CHILTON'S
AUTO REPAIR MANUAL
1964-1971

Publisher & Editor-In-Chief	Kerry A. Freeman, S.A.E.
Executive Editors	Dean F. Morgantini, W. Calvin Settle Jr., S.A.E.
Managing Editor	Nick D'Andrea
Assistant Managing Editor	Kenneth J. Grabowski, A.S.E., S.A.E.
Senior Editors	Eric O. Cole, Debra Gaffney, Jacques Gordon, Michael L. Grady, Kevin M. G. Maher, Richard J. Rivele, S.A.E., Richard T. Smith, Jim Taylor, Ron Webb
Project Managers	Larry Braun, S.A.E., A.S.C., Thomas P. Browne III, Joseph DeFrancesco, Robert E. Doughten, Benjamin E. Greisler, A.S.E., Martin J. Gunther, Craig P. Nangle, A.S.E., Richard Schwartz
Editorial Staff	Jaffer A. Ahmad, Bradley Bower, James Carr, Robert A. Chabot, William C. Cottman, A.S.E., Leonard Davis, A.S.E., Michael DiFurio Jr., S.A.E., Robert F. Dougherty Jr., John J. Ferraro, A.S.E., S.A.E., Sam Fiorani, Matthew E. Frederick, William C. Friedauer, Edward J. Giacomucci, A.S.E., S.A.E., Al Gibbs, Herbert Guie Jr, Dawn M. Hoch, David E. Jester, Lori Johnson, A.S.E., William Kessler, Kenneth F. Konzelman, Neil J. Leonard, A.S.E., James R. Marotta, Robert McAnally, Raymond K. Moore, Norman D. Norville, A.S.E., Christine L. Nuckowski, Eric S. Peterson, A.S.E., Ernest H. Ralph, Charles Ramsey, A.S.E., Roy Ripple, A.S.E., Don Schnell, A.S.E., S.A.E., Paul Shanahan, Larry E. Stiles, Gordon L. Tobias, Anthony Tortorici, A.S.E., S.A.E., Albert A. Wood, A.S.E.
Production Manager	Andrea M. Steiger
Assistant Production Manager	Marsha Park Herman
Production Specialists	Christina Davis, Kimberly T. Hayes, Joseph C. McGinty, Elizabeth E. Thompson
Director of Manufacturing	Mike D'Imperio
Manufacturing Manager	Robin Norman
OFFICERS	
Senior Vice President	Ronald A. Hoxter

CHILTON BOOK COMPANY

Manufactured in USA

© 1996 Chilton Book Company
Chilton Way, Radnor, PA 19089
ISBN 0-8019-5974-8

30 31 32 33 32109876

Contents

Car Section

Unit Repair Section

Troubleshooting Section

YEAR IDENTIFICATION

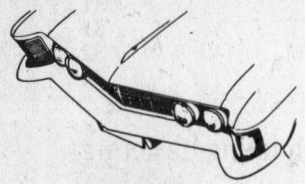

1964 Riviera

1964 Electra-Le Sabre

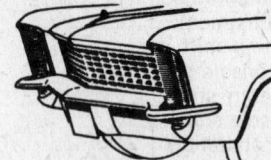

1965 Riviera

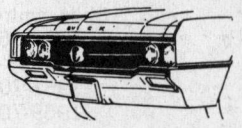

1964 Wildcat

1965 Electra-Le Sabre

1965 Wildcat

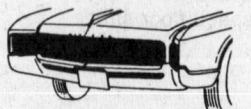

1966 Riviera

1966 Electra-Le Sabre

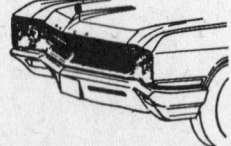

1966 Wildcat

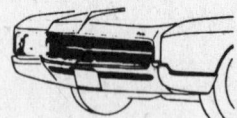

1967 Riviera

1967 Electra-Le Sabre

1967 Wildcat

1968-69 Riviera

1968 Le Sabre

1968 Electra

1968 Wildcat

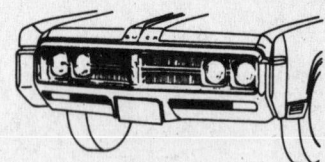

1969 Wildcat

1969 Le Sabre

1969 Electra

1970 Le Sabre

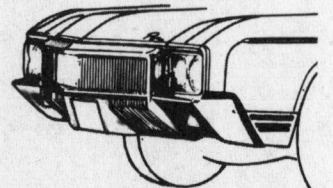

1970 Riviera

1971 Buick

1971 Riviera

FIRING ORDER

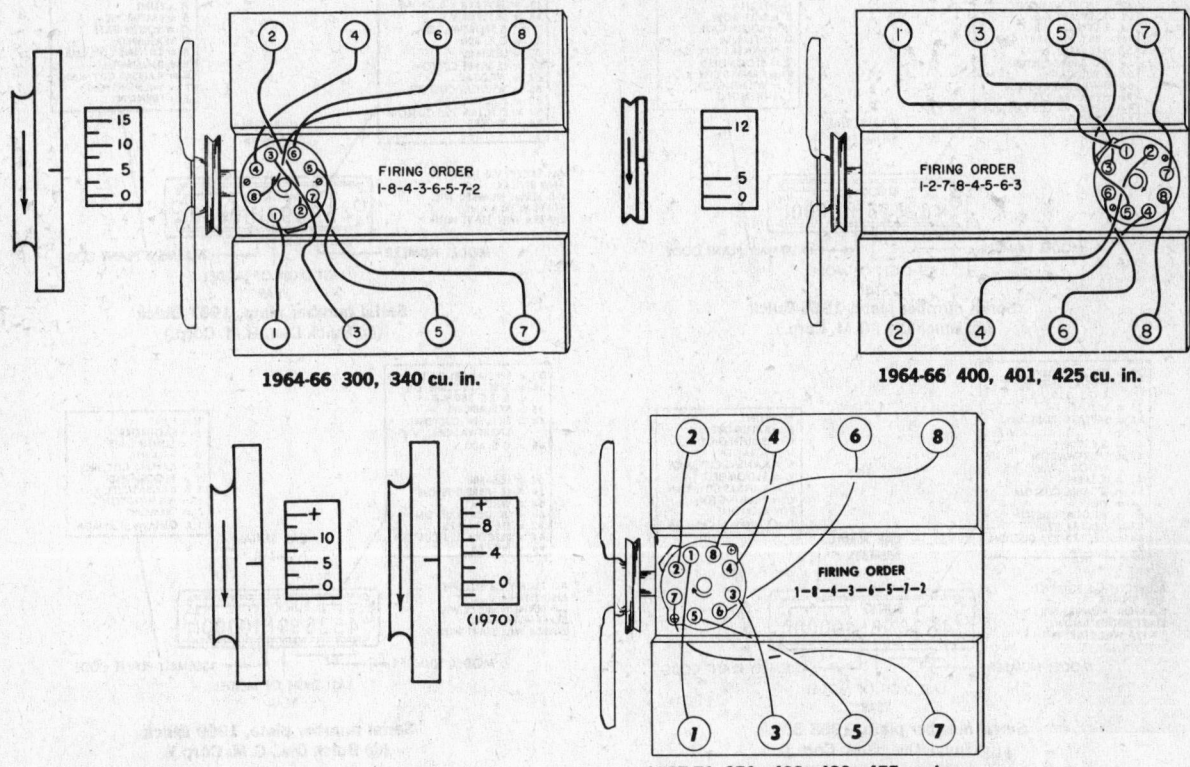

1964-66 300, 340 cu. in.

1964-66 400, 401, 425 cu. in.

(1970)

1967-71 350, 400, 430, 455 cu. in.

CAR SERIAL NUMBER LOCATION AND ENGINE IDENTIFICATION

The car serial number is used for registration and other legal records. This number is unique to the individual car. The production code number identifies the type of engine and its production date. The Engine Identification Code chart can be used to determine the type of engine in the particular vehicle. The engine number also appears on the vehicle identification plate following model and series identification. On 1964 models, the engine number and the car serial number are the same.

1964

The car serial number is located on the left front hinge pillar. The number is interpreted as follows:

1. The first digit is the series number

   4 for 4400 series
   6 for 4600 series
   7 for 4700 series
   8 for 4800 series
   9 for 4900 series

2. Next, a letter designating the year ("K" for 1964)
3. A digit (1-8) identifying the assembly plant
4. The remaining six digits are the individual car number.

The engine number is stamped on the top surface of the engine crankcase, just forward of the valve lifter cover on the left side. The engine number and the car serial number are the same.

The production code number is stamped to the left of the engine number.

1965

The serial number identification plate is attached to the left front body hinge pillar.

The engine serial and production code numbers on the 401 and 425 cu. in. engines are on the top surface of the engine crankcase, just forward of the valve lifter cover. The production code number is on the left and the serial number is on the right.

On the 300 cu. in. engine, the serial number is on the left front face of the engine crankcase to the side of the distributor cap. The production code

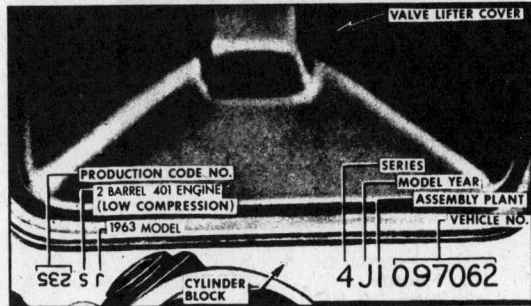

Engine numbers, 1964 Buick
(© Buick Div., G.M. Corp.)

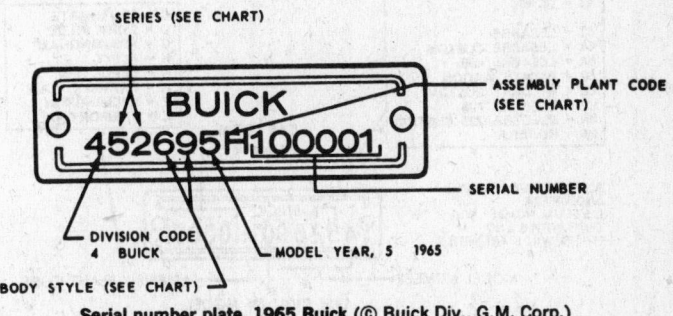

Serial number plate, 1965 Buick (© Buick Div., G.M. Corp.)

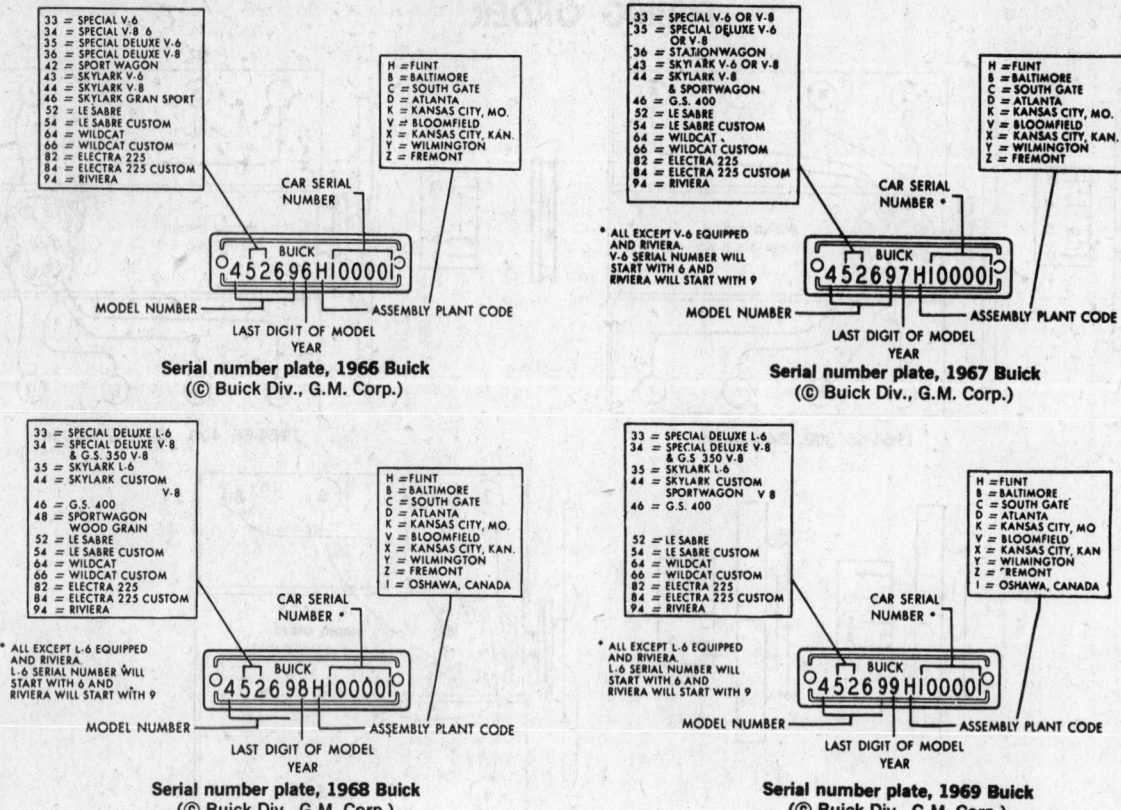

Serial number plate, 1966 Buick
(© Buick Div., G.M. Corp.)

Serial number plate, 1967 Buick
(© Buick Div., G.M. Corp.)

Serial number plate, 1968 Buick
(© Buick Div., G.M. Corp.)

Serial number plate, 1969 Buick
(© Buick Div., G.M. Corp.)

number is between the two middle branches of the right exhaust manifold.

1966

The serial number identification plate is attached to the left front body hinge pillar.

The engine serial and production code numbers on the 401 and 425 cu. in. engines are on the top surface of the crankcase, just forward of the valve lifter cover. The production code number is on the left and the serial number is on the right.

On the 300 and 340 cu. in. engines, the serial number is on the left front face of the engine crankcase to the side of the distributor cap. The production code number is between the

two middle branches of the right exhaust manifold.

1967

The car serial number identification plate is attached to the left front body hinge pillar.

On the 300 and 340 cu. in. engines, the serial number is on the left front face of the crankcase to the side of the distributor cap. The production code number is between the two middle branches of the right exhaust manifold.

On the 400 and 430 cu. in. engines, the production code number is between the two front branches of the right exhaust manifold, while the engine serial number is between the two rear branches.

1968

The serial number identification plate is attached to the left front body hinge pillar.

On the 350 cu. in. engine, the serial number is on the front of the left cylinder bank just below the cylinder head. The production code number is between the left exhaust manifold and the two front spark plugs on the left bank.

On the 400 and 430 cu. in. engines, the serial number is between the two front spark plugs and the exhaust manifold on the left side. The production code number is between the two rear spark plugs and the exhaust manifold, also on the left side.

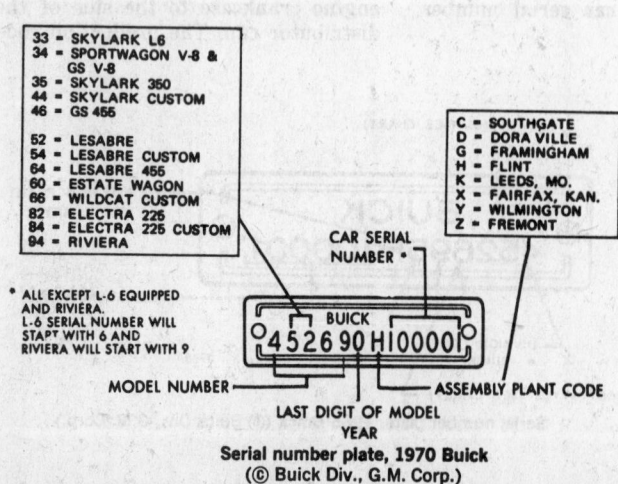

Serial number plate, 1970 Buick
(© Buick Div., G.M. Corp.)

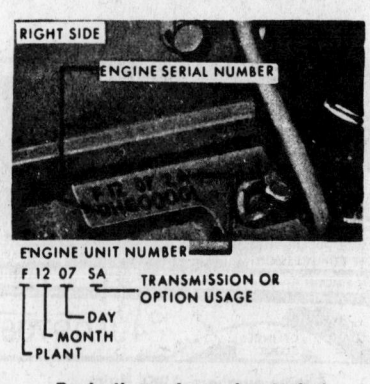

Production code number, typical
(© Buick Div., G.M. Corp.)

1969

The serial number identification plate is attached to the top of the instrument panel on the left side.

On the 350 cu. in. engine, the serial number is on the front of the left cylinder bank just below the cylinder head. The production code number is between the left exhaust manifold and the two front spark plugs on the left bank.

On the 400 and 430 cu. in. engines, the serial number is between the two front spark plugs and the exhaust manifold on the left side. The production code number is between the two rear spark plugs and the exhaust manifold, also on the left side.

1970-71

The serial number identification plate is attached to the top of the instrument panel on the left side.

On the 350 cu. in. engine, the serial number is on the front of the left cylinder bank just below the cylinder head. The production code number is between the left exhaust manifold and the two front spark plugs on the left bank.

On the 455 cu. in. engine, the serial number is between the two front spark plugs and the exhaust manifold on the left side. The production code number is between the two rear spark plugs and the exhaust manifold, also on the left side.

Engine numbers, 1968-70 350 cu. in. (© Buick Div., G.M. Corp.)

Engine numbers, 1967-70 400, 430, and 455 cu. in. (© Buick Div., G.M. Corp.)

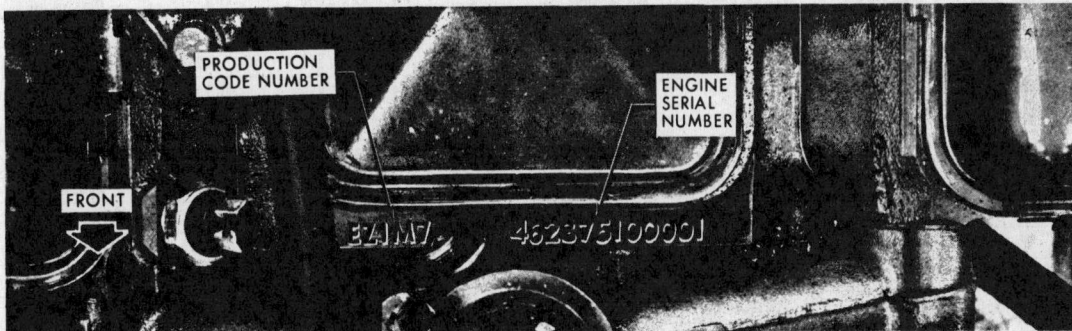

Engine numbers, 1964-66 401 and 425 cu. in. (© Buick Div., G.M. Corp.)

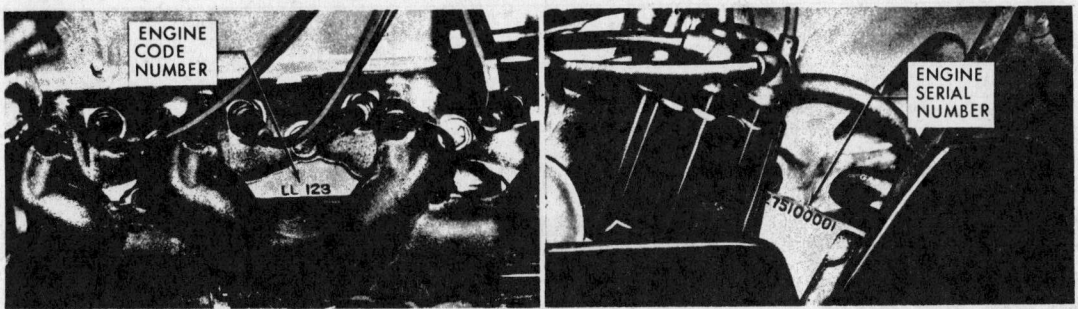

Engine numbers, 1965-67 300 and 340 cu. in. (© Buick Div., G.M. Corp.)

ENGINE IDENTIFICATION CODE

No. Cyls.	Cu. In. Displ.	Type	YEAR AND CODE						
			1964	1965	1966	1967	1968	1969	1970
8	300	Std. Eng.	KL						
8	300	HP	KP						
8	401		KT						
8	425		KW						
8	425	2 Bbl.	KX						
8	300	Std. Eng.		LL	ML	NL			
8	300	4 Bbl.		LP					
8	401	4 Bbl.		LT					
8	401	Riviera		LW					
8	425	2-4 Bbl.		LX					
8	340	2 Bbl.			MA	NA			
8	340	4 Bbl.			MB	NB			
8	425	4 Bbl.			MW				
8	401	4 Bbl.			MR, MT				
8	430	4 Bbl.				ND, MD	PD	RD	
8	350	2 Bbl.					PO	RO	SO
8	350	4 Bbl.					PP	RP	SR, SP
8	400	4 Bbl.				NR	PR	RR	
8	455	4 Bbl.							SP, SS, SF

GENERAL ENGINE SPECIFICATIONS

YEAR	CU. IN. DISPLACEMENT	CARBURETOR	DEVELOPED HORSEPOWER @ RPM	DEVELOPED TORQUE @ RPM (FT. LBS.)	A.M.A. HORSEPOWER	BORE & STROKE (IN.)	ADVERTISED COMPRESSION RATIO	VALVE LIFTER TYPE	NORMAL OIL PRESSURE (PSI)
1964–65	V8—300	2-BBL.	210 @ 4600	310 @ 2400	45.0	3.7500 x 3.40	9.0–1	Hyd.	33
	V8—300	4-BBL.	250 @ 4800	335 @ 3000	45.0	3.7500 x 3.40	11.0–1	Hyd.	33
	V8—401	4-BBL.	325 @ 4400	445 @ 2800	56.1	4.1875 x 3.64	10.25–1	Hyd.	40
	V8—425	4-BBL.	340 @ 4400	465 @ 2800	59.5	4.3125 x 3.64	10.25–1	Hyd.	40
	V8—425	2-4-BBL.	360 @ 4400	465 @ 2800	59.5	4.3125 x 3.64	10.25–1	Hyd.	40
1966	V8—340	2-BBL.	220 @ 4000	340 @ 2400	45.0	3.7500 x 3.850	9.0–1	Hyd.	33
	V8—340	4-BBL.	260 @ 4000	365 @ 2800	45.0	3.7500 x 3.850	10.25–1	Hyd.	33
	V8—401	4-BBL.	325 @ 4400	445 @ 2800	56.1	4.1875 x 3.640	10.25–1	Hyd.	40
	V8—425	4-BBL.	340 @ 4400	465 @ 2800	59.5	4.3125 x 3.640	10.25–1	Hyd.	40
	V8—425	2-4-BBL.	360 @ 4400	465 @ 2800	59.5	4.3125 x 3.640	10.25–1	Hyd.	40
1967	V8—340	2-BBL.	220 @ 4200	340 @ 2400	45.0	3.7500 x 3.850	9.0–1	Hyd.	33
	V8—340	4-BBL.	260 @ 4200	365 @ 2800	45.0	3.7500 x 3.850	10.25–1	Hyd.	33
	V8—430	4-BBL.	360 @ 5000	475 @ 3200	56.1	4.1875 x 3.900	10.25–1	Hyd.	40
1968–69	V8—350	2-BBL.	230 @ 4400	350 @ 2400	46.2	3.8000 x 3.850	9.0–1	Hyd.	37
	V8—350	4-BBL.	280 @ 4600	375 @ 3200	46.2	3.8000 x 3.850	10.25–1	Hyd.	37
	V8—430	4-BBL.	360 @ 5000	475 @ 3200	56.1	4.1875 x 3.900	10.25–1	Hyd.	40
1970	V8—350	2-BBL.	260 @ 4600	360 @ 2600	46.2	3.8000 x 3.850	9.0–1	Hyd.	37
	V8—350	4-BBL.	285 @ 4600	375 @ 3000	46.2	3.8000 x 3.850	9.0–1	Hyd.	37
	V8—350	4-BBL.	315 @ 4800	410 @ 3200	46.2	3.8000 x 3.850	10.25–1	Hyd.	37
	V8—455	4-BBL.	370 @ 4600	510 @ 2800	59.5	4.3125 x 3.900	10.0–1	Hyd.	40
1971	V8—350	2-BBL.	260 @ 4600	...	46.2	3.8000 x 3.850	9.0–1	Hyd.	37
	V8—350	4-BBL.	285 @ 4600	...	46.2	3.8000 x 3.850	9.0–1	Hyd.	37
	V8—350	4-BBL.	315 @ 4800	...	46.2	3.8000 x 3.850	...	Hyd.	37
	V8—455	4-BBL.	370 @ 4600	...	59.5	4.3125 x 3.900	9.0–1	Hyd.	40
	V8—455	4-BBL.	...	...	59.5	4.3125 x 3.900	...	Hyd.	40

CYLINDER HEAD BOLT TIGHTENING SEQUENCE

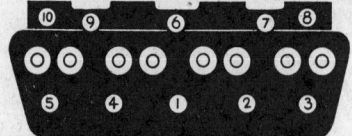

1964-67 300 cu. in.
1966-67 340 cu. in.
1967-69 400, 430 cu. in.
1968-69 350 cu. in.

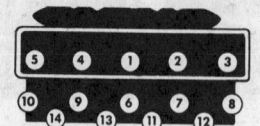

1964-66 400, 401, 425 cu. in.

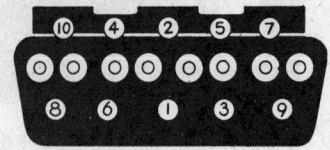

1970-71 350, 455 cu. in.

TUNE-UP SPECIFICATIONS

YEAR	MODEL	SPARK PLUGS		DISTRIBUTOR		IGNITION TIMING (Deg.) ▲	CRANKING COMP. PRESSURE (Psi)	VALVES			FUEL PUMP PRESSURE (Psi)	IDLE SPEED (Rpm) ★
		Type	Gap (In.)	Point Dwell (Deg.)	Point Gap (In.)			Tappet (Hot) Clearance (In.)		Intake Opens (Deg.)		
								Intake	Exhaust			
1964	V8—300 Cu. In.; 2-BBL.	44FFS	.032	30°	.016	2½ B	160	Zero	Zero	26B	5	550
	V8—300, 401 Cu. In.; 4-BBL.	44S	.032	30°	.016	2½ B	170	Zero	Zero	28B	6	500
	V8—425 Cu. In.; 4-BBL.	44S	.032	30°	.016	2½ B	170	Zero	Zero	31B	6	525
	V8—425 Cu. In.; 2-4-BBL.	44S	.032	30°	.016	12B	170	Zero	Zero	28B	6	525
1965	V8—300 Cu. In.; 2-BBL.	44FFS	.032	30°	.016	2½ B	160	Zero	Zero	26B	6	550
	V8—300, 401 Cu. In.; 4-BBL.	44S	.032	30°	.016	2½ B	170	Zero	Zero	28B	6	500
	V8—425 Cu. In.; 4-BBL.	44S	.032	30°	.016	2½ B	170	Zero	Zero	29B	6	525
	V8—425 Cu. In.; 2-4-BBL. M.T.	44S	.032	30°	.016	2½ B	170	Zero	Zero	28B	6	525
	V8—425 Cu. In.; 2-4-BBL. A.T.	44S	.032	30°	.016	12B	170	Zero	Zero	28B	6	525
1966	V8—340 Cu. In.	44TS	.032	30°	.016	2½ B	160	Zero	Zero	32B	6	550
	V8—401 Cu. In.	44S	.032	30°	.016	2½ B	170	Zero	Zero	28B	6	500
	V8—425 Cu. In.	44S	.032	30°	.016	2½ B	170	Zero	Zero	29B	6	525
	V8—425 (2-4-BBL.)	44S	.032	30°	.016	12B	170	Zero	Zero	28B	6	525
1967	V8—340 Cu. In.	44S	.032	30°	.016	2½ B	160	Zero	Zero	32B	5	550
	V8—430 Cu. In.	44TS	.032	30°	.016	2½ B	170	Zero	Zero	14B	6	550
1968	V8—350 Cu. In.; 2-BBL.	44TS	.030	30°	.016	2½ B	160	Zero	Zero	24B	5	550
	V8—350 Cu. In.; 4-BBL.	44TS	.030	30°	.016	2½ B	170	Zero	Zero	24B	5	550
	V8—430 Cu. In.; 4-BBL.	44TS	.030	30°	.016	2½ B	170	Zero	Zero	14B	6¼	550
1969	V8—350 Cu. In.; 2-BBL.	R45TS	.030	30°	.016	TDC	160	Zero	Zero	24B	5	550
	V8—350 Cu. In.; 4-BBL.	R45TS	.030	30°	.016	TDC	170	Zero	Zero	24B	5	550
	V8—430 Cu. In.; 4-BBL.	R44TS	.030	30°	.016	TDC	170	Zero	Zero	14B	6¼ ①	550
1970	V8—350 Cu. In.	R45TS	.030	30°	.016	6B	165	Zero	Zero	24B	5–8	550
	V8—455 Cu. In.	R44S	.030	30°	.016	6B	180	Zero	Zero	18B	5–8 ①	850
1971	V8—350 Cu. In.	R45TS	.030	30°	.016	★★	165	Zero	Zero	24B	5–8	550
	V8—455 Cu. In.	R44TS	.030	30°	.016	★★	165	Zero	Zero	18B	5–8	850

★—With synchromesh transmission in N and automatic in D. Add 50 rpm if air conditioned.

▲—With vacuum advance disconnected. NOTE: These settings are only approximate. Engine design, altitude, temperature, fuel octane rating and the condition of the individual engine are all factors which can influence timing. The limiting advance factor must, therefore, be the "knock point" of the individual engine.

①—Pump is in fuel tank—Riviera only.
B—Before top dead center.
TDC—Top dead center.
★★—See engine decal.

Caution

General adoption of anti-pollution laws has changed the design of almost all car engine production to effectively reduce crankcase emission and terminal exhaust products. It has been necessary to adopt stricter tune-up rules, especially timing and idle speed procedures. Both of these values are peculiar to the engine and to its application, rather than to the engine alone. With this in mind, car manufacturers supply idle speed data for the engine and application involved. This information is clearly displayed in the engine compartment of each vehicle.

TORQUE SPECIFICATIONS

YEAR	CYLINDER HEAD BOLTS (FT. LBS.)	ROD BEARING BOLTS (FT. LBS.)	MAIN BEARING BOLTS (FT. LBS.)	CRANKSHAFT BALANCER BOLT (FT. LBS.)	FLYWHEEL TO CRANKSHAFT BOLTS (FT. LBS.)	MANIFOLD (FT. LBS.)	
						Intake	Exhaust
1964–71	65–75 ①	40–45	100–110	200–220 ②	50–60	25–30 ③	10–15 ④

①—1967 400, 430 cu. in.—100–120 ft. lbs.
1968–71 400, 430, 455 cu. in.—100 ft. lbs.

②—All V8 except 455 cu. in.—120–140 ft. lbs. minimum.

③—1966–67—45–55 ft. lbs.
1968–71—55 ft. lbs.

④—1966–67—15–20 ft. lbs.
1968–71—18 ft. lbs.

VALVE SPECIFICATIONS

YEAR AND MODEL		SEAT ANGLE (DEG.)	FACE ANGLE (DEG.)	VALVE LIFT INTAKE (IN.)	VALVE LIFT EXHAUST (IN.)	VALVE SPRING PRESSURE (VALVE OPEN) LBS. @ IN.	VALVE SPRING INSTALLED HEIGHT (IN.)	STEM TO GUIDE CLEARANCE (IN.)		STEM DIAMETER (IN.)	
								INTAKE	EXHAUST	INTAKE	EXHAUST
1964-65	300 Cu. In.	45	45	.401	.401	168 @ 1.26	1.64	.0020-.0025①	.0025-.0030①	.3407-.3412②	.3402-.3407②
	401, 425 Cu. In.	45	45	.431★	.431★	101 @ 1.16	1.60	.0020-.0030①	.0025-.0035①	.3720-.3730②	.3715-.3725②
1966	340 Cu. In.	45	45	.401	.401	164 @ 1.34	1.72	.0012-.0032	.0025-.0030①	.3405-.3415②	.3402-.3407②
	401 Cu. In.	45	45	.431	.431	101 @ 1.16	1.60	.0020-.0030①	.0025-.0035①	.3720-.3730②	.3715-.3720②
	425 Cu. In.	45	45	.439	.441	101 @ 1.16	1.60	.0020-.0030①	.0025-.0035①	.3720-.3730②	.3715-.3720②
1967	340 Cu. In.	45	45	.399	.399	164 @ 1.34	1.72	.0012-.0032	.0015-.0035①	.3405-.3415	.3402-.3407
	430 Cu. In.	45	45	.421	.450	177 @ 1.45	1.89	.0015-.0035	.0015-.0035①	.3720-.3730	.3720-.3730
1968-69	350 Cu. In.	45	45	.371	.384	180 @ 1.34	1.72	.0015-.0035	.0015-.0035①	.3720-.3730	.3720-.3730②
	430 Cu. In.	45	45	.418	.448	177 @ 1.45	1.89	.0015-.0035	.0015-.0035①	.3720-.3730	.3720-.3730②
1970	350 Cu. In.	45	45	.381	.389	180 @ 1.34	1.72	.0015-.0035	.0015-.0035①	.3720-.3730	.3720-.3730②
	455 Cu. In.	45	45	.387	.458	177 @ 1.45	1.89	.0015-.0035	.0015-.0035①	.3720-.3730	.3720-.3730②
1971	350 Cu. In.	45	45	.381	.389	180 @ 1.34	1.72	.0015-.0035	.0015-.0035①	.3720-.3730	.3720-.3730②
	455 Cu. In.	45	45	.387	.458	177 @ 1.45	1.89	.0015-.0035	.0015-.0035①	.3720-.3730	.3720-.3730②

★—425 Cu. In.—intake .439, exhaust .441.
①—±.001 in. Guide Tapers with larger diameter at bottom.
②—±.0005 in. Guide Tapers with larger diameter at top.

CRANKSHAFT BEARING JOURNAL SPECIFICATIONS

YEAR AND MODEL		MAIN BEARING JOURNALS (IN.)				CONNECTING ROD BEARING JOURNALS (IN.)		
		Journal Diameter	Oil Clearance	Shaft End-Play	Thrust On No.	Journal Diameter	Oil Clearance	End-Play
1964-66	401 Cu. In.	2.4985	.0015	.006	3	2.2495	.0012	.009
	425 Cu. In.	2.4985	.0015	.006	3	2.2495	.0012	.009
	300 Cu. In.	2.2992	.0013	.006	3	2.0000	.0012	.009
	340 Cu. In.	2.9995	.0015	.006	3	2.0000	.0012	.010
1967	340 Cu. In.	2.9995	.0015	.006	3	2.0000	.0012	.010
	430 Cu. In.	3.2500	.0012	.006	3	2.2495	.0012	.010
1968-69	350 Cu. In.	2.9995	.0010	.006	3	2.0000	.0012	.008
	430 Cu. In.	3.2500	.0012	.006	3	2.2500	.0012	.008
1970-71	350 Cu. In.	2.9995	.0004-.0015	.003-.009	3	2.0000	.0002-.0023	.006-.014
	455 Cu. In.	3.2500	.0007-.0018	.003-.009	3	2.249-2.250	.0002-.0023	.005-.012

BATTERY AND STARTER SPECIFICATIONS

YEAR AND MODEL		BATTERY				STARTERS					
		Ampere Hour Capacity	Volts	Group Number	Terminal Grounded	Lock Test		No-Load Test			
						Amps.	Volts	Amps.	Volts	RPM	Brush Spring Tension (Oz.)
1964	4400	70	12	3SM	Neg.	330	3.5	85	10.6	4,350	35
	Exc. 4400	70	12	3SM	Neg.	330	2.0	100	10.6	4,450	35
1965-66	4500	61	12	28M	Neg.	330	3.0	85	10.6	4,350	35
	Exc. 4500	70	12	3SM	Neg.	330	2.0	110	10.6	4,450	35
1967	4500	61	12	2SMD	Neg.	Not Recommended		85	10.6	3,600	35
	Exc. 4500	70	12	2STA	Neg.	Not Recommended		120	10.6	4,700	35
1968	4500	61	12	9MJ3F	Neg.	Not Recommended		85	10.6	4,350	35
	Exc. 4500	70	12	9MJ6A	Neg.	Not Recommended		88	10.6	5,000	35
1969	4500	61	12	9MJ3F	Neg.	Not Recommended		70	9.0	4,000	35
	Exc. 4500	70	12	9MJ6A	Neg.	Not Recommended		61	9.0	5,200	35
1970-71	4500	61	12	R59	Neg.	Not Recommended		55-85	9.0	3,100-4,900	35
	4600 (455)	70	12	Y69	Neg.	Not Recommended		48-74	9.0	4,100-6,300	35
	46, 48, 49	70	12	Y71	Neg.	Not Recommended		48-74	9.0	4,100-6,300	35

CAPACITIES

YEAR	MODEL	ENGINE CRANKCASE ADD 1 Qt. FOR NEW FILTER	TRANSMISSIONS Pts. TO REFILL AFTER DRAINING			DRIVE AXLE (Pts.)	GASOLINE TANK (Gals.)	COOLING SYSTEM (Qts.) WITH HEATER
			Manual		Automatic			
			3-Speed	4-Speed				
1964	4400	4	2	$2\frac{1}{2}$	21	$4\frac{1}{2}$	20	$18\frac{1}{2}$
	4600, 4800	4	$3\frac{1}{2}$	$2\frac{1}{2}$	24	$4\frac{1}{2}$	20	$18\frac{1}{2}$
	4700	4	None	None	24	$4\frac{1}{2}$	20	$18\frac{1}{2}$
1965	45000	4	2	None	19	$4\frac{1}{2}$	25	$18\frac{1}{2}$
	46000	4	$3\frac{1}{2}$	$2\frac{1}{2}$	22	$4\frac{1}{2}$	25	$18\frac{1}{2}$
	48000	4	None	None	22	$4\frac{1}{2}$	25	$18\frac{1}{2}$
	49000	4	None	None	22	$4\frac{1}{2}$	20	$18\frac{1}{2}$
1966	45000	4	$3\frac{1}{3}$	None	19	$2\frac{1}{2}$	25	$14\frac{1}{2}$
	46000	4	$3\frac{1}{2}$	None	22	2	25	18
	48000	4	None	None	22	2	25	18
	49000	4	None	None	22	2	21	18
1967	4500	4	$3\frac{1}{3}$	None	19	$2\frac{3}{4}$	25	13
	46000, 48000, 49000	4	None	None	22	$4\frac{1}{4}$	25 ★	18
1968–69	45000	4	$3\frac{1}{2}$	$3\frac{1}{2}$	19.0	$4\frac{1}{4}$	25	$13\frac{1}{4}$
	46000, 48000, 49000	4	None	$3\frac{1}{2}$	22.0	$4\frac{1}{2}$	25 ★	$16\frac{3}{4}$
1970–71	45000	4	$3\frac{1}{2}$	None	20.0	3	25	16.2
	46000	4	$3\frac{1}{2}$	None	20.0	$4\frac{1}{4}$	25	19.7
	48000	4	None	None	20.0	$4\frac{1}{4}$	25	16.4
	49000	4	None	None	23.0	$4\frac{1}{4}$	21	19.7

★—Riviera (49000) 21 gals.

GENERAL CHASSIS AND BRAKE SPECIFICATIONS

YEAR AND MODEL		CHASSIS		BRAKE CYLINDER BORE				BRAKE DRUM	
		Overall Length (In.)	Tire Size	Master Cylinder (In.)		Wheel Cylinder Diameter (In.)		Diameter (In.)	
				Std.	Power	Front	Rear	Front	Rear
1964	V8, 4400	218.8	7.10 x 15	1.0	1.0	$1\frac{1}{8}$	1.0	12.002	12.002
	V8, 4600	218.8	7.60 x 15	1.0	1.0	$1\frac{1}{8}$	1.0	12.002	12.002
	V8, 4700	208.0	7.10 x 15	1.0	1.0	$1\frac{1}{8}$	1.0	12.002	12.002
	V8, 4800	222.8	8.00 x 15	1.0	1.0	$1\frac{1}{8}$	1.0	12.002	12.002
1965	V8, 45000, LeSabre	216.9	8.15 x 15	1.0	1.0	$1\frac{1}{8}$	1.0	12.002	12.002
	V8, 46000, Wildcat	219.9	8.45 x 15	1.0	1.0	$1\frac{1}{8}$	1.0	12.002	12.002
	V8, 48000, Electra	222.9	8.75 x 15	1.0	1.0	$1\frac{1}{8}$	1.0	12.002	12.002
	V8, 49000, Riviera	208.9	8.45 x 15	1.0	1.0	$1\frac{1}{8}$	1.0	12.002	12.002
1966	V8, 45000, LeSabre	217.0	8.15 x 15	1.0	1.0	$1\frac{1}{8}$	1.0	12.002	12.002
	V8, 46000, Wildcat	220.1	8.45 x 15	1.0	1.0	$1\frac{1}{8}$	1.0	12.002	12.002
	V8, 48000, Electra	223.5	8.85 x 15	1.0	1.0	$1\frac{1}{8}$	1.0	12.002	12.002
	V8, 49000, Riviera	211.2	8.45 x 15	1.0	1.0	$1\frac{1}{8}$	1.0	12.002	12.002
1967	V8, 45000, LeSabre	217.5	8.45 x 15	1.0①	1.0	$1\frac{3}{16}$	1.0	12.002②	12.002
	V8, 46000, Wildcat	220.5	8.45 x 15	1.0①	1.0	$1\frac{3}{16}$	1.0	12.002②	12.002
	V8, 48000, Electra	223.9	8.85 x 15	1.0①	1.0	$1\frac{3}{16}$	1.0	12.002②	12.002
	V8, 49000, Riviera	211.3	8.45 x 15	1.0①	1.0	$1\frac{3}{16}$	$\frac{15}{16}$	12.002②	12.002
1968–69	V8, 4500, LeSabre	217.5	8.45 x 15	1.0①	1.0	$1\frac{3}{16}$	1.0	12.002②	12.002
	V8, 4600, Wildcat	220.5	8.45 x 15	1.0①	1.0	$1\frac{3}{16}$	1.0	12.002②	12.002
	V8, 4800, Electra	224.9	8.85 x 15	1.0①	1.0	$1\frac{3}{16}$	1.0	12.002②	12.002
	V8, 4900, Riviera	215.2	8.45 x 15	1.0①	1.0	$1\frac{3}{16}$	$\frac{15}{16}$	12.002②	12.002
1970–71	V8, 4500, LeSabre	219.4	8.55 x 15	1.0①	1.0	$1\frac{3}{16}$	1.0	$12 \times 2\frac{1}{4}$②	12 x 2
	V8, 4600, Wildcat	219.4	8.55 x 15	1.0①	1.0	$1\frac{3}{16}$	1.0	$12 \times 2\frac{1}{4}$②	12 x 2
	V8, 4800, Electra	225.4	8.85 x 15	1.0①	1.0	$1\frac{3}{16}$	1.0	$12 \times 2\frac{1}{4}$②	12 x 2
	V8, 4900, Riviera	215.5	8.55 x 15	1.0①	1.0	$1\frac{3}{16}$	$\frac{15}{16}$	$12 \times 2\frac{1}{4}$②	12 x 2

① —Optional disc brakes—$1\frac{1}{8}$ in.
② —Optional disc brakes—$11\frac{25}{32}$ in. x 1.04 in.

AC GENERATOR AND REGULATOR SPECIFICATIONS

YEAR AND MODEL		ALTERNATOR			REGULATOR						
		Field Current Draw @ 12V.	Output @ Generator RPM		Model	Field Relay			Regulator		
			1100	6500		Air Gap (In.)	Point Gap (In.)	Volts to Close	Air Gap (In.)	Point Gap (In.)	Volts at 125°
1964	1100624	1.9-2.3	10	42	1119515	.015	.030	2.3-3.7	.060	.014	13.5-14.3
	1100679	2.2-2.6	10	65	1119515	.015	.030	2.3-3.7	.060	.014	13.5-14.3
	1100623	1.9-2.3	10	42	1119515	.015	.030	2.3-3.7	.060	.014	13.5-14.3
	1100661	2.2-2.6	10	65	1119515	.015	.030	2.3-3.7	.060	.014	13.5-14.3
1965-66	1100691	2.2-2.6	10	42	1119515	.015	.030	2.3-3.7	.060	.014	13.5-14.3
	1100705	2.2-2.6	10	37	1119515	.015	.030	2.3-3.7	.060	.014	13.5-14.3
	1100708	2.2-2.6	10	42	1119515	.015	.030	2.3-3.7	.060	.014	13.5-14.3
	1100709	2.2-2.6	10	55	1119515	.015	.030	2.3-3.7	.060	.014	13.5-14.3
	1100710	2.2-2.6	10	55	1119515	.015	.030	2.3-3.7	.060	.014	13.5-14.3
1967-69	1100691	2.2-2.6	10	42	1119515	.015	.030	2.3-3.7	.060	.014	13.6-14.4
	1100691	2.2-2.6	10	42	1119515	.015	.030	2.3-3.7	.060	.014	13.6-14.4
	1100774	2.2-2.6	10	55	1119515	.015	.030	2.3-3.7	.060	.014	13.6-14.4
1970	1100691	2.2-2.6	10	42	1119515	.015	.030	2.3-3.7	.060	.014	13.6-14.4
	1100691	2.2-2.6	10	42	1119515	.015	.030	2.3-3.7	.060	.014	13.6-14.4
	1100774	2.2-2.6	10	55	1119515	.015	.030	2.3-3.7	.060	.014	13.6-14.4
1971	1100693	2.2-2.6	10	42	...	Transistor type—no adjustment					
	1100693	2.2-2.6	10	42	...	Transistor type—no adjustment					
	1100776	2.2-2.6	10	55	...	Transistor type—no adjustment					

WHEEL ALIGNMENT

YEAR	MODEL	CASTER		CAMBER		TOE-IN (In.)	KING-PIN INCLINATION (Deg.)	WHEEL PIVOT RATIO	
		Range (Deg.)	Pref. Setting (Deg.)	Range (Deg.)	Pref. Setting (Deg.)			Inner Wheel	Outer Wheel
1964-66	All Series	$\frac{1}{2}$P to $1\frac{1}{2}$P	1P	$\frac{1}{4}$P to $\frac{3}{4}$P	$\frac{1}{2}$P	$\frac{7}{32}$ to $\frac{5}{16}$	10	$22\frac{1}{4}$	20
1967	All Series	$\frac{1}{2}$P to $1\frac{1}{2}$P	1P	$\frac{1}{4}$P to $\frac{3}{4}$P	$\frac{1}{2}$P	$\frac{7}{32}$ to $\frac{5}{16}$	10	20	18
1968	All Series	$\frac{1}{2}$P to $1\frac{1}{2}$P	1P	$\frac{1}{4}$N to $\frac{3}{4}$P	$\frac{1}{4}$P	$\frac{7}{32}$ to $\frac{5}{16}$	10	20	18
1969	LeSabre, Wildcat, Electra	$\frac{1}{4}$P to $1\frac{1}{4}$P	$\frac{3}{4}$P	$\frac{1}{2}$N to $\frac{1}{2}$P	0	$\frac{7}{32}$ to $\frac{5}{16}$	10	22	20
	Riviera	$\frac{1}{2}$P to $1\frac{1}{2}$P	1P	$\frac{1}{4}$N to $\frac{3}{4}$P	$\frac{1}{4}$P	$\frac{5}{32}$ to $\frac{1}{4}$	10	20	18
1970	LeSabre, Wildcat, Electra	$\frac{1}{4}$P to $1\frac{1}{4}$P	$\frac{3}{4}$P	$\frac{1}{2}$N to $\frac{1}{2}$P	0	$\frac{7}{32}$ to $\frac{5}{16}$	10	20	20
	Riviera	$\frac{1}{2}$P to $1\frac{1}{2}$P	1P	$\frac{1}{4}$N to $\frac{3}{4}$P	$\frac{1}{4}$P	$\frac{5}{32}$ to $\frac{1}{4}$	10	20	17
1971	LeSabre, Wildcat, Electra	$\frac{1}{4}$P to $1\frac{1}{2}$P	$\frac{3}{4}$P	$\frac{1}{2}$N to $\frac{1}{2}$P	0	$\frac{7}{32}$ to $\frac{5}{16}$	10	20	20
	Riviera	$\frac{1}{2}$P to $1\frac{1}{2}$P	1P	$\frac{1}{4}$N to $\frac{3}{4}$P	$\frac{1}{4}$P	$\frac{5}{32}$ to $\frac{1}{4}$	10	20	17

N—Negative
P—Positive

FUSES AND CIRCUIT BREAKERS

1964-65

- Back-up lights .... 10 AMP. $1\frac{1}{4}$ in.
- Clock .... 2 AMP. $\frac{5}{8}$ in.
- Directional signal & stop lights .... 10 AMP. $1\frac{1}{4}$ in.
- Dome & trunk light .... 20 AMP. $1\frac{1}{4}$ in.
- Heater, defroster, air conditioner .... 30 AMP. $1\frac{1}{4}$ in.
- Instrument lights .... 5 AMP. $1\frac{1}{4}$ in.
- Radio .... 7.5 AMP. $\frac{7}{8}$ in.
- Tail, license & instrument rheostat .... 10 AMP. $1\frac{1}{4}$ in.
- W/S washer & wiper .... 25 AMP. $1\frac{1}{4}$ in.
- Antennae motor .... 15 AMP $1\frac{1}{4}$ in.
- Head & front parking lights on H. L. switch . . 15 AMP. CB

1966

- Back-up lights .... 10 AGC
- Blowers .... 30 AGC
- Clock, radio .... 7.5 AGC
- Directional sig. & stop .... 10 AGC
- Dome, lighter .... 20 SFE
- License, tail, cornering .... 15 AGC
- Instrument lamps .... 4 AGC
- High beam indicator .... 15 AMP. CB
- Headlamps, parking .... 15 AMP. CB
- Trans. selector; wipers .... 25 AGC

1967-71

- Back-up lights-directional sig. .... 20 SAE
- Blowers .... 25 AGC
- Clock .... 20 AGC
- Dome light, lighter, trunk .... 20 SAE
- Dome light, lighter, trunk (4900) .... 15 AGW
- Hazard flasher & stop lamp .... 20 SAE
- License, tail, cornering .... 20 SAE
- Instrument lamps .... 4 AGC
- Wipers .... 25 AGC

LIGHT BULBS

1964–66

Ash tray	1445; .5 CP
Ash tray (4700)	53; 1 CP
Auto. trans. control dial (console)	1816; 3 CP
Auto. trans. control dial (instrument panel)	194; 2 CP
Back-up	1156; 32 CP
Clock dial	1893; 2 CP
Cornering	1195; 50 CP
Courtesy, console, rear seat side rail or arm rest	90; 6 CP
Cruise control dial	53; 1 CP
Dome, center roof	1004; 15 CP
Glove box	1893; 2 CP
Headlight high beam indicator	194; 2 CP
Headlight, 5¾ in. dia., type 1 (inner)	4001; 37.5 W
Headlight, 5¾ in. dia., type 2 (outer)	4002-L; 37.5–55 W
Heater-air conditioner control dial	1893; 2 CP
Ignition switch	1445; .5 CP
Indicator lights (hot, cold, oil and amp.)	194; 2 CP
Instrument cluster dials	161; 1 CP
License	1155; 4 CP
Parking brake warning	1816; 3 CP
Parking, lower (4700)	1155; 4 CP
Radio dial	1881; 1 CP
Turn signal and parking, front	1157A; 32-4 CP
Turn signal, tail and stop, rear	1157; 32-4 CP
Turn signal indicator	194; 2 CP
Trunk	89; 6 CP

1967–69

Ash tray	1445; .5 CP
Ash tray (4900)	53; 1 CP
Auto. trans. control dial	1816; 2 CP
Auto. trans. control dial (4900)	1893; 2 CP
A/C control	1893; 2 CP
Back-up	1157; 32-4 CP
Cornering	1195; 50 CP
Courtesy, console, rear seat side rail or arm rest	90; 6 CP
Cruise control dial	161; 1 CP
Dome, center roof	1004; 15 CP
Glove box	1893; 2 CP
Headlight high beam indicator	161; 1 CP
Headlight high beam indicator (4900)	194; 2 CP
Headlight, 5¾ in. dia., type 1 (inner)	4001; 375 CP
Headlight 5¾ in. dia., type 2 (outer)	4002L; 37.5–55 CP
Heater air cond. control dial	1893; 2 CP
Ignition switch	53; 1 CP
Indicator lights (hot, cold, oil & amp.)	194; 2 CP
Instrument cluster dials	194; 2 CP
License	97; 4 CP
Parking brake warning	194; 2 CP
Radio dial	1892; 1 CP
Turn signal and parking, front	1157A; 32-4 CP
Turn signal, tail & stop, rear	1157; 32-4 CP
Turn signal indicator	161; 1 CP
Trunk	89; 6 CP

1970–71

Ash tray (4900)	1445; .5CP
Ash tray (exc. 4900)	1455; 5CP
Auto. trans. control dial	1816; 2CP
Auto. trans. control dial (4900)	1893; 2CP
A/C control	1893; 2CP
Back-up	1157; 32-4CP
Cornering (exc. 4900)	1195; 50CP
Courtesy, console, rear seat (4900)	1295; 50CP
Courtesy, console, rear seat	212; 6CP
Courtesy lights under dash	89; 6CP
Cruise control dial	181; 3CP
Dome, center roof	211; 12CP
Glove compartment	1893; 2CP
Headlight high beam indicator	194; 2CP
Headlight, 5¾ in. type 1 inner	4001; 375CP
Headlight, 5¾ in. type 2 outer	4002L; 37.5–55CP
Heater, A/C control dial	53; 1CP
Ignition switch	53; 1CP
Indicator lights (hot, cold, oil, amps.)	194; 2CP
Instrument cluster	194; 2CP
License	97; 4CP
Parking brake warning	194; 2CP
Radio dial	1893; 2CP
Turn and parking (front)	1157A; 32-4CP
Turn, tail, stop (rear)	1157; 32-3CP
Turn indicator	194; 2CP
Trunk	89; 6CP
Water temperature indicator	194; 2CP

Distributor

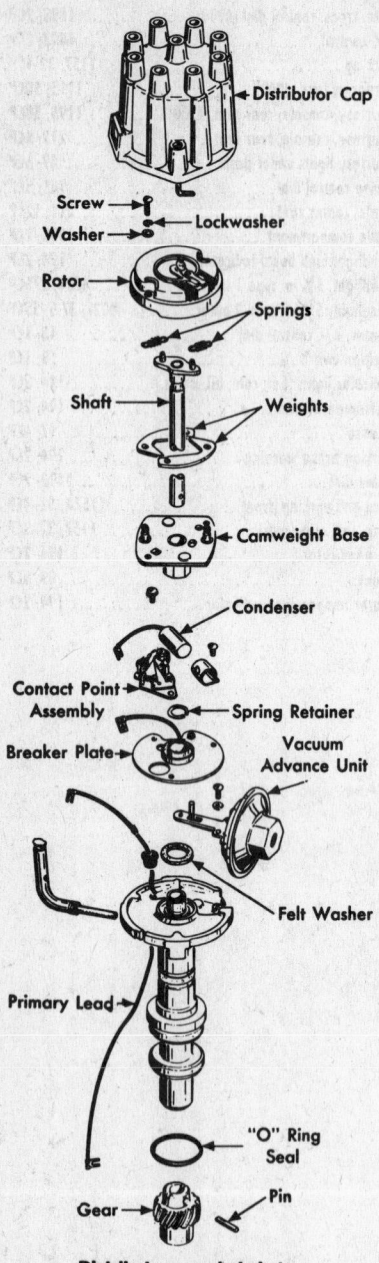

Distributor—exploded view
(© Buick Div. G.M. Corp.)

Distributor Removal

Disconnect the distributor primary wire from the coil and the tube from the vacuum unit. Remove distributor cap by inserting a screwdriver into upper slotted end of cap latches, pressing down and turning 90° counterclockwise.

Make a mark on the distributor body in line with the rotor. Match-mark position of vacuum unit to the engine.

Remove clamp to release distributor and remove from crankcase.

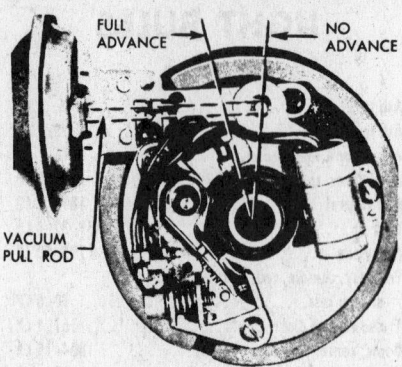

Vacuum advance mechanism 401 and 425 cu. in. engine
(© Buick Div. G.M. Corp.)

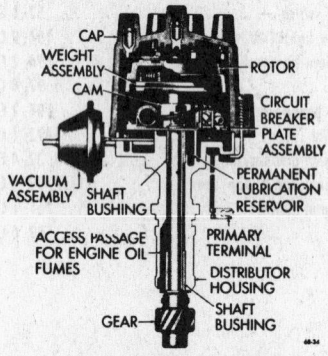

Distributor and cap assembly
(© Buick Div. G.M. Corp.)

Distributor Installation

If engine was inadvertently turned over while distributor was out, proceed as follows:

Remove right rocker arm cover. Using a wrench on the crankshaft pulley bolt, turn the engine over until both valves for No. 1 cylinder are closed. The timing mark on the harmonic balancer behind the crankshaft pulley should be aligned with the correct degree mark. No. 1 cylinder is now at firing point.

Install distributor in engine with rotor in position to fire No. 1 cylinder. The vacuum unit should align with the match-mark made when distributor was removed. Press down lightly

on distributor if it does not seat correctly. Use starter to turn engine until the tang on the distributor shaft slips into the slot in the oil pump shaft. This will not disturb the relationship between the distributor and the camshaft because the drive gear engages before the tang. However, it will be necessary to return the engine to the No. 1 firing point and check that rotor is also at No. 1 firing point. Reconnect vacuum tube and primary wire. Rotate the distributor body slightly until contacts just start to open. Install and tighten distributor clamp. Install distributor cap. Start engine and adjust point dwell.

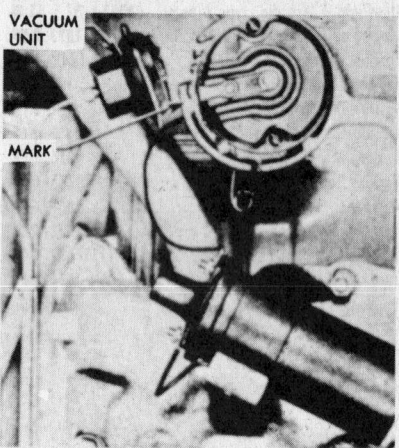

Installing distributor into engine
(© Buick Div. G.M. Corp.)

If the engine has not been disturbed since the distributor was removed proceed as follows:

Insert distributor into the block so that the rotor is pointing to the mark made on distributor housing and the vacuum advance unit is aligned with the match-mark made on the engine. Connect the vacuum tube, primary wire, and install the distributor cap. Install distributor clamp. Check that spark plug wires are correctly routed. Start engine and adjust point dwell and then adjust ignition timing. Rotate distributor body counterclockwise to advance the timing.

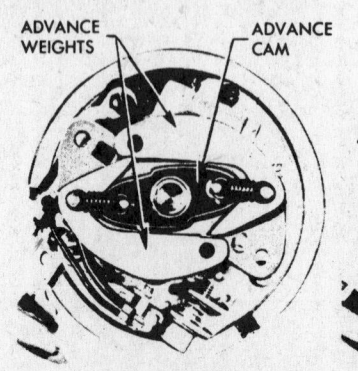 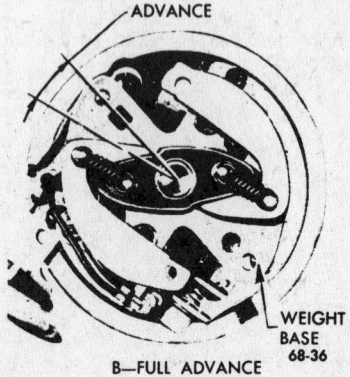

Centrifugal advance mechanism all except 401 and 425 cu. in.   (© Buick Div. G.M. Corp.)

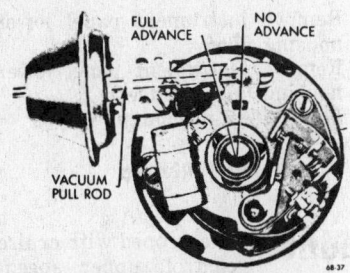

Vacuum advance mechanism—
all except 401 and 425 cu. in.
(© Buick Div. G.M. Corp.)

Generator, Regulator

Alternator Removal and Installation

Unfasten bolt holding tension bar to generator. Push generator in toward engine to release drive belt. Unfasten generator mounting bolt to release generator from engine.

When reinstalling, adjust generator drive belt to allow ½ in. play on the longest run between pulleys.

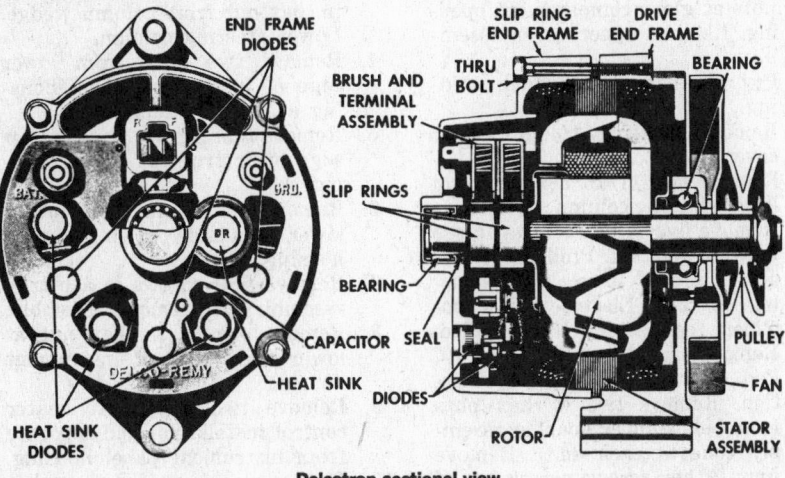

Delcotron sectional view

AC Generator (Delcotron)

Delcotron, the AC generator by Delco-Remy, is used on Buick passenger cars. These units are furnished with a voltage regulator to suit the application. A three unit regulator is used with vehicles having a 42, 55, or 61 Ampere Delcotron. A two unit regulator is used on vehicles equipped with an ammeter. Vehicles such as police equipment, etc. use a 60 or 62 Ampere Delcotron with a transistorized regulator.

Battery, Starter

Detailed information on the battery and starter will be found in the Battery and Starter Specifications table.

A more general discussion of starters and their troubles can be found in the Unit Repair Section.

Starter System

1964-70

The starter circuit is completed by a spring loaded ignition switch. To activate the starter, the ignition key is turned to the extreme right. When the engine fires, release the ignition key.

Instruments

Instrument Panel

1964-65

The instrument panel contains three separate assemblies which may be removed individually. These assemblies contain the speedometer, clock and warning lights, and gas gauge. On some models, the clock or gas gauge is mounted separately. Each of the three assemblies can be removed by removing one or two screws and pulling the assembly out of the panel housing. The speedometer cable may be disconnected after the speedometer assembly has been removed from the panel housing. The battery ground strap should be disconnected before removing any of the instrument assemblies.

Instrument Panel R & R

1966-67 Except Riviera

Before performing any instrument panel services, disconnect the battery ground cable.

1. Remove two windshield side garnish moldings.
2. Remove screws and pull instrument panel upper cover rearward. Disconnect radio speaker wire and remove cover.
3. Remove ash tray.
4. Remove one ⅜ in. hex nut

through hole in glove compartment.
5. Remove either radio bracket screw.
6. Remove two ⅜ in. hex head bolts from the outer ends of the instrument panel housing.
7. Remove air conditioner hose from center distribution duct and push out of way.
8. Remove light switch from instrument panel housing. Do not unplug connector.
9. Protect steering column against scratches, then tilt instrument panel back and position a ⅞ in. spacer block under each end of housing at attaching points.
10. Disconnect the following from instrument cluster:
    A. Shift indicator link.
    B. Printed circuit connector.
    C. Clock connector.
    D. Cruise switch connector.
    E. Cruise speedometer connector, and speedometer cable.
11. From below, remove ¼ in. hex screws from bottom edge of instrument cluster.
12. From above, remove ¼ in. hex screws from upper edge of cluster. Be careful not to lose two spacers.
13. Disconnect ground wire from upper edge of cluster.
14. Shift instrument cluster to the right and lift out.
15. Install by reversing removal procedure.

1969 Except Riviera

Caution If equipped with cruise control, upper speedometer cable must be disconnected from transducer before cluster housing is pulled back.

1. Disconnect battery ground cable.
2. Remove eight screws from instrument panel compartment body assembly and remove assembly.
3. Remove three ⅜ in. hex nuts at top underside of dash assembly and four screws at instrument panel housing assembly. Pull instrument panel upper cover rearward to remove.
4. Remove two screws from steering column filler and remove filler. Disconnect shift quadrant link wire at steering column. Remove two ⅜ in. hex nuts from steering column mounting bracket and one ⅜ in. hex bolt from column wedge. Lower steering column.
5. Remove two nuts from lower edge of instrument panel housing at steering column.
6. Remove four screws across upper edge of instrument panel housing.

7. Remove two screws at heater control installation and separate from instrument panel housing.
8. Remove four screws at ash tray assembly and remove assembly.
9. Remove one 3/8 in. hex nut at lower right side of instrument housing.
10. Remove headlight switch from instrument panel housing. Do not unplug connector.
11. Remove one 3/8 in. hex nut at lower left side of instrument housing.
12. Protect steering column so that instrument panel housing will not mar column when housing is tilted back.
13. Remove two screws at center air conditioning distribution duct (lower) and remove duct.
14. Disconnect from instrument cluster:
    A. Speedometer cable from below
    B. Printed circuit connector from above
    C. Wiring harness clip from below.
    D. Clock connector and two clock bulbs from above.
    E. Cruise switch connector from above
    F. Courtesy light connector from above
    G. Windshield wiper or washer switch connector from above
    H. Antenna and accessory switch connectors from above
    I. Cluster ground wire from above.
15. Remove complete instrument panel housing assembly to work bench.
16. Remove six 1/4 in. hex screws and remove instrument panel cluster from instrument panel housing.
17. Install instrument cluster by reversing above steps.

1969-70 Except Riviera

Caution If equipped with cruise control, upper speedometer cable must be disconnected from transducer before cluster housing is pulled back.
1. Disconnect battery ground cable.
2. Remove lower instrument panel filler by removing four screws, then sliding filler forward and down.
3. Remove glove compartment, leaving glove compartment door.
4. Remove instrument panel cover by removing two nuts above the glove compartment opening, removing three screws through the cluster housing, and removing all screws across the bottom edge of the cover. Remove both courtesy lights.

5. Remove two nuts at steering column. Disconnect shift indicator link. Lower steering column.
6. Protect steering column to avoid marring paint.
7. Remove eight screws from instrument cluster housing. Pull housing back on steering column and rotate so that back of housing is visible.
8. Disconnect from instrument cluster:
    A. Speedometer cable.
    B. Heater-air conditioner control panel.
    C. Cluster wiring connector.
    D. Buzzer connector
9. Remove six cluster to housing screws. Remove instrument panel cluster from instrument panel housing.
10. Install instrument cluster by reversing above steps.

1966 Riviera
1. Disconnect battery ground cable.
2. Remove ash tray assembly, center air outlet and duct.
3. Remove radio.
4. Remove three Phillips screws at cluster housing and two 3/8 hex nuts at glove compartment opening. Remove upper cover assembly.
5. Pry off instrument panel molding.
6. Remove steering column lower cover.
7. Remove two 11/16 hex nuts and lower steering column.
8. Remove five 1/4 hex screws across bottom and six Phillips screws across top of panel lower housing. Unplug Electro-Cruise amplifier connector if installed. Remove panel lower housing.
9. Pad steering column for protection. Remove two 3/8 hex nuts from below upper housing assembly (one at each end). Remove four 1/4 hex screws across top of housing. Pull upper housing assembly out to rest on steering column and knees.
10. Rotate upper housing assembly so that cluster retaining screws can be seen. Disconnect speedometer cable, unplug cluster connector, cruise connector, courtesy light connector and clock connector. Remove two 1/4 hex wiring harness clamp screws.
11. Remove five 1/4 hex screws across bottom and five 3/8 hex nuts across top of cluster. Remove instrument cluster assembly.
12. Install instrument cluster by reversing above steps.

1967 Riviera
This procedure is identical to that outlined above for the 1966 Riviera, with the substitution of the following steps:

6. Remove instrument panel lower housing filler.
7. Remove instrument panel lower housing (right and left).
8. Lower steering column.

1968 Riviera

Caution If equipped with cruise control, upper speedometer cable must be disconnected from transducer before cluster housing is pulled back.
1. Remove eight screws from instrument panel compartment body assembly and remove assembly.
2. Remove four 3/8 in. hex nuts at right underside of dash assembly and four Phillips screws at housing assembly. Pull instrument panel upper cover rearward to remove.
3. Remove two screws from steering column filler and remove filler. If column shift, disconnect shift quadrant link at steering column. Remove two 9/16 in. hex nuts from steering column mounting bracket and one 9/16 in. hex nut from column wedge. Lower steering column.
4. Remove two nuts from lower edge of instrument panel housing at steering column.
5. Remove four screws across upper edge of instrument panel housing.
6. Remove one 3/8 in. hex nut at lower left side of instrument housing.
7. Remove four screws at ash tray assembly and remove assembly.
8. Remove one 3/8 in. hex nut at lower right side of instrument housing.
9. Remove two screws at heater control installation and separate from instrument panel housing.
10. Protect steering column so that instrument panel housing will not mar column when housing is tilted back.
11. Disconnect from instrument cluster:
    A. Speedometer cable from above
    B. Two wiring harness clips from above
    C. Printed circuit connector from above.
12. Disconnect from instrument housing assembly:
    A. Clock connector and two clock bulbs from above
    B. Cruise control connector from above
    C. Courtesy light connector from above
    D. Windshield wiper or washer switch connector from above
    E. Antenna and accessory switch connectors from above

F. Cluster ground wire from above
G. Air conditioner hose from above
H. Headlight connector from above.

13. Remove instrument panel housing assembly.
14. Remove six ¼ in. hex screws and remove instrument panel cluster from instrument panel housing.
15. Install instrument cluster by reversing the above steps.

1969-70 Riviera

This procedure is identical to that outlined above for the 1968 Riviera, with the substitution of the following step:

1. Remove eight screws holding glove compartment. Remove glove compartment.

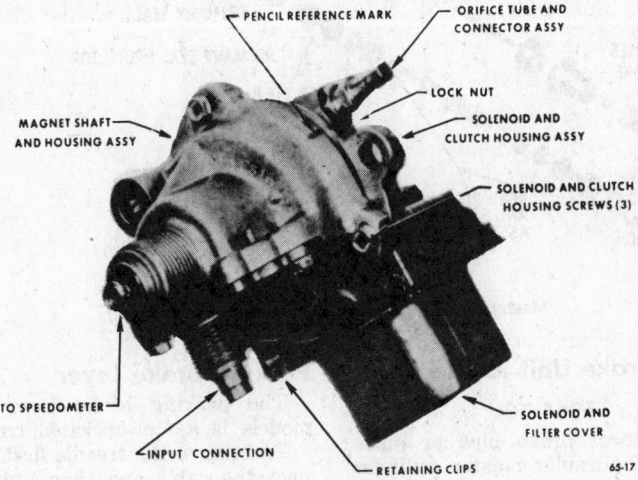

Speed transducer

Speedminder

1964-67

These models have the buzzer unit mounted under the dash, to the right of steering column. The buzzer circuit contains a 7.5 amp. fuse that has a dual purpose with the parking brake warning light. It is marked Bk and Bz on the fuse block.

1968-69

Speedminders for these models are fused at the clock fuse on the fuse block.

1970

This unit is not fused, but is connected at the starter solenoid and protected by the fusible links at the starter motor.

Charge Indicator

1964-70

An indicator light is used to show when system is not charging in all models without an ammeter.

Fuel Gauge—Dash Unit

1964-70

On 1964 and 1965 models access to the gas gauge is by removing the gauge unit mounting screws and pulling the assembly out of the panel housing. On 1966 and 1967 models the instrument panel upper cover must be removed. On 1968 models the center air-conditioning duct must be removed. 1969 and 1970 models have an access hole with a cover plate to the right of the gauge unit.

Information covering operation and troubles of the fuel gauge will be found in the Unit Repair Section.

Detailed information on the carburetor and how to adjust it will be found in the Unit Repair Section under the specific heading of the make of carburetor being used on the engine being worked on.

Dashpot adjustment can be found under Automatic Transmission Linkage Adjustment of this car section.

Temperature Gauge

1964-65 All Models, 1966-68 Except Riviera

A temperature switch located in the right cylinder head controls the operation of a cold indicator with a green lens and a hot indicator with a red lens. Le Sabre models have only the cold indicator.

1966-68 Riviera

A sending unit in the cylinder head controls the operation of a temperature gauge. When the engine is cold, the resistance of the sending unit is high, resulting in a cold reading on the gauge. As the engine approaches normal operating temperature, the resistance of the sending unit will become lower, resulting in a higher reading on the temperature gauge.

1969-70 All Models

A water temperature switch in the intake manifold controls the operation of a hot indicator with a red lens. A metal temperature switch in the left cylinder head controls operation of a stop engine indicator. Le Sabre models have only the hot indicator.

Ignition Switch

Lock Cylinder Replacement

1. Insert key and turn to Acc position.
2. With stiff wire (paper clip) in hole in face of switch depress lock pin and rotate cylinder counterclockwise and pull out.

Switch Replacement

1. Disconnect battery.
2. Remove cylinder (as above).
3. Remove ignition switch nut.
4. Install in reverse of above.

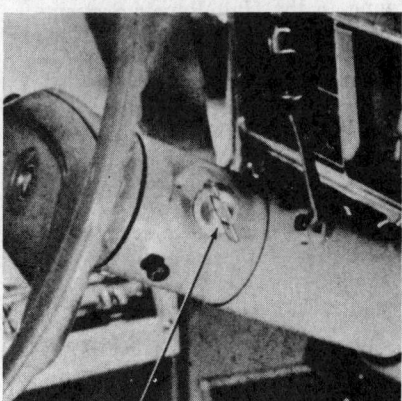

1969-70 ignition switch location
(© Buick Div. G.M. Corp.)

1969-70

The ignition switch has been relocated. It is no longer in its familiar place on the instrument panel, but it occupies a position on the steering column, just above the gear selector lever. This lock also prevents shifting the transmission and locks the steering.

Lighting Switch

Replacement

1. Disconnect battery.
2. Remove screws that retain vent control plate or access door to instrument panel.
3. Pull switch knob to last notch and depress spring loaded latch button on top of switch, while pulling knob and rod out of switch.
4. Remove escutcheon. Remove switch from cluster.
5. Disconnect multiple connector.
6. Install in reverse of above.

Rear Lights

The tail, stop, and turn-signal bulbs are accessible from inside the trunk compartment. On station wagon models it is necessary to remove the lamp lens.

Brakes

Specific information on brake cylinder sizes can be found in the General Chassis and Brake Specifications table of this section.

Since 1967 a dual master cylinder is used on all models. Information on the dual type system and brake adjustments, band replacement, bleeding procedure, master and wheel cylinder overhaul can be found in the Unit Repair Section.

Information on troubleshooting and overhauling power brakes can be found in the Unit Repair Section.

Information on the grease seals which may need replacement can be found in the Unit Repair Section.

Since 1967, some models are equipped with front wheel disc brakes. Beginning 1969 these disc brakes operate with single cylinder per wheel design. However, some 1969 models have the earlier four piston disc brakes. For details, consult the Unit Repair Section.

Self-adjusting brakes are standard equipment. Information on repairs and adjustments can be found in Unit Repair Section.

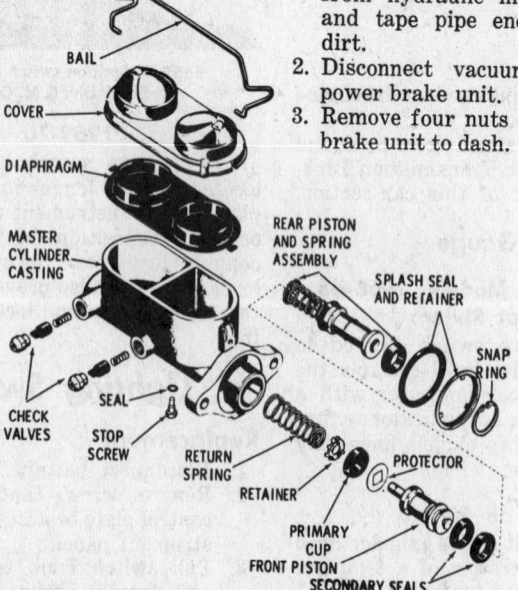

Dual-type master cylinder—Bendix (© Buick Div. G.M. Corp.)

Master Cylinder Removal

1964-70

1. Disconnect brake pipe or pipes from master cylinder and tape

end of pipe or pipes to prevent entrance of dirt.
2. Disconnect brake pedal from master cylinder at the pushrod.
3. Remove master cylinder-to-dash retaining bolts. Remove the master cylinder.

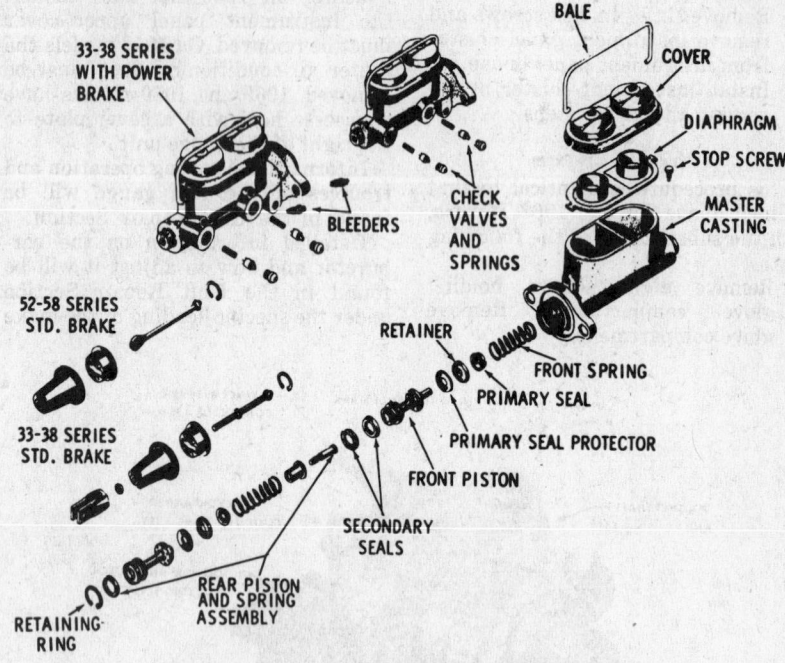

Master cylinder (Moraine) (© Buick Div. G.M. Corp.)

Power Brake Unit Removal

1964-70

1. Disconnect brake pipe or pipes from hydraulic master cylinder and tape pipe ends to exclude dirt.
2. Disconnect vacuum hose from power brake unit.
3. Remove four nuts holding power brake unit to dash.

4. Remove retainer and washer from brake pedal pin and disengage pushrod clevis.
5. Remove power brake unit from car.

Refilling Master Cylinder

1964-70

The master cylinder is located in the engine compartment and is filled from under the hood.

Parking Brake Lever

The parking brake lever on all models is a foot-operated treadle.

To remove the treadle first disconnect the cable and then unbolt the treadle frame from its mounting under the dash.

Parking Brake Cable Replacement—1964-70

Front Cable

1. Raise car.
2. Remove jam nut and adjusting nut from equalizer. Remove retainer clip from rear portion of front cable at frame. The retainer clip is not used on the Riviera.
3. At front of cable, bend snap-in retainer fingers in, so that retainer can be removed.
4. Disconnect cable from pedal assembly and remove cable.

NOTE: installation of a new cable can be eased by tying a cord to either end of the cable being removed and then pulling the new cable through the proper routing by use of the same cord. This is necessary since the cable is not long enough to follow a new path.

5. Install cable by reversing removal procedure.

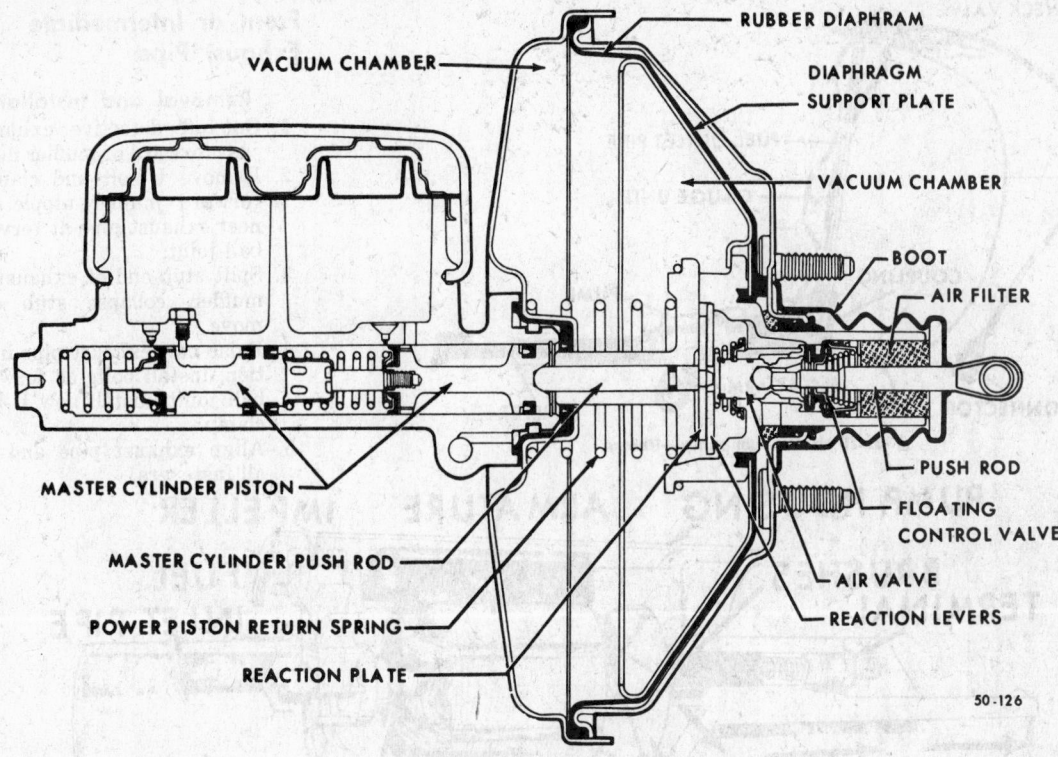

VACUUM CHAMBER

RUBBER DIAPHRAM

DIAPHRAGM
SUPPORT PLATE

VACUUM CHAMBER

BOOT

AIR FILTER

PUSH ROD

FLOATING
CONTROL VALVE

AIR VALVE

REACTION LEVERS

MASTER CYLINDER PISTON

MASTER CYLINDER PUSH ROD

POWER PISTON RETURN SPRING

REACTION PLATE

50-126

Power brake unit—1967-70 (© Buick Div. G.M. Corp.)

Center Cable

1. Raise car.
2. Remove jam nut and adjusting nut from equalizer.
3. Unhook connector at each end and disengage hooks and guides.
4. Install new cable by reversing removal procedure.

Rear Cable

1. Raise car.
2. Remove rear wheel and brake drum.
3. Loosen jam nut and adjusting nut at equalizer.
4. Disengage rear cable at connector.
5. Remove two bolts attaching cable assembly to backing plate. Disengage cable at brake shoe operating lever.
6. Install new cable by reversing removal procedure.

Parking Brake Adjustment

Adjustment of the parking brake is necessary whenever the rear brake cables have been disconnected or the parking brake pedal can be depressed more than eight rachet clicks under heavy foot pressure. The car should first be raised on a lift.

1. Make sure that service brakes are properly adjusted.
2. Depress parking brake pedal three rachet clicks.
3. Loosen jam nut on equalizer adjusting nut. Tighten adjusting nut until rear wheels can just be

turned rearward by hand but not forward.
4. Release rachet one click; the rear wheels should rotate rearward freely and forward with a slight drag.
5. Release rachet one more click; rear wheels should turn freely in either direction.

NOTE: be sure that the parking brake does not drag. An overtightened, dragging parking brake on a car with automatic brake adjusters will result in an extremely short life for rear brake linings.

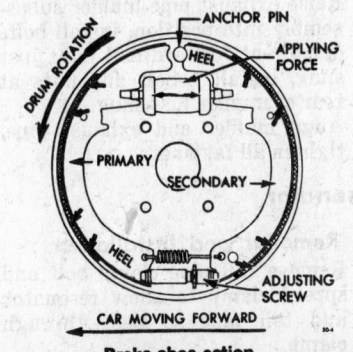

DRUM ROTATION

ANCHOR PIN

HEEL

APPLYING FORCE

PRIMARY

SECONDARY

HEEL

ADJUSTING SCREW

CAR MOVING FORWARD

Brake shoe action

Fuel System

Data on capacity of the gas tank will be found in the Capacities Table. Data on correct engine idle speed and

fuel pump pressure will be found in the Tune-Up Specifications Table.

Fuel Pump

1964-70

These models use a single action fuel pump mounted on the lower side of the engine front cover. Flexible type gas lines are used.

Electric wipers replace the vacuum type, therefore, no vacuum pump is needed.

Beginning 1966, the repairable fuel pump was discontinued. If the fuel pump is unsatisfactory, renew the unit.

1969-70 Riviera

These models have a turbine type electric fuel pump mounted at the bottom of the fuel tank. This pump maintains a steady pressure whenever the engine is running. The electrical circuit to the pump is completed by an oil pressure switch which closes at 3 psi oil pressure; the switch is bypassed for starting.

1967-70 All Engines with Air Conditioners, All 400, 430, 455 Cu. In.

All air-conditioner equipped cars have a special fuel pump with a metering outlet for a vapor return system. Hot fuel and fuel vapor is returned to the fuel tank. The fuel pump is continuously cooled by circulating fuel from the tank, thus

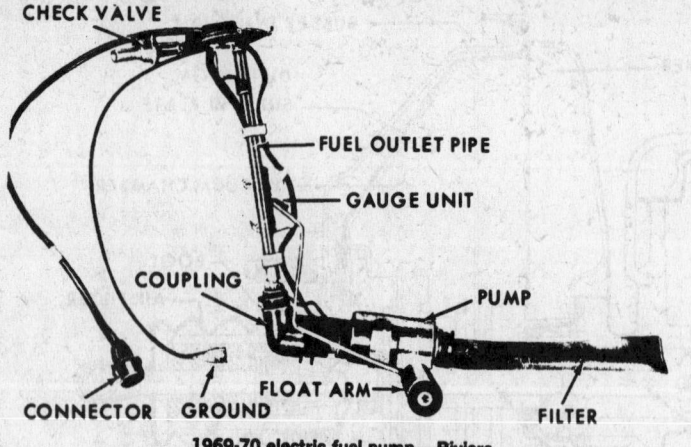

CHECK VALVE
FUEL OUTLET PIPE
GAUGE UNIT
COUPLING
PUMP
CONNECTOR   GROUND   FLOAT ARM   FILTER

1969-70 electric fuel pump—Riviera

Front or Intermediate Exhaust Pipe

Removal and Installation

1. Cut off defective exhaust pipe just forward of muffler nipple.
2. Remove U-bolt and clamp from forward muffler nipple. Disconnect exhaust pipe at forward end ball joint.
3. Split stub end of exhaust pipe in muffler, collapse stub and remove.
4. Raise new exhaust pipe into position, install bolts at forward end ball joint, install new U-bolt and clamp.
5. Align exhaust pipe and tighten all fasteners.

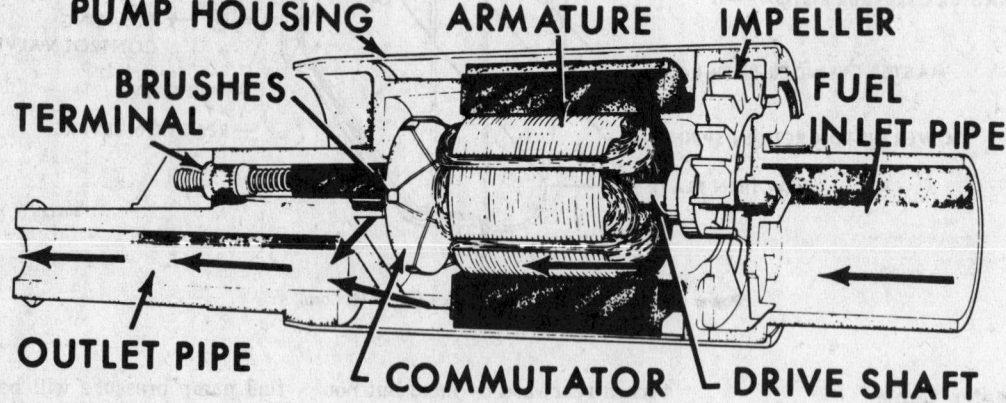

PUMP HOUSING   ARMATURE   IMPELLER
BRUSHES
TERMINAL
FUEL INLET PIPE
OUTLET PIPE   COMMUTATOR   DRIVE SHAFT

1969-70 electric pump—Riviera

greatly reducing the possibility of vapor lock.

Exhaust System

All Buick models, except Riviera, have a standard single exhaust system. The Riviera has a standard dual exhaust system. For 1964-66 models, a dual exhaust system is optional on all models except the Estate Wagon. For 1967-70, a dual exhaust system is optional on all models except the Le Sabre and the Estate Wagon. The dual exhaust system is also optional for 1970 Le Sabre models equipped with the 455 cu. in. engine.

Most exhaust system connections are of the ball joint type to allow easy alignment. The rest are slip joint connections. No gaskets are used in the entire system.

Muffler

Removal and Installation

1. Remove U-bolts and clamps from muffler inlet and outlet. Split muffler inlet and outlet nipples on opposite sides so that they are loose on inner pipes. Be careful not to damage inner pipes.

2. Disconnect front or intermediate exhaust pipe at forward end ball joint.
3. Pull exhaust pipe forward and twist to disengage from muffler. Lay exhaust pipe on floor. Remove old muffler by pulling forward and twisting.
4. Subassemble exhaust pipe and new muffler on floor, tightening U-bolt and nuts snug but not firm enough to prevent movement.
5. Raise exhaust pipe-muffler subassembly into position, install bolts or nuts at forward ball joint just snug, install U-bolt and nuts at rear of muffler just snug.
6. Align muffler and exhaust pipe, tighten all fasteners.

Resonator

Removal and Installation

1. Remove tail pipe clamp bolt and spread clamp to allow resonator and tail pipe to pass through clamp.
2. Split resonator inlet nipple on opposite sides, being careful not to damage exhaust pipe.
3. Slide old resonator to rear and then forward out of tail pipe clamp.
4. Install new resonator and tail-pipe assembly, tighten all fasteners.

Tail Pipe

Removal and Installation

1. Raise car so that rear axle and frame are separated as far as possible.
2. Cut off defective tail pipe behind muffler nipple.
3. Remove tail pipe clamp bolt and spread clamp. Remove old tail pipe.
4. Split stub end of tail pipe in muffler, collapse stub and remove.
5. Position new tail pipe over rear axle, install new U-bolt and clamp, install tail pipe hanger clamp bolt. Align tail pipe and tighten all fasteners.

Cooling System

Radiator Core Removal

1964-70

Remove the cap screws that hold the fan blades to the fan hub and take off the blades, spacer and pump pulley. Remove the top and bottom radiator hose and the two hoses which connect the oil cooler to the radiator. Remove the bolts that hold the radiator core to the cradle and lift the core straight up.

Water Pump Removal

It is possible to remove and replace the water pump on all Buicks without disturbing the radiator core. This is accomplished by removing the fan belt, fan blades, and pulley, disconnecting the hoses and removing the water pump attaching bolts.

Water Manifolds

1964-65 All Engines, 1966 401 and 425 Cu. In.

To remove the water manifold, detach the upper radiator hose and remove the two attaching bolts which hold the manifold to the front of each cylinder head. Lift the manifold straight up to free the neoprene seal in the water pump housing.

Caution On assembly use a new seal. Mounting gaskets may be coated with compound.

Thermostat Removal

1964-65 All Engines, 1966 401 and 425 Cu. In.

The thermostat is contained in the water outlet elbow mounted on the front of the water manifold.

To replace the thermostat, disconnect the upper radiator hose, remove the water outlet attaching bolts, lift off the outlet and take out the thermostat.

1966 300 and 340 Cu. In., 1967-70 All Engines

The thermostat is contained in the water outlet elbow mounted on the front of the intake manifold.

To replace the thermostat, disconnect the upper radiator hose, remove the water outlet attaching bolts, lift off the outlet and take out the thermostat.

Caution In replacing a thermostat, avoid installing the unit backwards.

Engine

In the specifications tables are listed all the available facts about the engines. Engine identification is covered prior to the Tune-Up Specifications table.

Exhaust Emission Control

1966-67

The California inspired Air Injection Reactor System (A.I.R.) is used on all General Motors cars (with a few design exceptions) for delivery into that state.

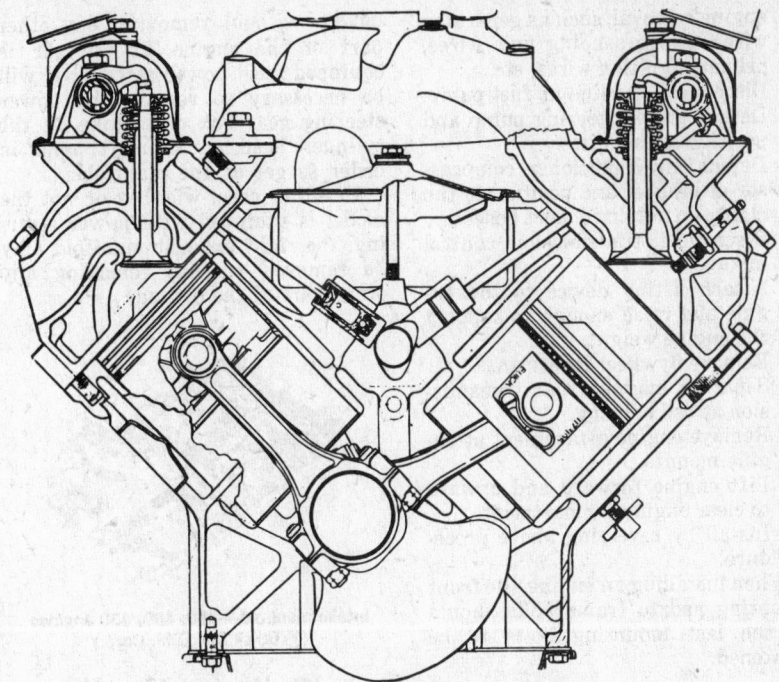

Cross-section of 401 and 425 V8 engine (© Buick Div. G.M. Corp.)

Beginning 1968

In compliance with anti-pollution laws involving all of the continental United States, the General Motors Corporation has adopted a special system of terminal exhaust treatment. This plan supersedes (in most cases) the method used to conform to 1966-67 California laws. The new system replaces (except with stick shift and special purpose engine applications) the use of the A.I.R. method previously used.

The new concept, Combustion Control System (C.C.S.) utilizes engine modification, with emphasis on carburetor and distributor changes. Any of the methods of terminal exhaust treatment requires close and frequent attention to tune-up factors of engine maintenance.

Since 1968, all car manufacturers post idle speeds and other pertinent data relative to the specific engine-car application, in a conspicuous place in the engine compartment. For details, consult the Unit Repair Section.

Valves

Detailed information on the valves and the type of valve guides used can be found in the Valve Specifications table.

A general discussion of valve clearance and a chart showing how to read pressure and vacuum gauges when using them to diagnose engine troubles will be found in the Troubleshooting Section.

Valve tappet clearance for each engine is given in the Tune-Up Specifications table.

Bearings

Detailed information on engine bearings will be found in the Crankshaft Bearing Journal Specifications table.

Engine crankcase capacities are listed in the Capacities table.

Approved torque wrench readings and head bolt tightening sequences are covered in the Torque Specifications table.

Information on the engine marking code will be found in the Engine Identification Code table.

Engine, Removal and Replacement

All Models

The reason for removal, degree of disassembly, and extent and type of shop equipment may all influence the following procedure.

1. Drain cooling system.
2. Scribe hinge outline on underside of hood. Remove hood attaching bolts and remove hood.
3. Disconnect battery cables.
4. Remove radiator and heater hoses.
5. Disconnect transmission oil cooler lines.
6. Remove attaching bolts and lift out radiator.
7. Disconnect exhaust pipe or pipes at the exhaust manifold/s.
8. Disconnect vacuum line to power brake unit.
9. Disconnect accelerator to carburetor linkage.
10. Disconnect all engine component wiring that would interfere with

engine removal, such as generator wires, gauge sending unit wires, primary ignition wires, etc.

11. Disconnect gas line at fuel pump.
12. Detach power steering pump and position to the left.
13. Detach air conditioner compressor at bracket and position to the right. Do not disconnect hoses.
14. Disconnect transmission control linkage.
15. Attach lifting device to the engine and raise enough to support the engine weight.
16. Remove flywheel cover pan.
17. Separate engine from transmission at bell housing.
18. Remove engine attachment at engine mounts.
19. Lift engine forward and upward to clear engine compartment.
20. Install by reversing above procedure.

When installing an engine, the front mounting pad to frame bolts should be the last mounting bolts to be tightened.

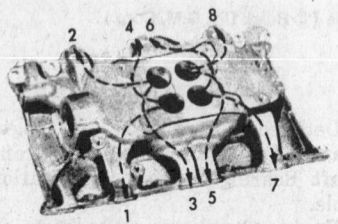

Intake manifold—300, 340 engine
(© Buick Div. G.M. Corp.)

Engine Manifolds

All Models

The intake manifold may be removed from the center of the engine block without removing any other part of the engine.

Take off the air cleaner and disconnect the vacuum and gas lines and accelerator linkage from the carburetor. Remove the air conditioning mounting bracket bolt, loosen the bracket to compressor bolts, and slide the bracket outboard. Drain the coolant. (This is unnecessary on engines with a water manifold mounted separately ahead of the intake manifold.)

The exhaust manifolds on each side may be removed, with some slight dif-

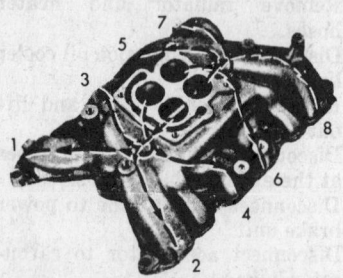

Intake manifold, 401-425 engines
(© Buick Div. G.M. Corp.)

ficulty, without removing any other part of the engine. If the car is equipped with power steering it will be necessary to remove the power steering gear box or to take off the cylinder head on the left bank in order to get at the manifold.

In either case, whether or not the model is equipped with power steering, the right exhaust manifold may be removed without removing any other part of the engine.

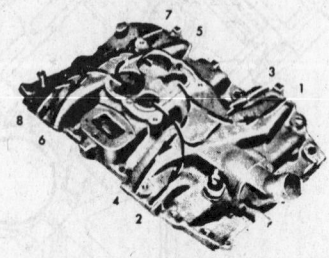

Intake manifold—350, 400, 430 engines
(© Buick Div. G.M. Corp.)

Cylinder Head

Removal

Remove the intake manifold as above.

Remove the rocker cover, then detach the exhaust manifold at the flange connection rather than at the head.

NOTE: the exhaust manifold can be disconnected from the head but this procedure takes somewhat longer than detaching it at the exhaust flange connection. Remove the Delcotron and the air conditioning compressor to remove the right head. Remove the dipstick and power steering pump to remove the left head.

Remove the rocker cover and the rocker assemblies. Mark them carefully for reassembly.

NOTE: there are no oil line connections to the rocker assemblies, since oil is fed through the rocker front bracket.

Detach the front water manifold, if applicable, from both cylinder heads, unbolt and lift off the head. It is important to prevent dirt from entering the engine. The hydraulic lifters, in particular, are very susceptible to dirt.

Valve System

Rocker Shaft and Pushrod Removal

To remove the rocker shafts, first remove the air cleaner, then the rocker cover. Then, take out the bolts which hold the rocker shaft brackets to the cylinder head.

Carefully mark the rocker shaft so that it will be returned to the same cylinder head from which it was removed.

If it is placed on the wrong head, the counter-bored bracket will not pass oil to the rocker shafts and they will very shortly wear out from lack of lubrication.

After the rocker shafts have been taken off, the pushrods can be removed from their bores without removing the cylinder head. Keep the pushrods in order, so that each can be returned to its original location.

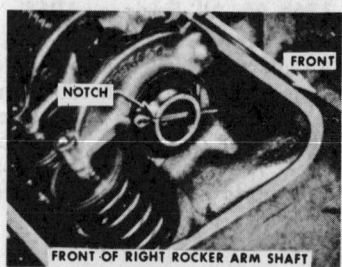

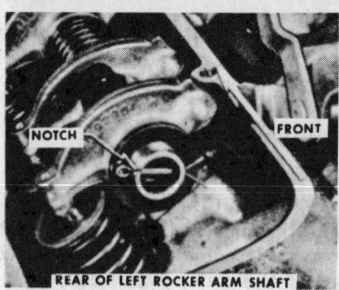

Rocker arm and shaft—300, 340, 1964-67
(© Buick Div. G.M. Corp.)

Rocker Shaft Lubrication

Oil is fed through the front rocker shaft bracket on both cylinder heads. The front bracket has an oversize bore which permits oil to pass around the outside of the bolt up to the hollowed out rocker shaft.

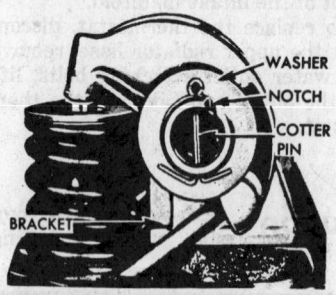

Rocker arm and shaft—401 and 425
(© Buick Div. G.M. Corp.)

Valve Guide Replacement

1964-65

Remove the cylinder head and the valves and valve spring assemblies. On engines equipped with removable guides carefully measure the amount the valve guide protrudes from the cylinder head before driving it out so that the new guide can be driven down exactly that amount. Make a stack of washers equivalent to the protrusion of the guide before removing the guide from the head.

When driving out the valve guides, support the cylinder head as near to the valve guide as is practical.

A pilot type driver should be used and the guide should be driven out from the bottom, or, it may be pressed out if an arbor press is available.

Place the new guide in the top of the head and tap it gently to insure that it is starting straight. Once started straight, it can be driven into position.

When the new valve guide has been driven in the correct distance, insert a new valve into the guide to make sure that the valve will operate freely up and down. The slightest sign of binding in the new valve guide means that the guide itself has become riveted over or slightly warped in the driving process and will have to be reamed.

Buick removable valve guides should be finish-reamed to size, after installation. Use a reamer that will give the valve stem to guide clearance listed in the Valve Specifications table.

NOTE: beginning 1965 the 300 cu. in. and the 340 cu. in. engines have valve guides that are cast integrally with the cylinder head. Since 1967, all guides are integral with the cylinder head.

In cases of excessive guide-to-valve stem clearance, the guides can be reamed and oversize valves installed. .004 in. oversize valves can be installed in the 300 and 340 cu. in. engines and .010 in. oversize valves can be installed in the 350, 400, 430, and 455 cu. in. engines.

Always reface the valve seat when new guides have been installed to be absolutely certain that the valve seat is in alignment with the new guide.

DRIVE OLD GUIDE OUT FROM COMBUSTION CHAMBER SIDE

DRIVE NEW GUIDE IN FROM TOP SIDE OF HEAD

Removing and installing valve guides
(© Buick Div. G.M. Corp.)

Valve Springs

To check the condition of the valve springs, line up the intake valve springs on a flat surface and, using a straightedge, compare the height of the springs. If all of the springs are the same height, as determined by the straightedge, it may be assumed that the springs are in good condition, since it is very unlikely that all of

the springs would collapse the same amount.

If one or more of the springs are lower than the rest it is advisable to procure at least one new spring and then compare the other springs with the new one for free length.

Replace all springs that do not come up to the standard established by the new one.

Repeat the operation on the exhaust valve springs.

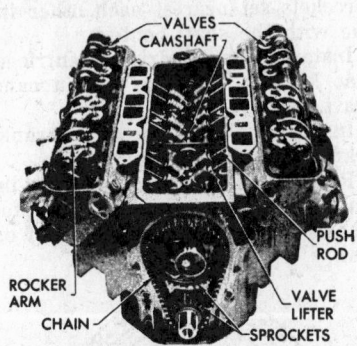

**Valve component locations—
1964-66 401 and 425 cu. in. engines**
(© Buick Div. G.M. Corp.)

The following is a method for replacing valve springs, oil seals or spring retainers without removing the cylinder head.

1. Entirely dismantel a spark plug and save the threaded shell.
2. To this shell, braze or weld an air chuck.
3. Remove the valve rocker cover. Remove the rocker arm from the affected valve.
4. Remove the spark plug from the affected cylinder.
5. Turn the crankshaft to bring the piston of this cylinder down, away from possible contact with the valve head. Sharply tap the valve retainer to loosen the valve lock.
6. Then, turn the crankshaft to bring the piston in this cylinder to the exact top of its compression stroke.
7. Screw in the chuck-equipped spark plug shell.
8. Hook an air hose to the chuck and turn on the pressure (about 200 psi).
9. With a strong and constant supply of air holding the valve closed, compress the valve spring and remove the lock and retainer.
10. Make the necessary replacements and reassemble.

NOTE: it is important that the operation be performed exactly as stated, in this order. The piston in the affected cylinder must be on exact TDC to prevent air pressure from turning the crankshaft.

Valve Removal

Remove the air cleaner, the rocker cover, the rockers and the intake manifold. Disconnect the exhaust manifolds at their flanges, leaving the manifold attached to the heads.

From the right bank, remove the generator mounting bracket, and from the left, the power steering pump. Disconnect the heat indicator, remove fuel and vacuum lines. Remove the bolts that hold the water manifolds, if applicable, to the cylinder heads, unbolt and remove the cylinder heads. Take the heads to a

bench and, using a C-type or lever type valve spring compressor, compress the valve springs, remove the keepers, release the valve springs, and push the valves to the combustion chamber side of the head.

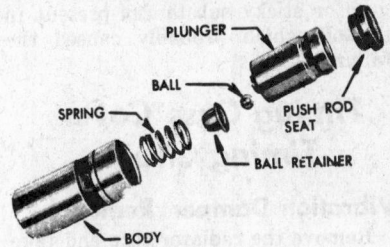

Hydraulic valve lifter parts
(© Buick Div. G.M. Corp.)

Hydraulic Lifters, Removal

To remove the lifters, remove the rocker cover and take off the rocker shaft assemblies and lift out the pushrods. Then remove the intake manifold.

The valve chamber cover plate can then be removed giving access to the lifters.

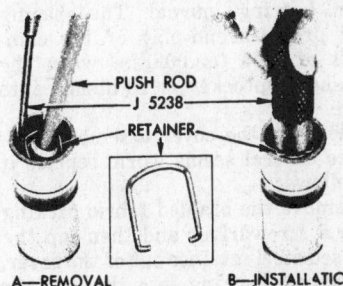

Removing and installing plunger retainer
(© Buick Div. G.M. Corp.)

The lifters are barrel type which come right up out of their bores requiring no other tools than the fingers.

If more effort is required than can be given by the fingers, it indicates gum or sticky substances present in the oil which probably caused the failure.

Timing Case Cover Timing Chain

Vibration Damper Removal

Remove the radiator core and take out the cap screws that hold the fan pulley to the vibration damper. Remove the large bolt from the center of the crankshaft and insert a bolt type puller into the holes which held the fan pulley. Pull off the vibration damper.

Timing Chain and Front Oil Seal Replacement

1964-66 401 and 425 Cu. In.

Drain cooling system and remove radiator core, shroud, fan belt, fan and pulley and vibration damper.

Remove all bolts holding timing gear cover and the water manifold to the engine block and cylinder heads. Do not remove the five small bolts holding the water pump to the chain cover.

Remove water-manifold-timing-chain-cover assembly, being careful not to tear the oil pan gasket.

Remove oil slinger from crankshaft and remove the bolt, lockwasher and plain washer holding fuel-pump-drive eccentric and camshaft sprocket to camshaft.

If there has been doubt about the valve timing, turn the crankshaft so that the camshaft sprocket keyway points down. The "O" marks on the sprockets shculd be set to be nearest each other inline with the shaft centers.

Remove camshaft sprocket and timing chain.

End thrust of the camshaft is taken up by a thrust plate fastened to the block behind the sprocket. End-play of the camshaft is controlled by a spacer ring just behind the thrust plate and in front of the camshaft front bearing journal. The spacing ring provides end-play of the camshaft of .004 to .008 in. when the camshaft sprocket is tightened into place.

Clean up the cover assembly, and, if the oil seal seems worn, replace it as follows:

Remove the braided fabric packing with a screwdriver and then tap the pressed steel retainer out of the cover. Work new packing into the retainer and drive the retainer into the recess. The packing should expand slightly as it seats. Install so that joint between ends of packing is toward top of engine. Smear the seal with vaseline.

When ready to install the chain turn the crankshaft, if it has been turned since chain was removed, so that pistons No. 1 and No. 4 are on top dead center. Turn camshaft so that sprocket keyway points down.

Place chain and sprocket back in place with the "O" marks on the sprockets set nearest each other inline with the shaft centers.

Install fuel pump drive eccentric so that keyway fits over key in camshaft. Fasten all in place.

Install oil slinger on end of crankshaft with hollow side outward.

Reverse procedure to complete installation. Keep engine speed low for a short while after installation of oil seal.

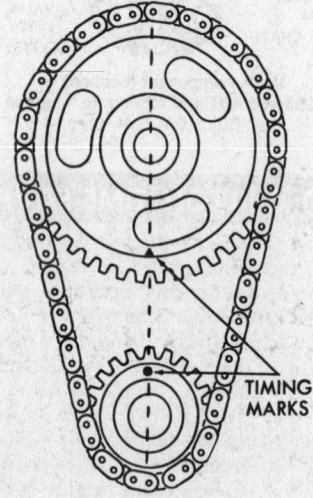

Timing chain and sprocket marks

1965-70 300, 340, 350 Cu. In.

1. Drain cooling system and remove radiator, shroud, fan, pulleys, and belts.
2. Remove crankshaft pulley, fuel pump and distributor.
3. Remove Delcotron and power steering pump, if necessary.
4. Loosen and slide rearward front clamp on thermostat by-pass hose. Remove harmonic balancer.
5. Remove bolts attaching timing chain cover to cylinder block and oil pan to timing chain cover bolts. Remove timing chain cover assembly and gasket. Clean cover thoroughly, being careful not to damage the gasket surface.
6. Turn the crankshaft so that the timing marks on the sprockets are adjacent to each other on a line with the shaft centers.
7. Remove crankshaft oil slinger.
8. Remove bolt, special washer, distributor drive gear, and fuel pump eccentric from camshaft.
9. Pry camshaft and crankshaft sprockets forward until camshaft sprocket is free. Then remove both sprockets and chain.

If oil seal appears worn or has been leaking, replace as follows:

10. Use a punch to drive out the old seal and retainer. Drive from front to rear of the timing chain cover.
11. Coil new packing around opening so that ends are at top. Drive in retainer. Stake the retainer in at least three places. Size the packing by rotating a hammer handle, etc. around the packing until the balancer hub fits through the packing.

If engine has been disturbed since chain and sprockets were removed:

12. Turn crankshaft until No. 1 piston is at top dead center.
13. Mount sprocket temporarily and turn camshaft so that timing mark is straight down.
14. Assemble chain and sprockets and mount on shafts with their timing marks closest to each other.
15. Mount slinger on sprocket with the concave side to the front.
16. Reinstall fuel pump eccentric, distributor drive gear, special washer, and bolt on camshaft. Reinstall Woodruff key with oil groove forward.
17. Remove oil pump cover and pack the space around the oil pump gears full of petroleum jelly, leaving no air spaces. Reinstall oil pump cover with new gasket. This step is very important. If it is not done the oil pump will not begin to pump oil as soon as the engine is started.
18. Reinstall timing chain cover with new gasket.

Keep engine speed low for a short time after installation of a new oil seal.

1967-70 400, 430, 455 Cu. In.

This procedure is identical to that outlined above for 1965-70 300, 340, and 350 cu. in. engines with the substitution of the following steps:

8. Remove oil pan. Remove camshaft sprocket bolts.
16. Reinstall oil pan. Reinstall camshaft sprocket bolts.

Engine Lubrication

Oil Pan Removal

The following procedures apply in general to all models:

1. Disconnect battery ground strap.
2. Raise car and support on stands.
3. Drain oil.
4. Disconnect or loosen shift linkage.
5. On manual transmission equipped models, loosen clutch equalizer bracket to frame mounting bolts

and disconnect exhaust crossover pipe at engine.

Depending on the individual car model, it may be necessary to do one or more of the following:

1. Remove lower flywheel housing.
2. Disconnect idler arm at frame and push steering linkage forward.
3. Remove fan shroud to radiator tie bar bolts.
4. Remove engine mounting bolts and raise engine by jacking under the crankshaft pulley mounting. When this is done on air conditioned cars, it will be necessary to support the right side of the engine-transmission as-

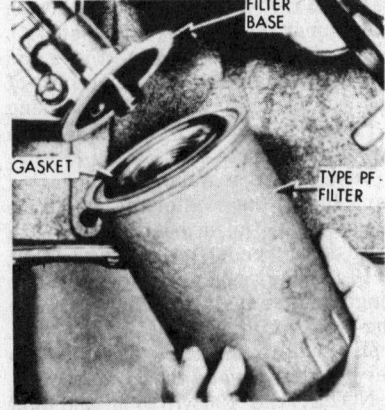

Oil filter installation (© Buick Div. G.M. Corp.)

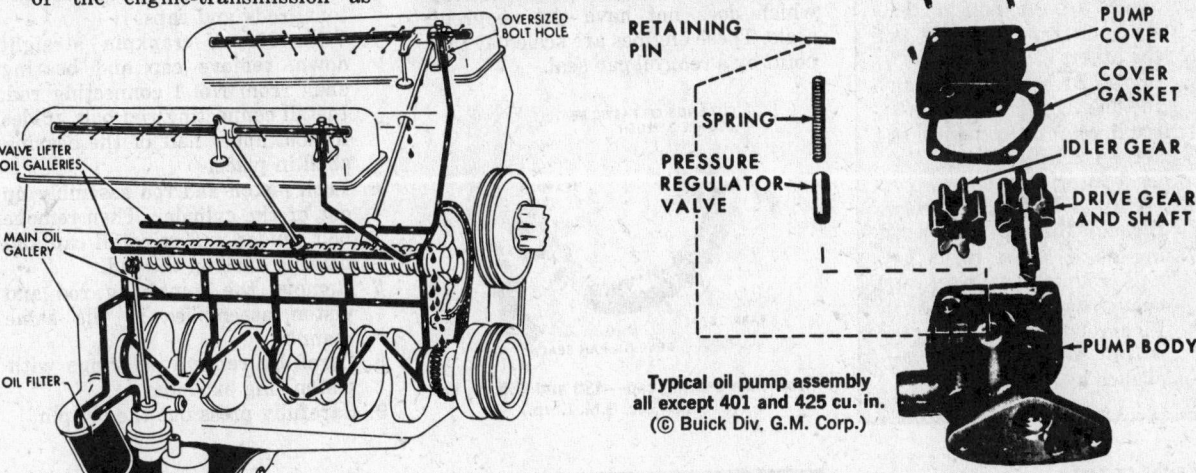

Engine lubrication—401 and 425 cu. in. engine (© Buick Div. G.M. Corp.)

Typical oil pump assembly all except 401 and 425 cu. in. (© Buick Div. G.M. Corp.)

sembly due to the off-center weight of the compressor.

5. Rotate crankshaft slightly while lowering pan to clear counterbalances.

Oil Pump Removal

With the exception of the 1964-66 401 and 425 cu. in. engines, the oil pump is located in the timing chain cover on the right-hand side. It is connected by a drilled passage in the crankcase to an oil screen housing and pipe assembly. The screen is submerged in the oil supply in the oil pan.

The 401 and 425 cu. in. engines have the oil pump bolted to the crankcase and contained in the oil pan. To gain access to the oil pump, the oil pan must be removed.

Oil Filter

1964-67

A screw-off, disposable element and can-type filter is used. The filter should be changed at 6,000 miles or six-month intervals, whichever comes first.

1968-70

The filter should be changed at every other oil change. Oil changes

should be made every four months or 6,000 miles, whichever occurs first.

Oil Filter Replacement

1. Coat the gasket on the new filter with oil.
2. Place the new filter in position on the block.
3. Hand tighten until contact is made between the filter gasket and the adapter face.
4. Tighten by further turning the filter two-thirds turn.
5. Run the engine at fast idle and check for oil leaks.
6. Check the oil and bring crankcase to level if necessary.

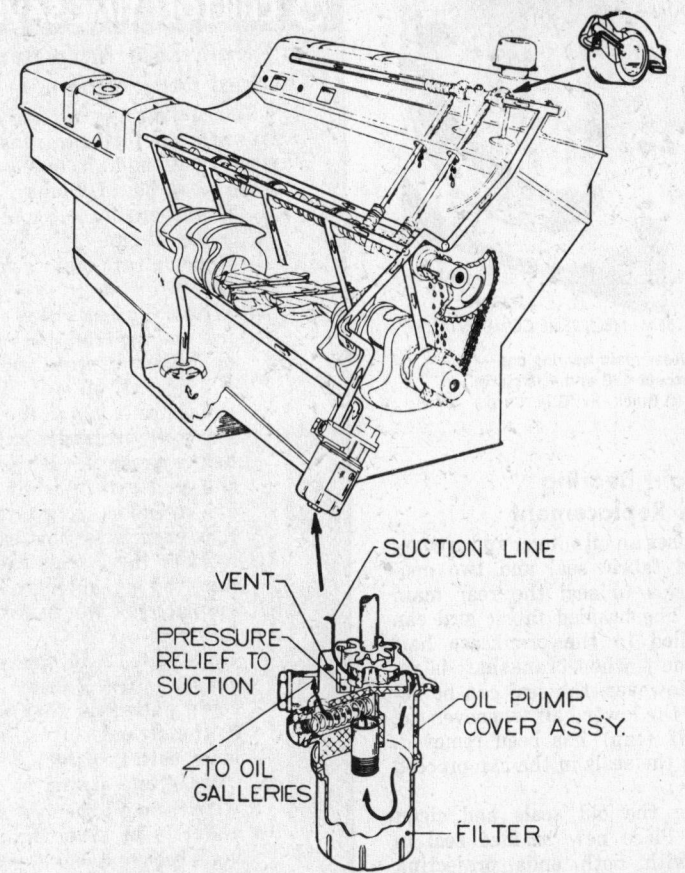

Oil flow—except 401 and 425 cu. in. engine (© Buick Div. G.M. Corp.)

⏱ CHILTON TIME-SAVER

A few cases of difficulty in oil filter replacement have come to our attention. The trouble starts with oil filter elements being turned on too tightly. The unit may be too tight to remove by hand and it may collapse in the grip of a tool that applies enough squeeze to grip the element hard enough to turn it.

1. Raise the car on a jack or hoist and place a drip pan under the filter.
2. With a 12 to 14 in. slender punch drive a hole in the element from one side to the other.
    NOTE: before punching the hole, consider the angle required for the punch to act as a lever, with the least interference.
3. With the drift all the way through the filter and acting as a lever, turn the unit counterclockwise far enough to break it loose.
4. Final loosening and removal can now be accomplished by hand.

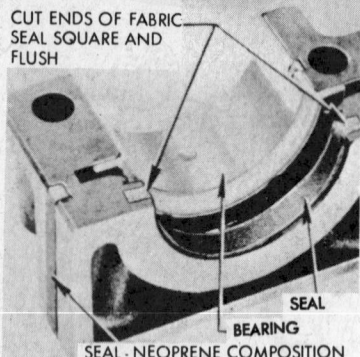

CUT ENDS OF FABRIC SEAL SQUARE AND FLUSH

SEAL

BEARING

SEAL - NEOPRENE COMPOSITION

Rear main bearing cap—except 430 and 455 cu. in.
(© Buick Div. G.M. Corp.)

Rear Main Bearing Oil Seal Replacement

Buick uses an oil slinger and groove, a braided fabric seal and two neoprene strips to seal the rear main bearing. The braided fabric seal can be installed in the crankcase half (upper) only when crankshaft is removed. However, the seal can be replaced in the lower half whenever the lower half (cap) has been removed. To renew the seals in the cap proceed as follows:

Remove the old seals and clean the cap. Place new braided seal in groove with both ends projecting above parting surface of cap. Force

seal into groove by rubbing down with a hammer handle or other smooth tool until seal is seated in groove and ends project above the parting face of the cap not more than 1/16 in. Using a razor blade, cut off ends flush with parting surface.

Just before installing the bearing cap, lightly lubricate the neoprene seals and install in bearing cap with the upper ends protruding about 1/16 in. The seals must not be cut to length. After installing the cap, force the seals up into the cap with a blunt instrument to insure a seal at the line between the cap and the case.

NOTE: the 1968-70 430 and 455 cu. in. engines use a rear bearing cap which does not have the neoprene seals. These engines are sealed at this point by a rear oil pan seal.

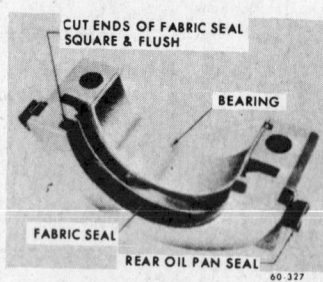

CUT ENDS OF FABRIC SEAL SQUARE & FLUSH

BEARING

FABRIC SEAL

REAR OIL PAN SEAL

60-327

Rear main bearing cap—430 and 455 cu. in.
(© Buick Div. G.M. Corp.)

⏱ CHILTON TIME-SAVER

Top Half, Rear Main Bearing Oil Seal Replacement

The following method has proven a distinct advantage in most cases and, if successful, saves many hours of labor.

1. Drain engine oil and remove oil pan.
2. Remove rear main bearing cap.
3. With a 6 in. length of 3/16 in. brazing rod, drive up on either exposed end of the top half oil seal. When the opposite end of the seal starts to protrude, have a helper grasp it with pliers and pull gently while the driven end is being tapped. It is surprising how easily most of these seals can be removed by this method.

To replace the woven fabric-type seal:

1. Obtain a 12 in. piece of copper wire (about the same gauge as that used in the strands of an insulated battery cable).
2. Thread one strand of this wire through the new seal, about ½ in. from the end, bend back and make secure.
3. Thoroughly saturate the

new seal with engine oil.
4. Push the copper wire up through the oil seal groove until it comes down on the opposite side of the bearing.
5. Pull (with pliers) on the protruding copper wire while the crankshaft is being turned and the new seal is slowly fed into place.

CAUTION: this snaking operation slightly reduces the diameter of the new seal and care will have to be used to keep the seal from slipping too far through the top half of the bearing.

6. When an equal amount of seal is extending from each side, cut off the copper wire close to the seal and tamp both ends of the seal up into the groove (this will tend to expand the seal again).

NOTE: don't worry about the copper wire left in the groove, it is too soft to cause damage.

7. Replace the seal in the cap in the usual way and replace the oil pan.

Caution The engine must be operated at slow speed when first started after installation of new braided seals.

Connecting Rods And Pistons

Piston Assembly Removal

1. Remove cylinder heads.
2. Remove oil pan.
3. Examine cylinder bores for top ridge. If ridge exists, remove it before taking pistons out.
4. Mark cylinder numbers on all pistons, rods and caps.
5. With No. 1 crankpin straight down, remove cap and bearing shell from No. 1 connecting rod. Install connecting rod bolt guides to hold upper half of the bearing shell in place.
6. Push piston and rod assembly up out of the cylinder. Then remove bolt guides and reinstall cap and bearing shell on the rod.
7. Remove the remaining rod and piston assemblies in the same manner.
8. Carefully remove old rings with piston ring expander.
9. Carefully press out the old pin.

NOTE: check the cylinder bores for out-of-round, taper or other damage. Any cylinders requiring attention may be bored or honed the same as any conventional cast iron cylinder block.

Fitting Rings and Pins

When new rings are installed without reboring the cylinders, cylinder wall glaze should be broken. This can be done by using the finest grade stones in a cylinder hone.

New piston rings must be checked for clearance in cylinder bores and for gap.

When fitting new rings to new pistons the side clearance for compression rings should be .003 to .005 in. Side clearance of the oil ring should be .0035 to .0095 in.

Check end gap of compression rings by placing them in the bore in which they will operate. Then, push them to the bottom of the bore with a piston. Now, measure the end-gap in each ring. The end-gap should be no less than .015 in. for all engines from 1964-67. For 1968-70 models end gap should be no less than .010 in. for the 350 cu. in. engine and .013 in. for the 400, 430, and 455 cu. in. engines. End gap for oil rings for 1968-70 engines should be no less than .015 in. Oil ring end gap need not be measured on 1964-67 engines.

If piston pin bosses are worn out of round or oversize, the piston and pin should be replaced. Oversize pins are not practical because the pin is a press fit in the connecting rod. Piston pins must fit the piston with an easy finger push at 70° F.

In assembling the piston to the connecting rod, a press is ideal. However, substitutes are available that will serve the purpose.

When the rod assemblies are replaced in the engine, the connecting rod bearing oil spurt hole must point up toward the camshaft.

Connecting Rod Bearings

1. Remove connecting rod cap with bearing shell. Wipe all oil from the bearing area.
2. Place a piece of Plastigage lengthwise along the bottom center of the lower bearing shell. Then install cap with bearing shell and torque the bolt nuts to the proper torque from Torque Specifications Chart.
   NOTE: do not turn crankshaft.
3. Remove the cap and shell. The gauging material will be flattened and adhering to either the bearing shell or the crankpin. Do not remove it.
4. Using the scale that comes with the gauge, measure the flattened gauge material at its widest point. The number within the gradua-

tion which comes closest to the width of the gauging material indicates the bearing to crankpin clearance in thousandths of an inch.
5. Desired clearance for a new bearing is .0002 to .0023 in. If the bearing has been in service, it is wise to install a new bearing if clearance exceeds .003 in.
6. If a new bearing is required, try a standard, then each undersize bearing in turn until one is found that is within specifications.
7. With the proper bearing selected, clean off the gauging material, reinstall the bearing cap and torque to specified torque.
8. After the bearing cap has been torqued, it should be possible to move the connecting rod back and forth on the crankpin, the extent of end clearance.

Piston Assembly Installation

1. Carefully assemble the piston to the connecting rod, press in the pin.
2. Remove piston and rod from the press. Rock the piston on the pin to be sure pin or piston boss was not damaged during the pressing operation.
3. Install ring expander in lower ring groove. Position the ends of the expander above the piston pin where groove is not slotted. The ends of the expander must butt together.
4. Install oil ring rails and rail spacer over expander with gaps up on same side of piston as oil spurt hole in connecting rod.
5. Install compression rings (with a ring expander) in top and center groove.
6. Coat all bearing surfaces, rings and piston skirt with engine oil.
7. Position the crankpin of the cylinder being worked on down.
8. Remove connecting rod bearing cap and with upper bearing shell correctly seated in the rod, install connecting rod bolt guides.
9. Make sure the gaps in the two oil ring rails are up toward the center of the engine. Make sure the gaps of the compression rings are not in line with each other or the oil ring rails. Be sure the ends of the oil ring spacer-expander are butted and not overlapped.
10. With a good ring compressor, install the piston and rod assembly into the cylinder bore and carefully tap down until the rod bearing is solidly seated on the crankpin.
11. Remove the connecting rod bolt guides and install cap and lower bearing shell. Torque to specified torque.
12. Install other piston and rod as-

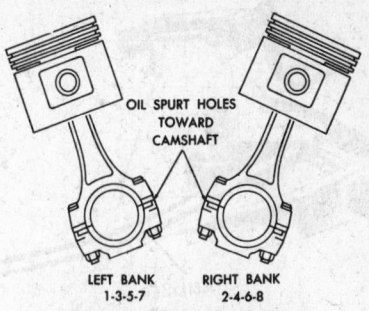

Piston and rod assembly— 300-340, 400-430-455

semblies in the same manner. When the assemblies are all installed, the oil spurt holes will be up. The rib on the edge of the rod cap will be on the same side as the conical boss on the connecting rod web. These marks will be toward the other connecting rod on the same crankpin.

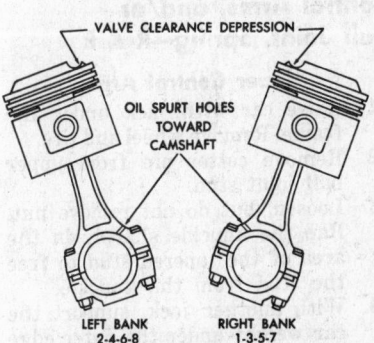

Piston and rod assembly—401-425

13. Accumulated end clearance between rod bearings on any crankpin should be .005 to .012 in.
14. Install oil screen and oil pan.
15. Install cylinder heads.

After starting the engine, avoid high speeds but do not run on slow idle for a while. A better break-in speed is about 800-1,000 rpm. for the first hour.

Front Suspension

General instructions covering the front suspension and how to repair and adjust it, together with information on installation of front wheel bearings and grease seals, are given in the Unit Repair Section.

Definitions of the points of steering geometry are covered in the Unit Repair Section. This article also covers troubleshooting front-end geometry and irregular tire wear.

Figures covering the caster, camber, toe-in, king pin inclination, and turning radius can be found in the Wheel Alignment table.

Tire size figures can be found in the General Chassis and Brake Specifications table.

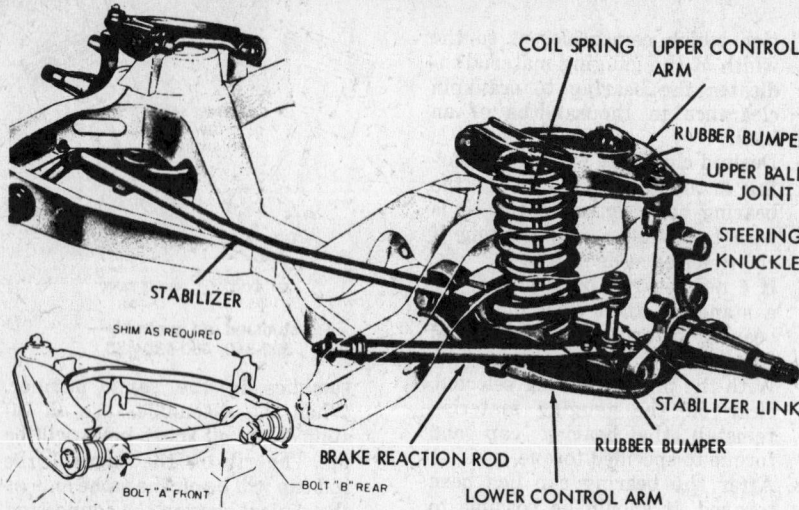

COIL SPRING   UPPER CONTROL ARM

RUBBER BUMPER

UPPER BALL JOINT

STEERING KNUCKLE

STABILIZER

SHIM AS REQUIRED

STABILIZER LINK

RUBBER BUMPER

BRAKE REACTION ROD

BOLT "A" FRONT

BOLT "B" REAR

LOWER CONTROL ARM

Typical front suspension—1964-70 (© Buick Div. G.M. Corp.)

Control Arms, and/or Ball Joint, Spring—R & R

Upper Control Arm

1. Raise car with jack under the frame. Remove wheel and tire.
2. Remove cotter pin from upper ball joint stud.
3. Loosen, but do not remove nut. Rap the knuckle sharply in the area of the tapered stud to free the stud from the knuckle.
4. With another jack, support the car weight under the outer edge of the lower control arm. Raise jack enough to free upper control arm from upper ball stud.
5. Wire brake and knuckle in place to prevent brake hose damage, then, lift upper arm from knuckle.
6. Remove the upper control arm shaft-to-bracket nuts and lock washers. Carefully note the number, thickness, and location of the adjusting shims. Remove control arm assembly.
7. The upper control arm is serviced only as an assembly. Therefore, if the arm is bent, bushings worn, or the control arm shaft bent, the entire assembly must be replaced.

Lower Control Arm, or Spring

1. Proceed as with upper control arm, Step 1.
2. Disconnect and remove shock absorber.
3. Remove front stabilizer rod link from lower control arm.
4. Disconnect brake reaction rod from lower control arm but leave it attached to the front frame crossmember.
5. Remove control arm bumper.
6. As a safety precaution and to gain maximum leverage, place a jack about ½ in. below the lower ball joint stud. Now, remove the ball stud cotter pin and loosen

the nut about ⅛ in. Do not remove the nut.
7. Rap the steering knuckle in the area of the stud to separate the stud from the knuckle.
8. After the stud has broken loose from the knuckle, raise the jack against the control arm. Remove nut and separate the steering knuckle from the tapered stud.
9. Carefully lower jack under the control arm and release the spring. With the jack entirely lowered, it may be necessary to pry the spring off its seat on the lower control arm with a pry bar.
10. After the spring is removed, the lower control arm may be removed by removing the lock nut attaching the control arm to the frame.
11. Install by reversing removal procedure.

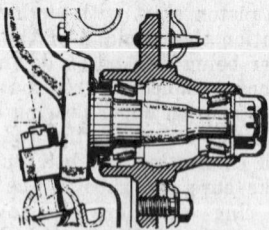

Front wheel bearing—1964-70 (© Buick Div. G.M. Corp.)

Front Wheel Bearing Adjustment

1964-70

Front wheels are now suspended upon tapered roller bearings. Adjustment of freshly cleaned and repacked roller bearings is as follows:

1. Torque spindle nut to 19 ft. lbs. while rotating the wheel.
2. Back off the nut until bearings are loose.
3. Retorque spindle nut to 11 ft. lbs. while rotating the wheel.

4. If either cotter pin hole in spindle lines up with nut castellations, back off the nut one-twelfth turn and install cotter pin. Otherwise, back off the nut to the first position that will accept an horizontal or vertical cotter pin.
5. Install cotter pin and lock spindle nut into position.

Jacking and Hoisting

Jack car at front spring seat of lower control arm or center of crossmember.

Jack car at rear, at axle housing.

To lift at frame, use side rails in front of body floor pan and at rear side rail at lower control arm front pivot.

Steering Wheel

Steering Wheel Removal

On all models of Buick the steering wheel can be removed readily after the horn actuator has been taken off. Be sure to disconnect wire at horn cable connector to prevent horn from blowing.

Caution In all cases a steering wheel puller should be used to remove the steering wheel since prying or driving at the wheel will result in damage.

Before the steering wheel is removed mark it carefully so that it can be assembled readily in the same position from which it was removed.

On most models a blind spline is used to insure that the wheel will only go on in one position.

NOTE: to align steering wheel, see Unit Repair Section.

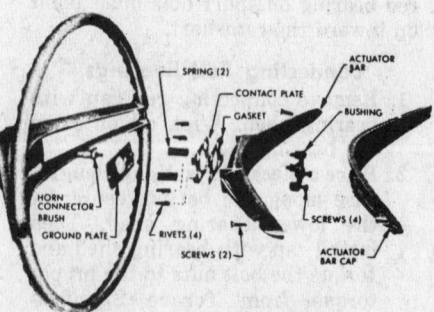

SPRING (2)

ACTUATOR BAR

CONTACT PLATE

GASKET

BUSHING

HORN CONNECTOR BRUSH

GROUND PLATE

RIVETS (4)

SCREWS (2)

SCREWS (4)

ACTUATOR BAR CAP

Steering wheel and horn contacts (© Buick Div. G.M. Corp.)

Horn Actuator Removal

1964-67

Remove screws securing horn actuator from back of steering wheel, remove actuator bar, springs, ground plate, and bushing and screw assembly.

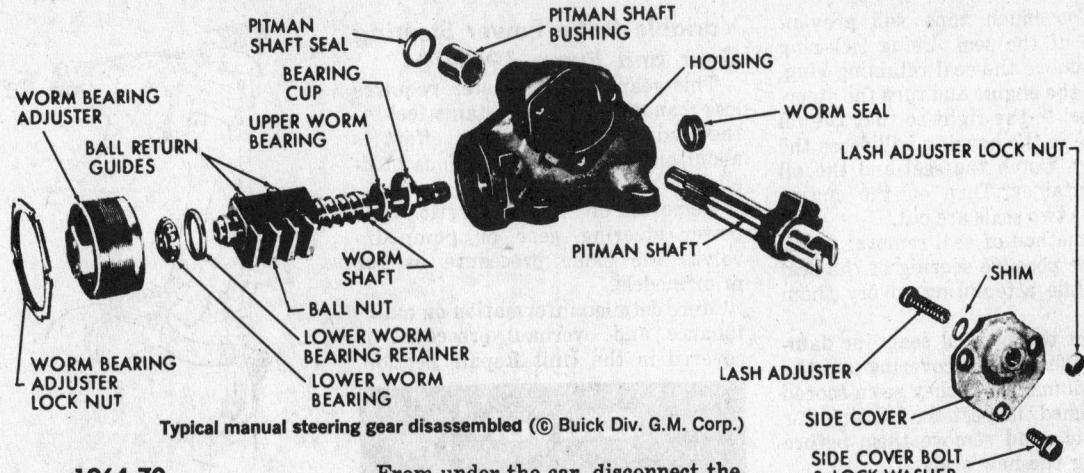

PITMAN SHAFT SEAL
BEARING CUP
UPPER WORM BEARING
PITMAN SHAFT BUSHING
HOUSING
WORM SEAL
WORM BEARING ADJUSTER
BALL RETURN GUIDES
LASH ADJUSTER LOCK NUT
PITMAN SHAFT
SHIM
WORM SHAFT
BALL NUT
LOWER WORM BEARING RETAINER
LOWER WORM BEARING
WORM BEARING ADJUSTER LOCK NUT
LASH ADJUSTER
SIDE COVER
SIDE COVER BOLT & LOCK WASHER
90-18

Typical manual steering gear disassembled (© Buick Div. G.M. Corp.)

1964-70

Remove screws securing horn actuator from back of steering wheel, partially lift off actuator, pull lead connector from cancelling cam, then fully lift off actuator.

Steering Gear

Manual

Instructions covering the overhaul of the steering gear will be found in the Unit Repair Section.

Manual Steering Gear Removal

1964-70

The steering gear mechanism is in two pieces coupled together with a flexible coupling.

To remove the lower gear box, disconnect the coupling by removing the pinch bolt securing the flexible coupling flange to the steering gear stub shaft.

From under the car, disconnect the steering pitman arm from the pitman shaft and remove the bolts at outside of left frame side rail which hold the gear housing to the frame, and lower the gear assembly to the floor.

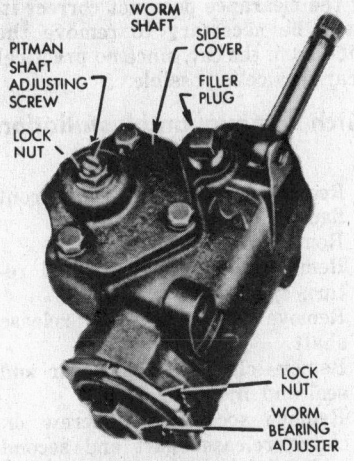

WORM SHAFT
SIDE COVER
FILLER PLUG
PITMAN SHAFT ADJUSTING SCREW
LOCK NUT
LOCK NUT
WORM BEARING ADJUSTER

Steering gear adjustment location (© Buick Div. G.M. Corp.)

When reinstalling, tighten pitman arm nut to 140 ft. lbs.

Steering Idler Arm

The idler arm bracket is held to the right frame side rail by two cap screws. Remove the cap screws and lower the tie rods. Unscrew the bracket from the bushing and then take the bushing out of the idler arm.

Bracket and bushing are usually replaced in sets.

Power

Troubleshooting and repair instructions covering power steering gear are given in the Unit Repair Section.

Replacement of Pitman Shaft Seals with Power Steering Gear Still in Place

Disconnect pitman arm from pitman shaft. Clean end of pitman shaft and housing. Tape the splines of the pitman shaft to keep them from cutting the seal. Use only one layer of

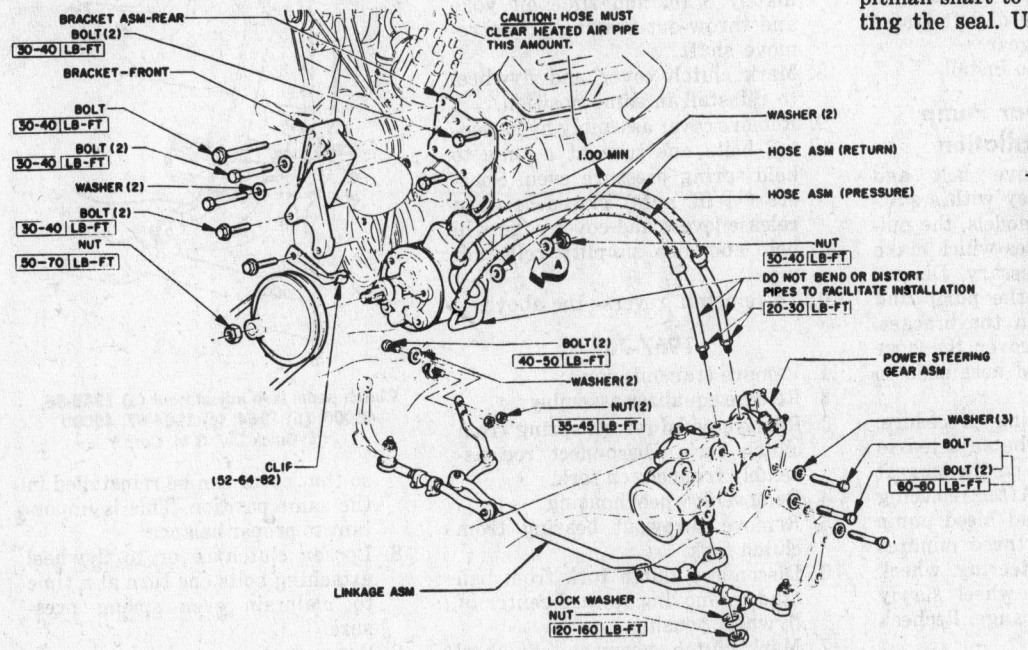

BRACKET ASM-REAR
BOLT (2)
30-40 LB-FT
BRACKET-FRONT
BOLT
30-40 LB-FT
BOLT (2)
30-40 LB-FT
WASHER (2)
BOLT (2)
30-40 LB-FT
NUT
50-70 LB-FT
CAUTION: HOSE MUST CLEAR HEATED AIR PIPE THIS AMOUNT.
1.00 MIN
WASHER (2)
HOSE ASM (RETURN)
HOSE ASM (PRESSURE)
NUT
30-40 LB-FT
DO NOT BEND OR DISTORT PIPES TO FACILITATE INSTALLATION
20-30 LB-FT
BOLT (2)
40-50 LB-FT
WASHER (2)
NUT (2)
35-45 LB-FT
CLIP
(52-64-82)
LINKAGE ASM
LOCK WASHER
NUT
120-160 LB-FT
POWER STEERING GEAR ASM
WASHER (3)
BOLT
BOLT (2)
60-60 LB-FT

STEERING LINKAGE
THE PITMAN SHAFT NUT MUST BE FLUSH OR ABOVE THE END OF THE PITMAN SHAFT WHEN TIGHTENED TO PROPER TORQUE.
CAUTION: TO PREVENT DAMAGE TO GEAR, USE PITMAN SHAFT NUT TO ASSEMBLE PITMAN ARM TO PITMAN SHAFT. USE PULLER FOR REMOVAL.
TIE ROD END NUTS MUST BE PULLED UP TO TORQUE & TIGHTENED TO NEAREST SLOT FOR INSERTION OF COTTER PIN. THE NUT MUST NEVER BE BACKED OFF TO INSERT COTTER PIN.

Power steering system (© Buick Div. G.M. Corp.)

tape. Too much tape will prevent passage of the seal. Using lock-ring pliers remove the seal retaining ring.

Start the engine and turn the steering wheel to the right so that the oil pressure in the housing will force the seals out. Catch the seal and the oil in a container. Turn off the engine when the two seals are out.

This method of seal removal eliminates the possible scoring of the seal seats while attempting to pry them out.

Inspect the two old seals for damage to the rubber covering on the outside diameter. If they seem scored or scratched, inspect the housing for burrs, etc., and remove them before installing the new seals.

Lubricate the two new seals with petroleum jelly. Put the one with a single lip in first, then insert a washer and drive seal in far enough to permit installation of double lip seal, washer and the seal retaining ring. The first seal is not supposed to bottom in its counterbore.

Fill reservoir to proper level, start engine, allow to idle for at least three minutes without turning steering wheel, turn wheel to right and check for leaks.

Remove the tape and reinstall the pitman arm. Tighten nut to 140 ft. lbs.

Power Steering Gear Removal

Disconnect the hoses at the steering gear and elevate their ends above the pump to prevent oil from draining out of the pump.

Remove pinch bolt securing flexible coupling to steering gear stub shaft.

Working under the car, remove pitman shaft nut and remove the pitman arm. Remove sheet metal baffle. Remove the four steering gear to frame bolts at outside of left frame rail. Remove steering gear.

Reverse procedure to install.

Power Steering Gear Pump Removal and Installation

Disconnect the drive belt and remove the pump pulley with a suitable puller. On some models, the pulley has bolt access holes which make pulley removal unnecessary. Disconnect the hoses from the pump and unbolt the pump from the bracket. Use caps or tape to cover the hose connectors, unions, and hose ends to keep out dirt.

Reinstall by reversing procedure. The drive belt should be adjusted to have about ½ in. play on the longest run between pulleys. After replacing pump, fill reservoir and bleed pump by idling engine for three minutes before moving the steering wheel. Then rotate steering wheel slowly throughout its entire range. Recheck oil level.

Variable Ratio Power Steering Gear and Pump 1969-70

This gear responds faster, requires less manual effort and retains feel-of-the-road steering. Parking effort is about one-half that of previous models.

Removal of variable ratio type power steering gear or pump involves the same procedure as for prior models.

More detailed information on maintenance and overhaul procedure is covered in the Unit Repair Section.

Clutch

1964-70

The only service adjustment that can be made on a Buick clutch is that of pedal clearance. If difficulty is experienced with the clutch and adjusting the clearance does not correct it, it will be necessary to remove the clutch from the car, since no practical in-car service is possible.

Clutch Removal and Installation

1964-66

1. Remove driveshaft from front flange.
2. Remove transmission.
3. Remove equalizer shaft and return spring.
4. Remove ball stud from release shaft.
5. Remove clutch release lever and seal and nylon bushing.
6. Remove socket head screw on clutch release shaft and second socket head screw with cone point from same hole.
7. Pull release shaft out approximately 3 in. and slide off yoke and throw-out bearing. Then remove shaft.
8. Mark clutch cover and flywheel to reinstall in same position.
9. Remove cover assembly by loosening bolts one turn at a time to hold spring pressure even. Spacers (¼ in. nuts) placed between release levers and cover edge will help and also simplify reinstalling.
10. To reinstall, reverse the above.

1967-70

1. Remove transmission.
2. Release equalizer assembly.
3. Remove pedal return spring from clutch fork. Disconnect rod assembly from clutch fork.
4. Remove flywheel housing.
5. Remove throwout bearing from clutch fork.
6. Disconnect clutch fork from ball stud by moving toward center of flywheel housing.
7. Mark clutch cover and flywheel

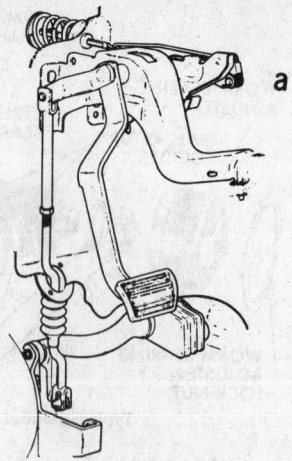

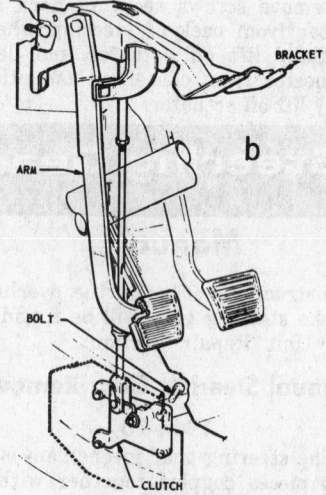

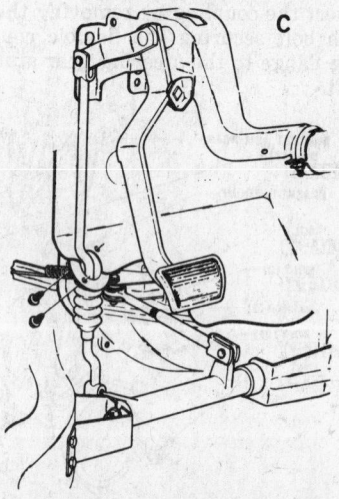

Clutch pedal lash adjustment (a) 1965-66, 46000; (b) 1964; (c) 1964-67, 45000
(© Buick Div. G.M. Corp.)

so that cover can be reinstalled in the same position. This is important to proper balance.

8. Loosen clutch cover to flywheel attaching bolts one turn at a time to maintain even spring pressure.
9. Support pressure plate and cover

assembly while removing bolts. Remove pressure plate and driven plate. Caution should be used to keep the driven plate clean.

10. Reinstall by reversing procedure.

Clutch Pedal Adjustment

1964-66, Except LeSabre

Clutch pedal clearance is adjusted under the dash at a turnbuckle on the rod between the clutch pedal and the equalizer. There should be $\frac{7}{8}$-1$\frac{1}{8}$ in. free-play of the clutch pedal before the throw out bearing strikes the diaphragm.

1964-66 LeSabre, 1967-70 All Models

Clutch pedal clearance is adjusted under the car at the link between the clutch throwout fork and the equalizer. There should be $\frac{5}{8}$-$\frac{7}{8}$ in. free-play of the clutch pedal before the throwout bearing strikes the fingers (or diaphragm).

Transmissions

Standard

Transmission refill capacities will be found in the Capacities table.

Shift Control Adjustment—Three-Speed Column Shift

1964-66

1. Place transmission levers in neutral.
2. Loosen shift rod adjusting clamp bolts.
3. Install a piece of $\frac{1}{4}$ in. round stock (The unthreaded part of a $\frac{1}{4}$ in. bolt is ideal.) into the bearing tab and first-reverse lever holes at the bottom of the steering column.
4. Push a large screwdriver blade between the selector plate and the second-third lever until the selector plate engages the tang on both shift levers.
5. Lift the column selector lever straight up toward the steering wheel several times to assure that the neutral positions in both the first-reverse and the second-third planes are aligned. If they are not aligned, shorten the first-reverse rod by pulling it through the swivel no more than 3/16 in.

1967-68

This procedure is identical to that described above for 1964-66 models with the substitutions of the following steps:
3. Install a 3/16 in. rod through the holes in the first-reverse lever, the selector plate, the second-

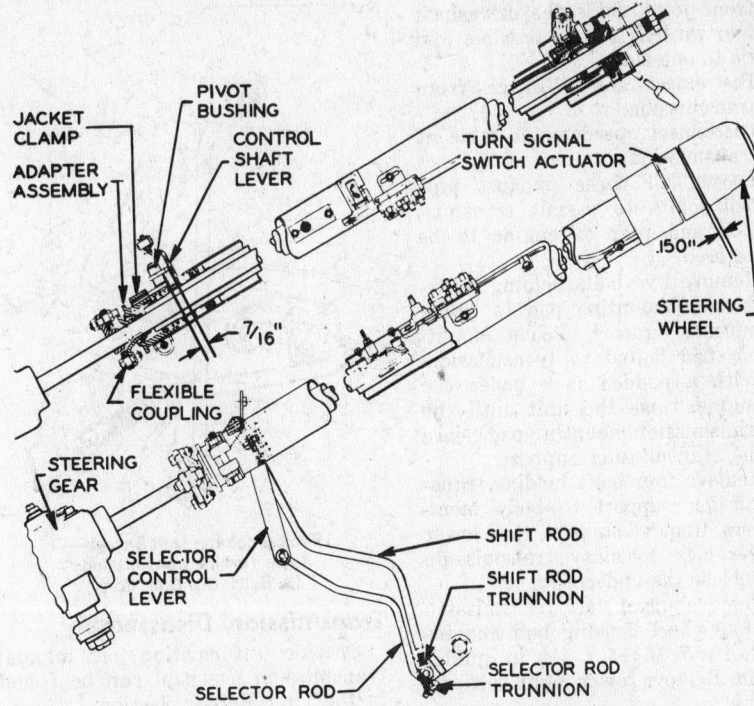

third lever, and the alignment plate at the bottom of the steering column.
4. Eliminate this step.

1969-70

1. Place column selector lever in reverse position. Loosen first-reverse adjusting clamp bolts.
2. Shift first-reverse transmission lever into reverse. Tighten first-reverse adjusting clamp bolt to 17-23 ft. lbs.
3. Shift transmission levers into neutral positions. Loosen second-third adjusting clamp.
4. Install 3/16 in. rod into alignment holes. Tighten second-third adjusting clamp bolt to 17-23 ft. lbs.

Synchromesh transmission shift mechanism (© Buick Div. G.M. Corp.)

WITH ALL TRANSMISSION & CONTROL LEVERS
IN NEUTRAL POSITION INSERT J-21196
TIGHTEN NUT & SPRING ASSY.
(AT TRANS.)

J-21196

Four-speed transmission controls with console—arrow points to $\frac{1}{4}$-in. gauge pin
(© Buick Div. G.M. Corp.)

1964-65 Four-Speed Floor Shift

The four-speed gearshift linkage uses three shift rods and levers. A simple gauge pin, a $\frac{1}{4}$ in. dia. rod, will help in making the proper adjustments. An alternative method is to have an assistant hold the manual shift lever in neutral.

1. Loosen shift rod adjusting clamps.
2. Place transmission in neutral and install gauge pin through shift levers (or have assistant hold shifter lever in neutral).
3. Tighten shift rod adjusting clamps to 17-23 ft. lbs.
4. Remove gauge pin and check for complete and easy shifting.
5. Hold shift control lever in fourth gear, then turn stop bolt until it contacts shift lever. Repeat operation at forward stop bolt for third gear.

NOTE: console equipped cars have a gauge pin access hole under the carpet.

Transmission Removal

NOTE: when removing any major part or assembly, always mark the parts so that the assemblies can be returned to the same position from which they were removed. This applies particularly to universal joint housings and covers, spring hangers, etc.

1964-70

1. Mark universal joint and transmission shaft companion flange for proper indexing at time of installation. Remove two U-bolts and disconnect driveshaft at the

front joint. Slide the driveshaft rearward as far as possible and tie to one side.

2. Disconnect shift linkage from transmission.

3. Disconnect speedometer cable at transmission.

4. Loosen all three exhaust pipe ball joints to permit transmission and rear of engine to be lowered.

5. Remove two bolts holding transmission mounting pad to transmission support. Leave mounting pad bolted to transmission.

6. With a padded jack under the engine, raise the unit until the transmission mounting pad clears the transmission support.

7. Remove four bolts holding transmission support to body members. Remove support, then lower the jack to allow transmission to clear the underbody.

8. Remove upper left transmission to flywheel housing bolt and install a 7/16 14 x 4½ in. guidepin. Remove lower right bolt and pin.

9. Remove the other two transmission attaching bolts. Slide the transmission back until the drive gear shaft disengages the clutch disc and clears the flywheel housing. Lower the transmission.

10. Install transmission by reversing the above procedure.

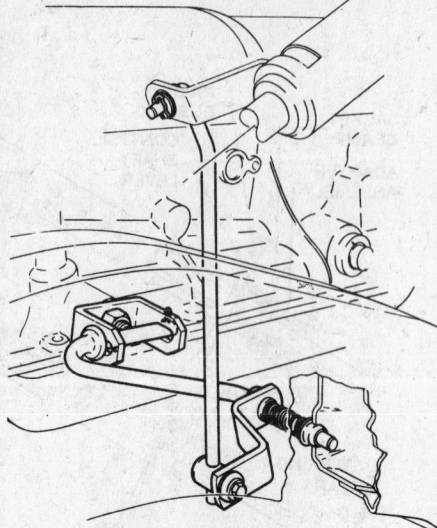

**Typical column shift linkage—
Super Turbine transmissions**
(© Buick Div. G.M. Corp.)

Transmission Disassembly

Specific information on manual transmission overhaul can be found in the Unit Repair Section.

Automatic

When automatic transmission trouble is reported, a road test and careful diagnosis are in order. Transmission Removal and Replacement and Linkage Adjustments are covered here in the following paragraphs. For test procedures, transmission overhaul and other detailed information, see Unit Repair Section.

The Super Turbine 300 transmission is standard equipment on Le Sabre models with the smallest engine, 1964-69. The Super Turbine 400 is optional on the small engine Le Sabres and standard on all other models, 1964-68. In 1969, the Super Turbine 400 is replaced by the virtually identical Turbo Hydra-Matic 400; it is optional on small engine Le Sabres and standard on all other models. In 1970, the Super Turbine 300 is eliminated; the Turbo Hydra-Matic 400 is the only automatic transmission used on 1970 Buicks.

Shift Control Adjustment

1964-66 Column Shift, 1964-68 Console Shift

1. Place manual control lever in Park position.

2. Loosen adjusting clamp bolt at transmission.

3. Place transmission lever in Park position.

4. Tighten adjusting clamp bolt to 17-23 ft. lbs. Overtightening will cause hard shifting.

5. Start engine. Check for proper shifting into all ranges.

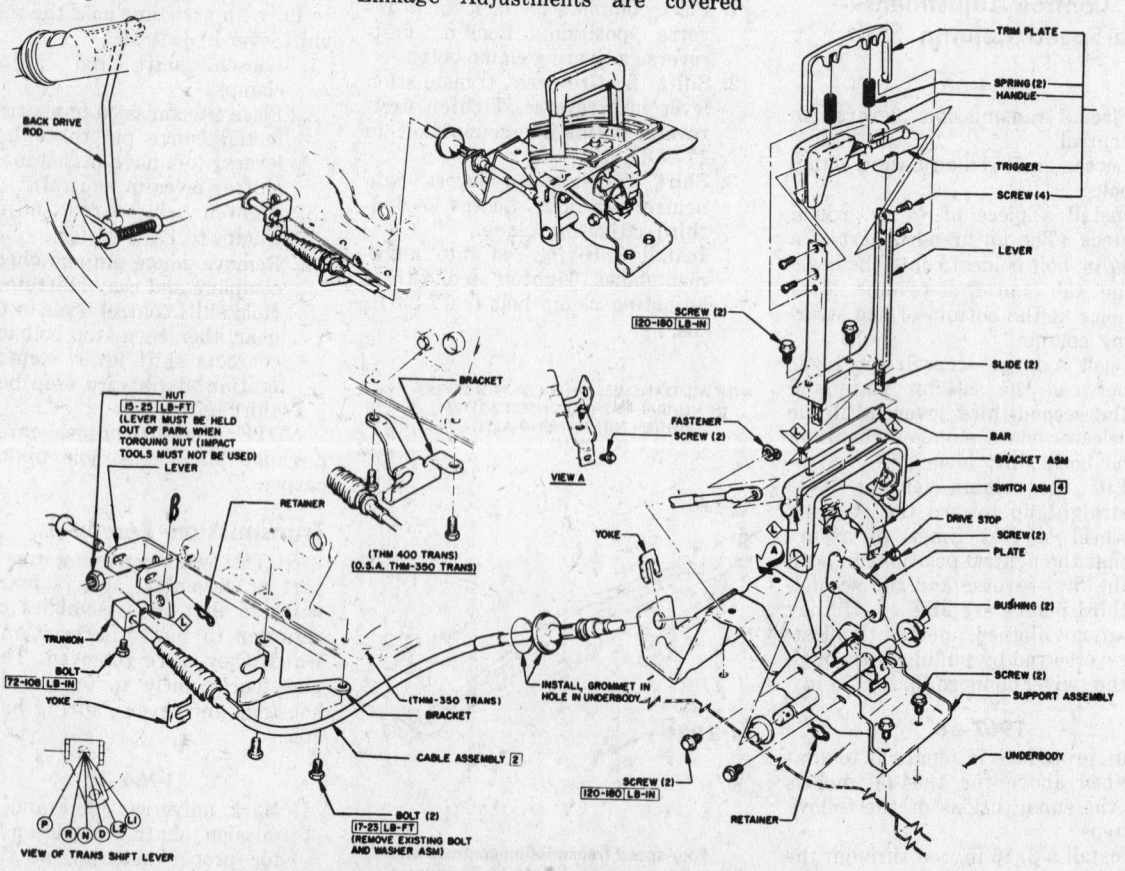

Cable type console shift linkage—1969-70 Turbo Hydra-Matic 400
(© Buick Div. G.M. Corp.)

1967-70 Column Shift

This procedure is identical to that described above for 1964-66 column shift and 1964-68 console shift with the substitution of the following step:

1. Place manual control lever against Drive stop.

1969-70 Console Shift

These units are operated by a cable linkage. Adjust as follows:

1. Loosen trunnion bolt at transmission end of cable.
2. Set manual control lever against Drive stop.
3. Place transmission bar assembly in Drive position.
4. Tighten trunnion bolt against cable end to 6-9 ft. lbs.
5. Place transmission bar assembly in Park position.
6. Loosen back drive rod clamp screw.
7. Push back drive rod (from linkage to steering column) up and hold lightly against stop.
8. Tighten screw in clamp at end of back drive rod to 17-23 ft. lbs.
9. Start engine. Check for proper shifting into all ranges.

Neutral Start Switch Adjustment—All Models

This safety switch prevents starting except in Neutral or Park positions. The switch combines function with the back-up light switch and is actuated by the transmission linkage. On column shift cars, the switch is under the instrument panel. On console shift cars, the switch is inside the console. To check switch adjustment:

1. Turn on ignition switch.
2. Place shift control lever in Reverse, and make sure back-up lights are on.
3. Set parking brake. Hold foot brake. Place shift control in Neutral and make sure engine will start. Repeat in Park, Drive, and Reverse. Engine must start only in Neutral or Park.
4. To adjust switch, loosen mounting screws and move switch on slotted mounting holes.

Idle Stator and Detent Switch Adjustments

1964-67 Super Turbine 300, 400

Refer to the accompanying illustrations for these adjustment procedures.

Detent Switch Adjustment

1968-69 Super Turbine 300
1968-70 Super Turbine 400, Turbo Hydra-Matic 400

Refer to the accompanying illustrations for these adjustment procedures.

Transmission Removal

1. Raise car and provide support for front and rear of car.
2. Disconnect front exhaust pipe bolts at the exhaust manifold and at the connection of the intermediate exhaust pipe location (single exhaust only). On dual exhaust the exhaust pipes need not be removed.
3. Remove pinion flange U-bolts and slide propeller shaft toward transmission as far as possible to separate universal joint from pinion flange. Remove propeller shaft from car.
4. Place suitable jack under transmission and fasten transmission securely to jack.
5. Remove vacuum line to vacuum modulator hose from vacuum modulator.
6. Loosen cooler line bolts and separate cooler lines from transmission.
7. Remove transmission mounting pad to crossmember bolts.
8. Remove transmission cross member support to frame rail bolts. Remove crossmember.
9. Disconnect speedometer cable and electrical wiring.
10. Disconnect shift linkage from transmission.
11. Disconnect transmission filler pipe at engine. Remove filler pipe from transmission.
12. Support engine at oil pan.

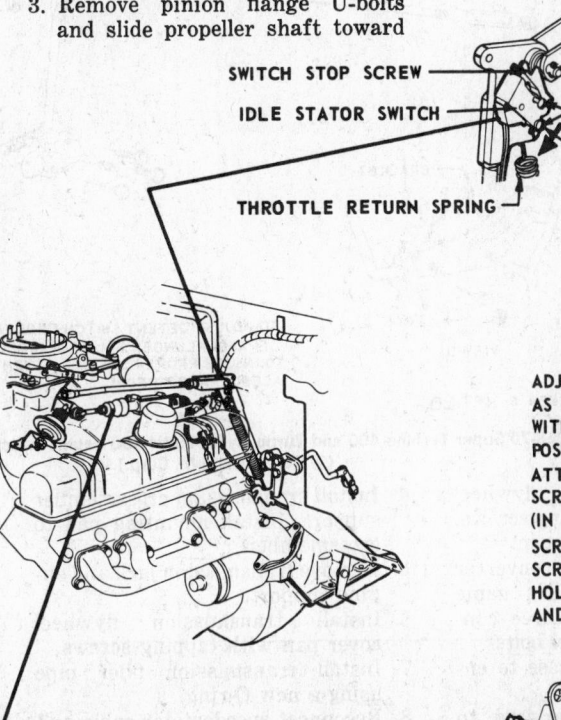

SWITCH STOP SCREW
IDLE STATOR SWITCH
ATTACHING SCREWS
SCREW "A"
THROTTLE RETURN SPRING

ADJUST IDLE STATOR SWITCH AS FOLLOWS: ADJUST SWITCH WITH THROTTLE AT CLOSED POSITION AND RETURN SPRING ATTACHED. WITH ATTACHING SCREWS LOOSE, ROTATE SWITCH (IN DIRECTION SHOWN) ABOUT SCREW "A" UNTIL SWITCH STOP SCREW BOTTOMS AGAINST CASE. HOLD SWITCH IN THIS POSITION AND TIGHTEN ATTACHING SCREWS.

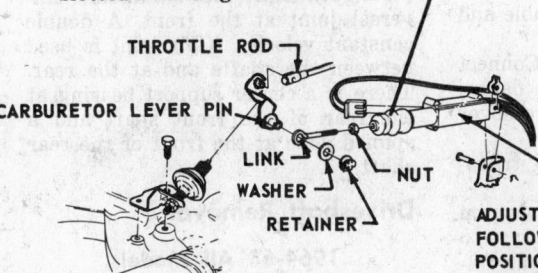

THROTTLE ROD
CARBURETOR LEVER PIN
LINK
WASHER
NUT
RETAINER
STATOR AND DETENT SWITCH

ADJUST STATOR AND DETENT SWITCH AS FOLLOWS WITH CARBURETOR IN WIDE OPEN POSITION AND SWITCH PLUNGER BOTTOMED, ADJUST LINK UNTIL IT WILL SLIP OVER CARBURETOR LEVER PIN, THEN SCREW LINK INTO PLUNGER 1 1/2 TURNS. INSTALL WASHER AND RETAINER.

74-2

1964-67 Super Turbine 300 and 400 stator and detent switch adjustment
(© Buick Div. G.M. Corp.)

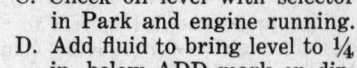

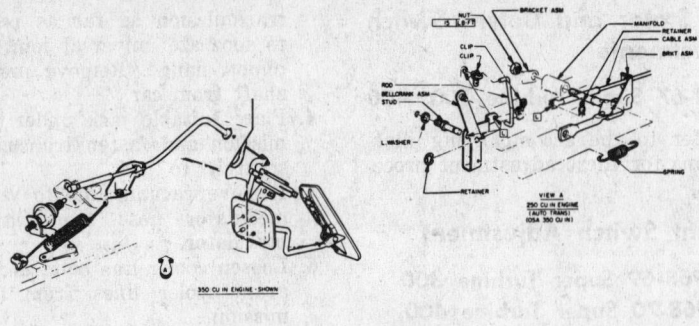

C. Check oil level with selector in Park and engine running.
D. Add fluid to bring level to ¼ in. below ADD mark on dipstick.
E. Bring fluid level up to full mark after transmission is up to normal operating temperature. About fifteen miles of highway driving is sufficient to attain normal temperature.

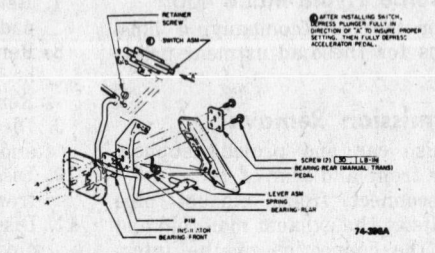

Drive Shaft, U Joints

1968-69 Super Turbine 300 detent switch adjustment (© Buick Div. G.M. Corp.)

1964-68 Except Riviera

The driveshaft consists of a two-piece open shaft with standard universal joints front and rear. A double constant velocity joint is used between the shafts with a center support bearing at rear of front shaft and splined front yoke at front of rear shaft.

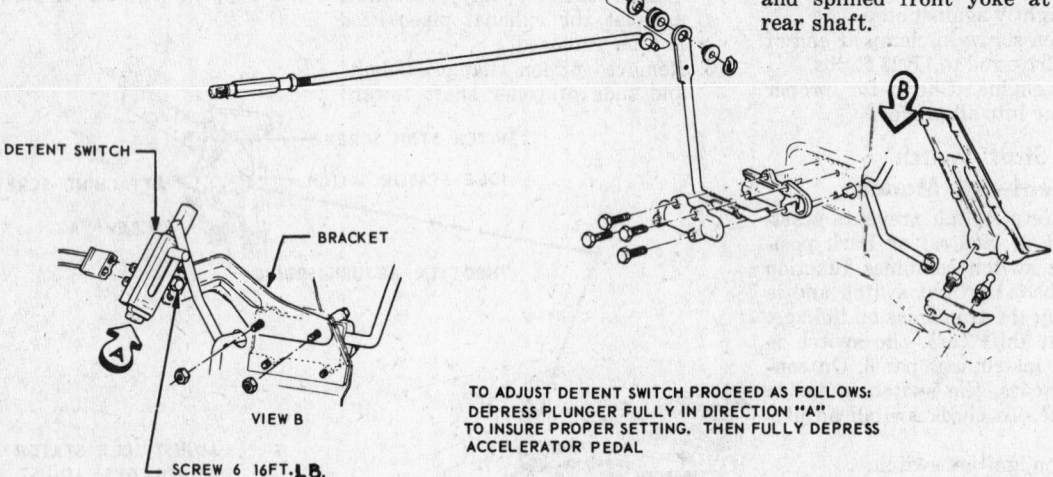

TO ADJUST DETENT SWITCH PROCEED AS FOLLOWS: DEPRESS PLUNGER FULLY IN DIRECTION "A" TO INSURE PROPER SETTING. THEN FULLY DEPRESS ACCELERATOR PEDAL

1968-70 Super Turbine 400 and Turbo Hydra-Matic 400 detent switch adjustment (© Buick Div. G.M. Corp.)

13. Remove transmission flywheel cover pan to case tapping screws. Remove flywheel cover pan.
14. Mark flywheel and converter pump for reassembly in same position, and remove three converter pump to flywheel bolts.
15. Remove transmission case to engine block bolts.
16. Move transmission rearward to provide clearance between converter pump and crankshaft. Lower transmission and move to bench.

Transmission Installation

1. Raise transmission into position using suitable jack. Rotate converter to permit coupling of flywheel and converter in original relationship.
2. Install transmission case to engine block bolts. Torque to 30-40 ft. lbs. Do not overtighten.
3. Install flywheel to converter pump bolts. Torque to 25-35 ft. lbs.

4. Install transmission crossmember support. Install mounting pad to crossmember.
5. Remove transmission jack and engine support.
6. Install transmission flywheel cover pan with tapping screws.
7. Install transmission filler pipe using a new O-ring.
8. Reconnect speedometer cable and wiring.
9. Install propeller shaft. Connect propeller shaft to pinion flange.
10. Reinstall front exhaust crossover pipe.
11. Install oil cooler lines to transmission.
12. Install vacuum line to vacuum modulator.
13. Fill transmission with oil as follows:
   A. Add five to six pints of oil.
   B. Start engine in neutral. Bring up to fast idle. Do not race engine. Shift through each range.

1969-70, Except Riviera

The driveshaft consists of a one-piece open shaft with universal joints front and rear. A splined slip yoke is located at the transmission end.

1964-70 Riviera

The driveshaft consists of a two-piece open shaft with a standard universal joint at the front. A double constant velocity (CV) joint is used between the shafts and at the rear. There is a center support bearing at the rear of the front shaft and a splined yoke at the front of the rear shaft.

Driveshaft Removal

1964-68 All Models, 1969-70 Riviera

1. Mark pinion flange and rear joint for reassembly. At rear pinion flange, remove U-bolt clamps from rear universal; on Riviera, remove four rear CV

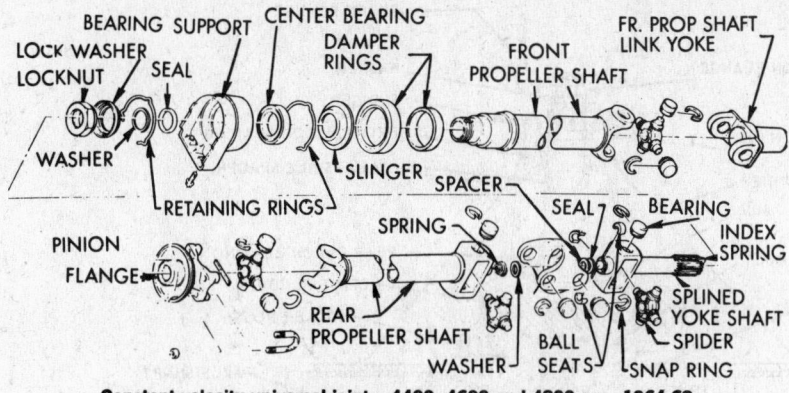

Constant velocity universal joint—4400, 4600 and 4800 ser., 1964-69
(© Buick Div. G.M. Corp.)

Lubrication of Slip Splines

1964-68 All Models, 1969-70 Riviera

Lubrication of driveshaft slip splines and constant velocity universal joint center ball is as follows:

The constant velocity universal joint center ball should be lubricated at 6,000 mile intervals.

1. Rotate driveshaft so fitting is visible through rear hole of frame tunnel.
2. Use special adapter held firmly against conical fitting to force grease into center ball socket.

The propeller shaft slip spline

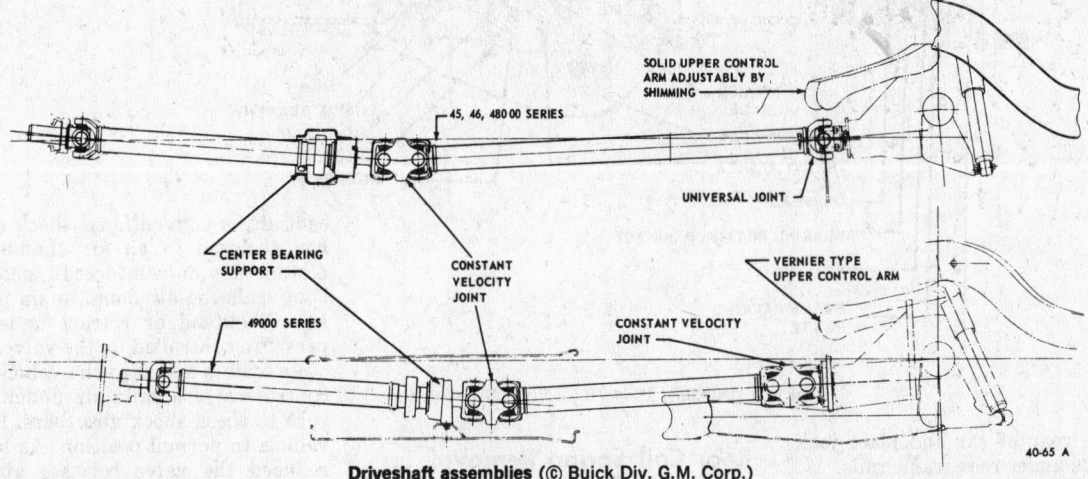

Driveshaft assemblies (© Buick Div. G.M. Corp.)

joint to pinion flange bolts. Use tape to secure bearings on the spider.
2. Remove four center bearing attaching bolts; two bolts on Riviera.
3. Support rear end of shaft. Slide assembly rearward until front yoke is free of transmission shaft splines. On Riviera, slide complete shaft assembly rearward through frame tunnel.
4. Protect the oil seal surface on the front yoke from dirt or marring.
CAUTION: do not bend CV joint to its extreme angle at any time.

1969-70, Except Riviera

1. Mark shaft and pinion flange for reassembly.
2. Remove U-bolts from rear pinion flange. Use tape to secure bearings on the spider.
3. Remove shaft assembly by sliding rearward to disengage splines on transmission shaft.

Driveshaft Disassembly

1964-68 All Models, 1969-70 Riviera

1. Loosen locknut and slide locknut and seal against the center CV joint. Slide the rear shaft from the front shaft.
2. Pull the center support and bear-

ing assembly from the shaft by use of a suitable puller.
3. Remove retainers and drive center bearing from support.
4. Disassemble CV joints:
   A. Mark all yokes for reassembly.
   B. Remove universal joint bearings from link yoke.
   C. Press two link yoke bearings out, one at a time.
   D. Carefully work universal joint and ball stud yoke assembly out of link yoke.
   E. Remove all remaining bearings from ball stud yoke. Slip out spider from yoke.
5. Pry out ball stud seat seal, spacer, ball seats, stop, spring, spring seat, and spring seat gasket.
6. Disassemble universal joints:
   A. Remove snap-rings.
   B. Press out shaft bearings, one at a time.
7. Reverse procedure for reassembly. Always install complete universal joint repair sets when repairs or replacements are necessary.

1969-70, Except Riviera

This procedure is identical to that described above for 1964-68 all models, and 1969-70 Riviera, with the exclusion of Steps 1-5.

should be lubricated every 6,000 miles.

NOTE: due to the position of the slip spline when the car is raised on some types of modern hoists, it is necessary to fill the slip spline lubricant cavity by forcing lubricant through the spline into the cavity. For this reason, either a fitting or an adapter is necessary in order to build up sufficient pressure to fill the slip spline cavity.

Drive Axle

Troubleshooting and Adjustment

General instructions covering the troubles of the drive axle and how to repair and adjust it, together with information on installation of drive axle bearings and grease seals, are given in the Unit Repair Section.

Capacities of the drive axles are given in the Capacities table.

Rear Axle Removal and Installation

1964-70

All repairs on the rear axle assembly can be made with the assembly mounted in the car. In the event of damage to the housing, remove as follows:

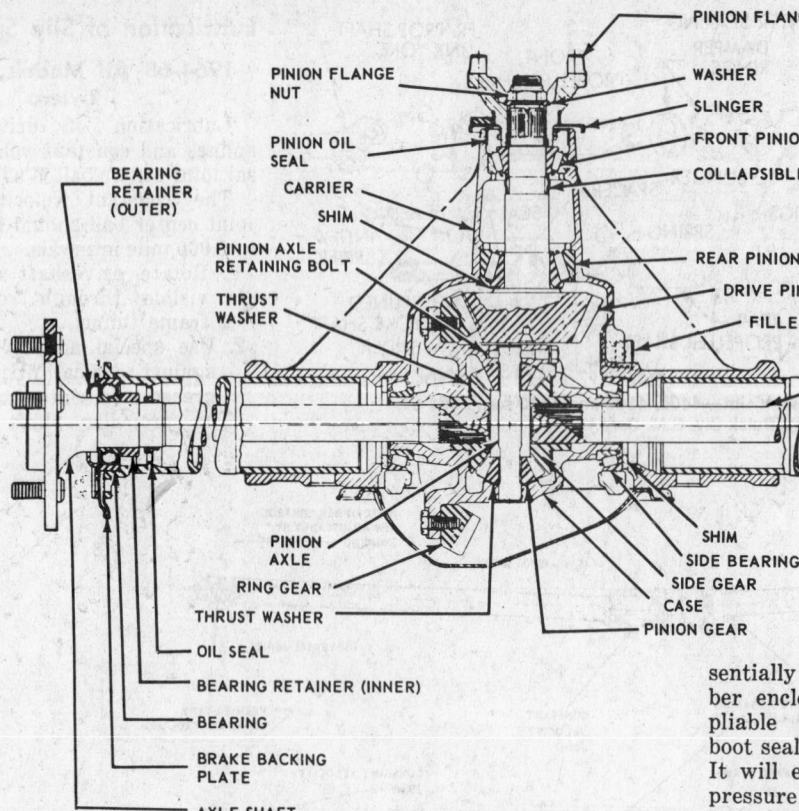

PINION FLANGE
PINION FLANGE NUT
PINION OIL SEAL
CARRIER
SHIM
PINION AXLE RETAINING BOLT
THRUST WASHER
BEARING RETAINER (OUTER)

WASHER
SLINGER
FRONT PINION BEARING
COLLAPSIBLE SPACER
REAR PINION BEARING
DRIVE PINION
FILLER PLUG
AXLE SHAFT

PINION AXLE
RING GEAR
THRUST WASHER
OIL SEAL
BEARING RETAINER (INNER)
BEARING
BRAKE BACKING PLATE
AXLE SHAFT

SHIM
SIDE BEARING
SIDE GEAR
CASE
PINION GEAR

Rear axle assembly, 1964-70 (© Buick Div. G.M. Corp.)

1. Raise rear of car and place jack stands under rear frame rails.
2. Mark rear universal and pinion flange for reassembly. Disconnect rear universal. On Riviera, mark flanged ball stud yoke and rear pinion flange for reassembly. Then disconnect rear CV joint.
3. Push driveshaft forward and support.
4. Detach parking brake cables. Remove shock absorbers. Support axle housing on jack.
5. Disconnect sway bar, and upper control arms.
6. Disconnect lower control arms.
7. Lower axle housing to floor.
8. To reinstall, reverse procedure.

Positive Traction Differential

No special attention is required in this area, except with the lubricant used.

Under no circumstances use anything but special positive traction lubricant.

Rear Coil Spring Removal

Disconnect the shock absorber link and the torsion bar. Place a jack at the frame in front of the rear spring and jack up the frame.

Remove the bolt that holds the spring to the lower perch.

NOTE: the left side of the car has a left-hand thread bolt holding the spring, and the right side of the car has a right-hand thread bolt holding the spring. Remove the upper and lower bolts and take out the coil spring. If necessary, jack the car up a sufficient amount so it will come out readily.

Automatic Level Control

The system consists of a vacuum-operated air compressor, connected to a control valve mounted at rear suspension cross member. The valve is then connected to Superlift rear shock absorbers.

The Superlift shock absorber is es-

sentially a conventional shock absorber enclosed in an air chamber. A pliable nylon-reinforced neoprene boot seals the air dome to air piston It will extend or retract under the pressure controlled by the valve.

As load is added to the vehicle, the control valve admits air under pressure to these shock absorbers, lifting vehicle to normal position. As load is reduced the valve releases air and lowers vehicle to the previous normal level.

The valve is connected by a link to the right rear upper control link. A deflection of at least ½ in. is required to make it operative.

A delay mechanism is built into the valve housing. This requires four to 22 seconds to cause valve to operate. It prevents operation during normal road motions.

Pressure at the shock absorber units is kept equal by the line connecting the two units, with only one unit connected directly to the control valve. This keeps approximately 8-15 psi on shock absorber units at all times. The pressure is released at the control valve and the equalizing pressure is maintained through a check valve at the release fitting.

The compressor is located in the engine compartment. It is operated by vacuum surge through a line connected just forward of the carburetor insulator connection. Air at atmospheric pressure is taken into the compressor through a line connected to the air cleaner. The compressed air from the compressor is supplied to a reservoir and then to the control valve.

Any service work on this system or other parts of the vehicle that may deflate the system will require reinflation to approximately 140 psi.

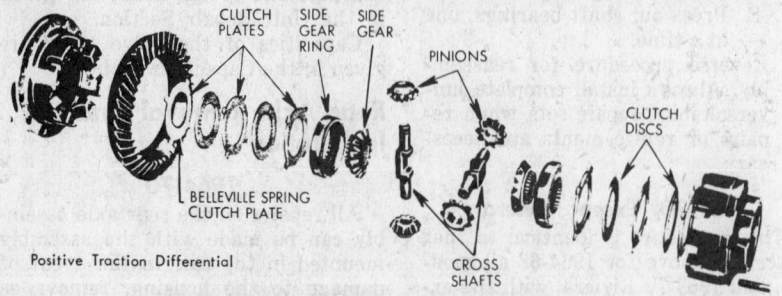

CLUTCH PLATES
SIDE GEAR RING
SIDE GEAR
PINIONS
CLUTCH DISCS
BELLEVILLE SPRING CLUTCH PLATE
CROSS SHAFTS

Positive Traction Differential

Positive traction differential (© Buick Div. G.M. Corp.)

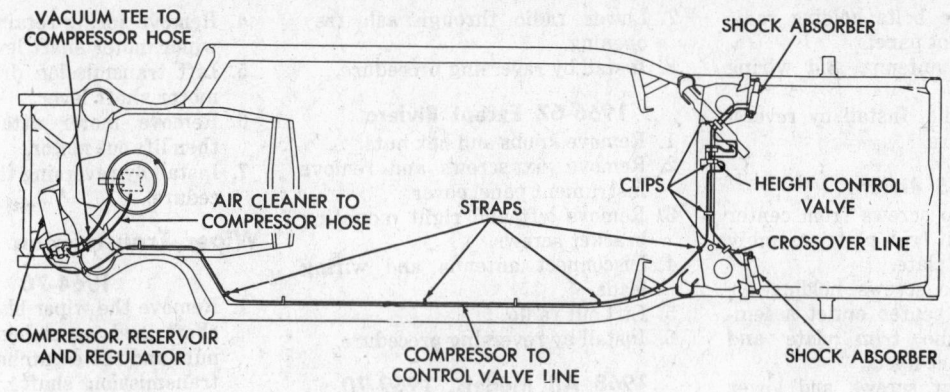

Automatic level control (© Buick Div. G.M. Corp.)

VACUUM TEE TO COMPRESSOR HOSE

AIR CLEANER TO COMPRESSOR HOSE

COMPRESSOR, RESERVOIR AND REGULATOR

STRAPS

COMPRESSOR TO CONTROL VALVE LINE

SHOCK ABSORBER

CLIPS

HEIGHT CONTROL VALVE

CROSSOVER LINE

SHOCK ABSORBER

Compressor, reservoir tank and regulator valve assembly (© Buick Div. G.M. Corp.)

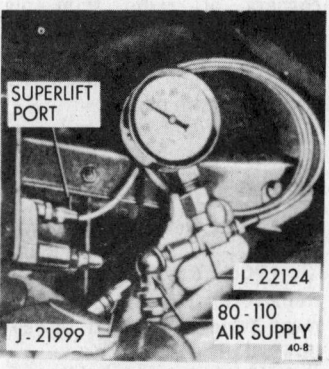

Connect test gauge to superlift port
(© Buick Div. G.M. Corp.)

SUPERLIFT PORT

J-22124

80-110 AIR SUPPLY

J-21999

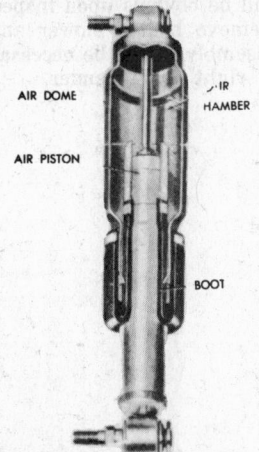

Superlift shock absorber
(© Buick Div. G.M. Corp.)

AIR DOME

AIR CHAMBER

AIR PISTON

BOOT

Remove superlift air supply line
(© Buick Div. G.M. Corp.)

SUPERLIFT PORT

J-21999

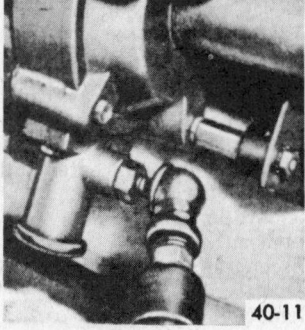

Filling system through service valve
(© Buick Div. G.M. Corp.)

All lines are ⅛ in. diameter flexible black tubing. In working on this system, use care not to kink the tubing. Keep tubing away from the exhaust system.

Radio

Always disconnect the battery ground cable before working on any part of the instrument panel.

Removal and Installation
1964 All Models

1. Remove knobs, felt washers, inner knobs, and hex nuts.
2. Remove trim plate.

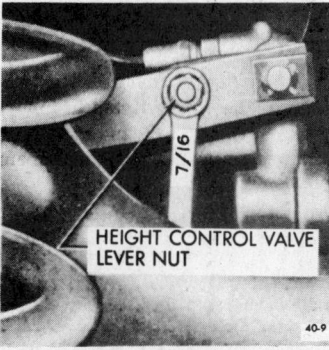

Loosen height control valve lever nut
(© Buick Div. G.M. Corp.)

HEIGHT CONTROL VALVE LEVER NUT

7/16

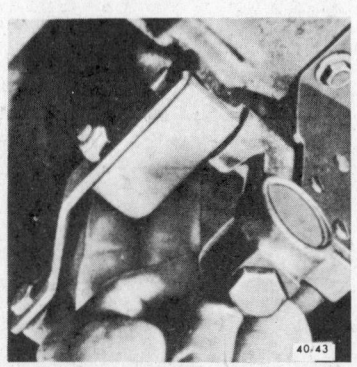

Hold overtravel body in exhaust position
(© Buick Div. G.M. Corp.)

3. Remove two bolts holding radio to instrument panel.
4. Disconnect antenna and wiring leads.
5. Remove radio. Install by reversing procedure.

1965 Riviera

1. Remove two screws from center console front trim plate assembly and remove plate.
2. Remove two screws holding air conditioner center outlet assembly to radio trim plate and remove outlet ducts.
3. Remove four screws and lower air conditioner control assembly.
4. Remove two screws from lower corners of radio trim plate. Remove two screws securing underside of receiver to cross support. Partially withdraw radio and radio trim plate.
5. Disconnect antenna and wiring leads.
6. Remove radio and radio trim plate.
7. Install by reversing procedure.

1965 Except Riviera

1. Remove six screws from instrument panel cover and tilt cover upward.
2. Remove four nuts holding cove molding to instrument panel and remove radio and molding as an assembly.
3. Install by reversing procedure.

1966-67 Riviera

1. Remove ash tray assembly.
2. Pry out chrome strip at center of instrument panel.
3. Remove two screws securing center outlet, lift off center outlet, and pull out plastic duct.
4. Remove knobs, two nuts, and escutcheon.
5. Disconnect antenna and wiring leads.
6. Remove radio support.

7. Lower radio through ash tray opening.
8. Install by reversing procedure.

1966-67 Except Riviera

1. Remove knobs and hex nuts.
2. Remove six screws and remove instrument panel cover.
3. Remove left and right mounting bracket screws.
4. Disconnect antenna and wiring leads.
5. Lift out radio.
6. Install by reversing procedure.

1968 All Models, 1969-70 Riviera

1. Remove ash tray assembly.
2. Remove knobs, escutcheons, and hex nuts.
3. Unplug antenna and wiring leads.
4. Remove radio downward.
5. Install by reversing procedure.

1969-70 Except Riviera

1. Remove center air-conditioning duct.
2. Remove right instrument trim panel and screw in bottom of radio.
3. Remove radio knobs, escutcheons, and two hex nuts.
4. Unplug antenna and wiring leads.
5. Remove radio downward.
6. Install by reversing procedure.

Windshield Wipers

Motor R & R

1964-70

1. Disconnect wire connectors from motor and pump.
2. Remove washer hoses from the pump.
3. Remove left side air intake grille.

4. Remove spring retainer clip from wiper motor shaft lever.
5. Lift transmission drive links off motor shaft lever.
6. Remove motor attaching bolts, then lift out motor.
7. Install by reversing the above procedure.

Wiper Transmission

1964-70

1. Remove the wiper blade and arm, shaft and escutcheon retaining nuts and the escutcheon from the transmission shaft.
2. Remove air intake grille.
3. Remove spring retainer clip from wiper motor shaft. Lift drive links off motor shaft.
4. Remove the transmission retaining screws.
5. Slide transmission and drive link toward opposite side of car. Lift transmission up at opening and remove.
6. Install by reversing the above procedure.

Heater System

To remove heater core, it is necessary to take out heater assembly. This will be obvious upon inspection.

To remove heater blower and air inlet assembly it will be necessary to remove right front fender.

Buick Special and Gran Sport

Index

YEAR IDENTIFICATION

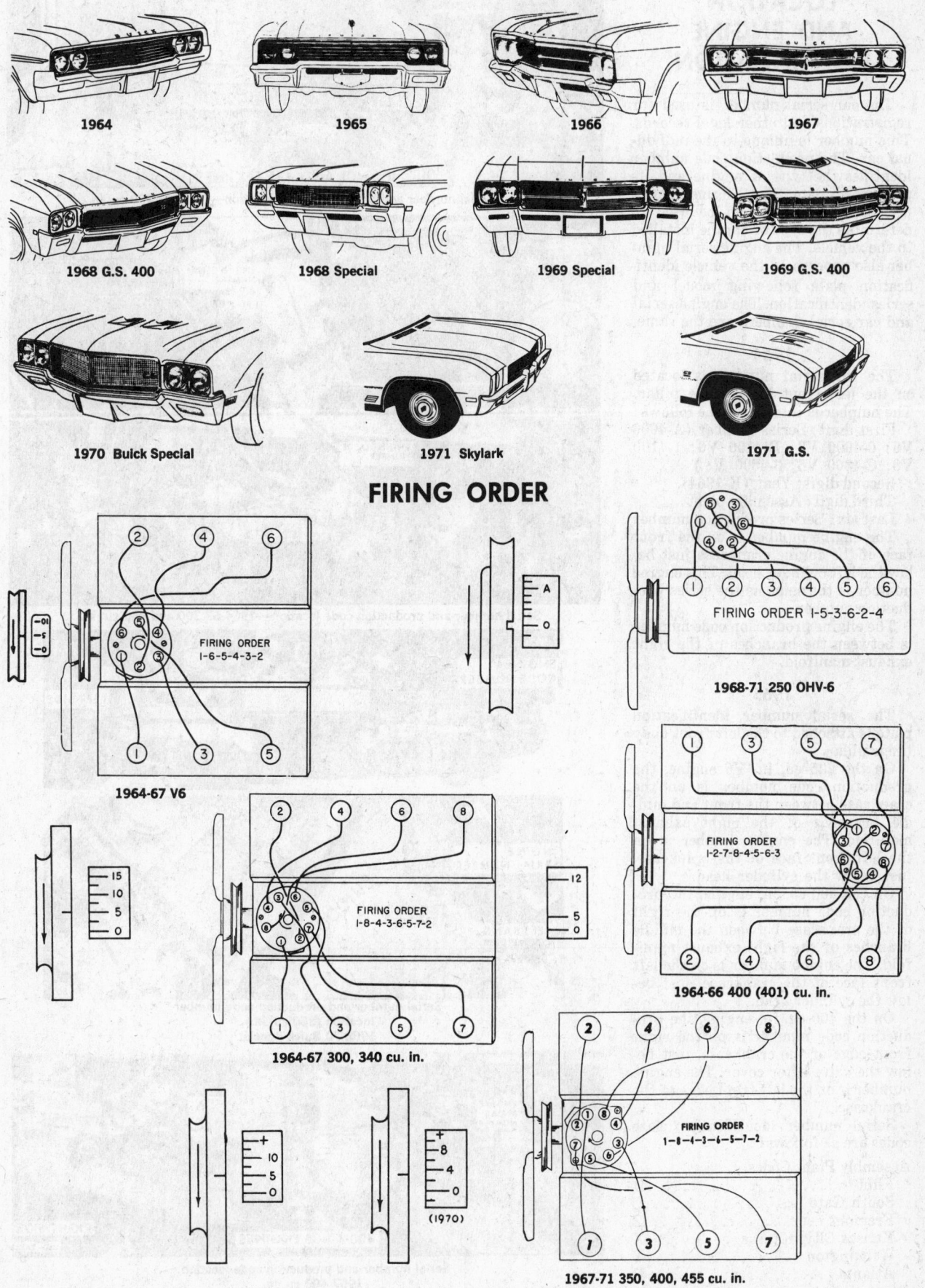

1964

1965

1966

1967

1968 G.S. 400

1968 Special

1969 Special

1969 G.S. 400

1970 Buick Special

1971 Skylark

1971 G.S.

FIRING ORDER

FIRING ORDER 1-6-5-4-3-2

1964-67 V6

FIRING ORDER 1-5-3-6-2-4

1968-71 250 OHV-6

FIRING ORDER 1-8-4-3-6-5-7-2

1964-67 300, 340 cu. in.

FIRING ORDER 1-2-7-8-4-5-6-3

1964-66 400 (401) cu. in.

(1970)

FIRING ORDER 1-8-4-3-6-5-7-2

1967-71 350, 400, 455 cu. in.

CAR SERIAL NUMBER LOCATION AND ENGINE IDENTIFICATION

The car serial number is used for registrations and other legal records. This number is unique to the individual car. The production code number identifies the type of engine and its production date. The Engine Identification Code chart can be used to determine the type of engine installed in the vehicle. The engine serial number also appears on the vehicle identification plate following model and series identification. The engine serial and car serial numbers are the same.

1964

The car serial number is located on the left front body hinge pillar. The number is interpreted as follows:

First digit: Series number (A-4000 V6; 0-4000 V8; B-4100 V6; 1-4100 V8; C-4300 V6; 3-4300 V8.)

Second digit: Year (K-1964)

Third digit: Assembly plant

Last six: Series production number

The engine number is on the front face of the engine crankcase just below the left cylinder head. This engine number is the same as the series production number.

The engine production code number is between the branches of the right exhaust manifold.

1965

The serial number identification plate is attached to the left front body hinge pillar.

On the 225 cu. in. V6 engine, the production code number is on the crankcase between the front and middle branches of the right exhaust manifold. The engine number is on the left front face of the crankcase, just below the cylinder head.

On the 300 cu. in. engine, the production code number is on the right of the crankcase between the middle branches of the right exhaust manifold. The engine number is on the left front face of the crankcase, just below the cylinder head.

On the 400 cu. in. engine, the production code number is on the right front edge of the crankcase, just below the valve lifter cover. The engine number is on the left front edge of the crankcase.

Serial number identification plate codes are as follows:

Assembly Plant Codes
```
Flint ...................... H
South Gate ................ C
Fremont ................... Z
Kansas City, Kansas ....... X
Wilmington ................ Y
Atlanta ................... D
```

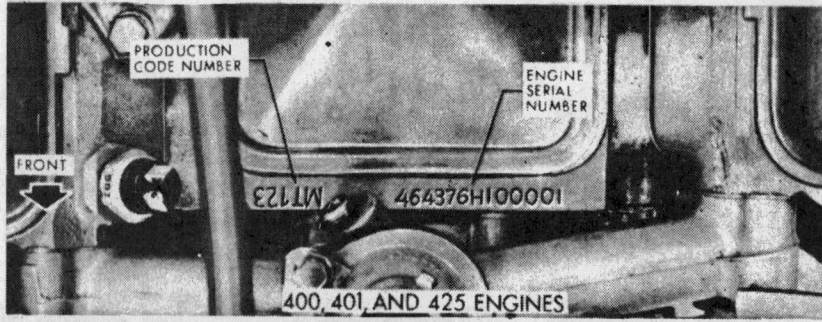

Serial number and production code location—1965-66 400 cu. in.

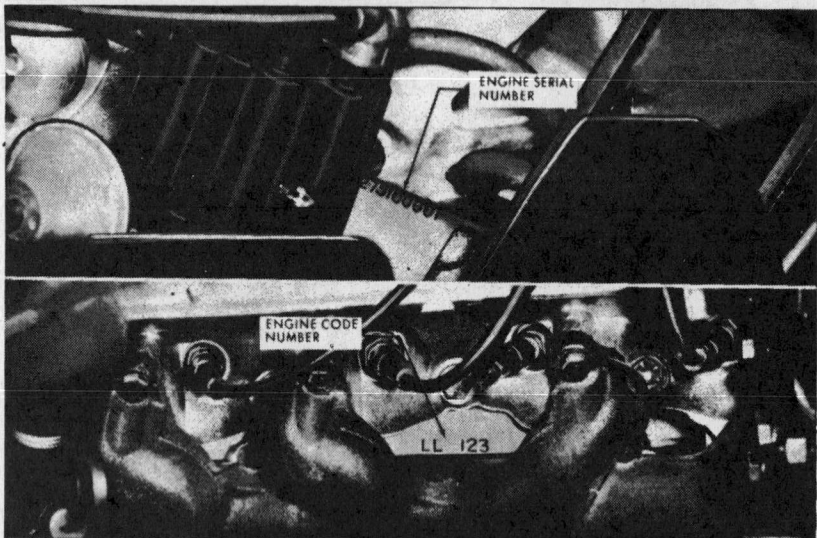

Serial number and production code location—1964-67 300 and 340 cu. in.

Serial number and production code number location (350 cu. in.), 1968-70 Buick Special

Serial number and production code location, 1967 400 cu. in.

Serial number and production code location (400 and 455 cu. in.)—1968-70 Buick Special

Baltimore ....B
Kansas City, Mo. ....K
Bloomfield V

Body Style Code No.
2-door coupe 27
4-door 2-seat station wag. 35
2-door hardtop coupe 37
4-door hardtop 39
2-door hardtop coupe 47
4-door 2-seat sportwagon 55
4-door 3-seat sportwagon 65
2-door convertible 67
4-door sedan 69

Series Identification	V6	V8
Special	33	34
Special Deluxe	35	36
Sportwagon		42
Skylark	43	44

1966

The car serial number identification plate is attached to the left front body hinge pillar. On the 225 cu. in. V6 engine, the production code number is between the front and middle branches of the right exhaust manifold. The engine number is on the left front face of the crankcase, just below the cylinder head.

On the 300 and 340 cu. in. engines, the production code number is between the middle branches of the right exhaust manifold. The engine number is just below the front of the left cylinder head.

On the 400 cu. in. engine, the production code number is on the right

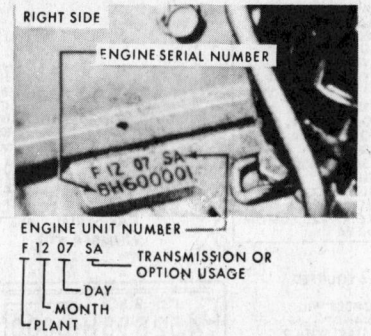

Serial number and production code location (OHV 6)—1968-70 Buick Special

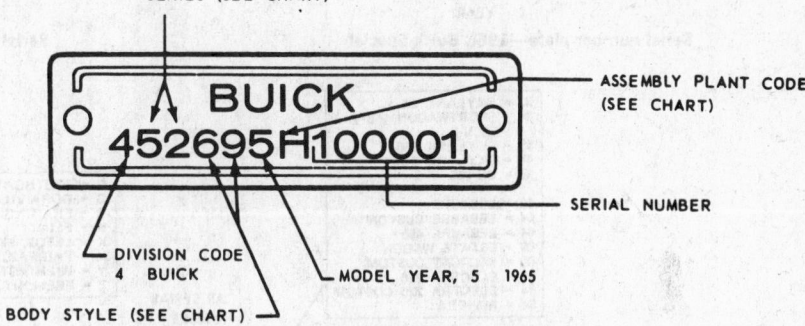

Serial number plate—1965 Buick Special

ENGINE IDENTIFICATION CODE

No. Cyls.	Cu. In. Displ.	Type	1964	1965	1966	1967	1968	1969	1970
6	225	1 & 2 BBL.	KH	LH	MH	NH			
8	300	Std. Eng.	KL	LP	ML	NL			
8	300	HP	KP	LL					
8	401	4-BBL.		LT					
8	340	2-BBL.			MA	NA			
8	340	4-BBL.			MB	NB			
8	455	4-BBL.							SR
6	250	1-BBL.							SS
8	455	4-BBL. St. 1							SS
8	350	2-BBL.					PO	RO	SO
8	400	4-BBL.			MR, MT	NR	PR	RR	
8	350	4-BBL.					PP	RP	SB, SP

2-BBL.—Two-barrel carburetor.
4-BBL.—Four-barrel carburetor.
HP—High Performance Engine.
St. 1—Stage 1.

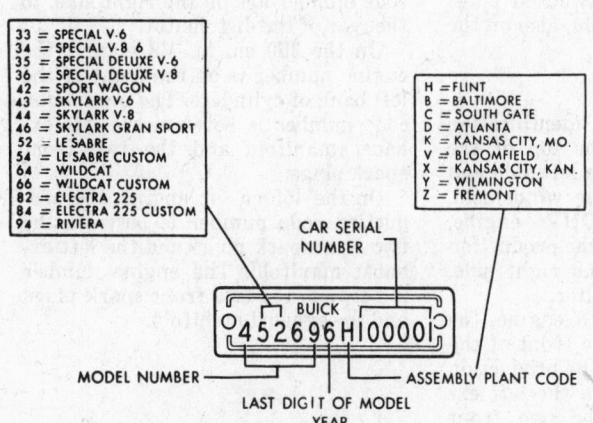

33 = SPECIAL V-6
34 = SPECIAL V-8 6
35 = SPECIAL DELUXE V-6
36 = SPECIAL DELUXE V-8
42 = SPORT WAGON
43 = SKYLARK V-6
44 = SKYLARK V-8
46 = SKYLARK GRAN SPORT
52 = LE SABRE
54 = LE SABRE CUSTOM
64 = WILDCAT
66 = WILDCAT CUSTOM
82 = ELECTRA 225
84 = ELECTRA 225 CUSTOM
94 = RIVIERA

H = FLINT
B = BALTIMORE
C = SOUTH GATE
D = ATLANTA
K = KANSAS CITY, MO.
V = BLOOMFIELD
X = KANSAS CITY, KAN.
Y = WILMINGTON
Z = FREMONT

Serial number plate—1966 Buick Special

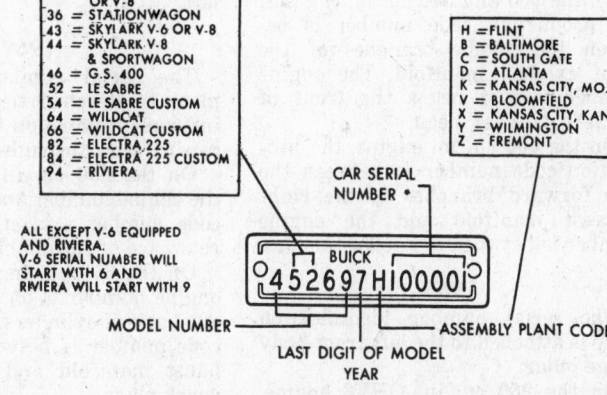

33 = SPECIAL V-6 OR V-8
35 = SPECIAL DELUXE V-6 OR V-8
36 = STATIONWAGON
43 = SKYLARK V-6 OR V-8
44 = SKYLARK V-8 & SPORTWAGON
46 = G.S. 400
52 = LE SABRE
54 = LE SABRE CUSTOM
64 = WILDCAT
66 = WILDCAT CUSTOM
82 = ELECTRA 225
84 = ELECTRA 225 CUSTOM
94 = RIVIERA

ALL EXCEPT V-6 EQUIPPED AND RIVIERA.
V-6 SERIAL NUMBER WILL START WITH 6 AND RIVIERA WILL START WITH 9

H = FLINT
B = BALTIMORE
C = SOUTH GATE
D = ATLANTA
K = KANSAS CITY, MO.
V = BLOOMFIELD
X = KANSAS CITY, KAN.
Y = WILMINGTON
Z = FREMONT

Serial number plate—1967 Buick Special

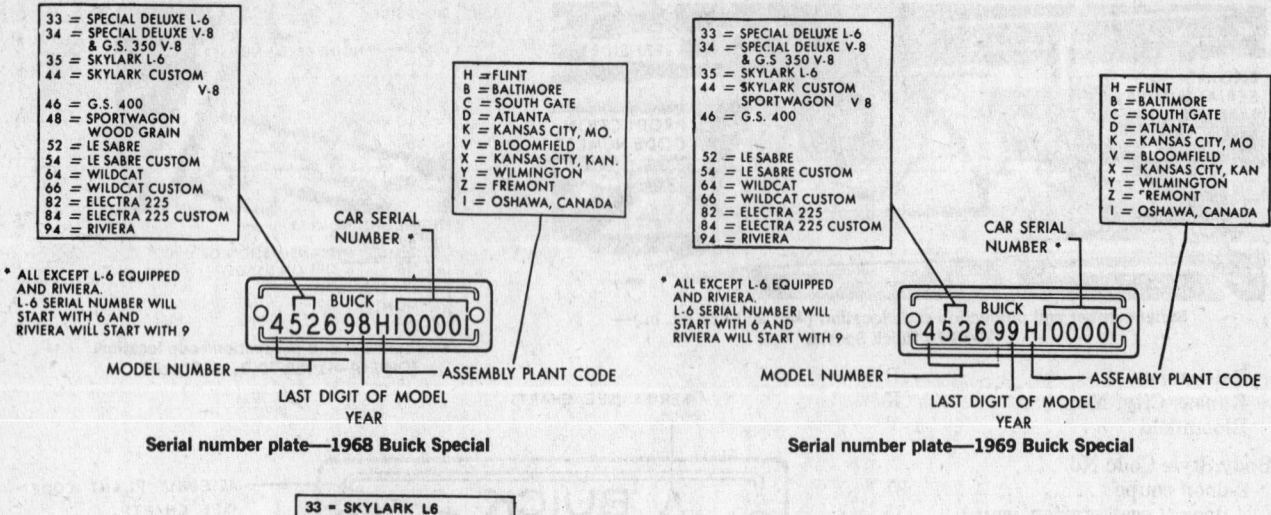

Serial number plate—1968 Buick Special

Serial number plate—1969 Buick Special

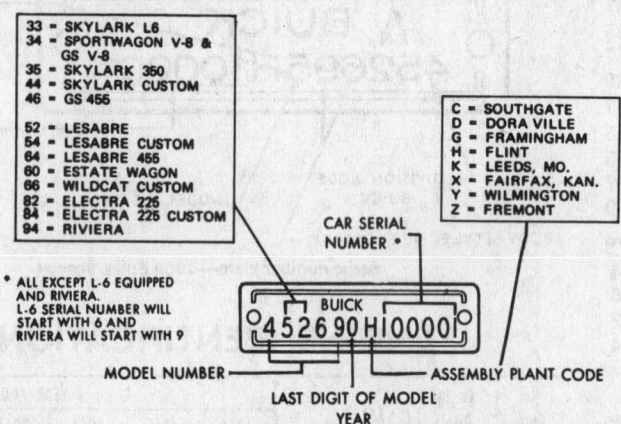

Serial number plate, 1970 Buick Special
(© Buick Div., G.M. Corp.)

front edge of the crankcase, just below the valve lifter cover. The engine number is on the left front edge of the same surface.

1967

The serial number identification plate is attached to the left front body hinge pillar.

On the 225 cu. in. V6 engine, the production code number is between the front and middle branches of the right exhaust manifold. The engine number is just below the front of the left cylinder head.

On the 300 and 340 cu. in. engines, the production code number is between the middle branches of the right exhaust manifold. The engine number is just below the front of the left cylinder head.

On the 400 cu. in. engine, the production code number is between the two forward branches of the right exhaust manifold and the engine number is between the rear branches.

1968

The serial number identification plate is attached to the left front body hinge pillar.

On the 250 cu. in. OHV6 engine, the engine number and the produc-

tion code number are on the right side of the engine, to the rear of the distributor.

On the 350 cu. in. V8 engine, the engine number is stamped on the front of the left cylinder bank. The production code number is between the left exhaust manifold and the two front spark plugs.

On the 400 cu. in. engine, the engine number is between the two front spark plugs and the left exhaust manifold, and the production code number is between the two rear plugs and the exhaust manifold, also on the left.

1969

The serial number identification plate is attached to the top of the instrument panel on the left side and can be seen through the windshield.

On the 250 cu. in. OHV6 engine, the engine number and the production code number are on the right side, rearward of the distributor.

On the 350 cu. in. V8 engine, the engine number is on the front of the left bank of cylinders. The production code number is between the left exhaust manifold and the two front spark plugs.

On the 400 cu. in. engine, the pro-

duction code number is between the two rear spark plugs and the left exhaust manifold. The engine number is between the two front plugs and the left exhaust manifold.

1970-71

The serial number identification plate is attached to the top left of the instrument panel. It can be seen through the windshield.

On the 250 cu. in. OHV6 engine, the engine number and production code number are on the right side, to the rear of the distributor.

On the 350 cu. in. V8 engine, the engine number is on the front of the left bank of cylinders. The production code number is between the left exhaust manifold and the two front spark plugs.

On the 455 cu. in. engine, the production code number is between the two rear spark plugs and the left exhaust manifold. The engine number is between the two front spark plugs and the exhaust manifold.

GENERAL ENGINE SPECIFICATIONS

YEAR	CU. IN. DISPLACEMENT	CARBURETOR	DEVELOPED HORSEPOWER @ RPM	DEVELOPED TORQUE @ RPM (FT. LBS.)	A.M.A. HORSEPOWER	BORE & STROKE (IN.)	ADVERTISED COMPRESSION RATIO	VALVE LIFTER TYPE	NORMAL OIL PRESSURE (PSI)
1964	V6—225	1-BBL.	155 @ 4600	225 @ 2400	33.8	3.750 x 3.400	9.0-1	Hyd.	33
	V8—300	2-BBL.	210 @ 4600	330 @ 2400	45.0	3.750 x 3.400	9.0-1	Hyd.	33
	V8—300	4-BBL.	250 @ 4800	352 @ 3000	45.0	3.750 x 3.400	11.0-1	Hyd.	33
1965	V6—225	1-BBL.	155 @ 4400	225 @ 2400	33.8	3.750 x 3.400	9.0-1	Hyd.	33
	V8—300	2-BBL.	210 @ 4600	310 @ 2400	45.0	3.750 x 3.400	9.0-1	Hyd.	33
	V8—300	4-BBL.	250 @ 4800	335 @ 3000	45.0	3.750 x 3.400	11.0-1	Hyd.	33
	V8—400	4-BBL.	325 @ 4400	445 @ 2800	56.1	4.1875 x 3.640	10.25-1	Hyd.	40
1966	V6—225	1-BBL.	160 @ 4200	235 @ 2400	33.8	3.750 x 3.400	9.0-1	Hyd.	33
	V8—300	2-BBL.	210 @ 4600	310 @ 2400	45.0	3.750 x 3.400	9.0-1	Hyd.	33
	V8—340	2-BBL.	220 @ 4000	365 @ 2800	45.0	3.750 x 3.850	9.0-1	Hyd.	33
	V8—340	4-BBL.	260 @ 4000	365 @ 2800	45.0	3.750 x 3.850	10.25-1	Hyd.	33
	V8—400	4-BBL.	325 @ 4400	445 @ 2800	56.1	4.187 x 3.640	10.25-1	Hyd.	40
1967	V6—225	2-BBL.	160 @ 4200	235 @ 2400	33.8	3.750 x 3.400	9.0-1	Hyd.	33
	V8—300	2-BBL.	210 @ 4600	310 @ 2400	45.0	3.750 x 3.400	9.0-1	Hyd.	33
	V8—340	2-BBL.	220 @ 4200	340 @ 2400	45.0	3.750 x 3.850	9.0-1	Hyd.	33
	V8—340	4-BBL.	260 @ 4200	365 @ 2800	45.0	3.750 x 3.850	10.25-1	Hyd.	33
	V8—400	4-BBL.	340 @ 5000	440 @ 3200	52.2	4.040 x 3.900	10.25-1	Hyd.	37
1968-69	L6—250	1-BBL.	155 @ 4200	235 @ 1600	36.0	3.875 x 3.530	8.5-1	Hyd.	38
	V8—350	2-BBL.	230 @ 4400	350 @ 2400	46.2	3.800 x 3.850	9.0-1	Hyd.	37
	V8—350	4-BBL.	280 @ 4600	375 @ 3200	46.2	3.800 x 3.850	10.25-1	Hyd.	37
	V8—400	4-BBL.	340 @ 5000	440 @ 3200	52.2	4.040 x 3.900	10.25-1	Hyd.	37
1970	L6—250	1-BBL.	155 @ 4200	235 @ 1600	36.0	3.875 x 3.530	8.5-1	Hyd.	37
	V8—350	2-BBL.	260 @ 4600	360 @ 2600	46.2	3.800 x 3.850	9.0-1	Hyd.	37
	V8—350	4-BBL.	285 @ 4600	375 @ 3000	46.2	3.800 x 3.850	9.0-1	Hyd.	37
	V8—350	4-BBL.	315 @ 4800	410 @ 3200	46.2	3.800 x 3.850	10.25-1	Hyd.	40
	V8—455	4-BBL.	350 @ 4600	510 @ 2800	59.5	4.3125 x 3.900	10.0-1	Hyd.	40
	V8—455	4-BBL.	360 @ 4600	510 @ 2800	59.5	4.3125 x 3.900	10.0-1	Hyd.	40
1971	L6—250	1-BBL.	N.A.	N.A.	36.0	3.875 x 3.530	8.5-1	Hyd.	37
	V8—350	2-BBL.	N.A.	N.A.	46.2	3.800 x 3.850	9.0-1	Hyd.	37
	V8—350	4-BBL.	N.A.	N.A.	46.2	3.800 x 3.850	9.0-1	Hyd.	37
	V8—350	4-BBL.	N.A.	N.A.	46.2	3.800 x 3.850	10.25-1	Hyd.	37
	V8—455	4-BBL.	N.A.	N.A.	59.5	4.3125 x 3.900	9.0-1	Hyd.	40
	V8—455	4-BBL.	N.A.	N.A.	59.5	4.3125 x 3.900	10.0-1	Hyd.	40

AC GENERATOR AND REGULATOR SPECIFICATIONS

YEAR	MODEL	ALTERNATOR			REGULATOR						
		Field Current Draw @ 12V.	Output @ Generator RPM		Model	Field Relay			Regulator		
			1100	6500		Air Gap (In.)	Point Gap (In.)	Volts to Close	Air Gap (In.)	Point Gap (In.)	Volts at 125°
1964	1100624	1.9-2.3	10A	42A	1119515	.015	.030	2.3-3.7	.060	.014	13.6-14.4
	1100663	1.9-2.3	10A	37A	1119515	.015	.030	2.3-3.7	.060	.014	13.6-14.4
	1100679	2.1-2.5	10A	55A	1119515	.015	.030	2.3-3.7	.060	.014	13.6-14.4
1965-66	1100705	2.2-2.6	10A	37A	1119515	.015	.030	2.3-3.7	.060	.014	13.6-14.4
	1100691	2.2-2.6	10A	42A	1119515	.015	.030	2.3-3.7	.060	.014	13.6-14.4
	1100708	2.2-2.6	10A	42A	1119515	.015	.030	2.3-3.7	.060	.014	13.6-14.4
	1100709	2.2-2.6	10A	55A	1119515	.015	.030	2.3-3.7	.060	.014	13.6-14.4
1967-69	1100761	2.2-2.6	10A	37A	1119515	.015	.030	2.3-3.7	.060	.014	13.6-14.4
	1100691	2.2-2.6	10A	42A	1119515	.015	.030	2.3-3.7	.060		13.6-14.4
1970	1100761	2.2-2.6	10A	37A	1119515	.015	.030	2.3-3.7	.060	.014	13.6-14.4
	1100691	2.2-2.6	10A	42A	1119515	.015	.030	2.3-3.7	.060	.014	13.6-14.4
1971	1100763	2.2-2.6	10A	37A	N.A.	Transistor type—no adjustment					
	1100694	2.2-2.6	10A	42A	N.A.	Transistor type—no adjustment					

TUNE-UP SPECIFICATIONS

YEAR	MODEL	SPARK PLUGS		DISTRIBUTOR		IGNITION TIMING (Deg.) ▲	CRANKING COMP. PRESSURE (Psi)	VALVES			FUEL PUMP PRESSURE (Psi)	IDLE SPEED (Rpm) ★
		Type	Gap (In.)	Point Dwell (Deg.)	Point Gap (In.)			Tappet (Hot) Clearance (In.) Intake	Exhaust	Intake Opens (Deg.)		
1964	V6—225 Cu. In.	44S	.035	30°	.016	5B	160	Zero	Zero	24B	4½–5¾	550
	V8—300 Cu. In.	44FFS	.035	30°	.016	5B	160	Zero	Zero	26B	5¼–6½	550
1965	V6—225 Cu. In.	44S	.035	30°	.016	5B	160	Zero	Zero	24B	4¼–5¾	550
	V8—300 Cu. In.	44FFS	.035	30°	.016	2½B	160	Zero	Zero	26B	4¼–5¾	550
	V8—400 Cu. In.	44S	.035	30°	.016	2½B	160	Zero	Zero	28B	4¾–6½	500
1966	V6—225 Cu. In.	44TS	.035	30°	.016	5B	160	Zero	Zero	24B	4¼–5¾	550
	V8—300 Cu. In.	44S	.035	30°	.016	2½B	160	Zero	Zero	26B	4–5¼	550
	V8—340 Cu. In.	44TS	.035	30°	.016	2½B	160	Zero	Zero	32B	4¼–5¾	550
	V8—400 Cu. In.	44S	.035	30°	.016	2½B	160	Zero	Zero	38B	5½–7	500
1967	V6—225 Cu. In.	44TS	.035	30°	.016	5B	160	Zero	Zero	24B	4¼–5¾	550
	V8—300 Cu. In.	44S	.035	30°	.016	2½B	160	Zero	Zero	30B	4¼–5¾	550
	V8—340 Cu. In.	44S	.035	30°	.016	2½B	160	Zero	Zero	32B	4¼–5¾	550
1968	L6—250 Cu. In.; 1-BBL.	46N	.035	32.5°	.019	4B	130	Zero	Zero	62B	3½–4½	500
	V8—350 Cu. In.	44TS	.030	30°	.016	TDC	160	Zero	Zero	24B	4½–5½	550
	V8—400 Cu. In.; 4-BBL.	44TS	.030	30°	.016	TDC	170	Zero	Zero	14B	5½–7	600
1969	L6—250 Cu. In.; M.T.	46N	.035	32.5°	.019	TDC	130	Zero	Zero	62B	3½–4½	700
	L6—250 Cu. In.; A.T.	46N	.035	32.5°	.019	4B	130	Zero	Zero	62B	3½–4½	550
	V8—350 Cu. In.; A.T.	R45TS	.030	30°	.016	TDC	160	Zero	Zero	24B	4½–5½	550
	V8—350 Cu. In.; M.T., G.S.	R45TS	.030	30°	.016	TDC	160	Zero	Zero	24B	4½–5½	700
	V8—400 Cu. In.; M.T., G.S.	R44TS	.030	30°	.016	2½A	170	Zero	Zero	14B	4½–7	700
	V8—400 Cu. In.; A.T.	R44TS	.030	30°	.016	TDC	170	Zero	Zero	14B	4½–7	550
1970–71	L6—250 Cu. In.; M.T.	R46N	.035	32°	.019	TDC†	130	Zero	Zero	16B	4–5	700
	L6—250 Cu. In.; A.T.	R46N	.035	32°	.019	4B†	130	Zero	Zero	16B	4–5	550
	V8—350 Cu. In.; M.T.	R45TS	.030	30°	.016	6B†	165	Zero	Zero	24B	4½–5½	700
	V8—350 Cu. In.; A.T.	R45TS	.030	30°	.016	6B†	165	Zero	Zero	24B	4½–5½	600
	V8—455 Cu. In.	R44TS	.030	30°	.016	6B†	180	Zero	Zero	18B	4½–5½	650
	V8—455 Cu. In., St. 1	R44TS	.030	30°	.016	10B†	180	Zero	Zero	18B	4½–5½	650

†—See engine decal for 1971 figures.

★—With synchromesh transmission in N and automatic in D. Add 50 rpm if air conditioned.

▲—With vacuum disconnected. NOTE: These settings are only approximate. Engine design, altitude, temperature, fuel octane rating and the condition of the individual engine are all factors which can influence timing. The limiting advance factor must, therefore, be the "knock point" of the individual engine.

B—Before top dead center.

A—After top dead center.

TDC—Top dead center.

GS--Gran Sport.

M.T.—Manual transmission.

A.T.—Automatic transmission.

Caution

General adoption of anti-pollution laws has changed the design of almost all car engine production to effectively reduce crankcase emission and terminal exhaust products. It has been necessary to adopt stricter tune-up rules, especially timing and idle speed procedures. Both of these values are peculiar to the engine and to its application, rather than to the engine alone. With this in mind, car manufacturers supply idle speed data for the engine and application involved. This information is clearly displayed in the engine compartment of each vehicle.

CYLINDER HEAD BOLT TIGHTENING SEQUENCE

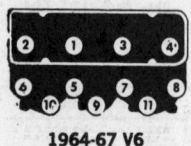

1964-67 V6

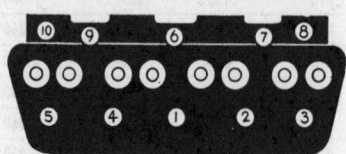

1964-67 300 cu. in.;
1966-67 340 cu. in.;
1967-69 400 cu. in.;
1968-69 350 cu. in.

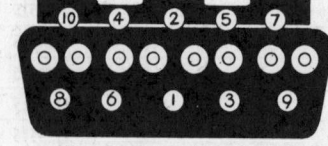

1970-71 350, 455 cu. in.

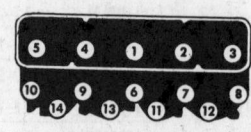

1964-66 400 (401)

GENERAL CHASSIS AND BRAKE SPECIFICATIONS

| YEAR | MODEL | CHASSIS | | | BRAKE CYLINDER BORE | | BRAKE DRUM | |
| | | Overall Length (In.) | Tire Size | Master Cylinder (In.) | Wheel Cylinder (In.) | | Diameter (In.) | |
					Front	Rear	Front	Rear
1964	All	203.5	6.50 x 14	1.00	1¹/₁₆	¹⁵/₁₆	9.5	9.5
1965	V6—4 dr. Sedans	203.4	6.95 x 14	1.00	1¹/₁₆	¹⁵/₁₆	9.5	9.5
	V6—Special, St. Wag.	203.4	7.35 x 14	1.00	1¹/₁₆	¹⁵/₁₆	9.5	9.5
	V8—Skylark	208.2	7.75 x 14	1.00	1¹/₁₆	¹⁵/₁₆	9.5	9.5
1966	Special Sdn.	204.0	6.95 x 14	1.00	1¹/₁₆	¹⁵/₁₆	9.5	9.5
	Special Wagon	204.0	7.75 x 14	1.00	1¹/₁₆	¹⁵/₁₆	9.5	9.5
	Sportswagon	209.0	8.25 x 4	1.00	1¹/₁₆	1.0	9.5	9.5
	Skylark	204.0	6.95 x 14	1.00	1¹/₁₆	¹⁵/₁₆	9.5	9.5
	Gran Sport	204.0	7.75 x 14	1.00	1¹/₈	¹⁵/₁₆	9.5	9.5
1967	Special, St. Wag.	209.3	7.75 x 14	1.00①	1¹/₈	¹⁵/₁₆	9.5	9.5
	Sedan	205.0	7.75 x 14	1.00①	1¹/₈	¹⁵/₁₆	9.5	9.5
	Sportswagon	214.0	8.25 x 14	1.00①	1¹/₈	1.0	9.5	9.5
1968–69	All Coupes (Exc. G.S.)	200.6	7.75 x 14★	1.00①	1¹/₈	⁷/₈	9.5	9.5
	All Sedans	204.6	7.75 x 14★	1.00①	1¹/₈	⁷/₈	9.5	9.5
	Sta. Wagon (Deluxe)	209.0	7.75 x 14★	1.00①	1¹/₈	⁷/₈	9.5	9.5
	Sta. Wagon (Spt. Wagon)	214.09	8.25 x 14#	1.00①	1¹/₈	1.00	9.5	9.5
	G.S. 350 & 400	200.6	7.75 x 14★★	1.00①	1¹/₈	⁷/₈	9.5	9.5
1970–71	All Coupes	202.2	8.24 x 14#	1.00①	1¹/₈	⁷/₈	9.495–9.505	9.495–9.505
	All Sedans	206.2	8.25 x 14#	1.00①	1¹/₈	⁷/₈	9.495–9.505	9.495–9.505
	Sport Wagon	212.7	8.55 x 14	1.00①	1¹/₈	⁷/₈	9.495–9.505	9.495–9.505
	Convertible	202.2	8.25 x 14#	1.00①	1¹/₈	⁷/₈	9.495–9.505	9.495–9.505

①—Optional disc brakes—1¹/₈ in.
★—Optional—8.25 x 14
#—Optional—8.55 x 14
★★—Optional—On 350, 8.25 x 14; Standard on 400, F70 x 14
G.S.—Gran Sport

WHEEL ALIGNMENT

| YEAR | MODEL | CASTER | | CAMBER | | TOE-IN (In.) | KING-PIN INCLINATION (Deg.) | WHEEL PIVOT RATIO | |
		Range (Deg.)	Pref. Setting (Deg.)	Range (Deg.)	Pref. Setting (Deg.)			Inner Wheel	Outer Wheel
1964–66	All	1N–0	½N	¼P–¾P	½P	⁷/₃₂–⁵/₁₆	8	21¹/₄°	20°
1967–69	All	1N–0	½N	0–1P	½P	¹/₈–¹/₄	8	20°	18¹/₂°
1970–71	All	1N–0	½N	0–1P	½P	¹/₈–¹/₄	8	20°	18¹/₂°

N—Negative
P—Positive

TORQUE SPECIFICATIONS

| YEAR | MODEL | CYLINDER HEAD BOLTS (FT. LBS.) | ROD BEARING BOLTS (FT. LBS.) | MAIN BEARING BOLTS (FT. LBS.) | CRANKSHAFT BALANCER BOLT (FT. LBS.) | FLYWHEEL TO CRANKSHAFT BOLTS (FT. LBS.) | MANIFOLD (FT. LBS.) | |
							Intake	Exhaust
1964	V6	65–70	30–35	100–125	140–160	50–60	25–30	10–15
	V8	65–70	30–35	50–55▲	140–160	50–60	25–30	10–15
1965–66	V6	65–70	30–35	100–125	140–160	50–60	25–30	10–15
	V8—300, V8—340	65–70	30–35	50–55▲	140–160	50–60	25–30	10–15
	V8—400	65–80	40–50	95–120	200–220	60–65	25–35	10–15
1967	V6	65–70	30–35	100–125	140–160	50–60	25–30	10–15
	V8	65–70①	30–35	50–55▲	140–160	50–60	25–30	10–15
1968–69	L6	90–95	35–45	60–70	Press Fit	55–65	25–35	■
	V8—350	65–70	30–35	50–55▲	140–160	50–60	25–30	10–15
	V8—400	100	45	110	200	50–60	55	18
1970–71	L6	95	35	60–70	Press Fit	55–65	25–35	■
	V8—350	75	35	95	120	60	55	18
	V8—455	100	45	110	200	58	55	18

①—V8—400 100–120 ft-lbs.
▲—Rear Main Bearing—65–75
■—Center Bolts 25–30; End Bolts 15–20.

CRANKSHAFT BEARING JOURNAL SPECIFICATIONS

YEAR	MODEL	MAIN BEARING JOURNALS (IN.)				CONNECTING ROD BEARING JOURNALS (IN.)		
		Journal Diameter	Oil Clearance	Shaft End-Play	Thrust On No.	Journal Diameter	Oil Clearance	End-Play
1964	V6	2.2992	.0005–.0021	.006	2	2.0000	.0020–.0022	.004–.008
	V8	2.2992	.0005–.0021	.006	3	2.0000	.0020–.0022	.004–.008
1965	V6	2.4995	.0005–.0021	.006	2	2.0000	.0020–.0023	.004–.008
	V8—300	2.4995	.0005–.0021	.006	3	2.0000	.0020–.0023	.004–.008
	V8—400	2.4985	.0005–.0019	.006	3	2.2495	.0020–.0023	.005–.012
1966	V6	2.4995	.0004–.0015	.006	2	2.0000	.0020–.0023	.006–.014
	V8—300	2.4995	.0004–.0015	.006	3	2.0000	.0020–.0023	.006–.014
	V8—340	2.9995	.0004–.0015	.006	3	2.0000	.0020–.0023	.006–.014
	V8—400	2.4985	.0004–.0015	.006	3	2.2495	.0020–.0023	.005–.012
1967	V6—225	2.4995	.0004–.0015	.006	2	2.0000	.0020–.0023	.006–.014
	V8—300	2.4995	.0004–.0015	.006	3	2.0000	.0020–.0023	.006–.014
	V8—340	2.9995	.0004–.0015	.006	3	2.0000	.0020–.0023	.006–.014
1968–69	L6—250	2.3004	.0003–.0029	.004	7	2.0000	.0007–.0027	.009–.013
	V8—350	2.9995	.0004–.0015	.006	3	2.0000	.0002–.0023	.006–.014
	V8—400	3.2500	.0007–.0018	.006	3	2.2500	.0002–.0023	.005–.012
1970–71	L6—250	2.3004	.0003–.0029	.002–.006	7	1.999–2.0000	.0007–.0027	.009–.013
	V8—350	2.9995	.0004–.0015	.003–.009	3	2.0000	.0002–.0023	.006–.014
	V8—455	3.2500	.0007–.0018	.003–.009	3	2.249–2.2500	.0002–.0023	.005–.012

VALVE SPECIFICATIONS

YEAR AND MODEL		SEAT ANGLE (DEG.)	FACE ANGLE (DEG.)	VALVE LIFT INTAKE (IN.)	VALVE LIFT EXHAUST (IN.)	VALE SPRING PRESSURE (VALVE OPEN) LBS.@ IN.	VALVE SPRING INSTALLED HEIGHT (IN.)	STEM TO GUIDE CLEARANCE (IN.)		STEM DIAMETER (IN.)	
								Intake	Exhaust	Intake	Exhaust
1964	V6	45	45	.391	.401	168 @ 1.26	1.64	.0020–.0025①	.0025–.0030①	.3407–.3412②	.3402–.3407②
	V8	45	45	.401	.401	168 @ 1.26	1.64	.0020–.0025①	.0025–.0030①	.3407–.3412②	.3402–.3407②
1965	V6	45	45	.391	.401	168 @ 1.26	1.64	.0020–.0025①	.0025–.0030①	.3407–.3412②	.3402–.3407②
	V8—300	45	45	.391	.401	168 @ 1.26	1.64	.0020–.0025①	.0025–.0030①	.3407–.3412②	.3402–.3407②
	V8—400	45	45	.431	.431	101 @ 1.16	1.60	.0020–.0030①	.0025–.0035①	.3720–.3730②	.3715–.3725②
1966	V6	45	45	.401	.401	164 @ 1.34	1.72	.0020–.0025①	.0025–.0030①	.3407–.3412②	.3402–.3407②
	V8—300	45	45	.393	.401	164 @ 1.34	1.72	.0012–.0032	.0025–.0030①	.3405–.3415②	.3402–.3407②
	V8—340	45	45	.399	.399	164 @ 1.34	1.72	.0012–.0032	.0025–.0030①	.3405–.3415②	.3402–.3407②
	V8—400	45	45	.431	.431	101 @ 1.16	1.60	.0020–.0030①	.0025–.0035①	.3720–.3730②	.3715–.3725②
1967	V6—225	45	45	.401	.401	168 @ 1.25	1.72	.0012–.0032	.0015–.0035①	.3405–.3415	.3402–.3407②
	V8—300	45	45	.393	.399	164 @ 1.34	1.72	.0012–.0032	.0015–.0035①	.3405–.3415	.3402–.3407②
	V8—340	45	45	.399	.399	164 @ 1.34	1.72	.0012–.0032	.0015–.0035①	.3405–.3415	.3402–.3407②
1968–69	L6—250	46	45	.388	.388	185 @ 1.27	1.66	.0010–.0027	.0010–.0027	.3410–.3417	.3410–.3417
	V8—350	45	45	.371	.384	180 @ 1.34	1.72	.0015–.0035	.0015–.0035①	.3720–.3730	.3720–.3730②
	V8—400	45	45	.418	.448	177 @ 1.45	1.89	.0015–.0035	.0015–.0035①	.3720–.3730	.3720–.3730②
1970–71	L6—250	46	45	.388	.388	185 @ 1.27	1.66	.0010–.0027	.0010–.0027	.3410–.3417	.3410–.3417
	V8—350	45	45	.381	.389	180 @ 1.34	1.72	.0015–.0035	.0015–.0035①	.3720–.3730	.3720–.3730②
	V8—455	45	45	.387	.458	177 @ 1.45	1.89	.0015–.0035	.0015–.0035①	.3720–.3730	.3720–.3730②

① Plus or minus 0.001 in. Guide tapers with larger diameter at bottom.
② Plus or minus 0.0005 in. Guide tapers with larger diameter at top.

LIGHT BULBS

1964–65

Automatic transmission control dial	1893; 2 CP
Back-up	1156; 32 CP
Clock dial	57; 2 CP
Courtesy light, instrument panel	89; 6 CP
Courtesy light, rear seat area	90; 6 CP
Dome, center roof	211; 15 CP
Glove box	1893; 2 CP

Headlight high beam indicator	158; 2 CP
Headlight, 5³/₄ in. dia., type 1 (inner)	4001; 37.5 W
Headlight, 5³/₄ in. dia., type 2 (outer)	4002-L; 37.5–55 W
Heater-air conditioner control dial	1893; 2 CP
Indicator lights (oil, temp. and gen.)	158; 2 CP
Instrument cluster dials	158; 2 CP
License	97; 4 CP
Radio dial	1881; 2 CP
Trunk	89; 6 CP

Turn signal and parking, front	1157; 32-4 CP
Turn signal, tail and stop	1157; 32-4 CP
Turn signal indicator	158; 2 CP

1966–71

Same as 1964–65 except the following:

Indicator lights	194; 2 CP
Instrument cluster	194; 2 CP

BATTERY AND STARTER SPECIFICATIONS

YEAR	MODEL	BATTERY: Ampere Hour Capacity	BATTERY: Volts	BATTERY: Group Number	BATTERY: Terminal Grounded	STARTERS: Lock Test Amps.	STARTERS: Lock Test Volts	STARTERS: Lock Test Torque	STARTERS: No-Load Test Amps.	STARTERS: No-Load Test Volts	STARTERS: No-Load Test RPM	STARTERS: Brush Spring Tension (Oz.)
1964-65	V6	61	12	28M	Neg.	300	4.0	N.A.	70	10.6	8,600	35
	V8	61	12	35M	Neg.	330	3.5	N.A.	85	10.6	4,350	35
1966	V6, V8—300	44	12	N.A.	Neg.	N.A.	N.A.	N.A.	62.5	10.6	6,200	35
	V8—340	61	12	28M	Neg.	N.A.	N.A.	N.A.	85	10.6	3,600	35
	V8—400	70	12	N.A.	Neg.	330	2.0	N.A.	120	10.6	4,700	35
1967	V6—225	44	12	17M2	Neg.	N.A.	N.A.	N.A.	58	10.6	6,200	35
	V8—300, 340	66	12	2SMD	Neg.	N.A.	N.A.	N.A.	85	10.6	3,600	35
1968	L6—250	45	12	17MJ1B	Neg.	N.A.	N.A.	N.A.	58	10.6	6,200	35
	V8—350	61	12	9MJ3F	Neg.	N.A.	N.A.	N.A.	85	10.6	3,600	35
	V8—400	70	12	9TJ3	Neg.	N.A.	N.A.	N.A.	85	10.6	3,600	35
1969	L6—250	44	12	Y54	Neg.	N.A.	N.A.	N.A.	68	10.6	6,700	35
	V8—350	61	12	9MJ3F	Neg.	N.A.	N.A.	N.A.	70	9.0	4,000	35
	V8—350 G.S.	61	12	9MJ3F	Neg.	N.A.	N.A.	N.A.	70	9.0	4,000	35
	V8—400 G.S.	70	12	9TJ3	Neg.	N.A.	N.A.	N.A.	61	9.0	5,200	35
1970-71	L6—250	44	12	Y55	Neg.	N.A.	N.A.	N.A.	49-87	10.6	6,200-10,700	35
	V8—350	61	12	R59	Neg.	N.A.	N.A.	N.A.	55-85	9.0	3,100-4,900	35
	V8—455 G.S.	70	12	R69	Neg.	N.A.	N.A.	N.A.	48-74	9.0	4,100-6,300	35

G.S.—Gran Sport

CAPACITIES

YEAR	MODEL	ENGINE CRANKCASE ADD 1 Qt. FOR NEW FILTER	TRANSMISSIONS Pts. TO REFILL AFTER DRAINING: Manual 3-Speed	Manual 4-Speed	Automatic	DRIVE AXLE (Pts.)	GASOLINE TANK (Gals.)	COOLING SYSTEM (Qts.) WITH HEATER
1964	V6	4	2	2 1/2	19	2	20	13
	V8	4	2	2 1/2	19	2	20	15
1965	V6	4	2	2 1/2	19	2 1/2	20	10.7
	V8—300	4	2	None	19	2 1/2	20	13.7
	V8—400	4	2	None	19	2 1/2	20	17.7
1966	V6	4	3 1/3	None	19	2 1/2	20	11.2
	V8—300	4	3 1/3	None	19	2 1/2	20	11.2
	V8—340	4	3 1/3	None	19	2 1/2	20	12.7
	V8—400	4	3 1/2	2 1/2	19	2 1/2	20	18.5
1967	V6	4	3 3/8	None	19	2 3/4	20	11.2
	V8—300, 340	4	3 3/8	None	19	2 3/4	20	12.7
1968-69	L6—250	4	3 3/8	None	20.0	2 3/4	20	11.3
	V8—350	4	3 3/8	None	20.0	2 3/4	20	13.5
	V8—400	4	3 3/8	3 1/2	23.0	2 3/4	20	16.2
1970-71	L6—250	4	3 1/2	None	20.0	3	20	16.4
	V8—350	4	3 1/2	None	20.0	3	20	16.4
	V8—350 G.S.	4	3 1/2	3	20.0	3	20	16.4
	V8—455 G.S.	4	3 1/2	3	23.0	3	20	19.1

FUSES AND CIRCUIT BREAKERS

1964–65

Back-up, stop, turn signal .... 15 AMP. 1 1/4 in.
Heater, air conditioner .... 30 AMP. 1 1/4 in.
Dome, trunk, cigar lighter.............. 15 AMP. 1 1/4 in.
Panel lights & rheostat .... 3 AMP. 1 1/4 in.
Radio .... 2.5 AMP. 7/8 in.
Tail, license, glove box, panel
light & clock .... 10 AMP. 1 1/4 in.
Transmission shift solenoid, wiper
& washer motor.................... 25 AMP. 1 1/4 in.
Headlamp and front parking .... 15 AMP. CB

1966

Same as 1964–65 except the following:
Instrument panel lights .... 4 AMP.
Radio.. 7.5 AMP.

1967–71

Radio, dir. signal .... 10 AMP. 1 1/4 in.
Stop lamps & hazard flasher............ 20 AMP. 1 1/4 in.
Heater & air conditioner .... 25 AMP. 1 1/4 in.
Instrument lights .... 4 AMP. 1 1/4 in.
Wiper, back-up & indicator lamps........ 20 AMP. 1 1/4 in.
Lighter & tail lamps.................... 20 AMP. 1 1/4 in.
Rear window defroster .... 5 AMP. 5/8 in.

Distributor

Detailed information on distributor drive, direction of distributor rotation, cylinder numbering, firing order, point gap, point dwell, timing mark location, spark plugs, spark advance, ignition resistor location, and idle speed, will be found in the Specifications tables of this section.

The distributor on the OHV 6 is located on the right side of the engine. On the V6, and most V8's, the distributor is located between the two cylinder banks, up front. On the 1964-66 400 cu. in. V8, the distributor is at the rear of the engine. The rotor turns clockwise, viewed from the top.

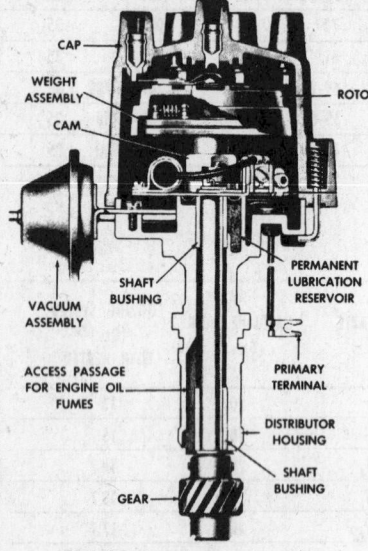

Distributor and cap assembly
(© Buick Div. G.M. Corp.)

Engine timing requirements are satisfied by centrifugal and vacuum advance mechanisms. Vacuum advance is controlled by the effort of a vacuum diaphragm working against spring tension. The diaphragm moves the breaker plate counterclockwise to advance the timing, and the springs move the plate clockwise to retard the timing. The degree of vacuum advance is determined by the amount of vacuum applied to the spring loaded diaphragm and breaker plate. Centrifugal advance is governed by engine speed. As speed increases, advance weights push against an advance cam which is integral with the distributor shaft. This causes the spark to be mechanically advanced. As speed decreases, the weights are returned by springs, returning the timing to the initial setting.

Caution Design of the V6-90° engine requires a special form of distributor cam. The dis-

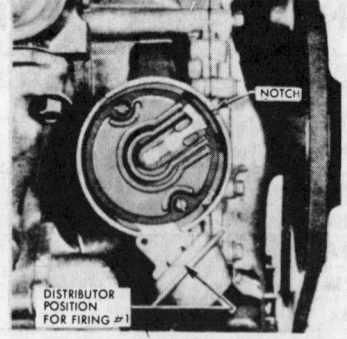

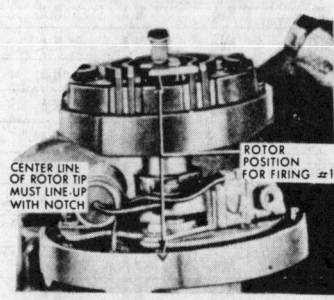

Installing distributor in V6 engine
(© Buick Div. G.M. Corp.)

tributor may be serviced in the regular way and should cause no more problems than any other distributor, if the firing plan is thoroughly understood.

The distributor cam is not ground to standard six cylinder indexing intervals. This particular form requires that the original pattern of spark plug wiring be used. The engine will not run in balance if No. 1 spark plug wire is inserted into No. 6 distributor cap tower, even though each wire in

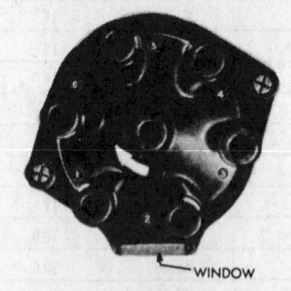

Spark plug wire installation in V6 engine
(© Buick Div. G.M. Corp.)

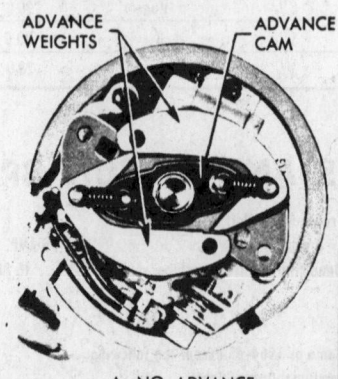

Centrifugal advance mechanism (© Buick Div. G.M. Corp.)

firing sequence is advanced to the next distributor tower. There is a difference between the firing intervals of each succeeding cylinder through the 720° engine cycle.

Distributor Removal

1. Remove distributor cap, primary wire and vacuum line at the distributor.
2. Scribe a mark on the distributor body, locating the position of the rotor and scribe another mark on the distributor body and en-

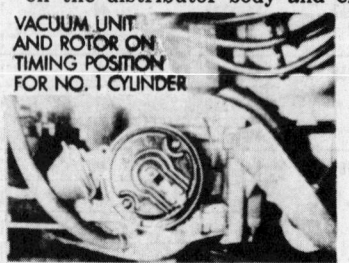

Installing distributor into V8 engine except 1964-66 400 cu. in.
(© Buick Div. G.M. Corp.)

gine block, showing the position of the body in the block.
3. Remove the hold-down screw and lift the distributor out of the block.

Distributor Installation

For firing order and cylinder numbering, see specifications.

1. If engine has been disturbed, rotate the crankshaft to bring the piston of No. 1 cylinder to the top of its compression stroke.
2. Position the distributor in the block with the rotor at No. 1 firing position. Make sure the oil pump intermediate drive shaft is properly seated in the oil pump.
3. Install the distributor lock but do not tighten.
4. Rotate the distributor body clockwise until the breaker points are just starting to open. Tighten the retaining screw.
5. Connect the primary wire and the vacuum line to the distributor, then install distributor cap.
6. Start the engine and check the timing with a timing light.

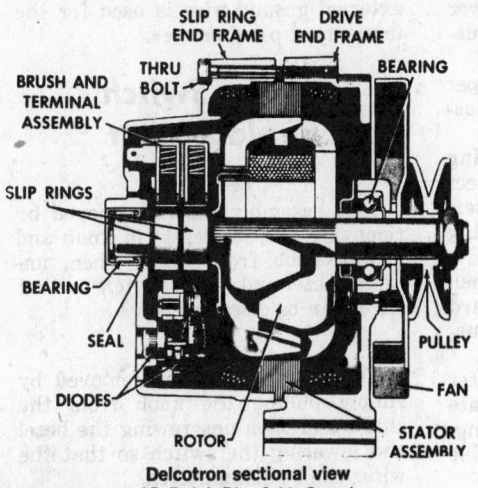

Delcotron sectional view
(© Buick Div. G.M. Corp.)

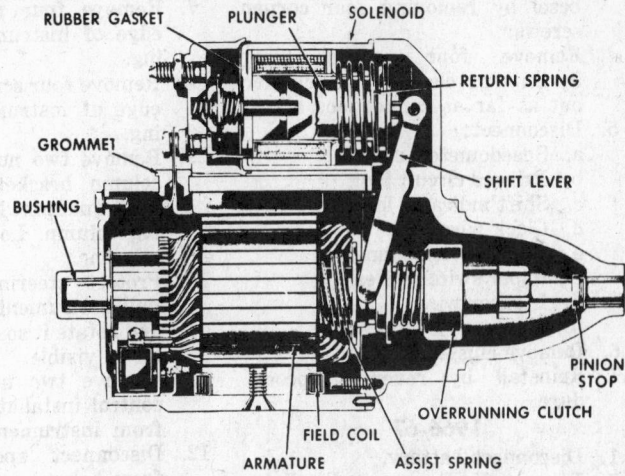

Starter motor and solenoid (© Buick Div. G.M. Corp.)

Generator, Regulator

Detailed facts on the generator and regulator can be found in the Specification tables.

General repair and troubleshooting can be found in the Unit Repair Section.

AC Generator (Delcotron)

An alternating current generator is used on Buick Special. This is to satisfy the increase in electrical loads imposed on the battery by conditions of traffic and driving patterns.

The Delcotron is covered in the Unit Repair Section.

Caution Since the Delcotron and regulator are designed for use on only one polarity system, the following precautions must be observed:

1. The polarity of the battery, generator and regulator must be matched and considered before making any electrical connections in the system.
2. When connecting a booster battery, be sure to connect the negative battery terminals together and the positive battery terminals together.
3. When connecting a charger to the battery, connect the charger positive lead to the battery positive terminal. Connect the charger negative lead to the battery negative terminal.
4. Never operate the Delcotron on open circuit. Be sure that all connections in the circuit are clean and tight.
5. Do not short across or ground any of the terminals on the Delcotron regulator.

6. Do not attempt to polarize the Delcotron.
7. Do not use test lamps of more than 12 volts for checking diode continuity.
8. Avoid long soldering times when replacing diodes or transistors. Prolonged heat is damaging to these units.
9. Disconnect the battery ground terminal when servicing any AC system. This will prevent the possibility of accidental reversing of polarity.

Delcotron R & R

Remove bolt holding tension bar to unit. Release drive belt. Unfasten mounting bolt to release Delcotron from engine. When reinstalling, adjust drive belt to allow 1/2 in. play on the longest run between pulleys.

Battery, Starter

Battery

A Delco 12 volt battery is used in all models. (See Battery and Starter Specifications.)

Starter

The starter circuit consists of the battery, battery cables, starting motor, starter motor solenoid switch, ignition-starter switch and the neutral safety switch, (used on cars with automatic transmission).

The starting motor and solenoid assembly is mounted on the flywheel upper housing, left side, on the 1965-66 400 cu. in. engine. On all other models the starting motor and solenoid assembly is on the right side.

The solenoid switch closes the circuit between the battery and the starting motor. It also operates the shift lever that moves the drive pin-

ion into mesh with the flywheel ring gear.

Starter R & R

Disconnect battery and solenoid wires. Remove attaching bolts and lift out starter.

Instruments

1964-69

The instrument cluster includes the speedometer head, the generator charge indicator, the oil pressure indicator, the temperature indicator, the fuel gauge, light switch, wiper and washer switch, starter and ignition switch and the cigarette lighter. On 1969 models, the ignition switch is on the right side of the steering column.

A printed circuit which is part of the speedometer housing is used to complete the circuit for the fuel gauge and the lights in the cluster.

1970

The instrument cluster assembly is comprised of three individual units. The left unit houses the fuel gauge, indicator lights and/or oil pressure gauge and ammeter. The center unit contains the speedometer. The right unit contains the optional clock or tachometer. Each unit contains its own printed circuit, fastened by three screws on the rear of the housing.

Cluster Removal and Installation

1964-65

1. Disconnect battery ground cable.
2. Remove cover extension assembly by removing four screws across bottom, then raising the entire extension to disengage four clips across top.
3. Remove heater control trim

bezel by removing four corner screws.

4. Remove four screws from instrument cluster. Pull cluster out as far as connections allow.
5. Disconnect:
   a. Speedometer cable
   b. Printed circuit plug
   c. Shift indicator lamp
   d. Clock wire
   e. Light switch connector
   f. Wiper switch wires
   g. Lighter wire
   h. Accessory switch wires
6. Remove cluster
7. Reinstall by reversing procedure.

1966-67

1. Disconnect battery.
2. Remove five screws and pull instrument panel upper cover rearward to remove.
3. Lower steering column and remove one 1/4 in. hex screw from lower edge of instrument panel housing.
4. Remove one 1/4 in. hex screw from lower edge of instrument panel housing through glove compartment hole.
5. Remove four remaining 1/4 in. hex screws from lower edge of instrument panel housing.
6. Remove six screws from across upper edge of instrument panel housing.
7. Disconnect speedometer cable.
8. Protect steering column from scratches, then pull instrument panel rearward and rotate it so back of cluster is accessible.
9. Disconnect from instrument cluster:
   A. Printed circuit connector.
   B. Clock connector.
   C. Shift quadrant light.
10. Remove four 1/4 in. hex screws and remove instrument panel cluster.
11. Install by reversing removal procedure.

1968

1. Disconnect battery.
2. Remove six screws from instrument panel compartment body and remove assembly.
3. Remove radio knobs and escutcheons.
4. Remove two 5/8 in. hex nuts and two screws from radio filler plate and remove plate. Do not remove radio.
5. Remove four 3/8 in. hex nuts at top underside of dash assembly and two screws at housing assembly. Pull instrument panel upper cover rearward to remove.
6. Remove four screws from steering column filler and remove filler. If car is equipped with air conditioner, remove one screw at left air conditioner plenum chamber and remove plenum from instrument panel housing.

7. Remove four nuts from lower edge of instrument panel housing.
8. Remove four screws across upper edge of instrument panel housing.
9. Remove two nuts from steering column bracket and disconnect shift quadrant link wire at steering column. Lower the steering column.
10. Protect steering column, then pull instrument panel rearward and rotate it so that back of cluster is visible.
11. Remove two nuts from heater control installation and separate from instrument panel housing.
12. Disconnect speedometer cable from below.
13. Disconnect from instrument cluster:
    A. Wiring harness clip.
    B. Printed circuit connector.
    C. Clock connector.
14. Disconnect accessory switch and cruise control switch assembly wires from top side of instrument panel housing.
15. Disconnect headlight switch connector.
16. Disconnect windshield wiper/washer switch connector.
17. Disconnect cigarette lighter connector.
18. Remove instrument panel housing assembly.
19. Remove four 1/4 in. hex screws and remove instrument panel cluster from instrument panel housing.
20. Install by reversing removal procedure.

1969-70

1. Disconnect battery ground cable.
2. Remove nine glove compartment screws and glove compartment.
3. Remove two screws through cluster housing. Remove two nuts above glove compartment opening. Pull instrument panel upper cover rearward to disengage three guide pins. Remove upper cover.
4. Remove steering column opening filler. Disconnect left air conditioning duct, if applicable.
5. Remove two nuts. Disconnect shift indicator link. Lower steering column.
6. Protect steering column.
7. Remove eight screws. Pull instrument cluster housing back. Rest housing on steering column and rotate so that back of cluster is visible.
8. Reinstall by reversing procedure.

Speedometer

To remove the speedometer head and printed circuit it is necessary to remove the instrument cluster. An

external ground wire is used for the instrument panel gauges.

Wiper Switch Replacement

1964-66

This assembly can be removed by removing the set screw in knob and sliding knob from shaft. Then, unscrew bezel and lower switch so that wires can be disconnected.

1967

The assembly can be removed by simply pulling the knob from the shaft, and then unscrewing the bezel and lowering the switch so that the wires can be disconnected.

1968-70

To remove the windshield wiper and washer switch assembly, pry the switch cover plate from the front of the cluster assembly. Remove the two retaining screws from the switch and pull the switch from the connector.

Ignition Switch Replacement

1964-68

1. Disconnect battery ground strap.
2. Turn key to Accessory position.
3. Insert stiff wire into hole in face of lock cylinder to depress lock pin. Rotate cylinder counterclockwise and pull out.
4. To remove ignition switch, remove retaining nut and lower switch.

To reinstall:

5. Install switch, tighten nut.
6. Insert key in cylinder, place cylinder in switch slightly counterclockwise from Accessory position, press inward and turn cylinder clockwise.

1969-70

The ignition switch and lock is located, not in the instrument panel, but in the steering column. The ignition lock also locks the steering and the transmission.

1. Disconnect battery ground cable.
2. Remove steering wheel, lock plate, and turn signal switch assembly.
3. Place switch in Accessory position.
4. Insert a thin screwdriver into the right hand slot next to the switch mounting screw boss, and depress spring latch at bottom of slot. Remove lock cylinder.
5. Using needle nose pliers, pull buzzer switch straight out of housing.
6. Remove two attaching screws and ignition switch.

To reinstall:

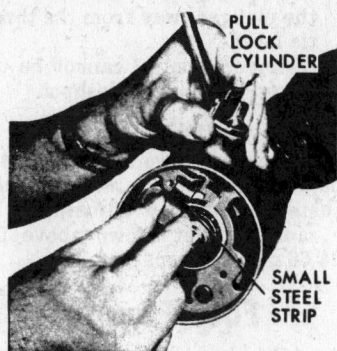

Removing ignition lock cylinder, 1969-70
(© Buick Div. G.M. Corp.)

7. Install ignition switch with two attaching screws. Install buzzer switch.
8. With key in lock cylinder assembly, hold cylinder sleeve and rotate lock knob on steering column clockwise against stop.
9. Insert cylinder. Push cylinder inward until end of lock cylinder touches end of lock sector.
10. Slowly rotate lock knob counterclockwise, pressing inward on cylinder until lock cylinder mates with lock sector in column.
11. Push lock cylinder inward until snap ring snaps into position. Check by pulling out on lock cylinder. If cylinder pulls out, start over at Step 4.
12. Reinstall steering wheel, lock plate, and turn signal switch assembly.

Lighting Switch Replacement

1. Disconnect battery.
2. Disconnect multiple connector from switch.
3. Pull switch knob to last notch and depress spring loaded latch button on top of switch while pulling knob and rod out of switch.
4. Remove escutcheon.
5. Install in reverse of above.

Rear Lights

The tail, stop and turn-signal bulbs are accessible from inside the trunk compartment on all except station wagons. On station wagons the bulbs are replaced by removing the lenses.

Brakes

The service brakes are of the conventional type, hydraulically operated. The lining is molded and attached to the shoes by tubular rivets. The primary shoe lining is shorter than the secondary lining and is of different composition.

Brake drum lining-contact-surfaces are cast iron, however, the drum proper is pressed steel with integral cooling fins.

The parking brake on all models is operated by a foot pedal and actuates the rear brakes only.

1964-69

Self adjusters have been added to the service brakes. The system is designed to react and progressively tighten the star wheel adjuster, a notch at a time, as required. The self adjusters only operate when the brakes are applied while the car is moving rearward.

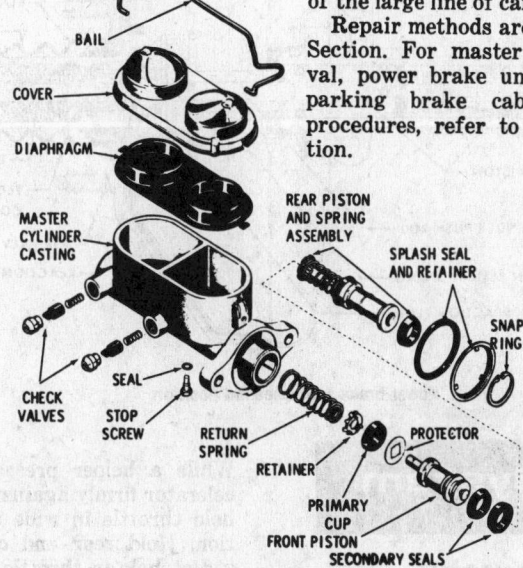

Dual master cylinder (Bendix)
(© Buick Div. G.M. Corp.)

For detailed service brake information, see Unit Repair Section.

Beginning 1967

A tandem master cylinder is standard equipment on all models. Information on this type system is covered in the Unit Repair Section.

1967-69

Disc brakes are optional on the front wheels of some models. Information on disc brakes is in the Unit Repair Section.

Power Brakes

1964-70

This installation is similar to that of the large line of cars.

Repair methods are in Unit Repair Section. For master cylinder removal, power brake unit removal, and parking brake cable replacement procedures, refer to the Buick Section.

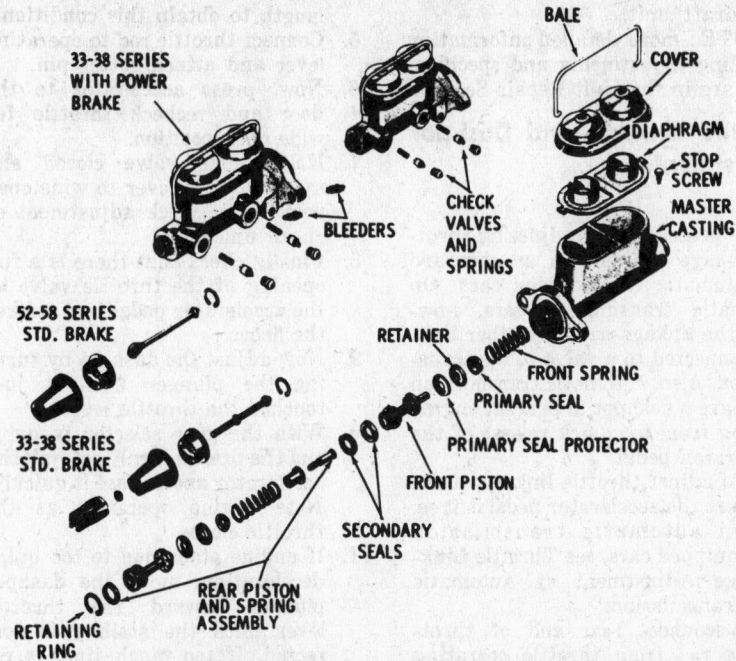

Dual master cylinder (Delco) (© Buick Div. G.M. Corp.)

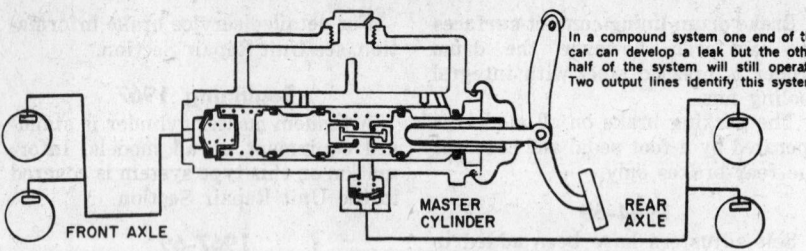

In a compound system one end of the car can develop a leak but the other half of the system will still operate. Two output lines identify this system.

FRONT AXLE　MASTER CYLINDER　REAR AXLE

Dual type master cylinder (© Buick Div. G.M. Corp.)

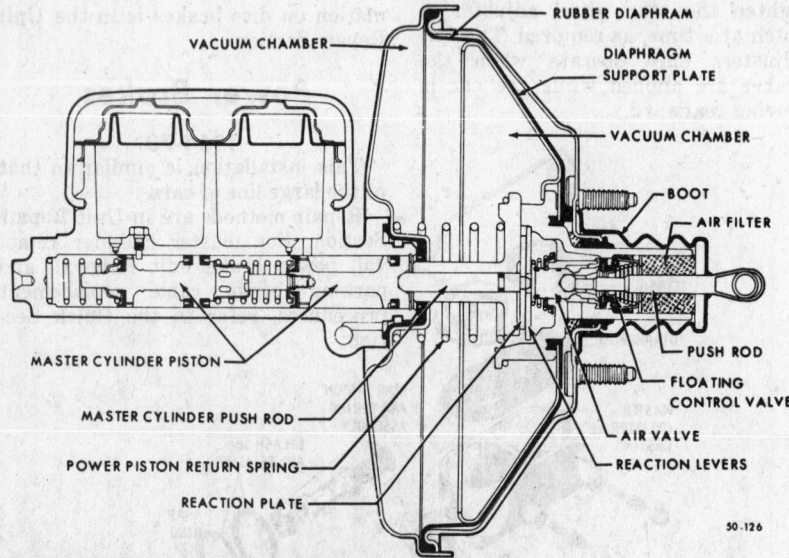

VACUUM CHAMBER　RUBBER DIAPHRAM　DIAPHRAGM SUPPORT PLATE　VACUUM CHAMBER　BOOT　AIR FILTER

MASTER CYLINDER PISTON　MASTER CYLINDER PUSH ROD　POWER PISTON RETURN SPRING　REACTION PLATE　PUSH ROD　FLOATING CONTROL VALVE　AIR VALVE　REACTION LEVERS

50-126

Power brake unit released position

Fuel System

Carburetor

The carburetor is a Carter or Rochester; one, two, or four-barrel downdraft unit.

NOTE: more detailed information including adjustments and specifications are in the Unit Repair Section.

Throttle Linkage and Dashpot Adjustments

1964-66

The procedure for adjusting throttle linkage is identical on standard or automatic transmission cars. On automatic transmission cars, however, the linkage actuates other linkage connected to a valve in the transmission. Also, automatic transmission cars have a dashpot to prevent engine stalling from too quick release of the accelerator pedal.

1. To adjust throttle linkage, make sure the accelerator pedal is free.
2. On automatic transmission equipped cars, see Throttle Linkage Adjustment of Automatic Transmission.
3. Disconnect rear end of throttle rod from throttle operating lever.
4. While a helper presses the accelerator firmly against the floor, hold throttle in wide open position. Hold rear end of throttle rod at hole in throttle operating lever. The rod end must be 1/16 in. short of entering the hole in the lever. Adjust throttle rod length to obtain this condition.
5. Connect throttle rod to operating lever and attach cotter pin.
6. Now, press accelerator to the floor and recheck throttle for wide open position.
7. Hold choke valve closed and move throttle lever to wide open position to check adjustment of choke unloader.
8. Finally check that there is a full opening of the throttle valve as the accelerator pedal just strikes the floor.
9. Now adjust the dashpot by turning the plunger until it just touches the throttle lever.
10. With the gear selector in drive and the brakes firmly set, jab the accelerator and release it quickly. Note engine operation as the throttle closes.
11. If engine stalls due to too quick deceleration, move the dashpot plunger toward the throttle lever until the stalling is corrected. If too much time is required for throttle to close, move

the plunger away from the throttle lever.
12. If correct control cannot be obtained, renew the dashpot.

1967-70

These models have a flexible cable type throttle linkage, which is not adjustable. Dashpot adjustment is the same as that shown above for 1964-66 models, Steps 9-12.

Fuel Pump

An AC fuel pump is used. The pump lever works from the underside of a camshaft eccentric. It is of the single-action diaphragm type and is equipped with a pulsation dampening chamber for stabilizing fuel flow. Beginning 1966, fuel pumps are sealed units. They are not to be repaired.

Beginning 1966, all air conditioned cars with V8 engines and all cars with 400 and 455 cu. in. engines have a special fuel pump. This pump has a vapor return line which returns hot fuel and fuel vapor to the fuel tank. The possibility of vapor lock is thus greatly reduced by keeping cool fuel circulating through the pump.

Fuel Filter

V6 Engine

The V6 engine gas filter is located in the carburetor fuel inlet. The element is sintered bronze and placed with the cupped end outward. The element is so spring loaded as to permit fuel by-pass in the event of element clogging.

The element should be removed and cleaned in a good solvent at 12,000 mile or 12 month periods.

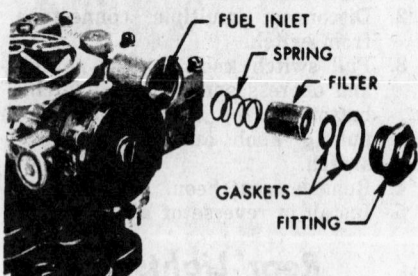

FUEL INLET　SPRING　FILTER　GASKETS　FITTING

Fuel filter (© Buick Div. G.M. Corp.)

1964-66 V8 Engine

On these models, the fuel filter is of the disposable, in-line type, mounted between the fuel pump and the carburetor. Air conditioned models have a fuel filter with a vapor return outlet. The filter should be replaced every 12,000 miles.

1967-70 OHV 6 & V8 Engines

These engines have a pleated paper fuel filter located in the carburetor inlet. The filter should be replaced every 12,000 miles.

Manifolds

OHV 6 Models

This engine uses a combined intake and exhaust manifold, equipped with thermostatic heat-riser valve.

To remove the manifold assembly disconnect the exhaust pipe flange and remove all connections to the carburetor. Take off the vacuum lines at the manifold and also at the carburetor.

Remove the carburetor. Unbolt the manifold from the side of the cylinder head.

Before reinstalling the manifold, thoroughly clean out the ports to prevent turbulence, particularly on the intake manifold.

Intake Manifold

V6 & V8 Engines

These engines have a low restriction, dual intake manifold. The manifold incorporates an exhaust heat passage to warm the carburetor throttle body. Engine coolant flows out of the engine through the water passages in the manifold and through the thermostat and water outlet elbow located at the front of the manifold. 1965-66 400 cu. in. engines do not have intake manifold water passages; they have a separate water manifold between the cylinder heads, forward of the intake manifold. The thermostat is contained in the water manifold. For manifold removal, see the Buick Section.

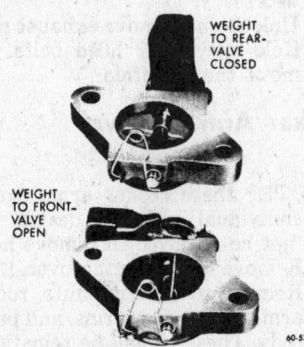

WEIGHT TO REAR-VALVE CLOSED

WEIGHT TO FRONT-VALVE OPEN

60-53

Manifold heat control valve

Exhaust Manifold

V6 & V8 Engines

The controlling source of exhaust heat is a heat control valve in one of the exhaust manifolds. This valve has a thermostat spring which tends to hold the valve closed under cold operating conditions.

This tension causes pressure build-up in the exhaust manifold, which forces exhaust through the crossover passage under the carburetor to the opposite exhaust manifold and out the pipe and muffler.

The 1964, 300 cu. in. V8 engine has no manifold exhaust heat valve. This engine has an aluminum intake manifold. See the Buick Section for general procedures on exhaust system and manifold removal.

Cooling System

The cooling system is pressurized to 15 psi. Coolant temperature is controlled by a thermostat housed in the forward (outlet) end of the intake manifold. In the 1965-66 400 cu. in. engine, the thermostat is housed in the water outlet manifold. In the OHV 6 engine, the thermostat is in the top front of the cylinder head. This thermostat controls circulation and temperature in the intake manifold as well as in the engine.

Caution

Be sure the thermostat is not reversed in its installed position. The temperature-sensitive side should extend toward the rear or down.

SPECIAL NOTE: it is advisable to use a highly inhibited ethylene-glycol antifreeze type coolant in the Buick Special, both winter and summer. The coolant should be changed every two years.

Water Pump

Removal

1. Drain cooling system.
2. Loosen belt or belts, then remove fan blades and pulley or pulleys from hub on water pump shaft. Remove belt or belts.
3. Disconnect hose from water pump inlet and heater hose from nipple. Remove bolts, then remove pump and gasket from the timing case cover.
4. Check pump shaft bearings for end-play or roughness. If bearings are not serviceable, the assembly must be replaced.

Installation

1. Install pump assembly with new gasket. Bolts and lock washers must be torqued evenly.
2. Connect radiator hose to pump inlet and heater hose to nipple. Fill cooling system and check all points of possible coolant leaks.
3. Install fan pulley or pulleys and fan blade. Install belt or belts and adjust for correct tension.

Engine Temperature and Oil Pressure Sending Unit

Temperature Gauge

A temperature switch controls the operation of the temperature indicator light located in the instrument cluster. The temperature switch is located in the thermostat housing on the OHV 6; in the right cylinder head on the 1965-66 400 cu. in. V8; and in the right front of the intake manifold on the V6 and all other V8 engines.

If the engine cooling system is not working properly and the coolant temperature reaches approximately 250°F, the temperature indicator light will burn in the instrument cluster.

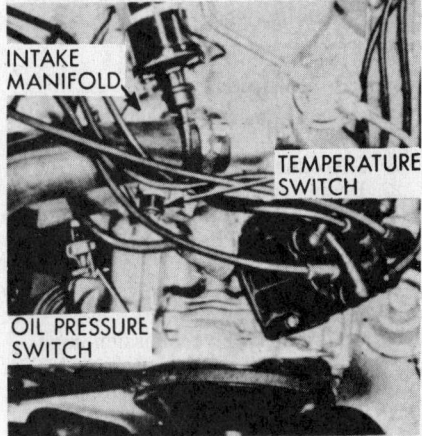

INTAKE MANIFOLD

TEMPERATURE SWITCH

OIL PRESSURE SWITCH

Oil pressure and temperature sending units—V8 except 1965-66 400 cu. in.
(© Buick Div. G.M. Corp.)

Oil Pressure

The oil pressure sending unit is located in the oil pump cover and operates an indicator light in the instrument cluster. On the 1965-66 400 cu. in. engine, the sending unit is on the right rear of the engine, in the main oil gallery.

If engine oil pressure drops below a safe level during operation, the circuit is completed through the sending unit to ground. This will cause the oil indicator light in the cluster to burn.

Engine

Exhaust Emission Control

1966-67

The Air Injection Reactor System (A.I.R.) is used on all General Motors cars (with a few design exceptions) for delivery into California.

Beginning 1968

In compliance with anti-pollution

laws involving all of the continental United States, the General Motors Corporation has adopted a special system of terminal exhaust treatment. This plan supersedes (in most cases) the method used to conform to 1966-67 California laws. The new system cancels out (except with stick shift and special purpose engine applications) the use of the A.I.R. method previously used.

The new concept, Combustion Control System (C.C.S.) utilizes engine modification, with emphasis on carburetor and distributor changes. Any of the methods of terminal exhaust treatment requires close and frequent attention to tune-up factors of engine maintenance.

Beginning 1968, all car manufacturers post idle speeds and other pertinent data relative to the specific engine-car application, in a conspicuous place in the engine compartment.

For details, consult the Unit Repair Section.

Engine Removal

This is a general procedure for all models.

1. Disconnect battery cables.
2. Drain cooling system and remove heater and radiator hoses.
3. Disconnect exhaust pipes at the manifolds.
4. Remove hood assembly.
5. Remove radiator.
6. On standard transmission models, disconnect the clutch control linkage. On automatic transmission equipped cars, disconnect the selector and throttle control linkage.
7. Disconnect the involved wiring such as generator, starter, sending units, etc.
8. Disconnect fuel line at fuel pump.
9. Disconnect accelerator linkage.
10. Disconnect vacuum line to power brake unit.
11. Remove the air conditioning compressor. Do not disconnect hoses.
12. Remove flywheel cover pan. Separate engine from transmission at bell housing.
13. Mark the front universal joint for reassembly, then disconnect the front joint by removing the U-bolts. Slide the driveshaft rearward as far as possible and tie to one side.
14. Disconnect both engine front mounts.
15. To afford clearance and prevent damage, it may be necessary to remove the oil filter unit.
16. Attach lifting device to engine.
17. Lift engine weight, then disconnect rear engine mount at transmission.
18. Lift the engine further and move it forward to clear the engine compartment.

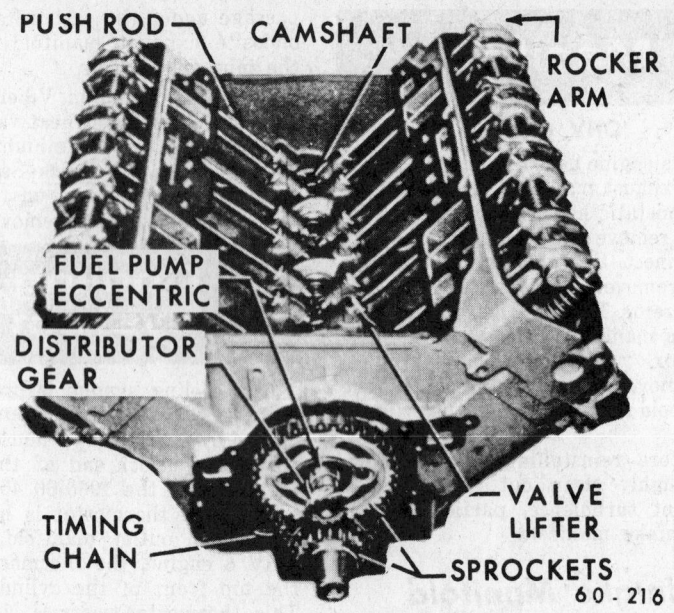

Engine valve mechanism—400 cu. in. 1967-70

19. Then raise fully and remove from the car.

Engine Installation

Install the engine in the reverse order of removal.

Cylinder Heads

Intake and Exhaust Manifold Removal

OHV 6

1. Disconnect accelerator rod, fuel and vacuum lines at carburetor.
2. Disconnect exhaust pipe at manifold flange, and remove manifold bolts and clamps. Remove intake and exhaust manifolds and carburetor as an assembly.

Intake Manifold Removal

V6 and V8

This is a general procedure for all models.

1. Drain cooling system.
2. Remove carburetor air cleaner. Disconnect all tubes and hoses from the carburetor. Disconnect and remove the coil.
3. Disconnect temperature indicator wire from sending unit.
4. Disconnect accelerator and transmission linkage at carburetor. Disconnect throttle return spring.
5. Slide front thermostat by-pass hose clamp back on the hose. Disconnect upper radiator hose at outlet.
6. Disconnect heater hose at the temperature control valve inlet. Force the end of the hose down to permit coolant to drain from intake manifold.
7. Remove manifold-to-head attaching bolts.

8. Remove intake manifold and carburetor as an assembly by sliding rearward to disengage the thermostat by-pass hose from the water pump. Remove intake manifold gasket.

Exhaust Manifold Removal

V6 and V8

This is a general procedure for all models.

1. Remove exhaust manifold-to-exhaust pipe attaching bolts.
2. On the right side, remove Delcotron, power steering pump, or air conditioning compressor as necessary.
3. Unlock and remove exhaust manifold-to-cylinder head bolts. Remove the manifold.

Rocker Arm Removal

OHV 6

NOTE: these rocker arms are of the individual pedestal design and need not be removed to remove head.

1. Remove rocker arm cover.
2. Remove rocker arm nuts, rocker arm bolts, rocker arms, and pushrods. These should be reinstalled in their original locations.

V6 and V8

1. Disconnect plug wires at the spark plugs and tie back out of the way.
2. Remove screws holding the rocker arm cover to the cylinder head. Remove rocker arm cover and gasket.
3. Remove rocker arm shaft bracket-to-cylinder head attaching bolts. Remove rocker arm and shaft assembly.
4. Remove the pushrods. These should be reinstalled in their original locations.

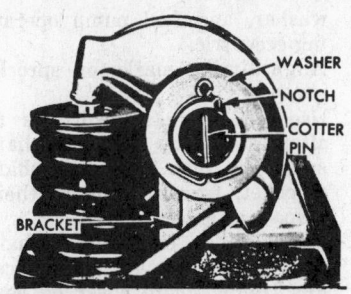

Rocker arm and shaft—1965-66 400 cu. in.
(© Buick Div. G.M. Corp.)

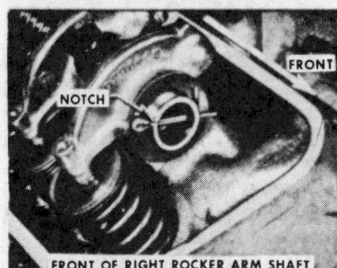

FRONT OF RIGHT ROCKER ARM SHAFT

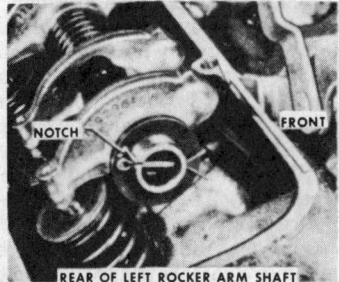

REAR OF LEFT ROCKER ARM SHAFT

Rocker arm and shaft—300 and 340 cu. in.
1964-67

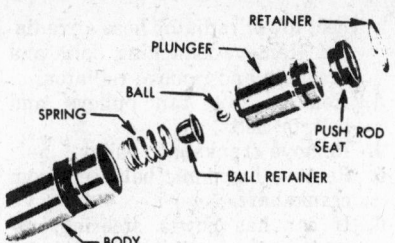

Valve lifter (© Buick Div. G.M. Corp.)

Delcotron and air conditioning compressor. Do not disconnect compressor hoses.

4. For left head only: Remove dipstick, power steering pump, and A.I.R. equipment. Do not disconnect steering unit hoses.

5. Disconnect exhaust pipe from manifold, remove manifold.

6. Remove rocker cover, rocker shaft assembly, and pushrods. The pushrods should be reinstalled in their original locations. It is important to protect the hydraulic lifters from dirt.

7. Loosen all head bolts, then remove all bolts and remove head and gasket.

1965-66 400 Cu. In.

1. Drain coolant and disconnect battery ground strap.
2. Remove intake manifold.
3. For right head only: Remove dipstick, automatic transmission filler pipe bracket, Delcotron

in the head during removal. Tape these items to the head to prevent loss. These same parts must be in the head during installation.

8. Loosen all head bolts, remove all bolts, and remove head and gasket.

V8 Except 1965-66 400 Cu. In.

Head removal procedure for these engines is the same as that detailed above for the 1965-66 400 cu. in. with the substitution of the following steps.

3. For right head only: Remove Delcotron and air conditioning compressor. Do not disconnect compressor hoses.

4. For left head only: Remove dipstick and power steering pump. Do not disconnect hoses.

5. Delete Step 5.

For valve guide replacement procedure, refer to the Buick Section.

Timing Case

Timing Gear Replacement

OHV 6

The timing gears are arranged so that (unless deliberately disturbed) the valve timing will remain as set at the factory when the engine was assembled. Unless the gears are badly worn or seriously damaged the valve timing will remain constant within reasonable limits.

If it becomes necessary to replace the timing gears due to wear or damage, remove the radiator core, disconnect the front motor mounts and jack up the front of the engine. Remove the fan belt, fan pulley, oil pan and timing case cover.

NOTE: it is recommended that the camshaft be removed from the car in order to remove and replace the gear in an arbor press.

Many successful mechanics prefer removing the camshaft in order to avoid possible risk when attempting to press a gear onto the shaft in place on the car. Sometimes when the gear is being pressed on in place on the car, damage results to the thrust washer behind the cam gear. Unfortunately, this damage is not noticed until the engine is started.

To replace the gear by removing the camshaft, remove the rocker arm assemblies and the distributor, all of the pushrods and all of the lifters. The camshaft may then be pulled out toward the front of the engine. It will be necessary to retime the ignition.

Runout of the camshaft timing gear should not exceed .004 in. Backlash between the two gears should not be less than .004 in. nor more than .006 in. End clearance of the thrust plate should be .001 to .005 in.

A different approach to this situa-

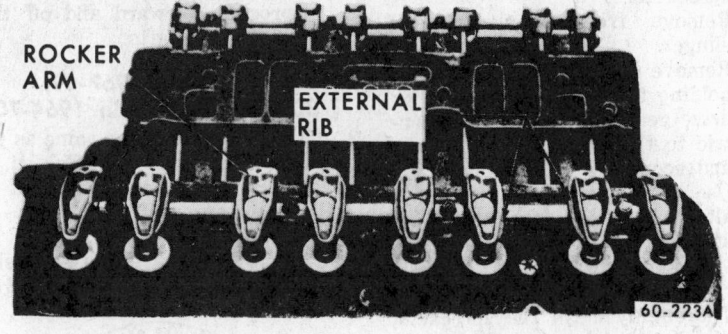

ROCKER ARM

EXTERNAL RIB

60-223A

Rocker arms and shaft, 455 cu. in. (© Buick Div. G.M. Corp.)

5. If lifters are to be serviced, remove them. If not, protect them with clean cloth. It is extremely important to protect the hydraulic lifters from dirt.

Cylinder Head Removal

OHV 6

Refer to the Chevrolet Section for this procedure. This engine is virtually identical to the Chevrolet 6 for the same model year.

V6

1. Drain coolant and disconnect battery ground strap.
2. Remove intake manifold.
3. For right head only: Remove

mounting bracket, and air conditioning compressor. Do not disconnect compressor hoses.

4. For left head only: Remove power steering pump. Do not disconnect hoses.

5. Remove water manifold.

6. Disconnect exhaust pipe from manifold, remove manifold.

7. Remove rocker arm cover, rocker arm and shaft assembly, and push rods. The pushrods should be reinstalled in their original locations. It is extremely important to protect the hydraulic lifters from dirt.

NOTE: due to lack of space in certain engine compartments, some of the bolts and pushrods must be left

tion, and certainly a quicker one, is as follows:

1. Very carefully center punch and drill two ¼ in. holes in opposite sides of the camshaft gear hub.
2. Break the fiber part of the camshaft gear away from the steel hub.
3. Split the steel hub with a cold chisel at the two newly drilled holes.
4. Remove broken camshaft gear and clean entire timing case area.
5. Place the new gear on the camshaft to line up with the keyway and the gear timing marks.

NOTE: be sure to allow for the helical cut on the gear when aligning the marks for timing.

6. Have a reliable helper, with the aid of a pinch bar, buck up against one of the camshaft lobes from underneath. (The success of the job depends upon this man's care in holding forward thrust on the camshaft. Failure on his part will allow the camshaft to be forced back and dislodge the oil sealing .expansion plug at the rear of the camshaft.)
7. With the aid of a 1¼ in. socket and a lead hammer, tap the new gear into place on the camshaft.

Caution The use of a dial indicator will reduce the possibility of driving the gear too far onto the camshaft. This would alter the desired camshaft thrust clearance of .001 to .005 in. Use care when approaching the final position of the gear on the shaft, because it is impossible to increase the thrust clearance without pulling the new gear. In the absence of a dial indicator, this end thrust can be measured with a feeler gauge. In this case the thrust clearance is to be measured between the camshaft gear hub and the thrust plate. A feeler gauge strip, inserted in either of the two large gear holes, will reach this point.

Timing Chain Removal

V6

1. Drain cooling system.
2. Disconnect radiator and heater hoses at water pump and discon-

nect lower radiator hose at radiator. Remove attaching bolts and brackets and remove radiator.
3. Remove fan, fan pulleys and belt, or belts.
4. Remove crankshaft pulley.
5. Remove harmonic balancer from crankshaft.
6. If car has power steering, remove the pump bracket bolts and move the steering pump out of the way.
7. Disconnect lines and remove the fuel pump.
8. Remove generator.
9. Remove distributor cap and spark plug wire retainers from brackets on rocker arm cover. Swing distributor cap, with wires, out of the way. Disconnect distributor primary wire.
10. Remove distributor.
11. Loosen and slide front clamp on thermostat by-pass hose rearward.
12. Remove bolts attaching timing cover to cylinder block. Remove two oil pan to timing cover bolts.
13. Lift off the timing case cover.
14. Temporarily install harmonic balancer bolt and washer to the end of crankshaft. Rotate crankshaft so sprockets are positioned as for timing, (shafts and sprocket O-marks on a centerline). Now remove harmonic balancer bolt with a sharp rap on the wrench handle to prevent changing the position of the sprockets.
15. Remove front crankshaft oil slinger.
16. Remove bolt and special washer holding the camshaft distributor drive gear and fuel pump eccentric to the camshaft. Slide gear and eccentric off the shaft.
17. Use two large . screwdrivers to alternately pry the camshaft sprocket then the crankshaft sprocket forward and off their respective shafts.
18. Thoroughly clean the sprockets, distributor drive gear, fuel pump eccentric and crankshaft oil slinger.

1965-66 400 Cu. In.

1. Drain cooling system.
2. Remove radiator core, fan belt, fan and pulley, and crankshaft balancer.
3. Remove all bolts attaching timing chain cover and water manifold to crankcase and cylinder heads. Do not remove five bolts attaching water pump to chain cover.
4. Remove chain cover and water manifold. Be careful not to damage oil pan gasket.
5. Remove crankshaft oil slinger. Remove bolt, lockwasher, plain

washer, and fuel pump operating eccentric.
6. Align timing marks on sprockets.
7. Use two large screwdrivers to alternately pry the camshaft sprocket, then the crankshaft sprocket, forward and off their respective shafts.
8. Thoroughly clean timing chain cover and crankcase surface. For timing chain cover oil seal replacement see Buick Section.

455 Cu. In.

1. Drain cooling system.
2. Remove radiator, fan, fan pulley and belt, and crankshaft pulley and pulley reinforcement.
3. Remove fuel pump and Delcotron.
4. Remove distributor.
5. Loosen clamp on thermostat bypass hose.
6. Remove harmonic balancer.
7. Remove timing chain cover to crankcase bolts. Remove oil pan to timing chain cover bolts. Thoroughly clean cover and crankcase surface. For timing chain cover oil seal replacement see Buick Section.
8. Align timing marks on sprockets.
9. Remove oil pan.
10. Remove crankshaft oil slinger. Remove camshaft sprocket bolts.
11. Use two large screwdrivers to alternately pry the camshaft sprocket, then the crankshaft sprocket, forward and off their respective shafts.

400 Cu. In. 1967-69, 300, 340, 350 Cu. In. 1964-70

This procedure is the same as that detailed above for the 455 cu. in. engine with the substitution of the following steps:

9. Delete Step 9.
10. Remove bolt, special washer, camshaft distributor drive gear,

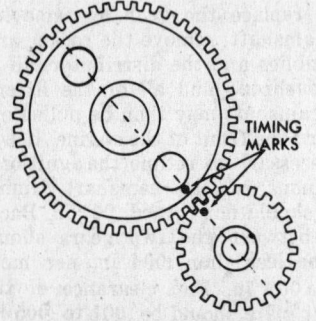

Timing marks—inline 6

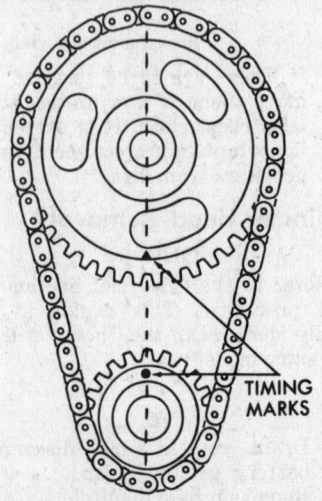

Valve timing mark V8

and fuel pump eccentric from camshaft. Remove crankshaft oil slinger.

Installation

V6 and V8, Except 1965-66 400 Cu. In. and 1970 455 Cu. In.

1. Make sure, with sprockets temporarily installed, that No. 1 piston is at top dead center and the camshaft sprocket O-mark straight down and on a centerline of both shafts.
2. Remove the camshaft sprocket and assemble the timing chain on both sprockets. Then slide the sprockets-and-chain assembly on the shafts with the O-marks in their closest together position and on a centerline with the sprocket hubs.
3. Assemble slinger on crankshaft with I.D. against the sprocket, (concave side toward front of engine).
4. Slide fuel pump eccentric on camshaft and Woodruff key with oil groove forward.
5. Install distributor drive gear.
6. Install drive gear and eccentric bolt and retaining washer. Torque to 40-55 ft. lbs.
7. Reinstall timing case cover by reversing removal procedure, paying particular attention to the following points.
   A. Remove oil pump cover and pack space around the oil pump gears completely full of petroleum jelly. There must be no air space left inside the pump. Reinstall the pump cover using new gasket.
   B. The gasket surface of the block and timing chain cover must be clean and smooth. Use a new gasket correctly positioned.
   C. Install chain cover being certain the dowel pins engage the dowel pin holes before starting the attaching bolts.
   D. Lube the bolt threads before installation and install them.
   E. If the car has power steering, the front pump bracket should be installed at this time.
   F. Lube the O.D. of the harmonic balancer hub before installation to prevent damage to the seal when starting the engine.

1965-66 400 Cu. In.

This procedure is similar to that outlined above for V6 and V8 except V8 1965-66 400 cu. in. and 1970 455 cu. in., with the substitution of the following steps:

7A. Delete this part of step 7. The oil pump on this engine is located in the oil pan and is thus not disturbed by this operation. Go on to parts B-F of Step 7.

455 Cu. In.

This procedure is similar to that above for V6 and V8, except V8 1965-66 400 cu. in. and 1970 455 cu. in., with the substitution of the following steps:

4. Delete Step 4.
5. Reinstall oil pan.
6. Install camshaft sprocket bolts. Torque to 22 ft. lbs.

Camshaft R & R

OHV 6 Cylinder

1. Drain cooling system.
2. Remove radiator, fan, and water pump pulley.
3. Remove grille.
4. Remove valve cover and gasket, then loosen rocker arm nuts and pivot rockers out of the way.
5. Remove pushrods.
6. Remove distributor, fuel pump, and spark plugs.
7. Remove coil, pushrod (tappet gallery) covers and gasket; reach in and remove tappets, keeping them in order.
8. Remove harmonic balancer, then loosen oil pan bolts and allow pan to drop.
9. Remove timing gear cover.
10. Remove two camshaft thrust plate bolts by rotating cam gear holes to gain clearance.
11. Remove the camshaft by pulling it straight forward.
NOTE: do not wiggle the camshaft; cam bearings could be dislodged.
12. If cam gear is to be replaced, press it from the shaft using an arbor press.
CAUTION: thrust plate must be positioned so that Woodruff key does not damage it during removal.

13. New cam gear must be pressed onto the shaft, with the shaft supported in back of the front bearing journal.
NOTE: the thrust plate end-play should be 0.001-0.005 in. If less than 0.001 in., replace spacer ring; if greater than 0.005 in., replace thrust plate.
14. Carefully install the camshaft into the engine, then turn crankshaft and camshaft so that timing marks coincide; tighten thrust plate bolts to 5-8 ft. lbs.
15. Check camshaft and crankshaft gear runout using a dial indicator. Cam gear runout should not exceed 0.004 in., crank gear runout should not exceed 0.003 in.
NOTE: if runout is excessive, remove gear and clean burrs from shaft.
16. Check gear backlash using a dial indicator; it should not exceed 0.006 in. and should not be less than 0.004 in.
17. To complete installation, reverse Steps 1-9.
NOTE: install distributor with No. 1 piston at TDC on compression stroke so that vacuum diaphragm faces forward and rotor points to No. 1 spark plug wire cap tower. Make sure oil pump drive shaft is properly indexed with distributor drive shaft.

Engine Lubrication

The engine lubrication system is the force feed type, in which oil is supplied under pressure to the crankshaft, connecting rods, camshaft bearings and valve lifters. Oil is supplied under controlled volume to the rocker arm bearings and pushrods. All other moving engine parts are lubricated by gravity flow or splash.

Oil Pump

On the V6 and most V8s, the oil pump is located in the timing chain cover, where it is connected by a drilled passage in the cylinder crankcase to an oil screen housing and standpipe assembly.

Oil is drawn into the pump through the screen and pipe and is discharged to the oil pump cover assembly. The cover assembly consists of an oil pressure relief valve, an oil filter by-pass valve and a nipple for installation of an oil filter. The oil pressure relief valve limits oil pressure to a maximum of 30 psi for most 1964-66 engines; 33 psi for the 1965-66 400 cu. in. engine; 40 psi for all 1967-70 V8s; and 45 psi for the OHV 6 engine. The oil filter bypass valve opens if the filter is clogged to the extent that a sufficient differential pressure exists between the filter inlet and outlet. This differential pressure is 4½ to 5 psi for 1964-66

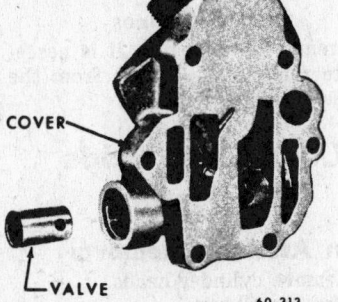

OIL PUMP COVER

SPRING

VALVE CAP

GASKET

VALVE

60-313

Pressure relief valve (© Buick Div. G.M. Corp.)

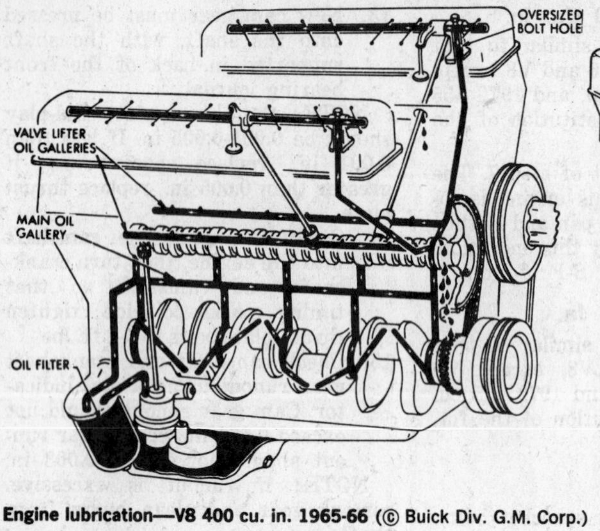

Engine lubrication—V8 400 cu. in. 1965-66 (© Buick Div. G.M. Corp.)

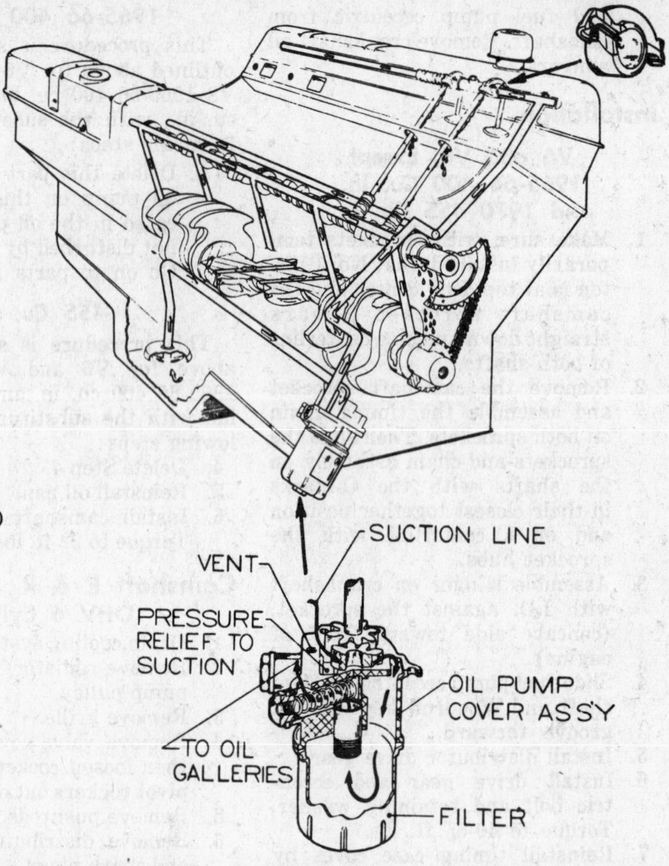

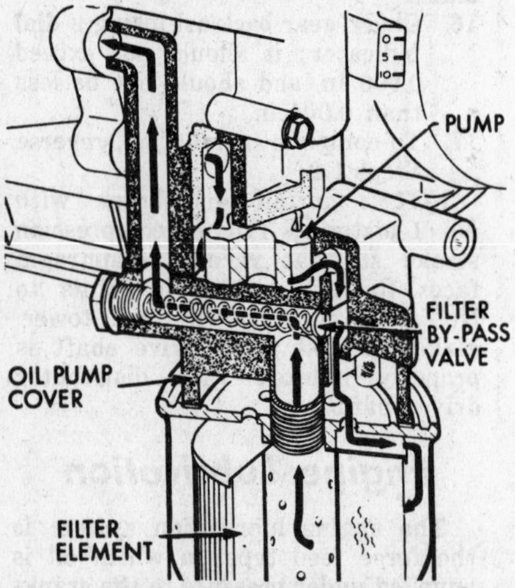

Oil flow through oil filter
(© Buick Div. G.M. Corp.)

Engine lubrication—all V8 except 1965-66 400 cu. in. (© Buick Div. G.M. Corp.)

engines and 15 psi for all 1967-70 engines. This is a safeguard for oil passage to the main engine oil galleries in case of filter stoppage.

On the OHV 6 and 1965-66 400 cu. in. V8 engine, the oil pump is located in the oil pan.

Oil Filter

An AC oil filter is mounted at the right front corner of the engine. On the 1965-66 400 cu. in. V8, the filter is mounted at the right rear. It requires no special tools and is completely disposable. For oil filter replacement, see the Buick Section.

Oil Pan Removal and Installation

V6 and V8 Engines

1. Raise car and support on stands.
2. Drain engine oil.
3. Disconnect exhaust pipe at crossover.
4. If standard transmission equipped, loosen clutch equalizer-to-frame attaching bolts.
5. Remove steering idler arm

bracket-to-suspension crossmember attaching bolts.
6. Support engine with a padded jack under the crankshaft pulley mounting.
7. Remove engine mounting bolts.
8. Raise engine.
9. Remove flywheel housing bolts. Then remove housing.
10. Remove oil pan bolts and lower the oil pan enough to remove oil pump pipe and screen-to-cylinder block attaching bolts.
11. Rotate crankshaft to provide maximum clearance at the front end of oil pan. Move the front of the pan to the right and lower the pan through opening between crossmember and steering linkage intermediate shaft.
12. Install by reversing removal procedure.

OHV 6 Engines

To remove the oil pan, it is necessary to remove the engine from the car. See Engine Removal.

Connecting Rods and Pistons

Piston Assembly Removal

1. Remove cylinder heads.
2. Remove oil pan.
3. Examine cylinder bores for top

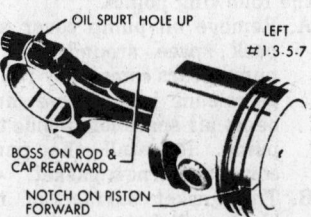

Piston and rod assembly, left bank—
300, 340, 350, 455, 1967-69 400
(© Buick Div. G.M. Corp.)

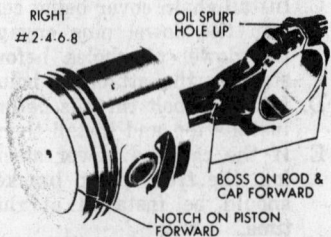

Piston and rod assembly, right bank—
300, 340, 350, 455, 1967-69 400
(© Buick Div. G.M. Corp.)

ridge. If ridge exists, remove it before taking pistons out.
4. Number all the pistons, connecting rods and caps.

With the V6 engine the right bank is numbered 2-4-6. The left bank, 1-3-5, from the front. The 1965-66 400 cu. in. V8 is numbered 2-4-6-8, left; and 1-3-5-7, right. All other V8 engines are numbered 1-3-5-7, left; and 2-4-6-8, right. The OHV 6 engine

is numbered 1-2-3-4-5-6, front to rear.

5. With No. 1 crankpin straight down, remove cap and bearing shell from No. 1 connecting rod. Install connecting rod bolt guides to hold upper half of the bearing shell in place.
6. Push piston and rod assembly up out of the cylinder. Then remove bolt guides and reinstall cap and bearing shell on the rod.
7. Remove the remaining rod and piston assemblies in the same manner.
8. Carefully remove old rings with piston ring expander.
9. Carefully press out the old pin.
NOTE: check the cylinder bores for distortion, taper or other damage. Any cylinders requiring attention may be bored or honed the same as any other conventional cast iron cylinder block.

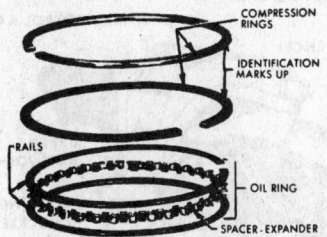

Piston rings, 1964-65 300 cu. in.
(© Buick Div. G.M. Corp.)

Fitting Rings and Pins

When new rings are installed without reboring the cylinders, cylinder wall glaze should be broken. This can be done by using the finest grade stones in a cylinder hone.

New piston rings must be checked for end gap and clearance in cylinder bores and ring grooves.

When fitting new rings to new pistons the side clearance for compression rings should be .003 to .005 in. for all engines except the OHV 6. For the OHV 6, top ring clearance should be .0012—.0027 in., and second ring, .0012—.0032 in. Side clearance of the oil ring should be .0035 to .0095 in., except for the OHV 6. OHV 6 oil ring side clearance should be .000—.005 in.

Check end gap of compression rings by placing them in the bore in which they will operate. Then push them to the bottom of the bore with a piston. Now measure the end gap in each ring. The end gap should be no less than .010 in. for 350 cu. in. engines, .013 in. for 1968-70 400 and 455 cu. in. engines, and .015 for all 1964-67 engines and the OHV 6 engine. Oil ring end gap need not be measured for all 1964-67 engines and the OHV 6 engine. For all 1968-70 V8 engines, oil end gap should be no less than .015 in.

If piston pin bosses are worn out of round or oversize, the piston and pin should be replaced. Oversize pins are not practical because the pin is a press fit in the connecting rod. Piston pins must fit the piston with an easy finger push at 70°F.

For V6 and V8 engines, the rod assemblies are correctly installed when the oil spurt holes are toward the camshaft, the boss on the rod is on the same side as the boss on the rod cap, and these bosses are toward the other connecting rod on the same crankpin. OHV 6 rod assemblies are correctly installed when all the oil spurt holes are toward the camshaft.

Connecting Rod Bearings

1. Remove connecting rod cap with bearing shell. Wipe all oil from the bearing area.
2. Place a piece of Plastigage lengthwise along the bottom center of the lower bearing shell. Then install cap with bearing shell and torque the bolt nuts to the specified figure from the Torque Specifications table.
NOTE: do not turn crankshaft.
3. Remove the cap and shell. The gauge material will be found flattened and adhering to either the bearing shell or the crankpin. Do not remove it.
4. Using the scale that comes with the gauge, measure the flattened gauge material at its widest point. The number within the graduation which comes closest to the width of the gauging material indicates the bearing to crankpin clearance in thousandths of an inch.
5. Desired clearance for a new bearing is .0002 in. to .0023 in. If the bearing has been in service, it is wise to install a new bearing if clearance exceeds .003 in.
6. If a new bearing is required, try a standard, then each undersize bearing in turn until one is found that is within specifications.
7. With the proper bearing selected, clean off the gauging material, reinstall the bearing cap and torque to the specified figure.
8. After the bearing cap has been torqued, it should be possible to move the connecting rod back and forth on the crankpin, the extent of end clearance.

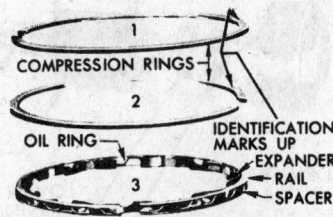

Piston rings, all V6, 1965-66 400, 1966-70 300-340-350
(© Buick Div. G.M. Corp.)

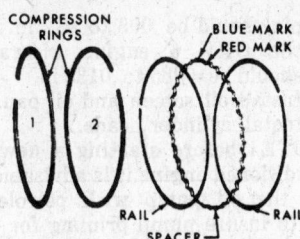

Piston rings, 1967-70 400, 455 cu. in.
(© Buick Div. G.M. Corp.)

Piston Assembly Installation

1. Carefully assemble the piston to the connecting rod (press in the pin).
2. Remove piston and rod from the press. Rock the piston on the pin to be sure pin or piston boss was not damaged during the pressing operation.
3. Install ring expander in lower ring groove. Position the ends of the expander above the piston pin where groove is not slotted. The ends of the expander must butt together.
4. Install oil ring rails over expander with gaps up on same side of piston as oil spurt hole in connecting rod.
5. Install compression rings, (with a ring expander) in top and center groove.
6. Coat all bearing surfaces, rings and piston skirt with engine oil.
7. Position the crankpin of the cylinder being worked on, down.
8. Remove connecting rod bearing cap and with upper bearing shell correctly seated in the rod, install connecting rod bolt guides.
9. Make sure the gaps in the two oil ring rails are up toward the center of the engine. Make sure the gaps of the compression rings are not in line with each other or the oil ring rails. Be sure the ends of the oil ring spacer-expander are butted and not overlapped.
10. With a good ring compressor, install the piston and rod assembly into the cylinder bore and carefully tap down until the rod bearing is soldidly seated on the crankpin.
11. Remove the connecting rod bolt guides and install cap and lower bearing shell. Torque to the specified figure.
12. Install other piston and rod assemblies in the same manner. When the assemblies are all installed, the oil spurt holes will be up. The rib on the edge of the rod cap will be on the same side as the conical boss on the connecting rod web. These marks will be toward the other connecting rod on the same crankpin.
13. Accumulated end clearance between rod bearings on any crank-

pin should be .005 to .012 in. For the OHV 6 engine, clearance should be .0085 to .0135 in.

14. Install oil screen and oil pan.
15. Install cylinder heads.

NOTE: before starting a new or reconditioned engine it is advisable to pack the oil pump with petroleum jelly to insure pump priming for immediate lubrication. See Timing Case Installation in this section.

After starting the engine, avoid high speeds but do not run on slow idle for a while. A better break-in speed is about 800-1,000 rpm for the first hour.

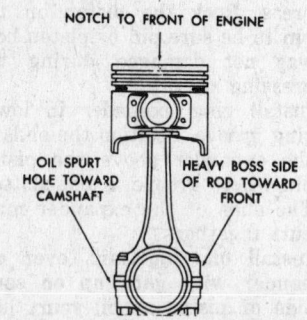

Piston and rod assembly—inline 6

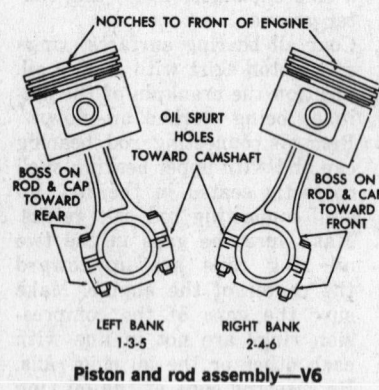

Piston and rod assembly—V6

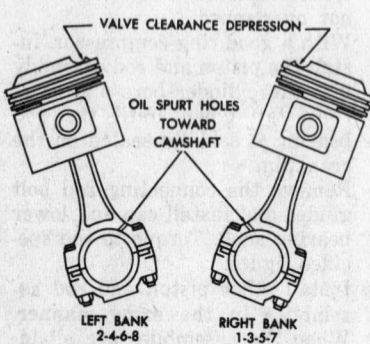

Piston and rod assembly—1965-66 400 cu. in.

Rear Main Bearing Oil Seal Replacement

For this procedure refer to the Buick Section. For the OHV 6 engine, refer to the Chevrolet Section.

Front Suspension

The Unit Repair Section covers front suspension repair, adjustment, installation of front wheel bearings and grease seals, and troubleshooting.

Figures for caster, camber, toe-in, king pin inclination, and turning radius can be found in the Wheel Alignment table.

Tire size figures can be found in the General Chassis and Brake Specifications table.

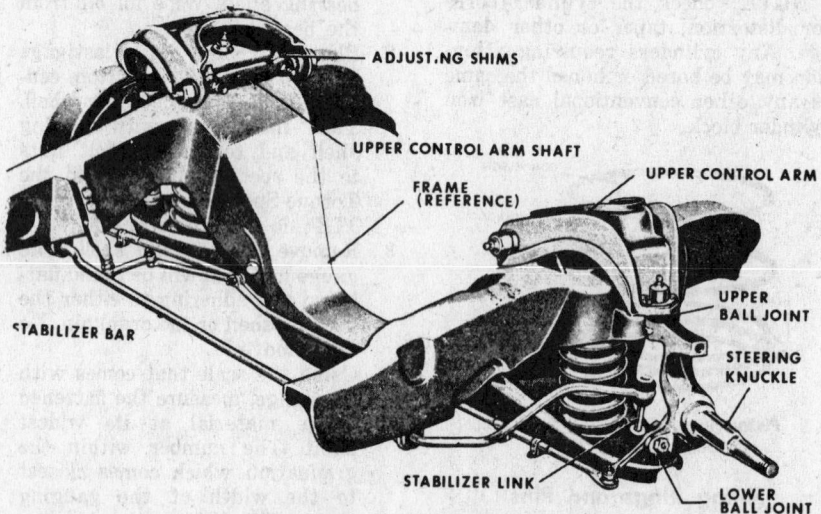

Front suspension (© Buick Div. G.M. Corp.)

For control arm, ball joint, and spring removal and replacement, and front wheel bearing adjustment, see the Buick Section.

Steering Wheel

Removal

1. Unplug the horn wire connector from the steering column.
2. On cars with standard wheel or optional wood-rim wheel, pull off cap, remove three screws and bushing spacer, receiver cup, and Belleville spring. On cars with horn actuator, remove screws securing actuator from underside of steering wheel, pull out lead connector plug, and remove actuator assembly.
3. Loosen steering wheel nut.
4. Apply steering wheel puller and pull wheel up to the nut. Now remove puller, nut and steering wheel.

Installation

NOTE: location marks are pro-

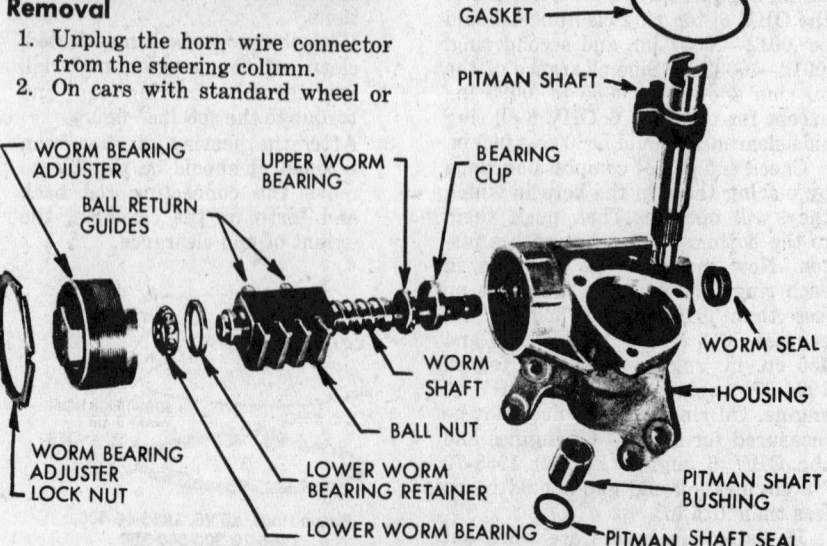

Manual steering gear, exploded view, typical (© Buick Div. G.M. Corp.)

vided on the steering wheel and shaft to simplify proper indexing at the time of installation.

1. Install wheel with the location mark aligned with that of the shaft.
2. Install the wheel nut and torque to 30 ft. lbs.
3. Reinstall horn button or actuator assembly.

Jacking, Hoisting

Jack car at front spring seat of lower control arm or center of cross member.

Jack car at rear at axle housing.

To lift at frame, use side rails in front of body floor pan and at rear side rail at lower control arm front pivot.

Steering Gear

Refer to the Unit Repair Section for adjustments and repairs to steering gear, both manual and power-assisted.

Manual

Removal and Installation

1. Raise and support front of car.
2. Remove pinch bolt securing flexible coupling flange to steering gear stub shaft.
3. Remove pitman arm retaining nut and pull off pitman arm.
CAUTION: do not hammer on puller. Hammering will damage the steering gear. If necessary, tapping on side of pitman arm may help in removal.

4. Remove steering gear to frame bolts and remove gear assembly.
5. Align mark on gear stub shaft with tab on coupling.
6. Install gear assembly to frame. Tighten bolts to 70 ft. lbs.
7. Install and tighten pinch bolt to 30 ft. lbs.
8. Reconnect pitman arm. Torque to 140 ft. lbs.

Power

Removal and Installation

This procedure is identical to that outlined above for the manual steering gear, with the substitution of the following step:

1. Raise and support front of car. Disconnect pressure and return hoses from steering gear. Elevate hose ends to prevent oil loss.

Clutch

A single plate, dry disc clutch is used in cars with manual transmissions. The unit is conventional in design with a diaphragm spring assembly. The clutch is not adjustable except for pedal clearance.

Clutch Removal

1964-70

1. Mark universal joint and transmission shaft companion flange for proper indexing at time of installation. Remove two U-bolts and disconnect driveshaft at the front joint. Slide the driveshaft rearward as far as possible and tie to one side.

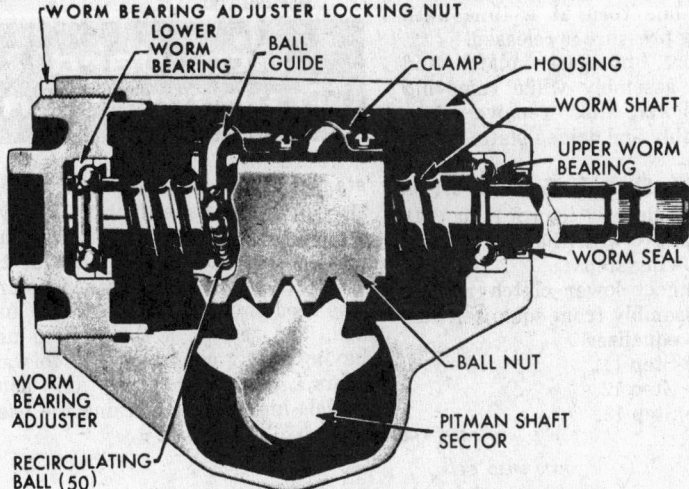

WORM BEARING ADJUSTER LOCKING NUT
LOWER WORM BEARING — BALL GUIDE — CLAMP — HOUSING — WORM SHAFT — UPPER WORM BEARING — WORM SEAL — BALL NUT — PITMAN SHAFT SECTOR — WORM BEARING ADJUSTER — RECIRCULATING BALL (50)

Manual steering gear cutaway (© Buick Div. G.M. Corp.)

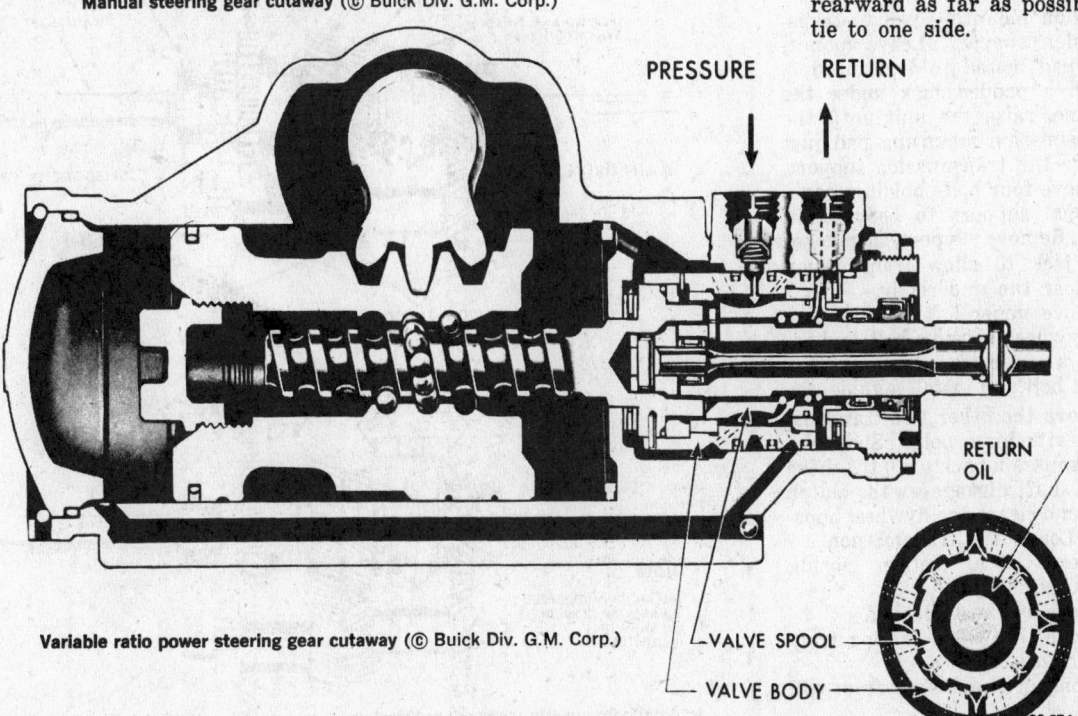

PRESSURE   RETURN

RETURN OIL

VALVE SPOOL — VALVE BODY

Variable ratio power steering gear cutaway (© Buick Div. G.M. Corp.)

90-27A

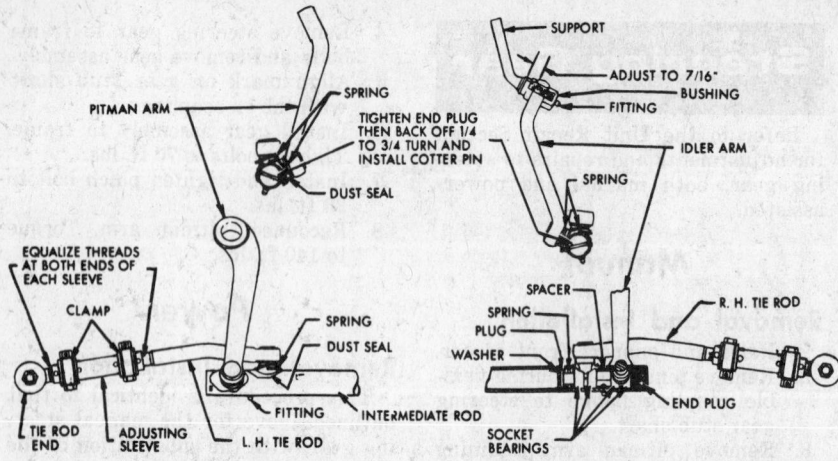

Manual and power steering linkage (© Buick Div. G.M. Corp.)

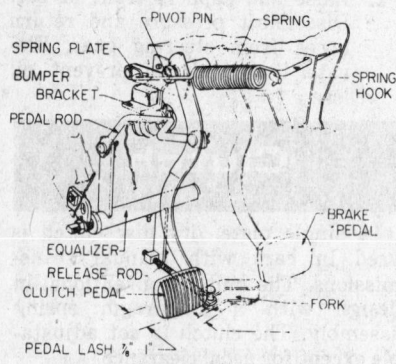

Clutch linkage

2. Disconnect shift linkage from transmission.
3. Disconnect speedometer cable at transmission.
4. Loosen all three exhaust pipe ball joints to permit transmission and rear of engine to be lowered.
5. Remove two bolts holding transmission mounting pad to transmission support. Leave mounting pad bolted to transmission.
6. With a padded jack under the engine, raise the unit until the transmission mounting pad just clears the transmission support.
7. Remove four bolts holding transmission support to body members. Remove support, then lower the jack to allow transmission to clear the underbody.
8. Remove upper left transmission to flywheel housing bolt and install a guide pin. Remove lower right bolt and install a guide pin.
9. Remove the other two transmission attaching bolts. Slide the transmission back until the drive gear shaft disengages the clutch disc and clears the flywheel housing. Lower the transmission.
10. Remove pedal return spring from clutch fork.
11. Remove flywheel housing.
12. Remove throw-out bearing from clutch fork.
13. Disconnect clutch fork from ball stud.

14. Mark clutch cover and flywheel to assure proper balance on reassembly.
15. Loosen clutch cover to flywheel bolts one turn at a time until spring pressure is released.
16. Support pressure plate and cover assembly while removing last bolts, then remove cover assembly and driven plate.

1965-66 Skylark Gran Sport
This procedure is identical to that described above with the substitution of the following steps:
10. Disconnect lower clutch release rod assembly from equalizer. Remove equalizer.
11. Delete Step 11.
12. Delete Step 12.
13. Delete Step 13.

Clutch Installation
Install clutch by reversing removal procedure. Use a clutch aligning pilot or a spare main drive gear through the hub of driven plate and into the pilot bushing. Be sure to align the clutch cover-to-flywheel index marks.

Clutch Linkage Adjustment
Check pedal lash (free-play) by pushing down on the pedal by hand. Lash should be approximately ¾ in. measured at the pedal pad.
1. Make sure the pedal is at full release position, contacting the rubber bumper stop.
2. Adjust clutch release rod underneath car to give zero lash at the clutch pedal.
3. Back off release rod adjustment 2-3 turns to give ¾ in. lash at pedal pad.
4. Tighten locknut on clutch release rod.

Standard Transmission

There are four basic types of standard transmissions used. First, the early three-speed unit with syncromesh on second and third gears only, which was discontinued at the end of the 1965 model year. This unit was used only with column shift linkage. The second, a three-speed unit having syncromesh on all forward gears, was introduced on some models in 1965 and became the basic

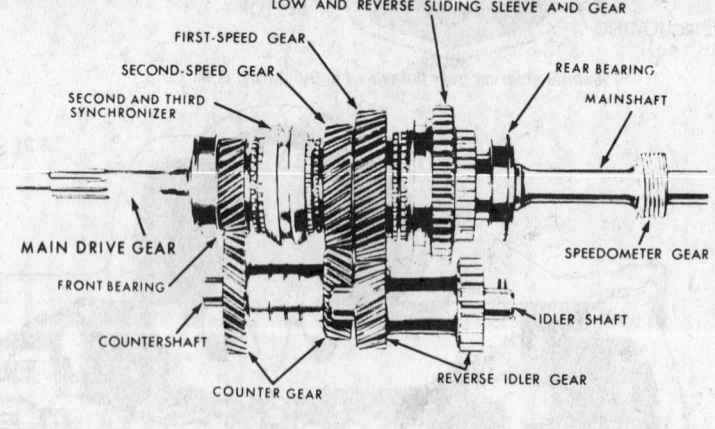

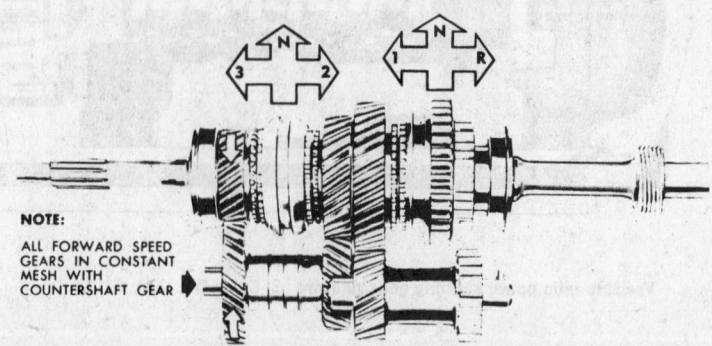

NOTE:
ALL FORWARD SPEED GEARS IN CONSTANT MESH WITH COUNTERSHAFT GEAR

Internal components, heavy-duty three-speed transmission (© Buick Div. G.M. Corp.)

three speed unit in 1966. This transmission was used with floorshift linkage only on 1965 Skylark Gran Sport models. The third is a heavy duty three-speed unit, available only with floorshift linkage. This unit is the same as that used on the larger series of Buicks. The fourth is a four-speed, all syncromesh unit, equipped only with floorshift linkage.

For repair procedures see Unit Repair Section.

Removal and Installation

See Clutch paragraphs.

Shift Linkage Adjustment

For information and adjustments on Hurst floorshift linkages, refer to the Unit Repair Section.

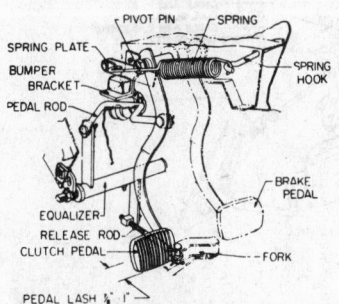

Clutch linkage

Adjustment Procedure —Three-Speed Column Shift
1964-66
1. Place column shift lever in Neutral.
2. Place transmission levers into neutral positions.
3. Loosen shift rod adjusting clamp bolts.
4. Install a 3/16 in. dia. rod through first-reverse lever and selector plate at bottom of steering column.
5. Push second-third shift rod through clamp until it sticks out from clamp 1/4 in.
6. Tighten shift rod adjusting clamps to 17-23 ft. lbs.
7. Lift column shift lever straight up toward steering wheel several times to assure a free neutral crossover. If neutral detents do not line up, loosen first-reverse rod clamp, pull first-reverse rod through clamp no more than 3/16 in., and tighten clamp.

1967
This procedure is similar to that outlined above for the 1964-66 models, with the substitution of the following steps:
4. Install a 3/16 in. dia. rod through first-reverse lever, selector plate,

and second-third lever at bottom of steering column.
5. Push second-third shift rod through clamp until it sticks out from clamp 1/4-1/2 in.

1968
This procedure is similar to that for 1964-66 models, with the substitution of the following step:
4. Install a 3/16 in. dia. rod through second-third lever, selector plate, first-reverse lever, and alignment plate.

1969-70
1. Place column shift lever in Reverse.
2. Loosen first-reverse clamp bolt.
3. Place transmission first-reverse lever into reverse position. Tighten clamp bolt to 17-23 ft. lbs.
4. Shift transmission levers into neutral positions.
5. Loosen second-third clamp bolt.
6. Install a 3/16 in. dia. rod through second-third lever, selector plate, first-reverse lever, and alignment plate.
7. Tighten second-third clamp bolt to 17-23 ft. lbs.

Adjustment Procedure —Three-Speed Floorshift
1. Place transmission levers into neutral.
2. Loosen shift rod adjusting clamp bolts.
3. Place a 5/16 in. dia. rod in notch in rear portion of shift bracket assembly.
4. Move both shift levers back against rod.
5. Tighten shift rod adjusting bolts to 17-23 ft. lbs.

Adjustment Procedure —Four-Speed Floorshift
1. Place transmission levers in neutral positions.
2. Place a 5/16 in. dia. rod in rear lower portion of shift bracket assembly.
3. Adjust all three shift levers back against rod.
4. Tighten adjusting clamp bolts to 17-23 ft. lbs.

Automatic Transmission

Only the Turbo Hydra-Matic 350 is covered in this section. For removal, installation, and stator and detent switch adjustments for the Super Turbine 300, 400, and Turbo Hydra-Matic 400, see the Buick Section.

Shift linkage adjustment for all models is covered in this section.

The transmission model number is usually located on the right side of

the transmission case. On the Turbo Hydra-Matic 350, the model designation and model year are stamped on the intermediate clutch accumulator cover on the middle right side.

Turbo Hydra-Matic 350

Removal
1. Raise and support car.
2. Disconnect exhaust pipe.
3. Remove driveshaft.
4. Place jack under transmission and fasten transmission to jack.
5. Remove vacuum line from vacuum modulator.
6. Disconnect oil cooler lines.
7. Remove detent cable from accelerator lever assembly and from detent valve link. Do not bend cable.
8. Remove transmission mounting pad to crossmember bolts. Remove crossmember.
9. Disconnect speedometer cable.
10. Disconnect shift linkage.
11. Remove transmission filler pipe.
12. Support engine at oil pan.
13. Remove flywheel cover pan. Mark flywheel and converter for reassembly. Remove converter to flywheel bolts.
14. Unbolt transmission and move rearward. Support converter and lower transmission.

Installation
1. Raise transmission into position.
2. Install transmission case to engine block bolts. Torque to 30-40 ft. lbs.
3. Install flywheel to converter pump bolts. Torque to 25-35 ft. lbs.
4. Install cross member support and mounting pad.
5. Remove transmission jack and engine support.
6. Install flywheel cover pan.
7. Install transmission filler pipe with new O-ring.
8. Reconnect speedometer cable.
9. Install driveshaft.
10. Reconnect exhaust pipe or pipes.
11. Install detent cable on detent valve link and on accelerator lever assembly. Do not bend cable.
12. Install oil cooler lines. Install vacuum line to modulator.
13. Add three pints of transmission fluid. With lever in Park, start engine. Do not race engine. Shift through each range. Check fluid level with lever in park, engine running, and vehicle on level surface. Add fluid to bring level to 1/4 in. below ADD mark. This allows for heat expansion

Detent Cable Adjustment
Refer to the accompanying illustration for this procedure.

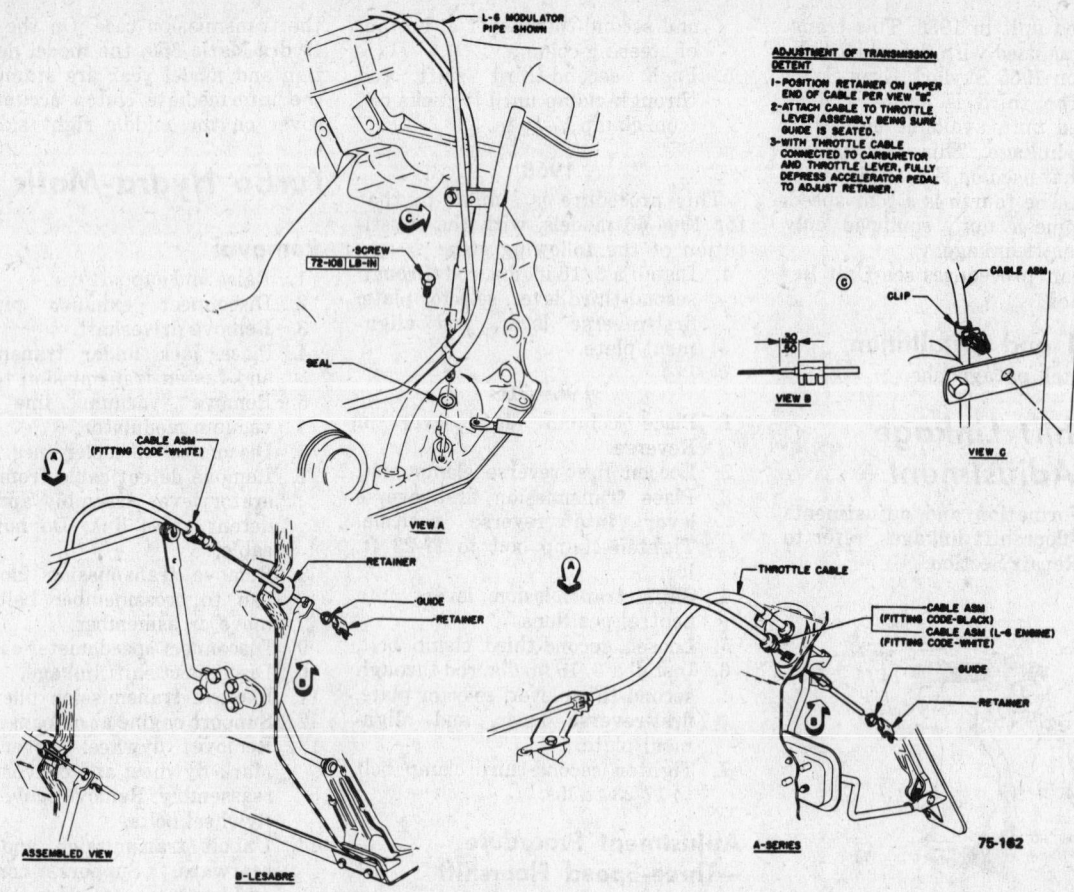

Transmission detent cable adjustment, Turbo Hydra-Matic 350
(© Buick Div. G.M. Corp.)

Shift Linkage Adjustment

Column Shift All Models
1. Place selector lever in Drive.
2. Loosen adjusting clamp bolt.
3. Place lever at transmission in drive position.
4. Tighten clamp bolt to 17-23 ft. lbs.

Console Shift 1965-66
1. Place selector lever in Park.
2. Loosen adjusting clamp bolt.
3. Place transmission in park position.
4. Tighten clamp bolt to 17-23 ft. lbs.

Console Shift 1967
1. Place selector lever in Drive.
2. Loosen adjusting clamp bolt.
3. Place transmission in drive position.
4. Tighten clamp bolt to 17-23 ft. lbs.

Console Shift 1968-70
This linkage uses a cable rather than a shifting rod.
1. Loosen trunnion bolt.
2. Set selector lever in Drive.
3. Place transmission in drive position.
4. Tighten trunnion bolt to 6-9 ft. lbs.

5. For 1969-70 models only: Place selector lever in Park. Place transmission in park position. Push back drive rod (to steering column) up to stop and hold lightly. Tighten back drive rod clamp screw.

Neutral Safety Switch
This switch prevents the engine from being started in any transmission position except Neutral or Park. The back-up light switch is combined with the neutral safety switch. On column shift cars, the switch is located on the steering column under the instrument panel. On console shift cars, the switch is located inside the console. When the neutral start portion of the switch is correctly adjusted, the back-up portion is adjusted automatically. Slotted mounting screw holes permit switch movement for adjustment.

CAUTION: when checking to see if engine will start in transmission positions other than Drive or Park, always hold the service brake firmly.

U Joints, Drive Lines

The driveshaft is a one piece unit with a splined slip yoke and a univer-sal joint at the transmission end, and a second universal joint at the differential end. This arrangement is similar to that used in the larger series of Buicks, 1969-70 except Riviera. For further information, consult the Buick Section.

Beginning 1969
Nylon-injected composite universal joints are used. To replace universal joints:
1. Mark driveshaft rear yoke and companion flange for correct alignment upon re-assembly.
2. Remove U-bolts from rear axle drive pinion companion flange.
3. If bearing tie wire has been removed, use band or tape to hold bearings onto the journals, to prevent loss of bearing rollers when joint is disconnected.
4. Slide assembly to the rear to clear splines at transmission output shaft.
5. By using a piece of pipe or similar tool, slightly larger than 1⅛ in. to encircle the bearing shell, apply force on the yoke until downward movement of the yoke and stationary position of journal force the bearing assembly almost out of the top of the yoke (the force applied on the yoke will shear nylon retainers which lock bearings in place).

6. Rotate propeller shaft 180° and repeat preceding step to partially remove the opposite bearing.
7. Complete removal of these bearings by tapping around the circumference of exposed portion of bearing.
8. Remove journal from driveshaft rear yoke.
9. Remove bearings and journal from splined yoke in the same way.

NOTE: new bearings and journal assembly kits must be used upon reassembly. The kit includes snap-rings and Delrin washers.

10. Install by inserting one bearing one-quarter way in one side of splined yoke, using brass hammer.
11. Insert journal into splined yoke (with dust shields installed).
12. Install opposite bearing, ensuring that the bearing rollers do not jam on journal. Check free rotary movement of journal in bearing.

13. Now, press both bearings into place (just far enough to install snap rings).
14. Assemble opposite end universal in the same way.

Rear Axles, Suspension

The rear axle assembly is of the semi-floating type in which the car weight is carried on the axle shafts through bearings in the rear axle tubes. Car drive is transmitted from the axle housing to body members through two lower and two upper control arms. Large rubber bushings at either end of these arms are designed to absorb vibration and noise. The arms are angle mounted to control sidewise movement of the suspension.

Repair instructions are given in the Unit Repair Section.

Rear Axle Assembly Removal

It is not necessary to remove the rear axle assembly for normal repairs. However, if the housing is damaged, the rear axle assembly may be removed and installed using the following procedure.

1. Raise rear of car high enough to permit working under the car.

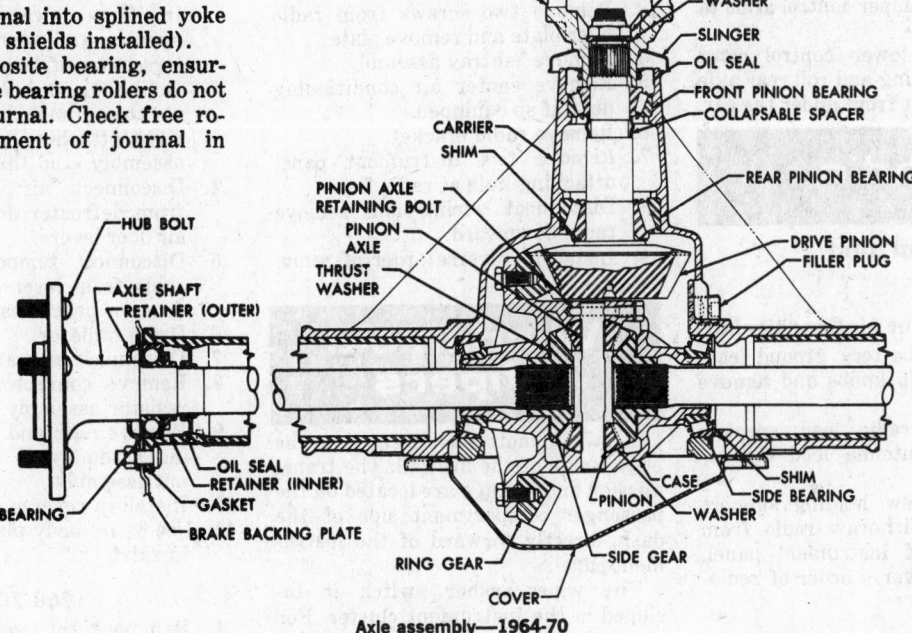

Axle assembly—1964-70

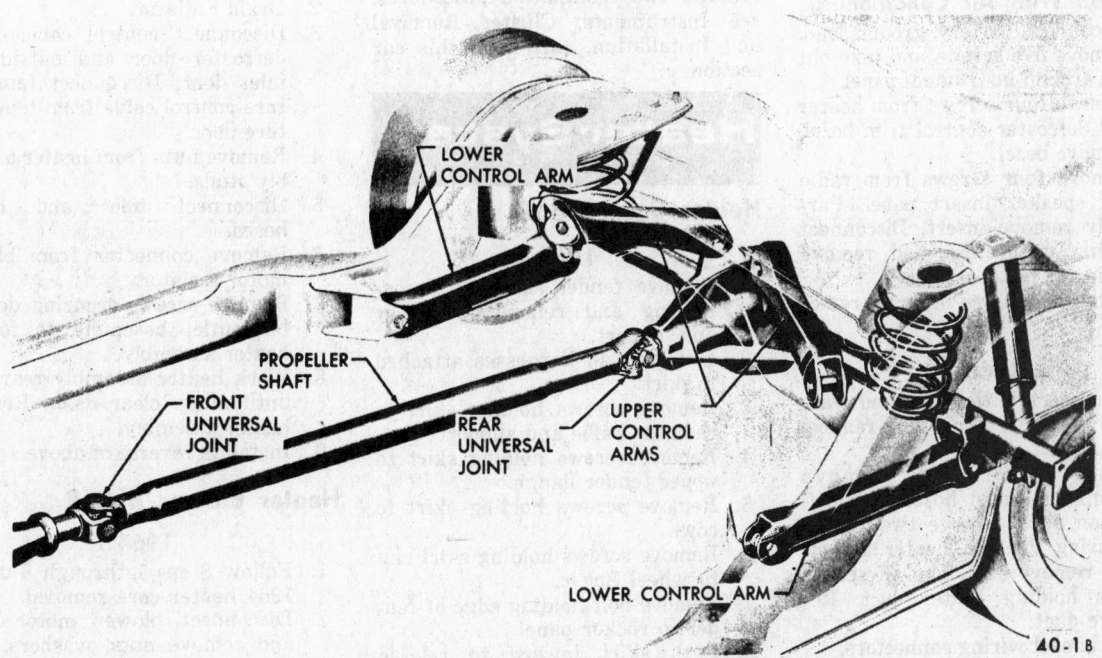

Rear suspension, 1964-70 (© Buick Div. G.M. Corp.)

40-1B

Place a floor jack under center of axle housing so it just starts to raise rear axle assembly. Place car stands solidly under body members on both sides.

2. Mark rear universal joint and pinion flange for proper reassembly. Then disconnect rear universal joint at pinion flange. Wire the driveshaft back out of the way.
3. Disconnect parking brake cables. Slide cable back until free of body.
4. Disconnect rear brake hose at floor pan.
5. Disconnect shock absorbers at axle housing. Lower jack under housing until rear springs can be removed.
6. Disconnect upper control arms at axle.
7. Disconnect lower control arms at axle housing and roll rear axle assembly out from under the car.

Radio

Radio Removal and Installation

1964-65 Without Air Conditioning
1. Disconnect battery ground lead.
2. Pull off radio knobs and remove hex nuts.
3. Disconnect radio lead, speaker lead, and antenna lead connectors.
4. Remove screw holding support to radio. Withdraw radio from underside of instrument panel.
5. Install in reverse order of removal.

1964-65 With Air Conditioning
1. Disconnect battery ground lead.
2. Remove five screws and take out rim around instrument panel.
3. Remove four screws from heater and defroster control trim bezel. Remove bezel.
4. Remove four screws from radio and speaker insert panel. Partially remove insert. Disconnect wiring connectors and remove radio and panel assembly.
5. Install in reverse order of removal.

1966
1. Disconnect battery ground lead.
2. Pull off radio knobs and remove hex nuts.
3. If air conditioned: Remove clamps on outlet hoses to distribution duct, remove two screws securing duct to heater assembly, remove duct. Pry off spring clips holding center duct. Remove duct.
4. Disconnect wiring connectors.
5. Remove screw holding support

to radio. Withdraw radio from underside of instrument panel.
6. Install in reverse order of removal.

1967
1. Disconnect battery ground lead.
2. Remove ashtray assembly.
3. Remove radio bracket to radio screw. Remove radio knobs, escutcheons, and hex nuts. Disconnect wiring connectors.
4. Remove radio downward.
5. Install in reverse order of removal.

1968-70
1. Disconnect battery ground lead.
2. Remove radio knobs, escutcheons, and hex nuts.
3. Remove two screws from radio filler plate and remove plate.
4. Remove ashtray assembly.
5. Remove center air conditioning duct, if so equipped.
6. Remove radio bracket.
7. Remove two instrument panel attaching nuts at radio face.
8. Disconnect wiring and remove radio downward.
9. Install in reverse order of removal.

Windshield Wipers

All wiper motors are located on the engine side of the firewall. The transmission and linkage are located on the passenger compartment side of the dash, directly forward of the instrument panel.

The wiper-washer switch is included in the instrument cluster. For removal and installation procedures, see Instruments, Cluster Removal and Installation, earlier in this car section.

Heater System

Heater Core R & R

1964
1. Remove fender skirt by disconnecting and removing battery and support.
2. Unfasten all harnesses attached to skirt.
3. Remove screws holding skirt to radiator baffle and support.
4. Remove screws holding skirt to upper fender flange.
5. Remove screws holding skirt to cowl.
6. Remove screws holding skirt rim to wheel house.
7. Remove bolt holding edge of fender to rocker panel.
8. Push skirt inward to release from lip of fender. Lift out.

9. Disconnect air control wires from levers of defroster door and outside air door.
10. Disconnect temperature control wire from lever of temperature door.
11. Drain radiator.
12. Disconnect inlet and outlet hoses.
13. Remove connector from blower resistor assembly.
14. Remove nuts and washers securing assembly to cowl, then lift out assembly.
15. Install in reverse of above.

1965-67
1. Remove right front wheel.
2. Draw an arc on inside of skirt, 11 in. from upper bolt of wheel opening. Draw another arc 16¾ in. from lower bolt of wheel opening and punch dimple at intersection of two arcs.
3. Drill a ¾ in. hole through skirt at this dimple. Remove lower right attaching nut from heater assembly stud through this hole.
4. Disconnect air control cables from defroster door and outside air door levers.
5. Disconnect temperature control cable from lever of temperature door on heater assembly.
6. Drain radiator.
7. Disconnect inlet and outlet hoses.
8. Remove connector from blower resistor assembly.
9. Remove nuts and washers securing assembly to cowl, then lift out assembly
10. Install in reverse of above.
11. Use ¾ in. body plug to close hole in skirt.

1968-70
1. Remove right front fender skirt.
2. Drain radiator.
3. Disconnect control cables from defroster door and outside air inlet door. Disconnect temperature control cable from temperature door.
4. Remove nuts from heater assembly studs.
5. Disconnect inlet and outlet hoses.
6. Remove connector from blower motor resistor.
7. Remove screws securing defroster outlet assembly to top of heater assembly.
8. Work heater assembly rearward until studs clear dash. Remove heater assembly.
9. Install in reverse of above.

Heater Blower R & R

1964
1. Follow Steps 1 through 8 under 1964 heater core removal.
2. Disconnect blower motor wire and remove nuts, washers, and screws holding blower and air inlet and remove assembly.

3. Install in reverse of above.

1965

1. Follow Steps 1 through 3 under 1965 heater core removal.
2. Remove lower right attaching nut from heater assembly through hole in skirt.
3. Remove remaining attaching nuts from heater assembly.
4. Remove screws holding blower and air inlet to cowl.

5. Disconnect blower wire and lift out blower and air inlet assembly.
6. Install in the reverse order and plug hole in fender skirt with ¾ in. body plug.

1966-67

1. Remove right front fender.
2. Remove nuts and screws securing blower and air inlet assembly to cowl.

3. Disconnect blower motor wire and remove assembly.
4. Install in reverse of above.

1968-70

Follow procedure above for 1966-67 models, substituting this step:

1. Remove right front fender skirt.

Cadillac

Index

YEAR IDENTIFICATION

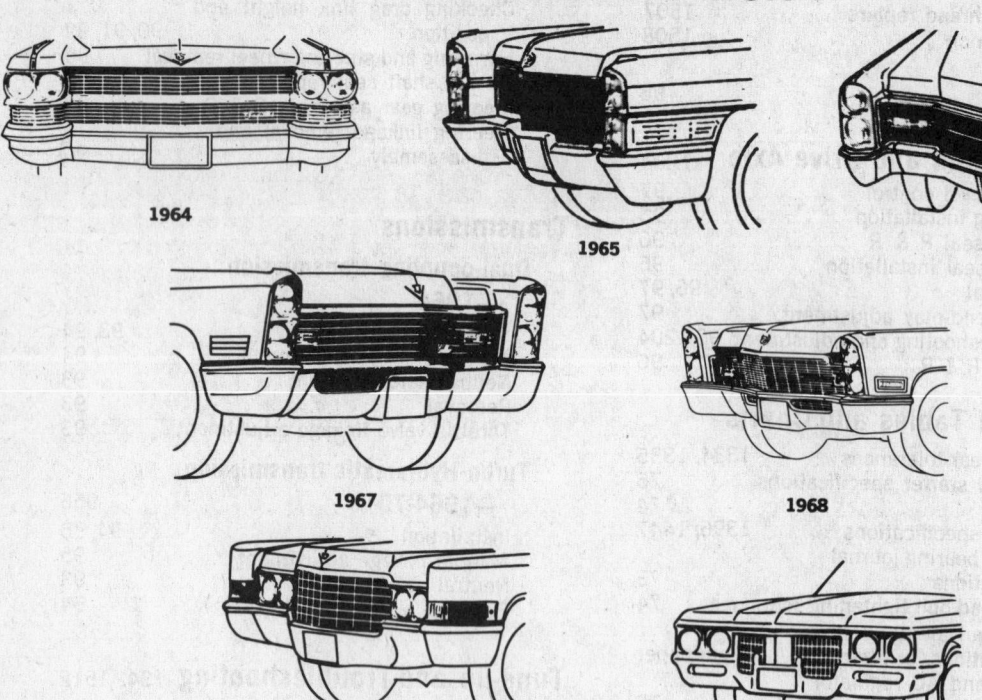

1964

1965

1966

1967

1968

1969-70

1971

FIRING ORDER

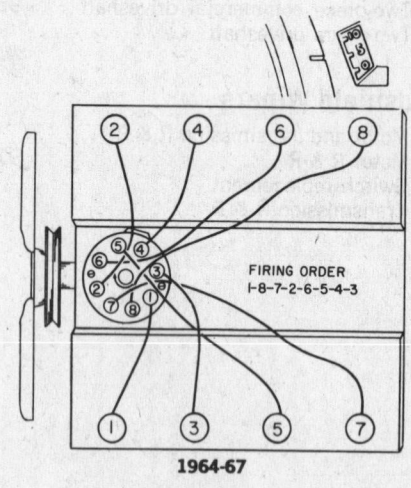

FIRING ORDER
1-8-7-2-6-5-4-3

1964-67

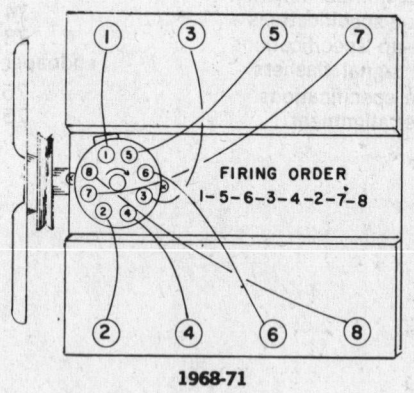

FIRING ORDER
1-5-6-3-4-2-7-8

1968-71

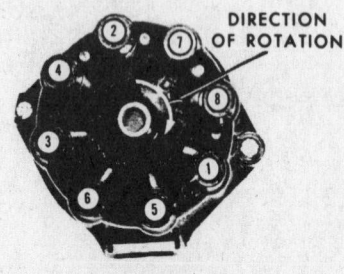

DIRECTION OF ROTATION

Distributor numbering— 1968 only

5° NOTCH ON TAB

NOTCH ON PULLEY

Timing marks— 1968-71

CAR SERIAL NUMBER LOCATION AND ENGINE IDENTIFICATION

1964-67

Each car has a serial number and an engine unit number. The car serial number is located on a plate attached to the left front door pillar post.

1968-71

Vehicle identification is located on a plate attached to the top of the left side of the dash, visible through the windshield.

Engine identification is stamped on a machined pad at the front, top edge of the right hand block. The engine number has nine digits. The first two indicate the model year; the third and fourth, the series and the succeeding numbers, the individual car serial number.

1964-67

A V8 engine of 429 cu. in. displacement is used. This engine can be identified by the distributor mounting; it is at the front instead of the rear.

1968-71

A new engine of 472 cu. in., developing 375 hp. is used. The distributor on this engine is also front mounted.

GENERAL ENGINE SPECIFICATIONS

YEAR	CU. IN. DISPLACEMENT	CARBURETOR	DEVELOPED HORSEPOWER @ RPM	DEVELOPED TORQUE @ RPM (FT. LBS.)	A.M.A. HORSEPOWER	BORE & STROKE (IN.)	ADVERTISED COMPRESSION RATIO	VALVE LIFTER TYPE	NORMAL OIL PRESSURE (PSI)
1964-67	V8—429	4-BBL.	340 @ 4600	480 @ 3000	54.6	4.125 x 4.000	10.50-1	Hyd.	33
1968/69	V8—472	4-BBL.	375 @ 4400	525 @ 3000	59.2	4.300 x 4.060	10.50-1	Hyd.	33
1970	V8—472	4-BBL.	375 @ 4400	525 @ 3000	59.2	4.300 x 4.060	10.00-1	Hyd.	35-40
1971	V8—472	4-BBL.	375 @ 4400	-----	59.2	4.300 x 4.060	8.8-1	Hyd.	35-40

TUNE-UP SPECIFICATIONS

YEAR	MODEL	SPARK PLUGS		DISTRIBUTOR		IGNITION TIMING (Deg.) ▲	CRANKING COMP. PRESSURE (Psi)	VALVES		Intake Opens (Deg.)	FUEL PUMP PRESSURE (Psi)	IDLE SPEED (Rpm) ★
		Type	Gap (In.)	Point Dwell (Deg.)	Point Gap (In.)			Tappet (Hot) Clearance (In.)				
								Intake	Exhaust			
1964-65	V8—All	44	.035	30°	.016	5B	180	Zero	Zero	39B	5³/₄	480
1966-67	V8—429	44	.035	30°	.016	5B	180	Zero	Zero	39B	5³/₄	500
1968-69	V8—472	44N	.035	30°	.019	5B	180	Zero	Zero	18B	5³/₄	550
1970	V8—472	R46N	.035	28-32°	.016	7¹/₂B	165-185	Zero	Zero	18B	5¹/₄-6¹/₂	600
1971	V8—472	R46N	.035	30°	.016	8B	160-170	Zero	Zero	18B	6	600

★—With automatic transmission in D. Add 50 rpm if equipped with air conditioning.

▲—With vacuum advance disconnected and hose plugged. NOTE: These settings are only approximate. Engine design, altitude, temperature, fuel octane rating and the condition of the individual engine are all factors which can influence timing. The limiting advance factor must, therefore, be the "knock point" of the individual engine.

B—Before top dead center.

Caution

General adoption of anti-pollution laws has changed the design of almost all car engine production to effectively reduce crankcase emission and terminal exhaust products. It has been necessary to adopt stricter tune-up rules, especially timing and idle speed procedures. Both of these values are peculiar to the engine and to its application, rather than to the engine alone. With this in mind, car manufacturers supply idle speed data for the engine and application involved. This information is clearly displayed in the engine compartment of each vehicle.

CRANKSHAFT BEARING JOURNAL SPECIFICATIONS

YEAR	MODEL	MAIN BEARING JOURNALS (IN.)				CONNECTING ROD BEARING JOURNALS (IN.)		
		Journal Diameter	Oil Clearance	Shaft End-Play	Thrust On No.	Journal Diameter	Oil Clearance	End-Play
1964-67	V8—429	3.000	.0003-.0026	.002-.012	3	2.2491	.0013	.008-.016
1968-69	V8—472	3.250	.0003-.0026	.002-.012	3	2.5000	.0005-.0028	.008-.016
1970-71	V8—472	3.250	.0003-.0026	.002-.012	3	2.5000	.0005-.0028	.008-.016

CYLINDER HEAD BOLT TIGHTENING SEQUENCE

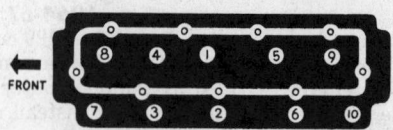

Cylinder head bolt
tightening sequence—1968-71

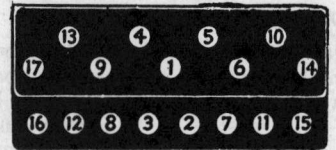

Cylinder head bolt
tightening sequence—1964-67

TORQUE SPECIFICATIONS

YEAR	MODEL	CYLINDER HEAD BOLTS (FT. LBS.)	ROD BEARING BOLTS (FT. LBS.)	MAIN BEARING BOLTS (FT. LBS.)	CRANKSHAFT BALANCER BOLTS (FT. LBS.)	FLYWHEEL TO CRANKSHAFT BOLTS (FT. LBS.)	MANIFOLD (FT. LBS.)	
							Intake	Exhaust
1964-67	All V8—429	60	40-45	90-100	65-70	75-80	25-30	25-30
1968-71	All V8—472	115	40	90	Press-fit	75	30	35

NOTE—Some bolts and nuts are marked on the heads to indicate the grade of steel used. Do not use bolts of a lower grade than those originally installed. The marks consist of lines: SAE5—3 lines; SAE7—5 lines; SAE8—6 lines.

GENERAL CHASSIS AND BRAKE SPECIFICATIONS

YEAR	MODEL	CHASSIS			BRAKE CYLINDER BORE			BRAKE DRUM	
		Overall Length (In.)	Tire Size	Master Cylinder (In.) Power	Wheel Cylinder Diameter (In.)			Diameter (In.)	
					Front	Rear		Front	Rear
1964-65	Series—All Exc. 75	223.5	8.00 x 15	1	1³/₁₆	1.0		12	12
	Series 75	243.8	8.20 x 15	1	1³/₁₆	1.0		12	12
1966-67	Fleetwood Brougham & 60 Spd. Sdn.	227.5	9.00 x 15	1	1³/₁₆	1.0		12	12
	75 Sdn. & Limousine	244.5	8.20 x 15	1	1³/₁₆	1.0		12	12
	All others	224	9.00 x 15	1	1³/₁₆	1.0		12	12
1968	Fleetwood 60, Brougham	228.2	9.00 x 15	1	1³/₁₆	15/₁₆		12	12
	Fleetwood 75	245.2	9.00 x 15	1	1³/₁₆	1.0		12	12
	All others	224.7	9.00 x 15	1	1³/₁₆	15/₁₆		12	12
	w/Disc. Brakes	—		1	2³/₄ ①	13/₁₆ ▲		12 (Disc.)	12
1969	Fleetwood 60, Brougham	228.2	9.00 x 15	1	2³/₄	7/₈		12	12
	Fleetwood 75	245.2	9.00 x 15	1	2³/₄	7/₈		12	12
	All others	224.7	9.00 x 15	1	2³/₄	13/₁₆		12	12
1970-71	Calais, DeVille	225.8	L78 x 15	1	2³/₄ Disc	13/₁₆		11.9 Disc	12
	Fleetwood 60, Brougham	228.8	L78 x 15	1	2³/₄ Disc	13/₁₆		11.9 Disc	12
	Fleetwood 75	247.3	L78 x 15	1	2³/₄ Disc	7/₈		11.9 Disc	12

▲—Fleetwood 75—7/₈
①—1968—2¹⁵/₁₆ single cylinder disc.

CAPACITIES

YEAR	MODEL	ENGINE CRANKCASE ADD 1 Qt. FOR NEW FILTER	TRANSMISSIONS Pts. TO REFILL AFTER DRAINING			DRIVE AXLE (Pts.)	GASOLINE TANK (Gals.)	COOLING SYSTEM (Qts.) WITH HEATER
			Manual		Automatic			
			3-Speed	4-Speed				
1964-65	All	4	None	None	18▲	5	26	17¼ ■
1966-67	All	4	None	Nonc	24	5	26	18 ■
1968-71	All	4	None	None	25	5	26	21.3 ●

▲—Turbo Hydramatic—24 Pts.
■—Series 67 & 75 use 1½ Qts. cadditional.
●—1968-71—75 Series—24.8 Qts.

VALVE SPECIFICATIONS

YEAR AND MODEL		SEAT ANGLE (DEG.)	FACE ANGLE (DEG.)	VALVE LIFT INTAKE (IN.)	VALVE LIFT EXHAUST (IN.)	VALVE SPRING PRESSURE (VALVE OPEN) LBS. @ IN.	VALVE SPRING INSTALLED HEIGHT (IN.)	STEM TO GUIDE CLEARANCE (IN.)		STEM DIAMETER (IN.)	
								INTAKE	EXHAUST	INTAKE	EXHAUST
1964–67	V8—429	45	44	.440	.440	160 @ 1½	1 15/16	.0005–.0025	.0010–.0025	.3420	.3420
1968–71	V8—472	45	44	.440	.454	160 @ 1½	1 15/16	.0005–.0025	.0010–.0025	.3420	.3420

WHEEL ALIGNMENT

YEAR	MODEL	CASTER Range (Deg.)	CASTER Pref. Setting (Deg.)	CAMBER Range (Deg.)	CAMBER Pref. Setting (Deg.)	TOE-IN (In.)	KING-PIN INCLINATION (Deg.)	WHEEL PIVOT RATIO Inner Wheel	WHEEL PIVOT RATIO Outer Wheel
1964–67	All Series	1½N–½N	1N	⅜N–⅜P	(2)	3/16–¼	6	22⅔	20
1968	All Series	1½N–½N	1N	(1)	(1)	3/16–¼	6	20	18
1969–71	All Series	1½N–½N	1N	⅜N–⅜P	0	⅛–¼	6	20	18

(1)—Left ⅜P–⅛N; zero preferred.
  Right ⅛P–⅜N; ¼N preferred.
(2)—¼–½, more on left than right.
P—Positive.
N—Negative.

AC GENERATOR AND REGULATOR SPECIFICATIONS

YEAR	MODEL	ALTERNATOR Field Current Draw @ 12V.	ALTERNATOR Output @ Generator RPM 1100	ALTERNATOR Output @ Generator RPM 6500	REGULATOR Model	Field Relay Air Gap (In.)	Field Relay Point Gap (In.)	Field Relay Volts to Close	Regulator Air Gap (In.)	Regulator Point Gap (In.)	Volts at 125°
1964	1100624	1.9–2.3	5A	42A	1119512	.015	.030	2.3–3.7	.067	.014	13.5–14.4
	1100617	1.9–2.3	4A	52A	1119512	.015	.030	2.3–3.7	.067	.014	13.5–14.4
1965	1100696	2.2–2.6	5A	42	1119515	.015	.030	2.3–3.7	.060	.014	13.5–14.4
	1100694	2.2–2.6	5A	55	1119515	.015	.030	2.3–3.7	.060	.014	13.5–14.4
1966–67	1100691	2.2–2.6	5A	42	1119515	.015	.030	2.3–3.7	.060	.014	13.5–14.4
	1100692	2.2–2.6	5A	55	1119515	.015	.030	2.3–3.7	.060	.014	13.5–14.4
1968–69	1100696	2.2–2.4	5A	42	1119515	.015	.030	1.5–3.2	.060	.014	13.5–14.4
	1100694	2.2–2.4	5A	55	1119515	.015	.030	1.5–3.2	.060	.014	13.5–14.4
	1100742	2.2–2.4	5A	63	1119519	.015	.030	1.5–3.2	.060	.014	13.5–14.4
1970	1100908	2.2–2.6	...	42	1119515	.015	.030	1.5–3.2	.060	.014	13.5–14.4
	1100694	2.2–2.4	...	55	1119515	.015	.030	1.5–3.2	.060	.014	13.5–14.4
	1100910	2.8–3.2	...	63	1119519	.015	.030	1.5–3.2	.060	.014	13.5–14.4
1971	1100911	2.8–3.2	...	...	1119602	Transistor Type—no adjustment					13.5–14.4

BATTERY AND STARTER SPECIFICATIONS

YEAR	MODEL	BATTERY Ampere Hour Capacity	BATTERY Volts	BATTERY Group Number	BATTERY Terminal Grounded	STARTERS Lock Test Amps.	STARTERS Lock Test Volts	STARTERS Lock Test Torque	STARTERS No-Load Test Amps.	STARTERS No-Load Test Volts	STARTERS No-Load Test RPM	Brush Spring Tension (Oz.)
1964–67	All Series	73	12	3KMB	Neg.	510	3.0	Locked	88	10.6	4,000	35
1968–71	All Series	74	12	—	Neg.	—	Not Recommended		70–99	10.6	7,800	35

LIGHT BULBS

1964

Ash tray, rear radio, turn signal indicator..... 1445; 1 CP
Back-up lamp .... 1156; 32 CP
Clock, glove comp., fuel, temp., radio,
    speedometer, gen. & oil ind. .... 1895; 2 CP
Cornering .... 1195; 50 CP
Courtesy and door warning................. 212; 6 CP
Dome lamp .... 1004; 15 CP
Headlamp (inner)....................... 4001; 37.5 W
Headlamp (outer)................... 4002; 37.5-50 W
License lamp .... 67; 4 CP
Stop, signal & turn.................... 1157; 32-4 CP

1965–66

Ash tray, cruise control .... 1445; 1 CP
Back-up lamps .... 1156; 32 CP
Cornering lamps .... 1195; 50 CP
Gen. & oil tell-tale, glove compartment,
    radio dial .... 1895; 2 CP
Headlamp (inner)....................... 4001; 37.5 CP
Headlamp (outer) .... 4002; 50/37.5 CP
License............................... 67; 4 CP
Parking & turn signal
    stop, tail & turn .... 1157; 32/4 CP
Trunk lamp .... 89; 6 CP

1967–69

Same as 1965–66 except the following:
Gen. oil tell-tale .... 161; 1 CP
Radio dial .... 1816; 3 CP

1970–71

FUNCTION	BULB NO.	C/P
Ash tray lamp	1445	.7
Back-up lamp	1156	32
Back-up lamp	1295	50
Clock	1816	3
Console lamp	57	2
Cornering lamp	1295	50
Courtesy lamp—rear quarter	90	6
Courtesy lamp-console	212/212-1	6
Courtesy lamp-instrument panel	89	6
Courtesy lamp—rear door	212/212-1	6
Courtesy lamp—rear quarter armrest	212/212-1	6
Cruise control speed selector illum. auto. lock lamps	1445	.7
Engine temp. warning light	161	1
Generator telltale lamp	161	1
Glove compartment lamp	1816	3
Headlamp—inner	L4001	37.5 Watts
Headlamp—outer	L4002	37.5W/55.0W
Headlamp switch lamp	1895	2
Heater control or climate control lamp	1816	3
High beam indicator	161	1
License lamp	67	4
Low brake telltale lamp	161	1
Low oil pressure telltale lamp	168	3
Map lamp	89	6
Marker lamp-front side	97A	4
Marker lamp-rear side	194	2
Panel lamp	168	3
Park-signal lamp	<u>1157 NA</u>	32/3
Radio dial lamp	1816	3
*Radio AM-FM band indicators	2182D	.4
*Radio AM-FM stereo indicators	2182D	.4
*Radio-rear control indicator	250	1
Spot lamp-front compartment	90	6
Spot lamp-reading	1004	15
Stop, tail and signal	<u>1157</u>	32/3
Trunk compartment lamp	89	6
Trunk lid tell tale	161	1
Turn signal indicator	168	3
Warning lamp-front door (combined with courtesy light)	212/212-1	6
Warning lamp-rear door (combined with courtesy light)	212/212-1	6
Water temperature tell tale	168	3

*Serviceable only by Radio Technician.

FUSES AND CIRCUIT BREAKERS

1964

Air conditioner and heater .... 25 AMP. FUSE
Antenna .... 14 AMP. FUSE
Body feed, cigar lighter, clock, map .... 25 AMP. FUSE
Cruise control....................... 6 AMP. FUSE
Guide-matic .... 4 AMP. FUSE
Head and parking lights.................. 25 AMP. CB
Horns .... 25 or 40 AMP. CB
Instrument and back-up lights .... 9 AMP. FUSE
Radio............................. 7½ AMP. FUSE
Seats and windows .... 40 AMP. CB
Tail and stop lights .... 25 AMP. FUSE
Cornering and courtesy lights .... 25 AMP. FUSE
Turn signal .... 14 AMP. FUSE
Windshield wipers .... 25 AMP. FUSE

1965–69

Same as 1964 except the following:
Head and parking lights .... 15 AMP. CB
Instrument & back-up lights .... 10 AMP. FUSE
Turn signal .... 15 AMP. FUSE

1970–71

Body feed .... 25 AMP.
    Cigar lighters

Clock
Courtesy lights
Glove box light
Map light
Trunk light
Cornering and Parking Lights .... 10 AMP.
    Ash tray light
    Cornering lights
    Front side marker lights
    Parking lights
Directional signal and back-up lights .... 20 AMP.
    Back-up lights
    Cruise control
    Rear window de-fogger
    Turn signals
Gages and transmission controls............... 10 AMP.
    Brake warning light
    CCS vacuum solenoid
    Downshift solenoid
    Fuel gauge
    Generator light
    Low oil pressure indicator
    Water temperature warning light
Headlights (integral with headlight switch).... 15 AMP. CB

Heater and accessories .... 25 AMP.
    Air conditioning amplifier
    Air conditioning blower relay
    Heater blower
    (On cars equipped with a heater only, the
      25 AMP fuse is replaced by a 15 AMP,
      fuse)
Horns................................... CB
    Convertible top
    Engine metal temperature light
    Horns
    Power seat
    Power windows
Instrument panel lights .... 4 AMP.
Low blower (air conditioning only) .... 10 AMP.
Radio and window control relay .... 7½ AMP.
Stop lights and hazard warning flasher .... 25 AMP.
Tail lights .... 25 AMP.
    License light
    Rear side marker lights
    Tail lights
Twilight sentinel (integral with headlight
    switch) .... 15 AMP. CB
Windshield wipers.......................... 25 AMP.

Distributor

Detailed information on distributor drive, direction of distributor rotation, cylinder numbering, firing order, point gap, cam dwell, timing mark location, spark plugs, spark advance, ignition resistor location, and idle speed will be found in the Specifications table.

Distributor Point Replacement

1964-70

1. Remove distributor cap by depressing and turning the retaining screws.
2. Remove two screws securing rotor cap and remove cap.
3. Remove condenser and primary leads from nylon insulated connection.
4. Loosen two screws holding base of contact assembly in place and remove points.
5. Inspect weight assembly, replace or lubricate as required.
6. Place new points under the two screws and tighten screws.
7. Connect the condenser and primary leads at the nylon insulated connection.
NOTE: be sure leads do not interfere with cap, weight base, or breaker advance.
8. Install rotor cap. Square and round lugs must be properly aligned.

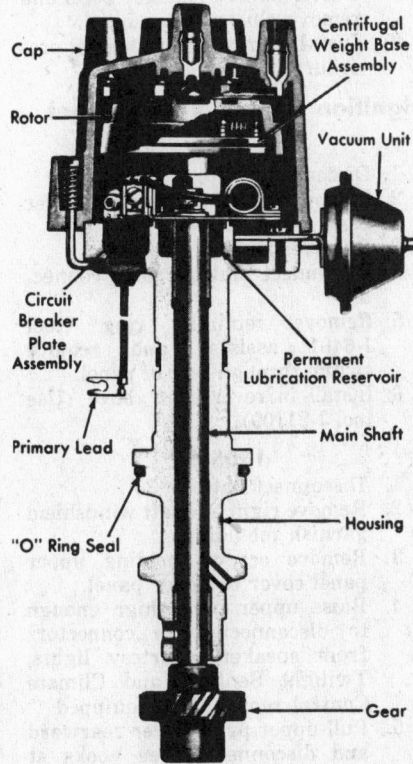

Distributor showing major components
(© Cadillac Div. G.M. Corp.)

Cap
Rotor
Circuit Breaker Plate Assembly
Primary Lead
"O" Ring Seal
Centrifugal Weight Base Assembly
Vacuum Unit
Permanent Lubrication Reservoir
Main Shaft
Housing
Gear

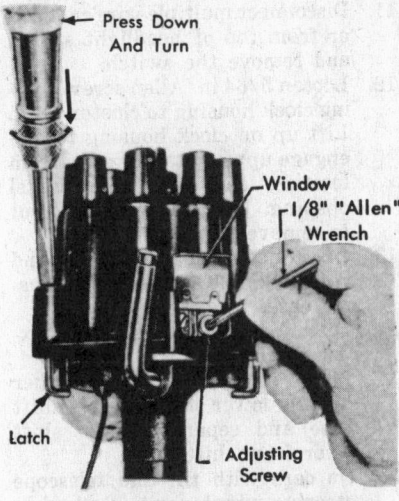

Press Down And Turn
Window
1/8" "Allen" Wrench
Latch
Adjusting Screw

Point adjustment with engine running

9. With ⅛ in. Allen wrench inserted, turn until points close while rubbing block is on high point of lobe. Then turn screw counterclockwise one-half turn.
10. Replace distributor cap.
11. With engine warmed up and off fast idle, set points to get proper dwell angle.

Distributor Removal

All Models

Remove distributor cap. Disconnect vacuum pipe. Disconnect primary lead at distributor.

Turn the engine to top dead center for No. 1 cylinder so that the rotor points to the No. 1 cylinder tower in the distributor cap and the pointer on the timing case cover points to the O-mark on the crankshaft pulley.

Using a scribe mark, index the vacuum advance unit to the cylinder block so that the distributor body will be correctly replaced at reassembly.

Distributor Replacement

All Models

Install the distributor so that the vacuum advance unit aligns with the match-mark made at removal. Turn the rotor slightly left of center so that as the gear engages the camshaft it will revolve into the proper position, pointing to the No. 1 contact in the cap.

NOTE: if the engine has been cranked, remove the No. 1 spark plug. Crank the engine until the No. 1 piston is in firing position with the pointer and the O-mark on the crank-shaft pulley aligned. Then proceed as above.

Install the hold-down clamp. Connect the primary lead and install the cap.

Fill the distributor oiler tube with 10W oil.

Plug the distributor vacuum line to the carburetor.

Insert an adapter pin alongside the No. 1 wire in the distributor cap and connect a timing light.

Clean the crankshaft pulley markings and the pointer.

The marks are at 5° and 10° B.T.D.C.

Set the timing.

Tighten clamp bolt to 18 ft. lbs.

Remove plug and adapter pin and reconnect the vacuum line to the advance unit.

Generator, Regulator

Detailed facts on the generator and the regulator can be found in the Delcotron Specifications table.

General information on generator and regulator repair can be found in the Unit Repair Section.

NOTE: under no circumstances should the A.C. generator be polarized.

Generator Removal

All Models

Disconnect the battery. Disconnect the wire leads at the generator. Remove generator adjusting strap clamp screw, mounting bolts, and drive belt. Remove generator from left exhaust manifold.

Battery, Starter

Detailed information on the battery and starter will be found in the Battery and Starter Specifications table.

A more general discussion of starters can be found in the Unit Repair Section.

Starter Removal

All Models

Remove coil feed wire, battery cable, and starter button wire, at starter solenoid.

Remove the two starter mounting bolts at flywheel housing. Pull starter forward, then down, and remove from car.

Instruments

Panel and Cluster Removal

1964-65

1. Disconnect negative cable at battery.
2. Remove right and left windshield garnish moldings. Remove five screws holding upper panel cover to lower panel. Three screws are located above instrument panel and two at inside top of glove compartment.

3. Raise upper panel high enough to disconnect three-way connector for map light and three-way connector for Guide-Matic phototube unit, if car is so equipped.
4. Pull cover rearward to disengage hooks at front of cover from retainers on cowl. Remove cover.
5. Remove three screws and one large screw from upper end of steering column lower cover.
6. Remove five screws from lower end of steering column. Remove cover.
7. Disconnect speedometer cable at the head.
8. Remove cluster bulb sockets.
9. Disconnect odometer reset cable from lower flange of instrument panel.
10. Remove four cluster retaining screws and remove cluster.
11. Install by reversing the above procedure.
NOTE: the above procedure is required when service or replacement is needed on the speedometer head or any of the instruments.

1966-70

Due to instrument panel and speedometer R & R involvement, the following procedure also includes the removal of:
Upper instrument panel cover
Headlight control switch
Clock
Steering column
Speedometer head
1. Disconnect battery.
2. Remove right and left windshield garnish moldings by removing four screws in each molding.
3. Remove six screws that hold upper panel cover to upper panel.
4. Raise upper panel high enough to gain access, and disconnect wire connectors for radio speaker/s, courtesy lights, and for Twilight Sentinel photocell and Automatic Climate Control sensor, if car is so equipped.
5. Pull upper panel cover rearward to disengage three hooks at front of cover from retainers on cowl, and remove cover.
6. Loosen set screw holding headlight switch housing to cluster bezel, using a 5/64 in. Allen wrench.
7. Carefully lift upward on bottom of headlight switch housing to disengage upper retainer clip from locating slot in cluster bezel opening. Pull headlight switch straight out to remove.
8. Disconnect headlight switch bulb socket from top of housing case.
9. Disconnect trunk warning lens dial bulb socket from bottom of housing case.
10. On cars equipped with Guide-Matic and/or Twilight Sentinel, disconnect control switch lead connectors.

11. Disconnect multiple wire connector from top of headlight switch and remove the switch.
12. Loosen 5/64 in. Allen screw holding clock housing to cluster bezel. Lift up on clock housing to disengage upper retaining clip from locating slot in cluster bezel opening. Pull clock straight out to remove.
13. Disconnect clock feed wire, and two clock bulb sockets from housing case.
14. Disconnect steering shaft at flexible coupling.
15. Remove screw and lockwasher holding lower shift lever to shift tube and separate lower shift lever from shift tube.
16. On cars with tilt and telescope steering wheel, position wheel in up position.
17. Remove four screws that hold lower end of steering column lower cover to lower instrument panel.
18. Remove one long screw holding upper end of steering column lower cover to clamp.
19. Disconnect front retainer flanges of steering column lower cover from front of lower instrument panel and lower steering column cover enough to gain access to cover.
20. Remove flasher unit from mounting clip and remove lower cover.
21. If car is a convertible, disconnect convertible top switch multiple connector from rear side of steering column lower cover.
22. On cars equipped with rear window defogger, disconnect blower switch single connector (light green wire and dark blue wire with white stripe) at accessory terminal on fuse panel. Disconnect blower switch T connector from wire assembly connector.
23. If the car has power windows or cruise control plus rear window defogger, disconnect override relay feed or cruise control wire from blower switch (6 in. long, dark blue wire with white stripe.)
24. If the car has a front seat warmer, detach connector leading from switch. Also remove two screws holding on-off switch to steering column lower cover. Remove the switch.
25. Remove transmission shift indicator pointer (use Allen wrench).
26. Disconnect horn wiring from chassis wiring harness. On tilt and telescope columns, disconnect connectors from turn signal and cornering switch on side of jacket. On standard column, disconnect multiple connectors.
27. Disconnect connectors and vacuum hoses from neutral safety and back-up light switch assembly.

28. Remove steering column lower cover clamp and remove two screws that hold clamp to instrument panel.
29. Remove two screws and washers that hold steering column upper clamp to support struts.
30. Slide rubber cover up on steering column and pull back carpet.
31. Remove steering column lower clamp screw, spacer and lockwasher, then loosen upper right support bolt at toe pan.
32. Pull steering column up and out of the car.
33. Disconnect seven cluster bulb sockets, fuel gauge connector and temperature gauge connector.
34. Disconnect speedometer cable at speedometer head.
35. Working through headlight switch opening, remove screw that holds cluster bezel left lower mounting bracket to left mounting bracket on instrument panel center brace.
36. Working through clock opening, remove screw that holds cluster bezel right lower mounting bracket on instrument panel center brace.
37. Remove two screws that hold cluster bezel to upper instrument panel center molding. Remove bezel and cluster assembly from instrument panel.
38. Loosen set screw that holds trip odometer reset shaft knob to reset shaft. Remove knob.
39. Remove four screws that hold cluster panel to cluster bezel and remove cluster from bezel.
40. Install by reversing removal procedure.

Ignition Switch Replacement
1964

1. Disconnect battery.
2. Remove steering column lower panel cover.
3. Remove lock cylinder.
4. Disconnect multiple wire connector.
5. Remove retaining ring (tool J-6481 assists) and remove switch through rear of panel.
6. Install in reverse of above (Use tool J-21109).

1965-66

1. Disconnect battery.
2. Remove right and left windshield garnish mouldings.
3. Remove screws holding upper panel cover to upper panel.
4. Raise upper panel high enough to disconnect wire connectors from speaker, courtesy lights, Twilight Sentinel and Climate Control sensor if so equipped.
5. Pull upper panel cover rearward and disconnect three hooks at front of cover and remove cover.
6. Remove lock cylinder assembly as described above.

7. Disconnect four-way connector at rear of ignition switch housing.
8. Remove switch nut. Tool J-21109 will assist.
9. Disconnect dial bulb socket at rear of housing and remove switch through rear.
10. Install in reverse of above.

1967-68

1. Disconnect battery.
2. Remove lock cylinder.
3. Remove steering column lower cover.
4. Remove two screws and clip that hold the switch assembly to the panel.
5. Disconnect dial bulb socket and wiring harness from back of switch assembly.
6. Pull switch rearward out of instrument panel.
7. Install by reversing removal procedure.

1969-70

1. Disconnect battery.
2. Position lock cylinder in "lock" position.
3. Remove steering column lower cover.
4. Loosen two nuts on upper steering column, allowing column to drop.
CAUTION: do not remove nuts, as column may bend under its own weight.
5. Disconnect ignition switch connector at switch.
6. Remove two screws securing ignition switch to steering column. Remove switch.
7. Install in reverse of above.

Lock Cylinder Replacement

1. Insert key and turn to Acc position.
2. With stiff wire in hole depress lock pin and rotate cylinder counterclockwise and pull out.

Rear Lights

Bumper Section—1964

1. Remove reflex attaching screws, and remove reflex.
2. Remove screws and retaining ornament insert bar.
3. Remove screws and retaining lens, and replace bulb.

Fender Section—1964

Remove lens retaining screws, and replace bulb.

Fender Section—1965

1. Remove bezel-to-housing screws, and remove bezel. Bezel is two-piece unit. Use caution that insert does not fall out.
2. Remove lens-to-housing screws and remove lens. Use caution that red filter does not fall out.

3. Replace bulb and reinstall parts in reverse of above.

Fender Section—1966-69

Working through opening behind rear bumper outer end in rear underbody, remove bulb socket by turning counterclockwise.
NOTE: back-up bulb is on the bottom, tail and stoplight bulbs are in the middle and top.

Fender Section—1970

1. Open trunk and pull carpet and liner back to gain access to lamp assemblies.
2. Remove bulbs and sockets by snapping out.

Brakes

Specific information on brake cylinder sizes can be found in the General Chassis and Brake Specifications table.
Information on overhauling power brakes can be found in the Unit Repair Section. All Cadillac cars are equipped with power brakes.
Information on the grease seals which may need replacement can be found in the Unit Repair Section.
To gain access to the front wheel star wheel adjustment, it will be necessary to remove the disc wheel (not the hub or drum).
The rear disc wheels must also be removed to adjust the rear star wheel adjustment.

Power Brake Assembly Removal

The master cylinder is part of the power unit on these models.
1. Disconnect output line from master cylinder of power unit.
2. Disconnect hoses from power units.
3. Remove clevis pin retaining power unit pushrod to brake pedal relay lever.
4. Remove four screws to release power unit from car.

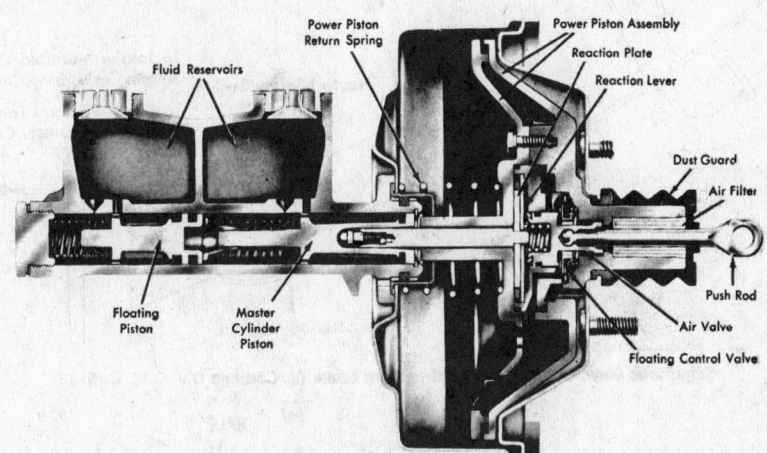

Moraine unit (© Cadillac Div. G.M. Corp.)

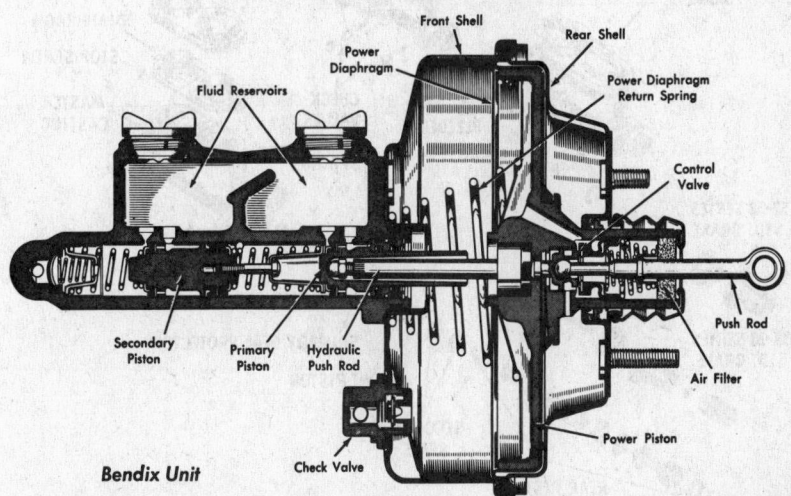

Bendix Unit

1963-66 Cadillac dual master cylinder. (It can be hooked up backwards; the proper hook-up is as follows: the rear master cylinder should operate the front brakes; the front master cylinder should operate the rear brakes.)
(© Cadillac Div. G.M. Corp.)

Self-Adjusting Service Brake

The rear braking system of all models consists of power-assisted, hydraulic service brakes. The front braking system is same as rear for models 1964-68. Front disc brakes are optional on the 1968 model and standard on all 1969-70 cars. All models use a foot-operated parking brake, which is applied at rear wheels through mechanical linkage.

The service brake has a self-adjusting brake shoe mechanism consisting of a link, actuator, pawl, and pawl return spring. The actuator is held against the secondary shoe by means of a hold-down cup and spring. The pawl is connected to the actuator and held in position by the pawl return spring.

The automatic adjustment takes place only when the brakes are applied when the car is moving rearward.

Over-adjustment is prevented by the shoe-to-drum clearance limiting secondary shoe travel to less than that required for the pawl to engage the next tooth of the star wheel.

Care must be used that the correct star wheel assembly is installed at the proper wheel, to insure that the self-adjuster work correctly.

Star wheel adjusters with left-hand threads have one groove on the long end of the star wheel adjuster and must be used on right-side drums. Those with right-hand threads have two grooves and must be used on the left-hand side of the car.

Caution
Fixed anchors are used.

Periodic wheel removal and lining inspection becomes more important to insure against drum and shoe damage due to neglect.

Parking Brake

Vacuum Release Parking Brake

A vacuum release assists the foot-operated parking brake. With the engine running, the brake automatically releases when the car is put into gear. This device eliminates the possibility of driving the car with the parking brake engaged.

Master Cylinder—1964-66

A dual master cylinder is used. The front reservoir supplying rear brakes and the rear one supplying the front brakes. This allows one pair of brakes to operate should there be a failure of the opposite pair. If the lines have been disconnected, be sure to reinstall them in their proper place, i.e. front to rear cylinder and rear to front.

Two different brake units were used on these models—Bendix and Delco Moraine. For identification purposes, the Bendix unit is painted all black, while the Delco Moraine vacuum cylinder is zinc plated.

With a pressure bleeder, the Bendix system can be bled from front reservoir by covering rear reservoir with solid cap. The Moraine-type must be bled separately, front and rear.

Without a pressure bleeder, keep both reservoirs nearly full.

Beginning 1967

The master cylinder design was revised. The master cylinder and reservoir responsible for front wheel brake application is now the front half of the cylinder instead of the rear, as in earlier models. Now, the front portion of the master cylinder serves the front wheels and the rear portion of the cylinder serves the rear wheels.

Front Wheel Disc Brakes

Front disc brakes were optional for 1968 models, standard for all 1969-70 models. Illustrations and procedures information may be found in the Unit Repair Section.

Fuel System

Data on capacity of the gas tank will be found in the Capacities table. Data on correct engine idle speed will be found in the Tune-Up Specifications table.

Information covering operation and troubles of the fuel gauge will be found in the Unit Repair Section.

Detailed information on the carburetor and how to adjust it will be found in the Unit Repair Section under the specific heading of the make of carburetor being used on the engine being worked on. Carter,

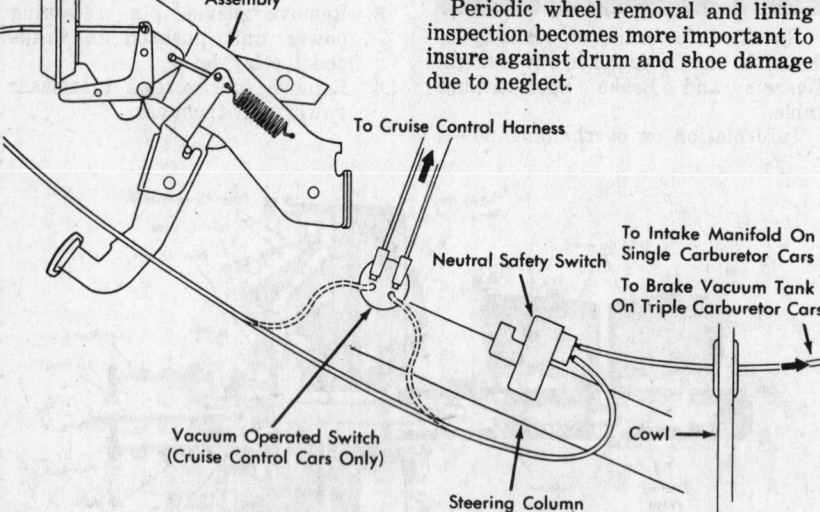

Schematic view, vacuum-operated parking brake (© Cadillac Div. G.M. Corp.)

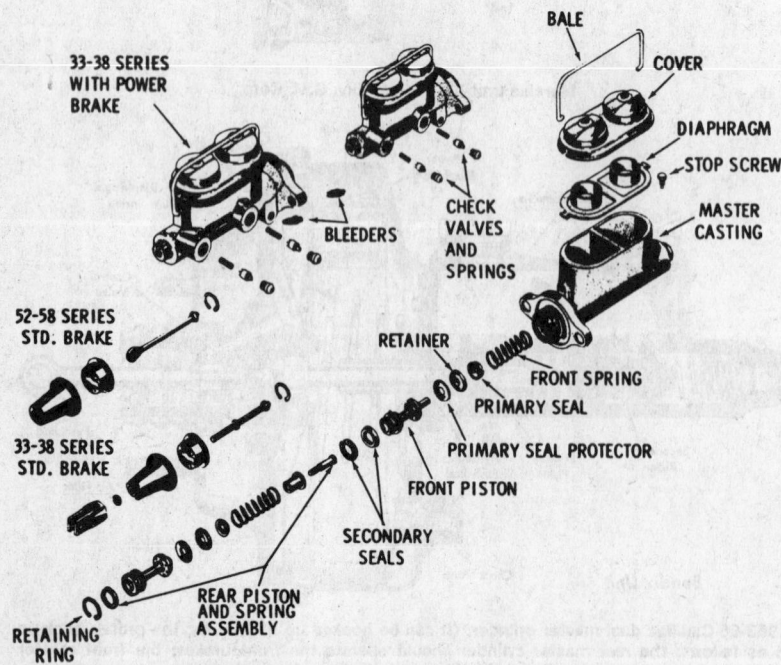

Master cylinder (Moraine) (© Cadillac Div. G.M. Corp.)

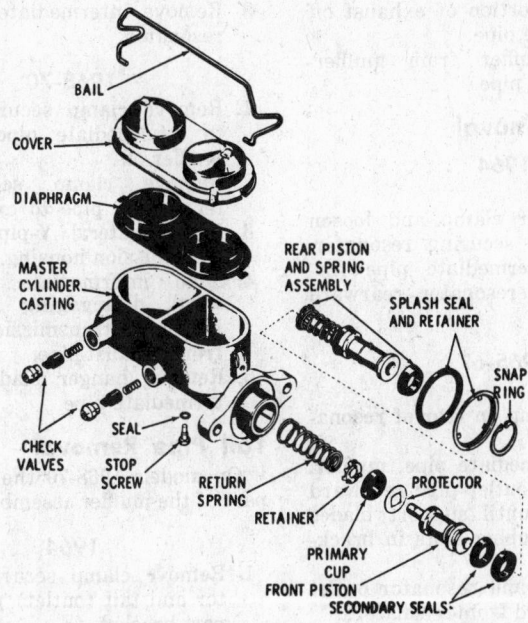

Master cylinder (Bendix) (© Cadillac Div. G.M. Corp.)

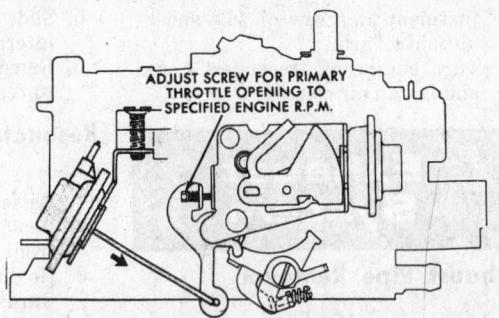

Speed up control adjustment (© Cadillac Div. G.M. Corp.)

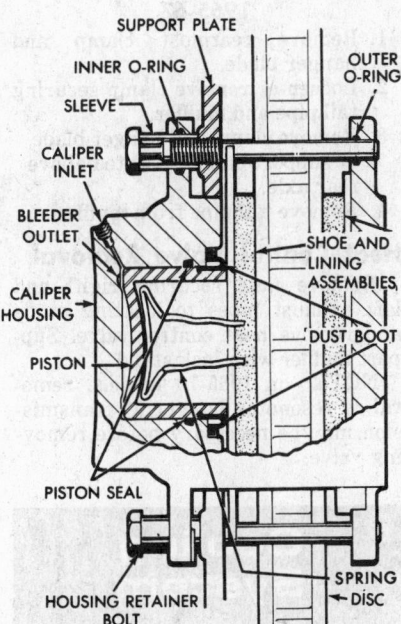

Disc brake caliper cross section

Holley, Rochester and Stromberg carburetors are covered.

Dashpot adjustment can be found in the Unit Repair Section.

Fuel Pump Removal

The fuel pump is mounted on the engine oil filler bracket. The pump is driven by an eccentric machined as an integral part of the camshaft. There is a fuel filter between the fuel pump and the carburetor. On air conditioned cars, the fuel filter has a passage and a connecting line to the fuel tank to return fuel vapors to the tank under high temperature conditions.

NOTE: on air conditioned cars, be sure to disconnect the flexible line connecting the fuel filter to the vapor return line from the tank.

1. Disconnect the flexible fuel line at the end near the front motor mount and at the fuel pump.
2. Disconnect line to carburetor at the filter.
3. Remove two screws and flat washers holding the pump to the oil filler housing.
4. Remove pump and filter as an assembly.
5. When reinstalling, use a new gasket on pump mounting flange.
6. The pump is best serviced by replacement.

Fuel Filter Removal

1. Clamp or plug rubber section of inlet hose.
2. Disconnect fuel pump outlet line at fuel pump.
3. Remove fuel outlet nut and remove filter.
NOTE: use two wrenches to prevent loosening of nut welded to pump cover.
4. Install in reverse of above.

Throttle Check

A vacuum-operated throttle check is used on some models. It operates by a combination of spring pressure and engine vacuum. Adjust length of plunger for correct operation: lengthen to prevent stalling; shorten to avoid racing.

Speed-Up Control Adjustment

Cars equipped with air conditioning have a vacuum-powered, solenoid-operated speed-up control attached to the carburetor.

This device increases the engine idle speed to 900 rpm when the transmission is in neutral and the air conditioner switch is on.

1964-66

1. Warm up engine.
2. Remove air cleaner.
3. Turn air conditioner on.
4. While in Neutral, adjust screw in speed-up unit plunger.
5. Shut off engine and replace air cleaner.

1967-68

1. In Park, warm up engine.
2. Remove air cleaner.
3. Remove and plug vacuum hose from Automatic Climate Control power servo vacuum actuator.
4. Set Automatic Climate Control selector on AUTO.
5. On 1967 model, adjust nuts on speed-up control rod to idle speed of 900. On 1968 model, the adjustment is made by turning the screw located on speed-up control arm.
6. Shut off engine, reconnect vacuum hose and install air cleaner.

1969-70

On these models, the speed-up control is actuated by water temperature, working only when radiator reaches 220°F. Air conditioner does not have to be on to have idle speed up. Curb and fast idle adjustments should be made before attempting the speedup idle adjustment. (See Carburetors in Unit Repair Section.)

1. Warm up engine.
2. With engine off, remove air cleaner.
3. Disconnect vacuum hose leading from thermo vacuum switch to reducing nipple near dash, at reducing nipple.
4. Disconnect vacuum hose at diverter valve and connect to reducing nipple.
5. Disconnect and plug distributor vacuum hose at vacuum unit.
6. Disconnect manifold vacuum hose from thermo vacuum switch nipple at "tee" and connect to distributor vacuum unit. This is the nipple closest to the block.
7. Turn air conditioner to HIGH and turn temperature dial to 65.
8. With transmission in Neutral or Park, start engine and make ad-

justment at screw on idle speed-up control arm.

9. Turn engine off, reconnect hoses and install air cleaner.

Exhaust System

Exhaust Pipe Removal

1964

1. Raise car.
2. Remove clamp at muffler inlet.
3. Remove two nuts securing both right and left exhaust pipes to locating studs.
4. Remove heat control valve.
5. Support muffler on jack stand.
6. Remove exhaust pipe from muffler inlet.

1965-67

1. Raise car.
2. Remove resonator.
3. Remove two nuts from both right and left exhaust manifold studs.
4. Lower exhaust pipe and remove heat control valve. Remove exhaust pipe.

1968-70

1. Raise car.
2. Loosen intermediate pipe hanger at transmission extension housing and remove lateral Y pipe.
3. Remove two nuts securing pipes to each exhaust manifold.
4. Allow exhaust pipe to drop and drive it forward out of intermediate pipe.

Muffler Removal

1964

1. Raise car.
2. Remove clamp securing exhaust pipe to muffler inlet.
3. Remove clamp and blade at muffler inlet.
4. Hold Y-pipe forward and drive muffler rearward.
5. Remove muffler from front intermediate pipe.

1965-67

1. Raise car.
2. Remove clamp at rear of muffler.
3. Pry muffler outlet pipe rearward.
4. Support system with jack.
5. Remove clamp at front of muffler.
6. Remove muffler from intermediate pipe.

1968-70

1. Raise car.
2. Loosen resonator clamp and remove hanger blade.
3. Remove two nuts securing front clamp to muffler. Remove clamp.
4. Remove two nuts securing rear clamp to muffler. Remove clamp.

5. Slide rear portion of exhaust off intermediate pipe.
6. Remove muffler from muffler-to-resonator pipe.

Resonator Removal

1964

1. Raise car.
2. Remove rear clamp and loosen front clamp securing resonator.
3. Holding intermediate pipe forward, drive resonator rearward to remove.

1965-67

1. Raise car.
2. Remove clamp on rear of resonator.
3. Work intermediate pipe, muffler, and muffler outlet pipe rearward as a unit, until support blades slide from rubber slots in brackets.
4. Pry muffler and resonator outlet pipe rearward from resonator.
5. Loosen clamp at front of resonator.
6. Pry resonator rearward.
7. Slide hanger blade out of slot in support bracket. Remove resonator.

1968-70

1. Raise car.
2. Remove clamp at joint of resonator to muffler-to-resonator pipe.
3. Remove rear exhaust hanger at resonator outlet.
4. Separate resonator from muffler-to-resonator pipe.
   CAUTION: do not use heat behind rear axle because of explosive fuel vapors.

Intermediate Pipe Removal

1964 Front

1. Raise car.
2. Remove clamp and blade at muffler outlet.
3. Separate front intermediate from muffler outlet by pulling Y-pipe forward. Support muffler with jack.
4. Remove clamp securing front to rear intermediate pipes.
5. Separate front and rear intermediate pipes.

1964 Rear

1. Remove resonator leaving outlet pipe attached.
2. Remove clamp securing front to rear intermediate pipes.
3. Remove clamp and blade from hanger securing rear intermediate pipe to support bracket.
4. Separate rear and front intermediate pipes.
5. Remove rear intermediate pipe

1965-67

1. Remove muffler.
2. Remove clamp at front end of intermediate pipe.

3. Remove intermediate pipe from resonator.

1968-70

1. Remove clamp securing muffler to intermediate pipe and slide muffler off.
2. Remove clamp securing intermediate pipe to exhaust pipe.
3. Loosen lateral Y-pipe brace at transmission housing.
4. Slide intermediate pipe rearward, disengaging blade from hanger at transmission, and pipe from exhaust pipe.
5. Remove hanger blade from intermediate pipe.

Tail Pipe Removal

On models 1968-70 the tail pipe is part of the muffler assembly.

1964

1. Remove clamp securing resonator and tail (outlet) pipe to support bracket.
2. Remove tail pipe.

1965-67

1. Remove rearmost clamp and hanger blade.
2. Loosen or remove clamp securing tail pipe and muffler.
3. Remove clamp and hanger blade at support bracket located above rear axle.
4. Remove tail pipe from muffler.

Heat Control Valve Removal

Remove nuts securing right and left exhaust pipes to locating studs and remove heat control valve. Support muffler with jack stand.

NOTE: on 1965-70 models, removal of resonator hanger at transmission may be necessary before removing valve.

Cooling System

Detailed information on cooling system capacity can be found in the Capacities table.

Information on the water temperature gauge can be found in the Unit Repair Section.

Radiator Core Removal

Drain the system and remove the upper and lower radiator hoses. Disconnect the hydramatic cooling lines. Remove the mounting bolts, or clamps, and lift the radiator up and out of the engine compartment.

All Models With Air Conditioning

1. Drain cooling system and disconnect radiator hoses.
2. Remove Freon hose retainer clip

from right hand upper side of radiator cradle.

3. Remove four screws holding condenser and dehydrator-receiver to brackets on sides of radiator. Access to the lower right is through the grille.
4. Remove left tie bar between radiator cradle and hood lock plate.
5. Lift out condenser and attachments and taking advantage of the flexible lines that do not have to be disconnected, swing the assembly out of the way.
6. Remove screws holding radiator assembly and condenser brackets to the radiator cradle. Remove the brackets.
7. Now pull radiator forward to clear the hoses and lift it up and out.

Thermostat Removal

The thermostat is located in the water manifold at the front of the block.

This thermostat is removed by disconnecting the upper radiator hose and taking off the four cap screws which hold the thermostat housing to the water manifold.

Thermostat Installation

Install thermostat in opening at top of water manifold with valve up. Be sure that the thermostatic spring strap is parallel to the centerline of the car. Install a new thermostat gasket with gasket cement. Install thermostat housing and tighten cap screws to 10-13 ft. lbs.

Water Pump Removal

Models Without Air Conditioning

1. Drain the cooling system, and remove the fan and drive belts.
2. On power steering cars, detach the power steering pump and bracket from the water pump.
3. Remove the radiator and heater hoses from the water pump, then take out the thermostat housing screws and remove the housing.
4. Take out the water pump flange to cylinder block screws, then remove the water pump and gaskets.

Models With Air Conditioning or Air Suspension

1. Drain cooling system and remove upper half of fan shroud.
2. Remove drive belts from power-steering-air suspension unit, Freon compressor, and generator.
3. Disconnect radiator and heater hoses at the water pump.
4. Exhaust air from accumulator tank and disconnect air inlet and outlet lines at air suspension compressor.

5. Unfasten air compressor, power steering pump assembly, and brackets. No need to disconnect power steering hoses. Swing assembly aside.
6. Disconnect throttle return spring.
7. Remove oil filter unit.
8. Remove the two water pump to block screws that hold the Freon compressor front mounting bracket.
9. Remove Freon compressor front mounting bracket to cylinder block screw, and screw holding generator adjusting link to Freon compressor mounting bracket screw.
10. Loosen front Freon compressor adjusting link to mounting bracket screw and swing lower portion of front mounting bracket to right side of car.
11. Remove four fan retaining screws to release fan and pulley.
12. Remove four remaining water pump to cylinder block screws to release water pump from engine.

Engine

References

In the Specifications table are listed all essential facts about the engines. When engines of different sizes are used, they can be identified by reference to Engine Identification.

Valves

Detailed information on the valves, and the type valve guide used, can be found in the Valve Specifications table.

Bearings

Detailed information on engine bearings can be found in the Crankshaft Bearing Journal Sizes table.

Exhaust Emission Control

In compliance with anti-pollution laws involving all of the continental United States, the Cadillac Division of General Motors has adopted as standard equipment, an integrated air injection emission control system. This method is designed and built into the engine castings and eliminates the need for some of the tubes and exterior air manifolding of previous plans. It does, however, use the same afterburner principle as that described in the Unit Repair Section.

Any of the present methods of terminal exhaust control requires close and frequent attention to tune-up factors of engine maintenance.

Since 1968, all car manufacturers post idle speeds and other pertinent data relative to the specific engine-car

application in a conspicuous place in the engine compartment.

Engine Assembly Removal

The engine is removed together with the transmission. Place the car on stand jacks and drain the cooling system, crankcase, and transmission. Disconnect the battery cables. Take a scribe and carefully mark the position of the hood hinges where they mount to the fender apron, and remove the hood complete with its hinge mechanism.

Disconnect the generator and remove the radiator core, fan and lower pulley. Disconnect the lines from the power steering pump at the pump. Disconnect the refrigerator lines on models with air conditioning, take out the heater hoses, the power brake vacuum line, the carburetor air cleaner, the carburetor and its linkage.

Remove the transmission gravel deflector and disconnect the levers and speedometer at the transmission, disconnect the fuel lines, take off the battery ground straps, the primary ignition wire, the oil pressure and cooling system temperature switch wires. Remove the ignition coil and take off the wires to the ignition resistor, disconnect the vacuum hoses to the manifold and windshield wipers.

Split the rear universal joint and slide the driveshaft from the back of the transmission. Remove the frame intermediate support, disconnect the starter and disconnect the exhaust pipe at the exhaust manifolds. Remove the bolts that hold the front motor supports at the frame and then take off the idler arm support screws and lower the idler arm and steering connecting link. Attach a lifting device and take up the slack until the lifting device has a little load on it. Disconnect and remove the rear engine support bracket from the frame. Carefully lift the engine with its transmission out of the car.

It may be necessary to support the transmission on some sort of movable floor jack so that it can be kept in a downward position and yet guided out easily.

Engine Manifolds

Exhaust Manifold Removal

To remove either of the exhaust manifolds, detach the manifold at the exhaust pipe flange and, in the case of the right manifold, remove the generator and then remove the bolts that hold the manifold to the cylinder head.

On some models, particularly those with heater ducts, access is easier from underneath the car.

However, they can be reached if the air intake ducts of the heater system are detached at both ends.

Intake Manifold Removal

Remove throttle linkage, gas and vacuum lines and the carburetor.

Take off the ignition wires, unbolt and lift off the manifold.

NOTE: on cars with air conditioning, partial removal of compressor is necessary. Do not disconnect compressor hoses.

Cylinder Head

Rocker Shaft Removal

The rocker shafts can be removed and serviced without disturbing the cylinder head or manifolds.

Get the spark plug wires out of the way, remove whatever heater or throttle linkage passes over the rocker cover, and then unbolt and remove the rocker cover.

The rocker shafts are held on brackets; the bolts for these do not pass through the cylinder head. These bracket bolts hold the brackets to the head but do not hold the head on.

Unbolt and remove the rocker shafts, being careful to replace them

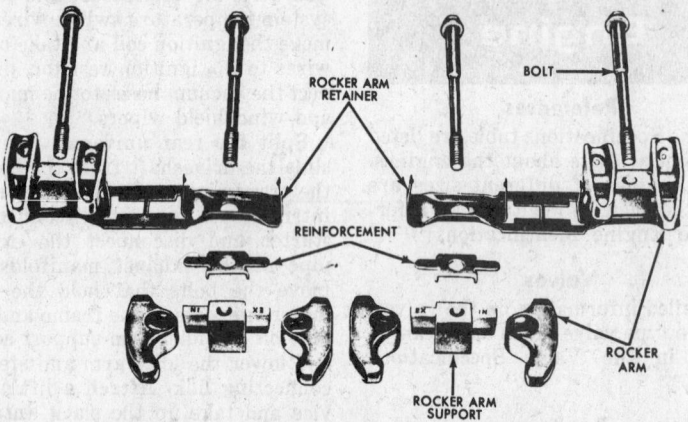

V8 472 cu. in. rocker arm (© Cadillac Div. G.M. Corp.)

on the head from which they were removed. If new rockers and/or shafts are to be installed, note the relative position the rocker occupies on the shaft. Then, if a new rocker is installed toward the center of the shaft, the balance of the rockers will be put on in the proper order, having the correct springs between the rockers.

Thoroughly, clean the rocker springs and shafts before reinstalling.

The push rods can be pulled directly up through the cylinder heads for examination to make sure that they are straight and not badly worn at either end.

Reinstall the rockers, reversing the removal procedure. The larger machined surface on the rocker bracket goes down. The little notch in the forward end of the rocker shaft points toward the camshaft.

Cylinder Head Removal and Installation

Service Note

Care must be used when replacing cylinder-head bolts. They are of different lengths.

1964-67

1. Disconnect the water manifold at the front of the cylinder head or heads. It is a good idea to remove the water pump and water manifold from the car. It is difficult to reinstall a cylinder head with the water pump in place on one head without damaging the water pump gasket.

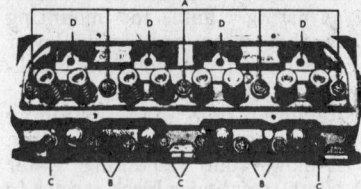

Cylinder head bolt location and length 1964-67
(© Cadillac Div. G.M. Corp.)

2. Remove all vacuum lines and carburetor connections; disconnect all ignition, throttle and battery connections.
3. Take off the intake manifold with the carburetor in place or if desired remove the carburetor.
4. Remove the rocker covers.

   NOTE: it is customary to remove the rocker covers together with the ignition wires and distributor cap as a unit unless service is to be done on the distributor.
5. Remove the generator if the right cylinder head is to be removed. The exhaust manifolds may be disconnected either from the head or from the flange connection to the exhaust pipe. It is better to leave them connected to the head.
6. Remove the head bolts that hold the rocker assemblies to the cyl-

inder head and lift off the rocker assemblies.
7. Remove the pushrods.
8. Remove the balance of the cylinder attaching bolts and lift the head off. It is very important that the head be handled carefully so as not to damage or mark the head gasket surface.
9. Installation is the reverse of the above.

Beginning 1968—472 Eng.

1. Remove intake manifold.
2. Drain engine coolant.
3. Disconnect ground strap at rear of cylinder heads from cowl. Disconnect wiring connector for high engine temperature warning system from sending unit at rear of left cylinder head.
4. Remove generator, if working on the right cylinder head, or partially remove the steering pump if working on the left head.
5. Disconnect A.I.R. injection pump tubes from cylinder heads.
6. Remove clamps holding the wire harness to the cylinder heads and tie harness back out of the way.
7. Remove screws holding exhaust manifold to cylinder heads.
8. Remove screws holding the rocker arm cover to the heads.
9. Remove cover.
10. Remove screws holding each rocker arm support to cylinder head, then remove rocker arm assemblies. Store these assemblies so that they may be reinstalled in their correct locations.
11. Remove pushrods and store them with their respective rocker arm assemblies.
12. Install two 7/16 x 6 in. screws to be used as lifting handles in two of the rocker arm support screw holes.
13. Remove ten cylinder-head bolts.
14. Lift cylinder head off the block.
15. Remove all gasket material from the cylinder head and block mating surfaces.
16. Install by reversing removal procedures.
17. Install ten cylinder-head bolts, finger tight as indicated in illustration.
18. Torque cylinder-head bolts to 100 ft. lbs. starting at the center of the head and working toward both ends.
19. Install pushrods and be sure that the rods seat in the valve lifter cups.
20. Install rocker arm assemblies and torque attaching screws to 60 ft. lbs.
21. Install new rocker arm cover gasket. Install the cover and torque attaching screws to 28 in. lbs.

FRONT OF ENGINE

FRONT OF ENGINE

Bolt Location	Length
A (Bolt)	4.36"
B (Bolt)	4.77"
C (Bolt)	3.02"
D (Bolt/Stud)	3.02"
E (Bolt/Stud)	4.77"

Cylinder head bolt location and length—1968

Bolt Location	Length
A (Bolt)	4.36"
B (Bolt)	4.77"
C (Bolt)	3.02"

Cylinder head bolt location and length—1969

22. Install exhaust manifold to cylinder head. Torque to 35 ft. lbs.
23. Position wiring harness along upper edge of rocker arm cover and secure with clamps.
24. Position A.I.R. tubes in cylinder heads and secure.
25. Install generator if working on right cylinder head, or power steering pump if working on left head.
26. Connect ground strap at rear of cylinder heads to cowl.
27. Connect high temperature warning wiring connector to sending unit at rear of left cylinder head.
28. Install intake manifold.
29. Refill cooling system.
30. Operate engine until normal running temperature is reached, then retorque rocker cover screws.
31. Make visual check of engine coolant and bring to required running attitude level.

Pistons, Connecting Rods and Main Bearings

Piston and Rod Removal

Rod and piston assemblies on all models are removed through the top of the block.

It is possible to replace any and all of the rod or main bearings from underneath the car without removing the crankshaft.

Clean out carbon from top of cylinder bore and ream off the ridge at the top of the bore. This will prevent breakage of the piston ring lands. Push the piston and rod assemblies up and out of the tops of the cylin-

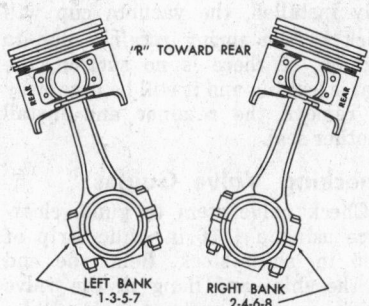

"R" TOWARD REAR

LEFT BANK 1-3-5-7   RIGHT BANK 2-4-6-8

Piston to connecting rod relationship

ders. Be careful not to nick the lower edge of the bores.

Assembling Pistons To Connecting Rods

Slipper-type pistons are used, but the piston has the word rear cast into the metal just beside the wrist-pin hole. With the connecting rod mounted in the vise, so that the number faces the operator, the word rear will go to his right hand on the odd numbered cylinders (left bank).

With the number on the connecting rod facing the operator, the word rear on the piston will go to his left hand on the even numbered pistons (right bank).

Since 1964 the oil spurt hole is discontinued.

Assembling Rod and Piston Assemblies to the Block

The numbers on the connecting rods face away from the camshaft; that is, the numbers on the left bank (odd numbers) face to the left; the numbers on the right bank (even numbers) face to the right. As a double check, the word rear, stamped on the piston, faces the rear of the engine on both banks.

Wick-Type Rear Main Bearing Oil Seal Replacement

This oil seal has a lip and cannot be interchanged with the old type. The two seal halves are identical.

1. Remove the oil pan, baffle, and combination oil and vacuum pump.
2. Remove the rear main bearing cap and loosen the bolts holding the other four bearings about three turns each. Remove the old rear main bearing seals.
3. Clean the groove in the cap and in the block.
4. Apply a slight coating of engine oil to the lips of the two seal halves.
5. Start the upper half into the groove in the block with the lip facing forward and rotate it into position, using care not to distort the seal. Press firmly on both ends to be sure it is protruding uniformly on each side.
6. Install the lower half of the seal into the bearing cap with the lip facing forward and one end of the seal over the ridge and flush

Rear Main Bearing Oil Seal Installer Tool No. J-6349

Cutting rear main bearing oil seal
(© Cadillac Div. G.M. Corp.)

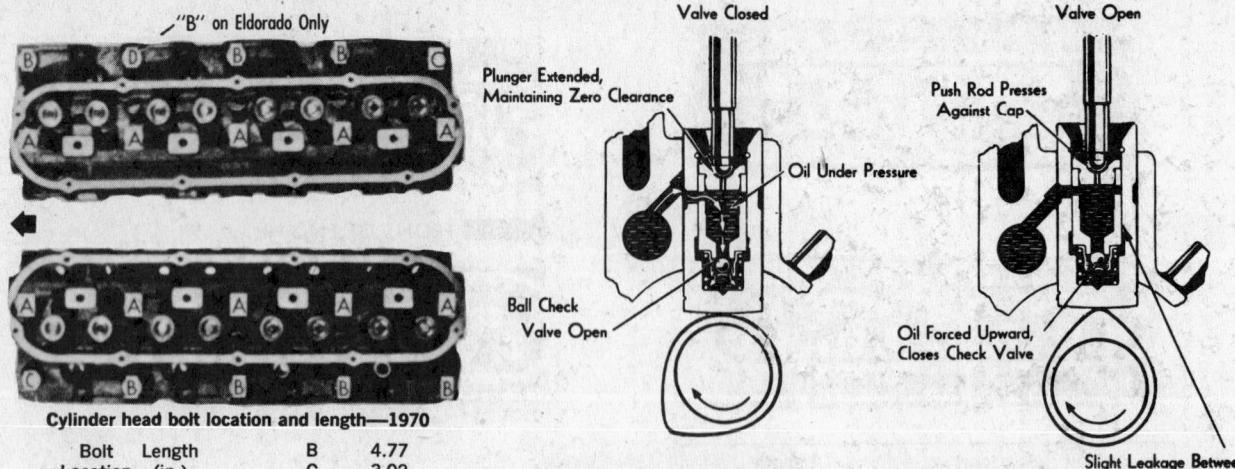

Cylinder head bolt location and length—1970

Bolt Location	Length (in.)		
		B	4.77
		C	3.02
A	4.36	D	4.77

Operation of hydraulic valve lifters (© Cadillac Div. G.M. Corp.)

with the split line. Hold one finger over this end to prevent it from slipping, and push the seal into seated position by applying pressure to the other end. Be sure the seal is firmly seated and protrudes evenly on each side. Do not apply pressure to the lip. This may damage the effectiveness of the seal.

7. Apply rubber cement to the mating surfaces of the block and cap being careful not to get any cement on the bearing, the crankshaft or the seal. The cement coating should be about .010 in. thick.
8. Install the bearing cap, tightening the bolts with the fingers only.
9. Move the crankshaft forward and rearward by pounding on the counterweight with a plastic hammer to assure alignment of the rear main bearing thrust surfaces.
10. Tighten the bearing bolts to 90-100 ft. lbs. Be sure to tighten the bolts of the other four bearings also.
11. Reinstall the oil pump, the baffle and the oil pan.

Valve System

Valve Removal

Cadillac uses a holding fixture for the cylinder head which incorporates a pedal-operated valve spring compressor.

With the head off the car and on the bench or in the holder, compress a valve spring and remove the valve keepers. This will release the valve spring retainer, valve spring, and rubber seal from lower groove in valve stem.

Repeat for the remaining valves, being sure to keep them in order so they can be reinstalled in the same position. When reinstalling, be sure that new rubber seal is seated in the groove closest to the valve head.

Check to see that the seal is properly installed. Strike the ends of the valve stems to seat the keepers. Compress a suction cup over the spring retainer and valve stem to test for leakage past the seal.

If the rubber oil seal has been properly installed, the vacuum cup will stick to the spring retainer due to suction. If there is no suction, the seal is leaking and it will be necessary to remove the retainer and install another seal.

Checking Valve Guides

Check valve stem to guide clearance using a 1/16 in. wide strip of .005 in. shim stock. Bend the end of the shim and hang in the valve guide on the pushrod side. Shim should not extend more than 1/4 in. into the guide. If the valve stem will enter the guide, the clearance is excessive.

Valve Guide Replacement

1964-67

Make a pile of washers equal to the projection of the valve guide, and set aside. Drive out valve guides from the bottom side of the cylinder head.

Using an installer or suitable driver, with the pile of washers, lubricate outer surface of the guide and start it into the head. Enter the longest taper first, pointing toward the rocker arm side.

Press guide into head until the installer contacts the plate or the piston end of the guide is flush with the pile of washers.

Beginning 1968—472 Eng.

The valve guides are cast integrally with the cylinder block. For excessive clearance between valve stem and guide, service valves are available with oversize stems (.003-.006-.013 in.). Guides must be reamed to compensate for these oversize stems.

Valve Lifter Removal

1964-67

Lifters may be removed without taking off the cylinder head.

Remove throttle and gas lines from the carburetor, disconnect hoses, vacuum lines and wires that pass over the rocker covers. Remove the distributor cap and disconnect the wires at the spark plugs. Remove the bolts that hold the rocker covers to the cylinder head and lift off the rocker covers leaving the spark plug wires attached to them. Remove the bolts that hold the intake manifold to the cylinder block and lift off the intake manifold. If desired, the carburetor can be detached from the manifold first, but this is not necessary. Remove the valve chamber cover plate. Remove the bolts that hold the rocker shafts to the cylinder head and lift off the rocker shafts.

Removing valve lifters
(© Cadillac Div. G.M. Corp.)

Pull the pushrods up through the holes in the cylinder heads, and the lifters can be pulled up out of their bores.

Sometimes gum residue forms on the bottom of the lifter, making it very difficult to pull the lifter up out of its bore. If this condition is suspected before the job is started, put a good solvent in the engine oil and run the engine for the time specified by the manufacturer of the solvent in order to dissolve the gum.

However, even when gum is present on the bottom of the lifter body, the lifter can be pulled up out of its bore using special pliers. These pliers are designed to grip the lifter firmly, without scoring or scratching it.

If a special tool isn't available, a good substitute can be made by grinding the teeth out of an ordinary pair of pliers and grinding a circle almost the size of the valve lifter body. When the pliers are squeezed down on the lifter body, it will contact a large surface, thus preventing scoring.

Beginning 1968—472 Eng.

Valve rocker arms are no longer fitted to one common shaft per head; they are mounted in pairs (four pairs to each cylinder head). They are of the modified pedestal-mounted type.

Rocker arms may be removed in pairs and do not require cylinder-head removal.

Torque rocker arm mounting screws to 60 ft. lbs.

Timing Case Cover— Chains and Sprockets

Timing Chain and Sprocket Removal

1964-70

1. Disconnect battery and remove carburetor air cleaner.
2. Drain coolant from engine cooling system.
3. Drain oil and remove engine oil pan.
4. Remove upper radiator hose.
5. Remove fan blade assembly, spacer and pulley.

NOTE: where air conditioning is involved, partially remove the compressor. Remove the compressor belt, then, remove the radiator fan shroud.

6. Remove power steering pump belt, generator belt and pulley.
7. Remove lower radiator hose.
8. Without disconnecting the hoses, remove the power steering pump bracket from the cylinder block. Position bracket out of the way.
9. Detach the generator support bracket from the cylinder-head water-outlet pipe and position the bracket out of the way.
10. Remove distributor assembly.
11. Remove fuel pump.

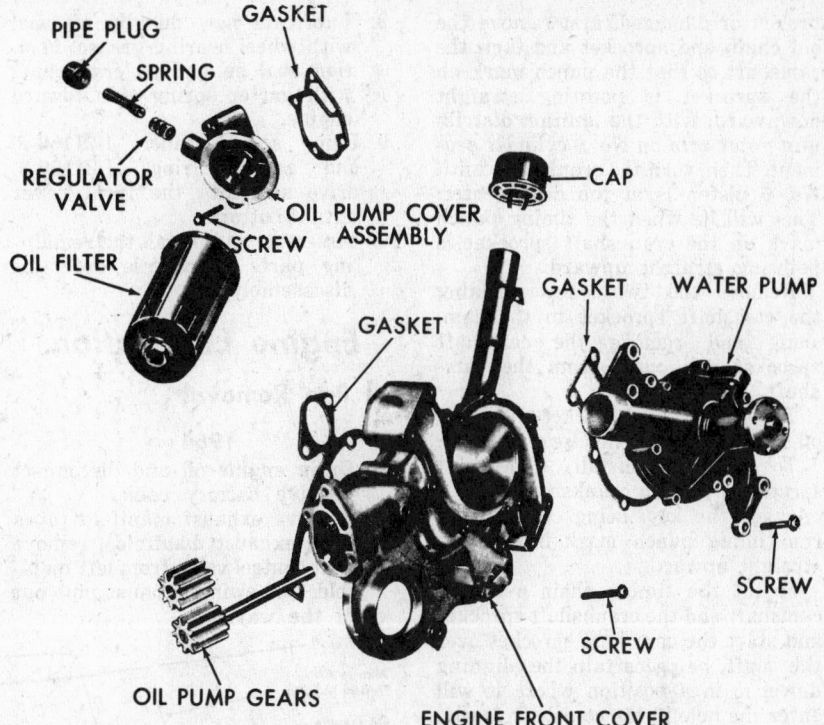

Engine front cover disassembled (© Cadillac Div. G.M. Corp.)

12. Remove four of the six cap screws that attach the crankshaft pulley to the harmonic balancer.
13. Remove cork plug from end of crankshaft, and install balancer puller pilot, J-21052-4 in the bore in the end of the crankshaft.
14. Install puller base, J-21052-1 on front of pulley, lining up index mark on puller base with key slot in harmonic balancer, and install attaching screws. Do not tighten screws.
15. Tighten puller screw to remove balancer. Remove pilot from end of crankshaft.

NOTE: on engines equipped with an Air Injector Reactor System, remove the air pump and bracket assembly and swing it out of the way.

16. Remove oil filter from oil pump cover assembly.

17. Remove the four cap screws that hold the cylinder-head water-outlet pipe to cylinder heads and remove the outlet pipe.
18. Remove remaining cap screws that attach the front cover to the cylinder block and remove the cover with water pump attached.
19. Align the two sprocket timing marks, then, remove the two camshaft sprocket attaching screws.
20. Remove camshaft sprocket, with chain, from camshaft.
21. Remove crankshaft sprocket.
22. To assemble, reverse the above procedure.

Valve Timing Procedure

The chain and sprocket assembly used on all Cadillac models is such that, unless deliberately disturbed, the valve timing will remain as set by the factory, unless the chain and sprockets or both are badly worn or damaged.

NOTE: it is necessary to lower the oil pan in order to remove the crankshaft sprocket.

Remove the timing case cover.

If the timing chain and/or sprockets are being replaced because they are loose or noisy but the car is still in good operating condition, turn the crankshaft until the No. 6 cylinder is in the firing position. The timing punch marks on the cam and crankshaft sprocket will be in line with each other between the shaft centers. This is done to avoid the necessity of having to reset the ignition timing.

If the chain has jumped or is

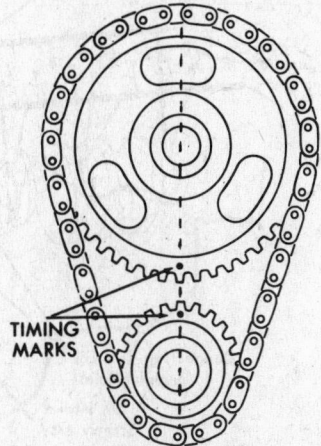

Timing location marks

broken or damaged, first remove the old chain and sprocket and turn the camshaft so that the punch mark on the sprocket is pointing straight downward, with the ignition distributor rotor arm on No. 6 cylinder segment. Then, turn the crankshaft until No. 6 piston is at top dead center. This will be when the timing punch mark on the crankshaft sprocket is pointing straight upward.

Remove the two screws holding the camshaft sprocket to the camshaft and remove the camshaft sprocket and chain from the camshaft.

The crankshaft sprocket will come off readily, without the use of a puller.

To replace, install crankshaft sprocket over the crankshaft until it engages the key, being certain that the timing punch mark is pointing straight upward.

Mount the timing chain over the camshaft and the crankshaft sprocket and start the camshaft sprocket over the shaft, being certain the aligning dowel is in a position where it will enter the hole in the camshaft freely. Make certain that the timing marks on the sprockets are in line between shaft centers.

Camshaft sprockets sometimes install a little stiffly. However, a comparatively easy way to install a tight-fitting sprocket is to draw it on carefully with two bolts somewhat longer than the regular mounting bolts. By drawing alternately against each bolt, and tapping gently with a plastic hammer, even a very tight camshaft gear sprocket can be installed.

When the camshaft is secured, turn the engine two full revolutions until the timing marks again assume the original position. Check to make certain that the punch marks, which are little round circles stamped into the front face of the sprockets, are in line between the shaft centers.

Timing Case Cover Oil Seal

1964-70

These cars are equipped with a molded-type front cover crankshaft oil seal. The seal may be replaced without removing the engine front cover.

1. Disconnect the battery and remove carburetor air cleaner.
2. Remove power steering pump drive belt.
3. Remove generator drive belt.
4. On air conditioned cars, and cars equipped with the A.I.R. system, remove the pump drive belts.
5. Raise and support the front of the car on stands.
6. Remove pulley and harmonic balancer, as outlined in Timing Chain and Sprocket Removal.
7. With a thin blade screwdriver, pry out front cover oil seal.

8. Lubricate new dual-lip oil seal with wheel bearing grease. Position seal on end of crankshaft with garter spring side toward engine.
9. Using seal installer, J-21150-2, and adapter ring, J-21150-3, drive seal into the front cover until it bottoms.
10. Assemble and install the remaining parts in reverse order of disassembly.

Engine Lubrication

Oil Pan Removal

1964

1. Drain engine oil and disconnect positive battery cable.
2. Remove exhaust manifold pipes from exhaust manifolds, remove heat control valve from left manifold, and swing exhaust pipe out of the way.

3. Remove idler arm support mounting screws from frame side member, and lower support.
4. On cars equipped with Hydramatic transmission, remove cap that holds transmission rear cooler pipe clamp to upper flywheel cover plate.
5. On cars equipped with Turbo-Hydramatic transmission, remove bolt that secures rear cooler pipe clamp to adapter ring plate.
6. Remove nut that holds transmission front cooler pipe clamp to right front oil pan locating stud, and remove clamp from stud.
7. Disconnect wires from starter solenoid, releasing wires from spring clip.
8. On cars equipped with Hydramatic transmission, remove two cap screws that hold starter motor assembly to flywheel housing and remove starter assembly.

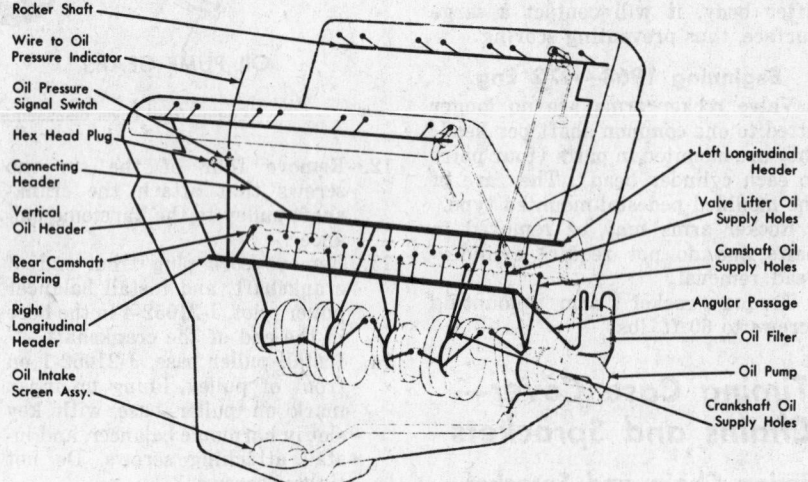

Engine oiling system—1964 (© Cadillac Div. G.M. Corp.)

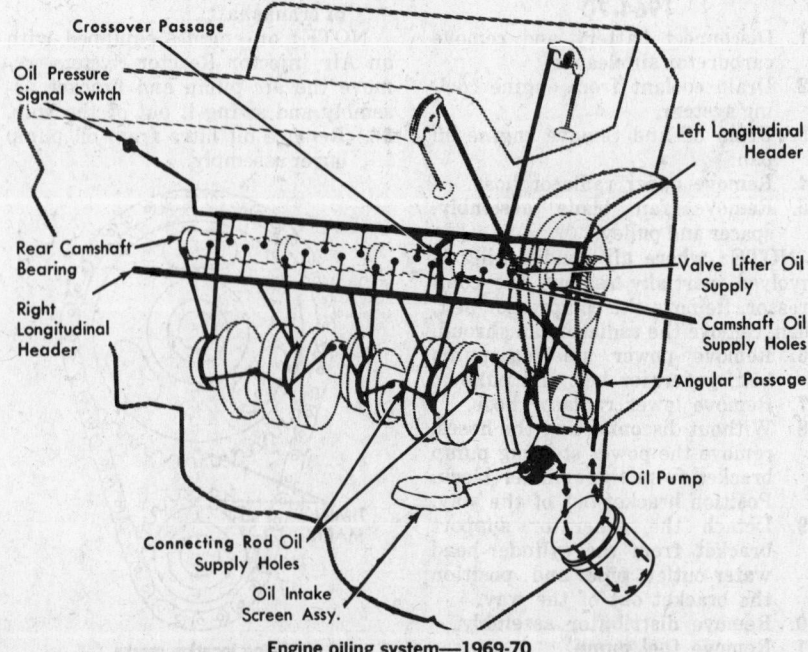

Engine oiling system—1969-70

9. On cars equipped with Turbo-Hydramatic transmission, remove two screws that secure bracket and spacer to crankcase. Remove three bolts that secure starter motor to adapter ring plate and remove starter motor.
10. On cars equipped with Hydramatic transmission, remove four cap screws and two nuts that hold upper flywheel cover plate to flywheel housing and engine oil pan, and remove cover plate.
11. On cars equipped with Turbo-Hydramatic transmission, remove three remaining bolts, two nuts and cap screw that hold front cover plate to adapter ring and engine oil pan, and remove front cover plate.
12. Remove oil pan.
13. When reinstalling, reverse above procedure and torque oil pan screws and nuts to 10 ft. lbs.

1965-70

1. Drain engine oil and disconnect positive battery cable.
2. Disconnect exhaust crossover pipe at exhaust manifold.
3. Disconnect exhaust support bracket at transmission extension housing, and position exhaust system to one side.
4. Remove starter motor.
5. Remove two idler arm support mounting screws from frame side member, and lower support.
6. Disconnect pitman arm at drag link, and lower steering linkage.
7. Remove transmission lower cover.
8. Remove engine oil pan.
9. When reinstalling, reverse above procedure and torque oil pan screws and nuts to 10 ft. lbs.

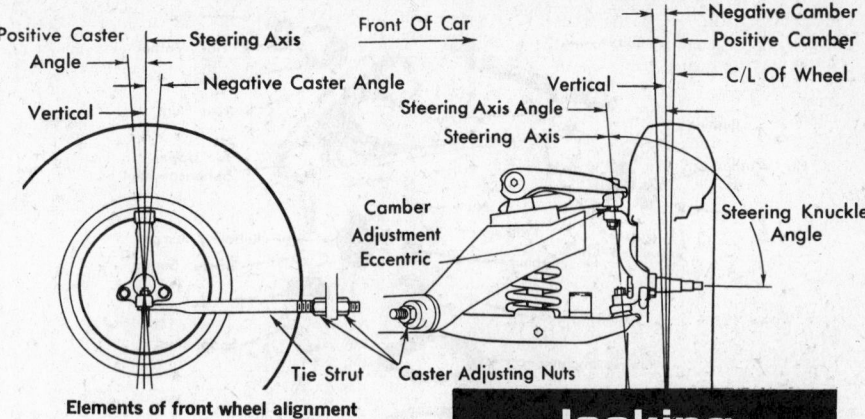

Elements of front wheel alignment

Oil Filter Element Replacement

A full flow filter is used that is disposable and is equipped with a by-pass safety valve.

Front Suspension

References

General instructions covering the front suspension and how to repair and adjust it, together with information on installation of front wheel bearings and grease seals, are given in the Unit Repair Section.

Definitions of the points of steering geometry are covered in the Unit Repair Section. This article also covers troubleshooting, front end alignment, and irregular tire wear.

Overall length and tire size figures can be found in the General Chassis and Brake Specifications table.

Jacking, Hoisting

1964

When jacking under front suspension arms, make sure lift is made from point outboard of the support plates on lower arms.

1965-70

Follow instructions on earlier types when jacking at front suspension.

When lifting on frame area, make sure of solid contact at the corners of the perimeter frame offset close to the bend at front and rear areas.

Lower Suspension Arm and Coil Spring Removal and Installation

1. Disconnect front shock at its upper mount.
2. Raise car and support under front frame side rails.
3. Disconnect stabilizer link from lower arm or spring to be removed.
4. Disconnect tie-strut at lower arm.
5. Remove bolt holding shock to lower arm, and remove shock from car.
6. Remove wheel and tire assembly.
7. Remove nut from pivot bolt in lower arm at frame mount.
8. Position jack under outboard end of lower suspension arm so that jack is supporting the arm.
9. Remove locknut from lower ball joint stud. Install standard nut on joint stud and run nut to within two threads of knuckle.
10. Strike knuckle with a hammer in area of ball joint stud to loosen the joint. Raising the opposite rear corner of the car will help compress the spring and assist in removing the joint stud from the knuckle.
11. Use jack to lift spring load from nut and remove nut from joint stud.
12. Slowly lower jack and remove spring.
13. Remove pivot bolt from lower

Engine oiling system—1965-68 (© Cadillac Div. G.M. Corp.)

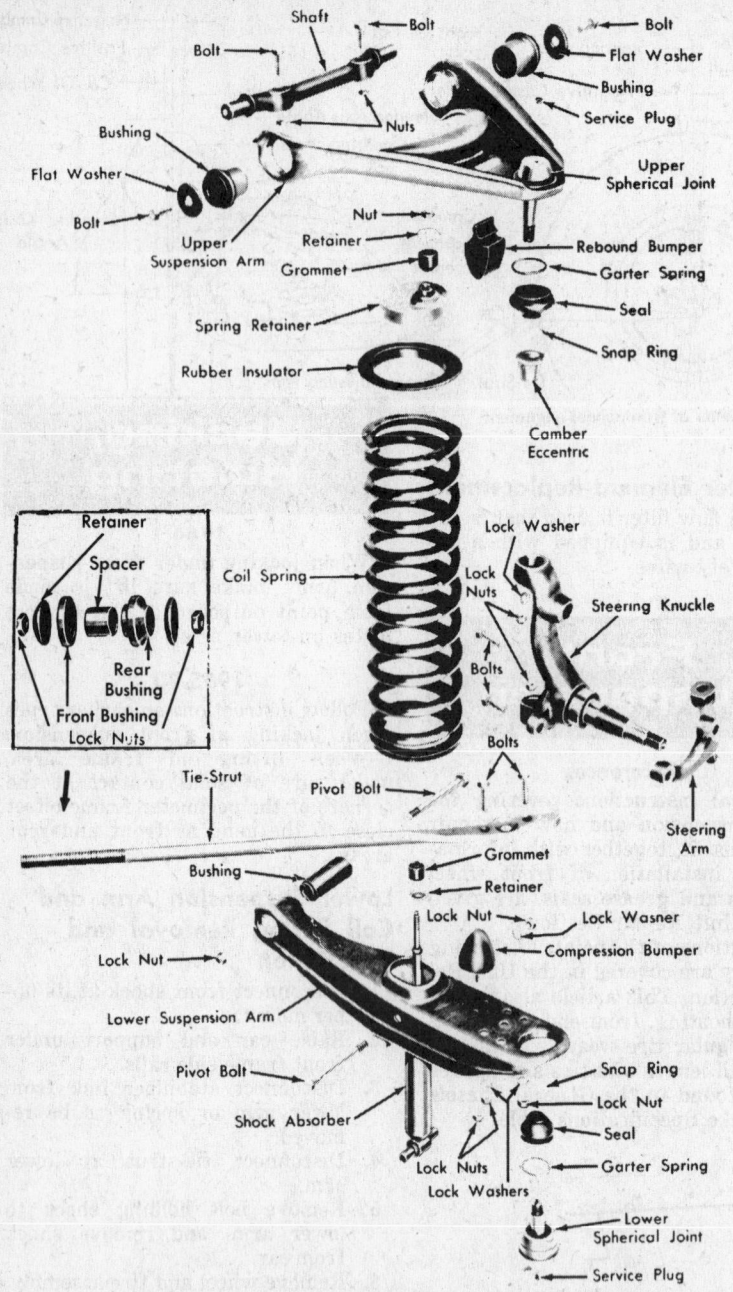

Typical front suspension—1964-70 (© Cadillac Div. G.M. Corp.)

When reinstalling, tighten nut to 45-50 ft. lbs. for 1964 model, 30-35 ft. lbs. for 1965-70 models.

Steering Linkage Removal and Disassembly

1. Remove cotter pins and nuts from outer tie-rod pivots.
2. Remove outer tie rod pivots from steering knuckles using tie-rod end puller.
3. Remove idler arm screws and lockwashers from side member.
4. Remove pitman arm cotter pin, nut and washer at steering linkage.
5. Remove steering linkage from pitman arm.
6. Remove drag link with tie-rods and idler arm attached.
7. Remove cotter pins and nuts from idler arm pivot and inner tie-rod pivots.
8. Remove tie-rod.
9. Remove idler arm from drag link.
10. Remove dust seals from pitman arm and idler arm pivot studs.
11. Remove outer tie-rod pivots by loosening nuts on outer clamp bolts and unscrewing the pivots from adjuster tubes.
12. To install, reverse removal procedure.

NOTE: all nuts should be tightened to 35-40 ft. lbs.

Checking Drag Link Height and Position

The distance between the lower edge of the drag link and the flat spot on the frame side bar, directly above the drag link at each end, should be checked in cases of steering wander and instability after normal corrective adjustments have been made. The procedure outlined below may be used to measure these distances.

1. Place a straight bar across two adjustable jacks, directly below the drag link.
2. Adjust the height of the jacks so that distance A (from top of bar to ½ in. outboard of side rail and directly above drag link) is equal on both sides.
    NOTE: adjusting jacks so that distance A is an even number of inches will simplify this measurement.
3. Measure distance B (from top of bar to bottom of drag link) on both sides.
4. Distance A minus B should be 4½ in. and equal on both ends of the drag link within ⅛ in. (drag link must be parallel to frame within ⅛ in). A tool to check drag link to frame parallelism may be made from any rigid material. Place the tool on the top of the drag link at the end, and check the distance between the

arm at frame mount and remove the arm.
14. Install by reversing the removal procedure.

Steering Gear

Power Steering Gear

Troubleshooting and repair instructions covering power steering gears are given in the Unit Repair Section.

NOTE: for power steering pump belt adjustment, loosen pump to mounting bracket screws, move pump upward until belt is tight. Tighten mounting bracket screws with car in neutral and engine running faster than idle speed, turn steering wheel full right or left. If belt squeals, it is too loose and should be tightened more.

Horn Ring and Steering Wheel Removal

Disconnect horn wire at neutral safety switch on the steering column. Using an Allen wrench, loosen the screws on the underside of the steering wheel spokes near the center and remove the horn ring assembly.

Remove the nut holding the steering wheel to the steering shaft.

Use a puller to remove the steering wheel. Note the match-marking of the shaft and wheel.

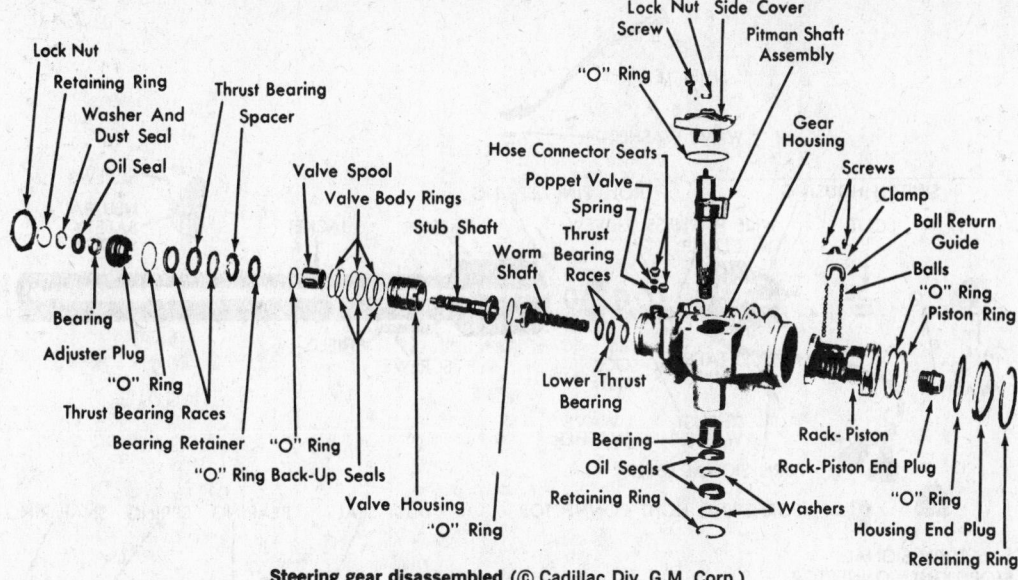

Steering gear disassembled (© Cadillac Div. G.M. Corp.)

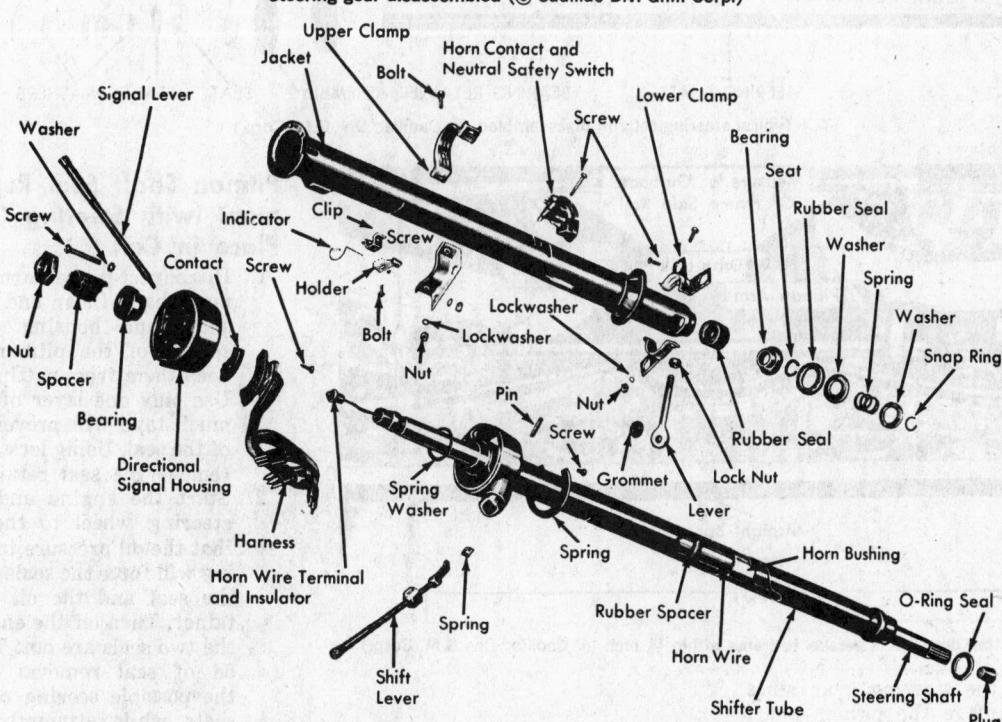

Steering column disassembled—1964 (© Cadillac Div. G.M. Corp.)

tip of the tool and a point located 1/2 in. outboard of the frame side rail with a 1/4 in. drill. If the tool plus the drill shank does not touch the frame, the drag link is too low, and if the tip of the tool will not fit in position, the drag link is too high. Check both ends to see that the clearance between the tool and frame is within the 1/8 in. allowed for parallelism to frame.

5. If the idler arm end of the drag link is not within limits, remove the idler-arm support mounting screws on the frame side bar, and screw the idler arm in or out of bushing until correct height is obtained.

Caution When turning the idler arm into the bushing to raise the drag link, be sure that the idler arm is at least one-half turn off of bottom to prevent interference on turns. When turning idler arm out of bushing, do not unscrew more than two and one-half turns from bottom or excessive play will result. If proper height cannot be obtained with this adjustment, it indicates a bent idler arm which should be replaced.

Play in the idler arm bushing which causes more than 1/8 in. vertical movement at the ball end of the idler arm may cause car wander or erratic steering. If this condition exists, parts should be replaced.

6. If the pitman arm end of drag link is not within limits, the pitman arm must be removed and bent as required.

Caution The bending operation on the pitman arm must be done very carefully to avoid internal stresses and must be performed cold, with a tool placed midway between the ball stud and the splined hole in the steering gear end. Do not bend unless drag link distance to frame at pitman arm end is not within limits. All adjustments for parallelism should be made at the idler arm end if possible. Do not attempt to bend the arm while it is

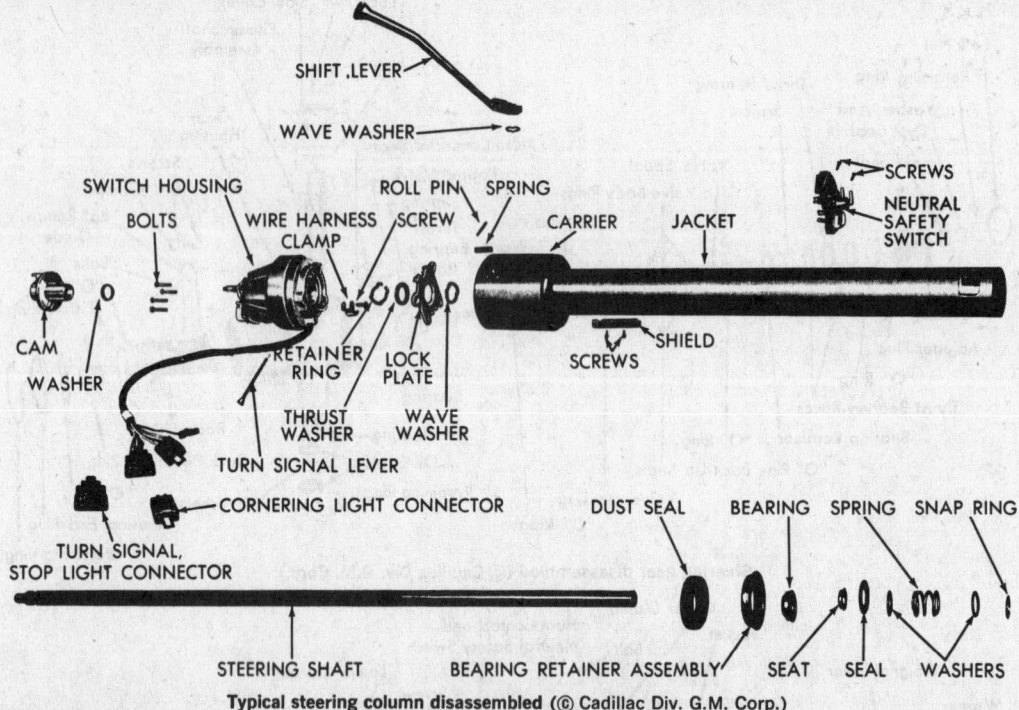

Typical steering column disassembled (© Cadillac Div. G.M. Corp.)

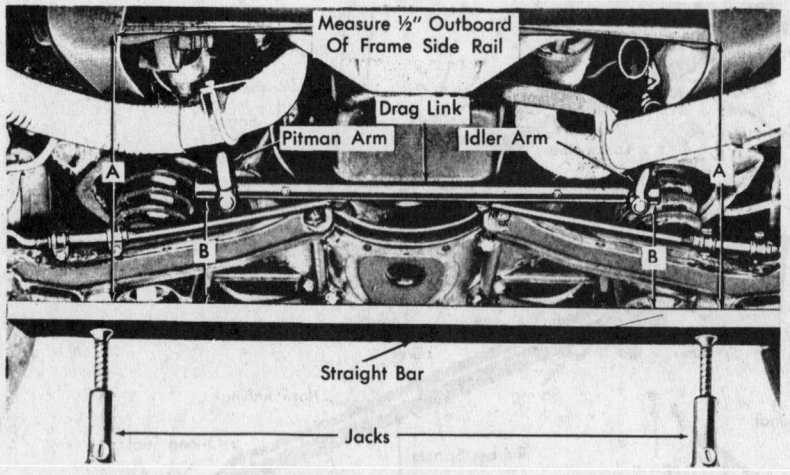

Checking that drag link is parallel to frame within ⅛ inch (© Cadillac Div. G.M. Corp.)

attached to the steering gear since this may damage the pitman shaft bearing.

Steering Gear Assembly Removal and Installation

1. Remove pump reservoir cover and siphon out all fluid.
2. Disconnect the hoses at the pump and cap.
3. Support front end of car on stand jacks near outer end of lower suspension arms. (At frame side members if air suspended.)
4. Remove clamp bolt from half of flexible coupling connecting gear to upper steering shaft.
5. Disconnect pitman arm from drag link.
6. Remove screws holding gear assembly to frame and so release gear assembly from car.

7. When reinstalling, tighten the clamp bolt in the upper half of the flexible coupling to 25-30 ft. lbs. Reconnect the pitman arm to the drag link as described in Steering Linkage Removal and Disassembly. Check that drag link is as described in Checking Drag Link Height And Position.
8. Reconnect hoses. Be sure that flexible coupling is in a flat plane with no visible bend. If distorted in any way, remove the lower steering column cover and lower clamp to jacket screw.
9. Loosen the steering jacket clamp screws at the instrument panel and slide the steering column jacket up or down to relieve all distortion in the flexible coupling.

Pitman Shaft Seal Replacement (with Steering Gear in Place in Car)

1. Disconnect pitman arm from pitman shaft. Clean end of pitman shaft and housing. Tape the splines of the pitman shaft to keep them from cutting the seal. Use only one layer of tape. Too much tape will prevent passage of the seal. Using lock ring pliers remove the seal retaining ring.
2. Start the engine and turn the steering wheel to the right so that the oil pressure in the housing will force the seals out. Catch the seal and the oil in a container. Turn off the engine when the two seals are out. This method of seal removal eliminates the possible scoring of the seal seats while attempting to pry them out.
3. Inspect the two old seals for damage to the rubber covering on the outside diameter. If they are scored or scratched, inspect the housing for burrs, etc. and remove them before installing the new seals.
4. Lubricate the two new seals with petroleum jelly. Put the one with a single lip in first, then put in a washer. Drive seal in far enough to permit installation of double lip seal, washer and the seal retaining ring. The first seal is not supposed to bottom in its counterbore.
5. Fill reservoir to proper level, start engine, turn wheel to right and check for leaks.
6. Remove the tape and reinstall

the pitman arm. Tighten nut to 100-125 ft. lbs.

Automatic Transmission

References

When automatic transmission trouble is reported, a road test and careful diagnosis are in order. Transmission Removal and Replacement and Linkage Adjustments are covered here. For test procedures, transmission overhaul and other detailed information, see Unit Repair Section.

Neutral Safety Switch, All Models

1. Check that the hand lever is correctly adjusted and that the neutral safety switch is properly positioned by this procedure.
2. Set the handbrake. Put the hand lever on the steering column in drive. Hold the ignition key (or starter button) on and slowly move the hand lever toward neutral until the starter cranks and the engine runs.
3. Without moving the lever farther, press the accelerator to determine whether the transmission is really in neutral.
4. If all is correct, the engine will have started when the hand lever got to the neutral position and the transmission will not be in gear.
5. Adjust the neutral safety switch by turning it and its mounting bracket until the above conditions have been met.

Dual-Coupling—1964

Throttle Valve Linkage Adjustment

1. Have engine and transmission at operating temperature. Adjust engine idle speed to 430-450 rpm.
2. At the transmission throttle valve lever (the outer one on the left side of the transmission) loosen the two nuts that hold the trunnion on the rod. Hold the lever back against its stop and adjust the trunnion to suit.
3. Tighten the locknuts.
4. Upshifts will be at too high a speed if the throttle linkage is too short. Upshifts will be at too low a speed if the throttle linkage is too long.

Transmission Removal

1. Remove oil level stick. Drain transmission oil pan. Remove oil filler tube. The first fluid coupling can be drained now or later.
2. Drain cooling system, discon-

nect battery, remove starter. Disconnect the controls at the transmission levers. Remove the levers so they won't be bent.
3. Remove two propeller shaft center - bearing - support - to - frame bolts, being careful to match-mark the support at the frame to facilitate alignment at reassembly. Be careful also to identify the shim packs under each side of the support so they can be returned to their original position. The bent ends of the shims go down.
4. Remove the U-bolts and locks at the rear axle pinion. Slide the propeller shaft front yoke to the rear and off the transmission output shaft.
5. Tie the rear end of the double shaft up to the frame. The front portion can rest on the frame crossmember.
6. Install engine support and transmission jack. Raise engine to take weight off rear engine mounts and remove mount-to-crossmember bolts.
7. Unbolt and remove the crossmember.
8. Lower engine enough to allow access to the upper bell housing bolts and remove them.
9. Remove the four flex plate-to-flywheel bolts. Match-mark flywheel to flex plate.
10. Remove remaining flywheel housing bolts.
11. Move transmission ¾ in. to rear to clear the dowels and lower assembly down and away.

Installation

1. Some models have a wick in the hole in the rear of the crankshaft, which acts as a pilot for the first fluid coupling. Lubricate this hole or the wick with synthetic oil seal lubricant.
2. Raise transmission to align with flex plate. Align match-marks on flywheel and flex plate and move transmission assembly forward to engage dowels.
3. Install two of the lower flywheel housing bolts. Check that first fluid coupling assembly has some end-play and that it is not binding in the pilot hole in the crankshaft. Push forward on the flywheel and measure clearance between a mounting pad on the flex plate and the front mounting face of the flywheel. This clearance should be between .013 and .024 in. If clearance is outside these limits install shims following one of the following procedures:
4. When clearance is less than .013 in., install a spacer over flywheel pilot. Then refasten the flywheel housing and again check that

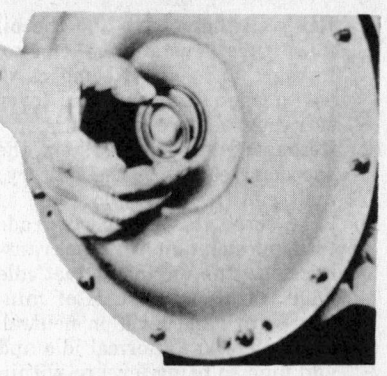
Installing spacer on flywheel pilot in order to increase clearance between flywheel and flexplate to .013-.024 in.
(© Cadillac Div. G.M. Corp.)

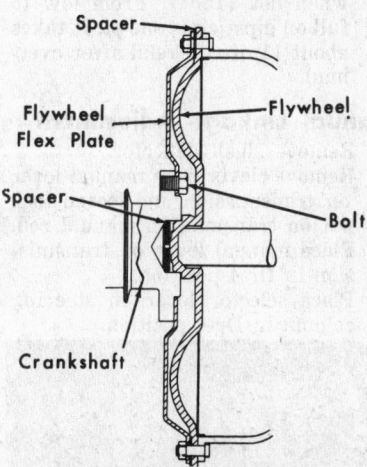
Dual-range transmission to crankshaft relationship
(© Cadillac Div. G.M. Corp.)

clearance between a mounting pad and the flywheel is between .013 and .024 in.
5. When clearance is more than .024 in., use vaseline coated shims on the flywheel mounting pads to reduce the clearance to between .013 and .024 in. Do not install the nuts but proceed to shim up the other mounting pad-to-flexplate surfaces to lie within the given limits. Try to keep the clearance equal.
6. Install the flywheel-to-flexplate nuts and tighten evenly to 15-20 ft. lbs.
7. Reinstall engine support crossmember. Install the engine rear supports.
8. Remove engine support device and transmission jack.
9. Slide the driveshaft into place on the output shaft. Refasten the rear universal joint at the rear axle. Align the match-marks at the driveshaft center support and install the two center bearing support-to-frame bolts.
10. Install the remaining flywheel housing-to-crankcase bolts and tighten evenly to 40-50 ft. lbs. Install the flywheel housing cover.

11. Reinstall the starter and the oil filler tube. Install the control linkages and levers, being careful to align the shaft and lever serrations before tightening.
12. Reconnect the cooling system, the speedometer cable, the battery; and refill the cooling system.
13. Pour eight qts. of a good grade transmission fluid into the transmission. Run engine at fast idle (800 rpm) for a couple of minutes with hand lever in neutral. Reduce speed to normal idle and add fluid to bring level to within ¼ in. of full mark. Fluid level should be within ¼ in. of full mark when cold; at full mark when hot (150°). From low to full on dipstick is one pt. It takes about 11 qts. to refill after overhaul.

Manual Linkage Adjustment

1. Remove slush deflector.
2. Remove clevis from manual lever on transmission and loosen lock nut on transmission manual rod.
3. Place manual lever on transmission in Dr-4 position.
4. Place selector lever on steering column in Dr-4 position.

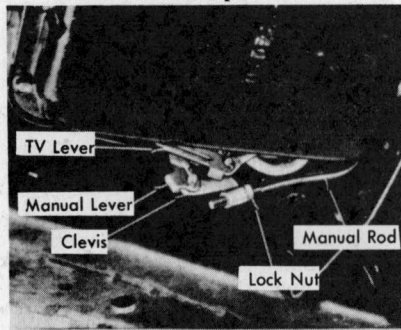

Manual linkage adjustment—1964
(© Cadillac Div. G.M. Corp.)

5. Adjust length of manual rod until hole in manual lever lines up with hole in clevis and then increase rod length one turn more.
6. Install clevis on manual lever and tighten lock nut.
7. Check selector lever in car. It should be free to enter Park and Dr-4 position on column should correspond with Dr-4 on transmission.

Turbo Hydramatic —1964-70

Transmission Removal

1. Raise car and provide support for front and rear of car.
2. Disconnect front exhaust pipe bolts at the exhaust manifold and at the connection of the intermediate exhaust pipe location (single exhaust only). On dual exhaust the exhaust pipes need not be removed.
3. Remove pinion flange U-bolts and slide propeller shaft toward transmission as far as possible to separate universal joint from pinion flange. Remove propeller shaft from car.
4. Place suitable jack under transmission and fasten transmission securely to jack.
5. Remove vacuum line to vacuum modulator hose from vacuum modulator.
6. Loosen cooler line bolts and separate cooler lines from transmission.
7. Remove transmission mounting pad to crossmember bolts.
8. Remove transmission crossmem-

ber support-to-frame rail bolts. Remove crossmember.
9. Disconnect speedometer cable.
10. Loosen shift linkage adjusting swivel clamp nut. Remove cotter key, spring, and washer attaching equalizer to outer range selector lever. Remove equalizer.
11. Disconnect transmission filler pipe at engine. Remove filler pipe from transmission.
12. Support engine at oil pan.
13. Remove transmission flywheel cover pan-to-case tapping screws. Remove flywheel cover pan.
14. Mark flywheel and converter pump for reassembly in same position, and remove three converter pump-to-flywheel bolts.
15. Remove transmission case to engine block bolts.
16. Move transmission rearward to provide clearance between converter pump and crankshaft. Lower transmission and move to bench.

Installation

1. Assemble transmission to suitable transmission jack and raise transmission into position. Rotate converter to permit coupling of flywheel and converter with original relationship.
2. Install transmission-case-to-engine-block bolts. Torque to 30-40 ft. lbs. Do not overtighten.
3. Install flywheel to converter pump bolts. Torque to 30-40 ft. lbs.
4. Install transmission crossmember support. Install mounting pad to crossmember.
5. Remove transmission jack and engine support.
6. Install transmission flywheel cover pan with tapping screws.

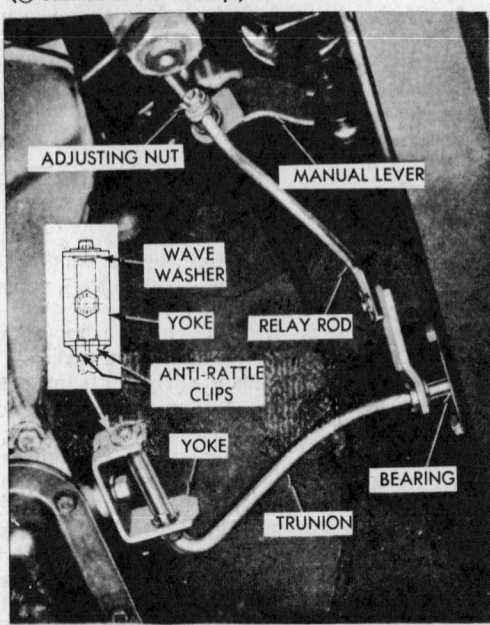

Manual linkage adjustment—1965-68
(© Cadillac Div. G.M. Corp.)

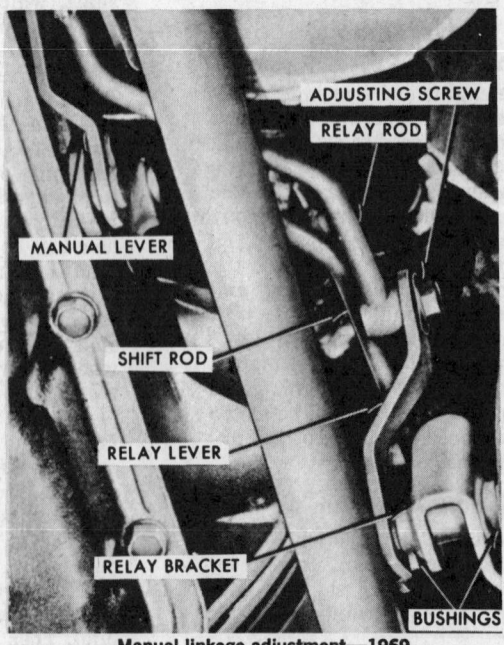

Manual linkage adjustment—1969
(© Cadillac Div. G.M. Corp.)

7. Install transmission filler pipe using a new O-ring.
8. Reconnect speedometer cable.
9. Install propeller shaft. Connect propeller shaft to pinion flange.
10. Reinstall front exhaust crossover pipe.
11. Install oil cooler lines to transmission.
12. Install vacuum line to vacuum modulator.
13. Fill transmission with oil as follows:
    A. Add four pts. of oil.
    B. Start engine in neutral. Do not race engine. Move manual control lever through each range.
    C. Check oil level, adjust oil level to full mark on dipstick.

Manual Linkage Adjustment

1. Loosen nut on steering column manual lever.
2. Pull relay rod up, positioning transmission shift valve in Park, then push rod down to the third or Neutral step.
3. Position selector lever in Neutral.
4. Tighten nut on steering column manual lever.
5. Check that positions selected on selector lever correspond with appropriate detents on transmission.

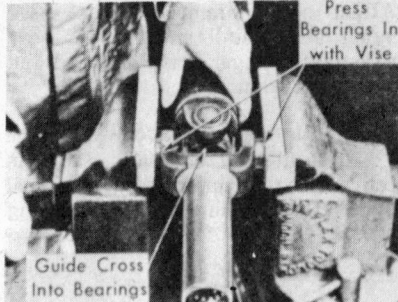

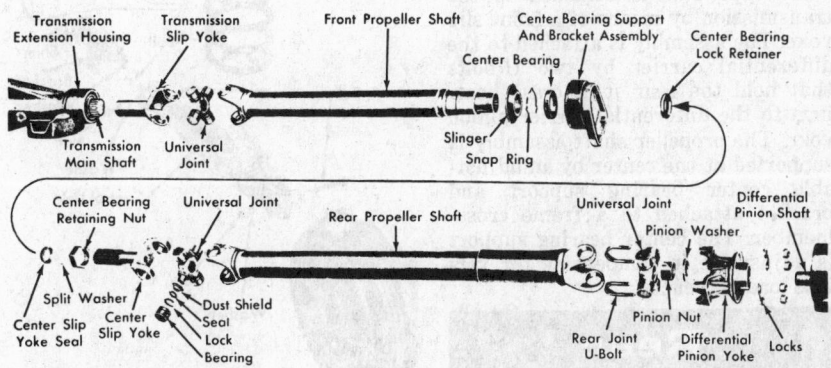
Driveshaft—1964 (© Cadillac Div. G.M. Corp.)

propeller shaft uses three constant velocity universal joints. These joints are located at each end and at the approximate center of the shaft assembly.

At the front end of the rear section of the propeller shaft is a splined male slip yoke that fits into a splined coupling in the rear end of the front section of the front shaft. This slip spline satisfies the normal lengthening and shortening of the propeller shaft due to road conditions and rear axle movement.

The propeller shaft assembly is attached to the transmission by means of a slip yoke, and to the differential drive pinion by a double flange connection. The propeller shaft assembly is supported midway by a bracket and bearing combination attached to a frame crossmember.

With the exception of the center bearing and support combination, the propeller shaft is serviced as an assembly.

Two-Piece-Shaft Type— Commercial Models

A two-piece propeller shaft assembly, using three standard universal joints, is used on commercial vehicles.

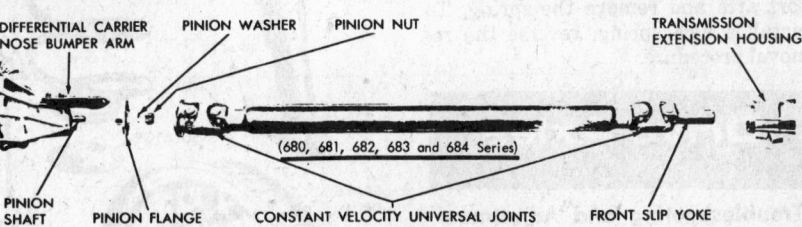

Manual linkage adjustment—1970 (© Cadillac Div. G.M. Corp.)

U Joints, Drive Lines

Universal joints and drive lines can be divided into three groups: single-shaft models, two-piece shaft models (except commercial), and two-piece shaft models (commercial).

Single-shaft Type

Two constant velocity universal joints are used on the single-shaft type: one at the front, and one at the rear. This type propeller shaft is serviced as a complete assembly.

Two-Piece-Shaft Type— Except Commercial Models

The non-commercial, two-piece

Driveshaft—1965-70 (© Cadillac Div. G.M. Corp.)

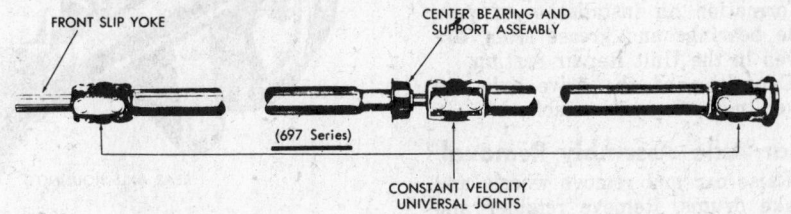
Commercial driveshaft—1966-70 (© Cadillac Div. G.M. Corp.)

Standard joints are used at each end of the shaft assembly and at the center support.

The universal joints are replaceable, but cannot be repacked. On original universal joints, the injected nylon ring that locks the bearing cup in the slip yoke will shear off when the bearing is removed. There are no provisions for replacing this nylon ring. When the joint becomes noisy or otherwise needs attention, renew the joint.

Front and rear sections of this shaft are splined together the same as in two-piece passenger car models.

The commercial vehicle two-piece-shaft type unit is attached to the transmission by means of a front slip yoke. The assembly is attached to the differential carrier by two U-bolts that hold the rear joint cross bearings to the differential carrier pinion yoke. The propeller shaft assembly is supported at the center by an adjustable center bearing support and bracket attached to a frame crossmember. The center bearing support is adjustable to compensate for various load influences.

Rear Suspension

Coil Spring R & R

1964-70

Jack up the rear of the car and place stands under the rear housing near the differential and also under the frame. Now disconnect the shock absorber from the bracket which is welded to the rear housing. Disconnect the emergency brake cables at the clip on the rear support arm. Now place an hydraulic jack under the rear support arm at the spring. Remove the bolt that holds the rear support arm to the bracket that is welded on the rear housing. Then lower the support arm and remove the spring. To install a new spring, reverse the removal procedure.

Drive Axles

Troubleshooting and Adjustment

General instructions covering the troubles of the drive axle and how to repair and adjust it, together with information on installation of rear axle bearings and grease seals, are given in the Unit Repair Section.

Capacities of the drive axle are given in the Capacities table.

Rear Axle Assembly Removal

Raise car and remove wheels and brake drums. Remove retainer and backing plate to rear axle housing.

Attach slide hammer to axle shaft puller and install on studs of rear axle shaft flange. Drive outward and remove axle shaft.

Inner Seal R & R

This seal may be replaced in the conventional manner of regular seal installation, while shaft and bearing assembly is out of housing.

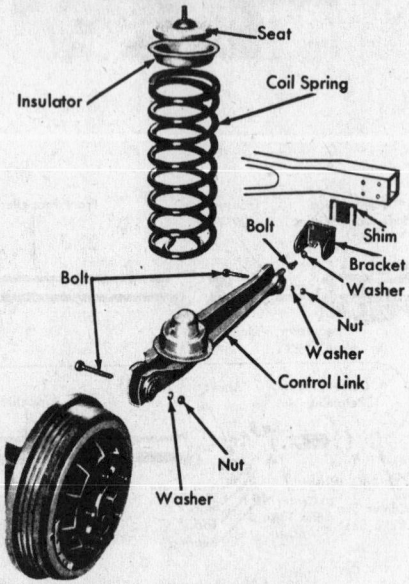

Rear lower control link
(© Cadillac Div. G.M. Corp.)

Outer Seal Installation

1. Adjuster ring seal is double-lipped. Install with short lip toward castellations. Coat lip with light lubricant.
2. Bearing retainer seal is double-lipped. Install with short lip toward axle shaft flange and open side toward wheel bearing. Coat lip with light lubricant.
3. With proper sized adapter, press bearing seal into adjusting ring or retainer. From face of seal to edge of bearing retainer or adjusting ring should be 3/16 in.

Bearing Installation

1. Thoroughly and carefully pack the bearing with lubricant.
2. Assemble parts in order on shaft in following sequence—adjuster ring with seal, retainer, cup cone and pressing sleeve. Press tightly against shaft shoulder. Tool—sleeve J-4704-1 is suggested.
3. Install clinch ring with chamfer toward bearing cone.

Shaft Assembly Installation

1. Thoroughly clean mating surfaces between housing flange and backing plate and between backing plate and bearing retainer.

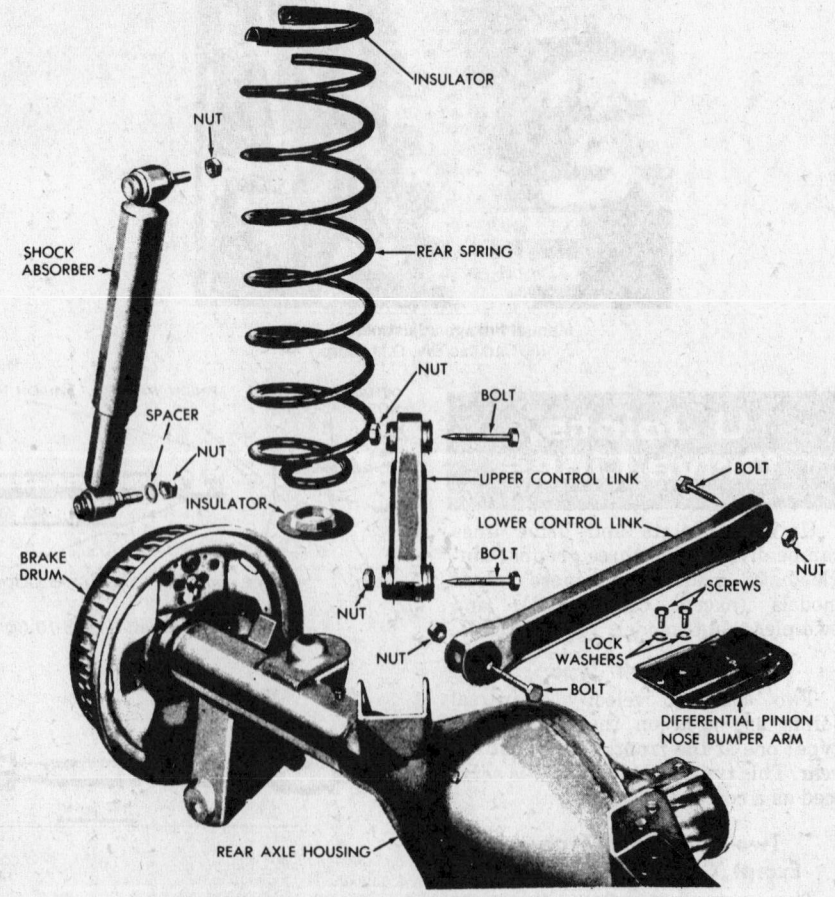

Rear suspension—1965-70 (© Cadillac Div. G.M. Corp.)

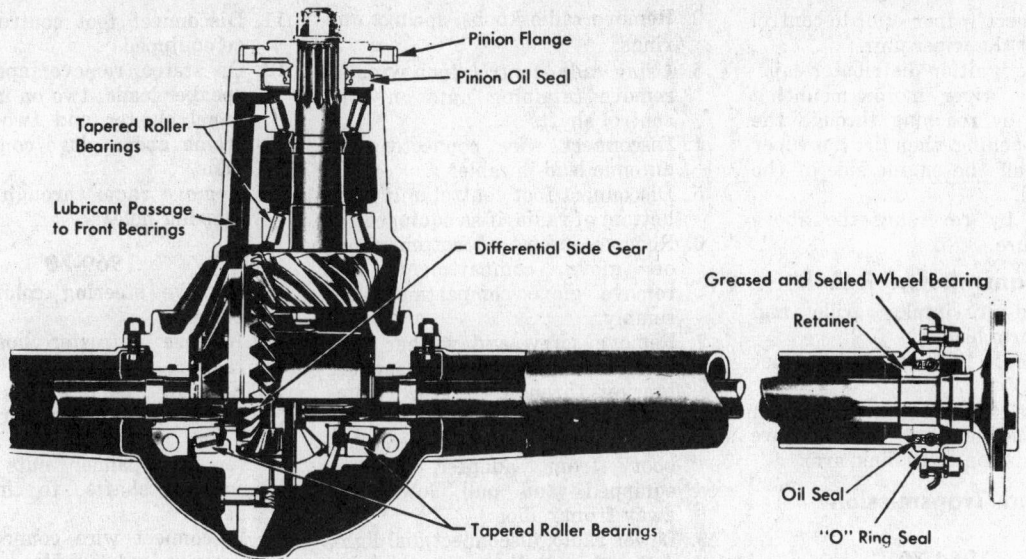

Rear axle assembly—1964-69 (© Cadillac Div. G.M. Corp.)

2. Install new gasket on each side of backing plate. Where shims are used on left side, they must be placed next to housing flange.
3. Straighten tang on adjusting ring lock, and, with approximately two threads of adjusting ring engaged in retainer, carefully insert axle shaft into housing. Use care not to damage inner axle oil seal.
4. Install the bearing retainer nuts and washers and progressively tighten to 40 ft. lbs. torque. Tap end of each axle shaft to be sure bearing cup is seated and center thrust block is not binding.

Shaft End-play Adjustment

1. Coat adjuster ring threads with waterproof sealer.
2. Tighten adjuster ring as far as possible by hand.
3. Insert screwdriver, or other suitable tool, through access hole in axle shaft flange. While holding axle shaft flange to prevent turning, tighten adjusting ring until all end-play has been removed. Then, back off ring two castellations. (Each castellation equals about .005 in.).
4. Tap on left axle shaft so that all end-play will be at the left axle shaft bearing.
5. Mount dial gauge at backing plate and measure shaft and flange assembly end-play. Turn adjusting ring to produce end-play of .001-.006 in.
6. Bend tang to lock adjusting ring.

Automatic Level Control—1965-70

The system consists of a vacuum-operated air compressor and a control valve mounted at rear suspension crossmember. The valve is then con-

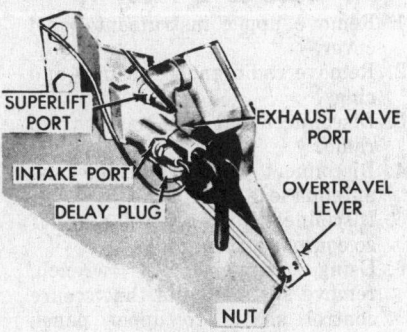

Height control valve

nected to Superlift rear shock absorbers.

The Superlift shock absorber is essentially a conventional shock absorber enclosed in an air chamber. A pliable nylon-reinforced neoprene boot seals the air dome to air piston. It will extend or retract under the pressure controlled by the valve.

As load is added to the vehicle, the control valve admits air under pressure to these shock absorbers, lifting vehicle to normal position. As load is reduced, the valve releases air and lowers vehicle to the previous normal level.

The valve is connected by a link to the right rear upper control link. A deflection of at least ½ in. is required to make it operative.

A delay mechanism is built into the valve housing. This requires that an attitude be assumed for four to 15 seconds in order for the valve to operate. It prevents operation during normal road motions.

Pressure at the shock absorber units is kept equal by means of the line connecting the two units, with only one unit connected directly to the control valve. This keeps approximately 8-15 psi. on shock absorber units at all times. The pressure is released at the control valve and the equalizing pressure is maintained

through a check valve at the release fitting.

The compressor is located in the engine compartment. It is operated by vacuum surge through a line connected just forward of the carburetor insulator connection. Air, at atmospheric pressure, is taken into the compressor through a line connected to the air cleaner. The compressed air from the compressor is supplied to a reservoir and then to the control valve.

Any service work on this system, or other parts of the vehicle, that may cause deflation will require system re-inflation to approximately 140 psi.

All lines are ⅛ in. diameter flexible black tubing. In working on this system, use care not to kink this tubing. Keep tubing away from the exhaust system.

Windshield Wipers

The wiper motor is an electrical unit and power is transmitted to the wiper arms through linkage.

Wiper Motor R & R

1. Disconnect battery cable.
2. Separate connections and washer hoses at the wiper.
3. Remove wiper arm and blade assemblies.
4. Remove cam spanner nut and remove the cam.
5. Remove escutcheon spanner nut and washer, then lift off escutcheon.
6. Raise the hood and remove screws holding front edge of grille to cowl.
7. Raise front of grille, slide forward, and remove the air intake grille.

8. Disconnect wiper unit-to-control arms at the wiper unit.
9. Remove ignition distributor cap.
10. Remove wiper motor mounting screws by reaching through the grille opening, then lift the wiper motor off the engine side of the firewall.
11. Install by reversing the above procedure.

Wiper Transmission R & R

1. Remove air intake grille and wiper arm blades.
2. Disconnect linkage at the wiper unit.
3. Remove the three transmission mounting screws and remove transmission with link arm.

Motor and Transmission R & R
1964-70

Motor and transmission R & R procedure is essentially the same as for earlier, electrically driven models. The exception is in the form of the drive links and their method of attachment.

Wiper Switch Replacement
1964

1. Disconnect battery.
2. If equipped with tilt wheel, place wheel in maximum up position.
3. Remove screws from upper end of column lower cover and remove cover.
4. Remove screws and washers from lower end of column lower cover and remove cover.
5. Disconnect three-way connector from switch and pry switch from panel.
6. Remove screws from switch and remove escutcheon.
7. Install in reverse of above.

1965-69

1. Remove steering column lower cover.
2. Remove two screws and clip that hold switch to panel.
3. Pull switch rearward to disengage from instrument panel.
4. Disconnect dial bulb socket and wiring harness connector from back of switch assembly.
5. Install in reverse of above.

1970

1. Open left front door to gain access to screw that holds control switch to instrument panel extension; loosen screw.
2. Pull switch out and disconnect electrical connector.

Radio

Removal
1964

1. Remove steering column lower cover.

2. Remove radio knobs, springs and rings.
3. Using radio control knob wrench, remove retaining nuts on both control shafts.
4. Disconnect wire connector and antenna lead-in cable.
5. Disconnect foot control unit from bottom of radio, if so equipped.
6. Remove screws at bottom front of glove compartment and remove glove compartment assembly.
7. Remove screw and washer that hold radio to rear support bracket through the glove compartment.
8. Disconnect air conditioner duct boot from adapter, if so equipped, and pull down and away from radio.
9. Lower radio, disconnect dial light and remove radio.

1965-66 & 1968

1. Remove upper instrument panel cover.
2. Remove radio knobs, springs and rings.
3. Disconnect dial bulb socket from radio.
4. Disconnect wire connector and antenna lead-in cable.
5. Disconnect foot control plug, if so equipped, from radio.
6. Using spanner nut wrench, remove spanner nuts that secure control shafts to upper panel, then remove escutcheons.
7. Remove locknut that holds radio front bracket to stud.
8. Loosen screw that holds rear bracket to radio and remove screw to upper panel.
9. Pull radio rearward, then remove through opening at top of panel.
NOTE: on AM-FM stereo radios, disconnect audio-amplifier unit connector after Step 1.

1967

1. Remove upper instrument panel cover.
2. Remove ash tray housing assembly.
3. Remove screws that secure ash tray frame to retaining plate, disconnect ash tray frame connector, and remove frame.
4. Remove knobs, springs and rings.
5. Using spanner nut wrench, remove spanner nuts that hold control shafts to instrument panel.
6. Remove screw on right side that secures radio bracket to frame.
7. Pull radio rearward and lower to gain access to wire connectors.
8. On AM-FM stereo radio, disconnect audio-amplifier connector.
9. Disconnect wire connector and antenna lead-in cable.
10. Disconnect dial bulb socket from radio.

11. Disconnect foot control plug, if so equipped.
12. On stereo, remove tape securing speaker leads; two on instrument panel cluster and two at panel frame above glove compartment door.
13. Remove radio through ash tray housing hole.

1969-70

1. Remove steering column lower cover.
2. Remove defroster hose behind radio.
3. Remove radio knobs, washers and rings by pulling straight out.
4. Using spanner nut wrench, remove spanner nuts securing control shafts to instrument panel.
5. Disconnect wire connectors and antenna lead-in cable.
6. Remove screws securing support bracket to radio and panel center support and remove radio.
7. Pull radio rearward and down.
8. Disconnect dial bulb socket and remove radio.

Heater System

Heater Blower— Air-Conditioned Cars

1. Drain cooling system.
2. Disconnect electric lead to blower.
3. Disconnect three vacuum hoses from diaphragms.
4. Disconnect rubber boot from blower.
5. Remove heater hoses from heater pipes.
6. Remove blower mounting screws and nuts, then remove assembly with rubber gasket.
7. Install in reverse of above.

Heater Blower— Non-Air-Conditioned Cars
1964

1. Lower antenna. Raise hood and remove lower antenna mounting screw and position antenna out of the way.
2. Disconnect motor ground wire and feed wire.
3. Remove vacuum hose.
4. Remove screws and nuts holding blower and inlet assembly and remove assembly through compartment.
5. Install in reverse of above.

Heater Blower— Air-Conditioned Cars
1964

1. Remove bolts holding right hood hinge to mounting brackets on cowl and dust shield.

2. Position hood panel out of way.
3. Remove nuts and bolts securing front and rear halves of duct from blower and remove front half of duct.
4. Remove electric lead and ground wire from blower.
5. Disconnect vacuum hoses from 20%-100% water valve and suction throttle valve diaphragms. Feed 6 in. of hose through rubber grommet on blower inlet assembly.
6. Remove screws holding blower inlet to cowl and nuts holding blower inlet to mounting studs.
7. Pull inlet assembly out from cowl and remove rear half of duct from assembly to evaporator case.
8. Reach inside of inlet assembly and pull vacuum hoses through grommet and remove assembly.
9. Install in reverse of above.

Heater Blower—
Non-Air-Conditioned Cars
1965-70

1. Drain cooling system.
2. Disconnect electrical connection to blower motor.
3. Remove blower motor mounting screws and remove blower.
4. Remove heater hoses from fittings on assembly, leaving clamps on fittings.
5. Disconnect cable to temperature valve at pivot point on assembly.
6. Remove screws holding bottom of assembly to cowl.
7. Remove screws holding top of assembly to cowl, then remove assembly.
8. Install in reverse of above.

Heater Blower—
Air-Conditioned Cars
1965-70

1. Disconnect battery.
2. Disconnect motor feed wire.
3. Disconnect air hose at motor.
4. Disconnect ground wire.
5. Remove screws holding motor to evaporator and remove motor.
6. Install in the reverse of above.

Heater Core—
Non-Air-Conditioned Cars
1964

1. Drain cooling system.
2. Open glove compartment door and remove screws at lower front of compartment, then remove assembly.
3. Disconnect electrical connectors from resistor panel on top right side of heater assembly.
4. Disconnect cables from defroster valve and temperature valve at pivot points on assembly.
5. Disconnect defroster hoses.
6. Remove both cowl kick panels and pull carpet away from front floor duct.

7. Remove screws holding floor duct connector.
8. Remove heater hoses from fittings on heater assembly, leaving clamps on fittings.
9. Remove blower assembly as described under 1963-64 Blower Removal.
10. Remove two spring retaining nuts from mounting studs and remove heater assembly.
11. Remove screws holding wire retaining clamps to assembly.
12. Slide core toward defroster outlets and lift out of heater assembly.
13. Remove retaining clamps from heater core ends.
14. Install in reverse of above.

Heater Core—
Air-Conditioned Cars
1964

1. Drain cooling system.
2. Open glove compartment door and remove screws at bottom front of compartment, then remove glove compartment.
3. Remove steering column lower cover.
4. Disconnect connectors from resistor panel on top right side of assembly.
5. Disconnect defroster hoses.
6. Disconnect hose from heater air-conditioner diaphragm.
7. Disconnect right and left outlet hoses from air conditioner cross-car duct.
8. Remove screws from center outlet to cross-car duct.
9. Remove cable from clip on top of duct.
10. Remove kick pads and pull front edge of carpet back from floor duct.
11. Remove screws securing duct connector to duct.
12. Remove clamp securing assembly to evaporator adapter.
13. Remove bolts holding right hood hinge to mounting brackets on cowl and dust shield.
14. Position hood panel out of way.
15. Remove nuts and bolts securing front and rear of duct from assembly and remove front half of duct.
16. Disconnect electrical lead and ground wire.
17. Disconnect vacuum hoses from 20%-100% water valve and suction throttle valve diaphragms. Feed 6 in. of hose through grommet on inlet assembly.
18. Remove screws holding blower inlet to cowl and nuts holding blower inlet assembly to mounting studs.
19. Pull blower inlet assembly out from cowl and remove rear half of duct.
20. Reach inside of inlet assembly, pull vacuum hose through grommet and remove assembly.

21. Disconnect water hoses from heater fittings, leaving clamps on fittings.
22. Remove spring retainer nut from assembly mounting stud, and remove assembly.
23. Install in reverse of above.

Heater Core—
Non-Air-Conditioned Cars
1965-70

1. Follow Steps 1 through 7 under Heater Blower—Non-Air-Conditioned Cars, 1965 models.
2. Remove screws from each side of heater core, securing wire retaining clamps to blower case, then remove clamps.
3. Pull core out of case and remove grommets from inlet and outlet fittings.
4. Install in reverse of above.

Heater Core—
Air-Conditioned Cars
1965-70

1. Drain cooling system.
2. Remove carburetor air cleaner.
3. Disconnect vacuum hose assembly connectors from servo vacuum valve and control vacuum valve on power servo unit.
4. Disconnect vacuum hose from power servo power unit.
5. Disconnect electrical connector from power servo unit.
6. Disconnect vacuum hoses and electrical connections from master switch.
7. Disconnect small diameter vacuum hose from center port of vacuum check valve.
8. Remove screw securing vacuum check valve mounting bracket to heater air-selector assembly, then position check valve and bracket out of way.
9. Disconnect vacuum hose from core door vacuum power unit.
10. Disconnect cable from control valve on power servo unit.
11. Remove hoses from heater inlet and outlet fittings.
12. Remove screws securing heater-air selector to cowl and remove assembly from engine compartment.
13. Remove screws securing heater core frame to heater-air selector case, then remove gasket from case.
14. Pull core frame, with core attached, away from heater-air selector case.
15. Remove grommets from inlet and outlet fittings.
16. Remove corner screws securing wire retaining clamps to core frame.
17. Remove clamps and core.
18. Install in reverse of above.

Cadillac Eldorado

Index

NOTE: For information not given in this section,
       refer to same year model in Cadillac section.

YEAR IDENTIFICATION

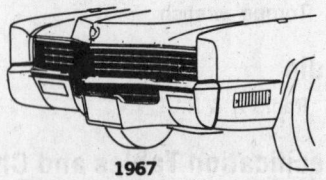
1967

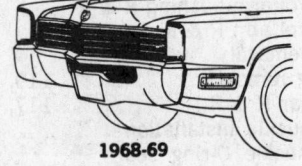
1968-69

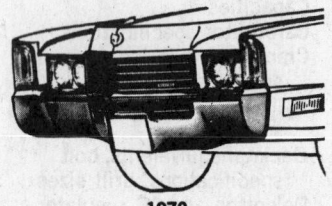
1970

1971

FIRING ORDER

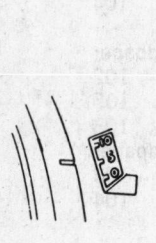

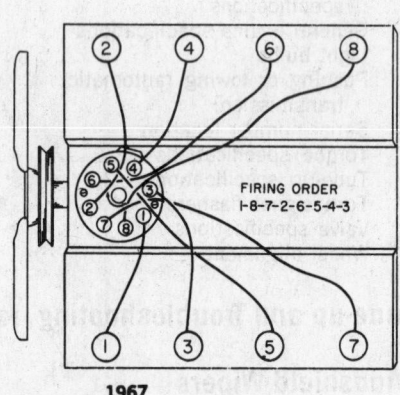

FIRING ORDER
1-8-7-2-6-5-4-3

1967

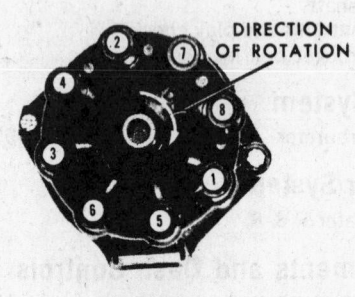

DIRECTION OF ROTATION

Distributor numbering— 1968 only

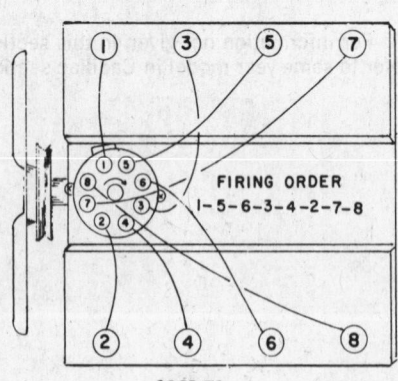

FIRING ORDER
1-5-6-3-4-2-7-8

1968-71

5° NOTCH ON TAB

NOTCH ON PULLEY

Timing marks— 1968-71

CAR SERIAL NUMBER LOCATION AND ENGINE IDENTIFICATION

1967

Each car has a serial number and an engine unit number. The car serial number is located on a plate attached to the left front door pillar post. A V8 engine of 429 cu. in. displacement is used.

1968-69

Vehicle identification number is lo-cated on the left side of the dash, visible through the windshield.

Engine identification is stamped on a machined pad at the front, top edge of the right hand block. The engine number has nine digits. The first two indicate the model year; the third and fourth, the series and the succeeding numbers, the individual car serial number.

A 472 cu. in. engine is used in all series.

1970-71

Vehicle identification number is located on the left side of the dash, visible through the windshield. A V8 engine of 500 cu. in. displacement is used.

GENERAL ENGINE SPECIFICATIONS

YEAR	CU. IN. DISPLACEMENT	CARBURETOR	DEVELOPED HORSEPOWER @ RPM	DEVELOPED TORQUE @ RPM (FT. LBS.)	A.M.A. HORSEPOWER	BORE & STROKE (IN.)	ADVERTISED COMPRESSION RATIO	VALVE LIFTER TYPE	NORMAL OIL PRESSURE (PSI)
1967	V8—429	4-BBL.	340 @ 4600	480 @ 3000	54.6	4.125 x 4.000	10.50-1	Hyd.	33
1968-69	V8—472	4-BBL.	375 @ 4400	525 @ 3000	59.2	4.300 x 4.060	10.50-1	Hyd.	33
1970	V8—500	4-BBL.	400 @ 4400	550 @ 3000	59.2	4.300 x 4.304	10.00-1	Hyd.	35-40
1971	V8—500	4-BBL.	N.A.	N.A.	59.2	4.300 x 4.304	9.00-1	Hyd.	35-40

TUNE-UP SPECIFICATIONS

YEAR	MODEL	SPARK PLUGS		DISTRIBUTOR		IGNITION TIMING (Deg.) ▲	CRANKING COMP. PRESSURE (Psi)	VALVES		Intake Opens (Deg.)	FUEL PUMP PRESSURE (Psi)	IDLE SPEED (Rpm) ★
		Type	Gap (In.)	Point Dwell (Deg.)	Point Gap (In.)			Tappet (Hot) Clearance (In.) Intake	Exhaust			
1967	V8—429	44	.035	30°	.016	5B	180	Zero	Zero	39B	5$\frac{3}{4}$	500
1968-69	V8—472	44N	.035	30°	.016	5B	180	Zero	Zero	18B	5$\frac{3}{4}$	550
1970	V8—500	R46N	.035	28-32°	.016	7$\frac{1}{2}$B	165-185	Zero	Zero	18B	5$\frac{1}{4}$-6$\frac{1}{2}$	600
1971	V8—500	R46N	.035	28-32°	.016	★★	160-175	Zero	Zero	18B	6	600

★—With automatic transmission in D. Add 50 rpm if equipped with air conditioning.

▲—With vacuum advance disconnected and hose plugged. NOTE: These settings are only approximate. Engine design, altitude, temperature, fuel octane rating and the condition of the individual engine are all factors which can influence timing. The limiting advance factor must, therefore, be the "knock point" of the individual engine.

B—Before top dead center.

★★—See engine decal.

Caution

General adoption of anti-pollution laws has changed the design of almost all car engine production to effectively reduce crankcase emission and terminal exhaust products. It has been necessary to adopt stricter tune-up rules, especially timing and idle speed procedures. Both of these values are peculiar to the engine and to its application, rather than to the engine alone. With this in mind, car manufacturers supply idle speed data for the engine and application involved. This information is clearly displayed in the engine compartment of each vehicle.

CRANKSHAFT BEARING JOURNAL SPECIFICATIONS

YEAR	MODEL	MAIN BEARING JOURNALS (IN.)				CONNECTING ROD BEARING JOURNALS (IN.)		
		Journal Diameter	Oil Clearance	Shaft End-Play	Thrust On No.	Journal Diameter	Oil Clearance	End-Play
1967	All	3.000	.0003-.0026	.003	3	2.2491	.0013	.011
1968-71	All	3.250	.0003-.0026	.002-.012	3	2.5000	.0005-.0028	.008-.016

VALVE SPECIFICATIONS

YEAR AND MODEL		SEAT ANGLE (DEG.)	FACE ANGLE (DEG.)	VALVE LIFT INTAKE (IN.)	VALVE LIFT EXHAUST (IN.)	VALVE SPRING PRESSURE (VALVE OPEN) LBS. @ IN.	VALVE SPRING INSTALLED HEIGHT (IN.)	STEM TO GUIDE CLEARANCE (IN.)		STEM DIAMETER (IN.)	
								INTAKE	EXHAUST	INTAKE	EXHAUST
1967	V8—429	45	44	.440	.440	160 @ 1$\frac{1}{2}$	1$\frac{15}{16}$	.0005-.0025	.0010-.0025	.3415-.3425	.3415-.3420
1968-69	V8—472	45	44	.440	.454	160 @ 1$\frac{1}{2}$	1$\frac{15}{16}$	.0005-.0025	.0010-.0025	.3415-.3425	.3415-.3420
1970-71	V8—500	45	44	.440	.454	160 @ 1$\frac{1}{2}$	1$\frac{15}{16}$	.0005-.0025	.0010-.0025	.3415-.3425	.3415-.3420

GENERAL CHASSIS AND BRAKE SPECIFICATIONS

YEAR	MODEL	CHASSIS			BRAKE CYLINDER BORE			BRAKE DRUM	
		Overall Length (In.)	Tire Size	Master Cylinder (In.) Power	Wheel Cylinder Diameter (In.) Front	Rear		Diameter (In.) Front	Rear
1967-69	V8—429,472	221.0	9.00 x 15	1	1$\frac{1}{8}$ ①	$\frac{7}{8}$ ②		11	11 Drum
1970	V8—500	221.0	L78 x 15	1.125	2$\frac{15}{16}$ Disc	$\frac{7}{8}$		11 Disc	11 Drum
1971	V8—500	221.0	L78 x 15	1.125	2$\frac{15}{16}$ Disc	$\frac{7}{8}$		11 Disc	11 Drum

① Disc Brakes—1967-68—1$\frac{15}{16}$—1969—2$\frac{15}{16}$

② Disc Brakes—1$\frac{3}{16}$

WHEEL ALIGNMENT

YEAR	MODEL	CASTER		CAMBER		TOE-IN (In.)	KING-PIN INCLINATION (Deg.)	WHEEL PIVOT RATIO	
		Range (Deg.)	Pref. Setting (Deg.)	Range (Deg.)	Pref. Setting (Deg.)			Inner Wheel	Outer Wheel
1967		$1^1/_2$ N to $2^1/_2$ N	2N	$^3/_8$ N to $^3/_8$ P	▲	0 to $^1/_8$	6	$22^2/_3$	20
1968–71		$1^1/_2$ N to $2^1/_2$ N	2N	$^3/_8$ N to $^3/_8$ P	0	0 to $^1/_8$	11	20	18

▲—$^1/_4$° to $^1/_2$° more on left than right.
N—Negative
P—Positive

LIGHT BULBS

1967–69

Ash tray, rear radio, turn signal indicator..... 1445; 1 CP
Back-up lamp .... 1156; 32 CP
Clock, glove comp., fuel, temp., radio,
   speedometer, gen. & oil ind. .... 1895; 2 CP

Cornering .... 1195; 50 CP
Courtesy and door warning................ 212; 6 CP
Dome lamp .... 1004; 15 CP
Headlamp (inner) .... 4001; 37.5 W

Headlamp (outer)..................... 4002; 37.5–50 W
License lamp .... 67; 4 CP
Stop, signal & turn .... 1157; 32–4 CP

1970–71

FUNCTION	BULB NO.	C/P
Ash tray lamp	1445	.7
Back-up lamp	1156	32
Back-up lamp	1295	50
Clock	1816	3
Console lamp	57	2
Cornering lamp	1295	50
Courtesy lamp—rear quarter	90	6
Courtesy lamp-console	212/212-1	6
Courtesy lamp-instrument panel	89	6
Courtesy lamp—rear door	212/212-1	6
Courtesy lamp—rear quarter armrest	212/212-1	6
Cruise control speed selector illum. auto. lock lamps	1445	.7
Engine temp. warning light	161	1
Generator telltale lamp	161	1

FUNCTION	BULB NO.	C/P
Glove compartment lamp	1816	3
Headlamp—inner	L4001	37.5 Watts
Headlamp—outer	L4002	37.5W/55.0W
Headlamp switch lamp	1895	2
Heater control or climate control lamp	1816	3
High beam indicator	161	1
License lamp	67	4
Low brake telltale lamp	161	1
Low oil pressure telltale lamp	168	3
Map lamp	89	6
Marker lamp-front side	97A	4
Marker lamp-rear side	194	2
Panel lamp	168	3
Park-signal lamp	1157 NA	32/3
Radio dial lamp	1816	3
Radio AM–FM band indicators	2182D	.4

FUNCTION	BULB NO.	C/P
Radio AM-FM stereo indicators	2182D	.4
Radio-rear control indicator	250	1
Spot lamp-front compartment	90	6
Spot lamp-reading	1004	15
Stop, tail and signal	1157	32/3
Trunk compartment lamp	89	6
Trunk lid tell tale	161	1
Trunk signal indicator	168	3
Warning lamp-front door (combined with courtesy light)	212/212-1	6
Warning lamp-rear door (combined with courtesy light)	212/212-1	6
Water temperature tell tale	168	3

FUSES AND CIRCUIT BREAKERS

1967–69

Air conditioner and heater............... 25 AMP. FUSE
Antenna .... 14 AMP. FUSE
Body feed, cigar lighter, clock, map .... 25 AMP. FUSE
Cruise control.......................... 6 AMP. FUSE
Guide-matic .... 4 AMP. FUSE

Head and parking lights .... 25 AMP. CB
Horns .... 25 or 40 AMP. CB
Instrument and back-up lights .... 9 AMP. FUSE
Radio .... $7^1/_2$ AMP. FUSE
Seats and windows .... 40 AMP. CB

Tail and stop lights .... 25 AMP. FUSE
Cornering and courtesy lights .... 25 AMP. FUSE
Turn signal .... 14 AMP. FUSE
Windshield wipers .... 25 AMP. FUSE

1970–71

Body feed .... 25 AMP.
  Cigar lighters
  Clock
  Courtesy lights
  Glove box light
  Map light
  Trunk light
Cornering and parking lights................. 10 AMP.
  Ash tray light
  Cornering lights
  Front side marker lights
  Parking lights
Directional signal and back-up lights .... 20 AMP.
  Back-up lights
  Cruise control
  Rear window de-fogger
  Turn signals

Gages and transmission controls.............. 10 AMP.
  Brake warning light
  CCS vacuum solenoid
  Downshift solenoid
  Fuel gage
  Generator light
  Low oil pressure indicator
  Water temperature warning light
Headlights (integral with headlight switch) .... 15 AMP.CB
Heater and accessories .... 25 AMP.
  Air conditioning amplifier
  Air conditioning blower relay
  Heater blower
  (On cars equipped with a heater only, the 25 AMP fuse is replaced by a 15 AMP. fuse)

Horns... CB
  Convertible top
  Engine metal temperature light
  Horns
  Power seat
  Power windows
Instrument panel lights .... 4 AMP.
Low blower (air conditioning only)........... 10 AMP.
Radio and window control relay .... $7^1/_2$ AMP.
Stop lights and hazard warning flasher .... 25 AMP.
Tail lights................................. 25 AMP.
  License light
  Rear side marker lights
  Tail lights
Twilight sentinel (integral with headlight switch)........................... 15 AMP. CB
Windshield wipers........................ 25 AMP.

TORQUE SPECIFICATIONS

YEAR	MODEL	CYLINDER HEAD BOLTS (FT. LBS.)	ROD BEARING BOLTS (FT. LBS.)	MAIN BEARING BOLTS (FT. LBS.)	CRANKSHAFT BALANCER BOLTS (FT. LBS.)	FLYWHEEL TO CRANKSHAFT BOLTS (FT. LBS.)	MANIFOLD (FT. LBS.) Intake	MANIFOLD (FT. LBS.) Exhaust
1967	All	70–80	40–45	90–100	65–70	75–80	25–30	25–30
1968–71	All	115	40	90	Press-Fit	75	30	35

NOTE—Some bolts and nuts are marked on the heads to indicate the grade of steel used.
Do not use bolts of a lower grade than those originally installed. The marks consist
of lines: SAE5—3 lines; SAE7—5 lines; SAE8—6 lines.

CYLINDER HEAD BOLT TIGHTENING SEQUENCE

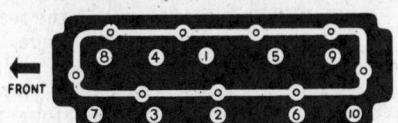

1968-71 cylinder head bolt
tightening sequence

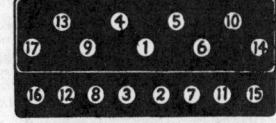

1967 cylinder head bolt
tightening sequence

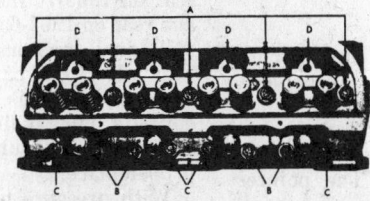

Cylinder head cap screw location and length

Bolt Location	Length
A	4.06"
B	2.75"
C	3.69"
D	5.94"

See text in Cadillac section for
1968-70 bolt location and length

BATTERY AND STARTER SPECIFICATIONS

YEAR	MODEL	BATTERY Ampere Hour Capacity	BATTERY Volts	BATTERY Group Number	BATTERY Terminal Grounded	STARTERS Lock Test Amps.	STARTERS Lock Test Volts	STARTERS Lock Test Torque	STARTERS No-Load Test Amps.	STARTERS No-Load Test Volts	STARTERS No-Load Test RPM	Brush Spring Tension (Oz.)
1967	V8—429	73	12	3KMB	Neg.	510	3.0	Locked	88	10.6	4,000	35
1968–71	V8—472,500	74	12	—	Neg.	Not Recommended			70–99	10.6	7,800	35

CAPACITIES

YEAR	MODEL	ENGINE CRANKCASE ADD 1 Qt. FOR NEW FILTER	TRANSMISSIONS Pts. TO REFILL AFTER DRAINING Manual 3-Speed	TRANSMISSIONS Pts. TO REFILL AFTER DRAINING Manual 4-Speed	Automatic	DRIVE AXLE (Pts.)	GASOLINE TANK (Gals.)	COOLING SYSTEM (Qts.) WITH HEATER
1967	V8—429	4	None	None	26	4½	24	17
1968–71	V8—472, 500	5	None	None	26	4½	24	20①

NOTE—① 1969-71—21.8 qts. & visual capacity tank

AC GENERATOR AND REGULATOR SPECIFICATIONS

YEAR	MODEL	ALTERNATOR Field Current Draw @ 12V.	ALTERNATOR Output @ Generator RPM 1100	ALTERNATOR Output @ Generator RPM 6500	REGULATOR Model	REGULATOR Field Relay Air Gap (In.)	REGULATOR Field Relay Point Gap (In.)	REGULATOR Field Relay Volts to Close	REGULATOR Regulator Air Gap (In.)	REGULATOR Regulator Point Gap (In.)	REGULATOR Regulator Volts at 125°
1967	1100760	2.2–2.6	5A	55	1119515	.015	.030	2.3–3.7	.060	.014	13.5–14.4
1968–69	1100696	2.2–2.4	5A	42	1119515	.015	.030	2.3–3.7	.060	.014	13.5–14.4
	1100694	2.2–2.6	5A	55	1119515	.015	.030	2.3–3.7	.060	.014	13.5–14.4
	1100742	2.2–2.4	5A	63	1119519	.015	.030	2.3–3.7	.060	.014	13.5–14.4
1970	1100909	2.2–2.6	—	42	1119515	.015	.030	1.5–3.2	.060	.014	13.5–14.4
	1100694	2.2–2.6	—	55	1119515	.015	.030	1.5–3.2	.060	.014	13.5–14.4
	1100910	2.8–3.2	—	63	1119519	.015	.030	1.5–3.2	.060	.014	13.5–14.4
1971	1100911	2.2–2.6	—	42	N.A.	Transistor type—no adjustment					
	1100908	2.2–2.6	—	55	N.A.	Transistor type—no adjustment					
	1100694	2.8–3.2	—	63	N.A.	Transistor type—no adjustment					

Distributor

Detailed information on distributor drive, direction of distributor rotation, cylinder numbering, firing order, point gap, cam dwell, timing mark location, spark plugs, spark advance, ignition resistor location, and idle speed, will be found in the Tune-up Specifications Table.

Breaker Point Installation and/or Adjustment

1. Remove distributor cap by depressing and turning the retaining screws.
2. Remove two screws securing cap and remove cap.
3. Remove condenser and primary leads from nylon insulated connection.

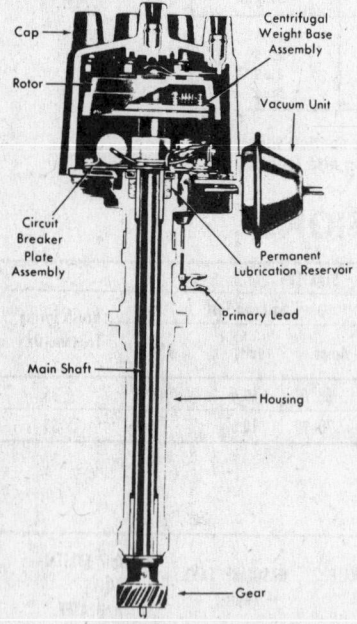

Distributor showing major components
(© Cadillac Div. G.M. Corp.)

4. Loosen two screws holding base of contact assembly in place and remove points.
5. Inspect weight assembly, replace or lubricate as required.
6. Place new points under the two screws and tighten screws.
7. Connect the condenser and primary leads at the nylon insulated connection.
   NOTE: be sure leads do not interfere with cap, weight base, or breaker advance.
8. Install rotor cap. Square and round lugs must be properly aligned.
9. With 1/8 in. Allen wrench, turn until points close while rubbing block is on high point of lobe. Then turn screw counterclockwise one-half turn.
10. Replace distributor cap.
11. With engine warmed up and off

Adjusting distributor points
(© Cadillac Div. G.M. Corp.)

fast idle, set points to get proper dwell angle.

Distributor Removal

1. Remove distributor cap. Disconnect vacuum pipe. Disconnect primary lead at distributor.
2. Turn the engine to top dead center for No. 1 cylinder, see firing order illustrations. The pointer on the timing case cover will point to the O-mark on the crankshaft pulley.
3. Match-mark the vacuum advance unit to the cylinder block so that the distributor body will be correctly replaced at reassembly.

Distributor Replacement

1. Install rubber seal-ring below distributor housing mounting flange.
2. Install the distributor so that the vacuum advance unit aligns with the match-mark made at removal. Turn the rotor slightly left of center so that as the gear engages the camshaft it will revolve into the proper position, pointing to No. 1 contact in the cap.
   NOTE: if the engine has been cranked, remove No. 1 spark plug. Crank the engine until No. 1 piston is in firing position with the pointer and the O-mark on the crankshaft pulley aligned.
3. Install the hold-down clamp. Connect the primary lead and install the cap.
4. Fill the distributor oiler tube with 10W oil.
5. Plug the distributor vacuum line to the carburetor.
6. Insert an adapter pin alongside No. 1 wire in the distributor cap and connect up a timing light.
7. Clean the crankshaft pulley markings and the pointer.
8. Insert an adapter pin alongside the wire from No. 6. Check relationship of pointer and chalk mark. If nearly the same as with the pin at No. 1, turn distributor body to divide the variance. If

the variance is excessive the distributor needs overhauling.
9. Tighten clamp bolt to 15-18 ft. lbs.
10. Remove plug and reconnect the vacuum line to the advance unit.

Brakes

The braking system used consists of power-assisted, hydraulic front and rear service brakes and a foot operated, vacuum released parking brake that works on the rear wheels through mechanical linkage.

Single piston sliding caliper front disc brakes are standard on 1969-70 models while four piston front disc brakes are optional on earlier models. Drum brakes are standard for the rear on 1969-70 models and for front and rear on 1967-68 Eldorados.

A dual master cylinder is used. This allows one pair of brakes to operate if there is a failure of the opposite pair. If lines have been disconnected, be sure to reinstall in proper place.

With pressure bleeder, the Bendix system can be bled from front reservoir by covering rear reservoir with solid cap. The Moraine type must be bled separately, front and rear.

Without pressure bleeder keep both reservoirs nearly full.

Information pertaining to repair procedures will be found in the Unit Repair Section.

Battery, Starter

Detailed information on the battery and starter will be found in the Battery and Starter Specifications Table.

A more general discussion of starters and their troubles can be found in the Unit Repair Section.

Starter R & R

1. Disconnect battery.
2. Raise front end of car.
3. Disconnect wires from starter solenoid.
4. Remove wire spring clip from solenoid housing.
5. Remove two starter mounting screws.
6. Remove starter motor by pulling it forward and lowering it.
7. Install by reversing removal procedure. Torque to 23 ft. lbs.

Instruments

Upper Instrument Panel Cover Panel R & R
1967-68

1. Disconnect battery.
2. Remove three Phillips head

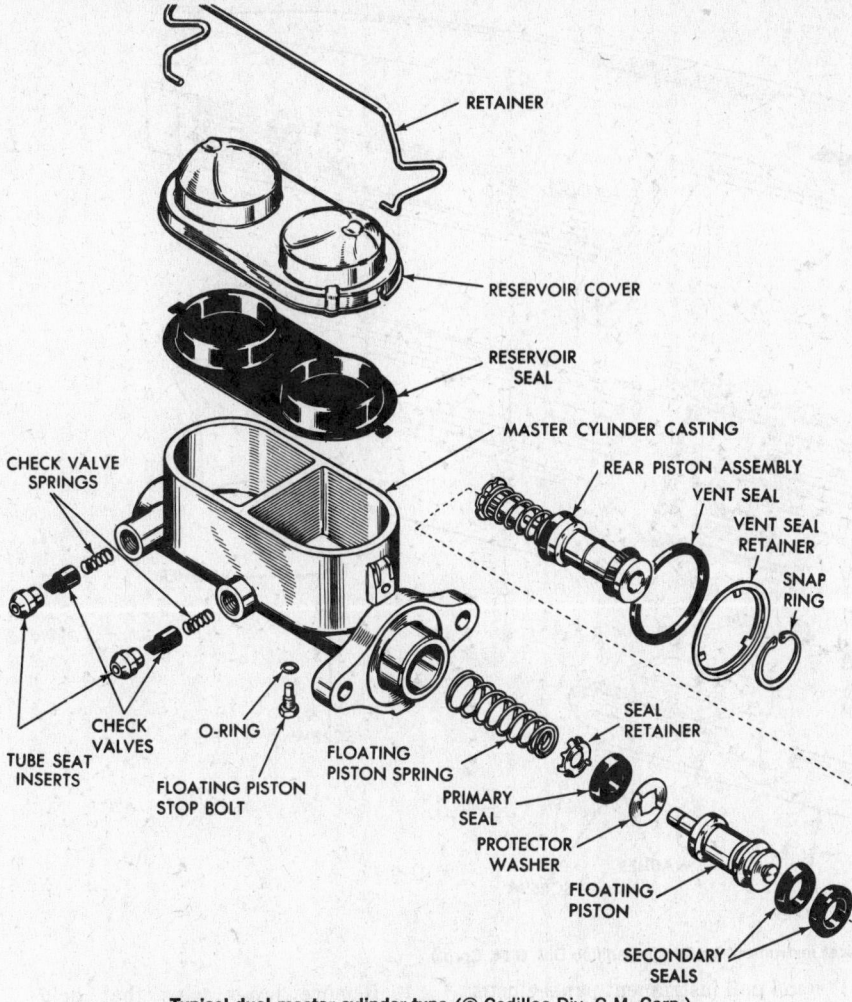

CHECK VALVE SPRINGS

TUBE SEAT INSERTS

CHECK VALVES

O-RING

FLOATING PISTON STOP BOLT

RETAINER

RESERVOIR COVER

RESERVOIR SEAL

MASTER CYLINDER CASTING

REAR PISTON ASSEMBLY

VENT SEAL

VENT SEAL RETAINER

SNAP RING

SEAL RETAINER

FLOATING PISTON SPRING

PRIMARY SEAL

PROTECTOR WASHER

FLOATING PISTON

SECONDARY SEALS

Typical dual master cylinder type (© Cadillac Div. G.M. Corp.)

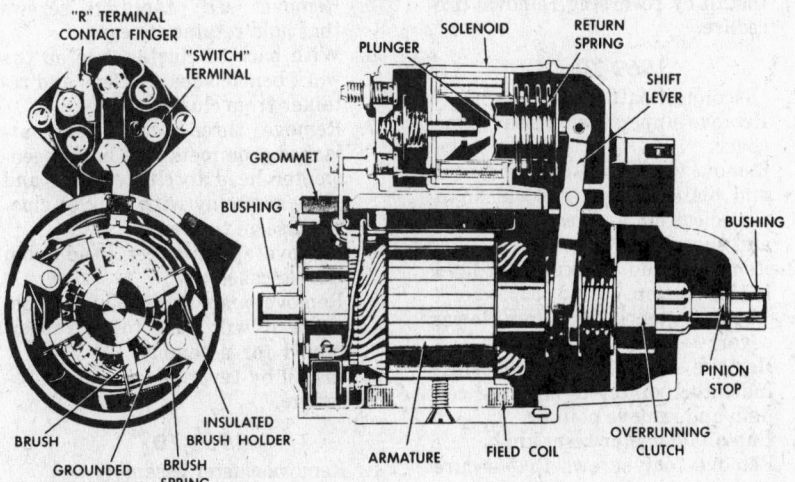

"R" TERMINAL CONTACT FINGER

"SWITCH" TERMINAL

GROMMET

BUSHING

BRUSH

GROUNDED BRUSH HOLDER

BRUSH SPRING

INSULATED BRUSH HOLDER

ARMATURE

FIELD COIL

PLUNGER

SOLENOID

RETURN SPRING

SHIFT LEVER

BUSHING

PINION STOP

OVERRUNNING CLUTCH

Starter cut-away view (© Cadillac Div. G.M. Corp.)

screws holding the instrument panel cluster bezel to the upper cover.

3. From inside the glove compartment door, remove two upper screws holding the upper cover to the right panel.

4. Raise upper panel high enough for clearance, then disconnect the following wire connectors: radio speaker/or speakers, Twilight Sentinel photocell and Automatic Climate Control sensor.

5. Pull upper cover rearward to disengage three hooks at the front of cover from retainers on cowl, and remove cover.

6. Install by reversing removal procedure.

1969-70

1. Disconnect battery.
2. Disconnect radio speaker connection near radio.
3. Remove three Phillips head screws that secure right and left garnish moldings and remove moldings.
4. Inside glove compartment, remove two screws that secure cover to instrument panel.
5. Remove three screws that secure cover to bezel assembly.
6. Lift cover up and rearward to disengage from cowl.
7. Disconnect Twilight Sentinel Photocell above cluster assembly. Remove cover.
8. Install in reverse of above.

Steering Column Lower Cover R & R

1967-68

1. Disconnect battery.
2. On cars equipped with tilt and telescope steering wheel, put wheel in up position.
3. Remove two Phillips head screws holding the lower end of the steering column lower cover to the lower instrument panel.
4. Remove one long special screw holding upper end of steering column lower cover to clamp.
5. Disengage steering column lower cover by pulling straight out to disengage two upper pins from cover and gain access to flasher unit on rear side of lower cover.
6. Remove flasher unit from mounting clip.
7. Install by reversing removal procedure.

1969-70

1. Disconnect battery.
2. Remove screws that secure lower cover to bezel assembly and loosen two screws that secure lower cover to lower instrument panel.
3. Pull lower cover up and out to disengage.
4. Disconnect ash tray wiring and Twilight Sentinel, if so equipped. Remove courtesy light bulb and socket.
5. Remove flasher units from clips on rear sides of cover.
6. Remove cover.
7. Install in reverse of above.

Instrument Cluster R & R

1967-68

1. Remove upper instrument panel cover as previously described.
2. Remove steering column lower cover as previously described.
3. Remove transmission shift indicator pointer with an Allen wrench.
4. Disconnect multiple connector at instrument panel cluster case and remove instrument panel

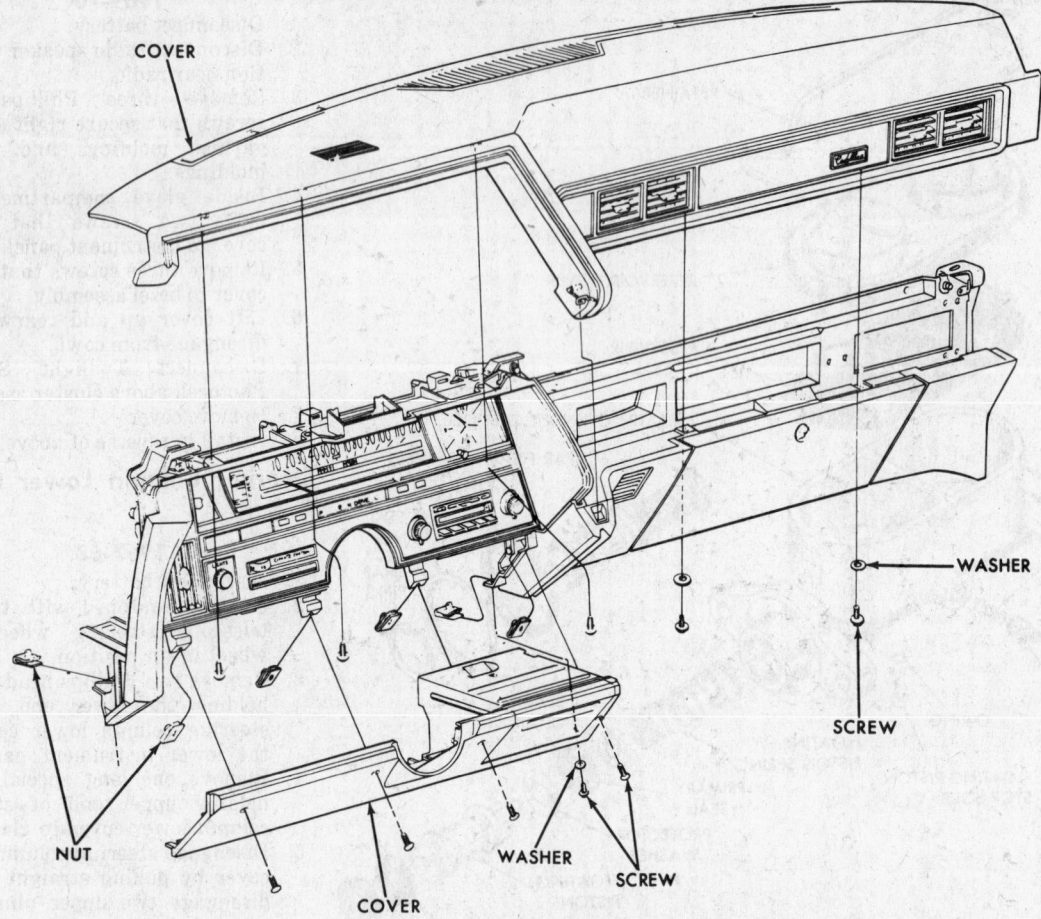

COVER

WASHER

SCREW

NUT

WASHER

SCREW

COVER

Typical instrument panel (© Cadillac Div. G.M. Corp.)

harness from cluster attachment.

5. Disconnect speedometer cable at speedometer head by depressing rises on wave washer. These rises are 180° apart.
6. Remove three upper screws that hold instrument panel cluster to bezel.
7. Remove right and center screws that hold instrument panel cluster to bezel. Center screw is located below the clock.
8. On cars equipped with Automatic Climate Control, remove center air outlet boot at both ends and remove the boot.
9. Remove screws that hold the center air outlet and remove the outlet.
10. If car is equipped with Automatic Climate Control, remove left Climate Control air outlet boot at inlet end.
11. By using a flexible hex driver, remove left instrument panel cluster to bezel screw.
12. If car is equipped with rear window defogger, seat warmer or convertible top, disconnect connectors.
13. Disconnect map light switch connector.
14. Loosen upper left and right bezel to bracket screws.
15. Pry left corner of bezel forward

and pull instrument panel cluster up to remove.
16. Install by reversing removal procedure.

1969-70

1. Disconnect battery.
2. Remove upper instrument panel cover.
3. Remove clock by removing screws and plate at top of clock, snapping clock out and removing bulbs and wires.
4. Remove radio as explained later in this section.
5. Remove steering column lower cover.
6. Remove screw that secures shift indicator pointer to steering column and remove pointer.
7. Pull off odometer reset knob.
8. Remove four screws that secure cluster to bezel assembly.
9. Move cluster forward and right, then downward to remove.

Speedometer Head R & R

1967-68

1. Remove cluster assembly as previously described.
2. Remove odometer reset knob retainer using a 1/16 in. Allen wrench.
3. Remove clock reset knob retainer using a 1/16 in. Allen wrench.

4. Remove lower clips that hold cluster lens to cluster case.
5. Remove self tapping screws that hold retainer to case.
6. With back of cluster case on the work bench, separate lens and retainer from cluster case.
7. Remove three screws and attached grommets that hold speedometer head to cluster case and place assembly with back of cluster case on the bench.
8. Remove speedometer head from cluster case.
9. Remove map light housing because it will have to be repositioned for assembly.
10. Install by reversing removal procedure.

1969-70

1. Remove cluster assembly.
2. Remove clips that secure cluster lens to case and remove lens.
3. Open staking along lower edges and remove sheet metal retainer and shift indicator dial.
4. Remove screws that secure speedometer head assembly to cluster case and remove assembly with back of cluster case.
5. Lift speedometer head assembly out of cluster case.
6. Install in reverse of above.

Printed Circuit R & R

1967-68

1. Remove instrument panel cluster as previously described.
2. Remove one nut and wave washer holding printed circuit to clock.
3. Remove 12 wedge-base sockets and bulbs from cluster case.
4. Remove two screws from fuel gauge.
5. Remove two screws from temperature gauge.
6. Remove four screws that hold the printed circuit to cluster case and remove printed circuit.

NOTE: do not attempt to repair this printed circuit.

7. Install by reversing removal procedure.

1969-70

1. Remove instrument panel cluster.
2. Remove 14 sockets and bulbs from cluster case.
3. Snap off fuel gauge cover and remove two screws that secure printed circuit flap to fuel gauge.
4. Remove screws that secure circuit to back of cluster case and remove circuit.
5. Install by reversing removal procedure.

Headlight Control Switch R & R

1967-68

1. Remove steering column lower cover as previously described.
2. Remove hoses at vacuum valve, which is integral with headlight switch.

3. Remove lower right screw holding headlight control switch housing to lower instrument panel.
4. On cars equipped with Automatic Climate Control, remove left outlet hose at inboard side to gain access to upper left screw.
5. Remove upper left screw that holds headlight control switch housing to lower instrument panel.
6. Pull headlight control switch assembly rearward, disconnect wiring harness connectors and two bulbs, then remove assembly.

7. Install by reversing removal procedure.

1969

1. Remove instrument panel upper cover.
2. Remove steering column lower cover.
3. Remove two screws that secure air conditioning duct to bezel assembly and remove duct.
4. Disconnect wiring harness that is below headlight switch assembly.
5. Depress spring loaded release button on top of headlight switch and remove switch, knob and rod assembly.
6. Remove screws that secure switch assembly to bezel.
7. Pull assembly rearward, disconnect wiring connectors, two bulbs and remove assembly.
8. Install in reverse of above.

1970

1. Remove steering column lower cover.
2. Disconnect wiring harness retainer below headlight switch assembly.
3. Depress spring loaded release button on top of headlight switch and remove switch, knob and rod assembly.

4. Remove screw with ground wire at bottom of switch housing.
5. Pull assembly down and rearward, disconnect wiring harness connectors, two bulbs and remove assembly.
6. Install in reverse of above.

Fuel System

Data on capacity of the gas tank will be found in the Capacities table. Data on correct engine idle speed and tune-up specifications can be found in this section.

Information covering operation and troubles of the fuel gauge will be found in the Unit Repair Section.

Detailed information on the carburetor and how to adjust it will be found in the Unit Repair Section.

Exhaust System

Exhaust Pipe Removal

1. Loosen clamp at muffler and drive exhaust pipe out of muffler.
2. Cut exhaust pipe at weld in pipe and remove rear portion.
3. Remove screws that secure left

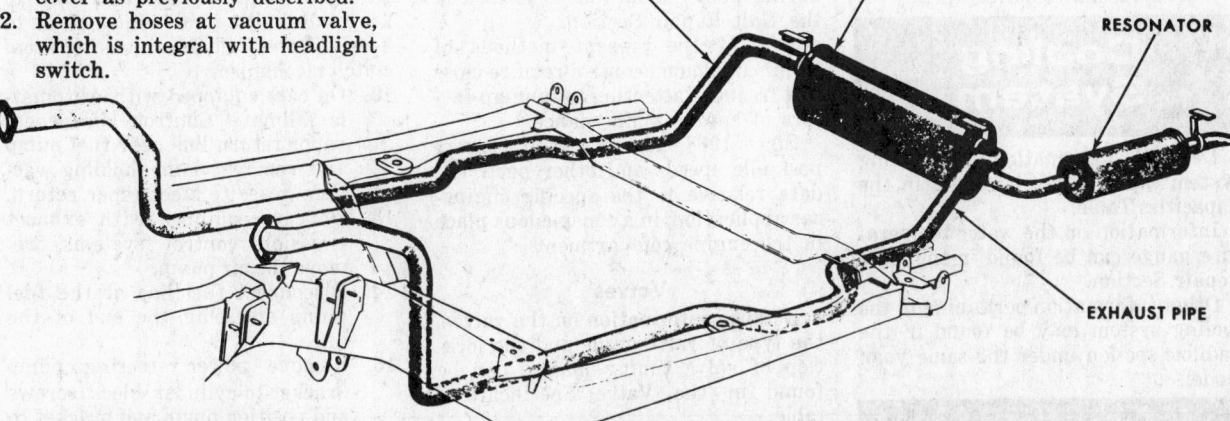

Eldorado exhaust system (© Cadillac Div. G.M. Corp.)

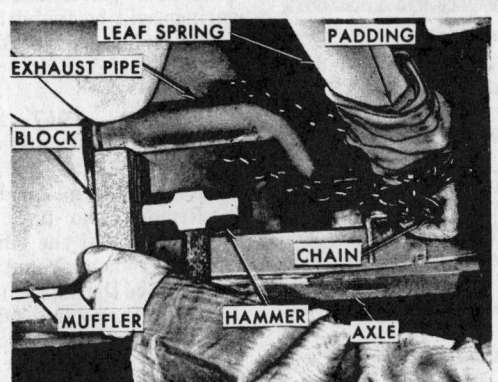

Muffler removal (© Cadillac Div. G.M. Corp.)

pipe flange to manifold or nuts that scure right pipe flange to manifold.

4. Remove forward portion of pipe.

NOTE: if right side is being removed, care must be exercised not to drop heat valve.

Muffler Removal

1. Loosen clamp that secures resonator inlet at muffler.
2. Remove screws that secure resonator hanger to body, remove hanger, then resonator assembly.
3. Remove right and left exhaust pipes to muffler clamps and hangers and lower muffler until exhaust pipes rest on rear axle.
4. Tie right exhaust pipe securely to leaf spring.

Caution The spring should be padded for protection.

5. Drive muffler off right exhaust pipe.
6. Repeat Steps 4 and 5 for left side.

Resonator Removal

1. Remove clamp that secures resonator clamp at muffler.
2. Remove screws that secure resonator outlet hanger to body. Remove hanger.
3. Remove resonator assembly by driving inlet out of muffler.
4. Remove resonator hanger blade from resonator outlet pipe.

Cooling System

Detailed information on cooling system capacity can be found in the Capacities Table.

Information on the water temperature gauge can be found in the Unit Repair Section.

Other information pertaining to the cooling system may be found in the Cadillac section under the same year model.

Engine

References

The engine in all Eldorado models is of the overhead valve 90°, V8 design. The 1967 model engine has a displacement of 429 cu. in. developing 340 horsepower at 4,600 rpm. The 1968-69 models have an engine of 472 cu. in. displacement developing 375 horsepower at 4,400 rpm. The stroke on the 1970 engine has been increased, giving it 500 cu. in. displacement and developing 400 horsepower at 4,400 rpm.

Specifications tables are found at the beginning of this section. Other information may be found by refer-

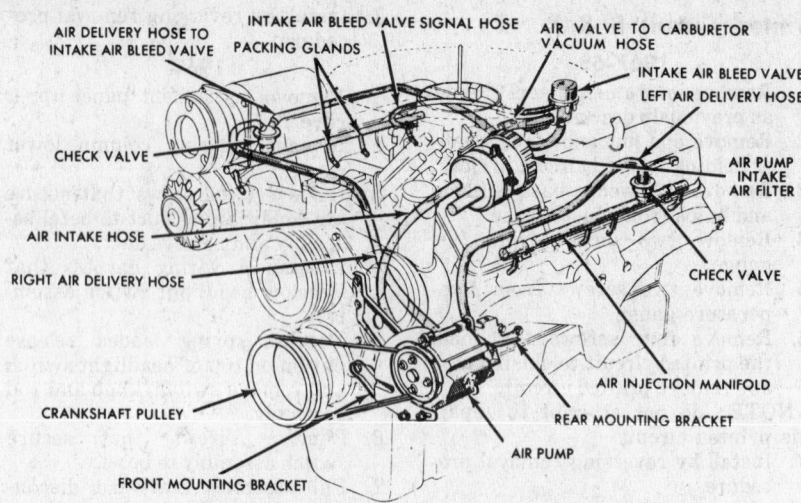

Air injection reactor system (© Cadillac Div. G.M. Corp.)

ring to procedures given for the same year model in the Cadillac section.

Exhaust Emission Control

In compliance with anti-pollution laws involving all of the continental United States, the Cadillac Division of General Motors has adopted as standard equipment, an integrated air injection emission control system. This method is designed and built into the engine castings and eliminates the need for some of the tubes and exterior air manifolding of previous plans. It does, however, use the same afterburner principle as that described in the Unit Repair Section.

Any of the present methods of terminal exhaust control require close and frequent attention to tune-up factors of engine maintenance.

Since 1968, all car manufacturers post idle speeds and other pertinent data relative to the specific engine-car application, in a conspicuous place in the engine compartment.

Valves

Detailed information on the valves, the type of valve guide and the location of valve timing marks, can be found in the Valve Specifications table.

Bearings

Detailed information on engine bearings will be found in the Crankshaft Bearing Journal Sizes Table.

Engine R & R

1967

1. Disconnect negative battery terminal.
2. Remove engine hood.
3. Remove two nuts holding the cowl rods at the wheel wells and pivot rods up from the cowl.
4. Remove the air cleaner.
5. Drain the cooling system.
6. Disconnect wires from generator.
7. On cars equipped with Automatic Climate Control, partially remove compressor.

8. Disconnect transmission cooler line at left front of final drive by removing the screw that holds attaching clip.
9. Remove heater hoses at engine and at water control valve.
10. Remove left and right shrouds by removing four screws that attach each shroud.
11. Remove four screws holding fan assembly to water pump pulley and remove as an assembly.

NOTE: fan clutches used on air-conditioned cars are always to be in an in-car position. When removed from car, support the assembly to keep clutch disc in a vertical plane to keep silicone fluid from leaking from clutch mechanism.

12. On cars equipped with Automatic Climate Control, disconnect vapor return line near fuel pump and remove clamp holding vacuum hoses to steel vapor return.
13. On cars equipped with exhaust emission control systems, remove the air pump.
14. Disconnect fuel line at the fuel pump and plug the end of the line.
15. Remove power steering pump bracket-to-cylinder block screws and position pump and bracket to one side. Remove pump belt. Do not disconnect power steering hoses.
16. Disconnect accelerator linkage at the carburetor.
17. Disconnect wiring connectors at transmission downshift switch.
18. Disconnect positive terminal wiring at coil and remove harness from two retaining clips on the left valve cover.
19. If car is equipped with Cruise Control, disconnect wiring connector at the power unit. Remove cotter pin securing accelerator linkage to exterior arm, remove washer and separate linkage from exterior arm. Also disconnect two cables at power unit.

20. Disconnect oil pressure switch connector at rear of engine.
21. Disconnect all vacuum hoses leading from intake manifold and carburetor.
22. Remove two nuts that hold left exhaust clamp to exhaust manifold and disconnect pipe.
23. Remove four upper transmission to adapter screws.
24. Remove right output shaft as described in later paragraph.
25. Disconnect starter motor retainer clamps by removing one screw at the bearing support and one screw at the engine mounting bracket.
26. Disconnect wiring at starter solenoid and remove two screws holding the starter motor to the engine and remove the starter.
27. Remove one screw at the brace at the final drive.
28. Remove two nuts holding the right exhaust pipe clamp at the exhaust manifold.
29. Remove four screws holding transmission front cover to transmission.
NOTE: the upper left screw is accessible with an extension and universal socket.
30. Remove three converter-to-flexplate attaching screws.
NOTE: this is done by removing the cork in the harmonic balancer and inserting a screw in the balancer. Rotate the screw to gain access to flexplate-to-converter screws. Do not pry on the flexplate ring gear to rotate the converter.
31. Remove vacuum modulator line at transmission and at engine.
32. Working through center crossmember, loosen, but do not remove, two transmission mounting nuts.
33. Remove two nuts and washers holding the engine mounting studs to front frame crossmember.
NOTE: there is one bolt left holding the final drive housing to the engine support bracket and two screws holding the transmission to the spacer to the engine. Do not proceed further until chain hoist is connected to the engine, because the engine may shift.
34. Attach chain hoist and take up slack.
35. Remove lower right and left transmission - to - adapter - to - cylinder-block screws.
36. Place small jack under final drive housing to support final drive and transmission.
37. Remove bolt and lockwasher holding final drive housing to engine support bracket.
38. Remove engine by pulling it slightly forward to disengage it from transmission and up from engine compartment. Turn engine slightly clockwise while removing to assist in clearing engine compartment.
39. Secure converter holding strap J-21366 to transmission case using a 5/16—18 nut, because the converter is now free.
40. Install by reversing removal procedure.

1968-70

Follow engine removal procedure given in Cadillac section with the following exception:

After step calling for removal of oil filter, remove right-hand output shaft by the following procedure.

1. Remove drive axle-to-output shaft screws and lock-washers.
2. Remove shaft support-to-engine bolts, and one support-to-brace screw.
3. Rotate inboard end of drive axle rearward toward starter motor.
4. Pull output shaft straight out, then lower and remove from underside of car. Proceed with engine removal procedure.

Cylinder Head

Rocker Shaft Removal

1967—429 Eng.

The rocker shafts can be removed and serviced without disturbing the cylinder head or manifolds.

Simply get the spark plug wires out of the way and remove whatever heater or throttle linkage that passes over the rocker cover and then unbolt and remove the rocker cover.

1. Remove the four mounting cap screws.
2. Remove each retainer and mark so that it can be reassembled correctly.
3. Remove rocker arm support assemblies from retainers and remove rocker arms from supports. Be sure to keep supports and rocker arms in order for correct assembly.

Rocker Shaft Installation

1. Install rocker arms on supports and place supports in retainers.
   NOTE: be sure that the EX on the support is positioned toward the exhaust valve and the IN toward the intake valve.
2. Put cap screws through the reinforcements, supports and retainers. Position assemblies on the cylinder head. Make sure that pushrods are properly seated in the lifter seats and in the rocker arms.
NOTE: lubricate rocker arm bearing surfaces before assembly to prevent wear.
3. Tighten four mounting screws and torque to 60 ft. lbs.

Beginning 1968

Valve rocker arms are no longer fitted to one common shaft per head, but are mounted in pairs (four pairs to a cylinder head). They are of the modified pedestal mounted type.

Rocker arms may be removed in pairs and do not require cylinder head removal.

Torque rocker mounting screws to 60 ft. lbs.

Cylinder Head Removal

1967—429 Eng.

With these engines the procedure is as follows:

1. Disconnect the water manifold at the front of the cylinder head or heads. It is a good idea to remove the water pump and water manifold from the car. It is difficult to reinstall a cylinder head with the water pump in place on one head without damaging the water pump gasket.
2. Remove all vacuum lines and carburetor connections, disconnect all ignition, throttle and battery connections.
3. Take off the intake manifold with the carburetor in place or if desired remove the carburetor.
4. Remove the rocker covers.

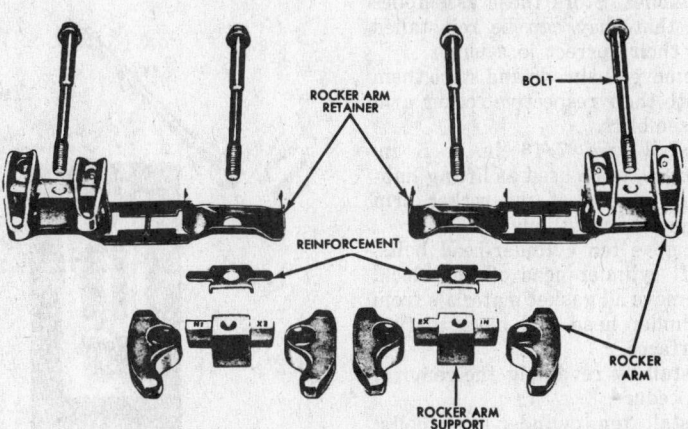

Disassembled rocker arm (© Cadillac Div. G.M. Corp.)

NOTE: it is customary to remove the rocker covers together with the ignition wires and distributor cap as a unit unless service is to be done on the distributor.

5. Remove the generator if the right cylinder head is to be removed.
6. The exhaust manifolds may be disconnected either from the head or from the flange connection to the exhaust pipe. It is better to leave them connected to the head.
7. Remove the head bolts which hold the rocker assemblies to the cylinder head and lift off the rocker assemblies.
8. Take out the valve pushrods.
9. Remove the balance of the cylinder attaching bolts and lift the head off. It is very important that the head be handled carefully so as not to damage or mark the head gasket surface.

1968-70

1. Remove intake manifold.
2. Drain coolant from engine cooling system.
3. Disconnect ground strap at rear of cylinder heads from cowl. Disconnect wiring connector for high engine temperature warning system from sending unit at rear of left cylinder head.
4. Remove generator, if working on the right cylinder head, or partially remove the power steering pump if working on the left cylinder head.
5. Remove clamps holding the A.I.R. tubes to both cylinder heads and remove the A.I.R. supply tubes.
6. Remove clamps holding the wire harness to the cylinder head and tie harness back out of the way.
7. Remove screws holding exhaust manifold to cylinder head.
8. Remove screws holding the rocker arm cover to the head.
9. Remove cover.
10. Remove screws securing each rocker arm support to cylinder head and remove rocker arm assemblies. Store these assemblies so that they can be reinstalled in their correct locations.
11. Remove pushrods and store them with their respective rocker arm assemblies.
12. Install two 7/16 in. x 6 in. screws, to be used as lifting handles, in two of the rocker arm support screw holes.
13. Remove ten cylinder-head bolts.
14. Lift cylinder head off the block.
15. Remove all gasket materials from cylinder head and block mating surfaces.
16. Install by reversing the removal procedure.
17. Install ten cylinder-head bolts, finger tight.

18. Torque cylinder-head bolts to 100 ft. lbs. Starting at the center of the head and working toward both ends.
19. Install pushrods and be sure that the rods seat in the valve lifter cups.
20. Install rocker arm assemblies and torque attaching screws to 60 ft. lbs.
21. Install new rocker arm lever gasket. Install the cover and torque attaching screws to 28 in. lbs.
22. Install exhaust manifold to cylinder head. Torque to 35 ft. lbs.
23. Position wiring harness along upper edge of rocker arm cover and secure with clamps.
24. Position A.I.R. manifold in openings in front of cylinder heads and secure. Right-hand (generator side) is installed with flanges facing cylinder head, while left-hand clamp is installed with flanges facing outward.
25. Install generator if working on right cylinder head, or power steering pump if working on the left head.
26. Connect ground strap at rear of cylinder heads to cowl.
27. Connect high temperature warning wiring connector to sending unit at rear of left cylinder head.
28. Install intake manifold.
29. Refill cooling system.
30. Operate engine to normal running temperature, then retorque rocker arm cover screws.
31. Make visual check of engine coolant and bring to required running attitude level.

Engine Lubrication

Oil Pan Removal

1967

1. Remove engine as previously de-

scribed in Engine R & R.
2. Drain engine oil.
3. Remove two brackets-to-block bolts on each side of engine front mounting support.
4. Remove nuts and cap screws that hold oil pan to cylinder block and engine front cover, then remove the oil pan.
5. Remove side gaskets and rubber front and rear seals from oil pan. Discard the gaskets and seals.
6. Install by reversing the removal procedure. Torque to 10 ft. lbs.

1968-70

1. Remove engine from chassis.
2. Remove nuts and screws that hold oil pan to cylinder block.
3. Remove oil pan.

Filter Element Replacement

A full flow filter is used. This is the disposable-type, equipped with a by-pass safety valve.

Valve System

Valve Removal

Eldorado uses a holding fixture for the cylinder head which incorporates a pedal-operated valve spring compressor.

With the head off the car and on the bench or in the holder, compress a valve spring and remove the valve keepers. This will release the valve spring retainer, valve spring, and rubber seal from lower groove in valve stem.

Repeat for the remaining valves being sure to keep them in order so they can be reinstalled in the same position. When reinstalling, be sure that new rubber seal is seated in the groove closest to the valve head.

Check to see that seal is properly installed. Strike the ends of the valve stems to seat the keepers. Compress a suction cup over the spring retainer

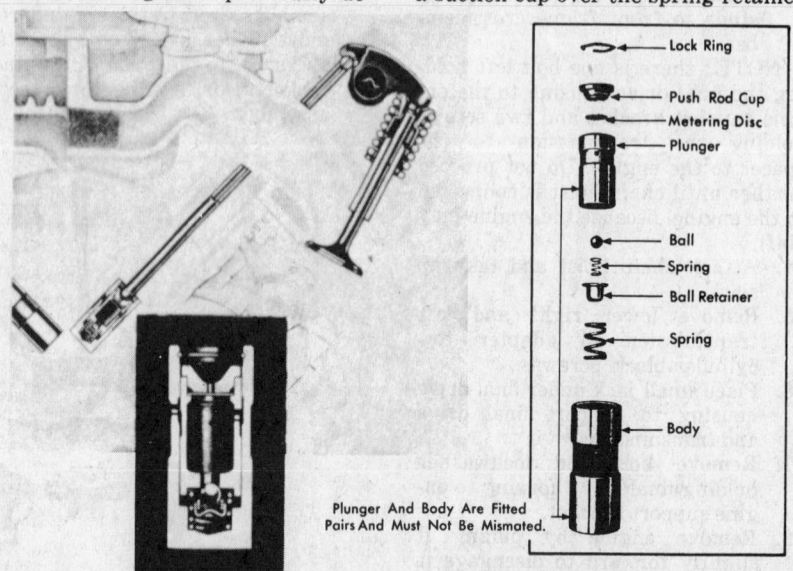

Lock Ring
Push Rod Cup
Metering Disc
Plunger

Ball
Spring
Ball Retainer
Spring

Body

Plunger And Body Are Fitted Pairs And Must Not Be Mismated.

Arrangement of valves and lifters (© Cadillac Div. G.M. Corp.)

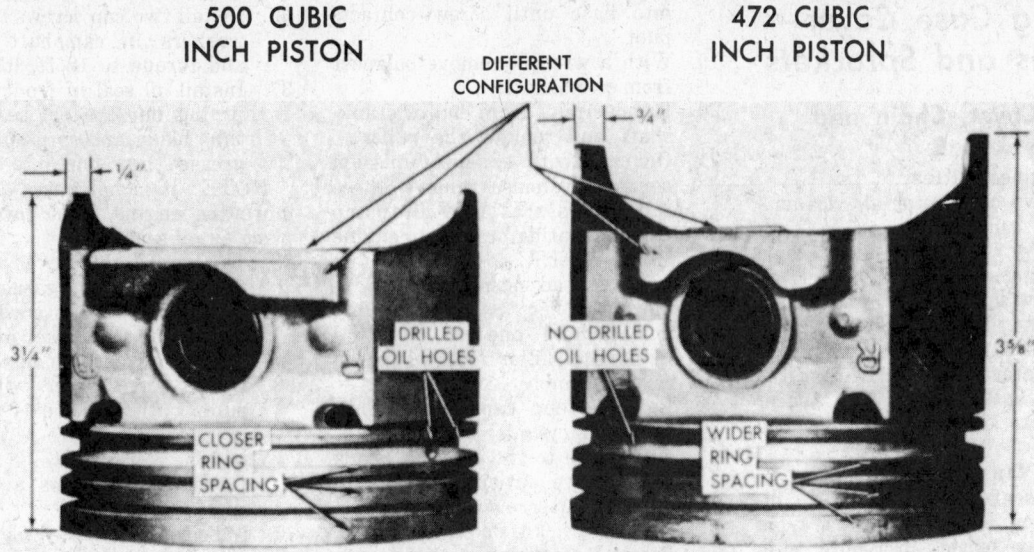

500 CUBIC INCH PISTON   DIFFERENT CONFIGURATION   **472 CUBIC INCH PISTON**

DRILLED OIL HOLES   NO DRILLED OIL HOLES

CLOSER RING SPACING   WIDER RING SPACING

Piston identification (© Cadillac Div. G.M. Corp.)

and valve stem to test for leakage past the seal.

If the rubber oil seal has been properly installed the vacuum cup will stick to the spring retainer due to suction. If there is no suction the seal is leaking and it will be necessary to remove the retainer and install another seal.

Checking Valve Guides

Check valve stem-to-guide clearance using a 1/16 in. wide strip of .005 in. shim stock. Bend the end of the shim and hang in the valve guide on the pushrod side. Shim should not extend more than ¼ in. into the guide. If now the valve stem will enter the guide the clearance is excessive.

Valve Lifter Removal

Lifters may be removed without taking off the cylinder head.

Remove throttle and gas lines from the carburetor, disconnect hoses, vacuum lines and wires that pass over the rocker covers. Remove the distributor cap and disconnect the wires at the spark plugs. Remove the bolts that hold the rocker covers to the cylinder head and lift off the rocker covers leaving the spark plug wires attached to them. Remove the bolts that hold the intake manifold to the cylinder block and lift off the intake manifold. If desired, the carburetor can be detached from the manifold first, but this is not necessary. Remove the valve chamber cover plate. Remove the bolts that hold the rockers or shafts to the cylinder head and lift off the rockers. Pull up the pushrods through the holes in the cylinder heads and the lifters can be pulled up out of their bores.

Sometimes gum residue forms on the bottom of the lifter, making it

very difficult to pull the lifter up out of its bore. If this condition is suspected before the job is started, put a good solvent in the engine oil and run the engine for the time specified by the manufacturer of the solvent in order to dissolve this gum.

However, even when gum is present on the bottom of the lifter body, the lifter can be pulled up out of its bore using special pliers. These pliers are designed to grip the lifter body firmly without scoring or scratching it.

If a special tool isn't available, a good substitute can be made by grinding the teeth out of an ordinary pair of pliers and grinding a circle almost the size of the valve lifter body so that when the pliers are squeezed down on the lifter body it will contact a large surface of the lifter body, thus preventing scoring.

Pistons, Connecting Rods and Main Bearings

Piston and Rod Removal
1967-68
Follow procedure given for same year model in the Cadillac section.

1969-70
1. Remove engine.
2. Remove cylinder heads.
3. Clean a carbon from top of bore and ream the upper ridge if necessary to prevent breakage of piston ring lands. Pack cylinder bore with cloth to catch shavings.
4. Remove oil pan.
5. Remove screws and nut that hold oil pick-up tube and strainer assembly to block. Discard O-ring.
6. Remove connecting rod nuts and remove connecting rod cap by sliding it off bolts.

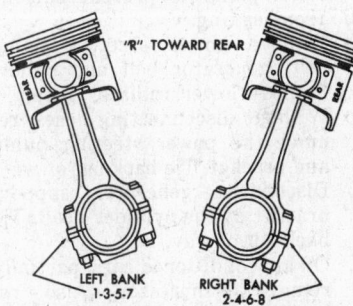

"R" TOWARD REAR

LEFT BANK 1-3-5-7   RIGHT BANK 2-4-6-8

Piston and connecting rod assembly

7. Install connecting rod guide set on bolts.
8. Push rod and piston assembly up and out.
9. Remove remaining rod and piston assemblies in same manner.

Assembling Pistons to Connecting Rods

Slipper-type pistons are used, but the piston has the word rear cast into the metal just beside the wrist-pin hole. With the connecting rod mounted in the vise so that the number faces the operator, the word rear will go to his right hand on the odd numbered cylinders (left bank).

With the number on the connecting rod facing the operator, the word rear on the piston will go to his left hand on the even numbered pistons (right bank).

Assembling Rod and Piston Assemblies to the Block

The numbers on the connecting rods face away from the crankshaft. The numbers on the left bank (odd numbers) face to the left; the numbers on the right bank (even numbers) face to the right. As a double check, the word rear stamped on the piston faces the rear of the engine on both banks.

Timing Case Cover— Chains and Sprockets

Timing Cover, Chain and Sprockets R & R

1. Disconnect battery.
2. Remove carburetor air cleaner.
3. Drain coolant.
4. Remove oil-pan-to-front cover nuts and studs.
5. Remove upper radiator hose.
6. Remove the cap screws that hold the fan blade assembly to the water pump and remove fan blade assembly, or hub spacer and fan on non-air conditioned cars.

NOTE: fan clutches on air-conditioned cars are to be kept in an in-car position. When removed from car for any service procedure, support assembly to keep clutch disc in a vertical plane to prevent silicone fluid from leaking.

7. Remove power steering pump belt, generator belt and pulley.
8. Remove lower radiator hose.
9. Without disconnecting hoses, remove the power steering pump and bracket. Tie back out of way.
10. Disconnect generator support bracket at the cylinder head. Tie back out of way.
11. On air-conditioned cars, partially remove compressor. Also remove compressor lower mounting bracket attached to engine front cover.
12. Remove distributor assembly.
13. Remove fuel pump as previously described.
14. Remove four of the six cap screws that hold the crankshaft pulley to the harmonic balancer.
15. Remove cork plug from end of crankshaft.
16. Install harmonic balancer puller

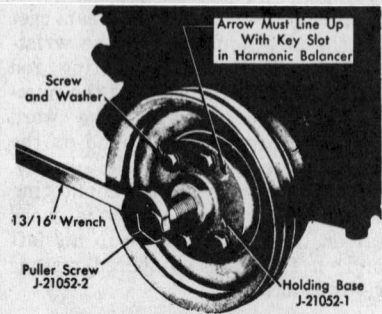

Balancer assembly removal
(© Cadillac Div. G.M. Corp.)

pilot, J-21052-4 in bore end of crankshaft.

17. Install holding base, J-21052-1, on front of pulley, lining up the scribe mark on base with key slot in harmonic balancer, and install four holding screws with washers finger tight. Do not use a wrench to tighten screws.
18. Thread puller screw, J-21052-2, into base until screw contacts pilot.
19. With a wrench, remove balancer from crankshaft.
20. Remove pilot from end of crankshaft and remove the puller.
21. On cars with exhaust emission control systems, remove three cap screws that hold air pump front mounting bracket to engine front cover. Loosen air pump drive belt adjusting bolt and remove the belt. Swing air pump and bracket to one side.
22. Remove oil filter from oil pump cover assembly.
23. Remove four cap screws that hold the cylinder-head coolant-outlet pipe to the cylinder heads and remove outlet pipe.
24. Remove cap screw that holds fuel filter to bracket on oil filler tube.
25. Remove remaining nine cap screws that hold engine front cover to cylinder block and remove the cover with water pump attached.
26. Remove the cap screws that hold the camshaft sprocket to the camshaft.
27. Remove camshaft sprocket with chain, from camshaft.
28. Remove crankshaft sprocket from crankshaft.
29. Remove Woodruff key from crankshaft key slot.
30. Install chain and sprockets by, first seating Woodruff key in the crankshaft key slot.
31. Install crankshaft sprocket on the crankshaft in line with the keyway.
32. Install camshaft sprocket in timing chain with timing mark toward the front.
33. Place chain over crankshaft sprocket.
34. Line up timing marks on both sprockets.
35. Hold camshaft sprocket in position against end of camshaft and press sprocket on camshaft by hand, being sure index hole in camshaft is lined up with index hole in sprocket.

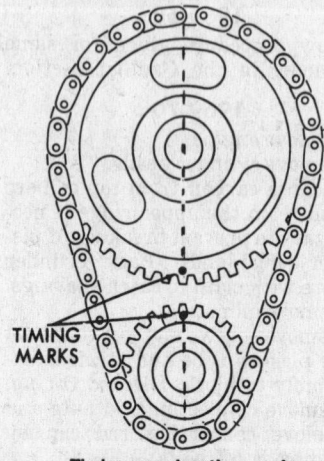

TIMING MARKS

Timing gear location marks

36. Install two cap screws with lock-washers in camshaft sprocket and torque to 18 ft. lbs.
37. Install oil seal in front cover by prying out the old seal with a thin blade and, pressing a well greased new one into place.

NOTE: it is not necessary to remove the engine front cover to replace an oil seal.

38. To install the cover, apply a small amount of gasket cement to the new front cover gasket and locate the gasket over locating dowels on cylinder block.
39. Install front cover with water pump attached, over end of crankshaft. Secure with 12 attaching screws. Refer to chart for screw locations and torque specifications.
40. Lubricate the new coolant outlet pipe to water pump O-ring seal with silicone, and install O-ring against shoulder in bore in pump body.
41. With cement on coolant outlet pipe flange surfaces, place new gaskets in position on coolant outlet pipe.
42. Install neck of coolant outlet pipe in bore in pump body, position flange surfaces against cylinder heads and install four attaching screws. Torque screws to 20 ft. lbs.
43. Install oil filter on oil pump cover assembly.
44. On cars with air-conditioning, install compressor lower mounting bracket to engine front cover. Also install compressor.
45. Install generator support bracket on coolant outlet pipe and secure with two attaching screws. Torque to 20 ft. lbs.
46. Secure fuel filter to bracket on oil filler tube with attaching screw and washer.
47. On cars with exhaust emission control system, position air pump and mounting bracket on engine front cover and secure with three attaching screws. See illustration for torque.
48. Install four pulley to harmonic balancer cap screws that were previously removed. Torque all six screws to 18 ft. lbs.
49. Lubricate bore of balancer with E.P. lubricant to prevent seizure.
50. Position balancer on the crankshaft, with key and keyway lined up.
51. Place holding base, J-21052-1, against front face of pulley and thread installer screw, J-21052-5, into end of crankshaft. Position thrust bearing with inner race forward, washer next, and installer nut, J-21052-6, last.
52. With a wrench, press harmonic balancer onto the crankshaft.
53. When balancer is in place on the

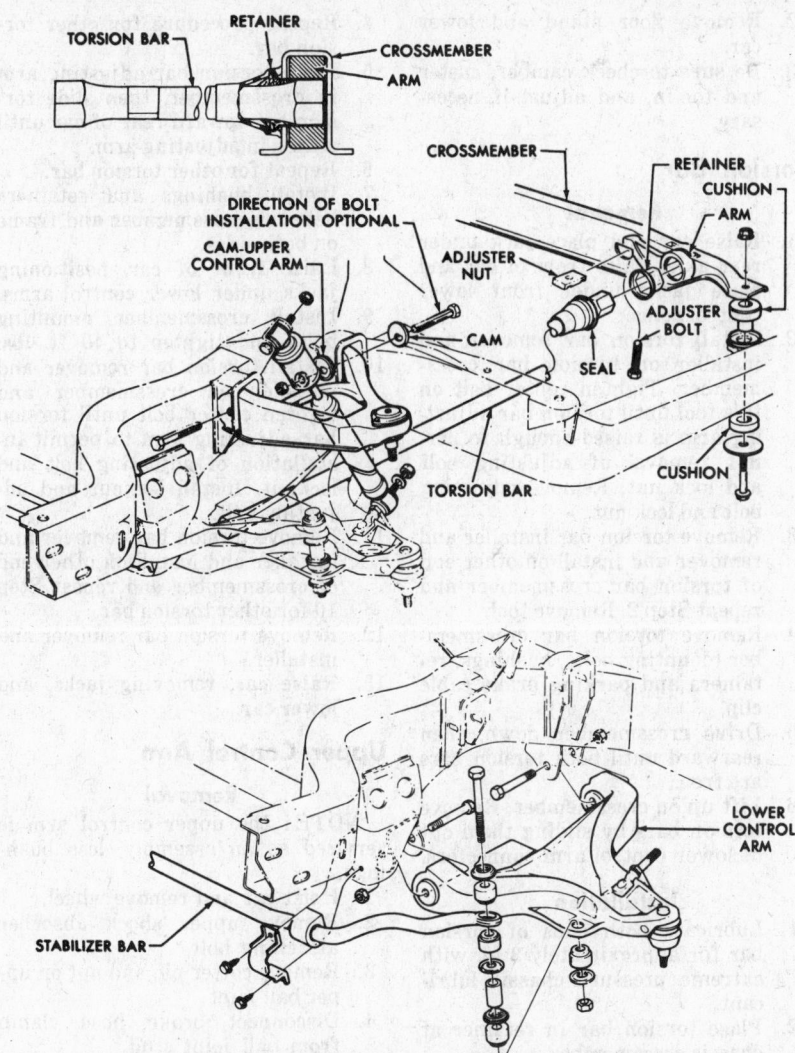

Front suspension disassembled (© Cadillac Div. G.M. Corp.)

65. Refill cooling system.
66. Reconnect battery.
67. Install distributor assembly and adjust timing to specifications.
68. Install carburetor air cleaner.
69. Run engine to check for coolant and oil leaks at all connections.

Front Suspension

References

The front suspension consists of control arms, stabilizer bar, shock absorbers and a right and left torsion bar. Torsion bars are used in place of conventional coil springs. The front end of the torsion bar is attached to the lower control arm. The rear of torsion bar is mounted into an adjustable arm at the torsion bar crossmember. The carrying height of the car is controlled by this adjustment.

Wheel Hub (Front) R & R

1. Carefully pull drum or disc from hub assembly.
2. Remove drive axle cotter pin, nut and washer.
3. Position access slot in hub assembly so each of the attaching bolts can be removed.
4. Position spacer tool J-22237 and install tool J-21579 and slide hammer J-2619.
5. Remove hub assembly.
6. To install, reverse removal procedure.

NOTE: O.D. of bearing must be lubricated with E.P. chassis lubricant. Use care when installing hub assembly over drive axle splines to assure spline alignment.

Steering Knuckle O-Ring Seal Replacement

Replacement of seal with wheel removed.

1. Remove hub and drum or disc.
2. Remove upper ball joint cotter pin and nut.

crankshaft, remove the installer tool. Finish positioning the balancer all the way on the crankshaft by, threading an appropriate bolt, (a 1962 Cadillac balancer-to-crankshaft screw will do) and washer in the end of the crankshaft. Tighten screw to 125 ft. lbs.

54. Remove this screw from the crankshaft, and install cork plug.
55. Install fuel pump.
56. Install power steering pump with bracket on the cylinder block and secure with two attaching screws.
57. Connect lower radiator hose at radiator outlet and water pump inlet pipe.
58. Install pulley and fan blade assembly on water pump and secure with four attaching screws. Torque screws to 18 ft. lbs.
59. On cars equipped with exhaust emission control system, install air pump drive belt and adjust.
60. Install generator belt and adjust.
61. Install upper radiator hose.
62. Install power steering pump belt and adjust.

63. On air-conditioned cars, install compressor, and adjust belt. Install fan shroud on radiator cradle and secure with eight cap screws. Torque screws to 12 ft. lbs.
64. Install two oil pan to engine front cover studs and nuts.

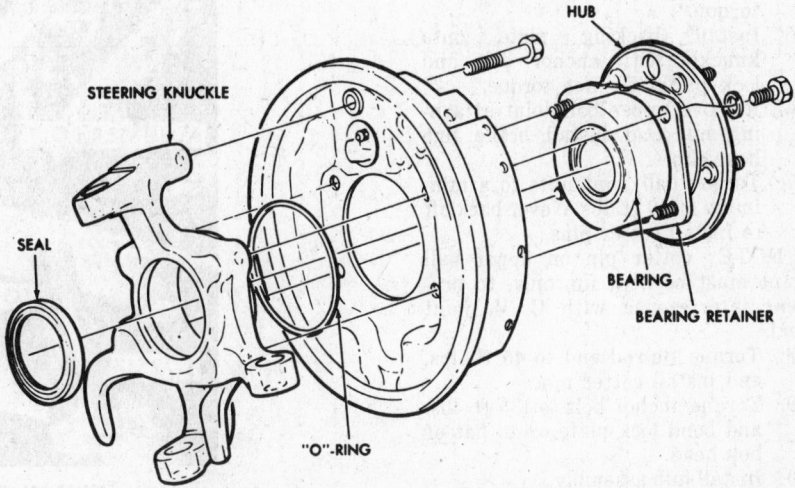

Front hub bearing and retainer (© Cadillac Div. G.M. Corp.)

3. Remove brake line hose clip from ball joint stud.
NOTE: do not loosen ball joint stud.
4. Bend lock plate on anchor bolt up and remove anchor bolt.
5. Carefully lift brake backing plate outboard over end of axle shaft and support brake hose so that it is not damaged.
6. Remove O-ring seal.
To install, reverse removal procedure.

Steering Knuckle

Removal

1. Raise car and support lower control arm with floor stands.
2. Remove wheel and drum or disc.
3. Remove hub assembly.
4. Follow Steps 2 through 6 on knuckle O-ring removal.
5. Place support block under drive axle to protect C. V. joint seal.
6. Using a brass drift and hammer loosen upper ball joint stud.
7. Remove cotter pin and nut from tie-rod end.
8. Using brass drift and hammer, remove tie-rod end from knuckle.
9. Remove cotter pin and nut from lower ball joint.
10. Carefully place ball joint puller adapter between ball joint seal and knuckle.
11. Remove lower ball joint from knuckle.
12. Remove knuckle.
13. Knuckle seat can be pried from the knuckle at this time.

Installation

1. Using seal installer, install seal into knuckle.
2. Install lower ball joint stud into knuckle and attach nut. Do not torque.
3. Install tie-rod and stud into knuckle and attach nut. Do not torque.
4. Install upper ball joint stud into knuckle and attach nut. Do not torque.
5. Install backing plate onto knuckle with anchor bolt and lock plate. Do not torque.
6. Remove upper ball joint attaching nut and install brake line hose clip.
7. Torque ball joint nuts to a minimum of 40 ft. lbs. Never back off to install cotter pins.
NOTE: cotter pin on upper ball joint must be bent up, only, to prevent interference with C. V. joint seal.
8. Torque tie-rod end to 45 ft. lbs. and install cotter pin.
9. Torque anchor bolt to 135 ft. lbs. and bend lock plate on to flat of bolt head.
10. Install hub assembly.
11. Install drum or disc and wheel.

12. Remove floor stand and lower car.
13. Be sure to check camber, caster and toe-in, and adjust if necessary.

Torsion Bar

Removal

1. Raise car and place jack under rear axle. Raise front of car and place jacks under front lower control arms.
2. Install torsion bar remover and installer on torsion bar crossmember. Tighten center bolt on this tool until torsion bar adjusting arm is raised enough to permit removal of adjusting bolt and lock nut. Remove adjusting bolt and lock nut.
3. Remove torsion bar installer and remover and install on other end of torsion bar crossmember and repeat Step 2. Remove tool.
4. Remove torsion bar crossmember mounting bolts, bushings, retainers and parking brake cable clip.
5. Drive crossmember down, then rearward until both torsion bars are free.
6. Lift up on crossmember. Remove torsion bars by sliding them out of lower control arm connectors.

Installation

1. Lubricate both ends of torsion bar for approximately 3 in. with extreme pressure chassis lubricant.
2. Place torsion bar in retainer at chassis crossmember.
3. Lubricate lower control arm torsion bar connector and slide torsion bar into connector.

4. Repeat procedure for other torsion bar.
5. Place torsion bar adjusting arm in crossmember, then slide torsion bar toward rear of car until seated in adjusting arm.
6. Repeat for other torsion bar.
7. Install bushings and retainers between crossmember and frame on both sides.
8. Raise front of car, positioning jacks under lower control arms.
9. Install crossmember mounting bolts and tighten to 40 ft. lbs.
10. Install torsion bar remover and installer on crossmember and tighten center bolt until torsion bar adjusting arm to permit installation of adjusting bolt and locknut. Install locknut and adjusting bolt.
11. Remove torsion bar remover and installer and install on other end of crossmember and repeat Step 10 for other torsion bar.
12. Remove torsion bar remover and installer.
13. Raise car, removing jacks, and lower car.

Upper Control Arm

Removal

NOTE: the upper control arm is serviced as an assembly, less bushings.
1. Hoist car and remove wheel.
2. Remove upper shock absorber attaching bolt.
3. Remove cotter pin and nut on upper ball joint.
4. Disconnect brake hose clamp from ball joint stud.
5. Using hammer and a drift, drive on spindle until upper ball joint stud is disengaged.

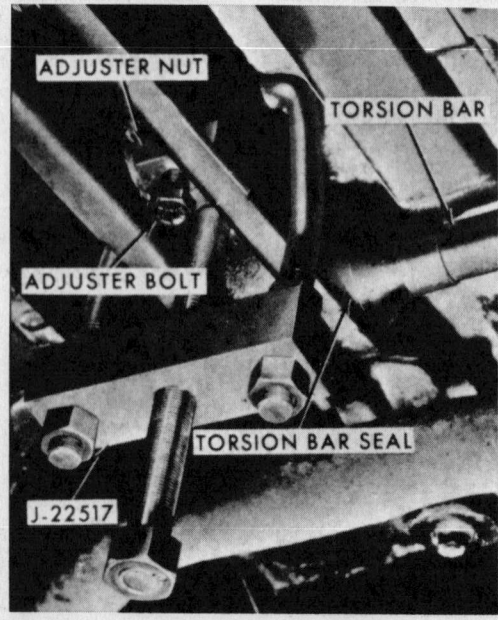

Torsion bar remover and installer
(© Cadillac Div. G.M. Corp.)

6. Remove upper control arm cam assemblies and remove control arm from car by guiding shock absorber through access hole in arm.

Installation

1. Guide upper control arm over shock absorber and install bushing ends into frame horns.
2. Install cam assemblies.
   NOTE: front cam is mounted up, rear cam is mounted down.
3. Install ball joint stud into knuckle.
4. Install brake hose clip on ball joint stud.
5. Install ball joint nut. Torque to 40 ft. lbs. and insert cotter pin, crimp.
   NOTE: cotter pin must be crimped toward upper control arm to prevent interference with outer C. V. joint seal.
6. Install upper shock attaching bolt and nut. Torque to 75 ft. lbs.
7. Install wheel.
8. Lower hoist.
9. Check camber, caster and toe-in, and adjust if necessary.

Upper Control Arm Bushing on the Car

The upper control arm bushings can be removed and installed on or off the car.

Removal

1. Hoist car and remove wheel.
2. Disconnect upper shock absorber attaching bolt.
3. Remove cam assemblies from control arms.
4. Move control arms out of frame horns and attach bushing removal tools.

Installation

1. Install tools and press bushings into control arm.
2. Move control arm into frame horns and install cam assemblies. Front cam is installed with the bolt in the lower position. Rear cam is installed with the bolt in the upper position.
3. Connect upper shock absorber attaching bolt. Torque to 75 ft. lbs.
4. Replace wheel and lower car.
5. Align front wheels.

Lower Control Arm

Removal

1. Remove wheel disc and loosen wheel mounting nuts.
2. Remove hub cotter pin. Loosen nut.
3. Raise car and remove wheel and tire.
4. Remove torsion bar.
5. Remove hub nut and washer, and brake line clip attached to frame.

6. Remove cotter pin, nut and brake line clip from upper spherical joint and remove joint from steering knuckle with a hammer and drift.
7. Disconnect shock absorber and remove.
8. Disconnect tie-rod end at steering knuckle with tie-rod end puller.
9. Disconnect stablizer bar and nut and link bolt.
10. Disconnect lower spherical joint with ball joint puller and adapter.
11. Disengage hub, knuckle and disc as an assembly and secure to upper control arm with wire.
12. Remove lower control arm to frame nuts and bolts and disengage arm from frame mounts.

Installation

1. Install hub, disc and knuckle assembly on drive axle.
2. Install lower control arms into mounts at chassis.
   NOTE: do not tighten nuts now.
3. Install lower control arm spherical joint into steering knuckle. Tighten nut to 85 ft. lbs. and install cotter pin.
4. Tighten lower control arm bolts to 75 ft. lbs.
5. Install shock absorber and tighten nut to 75 ft. lbs.
6. Install upper control arm spherical joint into steering knuckle and install brake line clip. Tighten nut to 85 ft. lbs. and install cotter pin.
7. Install brake line clip to chassis.
8. Install tie-rod end in steering knuckle, tightening nut to 30 ft. lbs.
9. Install stabilizer bar.
10. Install hub to drive axle washer and nut.
11. Install torsion bar.
12. Install wheel and tire.
13. Lower car.
14. Tighten hub to drive axle nut to 110 ft. lbs. and install cotter pin. Tighten wheel nuts to 105 ft. lbs.
15. Install wheel disc.

Lower Control Arm Bushings

Removal

Remove lower control arm and press bushings out of arm. To install, press bushings into arm and install arm. Check standing height and adjust if necessary.

Ball Joint Checks

Vertical Check

1. Raise the car and position floor stands under the left and right lower control arm, as near as possible to each lower ball joint. Car must be stable and should not rock on floor stands.
2. Position dial indicator to regis-

ter vertical movement at wheel hub.
3. Place a pry bar between the lower control arm and the outer race, and pry down on the bar. Care must be used so that the drive axle seal is not damaged. The vertical reading must not exceed .125 in.

Horizontal Check

1. Place car on floor stands as outlined in Step 1 in the Vertical Check.
2. Position dial indicator at the rim of the wheel, to indicate side play.
3. Grasp wheel, top and bottom, and push in on the bottom of the tire while pulling out at the top. Read gauge, then reverse the push-pull procedure. Horizontal deflection on the gauge should not exceed .125 in. at the wheel rim. This procedure checks both the upper and lower ball joints.

Lower Control Arm Ball Joint

Removal

1. Remove knuckle.
2. Using hack saw, saw the three rivet heads off.
3. By using a 7/32 in. drill bit, drill side rivets 3/16 in. deep.
4. Using hammer and punch, drive center rivet of joint, until joint is out of the control arm.

Installation

1. Install service ball joint into control arm and torque bolts and nut.
2. Reverse knuckle removal.

Lower Control Arm Ball Joint Seal

The lower ball joint seal can be installed with lower control arm either in or out of the car.

Removal

1. Remove steering knuckle.
2. Using hammer and chisel, drive seal from ball joint.
3. Wipe grease from ball joint and stud.

Installation

1. Position new seal over ball joint stud.
2. Lubricate jaws of camber adjusting wrench and carefully slide jaw between seal and retainer.
3. Tap lightly with hammer on center bolt of the wrench until retainer is fully seated.
4. Install knuckle.
5. Lubricate the ball joint fitting until grease is apparent in seal.

Stabilizer Bar

Removal

1. Remove link bolts, nuts, grom-

"C" Measured From Center Of
Bolt To Bolt On Shock

"D" Measured From Flat
On Bottom Of Stop Bracket
To Top Of Axle

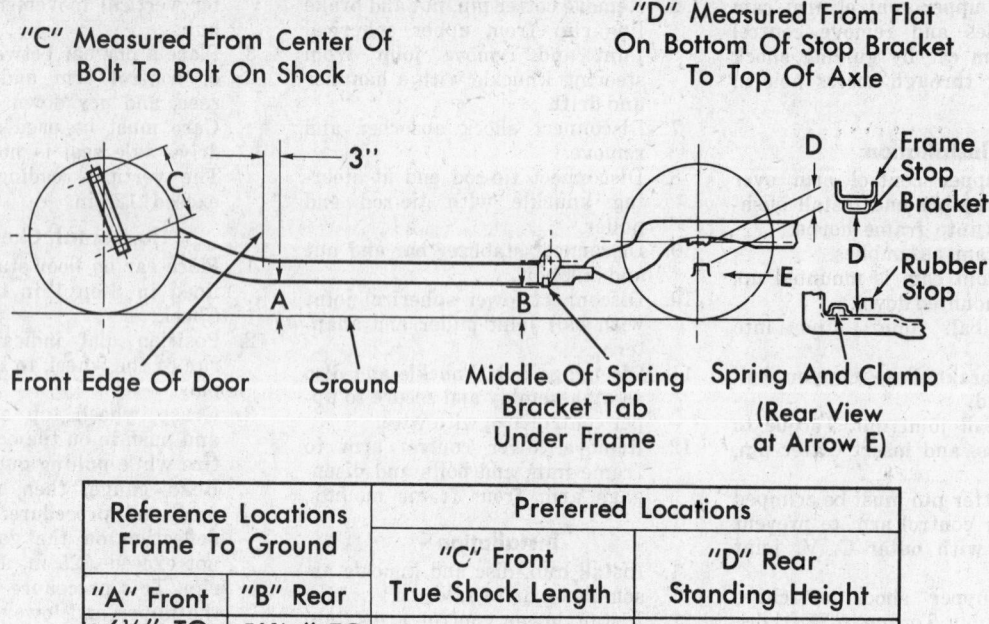

Reference Locations Frame To Ground		Preferred Locations	
"A" Front	"B" Rear	"C" Front True Shock Length	"D" Rear Standing Height
6¼" TO 6⁷⁄₁₆"	5¹⁵⁄₁₆" TO 6½"	14⅝" TO 14⅞"	4-23/32" To 5-15/32"

1969-70 Fleetwood Eldorado standing height chart (© Cadillac Div. G.M. Corp.)

mets, spacers and retainers from lower control arm. Discard bolts.
2. Remove two bolts attaching dust shield to frame, both sides.
3. Remove bracket to frame attaching bolts and remove stabilizer bar from front of car.

Installation

Reverse removal procedure.

NOTE: new link bolts are torqued to 14 ft. lbs., then cut off ¼ in. from nut.

Front Shock Absorber

Replacement

1. Remove wheel disc and loosen wheel mounting nuts.
2. Raise car, place on jacks, and remove wheel and tire.
3. Place a hydraulic jack under lower control arm and raise so that load is taken off shock absorber.
4. Disconnect shock absorber at upper and lower mount.
5. Compress shock absorber, working lower mount free from mount bolt.
6. Remove shock absorber.
7. Install by reversing procedure above, tightening shock absorber nuts to 75 ft. lbs. and wheel mounting nuts to 105 ft. lbs.

Front End Alignment

The standing heights are controlled by the adjustment setting of the torsion bar adjusting bolts. Clockwise rotation of the bolts increases the front height. It is very important that this height be considered and made correct before steering geometry is established.

For quick checks only, the locations at frame-to-ground points A and B, as illustrated, can be used. However, locations C and D, are preferred and should always be used for proper measurements of standing heights.

Quick Reference Locations

Location A is at a point 2 in. behind the front edge of the door. This is 39½ in. from the centerline of the front wheel.

Location B is at the middle of the spring bracket tab under the frame. This is 32¼ in. forward of the rear axle centerline.

Preferred Locations

Location C is from center to center of the shock absorber upper and lower mounting bolts.

Location D is from the flat on the bottom of the stop bracket to the top of the axle. The rear view of D, as illustrated, shows this location more clearly.

Caution Do not accidentally slip the measuring rule or device into the slot for the rubber bumper in the rear axle, because doing so would give a false reading.

Acceptable specifications are also given in illustration. Frame to ground dimensions must be within 1 in. from front to rear and within ⅝ in. from side to side.

If dimensions are not within tolerance, torsion bars must be adjusted.

Measuring Height

Before measuring standing height, check and correct the following items:
1. Car must be on a level surface.
2. Gas tank should be full or a compensating weight added. Estimate the amount of gasoline in the tank and add weights in the trunk in the space immediately above the float access hole area.
3. Front seat should be adjusted all the way to its rearmost position.
4. Front and rear tires should be inflated to 24 and 22 lbs. respectively.
5. Both doors closed.
6. No passengers or additional weight should be in the car or trunk (except as indicated above).

Checking Rear Height with Automatic Level Control Disconnected

Though rear height readings can be taken without disconnecting the A.L.C., it is preferred that measurements be taken with the leveling system disconnected—to eliminate the possibility of the A.L.C. system affecting the reading.

The A.L.C. can be disconnected at the black line connection at the leveling valve.

After the measurement is taken and the black line is reconnected, the A.L.C. system should be refilled by use of an air pressure hose to insure pressure reserve in the tank.

If rear of car raises when the automatic leveling system is reconnected, the leveling system should be checked and adjusted.

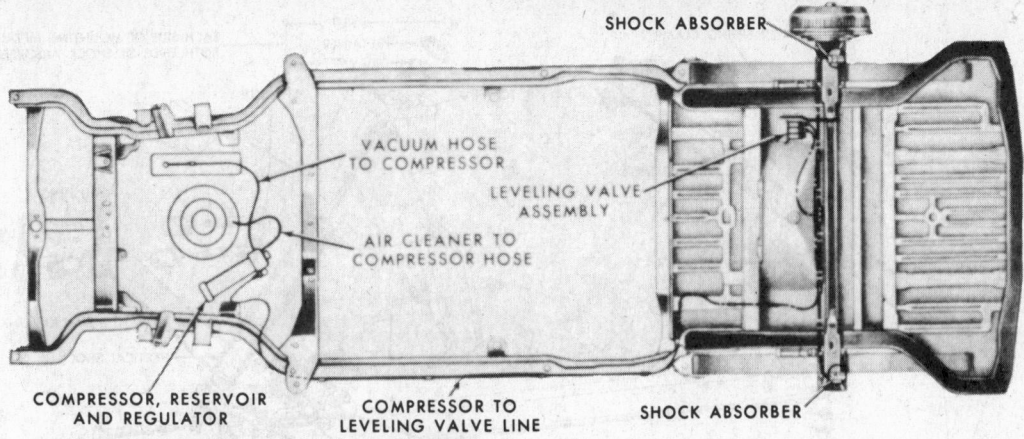

SHOCK ABSORBER

VACUUM HOSE TO COMPRESSOR

LEVELING VALVE ASSEMBLY

AIR CLEANER TO COMPRESSOR HOSE

COMPRESSOR, RESERVOIR AND REGULATOR

COMPRESSOR TO LEVELING VALVE LINE

SHOCK ABSORBER

Automatic leveling system (© Cadillac Div. G.M. Corp.)

CASTER CAMBER CAMS

Caster, camber cam locations (© Cadillac Div. G.M. Corp.)

Alignment Procedures Setting Camber & Caster

1. Check camber. The preferred setting for camber is 0° ± 3/8°. To adjust proceed as follows:
   A. Loosen nut on upper control arm front cam bolt.
   B. Note camber reading and rotate front bolt to correct for one-half of the incorrect reading or as near to that amount as possible. Tighten front nut.
   C. Loosen nut on upper control arm rear cam bolt and rotate rear cam bolt to bring camber reading to 0°. Tighten rear nut.
   D. Check caster. Preferred reading is 2° negative, ± 1/2°.
   NOTE: if caster requires adjustment, proceed with Step E; if not, move to Step I.
   E. Loosen front cam bolt nut.
   F. Using camber scale on alignment equipment, rotate front bolt so that the camber changes an amount equal to one-quarter of the desired caster change.
   NOTE: if adjusting to correct for excessive negative caster, rotate bolt

to increase positive camber. If adjusting to correct for excessive positive caster, rotate bolt to increase negative camber.
   G. Tighten front nut.
   H. Loosen nut on rear cam bolt and rotate the rear bolt until camber setting returns to 0°. This results in the correct caster setting.
   I. Tighten upper control arm cam nuts to 75 ft. lbs. Hold head of bolt securely; any movement of the cam will affect final setting and will require a recheck of the camber and caster adjustments.

Relationship of Front and Rear Cams

When setting camber and caster, remember this relationship:

Front cams. If turned for more positive camber, then caster also becomes more positive.

In other words, when turning the front cams to obtain a more positive setting for camber, caster will follow to a more positive setting.

The same is true when turning for more positive caster: more positive camber will follow.

Rear cams. If turned for more pos-

itive camber, then caster becomes more negative.

When turning the rear cams to obtain a more positive setting for camber, caster will advance in the opposite direction toward a more negative setting.

Toe-in Adjustment

Toe-in adjustment 0 in. to 1/8 in.
A. Center steering wheel, raise car and take wheel run out.
B. Loosen tie-rod and nuts, and adjust to proper setting.
C. Tighten tie-rod and nuts. Torque nuts 20 ft. lbs. Position tie-rod clamps so opening of clamps are facing down. This is a very necessary setting. Interference and a possible tie up of front end linkage could occur, if clamps snap anything while turning.

Rear Suspension

The rear suspension on the Cadillac Eldorado consists of two single leaf, semi-elliptical springs, two vertical and two horizontal shock absorbers.

Automatic Level Control

This system is basically the same as that used on other Cadillac models and functions identically. However, the on-car location of major components is different. Procedures will be found in the Cadillac section.

Rear Leaf Spring

Removal

1. Raise car.
2. Support rear axle at center with hydraulic jack.
3. Remove rear wheel from side being worked on.
4. Remove nut that secures Automatic Level Control link to axle bracket and remove link.
5. Remove nut that secures front of spring to frame bracket.

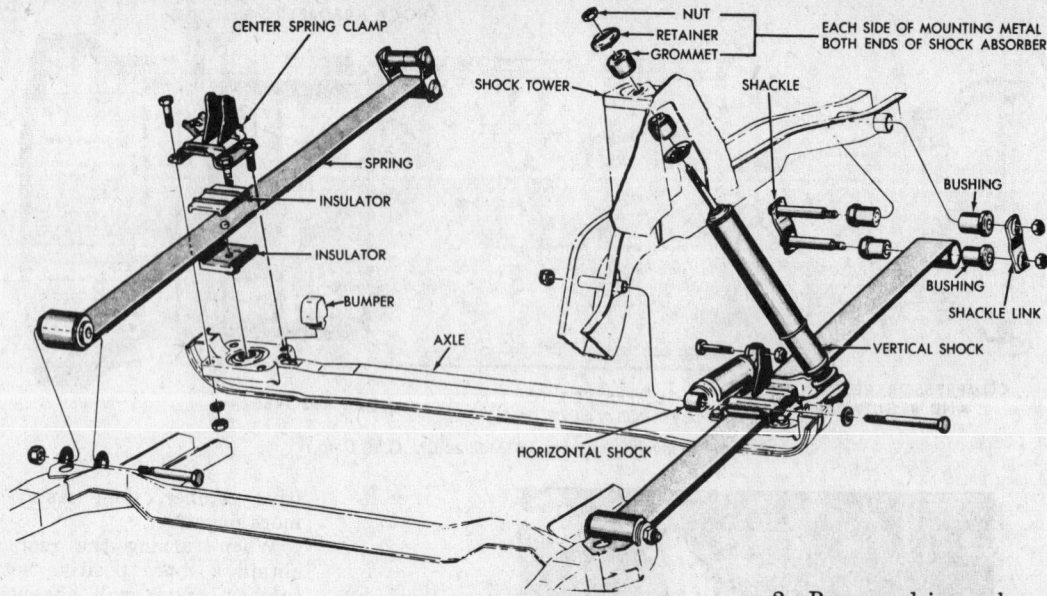

Rear suspension (© Cadillac Div. G.M. Corp.)

"C" Measured From Center Of Bolt To Bolt On Shock

"D" Measured From Flat On Bottom Of Stop Bracket To Top Of Axle

Reference Locations Frame To Ground		Preferred Locations	
		"C" Front True Shock Length	"D" Rear Standing Height
"A" Front	"B" Rear		
5-31/32" To 6-17/32"		14-15/32" To 14-23/32"	4-23/32" To 5-15/32"

1967-68 Fleetwood Eldorado standing height chart
(© Cadillac Div. G.M. Corp.)

NOTE: do not remove bolt now.
6. Remove two nuts at rear shackle outer link and remove link.
7. Remove four nuts and lockwashers that secure center spring clamp to rear axle and position out of the way.
8. Lower hydraulic jack until axle is free from spring.
9. Remove rear shackle assembly from spring and body.
10. Remove bolt from front of spring and remove spring.

Rear Axle
Removal
1. Raise car, and remove rear wheels.
2. Remove rear brake drum, then hub assembly.
3. Disconnect brake lines and hose and parking brake cable.
4. Disconnect overtravel lever link from bracket on rear axle.
5. Remove spring guides that hold parking brake cable to center spring clamp.

6. Remove brake backing plates.
7. Supporting rear axle at center with hydraulic jack, remove four nuts on each side of spring clamp assemblies.
8. Lower jack and remove rear axle.

Drive Axles

References
Drive axles are a complete flexible assembly and consist of an axle shaft and an inner and outer constant velocity joint. The inner constant velocity joint has complete flexibility, plus inward and outward movement. The outer constant velocity joint has complete flexibility only.

Drive Axle—Right Side
Removal
1. Hoist car under lower control arms.
NOTE: battery should be disconnected.

2. Remove drive axle, cotter pin, nut and washer.
3. Remove oil filter element.
4. Remove inner constant velocity joint attaching bolts.
5. Push inner constant velocity joint outward enough to disengage the right-hand final drive output shaft and move rearward.
6. Remove right-hand output shaft bracket bolts to engine and final drive.
7. Remove right-hand output shaft.
8. Drive axle assembly will be removed at this time.

Caution Care must be exercised so that constant velocity joints do not turn to full extremes, and that seals are not damaged against shock absorber or stabilizer bar.

Installation
1. Carefully place right-hand drive axle assembly into lower control arm and enter outer race splines into knuckle.
2. Lubricate final drive output shaft seal, with special seal lubricant (Part No. 1050169).
3. Install right-hand output shaft into final drive and attach the support bolts to engine and brakes. Torque the bolts to 50 ft. lbs.
4. Move right-hand drive axle assembly toward front of car and align with right-hand output shaft. Install attaching bolts and torque to 65 ft. lbs.
5. Install oil filter element.
6. Install washer and nut on drive axle. Torque to 60 ft. lbs. and insert cotter pin and crimp.
7. Remove floor stands and lower hoist.
8. Check engine oil. Add if necessary.

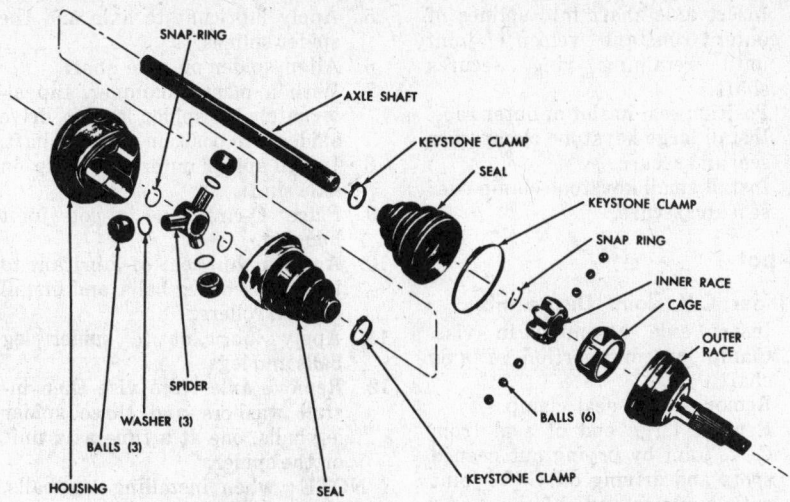

SNAP-RING
AXLE SHAFT
KEYSTONE CLAMP
SEAL
KEYSTONE CLAMP
SNAP RING
INNER RACE
CAGE
OUTER RACE
BALLS (6)
KEYSTONE CLAMP
SEAL
SPIDER
WASHER (3)
BALLS (3)
HOUSING
SEAL

Drive axle exploded view (© Cadillac Div. G.M. Corp.)

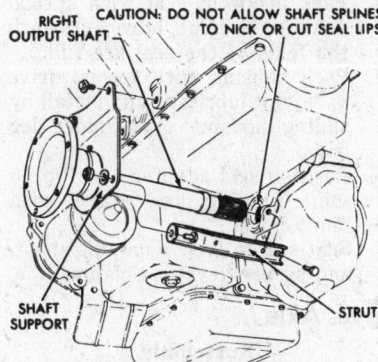

RIGHT OUTPUT SHAFT
CAUTION: DO NOT ALLOW SHAFT SPLINES TO NICK OR CUT SEAL LIPS
SHAFT SUPPORT
STRUT

Output shaft, right side
(© Cadillac Div. G.M. Corp.)

Drive Axle—Left Side

Removal

1. Hoist car under lower control arms.
2. Remove wheel and drum.
3. Remove drive axle cotter pin, nut and washer.
4. Position access slot in hub assembly so that each of the attaching bolts can be removed. It will be necessary to push aside adjuster lever to remove one of the bolts.
5. Remove hub assembly. It will again be necessary to push aside adjuster lever for clearance for hub assembly.
6. Remove tie-rod end cotter pin and nut.
7. Using hammer and brass drift, drive on knuckle until tie-rod end stud is free.
8. Remove bolts from drive axle assembly and left output shaft.
9. Remove upper control arm ball joint cotter pin and nut.
10. Using hammer and brass drift, drive on knuckle until upper ball joint stud is free.
11. Install adapter carefully between ball joint seal and knuckle.
12. Using ball joint puller remove lower ball joint from knuckle.

Care must be exercised so that the ball joint does not damage the drive axle seal.
13. Remove knuckle and support, so that brake hose is not damaged.
14. Carefully guide drive axle assembly outboard.

NOTE: care must be exercised so that constant velocity joints do not turn to full extremes and that seals are not damaged against shock absorber or stabilizer bar.

Installation

1. Carefully guide left-hand drive axle assembly onto lower control arm and into position on tool J-22193.
2. Insert lower control ball joint stud into knuckle and attach nut. Do not torque.
3. Center left-hand drive axle assembly in opening of knuckle and insert upper ball joint stud.
4. Place brake hose clip over upper ball joint stud and install nut. Do not torque.
5. Insert tie-rod and stud into knuckle and attach nut. Torque to 45 ft. lbs. Install cotter pin and crimp.
6. Lubricate hub assembly bearing O.D. with E.P. grease and install. Torque to 65 ft. lbs.
7. Align inner constant velocity joint with output shaft and install attaching bolts. Torque to 65 ft. lbs.
8. Torque upper and lower ball joint stud nuts to 40 ft. lbs. Install cotter pins and crimp.

NOTE: upper ball joint cotter pin must be crimped toward upper control arm to prevent interference with outer constant velocity joint seal.

9. Install drive axle washer and nut. Torque to 60 ft. lbs. Install cotter pin and crimp.
10. Install drum and wheel.
11. Remove floor stands and lower hoist.
12. Check camber, caster and toe-

in and adjust if necessary. Refer to Front End Alignment.

Constant Velocity Joint (Out of Car)

The constant velocity joints are to be replaced as a unit and are only disassembled for repacking and replacement of seals.

Outer C.V. Joint Disassembly

1. Insert axle assembly in vise. Hold by the mid-portion of the axle shaft.
2. Remove inner and outer seal clamps.
3. Slide seal down axle shaft to gain access to C. V. joint.
4. Using snap-ring pliers, spread retaining ring until C. V. joint can be removed from axle spline.
5. Remove retaining ring.
6. Slide seal from axle shaft.
7. Remove grease from constant velocity joint.
8. Holding constant velocity joint with one hand, tilt cage and inner race so that one ball can be removed. Continue until all six balls are removed.
9. Turn cage 90° and with large slot in cage aligned with land in inner race, lift out.
10. With cage and inner race assembly, turn inner race 90° to align with large hole in cage. Lift land on inner race up through large

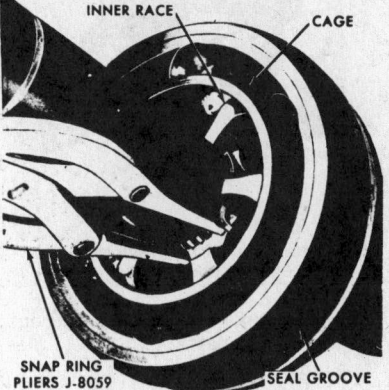

INNER RACE
CAGE
SNAP RING PLIERS J-8059
SEAL GROOVE

Removing outer axle joint
(© Cadillac Div. G.M. Corp.)

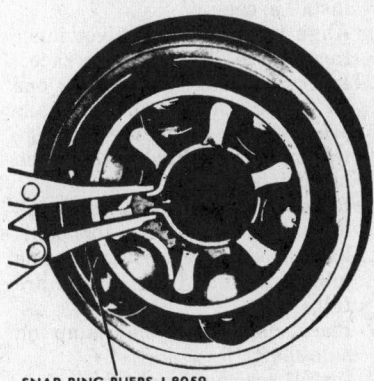

SNAP RING PLIERS J-8059

Removing inner race snap-ring
(© Cadillac Div. G.M. Corp.)

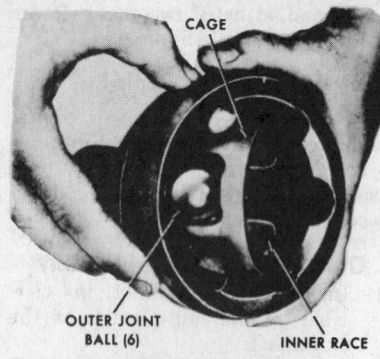

Removing balls from outer joint
(© Cadillac Div. G.M. Corp.)

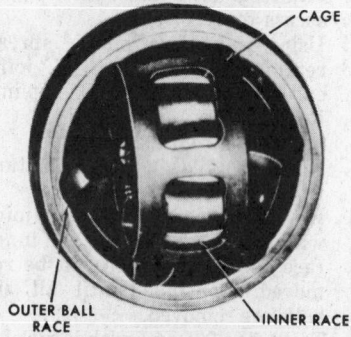

Removing cage and inner race
(© Cadillac Div. G.M. Corp.)

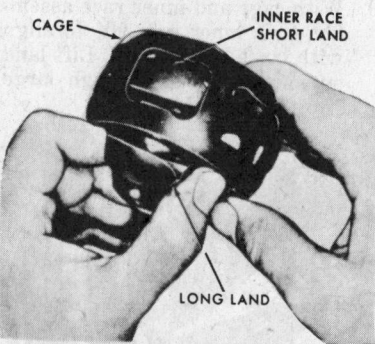

Removing inner race from cage
(© Cadillac Div. G.M. Corp.)

hole in cage and turn up and out to separate parts.

Assembly

1. Insert land of inner race into large hole in cage and pivot to install in cage.
2. Align inner race and pivot inner race 90° to align in outer race.
3. Insert balls into outer race one at a time until all six balls are installed. Inner race and cage will have to be tilted so that each ball can be inserted.
4. Pack constant velocity joint full of lubricant (Part No. 1050530).
5. Pack inside of seal with the same lubricant, until folds of seal are full.
6. Place small keystone clamp on axle shaft.
7. Install seal onto axle shaft.
8. Install retaining ring into inner race.

9. Insert axle shaft into splines of outer constant velocity joint until retaining ring secures shaft.
10. Position seal in slot of outer race.
11. Install large keystone clamp over seal and secure.
12. Install small keystone clamp over seal and secure.

Tri-pot

Inner C.V. Joint Disassembly

1. Insert axle assembly in vise. Clamp on mid-portion of axle shaft.
2. Remove small seal clamp.
3. Remove large end of seal from C. V. joint by prying out peened spots and driving off C. V. joint with hammer and chisel.
4. Slide the seal and adapter down the axle shaft until the tri-pot joint is exposed.
   NOTE: the tri-pot housing is now free to slide off the joint. Use care to prevent the spider leg balls from sliding off the spider legs. Each leg ball contains multiple bearing rollers.
5. Cup one hand under the tri-pot joint to prevent dropping spider leg balls and rollers while sliding housing off of joint.
6. Remove spider leg ball.
7. Remove O-ring seal from outer housing.
8. Wipe excess grease from outer housing to gain access to snap ring and remove spider outer snap-ring.
9. With a plastic hammer, tap alternately on spider legs to drive spider off shaft.
10. Remove spider inner snap-ring.
11. Slide seal off axle shaft.
12. Remove rollers from spider leg balls.

Assembly

1. Insert axle assembly in a vise. Hold by mid-portion of axle shaft.
2. Place small keystone clamp on axle shaft.
3. Position seal on shaft.
4. Place spider inner snap-ring in position on shaft.

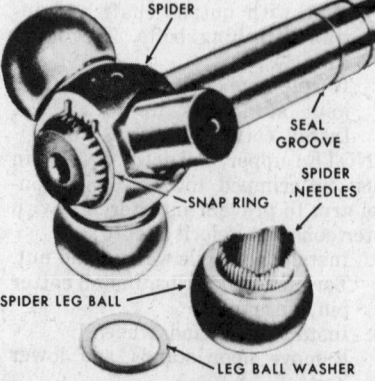

Spider assembly (© Cadillac Div. G.M. Corp.)

5. Apply lubricant to axle and the spider splines.
6. Align spider on axle shaft.
7. With a plastic hammer, tap alternately on spider legs to drive spider into position on axle shaft.
8. Install spider outer snap-ring on axle shaft.
9. Place O-ring on tri-pot joint housing.
10. Apply a thin coat of lubricant to inner race of leg balls, and install leg ball rollers.
11. Apply lubricant to spider leg balls and legs.
12. Remove axle from vise then install washers and three spider leg balls, one at a time as a unit on the spider.
    NOTE: when installing leg balls, use the leg ball washers as retainers for the spider rollers.
13. Pack inside of seal with special drive axle joint lubricant until the folds of the seal are full.
14. Pack housing with special drive axle joint lubricant and install by sliding housing over spider leg balls.
15. Position seal adapter over lip on joint housing and stake with blunt chisel.
16. Seat seal in groove on axle shaft, and secure keystone clamps.

Drive Axle

Disassembly

1. Remove drive axle assembly.
2. Remove outer C. V. joint seal clamps.
3. Remove inner C. V. joint seal by prying out peened spots and driving off seal.
4. Slide seals inboard on shaft.
5. Using snap-ring pliers, spread snap-rings until both C. V. joints can be removed from axle shaft.
6. Remove seals from axle.

Assembly

1. Pack seals with lubricant (Part No. 1050530).
2. Place outer C. V. joint seal on axle with keystone clamps in position on seal.
3. Insert axle into outer C. V. joint until retaining ring locks axle into position.
4. Position seal and clamps and secure keystone clamps.
5. Place inner C. V. joint seal on axle with keystone clamp in position on seal.
6. Insert axle into inner C. V. joint until retaining ring locks axle into position.
7. Place axle assembly into press using tool J-22247-1-2. Press seal into position and peen to secure.
8. Secure small keystone clamp with seal in position.
9. Install drive axle assembly.

Planetary-Type Differential

1967

This type differential replaces the conventional spider and beveled axle drive pinions with a planetary gear set to distribute torque to the respective drive axles.

Engine torque is transmitted from the power train, to the main drive pinion and ring gear, to the differential housing. Torque is then applied through the planetary and sun gear mechanism to both drive axles at variable speed requirements.

While the car is moving straight ahead, the planetary gears are fixed and rotate with the differential case and ring gear as a unit. However, when turning, the planetary gears revolve upon their individual axles with differential action, allowing the drive axles to rotate at different speeds.

This unit is not a controlled or limited slip differential.

Bevel Gear-Type Differential

Beginning 1968

Since 1968, a bevel gear-type differential is used on all front wheel drive models. This design supersedes the original planetary-type final drive. While unit removal and installation procedures are typical, the assembly, or its components are not interchangeable with the earlier design.

Overhauling the differential assembly is not encouraged. However, reconditioning procedures are found in later paragraphs of this final drive coverage. Differences in procedure are clearly indicated.

Output Shafts and Seals

Right Side Removal

1. Disconnect battery.
2. Hoist car.
3. Remove engine oil filter element.
4. Disconnect right-hand drive axle.
5. Disconnect support from engine and brace.
6. Remove output shaft assembly.
7. If seal is to be removed, install seal remover into seal and drive seal out with a hammer.
8. If output shaft bearing is to be removed, it can be removed with a press.

Installation

1. If output bearing was removed, assemble parts as illustrated.
2. Position assembly in a press and install bearing until seated.
3. Pack area between bearing and retainer with wheel bearing grease, then install slinger.

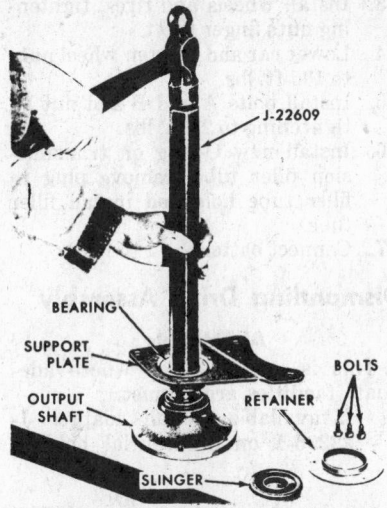

Installing output shaft support and bearing, right side
(© Cadillac Div. G.M. Corp.)

4. If seal was removed, it can now be installed.
5. Apply special seal lubricant to output shaft seal, then install output shaft into final drive, indexing the splines of both units.
6. Install support to engine and brace bolts.
7. Connect drive axle to output shaft.
8. Install engine oil filter element.
9. Connect battery, check engine oil level and check for oil leaks.

Left Side Removal

The left-hand output shaft can normally be removed only after removing the final drive assembly from the car. However, if the left-hand axle assembly has been removed for any reason, the output shaft and seal can be removed as follows:

1. Remove right-hand output shaft assembly, as in previous paragraphs.
2. Using a 9/16 in. socket, remove left-hand output shaft retaining bolt and remove left-hand shaft.
3. If seal is to be removed, insert seal remover into seal and drive seal out with a hammer.

Installation

1. If seal was removed, install new seal.
2. Apply special seal lubricant to seal, then insert output shaft into final drive assembly, indexing splines of output shaft with splines of final drive.
3. Install left-hand output shaft retaining bolt and torque to 45 ft. lbs.
4. Install right-hand output shaft, as described in previous paragraph.

Final Drive

Removal

1. Disconnect battery.
2. Pump about one gallon of transmission fluid out of filler tube. Remove bolt on the bracket that secures filler tube and remove filler tube, plugging the filler tube hole.
3. Remove bolts A and B and nut H. (See illustration.)
4. Remove support bracket bolts.
5. Raise car and remove wheels and tires.

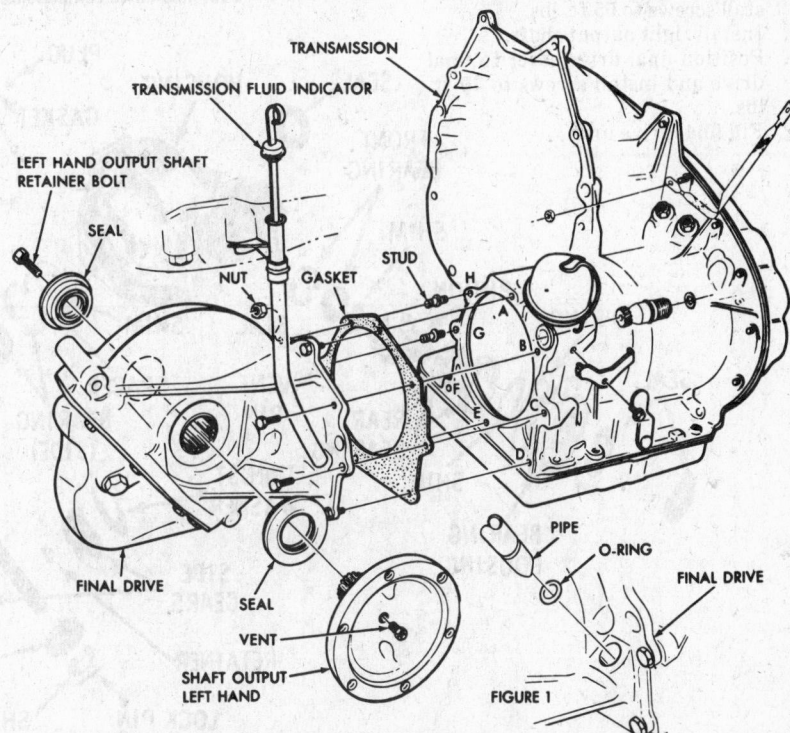

Final drive to transmission assembly (© Cadillac Div. G.M. Corp.)

6. Remove right hand output shaft.
7. Remove screws and washers that secure left drive axle to output shaft.
8. Loosen screws that secure final drive cover to final drive. Allow lubricant to drain then remove screws and cover.
9. Compress left drive axle inner constant velocity joint and secure drive axle to frame.
10. Remove final drive support bracket.
11. Remove bolts C, D, E, and F and nut G.
12. Disengage final drive splines from transmission.
13. Remove final drive unit, permitting ring gear to rotate up over steering gear.
14. Remove transmission to final drive gasket and discard.

Installation

1. Positioning new gasket on transmission, install final drive unit, permitting ring gear to rotate up over steering linkage.
2. Align final drive splines with splines in transmission.
3. Align bolt studs G and H on transmission with holes in final drive.
4. Install bolts C, D, E and F and nut G finger tight.
5. Install support bracket on final drive unit.
6. Install other support brackets.
7. Install bolt in oil cooler lines clamp and tighten to 8 ft. lbs.
8. Tighten bolts C, D, E and F and nut G to 25 ft. lbs.
9. Reposition left drive axle and install screws to 65 ft. lbs.
10. Install right output shaft.
11. Position final drive cover to final drive and install screws to 30 ft. lbs.
12. Fill final drive unit.

13. Install wheels and tires, tightening nuts finger tight.
14. Lower car and tighten wheel nuts to 105 ft. lbs.
15. Install bolts A and B and nut H, tightening to 25 ft. lbs.
16. Install new O-ring on transmission filler tube, remove plug in filler tube hole and install filler tube.
17. Connect battery.

Dismantling Drive Assembly

All Models

This is another area where adequate facilities are a must.
1. If available, install adapter J-22296-1 on differential holding fixture J-3289. Differential holding fixture must be modified to obtain clearance between fixture and final drive housing. Mount final drive in holding fixture.
2. Use a drain pan under the assembly. Remove the drain plug. Then, remove the cover attaching screws and remove the cover.
3. Rotate final drive until pinion points down, then check ring gear-to-pinion backlash with a dial indicator. Record backlash for reassembly. Check pinion and side bearing preload with tools, J-22208-1 and J-22208-2 with the help of a torque wrench. Record preload reading.
4. Remove side bearing caps.

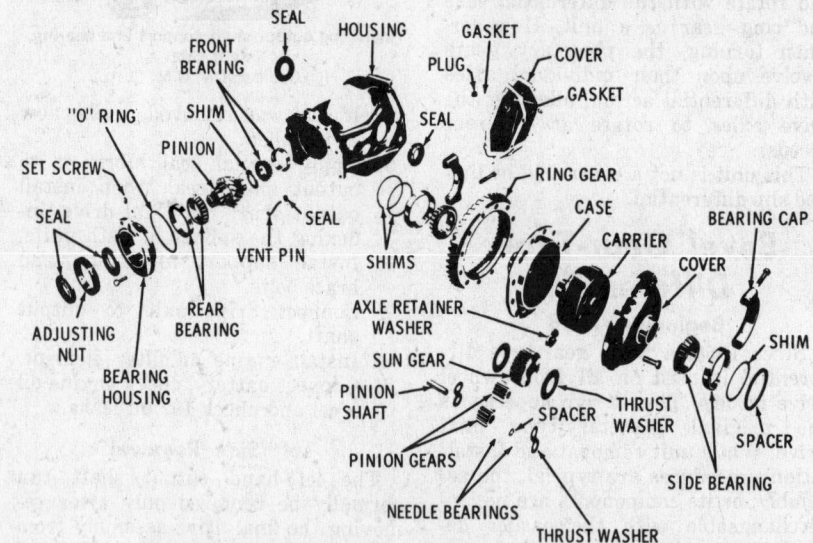

1967 final drive components (© Oldsmobile Div. G.M. Corp.)

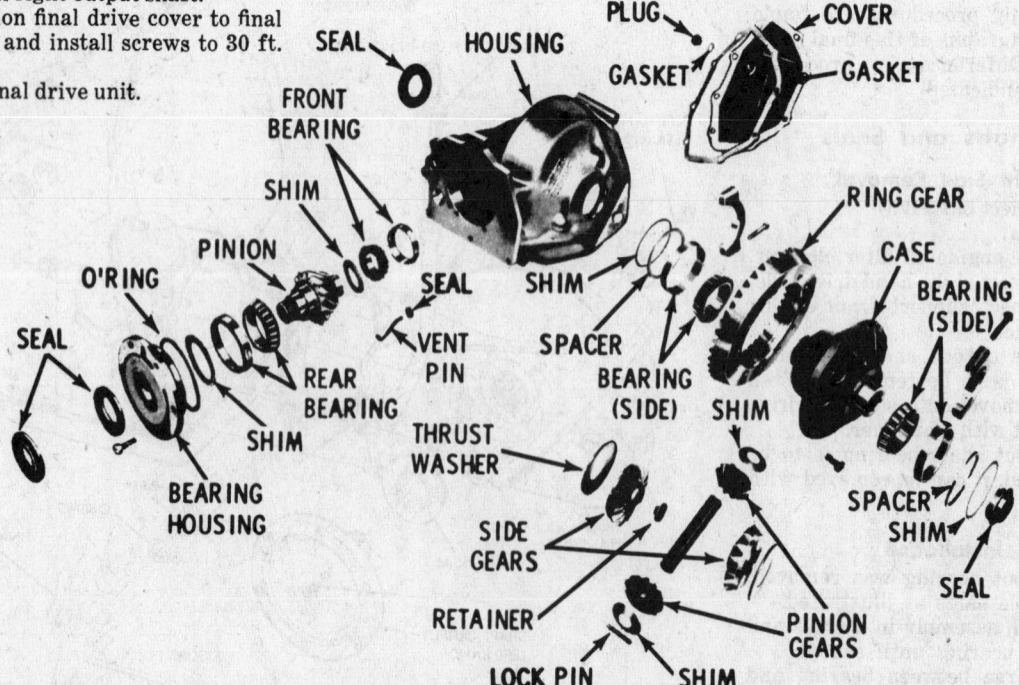

1968-70 final drive components (© Oldsmobile Div. G.M. Corp.)

NOTE: side bearing caps are of different size and can only be installed in one position.

5. Install spreader J-22196 on final drive, indexing the two guides on the spreader with the two holes on the carrier.
6. Turn the spreader screw to expand the spreader until the spacer and shims can be removed from between the small side of the bearing and the carrier.
7. Remove spreader from the carrier.
8. Remove the spacer and shims, then slide the case assembly to the left, away from the pinion gear. Remove case assembly from carrier. Check pinion bearing preload and record the reading.
9. Rotate carrier so the pinion is up.
10. Loosen set screw from adjusting nut.
11. Remove bearing housing bolts. Remove the drive pinion housing and remove the adjusting nut and housing from drive pinion. Remove rubber seal from bearing housing.
12. Remove rubber seal and vent wire from carrier.
13. With slide hammer J-2619 and tool J-22201, remove pinion front outer race.
14. Remove the output shaft oil seals.
15. Remove the two oil seals from the adjusting nut.
16. If necessary to remove pinion rear outer race, now is the time.

Pinion Bearings

All Models—Removal
1. Remove the pinion front bearing and selective shim. Bearing can be removed with a press.
2. Remove the pinion rear bearing.

Final Drive Case

All Models—Disassembly
1. If the side bearings are to be removed, it can best be done with tools J-22229-1, J-8433-1, and J-8416-1.
2. Mark ring gear, case and case cover, then remove all but two of the case cover to ring gear bolts. Leave two of the bolts 180° apart, loose.
3. Jar the assembly lightly on the bench to separate the halves of the case. Remove planet pinion carrier.
4. Clean all parts and examine all surfaces for wear or other damage.

Planetary-Type Pinion Gears

1967—Removal and Installation
1. Support the planetary pinion carrier assembly.

2. Press or drive the pinion pins out of the carrier.
3. Remove the pinion thrust washers, spacer, needle bearings, sun gear and thrust washers.
NOTE: the sun gear can be removed from only one opening of the carrier. This opening can be identified by the thinner wall at the carrier opening.
4. After removing the sun gear, the left axle retainer washer can be removed from the carrier.
5. Install by positioning loading tool J-22210 into planet pinion. Position a spacer washer over the loading tool, then install 24 needle bearings on each side of the spacer washer.
6. If the axle retainer washer was removed, install it at this time.
7. Position a thrust washer on each side of the sun gear, then insert the sun gear into carrier through large opening.
8. Position a thrust washer on each side of the planetary pinion, then insert planetary pinion into carrier.
9. Using a deep socket as a receiver, press pinion pin into carrier, until it bottoms.
10. Place a large punch in a vise, to be used as an anvil, and stake the opposite end of the pinion pin in three places.

Bevel-Type Pinion Gears

1968-70—Removal and Installation
1. After ring gear has been removed, drive lock pin from pinion shaft.
2. Push pinion shaft out of case.
3. Rotate one pinion gear and shim toward access hole in case, then remove.
NOTE: keep corresponding shims and pinion gear together for correct assembly.
4. Remove the other pinion and shim.
5. Remove side gears and thrust washers, keeping gears and washers in proper relationship for correct installation.
NOTE: the left-side gear has the threaded retainer which secures the (short) left output shaft. If threaded retainer is to be removed, use a brass drift to prevent trouble.
6. Upon assembling pinion and side gears into the case, lubricate components with a quality E. P. lubricant.
7. Place side gear thrust washers over the side gear hubs and install side gears in case. Gear with threaded retainer belongs in left side of case.
8. Position one pinion (without shims) between side gears, then rotate gears until pinion is directly opposite from loading opening in case. Place other pin-

ion between side gears so that the pinion shaft holes are in line; then rotate gears to make sure holes in pinions line up with holes in case.
9. If holes line up, rotate pinions back to loading opening to permit insertion of the pinion gear shims.
10. Install pinion shaft. Drive pinion shaft retaining lock pin into position.

Checking Pinion Depth

All Models
1. Install pinion front outer race. Drive race in until it bottoms.
2. Lubricate front bearing with final drive lubricant and install into front outer race.
3. Position tool J-21777-10 on the front bearing. Install tool J-21579 on final drive housing and retain with two bolts. Thread screw J-21777-13 into J-21579 until tip of screw engages tool J-21777-10. Torque tool J-21777-13 to 20 in. lbs. to preload the bearing.
4. Remove dial indicator post from tool J-21777-9 and install discs J-21777-11 and J-21777-12. Reinstall dial indicator post.
5. Place the gauging discs in the side bearing bores and install the side bearing caps.
6. Position the dial indicator on the mounting post of the gauge shaft with the contact button touching the indicator pad. Set dial indicator to zero, then depress the dial indicator until the needle rotates three-quarters of a turn clockwise. Tighten dial indicator.
7. Position the gauge shaft assembly in the carrier so that the dial indicator contact rod is directly over the gauging area of the gauge block, and the discs are seated fully in the side bearing bores.
8. Position gauge shaft so that the indicator rod contacts the gauging area. Rotate gauge rod back and forth until the indicator reads the greatest deflection. At the point of greatest deflection, set the indicator to zero. Repeat the rocking action to verify the zero setting.
9. After zero setting is obtained, rotate gauge shaft until the indicator rod does not touch the gauging area. Read the pinion depth directly from the dial indicator.
10. Select the correct pinion shim to be used during assembly on the following basis:
   A. If a service pinion is being used, or a production pinion with no marking, the correct shim will have a thickness equal to the indicator gauge reading found in step 9.

B. If a production pinion is used and marked +, the shim thickness indicated by the dial indicator on the pinion setting gauge must be increased by the amount etched on the pinion. If the pinion is marked −, the shim thickness indicated on the dial must be decreased by the amount etched on the pinion.

11. Remove pinion depth checking tools and front bearing from carrier.

Final Drive Case

All Models—Assembly

1. Install the planetary pinion carrier into the case.
2. With the case and cover alignment marks in index, insert four ring gear attaching bolts through case and cover. Align mark on ring gear with alignment marks on case and cover, then install ring and gear case. Alternately tighten the six attaching bolts to 85 ft. lbs.
3. If side bearings were removed, they can be installed now. Drive bearing on until it bottoms.
4. Install pinion rear bearing.
5. Position correct shim on drive pinion and install the drive pinion front bearing with tool J-21022 and a press.
6. Lubricate pinion bearings and install pinion into carrier.
7. Install seals into adjusting nut.
8. Install O-ring and vent pin on face of carrier. Torque attaching nuts to 35 ft. lbs.
9. Install seal protector J-22236 over drive pinion, then install the adjusting nut over the seal protector and thread into the housing.
10. Assemble tools as illustrated and adjust pinion bearing preload. The preload is 2 to 10 in. lbs. for new bearings, and 2 to 3 in. lbs. for used bearings. Adjust new bearing preload to 4 in. lbs. while rotating the pinion and checking preload until preload remains constant. When correct preload is obtained, tighten the set screw. Record preload reading because it will be used when making side bearing and preload adjustment. Leave the tools on pinion for side bearing preload adjustment.

Side Bearing Preload Adjustment

All Models

Differential side bearing preload is adjusted by means of shims located between the side bearings and the carrier. One spacer is used on the right side only. Shims are used on both sides and come in thickness increments of .002 in. from .036 to .070

in. By changing the thickness of both side shims equally, ring gear and pinion backlash will not change.

1. Lubricate the side bearings with final drive lubricant.
2. Place differential in position in the carrier.
3. If the original ring gear and pinion are being used, subtract the reading obtained in Step 8 from the reading obtained in Step 3 of the Dismantling Drive Assembly procedure. This determines the original side bearing preload and will aid in determining whether thicker or thinner shims are needed to bring the side bearing preload to specifications.
4. Install original shim on left side and spacer on the right side.
5. Install the carrier spreader and apply just enough tension to allow the shim to be installed between the spacer and the carrier.
6. Release tension on the spreader, install side bearing caps, then check preload. Preload should be 15 to 20 in. lbs. for new bearings and 5 to 7 in. lbs. for old bearings over the pinion bearing preload obtained in Step 11, Final Drive Case Assembly.
7. If preload is not within specifications, select thicker or thinner shims to bring preload within limits.

Backlash Adjustment

All Models

1. Rotate differential case a few times to seat bearings, then mount dial indicator in order to read movement at the outer edge of one of the ring gear teeth.
2. Check backlash at three points around the ring gear. Backlash must not vary more than .002 in.
3. Backlash at the minimum point should be between .006 in. and .008 in. for all new gears. If original ring gear and pinion was installed, backlash should be set at the same reading obtained in Step 3 of Dismantling Drive Assembly procedure, if reading was within specifications.
4. If backlash was not within limits, correct by increasing thickness of one differential shim and decreasing thickness of other side shim the same amount. This will not disturb differential side bearing preload. For each .001 in. change in backlash desired, transfer .002 in. shim thickness. To decrease backlash .001 in., decrease thickness of right shim .002 in. and increase thickness of left shim .002 in. To increase backlash .002 in. increase thickness of right shim .004 in. and decrease thickness of left shim .004 in.

5. When backlash is correct, remove spreader. Install bearing caps and bolts. Torque to 50 ft. lbs.
6. Install new output shaft seals.
7. Install new gasket on housing. Install cover, torque cover attaching bolts to 30 ft. lbs. Fill final drive to correct level with final drive lube.

Automatic Transmission

References

The Turbo Hydramatic transmission used on the Eldorado is a fully automatic transmission used for front wheel drive applications. It consists primarily of a three-element hydraulic torque converter, dual sprocket and link assembly, compound planetary gear set, three multiple-disc clutches, a sprag clutch, a roller clutch, two band assemblies, and an hydraulic control system. For major repair operations consult the Unit Repair Section.

Automatic Transmission, R & R

Removal

1. Disconnect the battery.
2. Remove the hood.
3. Remove the transmission dipstick, then remove the bolt holding the filler tube bracket to exhaust manifold and remove the filler tube. Discard the O-ring seal.
4. Remove bolts at locations A, B and C, holding the final drive case to the transmission. (See illustration).
5. Disconnect speedometer cable from the governor assembly.
6. Disconnect oil cooler pipes at the transmission and at the radiator. Cap the pipes and plug connector holes in transmission and radiator.
7. Remove bolt holding the cooler pipe bracket to final drive bracket and position pipes outboard of governor assembly.
8. Remove nut at location H, holding final drive to the transmission.
9. Remove bolts at locations I, J, K and L, holding transmission to engine and adapter plate.
10. Remove upper left bolt holding rear motor mount bracket to the transmission.
11. Remove bolt holding the ground strap to left side of cowl. Remove ground strap.
12. Remove upper left nut holding converter cover plate to transmission, (use 7/16 in. socket with universal and extension to

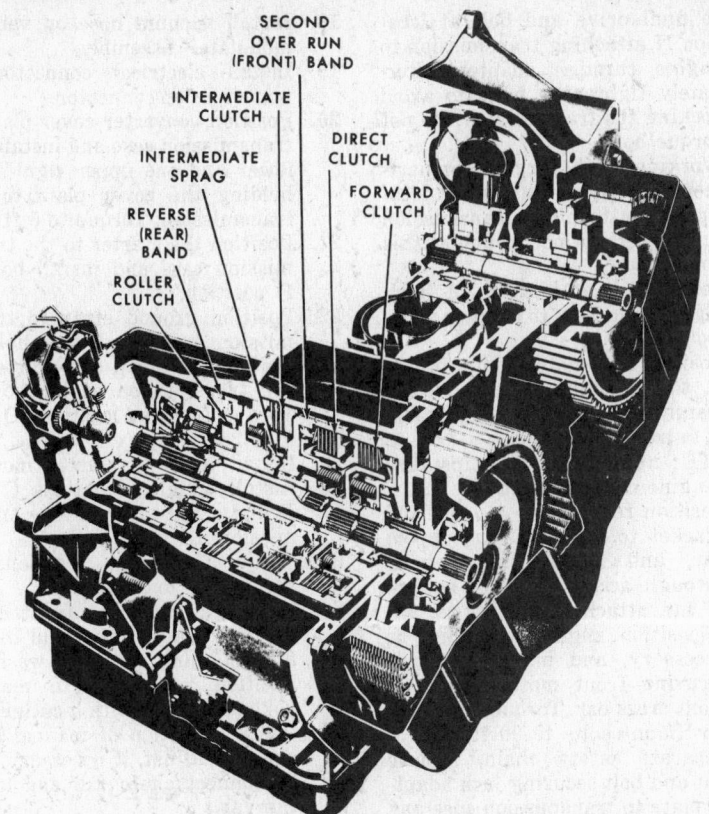

SECOND OVER RUN (FRONT) BAND
INTERMEDIATE CLUTCH
INTERMEDIATE SPRAG
REVERSE (REAR) BAND
ROLLER CLUTCH
DIRECT CLUTCH
FORWARD CLUTCH

Converter, chain drive and transmission parts layout (© Cadillac Div. G.M. Corp.)

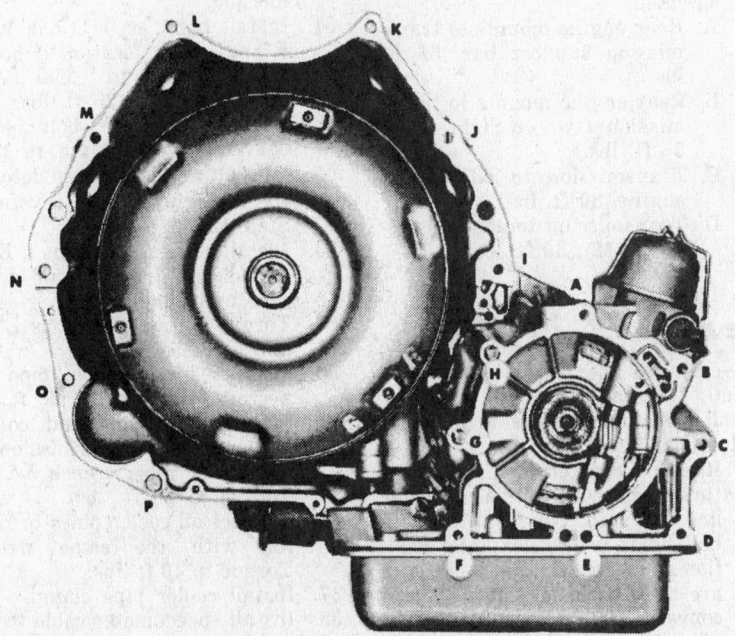

Transmission attaching bolt location (© Cadillac Div. G.M. Corp.)

19. Remove three remaining screws holding the converter cover plate to the transmission and remove the cover plate.
20. Position transmission jack, equipped with front end drive transmission adapter plate to transmission and install nut and bolt holding adapter brace to transmission at starter motor lower mounting bolt hole.
21. Disconnect electrical connector from transmission.
22. Remove vacuum line from vacuum modulator.
23. Secure transmission to transmission jack with safety chain.
24. Remove three flexplate-to-converter attaching bolts.
25. Remove bolts at locations M and N holding transmission to engine and adapter plate.
26. Remove cotter pin securing relay rod to manual yoke on left side of transmission and separate rod from yoke.
27. Remove bolts at locations D, E and F and nut at location G holding final drive to transmission.

NOTE: position a clean drain pan under a point where transmission and final drive meet to avoid leakage onto floor when the two units are separated.

28. Remove five bolts and washers holding the rear of acromat to front crossbar and frame horns and allow acromat to hang free.
29. Through access holes in bottom of front crossbar, remove left bolt and loosen right bolt holding front engine mount to front crossbar.
30. Have a helper, using a large pry bar, shift engine forward, while mechanic uses small pry bar to help separate transmission from engine and final drive.
31. Following initial separation, allow transmission to drain at the separation.
32. Remove two bolts on right side, holding the rear motor mount to the transmission.
33. Through access holes in the bottom of transmission support bar, remove two bolts, one on each side, holding the rear motor mounts to transmission support bar, and position motor mounts and bracket rearward to underbody.
34. While helper pries and holds engine forward, move transmission rearward to disengage transmission case from dowels on engine adapter and to disengage final drive from studs on transmission case. Top of transmission should be tilted slightly rearward.
35. Slowly lower transmission, until converter is about half-way exposed from flexplate.
36. Install converter holding clamp,

reach underneath the left exhaust manifold).
13. Position cable with looped ends under engine intake manifold and hook looped ends to chain fall and cable, putting engine mounts under tension.
14. Position safety chain over top of transmission.
15. Raise car and place on jack stands, adjusting chain fall as necessary.
16. Disconnect leads from starter.
17. Remove bolt at location O holding starter motor to transmission case and remove the ground strap.
18. While holding the starter, remove bolt at location P and remove starter.

J-21366, using a 5/16 in. x 18 nut to hold clamp screw to transmission case at location N.

Caution Converter holding clamp, J-21366, must be used to prevent the converter becoming disengaged when the transmission is removed.

37. Lower transmission from car.

Caution Rear motor mount bracket will follow transmission from car. To avoid damage or injury, remove bracket as soon as there is sufficient clearance.

38. Remove and discard final drive gasket and clean mounting surface of final drive.

Installation

1. Position transmission on jack, equipped with adapter plate, under the car.
2. Saturate new gasket with transmission fluid, then place gasket on final drive.
3. Position rear motor mount bracket on top of transmission support bar against underbody.
4. Raise transmission in place until converter is about half-way covered by flexplate, then remove converter holding clamp from transmission.
5. While helper pries engine forward with a pry bar, continue raising transmission, making sure the top of the transmission case clears splined input shaft of final drive, and position to engine.
6. Align the engine to the final drive, with the assistance of a helper by watching the following items:
   A. Studs on transmission case to mounting holes in final drive.
   B. Guide holes in transmission case to dowels on adapter.
   C. Internal flange on final drive to transmission.

Caution Since engagement of splined final drive input shaft to transmission is hidden, care must be taken to avoid damaging transmission or final drive assembly.

   D. To help engagement of final drive splines, rotate one front wheel while helper holds the other.

NOTE: when alignment is complete and correct, the gap between the final drive and transmission should not exceed ¼ in.

7. Loosely install bolts at locations D and F attaching transmission to final drive and bolt at location N attaching transmission to engine through adapter, alternately tightening bolts to avoid cocking the transmission. Do not torque bolts.
8. Working in the engine compartment, loosely install bolt at location J attaching transmission to adapter. Do not torque at this time.
9. Install bolt at location M holding transmission to adapter plate. Do not torque.
10. Position rear motor mount bracket to transmission and loosely install three bolts holding bracket to transmission.

NOTE: upper left bolt is installed from engine compartment.

11. Position rear engine mounts and bracket to transmission support bar, and loosely install bolts through access holes in bottom of bar, attaching mounts to bar.
12. Reposition engine assembly, as necessary, and install left bolt securing front motor mount to front cross bar. Torque front motor mount bolts to 90 ft. lbs.
13. Separate safety chain, remove nut and bolt securing jack adapter plate to transmission case and remove jack.
14. Torque the following bolts as specified:
   A. Rear engine mounts to transmission support bar, 55 ft. lbs.
   B. Rear engine mounts to transmission (two on right side), 55 ft. lbs.
   C. Transmission to adapter to engine, 30 ft. lbs.
   D. Transmission to adapter (location M), 30 ft. lbs.

Caution The following procedure for attaching the converter to the flexplate must be strictly followed to prevent improper installation and damage to flexplate and transmission.

15. Rotate converter until two of the three weld nuts on the converter line up with two of the three bolt holes in the flexplate. Position converter so that weld nuts are flush with flexplate. Be sure converter is not cocked and that pilot in center of converter is properly seated in crankshaft.
16. Install two flexplate to converter attaching bolts through access holes in flexplate and torque to 28 ft. lbs.

NOTE: bolts must be tightened at this time to assure proper alignment of converter.

17. Rotate flexplate and converter until third bolt hole is accessible. Install third bolt and torque to 28 ft. lbs.

18. Install vacuum hose on vacuum modulator assembly.
19. Install electrical connector to transmission connector.
20. Position converter cover plate to transmission case and install two lower and one upper right bolts holding the cover plate to the transmission. Torque to 5 ft. lbs.
21. Position the starter to the transmission case and install bolt at P position.
22. Position ground strap to transmission and install bolt holding the ground strap and starter to the transmission at location O. Torque bolts at locations O and P to 25 ft. lbs.
23. Install leads to starter motor.
24. Install bolts at locations C and E and a nut at G holding transmission to final drive.
25. Torque bolts at locations C through F to 25 ft. lbs.
26. Position acromat to front cross bar and frame horns and install five retaining bolts and washers.
27. Position relay rod to manual yoke and secure with a cotter pin.
28. Check operation of manual linkage and adjust, if necessary.
29. Disconnect chain fall and lower the car.
30. Remove cable from intake manifold and safety chain from transmission.
31. Install bolts at locations A and B and nut at location H holding transmission to final drive. Torque bolts to 25 ft. lbs.
32. Install upper left bolt holding converter cover plate to transmission in the manner described for removing it, the reversal of Step 12.
33. Install bolts at locations I, K and L, holding transmission to engine and adapter.
34. Torque bolts at locations I, J, K and L to 25 ft. lbs.
35. Tighten brass cooler pipe connectors at case to 28 ft. lbs. Clean connections and connect cooler pipes at transmission, using cooler pipe wrench J-21477. Torque to 28 ft. lbs.
36. Connect oil cooler pipes to radiator with the same wrench. Torque to 40 ft. lbs.
37. Install cooler pipe clamp.
38. Install speedometer cable to governor.
39. Install new O-ring on transmission filler tube and install filler tube through hole in final drive case.
40. Position transmission filler tube bracket to exhaust manifold and install retaining bolt.
41. Install body ground strap to firewall and secure it with a nut.
42. Connect battery.
43. Bring transmission to fluid level. Bring engine to operating tem-

perature, then recheck fluid level.

44. Thoroughly check entire power train for oil and coolant leaks.
45. Install and align hood assembly.

Drive and Driven Sprockets for the Transmission Drive

References

If it should be necessary to replace either the drive sprocket, chain, or driven sprocket, the three unit combination must be replaced as a set. They are matched and are not to be serviced separately.

Removal

1. Remove 18 cover housing attaching bolts.
2. Remove cover housing and gasket. Discard the gasket.
3. Install J-4646 snap-ring pliers into sprocket bearing retaining snap-rings located under the drive and driven sprockets and remove snap-rings from retaining grooves in support housings.
   NOTE: do not remove snap-rings from beneath the sprockets. Leave them in a loose position between the sprockets and the bearing assemblies.

Removing sprocket snap-ring

4. Remove drive and driven sprockets, link assemblies, bearings and shaft simultaneously by alternately pulling upward on the drive and driven sprockets until the bearings are out of the drive and driven support housings.
   NOTE: it may be necessary to pry up on the sprockets. Use care.

Caution Do not pry on the guide links or the aluminum case. Pry only on the sprockets.

5. Remove link assembly from drive and driven sprockets.
6. Remove two hook type oil seal rings from turbine shaft.
7. Inspect drive and driven sprocket bearing assemblies for rough or defective bearings.
   NOTE: do not remove bearing assemblies from drive or driven sprockets unless they need replacement.
8. If removal of bearing assembly from drive and/or driven sprockets is necessary, proceed as follows:
   A. Remove sprocket to bearing assembly retaining snap-ring using tool J-5589, snap-ring pliers.
   B. Mount sprocket with turbine or input shaft placed in hole

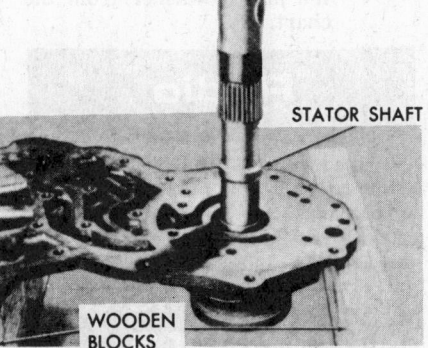

Removing drive sprocket support

in work bench on two 2 x 4 x 10 in. pieces of wood.

C. With a hammer and brass rod, drive the inner race alternately through each of the access openings until the bearing assembly is removed from the sprocket hub. Drive the sprocket, and turbine shaft and link assembly.

Inspection

1. Inspect drive sprocket teeth for nicks, burrs, scoring, gauling and excessive wear.
2. Inspect drive sprocket to ball bearing retaining snap-ring for damage.
3. Inspect drive sprocket ball bearing inner race mounting surface for damage.
4. Inspect turbine shaft for open lubrication passages. Run a tag wire through the passages to be sure they are open.
5. Inspect spline for damage.
6. Inspect the ground bushing journals for damage.
7. Inspect the two hook-type oil seal grooves for damage or excessive wear.
8. Inspect the turbine shaft for cracks or distortion.
9. Inspect the link assembly for damage or loose links.
   NOTE: take particular notice of the guide links. They are the wide outside links on each side of the link assembly.

Driven Sprocket at Input Shaft

Inspection

1. Inspect driven sprocket teeth for nicks, burrs, scoring, gauling and excessive wear.
2. Inspect sprocket to ball bearing retaining snapring for damage.
3. Inspect ball bearing inner race mounting surface for damage.
4. Inspect input shaft for open lubrication holes. Run a tag wire through the holes to be sure they are open.
5. Inspect spline for damage.
6. Inspect ground bushing journals for damage.

Removing sprockets and link assembly

Removing tight sprockets

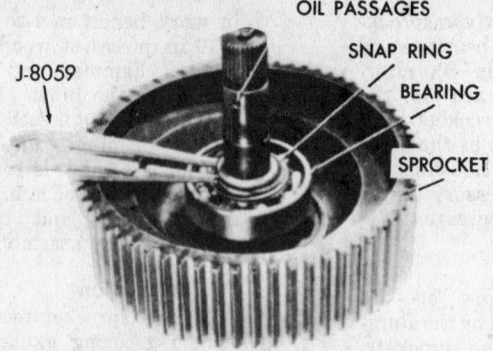

OIL PASSAGES
SNAP RING
BEARING
SPROCKET
J-8059

Removing sprocket bearing snap-ring

USE BOLTS FOR GUIDE PINS
WOODEN BLOCKS
Installing driven sprocket support

Sprocket Bearings

Installation

1. Turn sprocket so that turbine or input shaft is pointing upward.
2. Install new sprocket bearing as follows:
   A. Install snapring, letter side down on shaft.
   B. Assemble bearing assembly on turbine or input shaft.
   C. Using tool, J-6133-A, drive the bearing assembly onto the hub of the sprocket until it is resting on the bearing seat of the sprocket.
   D. Install sprocket to bearing assembly retaining snap-ring into groove sprocket hub.
3. Install two hook-type oil seal rings on turbine shaft.

Front Unit End-Play Check

Make front unit end-play check as follows:

A. Install front unit end-play checking tool J-22241 into driven sprocket housing so that the urethane on the tool can engage the splines and the forward clutch housing. Let the tool bottom on the main shaft and then withdraw it approximately 1/16-1/8 in.
B. Remove two of the 5/16—18 bolts from the driven support housing.
C. Install 5/16—18 threaded hammer bolt with jam nut into one bolt hole in driven support housing.
NOTE: do not thread slide hammer bolt deep enough to interfere with forward clutch travel.
D. Mount dial indicator on rod and index indicator to register with the forward clutch drum that can be reached through second bolt removed from driven support housing.
E. Push end-play tool down to remove slack.
F. Push and hold output flange outward. Place a screwdriver in case opening at parking area and push upward on output carrier.

G. Place another screwdriver between the metal lip of the end-play tool and the drive sprocket housing. Now push upward on the metal lip of the end-play tool and read the resulting end-play. This should be between .003-.024 in. The selective washer controlling this end play is the phenolic thrust washer located between the driven support housing and the forward clutch housing. If more or less washer thickness is required to bring the end-play within specifications, select the proper washer from the chart.

Radio

Procedures for removal and installation of radio may be found under instructions for same year model listed in the Cadillac section.

Windshield Wipers

The windshield wiper system consists of the wiper motor and transmission assembly. It is similar to that used on other Cadillac models.

Procedures will be found under instructions listed for same year model in the Cadillac section.

Heater System

Heater Blower

Removal

1. Drain cooling system.
2. Remove blower motor by removing cooling hose, disconnecting the wire connector and removing five screws that secure blower to heater case.
3. Remove left cowl - to - fender shield strut rod.

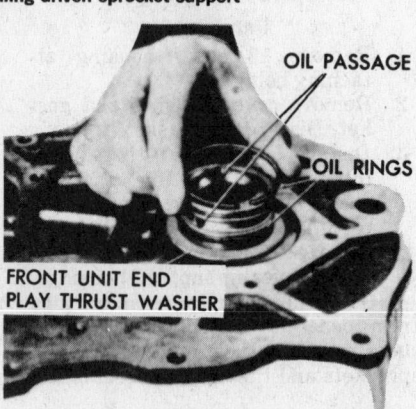

OIL PASSAGE
OIL RINGS
FRONT UNIT END PLAY THRUST WASHER

Installing oil rings on driven sprocket support

4. Remove heater hoses leaving clamps on fittings.
5. Disconnect temperature cable and position out of the way.
6. Disconnect wire connector from blower resistor.
7. Disconnect vacuum hoses and position out of the way.
8. Remove 12 screws that secure heater blower assembly to cowl.
9. Remove heater blower assembly by pulling away from cowl tipping blower end downward.
10. Install by reversing procedure above.

Heater Core

Removal

1. Remove heater blower assembly as described above.
2. Remove four screws, two on each side of the heater core, that secure the wire retaining clamps to the heater blower case. Remove the clamps.
3. Pull heater core out of heater blower case.
4. Install by reversing procedure above.

Index

YEAR IDENTIFICATION

1964 Chevy II

1965 Chevy II

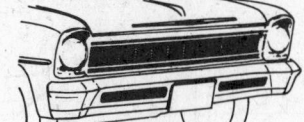

1966 Chevy II

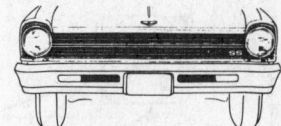

1967 Chevy II

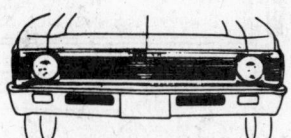

1968-69 Chevy II

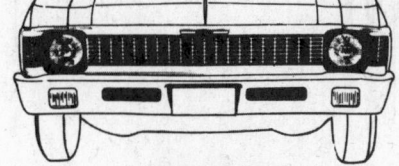

1970 Nova

1971 Nova

1964 Chevelle

1965 Chevelle

1966 Chevelle SS

1966 Chevelle

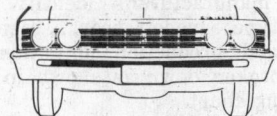

1967 Chevelle

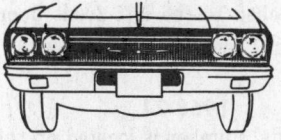

1968 Chevelle

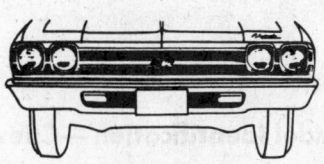

1969 Chevelle

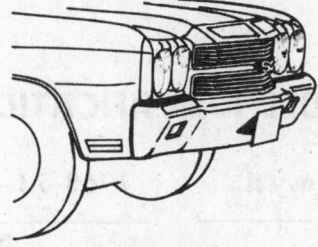

1970 Chevelle

1971 Chevelle

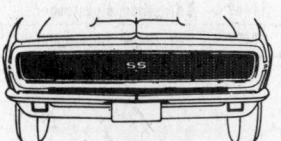

1967 Camaro SS

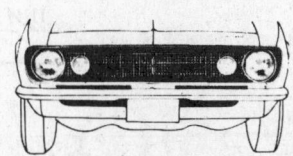

1967 Camaro

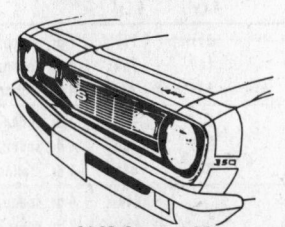

1968 Camaro SS

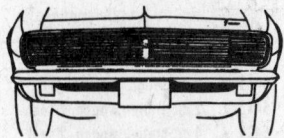

1968 Camaro

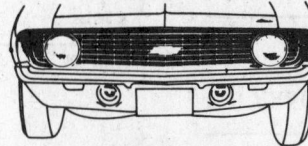

1969 Camaro

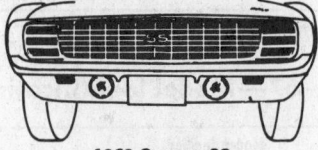

1969 Camaro SS

1970-71 Camaro

1970 Monte Carlo

1971 Monte Carlo

FIRING ORDER

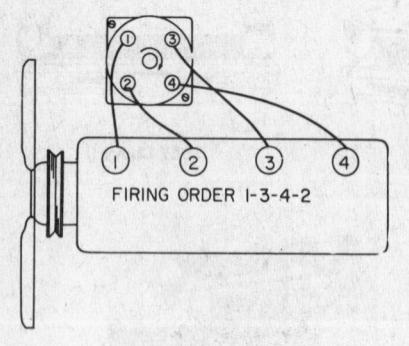

153 4 cyl.

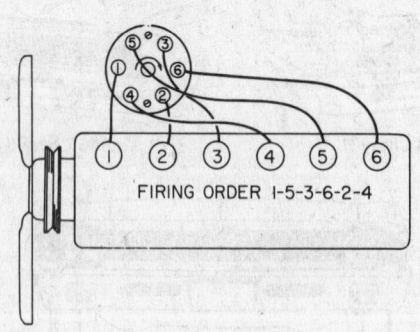

194, 230, 250 6 cyl.

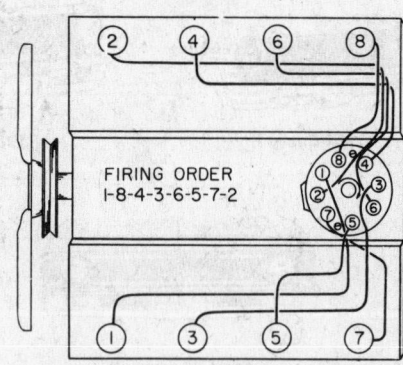

283, 327, 350, 396, 400, 454 V8

CAR SERIAL NUMBER LOCATION

1964-67
Car serial number is found on a plate attached to the left front door hinge pillar.

1968-71
Car serial number is located on the top left-hand side of the instrument panel, visible through the windshield.

Car Serial Number Interpretation
A typical vehicle serial number tag yields manufacturer's identity, vehicle type, model year, assembly plant and production unit number when broken down as shown in the following chart.

Mfr. Identity[1]	Body Style[2]	Model Year[3]	Assy. Plant[4]	Unit No.[5]
1	5645	8	F	100025

1. Manufacturer's identity number assigned to all Chevrolet built vehicles.
2. Model Identification
3. Last number of model year (1968).
4. F-Flint
5. Unit numbering will start at 100,001 at all plants.

MODEL IDENTIFICATION

1964—Model Identification—Chevy II

Series	Model Number		Description
	4 Cyl.	6 Cyl.	
Chevy II 100	111	211	2 dr. sedan, 6 passenger
	169	269	4 dr. sedan, 6 passenger
	135	235	4 dr. station wagon, 2 seat
Chevy II 300	311	411	2 dr. sedan, 6 passenger
	369	469	4 dr. sedan, 6 passenger
	345	445	4 dr. station wagon, 3 seat
Nova 400	—	449	4 dr. sedan, 6 passenger
	—	437★	2 dr. sport coupe, 5 passenger
	—	467★	2 dr. convertible, 5 passenger
	—	435	4 dr. station wagon 2 seat

★—Also available in Super Sport Models.

1965-67—Model Identification—Chevy II

Series	Model Number			Description
	4 Cyl.	6 Cyl.	V8	
100	11111	11311	11411	2 dr. sedan, 6 passenger
	11169	11369	11469	4 dr. sedan, 6 passenger
		11335	11435	4 dr. station wagon, 2 seat
Nova		11569	11669	4 dr. sedan, 6 passenger
		11537	11637	2 dr. sport coupe, 5 passenger
		11535	11635	4 dr. station wagon, 2 seat
Nova SS		11737	11837	2 dr. sport coupe, 4 passenger

1968-71—Model Identification—Chevy II

Series	Model Number			Description
	4 Cyl.	6 Cyl.	V8	
Nova	11127	11327	11427	2 dr. coupe, 5 passenger
	11169	11369	11469	4 dr. sedan, 6 passenger
Nova SS			11427	2 dr. coupe, 5 passenger

1964—Model Identification—Chevelle

Series	Model Number		Description
	6 Cyl.	V8	
Chevelle 300	5311	5411	2 dr. sedan, 6 passenger
	5369	5469	4 dr. sedan, 6 passenger
	5315	5415	2 dr. station wagon, 2 seat
	5335	5435	4 dr. station wagon, 2 seat
Malibu	5569	5669	4 dr. sedan, 6 passenger
	5537	5637	2 dr. sport coupe, 5 passenger
	5567	5667	2 dr. convertible, 5 passenger
	5535	5635	4 dr. station wagon, 2 seat
	5545	5645	4 dr. station wagon, 3 seat
Malibu SS	5737	5837	2 dr. sport coupe, 4 passenger
	5767	5867	2 dr. sedan convertible, 4 passenger
El Camino	5380	5480	2 dr. sedan pickup, 3 passenger reg.
	5580	5680	2 dr. sedan pickup, 3 passenger deluxe

1965–67—Model Identification—Chevelle

Series	Model Number		Description
	6 Cyl.	V8	
Chevelle 300	13111	13211	2 dr. sedan, 6 passenger
	13169	13269	4 dr. sedan, 6 passenger
	13115	13215	2 dr. station wagon, 2 seat
Chevelle 300 Deluxe	13311	13411	2 dr. sedan, 6 passenger
	13369	13469	4 dr. sedan, 6 passenger
	13335	13435	4 dr. station wagon, 2 seat
Malibu	13569	13669	4 dr. sedan, 6 passenger
	13537	13637	2 dr. sport coupe, 5 passenger
	13567	13667	2 dr. convertible, 5 passenger
	13535	13635	4 dr. station wagon, 2 seat
Malibu SS	13737	13837	2 dr. sport coupe, 4 passenger
	13767	13867	2 dr. convertible, 4 passenger
El Camino	13380	13480	2 dr. sedan pickup, 3 passenger reg.
	13580	13680	2 dr. sedan pickup, 3 passenger deluxe

1968—Model Identification—Chevelle

Series	Model Number		Description
	6 Cyl.	V8	
Chevelle 300	13127	13227	2 dr. coupe, 5 passenger
Chevelle 300 Deluxe	13327	13427	2 dr. coupe, 5 passenger
	13337	13437	2 dr. sport coupe, 5 passenger
	13369	13469	4 dr. sedan, 6 passenger
Malibu	13535	13635	4 dr. station wagon, 2 seat
	13537	13637	2 dr. sport coupe, 5 passenger
	13539	13639	4 dr. sport sedan, 6 passenger
	13567	13667	2 dr. convertible, 5 passenger
	13569	13669	4 dr. sedan, 6 passenger
Nomad	13135	13235	4 dr. station wagon, 2 seat
Nomad Custom	13335	13435	4 dr. station wagon, 2 seat
Concours Estate Wagon	13735	13835	4 dr. station wagon, 2 seat
SS 396		13837	2 dr. sport coupe, 5 passenger
		13867	2 dr. convertible, 5 passenger
El Camino	13380	13480	2 dr. sedan pickup, 3 passenger
	13580	13680	2 dr. sedan pickup, 3 passenger
		13880	2 dr. sedan pickup, 3 passenger

1969–71—Model Identification—Chevelle

Series	Model Number		Description
	6 Cyl.	V8	
300 Deluxe	*13327	*13427	2 dr. coupe, 6 passenger
	*13337	*13437	2 dr. sport coupe, 5 passenger
	*13369	*13469	4 dr. sedan, 6 passenger
Malibu	*13537	*13637	2 dr. sport coupe, 5 passenger
	*13539	*13639	4 dr. sport sedan, 6 passenger
	*13567	*13667	2 dr. convertible, 5 passenger
	*13569	*13669	4 dr. sedan, 6 passenger
Station Wagons	13135①	13235②	Nomad, 4 dr. 2 seat
	13335③	13435④	Greenbrier, 4 dr. 2 seat
	13346	13446	Greenbrier, 4 dr. 3 seat
	*13536	*13636	Concours, 4 dr. 2 seat
	13546	13646	Concours, 4 dr. 3 seat
	—	*13836	Concours Estate, 4 dr. 2 seat
	—	*13846	Concours Estate, 4 dr. 2 seat
El Camino	*13380	*13480	2 dr. sedan pickup, 3 passenger
	*13580	*13680	2 dr. sedan pickup, 3 passenger.

*—Applicable 1970 models.
① 13136—1970.　③ 13336—1970.
② 13236—1970.　④ 13436—1970.

1967–69—Model Identification—Camaro

Series	Model Number		Description
	6 Cyl.	V8	
Camaro	12337	12437	2 dr. sports coupe, 4 passenger
Camaro	12367	12467	2 dr. convertible, 4 passenger

1970–71—Model Identification—Camaro

Series	Model Number		Description
	6 Cyl.	V8	
Camaro	12387	12487	2 dr. sports coupe, 4 passenger

TORQUE SPECIFICATIONS

YEAR	MODEL	CYLINDER HEAD BOLTS (FT. LBS.)	ROD BEARING BOLTS (FT. LBS.)	MAIN BEARING BOLTS (FT. LBS.)	CRANKSHAFT BALANCER BOLT (FT. LBS.)	FLYWHEEL TO CRANKSHAFT BOLTS (FT. LBS.)	MANIFOLD (FT. LBS.)	
							Intake	Exhaust
1964–71	4 & 6 Cyl.	90–95	35–40	60–70	Press Fit	50–65	25–30	15–20
1964–71	V8—Excl. 400, E454	60–70	30–35	60–80	Press Fit	55–65	25–35	25–35
1966–71	V8—396(402)	60–70	35–45	95–105	Press Fit	55–65	25–35	25–35
1970–71	V8—400(402)	60–80	38–48	80–120①	100	85–95	28	22
1970–71	V8—454	60–80	38–48	80–120①	100	85–95	28	22

① — No. 5—120–160

CYLINDER HEAD BOLT TIGHTENING SEQUENCE

4 cylinder

6 cylinder

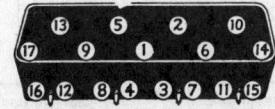

283, 302 (Z28), 307, 327, 350, 400 (small block) V8s

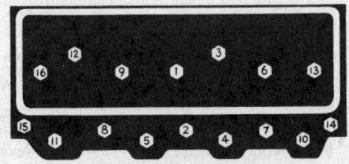

396, 400 (big block 402), 427, 454 V8s

ENGINE IDENTIFICATION

Chevelle Engine Identification Code Location

Engine identification code letter follows immediately after engine serial number.
6 Cyl.—pad at front right-hand side of cylinder block at rear of distributor.
V8—Pad at front right-hand side of cylinder block.

No. Cyls.	Cu. In. Displ.	Type	1964	1965	1966	1967	1968	1969	1970
6	194	M.T.	GF, G	AA	AA				
6	194	HDC	GG, GB	AC	AC				
6	194	AC	GM, GK	AG	AG				
6	194	HDC, AC	GN, GL	AH	AH				
6	194	Taxi	GH, GK	AK					
6	194	PG	K, KB	AL	AL				
6	194	PG, Taxi	KC, KD	AN					
6	194	PG, AC	KJ, KH	AR	AR				
6	194	w/ex. EM				AS			
6	194	w/ex. EM, AC				AT			
6	194	PG, w/ex. EM				AX			
6	194	PG, w/ex. EM, AC				AY			
6	230	HDC				BC	BC	BC	
6	230	HDC, AC				BB	BB	BB	
6	230	PG		BE			BF	BF	
6	230	PG, w/ex. EM		BN		BL		AN	
6	230	Hyd., AC						AR	
6	230	PG, w/ex. EM, AC				BM			
6	230	Hyd.						AD	
6	230	w/ex. EM				BN			
6	230	w/ex. EM, AC				BO			
6	230	PG, PCV, AC	BP						
6	230	M.T.			CA	CA	CA	BA	AM
6	230	M.T.	LM		CB	CB			
6	230	3 Spd. AC	LN, LL				CB		AP
6	230	PG			CC	CC	CC		
6	230	PG, AC	BM		CD	CD	CD	BH	AQ
6	250	3 Spd. or OD				CM	CM	BE	CCL
6	250	3 Spd. AC				CN	CN	BF	
6	250	3 Spd. or OD w/ex. EM				CO			
6	250	3 Spd. AC w/ex. EM					CP		
6	250	PG				CQ	CQ	BB	CCM
6	250	PG, AC				CR	CR	BC	
6	250	Hyd.						BD	CCK
6	250	PG, w/ex. EM					CS		
6	250	Hyd., AC						BH	
6	250	PG, AC w/ex. EM					CT		
8	283	3 Spd.	J	DA	DA	DA			
8	283	4 Spd.	JA	DB	DB	DB			
8	283	PG	JD	DE	DF	DE			
8	283	3 Spd., 4 Bbl.	JH	DG	DG				
8	283	PG, 4 Bbl.	JG	DH	DH				
8	283	w/ex. EM				DI	DI		
8	283	PG, w/ex. EM				DJ	DJ		
8	283	4 Spd., w/ex. EM				DK	DK		
8	283	4 Bbl., w/ex. EM				DL			
8	283	PG, 4 Bbl., w/ex. EM				DM			
8	283	HDC					DN		
8	307	Hyd.						DD	CNF
8	307	M.T.					DA	DA	CNC
8	307	4 Spd.					DE	DE	CND
8	307	PG					DB	DC	CNE
8	307	HDC					DN		
8	327	M.T.	JQ	EA	EA	EA	EA		
8	327	HP	JR	EB					
8	327	w/ex. EM			EB	EB			
8	327	SHP	JS	EC				ES	
8	327	PG, w/ex. EM			EC	EC			

No. Cyls.	Cu. In. Displ.	Type	1964	1965	1966	1967	1968	1969	1970
8	327	w/T. Ign.	JT	ED					
8	327	3 or 4 Spd. (325 H.P.)				EP			
8	327	HDC, 3 or 4 Spd. w/ex. EM (325 H.P.)				ER			
8	327	HDC (325 H.P.)				ES	ES		
8	327	HDC (275 H.P.)				ED	ED		
8	327	PG	SR	EE	EE	EE	EE		
8	327	PG, HP	SS	EF					
8	350	M.T.						HA	
8	350	Hyd.						HB	
8	350	2-BBL.						HC	
8	350	2-BBL., Hyd.						HD	
8	350	PG						HE	CNM(250)
8	350	PG, 2-BBL.						HF	
8	350	M.T.						HP	CNI(250)
8	350	M.T.						HR	CNJ(300)
8	350	PG						HR	CNK(300)
8	350	Hyd.						HS	CRE(300)
8	396	HDC			ED	ED	ED	ED	
8	396	HP			EF	EF	EF	JC	
8	396	SHP					EG	JD	
8	396	w/ex. EM				EH	EH		
8	396	HP, w/ex. EM				EJ	EJ		
8	396	PG				EK	EK	EK	EK
8	396	PG, HP				EL	EL	EL	EL
8	396	PG, w/ex. EM				EM	EM		
8	396	PG, HP, w/ex. EM				EN	EN		
8	396	Hyd. (325 H.P.)				ET	ET	ET	
8	396	Hyd. (350 H.P.)				EU	EU	EU	
8	396	w/ex. EM (325 H.P.)				EV			
8	396	w/ex. EM (350 H.P.)				EW			
8	396	M.T.						JA	CZX(265), CTX(350), CKT(375), CKO(375)
8	396	HP, 3-sp. Hyd. 400						JE	
8	396	Hyd. 400						JK	CTW(350)
8	396	SHP, Hyd. 400. (#—CKP only)						KF	CTY(375), CKP(375), CKU(375)
8	396	M.T.						KG	
8	396	Hyd. 400						KH	CKN(325)
8	396	M.T., HP						KB	
8	396	M.T.						JV	
8	396	SHP, M.T.						KD	
8	396	M.T.						KI	
8	396	M.T., HDC							CTZ(350), CKQ(375)
8	400	M.T. (330 H.P.)							CKR
8	400	M.T., HDC (330 H.P.)							CKS
8	454	M.T. (390 H.P.)							CRN, CRT
8	454	Hyd. 400 (390 H.P.)							CRQ
8	454	Hyd. 400 (450 H.P.)							CRR
8	454	Hyd. 400 #(450 H.P.)							CRS
8	454	M.T. (450 H.P.)							CRV

AC—air conditioned.
HDC—heavy duty clutch.
HP—high performance.
SHP—special high performance.
M.T.—manual transmission.
OD—overdrive.
PG—powerglide transmission.
PCV—positive crankcase ventilation.
w/ex. EM—with exhaust emission.
w/T. Ign.—with transistor ignition.
4 Bbl.—four barrel carburetor.
Hyd.—Hydramatic.
#—Aluminum heads.

Chevy II
Engine Identification Code, Location

4-6 Cyl.—Pad at front right-hand side of cylinder block at rear of distributor.
V8—pad at front right-hand side of cylinder block.

No. Cyls.	Cu. In. Displ.	Type	1964	1965	1966	1967	1968	1969	1970
4	153	M.T.	E	OA	OA		OA	AA	CCA
4	153	M.T., HDC	EB	OC	OC	OC	OC		
4	153	PG, Torque Dr.	EG	OH	OH	OH	OH	AB	CCB
4	153	M.T., Taxi	EK	OG					
4	153	PG, Taxi	EK	OG					
4	153	M.T., PCV	EP						
4	153	PG, PCV	EQ						
4	153	M.T., HDC, PCV	ER						
4	153	M.T., PCV, Taxi	ES						
4	153	PG, PCV, Taxi	ET						
6	194	M.T.	H	OK	OK				
6	194	M.T., HDC	HB	OM	OM	OM			
6	194	PG	HF	OR	OR	OR			
6	194	M.T., Taxi	HL	OQ					
6	194	PG, Taxi	HM	OT					
6	194	M.T., PCV, Taxi	HT						
6	194	PG, PCV, Taxi	HU		OS				
6	194	PG, w/ex. EM			ZX	ZX			
6	194	w/ex. EM			ZY	ZY			
6	230	PG	BT	PX	PX		BF		
6	230	PG, PCV	BU						
6	230	w/ex. EM			PC				
6	230	PG, w/ex. EM			PG			AN	
6	230	HDC, AC					BB		
6	230	HDC					BC		
6	230	PG, AC					BH	AQ	
6	230	Hyd. 350						AO	
6	230	AC						AP	
6	230	Hyd. 350, AC						AR	
6	230	M.T.	LR	LP	PV	PV		AM	
6	230	Torque Dr.						AN	CCD
6	250	M.T., or OD				PC, PV	CB	BE	CCG, CCI, CRF, CCL
6	250	PG, w/ex. EM				PI, PX	CQ	BB	CCM
6	250	PG, AC						BC	
6	250	Hyd. 350						BD	CCK
6	250	AC						BF	
6	250	Hyd. 350, AC						BH	
6	250	Torque Dr.						BB, BC	
8	283	M.T., 4-spd.	CF	PL	PL	PL			
8	283	4-spd., AC	CG	PM	PM	PM			
8	283	M.T., 3-spd.	CH	PD	PD	PD			
8	283	3-spd., AC	CJ						
8	283	PG	DE	PN	PN	PN			
8	283	PG, AC	DF	PP	PP	PP			
8	283	PG, 4-BBL., AC		PB					
8	283	4-BBL.		PE	QA				
8	283	w/ex. EM			PE	PE			
8	283	AC			PF	PF			
8	283	4-BBL., AC			PG	QB			
8	283	w/ex. EM, AC			PG				
8	283	PG, 4-BBL.		PK	PK				
8	283	PG, AC, w/ex. EM			PO				
8	283	4-spd., w/ex. EM			PQ	PQ			

No. Cyls.	Cu. In. Displ.	Type	1964	1965	1966	1967	1968	1969	1970
8	283	4-spd., AC, EM			PS				
8	283	PG, w/ex. EM			PU	PU			
8	283	4-BBL., EM			QC				
8	283	4-BBL., PG, EM			QD				
8	283	4-BBL., AC, EM			QE				
8	283	4-BBL., AC, PG, w/ex. EM			QF				
8	307	M.T.					MB	DA	CNC
8	307	SHP					ML		
8	307	PG					MM	DC	CNE
8	307	Hyd. 350						DD	CNF
8	307	4-spd.						DE	CND
8	327	M.T.		ZA	ZA	ZA	MK		
8	327	M.T., HP		ZB					
8	327	w/ex. EM			ZB	ZB			
8	327	w/ex. EM, AC			ZC				
8	327	w/ex. EM, PG			ZD	ZD			
8	327	M.T., AC		ZE	ZE	ZE			
8	327	M.T., AC, HP		ZF					
8	327	PG, ex. EM, AC			ZF				
8	327	SHP, w/ex. EM			ZG				
8	327	SHP, w/ex. EM, AC			ZH				
8	327	SHP			ZI		ML		
8	327	SHP, AC			ZJ				
8	327	PG		ZK	ZK	ZK	MM		
8	327	PG, HP		ZL					
8	327	PG, AC		ZM	ZM	ZM			
8	327	PG, HP, AC		ZN					
8	350	M.T.						HA, HQ	CNI(250), CNJ(300)
8	350	Hyd.						HB, HD	CNN(250), CRE(300)
8	350	2-BBL.						HC	
8	350	PG						HE, HR	CNK, CNM
8	350	PG, 2-BBL.						HF	
8	350	Hyd.						HS	
8	396	HP						JF	
8	396	SHP						JH, KA, KC	
8	396	HP, Hyd. 400						JI	
8	396	SHP. Hyd. 400						JL	
8	396	Hyd. 400						JM	CTW(350), CTY(375), CKN(325)
8	396	PG						JU	
8	396	M.T.						KE	CTX(350), CKO(375)
8	396	Hyd. 400#							CKP(375)
8	396	M.T., HDC							CTZ(350), CKQ(375)
8	396	M.T., HDC#							CKU(375)
8	396	M.T.#							CKT(375)
8	400	M.T.							CKR(330)
8	400	M.T., HDC							CKS(300)

AC—air conditioned.
HDC—heavy duty clutch.
HP—high performance engine.
SHP—special high performance engine.
PCV—positive crankcase ventilation.

OD—overdrive
w/ex. EM, or EM—with exhaust emission.
M.T.—manual transmission.
PG—Powerglide transmission.

3 Spd.—three speed transmission.
4 Spd.—four speed transmission.
Hyd.—Hydramatic
#—Aluminum heads

Camaro

Engine Identification Code Location

Engine identification code letter follows immediately after engine seal number.
6 Cyl.—pad at front right-hand side of cylinder block at rear of distributor.
V8—pad at front right-hand side of cylinder block.

No. Cyls.	Cu. In. Displ.	Type	YEAR AND CODE 1967	1968	1969	1970	No. Cyls.	Cu. In. Displ.	Type	YEAR AND CODE 1967	1968	1969	1970
6	230	3 or 4 Spd.	LA	BA	AM	CCC	8	327	3 or 4 Spd. (275 H.P.)	MK			
6	230	3 or 4 Spd. AC	LB	BB			8	327	3 or 4 Spd. w/ex. EM	ML			
6	230	3 or 4 Spd. w/ex. EM	LC				8	327	PG (275 H.P.)	MM			
6	230	3 or 4 Spd. AC,					8	327	PG, w/ex. EM	MN			
		w/ex. EM	LD				8	350	3 or 4 Spd.	MS	MS	HA,HQ	CNJ(300)
6	230	PG, Torque Dr.	LE	BF	AN	CCD	8	350	3 or 4 Spd. w/ex. EM	MT			
6	230	PG, AC	LH, LF	BH	AQ		8	350	PG	MU	MU	HE,HR	CNK(300)
6	230	Hyd. 350			AO		8	350	PG, w/ex. EM	MV			
6	230	AC			AP		8	350	Hyd.			HB,HS	CRE(300)
6	230	Hyd. 350, AC			AR		8	350	2-BBL.			HC	CNI(250)
6	230	PG, w/ex. EM	LG				8	350	Hyd., 2-BBL.			HD	CNM(250)
6	250	3 or 4 Spd.	LN	CM	BE		8	350	PG, 2-BBL.			HF	CNN(250)
6	250	3 or 4 Spd. AC	LO	CN			8	350	370 H.P. (new Z28)				N.A.
6	250	PG, w/ex. EM	LP				8	396	M.T. & P.G.	MW	MW		
6	250	3 or 4 Spd. w/ex. EM	LQ				8	396	Mt. & P.G. w/exh. EM.	MX			
6	250	AC			BF		8	396	Hyd. 400	MY	MY	JG	CJI(350)
6	250	PG, Torque Dr.	FM	CQ	BB		8	396	Hyd. w/exh. EM.	MZ			
6	250	PG, AC	FR	CR	BC		8	396	SHP		MQ	JH	
6	250	PG, w/ex. EM	GP				8	396	HP, Hyd.		MR	JI	
6	250	PG, AC, w/ex. EM	GQ		BC		8	396	SHP, ALUM. HEADS		MT		
6	250	Hyd. 350			BD		8	396	HP		MX	JF	
6	250	Hyd. 350, AC			BH		8	396	PG			JB	
8	302	Z28		DZ	DZ*		8	396	M.T., ALUM. HEADS			JJ,KE	
8	307	M.T.			DA	CNC	8	396	SHP, Hyd. 400			JL	CJL(375)
8	307	P.G.			DC	CNE	8	396	Hyd. 400, ALUM. HEADS			JM	
8	307	Hyd. 350			DD	CNF	8	396	M.T.,			JU	CJF(350)
8	307	4-spd.			DE	CND	8	396	M.T., SHP			KA,KC	CJH(375)
8	327	3 or 4-spd. (210)	MA	MA			8	400	Hyd. 400				CTW(350),
8	327	3 or 4-spd. w/ex. EM	MB										CKN(325),
8	327	PG, (210 H.P.)	ME	ME									CTY(375)
8	327	PG, w/ex. EM	MF				8	400	M.T.				CTX(350),
													CKO(375)

AC—air conditioned.
HDC—heavy duty clutch.
HP—high performance.
SHP—special high performance.
M.T.—manual transmission.
PG—Powerglide transmission.
w/ex. EM—with exhaust emission.
4 BBL.—four barrel carburetor.
2 BBL.—two barrel carburetor.
Hyd.—Hydromatic transmission (350 or 400).
*—CNA = late production.

WHEEL ALIGNMENT

YEAR	MODEL	CASTER Range (Deg.)	CASTER Pref. Setting (Deg.)	CAMBER Range (Deg.)	CAMBER Pref. Setting (Deg.)	TOE-IN (In.)	KING-PIN INCLINATION (Deg.)	WHEEL PIVOT RATIO Inner Wheel	WHEEL PIVOT RATIO Outer Wheel
1964	Chevy II	½P to 1½P	1P	0 to 1P	½P	¼ to ⅜	7	20	18¾
	Chevelle	¼N to ¾P	¼P	¼P to 1¼P	¾P	0 to ⅛	8¼	20	18¾
1965	Chevy II	½P to 1½P	1P	0 to 1P	½P	¼ to ⅜	7	20	18¾
	Chevelle	1½N to ½N	1N	¼N to ¾P	¼P	1/16 to 3/16	8	20	18¾
1966	Chevy II	½P to 1½P	1P	0 to 1P	½P	¼ to ⅜	7	20	18¾
	Chevelle	1½N to ½N	1N	0 to 1P	½P	⅛ to ¼	8	20	18¾
1967	Chevy II	½P to 1½P	1P	0 to 1P	½P	⅛ to ¼	7	20	18¾
	Chevelle	1½N to ½N①	1N	0 to 1P	½P	⅛ to ¼	8¼	20	18¾
	Camaro	0 to 1P	½P	¼N to ¾P	¼P	⅛ to ¼	8¾	20	18¾
1968–69	Chevy II	0 to 1P	½P	¼N to ¾P	½P	⅛ to ¼	8¾	20	N.A.
	Chevelle	1½N to ½N①	1N	0 to 1P	½P	⅛ to ¼	8¼	20	18½
	Camaro	0 to 1P	½P	¼N to ¾P	½P	⅛ to ¼	8¾	20	N.A.
1970–71	Chevy II	0 to 1P	½P	¼N to ¾P	½P	⅛ to ¼	8¼–9¼	20	N.A.
	Chevelle	1½N to ½N①	1N	0 to 1P	½P	⅛ to ¼	7¾–8¾	20	N.A.
	Camaro	0 to 2P	1P	¼N to 1¾P	¾P	⅛ to ¼	10-11	20	N.A.

①—SS396 & El Camino—0 to 1P
N—Negative
P—Positive

GENERAL CHASSIS AND BRAKE SPECIFICATIONS

YEAR	MODEL	CHASSIS		BRAKE CYLINDER SIZES				BRAKE DRUM	
		Overall Length (In.)	Tire Size	Master Cylinder (In.)		Wheel Cylinder—(In.)		Diameter (In.)	
				Standard	Metallic	Front	Rear	Front	Rear
1964	Chevy II, 4 & 6 Cyl.	182.9	6.00 x 13	1.0	.875	1.06	.875	9.5	9.5
	V8	182.9	6.50 x 14	1.0	.875	1.06	.875	9.5	9.5
	4 & 6 Cyl. Sta. Wag.	187.6	6.50 x 13	1.0	.875	1.06	.875	9.5	9.5
	V8 Sta. Wag.	187.6	6.50 x 14	1.0	.875	1.06	.875	9.5	9.5
	Chevelle, Exc. Wag.	193.9	6.50 x 14	1.0	.875	1.06	.875	9.5	9.5
	Station Wagon	198.8	7.00 x 14	1.0	.875	1.06	.875	9.5	9.5
1965	Chevy II, 4 Cyl.	182.9	6.00 x 13	1.0	.875	1.06	.875	9.5	9.5
	Chevy II, 6 Cyl.	182.9	6.50 x 13	1.0	.875	1.06	.875	9.5	9.5
	Chevy II, V8	182.9	6.95 x 14	1.0	.875	1.06	.875	9.5	9.5
	Chevy II, Sta. Wag.	182.9	7.00 x 13	1.0	.875	1.06	.875	9.5	9.5
	Chevelle, Exc. Wag.	196.6	6.95 x 14①	1.0	.875	1.06	.875	9.5	9.5
	Chevelle, Sta. Wag.	201.4	7.35 x 14①	1.0	.875	1.06	.875	9.5	9.5
1966	Chevy II, 4 Cyl.	183.0	6.50 x 13	1.0	.875	1.06	.875	9.5	9.5
	Chevy II, 6 Cyl.	183.0	6.50 x 13	1.0	.875	1.06	.875	9.5	9.5
	Chevy II, V8	183.0	6.95 x 14	1.0	.875	1.06	.875	9.5	9.5
	Chevy II, Wag.	187.4	6.95 x 14	1.0	.875	1.06	.875	9.5	9.5
	Chevelle, Exc. Wag.	197.0	6.95 x 14①	1.0	.875	1.06	.875	9.5	9.5
	Chevelle Wag.	197.6	7.75 x 14	1.0	.875	1.06	.875	9.5	9.5
1967	Chevy II, 4 Cyl. & 6 Cyl.	183.0	6.95 x 14	1.0	.875	1.06	.875	9.5	9.5
	Chevy II, V8	183.0	6.95 x 14	.875	.875	1.06	.875	9.5	9.5
	Chevy II, Wagon	187.4	6.95 x 14	1.0	.875	1.875	.875	9.5	9.5
	Chevelle, exc. Wag.	197.0	7.35 x 14②	1.0	.875	1.06	.938	9.5	9.5
	Chevelle, Wagon	199.9	7.75 x 14	1.125	.875	2.062	.938	9.5	9.5
	Camaro	184.6	7.35 x 14③	1.0	.875	1.125④	.875	9.5⑤	9.5
1968	Chevy II	187.7	7.35 x 14	1.00	.875	1.125	.875	9.5	9.5
	Chevy II w/Disc. Brakes	187.7	7.35 x 14	1.125	.875	2.062	.875	11 (Disc)	9.5
	Camaro	184.7	7.35 x 14	1.00	.875	1.125	.875	9.5	9.5
	Camaro w/Disc. Brakes	184.7	7.35 x 14	1.125	.875	2.062	.875	11 (Disc)	9.5
	Chevelle Coupes	196.8	7.35 x 14②	1.00	.875	1.125	.938	9.5	9.5
	Chevelle Sedans	200.8	7.35 x 14	1.00	.875	1.125	.938	9.5	9.5
	Chevelle Sta. Wag.	207.2	7.75 x 14	1.00	.875	1.125	.938	9.5	9.5
	Chevelle w/Disc. Brakes	—	—	1.125	—	2.062	.875	11 (Disc)	9.5
1969	Chevy II	189.4	7.35 x 14	1.0	—	1.125	.875	9.5	9.5
	Camaro	186.0	7.35 x 14	1.0	—	1.125	.875	9.5	9.5
	Chevelle Sedans	200.9	7.35 x 14	1.0	—	1.125	.875	9.5	9.5
	Chevelle, Coupe & Conv.	196.9	7.35 x 14	1.0	—	1.125	.875	9.5	9.5
	Chevelle Sta. Wag.	207.9	7.75 x 14	1.0	—	1.125	.875	9.5	9.5
	All w/Disc. Brakes	—	—	1.125	—	2.938	.875	11 (Disc)	9.5
1970–71	Chevy II	189.4	7.35 x 14	1.0	—	1.125	.875	9.5	9.5
	Camaro	188.0	7.35 x 14	1.125	—	2.938	.875	11 (Disc)	9.5
	Chevelle Sedans	201.2	7.35 x 14	1.0	—	1.125	.875	9.5	9.5
	Chevelle, Coupe & Conv.	197.2	7.35 x 14	1.0	—	1.125	.875	9.5	9.5
	Chevelle Sta. Wag.	206.5	7.75 x 14	1.0	—	1.125	.875	9.5	9.5
	All w/Disc. Brakes	—	—	1.125	—	2.938	.875	11 (Disc)	9.5

①—Models with 327 Engine—7.35 x 14.
With 396 Eng.—7.75 x 14.

②—396 cu. in. F70 x 14

③—Camaro with 350 cu. in. option—D70 x 14

④—Camaro with 350 cu. in. option 1⅛

⑤—Camaro with front disc option = 11.0

GENERAL ENGINE SPECIFICATIONS

YEAR	CU. IN. DISPLACEMENT	CARBURETOR	DEVELOPED HORSEPOWER @ RPM	DEVELOPED TORQUE @ RPM (FT. LBS.)	A.M.A. HORSEPOWER	BORE & STROKE (IN.)	ADVERTIZED COMPRESSION RATIO	VALVE LIFTER TYPE	NORMAL OIL PRESSURE (PSI)
1964	4 Cyl—153	1-BBL.	90 @ 4000	152 @ 2400	24.0	3.875 x 3.25	8.5-1	Hyd.	35
	6 Cyl—194	1-BBL.	120 @ 4400	177 @ 2400	30.5	3.563 x 3.25	8.5-1	Hyd.	35
	6 Cyl—230	1-BBL.	155 @ 4400	215 @ 2000	36.0	3.875 x 3.25	8.5-1	Hyd.	35
	V8—283	2-BBL.	195 @ 4800	285 @ 2400	48.0	3.875 x 3.00	9.25-1	Hyd.	35
	V8—283	4-BBL.	220 @ 4800	295 @ 3200	48.0	3.875 x 3.00	9.25-1	Hyd.	35
1965	4 Cyl—153	1-BBL.	90 @ 4000	152 @ 2400	24.0	3.875 x 3.25	8.5-1	Hyd.	35
	6 Cyl—194	1-BBL.	120 @ 4400	177 @ 2400	30.5	3.563 x 3.25	8.5-1	Hyd.	35
	6 Cyl—230	1-BBL.	140 @ 4400	215 @ 2000	36.0	3.875 x 3.25	8.5-1	Hyd.	35
	V8—283	2-BBL.	195 @ 4800	285 @ 2400	48.0	3.875 x 3.00	9.25-1	Hyd.	35
	V8—327	4-BBL.	250 @ 4400	350 @ 2800	51.2	4.000 x 3.25	10.5-1	Hyd.	35
	V8—327	4-BBL.	300 @ 5000	360 @ 3200	51.2	4.000 x 3.25	10.5-1	Hyd.	35
	V8—327	4-BBL.	350 @ 6000	360 @ 3200	51.2	4.000 x 3.25	11.0-1	Hyd.	35
1966	4 Cyl—153	1-BBL.	90 @ 4000	152 @ 2400	24.0	3.875 x 3.25	8.5-1	Hyd.	35
	6 Cyl—194	1-BBL.	120 @ 4400	177 @ 2400	30.5	3.563 x 3.25	8.5-1	Hyd.	35
	6 Cyl—230	1-BBL.	140 @ 4400	220 @1600	36.0	3.875 x 3.25	8.5-1	Hyd.	35
	V8—283	2-BBL.	195 @ 4800	285 @ 2400	48.0	3.875 x 3.00	9.25-1	Hyd.	45
	V8—283	4-BBL.	220 @ 4800	295 @ 3200	48.0	3.875 x 3.00	9.25-1	Hyd.	45
	V8—327	4-BBL.	275 @ 4800	355 @ 3200	51.2	4.000 x 3.25	10.5-1	Hyd.	45
	V8—327	4-BBL.	350 @ 5800	360 @ 3600	51.2	4.000 x 3.25	11.0-1	Hyd.	45
	V8—396	4-BBL.	325 @ 4800	410 @ 3200	53.6	4.094 x 3.76	10.25-1	Hyd.	60
	V8—396	4-BBL.	360 @ 5200	420 @ 3600	53.6	4.094 x 3.76	10.25-1	Hyd.	60
1967	4 Cyl—153	1-BBL.	90 @ 4000	152 @ 2400	24.0	3.875 x 3.25	8.5-1	Hyd.	35
	6 Cyl—194	1-BBL.	120 @ 4400	177 @ 2400	30.5	3.563 x 3.25	8.5-1	Hyd.	35
	6 Cyl—230	1-BBL.	140 @ 4400	220 @ 1600	36.0	3.875 x 3.25	8.5-1	Hyd.	35
	6 Cyl—250	1-BBL.	155 @ 4200	235 @ 1600	36.0	3.875 x 3.53	8.5-1	Hyd.	35
	V8—283	2-BBL.	195 @ 4800	285 @ 2800	48.0	3.875 x 3.00	9.25-1	Hyd.	45
	V8—327	4-BBL.	210 @ 4800	325 @ 3200	51.2	4.000 x 3.25	9.0-1	Hyd.	45
	V8—327	4-BBL.	275 @ 4800	355 @ 3200	51.2	4.000 x 3.25	10.25-1	Hyd.	45
	V8—327	4-BBL.	325 @ 4800	360 @ 3600	51.2	4.000 x 3.25	11.0-1	Hyd.	45
	V8—350	4-BBL.	295 @ 4800	380 @ 3200	51.2	4.000 x 3.48	10.25-1	Hyd.	45
	V8—396	4-BBL.	325 @ 4800	410 @ 3200	53.6	4.094 x 3.76	10.25-1	Hyd.	45
	V8—396	4-BBL.	350 @ 5200	420 @ 3600	53.6	4.094 x 3.76	10.25-1	Hyd.	45
1968	4 Cyl—153	1-BBL.	90 @ 4000	152 @ 2400	24.0	3.875 x 3.25	8.5-1	Hyd.	35
	6 Cyl—230	1-BBL.	140 @ 4400	220 @ 1600	36.0	3.875 x 3.25	8.5-1	Hyd.	35
	6 Cyl—250	1-BBL.	155 @ 4200	235 @ 1600	36.0	3.875 x 3.53	8.5-1	Hyd.	35
	V8—302	4-BBL.	290 @ 5800	290 @ 4200	51.2	4.000 x 3.00	11.0-1	Mech.	45
	V8—307	2-BBL.	200 @ 4600	300 @ 2400	48.0	3.875 x 3.25	9.0-1	Hyd.	45
	V8—327	2-BBL.	210 @ 4600	320 @ 2400	51.2	4.000 x 3.25	8.75-1	Hyd.	45
	V8—327	4-BBL.	275 @ 4800	355 @ 3200	51.2	4.000 x 3.25	10.0-1	Hyd.	45
	V8—327	4-BBL.	325 @ 5600	355 @ 3600	51.2	4.000 x 3.25	11.0-1	Hyd.	45
	V8—350	4-BBL.	295 @ 4800	380 @ 3200	51.2	4.000 x 3.48	10.25-1	Hyd.	45
	V8—396	4-BBL.	325 @ 4800	410 @ 3200	53.6	4.094 x 3.76	10.25-1	Hyd.	45
	V8—396	4-BBL.	350 @ 5200	415 @ 3400	53.6	4.094 x 3.76	10.25-1	Hyd.	45
	V8—396	4-BBL.	375 @ 5600	415 @ 3600	53.6	4.094 x 3.76	11.0-1	Mech.	45
1969	4 Cyl—153	1-BBL.	90 @ 4000	152 @ 2400	24.0	3.875 x 3.25	8.5-1	Hyd.	35
	6 Cyl—230	1-BBL.	140 @ 4400	220 @ 1600	36.0	3.875 x 3.25	8.5-1	Hyd.	35
	6 Cyl—250	1-BBL.	155 @ 4200	235 @ 1600	36.0	3.875 x 3.53	8.5-1	Hyd.	35
	V8—302	4-BBL.	290 @ 5800	290 @ 4200	51.2	4.000 x 3.00	11.0-1	Mech.	45
	V8—307	2-BBL.	200 @ 4600	300 @ 2400	48.0	3.875 x 3.25	9.0-1	Hyd.	45
	V8—327	2-BBL.	210 @ 4600	320 @ 2400	51.2	4.000 x 3.25	9.0-1	Hyd.	45
	V8—350	4-BBL.	255 @ 4800	365 @ 3200	51.2	4.000 x 3.48	9.0-1	Hyd.	45
	V8—350	4-BBL.	300 @ 5000	380 @ 3200	51.2	4.000 x 3.48	10.25-1	Hyd.	45
	V8—396	4-BBL.	325 @ 4800	410 @ 3200	53.6	4.094 x 3.76	10.25-1	Hyd.	45

GENERAL ENGINE SPECIFICATIONS, continued

YEAR	CU. IN. DISPLACEMENT	CARBURETOR	DEVELOPED HORSEPOWER @ RPM	DEVELOPED TORQUE @ RPM (FT. LBS.)	A.M.A. HORSEPOWER	BORE & STROKE (IN.)	ADVERTIZED COMPRESSION RATIO	VALVE LIFTER TYPE	NORMAL OIL PRESSURE (PSI)
	V8—396	4-BBL.	350 @ 5200	415 @ 3400	53.6	4.094 x 3.76	10.25-1	Hyd.	45
	V8—396	4-BBL.	375 @ 5600	415 @ 3600	53.6	4.094 x 3.76	11.0-1	Mech.	45
1970	4 Cyl—153	1-BBL.	90 @ 4000	152 @ 2400	24.0	3.875 x 3.250	8.5-1	Hyd.	30-45
	6 Cyl—230	1-BBL.	140 @ 4400	220 @ 1600	36.0	3.875 x 3.250	8.5-1	Hyd.	30-45
	6 Cyl—250	1-BBL.	155 @ 4200	235 @ 1600	36.0	3.875 x 3.530	8.5-1	Hyd.	30-45
	V8—307	2-BBL.	200 @ 4600	300 @ 2400	48.0	3.875 x 3.250	9.0-1	Hyd.	30-45
	V8—350	2-BBL.	250 @ 4800	345 @ 2800	51.2	4.001 x 3.480	9.0-1	Hyd.	30-45
	V8—350	4-BBL.	300 @ 4800	380 @ 3200	51.2	4.001 x 3.480	10.25-1	Hyd.	30-45
	V8—350	4-BBL.	360 @ 6000	380 @ 4000	51.2	4.001 x 3.480	11.0-1	Mech.	30-45
	V8—396(402)	4-BBL.	350 @ 5200	415 @ 3400	54.5	4.126 x 3.760	10.25-1	Hyd.	30-45
	V8—400(402)	4-BBL.	330 @ 4800	410 @ 3200	54.5	4.126 x 3.760	10.25-1	Hyd.	30-45
	V8—454	4-BBL.	360 @ 4400	500 @ 3200	57.8	4.251 x 4.000	10.25-1	Hyd.	30-45
1971	4 Cyl—153	1-BBL.	N.A.	N.A.	24.0	3.875 x 3.250	8.5-1	Hyd.	30-45
	6 Cyl—230	1-BBL.	N.A.	N.A.	36.0	3.875 x 3.250	8.5-1	Hyd.	30-45
	6 Cyl—250	1-BBL.	N.A.	N.A.	36.0	3.875 x 3.530	8.5-1	Hyd.	30-45
	V8—307	2-BBL.	N.A.	N.A.	48.0	3.875 x 3.250	9.0-1	Hyd.	30-45
	V8—350	2-BBL.	N.A.	N.A.	51.2	4.001 x 3.480	9.0-1	Hyd.	30-45
	V8—350	4-BBL.	N.A.	N.A.	51.2	4.001 x 3.480	9.25-1	Hyd.	30-45
	V8—350	4-BBL.	N.A.	N.A.	51.2	4.001 x 3.480	9.0-1	Mech.	30-45
	V8—396(402)	4-BBL.	N.A.	N.A.	54.5	4.126 x 3.760	9.25-1	Hyd.	30-45
	V8—400(402)	4-BBL.	N.A.	N.A.	54.5	4.126 x 3.760	9.25-1	Hyd.	30-45
	V8—454	4-BBL.	N.A.	N.A.	57.8	4.250 x 4.000	9.25-1	Hyd.	30-45

DELCOTRON AND AC REGULATOR SPECIFICATIONS

YEAR	MODEL	ALTERNATOR				REGULATOR					
		Field Current Draw @ 12V.	Output @ Generator RPM		Model	Field Relay			Regulator		
			2000	5000		Air Gap (In.)	Point Gap (In.)	Volts to Close	Air Gap (In.)	Point Gap (In.)	Volts at 125°
1964	1100668	1.9-2.3	5A	42A	1119515	.015	.030	2.3-3.7	.067	.014	13.5-14.4
	1100669	1.9-2.3	28A	40A	1119515	.015	.030	2.3-3.7	.067	.014	13.5-14.4
	1100670	1.9-2.3	5A	37A	1119515	.015	.030	2.3-3.7	.067	.014	13.5-14.4
	1117765	3.7-4.4	24A	62A	9000595	.015	.030	2.3-3.7	.067	.014	①
1965-67	1100693	2.2-2.6	27A	37A	1119515	.015	.030	2.3-3.7	.067	.014	13.5-14.4
	1100695	2.2-2.6	21A	32A	1119515	.015	.030	2.3-3.7	.067	.014	13.5-14.4
	1100794	2.2-2.6	27A	37A	1119515	.015	.030	2.3-3.7	.067	.014	13.5-14.4
1968	1100813	2.2-2.6	27A	37A	1119515	.015	.030	2.3-3.7	.067	.014	13.5-14.4
	1100693	2.2-2.6	27A	37A	1119515	.015	.030	2.3-3.7	.067	.014	13.5-14.4
1969	1100834	2.2-2.6	27A	37A	1119515	.015	.030	2.3-3.7	.067	.014	13.5-14.4
	1100836	2.2-2.6	27A	37A	1119515	.015	.030	2.3-3.7	.067	.014	13.5-14.4
1970	1100834	2.2-2.6	27A	37A	1119515	.015	.030	2.3-3.7	.067	.014	13.5-14.4
	1100837	2.2-2.6	27A	37A	1119515	.015	.030	2.3-3.7	.067	.014	13.5-14.4
1971	1100838	2.2-2.6	27A	37A	1119515	.015	.030	2.3-3.7	.067	.014	13.5-14.4
	1100839	2.2-2.6	27A	37A	1119515	.015	.030	2.3-3.7	.067	.014	13.5-14.4

①—13.0-13.6 @ 80°.

TUNE-UP SPECIFICATIONS

| YEAR | MODEL | SPARK PLUGS | | DISTRIBUTOR | | IGNITION TIMING (Deg.) ▲ | CRANKING COMP. PRESSURE (Psi) | VALVES Tappet (Hot) Clearance (In.) | | Intake Opens (Deg.) | FUEL PUMP PRESSURE (Psi) | IDLE SPEED (Rpm) ★ |
		Type	Gap (In.)	Point Dwell (Deg.)	Point Gap (In.)			Intake	Exhaust			
1964	4—153 Cu. In.	46N	.035	31°-34°	.019	4B	140	■	■	34B	3½-4½	575
	6—194 Cu. In.	46N	.035	31°-34°	.019	8B	140	■	■	34B	3½-4½	500
	6—230 Cu. In.	46N	.035	31°-34°	.019	4B	140	■	■	34B	5¼-6½	500
	V8—283 Cu. In.	45	.035	28°-32°	.019	4B	150	■	■	34B	5¼-6½	500
1965	4—153 Cu. In.	46N	.035	31°-34°	.019	4B	140	■	■	34B	3½-4½	575
	6—194 Cu. In.	46N	.035	31°-34°	.019	8B	140	■	■	34B	3½-4½	500
	6—230 Cu. In.	46N	.035	31°-34°	.019	4B	140	■	■	49B	5¼-6½	500
	V8—283 Cu. In., 2-BBL.	45	.035	28°-32°	.019	4B	150	■	■	34B	5¼-6½	500
	V8—327 Cu. In. (250 H.P.)	44	.035	28°-32°	.019	4B	160	■	■	32B	5¼-6½	550
	V8—327 Cu. In. (300 H.P.)	44	.035	28°-32°	.019	8B	160	■	■	32B	5¼-6½	550
	V8—327 Cu. In. (350 H.P.)	44	.035	28°-32°	.019	12B	160	■	■	35B	5¼-6½	800
1966	4 Cyl.—153 Cu. In.	46N	.035	31°-34°	.019	4B	140	■	■	33½B	3½-4½	575
	6 Cyl.—194 Cu. In.	46N	.035	31°-34°	.019	⊙	140	■	■	62B	3½-4½	500③
	6 Cyl.—230 Cu. In.	46N	.035	31°-34°	.019	4B	140	■	■	62B	3½-4½	500
	V8—283 Cu. In.	45	.035	28°-32°	.019	4B	150	■	■	32½B	5¼-6½	500
	V8—327 Cu. In. (275 H.P.)	44	.035	28°-32°	.019	⊙	160	■	■	32½B	5¼-6½	500③
	V8—327 Cu. In. (350 H.P.)	44	.035	28°-32°	.019	10B	160	■	■	54B	5¼-6½	700
	V8—396 Cu. In.	43N	.035	28°-32°	.019	4B	160	■	■	40B	5¼-6½	550
	V8—396 Cu. In. (360 H.P.)	43N	.035	28°-32°	.019	4B	150	■	■	58B	5¼-6½	550
1967	4 Cyl.—153 Cu. In. (90 H.P.)	46N	.035	31°-34°	.019	4B	130	■	■	33½B	3½-4½	500
	6 Cyl.—194 Cu. In. (120 H.P.)	45N	.035	31°-34°	.019	4B⊙	130	■	■	62B	3½-4½	500⊙
	6 Cyl.—230 Cu. In. (140 H.P.)	46N	.035	31°-34°	.019	4B⊙	130	■	■	62B	3½-4½	600⊙
	6 Cyl.—250 Cu. In. (155 H.P.)	46N	.035	31°-34°	.019	4B⊙	130	■	■	62B	3½-4½	500⊙
	V8—283 Cu. In. (195 H.P.)	45	.035	28°-32°	.019	4B⊙	150	■	■	36B	5¼-6½	550⊙
	V8—327 Cu. In. (210 H.P.)	44	.035	28°-32°	.019	4B⊙	160	■	■	36B	5¼-6½	600⊙
	V8—327 Cu. In. (275 H.P.)	44	.035	28°-32°	.019	8B⊙	160	■	■	36B	5¼-6½	600⊙
	V8—327 Cu. In. (325 H.P.)	44	.035	28°-32°	.019	10B⊙	160	■	■	54B	5¼-6½	600⊙
	V8—350 Cu. In. (295 H.P.)	44	.035	28°-32°	.019	4B⊙	160	■	■	36B	5¼-6½	600⊙
	V8—396 Cu. In. (325 H.P.)	43N	.035	28°-32°	.019	4B⊙	160	■	■	40B	5¼-6½	600⊙
	V8—396 Cu. In. (350 H.P.)	43N	.035	28°-32°	.019	4B	160	■	■	56B	5¼-6½	600
	V8—396 Cu. In. (375 H.P.)	43N	.035	28°-32°	.019	4B	160	.024	.028	N.A.	5-8½	750
1968	4 Cyl.—153 Cu. In. (90 H.P.)	46N	.035	31°-34°	.019	TDC①	130	■	■	17½B	3½-4½	750●
	6 Cyl.—230 Cu. In. (140 H.P.)	46N	.035	31°-34°	.019	TDC①	130	■	■	16B	3½-4½	700●
	6 Cyl.—250 Cu. In. (155 H.P.)	46N	.035	31°-34°	.019	TDC①	130	■	■	16B	3½-4½	700●
	V8—302 Cu. In. (290 H.P.)	43	.035	28°-32°	.019	4B	190	.030	.030	N.A.	5¼-6½	900
	V8—307 Cu. In. (200 H.P.)	45	.035	28°-32°	.019	2B	150	■	■	28B	5-6½	700●
	V8—327 Cu. In. (210 H.P.)	44	.035	28°-32°	.019	2A⑦	160	■	■	28B	5-6½	700●
	V8—327 Cu. In. (275 H.P.)	44	.035	28°-32°	.019	TDC①	160	■	■	28B	5-6½	700●
	V8—327 Cu. In. (325 H.P.)	44	.035	28°-32°	.019	4B	160	■	■	40B	5-6½	750
	V8—350 Cu. In. (295 H.P.)	44	.035	28°-32°	.019	TDC①	160	■	■	28B	5-6½	700●
	V8—396 Cu. In. (325 H.P.)	43N	.035	28°-32°	.019	4B	160	■	■	28B	5-6½	700●
	V8—396 Cu. In. (350 H.P.)	43N	.035	28°-32°	.019	TDC①	160	■	■	40B	7¼-8½	700●
	V8—396 Cu. In. (375 H.P.)	43N	.035	28°-32°	.019	4B	160	.024	.028	N.A.	5-8½	750
1969	4 Cyl. 153 Cu. In. (90 H.P.)	46N	.035	31°-34°	.019	TDC④	130	■	■	17½B	3½-4½	750⑦
	6 Cyl. 230 Cu. In. (140 H.P.)	46N	.035	31°-34°	.019	TDC④	130	■	■	16B	3½-4½	700⑥
	6 Cyl. 250 Cu. In. (155 H.P.)	46N	.035	31°-34°	.019	TDC④	130	■	■	16B	3½-4½	700⑥
	V8—302 Cu. In. (290 H.P.)	43	.035	28°-32°	.019	4B	190	.030	.030	N.A.	5-6½	900
	V8—307 Cu. In. (200 H.P.)	45S	.035	28°-32°	.019	2B	150	■	■	28B	5-6½	700⑦
	V8—327 Cu. In. (210 H.P.)	45S	.035	28°-32°	.019	2A⑤	160	■	■	28B	5-6½	700⑦
	V8—350 Cu. In. (255 H.P.)	44N	.035	28°-32°	.019	TDC④	160	■	■	28B	5-6½	700⑦
	V8—350 Cu. In. (300 H.P.)	44N	.035	28°-32°	.019	TDC④	160	■	■	28B	5-6½	700⑦
	V8—396 Cu. In. (325 H.P.)	43N	.035	28°-32°	.019	4B	160	■	■	28B	5-8½	800⑦

TUNE-UP SPECIFICATIONS, continued

| YEAR | MODEL | SPARK PLUGS | | DISTRIBUTOR | | IGNITION TIMING (Deg.) ▲ | CRANKING COMP. PRESSURE (Psi) | VALVES | | Intake Opens (Deg.) | FUEL PUMP PRESSURE (Psi) | IDLE SPEED (Rpm) ★ |
| | | Type | Gap (In.) | Point Dwell (Deg.) | Point Gap (In.) | | | Tappet (Hot) Clearance (In.) | | | | |
								Intake	Exhaust			
	V8—396 Cu. In. (350 H.P.)	43N	.035	28°–32°	.019	TDC④	160	■	■	56B	5–8½	800⑦
	V8—396 Cu. In. (375 H.P.)	43N	.035	28°–32°	.019	4B	160	.024	.028	N.A.	5–8½	750
1970	4 Cyl. 153 Cu. In. (90 H.P.)	R46N	.035	31°–34°	.019	TDC①	130	■	■	17½B	3–4½	750⑦
	6 Cyl. 230 Cu. In. (140 H.P.)	R46T	.035	31°–34°	.019	TDC④	130	■	■	16B	3–4½	750⑦
	6 Cyl. 250 Cu. In. (155 H.P.)	R46T	.035	31°–34°	.019	TDC④	130	■	■	16B	3–4½	750⑦
	V8—307 Cu. In. (200 H.P.)	R43	.035	28°–32°	.019	2B⑧	150	■	■	28B	5–6½	700⑦
	V8—350 Cu. In. (250 H.P.)	R44	.035	28°–32°	.019	TDC④	160	■	■	28B	5–6½	750⑦
	V8—350 Cu. In. (300 H.P.)	R44	.035	28°–32°	.019	TDC④	160	■	■	28B	5–6½	700⑦
	V8—350 Cu. In. (370 H.P.)	R43	.035	28°–32°	.019	14B	190	.030	.030	43B	5–6½	750
	V8—396 Cu. In. (350 H.P.)	R44T	.035	28°–32°	.019	4B	160	■	■	56B	5–8½	700⑦
	V8—400; 402 Cu. In. (330 H.P.)	R44T	.035	28°–32°	.019	4B	160	■	■	28B	5–8½	700⑦
	V8—454 Cu. In. (360 H.P.)	R43T	.035	28°–32°	.019	6B	160	■	■	56B	5–8½	700⑦
1971	4 Cyl.—153 Cu. In. (90 H.P.)	R46N	.035	31°–34°	.019	††	130	■	■	17½B	3–4½	750⑦
	6 Cyl.—230 Cu. In. (140 H.P.)	R46T	.035	31°–34°	.019	††	130	■	■	16B	3–4½	750⑦
	6 Cyl.—250 Cu. In. (155 H.P.)	R46T	.035	31°–34°	.019	††	130	■	■	16B	3–4½	750⑦
	V8—307 Cu. In. (200 H.P.)	R43	.035	28°–32°	.019	††	150	■	■	28B	5–6½	700⑦
	V8—350 Cu. In. (250 H.P.)	R44	.035	28°–32°	.019	††	160	■	■	28B	5–6½	750⑦
	V8—350 Cu. In. (300 H.P.)	R44	.035	28°–32°	.019	††	160	■	■	28B	5–6½	700⑦
	V8—350 Cu. In. (370 H.P.)	R43	.035	28°–32°	.019	††	190	.030	.030	43B	5–6½	750
	V8—396 Cu. In. (350 H.P.)	R44T	.035	28°–32°	.019	††	160	■	■	56B	5–8½	700⑦
	V8—400; 402 Cu. In.	R44T	.035	28°–32°	.019	††	160	■	■	28B	5–8½	700⑦
	V8—454 Cu. In. (360 H.P.)	R43T	.035	28°–32°	.019	††	160	■	■	56B	5–8½	700⑦

★—with manual transmission in N and automatic in D.
▲—with vacuum advanced disconnected and plugged. NOTE: These settings are only approximate. Engine design, altitude, temperature, fuel octane rating and the condition of the individual engine are all factors which can influence timing. The limiting advance factor must, therefore, be the "knock point" of the individual engine.
■—1 turn tighter than zero Lash.

⑩—1967—With California Air Injection:
194 (120 H.P.)—2° B.
230 (140 H.P.)—4° B.
250 (155 H.P.)—4° B.
283 (195 H.P.)—TDC.
327 (210 H.P.)—2° B.
327 (275 H.P.)—2° B.
327 (325 H.P.)—8° B.
350 (295 H.P.)—4° B.
396 (325 H.P.)—4° B.

⑨—1966 With California Air Injection System:
6 Cyl. 194, Dist. No. 1110360—8 B.
1110373—3 B.
V8—327 (275 H.P.) Dist. No. 1111152—8 B.
1111116—2 A.
①—w/A.T.—4B
②—w/A.T.—2B
③—With Calif. Air Injection—700 rpm.
●—4 Cyl. & V8 w/A.T., Idle 600 rpm.
6 Cyl. w/A.T., Idle 500 rpm.

④—w/A.T.—4B
⑤—w/A.T.—2B
⑥—w/A.T.—500 rpm.
⑦—w/A.T.—600 rpm.
⑧—w/A.T.—8B
A—After top dead center
B—Before top dead center
TDC—Top dead center
††—See engine decal

CAUTION

General adoption of anti-pollution laws has changed the design of almost all car engine production to effectively reduce crankcase emission and terminal exhaust products. It has been necessary to adopt stricter tune-up rules, especially timing and idle speed procedures. Both of these values are peculiar to the engine and to its application, rather than to the engine alone. With this in mind, car manufacturers supply idle speed data for the engine and application involved. This information is clearly displayed in the engine compartment of each vehicle.

CRANKSHAFT BEARING JOURNAL SPECIFICATIONS

YEAR	MODEL	MAIN BEARING JOURNALS (IN.)				CONNECTING ROD BEARING JOURNALS (IN.)		
		Journal Diameter	Oil Clearance	Shaft End-Play	Thrust on No.	Journal Diameter	Oil Clearance	End-Play
1964	4 Cyl.—153	2.2983–2.2993	.0008–.004	.002–.006	5	1.999–2.000	.0007–.0028	.008–.014
	6 Cyl.—194	2.2983–2.2993	.0008–.004	.002–.006	7	1.999–2.000	.0007–.0028	.008–.014
	6 Cyl.—230	2.2983–2.2993	.0008–.004	.002–.006	7	1.999–2.000	.0007–.0028	.008–.014
	V8—283	2.2978–2.2988	.0008–.004	.002–.006	5	1.999–2.000	.0007–.0028	.008–.014
1965	4 Cyl.—153	2.2983–2.2993	.0003–.0029	.002–.006	5	1.999–2.000	.0007–.0027	.009–.013
	6 Cyl.—194	2.2983–2.2993	.0003–.0029	.002–.006	7	1.999–2.000	.0007–.0027	.009–.013
	6 Cyl.—230	2.2983–2.2993	.0003–.0029	.002–.006	7	1.999–2.000	.0007–.0027	.009–.013
	V8—283	2.2978–2.2988	.0003–.0029 ①	.002–.006	5	1.999–2.000	.0007–.0027	.009–.013
	V8—327	2.2978–2.2988	.0008–.0034 ①	.002–.006	5	1.999–2.000	.0007–.0028	.009–.013
1966	4 Cyl.—153	2.2983–2.9993	.0003–.0029	.002–.006	5	1.999–2.000	.0007–.0027	.009–.013
	6 Cyl.—194	2.2983–2.2993	.0003–.0029	.002–.006	7	1.999–2.000	.0007–.0027	.009–.013
	6 Cyl.—230	2.2983–2.2993	.0003–.0029	.002–.006	7	1.999–2.000	.0007–.0027	.009–.013
	V8—283	②	.0003–.0029 ①	.003–.011	5	1.999–2.000	.0007–.0027	.009–.013
	V8—327	②	.0003–.0034 ①	.003–.011	5	1.999–2.000	.0007–.0028	.009–.013
	V8—396	③	④	.006–.010	5	2.199–2.200	.0007–.0028	.015–.021
1967	4 Cyl.—153	2.2983–2.2993	.0003–.0029	.002–.006	5	1.999–2.000	.0007–.0027	.009–.013
	6 Cyl.—194	2.2983–2.2993	.0003–.0029	.002–.006	7	1.999–2.000	.0007–.0027	.009–.013
	6 Cyl.—230	2.2983–2.2993	.0003–.0029	.002–.006	7	1.999–2.000	.0007–.0027	.009–.013
	6 Cyl.—250	2.2983–2.2993	.0003–.0029	.002–.006	7	1.999–2.000	.0007–.0027	.009–.013
	V8—283	⑤	⑦	.003–.011	5	1.999–2.000	.0007–.0027	.009–.013
	V8—327	⑤	⑦	.003–.011	5	1.999–2.000	.0007–.0028	.009–.013
	V8—350	2.24483–2.4493 ⑥	.0008–.002 ⑧	.003–.011	5	2.099–2.100	.0007–.0028	.009–.013
	V8—396	③	④	.006–.010	5	2.199–2.200	.0007–.0028	.015–.021
1968	4 Cyl.—153	2.2983–2.2993	.0003–.0029	.002–.006	5	1.999–2.000	.0007–.0027	.009–.013
	6 Cyl.—230	2.2983–2.2993	.0003–.0029	.002–.006	7	1.999–2.000	.0007–.0027	.009–.013
	6 Cyl.—250	2.2983–2.2993	.0003–.0029	.002–.006	7	1.999–2.000	.0007–.0027	.009–.013
	V8—302(Z28)	2.2984–2.2993	.0008–.003	.003–.011	5	1.999–2.000	.0007–.0028	.009–.013
	V8—307	2.4484–2.4493 ⑥	.0008–.002 ⑧	.003–.011	5	2.099–2.100	.0007–.0027	.009–.013
	V8—327	2.4484–2.4493 ⑥	.0008–.002 ⑤	.003–.011	5	2.099–2.100	.0007–.0028	.009–.013
	V8—350	2.4484–2.4493 ⑥	.0008–.002 ⑧	.003–.011	5	2.099–2.100	.0007–.0028	.009–.013
	V8—396	⑨	⑪	.006–.010	5	2.199–2.200	.0009–.0025	.015–.021
	V8—396(375 H.P.)	⑩	.0013–.0025 ⑫	.006–.010	5	2.1985–2.1995	.0014–.003	.019–.025
1969	4 Cyl.—153	2.2983–2.2993	.0003–.0029	.002–.006	5	1.999–2.000	.0007–.0027	.009–.013
	6 Cyl.—230	2.2983–2.2993	.0003–.0029	.002–.006	7	1.999–2.000	.0007–.0027	.009–.013
	6 Cyl.—250	2.2983–2.2993	.0003–.0029	.002–.006	7	1.999–2.000	.0007–.0027	.009–.013
	V8—302(Z28)	2.4479–2.4488	.0008–.003	.003–.011	5	1.999–2.000	.0007–.0028	.009–.013
	V8—307	2.4479–2.4488	.0008–.002 ⑧	.003–.011	5	2.099–2.100	.0007–.0027	.009–.013
	V8—327	2.4479–2.4488	.0008–.002 ⑧	.003–.011	5	2.099–2.100	.0007–.0028	.009–.013
	V8—350	2.4479–2.4488	.0008–.002 ⑧	.003–.011	5	2.099–2.100	.0007–.0028	.009–.013
	V8—396	⑨	⑪	.006–.010	5	2.199–2.200	.0009–.0025	.015–.021
	V8—396(375 H.P.)	⑩	.0013–.0025 ⑫	.006–.010	5	2.1985–2.1995	.0014–.003	.019–.025
1970–71	4 Cyl.—153	2.2983–2.2993	.0003–.0029	.002–.006	5	1.999–2.000	.0007–.0027	.009–.014
	6 Cyl.—230	2.2983–2.2993	.0003–.0029	.002–.006	7	1.999–2.000	.0007–.0027	.009–.014
	6 Cyl.—250	2.2983–2.2993	.0003–.0029	.002–.006	7	1.999–2.000	.0007–.0027	.009–.014
	V8—307	2.4484–2.4493 ⑥	⑮	.002–.006	5	2.099–2.100	.0007–.0028	.008–.014
	V8—350	2.4484–2.4493 ⑥	⑮	.002–.006	5	2.099–2.100	.0007–.0028	.008–.014
	V8—396(402)	⑬	⑯	.006–.010	5	2.199–2.200	.0009–.0025	.015–.021
	V8—400(402)	N.A.	⑯	.006–.010	5	2.199–2.200	.0009–.0025	.015–.023
	V8—454	⑭	.0013–.0025 ⑰	.006–.010	5	2.199–2.200	.0009–.0025	.015–.021

① No. 5—.0010–.0036
② No. 1—2.2987–2.2997
  Nos. 2-4—2.2983–2.2993
  No. 5—2.2978–2.2988
③ Nos. 1-2—2.7487–2.7497
  Nos. 3-4—2.7482–2.7492
  No. 5—2.7478–2.7488

④ Nos. 1-2—.0004–.002
  Nos. 3-4—.0009–.0025
  No. 5—.0013–.0029
⑤ No. 1—2.2984–2.2993
  Nos. 2-4—2.2983–2.2993
  No. 5—2.2978–2.2988
⑥ No. 5—2.4478–2.4488

⑦ No. 1—.0008–.002
  Nos. 2-4—.0018–.002
  No. 5—.0010–.0036
⑧ No. 5—.0018–.0034
⑨ Nos. 1-2—2.7484–2.7493
  Nos. 3-4—2.7481–2.7490
  No. 5—2.7478–2.7488

⑩ No. 1—2.7484–2.7493
  Nos. 2-4—2.7481–2.7490
  No. 5—2.7478–2.7488
⑪ Nos. 1-2—.0010–.0022
  Nos. 3-4—.0013–.0025
  No. 5—.0015–.0031
⑫ No. 5—.0015–.0031

⑬ Nos. 1-2—2.7487–2.7496
  Nos. 3-4—2.7481–2.7490
  No. 5—2.7478–2.7488
⑭ No. 1—2.7485–2.7494
  Nos. 2-4—2.7481–2.7490
  No. 5—2.7478–2.7488
⑮ No. 1—.0003–.0015

Nos. 2-4—.0006–.0018
No. 5—.0008–.0023
⑯ No. 1—.0007–.0019
  Nos. 2-4—.0013–.0025
  No. 5—.0024–.004
⑰ No. 5—.0024–.004

CAPACITIES

YEAR	MODEL	ENGINE CRANKCASE ADD 1 Qt. FOR NEW FILTER	TRANSMISSIONS Pts. TO REFILL AFTER DRAINING — Manual 3-Speed	Manual 4-Speed	Automatic	DRIVE AXLE (Pts.)	GASOLINE TANK (Gals.)	COOLING SYSTEM (Qts.) WITH HEATER
1964	4 Cyl.	3.5②	2	None	15.2	3.5	16	9
	6 Cyl.	4	2	None	15.2	3.5	16①	12
	V8—283	4	2	2.5	15.2	3.5	16①	17½
1965	4 Cyl.	3.5②	2	None	15.2	4	16	9
	6 Cyl.	4	2	None	15.2	4	16①	12
	V8—283	4	2	2.5	15.2	4	16①	17
	V8—327	4	2	2.5	15.2	4	16①	16③
1966	4 Cyl.	3.5②	2	None	15.2	3.5	16	9
	6 Cyl.	4	2	None	15.2	3.5	16①	12
	V8—283, 327	4	2	2.5	15.2	3.5	16①	17
	V8—396	4	2	2.5	15.2	3.5	20	22
1967	4 Cyl.	3.5②	3	None	17	3.5	16	9
	6 Cyl.	4	3	None	17	3.5	16①	11
	V8—283-327	4	3④	3	17	4	16①	16
	V8—350-396	4	3④	3	19	4	20①	23⑤
1968	4 Cyl.	3½②	3	None	17	3.5	18	9
	6 Cyl.	4	3	None	17	3.5	18⑥	12
	V8—302, 307, 327	4	3④	3	17	4	18⑥	16
	V8—350	4	3④	3	17	4	18	15
	V8—396	4	3④	3	19	4	18⑥	23
1969	4 Cyl.—153	3½②	3	None	17	⑧	18	9
	6 Cyl.—230, 250	4	3	None	17	⑧	18⑥	13
	V8—302, 307	4	3④	3	17⑦	⑧	18⑥	17
	V8—327	4	3④	3	19⑦	⑧	18⑥	17
	V8—350	4	3④	3	19	⑧	18⑥	16
	V8—396	4	3④	3	22	⑧	18⑥	23
1970-71	4 Cyl.—153	3½②	3	None	17	⑧	18	9
	6 Cyl.—230, 250	4	3	None	17⑦	⑧	18⑩	12
	V8—307	4	3	3	17⑦	⑧	18⑩	15
	V8—350	4	3	3	20⑨	⑧	18⑩	16
	V8—396(402)	4	None	3	22	⑧	⑩	23
	V8—400(402)	4	None	3	22	⑧	⑩	23
	V8—454	4	None	3	22	⑧	20	23

①—Chevy II—16 gal., Chevelle—20 gal., Camaro—18 gal.
②—Add 1 pint for new filter.
③—300 H.P.—18 qts., 350 H.P.—19 qts.
④—3 speed H.D. 3.5 pts.
⑤—Camaro 15 qts.
⑥—Chevelle 20 gals.
⑦—3 Speed Auto. 20 pts.
⑧—8.125 ring gear—3.5 pts. 8.875 ring gear—4.0 pts.
⑨—Turbohydramatic 400—22 pts.
⑩—Camaro with 250 and larger eng.—19 gals. Chevelle with 250 and larger eng.—20 gals.

BATTERY AND STARTER SPECIFICATIONS

Year	Model	BATTERY Amp. Hours Capacity	Volts	Terminal Grounded	STARTERS LOCK TEST Amp.	Volts	Torque	NO LOAD TEST Amps.	Volts	RPM	Brush Spring Tension (Oz.)
1964	All Series	44	12	Neg.	Not Recommended			49–76	10.6	7,800	35
1965	All Series	44	12	Neg.	Not Recommended			49–76	10.6	7,800	35
1966-67	4 & 6 Cyl., V8—283	44	12	Neg.	Not Recommended			49–76	10.6	7,800	35
	V8—327, 396	61	12	Neg.	Not Recommended			65–100	10.6	4,200①	35
1968-69	4 Cyl., 6 Cyl., V8—307	45	12	Neg.	Not Recommended			—	10.6	—	35
	V8—302, 327, 350, 396	61	12	Neg.	Not Recommended			—	9	—	35
1970-71	4 Cyl., 6 Cyl., V8—307	45	12	Neg.	Not Recommended			50–80	9	5,500–10,500	35
	V8—350	61	12	Neg.	Not Recommended			55–80	9	3,500–6,000	35
	V8—402(396)	61	12	Neg.	Not Recommended			65–95	9	7,500–10,500	35
	V8—454	62	12	Neg.	Not Recommended			65–95	9	7,500–10,500	35

①—Camaro—230 & 327 cu. in. = 9,750

VALVE SPECIFICATIONS

YEAR AND MODEL	SEAT ANGLE (DEG.)	FACE ANGLE (DEG.)	VALVE LIFT INTAKE (IN.)	VALVE LIFT EXHAUST (IN.)	VALVE SPRING PRESSURE (VALVE OPEN) LBS. @ IN.	VALVE SPRING INSTALLED HEIGHT (IN.)	STEM TO GUIDE CLEARANCE (IN.)		STEM DIAMETER (IN.)	
							Intake	Exhaust	Intake	Exhaust
1964-65 4 Cyl.—153	46	45	.3350	.3350	175 @ 1.26	1.66	.0010-.0027	.0015-.0033	.3404-.3417	.3410-.3417
6 Cyl.—194	46	45	.3350	.3350	170 @ 1.33	1.66	.0010-.0027	.0015-.0033	.3404-.3417	.3410-.3417
6 Cyl.—230	46	45	.4070	.4070	175 @ 1.26	1.66	.0010-.0027	.0015-.0033	.3404-.3417	.3410-.3417
V8—283	46	45	.3987	.3987	175 @ 1.26	1.66	.0010-.0027	.0015-.0033	.3404-.3417	.3410-.3417
V8—327	46	45	.3987	.3987	175 @ 1.26	1.66	.0010-.0027	.0010-.0027	.3404-.3417	.3410-.3417
1966 4 Cyl.—153	46	45	.3973	.3973	175 @ 1.26	1.66	.0010-.0027	.0010-.0027	.3404-.3417	.3410-.3417
6 Cyl.—194	46	45	.3318	.3318	175 @ 1.33	1.66	.0010-.0027	.0010-.0033	.3410-.3417	.3410-.3417
6 Cyl.—230	46	45	.3318	.3318	175 @ 1.33	1.66	.0010-.0027	.0010-.0033	.3410-.3417	.3410-.3417
V8—283	46	45	.3987	.3987	175 @ 1.26	1.66	.0010-.0027	.0010-.0027	.3410-.3417	.3410-.3417
V8—327	46	45	.3987	.3987	175 @ 1.26	1.66	.0010-.0027	.0015-.0033	.3410-.3417	.3410-.3417
V8—327 (350 H.P.)	46	45	.4472	.4472	182 @ 1.21	1.66	.0010-.0027	.0015-.0033	.3410-.3417	.3410-.3417
V8—396	46	45	.3983	.3983	220 @ 1.46	1.88	.0010-.0027	.0010-.0027	.3715-.3722	.3713-.3720
V8—396 (360 H.P.)	46	45	.4614	.4800	315 @ 1.38	1.88	.0010-.0027	.0010-.0027	.3715-.3722	.3713-.3720
1967 4 Cyl.—153	46	45	.3973	.3973	175 @ 1.26	1.66	.0010-.0027	.0010-.0027	.3410-.3417	.3410-.3417
6 Cyl.—194	46	45	.3318	.3318	177 @ 1.33	1.66	.0010-.0027	.0010-.0027	.3410-.3417	.3410-.3417
6 Cyl.—230	46	45	.3880	.3880	177 @ 1.33	1.66	.0010-.0027	.0010-.0027	.3410-.3417	.3410-.3417
6 Cyl.—250	46	45	.3880	.3880	186 @ 1.27	1.66	.0010-.0027	.0010-.0027	.3410-.3417	.3410-.3417
V8—283	46	45	.3900	.4100	200 @ 1.25	1.16	.0010-.0027	.0010-.0027	.3410-.3417	.3410-.3417
V8—327 (210 & 275 H.P.)	46	45	.3900	.4100	200 @ 1.25	1.16	.0010-.0027	.0010-.0027	.3410-.3417	.3410-.3417
V8—327 (325 H.P.)	46	45	.4472	.4472	200 @ 1.25	1.16	.0010-.0027	.0010-.0027	.3410-.3417	.3410-.3417
V8—350	46	45	.3900	.4100	200 @ 1.25	1.16	.0010-.0027	.0010-.0027	.3410-.3417	.3410-.3417
V8—396 (325 H.P.)	46	45	.3983	.3983	220 @ 1.46	1.88	.0010-.0027	.0010-.0027	.3715-.3722	.3713-.3720
V8—396 (350 H.P.)	46	45	.4614	.4800	315 @ 1.38	1.88	.0010-.0027	.0010-.0027	.3715-.3722	.3713-.3720
1968 4 Cyl.—153	46	45	.3973	.3973	175 @ 1.26	1.66	.0010-.0027	.0010-.0027	.3410-.3417	.3410-.3417
6 Cyl.—230	46	45	.3317	.3317	186 @ 1.27	1.66	.0010-.0027	.0017-.0027	.3410-.3417	.3410-.3417
6 Cyl.—250	46	45	.3880	.3880	186 @ 1.27	1.66	.0010-.0027	.0017-.0027	.3410-.3417	.3410-.3417
V8—307, 327 (210 & 275 H.P.)	46	45	.3900	.4100	198 @ 1.25	1.70	.0010-.0027	.0017-.0027	.3410-.3417	.3410-.3417
V8—302①	46	45	.3234	.3234	200 @ 1.25	1.56	.0010-.0027	.0010-.0027	.3410-.3417	.3410-.3417
V8—327 (325 H.P.)	46	45	.4471	.4471	198 @ 1.25	1.70	.0010-.0027	.0017-.0027	.3410-.3417	.3410-.3417
V8—350	46	45	.3900	.4100	198 @ 1.25	1.70	.0010-.0027	.0017-.0027	.3410-.3417	.3410-.3417
V8—396 (325 H.P.)	46	45	.3983	.3983	215 @ 1.48	1.88	.0010-.0027	.0017-.0027	.3715-.3722	.3713-.3720
V8—396 (350 H.P.)	46	45	.4614	.4800	315 @ 1.38	1.88	.0010-.0027	.0017-.0027	.3715-.3722	.3713-.3720
1969 4 Cyl.—153	46	45	.3973	.3973	175 @ 1.26	1.66	.0010-.0027	.0010-.0027	.3410-.3417	.3410-.3417
6 Cyl.—230	46	45	.3317	.3317	175 @ 1.33	1.66	.0010-.0027	.0010-.0027	.3410-.3417	.3410-.3417
6 Cyl.—250	46	45	.3880	.3880	186 @ 1.27	1.66	.0010-.0027	.0010-.0027	.3410-.3417	.3410-.3417
V8—302①	46	45	.3234	.3234	200 @ 1.25	1.70	.0010-.0027	.0010-.0027	.3410-.3417	.3410-.3417
V8—307, 350	46	45	.3945	.3945	200 @ 1.25	1.70	.0010-.0027	.0015-.0027	.3410-.3417	.3410-.3417
V8—396 (325 H.P.)	46	45	.3983	.3983	220 @ 1.46	1.88	.0010-.0027	.0010-.0027	.3715-.3722	.3713-.3722
V8—396 (350 H.P.)	46	45	.4614	.4800	312 @ 1.38	1.88	.0010-.0027	.0010-.0027	.3715-.3722	.3713-.3722
1970 4 Cyl.—153	46	45	.3973	.3973	175 @ 1.26	1.66	.0010-.0027	.0010-.0027	.3410-.3417	.3410-.3417
6 Cyl.—230	46	45	.3317	.3317	177 @ 1.33	1.66	.0010-.0027	.0010-.0027	.3410-.3417	.3410-.3417
6 Cyl.—250	46	45	.3880	.3880	186 @ 1.27	1.66	.0010-.0027	.0010-.0027	.3410-.3417	.3410-.3417

VALVE SPECIFICATIONS, continued

	YEAR AND MODEL	SEAT ANGLE (DEG.)	FACE ANGLE (DEG.)	VALVE LIFT INTAKE (IN.)	VALVE LIFT EXHAUST (IN.)	VALVE SPRING PRESSURE (VALVE OPEN) LBS. @ IN.	VALVE SPRING INSTALLED HEIGHT (IN.)	STEM TO GUIDE CLEARANCE (IN.) Intake	STEM TO GUIDE CLEARANCE (IN.) Exhaust	STEM DIAMETER (IN.) Intake	STEM DIAMETER (IN.) Exhaust
	V8—307	46	45	.3900	.4100	200 @ 1.25	1.70	.0010-.0027	.0010-.0027	.3410-.3417	.3410-.3417
	V8—350 (250, 300 H.P.)	46	45	.3900	4100	200 @ 1.25	1.70	.0010-.0027	.0010-.0027	.3410-.3417	.3410-.3417
	V8—350 (370 H.P.)	46	45	.4586	.4850	200 @ 1.25	1.70	.0010-.0027	.0010-.0027	.3410-.3417	.3410-.3417
	V8—396	46	45	.4614	.4800	②	③	.0010-.0027	.0010-.0027	.3715-.3722	.3715-.3722
	V8—400 (402)	46	45	.3983	.3983	②	③	.0010-.0027	.0010-.0027	.3715-.3722	.3715-.3722
	V8—454 (360 H.P.)	46	45	.4614	.4800	②	③	.0010-.0027	.0010-.0027	.3715-.3722	.3715-.3722
1971	4 Cyl.—153	46	45	.3973	.3973	175 @ 1.26	1.66	.0010-.0027	.0010-.0027	.3410-.3417	.3410-.3417
	6 Cyl.—230	46	45	.3317	.3317	177 @ 1.33	1.66	.0010-.0027	.0010-.0027	.3410-.3417	.3410-.3417
	6 Cyl.—250	46	45	.3880	.3880	186 @ 1.27	1.66	.0010-.0027	.0010-.0027	.3410-.3417	.3410-.3417
	V8—307	46	45	.3900	.4100	200 @ 1.25	1.70	.0010-.0027	.0010-.0027	.3410-.3417	.3410-.3417
	V8—350	46	45	.3900	.4100	200 @ 1.25	1.70	.0010-.0027	.0010-.0027	.3410-.3417	.3410-.3417
	V8—350 (370 H.P.)	46	45	.4586	.4850	200 @ 1.25	1.70	.0010-.0027	.0010-.0027	.3410-.3417	.3410-.3417
	V8—396	46	45	.4614	.4800	②	③	.0010-.0027	.0010-.0027	.3715-.3722	.3715-.3722
	V8—400 (402)	46	45	.3983	.3983	②	③	.0010-.0027	.0010-.0027	.3715-.3722	.3715-.3722
	V8—454 (360 H.P.)	46	45	.4614	.4800	②	③	.0010-.0027	.0010-.0027	.3715-.3722	.3715-.3722

①—Z28 engine

②—Outer—240 lbs. @ 1.38 in.
Inner—90 lbs. @ 1.28 in.

③—Outer—1.88 in.
Inner—1.78 in.

LIGHT BULBS

1964–68 Chevy II

Headlamp.................................. 6012; 40–50 W
Parking, tail, stop & turn signal .... 1157; 4–32 CP
Back-up lamps .... 1156; 32 CP
Gen., temp., oil, hi-beam indicator, clock ..... 1895; 2 CP
Dome lamp .... 211; 12 CP
License lamp.............................. 67; 4 CP
Radio dial .... 1893; 3 CP
Instrument panel .... 1816; 3 CP
Traffic hazard .... 1445; 1 CP

1964–68 Chevelle

Headlamp (outer)..................... 4002; 37½–55 W
Headlamp (inner)....................... 4001; 37½ W
Instrument panel .... 1895; 2 CP
Other lamps same as Chevy II

1967–68 Camaro

Front parking & turn signals............. 1034; 4–32 CP
Instrument panel, temp., gen., oil, hi-beam
 indicator.............................. 194; 2 CP

1969–70

LAMP USAGE	CANDLE POWER OR WATTAGE	BULB NUMBER
Headlamp unit		
Chevrolet, Chevelle & Corvette		
Outer—high beam	37½ Watts	4002
Outer—low beam	55 Watts	4002
Inner—high beam only	37½ Watts	4001
Nova & Camaro		
High beam	55 Watts	6012
Low beam	45 Watts	
Monte Carlo		
High beam	55 Watts	6012A
Low beam	45 Watts	
Parking lamp and directional signal		
Chevrolet & Chevelle	3–32	1157
Camaro, Corvette, Nova & Monte Carlo	3–32	1157NA
Tail, stop and directional signal	3–32	1157
Back-up lamp	32	1156
Instrument illumination lamps		
Chevrolet, Chevelle (1970), Monte Carlo, Camaro	2	194
Corvette, Chevelle (1969)	2	1895
Nova	3	168
Console instrument cluster (Nova)	2.5	1816
Temperature indicator (1969 Chevelle—2/1895)	2	194
Oil pressure indicator (1969 Chevelle—2/1895)	2	194
Generator indicator (1969 Chevelle—2/1895)	2	194
Hi-beam indicator		
All except Corvette	2	194
Corvette, 1969 Chevelle	2	1895
Directional indicator		
All except Corvette	2	194
Corvette, 1969 Chevelle	2	1895
Cigarette lighter lamp		
Corvette	1	1445
Warning lamps		
1969-70 Corvette—door ajar	2	1895
Headlamps up	2	1895
Seat belts	2	1895
1970 Chevrolet—low fuel	2	194
Check doors	2	194
Seat belt	2	194

FUSES AND CIRCUIT BREAKERS

1964–68

Headlamps	15 AMP. CB
Instrument Lights	3 AMP. FUSE
Tail, stop, courtesy, glove box, license plate, dome light, clock	15 AMP. FUSE
Radio	2½ AMP. FUSE
Heater, back-up light, brake signal light	10 AMP. FUSE
Air conditioning—with heater	20 AMP. FUSE
Air condition blower	30 AMP. FUSE
Overdrive	15 AMP. FUSE

1969–70

A circuit breaker in the light control switch protects the headlamp circuit, thus eliminating one fuse. A separate 30 amp. circuit breaker mounted on the firewall protects the power window, seat and top circuits. The under hood and spot lamp circuit is also protected by a separate 15 amp. inline fuse. Where current load is too heavy, the circuit breaker rapidly opens and closes, protecting the circuit until the cause is found and eliminated.

Fuses located in the fuse panel under the instrument panel are:

Wiper/washer, 3-spd. A/T downshift	SAE/SFE 25 AMP.
Back-up lamp, turn signal, cruise master, defogger, heater	SAE/SFE 25 AMP.
Air conditioning—transmission control spark solenoid (1970)	3AG/AGC 25 AMP.
Radio, power window	3AG/AGC 10 AMP.
Tail, marker and fender lamps	SAE/SFE 20 AMP.
Instrument lamps (Nova, Camaro and Chevelle)	SAE/SFE 4 AMP.
Instrument lamps (Corvette)	1AG/AGA 5 AMP.
Gauges and Tell-Tale Lamps	3AG/AGC 10 AMP.
Stop and Hazard	SAE/SFE 20 AMP.
Clock, lighter, courtesy lamps, dome and luggage lamps	SAE/SFE 20 AMP.

In-Line Fuses:

Air conditioning high blower speed fuse located in wire running from horn relay to A/C Relay	
Comfortron and Four Season	SAE 30 AMP.
Universal & All Weather (Nova)	SAE 20 AMP.

Fusible Links:

Pigtail lead at battery positive cable (except Corvette)	14 gauge brown wire
Molded splice at Solenoid "Bat" terminal (Corvette only)	14 gauge brown wire
Molded splice located at the horn relay	16 gauge black wire
Molded splice in voltage regulator #3 terminal wire	20 gauge orange wire
Molded splice in ammeter circuit (both sides of meter)	20 gauge orange wire

LIGHT BULBS

1969–70

LAMP USAGE	CANDLE POWER OR WATTAGE	BULB NUMBER
1970 Nova—low fuel	2	194
Heater or A/C Control Panel		
Chevelle and Monte Carlo	1	1445
Chevrolet, Nova, Camaro	2	1895
Corvette	2.5	1816
Glove compartment lamps		
Chevrolet, without A/C, Nova & Corvette (all 1969 Models use 1895)	2	1895
Chevelle, Monte Carlo & Chevrolet with A/C	2	1893
Dome and Courtesy Lamps		
Cartridge Type (All)	12	211
Bayonet Type (Exc. Corvette and 1970 Station Wagon third seat)	6	631
Corvette and 1970 Station Wagon third seat	6	90
Seat Separator-Courtesy Lamp	6	212
1969 Compartment Lamp	1	1445
Side Marker-Front	2	194
Side Marker-Rear	2	194
License Plate Lamp (Exc. Corvette)	4	67
(Corvette)	4	97
Radio Dial Lamp		
All AM Only Radios (all 1969 models use 2/1893)	2	293
All Tape Players and FM Radios	2	1893
Tape Player Lens Illumination Lamp (1970)	1	216
Stereo Indicator Lamp (1970)	.3	2182D
Automatic Transmission Control Indicator Lamp		
Chevrolet, Chevelle (without seat separator) & Monte Carlo (w/o seat separator)	2	194
Nova, Chevelle (with seat separator) & Monte Carlo (with seat separator)	1	1445
Brake Alarm Lamp		
All except Corvette	2	194
Corvette, 1969 Chevelle	2	1895
Luggage Compartment Lamp	15	1003
Map Lamp (Mirror) (1969 models—6/562)	4	563
Underhood	15	93
Rear Window Defogger Lamp (1970)	2	194

Distributor

Caution When using an auxillary starter switch for bumping the engine into position for timing or compression test, the primary distributor lead must be disconnected from the negative post of the ignition coil and the ignition switch must be on. Failure to do this may cause damage to the grounding circuit in the ignition switch.

4 and 6 Cylinder Models

Distributor design, (except for number of cam lobes and distributor cap) is similar for the unit(s) used

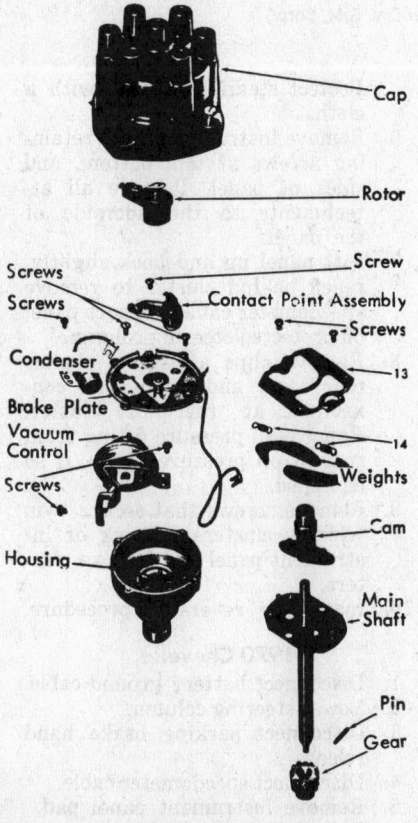

6 cylinder distributor exploded view—typical
(© Chevrolet Div. G.M. Corp.)

on the OHV 4 and 6 engines. Mounting is on the forward right side of the engine. Both units use centrifugal and vacuum controlled advance mechanism. Direction of rotation (as viewed from the top) is clockwise for both models. Other pertinent distributor specifications can be found in the Tune-Up Specifications charts.

V8 Models

The distributor is located between the two banks of cylinders at the back of the block.

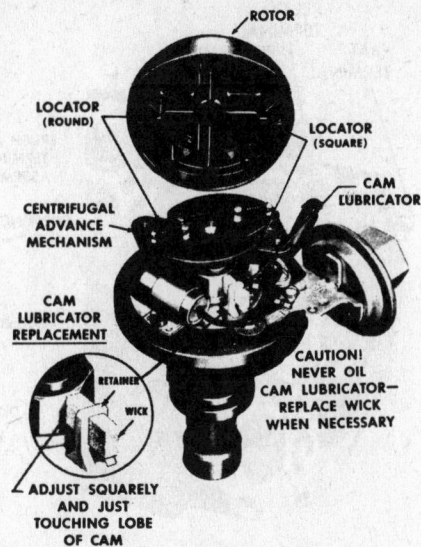

V8 distributor—typical (© G.M. Corp.)

Distributor Removal

4 and 6 Cylinder Models

1. Remove distributor cap, primary wire and vacuum line at distributor.
2. Scribe a mark on the distributor body, locating the position of the rotor. Scribe another mark on the distributor body and engine block, showing the position of the body in the block.
3. Remove the distributor hold-down screw and lift the distributor up and out of the engine.

V8 Models

The drive gear is attached to the distributor shaft. If it becomes necessary to remove the distributor, carefully mark the position of the rotor so that, if the engine is not turned after the distributor is taken out, the rotor can be returned to the position from which it was removed without difficulty.

To remove the distributor, take off the carburetor air cleaner, disconnect the coil primary wire and the vacuum line, remove the distributor cap, take out the single hold-down bolt located under the distributor body. With a pencil, mark the position of the body relative to the block, and then work the distributor up out of the block.

Distributor Installation

4 and 6 Cylinder Models

1. If the crankshaft was rotated, turn the engine until the piston of No. 1 cylinder is at the top of its compression stroke.
2. Position the distributor to the block so that the vacuum control unit is in its normal position.
3. Position the rotor to point toward the front of the engine (with distributor held out of the block, but in installed attitude). Turn rotor counterclockwise about one-eighth turn and push distributor down to engage camshaft drive. It may be necessary to move the rotor one way or the other to mesh the drive and driven gears properly.
4. While holding the distributor down in place, engage the starter a few times to make sure the oil pump shaft is engaged. Install hold-down clamp and bolt and snug up the bolt.
5. Once again, rotate the crankshaft until No. 1 cylinder is on the compression stroke and the harmonic balancer mark is on 0°.
6. Turn distributor body slightly until points open. Tighten distributor clamp bolt.
7. Place distributor cap in position and see that the rotor lines up with the terminal for the No. 1 spark plug.
8. Install cap, distributor primary wire, and double check plug wires in the cap towers.
9. Start engine and set timing according to the Tune-Up Specifications chart.
10. Reconnect vacuum hose to vacuum control assembly.

V8 Models

Remove No. 1 spark plug and, with finger on plug hole, crank the engine until compression is felt in No. 1 cylinder. Continue cranking until pointer lines up with the timing mark on the crankshaft pulley.

Position distributor in opening of the block in normal installed attitude; have rotor pointing to front of engine.

Turn the rotor counterclockwise about one-eighth of a turn (from straight front toward the left cylinder bank). Push the distributor down to engage the camshaft and while holding, turn the engine with the starter so that distributor shaft engages the oil pump shaft.

Return engine to compression stroke of No. 1 piston with timing mark on pulley aligned with the pointer. Adjust the distributor so that the points are opening. Install the cap being sure the rotor points to the contact for No. 1 spark plug. Connect the timing light and check that spark occurs as timing mark and pointer are aligned.

Generator, Regulator

AC generators and regulators are covered in the Unit Repair Section.

Delcotron Removal and Installation

1. Disconnect battery ground cable to prevent diode damage.
2. Disconnect Delcotron wiring.
3. Remove generator brace bolt. If power steering equipped, loosen pump brace and mount nuts. Detach drive belt(s).
4. Support generator and remove mount bolt(s). Remove unit from vehicle.
5. Reverse procedure to install. Adjust drive belt to have 1/4-1/2 in. play on longest run of belt.

AC Generator

The following are a few precautions to observe in servicing the Delcotron (AC) generator and the regulator.

1. When installing a battery, be certain that the ground polarity of the battery and the ground polarity of the generator and regulator are the same.
2. When connecting a booster battery, be sure to connect the correct battery terminals together.
3. When hooking up a charger, connect the correct charged leads to the battery terminals.
4. Never operate the generator on an open circuit. Be sure all connections in the charging circuit are tight.
5. Do not short across or ground any of the terminals on the generator or regulator.
6. Never polarize an AC system.

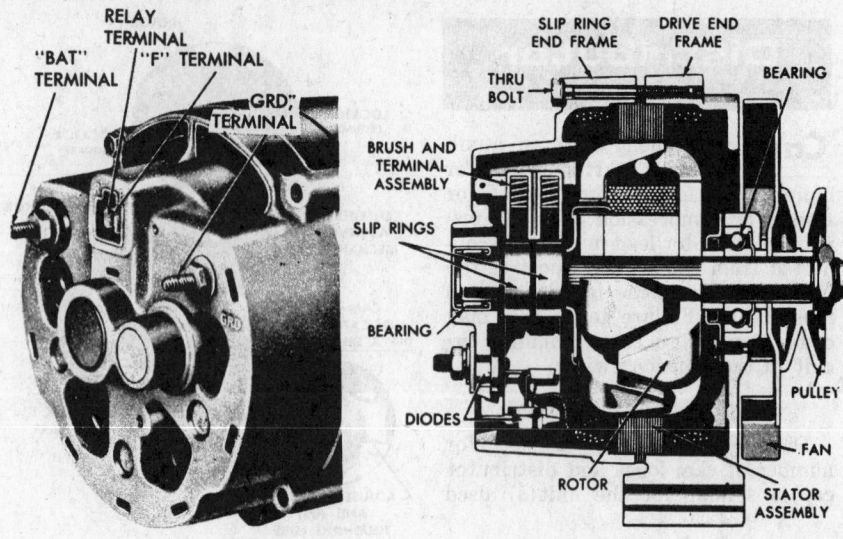

Delcotron (© Chevrolet Div. G.M. Corp.)

Instruments

Instrument Cluster Removal and Replacement

1964-65 Chevelle

1. Disconnect battery ground cable.
2. Remove upper mast jacket clamp bolt and bend clamp away from steering column.
3. Disconnect speedometer cable. Disconnect oil pressure line at gauge on SS model.
4. Remove the screws that attach the console to the instrument panel and lean console forward onto the mast jacket. Remove radio knobs before removing console from panel.
5. Disconnect all cluster lamps, harness connectors, and two harness retaining clips from rear of cluster.
6. Lift console forward and upward to remove.
7. Unscrew and remove cluster from console.
8. Install by reversing removal procedure.

1966-67 Chevelle

1. Disconnect battery.
2. Remove steering coupling bolt and disconnect steering shaft from coupling.
3. Loosen mast jacket lower clamp.
4. On air-conditioned cars, remove air-conditioning center distribution duct.
5. Remove radio rear support bracket screw.
6. Remove mast jacket trim cover and support clamp.
7. Loosen set screw and remove transmission dial indicator (if so equipped).
8. Disconnect speedometer shaft at speedometer head.
9. Remove the instrument panel attaching screws.
10. From under the console, remove the four lower retaining screws from the cluster housing.
11. With mast jacket padded, pull instrument panel from the console and lay forward on mast jacket.
12. Disconnect wiring harness, cluster lamps and wiring terminals from rear of cluster assembly.
13. Remove four screws holding the upper section of the cluster housing to panel and remove cluster from instrument panel.
14. Install by reversing removal procedure.

1968-69 Chevelle

1. Disconnect battery ground cable.
2. Remove ash tray and retainer.
3. Remove radio knobs, nuts, electrical connectors, and radio rear support. Remove radio.
4. Remove heater control screws, then push control head out of instrument panel.
5. Lower steering column. Remove automatic transmission indicator cable from steering column.

Protect steering column with a cloth.
6. Remove instrument panel retaining screws at top, bottom, and sides of panel. Remove all attachments to the underside of the panel.
7. Lift panel up and back slightly, reach behind cluster to remove speedometer cable, support panel on protected steering column.
8. Remove clips at top of cluster rear cover and remove all connectors at rear of cluster. Remove oil pressure fitting from rear of oil pressure gauge, if so equipped.
9. Remove screws that secure twin window clusters to back of instrument panel and remove clusters.
10. Install by reversing procedure.

1970 Chevelle

1. Disconnect battery ground cable.
2. Lower steering column.
3. Disconnect parking brake hand release.
4. Disconnect speedometer cable.
5. Remove instrument panel pad.
6. Disconnect radio speaker bracket from instrument panel. Disconnect speaker wire from radio.
7. Disconnect air conditioning center outlet and control head.
8. Remove radio knobs, washers, bezels, and wiring.
9. Unbolt radio braces. Roll radio out from under instrument panel.
10. Remove six instrument panel bolts and roll out instrument panel with the help of an assistant.
11. Reverse procedure for installation.

1964-65 Chevy II

1. Disconnect battery ground cable.

2. Loosen and lower mast jacket from the dash panel.
3. Disconnect speedometer cable.
4. Remove the four screws that attach the cluster to the dash panel. Cluster may now be pulled clear of dash.
5. Remove the two harness retaining clips from rear of cluster.
6. Disconnect all indicator and illuminating bulb sockets. Remove cluster from vehicle.
7. Reverse procedure to install.

1966-67 Chevy II

1. Disconnect battery.
2. Remove transmission selector dial indicator assembly (if so equipped) and mast jacket upper support clamp.
3. Disconnect speedometer cable at speedometer head.
4. Remove screws holding the cluster to the console.
5. With the mast jacket well padded, pull cluster forward of console opening and disconnect all wires and lamp connections. Remove the cluster.
6. Install by reversing removal procedure.

1968 Chevy II, 1969-70 Nova

1. Disconnect battery ground cable.
2. Lower steering column.
3. Remove three screws above heater control securing it to instrument cluster.
4. Remove radio knobs, washers, bezel nuts, and front support.
5. Remove screws at top, bottom, and sides of cluster.
6. Tilt cluster forward. Reach behind to disconnect speedometer cable, speedminder, and electrical connectors. Remove screws and lift instrument assembly out of instrument carrier.
7. Reverse procedure to install.

1967-68 Camaro

1. Disconnect battery ground cable.
2. Remove mast jacket lower support screws at toe pan.
3. Remove mast jacket upper support bolts and allow steering column to rest on seat cushion.
4. Remove cluster attaching screws from face of panel and partially remove assembly from instrument panel.
5. Reaching behind cluster assembly, disconnect speedometer cable, speed warning device, and harness connector.
6. Remove assembly from instrument panel.
7. Install by reversing removal procedure.

1969 Camaro

1. Disconnect battery ground cable.
2. Remove:
   a. instrument panel pad

b. air conditioning attachments
   c. radio brace attachments.
3. Lower steering column.
4. Remove cluster attaching screws.
5. Disconnect:
   a. speedometer cable
   b. speed warning device
   c. wiring connectors.
6. Remove cluster assembly.
7. Reverse procedure to install.

1970 Chevelle SS, 1970 Monte Carlo

1. Disconnect battery ground cable.
2. Remove instrument panel pad.
3. Disconnect air conditioning center outlet and control head.
4. Disconnect radio speaker and brackets from cluster.
5. Disconnect speaker leads from radio.
6. Remove radio knobs, washers, bezels, and wiring.
7. Remove bolts from radio braces and roll radio out from under instrument panel.
8. Remove steering column cover.
9. Remove two steering column attaching bolts. Note the placement of shims.
10. Disconnect automatic transmission indicator cable from steering column housing and lower steering column.
11. Disconnect parking brake hand release rod attachment.
12. Remove six instrument panel attaching bolts and roll out panel assembly. An assistant is required for this operation.
13. Disconnect speedometer cable.
14. Remove instrument cluster lamp sockets.
15. Remove clock, fuel gauge, power top, and rear window defogger connectors from rear of instrument panel.
16. Remove printed circuit.
17. Reverse procedure to install.

Ignition Switch Replacement

1964 Chevy II and Chevelle

1. Disconnect battery.
2. Remove lock cylinder by placing in Lock position and inserting a wire in small hole in cylinder face. While pushing on wire, continue to turn cylinder counterclockwise and pull cylinder from case.
3. Remove nut from passenger side of dash.
4. Pull switch from under dash and remove wiring connector.
5. To remove the theft resistant connector, the switch must be removed from under the dash. With screwdriver, depress tangs

and separate the connector.
6. Install in reverse of above.

1965-68 Chevy II, 1965 Chevelle

This procedure is identical to that for the 1964 Chevy II and Chevelle, with the substitution of the following step:
2. Remove lock cylinder by placing in Off position and inserting a wire in the small hole in cylinder face. While pushing on the wire, continue to turn cylinder counterclockwise. Pull cylinder from case.

1966-68 Chevelle

1. Disconnect battery ground cable.
2. Remove:
   a. ash tray
   b. ash tray retainer
   c. radio knobs, nuts, connectors, bracket, and radio.
3. Remove lock cylinder by positioning switch in Accessory position and inserting wire in hole in cylinder face. Push in wire and turn key counterclockwise to remove cylinder.
4. Remove bezel nut and pull out ignition switch.
5. Unsnap locking tangs on connector with a screwdriver. Unplug connector.
6. Reverse procedure to install.

1967-68 Camaro

This procedure is the same as that given above for the 1966-68 Chevelle, with the deletion of Step 2.

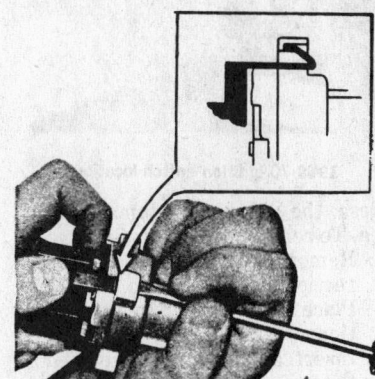

Unlocking ignition switch connector
(© G.M. Corp.)

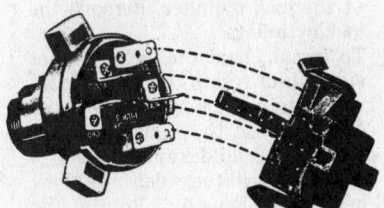

Separated ignition switch and connector
(© G.M. Corp.)

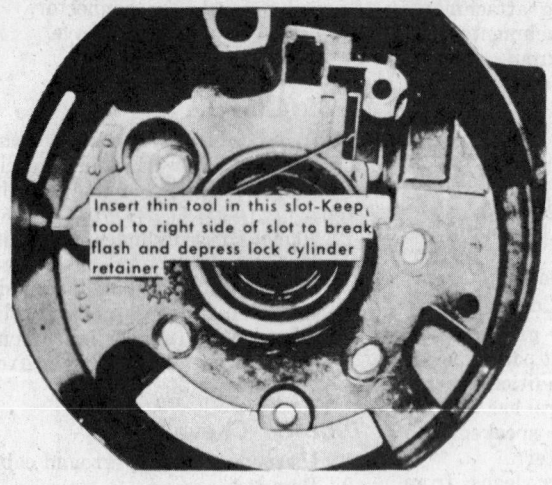

Removing 1969-70 lock cylinder
(© Chevrolet Div. G.M. Corp.)

SHAFT
RETAINER

Light switch, 1965-70
(© Chevrolet Div. G.M. Corp.)

1969-70 All Models

All 1969-70 models have the ignition lock cylinder located in the upper right side of the steering column. The ignition switch is inside the channel section of the brake pedal support. The switch is inaccessible

1969-70 ignition switch location

unless the steering column is lowered. To remove the lock cylinder:

1. Remove steering wheel and directional signal switch.
2. Place lock cylinder in Lock position.
3. Insert a small screwdriver into the 'turn signal housing slot. Keeping the screwdriver to the right side of the slot, break the housing flash loose and depress the spring latch at the lower end of the lock cylinder. Remove the lock cylinder.
4. To install, hold the lock cylinder sleeve and rotate the knob clockwise against the stop. Insert the cylinder into the housing, aligning the key and keyway. Hold a .070 in. drill between the lock bezel and housing. Rotate the cylinder counterclockwise, maintaining a light pressure until the drive section of the cylinder

mates with the sector. Push in until the snap ring pops into the grooves. Remove drill. Check cylinder operation.

To remove the ignition switch:
1. Lower steering column. The column must be carefully supported to prevent damage.
2. Remove lock cylinder as above.
3. Remove two switch screws and switch assembly.
4. When replacing switch, make sure switch and lock are in Lock position. Do not use switch screws longer than the originals, or the compressibility feature of the column may be lost.

Lighting Switch Replacement

1964-68 Chevy II, 1969-70 Nova, 1967-69 Camaro and Chevelle

1. Disconnect battery.
2. Pull knob out to on position.
3. Reach under instrument panel and depress the switch shaft retainer, and remove knob and shaft assembly.
4. Remove the retaining ferrule nut.
5. Remove switch from instrument panel.
6. Disconnect the multi-plug connector from the switch.
7. Install in reverse of above.

1970 Chevelle, Monte Carlo

1. Disconnect battery ground cable.
2. Remove six screws and instrument panel pad.
3. Remove left radio speaker.
4. Pull knob to on position.
5. Reach behind instrument panel and depress switch shaft retain-

er. Remove knob and shaft assembly.
6. Remove ferrule nut and switch assembly from instrument panel.
7. Reverse procedure to install.

Battery, Starter

Detailed information on the battery and starter will be found in the Battery and Starter Specifications table.

A more general discussion of batteries will be found in the Unit Repair Section.

A more general discussion of starters and their troubles can be found in the Unit Repair Section. The starter should require no lubrication or other maintenance between overhaul periods.

Starter Removal and Installation

1. Disconnect battery ground cable.
2. Raise and support vehicle.
3. Disconnect all wires at solenoid terminals. Note color coding of wires for reinstallation.
4. Remove starter front bracket and two mount bolts. On engines with solenoid heat shield, remove front bracket upper bolt and detach bracket from starter motor.
5. Remove front bracket bolt or nut. Rotate bracket clear. Lower starter front end first. Remove starter.
6. Reverse procedure to install. Torque mount bolts to 25-35 ft. lbs.

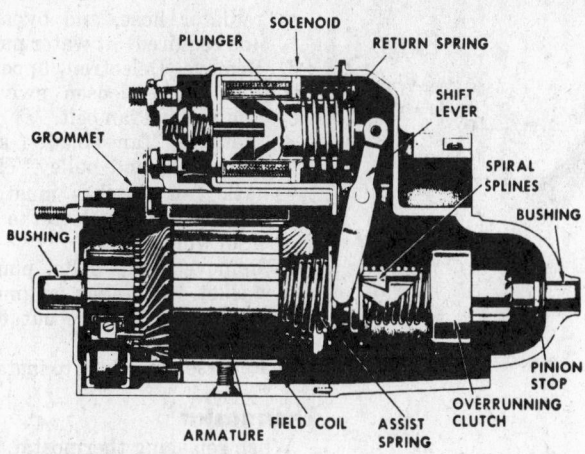

Starting motor cross-section (© Chevrolet Div. G.M. Corp.)

Brakes

References

General Motors cars have as standard equipment the duo-servo single anchor type service brake. Brake shoe linings are bonded and the shoes are self-adjusting. Segmented metallic brake linings are optional on some 1964-65 models. Drums are of cast iron.

Wheel cylinders are conventional double piston type.

Prior to 1967, the master cylinder consists of a single cylinder and reservoir mounted on the engine side of the firewall.

Since 1967, a dual type master cylinder is used. The front portion of the master cylinder supplies hydraulic pressure for the front wheels.

Pressure for rear wheel brake application is supplied from the rear portion of the master cylinder.

More detailed information on this dual type of hydraulic brake system can be found in the Unit Repair Section.

As an option, both Bendix and Moraine power brakes are available. Data on these two power brakes can be found in the Unit Repair Section.

Front wheel disc brakes are optional on some models starting 1966 and standard on the 1970 Camaro. Repair procedures for disc brakes are given in the Unit Repair Section.

Parking Brake

The parking brake is hand operated by a lever attached to the dash panel, just to the right of the steering column on the 1964-67 Chevy II and pedal operated on all other models. It functions through an equalizer and cables to the rear brake shoes.

Fuel System

Fuel Pump

The fuel pump is the single action AC diaphragm type. Two types of fuel pumps are used; serviceable and non-serviceable. The serviceable type is used on all 1964-65 engines, 1966 inline engines without the A.I.R. emission control system, and 1966 283 and 327 V8's. The non-serviceable type is used on all other engines.

The pump is actuated by an eccentric located on the engine camshaft.

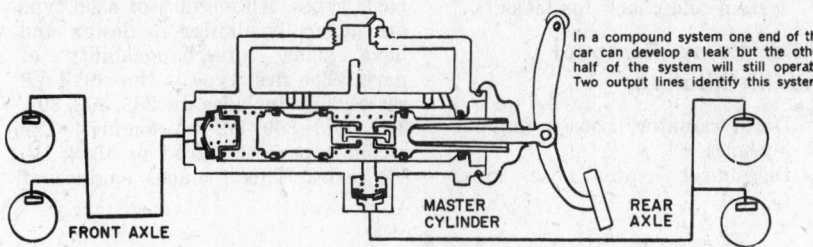

In a compound system one end of the car can develop a leak but the other half of the system will still operate. Two output lines identify this system.

Exploded view, master cylinder, 1967-70—typical (© Chevrolet Div. G.M. Corp.)

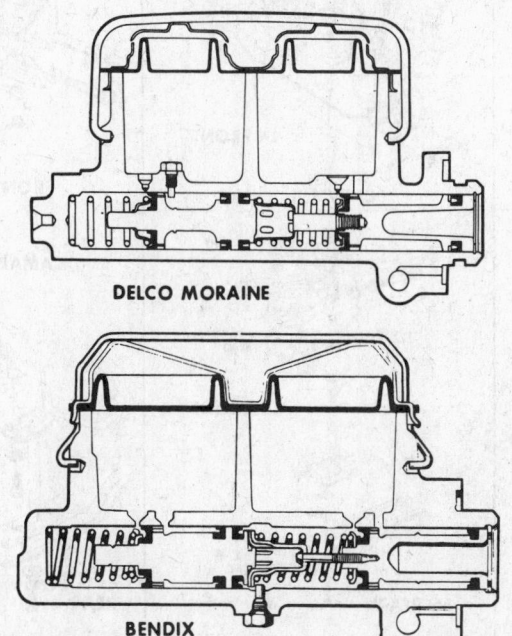

Dual type master cylinders, typical (© Chevrolet Div. G.M. Corp.)

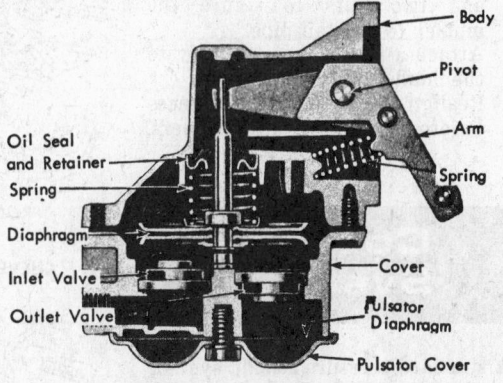

V8 fuel pump (serviceable) (© Chevrolet Div. G.M. Corp.)

On inline engines, the eccentric actuates a pump rocker arm. On V8 engines, a push rod between the camshaft eccentric and the fuel pump actuates the pump rocker arm.

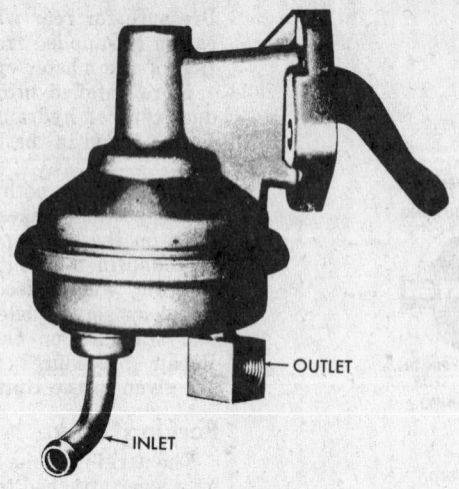

Fuel pump (non-serviceable)
(© Chevrolet Div. G.M. Corp.)

Exhaust System

Muffler Removal and Installation

1. Remove U bolt clamp at muffler mounting.
2. Remove muffler from exhaust pipe by cutting the pipe with a torch or hacksaw. Cut cleanly and close to the muffler inlet for ample surface for muffler replacement.
3. Remove the muffler from the tail pipe with a hammer.
4. Replace with a new muffler and or tail pipe and exhaust pipe, as required.
5. Use existing hardware for replacement, plus a new U-bolt and two nuts to secure the muffler to the tail pipe.
6. Attach a new clamp assembly to the muffler and exhaust pipe.
7. Realign and check all clearances before tightening all fasteners.

Cooling System

A standard pressure cooling system is used on all models. The radiator cap is designed to maintain a cooling system pressure of about 13 or 15 psi above atmospheric. The water pump requires no attention except to make certain the air vent at the top of the hosing and the drain holes in the bottom do not become clogged.

Radiator Removal

1. Drain radiator.
2. Disconnect hoses.

3. Remove radiator attaching bolts and lift radiator out of car.

Radiator Installation

1. Slide radiator into position.
2. Install attaching bolts.
3. Install hoses and close drain.
4. Fill cooling system, run engine until operating temperature has been reached. Again fill cooling system and check for leaks.

Water Pump Removal and Installation

1. Drain radiator. Loosen fan pulley bolts.
2. Disconnect heater hose, lower radiator hose, and bypass hose (as required) at water pump.
3. Remove Delcotron upper brace (V8 only), loosen swivel bolt, and remove fan belt.
4. Remove fan blade assembly bolts, fan, and pulley. Thermostatic fan clutches must not be tilted on removal or the silicone fluid will leak out.
5. Remove pump bolts, pump, and gasket. On inline engines, pull the pump straight out to avoid impeller damage.
6. Reverse procedure to install.

Thermostat

When replacing thermostat, be sure to install unit with the spring and body toward the engine.

Engine

Four, six, and eight cylinder engines are used. The four cylinder engine is a 153 cu. in. inline design with five main bearings. The six cylinder engines are also of the inline type, with seven main bearings. They are built in 230 and 250 cu. in. displacements. V8 engines are of two basic types. All engines of each type are generally similar in design and have some interchangeability of parts. The first type is the small V8 series. This includes the 283, 302, 307, 327, and 350 cu. in. engines. The second type is the large, or Mark IV, V8 series. This includes engines of

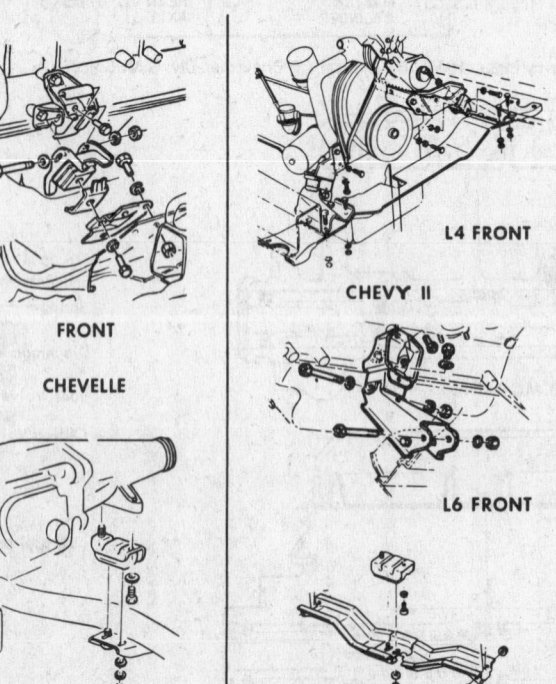

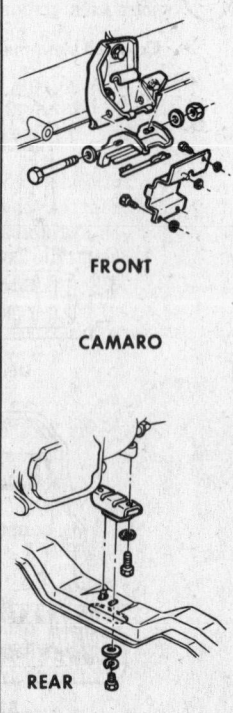

FRONT
CHEVELLE

REAR

CHEVY II

L4 FRONT

L6 FRONT

L6 REAR

FRONT
CAMARO

REAR

Engine mounts (© Chevrolet Div. G.M. Corp.)

396, 400, 427, and 454 cu. in. displacement.

Pertinent engine data can be found in the General Engine Specifications chart.

Exhaust Emission Control

The air injector reactor (A.I.R.) system is used on all vehicles, with certain design exceptions, sold in California in 1966-67, and later on virtually all vehicles. This system uses a belt driven air pump to inject air into the exhaust ports. The effect of this is to drastically reduce the percentage of pollutants in the vehicle exhaust.

Beginning with the 1970 models, all General Motors cars use a complete system of emission controls. Some of the features of this system were introduced as in previous years, particularly on vehicles sold in California. In most cases, the features are introduced to comply with the strict California anti-air pollution standards, then later added to all vehicles. Features of the emission control systems are Positive Crankcase Ventilation (P.C.V.), Controlled Combustion System (C.C.S.), Evaporation Emission Control (E.E.C.), and Transmission Controlled Spark (T.C.S.).

P.C.V. uses intake manifold vacuum to draw crankcase vapors through a P.C.V. metering valve into the combustion chambers for burning.

C.C.S. increases combustion efficiency through leaner carburetor adjustments and revised distributor calibration. Thermostatically controlled air intakes are also used on most models.

E.E.C. reduces fuel vapor emission from the fuel tank and the carburetor float bowl by use of an air filter canister.

T.C.S. eliminates distributor vacuum advance in the low forward gears of both standard and automatic transmissions. This is done by means of solenoid operated vacuum valves. Any of the methods of exhaust emission control requires close and frequent attention to tune-up factors of engine maintenance.

Since 1968, all car manufacturers post idle speeds and other pertinent data relative to the specific application in a conspicuous place in the engine compartment.

For details, consult the Unit Repair Section.

Engine R & R

NOTE: unless otherwise stated, the following operations cover the 4 cylinder, 6 cylinder and V8 engines.

Engine Removal

1. Raise car and place on jackstands.

V8 engine
(© Chevrolet Div. G.M. Corp.)

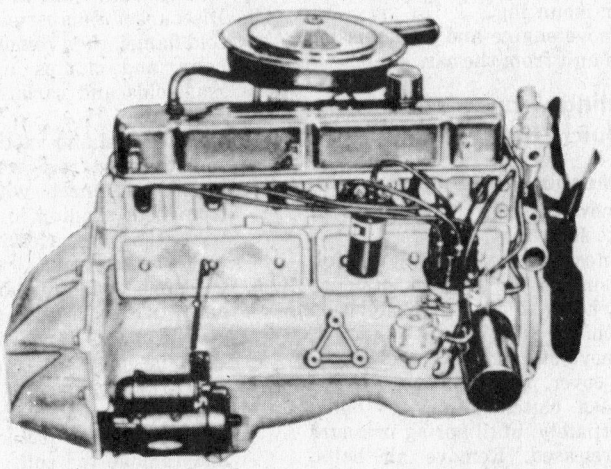

Inline engines (© Chevrolet Div. G.M. Corp.)

2. Drain cooling system, transmission, and crankcase.
3. Scribe alignment marks on underside of hood and around hood hinges, and remove hood from hinges.
4. Disconnect coolant and heater hoses at engine attachment.
5. Disconnect battery cables at battery.
6. Remove radiator and shroud assembly. Remove fan and pulley.
7. Remove air cleaner.
8. Disconnect coil, starter and Delcotron wires, engine-to-body ground strap, oil pressure and engine temperature sender wires.
9. Disconnect gas tank line at fuel pump.
10. Disconnect accelerator control linkage at firewall.
11. Disconnect hand choke linkage (4 cylinder).
12. Disconnect exhaust pipe from manifold. On V8, disconnect crossover pipe.
13. Disconnect clutch shaft bracket at frame and disconnect clutch linkage. On automatic transmission models, remove transmission oil filler tube and plug the opening.
14. Lower vehicle to floor.
15. Attach engine lifting tool. Attach to hoist and secure the engine.
16. Remove driveshaft.
17. Remove and set aside power steering pump and air conditioning compressor. Do not disconnect hoses.
18. Remove engine rear mounting bolts.
19. Disconnect speedometer cable and transmission control rod linkage lower ends.
20. Loosen front engine mounting bolts.
21. Raise engine slightly and remove bolts.
22. Remove transmission crossmember and free the transmission rear mounting.
23. Remove engine and transmission as a unit from the car.

Separating Transmission and Clutch from Engine

Manual Transmission

1. Remove clutch housing cover plate screws.
2. Remove bolts holding clutch housing to engine block. Remove clutch housing and transmission assembly.
3. Remove starter and clutch housing cover plate.
4. Loosen clutch-to-flywheel bolts, alternately, until spring pressure is released. Remove all bolts, clutch disc and pressure plate assembly.

5. Re-attach transmission by reversing above process.

Automatic Transmission

1. Lower the engine and support it on suitable blocks.
2. Remove starter and converter housing underpan.
3. Remove flywheel-to-converter assembly attaching bolts.
4. Support transmission on blocks.
5. Remove transmission-to-engine mounting bolts.
6. With engine hoist attached, remove blocks from engine only and slowly guide the engine from the transmission.
7. Re-attach automatic transmission by reversing above process.

Engine Installation

1. Bolt engine lifting tool to engine and lower engine and transmission into chassis as a unit. Guide engine to align front engine mounts with mounts on frame.
2. Install one rear transmission crossmember side bolt, swing crossmember up under transmission mount and install bolt in opposite side rail.
3. Align and install rear mount bolts.
4. Install engine front mount bolts and remove lifting tool from engine.
5. Install and connect all items in reverse order of engine removal procedure.

Cylinder Head

4 and 6 Cylinder Engines

Removal

1. Drain cooling system and remove air cleaner.
2. Disconnect choke cable (4 cylinder), accelerator pedal rod at bell crank on manifold, and fuel and vacuum lines at carburetor.
3. Disconnect exhaust pipe at manifold flange, then remove manifold bolts and clamps and remove manifolds and carburetor as an assembly.
4. Remove fuel and vacuum line retaining clip from water outlet. Then disconnect wire harness from heat sending unit and coil, leaving harness clear of clips on rocker arm cover.
5. Disconnect radiator hose at water outlet housing and battery ground strap at cylinder head.
6. Disconnect wires and remove spark plugs. On the 6 cylinder engine disconnect coil to distributor primary wire lead at coil and remove the coil.
7. Remove rocker arm cover. Back off rocker arm nuts, pivot rocker

arms to clear push rods and remove push rods.
8. Remove cylinder-head bolts, cylinder head and gasket.

Installation

1. Place a new cylinder-head gasket over dowel pins in cylinder block.
2. Guide and lower cylinder head into place over dowels and gasket.
3. Oil cylinder-head bolts, install and run them down snug.
4. Tighten the cylinder-head bolts a little at a time with a torque wrench in the correct sequence. Final torque should be 90 to 95 ft. lbs.
5. Install valve pushrods down through the cylinder-head openings and seat them in their lifter sockets.
6. Install rocker arms, balls and nuts and tighten rocker arm nuts until all pushrod play is taken up.
7. Install thermostat, thermostat housing and water outlet using new gaskets. Then connect radiator hose.
8. Install heat sending switch and torque to 15-20 ft. lbs.
9. Clean spark plugs or install new ones. Set gaps to .035 in.
10. Use new plug gaskets and torque to 20-25 ft. lbs.
11. Install coil (on six cylinder engine) then connect heat sending unit and coil primary wires, and connect battery ground cable at the cylinder head.
12. Clean surfaces and install new gasket over manifold studs. Install manifold. Install bolts and clamps and torque as specified.
13. Connect throttle linkage, and choke wire, (on four cylinder engine).
14. Connect fuel and vacuum lines to carburetor and secure lines in clip at water outlet.
15. Fill cooling system and check for leaks.
16. Adjust valve lash as outlined under Valves.
17. Install rocker arm cover and position wiring harness in clips.
18. Clean and install air cleaner.

V8 Engines

Removal and Installation

1. Drain coolant. Remove air cleaner.
2. Disconnect:
   a. battery
   b. radiator and heater hose from manifold
   c. throttle linkage
   d. fuel line
   e. coil wires
   f. temperature sending unit
   g. power brake hose, distributor vacuum hose, and crankcase vent hoses.

3. Remove:
   a. distributor, marking position
   b. Delcotron upper bracket
   c. coil and bracket
   d. manifold attaching bolts
   e. intake manifold and carburetor.
4. Remove:
   a. rocker arm covers
   b. rocker arm nuts, balls, rocker arms, and push rods. These items must be replaced in their original locations.
5. Remove cylinder head bolts, cylinder head, and gasket.
6. Reverse procedure to install. Tighten head bolts evenly to the specified torque.

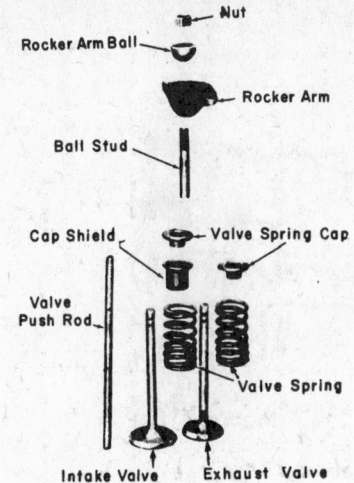

V8 valve assembly (© G.M. Corp.)

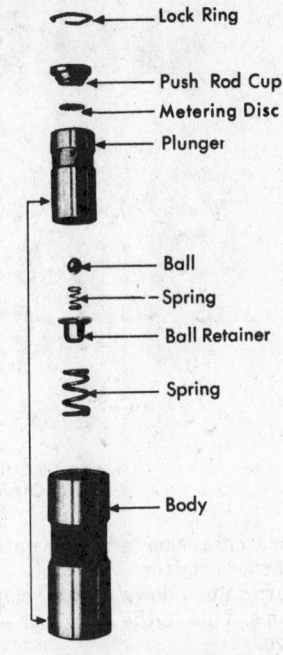

Plunger And Body Are Fitted Pairs And Must Not Be Mismated

Valve System

Chevrolet uses a hydraulically operated tappet system with adjustable rocker nuts to obtain zero lash.

Valve specifications may be obtained from the Valve Specifications chart and the Tune-up Specifications chart.

Hydraulic Valve Lifter Adjustment

In the case of disassembly, or any other cause for valve tappet adjustment, proceed as follows:

1. Adjust rocker arm nuts to eliminate lash. This must be done when lifter is on base of circle of cam.
2. Remove distributor cap and crank engine until distributor rotor points to No. 1 cylinder terminal, with points open.

The following valves can be adjusted with the engine in No. 1 firing position:

OHV 4—Intake No. 1, 2, Exhaust No. 1, 3

OHV 6—Intake No. 1, 2, 4, Exhaust No. 1, 3, 5

V8—Intake No. 1, 2, 5, 7, Exhaust No. 1, 3, 4, 8

3. Turn adjusting nut until all lash is removed from this particular valve train. This can be determined by checking pushrod side play while turning the adjustment. When all play has been removed, turn adjusting nut one more turn. This will place the lifter plunger in the center of its travel.
4. Follow steps 2 and 3 to adjust remaining valves.

The following valves can be adjusted with the engine in the No. 6 firing position (No. 4 on OHV 4):

OHV 4—Intake No. 3, 4, Exhaust No. 2, 4

OHV 6—Intake No. 3, 5, 6, Exhaust No. 2, 4, 6

V8—Intake No. 3, 4, 6, 8, Exhaust No. 2, 5, 6, 7

Readjust the lifters as follows with the engine hot and running.

1. Remove rocker arm covers and gaskets.
2. Place oil deflector clips on rocker arms.
3. With engine running at idle,

OPERATION OF VALVE LIFTER MECHANISM

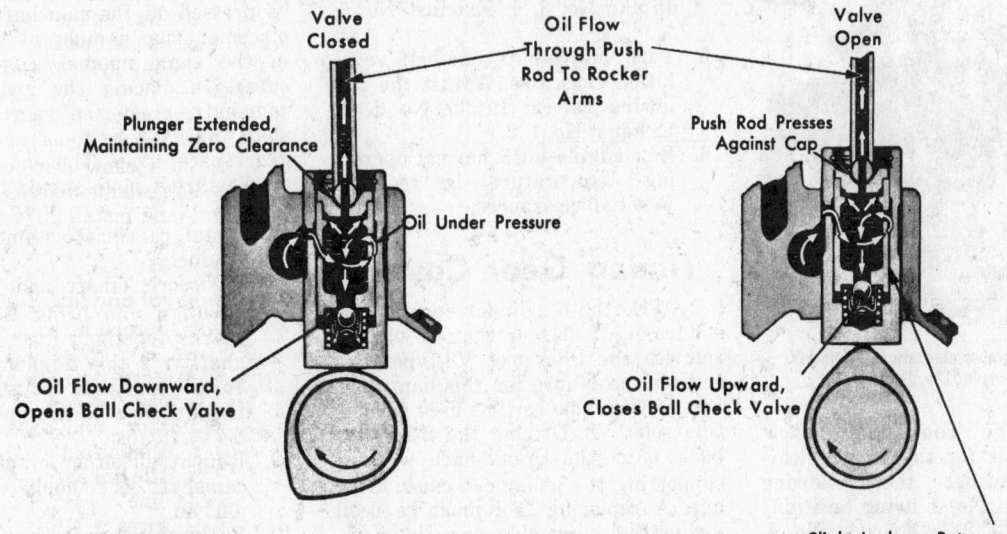

Valve lifter (© Chevrolet Div. G.M. Corp.)

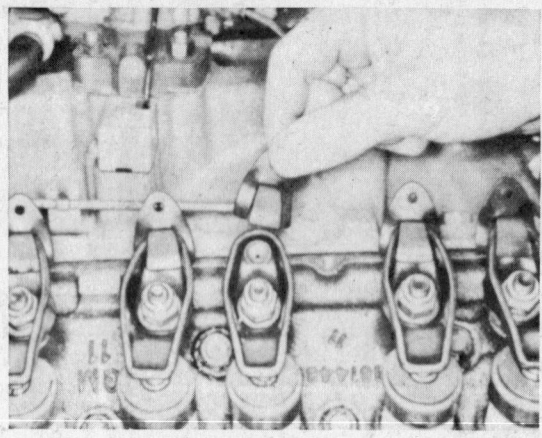

Oil deflector clips installed
(© Chevrolet Div. G.M. Corp.)

back off rocker arm nut until it starts to clatter.

4. Turn nut down until clatter stops. This is the zero lash position.

5. Tighten nut down one-quarter turn. Pause ten seconds. Repeat additional quarter turns and ten second pauses until nut has been tightened down one full turn from the zero lash position.

6. Repeat steps 3, 4, and 5 for all rocker arms.

7. Remove oil deflector clips and replace rocker arm covers.

Mechanical Valve Lifter Adjustment

1. Set engine in No. 1 firing position.

2. Adjust the clearance between the valve stems and the rocker arms using a feeler gauge.

Adjusting valve clearance 6 cylinder
(© Chevrolet Div. G.M. Corp.)

Check the Tune-Up Specifications table for the proper clearance. Adjust the following Valves in No. 1 firing position: Intake No. 2, 7, Exhaust No. 4, 8.

3. Turn crankshaft one-half revolution clockwise. Adjust the fol-

lowing valves: Intake No. 1, 8, Exhaust No. 3, 6.

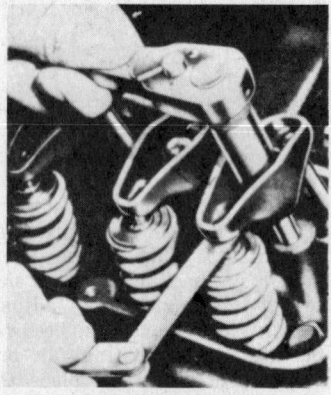

Adjusting valve clearance V8
with mechanical lifters
(© G.M. Corp.)

4. Turn crankshaft one-half revolution clockwise to No. 6 firing position. Adjust the following valves in No. 6 firing position: Intake No. 3, 4, Exhaust No. 5, 7.

5. Turn crankshaft one-half revolution clockwise. Adjust the following valves: Intake No. 5, 6, Exhaust No. 1, 2.

6. Run engine until normal operating temperature is reached. Reset all clearances.

Timing Gear Cover

NOTE: the 6 cylinder engine uses a harmonic balancer that closely resembles the Chevrolet V8-type. The removal procedure for this dampener will be the same as that used for the Chevrolet V8. Driving the dampener back onto the crankshaft without supporting the pulley can cause damage. A replacing tool must be used during the reassembly operation.

Cover Removal and Installation

1. Drain and remove radiator.

2. Remove harmonic balancer, (6 and 8 cylinder) or a crankshaft pulley, (4 cylinder) using a puller.

3. Drain engine oil and remove oil pan. Remove V8 water pump.

4. Remove timing gear cover attaching screws, and cover and gasket.

5. Reverse procedure to install.

Caution The 6 and 8 cylinder engines use a harmonic balancer. Breakage may occur where the balancer has been hammered back onto the crankshaft.

This balancer must be drawn back into place.

Oil Seal Replacement

1. After removing gear cover, pry oil seal out of front of cover with large screwdriver.

2. Install new lip seal with lip (open side of seal) inside and drive or press seal into place.

Inline Engine Camshaft Removal and Replacement

1. In addition to removing the timing gear cover, remove the grille assembly.

2. Remove valve cover and gasket, loosen all the valve rocker arm nuts and pivot the arms clear of the pushrods.

3. Remove distributor.

4. Remove coil, side cover and gasket. Remove pushrods and valve lifters.

5. Remove the two camshaft thrust plate retaining screws by working through holes in the camshaft gear.

6. Remove camshaft and gear assembly by pulling it out through the front of the block.

NOTE: if renewing either camshaft or camshaft gear, the gear must be pressed off the camshaft. The replacement parts must be assembled in the same manner (under pressure). In placing the gear on the camshaft, press the gear onto the shaft until it bottoms against the gear spacer ring. The end clearance of the thrust plate should be .001 to .005 in.

7. Install camshaft assembly in the engine.

8. Turn crankshaft and camshaft to align and bring the timing marks together. Push the camshaft into this aligned position. Install camshaft thrust plate-to-block screws and torque them to 6-7½ ft. lbs.

9. Runout on either crankshaft or camshaft gear should not exceed .003 in.

10. Backlash between the two gears should be between .004 and .006 in.

11. Install timing gear cover and

gasket.

12. Install oil pan and gaskets.
13. Install harmonic balancer.
14. Line up keyway in balancer with key on crankshaft and drive balancer onto shaft until it bottoms against crankshaft gear.
15. Install valve lifters and pushrods. Install side cover with new gasket. Attach coil wires.

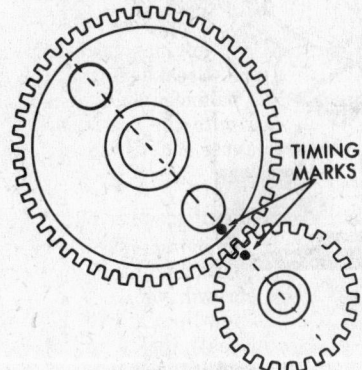

Timing mark, 4 and 6 cylinder

16. Install distributor and set timing as described under distributor at the beginning of the section.
17. Pivot rocker arms over pushrods and lash the valves as described in a previous paragraph, Valve Tappets Adjustment.
18. Add oil to the engine. Install and adjust fan belt.
19. Install radiator or shroud.
20. Install grille assembly.
21. Fill cooling system, start engine and check for leaks.
22. Check and adjust timing.

Timing Chain Replacement

V8 Models

V8 models are equipped with a timing chain. To replace the chain, remove the radiator core, water pump harmonic balancer, and the crankcase front cover. This will allow access to the timing chain. Crank the engine

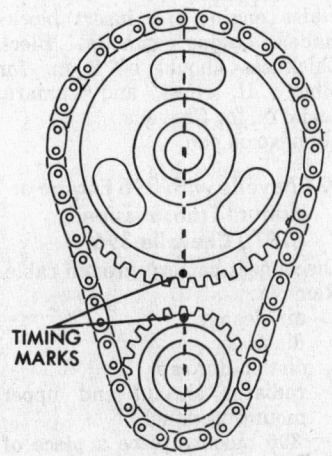

TIMING MARKS

Timing mark, V8 engine

until the zero marks punched on both sprockets are closest to one another and in line between the shaft centers. Then, take out the three bolts that hold the camshaft gear to the camshaft. This gear is a light press fit on the camshaft and will come off readily. It is located by a dowel.

The chain comes off with the camshaft gear.

A gear puller will be required to remove the crankshaft gear.

Without disturbing the position of the engine, mount the new crank gear on the shaft, then mount the chain over the camshaft gear. Arrange the camshaft gear in such a way that the timing marks will line up between the shaft centers and the camshaft locating dowel will enter the dowel hole in the cam sprocket.

Place the cam sprocket, with its chain mounted over it, in position on the front of the camshaft and pull up

with the three bolts that hold it to the camshaft.

After the gears are in place, turn the engine two full revolutions to make certain that the timing marks are in correct alignment between the shaft centers.

Engine Lubrication

Oil Pan Removal—Inline Engines

1964-67 Chevy II

1. Disconnect battery ground strap at battery.
2. Drain oil from engine.
3. Disconnect all wires from starter. Remove starter.
4. Disconnect steering idler arm bracket at right hand frame rail. Swing linkage down for pan clearance.

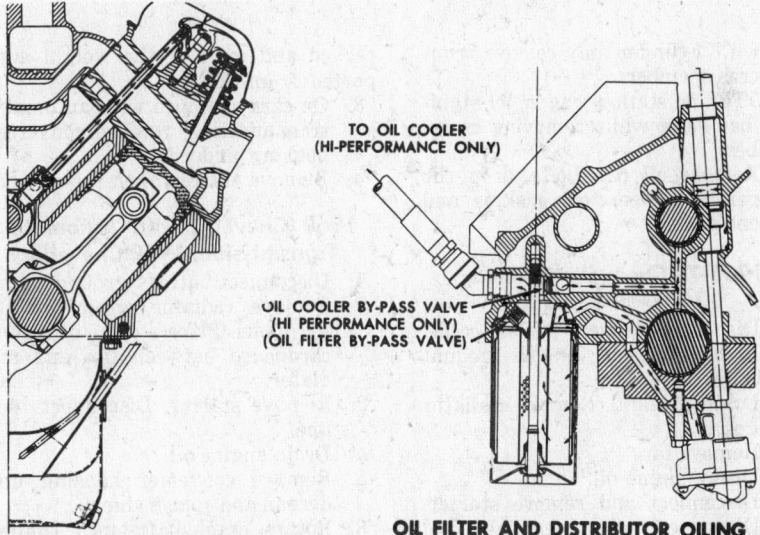

TO OIL COOLER
(HI-PERFORMANCE ONLY)

OIL COOLER BY-PASS VALVE
(HI PERFORMANCE ONLY)
(OIL FILTER BY-PASS VALVE)

OIL FILTER AND DISTRIBUTOR OILING

VALVE MECHANISM OILING

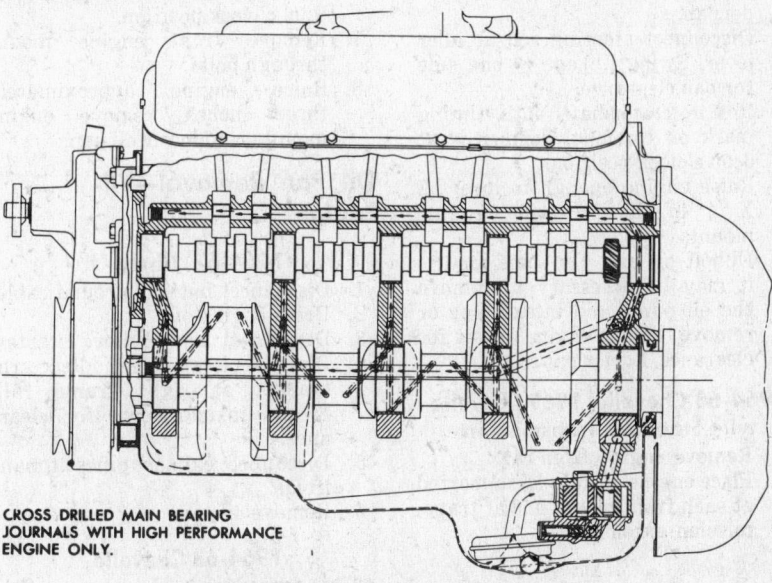

CROSS DRILLED MAIN BEARING
JOURNALS WITH HIGH PERFORMANCE
ENGINE ONLY.

CRANKCASE AND CRANKSHAFT OILING

Mark IV V8 lubrication (© G.M. Corp.)

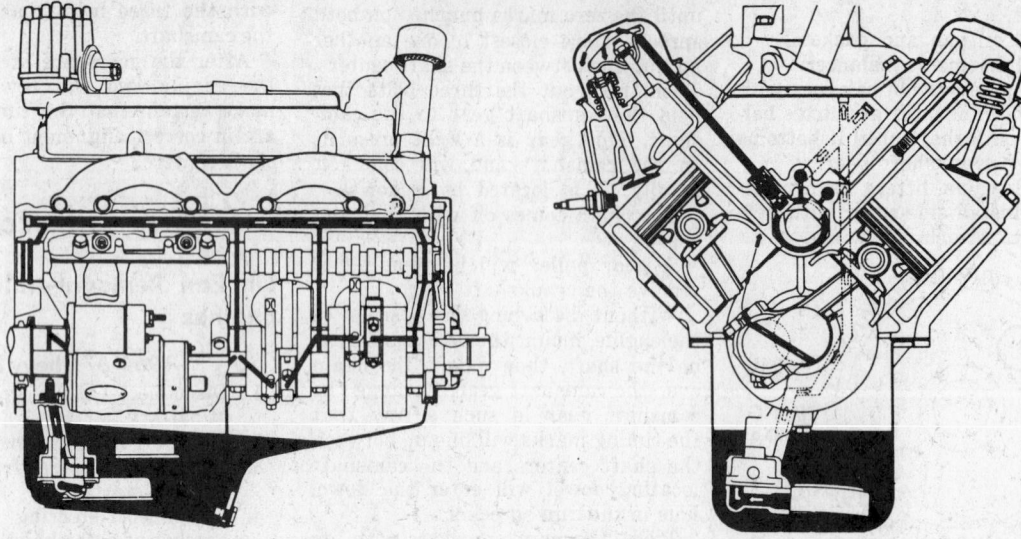

Oil passage diagram, small V8 engine (© G.M. Corp.)

5. On 6 cylinder only remove front crossmember.

NOTE: on station wagon, let stabilizer bar hang while removing crossmember.

6. Remove oil pan bolts, drop the pan and clean off gaskets and end seals.

1968-70 Chevy II and Nova, 1967-69 Camaro

1. Disconnect battery ground cable.
2. Remove front engine mount bolts.
3. Drain coolant. Remove radiator hoses.
4. Remove fan.
5. Drain engine oil.
6. Disconnect and remove starter.
7. Disconnect oil cooler lines and remove converter housing underpan.
8. Disconnect steering rod at idler lever. Swing linkage to one side for pan clearance.
9. Rotate crankshaft until timing mark on torsional damper is at 6:00 o'clock position.
10. Raise engine enough to insert 2 X 4 in. blocks under engine mounts.
11. Unbolt oil pan. On some models it may be necessary to remove the oil pump and intake pipe or remove the left engine mount for clearance. Lower pan.

1964-68 Chevelle, 1969 Chevelle with Standard Transmission

1. Remove engine from car.
2. Place engine on stands, supported at each front mount and at transmission extension.

Caution As a safety precaution, leave engine lift at-

tached and most of the weight supported from above.

3. On cars equipped with automatic transmissions, remove converter housing underpan.
4. Remove starter, then the oil pan.

1969 Chevelle with Automatic Transmission, 1970 Chevelle

1. Disconnect battery ground cable.
2. Remove radiator upper mounting panel. Place a piece of heavy cardboard between fan and radiator.
3. Remove starter. Disconnect fuel line.
4. Drain engine oil.
5. Remove converter housing underpan and splash shield.
6. Rotate crankshaft until timing mark on torsional damper is at 6:00 o'clock position.
7. Remove front engine mount through bolts.
8. Raise engine approximately three inches, remove engine mounts, and lower oil pan.

Oil Pan Removal—V8 Engines

1964-67 Chevy II

1. Disconnect battery ground cable.
2. Drain engine oil.
3. Disconnect and remove starter.
4. Disconnect steering idler arm bracket at right frame rail. Swing linkage down for clearance.
5. Disconnect exhaust pipes at manifolds.
6. Remove oil pan.

1964-68 Chevelle

This procedure is the same as that for 1964-68 Chevelle models with inline engines.

1967-69 Camaro, 1968-70 Chevy II and Nova, 1969-70 Chevelle, 1970 Monte Carlo

1. Disconnect battery ground cable.
2. Remove distributor cap.
3. Remove radiator upper mounting panel.
4. Remove fan. On 396 and 427 models, place a piece of heavy cardboard between the radiator and fan.
5. Drain engine oil.
6. Disconnect exhaust or crossover pipes.
7. Remove converter housing underpan and splash shield.
8. Disconnect steering idler lever at the frame. Swing linkage down.
9. Rotate crankshaft until timing mark on torsional damper is at 6:00 o'clock position.
10. Remove starter.
11. On small V8, remove fuel pump.
12. Remove front engine mount through bolts.
13. Raise engine and insert blocks under engine mounts. Block thickness should be 2 in. for Chevy II, Nova, and Camaro, and 3 in. for Chevelle.
14. Remove oil pan.

1969 Chevelle with 396 Engine or Manual Transmission, 1970 Chevelle 396

1. Disconnect battery ground cable.
2. Remove:
   a. air cleaner
   b. dipstick
   c. distributor cap
   d. radiator shroud and upper mounting panel.
3. On 396 models, place a piece of heavy cardboard between radiator and fan.

4. Disconnect engine ground straps. Remove fuel pump on 307 and 350 engines.
5. Disconnect accelerator control cable.
6. Drain oil. Remove filter on 307 and 350 engines.
7. Remove driveshaft and plug rear of transmission.
8. Remove starter.
9. Disconnect transmission linkage at transmission or remove floor-shift lever.
10. Disconnect speedometer cable and back-up switch connector.
11. On manual transmission vehicles disconnect clutch cordon shaft at frame. On automatic transmission vehicles, disconnect cooler lines, detent cable, rod or switch wire, and modulator pipe.
12. Remove crossmember bolts. Jack up engine. Move crossmember rearward.
13. Remove crossover or dual exhaust pipes.
14. Remove:
    a. flywheel housing cover
    b. transmission
    c. flywheel housing and throw-out bearing (manual transmission)
    d. front engine mount through bolts.
15. Raise rear of engine approximately 4 inches. Support engine by hoist.
16. Raise front of engine approximately 4 inches and insert 2 in. blocks under front engine mounts.
17. Rotate crankshaft until timing mark on torsional damper is at 6:00 o'clock position.
18. Unbolt and remove oil pan.

Connecting Rods and Pistons

Removal
1. Drain crankcase and remove oil pan.
2. Drain cooling system and remove cylinder heads.
3. Remove any ridge or deposits from the upper end of cylinder bores with a ridge reamer.
4. Check rods and pistons for identification numbers and, if necessary, number them.
5. Remove connecting rod cap nuts and caps. Push the rods away from the crankshaft and install caps and nuts loosely to their respective rods.
6. Push piston and rod assemblies up and out of the cylinders.

Piston Ring Installation
Before replacing rings, inspect cylinder bores.
1. Using internal micrometer measure bores both across thrust faces of cylinder and parallel to axis of crankshaft at a minimum of four locations equally spaced. The bore must not be out of round by more than 0.005 in. and it must not "taper" more than 0.005 in. "Taper" is the difference in wear between two bore measurements in any one cylinder. Bore any cylinder beyond limits of out of roundness or taper. Bore to diameter of next available oversize piston that will clean up wear. Compare piston diameter with cylinder bore size to determine piston to bore clearance. Check that clearance is within the specified limits.

Exploded view, 6 cylinder engine, typical (© G.M. Corp.)

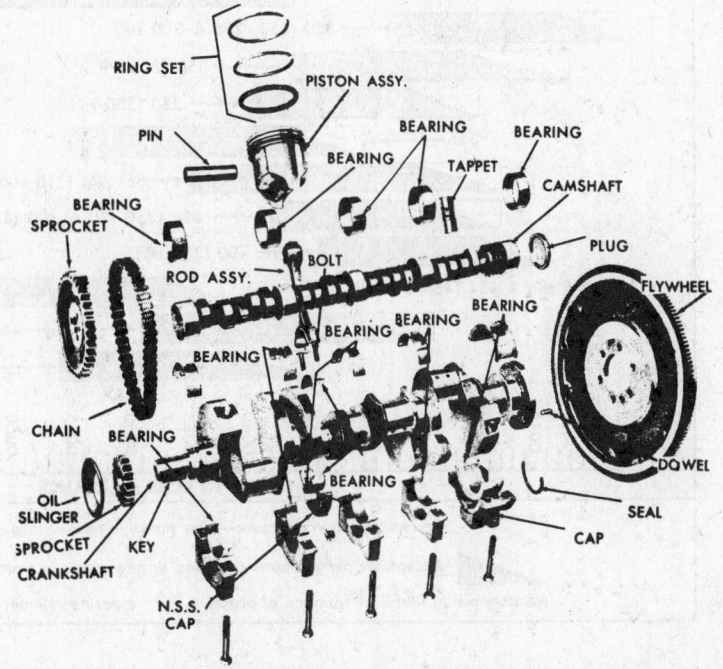

Exploded view, typical V8 (© G.M. Corp.)

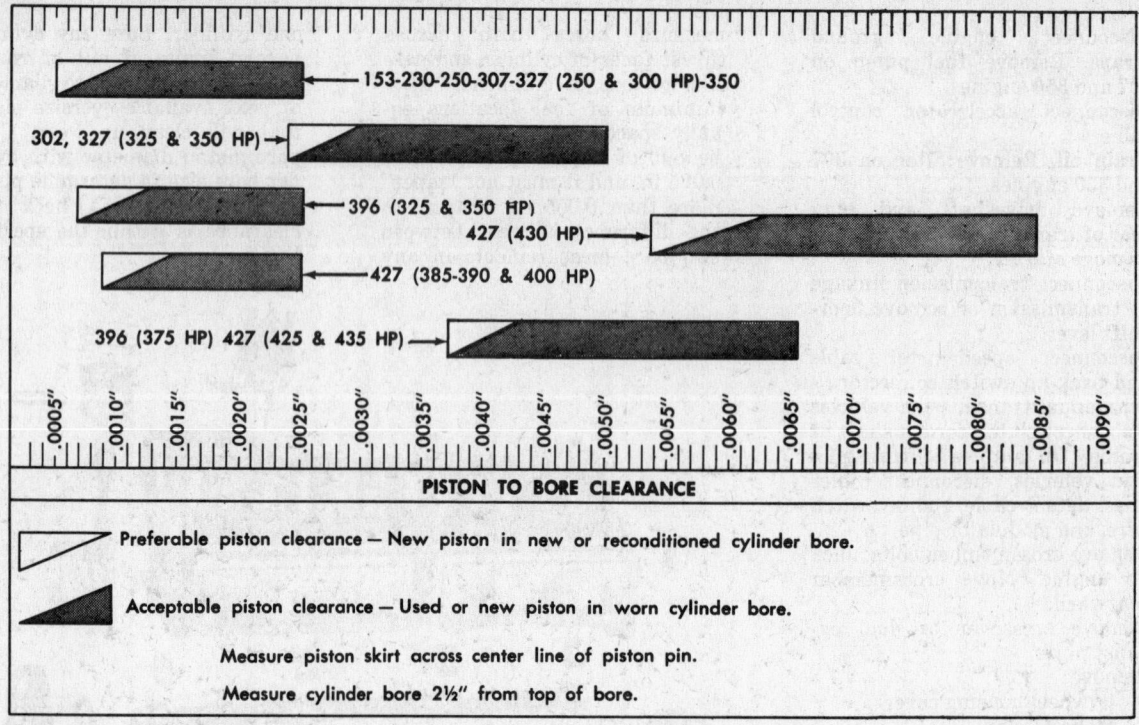

Piston—bore clearance, up to 1968 (© Chevrolet Div. G.M. Corp.)

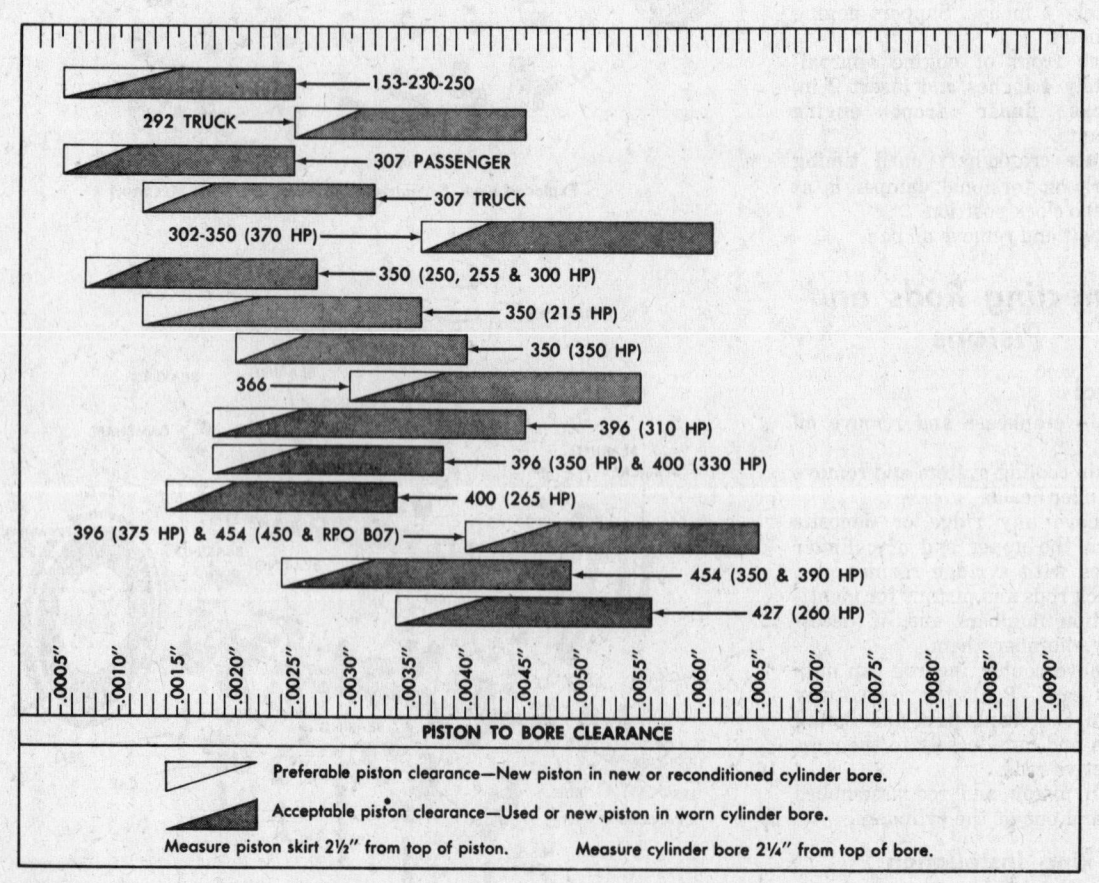

Piston—bore clearance, 1969-70 (© Chevrolet Div. G.M. Corp.)

2. If bore is within limits dimensionally, examine bore visually. It should be dull silver in color and exhibit a pattern of machining cross hatching intersecting at about 45-65 degrees. There should be no scratches, tool marks, nicks, or other damage. If any such damage exists, bore cylinder to clean up damage and then to next oversize piston diameter. Polished or shiny places in the bore are known as glazing. Glazing causes poor lubrication, high oil consumption and ring damage. Remove glazing by honing cylinders with clean, sharp stones of No. 180-220 grit to obtain a surface finish of 15-35 RMS. Use a hone to obtain correct piston clearance and surface finish in any cylinder that has been bored. This or any other machining operation should be done with the cylinder block completely disassembled. Hot tank cylinder block after honing or boring. To remove minor glazing when honing equipment is not available, run emery cloth back and forth across glazed area perpendicular to axis of bore. Scrub block and bores thoroughly with soap and water to remove all grit after using emery cloth.

NOTE: the emery cloth method should be used only as a last resort as it is a method much inferior to honing.

3. If cylinder bore is in satisfactory condition, place each ring in its bore in turn and square it in bore with head of piston. Measure ring end gap. If gap is greater than limit, get new ring. If gap is less than limit, file end of ring to obtain correct gap.

Ring End Gap

	Compression Rings (in.)	Oil Rings (in.)
all inline engines, 283, 307, 350, 400, 454	.010-.020	.015-.055
302, 327	.013-.025	.015-.055
396, 427	.010-.030	.010-.030

4. Check ring side clearance by installing rings on piston, and inserting feeler gauge of correct dimension between ring and lower land. Gauge should slide freely around ring circumference without binding. Any wear will form a step on lower land. Replace any pistons having high steps. Before checking ring side clearance be sure ring grooves are clean and free of carbon, sludge, or grit.

5. Space ring gaps at equal intervals around piston circumfer-

Ring Side Clearance

	Top Ring (in.)	Second Ring (in.)	Oil Ring (in.)
all inline engines, 283, 307, 400 (265 hp), 302, 327 (210, 250, 275 hp)	.0012-.0027	.0012-.0032	.000-.005
350 (215, 250 hp)	.0012-.0032	.0012-.0032	.002-.007
350 (300, 350, 370 hp), 327 (300, 325, 350 hp)	.0012-.0032	.0012-.0032	.000-.005
400 (330 hp), 454, 396, 427	.0017-.0032	.0017-.0032	.005-.0065

ence. Be sure to install piston in its original bore. Install piston and rod assembly with connecting rod bearing tang slots on the side opposite the camhaft on V8 engines. Inline engine pistons must have the piston notch fac-

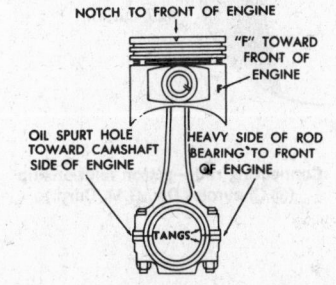

Piston and rod assembly, 4 and 6 cylinder

ing the front of the engine. Install short lengths of rubber tubing over connecting rod bolts to

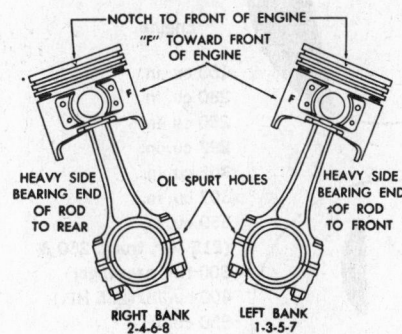

Piston and rod assembly, V8, 283 and 327

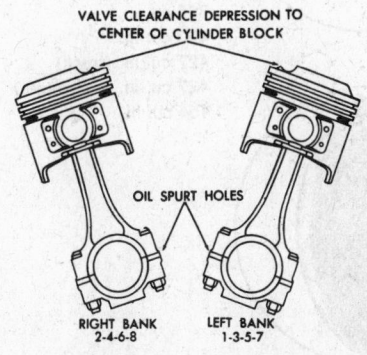

Piston and rod assembly, V8 396

prevent damage to the rod journals. Install ring compressor over rings on piston. Lower piston and rod assembly into bore until ring compressor contacts block. Using wooden handle of hammer push piston into bore while guiding rod onto journal.

Front Suspension

1964-67 Chevy II

Front suspension is an independent coil-spring, ball-joint-type with rubber bushed, pivoting upper and lower control arms. The coil springs are positioned at their lower ends on a pivoting spring seat bolted to the upper control arm. The upper end of the spring extends into spring towers formed in the front end sheet metal. Direct, double-acting shock absorbers are located inside the coil springs and are attached to the lower coil spring seat and to the upper bracket, accessible from the engine compartment.

Each lower control arm has a strut rod running diagonally forward to a brace attached between frame and radiator support. This strut rod provides for caster angle adjustment. Camber angle is adjusted by means of a cam-shaped lower control arm inner pivot bolt. A stabilizer rod, on station wagons, connects the two lower control arms and is rubber mounted to the front crossmember. Front wheel bearings are tapered roller bearings.

1964-70 Chevelle, 1968-70 Chevy II and Nova, 1967-69 Camaro, 1970 Monte Carlo

In these models, the springs ride on the lower control arms. Ball joints connect the upper and lower arms to the steering knuckle. Tapered roller wheel bearings are used.

Camber angle is adjusted by means of upper control arm inner support shaft shims.

Caster angle is adjusted by means of upper control arm inner support shaft shims.

A stabilizer bar is used on all Chevelles. 1968-70 Chevelles and 1970 Monte Carlo models have diagonal strut rods from the lower control arm to the frame.

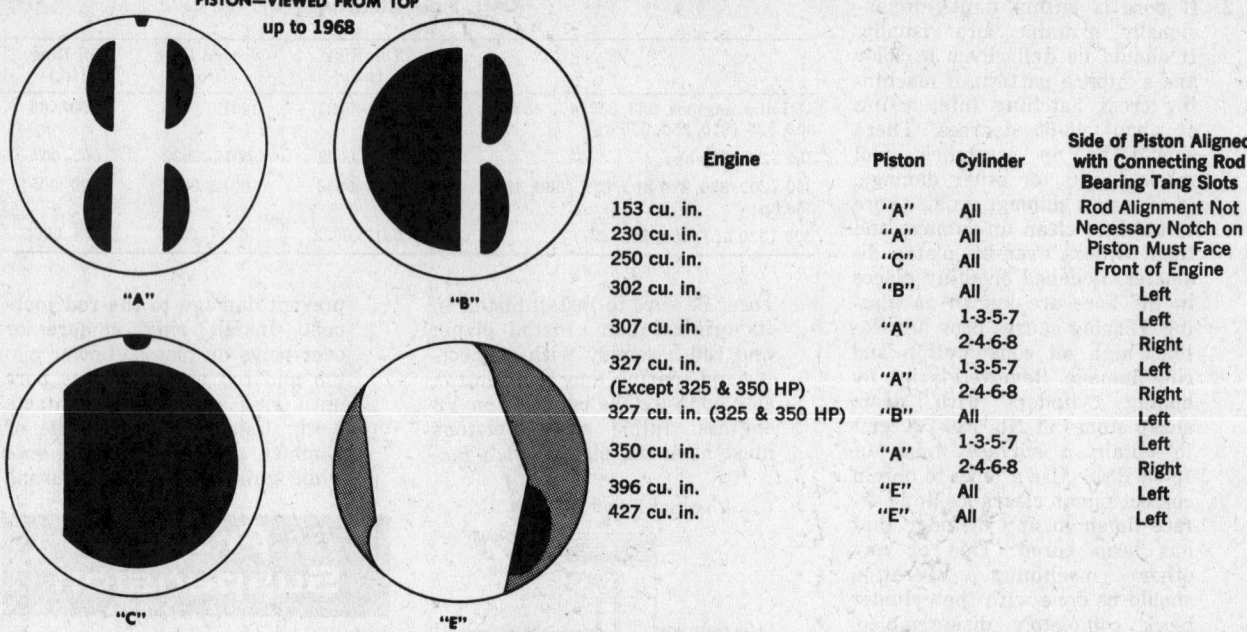

PISTON—VIEWED FROM TOP
up to 1968

"A"   "B"   "C"   "E"

Engine	Piston	Cylinder	Side of Piston Aligned with Connecting Rod Bearing Tang Slots
153 cu. in.	"A"	All	Rod Alignment Not Necessary Notch on Piston Must Face Front of Engine
230 cu. in.	"A"	All	
250 cu. in.	"C"	All	
302 cu. in.	"B"	All	Left
307 cu. in.	"A"	1-3-5-7	Left
		2-4-6-8	Right
327 cu. in. (Except 325 & 350 HP)	"A"	1-3-5-7	Left
		2-4-6-8	Right
327 cu. in. (325 & 350 HP)	"B"	All	Left
350 cu. in.	"A"	1-3-5-7	Left
		2-4-6-8	Right
396 cu. in.	"E"	All	Left
427 cu. in.	"E"	All	Left

Connecting rod—piston relationship
(© Chevrolet Div. G.M. Corp.)

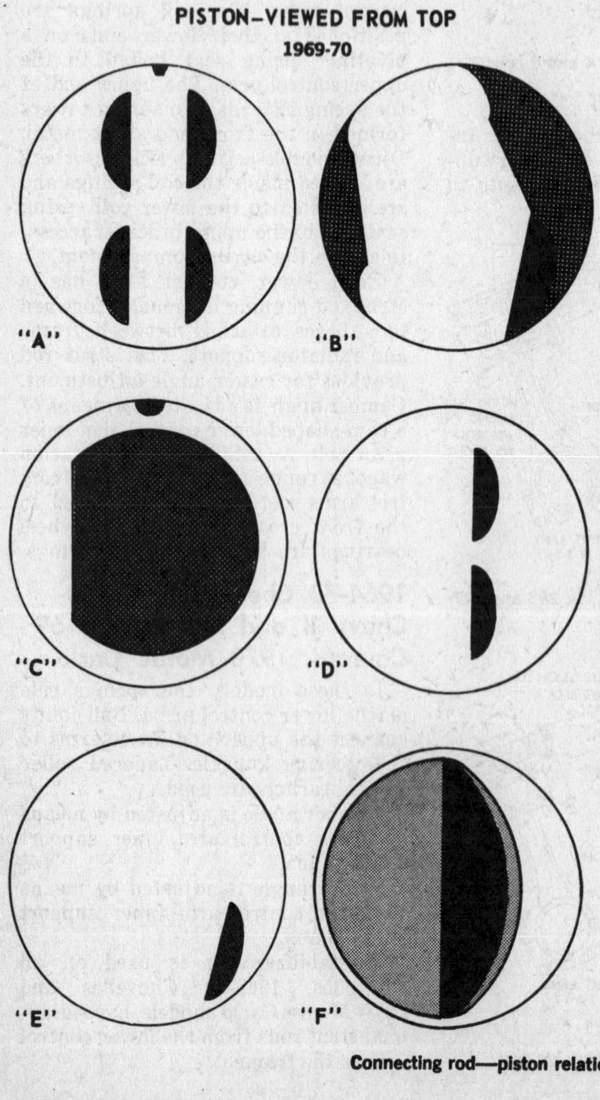

PISTON—VIEWED FROM TOP
1969-70

"A"   "B"   "C"   "D"   "E"   "F"

Engine	Piston	Cylinder	Side of Piston Aligned with Connecting Rod Bearing Tangs
153 cu. in.	"A"	All	Rod alignment not necessary—notch on piston must face front of engine.
230 cu. in.	"A"	All	
250 cu. in.	"C"	All	
292 cu. in.	"C"	All	
302 cu. in.	"F"	All	Left
307 cu. in.			
350 cu. in. (215 H.P. truck, 250 & 300 HP. passenger)	"A"	1-3-5-7	Left
		2-4-6-8	Right
400 cu. in. (265 HP.)			
350 cu. in. (255 HP. truck)	"D"	All	Left
350 cu. in. (350 & 370 HP. passenger)	"F"	All	Left
366 cu. in.	"E"	All	Left
396 cu. in.	"B"	All	Left
400 cu. in.			
427 cu. in. (truck)	"E"	All	Left
427 cu. in. (passenger)	"B"	All	Left
454 cu. in.	"B"	All	Left

Connecting rod—piston relationship (© Chevrolet Div. G.M. Corp.)

Periodic maintenance of the front suspension includes lubrication of the four ball joints, spring seat lower pivot shafts and adjustment and lubrication of the front wheel bearings.

Further data on front end alignment can be obtained from the Front Wheel Alignment chart and from the Unit Repair Section.

Coil Spring Replacement

1964-67 Chevy II—Coil Spring Removal

1. Raise car and remove wheel.
2. Support lower control arm with adjustable jackstand and raise slightly from full rebound position.
3. Remove shock absorber.
4. Insert spring compressor into upper spring tower so that lower U-bolt fits into shock absorber mounting holes in spring seat. Secure the two lower studs to the spring seat with nuts.
5. Fit tool upper pilot to top of spring and compress the spring by tightening the upper nut. Compress spring until the screw is bottomed out.
6. Remove lower spring seat retaining nuts, lift spring and seat assembly from control arm and guide it down and out through fender skirt.

1964-67 Chevy II— Coil Spring Installation

1. Install new spring into tool and compress spring until screw is bottomed out.
   NOTE: spring coil ends must be against spring stops in upper and lower seats.
2. Lift spring and tool assembly into place and position so that the upper spring stop is inboard.
3. Install lower spring seat to the control arm. Torque the nuts to 25-35 ft. lbs.
4. Loosen spring compressor until spring is properly seated in upper spring tower and remove the tool.
5. Install shock absorber.
6. Remove adjustable jackstand and

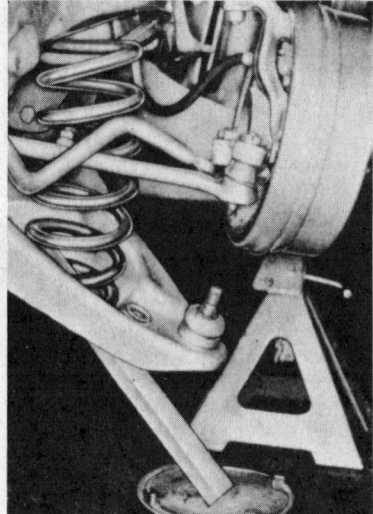

Front spring removal
(© Chevrolet Div. G.M. Corp.)

install wheel and tire. Lower car to the floor.

Coil Spring R & R—1964-70 Chevelle, 1968-70 Chevy II and Nova, 1967-69 Camaro, 1970 Monte Carlo

1. Hold the shock absorber upper stem from turning, then disconnect the shock absorber at the top.
2. Support the car by the frame, so the control arms hang free, remove wheel assembly (replace one wheel nut to hold the brake drum) shock absorber, and stabilizer bar to lower control arm link.
3. Place a steel bar through the shock absorber mounting hole in the lower control arm so that the notch seats over the bottom spring coil and the bar extends outboard beyond the end of the control arm and slightly toward the front of the car.
4. With a suitable jack, raise the end of the bar.
5. Remove lower ball stud cotter pin

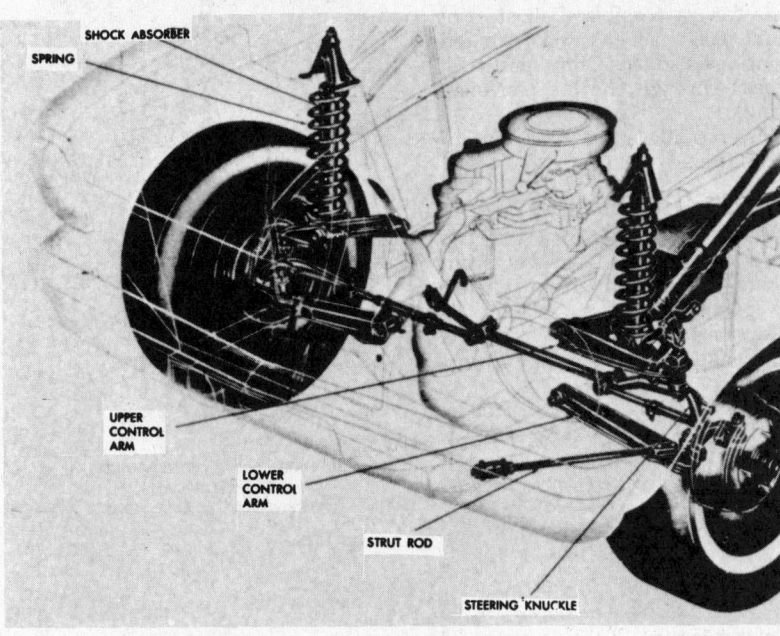

Front suspension, 1964-67 Chevy II (© G.M. Corp.)

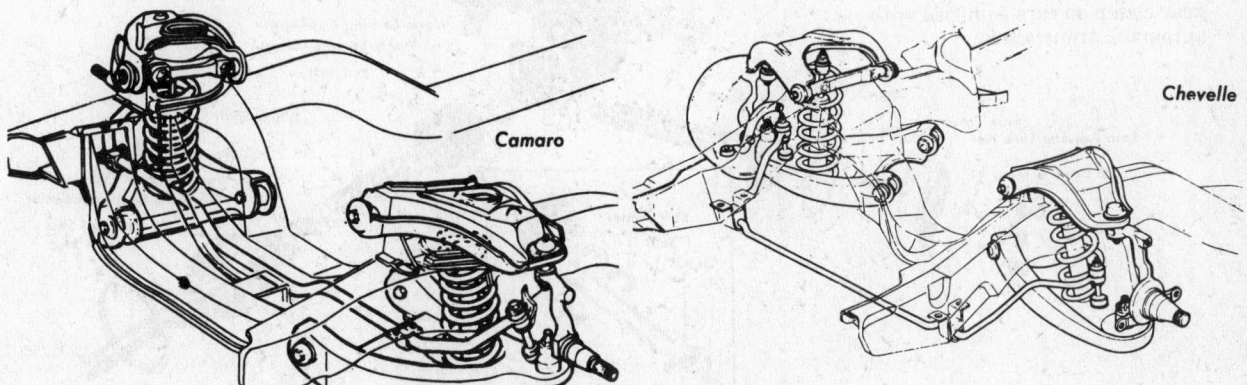

Front suspension (© Chevrolet Div. G.M. Corp.)

and nut, then remove the ball stud from the knuckle.

NOTE: do not damage the ball-joint seal during this operation.

6. Lower the jack supporting the steel bar and control arm until the spring can be removed.
7. Install by reversing.

Jacking, Hoisting

1. Jack car at front spring seat of lower control arm. Jack car at rear axle housing.
2. To lift at frame, use side rails in front of body floor pan and at rear corner at squared off corner of box ahead of rear wheel.

Steering Gear

Manual Steering Gear

Recirculating ball type gear is used on General Motors cars. Adjustment procedures are found in the Unit Repair Section.

Steering Gear R & R

1964-66 Chevy II

1. Remove pitman arm retaining nut. (Place an index mark on pitman arm and sector shaft so that the two parts can be reassembled in the same register.)
2. Remove pitman arm from sector shaft.
3. Disconnect transmission linkage from shift lever (S).
4. Remove the two lower steering gear-to-frame mounting bolts.
5. Disconnect external electrical wires from horn junction, directional switch and back-up lamp switch (if so equipped).
6. Remove steering wheel.
7. Remove screws holding mast jacket hole seal to toe panel.
8. Remove mast jacket-to-steering gear clamp on cars equipped with automatic transmission.

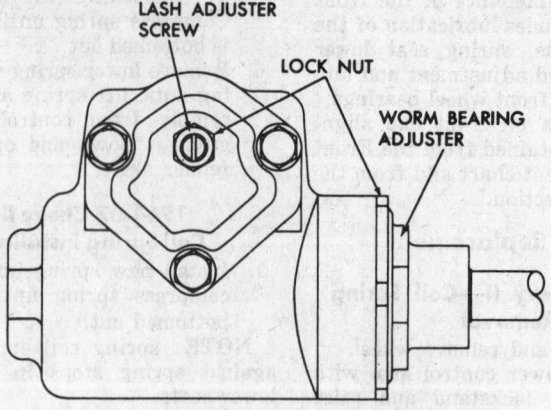

1967-70 Steering gear adjusting points
(© Chevrolet Div. G.M. Corp.)

9. Remove nuts from mast jacket-to-dash brace clamp.
10. Move mast jacket and steering gear assembly downward and away from dash, pivoting on remaining steering gear mounting bolt. Move front seat to the rear as far as possible. Pull steering and mast jacket toward you, rotating it so that shift levers will pass through the toe pan opening.
11. Reverse above procedure for installing.

All Models Except 1964-66 Chevy II

NOTE: on 1964-67 Chevelle, remove stabilizer to frame mounting brackets. Unbolt left front bumper bracket and brace from frame after marking location.

1. Disconnect steering shaft coupling.
2. Remove pitman arm with puller after marking relationship to shaft.
3. Remove bolts to frame. Remove steering gear.
4. Reverse procedure to install.

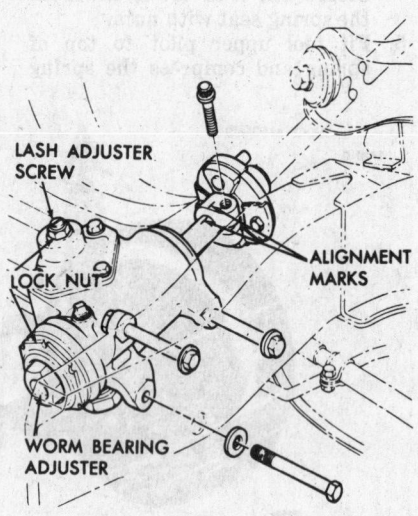

Steering gear adjusting points, 1964-67 models
(© Chevrolet Div. G.M. Corp.)

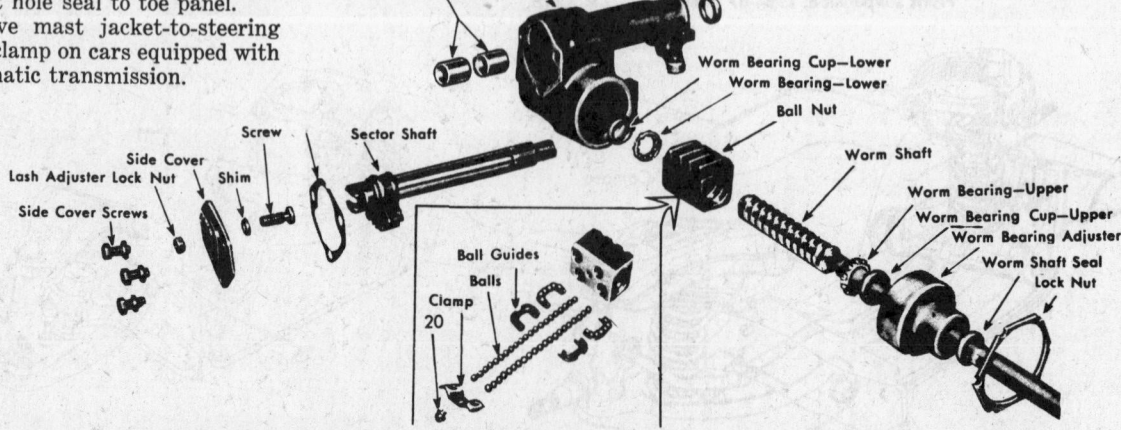

Exploded view of steering gear, Chevy II (© Chevrolet Div. G.M. Corp.)

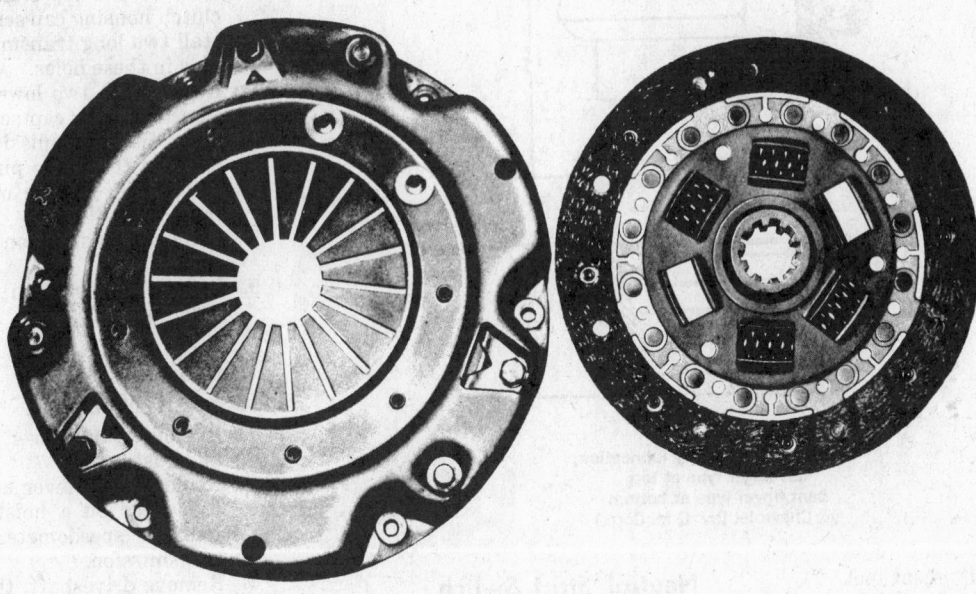

Diaphragm clutch assembly (© Chevrolet Div. G.M. Corp.)

Power Steering Gear

Two types of power steering are used. The 1964-67 Chevy II uses the linkage assist type of gear with a pump delivering an assist to a power cylinder attached to the steering linkage.

All other models use an integral type of power steering gear. A pump delivers hydraulic pressure through two hoses to the steering gear itself.

Detailed service coverage is found in the Unit Repair Section.

Clutch

A diaphragm type clutch assembly is used with all manual transmissions. A flat finger diaphragm clutch is used for normal service. V8 engines with four speed transmissions have a bent finger, centrifugal diaphragm clutch assembly. In this design the release fingers are bent back to gain a centrifugal boost and to insure quick re-engagement at high engine speeds. The centrifugal type clutch has the advantage of low pedal effort with high plate load. An optional heavy duty clutch assembly is used with some Mark IV engines. This unit is a dual plate bent finger diaphragm clutch.

The clutch release bearings used with the flat and bent finger diaphragms are not interchangeable. Using the flat finger release bearing with the bent finger clutch assembly will result in slippage and rapid wear.

The only service adjustment necessary on the clutch is to maintain the correct pedal free play. Clutch pedal free play, or throwout bearing lash, decreases with driven disc wear.

Further information on clutches may be found in the Unit Repair Section.

Removal
1. Support engine and remove transmission.
2. Disconnect clutch fork push rod and spring.
3. Remove flywheel housing.
4. Slide clutch fork from ball stud and remove fork from dust boot. Ball stud is threaded into clutch housing and may be replaced, if necessary.
5. Install an alignment tool to support the clutch assembly during removal. Mark flywheel and clutch cover for reinstallation, if they do not already have X marks.
6. Loosen clutch to flywheel attaching bolts evenly, one turn at a time, until spring pressure is released. Remove bolts and clutch assembly.

Installation
1. Clean pressure plate and flywheel face.

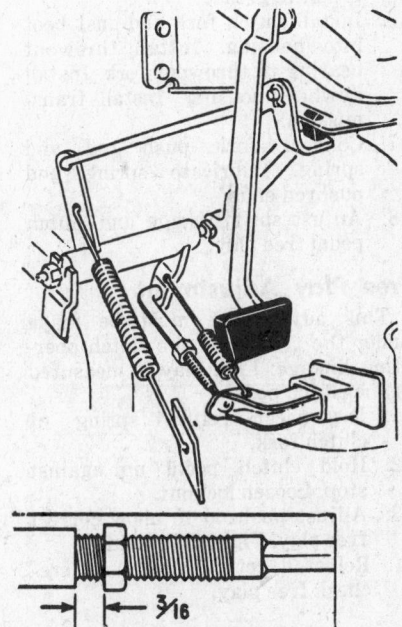

Clutch linkage, 1964-67 Chevelle (© Chevrolet Div. G.M. Corp.)

2. Support clutch disc and pressure plate with alignment tool. The driven disc is installed with the damper springs on the transmission side. The grease slinger is always on the transmission side.
3. Turn clutch assembly until mark on cover lines up with mark on flywheel, then install bolts. Tighten down evenly and gradually to avoid distortion.

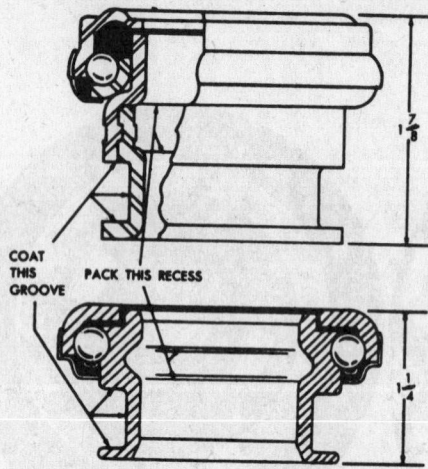

Clutch release bearing lubrication,
flat finger type at top,
bent finger type at bottom
(© Chevrolet Div. G.M. Corp.)

4. Remove alignment tool.
5. Lubricate ball socket and fork fingers at release bearing end with high melting point grease. Lubricate recess on inside of throwout bearing and throwout fork groove with a light coat of graphite grease.
6. Install clutch fork and dust boot into housing. Install throwout bearing to throwout fork. Install flywheel housing. Install transmission.
7. Connect fork push rod and spring. Lubricate spring and pushrod ends.
8. Adjust shift linkage and clutch pedal free play.

Free Play Adjustment

This adjustment must be made under the vehicle on the clutch operating linkage. Free play is measured at the clutch pedal.

1. Disconnect return spring at clutch fork.
2. Hold clutch pedal up against stop. Loosen locknut.
3. Adjust pushrod to allow correct free play. Tighten locknut.
4. Reinstall return spring and recheck free play.

Clutch Pedal Free Play

Vehicle	Free Play at Pedal Pad (in.)
1964 Chevy II	3/4-1
1964-65 Chevelle, 1965 Chevy II	3/4-1 1/8
1966 all models, 1967 Chevy II, 1967 Chevelle	1-1 1/2
1967-69 Camaro, 1968-70 Chevy II	1-1 1/8
1968-70 Chevelle	1 1/8-1 3/4

Neutral Start Switch

This switch, used on 1969-70 standard shift models, is operated by linkage from the clutch pedal arm, inside the vehicle. The function of the neutral start switch is to prevent the engine from being started unless the clutch pedal is fully depressed. There is no adjustment necessary for this switch.

Standard Transmission

Three and four speed transmissions are available on all models. Three speed transmissions from 1966-70 have all three forward gears synchronized; earlier models have only second and third gears synchronized. For 1966-69 models, a heavy duty three speed transmission is available.

All four speed transmissions have synchromesh in all forward gears. For 1966-70 models, a heavy duty four speed transmission is avilable.

A planetary overdrive, in combination with a three speed transmission, is available from 1966-70.

See the Unit Repair Section for transmission repairs, overdrive repairs, and Hurst shift linkage adjustments.

Transmission R & R Column Shift Models

1. Drain transmission.
2. Disconnect speedometer cable at transmission. Disconnect shift control rods from shift levers at the transmission.
3. Remove driveshaft.

4. Support rear of engine. Remove crossmember.
5. Remove two top transmission to clutch housing cap screws and install two long transmission guide pins in these holes.
6. Remove the two lower transmission mounting cap screws.
7. Slide the transmission straight back on the guide pins until the clutch gear is free of splines in the clutch disc.
8. Remove transmission from under car.
9. Install transmission in reverse order of removal.

Transmission R & R Floorshift Models

1. Remove shift lever trim plate and dust boot.
2. Remove shift lever assembly.
3. Raise car on a hoist, then disconnect speedometer cable at transmission.
4. Remove driveshaft, then support engine at the oil pan with a padded jack capable of supporting the engine weight when the transmission is removed.
5. Disconnect shift lever bracket. Remove all transmission shift levers and linkage.
6. Remove crossmember attaching bolts.
7. Loosen transmission crossmember and move rearward or remove.
8. Remove transmission-to-clutch housing retaining bolts and install two guide pins in top holes.
9. Slide transmission straight back to free the input shaft from the clutch hub.
10. When transmission has moved rearward enough, tilt front of transmission down and lower unit from car.
11. Install by reversing removal procedure.

Shift Linkage Adjustment

For adjustment on Hurst shift linkages, see the Unit Repair Section.

1964-68 Three Speed Column Shift

1. With transmission shifter rods disconnected at transmission levers, move both levers into neutral detents.
2. Move manual selector lever into neutral position.
3. Align first and reverse shifter tube lever with second and high shifter tube lever on the mast jacket. In some cases, a pin may be used to hold the levers in alignment.

NOTE: the key is engaged with the slot on the second and third shifter tube lever when selector lever is in the neutral position.

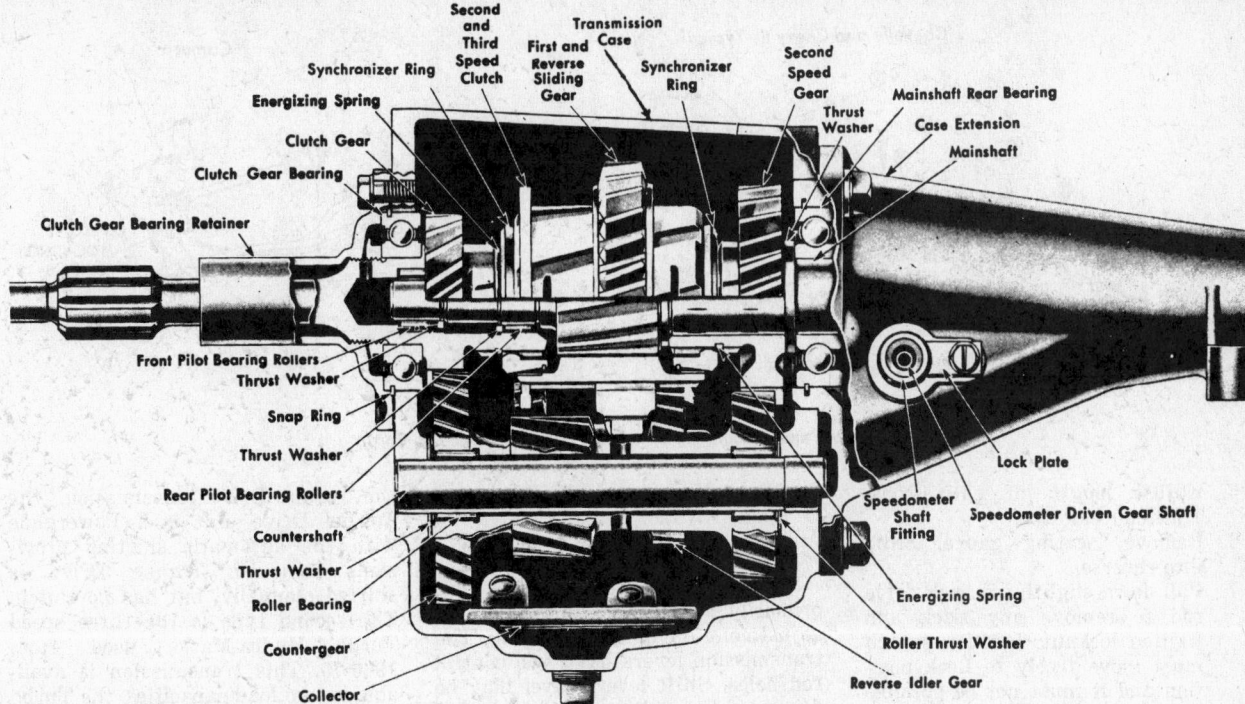

Typical 3-speed transmission (© G.M. Corp.)

4. Loosen control rod clamp bolts. Install control rods on mast jacket shifter levers and secure with retaining clips.
5. Adjust length of first-reverse rod. Tighten clamp bolt.
6. Adjust length of second-third control rod. Tighten clamp bolt.
7. Shift through all positions to check adjustment, and to insure positive and full gear engagement.

1969-70 Three Speed Column Shift

1. Place ignition switch in Off position.
2. Loosen shift rod lock nuts.
3. Set transmission first-reverse lever in reverse position. Push up on first-reverse control rod until column lever is in reverse detent position. Tighten first-reverse lock nut.
4. Shift column and transmission levers to neutral position. Insert a 3/16 in. dia. rod into alignment holes in levers and alignment plate.
5. Tighten second-third locknut.
6. Remove alignment rod. Shift column lever to reverse. Turn key to Lock. Ignition switch must move freely to Lock position and it must not be possible to turn key to Lock when in any transmission position other than reverse. If this interlock binds, leave switch in Lock position and readjust first-reverse rod.
7. Check shifting.

1966-68 Three Speed Floorshift

1. Loosen shift rod locknuts.
2. Set shift lever in neutral and in-

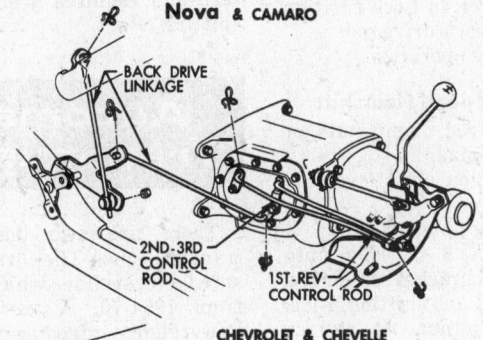

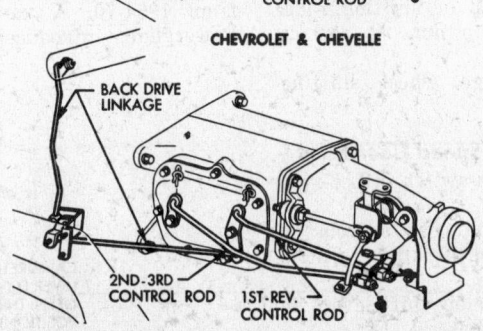

Typical three speed floorshift linkage, 1969-70
(© Chevrolet Div. G.M. Corp.)

stall locating pin into control lever bracket assembly. On some linkages, a flat locating gauge is used. This gauge is 1/8 thick X 41/64 wide X 3 in. long.
3. Place transmission shift levers in neutral positions.
4. Adjust length of control rods. Tighten locknuts.
5. Remove gauge or pin. Check shifting operation.

1969-70 Three Speed Floorshift

1. Turn ignition switch to Lock position.
2. Loosen locknuts on shift rods and back drive rod.
3. Set transmission levers in neutral positions.
4. Set floorshift lever in neutral. Install locating gauge, 1/8 thick X 41/64 wide X 3 in. long, into control lever bracket assembly.

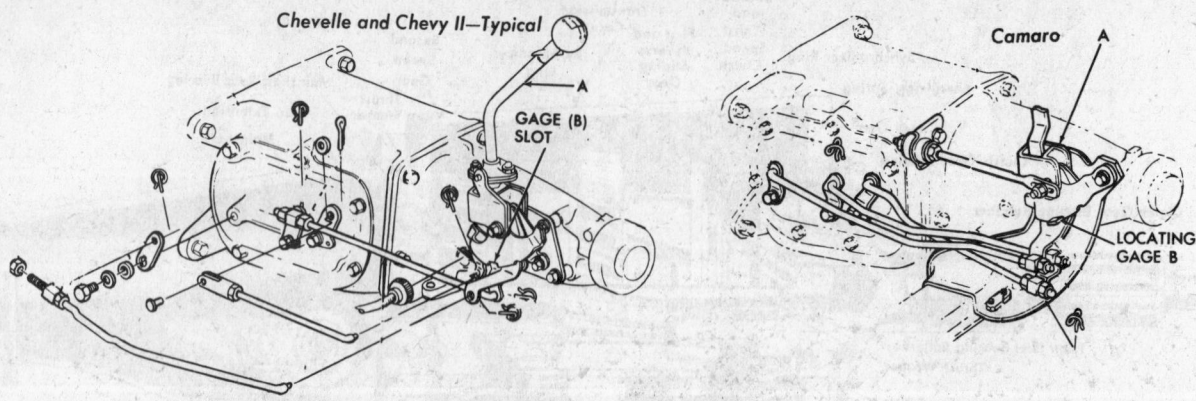

Chevelle and Chevy II—Typical

GAGE (B) SLOT

A

Camaro

A

LOCATING GAGE B

Typical four speed floorshift linkage (© Chevrolet Div. G.M. Corp.)

5. Adjust length of shift rods. Tighten locknuts.
6. Remove locating gauge. Shift into reverse.
7. Pull down slightly on back drive rod to remove any slack and tighten locknut. Ignition switch must move freely to Lock position and it must not be possible to turn key to Lock when in any transmission position other than reverse. If this interlock binds, leave the switch in Lock position and readjust back drive rod.
8. Check shifting operation.

1964-68 Four Speed Floorshift

1. Loosen shift rod clamp nuts or remove clevis pins.
2. Set transmission shift levers in neutral positions.
3. Insert locating gauge, $\frac{1}{8}$ thick X 41/64 wide X 3 in. long, into control lever bracket assembly.
4. Adjust length of shifting rods. Tighten clamp nuts or replace clevis pins.
5. Remove gauge. Check shifting operation.

1969-70 Four Speed Floorshift

1. Place ignition switch in Lock position.
2. Loosen locknuts at swivels on shift rods and back drive control rod.
3. Set transmission shift levers in neutral positions.
4. Shift lever into neutral. Insert locating gauge, $\frac{1}{8}$ thick X 41/64 wide X 3 in. long, into control lever bracket assembly.
5. Tighten shift rod locknuts and remove gauge.
6. Shift lever into reverse, then pull down slightly on back drive rod to remove slack. Tighten back drive rod locknut.
7. Ignition switch must move freely to Lock position and it must not be possible to turn key to Lock when in any transmission position other than reverse. Readjust back drive rod, if necessary.

8. Check for proper shifting operation.

Short Throw Shift Adjustment

Some four speed transmissions, primarily heavy duty units, have an adjustment for quicker shifting. The transmission levers have two control rod holes. Shift lever travel may be decreased by positioning the controls rods in the lower holes. This adjustment results in a tighter shift pattern and requires a slightly greater shifting effort.

Automatic Transmission

There are two basic automatic transmissions. The first is the two speed Powerglide, which is available from 1964-70. A variation on the Powerglide, introduced in 1969, is the Torque Drive transmission. The Torque Drive unit is a Powerglide with the automatic shifting provisions removed. Torque Drive is shifted manually, but has no clutch. The second type is the three speed Turbo Hydra-Matic, used from 1966-70. This transmission is available in two load capacities, the Turbo Hydra-Matic 350 and the Turbo Hydra-Matic 400. This transmission is used in most General Motors vehicles.

Powerglide Shift Linkage Adjustment

1964-70 Column Shift

Check adjustment as follows:
1. The shift tube and lever assembly must be free in the mast jacket.
2. Lift the selector lever toward the steering wheel. Allow the selector lever to be positioned in Drive by the transmission detent.

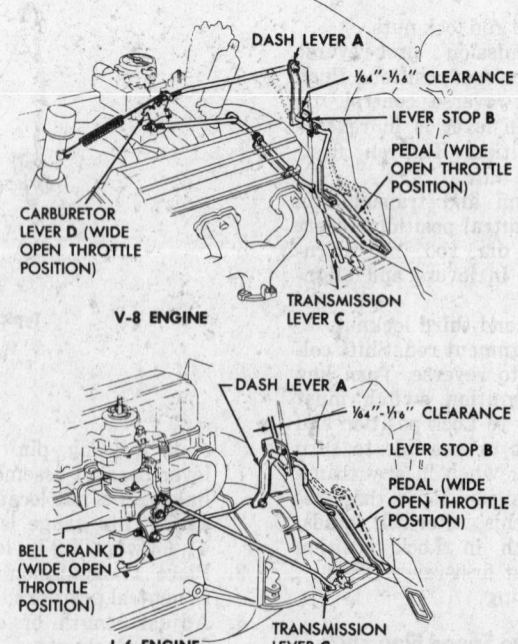

DASH LEVER A

$\frac{1}{64}$"-$\frac{1}{16}$" CLEARANCE

LEVER STOP B

PEDAL (WIDE OPEN THROTTLE POSITION)

CARBURETOR LEVER D (WIDE OPEN THROTTLE POSITION)

TRANSMISSION LEVER C

V-8 ENGINE

DASH LEVER A

$\frac{1}{64}$"-$\frac{1}{16}$" CLEARANCE

LEVER STOP B

PEDAL (WIDE OPEN THROTTLE POSITION)

BELL CRANK D (WIDE OPEN THROTTLE POSITION)

TRANSMISSION LEVER C

L-6 ENGINE

Powerglide throttle valve linkage adjustment, 1964-66 V8, 1964-70 inline engines (© G.M. Corp.)

3. Release selector lever. The selector lever should be prevented from engaging Low range unless the lever is lifted.

4. Lift the selector lever toward the steering wheel and allow the lever to be positioned in Neutral by the transmission detent.

5. Release the selector lever. The selector lever should now be prevented from engaging Reverse unless the lever is lifted. If the linkage is adjusted correctly, the selector lever should be prevented from moving beyond both the Neutral detent and the Drive detent unless the lever is lifted to pass over the mechanical stop in the steering column.

Adjust as follows:

6. Loosen adjustment clamp at cross-shaft. Set transmission lever in drive by rotating lever counterclockwise to low detent, then clockwise to drive.

7. Set selector lever in Drive. Remove any free play by holding cross-shaft upward and pulling shift rod downward.

8. Tighten clamp and recheck adjustment.

For 1969-70 models:

9. Place shift lever in Park and ignition switch in Lock. Loosen back drive rod clamp nut. Remove column lash and tighten clamp nut.

10. With selector lever in Park, the ignition key should move freely to Lock position. Lock position should be obtainable only when transmission is in Park.

1969-70 Torque Drive

1. Loosen swivel at idler lever.
2. Place transmission lever in hi position.

3. Set shift lever at lower end of column up against first position stop.

4. Adjust rod in swivel and tighten retaining nut.

5. Place shift lever in Park and ignition switch in Lock. Loosen back drive rod clamp nut. Remove column lash and tighten clamp nut.

6. With selector lever in Park, the ignition key should move freely to Lock position. Lock position should be obtainable only when transmission is in Park.

1964-67 Chevelle Floorshift

1. Loosen adjustment nuts at swivel. Set transmission lever in drive position by moving counterclockwise to low detent, then clockwise one detent position to drive.

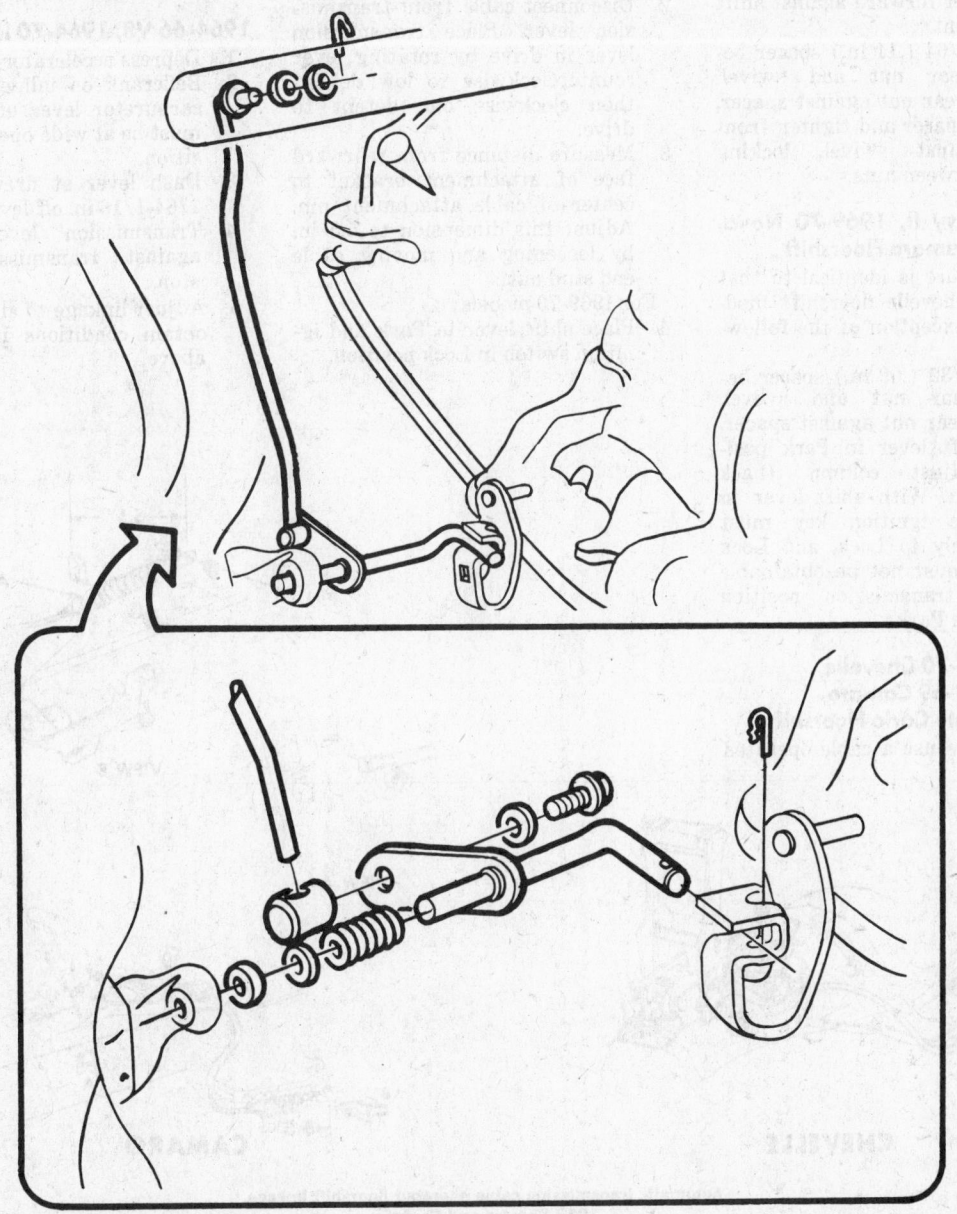

Automatic transmission column shift linkage, 1969 Chevelle (© Chevrolet Div. G.M. Corp.)

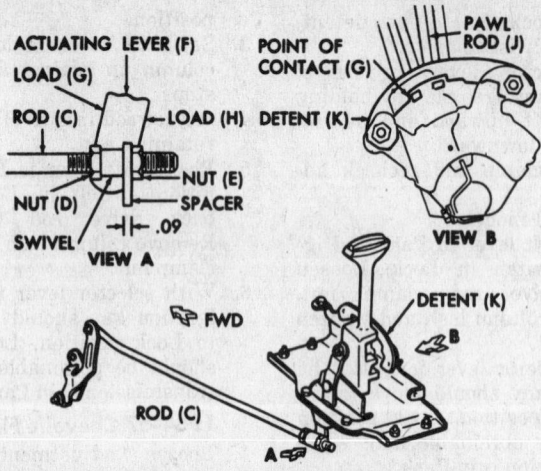

Automatic transmission rod operated floorshift linkage, 1967 Chevy II
(© Chevrolet Div. G.M. Corp.)

2. Set floorshift lever in Drive. Hold floorshift unit lower operating lever forward against shift lever detent.
3. Place a 7/64 (.11 in.) spacer between rear nut and swivel. Tighten rear nut against spacer.
4. Remove spacer and tighten front nut against swivel, locking swivel between nuts.

1965-68 Cheyy II, 1969-70 Nova, 1967 Camaro Floorshift

This procedure is identical to that for 1964-67 Chevelle floorshift models, with the exception of the following steps:

3. Place a 3/32 (.09 in.) spacer between rear nut and swivel. Tighten rear nut against spacer.
5. Place shift lever in Park position. Adjust column (back drive) rod. With shift lever in Park, the ignition key must move freely to Lock, and Lock position must not be obtainable in any transmission position other than Park.

1968-70 Chevelle, 1968-69 Camaro, 1970 Monte Carlo Floorshift

These models use a cable operated linkage.

1. Place shift lever in Drive position.
2. Disconnect cable from transmission lever. Place transmission lever in drive by rotating lever counterclockwise to low detent, then clockwise one detent to drive.
3. Measure distance from rearward face of attachment bracket to center of cable attachment pin. Adjust this dimension to 5.5 in. by loosening and moving cable end stud nut.

For 1969-70 models:

4. Place shift lever in Park and ignition switch in Lock position.

5. Loosen and adjust column (back drive) rod.
6. With selector lever in Park position, the ignition key should move freely to Lock position. Lock position should not be obtainable in any transmission position other than Park.

Turbo Hydra-Matic Shift Linkage Adjustment

The Turbo Hydra-Matic Linkages are the same as those used on Powerglide models. Adjustments are the same, except that the transmission lever is adjusted to drive position by moving the lever clockwise to the low detent, then counterclockwise two detent positions to drive.

Powerglide Throttle Valve Linkage Adjustment

1964-66 V8, 1964-70 Inline Engines

1. Depress accelerator pedal.
2. Bellcrank on inline engines and carburetor lever on V8 engines must be at wide open throttle position.
3. Dash lever at firewall must be 1/64-1/16 in. off lever stop.
4. Transmission lever must be against transmission internal stop.
5. Adjust linkage to simultaneously obtain conditions in Steps 1-4, above.

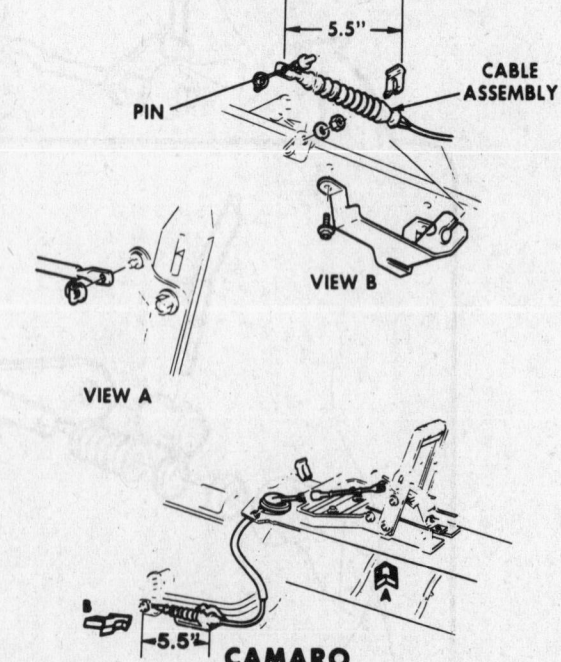

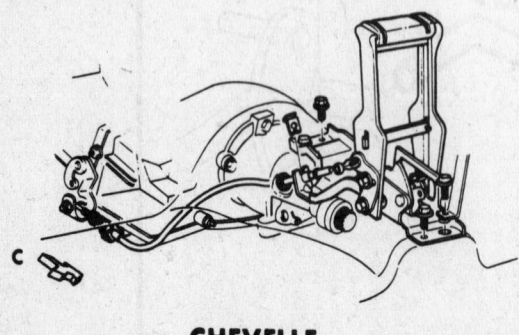

Automatic transmission cable operated floorshift linkage,
1969 Camaro and Chevelle
(© Chevrolet Div. G.M. Corp.)

1967-70 V8 Engines

1. Remove air cleaner.
2. Disconnect accelerator linkage at carburetor.
3. Disconnect both return springs.
4. Pull throttle valve upper rod forward until transmission is through detent.
5. Open carburetor to wide open throttle position. Adjust swivel on end of upper throttle valve rod so carburetor reaches wide open throttle position at the same time that the ball stud contacts the end of the slot in the upper throttle valve rod. A tolerance of 1/32 in. is allowable.

Turbo Hydra-Matic 350 Detent Cable Adjustment

The Turbo Hydra-Matic 350 has a detent, or downshift, cable between the caburetor linkage and the transmission.

1969-70—All models

1. Remove air cleaner.
2. Loosen detent cable screw or disengage snap lock.
3. Place carburetor lever in wide open throttle position. Make sure lever is against stop. On vehicles with Quadrajet carburetors, disengage the secondary lock out before placing lever in wide open throttle position.
   NOTE: detent cable must be pulled through detent position.
4. Engage snap lock or tighten detent screw.

Turbo Hydra-Matic 400 Detent Switch Adjustment

The Turbo Hydra-Matic 400 transmission has an electrical detent, or downshift, switch operated by the throttle linkage.

1968 Chevelle and Camaro

1. Place carburetor lever in wide open position.
2. Place automatic choke in off position.
3. Fully depress switch plunger.
4. Adjust switch mounting to ob-

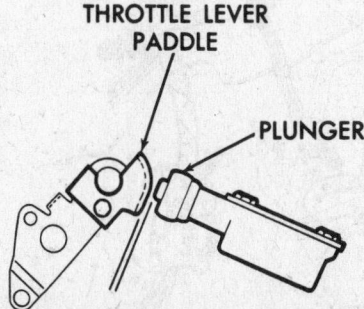

Turbo Hydra-Matic 400 detent switch adjustment, 1968 Chevelle, 1969-70 Nova, 1968-69 Camaro
(© Chevrolet Div. G.M. Corp.)

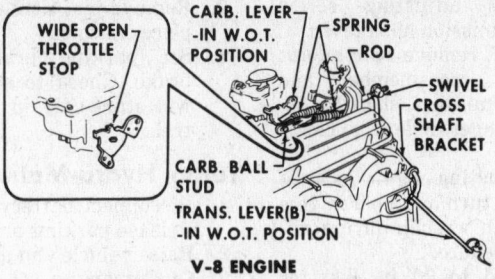

Powerglide throttle valve linkage, 1967-70 V8
(© Chevrolet Div. G.M. Corp.)

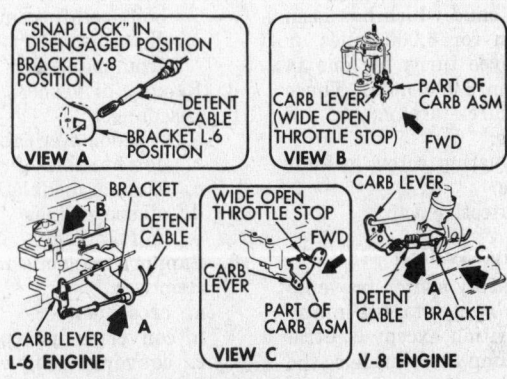

1969-70 Turbo Hydra-Matic 350 detent cable adjustment
(© Chevrolet Div. G.M. Corp.)

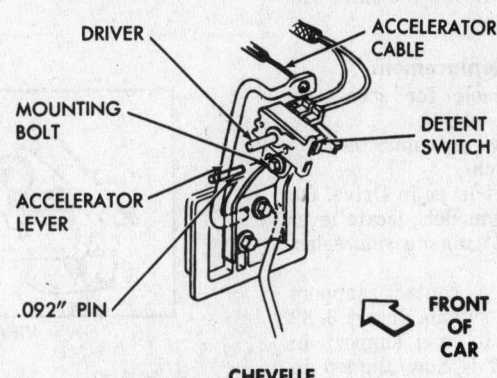

Turbo Hydra-Matic 400 detent switch adjustment, 1969-70 Chevelle, 1970 Monte Carlo
(© Chevrolet Div. G.M. Corp.)

tain distance between depressed switch plunger and throttle lever paddle of .05 in. for Chevelle and .20 in. for Camaro.

1969-70 Nova, 1969 Camaro

This procedure is the same as that for the 1968 Chevelle and Camaro, with the substitution of the following step:
4. Adjust switch mounting to obtain distance between depressed switch plunger and throttle lever paddle. of .22-.24 in.

1969-70 Chevelle, 1970 Monte Carlo

1. Pull detent switch driver rearward until hole in switch body aligns with hold in driver. Insert

a .092 in. dia. pin through the aligned holes to hold the driver in position.
2. Loosen mounting bolt.
3. Depress accelerator to wide open throttle position. Move switch forward until driver contacts accelerator lever.
4. Tighten mounting bolt. Remove pin.

Powerglide Low Band Adjustment

Adjustment should be performed at oil change intervals or whenever slippage becomes evident. See the Unit Repair Section for oil change intervals.

1. Raise vehicle and place selector lever in Neutral.
2. Remove protective cap from

transmission adjusting screw above transmission shift lever.

3. On Chevelle, remove rear mount bolts from crossmember and move transmission slightly toward passenger side for clearance.

4. Loosen adjusting screw locknut one-quarter turn and hold in this position with wrench during adjusting procedure.

5. Adjust band to 70 in. lbs. for 1967-70 models; 40 in. lbs. for 1964-66 models. Then back off adjusting screw four complete turns for a band which has been in operation for 6,000 miles or more, or three turns for one in use less than 6,000 miles. These back-off figures are exact, not approximate.

6. Tighten adjusting screw locknut to 15 ft. lbs.

7. Replace protective cap.

Neutral Safety Switch

The neutral safety switch prevents the engine from being started in any transmission position except Neutral or Park. On column shift models, the switch is located on the upper side of the steering column under the instrument panel. On floorshift models, the switch is located inside the shift console.

Switch Replacement

1. Remove console for access on floorshift models.
2. Disconnect wiring connectors.
3. Remove switch.
4. Position shift lever in Drive. On column shift models, locate lever tang against transmission selector plate.
5. Align slot in contact support with hole in switch. Insert 3/32 in. dia. pin to hold support in place. Switch is now aligned in drive position.
6. Place contact support drive slot over drive tang. Install screws.

7. Remove pin. Connect wiring. Replace console.
8. Set parking brake and footbrake. Check to see that engine will start only in Drive or Neutral.

Turbo Hydra-Matic 350 R & R

1. Disconnect battery ground cable. Release parking brake.
2. Raise vehicle and drain oil.
3. On Camaro:
   a. disconnect parking brake cables
   b. remove convertible underbody reinforcement plate
   c. disconnect left exhaust pipe from manifold.
4. Remove driveshaft.
5. Disconnect:
   a. speedometer cable
   b. detent cable
   c. vacuum modulator line
   d. oil cooler lines
   e. shift linkage.
6. Support transmission.
7. Remove:
   a. crossmember
   b. converter underpan
   c. converter to flywheel bolts.
8. Loosen exhaust pipe to manifold bolts.
9. Lower transmission until jack

is barely supporting it. Be careful that V8 distributor is not forced against firewall.

10. Remove:
    a. transmission to engine mounting bolts.
    b. oil filler tube.
11. Raise transmission to normal position. Support engine with a jack. Slide transmission to rear and then down. Use a strap to hold converter to transmission or keep rear of transmission down to avoid losing converter.
12. Reverse procedure to install.

Turbo Hydra-Matic 400 R & R

This procedure is the same as that given above for the Turbo Hydra-Matic 350, with the substitution of the following step:

5. Disconnect:
   a. speedometer cable
   b. electrical lead to case connector
   c. vacuum line modulator
   d. oil cooler lines
   e. shift linkage.

Powerglide R & R

1. Raise car on hoist and drain oil.
2. Disconnect oil cooler and vacuum modulator lines and the

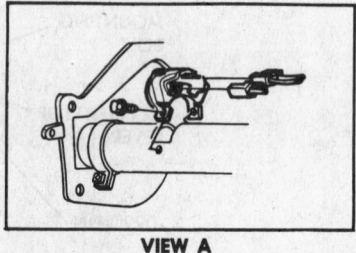

VIEW A

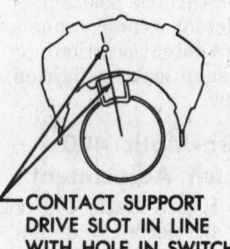

CONTACT SUPPORT
DRIVE SLOT IN LINE
WITH HOLE IN SWITCH

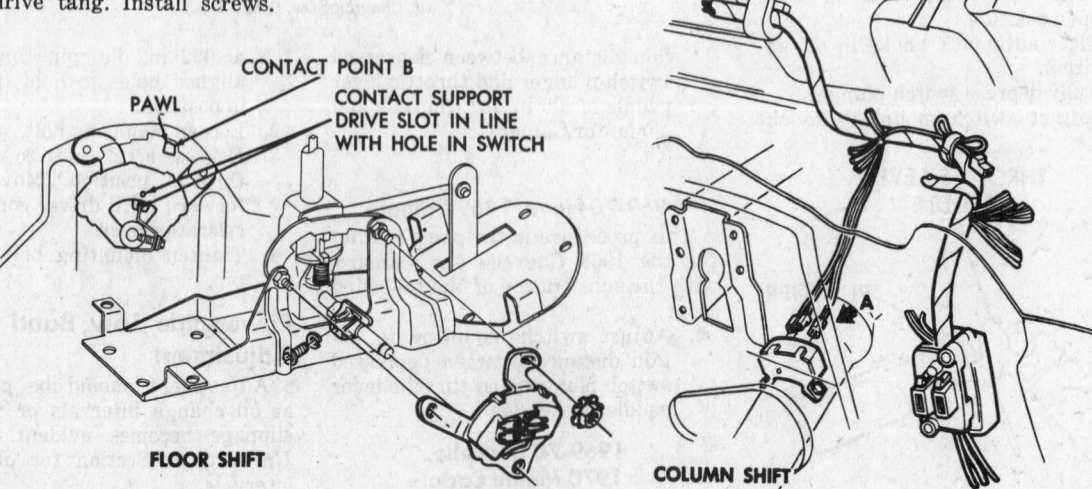

CONTACT POINT B

PAWL

CONTACT SUPPORT
DRIVE SLOT IN LINE
WITH HOLE IN SWITCH

FLOOR SHIFT

COLUMN SHIFT

Typical neutral safety switch installation (© Chevrolet Div. G.M. Corp.)

speedometer drive cable at transmission. Fasten lines out of way.

3. Disconnect manual and throttle valve control lever rods from transmission.
4. Disconnect driveshaft.
5. Install suitable transmission lift equipment.
6. Disconnect engine rear mount on transmission extension, then remove transmission support crossmember.
7. Remove converter underpan, scribe flywheel-converter relationship for assembly. Remove converter-to-flywheel attaching bolts.
8. Support the engine at the oil pan rail with a jack or other suitable brace capable of supporting the engine weight when transmission is removed.
9. Lower rear of transmission slightly so that the upper transmission housing-to-engine attaching bolts can be reached by using a universal socket and a long extension. Remove upper bolts.
10. Remove remainder of transmission-to-housing bolts.
11. Remove transmission by moving it slightly to the rear and downward, then remove from beneath the car and transfer it to a work bench.

NOTE: watch the converter when moving the transmission rearward. If it does not follow the transmission, pry it free of the flywheel before proceeding.

Caution Use some sort of holding strap to keep the converter from falling out of the transmission during transmission removal and handling.

12. Install transmission in the reverse order of removal.

Drive Shaft, U Joints

A one piece, exposed type, tubular driveshaft is used on all models. The driveshaft has two cross and roller universal joints and a splined slip joint. The cross and roller universal joints may be of two types. The first is the Cleveland type which uses external snap rings for trunnion retention. The second is the Saginaw design in which the trunnions are retained by a nylon material which is injected into a groove in the yoke.

See the Chevrolet Section for disassembly and repair of the universal joints.

Drive Axle, Suspension

The Chevelle and Monte Carlo have a coil spring rear suspension lo-

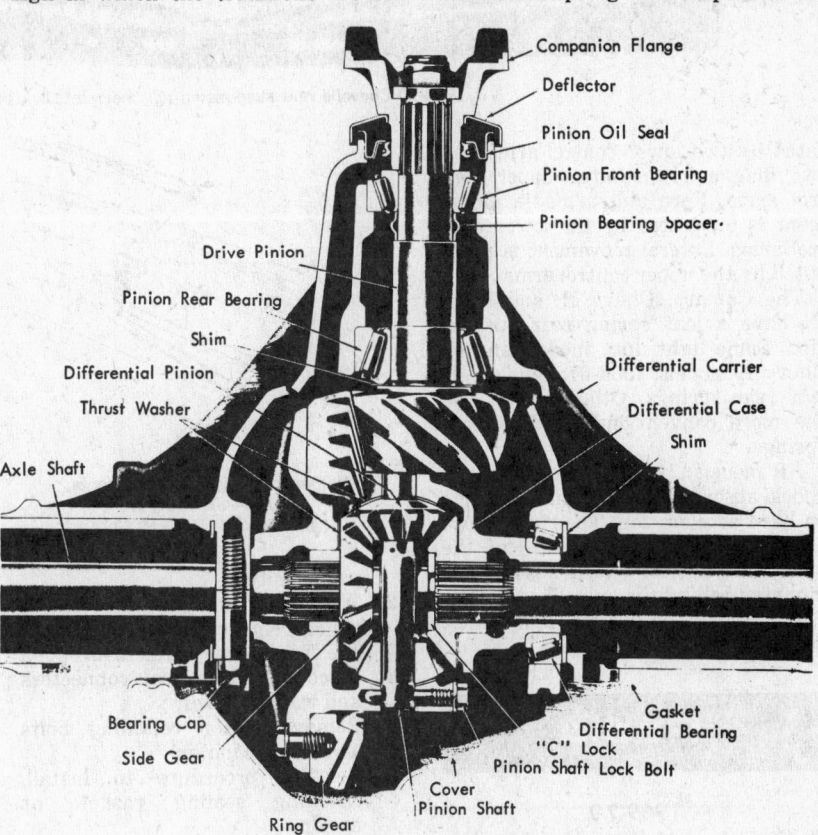

Differential, 1964-70 (© G.M. Corp.)

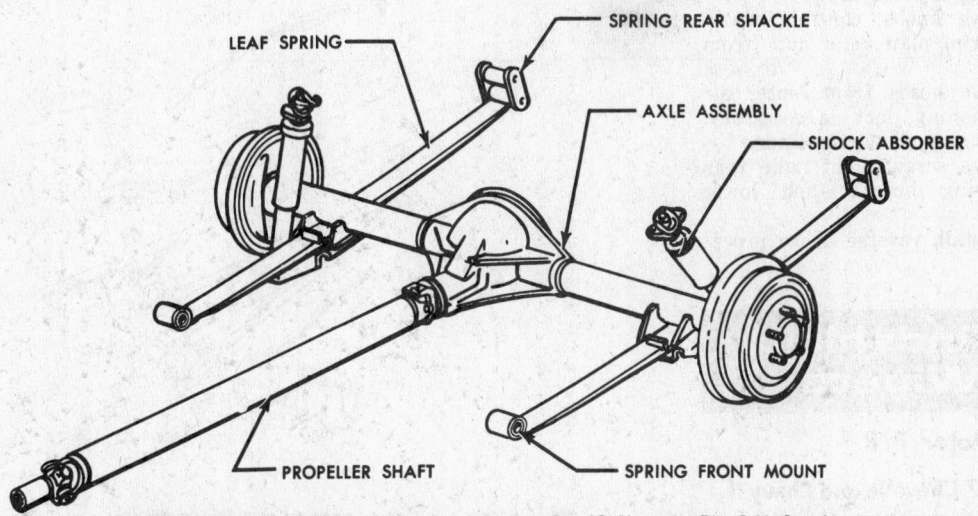

1968-70 Camaro and Chevy II (Nova) rear suspension (© Chevrolet Div. G.M. Corp.)

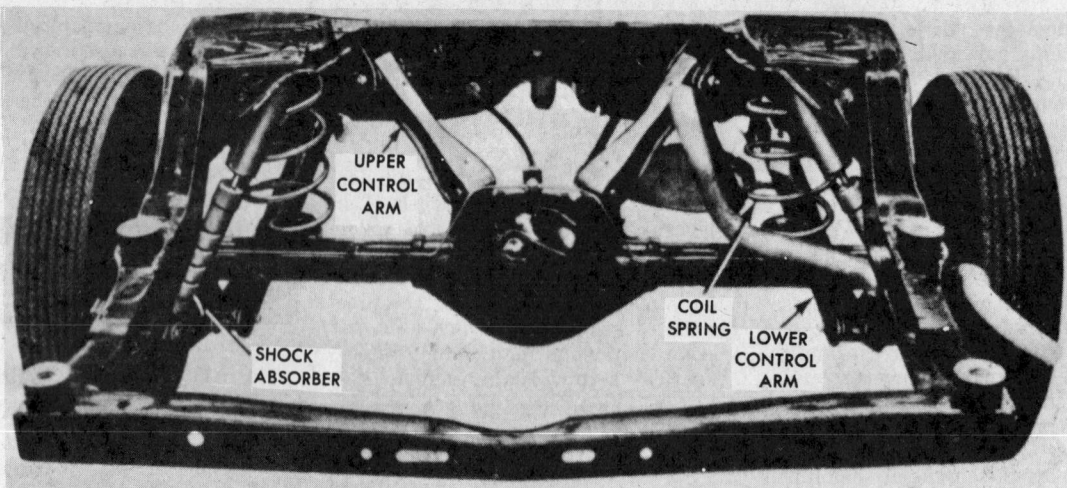

1968-70 Chevelle rear suspension (© Chevrolet Div. G.M. Corp.)

cated by two lower control arms and two diagonally mounted upper control arms. Fore and aft axle movement is prevented by the lower control arms. Lateral movement is prevented by the upper control arms.

The Camaro, Chevy II, and Nova all have a leaf spring rear suspension. Some light duty models and all Chevy II models, 1964-67, have single leaf rear springs. Other models use the more conventional multiple leaf springs.

All models, 1968-70, use staggered shock absorbers to prevent axle hop on hard acceleration. The right shock absorber is mounted forward of the axle and the left shock absorber is mounted behind the axle.

For repair details on rear axles, see the Unit Repair Section.

Radio

1969-70

1. Disconnect battery ground cable.
2. Remove ash tray and ash tray housing as necessary.
3. Remove knobs, controls, washers, trim plate, and nuts from radio.
4. Remove hoses from center air conditioning duct as necessary.
5. Disconnect all wiring leads.
6. Remove screw from radio rear mounting bracket and lower radio.
7. To install, reverse above procedure.

Windshield Wipers

Wiper Motor R R

1964-67 Chevelle and Chevy II

1. Make certain wiper motor is in

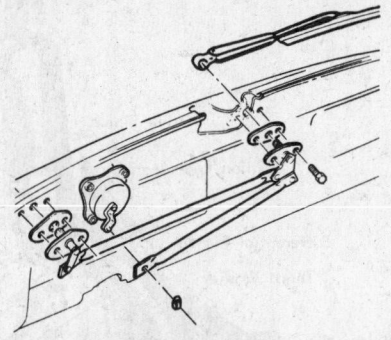

Wiper transmission components,
1964-67 Chevy II
(© Chevrolet Div. G.M. Corp.)

park position.
2. Working under instrument panel, remove transmission linkage from motor crank arm.
3. Disconnect electrical connectors and washer hoses.
4. Remove motor retaining bolts and remove motor.
5. Reverse procedure to install, checking sealing gaskets at motor.

1967 Camaro

1. Make certain wiper motor is in park position.
2. Disconnect washer hoses and electrical connectors.
3. Remove three motor bolts. Pull wiper motor assembly from cowl opening and loosen nuts retaining drive rod ball stud to crank arm.
4. Reverse procedure to install, checking sealing gaskets at motor.

1968-70 All Models

1. Make sure wiper motor is in park position.
2. Disconnect washer hoses and electrical connectors.
3. Remove the plenum chamber grille or access cover. Remove the nut retaining the crank arm to the motor assembly.
4. Remove the retaining screws or nuts and remove motor.
5. Reverse procedure to install, checking sealing gaskets at motor.

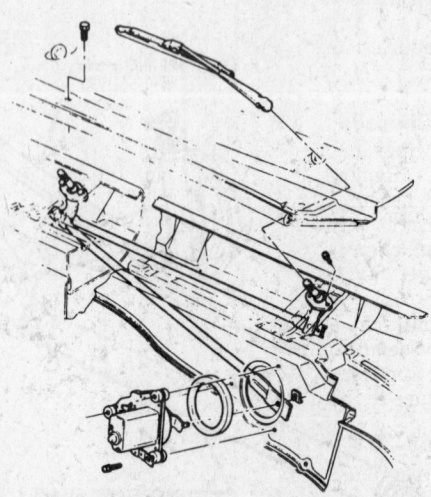

Wiper motor and linkage, 1964-67 Chevelle (© Chevrolet Div. G.M. Corp.)

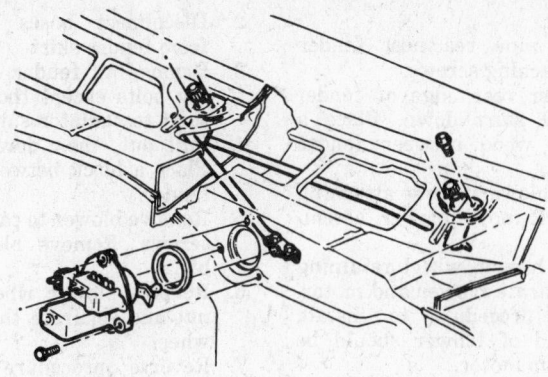

Wiper and motor linkage, 1967-69 Camaro (© Chevrolet Div. G.M. Corp.)

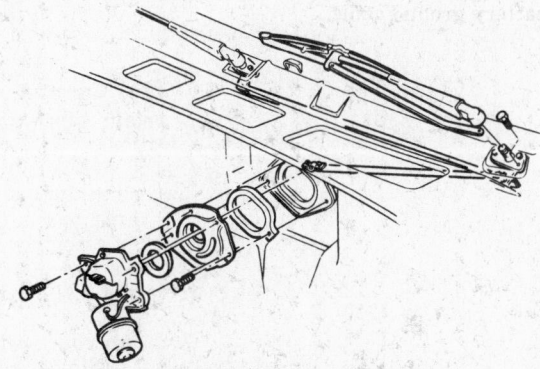

1968-70 Chevelle recessed wipers
(© Chevrolet Div. G.M. Corp.)

Wiper Transmission R & R

1964-67 Chevy II

1. Make certain wiper motor is in park position. Remove wiper arm and blade assemblies from transmission shaft.
2. Remove linkage from wiper crank arm. Remove left transmission link from right transmission.
3. Remove two screws securing transmission to cowl on one side. Remove transmission from under dash.
4. Reverse procedure to install, checking gasket.

1964-67 Chevelle

1. Make certain wiper motor is in park position. Remove wiper arm and blade assemblies from transmission shaft.
2. Remove plenum chamber grille.
3. Detach linkage from wiper crank arm.
4. Remove transmission retaining screws, lower assembly into plenum chamber, and remove unit.
5. Reverse procedure to install.

1967 Camaro, 1968-70 All models

1. Make sure wiper motor is in park position.
2. Disconnect battery ground cable.
3. Remove wiper arm and blade as-

semblies from transmission. On articulated left arm assemblies, remove carburetor type clip retaining pinned arm to blade arm.
4. Remove plenum chamber air intake grille or screen.
5. Loosen nuts retaining drive rod ball stud to crank arm and detach drive rod from crank arm.
6. Remove transmission retaining screws. Lower transmission and drive rod assemblies into plenum chamber.
7. Remove transmission and linkage from plenum chamber through cowl opening.
8. Reverse procedure to install, making sure wiper blade assemblies are installed in park position.

Heater System

Heater Core R & R

1964-67 Chevy II

1. Drain radiator.
2. From engine compartment, remove hoses from inlet and outlet connections.
3. Remove nuts around blower motor holding heater to dash panel.
4. From inside vehicle, remove glove

compartment and glove compartment door.
5. Remove screws attaching heater distributor bracket to dash.
6. Remove screw holding case bracket to adaptor bracket.
7. Detach heater assembly from dash panel and adaptor assembly, then lower toward floor.
8. Disconnect all cable connections, wire connector and defroster hoses.
9. Remove assembly from vehicle.
10. Remove screws attaching core cover to heater.
11. Remove core mounting screws and remove core from heater.
12. Install in reverse of above.

1964-67 Chevelle

1. Drain radiator.
2. Remove heater hoses at connections beside air inlet assembly.
3. Remove cable and electrical connectors from heater and defroster assembly.
4. On engine side of dash, remove screws and nuts holding air inlet to dash panel.
5. Inside vehicle, pull entire assembly from the firewall and remove assembly from vehicle.
6. Remove core assembly retaining springs and remove core.
7. Install in reverse of above.

1968-70 All Models

1. Disconnect battery ground cable.
2. Drain radiator.
3. Disconnect heater hoses. Plug core inlet and outlet.
4. Remove nuts from air distributor duct studs on firewall.
5. On Chevy II, remove glove compartment and door assembly.
6. From under dash, drill out lower right hand distributor duct stud with a ¼ in. drill.
7. Pull distributor duct from firewall mounting. Remove resistor wires. Lay duct on floor.
8. Remove core assembly from distributor duct.
9. Reverse procedure to install.

Heater Blower R & R

1964-67 Chevy II

1. Remove screws attaching the motor and blower to heater assembly.
2. Remove retainer attaching blower to motor shaft.
3. Install in reverse of above.

1964-67 Chevelle

1. Disconnect battery.
2. Unclip hoses from fender skirt.
3. Disconnect electrical feed from motor.
4. Turn vehicle front wheels to extreme right.
5. Remove right front fender skirt bolts and allow skirt to drop,

resting it on top of tire. It may be wedged away from fender lower flange with block of wood to provide better access to bolts.

6. Remove screws attaching motor mounting plate to air inlet housing.

7. Remove screws attaching motor to mounting plate.

8. Remove clip attaching cage to shaft and remove blower motor.

9. Install in reverse of above.

1967-69 Camaro

1. Disconnect battery ground cable.
2. Disconnect hoses and wiring from fender skirt.
3. Remove wheel opening trim.
4. Remove rocker panel molding.
5. Loosen rear lower fender to body bolt.
6. Remove nine rearmost fender skirt attaching screws.
7. Pull lower rear edge of fender out. Pull skirt down. Place a block of wood between fender and skirt.
8. Remove blower to case attaching screws. Remove blower assembly.
9. Remove blower wheel retaining nut. Separate blower and motor.
10. Reverse procedure to install. Open end of blower should be away from motor.

1968 Chevy II, 1969 Nova, 1968-69 Chevelle, 1970 All Models

1. Disconnect battery ground cable.
2. Disconnect hoses and wiring from fender skirt.
3. Remove all fender skirt attaching bolts except those attaching skirt to radiator support.
4. Pull out, then down, on skirt. Place a block between skirt and fender.
5. Remove blower to case attaching screws. Remove blower assembly.
6. Remove blower wheel retaining nut and separate the motor and wheel.
7. Reverse procedure to install. Open end of blower should be away from motor.

Index

YEAR IDENTIFICATION

CHEVROLET

1964

1965

1966

1967

CHEVROLET

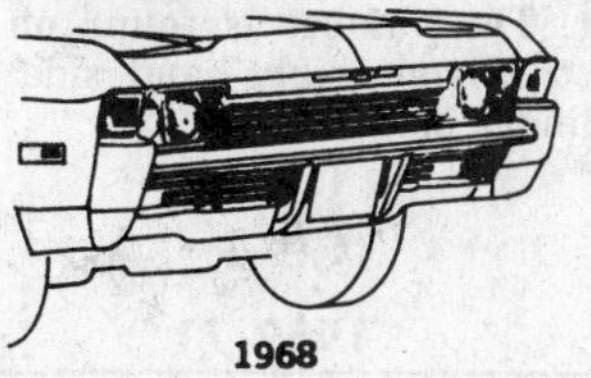

1968

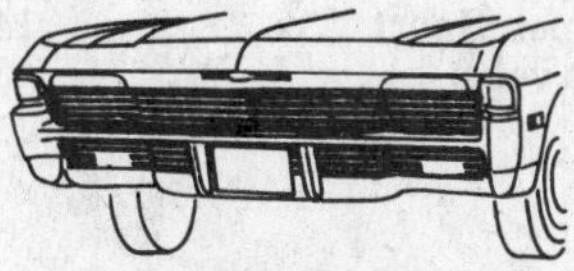

1968 Caprice

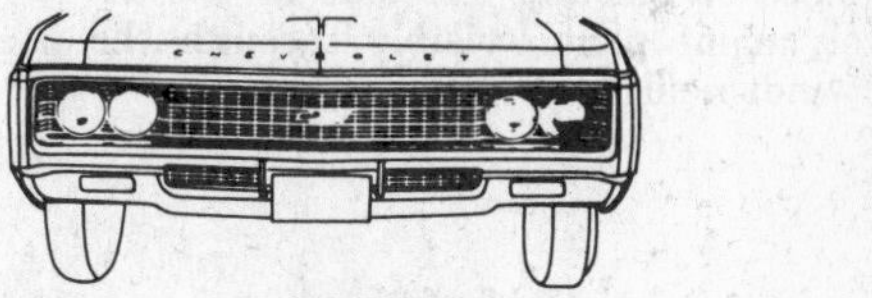

1969

CHEVROLET

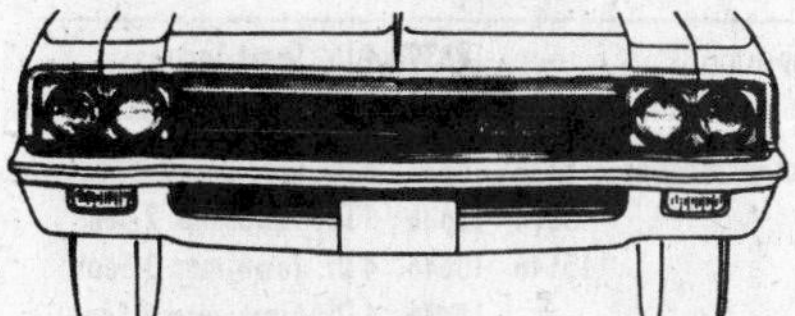

1970

1971

CORVETTE

1964-65

CORVETTE

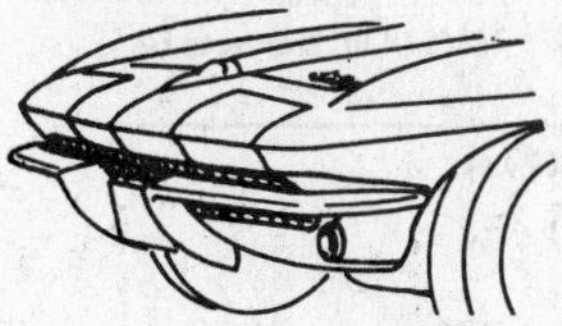

1966

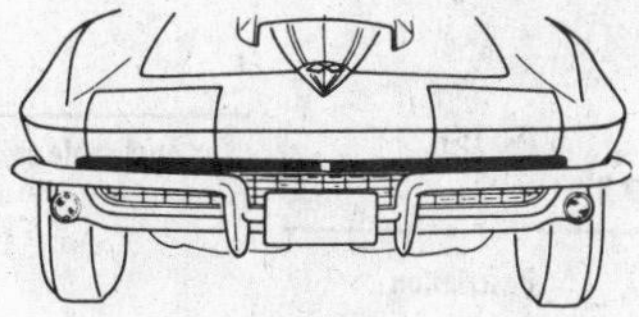

1967

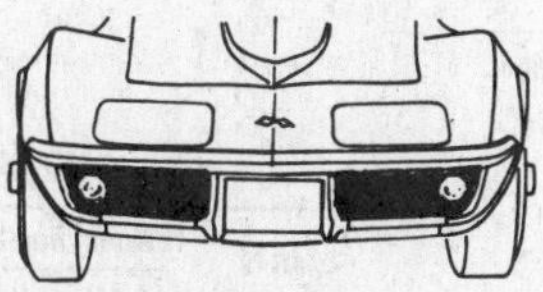

1968-69

1970-71

FIRING ORDER

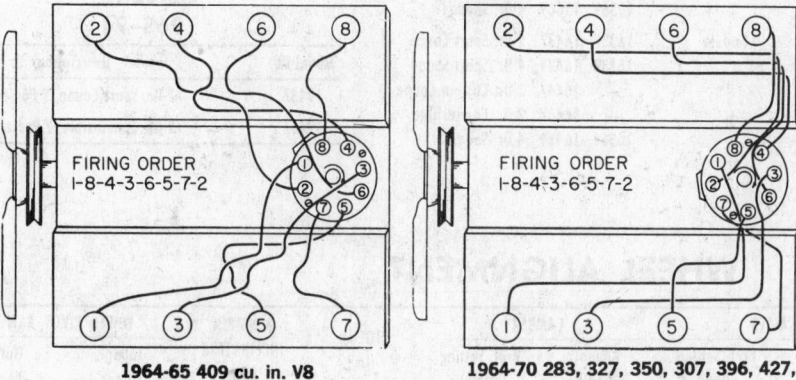

FIRING ORDER
1-8-4-3-6-5-7-2

1964-65 409 cu. in. V8

FIRING ORDER
1-8-4-3-6-5-7-2

1964-70 283, 327, 350, 307, 396, 427,
400, 454 cu. in. V8

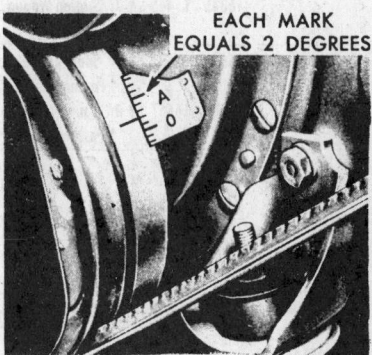

EACH MARK
EQUALS 2 DEGREES

Ignition timing marks V8 engine
(© Chevrolet Div., G.M. Corp.)

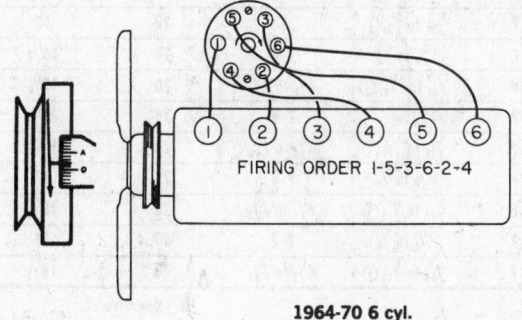

FIRING ORDER 1-5-3-6-2-4

1964-70 6 cyl.

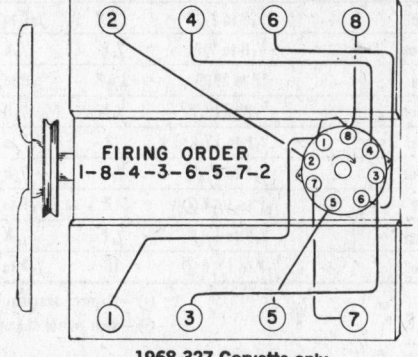

FIRING ORDER
1-8-4-3-6-5-7-2

1968 327 Corvette only

CAR SERIAL NUMBER LOCATION

1964-67

Vehicle serial number is found on a plate attached to left front door hinge pillar.

1968-71

Vehicle serial number is found on a plate on the top left side of the instrument panel, visible through the windshield.

Model Identification

1964

Six cylinder
Biscayne 1100
Bel Air 1500
Impala 1700
Super Sport 1300

V8
Biscayne 1200
Bel Air 1600
Impala 1800
Super Sport 1400
Corvette 800

Engine Identification

Six Cylinder Engines

The production code letters immediately follow the engine serial number. The number is found on a pad at the front right-hand side of the cylinder block, just to the rear of the distributor.

V8 Engines

The production code letters immediately follow the engine serial number. The number is found on a pad at the front right-hand side of the cylinder block.

1965-68

Series	Model Number 6 Cyl.	V-8	Description
Biscayne	15311	15411	2 Dr. Sedan, 6 Passenger
	15369	15469	4 Dr. Sedan, 6 Passenger
	15335	15435	4 Dr. Station Wagon, 2 Seats
Bel Air	15511	15611	2 Dr. Sedan, 6 Passenger
	15569	15669	4 Dr. Sedan, 6 Passenger
	15535	15635	4 Dr. Station Wagon, 2 Seats
	15545	15645	4 Dr. Station Wagon, 3 Seats
Impala	16369	16469	4 Dr. Sedan, 6 Passenger
	16339	16439	4 Dr. Sports Sedan, 6 Passenger
	16337	16487	2 Dr. Sport Coupe, 5 Passenger
	16367	16467	2 Dr. Convertible, 5 Passenger
	16335	16435	4 Dr. Station Wagon, 2 Seats
	16345	16445	4 Dr. Station Wagon, 3 Seats
Impala SS	16537	16637	2 Dr. Sport Coupe, 4 Passenger
	16567	16667	2 Dr. Convertible, 4 Passenger

1965-68

Series	Model Number 6 Cyl.	V-8	Description
Caprice	—	16635	4 Dr. Custom Wagon—2-Seat
	—	16639	4 Dr. Custom Wagon, 6 Passenger
	—	16645	4 Dr. Custom Wagon, 3-Seat
	—	16647	2 Dr. Custom Coupe, 4 Passenger

1969-71

Series	Model Number 6 Cyl.	V-8	Description
Biscayne	*15311	*15411	2 Dr. Sedan
	15369	15469	4 Dr. Sedan
Bel Air	*15511	*15611	2 Dr. Sedan
	15569	15669	4 Dr. Sedan
Impala	16337	16437	2 Dr. Sport Coupe
	16339	16439	4 Dr. Sport Sedan
	—	16447	2 Dr. Custom Coupe
	—	16467	2 Dr. Convertible
	16369	16469	4 Dr. Sedan

1969-71

Series	Model Number 6 Cyl.	V-8	Description
Caprice	—	16639	4 Dr. Sport Sedan
	—	16647	2 Dr. Custom Coupe
Station Wagons	15336	15436	4 Dr. Brookwood 2-Seat
	15536	15636	4 Dr. Townsman 2-Seat
	15546	15646	4 Dr. Townsman 3-Seat
	—	16436	4 Dr. Kingswood 2-Seat
	—	16446	4 Dr. Kingswood 3-Seat
	—	16636	4 Dr. Kingswood Est. 2-Seat
	—	16646	4 Dr. Kingswood Est. 3-Seat

*Applicable to 1969 only.

Corvette 1965-71

Model No.	Description
19437	2 Dr. Sport Coupe, 2 Passenger
19467	2 Dr. Convertible, 2 Passenger

WHEEL ALIGNMENT

YEAR	MODEL	CASTER Range (Deg.)	Pref. Setting (Deg.)	CAMBER Range (Deg.)	Pref. Setting (Deg.)	TOE-IN (In.)	KING-PIN INCLINATION (Deg.)	WHEEL PIVOT RATIO Inner Wheel	Outer Wheel
1964	Chevrolet	1/2 N to 1/2 P	0	0 to 1P	1/2 P	1/32 to 3/32	7 1/4	20	18
	Corvette	1 1/4 P to 2 1/4 P	1 3/4 P	1/4 P to 1 1/4 P ③	3/4 P	3/32 to 5/32 ①	7	20	18
1965-66	Chevrolet	1/4 N to 3/4 P	1/4 P	1/4 N to 3/4 P	1/4 P	1/8 to 1/4	7-8	20	20 1/4
	Corvette	1P to 2P ②	1 1/2 P	1/4 P to 1 1/4 P ③	3/4 P	1/32 to 11/32 ①	6 1/2 to 7 1/2	20	18 1/2
1967	Chevrolet	1/4 P to 1 1/4 P	3/4 P	1/4 P to 3/4 P	1/4 P	1/8 to 1/4	7-8	20	20 1/4
	Corvette	1/2 P to 1 1/2 P	1P	1/4 P to 1 1/4 P ③	3/4 P	3/16 to 5/16 ①	6 1/2 to 7 1/2	20	18 1/2
1968-69	Chevrolet	1/4 P to 1 1/4 P	3/4 P	1/4 N to 3/4 P	1/4 P	1/8 to 1/4	7-8	20	18
	Corvette	1/2 P to 1 1/2 P ④	1P	1/2 P to 1 1/4 P ⑤	3/4 P	3/16 to 5/16 ⑤	6 1/2 to 7 1/2	20	18 1/2
1970-71	Chevrolet	1/4 P to 1 1/4 P	3/4 P	1/4 N to 3/4 P	1/4 P	1/8 to 1/4	7-8	20	18
	Corvette	1/2 P to 1 1/2 P ④	1P	1/2 P to 1 1/4 P ⑤	3/4 P	3/16 to 5/16 ⑤	6 1/2 to 7 1/2	20	18 1/2

① —Rear wheels 1/16-3/16.
② —1966—1/2 P to 1 1/2 P.
③ —Rear wheels—1/2° ± 1/2°.
④ —W/pwr. steering: 1 1/4 P to 2 1/4 P.
⑤ —Rear wheel alignment: camber 1 3/8 N to 3/8 N toe-in 1/32-3/32.

N—Negative.
P—Positive.

Chevrolet Engine Identification Code

No. Cyls.	Cu. In. Displ.	Type	1964	1965	1966	1967	1968	1969	1970
6	230	M.T.	A	FA					
6	230	M.T., HDC	AE	FE					
6	230	M.T., AC	AF	FL					
6	230	M.T., AC, HDC	AG	FF					
6	230	PG	B	FM	FM				
6	230	PG, AC	BQ	FR					
6	230	Taxi, M.T.				FK			
6	230	Taxi, PG				FP			
6	250	PG, AC, w/ex. EM			GQ	GQ	CR	BO	
6	250	Taxi, PG, w/ex. EM			GR				
6	250	M.T.			FA	FA	CA	BA	CCG, CCH, CCZ, CRF, CRG
6	250	HDC, M.T.			FE	FE	CJ		
6	250	HDC, M.T., AC			FF	FF	CK		
6	250	M.T., AC			FL	FL	CN	BG	
6	250	Taxi, Police, M.T.			FK			BP	CCL
6	250	PG				FM	CQ		
6	250	PG, Taxi, Police				FP		BJ, BL	CCM
6	250	PG, AC			FR	FR	CR	BO	
6	250	AC			FV			BQ	
6	250	w/ex. EM			FY				
6	250	Taxi, w/ex. EM			FZ				
6	250	PG, w/ex. EM			GP	GP	CQ		
6	250	Hyd. 350, Police							CCK
8	283	M.T.	C	GA	GA	GA			
8	283	Police	CB						
8	283	PG	D	GF	GF	GF			
8	283	4 spd., w/ex. EM			GS	GS			
8	283	PG, w/ex. EM			GT	GT			
8	283	HDC				GU			
8	283	4-BBL.		GK	GW				
8	283	4-BBL., w/ex. EM			GX				
8	283	PG, 4-BBL., ex. EM			GZ				
8	283	4 spd.			GC	GC	GC		
8	283	w/ex. EM			GK	GK			
8	283	PG, 4-BBL.		GL	GL				
8	307	M.T.					DO		
8	307	4 spd. SS					DP		
8	307	HDC					DQ		
8	307	PG					DR		
8	327	M.T.	R	HA	HA	HA	HA	FA, FJ	
8	327	M.T., HP	RB	HB					
8	327	w/ex. EM			HB	HB			
8	327	PG	S	HC	HC	HC	HC	FB, FK	
8	327	PG, HP	SB	HD					
8	327	PG, w/ex. EM				HF	HF		
8	327	Hyd. 350				KL	HF	FC, FL	
8	327	HDC				KE			
8	327	Hyd. 400						FH	
8	327	Police						FG	
8	327	M.T., Taxi						FY	
8	327	PG, Taxi						FZ	
8	327	Hyd. 350, Taxi						GA	
8	327	Hyd. 400, Taxi						GB	
8	350	PG						GE, HK	
8	350	Hyd. 350, 2-BBL.						HD, HM	
8	350	PG, 2-BBL.						HF, HL	
8	350	M.T.						HG, HD	CND (250), CNQ (300)
8	350	Hyd. 400						HH, IA	
8	350	2-BBL.						HI	
8	350	Hyd. 400, 2-BBL.						HJ	
8	350	Hyd. 350						HN, HY	CNR (300), CNV (250)
8	350	M.T.						HP, HT	

No. Cyls.	Cu. In. Displ.	Type	1964	1965	1966	1967	1968	1969	1970
8	350	PG						HU	CNS (300), CNU (250)
8	350	M.T., Taxi, 2-BBL.						IL	
8	350	PG, Taxi, 2-BBL.						IM	
8	350	Hyd. 350, Taxi, 2-BBL.						IN	
8	350	Hyd. 400, Taxi, 2-BBL.						IP	
8	350	M.T., Taxi, Police						IQ, IR	CNP (250)
8	350	PG, Taxi, Police						IS, IX	CNW (250)
8	350	Hyd. 350, Taxi						IT, IY	CNT (300), CNX (250)
8	350	Hyd. 400, Taxi						IV, IZ	
8	350	M.T., Taxi						IW	
8	396	M.T.	IA	IA	IA	IA		JT	
8	396	w/ex. EM			IB	IB			
8	396	M.T., w/ex. EM	IC						
8	396	PG, w/ex. EM			IC	IC			
8	396	M.T., SHP	IE						
8	396	PG	IG	IG	IG	IG			
8	396	PG, Trans. Ign.	II						
8	396	Hyd., w/ex. EM			IN	IN			
8	396	Hyd.			IV	IV	IV		
8	396	Hyd., Trans. Ign.	IW						
8	396	M.T., 2-BBL.						JN	
8	396	Hyd. 400, Police, 2-BBL.						JO	
8	396	2-BBL., Police						JP	
8	396	Hyd. 400, 2-BBL.						JQ	
8	396	M.T., Police						JR	
8	400	M.T. (265 H.P.)							CGR
8	409	M.T., HP	QA	JA					
8	409	M.T.	QC	JB					
8	409	Trans. Ign.	QQ	JC					
8	409	Trans. Ign., HP	QN	JD					
8	409	PG	QG	JE					
8	409	PG, Trans. Ign.	QR	JF					
8	409	2-4-BBL.	QB						
8	409	2-4-BBL., Trans. Ign.	QP						
8	427	M.T., HDC			IE, IX	IE			
8	427	SHP		ID		ID			
8	427	Hyd., w/ex. EM			IS	IS			
8	427	w/ex. EM		II	II				
8	427	M.T.			IH	IH	IH	MA	
8	427	Hyd.			IJ	IJ	IJ		
8	427	Hyd., HP, w/ex. EM			IO	IO			
8	427	SS Hyd.				IF			
8	427	HP						LA	
8	427	4-BBL.						LB	
8	427	Hyd. 400, HP						LC	
8	427	M.T., SHP						LD	
8	427	Hyd. 400, 4-BBL.						LE	
8	427	Hyd., Police, HP						LF	
8	427	M.T., Police, HP						LG, LZ, MB	
8	427	M.T., HP						LH, MC	
8	427	Hyd. 400						LI	
8	427	Hyd. 400, Police, 4-BBL.						LJ	
8	427	M.T., Police, 4-BBL.						LK	
8	427	Hyd. 400, SHP						LS	
8	427	M.T., Police						LY	
8	427	M.T., SHP						MD	
8	454	M.T.							CGV (345), CGU (390)
8	454	M.T., Police							CGS (345), CGT (390)

AC—Air conditioned.
HDC—Heavy duty clutch.
HP—High performance.

SHP—Special high performance.
M.T.—Manual transmission.
PG—Powerglide transmission.

Hyd.—Turbo-Hydramatic transmission.
w/ex. EM—With exhaust emission controls.
Trans. Ign.—Transistorized ignition.

Corvette Engine Identification Code

No. Cyls.	Cu. In. Displ.	Type	1964	1965	1966	1967	1968	1969	1970
8	327	SHP, w/ex. EM			HD	HD			
8	327			HE	HE	HE	HE		
8	327	HP		HF					
8	327	w/F.I.		HG					
8	327	SHP		HH					
8	327	w/ex. EM			HH	HH			
8	327	AC		HI					
8	327	HP, AC		HJ					
8	327	SHP, AC		HK					
8	327	T. Ign.		HL					
8	327	T. Ign., AC		HM					
8	327	w/F.I., T. Ign.		HN					
8	327	PG			HO	HO			
8	327	T.H.					HO		
8	327	SHP, T.H.					HT		
8	327	PG, HP			HP				
8	327	PS, AC				HP			
8	327	4 Speed, AC, PS					HP		
8	327	PG, AC		HQ					
8	327	PG, HP, AC		HR					
8	327	PG, w/ex. EM			HR	HR			
8	327	SHP		HT	HT				
8	327	4 Speed				HT	HT		
8	327	SHP, AC		HU					
8	327	SHP, T. Ign.		HV					
8	327	SHP, T. Ign., AC		HW					
8	327	SHP, AC, W/ex. EM				KH			
8	327	M.T.	RC						
8	327	M.T., HP	RD						
8	327	M.T., SHP	RE						
8	327	M.T., w/F.I.	RF						
8	327	M.T., AC	RP						
8	327	M.T., HP, AC	RQ						
8	327	M.T., SHP, AC	RR						
8	327	M.T., T. Ign.	RT						
8	327	M.T., T. Ign., AC	RU						
8	327	W/F.I., T. Ign.	RX						
8	327	PG	SC						
8	327	PG, HP	SD						
8	327	PG, AC	SK						

No. Cyls.	Cu. In. Displ.	Type	1964	1965	1966	1967	1968	1969	1970
8	327	PG, HP, AC	SL						
8	350	HP						HW	CTN
8	350	HP, AC						HX	CTO
8	350	M.T.						HY	CTL
8	350	T.H. 400						HZ	CTM
8	350	HP, T. Ign.							CTP
8	350	HP, T. Ign., AC							CTQ
8	350	SHP							CTR
8	350	SHP, T. Ign.							CTU
8	350	SHP, T. Ign., M.T.							CTV
8	396	SHP		IF					
8	427	SHP			IK				
8	427	HP (390 H.P.), T.H.			IL	IL	IL	LL, LM	
8	427	w/ex. EM (390 H.P.)			IM	IM			
8	427	SHP			IP		IT	LO	
8	427	(390 H.P.), T.H.			IQ	IQ	IQ		
8	427	PG, w/ex. EM (390 H.P.)			IR	IR			
8	427	SHP (435 H.P.)				IT	IR	LR	
8	427	Aluminum Heads (435 H.P.)				IU	IU	LP	
8	427	4 Speed w/ex. EM (435 H.P.)				JA			
8	427	4 Speed (400 H.P.)				JC			
8	427	PG or T.H. (400 H.P.)				JD		IO	
8	427	4 Speed (435 H.P.)				JE			
8	427	4 Speed, w/ex. EM (400 H.P.)				JF	IM		
8	427	PG w/ex. EM (400 H.P.)				JG			
8	427	Aluminum Heads w/ex. EM (435 H.P.)				JH			
8	427	HP, T.H. 400, 3-2-BBL.						LN	
8	427	HP, 3-2-BBL.						LQ	
8	427	SHP, HDC, 3-2-BBL.						LT	
8	427	Aluminum Heads, HDC						LU	
8	427	Aluminum Heads, T.H. 400						LW	
8	427	T.H. 400 (heavy duty)						LV	
8	427	SHP, T.H. 400						LX	
8	454	HP, 4-BBL., T.H. 400							CGW
8	454	HP, 4-BBL.							CZU
8	454	Heavy duty, 4-BBL.							CZL
8	454	T.H. 400, 4-BBL.							CZN
8	454	HP, 4-BBL., T. Ign.							CRI

AC—Air conditioned.
HP—High performance.
SHP—Special high performance.

M.T.—Manual transmission.
PG—Powerglide Transmission.
HDC—Heavy duty clutch.

PS—Power steering.
w/ex. EM—With exhaust emission.
w/F.I.—With fuel injection.

T. Ign.—With transistor ignition.
4 BBL.—Four barrel carburetor.
T.H.—With Turbo-Hydramatic.

TORQUE SPECIFICATIONS

YEAR	MODEL	CYLINDER HEAD BOLTS (FT. LBS.)	ROD BEARING BOLTS (FT. LBS.)	MAIN BEARING BOLTS (FT. LBS.)	CRANKSHAFT BALANCER BOLT (FT. LBS.)	FLYWHEEL TO CRANKSHAFT BOLTS (FT. LBS.)	MANIFOLD (FT. LBS.) Intake	MANIFOLD (FT. LBS.) Exhaust
1964–71	6 Cyl.—230, 250 Cu. In.	90–95	35–45	60–70	Press Fit	55–65	25–35	①
1964–71	V8—283, 307, 327, 350 Cu. In.	60–70	30–35	60–70	Press Fit	55–65	25–35	①
1964–65	V8—409 Cu. In.	60–70	35–45	95–105	Press Fit	55–65	25–35	①
1965–69	V8—396 Cu. In. (Big Block)	60–70	35–45	95–105	Press Fit	55–65	25–35	①
1966–69	V8—427 Cu. In.	80	50	②	Press-Fit	55–65	30	20
1970–71	V8—400(402)	60–70 ③	50	95–105	Press-Fit	55–65	30	20
1970–71	V8—454	80	50	105	Press-Fit	65	30	20

① —Center bolts—25–30; end bolts—15–20.

② —2 Bolt Cap—95
4 Bolt Cap—115.

③ —330 H.P.—80.

DELCOTRON AND A.C. REGULATOR SPECIFICATIONS

YEAR	MODEL	ALTERNATInOR Field Current Draw @ 12V.	Output @ Generator RPM 2000	Output @ Generator RPM 5000	REGULATOR Model	Field Relay Air Gap (In.)	Field Relay Point Gap (In.)	Field Relay Volts to Close	Regulator Air Gap (In.)	Regulator Point Gap (In.)	Regulator Volts at 125°
1964	1100668	1.9–2.3	25A	35A	1119515	.015	.030	2.3–3.7	.067	.014	13.5–14.4
	1100669	1.9–2.3	28A	40A	1119515	.015	.030	2.3–3.7	.067	.014	13.5–14.4
	1100665	1.9–2.3	32A	50A	1119515	.015	.030	2.3–3.7	.067	.014	13.5–14.4
	1117765	3.7–4.4	18A①	62A①	9000595	.015	.030	2.3–3.7	.075	.014	13.2–14.2
	1100684	2.8–3.2	36A	58A	1116366	.015	.025	2.3–3.5	.075	.014	13.4–14.1
1965–67	1100693	2.2–2.6	27A	37A	1119515	.015	.030	2.3–3.7	.067	.014	13.5–14.4
	1100696	2.2–2.6	29A	42A	1119515	.015	.030	2.3–3.7	.067	.014	13.5–14.4
1968	1100693	2.2–2.6	27A	37A	1119515	.015	.030	2.3–3.7	.067	.014	13.5–14.4
	1100794	2.2–2.6	27A	37A	1119515	.015	.030	2.3–3.7	.067	.014	13.5–14.4
	1100696	2.2–2.6	29A	42A	1119515	.015	.030	2.3–3.7	.067	.014	13.5–14.4
1969	1100834	2.2–2.6	27A	37A	1119515	.015	.030	2.3–3.7	.067	.014	13.5–14.4
	1100836	2.2–2.6	27A	37A	1119515	.015	.030	2.3–3.7	.067	.014	13.5–14.4
	1100696	2.2–2.6	27A	37A	1119515	.015	.030	2.3–3.7	.067	.014	13.5–14.4
1970–71	1100834②	2.2–2.6	27A	37A	1119515	.015	.030	2.3–3.7	.067	.014	13.5–14.4
	1100900③	2.2–2.6	27A	37A	1119515	.015	.030	2.3–3.7	.067	.014	13.5–14.4
	1100901④	2.2–2.6	27A	37A	1119515	.015	.030	2.3–3.7	.067	.014	13.5–14.4

① — At 1100 & 6500 rpm
② — 1971 — 1100836
③ — 1971 — 1100902
④ — 1971 — 1100903

BATTERY AND STARTER SPECIFICATIONS

YEAR	MODEL	BATTERY Ampere Hour Capacity	BATTERY Volts	BATTERY Group Number	BATTERY Terminal Grounded	STARTERS Lock Test Amps.	STARTERS Lock Test Volts	STARTERS Lock Test Torque	STARTERS No-Load Test Amps.	STARTERS No-Load Test Volts	STARTERS No-Load Test RPM	Brush Spring Tension (Oz.)
1964–66	6 Cyl. & V8—283 Cu. In.	44	12	2SMR	Neg.	290	4.3	—	63	10.6	7,800	35
	V8—327 Cu. In.	61	12	2SMD	Neg.	330	3.5	—	83	10.6	4,350	35
	V8—396, 427 Cu. In.	61	12	2SMD	Neg.	Not Recommended			83	10.6	4,350	35
	V8—409 Cu. In.	70	12	2SMD	Neg.	330	2.0	—	83	10.6	4,350	35
1967	6 Cyl. & V8—283 Cu. In.	45	12	—	Neg.	Not Recommended			73	10.6	9,575	35
	V8—327 Cu. In.	61	12	—	Neg.	Not Recommended			83	10.6	4,250	35
	V8—396, 427 Cu. In.	61	12	—	Neg.	Not Recommended			85	10.6	9,900	35
1968	6 Cyl., V8—307, 327 (250 H.P.)	45	12	—	Neg.	Not Recommended			73	10.6	4,500	35
	V8, 327 (275 H.P.), 396, 427	61	12	—	Neg.	Not Recommended			85	10.6	10,000	35
	All Corvette Engs.	62	12	—	Neg.	Not Recommended			85	10.6	10,000	35
1969	6 Cyl., V8—327, 396	45	12	—	Neg.	Not Recommended			73	9	4,500	35
	V8—350, 427	61	12	—	Neg.	Not Recommended			85	9	10,000	35
	All Corvette Engs.	62	12	—	Neg.	Not Recommended			85	9	10,000	35
1970–71	6 Cyl.	45	12	—	Neg.	Not Recommended			50–80	9	5,500–10,500	35
	V8—350	61	12	—	Neg.	Not Recommended			55–80	9	3,500–6,000	35
	All Corvette Engs.	62	12	—	Neg.	Not Recommended			55–80	9	3,500–6,000	35
	V8—400	62	12	—	Neg.	Not Recommended			55–80	9	3,500–6,000	35
	V8—454	62	12	—	Neg.	Not Recommended			65–95	9	7,500–10,500	35

CYLINDER HEAD BOLT TIGHTENING SEQUENCE

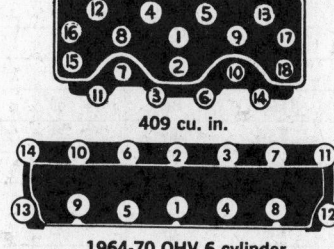

409 cu. in.

1964-70 OHV 6 cylinder

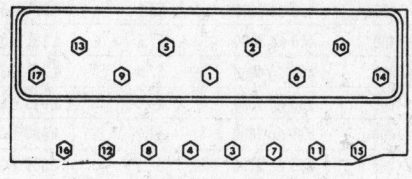

V8—283, 307, 327 and 350

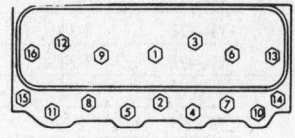

V8—396, 427, 454

GENERAL ENGINE SPECIFICATIONS

YEAR	CU. IN. DISPLACEMENT	CARBURETOR	DEVELOPED HORSEPOWER @ RPM	DEVELOPED TORQUE @ RPM (FT. LBS.)	A.M.A. HORSEPOWER	BORE AND STROKE (IN.)	ADVERTIZED COMPRESSION RATIO	VALVE LIFTER TYPE	NORMAL OIL PRESSURE (PSI)
1964	6 Cyl.—230	1-BBL.	140 @ 4400	220 @ 1600	36.0	3.875 x 3.250	8.50–1	Hyd.	35
	V8—283	2-BBL.	195 @ 4800	285 @ 2400	48.0	3.875 x 3.000	9.25–1	Hyd.	45
	V8—327	4-BBL.	250 @ 4400	350 @ 2800	51.2	4.000 x 3.250	10.50–1	Hyd.	45
	V8—327	4-BBL.①	300 @ 5000	360 @ 3200	51.2	4.000 x 3.250	10.50–1	Hyd.	45
	V8—327	Fuel Inj.	360 @ 6000	352 @ 5000	51.2	4.000 x 3.250	11.25–1	Mech.	45
	V8—409	4-BBL.	340 @ 5000	420 @ 3200	59.5	4.313 x 3.500	10.00–1	Hyd.	45
	V8—409	4-BBL.①	400 @ 5800	425 @ 3600	59.5	4.313 x 3.500	11.00–1	Mech.	45
	V8—409	2-4 BBL.①	425 @ 6000	425 @ 4200	59.5	4.313 x 3.500	11.00–1	Mech.	45
1965	6 Cyl.—230	1-BBL.	140 @ 4400	220 @ 1600	36.0	3.875 x 3.250	8.50–1	Hyd.	45
	V8—283	2-BBL.	195 @ 4800	285 @ 2400	48.0	3.875 x 3.000	9.25–1	Hyd.	45
	V8—327	4-BBL.	250 @ 4400	350 @ 2800	51.2	4.000 x 3.250	10.5–1	Hyd.	45
	V8—327	4-BBL.	300 @ 5000	360 @ 3200	51.2	4.000 x 3.250	10.5–1	Hyd.	45
	V8—327	4-BBL.	350 @ 5300	360 @ 3600	51.2	4.000 x 3.250	11.0–1	Hyd.	45
	V8—327	4-BBL.	365 @ 6200	350 @ 4000	51.2	4.000 x 3.250	11.0–1	Mech.	45
	V8—327	Fuel Inj.	375 @ 6200	350 @ 4600	51.2	4.000 x 3.250	11.0–1	Mech.	45
	V8—396	4-BBL.	325 @ 4800	410 @ 3200	53.6	4.094 x 3.760	10.25–1	Hyd.	45
	V8—396	4-BBL.	425 @ 6400	415 @ 4000	53.6	4.094 x 3.760	11.00–1	Mech.	45
	V8—409	4-BBL.	340 @ 5000	420 @ 3200	59.5	4.313 x 3.500	10.0–1	Hyd.	45
	V8—409	4-BBL.	400 @ 5800	425 @ 3600	59.5	4.313 x 3.500	11.0–1	Mech.	45
1966	6 Cyl.—250	1-BBL.	150 @ 4200	235 @ 1600	36.0	3.875 x 3.530	8.5–1	Hyd.	45
	V8—283	2-BBL.	195 @ 4800	285 @ 2400	48.0	3.875 x 3.000	9.25–1	Hyd.	45
	V8—283	4-BBL.	220 @ 4800	295 @ 3200	48.0	3.875 x 3.000	9.25–1	Hyd.	45
	V8—327	4-BBL.	275 @ 4800	355 @ 3200	51.2	4.000 x 3.250	10.5–1	Hyd.	45
	V8—327	4-BBL.	300 @ 5000	360 @ 3200	51.2	4.000 x 3.250	10.5–1	Hyd.	45
	V8—327	4-BBL.	350 @ 5800	360 @ 3600	51.2	4.000 x 3.250	11.0–1	Hyd.	45
	V8—396	4-BBL.	325 @ 4800	410 @ 3200	53.6	4.094 x 3.760	10.25–1	Hyd.	60
	V8—427	4-BBL.	390 @ 5200	460 @ 3600	57.8	4.251 x 3.760	10.25–1	Hyd.	60
	V8—427	4-BBL.	425 @ 5600	460 @ 3600	57.8	4.251 x 3.760	11.00–1	Mech.	60
1967	6 Cyl.—250	1-BBL.	155 @ 4200	235 @ 1600	36.0	3.875 x 3.530	8.5–1	Hyd.	45
	V8—283	2-BBL.	195 @ 4800	285 @ 2400	48.0	3.875 x 3.000	9.25–1	Hyd.	45
	V8—327	4-BBL.	275 @ 4800	355 @ 3200	51.2	4.000 x 3.250	10.25–1	Hyd.	45
	V8—327	4-BBL.	300 @ 5000	360 @ 3400	51.2	4.000 x 3.250	10.25–1	Hyd.	45
	V8—327	4-BBL.	350 @ 5800	360 @ 3600	51.2	4.000 x 3.250	11.00–1	Hyd.	45
	V8—396	4-BBL.	325 @ 4800	410 @ 3200	53.6	4.094 x 3.760	10.25–1	Hyd.	65
	V8—427	4-BBL.	385 @ 5200	460 @ 3400	57.8	4.251 x 3.760	10.25–1	Hyd.	65
	V8—427	4-BBL.	390 @ 5400	460 @ 3600	57.8	4.251 x 3.760	10.25–1	Hyd.	65
	V8—427	3-2-BBL.	400 @ 5400	460 @ 3600	57.8	4.251 x 3.760	10.25–1	Hyd.	65
	V8—427	4-BBL.	425 @ 5600	460 @ 3800	57.8	4.251 x 3.760	N.A.	Mech.	65
	V8--427	3-2-BBL.	435 @ 5800	460 @ 4000	57.8	4.251 x 3.760	11.00–1	Mech.	65
1968	6 Cyl.—250	1-BBL.	155 @ 4200	235 @ 1600	36.0	3.875 x 3.530	8.5–1	Hyd.	45
	V8—307	2-BBL.	200 @ 4600	300 @ 2400	48.0	3.875 x 3.250	9.0–1	Hyd.	45
	V8—327	4-BBL.	250 @ 4800	325 @ 3200	52.2	4.000 x 3.250	8.75–1	Hyd.	45
	V8—327	4-BBL.	275 @ 4800	355 @ 3200	51.2	4.000 x 3.250	10.0–1	Hyd.	45
	V8—327	4-BBL.	300 @ 5000	360 @ 3400	51.2	4.000 x 3.250	10.0–1	Hyd.	45
	V8—327	4-BBL.	350 @ 5800	360 @ 3600	51.2	4.000 x 3.250	11.0–1	Hyd.	45
	V8—396	4-BBL.	325 @ 4800	410 @ 3200	53.6	4.094 x 3.760	10.25–1	Hyd.	65
	V8—427	4-BBL.	385 @ 5200	460 @ 3400	57.8	4.251 x 3.760	10.25–1	Hyd.	65
	V8—427	4-BBL.	390 @ 5400	460 @ 3600	57.8	4.251 x 3.760	10.25–1	Hyd.	65
	V8—427	3-2-BBL.	400 @ 5400	460 @ 3600	57.8	4.251 x 3.760	10.25–1	Hyd.	65
	V8—427	3-2-BBL.	435 @ 5800	460 @ 4000	57.8	4.251 x 3.760	11.0–1	Mech.	65
1969	6 Cyl.—250	1-BBL.	155 @ 4200	235 @ 1600	36.0	3.875 x 3.530	8.5–1	Hyd.	35
	V8—327	2-BBL.	235 @ 4800	325 @ 2800	51.2	4.000 x 3.250	9.0–1	Hyd.	45
	V8—350	4-BBL.	255 @ 4800	365 @ 3200	51.2	4.000 x 3.480	9.0–1	Hyd.	45
	V8—350	4-BBL.	300 @ 4800	380 @ 3200	51.2	4.000 x 3.480	9.0–1	Hyd.	45
	V8—350	4-BBL.	350 @ 5600	380 @ 3600	51.2	4.000 x 3.480	11.0–1	Hyd.	45

GENERAL ENGINE SPECIFICATIONS, continued

YEAR	CU. IN. DISPLACEMENT	CARBURETOR	DEVELOPED HORSEPOWER @ RPM	DEVELOPED TORQUE @ RPM (FT. LBS.)	A.M.A. HORSEPOWER	BORE AND STROKE (IN.)	ADVERTIZED COMPRESSION RATIO	VALVE LIFTER TYPE	NORMAL OIL PRESSURE (PSI)
	V8—396	2-BBL.	265 @ 4800	400 @ 2800	53.6	4.094 x 3.760	9.0-1	Hyd.	45
	V8—427	4-BBL.	335 @ 4800	470 @ 3200	57.8	4.251 x 3.760	10.25-1	Hyd.	45
	V8—427	4-BBL.	390 @ 5400	460 @ 3600	57.8	4.251 x 3.760	10.25-1	Hyd.	45
	V8—427	3-2-BBL.	400 @ 5400	460 @ 3600	57.8	4.251 x 3.760	10.25-1	Hyd.	45
	V8—427	3-2-BBL.	435 @ 5800	460 @ 4000	57.8	4.251 x 3.760	11.0-1	Mech.	45
1970-71	6 Cyl.—250	1-BBL.	155 @ 4200	235 @ 1600	36.0	3.875 x 3.530	② 8.5-1	Hyd.	30-45
	V8—350	2-BBL.	250 @ 4800	345 @ 2800	51.2	4.000 x 3.480	② 9.0-1	Hyd.	30-45
	V8—350	4-BBL.	300 @ 4800	380 @ 3200	51.2	4.000 x 3.480	②10.25-1	Hyd.	30-45
	V8—350	4-BBL.	350 @ 5600	380 @ 3600	51.2	4.000 x 3.480	② 11.0-1	Hyd.	30-45
	V8—350	4-BBL.	370 @ 6000	380 @ 4000	51.2	4.000 x 3.480	② 11.0-1	Mech.	30-45
	V8—400	2-BBL.	265 @ 4400	400 @ 2400	54.4	4.125 x 3.750	② 9.0-1	Hyd.	30-45
	V8—454	4-BBL.	345 @ 4400	500 @ 3000	57.8	4.251 x 4.000	②10.25-1	Hyd.	30-45
	V8—454	4-BBL.	390 @ 4800	500 @ 3400	57.8	4.251 x 4.000	②10.25-1	Hyd.	30-45
	V8—454	4-BBL.	460 @ 5600	490 @ 3600	57.8	4.251 x 4.000	②11.25-1	Mech.	30-45

① —Special camshaft.
② —1971—slightly lower.

GENERAL CHASSIS AND BRAKE SPECIFICATIONS

YEAR	MODEL	CHASSIS Overall Length (In.)	Tire Size	Master Cylinder (In.)	BRAKE CYLINDER BORE Wheel Cylinder Diameter (In.) Front	Rear	BRAKE DRUM Diameter (In.) Front	Rear
CHEVROLET								
1964	Sedans	209.9	7.00 x 14	1.0①	1³/₁₆	1.0	11	11
	Conv.	209.9	7.50 x 14	1.0①	1³/₁₆	1.0	11	11
	Sta. Wag.	210.8	8.00 x 14	1.0①	1³/₁₆	1.0	11	11
1965	Sedans	213.0	7.35 x 14② ③	1.0①	1³/₁₆	1.0	11	11
	Sta. Wag.	213.3	8.25 x 14	1.0①	1³/₁₆	1.0	11	11
	Conv.	213.0	7.75 x 14③	1.0①	1³/₁₆	1.0	11	11
1966	Sedans	213.2	7.35 x 14② ③	1.0①	1³/₁₆	1.0	11	11
	Conv.	213.2	7.75 x 14③	1.0①	1³/₁₆	1.0	11	11
	Sta. Wag.	212.4	8.55 x 14	1.0①	1³/₁₆	1.0	11	11
1967	Exc. Sta. Wag.	213.2	8.25 x 14	1.0①	1³/₁₆ ④	1.0	11⑤	11
	Sta. Wag.	212.4	8.85 x 14	1.0①	1³/₁₆ ④	1.0	11⑤	11
1968	Exc. Sta. Wag.	214.7	8.25 x 14	1.0①	1³/₁₆ ④	1.0	11⑤	11
	Sta. Wag.	213.9	8.55 x 14	1.0①	1³/₁₆ ④	1.0	11⑤	11
1969	Exc. Sta. Wag.	215.9	8.25 x 14	1.0①	1³/₁₆ ④	1.0	11⑤	11
	Sta. Wag.	216.7	8.85 x 14	1.0①	1³/₁₆ ④	1.0	11⑤	11
1970-71	Exc. Sta. Wag.	215.9	8.25 x 14	1.0⑥	1³/₁₆ ④	1.0	11⑤	11
	Sta. Wag.	216.7	8.85 x 14	1.0⑥	1³/₁₆ ④	1.0	11⑤	11
CORVETTE								
1964		175.3	6.70 x 15	1.0①	1³/₁₆	1.0	11	11
1965-66	Conv.	175.1	7.75 x 15	1.0	1⅛	1³/₈	11.75 (Disc)	11.75 (Disc)
	Coupe	175.1	7.75 x 15	⁷/₈	1⅛	1.0	11.75	11.75
1967		175.1	7.75 x 15	1.0	1⅛	1³/₈	11.75 (Disc)	11.75 (Disc)
1968-69		182.1	F70 x 15	1.0	1⅛	1³/₈	11.75 (Disc)	11.75 (Disc)
1970-71		182.1	F70 x 15	1.0	1⅛	1³/₈	11.75 (Disc)	11.75 (Disc)

① —Metallic brakes—master cylinder bore—⁷/₈.
② —Models with 327 or 396 Eng.—7.75 x 14.
③ —Models with 409 or 427 Eng.—8.25 x 14.
④ —Disc 2¹⁵/₁₆.
⑤ —Disc 11.75.
⑥ —W/disc brakes—1⅛.

TUNE-UP SPECIFICATIONS

YEAR	MODEL	SPARK PLUGS		DISTRIBUTOR		IGNITION TIMING (Deg.) ▲	CRANKING COMP. PRESSURE (Psi)	VALVES		Intake Opens (Deg.)	FUEL PUMP PRESSURE (Psi)	IDLE SPEED (Rpm) ★
		Type	Gap (In.)	Point Dwell (Deg.)	Point Gap (In.)			Tappet (Hot) Clearance (In.)				
								Intake	Exhaust			
1964	6 cyl.—230 Cu. In.	46N	.035	32°	.019	4B	130	■	■	34B	4	500
	V8—283 Cu. In.	45	.035	30°	.019	4B	150	■	■	32½B	6	500
	V8—327 Cu. In. Std.	44	.035	30°	.019	4B	160	■	■	32½B	6	500
	V8—327 Cu. In. Hi. Perf.	44	.035	30°	.019	8B	160	.030	.030	32½B	6	500
	V8—409 Cu. In. Std.	43N	.035	30°	.019	6B	150	■	■	38½B	6	700
	V8—409 Cu. In. Hi. Perf.	43N	.035	30°	.019	12B	150	.012	.020	49½B	8	700
1965	6 cyl.—230 Cu. In.	46N	.035	32°	.019	4B	130	■	■	62B	4	500
	V8—283 Cu. In.	45	.035	30°	.019	4B	150	■	■	32½B	6	500
	V8—327 Cu. In. Std.	44	.035	30°	.019	4B	160	■	■	32½B	6	500
	V8—327 Cu. In. (300 H.P.)	44	.035	30°	.019	8B	160	■	■	32½B	6	500
	V8—327 Cu. In. (350 H.P.)	44	.035	30°	.019	12B	150	■	■	54B	6	800
	V8—327 Cu. In. (365 H.P.)	44	.035	30°	.019	10B	150	.030	.030	54B	6	800
	V8—327 Cu. In. (Fuel Inj.)	44	.035	30°	.019	12B	150	.030	.030	54B	6	800
	V8—396 Cu. In. Std.	43N	.035	30°	.019	4-6B	150	■	■	N.A.	5½	800
	V8—396 Cu. In. Hi. Perf.	43N	.035	30°	.019	10B	150	.020	.024	N.A.	7½	800
	V8—409 Cu. In. Std.	43N	.035	30°	.019	6B	150	■	■	38½B	8	500
	V8—409 Cu. In. Opt.	43N	.035	30°	.019	12B	150	.018	.030	50¾B	8	700
1966	6 Cyl.—250 Cu. In.	46N	.035	34°	.019	6B	130	■	■	62B	4	500
	V8—283 Cu. In.	45	.035	30°	.019	4B	150	■	■	32½B	6	500
	V8—327 Cu. In. Std.	44	.035	30°	.019	③	160	■	■	32½B	6	500
	V8—327 Cu. In. (350 H.P.)	44	.035	30°	.019	10B	160	■	■	54B	6	700
	V8—396 Cu. In.	43N	.035	30°	.019	4B	150	■	■	40B	6	500
	V8—427 Cu. In.	43N	.035	30°	.019	4B	150	■	■	58B	6	600
	V8—427 Cu. In. (425 H.P.)	43N	.035	30°	.019	8B	150	.020	.024	54B	6	800
1967	6 cyl.—250 Cu. In.	46N	.035	32°	.019	4B	130	■	■	62B	4	500
	V8—283 Cu. In. (195 H.P.)	45	.035	30°	.019	4B	150	■	■	36B	6	500
	V8—327 Cu. In. (275 H.P.)	44	.035	30°	.019	8B⑤	160	■	■	38B	6	500
	V8—327 Cu. In. (300 H.P.)	44	.035	30°	.019	6B⑥	160	■	■	38B	6	500
	V8—327 Cu. In. (350 H.P.)	44	.035	30°	.019	10B	150	■	■	54B	6	700
	V8—396 Cu. In. (325 H.P.)	43N	.035	30°	.019	4B	160	■	■	40B	6	500
	V8—427 Cu. In. (385 H.P.)	43N	.035	30°	.019	4B	160	■	■	56B	6	550
	V8—427 Cu. In. (390 H.P.)	43N	.035	30°	.019	4B	160	■	■	56B	6	550
	V8—427 Cu. In. (400 H.P.)	43N	.035	30°	.019	4B	160	■	■	56B	6	550
	V8—427 Cu. In. (425 H.P.)	43N	.035	30°	.019	10B	150	.022	.024	44B	6	1000
	V8—427 Cu. In. (435 H.P.)	43N	.035	30°	.019	5B	150	.024	.028	44B	6	750
1968	6 Cyl.—250 Cu. In.	46N	.035	32°	.019	TDC⑦	130	■	■	16B	4	700④
	V8—307 Cu. In. (200 H.P.)	45	.035	30°	.019	2B	150	■	■	28B	6	700①
	V8—327 Cu. In. (250 H.P.)	45	.035	30°	.019	4B	150	■	■	28B	6	700①
	V8—327 Cu. In. (275 H.P.)	44	.035	30°	.019	TDC⑦	160	■	■	28B	6	700①
	V8—327 Cu. In. (300 H.P.)	44	.035	30°	.019	4B	160	■	■	28B	6	700①
	V8—327 Cu. In. (350 H.P.)	44	.035	30°	.019	4B	160	■	■	40B	6	750
	V8—396 Cu. In. (325 H.P.)	43N	.035	30°	.019	4B	160	■	■	28B	6	700①
	V8—427 Cu. In. (385 H.P.)	43N	.035	30°	.019	4B	160	■	■	40B	8	700①
	V8—427 Cu. In. (390 H.P.)	43N	.035	30°	.019	4B	160	■	■	40B	8	700①
	V8—427 Cu. In. (400 H.P.)	43N	.035	30°	.019	4B	160	■	■	40B	8	750①
	V8—427 Cu. In. (435 H.P.)	43N	.035	30°	.019	4B	150	.024	.028	44B	8	750
1969	6 Cyl.—250 Cu. In.	46N	.035	32°	.019	TDC⑤	130	■	■	16B	4	700⑦
	V8—327 Cu. In. (235 H.P.)	45S	.035	30°	.019	2A④	150	■	■	28B	6	700①

TUNE-UP SPECIFICATIONS, continued

YEAR	MODEL	SPARK PLUGS Type	Gap (In.)	DISTRIBUTOR Point Dwell (Deg.)	Point Gap (In.)	IGNITION TIMING (Deg.) ▲	CRANKING COMP. PRESSURE (Psi)	VALVES Tappet (Hot) Clearance (In.) Intake	Exhaust	Intake Opens (Deg.)	FUEL PUMP PRESSURE (Psi)	IDLE SPEED (Rpm) ★
	V8—350 Cu. In. (255 H.P.)	44S	.035	30°	.019	TDC⑤	160	■	■	28B	6	700①
	V8—350 Cu. In. (300 H.P.)	44S	.035	30°	.019	TDC⑤	160	■	■	28B	6	700①
	V8—350 Cu. In. (350 H.P.)	44	.035	30°	.019	8B	160	■	■	52B	6	750
	V8—396 Cu. In. (265 H.P.)	44N	.035	30°	.019	TDC⑤	160	■	■	28B	6	700①
	V8—427 Cu. In. (335 H.P.)	43N	.035	30°	.019	4B	160	■	■	28B	8	800①
	V8—427 Cu. In. (390 H.P.)	43N	.035	30°	.019	4B	160	■	■	56B	8	800①
	V8—427 Cu. In. (400 H.P.)	43N	.035	30°	.019	4B	160	■	■	56B	8	800①
	V8—427 Cu. In. (435 H.P.)	43LX	.035	Transistor Ign.		4B	160	■	■	44B	8	750
1970	6 Cyl.—250 Cu. In.	R46T	.035	31–34	.019	TDC⑤	130	■	■	16B	4	750①
	V8—350 Cu. In. (250 H.P.)	R44	.035	29–31	.019	TDC⑤	160	■	■	28B	8	750①
	V8—350 Cu. In. (300 H.P.)	R44	.035	29–31	.019	TDC⑤	160	■	■	28B	8	750①
	V8—350 Cu. In. (350 H.P.)	R44	.035	29–31	.019	8B	160	■	■	52B	8	750
	V8—350 Cu. In. (370 H.P.)	R43	.035	Transistor Ign.		8B	160	.030	.030	43B	8	900
	V8—400 Cu. In. (265 H.P.)	R44	.035	29–31	.019	4B⑦	160	■	■	34B	8	700①
	V8—454 Cu. In. (345 H.P.)	R44T	.035	28–30	.019	6B	160	■	■	30B	8	600
	V8—454 Cu. In. (390 H.P.)	R43T	.035	28–30	.019	6B	160	■	■	56B	8	700①
	V8—454 Cu. In. (450 H.P.)	R43XL	.035	Transistor Ign.		8B	160	.024	.028	62B	8	700①
1971	6 Cyl.—250 Cu. In.	R46T	.035	31–34	.019	①	130	■	■	16B	4	750①
	V8—350 Cu. In. (250 H.P.)	R44	.035	29–31	.019	①	160	■	■	28B	8	750①
	V8—350 Cu. In. (300 H.P.)	R44	.035	29–31	.019	①	160	■	■	28B	8	750①
	V8—350 Cu. In. (350 H.P.)	R44	.035	29–31	.019	①	160	■	■	52B	8	750
	V8—350 Cu. In. (370 H.P.)	R43	.035	Transistor Ign.		①	160	.030	.030	43B	8	900
	V8—400 Cu. In. (265 H.P.)	R44	.035	29–31	.019	①	160	■	■	34B	8	700①
	V8—454 Cu. In. (345 H.P.)	R44T	.035	28–30	.019	①	160	■	■	30B	8	600
	V8—454 Cu. In. (390 H.P.)	R43T	.035	28–30	.019	①	160	■	■	56B	8	700①
	V8—454 Cu. In. (450 H.P.)	R43XL	.035	Transistor Ign.		①	160	.024	.028	62B	8	700①

★—with manual transmission in N and automatic in D. Add 50 rpm if equipped with air conditioning

▲—with vacuum advance disconnected. NOTE: These settings are only approximate. Engine design, altitude, temperature, fuel octane rating and the condition of the individual engine are all factors which can influence timing. The limiting advance factor must, therefore, be the "knock point" of the individual engine.

■—1 turn tighter than zero lash.

①—N.A. @ publication, see decal under hood for data.

⑩—if exhaust emission equipped:
  .283 Cu. In., w/M.T.—TDC

②—w/A.T.—4B

③—V8—327, Dist. No. 1111152—8° B.
  Dist. No. 1111116—2° A.
  Dist. No. 1111153—6° B.
  Dist. No. 1111117—4° A.

④—6 Cyl. w/A.T.—Idle 500 rpm.

⑤—w/Auto. Trans.—4°B.

⑥—w/Auto. Trans.—2°B.

⑦—w/Auto. Trans.—500 rpm.

⑧—w/Auto. Trans.—600 rpm

⑨—w/Auto. Trans.—8°B.

A—After top dead center.

B—Before top dead center.

TDC—Top dead center.

CAUTION

General adoption of anti-pollution laws has changed the design of almost all car engine production to effectively reduce crankcase emission and terminal exhaust products. It has been necessary to adopt stricter tune-up rules, especially timing and idle speed procedures. Both of these values are peculiar to the engine and to its application, rather than to the engine alone. With this in mind, car manufacturers supply idle speed data for the engine and application involved. This information is clearly displayed in the engine compartment of each vehicle.

VALVE SPECIFICATIONS

YEAR AND MODEL	SEAT ANGLE (DEG.)	FACE ANGLE (DEG.)	VALVE LIFT INTAKE (IN.)	VALVE LIFT EXHAUST (IN.)	VALVE SPRING PRESSURE (VALVE OPEN) LBS. @ IN.	VALVE SPRING INSTALLED HEIGHT (IN.)	STEM TO GUIDE CLEARANCE (IN.) Intake	STEM TO GUIDE CLEARANCE (IN.) Exhaust	STEM DIAMETER (IN.) Intake	STEM DIAMETER (IN.) Exhaust
1963-64 6 Cyl.—230 Cu. In.	46	45	.3349	.3349	175 @ 1.26	1.66	.0010–.0027	.0016–.0033	.3404–.3417	.3410–.3417
V8—283 Cu. In.	46	45	.3987	.3987	175 @ 1.26	1.66	.0010–.0027	.0016–.0033	.3404–.3417	.3410–.3417
V8—327 Cu. In.	46	45	.3987	.3987	175 @ 1.26	1.66	.0010–.0027	.0016–.0033	.3410–.3417	.3410–.3417
V8—409 Cu. In.	46	45	.4000	.4119	170 @ 1.33	1.66	.0010–.0027	.0025–.0042	.3715–.3722	.3710–.3717
V8—409 Cu. In.③	46	45	.5068	.5185	330 @ 1.20	1.66	.0010–.0027	.0025–.0042	.3715–.3722	.3710–.3717
1965-66 6 Cyl.—230 Cu. In.	46	45	.3318	.3318	175 @ 1.26	1.66	.0010–.0027	.0016–.0033	.3404–.3417	.3410–.3417
6 Cyl.—250 Cu. In.	46	45	.3880	.3880	185 @ 1.27	1.66	.0010–.0027	.0010–.0027	.3410–.3417	.3410–.3417
V8—283 Cu. In.	46	45	.3987	.3987	175 @ 1.26	1.66	.0010–.0027	.0016–.0033	.3404–.3417	.3410–.3417
V8—327 Cu. In.	46	45	.3987	.3987	175 @ 1.26	1.66	.0010–.0027	.0016–.0033	.3404–.3417	.3410–.3417
V8—327 Cu. In. (350 H.P.)	46	45	.4472	.4472	175 @ 1.26	1.66	.0010–.0027	.0016–.0033	.3410–.3417	.3410–.3417
V8—327 Cu. In. (Fuel Inj.)	46	45	.4850	.4850	175 @ 1.26	1.66	.0010–.0027	.0016–.0033	.3410–.3417	.3410–.3417
V8—396 Cu. In. Std.	46	45	.3980	.3980	220 @ 1.46	$1\frac{7}{8}$	.0005–.0024	.0012–.0029	.3715–.3722	.3713–.3720
V8—396 Cu. In. Hi. Perf.	46	45	.5000	.4960	315 @ 1.38	$1\frac{7}{8}$	.0005–.0024	.0012–.0029	.3715–.3722	.3713–.3720
V8—409 Cu. In.	46	45	.4005	.4119	170 @ 1.33	1.66	.0010–.0027	.0015–.0032	.3715–.3722	.3710–.3717
V8—409 Cu. In. (400 H.P.)	46	45	.5567	.5567	330 @ 1.20	1.68	.0010–.0027	.0015–.0032	.3715–.3722	.3710–.3717
V8—427 Cu. In.	46	45	.4614	.4800	315 @ 1.38	1.88	.0010–.0027	.0015–.0032	.3715–.3722	.3713–.3720
V8—427 Cu. In. (425 H.P.)	46	45	.5197	.5197	315 @ 1.38	1.88	.0010–.0027	.0015–.0032	.3715–.3722	.3713–.3720
1967 6 Cyl.—250 Cu. In.	46	45	.3880	.3880	186 @ 1.27	$1\frac{21}{32}$	.0010–.0027	.0010–.0027	.3410–.3417	.3410–.3417
V8—283 Cu. In.	46	45	.3900	.4100	200 @ 1.25	$1\frac{5}{32}$	.0010–.0027	.0010–.0027	.3410–.3417	.3410–.3417
V8—327 Cu. In.	46	45	.3900	.4100	200 @ 1.25	$1\frac{5}{32}$	.0010–.0027	.0010–.0027	.3410–.3417	.3410–.3417
V8—327 Cu. In. (300 H.P.)	46	45	.3900	.4100	200 @ 1.25	$1\frac{5}{32}$	.0010–.0027	.0015–.0032	.3410–.3417	.3410–.3417
V8—327 Cu. In. (350 H.P.)	46	45	.4472	.4472	200 @ 1.25	$1\frac{5}{32}$	.0010–.0027	.0015–.0032	.3410–.3417	.3410–.3417
V8—396 Cu. In.	46	45	.3983	.3983	220 @ 1.46	$1\frac{7}{8}$	.0010–.0027	.0015–.0032	.3715–.3722	.3713–.3720
V8—427 Cu. In. (385, 390, 400 H.P.)	46	45	.4614	.4800			.0010–.0027	.0015–.0032		
V8—427 Cu. In. (435 H.P.)	46	45	.5197	.5197	315 @ 1.38	$1\frac{7}{8}$	.0010–.0027	.0015–.0032	.3715–.3722	.3713–.3720
1968 6 Cyl.—250 Cu. In.	46	45	.3880	.3880	186 @ 1.25	1.66	.0010–.0027	.0017–.0027	.3410–.3417	.3410–.3417
V8—307 Cu. In.	46	45	.3900	.4100	198 @ 1.25	1.70	.0010–.0027	.0017–.0027	.3410–.3417	.3410–.3417
V8—327 Cu. In. (Exc. 350 H.P.)	46	45	.3900	.4100	198 @ 1.25	1.70	.0010–.0027	.0017–.0027	.3410–.3417	.3410–.3417
V8—327 (350 H.P.)	46	45	.4472	.4472	198 @ 1.25	1.70	.0010–.0027	.0010–.0027	.3410–.3417	.3410–.3417
V8—396 Cu. In.	46	45	.3983	.3983	220 @ 1.46	$1\frac{7}{8}$	.0010–.0027	.0015–.0032	.3715–.3722	.3713–.3720
V8—427 (Exc. 435 H.P.)	46	45	.4614	.4800	315 @ 1.38	1.88	.0010–.0027	.0015–.0032	.3715–.3722	.3713–.3722
V8—427 Cu. In. (435 H.P.)	46	45	.5197	.5197	315 @ 1.38	1.88	.0010–.0027	.0015–.0032	.3715–.3722	.3713–.3722
1969 6 Cyl. 250 Cu. In.	46	45	.3880	.3880	186 @ 1.27	1.66	.0010–.0027	.0010–.0027	.3410–.3417	.3410–.3417
V8—327, 350 Cu. In.	46	45	.3945	.3945	200 @ 1.25	1.70	.0010–.0027	.0010–.0027	.3410–.3417	.3410–.3417
V8—350 (350 H.P.)	46	45	.4500	.4600	200 @ 1.25	1.70	.0010–.0027	.0010–.0027	.3410–.3417	.3410–.3417
V8—396	46	45	.3983	.3983	220 @ 1.46	$1\frac{7}{8}$	.0010–.0027	.0010–.0027	.3410–.3417	.3410–.3417
V8—427 (335 H.P.)	46	45	.3983	.3983	198 @ 1.32	1.88	.0010–.0027	.0010–.0027	.3410–.3417	.3410–.3417
V8—427 (390 & 400 H.P.)	46	45	.4614	.4800	312 @ 1.38	1.88	.0010–.0027	.0010–.0027	.3715–.3722	.3713–.3722
V8—427 (435 H.P.)	46	45	.5197	.5197	312 @ 1.38	1.88	.0010–.0027	.0010–.0027	.3715–.3722	.3713–.3722
1970 6 Cyl.—250 Cu. In.	46	45	.3880	.3880	186 @ 1.27	1.66	.0010–.0027	.0015–.0032	.3410–.3417	.3410–.3417
V8—350 Cu. In. (250, 300 H.P.)	46	45	.3900	.4100	200 @ 1.25	1.70	.0010–.0027	.0012–.0029	.3410–.3417	.3410–.3417
V8—350 Cu. In. (350 H.P.)	46	45	.4500	.4600	200 @ 1.25	1.70	.0010–.0027	.0012–.0029	.3410–.3417	.3410–.3417
V8—350 Cu. In. (370 H.P.)	46	45	.4500	.4850	200 @ 1.25	1.70	.0010–.0027	.0012–.0029	.3410–.3417	.3410–.3417
V8—400 Cu. In. (265 H.P.)	46	45	.3983	.3983	200 @ 1.25	1.70	.0010–.0027	.0012–.0029	.3410–.3417	.3410–.3417
V8—454 Cu. In. (345 H.P.)	46	45	.3983	.4300	240 @ 1.38①	1.88①	.0010–.0027	.0012–.0027	.3715–.3722	.3715–.3722
V8—454 Cu. In. (390 H.P.)	46	45	.4614	.4800	240 @ 1.38①	1.88①	.0010–.0027	.0012–.0027	.3715–.3722	.3715–.3722
V8—454 Cu. In. (450 H.P.)	46	45	.5197	.5498	193 @ 1.32②	1.88②	.0010–.0027	.0012–.0027	.3715–.3722	.3715–.3722
1971 6 Cyl.—250 Cu. In.	46	45	.3880	.3880	186 @ 1.27	1.66	.0010–.0027	.0015–.0032	.3410–.3417	.3410–.3417
V8—350 Cu. In. (250-300 H.P.)	46	45	.3900	.4100	200 @ 1.25	1.70	.0010–.0027	.0012–.0029	.3410–.3417	.3410–.3417
V8—350 Cu. In. (350 H.P.)	46	45	.4500	.4600	200 @ 1.25	1.70	.0010–.0027	.0012–.0029	.3410–.3417	.3410–.3417
V8—350 Cu. In. (370 H.P.)	46	45	.4500	.4850	200 @ 1.25	1.70	.0010–.0027	.0012–.0029	.3410–.3417	.3410–.3417
V8—400 Cu. In. (265 H.P.)	46	45	.3983	.3983	200 @ 1.25	1.70	.0010–.0027	.0012–.0029	.3410–.3417	.3410–.3417

VALVE SPECIFICATIONS, continued

YEAR AND MODEL	SEAT ANGLE (DEG.)	FACE ANGLE (DEG.)	VALVE LIFT INTAKE (IN.)	VALVE LIFT EXHAUST (IN.)	VALVE SPRING PRESSURE (VALVE OPEN) LBS. @ IN.	VALVE SPRING INSTALLED HEIGHT (IN.)	STEM TO GUIDE CLEARANCE (IN.)		STEM DIAMETER (IN.)	
							Intake	Exhaust	Intake	Exhaust
V8—454 Cu. In. (345 H.P.)	46	45	.3983	.4300	240 @ 1.38 ①	1.88 ①	.0010–.0027	.0012–.0027	.3715–.3722	.3715–.3722
V8—454 Cu. In. (390 H.P.)	46	45	.4614	.4800	240 @ 1.38 ①	1.88 ①	.0010–.0027	.0012–.0027	.3715–.3722	.3715–.3722
V8—454 Cu. In. (450 H.P.)	46	45	.5197	.5498	193 @ 1.32 ②	1.88 ②	.0010–.0027	.0012–.0027	.3715–.3722	.3715–.3722

① —Inner spring 90 lbs. at 1.28 in., spring height 1.78 in.
② —Inner spring 101 lbs. at 1.22 in., spring height 1.78 in.
③ —with special camshaft.

CRANKSHAFT BEARING JOURNAL SPECIFICATIONS

YEAR	MODEL	MAIN BEARING JOURNALS (IN.)				CONNECTING ROD BEARING JOURNALS (IN.)		
		Journal Diameter	Oil Clearance	Shaft End-Play	Thrust On No.	Journal Diameter	Oil Clearance	End-Play
1964–65	6 Cyl.—230 Cu. In.	2.2983–2.2993	.0011②	.004	7	1.9990–2.0000	.0018	.0110
	V8—283/327 Cu. In.	2.2978–2.2988	.0016	.004	5	1.9990–2.0000	.0018	.0110
	V8—396 Cu. In. Std.	2.7487–2.7497③	.0014	.008	5	2.1990–2.2000	.0017	.0180
	V8—396 Cu. In. Hi. Perf.	2.7487–2.7497	.0021	.008	5	2.1985–2.1995	.0020	.0180
	V8—409 Cu. In.	2.4980–2.4990①	.0017	.008	5	2.1988–2.1998	.0020	.0180
1966–67	6 Cyl.—250 Cu. In.	2.3004	.0016	.004	7	1.9990–2.0000	.0017	.0110
	V8—283, 327 Cu. In.	2.2978–2.2988	.0016	.004	5	1.9990–2.0000	.0018	.0110
	V8—396 Cu. In.	2.7487–2.7497③	.0014	.008	5	2.1990–2.2000	.0017	.0180
	V8—427 Exc. Below	2.7487–2.7497③	.0014	.008	5	2.1990–2.2000	.0017	.0180
	V8—427 (425 & 435 H.P.)	2.7487–2.7497③	.0022	.008	5	2.1990–2.2000	.0021	.0180
1968–69	6 Cyl.—250 Cu. In.	2.3004	.0016	.004	7	1.9990–2.0000	.0017	.0110
	V8—307, 327, 350 Cu. In.	2.4505④	.0018	.008	5	2.0990–2.1000	.0018	.0110
	V8—396, 427 Cu. In.	2.7505	.0018	.008	5	2.1990–2.2000	.0018	.0180
1970–71	6 Cyl.—250 Cu. In.	2.2983–2.2993	.0003–.0029	.002–.006	7	1.9990–2.0000	.0007–.0027	.009–.014
	V8—350 Cu. In.	2.4484–2.4493⑤	⑦	.002–.006	5	2.0990–2.1000	.0013–.0035	.008–.014
	V8—400 Cu. In.	2.650	⑦	.002–.006	5	2.0990–2.1000	.0013–.0035	.008–.014
	V8—454 Cu. In.	2.7481–2.7490	⑧	.006–.010	5	2.1990–2.2000	.0009–.0025	.015–.023
	V8—454 Cu. In. (450 H.P.)	2.7481–2.7490	⑧	.006–.010	5	2.1990–2.2000	.0014–.0034	.015–.023

① —Rear main journal diameter—2.4977–2.4987.
② —Rear main oil clearance—.0016.
③ —No's. 3, 4—2.7482–2.7492, No. 5—2.7478–2.7488.
④ —No. 1—2.4502, No. 5—2.4507.
⑤ —No. 5—2.4779–2.4488.

⑥ —No. 1—2.7485–2.7494; No. 5—2.748–2.7488.
⑦ —No. 1—.0008–.0020; Nos. 2-4—.0013–.0025; No. 5—.0019–.0035.
⑧ —No. 1—.0007–.0019; Nos. 2-4—.0013–.0025; No. 5—.0019–.0035.
⑨ —No. 5—2.7478–2.7488.

FUSES AND CIRCUIT BREAKERS

1964–68

Headlamps, parking lamps.................. 15 Amp. CB
Panel and accessory lamps................ 3 Amp. Fuse
Tail, dome, glove box, clock, stop, back-up,
   brake indicator, courtesy, stop lamps.... 10 Amp. Fuse
Heater (no air conditioning) .... 10 Amp. Fuse
Air conditioning (including heater)........ 20 Amp. Fuse
Radio.................................... 7½ Amp. Fuse

1969–70

A circuit breaker in the light control switch protects the headlamp circuit, thus eliminating one fuse. A separate 30 Amp. circuit breaker mounted on the firewall protects the power window, seat and top circuits. The under hood and spot lamp circuit is also protected by a separate 15 Amp. inline fuse. Where current load is too heavy, the circuit breaker rapidly opens and closes, pro-tecting the circuit until the cause is found and eliminated.

Fuses located in the fuse panel under the instrument panel are:
Wiper/washer, 3-spd. A/T downshift... SAE/SFE 25 AMP.
Back-up lamp, turn signal, cruise
   master, defogger, heater .... SAE/SFE 25 AMP.
Air conditioning—transmission control
   spark solenoid .... 3AG/AGC 25 AMP.
Radio, power window........... 3AG/AGC 10 AMP.
Tail, marker and fender lamps....... SAE/SFE 20 AMP.
Instrument lamps (Nova, Chevrolet
   and Chevelle) .... SAE/SFE 4 AMP.
Instrument lamps (Corvette).......... 1AG/AGA 5 AMP.
Gauges and tell-tale lamps....... 3AG/AGC 10 AMP.
Stop and hazard.................. SAE/SFE 20 AMP.
Clock, lighter, courtesy lamps, dome
   and luggage lamps........... SAE/SFE 20 AMP.

In-Line Fuses
Air Conditioning Higher Blower Speed Fuse located in wire running from horn relay to A/C Relay
   Comfortron and Four Season............SAE 30 AMP.
   Universal & All Weather (Nova)..........SAE 20 AMP.
Fusible Links:
Pigtail lead at battery positive
   cable (except Corvette)...... 14 gauge brown wire
Molded splice at Solenoid "Bat"
   terminal (Corvette only)..... 14 gauge brown wire
Molded splice located at the horn
   relay .... 16 gauge black wire
Molded splice in voltage regulator
   #3 terminal wire........... 20 gauge orange wire
Molded splice in ammeter circuit
   (both sides of meter)........ 20 gauge orange wire

CAPACITIES

YEAR	MODEL	ENGINE CRANKCASE ADD 1 Qt. FOR NEW FILTER	TRANSMISSIONS Pts. TO REFILL AFTER DRAINING			DRIVE AXLE (Pts.)	GASOLINE TANK (Gals.)	COOLING SYSTEM (Qts.) WITH HEATER
			3-Speed★	4-Speed	Automatic			
CHEVROLET								
1964	6 Cyl.	4	2	None	18	4	20①	12
	V8—283 Cu. In.	4	2	2½	18	4	20①	18½
	V8—327 Cu. In.	4	2	2½	18	4	20①	18½
	V8—409 Cu. In.	5	2	2½	18	4	20①	22
1965	6 Cyl.	4	2	None	18	4	20②	12
	V8—283 Cu. In.	4	2	2½	18	4	20②	17
	V8—327 Cu. In. (250 HP)	4	2	2½	18	4	20②	16
	V8—327 Cu. In. (300 HP)	4	2	2½	18	4	20②	18
	V8—396 Cu. In. Std.	4	2	2½	22	4	20	22
	V8—396 Hi. Perf.	5	2	2½	19	4	20	22
	V8—409 Cu. In.	4	2	2½	18	4	20②	22
1966	6 Cyl.—250 Cu. In.	4	2	None	18	4	20②	12
	V8—283 Cu. In.	4	2	2½	18	4	20②	17
	V8—327 Cu. In.	4	2	2½	18	4	20②	16
	V8—396 Cu. In.	4	2	2½	18③	4	20②	23
	V8—427 Cu. In.	4	2	2½	19	4	20②	24
1967	6 Cyl.—250 Cu. In.	4	3	None	17	3½	24	12
	V8—283 Cu. In.	4	3	3	17	3½	24	17
	V8—327 Cu. In.	4	3	3	19	4	24	15
	V8—396 Cu. In.	4	3½	3	19③	4	24	23
	V8—427 Cu. In.	4	3½	3	22	4	24	22
1968	6 Cyl.—250 Cu. In.	4	3	None	17	3½	24	11
	V8—307 Cu. In.	4	3	3	17	3½	24	16
	V8—327 Cu. In.	4	3	3	19	4	24	14
	V8—396 Cu. In.	4	3½	3	19③	4	24	21
	V8—427 Cu. In.	4	3½	3	22	4	24	21
1969	6 Cyl.—250 Cu. In.	4	3	None	17	⑤	24	12
	V8—327 Cu. In.	4	3	3	19③	⑤	24	17
	V8—350 Cu. In.	4	3½	3	19③	⑤	24	15
	V8—396 Cu. In.	4	3½	3	22	⑤	24	23
	V8—427 Cu. In.	4	3½	3	22	⑤	24	22
1970–71	6 Cyl.—250 Cu. In.	4	3	None	6⑦	⑤	25⑥	12
	V8—350 Cu. In. (250 H.P.)	4	⑧	3	6½⑦	⑤	25⑥	16
	V8—400 Cu. In.	4	None	3	8⑦	⑤	25⑥	16
	V8—454 Cu. In.	4	None	None	8⑦	⑤	25⑥	22

LIGHT BULBS

1964–68

Headlamp (outer)	4002; 37½–55 W
Headlamp (inner)	4001; 37½ W
Parking and front directional, tail and stop and rear directional	1157; 4–32 CP
Tail lamp (belair)	1155; 4 CP
Back-up lamp	1156; 32 CP
Instrument lamps, panel compartment, temperature, oil pressure, generator, hi-beam indicator, clock lamp	1895; 2 CP
A.T. quadrant, directional signal, ignition lock, heater control panel	1445; 1 CP
Dome lamp	1004; 15 CP
License plate lamp	1155; 4 CP
Radio dial lamp	1893; 2 CP
Brake alarm lamp	257; 2 CP

1969–70

LAMP USAGE	CANDLE POWER	BULB NUMBER
Headlamp unit		
Chevrolet, Chevelle & Corvette		
Outer—high beam	37½ Watts	4002
Outer—low beam	55 Watts	4002
Inner—high beam only	37½ Watts	4001
Nova		
High beam	55 Watts	6012
Low beam	45 Watts	
Monte Carlo		
High beam	55 Watts	6012A
Low beam	45 Watts	
Parking lamp and directional signal		
Chevrolet & Chevelle	3–32	1157
Corvette, Nova & Monte Carlo	3–32	1157NA

CAPACITIES, continued

YEAR	MODEL	ENGINE CRANKCASE ADD 1 Qt. FOR NEW FILTER	TRANSMISSIONS Pts. TO REFILL AFTER DRAINING			DRIVE AXLE (Pts.)	GASOLINE TANK (Gals.)	COOLING SYSTEM (Qts.) WITH HEATER
			Manual		Automatic			
			3-Speed★	4-Speed				
CORVETTE								
1963	V8—327 Cu. In.	4	2	2½	18	4	20	16½
	V8—409 Cu. In.	5	2	2½	18	4	20	16½
1964-65	V8—327 Cu. In. Std.	4	2	2½	18	4④	20	16½
	V8—327 Cu. In. Hi. Perf.	5	2	2½	18	4④	20	16½
	V8—396 Cu. In. Std.	4	2	2½	22	4④	20	22
	V8—396 Cu. In. Hi. Perf.	5	2	2½	19	4④	20	22
1966	V8—327 Cu. In.	5	2	2½	18	3½	20	19
	V8—427 Cu. In.	5	None	2½	None	3½	20	22
1967	V8—327 Cu. In.	5	3	3	19	3½	20	19
	V8—427 Cu. In.	5	None	3	19	3½	20	22
1968	V8—327 Cu. In.	4	3	4	22	4	20	14
	V8—427 Cu. In.	5	3	4	22	4	20	21
1969	V8—350 Cu. In.	4	3	3	22	4	20	15
	V8—427 Cu. In.	5	None	3	22	4	20	22
1970-71	V8—350 Cu. In.	4	None	3	22	4	20	15**
	V8—454 Cu. In.	5	None	3	22	4	20	22

★—Add 1 pt. for Overdrive.
①—Station Wagon—19.
②—Station Wagon—23½.
③—With Turbo-Hydra-Matic—19 Pts. 1966, 22 Pts. 1967-70.
④—1965—3½ Pts.

⑤—8.125 ring gear 3.5 Pts.; 8.875 ring gear 4.0 Pts.
⑥—Station Wagon—22.
⑦—Case only—does not include converter.
⑧—300 H.P.—None.
**—370 H.P.—18 Qts.

1969-70 (continued)

LAMP USAGE	CANDLE POWER	BULB NUMBER
Tail, stop and directional signal	3-32	1157
Backing lamp	32	1156
Instrument illumination lamps		
Chevrolet, Chevelle & Monte Carlo	2	194
Corvette	2	1895
Nova	3	168
Console instrument cluster (Nova)	2.5	1816
Temperature indicator	2	194
Oil pressure indicator	2	194
Generator indicator	2	194
Hi-beam indicator		
All except Corvette	2	194
Corvette	2	1895
Directional indicator		
All except Corvette	2	194
Corvette	2	1895
Cigarette lighter lamp		
Corvette	1	1445
Warning lamps		
Corvette—Door ajar	2	1895
Headlamps up	2	1895
Seat belts	2	1895
Chevrolet—low fuel	2	194
check doors	2	194
seat belt	2	194
Nova—low fuel	2	194
Heater or A/C control panel		
Chevelle and Monte Carlo	1	1445
Chevrolet & Nova	2	1895
Corvette	2.5	1816
Glove box lamps		
Chevrolet, without A/C, Nova & Corvette	2	1895
Chevelle, Monte Carlo & Chevrolet with A/C	2	1893

LAMP USAGE	CANDLE POWER	BULB NUMBER
Dome and courtesy lamps		
Cartridge type (all)	12	211
Bayonet type (exc. Corvette and station wagon		
third seat)	6	631
Corvette and station wagon third seat	6	90
Seat separator-courtesy lamp	6	212
Side marker-front	2	194
Side marker-rear	2	194
License plate lamp (exc. Corvette)	4	67
(Corvette)	4	97
Radio dial lamps		
All AM only radios	2	293
All tape players and FM radios	2	1893
Tape player lens illumination lamp	1	216
Stereo indicator lamp	.3	2182D
Automatic transmission control indicator lamp		
Chevrolet, Chevelle (without seat separator) & Monte Carlo (w/o seat separator)	2	194
Nova, Chevelle (with seat separator) & Monte Carlo (with seat separator)	1	1445
Brake alarm lamp		
All except Corvette	2	194
Corvette	2	1895
Luggage compartment lamp	15	1003
Map lamp (mirror)	4	563
Underhood	15	93
Rear window defogger lamp	2	194

Distributor

References

Detailed information on distributor drive, direction of distributor rotation, cylinder numbering, firing order, point gap, cam dwell, timing mark location, spark plugs, spark advance, ignition resistor location and idle speed will be found in the Specifications tables.

A professional approach to engine diagnosis is treated in the Unit Repair Section.

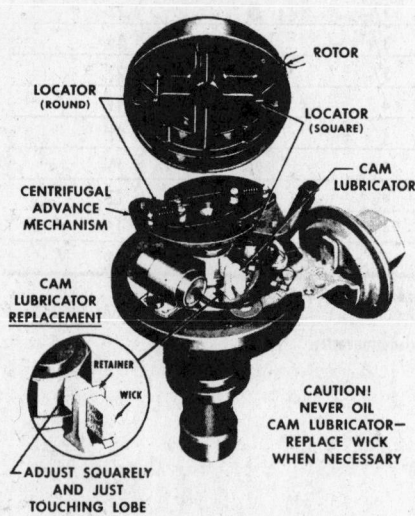

V8 distributor showing details of cam lubricator
(© Chevrolet Div. G.M. Corp.)

CAUTION!
NEVER OIL CAM LUBRICATOR—REPLACE WICK WHEN NECESSARY

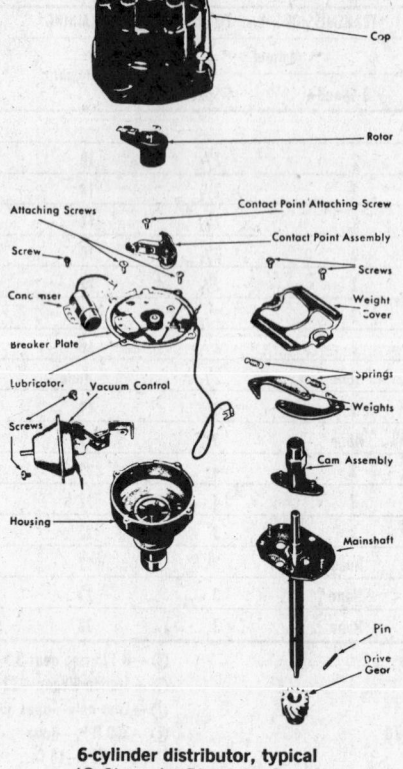

6-cylinder distributor, typical
(© Chevrolet Div. G.M. Corp.)

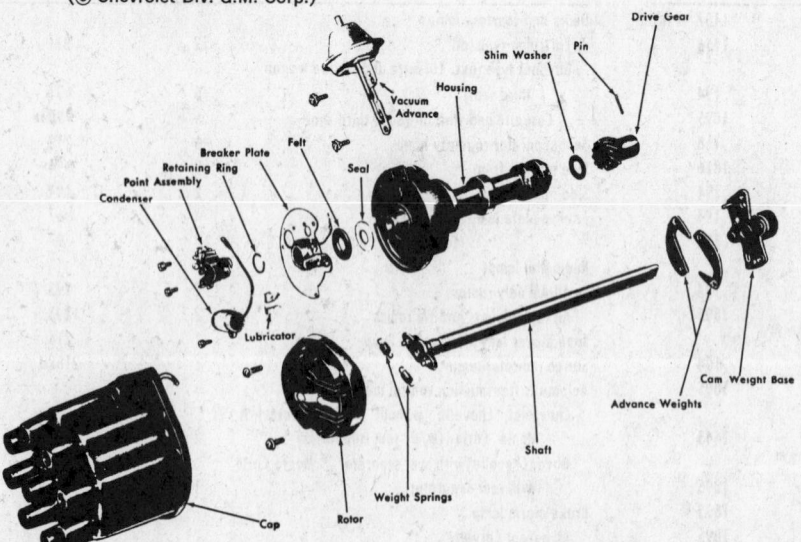

V8 distributor, typical (© Chevrolet Div. G.M. Corp.)

Caution When using an auxiliary starter switch for bumping the engine into position for timing, the primary distributor lead must be disconnected from the negative post of the ignition coil and the switch must be in the on position. Failure to do this may cause damage to the grounding circuit in the ignition switch.

Distributor Removal

6 Cylinder Models

The distributor assembly is mounted on the right side of the block and is driven directly from the camshaft.

To remove the distributor, first detach the vacuum lines from the vacuum advance unit and lift off the distributor cap.

The distributor body is fastened to the block by a single cap screw which holds the octane selector plate down against the block. Remove the retaining screw and lift the distributor out of the block.

V8 Models

The distributor is located between the two banks of cylinders at the back of the block.

The drive gear is attached to the distributor shaft; therefore, if it becomes necessary to remove the distributor, carefully mark the position of the rotor. Then, if the engine is not turned after the distributor is taken out, it can be returned to the position from which it was removed.

To remove the distributor, take off the carburetor air cleaner, disconnect the coil primary wire and the vacuum line, remove the distributor cap, take out the single hold-down bolt located under the distributor body, mark the position of the body relative to the block and then work the distributor up out of the block.

V8 Distributor Cam Lubricator Wick

If the car has gone 20,000 miles or more, the cam lubricator wick should be changed.

Take the distributor cap off. Using long-nose pliers, squeeze the wick assembly together at the base and lift it out. Wipe off the cam and install a new wick assembly so that the end of the wick touches the cam lobes. Overlubrication at this point results when the wick presses too hard against the cam surface. Do not put oil on the wick.

6 Cylinder Distributor Installation (After Engine Has Been Disturbed)

Remove No. 1 spark plug and, with finger on the plug hole, crank the engine until compression is felt. Slowly continue cranking until the timing ball on the flywheel, or the scribe mark on the vibration damper, lines up with the pointer or scale.

Place the distributor in its normal position. Turn the rotor clockwise, not quite one-eighth of a turn, and push the assembly down to engage the camshaft. It may be necessary to wiggle the rotor a bit. Holding the distributor in place, use starter to turn motor to make sure oil pump shaft is engaged with the tongue of the distributor shaft. With hold-down clamp at the octane selector drawn up tight, and the clamp screw loose, turn the distributor body

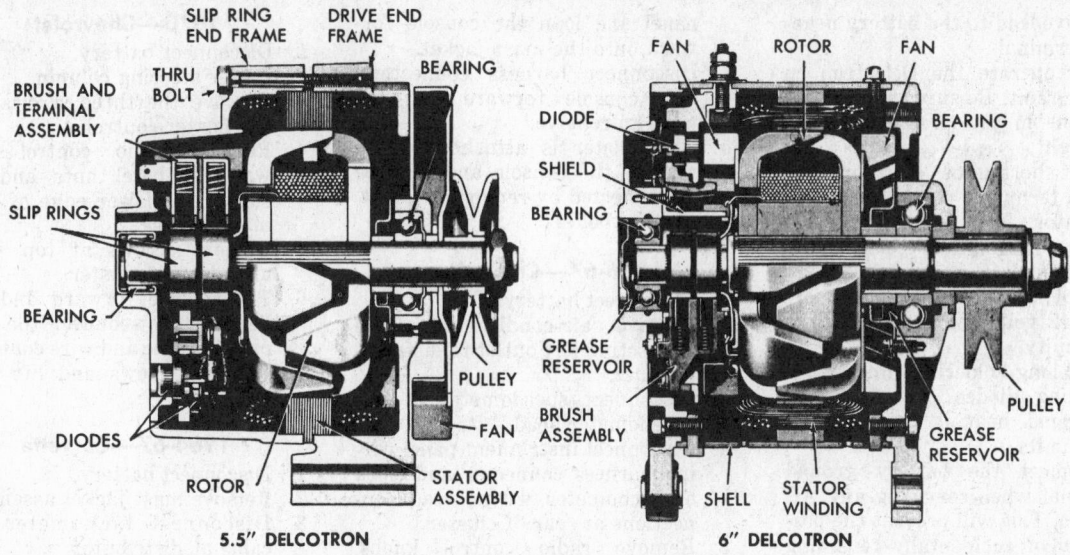

Cross-section of Delcotron units (© Chevrolet Div. G.M. Corp.)

until points are slightly open, and tighten the clamp screw.

Install the distributor cap, being sure that the rotor is pointing toward the terminal for No. 1 spark plug. Use a timing light to check spark timing.

V8 Distributor Installation (After Engine Has Been Disturbed)

1. Remove No. 1 spark plug and, with finger on plug hole, crank the engine until compression is felt in No. 1 cylinder. Continue cranking until pointer lines up with the timing mark on the crankshaft pulley.
2. Position distributor to opening in the block in normal, installed attitude with rotor pointing to front of engine.
3. Turn the rotor counterclockwise about one-eighth of a turn (from straight front toward the left cylinder bank). Push the distributor down to engage the camshaft and, while holding, turn engine with the starter so that distributor shaft engages the oil pump shaft.
4. Return engine to compression stroke of No. 1 piston with timing mark on pulley aligned with the pointer. Adjust the distributor so that the points are opening. Install the cap being sure the rotor points to the contact for No. 1 spark plug. Connect the timing light and check that spark occurs as timing mark and pointer are aligned.

Caution V8, the distributor body is involved in the engine lubricating system. The lubricating circuit can be interrupted to the right-bank valve train by misalignment of the distributor body.

This can cause serious trouble and may be hard to diagnose. See Firing Order and Timing illustrations for correct distributor positioning.

AC Generator (Delcotron)

References

Delcotron, the AC generator by Delco-Remy is available on Chevrolet cars and trucks. These units are furnished in two service types with companion voltage regulators. A three-unit regulator will be used with vehicles having a 42 or 52 ampere Delcotron. Vehicles such as police equipment, etc., will use a heavy duty Delcotron and be furnished with a two-unit transistorized regulator.

Repair and test details on the Delcotron and its regulators are covered in the Unit Repair Section.

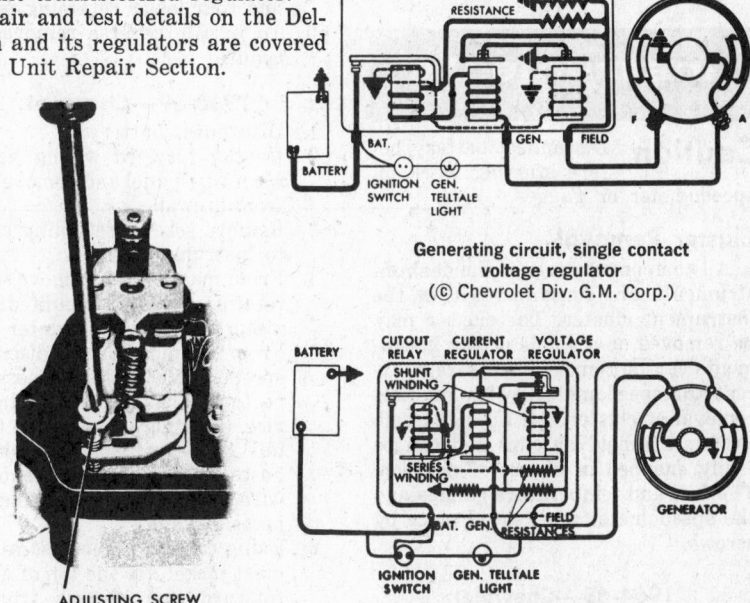

ADJUSTING SCREW (TURN TO ADJUST SETTING)

Adjusting voltage setting (© Chevrolet Div. G.M. Corp.)

Caution Since the Delcotron and regulator are designed for use on a single polarity system, the following precautions must be observed:

1. The polarity of the battery, generator, and regulator must be matched and considered before making any electrical connections in the system.
2. When connecting a booster battery, be sure to connect the negative battery terminals with one another, and the positive battery terminals with one another.
3. When connecting a charger to the battery, connect the charger positive lead to the battery positive terminal. Connect the charger

Generating circuit, single contact voltage regulator (© Chevrolet Div. G.M. Corp.)

Generating circuit, double contact voltage regulator (© Chevrolet Div. G.M. Corp.)

negative lead to the battery negative terminal.

4. Never operate the Delcotron on open circuit. Be sure that all connections in the circuit are clean and tight.
5. Do not short across or ground any of the terminals on the Delcotron regulator.
6. Do not attempt to polarize the Delcotron.
7. Do not use test lamps of more than 12 volts for checking diode continuity.
8. Avoid long soldering times when replacing diodes or transistors. Prolonged heat is damaging to these units.
9. Disconnect the battery ground terminal when servicing any AC system. This will prevent the possibility of accidentally reversing polarity.

Battery, Starter

References

Detailed information on the battery and starter will be found in the Battery and Starter Specifications table of this section.

A more general discussion of batteries and starters will be found in the Unit Repair Section.

Starter Removal

Disconnect the battery and the wires from the solenoid.

Remove the starter mounting bolt and lock washers. On V8s, a stud nut and lock washer are at the front of the starter.

Pull starter forward and out of car.

Instruments

Caution Disconnect battery before starting work on speedometer or gauges.

Cluster Removal

All component parts of the dash instruments are contained within the instrument cluster. The cluster may be removed as a whole or the instruments can be removed separately, except the speedometer which requires removal of cluster. All the light bulb sockets are held by clips and can be easily snapped in or out of position. The fuel and temperature gauges and the speedometer are held in place by screws.

1964-65—Chevrolet
1. Disconnect speedometer cable.
2. Remove the nine screws that attach the instrument console to

panel and lean the console forward onto the mast jacket.
3. Disconnect harness connectors.
4. Lift console forward and upward to remove.
5. The cluster is attached to the rear of the console and may be disconnected by removing the attaching bolts.

1966-67—Chevrolet
1. Disconnect battery.
2. Remove air-conditioning hose connecting left outlet to distributor duct.
3. Disconnect speedometer at the speedometer head.
4. Disconnect instrument panel wiring harness connector and clock or tachometer wiring lead connections at rear of cluster.
5. Remove radio control knobs, bezels and shaft retaining nuts. Push radio in to disengage shafts from panel openings.
NOTE: on rear-seat speaker models, disconnect fader control wiring from radio harness wiring.
6. Remove instrument panel compartment door and the compartment retaining screws.
7. Remove upper and lower instrument panel console retaining screws.
8. With adjacent painted surfaces protected, open right front door, roll console assembly forward and slide to the right to remove assembly from the car.
NOTE: at this point, all components (fuel gauge, speedometer, printed circuit, etc.) of the assembly may be serviced.
9. To separate instrument cluster assembly from the console, remove five screws in top of console and four lower retaining nuts.
10. To install, reverse removal procedure.

1968-69—Chevrolet
1. Disconnect battery.
2. Unplug forward wiring harness from fuse panel and remove panel from firewall.
3. Remove screws retaining cluster to instrument panel.
4. From mast jacket, remove screws retaining column-mounted automatic transmission pointer cable.
5. From behind cluster, disconnect speedometer cable, harness connector, clock, speed warning device, defogger, convertible top or tail gate switches and vacuum hoses, if so equipped. If equipped with gauge pack, disconnect oil pressure line.
6. Using care to prevent scratching mast jacket, tip the top of cluster forward and remove from vehicle.
7. To install, reverse the removal procedure.

1970—Chevrolet
1. Disconnect battery.
2. Lower steering column.
3. Remove the three screws above the heater control.
4. Remove radio control knobs, washers, bezel nuts and front support at lower edge of instrument cluster.
5. Remove screws at top, bottom and sides of cluster.
6. Tilt console forward and reach behind to disconnect the speedometer cable and wire connectors.
7. Remove screws and lift cluster out.

1964-67—Corvette
1. Disconnect battery.
2. Remove mast jacket assembly.
3. Disconnect tachometer drive cable at distributor.
4. Disconnect cowl vent control cable brackets and the headlamp panel control switch from the instrument cluster.
5. Remove lighting switch and ignition switch, and disconnect ignition switch lamp support at instrument panel.
6. Disconnect parking brake lever support bracket at cowl crossmember.
7. Disconnect oil pressure line at oil pressure gauge, then remove the lead wires from ammeter, wiper switch and cigarette lighter. Disconnect trip odometer at mast jacket support.
8. Remove instrument cluster-to-dash retaining screws and pull cluster assembly slightly forward to make clearance for removal of speedometer cable, tachometer cable, cluster ground wire, fuse gauge lead wires and the remaining indicator and cluster illuminating lamps.
9. Install by reversing the removal procedure.

1968-69 Corvette (Left-hand Section)
1. Disconnect battery.
2. Lower steering column.
3. Remove screws retaining left instrument panel to door opening, top of dash and left side of center instrument panel.
4. Unclip and remove floor console forward trim panel.
5. Pull cluster assembly forward to obtain clearance for removal of speedometer cable housing nut, tachometer cable housing nut, headlamp and ignition switch connectors and panel lamps.
6. To install, reverse removal procedure.

(Center Section)
1. Disconnect battery.
2. Remove screws from right side dash pad.

3. Remove clipped-in center floor console forward trim pads.
4. Remove radio knobs, bezel retaining nuts, aerial, speaker and electrical connections.
5. Remove radio, rear support bracket, and slide receiver out right side.
6. Remove upper center console trim plate screws, and tip trim plate forward for access to remove windshield wiper switch connector. Lift trim plate out.
7. Remove screws at left side of center console and nuts attached to underside of console studs.
8. Tilt console forward. Remove oil and electrical connections and lamps from rear of console.
9. Lift up and forward to remove console.
10. To install, reverse the removal procedure.

Ignition Switch Replacement

1964-70

1. Disconnect battery.
2. Remove cylinder by placing in lock position and insert stiff wire in small hole to depress

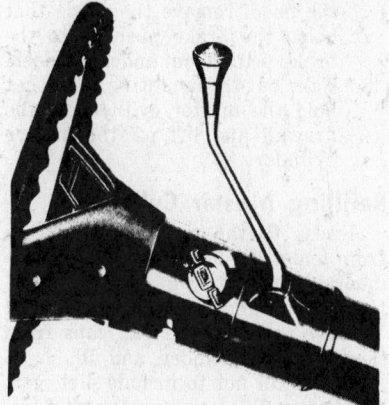

1969-70 ignition switch location

plunger. Turn cylinder counter-clockwise until cylinder can be removed.
3. Remove holding nut. (Tool J-7607 will assist.)
4. Pull switch from under dash and remove connector.
5. Using a screwdriver, unsnap the locking tangs.
6. Install in reverse of above.

Lighting Switch Replacement

1964-70

1. Disconnect battery.
2. Pull knob out to on position.
3. Reach under instrument panel and depress the switch shaft retainer. Remove knob and shaft assembly.
4. Remove the retaining ferrule nut. (Tool J-4880 will assist.)
5. Remove switch from instrument panel.

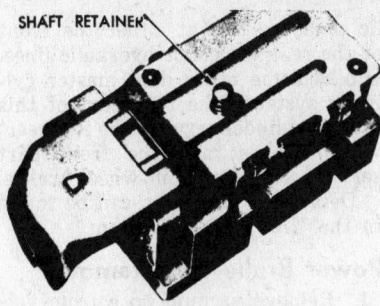

SHAFT RETAINER

Light switch shaft retainer
(© Chevrolet Div. G.M. Corp.)

6. Disconnect the multi-plug connector from the switch.
7. Replace in reverse of above. (In checking lights before installation, switch must be grounded to test dome light.)

Brakes

References

Specific information on brake cylinder sizes can be found in the General Chassis and Brake Specifications table.

Information on brake adjustments,

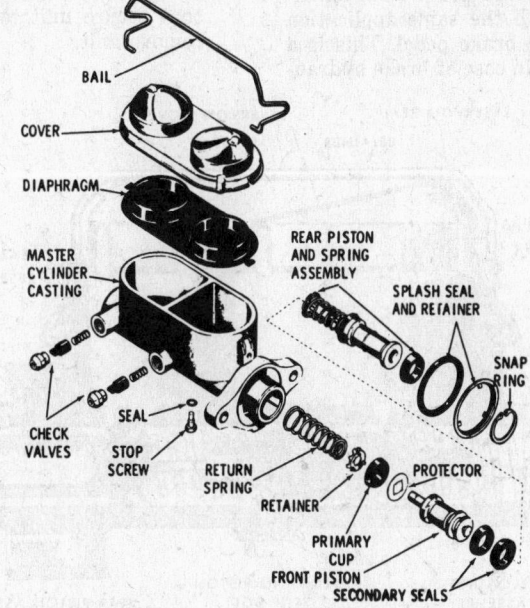

BAIL
COVER
DIAPHRAGM
MASTER CYLINDER CASTING
REAR PISTON AND SPRING ASSEMBLY
SPLASH SEAL AND RETAINER
SNAP RING
CHECK VALVES
SEAL
STOP SCREW
RETURN SPRING
RETAINER
PROTECTOR
PRIMARY CUP
FRONT PISTON
SECONDARY SEALS

Master cylinder (Bendix) (© Chevrolet Div. G.M. Corp.)

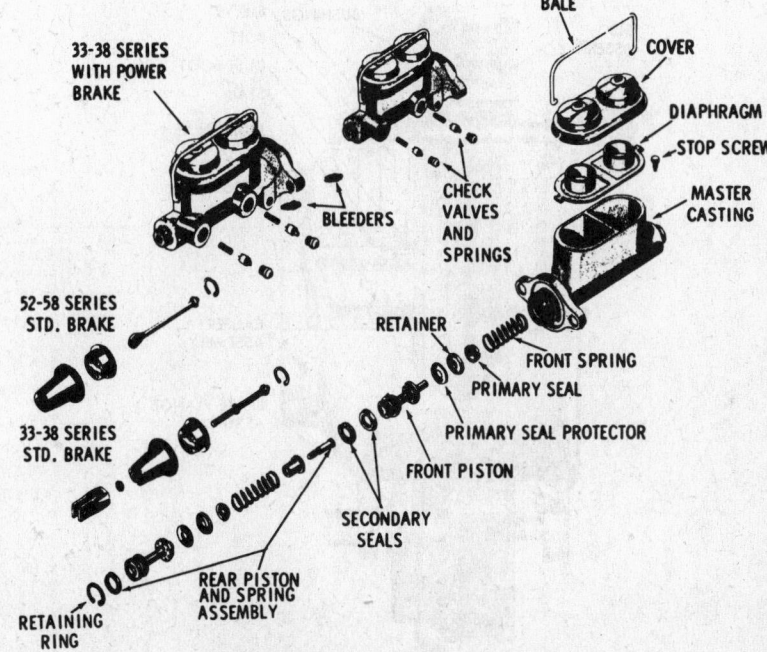

33-38 SERIES WITH POWER BRAKE
BLEEDERS
CHECK VALVES AND SPRINGS
52-58 SERIES STD. BRAKE
33-38 SERIES STD. BRAKE
RETAINING RING
REAR PISTON AND SPRING ASSEMBLY
SECONDARY SEALS
FRONT PISTON
PRIMARY SEAL PROTECTOR
PRIMARY SEAL
FRONT SPRING
RETAINER
BALE
COVER
DIAPHRAGM
STOP SCREW
MASTER CASTING

Master cylinder (Moraine) (© Chevrolet Div. G.M. Corp.)

band replacement, bleeding procedure, master and wheel cylinder overhaul can be found in the Unit Repair Section.

Information on troubleshooting and overhauling power brakes can be found in the Unit Repair Section.

Information on the grease seals which may need replacement can be found in the Unit Repair Section.

Beginning 1967

Since 1967, a dual master cylinder is used. The master cylinder is divided into two parts. This design supplies the vehicle with separate and independent hydraulic brake systems. Both systems are acted upon simultaneously with the same application of the service brake pedal. This is a safety factor in case of brake hydraulic pressure loss at either the front or the rear wheels or hydraulic lines. Unlike some other dual master cylinder systems, the rear part of this master cylinder services the two rear-wheel brakes; and the front part services the two front-wheel brakes.

Detailed information can be found in the Unit Repair Section.

Power Brake Unit Removal

1. Remove vacuum hose from vacuum check valve.
2. Disconnect hydraulic lines at unit.
3. Disconnect push rod at brake pedal assembly.
4. Remove nuts and lockwashers that secure unit to firewall and remove unit.

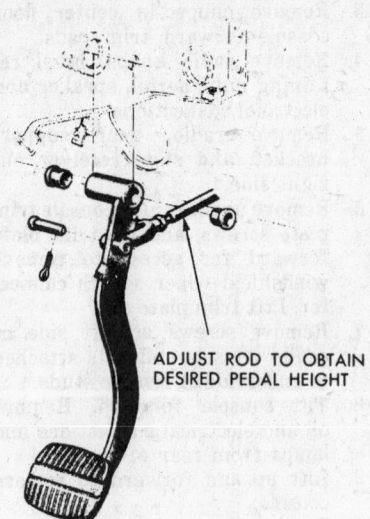

ADJUST ROD TO OBTAIN DESIRED PEDAL HEIGHT

Power brake pushrod adjustment
(© G.M. Corp.)

Brake Master Cylinder Removal

The pedals are pivoted from underneath the dash panel. The master cylinder is located on the engine side of the firewall.

1. To remove the master cylinder, disconnect the lines from under the hood, remove the clevis that holds the brake pushrod to the brake pedal from under the dash.
2. Take out the mounting bolts that hold the master cylinder to the firewall and lift off the master cylinder.

Refilling Master Cylinder

Access to the master cylinder is from under the hood. The master cylinder is bolted to the engine side of the firewall and is readily accessible. Remove the reservoir cap/caps from the top of the cylinder, and fill.

Be careful not to include dirt, grit or oil when filling the master cylinder.

Parking Brake

Chevrolet—Pedal

1. Disconnect battery.
2. Release the parking brake.
3. Remove equalizer check nut, and separate cable stud from equalizer.
4. Remove two attaching nuts from mounting studs on engine side of firewall.

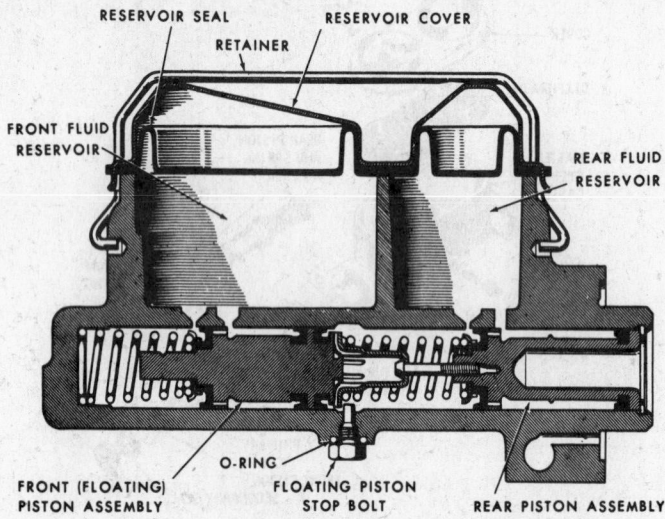

RESERVOIR SEAL — RESERVOIR COVER

RETAINER

FRONT FLUID RESERVOIR

REAR FLUID RESERVOIR

FRONT (FLOATING) PISTON ASSEMBLY — O-RING — FLOATING PISTON STOP BOLT — REAR PISTON ASSEMBLY

Bendix power disc brake master cylinder (© Chevrolet Div. G.M. Corp.)

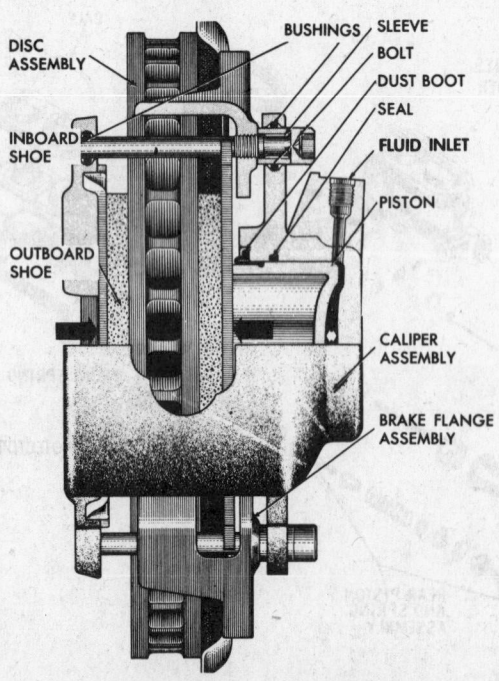

DISC ASSEMBLY

BUSHINGS   SLEEVE
BOLT
DUST BOOT
SEAL
FLUID INLET

INBOARD SHOE

PISTON

OUTBOARD SHOE

CALIPER ASSEMBLY

BRAKE FLANGE ASSEMBLY

1969 Corvette disc brake caliper assembly

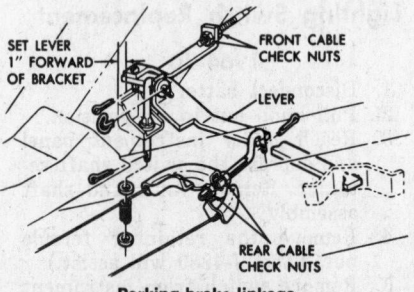

SET LEVER 1" FORWARD OF BRACKET

FRONT CABLE CHECK NUTS

LEVER

REAR CABLE CHECK NUTS

Parking brake linkage
(© Chevrolet Div. G.M. Corp.)

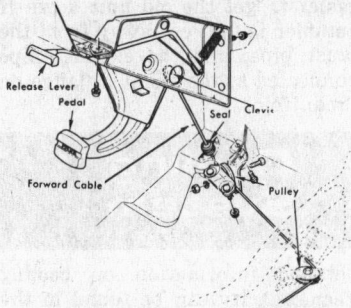

Parking brake pedal and release lever
(© Chevrolet Div. G.M. Corp.)

5. Remove front cable ball end from pedal assembly swivel.
6. Remove pedal assembly to dash brace attaching screw.
7. Remove pedal assembly by lowering rear slightly to avoid scratching dash when removing assembly.
8. Install by reversing removal procedure.

1964-67 Corvette—Lever

1. Disconnect battery.
2. Release parking brake.
3. Release equalizer rod at the idler lever under the car.
4. Remove pin from lever assembly near the firewall.
5. Rotate lever about 40° counterclockwise and remove cable from slot in lever.
6. Remove lever attaching nuts from engine side of firewall at the pulley.
7. Remove two lever to dash attaching screws, and remove lever assembly from car.
8. Install by reversing removal procedure.

1968-69 Corvette—Lever

1. Place lever in fully released position.
2. Raise car on hoist, unhook and remove return spring.
3. Remove rear nut from cable stud at equalizer and allow cable to hang down.
4. Lower car.

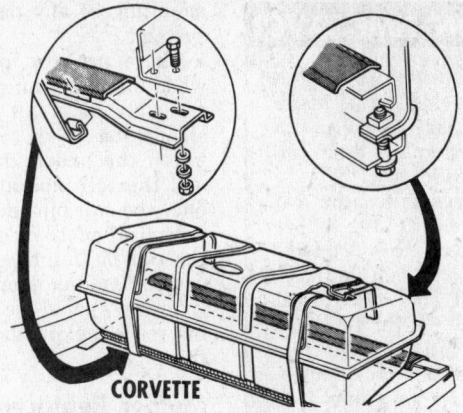

CORVETTE

Typical fuel tank installation—Corvette

5. Inside the car, remove the retaining screws and the lever case cover.
6. Remove trim seal from lever.
7. Remove the screw and washer that secure the parking brake switch to the side of the lever assembly.
8. Remove bolts that secure the lever assembly to the underbody and lift the lever assembly upward. Remove the lever forward mounting bracket.
9. Remove front cable from the lever assembly with long nose pliers and remove the lever.

Fuel System

References

Data on capacity of the gas tank will be found in the Capacities table. Data on correct engine idle speed and fuel pump pressure will be found in the Tune-up Specifications table. Both can be found at the start of this section.

Information covering operation and troubles of the fuel gauge is in the Unit Repair Section.

Detailed information on the carburetor and how to adjust it is in the Unit Repair Section.

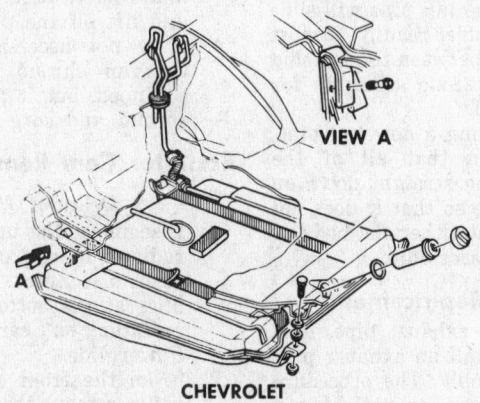

VIEW A

CHEVROLET

Typical fuel tank installation—Chevrolet

Dashpot adjustment is in the Unit Repair Section under the same heading as that of the specific automatic transmission, as well as under the make of carburetor.

Fuel Pump

NOTE: two types of fuel pumps are used; a serviceable type, used on inline engines (except on cars equipped with exhaust emission control) and on the V8 283 and the 327 cu. in. engines. The non-serviceable type is used on the V8 396 and 427 cu. in. engines, and on inline engines using the exhaust emission control system.

To remove the fuel pump, disconnect the input flex line and the output line to the carburetor. The fuel pump can then be detached from the side of the block and lifted off. On V8 models, the pump is actuated by a pushrod in the block.

Beginning 1967

All Chevrolet engines are equipped with the non-serviceable type fuel pump. Because of design, this pump is serviced as an assembly only.

Caution Fuel pump replacement V8. A fuel pump may fail to function at the time of replacement as a result of error in positioning or damage to the fuel pump pushrod of the V8 engine. Design characteristics require that a pushrod (or some other method of extension) be used to transmit camshaft lobe action to the pump activating arm. This pushrod can slip out of place during the process of pump replacement and result in no pump action from the newly replaced unit. Before tightening the fuel pump to the engine, have someone spin the engine using the starter while the mechanic feels the fuel pump body for movement. If the pump and pushrod are in correct position, movement will be felt in the pump as the pushrod pressure is applied and released from the pump arm.

283, 307, 327, and 350 cu. in.
When replacing a fuel pump on a 283, 307, 327 or 350 cu. in. engine, considerable time can be saved as follows:

1. Before removing the old pump, remove the upper bolt from the engine's right front mounting boss. This bolt hole is in direct alignment with the fuel pump pushrod. The threaded bolt hole continues into the pump pushrod bore. The bolt acts as an oil plug.
2. Temporarily insert a longer bolt, (about ⅜—16 x 2 in.) into the hole. Screw the bolt into the bore until it bottoms against the pump pushrod. (Don't tighten the bolt with a wrench or the rod can be damaged.)
3. The mechanic is now free to remove and install the fuel pump without worrying about fuel pump pushrod misalignment.
CAUTION: don't forget to reinstall original motor bolt.

396, 427 and 454 cu. in.
The design of these engines prevents the use of the method of simplifying fuel pump pushrod positioning while installing a fuel pump. However, to hold the pump pushrod in position while installing the fuel pump, the following has worked satisfactorily:

1. Clean oil from pushrod.
2. Pack a small quantity of nonfibrous grease in the area around the fuel pump pushrod to hold it in suspension long enough to position the fuel pump.
3. Install and check pump action, then torque attaching bolts.

Exhaust System

Muffler, Exhaust Pipe and Tail Pipe

When installing an exhaust pipe, muffler or tail pipe assembly, care should be taken to have these parts in proper relationship to one another and properly aligned.

Inaccurate alignment of the exhaust system is frequently the cause of annoying rattles because of incorrect clearances. Many unusual noises very hard to locate are sometimes due to a change or obstruction in the normal flow of gases caused by improper mounting of any part of the exhaust system.

There are two points to consider when installing an exhaust or muffler assembly or a tail pipe. First, there should be ½-¾ in. clearance between the underside of the floor pan and the tail pipe at the kickup. Second, the tail pipe support must be in a vertical position. If it is at an angle the tail pie might strike the bumper causing an annoying rattle.

There should be ¾ in. clearance between the bumper brace and the tail pipe.

Muffler Replacement

Chevrolet services the muffler and exhaust pipe as an assembly. However, if it is desired to replace the muffler only, cut the exhaust pipe as close to the muffler as is practical so that the new muffler will have plenty of purchase as it is slid over the exhaust pipe. Loosen the clamps that hold the exhaust pipe right at the muffler and the tail pipe at the back end of the muffler and cut the exhaust pipe out, removing the clamp at the back end so that the tail pipe can be slid out.

If any difficulty is experienced in getting the muffler off the tail pipe due to rust or corrosion, it is a good idea to let the joint soak in a good rust-dissolving fluid for about one-half hour or so before starting to disconnect the two pipes. Install the new muffler, being certain that the clamps are fitted tightly and evenly so that there are no gas leaks.

Tail Pipe Replacement

1. Use a good rust dissolving fluid at the joint between the tail pipe and the muffler so that they will come apart readily after the clamps are taken off.
2. Take off the clamps that hold the tail pipe to the body and frame and spread them enough so that the tail pipe can be dropped out of them.
3. Drop out the tail pipe and, if sufficient rust dissolving fluid has been used, the tail pipe will slide out of the muffler readily. It must be threaded between the housing and the rear axle assembly for easy removal.
4. When installing a new tail pipe, make certain that all of the brackets hang straight down on the new pipe so that it does not rattle against either the body or the bumper assembly.

Exhaust Pipe Replacement

To replace the exhaust pipe, it is customary to install an exhaust pipe and muffler assembly. The procedure is exactly the same as replacing a muffler, except that the pipe is not cut. However, in many instances it is easier to get the old unit down if the muffler is first cut away from the exhaust pipe, and the exhaust pipe disconnected at the exhaust flange on the manifold.

Cooling System

Detailed information on cooling system capacity can be found in the Capacities table.

Information on the water temperature gauge can be found in the Unit Repair Section.

Caution When replacing a thermostat, it is sometimes possible to install it in the reverse position. The spring-loaded end of the unit must be installed toward the engine.

Water Pump Removal

6 Cyl. Models
1. Disconnect the upper and lower radiator hoses, remove the bolts that hold the radiator core to its cradle, and remove the core.
2. Slack off on the generator belt and remove the bolts holding the fan pulley to the water pump hub. Disconnect the heater hoses and remove the bolts that hold the water pump to the block.
3. Carefully scrape the gasket from the block. If a new pump is to be installed, use a new gasket. It is good practice to put gasket compound on, since this is a thin paper gasket.

V8 Models
1. Slack off on the generator and remove the generator belt. Disconnect the heater hose and the lower radiator hose from the water pump manifold.
2. Remove the bolts that hold the fan blades and fan pulley to the hub, and take off the blades and pulley.
3. Remove the bolts that hold the manifold to both cylinder heads and lift off the pump assembly.
4. It is not necessary to remove the fan shroud on models so equipped, but it is easier if the shroud and core are removed.

Radiator Core Removal

6 Cylinder Models
1. Disconnect the upper and lower radiator hoses from the radiator. Also disconnect the oil cooler lines at the bottom tank of the radiator on cars fitted with Powerglide.
2. From the front of the radiator baffle, remove the bolts that hold the core to the baffle and pull the core straight forward and out.

V8 Models

1. Disconnect and remove the battery. Disconnect the upper and lower radiator hoses and the two oil cooler lines at the bottom of the core.
2. Remove the screws and/or bolts that hold the radiator core to its cradle and lift it straight up.
3. The new radiator core is reinstalled in reverse of the above procedure.

Engine

References

The specifications table lists most of the service references needed.

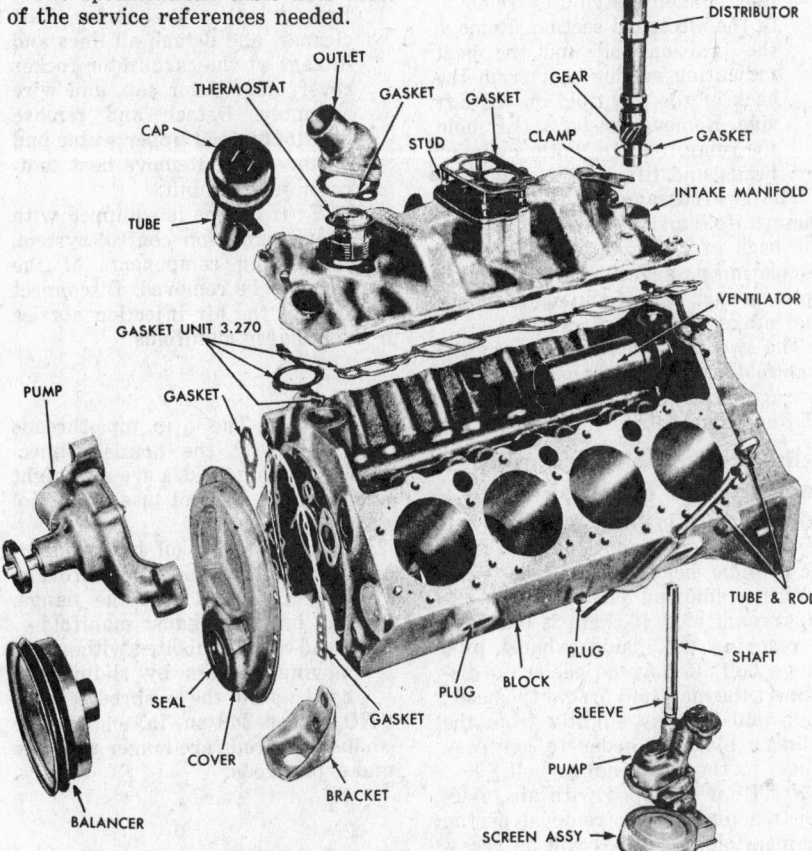

Exploded view, V8 engine external parts, typical (© Chevrolet Div. G.M. Corp.)

Where engines different in size are used, identification can be made by referring to Car Serial Number Location and Engine Identification, at the beginning of this car section.

Exhaust Emission Control

1966-67

The air injection system (K-19) is used on all General Motors cars (with a few design exceptions) for delivery to California.

Beginning 1968

In compliance with anti-pollution laws involving all of the continental United States, the General Motors Corporation has adopted a special system of terminal exhaust treatment. This plan supersedes (in most cases) the method used to conform to 1966-67 California laws. The new system replaces (except with stick shift and special purpose engine applications) the use of the A.I.R. method previously used.

The new concept, Combustion Control System (C.C.S.) utilizes engine modification, with emphasis on carburetor and distributor changes. Any of the methods of terminal exhaust treatment requires close and frequent attention to tune-up factors of engine maintenance.

Since 1968, all car manufacturers post idle speeds, and other pertinent data relative to the specific engine-car application, in a conspicuous place in the engine compartment.

For details, consult the Unit Repair Section.

Engine Removal and Installation

(Except 409 cu. in.)

1. Drain cooling system and engine oil.
2. Remove air cleaner and fuel pump.
3. Disconnect battery. It may aid if battery is removed in some cases.
4. Scribe around hood hinges and remove hood.
5. Remove radiator, shroud and fan blades.
6. Disconnect various wires from engine.
7. Disconnect various linkages, manifold pipes, power brake and power steering lines where used.
8. Raise vehicle and place on jack stands.
9. Remove driveshaft.
10. Disconnect shift linkage at transmission and speedometer cable at transmission.
11. On synchromesh vehicles, disconnect clutch linkage and remove cross-shaft from engine bracket.
12. Remove rocker arm covers and attach lifting bracket.
13. Remove front mount bolts.
14. Raise engine to relieve weight on front mounts, then remove rear mount bolts.
15. Raise engine to relieve weight on rear mount, then remove crossmember.
16. Remove engine and transmission as an assembly.
17. Install in reverse of above.

409 cu. in.

1. Follow Steps 1 through 15 above.
2. With Powerglide, disconnect transmission throttle valve rod at bellcrank.
3. With synchromesh, remove clutch cross-shaft.
4. Place floor jack under transmission.
5. Remove converter underpan or flywheel cover bolts.
6. Remove housing to engine bolts.
7. With Powerglide, install holding bracket or converter.
8. Move engine forward and upward alternately (raise transmission jack, also) and guide engine out of vehicle.
9. Install in reverse of above.

Manifolds

References

With the advent of exhaust emission control systems, interference may be encountered in either intake or exhaust manifold R & R procedures. In any event, especially with V8 engines where the air injection tubes enter the exhaust manifold, ex-

haust emission control parts involved must also be included in manifold R & R procedures.

Combination Manifold Used on 6 Cylinder Engines

All Chevrolet six cylinder engines are equipped with a combination intake and exhaust manifold. The exhaust manifold is equipped with a heat riser valve which, when the engine is cold, deflects the hot exhaust gases against the intake manifold to assist in rapid warm up.

If the engine doesn't seem to warm up properly or, when operated at a high speed, acts lean, it is a good idea to check this heat riser valve to be certain that it is functioning freely. Failure of the heat riser valve to open will increase the time required to warm the engine. Failure of the heat riser valve to close after the manifold is hot will cause the engine to run lean.

To remove the manifold assembly, disconnect the exhaust pipe flange and remove all connections to the carburetor. Take off the vacuum lines at the manifold and at the carburetor.

Remove the carburetor, and the manifold may be unbolted from the side of the cylinder head using socket wrenches and box wrenches. If necessary to remove either exhaust or intake manifolds they may be separ-

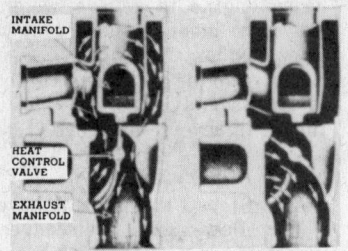

Manifold heat valve used on 6 cylinder engines
(© Chevrolet Div. G.M. Corp.)

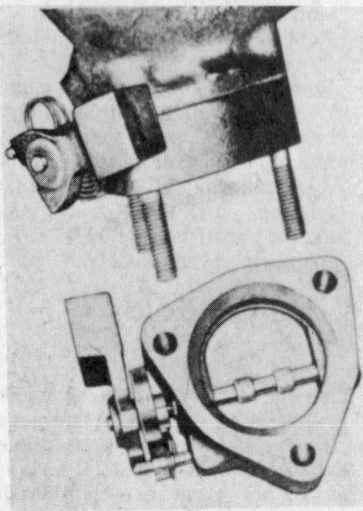

Manifold heat valve used on V8 engines
(© Chevrolet Div. G.M. Corp.)

ated by removing one bolt and two nuts at center of assembly.

Before reinstalling the manifold, thoroughly clean out the ports to avoid turbulence, particularly on the intake manifold.

Intake Manifold Assembly Removal (V8 Engines)

1. Disconnect the upper radiator hose and heater hose and then remove all of the rods, lines, and wires from the carburetor. While it is possible to take off the manifold with the carburetor in place, it is recommended that the carburetor be removed.
2. Remove the distributor as outlined under distributor removal in the electrical section. Remove the ignition coil and the heat indicating sender unit from the back of the manifold on the left side. Remove the bolts that hold the manifold to both cylinder heads and lift off the manifold.

The two passages at the front of the manifold and the two passages at the back of the manifold are water circulating passages. The four square holes on each side of the manifold are the intake manifold passages.

The two holes in the center of the manifold are the exhaust crossover passages that maintain manifold operating temperature.

Exhaust Manifold Assembly Removal (V8 Engines)

Two exhaust manifolds are used and are attached to the outside of each of the cylinder heads.

When removing either of the exhaust manifolds, if there is difficulty in reaching the inside exhaust pipe flange bolt, it may be easier to disconnect the manifold first at the head; then, pull it away slightly from the cylinder block in order to get easy access to the inner flange bolt.

NOTE: if equipped with air injection reactor system, removal of the air manifold and tubes will be necessary.

Cylinder Head Removal

6 Cylinder Models

To remove the cylinder head, detach the air cleaner and all rods, lines and vacuum tubes at the carburetor and manifold.

NOTE: if the engine is equipped with an exhaust emission control system, the injector connections must be disconnected at the cylinder head. Disconnect any interfering components and tie back out of the way.

Caution The ¼ in. pipe threads at the cylinder head air injection nozzles are a straight pipe thread. Do not use a ¼ in. tapered pipe tap. Hoses used in this

air injection system are of special material. Do not substitute.

1. Unbolt the manifold from the cylinder head, but not from the exhaust pipe flange. The manifold is simply pulled away from the head.
2. Remove the engine side plate covers and the gas lines at the fuel pump. Unbolt and lift off the rocker cover, disconnect the oil line leads to the rockers.
3. The rocker levers are supported separately and may be left intact until the head is removed.
4. Unbolt and lift off the cylinder head.

V8 Models

1. Remove the carburetor, air cleaner, and detach all lines and linkage at the carburetor rocker cover, distributor cap, and wire assembly. Detach and remove distributor and upper water and heater hoses. Remove heat indicator sending unit.

NOTE: if engine is equipped with an exhaust emission control system, the interfering components of the system must be removed. Disconnect the lines at the air injection nozzles in the exhaust manifolds.

Caution The ¼ in. pipe threads at the nozzle connections to the manifold/s are a straight pipe thread. Do not use a ¼ in. tapered pipe tap.

2. Unbolt and lift off intake manifold. Detach exhaust manifold at both the head and the flange, and lift off exhaust manifold.
3. Head can be unbolted without removing rockers by sliding the head up off the pushrods.

NOTE: on 348 cu. in. engine, the exhaust pushrods are longer than the intake pushrods.

Valve System

Where some engines have hydraulic valve lifters and others do not, a means of determining is given in the General Engine Specifications.

Detailed information on the valves, the type of valve guide and valve timing information, can be found in the Valve Specifications table of this section.

A general discussion of valve clearance, and a chart showing how to read pressure and vacuum gauges when using them to diagnose engine troubles, is in the Troubleshooting Section.

Valve tappet clearance for each engine is given in the Tune-up Specifications table.

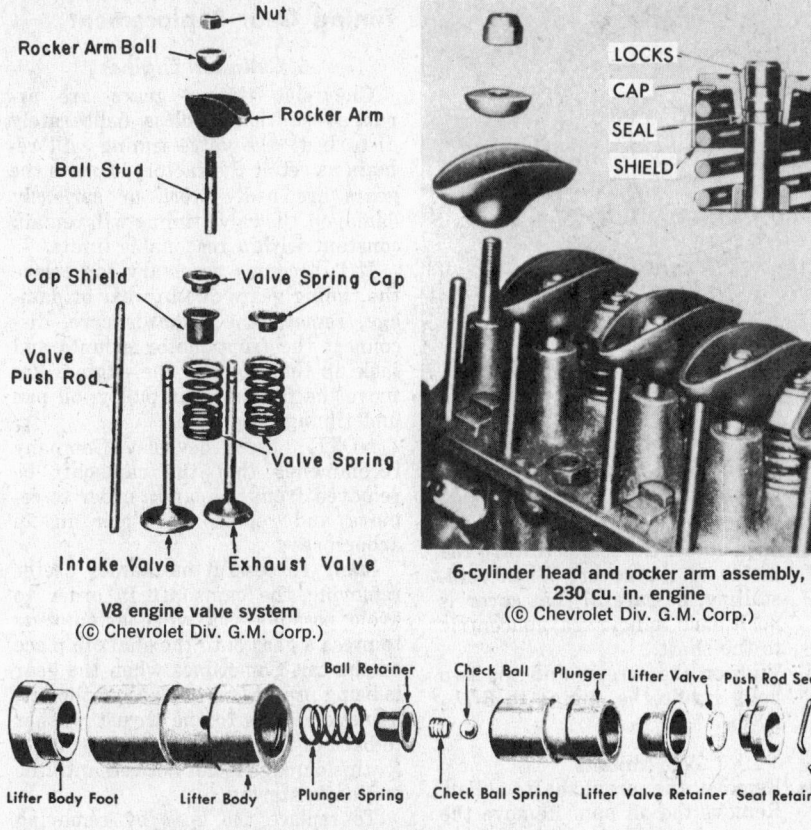

V8 engine valve system
(© Chevrolet Div. G.M. Corp.)

Typical hydraulic lifter exploded (© Chevrolet Div. G.M. Corp.)

6-cylinder head and rocker arm assembly,
230 cu. in. engine
(© Chevrolet Div. G.M. Corp.)

Rocker Arm Assemblies

6 Cylinder

The 6-cylinder engine is equipped with a valve train and rocker lubrication design similar to the V8 engines.

Adjusting valve clearance at the rocker arm
(© Chevrolet Div. G.M. Corp.)

The rocker arms are supported on individual pedestals with zero valve lash adjustment made at the rocker arm-to-pedestal ball seat and nut.

Rocker arm and cylinder head services are performed the same as on V8 engines.

V8 Models

The rocker arm on the V8 models is formed of pressed steel with an oval hole in the center that fits over a stud having a spherical type nut at the top for adjustment.

Oil is fed through the hollow pushrod and lies in the trough of the rocker arm providing adequate lubrication for the rocker arm spherical nut.

The rocker arm stud is a pressed fit in the cylinder head, and is available from Chevrolet stock in several oversizes for replacement.

Chevrolet supplies special reamers to accommodate oversize studs.

Valve Guides

V8 Models, 1963-69 and 6 Cylinder Models, 1963-69

Separate valve guides are not used on these engines. Instead, the valves are fitted directly into bores in the cylinder head.

Clearance of the valve stem to the bore should be .001 to .003 in. for intake valves and .002 to .004 in. for six cylinder exhaust valves and .0012 to .0029 for V8 engines. Clearance when worn should be within .002 in. of above limits.

This clearance should be checked by placing an indicator against the valve stem while gently pushing the valve from side to side, so that the total clearance can be read in thousandths of an inch on the indicator.

If the clearance is excessive, as checked with the dial indicator, try a new valve in the bore to determine whether the bore is worn or the valve is worn. If the valve is worn, it is necessary to install a new valve.

Valve Replacement

With cylinder head on bench and the rockers removed, compress the valve spring and remove the valve lock, seal, spring cap and spring.

Line the valve springs up on a flat surface. All should be the same height. Replace those that do not match with new.

When reinstalling the valves in the head, some asbestos washers with loading springs are available. These washers are placed over the intake valve stems onto the top of the valve guide before installation of the regular valve spring. These washers effect a good seal when the guides are not badly worn.

Check that the contact between the seal at the end of the valve stem and the spring cap is air tight. The closed coil portion of the valve springs contacts the cylinder head.

Quick Method for Valve Tappet Adjustment on Engines with Hydraulic Lifters

Adjust the rockers on both 6s and V8s until they touch the valve stem.

Start the engine and back off a rocker adjustment until it clatters, tighten until no noise is heard and tighten further one turn for 6s; three-quarters of a turn for V8s. Repeat the procedure on all valves.

Valve Tappet Adjustment (Engine Stopped)

On Engines with Hydraulic Lifters

For initial adjustment on these models: with valve closed and on base circle of cam, adjust nut to remove all freeplay in rocker, pushrod and hydraulic tappet. Now tighten nut one more complete turn.

Reaming stud hole, V8 engines
(© Chevrolet Div. G.M. Corp.)

Reaming should be done with extreme care in order to get a press fit on the oversize stud.

Timing Case

Crankshaft Pulley Replacement

NOTE: to prevent vibration damper damage, it is important that tool J-22197 be used on engines, except the 396 cu. in. engine, to install the damper. When installing the damper on the 396 engine, tool J-21058 should be used.

Removing torsional damper

6 Cylinder Models

1. Remove the radiator core and the fan belt. Remove accessory drive pulley and belt, if so equipped.
2. Use a screw-type puller to remove the balancer-pulley assembly.

V8 Models

1. Drain radiator and disconnect the hoses. Take off the fan belt, and the fan pulley assembly. Remove the battery.
2. Remove the fan shroud. Remove the radiator core. Unbolt the pulley portion of the balancer-pulley assembly.
3. Install screw-type puller and remove the balancer portion from the crankshaft.

Timing Case Cover and Front End Oil Seal Replacement

NOTE: the timing case cover oil seal may be replaced without removing the case cover on all Corvettes and Chevrolets.

After gaining access to the oil seal, pry the old seal out of the cover with a screwdriver. Then, lubricate the new seal and drive it into place with tool J-8340.

6 Cylinder Models

1. Remove the crankshaft pulley. Remove the oil pan.
2. Remove the timing case cover attaching screws and the two bolts that are installed from inside the engine through the front main bearing cap to hold the cover at the bottom.
3. Remove the cover and gasket. Pry the old seal out of the front

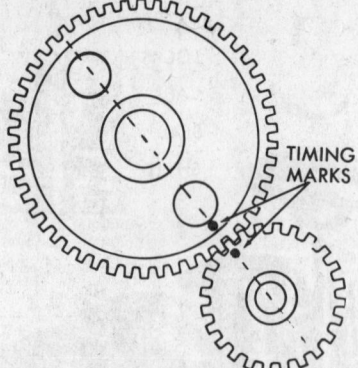

Timing mark alignment, 6 cylinder

side of the cover with a large screwdriver.
4. Install the new seal so that the open end of the seal is toward the inside of the cover. When reinstalling, be careful that cover is positioned to hold seal concentric to the shaft.
5. Tighten the screws and the two bolts inside the engine to 6-7½ ft. lbs.

V8 Models

1. Remove the crankshaft pulley. Remove the oil pan. Remove the water pump. Remove the screws holding the timing case cover to the block and remove the cover and gaskets.
2. Use a large screwdriver to pry the old seal out of the front face of the cover.

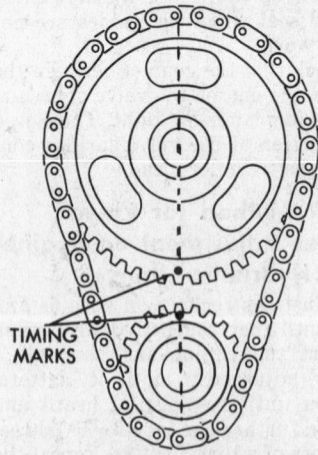

Timing mark alignment, V8

3. Install the new seal so that open end is toward the inside of the cover.
4. Check that the timing chain oil slinger is in place against the crankshaft sprocket.
5. Install the cover carefully onto the locating dowels.
6. Tighten the attaching screws to 6-8 ft. lbs.

Timing Gear Replacement

6 Cylinder Engines

Chevrolet timing gears are arranged so that (unless deliberately disturbed) the valve timing will remain as set at the factory. Unless the gears are badly worn or seriously damaged, the valve timing will remain constant within reasonable limits.

If it becomes necessary to replace the timing gears due to wear or damage, remove the radiator core, disconnect the front motor mounts and jack up the front of the engine. Remove the fan belt, fan pulley, oil pan and timing case cover.

NOTE: the Chevrolet Company recommends that the camshaft be removed from the car in order to remove and replace the gear in an arbor press.

Many successful mechanics prefer removing the camshaft in order to avoid possible risk when attempting to press a gear onto the shaft in place on the car. Sometimes when the gear is being pressed on in place on the car, damage results to the thrust washer in back of the cam gear. Unfortunately, this damage is not noticed until the engine is started.

To replace the gear by removing the camshaft, remove the rocker arm assemblies and the distributor, take out all of the pushrods and all of the lifters. The camshaft may then be pulled out toward the front of the engine. It will be necessary to retime the ignition.

Runout of the timing gear should not exceed .004 in. Backlash between the two gears should not be less than .004 in. nor more than .006 in. End clearance of the thrust plate should be .001 to .005 in.

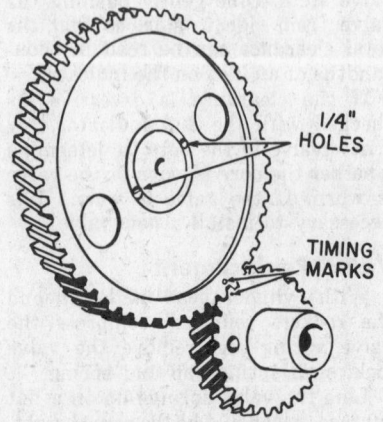

Time-saver for fast removal

A different approach to this situation, and certainly a quicker one, is as follows:

1. Very carefully center punch and drill two ¼ in. holes in opposite sides of the camshaft gear hub, as illustrated.
2. Break the fiber part of the camshaft gear away from the steel hub.
3. Split the steel hub with a cold chisel at the drilled holes.
4. Remove broken camshaft gear and clean entire timing case area.
5. Place the new gear on the camshaft to line up with the keyway and the gear timing marks.

NOTE: be sure to allow for the helical cut on the gear when aligning the marks for timing.

6. Have a reliable helper, with the aid of a pinch bar, buck up against one of the camshaft lobes from underneath. (The success of the job depends upon this man's care in holding forward thrust on the camshaft. Failure on his part will allow the camshaft to be forced back and dislodge the oil sealing expansion plug in the back at the rear of the camshaft.)
7. With the aid of a 1¼ in. socket and a lead hammer, tap the new gear into place on the camshaft.

CAUTION: the use of a dial indicator will reduce the possibility of driving the gear too far onto the camshaft. This would alter the desired camshaft thrust clearance of .001 to .005 in. Use care when approaching the final position of the gear on the shaft, because it is impossible to increase the thrust clearance without pulling the new gear. In the absence of a dial indicator, this end thrust can be measured with a feeler gauge. In this case, the thrust clearance is to be measured between the camshaft gear hub and the thrust plate. A feeler gauge strip, inserted in either of the two large gear holes, will reach this point.

Timing Chain Replacement

V8 Models

To replace the chain, remove the radiator core, water pump, the harmonic balancer and the crankcase front cover. This will allow access to the timing chain. Crank the engine until the timing marks on both sprockets are nearest each other and

in line between the shaft centers. Then take out the three bolts that hold the camshaft gear to the camshaft. This gear is a light press fit on the camshaft and will come off easily. It is located by a dowel.

The chain comes off with the camshaft gear.

A gear puller will be required to remove the crankshaft gear.

Without disturbing the position of the engine, mount the new crankshaft gear on the shaft, and mount the chain over the camshaft gear. Arrange the camshaft gear in such a way that the timing marks will line up between the shaft centers and the camshaft locating dowel will enter the dowel hole in the cam sprocket.

Place the cam sprocket, with its chain mounted over it, in position on the front of the car and pull up with the three bolts that hold it to the camshaft.

After the gears are in place, turn the engine two full revolutions to make certain that the timing marks are in correct alignment between the shaft centers.

End-play of the V8 camshaft is zero.

Pistons and Connecting Rods

Connecting Rod Bearings

Chevrolet does not recommend adjusting the slip-in type rod bearing. However, this bearing may be adjusted for normal wear by installing a taper or feather type shim between the lower bearing shell and the bearing cap.

Assembling Piston to Connecting Rod

6 Cylinder Engines

Where split skirt-type pistons are being installed, the split in the skirt of the piston should be placed opposite the clamp screw of the wristpin. This is also opposite the number on the bottom of the connecting rod.

Where solid skirt slipper-type pistons are being replaced, it is unimportant which way the piston is mounted onto the connecting rod. However, if

Correct relation of piston to rod, 6-cylinder 230 and 250 cu. in. engines

the old pistons are being reinstalled, the piston should be carefully marked before it is detached from the connecting rod in order that it may be replaced on the same side from which it was removed.

V8 Engines

Pistons are marked with a cast depression at the top of the piston and also the letter F on the piston strut. This depression and F always go toward the front.

For the left bank, pistons Nos. 1, 3, 5, and 7, the heavy flange at the bot-

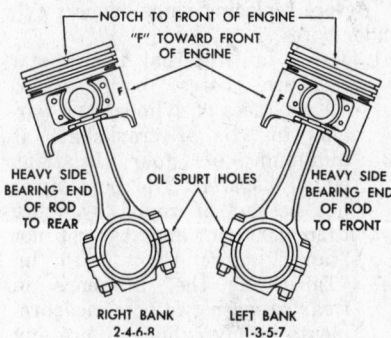

Correct relation of piston to rod (283, 307, 327 and 350 cu. in. engines)

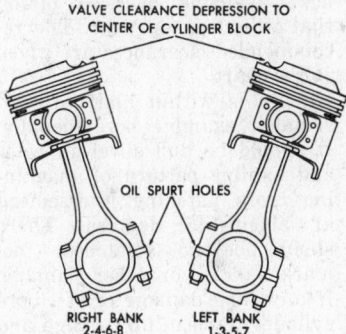

Correct relation of piston to rod (396, 427, and 454 cu. in. engines)

tom of the connecting rod goes on the side of the piston having the depression and F mark. For the right bank, cylinders Nos. 2, 4, 6, and 8, the heavy flange on the connecting rod goes to the side opposite the stamped letter F and the cast depression in the top of the piston.

Assembling Piston and Rod Assembly to the Engine

6 Cylinder Models

When assembling the rods to the pistons and installing the pistons in their respective bores, be sure that the flange, or heavy side of the rod at the bearing end, is toward the front of the piston (cast depression in top of piston head). The oil hole in the connecting rod goes toward the camshaft side of the engine.

V8 Models

Place the piston and rod assemblies into the cylinder so that the depres-

sion cast into the top of the piston (and the letter F stamped on the boss of the piston) face front. Double check that the pistons are in the correct bank by noting that on the left bank, pistons Nos. 1, 3, 5 and 7, the heavy flange on the connecting rod will also face forward, but on the right bank, cylinders Nos. 2, 4, 6 and 8, the heavy flange on the connecting rod will face toward the rear.

Piston Rings

Replacement

Before replacing rings, inspect cylinder bores.

1. Using an internal micrometer measure bores both across thrust faces of cylinder and parallel to axis of crankshaft at minimum of four locations equally spaced. The bore must not be out of round by more than 0.005 in. and it must not "taper" more than 0.010 in. "Taper" is the difference in wear between two bore measurements in any cylinder. Bore any cylinder beyond limits of out of roundness or taper to diameter of next available oversize piston that will clean up wear. The recommended clearances are given in the chart.

2. If bore is within limits dimensionally, examine bore visually. It should be dull silver in color and exhibit pattern of machining cross hatching intersecting at about 45 degrees. There should be no scratches, tool marks, nicks, or other damage. If any such damage exists, bore cylinder to clean up damage and then to next oversize piston diameter. Polished or shiny places in the bore are known as glazing. Glazing causes poor lubrica-

tion, high oil consumption and ring damage. Remove glazing by honing cylinders with clean, sharp stones of No. 180-220 grit to obtain surface finish of 15-35 RMS. Use a hone also to obtain correct piston clearance and surface finish in any cylinder that has been bored.

⏻ CHILTON TIME-SAVER

This or any other machining operation should be done with the cylinder block completely disassembled. Hot tank cylinder block after honing or boring. To remove minor glazing when honing equipment is not available, run emery cloth back and forth across glazed area perpendicular to axis of bore. Scrub block and bores thoroughly with soap and water to remove all grit after using emery cloth.

NOTE: the emery cloth method should be used only as a last resort as it is a method much inferior to honing.

3. If cylinder bore is in satisfactory condition, place each ring in bore in turn and square it in bore with head of piston. Measure ring gap. If ring gap is greater than limit, get new ring. If ring gap is less than limit, file end of ring to obtain correct gap.

4. Check ring side clearance by installing rings on piston, and inserting feeler gauge of correct dimension between ring and lower land. Gauge should slide freely around ring circumference without binding. Any wear will form a step on lower land. Replace any pistons having high steps. Before checking ring side clearance be sure ring grooves

are clean and free of carbon, sludge, or grit.

5. Space ring gaps at equidistant intervals around piston circumference. Be sure to install piston in its original bore. Install short lengths of rubber tubing over connecting rod bolts to prevent damage to rod journal. Install ring compressor over rings on piston. Lower piston rod assembly into bore until ring compressor contacts block. Using wooden handle of hammer push piston into bore while guiding rod onto journal.

Caution Some pistons, such as those used in the 454, 396 or 427 engines with high lift camshafts, have a peculiar head shape. Where the standard piston is equipped with a gabled, or double slanted top, the high torque piston top has only one slanted side. This is a means of raising compression ratio.

Use care in assembly because the piston head is clearance-bored to allow for the valve head.

Engine Lubrication

Oil Pan Removal

Chevrolet 6 Cylinder Manual Transmission

The oil pan can be removed, either after removing engine, or as follows:

1. Drain radiator and oil pan.
2. Disconnect gas tank line at fuel pump and upper and lower radiator hoses.
3. Remove clutch housing-to-engine block bolt above dowel on right side.
4. Raise vehicle on hoist or place on jack stands.

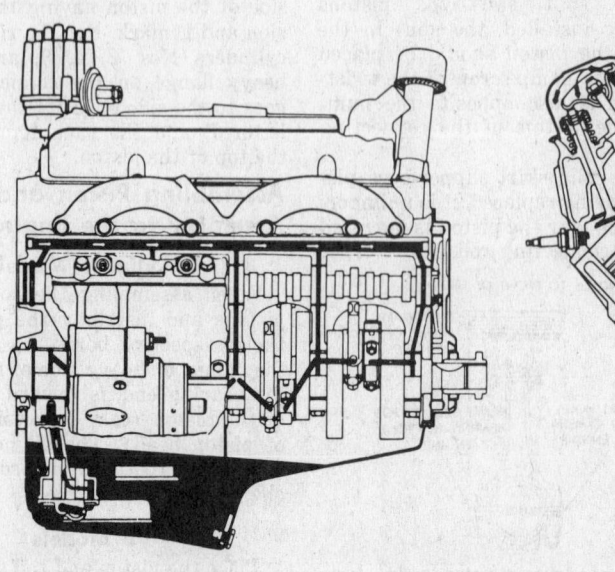
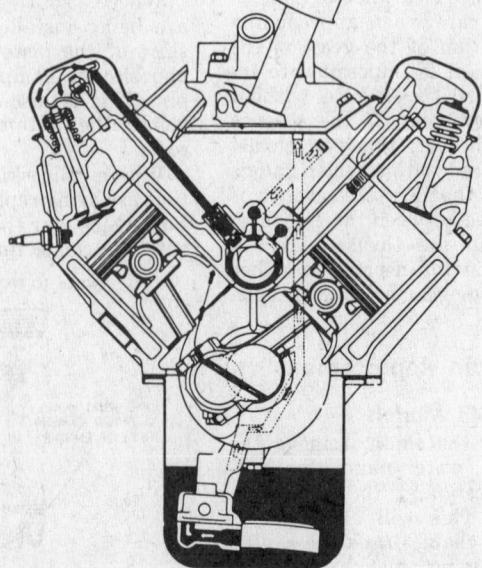

283, 307, 327 and 350 cu. in. V8 engine lubrication (© Chevrolet Div. G.M. Corp.)

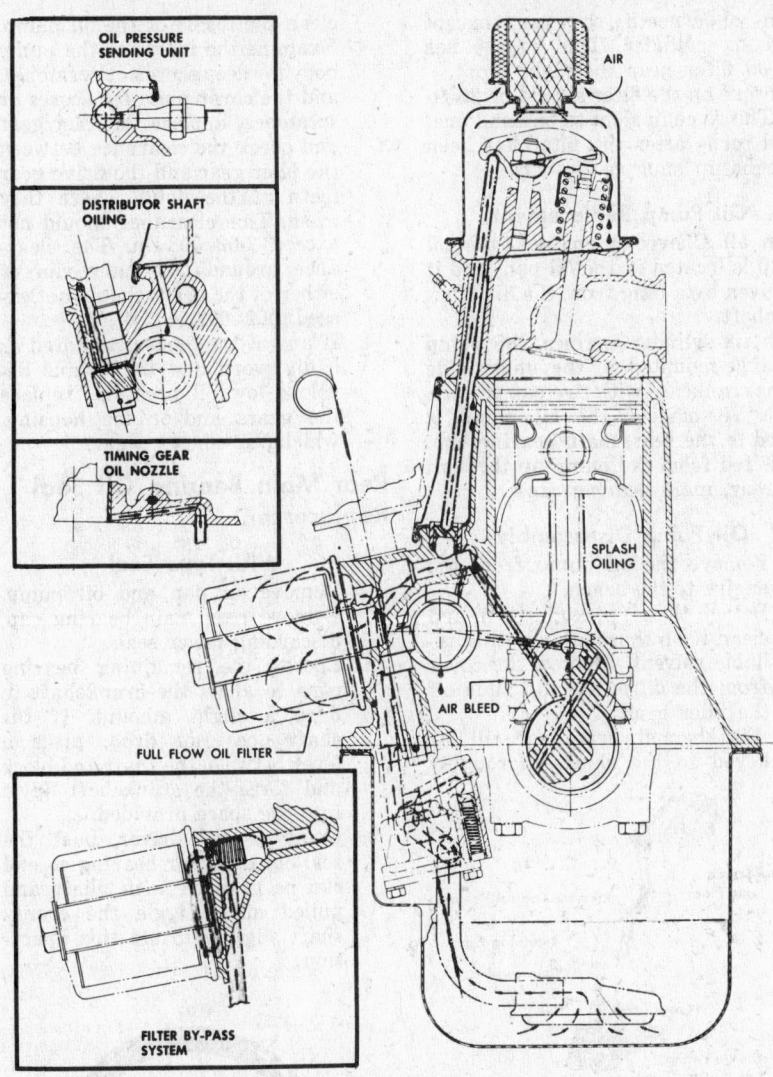

230 cu. in. 6-cylinder engine lubrication (© Chevrolet Div. G.M. Corp.)

5. Rotate engine to align distributor rotor No. 3 and No. 5 plug wires. (This locates No. 6 crank throw part way up.)
6. Remove starter and flywheel front cover plate.
7. Remove front mount through bolts.
8. Jack front of engine. (Tool J-6987 on balancer will aid.) Raise as far as possible always using care by checking various dash and body tunnel clearances.
9. Remove front engine mount frame bracket on right side and remove oil filter where necessary.
10. Remove oil pan screws and lower pan to frame.
11. Remove oil pump to gain clearance, then remove oil pan by sliding and rotating front to right and then to rear, and down at an angle. (On certain earlier models, these procedures may be varied in some self-evident areas.)
12. Install in reverse of above.
NOTE: gasket can be replaced by completely removing pan from vehicle.

Chevrolet 6 Cylinder Automatic Transmission

1. Drain radiator and crankcase.
2. Disconnect gas tank line at fuel pump, and radiator hoses at radiator.
3. Remove clutch housing-to-engine block bolt above dowel pin on each side.
4. Rotate engine to align distributor between No. 3 and No. 5 plug wires. (This locates No. 6 crank throw part way up.)
5. Raise vehicle on hoist or on jack stands.

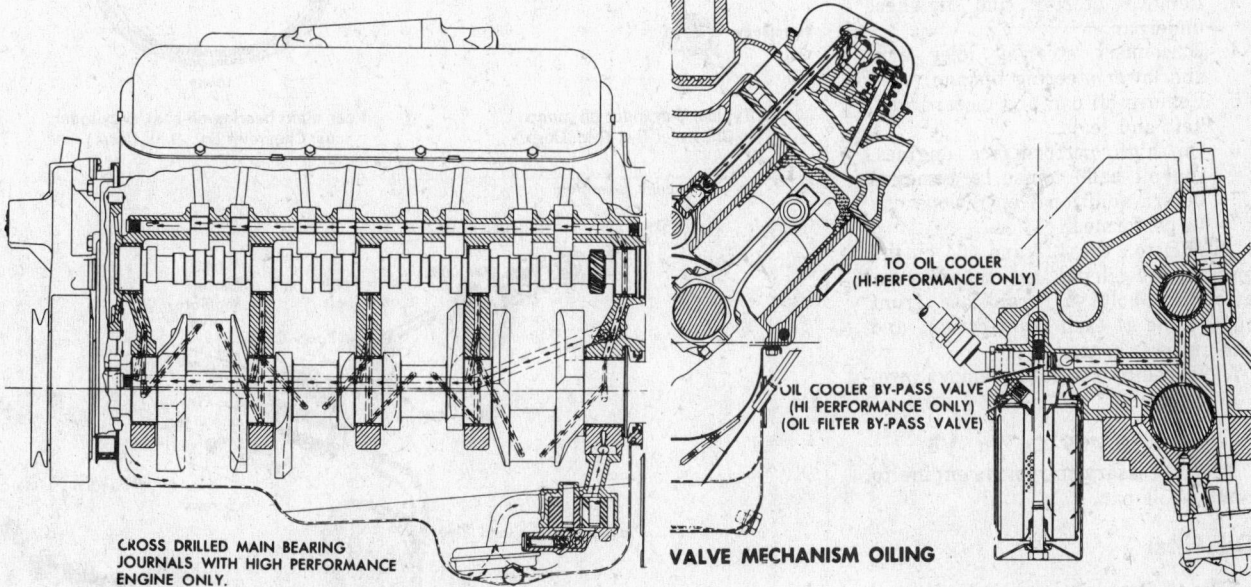

CROSS DRILLED MAIN BEARING JOURNALS WITH HIGH PERFORMANCE ENGINE ONLY.

CRANKCASE AND CRANKSHAFT OILING

VALVE MECHANISM OILING

OIL FILTER AND DISTRIBUTOR OILING

396 and 427 cu. in. V8 lubrication

6. Remove converter cover pan, and starter assembly.
7. Follow Steps 7 through 12, listed above.

Chevrolet—V8
(Except. 409 Cu. In. 1964-65)

1. Disconnect battery positive cable.
2. Remove distributor cap from distributor to prevent breakage against firewall.
3. Drain cooling system. Remove radiator hoses, and remove oil dipstick and tube, where necessary.
4. Remove fan blade assembly.
5. Raise car, and drain engine oil.
6. Remove thru bolts from engine front mounts. Disconnect and remove starter.
7. On cars with automatic transmissions, remove converter housing underpan.
8. Disconnect steering rod at idler lever and swing linkage down for pan clearance.
9. Rotate crankshaft until timing mark on the damper is at six o'clock position.
10. Using a block of wood and a suitable jack, raise engine enough to insert 2 x 4 in. wood blocks under engine mounts then lower engine onto blocks.
11. Remove engine oil pan.
12. Install by reversing removal procedures.

NOTE: the 396, 427 and 454 cu. in. engines use three ¼ in. attaching bolts at crankcase front cover; one at each corner, and one at the lower center.

Corvette

1. Disconnect battery, and remove dipstick and tube.
2. Raise car and support on stands. Drain engine oil.
3. Remove starter and flywheel underpan.
4. Disconnect steering idler arm and lower steering linkage.
5. Remove oil pan and discard gaskets and seals.
6. On high performance engines, the oil baffle must be removed before additional operations can be performed.

NOTE: on the 427 and 454 cu. in. engine, the oil pan has three ¼ in. attaching bolts at crankcase front cover; one at each front corner, and one at lower center.

7. Install by reversing removal procedure.

1964-65 409 Cu. In. V8

It is necessary to remove engine to remove oil pan.

Oil Filter

The oil filter is located under the engine at the left side just forward of the flywheel housing and is accessible from underneath the car, except 1964 six cylinder. This engine has the oil filter near the right front.

Torque on the filter should be 25 ft. lbs. This is equivalent to one and one-third turns after the filter has been brought up snug to the case.

Oil Pump Replacement

On all Chevrolet engines, the oil pump is located in the oil pan, and it is driven by a tang from the distributor shaft.

On six-cylinder engines, the pump is flange-mounted to the under side of the crankcase with two cap screws.

On V8 models, the oil pump is bolted to the rear, main bearing cap. Oil is fed from the pump up through the rear, main bearing cap.

Oil Pump Disassembly

1. Remove the oil pump from the engine to the bench.
2. Detach the oil pickup screen and clean it up thoroughly with a reliable solvent. Remove the cover from the oil pump and slide off the idler gear.
3. With the main drive gear still left keyed to the shaft, thoroughly

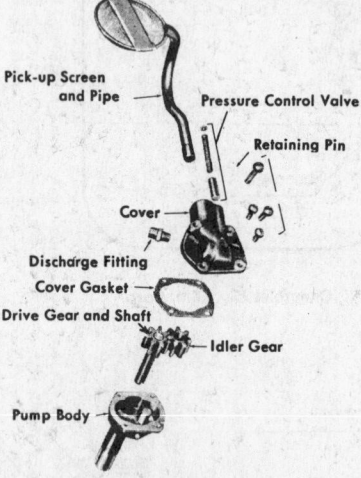

Pick-up Screen and Pipe
Pressure Control Valve
Retaining Pin
Cover
Discharge Fitting
Cover Gasket
Drive Gear and Shaft
Idler Gear
Pump Body

Typical 6-cylinder oil pump
(© Chevrolet Div. G.M. Corp.)

clean the inside of the oil pump. Examine the inside of the pump body for deep scores or scratches, and the cover plate for scores or scratches. Replace the idler gear and check the clearance between the idler gear and the drive gear teeth at the point where they mesh. This clearance should not exceed .002-.003 in. The clearance around the outer rim of either of the gears should not exceed .002-.003 in.

4. If any of the parts are scored or badly worn and the engine develops low oil pressure, replace the gears and/or the housing, whichever shows wear.

Rear Main Bearing Oil Seal Replacement

Wick Type Seal

1. Remove oil pan and oil pump. Remove rear main bearing cap, discarding lower seal.
2. Loosen the remaining bearing caps to allow the crankshaft to drop a slight amount. If the shaft does not drop, place a lever between the shaft and block and force the crankshaft down into the space provided.
3. Using a screwdriver, push the seal out of upper bearing so end can be grabbed with pliers and pulled out. Wiggle the crankshaft slightly to aid this operation.

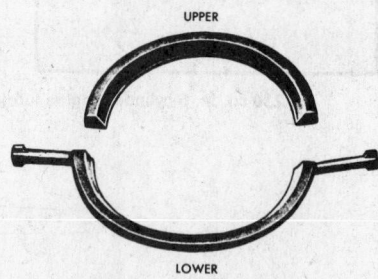

UPPER

LOWER

Rear main bearing oil seal, 6 cylinder
(© Chevrolet Div. G.M. Corp.)

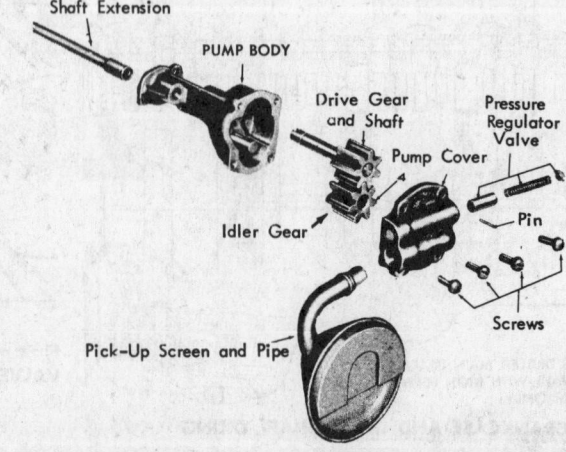

Shaft Extension
PUMP BODY
Drive Gear and Shaft
Pressure Regulator Valve
Pump Cover
Idler Gear
Pin
Screws
Pick-Up Screen and Pipe

Layout of typical V8 oil pump (© Chevrolet Div. G.M. Corp.)

4. Use the lower bearing cap to form the new upper seal into a semi-circle. Insert a wire (soft tag-type) through the seal about ¼ in. from the end. Wrap the wire around the end of the seal so it has a good grip.

5. Use a light coat of vaseline to lubricate the seal and insert the wire into the seal groove and up and over the crankshaft. Pull the seal gently into position. It may be necessary to rotate the crankshaft to aid in getting the seal placed. With seal centered in the opening, cut off the ends so that they stick out 1/64 in. beyond the parting surface.

6. Install the lower seal in the groove and roll into place. Cut the small portions of the ends that protrude from the groove flush with the surface of the bearing cap. Install bearing cap over the crankshaft onto the block.

Neoprene Seal

1. Remove rear main bearing cap and pry old seal from groove. Insert new seal with lubricant only on the lip. Do not get oil on the glue-treated parting line surfaces. Lip faces front of engine.

2. Using a hammer and small punch, revolve the upper half of the seal until it protrudes far enough to remove with pliers.

3. Oil the seal except at the glue-treated ends and, using a hammer handle, roll the seal into place in the block.

4. These seals are made to size and require no trimming. Install the lower half over the crankshaft and in place onto the block.

Front Suspension

References

General instructions covering the front suspension and how to repair and adjust it, together with information on installation of front wheel bearings and grease seals, are given in the Unit Repair Section.

Definitions of the points of steering geometry are covered in the Unit Repair Section. This article also covers troubleshooting front end geometry and irregular tire wear.

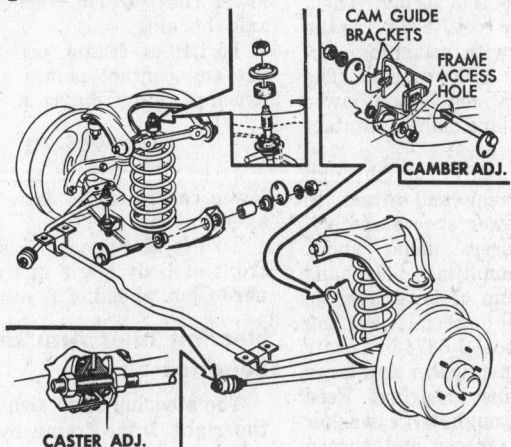

Front suspension, 1965-67 (© Chevrolet Div. G.M. Corp.)

Figures covering the caster, camber, toe-in, kingpin inclination, and turning radius can be found in the Wheel Alignment table of this section.

Tire sizes can be found in the General Chassis and Brake Specifications Table of this section.

Front Springs R & R

Chevrolet

1. Remove shock absorber upper stem retaining nut and grommet.

2. Support the car by the frame so that the control arms hang free. Remove the wheel assembly, shock absorber, stabilizer to lower control arm link, strut rod to lower control arm attaching nuts, bolts and lockwashers, and the tie-rod end.

3. Scribe the position of the inner pivot camber adjusting cam bolt and then remove the nut, lock washer and outer cam.

4. Install a steel bar through the shock absorber mounting hole in the lower control arm so that the notch in the bar seats over the bottom spring coil and the bar extends inboard and under the inner bushing. Fit a 5 in. wood block between the bar and the lower arm inner support bushing.

5. With a floor jack, raise the end of the steel bar enough to remove tension from the inner pivot cam bolt. The bolt can then be removed.

6. Carefully lower the inner end of the control arm. Tension on the

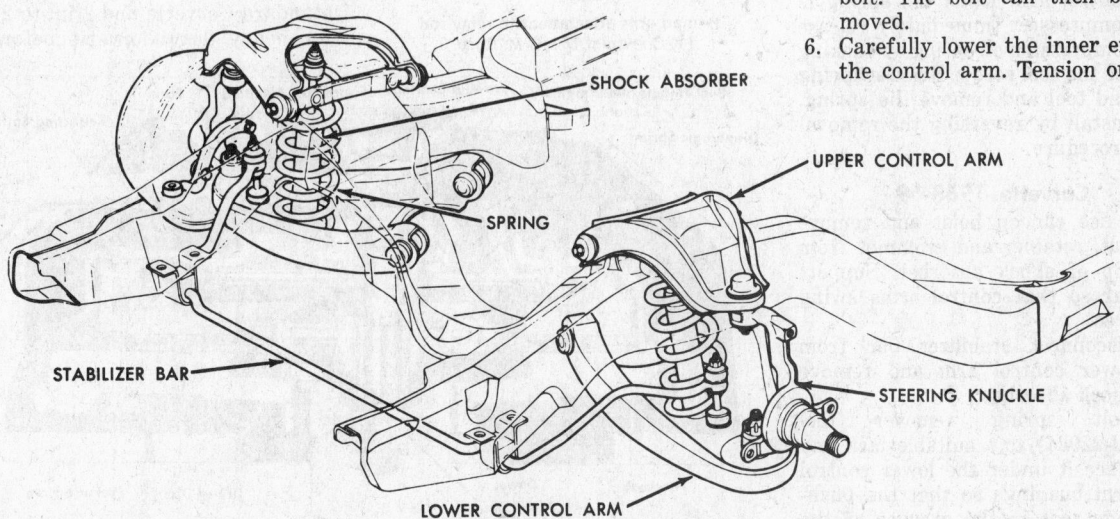

Front suspension, Chevrolet 1968-70

spring must be removed before the spring can be taken out of the car.

7. Remove the spring.
8. Install by reversing removal procedure.

Corvette—1964-67

1. Support car by the frame to allow control arms to swing free. Remove wheel assembly, stabilizer bar and shock absorber. Loosen lower ball joint to steering knuckle nut, and the two lower control arm cross-shaft bushing bolts.
2. Place tool J-6874-1 across top of sixth coil of the spring. Then, loosely secure tool J-6874-2 to the upper shoe, with attaching capscrews and lockwashers. The upper shoe V notch and lower shoe raised land should contact the spring.
3. Insert tool J-6874 up through center of spring and attach to upper and lower shoe assembly.
4. Position spacers under shock absorber mounting hole and against bottom of lower control arm. Install special bearing washer and tool J-6874-5. Locate bearing against spacer and large washer against bearing. Feed screw up through large washer bearing and spacer and thread into tool J-6874 and tighten.
5. Center shoe assembly on spring and tighten screw until a very slight compression is exerted on the spring. Then, firmly tighten the two capscrews holding the upper and lower shoes to lock these shoes to the spring.
6. Tighten the spring compressor enough to permit the spring to clear the spring tower, then remove the lower ball joint to steering knuckle nut.
7. Disconnect lower ball joint from the steering knuckle and lower control arm while the spring is compressed. Immediately release compression on spring by backing off the tool screw. Release spring and tool and remove the spring.
8. Install by reversing the removal procedure.

Corvette 1968-69

1. Raise car on hoist and remove nut, retainer and grommet from top of shock absorber. Support car so that control arms swing free.
2. Disconnect stabilizer bar from lower control arm and remove shock absorber.
3. Bolt spring remover tool (J-22944) to a suitable jack and place it under the lower control arm bushings so that the bushings seat in the grooves of the tool.

4. Remove cross shaft rear retaining nut and the two front retaining bolts.
5. Slowly release jack, swing control arm forward, then remove spring.
6. Install by reversing procedure above.

Jacking, Hoisting

1964

Jack car at front spring seats of lower control arm. Jack car at rear axle housing.

To lift at frame, use adapters on lift and contact points as shown in accompanying diagram.

1965-70

Jack car at front spring seats of lower control arm. Jack car at rear axle housing.

To lift at frame, use side rails at front of body angle and at rear corner of box ahead of rear wheel.

Steering Idler Arm and Bearings

The steering idler arm is bolted to the right front frame by two bolts.

It is connected to the center tie-rod by a nonadjustable tie-rod end which is an integral part of the intermediate steering arm.

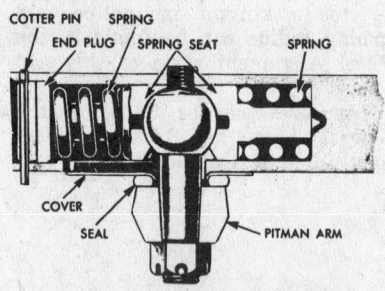

Pitman arm attachment to relay rod
(© Chevrolet Div. G.M. Corp.)

The tie-rod end portion is removed in the same manner as any tie-rod end.

Steering Reach Rod or Relay Rod Replacement

The relay rod reaches from the steering pitman arm to the idler arm attached to the right frame side member.

Remove the cotter pin and the large adjusting screw from the end of the rod. The seats and spring assemblies can then be taken out of the tube.

Examine them for pits, scratches or wear, and if they cannot be adjusted so that they operate smoothly and easily they will have to be replaced.

It is unusual to find that these units require replacement and, generally speaking, they can be smoothed up to function well unless a distinct flat spot has been worn on the ball stud on the pitman arm.

Clutch

On most Chevrolet models with synchromesh transmissions, a diaphragm spring type clutch is used.

A ball bearing-type throw-out bearing is used, and no provision is made for lubrication of this bearing, because it was sealed full of lubricant at assembly.

The throw-out fork pivots on a ball stud which is mounted in the rear face of the bell housing.

Clutch Adjustment

Free-travel is adjustable at one point on the clutch linkage. This free-travel adjustment will compensate for all normal clutch wear.

Clutch Pedal Free-Travel

The pedal should travel 1 in. to 1½ in. for Chevrolet, 1¼ in. to 2 in. for standard Corvette and 2 in. to 2½ in. for heavy duty Corvette before the

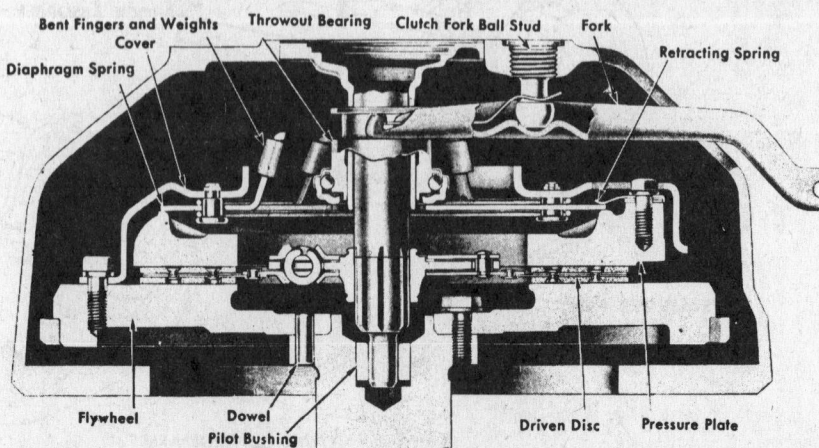

Typical V8 clutch cross-section (© Chevrolet Div. G.M. Corp.)

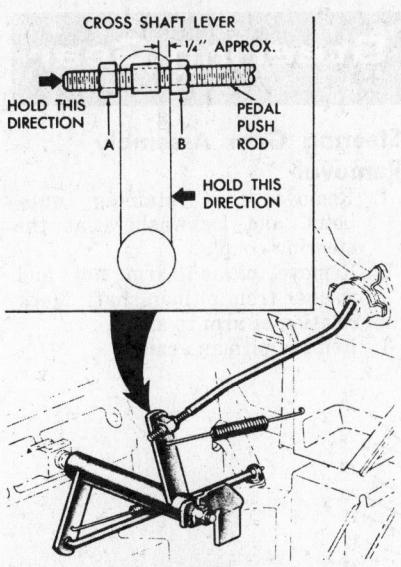

CROSS SHAFT LEVER

¼" APPROX.

HOLD THIS DIRECTION

A B

PEDAL PUSH ROD

HOLD THIS DIRECTION

Typical clutch linkage
(© Chevrolet Div. G.M. Corp.)

throw-out bearing engages the diaphragm spring.

This should be checked at the pedal by hand. The check should be made by hand, and not by foot, because the feel is sensitive; ¾ in. true free-travel of the bearing will approximate 1 in. feel at the pedal.

The adjustment is made on the fork pushrod running from the lever and shaft assembly to the clutch fork. On some models, the adjustment is made at the fork end by changing the position of two jam nuts. On other models, the adjustment is made at the front end of the rod by turning an adjustable swivel. On this type, one turn of the swivel equals approximately 3/16 in. at the pedal. Also, on this type, the adjustment can be made by holding the fork pushrod rearward to remove all lash, then adjusting the swivel to line up a conical point stamped on the swivel with a dimple stamped on the lever to which it attaches.

Clutch Pedal Height

The top of the clutch pedal pad should be at least 7 in. above the deadener felt glued to the metal floor pan. Do not measure to the floor mat.

If less than 7 in., cut off the rubber pedal stop to obtain proper pedal height. On some models, the rubber pedal stop is fastened to a metal piece held to the instrument panel brace by a bolt and nut. A slotted hole in the brace allows for adjustment of the bumper holding piece.

If more than 7 in. of pedal travel occurs, it may be that the diaphragm spring is being overstressed. At any rate, to reduce pedal travel add a pad under the floor mat.

Clutch Assembly Removal

The clutch is enclosed in a 360° aluminum housing. This housing must be removed for access to the clutch assembly.

1. For clutch removal, take out transmission. Disconnect linkage spring from fork and let pushrod hang free.
2. Remove throw-out bearing from fork. Remove housing cover plate screws and housing bolts. Remove housing. On vehicles with V8 engines, the screw at oil filter may be removed last, while supporting the housing.
3. Tool J-5824, or equivalent, should be used to support the clutch assembly during removal. Loosen the clutch attaching bolts evenly until diaphragm spring tension is released, then remove bolts, clutch assembly, and pilot tool.
4. Slide clutch fork from ball stud and remove fork from boot.

Clutch Disassembly

1. Remove three drive-strap-to-pressure-plate bolts and retracting springs.
2. Note position of grooves on edge of pressure plate and cover. These marks must be aligned in assembly to maintain balance. Remove the pressure plate from the cover.
3. The clutch diaphragm spring and two pivot rings are riveted to the clutch cover. Spring, rings, and cover should be inspected for excessive wear or damage. If there is a defect, it is necessary to replace the complete cover assembly.

Clutch Assembly

1. Install the pressure plate in the cover assembly, lining up the groove on the edge of the pressure plate with the groove on the edge of the cover.
2. Install pressure plate retracting springs and drive-strap-to-pressure-plate bolts. The drive-straps should not be loose in the rivets or at the bolts. Tighten to 10-15 ft. lbs.

Clutch Installation

1. Turn crankshaft so the X mark on the flywheel is at the bottom.

2. With clutch fork in the housing, but not on ball stud, install the clutch disc, pressure plate and cover assembly, and support on a pilot shaft.

NOTE: on 6 cylinder engines, the clutch driven disc is installed with the damper springs toward the flywheel. On V8s, the damper springs are toward the transmission; however, the grease slinger is always toward the transmission.

3. Turn the clutch assembly until the X mark on the clutch cover flange lines up with the X mark on the flywheel. Align nearest bolt holes in clutch and flywheel and install bolts in every other hole. They are marked L. Tighten gradually, and then install remaining three bolts. Remove the pilot shaft.
4. Pack the clutch fork ball seat with a small amount of high-melting-point grease.
5. Remove fork on ball stud. Lubricate the recess on the inside of the throw-out bearing collar and coat the throw-out fork groove with a small amount of graphite grease. Install throw-out bearing assembly to the fork. Hook up linkage. Reinstall transmission.

Horn Button

Horn Button Removal

The horn blowing ring is held to the steering wheel by three screws.

1. Disconnect turn signal switch harness from chassis wiring harness at the connector.
2. Pull out horn button on standard models. On other models, pull out center ornament from horn ring.
3. Remove three screws from the receiving cup or horn ring.
4. Remove the receiving cup or horn ring, belleville spring, bushing and on deluxe wheel, the pivot ring.

Horn Relay

The horn relay is located on the radiator core panel above the generator regulator.

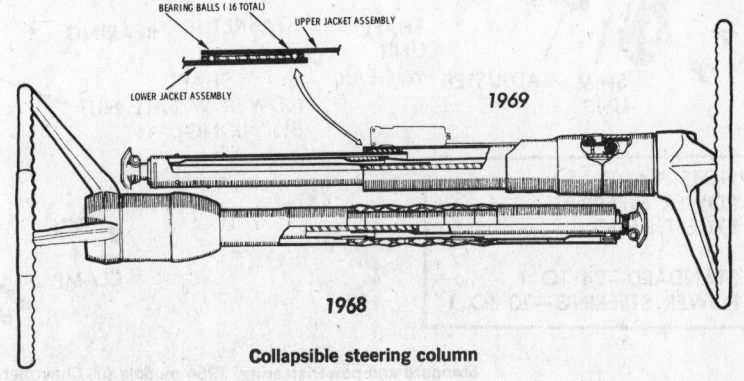

BEARING BALLS (16 TOTAL)

UPPER JACKET ASSEMBLY

LOWER JACKET ASSEMBLY

1969

1968

Collapsible steering column

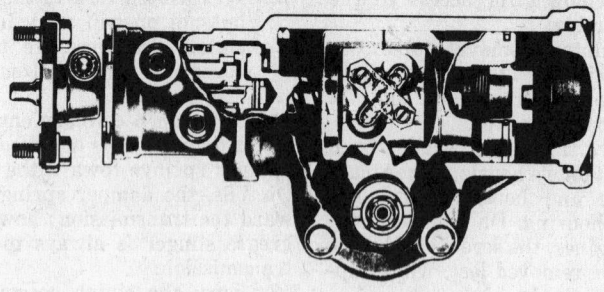

1969-70 steering gear cross-section

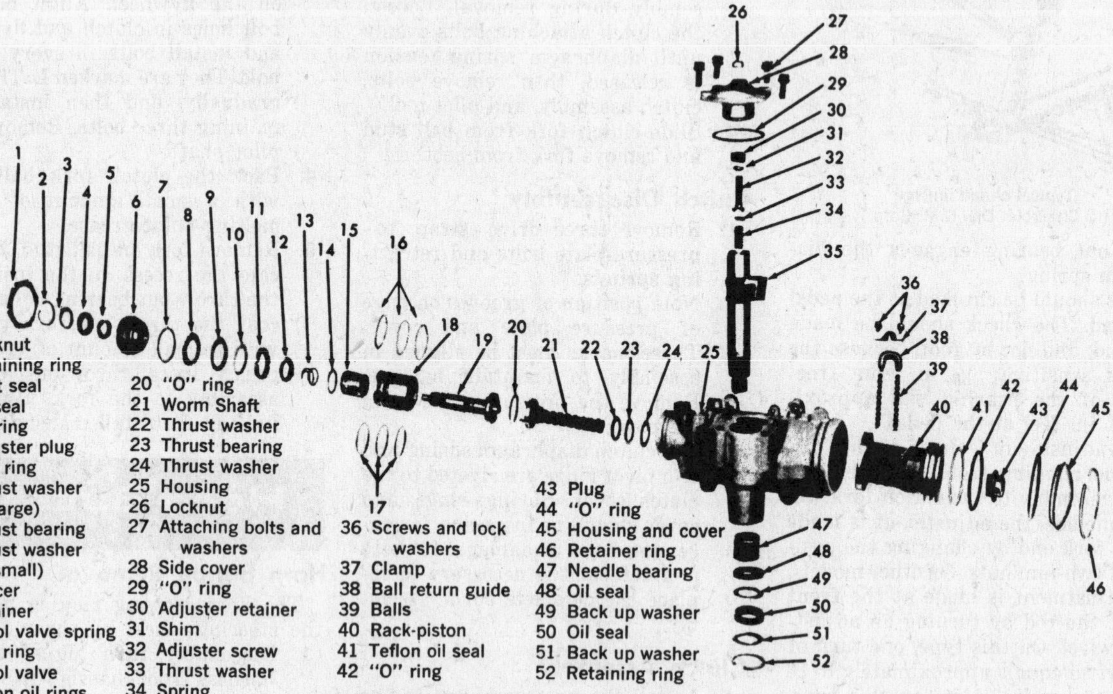

Steering Gear Assembly Removal

1. Remove the retaining nuts, bolts, and lockwashers at the steering couple.
2. Remove pitman arm nut and washer from pitman shaft. Mark relation of arm to shaft.
3. Remove pitman arm.

1 Locknut
2 Retaining ring
3 Dust seal
4 Oil seal
5 Bearing
6 Adjuster plug
7 "O" ring
8 Thrust washer (large)
9 Thrust bearing
10 Thrust washer (small)
11 Spacer
12 Retainer
13 Spool valve spring
14 "O" ring
15 Spool valve
16 Teflon oil rings
17 "O" rings
18 Valve body
19 Stud shaft

20 "O" ring
21 Worm Shaft
22 Thrust washer
23 Thrust bearing
24 Thrust washer
25 Housing
26 Locknut
27 Attaching bolts and washers
28 Side cover
29 "O" ring
30 Adjuster retainer
31 Shim
32 Adjuster screw
33 Thrust washer
34 Spring
35 Pitman shaft

36 Screws and lock washers
37 Clamp
38 Ball return guide
39 Balls
40 Rack-piston
41 Teflon oil seal
42 "O" ring

43 Plug
44 "O" ring
45 Housing and cover
46 Retainer ring
47 Needle bearing
48 Oil seal
49 Back up washer
50 Oil seal
51 Back up washer
52 Retaining ring

Power steering gear, 1965-67 (© Chevrolet Div. G.M. Corp.)

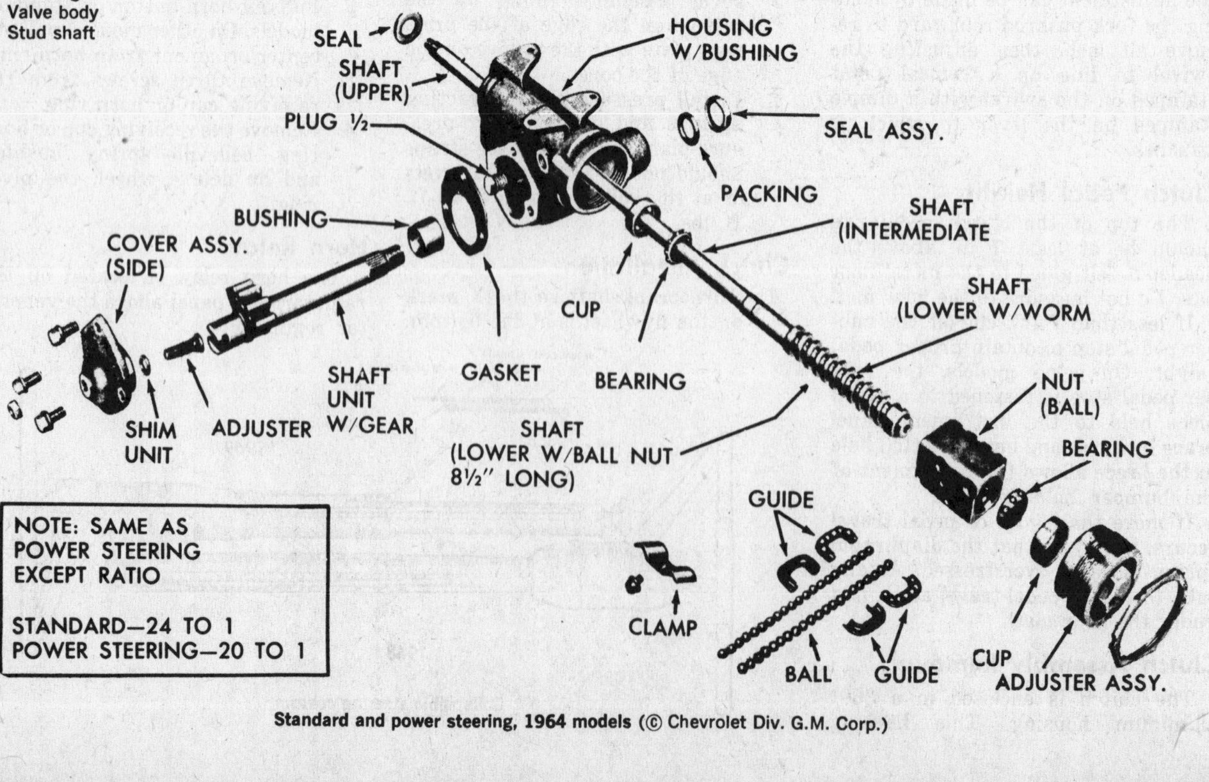

SEAL
SHAFT (UPPER)
PLUG ½
HOUSING W/BUSHING
SEAL ASSY.
PACKING
SHAFT (INTERMEDIATE
BUSHING
COVER ASSY. (SIDE)
SHAFT (LOWER W/WORM
SHIM UNIT
ADJUSTER
SHAFT UNIT W/GEAR
GASKET
CUP
BEARING
SHAFT (LOWER W/BALL NUT 8½" LONG)
NUT (BALL)
BEARING
GUIDE
CUP ADJUSTER ASSY.
CLAMP
BALL
GUIDE

NOTE: SAME AS POWER STEERING EXCEPT RATIO

STANDARD—24 TO 1
POWER STEERING—20 TO 1

Standard and power steering, 1964 models (© Chevrolet Div. G.M. Corp.)

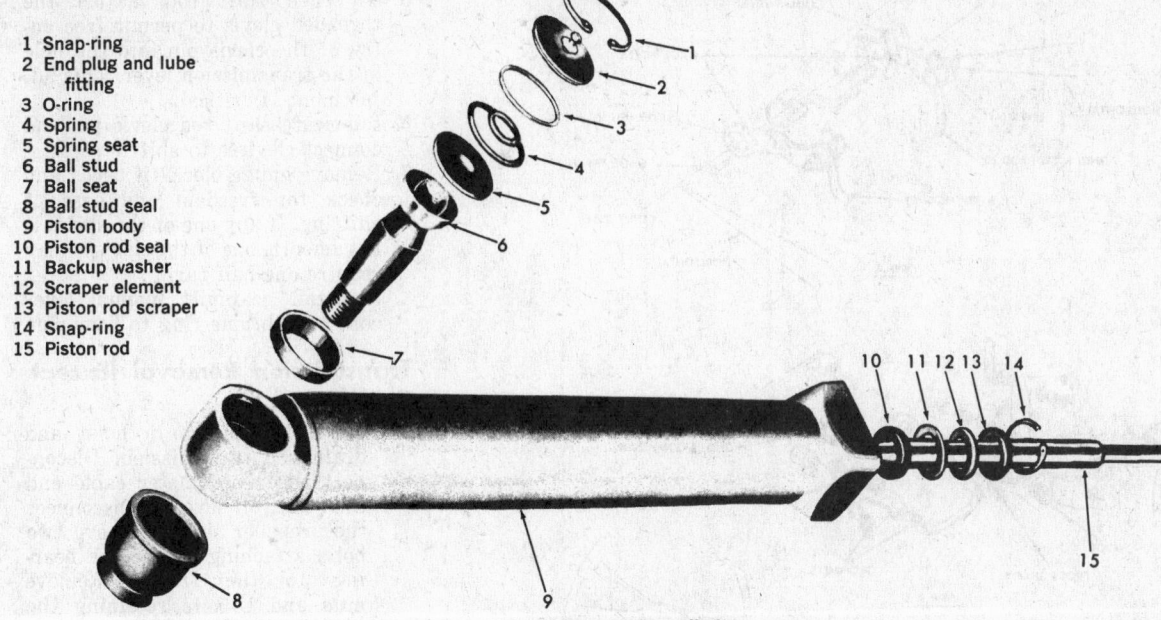

1 Snap-ring
2 End plug and lube fitting
3 O-ring
4 Spring
5 Spring seat
6 Ball stud
7 Ball seat
8 Ball stud seal
9 Piston body
10 Piston rod seal
11 Backup washer
12 Scraper element
13 Piston rod scraper
14 Snap-ring
15 Piston rod

Power steering cylinder

4. Remove screws that secure steering gear to frame and remove gear from car.

Standard Transmission

Transmission refill capacities are in the Capacities table of this section.

General information and exploded views, together with troubleshooting charts, are included in the articles on each automatic transmission.

Troubleshooting and repair of manual transmissions are covered in the Unit Repair Section.

Shift Linkage Adjustment

Three-Speed

1. Loosen swivel attaching nuts on shifter rod-to-lever attachments at bottom of steering column.
2. Move both control rods until transmission levers are in neutral.
3. Move selector lever to neutral.
4. Engage second and third shifter lever at bottom of steering column with relay lever.
5. Center the levers in the mast jacket by measuring from edge of slot in jacket to edge of slot in spacer at each side of lever.
6. Adjust swivel on end of second and third shifter control rod until swivel enters hole in lever. Install swivel and insert retaining clip.
7. Move first and reverse lever on tube until lug on lever lines up with slot in relay lever.
8. Adjust swivel on end of first and reverse control rod until swivel enters hole in lever. Install swivel and insert retaining clip.
9. Move selector lever through all positions of shift to check adjustment and smoothness of operation.

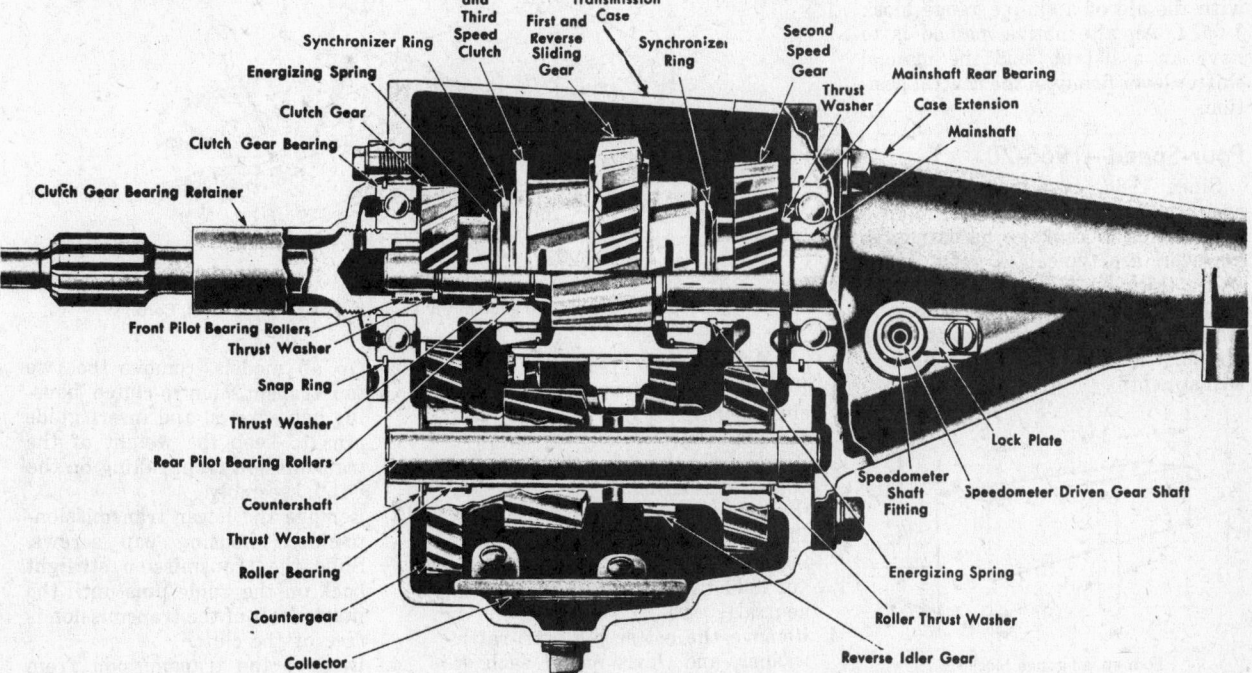

Typical 3-speed transmission (© Chevrolet Div. G.M. Corp.)

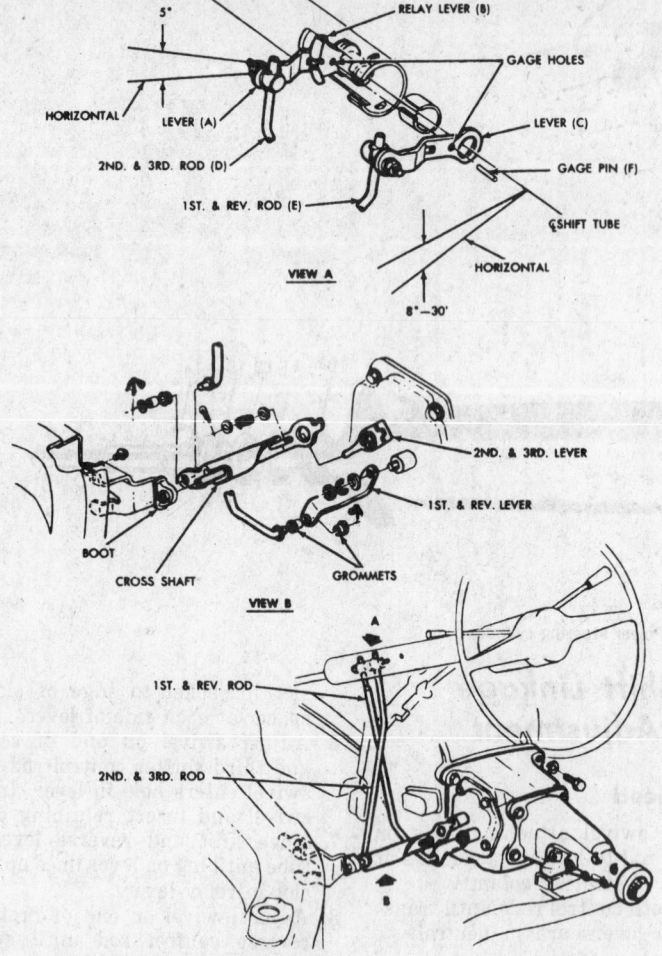

Heavy duty 3-speed linkage adjustment (© Chevrolet Div. G.M. Corp.)

Four-Speed—1964-65

The four-speed transmission gearshift linkage uses three shift rods and levers. Adjustment can be made with the aid of a simple gauge block J-9574. An alternative method is to have an assistant hold the manual shifter lever firmly in the neutral position.

Four-Speed—1966-70

Since 1966, two makes of four-speed transmission are used, Muncie and Saginaw. Linkage adjustments, however, are typical. A gauge block ⅛ in. thick by 41/64 in. wide and 3 in. long should be used to locate and maintain neutral detent position of the shift lever while making linkage adjustments.

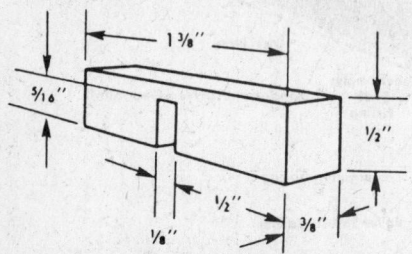

Four-speed gauge block
(© Chevrolet Div. G.M. Corp.)

Adjustment Procedure

1. Remove three screws holding the chrome ring to the floor pan and remove.
2. Remove three screws holding the boot and retainer to the floor pan, then slide boot up the shift rod.
3. Place transmission in neutral and install gauge block (or have assistant hold the manual shift lever in neutral).
4. Remove the cotter pin, anti-rattle washer, and clevis pin at each of the three shift levers.

5. On each shift rod, adjust the threaded clevis to permit free entry of the clevis pin into the hole in the transmission lever. This adjustment is critical.
6. Lubricate shift rod clevis pin and connect clevises to shift levers.
7. Remove gauge block (if used) and check for freedom and ease of shifting. If any one of the shifts is not smooth, one of the clevises may require one-half turn.
8. Reinstall gearshift manual lever boot and chrome ring to floor pan.

Transmission Removal (Except Corvette)

1. Raise the car on a hoist and drain the transmission. Disconnect the speedometer cable and the control levers. Disconnect the propeller shaft. Remove two bolts attaching the center bearing to the frame. Remove nuts and U-bolts retaining the rear universal joint bearing to the differential pinion drive flange. Move the propeller shaft rearward to the left and under the rear axle housing to withdraw the front universal joint from the transmission output shaft. Remove the transmission rear mounting pad bolts and unbolt the support member from the frame.

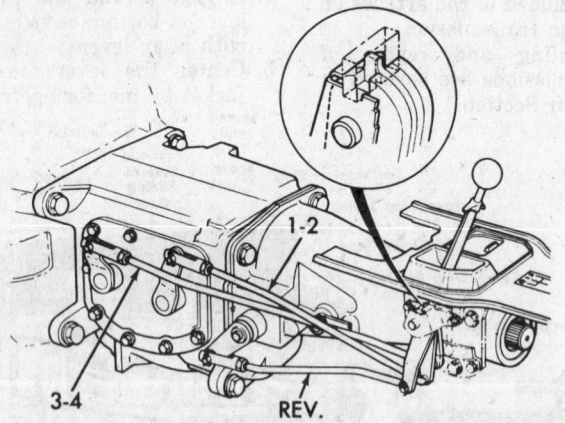

Four-speed transmission gearshift linkage (© Chevrolet Div. G.M. Corp.)

2. On all models, remove the two top transmission-to-clutch housing cap screws, and insert guide pins to keep the weight of the transmission from falling on the clutch assembly.
3. Remove the lower transmission-to-clutch housing cap screws. Slide the transmission straight back on the guide pins until the input shaft of the transmission is free of the clutch.
4. Remove the transmission from under the car.

Transmission Removal
(Corvette)
1. Disconnect battery.
2. Disassemble shift control lever assembly.
3. Raise car on a hoist.
4. Place a block of wood between the top of the differential housing and the underbody.
5. Disconnect differential carrier front support from the frame bracket at the biscuit mount.
6. Pry the carrier down, while removing the two center mounting bolts from the carrier front support.
7. Pivot carrier support downward for access to propeller shaft U-joint.
8. Disconnect propeller shaft U-joints, front and rear.
9. Disconnect parking brake cable from ball socket at idler lever near center of underbody.
10. Remove propeller shaft.
11. Remove heat deflectors from right and left exhaust pipes.
12. Remove left bank exhaust pipe. Remove right bank exhaust pipe and heat riser.
13. Disassemble transmission mount as follows:
    A. Remove two bolts that hold rear mount cushion to rear mount bracket.
    B. Support engine under the oil pan (with a well padded jack) and raise engine to take weight off rear mount cushion.
    C. Remove the three transmission mount bracket-to-crossmember bolts, and remove mount bracket.
    D. Remove the two bolts from mount pad to transmission case, and remove rubber mount cushion and exhaust pipe yoke.
14. Disconnect transmission linkage by removing shift levers at transmission.
15. Disconnect speedometer cable at transmission.
16. Remove two bolts to disconnect the transmission gearshift control lever and bracket assembly from its adapter plate on side of transmission.
17. Lower transmission assembly from the car, letting the gearshift lever slide down through the dust boot in the console.
18. Remove transmission-to-clutch housing attaching bolts.
19. Remove transmission rearward from the clutch and rotate the assembly to gain access to the three flathead machine screws in the control lever bracket adapter plate. Rotate transmission back to upright position.
20. Slowly lower rear of engine until tachometer drive cable at distributor clears the ledge across front of dash.
21. Slide transmission rearward out of clutch, then tip front of transmission down and lower the assembly out of the car.
22. Install by reversing removal procedure.

Powerglide Transmission

Quick Service Information
When automatic transmission trouble is reported, a road test and careful diagnosis are in order. Transmission Removal and Replacement and Linkage Adjustments are covered in the following paragraphs. For Test Procedures, Transmission Overhaul and other detailed information, see Unit Repair Section.

Neutral Safety Switch Adjustment
Place hand lever in neutral and loosen the two switch-mounting screws. Move the switch bracket so that a snug fitting pin can be inserted through the two locating holes. Tighten the screws to hold the switch securely, and remove the pin.
NOTE: late models use two pins.

6 Cylinder and V8 Models
Follow adjustment of rod E in lever A on V8s, and rod G in bell crank A on six cylinder cars to arrive at dimensions shown in accompanying diagram.

Powerglide Removal
1. Raise car on hoist. Remove oil and oil pan, then reattach pan with a few bolts.
2. Disconnect oil cooler lines, vacuum modulator line and the speedometer cable at the transmission.
3. Remove crankcase ventilator tube clamp bracket bolt, washer and nut from the transmission.
4. Disconnect manual and TV control lever rods from transmission.
5. Disconnect propeller shaft at transmission.
6. Install suitable transmission, lift and support the transmission.
7. Disconnect engine rear mount on transmission extension, then remove transmission support crossmember.

Caution Note any shims that may be installed between the extension mounting boss and the crossmember. It is important that the same number of shims be reinstalled because these influence drive line angles.

8. Remove converter underpan, scribe flywheel-to-converter relationship for reassembly, then remove flywheel-to-converter attaching bolts.
9. Support engine at oil pan rail with a jack, or other suitable brace, for engine support safety.
10. Lower rear of transmission slightly so that the upper transmission housing-to-engine attaching bolts can be reached with a universal socket and long extension. Remove the upper bolts.

Caution On V8 engines, use care not to lower the transmission too far because distributor-firewall interference may cause damage. Have a helper watch this area.

11. Remove the balance of the transmission attaching bolts.
12. Remove the transmission by lowering and moving the unit toward the rear. Be sure to use a conver-

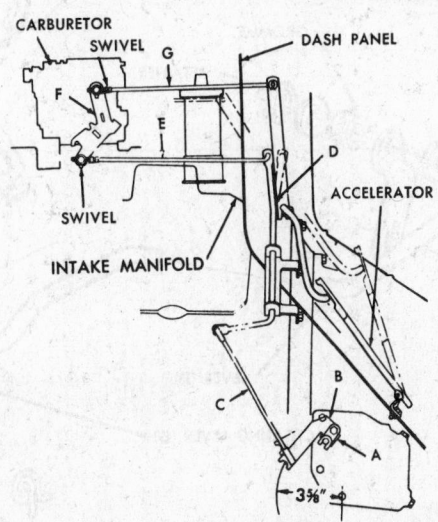

V8 throttle linkage; hold lever A counterclockwise. Measure distance from outer end of lever B to indicated bolt hole. It should be 3⅝".
(© Chevrolet Div. G.M. Corp.)

ter-holding-strap, or some improvisation, to keep converter from falling while removing the transmission unit.

13. Install transmission unit by reversing the above procedure.

Powerglide—Type B, Aluminum Case

All models, except those equipped with 409, 396 and 427 cu. in. engines, are available with Powerglide B (aluminum case) transmissions.

Turbo-Hydramatic

Manual Linkage Adjustment

Since 1965, Turbo-Hydramatic has been used with some V8 options.

1. Adjust as in illustrations.
2. When properly adjusted, the following condition must be met by manual operation of the shift selector.

Through the entire range of selector travel, the detent feel must be noted and in register with the correct position on the dial.

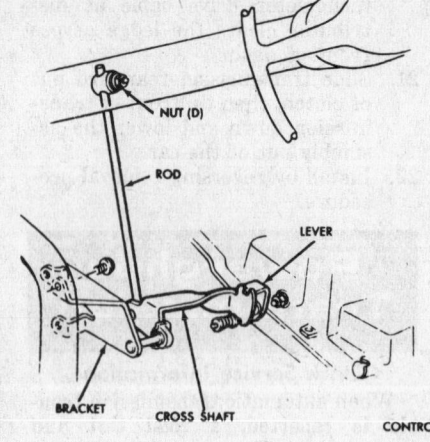

CONTROL ADJUSTMENT

1. Set transmission lever in "Drive" position.
2. Set shift tube & lever assby. in "Drive" position.
3. Tighten nut (D) to 10 ft. lbs.
4. Check shift pattern in all ranges. Readjust if necessary.

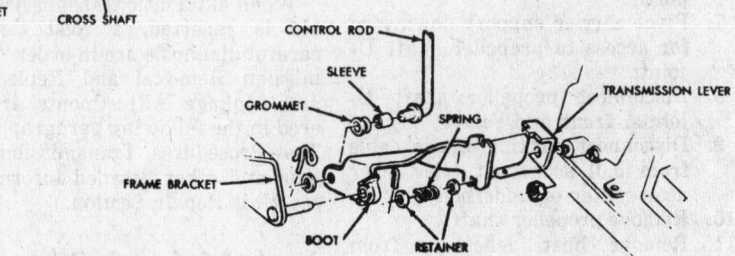

Turbo-hydramatic, column linkage adjustment (© Chevrolet Div. G.M. Corp.)

LINKAGE ADJUSTMENT

1. Set lever (D) & control lever (E) in "Drive" detent.

2. Apply Forward load (Y) on lever (G) to Fully seat lever (E) in "Drive".

3. Place a ⁷⁄₆₄" spacer (H) between nut (A) & Swivel (J), run nut (A) to spacer. Remove spacer & apply rearward load (X) until lever (G) touches nut (A). Tighten nut B.

4. Check shift pattern in all ranges, readjust if required.

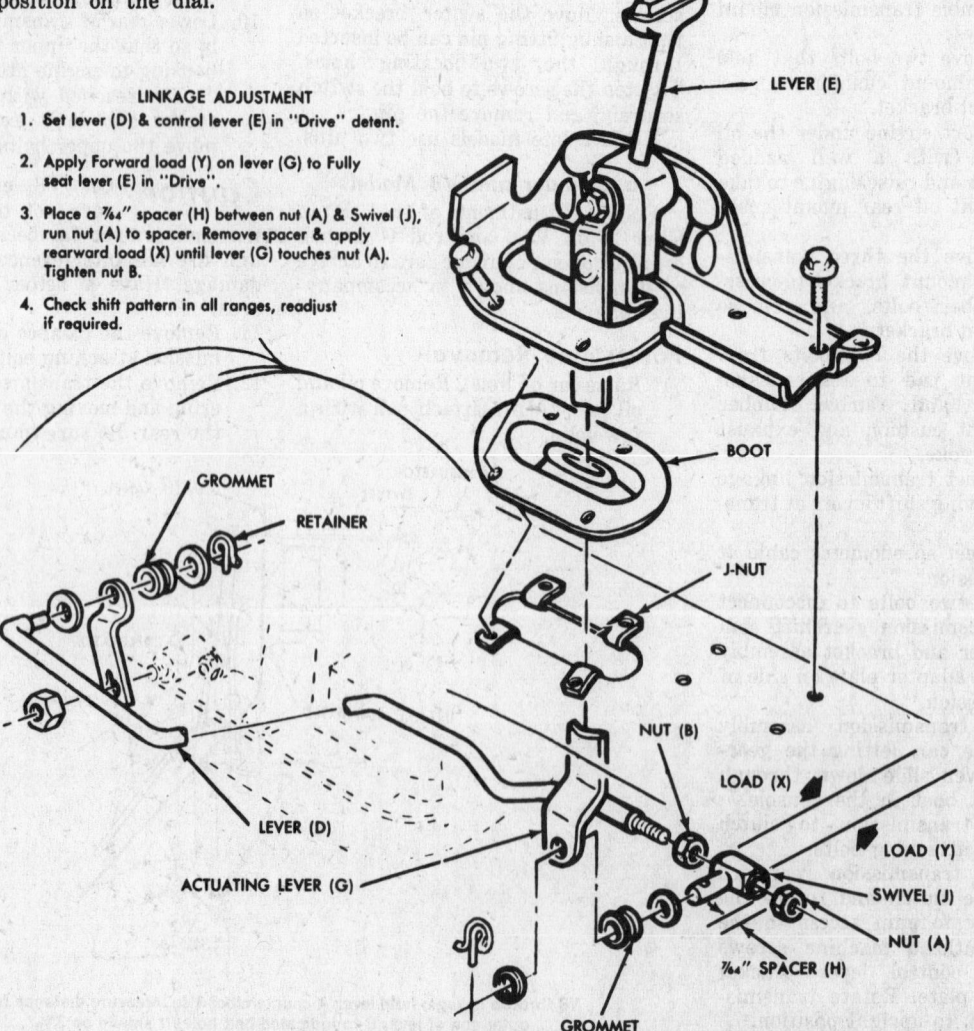

Turbo-hydramatic, console linkage adjustment (© Chevrolet Div. G.M. Corp.)

Transmission R & R

1. Disconnect battery, then raise the car.
2. Remove drive shaft.
3. Disconnect speedometer cable at transmission, electrical lead at transmission connector, vacuum line at modulator, and oil cooler pipes.
4. Disconnect shift control linkage.
5. Support transmission with suitable transmission jack.
6. Disconnect rear mount from frame crossmember.
7. Remove two bolts at each end of frame crossmember and remove crossmember.
8. Remove oil cooler lines, vacuum modulator line and detent solenoid connector wire at transmission.
9. Remove converter underpan.
10. Remove converter to flywheel bolts.
11. Loosen exhaust pipe to manifold bolts about ¼ in., then lower transmission until jack is just supporting it.
12. Remove transmission to engine mounting bolts and remove oil filler tube at transmission.
13. Raise transmission to its normal position, support engine with a jack and slide transmission rearward, then lower it away from the car.
   NOTE: use converter holding tool J-5384, or suitable substitute, to keep from losing converter from transmission while handling the assembly.
14. Install by reversing the removal procedure.

Drive Shaft, U Joints

Universal Joint Disassembly

1964—All Models

1. Three universal joints are used. To remove the shaft and/or joint assemblies, first remove the bolts attaching the center bearing support to the frame crossmember.
2. At the rear universal joint, split

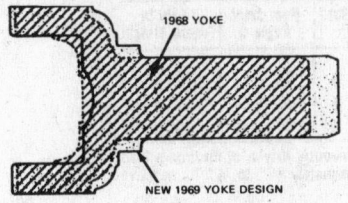

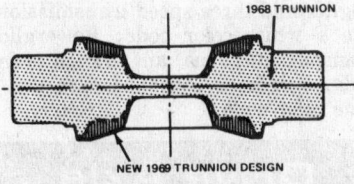

Universal joint yoke comparison, 1968-69

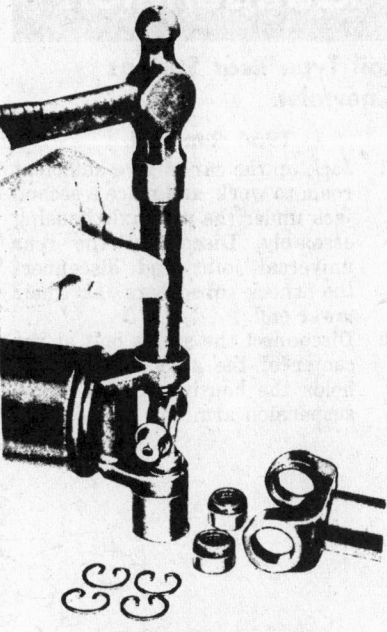

Disassembling universal joint
(© Chevrolet Div. G.M. Corp.)

the joint by removing the trunnion bearing U-clamps. Withdraw the drive shaft and bearing assembly by moving it rearward to the left under the axle housing assembly.
3. At the center universal joint, remove the bearing lock rings and drive the universal joint cross from one side to the other to

force out the bearing on the opposite side.
4. Repeat at all four bearings and then at the rear and front joint.
5. Reinstall by starting the driveshaft in up over the rear axle, down through the tunnel in the frame until the front universal joint spline slides up on the transmission output shaft.
6. Connect the rear universal joint by installing two U-bolts. Be sure the trunnions are properly seated in the rear axle drive flange. Torque should be 14 to 18 ft. lbs.
7. On all units, except those with air suspension, with car at normal height the center bearing mounting should be allowed to fall freely into place over the slotted holes in the frame crossmember and tightened at this position.
8. On air suspension models, with springs full of air the center mounting should be forced forward ⅛ to ¼ in. from its free position before being bolted tight.

Chevrolet, 1965-70

A one-piece driveshaft is used on these models. The front and rear joints are cardon-type with lube-for-life design. No periodic maintenance is required.

Corvette, 1965-69

Due to independent rear suspension, Corvette is equipped with a single, open, tubular drive shaft and two universal joints, one joint at either end.

Driveshaft Alignment

1964

The relative angles formed by the drive shaft sections with the transmission output shaft and the pinion gear shaft are critical.

Consequently, in cases of driveshaft vibration, collision rear control arm replacement and such, the angles should be checked.

On cars with coil springs the angles should be (in degrees) :

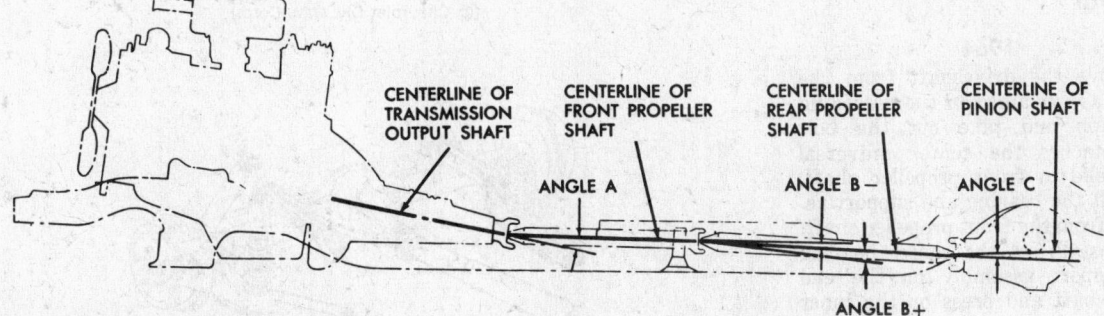

ANGLES POINTING DOWN ARE POSITIVE
ANGLES POINTING UP ARE NEGATIVE

Open driveline angles (© Chevrolet Div. G.M. Corp.)

	Front Joint Angle A	Center Joint Angle B*	Rear Joint Angle C	Axle to Frame Height
All Models Except SS and Sta. Wag.	2°40′ ± ½°	−1°29′ ± ½°	3°15′ ± ½°	6.52″ ± ¼°
Station Wagons	1°40′ ± ½°	−2°17′ ± ½°	3°15′ ± ½°	7.65″ ± ¼°
Super Sport	3°00′ ± ½°	−0°14′ ± ½°	3°15′ ± ½°	5.64″ ± ¼°

*For all vehicles which are normally and continuously driven at maximum load conditions, angle B may be increased approximately ¼° to ½° in negative direction for more satisfactory results.

Do not use a frame contact hoist. Support car by tires or axle. Support front of car 6 in. higher than the rear.

Before checking the angles, check the transmission support crossmember. For correct assembly, two conditions are important.

The taper in the top surface of the crossmember must be toward the rear of the vehicle.

For synchromesh and Turboglide transmissions the support must be attached to the frame by the upper holes in the frame brackets. For overdrive and Powerglide transmissions, the support must be attached by the two lower holes in the frame brackets.

Use a bubble protractor to measure the degrees of angle.

Angle A is obtained by subtracting the front propeller shaft angle from the engine angle.

Angle B is obtained by subtracting the rear propeller shaft angle from the front propeller shaft angle.

Angle C is obtained by subtracting the pinion shaft angle from the rear propeller shaft angle if the pinion shaft points up toward the engine. When the pinion shaft points forward toward the ground, the angle must be added to that of the rear propeller shaft to get angle C.

If any of the joint angles are not within the limits given, suitable shims should be added or removed at the transmission support and at the rear suspension upper control attaching point.

Driveshaft Vibration

Excessive vibration of the driveshaft may be due to improper driveline angle, worn U-bolts, bent shaft, pinion flange runout, balance weights missing from shafts.

Driveshaft Center Bearing Removal

1964

Remove the driveshaft from the vehicle as explained in the preceding paragraph and take out the bolt that attaches the center universal joint yoke to front propeller shaft. Press off the bearing and support assembly from the front propeller shaft.

To install, set the center bearing and support assembly on the end of the shaft and press on the inner race, pressing it on until the bearing bottoms on the machined shoulder on the propeller shaft. Depend-

ing on the transmission used, three front propeller shafts are used. Turboglide and three-speed transmissions use a white color code; Powerglide transmissions use an orange color code; overdrive equipped cars use a blue color code.

Drive Axle, Suspension

Coil Type Rear Springs (Chevrolet)

1964 Removal

1. Jack up the car to give sufficient room to work, and place a second jack under the rear axle housing assembly. Disconnect the rear universal joint and disconnect the shock absorbers at their lower end.
2. Disconnect the single bolt at the center of the axle housing that holds the housing to the center suspension arm.

3. Let the housing assembly come downward until the pressure is relieved from the coil spring, and lift off the coil spring.
4. Install in reverse order of renewal, being careful when installing the new spring that the end coil of the spring fits properly into the form plate both top and bottom.

1965-70 Removal

1. Raise rear of vehicle and place jack stands under frame. Support weight of vehicle at rear axle housing separately from above frame position by using either jacks of twin-post lift.
2. Remove both rear wheels.
3. With car supported as in Step 1, and springs compressed by weight of vehicle:
   A. Disconnect both rear shocks from anchor pin lower connection.
   B. Loosen the upper control arm(s) rear pivot bolt (do not remove the nut).
   C. Loosen both left and right lower control arm rear attachment (do not disconnect from axle brackets).
   D. Remove rear suspension tie rod from stud on axle tube.
4. Slightly loosen the nut on the bolt that retains the spring and seat to control arm at lower seat

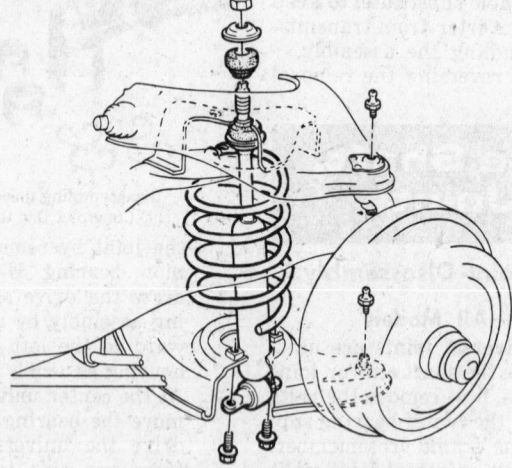

Installing shock absorber—typical
(© Chevrolet Div. G.M. Corp.)

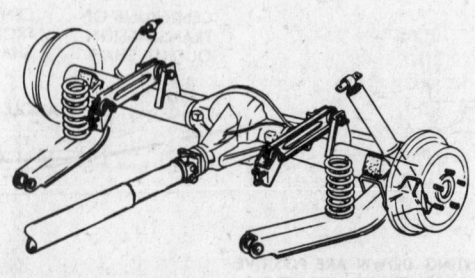

Chevrolet rear suspension (© Chevrolet Div. G.M. Corp.)

of both rear springs. When bolt has been backed off the maximum distance, all threads of the nut should still be engaged on the bolt.

Caution Under no condition should the nut, at this time, be removed from the bolt in the seat of either spring.

5. Slowly lower the rear axle assembly, allowing the axle to swing down, carrying the springs out of the upper seat. This provides access for spring removal.
6. Remove the lower seat attaching parts from each spring, then remove springs from vehicle.

1965-70 Installation

1. Position springs in upper seat and install lower seat parts on control arm. Install nut of spring retaining bolt finger-tight.
   NOTE: Omit lockwasher under the special high carbon bolt, so that sufficient threads will be available to start the nut. Lockwashers will be installed later.
2. Alternately raise the axle slightly and retighten the nut on each spring lower seat bolt. Continue in until the weight is fully supported on the jack or lift. With spring now completely compressed to approximate curb position, completely position the springs in the lower seats by torquing the nut on the lower seat bolt.
3. Reconnect shock absorbers, torque rear attachment of upper and lower control arms, and reconnect the axle tie-rod.
4. While still jacekd under axle, remove the nut from the lower seat bolt of one rear spring and install lockwasher and replace nut and tighten. Similarly install lockwasher at other spring.
5. Install rear wheels and lower car to floor.

Transverse Leaf Rear Spring (Corvette)

1964-69 R & R

1. Raise car and support it by the frame, slightly forward of torque control pivot points. Remove wheel assemblies.
2. Place floor jack under spring near link bolt, and raise spring until nearly flat.
3. Tie the end of the spring to the suspension crossmember to hold this flat attitude, with a ¼ in. or 5/16 in. chain and grab hook wrapped around the spring and crossmember. To prevent chain slipping, use a C-clamp on the spring adjacent to the chain.
4. Remove link bolt and rubber bushings.
5. Support and raise spring end, as before, and remove chain.

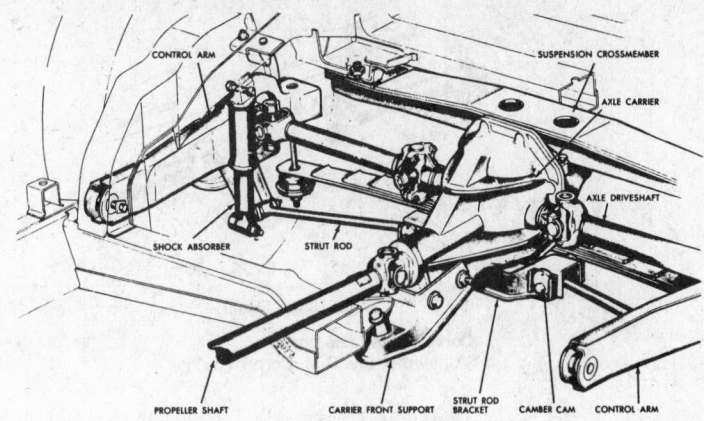

Corvette rear suspension (© Chevrolet Div. G.M. Corp.)

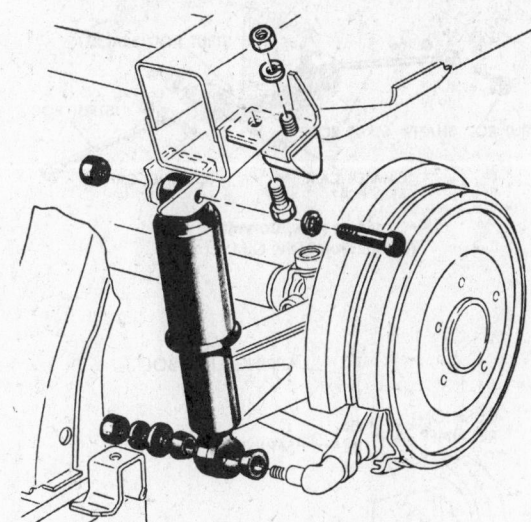

Shock absorber installation, Corvette (© Chevrolet Div. G.M. Corp.)

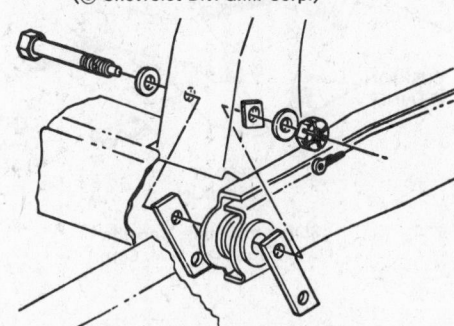

Toe-in adjustment shim location, Corvette

6. Carefully lower jack to completely relax spring.
7. Repeat foregoing procedure on the other side of car.
8. Remove four bolts and washers attaching the spring at the center.
9. Remove spring from car.
10. Install by reversing removal procedure.

Strut Rod and Bracket (Corvette)

Rod and Bracket—Removal

1. Raise car on a hoist.
2. Disconnect shock absorber lower eye from strut rod shaft.
3. Remove strut rod shaft cotter pin and nut. Withdraw shaft by pulling toward the front of the car.
4. Mark related position of camber adjustment, so that adjustment is maintained upon reassembly.
5. Loosen camber bolt and nut. Remove four bolts holding strut rod bracket to carrier and lower the bracket.
6. Remove cam bolt and cam bolt assembly. Pull strut down out of bracket and remove bushing caps.
7. Inspect strut rod bushings for

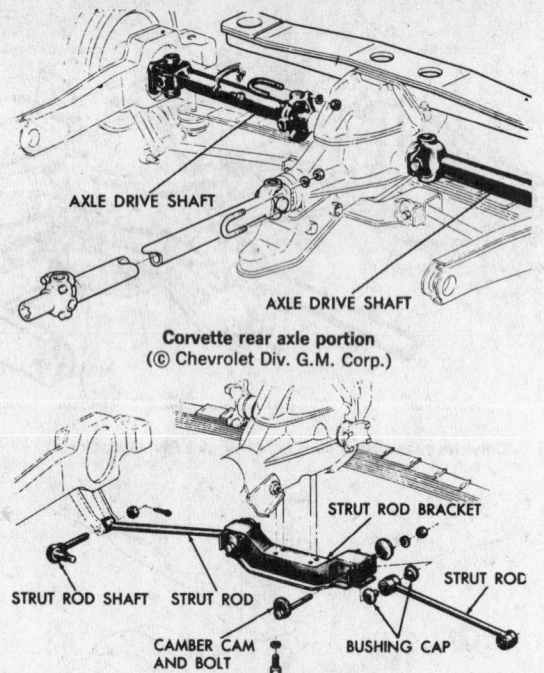

Corvette rear axle portion
(© Chevrolet Div. G.M. Corp.)

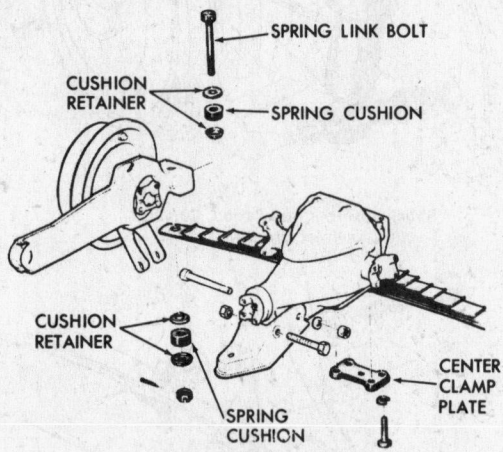

Strut rods, Corvette
(© Chevrolet Div. G.M. Corp.)

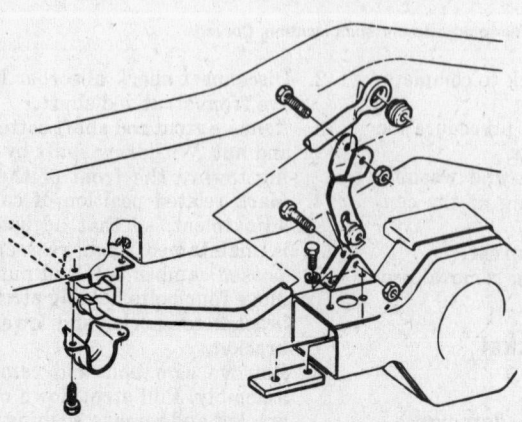

Spring mounting, Corvette
(© Chevrolet Div. G.M. Corp.)

wear and replace where necessary. Replace strut rod if it is bent or damaged in any way.

8. Install by reversing removal procedure.

9. Check rear wheel camber and adjust to specifications.

Torque Control Arm R & R (Corvette)

1. Disconnect spring on the side from which the torque arm is to be removed. Follow procedure for Springs R & R 1964-69, Corvette.

   NOTE: if so equipped, disconnect stabilizer rod from torque arm.

2. Remove shock absorber lower eye from strut rod shaft.

3. Disconnect and remove strut rod shaft and swing strut rod down.

4. Remove four bolts holding the axle driveshaft to spindle flange and disconnect drive shaft.

5. Disconnect brake line at wheel cylinder inlet or caliper and from torque arm. Disconnect parking brake cable.

6. Remove torque arm pivot bolt and toe-in shims, then pull torque arm out of frame. Tape shims together to assure relationship for reassembly.

7. To install, place torque arm in frame opening.

8. Position toe-in shims in original location on both sides of torque arm. Install pivot bolt and lightly tighten at this time.

9. Raise axle driveshaft into position and install to drive flange. Torque bolts to 75 ft. lbs.

10. Raise strut into position and insert strut rod shaft so that flat lines up with flat in spindle support fork. Install nut and torque to 80 ft. lbs.

11. Install shock absorber lower eye and tighten nut to 35 ft. lbs.

12. Connect spring end as outlined under Leaf Type Rear Springs—Corvette, 1964-69 R & R.

    NOTE: if car is so equipped, connect stabilizer shaft.

13. Install brake drum or disc and

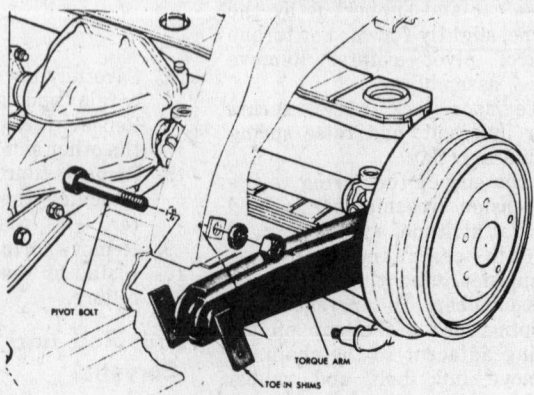

Stabilizer shaft installation, Corvette
(© Chevrolet Div. G.M. Corp.)

Torque control arm, Corvette
(© Chevrolet Div. G.M. Corp.)

caliper, and wheel. Then lower the car. Tighten torque pivot bolt to 50 ft. lbs.

14. Bleed brakes and check camber and toe-in.

Troubleshooting and Adjustment

General instructions covering the troubles of the rear axle and how to repair and adjust it, together with information on installation of rear axle bearings and grease seals, are given in the Unit Repair Section.

Capacities of the rear axle are given in the Capacities table.

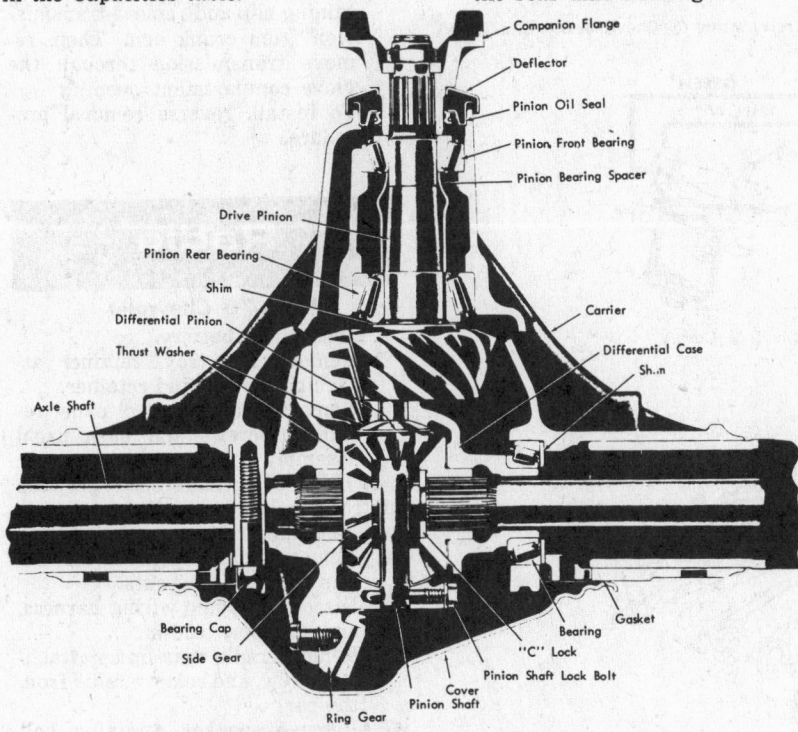

Chevrolet differential assembly, 1965-70 (© Chevrolet Div. G.M. Corp.)

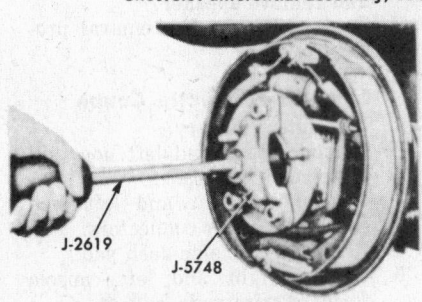

Removing rear axle shaft
(© Chevrolet Div. G.M. Corp.)

Rear Axle Assembly Removal

Chevrolet

1. Jack up the back of the car to allow sufficient room to work, then place another jack under the rear axle housing. Disconnect the driveshaft.
2. Remove the single bolt which holds the center support arm to the rear of the axle banjo housing.
3. Remove the wheel assemblies, disconnect the shocks at the lower end, and then slowly lower the rear axle housing.
4. Remove the bolt on each side which holds the torque arm to the rear axle housing, disconnect the hand brake cable and the hydraulic line and the T fitting over the rear axle housing. Then, slide the housing assembly out from underneath the vehicle.

Corvette, 1964-69

Corvette is equipped with an independent rear suspension. The differential is solidly attached to the car frame, the rear wheels being driven through tubular rear axles, each fitted with two universal joints. A transverse, multiple leaf rear spring provides rear suspension. Brake torque and driving forces are transmitted through radius arms to the frame. The spring supports vertical loads, while lateral forces, on turns etc., are taken by the axles and control rods to the fixed differential and to the frame.

1. Raise car on hoist.
2. Disconnect inboard driveshaft trunnion from side gear yoke.
3. Bend bolt lock tabs down.
4. Remove bolts that secure shaft flange to spindle drive flange.
5. Pry driveshaft out of outboard drive flange pilot and remove by withdrawing outboard end first.

Positraction Differential

No special attention is required in this area, except with the lubricant used.

Under no circumstances use anything but special G.M. 3758790 or 3758791 lubricant.

Failure to follow these instructions may result in permanent damage to the unit.

Windshield Wipers

Motor R & R

These models may be equipped with vacuum or electric motors. Power transmission may be through cable, or through levers and links.

Chevrolet, 1964-70

1. Make sure the battery is disconnected and the wiper motor is in parked position.
2. Remove washer hoses, if present, and all electrical connectors.
3. Remove plenum chamber ventilator grille.
4. Disconnect transmission drive linkage from wiper motor crank arm.
5. Remove motor retaining bolts, then remove the motor.
6. Install motor by reversing the removal procedure.

NOTE: make sure the wiper motor is properly grounded.

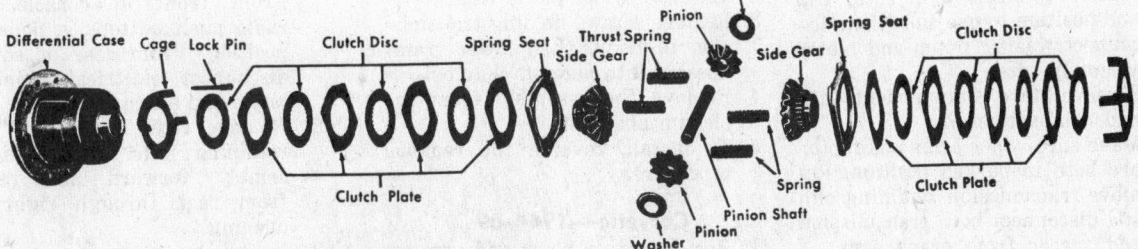

Positraction differential (© Chevrolet Div. G.M. Corp.)

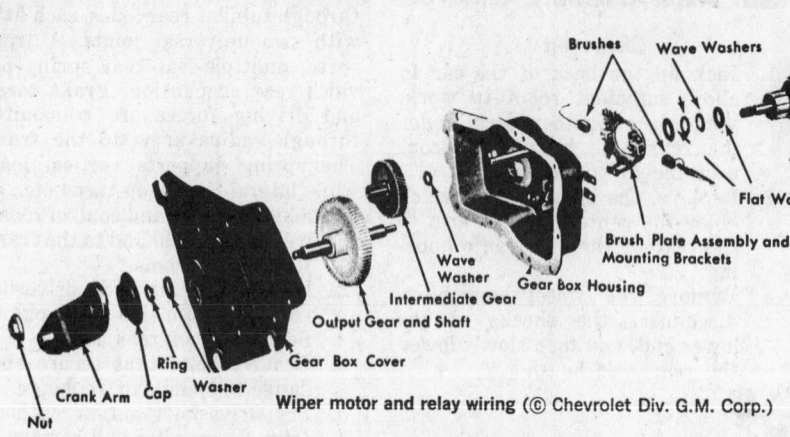

Wiper motor and relay wiring (© Chevrolet Div. G.M. Corp.)

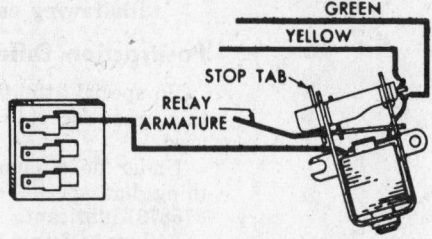

Relay wiring (© Chevrolet Div. G.M. Corp.)

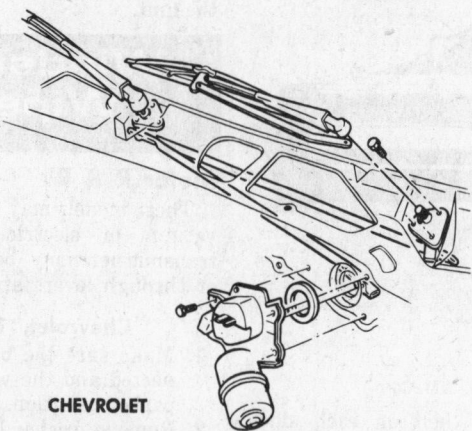

CHEVROLET

Wiper assembly installation

Corvette, 1964-69

1. Disconnect battery.
2. Remove ignition distributor shielding and left bank plug wire vertical shield.
3. Disconnect left bank plug wire bracket-to-manifold, position assembly to one side.
4. Disconnect ignition resistor at firewall, then remove washer pump inlet and outlet hose at pump valve assembly.
5. Remove ignition distributor cap and position to one side, then disconnect washer pump and motor assembly wires.
6. Remove glove compartment door and compartment.
7. Make sure wiper arms and motor are both in parked position. Remove transmission retaining clip and disconnect both transmission and spacer from crank arm.
8. Remove four wiper motor-to-dash

wall mounting bolts and remove wiper motor from the car.
9. Install motor by reversing removal procedure.

Transmission R & R

Chevrolet, 1964-70

1. With the wiper motor in parked position, remove shroud top ventilator grille.
2. Detach transmission drive linkage from wiper motor assembly.
3. Remove screws holding transmission to body, then lower transmission into plenum chamber.
4. Remove the assembly from the plenum chamber.
5. To install, reverse the removal procedure.

Corvette—1964-69

1. Remove wiper block and arm assembly from transmission.

2. Remove glove compartment door and compartment.
3. Remove three transmission-to-cowl retaining screws.
4. Remove wiper transmission retaining clip and remove transmission from crank arm. Then, remove transmission through the glove compartment opening.
5. To install, reverse removal procedure.

Radio

1966-70 Chevrolet

1. Disconnect battery.
2. Remove ash tray, retainer attaching screws and retainer.
3. Remove heater control panel retaining screws and push panel assembly from console.
NOTE: if interference between control panel and radio is met, loosen radio retaining nuts.
4. Remove radio control knobs, bezels and retaining nuts.
5. Disconnect radio wiring harness, and antenna lead-in.
6. Remove radio rear brace attaching screw, and remove radio from the car.
7. Remove speaker retaining bolt and remove speaker.
8. To install, reverse removal procedure.

1966-69—Corvette Coupe

1. Disconnect battery.
2. Remove right and left door sill plates and kick pads.
3. Disconnect right and left side radio-to-speaker connectors.
4. Remove right side dash pad.
5. Remove right and left console forward trim pads.
6. Remove bolt and remove the heater floor outlet duct by pulling it through left hand opening.
7. From front of console, tape radio push buttons in depressed position. From rear of console, disconnect electrical connector, brace and antenna lead-in.
8. Remove radio knobs and bezel retaining nuts. Push radio assembly forward and remove from rear through right side opening.
9. Install by reversing procedure above.

1966-69 Corvette Convertible

1. Disconnect batery.
2. Remove right instrument panel pad.
3. Disconnect speaker connectors.
4. Remove wiper switch trim plate screws to gain access to switch connector and remove connector and trim plate from cluster assembly.
5. Unclip and remove right and left console forward trim pads and remove forwardmost screw on right and left side of console.
6. Inserting a flexible drive socket between the console and metal horseshoe brace, remove the nuts from the two studs on the lower edge of the console cluster. Remove the remaining screws that retain the cluster assembly to the instrument panel.
7. From rear of console, disconnect electric connector, brace and antenna lead-in.
8. Remove radio knobs and bezel retaining nuts.
9. Pull radio assembly forward and remove through right side opening.
10. Install by reversing procedure.

Heater System

Heater Blower R & R

Chevrolet—1964-70

1. Disconnect battery.
2. Unclip hoses from fender skirt.
3. Disconnect electrical feed from motor.
4. Turn vehicle front wheels to extreme right.
5. Remove right front fender skirt bolts and allow skirt to drop, resting it on top of tire. It may be wedged away from fender lower flange with block of wood to provide better access to bolts.
6. Remove screws attaching motor mounting plate to air inlet housing.
7. Remove screws attaching motor to mounting plate.
8. Remove clip attaching cage to shaft and remove blower motor.
9. Install in reverse of above.

Corvette—1964-69

1. Remove the radiator supply tank from its retaining straps. Move it out of the way. Disconnect the battery.
2. Remove blower motor electrical connectors.
3. Scribe a reference mark on the blower motor mounting plate and the blower motor.
4. Remove the five screws that mount the blower mounting plate to the blower inlet assembly.
5. Withdraw the blower assembly from the inlet assembly.

Piston Fitting Specifications

Displacement (cu. in.)	Advertised Brake H.P.	Piston clearance (Prod.) (in.)	Piston clearance (Service Max.) (in.)	Compression Ring Groove clearance (in.)			Compression Ring Gap (in.)			Oil Ring Groove clearance (in.)		Oil Ring Gap (in.)	
				Top (Prod.)	Second (Prod.)	Service (high prod. limit)	Top (Prod.)	Second (Prod.)	Service (high prod. limit)	Prod.	Service	Prod.	Service (high prod. limit)
153	90	.0005-.0015	.0025	.0012-.0027	.0012-.0032	+.001	.010-.020	.010-.020	+.01	.000-.005	+.001	.015-.055	+.01
230	140	.0005-.0015	.0025	.0012-.0027	.0012-.0032	+.001	.010-.020	.010-.020	+.01	.000-.005	+.001	.015-.055	+.01
250	155	.0005-.0015	.0025	.0012-.0027	.0012-.0032	+.001	.010-.020	.010-.020	+.01	.000-.005	+.001	.015-.055	+.01
283, 307, 327	All	.0005-.0011	.0025	.0012-.0027	.0012-.0032	+.001	.010	.020	+.01	.000-.005	+.001	.015-.055	+.01
350	250	.0007-.0013	.0027	.0012	.0032	+.001	.010-.020	.013-.025	+.01	.002-.007	+.001	.015-.055	+.01
	300	.0007-.0013	.0027	.0012-.0032	.0012-.0027	+.001	.010-.020	.013-.025	+.01	.000-.005	+.001	.015-.055	+.01
	350	.0020-.0026	.0036	.0012-.0032	.0012-.0027	+.001	.010-.020	.013-.023	+.01	.000-.005	+.001	.015-.055	+.01
	370	.0036-.0042	.0061	.0012-.0032	.0012-.0027	+.001	.010-.020	.013-.023	+.01	.000-.005	+.001	.015-.055	+.01
396	350	.0018-.0026	.0038	.0017-.0032	.0017-.0032	+.001	.010-.020	.010-.020	+.01	.0005-.0065	+.001	.010-.030	+.01
	375	.0036-.0046	.0065	.0017-.0032	.0017-.0032	+.001	.010-.020	.010-.020	+.01	.0005-.0065	+.001	.010-.030	+.01
400	265	.0014-.0020	.0034	.0012-.0027	.0012-.0032	+.001	.010-.020	.010-.020	+.01	.000-.005	+.001	.015-.055	+.01
	330	.0018-.0026	.0038	.0017	.0032	+.001	.010-.020	.010-.020	+.01	.0005-.0065	+.001	.015-.055	+.01
409	340	.0009-.0015	.0045	.0017-.0032	.0017-.0032	+.001	.010-.020	.010-.020	+.01	.0012-.0050	+.001	.015-.055	+.01
	400	.0035-.0042	.0045	.0017-.0032	.0017-.0032	+.001	.010-.020	.010-.020	+.01	.0012-.0050	+.001	.015-.055	+.01
	425	.0035-.0042	.0045	.0017-.0032	.0017-.0032	+.001	.010-.020	.010-.020	+.01	.0012-.0050	+.001	.015-.055	+.01
427	390	.0024-.0034	.0045	.0017-.0032	.0017-.0032	+.001	.010-.020	.010-.020	+.01	.0005-.0065	+.001	.010-.030	+.01
	400	.0024-.0034	.0045	.0017-.0032	.0017-.0032	+.001	.010-.020	.010-.020	+.01	.0005-.0065	+.001	.010-.030	+.01
	430	.0058-.0068	.0080	.0017-.0032	.0017-.0032	+.001	.010-.020	.010-.020	+.01	.0005-.0065	+.001	.010-.030	+.01
	435	.0040-.0050	.0065	.0017-.0032	.0017-.0032	+.001	.010-.020	.010-.020	+.01	.0005-.0065	+.001	.010-.030	+.01
454	345	.0024-.0034	.0049	.0017-.0032	.0017-.0032	+.001	.010-.020	.010-.020	+.01	.0005-.0065	+.001	.015-.055	+.01
	360	.0024-.0034	.0049	.0017-.0032	.0017-.0032	+.001	.010-.020	.010-.020	+.01	.0005-.0065	+.001	.015-.055	+.01
	390	.0024-.0034	.0049	.0017-.0032	.0017-.0032	+.001	.010-.020	.010-.020	+.01	.0005-.0065	+.001	.015-.055	+.01
	450	.0040-.0050	.0065	.0017-.0032	.0017-.0032	+.001	.010-.020	.010-.020	+.01	.0005-.0065	+.001	.015-.055	+.01

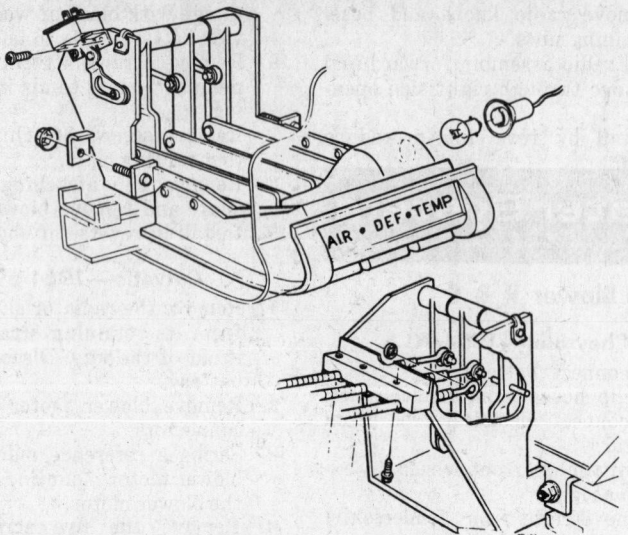

Heater control assembly (© Chevrolet Div. G.M. Corp.)

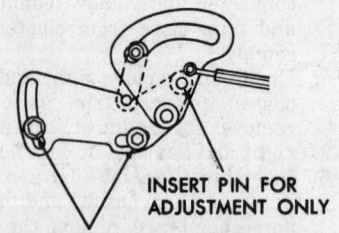

INSERT PIN FOR
ADJUSTMENT ONLY

LOOSEN BOLTS—APPLY APPROX. 3 LBS.
FORCE ON ENTIRE CAM ASSEMBLY
IN A CLOCKWISE DIRECTION
AND HOLD WHILE TIGHTENING BOLTS

Temperature door cam adjustment
(© Chevrolet Div. G.M. Corp.)

6. Install in reverse of removal procedure.

Heater Core R & R

1. Drain radiator.
2. Remove heater hoses at connections beside air inlet assembly.
3. Remove cable and electrical connectors from heater and defroster assembly.
4. On engine side of dash, remove screws and nuts holding air inlet to dash panel.
5. Inside vehicle, pull entire assembly from firewall and remove assembly from vehicle.
6. Remove core assembly retaining springs and remove core.
7. Install in reverse of above.

Index

YEAR IDENTIFICATION

CHRYSLER

1964 Newport

1964 300 & 300K

1964 New Yorker

1965 Newport

1965 "300"

1965 New Yorker

1966 Newport

1966 "300"

1966 New Yorker

1967 Newport

1967 "300"

1967 New Yorker

1968 Newport

1968 "300"

1968 New Yorker

1969 Newport

1969 "300"

1969 New Yorker

1970 Newport Custom

1970 300

1970 New Yorker

1971 Newport

IMPERIAL

1964

1965

1966

1967

1968

1969

1970 Imperial LeBaron

1971 Le Baron

CAR SERIAL NUMBER LOCATION

1964

The vehicle number is located on a metal plate attached to the left front door hinge pillar.

All vehicle numbers contain ten digits. They are interpreted as follows:

First: Make of vehicle.
Second: Model of vehicle.
Third: Year built.
Fourth: Assembly plant. (3 = Detroit; 6 = Delaware)
Last six: Sequence production number.

The starting serial numbers are as follows:

Chrysler Models
VC-1 Newport 81-43100001
VC-2 300 82-43100001
VC-3 New Yorker ..83-43100001
VC-1 Newport
  (T&C) 85-43100001
VC-3 New Yorker
  (T&C) 87-43100001
VC-3 New Yorker
  Salon 88-43100001
VC-4 300J ....—43100001

Imperial Models
VY-1 Crown 92-43100001
VY-1 LeBaron 93-43100001

1965

Vehicle number location and code same as 1964.

Chrysler Models
AC-1 Newport
  ....C1-5 (3 or 6) 100001
AC-2 300 ..C2-5 (3 or 6) 100001
AC-3 New Yorker
  ....C3-5 (3 or 6) 100001
AC-2 300L C4-5 (3 or 6) 100001
AC-3 New Yorker (sta. wgn.)
  ....C7-5 (3 or 6) 100001
AC-1 Newport (sta. wgn.)
  ....C5-5 (3 or 6) 100001
Police
  ....C9-5 (3 or 6) 100001

FIRING ORDER

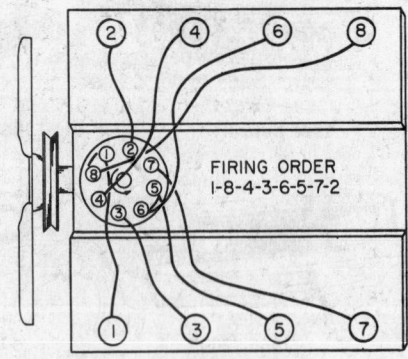

FIRING ORDER
1-8-4-3-6-5-7-2

1964-71—all V8

Imperial Models
AY-2 Crown Y2-53100001
AY-3 LeBaron Y3-53100001

1966

Vehicle number location same as 1964. All vehicle numbers now con-

GENERAL ENGINE SPECIFICATIONS

YEAR	CU. IN. DISPLACEMENT	CARBURETOR	DEVELOPED HORSEPOWER @ RPM	DEVELOPED TORQUE @ RPM (FT. LBS.)	A.M.A. HORSEPOWER	BORE AND STROKE (IN.)	ADVERTIZED COMPRESSION RATIO	VALVE LIFTER TYPE	NORMAL OIL PRESSURE (PSI)
1964	V8—361	2-BBL.	265 @ 4400	380 @ 2400	54.5	4.125 x 3.375	9.10-1	Hyd.	60
	V8—383	2-BBL.	305 @ 4600	410 @ 2400	57.8	4.250 x 3.375	10.10-1	Hyd.	60
	V8—413	4-BBL.	340 @ 4600	470 @ 2800	56.1	4.188 x 3.750	10.10-1	Hyd.	60
	V8—413	2-4-BBL.	390 @ 4800	485 @ 3600	56.1	4.188 x 3.750	9.60-1	Mech.	60
1965	V8—383	2-BBL.	270 @ 4400	390 @ 2800	57.8	4.250 x 3.375	9.2-1	Hyd.	60
	V8—383	4-BBL.	330 @ 4600	425 @ 2800	57.8	4.250 x 3.375	10.0-1	Hyd.	60
	V8—413	4-BBL.	340 @ 4600	470 @ 2800	56.1	4.188 x 3.750	10.1-1	Hyd.	60
	V8—413	2-4-BBL.	360 @ 4800	470 @ 3200	56.1	4.188 x 3.750	10.1-1	Mech.	60
1966	V8—383	2-BBL.	270 @ 4400	390 @ 2800	57.8	4.250 x 3.375	9.2-1	Hyd.	55
	V8—383	4-BBL.	325 @ 4800	425 @ 2800	57.8	4.250 x 3.375	10.0-1	Hyd.	55
	V8—440	4-BBL.	350 @ 4400	480 @ 2800	59.7	4.320 x 3.750	10.0-1	Hyd.	55
1967	V8—383	2-BBL.	270 @ 4400	390 @ 2800	57.8	4.250 x 3.375	9.2-1	Hyd.	55
	V8—383	4-BBL.	325 @ 4800	425 @ 2800	57.8	4.250 x 3.375	10.0-1	Hyd.	55
	V8—440	4-BBL.	350 @ 4400	480 @ 2800	59.7	4.320 x 3.750	10.0-1	Hyd.	55
	V8—440	4-BBL.	375 @ 4600	480 @ 3200	59.7	4.320 x 3.750	10.0-1	Hyd.	55
1968	V8—383	2-BBL.	290 @ 4400	380 @ 2400	57.8	4.250 x 3.375	9.2-1	Hyd.	55
	V8—383	4-BBL.	330 @ 5000	425 @ 3200	57.8	4.250 x 3.375	10.0-1	Hyd.	55
	V8—440	4-BBL.	350 @ 4400	480 @ 2800	59.7	4.320 x 3.750	10.1-1	Hyd.	55
	V8—440	4-BBL.	375 @ 4600	480 @ 3200	59.7	4.320 x 3.750	10.1-1	Hyd.	55
1969	V8—383	2-BBL.	290 @ 4400	380 @ 2400	57.8	4.250 x 3.375	9.2-1	Hyd.	55
	V8—383	4-BBL.	330 @ 5000	425 @ 3200	57.8	4.250 x 3.375	10.0-1	Hyd.	55
	V8—440	4-BBL.	350 @ 4400	480 @ 2800	59.7	4.320 x 3.375	10.1-1	Hyd.	55
	V8—440	4-BBL.	375 @ 4600	480 @ 3200	59.7	4.320 x 3.375	10.1-1	Hyd.	55
1970	V8—383	2-BBL.	290 @ 4400	390 @ 2800	57.8	4.250 x 3.375	8.7	Hyd.	45-65
	V8—383	4-BBL.	330 @ 5000	425 @ 3200	57.8	4.250 x 3.375	9.5	Hyd.	45-65
	V8—440	4-BBL.	350 @ 4400	480 @ 2800	59.7	4.320 x 3.375	9.7	Hyd.	45-65
	V8—440	4-BBL.	375 @ 4600	480 @ 3200	59.7	4.320 x 3.375	9.7	Hyd.	45-65
1971	V8—383	2-BBL.	290 @ 4400	390 @ 2800	57.8	4.250 x 3.375	8.7-1	Hyd.	45-65
	V8—383	4-BBL.	330 @ 5000	425 @ 3200	57.8	4.250 x 3.375	9.5-1	Hyd.	45-65
	V8—440	4-BBL.	350 @ 4400	480 @ 2800	59.7	4.320 x 3.750	9.7-1	Hyd.	45-65
	V8—440	4-BBL.	375 @ 4600	480 @ 3200	59.7	4.320 x 3.750	N.A.	Hyd.	45-65

tain thirteen digits. They are interpreted as follows:

First: Car line (make). (C = Chrysler; Y = Imperial)
Second: Price class. (see note 1)
Third: Body type. (see note 2)
Fourth: Body type. (see note 2)
Fifth: Engine displacement. (see note 3)
Sixth: Model year.
Seventh: Assembly plant. (3 = Jefferson, 6 = Newark, 8 = Export)
Last six: Production sequence number.
Note 1: E = economy; L = low; M = medium; H = high; P = premium.
Note 2: 23 = 2-dr. hardtop; 27 = convt.; 41 = 4-dr. sdn.; 42 = 4-dr. town sdn.; 43 = 4-dr. hardtop; 45 = 2-seat sta. wgn.; 46 = 3-seat sta. wgn.
Note 3: F = 383 cu. in.; G = 413 cu. in.; J = 440 cu. in.

Chrysler Models

Newport BC-1
300 ....BC-2
New Yorker BC-3

Imperial Models

Crown BY-3
LeBaron BY-3

1967

Vehicle number location same as 1964. Vehicle number interpretation code same as 1966, with the following exceptions:
(1) Body type 42 is deleted.
(2) G = 383 cu. in.; J = 440 cu. in.; K = Spec. Ord. 8.

1968

Vehicle number is located on a plate on the left side of the instrument panel, visible through the windshield. Vehicle number interpretation code same as 1966, with the following exceptions:
(1) Body type 42 is deleted.
(2) Premium price class is deleted.
(3) G = 383 cu. in.; H = 383 High perf.; K = 440 cu. in.; L = 440 cu. in. high perf.; M = Spec. Ord. 8.
(4) C = Jefferson assembly

plant; F = Newark assembly plant.

Chrysler Models

Newport DC-1
Newport Custom DC-1
300 ....DC-2
New Yorker DC-3

Imperial Models

Crown DY-1
LeBaron DY-1

1969

Vehicle number location same as 1968. Vehicle number interpretation code same as 1968.

Chrysler Models

Newport
Newport Custom
300
New Yorker

Imperial Models

Imperial

Engine Identification

On 1964-67 engines the engine number is stamped on a boss on the top of the engine block right in back of the water pump. On 1968 and later engines, the engine number is stamped on the engine block oil pan rail at the left rear corner below the starter opening. Engines are as follows:

1964

Newport, VC-1 361
Chrysler 300, VC-2 383
Chrysler 300, VC-2 413
New Yorker, VC-3 413
Chrysler Imp., VY-1 413
Chry. 300K, VC-2-300K 413

1965

Newport, AC-1 383
Chrysler 300, AC-2 383
Chrysler 300L, AC-2 413
New Yorker, AC-3 413
Imperial, AY-1 413

1966

Newport, BC-1 383
Chrysler 300, BC-2 383
New Yorker, BC-3 440
Imperial, BY-1 440

1967

Newport, CC-1 383
Newport Custom, CC-1 383
300, CC-2 440
New Yorker, CC-3 440
Imperial Crown, CY-1 440
Imperial LeBaron, CY-1 440

1968

Newport, DC-1 ....383, 440
Newport Custom, DC-1 .383, 440
300, DC-2 440
New Yorker, DC-3 440
Imperial Crown, DY-1 440
Imperial LeBaron, DY-1 ...440

1969

Newport, LB
 383 cu. in. (2-BBL., std. cam)
Newport, LB
 383 cu. in. (4-BBL., std. cam)
Newport, RB
 440 cu. in. (4-BBL., spl. cam)
300, LB
 440 cu. in. (4-BBL., std. cam)
300, RB
 440 cu. in. (4-BBL., spl. cam)
New Yorker, RB
 440 cu. in. (4-BBL., std. cam)
New Yorker, RB
 440 cu. in. (4-BBL., spl. cam)
Imperial, RB
 440 cu. in. (4-BBL., std. cam)

1970

Newport Town & Country
 Std. LB, 383, 2-BBL, std. cam, single exhaust
 Opt. LB, 383, 4-BBL., std. cam, dual exhaust
 Opt. RB, 440, 4-BBL., spec. cam, dual exhaust
300
 Std. LB, 440, 4-BBL., std. cam, single exhaust
 Opt. RB, 440, 4-BBL., spec. cam, dual exhaust
New Yorker
 Std. RB, 440, 4-BBL., std. cam, single exhaust
 Opt. RB, 440, 4-BBL., spec. cam, dual exhaust
Imperial
 RB, 440, 4-BBL., std. cam. single exhaust

AC GENERATOR AND REGULATOR SPECIFICATIONS

YEAR	MODEL	ALTERNATOR			REGULATOR			
		Field Current Draw @ 12 V.	Current Output @ 1250 Engine RPM	Model	Point Gap (in.)	Air Gap (In.)	Voltage at 140°F.	
1964–67	Standard	2.3–2.7	35 Amp.	2098300	.015	.050	13.4–14.0	
	Heavy Duty	2.3–2.7	40 Amp.	2098300	.015	.050	13.4–14.0	
1968–69	Standard	2.3–2.7	37 Amp.	2098300	.015	.050	13.4–14.0	
	W/Air Cond.	2.3–2.7	46 Amp.	2098300	.015	.050	13.4–14.0	
1970–71	Standard	2.38–2.75	34.5 ± 3 Amp.	3438150	.015	.050	13.3–14.0	
	Heavy Duty and/or Air. Cond.	2.38–2.75	44.5 ± 3 Amp.	3438150	.015	.050	13.3–14.0	

TUNE-UP SPECIFICATIONS

YEAR	MODEL	SPARK PLUGS Type	Gap (In.)	DISTRIBUTOR Point Dwell (Deg.)	Point Gap (In.)	IGNITION TIMING (Deg.) ▲	CRANKING COMP. PRESSURE (Psi)	VALVES Tappet (Hot) Clearance (In.) Intake	Exhaust	Intake Opens (Deg.)	FUEL PUMP PRESSURE (Psi)	IDLE SPEED (Rpm) ★
1964	V8—361 Cu. In.	J12Y	.035	30°	.017	10B	140	Zero	Zero	13B	4½	500
	V8—383 Cu. In.	J12Y	.035	30°	.017	10B	150	Zero	Zero	18B	4½	500
	V8—413 Cu. In.	J12Y	.035	30°	.017	10B	150	Zero	Zero	18B	4½	500
	V8—413 Cu. In.; Hi. Perf.	XJ10Y	.035	①	.017	12½B	150	.015	.024	24B	4½	500
1965	V8—383 Cu. In.; Std.	J14Y	.035	30°	.017	10B	140	Zero	Zero	13B	4½	500
	V8—383 Cu. In.; Hi. Perf.	J10Y	.035	30°	.017	10B	150	Zero	Zero	24B	4½	500
	V8—413 Cu. In.	J12Y	.035	30°	.017	10B	150	Zero	Zero	24B	4½	500
	V8—413 Cu. In.; Hi. Perf.	J10Y	.035	①	.017	10B	150	.015	.024	25B	4½	500
1966	V8—383 Cu. In.; Std., 2-BBL.	J14Y	.035	30°	.017	12½B ⑥	140	Zero	Zero	13B	4½	500②
	V8—383 Cu. In.; 4-BBL.	J13Y	.035	30°	.017	12½B ⑥	150	Zero	Zero	18B	4½	500②
	V8—440 Cu. In.; 4-BBL.	J13Y	.035	30°	.017	12½B ⑥	150	Zero	Zero	18B	4½	500②
1967	V8—383 Cu. In.; Std. 2-BBL.	J14Y	.035	30°	.017	12½B ⑥	145	Zero	Zero	16B	4½	500②
	V8—383 Cu. In.; 4-BBL.	J13Y	.035	30°	.017	12½B ⑥	145	Zero	Zero	16B	4½	500②
	V8—440 Cu. In.; 4-BBL.	J13Y	.035	30°	.017	12½B ⑥	145	Zero	Zero	18B	4½	500②
	V8—440 Cu. In.; Hi. Perf.	J11Y	.035	30°	.017	12½B ⑥	145	Zero	Zero	19B	4½	500②
1968	V8—383 Cu. In.; 2-BBL., M.T.	J14Y	.035	30°	.017	TDC	145	Zero	Zero	18B	4½	650
	V8—383 Cu. In.; 2-BBL., A.T.	J14Y	.035	30°	.017	7½B	145	Zero	Zero	18B	4½	650
	V8—383 Cu. In.; 4-BBL. M.T.	J11Y	.035	30°	.017	5B	145	Zero	Zero	18B	4½	650
	V8—383 Cu. In.; 4-BBL., A.T.	J11Y	.035	30°	.017	TDC	145	Zero	Zero	18B	4½	650
	V8—440 Cu. In., M.T.	J13Y	.035	30°	.017	TDC	145	Zero	Zero	18B	4½	650
	V8—440 Cu. In., A.T.	J13Y	.035	30°	.017	7½B	145	Zero	Zero	18B	4½	600
	V8—440 Cu. In.; Hi. Perf.	J11Y	.035	30°	.017	5B	145	Zero	Zero	21B	4½	650
1969	V8—383 Cu. In.; 2-BBL., M.T.	J14Y	.035	32°	.017	TDC	145	Zero	Zero	18B	4½	650
	V8—383 Cu. In.; 2-BBL., A.T.	J14Y	.035	32°	.017	7½B	145	Zero	Zero	18B	4½	650
	V8—383 Cu. In.; 4-BBL., A.T.	J11Y	.035	32°	.017	5B	145	Zero	Zero	18B	4½	650
	V8—440 Cu. In.	J13Y	.035	32°	.017	7½B	145	Zero	Zero	18B	4½	650
	V8—440 Cu. In.; Hi. Perf.	J11Y	.035	32°	.017	5B	145	Zero	Zero	21B	4½	650
1970	V8—383 Cu. In.; 2-BBL., M.T.	J14Y	.035	28.5-32.5⑤	.016-.021	10B⑥	100③	Zero	Zero	18B	3½-5	750
	V8—383 Cu. In.; 2-BBL., A.T.	J14Y	.035	28.5-32.5⑤	.016-.021	12½B⑥	100③	Zero	Zero	18B	3½-5	650
	V8—383 Cu. In.; 4-BBL., A.T.	J11Y	.035	28.5-32.5⑤	.016-.021	12½B④⑥	110③	Zero	Zero	18B	3½-5	700
	V8—440 Cu. In.	J13Y	.035	28.5-32.5⑤	.016-.021	12½B④⑥	110③	Zero	Zero	18B	3½-5	650
	V8—440 Cu. In.; Hi. Perf.	J11Y	.035	28.5-32.5⑤	.016-.021	12½B④⑥	110③	Zero	Zero	21B	3½-5	800
1971	V8—383 Cu. In.; 2-BBL., M.T.	J14Y	.035	28.5-32.5⑤	.016-.021	**	100-140	Zero	Zero	18B	3½-5	750
	V8—383 Cu. In.; 2-BBL., A.T.	J14Y	.035	28.5-32.5⑤	.016-.021	**	100-140	Zero	Zero	18B	3½-5	650
	V8—383 Cu. In.; 4-BBL., A.T.	J11Y	.035	28.5-32.5⑤	.016-.021	**	110-150	Zero	Zero	18B	3½-5	700
	V8—440 Cu. In. Std.	J13Y	.035	28.5-32.5⑤	.016-.021	**	110-150	Zero	Zero	18B	3½-5	650
	V8—440 Cu. In.; Hi. Perf.	J11Y	.035	28.5-32.5⑤	.016-.021	**	110-150	Zero	Zero	21B	3½-5	800

★—With sychromesh transmission in N and automatic in D. Add 50 rpm if air conditioned.

▲—With vacuum advance disconnected. NOTE: These settings are only approximate. Engine design, altitude, temperature, fuel octane rating and the condition of the individual engine are all factors which can influence timing. The limiting advance factor must, therefore, be the "knock point" of the individual engine.

⑥—1966 with Cleaner Air Package (California). Set at 5°A.

⑥—1967, manual transmission TDC; automatic transmission 5°B.

①—With double point set, total cam angle = 34°

②—With Cleaner Air Package, 625 rpm.

③—Minimum.

④—10B for M.T.

⑤—When setting dwell, disconnect vacuum advance line and distributor solenoid.

⑥—When timing engine, disconnect vacuum advance line only.

A—After top dead center.
B—Before top dead center.
TDC—Top dead center.
A.T.—Automatic transmission.
M.T.—Manual transmission.
**—See engine decal.

TORQUE SPECIFICATIONS

YEAR	MODEL	CYLINDER HEAD BOLTS (FT. LBS.)	ROD BEARING BOLTS (FT. LBS.)	MAIN BEARING BOLTS (FT. LBS.)	CRANKSHAFT BALANCER BOLT (FT. LBS.)	FLYWHEEL TO CRANKSHAFT BOLTS (FT. LBS.)	MANIFOLD (FT. LBS.) Intake	Exhaust
1964-71	V8—All	65-75	40-45	80-85	▲	55-65	50	30

▲—Crankshaft Pulley Bolt: Vibration damper bolts—15 ft. lbs.   Bolt in end of crankshaft—135 ft. lbs.

CRANKSHAFT BEARING JOURNAL SPECIFICATIONS

YEAR	MODEL	MAIN BEARING JOURNALS (IN.)				CONNECTING ROD BEARING JOURNALS (IN.)		
		Journal Diameter	Oil Clearance	Shaft End-Play	Thrust On No.	Journal Diameter	Oil Clearance	End-Play
1964–65	V8—361, 383 Cu. In.	2.625	.0010	.0045	3	2.375	.0010	.013
	V8—413 Cu. In.	2.750	.0010	.0045	3	2.375	.0010	.013
1966–71	V8—383 Cu. In.	2.625	.0005–.0020	.002–.007	3	2.38	.0005–.0020	.009–.017
	V8—440 Cu. In.	2.750	.0005–.0020	.002–.007	3	2.38	.0010–.0020	.009–.017

VALVE SPECIFICATIONS

YEAR AND MODEL	SEAT ANGLE (DEG.)	FACE ANGLE (DEG.)	VALVE LIFT INTAKE (IN.)	VALVE LIFT EXHAUST (IN.)	VALVE SPRING PRESSURE (VALVE OPEN) LBS. @ IN.	VALVE SPRING INSTALLED HEIGHT (IN.)	STEM TO GUIDE CLEARANCE (IN.) INTAKE	EXHAUST	STEM DIAMETER (IN.) INTAKE	EXHAUST
1964–65 V8—Exc. 300	45	45	.392	.390	195 @ 1$\frac{15}{32}$	1$\frac{7}{8}$	.001–.003	.002–.004	.372–.373	.371–.372
V8—300 Std.	45	45	.430	.430	195 @ 1$\frac{15}{32}$	1$\frac{7}{8}$	.001–.003	.002–.004	.372–.373	.371–.372
V8—300 Hi. Perf.	45	45	.444	.450	225 @ 1$\frac{7}{16}$	1$\frac{7}{8}$	.001–.003	.002–.004	.372–.373	.371–.372
1966 V8—383, 2-BBL.	45	45	.392	.390	195 @ 1$\frac{15}{32}$	1$\frac{55}{64}$	.001–.003	.002–.004	.372–.373	.371–.372
V8—383, 440, 4-BBL.	45	45	.425	.435	200 @ 1$\frac{7}{16}$	1$\frac{55}{64}$	.001–.003	.002–.004	.372–.373	.371–.372
1967 V8—383, 2-BBL.	45	45	.425	.435	200 @ 1$\frac{7}{16}$	1$\frac{55}{64}$	.001–.003	.002–.004	.372–.373	.371–.372
V8—383, 440, 4-BBL.	45	45	.425	.435	200 @ 1$\frac{7}{16}$	1$\frac{55}{64}$	.001–.003	.002–.004	.372–.373	.371–.372
V8—440, 4-BBL. Hi. Perf.	45	45	.450	.458	246 @ 1$\frac{23}{64}$	1$\frac{55}{64}$	.001–.003	.002–.004	.372–.373	.371–.372
1968–69 V8—383, 2-BBL.	45	45	.425	.435	200 @ 1$\frac{7}{16}$	1$\frac{55}{64}$	.001–.003	.002–.004	.372–.373	.371–.372
V8—383, 4-BBL.	45	45	.425	.435	200 @ 1$\frac{7}{16}$	1$\frac{55}{64}$	.001–.003	.002–.004	.372–.373	.371–.372
V8—440	45	45	.425	.435	200 @ 1$\frac{7}{16}$	1$\frac{55}{64}$	.001–.003	.002–.004	.372–.373	.371–.372
V8—440 Hi. Perf.	45	45	.450	.458	230 @ 1$\frac{13}{32}$	1$\frac{55}{64}$	.001–.003	.002–.004	.372–.373	.371–.372
1970 V8—383, 2-BBL.	45	45	.425	.437	200 @ 1$\frac{7}{16}$	1$\frac{55}{64}$	.001–.003	.002–.004	.372–.373	.371–.372
V8—383, 4-BBL.	45	45	.425	.437	234 @ 1$\frac{7}{16}$	1$\frac{55}{64}$	.001–.003	.002–.004	.372–.373	.371–.372
V8—440	45	45	.425	.437	234 @ 1$\frac{7}{16}$	1$\frac{55}{64}$	.001–.003	.002–.004	.372–.373	.371–.372
V8—440 Hi. Perf.	45	45	.450	.458	234 @ 1$\frac{7}{16}$	1$\frac{55}{64}$	.001–.003	.002–.004	.372–.373	.371–.372
1971 V8—383, 2-BBL.	45	45	.425	.435	200 @ 1$\frac{7}{16}$	1$\frac{55}{64}$	.001–.003	.002–.004	.372–.373	.371–.372
V8—383, 4-BBL.	45	45	.450	.458	246 @ 1$\frac{23}{64}$	1$\frac{55}{64}$	.001–.003	.002–.004	.372–.373	.371–.372
V8—440	45	45	.425	.435	246 @ 1$\frac{23}{64}$	1$\frac{55}{64}$	.001–.003	.002–.004	.372–.373	.371–.372
V8—440 Hi. Perf.	45	45	.450	.458	246 @ 1$\frac{23}{64}$	1$\frac{55}{64}$	.001–.003	.002–.004	.372–.373	.371–.372

BATTERY AND STARTER SPECIFICATIONS

YEAR	MODEL	BATTERY Ampere Hour Capacity	Volts	Terminal Grounded	STARTERS Lock Test Amps.	Volts	No-Load Test Torque	Amps.	Volts	RPM	Brush Spring Tension (Oz.)
1964–66	① ③	59	12	Neg.	400	4.0	8.5	78	11.0	3,800	34
	② ④	70	12	Neg.	450	4.0	24.0	90	11.0	2,175	34
1967–69	383, 440—W.O./Air Cond.④	59	12	Neg.	425	4.0	24.0	90	11.0	2,300	40
	All others④	70	12	Neg.	425	4.0	24.0	90	11.0	2,300	40
1970–71	383④	59	12	Neg.	400–450	4.0	—	90	11.0	1,925–2,600	32–36
	440④	70	12	Neg.	400–450	4.0	—	90	11.0	1,925–2,600	32–36

① —Except New Yorker, Imperial and 300 hi. perf.
② —New Yorker, Imperial and 300 hi. perf.
③ —Direct drive starter.
④ —Gear reduction starter.

CYLINDER HEAD BOLT TIGHTENING SEQUENCE

All V8—1964-71

GENERAL CHASSIS AND BRAKE SPECIFICATIONS

YEAR	MODEL	CHASSIS			BRAKE CYLINDER BORE			BRAKE DRUM	
		Overall Length (In.)	Tire Size	Master Cylinder (In.)	Wheel Cylinder Diameter (In.)			Diameter (In.)	
					Front	Rear		Front	Rear
1964	Exc. Sta. Wag. & Imp.	215.3	8.00 x 14	1.0	1.125	.9375		11	11
	Sta. Wagon	219.4	8.50 x 14	1.0	1.125	.9375		11	11
	Imperial	225.8	8.20 x 15	1.0	1.125	.9375		11	11
1965	Newport sed.	218.2	8.25 x 14	1.0	1.125	.9375		11	11
	Npt. St. W., 300 & N.Y. sdn.	218.2⑤	8.55 x 14	1.0	1.125	.9375		11	11
	New Yorker, sta. wag.	219.0	8.85 x 14	1.0	1.125	.9375		11	11
	Imperial	227.8	9.15 x 15	1.0	1.125	.9375		11	11
1966	Newport	218.2	8.25 x 14①	1.0	1.125	.9375		11	11
	"300"	218.2	8.55 x 14	1.0	1.125	.9375		11	11
	New Yorker	218.2	8.55 x 14	1.0	1.125	.9375		11	11
	Newport & N.Y. sta. wag.	219.0	9.00 x 14	1.0	1.125	.9375		11	11
	Imperial	227.0	9.15 x 15	1.0	1.125	.9375		11	11
	All with Disc. Brake	—	8.45 x 15①	1.125	2.375	.8750		11.87	11
1967	Newport	219.3	8.25 x 14	1.0	1.125④	.9375		11②	11
	New Yorker	219.3	8.55 x 14	1.0	1.125④	.9375		11②	11
	"300"	223.4	8.55 x 14	1.0	1.125④	.9375		11②	11
	Imperial	224.7	9.15 x 15	1.0③	1.125④	.9375		11②	11
	Sta. Wag., 2 Seat	219.5	8.85 x 15	1.0③	1.125④	.9375		11②	11
	Sta. Wag., 3 Seat	220.3	8.85 x 15	1.0③	1.125④	.9375		11②	11
1968	Newport	219.2	8.55 x 14	1.0	1.125	.9375		11	11
	New Yorker	219.2	8.55 x 14	1.0	1.125	.9375		11	11
	"300"	221.7	8.55 x 14	1.0	1.125	.9375		11	11
	Imperial (Disc.)	224.5	9.15 x 15	1.125	2.375	.9375		11.76	11
	Sta. Wag., 2 Seat	219.5	8.85 x 14	1.0	1.125	.9375		11	11
	Sta. Wag., 3 seat	220.3	8.85 x 14	1.0	1.125	.9375		11	11
	Disc. Brakes (exc. wag.)	—	8.45 x 15	1.125	2.375	.9375		11.76	11
	Disc. Brakes (sta. wag.)	—	8.85 x 15	1.125	2.375	.9375		11.76	11
1969	Newport	224.7	8.55 x 15	1.0	1.125	.9375		11	11
	New Yorker	224.7	8.55 x 15	1.0	1.125	.9375		11	11
	"300"	224.7	8.55 x 15	1.0	1.125	.9375		11	11
	Imperial (Disc.)	229.7	9.15 x 15	1.125	2.375	.9375		11.76	11
	Disc. Brakes exc. wag.	—	8.45 x 15	1.0	2.375	.9375		11.76	11
	Disc. Brakes sta. wag.	—	8.85 x 15	1.125	2.375	.9375		11.76	11
1970–71	Newport	224.7	H78 x 15	1.0	1.125	.9375		11	11
	New Yorker	224.7	J78 x 15	1.0	1.125	.9375		11	11
	"300"	224.7	H78 x 15	1.0	1.125	.9375		11	11
	Imperial (Disc.)	229.7	L84 x 15	1.125	2.75	.9375		11.75	11
	Disc. Brakes exc. wag.	—	—	1.125	2.75	.9375		11.75	11
	Disc. Brakes sta. wag.	—	—	1.125	2.75	.9375		11.75	11
	Sta. wagon	224.8	L84 x 15	1.0	1.125	.9375		11	11

①—Disc. Brakes—Sedans & Conv. 8.15 x 15, w/AC 8.45 x 15.
②—Disc. Brakes—11.87
③—Disc. Brakes—1⅛
④—Disc. Brakes—2.375
⑤—Newport Sta. Wag.—219.0

VOLTAGE TEMPERATURE INDEX

Temp. in Degrees	0°	25°	48°	70°	95°	118°	140°
Volt Reading—Min.	14.0	13.9	13.8	13.7	13.6	13.5	13.4
	to	to	to	to	to	to	to
Volt Reading—Max.	14.6	14.5	14.4	14.3	14.2	14.1	14.0

WHEEL ALIGNMENT

YEAR	MODEL	CASTER Range (Deg.)	CASTER Pref. Setting (Deg.)	CAMBER Range (Deg.)	CAMBER Pref. Setting (Deg.)	TOE-IN (In.)	KING-PIN INCLINATION (Deg.)	WHEEL PIVOT RATIO Inner Wheel	WHEEL PIVOT RATIO Outer Wheel
1964-65	Manual Steering	1N-0	$\frac{1}{2}$N	①	①	$\frac{3}{32}$-$\frac{5}{32}$	5-7	21$\frac{1}{2}$	20
	Power Steering	$\frac{1}{4}$P-1$\frac{1}{4}$P	$\frac{3}{4}$P	①	①	$\frac{3}{32}$-$\frac{5}{32}$	5-7	21$\frac{1}{2}$	20
1966	Manual Steering	1N-0	$\frac{1}{2}$N	②	②	$\frac{3}{32}$-$\frac{5}{32}$	7$\frac{1}{2}$③	20	18.8④
	Power Steering	$\frac{1}{4}$P-1$\frac{1}{4}$P	$\frac{3}{4}$P	②	②	$\frac{3}{32}$-$\frac{5}{32}$	7$\frac{1}{2}$③	20	18.8④
1967	Chrysler, Manual	1N-0	$\frac{1}{2}$N	②	②	$\frac{3}{32}$-$\frac{5}{32}$	7$\frac{1}{2}$	20	18.8
	Chrysler, Power	$\frac{1}{4}$P-1$\frac{1}{4}$P	$\frac{3}{4}$P	②	②	$\frac{3}{32}$-$\frac{5}{32}$	7$\frac{1}{2}$	20	18.8
	Imperial, Power	$\frac{1}{4}$P-1$\frac{1}{4}$P	$\frac{3}{4}$P	②	②	$\frac{3}{32}$-$\frac{5}{32}$	9	20	17.9
1968	Manual Steering	1N-0	$\frac{1}{2}$N	②	②	$\frac{3}{32}$-$\frac{5}{32}$	7$\frac{1}{2}$⑤	20	18.8
	Power Steering	$\frac{1}{4}$P-1$\frac{1}{4}$P	$\frac{3}{4}$P	②	②	$\frac{3}{32}$-$\frac{5}{32}$	7$\frac{1}{2}$⑤	20	18.8⑥
1969	Man, Pow. exc. Imperial	1N-0	$\frac{1}{2}$N	②	②	$\frac{3}{32}$-$\frac{5}{32}$	7$\frac{1}{2}$⑤	20	18.8
	Power Steering Imperial	$\frac{1}{4}$P-1$\frac{1}{4}$P	$\frac{3}{4}$P	②	②	$\frac{3}{32}$-$\frac{5}{32}$	9	20	17.9
1970-71	Manual Steering	$\frac{1}{2}$N ± $\frac{9}{16}$	$\frac{1}{2}$N	⑦	⑧	$\frac{1}{8}$ ± $\frac{1}{16}$	9	20	18.8
	Power Steering—Chrysler	$\frac{1}{2}$N ± $\frac{9}{16}$	$\frac{1}{2}$N	⑦	⑧	$\frac{1}{8}$ ± $\frac{1}{16}$	9	20	18.8
	Power Steering—Imperial	$\frac{3}{4}$P ± $\frac{9}{16}$	$\frac{3}{4}$P	⑦	⑧	$\frac{1}{8}$ ± $\frac{1}{16}$	9	20	17.9

①—Camber range and preferred setting: Set right side at $\frac{1}{8}$N-$\frac{1}{8}$P with 0 preferred. Set left side at $\frac{1}{8}$P to $\frac{3}{8}$P with $\frac{1}{4}$ preferred.

②—Left side—P$\frac{1}{4}$ to P$\frac{3}{4}$. Preferred P$\frac{1}{2}$. Right side—0 to P$\frac{1}{2}$. Preferred P$\frac{1}{4}$.

③—Imperial 6$\frac{1}{2}$ degrees.

④—Imperial 18.5 degrees.

⑤—Imperial 9 degrees.

⑥—Imperial 17.9.

⑦—Left side—$\frac{1}{2}$P ± $\frac{1}{4}$; Right side—$\frac{1}{4}$P ± $\frac{1}{4}$.

⑧—Left side—$\frac{1}{2}$P; Right side—$\frac{1}{4}$P.

N—Negative.

P—Positive.

CAPACITIES

YEAR	MODEL	ENGINE CRANKCASE ADD 1 Qt. FOR NEW FILTER	TRANSMISSIONS Pts. TO REFILL AFTER DRAINING Manual③ 3-Speed	TRANSMISSIONS Pts. TO REFILL AFTER DRAINING Manual③ 4-Speed	TRANSMISSIONS Pts. TO REFILL AFTER DRAINING Automatic	DRIVE AXLE (Pts.)	GASOLINE TANK (Gals.)	COOLING SYSTEM (Qts.) WITH HEATER
1964	Newport, 300	5	5	7$\frac{1}{2}$	20	4	23①	17
	Others	5	None	7$\frac{1}{2}$	20	4	23①	17
1965	Town & Country	4	5	7$\frac{1}{2}$	19$\frac{1}{2}$	4	22	17
	Imperial	5	None	None	19$\frac{1}{2}$	4	23	17
	Others	4	5	7$\frac{1}{2}$	19$\frac{1}{2}$	4	25	17
1966	Imperial	5	None	None	18$\frac{1}{2}$	4	23	18
	Town & Country	4	6	None	18$\frac{1}{2}$	4	22	17
	Others	4	6	None	18$\frac{1}{2}$	4	25	17
1967	Imperial	5	None	None	18$\frac{1}{2}$	4	25	18
	Town & Country	4	None	None	18$\frac{1}{2}$	4	22	17
	Others	4	6$\frac{1}{2}$	None	18$\frac{1}{2}$	4	25	17
1968	Imperial	4	None	None	18$\frac{1}{2}$	4	24	18②
	Town & Country	4	None	None	18$\frac{1}{2}$	4	22	17②
	Others	4	6	None	18$\frac{1}{2}$	4	24	17②
1969	Imperial	4	None	None	18$\frac{1}{2}$	4	24	19②
	Town & Country	4	None	None	18$\frac{1}{2}$	4	23	17②
	Others	4	6	None	18$\frac{1}{2}$	4	24	17②
1970-71	Imperial	4	None	None	19	4.4	24	⑥
	440 Hi. Perf.	6	None	None	19	4.4	⑤	⑥
	Others	4	4$\frac{3}{4}$	None	④	4.4	⑤	⑥

①—Town & Country = 21 gals.

②—Add 1$\frac{1}{2}$ qts. for rear seat heater.

③—Use SAE 80 or 90 gear oil in warm weather. For year-round use, it is permissible to use automatic transmission fluid, type A, suffix A.

④—383 2-BBL.—19; 383 4-BBL.—16; 440—19.

⑤—Exc. sta. wag.—24; sta. wag.—23.

⑥—383 2-BBL.—14.5; w/AC—16; Chrysler 383 4-BBL. and all 440—15.5; w/AC—17; Imperial—16.5.

FUSES AND CIRCUIT BREAKERS

1963–69

Lighting system	15 Amp. CB
Top, power windows, power seats	30 Amp. CB
Dome, stop, trunk, park, tail lamps	20 Amp. Fuse
Cigarette lighter, glove compartment	20 Amp. Fuse
Instrument panel lamps	2 Amp. Fuse
Air conditioner or heater	20 Amp. Fuse
Air conditioner, rear	20 Amp. CB
Windshield wiper (1-speed) Chrys.	5 Amp. CB

Windshield wiper (2-speed) Chrys.	7 Amp. CB
Windshield wiper (2-speed) Imp.	6 Amp. CB
Radio	7½ Amp. Fuse

1970–71

Accessory and turn signal	20 Amp. (Imperial)
Accessories	20 Amp. (Chrysler)
Headlight sentinel	20 Amp.
Cigarette lighter	20 Amp.

Console	20 Amp. (Chrysler)
Emergency flasher	20 Amp. (Chrysler)
Emergency flasher and stop lights	20 Amp. (Imperial)
Headlight dimmer	4 Amp.
Heater or A/C blower motor	20 Amp.
Instrument lights	3 Amp. (Chrysler)
Instrument lights	5 Amp. (Imperial)
Low fuel warning relay	5 Amp. (Imperial)
Radio and back-up lights	20 Amp.

LIGHT BULBS

1964–69

Electroluminescent lighting used on some of the instruments.

Headlamp (dual type) (inner)	4001; 37½ W
Headlamp (dual type) (outer)	4002; 37½–50 W
Beam indicator, glove box, map lamp	57; 1 CP
Hand brake, warning lamp	90; 1 CP
Front and rear parking, stop, front and rear turn signal lamps	1034; 32–4 CP
License plate lamp	67; 3 CP
Map and dome lamps	1004; 15 CP
Back-up lamps	1073; 32 CP
Transmission push-button, clock light	57; 2 CP
Transmission selector, auto-pilot lamps	1816; 2.85 CP
Radio dial	1893; 2 CP
Trunk lamp	1003; 15 CP
Tachometer w/console	1893; 2 CP

1970–71

	CHRYSLER	IMPERIAL
Arm rest lamp		1445
Ash receiver	1445(2)	*1445(2)
Auto-temp	*(168)	*(704)
Back-up lights	1156(2)	1156(2)

Brake system warning light	57	57
Clock	*(168)	*(704)
Cornering light	1293	1293
Dome and/or "C" pillar light	550	
Door and pocket panel and/or reading light	90	90
Fasten belts indicator	57	57
Fender mounted turn signal indicator	330(2)	1813(2)
Gear selector indicator (column)	*(168)	*(704)
Gear selector with console	57	—
Glove compartment	1891	1891
Heater and/or A/C Control	*168	*(704)
High beam indicator	57	57
Instrument cluster and speedometer illumination	*(168)	*(704)(4)
Ignition lamp	1445	1445
License light	67(1)-67(2)	67
	Station Wagon	
Lock doors indicator	158	
Map lamp	90	90
Oil pressure indicator	57	Gauge

Open door indicator		57
Panel rheostat valve	24 Ohms	12 Ohms
Park and turn signal (front)	1157(2)	1157NA
Portable reading light		89
Radio	*(168)	*(704)
Sealed beam—Hi-beam (No. 1)	4001	4001
Sealed beam—Hi-low beam (No. 2)	4002	4002
Sentry signal	57	57
Side marker	1895(4)	1895(4)
Stereo indicator	1445	1445
Switch lighting	*(168)	*(704)
Tail light (only)	1095(2)	
Tail, stop and turn signal	1157(2)	1157(6)
Temperature indicator	57(2)	—
Trunk and/or under hood light	1004	1004
Turn signal indicator (panel)	*57-168	—

‡Included in instrument cluster lighting.

**Optional.

NOTE: All of the above bulbs are brass base. Aluminum base bulbs are not approved and not to be used.

Distributor

Detailed information on drive, direction of rotation, cylinder numbering, firing order, point gap, point dwell, timing mark location, spark plugs, spark advance and idle speed will be found in the Specifications tables.

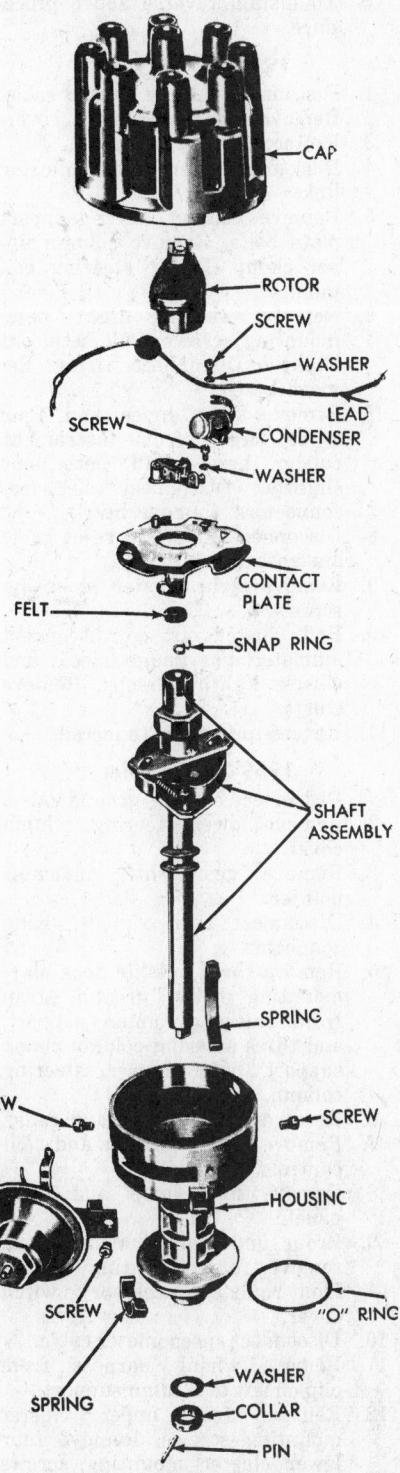

Distributor
(© Chrysler Corp.)

Spark Plug Cables

All Chrysler and Imperial engines incorporate conventional spark plugs (without resistors) along with resistance type spark plug cables to eliminate radio interference.

For identification purposes, this cable has "radio" or "electronic suppression" printed on it.

The cable uses a graphite- or composition-type conducting core replacing the copper wire in the center of conventional spark plug cable. Full contact is made between the core and terminals by a short wire pin pushed into the ends of the cable.

Precautions must be taken in handling the core to prevent damage. The cable should be removed from the spark plug by grasping the cable cover and pulling straight off with a steady, even pull. Pulling sideways could jam the terminal on the spark plug and cause the cable to separate from the terminal. The cable terminal should not be crimped to the point that excessive force is required to remove it from the spark plug.

The cables should never be removed by giving them a quick jerk. Doing this can stretch the core and cause a high resistance or open circuit. If a damaged core is suspected, a resistance check with an ohmmeter should be made.

If any cable has more resistance than 30,000 ohms, check to be sure the terminals are in contact with the pin and the pin is in full contact with the core. If the terminals and pins are properly installed and the cable resistance is still more than specified, the cable should be replaced with a new resistance type cable.

A new terminal should never be attached to the resistance core cables unless the wire pin is in place. Contact will not be maintained with the core. This will result in arcing and burning of the core which will cause engine malfunction and radio interference.

Spark Plug Caution

Resistor-type spark plugs are never to be used with the resistance-type cable. The added resistance of the spark plugs may cause malfunctioning of the ignition system. When replacing a spark plug, be sure to use the type specified for the particular engine.

Cable Troubles

If the radio develops excessive noise, or if there is a pronounced engine miss, check for faulty or broken cables.

Distributor Removal and Installation

Removal Procedure

Disconnect vacuum advance hose and primary wire. Lift off distributor cap. Mark edge of housing to aid in locating position of rotor for reinstalling. Remove the hold-down bolt, lock plate and distributor.

Installation Procedure

If engine has been turned after removal, make sure No. 1 piston is at top dead center and install the distributor so that the rotor is pointing to No. 1 firing position. Install the lock plate and screw, but not tightly. Rotate the crankshaft to align the specified degree mark on the crankshaft pulley with the pointer. Rotate distributor until contacts are opening. Tighten the hold-down bolt and reconnect the primary lead and the vacuum pipe. Check that timing is correct, using a timing light.

Generator, Regulator

Detailed facts on the alternator are in the Generator and Alternator Specifications table.

General information on generator and regulator repair and troubleshooting is in the Unit Repair Section.

Alternator R&R

1. Disconnect battery ground cable.
2. Disconnect alternator output "BATT" and field "FLD" leads. Disconnect ground wire.
3. Remove mounting bolts. Remove alternator.
4. Reverse procedure for installation. Adjust drive belt to have ¼ in. deflection at the center of the longest run between pulleys.

Caution Under no circumstances should the alternator circuit be polarized.

Battery, Starter

Detailed information on the battery and starter is found in the Battery and Starter Specifications table.

A more general discussion of starters and their troubles is found in the Unit Repair Section.

Starter R & R

Disconnect battery and starter wires. If equipped with automatic

transmission, slide cooler tube bracket off stud. Remove attaching bolts and lift out starter.

Instruments

Instrument Cluster Removal

1964-66 Imperial

The instrument cluster is serviced in three separate sections. When servicing the cluster, it is necessary to remove only the section containing the desired unit.

1. Disconnect the battery ground cable.
2. Remove screws attaching instrument cluster chrome bezel to instrument cluster. Remove bezel.
3. Remove trip odometer knob, clock reset knob, and temperature control level knob.
4. Remove lens from cluster.
5. Remove screws attaching cluster face plate to cluster. Remove face plate.
6. To remove speedometer: Disconnect speedometer and odometer reset cables from under instrument panel. Remove four speedometer to cluster screws. Remove speedometer.
7. To remove printed circuit assembly: Remove assembly to cluster screws. Pull assembly forward. Disconnect printed circuit connector. Remove assembly to service fuel, oil, temperature, and ammeter gauges.
8. To remove clock: Remove screws. Pull clock forward and disconnect wire. Remove clock.
9. To install reverse above procedure.

1967-68 Imperial

1. Disconnect battery ground cable.
2. Tape top of steering column to prevent scratching.
3. Remove four steering column trim plate screws. Remove cover.
4. Loosen allen screw on right underside of steering column. Push gear selector forward and rotate clockwise to remove.
5. Remove steering column upper clamp.
6. Remove air conditioning left spot cooler hose by releasing alligator clamp at the connection.
7. Remove four upper bezel screws. Remove four lower bezel screws from lower left corner of bezel, each side of steering column, and inside the ash tray.
8. Raise lower edge of bezel. Disconnect electrical connectors. Remove vacuum hose from rear air switch.
9. Remove bezel with spot cooler hose attached.

10. Remove odometer reset cable bezel nut. Push cable up into panel.
11. Remove eight cluster mounting screws. Pull cluster out, bottom edge first. Disconnect printed circuit connector, speedometer cable, and ammeter wires. Remove cluster.
12. Reverse procedure for installation.

1969-70 Imperial

1. Disconnect battery ground cable.
2. Remove left ash tray and radio.
3. Remove heater controls.
4. Disconnect vent control cables at fresh air doors.
5. Remove:
   a. vent control.
   b. map lamp.
   c. lamp panel assembly.
   d. cluster accessory bezel.
   e. steering column cover.
   f. gear shift indicator.
   g. steering column clamp.
   h. cover screws at floor panel.
6. Lower steering column.
7. Disconnect speedometer cable.
8. Remove five upper cluster mounting screws. Remove five lower cluster mounting screws through access holes in lower instrument panel.
9. Move cluster to the right. Push right end toward front and turn top down. Pull left end of cluster out of panel.
10. Disconnect all wiring and connectors. Remove cluster.
11. Reverse procedure to install.

1964 Chrysler

1. Disconnect battery ground cable.
2. Tape steering column to prevent scratching.
3. Remove upper and lower moulding screws at center of instrument panel.
4. Remove heater bezel. Disconnect bowden cable, vacuum lines, and wiring to heater control switch.
5. Remove instrument panel lower hood.
6. Remove wiper switch knob, retaining bezel, and wiper switch.
7. Remove three screws from across top of instrument cluster and three screws from cluster. lower chrome moulding.
8. Pull cluster forward. Disconnect speedometer cable.
9. Disconnect wiring connectors. Remove cluster.
10. Reverse procedure to install.

1965-66 Chrysler

1. Disconnect battery ground cable
2. Remove steering column cover.
3. Disconnect gear shift indicator link.
4. Remove column clamp. Loosen column lower support plate. Lower steering column.

5. Remove steering column upper filler.
6. Disconnect speedometer, odometer, and circuit board connector.
7. Remove ignition switch and accessory switch.
8. Remove lower and upper cluster trim bezels.
9. Remove cluster retaining screws. Lower cluster and disconnect ammeter wires. Remove cluster.
10. To install, reverse above procedure.

1967-68 Chrysler

1. Disconnect battery ground cable.
2. Remove steering column cover.
3. Protect steering column.
4. Disconnect gear shift indicator link.
5. Remove column lower support plate bolts. Remove column upper clamp. Lower steering column.
6. Remove warning light bezel mounting screws. Pull bezel out slightly. Disconnect wiring. Remove bezel.
7. Remove four upper and four lower screws from instrument cluster bezel. Pull bezel out slightly. Disconnect electrical connectors. Remove bezel.
8. Disconnect odometer reset cable and speedometer cable.
9. Remove eight cluster mounting screws.
10. Pull cluster out to disconnect ammeter, gas gauge, clock, and cluster lighting lamps. Remove cluster.
11. Reverse procedure to install.

1969-70 Chrysler

1. Disconnect battery ground cable.
2. Remove lower steering column cover.
3. Remove gear shift indicator pointer.
4. Disconnect turn signal wiring connector.
5. Remove three outside floor plate mounting bolts, ground strap from steering column support, and three steering column clamp support nuts. Lower steering column.
6. Remove left ash tray and radio.
7. Remove heater controls and vent controls.
8. Remove map lamp and lamp panel.
9. From under instrument panel, remove four mounting screws from right end accessory switch cover.
10. Disconnect speedometer cable.
11. Remove wiring harness from clip on left of column support.
12. Remove four upper cluster mounting screws. Remove four lower cluster mounting screws through access holes in the lower panel.

13. Move cluster right, rotating the right end of cluster toward front and down. Roll top of cluster down and rock slightly to the left. Disconnect wiring. Roll cluster out from instrument panel.
14. Reverse procedure for installation.

Ignition Switch Replacement

1964-66 Imperial

1. Remove lower steering column cover plate.
2. Remove accessory switch knobs.
3. Remove screw attaching left end of bezel to instrument panel. This screw may be reached from inside steering column opening.
4. Remove screw in switch well.
5. Lift off bezel.
6. Remove mounting nut from ignition switch.
7. From under panel, pull switch down, disconnect wiring and remove switch from under panel.
8. Install in reverse of above.

1964-66 Chrysler
1967-69 All Models

1. Remove switch bezel nut.
2. Push switch through panel.
3. Disconnect wiring connector.
4. Remove switch.
5. To install, reverse procedure. Align key on switch with slot in panel.

1970 All Models

On all 1970 models the ignition switch is mounted in the steering column. The switch also locks the steering and the transmission linkage.

1970 Standard Steering Column

1. Disconnect ground cable from battery.
2. Remove steering wheel.
3. Remove turn signal lever.
4. Remove turn signal switch and upper bearing retainer.
5. Remove ignition key lamp assembly.
6. Remove snap ring. Remove three bearing housing to lock housing screws.
7. Pull bearing housing from shaft. Remove bearing lower snapring.
8. Pry sleeve off steering shaft lock plate hub to expose pin.
9. Press, do not hammer, pin from shaft.
10. Remove lock plate from shaft.
11. Remove buzzer switch and lock lever guide plate.
12. Lock switch and remove key. Insert a stiff wire into lock cylinder release hole, push in to re-

lease lock retainer, and pull lock cylinder out.
13. Remove three retaining screws and ignition switch assembly.
14. Reverse procedure for installation.

1970 Tilting Steering Column

1. Disconnect ground cable from battery.
2. Remove steering wheel.
3. Remove turn signal lever.
4. Remove gearshift lever pivot pin and lever, tilt release lever, and turn signal switch lever.
5. Press down lock plate and carrier and remove C-ring. Remove lock plate, carrier and spring.
6. Remove three turn signal switch attaching screws, place shift bowl in low position, and remove switch and wiring. (The shift bowl is the section to which the shift lever is attached.)
7. Remove buzzer switch.
8. Lock switch and remove key.
9. Insert a small screwdriver into the slot next to the switch mounting screw boss (right-hand slot) and depress spring latch at bottom of slot. Remove lock.
10. Remove housing cover.
11. Install tilt release lever and place column in full up position.
12. Insert screwdriver into tilt spring retainer slot, press in approximately 3/16 in., turn approximately ⅛ turn counter-clockwise until ears align with housing grooves, and remove spring and guide.
13. Remove seat and upper bearing race.
14. Remove ignition switch mounting screws and switch.
15. Reverse procedure for installation. For further information on the tilting steering column, refer to the Unit Repair Section.

Lighting Switch Replacement

1964-66 Imperial

1. Remove lower steering column cover plate.
2. Remove headlight switch knob, stem assembly and windshield-wiper switch knob.
3. Remove screw that attaches right end of switch bezel. This screw can be reached from inside the steering column opening.
4. Remove headlight switch retaining nut, using a spanner wrench.
5. Lift off switch bezel.
6. Remove light switch and wiper switch stem light seals.
7. Remove switch mounting nut.

8. Pull switch down from under panel and disconnect wiring.
9. Install in reverse of above.

1964-66 Chrysler

All switches on the instrument panel or in the clusters can be serviced from under the panel by removing the knob, mounting nut and bezel, and disconnecting the wires.

1967-68 All Models

1. Remove the instrument cluster bezel as outlined above under instrument Cluster Removal.
2. Remove headlamp switch from rear of bezel.
3. Install in reverse of above.

1969-70 All Models

1. Remove instrument cluster as outlined above under Instrument Cluster Removal.
2. Remove headlamp switch from rear of cluster.
3. Install in reverse of above.

Brakes

Specific information on brake cylinder and drum sizes is in the General Chassis and Brake Specifications table.

Information on troubleshooting and overhauling power brakes is in the Unit Repair Section.

Information on the grease seals which may need replacement is in the Unit Repair Section.

Brake Types

A servo-contact, self-energizing brake is used. It uses a double-acting wheel cylinder at tops of shoes at each assembly.

It is also a self-adjusting brake. It operates through a link, cable and return spring so connected that when brake is applied during reverse stops, the link indexes the star wheel to maintain proper shoe clearance. The self-adjusting feature is not used on heavy duty brakes.

Beginning 1966

Front wheel disc brakes are used on front wheels of some models. Information on this type of brake is in the Unit Repair Section.

Beginning 1967

A compound brake system is used on all models. This system is, in effect, two independent hydraulic systems, one for the front brakes and another for the rear brakes. The master cylinder has two pistons in tandem and two fluid outlets. The front outlet tube is connected to the hydraulic system safety switch and to the rear brakes. The rear outlet

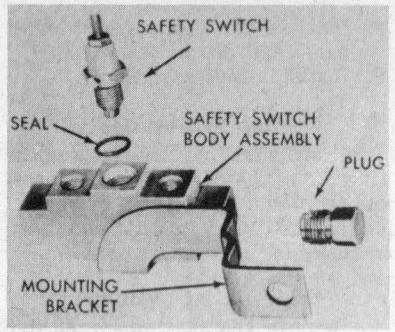

Hydraulic system safety switch
(© Chrysler Corp.)

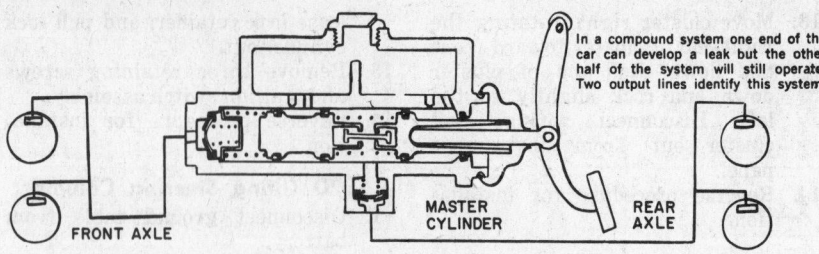

In a compound system one end of the car can develop a leak but the other half of the system will still operate. Two output lines identify this system.

FRONT AXLE   MASTER CYLINDER   REAR AXLE

Dual master cylinder, typical—1967-70

tube is also connected to the safety switch and the front brakes. In the event of a pressure loss in either branch of the system, the safety switch causes a warning light to be illuminated on the instrument panel.

Power Brakes

The power brake unit features a direct pedal connection to a vacuum unit mounted on the firewall with a master cylinder directly mounted to a vacuum booster.

This vacuum-suspended system utilizes engine intake manifold vacuum and atmospheric pressure for its power boost to the master cylinder.

Master Cylinder Removal

1. Disconnect fluid lines. On disc brake cylinders, plug front brake outlet to prevent leakage.
2. Remove nuts attaching master cylinder to cowl panel or to power brake unit.
3. Disconnect pedal push rod (non-power brakes) from brake pedal.
4. Remove master cylinder from vehicle.
5. Reverse procedure to install.

Parking Brake Adjustment

1. Raise and support vehicle. Release parking brake lever. Loosen cable adjusting nut.
2. Tighten cable adjusting nut

until a slight drag is felt while rotating wheel.
3. Loosen cable adjusting nut until both rear wheels can be rotated freely. Back off cable adjusting nut two full turns.
4. Apply parking brake several times. Check to see that rear wheels rotate freely without dragging.

Fuel System

Data on capacity of the gas tank is in the Capacities table. Data on correct engine idle speed and fuel pump pressure is in the Tune-Up Specifications table.

Information covering operation and troubles of the fuel gauge is in the Unit Repair Section.

Detailed information on the carburetor and how to adjust it is in the Unit Repair Section under the specific heading of the make of carburetor used on the engine involved.

Vapor Separator

The 1968 440 high performance engine is equipped with a vapor separator mounted between the carburetor and the fuel pump.

Serviced only as a unit, the vapor separator consists of a steel can containing a filter screen. It has an inlet from the fuel pump and an outlet to the carburetor. It also has a metered vapor bleed hole which is piped back to the fuel tank.

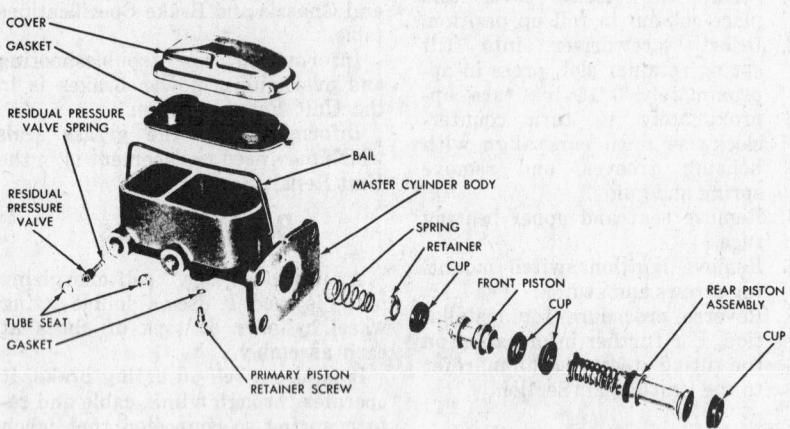

Dual type master cylinder (exploded view) with disc brakes

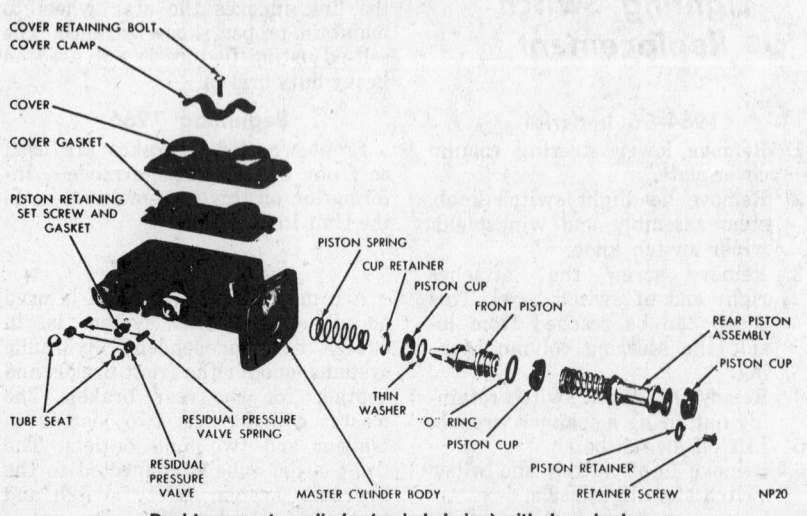

Dual type master cylinder (exploded view) with drum brakes

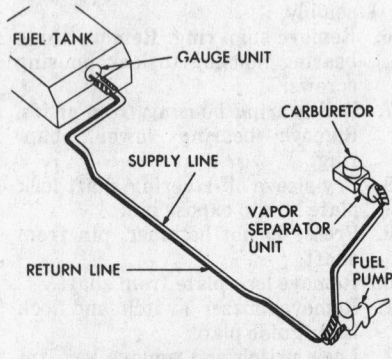

Schematic showing piping to vapor separator
(© Chrysler Corp.)

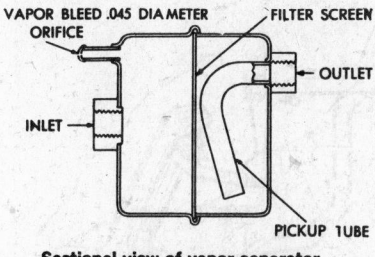

VAPOR BLEED .045 DIA METER
ORIFICE   FILTER SCREEN

OUTLET

INLET

PICKUP 1UBE

Sectional view of vapor separator
(© Chrysler Corp.)

When operating, the vapor separator is full of fuel. Any vapor that may form goes to the top of the can and, passing through the metered orifice, returns to the fuel tank.

The unit cannot be serviced. If vaporlock is evident, remove the return line at the top of the unit and pass a thin wire through the orifice to clear it.

Fuel Pump

The standard fuel pump on Chrysler and Imperial models is of pressed steel construction and cannot be disassembled for service. A repairable fuel pump is optional for 1968-70 models. This type is easily recognized by the assembling screws. The fuel pump is driven by an eccentric on the camshaft and a short push rod.

Fuel Filter

An in-line fuel filter is used near the carburetor. This filter is the throwaway type and cannot be cleaned. To assure an unrestricted fuel supply, the filter should be replaced at least every 24,000 miles, and more frequently if fuel contamination is suspected.

Throttle Linkage Adjustment

1964—361 and 383 Engines

Refer to the linkage illustrations for this procedure.

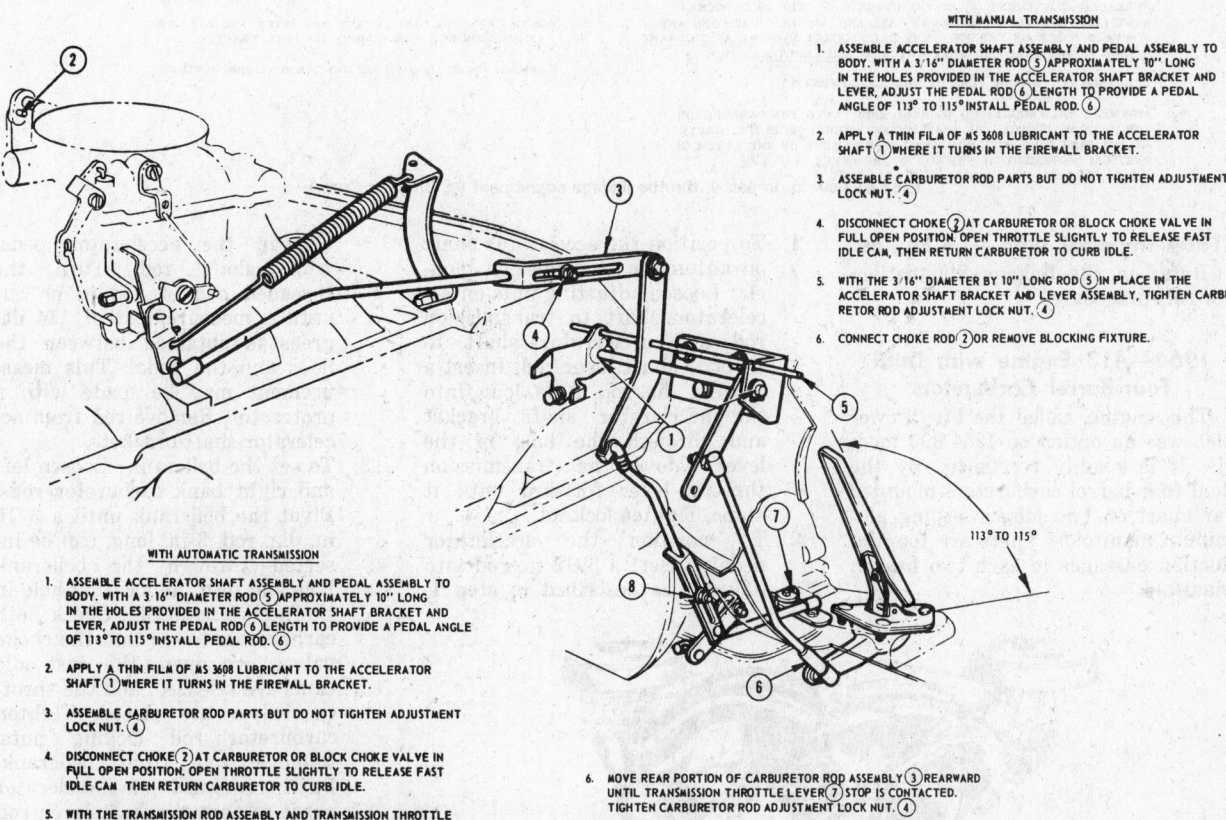

WITH MANUAL TRANSMISSION

1. ASSEMBLE ACCELERATOR SHAFT ASSEMBLY AND PEDAL ASSEMBLY TO BODY. WITH A 3/16" DIAMETER ROD (5) APPROXIMATELY 10" LONG IN THE HOLES PROVIDED IN THE ACCELERATOR SHAFT BRACKET AND LEVER, ADJUST THE PEDAL ROD (6) LENGTH TO PROVIDE A PEDAL ANGLE OF 113° TO 115°. INSTALL PEDAL ROD. (6)

2. APPLY A THIN FILM OF MS 3608 LUBRICANT TO THE ACCELERATOR SHAFT (1) WHERE IT TURNS IN THE FIREWALL BRACKET.

3. ASSEMBLE CARBURETOR ROD PARTS BUT DO NOT TIGHTEN ADJUSTMENT LOCK NUT. (4)

4. DISCONNECT CHOKE (2) AT CARBURETOR OR BLOCK CHOKE VALVE IN FULL OPEN POSITION. OPEN THROTTLE SLIGHTLY TO RELEASE FAST IDLE CAM, THEN RETURN CARBURETOR TO CURB IDLE.

5. WITH THE 3/16" DIAMETER BY 10" LONG ROD (5) IN PLACE IN THE ACCELERATOR SHAFT BRACKET AND LEVER ASSEMBLY, TIGHTEN CARBURETOR ROD ADJUSTMENT LOCK NUT. (4)

6. CONNECT CHOKE ROD (2) OR REMOVE BLOCKING FIXTURE.

113° TO 115°

WITH AUTOMATIC TRANSMISSION

1. ASSEMBLE ACCELERATOR SHAFT ASSEMBLY AND PEDAL ASSEMBLY TO BODY. WITH A 3/16" DIAMETER ROD (5) APPROXIMATELY 10" LONG IN THE HOLES PROVIDED IN THE ACCELERATOR SHAFT BRACKET AND LEVER, ADJUST THE PEDAL ROD (6) LENGTH TO PROVIDE A PEDAL ANGLE OF 113° TO 115°. INSTALL PEDAL ROD. (6)

2. APPLY A THIN FILM OF MS 3608 LUBRICANT TO THE ACCELERATOR SHAFT (1) WHERE IT TURNS IN THE FIREWALL BRACKET.

3. ASSEMBLE CARBURETOR ROD PARTS BUT DO NOT TIGHTEN ADJUSTMENT LOCK NUT. (4)

4. DISCONNECT CHOKE (2) AT CARBURETOR OR BLOCK CHOKE VALVE IN FULL OPEN POSITION. OPEN THROTTLE SLIGHTLY TO RELEASE FAST IDLE CAM, THEN RETURN CARBURETOR TO CURB IDLE.

5. WITH THE TRANSMISSION ROD ASSEMBLY AND TRANSMISSION THROTTLE LEVER IN PLACE, MOVE THE TRANSMISSION THROTTLE LEVER (7) FORWARD AGAINST THE STOP AND TIGHTEN TRANSMISSION ROD ADJUSTMENT LOCK NUT. (8) REMOVE 3/16" DIAMETER ROD (5) FROM ACCELERATOR SHAFT BRACKET.

6. MOVE REAR PORTION OF CARBURETOR ROD ASSEMBLY (3) REARWARD UNTIL TRANSMISSION THROTTLE LEVER (7) STOP IS CONTACTED. TIGHTEN CARBURETOR ROD ADJUSTMENT LOCK NUT. (4)

7. CONNECT CHOKE ROD (2) OR REMOVE BLOCKING FIXTURE.

1964 361 and 383 cu. in. Chrysler throttle linkage adjustment (© Chrysler Corp.)

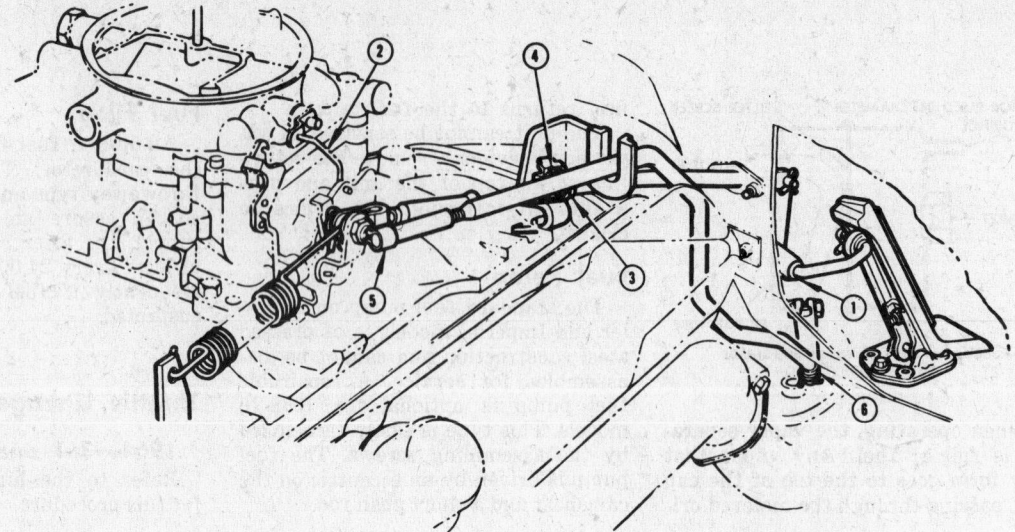

1. ASSEMBLE TRANSMISSION CONTROL LINKAGE PARTS IN PLACE BUT DO NOT ASSEMBLE TRANSMISSION ROD BALL SOCKET ⑤ TO BALL END.

2. APPLY A THIN FILM OF MS 3701 LUBRICANT TO THE ACCELERATOR SHAFT ① WHERE IT TURNS IN THE BRACKET.

3. DISCONNECT CHOKE ② AT CARBURETOR OR BLOCK CHOKE VALVE IN FULL OPEN POSITION, OPEN THROTTLE SLIGHTLY TO RELEASE FAST IDLE CAM, THEN RETURN CARBURETOR TO CURB IDLE.

4. HOLD THE TRANSMISSION LEVER ⑥ FORWARD AGAINST ITS STOP AND ADJUST THE LENGTH OF THE TRANSMISSION ROD BY MEANS OF THE THREADED ADJUSTMENT ⑤ AT THE UPPER END   THE BALL SOCKET MUST LINE UP DIRECTLY WITH THE BALL END WITHOUT EXERTING ANY FORWARD FORCE ON THE ROD. THE BALL SOCKET MUST BE AT THE SAME HEIGHT AS THE BALL END WHEN CHECKING ROD LENGTH.

5. LENGTHEN ROD BY ONE TURN OF THE ADJUSTMENT. ⑤

6. ASSEMBLE BALL SOCKET ⑤ TO BALL END. WHEN THE CARBURETOR THROTTLE IS OPENED, THE TRANSMISSION SHOULD BEGIN ITS TRAVEL AT THE SAME TIME WITH NO VERTICAL MOVEMENT OF THE LEVER OR VERTICAL MOVEMENT OF THE ROD IN THE LEVER.

7. ASSEMBLE REMAINDER OF THE LINKAGE PARTS IN PLACE. WITH THE CABLE CLAMP NUT ④ LOOSE, ADJUST THE POSITION OF THE CABLE HOUSING FERRULE ③ IN THE CLAMP SO THAT ALL SLACK IS REMOVED FROM THE CABLE WITH THE CARBURETOR AT CURB IDLE. TO REMOVE SLACK FROM THE CABLE, MOVE THE FERRULE ③ IN THE CLAMP IN THE DIRECTION AWAY FROM THE CARBURETOR LEVER.

8. BACK OFF FERRULE ③ 1/2". THIS PROVIDES 1/2" FREE PLAY BETWEEN THE REAR SURFACE OF THE ACCELERATOR BRKT. AND THE FRONT EDGE OF ACCELERATOR SHAFT LEVER. TIGHTEN CLAMP NUT ④

9. ROUTE CABLE SO THAT IT DOES NOT INTERFERE WITH THE TRANSMISSION ROD THROUGHOUT ITS FULL TRAVEL.

10. CONNECT CHOKE ROD ② OR REMOVE BLOCKING FIXTURE.

1964 413 cu. in. Imperial, throttle linkage adjustment (© Chrysler Corp.)

1964 Imperial with 413 Engine
Refer to the linkage illustrations for this procedure.

1964—413 Engine with Dual Four-Barrel Carburetors
This engine, called the Fire Power 390, was an option on 1964 300 models. It is readily recognized by the dual four-barrel carburetors mounted far apart on two long sweeping aluminum manifolds. There are four induction passages in each two branch manifold.

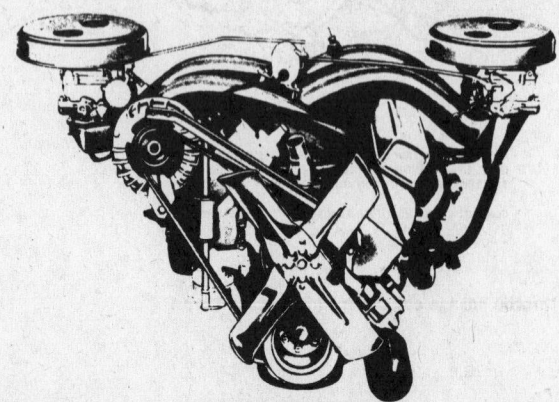

1964 Fire Power 390
(© Chrysler Corp.)

1. To position the accelerator shaft on automatic transmission models: Loosen adjusting nuts on accelerator shaft to transmission rod and accelerator shaft to throttle shaft lever rod. Insert a 3/16 in. dia. rod, 10 in. long, into the accelerator shaft bracket and through the hole in the lever. Move the transmission throttle lever forward until it stops. Tighten locknut.
2. To position the accelerator pedal: Insert a 3/16 in. rod into linkage, as described in step 1.

Unsnap the accelerator pedal from shaft rod. Turn the threaded end of rod in or out until a measurement of 114 degrees is obtained between the floor and the pedal. This measurement may be made with a protractor. Remove rod from accelerator shaft bracket.
3. To set the bellcrank: Loosen left and right bank carburetor rods. Pivot the bellcrank until a 3/16 in. dia. rod, 3 in. long, can be inserted through the bellcrank hole and into the locating hole in the intake manifold. Check both carburetors to see that the choke valves are open, the fast idle cams are released, and the throttle valves are closed. Tighten carburetor rod locking nuts. Remove rod from bellcrank. Push rearward on accelerator shaft to throttle shaft lever rod adjusting link until stop is felt. Tighten locking nut.
4. To set the slow idle speed: Warm engine to normal temperature. Be sure chokes are fully off and engine at slow idle. Turn idle mixture screws from one to two turns open. Set idle bypass air screws, between idle mixture screws, two turns open. Adjust

idle speed to 700 rpm; for air conditioned cars, 500 rpm with compressor operating. Adjust idle mixture screws on each carburetor for maximum rpm. Road test and readjust as necessary.

5. To make a fast idle adjustment: Make sure slow idle speed is correctly set. Turn air conditioning compressor off. Remove air cleaners and disconnect throttle rods. Position left carburetor at fast idle. Adjust fast idle screw for a 1,400 rpm idle. Return carburetor to slow idle position. Repeat procedure for right carburetor. It is important that the carburetors have identical fast idle speeds. Reconnect throttle rods.

1965-70 Manual Transmission Models

1. Apply multi-purpose grease lightly to all linkage friction points.

2. Disconnect choke at carburetor. Set carburetor on slow idle position.

3. Adjust position of cable housing ferrule in clamp to remove slack from cable. Back off ferrule ¼ in. to provide ¼ in. cable slack at idle.

4. Connect choke rod.

1965-66 Chrysler with Automatic Transmission
1967-68 All Automatic Transmission Models

1. Apply multi-purpose grease lightly to all linkage friction points.

2. Disconnect return spring and slotted transmission rod from carburetor lever pin. Disconnect transmission intermediate rod ball socket from upper bellcrank ball end.

3. Disconnect choke. Set carburetor on slow idle position.

4. Place a 3/16 in. dia. rod in holes in the engine mounted bellcrank and lever. Adjust intermediate transmission rod at upper end. The ball socket must line up with the ball end with rod held upward against the transmission stop. Assemble ball socket to ball end.

5. Remove 3/16 in. rod from upper bellcrank and lever.

6. Hold carburetor rod forward against transmission stop and adjust so that rear end of adjusting link slot just contacts the carburetor lever pin. Lengthen carburetor rod two full turns.

7. Assemble slotted link to carburetor.

8. Adjust cable housing ferrule in clamp to remove all cable slack. Back off ferrule ¼ in.

9. Connect choke rod.

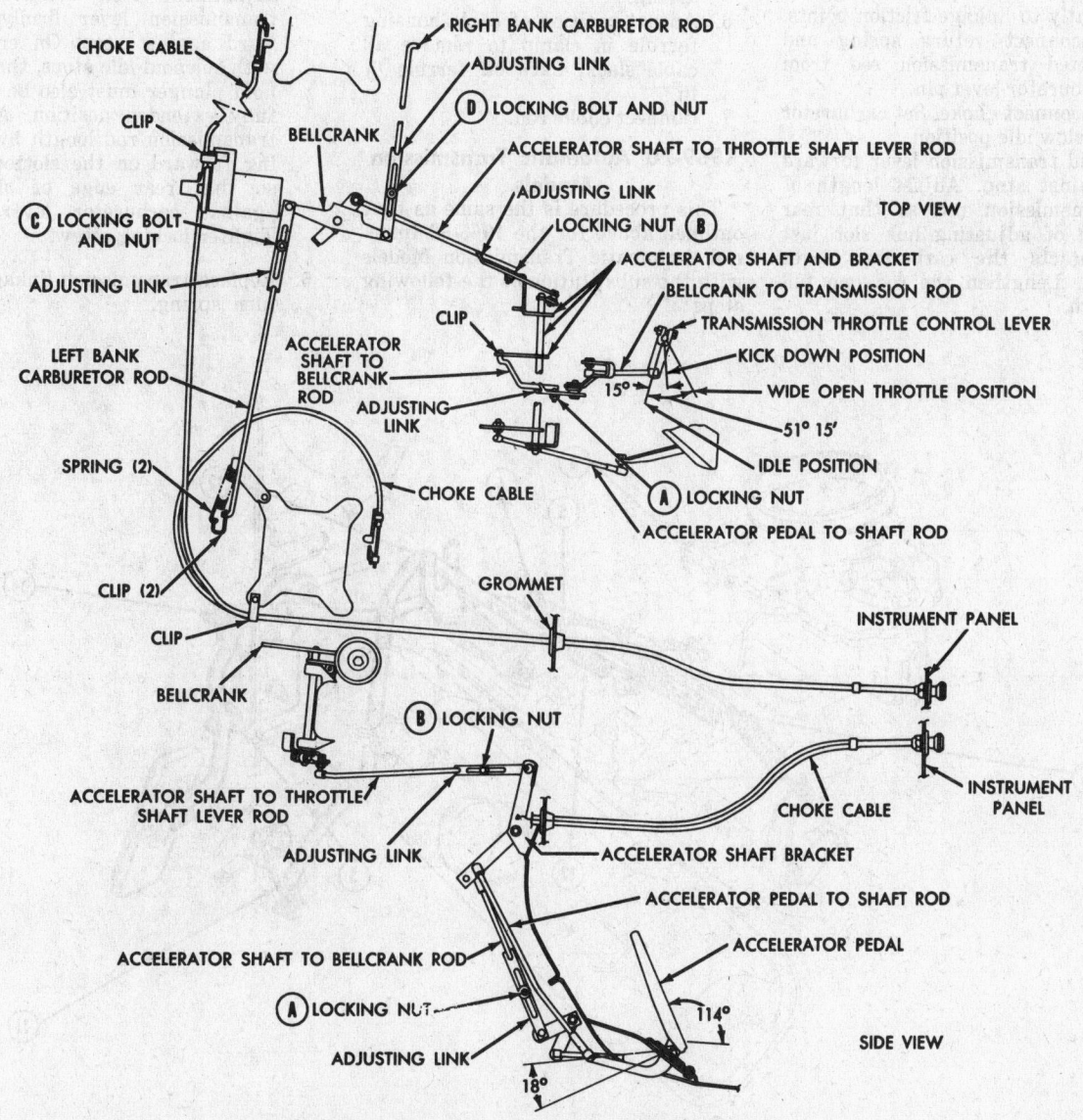

RIGHT BANK CARBURETOR ROD
CHOKE CABLE
ADJUSTING LINK
Ⓓ LOCKING BOLT AND NUT
CLIP
BELLCRANK
ACCELERATOR SHAFT TO THROTTLE SHAFT LEVER ROD
ADJUSTING LINK
Ⓒ LOCKING BOLT AND NUT
LOCKING NUT Ⓑ
TOP VIEW
ACCELERATOR SHAFT AND BRACKET
ADJUSTING LINK
BELLCRANK TO TRANSMISSION ROD
LEFT BANK CARBURETOR ROD
CLIP
TRANSMISSION THROTTLE CONTROL LEVER
ACCELERATOR SHAFT TO BELLCRANK ROD
KICK DOWN POSITION
ADJUSTING LINK
15°
WIDE OPEN THROTTLE POSITION
51° 15'
SPRING (2)
IDLE POSITION
CHOKE CABLE
Ⓐ LOCKING NUT
ACCELERATOR PEDAL TO SHAFT ROD
CLIP (2)
GROMMET
INSTRUMENT PANEL
CLIP
BELLCRANK
Ⓑ LOCKING NUT
INSTRUMENT PANEL
ACCELERATOR SHAFT TO THROTTLE SHAFT LEVER ROD
CHOKE CABLE
ADJUSTING LINK
ACCELERATOR SHAFT BRACKET
ACCELERATOR PEDAL TO SHAFT ROD
ACCELERATOR SHAFT TO BELLCRANK ROD
ACCELERATOR PEDAL
Ⓐ LOCKING NUT
114°
ADJUSTING LINK
SIDE VIEW
18°

1964 413 cu. in Fire Power 390 throttle linkage (© Chrysler Corp.)

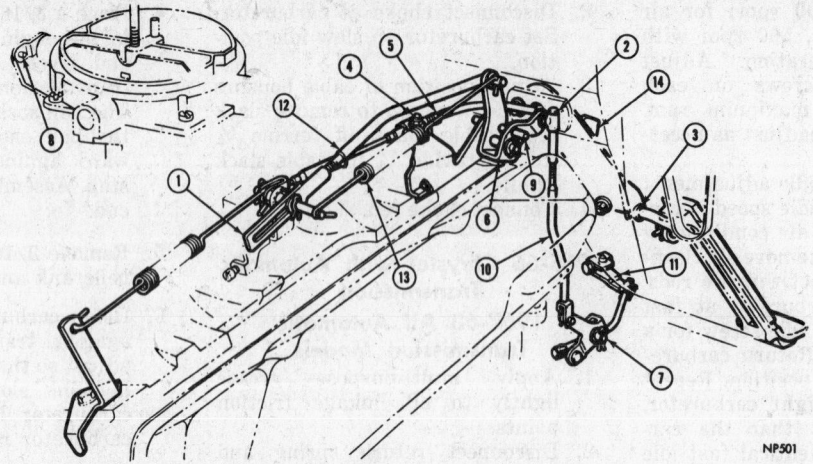

NP501

1967 throttle linkage—all automatic transmission (© Chrysler Corp.)

1965-66 Imperial Automatic Transmission Models

1. Apply multi-purpose grease lightly to linkage friction points.
2. Disconnect return spring and slotted transmission rod from carburetor lever pin.
3. Disconnect choke. Set carburetor on slow idle position.
4. Hold transmission lever forward against stop. Adjust length of transmission rod so that rear end of adjusting link slot just contacts the carburetor lever pin. Lengthen the rod one full turn.

5. Replace slotted adjustment rod on carburetor lever pin. Replace transmission linkage return spring.
6. Adjust position of cable housing ferrule in clamp to remove all cable slack. Back off ferrule ¼ in.
7. Connect choke rod.

1969-70 Automatic Transmission Models

This procedure is the same as that outlined above for the 1965-66 Imperial Automatic Transmission Models with the substitution of the following steps:

2. Disconnect return spring.

4. Loosen transmission throttle rod adjustment lock screw. Hold transmission lever firmly forward against stop. On engines with solenoid idle stops, the solenoid plunger must also be in its fully extended position. Adjust transmission rod length by pulling forward on the slotted link so that rear edge of slot is against carburetor lever pin. Tighten locking screw.

5. Replace transmission linkage return spring.

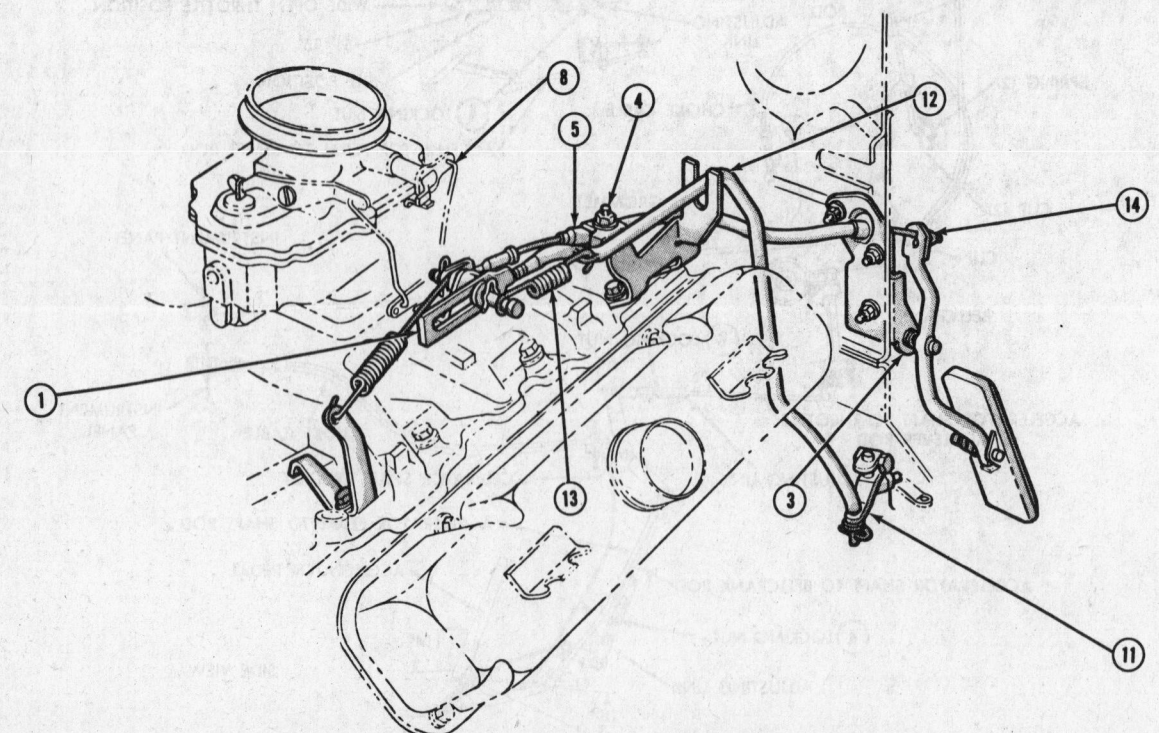

Throttle linkage, all 1969 automatic transmission models (© Chrysler Corp.)

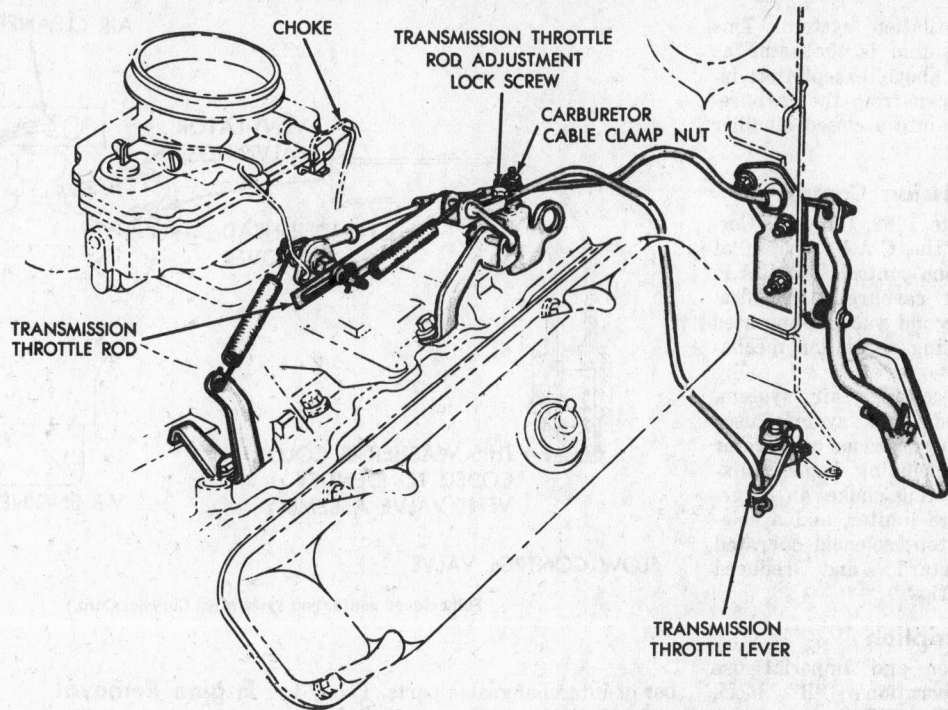

Throttle linkage, all 1970 automatic transmission models (© Chrysler Corp.)

Cooling System

Detailed information on cooling system capacity is in the Capacities table.

Information on the water temperature gauge is in the Unit Repair Section.

Water Pump R & R—1964-70

1. Drain cooling system. Remove upper half of fan shroud, if so equipped.
2. Loosen the power steering pump, idler pulley, and alternator. Remove all belts.
3. Remove fan, spacer or fluid drive, and pulley.
4. Remove pump attaching bolts.

5. Reverse procedure to install. Torque pump attaching bolts to 30 ft. lbs. and fan nuts to 15 ft. lbs. Adjust power steering belt to have 5/32 in. deflection at midpoint of belt; fan belt, 3/32 in.; and alternator belt, ¼ in.

Thermostat R & R—1964-70

1. Drain cooling system down to thermostat level or below.
2. Remove upper radiator hose from thermostat housing at top of water pump housing.
3. Remove thermostat housing and thermostat. Discard gasket.
4. Using a new gasket, install thermostat so pellet end is toward engine.
5. Replace upper hose and fill radiator with coolant to 1¼ in. below filler neck.

Radiator Core R & R—1964-70

1. Drain coolant.
2. On automatic transmission vehicles, disconnect oil cooler lines at radiator bottom tank.
3. Remove upper and lower hoses.
4. Remove shroud from radiator.
5. Remove radiator attaching screws and lift out radiator.
6. Reverse procedure to install.

Engine

The General Engine and Tune-Up specifications tables list information about the engines. The engine can be identified by the Engine Identification Code table.

Engine crankcase capacities are listed in the Capacities table.

Correct torque wrench readings and head bolt tightening sequences are in the Torque Specifications table.

Positive Crankcase Ventilation

All models are equipped with a positive crankcase ventilation system. Air is drawn into the engine through the oil filler cap and circulated through the engine. The air and vapors are drawn out of the engine through the rocker arm cover, a flow control valve, and the intake manifold, into the combustion chambers where they are consumed.

Vehicles sold in California from 1965 to 1967, and all vehicles from 1968 on, have a fully closed

BOLT — OUTLET ELBOW
THERMOSTAT
LOCK WASHER
TEMPERATURE SENDING UNIT
BOLT AND LOCKWASHER
GASKET
PLUGS
IMPELLER
GASKET
SEAL
SHAFT AND BEARING
HUB
BOLT AND LOCKWASHER
HOUSING
GASKET
HOUSING
SLINGER
KR233C

Water pump—383, 413 and 440 cu. in. (© Chrysler Corp.)

crankcase ventilation system. This fully closed system is the same as that described above, except that intake air is drawn from the carburetor air cleaner into a closed oil filler cap.

Exhaust Emission Control

From 1966 to 1969, Chrysler Corporation uses the C.A.P. system of exhaust emission control. The C.A.P. system utilizes carburetor modifications and a second vacuum-operated distributor timing regulation mechanism.

In 1970, a cleaner air system, C.A.S., is used. This system uses heated intake air; various carburetor modifications including leaner mixtures, a fast acting choke, an external idle mixture limiter, and a solenoid throttle stop; solenoid operated distributor retard; and reduced compression ratios.

Engine Description

The Chrysler and Imperial use Chrysler Corporation's "B" block type engines. The "B" block type engines are divided into two series, the low block series and the high block series. These series differ in block deck height, main bearing journal size, connecting rod length, and push rod length. In other respects, these engines are similar and have a number of interchangeable parts. The 361 and 383 cu. in. engines are low block engines. The 413 and 440 cu. in. engines are high block engines. These engines are all conventional overhead valve V8's with wedge shaped combustion chambers and deep blocks extending well below the crankshaft centerline.

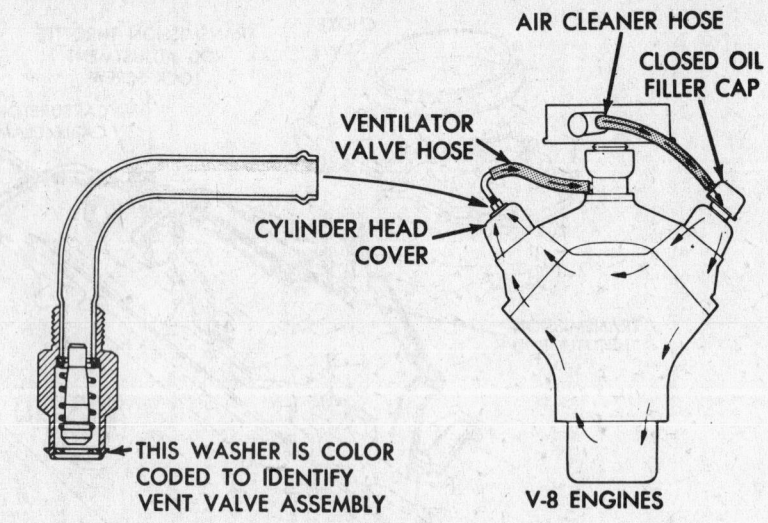

Fully closed ventilation system (© Chrysler Corp.)

Engine Removal
1964-70
1. Scribe outline of hood hinge brackets and remove hood.
2. Drain cooling system. Remove battery.
3. Remove all hoses, fan shroud, and disconnect oil cooler lines, then remove radiator.

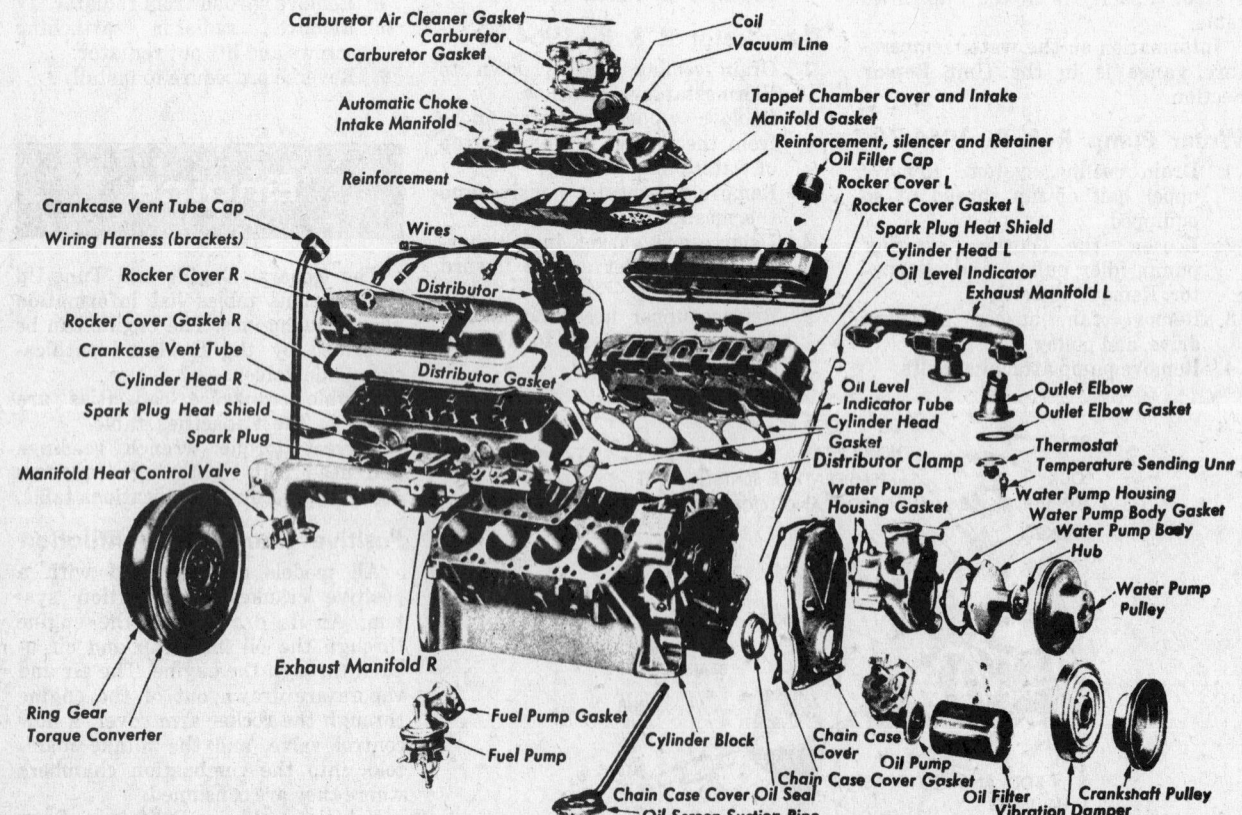

1964-70 engine external parts (© Chrysler Corp.)

4. Disconnect fuel lines and wires attached to engine units.
5. Remove air cleaner and carburetor.
6. Attach lifting fixture.
7. Raise vehicle on hoist and install engine support fixture on frame to support rear of engine.
8. Drain transmission and torque converter.
9. Disconnect exhaust pipes at manifolds, driveshaft, wires, linkage, cable, and oil cooler lines at transmission.
10. Remove engine rear support crossmember and transmission from vehicle.
11. Lower vehicle and attach chain hoist to fixture.
12. Remove engine front mounting bolts and, with chain hoist, work engine slowly out of the chassis.
13. Install in reverse of above.

Manifolds

Intake Manifold R & R —All Models

1. Drain cooling system.
2. Remove alternator, carburetor, air cleaner, and fuel line.
3. Disconnect accelerator linkage.
4. Remove vacuum control tube at carburetor and distributor.
5. Disconnect distributor cap, coil wires, and heater hose.
6. Disconnect heat indicator sending unit wire.
7. Remove intake manifold. Reverse procedure to install. Torque manifold bolts to correct figure from Torque Specifications Table.

Exhaust Manifold R & R— All Models

1. Remove spark plugs.
2. Remove alternator from right cylinder head.
3. Disconnect exhaust pipe from exhaust manifold.
4. Unbolt manifold from cylinder head. Slide manifold off studs and away from cylinder head.
5. Reverse procedure to install.

Torque manifold bolts to correct figure from Torque Specifications Table. Tighten exhaust pipe bolts to 40 ft. lbs.

Cylinder Head

All Engines—1964-70

1. Drain the cooling system. Remove alternator, carburetor air cleaner and fuel line. Disconnect the accelerator linkage. Remove closed ventilation system and evaporative control system (if so equipped). Remove the vacuum control tube at carburetor and distributor. Disconnect the distributor cap, coil wires, and heater hose. Disconnect the heat indicator sending-unit wire. Remove spark plugs, intake manifold, ignition coil, and carburetor as an assembly. Remove the tappet chamber cover, rocker covers, and gaskets.
NOTE: on air-conditioned cars, No. 8 cylinder exhaust valve must be open to allow clearance between the right bank cylinder-head cover and the heater housing.
2. Remove exhaust manifolds, the rocker arm and shaft assemblies. Remove the pushrods and place them in their respective slots in a holder. Remove the 17 head bolts from each cylinder head, then remove cylinder heads.
3. Reinstall in reverse of above.

Rocker Shaft Removal

1. Remove the carburetor air cleaner and pull the wires off the spark plugs. Remove the rocker covers.
NOTE: a new gasket must be used on reinstallation.
2. Loosen the rocker shaft retaining bolts, gradually and alternately, until all valve spring tension has been relieved. Remove rocker shaft bolts and retainers. Then, lift off rockers and shaft as an assembly.

Rocker Arm Installation

1. Install pushrods, small end first.
2. Install rocker shaft assembly, being sure to install the long stamped steel retainers in the number two and four positions. Install rocker shafts so that the 3/16 in. dia. rocker arm lubrication holes point downward into the rocker arm, toward the valve end of the rocker arms. This is necessary to provide proper lubrication.
3. Gradually tighten rocker shaft bolts to 25 ft. lbs., allowing time for tappets to bleed down to their operating length.
4. Install rocker covers with new gaskets. Torque to approximately 40 in. lbs.

Valve System

Valves

Detailed information on the valves and valve guides is in the Valve Specifications table.

A general discussion of valve clearance and a chart showing how to read pressure and vacuum gauges to diagnose engine troubles is in the Unit Repair Section.

Valve tappet clearance for each engine is given in the Tune-up Specifications table.

Valve Guides

Separate valve guides are not used. The guide is cast, integrally, with the cylinder head.

In cases of excessive stem-to-guide clearance, valves with oversize stems are available in the following increments: .005 in., .015 in. and .030 in. When installing valves with oversize stems, the guides must be reamed. Reamers to accommodate the oversize stems are available.

Caution Do not attempt to ream the valve guides in their present worn state to .030 in.,

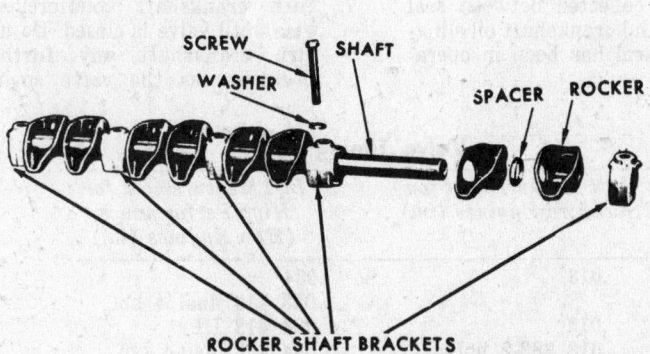

ROCKER SHAFT BRACKETS

Rocker shaft assembly (© Chrysler Corp.)

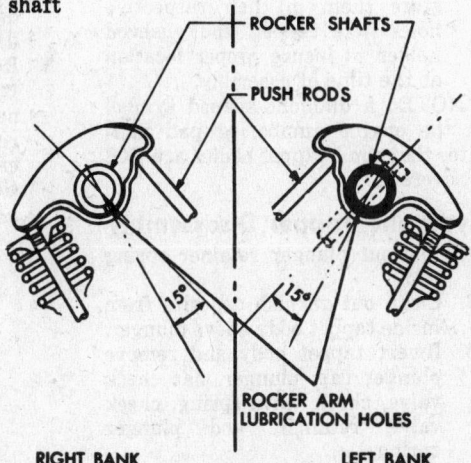

ROCKER SHAFTS

PUSH RODS

15°   15°

ROCKER ARM LUBRICATION HOLES

RIGHT BANK   LEFT BANK

Rocker arm positioning (© Chrysler Corp.)

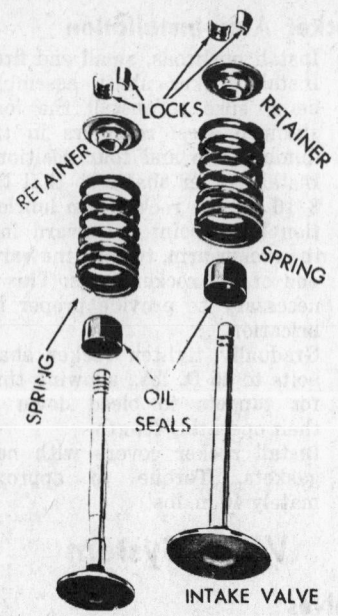

LOCKS

RETAINER   RETAINER

SPRING

SPRING

OIL SEALS

INTAKE VALVE

EXHAUST VALVE

Valve assembly
(© Chrysler Corp.)

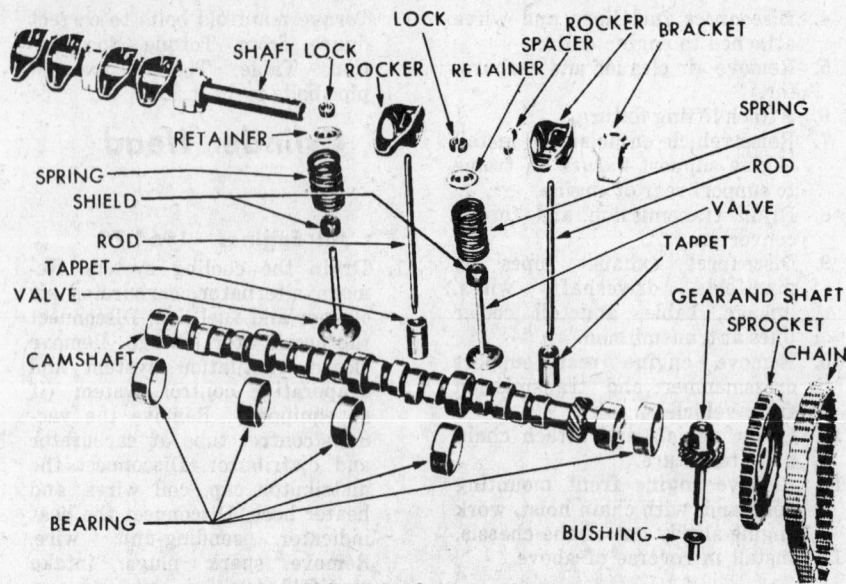

Valve train (© Chrysler Corp.)

or even .015 in., in one step. To maintain original bore angle, it is important to use step reaming procedures of .005 in., .015 in., then, if necessary, .030 in.

Hydraulic Tappet Removal

The tappet can be removed without removing the intake manifold or cylinder heads by following this procedure:

1. Remove rocker arm covers.
2. Remove rocker arms and shaft assembly.
3. Remove pushrods and place them in their respective places, on the bench, to retain their identity.
4. Slide tappet puller through pushrod opening in the cylinder head, and seat the tool firmly in the head of the tappet.
5. Pull tappet out of bore with a twisting motion.
   If all tappets are to be removed, store them in their respective holes in a tappet, and pushrod holder to insure proper location at the time of assembly.

NOTE: a diamond-shaped symbol on the engine numbering pad indicates that some tappet bodies are .008 in. oversize.

Hydraulic Tappet Disassembly

1. Pry out plunger retainer spring clip.
2. Clean out varnish deposits from inside tappet body above plunger.
3. Invert tappet body and remove plunger cap, plunger, flat check valve, check valve spring, check valve retainer. and plunger spring.
4. Clean tappet parts in solvent to remove all varnish and carbon.

NOTE: tappet parts are not interchangeable. Do not mix parts.

5. If tappet plunger shows scoring or wear, check valve is pitted, or plunger valve seat shows any condition which would prevent the valve from seating properly, replace tappet.

Stuck Hydraulic Lifters

If sticking has been experienced with hydraulic lifters, before attempting to take them out of the engine, first run a good solvent through the engine by mixing the solvent with the engine oil in order to remove the gum, tar, and oil residue from the bottom of the lifter. This will let it come up out of its bore easily. The residues of oil sometimes make it very difficult to pull a lifter body up out of its bore.

Timing Case Cover Removal

1. Drain coolant, remove radiator and water pump assembly.
2. Remove crankshaft vibration damper attaching bolt. Remove damper with a puller.
3. Remove chain cover and gasket. It is normal to find particles of neoprene collected between seal retainer and crankshaft oil slinger after seal has been in operation.

4. Slide crankshaft oil slinger off end of crankshaft.
5. Reverse procedure to install.

Checking Valve Timing

1. Turn crankshaft until No. 6 exhaust valve is closing and No. 6 intake valve is opening.
2. Insert a ¼ in. spacer between rocker arm pad and stem tip of No. 1 intake valve.
3. Install a dial indicator so plunger contacts valve spring retainer as nearly perpendicular as possible.
4. Allow spring load to bleed tappet down. Zero the indicator.
5. Turn the crankshaft clockwise until the intake valve has lifted the amount specified in the accompanying chart. The timing, read at the timing indicator on the chain case cover, should read from 10 degrees BTDC to 2 degrees ATDC.
6. If the reading is not within the specified limits: Inspect timing sprocket index marks, inspect timing chain for wear, and check accuracy of timing indicator marks.
7. Turn crankshaft counterclockwise until valve is closed. Do not turn crankshaft any further clockwise, as the valve spring

Valve Timing		
	Lift Measurement for Standard Engines (in.)	Lift Measurement for High Performance (HP) Engines (in.)
1964	.013	.034
		.033 413 dual 4 bbl.
1965	.013	.034 413 HP
1966	.013 383-2 bbl.	.034 383, 440-4 bbl.
1967-70	.025	.033 440 HP

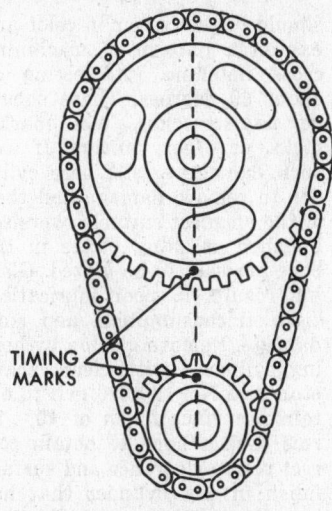

TIMING MARKS

Timing mark alignment

might bottom, causing serious damage. Remove indicator and spacer.

Engine Lubrication

Oil Pump

Removal

1. Remove oil pan and filter assembly.
2. Remove oil pump from bottom side of engine.

Disassembly

1. Remove filter base and oil seal ring.
2. Remove pump rotor and shaft. Lift out outer pump rotor.
3. Remove oil pressure relief valve plug. Lift out spring and relief valve plunger.

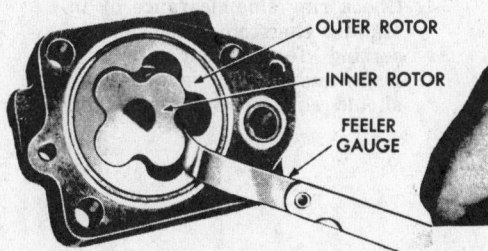

OUTER ROTOR
INNER ROTOR
FEELER GAUGE

Measuring clearance between oil pump rotors
(© Chrysler Corp.)

Inspection

1. Clean all parts thoroughly in solvent.
2. Inspect mating face of oil pump cover. Replace cover if face is scratched or grooved.
3. Lay straight edge across oil pump cover face. If .0015 feeler gauge can be inserted between cover and straight edge, replace cover.
4. Measure outer rotor. If rotor length is less than 0.943 in., re-

place rotor. If rotor diameter is less than 2.469 in., replace rotor.
5. Measure inner rotor. If rotor length is less than 0.942 in., replace rotor.
6. Install outer rotor in pump body. Holding rotor to one side, measure clearance between rotor and body. If clearance is greater than 0.014 in., replace pump body.
7. Install inner rotor into pump body. Place straight edge across body between bolt holes. If feeler gauge thicker than .004 in. can be inserted between rotors and body, replace body.
8. Measure clearance between tips of inner and outer rotors where they are opposed. If clearance exceeds .010 in., replace inner and outer rotors.

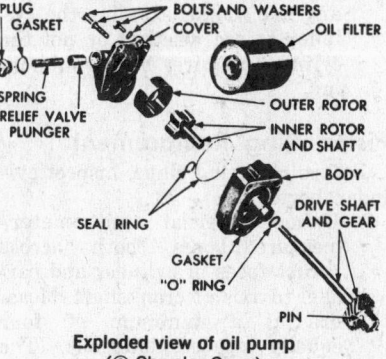

PLUG
GASKET
BOLTS AND WASHERS
COVER
OIL FILTER
SPRING
RELIEF VALVE PLUNGER
OUTER ROTOR
INNER ROTOR AND SHAFT
SEAL RING
BODY
DRIVE SHAFT AND GEAR
GASKET
"O" RING
PIN

Exploded view of oil pump
(© Chrysler Corp.)

Assembly and Installation

1. Assemble pump using new parts as required.
2. Install new seal rings between filter base and body. Torque bolts to 10 ft. lbs.
3. Install new O-ring seal on pilot of oil pump before attaching oil pump to cylinder block.
4. Install oil pump on engine using a new gasket. Tighten attaching bolts to 30 ft. lbs. Install oil filter assembly and oil pan.

Oil Pan Removal and Installation

From under the car, disconnect the exhaust pipes at the exhaust manifolds. Loosen the clamp at the Y connection where the two exhaust pipes come together, and remove the Y exhaust pipes. With dual exhaust, disconnect both pipes at manifolds. Disconnect the steering idler arm and let the steering linkage drop down out of the way. Remove the transmission (or converter) dust shield at the front of the flywheel housing.

Remove the bolts that hold the oil pan to the block. Turn the crankshaft in order to get the front counterweight out of the way and turn the pan counterclockwise to clear the oil pump screen. Then, slide the oil pan backward and down. Reverse procedure to install pan.

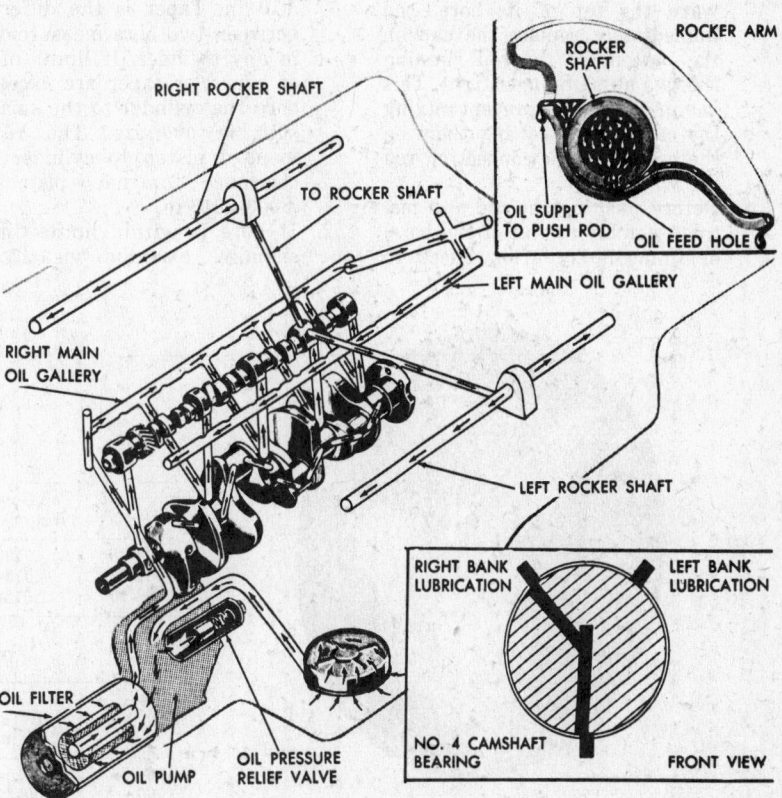

RIGHT ROCKER SHAFT
TO ROCKER SHAFT
ROCKER SHAFT
ROCKER ARM
OIL SUPPLY TO PUSH ROD
OIL FEED HOLE
LEFT MAIN OIL GALLERY
RIGHT MAIN OIL GALLERY
LEFT ROCKER SHAFT
RIGHT BANK LUBRICATION
LEFT BANK LUBRICATION
OIL FILTER
OIL PUMP
OIL PRESSURE RELIEF VALVE
NO. 4 CAMSHAFT BEARING
FRONT VIEW

Engine lubrication (© Chrysler Corp.)

Oil Filter Replacement

1964-70

The oil filter is at the lower left front corner of the engine. Unscrew the filter from its base. Wipe the base clean. Screw new filter on base until gasket contacts base. Tighten one-half turn more by hand or as indicated on new filter and check for leaks. The filter should be replaced at every other oil change.

Pistons, Connecting Rods and Main Bearings

Rod and Piston Assembly Removal

1. Remove the cylinder head and oil pan. Select two pistons in the down position, insert a good cylinder ridge reamer into those two cylinders and remove the ring ridge.
2. Turn the crankshaft until two other pistons are in the same down position and repeat the ridge-removing operation.
3. From underneath the car, select two connecting rods in the down position, take off the locking device (cotter pin or pal nut) and remove the two lower connecting rod nuts. Tap the cap of the connecting rod gently, and slide it off the two bolts. Push that rod and piston assembly up toward the top of its bore, and immediately replace the cap on the bottom of the rod running the two nuts up finger tight. This is a precaution to prevent mixing the caps up or getting them on the bottom of the connecting rod the wrong way.
4. Before pushing the rod and piston assembly up out of the bore, or immediately after, check to

ascertain if the number of the connecting rod is stamped on the bottom of the rod. If it is not, either file or mark the rod with the cylinder number so that it can be replaced in the cylinder from which it was removed.

Piston Inspection

1. Thoroughly clean the pistons and remove all traces of carbon from the ring grooves. Drill out the oil drain holes in the back of the lower ring grooves. Examine the ring grooves for bell mouth condition. If the groove is bell mouthed it will either have to be machined to a square corner and compensating rings installed, or new pistons and rings installed.
2. Examine the thrust face of the piston for scores or scratches. If any are found, examine the cylinder to see whether or not the cylinder requires boring or honing.

Piston Ring Replacement

Before replacing rings, inspect cylinder bores.

1. Using internal micrometer, measure bores, both across thrust faces of cylinder and parallel to axis of crankshaft. Measure at a minimum of four equally spaced locations. The bore must not be out of round by more than .005 in. The bore must not be tapered more than .010 in. Taper is the difference between two bore measurements in any cylinder. If limits of out of round or taper are exceeded, bore the cylinder to the smallest suitable oversize. The recommended piston to cylinder wall clearance for new pistons is .0003-0013 in.
2. If bore is within limits dimensionally, examine visually. It

should be dull silver in color and exhibit a pattern of machining cross hatching intersecting at about 60 degrees. There should be no scratches, tool marks, nicks, or other damage. If any such damage exists, bore cylinder to remove damage and then to the smallest suitable oversize. Polished or shiny places in the bore are said to be glazed. Glazing results in poor lubrication, high oil consumption, and ring damage. Remove glazing by honing cylinders with clean, sharp stones of No. 180-220 grit to obtain a surface finish of 15 - 35 rms. Use a hone to obtain correct piston clearance and surface finish in any cylinder that has been bored. Any machining operations should be done with the cylinder block completely disassembled. Hot tank the cylinder block after honing or boring. To remove minor glazing when honing equipment is not available, run emery cloth back and forth across glazed area perpendicular to the axis of the bore. Scrub the block and bores thoroughly with soap and water to remove all grit after using emery cloth.

NOTE: the emery cloth method should be used only as a last resort as it is a method much inferior to honing.

3. If cylinder bore is in satisfactory condition, place each ring in its bore in turn and square it in the bore with the head of a piston. Measure ring end gap. If ring gap exceeds limits, try another ring. If ring gap is too small, file end of ring to obtain correct gap.
4. Check ring side clearance by installing rings on piston, and inserting feeler gauge between ring and lower land. Gauge should slide freely around ring

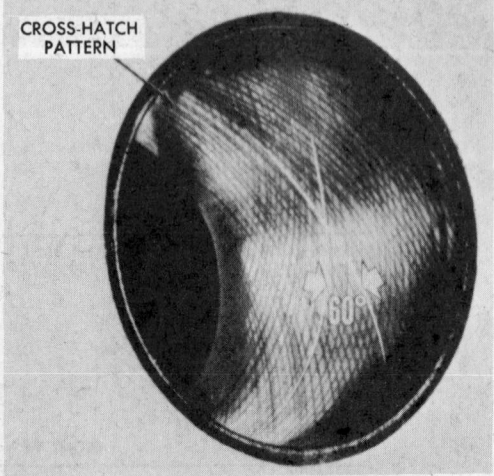

CROSS-HATCH PATTERN

Cylinder wall cross-hatching pattern
(© Chrysler Corp.)

Piston Ring End Gap

	Top Compression Ring Gap (in.)	Bottom Compression Ring Gap (in.)	Oil Control Ring Gap (in.)
1964-65	.013-.025	.013-.025	.013-.055
1966-67	.013-.025	.013-.025	.015-.055
1968-70	.013-.023	.013-.023	.015-.055

Piston Ring Side Clearance

	Top Compression Ring Side Clearance (in.)	Bottom Compression Ring Side Clearance (in.)	Oil Control Ring Side Clearance (in.)
1964-65	.0015-.0030	.0015-.0040	.009
1966	.0015-.0040	.0015-.0040	.0002-.0050
1967-70	.0015-.0040	.0015-.0040	.0000-.0050

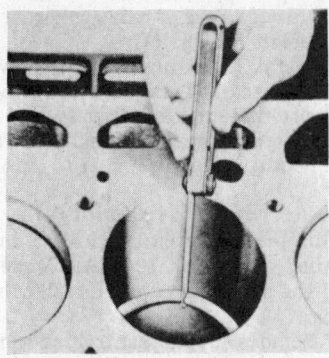

Measuring piston ring end gap
(© Ford Motor Company)

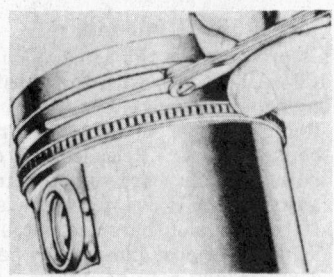

Measuring piston ring side clearance
(© Ford Motor Company)

circumference without binding. Any wear will show as a step on the lower land. Replace any pistons having high steps.

5. Space ring gaps at equidistant intervals around piston circumference. Be sure to install piston in its original bore. Install short lengths of rubber tubing over connecting rod bolts to prevent damage to rod journal. Install ring compressor over rings on piston. Lower piston and rod assembly into bore until ring compressor contacts block. Using wooden handle of hammer push piston into bore while guiding rod onto journal.

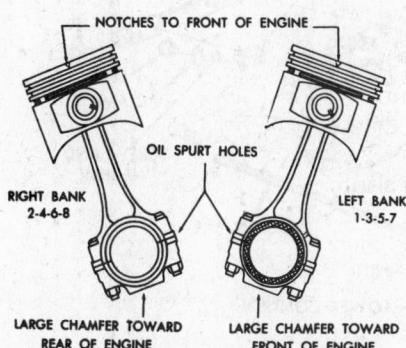

NOTCHES TO FRONT OF ENGINE

OIL SPURT HOLES

RIGHT BANK 2-4-6-8

LEFT BANK 1-3-5-7

LARGE CHAMFER TOWARD REAR OF ENGINE

LARGE CHAMFER TOWARD FRONT OF ENGINE

Piston and connecting rod assembly

Rod and Piston Assembly Installation

The pistons are assembled to the engine with the V notch in the head of the piston toward the front.

Bearings

Detailed information on engine bearings is in the Crankshaft Bearing Journal Specifications table.

Rear Main Bearing Oil Seal

To install the upper half of the rear main bearing seal, the crankshaft must be removed from the engine.

Upper Rear Main Seal Installation

1. Install a new rear main bearing oil seal in cylinder block so that both ends protrude.
2. Tap seal down into position, using installation tool with bridge removed.
3. Cut off portion of seal that extends above the block on both sides.

Lower Rear Main Seal Installation

1. Install a new seal in seal retainer so ends protrude.
2. Tap seal into position, using installation tool with bridge installed.
3. Trim off that portion of the seal that protrudes above the cap.

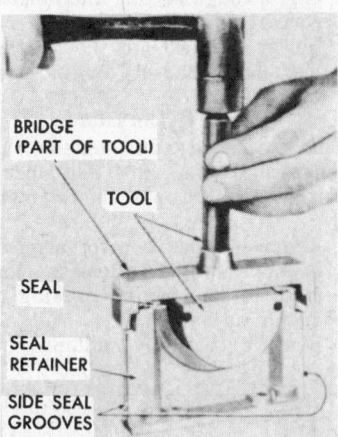

BRIDGE (PART OF TOOL)

TOOL

SEAL

SEAL RETAINER

SIDE SEAL GROOVES

Installing rear main bearing lower oil seal
(© Chrysler Corp.)

Side Seals Installation

Perform the following operations as rapidly as possible. The side seals are made from a material that expands quickly when oiled.

1. Apply mineral spirits or diesel fuel to the side seals. Failure to pre-oil the seals will result in an oil leak.
2. Install seals in the seal retainer grooves immediately.
3. Install seal retainer and tighten screws to 30 ft. lbs.

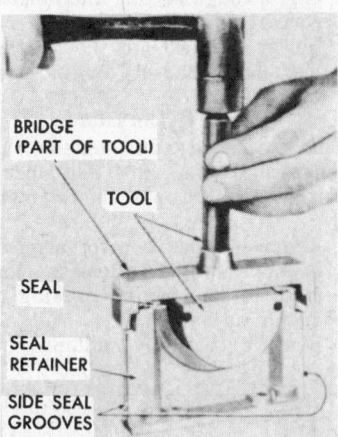

CHILTON TIME-SAVER

Top Half, Rear Main Bearing Oil Seal Replacement

The following method has proven a distinct advantage in most cases and, if successful, saves many hours of labor.

1. Drain engine oil and remove oil pan.
2. Remove rear main bearing cap.
3. With a 6 in. length of 3/16 in. brazing rod, drive up on either exposed end of the top half oil seal. When the opposite end of the seal starts to protrude, have a helper grasp it with pliers and pull gently while the driven end is being tapped. It is surprising how easily most of these seals can be removed by this method.

To replace the woven fabric-type seal:

1. Obtain a 12 in. piece of copper wire (about the same gauge as that used in the strands of an insulated battery cable).
2. Thread one strand of this wire through the new seal, about ½ in. from the end, bend back and make secure.
3. Thoroughly saturate the new seal with engine oil.
4. Push the copper wire up through the oil seal groove until it comes down on the opposite side of the bearing.
5. Pull (with pliers) on the protruding copper wire while the crankshaft is being turned and the new seal is slowly fed into place.

CAUTION: this snaking operation slightly reduces the diameter of the new seal and care will have to be used to keep the seal from slipping too far through the top half of the bearing.

6. When an equal amount of seal is extending from each side, cut off the copper wire close to the seal and tamp both ends of the seal up into the groove (this will tend to expand the seal again).

NOTE: don't worry about the copper wire left in the groove, it is too soft to cause damage.

7. Replace the seal in the cap in the usual way and replace the oil pan.

Front Suspension

General instructions covering the front suspension and how to repair and adjust it, together with information on installation of front wheel bearings and grease seals, are given in the Unit Repair Section.

Definitions of the points of steering geometry are covered in the Unit Repair Section. This article also covers troubleshooting front end geometry and irregular tire wear.

Figures covering the caster, camber, toe-in, kingpin inclination, and turning radius are in the Wheel Alignment table.

Overall car length and tire and brake drum sizes are in the General Chassis and Brake Specifications table.

Torsion Bar Springs

Contrary to appearance, the torsion bars are not interchangeable from right to left. They are marked with an R or an L, according to their location.

Torsion Bar Removal and Replacement

1964 All Models

1. Raise vehicle so that the front suspension is under no load.
2. Release load from torsion bar by backing off anchor adjusting bolt.
3. Remove balloon seal, lock ring, and plug from rear of torsion bar anchor.
4. Slide torsion bar toward rear to disengage forward end from lower control arm. Slide bar forward and down, disengaging it from the anchor. Remove torsion bar from vehicle.
5. Inspect and replace adjusting bolt, swivel, and balloon seal as necessary when installing tor-

sion bar. Be sure that the bar is being installed on the correct side.
6. Apply a liberal coating of a lithium base grease around each hex end of the torsion bar.
7. Install the torsion bar through the rear anchor.
8. Slide the balloon seal over the bar with the cupped side toward the rear.
9. Turn the bar until the anchor end is positioned about 120 degrees (at an eight or four o'clock position) down from the frame. If this is not done, the suspension cannot be adjusted to the correct height.
10. Engage the front end of the bar in the hex opening of the lower control arm.
11. Center the bar so that full contact of the hex ends is obtained at the anchor and control arm shaft. Install the lock ring. Make sure that the lock ring seats in its groove.
12. Pack the annular opening in the rear anchor full of lithium base grease. Position the lip of the seal in the groove in the anchor hub. Install the plastic plug into the rear end of the torsion bar anchor.
13. Install the adjusting bolt, swivel, and seat. Tighten bolt until about 1 in. of threads are showing above the swivel. This is only a rough setting and must be corrected later.
14. Lower vehicle to floor, then measure and adjust height.

1965-70 Chrysler

This procedure is the same as that detailed above for 1964 All Models, with the substitution of the following steps:
2. Remove upper control arm rebound bumper. Release load from torsion bar by backing off anchor adjusting bolt.
4. Remove bar by sliding out

through rear of rear anchor.
9. Delete Step 9.
13. Install the adjusting bolt, swivel, and seat. Turn adjusting bolt clockwise to load torsion bar.
15. Install upper control arm rebound bumper.

1965-66 Imperial

This procedure is identical to that detailed above for 1964 All Models, with the substitution of the following steps:
2. Remove upper control arm rebound bumper. Release load from torsion bar by backing off anchor adjusting bolt.
15. Install upper control arm rebound bumper.

1967-70 Imperial

This procedure is the same as that for 1964 All Models, with the substitution of the following steps:
2. Remove upper control arm rebound bumper. Release load on both torsion bars by turning adjusting bolts counterclockwise. If both torsion bars are not unloaded, serious damage or personal injury may result.
13. Install the adjusting bolt, swivel, and seat. Turn both adjusting bolts clockwise to place a load on both torsion bars.
15. Install upper control arm rebound bumper.

Front Height Adjustment

1. Jounce vehicle several times, releasing it on downward motion.
2. Measure distance A. For all 1964 Chrysler and 1964-66 Imperial models, this is measured from the lowest point of the lower control arm bushing housing to the floor. For 1965-70 Chrysler models, the measurement is taken from the lowest point of the adjusting blade. For 1967-70 Imperial models, measure from the lowest point of the front tor-

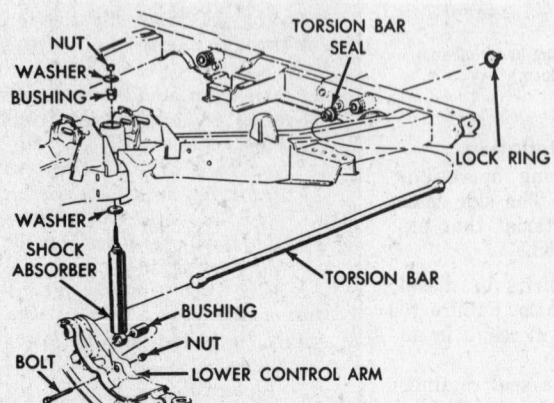

1965-70 Chrysler torsion bar
(© Chrysler Corp.)

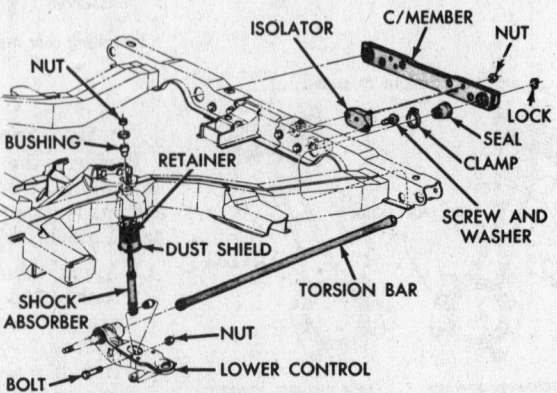

1967-70 Imperial torsion bar
(© Chrysler Corp.)

Front Suspension Height

	Suspension Type	Chrysler (in.)	Imperial (in.)
1964		$2 \pm \frac{1}{8}$	$2 \pm \frac{1}{8}$
1965-66	Standard	$1\frac{1}{8} \pm \frac{1}{8}$	$2 \pm \frac{1}{8}$
	Heavy Duty	$1\frac{1}{8} \pm \frac{1}{8}$	$2\frac{3}{8} \pm \frac{1}{8}$
	Limousine		$2\frac{3}{8} \pm \frac{1}{8}$
1967	Standard	$1\frac{1}{8} \pm \frac{1}{8}$	$1\frac{3}{4} \pm \frac{1}{8}$
	Heavy Duty	$1\frac{1}{8} \pm \frac{1}{8}$	
1968	Standard	$1\frac{1}{8} \pm \frac{1}{8}$	$1\frac{3}{4} \pm \frac{1}{8}$
	Heavy Duty	$1\frac{1}{8} \pm \frac{1}{8}$	$1\frac{3}{4} \pm \frac{1}{8}$
1969-70		$1\frac{1}{8} \pm \frac{1}{8}$	$1\frac{3}{4} \pm \frac{1}{8}$

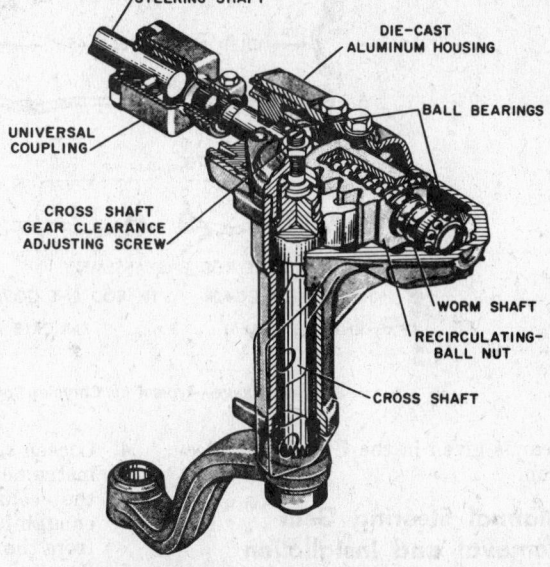

1964-70 manual steering gear (© Chrsyler Corp.)

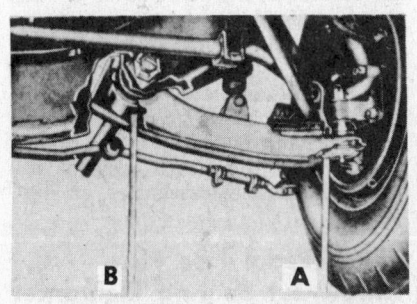

Measuring front suspension height—
1964 Chrysler and 1964-66 Imperial
(© Chrysler Corp.)

sion bar anchor at the rear of the lower control arm flange.
3. Measure distance B. This is the distance between the lowest point of the lower ball joint housing and the floor.
4. Subtract distance A from distance B to obtain front suspension height.
5. Measure the other side. There should be no more than ⅛ in. difference in height from one side to the other.
6. Adjust height, as necessary, by turning torsion bar adjusting bolt clockwise to increase height and counterclockwise to decrease height.

Jacking, Hoisting

Jack car at front under lower control arm and at rear under axle housing.

To lift at frame, use adapters so that contact will be made at points shown. Lifting pad must extend beyond sides of supporting structure.

Steering Gear

Manual Steering Gear
Instructions covering the adjustment of the steering gear are in the Unit Repair Section.

Power Steering Gear
Troubleshooting and repair instructions covering power steering

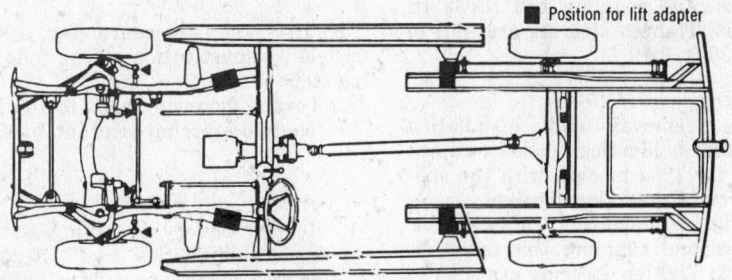

Positioning lift adapter (© Chrysler Corp.)

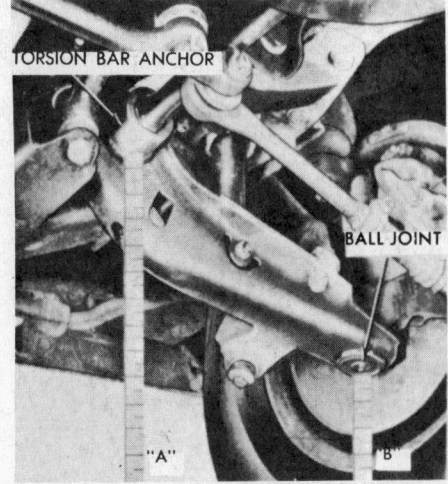

Measuring front suspension height, 1967-70 Imperial
(© Chrysler Corp.)

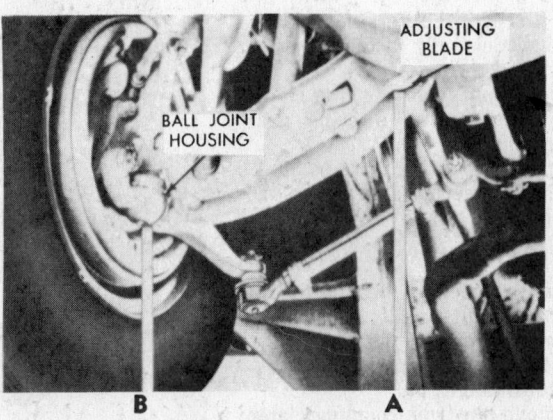

Measuring front suspension height, 1965-70 Chrysler
(© Chrysler Corp.)

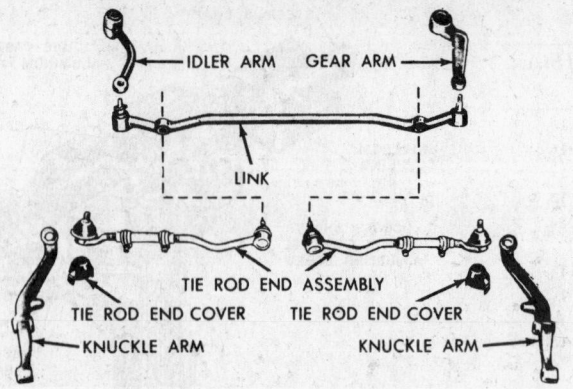

Steering linkage—typical (© Chrysler Corp.)

gear is given in the Unit Repair Section.

Manual Steering Gear Removal and Installation

1964

1. Remove steering arm retaining nut. Pull off steering arm.
2. Remove bolt from coupling clamp at upper end of steering gear wormshaft.
3. Loosen steering column jacket to instrument panel clamp. Slide the column assembly up far enough to disengage the coupling from the wormshaft.
4. Remove three steering gear housing bolts and remove steering gear from under vehicle.
5. Reverse procedure to install.
6. Tighten mounting bolts to 50 ft. lbs. and coupling bolt to 33 ft. lbs. Tighten steering arm nut to 120 ft. lbs.

1965

This removal and installation procedure is identical to that outlined above for 1964 models, with the substitution of the following step:
6. Tighten mounting bolts to 80 ft. lbs. and coupling bolt to 33 ft. lbs. Tighten steering arm nut to 120 ft. lbs.

1966

This procedure differs from that for 1964 models only in the substitution of the following steps:
2. Remove pin from coupling clamp at upper end of steering gear wormshaft.
6. Tighten mounting bolts to 80 ft. lbs. and steering arm nut to 120 ft. lbs.

1967

1. Remove steering gear arm retaining nut and lockwasher. Remove arm with puller.
2. Disconnect column shift gear selector linkage.
3. Remove pin from coupling clamp at upper end of steering gear wormshaft.

4. Loosen steering column jacket to instrument panel clamp. Slide the column assembly up far enough to disengage the coupling from the wormshaft.
5. Remove column lower support plate to floor pan bolts.
6. Remove housing bolts. Remove steering gear.
7. Reverse procedure to install.
8. Torque mounting bolts to 80 ft. lbs. and steering arm nut to 120 ft. lbs.

1968

Modify the procedure for 1967 models by the substitution of the following step:
8. Torque mounting bolts to 80 ft. lbs. and steering arm nut to 175 ft. lbs.

1969

Modify the procedure for 1967 models by substitution of the following step:
8. Torque mounting bolts to 100 ft. lbs. and steering arm nut to 175 ft. lbs.

1970

1. Remove energy absorbing steering column. See Unit Repair Section for this procedure.
2. Remove steering arm nut and lockwasher. Pull off steering arm.
3. Remove frame mounting bolts. Remove steering gear.
4. Reverse procedure to install. Torque mounting bolts to 100 ft.

lbs. and steering arm nut to 180 ft. lbs. When installing steering gear in vehicle, be sure that vehicle wheels are straight ahead and that steering gear is centered.

Clutch

The only practical service possible on the clutch assembly is adjusting the pedal free play. All other service requires the removal of the clutch assembly

If the clutch assembly is being removed because of chatter or malfunction, it is advisable to check to see if there is any oil leaking from the rear main bearing or from the transmission. Oil on the clutch facings will produce a noticeable chatter, or slipping.

Clutch Pedal Free Play (Throw-out Bearing Clearance)

This adjustment is done at the adjusting nut on the clutch release

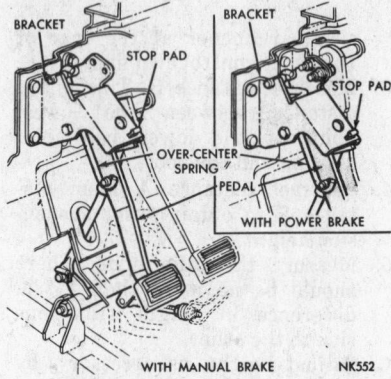

Clutch pedal and linkage—1965-70

rod. There should be 1 in. free-travel of the pedal before the throw-out bearing strikes the fingers of the clutch.

Clutch Assembly Removal

Remove the driveshaft. Disconnect the gear shift control rods and the speedometer cable. Support the back of the transmission and remove the bolts that hold the transmission to the bell housing.

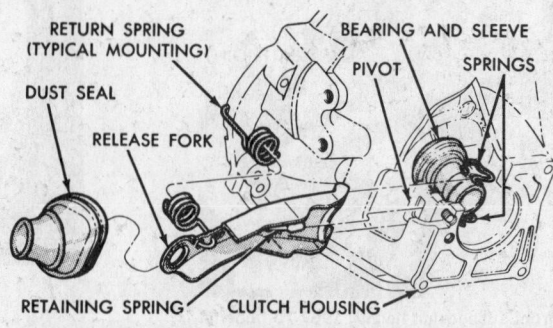

Clutch release fork, bearing and sleeve (© Chrysler Corp.)

Some shops prefer to replace the two upper bolts with two long pilots so the transmission can be slid straight back on the pilots. This eliminates the risk of bending the clutch disc.

On all models, remove the transmission assembly, the clutch housing lower pan and the clutch throw-out fork. The clutch cover assembly should be stamped, showing its relation to the flywheel so that it can be reassembled in the same position from which it was removed. The cover bolts should be removed, a few turns at a time, in order to avoid springing the clutch cover.

Disassembly service on the clutch requires special jigs and fixtures. Instructions in the use of the fixtures are supplied by the manufacturer.

Standard Transmission

Transmission refill capacities are in the Capacities table.

See Unit Repair Section for overhaul procedures.

Transmission Removal and Installation

1. Disconnect all attaching parts at the transmission.
2. Remove driveshaft.
3. Support the rear of the engine on a padded jack and take off the transmission crossmember.
4. Detach the transmission from the clutch housing and move down and out of the car.

5. Reverse procedure for installation.

Shift Linkage Adjustment

1964 Three Speed Floorshift

1. Remove screws, retaining ring and upper boot from the floor pan.
2. Disconnect first and reverse shift rod and disengage rod from lever.
3. Disconnect second and high shift rod and disengage rod from lever.
4. Place the transmission shift levers in neutral, then refer to illustration for adjustment details.

1964-65 Four Speed Floorshift

1. Remove the shift boot attaching screws, and slide the boot up on the shift lever.
2. Disconnect all the shift rods at the adjusting swivels.
3. Bend a ¼ in. diameter rod to a right angle. Insert through the aligning holes in the control rod levers. This locks all three levers in neutral positions.
4. Adjust the length of the three shift rods until the swivel stub shafts match the control rod lever holes. Install the stub shafts and secure with clips.
5. Remove the ¼ in. diameter aligning rod.
6. With transmission hand shift lever in third or fourth speed detent position, adjust the lever stop screws (front and rear) to provide from .020-.040 in. clearance between the lever and the stops. Tighten adjusting screw locknuts.
7. Test linkage adjustment for ease

of shifting and crossover operation.

Caution Accuracy of adjustment is very important because there is no reverse gear interlock to prevent engaging two gears at the same time.

8. Slide the boot down the shift lever shaft to the floor and tighten with attaching screws.

1965 Column Shift

1. Disconnect second-third rod from column lever and first-reverse rod from transmission lever. Position both transmission levers in neutral positions.
2. If ends of lower column levers can be moved more than 1/16 in. up and down, loosen the two upper bushing screws (just above the lever slot) and rotate the bushing downward to remove play. Tighten screws.
3. Use a screwdriver between the crossover blade and second-third lever, to engage the crossover blade with both lever pins.
4. Adjust the length of each control rod and connect the rods.
5. Remove screwdriver and check shifting.

1966-69 Column Shift

These models have an interlock to prevent engaging two gears at once. Add the following steps to the 1965 Column Shift adjustment procedure:
6. Shift transmission into neutral. Make sure clutch free play is correct.
7. Loosen clutch interlock rod swivel clamp bolt. Hold clutch

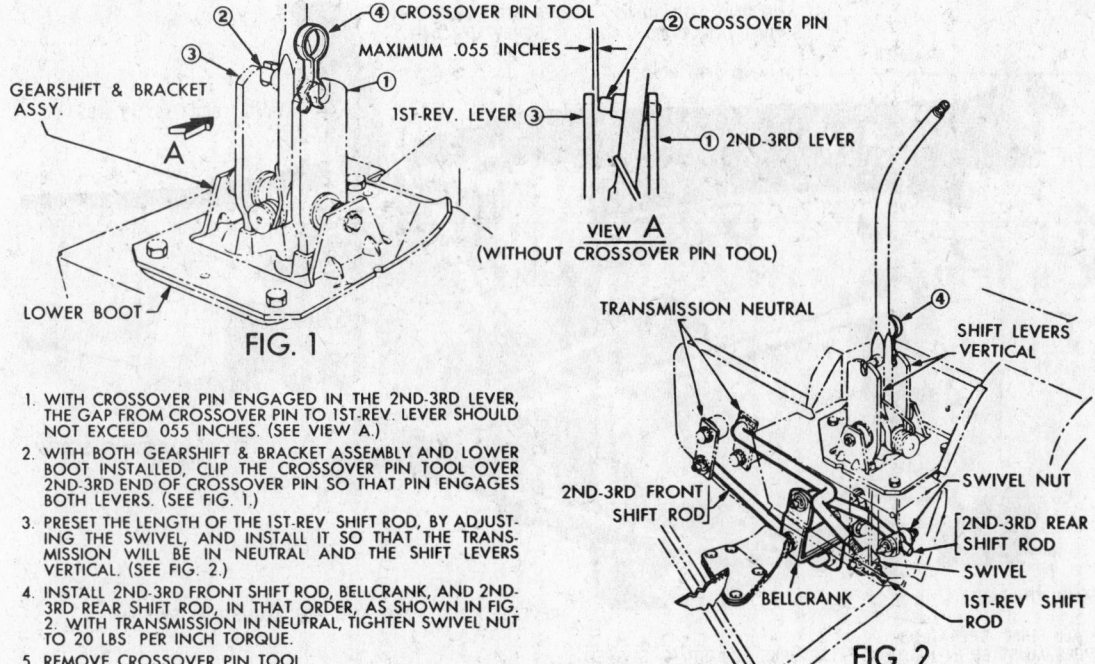

FIG. 1

1. WITH CROSSOVER PIN ENGAGED IN THE 2ND-3RD LEVER, THE GAP FROM CROSSOVER PIN TO 1ST-REV. LEVER SHOULD NOT EXCEED .055 INCHES. (SEE VIEW A.)
2. WITH BOTH GEARSHIFT & BRACKET ASSEMBLY AND LOWER BOOT INSTALLED, CLIP THE CROSSOVER PIN TOOL OVER 2ND-3RD END OF CROSSOVER PIN SO THAT PIN ENGAGES BOTH LEVERS. (SEE FIG. 1.)
3. PRESET THE LENGTH OF THE 1ST-REV SHIFT ROD, BY ADJUSTING THE SWIVEL, AND INSTALL IT SO THAT THE TRANSMISSION WILL BE IN NEUTRAL AND THE SHIFT LEVERS VERTICAL. (SEE FIG. 2.)
4. INSTALL 2ND-3RD FRONT SHIFT ROD, BELLCRANK, AND 2ND-3RD REAR SHIFT ROD, IN THAT ORDER, AS SHOWN IN FIG. 2. WITH TRANSMISSION IN NEUTRAL, TIGHTEN SWIVEL NUT TO 20 LBS PER INCH TORQUE.
5. REMOVE CROSSOVER PIN TOOL

FIG. 2

Three-speed floorshift unit—1964 (© Chrysler Corp.)

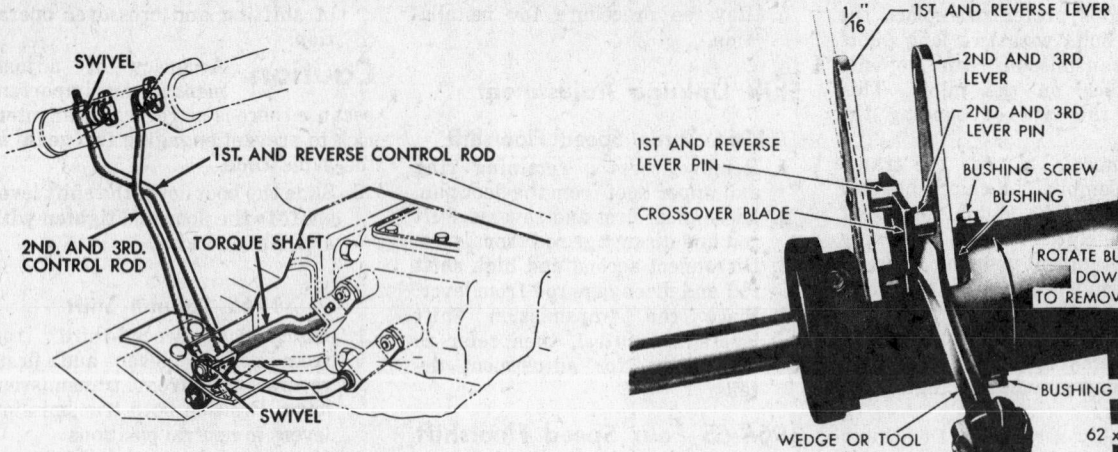

Column shift linkage, 1965-69
(© Chrysler Corp.)

Holding crossover blade in neutral position, 1964-70 column shift
(© Chrysler Corp.)

pedal to fully returned position. Slide swivel along interlock rod to engage pawl with lever. Tighten clamp bolt.

8. Disengage clutch. Shift transmission halfway between neutral and first. Let out clutch pedal. The interlock should only allow the pedal to return to one or two inches from the floor.

1970 Column Shift

1. Disconnect both shift rods from transmission levers. Check that transmission levers are in neutral positions.
2. Move shift lever to line up locat-ing slots in bottom of steering column shift housing and bear-

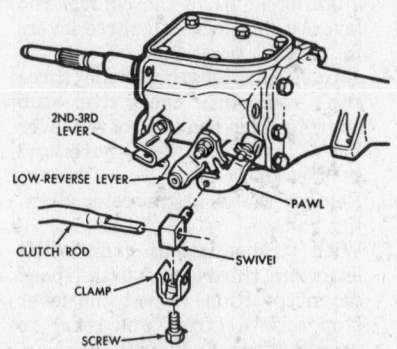

Three-speed interlock—1966-69
(© Chrysler Corp.)

ing housing. Install suitable tool in slot and lock ignition switch.

3. Place screwdriver between crossover blade and second-third lever so that both pins are engaged by crossover blade.
4. Set first-reverse transmission lever in reverse position. Adjust rod length and connect the rod.
5. Remove gearshift housing locating tool. Unlock ignition switch. Shift column lever to neutral position.
6. Install and adjust length of second-third rod.
7. Remove screwdriver. Steering column should lock in reverse but not in second position.

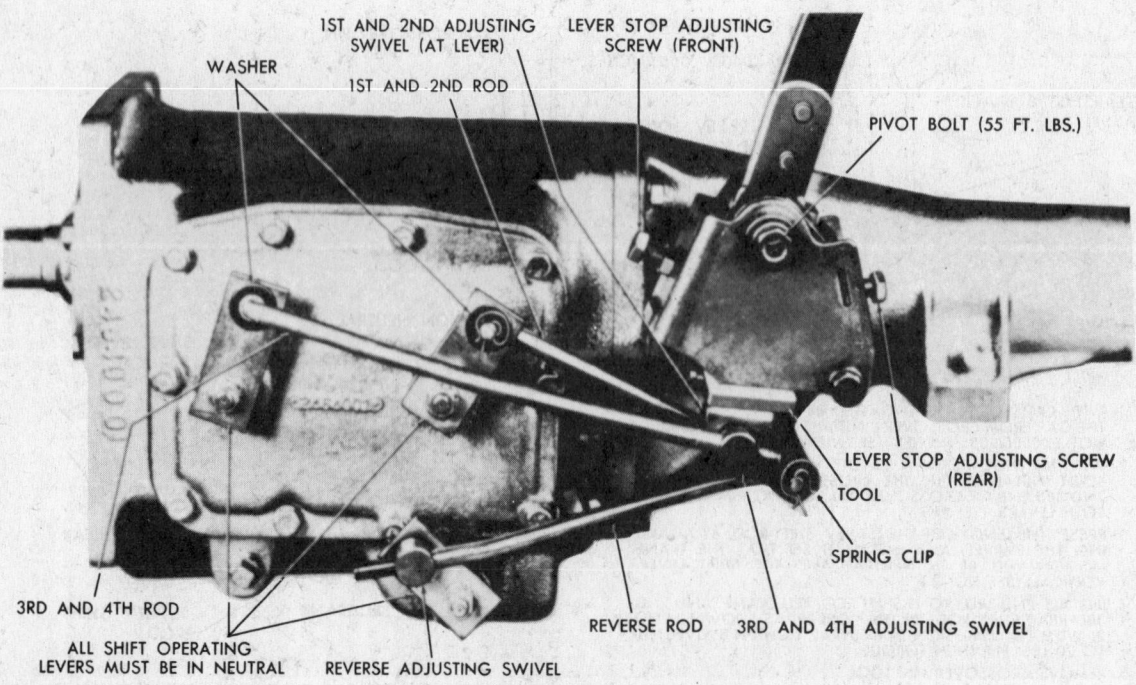

Four speed linkage, 1964-65 (© Chrysler Corp.)

Automatic Transmission

Anti-Stall Adjustment

The anti-stall consists of a diaphragm and a plunger with a small orifice. The air trapped behind the diaphragm is bled out at a specific rate by the orifice. The device acts to keep the throttle from snapping shut during the last ¼ in. of travel.

To check, open the throttle by hand and release. The closing of the throttle should be visibly slowed by the action of the anti-stall.

To adjust, have the engine at operating temperature and set adjusting screw so that the plunger has 1/16 in. travel after the throttle is fully closed.

Shift Linkage Adjustment

There are two basic types of Torqueflite linkages used in Chrysler and Imperial vehicles. A cable linkage is used on 1964-65 models, and a rod linkage is used on 1966-70 models. Adjustment procedures for both types are given below.

1964 Push Button Selector

1. Have an assistant hold the R button firmly depressed.
2. Hold the gearshift control cable guide centered in the hole of the transmission case and apply light pressure to bottom the assembly is the reverse detent. Rotate the adjustment wheel clockwise to line up an adjustment hole with the screw hole in the case. Counting this hole as number one, turn the wheel clockwise until the fifth hole lines up with the screw hole.
3. Install lock screw. Torque to 75 in. lbs.

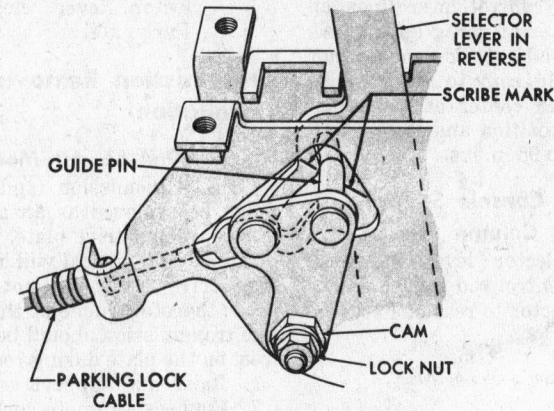

1965 Console selector adjustment
(© Chrysler Corp.)

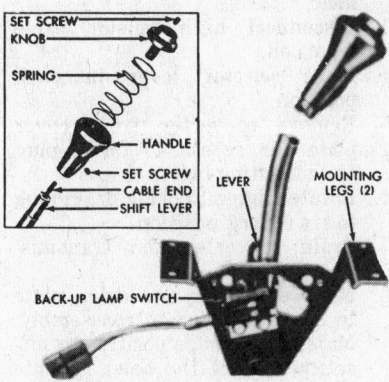

Automatic transmission console shifter—
1965-70
(© Chrysler Corp.)

4. Check to see that parking lock lever has clearance at both ends of instrument panel slot. To adjust the lever:
5. Loosen cable clamp bolt at transmission, making sure the cable housing is free to move in the parking sprag cover hole.
6. Move lever to OFF. Loosen two nuts attaching lock arm to push button housing.
7. Block lever 1/16 in. from OFF end of lever slot. Tighten the two nuts on the push button housing. Place lock lever in OFF position.
8. Loosen clamp bolt securing cable housing in parking lever cover on bottom of transmission extension housing. Tap end of clamp bolt lightly to release cable.
9. Pull gently outward on cable housing to limit of travel, release, and tighten clamp bolt to 10 in. lbs.
10. Check adjustment by allowing vehicle to roll on slight incline. The parking sprag should fully engage in ON position, and there should be no racheting noise in OFF.

1964 Console Selector, 1965 Column Selector

This adjustment procedure is similar to that for the 1964 Push Button Selector, with the substitution of the following steps:

1. Have an assistant hold the selector lever firmly in the 1 (Low) position.
2. Hold the gearshift control cable guide centered in the hole of the transmission case and apply light pressure to bottom the assembly in the low detent. Rotate the adjustment wheel clockwise until it contacts the case squarely. Turn the wheel counter-clockwise to line up an adjustment hole with the screw hole in the case. Counting this hole as number one, turn the wheel counter-clockwise until the fifth hole lines up with the screw hole.

4-7. Delete steps 4-7.

1965 Console Selector

Add the following to the procedure for the 1964 Console Selector:

11. With the selector lever in Reverse, look through the console clearance hole to see if scribe

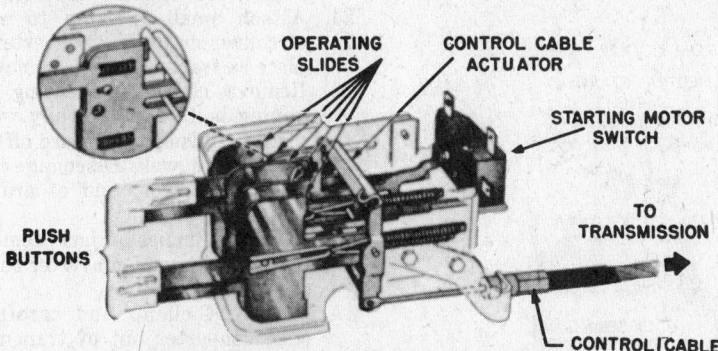

Typical push button mechanism (© Chrysler Corp.)

mark on control lever lines up with center of sprag lever guide pin. To adjust, loosen locknut and rotate cam to align scribe mark with center of guide pin. Hold in position and torque cam locknut to 95 in. lbs.

1966-69 Console Selector, 1966-69 Column Selector

1. Place selector lever in Park. Loosen control rod swivel clamp.
2. Move selector to rear of Park detent.

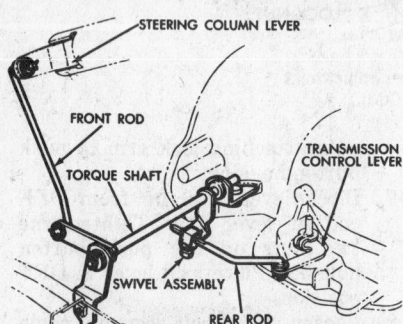

Automatic transmission columnshift linkage —1966 Chrysler and 1967-69 All (© Chrysler Corp.)

3. Check that transmission lever is in park position. Tighten swivel clamp screw to 100 in. lbs.

1970 Console Selector, 1970 Column Selector

1. Free adjustable rod ends.
2. Place selector lever in Park. Lock steering column. On console selector, line up locating slots in bottom of shift housing and bearing housing. Hold in place with a suitable tool.
3. Move selector to rear of Park detent.
4. Set adjustable rods to proper length.
5. Check adjustment:
   a. Detents and gate stops should be positive.
   b. Selector lever must not remain out of detent position when placed against gate and then released.
   c. Key start must occur with se-

lector lever held against Park gate.

Transmission Removal and Installation

1964-65 All Models

The transmission and converter must be removed as an assembly or the converter drive plate, front pump bushing, and oil seal will be damaged. The drive plate will not support a load; therefore, none of the weight of the transmission should be allowed to rest on the plate during removal.

1. Raise and support vehicle.
2. Connect a remote control starter switch to solenoid so that engine may be rotated from under vehicle.
3. Disconnect high tension cable from coil.
4. Place selector lever in park position.
5. Remove converter front cover plate for access to drain plug and mounting bolts.
6. Rotate engine to bring drain plug to six o'clock position.
7. Drain converter and transmission.
8. Mark converter and drive plate to locate position in reassembly. Matching is made positive by offsetting one of the holes in plate and in converter.
9. Rotate engine to five and seven o'clock positions and remove two drive plate bolts. Then rotate engine and remove remaining bolts. Do not rotate converter or drive plate by prying because this can distort the drive plate. Also, the starter should never be engaged if the drive plate is not attached to the converter with at least one bolt, or if the transmission case to engine block bolts have been loosened.
10. Disconnect battery ground cable.
11. Remove starting motor assembly.
12. Disconnect neutral starting switch wire.
13. Remove gearshift control cable to transmission adjusting wheel lock screw. Pull cable out of transmission case as far as pos-

sible. It may be necessary to back off on the adjusting wheel a few turns.
14. Insert small screwdriver above and slightly to right of gearshift cable. Disengage cable adapter lock spring by pushing screwdriver to right while pulling outward on cable.
15. Disconnect throttle rod from relay lever at left side of the transmission.
16. Disconnect oil cooler lines at transmission and remove oil filler tube.
17. Remove speedometer pinion and sleeve.
18. Loosen parking lock cable clamp bolt where cable enters the cover. Tap end of bolt lightly to release

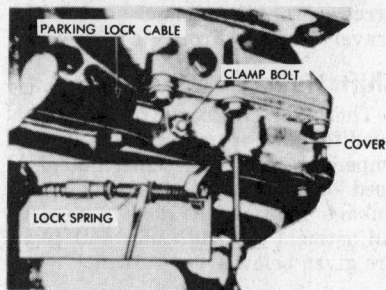

Removing parking lock cable (© Chrysler Corp.)

hold on cable. Remove housing cover lower plug. Insert screwdriver through hole and gently exert pressure against projecting portion of cable lock spring and withdraw lock cable.
19. Disconnect driveshaft at the rear universal joint and pull shaft assembly out of extension housing. (Except Imperial models.)
    Remove driveshaft center bearing housing bolts and slide shaft rearward to disengage front joint from front yoke. (Imperial models)
20. Remove rear engine mount insulator-to-extension housing bolts.
21. Install rear support fixture at engine and raise engine slightly.
22. Remove attaching bolts and the crossmember.
23. Place proper jack under transmission to support assembly.
24. Attach small C-clamp to edge of converter to hold converter in place as transmission is removed.
25. Remove converter housing retaining bolts and carefully work the transmission rearward off engine block dowels. Disengage converter hub from end of crankshaft.
26. Lower transmission jack, remove transmission and converter as an assembly.
27. Remove C-clamp and carefully slide converter out of transmission.

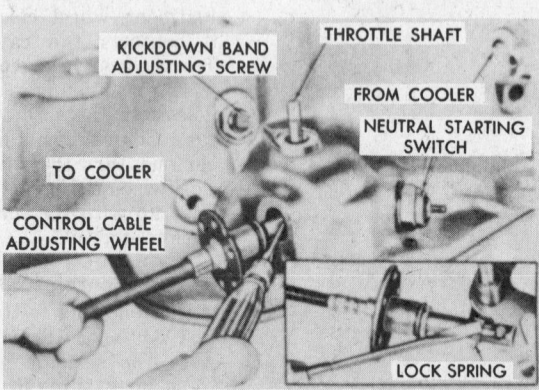

Removing gearshift control cable (© Chrysler Corp.)

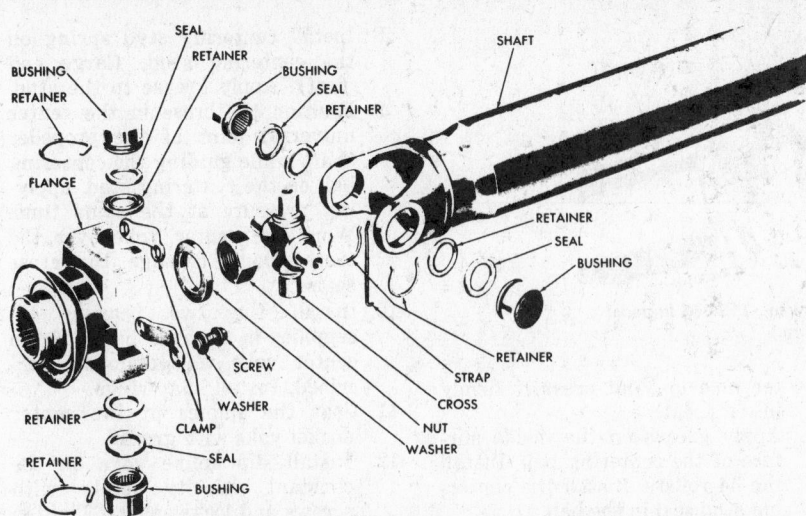

Rear cross and roller U-joint (© Chrysler Corp.)

28. Install in reverse of above. It is important that front pump rotors properly align to engage the front pump inner lugs. An aligning tool is necessary. The converter hub hole in crankshaft should be coated with wheel bearing lubricant.

1966 All Models, 1967-70 Chrysler

This procedure is the same as that for 1964-65 All Models outlined above, with the substitution of the following steps:
13. Disconnect gear selector rods from transmission.
14. Delete Step 14.
18. Delete Step 18.

1967-70 Imperial

This procedure is the same as that for 1964-65 All Models, with the substitution of the following steps:
13. Disconnect gear selector rods from transmission.
14. Delete Step 14.
18. Delete Step 18.
23. Place jack under transmission to support assembly. Through openings on rear side of torsion bar rear anchor crossmember, remove four large bolts securing rubber isolators to center crossmember. Remove six additional bolts securing center crossmem-

ber, then remove crossmember from stub frame. Do not remove rear anchor crossmember from the torsion bars.

U Joints, Drive Lines

All 1964-70 Chrysler and 1967-70 Imperial models use one piece driveshafts with two universal joints. All 1964 Chrysler models and 1965 Chrysler manual transmission models have a ball and trunnion joint at the front and a cross and roller joint at the rear. 1965 Chrysler automatic transmission models and all 1966-70 Chrysler models have two cross and roller joints, with a slip spline at the front universal joint. 1967-70 Imperials have two constant velocity universal joints with a sliding yoke at the front.

All 1964-66 Imperials have two piece driveshafts with three universal joints, a support bearing forward of the center joint, and a sliding spline. 1964 models have three cross and roller joints; 1965-66 models have two constant velocity joints and a cross and roller joint at the front.

Refer to the Dodge-Plymouth Section for disassembly and repair details on ball and trunnion and cross and roller universal joints.

Constant Velocity Universal Joint

Disassembly

Remove the driveshaft and, before disassembling any parts, mark the joints for proper indexing at the time of assembly.
1. Remove four screws and lockwashers. Remove spline yoke.
2. Remove two loose bearings from centering socket yoke.
3. Remove snap-rings holding the bearing assemblies in the center socket yoke shaft, and center yoke bores.
4. Press bearing assemblies from the yokes by using a 3/4 in. socket as a remover and a pipe or socket with an inside diameter of not less than 1 1/16 in. as a receiver on the opposite bearing. With the aid of a press or vise, press one of the rear yoke bearings about 3/8 in. out of the yoke.
5. Clamp the exposed bearing in the vise and drive the yoke from the bearing with a brass drift.
6. Using the same procedure, press the exposed end of the cross to force the bearing on the opposite end about 3/8 in. out of the yoke. Remove the bearing from the yoke as previously described in Step 5.

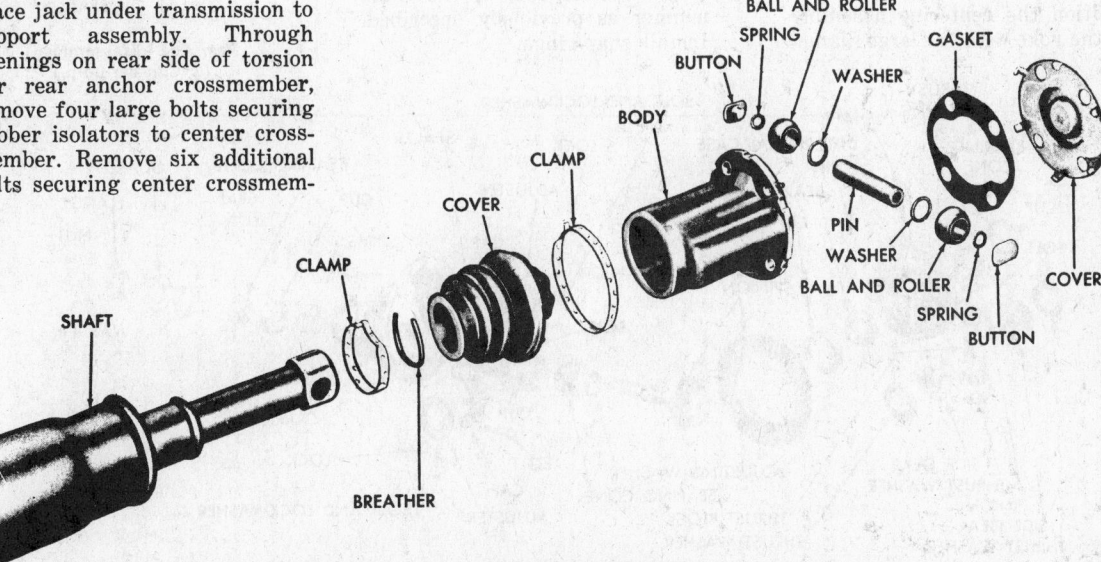

Ball and trunnion universal joint, 1964-65 Chrysler (© Chrysler Corp.)

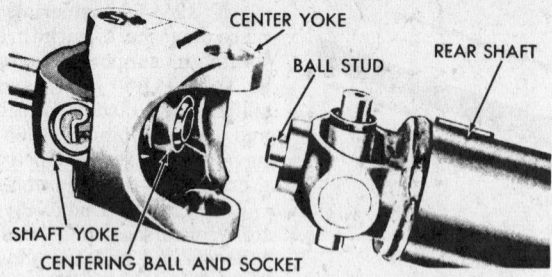

Center constant velocity universal joint, 1965-66 Imperial
(© Chrysler Corp.)

7. Remove the remaining set of bearings from the propeller shaft yoke in the same way.
8. With the shaft held in the vise, press in on the yoke shaft and work the center joint off the cross.
9. Remove the cross from the propeller shaft yoke. Remove centering stud spring from the propeller shaft.
10. Remove the four roller bearing assemblies to separate the yoke shaft from the center yoke, as previously described.

If it is necessary to remove the centering ball and socket assembly, proceed as follows:

11. Carefully pry the centering ball seal assembly from the yoke shaft.
12. Remove seal from the centering stud seal retainer and the bearing rollers from the centering ball.
13. Fill the cavity behind the centering ball and inside the ball with lithium base grease.
14. Insert a rod, slightly smaller than the inside diameter of the centering ball, into the ball, then strike it sharply with a hammer. The hydraulic force applied should force the ball and retainer from the yoke.

Assembly

1. Position the centering assembly in the yoke with the large diameter hole up, and press it firmly into its seat.
2. Apply grease on the inside surface of the centering ball. Install the 34 rollers. Install the centering stud seal in the ball.
3. Install centering ball seal assembly on the yoke and press firmly into place.
4. Coat the inside surfaces of the bearing races with the same grease, and install the 32 rollers. Also, pack the reservoirs in the ends of the cross with the same grease.
5. Place the cross in the shaft yoke. Insert one bearing assembly in the bearing bore of the shaft yoke. With the bar stock or socket used as a remover when disassembling, press the bearing into the bore. At the same time, guide the cross into the bearing. Press the bearing into the yoke far enough to install the snap-ring. Install the snap-ring. Reverse the position of the yoke and install the opposite bearing and snap-ring in the same manner.
6. Place the center yoke on the cross installed in the shaft yoke. Install the two bearings and snap-rings in the yoke, as previously described.
7. Install the cross and two bearings in the shaft yoke, in the same manner as previously described. Install snap-rings.

8. Install centering stud spring on the centering stud, (large end first). Apply grease to the stud.
9. Position the cross in the center universal joint of the propeller shaft while guiding the centering ball on the centering stud, applying pressure at the same time. Work the center yoke over the cross. Don't damage the cross seals.
10. Install the two bearing assemblies in the rear bores of the center yoke, as previously described. Install snap-rings.
11. Coat the splines of the center socket yoke with grease.
12. Install slip spline yoke on the constant velocity joints with screws and lockwashers. Tighten to 300 in. lbs.

Rear Axles, Suspension

Troubleshooting and Adjustment

General instructions covering the troubles of the drive axle and how to repair and adjust it, together with information on installation of rear axle bearings and grease seals, are given in the Unit Repair Section.

Capacities of the drive axle are given in the Capacities table.

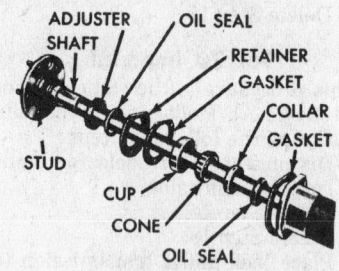

Rear axle (disassembled)
(© Chrysler Corp.)

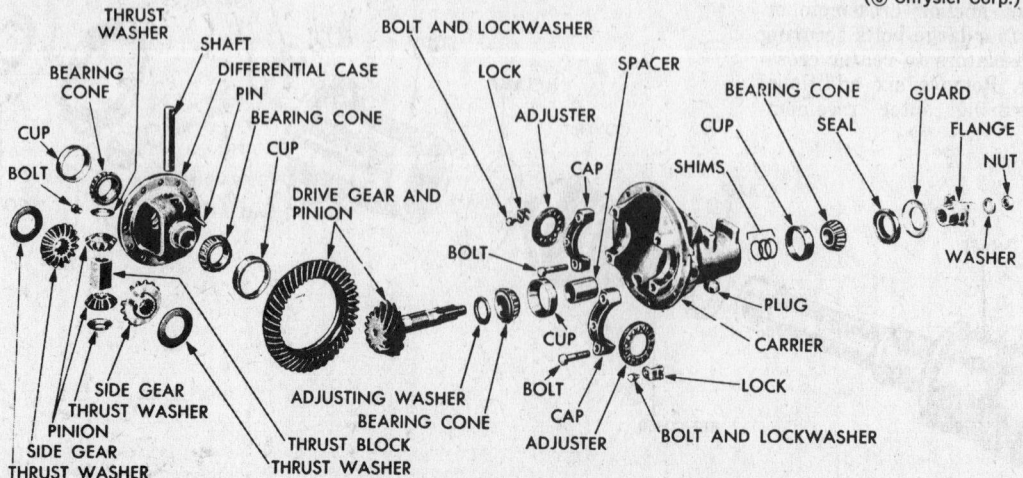

Differential (disassembled) (© Chrysler Corp.)

Rear Shock Absorbers

All Chrysler models are equipped with airplane-type direct-acting shock absorbers on the rear axle.

The shock absorbers can be removed readily by detaching the upper and lower mounting bolts and sliding the shock absorber out of the way.

If the shock absorbers are defective, they should be replaced.

Radio

Removal and Installation

1964-66 Imperial

1. Disconnect battery ground cable.
2. Remove ash tray screws. Lower ash tray assembly and disconnect turn signal flasher and ash tray light. Remove ash tray.
3. Remove lower radio bracket.
4. Disconnect radio wiring.
5. Remove instrument panel pencil brace, just left of radio.
6. Remove radio knobs and mounting nuts.
7. Pull radio out of panel and rotate so radio face is to right of vehicle. Remove radio from under instrument panel.
8. Reverse procedure to install.

1964 Chrysler

1. Disconnect battery ground cable.
2. Remove upper and lower instrument panel moldings. Remove speaker grille.
3. Remove speaker.
4. Disconnect radio wiring.
5. Remove radio knobs and mounting nuts.
6. Remove radio support brackets. Remove radio.
7. Reverse procedure to install.

1965-66 Chrysler

1. Disconnect battery ground cable.
2. On air conditioned models, remove center outlet hose.
3. Remove ash tray and housing assembly.
4. Remove radio knobs and mounting nuts.
5. Disconnect radio wiring.
6. Remove mounting bracket.
7. Remove radio through ash tray assembly opening.
8. Reverse procedure to install.

1967-68 Imperial

1. Disconnect battery ground cable.
2. Remove air conditioning duct and hoses.
3. Disconnect heater blower motor wire connectors from resistor.
4. Disconnect radio wiring.
5. Remove radio support bracket.
6. Remove radio knobs and mounting nuts. Slide radio down and to the right. Rotate front of radio up and remove.
7. Reverse procedure to install.

1967 Chrysler

1. Disconnect battery ground cable.
2. Remove ash tray and housing.
3. Remove air conditioning ducts and hoses.
4. Remove bezels next to map light.
5. Remove two upper radio bezel screws now exposed and remove screw in lower center edge of bezel.
6. Disconnect antenna cable. Remove radio mounting screws from instrument panel.
7. Remove support bracket screw at lower lip of instrument panel and loosen nut on mounting stud at back of radio.
8. Rotate rear edge of radio out and down enough to disconnect speaker and feed wires. Remove radio.
9. Reverse procedure to install.

1968 Chrysler

1. Disconnect battery ground cable.
2. Remove ash tray. Lower ash tray housing. Disconnect two ash tray lights and remove housing.
3. Remove heater temperature control knob.
4. Remove blower switch connector.
5. Remove heater control plate attaching nuts and drop controls down to ash tray opening.
6. Disconnect electrical connections, vacuum switch connector, and bowden cable.
7. Remove heater controls through ash tray opening.
8. Remove fader cover plate and reverberator cover plate.
9. Open glove compartment and remove center bezel (three screws).
10. Remove radio mounting nuts.
11. Remove radio mounting bracket.
12. Disconnect radio wiring.
13. Tilt radio toward instrument panel and slightly toward right to disconnect stereo plug, if so equipped. Remove radio through ash tray opening.
14. Reverse procedure to install.

1969-70 All Models

1. Disconnect battery ground cable.
2. Remove left ash tray.
3. Remove steering column cover.
4. Unscrew stereo tape reset knob, if so equipped.
5. Disconnect radio wiring.
6. Move defroster vacuum actuator to facilitate radio removal.
7. Remove two radio mounting screws through access openings in lower instrument panel. On search-tune and AM radios, remove knobs, bezels, and nuts.
8. Remove radio support bracket mounting screw from lower reinforcement. Support radio.
9. Remove radio support bracket. Remove radio from under instrument panel.
10. Reverse procedure to install.

Windshield Wipers

Wiper Motor Removal and Installation

1964-65 Imperial

1. Disconnect battery ground cable.
2. Remove left air conditioning spot cooler hose.
3. Remove right defroster hose.
4. Remove right instrument panel lower reinforcement to windshield wiper motor mounting bracket pencil brace.
5. Remove glove compartment on air conditioned models.
6. Pivot defroster control vacuum actuator down out of the way.
7. Disconnect motor wiring.
8. Disconnect right and left wiper links at the wiper pivots.
9. Remove nuts attaching wiper motor mounting bracket to cowl panel.
10. Remove wiper motor, motor mounting bracket, and wiper links as an assembly from under the instrument panel.
11. Reverse procedure to install.

1964 Chrysler

1. Disconnect battery ground cable.
2. Remove upper and lower moulding screws. Slip mouldings out from behind passenger assist handle and remove mouldings.
3. Loosen screws attaching heater bezel to instrument cluster. Do not remove bezel.
4. Remove speaker grille, speaker, and mounting plate.
5. Remove glove compartment.
6. Disconnect wiper links at pivots.
7. Remove panel support bracket from wiper motor mounting bracket.
8. Disconnect motor wiring.
9. Remove three nuts attaching wiper motor bracket to cowl panel.
10. Remove wiper motor assembly, with both links attached, through glove compartment opening.
11. Reverse procedure to install.

1965-66 Chrysler, 1967-68 All Models

1. Disconnect battery ground cable.
2. Remove wiper arm and blade assemblies.

3. Remove windshield lower moulding.
4. Remove cowl grille panel.
5. Remove drive crank arm retaining nut and drive crank. Disconnect motor wiring.
6. Remove three nuts mounting motor to bulkhead and remove motor.
7. Reverse procedure to install.

1966 Imperial

1. Disconnect battery ground cable.
2. Remove air conditioning right spot cooler hose and distribution duct.
3. Remove instrument panel lower reinforcement to windshield wiper motor mounting bracket pencil brace.
4. Disconnect motor wiring.
5. Remove both left and right link to pivot retainers.
6. Remove three motor bracket mounting nuts.
7. Work motor and link assembly out from under panel toward right side of vehicle.
8. Remove links and motor mounting bracket.
9. Reverse procedure to install.

1969-70 All Models

1. Disconnect battery ground cable.
2. Lift the wiper arm and insert a .090 pin or drill. Pull wiper arm from shaft with a rocking motion.
3. Remove windshield lower moulding.
4. Remove cowl screen.
5. Remove drive crank arm retaining nut and drive crank. Disconnect motor wiring.
6. Remove three mounting nuts. Remove motor.
7. Reverse procedure to install.

Heater System

Heater Blower R & R

1964 All Models,
1965-66 Imperial

1. Disconnect battery ground cable.
2. Disconnect heater ground wire at wiper motor mounting bracket.
3. Disconnect heater wires from harness connector.
4. Disconnect vacuum hoses from units (where used).

5. Remove hoses from attaching clips.
6. Remove valve capillary coil from opening in housing (driver's compartment).
7. Remove clips from housing.
8. Remove screws attaching heater duct to dash panel (left of vent door, below heater on passenger side, and wiper motor right link pivot).
NOTE: disconnect wiper right link at pivot to expose housing screws.
9. Remove housing and blower by pulling down and out.
10. Remove blower, mounting plate, and motor.
11. Install in reverse of above.

1965-66 Chrysler,
1967-68 All Models

1. Disconnect battery ground cable.
2. Disconnect water hoses at dash panel (engine side). Plug heater hose fittings to prevent spilling water in passenger area.
3. Under instrument panel, remove bracket from top of heater to dash.
4. Remove defroster hoses at heater, and disconnect vacuum lines at heater.
5. Disconnect wiring at heater blower motor resistor.
6. Remove glove compartment.
7. Disconnect control cable at heater end.
8. Unclamp flexible connector at right end of heater. Do not remove connector from cowl side.
9. Pull carpet or mat from under instrument panel.
10. From engine compartment, remove nuts mounting the assembly to the instrument panel.
11. Pull heater toward rear to clear mounting studs from dash. Rotate heater assembly until studs are down, then remove heater.
12. Disconnect wiring from heater assembly to blower motor.
13. Remove motor cooler tube.
14. Remove heater back plate.
15. Remove fan from motor shaft.
16. Remove blower motor from back plate.
17. Install in reverse of above.

1969-70 All Models

The blower motor is mounted to the engine side housing under the right front fender between the inner fender shield and the fender. The inner fender shield must be removed to service the blower motor.

Heater Core R & R

1964 All Models,
1965-66 Imperial

1. Disconnect battery ground cable.
2. Drain cooling system.
3. Disconnect heater hoses at heater.
4. Remove screws attaching heater housing to instrument panel.
5. Remove housing and core as an assembly.
6. Remove mastic to expose housing.
7. Remove core from outer housing.
8. Install in reverse of above.

1965-66 Chrysler,
1967-68 All Models

1. Follow Steps 1 through 11 under Heater Blower R&R, 1965-66 Chrysler, 1967-68 All Models.
2. Remove heater cover plate.
3. Remove screws attaching heater core to heater, and remove core.
4. Install in reverse of above.

1969-70 All Models

1. Disconnect battery ground cable. Drain coolant.
2. Disconnect heater hoses and plug fittings.
3. Slide front seat back. Unplug antenna from radio.
4. Remove vacuum hoses from trunk lock, if so equipped.
5. Disconnect blower motor resistor block.
6. Remove vacuum hoses from defroster actuator and heater shut off door actuator.
7. Swing support bracket up out of the way.
8. Remove four retaining nuts from studs on engine side housing.
9. Remove locating bolt from bottom center of passenger side housing.
10. Roll or tip housing out from under instrument panel.
11. Remove temperature control cable retaining clip and cable from heat shut off door crank.
12. From inside housing, remove two retaining nuts from right side of heater core and four screws from outside of housing.
13. Remove core tube locating metal screw from top of housing.
14. Carefully pull heater core out of housing.
15. Reverse procedure to install.

Corvair

Index

YEAR IDENTIFICATION

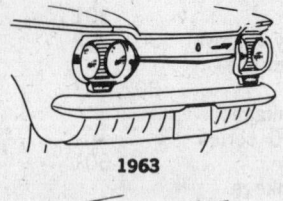

1963

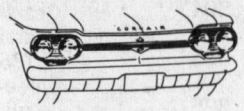

1964

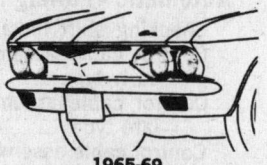

1965-69

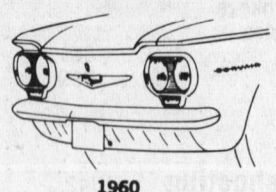

1960

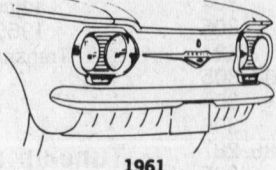

1961

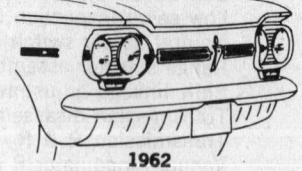

1962

FIRING ORDER

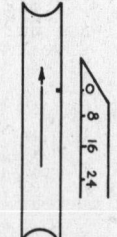

Timing marks—1966 140 and
180 H.P.; 1966-67 110 H.P.,
A.T., AC

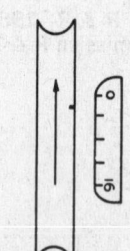

Timing marks—1965 95, 110,
140 H.P.; 1966 95, 110 H.P.,
and 1967-69 all

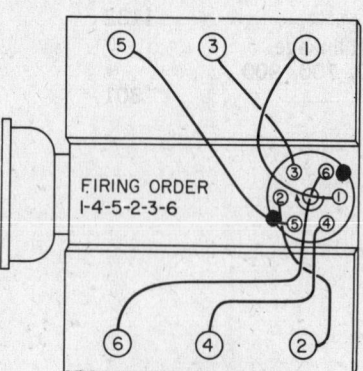

FIRING ORDER
1-4-5-2-3-6

Firing order—1961-69

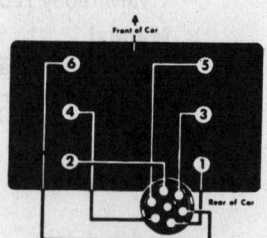

Firing order—1960

Timing marks—1962-65
turbo charged engines

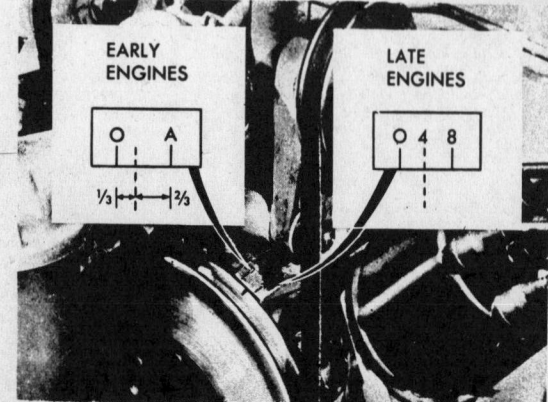

EARLY ENGINES   LATE ENGINES

Timing marks—1960

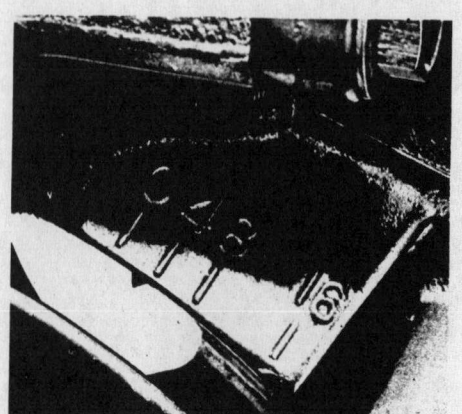

Timing marks—1961-64

CAR SERIAL NUMBER LOCATION

1960-64

The serial number is located on the left front door lock pillar. This number identifies the vehicle type, model year, assembly plant and the production unit number.

1965-69

The serial number is located on the left-hand top of frame side rail just to the rear of battery bolts.

Engine Identification And Number Location

1960-64

The number is located on top of the engine block, just forward of the generator oil filter adapter.

1965-69

The number can be found on top of the engine block behind the oil pressure sending unit.

The engine number consists of six units. The first unit is a letter and identifies the point of manufacture. The next two digits show the month of manufacture, and the next two digits the day of manufacture. The last unit is a letter, which identifies the type of transmission used.

Example:

T0430-Z

T—Manufacturing plant (Tonawanda)
04—Month of manufacture (April))
30—Day of manufacture (Thirtieth)

Z—Transmission type (Z = Powerglide; Y = manual)

Model Identification

1960

Standard 569	4-dr. sedan, 6 pass.
Standard 527	club coupe, 5 pass.
Deluxe 769	4-dr. sedan, 6 pass.
Deluxe 727	club coupe, 5 pass.

1961

Standard 527	2-dr. club coupe, 5 pass.
Standard 535	4-dr. Lakewood sta. wagon, 6 pass.
Standard 569	4-dr. sedan, 6 pass.
Deluxe 727	2-dr. club coupe, 5 pass.
Deluxe 735	4-dr. Lakewood sta. wagon, 6 pass.
Deluxe 769	4-dr. sedan, 6 pass.
Monza 927	2-dr. club coupe, 4 pass.

1962-63

Standard 527	2-dr. club coupe, 5 pass.
Deluxe 727	2-dr. club coupe, 5 pass.
Deluxe 735	4-dr. Lakewood sta. wagon, 6 pass.
Deluxe 769	4-dr. sedan, 6 pass.
Monza 927	2-dr. club coupe, 4 pass.
Monza 969	4-dr. sedan, 6 pass.

1964

Standard 527	2-dr. coupe, 5 pass.
Deluxe 769	4-dr. sedan, 6 pass.
Monza 927	2-dr. coupe, 4 pass.
Monza 967	2-dr. convt., 4 pass.
Monza 969	4-dr. sedan, 5 pass.
Monza Spyder 627	2-dr. coupe, 4 pass.
Monza Spyder 667	2-dr. convt., 4 pass.

1965-67

Standard	10139 (4-dr. sport sdn., 6 pass.)
	10137 (2-dr. sport cpe., 4 pass.)
Monza	10539 (4-dr. sport sdn., 6 pass.)
	10537 (2-dr. sport cpe., 4 pass.)
	10567 (2-dr. convt., 4 pass.)
Corsa	10737 (2-dr. sport cpe., 4 pass.)
	10767 (2-dr. convt., 4 pass.)

1968-69

Standard	10137 (2-dr. sport cpe., 5 pass.)
Monza	10537 2-dr. sport cpe., 4 pass.)
	10567 (2-dr. convt., 4 pass.)

GENERAL ENGINE SPECIFICATIONS

YEAR	CU. IN. DISPLACEMENT	CARBURETOR	DEVELOPED HORSEPOWER @ RPM	DEVELOPED TORQUE @ RPM (FT. LBS.)	A.M.A. HORSEPOWER	BORE AND STROKE (IN.)	COMPRESSION RATIO	VALVE LIFTER TYPE	NORMAL OIL PRESSURE (PSI)
1960	140 Std.	2-1-BBL.	80 @ 4400	125 @ 2400	27.3	3.375 x 2.600	8.0-1	Hyd.	35
1961	145 Turbo Air	2-1-BBL.	80 @ 4400	128 @ 2300	28.4	3.4375 x 2.600	8.0-1	Hyd.	35
	145 Super T.A.	2-1-BBL.	98 @ 4600	132 @ 2800	28.4	3.4375 x 2.600	8.0-1	Hyd.	35
1962-63	145 Turbo Air	2-1-BBL.	80 @ 4400	128 @ 2300	28.4	3.438 x 2.600	8.0-1	Hyd.	35
	145 Monza P.G.	2-1-BBL.	84 @ 4400	130 @ 2300	28.4	3.438 x 2.600	9.0-1	Hyd.	35
	145 Super T.A.	2-1-BBL.	102 @ 4400	134 @ 2900	28.4	3.438 x 2.600	9.0-1	Hyd.	35
	145 Super Chg.	1-1-BBL.	150 @ 4400	210 @ 3300	28.4	3.438 x 2.600	8.0-1	Hyd.	35
1964	164 Turbo Air	2-1-BBL.	95 @ 3600	154 @ 2400	28.4	3.438 x 2.937	8.25-1	Hyd.	35
	164 Super T.A.	2-1-BBL.	110 @ 4400	160 @ 2800	28.4	3.438 x 2.937	9.25-1	Hyd.	35
	164 Super Chg.	1-1-BBL.	150 @ 4000	232 @ 3200	28.4	3.438 x 2.937	8.25-1	Hyd.	35
1965-67	164 Turbo Air	2-1-BBL.	95 @ 3600	154 @ 2400	28.4	3.438 x 2.937	8.25-1	Hyd.	35
	164 Turbo Air	2-1-BBL.	110 @ 4400	160 @ 2800	28.4	3.438 x 2.937	9.0-1	Hyd.	35
	164 Turbo Air	4-1-BBL.	140 @ 5200	160 @ 3600	28.4	3.438 x 2.937	9.0-1	Hyd.	35
	164 Super Chg.	1-1-BBL.	180 @ 4000	265 @ 3200	28.4	3.438 x 2.937	8.0-1	Hyd.	35
1968-69	164 Turbo Air	2-1-BBL.	95 @ 3600	154 @ 2400	28.4	3.438 x 2.937	8.25-1	Hyd.	30
	164 Turbo Air	2-1-BBL.	110 @ 4400	160 @ 2800	28.4	3.438 x 2.937	9.25-1	Hyd.	30
	164 Turbo Air	4-1-BBL.	140 @ 5200	160 @ 3600	28.4	3.438 x 2.937	9.25-1	Hyd.	30

SERIAL NUMBER LOCATION
1960-64—Left front door hinge pillar.
1965-69—Left-hand top of frame side rail rearward of battery bolts.

ENGINE IDENTIFICATION CODE
1960-64—Top of engine block forward of generator oil filter adapter.
1965-69—Top of engine block behind oil pressure sending unit.

No. Cyls.	Cu. In. Displ.	Type	YEAR AND CODE					No. Cyls.	Cu. In. Displ.	Type	YEAR AND CODE				
			1960	1961	1962	1963	1964				1965	1966	1967	1968	1969
6	145	M.T.	YH, V	YH, V	YH, V	V		6	164	MT, AC, w/ex. EM			QM		
6	164	M.T.					V	6	164	PG, AC, w/ex. EM			QO		
6	145	A.T.	W, ZL	W, ZL	W, ZL	W		6	164	HP, PG, w/ex. EM			QP		
6	164	A.T.					W	6	164	HP, MT, AC, w/ex. EM			QS		
6	145	4 Spd., TC	YR	YR	YR	Y		6	164	M.T.	RA	RA	RA	RS	RS
6	145	M.T.	YC	YC	YC	YC		6	164		RB	RB			
6	164	M.T.					YC	6	164	HP	RD	RD	RD	RM	RM
6	145	M.T., AC	YL	YL	YL	YL		6	164	M.T., AC	RE	RE	RE		
6	164	M.T., AC					YL	6	164	HP, AC	RF	RF	RF	RF	RF
6	145	M.T., HP, AC	YM	YM	YM	YM		6	164	PG	RG	RG	RG		
6	164	4 Spd., TC					YM	6	164	PG, HP	RH	RH	RH	RW	RW
6	145	M.T., HP	Y, YN	Y, YN	Y, YN	YN		6	164	PG, AC	RJ	RJ	RJ	RJ	RJ
6	164	M.T., HP					YN	6	164	PG, HP, AC	RK	RK	RK	RK	RK
6	145	A.T.	ZB	ZB	ZB	Z		6	164	TC	RL	RL			
6	164	A.T.					Z	6	164	M.T., SHP	RM	RM			
6	145	A.T., AC	ZJ	ZJ	ZJ	ZD		6	164	PG, SHP	RN	RN			
6	164	A.T., AC					ZD	6	164	SHP, w/ex. EM		RQ			
6	145	A.T., HP	ZF, ZK	ZF, ZK	ZF, ZK	ZF		6	164	AC		RR		RE	RE
6	164	A.T., HP					ZF	6	164	M.T., HP, w/ex. EM	RS	RS	RS		
6	145	A.T., HP, AC	ZG	ZG	ZG	ZG		6	164	w/ex. EM		RT			
6	164	A.T., HP, AC					ZG	6	164	M.T., HP	RU	RU	RU		
6	145	A.T.	ZH	ZH	ZH	ZH		6	164	PG	RV			RV	RV
6	164	A.T.					ZH	6	164	PG, w/ex. EM		RV	RV		
								6	164	HP, w/ex. EM		RW	RW		
								6	164	PG, SHP, PG, AC		RY			
								6	164	HP, PG	RX				
								6	164	PG, SHP, w/ex. EM			RX		
								6	164	SHP, AC		RZ			

AC—Air conditioned.　　SHP—Special high performance.　　M.T.—Manual transmission.　　w/ex. EM—With exhaust emission.
HP—High performance.　　PG—Powerglide transmission.　　TC—Turbocharged.　　A.T.—Automatic transmission.

GENERATOR AND REGULATOR SPECIFICATIONS

YEAR	MODEL	GENERATORS				REGULATORS			
		Field Current in Amperes At 12 Volts	Brush Spring Tension (Oz.)	Cut-Out Relay		Current & Voltage Air Gaps (In.)	Current Regulator Setting (Amps.)	Voltage Regulator Setting (Volts)	
				Air Gap (In.)	Closing Voltage				
1960-61		1.50-1.62	28	.020	11.8-13.5	.075	27-33	13.8-14.8	
1962-63	1102226/30A.	1.69-1.79	28	.020	11.8-13.5	.075/.060	27-31	13.8-14.7	
	1102227/30A.	1.69-1.79	28	.023	11.8-13.0	.075/.067	31.0-35.5	13.8-14.7	
	1105135/35A.	2.73-3.00	28	.023	11.8-13.0	.075/.067	27.8-32.4	13.8-14.7	
	1105139/40A.	2.73-3.00	28	.020	11.8-13.0	.075/.067	31.0-35.5	13.8-14.6	
1964	1102336/35A.	1.69-1.79	28	.020	11.8-13.5	.075	31.0-35.5	13.8-14.7	
	1105135/35A.	2.73-3.00	28	.020	11.8-13.0	.075/.067	31.0-35.5	13.8-14.6	

DELCOTRON AND AC REGULATOR SPECIFICATIONS

YEAR	MODEL	ALTERNATOR			REGULATOR						
		Field Current Draw @ 12V.	Output @ Generator RPM		Model	Field Relay			Regulator		
			500	1500		Air Gap (In.)	Point Gap (In.)	Volts to Close	Air Gap (In.)	Point Gap (In.)	Volts at 125°
1965-69	1100639	2.2-2.6	25	35	1119515	.015	.030	2.3-3.7	.067	.014	13.5-14.4
	1100698*	2.8-3.2	35	45	1119515	.015	.030	2.3-3.7	.067	.014	13.8-14.8

*—with external field discharge diode.

TUNE-UP SPECIFICATIONS

YEAR	MODEL AND DISPLACEMENT	Type	Gap (In.)	Point Dwell (Deg.)	Point Gap (In.)	IGNITION TIMING (Deg.) ▲	CRANKING COMP. PRESSURE (Psi)	Tappet (Hot) Clearance (In.) Intake	Exhaust	Intake Opens (Deg.)	FUEL PUMP PRESSURE (Psi)	IDLE SPEED (Rpm) ★
1960	140 Cu. In.	44FF	.035	31–36°	.019	4B	130#	■	■	43B	4–5	450–500☆
1961	145 Cu. In., T.A., M.T.	46FF	.035	33°	.019	4B	130#	■	■	43B	4–5	500☆
	145 Cu. In., T.A., A.T.	46FF	.035	33°	.019	13B	130#	■	■	43B	4–5	500☆
	145 Cu. In., Super T.A.	46FF	.035	33°	.019	13B	130#	■	■	54B	4–5	500☆
1962–63	145 Cu. In.; Turbo Air M.T.	46FF	.035	33°	.019	4B	130#	■	■	43B	4½	475
	145 Cu. In.; Turbo Air A.T.	46FF	.035	33°	.019	13B	130#	■	■	43B	4½	475
	145 Cu. In.; Monza A.T.	44FF	.035	33°	.019	13B	130#	■	■	43B	4½	475
	145 Cu. In.; Super Turbo Air	44FF	.035	33°	.019	13B	130#	■	■	54B	4½	475
	145 Cu. In.; Super Charged	44FF†	.035	33°	.019	24B	130#	■	■	54B	4½	800
1964	164 Cu. In.; Turbo Air M.T.	46FF	.035	33°	.019	2B	130#	■	■	44B	4½	500
	164 Cu. In.; Turbo Air A.T.	46FF	.035	33°	.019	10B	130#	■	■	44B	4½	500
	164 Cu. In.; Super Turbo Air	44FF	.030	33°	.019	12B	130#	■	■	54B	4½	600☆
	164 Cu. In.; Super Charged	44FF	.030	33°	.019	24B	130#	■	■	54B	4½	850☆☆
1965–67	164 Cu. In.; Turbo Air (95 HP)	46FF	.035	33°	.019	6B●◎	130#	■	■	44B	4½	500
	164 Cu. In.; Turbo Air (110 HP)	46FF	.035	33°	.019	14B◎	130#	■	■	55B	4½	500
	164 Cu. In.; Turbo Air (140 HP)	44FF	.030	33°	.019	18B	130#	■	■	55B	4½	600
	164 Cu. In.; Super Chg. (180 HP)	44FF	.030	33°	.019	24B	130#	■	■	82B	4½	850
1968–69	164 Cu. In.; Turbo Air (95 HP) M.T.	46FF①	.035	33°	.019	6B	130#	■	■	26B	6	700
	164 Cu. In.; Turbo Air (95 HP) A.T.	46FF①	.035	33°	.019	14B	130#	■	■	26B	6	600
	164 Cu. In.; Turbo Air (110 HP) M.T.	44FF	.030	33°	.019	4B	130#	■	■	37B	6	700
	164 Cu. In.; Turbo Air (110 HP) A.T.	44FF	.030	33°	.019	12B	130#	■	■	37B	6	600
	164 Cu. In.; Turbo Air (140 HP)	44FF	.030	33°	.019	4B	130#	■	■	70B	6	650●●

★—With manual transmission in N and automatic in D.
☆—500 rpm with automatic in D.
☆☆—No automatic.
▲—With vacuum advance disconnected and plugged. NOTE: These settings are only approximate. Engine design, altitude, temperature, fuel octane rating and the condition of the individual engine are all factors which can influence timing. The limiting advance factor must, therefore, be the "knock point" of the individual engine.
■—1 turn tighter than zero lash.
●—Powerglide—14 B.

●●—W/A.T., 140 HP—550 rpm.
◎—W/exh. emission—95 HP—TDC, 110 HP—4 B.
①—1969—44FF.
B—Before top dead center.
#—Minimum.
†—42FF for racing.
A.T.—Automatic transmission.
M.T.—Manual transmission.
T.A.—Turbo Air.
††—New points; used points = .016.

CAUTION

General adoption of anti-pollution laws has changed the design of almost all car engine production to effectively reduce crankcase emission and terminal exhaust products. It has been necessary to adopt stricter tune-up rules, especially timing and idle speed procedures. Both of these values are peculiar to the engine and to its application, rather than to the engine alone. With this in mind, car manufacturers supply idle speed data for the engine and application involved. This information is clearly displayed in the engine compartment of each vehicle.

CRANKSHAFT BEARING JOURNAL SPECIFICATIONS

YEAR	MODEL	MAIN BEARING JOURNALS (IN.) Journal Diameter	Oil Clearance	Shaft End-Play	Thrust On No.	CONNECTING ROD BEARING JOURNALS (IN.) Journal Diameter	Oil Clearance	End-Play
1960–69	All Engines	▲	■	.002–.006	1	1.799–1.800	.0007–.0027	.005–.010

▲—#1 & 2, 2.0978–2.0988; #3 & 4, 2.0983–2.0993 (1960—all four same as 1 and 2).
■—#1 & 2, .0012–.0027; #3 & 4, .0007–.0022.

WHEEL ALIGNMENT

YEAR	MODEL	CASTER Range (Deg.)	CASTER Pref. Setting (Deg.)	CAMBER Range (Deg.)	CAMBER Pref. Setting (Deg.)	TOE-IN (In.)	KING-PIN INCLINATION (Deg.)	WHEEL PIVOT RATIO Inner Wheel	Outer Wheel
1960	All Models	2½P to 3P	2¾P	0–1P	½P	3/16 +0 −1/16 ●	7 ± ½	20	18
1961	500, 700, 900	1½P to 2P	2P	0–1P	½P	3/16 +0 −1/16 ●	7 ± ½	20	18
	1200 and 95	2¼P to 2¾P	2½P	0–½P	¼P	⅛●	—	—	—
1962–63	500, 700, 900	1½P to 2P	2P	0–1P	½P	¼–³/₈●●	7 ± ½	20	18
	1200 Panel, P.U.	1P to 1½P	1¼P	¼N–¼P	0	1/16–3/16#	—	—	—
	1200 Greenbrier	2P to 2½P	2¼P	¼P–¾P	½P	1/16–3/16##	—	—	—
1964	500, 600, 700, 900	1½P to 2P	2P	½N–½P	0	¼–³/₈●●	7 ± ½	20	18 *
	1200	-----Same as 1962–63 Specifications-----							
1965–66		1½ to 2½P ▲	2P	½P to 1⅓P ■	1P	¼ to ³/₈ ★	7	20	18
1967–69		1¾P to 2¾P	2¼P	½P to 1½P ■	1P	3/16 to 5/16 ★	6½	20	18

●—Rear toe-in 0 to ¼, with non-adjustable camber—2°N ± ½°.
●●—Rear toe-in ⅛ to ³/₈, with non-adjustable camber—1°N ± ½°.
★—Rear toe-in ⅛ to ³/₈, camber—1°N to 0.
▲—1966—2½ to 3½P.
■—Rear wheel camber 0 to 1N.

N—Negative.
P—Positive.
P.U.—Pickup truck.
#—Rear toe-in 1/16–3/16, rear camber 4⅝°N.
##—Rear toe-in 1/16–3/16, rear camber 3½°N ± ½°.

VALVE SPECIFICATIONS

YEAR AND MODEL	SEAT ANGLE (DEG.)	FACE ANGLE (DEG.)	VALVE LIFT INTAKE (IN.)	VALVE LIFT EXHAUST (IN.)	VALVE SPRING PRESSURE (VALVE OPEN) LBS. @ IN.	VALVE SPRING INSTALLED HEIGHT (IN.)	STEM TO GUIDE CLEARANCE (IN.) INTAKE	EXHAUST	STEM DIAMETER (IN.) INTAKE	EXHAUST
1960 140 Cu. In.	45 in. 45 ex.	45 in. 44 ex.	.314	.344	144 @ 1.148	1.508	.0010–.0027	.0015–.0032	.3415–.3422	.3410–.3417①
1961 145 Turbo Air	45 in. 45 ex.	45 in. 44 ex.	.314	.344	144 @ 1.148	1.508	.0010–.0027	.0015–.0032	.3415–.3422	.3410–.3417①
145 Super T.A.	45 in. 45 ex.	45 in. 44 ex.	.380	.380	165 @ 1.306	1.696	.0010–.0027	.0015–.0032	.3415–.3422	.3410–.3417①
1962–63 145 Turbo Air & Monza P.G.	45 in. 45 ex.	44 in. 44 ex.	.314	.344	144 @ 1.148	1.508	.0010–.0027	.0010–.0027	.3415–.3422	.3410–.3417①
145 Super T.A. & Spyder T.C.	45 in. 45 ex.	45 in. 44 ex.	.378	.378	165 @ 1.306	1.696	.0010–.0027	.0010–.0027	.3415–.3422	.3410–.3417①
1964 164 Turbo Air & Super T.A.	45 in. 45 ex.	45 in. 44 ex.	.385	.385	175 @ 1.260	1.656 ±.030	.0010–.0027	.0014–.0029	.3414–.3422	.3407–.3418
164 Spyder T.C.	45 in. ex.	45 in. ex.	.390	.390	175 @ 1.260	1.656 ±.030	.0010–.0027	.0014–.0029	.3414–.3422	.3407–.3418
1965 164 Turbo Air	45 in. 45 ex.	45 in. 44 ex.	.385	.385	175 @ 1.260	1.656 ±.030	.0010–.0027	.0014–.0029	.3414–.3422	.3407–.3418
164 (110 & 140 H.P.)	45 in. 45 ex.	45 in. 44 ex.	.390	.390	175 @ 1.260	1.656 ±.030	.0010–.0027	.0014–.0029	.3414–.3422	.3407–.3418
164 Spyder T.C.	45 in. ex.	45 in. ex.	.374	.374	175 @ 1.260	1.656 ±.030	.0010–.0027	.0014–.0029	.3414–.3422	.3407–.3418
1966–69 164 Turbo Air	46	45	.403	.403	175 @ 1.260	1.656 ±.030	.0010–.0028	.0014–.0029	.3414–.3422	.3407–.3418
164 (110 & 140 H.P.)	46	45	.409	.409	175 @ 1.260	1.656 ±.030	.0010–.0028	.0014–.0029	.3414–.3422	.3407–.3418
164 Spyder T.C.	46	45	.392	.392	175 @ 1.260	1.656 ±.030	.0010–.0028	.0014–.0029	.3414–.3422	.3407–.3418

① —Top; bottom = .3400–.3407.

GENERAL CHASSIS AND BRAKE SPECIFICATIONS

YEAR	MODEL	CHASSIS Overall Length (In.)	Tire Size	BRAKE CYLINDER BORE Master Cylinder (In.)	Wheel Cylinder (In.) Front	Rear	BRAKE DRUM Diameter (In.) Front	Rear
1960–64	All, exc. below	180.0	6.50 x 13★	1.0	⅞ ①	15/16 ①	9.0	9.0
1965–66	All, exc. below	183.3●	6.50 x 13	1.0	⅞	15/16	9.5	9.5
1967–69	All	183.3	7.00 x 13	1.0	⅞	15/16	9.5	9.5

★—1961–64—1200 Series and Sports Wagon—7.00 x 14 and 7.00 x 13.
●—Greenbrier—179.7.
① —1200 Series—1⅛ front, 1.0 rear.

CAPACITIES

| YEAR | MODEL | ENGINE CRANKCASE (Qts.) ADD 1 Pt. FOR NEW FILTER | TRANSMISSIONS Pts. TO REFILL AFTER DRAINING | | | DRIVE AXLE (Pts.) | GASOLINE TANK (Gals.) | COOLING SYSTEM (Qts.) WITH HEATER |
| | | | Manual | | Automatic | | | |
			3-Speed	4-Speed				
1960	All	4	2	—	8	3	11	Air Cooled
1961–63	500, 700, 900	4	3	$3\frac{3}{4}$	13	3★	14	Air Cooled
	1200	4	3	$3\frac{3}{4}$	13	3★	18.6	Air Cooled
1964	All except below	4	$2\frac{1}{4}$	$3\frac{3}{4}$	13	$4\frac{1}{2}$	14▲	Air Cooled
1965–68	All	4	3	$3\frac{3}{4}$	13	4	14	Air Cooled
1969	All	4	3.7	3.7	13	2.7	14	Air Cooled

▲—Station Wagon or truck; $18\frac{1}{2}$ gal.
★—1963—$4\frac{1}{2}$.

TORQUE SPECIFICATIONS

| YEAR | MODEL | CYLINDER HEAD BOLTS (FT. LBS.) | ROD BEARING BOLTS (FT. LBS.) | MAIN BEARING BOLTS (FT. LBS.) | CRANKSHAFT BALANCER BOLT (FT. LBS.) | FLYWHEEL TO CRANKSHAFT BOLTS (FT. LBS.) | MANIFOLD (FT. LBS.) | |
							Intake	Exhaust
1960–69	All	27–33	20–26	42–48	60–80	20–30 ①	None	12–27

①—All 1960–64 = 20–30 FT. LBS., same for 1965–69 Powerglide models.
1965–69 Standard transmission models = 40–50 FT. LBS.

BATTERY AND STARTER SPECIFICATIONS

| YEAR | MODEL | BATTERY | | | | STARTERS | | | | | |
| | | Ampere Hour Capacity | Volts | Terminal Grounded | Lock Test | | | No-Load Test | | | Brush Spring Tension (Oz.) |
					Amps.	Volts	Torque	Amps.	Volts	RPM	
1960–61	Standard	35	12	Neg.	280–320	4.0	—	58–80	10.6	6,750–8,600	35
	Heavy Duty	40	12	Neg.	—	—	—	—	—	—	—
1962–64	All	42	12	Neg.	280	4.0	—	58–80	10.6	6,750–10,500	35
1965–66	All	44	12	Neg.	280	4.0	—	58–80	10.6	6,750–10,500	35
1967–69	All	45	12	Neg.	Not Recommended			69	10.6	6,750–10,500	35

CYLINDER HEAD BOLT TIGHTENING SEQUENCE

1960-68

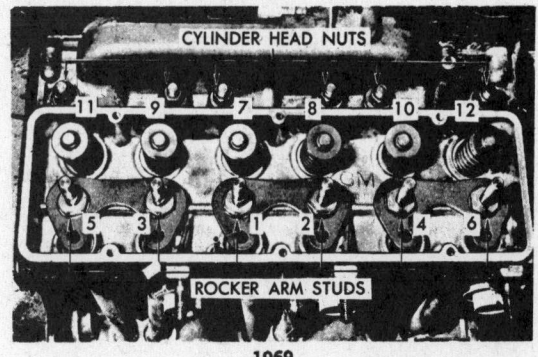

1969

FUSES AND CIRCUIT BREAKERS

1960–65

Light circuits (except following)............. 15 Amp. CB
   Tail, stop, dome lamp and cigar lighter .. 10 Amp. Fuse
   Glove comp. lamp and heater blower.... 10 Amp. Fuse
   Back-up lamp and total heater system... 20 Amp. Fuse
   Radio .... 4 Amp. Fuse
   Instrument lamp, radio panel lamp,
      heater control panel .... 3 Amp. Fuse
   Electric windshield wiper .... 20 Amp. Fuse
   Air conditioner .... 15 Amp. Fuse

1966–69

Light circuits (except following) .... CB
   Radio.............................. 10 Amp. Fuse
   Instrument lamps .... 4 Amp. Fuse
   Gauges & back-up lamps .... 10 Amp. Fuse
   Clock, lighter, courtesy lamps,
      and hazard flasher................. 20 Amp. Fuse
   Stop, tail/side marker and tail lights .... 20 Amp. Fuse
   Wiper motor .... 20 Amp. Fuse
   Heater & air conditioner .... 25 Amp. Fuse

LIGHT BULBS

1960–63

Headlamp (outer).................. 4002; $37\frac{1}{2}$–50 W
Headlamp (inner)................... 4001; $37\frac{1}{2}$ W
Parking, tail, stop and dir. signals .... 1034; 4–32 CP
Back-up lamps .... 1073; 32 CP
Instrument lamps...................... 1816; 3 CP
Radio dial lamp....................... 1891; 2 CP
Indicator lamps, glove compartment lamp ...... 57; 2 CP
High beam indicator, heater control panel ...... 53; 1 CP
Dome lamp .... 211; 15 CP
Courtesy lamp........................ 89; 6 CP
License plate lamp...................... 67; 4 CP

1964–69

Headlamp (outer).................. 4002; $37\frac{1}{2}$–50 W
Headlamp (inner)................... 4001; $37\frac{1}{2}$ W
Parking, tail, stop and dir. signals .... 1157; 4–32 CP
Back-up lamps .... 1156; 32 CP
Instrument lamps...................... 1816; 3 CP
Directional signal, headlamp beam, heater
   control panel indicator lamps............. 1445; 1 CP
Temperature, oil, generator indicator and
   glove compartment lamps .... 1895; 2 CP
Dome lamp .... 211; 12 CP
Courtesy lamp........................ 631; 6 CP
License plate lamp...................... 67; 4 CP
Radio dial lamp....................... 1893; 2 CP

Distributor

The distributor is located on the right side of the engine, at the rear. It is driven through a helical gear by the crankshaft. The distributor shaft is extended into the crankcase area to drive the oil pump.

Removal

1. Disconnect distributor primary wire.
2. Remove distributor cap.
3. Remove vacuum control line.
4. Mark position of rotor to simplify timing at the time of reinstallation.
5. Remove distributor clamp, then lift out distributor.
6. If removal of secondary wires from the cap is necessary, mark No. 1 location on the cap tower. Consult the Timing Location and Firing Order diagram for firing order and location data.

Disassembly

1. Remove cap and rotor.
2. Remove weight springs and centrifugal weights
3. Drive gear pin and thrust washer pin from shaft using a pin punch.
4. Remove gear and thrust washer, then file away any burrs around pin hole.
5. Remove cam and weight base, along with shaft, from housing.
6. Remove cam and weight base from shaft.
7. Remove condenser.
8. Remove spring retainer ground lead and breaker plate.
9. Remove vacuum advance unit and felt washer around housing bushing.

Inspection

1. Wash all parts in thinner except vacuum advance unit and electrical parts.

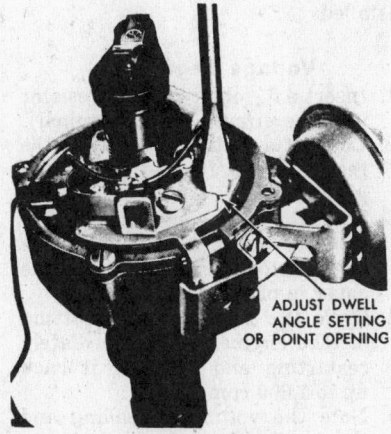

ADJUST DWELL ANGLE SETTING OR POINT OPENING

Point adjustment
(© G.M. Corp.)

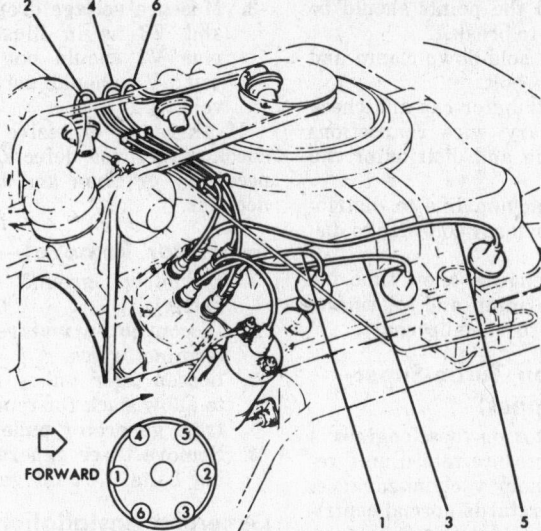

Distributor and position of spark plug wires—1961-69
(1960 slightly different configuration)
(© G.M. Corp.)

2. Inspect breaker plate for wear and replace if necessary.
3. Inspect shaft for wear. If bearings are worn, entire housing must be replaced.
4. Inspect shaft for runout (wobble); maximum permitted is 0.002 in.
5. Check governor weights and cam for burrs; remove any existing with emery paper.
6. Assemble distributor components in reverse order of removal. When replacing points, always replace condenser at the same time.

NOTE: do not oil cam lubricator—replace it every time points are replaced. A dry Corvair distributor cam makes a loud "knocking" noise similar to bad rod bearings.

Installation (If Engine has been Disturbed)

1. With No. 1 piston coming up on compression stroke, continue cranking the engine until the pulley timing notch lines up with the O-mark on the engine rear housing pad.
2. Position the distributor to the opening in the block with the vacuum control unit at about four o'clock.
3. Point the rotor toward the rear of the engine (six o'clock) then move the rotor counterclockwise, slightly. This movement should permit engagement of the distributor driven gear with the crankshaft.
4. Press down on the distributor housing while kicking the starter a few times to insure distributor shaft engagement with the oil pump driven member.
5. Bring No. 1 piston back up on

the compression stroke until the pulley notch lines up with the O-mark on the engine rear housing pad. The rotor should point toward the distributor cap No. 1

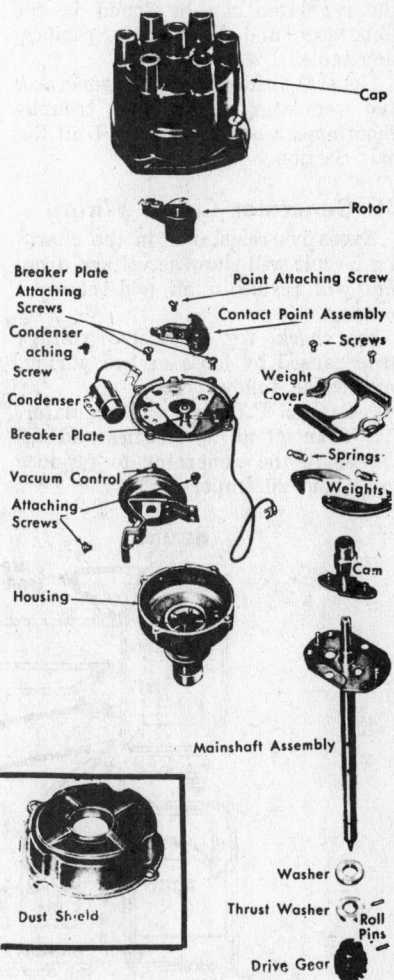

Cap

Rotor

Breaker Plate Attaching Screws

Point Attaching Screw

Contact Point Assembly

Condenser aching Screw

Screws

Condenser

Weight Cover

Breaker Plate

Springs

Vacuum Control Attaching Screws

Weights

Cam

Housing

Dust Shield

Mainshaft Assembly

Washer

Thrust Washer

Roll Pins

Drive Gear

Distributor—dust shield (inset)
is used on 1964-69
(© Chevrolet Div. G.M. Corp.)

tower, and the points should be just ready to break.

6. Install the hold-down clamp and tighten the bolt.

7. Install distributor cap and check all secondary wire connections (spark plug and distributor cap ends.)

8. Connect vacuum line to distributor and primary coil wire to distributor.

9. Start the engine, allow time for normal warm-up and set timing according to tune-up chart.

Distributor (on Turbo-Super-charged Engines)

The distributor on these engines is different. A pressure-retard-unit replaces the ordinary vacuum-advance-unit. This unit retards normal centrifugal advance at higher engine rpm. on regular distributors.

DC Generator, Regulator

Detailed facts on the generator and the regulator can be found in the Generator and Regulator Specification table.

General information on generator and regulator repair and troubleshooting can be found in the Unit Repair Section.

DC Generator Circuit Wiring

Excessive resistance in the charging circuit will show as voltage drop, and will result in an undercharged battery.

To check for excessive voltage drop caused by loose or bad wiring, proceed as follows:

1. Ground F terminal of regulator.

2. Turn off all accessories and operate the generator to produce about 20 amperes.

3. Measure voltage drop at V1, V2 and V3 as in illustration. V1 plus V2 should not exceed 0.5 volt; V3 should not exceed 0.3 volt.

If there is excessive resistance, check wiring for defects, replace as necessary or clean and tighten connections.

Generator Removal—1960-63

1. Disconnect ground cable from battery.

2. Disconnect armature and field terminal wires.

3. Loosen idler pulley on the belt to allow slack for removal of belt from generator pulley.

4. Remove three generator mounting bolts. Lift off generator.

Generator Installation— 1960-63

1. Lift generator into place and install attaching parts. Leave the assembly loose.

2. Place blower belt over generator drive pulley. Don't tighten.

3. Secure the generator by torquing the bolts to 15-22 ft. lbs. (Except mounting bracket-to-generator bolt. This torque should be 8-11 ft. lbs.)

4. Use a 16 in. screwdriver between the idler pulley and bracket. Locate the end of the screwdriver against the raised boss on the engine.

5. Exert a 25 lb. pull on the screwdriver handle and tighten the idler pulley bolt and nut.

NOTE: on cars with radio, connect the radio by-pass condenser to the armature terminal (A), not to the field terminal (F).

Removal—1964

1. Disconnect ground cable from battery.

2. Disconnect armature and field terminal wires.

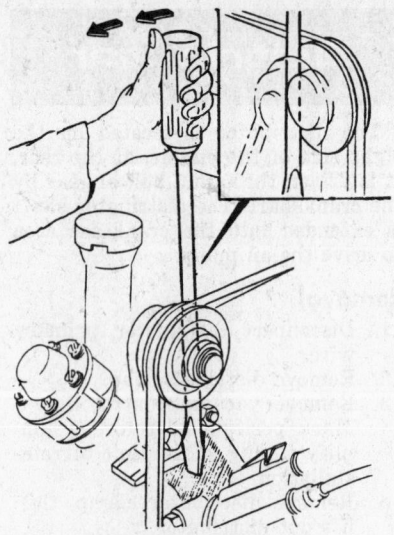

Generator installation—
adjusting blower belt tension
(© G.M. Corp.)

3. Loosen idler pulley bracket to permit belt removal.

4. Disconnect fuel inlet line at fuel pump.

5. Remove oil pressure sending unit from top of oil filter bracket.

6. Remove pulley nut and pulley.

NOTE: nut has a left-hand thread.

Installation—1964

1. Complete steps same as 1960-63, except that oil pressure sending unit, fuel inlet line and pulley must be replaced as well.

2. Install pulley nut and tighten to 50-60 ft. lbs.

Regulator Adjustment

When checking the temperature sensitive regulator, the cover must be in place and the unit must be at operating temperature. Normal operating temperature is achieved by 15 minutes charging at an 8-10 Amp. rate. If the regulator must be replaced, note that the Corvair regulator has a copper-colored ground strap; only regulators having this feature may be installed.

Voltage Regulator

1. Insert a ¼ ohm, 25-watt resistor into the circuit (BAT terminal).

2. Connect a voltmeter between the regulator BAT terminal and ground.

3. Start the engine and run at 1,600 rpm for about 15 minutes, with the resistor in the circuit and the cover in place.

4. Cycle the generator by shutting off the engine, then immediately restarting and bringing it back up to 1,600 rpm.

5. Note the voltmeter reading and the temperature reading (¼ in. from cover). The reading can be

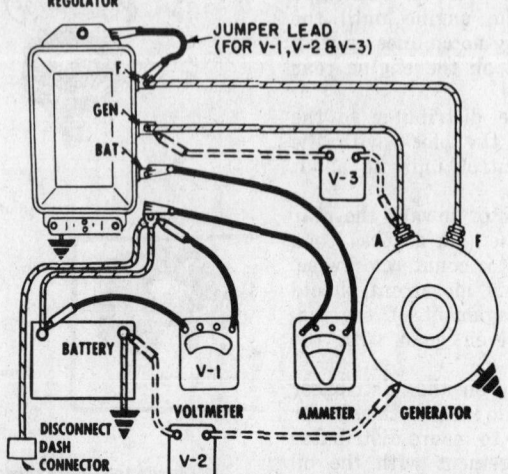

Checking generator circuit resistances—1960-64 (© G.M. Corp.)

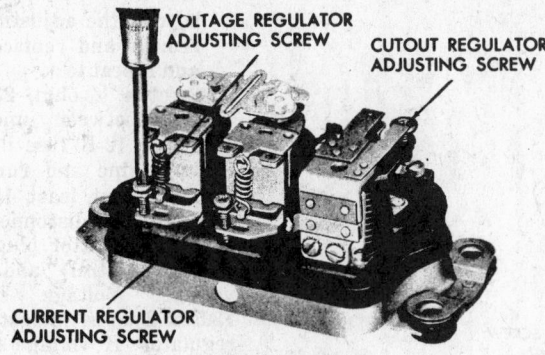

VOLTAGE REGULATOR ADJUSTING SCREW

CUTOUT REGULATOR ADJUSTING SCREW

CURRENT REGULATOR ADJUSTING SCREW

Adjusting regulator
(© G.M. Corp.)

AC Generator (Delcotron)

The Delcotron continuous output AC generator consists of two major parts—the stator and the rotor. The stator is composed of many turns of wire on the inside of a laminated core that is attached to the generator frame. The rotor is mounted on bearings at each end. Two brushes carry current through slip rings to the field coils, which are wound on the rotor shaft.

Six diodes, mounted on internal heat sinks, change the AC current output into DC current. This current is controlled by the regulator. The regulator is a double-contact unit combined with a field relay. When the optional, higher output, Delcotron is used the regulator incorporates an external field discharge diode.

Following are a few precautions to observe in servicing the electrical system of cars equipped with the Delcotron (AC) generator and the regulator.

1. When installing a battery, be certain that the ground polarity of the battery and the ground polarity of the generator and regulator are the same.
2. When connecting a booster battery, be sure to join the correct battery terminals together.
3. When hooking up a charger, connect the correct charged leads to the battery terminals.
4. Never operate the generator on open circuit. Be sure all connections in the charging circuit are tight.
5. Do not short across or ground any of the terminals on the generator or regulator.

compared to the following chart:

Temperature (°F)	Voltage (volts)
165	13.1-13.9
145	13.5-14.3
125	13.8-14.7*
105	14.0-14.9
85	14.2-15.2
65	14.4-15.4
45	14.5-15.6

* *Normal Range*

6. To adjust the voltage setting, remove the regulator cover and turn the *voltage* adjusting screw; increase spring tension to raise setting, decrease to lower.

NOTE: failure of the regulator to hold a setting usually is caused by the screw head not making good contact with the spring support. The regulator must be cycled, with cover on, before checking the setting.

Cutout Relay

It is seldom necessary to adjust this, so long as the relay functions to open and close the charging circuit. The cutout voltage should be at least 0.5 volt below the voltage regulator setting.

1. Connect a voltmeter between regulator GEN terminal and ground.
2. Slowly increase speed and note voltage at which relay closes; should be 11.8-13.5.

3. To adjust, turn illustrated screw clockwise to increase setting, counterclockwise to decrease.

Current Regulator

It is seldom necessary to adjust this, unless armature shows signs of overheating.

1. Disconnect battery ground cable.
2. Remove both power wires from regulator BAT terminal, then connect an ammeter between the removed wires and the BAT terminal.
3. Reconnect battery ground cable.
4. Turn on all lights and connect an additional load (carbon pile or bank of lights) across the battery to produce a system voltage of 12.5-13.0
5. Run engine for at least 15 minutes at 1,600 rpm, with regulator cover in place.
6. Cycle the generator and note current regulator setting, adjust by turning regulator center screw.

Polarizing the DC Generator

After reconnecting leads, momentarily connect a jumper lead between the Gen. and Bat. terminals of the regulator. Failure to do this may result in damage to the equipment, because reverse polarity causes vibration, arcing and burning of the relay points.

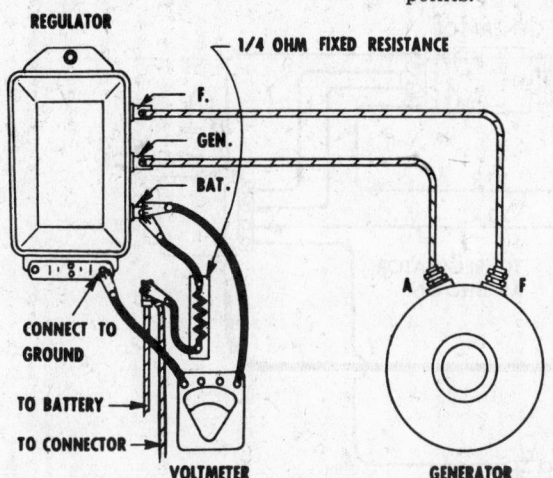

REGULATOR

1/4 OHM FIXED RESISTANCE

F.

GEN.

BAT.

CONNECT TO GROUND

TO BATTERY

TO CONNECTOR

VOLTMETER

GENERATOR

Checking voltage regulator setting—1960-64
(© G.M. Corp.)

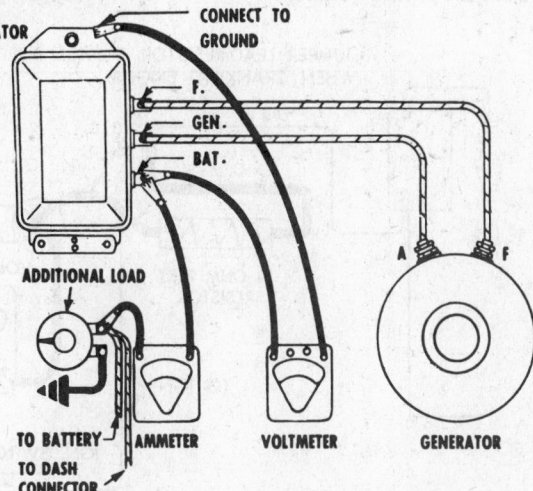

REGULATOR

CONNECT TO GROUND

F.

GEN.

BAT.

ADDITIONAL LOAD

TO BATTERY

TO DASH CONNECTOR

AMMETER

VOLTMETER

GENERATOR

A   F

Checking current regulator setting
(© G.M. Corp.)

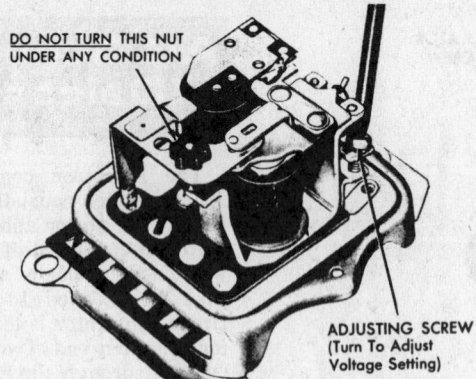

DO NOT TURN THIS NUT UNDER ANY CONDITION

ADJUSTING SCREW
(Turn To Adjust
Voltage Setting)

Adjusting voltage regulator—alternator system
(© G.M. Corp.)

Alternator Removal and Installation—1965-69

1. Disconnect battery ground cable at battery.
2. Disconnect Delcotron wiring.
3. Remove blower belt from pulley.
4. Remove three securing bolts and remove Delcotron.
5. Reverse removal procedure to install.

Troubleshooting

Problems in the AC charging system usually are indicated by one of the following conditions:

1. Faulty indicator light operation.
2. Undercharged battery; low cranking speed.
3. Overcharged battery; excessive battery water loss.
4. Excessive generator noise and/or vibration.

Before proceeding with further troubleshooting checks, check for a loose or broken blower belt, defective battery and loose connections in the system (especially at the push-on Delcotron connectors).

AC Generator Circuit Wiring

1. Place transmission in Neutral and set the parking brake.

2. Connect a voltmeter between junction block at horn relay and regulator ground (base).
3. Connect a tachometer between distributor to coil primary wire and ground.
4. Turn on ignition switch and look at indicator light; if it does not light, check circuit and the bulb itself. If it does light, start the engine and run at 1,500 rpm +. If light goes out, field circuit is O.K. and any trouble is in generator or regulator.
5. Turn on headlights (high beam) and heater motor blower (high), run engine at 1,500 rpm + and read voltage on the meter.
   a. If reading is 12.5 volts or greater, turn off electrical accessories and shut off engine. Proceed to Step 6.
   b. If reading is less than 12.5 volts, test Delcotron output as described later in this section. If Delcotron tests "bad," repair unit; if Delcotron tests "good," disconnect regulator plug, remove regulator cover and reconnect plug. Repeat Step 5 and turn voltage adjusting screw to raise setting to 12.5 volts. If 12.5 volts is

beyond the adjustment range, remove and replace regulator and repeat test.

6. Connect a 1/4-ohm, 25-watt test resister between junction block and wire 12-B (see illustration).
7. Start engine and run at 1,500 rpm+ for at least 15 minutes, then quickly disconnect and reconnect regulator plug (to cycle voltage control) and read the voltage. Voltage should be 13.5-15.2 volts, indicating a good regulator. If voltage is not correct, leave engine running at 1,500 rpm and follow the procedure below:
   a. Disconnect the four-terminal plug and remove regulator cover. Reconnect plug and adjust voltage to 14.2-14.6 volts.
   b. Disconnect plug and install cover, then plug.
   c. Run engine at 1,500 rpm+ for 10 minutes to bring internal regulator temperature up to par, recycle regulator as previously described and check voltage.

Alternator Output Testing

1. Disconnect regulator plug, then disconnect plug from alternator F and R terminals.
2. Connect a jumper wire between alternator BAT terminal and F terminal to give full field excitation.
3. Connect a voltmeter between alternator BAT terminal and GRD terminal.
4. Start engine, turn on headlights (high beam), heater blower (high) or air conditioner blower (medium). Run engine at 1,500 rpm+ and note voltage; it should exceed 12.5 volts within a few minutes. If not, Delcotron must be repaired—see Unit Repair Section of this manual.

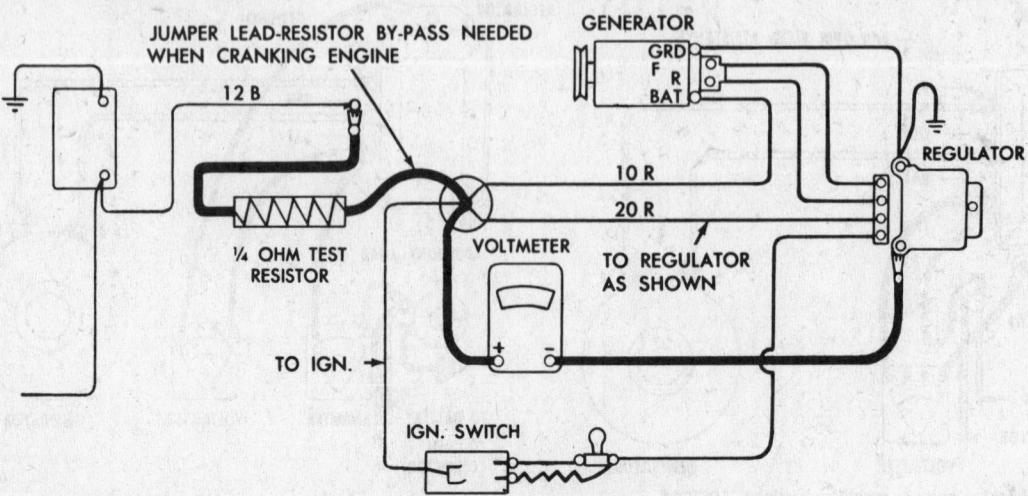

JUMPER LEAD-RESISTOR BY-PASS NEEDED WHEN CRANKING ENGINE

GENERATOR

GRD
F
R
BAT

12 B

REGULATOR

10 R

20 R

1/4 OHM TEST RESISTOR

VOLTMETER

TO REGULATOR AS SHOWN

TO IGN.

IGN. SWITCH

Testing alternator circuit (© G.M. Corp.)

Indicator Light Testing

The indicator light is important in AC charging systems, for it provides initial field excitation current to the alternator. The light goes out when the field relay closes, which applies battery current to both sides of the bulb. If the light does not go on when key is turned, the bulb could be faulty, there could be an open circuit in the wiring or an positive diode in the alternator could be shorted to ground.

1. Disconnect plug from regulator and connect a jumper wire between terminal No. 4 (in plug) and ground. Turn on ignition switch and observe the indicator light. If light does not go on, check bulb, socket or wiring between switch and regulator plug. If light goes on, check regulator (as (described previously), wiring between regulator F terminal and alternator, or Delcotron itself (see Unit Repair Section for service procedures).

2. Disconnect jumper wire at ground end and reconnect to F terminal in plug. Turn on ignition for a second and note light. If light goes on, problem is in regulator. If light does not go on, problem is in wire between F terminals (regulator and alternator).

3. Disconnect jumper at plug F terminal and reconnect the free end to F terminal at alternator. Turn on ignition switch for a second and note light. If light goes on, the problem is an open circuit in the wire connecting the regulator and alternator F terminals. If light does not go on, the alternator field windings are defective.

If the indicator light does not ex-

Adjusting field relay closing voltage
(© G.M. Corp.)

tinguish when engine is started, check for a loose drive belt, faulty field relay, faulty alternator, open parallel resistance wire (usually shows up at idle). If the light stays on with the key turned off, an alternator positive diode is shorted to ground.

Field Circuit Resistance Testing

The resistance wire is an integral part of the ignition wiring harness. The wire cannot be soldered; any connections must be made using crimp-type connectors. Resistance is 10 ohms, 6¼ watts.

1. Connect a voltmeter between the wiring harness terminal No. 4 and ground.
2. Turn on ignition switch, needle must indicate or resistor is open.

Field Relay Testing and Adjustment

1. Connect a voltmeter between No. 2 regulator terminal and ground.
2. Start engine and run at 1,500-2,000 rpm; observe voltmeter reading.

3. If voltmeter indicates zero volts at regulator, check the circuit between No. 2 regulator terminal and Delcotron R terminal.
4. To adjust, connect a 50 ohm rheostat between wiring harness terminal No. 3 and regulator terminal No. 2, after disconnecting the spade lug on the end of the No. 2 regulator terminal wire. Connect a voltmeter between regulator terminal No. 2 and ground, then turn the resistor to "open" position, turn off ignition switch and slowly decrease resistance until relay closes (noting voltage at this point). Voltage can be adjusted by bending heel iron as illustrated.

Instruments

Instrument Cluster R&R— 1960-64

1. Disconnect battery ground cable.
2. Remove connector from back of cluster.
3. Pull harness out of retaining clips; disconnect speedometer cable.
4. Remove light switch and cigarette lighter assembly.
5. Lower mast jacket and steering wheel assembly or remove wheel (on cars with 3-speed transmission) or steering wheel and turn signal housing (Powerglide).
6. Remove 11 screws that hold cluster to instrument panel; two screws must be removed from the rear.
7. Remove cluster by pulling straight out.

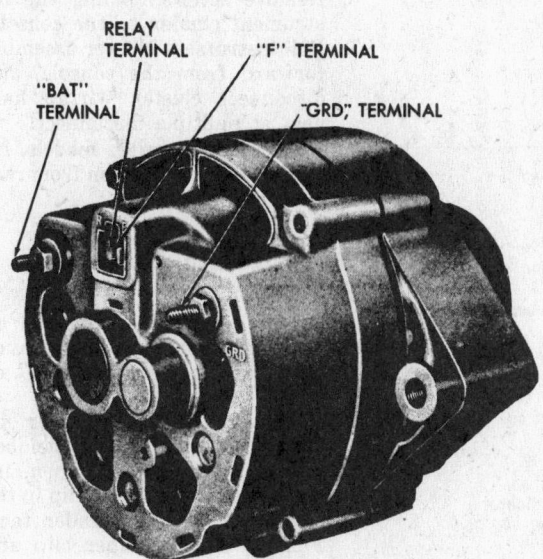

Delcotron alternator terminals
(© G.M. Corp.)

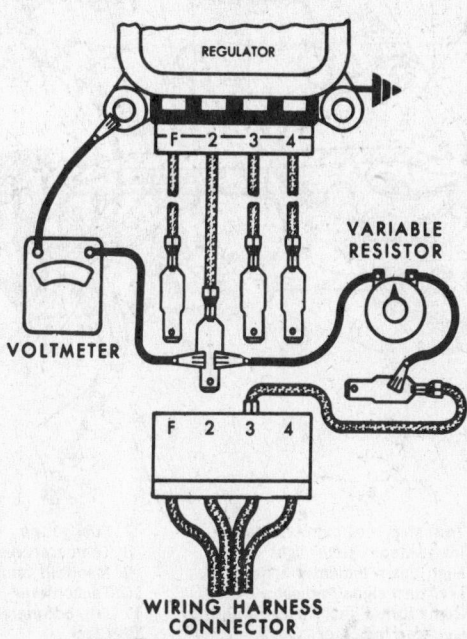

Checking field relay closing voltage
(© G.M. Corp.)

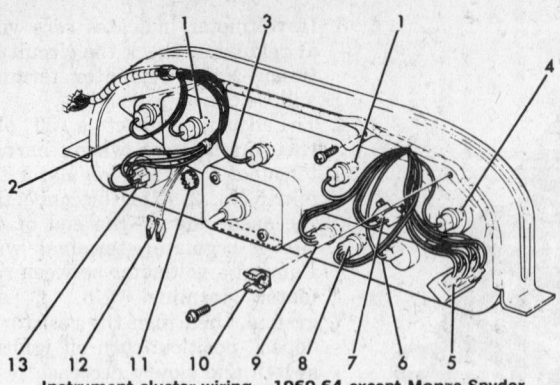

1　Instrument cluster lights (gray)
2　Right turn signal indicator (dark blue)
3　High beam indicator light (light green)
4　Left turn signal indicator (light blue)
5　Light switch
6　Windshield wiper switch (red)
7　Connector—gas gauge (brown and tan)
8　Oil pressure and temperature indicator light (dark blue and tan)
9　Generator charge light (brown and tan)
10　Harness clip
11　Cigarette lighter plug
12　Stop light connectors (orange and white)
13　Ignition switch

Instrument cluster wiring—1960-64 except Monza Spyder
(© G.M. Corp.)

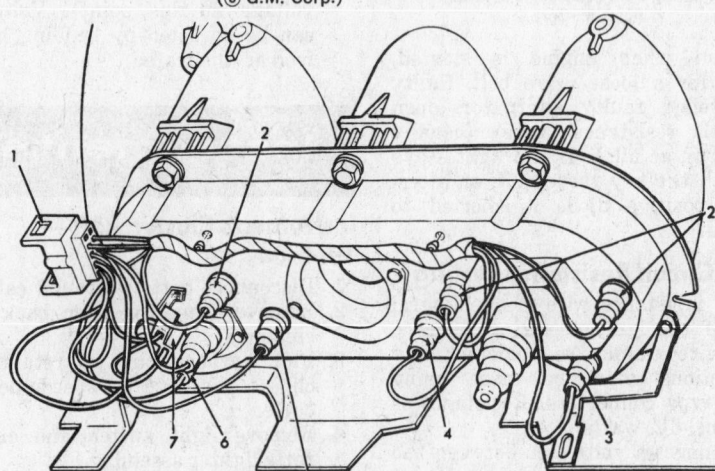

1　To instrument panel harness
2　Instrument cluster bulb
3　High-beam indicator
4　Left turn signal indicator
5　Right turn signal indicator
6　Gen/Fan indicator
7　Fuel gauge

Instrument cluster—1965-69 except Corsa
(© G.M. Corp.)

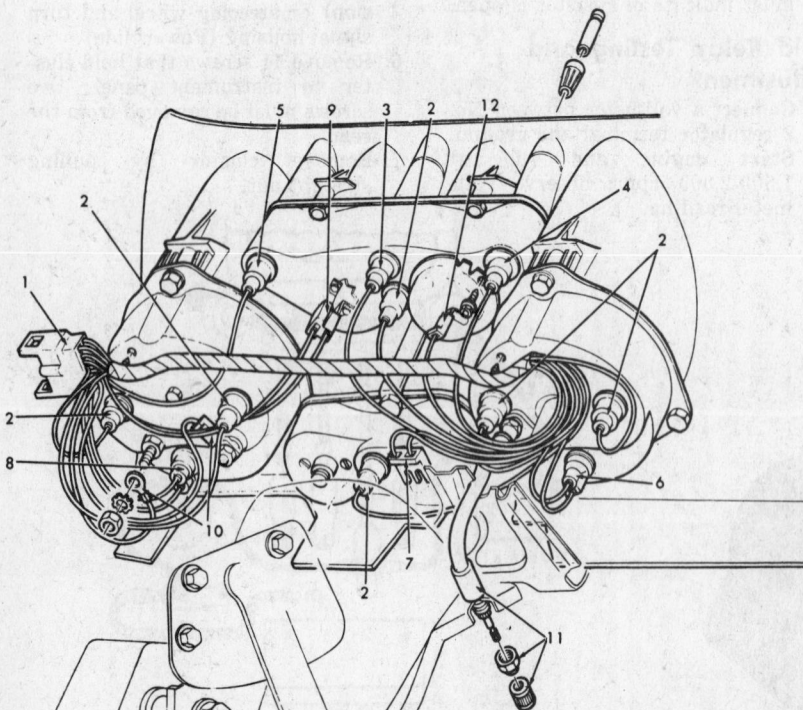

1　To instrument panel harness
2　Instrument cluster light
3　High beam indicator light
4　Left turn signal indicator light
5　Right turn signal indicator light
6　Gen/fan indicator
7　Fuel gauge
8　Temperature/oil pressure indicator
9　Manifold temperature gauge
10　Tachometer
11　Trip odometer
12　Clock

Instrument cluster wiring—1965-69 (© G.M. Corp.)

8. To install, reverse removal procedure.

Instrument Cluster R & R—1965-69

1. Disconnect battery.
2. Remove upper mast jacket support clamp (Powerglide models only).
3. Remove light and wiper switch bezel nuts.

NOTE: on Powerglide models, remove shift lever knob.

4. Remove heater or air-conditioning control retaining screws and allow control to hang below instrument console. On air-conditioned models, remove air outlet from panel.
5. Disconnect speedometer cable from rear of panel. If so equipped, disconnect trip odometer and speed warning unit.
6. Remove screws holding the instrument cluster to the console.
7. Pull instrument cluster assembly forward from the console, and disconnect cluster wiring harness at multiple disconnect.

NOTE: on Powerglide models, remove shift lever mechanism from rear of cluster.

8. Reverse procedure to install.

Ignition Switch Removal—1960-69

The ignition switch is located on the right side of the instrument panel.

1. Disconnect battery cable at battery.
2. Remove the lock cylinder by turning the switch to the lock position (Acc.-1969). Then, insert the end of a paper clip in the small hole of the cylinder face. Push in on the paper clip and turn the key counterclockwise until the lock cylinder can be removed.

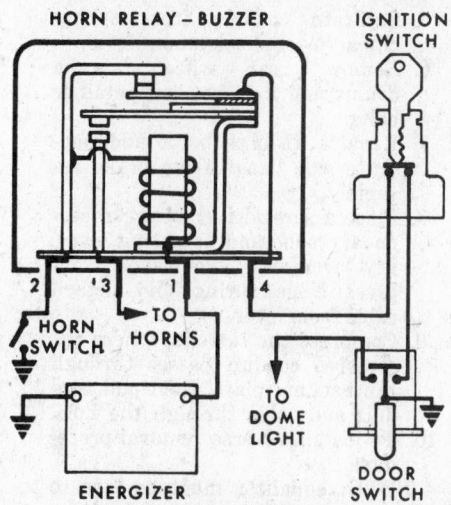

Key warning buzzer circuit
(© G.M. Corp.)

3. Using spanner wrench (tool J-7607), remove the front attaching nut.
4. Withdraw switch.
5. To remove the theft resistant connector:
   A. Make a suitable tool out of heavy wire stock and press on the lock clip within the slot in the connector.
   B. Holding the wire tool firmly in place, work the connector

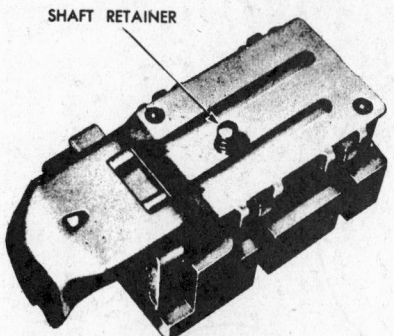

Light switch shaft retainer
(© G.M. Corp.)

away from the switch until the connector is released.

Lighting Switch R & R—1960-69

1. Disconnect battery.
2. Pull knob out to on position.
3. Reach under panel and depress the switch shaft retainer. Remove shaft and knob assembly.
4. Remove the retaining ferrule and bezel and lower the switch assembly.
5. Disconnect the multi-plug connector.
6. Install in reverse of above.

Key Warning Buzzer—1969

A buzzer, consisting of an upper contact in the horn relay, will sound if the driver's-side door is opened with the key in the off or accessory position. The buzzer will not operate if the key is slightly withdrawn.

Battery, Starter

Battery

Detailed information on the battery and starter will be found in the Battery and Starter Specifications table.

A general discussion of starters and batteries will be found in the Unit Repair Section.

Starter R & R—1960-69

1. Disconnect battery ground at battery.
2. Jack up and place stands under rear of car.
3. Disconnect rear throttle control rod from bellcrank at transmission, and from the cross-shaft in the engine compartment.
4. Disconnect the three wires at the starter solenoid.
5. With a ⅜ in. drive (9/16 in. shallow socket on a 9 in. extension

with ratchet) remove the two starter mounting bolts. Remove the upper bolt first. Then pull the starter forward and out of the engine.
6. Install starter in reverse order.

Brakes

Foot Brakes

The service brakes, both front and rear of all models, are the duo-servo single-anchor-type. This brake, except for size, is basically the same as that used on Chevrolet and should be serviced in the same way.

Brake adjusting and repair procedures can be found in the Unit Repair Section.

Beginning 1967

Since 1967, a tandem-type master cylinder has been used on all models. This design divides the brake hydraulic system into two independent and hydraulically separated halves.

Details and repair procedures on this tandem system may be found in the Unit Repair Section.

Master Cylinder Pushrod Clearance

The brake pedal has a non-adjustable rubber stop located at the release end of pedal travel. If the master cylinder pushrod is incorrectly adjusted, the pedal will contact the release stop before the master cylinder compensating port is open, resulting in locked-up or dragging brakes. Adjusting too far in the opposite direction will result in long pedal travel.

Adjustment—1960-69

1. Loosen the swivel locknut.
2. Adjust the pushrod by turning it until pedal pad movement, before pushrod touches master cylinder piston, is ⅛-⅜ in. for 1960-61

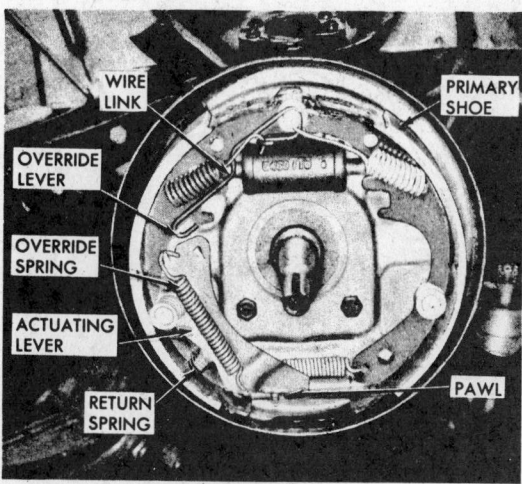

Self-adjusting brakes—1963-69
(© G.M. Corp.)

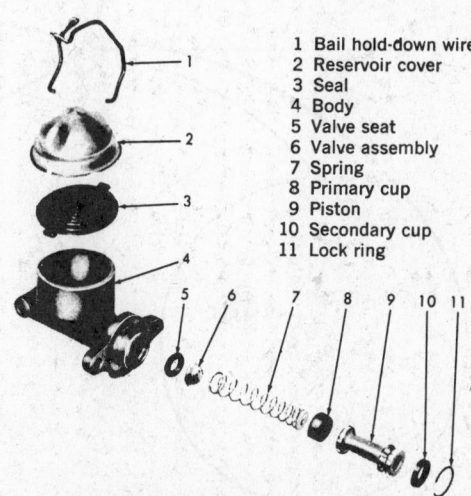

1 Bail hold-down wire
2 Reservoir cover
3 Seal
4 Body
5 Valve seat
6 Valve assembly
7 Spring
8 Primary cup
9 Piston
10 Secondary cup
11 Lock ring

Single circuit master cylinder—1960-66
(© G.M. Corp.)

cars, 1/16-1/4 in. for 1962-69 cars.

3. Tighten locknut.

Parking Brake

The parking brake apply lever is located on the dash, to the left of the steering column. A pull on the apply handle is transmitted through cables and an equalizer to the rear wheel brake shoes in the conventional manner.

Built into the parking brake mechanism is a self-adjusting feature. If the first full stroke of brake application is not enough to securely hold the car, a second and more solid stroke may be applied.

Release of the brake is by pulling a separate release handle that extends from under the dash to the left of the mast jacket. The release mechanism also provides a visual indication when the apply handle is pulled by causing the release handle to pop out about 2 in. during the first ½ in. of apply handle travel.

Parking Brake Adjustment— 1960-69

1. Jack up rear of car so that both rear wheels clear the floor.
2. Pull handbrake lever up three notches (1960), four notches (1961-64) or one notch (1965-69) from full release position.
3. Loosen forward locknut (at equalizer) and tighten rear nut until rear wheels drag when rotated.
4. Tighten nuts.
5. Fully release brakes and check that no drag exists.

NOTE: lubricate brake cables and pulleys every 12,000 miles, or once a year.

Forward Parking Brake Cable R&R—1960-69

1. Disconnect positive battery cable at battery.
2. Release parking brake.
3. Remove equalizer nuts and pull cable stud bolt from equalizer.
4. Remove underbody tunnel cover, then toe pan cover.
5. Remove cable pulley from upper toe board bracket.
6. Remove cable ball from lever clevis and withdraw cable from car.

NOTE: on 1960-64 cars having manual transmission, clutch cable must be disconnected to relieve tension on pulley before starting Step 5.

7. Reverse removal procedure to install, making sure that original length screws *only* are used to secure the toe pan cover. If longer screws are used, they could puncture the gas tank.

Rear Parking Brake Cable R&R —1960-64

1. Release parking brake.
2. Remove brake return spring.
3. Remove forward equalizer locknut and pull cable stud from equalizer.
4. Remove U-clips from support brackets (bolted to control arm pivots).
5. Separate conduit from spring clip at lower shock mount.
6. Remove rear wheels, brake drums and axle bearing retainer bolts.
7. Remove U-joint bolts and separate axle flange from brake assembly.
8. Place a screwdriver between secondary shoe and actuating lever, pry lever away from shoe, compress cable spring and detach cable from lever.
9. Compress the fastener where the armored conduit passes through the backing plate, then pull conduit and cable through the hole.
10. To install, reverse removal procedure.

NOTE: equalizer must be free to slide on rear cable.

Rear Parking Brake Cable R&R —1965-69

1. Release parking brake.
2. Remove cable return spring.
3. Remove rear equalizer locknut and separate forward cable stud from equalizer.
4. Remove U-clips and cable from support brackets on rear crossmember.
5. Remove the bolt that holds cable clip to control arm.
6. Remove rear wheels and brake drums.

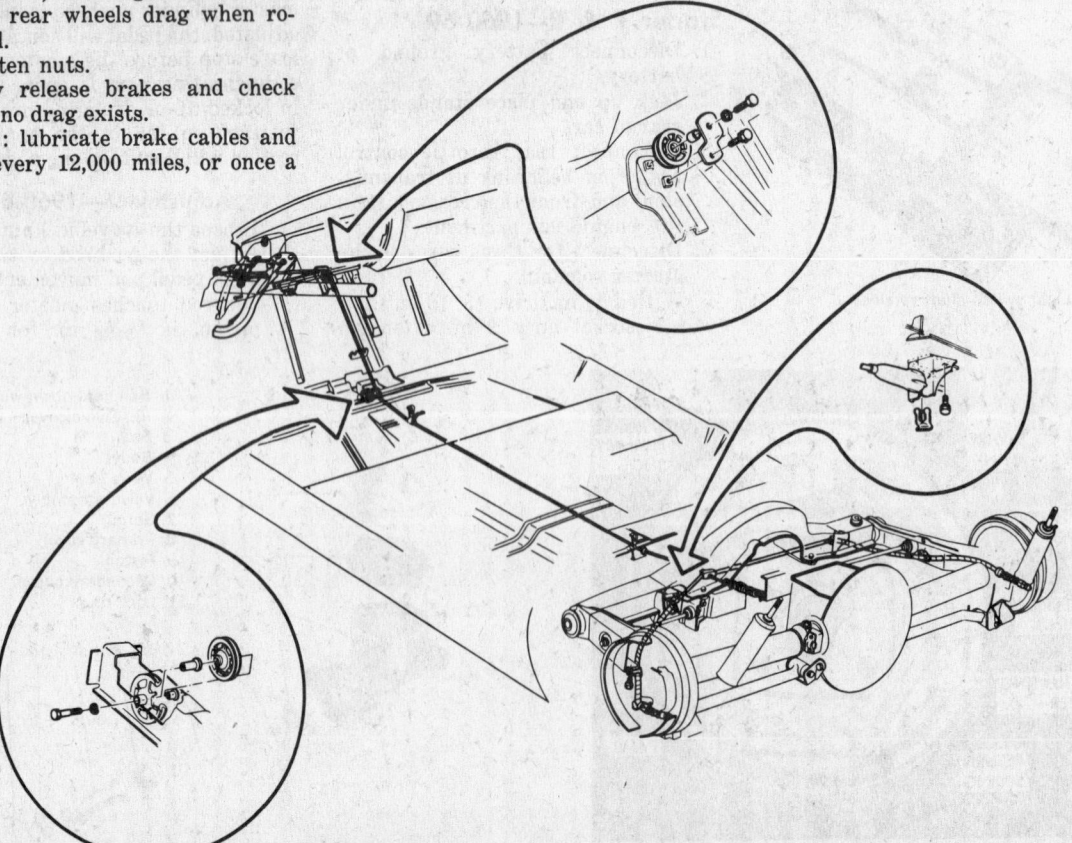

Parking brake cable configuration—typical 1960-69 (© G.M. Corp.)

7. Pry between actuating lever and secondary shoe with a screwdriver, compress cable spring and detach cable from lever.
8. Compress the fastener where the armored conduit (outer cable) passes through the backing plate, then pull conduit and cable through the hole.
9. To install, reverse removal procedure.

NOTE: equalizer must be free to slide on rear cable.

Fuel System

Carburetors

Corvair carburetion consists of either two carburetors—one per bank, of four carburetors—two per bank. The Rochester model H units have either a centrally-located automatic choke (on air cleaner, 1960), separate manual chokes (1961) or cylinder head-mounted automatic choke, (1962-64). Starting in 1965, Rochester model HV carburetors with automatic chokes are used, although the

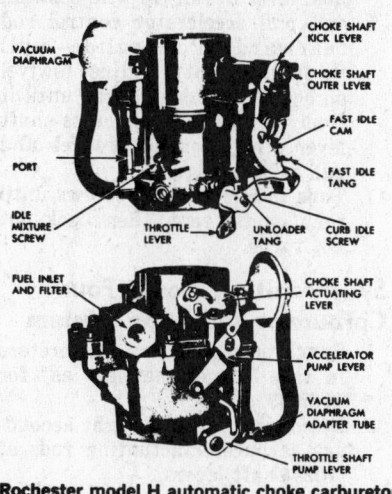

Rochester model H automatic choke carburetor
(© G.M. Corp.)

four-carburetor Corsa model has modified model H units as secondary carburetors. The Model H used on the four-carburetor engine is not equipped with choke, power enrichment or low-speed circuit.

The two carburetor throttle levers are connected by a cross-shaft, actuated by the accelerator linkage. Care must be used to insure synchronized adjustment.

The automatic choke vacuum source is located at the mid-point of the balance tube that extends between the two engine manifolds.

Accelerator Linkage Adjustment—1960-64

1. Remove three screws and the mud guard over idler lever A.
2. Disconnect swivel D from lever A.
3. Disconnect swivel B from the left carburetor cross-shaft E.
4. Pull rod C to wide-open throttle (through detent for Powerglide and against stop for standard transmission) and turn lever E in wide-open throttle position (carburetor lever against the top). Adjust swivel B to align with the hole in lever E. Now lengthen rod C by backing off swivel five full turns (seven-1960).
5. Position accelerator pedal by placing a block of wood between pedal end and floor mat. (Powerglide 1¼ in., Standard 1 in.)
6. Hold lever A in wide-open position (through detent) turn

Carburetor linkage
(© G.M. Corp.)

swivel D to align with hole in lever A. Install mud guard.
7. Remove the accelerator block.

Accelerator Linkage Adjustment—Turbo-Supercharged Engines

1. Remove air cleaner and block open choke.
2. Disconnect accelerator rod swivel (3) from cross-shaft lever (4).
3. Make sure throttle lever is against idle speed screw, then check linkage angle X. Angle must be 126°; it can be adjusted by changing length of rod (1).
4. Pull accelerator rod (5) rearward against bellcrank stop on transmission, then rotate lever (4) to full throttle position.
5. Adjust swivel (3) to just enter hole in lever (4), then connect.

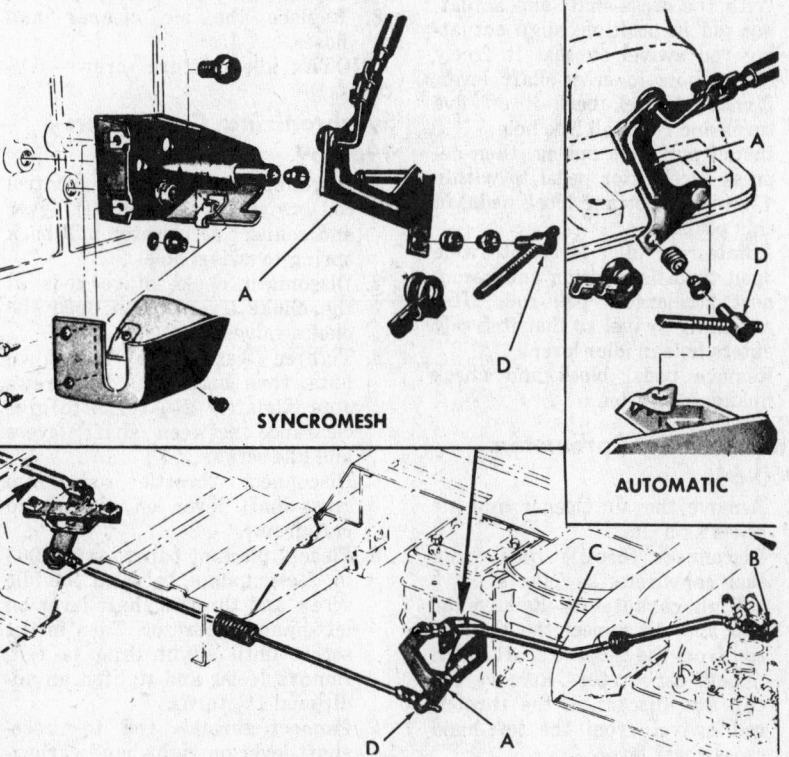

Accelerator linkage (© G.M. Corp.)

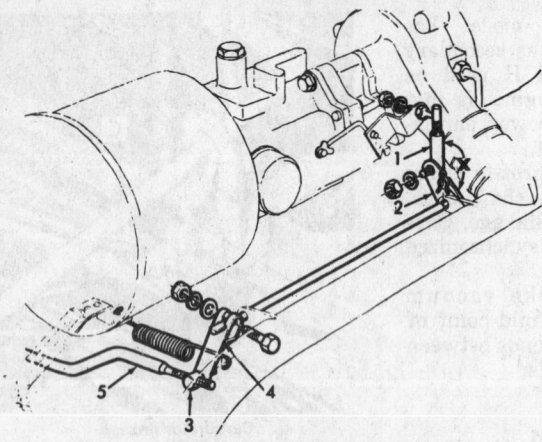

Accelerator linkage—Turbocharged engine
(© G.M. Corp.)

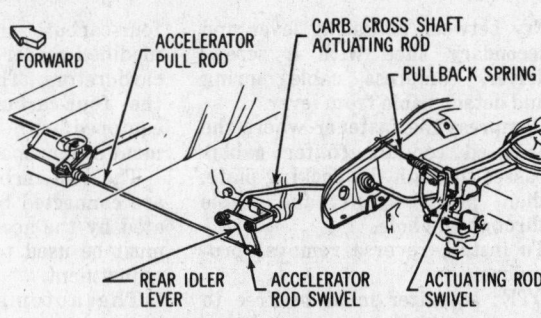

Accelerator linkage—1965-69
(© G.M. Corp.)

6. Check linkage action; when accelerator rod is moved to full throttle, throttle lever on carburetor also must move to full throttle position without binding.

Accelerator Linkage Adjustment—1965-69

1. Disconnect accelerator pull rod swivel from rear idler lever.
2. Disconnect pull back spring, then remove cross-shaft actuating rod swivel from cross-shaft.
3. Pull carburetor cross-shaft actuating rod to the rear until the rear idler lever hits the stop. Rotate cross-shaft to move the left-hand carburetor to wide open position (Powerglide—through detent).
4. With the cross-shaft and actuating rod in position, align actuating rod swivel so that it freely enters hole in cross-shaft lever. Remove swivel, back it off five turns and reinstall into hole.
5. Install pull back spring, then depress accelerator pedal to within 1⅛ in. of floor and block pedal in this position.
6. Rotate rear idler lever into wide open throttle position and reconnect accelerator pull rod, after adjusting swivel so that it freely enters hole in idler lever.
7. Remove pedal block and check linkage operation.

Synchronizing Carburetors—1960-64

1. Remove the air cleaner and air hoses as a unit.
2. Disconnect throttle rods from each carburetor as follows:
   Right carburetor—Remove the clip and disconnect the throttle rod from the throttle shaft lever.
   Left carburetor—Remove the clip and disconnect the throttle rod swivel from the left-hand cross-shaft lever.
3. Back off each carburetor idle screw until each throttle valve is fully closed.
4. With a .003 in. feeler between the idle screw and the throttle lever, turn the screw in until it holds the gauge. Now, turn the screw 1-1½ turns. Do this on each carburetor.
5. On the right-hand carburetor, connect and clip the throttle rod to the carburetor throttle lever.
6. With the left hand throttle lever held fully clockwise, adjust the swivel at the top of the rod so that it will freely enter the hole in the cross-shaft lever. Turn one more turn for 1962-64, then clip in place.
7. The carburetors are now synchronized and any change of idle adjustment must be in duplicate.
8. Replace the air cleaner and hoses.

NOTE: idle mixture screws—1½ turns.

Synchronizing Carburetors—1965-69

1. Disconnect accelerator control rod swivel at cross-shaft lever and connect accelerator pull back spring to swivel hole.
2. Disconnect choke stove rods at the choke levers, then open the choke valves.
3. Tighten carburetor hold-down nuts, then back out idle screws approximately 2½ turns to give clearance between shaft levers and idle screws.
4. Disconnect throttle rod from cross-shaft lever on right-hand carburetor.
5. Place a piece of paper, or a 0.003 in. feeler gauge, between the idle screw and throttle shaft lever on left-hand carburetor. Turn in the screw until slight drag is felt, remove feeler and tighten an additional 1½ turns.
6. Connect throttle rod to cross-shaft lever on right-hand carburetor and disconnect corresponding rod on left-hand carburetor.
7. Adjust idle screw on right-hand carburetor the same as in Step 5.
8. Adjust left-hand carburetor throttle rod by holding up on rod, so that throttle shaft lever is against idle screw, and turning rod lower swivel until the rod can freely enter the hole in the cross-shaft lever.
9. Connect the throttle rod on left-hand carburetor to cross-shaft lever.
10. Remove accelerator pull back spring from cross-shaft lever, hold lever in full throttle position and pull accelerator control rod rearward (Powerglide—pull through detent). Adjust swivel on accelerator control rod until it freely enters hole in cross-shaft lever, then connect swivel and pull back spring.
11. Turn in idle mixture screws until they gently seat, then back out 1½ turns.

Synchronizing Corsa Four-Carburetor Induction System

1. Synchronize primary carburetors in the same manner as for 1965-69.
2. Disconnect left and right secondary carburetor actuating rods at cross-shaft levers.
3. Disconnect accelerator return spring and rotate accelerator control lever on cross-shaft until primary carburetors are at full throttle position.
4. While holding primary carburetors at full throttle position, open secondary carburetor to full throttle and adjust swivel so that it can just enter the front of the slot in cross-shaft lever.
5. Return primary carburetors to idle position, then connect both secondary carburetor rods.
6. Check linkage action, making sure that all carburetors open at the same rate (primary before secondary) and that all carburetors are fully open at full throttle position.

NOTE: actuate throttles only at cross-shaft lever.

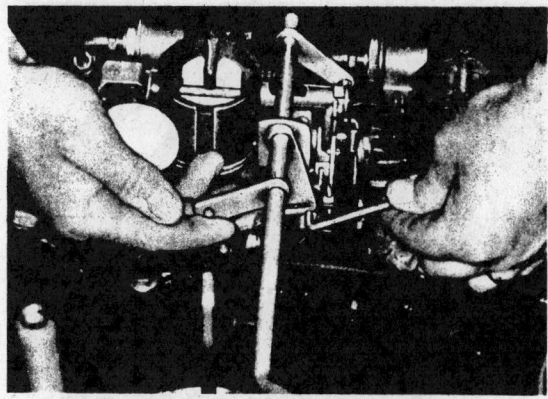

Corsa secondary carburetor synchronization
(© G.M. Corp.)

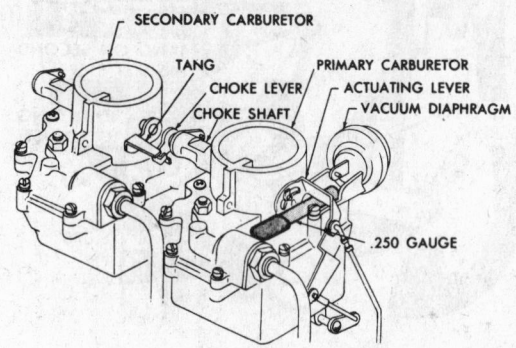

Corsa secondary carburetor lock-out adjustment
(© G.M. Corp.)

Corsa four-carburetor swivel adjustment
(© G.M. Corp.)

Fast Idle Cam Adjustment—1960

1. Synchronize carburetors and set idle speed.
2. Depress accelerator pedal and check carburetors for wide-open throttle.
3. Place a ⅛ in. spacer between the throttle lever and adjusting screw at the right-hand carburetor.
4. Remove the clip attaching the fast idle swivel to the fast idle lever.
5. Turn down fast idle screw until 5/16 in. of the screw threads projects beyond the lever.
6. Rotate the fast idle cam counterclockwise (viewed from left of vehicle) until the fast idle screw rests on the highest portion of the cam.
7. Holding the screw and lever in this position, adjust the swivel on the fast idle link so the pin will just enter the hole in the fast idle lever. Clip the swivel and link in this position.
8. Remove the ⅛ in. spacer. Linkage clearances will have reduced this clearance, so readjust the fast idle screw until the ⅛ in. spacer can be reinstalled. This adjustment will provide the correct fast idle speed.

Fast Idle Cam Adjustment—1961

1. With choke fully open, insert a 0.010 in. (automatic) or 0.030 in. (manual) feeler gauge between the fast idle screw and the tang on the throttle lever.
2. Adjust screw until the gauge can be pulled out with slight resistance.

Fast Idle Cam Adjustment—1962-69

1. Insert a 0.003 in. feeler gauge, or a strip of paper, between the idle screw and the throttle lever.
2. Hold the throttle lever closed with a rubber band, then turn in

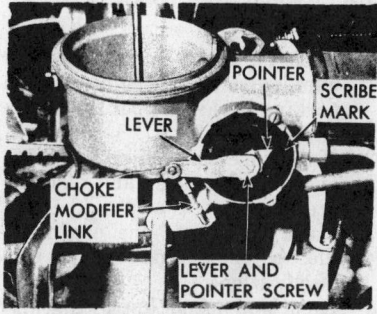

Automatic choke adjustment—1960
(© G.M. Corp.)

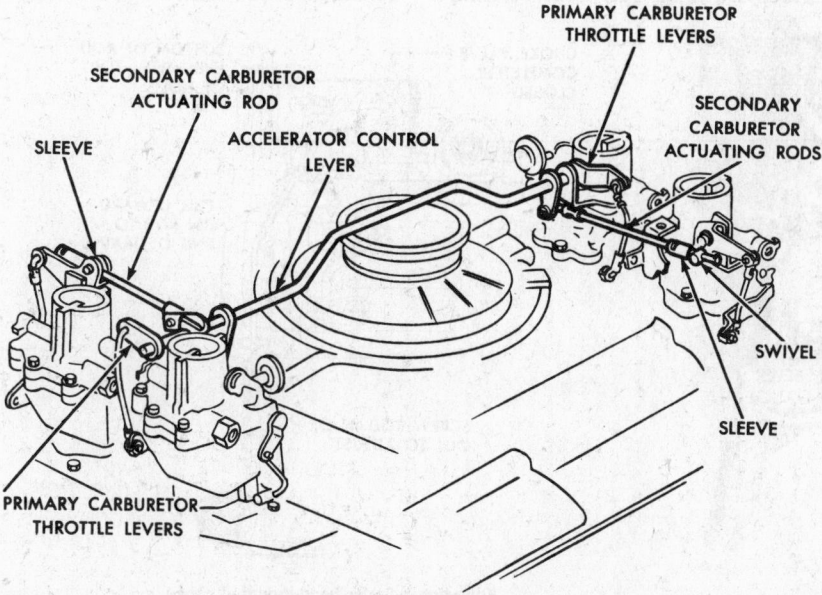

Corsa 4 carburetor assembly

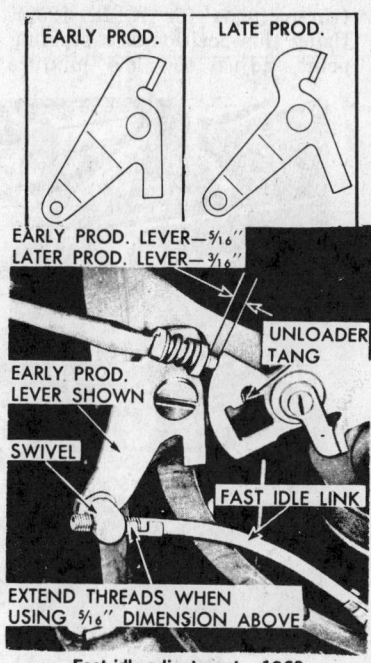

Fast idle adjustment—1960
(© G.M. Corp.)

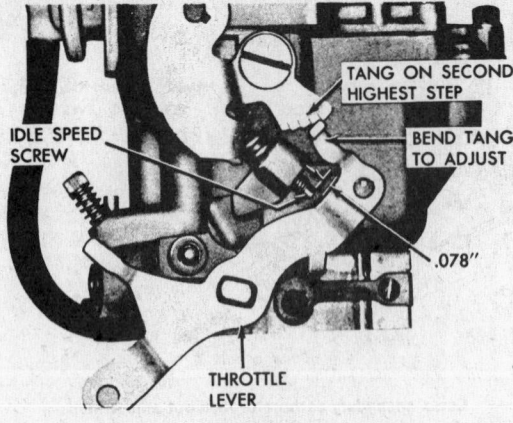

Fast idle cam adjustment—1962-69
(© G.M. Corp.)

Idle mixture adjustment
(© G.M. Corp.)

idle screw until a resistance is felt.
3. Remove gauge and turn in idle screw an additional 1½ turns.
4. With the throttle lever on the second highest step of the fast idle cam, bend tang to obtain 0.078 in. clearance between idle speed screw and throttle lever.

Idle Speed and Mixture Adjustment

1. Hook up a tachometer and vacuum gauge to the engine. Set the parking brake and put the transmission in neutral.
2. Make sure the engine is warm, and that the choke and fast idle are off.
3. Start and allow the engine to idle. Adjust engine idle speed to specifications.
4. Turn, but do not force, the idle mixture screws all the way in, then back out two full turns (three turns for A.I.R. 1969). Using this position as a starting point, adjust the idle mixture screw on each carburetor independently for smoothest idle, or highest vacuum reading in range 14-18 in. Hg.
NOTE: for A.I.R. 1969 engines, ¼ turn out from "lean roll," or 20-30 rpm drop point as mixture is leaned out.
5. It may be necessary, at this time, to readjust for correct rpm and then set the idle mixture again.
6. Remove tachometer and vacuum gauge, then road-test the car.

Turbo-Supercharger

The Monza Spyder is equipped with an exhaust driven supercharger. An exhaust-driven unit forces the air-fuel mixture into the intake manifold at higher-than-atmospheric pressure.

It is a precision balanced rotating group with a turbine at one end and a centrifugal impeller at the other, each in a contoured housing.

Hot exhaust gases directed against the turbine blades spin the turbine, shaft and impeller at high speeds. The impeller draws air-fuel mixture from the carburetor and passes it to the intake manifold at higher-than-atmospheric pressure.

This increases the air-fuel mixture available to the cylinders, resulting in greater horsepower.

Under heavy demands, the supercharger speed is increased because of increased exhaust gases, providing more air-fuel mixture to meet the demand.

The shaft rotates on a semi-floating sleeve bearing, lubricated by engine oil from the oil filter adapter. Excess oil drains through a tube to the right-hand bank rocker arms.

Carburetor Adjustment—Type YH

Idle Speed and Mixture Adjustment

1. Start engine and bring to operating temperature.
2. Shut off engine and back out idle speed screw, then adjust inwards until throttle plate is slightly open.
3. Screw in idle mixture screw until

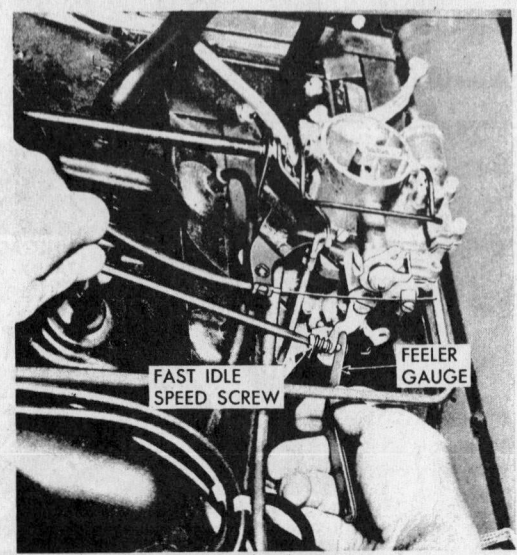

Fast idle cam adjustment—1961
(© G.M. Corp.)

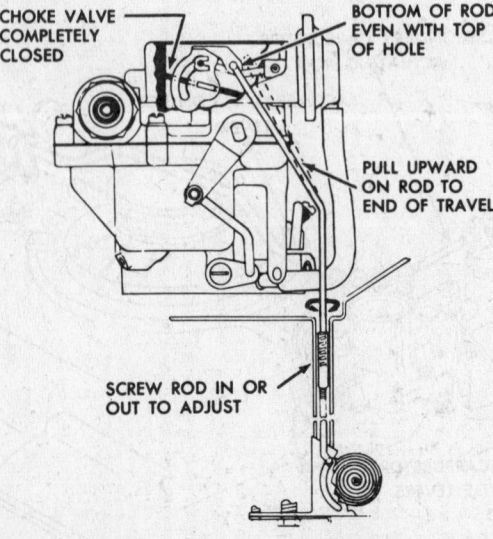

Adjusting cylinder head-mounted choke
(© G.M. Corp.)

it gently seats, then back out ¾ turn.

4. Connect a tachometer between the distributor-to-coil primary wire and ground, then connect a vacuum gauge into the manifold fitting.

5. Check the fast idle linkage and make sure it is not on. To check this, remove air cleaner and observe choke plate; it should be open.

6. Start engine and adjust idle speed screw to obtain 850 rpm idle, then adjust mixture screw and idle speed screw alternately to obtain the highest steady vacuum reading at 850 rpm.

7. Shut off engine, remove instruments and reconnect vacuum hose.

Dashpot Adjustment—1962-63

1. Adjust idle speed and mixture, as previously outlined.

2. With engine running at 850 rpm, choke open, measure the clearance between the throttle plate lever on the carburetor and the dashpot bolt head; clearance should be 0.030 in.

3. If necessary, adjust the clearance by turning the hex bolt, with a wrench on the flat for support.

Automatic Choke Adjustment

1. Loosen three housing screws, rotate housing to desired setting.

2. Tighten housing screws and check free operation of choke.

NOTE: must be set with engine cold; 1962-64—one notch lean, 1965 up—index.

Service

Carburetor Removal

1. Remove air cleaner.

2. Disconnect choke heat tube, fuel line and accelerator linkage at carburetor.

3. Remove carburetor hold-down nuts and carburetor, using a very short open-end wrench to loosen the front nut.

NOTE: remove dashpot, if so equipped, before Step 3.

Supercharger Removal

1. Remove spare tire and air cleaner assembly.

2. Disconnect choke heat tube and fuel line at carburetor.

3. Remove heat shield; disconnect oil feed and drain lines at housing.

4. Disconnect accelerator linkage at carburetor.

5. Loosen turbine housing V-clamp nut, support the supercharger assembly and remove the clamp.

6. Lift the supercharger carefully up and out of the car, being careful not to score the turbine wheel.

7. Remove carburetor hold-down nuts and carburetor.

8. Remove seven inlet and outlet flange bolts.

Supercharger Disassembly and Inspection

The supercharger must be disassembled in a clean, dust-free area, using clean tools and equipment. Any scratches, nicks or dirt on any of the rotating parts can cause premature failure of the unit. Be sure to remove *all* carbon build-up and replace any oil seals which show signs of leakage.

1. Remove the six bolts that secure the compressor housing to the bearing housing, then remove the housing and gasket.

2. Inspect the compressor housing for scoring, erosion or pit marks on its inner surface.

3. Inspect the impeller wheel for damaged blades or scuff marks.

NOTE: if the impeller must be cleaned, use only a nylon brush and kerosene.

4. Hold the turbine wheel blades with a rag and remove the self-locking nut by turning it clockwise (left-hand thread). Remove impeller washer.

5. Support supercharger in a press, with impeller wheel upward, and cover the area with rags to absorb the shock when the turbine wheel drops out.

6. Using a 0.25 in. diameter brass rod, press the turbine shaft from the impeller wheel.

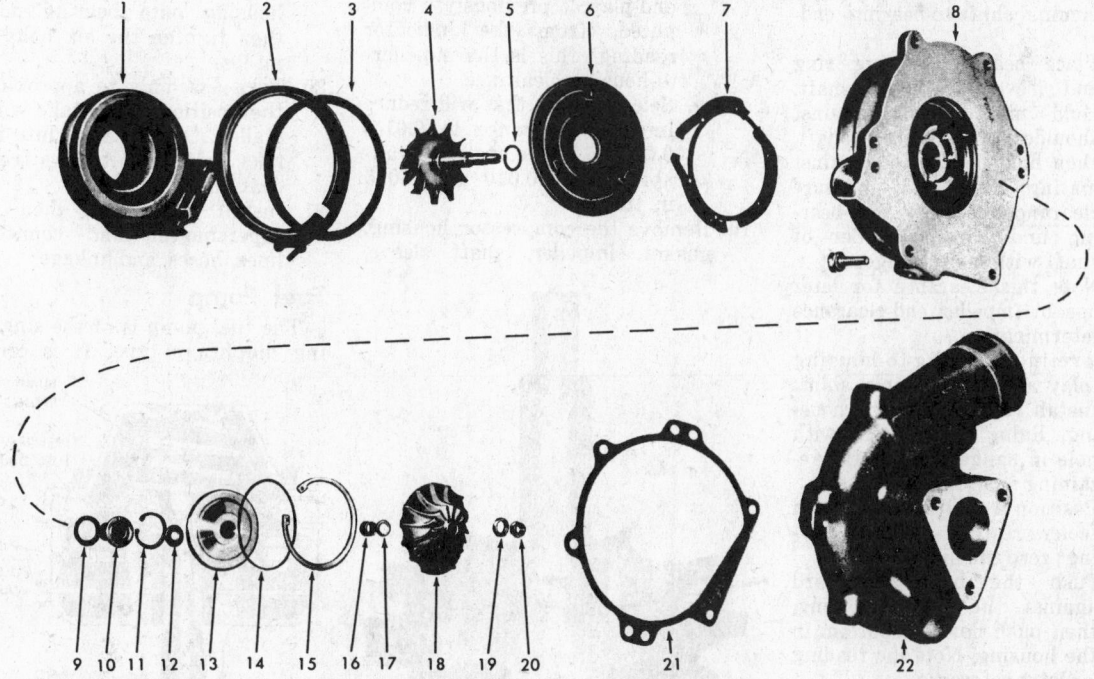

1 Turbine Housing	6 Shield Plate	12 Mating Ring (Washer)	18 Impeller
2 Turbine Housing Clamp	7 Spring Ring	13 Oil Seal Assembly	19 Impeller Special Washer
3 Gasket	8 Bearing Housing	14 O-Ring Seal	20 Impeller Nut
4 Turbine Wheel and Shaft	9 Bearing Shim	15 Seal Retaining Ring	21 Compressor Housing
5 Turbine Shaft Oil Seal Ring	10 Bearing	16 Shaft Sleeve	Gasket
	11 Bearing Retaining Ring	17 Impeller Shim	22 Compessor Housing

Turbosupercharger components (© G.M. Corp.)

7. Remove impeller wheel, shim/s, shaft sleeve, turbine wheel and shaft, turbine shield and shield spring ring.
8. Check impeller for nicks, scuffing on blades or back face and tight fit on turbine shaft.
9. Remove oil seal retainer from bearing housing using snap-ring pliers.
10. Turn bearing housing and push oil seal, O-ring and mating ring from housing using a ½ in. diameter punch.
11. Remove bearing retainer, bearing and shim.
12. Inspect bearing housing for scoring, wear in bearing bore, damaged flange face, tight bearing roll pin, damaged snap-ring grooves or oil ring seats, thread damage.
13. Inspect bearing for wear, foreign material or damage.
14. Inspect turbine shield for straightness, scoring or pitting.
15. Inspect spring ring for warpage or loss of tension.
16. Clean all parts with a nylon brush and kerosene, then wipe dry with a lint-free cloth.

Supercharger Assembly

1. Support the bearing housing with impeller upwards.
2. Install new roll pin into bearing housing with slot aligned radially inwards.
3. Determine shaft-to-bearing end-play:
   a. Place bearing, mating ring and sleeve onto turbine shaft.
   b. Hold mating ring against shoulder on turbine shaft, then hold bearing up against mating ring and measure clearance between the bearing and lower shoulder of shaft with feeler gauges.
   c. Note this clearance for later use in impeller end-clearance determination.
4. Determine bearing-to-housing end-play and select proper shim:
   a. Install bearing into the housing, lining up roll pin with hole in flange, then install retaining snap-ring.
   b. Position a dial indicator with feeler resting against bearing; zero the indicator.
   c. Push the bearing upward against the retaining ring, then push down to bottom in the housing. Note the reading for later reference.
   d. Remove the retaining snap-ring and bearing and select the shim that will reduce end-play to 0.001-0.002 in. Available shim sizes are 0.008, 0.009, 0.010, 0.011, 0.012 and 0.014 in.

e. The adjusted bearing-to-housing play plus the previously determined shaft-to-bearing play is the total shaft end-play.
5. Install selected shim, bearing and bearing retainer ring (beveled side up).
6. Position mating ring on the bearing flange (centered).
7. Lubricate O-ring, using silicone grease, and install.
8. Install oil seal assembly in housing by pressing by hand, then install retaining ring (beveled side up).
9. Determine impeller-to-housing clearance and select proper shim:
   a. Place shaft sleeve in the center of the oil seal assembly, then place impeller over the seal so that its center hub rests on the shaft sleeve.
   b. Install gasket and compressor housing in place on bearing housing and install every second bolt, tightening them to 80 in. lbs.
   c. Install turbine wheel and shaft assembly (without seal) into impeller just far enough to hold impeller to shaft.
   d. Position a dial indicator with the feeler resting against the turbine hub; zero indicator.
   e. Lift up turbine wheel and note reading.
   f. Subtract the total shaft end-play, previously computed, from the indicator reading; this is the impeller-to-housing clearance.
   g. Select a shim that will reduce impeller clearance to 0.015-0.020 in. There are two available shims, 0.010 and 0.015 in. thick.
10. Remove the compressor housing, gasket, impeller, shaft sleeve,

and turbine wheel and shaft assembly from the bearing housing.
11. Turn over housing and install spring ring, then position the turbine shield with the three projections lined up over flat areas on spring ring. Hold the spring ring compressed in this position with two small C-clamps.
12. Lubricate the turbine shaft seal ring groove with oil and install ring. Compress the ring into the groove using a length of thin wire.
13. Lubricate the bearing area of the shaft and install through the bearing. Remove the wire used in Step 12, holding wheel so that it does not pass ring.
14. Hold turbine wheel tightly against the shield, turn over and place in a press with turbine wheel hub down.
15. Install shaft sleeve, impeller shim and impeller.
16. Press impeller onto the shaft, using an arbor, until impeller bottoms, then remove the two C-clamps. If no press is available, heat impeller to 300°F., then install by hand.
17. Remove assembly from press and position special impeller washer with dished side up, then install the left-hand thread, self-locking nut.
18. Hold turbine wheel with a rag; tighten the nut to 80 in. lbs.
19. Place the gasket and compressor housing onto bearing housing, then tighten the six bolts to 80 in. lbs.
20. Turn assembly to approximately the position it will take when installed, then pour oil into the oil inlet hole until it flows from the drain hole.
21. Install carburetor, then install supercharger and connect all lines, hoses, and linkage.

Fuel Pump

The fuel pump is of the single-acting diaphragm-type. It is centrally

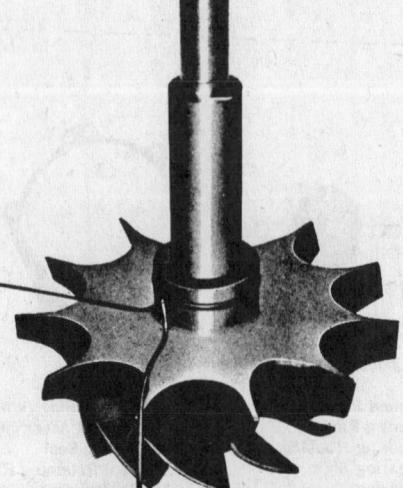

Installing supercharger turbine shaft seal ring using thin wire
(© G.M. Corp.)

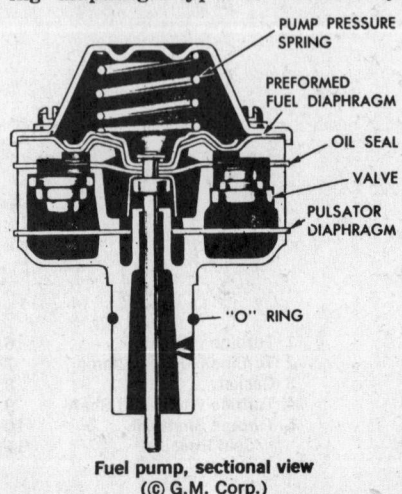

Fuel pump, sectional view
(© G.M. Corp.)

PUMP PRESSURE SPRING

PREFORMED FUEL DIAPHRAGM

OIL SEAL

VALVE

PULSATOR DIAPHRAGM

"O" RING

located at the rear of the engine, directly over the crankshaft pulley. The drive is through a crankshaft eccentric and a spring loaded push-rod.

Check fuel pump while on the car, in the conventional manner. The volume, at starter cranking speed, should be about one pint in 40 seconds. The pressure should be 5¼ psi minimum, and should remain constant at running speeds of 450 to 1,000 rpm.

The pump is held in place by a lock screw and jam nut.

Fuel Tank

The tank is mounted just to the rear of the front crossmember and in front of the toe-pan. It is held in place by a single metal strap attached to the underbody at each end by an adjustable hook.

The fuel pick-up is part of the tank gauge-sending unit located at the lower right rear of the tank. A fine mesh screen is located at the end of the pick-up pipe. The tank has no drain plug, draining to be accomplished by removing the tank gauge.

Exhaust System

The exhaust system is a single unit including exhaust pipes, muffler and tail pipe. The exhaust pipes are flange connected (and packed) to the exhaust manifolds. The packed flanges at the manifolds and the bracket at the engine comprise the entire exhaust mounting system.

Removal

1. Remove the four flange nuts, A.
2. Remove muffler clamp bolt, B.
3. Pull the muffler and exhaust pipe assembly from under the car.

Replacement

Replace in the reverse order of removal. Be sure the packing is in place at the two flanges. Torque nuts A to

Exhaust system and mounting bolts
(© G.M. Corp.)

15-20 ft. lbs., and bolt B to 8-10 ft. lbs.

NOTE: turbo charged muffler tips must be transferred to new muffler.

Cooling System

The engine is air-cooled by a blower mounted on top of the crankcase cover. The engine is entirely shrouded with sheet metal pieces that attach directly to the engine and form a plenum chamber. Air leaving the blower travels outward and downward over the cylinders and heads. The air then enters a duct under each bank from where it travels rearward to be exhausted at an opening at the rear of the engine.

Cooling is regulated by a bellows-type thermostat in the lower part of the plenum. This thermostat operates a cooling air valve that moves in and out of the eye of the blower to control air flow. In the event of thermostat failure, the air valve will remain open to prevent engine overheating.

The blower runs on a sealed, permanently lubricated ball bearing. It is belt-driven by a pulley mounted on the rear end of the crankshaft.

Cooling System Thermostat

In the event of thermostat failure, the cooling aid throttle valve will remain open to prevent engine overheating.

When installing a new thermostat, it is necessary to readjust the air throttle valve opening. This adjustment should be made with the engine warm so the thermostat rod can be easily pulled up against the pull of the thermostat bellows.

Adjustment

1. With the swivel inserted into the hinge lever, pull up on the thermostat rod until the bellows is stopped within its mounting bracket.

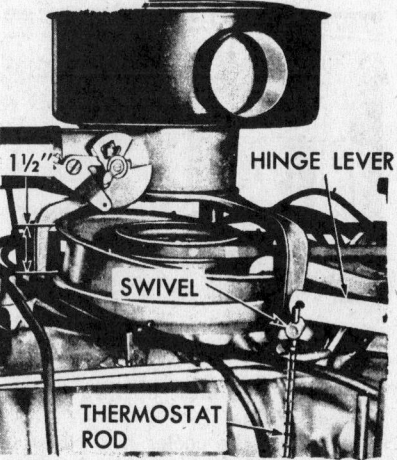

Cooling air throttle valve adjustment—1960
(© G.M. Corp.)

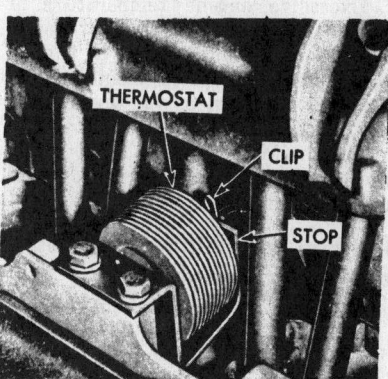

Cooling air valve thermostat
(© G.M. Corp.)

Removing foreign matter from oil cooler
(© G.M. Corp.)

2. Adjust the swivel to produce a 1½ in. opening of the cooling air valve (2 11/32 in.—1961-69). This measurement is to be made directly opposite the hinge at the maximum opening.

Oil Cooler

An oil cooler, through which a portion of the cooling air passes before discharge, is mounted above the air exhaust duct near the left rear corner of the engine.

It is important that the oil cooler be cleaned every 1,000 miles.

1. Remove oil cooler access hole cover and brush clean the cooler fins.
2. Insert an air hose under the

Oil cooler access hole cover
(© G.M. Corp.)

cooler and blow up through the cooler fins.

3. Replace the cooler hole cover.

Engine Temperature Indicator

Excessive engine temperature or low oil pressure is indicated through the use of a telltale light in the instrument cluster. If engine (oil) temperature is high, or oil pressure is low, the light will come on.

The indicator should light when the ignition switch is turned on, and before the engine is started.

Ignition on, Engine Stopped and Telltale Light off

1. Telltale light burned out, replace bulb.
2. Open circuit between light and ignition switch or between light and pressure switch.
3. Pressure switch stuck. Replace switch.
4. One of the switches is not grounded. Check for looseness or dirty threads.

Telltale Light on, Engine Running

1. Oil pressure is low, or oil temperature is high. First check oil level and apparent operating temperature of engine. If light is still on, remove the pressure sender and check oil pressure with a hydraulic gauge. If en-

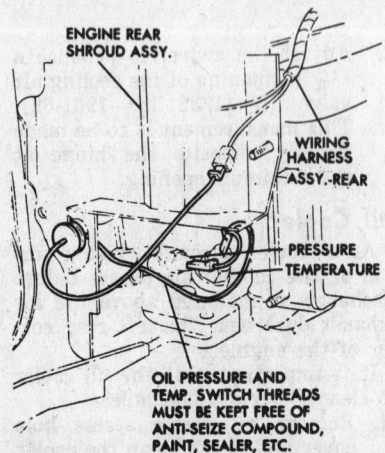

OIL PRESSURE AND
TEMP. SWITCH THREADS
MUST BE KEPT FREE OF
ANTI-SEIZE COMPOUND,
PAINT, SEALER, ETC.

Oil pressure and temperature senders
(© G.M. Corp.)

gine has cooled, and light remains on, temperature-sending unit is bad.
2. Electric circuit is grounded between senders and telltale light.
3. Sender calibration is wrong. To check oil pressure sender, increase engine speed and see if light goes out. To check oil temperature sender, disconnect wire at pressure sender and turn on ignition switch. With a jumper lead (7 ft. long) from temperature sender to the connector, remove the sender and immerse in container of hot oil. With a common kitchen cooking thermometer (400° scale) and the sending unit in the oil, heat the oil and note the temperature required to light the indicator. Temperature should be 280°-320°F.

Power Train

The power train comprises the engine, clutch, transmission, and differential. Some of the components of this train are more easily serviced with the entire power train removed from the vehicle. However, many operations can be performed without disturbing the assembly to any great extent. Following are some of the components that require the removal of the power train for easier servicing:

—Crankshaft
—Camshaft
—Connecting rod
—Flywheel
—Flywheel housing
—Fuel pump eccentric
—Rear housing
—Camshaft drive gear
—Crankshaft drive gear
—Distributor drive gear
—Piston, pins or rings
—Clutch driving plate
—Clutch driven plate
—Clutch housing
—Clutch release bearing
—Manual transmission
—Automatic transmission parts except those of the controls, the governor, the servo and valve body
—Torque converter
—Differential

Power train removal and installation stresses the importance of using proper equipment and recommended procedures. The preferred way to remove and install the power train is with special power train cradle, tool J-7894, mounted on a suitable transmission jack, such as tool J-8394, with the car on a hoist.

Equipment limitations may modify the approach to the operation, but basic principles and precautions remain the same.

The following precautions must be observed:

1. Do not support the complete power train except at the engine pan rail. Support cradle J-7894 is designed to correctly support and lock the power train in a balanced position.
2. No jacking or lifting tool should be used unless capable of supporting the weight of the power train assembly, about 460 lb.
3. No jacking device should be used unless sufficiently wide and stable to support the unit at its maximum height without tipping. The balance point of the complete power train is about 2/10 in. behind the front face of the cylinder block.
4. No jack should be used that will not permit gradual and steady lowering and raising.

Removal

NOTE: as a precaution, it is wise to insert a bolt through holes in the lid support to assure holding the lid open

1. Disconnect ground cable from battery and ground straps to the engine.
2. Disconnect wires from the coil, generator and oil pressure and the temperature sending units.
3. Disconnect accelerator return spring and rod from carburetor cross-shaft.
4. Raise the car on a hoist.
5. Remove engine side and rear shield seal retainers.
6. Remove both real wheels.
7. Remove real axle shaft universal joints from differential and remove axle shafts.
8. Disconnect the speedometer shaft and remove the wires from the starter solenoid.
9. Disconnect carburetor cross-shaft rod. Now push the rod up into the engine compartment. Remove the rear idler control lever rod.
10. Disconnect shift rod coupling on cars with standard transmission, or disconnect flex-control cable, TV and accelerator rod on Powerglide transmissions.
11. Disconnect and plug main and heater fuel lines.
12. On cars with standard transmission, disconnect the clutch fork return spring and unhook the clutch cable clevis. Disconnect clutch fork pull rod. Loosen ball stud at clutch control cable cross-shaft, and remove cross-shaft.
13. Remove parking brake tension spring.
14. Remove screws and rear engine grille.

15. Remove skid plate bolt.
    NOTE: on 1965-69 cars, match mark adjuster cam on outer end of rear strut rods, then loosen nut. Disconnect left and right rear strut rod bracket from differential carrier, then swing rods down.
16. Position and secure transmission jack and cradle under the assembly. Raise slightly.
17. Remove the two castellated nuts from the front engine mount, and one castellated nut from the rear mount.
    NOTE: if front engine mounting bracket and shims are disturbed, care must be used in replacement. Rear wheel toe-in will be influenced by any change at this point.
18. Carefully lower the power train while watching for interference at the rear mount and the left rear lower control arm.
19. Remove exhaust pipe and muffler assembly.

Installation

Installation procedure is the reverse of removal, keeping the following in mind:
1. As with any other major installation, keep all wires, lines, levers and rods out of the way.
2. Rear wheel toe-in will be changed if the shim value at the engine front mounting bracket is altered.

Separating Power Train Assembly

Standard Transaxle

1. With the power train out of the car and secured to a stable and substantial fixture, support the transaxle with a chain hoist and properly applied sling.
2. Drain transmission and differential.
3. Remove starter.
4. Remove the differential to clutch housing bolts and carefully withdraw the transaxle combination from the engine unit.
5. Remove the two screws holding the clutch pull rod dust seal, and the pull rod-to-clutch fork attaching pin. Remove the clutch shaft.
6. Complete the operation by taking the clutch fork and release bearing out of the differential carrier. The fork is attached to the carrier by a ball socket and spring retainer which is easily removed to permit the release bearing to be slipped off its shaft in the carrier.

Separation of Standard Transmission and Differential

To separate the standard transmission from the differential carrier

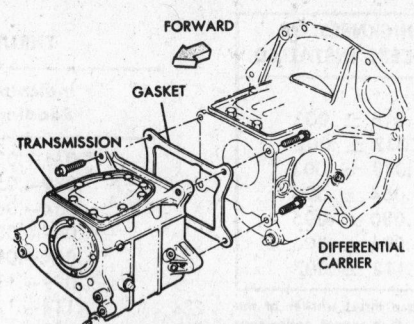

FORWARD

GASKET

TRANSMISSION

DIFFERENTIAL CARRIER

3-speed transmission differential carrier attachment
(© G.M. Corp.)

simply requires the removal of the four attaching bolts shown. Two of these attaching bolts are removed from the transmission, on the right-hand side, and two from the differential, on the left-hand side.

Powerglide Transaxle

1. With the power train out of the car and secured to a stable and substantial fixture, support the transaxle with a chain-hoist and properly applied sling.
2. Place a drain pan under the transmission and remove transmission oil by loosening the filler tube at the transmission oil pan.
3. Disconnect the hose from the vacuum modulator tube to the carburetor balance tube.
4. Remove engine front shield.
5. Remove starter.
6. Disconnect the converter from the engine flex plate by removing the three attaching bolts. The bolts may be reached through the access hole at the twelve o'clock location in the converter housing.
7. Apply a slight lifting effort on the chain hoist and sling.
8. Remove the differential carrier-to-converter housing bolts and pull the transaxle assembly away from the engine. Then remove converter and the turbine shaft.

Separation of Powerglide Transmission and Differential

1. To separate the Powerglide transmission from the differential carrier, place the transaxle on a flat surface.

2. Carefully pull the turbine shaft through the transmission and differential carrier. Be careful not to damage the turbine shaft bushings on the pump shaft splines.
3. Remove governor attaching screw and lift out governor.
4. Remove the three remaining transmission-to-differential attaching screws. Carefully pull the transmission straight away from the carrier to prevent pump shaft and bushing damage.
5. Remove transmission-to-differential carrier gasket, and governor drive gear and selective spacers from the pinion shaft of the differential carrier.

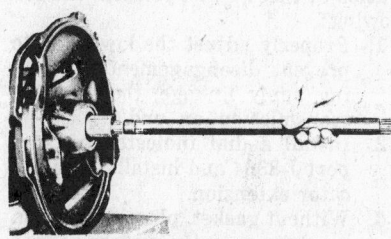

Removing turbine shaft
(© G.M. Corp.)

Powerglide Thrust Washer Determination

For proper operation of the Corvair Powerglide it is necessary that correct end-play be maintained within the unit. Selective thrust washers of various thickness are used at two locations to accomplish this.

A. At the front of the transmission between the front pump body and the clutch drum, and

B. At the rear of the transmission between the rear face of the planet carrier hub and the front face of the governor gear.

From a service standpoint, the use of selective washers at both front and rear is a great advantage. Final transmission end-play can be controlled from either end of the transmission, whichever is more advantageous.

If the repair requires axle and transmission separation, make the washer selection at the rear.

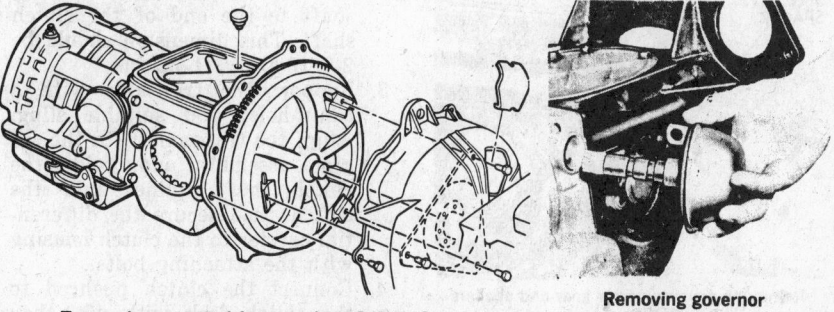

Transaxle separated from engine (© G.M. Corp.)

Removing governor
(© G.M. Corp.)

INDICATOR READING	NUMBER .016" SPACERS REQ'D	THICKNESS OF SPACERS INSTALLED
*.025—.046	None	
.047—.062	1	.016 ± .001
.063—.078	2	.032 ± .002
.079—.094	3	.048 ± .003
.095—.110	4	.064 ± .004
.111—.126	5	.080 ± .005
.127—.142	6	.096 ± .006
.143—.155	7	.112 ± .007

*If initial indicator reading is below .025", replace thrust washer at the clutch hub—front pump with .050" thrust washer, then repeat entire rear thrust spacer selection procedure.

Powerglide rear thrust washer determination chart—1961-69
(© G.M. Corp.)

POWERGLIDE REAR THRUST SPACER USAGE CHART	
Indicator Reading	Spacers to Be Used
*.011—.038	NONE
.039—.053	.016
.054—.068	.031
.069—.083	.046
.084—.098	.046 + .016
.099—.113	.046 + .031
.114—.128	.046 + .046
.129—.145	.046 + .046 + .016
.146—.155	.046 + .046 + .031

*If initial indicator reading is below .011", replace .088" thrust washer at the clutch hub—front pump with an .076" or .050" thrust washer, then repeat entire rear thrust spacer selection procedure.

Spacer	Part No.
.016"	6256827
.031"	6256828
.046"	6255664

1960 only

Prior to reassembling the differential carrier to the Powerglide transmission after repairs requiring separation of these units, perform the following:

1. Properly adjust the low band to prevent disengagement or cocking apply linkage, then tip the transmission on end.
2. Install a dial indicator on support J-8364 and install 3 in. indicator extension.
3. Without gasket, place support on rear pump cavity surface so that dial indicator tip rests on planet carrier hub. Set indicator dial to zero.
4. Slowly raise support J-8364 and indicator off rear pump cavity and note its range of dial reading. The deflection should not exceed .050 in.
5. Place J-8364 and dial indicator on governor gear of the differential pinion shaft, and lower the support slowly, so that revolutions of the indicator needle can

Measuring mounting difference with indicator
(© G.M. Corp.)

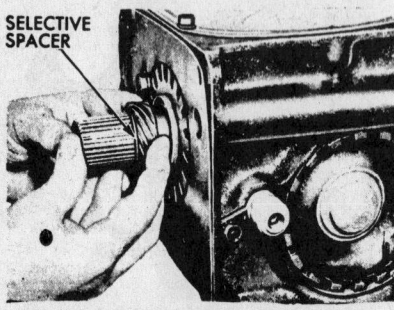

SELECTIVE SPACER

Removing governor drive gear and spacers
(© G.M. Corp.)

be counted. Measurement starts once the indicator needle again reaches zero. Fully depress support on governor gear, note reading and refer to chart for spacers to be installed on governor gear.
6. Install spacers on the governor gear as selected from chart.

Assembling the Power Train Assembly

Standard Transmission to Differential

To assemble the standard transmission to the differential carrier:
1. Apply petroleum jelly to a new gasket and couple the transmission to the differential carrier. Use care in starting the transmission mainshaft in the internal splines of the differential pinion.
2. Secure the transmission to the differential carrier with the four bolts, and torque to 24-32 ft. lbs.

Standard Transaxle to Engine

1. Install the clutch bearing and the release fork on the differential carrier.
2. Bottom the clutch shaft in the transaxle, then measure from the end of the clutch-release-bearing shaft to the end of the clutch shaft. This dimension should be 2-9/16 in. ± 1/32 in.
3. Support the transaxle with a chain-hoist and suitable sling. Align the transaxle with the engine assembly and guide the clutch shaft splines into the clutch, then secure the differential carrier to the clutch housing with the attaching bolts.
4. Connect the clutch pushrod to the clutch fork with pin, then

attach the pushrod dust seal assembly.
5. Install starter.
6. Fill transmission and differential carrier, as prescribed.

Powerglide Transmission to Differential

1. With petroleum jelly, apply a new gasket to the rear face of the transmission.
2. Align the transmission and carrier on a clean flat surface and guide the pump shaft through the differential carrier so as not to damage the bushing in the pinion. Then, engage the splines of the pinion with the internal splines in the transmission.
3. Install the governor, then secure the transmission to the differential and torque the bolts to 24-32 ft. lbs.
4. Install turbine shaft, using care to engage the forward splines, (clutch drum) and the rear splines (planet gear set).
5. Install converter, being sure to

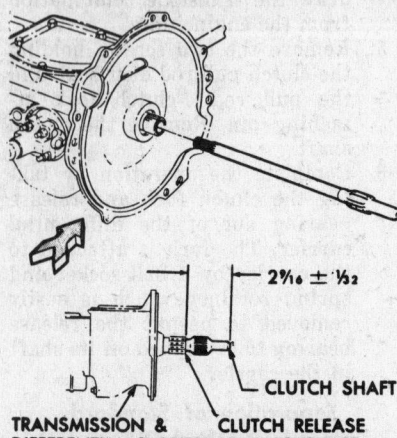

2⁹⁄₁₆ ± ¹⁄₃₂

CLUTCH SHAFT

TRANSMISSION & DIFFERENTIAL CARRIER   CLUTCH RELEASE BEARING SHAFT.

Clutch shaft dimensions
(© G.M. Corp.)

get full depth engagement of stator shaft, turbine shaft and front pump shaft.

NOTE: after the converter is in place, improvise a device for securing the converter to the transaxle during further handling.

Powerglide Transaxle to Engine

1. Locate the transaxle next to the engine on a chain hoist and suitable sling.
2. Remove the converter-to-transaxle securing device and align the converter with the flex plate.
3. Guide the converter hub into the crankshaft, then align the mounting bolt holes of the differential housing and converter housing. Secure the two housings by installing the top left bolt, eleven o'clock position.
4. Install the flex plate-to-converter bolt, accessible at the one o'clock position, as a temporary measure.
5. Install and finish tightening the housing-to-carrier mounting bolts.
6. Install the starter motor.
7. Install the remaining two, and tighten all three converter-to-flex plate bolts via the access hole in the converter housing.
8. Install the engine front shield.
9. If transmission filler tube has been removed from the engine front shield, reinsert it through the front shield at this time. Then connect the vacuum modulator to the carburetor balance tube with a short length of hose.
10. Connect filler tube to transmission oil pan and tighten.
11. Refill the transmission.

Power Train Installation

1. Position the power train jack and cradle with the power train attached under the engine compartment.
2. Raise power train until front and rear engine mounting brackets are in place on mounts, then install the nuts. Torque front mounts to 60-80 ft. lbs. and the rear mount to 50-60 ft. lbs. Install cotter pins at both mountings.

Standard Transmission

1. Install clutch control cable cross-shaft ball stud.
2. Attach clutch cable clevis to inboard lever.
3. Install transmission shift coupling and boot assembly.
4. Install clutch pull rod and adjust.
5. Hook up parking brake tension spring.

6. Connect and adjust throttle valve linkage.

Powerglide Transmission

1. Connect and adjust manual valve control cable.
2. Connect and adjust throttle valve linkage.
3. Disconnect cradle from power train, then lower and remove transmission jack.
4. Connect starter wires to solenoid and hook up speedometer cable.
5. Install axle shaft universal joints at differential.
6. Install rear wheel brake drums.
7. Install engine side and rear shield seal retainers. Place and tighten bolt in rear skid plate. Install exhaust system.
8. Install rear engine grill and retaining screws.
9. Lower the car, then connect carburetor control rod and return spring.
10. Connect multiple harness connector and wires to the coil, generator and the oil pressure and temperature-sending-units.
11. Connect battery cable and radio ground cable.
12. Check engine oil, transmission oil and differential lubricant.
13. Start the engine, check for leaks, and make necessary adjustments.

Engine

References

The engine is a horizontally-opposed, air-cooled, six cylinder unit.

It has a cast aluminum alloy crankcase, vertically divided into halves. These halves are held together by bolts. Each crankcase half is equipped with individual cast iron cylinders, positioned and held in place by four long studs for each cylinder. These studs extend and pass freely through the cooling fin structure and cylinder head, and serve to secure the cylinders and head to the crankcase.

Valves, Pushrods and Tappets

After removing the cylinder head, any service to valves, pushrods or tappets can be handled in a conventional manner.

NOTE: beginning with the Turbocharged engine, an extremely durable exhaust valve is used. It is recommended that it be used as replacement when unsatisfactory valve life is encountered in other Corvair engines (80, 84, 98 and 102 H.P.) installed

in 1963 cars of the 500-700-900 series.

A rotator is used with those in the 95 series.

It is also important that new valve keys be used in all such replacements.

Valve Lash Adjustment

1. Normalize engine temperature. Remove valve cover.
2. Use reworked cover (one with top

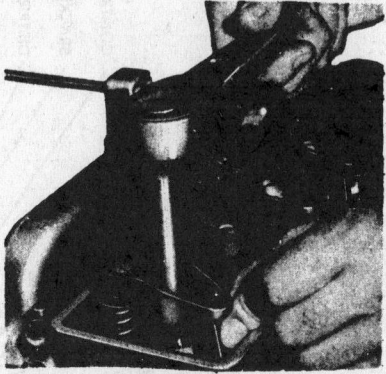

Adjusting valve lash
(© G.M. Corp.)

cut out) and gasket to stop oil from running out.
3. With engine at idle, back off rocker arm nut until arm starts to clatter.
4. Turn rocker arm nut down slowly until clatter just stops. This is zero lash position.
5. Turn nut down one-quarter additional turn and wait 10 seconds, then repeat one-quarter additional turn until one complete turn has been made beyond zero lash position.

NOTE: this one-turn preload adjustment must be done slowly, to allow lifter to adjust itself to prevent possibility of interference between the intake valve head and top of piston.
6. After each valve is adjusted, remove reworked cover and install regular cover with new gasket.

Engine Disassembly

1. Separate transmission from differential carrier.
2. Separate differential carrier from engine.
3. Separate clutch (if equipped) from engine.
4. Remove carburetor induction hoses, choke hose, and air cleaner. Disconnect carburetor and choke linkage. Remove air horn and choke assembly and air horn support.
5. Disconnect throttle linkage and remove carburetor cross-shaft.
6. Release belt tension and remove blower belt.
7. Disconnect fuel lines at carburetors and fuel pump. Loosen locknut at attaching screw and remove pump, pushrod return

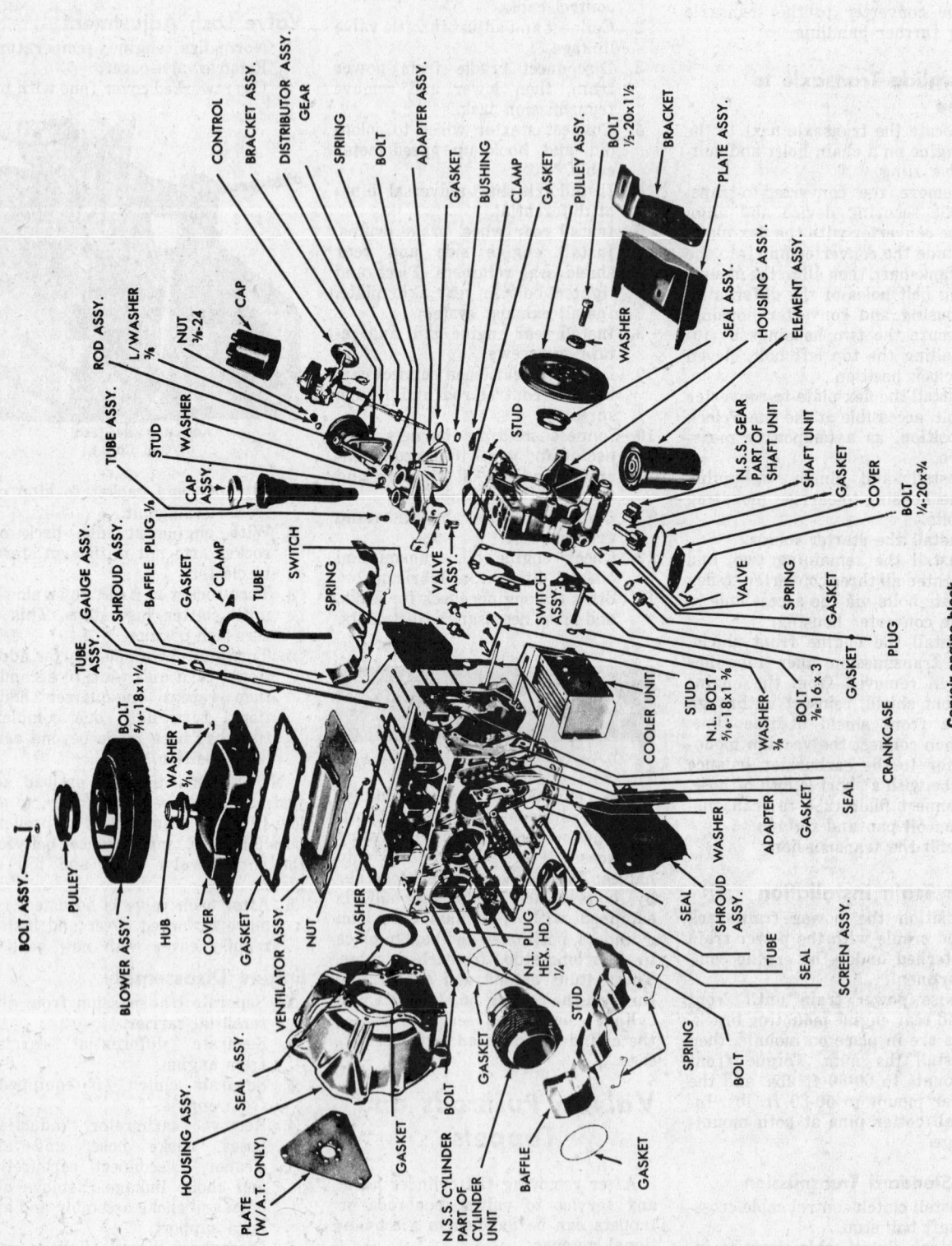

Corvair engine block

spring from oil filter and generator adapter.

8. Remove generator.
9. Disconnect vacuum balance tube at engine front shield and carburetor mounting pad. Remove engine front shield and vacuum balance tube.
10. Remove both carburetors.
11. Disconnect and remove cooling air throttle valve.
12. Remove fuel lines and oil level gauge.
13. Remove distributor cap and spark plug wiring harness.
14. Remove spark plugs.
15. Remove ignition coil and generator brace from cylinder head. Remove engine upper shroud, and left and right side shields.
16. Remove oil filter and generator adapter retaining bolts and adapter. Throw the gasket away.
17. Remove blower and pulley from crankcase cover. Remove crankcase vent tube and gasket.
18. Remove crankcase cover and gasket, blower bearing, and vent assembly.
19. Remove engine front and lower shrouds and exhaust ducts.
20. Turn engine upside down and remove oil pan and gasket.
21. Bend lock tabs back, remove nuts attaching nuts and clamps, and exhaust manifolds.
22. Remove manifold-to-cylinder head choke heat tube at right head.
23. Remove engine rear mounting bracket and engine skid plate at engine rear housing.
24. Remove valve rocker covers. Remove rocker arm nuts, balls and rocker arms.
25. Remove pushrods and identify pushrods for reinstallation (oil holes up).
26. Remove pushrod guide plates.
27. Remove valve rocker stubs, the chamfered washers, and discard the O-rings.
28. Remove cylinder-head nuts and flat washers.
29. Remove O-rings from pushrod drain tubes, and remove drain tubes.
30. Remove cylinder-head assemblies.

NOTE: the cast iron cylinders will need a holding device when rotating the crankshaft for engine disassembly. Six pieces of 3/8-1/2 in. iron or galvanized pipe, 4 1/2-4 5/8 in. long can be used on the long cylinder stubs (one to the cylinder). Six pieces of the same diameter pipe, 3 1/2-3 3/4 in. long to be used on the short studs (one to each cylinder).

31. Remove hydraulic lifters with a magnet or wire hook, and identify them for reassembly.
32. With engine upside down, install two 3/8–16 bolts in crankshaft pulley, exactly 1/4 in. deep.
33. Using a bar between the bolts, or a wrench on 3/4 in. pulley nut, rotate the crankshaft to give access to the connecting rod bearing being worked on.
34. Mark the cylinder numbers on each rod and piston for identity for reassembly.
35. Remove cylinder-holding device (long and short iron pipe) from one cylinder at a time.
36. Remove retainer springs on cylinder air baffle, and remove baffle.
37. Remove each cylinder with piston and rod assembly as a unit.
38. Remove the piston assembly from the cylinder bore with a hammer handle.
39. Remove cylinder ridge, or deposits, from the bore by mounting the cylinder barrel on an improvised jig (a block of wood and four studs will do).
40. Remove pulley with puller, then remove engine rear housing bolts and rear housing.
41. Support engine on a couple of 2 in. x 4 in. wood strips to protect the oil pump screen and pick-up tube.
42. Remove flywheel housing and gasket.
43. Remove front oil slinger snap ring and slinger.
44. Loosen crankshaft bolts (on side of crankcase), eight long and three short bolts. Support the crankcase at about a 15° angle (with a block of wood) so the crankshaft and camshaft will not fall out when the case halves are separated.
45. Remove left crankcase half.
46. Remove camshaft by turning while lifting.
47. Lift out crankshaft.
48. Slide main bearing shells out of crankcase halves.

NOTE: timing gears are best removed from the crankshaft and camshaft in an arbor press. Distributor drive gear and fuel pump eccentric may be removed from the crankshaft with a puller.

Engine Assembly

1. Lubricate the crankshaft, front end, and install Woodruff key. Press crankshaft gear in place, using arbor press.
2. Lubricate the crankshaft, rear end, and install two Woodruff keys. Locate and press the fuel pump eccentric and spacer into place, then press the distributor drive gear into place.
3. The crankshaft is serviced and the main bearings fitted in the conventional manner.
4. Place the bearing shells in their respective locations in the left half of the crankcase, lubricate with light engine oil.
5. Place the corresponding bearing shell halves in their locations in the right side of the crankcase, lubricate with light engine oil.
6. Carefully lay the crankshaft into place in the left crankcase half.
7. To install camshaft gear and thrust washer on camshaft, support the shaft at back of the front journal in an arbor press.
8. Lubricate gear hub of camshaft, and install thrust washer and Woodruff key.
9. Press gear onto shaft until it bottoms against the thrust washer.

Timing Gear Marks

1. Install the camshaft assembly into the left crankcase half while indexing the two timing gears so that the timing marks line up.

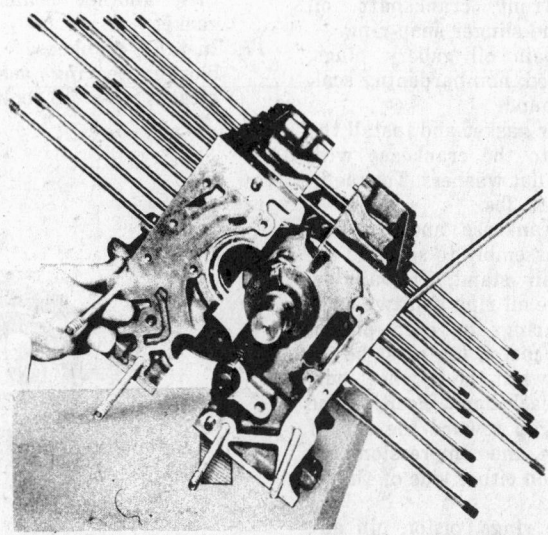

Removal and installation of crankcase sections (© G.M. Corp.)

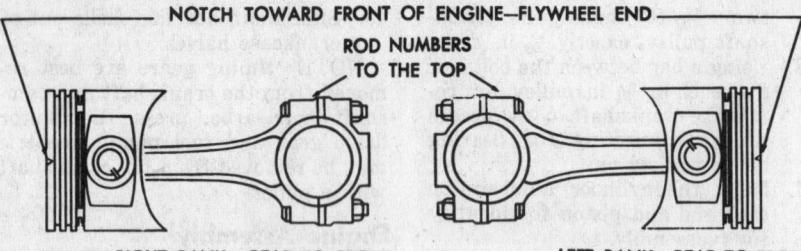

NOTCH TOWARD FRONT OF ENGINE—FLYWHEEL END
ROD NUMBERS TO THE TOP

RIGHT BANK — REAR TO FRONT 1-3-5

LEFT BANK — REAR TO FRONT 2-4-6

Rod and piston identification markings and relationship

Torquing connecting rods—
showing tubes holding opposite cylinder

Lubricate camshaft and journals with light engine oil.

2. Install the other half of crankcase onto the crankshaft and camshaft, and torque to proper tightness in correct order. Torque eight, 7/16 –20 bolts 42-48 ft. lbs. Torque three, small 5/16–18 bolts 7-13 ft. lbs. Camshaft end-play should be .003 in. to .007 in. End-play is influenced by wear at the thrust

TIMING MARKS

Timing marks

washer or crankcase groove. Normal timing gear backlash is .002 in. to .004 in.

3. Install front crankshaft oil slinger and slinger snap-ring.
4. Install main oil gallery plugs, with a good, non-hardening sealing compound.
5. Use a new gasket and install the flywheel to the crankcase with bolts and flat washers. Torque to 20 to 30 ft. lbs.
6. Mount crankcase and flywheel housing assembly to suitable engine repair stand, if available.
7. Install one oil ring and two compression rings on each piston. Notch on piston top must be installed toward the flywheel end, front of each bank. Position the oil ring gap toward the top of the engine, and compression ring gaps 45° on either side of the oil ring gap.
8. Lubricate rings, piston pin and skirt and install ring compressor.

9. Push piston assembly into the bore with a hammer handle until the piston head is slightly below the top of the cylinder bore.
10. With piston assemblies in the bores, put rod bearing inserts in place in the rods and caps.
11. Rotate the crankshaft to bring the crankpin in line with the piston and rod to be installed.
12. Protect the crankshaft journals by taping the rod bolt threads or sliding a piece of 5/16 in. I.D. plastic hose over the exposed rod bolts.
13. With a new copper cylinder gasket in place, start the cylinder and piston assembly on the long pilot bolts. With hammer handle pressure on the head of the piston, guide the cylinder into place on the crankcase. Tap the piston until firm rod-to-crankshaft journal contact is felt. Then, install rod cap and torque to 20 to 26 ft. lbs.
14. Install cylinder holding fixture tubes. Repeat this procedure on the remaining five cylinders. Side clearance at the connecting rod bearings should be .005 in. to .010 in.
15. Install cylinder air baffles with retainer springs.
16. Install new crankcase cover gasket. Install crankcase vent plate, then another crankcase cover gasket.
17. Install crankcase cover and blower bearing assembly and attach with 16 bolts and flat

washers. Torque to 7 to 13 ft. lbs. Attach crankcase vent tube and gasket.

NOTE: if either a new or the original pump screen and tube assembly is to be installed in the original crankcase, the outside of the end of the tube must be tinned with solder before installation.

18. Drive oil pickup screen and tube assembly into the crankcase with tool J-8369 (or substitute clamp) and a hammer. Do not drive on the screen. Align screen parallel to the oil pan rails. Install retaining clamp.
19. Coat threads on temperature and oil pressure sending units with non-hardening sealing compound and install sending units.
20. Remove cylinder holding fixture tubes from all cylinders of the bank to which the head is to be installed.
21. Install the three cylinder-head gasket rings. Position cylinder head on studs and carefully lower head into place.

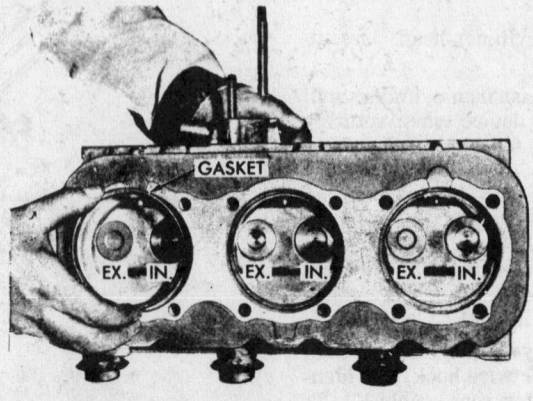

GASKET

EX. — IN.   EX. — IN.   EX. — IN.

Installation of cylinder head gasket (© G.M. Corp.)

Installation of cylinder head
(© G.M. Corp.)

22. Install flat washers and nuts on the six long studs, next to the intake manifold. Lubricate and install six new O-rings in the valve rocker stud counterbores. Coat rocker stud bore with anti-seize compound and install special flat washers and valve rocker studs.
23. Torque cylinder-head nuts and valve rocker studs, in proper sequence, to 27-33 ft. lbs.
24. With light oil, lubricate hydraulic lifters and install them in their proper bores.
25. Install pushrod oil drain tubes through cylinder head. Place O-rings, one on each end of the drain tube. Lubricate O-rings and push tubes into place at valve lifter bore and cylinder head. Install tube with long end toward cylinder head.
26. Install pushrods with the side oil hole up, at the rocker arm socket. Install pushrod guides in

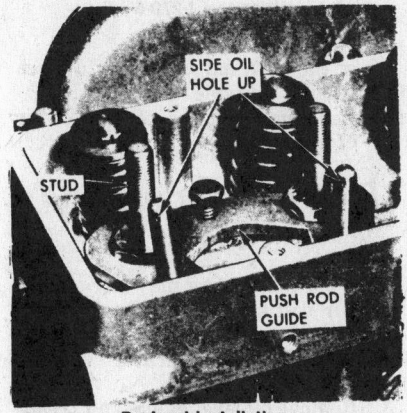

Pushrod installation
(© G.M. Corp.)

place over the rocker studs and pushrods. Rod guide retaining bolts, torque to 60 to 80 in. lbs.
27. Install rocker arms, balls and nuts loosely in place.
28. With a new gasket in place, position and attach the engine rear housing. Use flat washers and anti-seize compound on the attaching bolts. Torque to 7 to 13 ft. lbs.
29. Block the crankshaft from turning with a wood wedge. Lubricate and install pulley on the crankshaft wth heavy washer and retaining bolt. Draw the pulley into place with retaining

bolt. After bottoming the pulley, back screw off one turn, then torque to 50-60 ft. lbs.

Caution Do not drive pulley onto shaft.

30. With new gasket in place, install oil cooler adapter with anti-seize compound on bolts and flat washers. Torque adapter bolts to 7-13 ft. lbs. Install engine rear

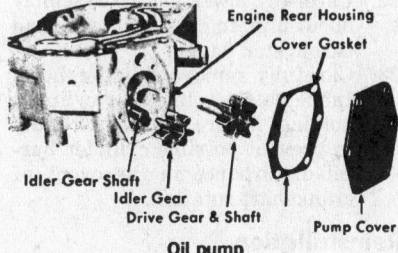

Oil pump
(© G.M. Corp.)

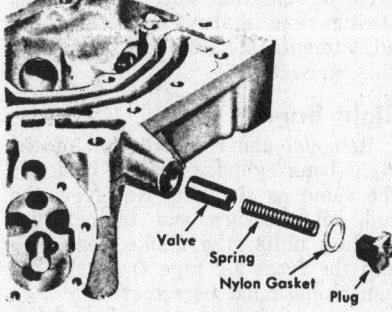

Pressure regulator
(© G.M. Corp.)

shroud. Install new seals in oil cooler adapter.
31. Install oil cooler, with anti-seize compound on the bolt and flat washer. Torque to 8-12 ft. lbs.
32. With new gaskets in place, install exhaust manifolds. On the right bank, install heat tube in manifold. Install clamps, torque nuts to 10-20 ft. lbs.
33. With new gasket, install oil filter and generator adapter on the engine rear housing. With anti-seize compound on bolts and washers, secure and torque the adapter to 7-13 ft. lbs. Install new oil filter cartridge and torque the attaching bolt, 9-15 ft. lbs. Install engine right rear shroud.
34. Install engine skid plate and rear mounting bracket, torque nuts, 20-30 ft. lbs. Attach lifting adapter to rear engine mount.
35. Install blower and blower pulley to crankcase cover blower bearing hub assembly.
36. Install front shroud and exhaust ducts. Install left and right side shields.
37. Install upper shroud assembly. Turn blower and check clearance while tightening upper shroud retaining screws.
38. Install fuel lines and oil level

gauge. Install thermostat and lower engine shrouds. Install cooling air throttle valve assembly.
39. Install idler bracket and pulley in place on the generator adapter. Be sure the adjusting slot is toward the front (flywheel) end of the engine.
40. Install coil bracket, coil and generator brace on cylinder head. Install blower-generator belt and adjust according to procedure outlined in the Generator Removal and Replacement paragraph.
41. Install spark plugs, and torque to 20-25 ft. lbs. Install plug and distributor wires, coil wires, and carburetors.
42. Turn engine bottom up, and install oil pan with new gasket. Check crankcase-to-flywheel parting line to see if surface is in condition for good sealing. Torque oil pan bolts 40-60 in. lbs.
43. Lift the engine off the stand or bench and secure it to the cradle, tool J-7894, in preparation for power train assembly and car installation.
44. Install fuel pump pushrod and spring into the oil filter and generator adapter. Install fuel pump with new O-ring seal. Secure pump with set screw and lock nut, both torqued 9-15 ft. lbs.
45. Connect fuel lines, install engine front shield and support strap, connect vacuum balance tube to both carburetors. Connect choke heat tube and fresh air tube.
46. Install carburetor cross-shaft and air horn support. Install air horn, air cleaner and air inlet hoses.
47. Refer to Cooling System for data on adjusting the cooling air throttle valve and thermostat. Refer to Corvair Powerglide, Linkage Adjustments.

Engine Rear Housing— Renewal

The oil pump and pressure regulator are contained in the engine rear housing. The pump is driven by the extended distributor shaft. Any major service to the oil pump (including replacement of early model engine rear oil seals) requires engine removal.

Cylinder Head R & R

(In the Car)

Valve conditioning, head gasket replacement, or any other service directly connected to the cylinder head,

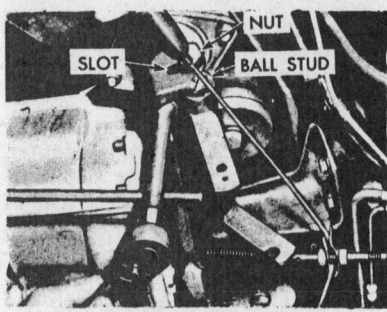

Lowering front engine mount
(© G.M. Corp.)

may be performed with the engine in the car.

Left Bank Removal

1. Drain engine oil.
2. Disconnect battery and ground connection to the engine. Disconnect radio ground strap.
3. Remove air intake hose connected to air cleaner and carburetor.
4. Remove carburetor accelerator return spring and disconnect accelerator rod from the carburetor.
5. Disconnect left side carburetor cross-shaft support from carburetor.
6. Disconnect fuel line and remove left carburetor. Remove the long, outboard stud from carburetor mounting.
7. Disconnect vacuum balance tube at carburetor mounting flange. Remove generator mounting bolts and swivel bracket up and away from the engine upper shroud.
8. Remove spark plugs and wires from cylinder head, throw away plug gaskets. Remove A.I.R. pump, if fitted.
9. Loosen all engine side shield retaining screws. Remove screw from engine side shield under carburetor in engine compartment, attached to cylinder head. Remove engine side shield.
10. Remove oil cooler access hole cover and remove oil cooler.
11. Raise car on a hoist and attach lifting cradle and jack. Support the engine.
12. Remove both engine side seal retainers and engine rear seal retainers.
13. Remove engine rear center shield and seal assembly.
14. Remove lower engine shroud. Remove exhaust pipe to manifold nuts.
15. Remove clamps and exhaust manifold.
16. Remove engine-to-body rear mounting bracket.
17. Remove screws and lift off rocker arm cover. Have oil drain pan in position to catch oil from head.

18. Remove rocker arm nuts, balls and rocker arms. Remove pushrods and rod guides.
19. Remove rocker stubs, washers and O-rings from cylinder head.
20. Remove O-rings from bottom of pushrod drain tubes with hooked tweezers, then remove tubes from cylinder head.
21. Remove nuts and washers from long cylinder head studs.
22. Carefully lower engine assembly about 3 in. to clear cylinder-head carburetor flange.
23. Carefully remove cylinder head. Immediately install cylinder holding fixture tubes and nuts to prevent possible cylinder barrel disturbance in the event of crankshaft rotation.

Reinstallation

Install cylinder head in the reverse order of removal, using new gaskets, O-rings and seals. For details and adjustments, see Engine Assembly in this car section.

Right Bank

Removal and reinstallation of the right bank cylinder head is basically the same as the left. However, the coil, oil pressure and temperature sending units, the choke heat pipe and the fresh-air pipe from the exhaust pipe must be removed.

Engine Rear Oil Seal

Unless the engine is of very early production, the rear oil seal can be changed without removing the engine. Early model engine rear housing seals were designed and installed with the seal flange located inside the housing, requiring engine removal. Later design provides for replacement from the outside by the following procedure:

1. Disconnect battery and drain engine oil.
2. Remove all side shield seal retainers.
3. Remove engine rear center shield. Remove engine skid bolt and attach cradle J-7894 to the underside of the engine, car raised on a hoist.
4. Remove engine rear body grille and engine rear mount.
5. On cars equipped with standard transmission, refer to Clutch Linkage Adjustment description. Remove clutch return spring and disconnect control cable. Disconnect clutch pull rod.
6. Loosen the outboard stud nut and slide part way out of the engine front mounting bracket slot. Remove shift rod coupling.
7. Loosen front engine mounting nuts and lower engine until nuts are flush with the front engine mount studs.

NOTE: do not remove front mounting nuts.

8. Lower engine enough to remove engine rear mounting bracket.
9. Remove oil filter and belt from the pulley.
10. Remove crankshaft pulley with puller.
11. Remove engine rear oil seal by prying on the outer edge of the seal with a couple of screw drivers.
12. Clean rear housing seal seat with solvent and check for surface damage.
13. Lubricate the outside diameter

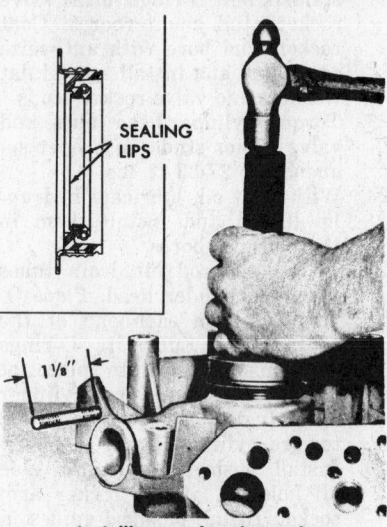

Installing rear housing seal, using tool J-270-6
(© G.M. Corp.)

of the new seal and tap into place with a suitable tool.
14. Reassemble in the reverse order of removal.

Blower Bearing and Cover

The blower bearing may be replaced without disturbing the power train.
1. Disconnect battery.
2. Disconnect accelerator linkage and fuel lines.
3. Remove oil level gauge and fresh-air choke at air cleaner. Remove choke pipe.
4. Remove air cleaner, air horn and support assembly.

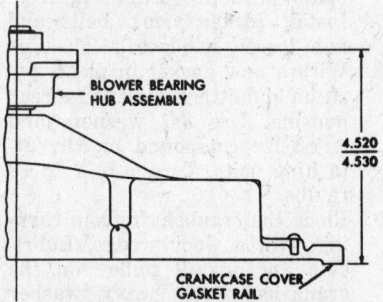

Blower bearing and crankcase cover
(© G.M. Corp.)

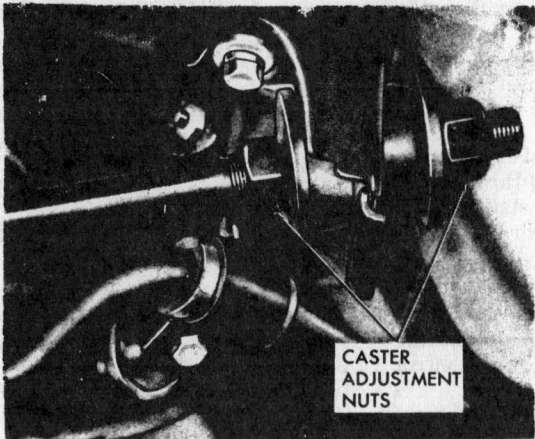

Caster adjustment—1963-69 (© G.M. Corp.)

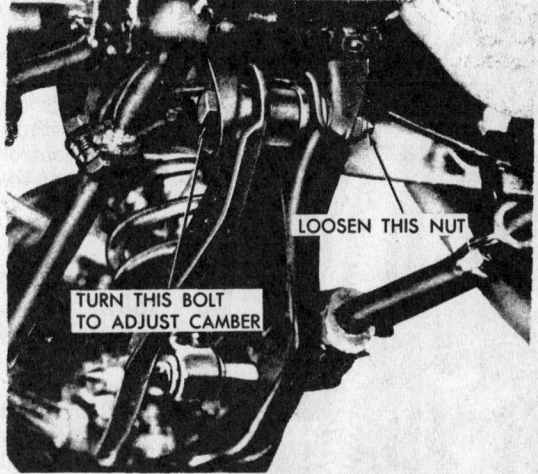

LOOSEN THIS NUT

TURN THIS BOLT TO ADJUST CAMBER

Camber adjustment—1965-69
(© G.M. Corp.)

5. Remove carburetor vacuum balance tube and retaining strap at engine upper shroud.
6. Remove blower belt and wire harness.
7. Disconnect cooling air throttle valve lever swivel.
8. Remove engine upper shroud and thermostat as an assembly.
9. Remove four bolts from blower pulley and lift off pulley and blower from blower bearing hub assembly.
10. Remove crankcase cover bolts and lift off the cover and bearing assembly.
11. Remove crankcase vent and gasket.

Blower Bearing Replacement

1. Carefully support the crankcase cover in an inverted position and press out the old blower bearing shaft.
2. Install a new bearing hub assembly in the crankcase cover (cover in its normal operating position), using hypoid lubricant on the bearing shaft.
3. Press on the shaft (only) of the

blower bearing until a height of 4.475 in. ± .015 in. for 1960-66, from crankcase cover rail to the top of the blower bearing flange, is reached.

4. Replace crankcase cover, crankcase vent and new gaskets by reversing the removal procedure. Torque blower pulley bolts to 20-25 ft. lbs., crankcase cover bolts to 7-13 ft. lbs.

Oil Pan

Drain oil and remove in the conventional manner. Retorque attaching bolts to 40-60 in. lbs.

Front Suspension

Any major operations on the front suspension will be simplified by use of a hoist. The suspension should be permitted to swing free.

Front Wheel Alignment

References

With the exception of rear wheel toe-in, wheel alignment and steering geometry are covered in the Unit Repair Section.

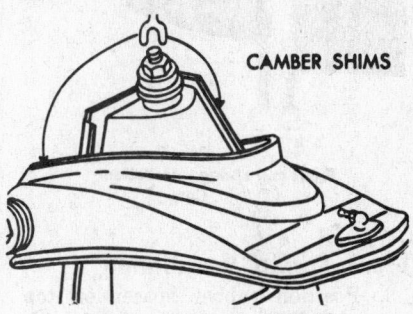

CAMBER SHIMS

Camber adjustment—1960-64
(© G.M. Corp.)

Control Arms, Ball Joints and Springs— Front Replacement

Coil Spring Removal.

1. Place car on hoist so that the control arms can swing free.
2. Remove shock absorber.
3. Remove the two strut rod-to-control arms, nuts and lockwashers.
4. Loosen the lower control arm inner pivot nut.
5. Place jack stand under lower control arm and take up slightly on spring pressure.
6. Remove control arm pivot nut and washer. Tap out pivot pin.
7. Carefully lower jackstand until spring is free. Lift out spring.

NOTE: a bar placed through the control arm and into the spring tower will keep the spring from slipping until it is free.

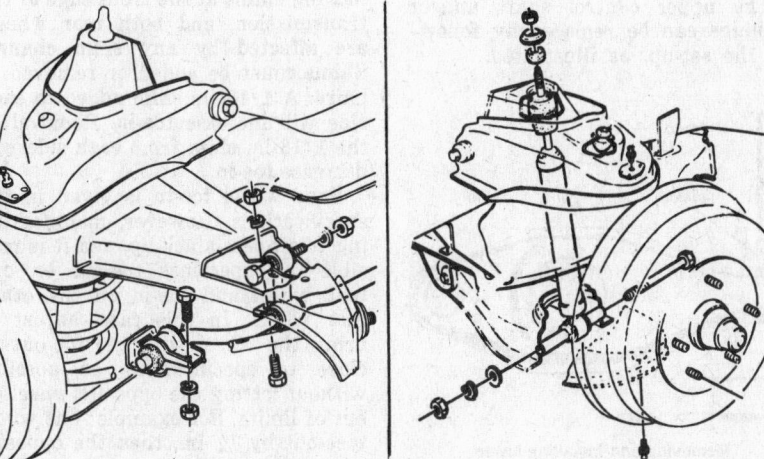

Front suspension

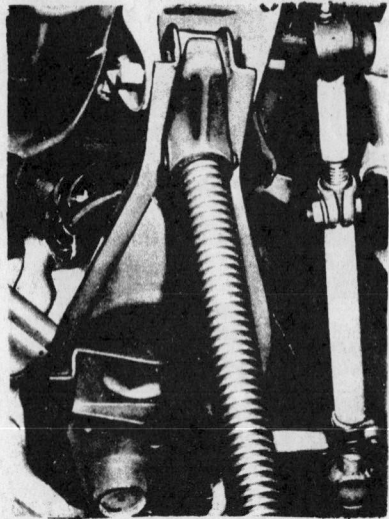

Position of jackstand for spring removal
(© G.M. Corp.)

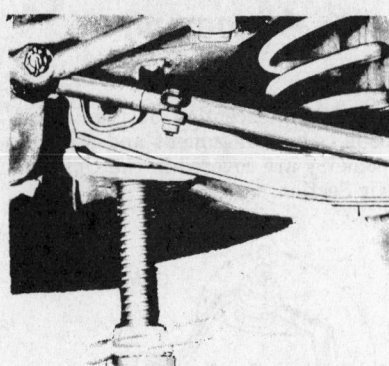

Front coil spring installation
(© G.M. Corp.)

Front Spring Installation

8. Position rubber spacer on top of coil spring and secure with friction tape. Contour of spacer and spring must match.
9. Place spring on control arm with spacer up. Rotate spring until upper end of spring and spacer finds its proper seat.
10. Using jack stand, raise control arm and install the lower control arm inner (pivot) bolt.
11. Attach strut rod to control arm with two attaching nuts and lock washers.
12. Install shock absorber.
13. Lower car to the floor. Neutralize by bouncing the car a few times. Then, tighten the lower control arm inner pivot bolt and nut.

Lower Control Arm Bushings

The control arm bushings can be replaced by following the set-up illustrated.

Lower Ball Joint Installation

1. Start the new ball joint into the control arm carefully and straight, (it is self-tapping).

Torque it up solidly with a 1⅝ in., 6-point wrench.
2. Install lubrication fitting in bottom of joint, then service with chassis lube. Place rubber cap over stud.
3. Set control arm in place with the stud of the joint up through the steering knuckle. Install, tighten the nut and lock with cotter pin.

Upper Control Arm and Ball Joint

1. Support the car weight at the outer end of the lower control arm.
2. Remove wheel assembly.
3. Remove cotter pin and nut from upper ball joint stud. Remove stud.
4. Remove two nuts holding the upper control arm cross-shaft-to-front crossmember. Count number of shims at each bolt.

Ball Joint Removal

1. Center-punch the four attaching rivets.
2. Drill through these rivets.
3. Chisel the heads off and remove these rivets. Discard the old ball joint.

Lower Control Arm and Ball Joint

To remove control arm, ball joint and/or bushing, perform Steps 1 through 7 of Coil Spring Removal. Then proceed as follows:
1. Remove cotter pin and nut from lower ball joint. Tap on the control arm to remove it and the ball joint from the steering knuckle.
2. If ball joint is to be replaced, grip the control arm in a vise and use a 1⅝ in., (six-point) wrench to remove the self-tapping joint from the lower control arm.

Upper Control Arm Bushings

The upper control shaft and/or bushings can be replaced by following the set-up, as illustrated.

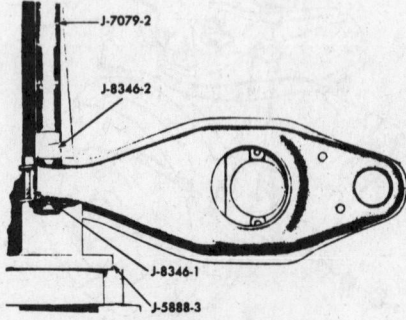

Removing and installing lower
control arm bushings
(© G.M. Corp.)

Upper Ball Joint Installation

1. Install new ball joint against the top side of the upper control arm. Secure the joint to the control arm with the four special alloy bolts and nuts that are furnished with the replacement joint.
2. Torque these nuts to 20-25 ft. lbs.

Upper Control Arm Installation

1. Install upper control arm to the car.
2. Install two attaching nuts and lockwashers to the studs holding the upper control arm shaft to front crossmember. Install same number of shims as removed at each bolt.
3. Install new rubber seal, then the bar stud through the knuckle, and the nut and cotter pin.
4. Install wheel assembly.
5. Lower car to the floor and bounce the car to neutralize suspension.
6. Torque cross-shaft bolts to 35-40 ft. lbs.
7. Recheck caster and camber.

Rear Suspension

Rear Wheel Camber

Since 1965, camber angle has been adjusted by turning the eccentric cam and bolt assembly located at the outboard mounting of the rear strut rod.
1. Place the rear wheels on an alignment machine.
2. Loosen cam bolt and rotate the cam to obtain camber specifications.
3. Torque cam bolt to 75-90 ft. lbs.

Rear Wheel Toe-in

Due to Corvair suspension design it is necessary to establish and maintain rear wheel toe-in.

This is controlled by adding or removing shims at the front edge of the transmission, and both rear wheels are affected by any shim change. Shims must be added or removed in pairs. A 1/16 in. shim added to each side will increase toe-in. Removal of the 1/16 in. shim from each side will decrease toe-in.

Rear wheel toe-in is given in the specifications. However, manufacturing tolerances stack-up, and it is possible to experience toe-out on one rear wheel and toe-in on the other rear wheel. In this case, adjust to bring the wheel with the toe-out as close to specifications as possible without letting the opposite wheel go out of limits. For example: one wheel toes-out by ¼ in., then the opposite wheel must toe-in enough to result in 0 in. to ¼ in. overall toe-in.

Caution If the case exceeds the above limits, look for bent or damaged parts or an uneven distribution of shims.

Shock Absorbers

Rear shock absorber mounting and action is functionally the same as the front. But, the rear shock absorber holds all of the rear spring compression. When removing or installing rear shock absorbers, the weight of the car must be resting on the tires.

1. Place car on a drive-on ramp type hoist, frame contact hoist or place jack stands under the body at each side rail just forward of the rear wheel openings.
2. Raise the body high enough to allow the rear wheels to hang free.
3. With a jack placed under the tire, raise the wheel to its normal position.
4. Now the shock absorber can be repaired without interference with the floor.

Riding Height and Front and Rear Spring Sag—1960-64

1. Place the car on a smooth, level floor. (The car body should be empty except for spare tire.)

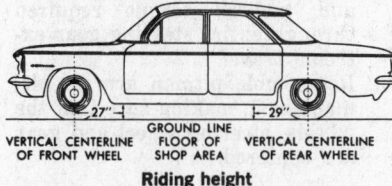

Riding height
(© G.M. Corp.)

2. Bounce the car to normalize shock absorbers and springs.
3. Measure the height from the shop floor to the bottom of the rocker panel 27 in. back of the centerline of the front wheel.

This measurement indicates front spring condition and should be 9⅜ in. ± ½ in.

4. Measure the height from the shop floor to the bottom of the rocker panel 29 in. forward of the centerline of the rear wheel. This measurement indicates rear spring condition and should be 9 in. ± ½ in.
5. This check should be made on all four springs, and replacements made where necessary.

Riding Height and Front Spring Sag—1965-68, Rear Spring Sag—1965-69

1. Follow Steps 1 and 2 in 1960-64 procedure.
2. Measure the distance between floor and lower edge of front wheelwell; the distance should be 26.2 in. ± ½ in.
3. To check rear height, measure the distance between floor and bottom of rocker panel 29 in. forward of the centerline of the rear wheel; the distance should be 8½ in. ± ½ in.
4. This check should be made on all four springs and replacements made if measurements for both front (and both rear) wheels vary more than ½ in.

Riding Height and Front Spring Sag—1969

1. Place car on a smooth, level floor with full gas tank and empty trunk (except for spare tire).
2. Measure the distance between the floor and the center of the inner lower control arm pivot.
3. Measure the distance between the floor and the bottom surface of the lower ball joint.
4. The difference between these two measurements should be 2½ in. ± ½ in. for two door models, 2¼

in. ± ½ in. for convertible models.

5. Measure the other side; difference should be ½ in. or less between left and right sides.

Rear Spring Replacement

1. Raise car by side rails, high enough for rolling floor jack to be placed under the brake drum.
2. Loosen control arm cross-shaft bolts.
3. Remove the bolt that holds the brake hose bracket to the underbody.
4. Remove the wheel assembly.
5. Replace the wheel lug nuts to hold the drum in place.
6. Position the axle shaft as illustrated. The side surface of the universal joint yoke must be at 45° to the centerline of the axle case.
7. Place a rolling floor jack under the drum and backing plate.
8. Raise the floor jack slightly, detach and remove the shock absorber.
9. Carefully lower the floor jack until the spring is free. Do not remove or lower the jack too far, because this will apply strain to the axle shaft and brake hose.
10. Install in the reverse order of removal.

Lower Control Arm R & R

1. Remove rear springs, as outlined.
2. Support the control arm with a suitable stand high enough to prevent a strain on the brake hose.
3. Remove brake drum.
4. Pry the upper end of the brake shoe up and onto the brake anchor pin. This will permit the axle flange plate and shaft to be pulled out past the parking brake strut.
5. Remove the four backing plate attaching nuts and washers.
6. Pry between the axle flange plate and backing plate until the axle shaft and universal joint assembly can be pulled out of the axle case.
7. Disconnect the universal joint at the U-bolts.
8. Remove the universal joint flange retaining bolt and washers from the end of the axle and pull the joint flange.
9. Pull out the axle shaft.
10. Disconnect the backing plate and brake shoe assembly from the control arm and tie the assembly up, out of the way.
11. Remove the four bolts and nuts from the control arm-to-crossmember. Remove the control arm.
12. Reinstall in the reverse order of removal.

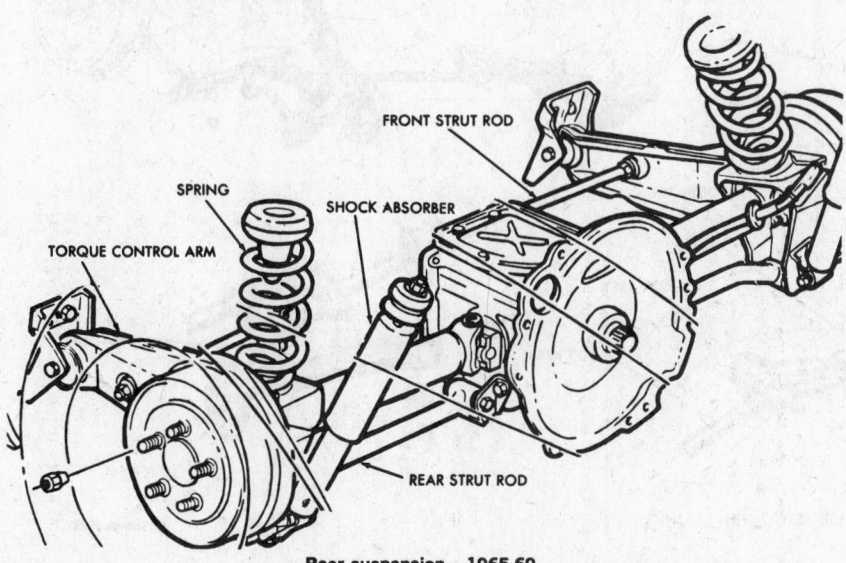

Rear suspension—1965-69

Jacking, Hoisting

For individual front wheel jacking, the lower support under front spring may be used.

Steering Gear

The steering gear is the recirculating ball type, basically the same as that used in Chevrolet. Linkage is the relay type. The tie-rod ends have self-adjusting socket joints.

Steering Gear Adjustment

Only two adjustments are possible and they must be made in the following order:
1. Remove the pitman arm.
2. Loosen pitman shaft lash adjuster screw locknut. Turn adjusting screw counterclockwise to unload the gear.

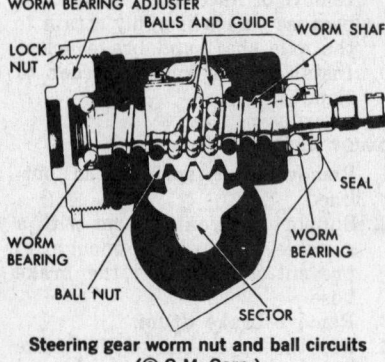

Steering gear worm nut and ball circuits
(© G.M. Corp.)

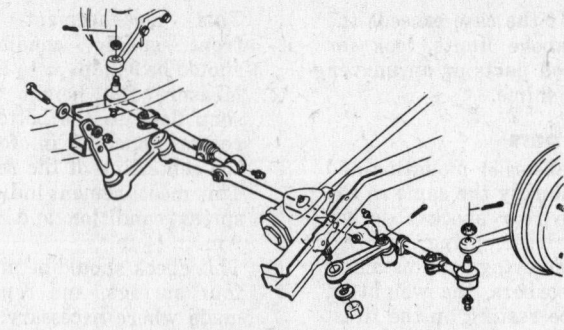

Steering linkage (© G.M. Corp.)

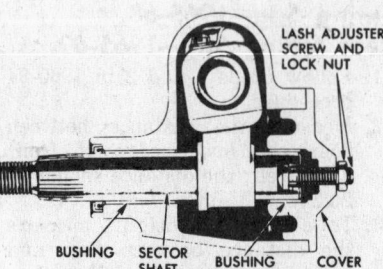

Steering gear pitman shaft
(© G.M. Corp.)

3. Gently turn the steering wheel in one direction to the end of its travel. Now back up one turn.
4. Pry off the horn button. With a suitable socket and an inch-pounds torque wrench, determine the torque required to keep the wheel rotating, (one full turn). This is between 2 and 6 in. lbs. If this is not the case, adjustment of the worm shaft endplay is in order.
5. To adjust play out of worm shaft bearings and obtain proper load, loosen worm bearing adjuster lock nut and turn the adjuster until there is no play. Check

shaft rotation pull with the torque wrench and turn the bearing adjuster until the 2 to 6 in. lbs. torque load is obtained.
6. Turn the steering wheel gently from one extreme to the other. Now turn the wheel back half way. This should be the steering gear high spot and the straight ahead gear and wheel position.
7. Turn the lash adjusting screw clockwise to take out all the lash between the gear teeth. Now adjust the lash adjuster, in or out, to produce 7 to 12 in. lbs., torque reading on the wrench applied to the steering wheel (worm shaft) nut.
8. Secure both adjustment locknuts and recheck torque required through entire steering gear extreme travel.
9. Reassemble pitman arm to pitman shaft, making sure that the wheels, steering wheel and gear are centered.

Steering Wheel R & R
1. Pry off horn button.
2. Remove three screws holding the

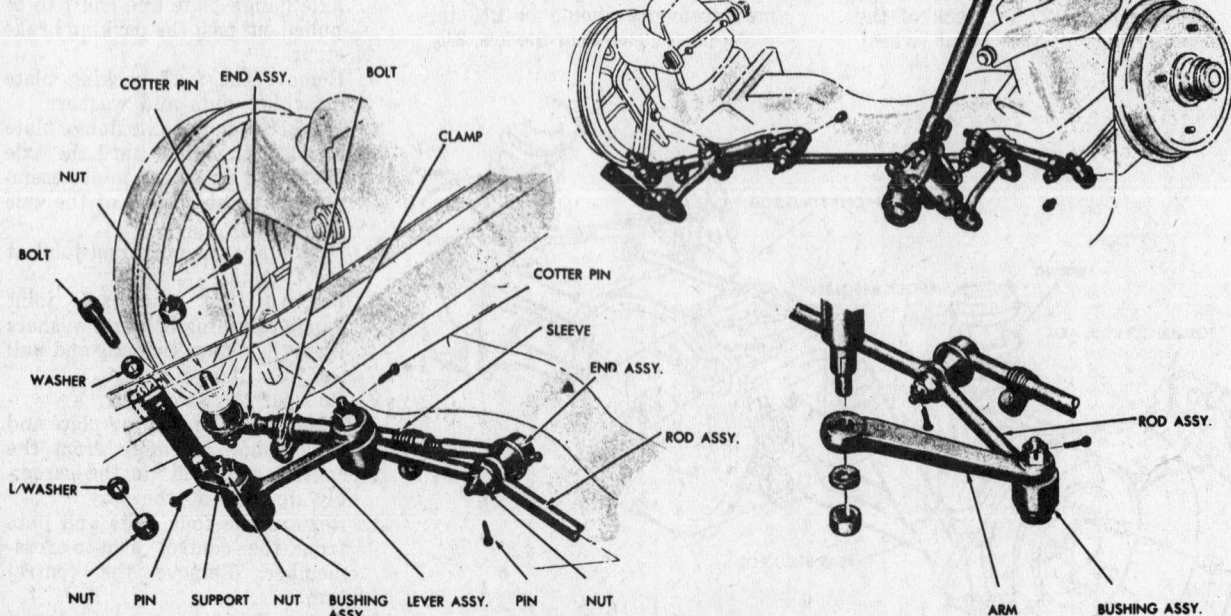

Steering linkage

Exploded view of mast jacket and steering wheel (© G.M. Corp.)

receiver cup and bushing spacer to the steering wheel, then remove the Belleville spring.

3. Remove steering wheel nut and washer from the worm shaft.
4. Reinstall in reverse order.

Steering Gear Assembly R & R

1. Lift car on suitable hoist. Disconnect pitman arm from sector shaft.
2. Remove three steering gear-to-frame mounting bolts.
3. Remove nuts and bolts from the steering gear shaft coupling and slide the gear assembly forward and down, removing it from the car.
4. Reinstall in reverse order.

Clutch

The clutch for the standard transmission is the diaphragm spring type. Due to the torsional flexibility of the input shaft, the driven disc is solidly mounted.

The clutch is operated through a conventional clutch fork by pulling instead of pushing. The clutch fork is pivoted on the axle housing and is operated by a cable, bellcrank, and pulley. In 1964, a bent-finger diaphragm clutch was introduced. This clutch takes a shorter ($1\frac{1}{4}$ in. vs. $1\frac{7}{8}$ in.) throwout bearing.

Clutch Linkage Adjustment —1960-63

1. Disconnect the clutch fork pull rod from the cross-shaft lever.
2. Attach the return spring to lower hole in the cross-shaft lever, #3 in illustration (for 1960-61).
3. Adjust clevis #2, in illustration until the outboard lever #3, on the clutch lever control cable cross-shaft has a clearance of $\frac{1}{8}$-$\frac{3}{8}$ in. for 1960-61, $\frac{3}{8}$-$\frac{5}{8}$ in. for 1962-63.
4. Lock up the clevis jam nut.
5. Manually pull the fork pull rod #4, in illustration until slack is out of the clutch fork, the release bearing touching the diaphragm fingers.
6. With the fork pull rod in this position, align the swivel #3, with upper hole in outboard lever #2. Back off the swivel three full turns and assemble to lever #2.

Clutch Linkage Adjustment— 1964

1. Check clutch control cable engagement at pedal and pulleys.
2. Disconnect clutch fork pull rod swivel from cross-shaft lever.
3. Check clearance between cross-shaft outboard lever and transmission mount bracket; should be $\frac{1}{2}$-$\frac{5}{8}$ in. If necessary, disconnect

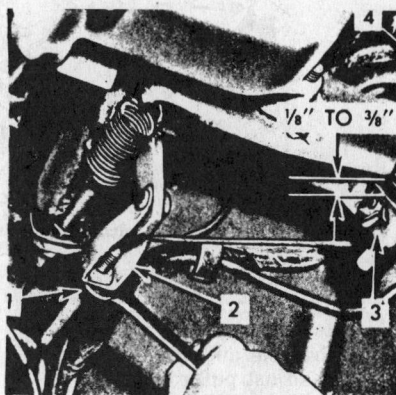

Adjusting clutch cable
(© G.M. Corp.)

return spring and adjust control cable clevis.
4. Pull clutch fork pull rod until slack is taken up at clutch fork and the throwout bearing is touching the diaphragm fingers, then turn swivel to align with hole in cross-shaft lever.
5. Lengthen pull rod 3 turns, connect swivel to cross-shaft lever, and install retainer clip.

Clutch Linkage Adjustment— 1961-64 Greenbrier and 95

1. Check front end clutch controls for proper location of bumper stop, proper return spring hookup and front cable housing clamp tightness.
2. Disconnect rear clutch return spring (1) and loosen clutch pull rod clevis locknut (2).
3. Disconnect clutch fork pull rod from control cable by removing clevis pin (3).
4. Pull rearward on cable clevis (5) to make sure all slack is out of system from clevis forward.

High-performance nodular cast-iron clutch identification— arrows point to cast-in lugs
(© G.M. Corp.)

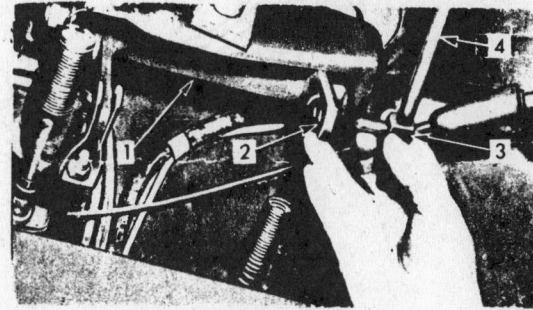

Adjusting clutch fork pull rod (© G.M. Corp.)

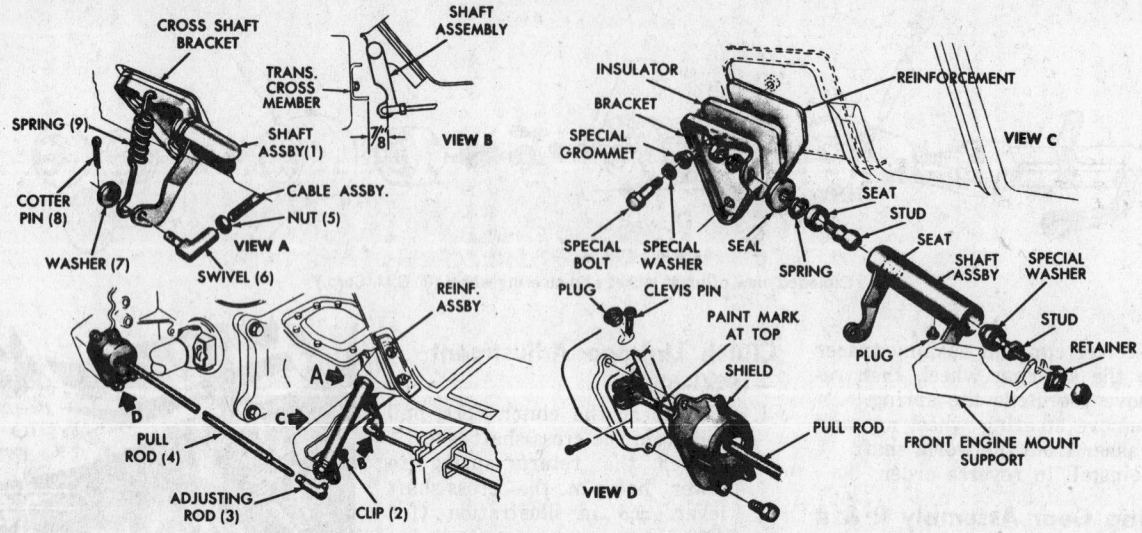

Clutch linkage—1965-69. View 'B' should read ⅜ in. for 1969 models (© G.M. Corp.)

5. Pull forward on clutch fork pull rod until throwout bearing contacts diaphragm fingers.

6. With cable and rod in this position, adjust pull rod clevis (4) to align holes, then lengthen pull rod 3½ turns and tighten locknut.

7. Install clevis pin and pin retainer, install pull rod return spring, and check pedal free travel (¾-1¼ in.).

Clutch Linkage Adjustment— 1965-69

1. Drive nut (5) to within ⅛ in. of end of threads.

2. Tension clutch cable rod assembly to 15 lbs. and thread swivel (6) to line up with hole in shaft assembly (1) inboard lever, with lever positioned ⅞ in. (for 1965-68) or ⅜ in. (1969) from transmission crossmember.

3. Torque nut (5) to swivel (6) to 8-12 ft. lbs.

4. Install spring (9), washer (7), and pin (8).

5. Manually pull clutch pull rod (4) until slack is taken up at clutch fork, then, with pull rod in this position, align adjusting rod (3) with hole in outboard shaft assembly lever.

6. Back off adjusting rod (3) 2 turns (1969) or 3 turns

(1965-68), then secure assembly to lever with clip (2).

Clutch Cable R&R—1960-69

1. Remove return spring from clutch cross-shaft bracket; remove rear cable swivel at cross-shaft inboard lever.

2. Remove cotter pin and washer (or other fastener) from swivel and remove swivel from cross-shaft.

3. Loosen swivel locknut and remove both swivel and nut from cable.

4. Remove underbody tunnel covers.

5. Unhook cable from clutch pedal

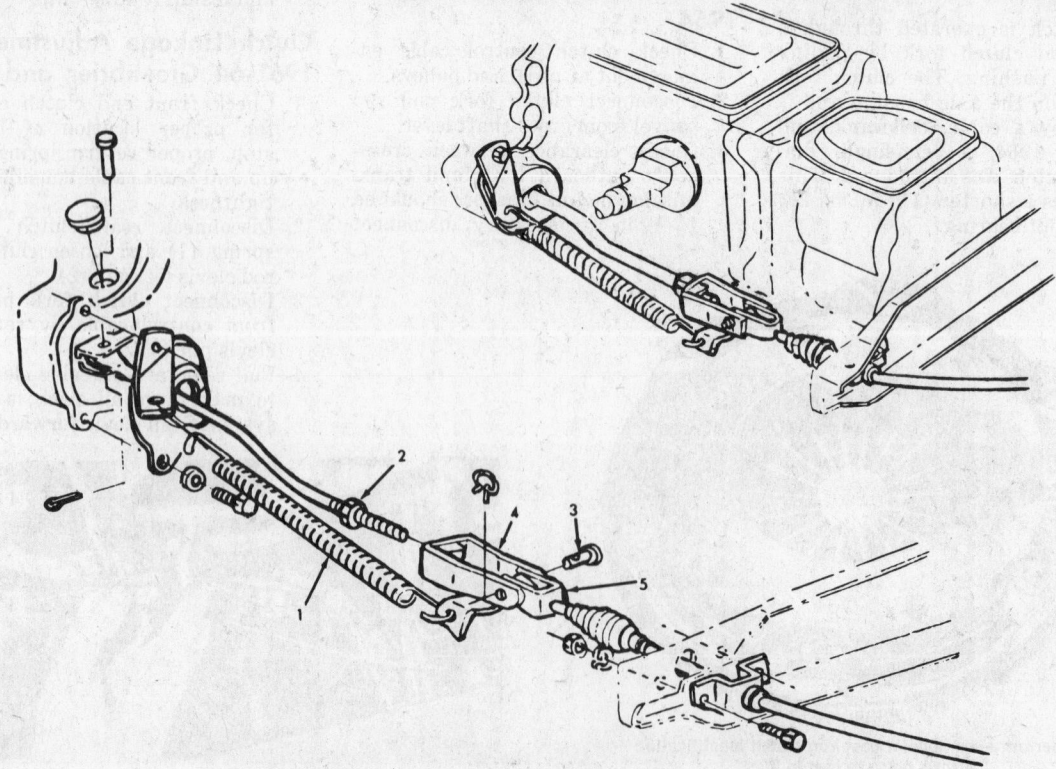

Clutch linkage—1964 Greenbrier and 95 (1961-63 similar) (© G.M. Corp.)

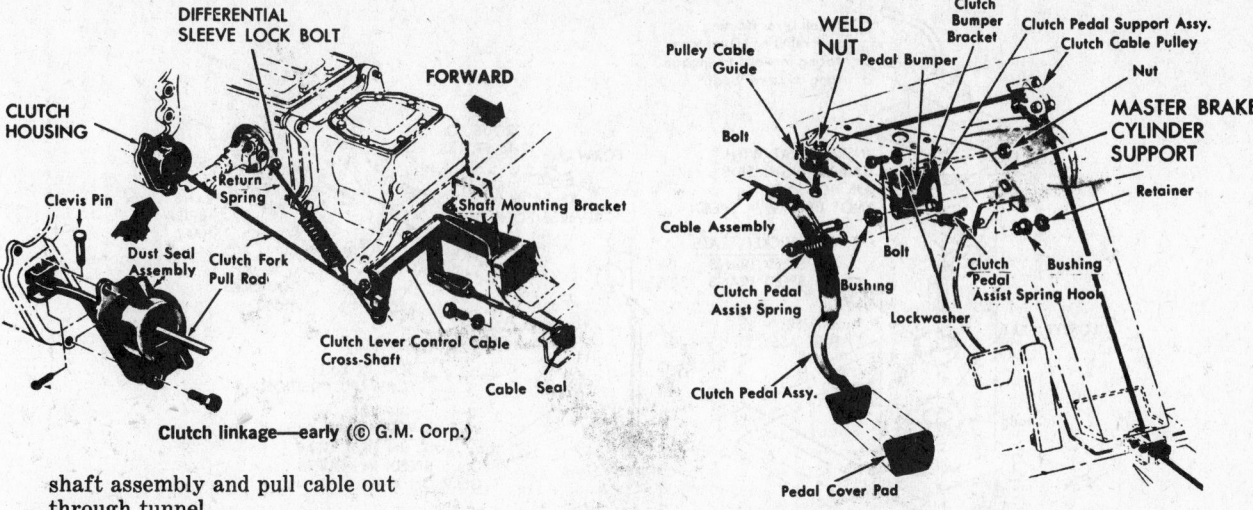

Clutch linkage—early (© G.M. Corp.)

Clutch pedal and cable—1960 (© G.M. Corp.)

shaft assembly and pull cable out through tunnel.

NOTE: it may be necessary to remove the pulley to unhook the front cable end.

6. To install, reverse removal procedure.

NOTE: often, it is easier to replace just the inner cable. Unhook both cable ends, fasten a wire to one end, new cable attached to wire, and pull the old cable from the sheath. This procedure, at the same time, pulls the new cable into place. Most cables break near the end, allowing the wire to be pulled completely through the outer cable sheath. If the cable breaks inside the sheath, however, it is best to replace the entire assembly due to roughness of the inner sheath surface near the break.

Clutch Assembly R & R

1. Remove power train.
2. Separate the transmission and axle units from the engine.

NOTE: the clutch fork, ball stud and clutch release bearing are removed with the axle housing.

3. Disconnect clutch fork from ball stud, and remove the release bearing from the clutch release shaft.
4. Loosen the six clutch attaching bolts, one turn at a time, until clutch spring pressure is zero.
5. Lift the clutch from the engine. The pilot bearing is an oil-impregnated-type pressed into the crankshaft.
6. Reinstall the clutch disc in reverse order of removal, with short hub towards flywheel.

Standard Transmission

Step-by-step repair procedures are covered in the Unit Repair Section. These transmissions are grouped according to type.

Three-Speed Gearshift Linkage Adjustment—1960-63

Series 500 and 700

NOTE: before making adjustment, play in the linkage is to be taken up by moving the shift control rod to the rear.

1. With seat in full forward position, shift transmission into first gear, then loosen the coupling nut on the transmission shift control rod.
2. Adjust gearshift lever to 1/2 in.

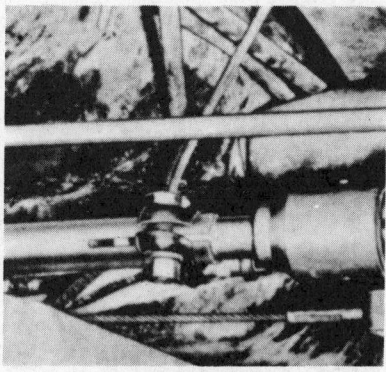

Shift rod clamp
(© G.M. Corp.)

from edge of seat, then tighten the coupling clamp nut.
3. Test shift in all ranges.

Series 900

NOTE: shift adjustment procedure for the 900 model is similar to other three-speed models except:

Shift transmission into reverse gear, then adjust the gearshift lever to 2¼ in. from the center of the gearshift lever housing, rearward, to the center of the gearshift lever knob.

Four-Speed Gearshift Linkage Adjustment—1960-63

Series 500, 700 and 900

1. With the seat fully forward, shift transmission to fourth gear (500 and 700 models), then loosen the coupling clamp nut on transmission shift control rod.
2. Adjust gearshift lever to 1/2 in. from the edge of the seat, then tighten the coupling clamp nut.
3. Test shift in all ranges.

NOTE: adjustment procedure is the same for the 900 model except:

4. Shift the transmission into reverse, then adjust the gearshift lever a distance of 3¼ in. from the center of the gearshift lever housing, rearward, to the center of the gearshift lever knob.

Series 1200 and 95

1. Remove spring from the gearshift lever shaft.
2. Rotate the shift control rod to the left of the vehicle (clockwise, when viewed from the front of the vehicle) and pull out until high gear is engaged.
3. Loosen shift control rod coupling nut. Then insert a 1½ in. block of wood (1⅞ in.—1964) between the rod end and the front flange of the front cross-member.
4. Remove the block, tighten clamp nut, and assemble spring.

Gearshift Linkage Adjustment —1964-69

See illustrations.

Gearshift Lever Assembly Removal

1. Remove the tunnel front plate.
2. Disconnect and remove the gearshift lever assembly from the

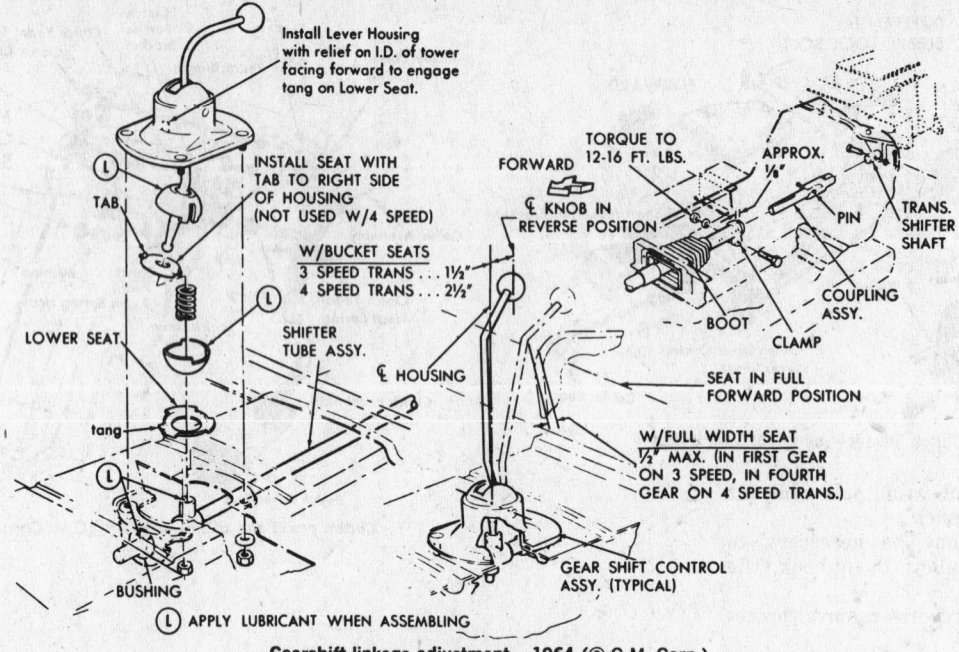

Gearshift linkage adjustment—1964 (© G.M. Corp.)

floor pan. The two rear (1960-69) or front (late 1961) nuts also hold the shift control shaft front mounting bracket.
3. Lift the assembly up until its studs clear the floor, then remove the unit by lifting the floor mat at the center of the seat.

4. Unscrew the gearshift knob, then invert and gently clamp the gearshift housing in a vise.
5. Using a length of 1½ in. pipe, depress the retainer plate and rotate one-third turn.
6. Remove the lower ball joint, spring and seat, then lift the lever out of the housing.
7. Assemble and reinstall in reverse order.

Transmission Removal and Installation

See Separating Power Train Assembly.

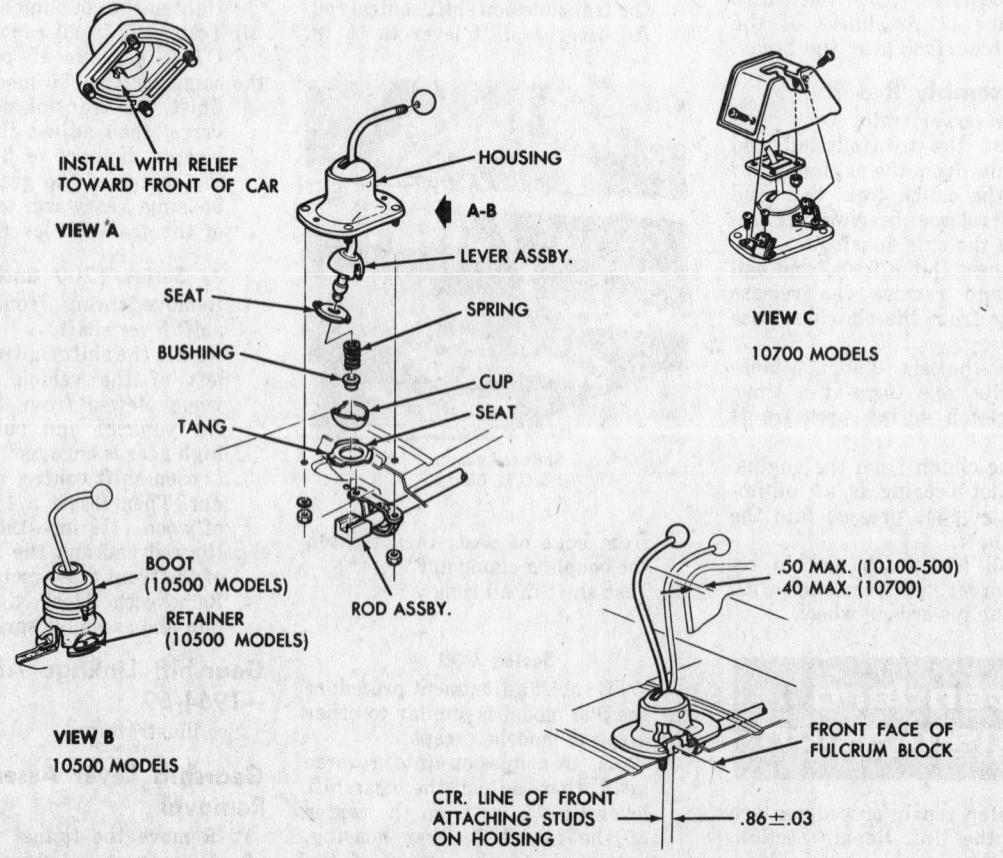

Gearshift linkage adjustment—1965-68 (© G.M. Corp.)

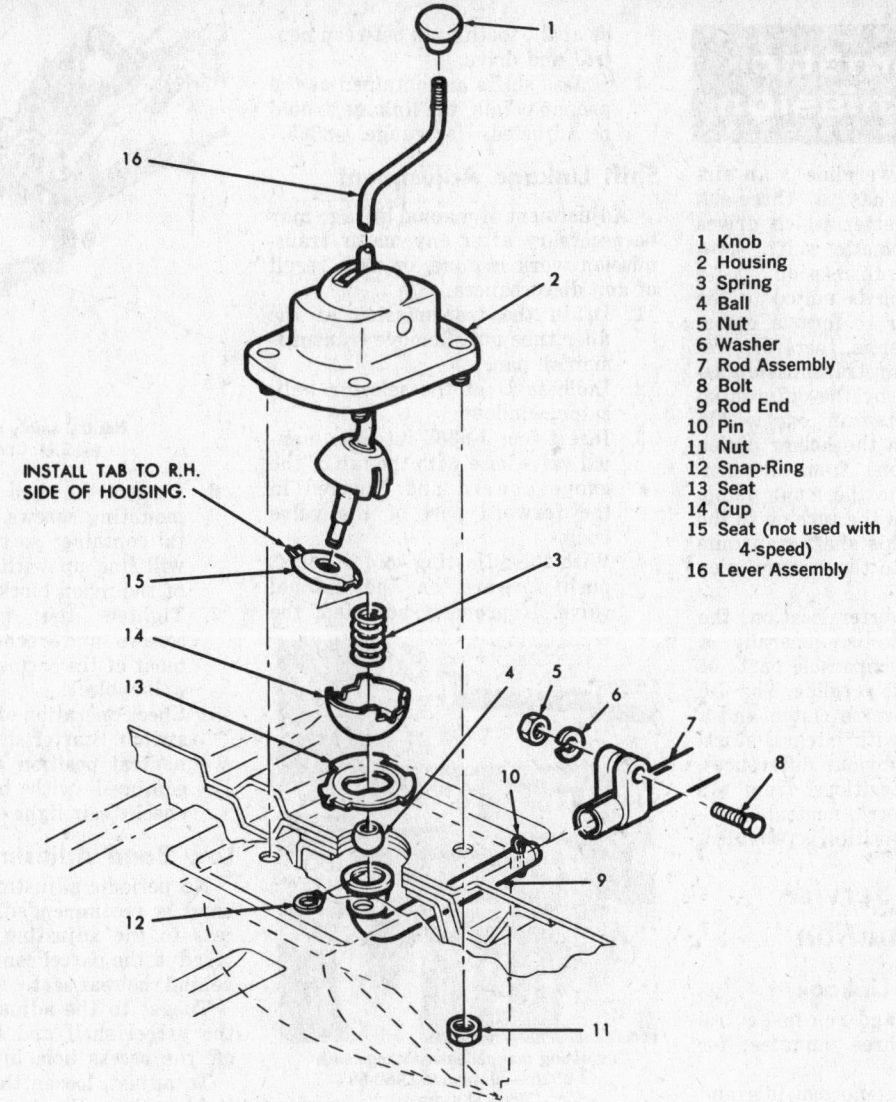

INSTALL TAB TO R.H.
SIDE OF HOUSING.

1 Knob
2 Housing
3 Spring
4 Ball
5 Nut
6 Washer
7 Rod Assembly
8 Bolt
9 Rod End
10 Pin
11 Nut
12 Snap-Ring
13 Seat
14 Cup
15 Seat (not used with 4-speed)
16 Lever Assembly

Gearshift linkage—1964 L.D.F.C. (95) (© G.M. Corp.)

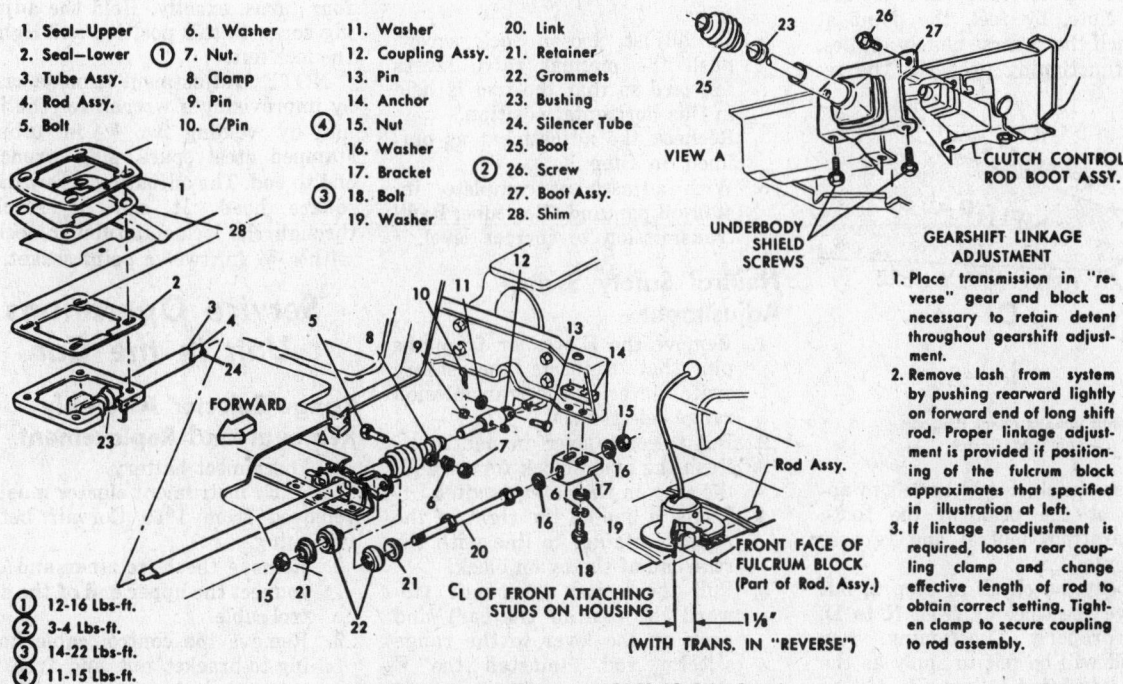

1. Seal—Upper
2. Seal—Lower
3. Tube Assy.
4. Rod Assy.
5. Bolt
⓵ 7. Nut.
6. L Washer
8. Clamp
9. Pin
10. C/Pin
11. Washer
12. Coupling Assy.
13. Pin
14. Anchor
⓸ 15. Nut
16. Washer
17. Bracket
⓷ 18. Bolt
19. Washer
20. Link
21. Retainer
22. Grommets
23. Bushing
24. Silencer Tube
25. Boot
⓶ 26. Screw
27. Boot Assy.
28. Shim

VIEW A

CLUTCH CONTROL
ROD BOOT ASSY.

UNDERBODY
SHIELD
SCREWS

FORWARD

Rod Assy.

FRONT FACE OF
FULCRUM BLOCK
(Part of Rod Assy.)

C_L OF FRONT ATTACHING
STUDS ON HOUSING

1⅛"

(WITH TRANS. IN "REVERSE")

⓵ 12-16 Lbs-ft.
⓶ 3-4 Lbs-ft.
⓷ 14-22 Lbs-ft.
⓸ 11-15 Lbs-ft.

GEARSHIFT LINKAGE ADJUSTMENT

1. Place transmission in "reverse" gear and block as necessary to retain detent throughout gearshift adjustment.

2. Remove lash from system by pushing rearward lightly on forward end of long shift rod. Proper linkage adjustment is provided if positioning of the fulcrum block approximates that specified in illustration at left.

3. If linkage readjustment is required, loosen rear coupling clamp and change effective length of rod to obtain correct setting. Tighten clamp to secure coupling to rod assembly.

Gearshift linkage adjustment—1969 (© G.M. Corp.)

Automatic Transmission

The Corvair Powerglide is an air-cooled unit. It has a three-element torque converter which drives through an automatic shift, two-speed planetary transmission.

The transmission is united to the differential carrier to form a transaxle. The converter is, therefore, remote from the main transmission assembly, separated by the differential carrier. Two shafts run, one within the other, through the hollow pinion shaft. There is one from the converter cover hub to the front pump, and the other from the turbine to the input sun gear. This shaft transmits converter torque to the transmission gear box.

Except for converter location, the Corvair Powerglide is, generally, a small version of comparable parts of the conventional Powerglide. The use of a plate-type reverse clutch and a welded converter with integral starter ring gear are obvious differences.

Selector lever positions, from top to bottom are reverse, neutral, drive and low. No park position is provided.

Quick Service Information

Checking Shift Linkage

1. Start engine and run in neutral for about three minutes for warm-up.
2. With engine at normal idle and the parking brake set, slowly move the selector from N toward R. Note, by feel, the point at which the reverse clutch applies. If functioning properly, the re-

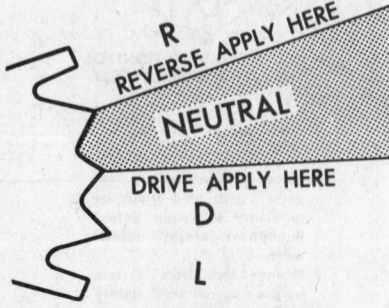

Shift linkage check diagram
(© G.M. Corp.)

verse clutch should be felt to apply at the peak of the tooth separating neutral and reverse detents.
3. Repeat the check in Step 2, but move the selector from N to D. If properly functioning, low band will be felt to apply as the selector lever follower is felt to

be at the tooth peak between neutral and drive.
4. Unless shifts are obtained at the proper points, the linkage should be adjusted. Use gauge J-8365.

Shift Linkage Adjustment

Adjustment of manual linkage may be necessary after any major transmission work is done, or as a result of unit disturbances.

1. Drain the transmission at the filler tube nut. Remove transmission oil pan.
2. Indicate D at the selector indicator window.
3. Insert tool J-8365 into the manual valve bore with the tab of the gauge upward and engaged in the forward port of the valve body.
4. With the adjusting tool in place, push forward on the manual valve. If properly adjusted, the

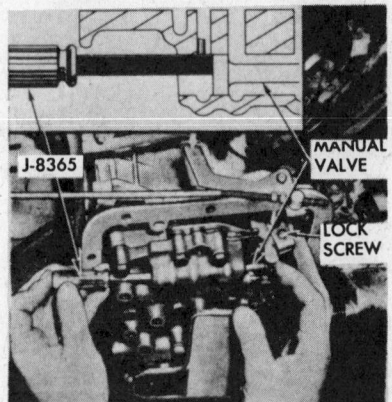

Adjusting manual valve linkage with J-8365—all models 1960-69
(© G.M. Corp.)

tool will be held in place horizontally.
5. To adjust, loosen lock screw, push the manual valve levers forward so that the tool is held in this horizontal position. Recheck the adjustment as outlined in Step 4.
6. With adjustment complete, install oil pan and filler tube. Refill transmission to correct level.

Neutral Safety Switch Adjustment

1. Remove the E washer from the pin that connects the safety switch lever to the transmission range selector rod.
2. Put selector lever in neutral.
3. Push the nylon block forward all the way in the safety switch.
4. Scribe a line on the right of the metal container in line with the rear end of the nylon block.
5. Pull the switch lever out (toward the rear of the car) and hook up the lever to the range selector rod. Reinstall the E washer.

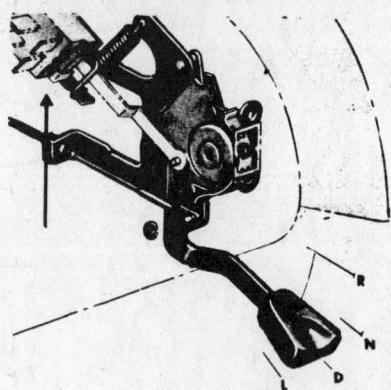

Neutral safety switch
(© G.M. Corp.)

6. Loosen the two safety switch mounting screws. Move the metal container so the scribed line will line up with the front end of the nylon block.
7. Tighten the two mounting screws and recheck the alignment of the scribed line with the nylon block.
8. Check operation of neutral safety switch. Starter should operate in neutral position only. If car is equipped with back-up lights, check their light operation.

Low Band Adjustment

No periodic adjustment of the low band is recommended. However, access to the adjusting screw is provided in the parcel compartment area behind the rear seat.

To get to the adjustment, remove the parcel shelf and take the cover off the access hole in the floor pan.

To adjust, loosen the lock nut and tighten the adjusting screw until it bottoms, finger tight. Then back off four turns, exactly. Hold the adjusting screw in this position and tighten the lock nut.

NOTE: Adjustment is made easier by improvising a wrench for the lock nut, by welding two ¾ in. tubular stamped steel spark plug wrenches end to end. The adjusting screw has a square head. It can be reached through the lock nut tubular socket, with a ⅜ in. twelve point socket.

Service Operations Unit in the Car

Range Selector Assembly Removal and Replacement

1. Disconnect battery.
   NOTE: instrument cluster must be removed from 1969 Corvair before beginning.
2. Remove the E retainer and disconnect the upper end of the control cable.
3. Remove the control cable housing-to-bracket nut and free the cable from the selector.

4. Remove the instrument cluster.
5. Disconnect the wires to the safety switch.
6. Remove two screws that hold the range selector to the instrument cluster, and remove the quadrant light from its clip on the selector.

NOTE: the range selector is serviced only as an assembly.

Reverse above procedure for replacement.

Control Cable Assembly
Removal

NOTE: instrument cluster must be removed from 1969 Corvair before beginning.

1. Disconnect the control cable and housing at the upper end.
2. Remove tunnel covers.
3. At the front of the car, remove cable housing from multiple clip at the toe-pan and remove upper toe-pan clip.
4. Remove cable housing from the three clips in the tunnel.
5. Remove grommet plate at the rear of the tunnel, free the cable sheath from the plate, and remove the clip in the underbody kick-up area.
6. Disconnect the throttle rods from the throttle valve lever on the transmission.
7. Complete cable removal by rotating the transmission TV lever counterclockwise to its full limit. Free the cable ball from the inner manual valve lever slot in the transmission and withdraw the cable.

NOTE: the cable assembly with its two captive grommets is serviced only as an assembly.

Replacement

1. With tunnel covers still removed, lay the cable out beneath the car in its proper form.
2. Insert the front end of the cable up and into the driver's compartment. The cable must then be snaked under the parking brake cable, then over the brake pipe to prevent chafing.
3. Connect the shift cable to the range selector.
4. Put selector in D, attach the cable with the upper clip in the toe-pan, and secure the cable in

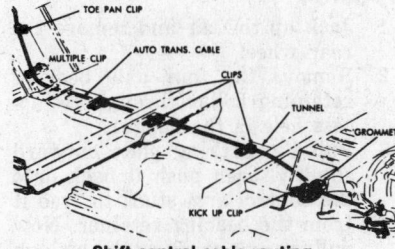

Shift control cable routing
(© G.M. Corp.)

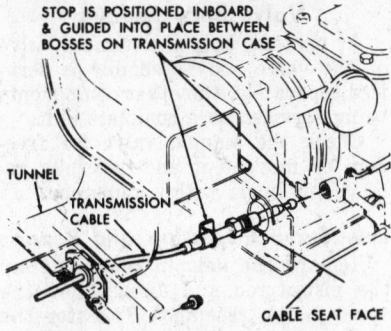

STOP IS POSITIONED INBOARD & GUIDED INTO PLACE BETWEEN BOSSES ON TRANSMISSION CASE

TUNNEL

TRANSMISSION CABLE

CABLE SEAT FACE

Cable to case installation—early design
(© G.M. Corp.)

the multiple clip at the base of the toe-pan.
5. Secure the cable with all three clips in the tunnel area. Guide the cable through the hole in the engine front support.

Early Design

1. With throttle rods disconnected from the TV lever on the transmission, rotate the lever fully counterclockwise and insert the cable ball into the slot of the transmission manual valve lever while guiding the cable sheath into the slot.
2. Insert and torque the cable sheath nut to 8-10 ft. lbs.

Late Design

1. Install O-ring seal on the cable. Lubricate the O-ring with transmission fluid.
2. With throttle rods disconnected from the TV lever on the trans-

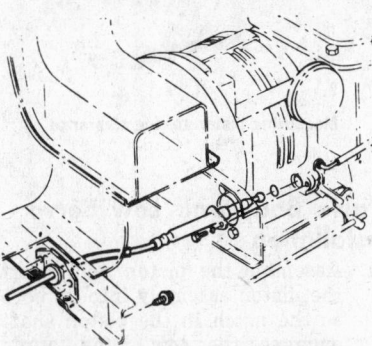

Cable to case installation—late design
(© G.M. Corp.)

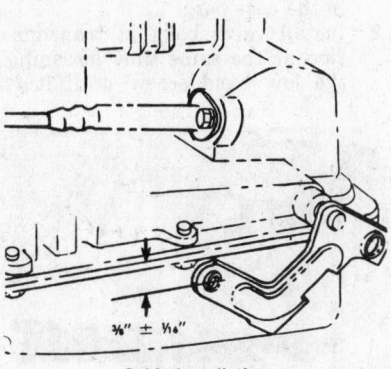

3/8" ± 1/16"

Cable installation
(© G.M. Corp.)

mission, rotate the lever fully counterclockwise, and insert the cable ball into the slot of the transmission manual valve lever.
3. Fully seat O-ring and secure installation by positioning cap screw and lock washer.
4. To check, exert a slight hand pressure, counterclockwise, and see that the hole in the notched arm of the TV lever is below the transmission oil pan rail. If hole is above the rail, installation is faulty.
5. Install cable rear grommet and plate in the rear of the tunnel.
6. Install clip on cable in rear kick-up area.
7. Check shift linkage for proper operation.

Vacuum Modulator

The vacuum modulator is mounted on the right side of the transmission and can be serviced from underneath the car.

Removal

1. Remove hose at vacuum modulator and unscrew the modulator with a thin, 1-in. end wrench.
2. Check modulator valve for nicks and other visual damage. A vac-

Removing vacuum modulator and valve
(© G.M. Corp.)

uum leak within the modulator can be troublesome and may cause loss of transmission fluid and smoky exhaust. The vacuum gauge, or some distributor test equipment, may be used to check for modulator leak trouble, or a thorough look at the inside of the modulator hose should detect oil if the modulator is leaking. If a leak is found, replace the modulator valve assembly.

Replacement

Reverse above procedure for replacement.

Governor

The governor can be reached from under the car. It is mounted on the left side.

Remove by taking out the lock screw from the governor retaining

Removing valve body
(© G.M. Corp.)

tab and lifting out the governor.

The only recommended service to the governor is the replacement of a driven gear, if necessary.

Valve Body and Low Servo

If service is required on the valve body or low servo piston while the transmission is still in the car, proceed as follows:

1. Remove the parcel compartment shelf, back of the rear seat.
2. Remove band adjusting hole cover and loosen the low band adjusting screw locknut.
3. Drain transmission oil pan and remove filler pipe.
4. Disconnect control rods from TV lever on transmission.
5. Remove oil pan, gasket and pick-up pipes.
6. Remove valve body attaching bolts and jar the valve body assembly lightly with a soft hammer to loosen it from its locating dowels.

Caution Carefully lower the valve body while tightening the low band adjusting screw. This must be done simultaneously with lowering the valve body, until the screw is fully tightened.

7. Remove the low servo piston by pulling downward on the hub of the piston shaft with a screwdriver.

Valve body installed
(© G.M. Corp.)

Valve Body Repairs

At present, only the manual valve of the valve body assembly is serviced separately, the other components being serviced as a complete unit.

Check the manual valve for freedom of movement and carefully remove any burrs with a slip stone.

Low Servo Inspection and Repairs

Remove the hairpin retainer from the piston rod and disassemble the servo piston assembly. Transfer the piston ring from the piston to the servo bore and measure gap clearance. The gap should be .002-.012 in. With the ring on the piston, ring groove clearance should be .0005-.005 in.

Measuring low servo piston ring gap
(© G.M. Corp.)

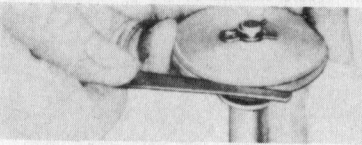

Measuring piston-to-ring clearance
(© G.M. Corp.)

Valve Body and Low Servo Installation

1. Assemble the piston and insert the piston assembly into its bore so the notch in the piston shaft engages the low band apply strut. Loosen the low band screw to permit the piston ring to seat in the case bore.
2. Install valve body in transmission at the same time loosening the low band screw until it is

Installing low servo piston and return spring
(© G.M. Corp.)

Installing oil pick-up in valve body
(© G.M. Corp.)

possible to align the valve body on the locating dowels. If manual valve is installed, index it with the manual valve lever in the case. Install the 20 attaching bolts and torque to 9-11 ft. lbs.
3. Lubricate and install O-ring seal in valve body and install oil pick-up pipe.
4. Install oil pan, torque to 3-4 ft. lbs.
5. Tighten filler tube connection, refill transmission and adjust low range band.

Automatic Transmission Disassembly

For all practical purposes, this transmission is almost identical to the Chevrolet Powerglide transmission included in the Unit Repair Section.

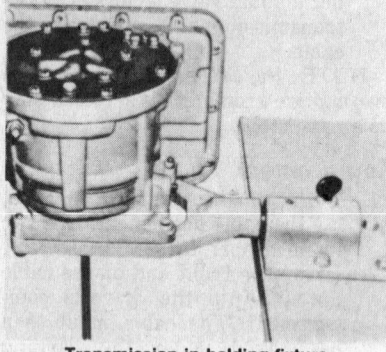

Transmission in holding fixture
(© G.M. Corp.)

Drive Axle Assembly

Axle and Universal Joint— 1960-64

1. Jack up the car and remove the rear wheel.
2. Remove the four axle bearing retaining bolts (through the access hole in the flange).
3. Pull the backing plate outward slightly, then push it back onto the control arm studs to free it from the bearing retainer. Now pull the axle shaft out far enough to free the U-joint splines from

Removing axle bearing retainer nuts
(© G.M. Corp.)

the side gears in the rear axle.

4. Remove the four nuts from the U-bolts holding the U-joint.
5. Remove bolt and two washers attaching the yoke to the axle shaft and remove the yoke. Now, withdraw the axle shaft from the lower control arm.

Axle and Universal Joint—1965-69

1. Jack up car and support with axle stands.
2. Place a jack under the torque arm or rear strut rod torque arm bracket, then raise jack until driveshaft is level.
3. Disconnect inboard driveshaft trunnion from side gear yoke by removing four bolts and retaining straps.
4. Remove the four bolts and straps which secure the outboard trunnion to drive spindle flange.
5. Pry shaft from either end and remove from vehicle.

Axle Shaft Bearing—1960-64

1. Place the axle shaft and bearing in a press. Attach a split ring

type puller between the axle flange and the bearing puller ring.
2. Pull oil deflector, bearing and puller ring.
3. To avoid damage in pressing the new bearing assembly into place, put a new puller ring and bearing on the axle. Now place the old puller ring, with its flat face against the bearing inner race. Press the puller and bearing onto the axle.
4. Remove the old puller ring. Then press on the oil deflector.

Side Bearing Adjuster Seal—1960-69

The differential side bearing adjusting sleeve seal may be replaced while the axle shaft is out by:
1. Removing the universal joint from the side bearing adjusting sleeve.
2. Prying out the old seal, then installing a new seal using a flat faced object as a driver, the seal mounts flush. The seal lips must face inward.

Carrier Disassembly and Assembly

For removal and installation procedures, see Power Train in this section.
1. Drain the differential carrier.
2. Remove speedometer-driven gear assembly.
3. Remove six cover bolts and lift off the cover.
4. Remove the tab-locks and screw out the side bearing adjusting sleeves.

5. Unlock, then unscrew the pinion bearing adjusting sleeve.
6. Remove pinion with bearings by lifting the pinion shaft toward the transmission end of the carrier. Remove the shaft and bearings through the carrier cover hole.
7. Remove the differential assembly from the carrier by angling the assembly through the carrier cover hole.

Further dismantling of components is done in the conventional manner.

NOTE: there is a difference between some of the differential components used in cars equipped with standard transmissions and those used in automatic transmissions.

Pinion or Bearing Replacement

In this area, special tools are required and should be used for best results.

To determine shim thickness to be used between the pinion bearing and pinion gear, proceed as follows:
1. Mount the differential carrier as illustrated. Place pinion rear bearing to be used in assembly in carrier and rotate several times to seat it.
2. Insert adapter pilot J-6266-25 into bore of stator shaft or clutch bearing shaft. Place depth setting gauge plate J-6266-5 on rear bearing. Insert clamp bolt through the gauge plate and the pilot, then lightly tighten the nut. Work the plate and the bearing while tightening the nut to ensure centering in the bore. Now tighten the clamp nut to 6 ft. lbs.
3. Place gauge cylinder adapter, J-6266-18 in the unthreaded portion of the side bearing adjusting sleeve bore. Now insert gauge cylinder J-6266-01 in adapter with plunger and mounting post horizontal. Rotate the gauge body back and forth to insure that the adapter crescents and body are fully seated in the side bearing bores.
4. Place gauge J-6266-19 on the

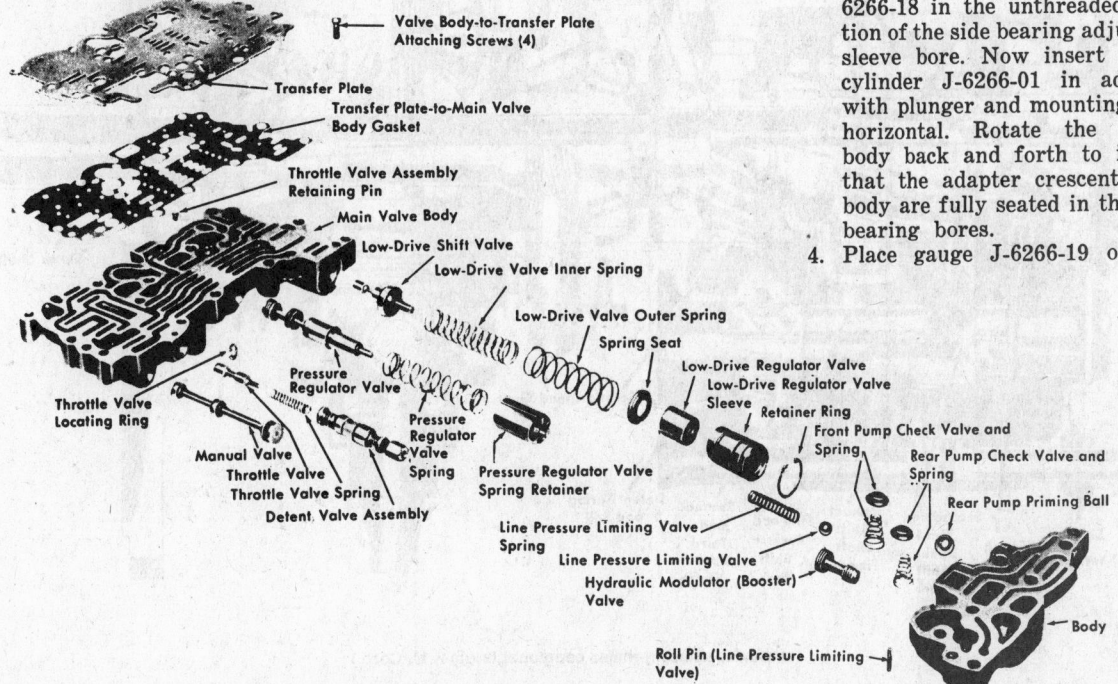

Corvair Powerglide valve body (© G.M. Corp.)

gauge plate so that it is centered beneath the gauge body. Loosen clamp screw in gauge and slide the plunger back and forth to obtain exact center between the low point of the gauge cylinder and the gauge plate. When this position is obtained, tighten the screw in the plunger and remove the plunger.

5. Using a 2 in. micrometer, measure the gauge plunger.

6. Check the pinion marking stamped on the front face of the pinion gear and the gauge plunger measurement obtained in procedure E with the following chart.

For example: the gauge reading is 1.255 in. and the pinion marking is 15. Following the micrometer reading of 1.255 in., across the vertical pinion marking columns to the column headed 15, see indication 15. This indicates that, for this particular pinion and gauge setting, one .015 in. shim should be installed between the pinion rear bearing and the pinion gear.

If the gauge reading were the same, but the pinion marking were 19, the chart would indicate 15 + 6, therefore one .015 in. shim and one .006 in. shim are required.

7. Assemble shim or shims on the rear face of the pinion gear, then install pinion rear bearing.

8. Install pinion front bearing.

Pinion Front Bearing Race

1. Remove the old bearing race from the pinion adjusting sleeve with a punch. On automatic transmission models, it is necessary to remove the seal, also.

2. Install new race in pinion adjusting sleeve, using driver J-7137 and handle, if available.

3. On automatic transmission units, install new seal with tool J-8340.

Side Bearing Adjusting Sleeve Race

1. Punch mark the side bearing adjusting sleeve at two places, 180° apart and 9/16 in. outboard from the seal bore.

2. With a 1/8 in. or 3/16 in. drill, bore through the adjusting sleeve at the punch marks until the race is encountered.

3. Drive out the bearing race with a pin punch through the two drilled holes.

4. Deburr the race seat of the adjusting sleeve to insure positive race seating. Drive a new bearing race into the sleeve until it is home and solidly seated.

5. Seal the drilled holes by using lead balls swaged into place with a punch. (These lead balls are available from your carburetor parts supplier.)

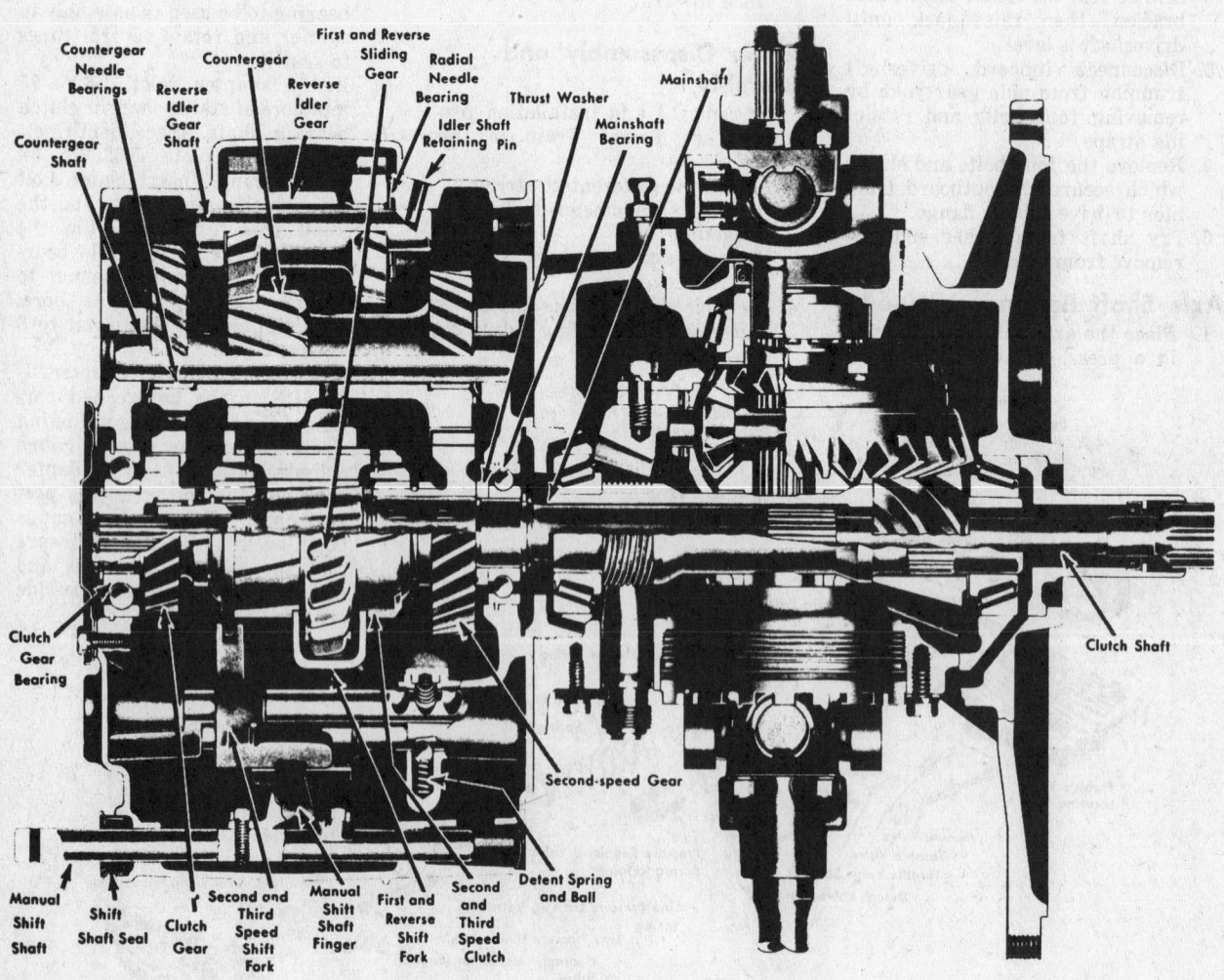

Transaxle, showing major components (© G.M. Corp.)

Differential Assembly

Reconditioning

Differential reconditioning is conducted in the traditional way with the exception of the following:

Three-Speed Transmission

Oil Seal

1. Remove the split ring and old seal from clutch release bearing shaft.
2. Drive a new seal, open side inward in shaft, using an old ¾ in. socket and soft hammer. Bottom the seal, then install split ring in clutch release bearing shaft.

Clutch Release Bearing Shaft or Pinion Bearing Race Replacement

1. Press out both clutch release bearing shaft and pinion rear bearing race.
2. If a new clutch bearing release shaft is being installed, first position the inner seal. Install a new seal ring in groove on outer diameter of bearing shaft and lubricate with petroleum jelly.
3. Support the differential carrier (only on boss) at clutch release bearing location with something cylindrical. Then, place bearing race on clutch release bearing shaft and press both into differential carrier. Press until cup is flush with adjacent surface inside the carrier.

Automatic Transmission

Pinion Shaft or Converter Hub Oil Seal Replacement

The pinion shaft front oil seal and converter hub oil seal are opposite each other (fore and aft) in the carrier.

Their replacement involves the same basic operation with the same tool, J-8340. This tool is designed to install the pinion front seal. The seal fits into the inner diameter of the tool for installation to a predetermined depth. When used to install the converter hub seal, the stop surface of J-8340 is used to drive the seal, which is mounted flush.

1. Pry out the old seal.
   Coat the outer diameter of the seal with non-hardening sealer, then install the new seal with tool J-8340.

Pinion Shaft Rear Oil Seal Replacement

1. Drive out the old seal with a punch inserted in the access hole in the stator shaft.
2. Install new seal by tapping until the seal bottoms.

Pinion Shaft Bushing Replacement

1. Remove the bushing with a chisel or other suitable tool.
2. Install new bushing to prescribed depth using tool J-8333.

Stator Shaft and Pinion Rear Bearing Race

1. Remove stator shaft and pinion bearing cup from carrier by using a press. Press downward on the end of the stator shaft.
2. Install seal ring in groove on outside diameter of stator shaft and lubricate with petroleum jelly. This seal is not used on later models.
   NOTE: when installing a new stator shaft it will be necessary to install a new pinion rear oil seal.
3. Align notch in stator shaft with drain passage in carrier. Place bearing race on the shaft and press race and stator shaft into housing carrier, using tool J-7137.

Assembling the Carrier Assembly

1. Place differential assembly into carrier with side bearing cones in place on differential hubs.
2. Before any parts have been attached to the carrier, insert pinion through the cover hole.

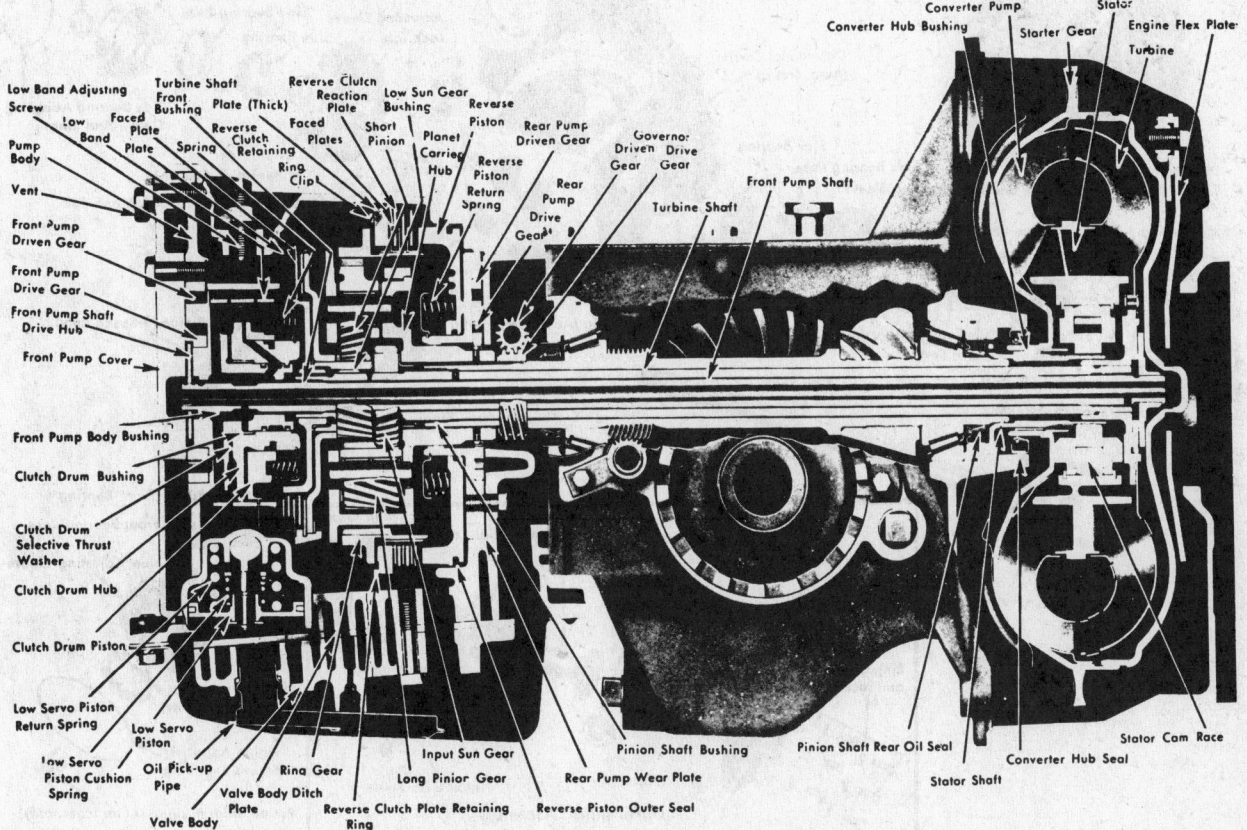

Automatic transmission rear axle, sectional side view (© G.M. Corp.)

3. Engage pinion with ring gear and carefully position the pinion rear bearing in the race. On automatic transmission models, care must be used to prevent damage to the seal at this location.

4. Install new O-ring seals in side bearing adjusting sleeves, coat adjusting sleeves with non-hardening type sealer. Loosely install adjusting sleeves in the carrier with the side bearing in position.

5. On automatic transmission models, install a new O-ring in the pinion adjusting sleeve, position pinion so its front bearing will pick up the bearing race in the adjusting sleeve. Then, loosely install pinion adjusting sleeve in the carrier.

6. Tighten both side bearing adjusting sleeves and the pinion adjusting sleeve to the point of contact between bearings and races. At this point, there should be no preload on any of the bearings and the gear lash should be just enough to permit the differential to rotate freely and smoothly.

Ring Gear and Pinion Bearing Adjustment

1. Tighten right side bearing adjusting sleeve while rocking the differential assembly with one hand until there is zero backlash. Mark this point, then back off the adjustment three notches to neutralize the O-ring. Retighten the sleeve two notches and lock tab.

2. Tighten the left side adjusting sleeve while chucking the differential side ways until all lash is gone. Mark the left side adjusting sleeve and carrier, then back off the sleeve adjustment three notches to neutralize the O-ring. Retighten adjustment sleeve to the no-lash mark, then two to three additional notches and lock it up with the lock tab. This operation preloads the differential side bearings.

3. Tighten pinion bearing sleeve with spanner wrench until bearings are in good contact with their races. Then tighten pinion sleeve two additional notches. Measure pinion turning torque, using tool J-8362 adapter and inch-pound torque wrench. The finally adjusted pinion turning torque should be 5-15 in. lbs. for used bearings, or 15-30 in. lbs. for new bearings.

4. When satisfied with preload, tab lock the adjusting sleeve.

5. Engage the speedometer driven gear with the drive gear, then secure the driven gear assembly in the carrier by tightening the lock tab.

Ring Gear-to-Pinion Lash

Ring gear-to-pinion lash should be .003 in. to .010 in. (.005-.008 in.

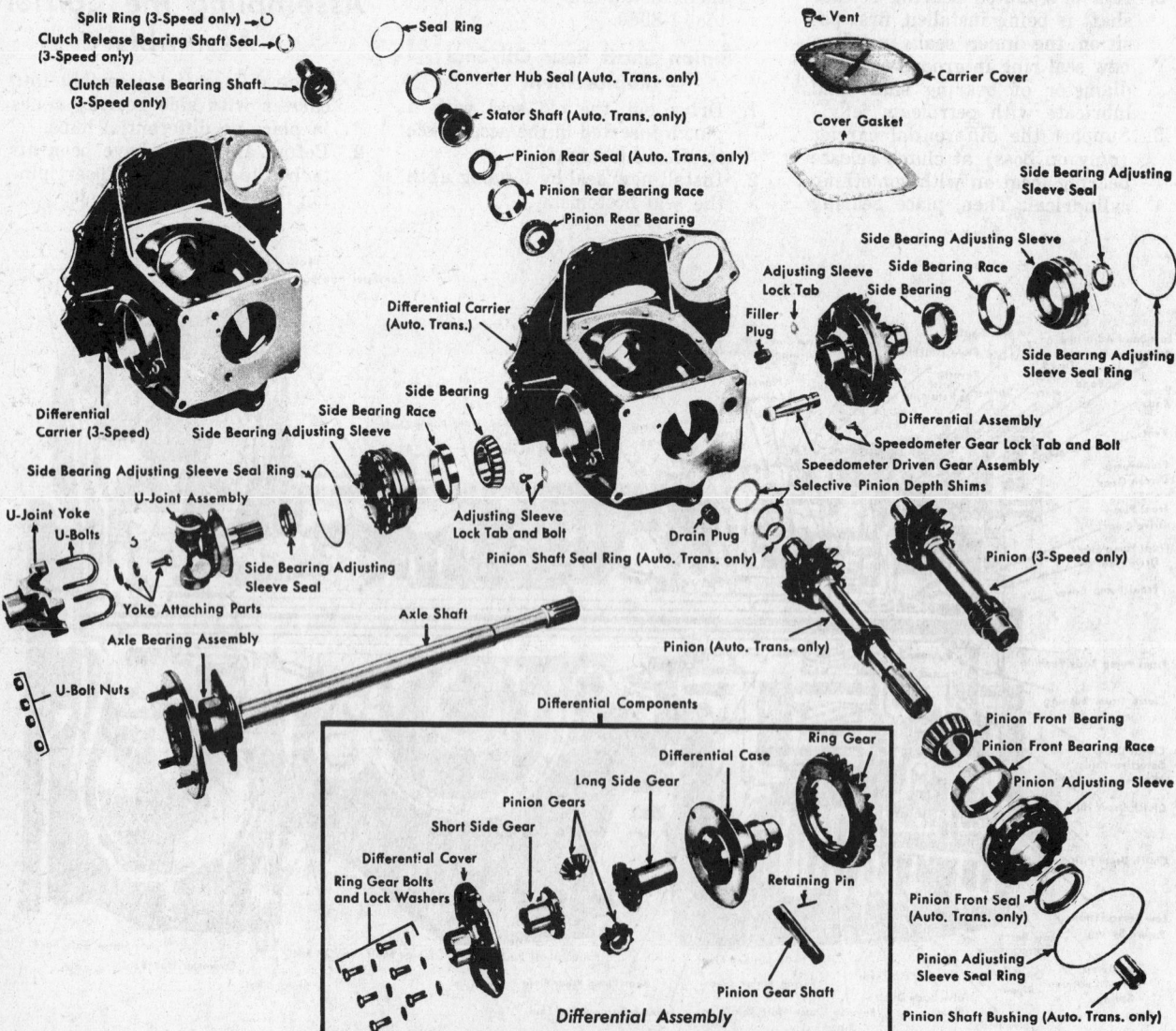

Differential carrier exploded view (© G.M. Corp.)

Micrometer Reading	Pinion Marking										
	10	11	12	13	14	15	16	17	18	19	20
1.250	6	6	9	9	9	12	12	12	15	15	15
1.251	6	9	9	9	12	12	12	15	15	15	18
1.252	9	9	9	12	12	12	15	15	15	18	18
1.253	9	9	12	12	12	15	15	15	18	18	18
1.254	9	12	12	12	15	15	15	18	18	18	15+6
1.255	12	12	12	15	15	15	18	18	18	15+6	15+6
1.256	12	12	15	15	15	18	18	18	15+6	15+6	15+6
1.257	12	15	15	15	18	18	18	15+6	15+6	15+6	18+6
1.258	15	15	15	18	18	18	15+6	15+6	15+6	18+6	18+6
1.259	15	15	18	18	18	15+6	15+6	15+6	18+6	18+6	18+6
1.260	15	18	18	18	15+6	15+6	15+6	18+6	18+6	18+6	18+9
1.261	18	18	18	15+6	15+6	15+6	18+6	18+6	18+6	18+9	18+9

Pinion bearing shim selection chart

preferred). This critical measurement should be read with a dial indicator.

To reduce backlash, the ring gear and differential must be moved toward the pinion. To increase backlash, the ring gear and differential must be moved away from the pinion. One adjustment notch is equivalent to about .003 in. backlash change.

If the backlash is zero, turn the differential side bearing adjustment (the side away from the pinion) counterclockwise two notches. Turn the side bearing adjustment, (the pinion side of the carrier) clockwise two notches.

By following this procedure, the preload is maintained and the lash should now be .006 in.

Positraction Differential

The Positraction differential is a multi-plate clutch unit incorporated into the right-hand side of the differential case. This unit affords better traction by allowing the wheel with the most traction to turn. A Belleville clutch plate and disc are compressed during assembly so as to provide a preload on the clutch pack. This preload, in addition to the normal load resulting from the differential side

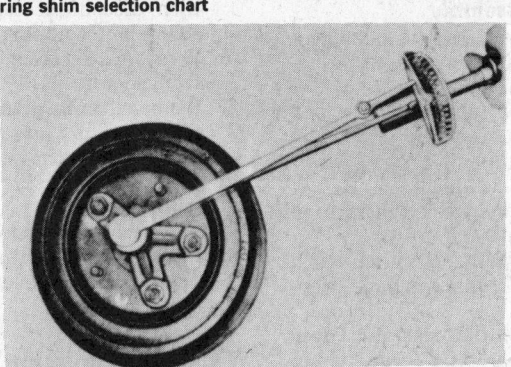

Checking preload of Positraction Belleville clutch
(© G.M. Corp.)

gear separating forces, provides the non-slip action of the differential. Repair procedures are identical with those for standard differential units, with the following exceptions:

Preload Check—Unit in Car

1. Jack up the rear of the car and place on axle stands.
2. Remove one wheel.
3. Back off brake adjustment to eliminate brake drag.
4. Attach an adapter for a torque wrench to the brake drum, as illustrated.
5. With transmission in Neutral, rotate axle shaft with a torque wrench and note reading; if torque is below 50 ft. lbs., remove differential assembly and examine clutch pack.

Disassembly

1. Remove differential side bearings.
2. If ring gear is to be replaced, remove it now.
3. Matchmark differential case and cover, then remove the two factory-installed 5/16-18 bolts (flat head) and separate case and cover.

NOTE: it is not necessary to replace these bolts.

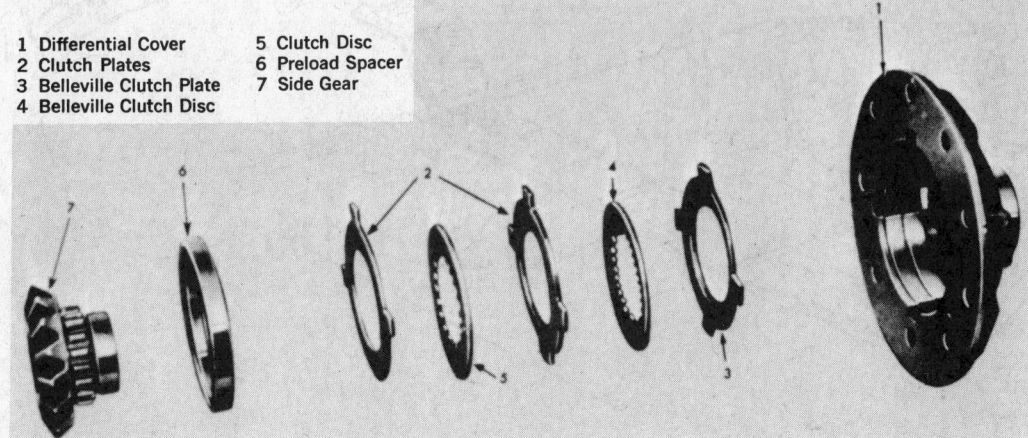

1 Differential Cover
2 Clutch Plates
3 Belleville Clutch Plate
4 Belleville Clutch Disc
5 Clutch Disc
6 Preload Spacer
7 Side Gear

Positraction differential components (© G.M. Corp.)

4. Remove the left-hand side gear, pinion shaft and pinion.
5. Remove right-hand side gear, clutch pack preload spacer, and clutch pack.
NOTE: clutch pack is replaced as a unit if defective.

Inspection

1. Clean and inspect clutch plates and discs, preload spacer and side gear thrust surfaces for excessive wear, scoring or cracks.
2. Clean and inspect preload spacer surface on differential case.
3. Using new parts where necessary, begin assembly.

Assembly

1. Oil clutch plates and side gear, then install in this order:
   Belleville clutch plate
   Belleville clutch disc
   Flat clutch plate
   Flat clutch disc
   Flat clutch plate
   Preload spacer (chamfered edge up)
2. Install side gear into clutch pack, making sure splines engage properly.
3. Align differential case and cover and assemble ring gear, using homemade guide pins. Remove pins and install bolts, tightening them to 40-60 ft. lbs.
4. Install differential side bearing assemblies.
5. Assemble differential into case in the same manner as standard differential.

Windshield Wipers

Motor R & R

1960-69

1. Remove the retainer holding the drive link to the motor drive arm.
NOTE: on 1965-69 cars, remove plenum chamber grill.
2. If equipped with a windshield washer, note the locations of washer hoses to wiper motor, then disconnect hoses from inside front compartment.
3. Remove electrical connections from motor.
4. Remove three attaching screws from motor assembly-to-body, then remove the motor.
5. Install by reversing removal procedure.
NOTE: when tightening the motor mounting screws, torque solidly to seat the spacing sleeves, preventing wiper motor float.

Wiper Transmission R & R

1960-69

1. Position wiper in park position.
NOTE: on 1965-69 cars, remove plenum chamber grill.
2. Remove a wiper arm and blade.
3. Remove retainer holding drive link to wiper motor drive arm.

4. Remove retainers holding the ends of the link to each wiper transmission and remove the link.
5. Remove the three mounting screws at each transmission, then remove each transmission.
6. Install by reversing removal procedure. The tips of the wiper blades should be 1½ in. above the lower windshield opening with wiper motor in park position.

Heater System

Blower R & R

1. From beneath vehicle, remove large diameter heater air hose from right side of air inlet assembly.
2. Remove blower motor wire at connector.
3. Remove blower mounting plate screws attaching blower to housing.
4. Work motor and mounting plate down and out of housing.
5. Install in reverse of above.

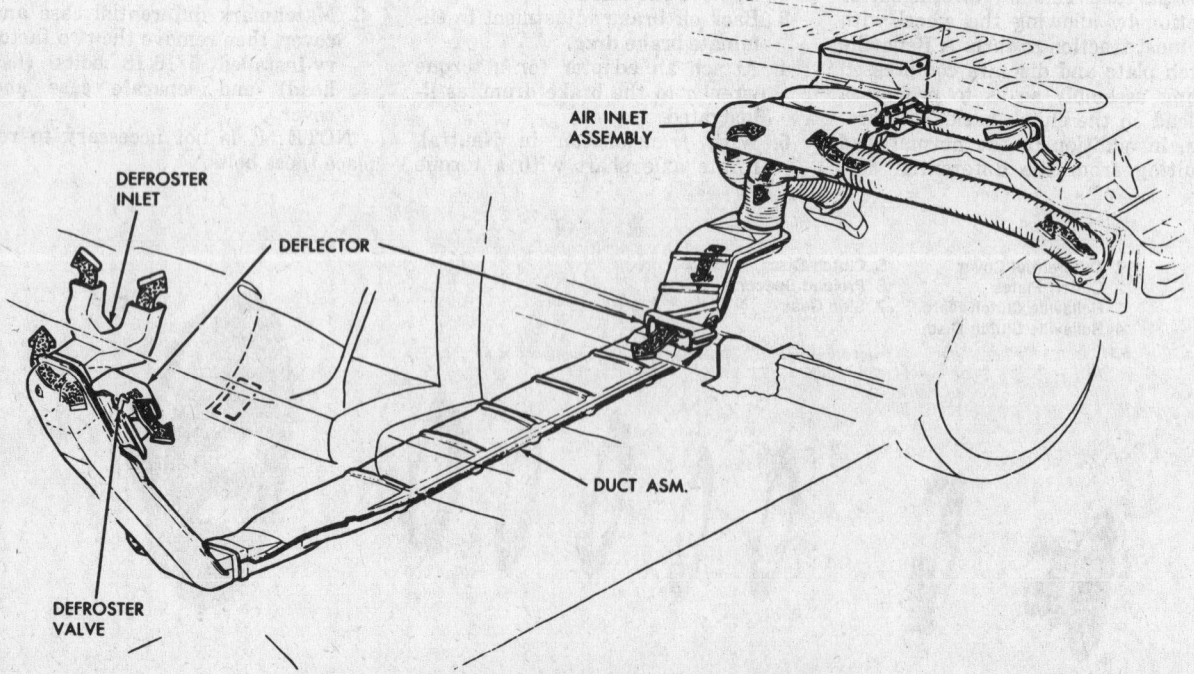

Heater ducting and blower location—typical 1960-69 © G.M. Corp.)

YEAR IDENTIFICATION
DODGE

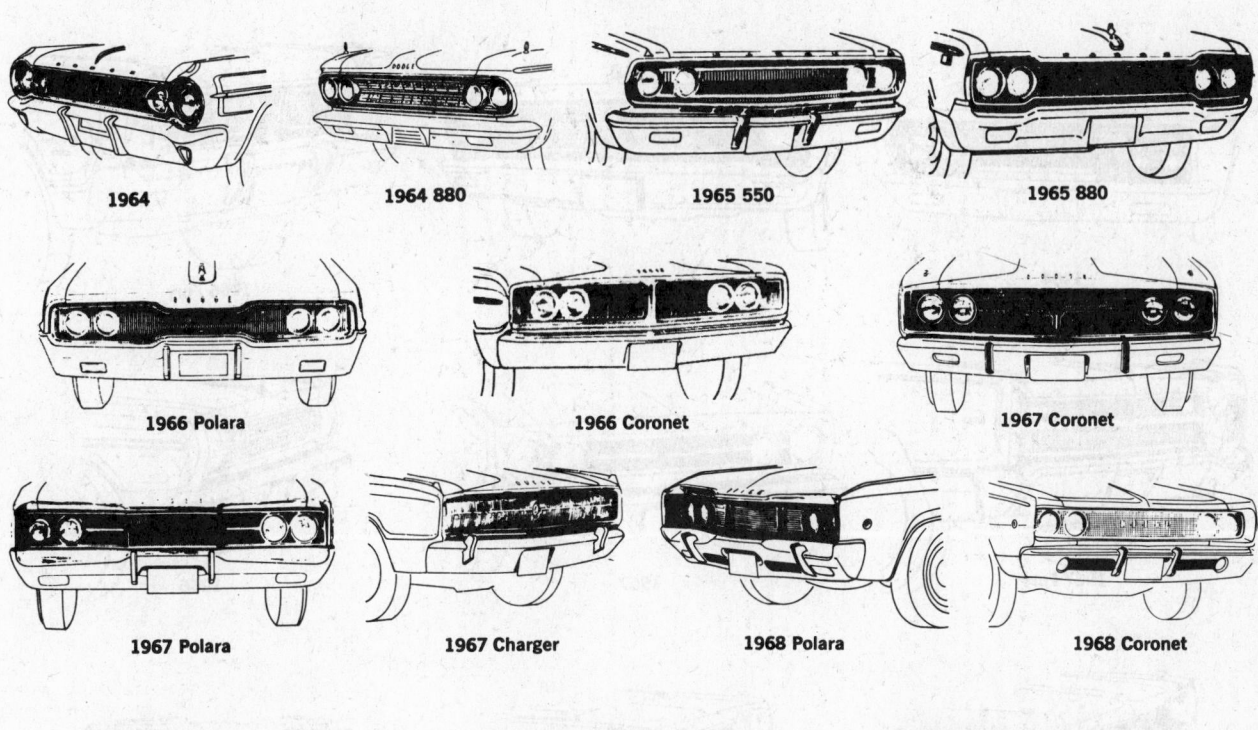

1964

1964 880

1965 550

1965 880

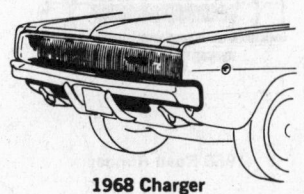
1966 Polara

1966 Coronet

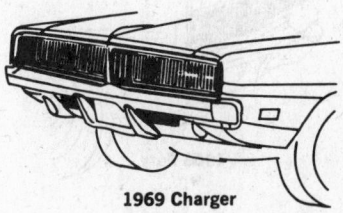
1967 Coronet

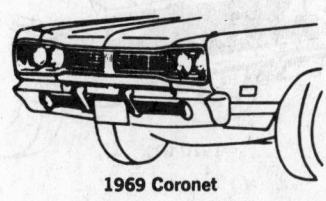
1967 Polara

1967 Charger

1968 Polara

1968 Coronet

1968 Charger

1969 Polara

1969 Charger

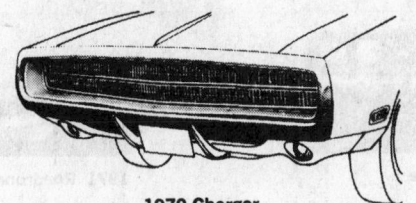
1969 Coronet

1970 Coronet 500

1970 Polara

1970 Charger

1970 Coronet 440

1971 Coronet

1971 Charger

1971 Polara

YEAR IDENTIFICATION

PLYMOUTH

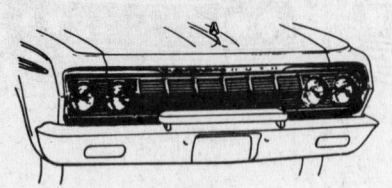

1964

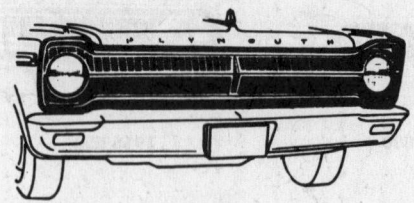

1965

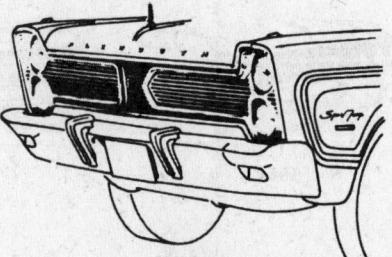

1966 Fury

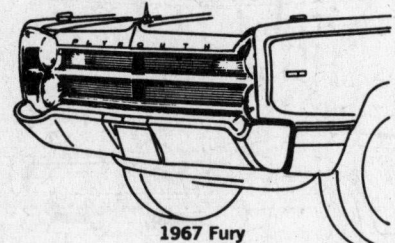

1967 Fury

1967

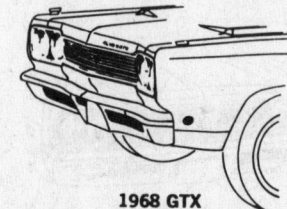

1966

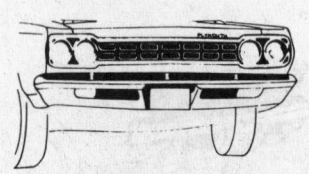

1968 Fury

1968 GTX

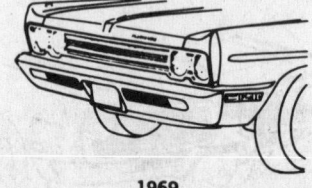

1968 Road Runner

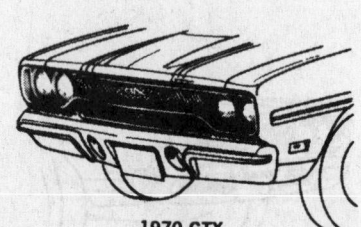

1969

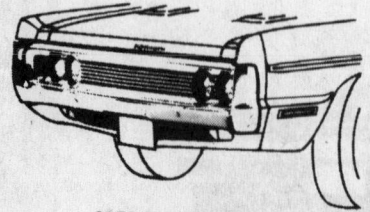

1969

1970 GTX

1970 Plymouth Fury

1971 Satellite

1971 Roadrunner

1971 Fury

FIRING ORDER

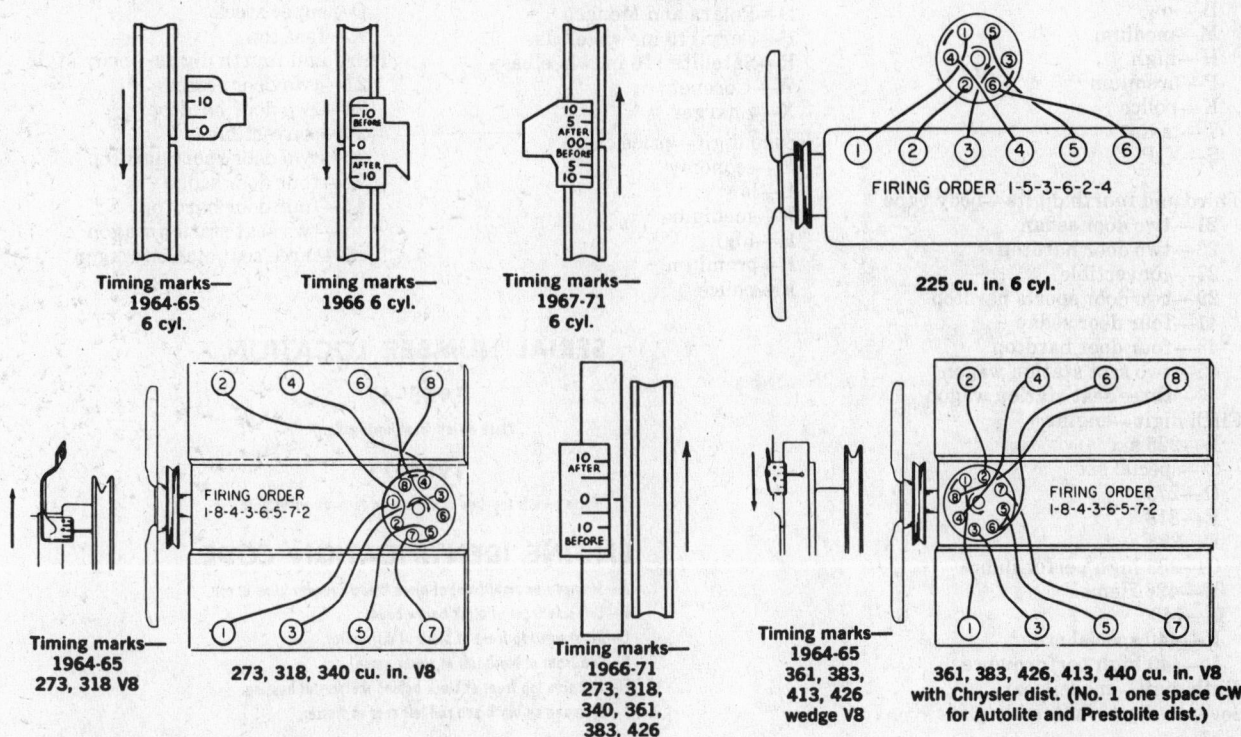

Timing marks—1964-65 6 cyl.

Timing marks—1966 6 cyl.

Timing marks—1967-71 6 cyl.

225 cu. in. 6 cyl.

FIRING ORDER 1-5-3-6-2-4

Timing marks—1964-65 273, 318 V8

273, 318, 340 cu. in. V8

FIRING ORDER 1-8-4-3-6-5-7-2

Timing marks—1966-71 273, 318, 340, 361, 383, 426 Hemi, 440 V8

Timing marks—1964-65 361, 383, 413, 426 wedge V8

361, 383, 426, 413, 440 cu. in. V8 with Chrysler dist. (No. 1 one space CW for Autolite and Prestolite dist.)

FIRING ORDER 1-8-4-3-6-5-7-2

CAR SERIAL NUMBER LOCATION AND ENGINE IDENTIFICATION

Car Serial Number Location

1964-67

Plate on left front door hinge post.

1968-71

Top of instrument panel, visible through windshield.

Model Identification

1964

Serial numbers contain ten digits as follows:

First digit—car line
2—Plymouth six
3—Plymouth eight
4—Dodge six
5—880 and 880 Custom
6—Dodge eight

Second digit—price class
1—Savoy, low priced 880, low priced Dodge
2—Belvedere, high priced 880, medium priced Dodge
3—Fury, high priced Dodge
4—Sport Fury, premium Dodge
5—Savoy station wagon, 880 station wagon, low priced Suburban
6—880 Custom station wagon, medium priced Suburban
7—Fury station wagon, high priced suburban

8—Taxi
9—special
Third digit—model year
Fourth digit—assembly plant

1965

First digit—car line
D—Polara, 880, and Monaco
P—Fury
R—Belvedere
W—Coronet eight cyl.
4—Coronet six cyl.

Second digit—price class
1—Polara, Belvedere I, Fury I, and low priced Coronet
2—Fury II
3—880, Belvedere II, Fury III, and high priced Coronet
4—Monaco, Satellite, Sport Fury, and premium Coronet
5—Polara station wagon, Belvedere I station wagon, Fury I station wagon, and low priced Coronet station wagon
6—Fury II station wagon
7—880 station wagon, Belvedere II station wagon, Fury III station wagon and high priced Coronet station wagon.
8—taxi
9—police
Third digit—model year
Fourth digit—assembly plant

1966

First digit—car line
D—Polara and Monaco
P—Fury

R—Belvedere
W—Coronet
Second digit—price class
E—economy
L—low
H—high
P—premium
K—police
T—taxi
S—VIP
M—medium
Third and fourth digits—body type
21—two door sedan
23—two door hardtop
27—convertible
29—two door sports hardtop
41—four door sedan
42—four door hardtop
43—four door hardtop
45—station wagon
46—station wagon
Fifth digit—engine
B—225 six
C—273
D—318
E—361
F—383
H—426
Sixth digit—model year
Seventh digit—assembly plant

1967

First digit—car line
D—Polara and Monaco
P—119 in. wheelbase Plymouth
R—116 in. wheelbase Plymouth
W—Coronet
X—Charger

Second digit
  E—economy
  L—low
  M—medium
  H—high
  P—premium
  K—police
  T—taxi
  S—VIP

Third and fourth digits—body type
  21—two door sedan
  23—two door hardtop
  27—convertible
  29—two door sports hardtop
  41—four door sedan
  43—four door hardtop
  45—two seat station wagon
  46—three seat station wagon

Fifth digit—engine
  B—225 six
  C—special six
  D—273
  E—318
  F—383
  G—383 high performance
  H—426 Hemi
  J—440
  K—440 special order
  L—440 high performance
Sixth digit—model year
Seventh digit—assembly plant

1968
First digit—car line
  D—Polara and Monaco
  P—119 in. wheelbase Plymouth
  R—116 in. Plymouth
  W—Coronet
  X—Charger
Second digit—price class
  E—economy
  L—low
  H—high
  M—medium
  P—premium
  K—police
  T—taxi
  S—special
  O—super stock
Third and fourth digits—body type
  21—two door sedan
  23—two door hardtop
  27—convertible
  29—two door sports hardtop
  41—four door sedan
  43—four door hardtop
  45—two seat station wagon
  46—three seat station wagon
Fifth digit—engine
  B—225 six
  C—special six
  D—273
  F—318
  G—383
  H—383 high performance
  J—426 Hemi
  K—440
  L—440 high performance
  M—special V8
Sixth digit—model year
Seventh digit—assembly plant

1969
First digit—car line
  D—Polara and Monaco
  P—Fury 119 in. wheelbase
  R—Satellite 116 in. wheelbase
  W—Coronet
  X—Charger
Second digit—price class
  E—economy
  L—low
  M—medium
  H—high
  P—premium
  K—police
  T—taxi
  S—special
  O—super stock
  X—fast top
Third and fourth digits—body style
  21—two door sedan
  23—two door hardtop
  27—convertible
  29—two door sport hardtop
  41—four door sedan
  43—four door hardtop
  45—two seat station wagon
  46—three seat station wagon

SERIAL NUMBER LOCATION

1963–67
Plate on left front hinge pillar.

1968–71
Plate on left top side of dash seen through windshield.

ENGINE IDENTIFICATION CODE
A—Stamped on right front of block below cylinder head at coil.
B—Left side front of block below head.
C—Right side top front of block at distributor.
D—Top front of block left of water pump.
E—Left side top front of block behind thermostat housing.
F—Stamped on block pan rail left rear at starter.

YEAR	ENGINE	CODE	LOCATION
1964	6 Cyl.—225	V22	A
	8 Cyl.—318	V318	B
	8 Cyl.—361	V36	C
	8 Cyl.—383	V38	D
	8 Cyl.—413 (police)	V41	D
	8 Cyl.—426 Wedge	TMP426	E
1965	6 Cyl.—225	A225	A
	8 Cyl.—273	A273	B
	8 Cyl.—318	A318	B
	8 Cyl.—361	A361	C
	8 Cyl.—383	A383	C
	8 Cyl.—426 Wedge	A426	E
1966	6 Cyl.—225	B225	A
	8 Cyl.—273	B273	B
	8 Cyl.—318	B318	B
	8 Cyl.—383	B383	C
	8 Cyl.—426 Hemi	BH426	D
	8 Cyl.—440	B440	E
1967	6 Cyl.—225	C225	A
	8 Cyl.—273	C273	B
	8 Cyl.—318	C318	B
	8 Cyl.—383	C383	C
	8 Cyl.—426 Hemi	CH426	D
	8 Cyl.—440	C440	E
1968-70	6 Cyl.—225	225	A
	8 Cyl.—273 (68–69)	273	B
	8 Cyl.—318	318	B
	8 Cyl.—383	383	F
	8 Cyl.—383 Hi. Perf.	383	F
	8 Cyl.—426 Hemi	426	F
	8 Cyl.—440	440	F
	8 Cyl.—440 Hi. Perf.	440	F
1971	6 Cyl.—225	225	A
	8 Cyl.—318	318	B
	8 Cyl.—383	383	F
	8 Cyl.—383 Hi. Perf.	383	F
	8 Cyl.—426 Hemi	426	F
	8 Cyl.—440	440	F
	8 Cyl.—440 Hi. Perf.	440	F

Fifth digit—engine
  B—225 six
  C—special six
  D—273
  F—318
  G—383
  H—383 high performance
  J—426 Hemi
  K—440
  L—440 high performance
  M—special V8
Sixth digit—model year
Seventh digit—assembly plant

1970
First digit—car line
  D—Polara and Monaco
  P—Fury
  R—Satellite

O—super stock
Third and fourth digits—body style
  21—two door sedan
  23—two door hardtop
  27—convertible
  29—Charger
  41—four door sedan
  43—four door hardtop
  W—Coronet
  X—Charger
Second digit—price class
  L—low
  M—medium
  H—high
  P—premium
  K—police
  N—N.Y. taxi
  T—taxi
  S—special

45—six passenger station wagon
46—nine passenger station wagon
Fifth digit—engine
  C—225 six
  E—special six
  G—318
  L—383
  N—383 high performance
  R—426 Hemi
  T—440
  U—440 high performance
  V—440 six pack
  Z—special order V8
Sixth digit—model year
Seventh digit—assembly plant

CRANKSHAFT BEARING JOURNAL SPECIFICATIONS

YEAR	MODEL	MAIN BEARING JOURNALS (IN.)				CONNECTING ROD BEARING JOURNALS (IN.)		
		Journal Diameter	Oil Clearance	Shaft End-Play	Thrust On No.	Journal Diameter	Oil Clearance	End-Play
1964-65	6 Cyl.—225 Cu. In.	2.7495-2.7505	.0005-.0015	.002-.007	3	2.1865-2.8175	.0005-.0015	.006-.012
	V8—273, 318 Cu. In.	2.4495-2.5005	.0005-.0015	.002-.007	3	2.1240-2.1250	.0005-.0015	.009-.017
	V8—361, 383 Cu. In.	2.6245-2.6255	.0005-.0015	.002-.007	3	2.3740-2.3750	.0005-.0015	.009-.017
	V8—413, 426 Cu. In.	2.7495-2.7505	.0005-.0015	.002-.007	3	2.3740-2.3750	.001-.002	.006-.012
1966-71	6 Cyl.—225 Cu. In.	2.7495-2.7505	.0005-.0015	.002-.007	3	2.1865-2.1875	.0005-.0015	.006-.012
	V8—318 Cu. In.	2.4995-2.5005	.0005-.0015	.002-.007	3	2.1240-2.1250	.0005-.0015	.009-.017
	V8—383 Cu. In.	2.6245-2.6255	.0005-.0015	.002-.007	3	2.3740-2.3750	.0005-.0015	.009-.017
	V8—426 Cu. In. Hemi	2.7495-2.7505	.0015-.0030	.002-.007	3	2.3740-2.3750	.0015-.0025	.013-.017
	V8—440 Cu. In.	2.7495-2.7505	.0005-.0020	.002-.007	3	2.3740-2.3750	.0010-.0020	.009-.019
	V8—440 Cu. In. Six Pack	2.7495-2.7505	.0015-.0025	.002-.007	3	2.3740-2.3750	.0010-.0020	.009-.019

TORQUE SPECIFICATIONS

YEAR	MODEL	CYLINDER HEAD BOLTS (FT. LBS.)	ROD BEARING BOLTS (FT. LBS.)	MAIN BEARING BOLTS (FT. LBS.)	CRANKSHAFT BALANCER BOLT (FT. LBS.)	FLYWHEEL TO CRANKSHAFT BOLTS (FT. LBS.)	MANIFOLD (FT. LBS.)	
							Intake	Exhaust
1964-71	Slant 6	65	45	85	Press	55	10●	10
1964-71	273, 318 V8	85	45	85	▲	55-65	40	25
1964-71	361, 383, 413, 426 Wedge, 440 V8	70	45	85	▲	55	40-50	30
1966-71	426 Hemi	70-75	75	100●●	135	70	See Text	35

● —Intake to exhaust—17-20 ft. lbs.
●● —Crossbolts—45 ft. lbs.
▲ —Balancer bolts—15 ft. lbs.
  End of crankshaft bolt—135 ft. lbs.

CYLINDER HEAD BOLT TIGHTENING SEQUENCE

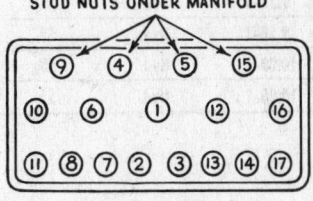

426 Hemi V8

361, 383, 413, 426 wedge, 440 V8

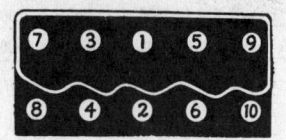

273, 318, 340 V8

225 6 cyl.

GENERAL ENGINE SPECIFICATIONS

YEAR	CU. IN. DISPLACEMENT	CARBURETOR	DEVELOPED HORSEPOWER @ RPM	DEVELOPED TORQUE @ RPM (FT. LBS.)	A.M.A. HORSEPOWER	BORE & STROKE (IN.)	ADVERTIZED COMPRESSION RATIO	VALVE LIFTER TYPE	NORMAL OIL PRESSURE (PSI)
1964	6 Cyl. —225	1-BBL.	145 @ 4000	215 @ 2400	27.7	3.400 x 4.125	8.40–1	Mech.	55
	V8—318	2-BBL.	230 @ 4400	340 @ 2400	48.9	3.910 x 3.310	9.00–1	Mech.	55
	V8—361	2-BBL.	265 @ 4400	380 @ 2400	54.0	4.125 x 3.375	9.00–1	Hyd.	55
	V8—383	2-BBL.	305 @ 4600	410 @ 2400	57.8	4.250 x 3.375	10.00–1	Hyd.	55
	V8—383	4-BBL.	330 @ 4600	425 @ 2800	57.8	4.250 x 3.375	10.00–1	Hyd.	55
	V8—413	4-BBL.	360 @ 4800	470 @ 3200	57.1	4.188 x 3.750	10.00–1	Mech.	55
	V8—426 Wedge	4-BBL.	375 @ 4600	465 @ 2800	57.8	4.250 x 3.750	10.50–1	Mech.	55
	V8—426 Wedge	2-4-BBL.	425 @ 6000	480 @ 4600	57.8	4.250 x 3.750	12.50–1	Mech.	55
1965	6 Cyl. —225	1-BBL.	145 @ 4000	215 @ 2400	27.7	3.400 x 4.125	8.40–1	Mech.	55
	V8—273	2-BBL.	180 @ 4200	260 @ 1600	42.0	3.625 x 3.310	8.80–1	Mech.	55
	V8—273	4-BBL.	235 @ 5200	280 @ 4000	42.0	3.625 x 3.310	10.50–1	Mech.	55
	V8—318	2-BBL.	230 @ 4400	340 @ 2400	48.9	3.910 x 3.310	9.00–1	Mech.	55
	V8—361	2-BBL.	265 @ 4400	380 @ 2400	54.0	4.125 x 3.375	9.00–1	Hyd.	55
	V8—383	2-BBL.	270 @ 4400	390 @ 2800	57.8	4.250 x 3.375	9.20–1	Hyd.	55
	V8—383	4-BBL.	330 @ 4600	425 @ 2800	57.8	4.250 x 3.375	10.00–1	Hyd.	55
	V8—413	4-BBL.	340 @ 4600	470 @ 2800	57.1	4.188 x 3.750	10.10–1	Hyd.	55
	V8—413	4-BBL.	360 @ 4800	470 @ 3200	57.1	4.188 x 3.750	10.30–1	Mech.	55
	V8—426 Wedge	2-4-BBL.	365 @ 4800	470 @ 3200	57.8	4.250 x 3.750	10.30–1	Hyd.	55
	V8—426 Wedge	2-4-BBL.	425 @ 6000	480 @ 4600	57.8	4.250 x 3.750	12.50–1	Mech.	55
1966	6 Cyl. —225	1-BBL.	145 @ 4000	215 @ 2400	27.7	3.400 x 4.125	8.40–1	Mech.	55
	V8—273	2-BBL.	180 @ 4200	260 @ 1600	42.0	3.625 x 3.310	9.00–1	Mech.	55
	V8—273	4-BBL.	235 @ 5200	280 @ 4000	42.0	3.625 x 3.310	10.50–1	Mech.	55
	V8—318	2-BBL.	230 @ 4400	340 @ 2400	48.9	3.910 x 3.310	9.00–1	Mech.	55
	V8—361	2-BBL.	265 @ 4400	380 @ 2400	54.0	4.125 x 3.375	9.00–1	Hyd.	55
	V8—383	2-BBL.	270 @ 4400	390 @ 2800	57.8	4.250 x 3.375	9.20–1	Hyd.	55
	V8—383	4-BBL.	325 @ 4800	425 @ 2800	57.8	4.250 x 3.375	10.0–1	Hyd.	55
	V8—426 Hemi	4-BBL.	425 @ 5000	490 @ 4000	57.8	4.250 x 3.750	10.25–1	Hyd.	55
	V8—440	4-BBL.	350 @ 4400	480 @ 2800	59.7	4.320 x 3.750	10.0–1	Hyd.	55
1967	6 Cyl. —225	1-BBL.	145 @ 4000	215 @ 2400	27.7	3.400 x 4.125	8.40–1	Mech.	55
	V8—273	2-BBL.	180 @ 4200	260 @ 1600	42.0	3.625 x 3.310	9.20–1	Mech.	55
	V8—318	2-BBL.	230 @ 4400	340 @ 2400	48.9	3.910 x 3.310	9.20–1	Hyd.	55
	V8—383	2-BBL.	270 @ 4400	390 @ 2800	57.8	4.250 x 3.375	9.20–1	Hyd.	55
	V8—383	4-BBL.	325 @ 4800	425 @ 2800	57.8	4.250 x 3.375	10.00–1	Hyd.	55
	V8—426 Hemi	2-4-BBL.	425 @ 5000	490 @ 4000	57.8	4.250 x 3.750	10.25–1	Mech.	55
	V8—440	4-BBL.	350 @ 4400	480 @ 2800	59.7	4.320 x 3.750	10.00–1	Hyd.	55
	V8—440	4-BBL.	375 @ 4600	480 @ 3200	59.7	4.320 x 3.750	10.00–1	Hyd.	55
1968	6 Cyl. —225	1-BBL.	145 @ 4000	215 @ 2400	27.7	3.400 x 4.125	8.40–1	Mech.	55
	V8—273	2-BBL.	190 @ 4400	260 @ 2000	42.0	3.625 x 3.310	9.00–1	Hyd.	55
	V8—318	2-BBL.	230 @ 4400	340 @ 2400	48.9	3.910 x 3.310	9.20–1	Hyd.	55
	V8—383	2-BBL.	290 @ 4400	390 @ 2800	57.8	4.250 x 3.375	9.20–1	Hyd.	55
	V8—383	4-BBL.	330 @ 5000	425 @ 3200	57.8	4.250 x 3.375	10.00–1	Hyd.	55
	V8—426 Hemi	2-4-BBL.	425 @ 5000	490 @ 5000	57.8	4.250 x 3.750	10.25–1	Mech.	55
	V8—440	4-BBL.	350 @ 4400	480 @ 2800	59.7	4.320 x 3.750	10.10–1	Hyd.	55
	V8—440	4-BBL.	375 @ 4600	480 @ 3200	59.7	4.320 x 3.750	10.10–1	Hyd.	55
1969	6 Cyl. —225	1-BBL.	145 @ 4000	215 @ 2400	27.7	3.400 x 4.125	8.40–1	Mech.	55
	V8—318	2-BBL.	230 @ 4400	340 @ 2400	48.9	3.910 x 3.310	9.20–1	Hyd.	55
	V8—383	2-BBL.	290 @ 4400	390 @ 2800	57.8	4.250 x 3.375	9.20–1	Hyd.	55
	V8—383	4-BBL.	330 @ 5000	425 @ 3200	57.8	4.250 x 3.375	10.00–1	Hyd.	55
	V8—383	4-BBL.	335 @ 5200	425 @ 3400	57.8	4.250 x 3.375	10.00–1	Hyd.	55

GENERAL ENGINE SPECIFICATIONS, continued

YEAR	CU. IN. DISPLACEMENT	CARBURETOR	DEVELOPED HORSEPOWER @ RPM	DEVELOPED TORQUE @ RPM (FT. LBS.)	A.M.A. HORSEPOWER	BORE & STROKE (IN.)	ADVERTIZED COMPRESSION RATIO	VALVE LIFTER TYPE	NORMAL OIL PRESSURE (PSI)
	V8—426 Hemi	2-4-BBL.	425 @ 5000	490 @ 4000	57.8	4.250 x 3.750	10.25-1	Mech.	55
	V8—440	4-BBL.	350 @ 4400	480 @ 2800	59.7	4.320 x 3.750	10.10-1	Hyd.	55
	V8—440	4-BBL.	375 @ 4600	480 @ 3200	59.7	4.320 x 3.750	10.10-1	Hyd.	55
1970	6 Cyl.—225	1-BBL.	145 @ 4000	215 @ 2400	27.7	3.400 x 4.125	8.40-1	Mech.	45-60
	V8—318	2-BBL.	230 @ 4400	320 @ 2000	48.9	3.910 x 3.310	8.8-1	Hyd.	45-65
	V8—383	2-BBL.	290 @ 4400	390 @ 2800	57.8	4.250 x 3.375	8.7-1	Hyd.	45-65
	V8—383	4-BBL.	330 @ 5000	425 @ 3200	57.8	4.250 x 3.375	9.5-1	Hyd.	45-65
	V8—383 Hi Perf.	4-BBL.	335 @ 5200	425 @ 3400	57.8	4.250 x 3.375	9.5-1	Hyd.	45-65
	V8—426 Hemi	2-4-BBL.	425 @ 5000	490 @ 4000	57.8	4.250 x 3.750	10.25-1	Hyd.	45-65
	V8—440	4-BBL.	350 @ 4400	480 @ 2800	59.7	4.320 x 3.750	9.7-1	Hyd.	45-65
	V8—440	3-2-BBL.	390 @ 4700	490 @ 3200	59.7	4.320 x 3.750	10.5-1	Hyd.	45-65
1971	6 Cyl.—225	1-BBL.	145 @ 4000	215 @ 2400	27.7	3.400 x 4.125	8.4-1	Mech.	45-60
	V8—318	2-BBL.	N.A.	N.A.	48.9	3.910 x 3.310	8.8-1	Hyd.	45-65
	V8—383	2-BBL.	N.A.	N.A.	57.8	4.250 x 3.375	8.7-1	Hyd.	45-65
	V8—383 Hi. Perf.	4-BBL.	N.A.	N.A.	57.8	4.250 x 3.375	9.5-1	Hyd.	45-65
	V8—426 Hemi	2-4-BBL.	N.A.	N.A.	57.8	4.250 x 3.750	N.A.	Hyd.	45-65
	V8—440	4-BBL.	N.A.	N.A.	59.7	4.320 x 3.750	9.7-1	Hyd.	45-65
	V8—440 Hi. Perf.	4-BBL.	N.A.	N.A.	59.7	4.320 x 3.750	N.A.	Hyd.	45-65
	V8—440	3-2-BBL.	N.A.	N.A.	59.7	4.320 x 3.750	N.A.	Hyd.	45-65

ALTERNATOR AND AC REGULATOR SPECIFICATIONS

YEAR		ALTERNATOR			REGULATOR			
	Model	Field Current Draw @ 12 V.	Current Output @ 1250 Engine RPM	Model	Point Gap (In.)	Air Gap (In.)	Voltage at 140°F.
1964-69	6 Cyl.—225	2.38-2.75	26 ± 3 Amps.	2098300①	.015	.050	13.4-14.0
	V8 Std.—All	2.38-2.75	34.5 ± 3 Amps.				
	Heavy Duty, A/C	2.38-2.75	44 ± 3 Amps.	2444980②	.015	.032-.042	13.2-14.2
	Special Equip.	2.38-2.75	51 ± 3 Amps.				
1970-71	6 Cyl.—225	2.38-2.75	26 ± 3 Amps.	3438150③	.015	.050	13.3-14.0
	V8 Std.—All	2.38-2.75	34.5 ± 3 Amps.				
	Heavy Duty, A/C	2.38-2.75	44.5 ± 3 Amps. ④				

① Chrysler built—used interchangeably with 2444980.
② Essex wire built—used interchaneably with 2098300.
③ Electronic type.
④ 51 ± 3 Amps.—Special equipment.

BATTERY AND STARTER SPECIFICATIONS

YEAR	MODEL	BATTERY				STARTERS							Brush Spring Tension (oz.)
		Ampere Hour Capacity	Volts	Terminal Grounded	Model Number	Lock Test			No-Load Test				
						Amps.	Volts	Torque	Amps.	Volts	RPM		
1964-69	225, 318 std., 273 std.,	48	12	Neg.	2095753①	350	4.0	—	78	11.0	3,800		32-48
	361, 383 std.,	59	12	Neg.	2095150②	400-450	4.0	—	90	11.0	1,925-2,400		32-48
	426, 440 std., opt. all others	70	12	Neg.	1889100①③	350	4.0	8.5	78	11.0	3,800		32-36
	426 Hemi	70	12	Neg.	2642930④①	310-445	4.0	—	78	11.0	3,800		32-36
1970-71	225, 318 std.,	46	12	Neg.	2875560⑦⑤	400-450	4.0	—	90	11.0	1,925-2,600		32-36
	383 std.	59	12	Neg.									
	426, 440 std., opt. all others	70	12	Neg.									

① Direct drive—1964.
② Gear reduction—All engines except 1964-69, 225 Taxi six and Hemi.
③ Taxi 225 cu. in. with 11 in. clutch.
④ 426 Hemi.
⑤ All engines—1970-71.

TUNE-UP SPECIFICATIONS

| YEAR | MODEL AND CU. IN. DISPLACEMENT | SPARK PLUGS | | DISTRIBUTOR | | IGNITION TIMING (Deg.) ▲ | CRANKING COMP. PRESSURE (Psi) | VALVES | | | FUEL PUMP PRESSURE (Psi) | IDLE SPEED (Rpm) ★ |
| | | Type | Gap (In.) | Point Dwell (Deg.) | Point Gap (In.) | | | Tappet Clearance (In.) | | Intake Opens (Deg.) | | |
								Intake	Exhaust			
1964	6 Cyl.—225	N14Y	.035	42	.020	2½B	125	.010H	.020H	8B	4½	550
	V8—273 M.T.	N14Y	.035	30	.017	5B	135	.013H	.021H	14B	6½	500
	V8—273 A.T.	N14Y	.035	30	.017	10B	135	.013H	.021H	14B	6½	500
	V8—318 M.T.	J12Y	.035	30	.017	5B	135	.013H	.021H	19B	6	550
	V8—318 S.T.	J12Y	.035	30	.017	10B	135	.013H	.021H	19B	6	475
	V8—361	J12Y	.035	30	.017	10B	140	Zero	Zero	13B	4½	500
	V8—383 2-BBL.	J12Y	.035	30	.017	10B	145	Zero	Zero	13B	4½	500
	V8—383 4-BBL.	J10Y	.035	30③	.017	10B	145	Zero	Zero	24B	4½	500
	V8—413	J10Y	.035	30	.017	10B	145	.028C	.032C	24B	7	500
	V8—426	N14Y	.035	30③	.017	10B	145	.028C	.032C	24B	7	500
	V8—426	N61Y	.020	36③	.012	31B⑤	145	.028C	.032C	36B	7	1400
1965	6 Cyl.—225	N14Y	.035	42	.020	2½B	125	.010H	.020H	10B	4½	550
	V8—273 M.T.	N14Y	.035	30	.017	5B	135	.013H	.021H	14B	6½	500
	V8—273 A.T.	N14Y	.035	30	.017	10B	135	.013H	.021H	14B	6½	500
	V8—273 4-BBL.	J10Y	.035	30③	.017	10B	135	.013H	.021H	14B	6½	600
	V8—318 M.T.	J14Y	.035	30	.017	5B	140	.013H	.021H	19B	6	550
	V8—318 A.T.	J14Y	.035	30	.017	10B	140	.013H	.021H	19B	6	475
	V8—361	J14Y	.035	30	.017	10B	140	Zero	Zero	13B	4½	500
	V8—383 2-BBL.	J14Y	.035	30	.017	10B	140	Zero	Zero	13B	4½	500
	V8—383 4-BBL.	J10Y	.035	30③	.017	10B	145	Zero	Zero	24B	4½	550
	V8—413 Std.	J14Y	.035	30	.017	12½B	145	Zero	Zero	14B	4½	500
	V8—413 Spec. Cam	J10Y	.035	30③	.017	10B	155	Zero	Zero	14B	4½	500
	V8—426	J10Y	.035	30③	.017	10B	145	.028C	.032C	24B	4½	500
1966	6 Cyl.—225	N14Y	.035	42½	.020	2½B⑥	125	.010H	.020H	10B	4½	550②
	V8—273 M.T. 2-BBL.	N14Y	.035	30	.017	5B⑥	135	.013H	.021H	14B	6	500①
	V8—273 A.T. 2-BBL.	N14Y	.035	30	.017	10B⑥	135	.013H	.021H	14B	6	500①
	V8—273 4-BBL.	N10Y	.035	29③	.017	10B⑥	150	.013H	.021H	14B	6	500①
	V8—318 M.T.	J14Y	.035	30	.017	5B⑥	140	.013H	.021H	19B	6	500②
	V8—318 A.T.	J14Y	.035	30	.017	10B⑥	140	.013H	.021H	19B	6	500②
	V8—361, 383 2-BBL.	J14Y	.035	30	.017	12½B⑥	140	Zero	Zero	13B	4½	500②
	V8—383 4-BBL.	J13Y	.035	30	.017	12½B⑥	150	Zero	Zero	18B	4½	500②
	V8—426 Hemi	J13Y	.035	30	.017	12½B⑥	150	.028C	.032C	30B	4½	500②
	V8—440 4-BBL.	J13Y	.035	30	.017	12½B⑥	150	Zero	Zero	18B	4½	500②
1967	6 Cyl.—225	N14Y	.035	42½	.020	5B	125	.010H	.020H	10B	4½	550
	6 Cyl.—225 Exh. Em.	N14Y	.035	42½	.020	TDC	135	.013H	.021H	14B	6	650
	V8—273 M.T. 2-BBL.	N14Y	.035	30	.017	5B⑥	135	.013H	.021H	14B	6	500①
	V8—273 A.T. 2-BBL.	N14Y	.035	30	.017	10B⑥	135	.013H	.021H	14B	6	650
	V8—318 M.T. 2-BBL.	N14Y	.035	30	.017	5B⑥	140	Zero	Zero	14B	6	500①
	V8—318 A.T. 2-BBL.	N14Y	.035	30	.017	10B⑥	140	Zero	Zero	14B	6	500①
	V8—383 M.T. 2-BBL.	J14Y	.035	30	.017	12½B	140	Zero	Zero	16B	4½	550
	V8—383 M.T. Exh. Em. 2-BBL.	J14Y	.035	30	.017	TDC	140	Zero	Zero	16B	4½	650
	V8—383 A.T. 4-BBL.	J13Y	.035	30	.017	12½B	150	Zero	Zero	16B	4½	550
	V8—383 A.T. Exh. Em. 4-BBL.	J13Y	.035	30	.017	5B	150	Zero	Zero	16B	4½	650
	V8—426 Hemi	N10Y	.035	30	.017	12½B	150	.028C	.032C	30B	7¼	750
	V8—426 Hemi Exh. Em.	N10Y	.035	30	.017	TDC	150	.028C	.032C	30B	7¼	750
	V8—440	J11Y	.035	30	.017	12½B	150	Zero	Zero	19B	4½	650
	V8—440 M.T. Exh. Em.	J11Y	.035	30	.017	TDC	150	Zero	Zero	16B	4½	650
	V8—440 A.T. Exh. Em.	J11Y	.035	30	.017	5B	150	Zero	Zero	16B	4½	650

TUNE-UP SPECIFICATIONS, continued

YEAR	MODEL AND CU. IN. DISPLACEMENT	SPARK PLUGS Type	Gap (In.)	DISTRIBUTOR Point Dwell (Deg.)	Point Gap (In.)	IGNITION TIMING (Deg.) ▲	CRANKING COMP. PRESSURE (Psi)	VALVES Tappet Clearance (In.) Intake	Exhaust	Intake Opens (Deg.)	FUEL PUMP PRESSURE (Psi)	IDLE SPEED (Rpm) ★
1968	6 Cyl.—225	N14Y	.035	42	.020	TDC	135	.010H	.020H	10B	4½	650
	V8—273 M.T.	N14Y	.035	30	.017	5A	135	Zero	Zero	10B	6	700
	V8—273 A.T.	N14Y	.035	30	.017	2½A	135	Zero	Zero	10B	6	650
	V8—318 M.T.	N14Y	.035	30	.017	5A	140	Zero	Zero	10B	6	650
	V8—318 A.T.	N14Y	.035	30	.017	2½A	140	Zero	Zero	10B	6	600
	V8—383 2-BBL. M.T.	J14Y	.035	30	.017	TDC	150	Zero	Zero	18B	4½	650
	V8—383 2-BBL. A.T.	J14Y	.035	30	.017	7½B	150	Zero	Zero	18B	4½	600
	V8—383 4-BBL. M.T.	J11Y	.035	30	.017	TDC④	150	Zero	Zero	18B	4½	650
	V8—383 4-BBL. A.T.	J11Y	.035	30	.017	5B	150	Zero	Zero	18B	4½	650
	V8—426	N10Y	.035	⑥	.017	TDC	175	.078C	.032C	36B	7½	750
	V8—440 (HP) M.T.	J11Y	.035	⑥	.017	TDC	150	Zero	Zero	21B	6¾	650
	V8—440 A.T.	J13Y	.035	30	.017	7½	150	Zero	Zero	18B	4½	600
	V8—440 (HP) A.T.	J11Y	.035	⑥	.017	5B	150	Zero	Zero	21B	6¾	650
1969	6 Cyl. 225	N14Y	.035	44	.020	TDC	135	.010H	.020H	10B	4½	650
	V8—318	N14Y	.035	32	.017	TDC	140	Zero	Zero	10B	6	650
	V8—383, 2-BBL. M.T.	J14Y	.035	32	.017	TDC	150	Zero	Zero	18B	4½	650
	V8—383, 2-BBL. A.T.	J11Y	.035	32	.017	7½B	150	Zero	Zero	18B	4½	600
	V8—383, 4-BBL. M.T.	J11Y	.035	32	.017	TDC	150	Zero	Zero	18B	4½	650
	V8—383, 4-BBL. A.T.	J11Y	.035	32	.017	5B	150	Zero	Zero	18B	4½	600
	V8—383, Hi. Perf. M.T.	J11Y	.035	⑥	.017	TDC	150	Zero	Zero	21B	4½	650
	V8—383, Hi. Perf. A.T.	J11Y	.035	⑥	.017	5B	150	Zero	Zero	21B	4½	600
	V8—426	N10Y	.035	⑥	.017	TDC	175	.028C	.028C	36B	7½	750
	V8—440 Std. Eng.	J13Y	.035	32	.017	7½B	150	Zero	Zero	18B	4½	650
	V8—440 Hi. Perf. M.T.	J11Y	.035	⑥	.017	TDC	150	Zero	Zero	21B	4½	650
	V8—440 Hi. Perf. A.T.	J11Y	.035	32	.017	5B	150	Zero	Zero	21B	4½	650
1970	6 Cyl.—225	N14Y	.035	41–46	.017–.023	TDC ± 2½	100⑨	010	.020	10B	3½–5	⑦
	V8—318	N14Y	.035	30–34	.014–.019	TDC ± 2½	100⑨	Zero	Zero	10B	5–7	⑧
	V8—383, 2-BBL., M.T.	J14Y	.035	28½–32½	.016–.021	TDC ± 2½	100⑨	Zero	Zero	18B⑩	3½–5	650
	V8—383, 4-BBL., A.T.	J11Y	.035	28½–32½	.016–.021	2½B	110⑨	Zero	Zero	18B⑩	3½–5	750
	V8—383, 2-BBL., A.T.	J14Y	.035	28½–32½	.016–.021	2½B ± 2½	100⑨	Zero	Zero	18B⑩	3½–5	750
	V8—383, 4-BBL., M.T.	J11Y	.035	28½–32½	.016–.021	TDC ± 2½	110⑨	Zero	Zero	18B⑩	3½–5	750
	V8—426 Hemi, M.T.	N10Y	.035	27–32	.014–.019	TDC ± 2½	110⑨	⑪	⑪	36B	7–8½	900
	V8—426 Hemi, A.T.	N10Y	.035	27–32	.014–.019	5B ± 2½	110⑨	⑪	⑪	36B	7–8½	900
	V8—440, M.T.	J11Y	.035	28½–32½	.016–.021	TDC ± 2½	110⑨	Zero	Zero	21B	⑫	900
	V8—440, A.T.	J11Y	.035	28½–32½	.016–.021	2½B ± 2½	110⑨	Zero	Zero	21B	⑫	800
	V8—440 Six Pack	J11Y	.035	③	.014–.019	5B ± 2½	110⑨	Zero	Zero	21B	⑫	900 ⑬

★—With manual transmission in N and automatic in D. Add 50 rpm if equipped with air conditioning.

▲—With vacuum advance disconnected. NOTE: These settings are only approximate. Engine design, altitude, temperature, fuel octane rating and the condition of the individual engine are all factors which can influence timing. The limiting advance factor must, therefore, be the "knock point" of the individual engine.

⓪ —With Cleaner Air Package (California). Set at 5° ATDC.
① —With Cleaner Air Package. manual trans. 700, auto. 650.
② —With Cleaner Air Package. manual trans. 650, auto. 650.
③ For distributor with double contact points set at 37–42° combined, 27–32° each set.
④ —383 Cu. In. high performance—Ign. 7½°, and mechanical lifters at .016 and .028 in. cold.

⑤ —At 3000 rpm.
⑥ —Dual points, 27°–31°—both sets, 36°–40°.
⑦ —M.T.—700, A.T.—650.
⑧ —M.T.—750, A.T.—700.
⑨ —Minimum.
⑩ —Super Bee—21B.
⑪ —See text.
⑫ —Except high perf.—3½–5; high perf.—6–7½.
⑬ —With solenoid throttle stop connected.
B—Before top dead center.
A—After top dead center.
A.T.—Automatic transmission.
M.T.—Manual transmission.

CAUTION

General adoption of anti-pollution laws has changed the design of almost all car engine production to effectively reduce crankcase emission and terminal exhaust products. It has been necessary to adopt stricter tune-up rules, especially timing and idle speed procedures. Both of these values are peculiar to the engine and to its application, rather than to the engine alone. With this in mind, car manufacturers supply idle speed data for the engine and application involved. This information is clearly displayed in the engine compartment of each vehicle.

TUNE-UP SPECIFICATIONS, continued

YEAR	MODEL	SPARK PLUGS Type	Gap (In.)	DISTRIBUTOR Point Dwell (Deg.)	Point Gap (In.)	IGNITION TIMING (Deg.) ▲	CRANKING COMP. PRESSURE (Psi)	VALVES Tappet (Hot) Clearance (In.) Intake	Exhaust	Intake Opens (Deg.)	FUEL PUMP PRESSURE (Psi)	IDLE SPEED (Rpm) ★
1971	6 Cyl.—225	N14Y	.035	41–46	.017–.023	**	100–125	.010	.020	10B	3½–5	⑦
	V8—318	N14Y	.035	30–34	.014–.019	**	100–140	Zero	Zero	10B	5–7	①
	V8—383, 2-BBL., M.T.	J14Y	.035	28½–32½	.014–.019	**	100–140	Zero	Zero	18B ⑩	3½–5	650
	V8—383, 2-BBL., A.T.	J14Y	.035	28½–32½	.014–.019	**	100–140	Zero	Zero	18B ⑩	3½–5	750
	V8—383, 4-BBL., M.T.	J11Y	.035	28½–32½	.014–.019	**	110–150	Zero	Zero	18B ⑩	3½–5	700
	V8—383, 4-BBL., A.T.	J11Y	.035	28½–32½	.014–.019	**	110–150	Zero	Zero	18B ⑩	3½–5	650
	V8—426 Hemi, M.T.	N10Y	.035	28–32	.014–.019	**	110–150	⑪	⑪	36B	7–8½	900
	V8—426 Hemi, A.T.	N10Y	.035	28–32	.014–.019	**	110–150	⑪	⑪	36B	7–8½	900
	V8—440, M.T.	J11Y	.035	28½–32½	.016–.021	**	110–150	Zero	Zero	18B	5–7	**
	V8—440, A.T.	J11Y	.035	28½–32½	.016–.021	**	110–150	Zero	Zero	18B	5–7	**
	V8—440, Hi. Perf., M.T.	J11Y	.035	28½–32½	.016–.021	**	110–150	Zero	Zero	21B	⑫	900
	V8—440, Hi. Perf., A.T.	J11Y *	.035	28½–32½	.016–.021	**	110–150	Zero	Zero	21B	⑫	800
	V8—440 Six Pack	J11Y	.035	③	.014–.019	**	110–150	Zero	Zero	21B	⑫	900

VALVE SPECIFICATIONS

YEAR AND MODEL	SEAT ANGLE (DEG.)	FACE ANGLE (DEG.)	VALVE LIFT INTAKE (IN.)	VALVE LIFT EXHAUST (IN.)	VALVE SPRING PRESSURE (VALVE OPEN) LBS. @ IN.	VALVE SPRING INSTALLED HEIGHT (IN.)	STEM TO GUIDE CLEARANCE (IN.) Intake	Exhaust	STEM DIAMETER (IN.) Intake	Exhaust
1964 6 Cyl.—225 Cu. In.	45	①	.375	.360	177 @ 1 5/16	1 11/16	.001–.003	.002–.004	.372–.373	.371–.372
V8—273 Cu. In.	45	45	.395	.395	145 @ 1 5/16	1 11/16	.001–.003	.002–.004	.372–.373	.371–.372
V8—318 Cu. In.	45	45	.403	.389	177 @ 1 5/16	1 11/16	.001–.003	.002–.004	.372–.373	.371–.372
V8—361 Cu. In.	45	45	.392	.389	195 @ 1 15/32	1 55/64	.001–.003	.002–.004	.372–.373	.371–.372
V8—383 Cu. In.	45	45	.389	.389	195 @ 1 15/32	1 55/64	.001–.003	.002–.004	.372–.373	.371–.372
V8—426 Cu. In.	45	45	.430	.430	195 @ 1 15/32	1 55/64	.001–.003	.002–.004	.372–.373	.371–.372
1965 6 Cyl.—225 Cu. In.	45	①	.375	.360	145 @ 1 15/16	1 11/16	.001–.003	.002–.004	.372–.373	.371–.372
V8—273 2-BBL.	45	45	.395	.405	145 @ 1 5/16	1 11/16	.001–.003	.002–.004	.372–.373	.371–.372
V8—273 4-BBL.	45	45	.415	.425	177 @ 1 5/16	1 11/16	.001–.003	.002–.004	.372–.373	.371–.372
V8—318 Cu. In.	45	45	.397	.403	145 @ 1 5/16	1 11/16	.001–.003	.002–.004	.372–.373	.371–.372
V8—361 Cu. In.	45	45	.392	.390	195 @ 1 15/32	1 55/64	.001–.003	.002–.004	.372–.373	.371–.372
V8—383 Cu. In.	45	45	.430	.430	195 @ 1 15/32	1 55/64	.001–.003	.002–.004	.372–.373	.371–.372
V8—426 Cu. In.	45	45	.430	.430	195 @ 1 15/32	1 55/64	.001–.003	.002–.004	.372–.373	.371–.372
1966 6 Cyl.—225 Cu. In.	45	⑦	.395	.395	145 @ 1 15/16	1 11/16	.001–.003	.002–.004	.372–.373	.371–.372
V8—273 2-BBL.	45	45	.395	.405	145 @ 1 5/16	1 11/16	.001–.003	.002–.004	.372–.373	.371–.372
V8—273 4-BBL.	45	45	.415	.425	177 @ 1 5/16	1 11/16	.001–.003	.002–.004	.372–.373	.371–.372
V8—318 Cu. In.	45	45	.397	.403	145 @ 1 5/16	1 11/16	.001–.003	.002–.004	.372–.373	.371–.372
V8—361 Cu. In.	45	45	.392	.390	195 @ 1 15/32	1 55/64	.001–.003	.002–.004	.372–.373	.371–.372
V8—383 2-BBL.	45	45	.392	.390	195 @ 1 15/32	1 55/64	.001–.003	.002–.004	.372–.373	.371–.372
V8—383 4-BBL.	45	45	.425	.435	195 @ 1 15/32	1 55/64	.001–.003	.002–.004	.372–.373	.371–.372
V8—426 Hemi	45	45	.480	.460	184 @ 1 13/32	1 55/64	.002–.004	.003–.005	.309	.308
V8—440 Cu. In.	45	45	.425	.437	200 @ 1 7/16	1 55/64	.001–.003	.002–.004	.372–.373	.371–.372
1967 6 Cyl.—225 Cu. In.	45	⑦	.395	.395	145 @ 1 7/16	1 11/16	.001–.003	.002–.004	.372–.373	.371–.372
V8—273 2-BBL.	45	45	.395	.405	145 @ 1 5/16	1 11/16	.001–.003	.002–.004	.372–.373	.371–.372
V8—273 4-BBL.	45	45	.415	.425	177 @ 1 5/16	1 11/16	.001–.003	.002–.004	.372–.373	.371–.372
V8—318 Cu. In.	45	45	.390	.390	177 @ 1 5/16	1 11/16	.001–.003	.002–.004	.372–.373	.371–.372
V8—383 2-BBL.	45	45	.425	.437	195 @ 1 15/32	1 55/64	.001–.003	.002–.004	.372–.373	.371–.372
V8—383 4-BBL.	45	45	.425	.437	200 @ 1 7/16	1 55/64	.001–.003	.002–.004	.372–.373	.371–.372
V8—426 Hemi	45	45	.480	.460	184 @ 1 13/32	1 55/64	.002–.004	.003–.005	.309	.308
V8—440 4-BBL.	45	45	.425	.437	200 @ 1 7/16	1 55/64	.001–.003	.002–.004	.372–.373	.371–.372
V8—440 Hi. Perf.	45	45	.450	.465	246 @ 1 23/64	1 55/64	.001–.003	.002–.004	.372–.373	.371–.372

VALVE SPECIFICATIONS, continued

YEAR AND MODEL	SEAT ANGLE (DEG.)	FACE ANGLE (DEG.)	VALVE LIFT INTAKE (IN.)	VALVE LIFT EXHAUST (IN.)	VALVE SPRING PRESSURE (VALVE OPEN) LBS. @ IN.	VALVE SPRING INSTALLED HEIGHT (IN.)	STEM TO GUIDE CLEARANCE (IN.)		STEM DIAMETER (IN.)	
							Intake	Exhaust	Intake	Exhaust
1968 6 Cyl.—225 Cu. In.	45	①	.395	.395	145 @ 1 5/16	1 11/16	.001–.003	.002–.004	.372–.373	.371–.372
V8—273 Cu. In.	45	45	.372	.400	177 @ 1 5/16	1 11/16	.001–.003	.002–.004	.372–.373	.371–.372
V8—318 Cu. In.	45	45	.372	.400	177 @ 1 5/16	1 11/16	.001–.003	.002–.004	.372–.373	.371–.372
V8—383 2-BBL.	45	45	.425	.437	200 @ 1 7/16	1 55/64	.001–.003	.002–.004	.372–.373	.371–.372
V8—383 4-BBL.	45	45	.450	.465	225 @ 1 7/16	1 55/64	.001–.003	.002–.004	.372–.373	.371–.372
V8—383 Hi. Perf.	45	45	.450	.465	230 @ 1 13/32	1 55/64	.001–.003	.002–.004	.372–.373	.371–.372
V8—426 Hemi	45	45	.490	.480	280 @ 1 3/8	1 55/64	.002–.004	.003–.005	.309	.308
V8—440 4-BBL.	45	45	.425	.435	200 @ 1 7/16	1 55/64	.001–.003	.002–.004	.372–.373	.371–.372
V8—440 Hi. Perf.	45	45	.450	.458	230 @ 1 13/32	1 55/64	.001–.003	.002–.004	.372–.373	.371–.372
1969 6 Cyl.—225 Cu. In.	45	①	.395	.395	145 @ 1 5/16	1 11/16	.001–.003	.002–.004	.372–.373	.371–.372
V8—318 Cu. In.	45	①	.372	.400	177 @ 1 5/16	1 11/16	.001–.003	.002–.004	.372–.373	.371–.372
V8—383 2-BBL.	45	45	.425	.437	200 @ 1 7/16	1 55/64	.001–.003	.002–.004	.372–.373	.371–.372
V8—383 4-BBL.	45	45	.425	.437	246 @ 1 23/64	1 55/64	.001–.003	.002–.004	.372–.373	.371–.372
V8—383 Hi Perf.	45	45	.450	.465	246 @ 1 23/64	1 55/64	.001–.003	.002–.004	.372–.373	.371–.372
V8—426 Hemi	45	45	.490	.480	280 @ 1 3/8	1 55/64	.002–.004	.003–.005	.309	.308
V8—440 4-BBL.	45	45	.425	.435	200 @ 1 7/16	1 55/64	.001–.003	.002–.004	.372–.373	.371–.372
V8—440 Hi. Perf.	45	45	.450	.458	246 @ 1 23/64	1 55/64	.001–.003	.002–.004	.372–.373	.371–.372
1970 6 Cyl.—225 Cu. In.	45	①	.395	.395	145 @ 1 5/16	1 11/16	.001–.003	.002–.004	.372–.373	.371–.372
V8—318 Cu. In.	45	①	.372	.400	177 @ 1 5/16	1 11/16	.001–.003	.002–.004	.372–.373	.371–.372
V8—383 2-BBL.	45	45	.425	.437	200 @ 1 7/16	1 57/64	.001–.003	.002–.004	.372–.373	.371–.372
V8—383 4-BBL.	45	45	.425	.437	246 @ 1 23/64	1 57/64	.001–.003	.002–.004	.372–.373	.371–.372
V8—383 Hi Perf.	45	45	.450	.465	246 @ 1 23/64	1 57/64	.001–.003	.002–.004	.372–.373	.371–.372
V8—426 Hemi	45	45	.490	.480	310 @ 1 3/8	1 55/64	.002–.004	.003–.005	.309	.308
V8—440 4-BBL.	45	45	.425	.437	200 @ 1 7/16	1 55/64	.001–.003	.002–.004	.372–.373	.371–.372
V8—440 Hi Perf.	45	45	.450	.465	246 @ 1 23/64	1 55/64	.001–.003	.002–.004	.372–.373	.371–.372
V8—440 3-2-BBL.	45	45	.450	.465	310 @ 1 3/8	1 55/64	.001–.003	.002–.004	.372–.373	.371–.372
1971 6 Cyl.—225 Cu. In.	45	①	.406	.414	144 @ 1 5/16	1 11/16	.001–.003	.002–.004	.372–.373	.371–.372
V8—318 Cu. In.	45	①	.373	.399	177 @ 1 5/16	1 11/16	.001–.003	.002–.004	.372–.373	.371–.372
V8—383 2-BBL.	45	45	.425	.435	200 @ 1 7/16	1 55/64	.001–.003	.002–.004	.372–.373	.371–.372
V8—383 Hi Perf.	45	45	.450	.458	246 @ 1 23/64	1 57/64	.001–.003	.002–.004	.372–.373	.371–.372
V8—426 Hemi	45	45	.490	.481	310 @ 1 3/8	1 55/64	.002–.004	.003–.005	.310	.309
V8—440 4-BBL.	45	45	.425	.435	200 @ 1 7/16	1 55/64	.001–.003	.002–.004	.372–.373	.371–.372
V8—440 Hi Perf.	45	45	.450	.458	246 @ 1 23/64	1 55/64	.001–.003	.002–.004	.372–.373	.371–.372
V8—440 3-2-BBL.	45	45	.450	.458	246 @ 1 23/64	1 55/64	.001–.003	.002–.004	.372–.373	.371–.372

① —Intake 45; Exhaust 43.
② —Intake 45; Exhaust 47.

WHEEL ALIGNMENT

YEAR	MODEL		FRONT END HEIGHT ▲ (In.)	CASTER Range (Deg.)	CASTER Pref. Setting (Deg.)	CAMBER Range (Deg.)	CAMBER Pref. Setting (Deg.)	TOE-IN (In.)	KING PIN INCLINATION (Deg.)	WHEEL PIVOT RATIO Inner Wheel	Outer Wheel
1964	Dodge, Plymouth	M.S.	$1\frac{3}{4} \pm \frac{1}{8}$②	0–1N	$\frac{1}{2}$N	①	①	$\frac{3}{32}$–$\frac{5}{32}$	$7\frac{1}{2}$	20	17.5
	Dodge, Plymouth	P.S.	$1\frac{3}{4} \pm \frac{1}{8}$②	$\frac{1}{4}$P–$1\frac{1}{4}$P	$\frac{3}{4}$P	①	①	$\frac{3}{32}$–$\frac{5}{32}$	$7\frac{1}{2}$	20	17.5
	Dodge 880	M.S.	$2 \pm \frac{1}{8}$	0–1N	$\frac{1}{2}$N	①	①	$\frac{3}{32}$–$\frac{5}{32}$	$5\frac{1}{2}$–$7\frac{1}{2}$	20	21.5
	Dodge 880	P.S.	$2 \pm \frac{1}{8}$	$\frac{1}{4}$P–$1\frac{1}{4}$P	$\frac{3}{4}$P	①	①	$\frac{3}{32}$–$\frac{5}{32}$	$5\frac{1}{2}$–$7\frac{1}{2}$	20	21.5
1965	Belvedere Satellite	M.S.	$1\frac{3}{4} \pm \frac{1}{8}$②	0–1N	$\frac{1}{2}$N	①	①	$\frac{3}{32}$–$\frac{5}{32}$	$6\frac{1}{2}$	20	17.8
	Belvedere Satellite	P.S.	$1\frac{3}{4} \pm \frac{1}{8}$②	$\frac{1}{4}$P–$1\frac{1}{4}$P	$\frac{3}{4}$P	①	①	$\frac{3}{32}$–$\frac{5}{32}$	$6\frac{1}{2}$	20	17.8
	Fury, 880 Monaco	M.S.	$1\frac{3}{8} \pm \frac{1}{8}$③	0–1N	$\frac{1}{2}$N	①	①	$\frac{3}{32}$–$\frac{5}{32}$	9	20	18.8
	Fury, 880 Monaco	P.S.	$1\frac{3}{8} \pm \frac{1}{8}$③	$\frac{1}{4}$P–$1\frac{1}{4}$P	$\frac{3}{4}$P	①	①	$\frac{3}{32}$–$\frac{5}{32}$	9	20	18.8
	Coronet	M.S.	$1\frac{3}{4} \pm \frac{1}{8}$② ④	0–1N	$\frac{1}{2}$N	①	①	$\frac{3}{32}$–$\frac{5}{32}$	$7\frac{1}{2}$	20	17.8
	Coronet	P.S.	$1\frac{3}{4} \pm \frac{1}{8}$② ④	$\frac{1}{4}$P–$1\frac{1}{4}$P	$\frac{3}{4}$P	①	①	$\frac{3}{32}$–$\frac{5}{32}$	$7\frac{1}{2}$	20	17.8
1966	Coronet, Belvedere, Satellite	M.S.	$1\frac{1}{8} \pm \frac{1}{8}$	0–1N	$\frac{1}{2}$N	①	①	$\frac{3}{32}$–$\frac{5}{32}$	$7\frac{1}{2}$	20	17.8
	Coronet, Belvedere, Satellite	P.S.	$1\frac{1}{8} \pm \frac{1}{8}$	$\frac{1}{4}$P–$1\frac{1}{4}$P	$\frac{3}{4}$P	①	①	$\frac{3}{32}$–$\frac{5}{32}$	$7\frac{1}{2}$	20	17.8
	Fury, 880, Monaco	M.S.	$1\frac{1}{8} \pm \frac{1}{8}$	0–1N	$\frac{1}{2}$N	①	①	$\frac{3}{32}$–$\frac{5}{32}$	$7\frac{1}{2}$	20	18.8
	Fury, 880, Monaco	P.S.	$1\frac{1}{8} \pm \frac{1}{8}$	$\frac{1}{4}$P–$1\frac{1}{4}$P	$\frac{3}{4}$P	①	①	$\frac{3}{32}$–$\frac{5}{32}$	$7\frac{1}{2}$	20	18.8
1967–69	Coronet, Charger, Belvedere, Satellite	M.S.	$1\frac{7}{8} \pm \frac{1}{8}$	0–1N	$\frac{1}{2}$N	①	①	$\frac{3}{32}$–$\frac{5}{32}$	$7\frac{1}{2}$	20	17.8
	Coronet, Charger, Belvedere, Satellite	P.S.	$1\frac{7}{8} \pm \frac{1}{8}$	$\frac{1}{4}$P–$1\frac{1}{4}$P	$\frac{3}{4}$P	①	①	$\frac{3}{32}$–$\frac{5}{32}$	$7\frac{1}{2}$	20	17.8
	Fury, Monaco, Polara	M.S.	$1\frac{3}{8} \pm \frac{1}{8}$⑤	0–1N	$\frac{1}{2}$N	①	①	$\frac{3}{32}$–$\frac{5}{32}$	$7\frac{1}{2}$	20	18.8
	Fury, Monaco, Polara	P.S.	$1\frac{3}{8} \pm \frac{1}{8}$⑤	$\frac{1}{4}$P–$1\frac{1}{4}$P	$\frac{3}{4}$P	①	①	$\frac{3}{32}$–$\frac{5}{32}$	$7\frac{1}{2}$	20	18.8
1970–71	Coronet, Charger, Satellite, Belvedere	M.S.	$1\frac{7}{8} \pm \frac{1}{8}$	$\frac{1}{2}$N $\pm \frac{1}{2}$	$\frac{1}{2}$N	①	①	$\frac{3}{32}$–$\frac{5}{32}$	$7\frac{1}{2}$	20	17.8
	Coronet, Charger, Satellite, Belvedere	P.S.	$1\frac{7}{8} \pm \frac{1}{8}$	$\frac{3}{4}$P $\pm \frac{1}{2}$	$\frac{3}{4}$P	①	①	$\frac{3}{32}$–$\frac{5}{32}$	$7\frac{1}{2}$	20	17.8
	Polara, Monaco, Fury	M.S.	$1\frac{3}{4} \pm \frac{1}{8}$⑤	$\frac{1}{2}$N $\pm \frac{1}{2}$	$\frac{1}{2}$N	①	①	$\frac{3}{32}$–$\frac{5}{32}$	9.0⑥	20	18.8
	Polara, Monaco, Fury	P.S.	$1\frac{3}{4} \pm \frac{1}{8}$⑤	$\frac{1}{2}$N $\pm \frac{1}{2}$	$\frac{1}{2}$N	①	①	$\frac{3}{32}$–$\frac{5}{32}$	9.0⑥	20	18.8

▲—For procedure, see text.
M.S.—Manual steering.
P.S.—Power steering.
N—Negative.
P—Positive.
①—Left—$\frac{1}{4}$P–$\frac{3}{4}$P; $\frac{1}{2}$P preferred.

Right—0–$\frac{1}{2}$P; $\frac{1}{4}$P preferred.
②—Heavy duty—$2\frac{1}{8} \pm \frac{1}{8}$.
③—Fury sta. wag., Monaco, Polara, Custom 880—$1\frac{1}{8} \pm \frac{1}{8}$.
④—Sta. wag.—$1\frac{3}{4} \pm \frac{1}{8}$.
⑤—Monaco, Polara—$1\frac{1}{8} \pm \frac{1}{8}$.
⑥—Fury—$7\frac{1}{2}$.

CAPACITIES

YEAR	MODEL	ENGINE CRANKCASE ADD 1 Qt. FOR NEW FILTER	TRANSMISSIONS Pts. TO REFILL AFTER DRAINING			DRIVE AXLE (Pts.)	GASOLINE TANK (Gals.)	COOLING SYSTEM (Qts.) WITH HEATER
			Manual		Automatic			
			3-Speed	4-Speed				
1964	Dodge 330, 440 6 Cyl.	4	5	None	18	4.0	19①	13
	Dodge 330, 400 V8	4	5	7½	18	4.0	19①	21
	Dodge Polara V8	4	5	7½	18	4.0	19	21
	Dodge 880 V8	4	5	7½	20	4.0	23①	16
	Plymouth 6 Cyl.	4	6	None	18	4.0	19①	13
	Plymouth V8	4	4½②	7½	20	4.0	19	20③
1965	6 Cyl.—225	4	6	None	17	4.0	19①	13
	V8—273	4	4½	7½	19½	4.0	19①	18
	V8—318	4	4½	7½	19½	4.0	19①	21
	V8—361	4	4½	7½	19½	4.0	19①	17
	V8—383	4	4½	7½	19½	4.0	19①	17
	V8—413, 426	4	4½	7½	19½	4.0	19①	17
1966	6 Cyl.—225	4	6½	None	16	2.0⑤	19①	13④
	V8—318	4	6	9	18½	4.0	25⑥	22⑦
	V8—383	4	6	9	18½	4.0	25⑥	18⑧
	V8—426 Hemi	6	None	9	18½	4.0	25⑥	18⑧
1967	6 Cyl.—225	4	6½	None	16	2.0⑤	19	13④
	V8—273	4	6½	8	16	⑨	19	19⑩
	V8—318	4	6½	8½	18½	⑨	19	18⑪
	V8—318 Police	4	6½	8½	18½	⑨	19	19⑩
	V8—383	4	6½	8½⑫	18½	⑨	19, 25⑥	17⑪
	V8—440	4	6½	8½⑫	18½	⑨	19, 25⑥	17⑪
	V8—426 Hemi	6	None	8½⑫	18½	⑨	19, 25⑥	18
1968	6 Cyl.—225	4	6½	None	15½	⑨	19	13④
	V8—273	4	5¾	8	15½	⑨	19	19⑪
	V8—318	4	5¾	8	15½	⑨	19⑮	18⑪
	V8—318 Police	4	5¾	8	18½	⑨	19	19⑪
	V8—383	4	5¾	9	18½⑬	⑨	19⑮	17⑪
	V8—426 Hemi	6	None	9	16	⑨	19	18
	V8—440	4	5¾	9	18½⑬	⑨	19⑮	17⑭⑪
1969	6 Cyl.—225	4	6½	None	15½	⑨	19	13⑯
	V8—318	4	6	None	15½⑲	⑨	19⑮	16⑰
	V8—318 Police	4	6	7½	18½	⑨	19	19
	V8—383	4	6	7½⑱	18½⑬	⑨	19⑮	16⑪
	V8—383 Police	4	6	7½	18½	⑨	19	17
	T8—426 Hemi	6	None	7½⑱	16	⑨	19	18
	V8—440	4	6	7½⑱	18½⑬	⑨	19⑮	17⑭⑪
1970-71	6 Cyl.—225	4	4¾㉑	None	17	⑨	19	13⑯
	V8—318	4	4¾㉑	None	17	⑨	19⑮	16⑰
	V8—383	4	4¾㉑	7½	19⑳	⑨	19⑮	14½⑯
	V8—426 Hemi	6	None	7½	16.8	⑨	19	17
	V8—440	4	4¾㉑	7½	19	⑨	19	17㉒
	V8—440 Six Pack	6	None	7½	19	⑨	19⑮	17

① Station wagon—21.
② High perf.—3.
③ Commando—16.
④ With air cond. or heavy duty rad.—14.
⑤ Station wagon and 1967 Sure-Grip—4.0.
⑥ Station wagon—22.
⑦ With air cond., trailer pack, or heavy duty rad.—23.
⑧ With air cond., trailer pack, or heavy duty rad.—19.
⑨ 7¼ in. axle, 2.0; 8¼ in. axle, 4.4; 8¾ in. axle, 4.4; 9¾ in. axle, 5.5.

Axle	Filler Location	Cover fastening
7¼ in.	Cover	9 bolts
8¼ in.	Carrier, right side	10 bolts
8¾ in.	Carrier, right side	Welded
9¾ in.	Cover	10 bolts

⑩ With air cond., trailer pack, or heavy duty rad.—20.
⑪ With air cond., trailer pack, or heavy duty rad.—add 1 Qt.
⑫ All Plymouth Fury—9.0.
⑬ High performance—15½.
⑭ With manual transmission—18.
⑮ Polara, Monaco, Fury—24 except station wagon. Station wagon—22 (23—1970-71).
⑯ Add 2 Qts. for air cond. or heavy duty rad.
⑰ Add 3 Qts. for air cond. or heavy duty rad.
⑱ All Plymouth Fury—7¾.
⑲ A904 transmission—17.
⑳ High performance—16.3.
㉑ All synchro 3-sp.—use Dexron A.T. Fluid.
㉒ Fury—15½.

GENERAL CHASSIS AND BRAKE SPECIFICATIONS, DODGE

YEAR	MODEL	CHASSIS		BRAKE CYLINDER BORE			BRAKE DRUM	
		Overall Length (In.)	Tire Size	Master Cylinder (In.)	Wheel Cylinder Diameter (In.)		Diameter (In.)	
					Front	Rear	Front	Rear
1964	All Exc. Sta. Wag.	209.8	7.00 x 14	1	1⅛	¹⁵/₁₆	10	10
	Station Wagon	212.8	7.50 x 14	1	1⅛	¹⁵/₁₆	10	10
	Custom 880, Exc. Sta. Wag.	214.8	8.00 x 14	1	1⅛	¹⁵/₁₆	11	11
	Custom 880 Sta. Wag.	216.3	8.00 x 14	1	1⅛	¹⁵/₁₆	11	11
1965	Coronet Pass.	204.3	7.35 x 14	1	1⅛	¹⁵/₁₆	10	10
	Coronet Sta. Wag.	209.3	7.75 x 14	1	1⅛	¹⁵/₁₆	10	10
	Polara Exc. Sta. Wag.	212.3	8.25 x 14	1	1⅛	¹⁵/₁₆	11	11
	Polara Sta. Wag.	217.1	8.55 x 14	1	1⅛	¹⁵/₁₆	11	11
	Custom 880 and Monaco	212.3	8.25 x 14	1	1⅛	¹⁵/₁₆	11	11
	Custom 880 Sta. Wag.	217.1	8.55 x 14	1	1⅛	¹⁵/₁₆	11	11
1966	Coronet, 6 Cyl. Sed.	203.2	6.95 x 14	1	1⅛	¹⁵/₁₆	10	10
	Coronet, 6 Cyl. Conv.	203.2	7.35 x 14	1	1⅛	¹⁵/₁₆	10	10
	Coronet, V8 Sed. and Conv.	203.2	7.35 x 14	1	1⅛	¹⁵/₁₆	10	10
	Coronet, Sta. Wag., 2 Seat	207.9	7.75 x 14	1	1⅛	¹⁵/₁₆	10	10
	Coronet, Sta. Wag., 3 Seat	209.0	8.25 x 14	1	1⅛	¹⁵/₁₆	10	10
	Polara and Monaco, Exc. Sta. Wag.	213.3	8.25 x 14	1①	1⅛②	¹⁵/₁₆③	11	11
	Station Wagon	217.1	8.55 x 14	1①	1⅛②	¹⁵/₁₆③	11	11
1967	Coronet, Exc. Sta. Wag.	219.6	7.35 x 14	1①	1⅛②	¹⁵/₁₆	10	10
	Coronet, Sta. Wag., 2 Seat	207.9	8.25 x 14	1①	1⅛②	¹⁵/₁₆	10	10
	Coronet, Sta. Wag., 3 Seat	210.4	8.25 x 14	1①	1⅛②	¹⁵/₁₆	10	10
	Charger	203.6	7.35 x 14	1①	1⅛②	¹⁵/₁₆	10	10
	Polara and Monaco, Exc. Sta. Wag.	219.6	8.25 x 14	1①	1⅛②	¹⁵/₁₆③	11	11
	Polara and Monaco, Sta. Wag.	221.3	8.45 x 15	1①	1⅛②	¹⁵/₁₆③	11	11
1968–69	Coronet, Exc. Sta. Wag.	206.6	7.35 x 14	1	1⅛	¹⁵/₁₆	10⑤	10⑤
	Coronet, Sta. Wag.	210.0	8.25 x 14	1	1⅛	¹⁵/₁₆	10⑤	10⑤
	Charger	208.0	F-70 x 14	1	1⅛	¹⁵/₁₆	10⑤	10⑤
	Polara and Monaco, Exc. Sta. Wag.	219.0	8.25 x 14	1	1⅛	¹⁵/₁₆	11	11
	Polara and Monaco, Sta. Wag., 2 Seat	219.6	8.55 x 14	1	1⅛	¹⁵/₁₆	11	11
	Polara and Monaco, Sta. Wag., 3 Seat	220.0	8.55 x 14	1	1⅛	¹⁵/₁₆	11	11
	Coronet and Charger (Disc Brakes)	—	—	1⅛	2	¹⁵/₁₆	11.04	10
	Polara and Monaco (Disc Brakes)	—	8.15 x 15④	1⅛	2⅜⑥	¹⁵/₁₆	11.76	11
1970–71	Coronet, 6 Cyl., Exc. Sta. Wag.	209.2	F78 x 14	1.00	1.125	.9375	10	10
	Coronet, V8 318 & 383, Exc. Sta. Wag.	209.2	G78 x 14	1.00	1.125	.9375	10⑦	10⑦
	Coronet, V8 426 & 440, Exc. Sta. Wag.	209.2	F70 x 14	1.00	1.125	.9375	11	11
	Coronet, Sta. Wag.	211.7	G78 x 14	1.00	1.125	.9375	11	11
	Charger, 6 Cyl.	208.0	F78 x 14	1.00	1.125	.9375	10	10
	Charger, V8 318	208.0	G78 x 14	1.00	1.125	.9375	10	10
	Charger, V8 426 & 440	208.0	F70 x 14	1.00	1.125	.9375	11	11
	Polara & Monaco, Exc. Sta. Wag.	219.9	H78 x 15	1.00	1.125	.9375	11	11
	Polara & Monaco, Sta. Wag., 2-seat	223.9	J78 x 15	1.00	1.125	.9375	11	11
	Polara & Monaco, Sta. Wag., 3-seat Exc. 318	223.9	L78 x 15	1.00	1.125	.9375	11	11
	Polara & Monaco, Sta. Wag., 3-seat 318	223.9	J78 x 15	1.00	1.125	.9375	11	11
	Coronet & Charger (Disc Brakes)	—	—	1.125	2.75	.9375	10.72	11
	Polara & Monaco (Disc Brakes)	—	—	1.125	2.75	.3975	11.75	11

① —Disc brakes—Budd type—1⅛ in.
② —Disc brakes—Budd type—2⅜ in., Kelsey-Hayes 1.638 in.—Bendix 2.000.
③ —When car is equipped with front discs—⅞ in.
④ —Station Wagons—8.45 x 15 in.

⑤ —11 in. STD. on Coronet R/T; OPT. W/V8—426 and 440.
⑥ —1969—2¾ in.
⑦ —V8 383 high perf.—11 in.

GENERAL CHASSIS AND BRAKE SPECIFICATIONS
PLYMOUTH

YEAR	MODEL	CHASSIS		BRAKE CYLINDER BORE			BRAKE DRUM	
		Overall Length (In.)	Tire Size	Master Cylinder (In.)	Wheel Cylinder Diameter (In.) Front*	Rear	Diameter (In.) Front	Rear
1964	All, exc. sta. wagon	206.5	7.00 x 14	1	1	15/16	10**	10**
	Station wagons	211.5	7.50 x 14	1	1	15/16	10**	10**
1965	Belvedere, Satellite	203.4	7.35 x 14▲	1	1 1/8	15/16	10	10
	Fury	209.4	7.35 x 14●	1	1 1/8	15/16	10	10
	Sports Fury	209.4	7.75 x 14●	1	1 1/8	15/16	10	10
1966	6 Cyl. Belvedere, Satellite	200.5	6.95 x 14†	1	1 1/8	15/16	10	10
	V8	200.5	7.35 x 14†	1	1 1/8	15/16	10	10
	6 Cyl. Fury	209.8	7.35 x 14‡	1	1 1/8	15/16	11	11
	V8	209.8	7.75 x 14‡	1	1 1/8	15/16	11	11
	Disc Brakes, Kelsey-Hayes	—	—	1	1 5/8	15/16★	11	11
	Disc. Brakes, Budd	—	—	1 1/8	2 3/8	15/16★	11	11
1967	6 Cyl. Belvedere, Satellite	200.5†	7.35 x 14†	1②	1 1/8③	15/16	10⑤	10
	V8	200.5①	7.35 x 14①	1②	1 1/8③	15/16	10⑤	10
	Fury	213.1‡	7.75 x 14‡	1②	1 1/8④	15/16	11⑥	11
	V8 Belvedere—GTX	200.5	7.75 x 14	1②	1 1/8③	15/16	11⑥	11
1968	Belvedere, Satellite (STD.)	202.7	8.25 x 14	1	1 1/8	15/16	10	10
	GTX, Opt. 426 & 440	202.7	8.55 x 14	1	1 1/8	15/16	11	11
	Belvedere & Satellite sta. wag.	208.0	8.15 x 14	1	1 1/8	15/16	10	10
	Belvedere & Satellite (disc brakes)	202.7	—	1 1/8	2.0	15/16	11.04 (Disc.)	10
	Fury & V.I.P.	213	8.25 x 14	1	1 1/8	15/16	11	11
	Suburban (two-seat)	216	8.55 x 14	1	1 1/8	15/16	11	11
	Suburban (three-seat)	217	8.55 x 14	1	1 1/8	15/16	11	11
	Suburban (disc brakes)	—	8.45 x 15	1 1/8	2 3/8	15/16	11.76 (Disc.)	11
	Fury models (disc brakes)	213	8.15 x 15	1 1/8	2 3/8	15/16	11.76 (Disc.)	11
1969	Satellites							
	6 Cyl. exc. sta. wag.	202.7	7.35 x 14	1	1 1/8	15/16	10	10
	Sta. wag.	208.0	8.25 x 14	1	1 1/8	15/16	10	10
	V8, exc. below	202.7	7.35 x 14	1	1 1/8	15/16	10⑦	10⑦
	V8, sta. wag.	208.0	8.25 x 14	1	1 1/8	15/16	10	10
	V8, Road Runner	202.7	F70 x 14	1	1 1/8	15/16	11	11
	V8, GTX	202.7	F70 x 14	1	1 1/8	15/16	11	11
	Disc brake	—	—	1 1/8	2	15/16	(Disc.)	10
	Fury							
	6 Cyl. exc. sta. wag.	214.5	7.75 x 15	1	1 1/8	15/16	11	11
	Sta. wag.	219.1	8.85 x 15	1	1 1/8	15/16	11	11
	V8, exc. below	214.5	7.75 x 15	1	1 1/8	15/16	11	11
	V8, sta. wag.	219.1	8.85 x 15	1	1 1/8	15/16	11	11
	Disc. brake	—	—	1 1/8	2 3/4	15/16	(Disc.)	11
1970–71	Belvedere & Satellite, Exc. Sta. Wag.	204.0	⑧	1	1 1/8	15/16	⑩	⑩
	Belvedere & Satellite, Sta. Wag.	209.1	G78 x 14	1	1 1/8	15/16	11	11
	Disc brakes	—	—	1 1/8	2 3/4	15/16	10.72	11
	Fury, Exc. Sta. Wag.	214.9	⑨	1	1 3/16	15/16	11	11
	Fury, Sta. Wag.	220.6	J78 x 14	1	1 3/16	15/16	11	11
	Disc brakes	—	—	1 1/8	2 3/4	15/16	11.75	11

**—Optional with 11 in. drums and 13/16 in. rear wheel cylinders.
▲—Option & station wagon—7.75 x 14.
●—Optional—8.25 x 14.
Station wagon—8.55 x 14.
†—Sta. Wag. two-seat 7.75 x 14, three-seat 8.25 x 14, Overall Length 207.
‡—Sta. Wag. 8.55 x 14, overall length, two-seat 216.1, three-seat 217.4.
★—6 Cyl. application.
①—Sta. wag. 8.25 x 14, overall length, two-seat 207.1, three-seat 208.8.
②—Disc brakes—1 1/8 in.
③—Disc brakes—2 in.
④—Disc brakes 2 3/8 in.
⑤—Disc brakes 11 3/16 in.
⑥—Disc brakes 11 7/8 in.
⑦—383 hi. perf. 426, 440—11 inches.
⑧—225 and 318—F78 x 14; 426 and 440—F70 x 14.
⑨—Except Sport Fury GT—F78 x 15; Sport Fury GT—H70 x 15.
⑩—225, 318 and 383 except high perf.—10 in.; 383 high perf. 426 and 440—11 in.

LIGHT BULBS
DODGE

1964

	Dodge	Dodge 880
Sealed beam—low beam	4002	4002
Sealed beam—high beam	4001	4001
Tail, stop & turn signal	1034	1034
Park & turn signals	1034	1034
Back up lamps	1073	1073
License lamp	67	67
Trunk and/or under hood	1004	1004
Glove compartment	1891	1891
Radio	1892-AM 1893-AM 57-AM-FM	57
Trans. control push buttons	53X	1816
Handbrake indicator	57	57
Dome lamp	1004	1004
Map lamp	90	1004
Heater and/or AC cont. P/B	1892	1816
Turn signal indicator	57	57
High beam indicator	57	1445
Oil pressure warning light	—	—
Instrument cluster illumination	57	57
Ash tray	—	53
Auto pilot	—	1816
Door and/or pocket panel	—	90

1965

	Coronet	Polara, Monaco, Custom 880
Sealed beam—low beam	4002	4002
Sealed beam—high beam	4001	4001
Tail, stop & turn signal	1034	1034
Tail light	—	—
Park & turn signal	1034	1034
Back up lights	1073	1073
License light	67	67
Trunk and/or under hood	1004	1004
Glove compartment	1891	1891
Radio	1893	1893
Gear shift indicator	1445	53X
Handbrake indicator	57	257
Dome lamp	1004	1004
Map lamp	90	1004
Heater and/or AC cont. P/B	1892	53X
Turn signal indicator	158	158
High beam indicator	158	57
Oil pressure warning light	158	158
Instrument cluster illumination	158	158
Ignition switch	—	1445
Ash tray	—	53X
Auto pilot	—	57
Door and/or pocket panel	—	90

1966

	Polara, Monaco	Coronet
Ash tray	53X	—
Auto pilot	57	—
Back-up lights	1073	1073
Clock	158	(a)
Dome lights	1004	1004
Door and/or pocket	90	90
Emergency flasher	57	57
Fender mounted turn signals	1893	—
Gear selector indicator	53X	1445
Gear selector with console	57	57
Glove compartment	1891	1891
Handbrake indicator	257	57
Heater and/or A. C. control	53X	—
High beam indicator	57	158
Instrument cluster illumination	158	158-57
Ignition switch	1445	—
License light	67	67
Map light	1004	90
Oil pressure indicator	158	158
Panel and/or ridge light	90	90
Park and turn signal	1034A	—
Radio	1893	1893
Sealed beam—hi-beam (No. 1)	4001	4001
Sealed beam—lo-beam sealed beam —hi-lo beam (No. 2)	4002	4002
Tachometer with console	57	57
Tail lights	67	—
Tail stop and turn signal	1034	1034
Trunk and/or under hood light	1004	1004
Turn signal indicator	158	158

(a) In instrument cluster lighting.

1967

	Polara, Monaco	Coronet	Charger
Air conditioning control	*	1445	EL
Back-up lights	1073	1141	1141
Brake system warning light	158	257	257
Clock	*	*	57
Dome and/or "C" pillar light	1004	1004	1004
Door, pocket panel and/or reading light	90	90	90
Fender mounted turn signal indicator	1893	1893	330
Gear selector indicator	*	1445	1445
Gear selector with console	57	57	53X
Glove compartment	1891	1891	1891
Handbrake indicator	158	257	257
High beam indicator	150	158	57
Instrument cluster illumination	158	158	EL
License light	67	67	67
Map light	90	90	90
Oil pressure indicator	158	158	Gauge
Park and turn signal	1034	1034	1034
Portable reading light	90	—	—
Radio	1893	1893	EL
Sealed beam—hi-beam (No. 1)	4001	4001	4001
Sealed beam—hi-lo beam (No. 2)	4002	4002	4002
Tachometer	1816	1816	EL
Tail light	67	67	—
Tail, stop and turn signal	1034	1034	1034
Trunk and/or under hood light	1004	1004	1004
Turn signal indicator (panel)	158	158	57
Ash tray	53X	—	—
Auto pilot	1445, 1892	—	—
Ignition switch	53X	—	—

*Included in instrument cluster lighting.
EL—Electroluminescent lighting.

1968

	Polara, Monaco	Coronet	Charger
Air conditioning indicator	1445	1892	1892
Ash tray	1445	1445	1445
Back-up lights	1073 (Sta. Wagon—1141)	1073 (1141—Sta. Wagon)	1073
Brake system warning light	158	158	57
Clock	*	57	***
Courtesy lamp	—	89	89
Dome and/or "C" pillar light	1004	1004	1004
Door, pocket panel and/or reading light	90	90	90
Fender mounted turn signal indicator	330, 1893	330	**
Gear selector indicator	*	1445	—
Gear selector with console	57	57	57
Glove compartment	1891	1891	1891
High beam indicator	158	158	57
Ignition lamp	53X	1445	1445
Instrument cluster illumination	57	158	57-158
License light	67	67	67
Map light	90	90	90
Oil pressure indicator	158	158	Gauge
Park and turn signal	1034	1034	1034
Radio	1816	1816	1816
Sealed beam—hi-beam (No. 1)	4001	4001	4001
Sealed beam—hi-lo beam (No. 2)	4002	4002	4002
Side marker	1895	1895	1895
Tachometer	1816	57	*
Tail, stop and turn signal	1034	1034	1034
Trunk and/or under hood light	1004	1004	1004
Turn signal Indicator (panel)	158	158	57
Auto. temp	53X	—	—
Cornering lights	1195P	—	—
Tail light	1095	—	—

*Included in instrument cluster lighting.
***Hood mounted.
***Not lighted.

1969

	Polara, Monaco	Coronet	Charger
Air conditioning indicator	—	**1892(2)	**1892(2)
Ash tray	**1445	**1892	**53
Back-up lights	1156(2)	1156(2)	1156(2)
Brake system warning light	158	158	57
Clock	*	**57	*
Courtesy lamp	90	89	89
Dome and/or "C" pillar light	551	1004	1004

LIGHT BULBS
DODGE

Door, pocket panel and/or reading light ..	90	90	90
Fender mounted turn signal indicator	330(2)	330(2)	1816(2)***
(Tail lamp only) ..	—	—	1095(2)
Gear selector indicator	**57	**1445	**1445
Gear selector with console ..	**57	**57	**57
Glove compartment........	1891	1891	1891
High beam indicator	158	158	57
Ignition lamp....	1445	1445	1445
Instrument cluster illumination...	**158(4)	**158(4)	**57(3) 158(3)
License light.....	67	67	67
Map light.....	90	90	90
Oil pressure indicator	158	158	Gauge
Park and turn signal........	1157(2)	1157(2)	1157A(2)
Radio	**1816	**1816	**1816
Radio with tape ..	—	**1815	—
Reverse 4-speed transmission indicator	—	53	53
Sealed beam—hi-beam (No. 1)	4001	4001	4001
Sealed beam—hi-lo beam (No. 2)	4002	4002	4002
Tachometer	—	**57	*
Tail, stop and turn signal....	1157(2)	1157(2)	1157(4)

Trunk and/or under hood light	1004	1004	1004
Turn signal indicator (panel) ..	158(2)	158(2)	57(2)
Air cond. controls	**1892(2)	—	—
Auto. temp. buttons.......	**1893(2)	—	—
Auto. temp. thumbwheel ..	**53		
Cornering lights..	1195(2)	—	—
Heater controls ..	**1892(2)	—	—
Side marker.....	67(2)	—	—
Stereo indicator..	1445	—	—
Super lite indicator	1445	—	—
Tail light ...	1095(2)		
Switch lighting ...	1892	—	—

*Included in instrument cluster lighting.
**Headlamp rheostat dimming.
***Hood mounted.

1970–71

	Polara, Monaco	Coronet	Charger
Air conditioning indicator	—	1892	1892
Ash tray.................	**1445	1445	1445
Back-up lights...........	1156(2)	1156(2)	1156(2)
Brake system warning light.................	158	158	57
Clock	*	57	57
Courtesy lamp...........	90	89	89
Dome and/or "C" pillar light...........	551	1004	1004
Door, pocket panel and/or reading light...	90	90	90
Fender mounted turn signal indicator........	330(2)	330(2)	330(2)
(Tail lamp only)..........	—	—	1095
Gear selector indicator....	**57	161	161
Gear selector with console	**57	57	57
Glove compartment	1891	1891	1891
High beam indicator......	158	158	57
Ignition lamp............	1445	1445	1445
Instrument cluster illumination	**158(4)	158	57,158
License light	67	67	67
Map light	90	90	90
Oil pressure indicator.....	158	158	158
Park and turn signal......	1157(2)	1157(2)	1157(2)
Radio	**1816	1816	1816
Radio with tape..........	—	1815	—
Reverse 4-speed transmission indicator	—	53	53
Sealed beam—hi-beam (No. 1).............	4001	4001	4001
Sealed beam—hi-lo beam (No. 2).............	4002	4002	4002
Seat belt indicator	—	53	53
Tachometer.............	—	57	—
Tail, stop and turn signal..	1157(2)	1157(2)	1157(2)
Trunk and/or under hood light.............	1004	1004	1004
Turn signal indicator (panel)...............	158(2)	158	158
Air cond. controls	**1892(2)	—	—
Auto. temp. buttons.....	**1893(2)	—	—
Auto. temp. thumbwheel ..	**53	—	—
Cornering lights	1195(2)	—	—
Heater controls..........	**1892(2)	—	—
Side marker	67(2)	—	—
Stereo indicator	1445	—	—
Super lite indicator	1445	—	—
Tail light	1095(2)	—	—
Switch lighting	1892	—	—

*Included in instrument cluster lighting.
**Headlamp rheostat dimming.
***Hood mounted.

LIGHT BULBS
Plymouth

1964

	Plymouth
Sealed beam—lo-beam	4002
Sealed beam—hi-beam	4001
Tail, stop & turn signal	1034
Park & turn signal	1034
Back-up lamps	1073
License lamp	67
Trunk and/or under hood lamp	1004
Glove compartment	1891
Radio	1892-AM 57-FM
Transmission control push buttons	(a)
Handbrake indicator	57
Dome lamp	1004
Map lamp	90
Clock	(a)
Heater and/or A.C. control P/B	(a)
Turn signal indicator	57
High beam indicator	57
Oil pressure indicator	57
Instrument cluster illumination	57

(a) Included in instrument cluster lighting.

1965–66

	Belvedere, Satellite	Fury I, II, III
Sealed beam—lo-beam	—	4002
Sealed beam—hi-beam	—	4001
Single beam 2 filament	6012	—
Tail, stop & turn signal	1034	1034
Park & turn signal	1034A	1034A
Back-up lamps	1073	1073
License lamp	67	67
Trunk and/or under hood lamp	1004	1004
Glove compartment	1891	1891
Radio	1893	1893
Transmission gear shift control	1445	—
Handbrake indicator	57	256
Dome lamp	1004	1004
Map lamp	90	90
Clock	—	57
Heater and/or A.C. control	—	57
Turn signal indicator	158	57
High beam indicator	158	158
Oil pressure indicator	158	158
Instrument and speedometer cluster illumination	57, 158	158
Emergency flasher	57	57
Auto pilot	—	1816
Gear selector with console	57	57

1967

	Belvedere, Satellite	Fury, V.I.P.
Air conditioning indicator	1893	FL
Auto pilot	—	1445 & 1892
Brake system warning light	257	57
Clock	57	—
Dome and/or "C" pillar light	1004	1004
Door, pocket panel and/or reading light	90	90
Fender mounted turn signal indicator	330	330
Gear selector indicator	1445	FL
Gear selector with console	57	57
Glove compartment	1891	1891
Handbrake indicator	57	57
High beam indicator	158	57
Instrument cluster illumination	158	1893
Ignition switch	—	1445
License light	67	67
Map light	90	90
Oil pressure indicator	158	57
Park and turn signal	1034A	1034
Portable reading light	—	90
Radio	1893	FL
Sealed beam—hi-beam (No. 1)	4001	4001
Sealed beam—hi-lo beam (No. 2)	4002	4002
Tail, stop and turn signal	1034	1034
Trunk and/or under hood light	1004	1004
Turn signal indicator (panel)	158	57

FL—Floodlighted

1968

	Belvedere, Satellite	Fury, V.I.P.
Air conditioning indicator	1892*	FL
Ash receiver	1445*	FL
Back-up lights	1073	1073
Brake system warning light	158	57
Clock	57	—
Courtesy lamp	89	—
Dome and/or "C" pillar light	1004	1004
Door, pocket panel and/or reading light	90	90
Fender mounted turn signal indicator	330	330M
Gear selector indicator	1445	FL
Gear selector with console	57	57
Glove compartment	1891	1891
High beam indicator	158	57
Instrument cluster illumination	158	1893
Ignition switch	1445	1445
License light	67	67
Map light	90	1445
Oil pressure indicator	158	57
Park and turn signal	1034	1034NA
Radio	1816	FL
Sealed beam—hi-beam (No. 1)	4001	4001
Sealed beam—hi-lo beam (No. 2)	4002	4002
Side Marker	1895	1895
Tachometer	57	FL
Tachometer with console	—	1816
Tail, stop and turn signal	1034	1034
Trunk and/or under hood light	1004	1004
Turn signal indicator (panel)	158	57

FL—Floodlighted
*Included in instrument cluster.

1969

	Belvedere, Satellite	Fury, V.I.P.
Air conditioner control and auto-temp	**1892 (2)	**1893 (FL)
Ash receiver	**1892	**1445
Back-up lights	1156 (2)	1156 (2)
Brake system warning indicator	158	57
Clock	** 57	**1893 (FL)
Cornering lights	—	1293 (2)
Courtesy lamp	89	—
Dome lamp	1004	551
Pocket panel lamp	90	90
Fender mounted turn signal indicator	330 (2)	330 (2)
Gear selector indicator	**1445	**1893 (FL)
Gear selector with console	** 57	** 57
Glove compartment	1891	1891
High beam indicator	158	1892
Instrument cluster and speedometer illumination	** 158 (4)	**1893 (FL)
Ignition lamp	1445	1445
License light	67	704
Map and courtesy lamp	90	90
Oil pressure indicator	158	57
Park and turn signal	1157 (2)	1157 (2)
Radio	**1816	**1893 (FL)
Radio with tape	**1815	—
Sealed beam—hi-beam (No. 1)	4001	4001
Sealed beam—hi-lo beam (No. 2)	4002	4002
Stereo indicator	—	1445
Switch lighting	—	**1893 (FL)
Tachometer	** 57	**1893 (FL)
Tail, stop and turn signal	1157 (2)	1157 (2)
Trunk and/or under hood lamp	1004	1004
Turn signal indicator (panel)	158 (2)	57 (2)
Reverse 4-speed transmission indicator	53	53
Heat control	—	1893

FL—Floodlighted
NA—Not Available
*—Included in instrument cluster lighting
**—Headlamp rheostat dimming

1970–71

	Belvedere, Satellite	Fury
Air conditioner control and auto-temp	1892	1893
Ash receiver	1445	1445
Back-up lights	1156	1156
Brake system warning indicator	158	57
Clock	57	1893
Cornering lights	—	1293
Courtesy lamp	89	—
Dome lamp	1004	550
Pocket panel lamp	90	90
Fender mounted turn signal indicator	330	330
Gear selector indicator	161	1893
Gear selector with console	57	57
Glove compartment	1891	1891
High beam indicator	158	1892
Instrument cluster and speedometer illumination	158	1893
Ignition lamp	1445	1445
License light	67	67
Map and courtesy lamp	90	90
Oil pressure indicator	158	57
Park and turn signal	1157	1157
Radio	1816	1893
Radio with tape	1815	—
Sealed beam—hi-beam (No. 1)	4001	4001
Sealed beam—hi-lo beam (No. 2)	4002	4002
Switch lighting	—	1893
Tachometer	57	1893
Tail, stop and turn signal	1157	1157
Trunk and/or under hood lamp	1004	1004
Turn signal indicator (panel)	158	57
Reverse 4-speed transmission indicator	53	53
Heat control	—	1893

FUSES AND CIRCUIT BREAKERS
DODGE

1964

FUSES

Radio	3 AG/AGC; 7.5 AMP
Heater or air conditioning	3 AG/AGC; 20 AMP
Accessories	3 AG/AGC; 15 AMP
Rear air conditioning (880 only)	3 AG/AGC; 20 AMP
Cigar lighter	3 AG/AGC; 20 AMP
Tail, stop, dome	3 AG/AGC; 20 AMP
Instrument lamps	3 AG/AGC; 2 AMP

CIRCUIT BREAKERS

Windshield wiper—variable speed (back of wiper switch)	6 AMP
Windshield wiper—single speed (integral with wiper switch)	5 AMP
Lighting system (integral with headlamp switch)	15 AMP
Elec. window lifts, six-way seat, top lift (behind left front kick panel)	30 AMP

1965

FUSES

Radio	3 AG/AGC; 5 AMP
Heater or air cond.	3 AG/AGC; 20 AMP
Accessories	3 AG/AGC; 20 AMP
Cigar lighter	3 AG/AGC; 20 AMP
Tail, stop, dome	3 AG/AGC; 20 AMP
Instrument lights	3 AG/AGC; 2 AMP—Dart
	3 AMP—Coronet
	4 AMP—Polara, Monaco, 880
Rear Air conditioning (Polara, Monaco, 880 only)	3 AG/AGC; 20 AMP

CIRCUIT BREAKERS

Winshield wiper—variable speed (integral with wiper switch)	7½ AMP
	(6—Coronet)
Windshield wiper—single speed (integral with wiper switch	5 AMP
Lighting system (integral with headlight)	15 AMP
	(20—Coronet)
Power windows, power seats, top lift (behind left front kick panel)	30 AMP
Door locks (Polara, Monaco, 880 only) (behind left front kick panel)	15 AMP

1966

FUSES

Accessories	3 AG/AGC; 20 AMP
Cigar lighter	3 AG/AGC; 20 AMP
Heater or air cond.	3 AG/AGC; 20 AMP
Instrument lights	3 AG/AGC; 4 AMP—Polara, Monaco
	3 AMP—Coronet
Radio	3 AG/AGC; 5 AMP
Tail, stop, dome	3 AG/AGC; 20 AMP
Rear air cond. (Monaco, Polara only)	3 AG/AGC; 20 AMP

CIRCUIT BREAKERS

Door locks (behind left front kick panel)	15 AMP

Lighting system (integral with headlight)	20 AMP
Power windows, power seats, top lift and tail gate (behind left front kick panel)	30 AMP
Windshield wiper—single speed (integral with wiper switch)	5 AMP
Variable speed (integral with wiper switch)	7½ AMP
	(6—Coronet)

1967

FUSES

Accessories	20 AMP
Cigar lighter (front)	20 AMP
Console*	20 AMP
Emergency flasher**	20 AMP
Heater or air conditioning	20 AMP
Instruments	4 AMP
Radio	5 AMP
Tail, stop, dome**	20 AMP

*Inline fuse.

**Emergency flasher and stoplight use same inline fuse.

***Dome light and front cigar lighther on same fuse.

CIRCUIT BREAKERS

Cigar lighter—rear (behind left front cowl trim panel)	15 AMP
Convertible top (behind left front cowl trim panel)	30 AMP
Door locks (behind left front cowl trim panel)	15 AMP
Headlights (integral with headlight swtich)	20 AMP
Power seats (behind left front cowl trim panel)	30 AMP
Power tailgate (behind left front cowl trim panel)	30 AMP
Power windows (behind left front cowl trim panel)	30 AMP
Windshield wiper (integral with wiper switch)	7.5 AMP

1968

FUSES

	Monaco, Polara	Coronet, Charger
Accessories	20 AMP	20 AMP
Cigar lighter (front) and dome lamp	20 AMP	20 AMP
Console (inline fuse; Monaco, Polara)	20 AMP	20 AMP
Emergency flasher	20 AMP	20 AMP
Heater or air conditioning	20 AMP	20 AMP
Instruments	4 AMP	3 AMP
Radio and back-up lamps	5 AMP	5 AMP
Tail and stop lamps	20 AMP	20 AMP

CIRCUIT BREAKERS

	Monaco	Polara
Cigar lighter—rear (behind left front cowl trim panel)	15 AMP	15 AMP
Convertible top (behind left front cowl trim panel)	30 AMP	30 AMP
Door locks (behind left front cowl trim panel)	15 AMP	15 AMP
Headlights (integral with headlight switch)	20 AMP	20 AMP

	Coronet	Charger
Power seats (behind left front cowl trim panel)	30 AMP	30 AMP
Power tailgate (behind left front cowl trim panel)	30 AMP	30 AMP
Power windows (behind left front cowl trim panel)	30 AMP	30 AMP
Windshield wiper (integral with wiper switch)	7.5 AMP	7.5 AMP

	Coronet	Charger
Convertible top (instrument panel cluster behind ammeter)	30 AMP	30 AMP
Headlights (integral with headlight switch)	20 AMP	15 AMP
Power tailgate (instrument panel cluster behind ammeter)	30 AMP	—
Power windows (instrument panel cluster behind ammeter)	30 AMP	30 AMP
Windshield wipers (integral with wiper switch)	6 AMP	6 AMP

1969–71

FUSES

Accessories	20 AMP
Console	20 AMP
Emergency flasher	20 AMP
Heater or air conditioning	20 AMP
Instrument lamps	3 AMP
Radio and back-up lamps	7.5 AMP
Stop and dome	20 AMP
Tail and cigar lighter	20 AMP

CIRCUIT BREAKERS

	Monaco	Polara
Cigar lighter—rear (on fuse block)	30 AMP	30 AMP
Convertible top (on fuse block)	30 AMP	30 AMP
Door locks (behind right front cowl trim panel)	15 AMP	15 AMP
Headlights (integral with headlight switch)	20 AMP	20 AMP
Power seats (on fuse block)	30 AMP	30 AMP
Power tailgate (on fuse block)	30 AMP	30 AMP
Power windows (on fuse block)	30 AMP	30 AMP
Windshield wiper (integral with wiper switch)	7.5 AMP	7.5 AMP

	Coronet	Charger
Convertible top (instrument panel cluster behind ammeter)	30 AMP	30 AMP
Headlights (integral with headlight switch)	20 AMP	20 AMP
Power tail gate (instrument panel cluster behind ammeter)	30 AMP	—
Power windows (instrument panel cluster behind ammeter)	30 AMP	30 AMP
Windshield wipers (integral with wiper switch (3-speed)	7.5 AMP	7.5 AMP
(2-speed)	6.0 AMP	6.0 AMP

FUSES AND CIRCUIT BREAKERS
PLYMOUTH

1964
FUSES

Radio	3 AG/AGC; 7.5 AMP
Heater or air conditioning	3 AG/AGC; 20 AMP
Accessories	3 AG/AGC; 15 AMP
Cigar lighter	3 AG/AGC; 20 AMP
Tail, stop, dome	3 AG/AGC; 20 AMP
Instrument lamps	3 AG/AGC; 2 AMP

CIRCUIT BREAKERS

Winshield wiper—variable speed (back of wiper switch)	6 AMP
Windshield wiper—single speed (integral with wiper switch)	5 AMP
Lighting system (integral with headlamp switch)	15 AMP
Elec. window lifts, power seats, top lift (Electric window lifts and power seats—Plymouth only) (behind left front kick panel)	30 AMP

1965–66
FUSES

Radio	5 AMP
Heater or air conditioning	20 AMP
Accessories	20 AMP
Cigar lighter	20 AMP
Tail, stop, dome	20 AMP
Instrument lamps	*

*AV-1 and AV-2—2 ampere; AR-1 and AR-2—3 ampere; AP-1 and AP-2—4 ampere.

CIRCUIT BREAKERS

Winshield wiper—variable speed (integral with wiper switch)	*
Windshield wiper—single speed (integral with wiper switch)	5 AMP
Lighting system (integral with headlamp switch)	15 AMP
Power windows, power seats, top lift (behind left front kick panel)	30 AMP
Electric door locks (behind left front kick panel)	15 AMP

*AR-1 and AR-2—6 ampere; AP-1 and AP-2—7½ ampere.

1967
FUSES

Accessories	20 AMP
Cigar lighter (front)	20 AMP
Console*	20—All

Emergency flasher**	20 AMP
Heater or air conditioner	20 AMP
Instrument lights	3—Belvedere and Satellite
	4—Fury and V.I.P.
Radio	5 AMP
Tail, stop, dome	20 AMP

*Inline fuse.

**Stop light and emergency flasher use same inline fuse on Fury and V.I.P. models only.

CIRCUIT BREAKERS

	Belvedere Satellite	Fury V.I.P.
Cigar lighter—rear (behind left front cowl trim panel)	—	15 AMP
Convertible top (*integral with top lift switch)	30 AMP	30 AMP
Door locks (behind left front cowl trim panel)	—	15 AMP
Headlights (integral with headlight switch)	20 AMP	20 AMP
Power seats (behind left front cowl trim panel)	—	30 AMP
Power tailgate (*integral with tail-gate switch)	30 AMP	30 AMP
Power windows (**behind left front cowl trim panel)	30 AMP	30 AMP
Windshield wipers (integral with wiper switch)	6 AMP	7.5 AMP

*Behind left front cowl trim panel in Fury and V.I.P. models.

**Integral with top lift switch on Belvedere or Satellite convertible models.

1968
FUSES

	Belvedere Satellite	Fury V.I.P.
Accessories	20 AMP	20 AMP
Cigar lighter (front) and dome light	20 AMP	20 AMP
Console (inline fuse)	—	20 AMP
Emergency flasher	—	20 AMP
Heater or air conditioner	20 AMP	20 AMP
Instrument lights	3 AMP	4 AMP
Radio and back-up lamps	5 AMP	5 AMP
Tail stop (and emergency flasher—except Fury and VIP)	20 AMP	20 AMP

CIRCUIT BREAKERS

	Belvedere Satellite	Fury V.I.P.
Cigar lighter—rear (behind left front cowl trim panel)	—	15 AMP
Convertible top (integral with top lift switch, behind left front cowl trim panel in Fury and V.I.P. models.)	—	30 AMP
Convertible top (instrument panel cluster behind ammeter)	30 AMP	—
Door locks (behind left front cowl trim panel)	—	15 AMP
Headlights (integral with headlight switch)	20 AMP	20 AMP
Power seats (behind left front cowl trim panel)	—	30 AMP
Power tailgate (instrument panel cluster behind ammeter)	30 AMP	30 AMP
Power windows (instrument panel cluster behind ammeter)	30 AMP	30 AMP
Windshield wipers (integral with wiper switch)	6 AMP	7.5 AMP

1969–71
FUSES

	Belvedere Satellite	Fury VIP
Accessory	20 AMP	20 AMP
Console	—	20 AMP
Emergency flasher	20 AMP	20 AMP
Heater and air conditioner	20 AMP	20 AMP
Instrument lamps	3 AMP	3 AMP
Radio and back-up lamps	7.5 AMP*	7.5 AMP*
Stop and dome lamps	20 AMP	20 AMP
Tail lamps and cigar lighter	20 AMP	20 AMP

*—1970-71—20

CIRCUIT BREAKERS

	Belvedere Satellite	Fury VIP
Convertible (on fuse block integral with switch)	30 AMP	30 AMP
Door locks (behind right front cowl trim panel)	15 AMP	15 AMP
Headlights (integral with headlamp switch)	20 AMP	20 AMP
Power seats (on fuse block)	—	30 AMP
Power tail gate (on fuse block)	30 AMP	30 AMP
Power windows (on fuse block)	30 AMP	30 AMP
Windshield wipers (integral with wiper switch)	7.5 AMP	7.5 AMP

Distributor

Detailed information on distributor drive, direction of distributor rotation, cylinder numbering, firing order, point gap, point dwell, timing mark location, spark plugs, and spark advance will be found in the Specification Tables.

The 1970 440 Magnum, 440 six-pack, and Hemi 426 require idle speeds between 800 and 1000 rpm to obtain an acceptable level of hydrocarbon emission. To prevent "running on" with this high idle speed, these engines have a solenoid throttle stop which holds throttle at correct idle position, but allows throttle blades to close completely when ignition is turned off.

Distributor Assembly Removal

1. Take off the cap and wire assembly.
2. Disconnect the primary coil wire and vacuum control tube.
3. Mark the distributor and rotor relative positions.
4. Loosen the distributor mounting and lift out the distributor.
   NOTE: to simplify reinstallation, do not disturb the engine while the distributor is out.
5. Reinstall by reversing the above procedure.

Ignition Change

In compliance with anti-pollution laws involving all of the continental United States, Chrysler Motors Corporation has continued to use their

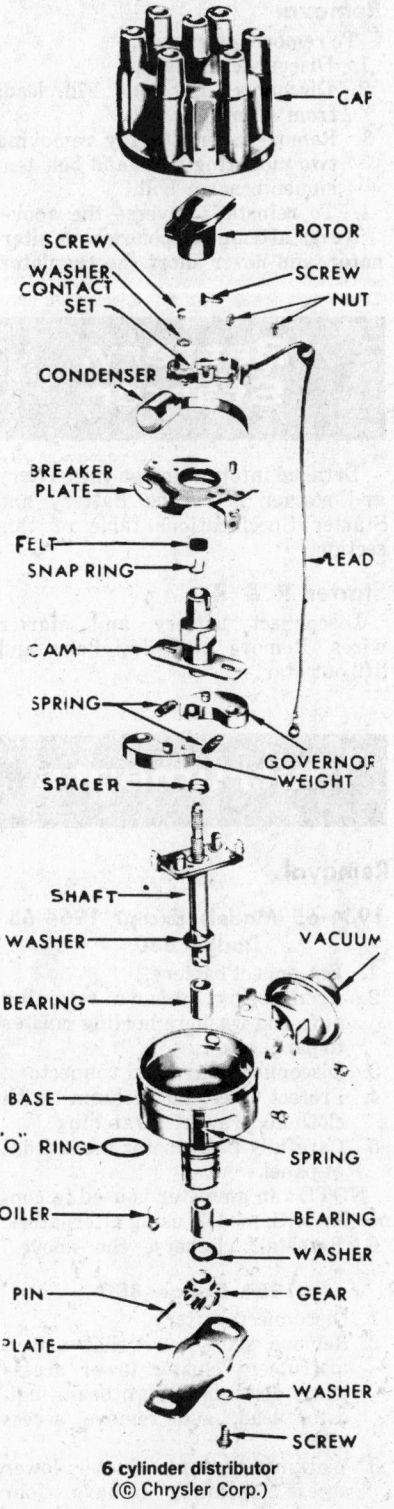

6 cylinder distributor
(© Chrysler Corp.)

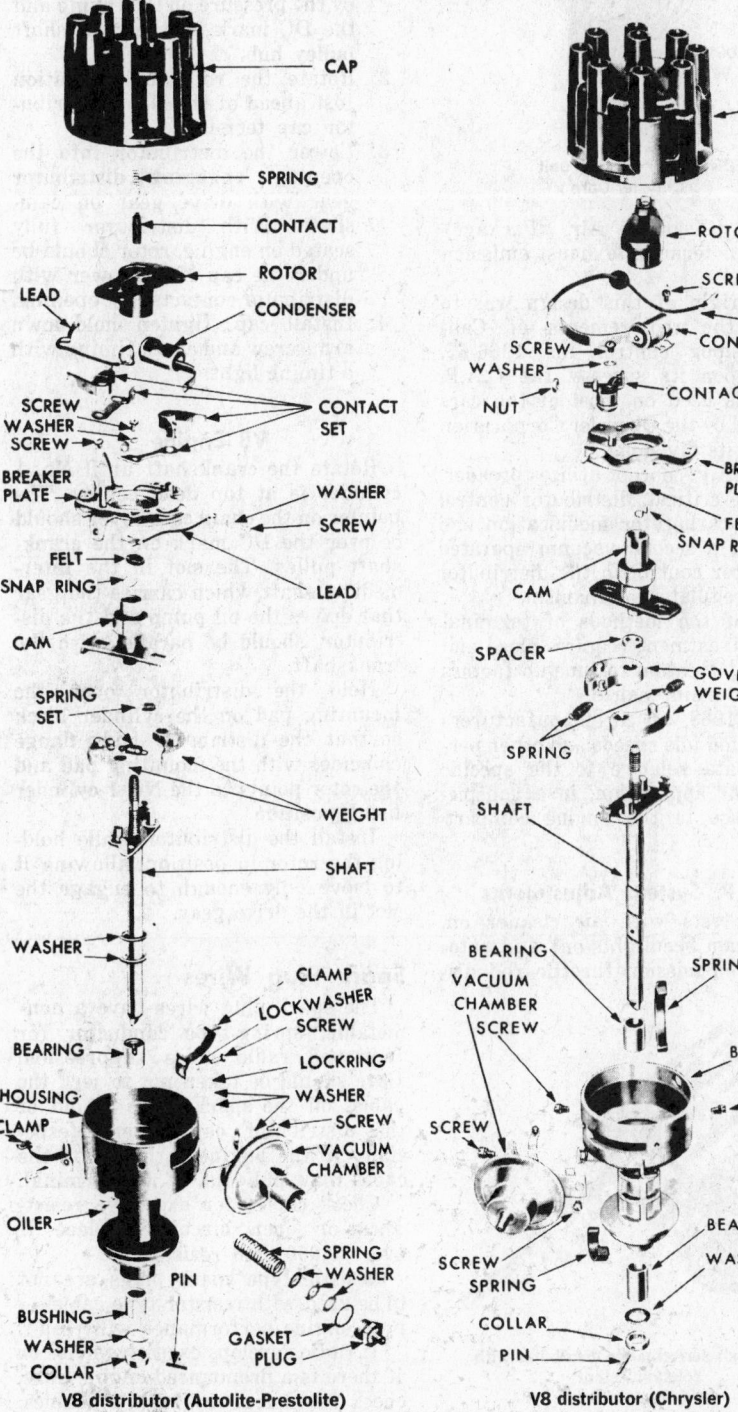

V8 distributor (Autolite-Prestolite)

V8 distributor (Chrysler)

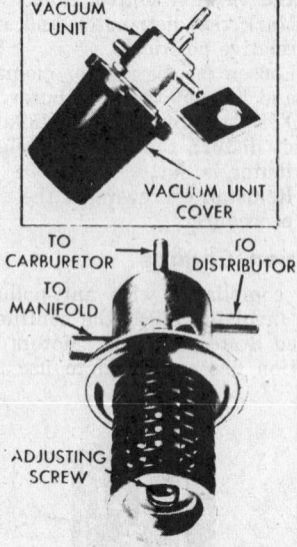

Distributor vacuum unit
(© Chrysler Corp.)

C.A.P. (Cleaner Air Package) method of terminal exhaust emission control.

The origin of this design was to satisfy the requirements of California smog control for 1966-67. Based upon its success, the C.A.P. system is used on most of the cars produced by the Chrysler Corporation in all of its divisions.

The C.A.P. concept utilizes broader, yet more critical, distributor control through carburetor modification and by using a second vacuum operated distributor control in the distributor timing regulator mechanism.

Any of the methods of terminal exhaust treatment requires close and frequent attention to tune-up factors of engine maintenance.

Since 1968, all car manufacturers have posted idle speeds and other pertinent data relative to the specific engine-car application, in a conspicuous place in the engine compartment.

C.A.P. System Adjustments

Make tests with air cleaner on, (high-beam headlights on) and automatic transmission throttle rod dis-

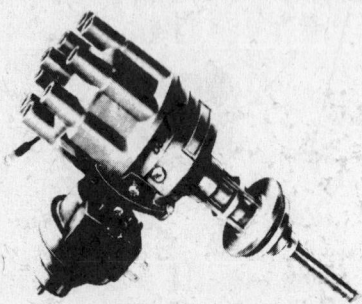

1970 High performance distributor with solenoid retard
(© Chrysler Corp.)

connected. When setting timing, disconnect and plug distributor vacuum line.

More detailed information on the C.A.P. system can be found in the Unit Repair Section.

Distributor Replacement (When Engine has been Disturbed)

Slant 6 Engine

1. Remove No. 1 spark plug and, with the thumb closing the hole, rotate the engine until No. 1 piston is up on compression at top dead center. This is determined by the pressure on the thumb and the DC mark on the crankshaft pulley hub.
2. Rotate the rotor to a position just ahead of the No. 1 distributor cap terminal.
3. Lower the distributor into the opening, engaging distributor gear with drive gear on camshaft. With distributor fully seated on engine, rotor should be under the cap No. 1 tower with distributor contacts just opening.
4. Install cap, tighten hold-down arm, screw and check timing with a timing light.

V8 Engine

Rotate the crankshaft until No. 1 cylinder is at top dead center. The pointer on the chain case cover should be over the DC mark on the crankshaft pulley. The slot in the intermediate shaft which carries the gear that drives the oil pump and the distributor, should be parallel with the crankshaft.

Hold the distributor over the mounting pad on the cylinder block so that the distributor body flange coincides with the mounting pad and the rotor points to the No. 1 cylinder firing position.

Install the distributor while holding the rotor in position, allowing it to move only enough to engage the slot in the drive gear.

Spark Plug Wires

The spark plug wires have a non-metallic, spring-type conductor for improved radio noise suppression. Care should be taken not to jerk the cables off the spark plugs or out of the distributor cap towers (especially if the engine is hot) for the cable may pull out of its terminal.

Check cables for excessive resistance or open circuit. Replace if over 30,000 ohm resistance.

Resistor-type spark plugs are not to be used with resistor-type cables—poor engine performance will result.

If radio develops excessive noise or if there is a pronounced engine miss, check for defective (broken) cables.

Alternator

References

Details on alternator and regulator are in Specification table of this section.

General information on alternator and regulator repair and troubleshooting is in the Unit Repair Section.

Removal

To remove alternator:
1. Disconnect battery.
2. Disconnect Bat. and Fld. leads from alternator.
3. Remove alternator by removing two mounting bolts and belt tensioner bracket bolt.
4. To reinstall: reverse the above.

Never attempt to polarize an alternator, and never short the regulator.

Battery, Starter

Detailed information on the battery and starter is in the Battery and Starter Specifications table of this section.

Starter R & R

Disconnect battery and starter wires. Remove attaching bolts and lift out starter.

Instruments

Removal

1964-65 Models Except 1964-65 Dodge 880

1. Disconnect battery.
2. From below, remove retaining nuts and washers holding cluster to panel.
3. Disconnect wires and connectors.
4. Protect steering column with cloth to prevent scratching.
5. Carefully pull cluster from front of panel.

NOTE: an ammeter is used in connection with models using alternators.
6. Reinstall: Reverse the above.

1964 Dodge 880

1. Disconnect battery.
2. Remove six screws that attach instrument cluster lower access plate, disconnect high beam indicator lead, and remove access plate.
3. Working through cluster lower access opening, remove four

screws that attach cluster bezel to cluster housing.

4. Remove parking sprag, temperature control lever, and clock reset knobs.
5. Remove instrument cluster bezel.
6. Disconnect wiring terminals to gauges and clock.
7. Remove four screws that attach instrument cluster to instrument cluster housing.
8. Remove cluster through lower access opening.
9. Remove two screws that attach desired gauge to cluster and remove. Reverse procedure to install.

1966-67 Belvedere and Satellite

1. Disconnect battery and tape steering column.
2. Disconnect speedometer cable, remove steering column cover and steering column clamp.
3. Remove six screws in upper and lower face of cluster bezel. Pull cluster out far enough to remove multiple connectors from headlamp and wiper switches and printed circuit board. Remove ammeter gauge wires, heater switch wires and control cables.
4. Remove cluster. Reverse procedure to install.

1966-68 Fury and VIP

1. Disconnect battery.
2. Remove eight instrument cluster light panel retaining screws, remove panel and rest panel on top of trim pad. It is not necessary to disconnect wire.
3. Remove heater or air conditioning control knobs, and check reset knob.
4. Remove six bezel retaining screws and remove bezel.
5. Remove four screws from steering column cover and drop cover with vent controls attached.
6. In vehicles with automatic transmission and column shift, remove gear selector link nut, spring washer and bolt.
7. Remove four stereo speaker grille screws and place speaker on top of instrument panel.
8. Disconnect speedometer cable, and remove five cluster mounting screws. Raise cluster slightly, roll upper edge out and disconnect ammeter leads.
9. With cluster face down disconnect fuel and temperature gauge wires and high beam, oil pressure, and turn signal light sockets.
10. Remove cluster. Reverse procedure to install.

Polara, Custom 880, Monaco—1965-66

1. Disconnect battery.
2. Disconnect speedometer cable.

3. Remove cluster mounting screws and roll cluster out.
4. Disconnect printed circuit plug and remove cluster.
5. Reinstall in reverse of above.

1966 Coronet

1. Disconnect battery, and tape top of steering column for scratch protection.
2. If air-conditioned, remove left spot cooler, duct and hose, and the fuse block.
3. Disconnect speedometer cable at head.
4. Remove steering column support bracket, then lower column support plate at firewall.
5. Remove radio knobs, mounting nuts, ash tray housing and the lighter.
6. From under the dash, remove mounting nut next to the heater blower switch.
7. Remove headlight switch knob and bezel by depressing release button on headlight switch and pulling out on headlight switch knob. Use tool C-3824 to remove switch bezel. Do not remove switch from the panel.
8. Loosen set screw in wiper switch knob and remove the switch bezel with tool C-3824. Do not remove switch from panel.
9. Remove four instrument cluster retaining screws and pull cluster out far enough to disconnect printed circuit board and ignition switch multiple connectors. Disconnect the two ammeter wires and remove from car.
10. Install by reversing removal procedure.

NOTE: all instruments may be serviced after removing the four screws holding the cluster bezel to the cluster housing and the ignition switch. Remove the temperature or fuel gauge terminal nuts from the printed circuit board. The alternator gauge is serviced in the same way from the cluster housing. The printed circuit board is serviced after the instruments, voltage limiter, light bulbs and four retaining screws are removed.

1966-67 Charger and 1967 Coronet

1. Disconnect battery.
2. Tape steering column to protect paint.
3. Remove heater control knobs and radio knobs and mounting nuts.
4. Open glove box door.
5. Disconnect speedometer cable.
6. Remove wiring harness from clips at steering column bracket.
7. Remove eight Phillips head screws from upper and lower lips of cluster bezel on Coronets or seven Phillips head screws on Chargers.

8. Carefully pull cluster out and to the right far enough to reach around left end of cluster and disconnect printed circuit multiple connector.
9. Remove ammeter wires and clock light on Coronets. On Chargers, remove wires from each gauge including tachometer. Disconnect panel lighting (white wire).
10. Roll top of cluster down while working to right over open glove box door. Remove Cluster. Reverse procedure to install.

1967-68 Polara and Monaco, 1968-70 Coronet, Belvedere, and Satellite

1. Disconnect battery.
2. Remove steering column trim cover by removing four screws.
3. Roll carpet back and remove lower mounting plate.
4. Remove steering column clamp.
5. Remove upper trim moulding, if so equipped, and left side trim moulding and left side trim plate, if so equipped.
6. Remove radio trim plate, and four switch bezel mounting screws, and switch bezel.
7. Remove ignition switch.
8. Remove center air conditioning register cover, if so equipped.
9. Remove lower trim pad by removing six screws from under panel and four screws from front panel.
10. Disconnect speedometer cable.
11. Remove six cluster mounting screws, and rock cluster out.
12. Disconnect wiring and remove cluster. Reverse procedure to install.

1968-70 Charger

1. Disconnect battery.
2. Remove steering column opening cover, steering column lower support plate, and steering column upper clamp.
3. Disconnect speedometer cable.
4. Remove five screws mounting cluster to panel.
5. Release wire harness from three retaining clips, and rock cluster out of panel far enough to disconnect wiring at ammeter, switches, tachometer or clock, light bulbs, and printed circuit. Roll out instrument panel. Reverse procedure to install.

1969-70 Polara and Monaco

1. Disconnect battery.
2. Remove steering column cover.
3. Remove gear shift indicator from column.
4. Remove lower column floor plate by removing three bolts.
5. Remove three upper column mounting nuts.
6. Lower steering and let wheel rest on seat.

7. Remove switch bezel, radio bezel.
8. Remove cluster trim pad and trim pad.
9. Disconnect clock reset cable at instrument panel lower reinforcement.
10. Disconnect clock electrical lead.
11. Remove five cluster assembly mounting screws.
12. Roll cluster out slightly and disconnect speedometer cable.
13. Disconnect gear shift indicator lamp.
14. Disconnect main harness from cluster.
15. Disconnect alternator gauge electrical leads and remove cluster. Reverse procedure to install.

1969-70 Fury and VIP

1. Disconnect battery.
2. Remove nine panel screws, and disconnect and remove light panel.
3. Remove steering column cover.
4. Remove radio trim bezel.
5. Remove left trim bezel and/or spot cooler.
6. Remove left and center air conditioner ducts, if so equipped.
7. From under panel disconnect leads to switches and clock, lamp assemblies, and speedometer cable.
8. Remove gear shift indicator pointer from column.
9. Tape column to protect paint.
10. Remove column upper clamp and lower support.
11. Lower column and rest on seat cushion.
12. Remove eight screws mounting cluster to instrument panel. Roll cluster out. Disconnect electrical connections to high beam indicator, fuel gauge, ammeter, and temperature gauge. Remove cluster. Reverse procedure to install.

Speedometer

Removal

1964-65 Plymouth, 1964-66 Coronet, and 1969-70 Belvedere and Satellite

After removing instrument cluster, remove speedometer by taking out two mounting screws.

1966-68 Plymouth, 1966-70 Charger, 1967-70 Coronet, and 1967-68 Monaco and Polara

1. Remove instrument cluster.
2. With cluster face down remove eight cluster bezel to cluster housing screws and separate the two.
3. On some models remove temperature, fuel and ammeter gauges to gain clearance.
4. Remove two screws and rubber washers, and withdraw speedom-

eter. Reverse procedure to install.

1969-70 Fury and VIP

1. Remove instrument cluster.
2. Remove gearshift indicator housing from cluster by removing two screws.
3. Remove cluster assembly from bezel assembly by removing four screws.
4. Remove cluster lens from cluster housing by turning up two upper cluster mounting pads.
5. Remove two mounting screws and remove speedometer. Reverse procedure to install.

1964 Dodge 880

1. Remove instrument cluster.
2. Disconnect speedometer cable.
3. Remove heater and transmission control buttons by pulling out on buttons. Remove neutral button by turning button out of metal slide of the control box after control box has been removed from the die cast mounting box.
4. Remove two nuts that attach transmission control box to die cast mounting box, and push transmission control box clear of mounting box. Remove neutral button.
5. Remove two screws that attach heater control assembly to instrument cluster housing, and push heater control assembly clear of speedometer.
6. Unplug three speedometer illumination bulbs from speedometer.
7. Remove four screws that attach speedometer to instrument cluster housing.
8. Remove instrument panel support pencil brace from left side of steering column opening.
9. Remove speedometer by pulling it down and out through area on left side of steering column.

1965-66 Dodge 880, Monaco, and Polara

1. Disconnect battery.
2. Disconnect speedometer cable.
3. Remove three screws that attach speedometer cluster to housing.
4. Roll cluster out and disconnect printed circuit plug.
5. Remove cluster.

1969-70 Monaco and Polara

On 1969-70 Polaras and Monacos the speedometer is integral with cluster.

Tachometer—Charger

The tachometer is removed from the instrument cluster housing after bezel and housing are separated. The electroluminescent light wire is disconnected from the tachometer dial and the four mounting screws are removed from the back of the housing.

Brakes

Brake Information

Specific information on brake cylinder sizes can be found in the General Chassis and Brake Specifications table.

Information on brake adjustments, band replacement, disc brakes, bleeding procedure, master and wheel cylinder overhaul can be found in the Unit Repair Section.

Information on troubleshooting and overhauling power brakes can be found in the Unit Repair Section.

Information on the grease seals which may need replacement can be found in the Unit Repair Section.

Beginning 1963

A new servo-contact, self-energizing brake is used. It uses a double-acting wheel cylinder at tops of shoes at each assembly.

It is also a self-adjusting brake. It operates through a link, cable and return-spring connected so that, when brake is applied during reverse stops, link indexes the star wheel to maintain proper shoe clearance.

Beginning 1967

Since 1967, a tandem-type master cylinder has been used on all models. This design divides the brake hydraulic system into two independent and hydraulically separated halves.

Details and repair procedures on this tandem system may be found in the Unit Repair Section.

Brake Pedal Clearance

1963 Through 1966

Cars with conventional brakes have a change incorporated in the brake master cylinder. With this change, a pedal stop is built into the master cylinder. This eliminates the necessity of a free-play adjustment on the master cylinder pushrod.

Since the master cylinder piston determines the position of the brake pedal when the brakes are released, a free-play at the pedal is no longer needed. However, if the pedal is pulled back by hand, some play may be noted, due to clearance allowed at the pushrod eye for self-alignment of the pushrod.

Pulling on the brake pedal in a direction away from the master cylinder (with a force of 50 lbs. or more) will result in deflection of the piston stop. Since this would result in excess pedal travel before any braking effort would be realized, don't do it.

During disassembly, the master cylinder pushrod cannot be removed from the piston; therefore, these

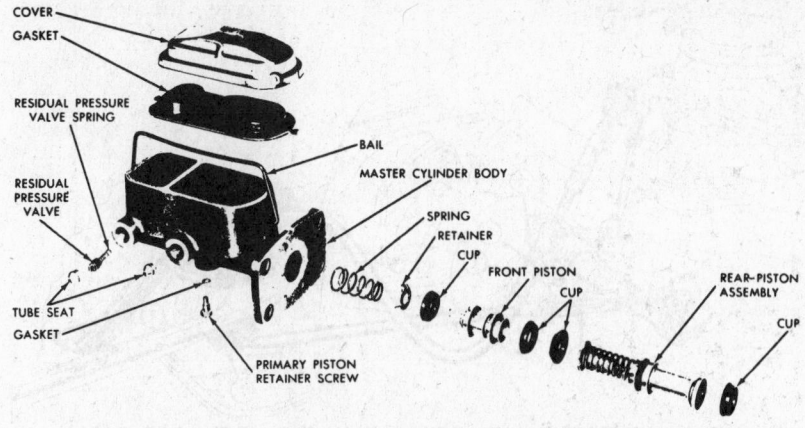

Dual type master cylinder used with disc brakes (© Chrysler Corp.)

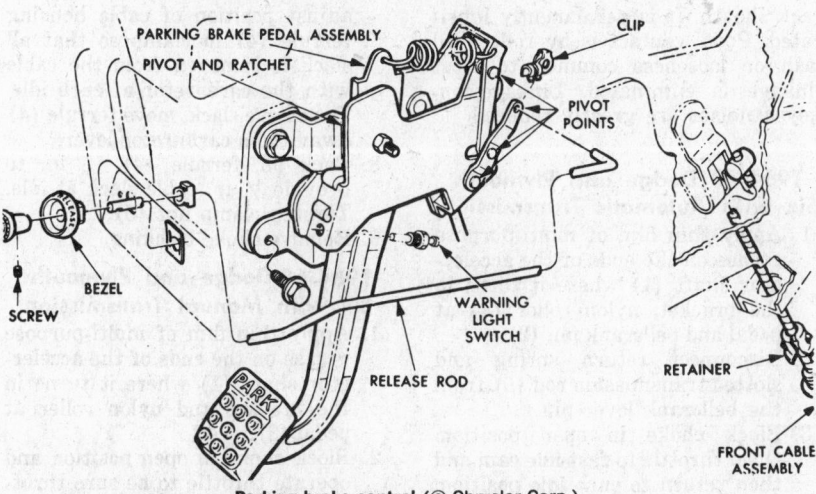

Parking brake control (© Chrysler Corp.)

parts along with the piston stop and boot retainer, are removed from the cylinder as an assembly and are serviced as an assembly.

The new- and old-type master cylinders are not interchangeable, because the old type necessitates installing a pedal stop and a pedal return spring, as well as the adjustable pushrod and nut.

Master Cylinder Removal— 1964-66

The master cylinder is mounted on the front side of the firewall under the hood.

1. Disconnect the brake pushrod and stop light wires from under the dash, remove the brake line, unbolt the master cylinder from the dash panel.

Master Cylinder Removal— 1967-70

1. Disconnect front and rear brake tubes from master cylinder.

NOTE: on drum brake master cylinders, residual pressure valves will keep cylinder from draining, but front brake outlet (rearmost) must be plugged on disc brake master cylinders.

2. Remove nuts that attach master cylinder to cowl panel or power brake unit.
3. On manual drum brakes, disconnect pedal pushrod from brake pedal.
4. Slide master cylinder straight out from cowl or power brake unit.

Master Cylinder Refilling

The master cylinder is located on the engine side of the firewall and can be filled by raising the hood and removing the top from the master cylinder body.

Disc Brakes

Since 1965, disc brakes have been available on front wheels of some models. Complete service procedures are covered in the Unit Repair Section.

Parking Brake Adjustment

1. Release parking brake lever and loosen cable adjusting nut to ensure cable is slack. Before loosening cable adjusting nut, clean threads with wire brush and lubricate.
2. Tighten cable adjusting nut until a slight drag is felt while rotating wheel. Loosen cable enough to allow both wheels to rotate freely. Back off cable adjusting nut two full turns.
3. Apply and release parking brake several times. Test to see that rear wheels rotate freely without grabbing.

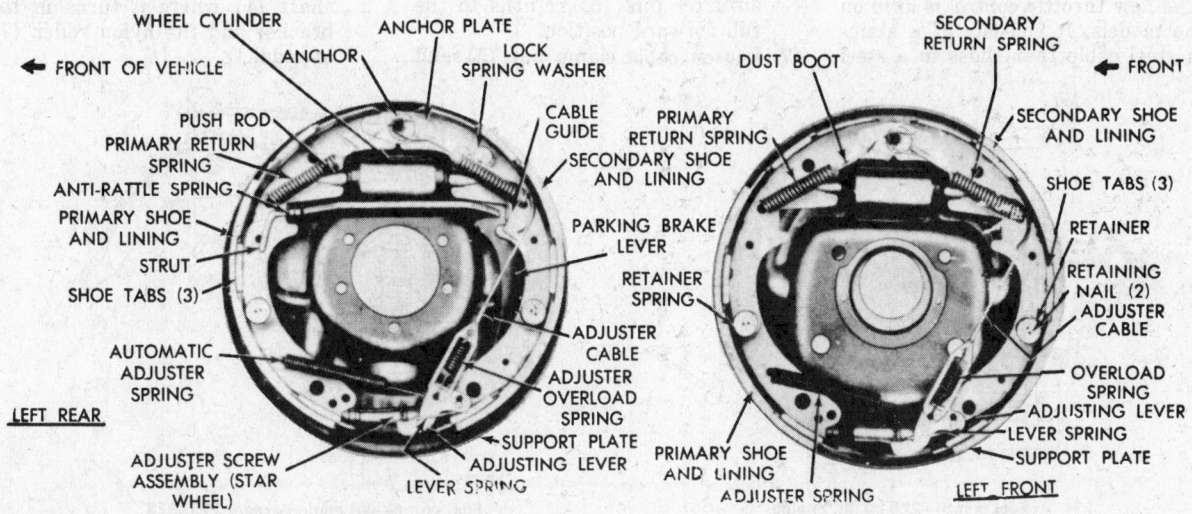

Drum brakes, showing major components (© Chrysler Corp.)

Power Brake

1965-70

Various types of power brakes are used. One is the vacuum-type, using a master cylinder and pedal linkage of the reactionary type. A vacuum cylinder is combined with a conventional master cylinder.

Another is a tandem-diaphragm-type, consisting of a self-contained vacuum hydraulic power unit.

The basic elements are vacuum power chamber with a front and rear shell, a center plate, front and rear diaphragm, pushrod and diaphragm return spring.

A control valve integral with the diaphragms regulates the amount of application.

For information on reconditioning see Unit Repair Section.

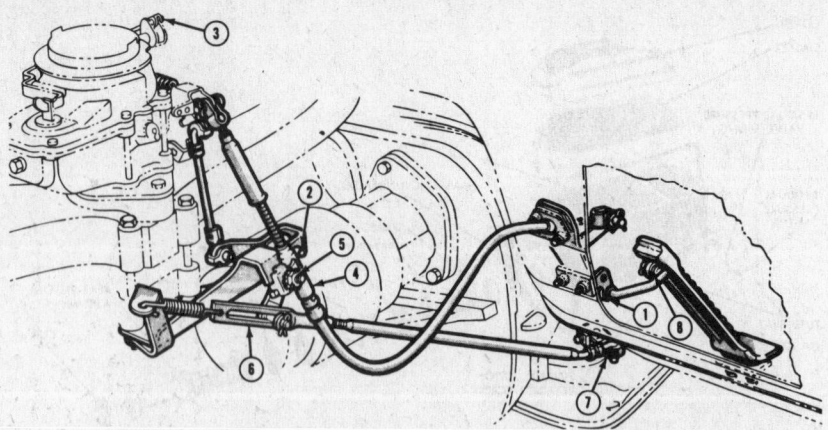

Linkage—1964-68 Coronet and Plymouth six-cylinder (© Chrysler Corp.)

Fuel System

Data on capacity of the gas tank will be found in the Capacities table. Data on correct engine idle speed and fuel pump pressure will be found in the Tune-Up Specifications table.

Information covering operation and troubles of the fuel gauge will be found in the Unit Repair Section.

Detailed information on the carburetor, and how to adjust it, will be found in the Unit Repair Section. Carter, Holley, Rochester and Stromberg carburetors are covered.

Fuel Pump Removal

All Models

Remove all lines at the fuel pump, and the pump-to-block mounting screws. Remove the pump.

Throttle Linkage Adjustment

1963-64

The new throttle control is used on some models. It consists of a stainless steel cable that slides in a steel rack sheath. It is permanently lubricated. Pedal contact is by roller. All lash or looseness common to most linkage is eliminated. Linkage-conveyed noises are greatly reduced.

1964-70 Dodge and Plymouth Six with Automatic Transmission

1. Apply thin film of multi-purpose grease on the ends of the accelerator shaft (1) where it turns in the bracket, nylon roller (8) at pedal and bellcrank pin (2).
2. Disconnect return spring and slotted transmission rod (6) from the bellcrank lever pin.
3. Block choke in open position. Open throttle to fast idle cam and then return to curb idle position.
4. Hold transmission lever forward (7) against its stop (rod or lever must not be moved vertically) and adjust transmission rod at threaded adjustment (6) at upper end. Rear of slot should contact bellcrank lever pin without exerting any force.
5. Lengthen rod by two full turns.
6. Assemble slotted adjustment (6) to bellcrank pin with washer and retainer pin and reconnect return spring. Check to be sure slotted adjuster link (6) returns to the full forward position.
7. Loosen cable clamp nut (5) and

adjust position of cable housing ferrule (4) in clamp so that all slack is removed from the cable with the carburetor at curb idle. To remove slack. move ferrule (4) away from carburetor lever.
8. Back off ferrule (4) ¼ in. to provide ¼ in. cable slack at idle. Tighten clamp nut (5).
9. Remove choke blocking.

1964-68 Dodge and Plymouth Six with Manual Transmission

1. Apply thin film of multi-purpose grease on the ends of the accelerator shaft (1) where it turns in the bracket and nylon roller at pedal (8).
2. Block choke in open position and operate throttle to be sure throttle returns to curb idle.
3. Follow steps 7, 8, and 9 above, under Automatic Transmission. (See illustration.)

1965 Dodge and Plymouth 273 Engine with Automatic Transmission and 1964 and some 1965 Dodge and Plymouth 318 Engines with Automatic Transmission

1. Apply thin film of multi-purpose grease on both ends of accelerator shaft (1) where it turns in the bracket and the nylon roller (7) at pedal.

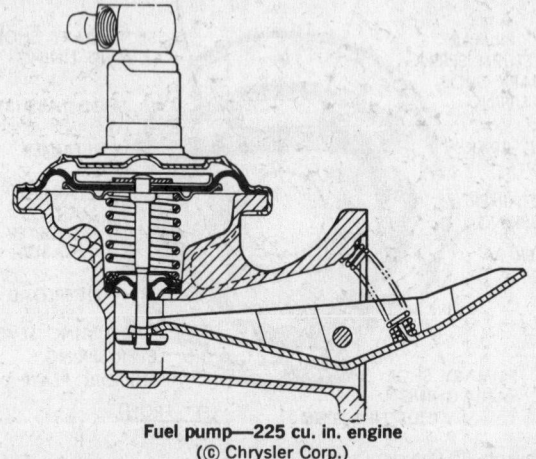

Fuel pump—225 cu. in. engine
(© Chrysler Corp.)

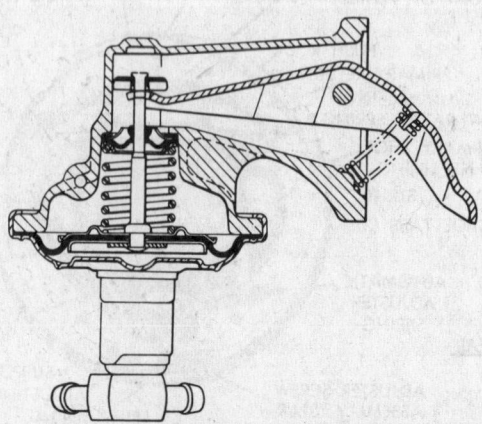

Fuel pump—V8 engine except 273, 318
(© Chrysler Corp.)

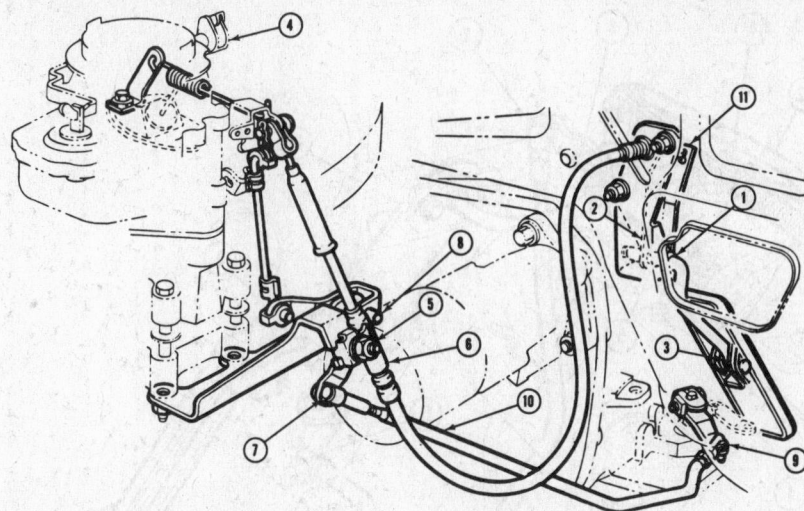

Linkage—1967 six cylinder (© Chrysler Corp.)

2. Disconnect return spring and slotted transmission rod (5) from carburetor lever pin.

3. Block choke in open position and operate throttle to be sure throttle returns to curb idle.

4. Hold transmission lever (6) forward against its stop (rod or lever must not be moved vertically) and adjust length of transmission rod at threaded adjustment (5). Rear of slot should contact Carburetor lever pin without exerting any force.

5. Lengthen rod by two full turns.

6. Assemble slotted adjustment (5) to carburetor lever pin and install

washer and retainer pin. Connect linkage return spring. Check to be sure slotted adjuster link (5) returns to full forward position.

7. Loosen cable clamp nut (4) and adjust ferrule in clamp to remove all slack with the carburetor at curb idle.

8. Back off ferrule (3) ¼ in. This provides ¼ in. cable slack at idle. Tighten cable clamp nut (4).

9. Remove choke blocking.

1965 Dodge and Plymouth 273 Engines with Automatic, and 1964 and some 1965 318 Engines with Automatic

1 Apply thin film of multi-purpose grease on both ends of accelerator shaft (1) where it turns in the

bracket; and the nylon roller (7) at pedal.

2. Block choke open and operate throttle to be sure throttle returns to curb idle.

3. Follow steps 7, 8 and 9 under Automatic Transmission. (See illustration.)

1964-65 Coronet and Plymouth with 361, 383 or 426 cu. in. Engine and Automatic Transmission

Follow procedure listed for 273 and 318 cu. in. Engines and Automatic Transmissions.

1964-65 Coronet and Plymouth with 361, 383 or 426 cu. in. Engine and Manual Transmission

Follow procedure listed for 273 and 318 cu. in. Engines and Manual Transmissions.

1964 Dodge Custom 880 with Automatic Transmission

1. Assemble accelerator shaft assembly and pedal to body. With 3/16 in. diameter rod (5) approximately ten in. long in holes provided in accelerator shaft bracket and lever, adjust pedal rod (6) length to provide pedal angle of 113 to 115 degrees. Install pedal-rod (6).

2. Apply thin film of MS 3608 lubricant to accelerator shaft (1) where it turns in firewall bracket.

3. Assemble carburetor rod parts but do not tighten adjustment lock nut (4).

4. Disconnect choke (2) at carburetor or block choke valve in open position. Open throttle slightly to release fast idle cam. Return carburetor to idle.

5. With transmission rod assembly and transmission throttle lever in place, move transmission throttle lever forward against stop and tighten transmission rod adjustment lock nut (8). Remove 3/16 in. rod (5) from accelerator shaft bracket.

6. Move rear portion of carubretor rod assembly (3) rearward until transmission throttle lever (7) is contacted.

7. Connect choke rod or remove blocking fixture.

1964 Dodge Custom 880 with Manual Transmission

1. Follow steps 1 through 4 of automatic transmission procedure.

2. With 3/16 in. diameter by ten in. long rod (5) in place in the accelerator shaft bracket and lever assembly, tighten carburetor rod adjustment lock nut (4).

3. Connect choke rod (2) or remove blocking fixture.

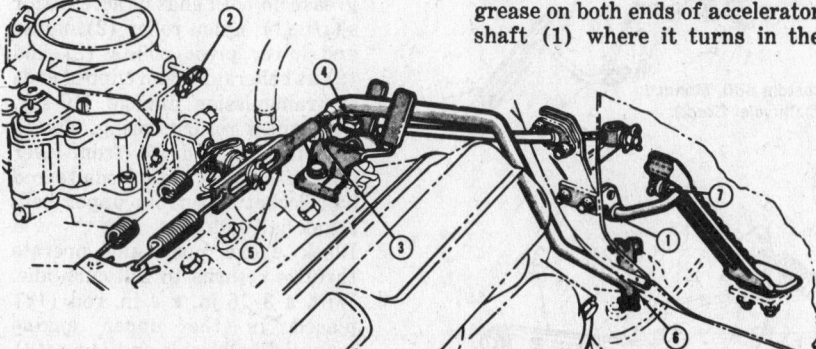

Linkage—1964-65 Coronet and Plymouth 273-318 (© Chrysler Corp.)

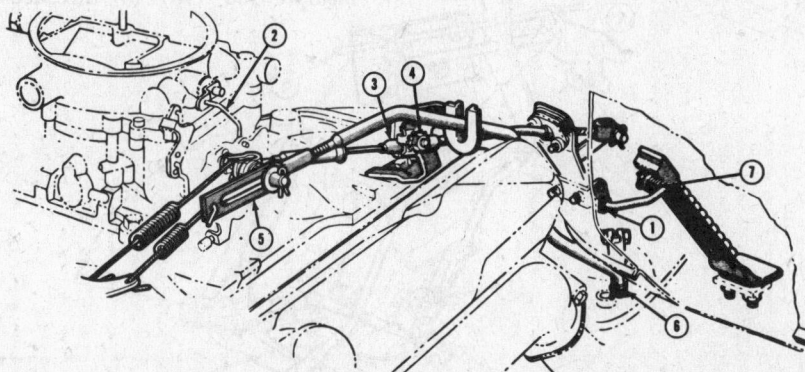

Linkage—1964 Dodge and 1965 Plymouth (© Chrysler Corp.)

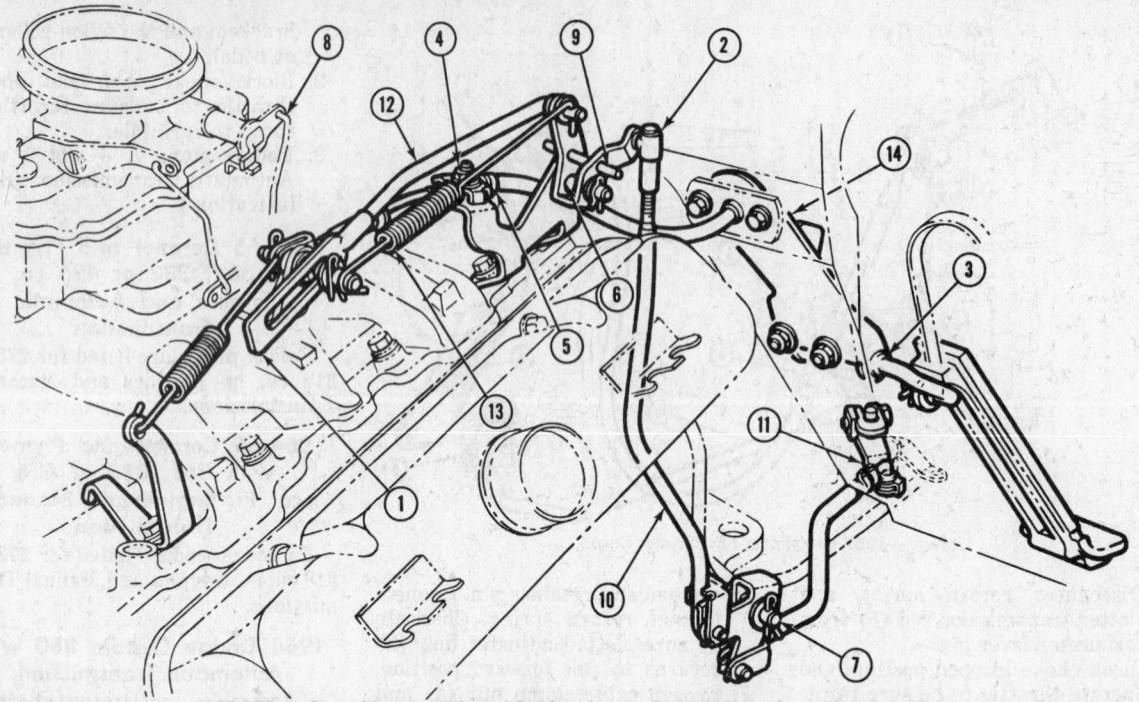

1965 273 engines with automatic and 1964 and some 1965 318 engines with automatic
(© Chrysler Corp.)

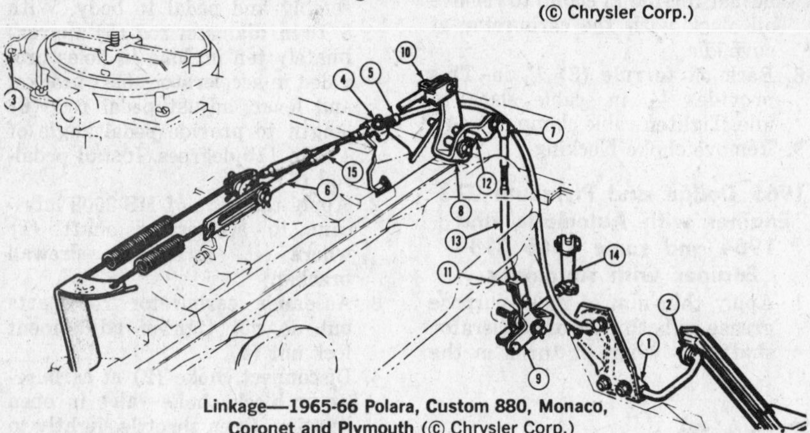

Linkage—1965-66 Polara, Custom 880, Monaco,
Coronet and Plymouth (© Chrysler Corp.)

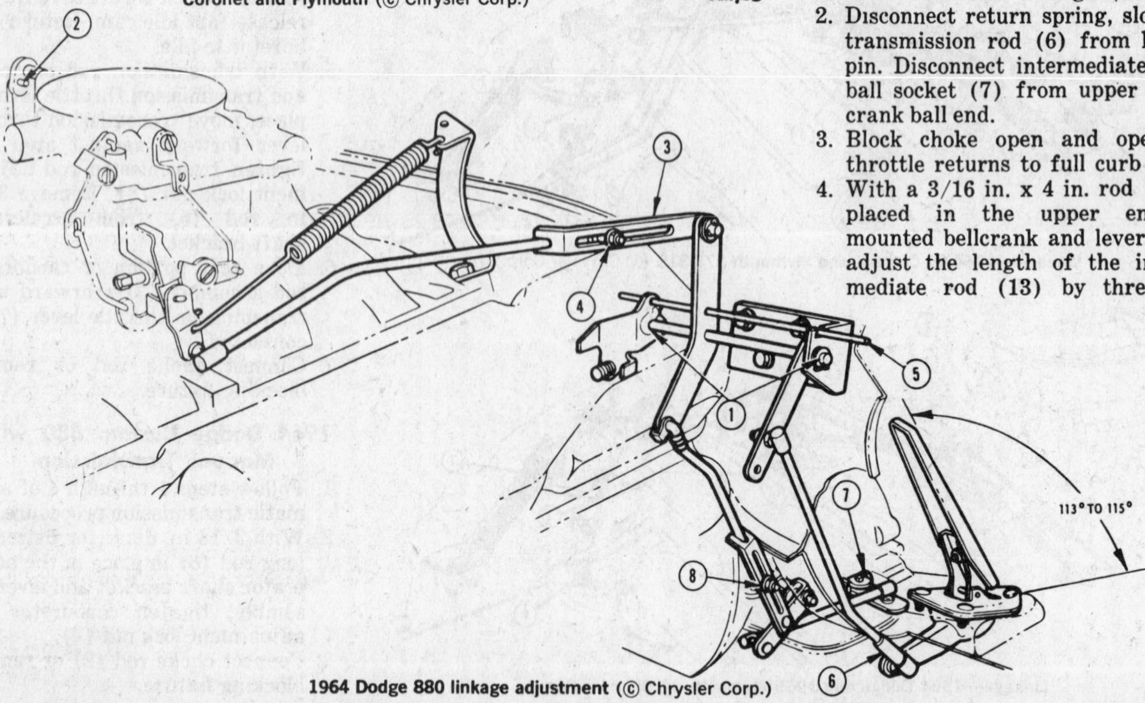

1965-66 Polara, Custom 880, Monaco with all Engines and Automatic Transmission and 1966 Coronet and Plymouth 318, 361, 383 Engines with Automatic, and some 1965 Coronet and Plymouth 318, 361, 383 and 426 Engines with Automatic

1. Apply thin film of multi-purpose grease on both ends of accelerator shaft (1), nylon roller (2), upper and lower pivot points (8) and (9) at bellcranks, and clipped ends of transmission linkage (10-11).
2. Disconnect return spring, slotted transmission rod (6) from lever pin. Disconnect intermediate rod ball socket (7) from upper bell-crank ball end.
3. Block choke open and operate throttle returns to full curb idle.
4. With a 3/16 in. x 4 in. rod (12) placed in the upper engine mounted bellcrank and lever (8) adjust the length of the inter-mediate rod (13) by threaded

113° TO 115°

1964 Dodge 880 linkage adjustment (© Chrysler Corp.)

upper end adjustment. The socket (7) must line up with the ball, with the rod held against the transmission stop (14).

5. Assemble ball socket (7) to ball end and remove gauge rod (12).
6. Hold carburetor rod (15) forward against transmission stop (14) and adjust its length at the slotted link (6) so that rear of slot (6) just contacts the carburetor lever pin.
7. Lengthen carburetor rod (15) two full turns at link (6).
8. Assemble link (6) to carburetor.
9. Loosen cable clamp nut (4) and adjust cable position in clamp so that all slack is removed from cable at curb idle.
10. Back off ferrule (5) ¼ in. This provides ¼ in. cable slack at idle and retighten cable clamp nut (4).
11. Remove choke block.

1965-66 Polara, 880, and Monaco with all Engines and Manual Transmissions, and 1966 Coronet and Plymouth 318, 361, and 383 Engines with Manual Transmissions, and some 1965 Coronet and Plymouth 318, 361, and 383 Engines with Manual Transmission

1. Follow steps 1, 3, 9, 10 and 11 listed under Automatic Transmission.

1967-70 Dodge and Plymouth 318, 383, and 440 Engines with Automatic Transmission

1. Apply a thin film of multi-purpose grease at all points of bell crank, throttle cable and linkage movement.
2. Disconnect choke at the carburetor, or block the choke valve in fully open position. Open throttle slightly to release fast idle cam, then return carburetor to curb idle.
3. Hold, or wire, the transmission lever (11) firmly forward

against its stop, while performing the next four steps of adjustment.
4. With a 3/16 in. diameter rod (9) placed in the holes provided in upper bell crank (6) and lever, adjust transmission rod (10) by means of adjustment (2) at upper end. The ball socket (2) must line up with the ball end with a slight downward effort on the rod.
5. Assemble ball socket (2) to ball end and remove the 3/16 in. locating rod from upper bell crank lever (6).
6. Disconnect return spring (13), then adjust length of carburetor rod (12) by pushing rearward on the rod lightly and turning the threaded adjustment (1). The rear end of slot should contact carburetor pin without exerting forward force on pin when slotted adjuster link (1) is in its normal operating position against lever pin nut.
7. Assemble slotted adjustment (1) to carburetor lever pin and install washer and retainer pin. Assemble transmission linkage return spring (13) in place.
8. Remove wire holding the transmission lever, then check transmission linkage for freedom of operation. Move slotted adjuster link (1) to full rearward position, then be sure it returns to its full forward position.
9. Loosen cable clamp nut (4), adjust position of cable housing ferrule (5) in the clamp so that all slack is removed from the cable at curb idle. To remove slack from cable, move cable (5) in the clamp in a direction away from the carburetor lever.
10. Back off ferrule (5) ¼ in. This provides free play between front edge of accelerator shaft lever and the dash bracket. Tighten cable clamp nut (4) to 45 in. lb.
11. Connect choke (8) rod or remove blocking fixture.

1967-70 Coronet and Belvedere 273 and 318 Engines with Automatic Transmission

1. Apply thin film of multi purpose grease on accelerator shaft (3) where it turns in bracket, ball-end, and support (14) at rear end of throttle cable.
2. Disconnect choke (8) at carburetor or block choke valve in full open position. Open throttle slightly to release fast idle cam, then return carburetor to idle.
3. Hold or wire transmission lever (11) firmly forward against stop while performing next four steps.
4. With 3/16 in. rod (9) placed in holes provided in upper bellcrank and lever, adjust length of intermediate transmission rod (10) by means of threaded adjustment at upper end. The ball socket (2) must line up with ball end with a downward effort on rod.
5. Assemble ball socket (2) to ball end and remove 3/16 in. rod from upper bell crank and lever.
6. Disconnect return spring (13), then adjust length of carburetor rod (12) by pushing rearward on rod with slight effort and turning threaded adjustment (1). The rear end of the slot should contact carburetor lever pin without exerting any forward force on pin when slotted adjuster link (1) is in its normal operating position against lever pin nut.
7. Assemble slotted adjustment (1) to carburetor lever pin and install washer and retainer pin. Assemble transmission linkage return spring (13) in place.
8. Remove wire securing transmission lever, then check transmission linkage freedom of operation and move slotted adjuster link (1) to full rearward position.
9. Loosen cable clamp nut (4). Adjust position of cable housing ferrule (5) in clamp so that all slack is removed from cable with carburetor at an idle. To remove slack from cable, move ferrule (5) in the clamp away from carburetor levers.
10. Back off ferrule (5) ¼ in. This provides ¼ in. free play between front edge of accelerator shaft lever and dash bracket. Tighten cable clamp nut (4) to 45 in. lbs.
11. Connect choke (8) or remove blocking fixture.

1967-68 Coronet and Belvedere 273 and 318 Engines with Manual Transmission

1. Apply light film of multi-purpose grease on accelerator shaft (3) where it turns in bracket, ball-

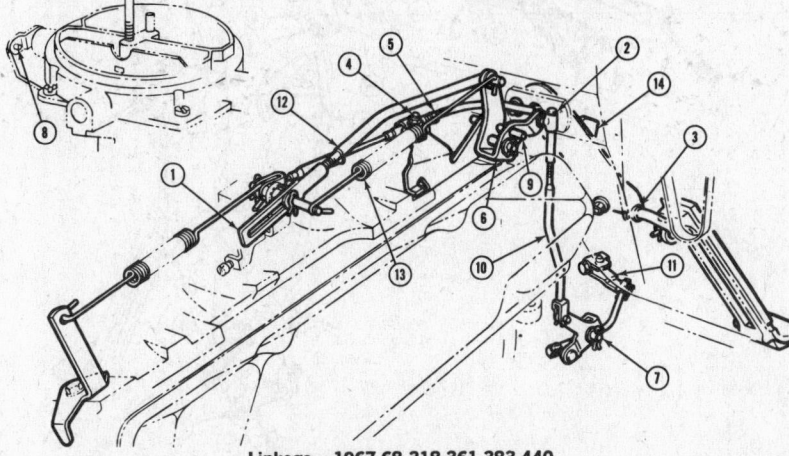

Linkage—1967-68 318-361-383-440
(© Chrysler Corp.)

end, and support (14) at rear end of throttle.

2. Disconnect choke (8) at carburetor or block choke valve in full open position. Open throttle slightly to release fast idle cam, then return carburetor to curb idle.

3. Loosen cable clamp nut (4). Adjust position of cable housing ferrule (5) in the clamp so that all slack is removed from cable with carburetor at an idle. To remove slack from cable, move ferrule (5) in the clamp away from carburetor lever.

4. Back off ferrule (5) ¼ in. This provides ¼ in. cable slack at idle. Tighten cable clamp nut (4) to 45 in. lbs.

5. Connect choke (8) rod or remove blocking fixture.

1967-68—All Models, 318, 383, 440—Manual Transmission

1. Apply a thin film of multi-purpose grease to all points of bell crank, throttle cable and linkage movement.

2. Disconnect choke at carburetor, or block the choke valve in fully open position. Open throttle slightly to release fast idle cam, then return carburetor to curb idle.

3. Loosen cable clamp (4), adjust position of cable housing ferrule (5) in the clamp so that all slack is removed from cable with carburetor at curb idle. To remove slack from cable, move ferrule (5) in the clamp in the direction away from the carburetor lever.

4. Back off ferrule (5) ¼ in. This provides cable slack at idle. Tighten cable clamp nut (4) to 45 in. lb.

5. Connect choke rod at carburetor, or remove locking fixture.

1967-70—All Models, 426 (Hemi.) Engine—Automatic Transmission

1. Apply a thin film of multi-purpose grease to all points of bell crank and throttle linkage.

2. Block choke valve in open position. Open throttle to release fast idle cam, then return carburetor to curb idle.

3. Hold or wire transmission lever (10) forward against its stop, while performing the next four steps of adjustment.

4. With a 3/16 in. diameter rod (8) placed in holes provided in bell crank and lever (15), adjust length of transmission rod (9) by means of threaded upper end. The ball socket must line up with the ball end while the rod is held upward against the transmission stop (10).

5. Assemble ball socket to ball end and remove the 3/16 in. positioning rod (8) from upper bell crank and lever (15).

6. Disconnect return spring (11), adjust length of rod (20) by pushing rearward on rod and turning threaded adjuster link (2). The rear end of the slot should contact carburetor lever stud without exerting any forward force on the stud when slotted adjuster link is in its normal operating position.

7. Assemble slotted adjuster link (2) to carburetor lever stud and install washer and retainer pin. Assemble transmission linkage return spring (11) in place.

8. Remove wire holding transmission lever, then check transmission linkage for freedom of operation. Move slotted adjuster link (2) to full rearward position, then allow it to return slowly. Be sure it returns to full forward position against the stud.

9. Loosen cable clamp nut (12), adjust position of cable housing ferrule (13) in the clamp (14) so that all slack is removed from the cable with rear carburetor at curb idle. To remove slack from cable, move ferrule (13) in clamp (14) in direction away from carburetor lever.

10. Back off ferrule (13) ¼ in. This will provide free play between front edge of accelerator shaft lever and dash bracket. Tighten clamp (14) to 45 in. lbs.

11. Route cable so it does not interfere with carburetor rod (20) or upper bell crank (15) through full throttle travel.

12. Attach carburetor rod assembly (4) between the carburetors with slotted rod end (16) attached to outboard side of inboard lever on rear carburetor. With rear carburetor at wide-open throttle, adjust length of connector rod (4) so that front carburetor is also at wide open throttle. To lengthen rod (4), turn adjusting stud (17) clockwise as viewed from front of engine. Tighten lock nut (18).

13. Remove choke valve blocking fixture.

1969-70 Dodge and Plymouth —All Engines with Manual Transmission, Except 426 Hemi

1. Apply thin film of multi-purpose grease on accelerator shaft where it turns in the bracket, ball end, and socket at rear end of throttle cable.

2. Disconnect choke at carburetor or block choke valve in full open position. Open throttle slightly to release fast idle cam, then return carburetor to idle.

3. Loosen cable clamp nut (1). Adjust position of cable housing ferrule (2) in clamp so that all slack is removed from cable with carburetor at idle. To remove slack from cable, move ferrule (2) in clamp away from carburetor lever.

4. Back off ferrule (2) ¼ in. to provide ¼ in. cable slack at idle. Tighten cable clamp nut to 45 in. lbs.

5. Connect choke rod or remove blocking fixture.

1967-70—All Models, 426 (Hemi) Engine—Manual Transmission

1. Apply a thin film of multi-purpose grease to all points of bell crank and throttle linkage.

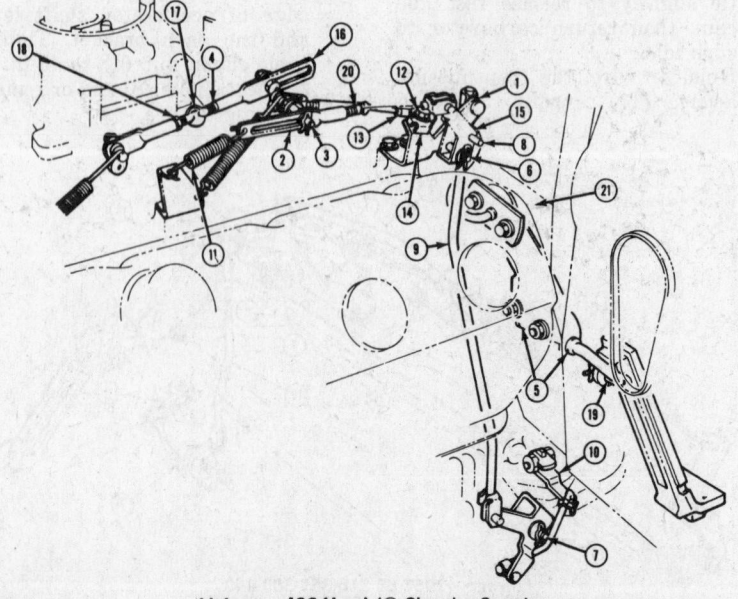

Linkage—426 Hemi (© Chrysler Corp.)

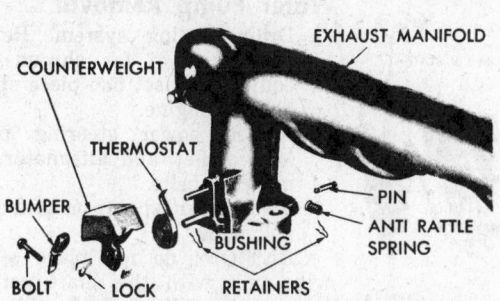

V8 manifold heat valve (© Chrysler Corp.)

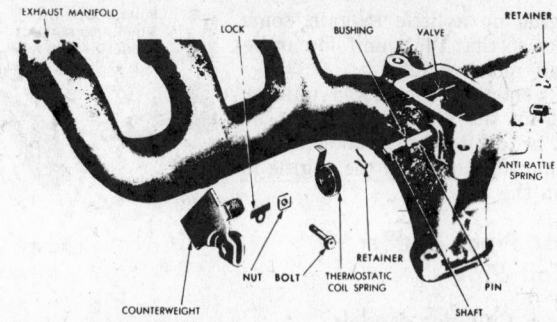

Slant 6 manifold heat valve (© Chrysler Corp.)

2. Block choke valve open, then open throttle slightly to release fast idle cam. Return carburetor to curb idle.

3. Loosen cable clamp nut (12), adjust position of cable housing ferrule (13) in clamp (14) so that all slack is removed from cable with rear carburetor at curb idle. To remove slack, move ferrule (13) in clamp (14) in the direction away from the carburetor lever.

4. Back off ferrule (13) ¼ in. This provides free play between front edge of accelerator shaft lever and the dash bracket. Tighten clamp (14) to 45 in. lbs.

5. Attach carburetor rod (4) assembly between the carburetors with slotted rod end (16) attached to outboard side of inboard lever on rear carburetor. With rear carburetor at wide-open throttle, adjust length of connector rod (4) so that front carburetor is also at wide-open throttle. To lengthen rod (4),

turn adjusting stud (17) clockwise as viewed from front of engine. Tighten lock nut (18).

6. Remove choke valve blocking fixture.

Exhaust System

Manifold Heat Valve

A heat riser control is incorporated in the exhaust manifold to regulate the amount of heat bypassing the intake manifold heat chamber.

The most common service required by the heat riser control is to keep it free to turn against its thermostat spring.

If difficulty is noticed in the warm-up period, or if after the car has become warm it seems to run lean, check the heat riser valve to make certain that it is turning freely on its shaft. If it is not, before removing the manifold, try to loosen it up with the

use of a good penetrating oil. If this fails to loosen it, it may be necessary to remove the manifold in order to free the heat riser valve.

Combination Manifold Removal

6 Cylinder Models

Remove all leads to the carburetor—vacuum, gasoline and throttle. Detach the exhaust manifold at the flange and, using socket and box wrenches, unbolt the manifold from the side of the block.

Exhaust Manifolds

V8 Models

Disconnect the exhaust manifold at the pipe flange. Access to these bolts is underneath the car.

The exhaust manifold mounting bolts are very difficult to reach. Unless the operator is particularly adept at working in close spaces, it might be an excellent idea to loosen the front engine mounting bolts and jack

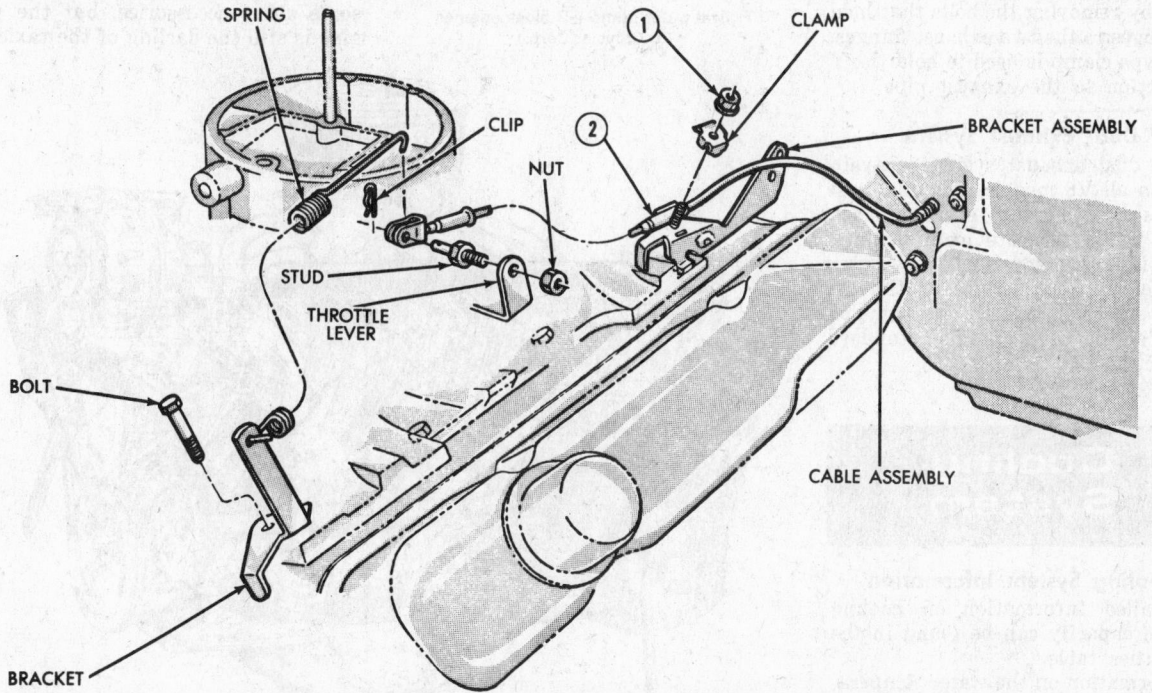

Linkage adjustment 1969-70—all engines with manual transmission except 426 Hemi
(© Chrysler Corp.)

the engine up a little to gain some clearance so that the manifold can be taken off more readily.

It is generally considered quicker to jack up the engine about an in. than it is to attempt to take the exhaust manifold off with the engine in place on the car.

Exhaust Pipe, Muffler and Tail Pipe

6 Cylinder Models

The oval muffler used on all models is of the straight-through type. When installing a new muffler, the word, front, stamped at one end is installed toward the front of the car.

If difficulty is experienced in separating the muffler from the exhaust and/or tail pipe, soak the joint for a few minutes with a good penetrating oil or a rust dissolving fluid.

The exhaust pipe can be removed by detaching it at the manifold and at the exhaust pipe flange, and it can be threaded out through the back.

Sometimes this is a little difficult since it requires careful threading to get it through.

The tail pipe can be removed by detaching it from its hangers and removing the rear muffler clamp.

Access to the exhaust flange bolt is either from under the hood or under the car, using a long extension on a socket wrench.

V8 Models With Single Exhaust

A Y-type exhaust pipe is used to connect the two manifolds to a single exhaust line.

The Y connection can be taken down by removing the bolts that hold its flanges to the two exhaust flanges. A U-type clamp is used to hold the Y connection to the exhaust pipe.

Dual Exhaust System

The dual exhaust system is available on all V8 models. The dual system is two separate systems, each going to its separate manifold, and there is no crossover pipe.

Otherwise, the service on the exhaust system is exactly the same as it is for the single muffler standard production car.

Cooling System

Cooling System Information

Detailed information on cooling system capacity can be found in the Capacities table.

Information on the water temperature gauge can be found in the Unit Repair section.

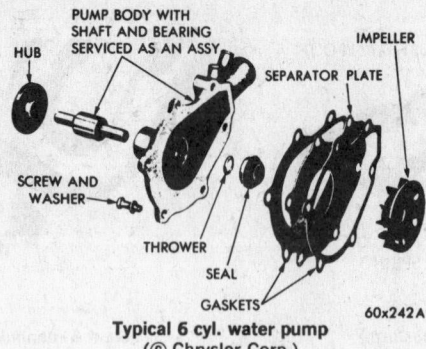

Typical 6 cyl. water pump
(© Chrysler Corp.)

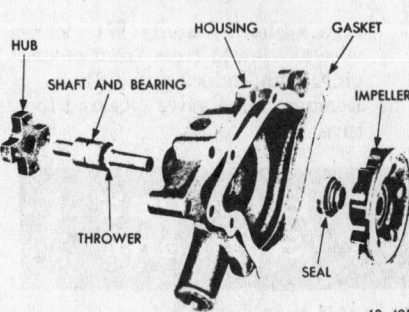

Typical V8 318 water pump
(© Chrysler Corp.)

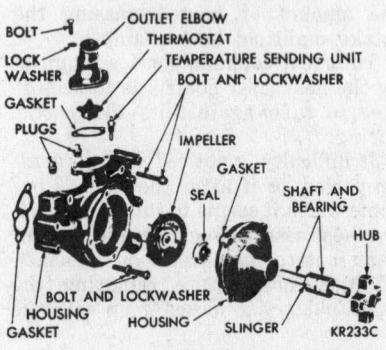

Typical water pump—B block engines
(© Chrysler Corp.)

Water Pump Removal

1. Drain cooling system. Remove upper half of fan shroud if so equipped or set one piece shroud back on engine.
2. Loosen power steering pump, idler pulley and alternator. Remove all belts.
3. Remove fan, spacer or fluid drive, and pulley.
   CAUTION: do not place a fluid drive unit with the shaft pointing downwards. Silicon fluid will drain into fan drive bearing and ruin grease.
4. Remove bolts attaching water pump to housing. Remove water pump and discard gasket.

Thermostat

On all models, the thermostat is located in the water outlet elbow just under the upper radiator hose connection.

Caution Be sure to install thermostat with the bellows or spring toward the engine.

Engine

Dodge and Plymouth Engines 1964-70

The standard equipment engine in most Dodge and Plymouth car models is the 225 cu. in. slant six. Although this engine has a very long stroke by modern standards, it presents a low profile because the entire block is canted 30 degrees to the right. This restricts access to the distributor and some other accessories, but the engine is still the darling of the taxicab

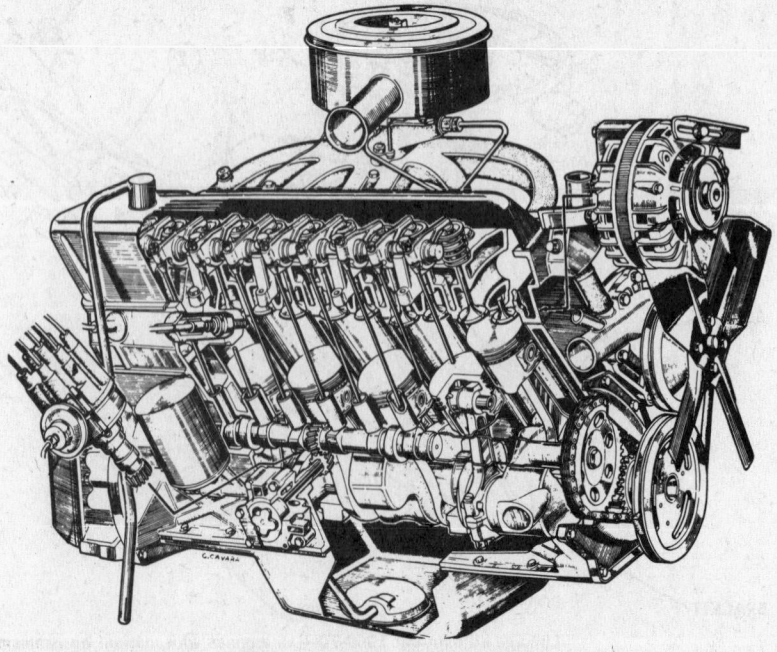

225 cu. in. engine (© Chrysler Corp.)

fleets for its ruggedness and long life.

The 273 cu. in. and 318 cu. in. engines are Chrysler Corporation's "A" block series of V8s. The oldest of the current "A" blocks, the 318, originally used polyspherical combustion chambers which were reckoned to be virtually as good as hemispherical combustion chambers. Virtually as good was evidently not good enough, for when the 273 was introduced in 1964, it had the simpler wedged shaped combustion chambers, and in 1967 the 318 adopted them also.

Chrysler Corporation's "B" block series is really two series of engines, the low-block series and the raised-block series. These series differ in block deck height, main journal diameter, connecting rod length, and pushrod length. Otherwise these engines are similar and many parts interchange. The 361 and 383 cu. in. engines are low-block engines, and the 413, 426 wedge head, and the 440 cu. in. V8s are raised-block engines. All these engines are conventional V8s with wedge shaped combustion chambers and deep blocks that extend well below crankshaft center-line.

The 426 Hemi is Dodge and Plymouth's largest, heaviest, most complicated, and most powerful engine. It is basically a "B" series, raised-block engine, but with so many differences that it must be treated as a completely separate engine. It has hemispherical combustion chambers with 2.25 in. intake and 1.95 in. exhaust valves actuated by rocker arms mounted on separate intake and exhaust rocker shafts. The spark plugs are centrally located in the combustion chambers, and aluminum tubes protect the plugs and wires from oil where they pass through the rocker covers. Because of the huge intake ports, there is no room for head bolts on the intake side. Instead studs are mounted in the head which extend down into the valley between the cylinder heads. To reduce piston side thrust, Hemis use longer connecting rods than other raised-block "B" engines, and to strengthen the lower end, the main caps are crossbolted.

Valves

Valve tappet clearance for each engine is given in the Tune-Up Specifications table.

Bearings

Detailed information on engine bearings will be found in the Crankshaft Bearing Journal Sizes table.

C.A.P. System

To comply with California air polution laws, the Chrysler Corporation has modified the treatment of engine exhaust on cars in present production for delivery into that state. This system is known as the C.A.P. (Cleaner Air Package.) Detailed information on all popular emission control systems may be found in the Unit Repair Section.

Engine Reassembly Notes

Engine crankcase capacities are listed in the Capacities table.

Approved torque wrench readings and head bolt tightening sequences are covered in the Torque Specifications table.

Information on the engine mark-ing code will be found in the Model Year Identification table.

Engine Removal

V8 Models, 1963-70

1. Scribe a line where the hood hinge connects to the hood so that reinstallation will be simplified. Remove the bolts and take off the hood. Remove the carburetor air cleaner and all lines to the carburetor. Disconnect the battery and remove the heat indicater and engine ground lines. Disconnect the ignition primary wire and remove the radiator core. Detach the two exhaust pipes at the exhaust manifold flanges. Remove all connections to the clutch and/or transmission, such as the clutch idler rod and throwout rod, shift levers, speedometer cable, handbrake cable, and back-up light wires.

2. Place a jack under the transmission and remove the bolts that hold the transmission to the crossmember. Take a little load on the jack, and remove the bolts that hold the crossmember to the frame, then let the crossmember come down. Disconnect the front universal joints. Place the lifting device on the engine and take a load on the lifter.

3. Remove the bolts that hold the front engine mounting bracket to the side of the cylinder block and take out the bolts. Crank the lifting device and pull the engine upward and forward. This operation is easier if a roller jack is placed under the transmission so that, as the engine comes forward, the roller jack will roll, still supporting the transmission.

6-Cylinder Engines

1. Carefully scribe the hood hinges where they contact the hood so that the hood can be reinstalled promptly without difficulty in alignment. Remove the bolts and take off the hood.

2. Remove the radiator core and take off the carburetor and generator.

3. Disconnect all fuel lines and heat indicator lines, battery ground straps, ignition primary wire and vacuum lines.

4. At the transmission, disconnect the clutch throwout rod and the transmission shift links. Remove the speedometer cable and hand brake cable. Disconnect the front universal joint. Disconnect the exhaust pipe at the flange. Place a roller jack under the transmission and take a slight load on the roller jack. Remove the bolts that hold the transmission to the

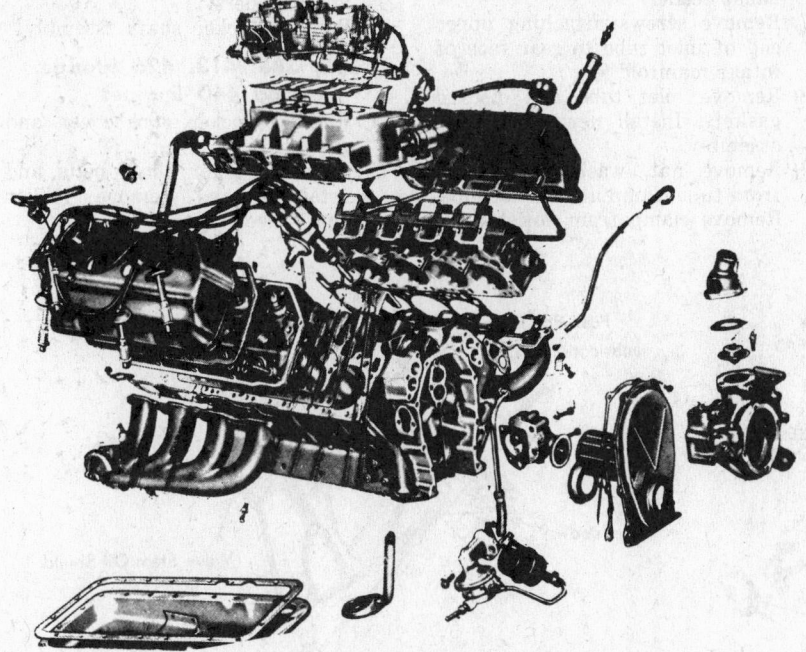

426 Hemi V8 engine (© Chrysler Corp.)

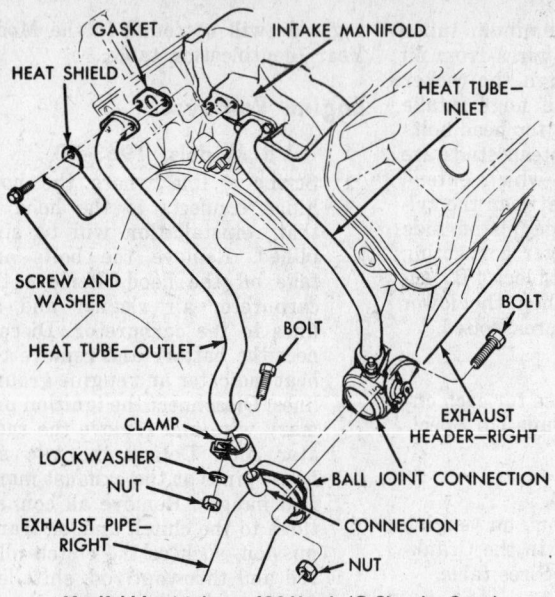

Manifold heat tubes—426 Hemi (© Chrysler Corp.)

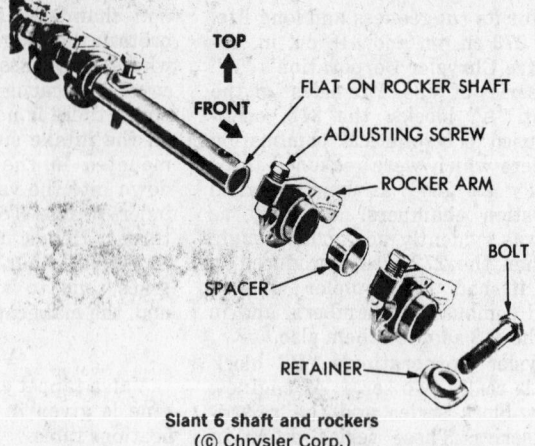

Slant 6 shaft and rockers
(© Chrysler Corp.)

crossmember and then take a slight load on the jack and remove the bolts that hold the crossmember to the frame and let the crossmember come down, leaving the weight of the transmission on the jack.

5. Attach the lifting device to the front part of the engine and take a slight load. Remove the bolts that hold the engine front motor mounts to the block and lift the engine upward and forward.

The back of the engine will ride on the roller jack until it is clear.

Intake Manifold Removal

All Engines Except 426 Hemi

1. Drain cooling system and disconnect battery.
2. Remove alternator, carburetor air cleaner, and fuel line.
3. Disconnect accelerator linkage.
4. Remove vacuum control between carburetor and distributor.
5. Remove distributor cap and wires.
6. Disconnect coil wires, tempera-

ture sending unit wire, heater hoses, and bypass hose.

7. Remove intake manifold, ignition coil, and carburetor as an assembly.

426 Hemi

1. Drain cooling system and disconnect battery.
2. Remove alternator, air cleaner, and fuel lines.
3. Disconnect accelerator linkage.
4. Remove vacuum control between carburetor and distributor. Remove distributor cap and wires.
5. Disconnect coil wires, heater hoses and bypass hose.
6. Remove two stud nuts and washers which retain intake manifold inlet heat tube to right hand exhaust header.
7. Remove screws attaching upper end of inlet tube to rear face of intake manifold.
8. Remove inlet tube and discard gaskets. Install new gaskets at assembly.
9. Remove nut, washer, and bolt from tube clamp at exhaust pipe. Remove clamp from outlet tube.

10. Remove screws attaching heat shield and outlet tube to rear face of intake manifold and remove tube and shield.
11. Remove intake manifold, coil and carubretors as an assembly.

Rocker Shaft Removal

Six Cylinder

1. Remove closed ventilation system (PCV).
2. Remove rocker arm cover and gasket.
3. Remove rocker shaft bolts and retainers.
4. Remove rocker shaft assembly.

273 and 318 Engines

1. Disconnect spark plug wires.
NOTE: when removing heads of any engine in the car, first remove spark plugs to avoid damage to insulators.
2. Disconnect closed ventilation system (PCV).
3. Remove rocker cover and gasket.
4. Remove five rocker shaft bolts and retainers.
5. Remove rocker shaft assembly.

361, 383, 413, 426 Wedge, and 440 Engines

1. Remove rocker arm cover and gasket.
2. Remove rocker shaft bolts and retainers and remove rocker shaft assembly.

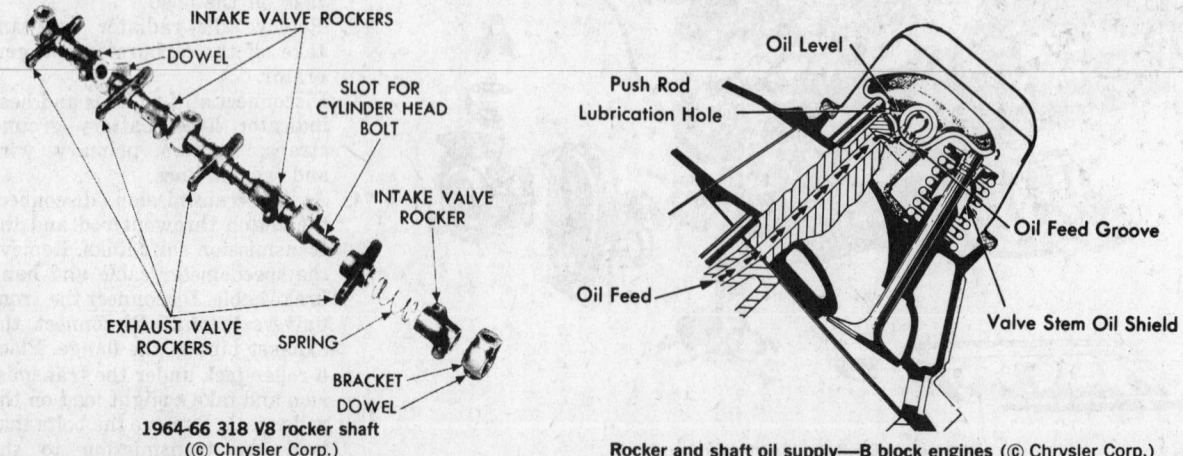

1964-66 318 V8 rocker shaft
(© Chrysler Corp.)

Rocker and shaft oil supply—B block engines (© Chrysler Corp.)

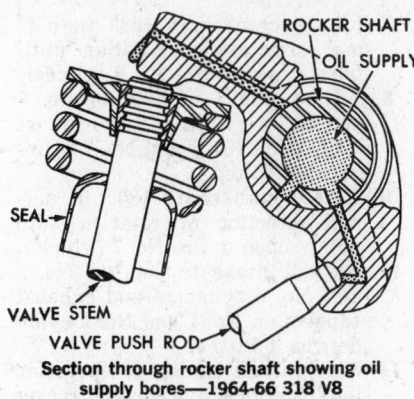

Section through rocker shaft showing oil supply bores—1964-66 318 V8
(© Chrysler Corp.)

426 Hemi

1. Remove air cleaner, and distributor cap with spark plug cables and secondry coil cable as an assembly.
2. Grasp secondary cables at plastic spark covers and pull covers straight out.
3. Remove spark plugs.
4. On left bank, disconnect brake lines at master cylinder, and remove cotter pin and clevis pin from linkage in back of power brake.
5. Remove four nuts attaching booster to mounting bracket and remove power brake and master cylinder assembly.
6. Remove rocker covers and gaskets.
7. Remove five bolts that attach rocker shafts assembly on each head.

NOTE: these rocker shaft assembly bolts pass through the head and into the block. Anytime rocker shaft assembly is removed, remove that

head, fit a new gasket, reassemble and torque.

8. Lift off rocker shafts assembly.

Cylinder Head

Removal

Six Cylinder

1. Drain cooling system.
2. Remove air cleaner and fuel line.
3. Remove vacuum line at carburetor and distributor.
4. Disconnect accelerator linkage.
5. Disconnect spark plug wires by pulling straight out in line with plugs.
6. Disconnect heater hose and by-pass hose clamp.
7. Disconnect temperature sending wire.
8. Disconnect exhaust pipe at exhaust manifold flange.
9. Remove intake and exhaust manifold as an assembly.
10. Remove closed vent system (PCV) and rocker cover.
11. Remove rocker shaft assembly.
12. Remove pushrods in sequence and save them to re-install in original bores.
13. Remove 14 head bolts.
14. Remove head.
15. Remove spark plugs and tubes.

273 and 318 Engines

1. Drain cooling system and disconnect battery.
2. Remove intake manifold.
3. Remove exhaust manifolds.
4. Remove rocker shaft assemblies.
5. Remove pushrods in sequence and save them to install in their original bores.
6. Remove ten head bolts from each cylinder head.

361, 383, 413, 426, and 440 Engines

1. Drain cooling system and disconnect battery.
2. Remove alternator, air cleaner, and fuel line.
3. Remove intake manifold.
4. Remove tappet chamber cover.
5. Remove rocker covers and gaskets.
6. Remove exhaust manifolds.
7. Remove rocker shaft assemblies.
8. Remove pushrods in sequence and save them to install in their

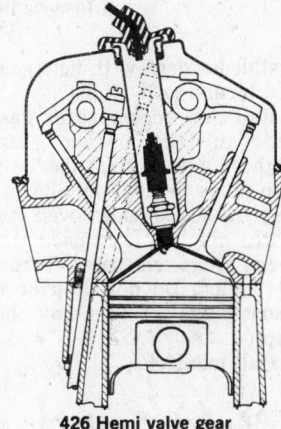

426 Hemi valve gear
(© Chrysler Corp.)

original bore.

9. Remove 17 head bolts from each cylinder head and remove heads.

426 Hemi—Removal and Installation

1. Remove rocker covers.
2. Remove rocker shaft assemblies.
3. Remove intake manifold.
4. Disconnect exhaust headers, and tie out of way.
5. Remove eight lower head bolts and four stud nuts from each head.
6. Remove heads. Do not set heads on studs at any time. Because of the unusual use of rocker shaft bolts as head bolts follow installation procedure carefully.
7. Coat new head gasket with sealer and install with raised bead towards block.
8. Install cylinder heads taking care not to damage studs.
9. Install nuts on cylinder head studs and short cylinder head bolts in outer bolt holes, but do not tighten either.
10. Install pushrods in their original bores. The short rods go in the upper holes and the long rods go in the lower holes.
11. Position rocker shafts assemblies on heads and install five long head bolts in each after lining up pushrods with rockers.
12. Torque bolts and stud nuts in sequence given at front of section.
13. Adjust valve lash on 1965-69 Hemis.

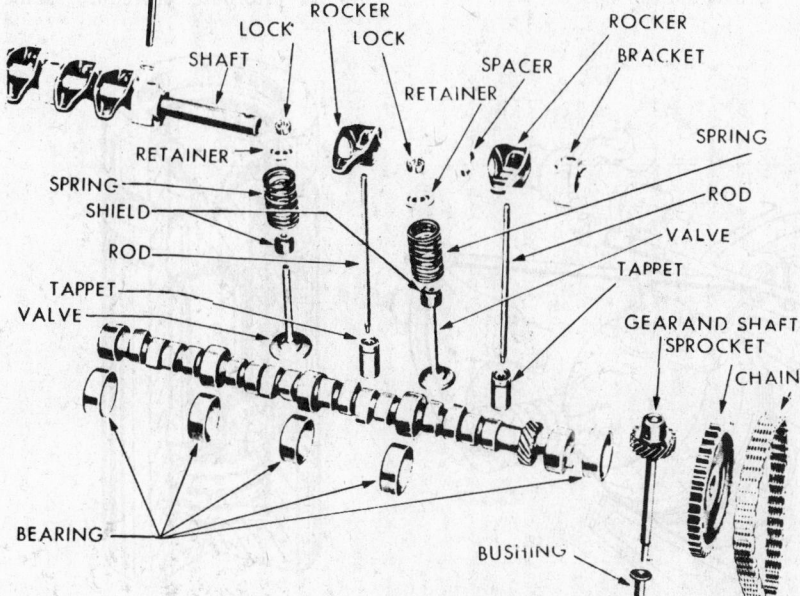

Valve train components—B block engines (© Chrysler Corp.)

Torquing Hemi head stud nuts
(© Chrysler Corp.)

14. Install headers with new gaskets and torque to 35 ft. lbs.
15. Install new rocker cover gaskets and install rocker covers. Tighten nuts to 10 ft. lbs.
16. Gap plugs to 0.035 in. Slide spark plug tube shields over tubes. With six in. extension install spark plugs and tubes. Torque to 30 ft. lbs. Do not drop or bang spark plugs for this may change gap.
17. Install manifold.

Valve System

Hydraulic Lifter Adjustment Except 426 Hemi

To be certain the hydraulic lifter is operating some place near the middle of its stroke, the Chrysler Corporation recommends that the length of the valve stem protruding from the cylinder head be checked with gauge C-3061 (engines with two rocker shafts) or C-3684 (engines with one rocker shaft.) However, if this gauge is not available an emergency check can be made by turning the engine until the valve be-

ing checked is in the fully closed position and the lifter is on the bottom of the cam. Depress the pushrod to force the lifter to leak down. While held in this position, there should be from .060 in. to .210 in. between the rocker arm pad and the end of the valve stem.

Hydraulic Tappet Adjustment —426 Hemi

1. Adjust ignition timing to TDC.
2. Mark crankshaft damper with chalk at TDC and 180° opposite TDC.
3. Rotate crankshaft until No. 1 cylinder is at TDC and points are just opening.
4. Adjust intake tappets on No. 2 and No. 7 cylinders and exhaust tappets on No. 4 and No. 8 cylinders to zero lash. Tighten adjustment 1½ turns more, then torque locknuts to 25 ft. lbs.
5. Rotate crankshaft 180° in normal direction of rotation until points open to fire No. 4 cylinder.
6. Adjust intake tappets on No. 1 and No. 8 cylinders and exhaust tappets on No. 3 and No. 6 cylinders as in Step 4.

7. Rotate crankshaft 180° in normal direction of rotation until points open to fire No. 6 cylinder.
8. Adjust intake tappets on No. 3 and No. 4 cylinders and exhaust tappets on No. 5 and No. 7 cylinders as in Step 4.
9. Rotate crankshaft 180° in normal direction of rotation until points open to fire No. 7 cylinder.
10. Adjust intake tappets on No. 5 and No. 6 cylinders and exhaust tappets on No. 1 and No. 2 cylinders as in Step 4.
11. Set ignition timing to operating specifications and install rocker covers.

Cylinder Head Disassembly

1. Remove cylinder heads.
2. Compress valve springs, using valve spring compressor.
3. Remove valve locks or keys.
4. Release valve springs.
5. Remove valve springs, retainers, oil seals, and valves.

NOTE: if a valve does not slide out of the guide easily, check end of stem for mushrooming or "heading over." If condition exists, file off excess, remove and discard valve. If stem is good, lubricate and remove valve, then check for guide and/or stem damage.

Valve Guides

Dodge and Plymouth engines do not have separate valve guides. They do have, however, 0.005, 0.015, and 0.030 in. oversize valves (stem diameter). To use these, ream the worn guides to the smallest oversize that will clean up wear. Always start with the smallest reamer and proceed in steps to the largest, as this maintains the concentricity of the guide with the valve seat.

As an alternate procedure, some

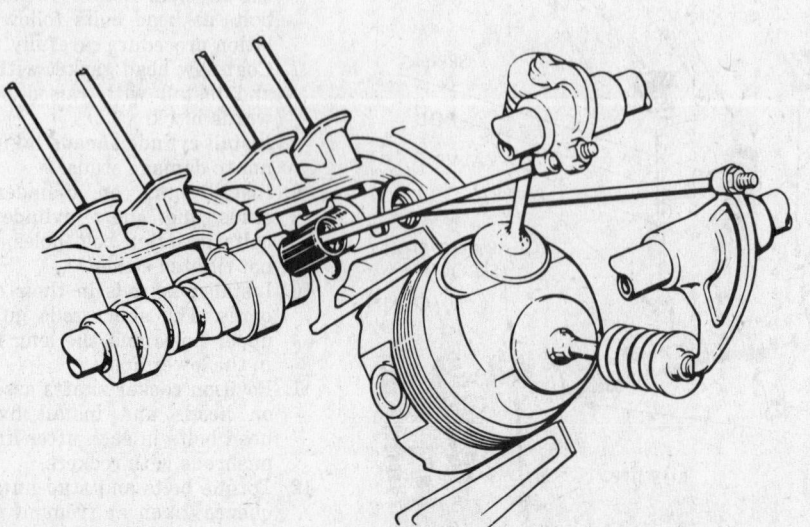

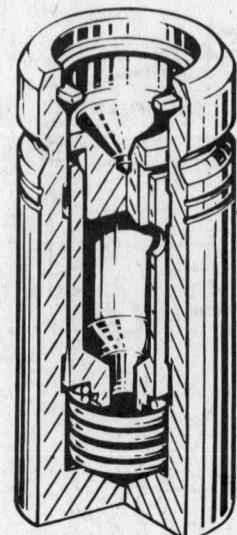

Hemi hydraulic tappet (© Chrysler Corp.)

local automotive machine shops bore out the stock guides and replace them with bronze or cast iron guides which are of stock internal dimensions.

Valves

Inspect valve stems for wear. Valve stems worn more than 0.002 in. must be replaced. Inspect valves for burning or warping and replace damaged valves. To find a bent valve, install valve in a valve grinder or a drill, rotate chuck and watch for head oscillation.

Valve Springs

It is recommended that the valve springs be checked for free height and squareness rather than for pressure.

Perhaps the easiest way to do this is to lay the valve springs on a straight flat surface and compare one with the other. If all of the valve springs are of the same height, it may be assumed that all are usable.

Hydraulic Tappets

Removal—All Except 426 Hemi

Tappets may be removed without removing manifold or cylinder heads, as follows:

1. Remove rocker covers and rocker shaft assembly.
2. Remove pushrods and identify to insure correct installation into original bore.
3. Slide magnetic or claw tool through opening in cylinder head and seat tool firmly in head of tappet.
4. Pull tappet out of bore with a twisting motion.

NOTE: all tappets must be installed in their original bores.

Disassembly—All Except 426 Hemi

The 426 Hemi has had mechanical tappets until 1970. The hydraulic

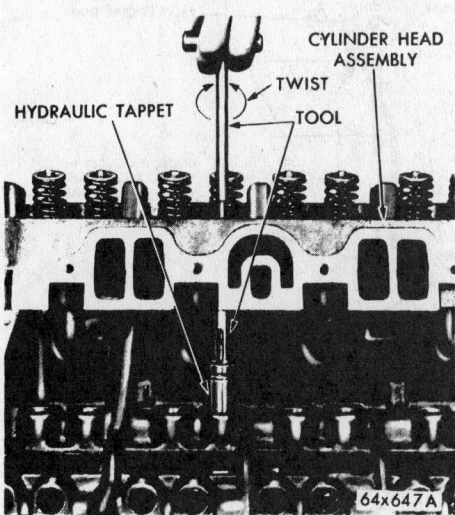

Removal of hydraulic tappets
(© Chrysler Corp.)

⏻ CHILTON TIME-SAVER

The following is a method for replacing valve springs, oil seals or spring retainers without removing the cylinder head.

1. Entirely dismantle a spark plug and save the threaded shell.
2. To this shell, braze or weld an air chuck.
3. Remove the valve rocker cover. Remove the rocker arm from the affected valve.
4. Remove the spark plug from the affected cylinder.
5. Turn the crankshaft to bring the piston of this cylinder down, away from possible contact with the valve head. Sharply tap the valve retainer to loosen the valve lock.
6. Turn the crankshaft to bring the piston in this cylinder to the exact top of its compression stroke.
7. Screw in the chuck-equipped spark plug shell.
8. Hook up an air hose to the chuck and turn on the pressure (about 200 lbs.).
9. With a strong and constant supply of air holding the valve closed, compress the valve spring and remove the lock and retainer.
10. Make the necessary replacements and reassemble.

NOTE: it is important that the operation be performed exactly as stated, in this order. The piston in the affected cylinder must be on exact top center to prevent air pressure from turning the crankshaft.

---

tappets in the 1970 Hemi are special high performance, high speed units which should not be disassembled.

1. Pry out plunger retainer spring clip.
2. Clean varnish deposits from tappet body above plunger.
3. Invert tappet body and remove plunger cap, plunger, flat check valve, check valve spring, check valve retainer, and plunger spring.
4. Clean tappet components in solvent (such as laquer thinner).
5. If tappet plunger shows signs of scoring or wear, or if check valve is pitted, replace tappet as a unit.

NOTE: tappet components are not interchangeable—do not mix.

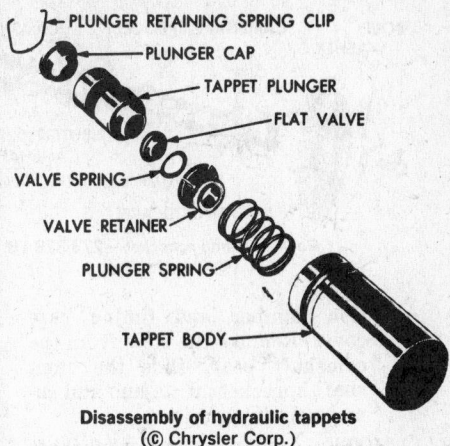

Disassembly of hydraulic tappets
(© Chrysler Corp.)

Timing Case Cover Chains and Gears

Vibration Damper Removal

It is necessary to remove radiator to prevent core damage on all models.

Timing Case Cover Removal

1. Drain cooling system, remove radiator and water pump.
2. Remove crankshaft vibration damper attaching bolt, then pull off the damper using a puller.
3. Remove timing chain cover and gasket. It is normal to find particles of neoprene collected between seal retainer and crankshaft oil slinger after seal has been in service.

NOTE: on six-cylinder engine, loosen oil pan bolts for clearance before removing chain cover. On 426 Hemi engine, remove the two front pan bolts.

4. Slide crankshaft oil slinger from end of crankshaft.

Timing Chain and Sprocket Removal

6-Cylinder Engines

1. To replace a timing chain, cam or crankshaft sprocket, or to retime valves where the timing has jumped, proceed as follows: take off the radiator core, vibra-

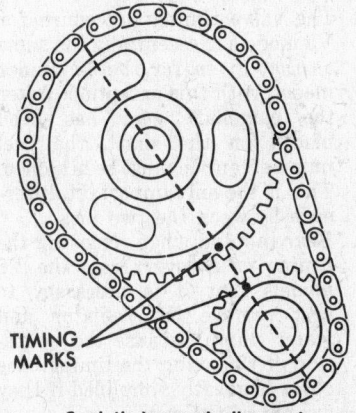

6 cyl. timing mark alignment

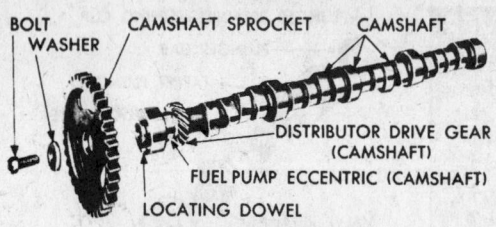

Camshaft and sprocket—273-318 V8
(© Chrysler Corp.)

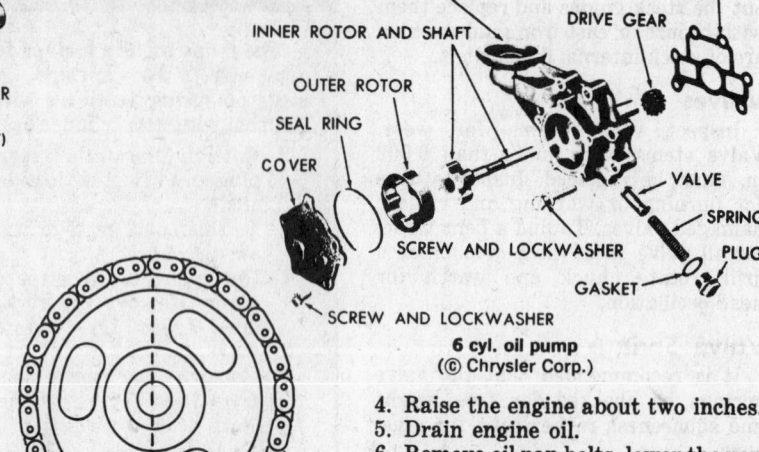

6 cyl. oil pump
(© Chrysler Corp.)

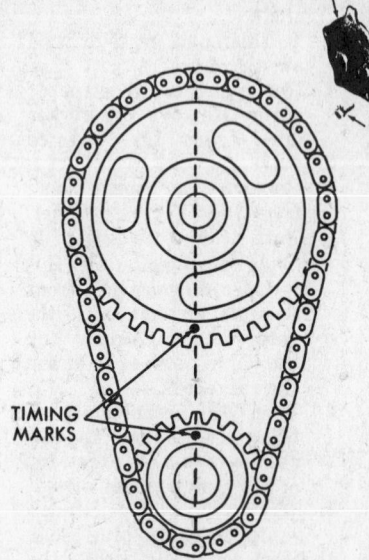

V8 timing mark alignment

tion damper and timing case cover. Remove the bolts from the camshaft gear, slide the camshaft sprocket off its hub and remove timing chain.

NOTE: unless the crankshaft sprocket is to be replaced it will not be necessary to remove it from the shaft.

2. Rotate the crankshaft so that the mark in the crankshaft sprocket is toward the camshaft and in exact alignment between the two shaft centers.
3. Now, install the timing chain over the camshaft sprocket so that the marks in the cam and crankshaft sprocket are nearest each other.

The timing bolt holes are staggered in such a way that the camshaft sprocket will attach only one way and permit the bolt to enter through the threaded holes in the hub.
4. Rotate the camshaft so that the holes align and the timing marks are still in line between the shaft centers. Mount the gear on the hub and draw up the bolts.
5. Turn the crankshaft two full revolutions and check to see that the marks are still in alignment between the shaft centers.
6. When set in the manner described, with the marks aligned between the two shaft centers, it is immaterial which piston is at dead center. It is necessary, however, to retime the ignition at any time the camshaft setting is disturbed.

V8 Engines

1. The valve timing procedure for V8 models is essentially the same as that given for the 6-cylinder model. with this exception. After the camshaft gear has been placed on the shaft. the fuel pump eccentric must be attached. This is the only important difference between the two jobs.
2. Bear in mind when removing the timing case cover from the V8 models that it is necessary to first remove the radiator and water pump because the operation of removing the timing case cover is greatly simplified if they are out of the way.

The camshaft gear is located on the camshaft by a very tight-fitting dowel pin.

Engine Lubrication

Oil Pan Removal

Slant 6 OHV

1. Remove the tie-rod at the steering and idler arms.
2. Remove the two front engine mounting bolts.
3. Remove left side support, connecting converter housing and cylinder block.

4. Raise the engine about two inches.
5. Drain engine oil.
6. Remove oil pan bolts, lower the pan to the rear.

NOTE: do not turn the oil pick-up out of position.

V8 Engines

1. Rotate the crankshaft until the timing marks are in the 5 o'clock position, as viewed from the front.
2. Disconnect the steering linkage at the idle arm bracket. On single exhaust systems, remove the crossover pipe. Remove the starting motor. Disconnect the front motor mount brackets and jack up the front of the engine about ½ in.
3. Remove the oil pan bolts and work the pan to clear the crossmember.

Oil Pump Removal

Six Cylinder

1. Drain radiator, disconnect upper and lower hoses, and remove fan shroud.
2. Raise vehicle on hoist, support front of engine with jackstand placed under right front corner

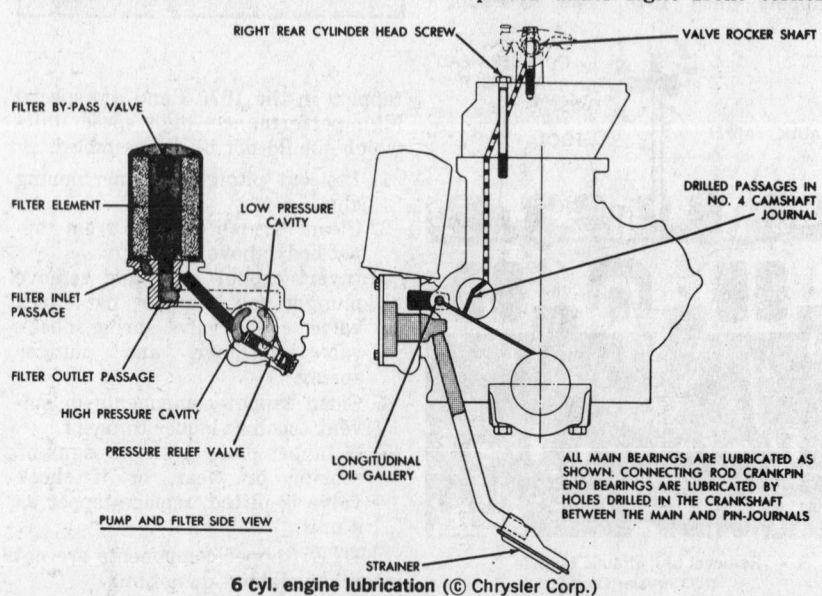

6 cyl. engine lubrication (© Chrysler Corp.)

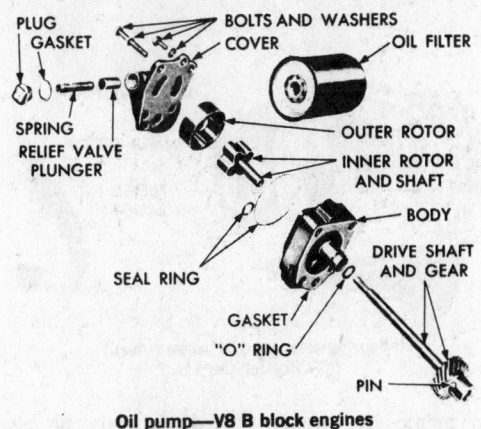

Oil pump—V8 B block engines
(© Chrysler Corp.)

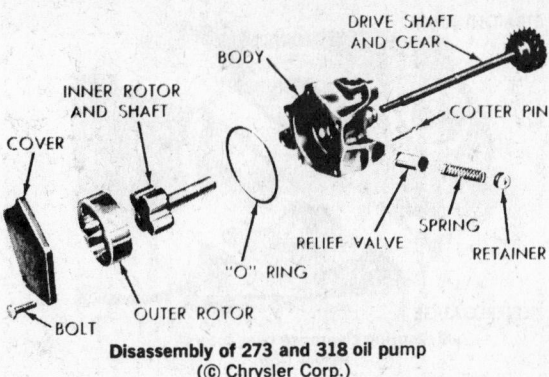

Disassembly of 273 and 318 oil pump
(© Chrysler Corp.)

of oil pan, and remove engine mount bolts. Do not support engine at crankshaft pulley or vibration damper.
3. Raise engine approximately 1½ to 2 in.
4. Remove oil filter, oil pump attaching bolts, and pump assembly.

273 and 318 Engines
1. Remove oil pan.
2. Remove oil pump from rear main bearing cap.

361, 383, 413, 426 wedge, 440, and 426 Hemi Engines
1. Remove oil pan and filter assembly.
2. Remove oil pump from bottom side of engine.

Oil Pump Disassembly
Six Cylinder
1. Remove pump cover and seal ring.

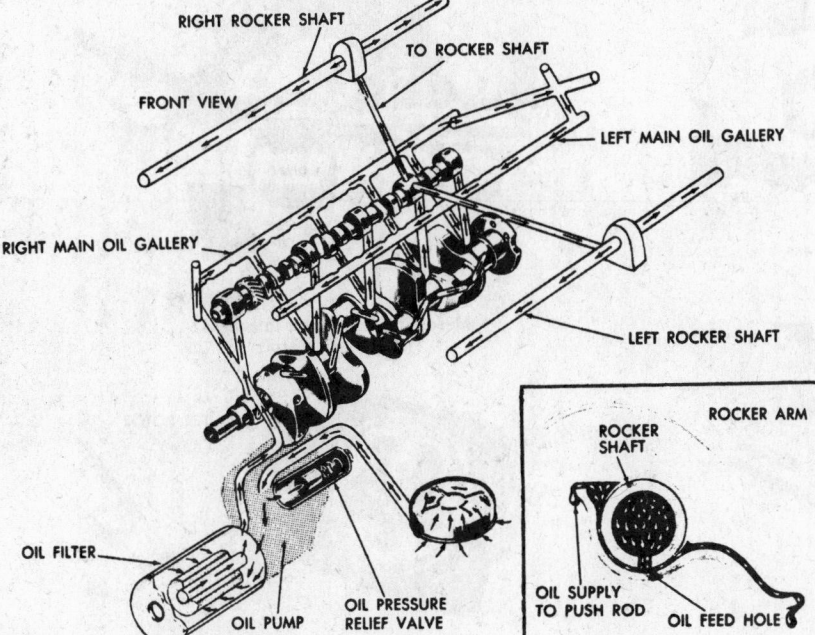

Engine lubrication—B block engines (© Chrysler Corp.)

2. Press off drive gear. Support gear to keep load off aluminum body.
3. Remove pump rotor and shaft and lift out outer pump rotor.
4. Remove oil pressure relief valve plug and lift out spring and retainer.

273 and 318 Engines
1. Remove cotter pin from relief valve, drill ⅛ in. hole into relief valve cap, and insert self threading sheet metal screw into cap.
2. Clamp sheet metal screw in vise, and, supporting oil pump body, tap body with soft-headed hammer until cap comes out. Discard cap and remove spring and relief valve.
3. Remove oil pump cover bolts and lock washers and lift off cover.
4. Discard oil seal ring.
5. Remove pump motor and shaft and remove outer rotor.

361, 383, 413, 426 Wedge, 440, and 426 Hemi
1. Remove filter base and oil seal ring.
2. Remove pump rotor and shaft and lift out outer pump rotor.
3. Remove oil pressure relief valve plug and lift out spring and relief valve plunger.

Oil Pump Inspection— All Engines
1. Clean all parts thoroughly in solvent.
2. Inspect mating surface of oil pump cover. Mating face should be smooth with no scratches or grooving. Replace if scratched or grooved.
3. Lay straight edge across oil pump cover face. If 0.0015 in. feeler gauge can be inserted between cover and straight edge, replace cover.
4. Measure outer rotor. If rotor length is less than 0.649 in. for six cylinders, 0.825 in. for 273 and 318 V8s, or 0.943 in. for larger V8s, replace rotor. If rotor diameter is less than 2.469 in. for either sixes or eights replace rotor.
5. Measure inner rotor. If rotor length is less than 0.649 in. for six cylinders, 0.825 in. for 273 and 318 V8s, or 0.942 in. for larger V8s, replace rotor.
6. Install outer rotor in pump body and holding against one side of body measure clearance between rotor and body. If clearance is greater than 0.014 in., replace oil pump body.
7. Install inner rotor into pump body and place straight edge across pump body between bolt holes. If feeler gauge greater than 0.004 in. can be inserted between rotors and body, replace oil pump body.
8. Measure clearance between tips of inner and outer rotor where they are opposed. If clearance exceeds 0.010 in., replace inner and outer rotors.

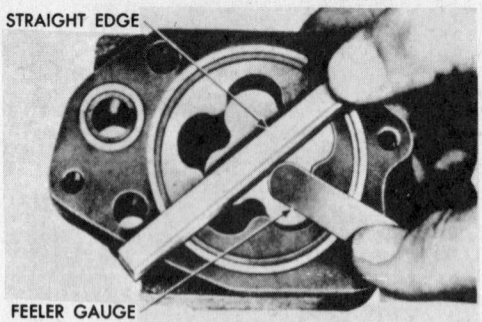

Measuring clearance over rotors
(© Chrysler Corp.)

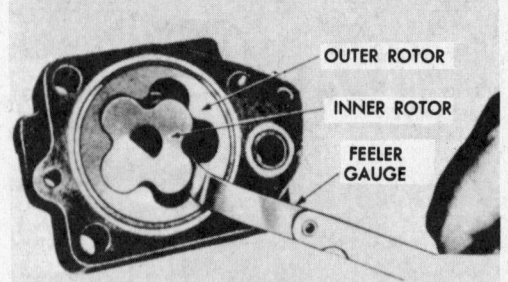

Measuring clearance between rotors
(© Chrysler Corp.)

Oil Pump Assembly and Installation

Six Cylinder

1. Assemble pump using new parts as required.
2. Install new seal rings between cover and body. Tighten cover bolts to 95 in. lbs.
3. Install pump on engine. Tighten attaching bolts to 200 in. lbs.
4. Lower engine and install engine mount bolts.
5. Replace fan shroud, connect upper and lower hoses, and fill radiator.

273 and 318 V8s

1. Assemble pump using new parts as required.
2. Install new seal rings between cover and body. Tighten cover bolts to 95 in. lbs.

3. Fill pump with oil to prime it and install pump and strainer on rear main bearing cap. Tighten bolts to 35 ft. lbs.
4. Replace oil pan.

361, 383, 413, 426 Wedge, 440, and 426 Hemi V8s

1. Assemble pump using new parts as required.
2. Install new seal rings between filter base and body. Torque bolts to 10 ft. lbs.
3. Install new "O" ring seal on pilot of oil pump before attaching oil pump to cylinder block.
4. Install oil pump on engine using new gasket and tightening bolts to 30 ft. lbs. Install oil filter assembly and oil pan.

Oil Filter

The oil filter is located at the left side of the engine and is the throw-away-type. It is only necessary to unscrew the filter from the base by hand and discard.

Screw on the new filter until the gasket contacts the base, then tighten at least ½ turn more. Run the engine and check for leaks.

Pistons, Connecting Rods and Main Bearings

Rod and Piston Assembly Removal

All Models

1. Remove the cylinder head and oil pan.
2. Insert a good cylinder ridge reamer into the top of the bores accessible without turning the crankshaft, and remove the

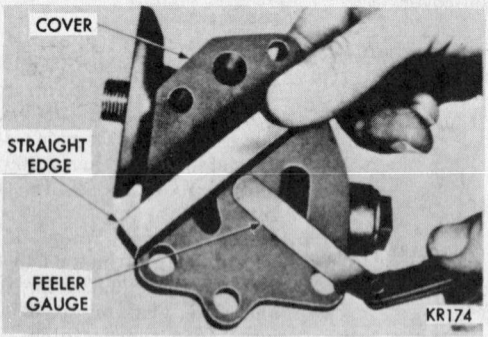

Measuring oil pump cover flatness
(© Chrysler Corp.)

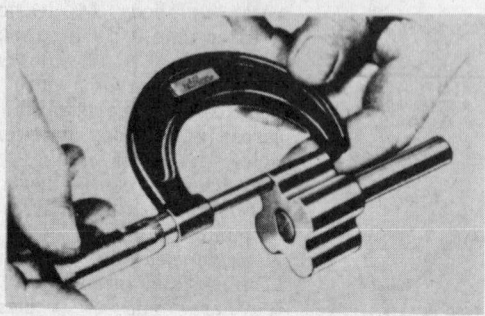

Measuring inner rotor thickness
(© Chrysler Corp.)

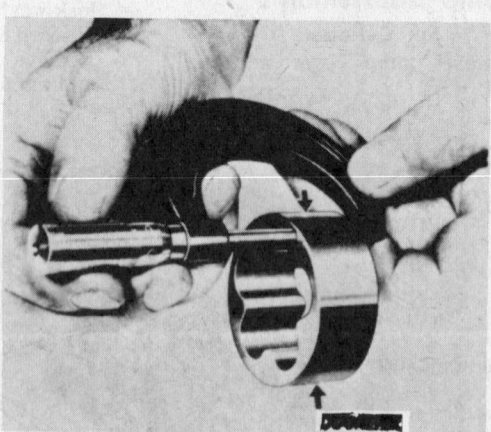

Measuring outer rotor thickness
(© Chrysler Corp.)

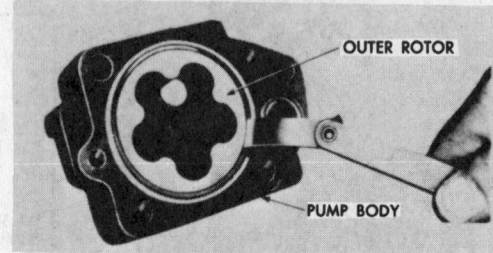

Measuring outer rotor clearance
(© Chrysler Corp.)

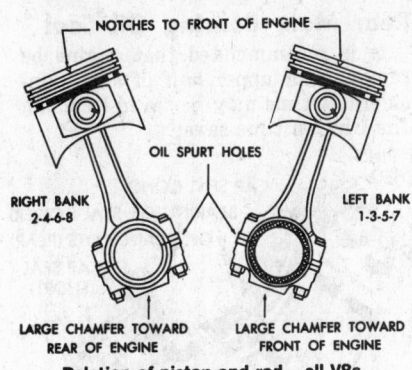

NOTCHES TO FRONT OF ENGINE

OIL SPURT HOLES

RIGHT BANK 2-4-6-8

LEFT BANK 1-3-5-7

LARGE CHAMFER TOWARD REAR OF ENGINE

LARGE CHAMFER TOWARD FRONT OF ENGINE

Relation of piston and rod—all V8s

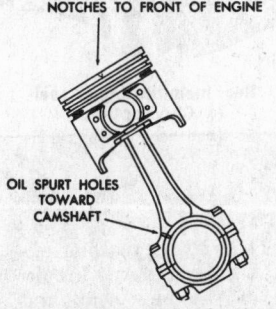

NOTCHES TO FRONT OF ENGINE

OIL SPURT HOLES TOWARD CAMSHAFT

Relation of piston and rod—slant 6

ridge. Detach the tool, turn the crankshaft, reattach the tool and remove the ridge on the next cylinder. Continue this process until all cylinder ridges have been removed.

Caution This is not a boring bar, so merely remove the ridge.

3. From underneath the car, select the connecting rods in the down position, and remove the locking device (pawl nut or cotter pin). Take off the two nuts that hold the cap to the lower end of the connecting rod. Tap the cap gently and slide it off the end of the bolts. Be careful not to lose the lower half of the rod bearing.
4. Start the connecting rod and piston assembly up toward the top of the bore, but, before pushing it out, replace the cap so that there isn't the slightest chance of it getting mixed up or put on in the wrong way.
5. At this point, note whether the number of the cylinder is stamped on the connecting rod, and, if it is not, some provision will have to be made to mark the rod, such as a file mark or a punch mark. Push the rod and piston assembly up until the rings snap out of the cylinder.
6. When assembling pistons to connecting rods, and the assemblies to the engine, on the slant 6, be sure to locate the squirt hole to the proper side. The 1963-66 engine using cast iron block has the

piston head notch at the front, with the oil squirt hole to the right side of the engine. Where aluminum blocks are used, the piston head notch is front with the oil squirt hole to the left side of the engine.

Piston Rings

Replacement

NOTE: before replacing rings, inspect cylinder bores.

1. Using internal micrometer, measure bores both across thrust faces of cylinder and parallel to crankshaft axis at a minimum of four, equally spaced locations.
2. The bore must not be out of round by more than 0.005 in. and it must not taper more than 0.010 in. Taper is the difference in wear between two bore measurements in any cylinder. Bore any out-of-tolerance cylinder to the next available oversize that will clean up cylinder wear.
3. Recommended piston clearances (for new pistons are 0.0005-0.0015 in. for six-cylinder engines, 0.0005-0.0015 in. for 273 and 318 V8 engines, 0.0003-0.0013 in. for 361, 383, 413, 426 wedge, and 440 V8 engines, and 0.0025-0.0035 in. for 426 Hemi engines.
4. If cylinder bore is within limits dimensionally, examine bore visually. It should be dull silver in color and exhibit cross-hatch machining pattern intersecting at about 60°. There should be no scratches, tool marks, nicks, or other damage. If any such damage exists, bore cylinder to next possible oversize that will clean up bore defects. Polished or shiny places in the bore are

CROSS-HATCH PATTERN

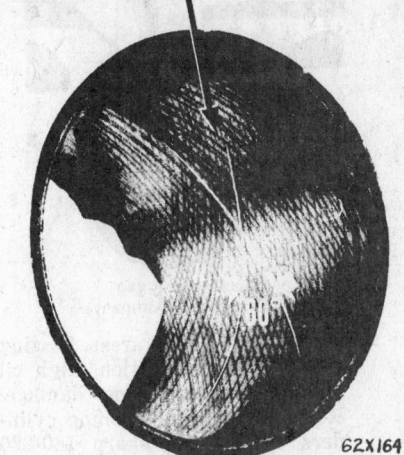

62X164

Cross hatch pattern
(© Chrysler Corp.)

⏻ CHILTON TIME-SAVER

This or any other machining operation should be done with the cylinder block completely disassembled. Hot tank cylinder block after honing or boring to remove all metal particles. To remove minor glazing when honing equipment is not available, run emery cloth back and forth across glazed area perpendicular to bore axis. Scrub block and bores thoroughly with soap and water to remove all grit particles.

NOTE: the emery cloth method should be used only as a last resort as it is a method much inferior to honing.

Ring end gap (in.)

Year and Engine	Top compression	Bottom compression	Oil control
1964-65 six	0.010-0.020	0.010-0.020	0.015-0.062
1964 318	0.010-0.020	0.010-0.020	0.010-0.020
1966-70 six	0.010-0.020	0.010-0.020	0.015-0.055
1965-70 273, 318	0.010-0.020	0.010-0.020	0.015-0.055
1964-65 big V8s	0.013-0.025	0.013-0.025	0.013-0.025
1966-67 big V8s	0.013-0.025	0.013-0.025	0.015-0.055
1968-70 big V8s	0.013-0.023	0.013-0.023	0.015-0.055

Ring side clearance (in.)

Year and Engine	Top compression	Bottom compression	Oil control
1964-65 six	0.0015-0.0040	0.0015-0.0040	0.009 max.
1964-65 all V8	0.0015-0.0030	0.0015-0.0040	0.009 max.
1966-70 six and 273, 318 V8s	0.0015-0.0040	0.0015-0.0040	0.0002-0.0050
1966 361, 383, 413, 426 wedge	0.0015-0.0040	0.0015-0.0040	0.0002-0.0050
1967-70 383, 440 V8s	0.0015-0.0040	0.0015-0.0040	0.0000-0.0050
1966-70 Hemi	0.0010-0.0030	0.0010-0.0030	0.0002-0.0050

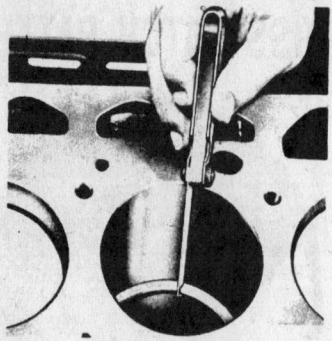

Checking ring gap
(© Ford Motor Company)

known as "glazed" areas. Glazing causes poor lubrication, high oil consumption and ring damage. Remove glazing by honing cylinders with clean, sharp 180-220 grit stones to obtain a surface finish of 15-35 RMS. The hone is also used to obtain correct piston clearance and surface finish in an overbored engine.

5. If cylinder bore is in satisfactory condition, place each ring in bore (in turn) and align using the head of a piston. Measure the ring end gap with feeler gauges; if gap is greater than limit, use new ring; if gap is less than limit, file end of ring to obtain correct gap. The correct ring gaps are found in the chart.

6. Check ring side clearance by installing rings onto piston and inserting feeler gauge of correct thickness between ring and lower land. Gauge should slide freely

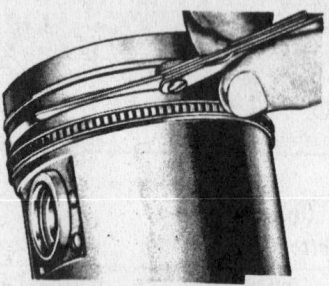

Checking ring side clearance
(© Ford Motor Company)

Piston Ring Compressor

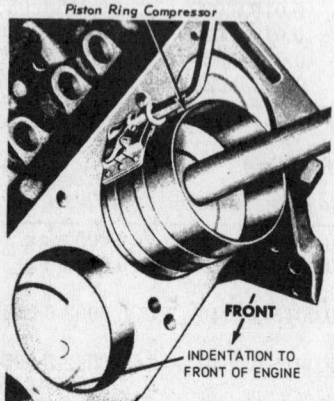

FRONT

INDENTATION TO
FRONT OF ENGINE

Piston installation
(© Ford Motor Company)

around ring circumference without binding. Any wear will usually show up as a step on the lower land; replace any pistons having high steps. Ring side clearance is found in the chart.

7. Space ring gaps at equidistant intervals around piston circumference. Be sure to install piston into its original bore. Install short lengths of rubber tubing over connecting rod bolts to prevent rod journal damage, then install ring compressor over rings on piston and lower piston-rod assembly into bore until ring compressor contacts block. Using a woden hammer handle, push piston into bore while guiding rod onto journal.

Rear Main Bearing Oil Seal

It is recommended that engine be removed for upper half of seal. Considerable time may be saved by using the Chilton time-saver.

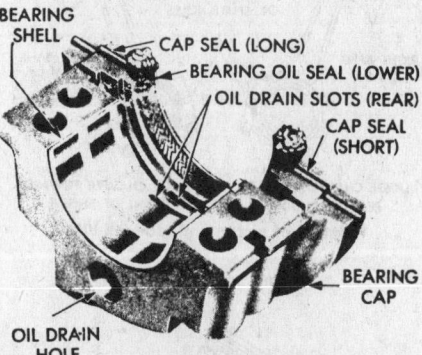

BEARING SHELL — CAP SEAL (LONG) — BEARING OIL SEAL (LOWER) — OIL DRAIN SLOTS (REAR) — CAP SEAL (SHORT) — BEARING CAP — OIL DRAIN HOLE

Rear main bearing oil seal
(© Chrysler Corp.)

⏻ CHILTON TIME-SAVER

Top Half, Rear Main Bearing Oil Seal Replacement

The following method has proven a distinct advantage in most cases and, if successful, saves many hours of labor.

1. Drain engine oil and remove oil pan.
2. Remove rear main bearing cap.
3. With a 6 in. length of 3/16 in. brazing rod, drive up on either exposed end of the top-half oil seal. When the opposite end of the seal starts to protrude, have a helper grasp it with pliers and pull gently while the driven end is being tapped. It is surprising how easily most of these seals can be removed by this method.

Woven Fabric Type Seal Installation

1. Obtain a 12 in. piece of copper wire about the same gauge as that used in the strands of an insulated battery cable.
2. Thread one strand of this wire through the new seal, about ½ in. from the end, bend back and make secure.
3. Thoroughly saturate the new seal with engine oil.
4. Push the copper wire up through the oil seal groove until it comes down on the opposite side of the bearing.
5. Pull (with pliers) on the protruding copper wire while the crankshaft is being turned and the new

seal is slowly fed into place.

CAUTION: this snaking operation slightly reduces the diameter of the new seal, and care will have to be used to keep the seal from slipping too far through the top half of the bearing.

6. When an equal amount of seal is extending from each side, cut off the copper wire close to the seal and tamp both ends up into the groove (this will tend to expand the seal again).

NOTE: don't worry about the copper wire left in the groove; it is too soft to cause damage.

7. Replace the seal in the cap in the usual way and replace the oil pan.

It is sometimes possible to seal up a small leak in the rear main bearing oil seal by installing a new lower half and letting it project slightly above the main bearing cap. The main bearing cap is then bolted tightly into place and immediately taken down. The riveted over-portion of the projecting lower half of the seal should be cleaned off, and the cap again bolted up into place until the cap seats firmly in the block without riveting over the oil seal.

The purpose of this is to squeeze the upper half of the oil seal more tightly into its groove, causing it to compress somewhat down onto the crankshaft.

This particular method is not always successful but has been used in many instances to prevent minor leaks.

Front Suspension

Adjustment

General instructions covering the front suspension, and how to repair and adjust it, together with information on installation of front wheel bearings and grease seals, are given in the Unit Repair Section.

Definitions of the points of steering geometry are covered in the Unit Repair Section. This article also covers troubleshooting front end geometry and irregular tire wear.

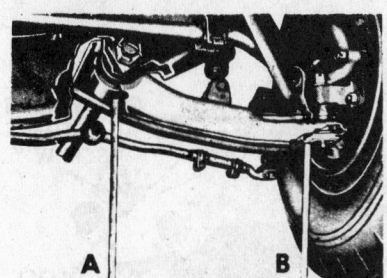

Checking front suspension height at ball joint and lower control arm
(© Chrysler Corp.)

Figures covering the caster, camber, toe-in, king pin inclination, and turning radius can be found in the Front Wheel Alignment table.

Car length and tire size figures can be found in the General Chassis and Brake Specifications table of this car section.

Ball Joints

1964-70

Balloon type and semi-permanently lubricated steering linkage, and front suspension ball joints are used. Relubrication at these points is required at about 36,000 miles, or three-year periods, (whichever comes first). However, the balloon seals should be inspected for leaks or other damage, two or three times a year.

When lubricating these points, use only multi-mileage long-life chassis grease. Remove the threaded plug from each joint to be lubricated, and temporarily install lube fittings. Inject lubricant, while feeling the seal with the fingers. Stop just before the seal starts to balloon. Remove lube fittings and reinstall plugs.

Front Height

Adjustment (Without Gauge)

1. Jounce the car and measure from the lower ball joint to the floor (measurement A).
2. Measure from the control arm torsion bar spring anchor housing to the floor, (measurement B).
3. Subtract A from B. The difference should be as shown in specification table.
4. Measure the other side in the same way.
5. Adjust by turning the torsion bar anchor adjusting nut, clockwise to raise, and counter-clockwise to lower.

Torsion Bar Springs

Contrary to appearance, the torsion bars are not interchangeable from right to left. They are marked with an R or an L, according to their location.

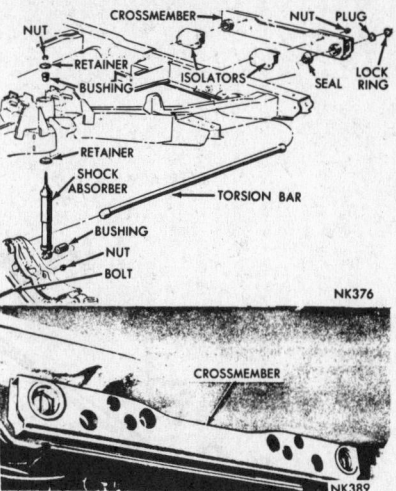

Torsion bar suspension—1965-70
(© Chrysler Corp.)

Removal

1. Lift the car by the body only so that the front suspension is free of all load. If the car is to be raised with jacks, place jack under center of frame crossmember and raise until suspension is free of all load.
2. Release load from torsion bar by backing off anchor adjusting nuts. Remove the adjusting nut and swivel bolt.
3. Remove the lower control arm strut.
4. Remove the lock spring from the rear of torsion bar rear anchor.
5. Install tool C-3728, or other suitable clamp, and remove tor-

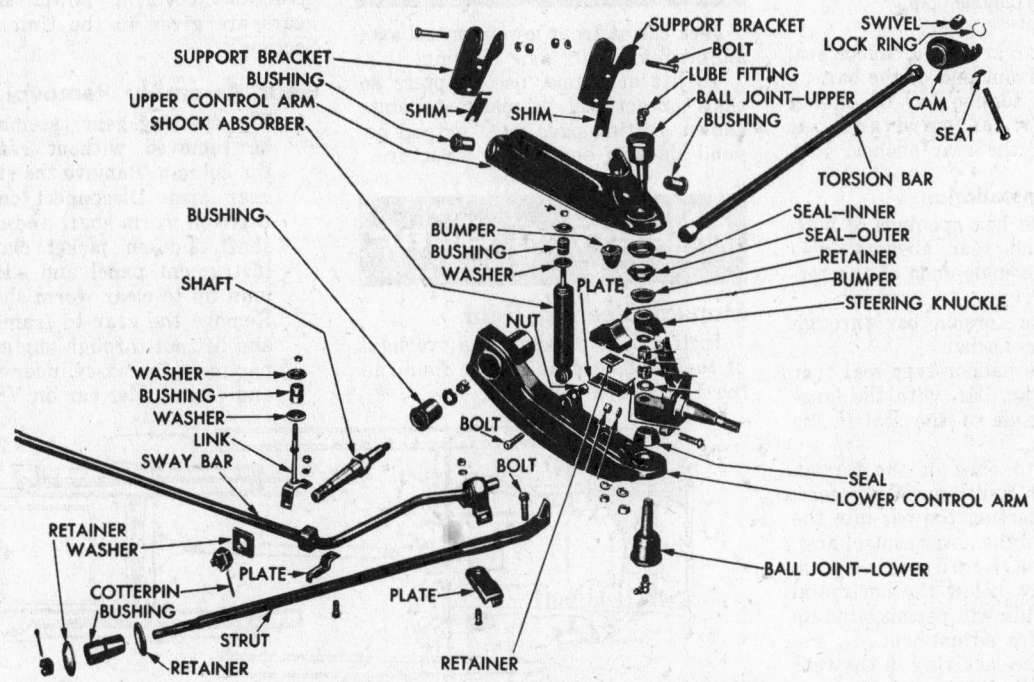

Exploded view of front suspension (© Chrysler Corp.)

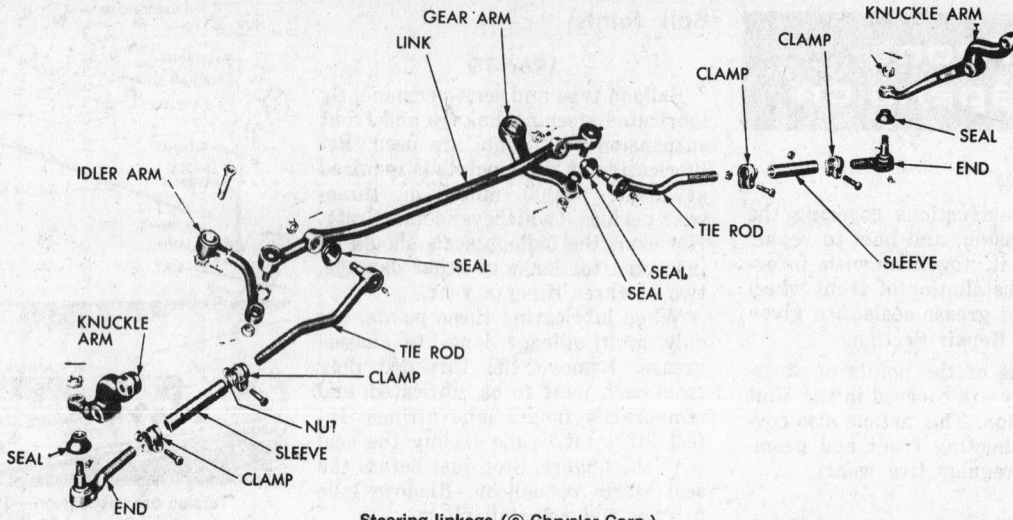

Steering linkage (© Chrysler Corp.)

sion bar rearward by striking the clamping tool with a hammer.

Do not apply heat to the front or rear anchors. Do not scratch or otherwise mar the skin of the torsion bar during removal or installation.

6. Remove the clamping tool and

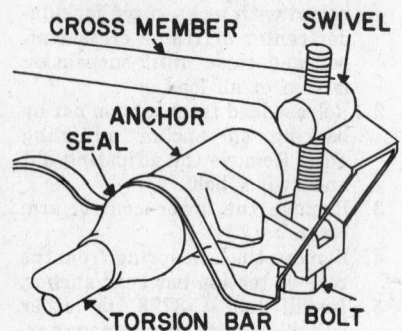

Torsion bar adjustment bolt
(© Chrysler Corp.)

slide the rear anchor balloon seal off the front end of the bar.

7. Remove torsion bar by sliding the bar rearward and out through the rear anchor.

Installation

1. Clean the hex openings of both front and rear anchors, also clean the male ends of the torsion bar.
2. Feed the torsion bar through the rear anchor.
3. Slide the balloon-type seal over the torsion bar, with the large cupped side of the seal facing the rear.
4. Coat both ends of the torsion bar with multi-purpose grease.
5. When starting the bar into the anchor in the lower control arm, position the adjusting arm about 60° below the horizontal plane. This will permit wind-up for future adjustment.
6. Install the lock ring in the rear anchor, then move torsion bar

rearward until the bar contacts the lock ring.

7. Position swivel bolt on the control arm and hold in place while installing the adjusting nut and seat. Tighten the adjustment about 10 turns before lowering car to the floor.
8. Pack the annular opening in the rear anchor with multi-purpose grease. Slide the rear anchor balloon-type seal into position over the rear anchor until the lip of the seal fits in the groove.
9. Install lower control arm strut.
10. Lower car to the floor and adjust front suspension height.

Jacking, Hoisting

Jack car at front lower control arm and at rear under axle housing.

To lift at frame, use adapters so that contact will be made at points shown. Lifting pads must extend beyond sides of supporting structure.

Steering Gear

Manual Steering Gear

Instructions covering the overhaul of the steering gear will be found in the Unit Repair Section.

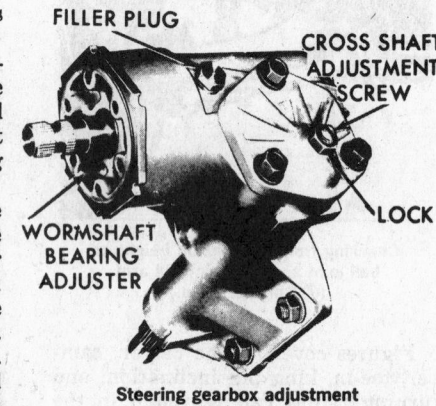

Steering gearbox adjustment
(© Chrysler Corp.)

Power Steering Gears

Troubleshooting and repair instructions covering power steering gears are given in the Unit Repair Section.

Gear Assembly Removal

1. The steering gear assembly can be removed without removing the column. Remove the steering gear arm. Disconnect coupling between worm shaft and column shaft. Loosen jacket clamp at instrument panel and slide column up to clear worm shaft.
2. Remove the gear to frame bolts and lift out through engine compartment on six-cylinder models and from under car on V8s.

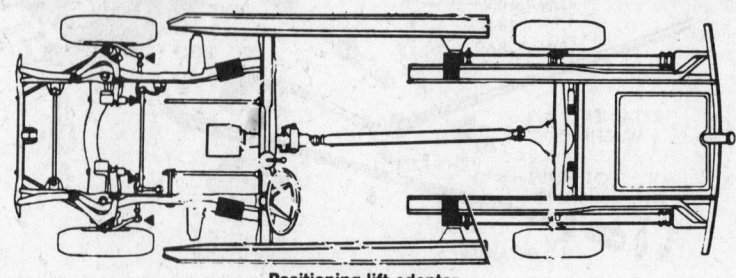

Positioning lift adapter

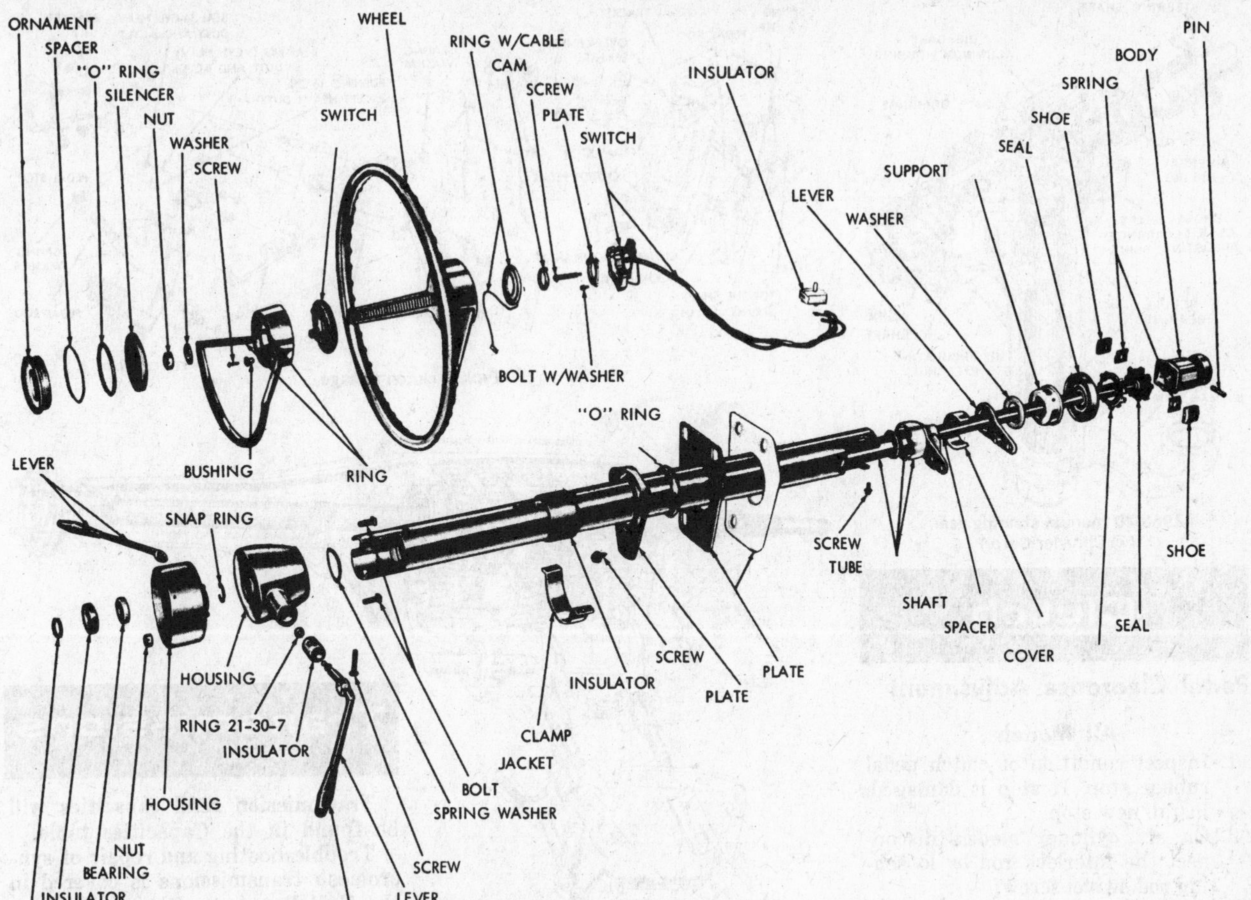

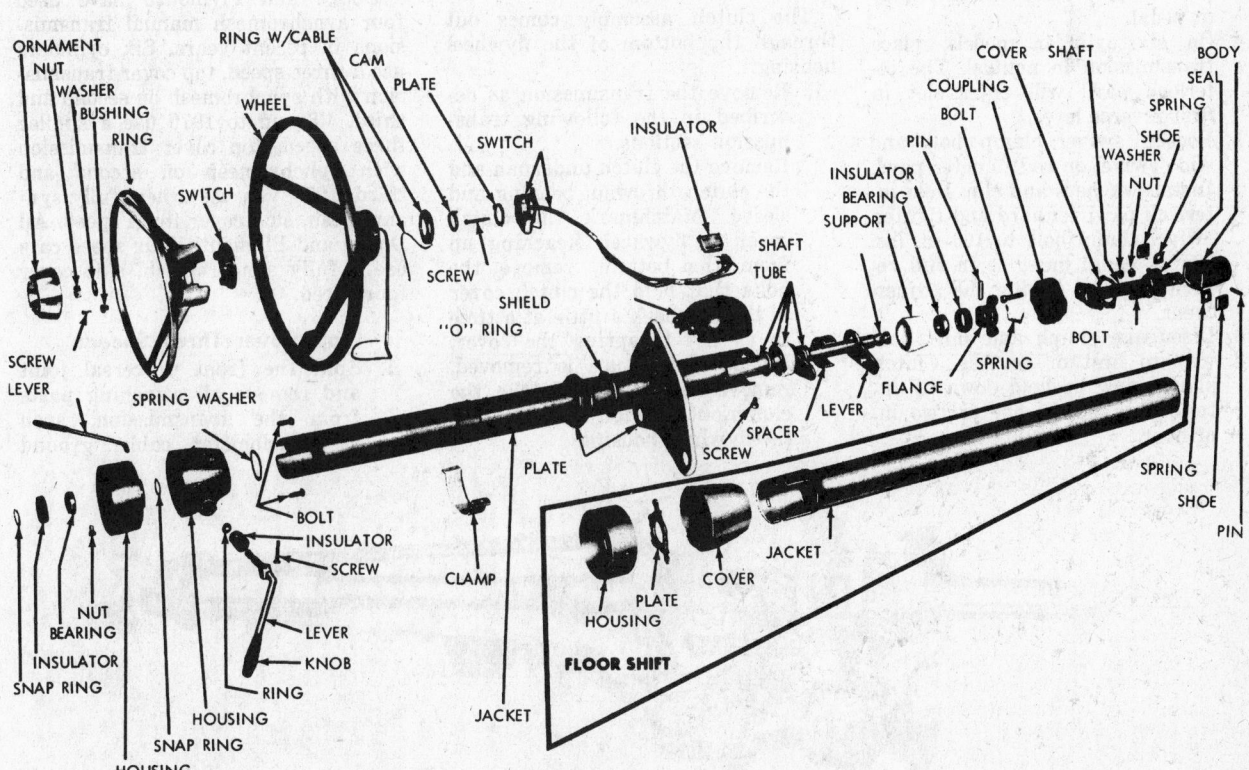

1965-66 column shift (© Chrysler Corp.)

ORNAMENT
SPACER
"O" RING
SILENCER
NUT
WASHER
SCREW
WHEEL
SWITCH
RING W/CABLE
CAM
SCREW
PLATE
SWITCH
INSULATOR
LEVER
WASHER
SUPPORT
SEAL
SHOE
SPRING
BODY
PIN

BOLT W/WASHER
"O" RING
SCREW
TUBE
SHAFT
SPACER
COVER
SEAL
SHOE

LEVER
SNAP RING
BUSHING
RING

HOUSING
RING 21-30-7
INSULATOR
HOUSING
NUT
BEARING
INSULATOR
SNAP RING
KNOB
SCREW
LEVER
SPRING WASHER
BOLT
CLAMP
JACKET
INSULATOR
SCREW
PLATE
PLATE

1965-66 column shift—Dodge Polara and Monaco (© Chrysler Corp.)

ORNAMENT
NUT
WASHER
BUSHING
RING
SWITCH
RING W/CABLE
CAM
PLATE
WHEEL
SWITCH
SCREW
INSULATOR
COVER
SHAFT
COUPLING
BOLT
PIN
INSULATOR
BEARING
SUPPORT
SHAFT
TUBE
BODY
SEAL
SPRING
SHOE
WASHER
NUT

SCREW
LEVER
SPRING WASHER
SHIELD
"O" RING
FLANGE
LEVER
SPRING
BOLT
SPRING
SHOE
PIN

NUT
BEARING
INSULATOR
SNAP RING
BOLT
INSULATOR
SCREW
LEVER
KNOB
RING
HOUSING
SNAP RING
HOUSING
CLAMP
SPACER
SCREW
PLATE
JACKET
PLATE
HOUSING
COVER
JACKET

FLOOR SHIFT

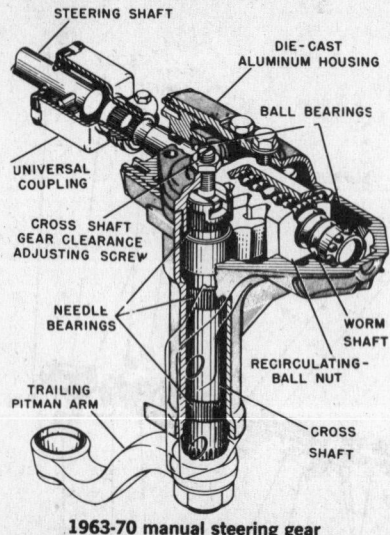

1963-70 manual steering gear
(© Chrysler Corp.)

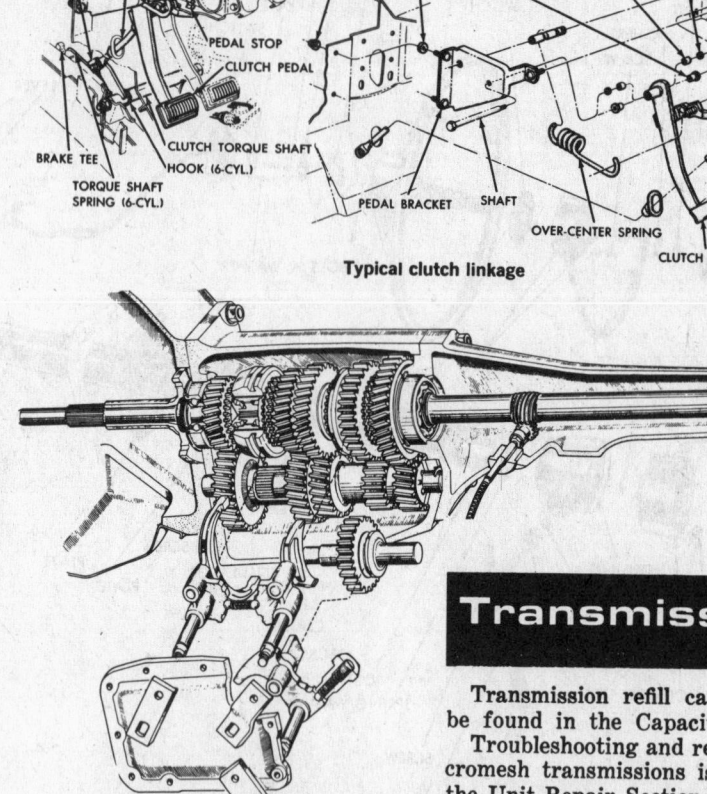

Typical clutch linkage

4-speed transmission
(© Chrysler Corp.)

Clutch

Pedal Clearance Adjustment

All Models

1. Inspect condition of clutch pedal rubber stop. If stop is damaged, install new stop.
2. On six cylinder models disconnect the interlock rod by loosening rod swivel screw.
3. Adjust fork rod by turning self-locking self-adjusting nut to provide 5/32 in. free movement at end of fork. This movement will provide prescribed one inch play at pedal.
4. On six cylinder models, place transmission in neutral. The interlock pawl will enter slot in first-reverse lever.
5. Loosen swivel clamp bolt and slide swivel on rod to enter pawl. Install washers and clip. Hold interlock pawl forward and tighten swivel clamp bolt to 100 in. lbs. Clutch pedal must be in full return position during the adjustment.
6. Disengage clutch and shift half way to first or reverse. Clutch should now be held down by interlock to within one or two in. of floor.

Clutch Assembly Removal

The clutch assembly comes out through the bottom of the flywheel housing.

1. Remove the transmission as described in the following transmission sections.
2. Remove the clutch underpan and the clutch throwout bearing and sleeve. Matchmark the clutch cover to flywheel. Reaching up from the bottom, remove the bolts that hold the clutch cover to the flywheel, a little at a time so as not to spring the cover. When all pressure is removed, remove the bolts and take the clutch out through the bottom of the flywheel housing.

Transmissions

Transmission refill capacities will be found in the Capacities table.

Troubleshooting and repair of syncromesh transmissions is covered in the Unit Repair Section.

Synchromesh Transmission Removal

Dodge and Plymouth have used four synchromesh manual transmissions in recent years. Six cylinders use a three speed, top cover transmission with synchromesh on second and third. V8s up to 1970 use a similar three speed, top cover transmission with synchromesh on second and third. 1970 V8s use a new fully synchromesh, side cover three speed. All Dodge and Plymouth four speed cars use a fully synchromesh, side cover four speed.

Top Cover Three Speeds

1. Split the front universal joint and remove all attaching parts from the transmission, such as speedometer cable, ground

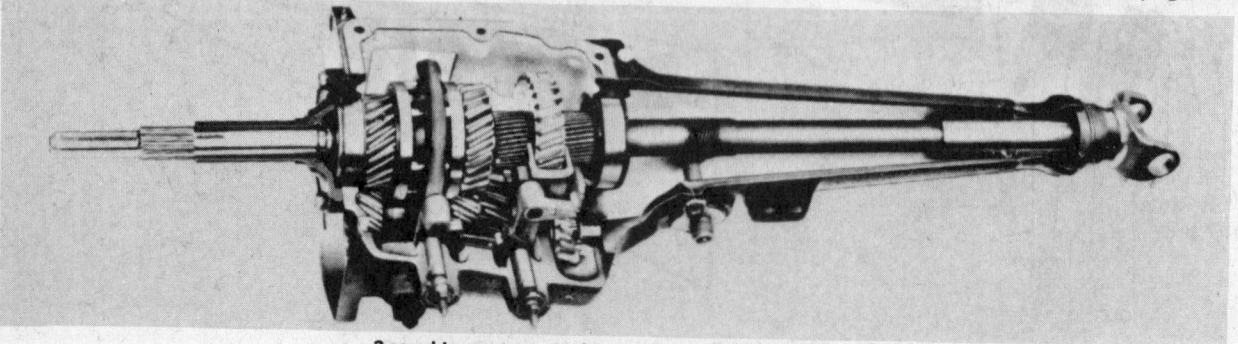

3-speed top cover manual transmission (© Chrysler Corp.)

cables, shift levers and rods, hand brake cables, etc.

2. Place a jack under the back of the engine and take a slight load on the jack. Remove the bolts that hold the transmission to the frame crossmember, then take out the bolts that hold the crossmember to the frame. Let the crossmember come down.

3. Remove the two upper bolts that hold the transmission to the bell housing and replace them with two long pilot studs. Remove the two bottom bolts and slide the transmission assembly back along the two upper pilot studs until the clutch shaft clears the clutch hub. Slide off the end of the pilot shafts and lower to the floor.

4. Reverse the above for installation. Be sure not to allow the transmission to hang after the pinion has entered the clutch disc.

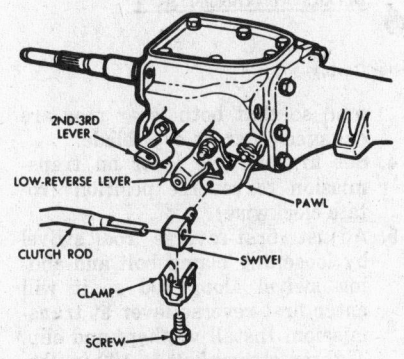

3-speed shift linkage
(© Chrysler Corp.)

Fully Synchromesh, Side Cover Three Speed

1. Remove shift rods from transmission levers.
2. Drain transmission fluid.
3. Disconnect drive shaft at rear universal joint. Mark both parts for reassembly.
4. Carefully pull yoke out of transmission extension.
5. Disconnect speedometer and backup lights.
6. Remove part of exhaust if it blocks transmission.
7. Raise engine slightly and block in place.
8. Support transmission with jack, and remove crossmember.
9. Remove transmission to clutch housing bolts.
10. Slide transmission to rear until drive pinion shaft clears. Clear clutch disc, lower transmission, and remove from vehicle.

Four Speed

1. Raise vehicle on a hoist and drain transmission.
2. Disconnect all shift controls from transmission levers. Remove

three bolts securing shift unit to extension housing.

3. Disconnect propeller shaft at rear universal joint. Carefully pull yoke out of transmission extension.
4. Disconnect speedometer cable and backup light switch leads.
5. Disconnect left exhaust pipe or dual exhausts. Disconnect parking brake cable.
6. Raise engine slightly and block in place.
7. Disconnect transmission extension from crossmember.
8. Remove crossmember.
9. Support transmission with jack. Remove clutch housing to transmission bolts.
10. Slide transmission to rear until drive pinion shaft clears clutch disc.
11. Lower transmission and remove from vehicle.

Manual Shift Adjustments

Three Speed 1964-67 Dodge and Plymouth Except 1964 Dodge 880

1. With the second and third control rod disconnected from the lever on the column, and first and reverse control rod disconnected at the transmission lever, position both transmission levers in neutral.
2. Check for axial freedom of the shift levers in the column. If the outer ends of the levers move up or down along the column axis over 1/16 in., loosen the two upper bushing screws and rotate the plastic bushing, downward, until all of the axial play is eliminated. Retighten bushing screws.
3. Wedge a screwdriver between the crossover blade and the second and third lever, so that the crossover blade is engaged with both lever crossover pins.
4. Adjust the swivel on the end of second and third rod until the stub shaft of the swivel enters the hole in the column lever. Install washers and clip. Tighten swivel lock nut to 70 in. lbs.
5. Slide the clamp and swivel on the end of the first and reverse control rod until the swivel stub shaft enters the hole in the transmission lever. Install washers and clip. Tighten the swivel clamp bolt to 100 in. lbs.

Three Speed Floor Shift 1964 Dodge 880

1. Remove the screws that hold the upper boot and retaining ring to the floor pan.
2. Remove the retaining ring and slide the boot up on the gearshift

lever far enough to expose the shift mechanism.

3. Disconnect the first and reverse shift rod. Disengage rod from lever.
4. Disconnect the second and high shift rod. Disengage rod from lever.
5. Place transmission levers in neutral, then follow the procedure in the illustration.

Three Speed 1968-69 Coronet, Charger and Belvedere

1. Remove second-third swivel from steering column lever and first-reverse swivel from transmission lever.
2. Make sure transmission shift levers are in neutral (middle detent) position.
3. Loosen lock nut and adjust second-third swivel so it will enter second-third lever at steering column while hand lever on steering column is held 12 degrees above horizontal position. Install washers and clip. Tighten swivel nut to 70 in. lbs.
4. Place screwdriver or suitable tool between cross-over blade and second-third lever at steering column so that both lever pins are engaged by cross-over blade.
5. Adjust first reverse rod swivel by loosening clamp bolt and sliding swivel along rod so it will enter first-reverse lever at transmission. Install washers and slip. Tighten swivel bolt to 100 in. lbs.
6. Remove tool from cross-over blade at steering column and shift through all gears to check adjustment and cross over smoothness.

Three Speed 1968-69 Polara, Monaco, and Fury

1. Remove first-reverse rod swivel from steering column and second-third rod swivel from torque shaft lever.
2. Make sure transmission shift levers are in neutral (middle detent) position.
3. Adjust second-third rod swivel by loosening clamp bolt and sliding swivel along rod so it will enter torque shaft lever while hand lever on steering column is held 12 degrees above horizontal position. Install washers and clip. Tighten swivel clamp bolt to 100 in. lbs.
4. Place screwdriver or suitable tool between cross-over blade and second-third lever at steering column so that both lever pins are engaged by cross-over blade.
5. Adjust first-reverse rod swivel by loosening lock nut and turning swivel so it will enter first-reverse lever at steering column. Install washers and clip. Tighten

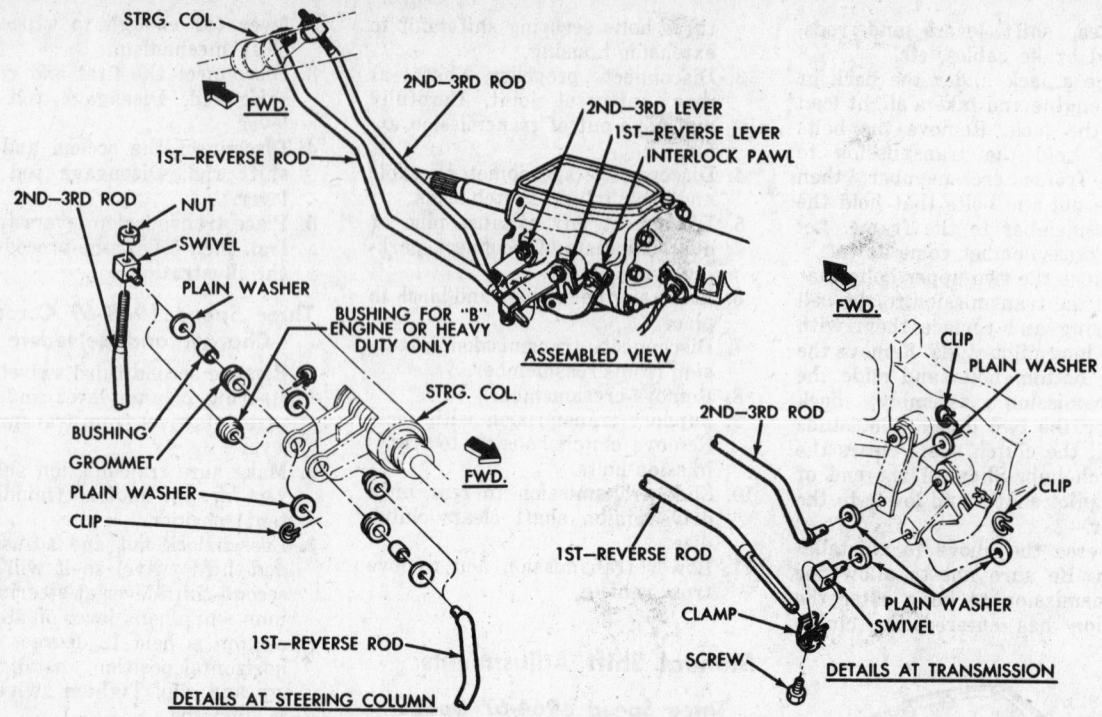

1968-69 Coronet, Charger, and Belvedere gearshift linkage (© Chrysler Corp.)

swivel lock nut to 70 in. lbs.

6. Remove tool from cross-over blade at steering column and shift through all gears to check adjustment and cross-over smoothness.

Three Speed 1970

1. Remove both shift rod swivels from transmission shift levers. Make sure transmission shift le-

vers are in neutral (middle detent) position.

2. Move shift lever to line up locating slots in bottom of steering column shift housing and bearing housing. Install suitable tool in slot and lock ignition switch.

3. Place screwdriver or suitable tool between crossover blade and second-third lever at steering col-

umn so that both lever pins are engaged by cross-over blade.

4. Set first-reverse lever on transmission to reverse position (rotate clockwise).

5. Adjust first-reverse rod swivel by loosening clamp bolt and sliding swivel along rod so it will enter first-reverse lever at transmission. Install washers and clip. Tighten swivel bolt to 100 in. lbs.

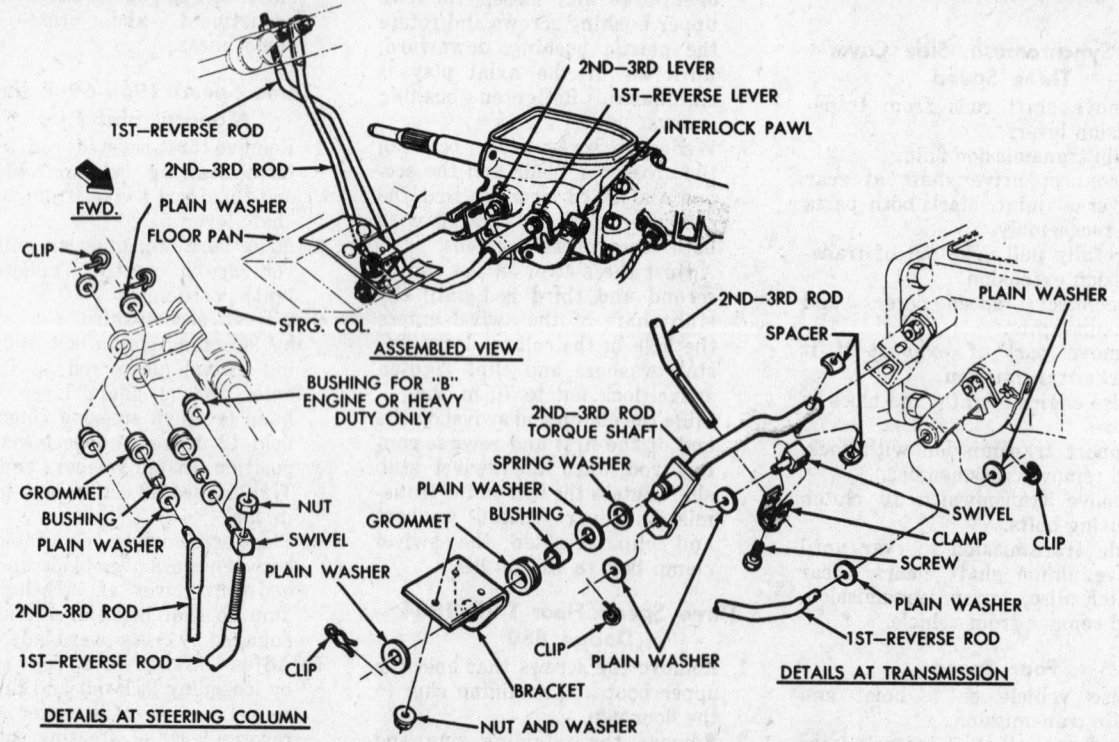

1968-69 Polara, Monaco, Fury and VIP gearshift linkage (© Chrysler Corp.)

6. Remove gearshift housing locating tool, unlock ignition switch, and shift column lever into neutral position.
7. Adjust second-third rod swivel by loosening clamp bolt and sliding swivel along rod so it will enter second-third lever at transmission. Install washers and clip. Tighten swivel bolt to 100 in. lbs.
8. Remove tool from crossover blade at steering column, and shift through all gears to check adjustment and cross-over smoothness.

Four-Speed

Many Dodge and Plymouth four speed transmissions use Hurst shift linkages. To adjust these linkages, see Hurst Shift Linkage section. Adjust Chrysler Corporation linkages as follows.
1. Remove the shift boot attaching screws, and slide the boot up on the shift lever.
2. Disconnect all the shift rods at the adjusting swivels.
3. Bend a ¼ in. diameter rod to a right angle, and insert through the aligning holes in the control rod levers and the slots provided in the gear shift support.
4. Adjust the length of the three shift rods until the swivel stub shafts match the control rod lever holes. Install the stub shafts and secure with clips.
5. Remove the ¼ in. diameter aligning rod.
6. With transmission hand shift lever in third or fourth speed

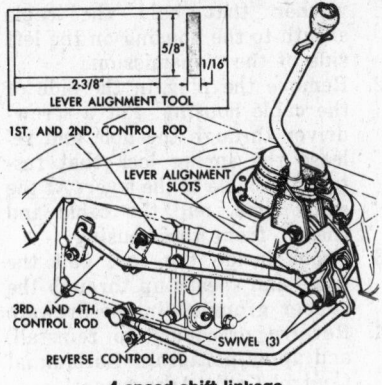

4-speed shift linkage
(© Chrysler Corp.)

detent position, adjust the lever stop screws (front and rear) to provide from .020 to .040 in. clearance between the lever and the stops. Tighten adjusting screw locknuts.
7. Test linkage adjustment for ease of shifting and crossover operation.

Caution
Accuracy of adjustment is very important because there is no reverse gear interlock to prevent engaging two gears at the same time.

8. Slide the boot down the shift lever shaft to the floor and secure with attaching screws.

Beginning 1966

1. Make up a lever aligning tool from 1/16 in. thick metal as in illustration.
2. With transmission in neutral,

disconnect all control rods from the transmission levers.
3. Insert lever aligning tool through the slots in the levers and against the back plate. This locks the three levers in neutral.
4. With all transmission levers in neutral, adjust the length of the control rods so they enter the transmission levers freely without rearward or forward movement.
5. Install control rod flat washers and retainers. Remove the aligning tool.
6. Check linkage for ease of shifting into all gears and for ease of crossover.

Automatic Transmission

TORQUEFLITE B (aluminum case) transmissions are used with the Slant 6 and V8 models up to 1965.
TORQUEFLITE (A-727-A and B) is a version of the Torqueflite B, but with no rear pump, and is used in all models beginning 1966.

Quick Service Information Reference

When automatic transmission trouble is reported, a road test and careful diagnosis are in order. Transmission Removal and Replacement and Linkage Adjustments are covered here. For Test Procedures, Transmission Overhaul and other detailed information, see Unit Repair Section.

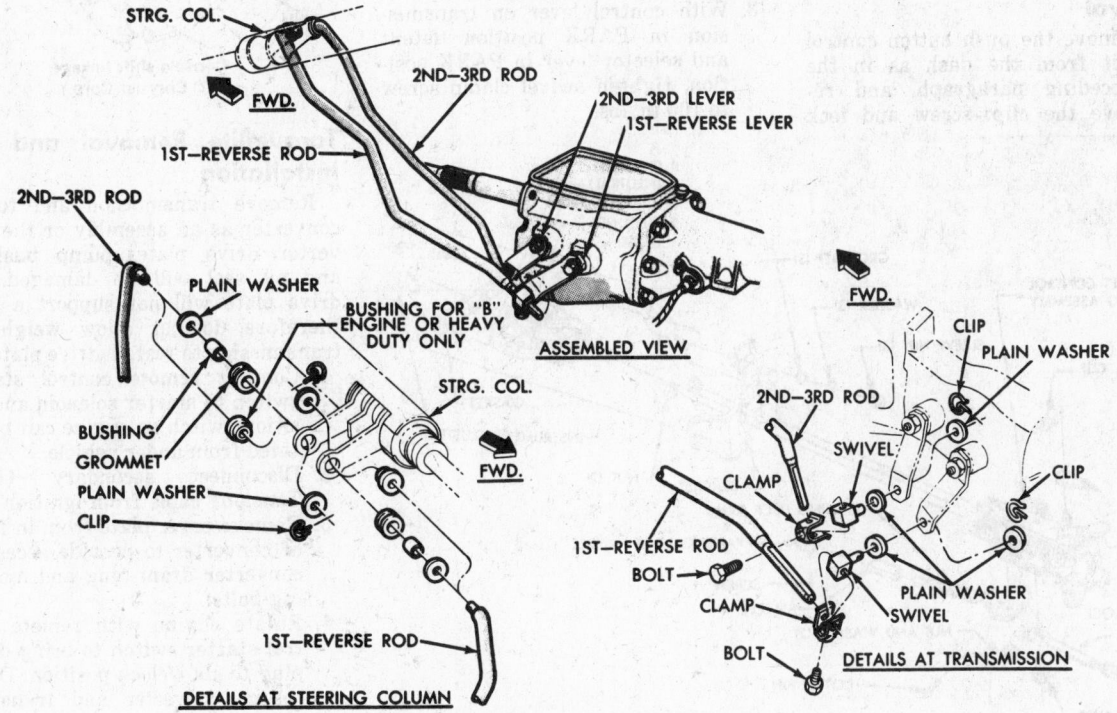

1970 Gearshift linkage (© Chrysler Corp.)

Torqueflite

Push Button Unit Light Bulb Renewal

Remove the three bezel retaining screws (one screw is located on underside of unit) and the bezel. Pull the D button off its slider, and gain clearance to remove the bulb. Install new bulb. Compress ends of slider and push D button back in place. Use a small screwdriver to assure that ends of slider are firmly seated in rear of plastic D button.

Back-up Light Switch Renewal

The back-up light switch is fastened to the push button unit by four tabs. Remove the push button control unit from the dash. Straighten the tabs and remove the switch.

Push Button Unit Removal

Remove the three bezel retaining screws (one screw is located on underside) and remove bezel. This will expose the two hex nuts and washers that hold the unit to the dash. Move the unit to the back of the dash. Remove the two screws holding cable assembly bracket to the unit, and remove the clip that holds the cable to the unit. Unfasten the wires from the back-up light switch. Note that the wire to the light in the unit comes over the upper stud and down between the sliders. The bracket that holds the light fits over the upper stud. Reverse the procedure to reinstall.

Manual Control Cable Removal

1. Remove the push button control unit from the dash as in the preceding paragraph, and remove the clip, screw and lock

washer that hold the cable sheath to the housing on the left side of the transmission.
2. Remove the plug in the side of the cable housing. Put a screwdriver through the hole and release the spring lock that fastens the cable to the lever. At the same time, pull the cable and sheath from the housing.
3. From inside the car, pull the cable and sheath up through the rubber grommet in the firewall.
4. Reverse procedure to reinstall, and apply paragraph on Manual Control Cable Adjustment.

Gearshift Linkage Adjustment Column Shift 1965-70

1. Place gearshift selector lever in PARK position and loosen con-

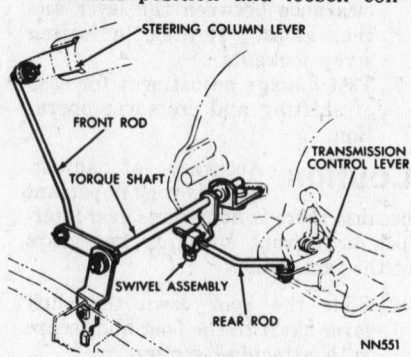

Automatic transmission column shift linkage
(© Chrysler Corp.)

trol rod swivel clamp screw a few turns.
2. Move transmission lever all the way to rear (in PARK detent).
3. With control lever on transmission in PARK position detent and selector lever in PARK position, tighten swivel clamp screw to 100 in. lbs.

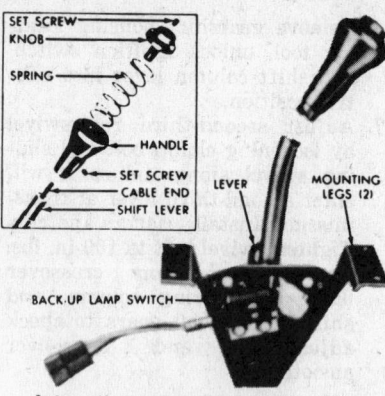

Automatic transmission console shifter
(© Chrysler Corp.)

Gearshift Linkage Adjustment Console Shift 1965-70

1. Place gearshift selector lever in PARK position and loosen lower rod swivel clamp screw a few turns.
2. Move transmission all the way to rear (in PARK detent).
3. With control lever on transmission in PARK position detent, and selector lever in PARK position, tighten swivel clamp screw or adjusting lever bolt securely.

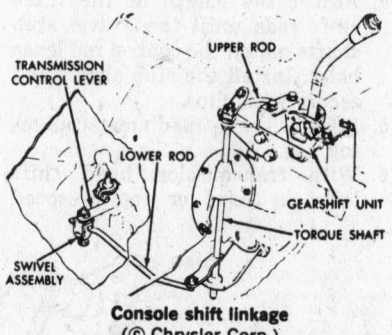

Console shift linkage
(© Chrysler Corp.)

Torqueflite Removal and Installation

Remove transmission and torque converter as an assembly or the converter drive plate, pump bushing, and oil seal will be damaged. The drive plate will not support a load, therefore, do not allow weight of transmission to rest on drive plate.

1. Connect remote control starter switch to starter solenoid and position switch so engine can be rotated from under vehicle.
2. Disconnect secondary (High Tension) cable from ignition coil.
3. Remove cover plate from in front of converter to provide access to converter dram plug and mounting bolts.
4. Rotate engine with remote control starter switch to bring drain plug to six o'clock position. Drain torque converter and transmission.
5. Mark converter and drive plate

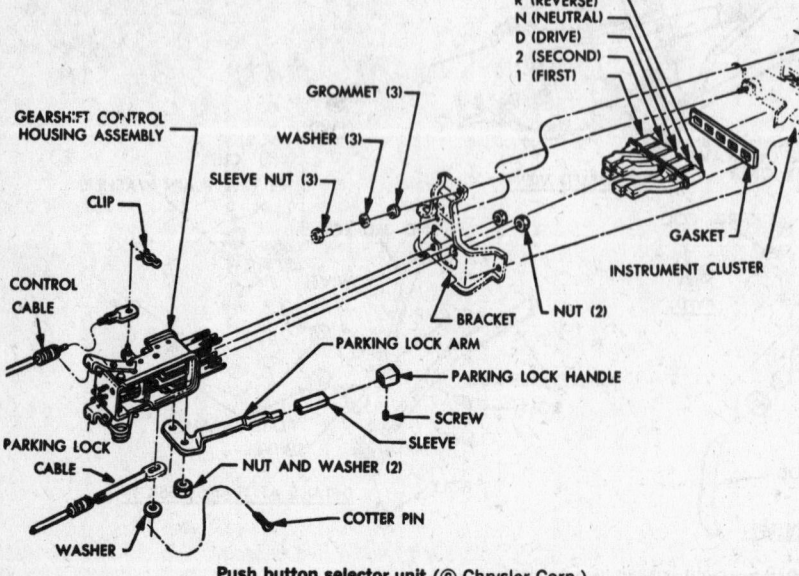

Push button selector unit (© Chrysler Corp.)

to aid in reassembly. There is an offset hole in crankshaft flange bolt circle, inner and outer circle of holes in drive plate, and four tapped holes in front face of converter so their parts will be installed in original position.

6. Rotate engine with remote control switch to locate two converter to drive plate bolts at five and seven o'clock positions. Remove bolts, rotate engine and remove two more bolts. Do not rotate converter by prying as this will distort drive plate. Do not engage starter if drive plate is not attached to converter with at least one bolt or if transmission to block bolts have been loosened.

7. Disconnect battery.
8. Remove starter.
9. Disconnect wire from neutral start switch.
10. Disconnect gearshift rod from transmission lever. Remove gearshift torque shaft from transmission housing and left side rail. On console shifts, remove two bolts securing gearshift torque shaft lower bracket to extension housing. Swing bracket out of way for transmission removal. Disconnect gearshift rod from transmission lever.
11. Disconnect throttle rod from bellcrank at left side of transmission bell housing.
12. Disconnect oil cooler lines at transmission and remove oil filler tube. Disconnect speedometer cable.
13. Disconnect drive shaft at rear universal joint. Carefully pull shaft assembly out of extension housing.
14. Remove transmission mount to extension housing bolts.
15. Raise engine slightly and block in place.
16. Remove crossmember attaching bolts and remove crossmember.
17. Support transmission with jack.
18. Attach a small C clamp to edge of converter housing to hold converter in place during removal of transmission.
19. Remove converter housing retaining bolts. Carefully work transmission to rear off engine dowels and disengage converter hub from end of crankshaft.
20. Lower and remove transmission. Remove C clamp and remove converter. Reverse procedure to install.

U Joints, Drive Lines

Cross- and Bearing-Type Joint Disassembly

The cross- and bearing-type joint can be identified easily since the joint is not covered.

1. To disassemble the joint, remove the four bolts that hold the two bearing assemblies to the companion flange and knock the bearings off the flange.
2. To remove the bearings from the yoke, first remove the bearing retainer lock washers or C-washers, then pressing on one of the bearings, drive the bearing in toward the center of the joint. This will force the cross to push the opposite bearing out of the universal joint yoke. After it has been pushed all the way out of the yoke, pull up the cross slightly and pack some washers under it. Then press on the end of the cross from which the bearing was just removed to force the first bearing out of the yoke.
3. Perhaps the easiest way to reassemble is to start both bearing retainers into the yoke at the same time, hold the cross carefully in the fingers and squeeze both bearings in a vise or heavy C-clamp. Driving the bearings into place usually cocks the little rollers, greatly reducing the life of the bearings.
4. Reinstall the locking devices.

Ball- and Trunnion-Type Joint Disassembly

The housing of the ball- and trunnion-type joint is held to its companion flange by four bolts. Remove the four bolts and pry the cover assembly backward away from the companion flange so that the shaft can be lowered. If two ball and trunnion type joints are used, both must be disconnected from their companion flanges in order to get the driveshaft over to the bench.

Remove the grease cover which will release the centering spring.

Remove the centering button spring and the ball and roller assemblies from the cross pin. Supporting the propeller shaft ball, press out the cross pin.

The cover and boot assembly can then be slid off the end of the propeller shaft.

Some models use a cross- and bearing-type joint at the rear joint, and a ball- and trunnion-type at the front.

Ball- and trunnion-type universal joints do not require a slip yoke since the driveshaft can work back and forth in the universal joint housing.

Propeller Shaft Center Bearing

Some models use a three universal joint drive line having two drive shafts and a center support bearing. The center support bearing and housing assembly are removed with the front driveshaft.

Split the rear universal joint and remove the rear propeller shaft.

Disconnect the front propeller shaft at the transmission flange and remove the bolts that hold the center bearing housing to the frame. Take off the center bearing housing, together with the front propeller shaft.

On the bench, remove the nut that holds the center universal joint flange to the driveshaft and, with a puller, draw off the flange. The bearing and housing assembly can then be pulled off the front of the shaft.

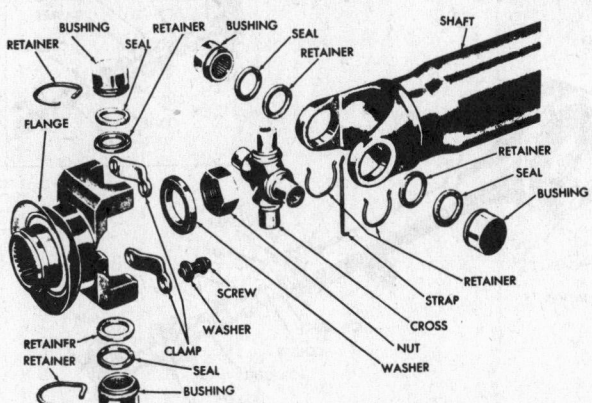

Exploded view—cross and bearing type universal joint
(© Chrysler Corp.)

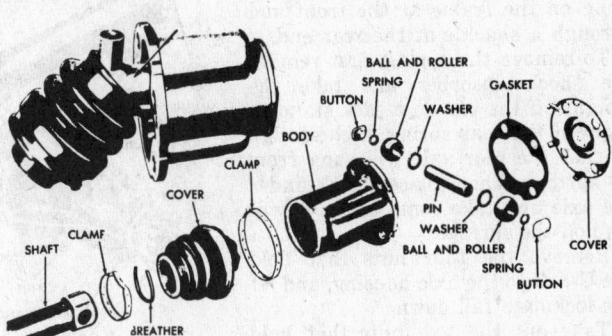

Exploded view—ball and trunnion type universal joint
(© Chrysler Corp.)

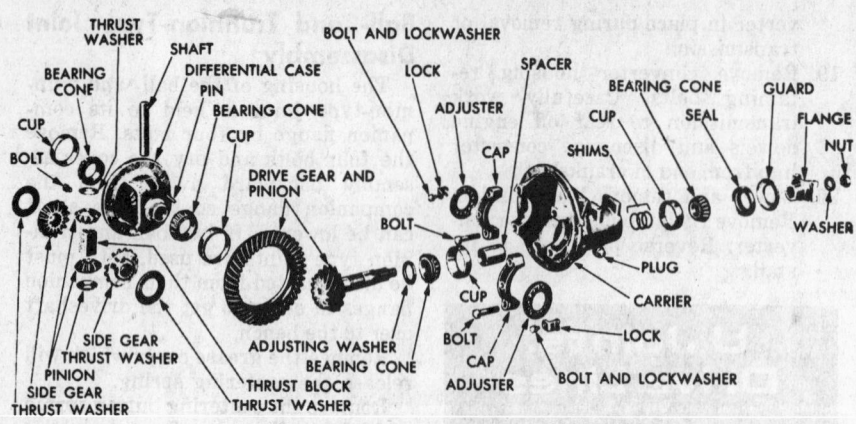

Differential—1963-64 (© Chrysler Corp.)

Radio

Radio Removal

1964 Dodge 880

1. Disconnect battery.
2. From under instrument panel, disconnect antenna lead, radio feed wire, radio illumination wire, and radio speaker leads.
3. Remove ash tray assembly to gain access to radio mounting screws.
4. Remove radio mounting screws.
5. From under instrument panel remove radio to instrument panel lower reinforcement mounting bracket.
6. Remove radio from under instrument panel. Reverse procedure to install.

1964-65 Coronet and Belvedere, and 1964 Fury

On vehicles equipped with air conditioning, service radio through glove box opening in instrument panel.

1. Disconnect battery.
2. Remove radio control knobs.
3. From under instrument panel disconnect radio feed wire at fuse block.
4. Disconnect antenna and radio speaker leads.
5. Remove screw attaching radio mounting bracket to instrument panel lower reinforcement.
6. Loosen top screw on radio mounting bracket and remove mounting bracket.
7. From front of radio remove two nuts attaching radio to the instrument panel and remove radio. Reverse procedure to install.

1965 Dodge 880, Polara, Monaco, and Plymouth Fury

1. Disconnect battery.
2. Remove radio knobs.
3. Remove radio mounting nuts.

Drive Axle

Troubleshooting and Adjustment

General instructions covering the troubles of the drive axle and how to repair and adjust it, together with information on installation of rear axle bearings and grease seals, are given in the Unit Repair Section.

Capacities of the drive axle are given in the Capacities tables.

Rear Shock Absorber Replacement

On all models, a direct acting shock absorber is used. To remove it, simply detach at the top and bottom and lift off the car.

Shock Absorber Service

Service on shock absorbers is a highly specialized job, requiring specific equipment for each type of shock absorber and this type of work should not be attempted in the average shop. If the shock absorber is defective or inefficient, it should be replaced with a new or rebuilt one.

Rear Spring Replacement

On all models, the rear spring is hung on the frame at the front and through a shackle at the rear end.

To remove the spring, first remove the shock absorber and take the weight off the car on a jack stand in front of the rear spring, high enough so that the rear axle will hang from its springs. Then, place a jack under the axle and take some of the download off the spring.

Remove the four nuts that hold the U-bolt to the axle housing, and let the lockplate fall down.

Take out the two nuts that hold the rear shackle, the top one to the frame, the bottom one to the spring, and drive off the rear shackle.

Remove the single bolt that holds the spring at the front, and lower the spring to the floor.

Rear Axle Assembly Removal

To remove the rear axle assembly on all models, detach the brake line at the T-fitting, detach the rear universal joint, remove the rear shock absorbers, remove the nuts that hold the U-bolts to the rear springs and rear axle housing and disconnect the spring at the back link. Let the spring drop to the floor.

Jack stand should be placed on the frame in front of the front spring, or the body should be raised with a chain block attached to the back bumper.

If it is difficult or impossible to raise the car sufficiently high to let the rear wheels pass under the fenders, the rear wheels can be removed. Roll the rear axle assembly out from underneath the car.

On models that use a rear torsion bar, it will be necessary to detach the torsion bar before removing the axle.

Replace the rear axle assembly by reversing the removal procedure.

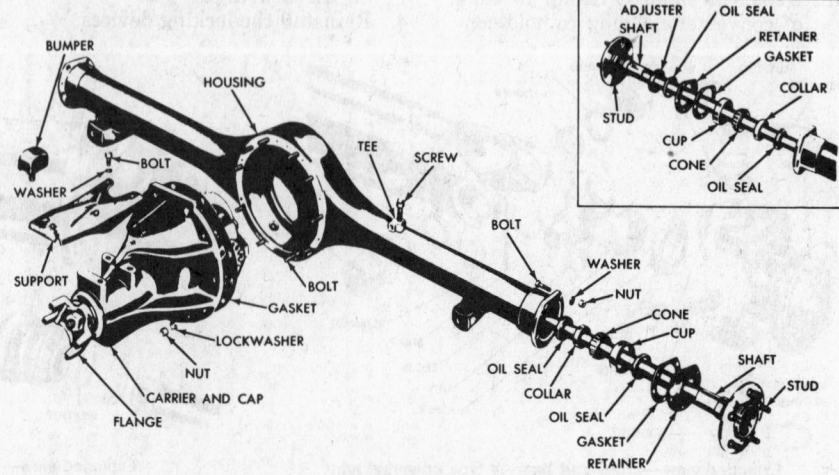

Rear axle—1965-70 (© Chrysler Corp.)

4. From under instrument panel disconnect speaker leads, radio feed wire, and antenna lead cable at radio.
5. Remove radio to instrument panel bracket and remove radio. Reverse procedure to install.

1966 Dodge Polara and Monaco and Plymouth Fury
1. Disconnect battery.
2. Remove control knobs and two mounting nuts.
3. On air-conditioned cars, remove spot cooler hoses and distribution duct, then disconnect speaker, power supply and antenna leads.
4. Remove radio support bracket.
5. Rotate front end of radio down and remove radio from under the dash panel.

1966-67 Coronet and Charger and 1966 Belvedere
1. Disconnect battery.
2. Remove upper half of glove compartment and disconnect wires from speaker.
3. Remove radio knobs and two mounting nuts.
4. Disconnect both defroster hoses at the heater.
5. Disconnect antenna cable and radio feed wires at connector.
6. Loosen radio support bracket retaining nut at radio and remove support bracket mounting screw from lower edge of instrument panel.
7. Remove radio from under instrument panel.
8. To install, reverse removal procedure.

1967 Polara, Monaco and Fury
1. Disconnect battery.
2. Remove heater or air-conditioner knobs.
3. Remove five center bezel retaining screws (3 in underside of lip and 2 in face of bezel.) Remove the bezel.
4. Remove ash tray and housing.
5. Disconnect right defroster hose at heater outlet. Tie hose out of way.
6. From ash tray opening, remove two heater or air-conditioner control mounting nuts and move the control assembly out of the way. It is not necessary to disconnect control cables.
7. Disconnect radio feed wires and antenna cable from the radio.
8. Remove two radio mounting screws from front of panel. Remove radio.
NOTE: on air-conditioned cars, the distribution duct must be removed.

9. To install, reverse removal procedure.
10. Remove two radio mounting screws.
11. Reach through ash tray opening and disconnect antenna leads and electrical leads.
12. Remove radio from panel by tipping radio down and lowering through ash tray opening. Reverse procedure to install.

1968-70 Charger
1. Disconnect battery.
2. Remove radio finish plate.
3. On air conditioned vehicles, remove lower center air duct, left air duct, and upper center duct.
4. Remove radio mounting bracket.
5. Remove two screws mounting radio to front of instrument panel.
6. Disconnect antenna and speaker leads.
7. Remove radio from under panel. Reverse procedure to install.

1967-68 Fury and VIP
1. Remove instrument cluster bezel. See Instrument Cluster Removal in this section.
2. From under panel, loosen radio support bracket nut at upper end.
3. Disconnect feed wires, speaker wires, and antenna cable at radio.
4. From front of instrument panel, remove three radio mounting screws and lift radio out of panel.
5. With radio on bench, remove knobs and four lens retaining screws. Remove lens. Reverse procedure to install.

1968 Polara and Monaco
1. Disconnect battery.
2. Disconnect cigar lighter lead and remove ash tray and housing.
3. Remove automatic temperature control, if so equipped.
4. Remove center air outlets, if so equipped.
5. Remove radio mounting bracket. (Loosen one nut at radio, remove one screw in lower reinforcement, and swing bracket toward glove box to clear area for radio removal.)
6. Remove eight bezel mounting screws (three in upper center trim bezel, three in upper right trim bezel, and two in lower center trim bezel).
7. Pull bezel out slightly and disconnect fader control harness and remove two fader control mounting screws from rear of fader and reverberator, if so equipped.

8. Remove reverberator knob, if so equipped.
9. Slide center trim bezel out of upper moulding toward cluster.

1968-70 Coronet and 1969-70 Belvedere and Satellite
1. Disconnect battery.
2. Remove radio upper trim panel.
3. Remove radio finish plate.
4. Remove radio rear mounting nut from mounting bracket.
5. Disconnect electrical wiring and antenna lead.
6. Remove two mounting screws from front of instrument panel.
7. Remove radio from instrument panel.

1968 Belvedere and Satellite and 1969-70 Polara and Monaco
1. Disconnect battery.
2. Remove automatic temperature control, if so equipped.
3. Remove radio bezel.
4. Remove two radio mounting bolts at front of instrument panel.
5. Remove air conditioner duct, if so equipped.
6. Disconnect electrical leads and antenna lead.
7. Loosen radio mounting bracket stud nut and slide radio and stud towards front of car from mounting bracket.
8. Carefully remove radio from under panel to avoid damaging electrical leads from main harness or automatic temperature control asperator tube. Reverse procedure to install.

1969-70 Fury and VIP
1. Disconnect battery.
2. Remove nine lamp panel mounting screws, lower lamp panel assembly slightly, disconnect lamp harness from main harness, and remove lamp panel from instrument panel.
3. Remove steering column cover.
4. Remove radio trim bezel mounting screws and bezel.
5. Remove center lower air conditioner duct, if so equipped.
6. Disconnect electrical leads and antenna lead at radio.
7. Remove radio support mounting bracket.
8. Remove two radio mounting bolts.
9. Move radio down through bottom of instrument panel carefully to avoid damage to vacuum hoses and electrical leads. Reverse procedure to install.

Windshield Wipers

Motor and Linkage R & R

1964-65

1. Disconnect battery at negative post.
2. Remove speaker grille and speaker. (Padded dash models require the removal of the padding to permit removal of the speaker grille.)
3. Remove the radio.
4. Remove bolts holding the motor bracket to the cowl and the instrument panel brace.
5. Disconnect wires at wiper motor.
6. Disconnect links at the pivot cranks. Clips are removed by lifting the top tab and sliding it sideways out of engagement with the groove in the pilot crank pin. The right pivot may be reached through the glove compartment opening.
7. Tilt the wiper motor and bracket on its side and remove the left link.
8. Remove the right defroster tube, the pencil brace between the instrument panel reinforcement and wiper motor mounting.
9. Remove the motor and bracket through the speaker opening.
10. Install by reversing the removal procedure.

1966-70

1. Disconnect battery.
2. Disconnect wiper motor multiple connector from engine side of firewall.
3. Remove wiper motor mounting nuts and pull motor out far enough to gain access to drive crank.
4. Rotate crank until drive link retainer is accessible.
5. With a short screwdriver, pry lip of retainer over drive link pivot pin and remove retainer and spring washer and the drive link.
6. Remove the wiper motor.

Linkage

1964-70

If car is air-conditioned, remove glove compartment to gain access to right pivot retainer.
1. Disconnect battery.
2. Loosen fuse block and move out of the way.
3. Remove wiper motor mounting nuts and pull motor out far enough to remove drive link from drive crank arm.
4. From under the panel, remove left pivot mounting nuts and right pivot retainer.
5. Remove the two links and left pivot as an assembly from under the panel or through the glove compartment.

Heater System

1964 All Except Dodge 880
1965 Dodge Coronet

Heater Blower R & R

1. Disconnect battery.
2. Disconnect heater ground wire at windshield.
3. Loosen and remove fresh air duct from the blower.
4. Remove screws holding blower to plenum.
5. Remove blower from housing and disconnect blower assembly wires.
6. Install in reverse of above.

1964 (880)

Heater Blower R & R

1. Disconnect battery.
2. Disconnect heater blower ground wire at wiper motor mounting bracket.
3. Disconnect heater wires from harness connectors.
4. Disconnect vacuum hoses at units. Remove hoses from clips.
5. Remove valve capillary coil from opening in heater housing in driver's compartment, and remove clip from housing.
6. Remove screws holding duct to dash panel, at left of vent door, below heater and at right wiper link pivot.
7. Remove housing and blower by pulling down and out of compartment.
8. Remove the blower, mounting plate and motor.
9. Install in reverse of above.

1964 All Except Dodge 880,
1965 Dodge Coronet

Heater Core R & R

1. Disconnect battery.
2. Drain cooling system below heater hose level and disconnect heater hoses.
3. Remove air intake duct from blower.
4. Remove screw holding blower to plenum bracket.
5. Disconnect actuator lines.
6. Disconnect heater wires from left side of heater.
7. Disconnect heater ground wire.
8. Remove control lever cable from control valve.

9. Remove defroster tubes from heater housing.
10. From engine compartment, remove nuts and washers attaching heater assembly to dash panel.
11. Remove heater to dash panel attaching screws.
12. Remove assembly from car.
13. Remove control valve screws and remove valve by pulling it straight out from housing. The valve is held to core by use of an O-ring.
14. Remove heater to core screws and remove core assembly.
15. Install in reverse of above.

1964 (880)

Heater Core R & R

1. Disconnect battery.
2. Drain cooling system below heater hose level and disconnect heater hoses.
3. Remove screws attaching core housing to dash.
4. Remove mastic to expose plastic rivets (if used).
5. Remove core from outer housing. In removing plastic rivets, use care not to damage rivets or housing when separating.
6. Install in reverse of above.

Polara, Monaco, Fury 1966-68

Heater R & R

1. Disconnect battery.
2. Disconnect hoses from heater and plug fittings to prevent coolant from leaking and spilling on inside of body when removed.
3. Remove bracket from top of heater to dash panel.
4. Remove defroster hoses at heater and vacuum actuator hose.
5. Disconnect wire at blower motor resistor.
6. Remove glove compartment.
7. Disconnect control cable at heater end.
8. Unclamp connector at right end of heater. Do not remove connector.
9. Pull carpet or mat from under instrument panel.
10. From inside engine compartment, remove nuts that attach heater assembly to dash panel.
11. Pull heater toward rear to clear mounting studs from dash panel and rotate heater until studs are down, then remove heater from panel.
12. Disconnect wiring, heater assembly-to-blower motor.
13. Remove back plate from heater assembly.
14. Remove fan from motor shaft.
15. Remove motor from heater back plate.
16. Install in reverse of above.

1965 Dodge 880, Monaco and Polara; 1966-70 Coronet, Charger, Belvedere and Satellite

Heater Assembly R & R

1. Drain radiator and disconnect battery.
2. Remove upper half of glove box.
3. Disconnect heater hoses at bulkhead. Plug hose fittings on heater to prevent spilling coolant on trim when removing heater.
4. From under instrument panel remove heater to cowl support bracket.
5. Remove defroster hoses and disconnect wiring from heater motor resistor.
6. Disconnect fresh air vent control and shut off door cables at heater from under instrument panel. Reaching through glove box, disconnect temperature control door cable.
7. From inside engine compartment, remove three nuts that mount heater to bulkhead.
8. Rotate heater assembly until mounting studs are up and carefully remove heater from under instrument panel.
9. Reverse procedure to install.

1969-70 Polara, Monaco, Fury, and VIP

Heater Assembly R & R

1. Disconnect battery and drain radiator.
2. Disconnect heater hoses at dash panel. Plug hose fittings on heater to prevent spilling coolant on trim.
3. Slide front seat back to allow room.
4. Disconnect radio antenna.
5. Disconnect electrical conductors from blower motor resistor block on face of housing.
6. Remove vacuum hoses from trunk lock if so equipped.
7. Remove control cables from defroster door crank and heat shut off door crank.

8. Remove bottom retaining nut from support bracket and swing bracket up and out of way.
9. In engine compartment remove four retaining nuts from studs on engine side housing.
10. Remove locating bolt from bottom center of passenger side housing.
11. Roll or tip housing out from under instrument panel.
12. Remove temperature control cable retaining clip and cable from heat shut off door crank.

All Models 1966-70

Heater Core R & R

1. Follow items 1 through 11 under 1965 Polara blower removal.
2. Remove heater cover plate.
3. Remove screws attaching heater core to assembly and remove core.

Heater Blower R & R

1. Remove heater as outlined in Heater Removal.
2. Disconnect wiring from blower motor to heater assembly.
3. Remove motor cooler tube.
4. Remove heater back plate assembly from heater.
5. Remove fan from motor shaft.
6. Remove blower motor from back plates.

1965 Belvedere and Satellite

Heater Blower R & R

1. Disconnect battery.
2. Disconnect heater ground wire.
3. Loosen air intake duct clamp at blower end and remove duct from blower assembly.
4. Remove screw which holds blower to plenum.
5. Remove blower assembly from housing.
6. From inside housing, disconnect blower assembly wires.
7. Install in reverse of above.

Heater Core R & R

1. Disconnect battery.
2. Drain cooling system.
3. Disconnect hoses at heater.

4. Remove fresh air duct from blower.
5. Remove blower plenum bracket screw.
6. Disconnect actuator lines.
7. Disconnect heater wire receptacle from left side of heater.
8. Disconnect heater ground wire.
9. Remove control lever cable from control valve.
10. Remove defroster tubes from heater housing.
11. From engine compartment, remove screws and washers which hold heater assembly to dash.
12. Remove heater to dash panel attaching screws.
13. Remove heater assembly.
14. Remove control valve screws and pull valve straight out from housing. The valve is held into core by an O-ring only.
15. Remove core attaching screws and remove core.
16. Install in reverse of above.

1965 (Fury only)

Heater Assembly R & R

1. Disconnect battery.
2. Disconnect heater hoses at dash panel. Plug heater hose fittings to prevent coolant spillage.
3. Under instrument panel, remove bracket, top of heater to dash.
4. Remove defroster hoses at heater.
5. Disconnect wires at blower motor resistor.
6. Remove glove compartment.
7. Disconnect cables at heater end.
8. Unclamp flexible connector at right end of heater. Do not remove connector.
9. Pull carpet or mat from under instrument panel.
10. From engine compartment, remove nuts attaching heater assembly to dash panel.
11. Pull assembly toward rear of vehicle until studs are clear of panel. Rotate heater assembly until studs are down and remove heater from under instrument panel.
12. Install in reverse of above.

Fairlane, Falcon, Mustang, Comet, Cougar, Montego, Maverick, Torino

Index

YEAR IDENTIFICATION

FAIRLANE AND TORINO

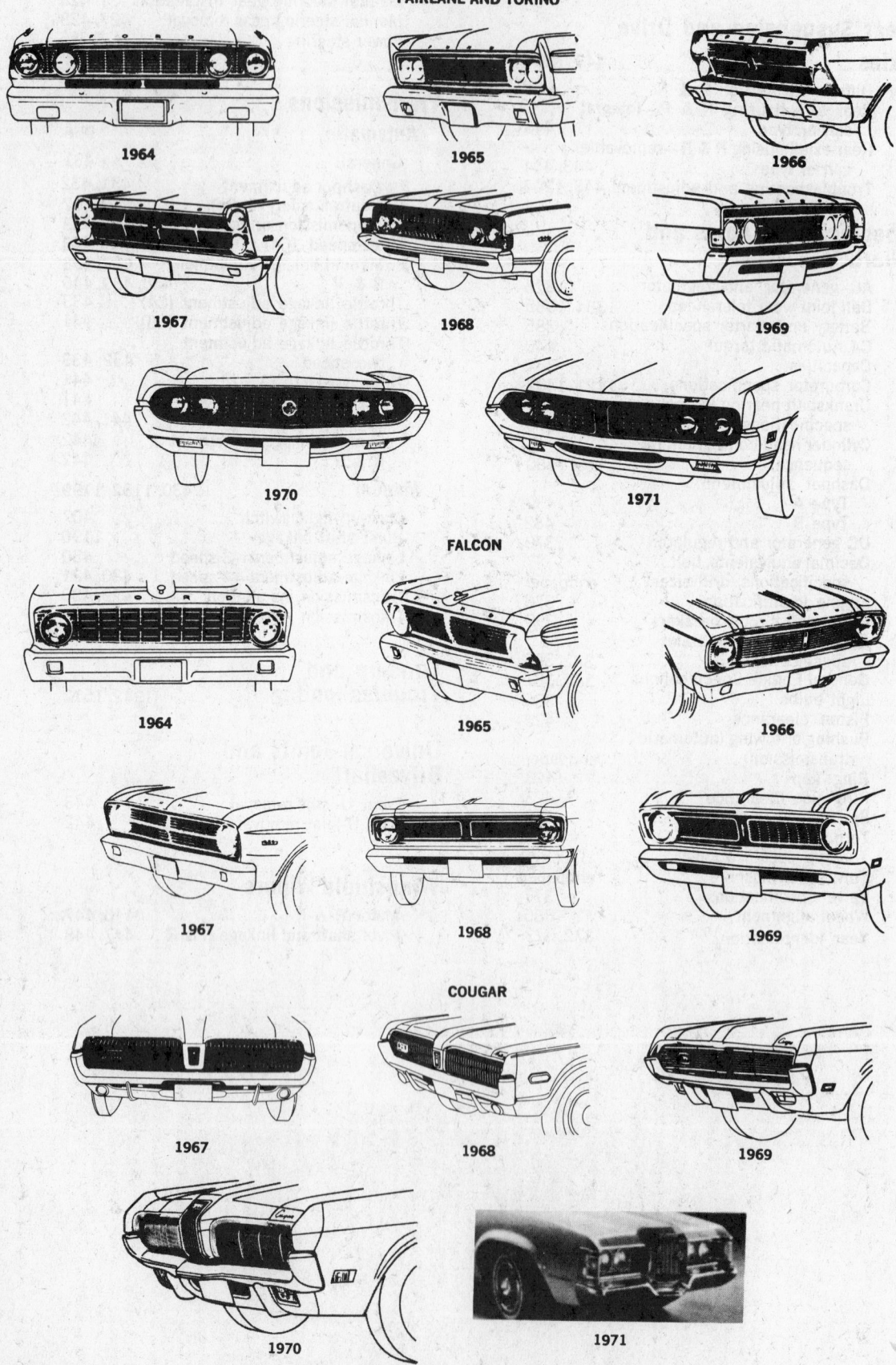

1964

1965

1966

1967

1968

1969

1970

1971

FALCON

1964

1965

1966

1967

1968

1969

COUGAR

1967

1968

1969

1970

1971

YEAR IDENTIFICATION

MUSTANG

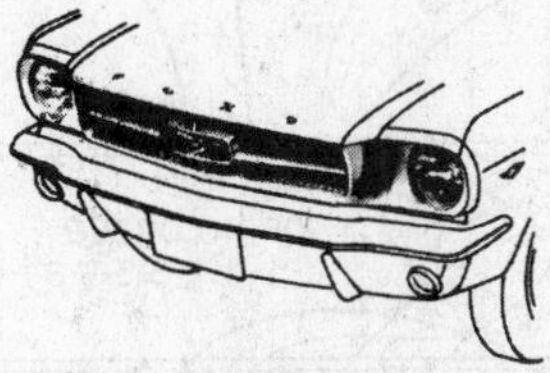

1964-65 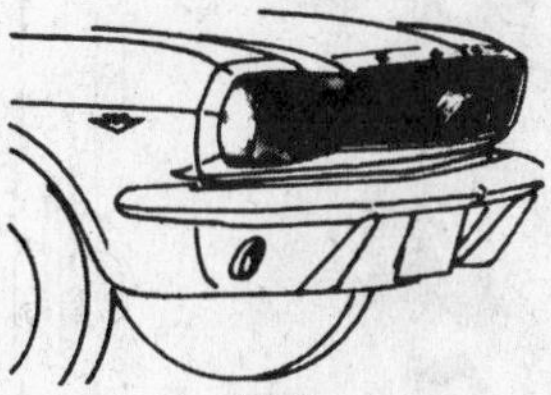   1966    1967 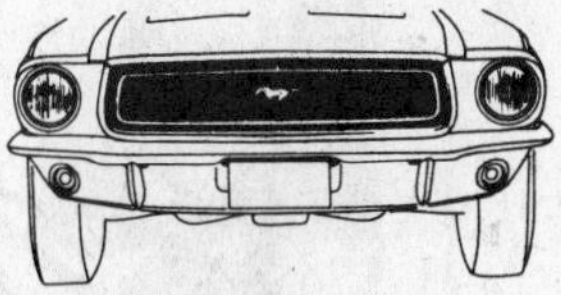   1968

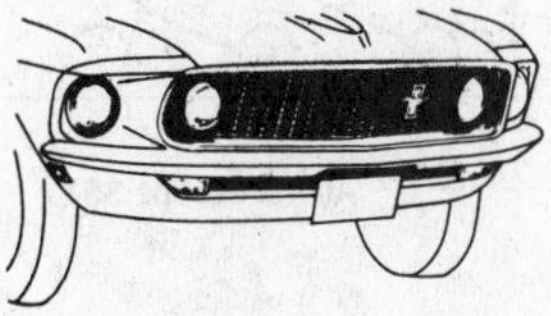

1969 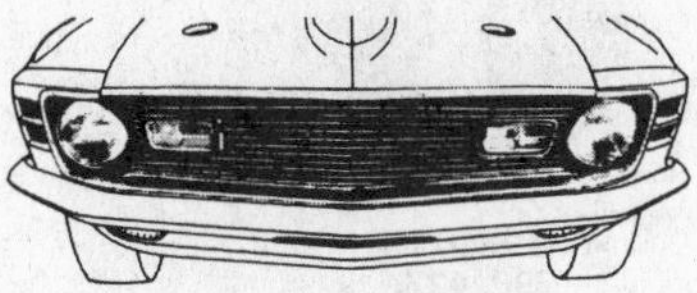   1970 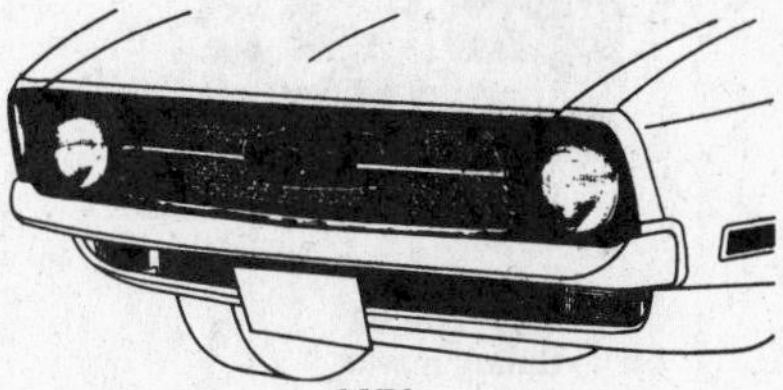   1971

COMET

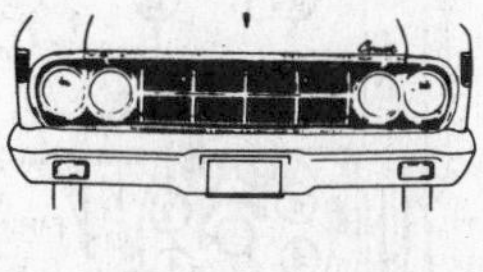

1964 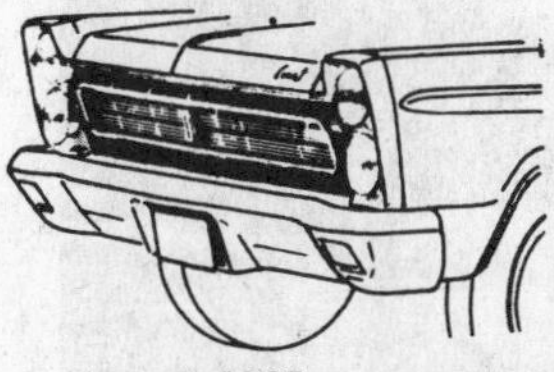   1965 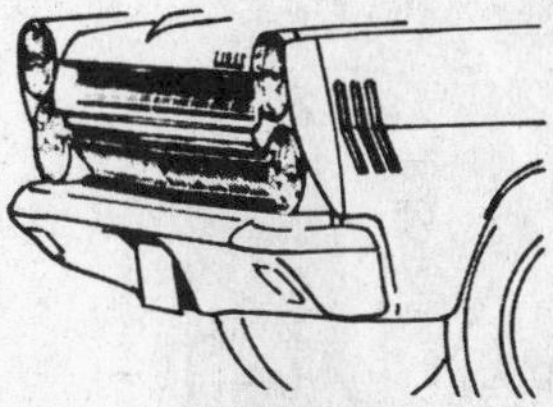   1966

1967    1971

MONTEGO

1968    1969    1970    1971

MAVERICK

1970    1971

FIRING ORDER

1968-71
289, 302 V8
timing marks

1968-71
390, 427,
428, 429 V8
timing marks

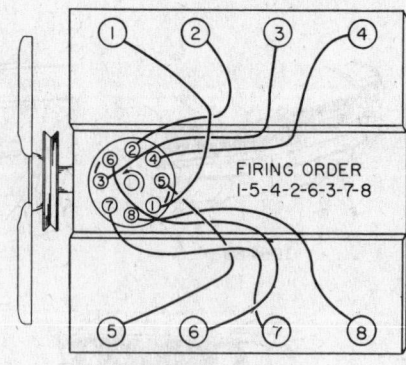

FIRING ORDER
1-5-4-2-6-3-7-8

All V8 except 351

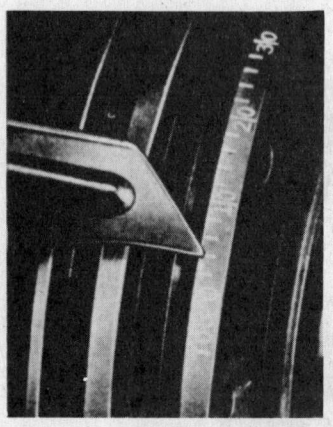

1964-67
390, 427 V8
timing marks

1964-67
260, 289 V8
timing marks

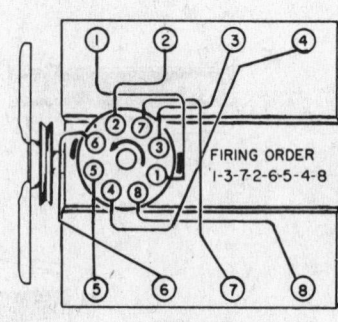

FIRING ORDER
1-3-7-2-6-5-4-8

351 V8

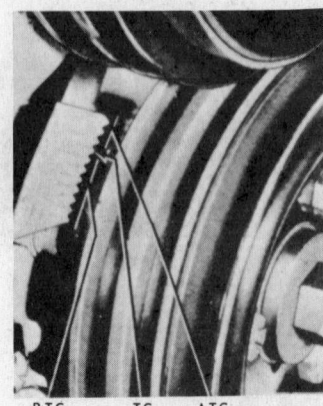

BTC   TC   ATC

1968-71
6 cyl.
timing marks

1964-67
6 cyl.
timing marks

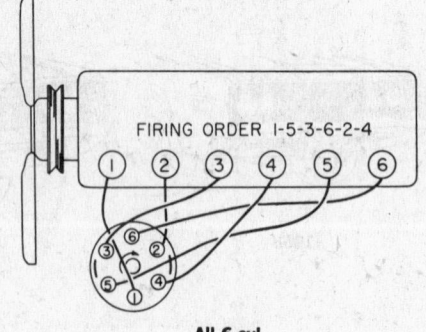

FIRING ORDER 1-5-3-6-2-4

All 6 cyl.

CAR SERIAL NUMBER LOCATION AND ENGINE IDENTIFICATION

1964-67

Engine is identified through car serial number. Serial numbers, and other pertinent information, are to be found on a plate riveted to the rear edge of the left front door.

The engine number is stamped on the top surface of the engine block near the crankcase breather pipe, front left side.

The car serial number is composed of two sections, the first five units giving the year, assembly plant, model and engine type. The second section gives consecutive order of production.

1968-71

The serial number is found on a plate attached to the top of the instrument panel, visible through the windshield. The plate is interpreted as per the illustrations.

Vehicle Certification Label 1970-71

The vehicle certification label is located on the rear face of the driver's door. The upper half of the label contains the name of the manufacturer, the month and year of manufacture, and the certification statement. For interpretation of the lower half of the label, see the illustration.

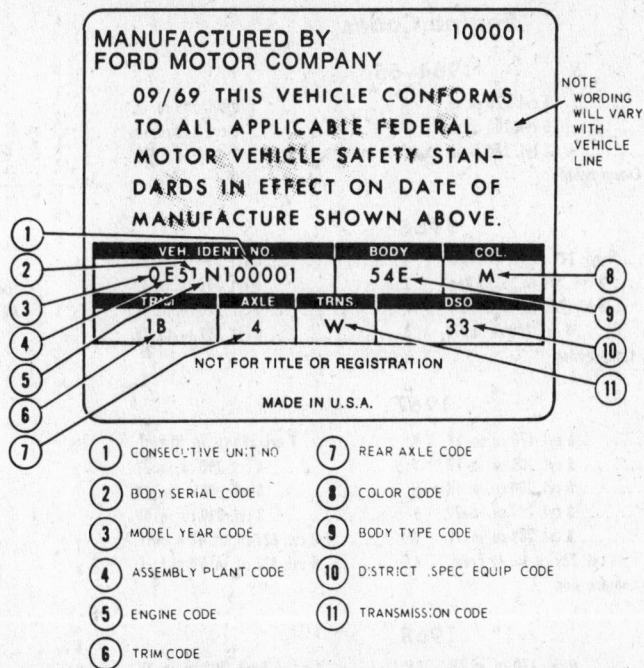

Vehicle certification label—1970-71

Typical vehicle identification number (VIN) tab

1 Model year code
2 Assembly plant code
3 Body serial code
4 Engine code
5 Consecutive unit number
6 Body type code
7 Color code
8 Trim code
9 Date code
10 District—special equipment code
11 Rear axle code
12 Transmission code

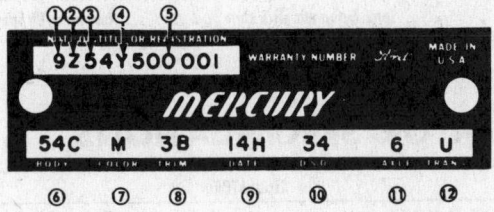
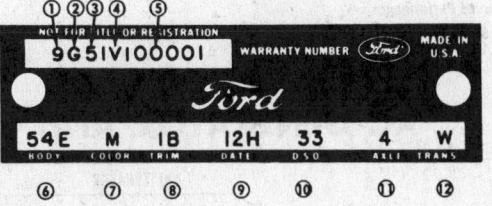

Engine number and code location, V8 engines

CRANKSHAFT BEARING JOURNAL SPECIFICATIONS

YEAR	MODEL	MAIN BEARING JOURNALS (IN.)				CONNECTING ROD BEARING JOURNALS (IN.)		
		Journal Diameter	Oil Clearance	Shaft End-Play	Thrust On No.	Journal Diameter	Oil Clearance	End-Play
1964-71	6 Cyl.—170, 144	2.2482-2.2490	.0008-.0015	.004-.008	3	2.1232-2.1240	.0008-.0015	.003-.010
	6 Cyl.—200	2.2482-2.2490	.0008-.0015	.004-.008	5	2.1232-2.1240	.0008-.0015	.003-.010
	6 Cyl.—250	2.3982-2.3990	.0005-.0015	.004-.008	5	2.1232-2.1240	.0008-.0015	.003-.010
	V8-302, 260, 289	2.2482-2.2490	.0005-.0015	.004-.008	3	2.1228-2.1236①	.0008-.0015②	.010-.020
	V8-351W	2.2994-3.0002	.0013-.0025	.004-.008	3	2.3103-2.3111	.0008-.0015	.010-.020
	V8-351C	2.7484-2.7492	.0005-.0015	.004-.010	3	2.3103-2.3111	.0010-.0015	.010-.020
	V8-390	2.7484-2.7492	.0005-.0015	.004-.010	3	2.4380-2.4388	.0008-.0015	.010-.020
	V8-427	2.7484-2.7492	.0010-.0020	.004-.010	3	2.4380-2.4388	.0020-.0030	.010-.020
	V8-428	2.7484-2.7492	.0010-.0020	.004-.010	3	2.4380-2.4388	.0020-.0030	.010-.020
	V8-429	2.9994-3.0002	.0005-.0015③	.004-.008	3	2.4992-2.5000	.0008-.0015④	.010-.020

① —Boss 302—2.1222-2.1230.   ② —Boss 302—.0015-.0025.   ③ —Boss 429—.0010-.0025.   ④ —Boss 429—.0015-.0025.

Engine Codes

1964-65

U	6 cyl. 170 cu. in.	K	8 cyl. 289 cu. in.
T	6 cyl. 200 cu. in.	*4	6 cyl. 170 cu. in.
F	8 cyl. 260 cu. in.	*6	8 cyl. 260 cu. in.

*Low Compression

1966

A	8 cyl. 289 cu. in. 4V Prem.	Y	8 cyl. 390 cu. in. 2V
C	8 cyl. 289 cu. in. 2V Reg.	Z	8 cyl. 390 cu. in. 4V
K	8 cyl. 289 cu. in. Hi-Perf.	*2	6 cyl. 200 cu. in. 1V
T	6 cyl. 200 cu. in. 1V	*3	8 cyl. 289 cu. in. 2V

*Low Compression

1967

U	6 cyl. 170 cu. in. 1V	K	8 cyl. 289 cu. in. Hi-Perf.
T	6 cyl. 200 cu. in. 1V	Y	8 cyl. 390 cu. in. 2V
*2	6 cyl. 200 cu. in. 1V	H	8 cyl. 390 cu. in. 2V
*3	8 cyl. 289 cu. in. 2V	S	8 cyl. 390 cu. in. 4V
C	8 cyl. 289 cu. in. 2V	W	8 cyl. 427 cu. in. 4V Hi-Perf.
A	8 cyl. 289 cu. in. 4V Prem.	R	8 cyl. 427 cu. in. 8V Hi-Perf.

*Low Compression

1968

U	6 cyl. 170 cu. in. 1V	*6	8 cyl. 302 cu. in. 2V
T	6 cyl. 200 cu. in. 1V	J	8 cyl. 302 cu. in. 4V
*2	6 cyl. 200 cu. in. 1V	Y	8 cyl. 390 cu. in. 2V
C	8 cyl. 289 cu. in. 2V	X	8 cyl. 390 cu. in. 2V Prem.
F	8 cyl. 302 cu. in. 2V	S	8 cyl. 390 cu. in. 4V GT
		W	8 cyl. 427 cu. in. 4V Hi-Perf.

*Low Compression

1969

U	6 cyl. 170 cu. in. 1V	F	8 cyl. 302 cu. in. 2V
T	6 cyl. 200 cu. in. 1V	*6	8 cyl. 302 cu. in. 2V
*2	6 cyl. 200 cu. in. 1V	D	8 cyl. 302 cu. in. 2V Police
L	6 cyl. 250 cu. in. 1V	H	8 cyl. 351 cu. in. 2V
*3	6 cyl. 250 cu. in. 1V	M	8 cyl. 351 cu. in. 4V
		#S	8 cyl. 390 cu. in. 4V
		Q	8 cyl. 428 cu. in. 4V CJ
		SR	8 cyl. 428 cu. in. 4V CJ

*Low Compression
#Improved Performance
S Ram Air Induction

1970

U	6 cyl. 170 cu. in. 1V	G	8 cyl. 302 cu. in. 4V Boss
T	6 cyl. 200 cu. in. 1V	H	8 cyl. 351 cu. in. 2V
*2	6 cyl. 200 cu. in. 1V	M	8 cyl. 351 cu. in. 4V
L	6 cyl. 250 cu. in. 1V	Q	8 cyl. 428 cu. in. 4V CJ
*3	6 cyl. 250 cu. in. 1V	#R	8 cyl. 428 cu. in. 4V CJ
F	8 cyl. 302 cu. in. 2V	N	8 cyl. 429 cu. in. 4V
*6	8 cyl. 302 cu. in. 2V	C	8 cyl. 429 cu. in. 4V CJ
D	8 cyl. 302 cu. in. 2V Taxi	#J	8 cyl. 429 cu. in. 4V CJ
		Z	8 cyl. 429 cu. in. 4V Boss

*Low Compression
#Ram Air Induction

Transmission Codes

1964-65

1	Three speed manual	4	Three speed automatic
3	Two speed automatic	5	Four speed manual

1966

1	Three speed manual	5	Four speed manual
4	Three speed automatic type C6	6	Three speed automatic type C4

1967

1	Three speed manual	5	Four speed manual
2	Overdrive	W	Automatic C4
3	Three speed manual	U	Automatic C6

1968

1	Three speed manual	W	Automatic C4
5	Four speed manual	U	Automatic C6

1969

1	Three speed manual	Y	Automatic MX
5	Four speed manual-wide ratio	X	Automatic FMX
6	Four speed manual-close ratio	Z	Automatic C6 Special for Police
W	Automatic C4		and trailer towing
U	Automatic C6		

1970

1	Three speed manual	W	Automatic C4
5	Four speed manual-wide ratio	U	Automatic C6
6	Four speed manual-close ratio	X	Automatic FMX
V	Semi-Automatic stick shift	Z	Automatic C6 Special for Police
			and trailer towing

AC GENERATOR AND REGULATOR SPECIFICATIONS

YEAR	MODEL	ALTERNATOR		REGULATOR						
				Field Relay			Regulator		
		Field Current Draw @ 12V.	Output @ 6500 Generator RPM	Air Gap (In.)	Point Gap (In.)	Volts to Close	Air Gap (In.)	Point Gap (In.)	Volts at 125°
1964	Standard	2.5	42A	.017	.020	2.5	.036	.017	14.1-14.9
	Heavy Duty	3.0	60A	.017	.020	2.5	.036	.017	14.1-14.9
1965-69	Autolite	2.8-3.3	38A	—	—	2.5	.049-.056	.017-.022	13.8-14.4
	Autolite	2.9-3.1	42A	.012-.022	.015-.022	2.5	.049-.056	.017-.022	13.8-14.4
	Autolite	2.9	45A	.015	—	2.5-4	.052	.019	13.8-14.6
	Autolite	2.9	55A	.015	—	2.5-4	.052	.019	13.8-14.6
	Leece-Neville	2.9	53A	.012	.025	7.0	.047	.019	14.1-14.9
1970-71	Autolite Purple	2.4	38A	—Transistor type—not adjustable—					
	Autolite Orange	2.9	42A	—Transistor type—not adjustable—					
	Autolite Red	2.9	55A	—Transistor type—not adjustable—					
	Autolite Black	2.9	65A	—Transistor type—not adjustable—					

VALVE SPECIFICATIONS

YEAR AND MODEL		SEAT ANGLE (DEG.)	FACE ANGLE (DEG.)	VALVE LIFT INTAKE (IN.)	VALVE LIFT EXHAUST (IN.)	VALVE SPRING PRESSURE (VALVE OPEN) LBS. @ IN.	VALVE SPRING INSTALLED HEIGHT (IN.)	STEM TO GUIDE CLEARANCE (IN.) Intake	Exhaust	STEM DIAMETER (IN.) Intake	Exhaust
1964–65	6 Cyl.—144, 200	45	44	.348	.348	142-158 @ 1.222	1 9/16-1 39/64	.0008-.0025	.0018-.0035	.3100-.3107	.3090-.3097
	6 Cyl.—170	45	44	.348	.348	112-122 @ 1.222	1 9/16-1 39/64	.0008-.0025	.0018-.0035	.3100-.3107	.3090-.3097
	V8—289	45	44	.380	.380	161-178 @ 1.380	1 3/4-1 25/32	.0010-.0027	.0020-.0037	.3416-.3423	.3406-.3413
	V8—260	45	44	.368	.380	161-177 @ 1.390	1 3/4-1 25/32	.0008-.0025	.0018-.0035	.3100-.3107	.3090-.3097
1966	6 Cyl.—170, 200	45	44	.348	.348	142-158 @ 1.222	1 9/16-1 39/64	.0008-.0025	.0010-.0027	.3100-.3107	.3098-.3105
	V8—289	45	44	.368	.380	161-177 @ 1.390	1 3/4-1 25/32	.0010-.0027	.0020-.0037	.3416-.3423	.3406-.3413
	V8—289 H.P.	45	44	.471	.457	235-260 @ 1.320	1 3/4-1 25/32	.0010-.0027	.0020-.0037	.3416-.3423	.3406-.3413
	V8—390	45	44	.408	.408	233-257 @ 1.380	1 13/16-1 27/32	.0010-.0024	.0010-.0024	.3711-.3718	.3711-.3718
1967	6 Cyl.—170	45	44	.348	.348	142-158 @ 1.222	1 9/16-1 39/64	.0008-.0025	.0010-.0027	.3100-.3107	.3098-.3105
	6 Cyl.—200	45	44	.367	.367	142-158 @ 1.222	1 9/16-1 39/64	.0008-.0025	.0010-.0027	.3100-.3107	.3098-.3105
	V8—289①	45	44	.368	.381	174-192 @ 1.270	1 5/8-1 11/16	.0010-.0027	.0010-.0027	.3416-.3423	.3416-.3423
	V8—289②	45	44	.426	.425	171-189 @ 1.230	1 5/8-1 11/16	.0010-.0027	.0010-.0027	.3416-.3423	.3416-.3423
	V8—289 H.P.	45	44	.477	.477	235-260 @ 1.320	1 3/4-1 25/32	.0010-.0027	.0010-.0027	.3416-.3423	.3416-.3423
	V8—390	45	44	.440	.440	209-231 @ 1.380	1 13/16-1 27/32	.0010-.0024	.0010-.0024	.3711-.3718	.3711-.3718
	V8—390GT	45	44	.481	.490	255-280 @ 1.320	1 51/64-1 53/64	.0010-.0024	.0010-.0024	.3711-.3718	.3711-.3718
	V8—427	30 int. 45 exh.	29 int. 44 exh.	.500	.500	255-280 @ 1.320	1 51/64-1 53/64	.0010-.0024	.0020-.0034	.3711-.3718	.3701-.3708
1968	6 Cyl.—170, 200	45	44	.348	.348	142-158 @ 1.222	1 9/16-1 39/64	.0008-.0025	.0010-.0027	.3100-.3107	.3098-.3105
	V8—289, 302	45	44	.368	.381	171-189 @ 1.230	1 5/8-1 11/16	.0010-.0027	.0015-.0032	.3416-.3423	.3411-.3418
	V8—390 2-V	45	44	.427	.430	209-231 @ 1.380	1 13/16-1 27/32	.0010-.0027	.0015-.0032	.3711-.3718	.3706-.3713
	V8—390GT	45	44	.481	.490	255-280 @ 1.320	1 51/64-1 53/64	.0010-.0024	.0015-.0032	.3711-.3718	.3706-.3713
	V8—427	30 int. 45 exh.	29 int. 44 exh.	.481	.490	255-280 @ 1.320	1 51/64-1 53/64	.0010-.0024	.0020-.0034	.3711-.3718	.3701-.3708
1969	6 Cyl.—170, 200	45	44	.348	.348	142-158 @ 1.222	1 9/16-1 39/64	.0008-.0025	.0010-.0027	.3100-.3107	.3098-.3105
	6 Cyl.—250	45	44	.368	.368	142-158 @ 1.222	1 9/16-1 39/64	.0008-.0025	.0010-.0027	.3100-.3107	.3098-.3105
	V8—302	45	44	.368	.381	171-189 @ 1.230	1 5/8-1 11/16	.0010-.0027	.0010-.0027	.3416-.3423	.3411-.3418
	V8—351W	45	44	.418	.448	204-226 @ 1.340	1 25/32-1 13/16	.0010-.0027	.0010-.0027	.3416-.3423	.3706-.3713
	V8—390 H.P.	45	44	.481	.490	209-231 @ 1.380	1 13/16-1 27/32	.0010-.0027	.0015-.0032	.3711-.3718	.3706-.3713
	V8—v28CJ	30 int. 45 exh.	29 int. 44 exh.	.481	.490	255-280 @ 1.320	1 51/64-1 53/64	.0010-.0027	.0015-.0032	.3711-.3718	.3706-.3713
	V8—428CJ③	30 int. 45 exh.	29 int. 44 exh.	.481	.490	255-280 @ 1.320	1 51/64-1 53/64	.0010-.0027	.0015-.0032	.3711-.3718	.3706-.3713
1970–71	6 Cyl.—170, 200	45	44	.348	.348	142-158 @ 1.222	1 9/16-1 5/8	.0008-.0025	.0010-.0027	.3100-.3107	.3098-.3105
	6 Cyl.—250	45	44	.368	.368	142-158 @ 1.222	1 9/16-1 5/8	.0008-.0025	.0010-.0027	.3100-.3107	.3098-.3105
	V8—302	45	44	.368	.381	171-189 @ 1.230	1 5/8-1 21/32	.0010-.0027	.0010-.0027	.3416-.3423	.3411-.3418
	V8—302 Boss	45	44	.477	.477	299-331 @ 1.320	1 13/16-1 27/32	.0010-.0027	.0015-.0032	.3416-.3423	.3411-.3418
	V8—351W	45	44	.418	.448	204-226 @ 1.340	1 3/4-1 13/16	.0010-.0027	.0010-.0027	.3416-.3423	.3411-.3418
	V8—351C 2-V	45	44	.407	.407	199-221 @ 1.420	1 13/16-1 27/32	.0010-.0027	.0015-.0032	.3416-.3423	.3411-.3418
	V8—351C 4-V	45	44	.427	.427	271-299 @ 1.320	1 13/16-1 27/32	.0010-.0027	.0015-.0032	.3416-.3423	.3411-.3418
	V8—428CJ	30 int. 45 exh.	29 int. 44 exh.	.481	.490	255-280 @ 1.320	1 13/16-1 27/32	.0015-.0032	.0015-.0032	.3711-.3718	.3706-.3713
	V8—428CJ③	30 int. 45 exh.	29 int. 44 exh.	.481	.490	255-280 @ 1.320	1 13/16-1 27/32	.0015-.0032	.0015-.0032	.3711-.3718	.3706-.3713
	V8—429CJ	30 int. 45 exh.	29 int. 44 exh.	.515	.515	294-318 @ 1.360	1 13/16-1 27/32	.0010-.0024	.0020-.0034	.3416-.3423	.3416-.3418
	V8—429SCJ	30 int. 45 exh.	29 int. 44 exh.	.500	.500	294-318 @ 1.360	1 13/16-1 27/32	.0010-.0024	.0020-.0034	.3416-.3423	.3416-.3418
	V8—429 Boss	30 int. 45 exh.	29 int. 44 exh.	.478	.505	300-330 @ 1.320	1 13/32-1 27/32	.0010-.0024	.0020-.0034	.3711-.3718	.3701-.3708

① —Some engines with early Thermactor exhaust emission and all engines not having Thermactor.
② —Late production with Thermactor.
③ —Ram Air.

TUNE-UP SPECIFICATIONS

YEAR	MODEL	SPARK PLUGS Type	Gap (In.)	DISTRIBUTOR Point Dwell (Deg.)	Point Gap (In.)	IGNITION TIMING (Deg.) ▲	CRANKING COMP. PRESSURE (Psi)	VALVES Tappet (Hot) Clearance (In.) Intake	Exhaust	Intake Opens (Deg.)	FUEL PUMP PRESSURE (Psi)	IDLE SPEED (Rpm) ★
1964-65	6 Cyl.—Exc. 200	BF82	.034	37.5°	.025	8B	170	Hyd.	Hyd.	13B	4½	500
	6 Cyl.—Exc. 200 Auto.	BF82	.034	37.5°	.025	12B	170	Hyd.	Hyd.	13B	4½	485
	6 Cyl.—200 Auto. Trans.	BF82	.034	37.5°	.025	12B	150	Hyd.	Hyd.	6B	4½	485
	V8—260 Std. Trans.	BF42	.034	27.0°	.015	6B①	150	Hyd.	Hyd.	21B	5	575
	V8—260 Auto. Trans.	BF42	.034	27.0°	.015	10B①	150	Hyd.	Hyd.	21B	5	500
	V8—289 Std. Trans.	BF42	.034	27.0°	.015	6B②	150	Hyd.	Hyd.	20B	5	575
	V8—289 Auto. Trans.	BF42	.034	27.0°	.015	8B	150	Hyd.	Hyd.	20B	5	500
	V8—289 Hi Perf.	BF32	.034	27.0°	.020	10B	200	.018	.018	46B	5	750
1966	6 Cyl. 170, 200 M.T.	BF82	.034	39.0°	.025	6B⊙	170	Hyd.	Hyd.	13B	4½	500
	6 Cyl. 170, 200 A.T.	BF82	.034	39.0°	.025	12B⊙	170	Hyd.	Hyd.	13B	4½	500
	V8—289 M.T.	BF42	.034	27.0°	.017	6B⊙	150	Hyd.	Hyd.	20B	5	600
	V-8, 289 A.T.	BF42	.034	27.0°	.017	6B⊙	150	Hyd.	Hyd.	20B	5	500
	V-8, 289 Hi Perf.	BF32	.030	31.0°	.020	12B	200	.018	.018	46B	5	750
	V-8, 390 M.T.	BF42	.034	27.0°	.017	10B⊙	180	Hyd.	Hyd.	26B	5	600
	V-8, 390 A.T.	BF42	.034	28.0°	.017	10B⊙	180	Hyd.	Hyd.	26B	5	500
1967	6 Cyl.—170, 200 M.T.	BF82	.034	37.0°	.025	6B⊙	170	Hyd.	Hyd.	9B	4½	600
	6 Cyl.—170, 200 A.T.	BF82	.034	37.0°	.025	12B⊙	170	Hyd.	Hyd.	9B	4½	525
	V8—289, 2-BBL.	BF42	.034	27.0°	.017	6B⊙	150	Hyd.	Hyd.	16B	5	600
	V8—289, 4-BBL.	BF42	.034	27.0°	.017	6B⊙	150	Hyd.	Hyd.	16B	5	600
	V8—289 Hi-Perf.	BF32	.034	31.0°	.020	12B⊙	200	.019	.021	46B	5	750
	V8—390, 2-BBL. M.T.	BF42	.034	26.0°	.017	10B⊙	180	Hyd.	Hyd.	16B	5	600
	V8—390, 2-BBL. A.T.	BF42	.034	26.0°	.017	10B⊙	180	Hyd.	Hyd.	16B	5	500
	V8—390, 4-BBL. M.T.	BF32	.034	26.0°	.017	12B⊙	180	Hyd.	Hyd.	20B	5	575
	V8—390, 4-BBL. A.T.	BF32	.034	26.0°	.017	12B⊙	180	Hyd.	Hyd.	20B	5	500
	V8—427, Transistor—All	BF32	.030	23.0°	.020	8B	180	.025	.025	48B	5	800
1968	6 Cyl. 170, 200	BF82	.034	37.0°	.027	6B	170	Hyd.	Hyd.	9B	4½	600
	V8—289 M.T.	BF42	.034	27.0°	.017	6B	150	Hyd.④	Hyd.④	9B	5	625
	V8—289 A.T.	BF42	.034	27.0°	.017	6B	150	Hyd.	Hyd.	9B	5	525
	V8—302	BF42	.034	27.0°	.017	6B	150	Hyd.	Hyd.	15B	5	650③
	V8—390 2-BBL.	BF32	.034	27.0°	.020	6B	180	Hyd.	Hyd.	13B	5	650
	V8—390 4-BBL.	BF32	.034	27.0°	.020	6B	180	Hyd.	Hyd.	18B	5	650
	V8—427	BF32	.034	27.0°	.020	6B	180	Hyd.	Hyd.	18B	5	650
1969	6 Cyl. 170, 200	BF82	.034	37.0°	.027	6B	170	Hyd.	Hyd.	9B	4½	⑪
	6 Cyl. 250	BF82	.034	40.0°	.025	6B	170	Hyd.	Hyd.	10B	5	⑤ ⑯
	V8—302, 2-BBL. A.T.	BF42	.034	29.0°	.017	6B	150	Hyd.	Hyd.	16B	5	550
	V8—302, 2-BBL. M.T.	BF42	.034	27.0°	.021	6B	150	Hyd.	Hyd.	16B	5	650
	V8—351, 2-BBL.	BF42	.034	29.0°	.017	6B	170	Hyd.	Hyd.	11B	5	⑰
	V8—351, 4-BBL.	BF32	.034	29.0°	.017	6B	170	Hyd.	Hyd.	11B	5	⑱
	V8—390, 4-BBL. A.T.	BF32	.034	29.0°	.017	6B	170	Hyd.	Hyd.	16B	5	550
	V8—390, 4-BBL. M.T.	BF42	.034	29.0°	.017	6B	170	Hyd.	Hyd.	16B	5	650-700
	V8—428, 4-BBL. A.T. CJ	BF32	.034	29.0°	.017	6B	180	Hyd.	Hyd.	18B	5	650
	V8—428, 4-BBL. M.T. CJ	BF32	.034	29.0°	.017	6B	180	Hyd.	Hyd.	18B	5	700

CAUTION

General adoption of anti-pollution laws has changed the design of almost all car engine production to effectively reduce crankcase emission and terminal exhaust products. It has been necessary to adopt stricter tune-up rules, especially timing and idle speed procedures. Both of these values are peculiar to the engine and to its application, rather than to the engine alone. With this in mind, car manufacturers supply idle speed data for the engine and application involved. This information is clearly displayed in the engine compartment of each vehicle.

TUNE-UP SPECIFICATIONS, continued

YEAR	MODEL	SPARK PLUGS Type	Gap (In.)	DISTRIBUTOR Point Dwell (Deg.)	Point Gap (In.)	IGNITION TIMING (Deg.) ▲	CRANKING COMP. PRESSURE (Psi)	VALVES Tappet (Hot) Clearance (In.) Intake	Exhaust	Intake Opens (Deg.)	FUEL PUMP PRESSURE (Psi)	IDLE SPEED (Rpm) ★
1970-71	6 Cyl. 170, 200	BF82	.035	37.0°	.027	6B	170	Hyd.	Hyd.	9B	4½	⑪
	6 Cyl. 250	BF82	.035	40.0°	.025	6B	170	Hyd.	Hyd.	10B	5	⑫
	V8—302, 2-BBL. A.T.	BF42	.035	⑨	⑨	6B	150	Hyd.	Hyd.	16B	5	600/500⑤ ⑧
	V8—302, 2-BBL. M.T.	BF42	.035	27.0°⑨	⑨	6B	150	Hyd.	Hyd.	16B	5	800/500⑤
	V8—302 Boss	AF32	.035	30-33°	.020⑩	16B	180	.025	.025	40B	5-6	800/500⑤
	V8—351, 2-BBL. A.T.⑥	BF42	.035	⑨	⑨	10B	170	Hyd.	Hyd.	11B	5	600/500⑤ ⑧
	V8—351, 2-BBL. M.T.⑥	BF42	.034	⑨	⑨	10B	170	Hyd.	Hyd.	11B	5	700/500⑤
	V8—351, 2-BBL. A.T.⑦	AF42	.035	⑨	⑨	6B	170	Hyd.	Hyd.	12B	5	600/500⑤ ⑧
	V8—351, 2-BBL. M.T.⑦	AF42	.035	⑨	⑨	6B	170	Hyd.	Hyd.	12B	5	700/500⑤
	V8—351, 4-BBL. A.T.⑦	AF32	.035	⑨	⑨	6B	170	Hyd.	Hyd.	14B	5	600/500⑤ ⑧
	V8—351, 4-BBL. M.T.⑦	AF32	.035	⑨	⑨	6B	170	Hyd.	Hyd.	14B	5	700/500⑤
	V8—428, 4-BBL. A.T. CJ	BF32	.035	⑨	⑨	6B	180	Hyd.	Hyd.	18B	5	⑭ ⑤
	V8—428, 4-BBL. M.T. CJ	BF32	.035	⑨	⑨	6B	180	Hyd.	Hyd.	18B	5	⑭ ⑤
	V8—429 Boss	AF32	.035	30-33°	.020⑩	10B	180	.013C	.013C	40B	6-8	700/500⑤
	V8—429, 4-BBL. CJ	AF32	.035	⑨	⑨	10B	180	Hyd.	Hyd.	32B	6-8	850/500⑤ ⑬
	V8—429, 4-BBL. SCJ	AF32	.035	⑨	⑨	10B	180	.019	.019	40½B	6-8	⑮ ⑤

★—With manual transmission in N and automatic in D. Add 50 rpm if air conditioned.
▲—With vacuum advance disconnected and plugged. NOTE: These settings are only approximate. Engine design, altitude, temperature, fuel octane rating and the condition of the individual engine are all factors which can influence timing. The limiting advance factor must, therefore, be the "knock point" of the individual engine.
B—Before top dead center.
CJ—Cobra Jet.
A.T.—Automatic transmission.
M.T.—Manual transmission.
①—Mustang—4B.
②—Mustang—8B.
③—550 rpm with automatic.
④—High perf. engine—Intake = .019 in.; Exhaust = .021 in.
⑤—Higher idle speed with solenoid throttle positioner energized, lower with it disconnected.
⑥—Windsor-built engine.
⑦—Cleveland-built engine.
⑧—575 rpm (600 for 351C) with no solenoid throttle positioner installed.
⑨—Dual diaphragm unit—one set of points = 24-29° dwell, .021 point gap. Single diaphragm unit—one set of points = 26-31° dwell, .017 point gap.
⑩—Dual points.
⑪—M.T.—750 A.T.—550
⑫—M.T.—750/500 A.T.—600/500

⑬—A.T.—600
⑭—M.T.—725/500 725 w/o A/C. A.T.—675/500 675 w/o A/C.
⑮—M.T.—650/500 A.T.—700/500
⑯—M.T.—700 w/o A/C 700/500 with A/C A.T.—550 w/o A/C 550/450 with A/C
⑰—M.T.—650 A.T.—550
⑱—M.T.—675 A.T.—575
⊙—1966 with Thermactor system: Timing—170,200,289 = TDC 390 = 6B Idle speed—M.T. = 625 rpm A.T. = 550-575 for 170 & 200 475-500 for 289 & 390
⊙—1967 with Thermactor system: Timing—170 = TDC 200 = 5B 289 = TDC 289 High perf. & 390 = 6B

GENERATOR AND REGULATOR SPECIFICATIONS

YEAR	MODEL	GENERATORS Field Current in Amperes At 12 Volts	Brush Spring Tension (Oz.)	Cut-Out Relay Air Gap (In.)	Closing Voltage	REGULATORS Current & Voltage Air Gaps (In.)	Current Regulator Setting	Voltage Regulator Setting
1964	Standard	1.2-1.8	32-40	①	12.4-13.2	①	25	13.9-14.7
	Optional	1.2-1.8	20-26	①	12.0-12.8	①	30	13.9-14.7

①—Not adjustable.

GENERAL ENGINE SPECIFICATIONS

YEAR	CU. IN. DISPLACEMENT	CARBURETOR	DEVELOPED HORSEPOWER @ RPM	DEVELOPED TORQUE @ RPM (FT. LBS.)	A.M.A. HORSEPOWER	BORE AND STROKE (IN.)	ADVERTISED COMPRESSION RATIO	VALVE LIFTER TYPE	NORMAL OIL PRESSURE (PSI)★
1964	6 Cyl.—144	1-BBL.	85 @ 4200	134 @ 2000	29.4	3.500 x 2.500	8.7-1	Hyd.	45
	6 Cyl.-170	1-BBL.	101 @ 4400	156 @ 2400	29.4	3.500 x 2.940	8.7-1	Hyd.	45
	6 Cyl.—200	1-BBL.	116 @ 4400	175 @ 2400	32.5	3.680 x 3.126	8.8-1	Hyd.	45
	V8—260	2-BBL.	164 @ 4400	258 @ 2200	46.2	3.800 x 2.870	9.0-1	Hyd.	45
	V8—289	4-BBL.	210 @ 4400	300 @ 2800	51.2	4.000 x 2.870	9.0-1	Hyd.	45
	V8—289	4-BBL.	220 @ 4400	304 @ 2800	51.2	4.000 x 2.870	9.2-1	Hyd.	45
	V8—289	4-BBL.	271 @ 6000	312 @ 3400	51.2	4.000 x 2.870	10.5-1	Mech.	45
1965-66	6 Cyl.—170	1-BBL.	105 @ 4400	158 @ 2400	29.4	3.500 x 2.940	9.1-1	Hyd.	45
	6 Cyl.—200	1-BBL.	120 @ 4400	185 @ 2400	32.5	3.680 x 3.126	9.2-1	Hyd.	45
	V8—289	2-BBL.	200 @ 4400	282 @ 2400	51.2	4.000 x 2.870	9.3-1	Hyd.	45
	V8—289	4-BBL.	225 @ 4800	305 @ 3200	51.2	4.000 x 2.870	10.0-1	Hyd.	45
	V8—289	4-BBL.	271 @ 6000	312 @ 3400	51.2	4.000 x 2.870	10.5-1	Mech.	45
	V8—390	2-BBL.	265 @ 4400	401 @ 2600	52.49	4.050 x 3.780	9.5-1	Hyd.	45
	V8—390	4-BBL.	315 @ 4600	427 @ 2800	52.49	4.050 x 3.780	10.5-1	Hyd.	45
	V8—390	4-BBL.	335 @ 4800	427 @ 3200	52.49	4.050 x 3.780	10.5-1	Hyd.	45
1967	6 Cyl.—170	1-BBL.	105 @ 4400	158 @ 2400	29.4	3.500 x 2.940	9.1-1	Hyd.	50
	6 Cyl.—200	1-BBL.	120 @ 4400	190 @ 2400	32.5	3.680 x 3.130	9.2-1	Hyd.	50
	V8—289	2-BBL.	200 @ 4400	282 @ 2400	51.2	4.000 x 2.870	9.3-1	Hyd.	50
	V8—289	4-BBL.	225 @ 4800	305 @ 3200	51.2	4.000 x 2.870	9.8-1	Hyd.	50
	V8—289	4-BBL.	271 @ 6000	312 @ 3400	51.2	4.000 x 2.870	10.0-1	Mech.	50
	V8—390	2-BBL.	265 @ 4400	401 @ 2600	52.49	4.050 x 3.780	9.5-1	Hyd.	62
	V8—390	2-BBL.	275 @ 4400	405 @ 2600	52.49	4.050 x 3.780	9.5-1	Hyd.	62
	V8—390	4-BBL.	315 @ 4600	427 @ 2800	52.49	4.050 x 3.780	10.5-1	Hyd.	62
	V8—390	4-BBL.	335 @ 4800	427 @ 3200	52.49	4.050 x 3.780	10.5-1	Hyd.	62
	V8—427	4-BBL.	410 @ 5600	476 @ 3400	57.33	4.233 x 3.781	11.1-1	Mech.	65
	V8—427	2-4-BBL.	425 @ 6000	480 @ 3700	57.33	4.233 x 3.781	11.1-1	Mech.	65
1968	6 Cyl.—170	1-BBL.	100 @ 4400	156 @ 2400	29.40	3.500 x 2.940	8.7-1	Hyd.	50
	6 Cyl.—200	1-BBL.	115 @ 3800	190 @ 2200	32.50	3.680 x 3.130	8.8-1	Hyd.	50
	V8—289	2-BBL.	195 @ 4400	280 @ 2400	51.20	4.000 x 2.870	8.7-1	Hyd.	50
	V8—302	2-BBL.	210 @ 4600	295 @ 2600	51.20	4.000 x 3.000	9.0-1	Hyd.	50
	V8—302	4-BBL.	230 @ 4800	310 @ 2800	51.20	4.000 x 3.000	10.0-1	Hyd.	50
	V8—390	2-BBL.	265 @ 4400	390 @ 2600	52.49	4.050 x 3.780	9.5-1	Hyd.	50
	V8—390	2-BBL.	280 @ 4400	403 @ 2600	52.49	4.050 x 3.780	10.5-1	Hyd.	50
	V8—390	4-BBL.	325 @ 4800	427 @ 3200	52.49	4.050 x 3.780	10.5-1	Hyd.	50
	V8—427	4-BBL.	390 @ 5600	460 @ 3200	57.30	4.236 x 3.780	10.9-1	Hyd.	50
1969	6 Cyl.—170	1-BBL.	105 @ 4400	158 @ 2400	29.40	3.500 x 2.940	9.1-1	Hyd.	50
	6 Cyl.—200	1-BBL.	120 @ 4400	190 @ 2400	32.50	3.680 x 3.130	8.8-1	Hyd.	50
	6 Cyl.—250	1-BBL.	155 @ 4000	240 @ 1600	32.50	3.680 x 3.910	9.0-1	Hyd.	50
	V8—302	2-BBL.	210 @ 4400	275 @ 2400	51.20	4.000 x 3.000	9.5-1	Hyd.	50
	V8—351 ①	2-BBL.	250 @ 4600	355 @ 2600	51.20	4.000 x 3.500	9.5-1	Hyd.	50
	V8—351 ①	4-BBL.	290 @ 4800	385 @ 3200	51.20	4.000 x 3.500	10.7-1	Hyd.	50
	V8—390 A.T.	4-BBL.	270 @ 4400	390 @ 2600	52.49	4.050 x 3.780	10.5-1	Hyd.	50
	V8—390 M.T.	4-BBL.	320 @ 4800	427 @ 3200	52.49	4.050 x 3.780	10.5-1	Hyd.	50
	V8—428 A.T.	4-BBL.	360 @ 5400	459 @ 3200	54.59	4.130 x 3.980	10.5-1	Hyd.	50
	V8—428 M.T.	4-BBL.	335 @ 5200	440 @ 3400	54.59	4.130 x 3.980	10.6-1	Hyd.	50

CYLINDER HEAT BOLT TIGHTENING SEQUENCE

289, 302, 302 Boss, 351W,
351C (similar bolt pattern),
390, 427, 428, 429, 429 Boss
(similar bolt pattern) cu. in. V8

170, 200, 250 cu. in. 6 cyl.

GENERAL ENGINE SPECIFICATIONS, continued

YEAR	CU. IN. DISPLACEMENT	CARBURETOR	DEVELOPED HORSEPOWER @ RPM	DEVELOPED TORQUE @ RPM (FT. LBS.)	A.M.A. HORSEPOWER	BORE STROKE (IN.)	COMPRESSION RATIO	VALVE LIFTER TYPE	NORMAL OIL PRESSURE (PSI)
1970	6 Cyl.—170	1-BBL.	105 @ 4400	156 @ 2200	29.40	3.502 x 2.940	8.7-1	Hyd.	50
	6 Cyl.—200	1-BBL.	120 @ 4000	190 @ 2200	32.50	3.682 x 3.126	8.7-1 ①	Hyd.	50
	6 Cyl.—250	1-BBL.	155 @ 4000	240 @ 1600	32.50	3.682 x 3.910	9.0-1	Hyd.	50
	V8—302	2-BBL.	220 @ 4600	300 @ 2600	51.20	4.002 x 3.000	9.5-1	Hyd.	50
	V8—302 Boss	4-BBL.	290 @ 5800	290 @ 4300	51.20	4.000 x 3.000	10.5-1	Mech.	50
	V8—351 ②	2-BBL.	250 @ 4600	355 @ 2600	51.20	4.002 x 3.500	9.5-1	Hyd.	50
	V8—351 ③	2-BBL.	250 @ 4600	355 @ 2600	51.20	4.002 x 3.500	9.5-1	Hyd.	50
	V8—351 ③	4-BBL.	300 @ 5400	380 @ 3400	51.2	4.002 x 3.500	11.0-1	Hyd.	50
	V8—351 ③ Ram Air	4-BBL.	300 @ 5400	380 @ 3400	51.2	4.002 x 3.500	11.0-1	Hyd.	50
	V8—428 CJ	4-BBL.	335 @ 5200	440 @ 3400	54.59	4.132 x 3.984	10.6-1	Hyd.	50
	V8—429	4-BBL.	360 @ 4600	480 @ 2800	60.83	4.362 x 3.590	10.5-1	Hyd.	55
	V8—429 CJ	4-BBL.	370 @ 5400	450 @ 3400	60.83	4.362 x 3.590	11.3-1	Hyd.	55
	V8—429 Nascar	4-BBL.	375 @ 5600	450 @ 3400	60.83	4.362 x 3.590	11.3-1	Mech.	60
1971	6 Cyl.—200	1-BBL.	120 @ 4000	190 @ 2200	32.5	3.682 x 3.126	8.7-1	Hyd.	50
	6 Cyl.—250	1-BBL.	155 @ 4000	240 @ 1600	32.5	3.682 x 3.910	9.0-1	Hyd.	50
	V8—302	2-BBL.	220 @ 4600	300 @ 2600	51.2	4.002 x 3.000	9.5-1	Hyd.	50
	V8—302 Boss	4-BBL.	290 @ 5800	290 @ 4300	51.2	4.000 x 3.000	N.A.	Mech.	50
	V8—351 ③	2-BBL.	250 @ 4600	355 @ 2600	51.2	4.002 x 3.500	9.5-1	Hyd.	50
	V8—351 ③	4-BBL.	300 @ 5400	380 @ 3400	51.2	4.002 x 3.500	N.A.	Hyd.	50
	V8—429	4-BBL.	360 @ 4600	480 @ 2800	60.83	4.362 x 3.590	N.A.	Hyd.	55
	V8—429 CJ	4-BBL.	370 @ 5400	450 @ 3400	60.83	4.362 x 3.590	N.A.	Hyd.	55

① —Maverick 9.2-1.
② —Windsor-built engine.
★ —Oil pressure approx. at 2,000 rpm.
③ —Cleveland-built engine.

TORQUE SPECIFICATIONS

YEAR	MODEL	CYLINDER HEAD BOLTS (FT. LBS.)	ROD BEARING BOLTS (FT. LBS.)	MAIN BEARING BOLTS (FT. LBS.)	CRANKSHAFT BALANCER BOLTS (FT. LBS.)	FLYWHEEL TO CRANKSHAFT BOLTS (FT. LBS.)	MANIFOLD (FT. LBS.) Intake	Exhaust
1964-68	6 Cyl.	70-75	19-24	60-70	85-100	75-85	None	13-18
	V8—260, 289, 302	65-70	19-24①	60-70	70-90	75-85	12-15②	15-20
	V8—390, 428	80-90	40-45	95-105	70-90	75-85	32-35	18-24
	V8—427	100-110	53-58	95-105	70-90	75-85	32-35	18-24
1969	6 Cyl.	70-75	19-24	60-70	85-100	75-85	—	13-18
	V8—302	65-72	19-24	60-70	70-90	75-85	23-25	12-16
	V8—351W	95-100	40-45	95-105	70-90	75-85	23-25	18-24
	V8—390	80-90	40-45	95-105	70-90	75-85	32-35	18-24
	V8—428	80-90	53-58	95-105	70-90	75-85	32-35	18-24
1970-71	6 Cyl.	70-75	19-24③	60-70	85-100	75-85	—	13-18
	V8—302	65-72	19-24④	60-70⑤	70-90	75-85	23-25	12-16
	V8—351W	95-100	40-45	95-105	70-90	75-85	23-25	18-24
	V8—351C	95-100	40-45	60-70	70-90	75-85	23-25 (⁵⁄₁₆) 28-32 (⅜)	18-24
	V8—428	80-90	53-58	95-105	70-90	75-85	32-35	18-24
	V8—429	130-140	40-45	95-105	70-90	75-85	25-30	28-33

① —289 High perf. 40-45.
② —1966-68, 20-22 FT. LBS.
③ —250-21-26.
④ —302 Boss-40-45.
⑤ —302 Boss-outer bolts 35-40.

GENERAL CHASSIS AND BRAKE SPECIFICATIONS

YEAR	MODEL	CHASSIS		BRAKE CYLINDER BORE			BRAKE DRUM	
		Overall Length (In.)	Tire Size	Master Cylinder (In.)	Wheel Cylinder Diameter (In.)		Diameter (In.)	
					Front	Rear	Front	Rear
1964	6 Cyl.—Falcon; Sed.	181.6	6.00 x 13	1.0	$1\frac{1}{16}$	$\frac{13}{16}$	9	9
	Falcon; Sta. Wag.	190.0	6.50 x 13	1.0	$1\frac{1}{8}$	$\frac{7}{8}$	9	9
	Comet; Sed.	195.1	6.50 x 13	1.0	$1\frac{1}{16}$	$\frac{13}{16}$	9	9
	Comet; Sta. Wag.	192.1	7.00 x 13	1.0	$1\frac{1}{8}$	$\frac{7}{8}$	9	9
	V8—Falcon; Sed.	181.6	6.50 x 13	1.0	$1\frac{1}{8}$	$\frac{29}{32}$	10	10
	Falcon; Sta. Wag.	190.0	7.00 x 13	1.0	$1\frac{1}{8}$	$\frac{15}{16}$	10	10
	Comet; Sed., exc. Cal.	195.1	6.50 x 13	1.0	$1\frac{1}{8}$	$\frac{29}{32}$	10	10
	Comet; Caliente	195.1	6.50 x 14	1.0	$1\frac{1}{8}$	$\frac{29}{32}$	10	10
	Comet; Sta. Wag.	192.1	7.00 x 13	1.0	$1\frac{1}{8}$	$\frac{15}{16}$	10	10
	Mustang; All	181.6	6.50 x 13	1.0	$1\frac{1}{8}$	$\frac{29}{32}$	10	10
1965	6 Cyl.—Falcon Sedan	181.6	6.00 x 13	1.0	$1\frac{1}{16}$	$\frac{13}{16}$	9	9
	Falcon Sta. Wag.	190.0	6.45 x 14	1.0	$1\frac{1}{8}$	$\frac{7}{8}$	9	9
	Comet Sedan	195.3	6.95 x 14	1.0	$1\frac{1}{16}$	$\frac{13}{16}$	9	9
	Comet Sta. Wag.	195.3	6.95 x 14	1.0	$1\frac{1}{8}$	$\frac{7}{8}$	9	9
	V8—Falcon Sedan	181.6	6.45 x 14	1.0	$1\frac{1}{8}$	$\frac{29}{32}$	10	10
	Falcon Sta. Wag.	190.0	6.95 x 14	1.0	$1\frac{1}{8}$	$\frac{15}{16}$	10	10
	Comet Sd., Ex. Cali.	195.3	6.95 x 14	1.0	$1\frac{1}{8}$	$\frac{29}{32}$	10	10
	Comet Caliente	195.3	6.95 x 14	1.0	$1\frac{1}{8}$	$\frac{29}{32}$	10	10
	Comet Sta. Wag.	195.3	6.95 x 14	1.0	$1\frac{1}{8}$	$\frac{15}{16}$	10	10
	Mustang, All	181.6	6.50 x 13†	1.0	$1\frac{1}{8}$■	$\frac{29}{32}$	10■	10
	Fairlane, Exc. Sta. Wag.	198.4	6.95 x 14☆	1.0	$1\frac{1}{8}$	$\frac{29}{32}$	10	10
	Fairlane, Sta. Wag.	203.2	7.35 x 14	1.0	$1\frac{3}{32}$	$\frac{15}{16}$	10	10
1966	6 Cyl.—Falcon; Sedan	184.3	6.50 x 13	1.0	$1\frac{1}{16}$	$\frac{27}{32}$	9	9
	Falcon; Sta. Wag.	198.7	7.35 x 14	1.0	$1\frac{1}{16}$	$\frac{27}{32}$	9	9
	Comet; Sedan	203 *	6.95 x 14	1.0	$1\frac{1}{16}$	$\frac{27}{32}$	9	9
	Comet; Sta. Wag.	199.9	7.75 x 14	1.0	$1\frac{1}{16}$	$\frac{27}{32}$	9	9
	V8—Falcon; Sedan	184.3	6.95 x 14	1.0	$1\frac{1}{8}$	$\frac{29}{32}$	10	10
	Falcon; Sta. Wag.	198.7	7.35 x 14	1.0	$1\frac{1}{8}$	$\frac{29}{32}$	10	10
	Comet; Sedan	203 *	6.95 x 14●	1.0	$1\frac{1}{8}$	$\frac{29}{32}$	10	10
	Comet; Sta. Wag.	199.9	7.75 x 14	1.0	$1\frac{1}{8}$	$\frac{29}{32}$	10	10
	Mustang; All	181.6	6.95 x 14	1.0▲	$1\frac{1}{8}$■	$\frac{29}{32}$	10■●●	10●●
	Fairlane 6 & 8; Exc. Wag.	197.0	6.95 x 14	1.0	$1\frac{1}{8}$	$\frac{29}{32}$	10	10
	Fairlane, Sta. Wag.	199.8	6.95 x 14	1.0	$1\frac{1}{8}$	$\frac{29}{32}$	10	10
	Fairlane V8—390	197.0	7.35 x 14	1.0	$1\frac{1}{8}$	$\frac{29}{32}$	10	10
1967-68	6 Cyl.—Falcon (Sedan)	184.3	6.50 x 13	1.0②	$1\frac{1}{16}$③	$\frac{27}{32}$	9	9
	6 Cyl.—Falcon (Sta. Wag.)	198.7	7.75 x 14	1.0②	$1\frac{3}{32}$	$\frac{15}{16}$	10	10
	V8—Falcon (Sedan)	184.3	6.95 x 14	1.0②	$1\frac{1}{8}$	$\frac{29}{32}$	10	10
	V8—Falcon (Sta. Wag.)	198.7	7.75 x 14	1.0②	$1\frac{3}{32}$	$\frac{15}{16}$	10	10
	6 Cyl.—Fairlane (Sedan)	197.0	7.35 x 14	1.0②	$1\frac{1}{8}$	$\frac{29}{32}$	10	10

GENERAL CHASSIS AND BRAKE SPECIFICATIONS, continued

YEAR	MODEL	CHASSIS		Master Cylinder (In.)	BRAKE CYLINDER BORE		BRAKE DRUM	
		Overall Length (In.)	Tire Size		Wheel Cylinder Diameter (In.)		Diameter (In.)	
					Front	Rear	Front	Rear
	6 Cyl.—Fairlane (Sta. Wag.)	199.9	7.75 x 14	1.0⑦	$1^3/_{32}$	$^{15}/_{16}$	10	10
	V8—Fairlane (Sedan)	197.0	7.35 x 14	1.0②	$1^1/_8$	$^{29}/_{32}$	10	10
	V8—Fairlane (Sta. Wag.)	199.9	7.75 x 14	1.0⑦	$1^3/_{32}$	$^{15}/_{16}$	10	10
	6 Cyl.—Mustang	183.6	7.35 x 14	1.0	$1^1/_{16}$ ③	$^{27}/_{32}$	10	10
	V8—Mustang	183.6	7.35 x 14	1.0	$1^1/_8$ ④	$^7/_8$	10■‡	10
	6 Cyl.—Comet (Exc. 2 Door Sedan)	203.5	7.35 x 14	1.0⑦	$1^1/_8$	$^{29}/_{32}$	10	10
	6 Cyl.—Comet (2 Door Sedan)	196.4	7.35 x 14	1.0⑦	$1^1/_8$	$^{29}/_{32}$	10	10
	6 Cyl.—Comet (Sta. Wag.)	199.9	7.75 x 14	1.0⑦	$1^3/_{32}$	$^{15}/_{16}$	10	10
	V8—Comet (Exc. 2 Door Sedan)	203.5	7.35 x 14	1.0②	$1^1/_8$	$^{29}/_{32}$	10	10
	V8—Comet (2 Door Sedan)	196.4	7.35 x 14	1.0⑦	$1^1/_8$	$^{29}/_{32}$	10	10
	V8—Comet (Sta. Wag.)	199.9	7.75 x 14	1.0②	$1^3/_{32}$	$^{15}/_{16}$	10	10
	V8—Cougar (289, 302 Engine)	190.3	7.35 x 14⑥	1.0	$1^1/_8$ ④	$^7/_8$ ⑤	11.4①	10
	V8—Cougar (390, 427 Engine)	190.3	7.35 x 14⑥	1.0	$1^3/_{32}$ ④	$^{13}/_{16}$ ⑤	11.4①	10
	6 Cyl. Montego	206.1	7.35 x 14	1.0	N.A.	N.A.	N.A.	10
	V8—Montego	206.1	7.35 x 14⑥	1.0	N.A.	N.A.	N.A.	10
1969	Cougar	193.8	⑦	1.0	$1^3/_{32}$	$^7/_8$	10	10
	Falcon 6 Cyl.—200	184.3	⑦	1.0	$1^1/_{16}$	$^{27}/_{32}$	9	9
	Falcon Sta. Wag.	198.7	⑦	1.0	$1^3/_{32}$	$^{15}/_{16}$	10	10
	Falcon Pass. Car, Exc. 6 Cyl.—200	184.3	⑦	1.0	$1^1/_8$	$^7/_8$	10	10
	Fairlane & Montego, Pass. 250, 302	201.1⑧	⑦	1.0	$1^1/_8$	$^7/_8$	10	10
	Fairlane & Montego, P & C, 351, 390	201.1⑧	⑦	1.0	$1^3/_{32}$	$^7/_8$	10	10
	Fairlane & Montego, Conv. exc. above	201.1⑧	⑦	1.0	$1^3/_{32}$	$^7/_8$	10	10
	Fairlane Sta. Wagons	203.9	⑦	1.0	$1^3/_{32}$	$^{15}/_{16}$	10	10
	Mustang 6 Cyl.	187.4	⑦	1.0	$1^1/_{16}$	$^{27}/_{32}$	9	9
	Mustang V8—302	187.4	⑦	1.0	$1^1/_8$	$^7/_8$	10	10
	Mustang V8 exc. 302	187.4	⑦	1.0	$1^3/_{32}$	$^7/_8$	10	10
1970	Cougar	196.1	⑦	1.0	$1^1/_8$	$^{29}/_{32}$	10 x $2^1/_2$■	10 x 2
	Falcon 6 Cyl. Sedan	206.2	⑦	1.0	$1^1/_{16}$	$^{27}/_{32}$	9 x $2^1/_4$	9 x $1^1/_2$
	Falcon Sta. Wag.	209.0	⑦	1.0	$1^1/_8$	$^{31}/_{32}$	10 x $2^1/_2$■	10 x 2
	Falcon V8 Sedan	206.2	⑦	1.0	$1^1/_8$	$^{29}/_{32}$	10 x $2^1/_4$■	10 x 2
	Torino & Montego Sedan	206.2⑨	⑦	1.0	$1^1/_8$	$^{29}/_{32}$	10■**	10 x 2
	Torino & Montego Sta. Wagon	209.0⑩	⑦	1.0	$1^1/_8$	$^{31}/_{32}$	10 x $2^1/_2$■	10 x $2^1/_2$***
	Maverick 6 Cyl.	179.4	⑦	1.0	$1^1/_{16}$	$^{27}/_{32}$	9 x $2^1/_4$	9 x $1^1/_2$
	Mustang 6 Cyl.	187.4	⑦	1.0	$1^1/_{16}$	$^{27}/_{32}$	9	9
	Mustang V8	187.4	⑦	1.0	$1^1/_8$	$^7/_8$	10■	10
1971	Cougar	196.9	⑦	1.0	$1^1/_8$	$^{29}/_{32}$	10 x $2^1/_2$■	10 x 2
	Mustang	189.5	⑦	1.0	$1^1/_8$	$^{29}/_{32}$	10 x $2^1/_2$■	10 x 2
	Maverick 6 Cyl.	179.4	⑦	1.0	$1^1/_{16}$	$^{27}/_{32}$	9 x $2^1/_4$	9 x $1^1/_2$
	Maverick V8	179.4	⑦	1.0	$1^1/_8$	$^7/_8$	10	10
	Comet (UTAH)—200 6 Cyl.	181.8⑪	⑦	1.0	$1^1/_{16}$	$^{27}/_{32}$	9 x $2^1/_4$	9 x $1^1/_2$
	Comet (UTAH)—250, 302	181.8⑪	⑦	1.0	$1^1/_8$	$^7/_8$	10 x $2^1/_2$■	10 x $1^3/_4$
	Montego, Torino, Falcon Sedans	206.2⑨	⑦	1.0	$1^1/_8$	$^{29}/_{32}$	■**	10 x 2
	Montego, Torino, Falcon Sta. Wagons	209.0⑩	⑦	1.0	$1^1/_8$	$^{31}/_{32}$	10 x $2^1/_2$■	10 x $2^1/_2$***

†—Optional—6.95 x 14.
■—Optional—disc brakes.
*—202 Series—195.9
●—With V8—390, 7.35 x 14 & 7.75 x 14
▲—With Pwr. Brakes—.875
● ●—6 Cyl. 9 in. Drums
☆—V8 Models—7.35 x 14
■‡—Disc 11.38 in.
①—Disc front.
②—0.9375 in. when equipped with power brakes.
③—When equipped with 9 in. drums.

④—Disc Brakes 1.625 in.
⑤—Disc Brakes $1^3/_{16}$ in.
⑥—GT Models F70 x 14
⑦—All tire information posted in each individual car.
⑧—Montego—206.2
⑨—Montego—209.9
⑩—Montego—211.8
⑪—4-dr. sedan—188.7
**—All 10 x $2^1/_2$ exc. 250 and 302 sedans and hardtops—these models 10 x $2^1/_4$
***—All exc. 250 and 302 station wagons and Ranchero—these models 10 x 2

CAPACITIES

| YEAR | MODEL | ENGINE CRANKCASE ADD 1 Qt. FOR NEW FILTER | TRANSMISSIONS Pts. TO REFILL AFTER DRAINING | | | DRIVE AXLE (Pts.) | GASOLINE TANK (Gals.) | COOLING SYSTEM (Qts.) WITH HEATER |
| | | | Manual | | Automatic | | | |
			3-Speed	4-Speed				
1964–65	6 Cyl.—144 & 170 Cu. In.	$3\frac{1}{2}$	$2\frac{1}{2}$	$4\frac{1}{2}$	15	$2\frac{1}{2}$	③	$10\frac{1}{2}$
	6 Cyl.—200 Cu. In.	$3\frac{1}{2}$	$2\frac{1}{2}$	$4\frac{1}{2}$	15	$2\frac{1}{2}$	③	17
	V8—260 Cu. In.	4	$3\frac{1}{2}$	$3\frac{1}{2}$	17	$4\frac{1}{2}$	③	$15\frac{1}{2}$
	V8—289 Cu. In.	4	$3\frac{1}{2}$	$3\frac{1}{2}$	17	$4\frac{1}{2}$	③	16
	Fairlane 6 Cyl.	$3\frac{1}{2}$	$2\frac{1}{2}$	$4\frac{1}{2}$	17	$4\frac{1}{2}$	16	$9\frac{1}{2}$
	Fairlane V8—289	4	$3\frac{1}{2}$	$3\frac{1}{2}$	17	$4\frac{1}{2}$	16	15
	Fairlane V8—289 Hi Perf.	4	None	$3\frac{1}{2}$	17	5	16	15
	Mustang, 6 Cyl.	$3\frac{1}{2}$	$2\frac{1}{2}$	$4\frac{1}{2}$	15	$2\frac{1}{2}$	16	$9\frac{1}{2}$
	Mustang, V8	4	$3\frac{1}{2}$	4	17	$4\frac{1}{2}$	16	15
1966	Falcon 6 Cyl.	$3\frac{1}{2}$	2	None	15	$2\frac{1}{2}$	16①	$9\frac{1}{2}$
	Falcon V8—289	4	$3\frac{1}{2}$	4	$17\frac{3}{4}$	$4\frac{1}{2}$	16①	15
	Comet & Fairlane 6 Cyl.	$3\frac{1}{2}$	2	None	15	$4\frac{1}{2}$	20	$9\frac{1}{2}$
	Comet & Fairlane V8—289	4	$3\frac{1}{2}$	4	$17\frac{3}{4}$	$4\frac{1}{2}$	20	15
	Comet & Fairlane V8—390	4	$3\frac{1}{2}$	4	26	$4\frac{1}{2}$	20	$20\frac{1}{2}$
	Mustang 6 Cyl.—200	$3\frac{1}{2}$	2	4	15	$2\frac{1}{2}$	16	$9\frac{1}{2}$
	Mustang V8—289	4	$3\frac{1}{2}$	4	$17\frac{3}{4}$	$4\frac{1}{2}$	16	15
1967	Falcon 6 Cyl.	$3\frac{1}{2}$	2	None	$15\frac{3}{4}$	$2\frac{1}{2}$	16①	$9\frac{1}{2}$
	Falcon V8—289	4	$3\frac{1}{2}$	4	$17\frac{3}{4}$	$4\frac{1}{2}$	16①	15
	Comet & Fairlane 6 Cyl.	$3\frac{1}{2}$	2	None	$15\frac{3}{4}$	$4\frac{1}{2}$	20	$9\frac{1}{2}$
	Comet & Fairlane V8—289	4	$3\frac{1}{2}$	4	$17\frac{3}{4}$	$4\frac{1}{2}$	20	15
	Comet & Fairlane V8—390	4	$3\frac{1}{2}$	4	26	$4\frac{1}{2}$	20	$20\frac{1}{2}$
	Mustang V8—289	4	None	4	$17\frac{3}{4}$	5	17	15
	Cougar V8—289	4	$3\frac{1}{2}$	4	$17\frac{3}{4}$	4	17	15
	Cougar V8—390	4	$3\frac{1}{2}$	4	26	5	17	$20\frac{1}{2}$
	All V8—427	5	None	4	None	5	20	$19\frac{1}{2}$
1968	Comet & Fairlane 6 Cyl.	$3\frac{1}{2}$	$3\frac{1}{2}$	None	16	4	20	$9\frac{1}{2}$
	Comet & Fairlane—289, 302	4	$3\frac{1}{2}$	4	18	4	20	15
	Comet & Fairlane—390, 427	4	$3\frac{1}{2}$	4	26	5	20	$20\frac{1}{2}$
	Falcon 6 Cyl.	$3\frac{1}{2}$	$3\frac{1}{2}$	None	16	2	16	$9\frac{1}{2}$
	Falcon V8	4	$3\frac{1}{2}$	4	26	4	16①	15
	Mustang 6 Cyl.	$3\frac{1}{2}$	$3\frac{1}{2}$	None	16	$2\frac{1}{2}$	16	$9\frac{1}{2}$
	Mustang V8	4	$3\frac{1}{2}$	4	26④	5⑤	16	$20\frac{1}{2}$⑥
	Cougar V8—302	4	$3\frac{1}{2}$	4	18	4	17	15
	Cougar V8—390, 427	4	$3\frac{1}{2}$	4	26	5	17	$20\frac{1}{2}$
	Montego 6 Cyl.	$3\frac{1}{2}$	$3\frac{1}{2}$	None	16	$2\frac{1}{2}$	20	$9\frac{1}{2}$
	Montego V8	4	$3\frac{1}{2}$	4	26④	5⑤	20	$20\frac{1}{2}$⑥
1969	6 Cyl. 170	$3\frac{1}{2}$	$3\frac{1}{2}$	None	16	$2\frac{1}{2}$	16	$9\frac{1}{2}$
	6 Cyl. 200	$3\frac{1}{2}$	$3\frac{1}{2}$	None	16	$2\frac{1}{2}$	⑦	$9\frac{1}{2}$
	6 Cyl. 250	$3\frac{1}{2}$	$3\frac{1}{2}$	None	18	4	⑦	10
	V8—302	4	$3\frac{1}{2}$	4	18	4	⑦	15
	V8—351W	4	$3\frac{1}{2}$	4	22	5	⑦	15
	V8—390	4	$3\frac{1}{2}$	4	$25\frac{1}{2}$	5	⑦	$20\frac{1}{2}$
	V8—428	4	$3\frac{1}{2}$	4	$25\frac{1}{2}$	5	⑦	$20\frac{1}{2}$
1970–71	6 Cyl. 170 ⑪	$3\frac{1}{2}$	$3\frac{1}{2}$	None	16	$2\frac{1}{2}$	16	$9\frac{1}{2}$
	6 Cyl. 200	$3\frac{1}{2}$	$3\frac{1}{2}$	None	16	$2\frac{1}{2}$	⑦	$9\frac{1}{2}$
	6 Cyl. 250	$3\frac{1}{2}$	$3\frac{1}{2}$	None	18	4	⑦	9.8
	V8—302 ⑬	4	$3\frac{1}{2}$	4	18	4	⑦	13.5*
	V8—351W and 351C	4	$3\frac{1}{2}$	4	22	5	⑦	14.6*⑦
	V8—428	4	$3\frac{1}{2}$	4	$25\frac{1}{2}$	5	⑦	19.3
	V8—429	4⑩ ⑫	None	None	$26\frac{1}{2}$	5	⑦	19⑦

① Ranchero and wagon—20 gal.
② Fairlane—16 gal.
③ 1964 Falcon and Comet: 6 cyl.—14 gal.; V8—20 gal.; 1965 Falcon—16 gal.; 1965 Comet—20 gal.
④ V8 289 and 302—18 pts.
⑤ V8 289 and 302—4 pts.
⑥ V8 289 and 302—15 qts.
⑦ Falcon and Maverick—16 gal. Mustang and Falcon wagon—20 gal. Fairlane, Montego and Cougar—20 gal.
⑧ —Montego—15.4.
⑨ —CJ and SCJ—19.6.
⑩ —SCJ—6 (includes 1 qt. for filter and 1 qt. for oil cooler).
⑪ —Not available in 1971.
⑫ —Boss 429—8 (incl. 2 qts. for filter and cooler.)
⑬ —Add an additional qt. for Boss 302 oil cooler.
*—With A/C, add $1\frac{1}{2}$.

WHEEL ALIGNMENT

YEAR	MODEL	CASTER		CAMBER		TOE-IN (In.)	KING-PIN INCLINATION (Deg.)	WHEEL PIVOT RATIO	
		Range (Deg.)	Pref. Setting (Deg.)	Range (Deg.)	Pref. Setting (Deg.)			Inner Wheel	Outer Wheel
1964-65	Falcon & Comet	0 to 1P	1/2 P	1/4 N to 1 1/4 P	1/2 P	1/4 to 3/16	7	20 3/4	20
	Fairlane (1964)	1N to 1P	0	1/4 N to 1 1/4 P	1/2 P	3/16 to 5/16	7 1/4	20	17 1/2
	Fairlane (1965)	1N to 1P	0	1/2 N to 1P	1/4 P	3/32 to 11/32	7 3/4	20	17 1/2
	6 Cyl.—Mustang	3/4 P to 1 3/4 P	1 1/4 P	0 to 1P	1/2 P	1/4 to 5/16	7	20	19
	V8—Mustang	1/4 N to 3/4 P	1/4 P	0 to 1P	1/2 P	1/4 to 5/16	7	20	18 3/4
1966	Falcon	1N to 1P	0	1/2 N to 1P	1/4 P	1/8 to 3/8	7 3/4	20	17 3/4
	Fairlane & Comet	1N to 1P	0	1/2 N to 1P	1/4 P	3/32 to 11/32	7 1/2	20	17 1/2
	6 Cyl. Mustang	0 to 2P	1P	1/4 N to 1 1/4 P	1/2 P	1/8 to 3/8	7	20	18 7/8 ①
	V8 Mustang	1N to 1P	0	1/4 N to 1 1/4 P	1/2 P	1/8 to 3/8	6 7/8	20	19 1/8 ①
1967	Comet, Falcon, Fairlane	1N to 0	1/2 N	1/4 N to 3/4 P	1/4 P	1/4 ± 1/16	7 1/2	20	17 3/4
	Cougar, Mustang	1/4 N to 3/4 P	1/4 P	1/2 P to 1 1/2 P	1P	3/16 ± 1/16	6 3/4	20	18 3/4
1968	Comet, Falcon, Fairlane, Montego	1 1/2 N to 1/2 P	1/2 N	1/2 N to 1P	1/4 P	1/4 ± 1/16	7	20	18 1/8
	Cougar & Mustang	3/4 N to 1 1/4 P	1/4 P	1/4 P to 1 3/4 P	1P	1/4 ± 1/16	6 3/4	20	18 3/4
1969	Falcon, Torino, Montego	1 3/4 N to 1/4 P	3/4 N	1/2 N to 1P	1/4 P	3/16 ± 1/16	7	20	18 1/8
	Cougar & Mustang	3/4 N to 1 1/4 P	1/4 P	1/4 P to 1 3/4 P	3/4 P	3/16 ± 1/16	6 3/4	20	18 3/4
1970-71	Montego, Falcon, Torino	1 1/4 N to 1/4 N	3/4 N	1/4 N to 3/4 P	1/4 P	1/4 ± 1/16	7 2/3 ②	20	③
	Cougar & Mustang	1/2 N to 1/2 P	0	1/4 P to 1 3/4 P	1P	3/16 ± 1/16	6 3/4	20	18°40′
	Maverick	1/4 N to 3/4 P	1/4 P	1/4 N to 3/4 P	1/4 P	3/16 ± 1/16	6 3/4	20	18 3/4 ④

① —V8 power steering—18 3/4°; 6 cyl. power steering—20 1/8°.

② —Falcon—6 2/3.

③ —Falcon—18°6′; others with manual steering—17°19′; power steering—17°49′.

④ —18.2° for power steering.

N—Negative.

P—Positive.

BATTERY AND STARTER SPECIFICATIONS

YEAR	MODEL	BATTERY			STARTERS						
		Ampere Hour Capacity	Volts	Terminal Grounded	Lock Test			No-Load Test			Brush Spring Tension (Oz.)
					Amps.	Volts	Torque	Amps.	Volts	RPM	
1964-65	6 Cyl.—All	40	12	Neg.	540	4.2	14.0	50	12	9,500	45
	V8—All	55	12	Neg.	670	6.0	15.5	70	12	9,500	45
1966	6 Cyl.	45	12	Neg.	460	5	9.0	70	12	9,500	40
	V8	55	12	Neg.	670	5	15.5	70	12	9,500	40
	Option	70	12	Neg.	670	5	15.5	70	12	9,500	40
	Option	80	12	Neg.	670	5	N.A.	70	12	9,500	40
1967-71	6 Cyl.—170 Cu. In.	45	12	Neg.	460	5	9.0	50	12	9,500	40
	6 Cyl.—200 Cu. In.	45	12	Neg.	670	5	15.5	70	12	9,500	40
	V8—exc. below	55	12	Neg.	460①	5	9.0②	70	12	9,500	40
	V8—351 4-BBL.	70	12	Neg.	460①	5	9.0②	70	12	9,500	40
	V8—428, 429	80	12	Neg.	460①	5	9.0②	70	12	9,500	40

① —4 1/2 in. starter—670.

② —4 1/2 in. starter—15.5.

LIGHT BULBS

1964

FALCON			COMET		
Unit	Candle Power or Wattage	Trade Number	Unit	Candle Power or Wattage	Trade Number
Headlight	50/40 W.	6012	Headlight no. 1 (inner)	37.5 W.	4001
Front turn signal/parking	4/32 C.P.	1157 (clear)	Headlight no. 2 (outer)	37.5/50 W.	4002
Rear turn signal and stop/tail	4/32 C.P.	1157	Front turn signal/parking	4/32 C.P.	1157
Stop/tail only			Rear turn signal and stop/tail	4/32 C.P.	1157
License plate	4 C.P.	1155	License plate	4 C.P.	1155
Back-up lights	32 C.P.	1156	Back-up lights	21 C.P.	1141
Dome light	15 C.P.	1003	Spot light	30 W.	4405
Instrument panel indicators:			Luggage compartment	6 C.P.	631
Hi beam	2 C.P.	1895	Dome light	15 C.P.	1003
Oil pressure	2 C.P.	1895	Instrument panel indicators:		
Generator	2 C.P.	1895	Hi beam	2 C.P.	1895
Turn signal	2 C.P.	1895	Oil pressure	2 C.P.	1895
Parking brake warning	2 C.P.	1895	Generator	2 C.P.	1895
Illumination			Turn signal	2 C.P.	1895
Speedometer	2 C.P.	1895	Parking brake warning	2 C.P.	1895
Cluster	2 C.P.	1895	Illumination		
Heater control	2 C.P.	1895	Speedometer	2 C.P.	1895
Clock	2 C.P.	1895	Cluster	2 C.P.	1895
Radio dial	2 C.P.	1891	Heater control	2 C.P.	1895
Courtesy and/or map (door mounted)	6 C.P.	631	Clock	2 C.P.	1895
Automatic transmission control	2 C.P.	1895	Radio dial	2 C.P.	1891
			Courtesy and/or map (door mounted)	6 C.P.	631
			Automatic transmission control	2 C.P.	1895

1965–66

Unit	Candela[1] or Wattage	Trade Number	Unit	Candela[1] or Wattage	Trade Number
Headlight no. 1 (inner or lower)	37.5 W.	4001	Turn signal (inst. panel)	2 C.	1895
Headlight no. 2 (outer or upper)	37.5/50 W.	4002	Illumination		
Headlight (Falcon-Mustang)	40/50 W.	6012	Instruments	2 C.	1895
Fog light (Mustang)	35 W.	4415	Clock	2 C.	1895
Front turn signal/parking	32 C.	1157	Heater control	2 C.	1895
Rear turn signal and stop/tail	32 C.	1157	Hi-beam indicator	2 C.	1895
License plate	4 C.	1155	Speedometer	2 C.	1895
Back-up lights (Falcon sdn)	32 C.	1156	Glove compartment	2 C.	1895
Back-up lights (Falcon sta. wag.)	32 C.	1076	Glove compartment (Mustang console)	1.5 C.	1445
Back-up lights (Mustang)	21 C.	1142	Courtesy light (ins. panel)	6 C.	631
Spot light	30 W.	4405	Radio pilot light (Fairlane)	2 C.	1892
Luggage compartment	6 C.	631	Radio pilot light (Mustang)	1.9 C.	1891
Luggage compartment (Fairlane)	15 C.	93	Radio pilot light (Falcon)	2 C.	1891
Cargo light (wagon—Comet)	15 C.	1003	Radio dial (Comet)	2 C.	1895
Engine compartment (Comet)	15 C.	93	Arm rest courtesy (Fairlane)	6 C.	631
Dome light	15 C.	1003	Courtesy light (front door—Comet)	4 C.	1155
Warning lights			Courtesy light (fast back—Mustang)	15 C.	1003
Oil and generator	2 C.	1895	Courtesy light (Falcon conv.)	6 C.	631
Park brake (Comet—Mustang)	1 C.	257	Ash receptacle (Fairlane)	1.5 C.	1445
Park brake	2 C.	1895	Courtesy light (Mustang console)	3 C.	1816
Seat belt (Fairlane)	1.6 C.	257	Clock (Mustang rally pac)	3 C.	1816
Seat belt (Comet—Falcon)	2 C.	1895	Tachometer (Mustang—rally pac)	2 C.	1895
Emergency flasher	2 C.	1895	Automatic transmission control	2 C.	1895
Alternator	2 C.	1895	Map light	6 C.	631

[1]—Candela is the new international term for candlepower

1967
COUGAR

Light Description	Bulb No.	Candela or Wattage	Light Description	Bulb No.	Candela or Wattage
Headlight-hi-lo beam (outer)	4002	37.5 & 50 Watts	Gages	1895	2 C.
Headlight-hi beam (inner)	4001	37.5 Watts	Hi-beam	1895	2 C.
Front parking and turn signal	1157A	4-32C.	Turn signal	1895	2 C.
Rear and stop and turn signal	1157	4-32C.	Clock	1895	2 C.
License plate	97	4 C.	Turn signal indicator ①	1895-G	2 C.
Map light	631	6 C.	Courtesy light	631	6 C.
"C" pillar light	1003	15 C.	Hand brake signal	1895	2 C.
Auto-trans. quadrant	158	2 C.	Radio AM	1893	1.9 C.
Turn signal indicator	53X	1 C.	Spotlight	4405	30 Watts
Door courtesy	1816	3 C.	Console light	1816	3 C.
Luggage compartment	631	6 C.	Low fuel warning	1445	1.5 C.
Glove compartment	631	6 C.	Emergency flasher warning	1445	1.5 C.
Back-up light	1156	32 C.	Door lock warning	256	1.6 C.
Dome	1816	3 C.	Seat belt warning	1445	1.5 C.
Speedometer	1445	1.5 C.			

① —For States of Minnesota and Wisconsin

MERCURY INTERMEDIATE, FALCON, FAIRLANE, AND MUSTANG

Light Description	Mercury Intermediate		Falcon		Fairlane		Mustang	
	Bulb No.	Candela or Wattage	Bulb No.	Candela or Wattage	Bulb No.	Candela or Wattage	Bulb No.	Candela or Wattage
Ash receptacle or cigar lighter					1445	1.5 C.	1895	2 C.
Backup lights—sedan	1156	32 C.	1156	32 C.	1156	32 C.	1142	21 C.
station wagon	1156	32 C.	1076	32 C.	1156	32 C.		
Courtesy lamp								
(console)							1895	2 C.
(convertible and inst. panel)	631	6 C.	631	6 C.	631	6 C.	631	6 C.
(door mounting or armrest)	631	6 C.			631	6 C.	1004⑤	15 C.
Cargo lamp (station wagon)	1003	15 C.	1003	15 C.	1003	15 C.		
Dome lamp	1003	15 C.	1003	15 C.	1003	15 C.	1003	15 C.
Engine compartment	631	6 C.	631	6 C.	631	6 C.	631	6 C.
Glove compartment or console	1895	2 C.	1895	2 C.	1895	2 C.	1445	1.5 C.
Headlamps hi-lo beam (outer or upper)	4002	37.5-50 W	6012	40-50 W	4002	37.5-50 W	6012④	40-50 W
hi beam (inner or lower)	4001	37.5 W			4001	37.5 W		
License lamp	97	4 C.	97	4 C.	97	4 C.	97	4 C.
Luggage compartment	631	6 C.	631	6 C.	631	6 C.	631	6 C.
Map light	631	6 C.	631	6 C.	631	6 C.	631	6 C.
Park and turn signal lamp	1157-A	4-32 C.	1157	4-32 C.	1157	4-32 C.	1157-A	4-32 C.
Spotlight (4.4 in. dia.)	4405	30 W	4405	30 W	4405	30 W	4405	30 W
Taillight, stop and turn signal lamp	1157	4-32 C.	1157	4-32 C.	1157	4-32 C.	1157	4-32 C.
Instrument panel illumination								
Clock & ignition key	1895	2 C.	1816	3 C.	1895	2 C.	1895①	2 C.
Control nomenclature	1895	2 C.	1895	2 C.	1895	2 C.	1895	2 C.
Gages and speedometer	1895	2 C.	1895	2 C.	1895	2 C.	1895	2 C.
Radio dial light	1893	1.9 C.	1445	1.5 C.	1895	2 C.	1893	1.9 C.
Tachometer	1895	2 C.	1895	2 C.	1895	2 C.	1895	2 C.
Trans control selector indicator	1445	1.5 C.	1445	1.5 C.	1445	1.5 C.	1445	1.5 C.
Turn signal indicators (L and R)③	1895	2 C.	1895	2 C.	1895	2 C.	1895	2 C.
Warning lights, panel or lamp kits								
Emergency flasher	1895	2 C.	1895	2 C.	1895	2 C.	1895	2 C.
Hi-beam indicator	1895	2 C.	1895	2 C.	1895	2 C.	1895	2 C.
Oil press., alt., and temp warning	1895	2 C.	1895	2 C.	1895	2 C.	1895	2 C.
Parking brake warning	1895	2 C.	1895	2 C.	1895	2 C.	257	1.6 C.
Seat belt warning	1895	2 C.	1895	2C②	257	1.6 C.	1891	2 C.

① —1816 (3C.) for Rally PAC   ③ —1895G bulb for States of Minnesota and Wisconsin   ⑤ —Fastback 1004 15C. for door mounted light
② —R.P.O.   ④ —Fog light 4415 (35W)

1968
FALCON, FAIRLANE, MONTEGO, AND MUSTANG

Light Description	Montego		Falcon		Fairlane		Mustang	
	Bulb No.	Candela or Wattage	Bulb No.	Candela or Wattage	Bulb No.	Candela or Wattage	Bulb No.	Candela or Wattage
Ash receptacle or cigar lighter					1445	1.5C.	1895	2C.
Backup lights—sedan	1156	32C.	1156	32C.	1156	32C.	1142	21C.
station wagon	1156	32C.	1076	32C.	1156	32C.	—	—
Courtesy lamp (console)							1816	3C.
(convertible and inst. panel)	631	6C.	631	6C.	631	6C.	631	6C.
(door mounting or armrest)	631	6C.	—	—	631	6C.	631	6C.
Cargo lamp (station wagon)	1003	15C.	1003	15C.	1003	15C.	—	—
Dome lamp	1003	15C.	1003	15C.	1003	15C.	1003	15C.
Engine compartment	631	6C.	631	6C.	631	6C.	631	6C.
Glove compartment or console	1895	2C.	1895	2C.	1895	2C.	1895	2C.
Headlamps—hi-lo beam (outer)	4002	37.5-50W	6012	40-50W	4002	37.5-50W	6012①	40-50W
hi beam (inner)	4001	37.5W	—	—	4001	37.5W	—	—
License lamp	97	4C.	97	4C.	97	4C.	97	4C.
Luggage compartment	631	6C.	631	6C.	631	6C.	631	6C.
Map light	631	6C.	631	6C.	631	6C.	631	6C.
Park and turn signal lamp	1157-A	4-32C.	1157-NA	4-32C.	1157-NA	4-32C.	1157-A	4-32C.
Spotlight (4.4 in. dia.)	4405	30W	4405	30W	4405	30W	4405	30W
Taillight, stop and turn signal lamp	1157	4-32C.	1157	4-32C.	1157	4-32C.	1157	4-32C.
Instrument panel illumination								
Clock & ignition key	1895	2C.	1816	3C.	1895	2C.	1895②	2C.
Control nomenclature	1895	2C.	1895	2C.	1895	2C.	1895	2C.
Gages and speedometer	194	2C.	194	2C.	194	2C.	1895	2C.
Radio dial light	1893	1.9C.	1893	1.9C.	1893	1.9C.	1893	1.9C.
Tachometer	1895	2C.	1895	2C.	1895	2C.	1895	2C.
Trans. control selector indicator	161	1C.	161	1C.	161	1C.	1893	1.9C.
Turn signal indicators (L and R)	194	2C.	194	2C.	194	2C.	1895	2C.
Warning lights, panel or lamp kits								
Emergency flasher	194	2C.	194	2C.	194	2C.	1895	2C.
Hi-beam indicator	194	2C.	194	2C.	194	2C.	1895	2C.
Oil press., alt., and temp warning	194	2C.	194	2C.	194	2C.	1895	2C.
Parking brake warning ③	194	2C.	1895	2C.	1895	2C.	256	1.6C.
Seat belt warning	1895	2C.	1895	2C.	1895	2C.	—	—
Ignition key	—	—	1895	2C.	—	—	—	—
Front side marker	1178-A	4C.	1178-A	4C.	—	—	1178-A	4C.
Turn signal indicator (mounted on hood) ③	—	—	—	—	—	—	53X	1C.

① —Fog light 4415 (35W)　　② —1816 (3C.) for rally pac　　③ —R.P.O.

COUGAR

Light Description	Bulb No.	Candela or Wattage	Light Description	Bulb No.	Candela or Wattage
Headlight—hi-lo beam (outer)	4002	37.5 & 50 Watts	Gages	1895	2C.
Headlight—hi beam (inner)	4001	37.5 Watts	Hi-beam	1895	2C.
Front parking and turn signal	1157A	4-32C.	Turn signal	1895	2C.
Rear and stop and turn signal	1157	4-32C.	Clock	1895	2C.
License plate	97	4C.	Turn signal indicator	1895	2C.
Map light	631	6C.	Courtesy light	631	6C.
"C" pillar light	1003	15C.	Hand brake signal	1895	2C.
Auto-trans. quadrant	1445	1.5C.	Radio AM	1893	1.9C.
Door courtesy	1816	3C.	Spotlight	4405	30 Watts
Luggage compartment	631	6C.	Console light	1816	3C.
Glove compartment	631	6C.	Low fuel warning	1895	2C.
Back-up light	1156	32C.	Door lock warning	257	1.6C.
Front side marker	97N.A.	4C.	Seat belt warning	1895	2C.
Speedometer	1895	2C.			

1969

Light Description	Montego Bulb No.	Montego Candela or ① Wattage	Falcon Bulb No.	Falcon Candela or ① Wattage	Fairlane Bulb No.	Fairlane Candela or ① Wattage	Mustang Bulb No.	Mustang Candela or ① Wattage	Cougar Bulb No.	Cougar Candela or ① Wattage
Ash receptacle or cigar lighter			1445	1.5C.	1445	1.5C.			1445	1 5C.
Auxiliary instrument cluster					1895	2C.				
Back-up lights										
sedan	1156	32C.	1156	32C.	1156	32C.	1142	21C.	1156	32C.
station wagon	1157	32C. 4C.	1076	32C.	1156	32C.				
Cargo light (station wagon)	1003	15C.	1003	15C.	1003	15C.				
Courtesy light										
C-pillar							1003	15C.	1003	15C.
floor console							1445	1.5C.	1816	3C.
convertible or inst. panel	631	6C.	631	6C.	631	6C.	631	6C.	631	6C.
door mounted							212	6C.	212	6C.
Cruise light (speed control)									1895	2C.
Dome light	1003	15C.	1003	15C.	1003	15C.	1003	15C.		
Door ajar indicator	1895	2C.	1895	2C.	1895	2C.			194	2C.
Door ajar indicator (convenience pkg.)									257	1.6C.
Engine compartment	631	6C.	631	6C.	631	6C.	631	6C.	631	6C.
Fog light					4415	35W	4415	35W		
Fog light switch					53X	1C.	53X	1C.		
Front park and turn signal light	1157A	3-32C.	1157	3-32C.	1157NA②	3-32C.	1157A	3-32C.	1157A	3-32C.
Front side marker	1178A	4C.	1178A	4C.			1178A	4C.	194A	2C.
Glove compartment	1891	2C.	1891	2C.	1891	2C.	1891	2C.	1891	2C.
Headlights										
hi-lo beam	4002	37.5-50W	6012	40-50W	4002	37.5-50W	4002	37.5-50W	4002	37.5-50W
hi-beam	4001	37.5W			4001	37.5-5W			4001	37.5W
Hi-beam indicator	194	2C.	194	2C.	194	2C.	194	2C.	194	2C.
Ignition key			1895	2C.	1895	2C.			1895	2C.
Instrument panel illumination										
clock	1895	2C.			1895	2C.	1895	2C.	1895	2C
control nomenclature heater and A/C			1895	2C.	1895	2C.	1895	2C.	1895	2C.
gauges and speedometer	194	2C.	194	2C.	194	2C.	194	2C.	194	2C.
tachometer							1895	2C.		
License light	97	4C.	97	4C.	97	4C.	97	4C.	97	4C.
Low fuel indicator	1895	2C.	1895	2C.	1895	2C.			194	2C.
Low fuel indicator (convenience pkg.)									1895	2C.
Luggage compartment	631	6C.	631	6C.	631	6C.	631	6C.	631	6C.
Map light	631	6C.	631	6C.	631	6C.			1816	3C.
Oil press., service brakes, alternator and temp. indicator	194	2C.	194	2C.	194	2C.	194	2C.	194②	2C.
Parking brake indicator					257	1.6C.	256	1.6C.	194④	2C.
Parking brake indicator (convenience pkg.)	1895	2C.	257	1.6C.	1895	2C.			256③	1.6C.
Radio dial light	1893	1.9C.	1893	1.9C.	1893	1.9C.	1893	1.9C.	1893	1.9C.
AM/FM stereo dial light	1892	1.3C.			1892	1.3C.	1892	1.3C.	1892	1.3C.
Rear side marker	194	2C.	194	2C.	194	2C.	194	2C.	194	2C.
Roof console							631	6C.		
Seat belt indicator	1895	2C.	1895	2C.	1895	2C.	1895	2C.	194	2C.
Seat belt indicator (convenience pkg.)							1891	2C.	1891	2C.
Spotlight (4.4 inch diameter)	4405	30W	4405	30W	4405	30W	4405	30W	4405	30W
Taillight, stop and turn signal light	1157	3-32C.	1157	3-32C.	1157	3-32C.	1157A	3-32C.	1157	3-32C.
Trans. control selector indicator (column mounted)	161	1C.	161	1C.	161	1C.	1893	1.9C.	1445	1.5C.
Trunk light (portable)	1003	15C.	1003	15C.	1003	15C.	1003	15C.	1003	15C.
Turn signal indicators (L and R)	194	2C.	194	2C.	194	2C.	194	2C.	194	2C.
Turn signal indicator (mounted on hood)					1895	2C.	1895	2C.		

① —Candela is the international term for candle power.
② —Service brake indicator light only.
③ —Except XR7.
④ —XR7.

1970–71

FAIRLANE

Light Description	Candle Power or Wattage	Trade No.	Light Description	Candle Power or Wattage	Trade No.
Standard Equipment			Emergency flasher indicators—		
Headlamps—hi & lo	37.5 & 50 Watts	4002	included in turn signal indicators		
			Optional Equipment		
Headlamps—hi-beams	37.5 Watts	4001	Courtesy lamp	6 c.p.	631
Front park/turn/side marker	3-32 c.p.	1157A①	Fog lamps—clear	35 Watts	4415
Rear tail/stop/turn sedan	3-32 c.p.	1157	Fog lamp switch	1 c.p.	53X
sta. wagon	3-32 c.p.	1157	Glove compartment	2 c.p.	1891
Back-up lamps	32 c.p.	1156	Spotlight—4.4″ dia.	30 Watts	4405
License plate lamp	4 c.p.	97	Radio pilot light (AM radio)	1.9 c.p.	1893
sta. wagon	4 c.p.	97	Tachometer	2 c.p.	1895
Dome lamp	12 c.p.	105	Auto. trans. quadrant	1.5 c.p.	1445
Hoot mtd. turn signals (GT)	2 c.p.	1895	Console lamp	3 c.p.	1816
Emergency flashers—included in front & rear turn signals			Luggage comp. light	12 c.p.	105
			Floor shift quadrant	1.5 c.p.	1445
Front side marker	2 c.p.	194	Parking brake—warning	2 c.p.	1895
Rear side marker	2 c.p.	194	Engine compartment	6 c.p.	631
Instrument Panel Cluster	2 c.p.	194	Portable trunk lamp	15 c.p.	1003
Courtesy lamp (conv. only)	Fuse type	562	Aux. inst. cluster	2 c.p.	1895
Hi-beam indicator	2 c.p.	194	Map lamp	6 c.p.	631
Turn signal indicators	2 c.p.	194	Cargo lamp—sta. wag.	12 c.p.	105
Warning lights (oil, alt., brake system)	2 c.p.	194	AM/FM (MPX) radio	1.9 c.p.	1893
Fuel gauge—speedometer	2 c.p.	194	Hood mtd. turn signals (optional with hi perf. eng.)	2 c.p.	1895
Heater (or A/C opt.) bar.	2 c.p.	1895			
Clock (optional)	2 c.p.	1895	Ash tray	1.5 c.p.	1445
Deluxe seat belt (option)	2 c.p.	1895	Headlights—on	2 c.p.	1859

① —Amber.

MONTEGO

Lamp Description	Candle Power or Wattage	Trade No.	Lamp Description	Candle Power or Wattage	Trade No.
Standard Equipment			Radio dial	2 c.p.	1893
Headlamps			Turn signal	2 c.p.	194
hi-lo beam	37.5 & 50 Watts	4002	Clock	2 c.p.	1895
			Instrumentation package	2 c.p.	194
hi-beam	37.5 Watts	4001	Ashtray	1.5 c.p.	1445
Front park and turn	3-32 c.p.	1157A	PRND21 (column)	1.5 c.p.	1445
Rear, stop and turn (passenger)	3-32 c.p.	1157	**Accessory Equipment**		
Rear, stop and turn (station wagon)	3-32 c.p.	1157	Glove compartment	2 c.p.	1891
License plate lamp	4 c.p.	97	Luggage compartment	12 c.p.	105
Dome lamp	12 c.p.	105	Engine compartment	6 c.p.	631
Back-up lamp	32 c.p.	1156	Spot lamp	30 Watt	4405
Front side marker	2 c.p.	194	Courtesy lamp (in instr. panel)	6 c.p.	562
Rear side marker	2 c.p.	194	Cargo lamp (wagon)	12 c.p.	105
Instrument Panel			Tachometer	2 c.p.	1895
Warning lights	2 c.p.	194	Cluster	2 c.p.	1895
Instrument (speed-o & gauges)	2 c.p.	194			
Hi-beam	2 c.p.	194			

FALCON AND MAVERICK

Lamp Description	Candle Power or Wattage	Trade No.	Lamp Description	Candle Power or Wattage	Trade No.
Standard Equipment			**Optional Equipment**		
Headlamps	40-50 Watts	6012	Spotlight—4.4" dia.	30 Watts	4405
Front part/turn signal (Maverick)	3-32 c.p.	1157A	Air cond. controls	2 c.p.	1895
Front park/turn signal (Falcon)	3-32 c.p.	1157	Radio pilot light	1.9 c.p.	1893
Rear tail/stop/turn	3-32 c.p.	1157	Auto. trans. quadrant	1.5 c.p.	1445
Back-up light	32 c.p.	1156	Tachometer light (round)	2 c.p.	1895
Back-up light (Falcon station wagon)	32 c.p.	1076	Hood mounted (Maverick)	2 c.p.	57
License plate lamp	4 c.p.	97	Luggage comp. light (Maverick)	6 c.p.	631
Dome lamp	12 c.p.	105	Seat belt warning (Maverick)	2 c.p.	1895
Front & rear side marker lamp	2 c.p.	194	Falcon only courtesy light (inst. panel)	6 c.p.	562
Instrument Panel			Clock light	3 c.p.	1816
Hi-beam indicator	2 c.p.	194	Luggage comp. light	12 c.p.	105
Turn signal indicators	2 c.p.	194	Portable trunk light	15 c.p.	1003
Warning lights (oil, alt., hot, brakes)	2 c.p.	194	Parking brake warning	16 c.p.	257
Speedometer & fuel gauge	2 c.p.	194	Engine compt.	6 c.p.	631
Heater controls	2 c.p.	1895	Map light	6 c.p.	631
Ash tray light (Maverick)	1.5 c.p.	1445	Cargo light	12 c.p.	105

COUGAR

Lamp Description	Candle Power or Wattage	Trade No.	Lamp Description	Candle Power or Wattage	Trade No.
Headlamps—hi & lo	37.5 & 50 Watts	4002	Glove compartment	3 c.p.	1816
			Warning Lamps		
Headlamps—highbeam	37.5 Watts	4001	**All Except XR-7 Models**		
Front park and turn signal	3-32 c.p.	1157A	Low "Fuel"	2 c.p.	1895
Rear tail, stop and turn signal	3-32 c.p.	1157	"Door" ajar	1.6 c.p.	257
Front side markers	4 c.p.	97NA	Seat belts	2 c.p.	1895
Rear side markers	2 c.p.	194	"Park" brake	2 c.p.	194
License plate lamp	4 c.p.	97	"Brakes"	2 c.p.	194
Auto. trans. quadrant	1.5 c.p.	1445	"Lights"	2 c.p.	194
Luggage compartment	6 c.p.	631	"Alternator"	2 c.p.	194
Back-up lamp	32 c.p.	1156	"Oil"	2 c.p.	194
Courtesy lamps—under panel	6 c.p.	562	**XR-7 Models**		
Courtesy lamps—pillar	12 c.p.	105	Low "Fuel"	2 c.p.	194
Courtesy lamps—door	6 c.p.	212	"Door" ajar	2 c.p.	194
Console lamp	3 c.p.	1816	Seat "belts"	2 c.p.	194
Instrument Panel			"Park" brake	2 c.p.	194
Speedometer	2 c.p.	194	"Brakes"	2 c.p.	194
Gauges	2 c.p.	194	"Lights"	2 c.p.	194
Hi-beam	2 c.p.	194	**Optional Equipment**		
Turn signal indicators	2 c.p.	194	Heater—air conditioner control	1.5 c.p.	1445
Clock	2 c.p.	1895	Radio AM	1.9 c.p.	1893
Cluster illumination	2 c.p.	194	Radio AM-FM	1.9 c.p.	1893
Ash tray	1.5 c.p.	1445	Radio AM-tape player	1.9 c.p.	1893
Ash tray—console	1.3 c.p.	1892	Engine compartment	6 c.p.	631
Map lamp	3 c.p.	1816	Spotlight	30 Watt	4405

MUSTANG

Light Description	Candle Power or Wattage	Trade No.	Light Description	Candle Power or Wattage	Trade No.
Standard Equipment			**Optional Equipment**		
Headlamps—hi & lo	40-50 Watts	6012	Ash tray (console)	1.3 c.p.	1892
Front park & turn signal	3-32 c.p.	1157②	Ash tray (instrument panel)	1.5 c.p.	1445
Front side marker	4 c.p.	97②	Courtesy lamp	6 c.p.	631
Rear tail stop/turn signal	3-32 c.p.	1157	Fog lamps	35 Watts	4415
Four way emergency flashers— included in front & rear turn signals			Fog lamp—switch	1 c.p.	53X
			Map lamp	15 c.p.	1004
Back-up lamp	32 c.p.	1156	Spotlight 4.4" dia.	30 Watts	4405
License plate lamp	4 c.p.	97	R & L turn signal indicators (outside— in hood scoop) all except Grande hardtop	1 c.p.	53X
Courtesy lamp—Fastback 64 "C" pillar	12 c.p.	105			
Courtesy lamp—under inst. pnl. (63 & 76)	Fuse type	562	Radio pilot light (AM radio)	1.9 c.p.	1893
Dome courtesy (65 only)	12 c.p.	105	Warning light—brake	1.6 c.p.	256
Hood mtd. turn signals (Mach I)	1 c.p.	53X	Conv. pkg. seat belt	2 c.p.	1891
Rear side marker lamp	2 c.p.	194	Trans. control selector indicator	1.5 c.p.	1445
Instrument Panel			Headlights on	1.6 c.p.	0000
Hi-beam indicator	2 c.p.	194	Luggage compt. lamp	6 c.p.	631
Turn signal indicators	2 c.p.	194	Engine compt. lamp	6 c.p.	631
Turn signal indicators are also emergency flasher indicators			Portable trunk lamp	15 c.p.	1003
			Clock	2 c.p.	1895
*Warning lights—oil, alt. & Brakes①	2 c.p.	194	AM/FM (MPX) radio	1.9 c.p.	1893
Glove compt./light	2 c.p.	1891	Warning light belts	2 c.p.	1891
Instruments	2 c.p.	194	Door courtesy r.p.o.	6 c.p.	212
Seat belt warning light r.p.o.	1.5 c.p.	1445	Radio AM-stereo type	9 c.p.	1893
Heater control	2 c.p.	1895			

① —Oil and alternator warning lights used with tachometer installation only.
② —Natural amber.

FUSES AND CIRCUIT BREAKERS

1964

FALCON

COMET

Circuit	Protective Device	Location	Circuit	Protective Device	Location
Headlights	Circuit breaker	Incorporated in lighting switch	Headlights	Circuit breaker	Incorporated in lighting switch
Instrument panel, dome, and all exterior lights, except headlight	3AG-15 or AGC-15 fuse	Fuse panel on lighting switch	Instrument panel, dome, and all exterior lights, except headlight	3AG-15 or AGC-15 fuse	Fuse panel on lighting switch
Turn signals and back-up	SFE-14 fuse	Fuse panel on lighting switch	Turn signals and back-up	SFE-14 fuse	Fuse panel on lighting switch
Radio	SFE-7.5 fuse	Fuse panel on lighting switch	Radio	SFE-7.5 fuse	Fuse panel on lighting switch
Heater blower	SFE-14 fuse	Fuse panel on lighting switch	Heater blower	SFE-14 fuse	Fuse panel on lighting switch
Electric windshield wiper (single speed) (dual speed)	Circuit breaker Circuit breaker	Integral with motor	Electric windshield wiper (single speed) (dual speed)	Circuit breaker Circuit breaker	Integral with motor
Cigar lighter	Reset circuit breaker	On back of cigar lighter socket	Cigar lighter	Reset circuit breaker	On back of cigar lighter socket
Air conditioner	3AG-15 or AGC-15 fuse	Cartridge in feed wire	Air conditioner	3AG-15 or AGC-15 fuse	Cartridge in feed wire
Spot light	SFE-7.5 fuse	Cartridge in feed wire	Spot light	SFE-7.5 fuse	Cartridge in feed wire
Instrument panel light rheostat	1AG-1 or AGA-1 fuse	Cartridge in feed wire	Instrument panel light rheostat	1AG-1 or AGA-1 fuse	Cartridge in feed wire
Clock	1AG-2 or AGA-2 fuse	Fuse panel on lighting switch	Clock	1AG-2 or AGA-2 fuse	Fuse panel on lighting switch

1965–66

Function	Location	Falcon Comet	Fairlane	Mustang
		Rating Type	Rating Type	Rating Type
Dome courtesy-map cargo	Fuse panel	7½ SFE	7½ SFE	7½ SFE
Tail-park license	Light switch	15 C.B.	15 C.B.	15 C.B.
Stop light	Light switch	15 C.B.	15 C.B.	15 C.B.
Clock	Fuse panel	7½ SFE	7½ SFE	7½ SFE
Back-up	Fuse panel	14 SFE	14 SFE	14 SFE
Turn signals (act as a circuit breaker)				
Radio	Fuse panel (acc. socket)	14 SFE	14 SFE	14 SFE
Heater	Fuse panel	14 SFE	14 SFE	14 SFE
Heater and PRNDL dial	Fuse panel heater socket			14 SFE
Cigar lighter	Fuse panel	14 SFE	14 SFE	14 SFE
Cigar lighter and emergency warning flasher	Fuse panel cigar lighter socket	20 SFE	20 SFE	20 SFE
Tachometer	Fuse panel (dome socket)	7½ SFE	7½ SFE	7½ SFE
Convertible top	Between starter relay & junction block	Safety link	Safety link	Safety link
Power window—power seat	On starter relay	20 C.B.	20 C.B.	20 C.B.
Overdrive	Clips to overdrive relay	20 SFE	20 SFE	20 SFE
Seat belt warning	Fuse panel acc. socket	14 SFE	14 SFE	14 SFE
Windshield washer (two speed wiper)		14 SFE	14 SFE	
Windwhield washer (single speed wiper)		14 SFE	14 SFE	14 SFE
Windshield washer (two speed wiper)	Wiper switch			12 C.B.
Windshield wiper	Wiper switch	6 (S. Sp.) C.B. 7 (2 Sp.) C.B. (Falcon) 7 Intermittent	6 (S. Sp.) C.B. 7 (2 Sp.) C.B.	5 (S. Sp.) C.B. 12 (2 Sp.) C.B.
Light-instrument panel	Fuse panel instrument LP socket	2½ AGA	2½ AGA	2½ AGA
Light-instrument cluster	connected into 15 C.B. light switch	2½ AGA	2½ AGA	2½ AGA
Light-clock		2½ AGA	2½ AGA	2½ AGA
Light-tachometer		2½ AGA	2½ AGA	2½ AGA
Light-ash receptacle		2½ AGA	2½ AGA	2½ AGA
Light-PRNDL dial		2½ AGA	2½ AGA	
Light-PRNDL dial (console only)	Fuse panel acc. socket	14 SFE	14 SFE	
Light-luggage compartment	Fuse panel	7½ SFE	7½ SFE	7½ SFE
Light-door open warning	Fuse panel		7½ SFE	
Light-glove box	Fuse panel	7½ SFE	7½ SFE	7½ SFE
Light-spotlight	In line	7½ SFE	7½ SFE	7½ SFE
Light-headlights	Light switch	18 Comet C.B. 12 Falcon C.B.		
Emergency warning flasher	Fuse panel	14 SFE	14 SFE	14 SFE
Horns	Light switch	15 C.B.	15 C.B.	15 C.B.
Air conditioner (integrated)	On ign. switch	25 C.B.	25 C.B.	25 C.B.
Back light control	At starter relay	20 C.B.	20 C.B.	
Economy air conditioner	In line	15 AGC	15 AGC	15 AGC
Motor windshield wiper	Circuit breaker integral with motor	C.B.	C.B.	C.B.
Motor convertible top		C.B.	C.B.	C.B.
Motor power window		C.B.	C.B.	C.B.
Motor power seat		C.B.	C.B.	C.B.

1967
MERCURY INTERMEDIATE, FALCON, FAIRLANE, COUGAR AND MUSTANG

Fuse Panel Circuits	Protective Device—Fuses in Amperes				
	Mercury Intermediate	Falcon	Fairlane	Cougar	Mustang
Ash tray light	AGA-2.5	AGA-2.5	AGA-2.5	AGA-2.5	AGA-2.5
Back up light	SFE-14	SFE-14	SFE-14	SFE-14	SFE-14
Cargo light	SFE-7.5	SFE-7.5	SFE-7.5	SFE-7.5	SFE-7.5
Cigar lighter	SFE-20	SFE-20	SFE-20	SFE-20	SFE-20
Clock	SFE-20	—	SFE-20	SFE-7.5	SFE-7.5
Clock light	AGA-2.5	—	AGA-2.5	AGA-2.5	AGA-2.5
Courtesy lights	SFE-7.5	SFE-7.5	SFE-7.5	SFE-7.5	SFE-7.5
Dome lights	SFE-7.5	SFE-7.5	SFE-7.5	SFE-7.5	SFE-7.5
Door ajar warning (door open warning, Fairlane taxi)	—	—	SFE-7.5	SFE-14	SFE-7.5
Emergency warning flasher	SFE-20	SFE-20	SFE-20	SFE-14	SFE-20
Glove box light	SFE-7.5	SFE-7.5	SFE-7.5	SFE-7.5	SFE-7.5
Heater and defroster	SFE-14	SFE-14	SFE-14	SFE-14	SFE-14
Instrument panel and cluster lights	AGA-2.5	AGA-2.5	AGA-2.5	AGA-2.5	AGA-2.5
Luggage compartment light	SFE-7.5	SFE-7.5	SFE-7.5	SFE-7.5	SFE-7.5
Map light	SFE-7.5	SFE-7.5	SFE-7.5	SFE-7.5	SFE-7.5
Radio	SFE-14	SFE-14	SFE-14	SFE-14	SFE-14
Radio light	AGA-2.5	AGA-2.5	AGA-2.5	AGA-2.5	AGA-2.5
Seat belt warning light (inst. or safety conv. panel)	SFE-14	SFE-14	SFE-14	SFE-7.5	SFE-7.5
Seat belt warning—Cougar model 65B only	—	—	—	SFE-14	—
Tachometer light	AGA-2.5	—	AGA-2.5	AGA-2.5	AGA-2.5
Transmission selector light	AGA-2.5	AGA-2.5	AGA-2.5	SFE-14	SFE-14
Turn signals	SFE-14	SFE-14	SFE-14	SFE-14	SFE-14
Windshield washers	SFE-14	SFE-14	SFE-14	—	—

Miscellaneous Circuits	Location	Protective Device—Fuse or Circuit Breaker (C.B.) in Amperes				
		Mercury Intermediate	Falcon	Fairlane	Cougar	Mustang
Headlights	In headlight switch	18 C.B.	12 C.B.	18 C.B.	18 C.B.	12 C.B.
Tail lights, stop lights, license lights, parking lights and horns	In headlight switch	15 C.B.	15 C.B.	15 C.B.	15 C.B.	15 C.B.
Tail lights (Cougar)	Near left rear light assy.	—	—	—	5 C.B.	—
Emergency warning flasher	Windshield wiper bracket	—	—	—	15 C.B.	—
Windshield wipers	Windshield wiper switch	6 C.B.	6 C.B.	6 C.B.	6 C.B.	6 C.B.
Intermittent windshield wipers	Windshield wiper switch	7 C.B.	—	—	—	—
Convertible top	Between starter relay and junction block	Safety link	—	Safety link	—	Safety link
Power windows, power seats and back window control	On starter relay	20 C.B.	20 C.B.	20 C.B.	—	—
Overdrive	Clip to overdrive relay	—	—	SFE-20	—	—
Air conditioner (integrated)	Acc. terminal ignition switch	25 C.B.	25 C.B.	25 C.B.	25 C.B.	25 C.B.
Air conditioner (economy)	In-line from acc. terminal of ignition switch	AGC-15	AGC-15	AGC-15	AGC-15	AGC-15
Speed control	In-line from acc. terminal of ignition switch	—	—	—	SFE-7.5	SFE-7.5
Motors: windshield wiper, convertible top, power windows and power seats	Integral part of motor	C.B.	C.B.	C.B.	C.B.	C.B.
Transmission selector light (console)	In-line	AGW-4	AWG-4	AGW-4	—	—
Parking brake warning	In-line	AGW-4	AGW-4	AGW-4	SFE-7.5	SFE-7.5

1968
COUGAR, FAIRLANE, FALCON, MONTEGO AND MUSTANG

Fuse Panel Circuits	Protective Device—Fuses in Amperes				
	Cougar	Fairlane	Falcon	Montego	Mustang
Ash tray light	AGA-2.5	SFE-4	SFE-4	SFE-4	AGA-2.5
Back up light	SFE-14	SFE-20	SFE-20	SFE-20	SFE-14
Cargo light	SFE-7.5	SFE-14	SFE-14	SFE-14	SFE-7.5
Cigar lighter	SFE-14	SFE-20	SFE-20	SFE-20	SFE-14
Clock	SFE-7.5	SFE-20	—	SFE-20	SFE-7.5
Clock light	AGA-2.5	SFE-4	—	SFE-4	AGA-2.5
Courtesy light	SFE-7.5	SFE-14	SFE-14	SFE-14	SFE-7.5
Defogger (cartridge in feed line for Cougar & Mustang)	SFE-7.5	SFE-20	SFE-20	SFE-20	SFE-7.5
Door ajar warning (door open warning, Fairlane taxi)	SFE-14	SFE-14	SFE-14	SFE-14	SFE-7.5
Dome light	SFE-7.5	SFE-14	SFE-14	SFE-14	SFE-7.5
Emergency warning flasher	—	SFE-20	SFE-20	SFE-20	SFE-20
Engine compartment light	—	—	—	—	SFE-7.5
Fog lights (bracket at fuse panel)	—	—	—	—	10 Amp. C.B.
Glove compartment light	SFE-7.5	SFE-14	SFE-14	SFE-14	SFE-7.5
Heater and defroster (30 AMP. req'd for A/C) ①	SFE-14	SFE-14	SFE-14	SFE-14	SFE-14
Instrument panel and cluster lights	AGA-2.5	SFE-4	SFE-4	SFE-4	AGA-2.5
Luggage compartment light	SFE-7.5	SFE-14	SFE-14	SFE-14	SFE-7.5
Map light	SFE-7.5	SFE-14	SFE-14	SFE-14	SFE-7.5
Radio and stereo tape	SFE-14	SFE-20	SFE-20	SFE-20	SFE-14
Radio light	AGA-2.5	SFE-4	SFE-4	SFE-4	AGA-2.5
Seat belt warning light (inst. or conv. control panel)	SFE-7.5	SFE-14	SFE-14	SFE-14	SFE-7.5
Seat belt warning light (inst. or conv. control panel)	SFE-14	—	—	—	—
Swing-tilt steering wheel (cartridge in feed line)	SFE-7.5	—	—	—	SFE-7.5
Tachometer light	AGA-2.5	SFE-4	—	SFE-4	AGA-2.5
Tilt steering wheel (cartridge in feed line)	SFE-7.5	—	—	—	SFE-7.5
Transmission selector light	SFE-14	SFE-4	SFE-4	SFE-4	SFE-14
Turn signals (cartridge in feed line)	SFE-15	SFE-20	SFE-20	SFE-20	SFE-14
Windshield washers	—	SFE-20	SFE-20	SFE-20	—

① —Located in fuse panel.

Miscellaneous Circuits	Location	Protective Device—Fuse or Circuit Breaker (C.B.) in Amperes				
		Montego	Falcon	Fairlane	Cougar	Mustang
Headlights	In headlight switch	18 C.B.	12 C.B.	18 C.B.	18 C.B.	12 C.B.
Tail lights, stop lights, license lights, parking lights, horns & marker lights	In headlight switch	15 C.B.	15 C.B.	15 C.B.	15 C.B.	15 C.B.
Tail lights (Cougar)	Near left rear light assy.	—	—	—	5 C.B.	—
Emergency warning flasher	Bracket at fuse panel	—	—	—	15 C.B.	—
Windshield wipers	Windshield wiper switch	6 C.B.	6 C.B.	6 C.B.	6 C.B.	6 C.B.
Intermittent windshield wipers	Windshield wiper switch	7 C.B.	—	—	—	—
Convertible top	Between starter relay and junction block	Safety Link	—	Safety Link	—	Safety Link
Power windows, power seats and back window control	On starter relay	20 C.B.	20 C.B.	20 C.B.	—	—
Air conditioner (integrated)	① ②	①	①	①	②	②
Speed control	In-line from acc. terminal of ignition switch	—	—	—	SFE-7.5	SFE-7.5
Motors: windshield wiper, convertible top, power windows and power seats	Integral part of motor	C.B.	C.B.	C.B.	C.B.	C.B.
Transmission selector light (console)	In-line	AGW-4	AGW-4	AGW-4	—	—
Parking brake warning	In-line	AGW-4	AGW-4	AGW-4	SFE-7.5	SFE-7.5

① —30 amp. fuse in fuse panel.   ② —30 amp. fuse in bracket attached to fuse panel.

1969

COUGAR, FALCON, FAIRLANE, MONTEGO AND MUSTANG

Circuits	Location	Protective Device Fuse or Circuit Breaker (C.B.) in Amperes				
		Cougar	Fairlane	Falcon	Montego	Mustang
Ash tray light		SFE 4	SFE 14	SFE 14	SFE 14	SFE 4
Back-up light		AGC or SFE 20	AGC or SFE 20	AGC or SFE 20	AGC or SFE 20	AGC or SFE 20
Cigar lighter	Fuse panel	SFE 14	AGC or SFE 20	AGC or SFE 20	AGC or SFE 20	AGC or SFE 20
Clock feed		SFE 14	AGC or SFE 20	AGC or SFE 20	AGC or SFE 20	SFE 14
Convertible top feed	Between starter relay and junction block	Fuse Link	20 C.B. ①	—	20 C.B. ①	Fuse Link
Convertible top motor	Integral part of motor	C.B.	C.B.	C.B.	C.B.	C.B.

Courtesy lights
  C pillar
  Cargo
  Console (Fairlane and Montego only) door
  Dome
  Glove compartment
  Luggage compartment
  Map light
	Fuse panel	SFE 14	SFE 14	SFE 14	SFE 14	SFE 14

Emergency flashers		15 C.B. ②	AGC or SFE 20	AGC or SFE 20	AGC or SFE 20	AGC or SFE 20
Engine compartment light	Cartridge in feed line	SFE 7.5	SFE 7.5	SFE 7.5	SFE 7.5	SFE 7.5
Fog lights	Bracket at fuse panel	—	10 C.B.	—	—	10 C.B.
Headlights	In headlight switch	18 C.B.	18 C.B.	18 C.B.	18 C.B.	18 C.B.
Heater and defroster	Fuse panel	SFE 14	SFE 14	SFE 14	SFE 14	SFE 14
Heater and defroster with air conditioner	Fuse panel	8 AG or AGX 30	8 AG or AGX 30	8 AG or AGX 30	8 AG or AGX 30	8 AG or AGX 30
Horns	In headlight switch	15 C.B.	15 C.B.	15 C.B.	15 C.B.	15 C.B.
Indicator lights (convenience control panel) Door ajar		AGC or SFE 20	—	—	—	AGC or SFE 20
Dual brake		—	SFE 14 ③	SFE 14 ③	SFE 14 ③	—
Engine temperature		—	SFE 14 ③	SFE 14 ③	SFE 14 ③	—
Low fuel	Fuse panel	AGC or SFE 20	—	—	—	AGC or SFE 20
Oil		—	SFE 14 ③	SFE 14 ③	SFE 14 ③	—
Seat belt (instrument or convenience control panel)		AGC or SFE 20	SFE 14 ③	SFE 14 ③	SFE 14 ③	AGC or SFE 20

Instrumental panel lights
  Clock light
  Heater controls
  Radio
	Fuse panel	SFE 4	SFE 14	SFE 14	SFE 4	SFE 4

1969
COUGAR, FALCON, FAIRLANE, MONTEGO AND MUSTANG

Circuits	Location	Protective Device Fuse or Circuit Breaker (C.B.) in Amperes				
		Cougar	Fairlane	Falcon	Montego	Mustang
License lights						
Marker lights	In headlight switch	15 C.B. ②	15 C.B.	15 C.B.	15 C.B.	15 C.B.
Parking lights						
Power backlight	Attached to starting motor relay	—	20 C.B.	20 C.B.	20 C.B.	—
Power backlight motor	Integral with motor	—	C.B.	C.B.	C.B.	—
Power backlight relay feed	Fuse panel	—	AGC or SFE 20	AGC or SFE 20	—	—
Power seats	On starter relay	20 C.B.	AGC or SFE 20 in Fuse panel	20 C.B.	20 C.B.	20 C.B.
Power seats motors	Integral part of motor	C.B.	C.B.	C.B.	C.B.	C.B.
Power window	Attached to starting motor relay	20 C.B.	20 C.B.	—	20 C.B.	20 C.B.
Power window motor	Integral with motor	C.B.	C.B.	—	C.B.	C.B.
Power window relay feed	Fuse panel	AGC or SFE 20	AGC or SFE 20	—	AGC or SFE 20	AGC or SFE 20
PRNDL (auto. trans.)	Fuse panel	—	SFE 14	SFE 14	SFE 14	—
Radio and/or stereo tape feed	Fuse panel	—	AGC or SFE 20	—	AGC or SFE 20	—
Speed control		AGC or SFE 20	—	—	—	AGC or SFE 20
Stop lights	In headlight switch	15 C.B. ②	15 C.B.	15 C.B.	15 C.B.	15 C.B.
Swing-tilt steering wheel	Fuse panel	AGC or SFE 20	—	—	—	AGC or SFE 20
Taillights	In headlight switch	15 C.B.	15 C.B.	15 C.B.	15 C.B.	15 C.B.
Transmission selector light	Fuse panel	AGC or SFE 20	SFE 4	SFE 4	SFE 4	AGC or SFE 20
Turn signals	Fuse panel	AGC or SFE 20	AGC or SFE 20	AGC or SFE 20	AGC or SFE 20	AGC or SFE 20
Windshield washers	Fuse panel					
Windshield wipers	Windshield wiper switch	6 C.B.	8 C.B.	8 C.B.	8 C.B.	6 C.B.
Windshield wiper motor	Integral part of motor	C.B.	C.B.	C.B.	C.B.	C.B.

① —Attached to starting motor relay   ② —Bracket attached to L.H. lower instrument panel   ③ —Instrument panel mounted light uses AGC or SFE 20

1970–71
MUSTANG, COUGAR, FALCON, MAVERICK, FAIRLANE and MONTEGO

Circuits	Location	Protective Device Fuse or Circuit Breaker (C.B.) in Amperes				
		Cougar	Fairlane	Falcon/Maverick	Montego	Mustang
Power window motor	Integral with motor	C.B.	C.B.		C.B.	
Power window relay feed	Fuse panel	AGC or SFE 15	AGC or SFE 20	AGC or ⑤ SFE 20	AGC or SFE 20	
PRNDL (auto. trans.)	Fuse panel	15 C.B. ①	SFE 4	SFE 4	SFE 4	15 C.B. ①
Radio and/or stereo tape feed	Fuse panel	AGC or SFE 15	AGC or SFE 15	AGC or SFE 15	AGC or SFE 20	AGC or SFE 15
Rear window defogger	Fuse panel	AGC or SFE 15	AGC or SFE 20	AGC or SFE 20	AGC or SFE 20	
Spotlight	Cartridge in feed wire	SFE 7.5	SFE 7.5			
Stop lights	In headlight switch	15 C.B. ②	15 C.B.	15 C.B.	15 C.B.	15 C.B.
Taillights	In headlight switch	15 C.B.	15 C.B.	15 C.B.	15 C.B.	15 C.B.
Transmission selector light	Fuse panel	15 C.B. ①	SFE 4	SFE 4	SFE 4	15 C.B. ①
Turn signals / Windshield washers	Fuse panel	AGC or SFE 20	AGC or SFE 20	AGC or SFE 20	AGC or SFE 20	AGC or SFE 15
Windshield wipers	Windshield wiper switch	6 C.B.	8 C.B.	8 C.B.	8 C.B.	6 C.B.
Windshield wiper motor	Integral part of motor	C.B.	C.B.	C.B.	C.B.	C.B.

① —Integral with headlight switch   ③ —Instrument panel mounted light used AGC or SFE 20   ⑤ —Falcon only
② —Bracket attached to L.H. lower instrument panel   ④ —Economy dealer-installed A/C used AGC or SFE 20

1970–71
MUSTANG, COUGAR, FALCON, MAVERICK, FAIRLANE and MONTEGO

Circuits	Location	Protective Device Fuse or Circuit Breaker (C.B.) in Amperes				
		Cougar	Fairlane	Falcon/Maverick	Montego	Mustang
Ash tray light	Fuse panel	SFE 4	SFE 4	SFE 4	SFE 4	SFE 4
Automatic seat back latch release			AGC or SFE 20			
Back-up light		AGC or SFE 15	AGC or SFE 15	AGC or SFE 15	AGC or SFE 20	AGC or SFE 15
Cigar lighter		SFE 14	AGC or SFE 20	AGC or SFE 20	AGC or SFE 20	SFE 14
Clock feed		SFE 14	AGC or SFE 20	AGC or SFE 20	AGC or SFE 20	SFE 14
Convertible top feed	Attached to starting motor relay	20 C.B.	20 C.B.		20 C.B.	20 C.B.
Convertible top motor	Integral part of motor	C.B.	C.B.	C.B.	C.B.	C.B.
Courtesy lights C pillar						
Cargo						
Console (Fairlane and Montego only) door	Fuse panel	SFE 14	SFE 14	SFE 14	SFE 14	SFE 14
Dome						
Glove compartment						
Luggage compartment						
Map light						
Emergency flashers	Fuse panel	15 C.B. ⑦	AGC or SFE 20	AGC or SFE 20	AGC or SFE 20	AGC or SFE 20
Emission control and/or throttle solenoid		SFE 14	SFE 14		SFE 14	SFE 14
Engine compartment light	Cartridge in feed line	SFE 7.5			SFE 7.5	SFE 7.5
Headlights	In headlight switch	18 C.B.	18 C.B.	12 C.B.	18 C.B.	12 C.B.
Heater and defroster	Fuse panel	SFE 14	SFE 14	SFE 14	SFE 14	SFE 14
Heater and defroster with air conditioner	Fuse panel	8 AG or AGX 30	8 AG or AGX 30 ④	8 AG or AGX 30	8 AG or AGX 30	8 AG or AGX 30
Horns	In headlight switch	15 C.B.	15 C.B.	15 C.B.	15 C.B.	15 C.B.
Indicator lights (convenience control panel) door ajar	Fuse panel	AGC or SFE 15				
Dual brake	Fuse panel		SFE 14	SFE 14⑤	SFE 14	
Engine temperature	Fuse panel		SFE 14	SFE 14⑤	SFE 14	
Low fuel	Fuse panel	AGC or SFE 15				
Oil	Fuse panel		SFE 14	SFE 14⑤	SFE 14	
Seat belt (instrument or convenience control panel)	Fuse panel	AGC or SFE 15	SFE 14③	SFE 14①	AGC or SFE 20	AGC or SFE 15
Instrument panel lights	Fuse panel					
Clock light	Fuse panel	SFE 4	SFE 4	SFE 4	SFE 4	SFE 4
Heater and A/C controls	Fuse panel					
Radio	Fuse panel					
Windshield wiper/washer controls	Fuse panel		SFE 4		SFE 4	
License lights						
Marker lights	In headlight switch	15 C.B.	15 C.B.	15 C.B.	15 C.B.	15 C.B.
Parking lights						
Parking brake indicator light	Fuse panel					AGC or SFE 15
Power backlight	Attached to starting motor relay		20 C.B.	20 C.B.⑤	20 C.B.	
Power backlight motor	Integral with motor		C.B.	C.B. ⑤	C.B.	
Power backlight relay feed	Fuse panel		AGC or SFE 20	AGC or SFE 20⑤		
Power seats	On starter relay				20 C.B.	
Power seats motors	Integral part of motor				C.B.	
Power tailgate	Attached to starting motor relay		20 C.B.			
Power tailgate motor	Integral part of motor		C.B.			
Power window	Attached to starting motor relay	20 C.B.	20 C.B.		20 C.B.	

Distributor

References

There are three different types of distributor used, the vacuum advance, as used with 6-cylinder engines, the dual advance distributor, as used with the V8 standard production engines and the centrifugal advance distributor, used on some of the high performance engines.

In 1968 all models adopted the dual advance distributor to provide more accurately timed ignition and cleaner, smog-free exhaust. Some of these distributors have dual diaphragm vacuum advance mechanisms. These have two vacuum lines to the distributor bellows, one to advance the timing during high speed road use, and one to retard the timing at idle. On engines using dual advance distributors before 1967, it was customary to check initial advance with a timing light and with the vacuum line connected. Since 1967 any attempt to check the timing without disconnecting and plugging the vacuum lines will give inaccurate readings.

Distributor installation—6 cyl.
(© Ford Motor Co.)

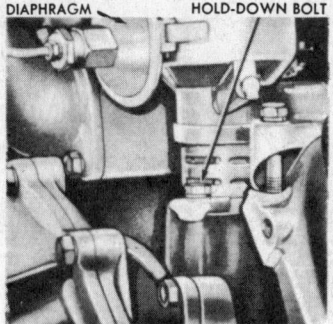

Distributor installation—V8
(© Ford Motor Co.)

Detailed information on distributor drive, direction of distributor rotation, cylinder numbering, firing order, point gap, point dwell, timing mark location, spark plugs, spark advance, ignition resistor location, and idle speed will be found in the Specifications table.

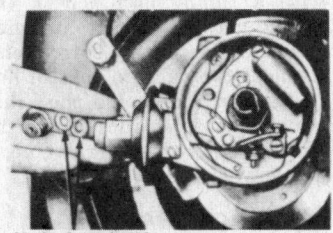

Vacuum advance unit
(© Ford Motor Co.)

6-Cylinder Vacuum Advance

Ignition timing changes are entirely satisfied by the action of the breaker plate. The position of the plate is controlled by a vacuum-actuated diaphragm working against the tension of two accurately calibrated breaker plate springs. The diaphragm moves the breaker plate counterclockwise to advance the spark. The springs tend to counteract this movement to return timing to a retarded position. Cam and rotor rotation are clockwise as viewed from the top.

Dual Advance

The dual advance distributor has two independently operated spark-timing control systems. A governor-type and a vacuum-type control are used on each distributor of standard production engines. Centrifugal weights cause the cam to advance or rotate ahead, relative to the distributor shaft.

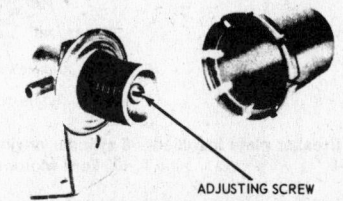

Distributor vacuum advance control valve—cover removed
(© Ford Motor Co.)

The vacuum control mechanism operates through a spring loaded diaphragm and movable breaker plate, about the same as the vacuum advance distributor.

The 289, high performance (4-bbl.

Distributor vacuum control valve
(© Ford Motor Co.)

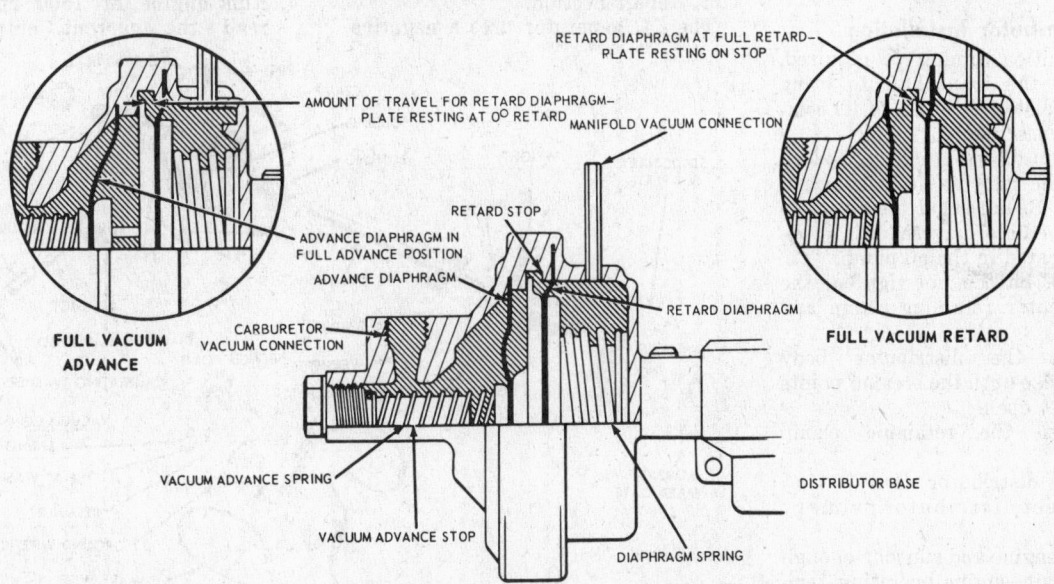

Dual diaphragm vacuum advance mechanism (© Ford Motor Co.)

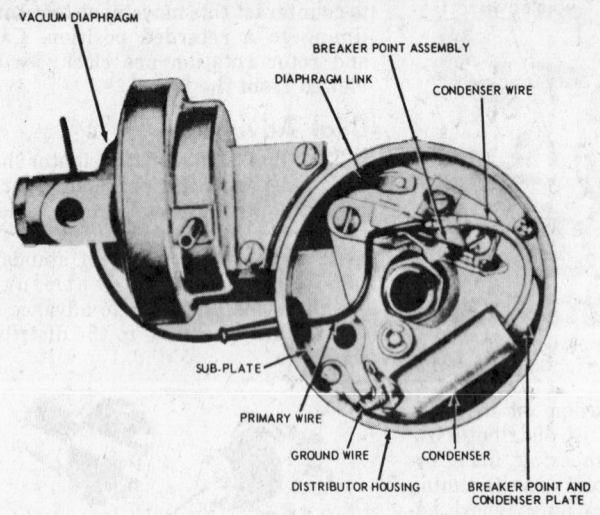

Breaker plate installed—6 cylinder engine, dual diaphragm distributor (© Ford Motor Co.)

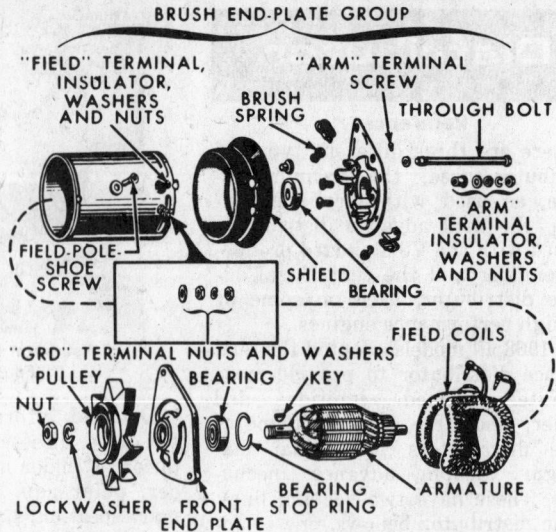

Exploded view of generator (© Ford Motor Co.)

carb.) engine is equipped with centrifugal advance only.

Distributor Removal

1. Remove distributor cap. Disconnect the primary wire at the coil and the vacuum control line at the distributor.
2. Scribe a mark on the distributor body, showing position of the rotor. Then, scribe another mark on the distributor body and engine block, showing the position of the body in the block. These marks can be used to advantage when reassembling the distributor in an undisturbed engine.
3. Remove the screw, lockwasher and hold-down clamp. Pull the distributor out of the block. Do not rotate crankshaft while distributor is out of block because it will then be necessary to re-time ignition.

Distributor Installation

1. If ignition timing is required, rotate the crankshaft to bring No. 1 piston to T.D.C. of its compression stroke.
2. Position distributor in the block with the rotor at No. 1 firing position. Be sure that the oil pump intermediate driveshaft is properly seated in the oil pump.
3. Install, but do not tighten, the distributor retaining clamp and screw.
4. Rotate the distributor body clockwise until the breaker points start to open.
5. Tighten the retaining clamp screw.
6. Install distributor cap.
7. Connect distributor primary wire.
8. Start engine and run long enough to obtain engine operating temperature.
9. Idle engine to 500 rpm. Then,

with a timing light, check the timing marks at the front pulley and make necessary corrections.
10. Connect the vacuum control line to the distributor and check advance characteristics with the timing light when the engine is accelerated.

V8 Centrifugal Advance

Follow procedure as in the Dual Advance type, eliminating the vacuum control items.

Generator, Regulator

The charging system may consist of a DC generator or the newer AC unit. More detailed information may be obtained on these systems from the Unit Repair Section.

The DC generator uses a negative ground system. Output is controlled by a regulator that is connected between the generator armature and the field. The field is grounded internally.

The armature shaft is supported on both ends by permanently lubricated ball bearings that fit into the end plates.

DC Generator Test (on the Car)

1. Disconnect regulator armature and field wires at the generator.
2. Connect a jumper wire from the generator armature terminal to the generator field terminal and the positive lead of a 0-50 ammeter to the generator armature terminal.
3. Start engine. While it is idling, connect the ammeter negative lead to the positive terminal of the battery.
4. Run engine at 1500 rpm and read the current output on

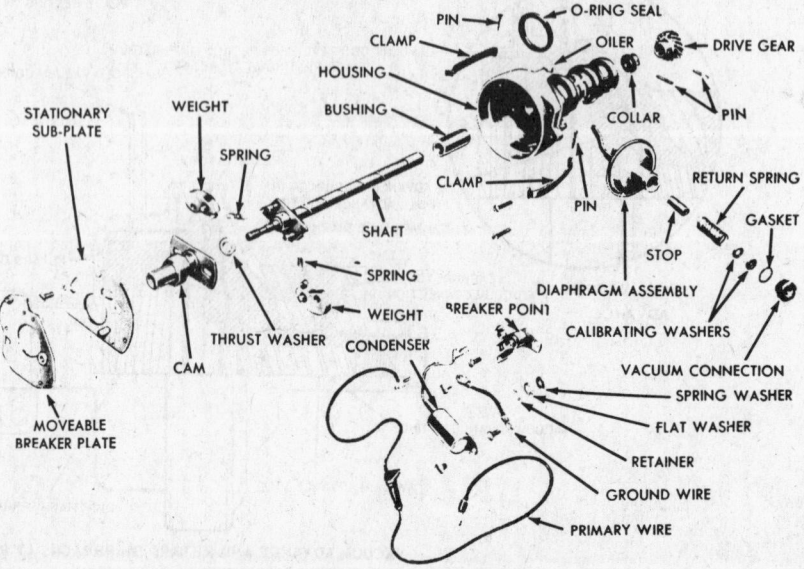

Typical dual advance distributor (© Ford Motor Co.)

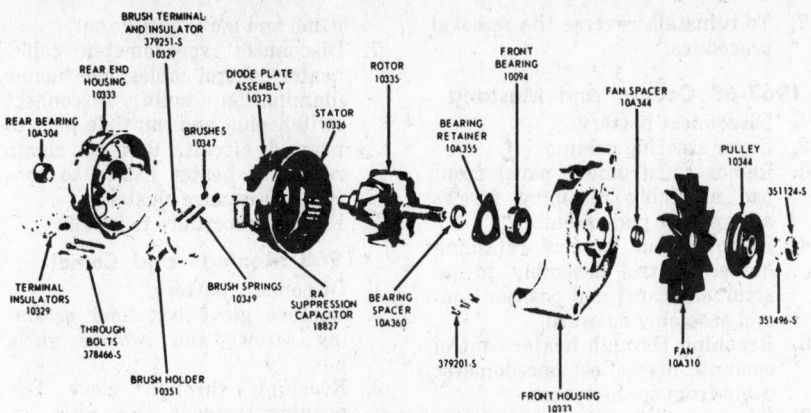

Exploded view of alternator (© Ford Motor Co.)

the ammeter. Generator output should reach or exceed 30 amperes.

NOTE: disconnect test leads as soon as test is completed to prevent overheating the generator. Then stop the engine.

Generator Removal

1. Disconnect all wires from generator.
2. Remove generator attaching bolts, then remove the generator.

Generator Installation

1. Clean mating surfaces of generator frame and mounting bracket.
2. Install generator in the mounting bracket with the two pivot bolts and lock washers.
3. Install the generator belt, and the adjustment arm to generator bolt. Adjust the belt tension and tighten all bolts securely.
4. Connect the armature, field, and ground leads to the generator terminals.
5. Start the engine and check generator operation.

Polarizing the DC Generator

To polarize a DC generator on the car, disconnect the field wire and the battery wire from the regulator. With the engine turned off, momentarily connect the two wires.

NOTE: do not polarize a generator by any method that applies battery voltage to the field terminal of the regulator, such as shorting from the battery terminal to the field terminal of the regulator. Connecting a jumper wire directly from the battery to the generator field terminal will cause regulator damage.

Caution Under no circumstances should the AC generator be polarized.

AC Generator

The AC generator is covered in the Unit Repair Section.

Caution Since the AC generator and regulator are designed for use on only one polarity system, the following precautions must be observed:

1. The polarity of the battery, generator and regulator must be matched and considered before making any electrical connections in the system.
2. When connecting a booster battery, be sure to join the negative battery terminals together and the positive battery terminals together.
3. When connecting a charger to the battery, connect the charger positive lead to the battery positive terminal. Connect the charger negative lead to the battery negative terminal.
4. Never operate the AC generator on open circuit. Be sure that all connections in the circuit are clean and tight.
5. Do not short across or ground any of the terminals on the AC generator.
6. Do not attempt to polarize the AC generator.
7. Do not use test lamps of more than 12 volts for checking diode continuity.
8. Avoid long soldering times when replacing diodes or transistors.

Prolonged heat is damaging to these units.

9. Disconnect the battery ground terminal when servicing any AC system. This will prevent the possibility of accidental reversing of polarity.

Starter

The starter is a four-brush, series-parallel wound unit. The circuit is completed by means of a relay controlled switch which is part of the ignition switch.

Removal and Installation

Due to interference of the exhaust inlet pipe on some models, the steering idler arm must be lowered to provide clearance for starter removal.

1. Disconnect the starter cable at the starter terminal, remove the flywheel housing to starter retaining screws. Remove the starter assembly and the rubber dust ring.
2. Position the rubber dust ring on the flywheel housing.
3. Position the starter assembly to the flywheel housing, and begin on the starter retaining screws. On a car with an automatic transmission, the transmission dipstick tube bracket is mounted under the starter side mounting bolt. Snug all bolts, then tighten to 15 ft. lbs. tightening the middle bolt first.

Instruments

References

For detailed information on gauges and indicator lights, see the Unit Repair Section.

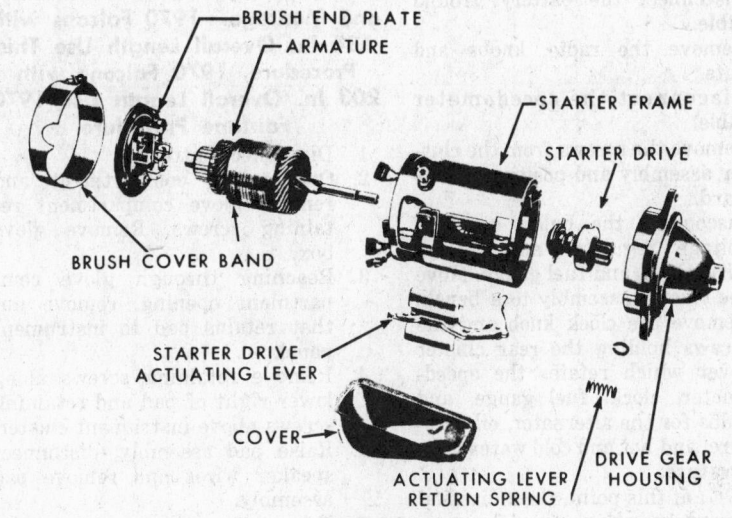

Starter motor
(© Ford Motor Co.)

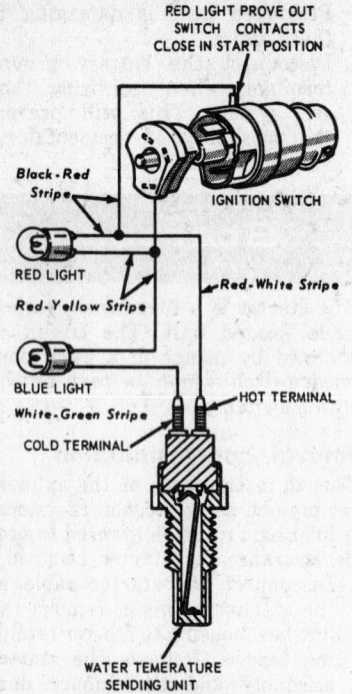

RED LIGHT PROVE OUT
SWITCH CONTACTS
CLOSE IN START POSITION

Black-Red
Stripe

IGNITION SWITCH

RED LIGHT

Red-Yellow Stripe

Red-White Stripe

BLUE LIGHT

White-Green Stripe

HOT TERMINAL

COLD TERMINAL

WATER TEMERATURE
SENDING UNIT

Temperature gauge circuit
(© Ford Motor Co.)

Cluster R & R

1964-67 Falcon and Comet and 1966 Mustang

1. Disconnect the battery cable.
2. Disconnect speedometer cable.
3. Remove the screws retaining the instrument cluster assembly to the instrument panel and tilt the cluster forward.
4. Disconnect the wiring and the bulb sockets. Remove the cluster assembly.

NOTE: all individual instruments may be removed and serviced at this time.

5. To reinstall, reverse the removal procedure.

1966-67 Fairlane

1. Disconnect the battery ground cable.
2. Remove the radio, knobs and nuts.
3. Disconnect the speedometer cable.
4. Remove the screws from the cluster assembly and position it outward.
5. Disconnect the bulbs, constant voltage regulator and ground wire, clock, and fuel gauge. Move the cluster assembly to a bench.
6. Remove the clock knob and the screws holding the rear cluster cover which retains the speedometer, clock, fuel gauge, and bulbs for the alternator, oil pressure, and hot and cold water temperature.

NOTE: at this point, the individual instruments may be removed for servicing.

7. To reinstall, reverse the removal procedure.

1967-68 Cougar and Mustang

1. Disconnect battery.
2. Cover steering column.
3. Remove instrument panel front pad assembly retaining screws and remove pad assembly.
4. Remove four screws retaining heater control assembly to instrument panel and position control assembly outward.
5. Reaching through heater control opening, disconnect speedometer cable from speedometer.
6. Remove three ash tray receptacle retaining screws.
7. Disconnect cigar lighter element wiring connector and remove ash tray receptacle.
8. Reaching through ash tray opening, remove nut and washer which retain inboard end of instrument cluster to instrument panel.
9. Remove seven external screws retaining instrument cluster to instrument panel.
10. Position cluster assembly to outward, disconnect two multiple connectors and remove instrument cluster. Reverse procedure to install.

1968 Fairlane

1. Disconnect battery.
2. Remove instrument panel cover assembly.
3. Remove right instrument panel shield assembly retaining screws and remove shield.
4. Remove five screws retaining cluster assembly and move cluster out.
5. Disconnect speedometer, tachometer, if so equipped, and multiple plug from printed circuit and remove cluster.
6. Reverse procedure to install.

1968 Falcon and 1969 Fairlane and Montego—1970 Falcons with 184 In. Overall Length Use This Procedure, 1970 Falcons with 203 In. Overall Length Use 1970 Fairlane Procedure

1. Disconnect battery.
2. Open glove compartment and remove glove compartment retaining screws. Remove glove box.
3. Reaching through glove compartment opening, remove nut that retains pad to instrument panel.
4. Remove retaining screws along lower right of pad and retaining screws above instrument cluster.
5. Raise pad assembly, disconnect speaker wires and remove pad assembly.
6. Remove five screws retaining instrument cluster to instrument

panel and move cluster out.
7. Disconnect speedometer cable, heater control cables, and heater illumination bulb. Disconnect switch plug and multiple plug to printed circuit. Remove clamp retaining heater cables to control and remove cluster.
8. Reverse procedure to install.

1968 Montego and Comet

1. Disconnect battery.
2. Remove glove box liner retaining screws and remove glove box.
3. Reaching through glove box opening, remove two nuts retaining pad to panel.
4. Remove two pad retaining screws from top of instrument cluster.
5. Remove seven pad retaining screws from along bottom of pad.
6. Lift up pad and disconnect radio speaker wires and clock wires. Remove pad.
7. Remove eight screws retaining cluster to instrument panel.
8. Move cluster out and disconnect speedometer cable, multiple plug to cluster, multiple plug to convenience lights, if so equipped and heater control cables and switch.
9. Disconnect clock, if so equipped, and remove cluster. Reverse procedure to install.

1969-70 Cougar and Mustang

1. Disconnect battery.
2. Remove cluster opening finish panel from pad assembly below instrument cluster (two screws).
3. Remove right and left lower end mouldings for access to pad retaining screws at lower ends of instrument panel.
4. Remove three pad retaining screws from top inner edge of pad to pad support.
5. Remove two retaining screws from right and left lower pad end.
6. Remove three retaining screws from lower right pad to instrument panel.
7. Pull pad assembly back, disconnect clock and courtesy light wires behind right side of pad and remove pad assembly. On XR7 models, disconnect multiple connector behind center of pad before removing pad.
8. Remove six screws retaining cluster to instrument panel and withdraw cluster slightly.
9. Disconnect multiple plug to printed circuit, and tachometer plug if so equipped.
10. Disconnect speedometer cable by pressing on flat surface of plastic connector and pulling cable away from speedometer.

11. Remove cluster. Reverse procedure to install.

1969-70 Maverick

1. Disconnect battery.
2. From under instrument panel disconnect speedometer cable by pressing flat surface of plastic connector and pulling cable away from head.
3. Remove two retaining screws at top of cluster and swing cluster down from panel.

4. Disconnect multiple connector plug from printed circuit at back of cluster.
5. Remove cluster by disengaging brackets on cluster lower edge from slots in panel.

1970 Fairlane

1. Disconnect battery.
2. Remove one retaining screw from lower edge of dash pad at left of instrument cluster.
3. Remove five pad to instrument panel screws across bottom edge of pad to right of cluster.
4. Pull pad and retainer assembly free from clips on instrument panel.
5. Disconnect speaker and remove pad.
6. Remove four screws that retain cluster to panel and move cluster part way out of panel.
7. Disconnect speedometer cable, multiple plug to printed circuit and feed plug to tachometer or clock, if so equipped.
8. Disengage three wiring harness connected light bulb and socket assemblies from their receptacles.
9. Disconnect cable and five vacuum hoses from heater control, and feed plug to heater control switch and connector to heater control light.
10. Remove cluster. Reverse procedure to install.

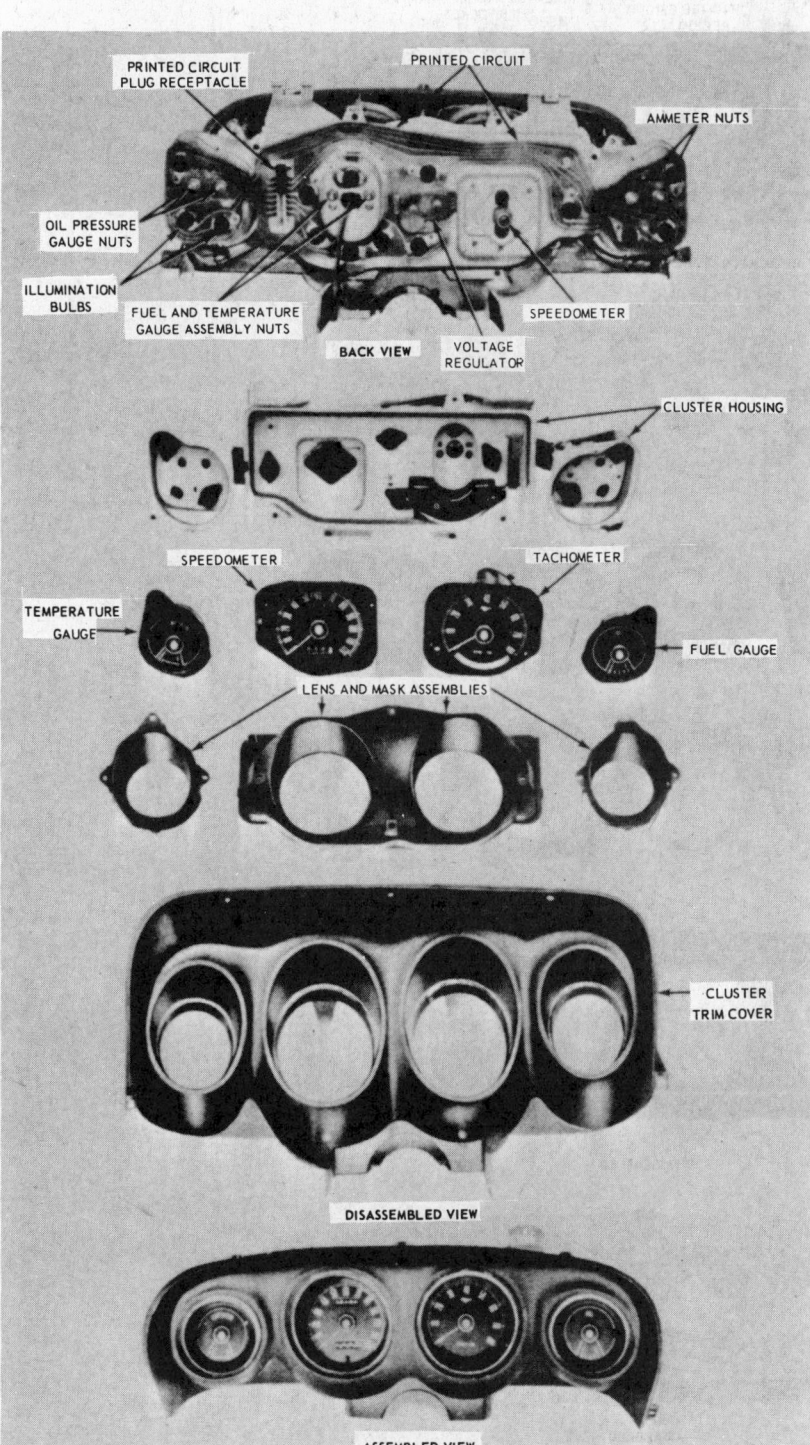

1969-70 Mustang instrument cluster—typical (© Ford Motor Co.)

1970 Montego and Cyclone

1. Disconnect battery.
2. Remove glove box for access and remove heater control panel by removing two retaining screws.
3. Disconnect two wires to light bulbs in heater control.
4. Remove two screws and left finish panel from pad.
5. Remove four retaining screws at lower right side and one screw at far left side of instrument panel pad.
6. Pull pad and retainer assembly free from clips on instrument panel.
7. On Cyclone Spoilers and other models with auxiliary instrument clusters, disconnect feed plugs to auxiliary cluster and tachometer.
8. Disconnect radio speaker and remove pad.
9. Remove four cluster to instrument panel retaining screws and move cluster part way out of panel.
10. Disconnect speedometer cable and cluster feed plug from its receptacle in printed circuit.
11. Disconnect electrical plug to clock and remove cluster. Reverse procedure to install.

1970 Cyclone Spoiler Auxiliary Instrument Cluster

This item is also optional in other models.

1. Perform steps 1-9 in Instrument Cluster removal.
2. Remove four cluster to pad retaining nuts from mounting studs.
3. Remove auxiliary cluster. Reverse procedure to install.

Speedometer Head R & R

1964-67 Falcon and Comet and 1966 Mustang

1. Disconnect battery.
2. Disconnect speedometer cable at the head.
3. Remove instrument cluster assembly.
4. Remove six screws holding the instrument cluster back plate assembly and remove the back plate.
5. Remove the speedometer head attaching screws and remove head.
6. Install in reverse of removal procedure.

1966-67 Fairlane

1. Disconnect battery.
2. Remove radio knobs, mounting nuts and radio.
3. Disconnect speedometer cable.
4. Remove the screws holding the cluster assembly to the instrument panel. Position the cluster assembly outward.
5. Disconnect bulbs, constant voltage regulator and its ground wire, the clock and fuel gauge.
6. Remove the cluster assembly.
7. Remove the clock knob and the screws holding the rear cluster cover.

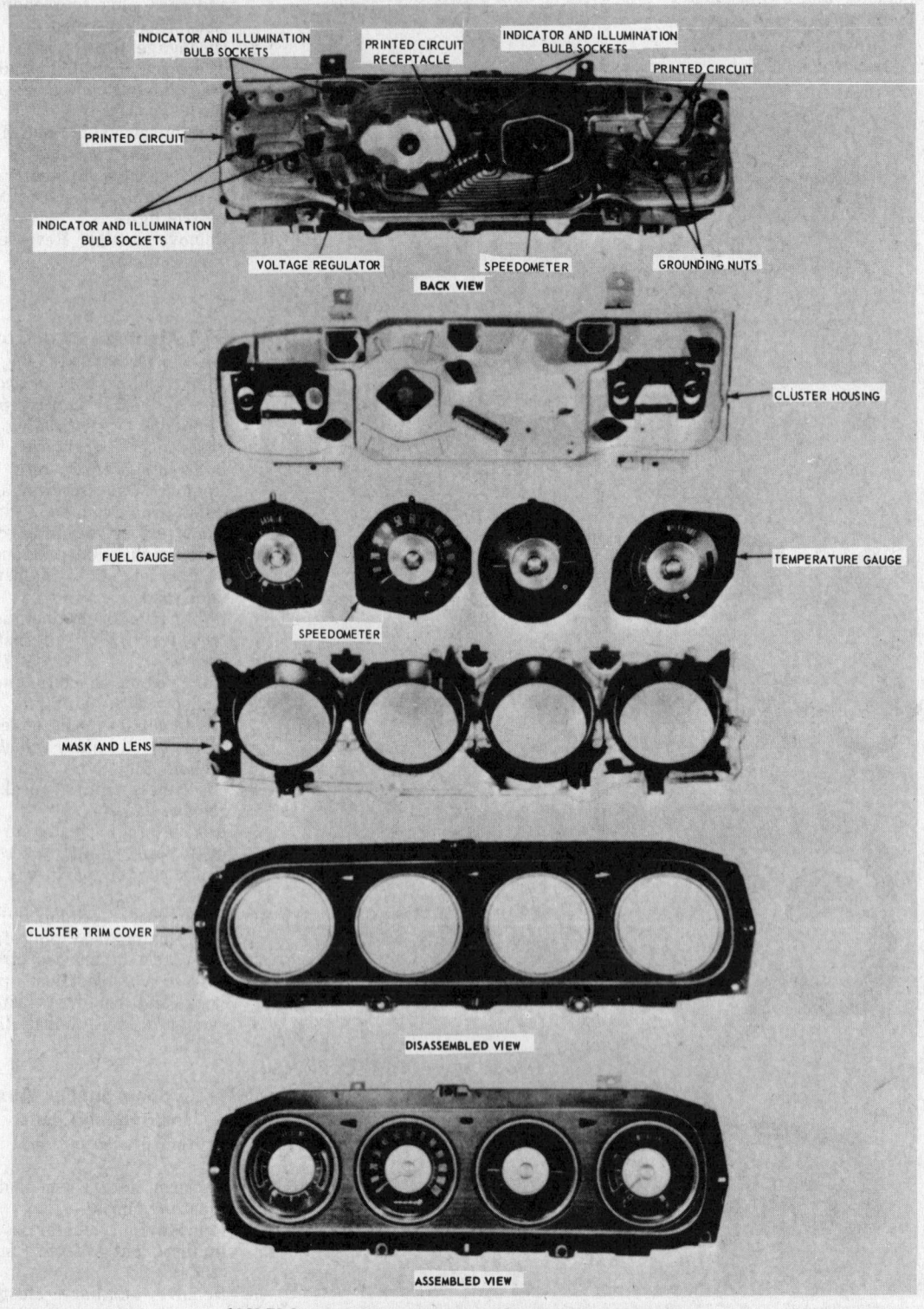

1969-70 Cougar instrument cluster—typical (© Ford Motor Co.)

8. Remove the screws holding the speedometer head to the cover and the four rubber insulators. Remove the speedometer head.
9. Install by reversing above procedure.

1967-68 Cougar and Mustang

1. Disconnect battery.
2. Remove instrument cluster.
3. Remove nine rear cluster housing retaining screws and remove rear cluster housing and gauges.
4. Remove two speedometer retaining nuts and remove speedometer. Reverse procedure to install.

1968-69 Fairlane

1. Disconnect battery.
2. Remove instrument cluster.
3. Remove eight button clips retaining lens and mask to cluster. Oil and alternator finish covers and four rubber spacers will come off with mask and lens.
4. Remove two screws retaining speedometer and remove speedometer. Reverse procedure to install.

1968-69 Falcon

1. Disconnect battery.
2. Remove instrument cluster.
3. Remove heater control knobs and remove eight screws retaining instrument cluster rear housing and remove rear housing.
4. Remove three screws retaining speedometer and remove speedometer. Reverse procedure to install.

1968-69 Montego and Comet

1. Disconnect battery.
2. Remove instrument cluster.
3. Remove heater control and switch knobs.
4. Remove seven screws retaining back of cluster and remove cluster back.
5. Remove two screws and remove speedometer. Reverse procedure to install.

1969-70 Mustang

1. Disconnect battery.
2. Remove instrument cluster.
3. Separate left and right instrument rear housings from printed circuit by removing three bulbs and two retaining nuts in each housing.
4. Remove center instrument rear housing from cluster by removing four screws. Printed circuit remains attached to housing.
5. Remove two retaining screws and remove speedometer. Reverse procedure to install.

1969-70 Cougar

1. Disconnect battery and remove instrument cluster.

2. Remove mask and lens from cluster.
3. Separate front and rear cluster housing assemblies. Reverse procedure to install.

1969-70 Maverick

1. Disconnect battery.
2. Remove instrument cluster.
3. Remove four retaining screws and separate cluster housing from finish panel.
4. Remove two retaining screws and remove speedometer.
5. Reverse procedure to install.

1970 Fairlane

1. Disconnect battery.
2. Remove instrument cluster.
3. Remove retaining screws at back of cluster.

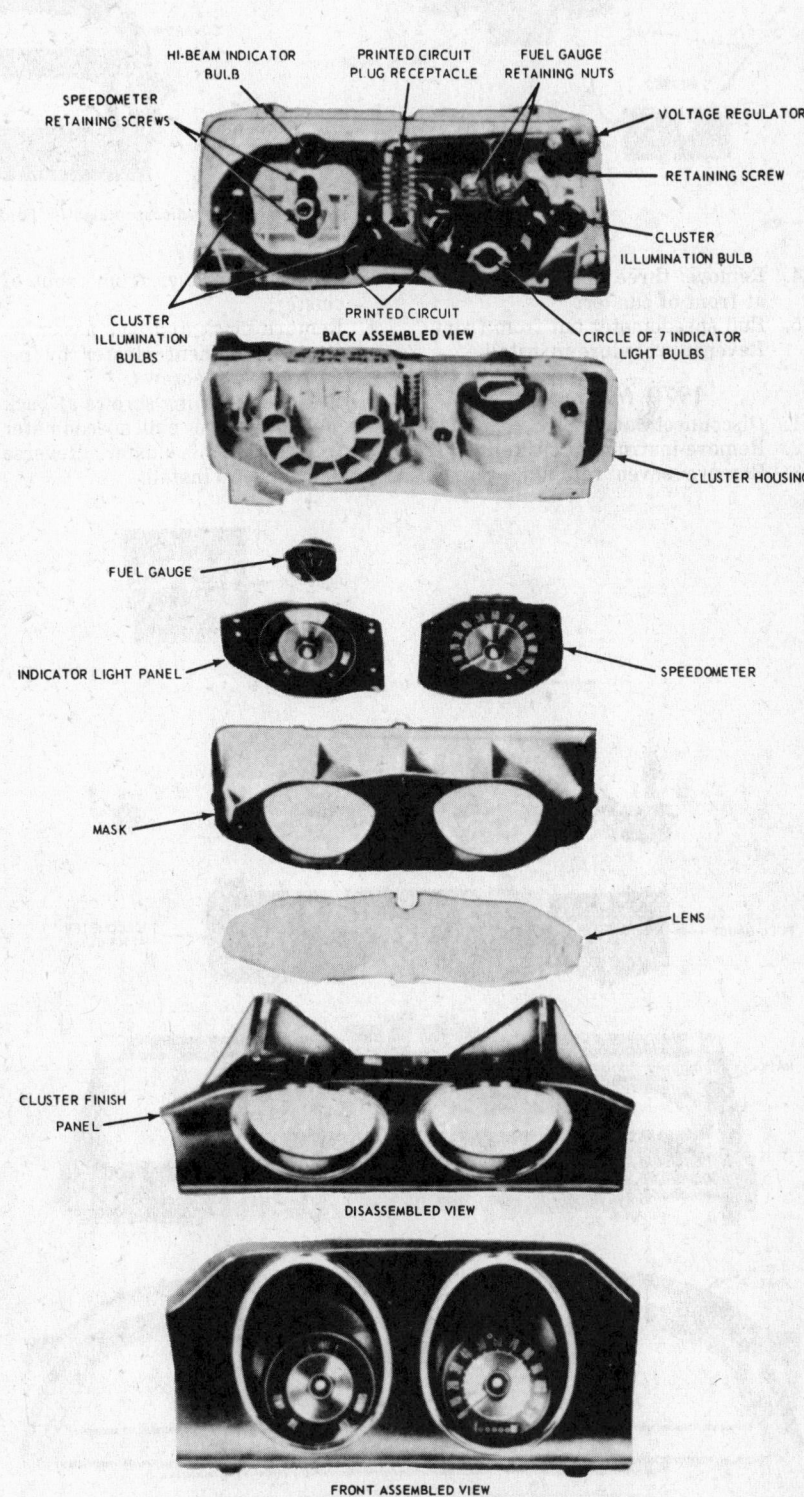

Maverick instrument cluster (© Ford Motor Co.)

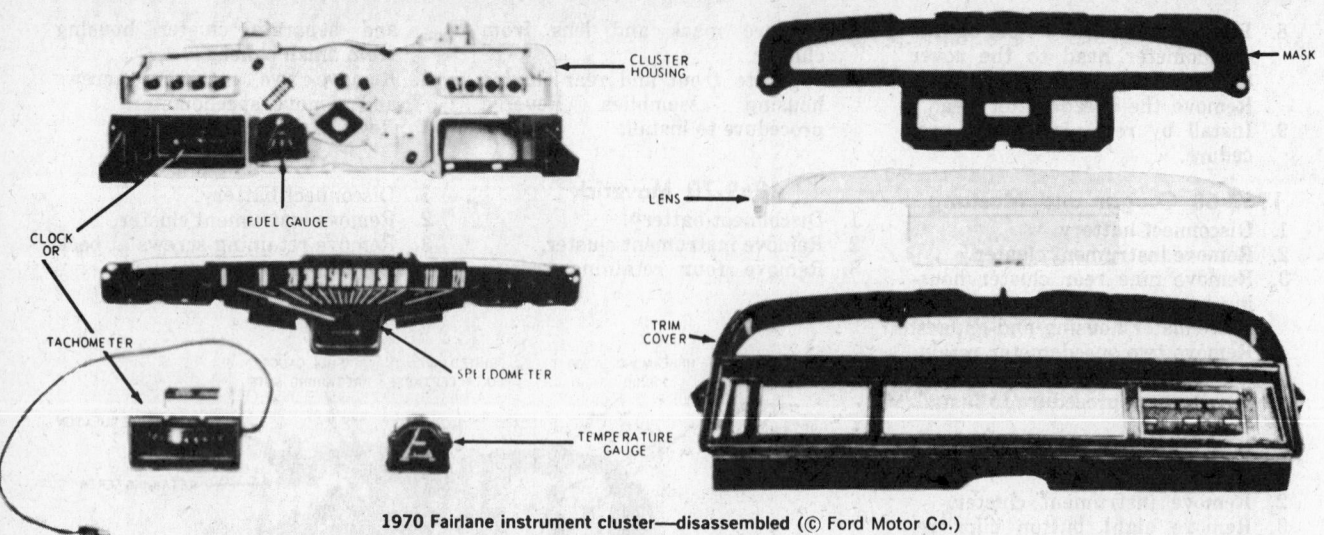

1970 Fairlane instrument cluster—disassembled (© Ford Motor Co.)

4. Remove three retaining screws at front of cluster.
5. Pull speedometer out from front. Reverse procedure to install.

1970 Montego
1. Disconnect battery.
2. Remove instrument cluster.
3. Remove seven retaining screws

and remove cover from front of cluster.
4. Remove mask and lens assembly from instrument cluster by removing eight screws.
5. Remove retaining screws at back of cluster and pull speedometer from front of cluster. Reverse procedure to install.

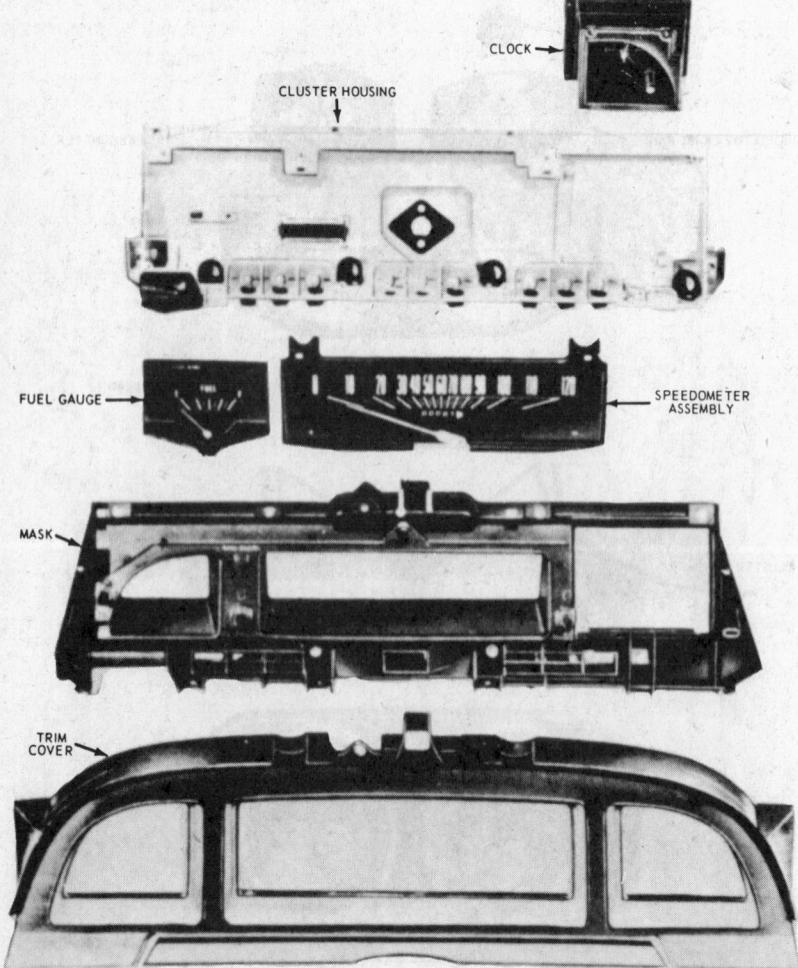

1970 Montego and Cyclone instrument cluster—disassembled (© Ford Motor Co.)

Oil Pressure Gauge
All models except XR7 and Cyclone Spoiler are equipped with a red indicator light on the dash.

To test the oil pressure switch (sending unit) on the engine, turn the ignition on, engine not running. The light should come on. If the indicator does not come on, short the terminal of the oil pressure switch to ground. If the light now comes on, the oil pressure switch is probably defective. If the light still does not come on, the bulb is burned out or the wires from the bulb to the ignition switch and oil pressure switch are bad.

Lock Cylinder Replacement 1964-69
1. Insert key and turn to Acc. position.
2. With stiff wire in hole, depress lock pin and rotate cylinder counterclockwise, then pull out cylinder.

Ignition Switch Replacement 1964-69
1. Remove cylinder as above.
2. Press in on rear of switch and rotate the switch one-eighth turn counterclockwise. Remove the bezel, switch and spacer.
3. Remove nut from back of switch. Remove the accessory and gauge feed wires from accessory terminal. Pull insulated plug from rear of switch.
4. Install in reverse of above.

1970 Ignition Switch
See Lincoln Continental section.

Lighting Switch Replacement
1. Disconnect battery.
2. Remove knob and shaft by pressing release knob button on switch housing and with knob in full on position.

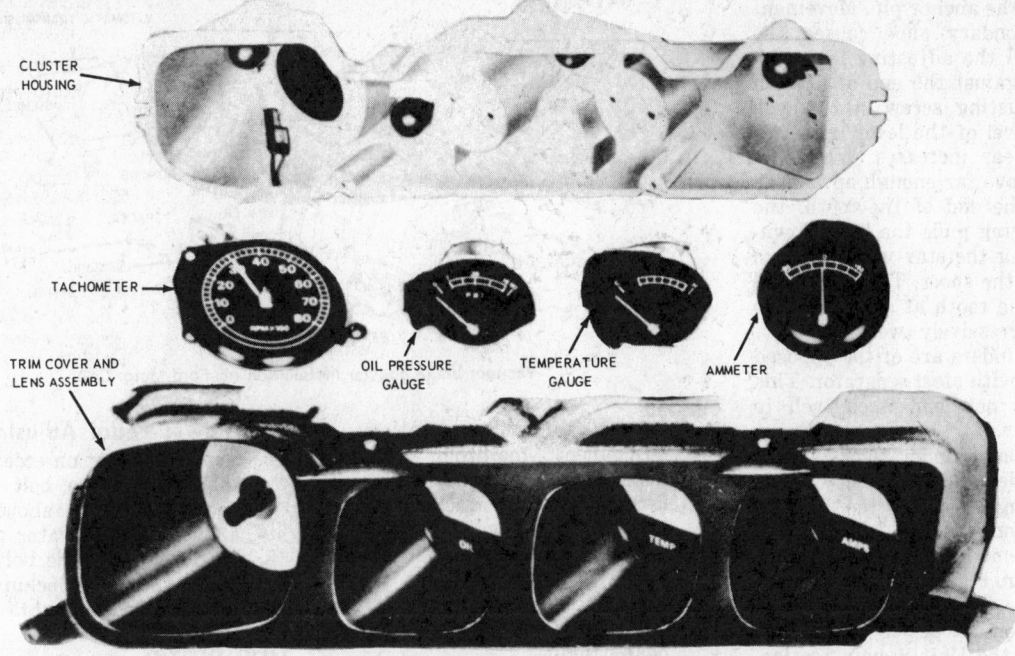

CLUSTER HOUSING

TACHOMETER

TRIM COVER AND LENS ASSEMBLY

OIL PRESSURE GAUGE

TEMPERATURE GAUGE

AMMETER

1970 Cyclone spoiler instrument cluster—disassembled (© Ford Motor Co.)

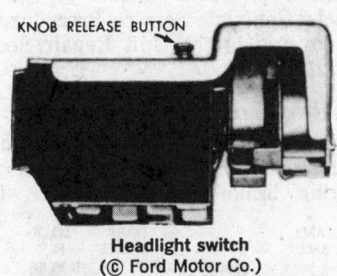

KNOB RELEASE BUTTON

Headlight switch
(© Ford Motor Co.)

3. Remove mounting nut, then switch.
4. Remove junction box from switch.
5. Install in reverse of above.

Neutral Safety and Back-Up Light Switch Assembly
1964-70

Ford small cars throughout this period have used the same neutral switches and back up switches as full sized Fords.

See Ford section for a complete breakdown of years, transmissions, and adjustments.

Back-Up Light Switch— Manual Transmission

The back-up light switch may be located in either one of two places. The back-up light switch location, on cars with column shift selector and linkage controls, is at the bottom of the column.

The back-up light switch location, on cars with consoles and floor shift selector, is on the left side of the transmission back at the shift control bracket.

Brakes

Single-anchor, internal-expanding hydraulic brakes are standard on all models.

An independent, hand-operated parking brake operates the rear wheel brake shoes through a mechanical cable linkage.

Self-Adjusting Brakes

The self-adjusting brake mechanism consists of a cable, cable guide, adjuster lever, and adjuster spring.

The cable is hooked over the anchor pin at the top and is connected to the lever at the bottom. The cable is connected to the secondary brake shoe by means of the cable guide. The adjuster spring is hooked to the primary brake shoe and to the lever.

The automatic adjuster operates only when the brakes are applied while the car is moving rearward.

With the car moving rearward and the brakes applied, the wrapping action of the shoes following the drum forces the upper end of the primary shoe against the anchor pin. Action of the wheel cylinder moves the upper end of the secondary shoe

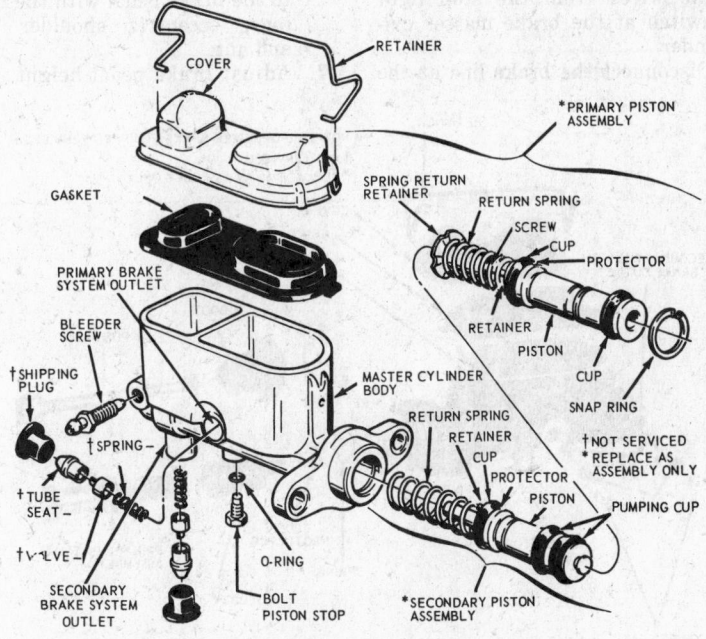

Dual master cylinder—disc brakes (© Ford Motor Co.)

away from the anchor pin. Movement of the secondary shoe causes the cable to pull the adjusting lever upward and against the end of a tooth on the adjusting screw star wheel. Upward travel of the lever increases as lining wear increases. When the lever can move far enough upward to pass over the end of the tooth, the adjuster spring pulls the lever downward causing the star wheel to turn and expand the shoes. The star wheel is turned one tooth at a time as the linings progressively wear.

Wheel cylinders are of the opposed piston type with steel separator. This design does not lend itself well to honing.

The master cylinder consists of a single cylinder and reservoir, mounted on the engine side of the firewall.

Information on brake adjustments, band replacement, bleeding procedure and cylinder reconditioning can be found in the Unit Repair Section.

Information on power brakes can be found in the Unit Repair Section.

Beginning 1967

Since 1967, a tandem-type master cylinder has been used on all models. This design divides the brake hydraulic system into two independent and hydraulically separated halves.

Details and repair procedures on this tandem system may be found in the Unit Repair Section.

Power Brakes

Power Unit Removal

1. Working inside the car below the instrument panel, disconnect booster valve operating rod from the brake pedal assembly.
2. Open the hood, and disconnect the wires from the stop light switch at the brake master cylinder.
3. Disconnect the brake line at the

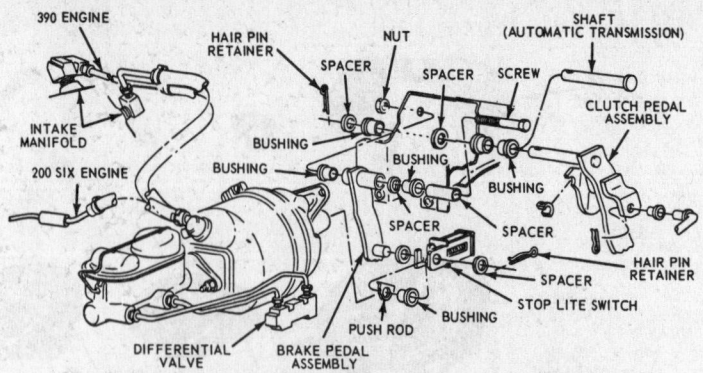

Vacuum brake booster installation (© Ford Motor Co.)

master cylinder outlet fitting.
4. Disconnect manifold vacuum hose from the booster unit.
5. Remove the four bracket-to-dash panel attaching bolts.
6. Remove the booster and bracket assembly from the dash panel, sliding the valve operating rod out from the engine side of the dash panel.

Power Unit Installation

1. Mount the booster and bracket assembly to the dash panel by sliding the valve operating rod in through the hole in the dash panel, and installing the attaching bolts.
2. Connect manifold vacuum hose to the booster.
3. Connect the brake line to the master cylinder outlet fitting.
4. Connect stop light switch wires.
5. Working inside the car below the instrument panel, install the rubber boot on the valve operating rod at the passenger side of the dash panel.
6. Connect the valve operating rod to the brake pedal with the bushings, eccentric shoulder bolt, and nut.
7. Adjust brake pedal height.

Power Pedal Adjustment

1. Loosen locknut on eccentric bolt, then rotate the bolt until the pedal height is about $1\frac{7}{8}$ in. above the accelerator pedal.
2. Hold the eccentric bolt securely and tighten the locknut.
3. Recheck pedal height.

Disc Brakes

Since 1965, disc brakes have been available on front wheels of some models. Complete Service Procedures are covered in the Unit Repair Section.

Parking Brake Adjustment

In most cases, a rear brake shoe adjustment will provide satisfactory parking brake action. However, if

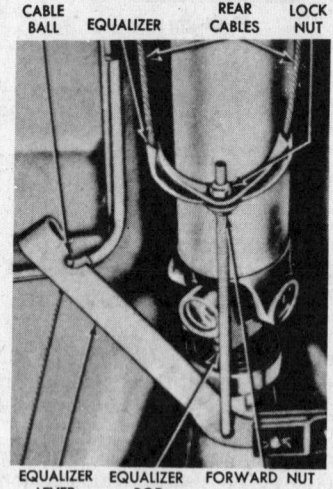

Parking brake linkage
(© Ford Motor Co.)

parking brake cables are excessively loose after releasing the handbrake, proceed as follows:

1. Pull up the handle to the third notch.
2. Loosen locknut on equalizer rod under the car. Then loosen the nut in front of the equalizer, several turns.
3. Turn the locknut forward against the equalizer until the cables are just tight enough to stop forward rotation of the wheels.

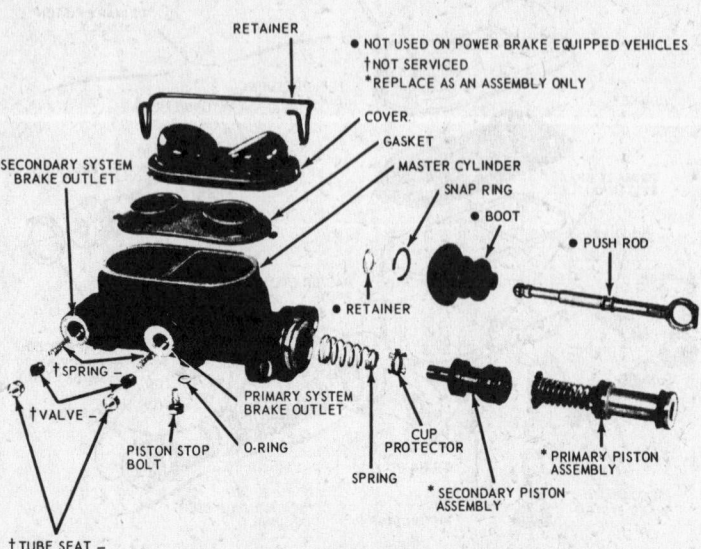

Dual master cylinder—drum brakes (© Ford Motor Co.)

4. When cables are properly adjusted, tighten both nuts against the equalizer.
5. Release the handle and feel for freeness of rear wheels.

Fuel System

References

Data on capacity of the gas tank is in the Capacities table.

Data on fuel pump pressure will be found in the Tune-up Specifications table. Both the above tables can be found in this car section.

Information covering operation and troubles of the fuel gauge will be found in the Unit Repair Section.

Detailed information on the carburetor and how to adjust it as well as the specific heading of the make of carburetor being used on the engine involved is in the Unit Repair Section.

Fuel Pump

A disposable-type, in-line filter is attached to the outlet side of the pump to clean the fuel before entering the carburetor.

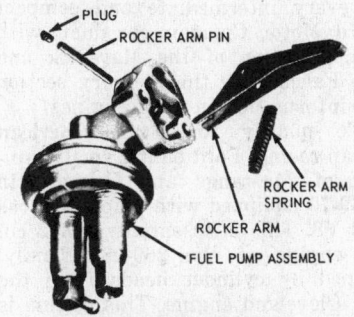

6 cylinder fuel pump
(© Ford Motor Co.)

Exhaust System

References

The exhaust system consists of a muffler inlet pipe, an inlet extension pipe, and a muffler (with integral outlet pipe). The muffler inlet pipe used with V8 engines is the Y-type, in which the inlet pipe from the left exhaust manifold crosses over beneath the transmission. It is welded to the inlet pipe from the right exhaust manifold.

Muffler and Outlet Pipe

Removal

1. Remove inlet extension pipe clamp at muffler.
2. Remove bolts that attach the

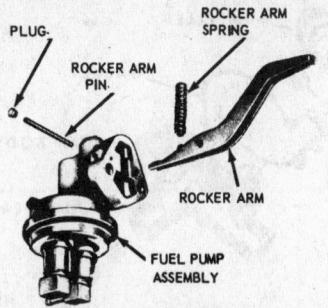

V8 fuel pump
(© Ford Motor Co.)

rear end of the muffler to the frame-mounted bracket.
3. Separate the muffler from the inlet extension pipe and remove the muffler and outlet pipe assembly.

Installation

1. Slide the new muffler and outlet pipe assembly on the inlet extension pipe. Position the inlet extension pipe clamp.
2. Position the muffler and outlet pipe assembly to the frame-mounted bracket and install the retaining bolts. Tighten inlet extension pipe clamp.
3. Start engine and check exhaust system for leaks.

Inlet Pipe

Removal

1. Remove inlet pipe clamp at inlet extension pipe.
2. Remove two nuts and lock washers holding the inlet pipe to the exhaust manifold (both exhaust manifolds on the V8 engines).
3. Pull the inlet pipe/s down and remove the inlet pipe from the inlet extension pipe.

Installation

1. Clean the gasket surfaces of the exhaust manifold/s.

2. Install a new gasket over the studs of the exhaust manifold/s.
3. Slide the new inlet pipe into the inlet extension pipe. Then position the inlet pipe clamp.
4. Position the inlet pipe on the studs of the exhaust manifold/s. Then, install the lock washers and nuts. Tighten the nuts.
5. Tighten inlet pipe clamp.
6. Start engine and check exhaust system for leaks.

Inlet Extension Pipe

Removal

1. Remove muffler and outlet pipe assembly by following directions given in Muffler and Outlet Pipe, Removal.
2. Remove the clamps at the inlet pipe and at the frame-mounted bracket. Remove the inlet extension pipe.

Installation

1. Slide the new inlet extension pipe on the inlet pipe.
2. Position the clamps at the inlet pipe and at the frame mounted bracket. Tighten the clamps.
3. Install the muffler and outlet pipe assembly by following Steps 1 through 3 of Muffler and Outlet Pipe, Installation.

Cooling System

Both the 6 cylinder and V8 engines employ cooling systems that are basically similar.

In the 6-cylinder engine, coolant flows from the cylinder head, past the thermostat (if it is open) and into the radiator upper tank. In the V8 engine, coolant from each cylinder head flows through water passages in the intake manifold, then past the

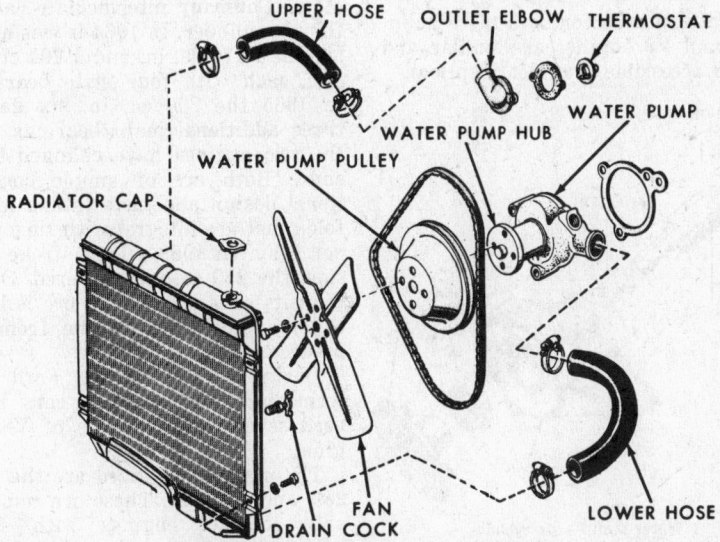

Radiator and related parts (© Ford Motor Co.)

thermostat (if it is open) and into the radiator upper tank.

The standard thermostat operating temperature is 185°-192°F. However, a low reading thermostat of 157°-162°F is available for use with non-permanent-type anti freeze solutions.

A single water pump assembly is used. The pump has a sealed bearing integral with the water pump shaft. The bearing requires no lubrication. There is a bleed hole in the water pump housing. This is not a lubrication hole.

Radiator

Removal

1. Drain cooling system.
2. Disconnect upper and lower hoses at the radiator.
3. On automatic transmission-equipped cars, disconnect oil cooler lines at radiator.
4. Remove radiator attaching bolts and lift out the radiator.

Installation

1. If a new radiator is to be installed, transfer the petcock from the old radiator to the new one. On cars equipped with automatic transmissions, transfer the oil cooler line fittings from the old radiator to the new one.
2. Position the radiator and install, but do not tighten, the radiator support bolts. On cars equipped with automatic transmissions, connect the oil cooler lines. Then tighten the radiator support bolts.
3. Connect the radiator hoses. Close the radiator petcock. Then fill and bleed the cooling system.
4. Start the engine and bring to operating temperature. Check for leaks.
5. On cars equipped with automatic transmissions, check the cooler lines for leaks and interference. Check transmission fluid level.

Water Pump

Water pumps for both the 6 cylinder and V8 engines are similar and quite accessible, but not identical.

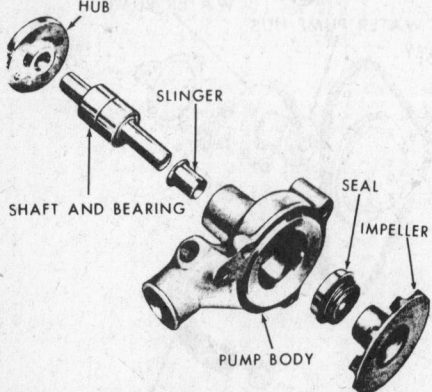

Water pump—six cylinder
(© Ford Motor Co.)

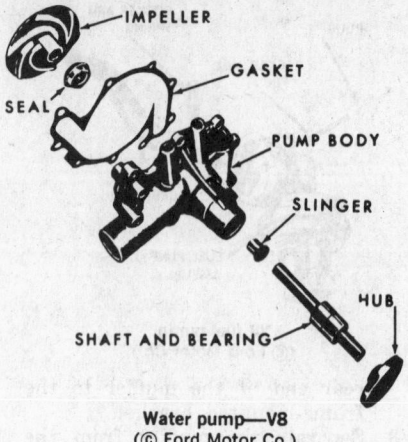

Water pump—V8
(© Ford Motor Co.)

Removal

1. Drain the cooling system.
2. Disconnect lower hose and heater hose at the water pump.
3. Remove the drive belt, fan, fan spacer (if so equipped) and the pulley.
4. Remove the attaching bolts and lift off the pump, gasket, and the timing pointer (from the V8 engine).

Installation

1. Clean mounting surfaces of both cylinder front cover and water pump.
2. Coat a new gasket on both sides, then position the gasket on the cylinder front cover and install the pump. (Position the timing pointer at the two lower mounting holes of V8 engines.)
3. Install pump attaching bolts and torque to 12-15 ft. lbs.

Engine

1964-70

The basic engine in most Ford Motor Company intermediate cars is the six cylinder. In 1964 it was available in a 170 cu. in. and a 200 cu. in. size, each with four main bearings. In 1965 the 200 cu. in. six gained three additional main bearings, and the two engines have changed little since. Both are of simple conventional design and have intake manifolds that are integral with the cylinder head. In 1969 a long stroke version, the 250 cu. in., appeared. Other than its greater power there is little to distinguish this engine from its predecessors.

During this period, Ford intermediate and compact cars have used seven general types of V8 engine.

The most widely used are the 260, 289, and 302 V8s. These are remarkably compact engines with stud-mounted rockers and wedge-shaped combustion chambers. Complete information on repairing these engines is in the Ford-Thunderbird section.

In 1969, Ford Motor Company introduced a longer stroke, higher block version of the 260, 289, 302 engine. This engine is the 351 Windsor engine and features the wedge-shaped combustion chambers and stud-mounted rockers of the small block engine in a new intermediate sized block. See the Ford-Thunderbird section for information on this engine.

In 1970, Ford Motor Company added the 351 Cleveland engine. The 351 Cleveland engine has the same bore and stroke as the Windsor engine, and there most of the resemblance ends. It has different main bearing size, larger valves, smaller plugs, semi-hemispherical combustion chambers and more power. It is used concurrently with the Windsor engine and is found in many of the same models. The Cleveland engine is treated in this section.

In late 1966, Ford installed the 390 cu. in. V8 in The Cyclone GT and Fairlane GT. This V8 was soon joined by other large Y block V8s, the 427 and the 428 Cobra Jet. These big V8s were installed in small numbers in every intermediate and compact Ford Motor Company product, with the exception of the Maverick and the Falcon. See the Mercury section for information on those engines.

To qualify for Trans-American sedan racing, Ford built a small number of Mustangs and Cougars in 1969-70 equipped with a special Boss 302 V8. This is essentially a 302 cu. in. engine of the 260-289 family, topped by cylinder heads from the 351 Cleveland engine. This engine is also treated in this section.

In 1970 some Mercury Montegos and Cyclones and Fairlanes use the 429 V8. This V8 comes in three forms. The first is the 429 4V engine, which is the same as is used in full sized Mercury and Ford cars. The second is the 429 CJ engine which uses stronger rods, big valve heads, smaller 14 mm. plugs, and an ignition governor set at 5,800 rpm. The third is the 429 Super CJ, which is similar to the 429 CJ except for forged pistons, four bolt main caps, solid lifters, and a 6,000 rpm governor. Despite the substantial differences between these engines, they are serviced in the same manner—see the Mercury section.

In 1969 and 1970 Ford released about one thousand Mustangs and Cougars powered by the Boss 429 engine. This engine is based loosely on the 429 engines discussed above, but its features are so individual it must be covered separately. It has modified hemispherical combustion chambers, very large valves and ports, valve

seats canted in two planes, rockers with individual rocker shafts, and O-rings and chevron seals in lieu of head gaskets. The heads are of aluminum, the valves take special valve seals, the main bearings have four bolt main caps, and the spark plugs pass through the rocker covers. In short, this is a very special engine that demands special procedures found in this section.

Exhaust Emission Control

In compliance with anti-pollution laws, the Ford Motor Company has adopted a distributor and a modified carburetor with some engine changes, to reduce terminal exhaust fumes to an acceptable level. This method is known as Ford's Improved Combustion (IMCO) System.

The plan supersedes (in most cases) the previous method used to conform to 1966-67 California laws. The new system phases out (except with stick shift and special purpose engine applications) the thermactor, or afterburner type of exhaust treatment.

The IMCO concept utilizes broader, yet more critical, distributor control through carburetor modification.

Since 1968, all car makers have posted idle speeds and other pertinent data, relative to the specific engine car application in a conspicuous place in the engine compartment.

For details on the IMCO system, consult the Unit Repair Section.

Engine Removal

1. Scribe the hood hinge outline on the under-hood, disconnect the hood and remove.

6 cyl. engine lifting hook
(© Ford Motor Co.)

2. Drain the entire cooling system and crankcase.
3. Remove the air cleaner, disconnect the battery at the cylinder head. On automatic transmission equipped cars, disconnect oil cooler lines at the radiator.
4. Remove upper and lower radiator hoses and remove radiator. If equipped with air conditioning, unbolt compressor and position compressor out of way with refrigerant lines intact. Unbolt and lay refrigerant radiator for-

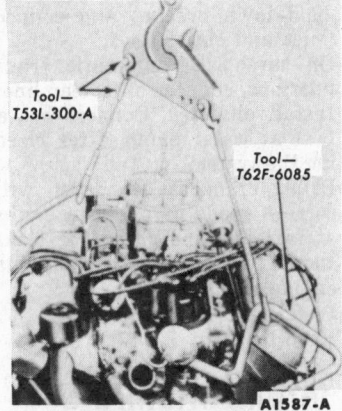

V8 engine lifting brackets and sling
(© Ford Motor Co.)

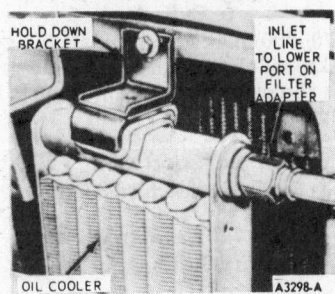

Engine oil cooler
(© Ford Motor Co.)

ward without disconnecting refrigerant lines. On some 428 CJ engines and all 429 Super CJ, Boss 302, and Boss 429 engines disconnect inlet and outlet lines from engine oil cooler, remove hold-down bracket and remove cooler.

5. Remove fan, fan belt and upper pulley.
6. Disconnect the heater hoses at the water pump and the carburetor spacer.
7. Disconnect the generator wires at the generator, the starter cable at the starter, the accelerator rod at the carburetor and, on the 6-cylinder engine, the choke control cable at the carburetor.
8. Disconnect fuel tank line at the fuel pump and plug the line.
9. Disconnect the coil primary wire at the coil. Disconnect wires at the oil pressure and water temperature sending units.
10. Remove the starter and dust seal.
11. On a car equipped with a manual-shift transmission, remove the clutch retracting spring. Disconnect the clutch equalizer shaft and arm bracket at the underbody rail and remove the arm bracket and equalizer shaft.
12. Raise the car. Remove the flywheel or converter housing upper retaining bolts through the access holes in the floor pan.
13. Disconnect the exhaust pipe or pipes at the exhaust manifold.

Disconnect the right and left motor mount at the underbody bracket. Remove the flywheel or converter housing cover.

14. On a car with manual shift, remove the flywheel housing lower retaining bolts.
15. On a car equipped with automatic transmission, disconnect throttle valve vacuum line at the intake manifold, disconnect the converter from the flywheel. Remove the converter housing lower retaining bolts. On a car with power steering, disconnect power steering pump from cylinder head. Put drive belt and wire steering pump out of the way.
16. Lower the car. Support the transmission and flywheel or converter housing with a jack.
17. Attach an engine lifting hook. Lift the engine up and out of the compartment and onto an adequate workstand.

Engine Installation

1. Place a new gasket over the studs of the exhaust manifold/s.
2. Attach engine sling and lifting device. Lift engine from workstand.
3. Lower the engine into the engine compartment. Be sure the exhaust manifold/s is in proper alignment with the muffler inlet pipe/s, and the dowels in the block engage the holes in the flywheel housing.

On a car with automatic transmission, start the converter pilot into the crankshaft.

On a car with manual-shift transmission, start the transmission main drive gear into the clutch disc. If the engine hangs up after the shaft enters, rotate the crankshaft slowly (with transmission in gear) until the shaft and clutch disc splines mesh.

4. Install the flywheel or converter housing upper bolts.
5. Install engine support insulator to bracket retaining nuts. Disconnect engine lifting sling and remove lifting brackets.
6. Raise front of car. Connect exhaust line/s and tighten attachments.
7. Position dust seal and install starter.
8. On cars with manual-shift transmissions, install remaining flywheel housing-to-engine bolts. Connect clutch release rod. Position the clutch equalizer bar and bracket, and install retaining bolts. Install clutch pedal retracting spring.
9. On cars with automatic transmissions, remove the retainer holding the converter in the housing. Attach the converter to the

flywheel. Install the converter housing inspection cover and the remaining converter housing retaining bolts.

10. Remove the support from the transmission and lower the car.

11. Connect engine ground strap and coil primary wire.

12. Connect water temperature gauge wire and the heater hose at coolant outlet housing. Connect accelerator rod at the bellcrank.

13. On cars with automatic transmission, connect the transmission filler tube bracket. Connect the throttle valve vacuum line.

14. On cars with power steering, install the drive belt and power steering pump bracket. Install the bracket retaining bolts. Adjust drive belt to proper tension.

15. Remove plug from the fuel tank line. Connect the flexible fuel line and the oil pressure sending unit wire.

16. Install the pulley, belt, spacer, and fan. Adjust belt tension.

17. Tighten generator adjusting bolts. Connect generator wires and the battery ground cable.

18. Install radiator. Connect radiator hoses. On air conditioned cars, install compressor and refrigerant radiator. On some 428 CJ engines, and all 429 Super CJ, Boss 302, and Boss 429 engines, install engine oil cooler and hold-down bracket and connect inlet and outlet lines.

19. On cars with automatic transmissions, connect oil cooler lines.

20. Install oil filter. Connect heater hose at water pump, after bleeding the system.

21. Bring crankcase to level with correct grade of oil. Run engine at fast idle and check for leaks. Install air cleaner and make final engine adjustments.

22. Install and adjust hood.

23. Road-test car.

Intake Manifold Removal

6 Cylinder

170, 200 and 250 cu. in. sixes have intake manifolds that are integral with the cylinder head and cannot be removed.

260, 289 and 302 V8

See Ford–Thunderbird Section.

390, 427, and 428 V8

See Mercury Section.

429, 429 CJ, and 429 Super CJ V8s

See Mercury Section.

351 Windsor V8

See Ford–Thunderbird Section.

302 Boss and 351 Cleveland V8s

1. Drain cooling system and remove air cleaner. On Boss 302 engine, disconnect Thermactor air hose from check valve at rear of intake manifold and loosen hose clamp at hose bracket. Remove air hose and Thermactor air by-pass valve from bracket and position out of way.

2. Disconnect accelerator linkage and accelerator downshift linkage, if so equipped, and position out of way. On Boss 302, disconnect choke cable from carburetor.

3. Disconnect high tension lead and wires from coil. Disconnect engine wire loom and position out of way.

4. Disconnect spark plug wires from spark plugs by grasping, twisting, and pulling molded cap only. Remove distributor cap and wire assembly.

5. Remove carburetor fuel inlet line.

6. Disconnect distributor vacuum hoses from distributor. Remove hold-down bolt and remove distributor.

7. Disconnect radiator upper hose from coolant outlet housing and disconnect temperature sender wire.

8. Loosen clamp on water pump by-pass hose at coolant outlet housing and slide hose off outlet housing.

9. Disconnect crankcase vent hose (PCV) at rocker cover.

10. If vehicle is air conditioned, remove compressor to intake manifold brackets.

11. Remove intake manifold and carburetor as an assembly. Discard all used gaskets and clean all mating surfaces.

12. Reverse procedure to install.

Boss 429 V8

1. Disconnect battery.

2. Drain cooling system.

3. Disconnect heater hose from manifold.

4. Disconnect positive crank case ventilation (PCV) hose from right-hand rocker cover. Disconnect and tag all vacuum lines from rear of intake manifold.

5. Twist and pull the molded spark plug wire cap from each plug. Remove plug wires from brackets on rocker covers.

6. Disconnect high tension lead from coil and remove distributor cap and wires from distributor as an assembly.

7. Disconnect accelerator linkage from carburetor. Remove bolts that attach accelerator linkage bellcrank. Disconnect linkage spring and position linkage to one side.

8. Disconnect all distributor vac-

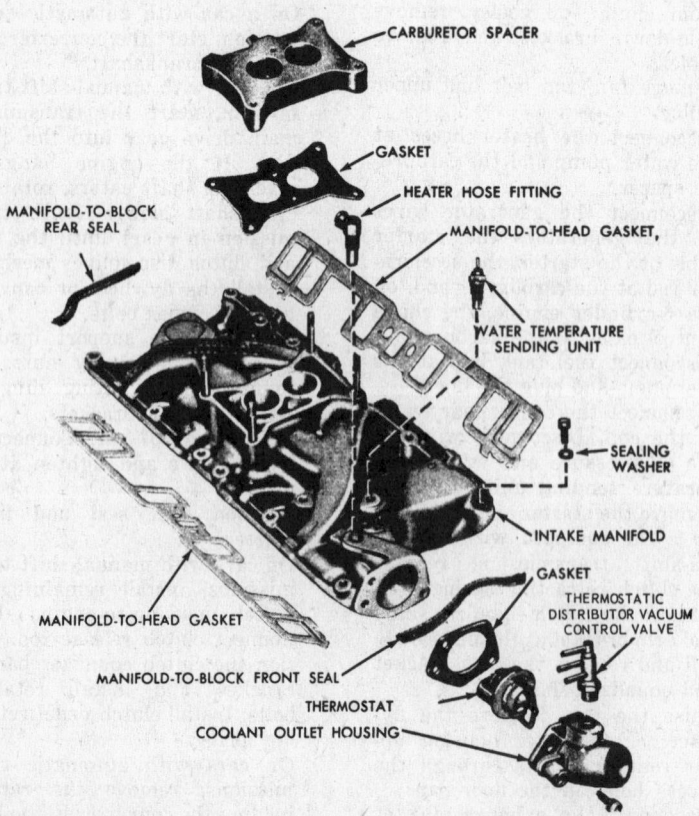

CARBURETOR SPACER

GASKET

HEATER HOSE FITTING

MANIFOLD-TO-HEAD GASKET,

MANIFOLD-TO-BLOCK REAR SEAL

WATER TEMPERATURE SENDING UNIT

SEALING WASHER

INTAKE MANIFOLD

GASKET

THERMOSTATIC DISTRIBUTOR VACUUM CONTROL VALVE

MANIFOLD-TO-HEAD GASKET

MANIFOLD-TO-BLOCK FRONT SEAL

THERMOSTAT

COOLANT OUTLET HOUSING

Intake manifold assembly—260, 289, 302, 351 Windsor V8s (© Ford Motor Co.)

uum lines from carburetor and vacuum control valves and tag them.

9. Disconnect carburetor fuel line.
10. Disconnect wiring harness from coil battery terminal, temperature sender unit, oil pressure sending unit, and other connections as necessary. Disengage wiring harness from retaining clips at left rocker cover bolts. Move harness out of way.
11. Disconnect Thermactor air by-pass valve from mounting bracket and place it to one side.
12. Remove coil and bracket assembly.
13. Disconnect manifold heat inlet and outlet tubes from rear of manifold and from exhaust pipe.
14. Remove distributor from engine.
15. Remove intake manifold attaching bolts.
16. Remove manifold and carburetor as an assembly. Discard used gaskets.
17. To install intake manifold, reverse above procedure—the manifold should be torqued in place as shown in the illustration.

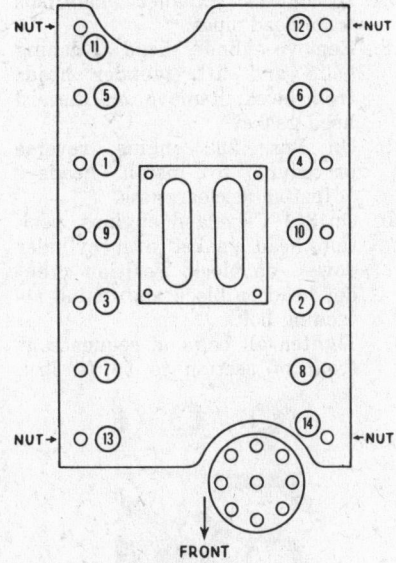

Tightening procedure

1 Torque bolts 1-10 to 4-6 ft. lbs. in sequence
2 Torque stud nuts 11-14 to 8-10 ft. lbs. in sequence
3 Torque all bolts and nuts to 15-20 ft. lbs. in sequence
4 Torque all bolts and nuts to 25-30 ft. lbs. in sequence
5 Retorque all bolts and nuts to 25-30 ft. lbs. in sequence

Boss 429 intake manifold torque sequence
(© Ford Motor Co.)

Cylinder Head

6 Cylinder Removal

1. Drain cooling system, remove the air cleaner and disconnect the battery cable at the cylinder head.

2. Disconnect exhaust pipe at the manifold end, spring the exhaust pipe down and remove the flange gasket.
3. Disconnect the fuel and vacuum lines from the carburetor. Disconnect the intake manifold line at the intake manifold.
4. Disconnect the accelerator and retracting spring at the carburetor. Disconnect the manually operated choke cable (if so equipped).
5. Disconnect the carburetor spacer outlet line at the spacer. Disconnect the radiator upper hose and the heater hose at the water outlet elbow. Disconnect the radiator lower hose and the heater hose at the water pump.
6. Disconnect the distributor vacuum control line at the distributor. Disconnect the gas filter line on the inlet side of the filter and the vacuum line at the fuel pump. Remove these three lines as an assembly, then remove the windshield wiper line at the vacuum pump, (if so equipped).
7. Disconnect the spark plug wires and remove the plugs.
8. Remove the rocker arm cover.
9. Back off all of the tappet adjusting screws to relieve tension on the rocker shaft. Loosen the rocker arm shaft attaching bolts and remove the rocker arm and shaft assembly. Remove the valve pushrods, in order, and keep them that way.
10. Remove one cylinder-head bolt from each end of the head (at opposite corners) and install cylinder head guide studs. Remove the remaining cylinder head bolts and lift off the cylinder head.

To help in removal and installation of cylinder head, two 6 in. x 7/16 = 14 bolts with heads cut off and the head end slightly tapered and slotted for installation, and removal with a screwdriver, will reduce the possibility of damage during head replacement. These guide studs make a handy tool during head removal and gasket and head replacement.

6-Cylinder Installation

1. Clean the cylinder head and block surfaces. Be sure of flatness and no surface damage.
2. Apply cylinder head gasket sealer to both sides of the new gasket and slide the gasket down over the two guide studs in the cylinder block.
3. Carefully lower the cylinder head over the guide studs. Place the exhaust pipe flange on the manifold studs (new gasket).
4. Coat the threads of the end bolts for the right side of the cylinder head with a small amount of water-resistant sealer. Install, but do not tighten, two head bolts at opposite ends to hold the head gasket in place. Remove the guide studs and install the remaining bolts.
5. Cylinder head torquing should proceed in three steps and in prescribed order. Tighten to 55 ft. lbs., then give them a second tightening to 65 ft. lbs. The final step is to 75 ft. lbs., at which they should remain undisturbed.
6. Lubricate both ends of the pushrods and install them in their original locations.
7. Apply a petroleum jelly-type lubricant to the rocker arm pads and the valve stem tips and position the rocker arm shaft assembly on the head. Be sure the oil holes in the shaft are in a down position.
8. Tighten all the rocker shaft retaining bolts to 30-35 ft. lbs. and do a preliminary valve adjustment (make sure there are no tight valve adjustments).
9. Hook up the exhaust pipe.
10. Reconnect the heater and radiator hoses.
11. Reposition the distributor vacuum line, the carburetor gas line and the intake manifold vacuum line on the engine. Hook them up to their respective connections and reconnect the battery cable to the cylinder head.
12. Connect the accelerator rod and retracting spring. Connect the choke control cable and adjust the choke.

Removing rocker arm assembly—6 cyl. (© Ford Motor Co.)

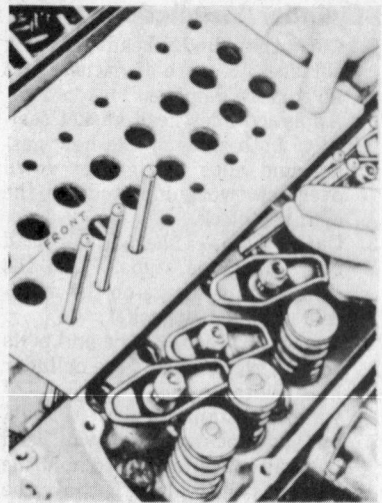

Pushrod removal—260, 289, 302, and 351 Windsor V8s
(© Ford Motor Co.)

13. Reconnect the vacuum line at the distributor. Connect the fuel inlet line at the fuel filter and the intake manifold vacuum line at the vacuum pump. Connect the windshield wiper vacuum line to the other side of the vacuum pump.
14. Lightly lubricate the spark plug threads, install them and torque to 25 ft. lbs. Connect spark plug wires and be sure the wires are all the way down in their sockets.
15. Fill the cooling system and bleed. Run the engine for about ½ hour at a good fast idle to stabilize all engine parts temperatures.
16. Adjust engine idle speed and idle fuel-air adjustment.
17. Reset valve tappet adjustment to .016 in. for a hot adjustment of both intake and exhaust valves on those engines not using hydraulic lifters.
18. Coat one side of a new rocker cover gasket with oil-resistant sealer. Lay the treated side of the gasket on the cover and install the cover. Be sure the gasket seals evenly all around the cylinder head.

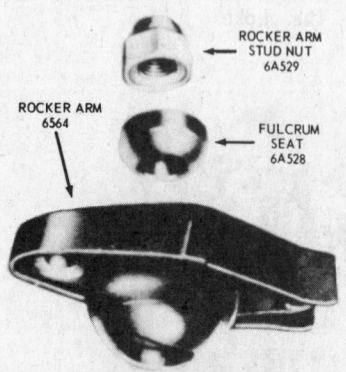

289, 302, and 351 Windsor rocker arm assembly
(© Ford Motor Co.)

260, 289, 302 and 351 Windsor V8
See Ford–Thunderbird Section.

390, 427, 428, 429, 429CJ, and 429 Super CJ V8s
See Mercury Section.

302 Boss and 351 Cleveland V8s

1. Remove intake manifold.
2. Remove attaching bolts and remove rocker covers.
3. Remove rocker arm lock nuts, stud nuts, fulcrum seats and rockers from Boss 302. Remove rocker arm fulcrum bolts, fulcrum seats and rockers from 351 Cleveland engines.
   NOTE: if cylinder head is not to be disassembled, loosen rocker arms

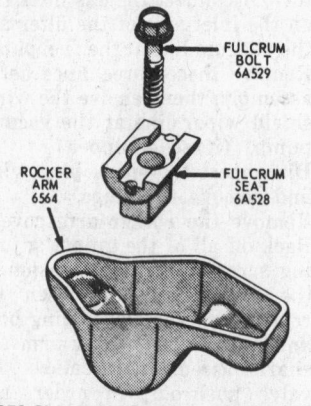

351 Cleveland V8 rocker arm assembly
(© Ford Motor Co.)

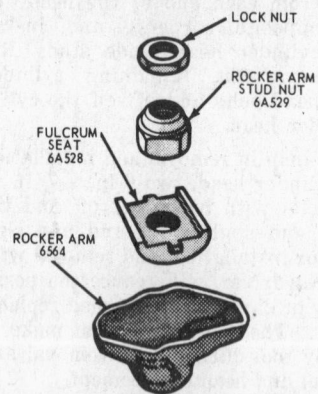

Boss 302 rocker arm assembly
(© Ford Motor Co.)

and pivot them sideways to clear pushrods.
4. Remove pushrods in sequence and save to replace in original bores.
5. If left cylinder head is to be removed, unbolt power steering pump, if any, from cylinder head. On Boss 302, remove ignition coil.
6. If right cylinder head is to be removed, remove alternator mounting bolt bracket and spacer. On 351 Cleveland,

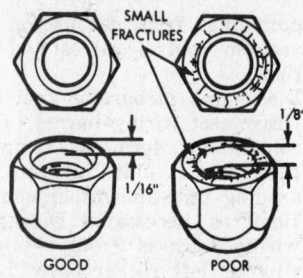

Rocker arm stud nut inspection
(© Ford Motor Co.)

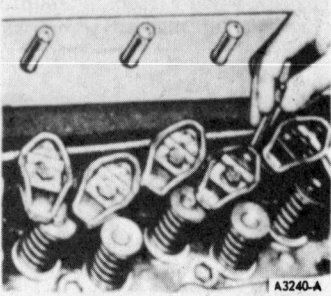

Removing push rods—351 Cleveland V8
(© Ford Motor Co.)

remove ground wire at back of cylinder head.
7. Disconnect exhaust manifolds from head pipes.
8. Remove cylinder head attaching bolts and lift cylinder heads from block. Remove and discard head gasket.
9. On Boss 302 engine, reverse procedure to install heads—adjust valve clearances.
10. On 351 Cleveland engines, position head gasket over cylinder dowels on block. Position cylinder head on block and install attaching bolts.
11. Tighten all bolts in sequence at front of section to 50 ft. lbs.,

Tappet removal
(© Ford Motor Co.)

then tighten bolts in sequence to 60 ft. lbs. Finally torque to specification.
12. Clean pushrods in solvent and blow out oil passage in pushrod with compressed air if available. Roll pushrods across flat surface to check for straightness. Discard bent pushrods, do not attempt to straighten them.
13. Lubricate and install pushrods into their original bores.
14. Lubricate top of valve stem, rocker arm, and fulcrum seat.
15. Position No. 1 piston on TDC at end of compression stroke. Install rocker arms, fulcrum seats,

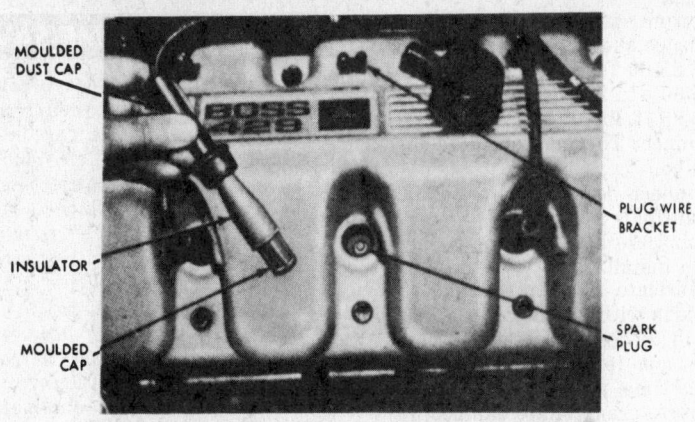

Removing spark plug wires on Boss 429 (© Ford Motor Co.)

MOULDED DUST CAP
INSULATOR
MOULDED CAP
PLUG WIRE BRACKET
SPARK PLUG

and fulcrum bolts on following valves: No. 1 intake, No. 1 exhaust, No. 4 intake, No. 3 exhaust, No. 8 intake, and No. 7 exhaust. Torque fulcrum bolts to 17-23 ft. lbs.

16. Rotate engine 180° clockwise and install rocker arms, fulcrum seats, and fulcrum bolts on No. 3 intake, No. 2 exhaust, No. 7 intake, and No. 6 exhaust.

17. Rotate crankshaft 270° clockwise and install rocker arms, fulcrum seats and fulcrum bolts on No. 2 intake, No. 4 exhaust, No. 5 intake, No. 5 exhaust, No. 6 intake, and No. 8 exhaust.

NOTE: be sure fulcrum seat base is inserted in its slot on cylinder head before torquing fulcrum bolt.

18. Connect exhaust manifolds at muffler inlet pipes. Torque nuts to 18-24 ft. lbs.

19. If right cylinder head was removed, install alternator mounting bracket through bolt and air cleaner inlet duct on right head assembly. Connect ground wire at rear of head. Adjust belt tension.

20. Apply oil resistant sealer to one side of new rocker cover gasket. Lay cemented side of gaskets in place in covers. Install rocker covers.

21. If left cylinder head was removed on power steering equipped vehicle, install drive belt and power steering pump bracket. Install bracket attaching bolts. Adjust drive belts.

22. Install intake manifold.

Boss 429 Removal

1. Disconnect battery.
2. Remove cap that connects crankcase ventilation hose (PCV) to left rocker cover.
3. Remove air cleaner.
4. If removing right head, remove crankcase ventilation hose.
5. Lift each plug wire from bracket.
6. Disconnect wires from spark plugs by twisting and pulling on molded duct caps.
7. If removing left head, disconnect brake master cylinder from booster and move it to one side to provide clearance.

8. Remove rocker cover attaching nuts and bolts.
9. Lift rocker covers from heads.
10. Clean gasket material from covers and heads.
11. Remove intake manifold.
12. Back off all rocker arm adjusting screws.
13. Remove all rocker shaft attaching nuts from rocker shafts.

NOTE: each rocker on a Boss 429 engine has its own individual rocker shaft.

14. Remove rocker arms, shafts and pedestals. Keep them in sequence so that they can be installed in their original position.
15. Lift pushrods from cylinder head. Keep them in sequence to install into their original bores.
16. Disconnect exhaust head pipe from exhaust manifold.
17. Disconnect air hose from thermactor check valve on head being removed.

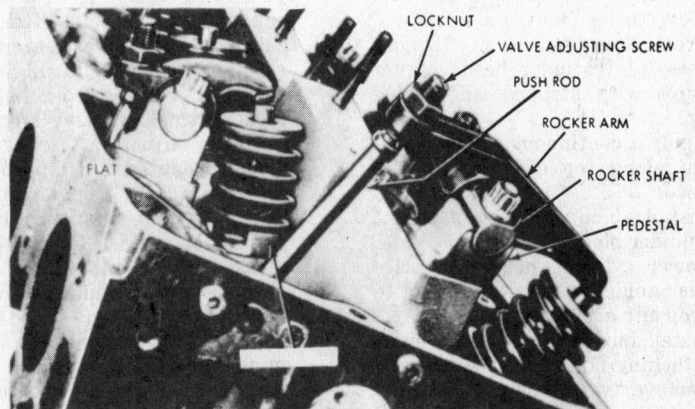

Boss 429 valve gear (© Ford Motor Co.)

LOCKNUT
VALVE ADJUSTING SCREW
PUSH ROD
ROCKER ARM
ROCKER SHAFT
PEDESTAL
FLAT

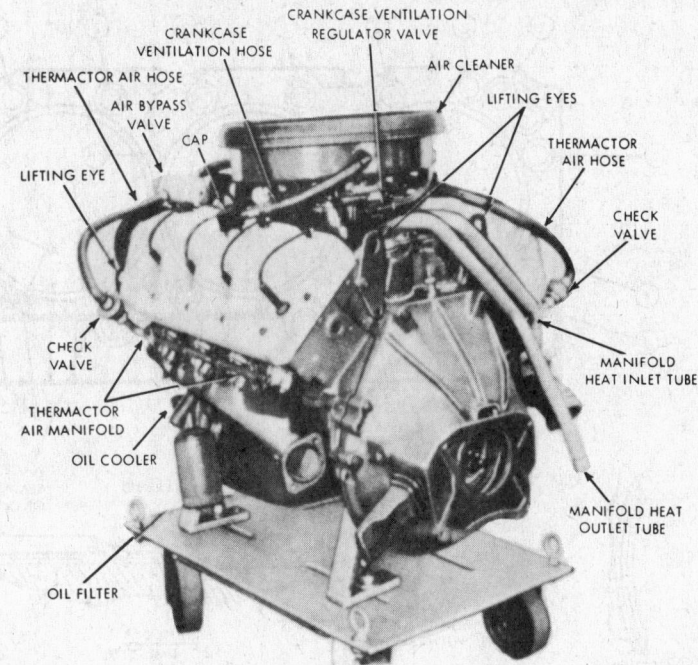

Boss 429 V8 (© Ford Motor Co.)

CRANKCASE VENTILATION HOSE
CRANKCASE VENTILATION REGULATOR VALVE
THERMACTOR AIR HOSE
AIR BYPASS VALVE
CAP
AIR CLEANER
LIFTING EYES
THERMACTOR AIR HOSE
LIFTING EYE
CHECK VALVE
MANIFOLD HEAT INLET TUBE
CHECK VALVE
THERMACTOR AIR MANIFOLD
OIL COOLER
MANIFOLD HEAT OUTLET TUBE
OIL FILTER

18. Remove ten cylinder head bolts. Connect lifting sling to lifting eye at each end of cylinder head, and lift cylinder head from block with hoist.
19. Remove all rubber and steel gaskets from head and block.
20. Clean cylinder block and head mating surfaces.

Boss 429 Installation

1. Wipe head and block surfaces with chlorathane.
2. Coat upper end of cylinder head and block with silicone rubber primer (Dow Corning A-4094 or equivalent). Coat gasket counter bores with quick drying adhesive sealer to prevent dropping gaskets while installing head.
3. Position four combustion chamber gaskets in counterbores with tabs seated down. Locate tabs by rotating gasket between finger and thumb to feel tabs.
4. Press four 1/4 in. ID gaskets into cylinder head counter bores with stepped side facing up.
5. Press seventeen 1/2 in. ID gaskets into cylinder head counter bores with stepped side facing up.
6. Apply a continuous strip of sealant along top edge of cylinder head.
7. Install guide pin at each end of cylinder block.
8. Lower cylinder head into place over guide pins. Take care not to drop any gaskets.
9. Install but do not tighten eight attaching bolts and flat washers.
10. Remove two guide pins and install two remaining bolts and washers.

11. Torque attaching bolts in sequence shown at front of section to 55-60 ft. lbs. Then torque to 75-80 ft. lbs. Finally torque to 90-95 ft. lbs.
12. Connect Thermactor air to check valve.
13. Connect lead pipe to exhaust manifold.
14. Lubricate both ends of pushrods and install.
15. Lubricate rocker arms and shafts with engine oil and install with loosened adjusting screws. Do not torque shafts down at this time.
16. Rotate crank shaft damper until No. 1 piston is at TDC at end of compression stroke.
17. Install distributor in cylinder block with rotor at No. 1 firing position and points just beginning to open. Install hold-down clamp and bolt.
18. Torque rocker shaft nuts on No. 1 cylinder intake and exhaust to 12-15 ft. lbs. If engine is equipped with a solid lifter camshaft, adjust valve clearance to specification (cold) using feeler gauge or valve gapper between rocker arm and valve stem tip. Torque adjusting screws in place. If engine is equipped with an hydraulic lifter camshaft, loosen locknut and turn in adjusting screw on No. 1 cylinder intake and exhaust rocker until all clearance is removed. Rotate pushrod with fingers while tightening adjusting screw to determine point when clearance is removed. Tighten adjusting screws 1/16 turn further. Hold adjusting screws in place and

torque locknuts to 20-30 ft. lbs.
19. Rotate crankshaft 90° to position No. 5 piston at TDC and repeat step 18 for No. 5 intake and exhaust rockers.
20. Rotate crank shaft 90° and repeat procedure in Step 18 for each cylinder in firing order (1-5-4-2-6-3-7-8).
21. Remove distributor.
22. Coat one side of new rocker cover gasket with oil resistant sealer and lay cemented side in place on cover.

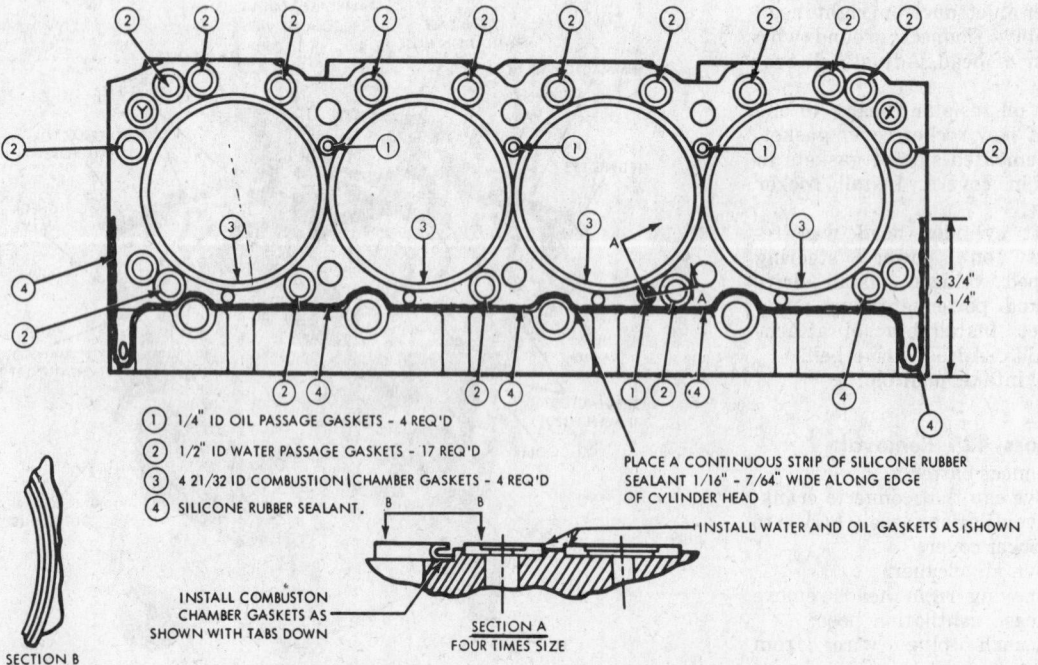

① 1/4" ID OIL PASSAGE GASKETS - 4 REQ'D
② 1/2" ID WATER PASSAGE GASKETS - 17 REQ'D
③ 4 21/32 ID COMBUSTION CHAMBER GASKETS - 4 REQ'D
④ SILICONE RUBBER SEALANT.

PLACE A CONTINUOUS STRIP OF SILICONE RUBBER SEALANT 1/16" - 7/64" WIDE ALONG EDGE OF CYLINDER HEAD

INSTALL WATER AND OIL GASKETS AS SHOWN

INSTALL COMBUSTION CHAMBER GASKETS AS SHOWN WITH TABS DOWN

SECTION A FOUR TIMES SIZE

SECTION B

Boss 429 cylinder head gasket location (© Ford Motor Co.)

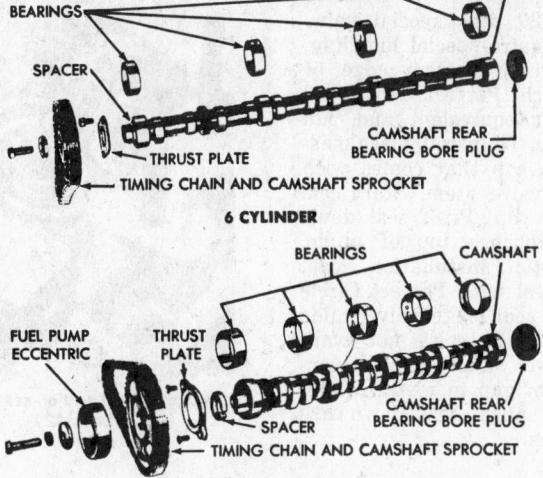

6 CYLINDER

8 CYLINDER
Camshaft and related parts (© Ford Motor Co.)

23. Install cover. Make sure gasket seats evenly all around cover.
24. Tighten cover attaching bolts evenly and alternately in two steps. Then torque cover bolts to 12-15 ft. lbs. Wait two minutes and retorque to 12-15 ft. lbs.
25. Install intake manifold.
26. Connect each spark plug wire to its respective plug. Insert plug wires into brackets on valve cover.
27. Install cap and crank case ventilation hose (PCV) on valve cover.
28. If installing left head, install master cylinder on booster.
29. Install air cleaner and connect battery.

Valve System

The 6-cylinder engines are equipped with tubular pushrods and barrel type tappets. Valve lash is controlled by self locking adjusting screws.

Early 1963, 6-cylinder engines use mechanical adjusters. However, later production engines incorporate hydraulic tappets with zero lash running clearance.

V8 engines, except the 289 and 427 high performance, use hydraulic tappets. The pushrods in the V8s also transfer oil under pressure to the friction areas of the rocker arms.

Dismantling 6 Cylinder

1. Remove cylinder head as described in previous paragraph.
2. Reach down through the pushrod area with a magnet, or other suitable extractor, and remove the tappets. Keep the tappets in order, so that they can be installed in their original location.
3. Loosen all valve rocker arm adjusting screws two turns at a time, and in sequence.
4. Remove rocker arm shaft pedestal bolts and lift off rocker arm shaft assembly.
5. Dismantle shaft and rockers by removing pin and spring washer from each end of the shaft.
6. Slide rocker arms, springs and pedestals off the shaft. (Be sure to identify the parts for proper assembly sequence.)

Assembly 6 Cylinder

1. Lubricate and assemble all rocker shaft components.
2. If the end plugs were removed from the rocker shaft, install new plugs, cup side out, to each end of the shaft.
3. Do valve grinding job, or any other service to the cylinder head, check valve spring condition, install new oil seals on valve stems, check springs and install valve assembly in place. Compress springs and keepers.
4. Install valve tappets in their proper bores.
5. Apply cylinder head gasket sealer to both sides of a new gasket and locate gasket on cylinder block. Install cylinder head and torque, in three progressive steps, to 75 ft. lbs.
6. Apply Lubriplate, or suitable substitute, to both ends of pushrods, then install pushrods.
7. Install rocker arm and shaft. Tighten rocker shaft pedestal bolts in progressive steps to 30-35 ft. lbs.

6-Cylinder Hydraulic Valve Adjustment

The following procedure is performed with the engine running.
1. After the engine has been brought to operating temperature, remove the valve cover.
2. With engine at normal idle speed, back off the valve rocker arm adjusters, one at a time, until the rocker arm starts to clatter.

3. Turn the arm adjuster down until the clatter stops.
4. Continue to turn down the adjuster exactly one turn. This will force the hydraulic lifter piston into the approximate center of its travel.
5. Install valve rocker cover.

On models with non-adjustable rockers, the pushrods must be changed to secure proper lifter position.

Disassembly of Cylinder Heads

1. Remove cylinder heads.
2. Compress valve springs using valve spring compressor.
3. Remove valve locks or keys.
4. Release valve springs.
5. Remove valve springs, retainers, oil seals, and valves. On Boss 302 and Boss 429 engines, remove valve spring seals also.

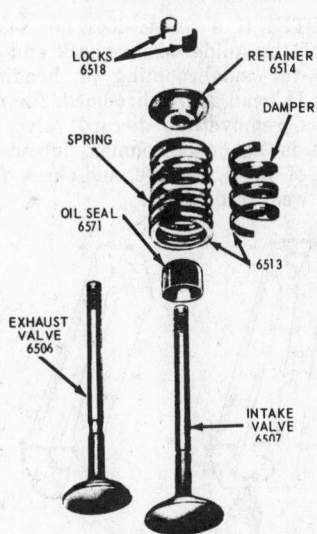

Valve assembly—351 Cleveland V8
(© Ford Motor Co.)

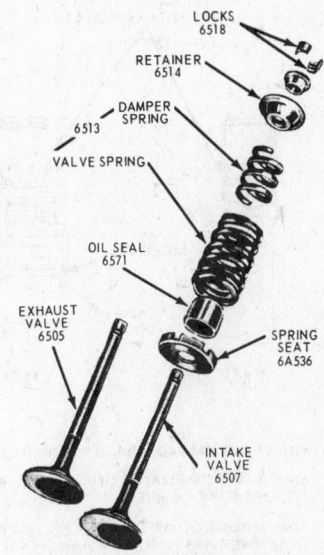

Valve assembly—Boss 302 V8
(© Ford Motor Co.)

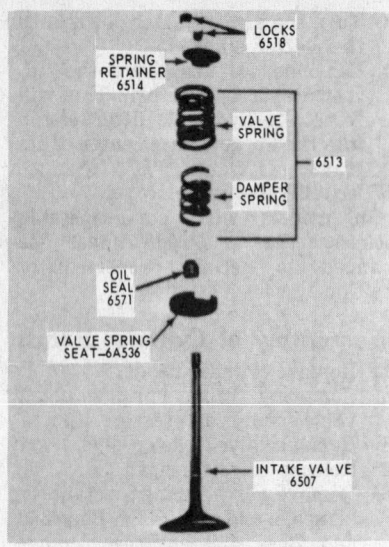

Valve assembly—Boss 429 V8
(© Ford Motor Co.)

NOTE: if a valve does not slide out of the guide easily check end of stem for mushrooming or heading over. If head is mushroomed, file off excess, remove and discard valve. If valve is not mushroomed, lubricate stem of valve, remove, and check for stem wear or damage.

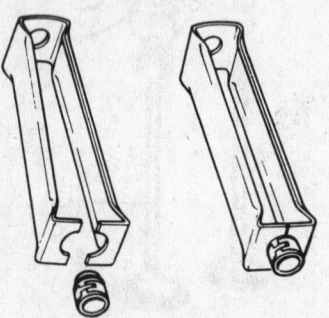

Boss 429 valve stem seal tool
(© Ford Motor Co.)

Valve Seals—Boss 429

The Boss 429 uses special valve seals which require special handling. To remove, grasp bottom edge of valve seal with Perfect Circle tool No. VSIT-1 or equivalent and pull seal from valve. To install, place plastic installation cap that comes with seal kit over valve stem. Start seal carefully over cap. Push seal down until jacket touches top of guide. Remove plastic installation cap. Grasp valve seal with Perfect Circle tool and push seal on to valve guide until it bottoms. If tool is not available, place two small screwdrivers about 90° from gap in metal retaining ring and push seal downward until it bottoms.

Valve Guides

Ford Motor Company engines use integral valve guides. Mercury and Ford dealers offer valves with oversize stems for worn guides. To fit these, enlarge valve guide bores with valve guide reamers to an oversize that cleans up wear. If a large oversize is required it is best to approach that size in stages by using a series of reamers of increasing diameter. This helps to maintain the concentricity of the guide bore with the valve seat. The correct valve guide to stem clearance is at front of this section. As an alternative, some local automotive machine shops will fit replacement guides that use standard stem valves.

Hydraulic Tappet Service

All Models

1. Remove lock ring from tappet body.
2. Remove pushrod cup and metering valve disc.

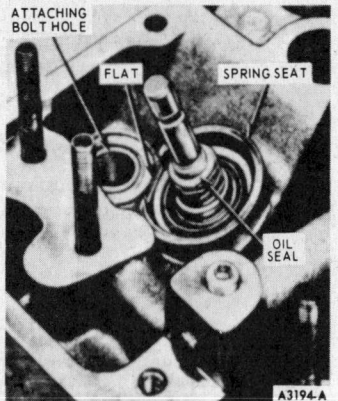

Valve spring seal and oil seal location—Boss 429
(© Ford Motor Co.)

3. Remove plunger from tappet body.
4. Invert the plunger and remove the check valve retainer by carefully prying up on it with a screwdriver.
5. Remove the check valve and the spring.
6. Remove the plunger spring from the tappet body.
7. Soak all tappet components in solvent (lacquer thinner works well on these tar and varnish substances).
8. After tappet parts have been thoroughly cleaned and blown dry, place the plunger upside down on a clean surface.
9. Place the check valve over the hole in the bottom of the plunger. Place check valve spring on top of the check valve.
10. Position check valve retainer over the spring and check valve, then push the retainer down into place on the plunger.
11. Place plunger spring and plunger into the tappet body.

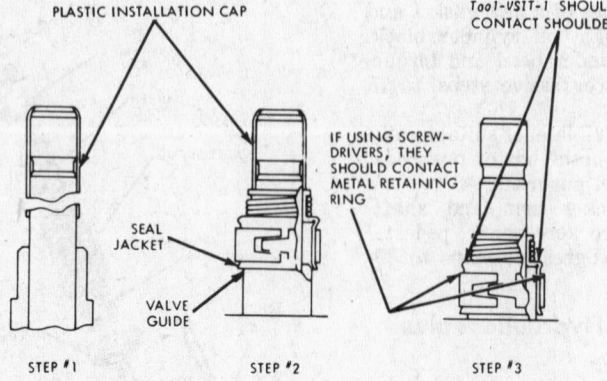

STEP #1 WITH VALVES IN HEAD. PLACE PLASTIC INSTALLATION CAP OVER END OF VALVE STEM.

STEP #2 START VALVE STEM SEAL CAREFULLY OVER CAP. PUSH SEAL DOWN UNTIL JACKET TOUCHES TOP OF GUIDE.

STEP #3 REMOVE PLASTIC INSTALLATION CAP. USE INSTALLATION TOOL-VSIT-1 OR SCREWDRIVERS TO BOTTOM SEAL ON VALVE GUIDE.

Installing valve stem seals—Boss 429 (© Ford Motor Co.)

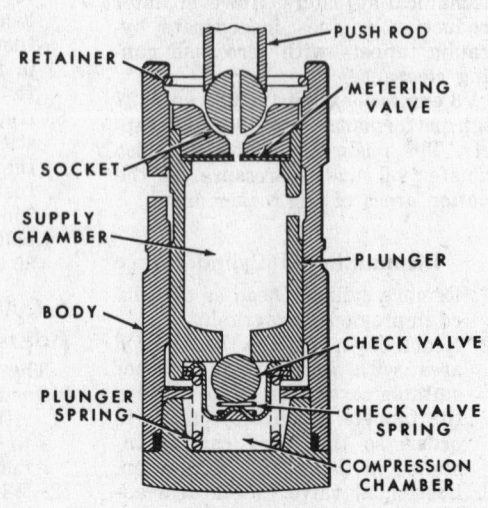

Hydraulic tappet
(© Ford Motor Co.)

12. Place metering valve disc and the pushrod cup in the plunger. Depress plunger and install the lock ring.
13. Test all hydraulic tappets (both new and used) before installing them in an engine. The factory quotes a bleed-down time of from 10 to 80 seconds, using special test oil and their approved tester.

Timing Cover and Chain

6 Cylinder Cover and Chain Removal

1. Drain cooling system, and disconnect radiator hoses.
2. Remove radiator.
3. Remove drive belt, fan and pulley.
4. Remove the crankshaft damper.
5. Remove timing chain cover and crankshaft front oil slinger.
6. Establish a reference point on the block. Now measure extreme slack in the loose side of the chain and record this measurement. Now, eliminate the slack by moving the camshaft sprocket within the chain slack limits and again take a measurement. The difference should not exceed ½ in. More than ½ in. slack justifies renewal of chain and sprockets.
7. Crank the engine until timing marks are aligned, as shown in Valve Timing illustration.
8. Remove camshaft sprocket retaining bolt and washer. Then, slide both sprockets and chain forward and remove as an assembly.

Cover Seal R & R (6 Cylinder & V8)

1. Drive out the old seal with a pin punch. Then, clean out the recess in the cover.

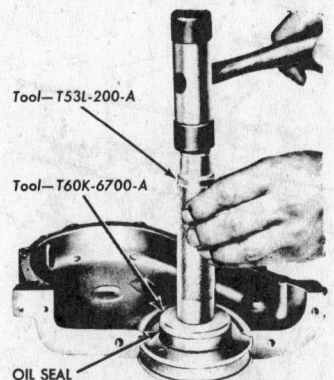

Tool—T53L-200-A
Tool—T60K-6700-A
OIL SEAL

Typical 6 cyl. crankshaft front oil seal replacement
(© Ford Motor Co.)

2. Coat a new seal with grease and drive it into place in the chain cover. Check that the spring is still properly positioned in the seal before installing cover.

6 Cylinder Cover and Chain Installation

1. Position sprockets and chain on the camshaft and crankshaft with both timing marks on a centerline. Install camshaft sprocket retain-

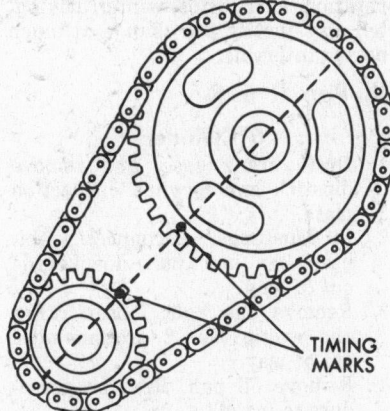

TIMING MARKS

Timing mark alignment—6 cyl.

ing bolt and washer. Torque attaching bolt to 35-45 ft. lbs.
2. Install the oil slinger so that the pointer on the slinger is aligned with the camshaft sprocket timing mark.
3. Clean front cover and cylinder block of old gasket material. Apply sealer to a new cover gasket and position the gasket to the cover.
4. Install the front cover, using a crankshaft - to - cover alignment tool. Torque attaching bolts to 7-9 ft. lbs.
5. Install crankshaft damper and torque to 45-55 ft. lbs.
6. Install fan, fan pulley and drive belt. Adjust the belt.
7. Install radiator. Then, connect the hoses.
8. Fill and bleed cooling system.
9. Start engine and check for leaks and final adjustments.

V8 Cover and Chain Removal

1. Drain cooling system, remove air cleaner and disconnect the battery.
2. Disconnect radiator hoses and remove the radiator.
3. Disconnect heater hose at water pump. Slide water pump by-pass hose clamp toward the pump.
4. Loosen generator mounting bolts

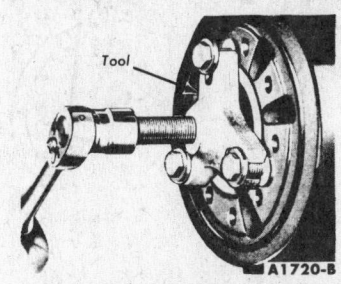

Tool

V8 crankshaft damper removal
(© Ford Motor Co.)

at the generator. Remove the generator support bolt at the water pump. Remove Thermactor pump on 428 CJ, 429 Super CJ, Boss 302, and Boss 429 engines.
5. Remove the fan, spacer, pulley, and drive belt.
6. Remove pulley from crankshaft pulley adapter. Remove cap screw and washer from front end of crankshaft. Remove crankshaft pulley adapter with a puller.
7. Disconnect fuel pump outlet line at the pump. Remove fuel pump retaining bolts and lay the pump to the side.
8. Remove front cover attaching bolts, and remove front cover and water pump as an assembly.
9. Remove crankshaft front oil slinger.
10. Rotate crankshaft in normal direction to remove slack from the chain on the fuel pump side of the engine.
11. Establish a reference point on the block and measure from this point to the chain.
12. Back up on crankshaft rotation to remove slack from the chain on the generator side of the engine. Force the fuel pump side of the chain out with the fingers and again take a measurement from chain to reference mark. Deflection is the difference between the two measurements. If deflection exceeds ½ in., a new chain and/or sprockets is warranted.
13. Crank engine until sprocket timing marks are aligned as shown in Valve Timing illustration.
14. Remove camshaft sprocket cap screw, washers, and fuel pump eccentric. Slide both sprockets and chain forward and off as an assembly.

V8 Cover and Chain Installation

1. Position sprockets and chain on the camshaft and crankshaft with both timing marks on a centerline. Install fuel pump eccentric, washers and sprocket attaching bolt. Torque the sprocket attaching bolt to 30-35 ft. lbs.
2. Install crankshaft front oil slinger.
3. Clean front cover and mating surfaces of old gasket material.
4. Coat a new cover gasket with sealer and position it on the block.
5. Install front cover, using a crankshaft-to-cover alignment tool. Torque attaching bolts to 12-15 ft. lbs.
6. Install fuel pump, torque attaching bolts to 23-28 ft. lbs., connect fuel pump outlet tube.
7. Install crankshaft pulley adapter and torque attaching bolt to

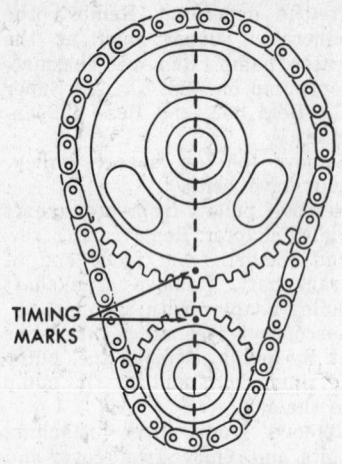

TIMING MARKS

Timing mark alignment—V8

70-90 ft. lbs. Install crankshaft pulley.

8. Install water pump pulley, drive belt, spacer and fan.
9. Install generator support bolt at the water pump. Tighten generator mounting bolts. Adjust drive belt tension. Install Thermactor pump if so equipped.
10. Install radiator and connect all coolant and heater hoses. Connect battery cables.
11. Refill and bleed cooling system.
12. Start engine and operate at fast idle to operating temperature.
13. Check for leaks, install air cleaner. Adjust ignition timing and make all final adjustments.

Engine Lubrication

All engines are equipped with full-flow-type oil filters to condition the oil before it reaches the main bearings. The filter is equipped with an internal, relief, by-pass valve as a safety precaution. The system of lubrication is best shown in the illustrations.

Under normal driving conditions, engine oil and oil filter should be changed at 6,000 mile intervals. However, adverse driving conditions, dusty operation, short trips, winter driving, etc., may justify the change at much shorter intervals.

Oil Pan R & R

6 Cylinder

1. Drain crank case, and remove dipstick and flywheel inspection plate.
2. In Mustangs, disconnect stabilizer bar and pull downwards out of way.
3. Remove one bolt, loosen other and swing No. 2 crossmember out of way.
4. Remove oil pan. Reverse procedure to install.

260, 289, 302, 351 Windsor, and 302 Boss V8s
See Ford–Thunderbird Section.

390, 427 and 428 Engines
See Ford–Thunderbird Section.

429 4V, 429 CJ, 429 Super CJ, and Boss 429
See Mercury Section.

351 Cleveland V8 in Cougar and Mustang

1. Remove dip stick, raise vehicle, and drain crank case.
2. Disconnect starter cable, remove starter.
3. Remove stabilizer bar.
4. Remove two bolts retaining No. 2 crossmember and remove crossmember.

5. Remove pan bolts, turn crank shaft for maximum clearance, and remove pan. Reverse procedure to install.

351 Cleveland V8 in Fairlane and Montego

1. Remove dipstick.
2. Remove fan shroud bolts and position fan shroud over pan.
3. Raise vehicle, drain crank case, disconnect starter cable, and remove starter.
4. Remove stabilizer bar attaching bolts, and lower sway bar for clearance.
5. Remove engine front support bolts.
6. Raise engine and place wood block between engine supports and chassis brackets.
7. Remove oil pan attaching bolts.
8. Move automatic transmission oil cooler lines, if any, out of way and remove pan. Reverse procedure to install.

Oil Pump

Removal—6 Cylinder

1. Remove oil pan.
2. Remove oil pump inlet tube and screen assembly.
3. Remove oil pump attaching bolts and remove oil pump gasket and intermediate shaft.

Removal—260, 289, 302, 351 Windsor and Boss 302 V8s

1. Remove oil pan.
2. Remove oil pump pickup tube and screen from oil pump.

NOTE: it is not necessary to remove oil baffle tray on Boss 302 engines to do this job.

3. Remove oil pump attaching bolts

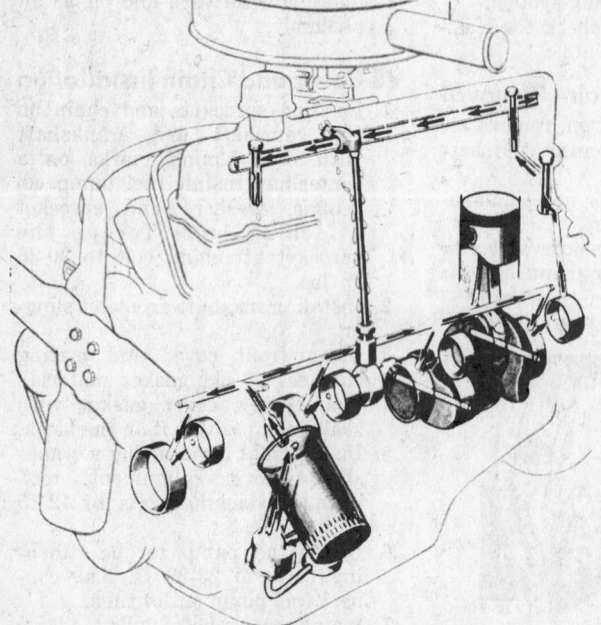

6 cyl. engine lubrication (© Ford Motor Co.)

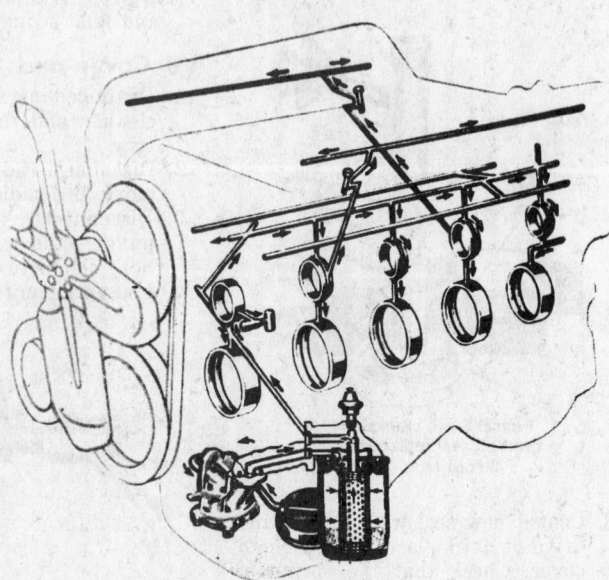

390, 427 and 428CJ V8s engine lubrication (© Ford Motor Co.)

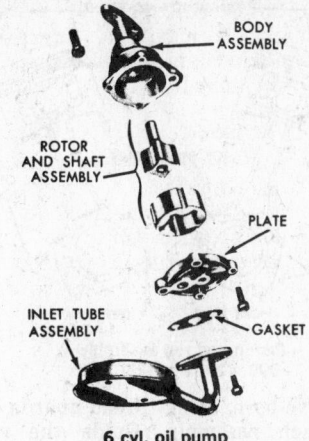

6 cyl. oil pump
(© Ford Motor Co.)

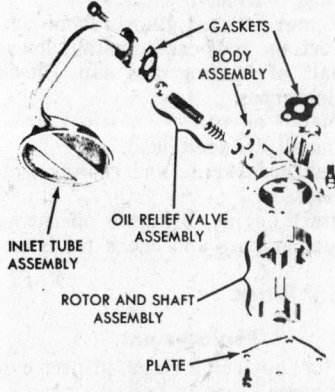

260, 289, 302, Boss 302, and 351 Windsor V8s oil pump
(© Ford Motor Co.)

and remove oil pump, gasket, and intermediate drive shaft.

Removal—390, 427, and 428 V8
See Mercury Section.

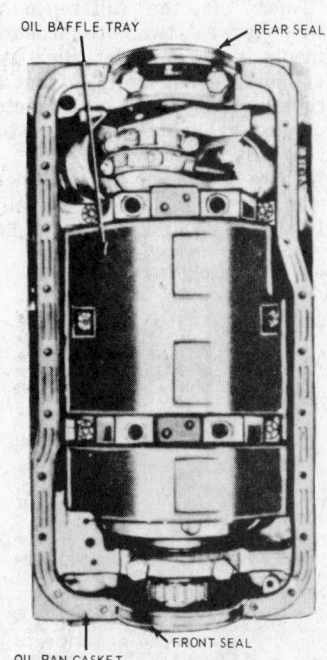

Oil baffle tray—Boss 302
(© Ford Motor Co.)

Removal—429 4V, 429 CJ, 429 Super CJ and Boss 429
See Mercury Section.

Removal—351 Cleveland V8
1. Remove oil pan.
2. Remove oil pump attaching bolts and remove oil pump with pickup tube and screen, gasket, and intermediate shaft.

Oil Pump Disassembly Except 429 4V, 429 CJ, 429 Super CJ, and Boss 429 V8s
1. Remove oil pump inlet tube and gasket.
2. Remove cover attaching screws and remove cover.
3. Remove inner rotor and shaft assembly and remove outer rotor.
4. Insert a self threading sheet metal screw into relief valve chamber cap and pull cap out of chamber.

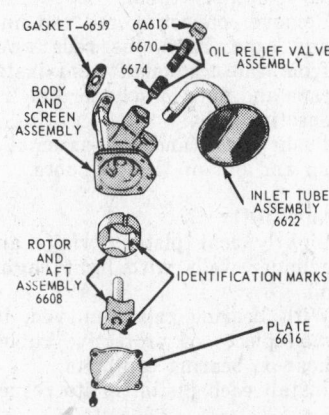

Oil pump—351 Cleveland V8
(© Ford Motor Co.)

5. Remove relief valve spring and plunger.

Oil Pump Disassembly 429 4V, 429 CJ, 429 Super CJ, and Boss 429 V8s
1. Remove four screws and washer securing pump cover and remove cover.
2. Remove inner rotor and shaft assembly and outer rotor.
3. Remove stake marks which secure relief valve plug.
4. Insert self tapping metal screw in relief valve plug.
5. Use pliers to remove plug and remove spring and relief valve.

Removing oil pump relief valve—429 V8s
(© Ford Motor Co.)

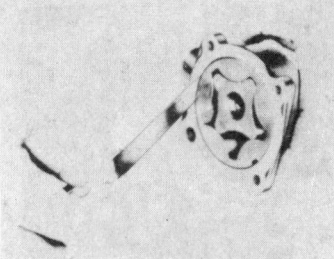

Measuring clearance between outer rotor and housing
(© Ford Motor Co.)

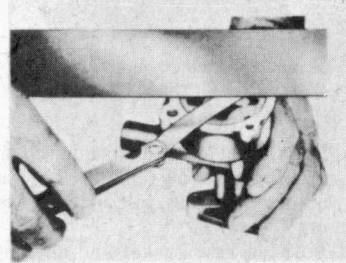

Measuring rotor end play
(© Ford Motor Co.)

Inspection—All Engines
1. Examine inner rotor, outer rotor, and pump body for wear or damage.
2. Examine mating surface of cover for wear, scoring, grooving, or warping. Replace damaged cover.
3. Install outer rotor in pump body, and holding rotor against one side, measure clearance between rotor and body. Clearance should be 0.006-0.013 in. If clearance is excessive, replace worn or damaged part or parts.
4. Install inner rotor and shaft in pump body. Place straight edge over pump body and rotors, and measure clearance between straight edge and rotors. Clearance should be 0.0011-0.0041 in. If clearance is excessive replace rotors.
NOTE: the inner rotor and shaft, and the outer rotor are replaceable only as an assembly.
5. Check drive shaft to housing bearing clearance by measuring OD of drive shaft journal area and ID of bearing. Clearance should be 0.0015-0.0029 in.
6. Inspect relief valve for collapsed or worn condition. Check relief valve for scoring and for free operation in bore.

Oil Pump Assembly Except 429 4V, 429 CJ, 429 Super CJ, Boss 429 V8s
1. Clean and oil all parts.
2. Install oil pressure relief valve plunger, spring, and new cap.
3. Install outer rotor and inner rotor and shaft assembly. Be sure identification marks on inner and outer rotors are aligned.

Oil pump and inlet tube installed—
351 Cleveland V8
(© Ford Motor Co.)

4. Install cover and torque cover attaching screws to 6-9 ft. lbs. on six cylinders, and 9-12 ft. lbs. on V8s.
5. Position new gasket and oil inlet tube on oil pump and install attaching bolts.

Oil Pump Assembly — 429 4V, 429 CJ, 429 Super CJ, and Boss 429

1. Install relief valve, spring, and plug in pump body. Press plug inward until it seats, then stake in place.
NOTE: the relief hole in plug must not be covered or obstructed.
2. Install outer rotor and inner rotor and shaft in body.
3. Install cover and four screws and washers.
4. Torque cover screws to 6-9 ft. lbs.

Installation—All Engines

1. Prime oil pump by filling inlet on outlet port with engine oil and rotating shaft of pump to distribute it.
2. Position intermediate drive shaft into distributor socket.
3. Position new gasket on pump body and insert intermediate drive shaft into pump body.
4. Install pump and intermediate shaft as an assembly.
NOTE: do not force pump if it does not seat readily. The drive shaft

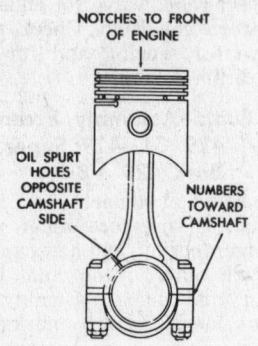

NOTCHES TO FRONT OF ENGINE

OIL SPURT HOLES OPPOSITE CAMSHAFT SIDE

NUMBERS TOWARD CAMSHAFT

Piston and rod assembly—6 cyl.

---

may be misaligned with the distributor shaft. To align rotate intermediate drive shaft into a new position.
5. Install and torque oil pump attaching screws to 12-15 ft. lbs. on six cylinder, 20-25 ft. lbs. on V8s.
6. Install oil pan.

Connecting Rods and Pistons

Removal

1. Drain crankcase and remove oil pan. Remove oil baffle tray if so equipped.
2. Drain cooling system and remove cylinder head or heads.
3. Remove any ridge and/or deposits from the upper end of cylinder bores with a ridge reamer.
4. Check rods and pistons for identification numbers and, if necessary, number them.
5. Remove connecting rod cap nuts and caps. Push the rods away from the crankshaft and install caps and nuts loosely to their respective rods.
6. Push piston and rod assemblies up and out of the cylinders.

Installation

1. Lightly coat pistons, rings and cylinder walls with light engine oil.
2. With bearing caps removed, install pieces of protective rubber hose on bearing cap bolts.
3. Install each piston in its respec-

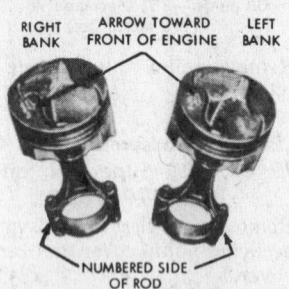

RIGHT BANK · ARROW TOWARD FRONT OF ENGINE · LEFT BANK

NUMBERED SIDE OF ROD

Piston and rod assembly—
351 Cleveland and Boss 302 V8s
(© Ford Motor Co.)

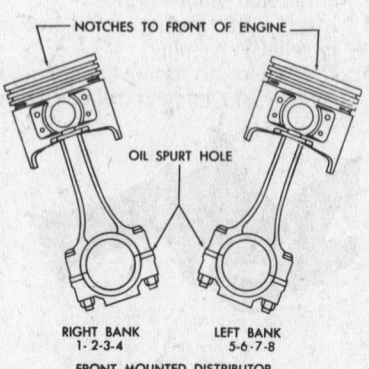

NOTCHES TO FRONT OF ENGINE

OIL SPURT HOLE

RIGHT BANK 1-2-3-4   LEFT BANK 5-6-7-8

FRONT MOUNTED DISTRIBUTOR

Piston and rod assembly—V8 289 cu. in.

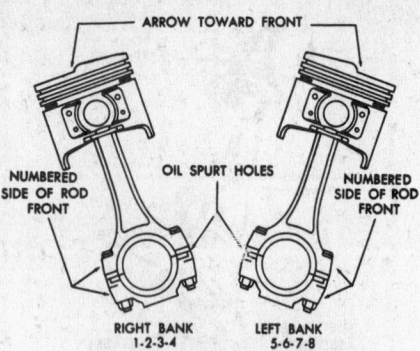

ARROW TOWARD FRONT

NUMBERED SIDE OF ROD FRONT   OIL SPURT HOLES   NUMBERED SIDE OF ROD FRONT

RIGHT BANK 1-2-3-4   LEFT BANK 5-6-7-8

Piston and rod assembly—
390, 427, and 428CJ V8s

tive bore, using thread guards on each assembly. Guide the rod bearing into place on the crankcase journal.
4. Remove thread guards from connecting rods and install lower half of bearing and cap. Check clearances.
5. Install oil pan.
6. Install cylinder head.
7. Refill crankcase and cooling system.
8. Start engine, bring to operating temperature and check for leaks.

Piston Rings

Replacement

Before replacing rings, inspect cylinder bores.

1. Using internal micrometer measure bores both across thrust faces of cylinder and parallel to axis of crankshaft at minimum of four locations equally spaced. The bore must not be out of round by more than 0.005 in. and it must not "taper" more than 0.010 in. "Taper" is the difference in wear between two bore measurements in any cylinder. Bore any cylinder beyond limits of out of roundness or taper to diameter of next available oversize piston that will clean up wear.
2. If bore is within limits dimensionally, examine bore visually. It should be dull silver in color

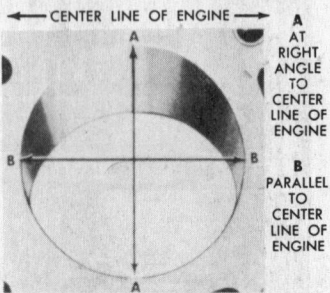

CENTER LINE OF ENGINE

A AT RIGHT ANGLE TO CENTER LINE OF ENGINE

B PARALLEL TO CENTER LINE OF ENGINE

1. OUT-OF-ROUND = DIFFERENCE BETWEEN A AND B
2. TAPER = DIFFERENCE BETWEEN THE A MEASUREMENT AT TOP OF CYLINDER BORE AND THE A MEASUREMENT AT BOTTOM OF CYLINDER BORE   A1025-A

Taper and out of roundness
(© Ford Motor Co.)

Measuring ring side clearance
(© Ford Motor Co.)

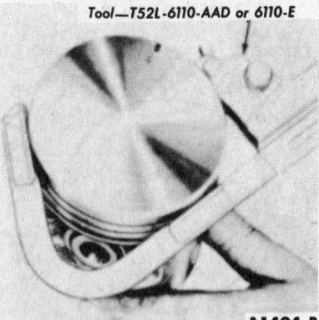

Tool—T52L-6110-AAD or 6110-E

A1404-B

Cleaning ring grooves
(© Ford Motor Co.)

and exhibit pattern of machining cross hatching intersecting at about 45 degrees. There should be no scratches, tool marks, nicks, or other damage. If any such damage exists, bore cylinder to clean up damage and then to next oversize piston diameter. Polished or shiny places in the bore are known as glazing. Glazing causes poor lubrication, high oil consumption, and ring damage. Remove glazing by honing cylinders with clean, sharp stones of No. 180-220 grit to obtain surface finish of 15-35 RMS. Use a hone also to obtain correct piston clearance and surface finish in any cylinder that has been bored.

3. If cylinder bore is in satisfactory condition, place each ring in bore in turn and square it in bore with head of piston. Measure ring gap. If ring gap is greater than limit, get new ring.

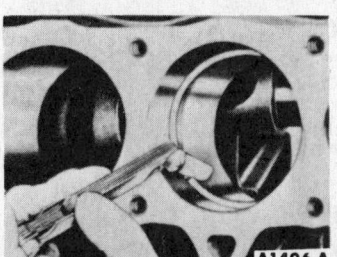

A1406-A

Measuring piston ring gap
(© Ford Motor Co.)

If ring gap is less than limit, file end of ring to obtain correct gap.

4. Check ring side clearance by installing rings on piston, and inserting feeler gauge of correct dimension between ring and lower land. Gauge should slide freely around ring circumference without binding. Any wear will form a step on lower land. Replace any pistons having high

steps. Before checking ring side clearance be sure ring grooves are clean and free of carbon, sludge, or grit.

5. Space ring gaps at equidistant intervals around piston circumference. Be sure to install piston in its original bore. Install short

Piston Clearance

Year and Engine	Minimum (in.)	Maximum (in.)	Replace (in.)
1964 144 and 170 Sixes	0.0018	0.0036	0.006
1964 200 Sixes	0.0016	0.0034	0.006
1964 260 V8	0.0021	0.0039	0.006
1964 289 V8	0.0020	0.0038	0.006
1965-69 Sixes	0.0014	0.0020	
1965-66 260 and 289 V8s	0.0014	0.0022	
1965-67 289 High Performance	0.0030	0.0038	
1967-70 289, 302 and 351 Windsor V8s	0.0018	0.0026	
1966-70 390 and 428 CJ V8s	0.0015	0.0023	
1967 427 V8①	0.0042	0.0066	
1968 427 V8②	0.0030	0.0038	
1969-70 Boss 302	0.0034	0.0042	
1970 351 Cleveland	0.0014	0.0022	
1970 429 4V V8	0.0014	0.0022	
1970 429 CJ, 429 Super CJ, and Boss 429 V8s	0.0030	0.0038	

① Solid lifter high performance version.
② Hydraulic lifter street version

Ring Gaps

Year and Engine	Top Compression (in.)		Bottom Compression (in.)		Oil Control (in.)	
	Min.	Max.	Min.	Max.	Min.	Max.
1964-65 Six	0.010	0.031	0.010	0.031	0.015	0.066
1964-65 260 V8	0.010	0.032	0.010	0.032	0.015	0.067
1964 289 V8	0.015	0.039	0.015	0.039	0.015	0.069
1965-70 289, 302, Boss 302, 351 Windsor, and 351 Cleveland V8s	0.010	0.020	0.010	0.020	0.015	0.069
1966-70 Six	0.010	0.020	0.010	0.020	0.015	0.055
1966-68 390 V8	0.010	0.031	0.010	0.020	0.015	0.066
1967 427 V8	0.010	0.031	0.010	0.020	0.015	0.066
1968 427 V8	0.018	0.028	0.010	0.025	0.015	0.055
1968-70 428 CJ, 429 4 V, 429 CJ, 429 Super CJ, and Boss 429 V8s	0.010	0.020	0.010	0.020	0.010	0.035
1969 390	0.010	0.020	0.010	0.020	0.015	0.055

Ring, Side Clearance

Year and Engine	Top Compression (in.)			Bottom Compression (in.)			Oil Control
	Min.	Max.	Replace	Min.	Max.	Replace	
1964-65 Sixes	0.0019	0.0036	0.006	0.0020	0.0040	0.006	Snug
1964-65 260 and 289 V8s	0.0019	0.0036	0.006	0.0010	0.0040	0.006	Snug
1966 170 Six	0.0009	0.0026	0.006	0.0020	0.0040	0.006	Snug
1966 200 Six	0.0019	0.0036	0.006	0.0020	0.0040	0.006	Snug
1966-67 289 V8	0.0019	0.0036	0.006	0.0020	0.0040	0.006	Snug
1966-67 390 V8	0.0020	0.0040	0.006	0.0020	0.0040	0.006	Snug
1967 427 V8	0.0024	0.0041	0.006	0.0020	0.0040	0.006	Snug
1968-70 All	0.0020	0.0040	0.006	0.0020	0.0040	0.006	Snug

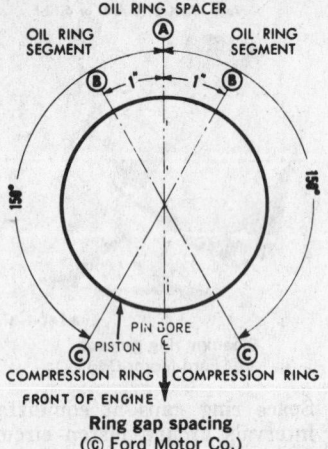

Ring gap spacing
(© Ford Motor Co.)

lengths of rubber tubing over connecting rod bolts to prevent damage to rod journal. Install ring compressor over rings on piston. Lower piston rod assembly into bore until ring compressor contacts block. Using wooden

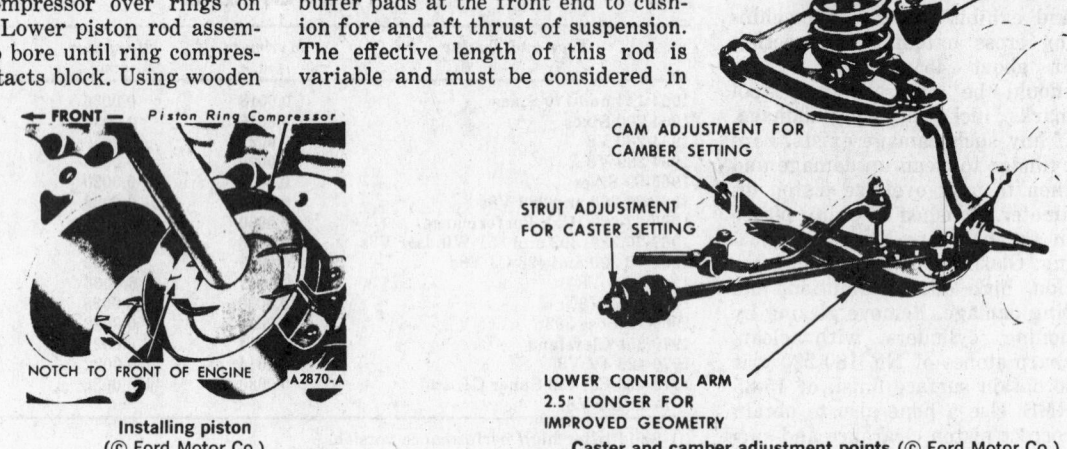

Installing piston
(© Ford Motor Co.)

handle of hammer, push piston into bore while guiding rod onto journal.

Front Suspension

References

The front coil springs are mounted on top of the upper control arm to a tower in the sheet metal of the body. This type of mounting provides good stability. The lower arm and stabilizing strut substitute for the conventional A frame and serve to guide the lower part of the spindle through its cycle of up-and-down movement. The rod-type stabilizing strut is mounted between two rubber buffer pads at the front end to cushion fore and aft thrust of suspension. The effective length of this rod is variable and must be considered in

maintenance. Ball joints are of the usual steel construction.

General instructions covering the front suspension, and how to repair and adjust it, together with information on installation of front wheel bearings and grease seals, are given in the Unit Repair Section.

Definitions of the points of steering geometry are covered in the Unit Repair Section. This article also covers troubleshooting front end geometry and irregular tire wear.

Figures covering the caster, camber, toe-in, kingpin inclination, and

Caster and camber adjustment points (© Ford Motor Co.)

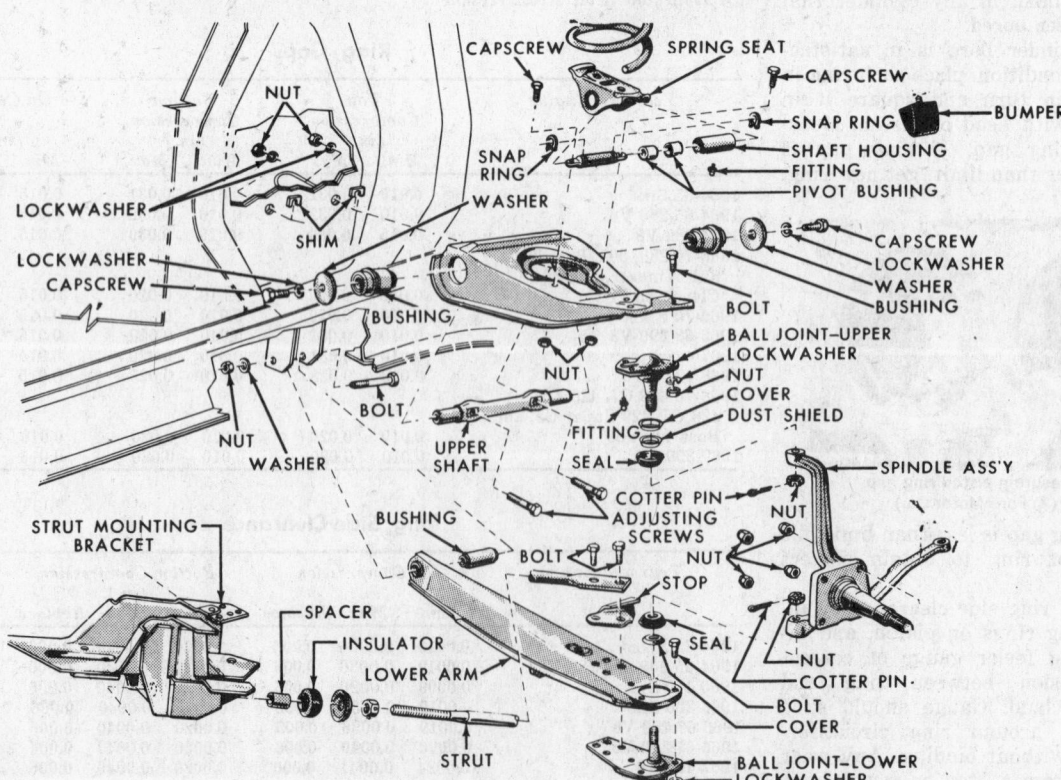

Typical front suspension (© Ford Motor Co.)

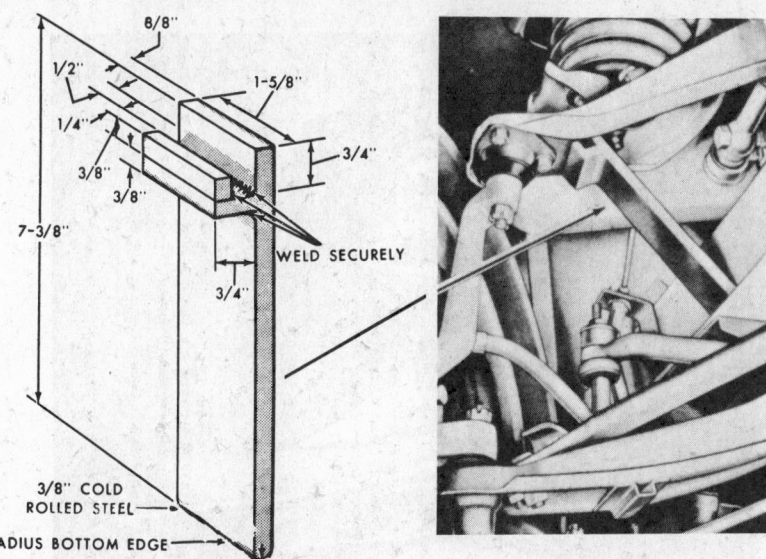

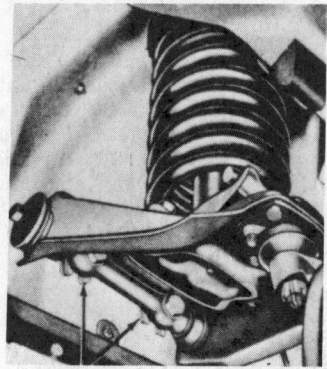

Upper control arm support (© Ford Motor Co.)

turning radius can be found in the Front Wheel Alignment table of this section.

Tire size figures can be found in the General Chassis and Brake Specifications table of this section.

Wheel Alignment

Prior To 1966

Caster and camber values are both maintained by changing the location, or by the addition or subtraction, of shims between the inner shaft of the front suspension upper arm and the underbody. Shims are available in 1/32 in. and ⅛ in. thicknesses.

Upper control arm assembly
(© Ford Motor Co.)

Beginning 1966
See Unit Repair Section.

Toe-In and Steering Wheel Alignment

Check steering wheel spoke position with the front wheels straight ahead. If the spoke position is not normal, the steering wheel can be properly adjusted while toe-in is being adjusted. See Unit Repair section for complete toe-in to steering wheel relation adjustment.

Front Spring

Removal

1. Raise hood and remove shock absorber upper mounting bracket bolts.
2. Raise front of vehicle, and place safety stands under inboard ends of lower control arms.
3. Remove shock absorber lower attaching nuts, washers and insulators.
4. Lift shock absorber and upper bracket from spring tower.
5. Remove wheel cover on hub cap.
6. Remove grease cap, cotter pin, nut lock, adjusting nut, and outer bearing.
7. Pull wheel, tire and hub and drum off spindle as an assembly.
8. Install spring compressor as shown in figures.
9. Compress spring until all tension is removed from control arms.
10. Remove two upper control arm attaching nuts and swing control arm out board.
11. Release spring compressor and remove.
12. Remove spring.

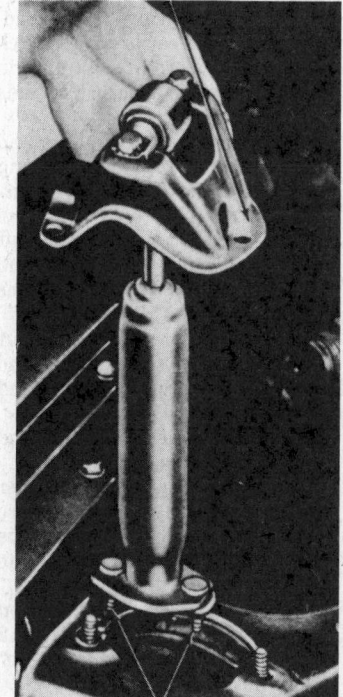

Removing shock absorber and bracket assembly
(© Ford Motor Co.)

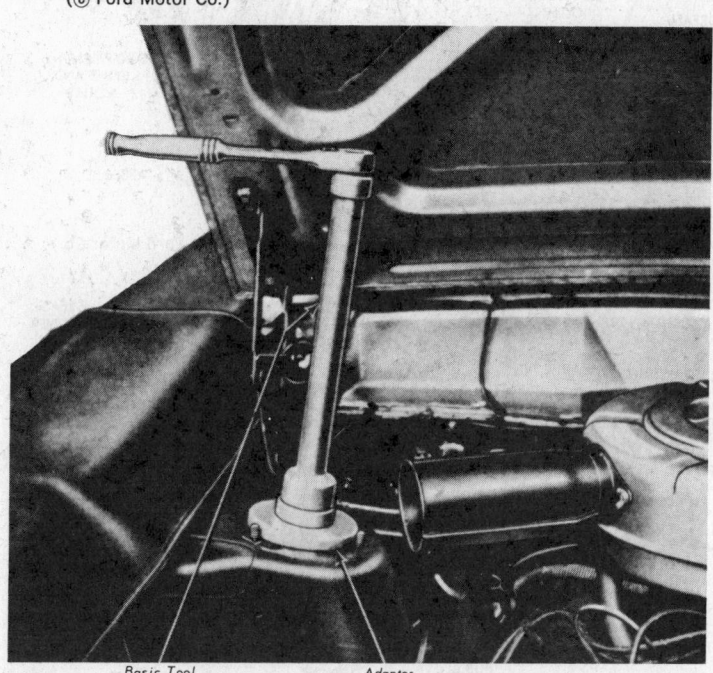

Spring compressor installed (© Ford Motor Co.)

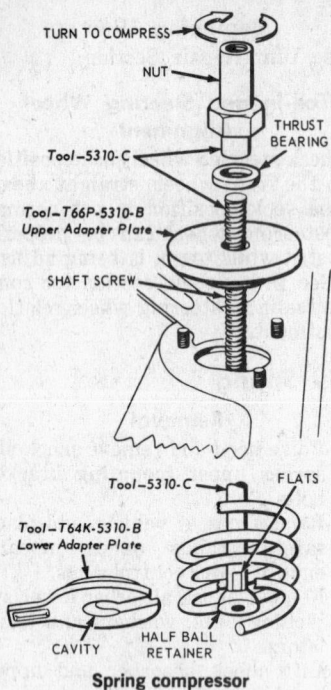

Spring compressor
(© Ford Motor Co.)

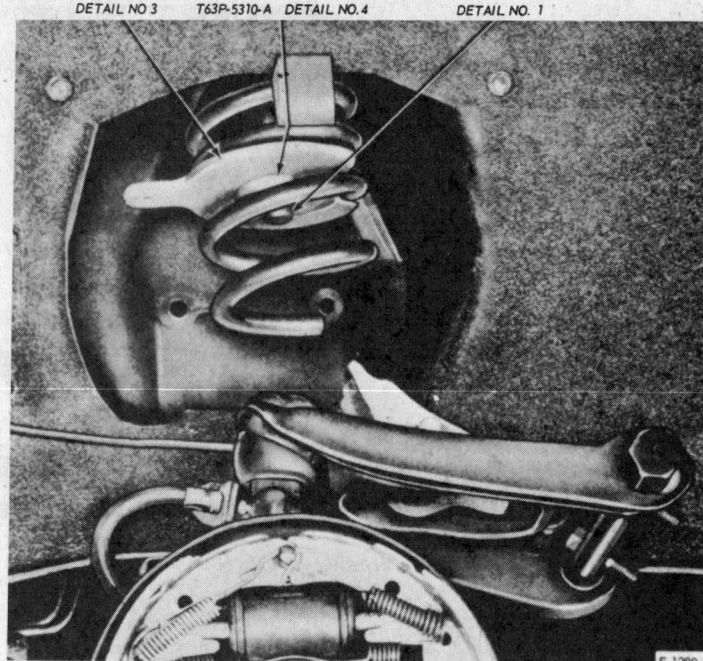

Compressing spring (© Ford Motor Co.)

Installation

1. Place upper spring insulator on spring and secure in place with tape.
2. Position spring in spring tower and compress with spring compressor.
3. Swing upper control arm in board and install attaching nuts. Torque nuts to 75-100 ft. lbs. on 1966-70 Montego, Fairlane, Comet, Falcon, and Maverick and 1967-70 Cougar and Mustang, 55-75 ft. lbs. on 1964-66 Mustangs, and 65-90 ft. lbs. on 1964-65 Comets and Falcons.
4. Release spring pressure and guide spring into upper arm spring seat. The end of the spring must be not more than ½ in. from tab on spring seat.
5. Remove spring compressor and position wheel, tire, and hub and drum on spindle.
6. Install bearing, washer and adjusting nut.
7. On disc brake cars, loosen adjusting nut three turns, and rock wheel hub and rotor assembly in and out to push disc brake pads away from rotor.
8. While rotating wheel, hub and drum assembly, torque adjusting nut to 17-25 ft. lbs. to seat bearing.
9. With 1⅛ in. box wrench back off adjusting nut ½ turn, and tighten nut to 10-15 in. lbs. or finger tight.
10. Position lock on adjusting nut and install new cotter pin. Bend ends of pin around castellated flange of nut lock.
11. Check front wheel rotation and install grease cap and hub cap.
12. Install shock absorber and upper bracket assembly, making sure shock absorber lower studs have insulators and are in pivot plate holes.
13. Install nuts and washers on lower studs and torque to 8-12 ft. lbs. on 1970 models and 12-17 ft. lbs. on 1964-69 models.
14. Install nuts on shock absorber upper bracket and torque to 10-15 ft. lbs. on Cougars, Mustangs, and Mavericks, 20-28 ft.

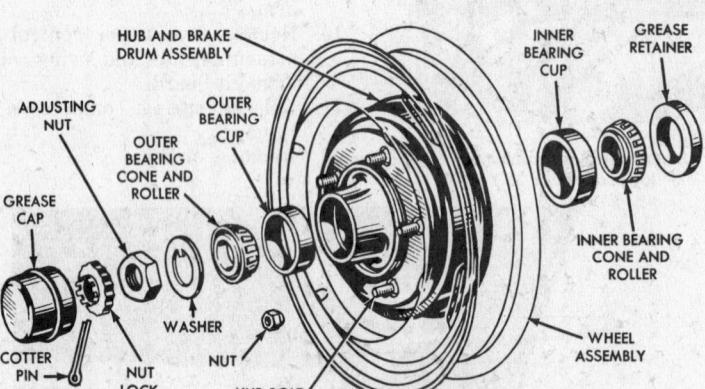

Front hub, bearings, and grease retainers—drum brakes (© Ford Motor Co.)

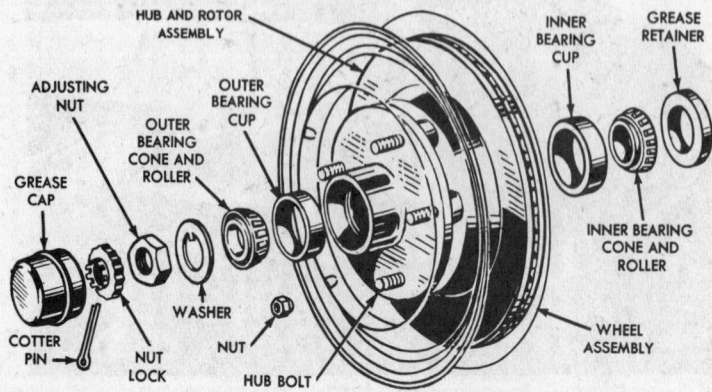

Front hub, bearings, and grease retainers—disc brakes (© Ford Motor Co.)

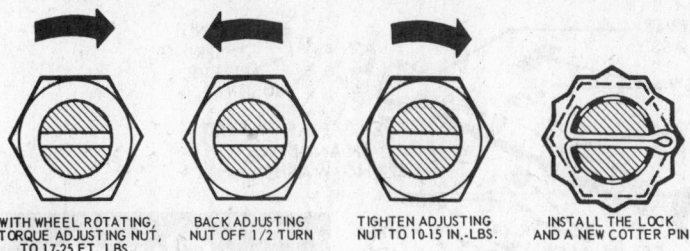

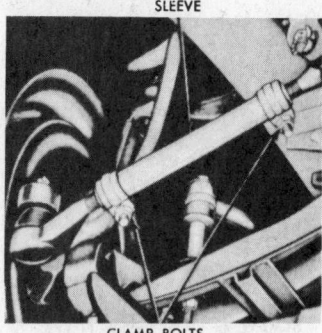

SLEEVE

CLAMP BOLTS

Tie-rod connecting sleeve
(© Ford Motor Co.)

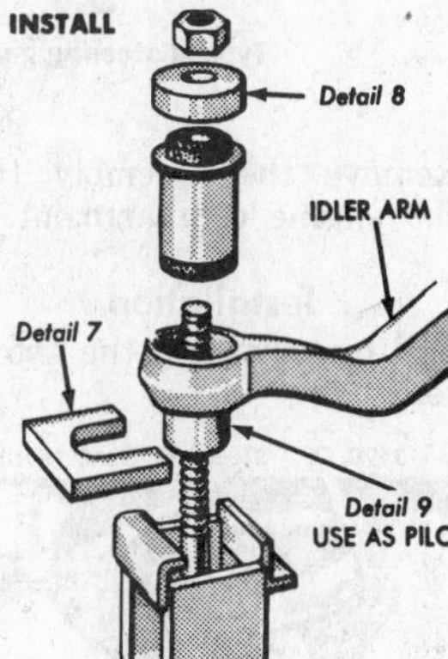

WITH WHEEL ROTATING, TORQUE ADJUSTING NUT, TO 17-25 FT. LBS.	BACK ADJUSTING NUT OFF 1/2 TURN	TIGHTEN ADJUSTING NUT TO 10-15 IN.-LBS.	INSTALL THE LOCK AND A NEW COTTER PIN

Adjusting wheel bearings (© Ford Motor Co.)

lbs. on 1966-1970 Montegos, Fairlanes, Comets, and Falcons, and 15-25 ft. lbs. on 1964-65 Comets and Falcons.

15. Lower car.

Steering Gear

A recirculating ball-type steering gear is used. Instructions for adjusting this assembly can be found in the Unit Repair Section.

Since late 1963, power steering has been available on all models. For overhaul procedures, see Unit Repair Section.

Gear Assembly

Removal

1. Raise front of car and place on safety stands.
2. Remove sector shaft.
3. Remove steering gear bolts from the underbody.
4. Disconnect shift rods from the shift levers.
5. Pull up the rubber seal on the column, fold back the floor mat and move the dash insulation out of the way. Then, take out weather seal retaining screws and the column cover plates and gasket.
6. Disconnect the horn and turn-signal wires under the instru-

ment panel. Remove the steering wheel.

7. Remove the upper bearing sleeve and spring and turn indicator lever.
8. Remove the clamp to instrument column bolts, then the clamp and insulator, and slide the tube from the steering gear shaft.

INSTALL

Detail 8

IDLER ARM

Detail 7

Detail 9 USE AS PILOT

Tool—T61P-3355-A

Idler arm bushing replacement
(© Ford Motor Co.)

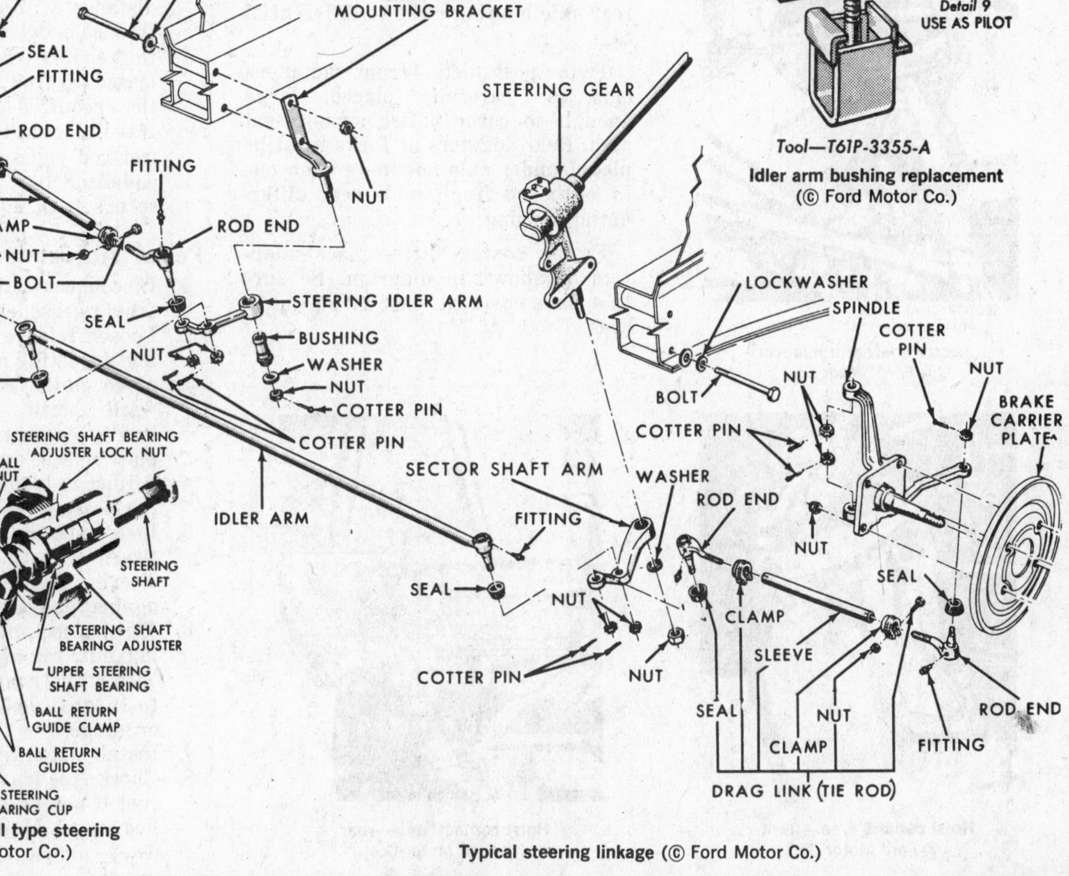

SPINDLE NUT COTTER PIN BOLT WASHER
COTTER PIN
NUT
SEAL
FITTING
NUT ROD END
IDLER ARM MOUNTING BRACKET
CLAMP
SLEEVE
CLAMP
NUT
BOLT
FITTING
NUT
ROD END
STEERING GEAR
DRAG LINK (TIE ROD)
FITTING
SEAL
SEAL
STEERING IDLER ARM
BUSHING
WASHER
NUT
COTTER PIN
NUT
COTTER PIN
LOCKWASHER
SPINDLE
COTTER PIN
NUT
NUT
BRAKE CARRIER PLATE
BOLT
COTTER PIN
SECTOR SHAFT ARM
WASHER
ROD END
NUT
FITTING
SEAL
IDLER ARM
SEAL
NUT
CLAMP
SLEEVE
NUT
ROD END
COTTER PIN
NUT
SEAL
CLAMP
FITTING

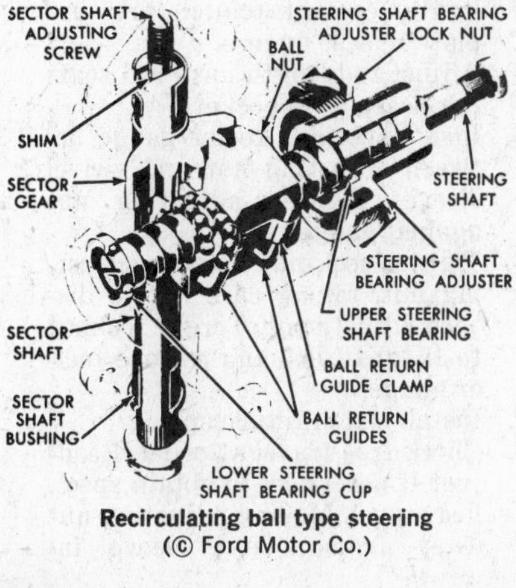

SECTOR SHAFT ADJUSTING SCREW
STEERING SHAFT BEARING ADJUSTER LOCK NUT
BALL NUT
SHIM
SECTOR GEAR
STEERING SHAFT
SECTOR SHAFT
STEERING SHAFT BEARING ADJUSTER
UPPER STEERING SHAFT BEARING
BALL RETURN GUIDE CLAMP
BALL RETURN GUIDES
SECTOR SHAFT BUSHING
LOWER STEERING SHAFT BEARING CUP

Recirculating ball type steering
(© Ford Motor Co.)

DRAG LINK (TIE ROD)

Typical steering linkage (© Ford Motor Co.)

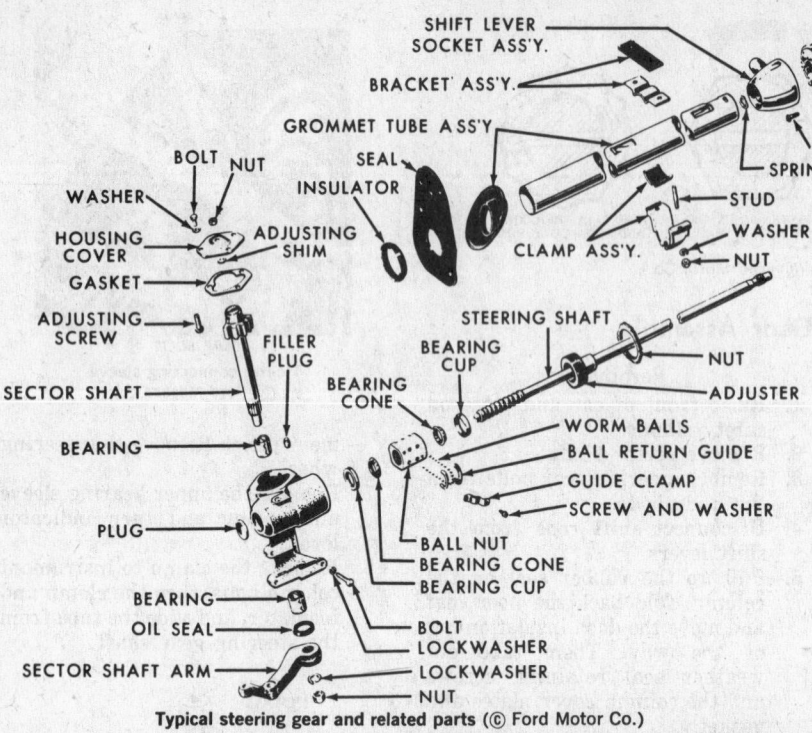

Typical steering gear and related parts (© Ford Motor Co.)

9. Remove the assembly through the engine compartment.

Installation

Install by reversing the above procedure.

Tool—3590-FC STEERING GEAR HOUSING

SECTOR SHAFT ARM
(PITMAN ARM)

Sector shaft arm removal
(© Ford Motor Co.)

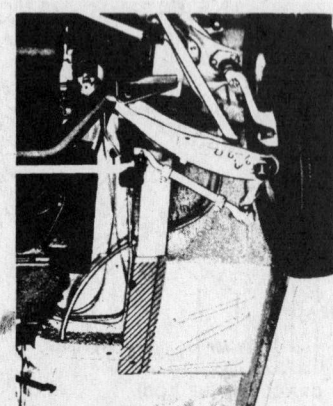

Hoist contact area—front
(© Ford Motor Co.)

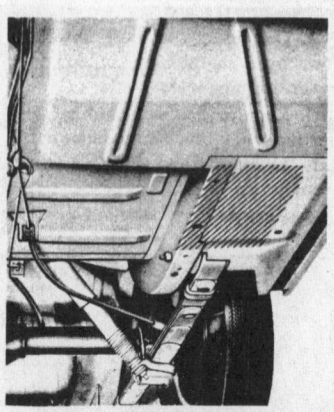

Hoist contact area—rear
(© Ford Motor Co.)

Jacking, Hoisting

Jack car at front under spring seat of lower control arm. Jack car at rear axle housing close to differential case.

Twin post lifts—front adapters must be carefully placed, large enough to cover entire spring seat area. Rear adapters or forks must be placed under axle not more than one in. outboard from welds near differential housing.

Frame contact lifts—place adapters as shown in diagram. Be sure that pads cover at least 12 sq. in. in area.

Clutch

The clutch cover and pressure plate, and the clutch disc plan is dependent upon power and transmission application.

Pedal Adjustment—1964-65

1. To check pedal assist spring tension, measure the distance between the inside radius of the spring hook and the front face of the link. This distance should be 1 3/16 in. Turn the nut on the retainer to get proper pedal assist.
2. Measure the total pedal travel. If the total travel is not within 6-6½ in., move the pedal bumper and bracket up or down, as necessary.

NOTE: Always check and adjust total travel before checking free travel.

3. To check pedal free travel, start to depress the pedal slowly until the release fingers contact the release bearing. Measure this distance with a rule. The difference between this reading and the reading when the pedal is released is free travel. To obtain the required ⅞-1⅛ in. free travel, loosen the pedal-to-equalizer-rod nuts and rotate the equalizer bar, as needed. Then secure both equalizer bar nuts.

Pedal Adjustment—1966-70

1. Disconnect clutch return spring from release lever.
2. Loosen release lever rod locknut and adjusting nut.
3. Move clutch release lever rearward until release bearing lightly contacts clutch pressure plate release fingers.
4. Adjust rod length until rod seats in release lever pocket.
5. Insert specified feeler gauge between adjusting nut and swivel sleeve. Tighten adjusting nut against gauge.
6. Tighten lock nut against adjusting nut, taking care not to disturb adjustment. Torque locknut to 15-20 ft. lbs. and remove feeler gauge.
7. Install clutch return spring.
8. Check free travel at pedal. Readjust if necessary to obtain specified travel. Moving adjusting nut away from swivel sleeve in-

creases travel. Moving adjusting nut toward swivel sleeve decreases travel.

9. As final check, measure pedal free travel with transmission in neutral and engine running at 3,000 rpm. If pedal travel is not minimum of ½ in., readjust free travel.

Clutch Pedal Adjustment

Year and Engine	Clearance* (in.)	Free Travel (in.)
1966 Six	0.178	⅞-1⅛
1966 V8	0.128	⅞-1⅛
1967 Six	0.178	¾-1⅛
1967 V8	0.128	¾-1⅛
1968 except 390, 427, 428	0.136	¾-1⅛
1968 390, 427 and 428	0.178	¾-1⅛
1969-70 except 390 and 428	0.136	⅞-1⅛
1969-70 390 and 428	0.178	⅞-1⅛

* Between adjusting nut and swivel sleeve

Clutch and/or Transmission Removal

1. Disconnect and remove starter and dust ring, if the clutch is to be removed.
2. Raise the car.
3. Disconnect the driveshaft at the rear universal joint and remove the driveshaft.
4. Disconnect the speedometer cable at the transmission extension.

5. Disconnect the gear shift rods from the transmission shift levers. If car is equipped with four speed, remove bolts that secure shift control bracket to extension housing.
6. Remove the bolt holding the extension housing to the rear support, and remove the muffler inlet pipe bracket to housing bolt.
7. Remove the two rear support bracket insulator nuts from the underside of the crossmember. Remove crossmember.
8. Place a jack (equipped with a protective piece of wood) under the rear of the engine oil pan. Raise the engine, slightly.
9. Remove transmission - to - flywheel-housing bolts. Thread two guide studs into the bottom attaching bolt holes.
10. Slide the transmission back and out of the car.
11. Remove release lever retracting spring and disconnect pedal at the equalizer bar.
12. Remove bolts that secure engine rear plate to front lower part of bellhousing.
13. Remove bolts that attach bell housing to cylinder block and remove housing and release lever as a unit.
14. Loosen six pressure plate cover attaching bolts evenly to release spring pressure. Mark cover and flywheel to facilitate reassembly in same position.
15. Remove six attaching bolts while holding pressure plate cover. Remove pressure plate and clutch disc.

Clutch and/or Transmission Installation

1. Wash flywheel surface with alcohol.
2. Attach the clutch disc and pressure plate assembly to the flywheel with the bolts finger tight.
3. Align the clutch disc with the pilot bushing. Torque cover bolts to 23-28 ft. lbs. on 1964-66 vehicles and to 12-20 ft. lbs. on 1967-70 vehicles.
4. Lightly lubricate the release lever fulcrum ends. Install the release lever in the flywheel housing and install the dust shield.
5. Apply very little lubricant on the release bearing retainer journal. Attach the release bearing and hub on the release lever.
6. Install the flywheel housing and torque the attaching bolts to 40-50 ft. lbs. on all 1964-65 vehicles and on 1966-70 V8s. Torque 1966-70 sixes to 23-33 ft. lbs. Install the dust cover and torque the bolts to 17-20 ft. lbs.
7. Connect the release rod and the retracting spring. Connect the pedal - to - equalizer - rod at the equalizer bar.
8. Install starter and dust ring.
9. Start the transmission extension housing up and over the rear support. After moving the transmission back just far enough for the pilot shaft to clear the clutch housing, move it upward and into position on the transmission guide studs.

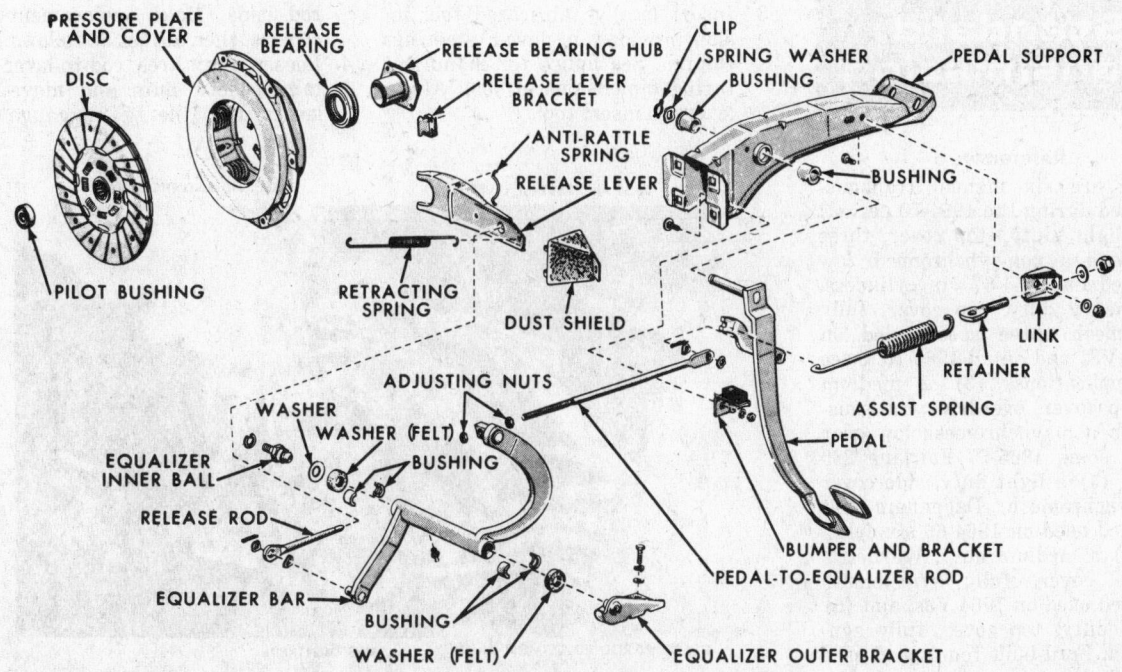

Typical clutch pedal mounting and linkage (© Ford Motor Co.)

3-speed synchromesh transmission (© Ford Motor Co.)

10. Move the transmission forward and into place against the flywheel housing.
11. Remove guide studs and attach the transmission with a torque of 37-42 ft. lbs. on all cars except 1964-66 six cylinder four speed. Torque 1964-66 six cylinder four speed to 40-45 ft. lbs.
12. Slowly lower the engine onto the crossmember.
13. Install and torque the insulator-to-crossmember nuts to 25-35 ft. lbs. on all 1964-69 vehicles except 1968-69 390, 427 and 428 CJ Cougars and Mustangs. Torque 1968-69 390, 427 and 428 CJ Cougars and Mustangs to 30-42 ft. lbs. Torque 1970 Montegos, Fairlanes and Mavericks to 30-50 ft. lbs. and 1970 Cougars and Mustangs to 25-35 ft. lbs.
14. Connect gear shift rods and the speedometer cable.
15. Hook up the drive shaft.
16. Refill transmission to proper level.

Standard Transmission

References

There are six manual transmissions used during the 1964-70 period: (1) a light duty, top cover, three speed with a non-synchromesh low gear used on 1964-67 six cylinders, (2) a heavy duty, top cover, fully synchromesh three speed used on 1964-67 V8s and on all 1968-70 three speed applications, (3) a medium duty, top cover, overdrive transmission with non-synchromesh low gear used on some 1965-67 Fairlane 289 2V V8s, (4) a light duty, side cover fully synchromesh, Dagenham-built four speed used on 1964-66 six cylinders, (5) a medium duty, Warner T 10, side cover, fully synchomesh four speed used on 1964 V8s, and (6) a heavy duty, top cover, fully synchromesh, Ford-built four speed used on 1965-70 V8s.

Three-Speed Column Shift Linkage Adjustment

With the transmission in neutral, the shift lever should be in a horizontal plane and parallel to the instrument panel line. Corrective adjustments should be made at the gear shift rods.

1964-65 and 1968-70

1. Place lever in neutral.
2. Loosen two gear shift rod adjustment nuts.
3. Insert 3/16 in. diameter alignment pin through first and reverse gear shift lever and second and third gear shift lever. Align levers to insert pin.
4. Tighten gear shift rod adjustment nuts, and remove pin.
5. Check gear lever for smooth crossover.

1966-67

1. Place lever in neutral.
2. Loosen two gear shift rod adjustment nuts.
3. Insert locally fabricated tool in slot provided in lower steering column. See figure for manufacturing dimensions of tool. Align levers to insert tool.

4. Tighten gear shift rod adjustment nuts, and remove tool.
5. Check gear lever for smooth crossover.

Three Speed Floor and Console Shift Linkage

1. Loosen three shift linkage adjustment nuts.
2. Install a ¼ in. diameter alignment pin through control bracket and levers.
3. Tighten three shift linkage adjustment nuts and remove alignment pin.
4. Check gear lever for smooth crossover.

Four-Speed Linkage— Dagenham Four Speed

1. Place shifter lever in neutral position, then raise car on a hoist.
2. Insert a ¼ in. rod into the alignment holes of the shift levers.
3. If the holes are not in exact alignment, check for bent connecting rods or loose lever lock nuts at the rod ends. Make replacements or repairs, then adjust as follows.
4. Loosen the three rod-to-lever retaining lock nuts and move the levers until the ¼ in. gauge rod

FIRST AND REVERSE LEVER   SECOND AND THIRD LEVER

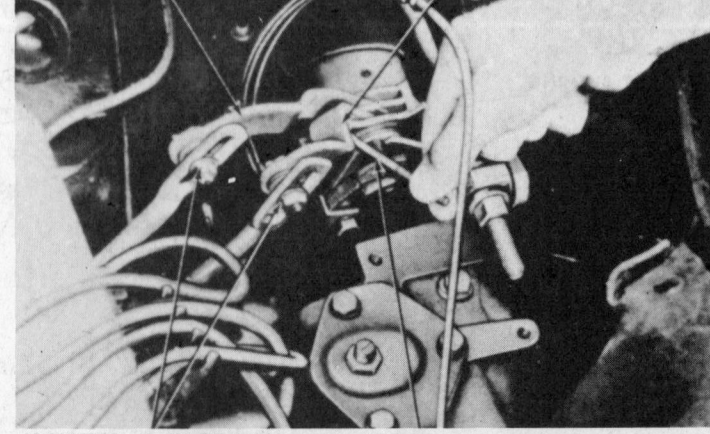

GEARSHIFT ROD ADJUSTMENT NUTS   ALIGNMENT PIN

Manual transmission column shift adjustment 1964-65 and 1968-70 (© Ford Motor Co.)

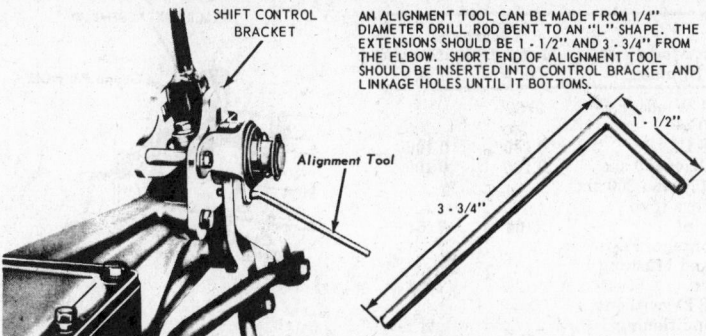

AN ALIGNMENT TOOL CAN BE MADE FROM 1/4" DIAMETER DRILL ROD BENT TO AN "L" SHAPE. THE EXTENSIONS SHOULD BE 1 - 1/2" AND 3 - 3/4" FROM THE ELBOW. SHORT END OF ALIGNMENT TOOL SHOULD BE INSERTED INTO CONTROL BRACKET AND LINKAGE HOLES UNTIL IT BOTTOMS.

Manual transmission floor or console shift adjustment (© Ford Motor Co.)

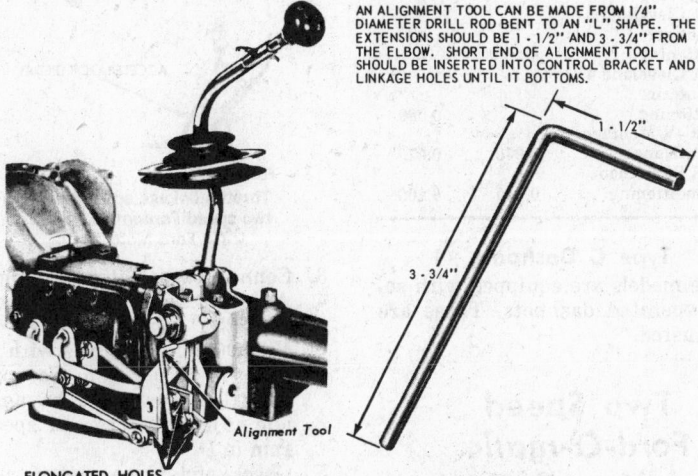

AN ALIGNMENT TOOL CAN BE MADE FROM 1/4" DIAMETER DRILL ROD BENT TO AN "L" SHAPE. THE EXTENSIONS SHOULD BE 1 - 1/2" AND 3 - 3/4" FROM THE ELBOW. SHORT END OF ALIGNMENT TOOL SHOULD BE INSERTED INTO CONTROL BRACKET AND LINKAGE HOLES UNTIL IT BOTTOMS.

Typical floor shift linkage adjustment. Dagenham four speed (© Ford Motor Co.)

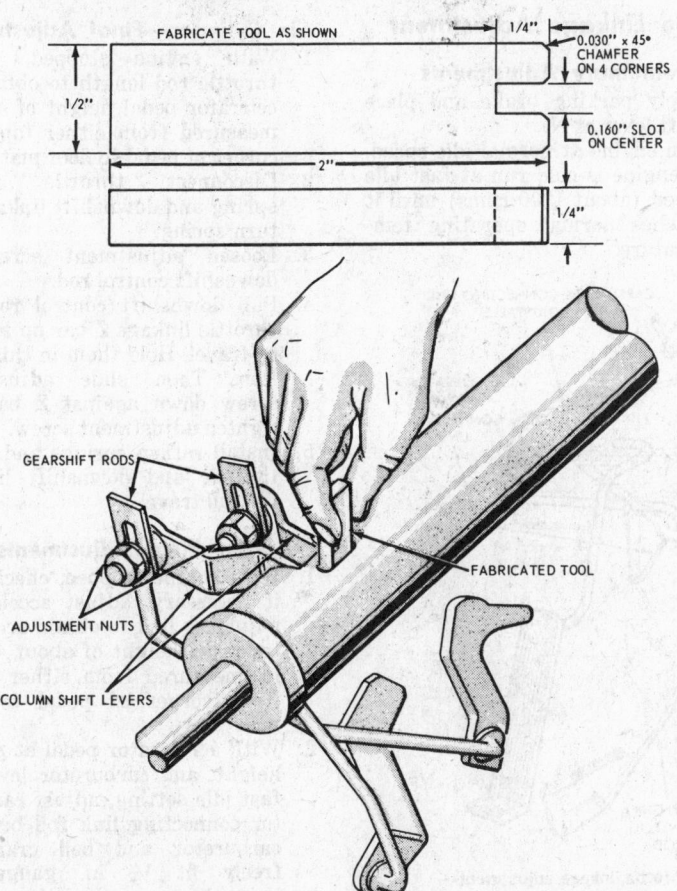

Manual transmission column shift adjustment 1966-67 (© Ford Motor Co.)

will enter the alignment holes. Be sure that the transmission shift levers are in neutral and the reverse shifter lever is in the neutral detent.

5. Install the shift rods and torque the lock nuts to 15 to 20 ft. lbs.
6. Remove the ¼ in. gauge rod.
7. Operate the shift levers to assure correct shifting.
8. Lower the car and road test.

Four Speed Linkage—Warner and Ford Transmissions

See Ford–Thunderbird Section.

Automatic Transmission

There are three automatic transmission designs used in compact and intermediate Ford and Mercury automobiles: the two speed Ford-O-Matic or Merc-O-Matic used on some 1964 models; the three speed, medium duty C4 automatic used on some 1964 models and all 1965 and later sixes and V8s up to 351 cu. in.; and the three speed, heavy duty, C6 transmission used on 1966 and later V8s of 351 cu. in. and larger.

Quick Service Information

When automatic transmission trouble is reported, a road test and careful diagnosis are in order. Transmission Removal and Replacement and Linkage Adjustments are covered in the following paragraphs. For test procedures, transmission reconditioning and other detailed information, see Unit Repair Section.

Dashpot Adjustment

Type A Dashpots

1. Adjust throttle to fast idle position and turn dashpot adjusting

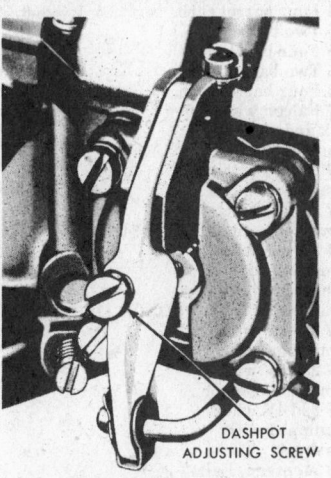

ADJUST THROTTLE TO HOT IDLE POSITION PRIOR TO ADJUSTING DASHPOT

Type A dashpot adjustment (© Ford Motor Co.)

screw out until it is clear of dashpot plunger assembly.

2. Turn screw in until it contacts plunger. Then turn dashpot adjusting screw in specified number of turns against plunger.

NOTE: Not all engines are equipped with dashpots, and not all are adjusted in this manner. This chart applies only to models mentioned.

Year, Engines and Models	No. of Turns
1964-66 six cylinder automatic vehicles	3½
1967 Falcon with 170 six cylinder, automatic and without emission control	3½
1967 Comet, Fairlane, Falcon, and Mustang with 200 six cylinder and automatic and without emission control	3½
1967 Comet, Fairlane, Falcon and Mustang with six cylinder, auto transmission, and emission control	2
1969 Falcon and Mustang 200 six cylinder with automatic	2¼
1969 Falcon and Mustang 200 six cylinder with manual transmission	3¼

Type B Dashpots

1. With engine idle speed and mixture properly adjusted and with engine at operating temperature, loosen dashpot lock nut.
2. Hold throttle in closed position and depress dashpot plunger. Measure clearance between plunger and cam. Adjust dashpot nut to give proper clearance.

NOTE: Not all engines have dashpots and not all are adjusted in this manner. This chart applies only to models mentioned.

Year, Model and Engine	Clearance Manual (in.)	Automatic (in.)
1964 Two barrel carb.		⅛-3/32
1964 Four barrel carb.		⅛-3/32
1965 Two barrel carb.		5/64
1965 Four barrel carb.		5/16
1966 Two barrel carb.		0.060-0.090
1966 Four barrel carb.		0.060-0.090
1967 Falcon with 170 six and emission control		⅛
1967 289 2 V without emission control		0.060-0.090
1967 289 2 V with emission control	0.110-0.140	0.110-0.140
1967 390 2V with emission control when equipped with dashpot	0.080-0.110	0.110-0.140
1967 390 2V without emission control	⅛	⅛
1967 289 4V with emission control	⅛	⅛
1967 390 4V when equipped with dashpot		⅛
1969 Montego, Fairlane and Mustang 250 six except when solenoid equipped	0.080	
1969 302 2V	⅛	⅛
1969 351 Windsor 2V	7/64	

Year, Model and Engine	Clearance Manual (in.)	Automatic (in.)
1969 351 Windsor 4V	3/32	
1969 390 4V	⅛	
1969 428 CJ	0.100	0.100
1969 Falcon 170 six	0.100	0.100
1970 170 and 200 six cylinders if so equipped	7/64	7/64
1970 Montego, Fairlane and Mustang 250 six		7/32
1970 302 2V without air conditioning		⅛
1970 351 Windsor 2V without air conditioning		⅛
1970 351 Cleveland 2V without air conditioning		⅛
1970 351 Cleveland 4V without air conditioning		0.080
1970 429 4V Montego and Fairlane	0.070	0.070
1970 428 CJ without air conditioning	0.140	0.200

Type C Dashpots

Some models are equipped with solenoid operated dashpots. These are not adjusted.

Two Speed Ford-O-matic and Merc-O-matic

Throttle Linkage Adjustment

Preliminary Adjustments

1. Apply parking brake and place shift lever at N.
2. Run engine at normal idle speed. If engine is cold run at fast idle speed (about 1200 rpms) until it reaches normal operating temperature.

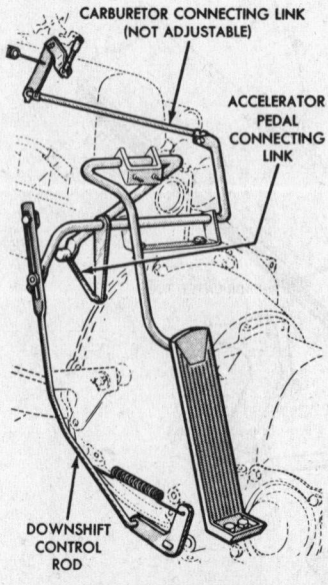

Throttle linkage adjustment—two speed Fordomatic with six cylinder
(© Ford Motor Co.)

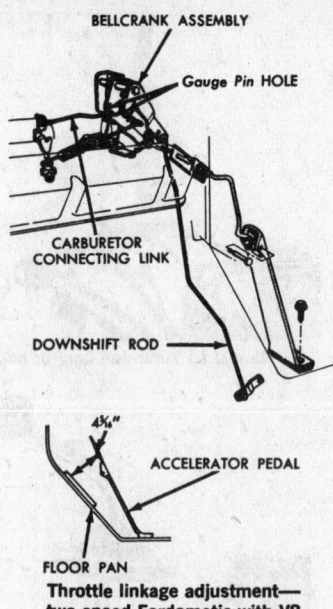

Throttle linkage adjustment—two speed Fordomatic with V8
(© Ford Motor Co.)

3. Connect tachometer to engine.
4. Adjust engine idle speed to 500-525 rpm for V8s and 475-500 rpm for sixes with shift lever in D position. Carburetor throttle lever must be against idle adjusting screw at specified rpm in D.
5. Check and if necessary, adjust dashpot plunger.

Six Cylinder—Final Adjustment

1. With engine stopped, adjust throttle rod length to obtain accelerator pedal height of 4¼ in. measured from either top front corner of pedal to floor mat.
2. Disconnect throttle return spring and downshift linkage return spring.
3. Loosen adjustment screw on downshift control rod.
4. Pull downshift control rod and throttle linkage Z bar up to limit of travel. Hold them in this position. Then slide adjustment screw down against Z bar and tighten adjustment screw.
5. Install return springs and check throttle and downshift linkage for full travel.

V8s—Final Adjustments

1. With engine stopped, check and, if necessary, adjust accelerator adjusting link to obtain accelerator pedal height of about 4 5/16 in. measured from either upper front corner of pedal to floor mat.
2. With accelerator pedal at proper height, and carburetor lever off fast idle setting, adjust carburetor connecting link rod between carburetor and bell crank to freely fit ¼ in. gauge pin through gauge pin holes of bell crank.

3. The down shift control rod is not adjustable.

Manual Linkage Adjustment

1. With engine stopped, loosen clamp at the shift lever so that the shift lever is free to slide in the clamp.
2. Put the selector in D position.
3. Shift the manual lever at the transmission into the D detent.
4. Tighten clamp on shift rod.
5. Recheck pointer alignment for all selector lever detent positions.

Ford-O-matic R & R

1. Drive car onto hoist, but do not raise at this time.
2. Disconnect starter cable.
3. Raise car on hoist.
4. Drain transmission fluid.
5. If transmission oil is water cooled, disconnect oil cooler lines from transmission.
6. Disconnect drive shaft at rear U-joint and remove it. Plug tail shaft to prevent loss of fluid.
7. Disconnect manual and downshift linkage.
8. Remove starter.
9. On some models it is necessary to remove exhaust pipe from manifolds and muffler.
10. Disconnect speedometer cable from extension housing and vacuum line from diaphragm.
11. Disconnect oil filler tube from case.
12. Remove parking brake front cable from equalizer bar.
13. Place jack under transmission and support transmission.
14. Remove bolts that retain transmission mount to transmission extension housing.
15. Lower transmission and support rear of engine.
16. Remove converter lower cover and remove four stud nuts that retain converter to flywheel.
CAUTION: do not turn flywheel with wrench on one of these nuts.
17. Remove converter housing to engine block bolts.
18. Secure transmission to jack and remove from car. Reverse procedure to install.

C4 Three-Speed Automatic

Throttle Linkage Adjustment

Initial Adjustments

1. Apply parking brake and place selector lever at N.
2. Run engine at normal idle speed. If engine is cold, run engine at fast idle speed (about 1200 rpm) until it reaches normal operating temperature. When en-

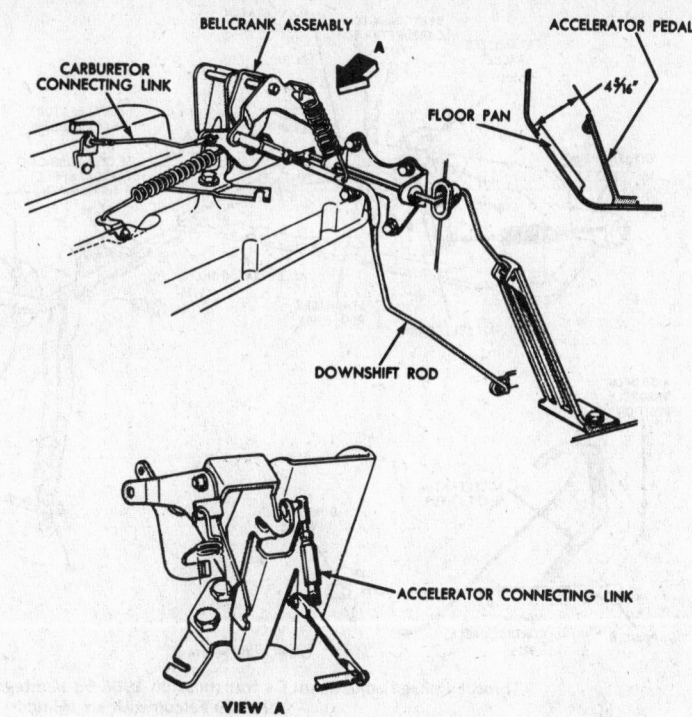

VIEW A

Throttle linkage adjustment C4 transmission 1964-65 Comet and Falcon
(© Ford Motor Co.)

gine is warm, slow it down to normal idle speed.
3. Connect tachometer to engine.
4. Adjust engine idle speed to specified rpm with transmission selector lever at D or D_1 or D_2.
5. The carburetor throttle lever must be against hot idle speed adjusting screw at specified idle speed in D or D_1 or D_2.

1964-65 Comet and Falcon— Final Adjustments

1. With engine off, check accelerator pedal for height of $4\frac{1}{4}$ in. measured from top of pedal to floor pan. To obtain correct height, adjust accelerator connecting link.
2. With engine off, loosen lock nut at adjustable upper end of downshift rod.
3. With carburetor choke in off position, depress accelerator pedal to the floor. Block pedal to hold it in wide open position.
4. Adjust downshift rod to place rod in kickdown detent position.
5. Back off adjustment to allow about 1/16 in. of free travel in bell crank assembly. Tighten lock nut and release accelerator pedal.

1967-69 Mustang and Cougar Sixes and V8s, 1966-68 Comet, Montego, and Fairlane Sixes and 1966-69 Falcon Sixes and V8s—Final Adjustments

1. With engine off, check accelerator pedal for height of $4\frac{1}{2}$ in. measured from top of pedal at

pivot point to floor pan. To obtain correct pedal height, adjust accelerator connecting link at point A in figure.
2. With engine off disconnect downshift control cable at point B from accelerator shaft lever.
3. With carburetor choke in off position, depress accelerator to floor. Block pedal to hold it in wide open position.
4. Rotate downshift lever C counter clockwise to place it against internal stop.
5. With lever held in this position, and all slack removed from cable, adjust trunnion so that it will slide into accelerator shaft lever. Turn one additional turn clockwise, then secure it to lever with retaining clip.
6. Remove block to release carburetor linkage.

1964-66 Mustang Sixes and V8s—Final Adjustments

1. With engine stopped and accelerator pedal in normal idle position, check pedal for height of $3\frac{7}{8}$ in. Be sure fast idle cam is not contacting fast idle screw of carburetor.
2. To check for free pedal travel, depress accelerator pedal to full throttle position (carburetor throttle lever against full throttle stop). Release pedal and recheck pedal height.
3. If necessary, adjust pedal height. On six cylinder engines, disconnect carburetor return spring and carburetor rod. Ad-

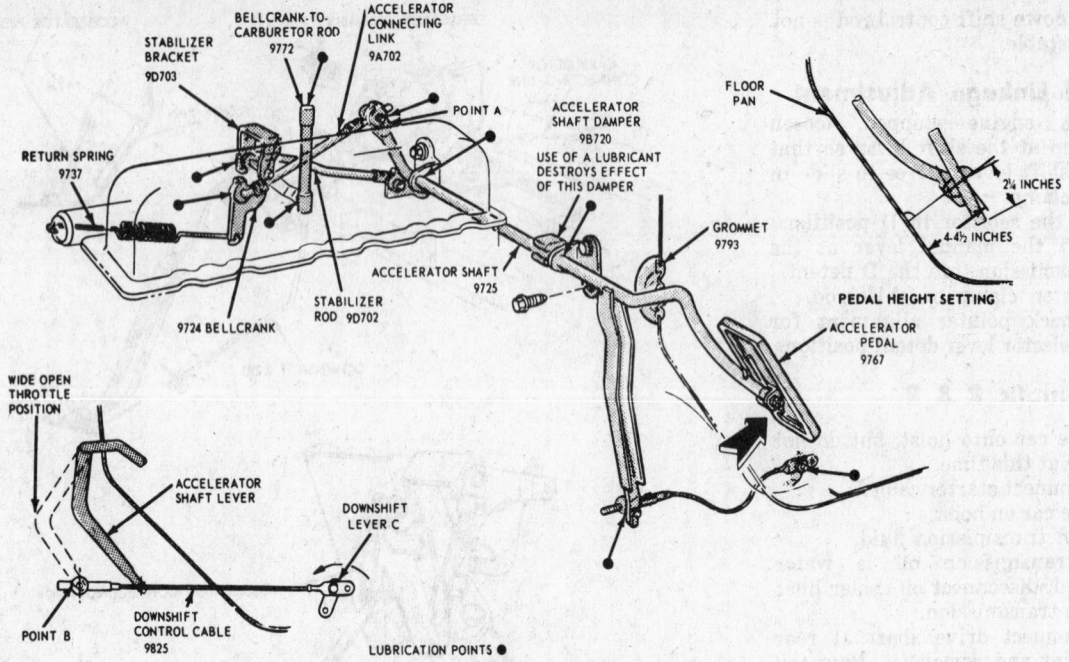

**Throttle linkage adjustment C4 transmission 1966-68 Montego, Comet, and Fairlane, and
1966-69 Falcon with six cylinder**
(© Ford Motor Co.)

just length of rod to bring pedal height within specifications. Connect carburetor rod, tighten jam nut and install return spring. On V8 engines, disconnect carburetor return springs and carburetor rod at point C. Adjust length of rod to bring pedal height within specifications. Connect carburetor rod and return spring.

4. On six cylinder engines disconnect downshift cable return spring at transmission, carburetor return spring at manifold

and downshift cable at point A.

5. Position downshift lever in downshift position (carburetor wide open).

6. Hold downshift lever on transmission against stop in counter clockwise direction (downshift position).

7. Adjust trunnion at point A on downshift cable so that it aligns with hole in downshift lever and install attaching clip.

8. Install return springs.

9. On V8 engines, disconnect downshift return spring at bell crank,

carburetor return spring, and downshift lever at point B.

10. Hold carburetor rod in wide open position. The step in rod should place bell crank in downshift position.

11. Hold downshift lever rod in downward position. This places transmission lever in downshift position.

12. Adjust downshift lever trunnion at point B so that it aligns with hole in bell crank. Install trunnion and retaining clip.

13. Release levers and install carburetor rod and bell crank.

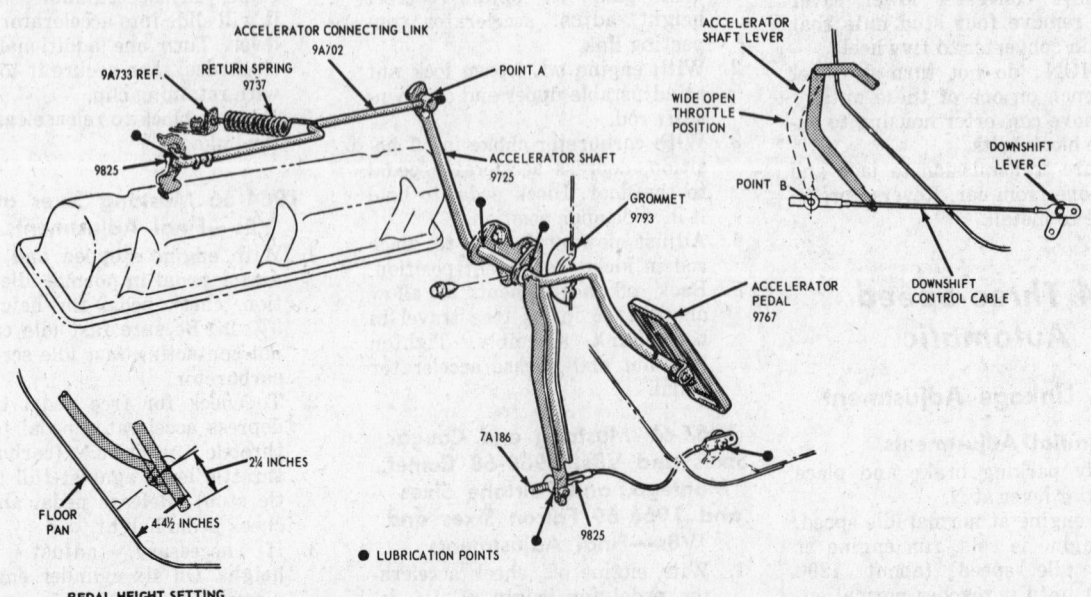

Throttle linkage adjustment C4 transmission 1966-69 Falcon V8 (© Ford Motor Co.)

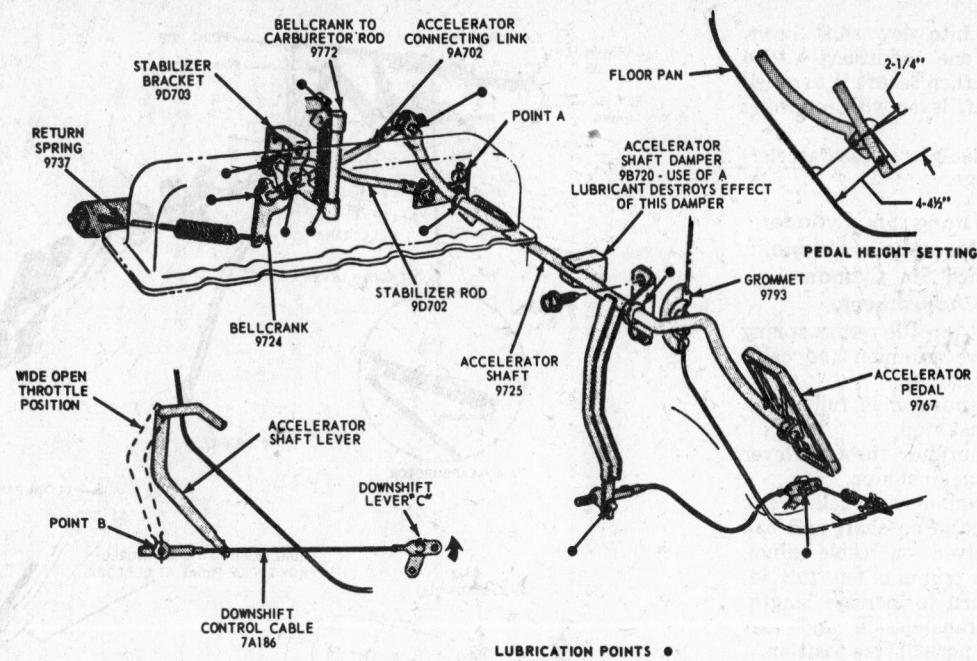

Throttle linkage adjustment C4 transmission 1967-68 Mustang six (© Ford Motor Co.)

1966-69 Comet, Montego, and Fairlane with V8 Engine, 1969 Montego and Fairlane with Six Cylinder Engine—Final Adjustment

1. Disconnect bell crank to carburetor rod at point C and accelerator connecting link from throttle shaft at point B.
2. Disconnect stabilizer rod from stabilizer at point B.
3. Insert 1/4 inch diameter pin through stabilizer and bracket.
4. Adjust length of stabilizer rod so that trunnion enters stabilizer freely. Secure stabilizer rod with retaining clip.
5. Secure carburetor to bell crank rod to bell crank with attaching clip at point C.
6. Adjust length of accelerator rod connecting link to obtain accelerator pedal height of 4-4 1/2 in. measured from top of pedal at pivot point. Connect accelerator connecting link to accelerator shaft with retaining clip after proper accelerator pedal height is obtained.
7. With engine off, disconnect downshift control cable at point D from accelerator shaft lever.
8. Rotate downshift lever E counter clockwise to place it against internal stop.
9. With lever held in this position, and all slack removed from cable, adjust trunnion so that it

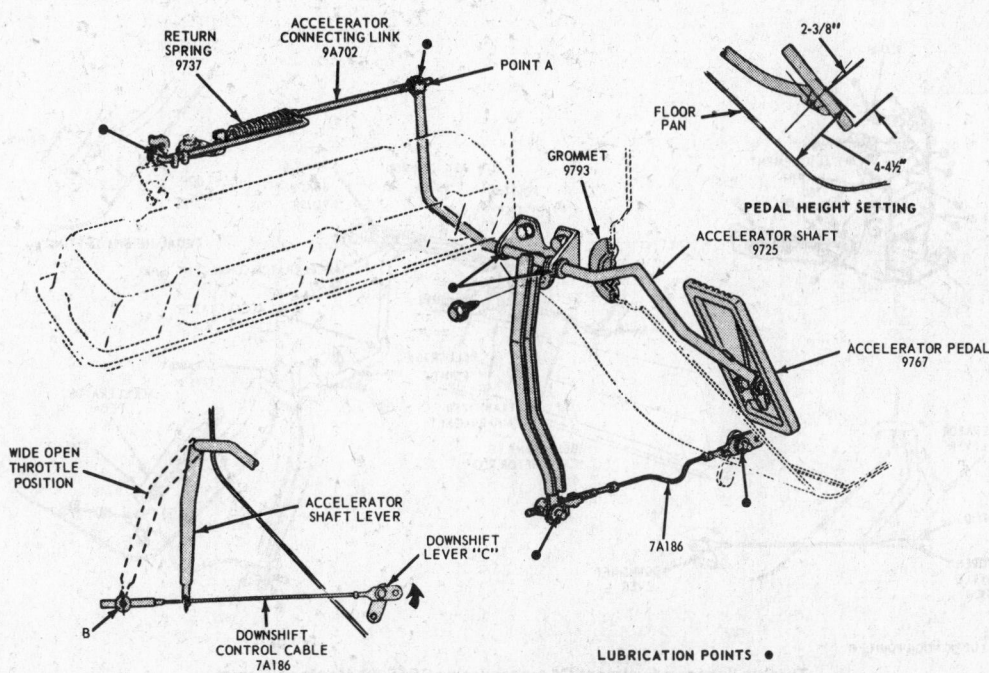

Throttle linkage adjustment C4 transmission 1967-68 Cougar and Mustang V8
(© Ford Motor Co.)

will slide into downshift lever. Turn it one additional turn clockwise, then secure it to accelerator shaft lever with retaining clip.

10. Remove block to release accelerator linkage.

1969-70 Mustang Six Cylinder, and 1970 Montego, Fairlane, and Maverick Six Cylinder— Final Adjustments

1. Disconnect throttle return spring an dremove trunnion and cable at bell crank.
2. Hold transmission in full downshift against stop.
3. Hold carburetor throttle lever wide open against stop.
4. Adjust trunnion at bell crank until ball stud on shaft and ball stud receiver on cable align. Then turn trunnion one full additional turn to increase length.
5. Release transmission and carburetor to normal free position.
6. Install throttle return spring.

1969-70 Cougar and Mustang V8s, and 1970 Montego and Fairlane V8s—Final Adjustments

1. Disconnect throttle and downshift return springs.
2. Hold carburetor throttle lever in wide open position against stop.
3. Hold transmission in full downshift position against internal stop.
4. Turn adjustment screw on carburetor downshift lever to within 0.040-0.080 in. of contacting pickup surface of carburetor throttle lever.

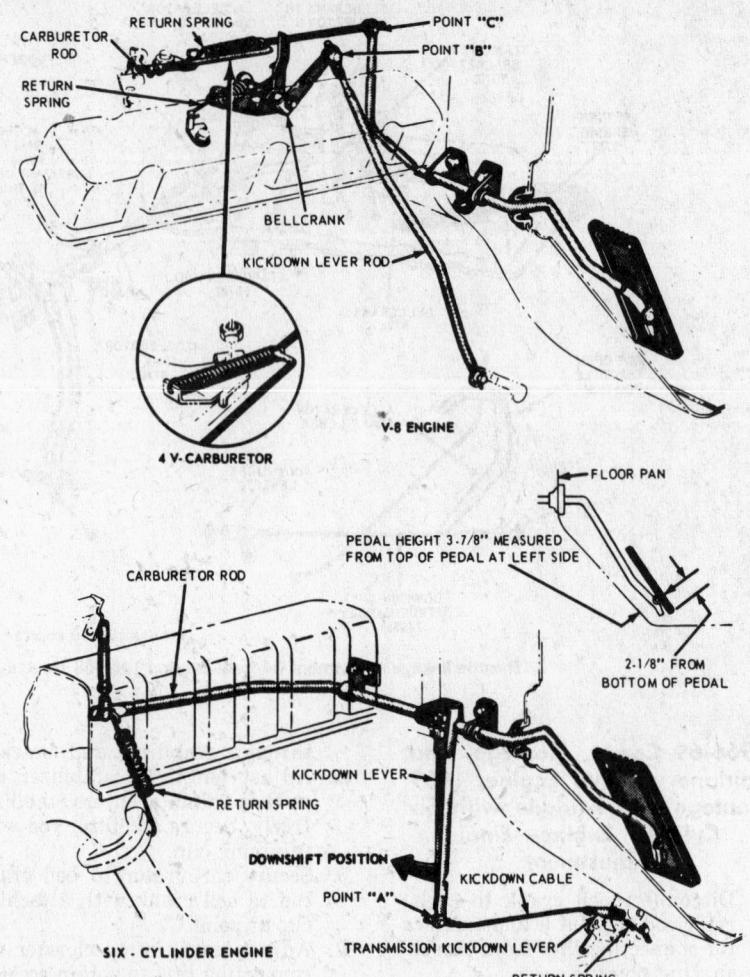

Throttle linkage adjustment C4 transmission 1964-66 Mustang (© Ford Motor Co.)

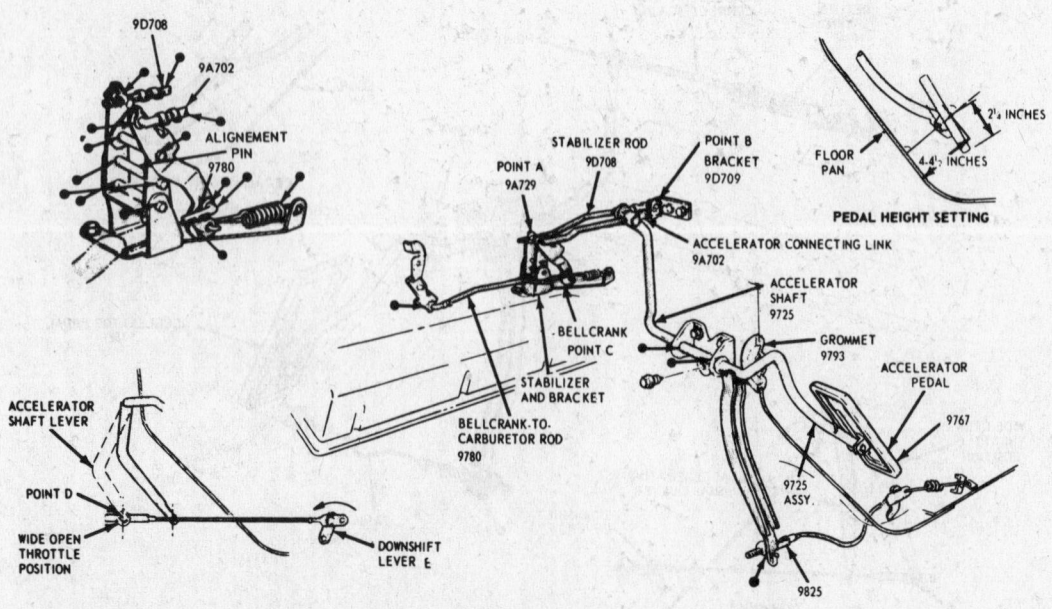

Throttle linkage adjustment C4 transmission 1966-69 Montego, Comet, Falcon, and Fairlane
(© Ford Motor Co.)

5. Release transmission and carburetor to normal free positions.
6. Install throttle and downshift return springs.

Manual Linkage Adjustment

1964-70 Column Shift

1. With engine stopped, loosen clamp at shift lever at point A so that shift rod is free to slide in clamp.
2. Place transmission shift lever into D or D_1 (large dot) position. On Maverick with semi-automatic transmission, place lever in Hi.
3. Shift manual lever at transmission into D, D_1, or Hi. Detent position. D on two speed transmission and D_1 on 1964-66 three speed transmission is second from rear. On 1967 and later transmissions, D or Hi is third detent from rear.
4. Tighten clamp on shift rod at point A to 10-20 ft. lbs.
5. Check pointer alignment and transmission operation for all selector lever positions.

Floor or Console Shift

1. Place transmission shift lever in D, (large dot) on most 1966 cars

POINT "A"
TRANSMISSION MANUAL LEVER
LOWER END OF STEERING COLUMN

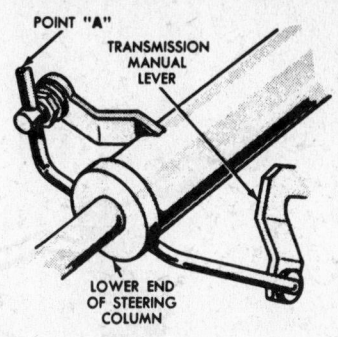

Typical column shift 1964-68
(© Ford Motor Co.)

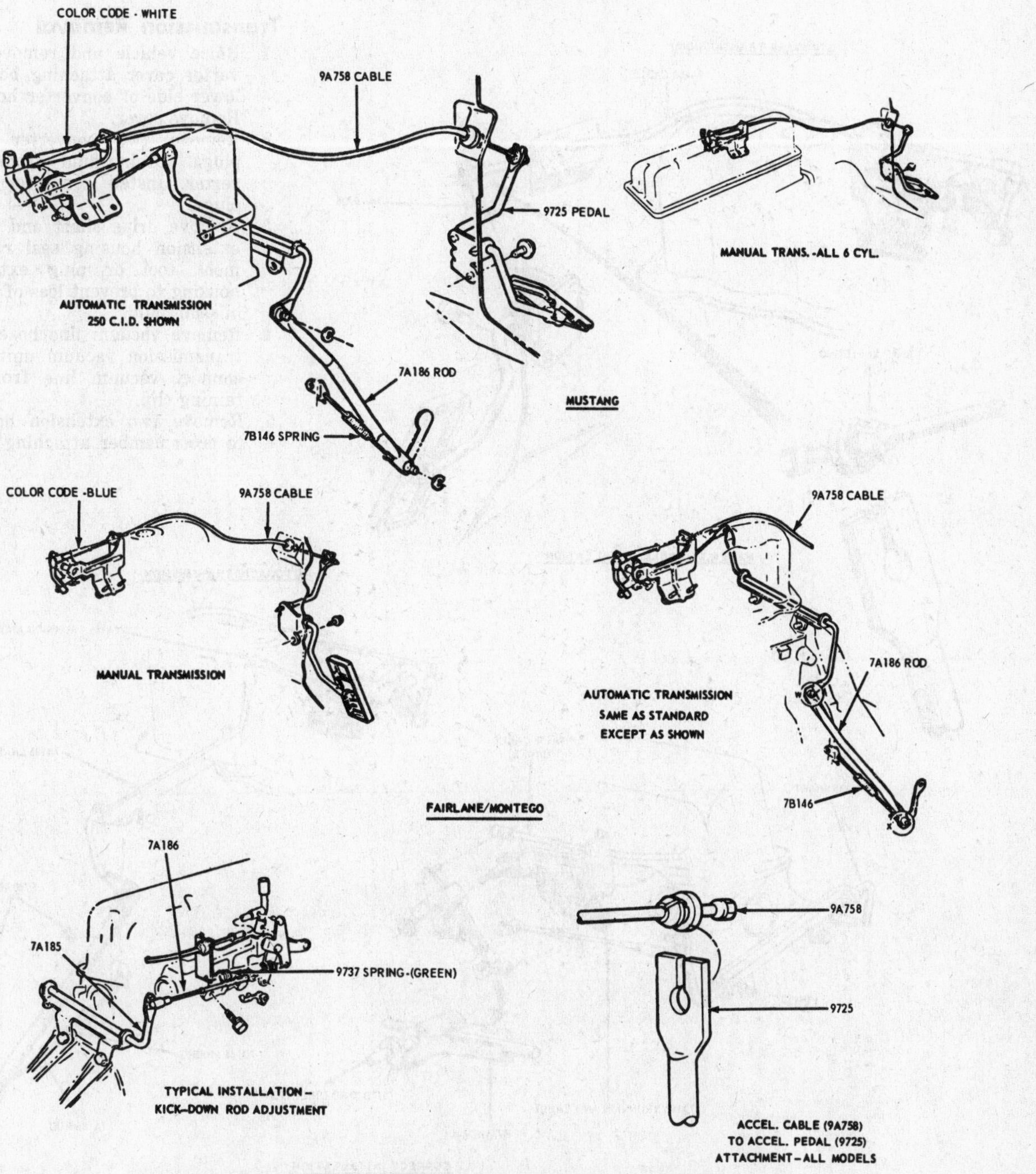

COLOR CODE - WHITE
9A758 CABLE
9725 PEDAL
AUTOMATIC TRANSMISSION
250 C.I.D. SHOWN
7A186 ROD
7B146 SPRING
MUSTANG
MANUAL TRANS. - ALL 6 CYL.
COLOR CODE - BLUE
9A758 CABLE
MANUAL TRANSMISSION
9A758 CABLE
7A186 ROD
AUTOMATIC TRANSMISSION
SAME AS STANDARD
EXCEPT AS SHOWN
7B146
FAIRLANE/MONTEGO
7A186
7A185
9737 SPRING - (GREEN)
TYPICAL INSTALLATION –
KICK-DOWN ROD ADJUSTMENT
9A758
9725
ACCEL. CABLE (9A758)
TO ACCEL. PEDAL (9725)
ATTACHMENT – ALL MODELS

Throttle linkage adjustment 1969-70 Mustang six and 1970 Montego and Fairlane six
(© Ford Motor Co.)

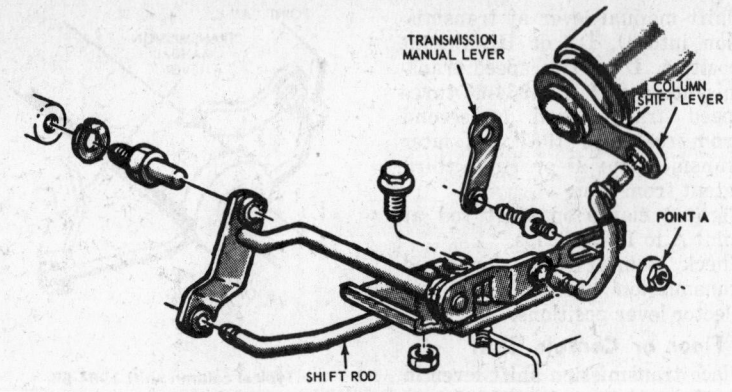

Typical column shift 1969-70 (© Ford Motor Co.)

and D on some 1966 and all 1967 and later cars.

2. Raise vehicle and loosen manual lever shift rod retaining nut. Move transmission lever to D_1 or D position. On most 1966 transmissions, D_1 is fifth detent from rear. On 1966 Cyclone and Fairlane GTs with select shift and on all 1967 and later cars, D is fourth detent from rear.

3. With transmission shift lever and transmission manual lever in position, tighten nut at point A to 10-20 ft. lbs.

4. Check transmission operation for all selector lever detent positions.

Transmission Removal

1. Raise vehicle and remove converter cover attaching bolts, at lower side of converter housing. Remove cover.

2. Remove two converter drain plugs. Drain fluid from converter. Install converter drain plugs.

3. Remove drive shaft and install extension housing seal replacement tool or plug extension housing to prevent loss of transmission fluid.

4. Remove vacuum line hose from transmission vacuum unit. Disconnect vacuum line from retaining clip.

5. Remove two extension housing to crossmember attaching bolts.

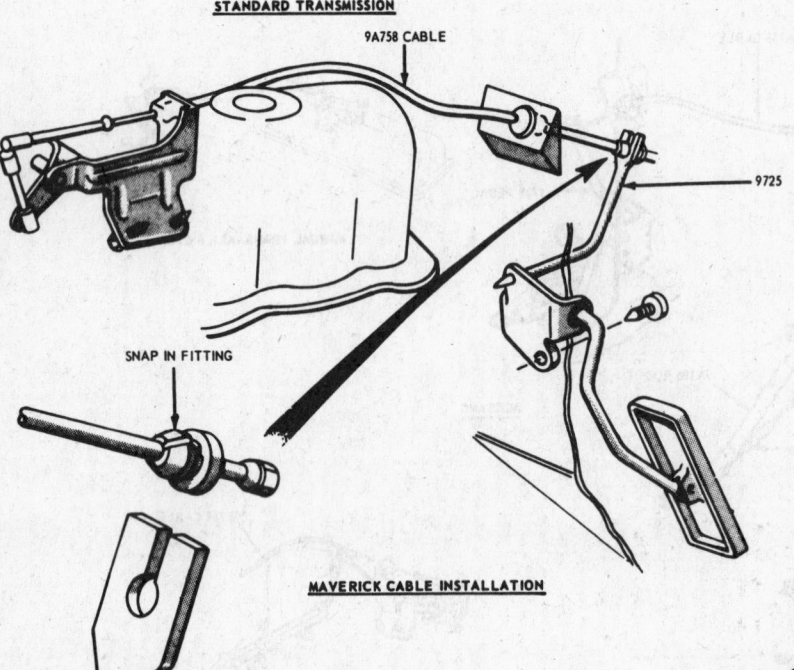

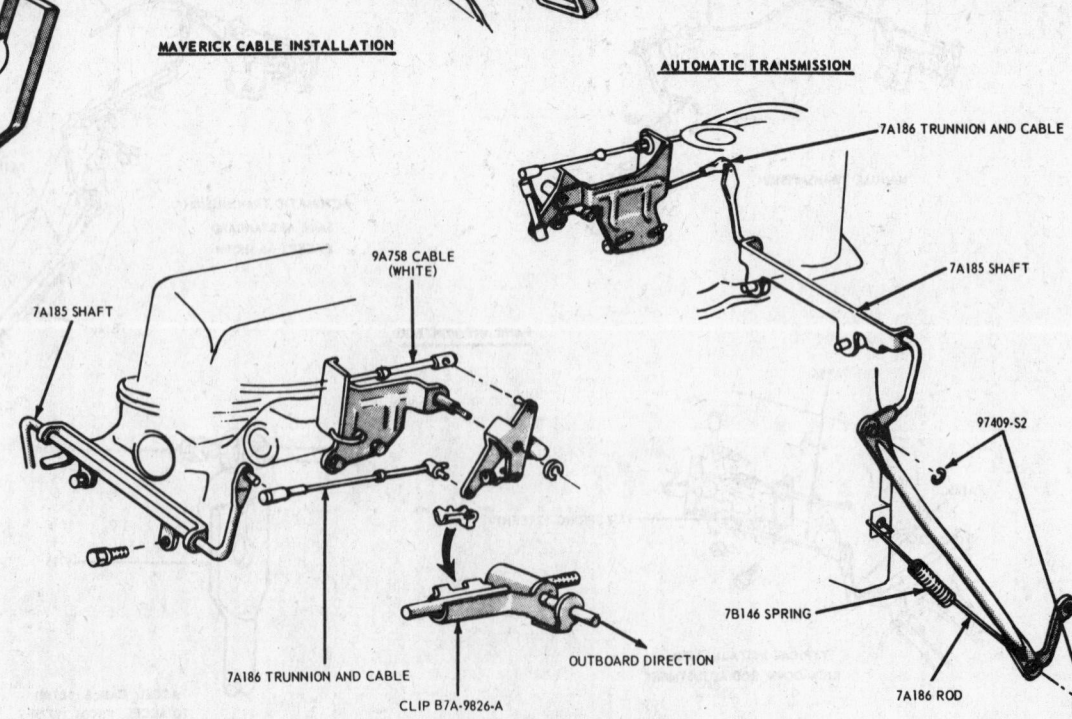

Throttle linkage adjustment—1969-70 Maverick (© Ford Motor Co.)

6. Remove speedometer cable from extension housing.
7. Disconnect exhaust pipe flange from manifolds.
8. Remove parking brake cable from equalizer lever.
9. Loosen transmission pan bolts and drain fluid at one corner of pan. Tighten attaching bolts after fluid has drained.
10. Disconnect fluid cooler lines from transmission case.
11. Remove manual and downshift rods from transmission control levers.
12. On 1969-70 Mustangs, disconnect neutral start switch wires from retaining clamps and connectors.
13. Disconnect starter cable, remove starter attaching bolts, and remove starter from converter housing.

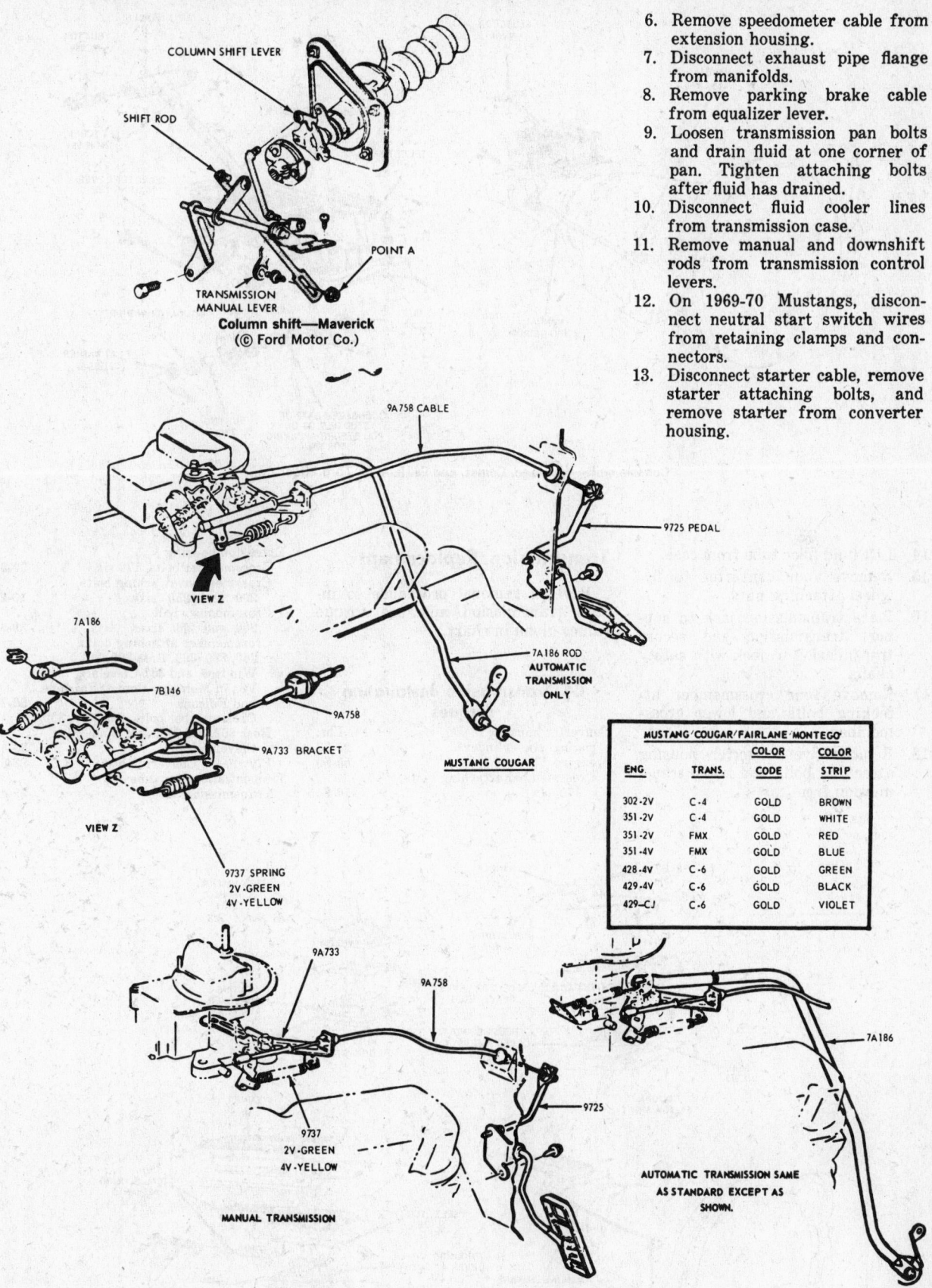

Column shift—Maverick
(© Ford Motor Co.)

MUSTANG/COUGAR/FAIRLANE/MONTEGO			
ENG.	TRANS.	COLOR CODE	COLOR STRIP
302-2V	C-4	GOLD	BROWN
351-2V	C-4	GOLD	WHITE
351-2V	FMX	GOLD	RED
351-4V	FMX	GOLD	BLUE
428-4V	C-6	GOLD	GREEN
429-4V	C-6	GOLD	BLACK
429-CJ	C-6	GOLD	VIOLET

AUTOMATIC TRANSMISSION SAME AS STANDARD EXCEPT AS SHOWN.

MANUAL TRANSMISSION

FAIRLANE/MONTEGO

Throttle linkage adjustment—1969-70 Cougar and Mustang V8s, and 1970 Montego and Fairlane V8s
(© Ford Motor Co.)

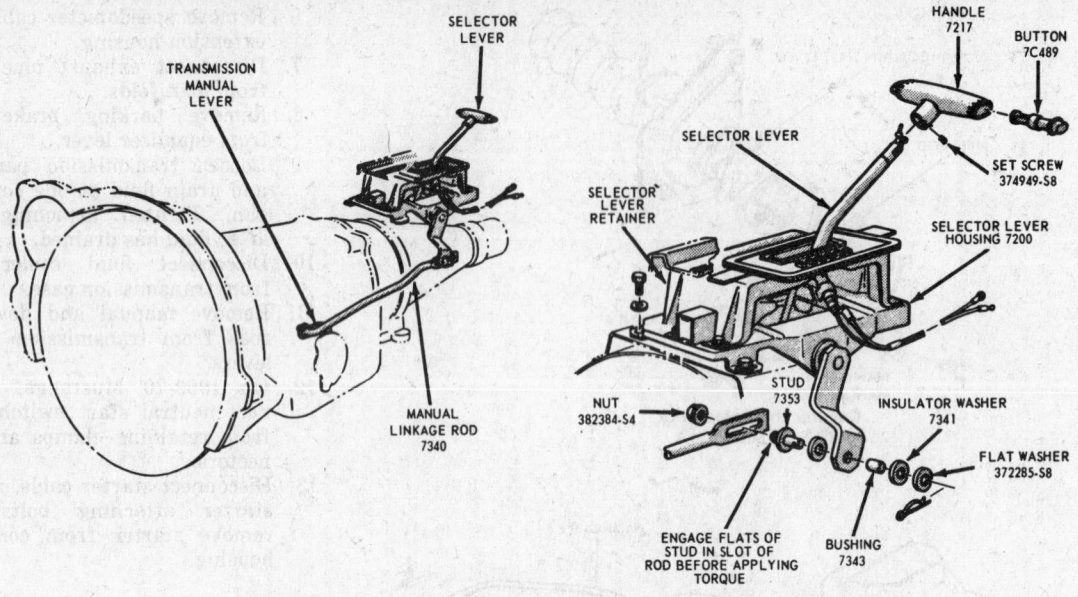

Console shift—Montego, Comet, and Fairlane (© Ford Motor Co.)

14. Lift fluid filler tube from case.
15. Remove four converter to flywheel attaching nuts.
16. Place transmission jack to support transmission and secure transmission to jack with safety chain.
17. Remove four crossmember attaching bolts and lower crossmember.
18. Remove five converter housing attaching bolts and lower transmission from car.

Transmission Replacement

Reverse removal procedure to install transmission and use torque values given in chart.

C4 Transmission Installation Torques

	Ft. Lbs.
Converter housing	
engine six cylinder	23-33
engine V8	40-50
Crossmember attaching bolts,	
170 six	14-24
Extension housing	
Crossmember bolts, 170 six	30-35
Crossmember attaching bolts,	
200 and 250 sixes	10-20
Crossmember bolts,	
200 and 250 sixes	30-35
Crossmember attaching bolts,	
260, 289, 302, Boss 302, 351	
Windsor, and 351 Cleveland	
V8s in Montego, Comet, Fairlane	
and Falcons	50-70
Crossmember bolts, 260, 289, 302,	
Boss 302, 351 Windsor, and 351	
Cleveland V8s	30-45
Flywheel—Converter bolts	23-33
Transmission filler tube	
Transmission case	32-42

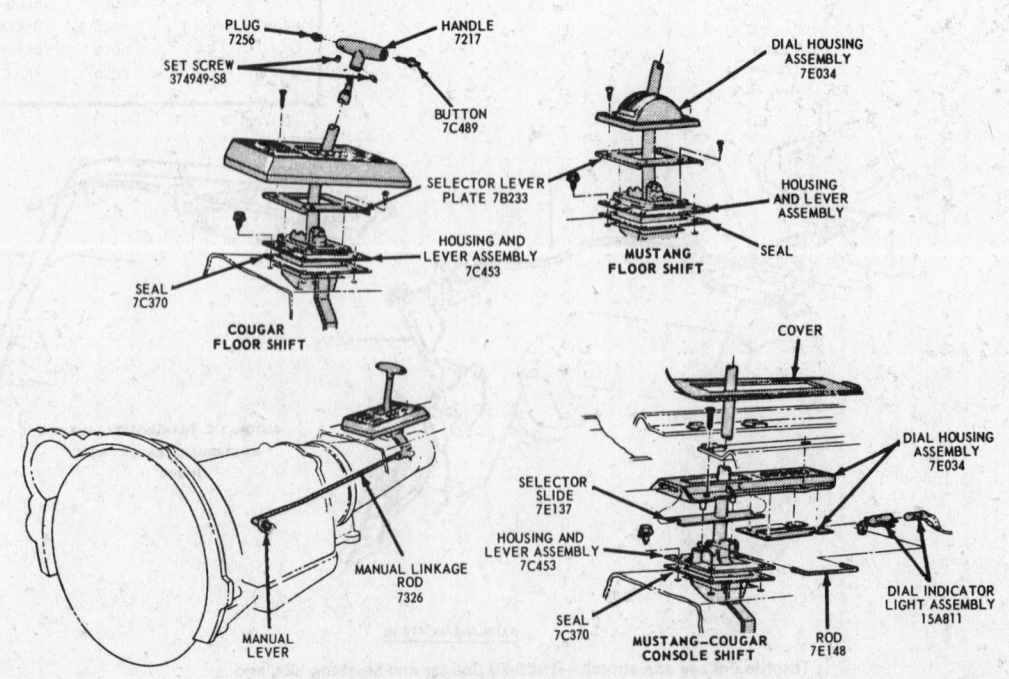

Floor or console shift—Cougar and Mustang (© Ford Motor Co.)

C6 Three-Speed Automatic

Throttle and Downshift Linkage

Initial Adjustments

See C4 three speed automatic.

1966-68 Montego, Comet, and Fairlane with 390, 427, and 428 CJ engines—Final Adjustments

1. Disconnect bellcrank to carburetor rod at point C and accelerator rod from throttle shaft at point B.
2. Disconnect stabilizer rod from stabilizer at point A.
3. Insert ¼ in. diameter pin through stabilizer and bracket.
4. Adjust length of stabilizer rod so that trunnion enters stabilizer freely. Secure stabilizer rod with retaining clip.
5. Secure carburetor to bellcrank rod to bellcrank with attaching clip at point C.
6. Adjust length of accelerator rod to obtain accelerator pedal height of 4-4½ in. measured at pedal.
7. Connect accelerator rod to accelerator shaft with retaining clip after proper accelerator pedal height has been established.
8. With engine off, disconnect downshift rod from lever at point D.
9. With carburetor choke in off position, depress accelerator pedal

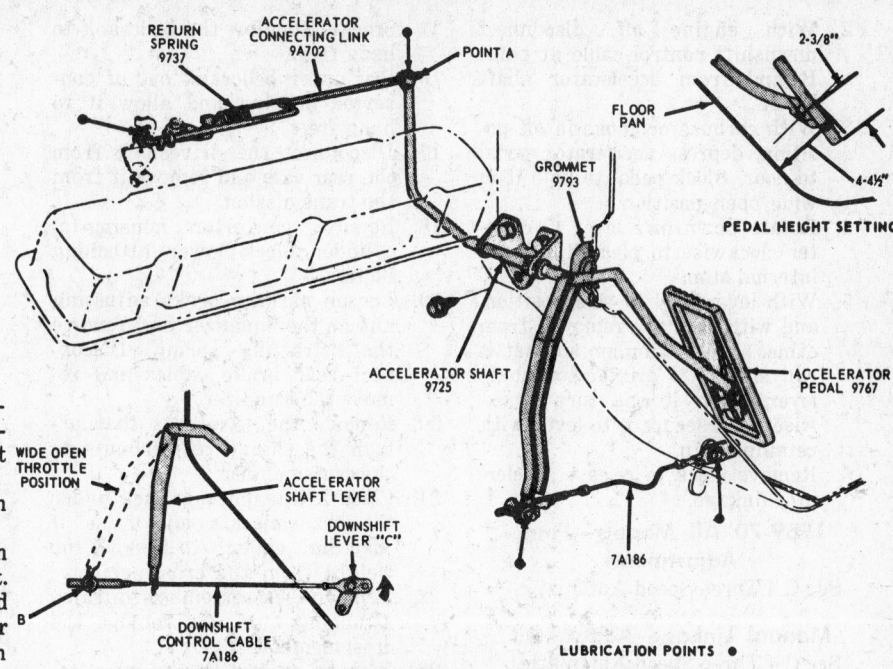

Throttle linkage adjustment—C6 transmission 1967-68 Cougar and Mustang with 390, 427, and 428CJ V8s
(© Ford Motor Co.)

to floor; block pedal to hold it in open position.
10. Rotate downshift lever on transmission in counter clockwise direction to place it against internal stop.
11. Adjust trunnion at point D so that it enters downshift lever freely.
12. Turn it one additional turn counter clockwise to lengthen rod. Secure it to lever with retaining clip.

13. Remove block from accelerator pedal.

1967-68 Cougar and Mustang with 390, 427, and 428 CJ Engines—Final Adjustments

1. With engine off, check accelerator pedal for height of 4½ in. measured from top of pedal at pivot point to floor pan. To obtain correct pedal height, adjust accelerator connecting link at point A.

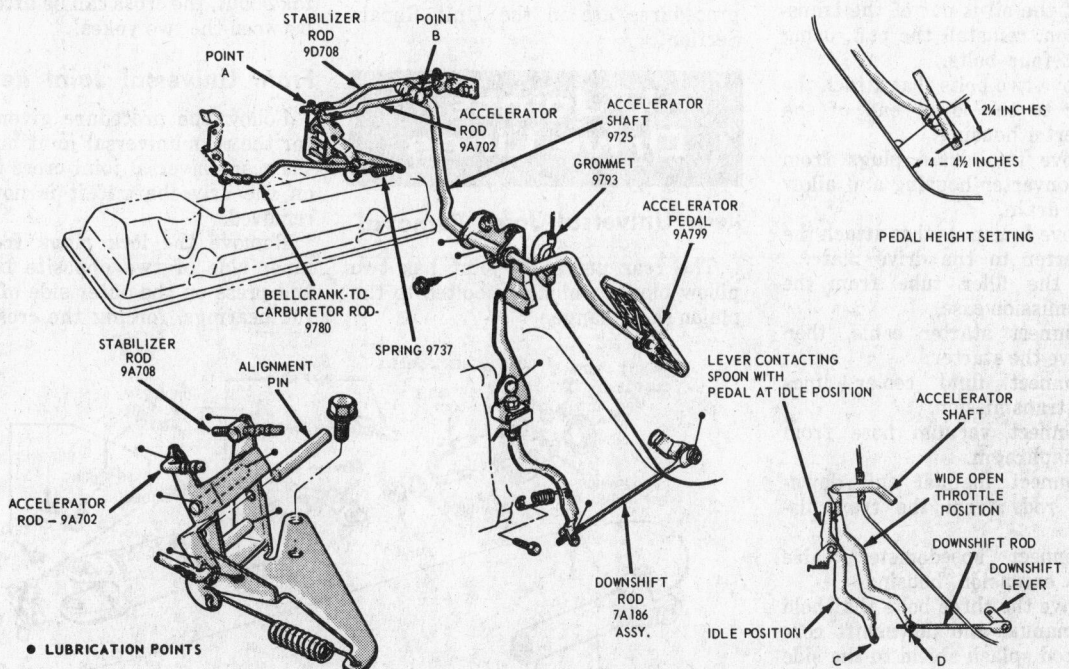

Throttle linkage adjustment—C6 transmission 1966-68 Montego, Comet and Fairlane with 390, 427 and 428CJ V8s
(© Ford Motor Co.)

2. With engine off, disconnect downshift control cable at point B and from accelerator shaft lever.

3. With carburetor choke in off position, depress accelerator pedal to floor. Block pedal to hold it in wide open position.

4. Rotate downshift lever C counter clockwise to place it against internal stop.

5. With lever held in this position, and with all slack removed from cable, adjust trunnion so that it will slide into accelerator shaft lever. Turn it one turn clockwise, then secure it to lever with retaining clip.

6. Remove block to release accelerator linkage.

1969-70 All Models—Final Adjustments

See C 4 Three Speed Automatic.

Manual Linkage Adjustment

See C 4 Three Speed Automatic.

Transmission Removal and Replacement—Beginning 1966

1. Raise hood and disconnect starter neutral switch wires.

2. Disconnect the transmission oil filler tube from the manifold.

3. Raise the car and remove the bolts that attach the reinforcement plate at the rear of the transmission oil pan. Remove the plate.

4. With a drain pan under the transmission, loosen the transmission oil pan bolts and slowly drain and remove the pan. After all of the oil is out of the transmission, reinstall the pan, using about four bolts.

5. Remove two bolts that attach the cover to the lower end of the converter housing.

6. Remove two drain plugs from the converter housing and allow it to drain.

7. Remove four nuts that attach the converter to the drive plate.

8. Lift the filler tube from the transmission case.

9. Disconnect starter cable, then remove the starter.

10. Disconnect fluid cooler lines from transmission.

11. Disconnect vacuum hose from the diaphragm.

12. Disconnect manual and downshift rods from the transmission.

13. Disconnect speedometer cable from extension housing.

14. Remove the three bolts that hold the manual and downshift control rod splash shield to the side rail and remove the shield.

15. Remove lower bellcrank bracket lower attaching bolt. Pivot the

bracket to allow the bellcrank to hang free.

16. Pry upper bellcrank out of converter housing and allow it to hang free.

17. Disconnect the driveshaft from the rear axle and remove it from the transmission.

18. Remove converter housing-to-cylinder block lower attaching bolts.

19. Loosen parking brake adjusting nut at the equalizer and remove the retracting spring. Disconnect rear brake cables and remove the equalizer.

20. Remove the two nuts that attach the engine rear mounts to the crossmember.

21. Place a transmission jack under the transmission and raise it just high enough to remove the weight from the crossmember.

22. Remove crossmember-to-frame attaching nuts and remove the crossmember.

23. Remove engine rear support-to-extension housing attaching bolts and remove support.

24. Secure the transmission to the jack with a safety chain. Lower the transmission and remove the upper converter housing-to-cylinder block attaching bolts.

25. Move the transmission away from the cylinder block. Lower it and remove it from under the car.

26. Remove the converter and mount the transmission in a holding fixture.

27. Replace the transmission by reversing removal procedure.

NOTE: transmission and servicing procedures are in the Unit Repair Section.

U Joints, Drive Lines

Rear Universal Joint Removal

The rear universal joint has two pillow blocks which are bolted to the pinion shaft flange.

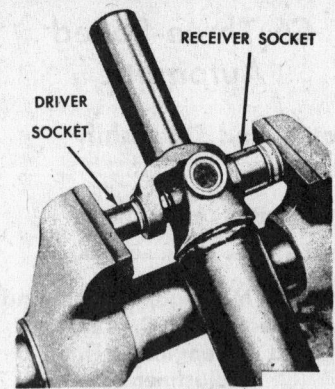

U-joint removal
(© Ford Motor Co.)

Take out the four bolts that hold the bearing blocks to the pinion shaft and gently tap off the bearing blocks.

Lower the back end of the drive shaft and the front end can be slid out of the back of the transmission together with the transmission yoke portion of the front universal joint.

Carry the assembly—the front universal joint complete, the driveshaft and the rear universal joint—to the bench and remove the cross from the rear universal joint by taking out the lock rings from the inner side of the bearings. Using a large punch or an arbor press, drive one of the bearings in toward the center, which will force out the opposite bearing.

When it is pressed out far enough to grip it with a pair of pliers, grip it and pull it out of the driveshaft yoke.

Now drive the cross in the opposite direction until the opposite bearing has been driven far enough out for gripping with a pair of pliers.

When both bearings have been taken out, the cross can be lifted from between the two yokes.

Front Universal Joint Removal

Follow the procedure given above for the rear universal joint but leave the rear universal joint cross in place on the driveshaft if it is not to be removed.

Remove the lock rings from the inner side of two opposite bearings and press on the outer side of one of the bearings, forcing the cross over,

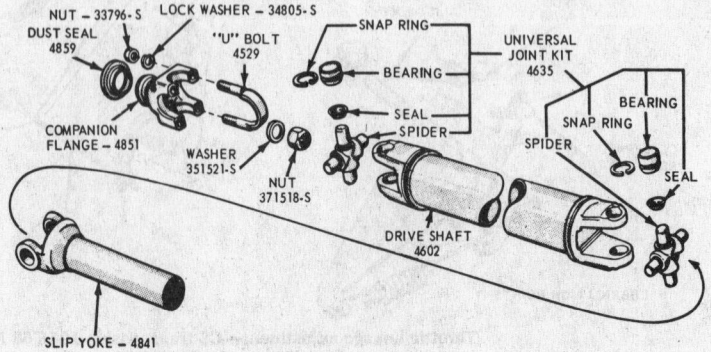

Driveshaft and universal joint assembly (© Ford Motor Co.)

which will force the bearing on the opposite side out of its yoke.

Remove the bearing which was forced out of the yoke and then press the cross in the opposite direction to press the other bearing out.

Repeat this procedure on the third and fourth bearings.

When installing the new bearings in the universal joint yoke, it is possible to put them in with a driver of some type, but it is recommended that this work be done in an arbor press since a heavy jolt on the needle bearings can very easily misalign them, which will greatly shorten their life.

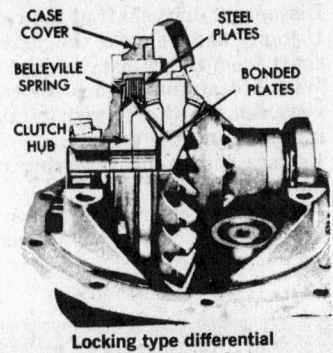

Locking type differential
(© Ford Motor Co.)

Drive Axle, Suspension

Types

There are three types of Ford axles used in compact and intermediate cars. The most prevalent is the integral carrier axle which uses a 7¼ in. ring and pinion gear. This is found on six cylinder models exclusively. All work on the differential or ring gear is done through an inspection plate on the back of the housing. The easiest way to work on this type is to remove the entire housing.

Most V8 models use a medium duty axle with an 8 in. ring gear. Large V8s use a heavy duty axle with 8¾ and 9 in. ring gears. These two axles have removable carrier assemblies which contain the ring gear and pinion and differential. Therefore most repairs can be done without remov-

ing the axle housings from the car. One unique Ford Motor Company feature of these two larger axles is the straddle mounted pinion in which the pinion gear is supported on both sides by a bearing. This has been a Ford feature since Model T days.

Troubleshooting and Adjustments

General instructions covering troubles of the drive axle with repair methods and adjustments are in the Unit Repair Section.

Integral Carrier Rear Axle
Removal

1. Raise the car and support it under the rear frame member.
2. Drain lubricant from the axle.
3. Disconnect driveshaft at pinion flange.

4. Disconnect lower end of shock absorbers.
5. Remove wheels, brake drums and both axle shafts.
6. Remove vent hose from rear axle vent tube and remove the tube from brake tube connection and axle housing.
7. Without opening the brake hydraulic system, remove the T-fitting from the axle housing. Remove the brake line clip from the axle housing.
8. Remove axle shaft oil seals.
9. Remove both brake backing plates and tie them back and out of the way. The brake lines and parking brake cables are still attached to the backing plates.
10. Support the rear axle housing on a jack, then remove the spring clip nuts. Remove spring clip plates.
11. Lower the axle housing and remove it from under the car.
12. If the axle housing is being replaced, transfer all the differential and pinion parts to the new housing. See Unit Repair Section.
13. Replace in reverse order of removal procedure.

Removable Carrier Rear Axle
Removal

1. Remove the carrier assembly from the axle housing as outlined in following paragraph.

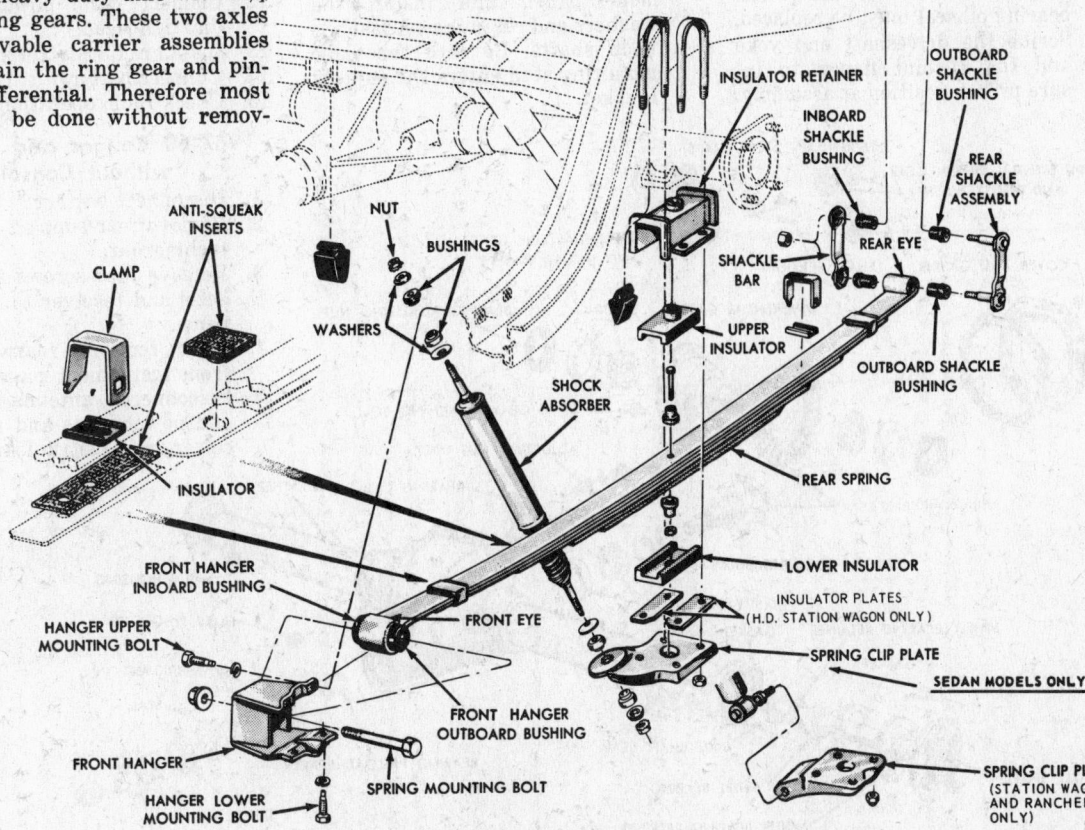

Typical rear suspension (© Ford Motor Co.)

2. With safety stands under the rear frame members, disengage brake line from the axle housing and keep clear.
3. Disconnect vent tube from axle housing.
4. Remove the brake carrier plate assemblies from the axle housing and support them out of the way. Do not open the line.
5. Disconnect each rear shock absorber from spring clip plate and position out of the way.
6. Lower rear axle slightly to reduce some tension. Disconnect the axle from the springs.
7. Remove rear axle from under the car.
8. Replace in reverse order of removal.

Differential Carrier R & R

1. Raise the car and support on stands. Remove rear wheel and tire assemblies.
2. Remove the two rear brake drums from the axle shaft flange studs.
3. Working through the hole provided in each axle flange, remove the rear wheel bearing retaining plate attaching nuts. Pull both axle shafts out of the axle housing. Install a nut on one of the brake carrier plate retaining bolts to hold the plate to the axle housing after the shaft has been removed. Whenever an axle shaft has been removed, the wheel bearing oil seal must be replaced.
4. Scribe the driveshaft end yoke and the U-joint flange to insure proper position at assembly.

Disconnect driveshaft at the rear U-joint, and remove the driveshaft from the transmission.
5. With a drain pan under the carrier, remove the carrier retaining nuts and the carrier.
6. Replace carrier by reversing removal procedure.
   NOTE: for differential carrier service procedures, see Unit Repair Section.

Radio

1964-66 All Models
1. Pull the radio and control knobs off. Remove the nuts and washers retaining the radio to the instrument panel.
2. Disconnect the antenna lead at the right side of the radio.
3. Disconnect the speaker lead.
4. Disconnect the radio lead wire at the fuse panel, and disconnect the pilot light wire. Remove the lead wire from the retaining clips.
5. Remove the radio right support bracket to radio retaining bolt. Remove the radio left support bracket to radio retaining nut (one bracket only on the Bendix radio).
6. Remove the radio assembly from the instrument panel.
7. Position the radio to the instrument panel, and install the washers and retaining nuts at the knob shafts. Be sure the radio mounting stud enters the support bracket.

8. Install the radio support bracket retaining nut and bolt. Tighten all mounting nuts to 25 in. lbs. torque.
9. Connect the antenna lead to the radio.
10. Connect the radio speaker lead and the pilot light lead.
11. Connect the radio power lead and the pilot light lead.
12. Install the radio control knobs.
13. Check the radio operation and adjust the antenna trimmer.

1967 Comet, Falcon and Fairlane
1. Disconnect negative cable from battery.
2. Pull radio control knobs off and remove nuts and washers that attach radio to instrument panel.
3. Disconnect antenna lead at right side of radio (at back of AM-FM radio).
4. Disconnect speaker lead.
5. Disconnect radio lead wire and dial light wire from quick disconnects.
6. Remove radio support bracket.
7. Remove radio from instrument panel.
8. To install, position radio in instrument panel and install washers and attaching nuts at knob shafts. Be sure radio mounting stud enters support bracket.
9. Install radio support bracket.
10. Connect antenna lead to radio.
11. Connect radio speaker lead.
12. Connect radio power lead and dial light lead.
13. Install radio control knobs.
14. Connect battery.
15. Check radio operation.

1967-68 Cougar and Mustang without Console
1. Disconnect battery.
2. Remove rear support bracket attaching nut.
3. Remove four screws that attach bezel and receiver to instrument panel.
4. Move receiver rearward away from instrument panel.
5. Disconnect antenna, speaker, and power leads and remove receiver from instrument panel.

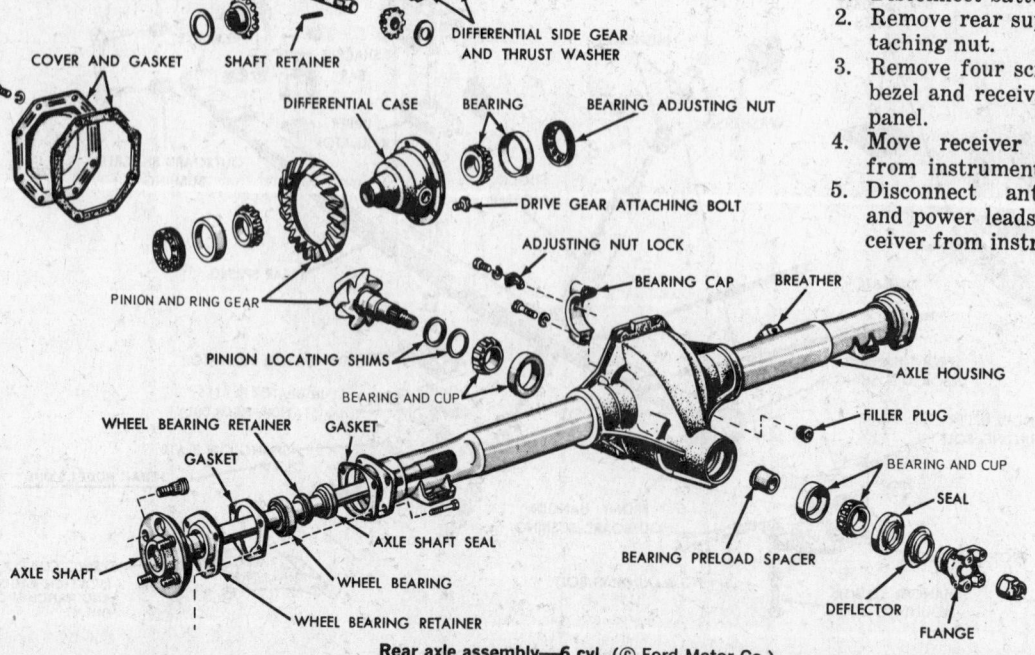

Rear axle assembly—6 cyl. (© Ford Motor Co.)

6. To install, position radio under instrument panel and connect speaker, antenna, and power leads.
7. Secure receiver to instrument panel with attaching screws.
8. Secure rear support bracket to receiver with attaching nut.
9. Connect battery.
10. Check operation of radio.

1967-68 Cougar and Mustang with Console

1. Disconnect battery.
2. Remove two screws attaching right and left supports to support bracket.
3. Remove console assembly.
4. Disconnect radio wiring and antenna lead.
5. Remove control knobs from radio.
6. Remove two nuts and washers from radio shafts and remove radio.
7. To install, position radio in opening and install nuts and washers on control shafts.
8. Install control knobs.
9. Connect radio wires and antenna lead cable.
10. Install console assembly.
11. Install two screws attaching right and left support to support bracket.
12. Connect battery.

1968-70 Montego, Comet, Fairlane, and Falcon

1. Disconnect battery.
2. Pull radio control knobs off shafts.
3. Remove radio support to instrument panel attaching screw.
4. Remove two bezel nuts from radio control shafts.
5. Lower radio and disconnect an-

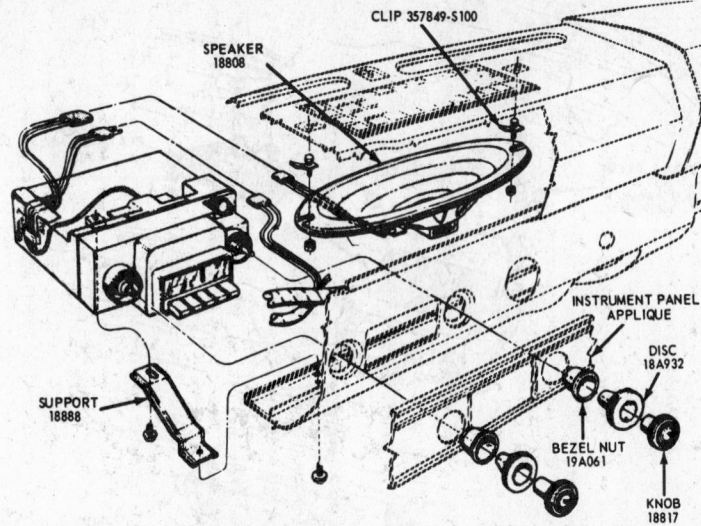

Radio removal—1968-70 Montego, Comet, Fairlane, and Falcon (© Ford Motor Co.)

tenna, speaker, and power leads. Remove radio.
6. To install, connect antenna, speaker and power leads to radio.
7. Position radio in instrument panel and install two bezel nuts. Torque bezel nuts to 30-35 in. lbs.
8. Install radio support bracket to instrument panel attaching screw and torque to 30-35 in. lbs.
9. Connect battery.
10. Adjust antenna trimmer, if necessary.
11. Install radio control knobs and set push buttons for desired stations.

1969-70 Cougar and Mustang

1. Disconnect battery.
2. Pull control knobs, discs, and sleeve from radio control shafts.
3. Remove radio applique from instrument panel.

4. Remove right and left finish panels.
5. Remove two mounting plate attaching screws.
6. Pull radio out of instrument panel and disconnect wires from radio.
7. Remove mounting plate and rear support from radio.
8. Remove radio.
9. To replace, install mounting plate and rear support on radio.
10. Position radio near opening and connect wires to radio.
11. Install jumper wire to ground radio to instrument panel.
12. Connect battery and check operation of radio.
13. Adjust antenna trimmer.
14. Disconnect battery and remove jumper cable.
15. Insert radio and wires into panel opening. Be sure radio rear support slips over instrument panel reinforcement.
16. Install mounting plate attaching screws.
17. Install left and right finish panels.
18. Install radio applique, sleeve, discs, and control knobs.
19. Connect radio ground cable and set push buttons.

1969-70 Maverick

1. Disconnect battery.
2. Remove radio rear support nut and lock washer.
3. Remove four radio to instrument panel retaining screws.
4. Pull radio from instrument panel and disconnect antenna, speaker, and power leads.
5. Remove radio.
6. Remove knob and disc assemblies from radio shafts.
7. Remove two bezel retaining nuts and remove bezel.
8. To install radio, position bezel

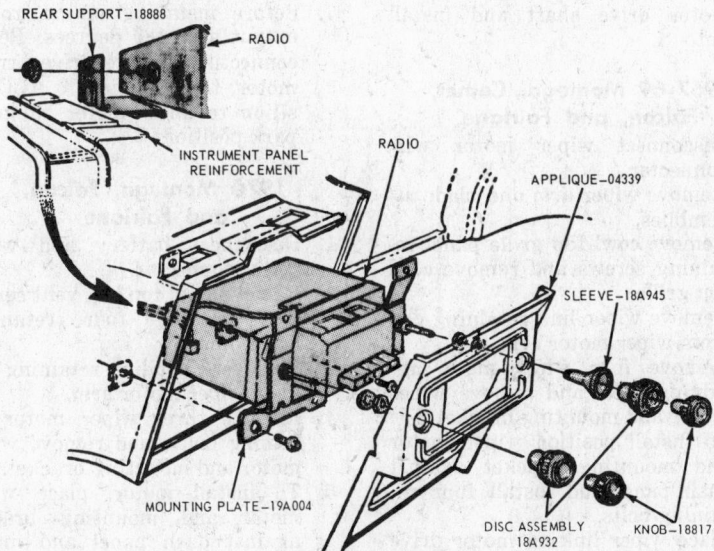

Radio removal—1969-70 Cougar and Mustang (© Ford Motor Co.)

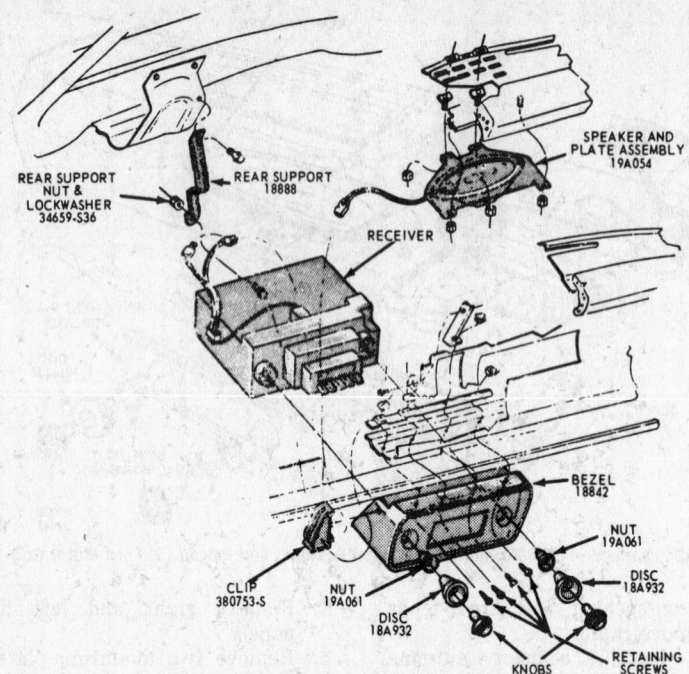

REAR SUPPORT NUT & LOCKWASHER 34659-S36

REAR SUPPORT 18888

SPEAKER AND PLATE ASSEMBLY 19A054

RECEIVER

BEZEL 18842

NUT 19A061

DISC 18A932

CLIP 380753-S

NUT 19A061

DISC 18A932

KNOBS

RETAINING SCREWS

Radio removal—Maverick (© Ford Motor Co.)

on radio and install two bezel retaining nuts.
9. Install disc and knob assemblies on radio shafts.
10. Connect antenna, speaker, and power connectors.
11. Position radio so that rear support mounting bolt enters hole in rear support mounting bracket.
12. Install four radio to instrument panel retaining screws.
13. Install radio rear support nut and lock washer.
14. Place speaker and power wire harnesses in clip on bezel.
15. Connect battery and check operation of radio.
16. Adjust selector buttons for desired stations.

Windshield Wipers

Motor R & R

1964-66 Comet, Falcon, and Mustang

1. Disconnect harness from wiper motor.
2. Remove three bolts retaining wiper motor and mounting bracket assembly to dash panel.
3. Lower assembly and disconnect wiper links at motor.
4. Remove motor and bracket assembly.
5. To install, assemble motor and bracket assembly and connect wiper links to motor.
6. Install motor and bracket assembly on dash by installing three retaining bolts.

7. Connect harness connector to wiper motor.
8. Connect battery and check operation of wiper motor.

1966 Fairlane

1. Disconnect wiper links drive arm from wiper motor drive shaft under instrument panel.
2. Disconnect wiring leads from motor.
3. Remove wiper motor mounting bolts and remove motor.
4. To install, apply sealer around edge of motor mounting bracket and around each bracket bolt hole. Place motor against dash and install mounting bolts.
5. Attach wiring leads.
6. Place wiper link drive arm on motor drive shaft and install nut.

1967-69 Montego, Comet, Falcon, and Fairlane

1. Disconnect wiper motor wire connector.
2. Remove wiper arm and blade assemblies.
3. Remove cowl top grille panel retaining screws and remove cowl top grille.
4. Remove wiper link retaining clip from wiper motor arm.
5. Remove four wiper motor retaining bolts and remove wiper motor and mounting bracket.
6. To install position wiper motor and mounting bracket against dash panel and install four retaining bolts.
7. Place wiper link on motor drive arm and install retaining clip.
8. Install cowl top grille panel.

9. Connect wiper motor wiring connectors.
10. Check motor operation.

1967-68 Cougar and Mustang

1. Disconnect battery.
2. Remove courtesy light. If car is air conditioned, lower air conditioner to floor.
3. Disconnect wiper motor plug connector.
4. Remove nut retaining pivot arm and wiper arms to motor.
5. Remove bolts and star washers retaining motor to mounting bracket, and remove motor.
6. To install motor, attach motor to mounting bracket with bolts and star washers.
7. Position pivot arm and wiper arms on motor and install retaining nut.
8. Connect motor wire plug and battery.
9. Check motor operation and install courtesy light and air conditioner.

1969-70 Cougar and Mustang

1. Remove wiper arm and blade assemblies from pivot shafts and disconnect left side washer hose at T fitting on cowl grille.
2. Remove eight screws and remove cowl top grille.
3. Motor is located inside left fresh air plenum chamber. Disconnect motor ground by removing one screw at forward edge of plenum chamber.
4. Disconnect motor wire at plug and push it back into plenum chamber.
5. Disconnect linkage drive arm from motor output arm crank pin by removing retaining clip.
6. Remove three bolts that retain motor to mounting bracket, rotate motor output arm 180 degrees, and remove motor.
7. Before installing motor, rotate output arm 180 degrees. Before connecting linkage drive arm to motor, turn ignition to ACC position to allow motor to go to park position.

1970 Montego, Falcon, and Fairlane

1. Disconnect battery and wiper motor connector.
2. Remove cowl top left vent screen by removing four retaining drive pins.
3. Remove wiper link retaining clip from wiper motor arm.
4. Remove three wiper motor retaining bolts, and remove wiper motor and mounting bracket.
5. To install motor, place wiper motor and mounting bracket against dash panel and install three retaining bolts.
6. Position wiper link on motor

drive arm, and install connecting clip. Be sure to force clip locking flange into locked position as shown in figure.

7. Install cowl top vent screen and secure with four drive pins.
8. Check motor operation and connect wiring plugs.

1969-70 Maverick

1. Remove instrument cluster.
2. If air conditioned, remove center connector and duct assembly. Remove mounting bracket screw behind center duct, disconnect assembly from plenum chamber and left duct, and pull center connector and duct assembly out through cluster opening.
3. Working through cluster opening, disconnect two pivot shaft links from motor drive arm by removing retaining clip.
4. Disconnect wiring plug at motor, remove three retaining bolts, and remove motor through cluster opening.
5. To install motor, bolt motor to mounting plate with three retaining bolts.
6. Connect right pivot shaft link to motor and then connect left pivot shaft link. Lock clip as shown.
7. On air conditioned vehicles, insert end of center connector and duct assembly near mounting bracket into left duct and opposite end into plenum chamber.
8. Secure assembly with mounting bracket screw.
9. Install instrument cluster, and check operation of wiper motor.

Pivot Shaft and Linkage R & R

1964-66 Comet, Falcon, Fairlane, and Mustang

1. Remove windshield wiper blade and arm assembly.
2. Remove pivot shaft retaining nut, bezel, and gasket.
3. Disconnect wiper link from motor, and remove link and pivot shaft assembly.
4. To install, place pivot shaft and link assembly on cowl and wiper motor. Connect link to motor.
5. Install pivot shaft through cowl and install gasket, bezel, and retaining nut.
6. Install wiper arm and blade assembly.

1967-70 Montego, Comet, Falcon, and Fairlane

1. Remove wiper arms and blades.
2. Remove cowl top grille retaining screws and remove cowl top grille.
3. Remove clip retaining drive arm to pivot.
4. Remove three retaining screws from each pivot and remove pivot shaft and link assembly.
5. To install, position pivot shaft and link assembly in the cowl and install pivot shaft retaining screws.
6. Place left link on motor drive arm and install retaining clip.
7. Install wiper arms and blades, and check wiper operation.
8. Install cowl top grille panel.

1967-68 Cougar and Mustang— Left Side

1. Disconnect battery.
2. Remove wiper arm and blade assembly.
3. Remove four screws that retain heater control to instrument panel and move heater control outward.
4. Remove clip that retains link to motor drive.
5. Working through heater control opening, remove three pivot shaft retaining bolts.
6. Remove pivot and link out through heater control opening.
7. To install, put a new gasket on pivot.
8. Install pivot through heater control opening, and install three retaining bolts.
9. Install clip that retains link to motor drive.
10. Install heater control assembly.
11. Install arm and blade assembly.

1967-68 Cougar and Mustang— Right Side

1. Disconnect battery.
2. Remove wiper arm and blade assembly.
3. Remove glove box liner retaining screws and remove glove box liner.
4. Working through glove box opening, remove three bolts which retain pivot assembly to cowl panel.
5. Remove clip which retains link to wiper motor drive and remove pivot and link assembly out through glove box opening.

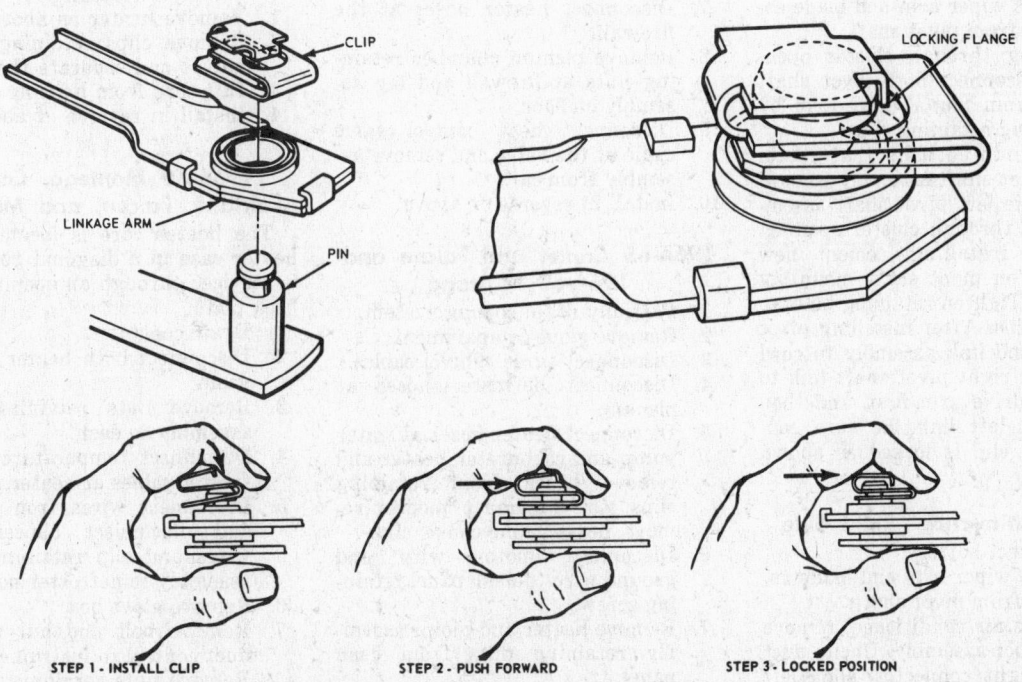

STEP 1 - INSTALL CLIP   STEP 2 - PUSH FORWARD   STEP 3 - LOCKED POSITION

Installation of windshield wiper connecting clips (© Ford Motor Co.)

6. To install, put new gasket on pivot.
7. Install pivot and link assembly through glove box opening and install three pivot retaining bolts.
8. Install link retaining clip on wiper motor drive.
9. Install glove box liner.
10. Connect battery and install wiper arm and blade.

1969-70 Cougar and Mustang
1. Disconnect battery.
2. Remove wiper arm and blade assemblies from pivot shafts.
3. Disconnect washer hose at T fitting on cowl grille.
4. Remove eight screws and remove cowl top grille.
5. Disconnect linkage drive arm from motor output arm crank pin by removing retaining clip.
6. Remove clip and disconnect right link from right arm and pivot shaft assembly.
7. Remove three retaining screws and remove right arm and pivot shaft assembly.
8. Remove three screws retaining left arm and pivot shaft.
9. Lift out pivot shaft and arm, left link, and linkage drive arm as one assembly. Assembly comes out to right.
10. When installing pivot shaft assemblies, tighten retaining bolts to 3-7 ft. lbs. Install left pivot shaft and linkage first. Be sure linkage connecting clips are forced into locked position as shown in figure.

1969-70 Maverick—Left Side
1. Remove instrument cluster.
2. Remove wiper arm and blade assembly from pivot shaft.
3. Working through cluster opening, disconnect both pivot shaft links from motor drive arm by removing retaining clip.
4. Remove three bolts that retain left pivot shaft assembly to cowl and take left pivot shaft assembly out through cluster opening.
5. Before installing, cement new gasket on pivot shaft mounting flange. Tighten retaining bolts to 3-7 ft. lbs. After installing pivot shaft and link assembly to cowl connect right pivot shaft link to motor drive arm first, and then connect left link. Be sure connecting clip is locked as shown in figure.

1969-70 Maverick—Right Side
1. Disconnect battery.
2. Remove wiper arm and blade assembly from pivot shaft.
3. If car is air conditioned, remove right duct assembly. Unclip duct from right connector, slide left end out of plenum chamber, and

lower duct assembly out from under instrument panel.
4. From under instrument panel, disconnect first left and then right pivot shaft link from motor drive arm by removing remaining clip.
5. Reaching between utility shelf and instrument panel, remove three bolts that retain right pivot shaft and link assembly to cowl. Lower assembly out from under instrument panel.
6. Before installing, cement new gasket to pivot shaft mounting flange. After installing pivot shaft and link assembly to cowl, be sure right pivot shaft link is connected to motor drive arm before left pivot shaft link. Be sure connecting clip is in locked position as shown in figure.

Heater System

Heater R & R
1964-65 Fairlane
1. Partially drain cooling system.
2. Disconnect the fresh-air inlet cable at instrument panel.
3. Disconnect right-air inlet boot at blower housing.
4. Disconnect defroster valve control cable at plenum chamber.
5. Disconnect defroster hoses at the outlets.
6. Disconnect heater wires from resistor and disconnect the motor wire.
7. Disconnect heater hoses at the firewall.
8. Remove plenum chamber retaining nuts at firewall and lay assembly on floor.
9. Disconnect heat control valve cable at the valve and remove assembly from car.
10. Install in reverse of above.

1964-65 Comet and Falcon and 1964-66 Mustang
1. Partially drain cooling system.
2. Remove glove compartment.
3. Disconnect three control cables.
4. Disconnect defroster hoses at plenum.
5. Disconnect heater hoses at water pump and carburetor heater and remove hoses from retaining clips. On 8-cylinder models remove hoses from choke clip.
6. Disconnect motor wire and ground wire-to-dash panel retaining screw.
7. Remove heater and motor assembly retaining nuts from dash panel.
8. Disconnect fresh-air inlet boot

and pull heater assembly from panel.
9. Install in reverse of above.

1969-70 Cougar and Mustang
1. Disconnect battery and drain coolant.
2. Remove instrument panel pad.
3. Remove glove compartment liner and door.
4. Remove air distribution duct from heater.
5. Disconnect control cables from heater assembly.
6. Disconnect wires from blower motor resistor.
7. Remove right courtesy light located on underside of instrument panel, if so equipped.
8. Remove heater support to dash panel retaining screw.
9. Disconnect vacuum hoses and remove power air vent duct, if so equipped.
10. Remove blower motor ground wire grounding screw.
11. Disconnect heater hoses from heater at dash panel.
12. Working in engine compartment, remove five heater assembly retaining nuts.
13. Remove instrument panel to cowl attaching screws.
14. Remove instrument panel right side brace.
15. Pull heater assembly and right side of instrument panel rearward, and remove heater assembly. Reverse procedure to install.

Heater Core R & R
1964-65 Comet, Fairlane, and Falcon and 1964-66 Mustang
1. Remove heater as above.
2. Remove clips retaining housing halves and separate the halves.
3. Lift core from housing chamber.
4. Install in reverse of above.

1966-70 Montego, Comet, Fairlane, Falcon, and Maverick
The heater core is located in the heater case in a diagonal position. It is serviced through an opening in the back plate.
1. Drain coolant.
2. Disconnect both heater hoses at dash.
3. Remove nuts retaining heater assembly to dash.
4. Disconnect temperature and defroster cables at heater.
5. Disconnect wires from resistor, and disconnect blower motor wires and clip retaining heater assembly to defroster nozzle.
6. Remove glove box.
7. Remove bolt and nut right air duct control to instrument panel. Remove nuts retaining right air duct and remove duct assembly.

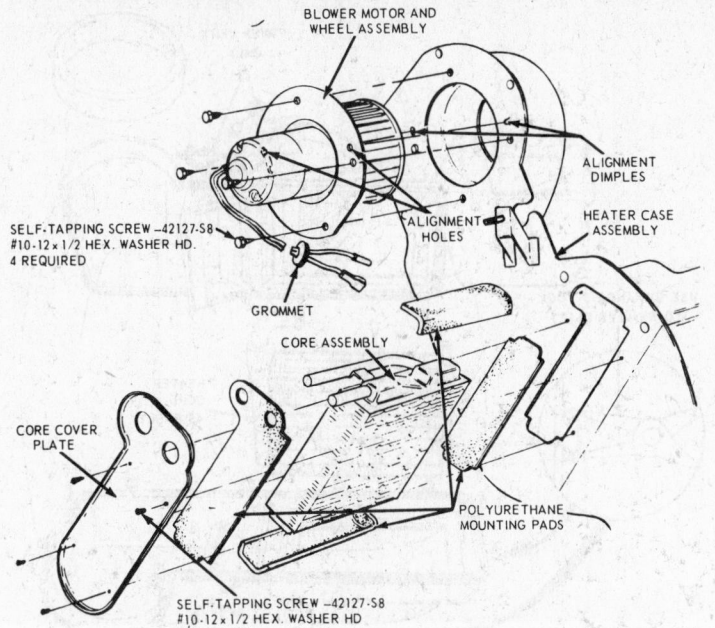

Heater, blower motor and core assemblies—1966-70 Montego, Comet, Fairlane, Falcon, and Maverick
(© Ford Motor Co.)

8. Remove heater assembly to bench.
9. Remove heater core cover and pad and remove core. Reverse procedure to install.

1967-68 Cougar and Mustang

1. Disconnect battery and drain coolant.
2. Disconnect heater hoses at engine.
3. Loosen screws at choke housing and position hose out.
4. Remove nuts retaining heater to dash.
5. Remove screw retaining ground wire at dash and disconnect two wires. Remove glove box liner.
6. Disconnect defroster hoses, temperature control cable, defroster control cable, and heat control cable.
7. Remove screw retaining heater to air intake.
8. Remove heater assembly from vehicle pulling hose through dash.
9. Remove both heater hoses at heater core and remove rubber boot from air intake and clips holding both halves of heater together.
10. Separate halves and remove heater core. Reverse procedure to install.

1969-70 Cougar and Mustang

1. Remove heater assembly.
2. Remove air inlet seal from heater assembly.
3. Remove eleven clips from heater assembly flange and separate heater assembly housing.
4. Remove heater core from heater assembly housing. Reverse procedure to install.

Blower Motor R & R

1964-65 Comet, Falcon, and Fairlane and 1964-66 Mustang

1. Remove heater assembly and lay assembly on floor.
2. Remove blower motor and bracket to blower housing retaining screws and remove blower assembly.
3. Loosen blower cage set screw and remove blower cage from motor.
4. Remove blower motor mounting plate from motor. Reverse procedure to install.

1966-70 Montego, Comet, Fairlane, Falcon, and Maverick

1. Disconnect battery and drain cooling system.
2. Disconnect heater hoses at engine.
3. Remove nuts retaining heater assembly to dash.
4. Disconnect temperature and defroster cables at heater.

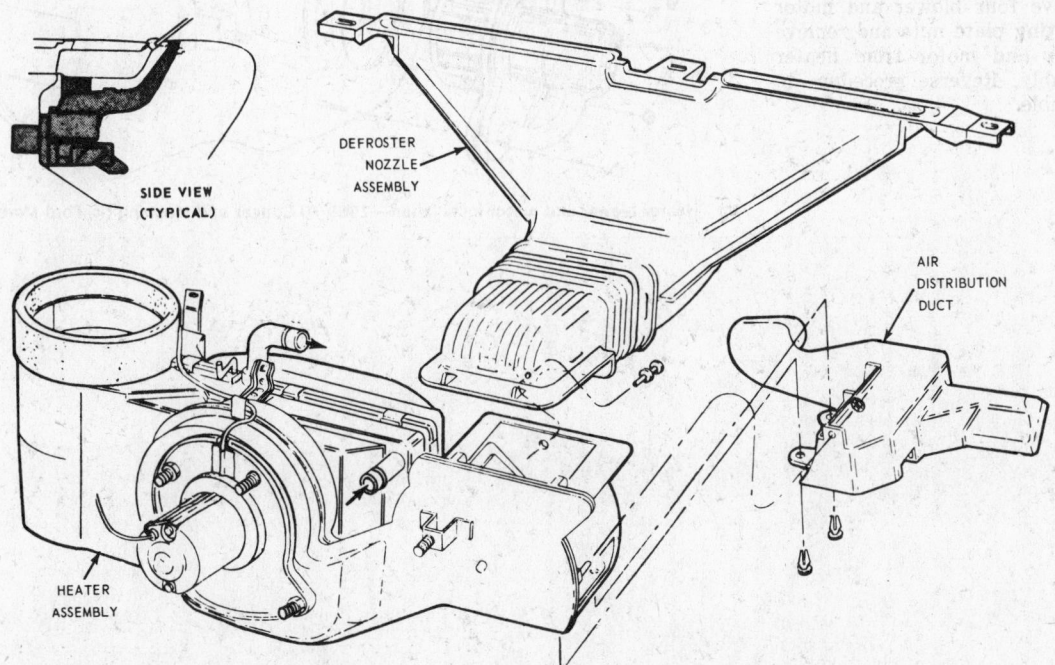

1969 Cougar and Mustang heater (© Ford Motor Co.)

5. Disconnect wires from resistor, blower motor wires, and clip retaining heater assembly to defroster nozzle.
6. Remove glove box.
7. Remove bolt and nut retaining right air duct control to instrument panel. Remove nuts retaining right air duct and remove duct assembly.
8. Remove heater assembly to bench.
9. Remove blower mounting screws and remove motor and wheel assembly. Reverse procedure to assemble.

1967-68 Cougar and Mustang

1. Disconnect battery and drain cooling system.
2. Disconnect heater hoses at engine.
3. Loosen screws at choke housing and move hose out.
4. Remove nuts retaining heater to dash.
5. Remove screw retaining ground wire at dash and disconnect two wires.
6. Remove screw retaining heater to air intake.
7. Lower heater to floor, pulling hoses through dash.
8. Remove nuts retaining motor mounting plate to heater and remove motor and blower assembly. Reverse procedure to install.

1969-70 Cougar and Mustang

1. Remove heater assembly.
2. Disconnect blower motor wire from resistor.
3. Remove four blower and motor mounting plate nuts and remove blower and motor from heater assembly. Reverse procedure to assemble.

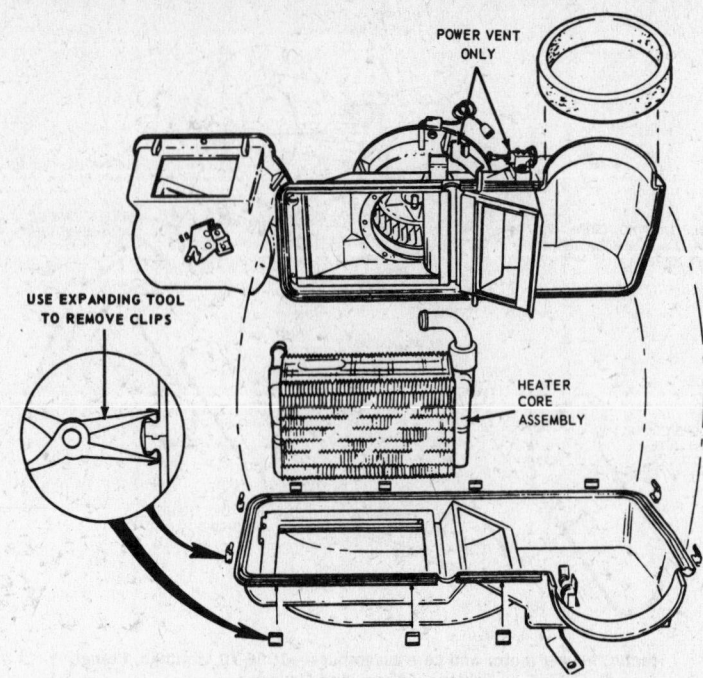

Heater core removal—1969-70 Cougar and Mustang (© Ford Motor Co.)

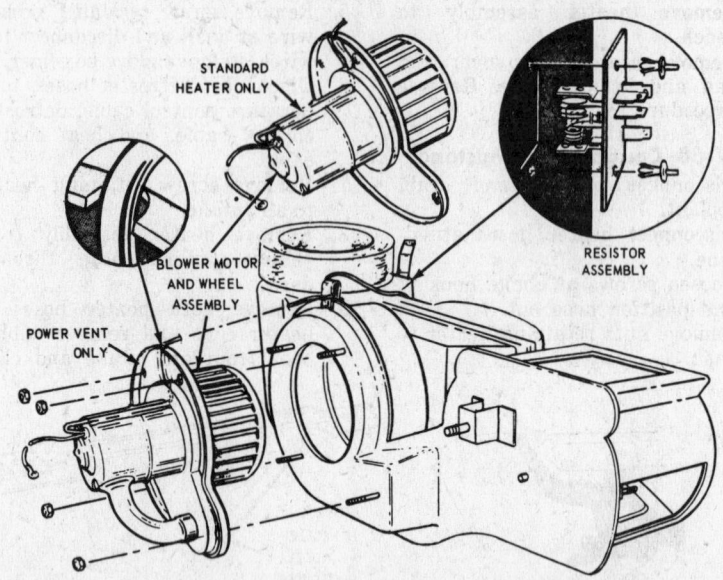

Heater blower and motor installation—1969-70 Cougar and Mustang (© Ford Motor Co.)

Index

451

YEAR IDENTIFICATION

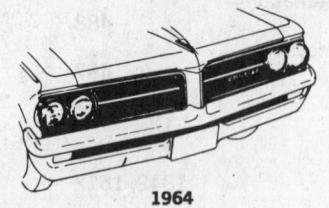

1964

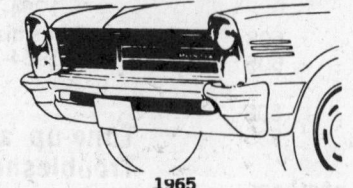

1965

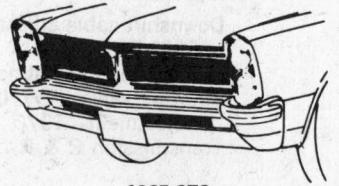

1965 GTO

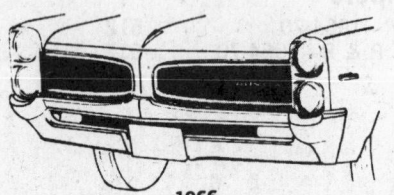

1966

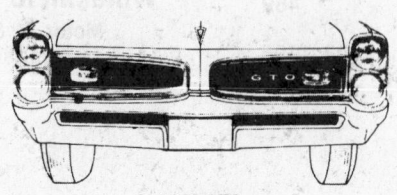

1966 GTO

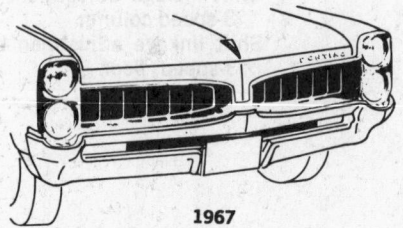

1967

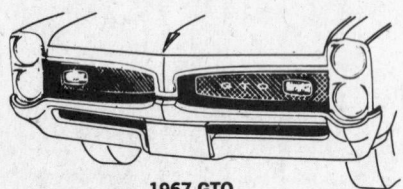

1967 GTO

1967 Firebird

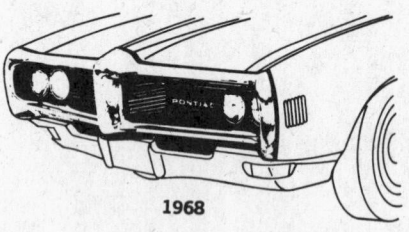

1968

1968 Firebird

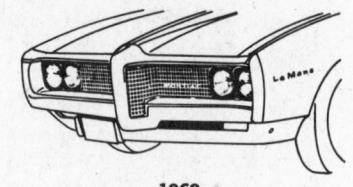

1969

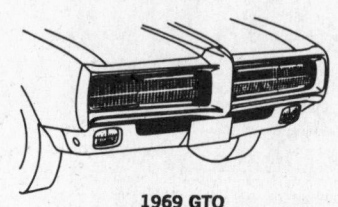

1969 GTO

1969 Firebird

1970-71 Firebird

FIRING ORDER

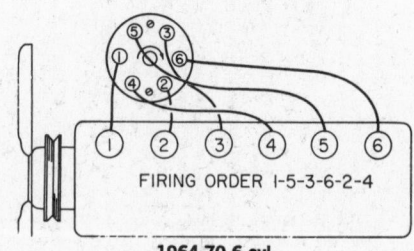

FIRING ORDER 1-5-3-6-2-4

1964-70 6 cyl.

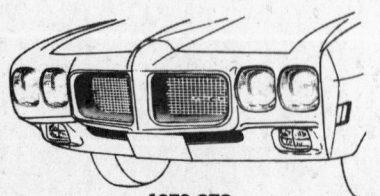

1970 GTO

1971 GTO

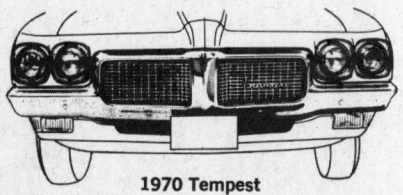

1970 Tempest

1971 Tempest

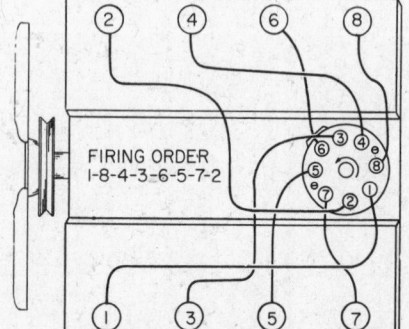

FIRING ORDER
1-8-4-3-6-5-7-2

1964-70 V8

CAR SERIAL NUMBER LOCATION

1964-67

The car serial number is found on a plate attached to the left front hinge pillar. The number is interpreted as follows:

1968-71

The car serial number is located on a plate attached to the top of the instrument panel, left-hand side, visible through the windshield. The number is interpreted in the same manner as 1965-71.

1964

First digit: Number of cylinders
Second digit: Series number
Third digit: Year manufactured
Fourth digit: Assembly plant
Last digits: Series production number

1965-71

First digit: Car division
Second and third digits: Series number
Fourth and fifth digits: Body style code
Sixth digit: Year manufactured
Seventh digit: Plant
Eighth digit: Engine used (1 = V8; 6 = 6 cyl.)

Engine Identification

The engine number, on V8 engines, is located on a machined pad on the right-hand bank of the engine block.

The engine production code is stamped immediately below this number. Use the following production code charts to find your engine and, for 1968-70, its special equipment.

On six cylinder engines, the engine code is stamped on the cylinder head-to-block contact surface behind the oil filler pipe. Use the following charts to find your engine and, for 1968-70, its special equipment.

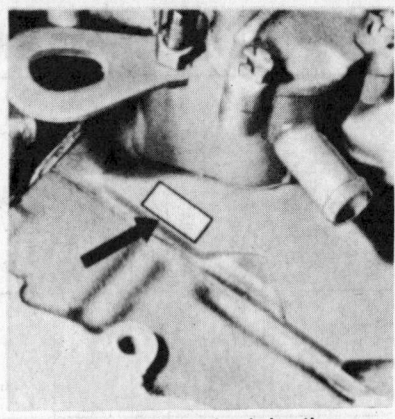
Engine number and code location—1964-69 V8 engines

6-cyl. engine serial number location

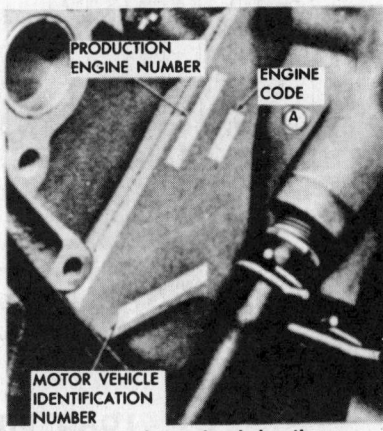
Engine number and code location—1970-71 V8 engines

1964–47 ENGINE IDENTIFICATION CODES

No. Cyls.	Cu. In. Displ.	Type	1964	1965	1966	1967	No. Cyls.	Cu. In. Displ.	Type	1964	1965	1966	1967
6	230	A.T., 1 BBL.			ZG	ZG	8	400	3 Spd., M.T., 4 BBL.				WT
6	215	3 Spd., M.T., 1 BBL.		ZK			8	389	3 Spd., M.T., 3-2 BBL.			WV	
6	230	3 Spd., M.T., 1 BBL.			ZK	ZK	8	400	4 BBL. w/Ex. Em.				WV
6	215	A.T., 1 BBL.		ZL			8	389	3 Spd., M.T., 4 BBL.			WW	
6	230	A.T., 4 BBL. w/Ex. Em.				ZL	8	400	M.T., 4 BBL. w/Ex. Em.				WW
6	215	A.T., 1 BBL.		ZM			8	326	3 Spd., M.T., 2 BBL.			WX	
6	230	A.T., 1 BBL.			ZM	ZM	8	326	3 Spd., M.T., 2 BBL. w/Ex. Em.				WX
6	215	3 Spd., M.T., 1 BBL.		ZN			8	389	A.T., 4 BBL.			XE	
6	250	A.T., 1 BBL.			ZN	ZN	8	400	A.T., 4 BBL. w/Ex. Em.				XE
6	215	3 Spd., M.T., 1 BBL.		ZR			8	326	A.T., 2 BBL. w/Ex. Em.			XF	XF
6	230	M.T., 4 BBL. w/Ex. Em.				ZR	8	326	A.T., 4 BBL. w/Ex. Em.			XG	XG
6	215	3 Spd., M.T., 1 BBL.		ZS			8	400	A.T., 2 BBL. w/Ex. Em.				XL
6	230	3 Spd., M.T., 1 BBL.			ZS	ZS	8	400	A.T., 2 BBL.				XM
8	389	3 Spd., M.T., 3-2 BBL.	76X				8	400	A.T., 4 BBL.				XP
8	389	A.T., 3-2 BBL.	77J				8	326	M.T., 4 BBL. w/Ex. Em.				XR
8	389	A.T., 4 BBL.	78X				8	400	M.T., 4 BBL.				XS
8	389	A.T., 4 BBL.	79J				8	400	A.T., 4 BBL. w/Ex. Em.				YI
6	215	3 Spd., M.T., 1 BBL.	80Z				8	326	A.T., 2 BBL.		YN	YN	
6	215	3 Spd., M.T., 1 BBL.	81Z				8	326	A.T., 4 BBL.		YP	YP	
6	215	3 Spd., M.T., 1 BBL.	84Z				8	400	A.T., 4 BBL. w/Ex. Em.				YQ
6	215	3 Spd., M.T., 1 BBL.	85Z				8	389	A.T., 3-2 BBL.		YR	YR	
6	215	A.T., 1 BBL.	88Y				8	400	M.T., 4 BBL. w/Ex. Em.				YR
8	326	3 Spd., M.T., 2 BBL.	925				8	389	A.T., 4 BBL.		YS	YS	
8	326	3 Spd., M.T., 4 BBL.	945				8	400	A.T., 4 BBL. (335 HP)				YS
8	326	A.T., 2 BBL.	960				8	400	A.T., 4 BBL. (360 HP)				YZ
8	326	A.T., 4 BBL.	971				6	215	3 Spd., M.T., 1 BBL.		ZD		
8	326	3 Spd., M.T., 2 BBL.		WP	WP	WP	6	230	3 Spd., M.T., 4 BBL.			ZD	ZD
8	326	3 Spd., M.T., 4 BBL.		WR	WR	WR	6	215	A.T., 1 BBL.		ZE		
8	389	3 Spd., M.T., 3-2 BBL.		WS	WS		6	230	A.T., 4 BBL.			ZE	ZE
8	400	3 Spd., M.T., 4 BBL.				WS	6	230	3 Spd., M.T., 1 BBL.			ZF	ZF
8	389	3 Spd., M.T., 4 BBL.		WT	WT								

1968 FIREBIRD ENGINE IDENTIFICATION

DISPLACEMENT	HORSEPOWER	ENGINE CODE	TRANS. MANUAL	TRANS. AUTOMATIC	CARB. 1-BBL	CARB. 2-BBL	CARB. QUADRAJET	COMP. RATIO 9.0	COMP. RATIO 9.2	COMP. RATIO 10.5	COMP. RATIO 10.75	CAM 9777254	CAM 9779066	CAM 9779067	CAM 9779068	CAM 9785744	CAM 9790826	CAM 9792539	DIST 1110430	DIST 1110431	DIST 1111281	DIST 1111282	DIST 1111447	DIST 1111270	DIST 1111449	VS SINGLE	VS STD. TWO	VS H.D. TWO (SPEC.)	VS RAM AIR
250 Cu. In. (six)	175	ZK	X		X			X									X		X							X			
	175	ZN		X	X			X									X		X							X			
	215	ZD	X				X		X									X		X								X		
	215	ZE		X			X		X										X		X							X		
350 Cu. In. (V8)	265	WC	X			X		X				X										X						X		
	265	YJ		X		X		X				X										X						X		
	320	WK	X				X			X			X											X				X		
	320	YM		X			X		X					X									X					X		
400 Cu. In. (V8)	335	XN		X			X				X				X													•	X
	335	WQ	X				X				X				X													•	X
	335	WI	X				X				X						X											•	X
	335	WZ	X				X				X			X													X	•	
	330	YW		X			X				X				X												X	•	
	330	YT		X			X				X				X												X	•	

• With 60 PSI Oil Pump Spring
All Cars Use CCS

1969 FIREBIRD ENGINE IDENTIFICATION

HORSEPOWER	ENGINE CODE	DISP 250	DISP 350	DISP 400	TRANS MANUAL	TRANS AUTOMATIC	CARB 1 BBL (MV)	CARB 2 BBL (2GV)	CARB 4 BBL (4MV)	CR 9.0:1	CR 9.2:1	CR 10.5:1	CR 10.75:1	CAM 9790826	CAM 9792539	CAM 9796327	CAM 9777254	CAM 9779067	CAM 9779068	CAM 9794041	DIST 1110474	DIST 1110475	DIST 1111941(b)	DIST 1111942	DIST 1111945	DIST 1111946(b)	DIST 1111952(b)	DIST 1111960	VS SINGLE	VS STD.-DUAL	VS H.D.-DUAL	VS H.D. SPEC.-DUAL	VS RAM AIR IV-DUAL	CH SMALL VALVE	CH LARGE VALVE
175	(a)ZK ZC	X			X		X			X				X									X						X					X	
175	(a)ZN ZF	X				X	X			X				X									X						X					X	
230	(a)ZD ZH	X			X			X			X						X				X									X				X	
215	(a)ZE ZL	X				X		X			X			X						X	X									X				X	
265	(a)WC WM		X		X			X									X								X					X	X			X	
265	(a)XL XB		X			X	X				X						X									X					X			X	
265	(a)YJ YE		X			X	X			X							X									X					X			X	
325	(a)WN WN		X		X				X	X								X						X			(e)				X		X		X
325	(a)XC XC		X			X			X	X								X								(d)				X		X		X
330	WZ			X	X				X			X						X						X							X		X		X
330	YT			X		X			X			X						X								X				X			X		X
335	WQ			X	X				X				X						X							X					X		X		X
335	YW			X		X			X				X						X							X				X			X		X
345	WH			X	X				X				X						X						X							X	X		X
345	XN			X		X			X				X						X						X							X	X		X

(a) Early production (small valve) engines with 30° intake valve seat angle. Later production (small valve) engines use 45° intake valve seat. NOTE: all large valve engines use 30° intake valve seat.

(b) Uses hardened drive gear for use with 60 psi oil pump and high tension distributor points.

(c) Two speed (M31) if equipped with A/C; Turbo-Hydramatic (M38) optional without A/C.

(d) Uses distributor 1111965

(e) Uses distributor 1111966

1968 TEMPEST ENGINE IDENTIFICATION

DISPLACEMENT	HORSEPOWER	ENGINE CODE	MANUAL	AUTOMATIC	1-BBL	2-BBL	QUADRAJET	8.6	9.0	9.2	10.5	10.75	9777254	9779067	9779068	9785744	9790826	9792539	1110430	1110431	1111281	1111282	1111447	1111272	1111270	1111449	SINGLE	STD. TWO	H.D. TWO (SPEC.)	RAM AIR
250 Cu. in. (6 cyl)	175	ZK	X		X				X								X		X									X	X		
	175	ZN		X	X				X								X			X								X	X		
	215	ZO	X				X				X							X		X									X		
	215	ZE		X			X				X							X			X								X		
350 Cu. in. (8A)	265	WD	X			X				X			X								X								X		
	265	YN		X		X				X			X								X								X		
	320	WR	X				X				X		X											X					X		
	320	YP		X			X				X		X									X							X		
400 Cu. in. (8A)	265	XM		X		X		X					X											X					X		
	360	WT	X				X					X		X			X								•					X	
	350	YS		X			X					X		X			X								•				X		
	360	WS	X				X					X				X									•					X	
	350	YZ		X			X					X		X											•				X		
	360	XS	X				X					X					X								•				X		X
	360	XP		X			X					X					X								•				X		X

● With 60 PSI Oil Pump Spring
All Cars Use CCS

1969 TEMPEST ENGINE IDENTIFICATION

HORSEPOWER	ENGINE CODE	250	350	400	MANUAL	AUTOMATIC	1 BBL (MV)	2 BBL (2GV)	4 BBL (4MV)	8.6:1	9.0:1	9.2:1	10.5:1	10.75:1	9790826	9792539	9796327	9777254	9785744	9779067	9779068	9794041	1110474	1110475	1111940	1111946 (b)	1111942	1111952 (b)	1111941 (b)	1111960	SINGLE	STD.-DUAL	H.D.-DUAL	H.D. SPEC.-DUAL	RAM AIR IV-DUAL	SMALL VALVE	LARGE VALVE
175	ZC [a]ZK	X			X		X				X				X									X							X					X	
175	ZF [a]ZN	X				X	X				X				X									X							X					X	
230	ZH [a]ZD	X			X				X				X				X						X									X				X	
215	ZL [a]ZE	X				X			X			X			X								X									X				X	
265	XS [a]XR		X			X		X				X						X							X							X				X	
(c) 265	YU [a]YN		X			X		X				X						X							X							X				X	X
265	WU [a]WP		X		X			X				X						X							X										X	X	
330	XU		X		X				X				X							X				X		(d)						X				X	X
330	WV		X			X			X				X							X				X		(e)						X				X	X
350	YS			X	X				X					X					X					X								X					X
350	WT			X	X	X			X					X					X						X								X				X
366	YZ			X	X				X					X					X						X								X				X
366	WS			X	X	X			X					X						X					X								X				X
265	XX [a]XM			X	X	X		X		X					X								X							X					X	X
370	XP			X	X				X					X							X	X											X		X
370	WW			X	X	X			X					X							X	X											X		X

(a) Early production (small valve) engines with 30° intake valve seat angle. Later production (small valve) engines use 45° intake valve seat. NOTE: all large valve engines use 30° intake valve seat.

(b) Uses hardened drive gear for use with 60 psi oil pump and high tension distributor points.

(c) Two speed (M31) if equipped with A/C; Turbo-Hydramatic (M38) optional without A/C.

(d) Uses distributor 1111965

(e) Uses distributor 1111966

1970 TEMPEST ENGINE IDENTIFICATION

HORSEPOWER	ENGINE CODE	250	350	400	455	MANUAL	AUTOMATIC	1 BBL. (MV)	2 BBL. (GV)	4 BBL. (4MV)	8.5:1	8.8:1	10.0:1	10.25:1	10.5:1	9777254 (U)	9779066 (N)	9779067 (P)	9779068 (S)	9794041 (T)	1110463	1110464	1112008	1112007	1111148④	1111176④	1112009④	1112012④	1112011*	SINGLE	STD.—DUAL	H.D. SPEC.—DUAL	RAM-AIR IV—DUAL	SMALL VALVE	LARGE VALVE
155	ZB①	X				X		X			X										X									X				X	
155	ZG①	X					X	X			X											X								X				X	
255	WU		X			X			X		X					X							X								X			X	
255	YU		X				X		X		X					X							X								X			X	
366	WS②			X		X				X				X				X										†				X			X
350	WT			X		X				X			X				X										X				X				X
370	WW③			X		X				X					X				X									X						X		X
370	XP③			X		X				X					X				X										X					X		X
265	XX		X			X			X		X					X								X							X			X	
330	XV		X			X				X		X					X							X							X			X	
350	YS		X			X				X		X						X							X						X				X
366	YZ②			X		X				X				X				X								X						X			X
370	WA				X	X				X				X					X								X				X				X
370	YC				X		X			X				X					X								X				X				X

① L-6 camshaft usage is 3864897 for both manual and automatic transmissions.
† WS Engine uses 1112024 Distributor.
② Ram Air III
③ Ram Air IV
④ Uses hardened drive gear for use with 60 psi oil pump and high tension distributor points.
* Uses cadmium gear for use with R.A. IV only.

GENERAL CHASSIS AND BRAKE SPECIFICATIONS

YEAR	MODEL	CHASSIS Overall Length (In.)	CHASSIS Tire Size	Master Cylinder Bore (In.)	Wheel Cylinder Diameter (In.) Front	Wheel Cylinder Diameter (In.) Rear	Brake Drum Diameter (In.) Front	Brake Drum Diameter (In.) Rear
1964	6 Cyl., Exc. Safari	203.0	6.50 x 14	1.0	1 1/8	15/16	9.5	9.5
	8 Cyl., and Safari	203.0	7.00 x 14	1.0	1 1/8	15/16	9.5	9.5
1965	All Exc. Safari & GTO	206.1	6.95 x 14▲	1.0	1 1/8	15/16	9.5	9.5
	Safari	204.4	7.35 x 14■	1.0	1 1/8	15/16	9.5	9.5
	GTO	206.1	7.75 x 14	1.0	1 1/8	15/16	9.5	9.5
1966	All Exc. Sta. Wag. & G.T.O.	206.4	6.95 x 14	7/8	1 1/8	15/16	9.5	9.5
	Station Wagon	203.6	7.75 x 14	7/8	1 1/8	15/16	9.5	9.5
	G.T.O.	206.4	7.75 x 14	7/8	1 1/8	15/16	9.5	9.5
1967	All Exc. Sta. Wag. & G.T.O.	206.6	7.75 x 14	1.0	1 1/8	7/8	9.5	9.5
	Station Wagon	203.4	7.75 x 14	1.0	1 1/8	7/8	9.5	9.5
	G.T.O.	206.6	F-70 x 14	1.125●	2 1/16●	7/8	11.0●	9.5
	Firebird	188.8	E-70 x 14	1.0	1 1/8	7/8	9.5	9.5
1968	Tempest, Custom, LeMans	204.7	8.25 x 14*	1.0	1 1/8	7/8	9.5	9.5
	Safari Wagon	211.0	8.25 x 14*	1.0	1 1/8	7/8	9.5	9.5
	GTO	200.7	G-70 x 14	1.0	1 1/8	7/8	9.5	9.5
	Firebird	188.8	F-70 x 14*	1.0	1 1/8	7/8	9.5	9.5
	All w/Disc Brakes	—	—	1.125	2 1/16	7/8	11.0	9.5
1969	Tempest, Custom, LeMans	205.5	8.25 x 14*	1.0	1 1/8	7/8	9.5	9.5
	Safari Wagon	211.0	8.25 x 14*	1.0	1 1/8	7/8	9.5	9.5
	GTO	201.2	G-70 x 14	1.0	1 1/8	7/8	9.5	9.5
	Firebird	191.1	F-70 x 14*	1.0	1 1/8	7/8	9.5	9.5
	All w/Disc Brakes	—	—	1.125	2 15/16	7/8	11.0	9.5
1970-71	Tempest, Custom, LeMans	205.5	8.25 x 14*	1.0	1 1/8	7/8	9.5	9.5
	Safari Wagon	210.6	8.25 x 14*	1.0	1 1/8	7/8	9.5	9.5
	GTO, Judge	201.2	G-70 x 14	1.0	1 1/8	7/8	9.5	9.5
	Firebird, Trans Am	192.6	E-70 x 14*	1.125●	2 11/16●	7/8	11.0●	9.5
	All w/Disc Brakes	—	—	1.125	2 15/16	7/8	11.0	9.5

▲—Optional—7.35 x 14.
■—Optional—7.75 x 14.
●—Disc Brakes.

*—Tempest 6 Cyl.—7.75 x 14
Firebird 6 Cyl. 1-BBL. E-70 x 14

GENERAL ENGINE SPECIFICATIONS

YEAR	CU. IN. DISPLACEMENT	CARBURETOR	DEVELOPED HORSEPOWER @ RPM	DEVELOPED TORQUE @ RPM (FT. LBS.)	A.M.A. HORSEPOWER	BORE AND STROKE (IN.)	ADVERTISED COMPRESSION RATIO	VALVE LIFTER TYPE	NORMAL OIL PRESSURE (PSI)
1964	215 (OHV)6	1-BBL.	140 @ 4200	206 @ 2000	33.7	3.75 x 3.25	8.6-1	Hyd.	35
	326 V8	2-BBL.	250 @ 4600	333 @ 2800	44.3	3.72 x 3.75	8.6-1	Hyd.	35
	326 V8	4-BBL.	280 @ 4800	355 @ 3200	44.3	3.72 x 3.75	10.5-1	Hyd.	35
	389 V8	4-BBL.	325 @ 4800	428 @ 3200	52.8	4.09 x 3.75	10.75-1	Hyd.	35
	389 V8	3-2-BBL.	348 @ 4900	428 @ 3600	52.8	4.09 x 3.75	10.75-1	Hyd.	35
1965	215(OHV)6	1-BBL.	140 @ 4200	206 @ 2000	33.7	3.75 x 3.25	8.6-1	Hyd.	35
	326 V8	2-BBL.	250 @ 4600	333 @ 2800	44.3	3.72 x 3.75	9.2-1	Hyd.	35
	326 V8	4-BBL.	285 @ 5000	359 @ 3200	44.3	3.72 x 3.75	10.5-1	Hyd.	35
	389 V8	4-BBL.	335 @ 5000	431 @ 3200	52.8	4.06 x 3.75	10.75-1	Hyd.	35
	389 V8	3-2-BBL.	360 @ 5200	424 @ 3600	52.8	4.06 x 3.75	10.75-1	Hyd.	35
1966	230 (OHC)6	1-BBL.	165 @ 4700	216 @ 2600	36.0	3.88 x 3.25	9.0-1	Hyd.	35
	230 (OHC)6	4-BBL.	207 @ 5200	228 @ 3800	36.0	3.88 x 3.25	10.5-1	Hyd.	35
	326 V8	2-BBL.	250 @ 4600	333 @ 2800	44.3	3.72 x 3.75	9.2-1	Hyd.	35
	326 V8	4-BBL.	285 @ 5000	359 @ 3200	44.3	3.72 x 3.75	10.5-1	Hyd.	35
	389 V8	4-BBL.	335 @ 5000	431 @ 3200	52.8	4.06 x 3.75	10.75-1	Hyd.	35
	389 V8	3-2-BBL.	360 @ 5200	424 @ 3600	52.8	4.06 x 3.75	10.75-1	Hyd.	35
1967	230 (OHC)6	1-BBL.	165 @ 4700	216 @ 2600	36.0	3.88 x 3.25	9.0-1	Hyd.	35
	230 (OHC)6	4-BBL.	215 @ 5200	240 @ 3800	36.0	3.88 x 3.25	10.5-1	Hyd.	35
	326 V8	2-BBL.	250 @ 4600	333 @ 2800	44.3	3.72 x 3.75	9.2-1	Hyd.	35
	326 V8	4-BBL.	285 @ 5000	359 @ 3200	44.3	3.72 x 3.75	10.5-1	Hyd.	35
	400 V8	2-BBL.	255 @ 4400	397 @ 2400	54.3	4.12 x 3.75	8.6-1	Hyd.	35
	400 V8	4-BBL.	360 @ 5100	438 @ 3600	54.3	4.12 x 3.75	10.75-1	Hyd.	35
1968–69	250 (OHC)6	1-BBL.	175 @ 4800	240 @ 2600	36.0	3.88 x 3.53	9.0-1	Hyd.	31
	250 (OHC)6	4-BBL.	215 @ 5200	255 @ 3800	36.0	3.88 x 3.53	10.5-1	Hyd.	31
	350 V8	2-BBL.	265 @ 4600	355 @ 2800	48.0	3.88 x 3.75	9.2-1	Hyd.	35
	350 H.O. V8	4-BBL.	320 @ 5100	380 @ 3200	48.0	3.88 x 3.75	10.5-1	Hyd.	35
	400 V8	2-BBL.	265 @ 4600	397 @ 2400	54.3	4.12 x 3.75	8.6-1	Hyd.	35
	400 H. O. V8	4-BBL.	335 @ 5000	430 @ 3400	54.3	4.12 x 3.75	10.75-1	Hyd.	35
	400 (Ram Air)V8	4-BBL.	335 @ 5300	430 @ 3600	54.3	4.12 x 3.75	10.75-1	Hyd.	35
	400 H. O. V8	4-BBL.	360 @ 5100	445 @ 3600	54.3	4.12 x 3.75	10.75-1	Hyd.	35
	400 (Ram Air)V8	4-BBL.	360 @ 5400	445 @ 3600	54.3	4.12 x 3.75	10.75-1	Hyd.	35
1970–71	250(OHV)6	1-BBL.	155 @ 4200	235 @ 1600	36.0	3.87 x 3.53	8.5-1 ①	Hyd.	35
	350 V8	2-BBL.	255 @ 4600	355 @ 2800	48.0	3.88 x 3.75	8.8-1 ①	Hyd.	35
	350 V8	4-BBL.	325 @ 5100	380 @ 3200	48.0	3.88 x 3.75	10.5-1 ①	Hyd.	35
	350 V8	4-BBL.	330 @ 5100	380 @ 3200	48.0	3.88 x 3.75	10.5-1 ①	Hyd.	35
	400 V8	2-BBL.	265 @ 4600	397 @ 2400	54.3	4.12 x 3.75	8.8-1 ①	Hyd.	35
	400 V8	4-BBL.	330 @ 4800	445 @ 2900	54.3	4.12 x 3.75	10.00-1 ①	Hyd.	35
	400 Trans Am	4-BBL.	335 @ 5000	430 @ 3400	54.3	4.12 x 3.75	10.75-1 ①	Hyd.	35
	400 Ram Air	4-BBL.	345 @ 5400	430 @ 3700	54.3	4.12 x 3.75	10.75-1 ①	Hyd.	35
	400 V8	4-BBL.	350 @ 5000	445 @ 3000	54.3	4.12 x 3.75	10.25-1 ①	Hyd.	35
	400 Judge	4-BBL.	366 @ 5100	445 @ 3600	54.3	4.12 x 3.75	10.50-1 ①	Hyd.	35
	400 Ram Air IV	4-BBL.	370 @ 5500	445 @ 3900	54.3	4.12 x 3.75	10.50-1 ①	Hyd.	35
	455 V8	4-BBL.	360 @ 4300	500 @ 2700	55.2	4.15 x 4.21	10.25-1 ①	Hyd.	35

① —1971—Slightly lower.

TUNE-UP SPECIFICATIONS

YEAR	MODEL	SPARK PLUGS		DISTRIBUTOR		IGNITION TIMING (Deg.) ▲	CRANKING COMP. PRESSURE (Psi)	VALVES Tappet (Hot) Clearance (In.)		Intake Opens (Deg.)	FUEL PUMP PRESSURE (Psi)	IDLE SPEED (Rpm) ★
		Type	Gap (In.)	Point Dwell (Deg.)	Point Gap (In.)			Intake	Exhaust			
1964-65	6 Cyl.	46N	.035	32	.016	4B	145	Zero	Zero	18B	5-6½	500
	V8—326 Cu. In.	45S	.035	30	.016	6B	145	Zero	Zero	22B	5¼-6¾	500
	V8—389 Cu. In.	45S	.035	30	.016	6B	160	Zero	Zero	23B	5¼-6½	500
1966	6 Cyl.; 1-BBL., OHC	44S	.035	32	.016	5B	145	Zero	Zero	12B	4-5½	500 ⊙
	6 Cyl.; 4-BBL., OHC	45S	.035	32	.016	5B	145	Zero	Zero	20B	4-5½	500 ⊙
	V8—326 Cu. In.	45S	.035	30	.016	6B	160	Zero	Zero	22B	5-6½	500 ⊙
	V8—389 Cu. In., 4-BBL.	45S	.035	30	.016	6B	160	Zero	Zero	23B	5-6½	500 ⊙
	V8—389 Cu. In., 3-2-BBL.	45S	.035	30	.016	6B	160	Zero	Zero	31B	5-6½	500 ⊙
1967	6 Cyl.; 1-BBL., OHC	44N	.035	32	.016	5B ⊙	145	Zero	Zero	7B	4-5½	500 ⊙
	6 Cyl.; 4-BBL., OHC	45S	.035	32	.016	5B ⊙	145	Zero	Zero	14B	4-5½	500 ⊙
	V8—326 Cu. In., 2-BBL.	45S	.035	30	.016	6B ⊙	160	Zero	Zero	22B	5-6½	500 ⊙
	V8—326 Cu. In., 4-BBL.	45S	.035	30	.016	6B ⊙	160	Zero	Zero	22B	5-6½	500 ⊙
	V8—400 Cu. In., 2-BBL.	44S	.035	30	.016	6B ⊙	160	Zero	Zero	31B	5-6½	500 ⊙
	V8—400 Cu. In., 4-BBL.	44S	.035	30	.016	6B ⊙	160	Zero	Zero	38B	5-6½	500 ⊙
1968	6 Cyl.; OHC, 1-BBL.	44N	.035	32	.016	TDC	145	Zero	Zero	14B	4-5½	700 ■
	6 Cyl.; OHC, 4-BBL.	44N	.035	32	.016	5B	145	Zero	Zero	14B	4-5½	800 ■
	V8—350; 2-BBL.	45S	.035	30	.016	9B	160	Zero	Zero	22B	5-6½	700 ■
	V8—350; 4-BBL., M.T.	45S	.035	30	.016	9B	160	Zero	Zero	23B	5-6½	850
	V8—350; 4-BBL., A.T.	45S	.035	30	.016	9B	160	Zero	Zero	30B	5-6½	650
	V8—400; 2-BBL.	44S	.035	30	.016	9B	160	Zero	Zero	22B	5-6½	700 ■
	V8—400; 4-BBL.	44S	035	30	016	9B	160	Zero	Zero	23B	5-6½	850 ■
	V8—400; H.O., M.T.	44S	.035	30	.016	9B	160	Zero	Zero	31B	5-6½	850
	V8—400; H.O., A.T.	44S	.035	30	.016	9B	160	Zero	Zero	23B	5-6½	650
	V8—400; RAM AIR, M.T.	44S	.035	30	.016	9B	160	Zero	Zero	38B	5-6½	1000
	V8—400; RAM AIR, A.T.	44S	.035	30	.016	9B	160	Zero	Zero	31B	5-6½	650
1969	6 Cyl., OHC, 1-BBL.	44N	.035	32	.016	TDC	145	Zero	Zero	14B	4-5½	700 ■
	6 Cyl., OHC, 4-BBL.	44N	.035	32	.016	5B	145	Zero	Zero	14B	4-5½	800 ■
	V8—350; 2-BBL., A.T.	46S	.035	30	.016	9B	160	Zero	Zero	22B	5-6½	650
	V8—350; 2-BBL., M.T.	46S	.035	30	.016	9B	160	Zero	Zero	22B	5-6½	850
	V8—350; 4-BBL., A.T.	45S	.035	30	.016	9B	160	Zero	Zero	30B	5-6½	650
	V8—350; 4-BBL., M.T.	45S	.035	30	.016	9B	160	Zero	Zero	23B	5-6½	850
	V8—400; 2-BBL. R. Fuel	46S	.035	30	.016	9B	160	Zero	Zero	22B	5-6½	700 ■
	V8—400; 4-BBL. A.T.	45S	.035	30	.016	9B	160	Zero	Zero	31B	5-6½	650
	V8—400; 4-BBL. M.T.	45S	.035	30	.016	9B	160	Zero	Zero	23B	5-6½	850
	V8—400, RAM Air	44S	.035	30	.016	15B	160	Zero	Zero	31B	5-6½	1000
1970-71	6 Cyl., 250 Cu. In. M.T.	R46T	.035	32	.019	TDC	130	Zero	Exhaust	16B	4-5	700
	6 Cyl., 250 Cu. In. A.T.	R46T	.035	32	.019	4B	130	Zero	Exhaust	16B	4-5	550
	V8—350; 2-BBL., A.T.	R46S	.035	30	.016	9B	160	Zero	Zero	22B	5-6½	650
	V8—350; 2-BBL., M.T.	R46S	.035	30	.016	9B	160	Zero	Zero	22B	5-6½	800
	V8—350; 4-BBL., A.T.	R45S	.035	30	.016	9B	160	Zero	Zero	30B	5-6½	650
	V8—350; 4-BBL., M.T.	R45S	.035	30	.016	9B	160	Zero	Zero	23B	5-6½	800
	V8—400; 2-BBL. R. Fuel	R46S	.035	30	.016	9B	160	Zero	Zero	22B	5-6½	700 ■
	V8—400; 4-BBL. A.T.	R45S	.035	30	.016	9B	160	Zero	Zero	31B	5-6½	650
	V8—400; 4-BBL. M.T.	R45S	.035	30	.016	9B	160	Zero	Zero	23B	5-6½	850
	V8—400, RAM AIR	R44S	.035	30	.016	9B	160	Zero	Zero	②	5-6½	1000
	V8—455	R44S	.035	30	.016	9B	160	Zero	Zero	23B ①	5-6½	850

▲—With vacuum advance disconnected. NOTE: These settings are only approximate. Engine design, altitude, temperature, fuel octane rating and the condition of the individual engine are all factors which can influence timing. The limiting advance factor must, therefore, be the "knock point" of the individual engine.

★—With manual transmission in N and automatic in D. Add 50 rpm if equipped with air conditioning.

⊙—California Air Injection System, 600-650.

B—Before top dead center.

TDC—Top dead center.

⊙—6 Cyl., 1-BBL.—TDC—All others same as non-exhaust emission type.

■—w/auto. trans.:
6 cyl., V8-2 BBL., 600 rpm.
V8-400, 4 BBL. 650 rpm.

①—M.T. models = 31B.

②—RAM AIR IV—42 B; RAM AIR III—31 B.

VALVE SPECIFICATIONS

YEAR AND MODEL		SEAT ANGLE (DEG.)	FACE ANGLE (DEG.)	VALVE LIFT INTAKE (IN.)	VALVE LIFT EXHAUST (IN.)	VALVE SPRING PRESSURE (VALVE OPEN) LBS. @ IN.	VALVE SPRING INSTALLED HEIGHT (IN.)	STEM TO GUIDE CLEARANCE (IN.)		STEM DIAMETER (IN.)	
								Intake	Exhaust	Intake	Exhaust
1964	6 Cyl.—215	46	45	.335	.335	171 @ 1.33	1 21/32	.0010–.0033	.0020–.0037	.341	.341
	V8—326	①	②	.370	.406	⑤	1 17/32	.0021–.0038	.0026–.0043	.341	.341
	V8—389	①	②	.405	.408	⑤	1 17/32	.0021–.0038	.0026–.0043	.341	.341
1965	6 Cyl.—215	46	45	.335	.335	171 @ 1.33	1 21/32	.0010–.0033	.0020–.0037	.341	.341
	V8—326	①	②	.370	.406	⑤	1 37/64	.0021–.0038	.0026–.0043	.341	.341
	V8—389	①	②	.406	.408	⑥	1 37/64	.0021–.0038	.0026–.0043	.341	.341
1966	6 Cyl.—1 BBL.—OHC	①	②	.400	.400	192 @ 1.18	1 37/64	.0021–.0038	.0026–.0043	.341	.341
	6 Cyl.—4 BBL.—OHC	①	②	.438	.438	⑤	1 37/64	.0021–.0038	.0026–.0043	.341	.341
	326—2-BBL.	①	②	.370	.406	⑤	1 37/64	.0021–.0038	.0026–.0043	.341	.341
	326—4-BBL.	①	②	.370	.406	⑤	1 37/64	.0021–.0038	.0026–.0043	.341	.341
	389—4-BBL.	①	②	.406	.408	⑥	1 37/64	.0021–.0038	.0026–.0043	.341	.341
	389—3-2-BBL.	①	②	.409	.409	⑥	1 37/64	.0021–.0038	.0026–.0043	.341	.341
1967	6 Cyl.—1 BBL.—OHC	①	②	.400	.400	192 @ 1.18	1 37/64	.0016–.0033	.0016–.0038	.341	.341
	6 Cyl.—4-BBL.—OHC	①	②	.438	.438	⑥	1 37/64	.0016–.0033	.0016–.0038	.341	.341
	V8—326—2-BBL.	①	②	.375	.410	⑥	1 37/64	.0016–.0033	.0016–.0038	.341	.341
	V8—326—4-BBL.	①	②	.375	.410	⑥	1 37/64	.0016–.0033	.0016–.0038	.341	.341
	V8—400—2-BBL.	①	②	.414	.413	⑥	1 37/64	.0016–.0033	.0021–.0038	.341	.341
	V8—400—4-BBL.	①	②	.414	.413	252 @ 1.34	1 37/64	.0016–.0033	.0021–.0038	.341	.341
1968	6 Cyl.; O.H.C., 1-BBL.	①	②	.400	.400	170 @ 1.23	1 21/32	.0016–.0033	.0021–.0038	.341	.341
	6 Cyl.; O.H.C., 4-BBL.	①	②	.438	.438	⑥	1 21/32	.0016–.0033	.0021–.0038	.341	.341
	V8—350; 2-BBL.	①	②	.376	.412	⑥	1 37/64	.0016–.0033	.0021–.0038	.341	.341
	V8—350, 4-BBL.	①	②	.410	.413	⑥	1 37/64	.0016–.0033	.0021–.0038	.341	.341
	V8—400; H.O., M.T.	①	②	.414	.413	⑥	1 37/64	.0016–.0033	.0021–.0038	.341	.341
	V8—400; H.O., A.T.	①	②	.410	.413	252 @ 1.34	1 37/64	.0016–.0033	.0021–.0038	.341	.341
	V8—400; Ram Air	①	②	.414	.413	252 @ 1.34	1 37/64	.0016–.0033	.0021–.0038	.341	.341
1969	6 Cyl.; O.H.C., 1-BBL.	46	45	.400	.400	170 @ 1.23	1 21/32	.0016–.0033	.0021–.0038	.341	.341
	6 Cyl.; O.H.C., 4-BBL.	46	45	.438	.438	⑦	1 21/32	.0016–.0033	.0021–.0038	.341	.341
	V8—350; 2-BBL.	③	④	.376	.412	⑥	1 37/64	.0016–.0033	.0021–.0038	.341	.341
	V8—350, 4-BBL.	③	④	.410	.413	⑥	1 37/64	.0016–.0033	.0021–.0038	.341	.341
	V8—400; H.O., M.T.	③	④	.414	.413	⑥	1 37/64	.0016–.0033	.0021–.0038	.341	.341
	V8—400; H.O., A.T.	③	④	.410	.413	185 @ 1.299	1 23/32	.0016–.0033	.0021–.0038	.341	.341
	V8—400; Ram Air	③	④	.414	.413	185 @ 1.299	1 23/32	.0016–.0033	.0021–.0038	.341	.341
1970	6 Cyl. 250	46	45	.388	.388	186 @ 1.27	1 21/32	.0010–.0027	.0010–.0027	.341	.341
	V8—350	45	44	.376	.412	⑥	1 37/64	.0016–.0033	.0021–.0038	.341	.341
	V8—400 2-BBL.	45	44	.376	.412	⑥	1 37/64	.0016–.0033	.0021–.0038	.341	.341
	V8—400 4-BBL.	③	④	.410	.413	⑥	1 37/64	.0016–.0033	.0021–.0038	.341	.341
	V8—400 Ram Air	③	④	.527	.527	⑨	—	.0016–.0033	.0021–.0038	.341	.341
	V8—455	③	④	.414	.413	⑧	1 9/16	.0016–.0033	.0021–.0038	.341	.341
1971	6 Cyl. 250	46	45	.388	.388	186 @ 1.27	1 21/32	.0010–.0027	.0010–.0027	.341	.341
	V8—350	45	44	.376	.412	⑥	1 37/64	.0016–.0033	.0021–.0038	.341	.341
	V8—400 2-BBL.	45	44	.376	.412	⑥	1 37/64	.0016–.0033	.0021–.0038	.341	.341
	V8—400 4-BBL.	③	④	.410	.413	⑥	1 37/64	.0016–.0033	.0021–.0038	.341	.341
	V8—400 Ram Air	③	④	.527	.527	⑨	—	.0016–.0033	.0021–.0038	.341	.341
	V8—455	③	④	.414	.413	⑧	1 9/16	.0016–.0033	.0021–.0038	.341	.341

①—Intake 30°; Exhaust 45°.
②—Intake 29°; Exhaust 44°.
③—Large Intake—30°, Small Intake—45°; Exhaust—45°.
④—Large Intake—29°; Small Intake—46°; Exhaust—44°.
⑤—Outer—114 @ 1.12; Inner—65 @ 1.07.
⑥—Outer—132 @ 1.178; Inner—90 @ 1.196.
⑦—Outer—120 @ 1.103; Inner—62 @ 1.17.
⑧—Outer—137 @ 1.1513; Inner—103 @ 1.107.
⑨—Outer 233 @ 1.291; Inner—111 @ 1.221.

AC GENERATOR AND REGULATOR SPECIFICATIONS

YEAR	ALTERNATOR			REGULATOR						
	Part Number	Field Current Draw @ 12V.	Output at Generator RPM 5000	Field Relay				Regulator		
				Part Number	Air Gap (In.)	Point Gap (In.)	Volts to Close	Air Gap (In.)	Point Gap (In.)	Volts at 125°
1964	1100668	1.9–2.3	37	1119515	.015	.030	2.3–3.7	.057	.015	13.5–14.4
	1100683	1.9–2.3	50	1119515	.015	.030	2.3–3.7	.057	.015	13.5–14.4
	1107294	1.9–2.3	50	1119515	.015	.030	2.3–3.7	.057	.015	13.5–14.4
	1100627	1.9–2.3	50	1119515	.015	.030	2.3–3.7	.057	.015	13.5–14.4
1965	1100714▲	1.9–2.3	37	1119515●	.015	.030	2.3–3.7	.057	.015	13.5–14.4
	1100704■	1.9–2.3	50	1119515●	.015	.030	2.3–3.7	.057	.015	13.5–14.4
1966	1100705●●	—	37	1119515●	.015	.030	2.3–3.7	.057	.015	13.8
	1100736●●●	—	37	1119515●	.015	.030	2.3–3.7	.057	.015	13.8
1967–69	1100761①	—	37	1119515●	.015	.030	2.3–3.7	.057	.015	13.8
	1100704②	—	37	1119515●	.015	.030	2.3–3.7	.057	.015	13.8
1970	1100905③	2.2–2.6	37	1119515	.015	.030	1.5–3.2	—	.014	13.8
	1100704②	2.2–2.6	37	1119515	.015	.030	1.5–3.2	—	.014	13.8
1971	1100907③	2.2–2.6	37	1119515	.015	.030	1.5–3.2	—	.014	13.8
	1100706②	2.2–2.6	37	1119515	.015	.030	1.5–3.2	—	.014	13.8

▲—1100701 or 1100707 with air conditioning.
■—1100700 or 1100692 with air conditioning (exc. with trans. ign.).
●—Exc. with transistor ign.
●●—with a/cond. 1100744 (55 Amp.); With a/cond. & tran. ign. 1100741 (60 Amp.).

●●●—With a/cond. 1100700 (55 Amp.); With a/cond. & tran. ign. 1100702 (60 Amp.).
①—1100760 (55 Amp.) with A/C
②—1100700 (55 Amp.) with A/C
③—1100906 (55 Amp.) with A/C

TORQUE SPECIFICATIONS

YEAR	MODEL	SPARK PLUGS (FT. LBS.)	CYLINDER HEAD BOLTS (FT. LBS.)	ROD BEARING BOLTS (FT. LBS.)	MAIN BEARING BOLTS (FT. LBS.)	CRANKSHAFT BALANCER BOLT (FT. LBS.)	FLYWHEEL TO CRANKSHAFT BOLTS (FT. LBS.)	MANIFOLD (FT. LBS.)	
								Intake	Exhaust
1964–69	6 Cyl.	15–25	85–100	30–35	60–70	PRESSED ON	60–70	25–40	15–25
	V8	15–25	85–100	40–46	90–110▲	130–190	85–100	20–35	30–45
1970–71	6 Cyl.	15	95	35	65	PRESSED ON	60	30	25
	V8	25	95	43	90–110	160	95	40	30

▲Rear Main 120 ft. lbs.

CYLINDER HEAD BOLT TIGHTENING SEQUENCE

V8

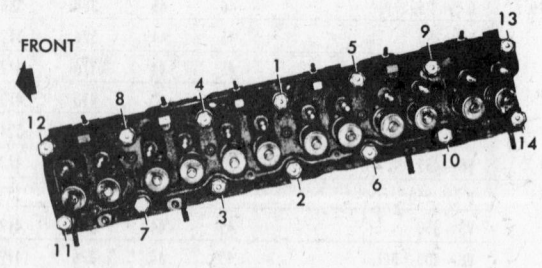

OHV 6 cyl.

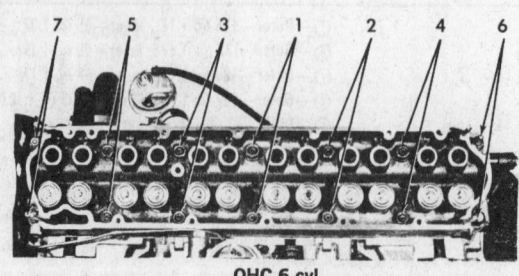

OHC 6 cyl.

BATTERY AND STARTER SPECIFICATIONS

YEAR	MODEL	BATTERY				STARTERS						Brush Spring Tension (Oz.)
		Ampere Hour Capacity	Volts	Group Number	Terminal Grounded	Lock Test			No-Load Test			
						Amps.	Volts	Torque	Amps.	Volts	RPM	
1964-65	4 Cyl. & 6 Cyl.	44	12	17MI	Neg.	Not Recommended			49-76	10.6	6,200-9,400	35
	V8	53	12	2SMB	Neg.	Not Recommended			65-120	10.6	3,600-5,400	35
1966	6 Cyl.	44	12	17MI	Neg.	Not Recommended			49-76	10.6	6,200-9,600	35
	326-V8	53	12	2SMB	Neg.	Not Recommended			65-100	10.6	3,600-5,100	35
	389-V8	66	12	2SMB	Neg.	Not Recommended			Not Recommended			35
1967-69	6 Cyl.	44	12	17MI	Neg.	Not Recommended			49-76			35
	326, 350-V8	53	12	2SM	Neg.	Not Recommended			65-100			35
	400-V8	61	12	2SM	Neg.	Not Recommended			Not Recommended			35
1970-71	6 Cyl.	45	12	17MI	Neg.	Not Recommended			41-47	10	—	35
	V8 350	53	12	25M	Neg.	Not Recommended			41-47	10	—	35
	V8 400, 455	61	12	2SM	Neg.	Not Recommended			41-47	10	—	35

CRANKSHAFT BEARING JOURNAL SPECIFICATIONS

YEAR	MODEL	MAIN BEARING JOURNALS (IN.)				CONNECTING ROD BEARING JOURNALS (IN.)		
		Journal Diameter	Oil Clearance	Shaft End-Play	Thrust On No.	Journal Diameter	Oil Clearance	End-Play
1964-65	6 Cyl.	2.30	.0003-.0029	.004	7	2.000	.0007-.0027	.0110
	V8	3.00	.0005-.0020	.006	4	2.250	.0005-.0025	.0085
1966-69	6 Cyl.	2.30	.0003-.0019	.004	7	2.000	.0007-.0027	.0110
	V8 350, 400	3.00	.0002-.0017	.006	4	2.250	.0005-.0025	.009
	V8 455	3.25	.0005-.0021	.006	4	2.250	.0005-.0021	.009
1970-71	6 Cyl.	2.299	.0003-.0029	.002-.006	7	2.000	.0007-.0027	.009
	V8 350, 400	3.00	.0002-.0017	.003-.009	4	2.250	.0005-.0025	.012
	V8 455	3.25	.0005-.0021	.003-.009	4	2.250	.0005-.0026	.012

CAPACITIES

YEAR	MODEL	ENGINE CRANKCASE ADD 1 Qt. FOR NEW FILTER	TRANSMISSIONS Pts. TO REFILL AFTER DRAINING			DRIVE AXLE (Pts.)	GASOLINE TANK (Gals.)	COOLING SYSTEM (Qts.) WITH HEATER
			Manual		Automatic			
			3-Speed	4-Speed				
1964-65	6 Cyl.—215 Cu. In.	4	3	3¾	15	4½ ▲	21½	13½
	V8—326 Cu. In.	4	3	3¾	15	4½ ▲	21½	20½
	V8—389 Cu. In.	5	3	3¾	15	4½ ▲	21½	20½
1966	6 Cyl.—230 Cu. In.	5	2¾	7½	15	3	21½	13½
	V8—326 Cu. In.	5	2¾	2½	15	3	21½	20½
	V8—389 Cu. In.	5	2¾	2½	15	3	21½	20
1967	6 Cyl.—230 Cu. In.	5	2¾	2½	15	3	21½	12¼
	V8—326 Cu. In.	5	2¾	2½	15	3	21½	18½
	V8—400 Cu. In.	6	2¾	2½	19	3	21½	17¾
1968	6 Cyl. 250 Cu. In.	5	3½	3½	15	3	21½ ●	12
	V8—350 Cu. In.	5	3½ ■	2½	15	3	21½ ●	19½
	V8—400 Cu. In.	5	2¾	2½	19①	3	21½ ●	17¾
1969	6 Cyl. 250 Cu. In.	4½	3½	3½	15	3	21½ ●	12
	V8—350 Cu. In.	5	3½ ■	2½	15	3	21½ ●	19½
	V8—400 Cu. In.	5	2¾	2½	19①	3	21½ ●	17¾
1970-71	6 Cyl. 250 Cu. In.	4	3	—	6	3②	21½ ●	12
	V8—350 Cu. In.	5	3½	2½	6	3②	21½ ●	19½
	V8—400 Cu. In.	5	2¾	2½	7½	3②	21½ ●	18½
	V8—455 Cu. In.	5	2¾	2½	7½	3②	21½	17¼

▲—1965, 3 Pts.
■—Lemans, 2¾ Pts.
●—Firebird, Gasoline Tank—18½ Gal.
①—"350" Trans.-20
②—5 Pts. w/8.875 in. ring gear.

WHEEL ALIGNMENT

YEAR	MODEL	CASTER Range (Deg.)	CASTER Pref. Setting (Deg.)	CAMBER Range (Deg.)	CAMBER Pref. Setting (Deg.)	TOE-IN (In.)	KING-PIN INCLINATION (Deg.)	WHEEL PIVOT RATIO Inner Wheel	WHEEL PIVOT RATIO Outer Wheel
1964–67	Exc. Firebird	2N to 1N	$1\frac{1}{2}$N	$\frac{1}{4}$N to $\frac{3}{4}$P	$\frac{1}{4}$P	0 to $\frac{1}{8}$	9	20	$18\frac{1}{4}$
1967	Firebird	0 to 1P	$\frac{1}{2}$P	$\frac{1}{4}$N to $\frac{3}{4}$P	$\frac{1}{4}$P	$\frac{1}{8}$ to $\frac{1}{4}$	$8\frac{3}{4}$	20	—
1968	Exc. Firebird	2N to 1N	$1\frac{1}{2}$N	$\frac{1}{4}$N to $\frac{3}{4}$P	$\frac{1}{4}$P	0 to $\frac{1}{8}$	9	20	$18\frac{1}{2}$
	Firebird	0 to 1P	$\frac{1}{2}$P	$\frac{1}{4}$N to $\frac{3}{4}$P	$\frac{1}{4}$P	$\frac{1}{8}$ to $\frac{1}{4}$	$8\frac{3}{4}$	20	$18\frac{1}{2}$
1969	Exc. Firebird	2N to 1N	$1\frac{1}{2}$N	$\frac{1}{4}$N to $\frac{3}{4}$P	$\frac{1}{4}$P	0 to $\frac{1}{8}$	9	20	$18\frac{1}{2}$
	Firebird	0 to 1P	$\frac{1}{2}$P	$\frac{1}{4}$N to $\frac{3}{4}$P	$\frac{1}{4}$P	$\frac{1}{8}$ to $\frac{1}{4}$	$8\frac{3}{4}$	20	$18\frac{1}{2}$
1970–71	Tempest, LeMans	1N to 2N	$1\frac{1}{2}$N	$\frac{1}{4}$N to $\frac{3}{4}$P	$\frac{1}{4}$P	0 to $\frac{1}{8}$	9	20	22
	Station Wagon	$1\frac{1}{2}$N to $2\frac{1}{2}$N	2N	$\frac{1}{4}$N to $\frac{3}{4}$P	$\frac{1}{4}$P	0 to $\frac{1}{8}$	9	20	22
	Firebird	$\frac{1}{2}$N to $1\frac{1}{2}$N	1N	$\frac{1}{4}$P to $1\frac{1}{4}$P	$\frac{3}{4}$P	$\frac{1}{8}$ to $\frac{1}{4}$	$8\frac{3}{4} \pm \frac{1}{2}$	20	22

N—Negative.
P—Positive.

FUSES AND CIRCUIT BREAKERS

1964–69

Headlamps, parking lights .... 15 Amp. CB
Power seat, power windows .... 40 Amp. CB
Tail lights, license, courtesy .... 10 Amp. Fuse
Dome, clock power .... 10 Amp. Fuse
Radio dial, instrument lamps .... 4 Amp. Fuse
Heater control lamps, clock lamp.......... 4 Amp. Fuse
Radio power .... 2.5 Amp. Fuse
Turn signals .... 15 Amp. Fuse
Back-up, heater blower, power .... 20 Amp. Fuse
Antenna, windshield washer & wiper ...... 20 Amp. Fuse

1970–71

	Temp.	F.B.
Air cond. controls lamp (custom)	SFE-4	SFE-4
Air cond. power & blower mtr. (custom)........................	AGC-30	AGC-30
Ash tray lamp	SFE-4	SFE-4
Back-up lamp......................	SFE-20	SFE-20
Cigar lighter......................	SFE-20	SFE-20
Cigar lighter lamp..................	SFE-4	SFE-4
Clock lamp	SFE-4	SFE-4
Clock power	SFE-20	SFE-20
Console compartment lamp	—	—
Cornering lamps	—	—
Deck lid release...................	SFE-20	SFE-20
Direction signal & stop lamp	SFE-20	SFE-20
Dome, rear qtr. or rear courtesy lamp	SFE-20	SFE-20
Electrocruise & low fuel lamp	AGC-10	—
Hazard flasher....................	SFE-20	SFE-20
Heater cont. lamp	SFE-4	SFE-4
Heater blower motor	AGC-25	AGC-25
Instrument lamps	SFE-4	SFE-4
Instrument panel compartment lamps.......................	SFE-20	SFE-20
Instrument panel courtesy lamp	SFE-20	SFE-20
License lamp	SFE-20	SFE-20
Lugg. compt. lamp.................	SFE-20	SFE-20
P.B. warning lamp.................	AGC-10	AGC-10
Power seat circuit breaker..........	40	40
Power tail gate wind. circuit breaker	40	40
Power side window circuit breaker...	40	40

	Temp.	F.B.
Radio dial lamp	SFE-4	SFE-4
Radio power......................	AGC-10	AGC-10
Rear wind. defogger	AGC-25	AGC-25
Safeguard speedo.................	AGC-25	—
Shift indicator lamp (column & console)	SFE-4	SFE-4
Side marker lamps frt. & rear	SFE-20	SFE-20
Tachometer lamp..................	SFE-4	SFE-4
Tail lamp	SFE-20	SFE-20
Under hood or utility	SFE-20	SFE-20
W/S washer pump.................	AGC-25	AGC-25
W/S wiper or washer motor........	AGC-25	AGC-25

LIGHT BULBS

1964–69

Headlamps (low beam) .... 4002; 50/37.5 W
Headlamps (high beam)................. 4001; 37.5 W
Park, turn signals, stop, tail .... 1157; 32 CP
License................................ 1155; N.A.
Panel indicators, tach., radio............... 1895; N.A.
Heater control, panel illum............... 1895; N.A.
Dome .... 1004; 15 CP
Courtesy .... 89; 6 CP

1970–71

	Temp.	F.B.
Air cond. controls	1895	1895
Ash tray	—	—
Back-up lamp	1156	1157
Brake warning	194	194
Clock	1895	—
Cigar lighter	1445	1445
Cornering lamps.................	—	—
Courtesy lamp	89	89
Courtesy (sta. wag. rear compt.)....	211-1	—
Dir. signal ind.	194	194
Dome lamp.....................	211-1	211-1

	Temp.	F.B.
Head lamp (lower) or (inner)......................	Type 1	—
Head lamp (upper) or (outer)......................	Type 2	Type 2
Head lamp beam ind..............	194	194
Heater cont.	1895	1895
Inst. panel compt.	1895	1895
License lamp.....................	67	67
Low fuel warning	159	194
Luggage compt.	89	1003
Oil pressure	194	194
Park & dir. signal................	1157 NA	1157 NA
Radio dial	1895	1895
Reading lamp	1004	—
Rr. seat arm rest (conv.)	68	—
Side marker-front	194A	194A
Side marker-rear	194	194
Side marker-rear	—	—
Shift ind. (exc. console)	1893①	1893①
Shift ind. (console)	1445	1445
Side roof rail	—	—
Speedo. Ill. (exc. aux. ga.)	194	194
Tail stop & signal	1157	1157
Tachometer (hood)	194	194
Tachometer (I.P. mtg.)............	1895	—
Temp. ga. (exc. aux. ga.).........	194	194
Temp. ga. (aux. ga. panel)........	—	1895
Under hood or utility	1003	1003

① —Exc. Tilt Wheel

Distributor

Removal

1. Disconnect distributor primary wire.
2. Remove distributor cap. (Unlatch the cap by using a screwdriver to disengage the latches.)
3. Crank engine so the rotor is in No. 1 position and the crankshaft pulley timing mark in line with the pointer.
4. Disconnect vacuum line at distributor.
5. Remove distributor clamp screw and hold-down clamp.
6. Lift out distributor and distributor-to-block gasket. Notice the slight rotation of the rotor as the distributor is removed from the block.

Installation—If Engine Has Been Disturbed

1. With No. 1 piston coming up on compression stroke, continue cranking the engine until the pulley timing mark indexes with the stationary mark.
2. Position the distributor to the opening in the block with reference to the firing order sequence shown in the firing order illustrations of this section.
3. Point the rotor toward distributor cap No. 1 tower location. Then, slightly retard the distributor rotor position.
4. Press down on distributor housing until seated, clamp unit in place.
5. Check initial timing, centrifugal and vacuum advance characteristics and engine vacuum.

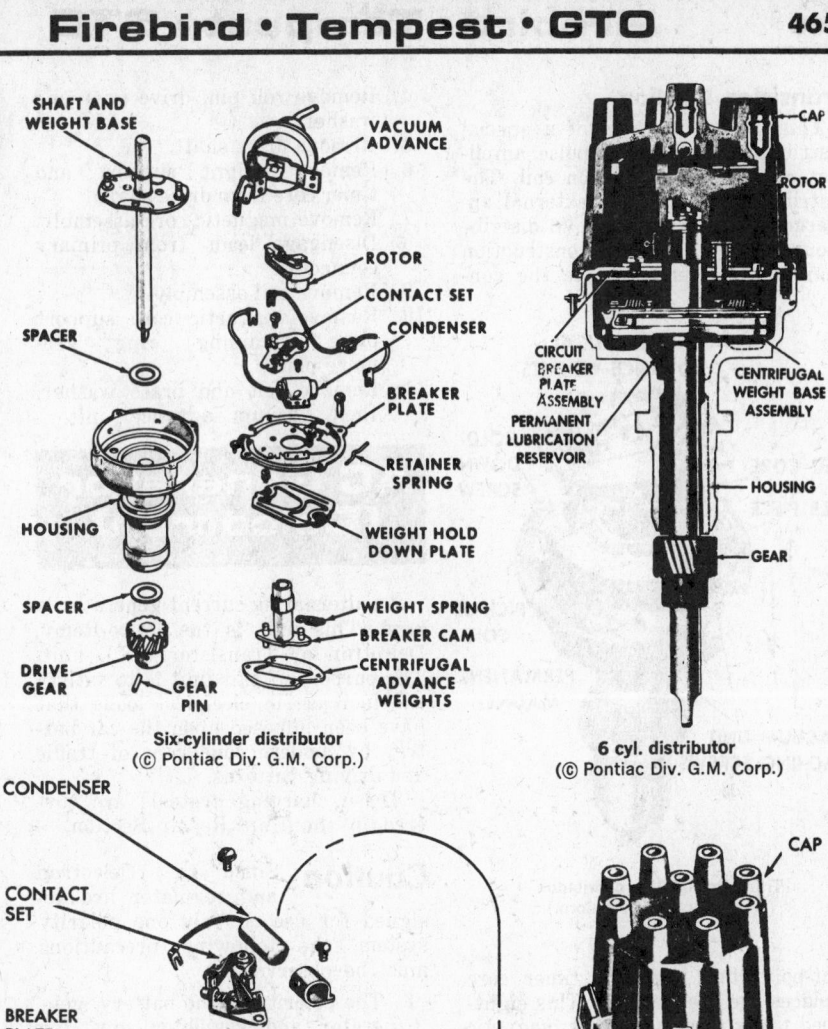

Six-cylinder distributor
(© Pontiac Div. G.M. Corp.)

6 cyl. distributor
(© Pontiac Div. G.M. Corp.)

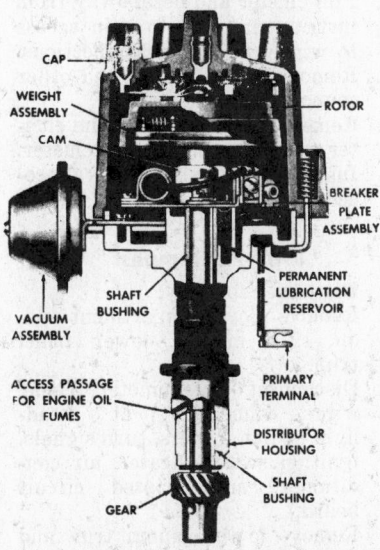

V8 distributor (radio interference shield is on 1970 distributors)
(© Pontiac Div. G.M. Corp.)

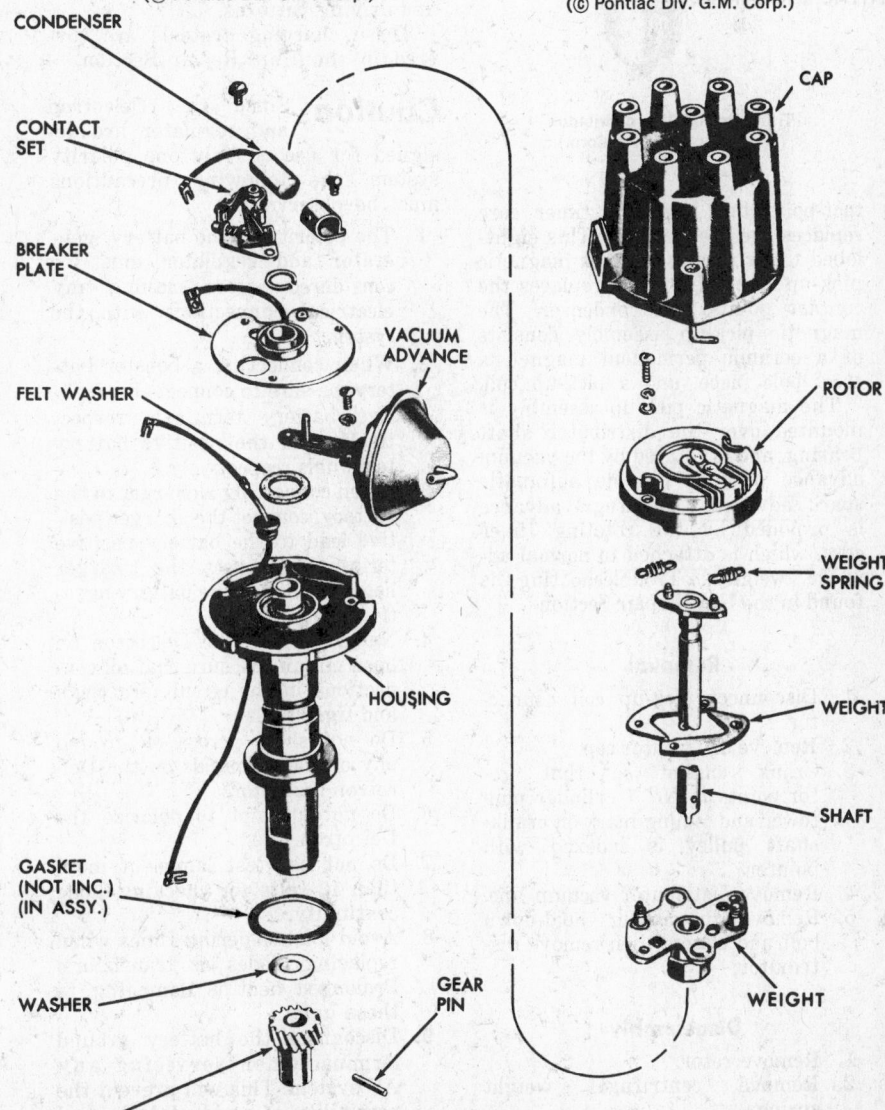

V8 distributor (© Pontiac Div. G.M. Corp.)

Transistor Ignition

This system consists of a special distributor, an ignition pulse amplifier, and a special ignition coil. The distributor is similar in external appearance to the standard V8 distributor, but the internal construction bears little resemblance to the con-

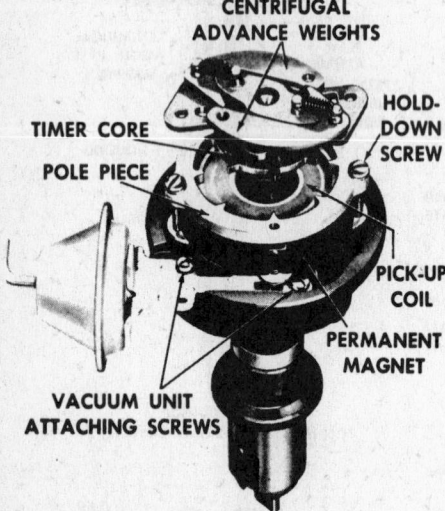

CENTRIFUGAL ADVANCE WEIGHTS
TIMER CORE
POLE PIECE
HOLD-DOWN SCREW
PICK-UP COIL
PERMANENT MAGNET
VACUUM UNIT ATTACHING SCREWS

Transistor ignition distributor
(© Pontiac Div. G.M. Corp.)

tact-point unit. An iron timer core replaces the breaker cam. This eight-lobed timer rotates inside a magnetic pick-up assembly, which replaces the contact points and condenser. The magnetic pick-up assembly consists of a ceramic permanent magnet, a steel pole piece, and a pick-up coil.

The magnetic pick-up assembly is mounted over the distributor shaft bearing, and is rotated by the vacuum advance unit to provide automatic spark advance. Centrifugal advance is provided by the rotating timer core, which is attached to normal advance weights. Troubleshooting is found in the Unit Repair Section.

Removal

1. Disconnect pick-up coil connector.
2. Remove distributor cap.
3. Crank engine so that rotor points to No. 1 cylinder plug tower and timing mark on crankshaft pulley is indexed with pointer.
4. Remove distributor vacuum line.
5. Remove distributor hold-down bolt and clamp, then remove distributor.

Disassembly

1. Remove rotor.
2. Remove centrifugal weight springs.
3. Remove weights.

4. Remove roll pin, drive gear and washer.
5. Remove drive shaft.
6. Remove weight support and timer core from drive shaft.
7. Remove magnetic core assembly.
8. Disengage leads from primary connector.
9. Remove coil assembly.
10. Remove magnetic core support plate retaining ring; remove plate.
11. Remove felt and brass washer, then vacuum advance unit.

AC Generator (Delcotron)

An alternating current generator is used. This unit is the Delco-Remy, Delcotron or Transistor C.S.I. unit. The purpose of this unit is to satisfy the increase in electrical loads that have been imposed upon the car battery by modern conditions of traffic and driving patterns.

These charging systems are covered in the Unit Repair Section.

Caution Since the Delcotron and regulator are designed for use on only one polarity system, the following precautions must be observed:

1. The polarity of the battery, generator and regulator must be considered before making any electrical connections with the system.
2. When connecting a booster battery, be sure to connect the negative battery terminals respectively, and the positive battery terminals respectively.
3. When connecting a charger to the battery, connect the charger positive lead to the battery positive terminal. Connect the charger negative lead to the battery negative terminal.
4. Never operate the Delcotron on open circuit. Be sure that all connections in the circuit are clean and tight.
5. Do not short across or ground any of the terminals on the Delcotron regulator.
6. Do not attempt to polarize the Delcotron.
7. Do not use test lamps of more than 12 volts for checking diode continuity.
8. Avoid long soldering times when replacing diodes or transistors. Prolonged heat is damaging to these units.
9. Disconnect the battery ground terminal when servicing any AC system. This will prevent the possibility of accidental reversal of polarity.

Instruments

The instrument cluster includes the speedometer head, the generator charge indicator, the oil pressure indicator and the temperature indicator. Also, the fuel gauge, light switch, wiper and washer switch, starter and ignition switch and the cigarette lighter.

Cluster Removal and Installation

1964-65 Tempest

1. Disconnect battery.
2. Remove upper retaining nuts.
3. Remove screws at lower edge of cluster.
4. Remove screws from lower steering column bezel.
5. Disconnect speedometer cable.
6. Pull cluster and housing out from instrument panel to gain access to wiring.
7. Starting at the top, remove bulbs and wiring.
8. Remove screws holding cluster to housing and remove the cluster.
9. To install, reverse above procedure.

1966-68 Tempest

1. Disconnect battery.
2. Remove screws holding bezel and cluster assembly to instrument panel.
3. Remove speedometer cable.
4. Disconnect heater control cables.
5. Lower steering column by removing trim plate and loosening nuts on column bracket.
6. Pull cluster and bezel away from instrument panel to gain access to wiring and other connections.
7. Remove bulbs, wiring and other connections, as necessary.
8. Remove screws holding the cluster to the bezel. Remove cluster.
9. Install by reversing removal procedure.

1969-70 Tempest

1. Disconnect battery.
2. Remove glove compartment and, on 1970 models, lower panel trim.
3. Disconnect speedometer cable, wire connectors at headlight switch, wipers, turn signals, ignition switch, heater, air conditioner, and printed circuit board.
4. Remove lower column trim and disconnect heater cable; remove lower air conditioner duct, if necessary.

5. Lower steering column.
6. Remove three instrument panel screws at gauge clusters.
7. Remove three upper right instrument panel nuts.
8. Remove ground strap retaining screws, if present, then disconnect harness conduit.
9. Remove cluster retaining screws, then cluster.
10. To install, reverse removal procedure.

NOTE: it may be necessary to shift the instrument panel around to gain access to some wiring.

1967-69 Firebird

1. Disconnect battery.
2. Remove lower instrument panel cover.
3. Remove ashtray bracket screws, radio retaining nuts, and glove compartment.
4. Disconnect heater control cables and wires.
5. Disconnect speedometer cable, then remove upper left-hand vent duct connector.
6. Disconnect headlight switch shaft.
7. Remove screws across top and bottom of instrument panel and the nut on the right-hand side (stud through dash).
8. Loosen toe plate screws, remove lower column support nuts, and drop steering column.
9. Pull panel out far enough to reach behind it, then disconnect printed circuit board, windshield wipers, and cigarette lighter.
10. Remove ground straps and cluster retaining screws.
11. Carefully remove cluster.
12. To install, reverse removal procedure, making sure steering column is properly aligned.

1964-68 Ignition Switch R & R

1. Remove switch from the dash by unscrewing the switch ferrule with a special spanner wrench, tool J-5893-A.
2. Remove switch from back of instrument panel and disconnect wires.
3. Replace switch by reversing above method.

Ignition and Locking Switch R & R—1969-70

The ignition and steering wheel locking switch is located just below the gear selector lever on the steering column.

1. Disconnect battery.
2. Loosen toe pan screws.
3. Remove column to panel nuts, lower steering column, and disconnect switch wire connectors.
4. Remove switch attaching screws and switch.
5. To install, move key lock to OFF-LOCK position.

6. Move actuator rod hole in switch to OFF-LOCK position.
7. Install switch, with rod in hole, then reverse removal procedure.

Lock Cylinder Replacement —1964-68

1. Disconnect battery, then insert key.
2. Remove lock cylinder by placing in off position and inserting wire into small hole in cylinder face. While pushing in on wire, continue to turn cylinder counterclockwise, then pull cylinder from case.

Lighting Switch Replacement —1964-70

1. Disconnect battery.
2. Pull knob to on position.
3. Reach under instrument panel and depress the switch shaft retainer (see illustration), then remove knob and shaft assembly.

NOTE: disconnect vacuum hose on vacuum-operated headlamp models.

4. Remove retaining ferrule nut.
5. Remove switch from instrument panel.
6. Disconnect multi-plug connector from switch.
7. Install in reverse of above. (In checking lights before installation, switch must be grounded to test dome lights on some models).

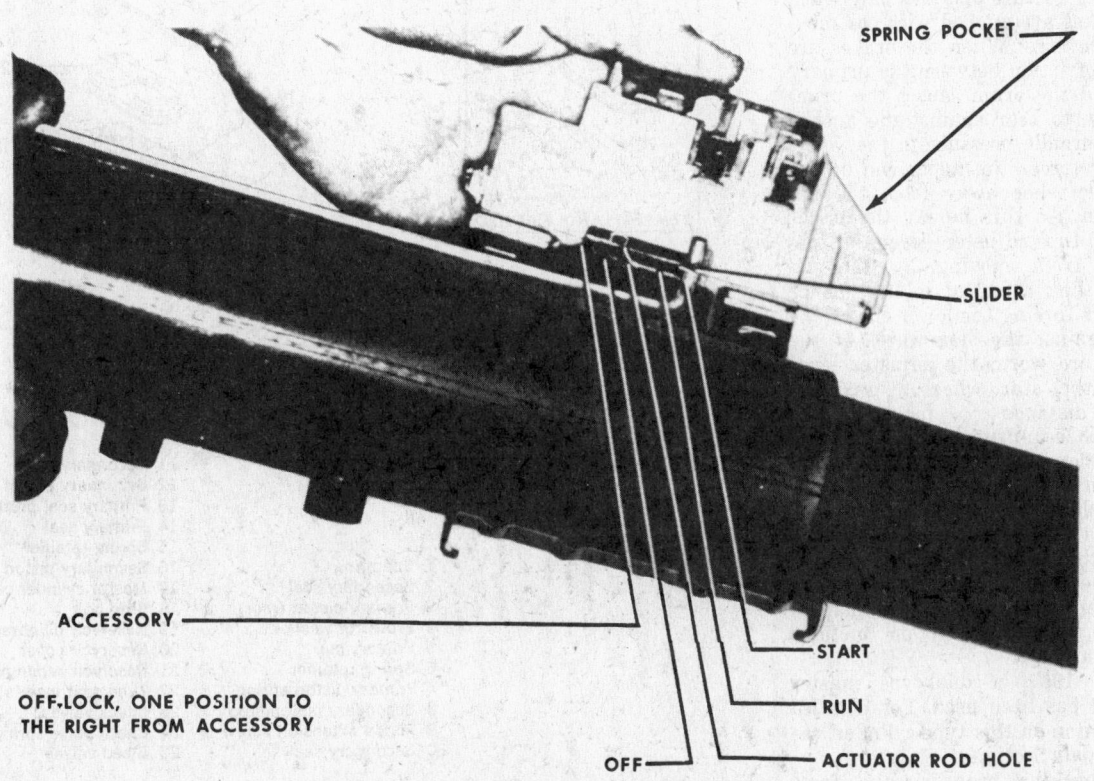

SPRING POCKET

SLIDER

ACCESSORY

OFF-LOCK, ONE POSITION TO THE RIGHT FROM ACCESSORY

START

RUN

OFF

ACTUATOR ROD HOLE

1969-70 ignition lock switch (© Pontiac Div. G.M. Corp.)

Battery, Starter

Battery

A Delco 12 volt battery is used in all models.

Starter

The starter circuit consists of the battery, battery cables, starting motor, starter motor solenoid switch, ignition-starter switch and the neutral safety switch, (used on cars with automatic transmission).

The starting motor and solenoid assembly is mounted on the flywheel housing.

The solenoid switch closes the circuit between the battery and the starting motor. It also operates the shift lever that moves the drive pinion into mesh with the flywheel ring gear.

Starter Removal

Disconnect battery and starter wires. Remove attaching bolts and lift out starter.

Brakes

Standard brakes are of the duo-servo, self-adjusting type. The self-adjusting feature operates only when the brakes are applied with car moving in reverse. When the brakes are applied, friction between the primary shoe and the drum causes the primary shoe to bear against the anchor pin. Hydraulic pressure in the wheel cylinder forces the upper end of the secondary shoe away from the anchor pin. As this moves, the upper end of the adjuster lever is prevented from moving by the actuating link attached to the anchor pin, thus forcing the lower end of the lever against the star wheel. If the linings are worn, the adjuster lever moves the star wheel a predetermined distance to maintain the proper shoe-to-drum clearance.

Metallic brake linings, used on some early high-performance models, never should be installed on cars equipped with standard brake drums unless the drums are radius ground and honed to a special finish. See the Unit Repair Section of this manual for more information on metallic brake linings.

Since 1967, a dual-type master cylinder has been used. For detailed information on this type cylinder, see Unit Repair Section.

The parking brake uses a foot-operated control lever, enclosed ca-

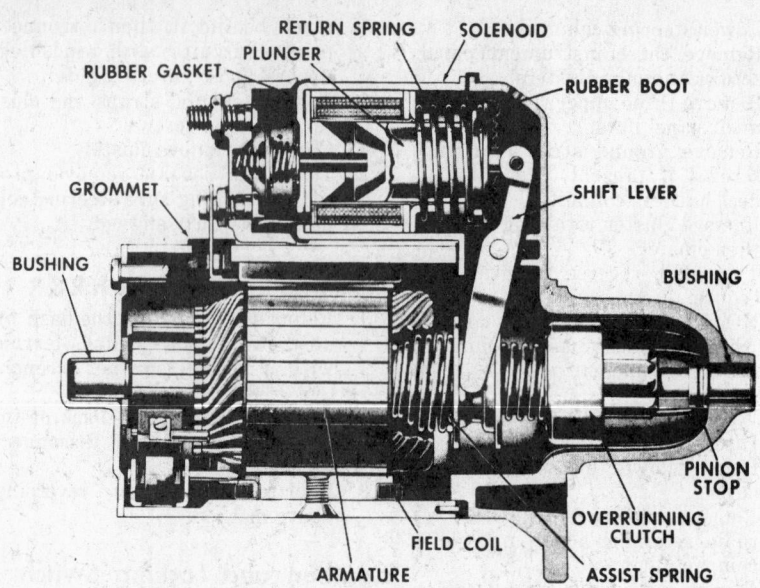

Typical V8 starter motor and solenoid (© Pontiac Div. G.M. Corp.)

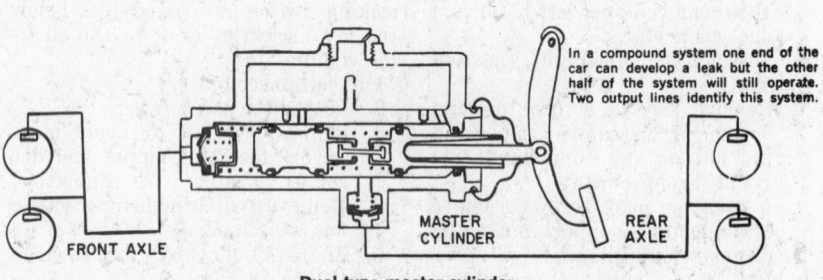

In a compound system one end of the car can develop a leak but the other half of the system will still operate. Two output lines identify this system.

Dual type master cylinder

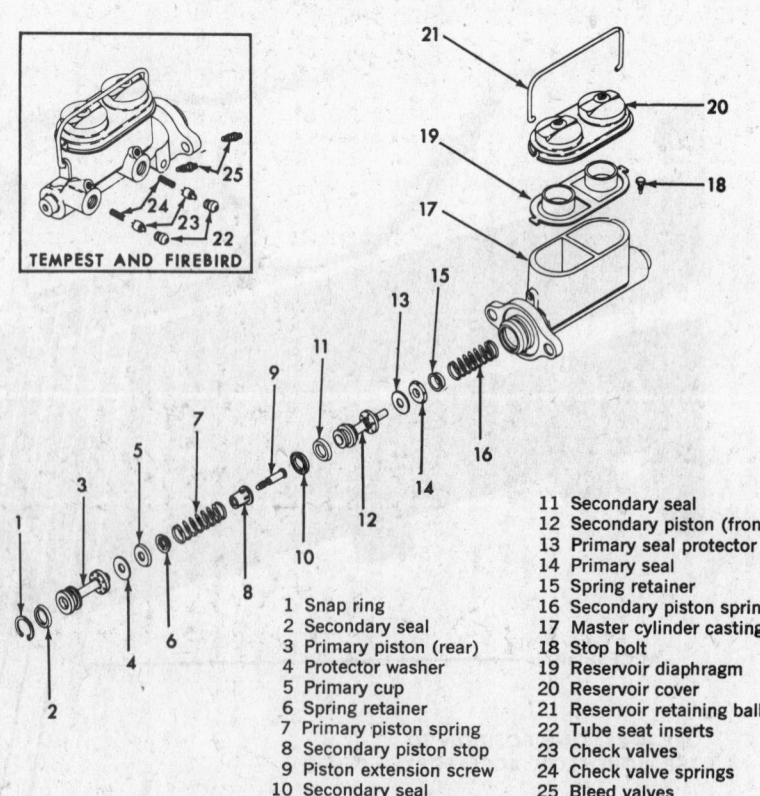

TEMPEST AND FIREBIRD

1 Snap ring
2 Secondary seal
3 Primary piston (rear)
4 Protector washer
5 Primary cup
6 Spring retainer
7 Primary piston spring
8 Secondary piston stop
9 Piston extension screw
10 Secondary seal
11 Secondary seal
12 Secondary piston (front)
13 Primary seal protector
14 Primary seal
15 Spring retainer
16 Secondary piston spring
17 Master cylinder casting
18 Stop bolt
19 Reservoir diaphragm
20 Reservoir cover
21 Reservoir retaining ball
22 Tube seat inserts
23 Check valves
24 Check valve springs
25 Bleed valves

Dual master cylinder (© Pontiac Div. G.M. Corp.)

bles, rear wheel brake shoe levers and struts to the rear wheel shoes. The parking brake is released by pulling the release lever.

Information on brake adjustments, lining replacement, bleeding procedure, master and wheel cylinder overhaul can be found in the Unit Repair Section.

Information on the grease seals which may need replacement can be found in the Unit Repair Section.

Disc Brakes

From 1967, single-piston, sliding-caliper disc brakes have been available as optional equipment on most models (standard with high performance packages). These brakes have a vented, cast-iron rotor with two braking surfaces.

Disc brakes need no adjustment because, during operation, the application and release of hydraulic pressure causes the piston and caliper to move only slightly. In the released position, the pads do not move very far from the rotor thus, as pads wear down, the piston simply moves farther out of the caliper bore and the caliper repositions itself on its mounting bolts to maintain proper pad-to-rotor clearance.

A metering valve in the front brake circuit prevents the discs from operating until about 75 psi exists in the system. This enables the rear drum brakes to operate in synchronization with the front discs and reduces the possibility of unequal brake application and premature lock-up. A proportioning valve in the rear brake circuit of some models accomplishes the same purpose. Disc brake pads should be examined for wear every 12,000 miles. See the Unit Repair Section of this manual for service procedures.

Parking Brake Adjustment —1964-70

The automatic self-adjusting feature incorporated in the rear brake mechanism normally maintains proper parking brake adjustment. For this reason, the rear brake adjustment must be checked before any adjustment of the parking brake cables is done. Check the parking brake mechanism and cables for free movement and lubricate all working surfaces before proceeding.

Caution It is very important that the parking brake cables are not too tight. If the cables are too tight, they create a drag and position the secondary shoes so that the self-adjusters continue to operate in compensation for drag wear. The result is rapidly worn rear brake linings.

1. Jack up both rear wheels.
2. Push parking brake pedal 5-7 notches from full release position (for Tempest and GTO) or 2 notches (Firebird).
3. Loosen rear equalizer locknut and adjust forward nut until light rear brake drag is felt as wheels are rotated by hand.
4. Tighten locknut and release parking brake pedal; no drag should be felt.

Fuel System

Fuel System Information

Data on capacity of the gas tank will be found in the Capacities table. Data on correct engine idle speed and fuel pump pressure will be found in the Tune-Up Specifications table. Both tables can be found in this section.

Information covering operation and troubles of the fuel gauge will be found in the Unit Repair Section.

The carburetor is Rochester or Carter, but varies with the application. See Unit Repair Section.

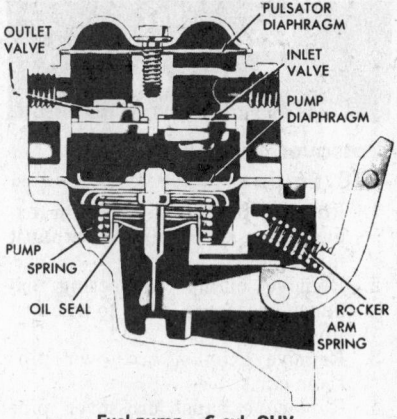

Fuel pump—6 cyl. OHV
(© Pontiac Div. G.M. Corp.)

Idle Stop Adjustment

Adjust plunger to obtain specified idle speed. Disconnect wires and observe operation of solenoid. The plunger should drop back to allow the carburetor idle screw to contact the idle cam; engine speed should drop to lower, "solenoid inactive" idle speed.

NOTE: idle stop unit must be disconnected when setting ignition timing on six-cylinder and Ram Air engines.

Fuel Pump

The fuel pump is of the single action diaphragm-type and is equipped with a pulsation dampening chamber for stabilizing fuel flow.

A vapor diverter is incorporated into the fuel pumps used on air conditioned V8 Models. The fuel pump is not repairable and must be replaced as a unit if defective.

Fuel Filter

The filter is contained in the fuel tank and is integral with the stand pipe and gas gauge tank unit.

The fuel tank must be dropped to gain access to the fuel filter, gas gauge tank unit or the stand pipe.

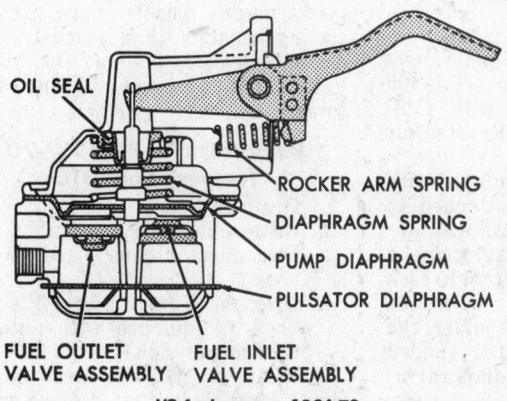

V8 fuel pump—1964-70
(© Pontiac Div. G.M. Corp.)

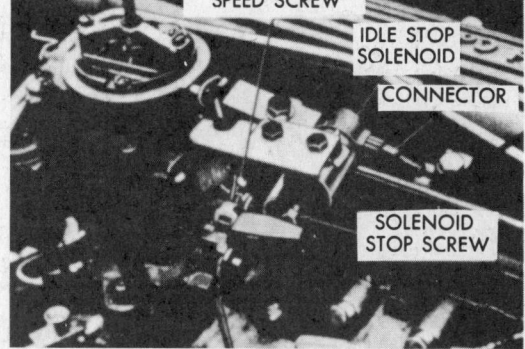

Idle stop solenoid—six-cylinder type shown
(© Pontiac Div. G.M. Corp.)

Exhaust System

Crossover Pipe R & R —V8 Engine

1. Remove four bolts holding exhaust crossover pipe to exhaust manifold.
2. Remove clamp connecting exhaust crossover pipe to exhaust pipe.
3. Remove exhaust crossover pipe from car.
4. Replace exhaust crossover pipe by reversing the above steps. Torque bolts connecting the crossover pipe to the manifold to 15-25 ft. lbs. Torque nuts on clamp to 33 ft. lbs.

Exhaust Pipe R & R

OHV Six—1964-65, 1970

1. Remove two bolts from exhaust manifold flange.
2. Cut pipe in front of muffler.
3. Remove exhaust pipe.
4. Replace pipe by reversing the above steps. Clamp new pipe in front of muffler; tighten flange bolts to 22-30 ft. lbs.

Standard OHC Six—1966-69

1. Remove two bolts from manifold flange.
2. Loosen exhaust pipe to muffler U-clamp on Tempest, intermediate pipe on Firebird.
3. Remove exhaust pipe from car.
4. To install, reverse removal procedure, tightening flange bolts to 35 ft. lbs., exhaust pipe U-clamp nuts to 18 ft. lbs.
NOTE: coat all slip joints with sealer before installation.

High Performance OHC Six —1966-69

1. Loosen front (behind muffler) and rear tailpipe supports.
2. Loosen exhaust pipe to muffler U-clamp.
3. Disconnect muffler from exhaust pipe by pulling muffler and tailpipe to the rear.

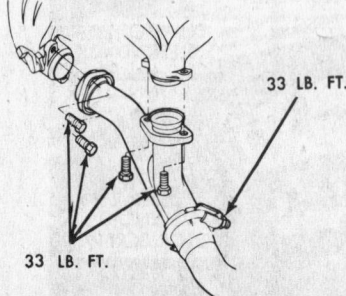

High performance OHC 6 exhaust manifold bolts. Bolts should be tightened to 25 ft. lbs. for 1969 engines.
(© Pontiac Div. G.M. Corp.)

4. Loosen U-clamp that secures exhaust pipe to manifold Y-pipe.
5. Remove exhaust pipe.
6. Remove four bolts that hold Y-pipe to exhaust manifold; remove Y-pipe.
7. To install, reverse removal procedure, tightening Y-pipe to manifold bolts to 33 ft. lbs. (25 ft. lbs.—1969), Y-pipe to exhaust pipe U-clamp bolts to 33 ft. lbs., front muffler U-clamp and tailpipe U-clamp to 18 ft. lbs., and resonator U-clamp to 33 ft. lbs.
NOTE: coat all slip joints with sealer before installation.

All V8 Single Exhaust—1964-70

1. Loosen crossover to exhaust pipe U-clamp.
2. Disconnect crossover pipe from exhaust pipe by pulling exhaust pipe to the rear.
3. Loosen front U-clamp on muffler and remove exhaust pipe.
4. To install, reverse removal procedure, tightening muffler U-clamp to 18 ft. lbs., crossover to exhaust pipe U-clamp to 33 ft. lbs., manifold flange bolts to 25 ft. lbs.
NOTE: coat all slip joints with sealer before installation. Tighten one manifold flange bolt finger-tight before tightening the other bolt with a wrench (each side).

All V8 Dual Exhaust—1964-70

1. Remove two or three nuts or bolts (depending on engine) from manifold flange.
2. Remove U-clamp at front of muffler.
3. Remove exhaust pipe/s.
4. To install, reverse removal procedure, tightening manifold flange bolts to 25-35 ft. lbs. and muffler front U-clamp nuts to 15-20 ft. lbs.
NOTE: it may be necessary to loosen resonator (if so equipped), remove tailpipe, and remove front muffler on some models.

Vacuum-Operated Exhaust —1970 GTO 400

A vacuum-operated exhaust system is available on 1970 GTO 400 models, unless equipped with California evaporative control system or Ram Air.

The dual mufflers are equipped with vacuum-operated servos attached to the front of each muffler. An actuator rod is connected to the diaphragm of each servo and passes through each muffler to operate spring-loaded valves. When the switch under the steering column is actuated (pulled out), vacuum on the servo diaphragms pulls the spring-loaded valves away from the muffler end baffle pipes, thus allowing exhaust to pass directly into the rear muffler chambers.

A vacuum reservoir tank, located on the left front inner fender panel, supplies vacuum to the system when engine vacuum is not available and during acceleration.

GTO Exhaust Extensions —1970

The dual-outlet exhaust extensions used on this model must be properly aligned to prevent rattling. (See illustrations.)

Cooling System

Radiator—1964-66 Tempest and GTO

A top-tank, down-flow unit is used. A drain cock is located at the inside lower left-hand corner of the radiator. The core is of the down-flow-tube and center-type, and is constructed of copper.

Radiator R & R—1964-66 Tempest and GTO

1. Drain cooling system.
2. Disconnect overflow, upper and lower radiator hoses.
3. Remove radiator fan shield.
4. Remove radiator.
5. To replace, reverse removal steps.

Radiator—1967-70 Tempest and GTO

A cross-flow radiator is used instead of a conventional down-flow and center type. With the cross-flow design, coolant flows horizontally through the core and the tanks are located on each side.

Advantages of the cross-flow radiator are improved cooling capability, more effective cooling surface area, and a low silhouette.

Automatic transmission radiators have oil coolers built into the right-hand tank, air-conditioned and high-performance models have greater cooling capacity than standard. The drain cock is located at the inside, lower left-hand corner of the radiator.

Radiator R & R—1967-70 Tempest and GTO

1. Drain coolant.
2. Remove fan shield assembly.
3. Disconnect upper and lower hoses.
4. Disconnect and plug oil cooler lines, if equipped with automatic transmission.
NOTE: on 1969 Tempest models remove fan blade, then remove entire radiator and shroud as an assembly.

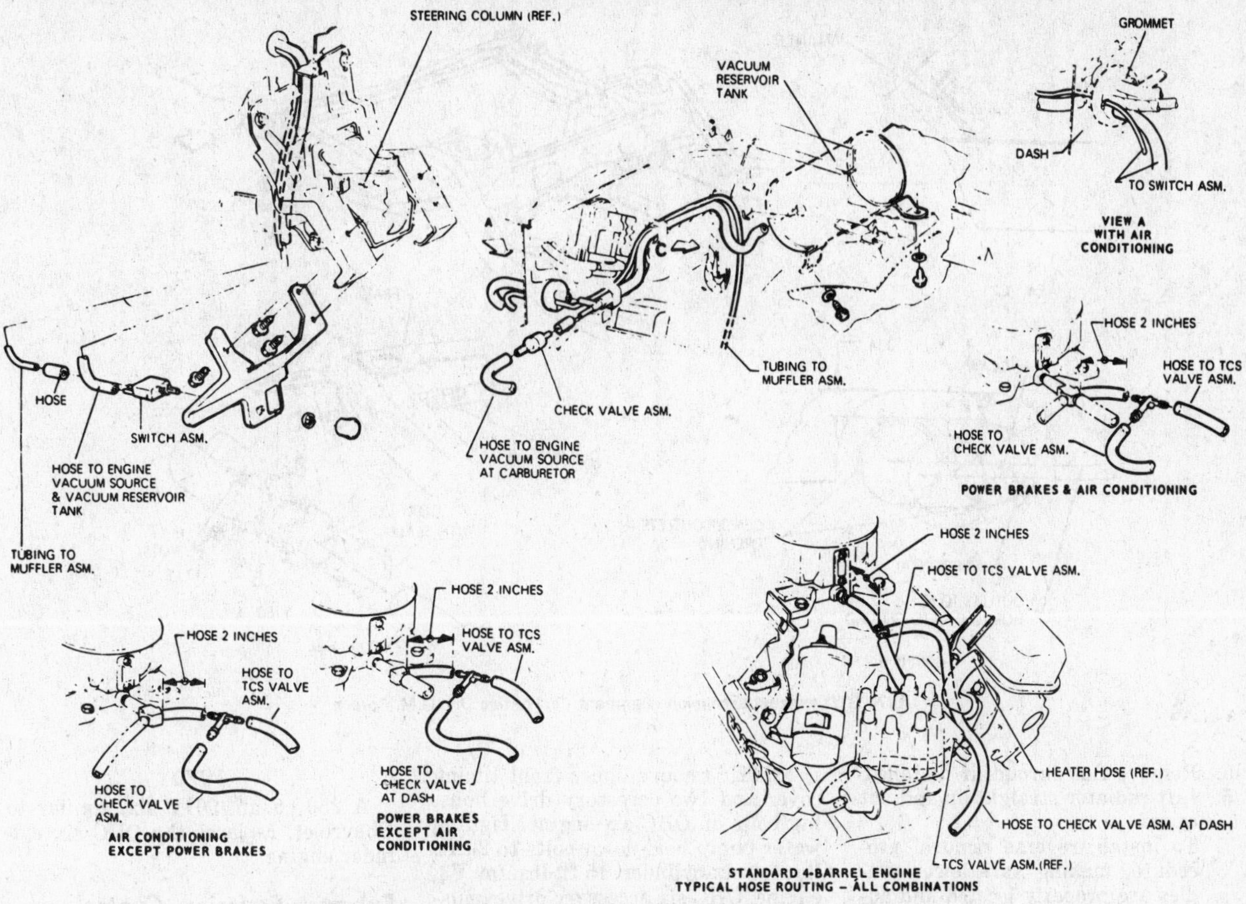

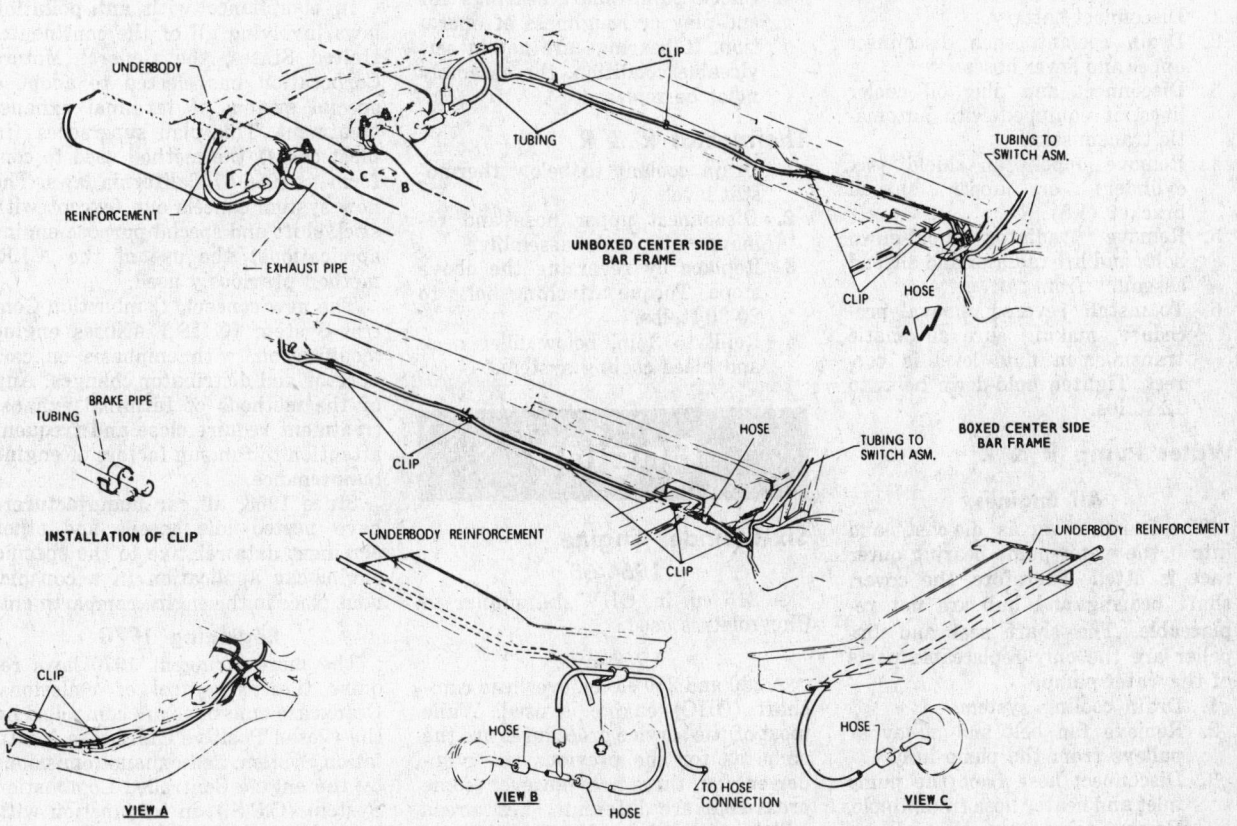

STEERING COLUMN (REF.)

VACUUM RESERVOIR TANK

GROMMET

DASH

TO SWITCH ASM.

VIEW A
WITH AIR
CONDITIONING

HOSE

SWITCH ASM.

HOSE TO ENGINE
VACUUM SOURCE
& VACUUM RESERVOIR
TANK

TUBING TO
MUFFLER ASM.

CHECK VALVE ASM.

HOSE TO ENGINE
VACUUM SOURCE
AT CARBURETOR

TUBING TO
MUFFLER ASM.

HOSE 2 INCHES

HOSE TO TCS
VALVE ASM.

HOSE TO
CHECK VALVE ASM.

POWER BRAKES & AIR CONDITIONING

HOSE 2 INCHES

HOSE TO TCS
VALVE ASM.

HOSE TO
CHECK VALVE
ASM.

AIR CONDITIONING
EXCEPT POWER BRAKES

HOSE 2 INCHES

HOSE TO TCS
VALVE ASM.

HOSE TO
CHECK VALVE
AT DASH

POWER BRAKES
EXCEPT AIR
CONDITIONING

HOSE 2 INCHES

HOSE TO TCS VALVE ASM.

HEATER HOSE (REF.)

HOSE TO CHECK VALVE ASM. AT DASH

TCS VALVE ASM. (REF.)

STANDARD 4-BARREL ENGINE
TYPICAL HOSE ROUTING – ALL COMBINATIONS

1970 GTO 400 vacuum-operated exhaust—engine line routing (© Pontiac Div. G.M. Corp.)

UNDERBODY

CLIP

TUBING

TUBING TO
SWITCH ASM.

REINFORCEMENT

EXHAUST PIPE

UNBOXED CENTER SIDE
BAR FRAME

CLIP

HOSE

A

TUBING

BRAKE PIPE

HOSE

CLIP

HOSE

TUBING TO
SWITCH ASM.

BOXED CENTER SIDE
BAR FRAME

INSTALLATION OF CLIP

UNDERBODY REINFORCEMENT

CLIP

UNDERBODY REINFORCEMENT

CLIP

HOSE

HOSE

TO HOSE
CONNECTION

HOSE

VIEW A

VIEW B

VIEW C

1970 GTO 400 vacuum-operated exhaust—chassis line routing (© Pontiac Div. G.M. Corp.)

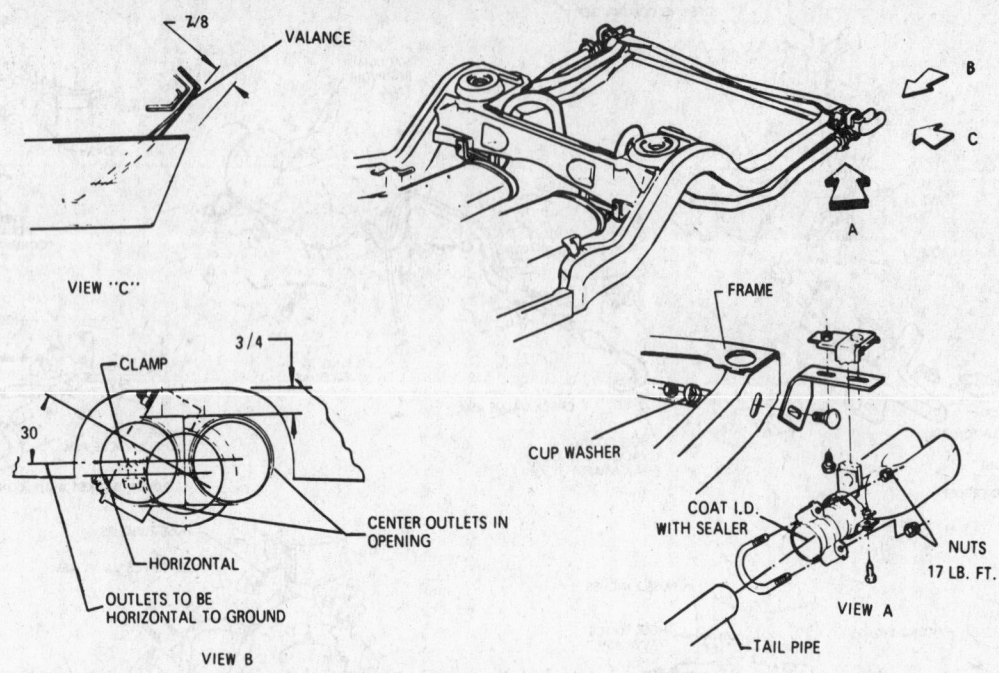

1970 GTO exhaust extension alignment (© Pontiac Div. G.M. Corp.)

5. Remove fan shroud, if installed.
6. Lift radiator straight up and out of car.
7. To install, reverse removal procedure, making sure lower cradles are properly located and automatic transmission is full.

Radiator R & R—1967-69 Firebird

1. Disconnect battery.
2. Drain coolant, then disconnect upper and lower hoses.
3. Disconnect and plug oil cooler lines, if equipped with automatic transmission.
4. Remove upper fan shield (six cylinder) or upper shroud bracket (V8).
5. Remove radiator hold-down bolts and lift radiator and shroud assembly from car.
6. To install, reverse removal procedure, making sure automatic transmission fluid level is correct. Tighten hold-down bolts to 12 ft. lbs.

Water Pump R & R

All Engines

The pump cover is die-cast, and into it the water pump bearing outer race is fitted. Therefore, the cover, shaft bearing and hub are not replaceable. The shaft seal and impeller are the only replaceable parts of the water pump.

1. Drain cooling system.
2. Remove fan belt and pulley or pulleys from the pump hub.
3. Disconnect hose from the pump inlet and heater hose from nipple. Remove pump assembly and gasket from timing chain cover.

NOTE: remove upper front timing cover and two accessory drive housing bolts on OHC six engine. Tighten water pump hold-down bolts to 20 ft. lbs. for six-cylinder, 15 ft. lbs. for V8, engine. OHC six accessory drive housing and upper timing cover bolts must be tightened to 15 ft. lbs. only.

4. Check pump shaft bearings for end-play or roughness of operation. If bearings are not in serviceable condition, the assembly must be replaced.

Thermostat R & R

1. Drain coolant to below thermostat level.
2. Disconnect upper hose and remove water outlet assembly.
3. Replace by reversing the above steps. Torque attaching bolts to 20-30 ft. lbs.
4. Refill to 3 in. below filler neck and bleed cooling system.

Engine

Six-Cylinder Engine

1964-65

A 215 cu. in. OHV six, similar to Chevrolet, is used.

1966-69

A 230 and 250 cu. in. overhead camshaft (OHC) engine is used. While most of the service procedures are the same as for the previous six cylinder engine, there are, however, some areas that are different. These areas will be covered in their regular departments.

1970

A 250 cu. in. OHV six, similar to Chevrolet, replaces the OHC six cylinder engine.

Exhaust Emission Control

Beginning 1968

In compliance with anti-pollution laws involving all of the continental United States, the General Motors Corporation has elected to adopt a special system of terminal exhaust treatment. This plan supersedes (in most cases) the method used to conform to 1966-67 California laws. The new system cancels out (except with stick shift and special purpose engine applications) the use of the A.I.R. method previously used.

The new concept, Combustion Control System (C.C.S.) utilizes engine modification, with emphasis on carburetor and distributor changes. Any of the methods of terminal exhaust treatment require close and frequent attention to tune-up factors of engine maintenance.

Since 1968, all car manufacturers have posted idle speeds and other pertinent data relative to the specific engine-car application in a conspicuous place in the engine compartment.

Beginning 1970

The more stringent 1970 laws require tighter control of emissions. Crankcase emissions are controlled by the Closed Positive Crankcase Ventilation System, and exhaust emissions by the engine Controlled Combustion System (C.C.S.), in conjunction with the new Transmission Controlled Spark System (T.C.S.).

In addition, cars sold in California are equipped with an Evaporation Control System that limits the amount of gasoline vapor discharged into the atmosphere (usually from the carburetor and fuel tank).

The T.C.S. system consists of a transmission switch, a solenoid valve, and a temperature switch. Under normal conditions, the system permits the vacuum distributor (spark) advance to operate only in high gear (both manual and automatic transmissions). When the engine temperature is below 85°F., or above 220°F., however, the system allows the vacuum advance to operate normally.

For details, consult the Unit Repair Section.

Engine R & R

1964-65 6 Cylinder and 1964-66 V8

1. Remove hood.
2. Drain cooling system and remove radiator.
3. Disconnect heater hoses at the engine.
4. Disconnect wiring harness at generator, ignition coil, starter solenoid, heater blower, thermogauge and oil pressure switch.
5. Disconnect ground strap at both sides of the engine.
6. Disconnect fuel line at fuel pump.
7. Disconnect vacuum modulator line at automatic transmission and power brake vacuum line.
8. Remove front fender cross brace.
9. Remove fan and fan pulley.
10. Disconnect accelerator rod at firewall.
11. Raise front of car.
12. Disconnect exhaust pipe at manifold.
13. Disconnect clutch and shift linkage on synchromesh, oil filler tube on automatic.
14. Disconnect rear U-joint bolts and remove driveshaft. Plug the end of the extension housing with rags to prevent oil loss. Disconnect speedometer cable at transmission then remove starter on 1965 models.
15. Disconnect engine support at crossmember, then lower car.
16. Raise engine with chain hoist, then remove transmission rear mount from crossmember. Move forward to clear the firewall and heater.
17. Lift and remove engine and transmission.
18. Replace by reversing the removal procedure.

1966-70 6 Cylinder and 1967-70 V8

1. Disconnect battery.
2. Drain cooling system.
3. Scribe alignment marks on hood and remove hood from hinges.
4. Disconnect engine wiring harness and ground straps.
5. Remove air cleaner and fan shield or shroud.
6. Disconnect radiator and heater hoses.
7. If equipped with manual transmission, remove radiator.
8. Remove fan and fan pulley.
NOTE: if equipped with power steering and/or air conditioning, disconnect and swing aside pump/compressor **without** disconnecting hoses.
9. Disconnect accelerator linkage and support bracket.
10. Disconnect automatic transmission vacuum modulator line and power brake vacuum line at carburetor.
NOTE: on Firebird models with air conditioning, remove wiper motor.
11. Jack up front of car and drain engine oil.
12. Disconnect fuel lines at pump.
13. Disconnect exhaust pipes.
14. Disconnect starter wires.
15. If equipped with automatic transmission, remove converter cover and three converter retaining bolts, then slide converter to the rear.
16. If equipped with manual transmission, disconnect clutch linkage and remove clutch cross-shaft.
NOTE: remove starter and lower flywheel cover on 1970 V8s.
17. Remove four lower bellhousing bolts (two per side).
18. Disconnect transmission filler tube support (automatic) and starter wire shield from cylinder heads.
19. Remove two front motor mount-to-frame bolts.
20. Lower car to floor then, using a jack and a wood block, support the transmission.
21. Remove two remaining bellhousing bolts.
22. Raise transmission slightly, using the jack and wood block, then, using a chain hoist, remove the engine.
23. To install, reverse removal procedure. Install the two upper bellhousing bolts first (with jack still under transmission).
NOTE: do not lower engine completely until jack and wood block are removed.

Cylinder Head

1964-65 and 1970 6 Cylinder

Removal

1. Drain cooling system, remove air cleaner. Disconnect radiator hoses.
2. Disconnect accelerator pedal rod at bellcrank, fuel and vacuum lines at carburetor. Disconnect exhaust pipe at manifold flange.
3. Remove manifold-to-cylinder

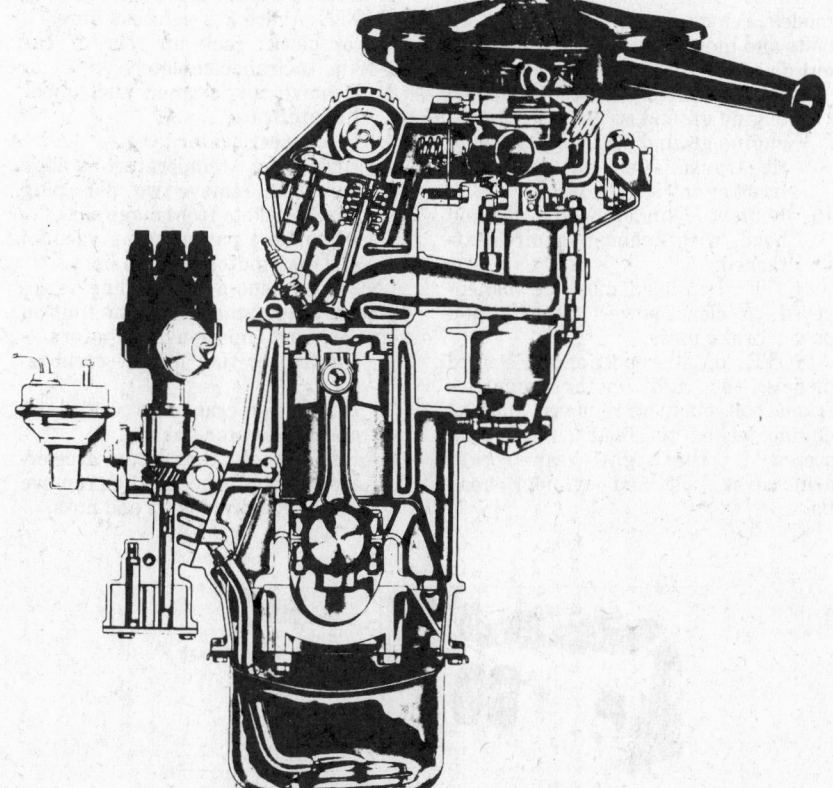

1966-69 6 cyl. OHC engine (© Pontiac Div. G.M. Corp.)

head attaching bolts and manifolds.

4. Remove rocker arm cover assembly, temperature sender and coil
wires.
5. Loosen rocker arm nuts and rotate rocker arms so the pushrods can be removed.
6. Remove pushrods and store them
so they can be installed in their
original locations.
7. Disconnect spark plug wires.
8. Remove cylinder head bolts.
9. Lift off the head.
10. Remove cylinder head gasket.

Installation

1. Position new cylinder head gasket on block, on locating dowels.
2. Place cylinder head in position.
3. Install cylinder head attaching
bolts. Torque to 95 ft. lbs.
4. Install pushrods in original location and position.
5. Position rocker arms and torque
rocker arm nuts to 15-25 ft. lbs.,
further tighten until valve train
play is removed, with lifter on
base circle of cam, then tighten
one more turn.
6. Install rocker arm cover.
7. Install manifold-to-cylinder head
bolts and torque to 30 ft. lbs.
(center) and 15-20 ft. lbs. (end).
8. Install pushrod cover and crankcase breather outlet pipe.
9. Connect all wires, hoses and linkage; fill cooling system and check
for leaks.
10. Connect spark plug wires.

1966-69 6 Cylinder OHC

Removal

1. Drain cooling system and remove
air cleaner.
2. Disconnect accelerator pedal
cable at bellcrank on manifold,
and fuel and vacuum lines at
carburetor.
3. Disconnect exhaust pipe at manifold flange, then remove manifold
bolts and clamps and remove
manifolds and carburetor as an
assembly.
4. Remove timing belt top front
cover.
5. Align timing marks, remove belt
tension then remove belt from
camshaft sprocket.
6. Remove rocker arm cover assembly.
7. Remove timing belt upper front
cover mounting support bracket
and rear lower cover.
8. Disconnect spark plug wires.
9. Remove rocker arms and hydraulic valve lash adjusters. Keep
rocker arms and hydraulic lash
adjusters in proper sequence for
exact location for installation.
10. Remove cylinder head bolts and
gasket.

11. Clean gasket surfaces and carbon
from cylinder head and block.

Installation

When installing new head, transfer
all serviceable parts to new head
using new seals on intake and exhaust
manifold gaskets.

1. Place new cylinder head gasket
in position over dowels in cylinder block.
2. Guide cylinder head into place
over dowels and gasket.
3. Start all cylinder head bolts in
threads.

NOTE: bolts are of two different
lengths. When inserted into proper
holes, all bolts will project an equal
distance from the head. Do not use
sealer of any kind on threads.

4. Tighten cylinder head a little at
a time with a torque wrench.
Tighten center bolts and then the
end bolts. Final torque should be
90-100 ft. lbs.
5. Reverse Steps 1-9 of removal to
complete installation procedure.

1964-70 V8

Removal

1. Remove intake manifold, valley
cover, and rocker arm cover.
2. Loosen all rocker arm retaining
nuts and pivot rockers off pushrods.
3. Remove pushrods and place in
order.
4. Remove exhaust pipe flange bolts.

NOTE: on air-conditioned Firebird
models, remove compressor hold-down
bolts and move compressor aside without disconnecting hoses.

5. Remove battery ground strap and
engine ground strap on left head;
engine ground strap and automatic transmission oil filler tube
bracket on right head.
6. Remove cylinder head bolts and
head, with exhaust manifold attached.

NOTE: left head must be maneuvered to clear power steering and
power brake units.

NOTE: on air-conditioned Firebird
models, the right motor mount-to-
frame bolt must be removed and the
engine jacked up about 2 in. to gain
access to the right rear rocker
arm cover bolt and cylinder head
bolt.

Installation

1. Check head surface for straightness, then place a new head gasket on block.

CAUTION: on air-conditioned Firebird models, install right rear head
bolt into head **before** placing head on
block.

NOTE: bolts are of three different
lengths on all V8s. When bolts are
properly installed, they will project an
equal distance from head.

2. Install all bolts and tighten
evenly to specified torque.
3. Install pushrods in original positions.
4. Position rocker arms over pushrods and tighten ball retaining
nuts to 20 ft. lbs. (except Ram
Air IV engines; see special procedure).
5. Replace rocker arm cover.
6. Replace valley cover.
7. Replace ground straps, oil filler
tube bracket, intake manifold,
and right motor mount bolt (on
A/C Firebird models).
8. Install exhaust pipe flange nuts.

NOTE: most left and right cylinder
heads are interchangeable within a
single year. Large- and small-valve
heads should not be used on the same
engine, nor should 1964 389 cu. in.
heads (standard) be interchanged
with 1964 GTO.

Intake Manifold R & R
—1964-70 V8

1. Drain radiator and block.

NOTE: there are petcocks on each
side of block; jack up rear of car
15-18 in. to drain completely.

2. Remove air cleaner and upper
radiator hose.
3. Disconnect heater hose.
4. Disconnect temperature gauge
wire, then remove two spark plug
wire brackets from manifold.
5. Disconnect power brake vacuum
and distributor vacuum lines.

NOTE: vacuum retard line is located at lower rear of vacuum unit on
some exhaust emission distributors.

6. Disconnect fuel line at carburetor.
7. Disconnect crankcase vent hose
and accelerator linkage.
8. Remove bolts that secure accelerator linkage bracket, then remove
intake manifold bolts and nuts.

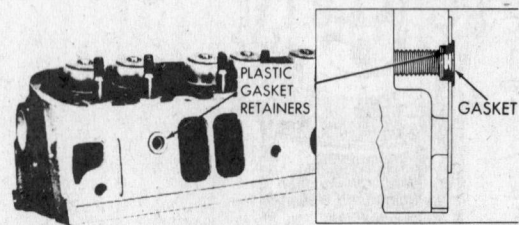

Intake manifold gaskets can be held in place
using plastic retainers, available at Pontiac dealers
(© Pontiac Div. G.M. Corp.)

9. Remove manifold and gasket.
CAUTION: make sure O-ring between intake manifold and timing chain cover is in place.
10. To install, reverse removal procedure, tightening timing chain cover to manifold bolts to 10-20 ft. lbs., manifold hold-down bolts and nuts evenly to 40-45 ft. lbs.

Intake and Exhaust Manifold R & R—1964-65 and 1970 OHV 6 Cylinder, 1966-69 OHC 6 Cylinder

1. Remove air cleaner.
2. Disconnect accelerator linkage and return spring.
3. Disconnect fuel and vacuum lines at carburetor; disconnect choke rod.
4. Disconnect exhaust pipe at manifold flange.
5. Remove manifold bolts and clamps, then remove manifolds.
NOTE: intake manifold can be separated from exhaust manifold by removing one bolt and two nuts. These fasteners should be tightened to 15-30 ft. lbs. after the manifolds are bolted to the engine.
6. To install, reverse removal procedure, tightening center clamp bolts to 25-30 ft. lbs., end bolts to 15-20 ft. lbs. (for OHV engines), or all bolts to 30 ft. lbs. (for OHC engines).

Rocker Arm Stud R & R

1964-65 and 1970 OHV 6 Cylinder

1. Remove rocker cover and rocker arm.

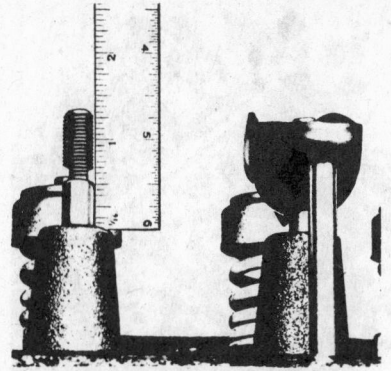

Rocker stud height—OHV 6 cyl.
(© Pontiac Div. G.M. Corp.)

2. File two slots 3/32-1/8 in. deep on opposite sides of stud. Bottom of slots should be ½ in. from top of stud hole.
3. Place spacer washer (or tool J-6392-3) over stud, then position a stud remover (or tool J-6392-1) on stud and tighten securely.
4. Place a spacer (socket or J-6392-2) over the stud remover, then thread a 7/8 in. nut on stud remover and turn in until stud pulls from head.
5. If an oversize stud is to be used (0.003 and 0.013 in. oversize studs are available), ream stud hole to proper size.
6. To install, coat press-fit area of stud with axle lube, then press or hammer into place.
NOTE: the factory recommends that tool J-6880 be used for this job. This tool is simply a sleeve that is held in place with an Allen screw—it protects the threads from damage. Any homemade tool similar to the one

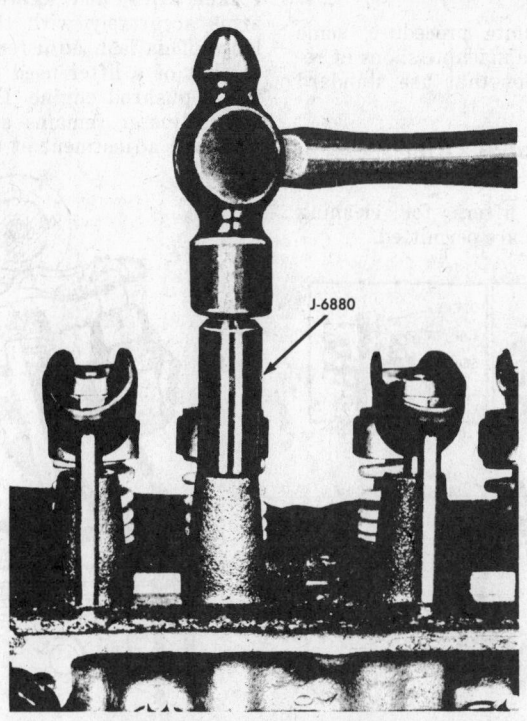

Installing rocker arm stud—OHV 6 cyl.
(© Chevrolet Div. G.M. Corp.)

illustrated will work if care is exercised. Do not hammer directly on the stud, because it is hardened to the point where it will fracture if subjected to shock.

1964-70 V8 Except GTO

Caution This procedure can be used **only** on engines with pressed-in rocker studs. GTO and some special high performance engines have screwed-in rocker studs which are easily identified by their hex head lower portion. Another common stud-securing procedure on standard engines is "pinning" pressed-in studs by drilling through the stud boss and stud and inserting an interference-fit roll pin. Make sure any such pins are removed before attempting the following procedure.

1. Disconnect battery and drain cooling system.
2. Remove rocker cover.
3. Pack oily rags around stud holes and engine openings.
4. Remove rocker arm and pushrod,

Removing rocker arm stud—V8
(© Pontiac Div. G.M. Corp.)

then file two slots 3/32-1/8 in. deep on opposite sides of the stud. The top of the slots should be ¼-3/8 in. below thread travel.
5. Place a spacer washer (or tool J-8934-3) over the stud, then position stud remover (or tool J-8934-1) on stud and tighten Allen screws.
6. Place a spacer (socket or J-8934-2) over the remover, then thread a 7/8 in. nut on stud remover and turn in until stud pulls from head.
7. If an oversize stud is to be used (0.003 in. oversize studs are available), ream stud hole to the proper size, then clean chips from area.
8. To install, refer to Step 6 of OHV 6 cylinder stud replacement procedure, substituting factory tool number J-23342 for J-6880.

NOTE: valve adjustment for Ram Air IV engines is covered later in this section.

GTO Screwed-In Rocker Studs

1. Remove rocker cover.
2. Remove rocker arm and nut.
3. Remove stud, using a deep socket.
4. Install new stud, tightening to 50 ft. lbs.

Cylinder Head Disassembly

1. Remove cylinder heads, as previously described.
2. Compress valve springs, using valve spring compressor.
3. Remove valve locks or keys.
4. Release valve springs.
5. Remove valve springs, retainers, oil seals, and valves.

NOTE: if a valve does not slide out of the guide easily, check end of stem for mushrooming or heading over. If head is mushroomed, file off excess material, remove and discard valve. If

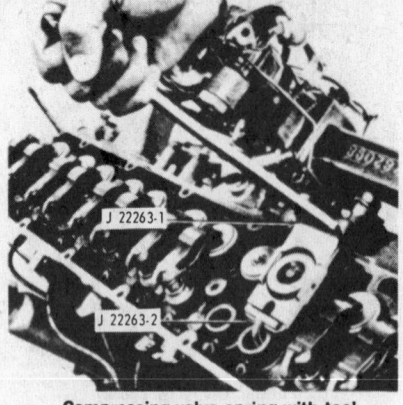

Compressing valve spring with tool
J22263.1—OHC 6 cyl.
(© Pontiac Div. G.M. Corp.)

valve is not mushroomed, lubricate stem, remove valve and check guide for galling.

Valves

Valve Guides

Pontiac engines have integral valve guides. Pontiac offers valves with oversize stems for worn guides (0.001, 0.003 and 0.005 in. being available for most engines). To fit these, enlarge valve guide bores with valve guide reamers to an oversize that cleans up wear. If a large oversize is required, it is best to approach that size in stages by using a series of reamers of increasing diameter. This helps to maintain the concentricity of the guide bores with the valve seats. The correct valve stem to guide clearance is given in the Valve Specifications table at the beginning of this section.

As an alternate procedure, some local automotive machine shops fit replacement guides that use standard stem valves.

Hydraulic Valve Lifter Disassembly

Disassemble lifters for cleaning only; no repairs are permitted.

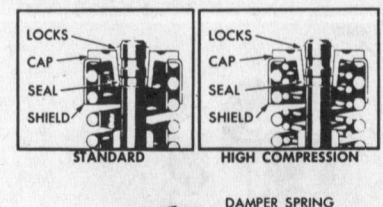

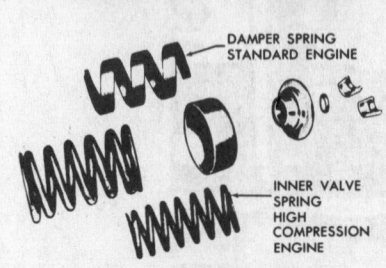

Typical valve spring assemblies
(© Pontiac Div. G.M. Corp.)

Checking valve spring installed height—6 cyl.
(© Pontiac Div. G.M. Corp.)

1. Grasp lock ring with needle nose pliers and remove. (Depress plunger to gain clearance.)
2. Remove pushrod cup, metering valve disc, and upper metering disc (if any). Do not bend metering disc.
3. Remove plunger assembly and plunger spring.
4. Remove spring, check valve retainer and check valve from plunger.
5. Clean all parts in solvent (lacquer thinner) and reassemble.

NOTE: internal parts are **not** interchangeable between lifters.

1966-69 6 Cylinder OHC Lash Adjuster (Valve Tappet)

This engine is equipped with hydraulic valve lash adjusters. These adjusters are located in the cylinder head and serve as a fulcrum of the rocker arms, and locate the rocker arms accurately with the camshaft lobes. This lash adjuster is identical to that of a lifter used in a conventional pushrod engine. However, the lash adjuster remains stationary to maintain adjustment at all times.

6 cyl. OHC valve train
(© Pontiac Div. G.M. Corp.)

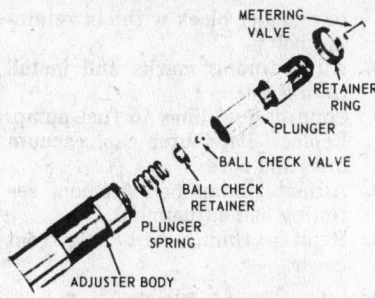

6 cyl. OHC valve lash adjuster
(© Pontiac Div. G.M. Corp.)

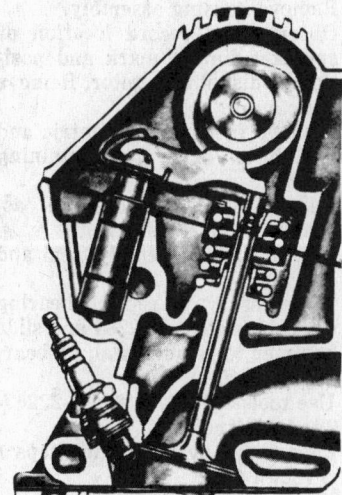

Details of automatic zero valve lash adjustment

These adjusters are to be serviced in the same manner as conventional hydraulic tappets.

R & R

1. Remove rocker cover assembly.
2. Remove rocker arm and hydraulic lash adjuster assemblies, keeping them in proper order for correct installation in original positions.
3. If lash adjuster sticks in its bore, proceed as follows:
   a. Remove rocker arm.
   b. Fill vent hole adjacent to lifter with SAE 30 oil.
   c. Insert a 4 in. length of 3/16 in. diameter rod into the vent hole and strike the end of the rod sharply with a hammer.
   NOTE: the hydraulic pressure generated in this operation should be sufficient to dislodge even the most stubborn adjuster.
4. To install, reverse removal procedure, with the exception of Step 3.

Valve Adjustment—Ram Air IV Engines

With this engine, it is not possible to adjust valves by tightening the rocker arm adjusting nut until it seats on the shoulder of the rocker arm. To adjust these limited travel lifters with the engine installed in the car, proceed as follows: (If engine has been removed, use other procedure.)

Engine In Car

1. Tighten rocker arm adjusting nuts so that pushrods will not jump out of place when engine is started.
2. Start engine and retighten rocker arm on any valve that is clattering. Tighten just until noise disappears.
   NOTE: oil deflector clips are a help in this operation; they are available through automotive parts jobbers.
3. Allow engine to run until normal operating temperature is achieved, then loosen each rocker arm adjusting nut until clattering begins. Retighten nut until noise disappears (this brings pushrod slightly into top of lifter travel) and, with adjusting nut in this position, tighten locknut to 30-40 ft. lbs.

Engine Out of Car

1. Rotate crankshaft until No. 1 piston is at TDC on compression stroke and distributor rotor points to No. 1 spark plug wire cap tower. Timing mark should be aligned with "O" on timing cover.
2. Tighten rocker arm adjusting nuts on No. 1 cylinder rockers to obtain 0.008 in. clearance between rocker arms and valve stems.
3. Tighten adjusting nuts an additional 1/8 turn ±5°, then tighten locknuts to 30-40 ft. lbs.
4. Rotate crankshaft 90°, in normal direction of rotation, to bring next piston in firing order (No. 8) to TDC on compression stroke, then complete Steps 2 and 3.
5. Continue as in Step 4 for the rest of the cylinders; firing order is 1-8-4-3-6-5-7-2.

Timing Case

Timing Gear Cover and Oil Seal R & R

1964-65 and 1970 OHV 6 Cylinder

1. Drain cooling system and disconnect radiator hoses at radiator.
2. Remove fan and water pump pulley.
3. Remove radiator and fan belt.
4. Remove harmonic balancer, using a puller.
5. Loosen oil pan bolts and allow pan to rest against front crossmember.
6. Remove timing gear cover bolts, then remove cover and gasket.
7. Pry out oil seal using a screwdriver.

NOTE: seal can be replaced with cover installed.
8. Install new seal, with lip toward inside of cover. Drive it into place, using proper seal installer

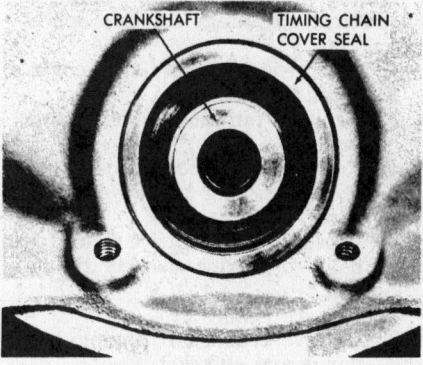

Timing chain cover oil seal
(© Pontiac Div. G.M. Corp.)

or an old wheel bearing outer race.
9. Inspect oil nozzle for damage and replace if necessary, then clean all gasket surfaces.
10. Install cover and gasket (stick gasket to block with Vaseline or wheel bearing grease), making sure cover is centered properly on crankshaft end.
   NOTE: the factory uses a centering tool (J-21742) for this job.
11. Tighten cover bolts to 7 ft. lbs., then install oil pan and harmonic balancer.

1964-70 V8

1. Drain radiator and cylinder block.
2. Loosen alternator adjusting bolts.
3. Remove fan, fan pulley, and accessory drive belts.
4. Disconnect radiator hoses.
5. Remove fuel pump.
6. Remove harmonic balancer bolt and washer.
7. Remove harmonic balancer.
   NOTE: do not pry on rubber-mounted balancers.
8. Remove front four oil pan to timing cover bolts.
9. Remove timing cover bolts and nuts and cover to intake manifold bolt.
10. Pull cover forward and remove.
11. Remove O-ring from recess in intake manifold, then clean all gasket surfaces.
12. To replace seal, pry it out of the cover using a screwdriver. Install the new seal with lip inwards.
   NOTE: seal can be replaced with cover installed.
13. To install, reverse removal procedure, making sure all gaskets are replaced. Tighten four oil pan bolts to 12 ft. lbs., harmonic balancer bolt to 160 ft. lbs., and fan pulley bolts to 20 ft. lbs.

Timing Belt, Crankshaft Sprocket, or Lower Crankcase Cover Seal R & R—1966-69 OHC 6 Cylinder

Radiator removal, at this point, is a distinct advantage for this operation.

1. Remove upper front timing cover.
2. Align timing marks.
3. Remove fan and water pump pulley.
4. Remove harmonic balancer.
5. Remove timing belt lower front cover.
6. Loosen accessory drive mounting bolts to provide slack in timing belt.
7. Remove timing belt.
8. Remove crankshaft timing belt flange and sprocket.
9. Carefully remove seal from crankcase cover.
10. Install new seal, with lip of seal inward, using seal installer J-22260.
11. Replace crankshaft timing belt sprocket and flange.
12. Align timing marks and replace timing belt.
13. Replace timing belt lower cover and harmonic balancer.
14. Adjust timing belt tension.
15. Replace water pump pulley and fan.
16. Replace timing belt upper front cover.

Front Crankcase Cover and Gasket R & R—1966-69 OHC 6 Cylinder

1. Remove timing belt sprocket, as described above.
2. Remove four front oil pan-to-crankcase cover retaining bolts.
3. Loosen remaining oil pan bolts, as necessary, to provide clearance between crankcase cover and oil pan.
4. Remove five front crankcase cover attaching bolts.
5. Remove front crankcase cover and gasket, clean off the old gasket.
6. Inspect cover seal for wear or distortion.
7. Using new gasket installed over dowels and, if necessary, new seal, reverse removal procedures, torque oil pan and crankcase cover bolts to 10-15 ft. lbs.

Housing Assembly, Oil Pump, Distributor and Fuel Pump—1966-69 OHC 6 Cylinder

The housing is unique, and consists of the oil pump, distributor and the fuel pump. The oil filter is also attached to this housing. The housing carries the drive sprocket for the above units and is used as a tensioner for the timing belt.

Oil Pressure Regulator R & R

1. Remove cap washer and spring from housing assembly.
2. Using magnet, remove valve from housing assembly.
3. Install valve on spring and install as an assembly.
4. Install cap washer.

Oil Pump R & R

1. Remove oil pump cover and gasket.
2. Remove drive gear and driven gear.
3. Install gears.
4. Replace cover using new gasket. Torque attaching bolts to 15-25 ft. lbs.

Housing Assembly R & R

1. Remove timing belt top front cover.
2. Align timing marks.
3. Loosen six housing assembly from cylinder block retaining bolts.
4. Remove timing belt from camshaft sprocket and distributor drive.
5. Disconnect fuel lines from fuel pump.
6. Remove distributor cap, vacuum lines and wires from distributor.
7. Remove housing by removing six retaining bolts.
8. Install, using a new gasket, and loosely install housing assembly to cylinder block with six retaining bolts.
9. Align timing marks and install timing belt.
10. Connect fuel lines to fuel pump.
11. Replace distributor cap, vacuum lines and wires.
12. Adjust timing belt tension, see timing belt adjustment.
13. Replace timing belt top front cover.

Distributor and Oil Pump Drive Housing (Except Oil Pump) Disassembly and Assembly

1. Remove housing assembly.
2. Observe and record location of sprocket timing mark and position of distributor rotor. Remove distributor.
3. Remove fuel pump eccentric and distributor drive gear retaining pin.
4. Remove shaft and sprocket assembly from housing.
5. Inspect shaft assembly, seal and bearing.
6. If necessary to replace bearing or seal, use tool J-22264 and slide hammer to remove seal, or bearing and seal together.
7. Use tools J-22267-1 and J-22267-2 to install seal.
8. Reassemble by reversing Steps 1 through 4.

NOTE: substitute tools can be made by duplicating tools in illustrations.

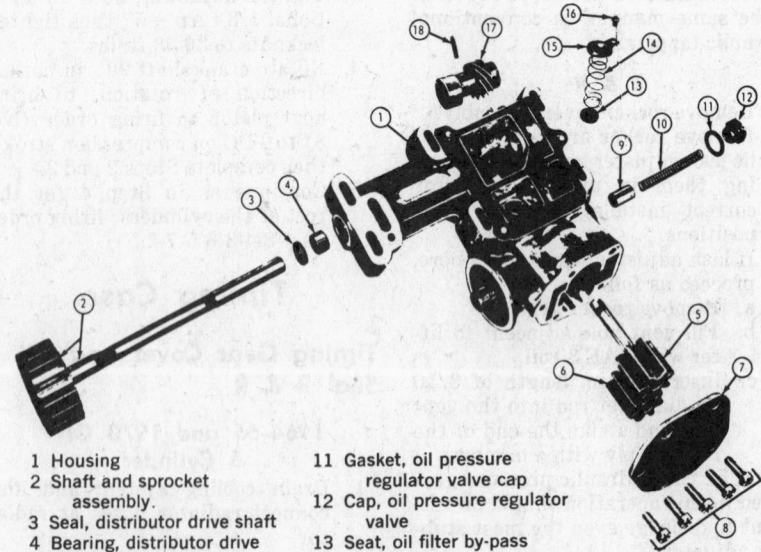

1 Housing	11 Gasket, oil pressure regulator valve cap
2 Shaft and sprocket assembly.	12 Cap, oil pressure regulator valve
3 Seal, distributor drive shaft	13 Seat, oil filter by-pass valve
4 Bearing, distributor drive	14 Spring, oil filter by-pass valve
5 Gear and shaft assembly oil pump drive	15 Retainer, oil filter by-pass valve
6 Gear, oil pump driven cover, oil pump	16 Screw, oil filter by-pass valve retainer
7 Cover, oil pump	17 Gear and eccentric
8 Bolt, oil pump cover to housing	18 Pin, distributor oil and fuel pump gear and eccentric
9 Valve, oil pressure regulator	
10 Spring, oil pressure regulator	

OHC 6 cyl. oil pump housing and distributor drive assembly
(© Pontiac Div. G.M. Corp.)

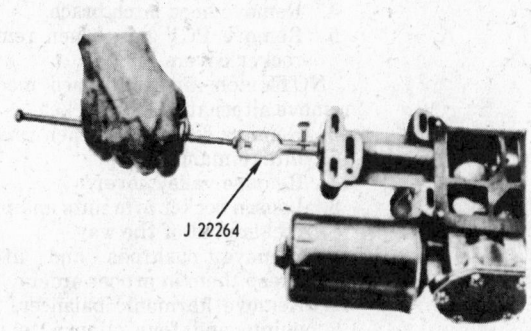

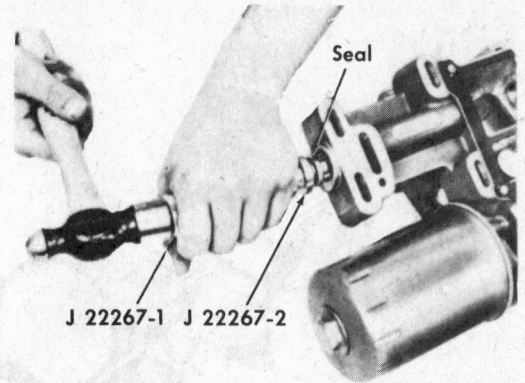

Removing seal from OHC 6 distributor and oil pump drive housing
(© Pontiac Div. G.M. Corp.)

Installing seal into OHC 6 distributor and oil pump drive housing
(© Pontiac Div. G.M. Corp.)

1966-69 OHC Timing Belt Adjustment

1. Remove timing belt top front cover.
2. Using J-22232-2 calibration bar, set the pointer of timing belt tension fixture J-22232-1 to zero.
NOTE: this calibration must be performed before each use of J-22232 fixture to insure an accurate timing belt adjustment.
3. Remove camshaft sprocket to camshaft bolt and install J-22232-1 (tension fixture) on the belt with the rollers on the outside (smooth) surface of belt. Thread the fixture mounting bolt into camshaft sprocket bolt location, finger-tight.
4. Squeeze indicator end (upper) of fixture and quickly release so the fixture assumes released or relaxed position.
5. With J-22232-1 installed, as above, adjust accessory drive housing up or down, as required, to obtain a tension adjustment indicator reading centered in the green range, with drive housing mounting bolts torqued to 15 ± 3 ft. lbs.
6. Remove tension fixture and install sprocket retaining bolt, making sure bolt threads and washers are free of dirt.
7. Install upper front timing belt cover.

Camshaft R & R

1964-65 and 1970 OHV 6 Cylinder

1. Drain cooling system.
2. Remove radiator, fan, and water pump pulley.
3. Remove grill.
4. Remove valve cover and gasket, then loosen rocker arm nuts and pivot rockers out of the way.
5. Remove pushrods.
6. Remove distributor, fuel pump, and spark plugs.

7. Remove coil, pushrod (tappet gallery) covers and gasket; reach in and remove tappets, keeping them in order.
8. Remove harmonic balancer, then loosen oil pan bolts and allow pan to drop.
9. Remove timing gear cover.
10. Remove two camshaft thrust plate bolts by rotating cam gear holes to gain clearance.
11. Remove the camshift by pulling it straight forward.
NOTE: do not wiggle the camshaft; cam bearings could be dislodged.
12. If cam gear is to be replaced, press it from the shaft using an arbor press.
CAUTION: thrust plate must be positioned so that Woodruff key does not damage it during removal.
13. New cam gear must be pressed onto the shaft, with the shaft supported in back of the front bearing journal.
NOTE: the thrust plate end-play

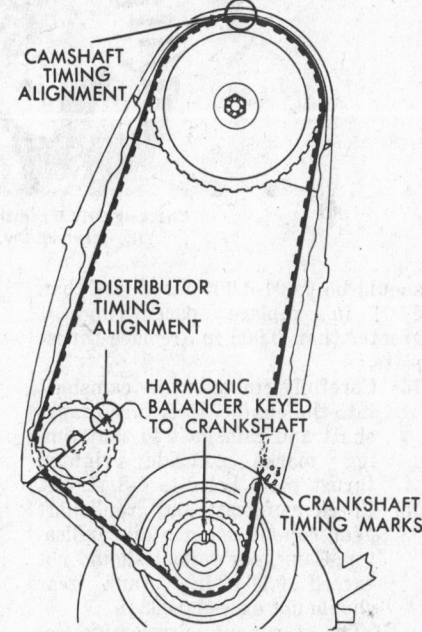

OHC 6 cyl. timing mark alignment

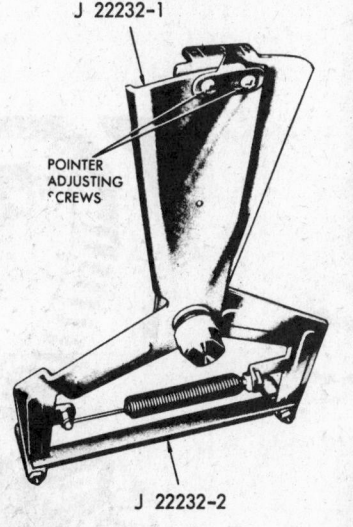

OHC 6 cyl. timing belt adjustment with tool J-22232
(© Pontiac Div. G.M. Corp.)

Checking OHV 6 camshaft gear backlash
(© Chevrolet Div. G.M. Corp.)

temperature gauge wire, and fuel lines.
4. Remove hood latch brace.
5. Remove PCV hose, then remove rocker covers.
NOTE: on air-conditioned models, remove alternator and bracket.
6. Remove distributor, then remove intake manifold.
7. Remove valley cover.
8. Loosen rocker arm nuts and pivot rockers out of the way.
9. Remove pushrods and lifters (keep them in proper order).
10. Remove harmonic balancer, fuel pump, and four oil pan to timing cover bolts.
11. Remove timing cover and gasket, then remove fuel pump eccentric and bushing.
12. Align timing marks, then remove timing chain and sprockets.
13. Remove camshaft thrust plate.
14. Remove camshaft by pulling straight forward, being careful not to damage cam bearings in the process.
NOTE: it may be necessary to jack up the engine slightly to gain clear-

should be 0.001-0.005 in. If less than 0.001 in., replace spacer ring; if greater than 0.005 in., replace thrust plate.
14. Carefully install the camshaft into the engine, then turn crankshaft and camshaft so that timing marks coincide; tighten thrust plate bolts to 5-8 ft. lbs.
15. Check camshaft and crankshaft gear runout using a dial indicator. Cam gear runout should not exceed 0.004 in., crank gear should not exceed 0.003 in.
NOTE: if runout is excessive, remove gear and clean burrs from shaft.
16. Check gear backlash using a dial indicator; it should not ex-

ceed 0.006 in. and should not be less than 0.004 in.
17. To complete installation, reverse Steps 1-9.
NOTE: install distributor with No. 1 piston at TDC on compression stroke so that vacuum diaphragm faces forward and rotor points to No. 1 spark plug wire cap tower. Make sure oil pump drive shaft is properly indexed with distributor drive shaft.

1964-70 V8

1. Drain cooling system and remove air cleaner.
2. Disconnect all water hoses, vacuum lines and spark plug wires.
3. Disconnect accelerator linkage,

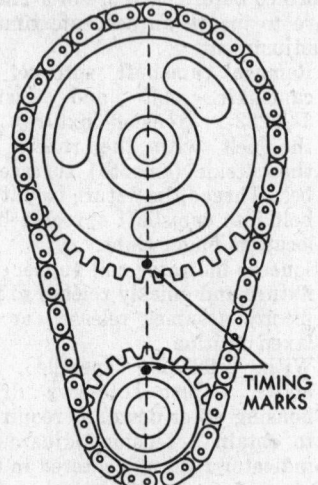

V8 timing mark alignment

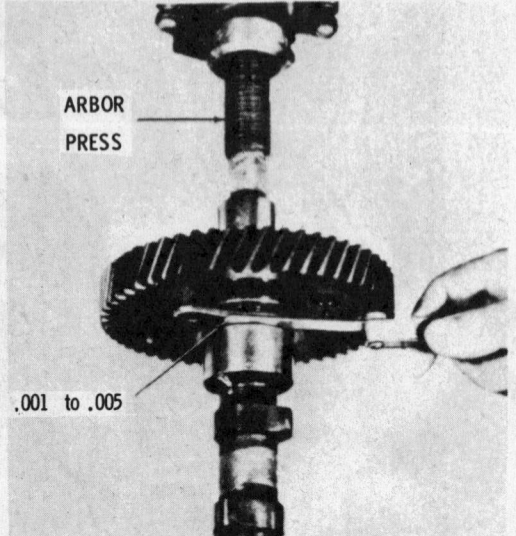

Installing OHV 6 camshaft gear and checking thrust plate end-play
(© Chevrolet Div. G.M. Corp.)

Checking OHV 6 camshaft gear runout
(© Chevrolet Div. G.M. Corp.)

ance, especially if motor mounts are worn.

15. Install new camshaft, with lobes and journals coated with heavy (SAE 50-60) oil, into the engine, being careful not to damage cam bearings.

NOTE: most specialty cams come with a special "break-in" lubricant for the lobes and journals; if such lubricant is available, use it instead of heavy oil.

16. Install camshaft thrust plate and tighten bolts to 20 ft. lbs.
17. To install, reverse Steps 1-12, tightening sprocket bolts to 40 ft. lbs., timing cover bolts and nuts to 30 ft. lbs., oil pan bolts to 12 ft. lbs., and harmonic balancer bolt to 160 ft. lbs.

1966-69 OHC 6 Cylinder

1. Remove camshaft sprocket and seal.
2. Remove rocker cover assembly.

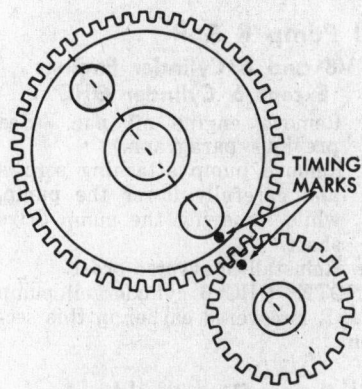

OHV 6 cyl. timing mark alignment

3. Using an adapter and a slide hammer, drive camshaft to the rear. Make sure bearing surfaces are not damaged during this operation.
4. Disconnect slide hammer and remove camshaft from rear of rocker cover.
5. Remove thrust washer, retaining washer, and bolt from rear of camshaft.
6. Clean and inspect all parts for

wear or damage, then inspect bearing surfaces for wear or scoring.
7. Clean camshaft oil passages.
8. To install, reverse removal procedure making sure thrust washer is installed as illustrated. Tighten retaining bolt to 40 ft. lbs.
9. Check camshaft end-play, using a dial indicator on the front sprocket; end-play should be 0.003-0.009 in. and is controlled by the camshaft bore plug.

NOTE: lubricate camshaft lobes and rockers with special lubricant, available at Pontiac dealers. Tighten rocker cover bolts and nuts to 15 ft. lbs. from center outward.

Engine Lubrication

Oil Pan R & R

1964-65 OHV 6 Cylinder and 1964-66 V8

1. Drain crankcase and cooling system, then remove engine and transmission from vehicle as a complete assembly.
2. Remove the oil pan bolts, then the pan.
3. To install, reverse removal procedure.

1967-70 V8

1. Remove engine from car, as previously described.
2. Remove oil pan bolts.
3. Remove oil pan.

NOTE: 1970 Tempest V8 oil pan can be removed, in some cases, in a manner similar to 1968-69 Firebird V8.

1968-69 Firebird V8

1. Disconnect battery cable at battery.
2. Remove distributor cap and fan shield.

3. Remove fan and fan pulley on air-conditioned models.
4. Disconnect engine ground straps.
5. On air-conditioned models, remove compressor and swing it out of the way without disconnecting hoses.
6. Jack up front of car and drain engine oil.
7. Disconnect steering idler arm from frame.
8. Remove exhaust crossover pipe on single exhaust cars; disconnect exhaust pipes at manifold flanges on dual exhaust cars.
9. Remove starter motor, starter motor bracket, and flywheel cover.
10. Support engine with a chain hoist, then remove motor mounts and loosen rear transmission mount.

NOTE: it may be necessary, in individual cases, to remove the rear transmission mount.

11. Remove oil pan bolts, raise engine about 4½ in., and move engine forward about 1½ in.
12. Remove oil pan by rotating clockwise (to clear oil pump) and pulling down.
13. To install, reverse removal procedure.

1970 OHV 6 Cylinder

1. Remove upper radiator shield assembly.
2. Disconnect negative battery cable.
3. Jack up front of car and drain engine oil.
4. Disconnect exhaust pipe at manifold flange.
5. Remove starter motor and flywheel cover.
6. Raise engine slightly, using a chain hoist, then remove both front motor mount to frame bolts and right motor mount.

INDEX THRUST WASHER TANG IN HOLE IN ROCKER ARM COVER

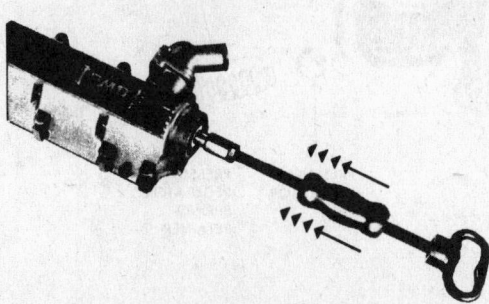

Removing camshaft with slidehammer—OHC 6
(© Pontiac Div. G.M. Corp.)

Camshaft thrust washer position—OHC 6
(© Pontiac Div. G.M. Corp.)

7. Remove oil pan bolts, then raise engine and remove oil pan.
8. To install, reverse removal procedure.

NOTE: bolts into timing gear cover should be installed last. They are installed at an angle and holes line up after rest of oil pan bolts are tightened finger-tight.

1966-69 6 Cylinder OHC Removal

1. Disconnect battery.
2. Remove air cleaner assembly.
3. On air conditioned cars, remove compressor from mounting brackets and position to one side.
4. Inspect all water hoses and wiring harness for routing and possible interference. (Engine is raised at least 4½ in. on Tempest, 2 in. on Firebird.)

NOTE: before raising the car, prop the hood open at least 6 in. to ensure enough clearance between timing belt cover and inner hood panel.

5. Raise car and drain crankcase.
6. Remove starter assembly and flywheel cover.
7. Reroute or disconnect any wiring between bellhousing and floor pan to insure against damage when bellhousing contacts pan.
8. Loosen transmission insulator to crossmember retaining bolts.
9. Remove right and left engine insulator to frame bracket through-bolts.
10. Rotate harmonic balancer until timing mark is at bottom. (This properly positions crankshaft counterweights.)
11. Bolt engine support bracket, tool J-22345 to front of harmonic balancer.
12. With suitable equipment, raise engine at J-22345 until insulators clear frame brackets.
13. Remove oil pan bolts.

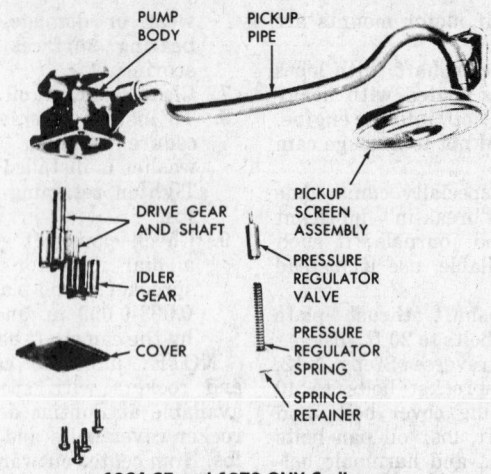

1964-65 and 1970 OHV 6 oil pump
(© Chevrolet Div. G.M. Corp.)

14. Raise engine. Apply a rearward force on the engine-transmission assembly until oil pan clears the flywheel housing. Then, remove the oil pan.

1966-69 6 Cylinder OHC Installation

1. Install new gasket on oil pan.
2. Apply enough rearward force on engine-transmission assembly to allow oil pan to clear flywheel housing.
3. Install oil pan and torque retaining bolts to 10-15 ft. lbs.
4. Lower engine, remove engine support bracket and install engine insulator bracket to frame through-bolts.
5. Tighten transmission insulator to crossmember bolts to 25-35 ft. lbs.
6. Replace flywheel cover and starter assembly.
7. Lower the car.
8. On air-conditioned cars, install compressor and adjust belt tension.
9. Replace air cleaner assembly.
10. Refill crankcase.
11. Connect battery.

Oil Pump R & R
V8 and 6 Cylinder Engines Except 6 Cylinder OHC

1. Remove engine oil pan. (See previous paragraph.)
2. Remove pump attaching screws and carefully lower the pump, while removing the pump drive shaft.
3. Reinstall in reverse order.

NOTE: OHC 6 cylinder oil pump R & R is covered earlier in this section.

Oil Pump Disassembly, Inspection and Assembly

1. Remove pressure regulator spring.
2. Remove cover hold-down bolts and cover.
3. Remove driven gear and drive gear, then remove drive shaft.

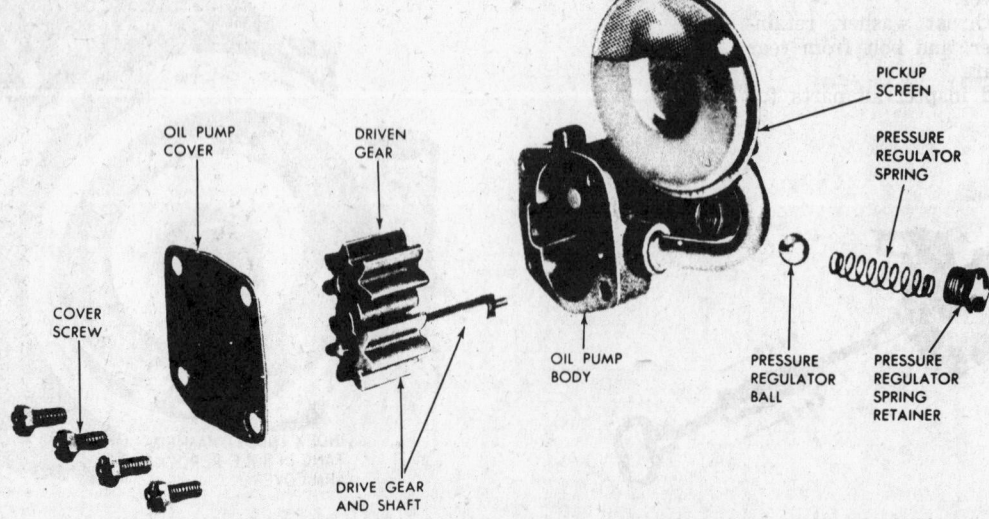

1964-70 V8 oil pump (© Pontiac Div. G.M. Corp.)

4. Clean all parts in solvent, especially pump pickup screen.
   CAUTION: do not remove or loosen oil pickup tube or screen.
5. Inspect regulator spring for distortion, wear, or cracks.
6. Inspect regulator ball for damage.
7. Inspect pump components for wear.
8. To assemble, first install drive and driven gears, then install cover.
   NOTE: it is a good idea to pack pump with Vaseline for proper priming.
9. Turn drive shaft by hand to make sure it turns freely.
10. Install regulator ball, spring, and retainer.
   NOTE: pressure regulator springs are normally 50 psi and are used with unplated gears. GTO oil pumps up to and including 1969, and all Ram Air IV oil pumps to 1970, use 60 psi springs in conjunction with cadmium plated gears. In 1970, the GTO uses a 60 psi spring and phosphate coated iron gears. Never use higher pressure springs with unplated gears, and never try to increase oil pressure by changing spring length.

Connecting Rods, Bearings and Pistons

Main Bearing Replacement

1. Remove oil pan and, on V8 engines, oil baffle plate.
2. Remove one main bearing cap and lower bearing insert.
3. With a "roll out" pin, or a cotter pin bent as illustrated, inserted

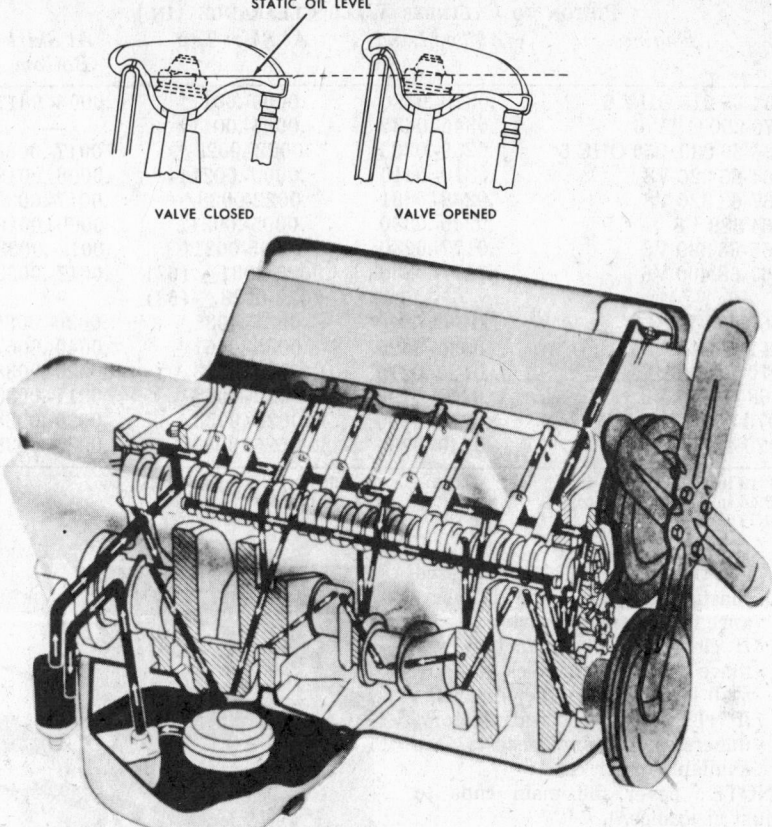

V8 oil flow (© Pontiac Div. G.M. Corp.)

into the oil hole in the crankshaft, rotate crankshaft, in normal direction of rotation (clockwise seen from front) to remove top bearing insert.
4. Oil the new upper bearing insert and place un-notched end of shell between crankshaft and notched upper bearing web.
5. With "roll out" pin positioned in oil hole, rotate new upper bearing insert into place.
6. Install new lower bearing insert into bearing cap, with a 0.002 in. strip of brass shim stock between insert and cap. Do not oil this bearing.
7. Place a strip of Plastigage (available from automotive parts jobbers) on the lower bearing insert, then install main cap, tightening bolts to specification.
   NOTE: do not rotate crankshaft.
8. Remove bearing cap and, using the scale on the envelope, meas-

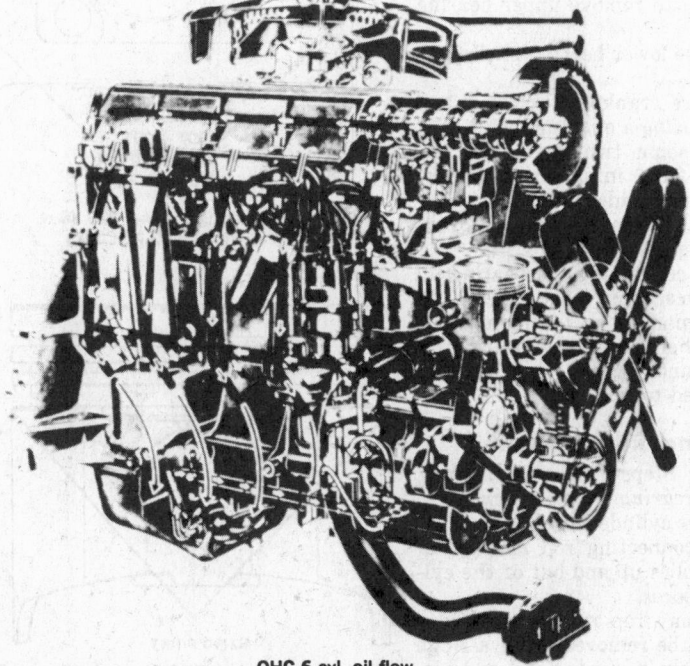

OHC 6 cyl. oil flow
(© Pontiac Div. G.M. Corp.)

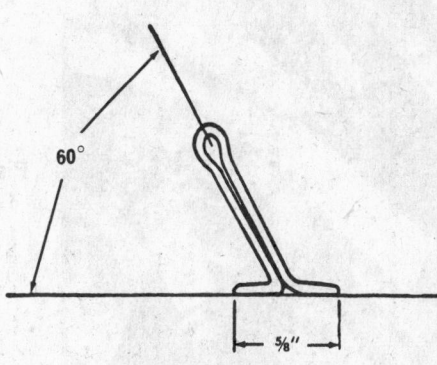

Homemade bearing "roll out" pin
(© Pontiac Div. G.M. Corp.)

PISTON TO CYLINDER WALL CLEARANCE (IN.)			
Engine	At Top Land	At Skirt Top	At Skirt Bottom
1964-65 215 OHV 6	.0320-.0420	.0005-.0011*	.0004-.0017
1970 250 OHV 6	.0345-.0435	.0005-.0011†	—
1966-69 230/250 OHC 6	.0250-.0303	.0022-.0028△	.0017-.0033
1964-66 326 V8	.0310-.0410	.0005-.0021‡	.0006-.0018
1967-68 326 V8	.0248-.0301	.0022-0028△	.0017-.0033
1964 389 V8	.0240-.0330	.0005-.0021‡	.0000-.0018
1965-66 389 V8	.0177-.0230	.0005-.0021‡	.0017-.0033
1967-68 400 V8	.0177-.0230	{.0025-0031△ (67) {.0022-0028△ (68)	.0017-.0033
1969 400 V8	.0170-.0210	.0025-.0031△	.0020-.0036
All Ram Air IV V8	.0330-.0420	.0055-.0061△	.0040-.0057
1970 400 V8	.0170-.0210	.0025-.0033△	.0020-.0038
1968-69 350 V8	.0240-.0290	.0022-.0028△	.0017-.0033
1970 350 V8	.0240-.0290	.0025-.0033△	.0020-.0038
1970 455 V8	.0240-.0290	.0025-.0033◎	.0020-.0038

* 2.16 in. below piston top.   ◎ 1.08 in. below piston top
† 2.44 in. below piston top.   ‡ 1.18 in. below piston top
△ 1.11 in. below piston top.   (.0007-.0013 in. preferred).

Measuring cylinder bore

ure the width of the compressed Plastigage; this is the oil clearance for this main bearing.

9. If clearance is satisfactory, remove brass shim stock and install lower cap and new bearing; if clearance is unsatisfactory, undersize bearing inserts are available to correct.

NOTE: never file main caps to adjust clearance.

10. Replace the remaining main bearings in the engine by following Steps 2-9, then install oil pan.

NOTE: it may be necessary to remove oil pump to gain clearance.

Connecting Rod Bearing Replacement

1. Remove oil pan and, on V8 engines, oil baffle.
2. Remove oil pump.
3. Rotate crankshaft to bring bearing caps, in turn, into position for removal.
4. Remove bearing cap.

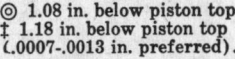

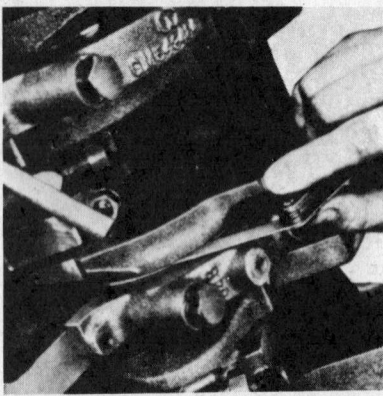
Measuring crankshaft endplay

5. Push sections of rubber tubing over connecting rod bolts (to protect crankpin), then push rod and piston assembly up **far** enough to remove upper bearing insert.
6. Remove lower bearing shell from cap.
7. Measure crankpin for taper and wear, using a micrometer.

NOTE: some two-barrel, V8 engines have 0.010 in. undersize crankpins. These are identified by a .010 U.S. stamp on front of No. 1, and rear of No. 8, counterweight.

8. Check connecting rod bearing oil clearance, using Plastigage, in the same manner as for main bearings, then install new bearing inserts and tighten rod bolts to specified torque.

Piston Ring Replacement

1. Follow Steps 1-6 in *Connecting Rod Bearing Replacement*.
2. Remove cylinder head/s.
3. Push connecting rod and piston assemblies up and out of the cylinder bores.

NOTE: any top ridge on cylinder wall should be removed using a ridge reamer before removing rod and piston assemblies.

4. Using an internal micrometer, measure the cylinder bores both across the thrust faces of the cylinder and parallel to the crankshaft axis. This should be done in at least four locations, equally spaced, to get a representative reading. The bore should not be out of round more than 0.005 in. and must not taper (top to bottom) more than 0.010 in. Cylinders out of tolerance should be bored to the next available oversize and new pistons fitted. If one cylinder is bored, the rest must be bored to the same oversize to preserve balance.

5. If bore is within tolerance, examine for visible damage. It should be dull silver in color and exhibit a pattern of machining cross hatching. There should be no

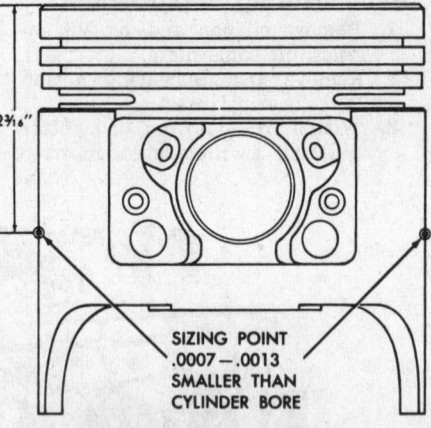
2³⁄₁₆"

SIZING POINT
.0007 — .0013
SMALLER THAN
CYLINDER BORE

Piston sizing points—326 V8
(© Pontiac Div. G.M. Corp.)

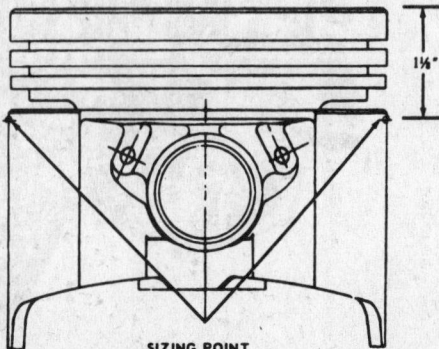
1¹⁄₂"

SIZING POINT

Piston sizing points—OHC 6 and all V8 except 326
(© Pontiac Div. G.M. Corp.)

Measuring bearing clearance using Plastigage
(© Pontiac Div. G.M. Corp.)

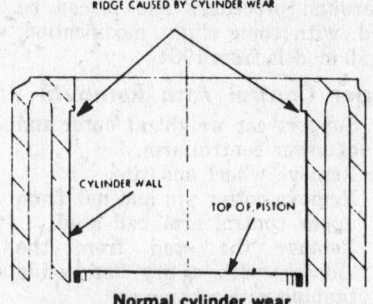

Normal cylinder wear

V8 piston and road assembly

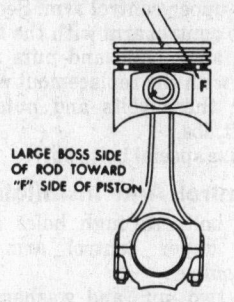

Piston and rod assembly—6 cyl.

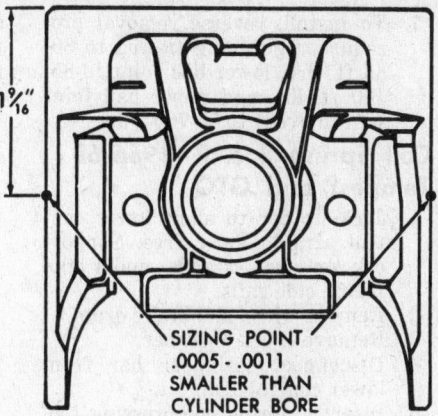

SIZING POINT
.0005 - .0011
SMALLER THAN
CYLINDER BORE

Piston sizing points

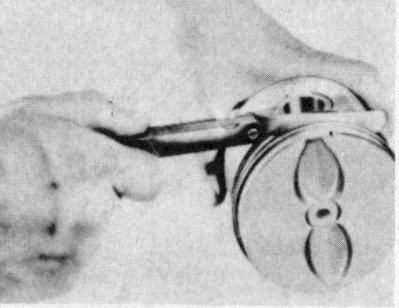

Measuring piston ring groove clearance
(© Pontiac Div. G.M. Corp.)

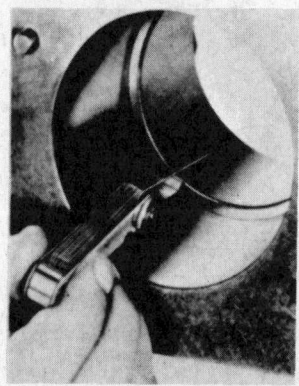

Measuring ring gap—V8

scratches, tool marks, nicks, or other damage. If any such damage exists, bore cylinder to clean up damage, then bore to next available oversize. Polished or shiny places in the bores cause poor lubrication, high oil consumption and ring damage. Re-

(!) CHILTON TIME-SAVER

This or any other machining operation should be done with cylinder block disassembled. Hot tank cylinder block after boring or honing. To remove minor glazing when honing equipment is not available, run emery cloth back and forth across glazed area perpendicular to bore axis. Scrub block and bores thoroughly with soap and water to remove all grit after using emery cloth.

Since this method is much inferior to honing, it is not recommended as a regular practice—only as a last resort.

move this glazing by honing cylinders with clean, sharp stones of No. 180-220 grit to obtain a surface finish of 15-35 RMS. Use a hone to correct piston clearance and surface finish any cylinder that has been bored.

6. If cylinder bore is in satisfactory condition, place each ring in bore, in turn, and square it in bore using a piston top. Measure the ring end gap with feeler gauges. If gap is greater than that specified, get new ring; if gap is less than that specified, file end of ring to bring within tolerance. Clearances are as follows:

1964-70 6 cylinder engines,
  including OHC   0.015 in. compression
                  0.035 in. oil
1964-67 326 V8 and 1964-66 389
V8 engines   No. 1   0.021 in.
                  No. 2   0.019 in.
                  oil   0.035 in.
1967 326 V8, 1967-69 400 V8 and
  1968-70 350 V8 engines
                  0.019 in. compression
                  0.035 in. oil
1970 400 V8 engine   No. 1   0.019 in.
                     No. 2   0.015 in.
                     oil   0.035 in.
1970 455 V8 engine   No. 1   0.021 in.
                     No. 2   0.015 in.
                     oil   0.035 in.

7. Clean ring grooves, using a commercial groove cleaner or a broken ring, then install rings. Space ring gaps equally around piston circumference and check ring to groove clearance. The following figures apply to all V8 and six-cylinder engines:

No. 1   0.0015-0.0030 in.
No. 2   0.0015-0.0035 in.
oil   0.0005-0.0055 in.

8. Lower rod and piston assemblies, one at a time, into bore until ring compressor contacts block. Using wooden hammer handle, push piston into bore while guiding rod onto journal.
9. Squirt oil on cylinder walls, then install rod bolts and tighten to specification.
10. Install cylinder head.
11. Reverse Steps 1-2 in *Connecting Rod Bearing Replacement*.

Front Suspension

Description

Ball joints, located at the outer ends of the upper and lower control arms, act as pivot points for both the vertical movement of the wheel and rotation of the steering knuckle. The spherical joints have a fixed boot grease seal to protect against dirt and water. Steering knuckles and spindles are one-piece forgings.

Rubber bushings at the upper inner control arm ends pivot on shafts attached to the frame. By varying shim thickness at this point, caster and camber adjustments are accomplished. The inner ends of the lower control arms are also rubber mounted, and are attached to the front crossmember through brackets.

The upper ends of the coil springs are seated in the frame, while the lower ends rest on the lower control arms. Double-action shock absorbers are located inside the coil springs, the rubber insulated upper end of each unit being fastened to the frame, the similarly insulated lower end to the lower control arm.

For increased roll stability, a stabilizer bar is rubber mounted to the frame and is connected to the lower control arms via links at each end. See the Unit Repair Section for wheel alignment procedures.

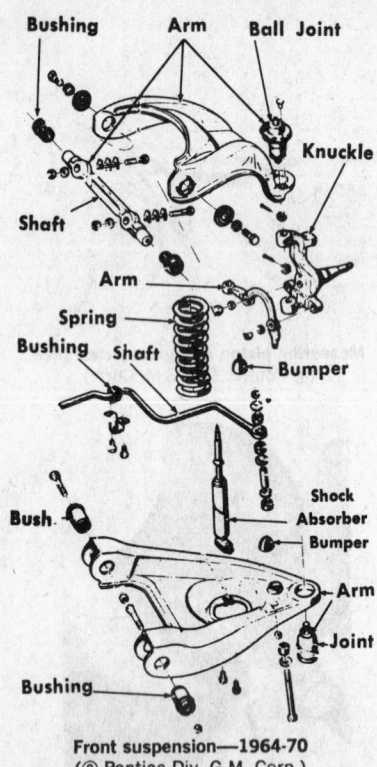

Front suspension—1964-70
(© Pontiac Div. G.M. Corp.)

Coil Spring R & R—1964-65 Tempest and GTO

1. Place car on a hoist which supports car at side rails. The front control arms must be allowed to swing free, and must be positioned so that the control arms may be raised or lowered with the hoist.
2. Remove shock absorber and wheel.
3. Disconnect stabilizer bar from control arm, then disconnect tie-rod.
4. Place stand under control arm and take up slightly on spring compression.
5. Remove lower ball stud from steering knuckle.
6. Carefully raise hoist until spring is free. Remove the spring.

7. To install, reverse removal procedure, tightening tie-rod to 30-45 ft. lbs., lower ball joint to 85-100 ft. lbs, and upper ball joint (if removed) to 55-70 ft. lbs.

Coil Spring R & R—1966-68 Tempest and GTO

1. Jack up car to allow lower control arm to hang free. Support car on jack stands under the frame side rails.
2. Remove wheel and brake drum.
3. Remove shock absorber.
4. Disconnect stabilizer bar from lower control arm.
5. Insert a spring compressing tool through the shock absorber mounting holes and compress the coil spring until it lifts from its seat.
NOTE: a spring compressor can be fabricated using a length of threaded rod, a support plate and a support hook.
6. Remove backing plate and swing it out of the way.
7. Disconnect lower ball joint stud and swing steering knuckle out of the way.
8. Pull lower control arm down far enough to remove spring.
9. To install, reverse removal procedure.
NOTE: spring must be compressed before installation.

Coil Spring R & R—1967-69 Firebird, 1969-70 Tempest and GTO

1. Jack up car and support on jack stands at frame side rails.
2. Remove shock absorber.
3. Disconnect stabilizer bar at lower control arm.
4. Support lower control arm with a hydraulic floor jack, then remove the two inner control arm to front crossmember bolts.
5. Carefully lower the control arm, allowing the spring to relax.
6. Reach in and remove spring.
NOTE: this is probably the best

all-around procedure and it can be used, with some slight modification, for all models from 1964.

Upper Control Arm Removal

1. Support car weight at outer end of lower control arm.
2. Remove wheel and tire.
3. Remove cotter pin and nut from upper control arm ball stud.
4. Remove the stud from the knuckle with a pry bar, while tapping with a hammer.
5. Remove two nuts that hold the upper control arm cross-shaft to front crossmember. Count number of shims at each bolt.
NOTE: on V8 Firebird models with air conditioning, swing compressor out of the way.

Upper Ball Joint Removal

1. Prickpunch the center of the four rivets.
2. Drill through the heads of these rivets.
3. Chisel off rivet heads and tap out rivets with a punch.

Upper Ball Joint Installation

1. Install new ball joint against top side of upper control arm. Secure joint to control arm with the four special alloy bolts and nuts furnished with the replacement part.
2. Torque these bolts and nuts to 10-12 ft. lbs.
NOTE: use special bolts only.

Upper Control Arm Installation

1. Install bolts through holes and install upper control arm to crossmember.
2. Secure two nuts and washers to the bolts holding the upper control arm shaft to front crossmember. Install same number of shims as removed at each bolt. Torque bolts to 50 ft. lbs.
3. Lubricate ball joint with chassis lube.
4. Install ball joint stud through knuckle. Install nut, and torque to 50 ft. lbs.
5. Install wheel and tire assembly.
6. Lower car to floor.
7. Bounce car to neutralize front end suspension and torque pivot shaft nuts to 50 ft. lbs. on Tempest, 35 ft. lbs. on Firebird.

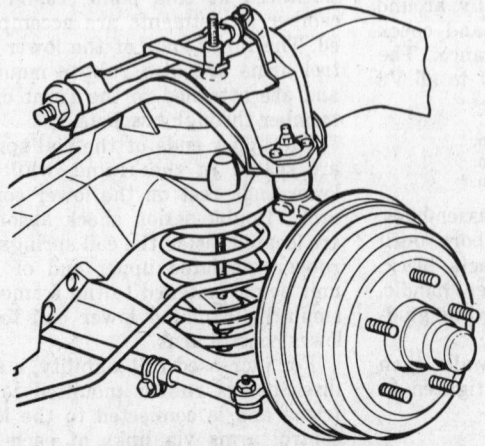

Typical front suspension—1964-70
(© Pontiac Div. G.M. Corp.)

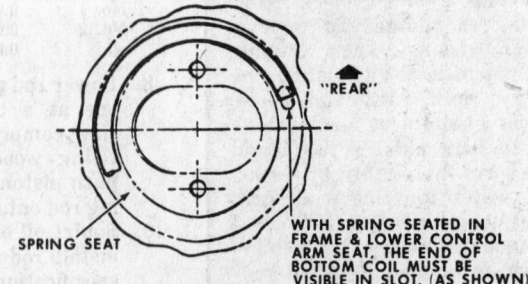

Front coil spring position on lower control arm—all except Firebird
(© Pontiac Div. G.M. Corp.)

8. Be sure to recheck caster and camber.

Lower Control Arm and Ball Joint R & R

1. Remove coil spring and lower control arm inner bolts.
2. Separate lower ball joint from steering knuckle by prying, while hammering sharply on steering knuckle.
3. Press lower ball joint from lower control arm using suitable arbors and a large bench vise.
4. To install, reverse removal procedure, tightening lower ball joint stud nut to 85-90 ft. lbs.

Jacking, Hoisting

Jack car at front spring seats of lower control arms. Jack car (1964-70) at rear under axle housing.

Steering Gear

Manual Steering Gear

NOTE: see Unit Repair Section.

Steering Gear Box R & R

1. Disconnect pitman arm from pitman shaft.
NOTE: the factory recommends a special puller be used, although the arm can be removed, in some cases, by using a pry bar and a hammer.
2. Matchmark the worm shaft flange and steering shaft, then disconnect lower flange.
3. Remove the three steering gear box to frame bolts, then the gear box.
4. To install, reverse removal procedure, tightening frame bolts to 70-90 ft. lbs.

In-Car Adjustment

There are two adjustments on the steering gear, worm bearing preload and pitman shaft overcenter adjustment.

The wheel should turn smoothly through its entire range. Roughness indicates internal trouble requiring disassembly. Binding (especially in straight ahead position) indicates too tight an adjustment.

1. Be sure the steering-gear-to-frame bolts are torqued to 70-90 ft. lbs.
2. Disconnect intermediate rod from pitman arm and loosen adjustment a few turns.
3. Turn steering wheel slowly from one extreme to the other. Never turn the wheel hard against the stopping point.
4. Remove emblem or cap from steering wheel.

Worm Bearing Preload

1. Check worm bearing preload by turning the steering wheel gently in one direction until it stops. This positions the gear away from the high point load.
2. Attach a socket and 25 in. lb. torque wrench to the steering wheel nut. Turn the worm shaft with the wrench, through a one-revolution range from either extreme. Torque required to keep the wheel moving through either revolution extreme should be 7 in. lbs.
3. Be sure the gear case side cover bolts are torqued to 30 ft. lbs. and the locknut to 85 ft. lbs.

Overcenter Adjustment

1. Turn steering wheel from one extreme to the other while counting total turns. Then, turn the wheel back exactly midway. This positions the steering gear on the high spot or straight ahead position. A slight drag should exist at this point.

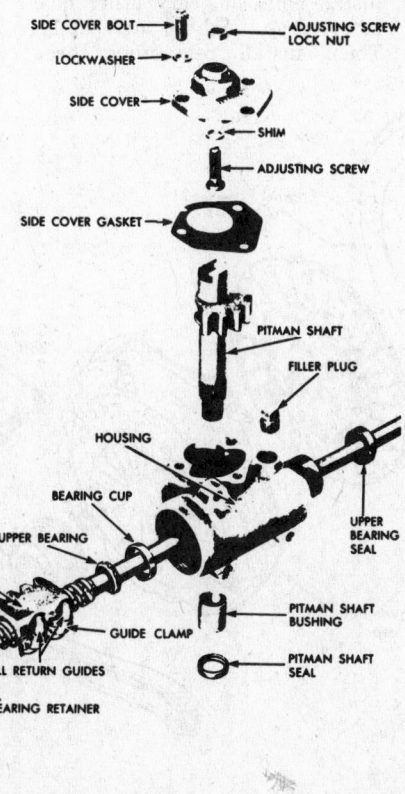

Manual steering gear (© Pontiac Div. G.M. Corp.)

2. Check torque used to rotate the wheel through the high point range. Torque should be 7 in. lbs. higher than worm bearing preload. The total over-center pull should not exceed 16 in. lbs.
3. Adjust pitman shaft overcenter preload by loosening locknut and turning pitman shaft adjuster screw to obtain 7 in. lbs. higher than worm bearing preload.
4. Tighten locknut. Rotate steering wheel through its entire range. Recheck for the maximum 16 in. lbs. torque while passing through the straight-ahead position.
5. With steering wheel spokes lined up, wheels should be straight. If not, adjust tie-rods.

Power Steering Gear

For detailed information on the power steering and pump, see Unit Repair Section.

Clutch

A single-plate, dry-disc, diaphragm-spring clutch is used on all Tempest, Firebird and GTO models. The clutch assembly consists of the driven plate, the pressure plate, and the release mechanism. Grooves on both sides of the driven plate prevent the plate's sticking to the flywheel and pressure plate due to vacuum.

Two types of diaphragm type pressure plates are used—a bent finger type, for the high performance OHC six-cylinder and V8s of more than 350 cu. in. displacement, and a flat finger type, for low performance V8s and six-cylinder standard engines. The diaphragm spring design is such that no overcenter spring is required.

The clutch release mechanism consists of a ball thrust (throwout) bearing and various linkage configurations (for the various models) to control this bearing. The throwout bearing slides on the front transmission extension housing (nose piece), which is concentric with and encloses the transmission main drive gear. When pedal pressure is applied, the clutch fork pivots on its ball socket, through linkage action, and the inner end of the fork forces the throwout bearing against the release levers.

A clutch safety switch prevents engine cranking unless the clutch is disengaged (on 1969-70 models). The only periodic clutch service required, other than adjustment for normal wear, is the lubrication of all linkage pivot points every 6,000 miles.

Removal—1964-70

1. Raise car and support on jackstands.

2. Support rear of engine with jack-stand.
3. Remove driveshaft.
4. Remove rear crossmember bolts from frame and transmission mounts, and remove crossmember.

NOTE: see transmission removal procedure for procedure variations.

5. Disconnect transmission shift linkage, speedometer cable and clutch return spring. Clutch fork pushrod will now hang free.
6. Remove clutch housing cover plate screws and let plate hang from starter gear housing.
7. Lower engine enough to gain access to clutch housing bolts at engine block, then remove all but uppermost bolt.
8. Hold transmission and clutch housing assembly against block over dowel pins while removing last bolt. Remove transmission and clutch housing as an assembly.
9. Matchmark pressure plate and flywheel with paint to make sure correct balance is maintained.
10. Loosen the six cover plate attaching screws, a little at a time, until clutch diaphragm spring tension is released. Remove bolts, clutch assembly and pilot tool.

Installation—1964-70

1. The clutch bearing is an oil-impregnated type bearing pressed into the crankshaft. Inspect and renew, if necessary.
2. Install clutch disc with long hub forward (toward flywheel).
3. Install pressure plate and cover

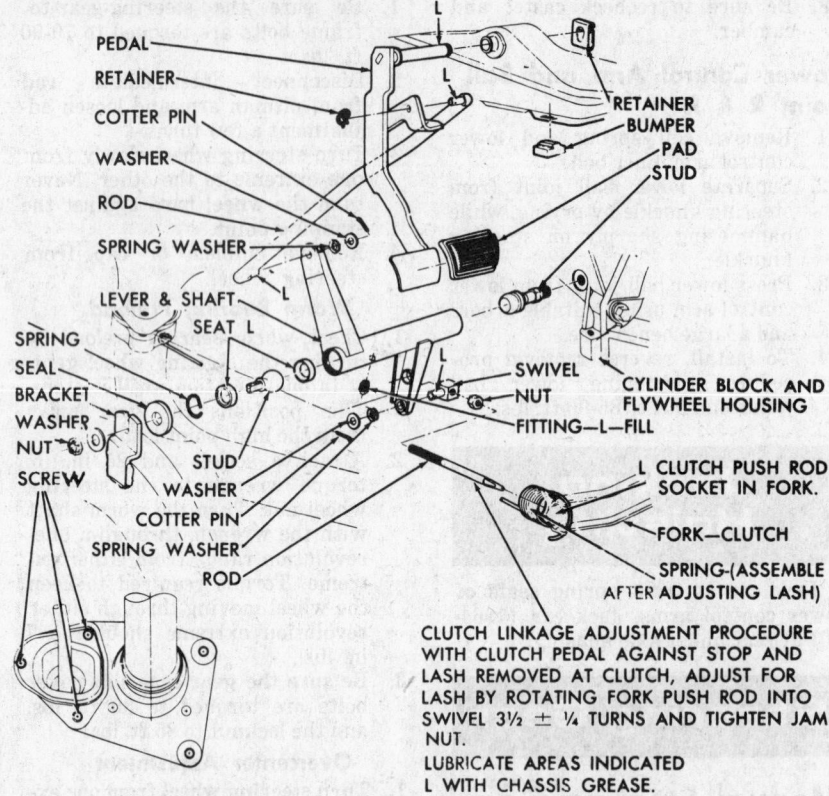

Clutch linkage—1964-70 (© Pontiac Div. G.M. Corp.)

CLUTCH LINKAGE ADJUSTMENT PROCEDURE WITH CLUTCH PEDAL AGAINST STOP AND LASH REMOVED AT CLUTCH, ADJUST FOR LASH BY ROTATING FORK PUSH ROD INTO SWIVEL 3½ ± ¼ TURNS AND TIGHTEN JAM NUT.
LUBRICATE AREAS INDICATED L WITH CHASSIS GREASE.

assembly, then align clutch disc by inserting pilot tool, or old transmission mainshaft, into splines. Align mark on clutch cover with mark on flywheel, then align nearest bolt holes.
4. Install bolts in every other hole in cover and tighten alternately. Then, install remaining three

bolts, tighten all six to 25 ft. lbs.
5. Remove clutch pilot tool and check to see that it can be reinserted and moved freely.
6. Install clutch fork and dust boot into clutch housing. Lubricate throwout bearing with graphite grease.

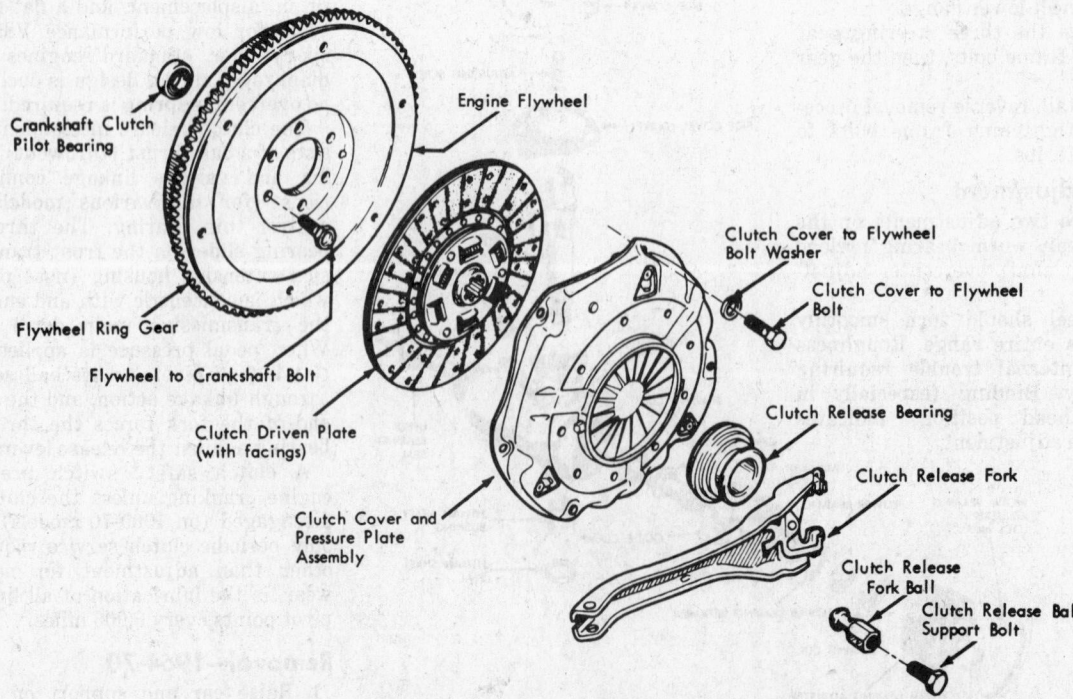

Typical diaphragm spring clutch assembly—1964-70 (© Pontiac Div. G.M. Corp.)

7. Complete the reassembly of clutch housing and transmission by reversing removal method. Tighten housing bolts to 40 ft. lbs.

8. Adjust shifter and clutch release linkage.

Clutch Pedal Adjustment —1964-70

1. Disconnect return spring.
2. With pedal against stop, loosen locknut to allow adjusting rod to be turned out of swivel (V8), or pushrod (6 cyl.), until the throwout bearing contacts the release fingers in the pressure plate.
3. Turn adjusting rod into swivel or pushrod 3½ turns; tighten locknut to 8-12 ft. lbs. for 1964-68, 30 ft. lbs. for 1969-70.
4. Install return spring and check pedal lash; it should be approximately 1 in.

Pilot Bearing and Flywheel Replacement—1964-70

1. Remove transmission and clutch.
2. Using a small cold chisel, remove stake marks which hold pilot bearing in the flywheel.
3. Pull the old bearing out of the flywheel, using a slide hammer if necessary.
4. With new bearing held in place, shielded side toward transmission, gently tap on bearing until

it enters the flywheel until flush. Stake in at least two places, using a prick punch.
5. If flywheel is removed, make sure to matchmark the flywheel and crankshaft flange.
6. To install, reverse removal procedure, tightening flywheel bolts to 95 ft. lbs.
   NOTE: flywheel bolts do not need lockwashers.

Transmissions

Three-Speed Manual Transmission

Linkage Adjustment—Column 1964

1. Place gearshift lever in Neutral.
2. Make sure levers at lower end of column are in Neutral.
3. Loosen two swivel nut assemblies and adjust until no binding exists with gearshift lever and lower selector levers in Neutral.
4. Tighten swivel nuts to 8-12 ft. lbs. and check shift pattern.
5. Lubricate all pivot points with chassis grease.

1965-66
See illustrations.

1967-68 Saginaw Transmission
See illustrations.

1969-70 Saginaw Transmission

1. Place gearshift lever in Reverse and lock ignition.
2. On Tempest, loosen swivel clamp bolt (C) at rear transmission shift lever (1st and Reverse) and bolt (D) at equalizer shaft and lever assembly. (See view B in illustration)
3. On Firebird, loosen swivel clamp nut (C) at rear transmission shift lever (1st and Reverse) as illustrated in view C in illustration, then loosen nut (D) at idler lever.
4. Position front transmission shift lever (2nd and 3rd) in Neutral and rear transmission shift lever (1st and Reverse) in Reverse.
5. Tighten swivel clamp bolt (C) or nut (C) to 20 ft. lbs., then unlock steering column and shift into Neutral.
6. Align lower gearshift levers (on column) (E and F) in Neutral position, then insert a 0.185 in. diameter gauge pin through hole in lower control levers.
7. Tighten swivel clamp bolt (D) or nut (D) to 20 ft. lbs., then remove gauge pin and check shift pattern.

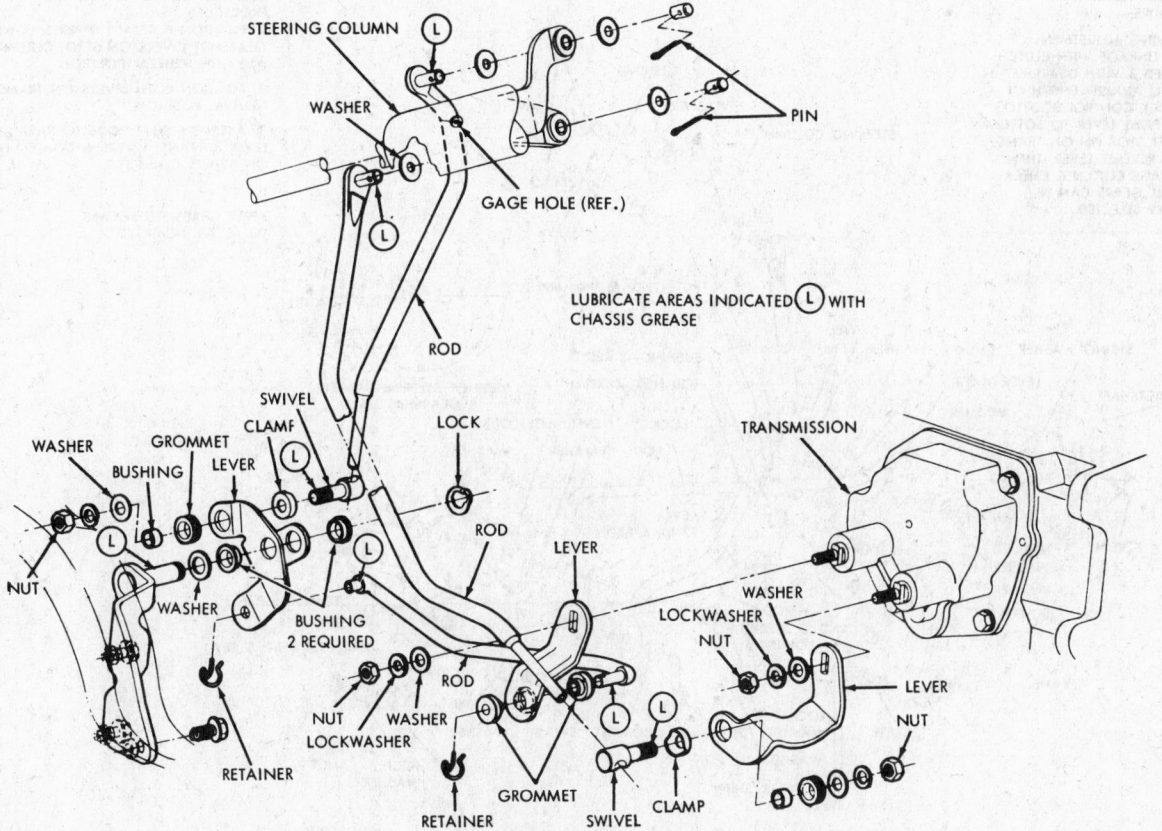

Gearshift column linkage—1964 V8 (© Pontiac Div. G.M. Corp.)

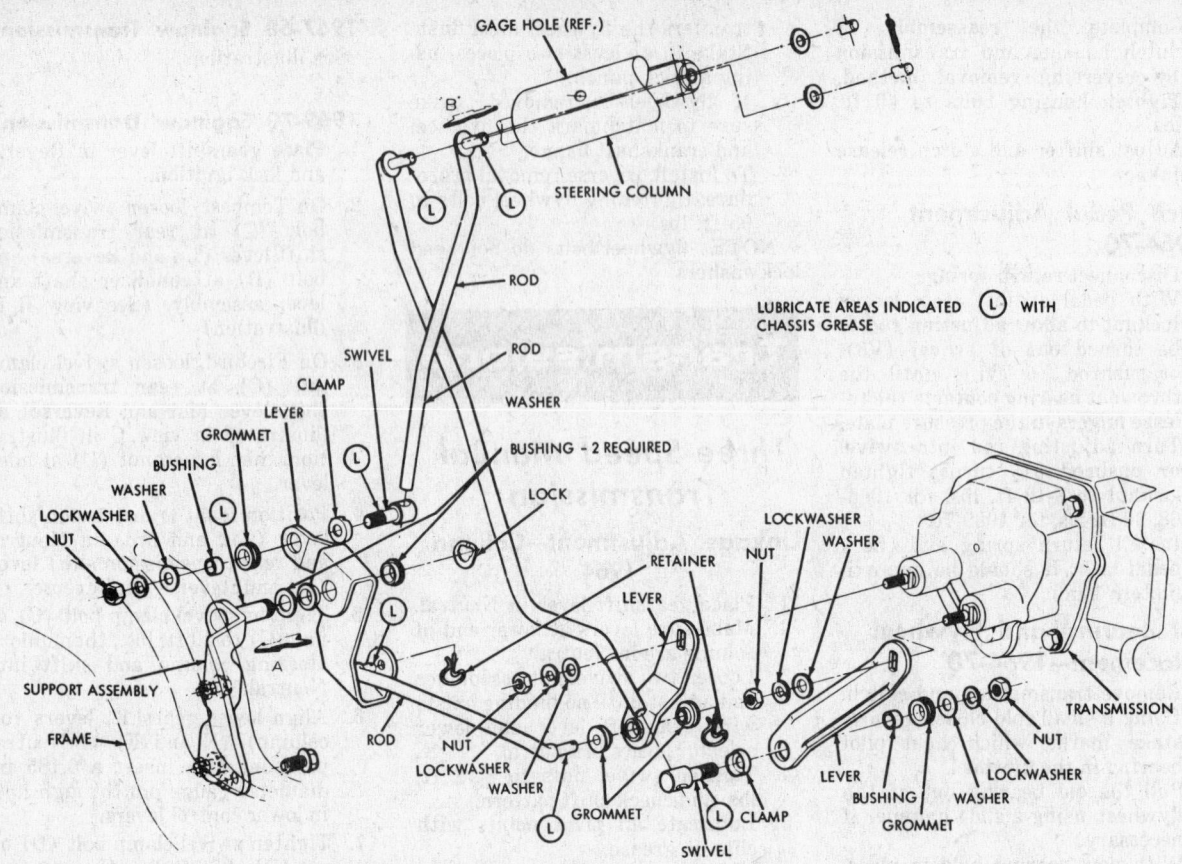

Gearshift column linkage—1964 6 cyl. (© Pontiac Div. G.M. Corp.)

INTERLOCK ADJUSTMENT PROCEDURE:—

FOLLOWING ADJUSTMENT OF CLUTCT LINKAGE WITH CLUTCH ENGAGED & WITH GEARSHIFT IN NEUTRAL, ADJUST LENGTH OF INTERLOCK CONTROL ROD TO PERMIT PAWL LEVER TO BOTTOM AGAINST STOP PIN ON TRANS. FIRST & REVERSE LEVER. THEN DISENGAGE CLUTCH & CHECK THAT ALL GEARS CAN BE PROPERLY SELECTED.

GEARSHIFT LEVERS ADJUSTMENT PROCEDURE:—

1. POSITION & RETAIN UPPER & LOWER GEARSHIFT LEVERS ON STRG. COLUMN ASS'Y. IN NEUTRAL POSITION.

2. POSITION BOTH LEVERS ON TRANS. IN NEUTRAL PUSITION.

3. ASSEMBLE SHIFT RODS TO IDLER LEVER & TRANS. LEVERS & TIGHTEN NUTS ON SWIVEL CLAMPS.

APPLY CHASSIS LUBRICANT TO AREAS INDICATED A

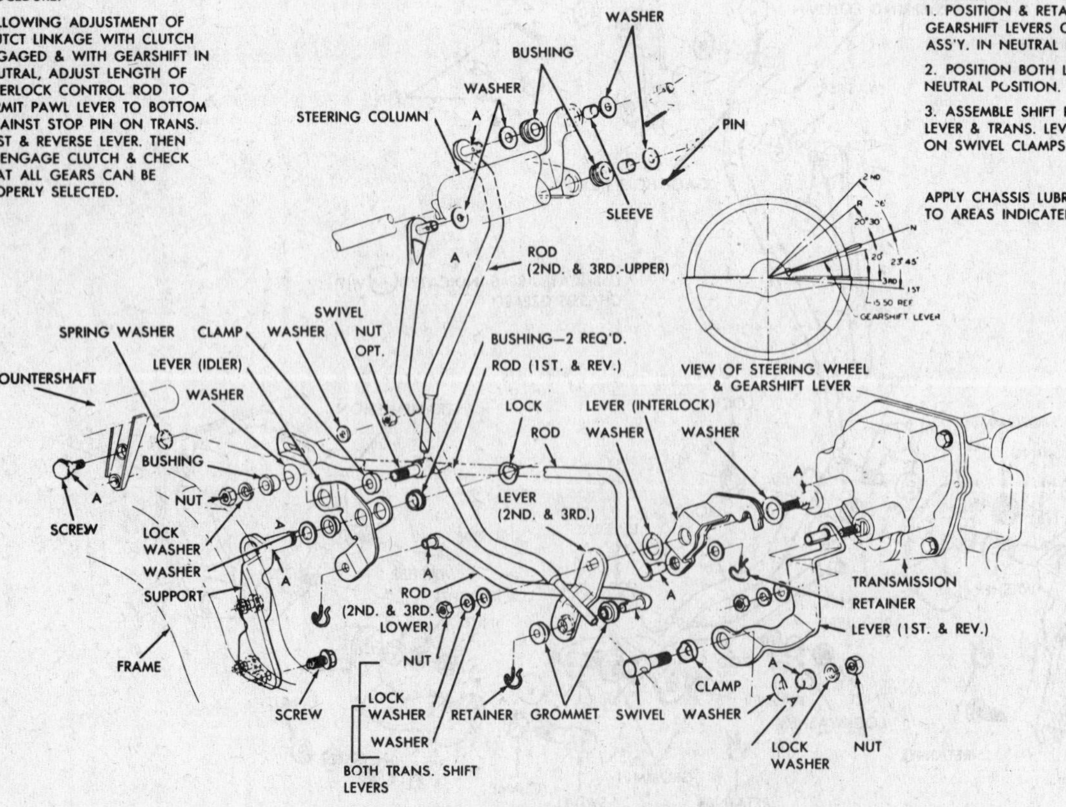

Gearshift column linkage—1965-66 V8 (© Pontiac Div. G.M. Corp.)

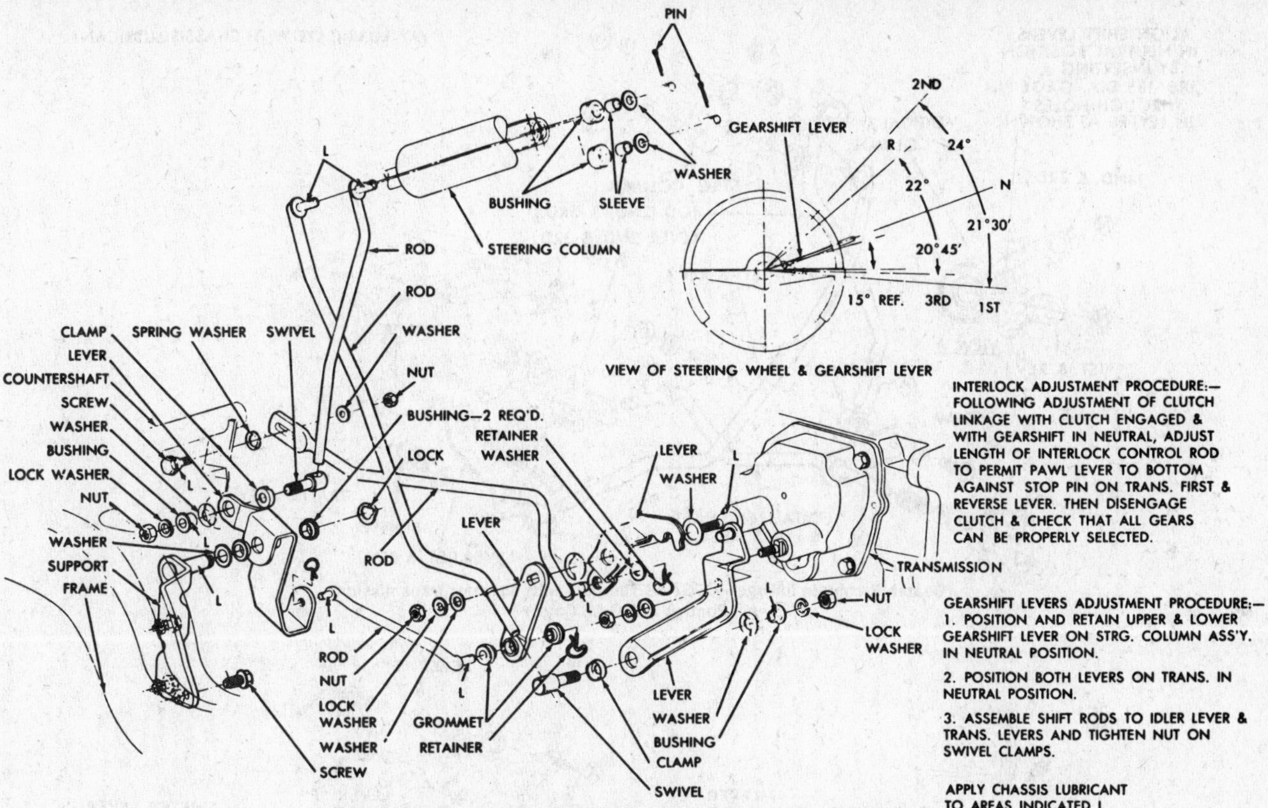

PIN

GEARSHIFT LEVER

WASHER

2ND

R 24°

22°

N

21°30'

20°45'

15° REF. 3RD 1ST

VIEW OF STEERING WHEEL & GEARSHIFT LEVER

L

BUSHING SLEEVE

STEERING COLUMN

ROD

ROD

WASHER

NUT

BUSHING—2 REQ'D.

RETAINER
WASHER

LOCK

LEVER

ROD

LEVER
WASHER

LEVER
WASHER

L

TRANSMISSION

NUT

LOCK
WASHER

ROD
NUT
LOCK
WASHER
WASHER
SCREW

GROMMET
RETAINER

LEVER
WASHER
BUSHING
CLAMP

SWIVEL

CLAMP
LEVER
COUNTERSHAFT
SCREW
WASHER
BUSHING
LOCK WASHER
NUT
WASHER
SUPPORT
FRAME

SPRING WASHER SWIVEL

L

INTERLOCK ADJUSTMENT PROCEDURE:—
FOLLOWING ADJUSTMENT OF CLUTCH
LINKAGE WITH CLUTCH ENGAGED &
WITH GEARSHIFT IN NEUTRAL, ADJUST
LENGTH OF INTERLOCK CONTROL ROD
TO PERMIT PAWL LEVER TO BOTTOM
AGAINST STOP PIN ON TRANS. FIRST &
REVERSE LEVER. THEN DISENGAGE
CLUTCH & CHECK THAT ALL GEARS
CAN BE PROPERLY SELECTED.

GEARSHIFT LEVERS ADJUSTMENT PROCEDURE:—
1. POSITION AND RETAIN UPPER & LOWER
GEARSHIFT LEVER ON STRG. COLUMN ASS'Y.
IN NEUTRAL POSITION.

2. POSITION BOTH LEVERS ON TRANS. IN
NEUTRAL POSITION.

3. ASSEMBLE SHIFT RODS TO IDLER LEVER &
TRANS. LEVERS AND TIGHTEN NUT ON
SWIVEL CLAMPS.

APPLY CHASSIS LUBRICANT
TO AREAS INDICATED L

Gearshift column linkage—1965-66 6 cyl. (© Pontiac Div. G.M. Corp.)

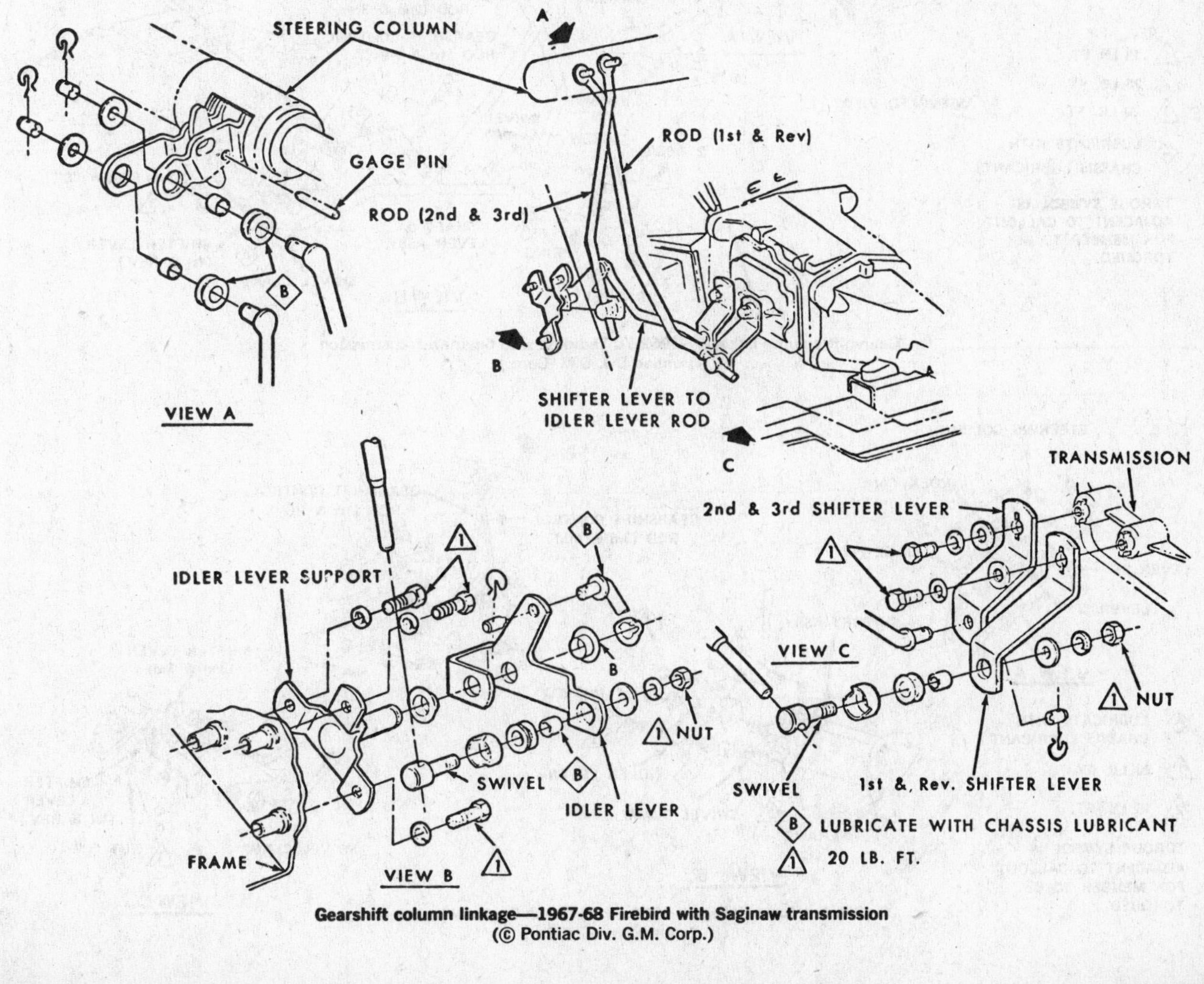

STEERING COLUMN

A

GAGE PIN

ROD (2nd & 3rd)

ROD (1st & Rev)

VIEW A

B

SHIFTER LEVER TO
IDLER LEVER ROD

C

TRANSMISSION

2nd & 3rd SHIFTER LEVER

A

B

IDLER LEVER SUPPORT

B

VIEW C

NUT

NUT

SWIVEL

B

IDLER LEVER

SWIVEL

1st &. Rev. SHIFTER LEVER

B LUBRICATE WITH CHASSIS LUBRICANT

FRAME

VIEW B

20 LB. FT.

Gearshift column linkage—1967-68 Firebird with Saginaw transmission
(© Pontiac Div. G.M. Corp.)

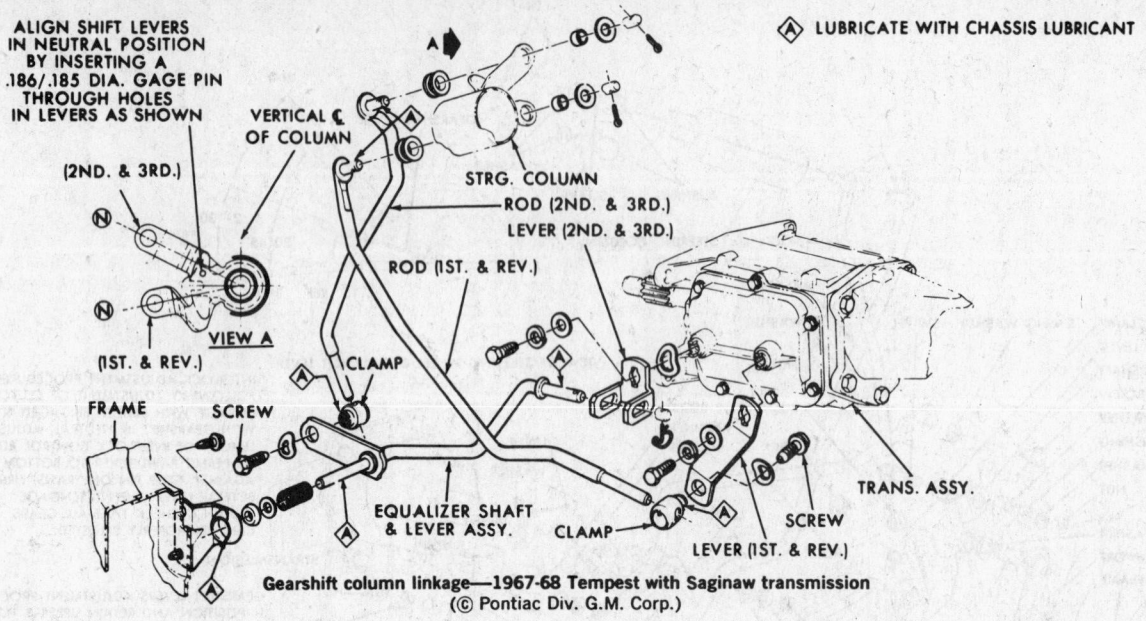

Gearshift column linkage—1967-68 Tempest with Saginaw transmission
(© Pontiac Div. G.M. Corp.)

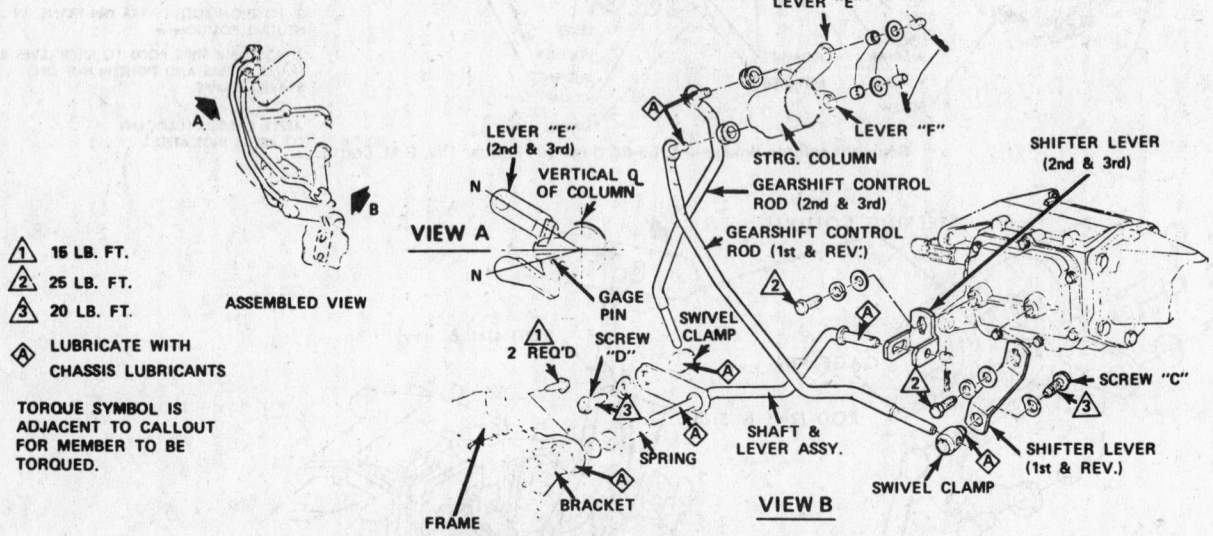

Gearshift column linkage—1969-70 Tempest with Saginaw transmission
(© Pontiac Div. G.M. Corp.)

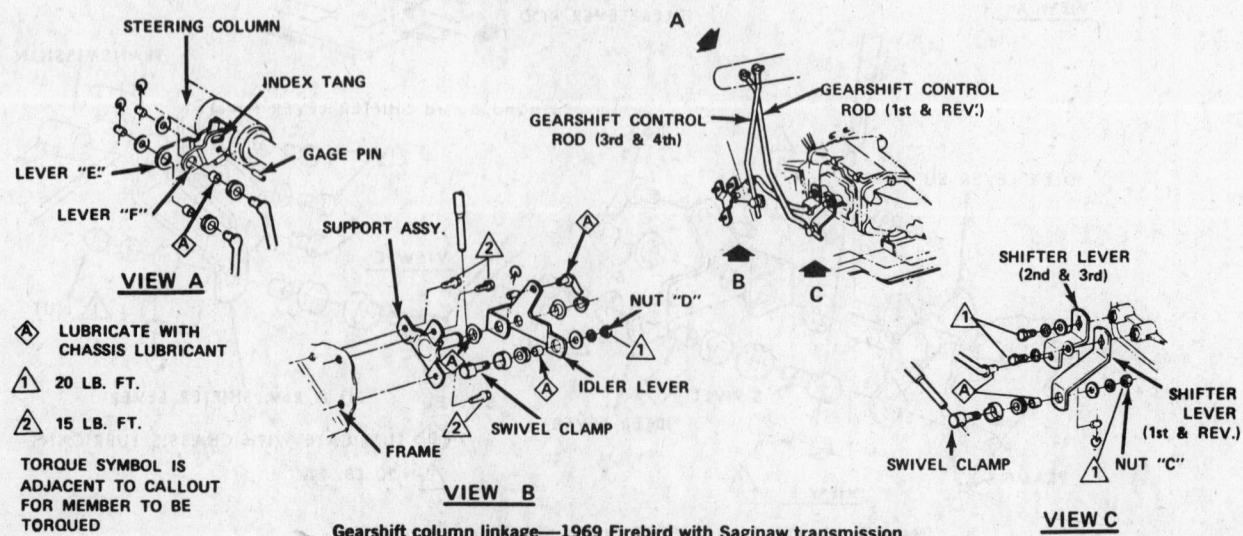

Gearshift column linkage—1969 Firebird with Saginaw transmission
(© Pontiac Div. G.M. Corp.)

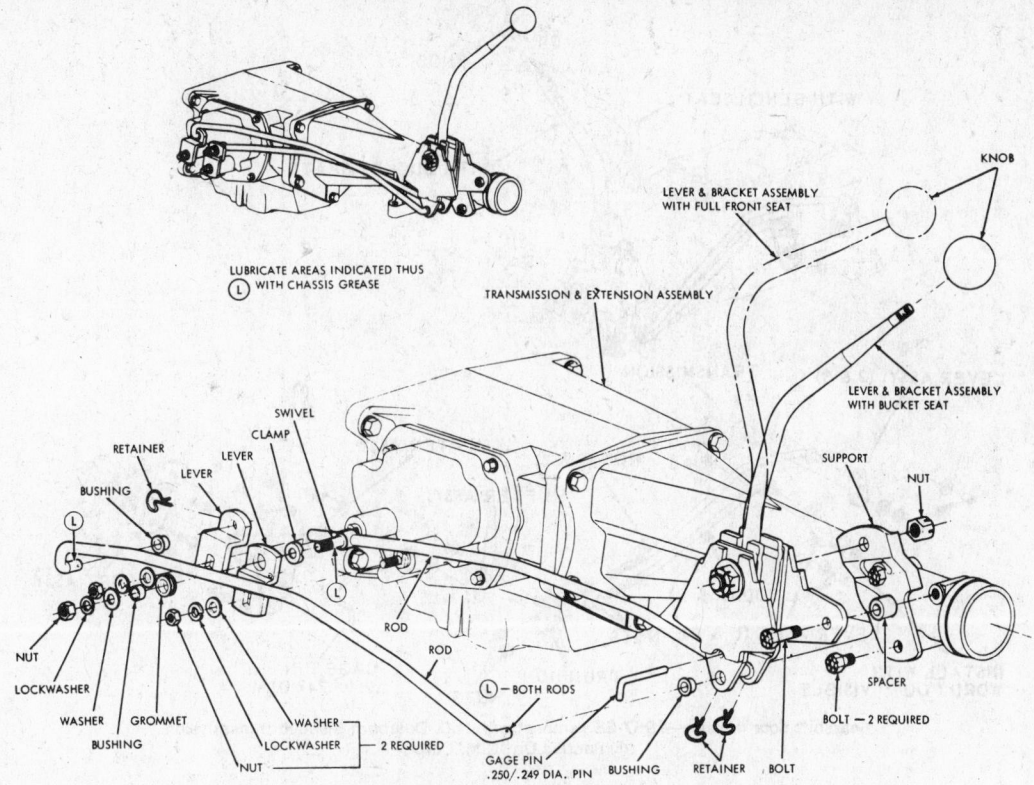

Gearshift floor linkage—1964 Tempest with 3-speed transmission
(© Pontiac Div. G.M. Corp.)

Linkage Adjustment—Floor

1964-66

1. Place gearshift lever in Neutral.
2. Loosen two swivel nut assemblies.
3. Insert a 1/4 in. drill rod into bracket and lever assembly and align shift levers in Neutral position.
4. Position transmission shift levers in Neutral position.
5. Tighten swivel nut assemblies to 8-12 ft. lbs.
6. Remove gauge pin and check shift pattern.

1967-68 H.D. Dearborn Transmission

See illustrations. Procedure same as 1964-66 except that locknuts must be tightened to 30 ft. lbs.

SHIFT CONTROL ADJUSTMENT

1. WITH CONSOLE INSTALLATION: INSERT GAGE PIN INTO LEVER & BRACKET ASSEMBLY. ADJUST LEVER & BRACKET ASSEMBLY AS SHOWN IN VIEW "A". TIGHTEN ATTACHING BOLTS TO SPECIFIED TORQUE.

2. EXCEPT CONSOLE INSTALLATION: INSERT GAGE PIN INTO LEVER & BRACKET ASSEMBLY AND ADJUST ASSEMBLY CENTRALLY IN SLOT. TIGHTEN ATTACHING BOLTS TO SPECIFIED TORQUE.

3. POSITION BOTH LEVERS ON TRANSMISSION IN NEUTRAL POSITION WITH PIN IN PLACE IN LEVER & BRACKET ASSEMBLY. ASSEMBLE SHIFT RODS TO LEVER & BRACKET ASSEMBLY CONTROL LEVERS. TIGHTEN JAM NUTS AT ROD & TRUNNION ASSEMBLIES TO SPECIFIED TORQUE.

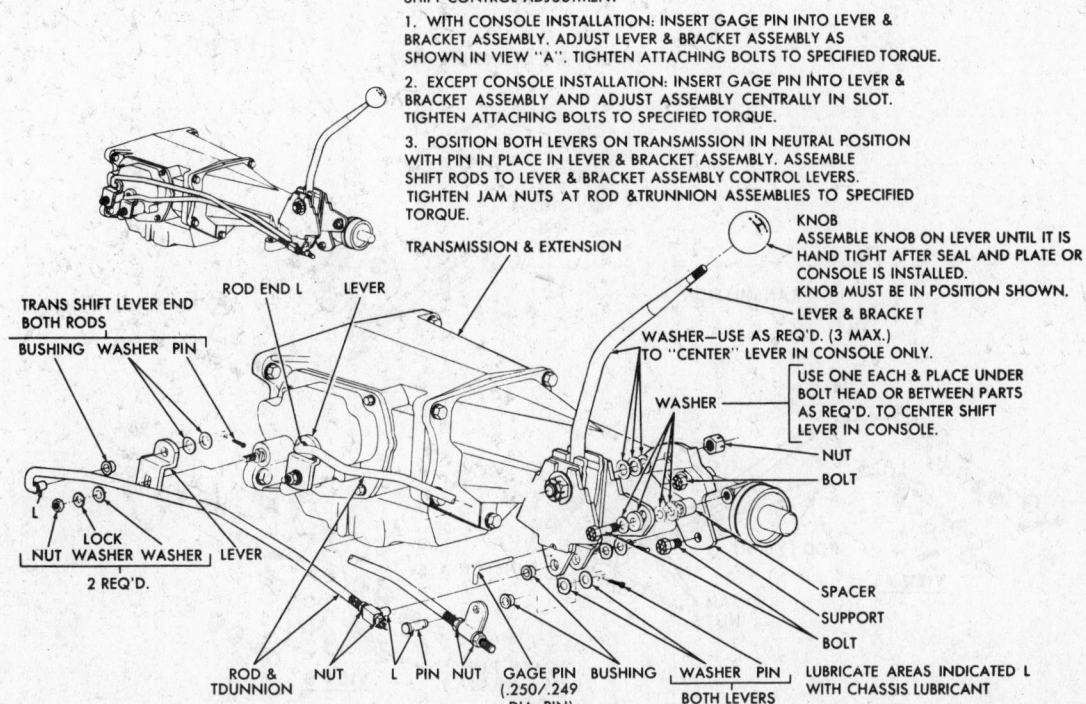

Gearshift floor linkage—1964-66 Tempest 3-speed transmission (© Pontiac Div. G.M. Corp.)

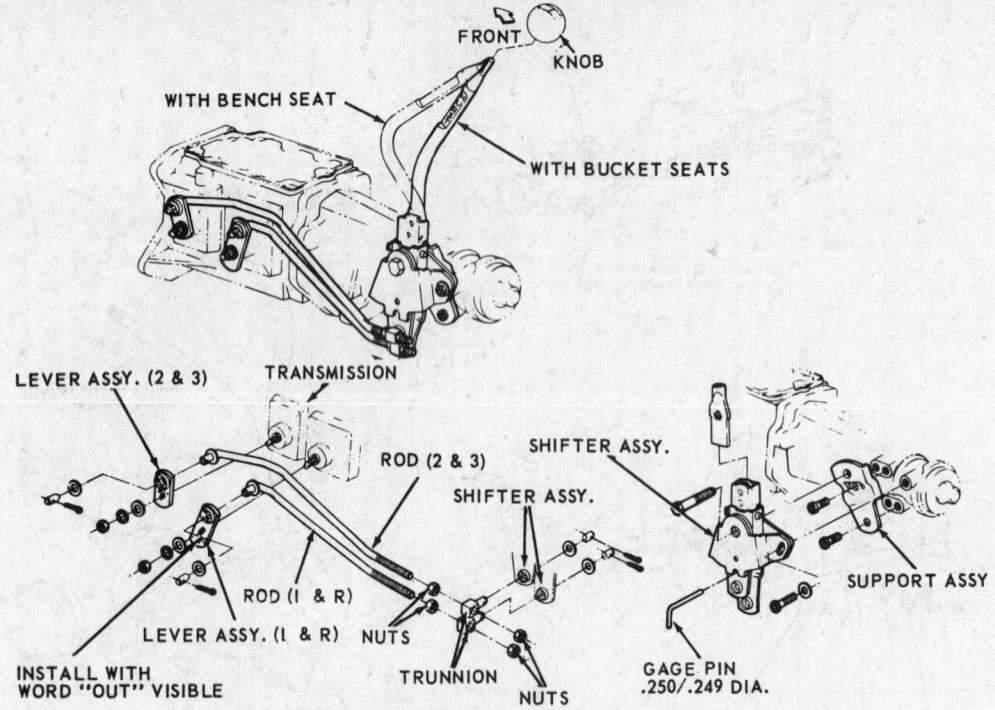

Gearshift floor linkage—1967-68 Tempest with H.D. Dearborn 3-speed transmission
(© Pontiac Div. G.M. Corp.)

1967-68 Saginaw Transmission

See illustrations. Procedure same as 1964-66 except that locknuts must be tightened to 30 ft. lbs.

1969 Saginaw and H.D. Dearborn Transmission

1. Place gearshift lever in Neutral.
2. Loosen swivel clamp on gearshift control rod (see Back Drive Link-age illustrations for Tempest and Firebird variations).
3. Loosen trunnion locknuts on 1st-Reverse and 2nd-3rd transmission control rods.
4. Insert a 1/4 in. drill rod into shifter assembly (view B in floor-shift linkage illustrations for Firebird and Tempest).
5. If gearshift lever is not properly aligned with floor opening (view C):
   a. *Console*—loosen two shifter to support bolts and align shift-er as in view C in illustration; tighten bolts.
   b. *Without console*—loosen two shifter to support bolts and center shifter in boot; tighten bolts.

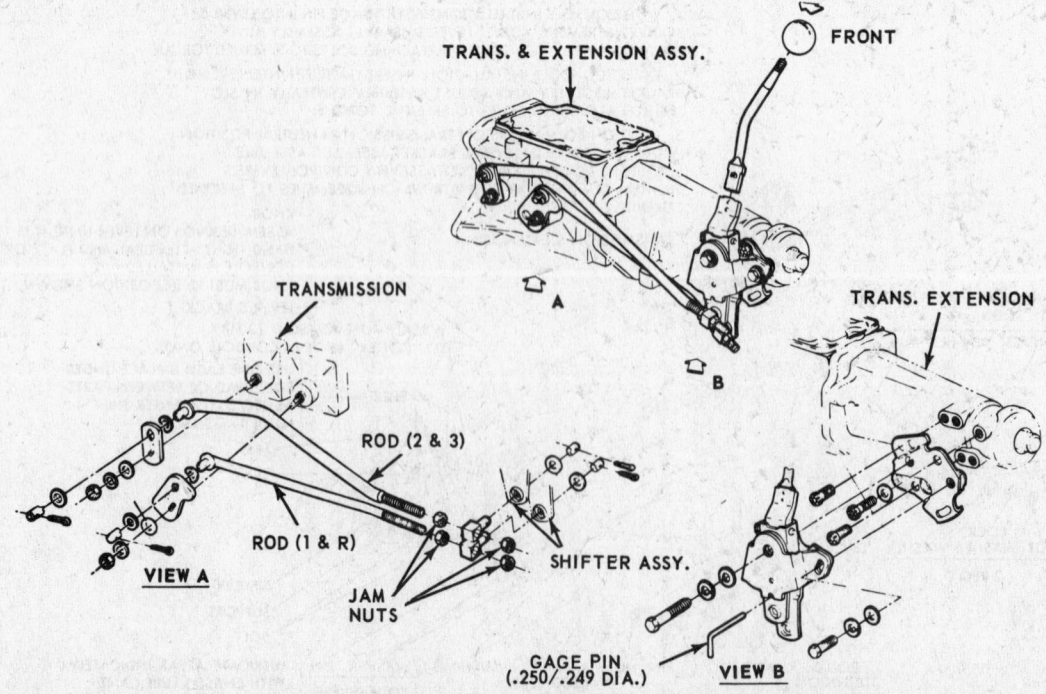

Gearshift floor linkage—1967-68 Firebird with H.D. Dearborn 3-speed transmission
(© Pontiac Div. G.M. Corp.)

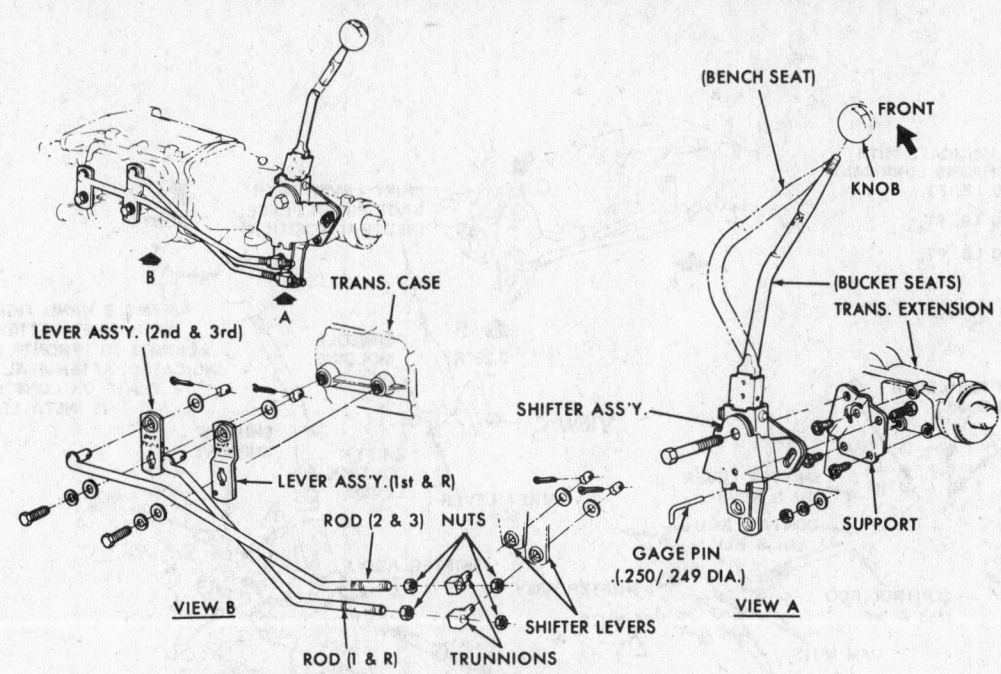

Gearshift floor linkage—1967-68 Tempest with Saginaw 3-speed transmission
(© Pontiac Div. G.M. Corp.)

6. Position both transmission shift levers in Neutral and tighten locknuts to 30 ft. lbs.
7. Remove gauge pin and check shift pattern.
8. Place gearshift lever in Reverse, then place steering column lower

lever in Lock position and lock ignition.
9. Push up on gearshift control rod to take up lash in column lock mechanism, then tighten adjusting swivel clamp to 20 ft. lbs.

1970 H.D. Muncie Transmission

See illustrations. Procedure is the same as 1969 Saginaw and H.D. Dearborn transmission linkage adjustment.
NOTE: the 1970 Saginaw transmission is not available with floorshift.

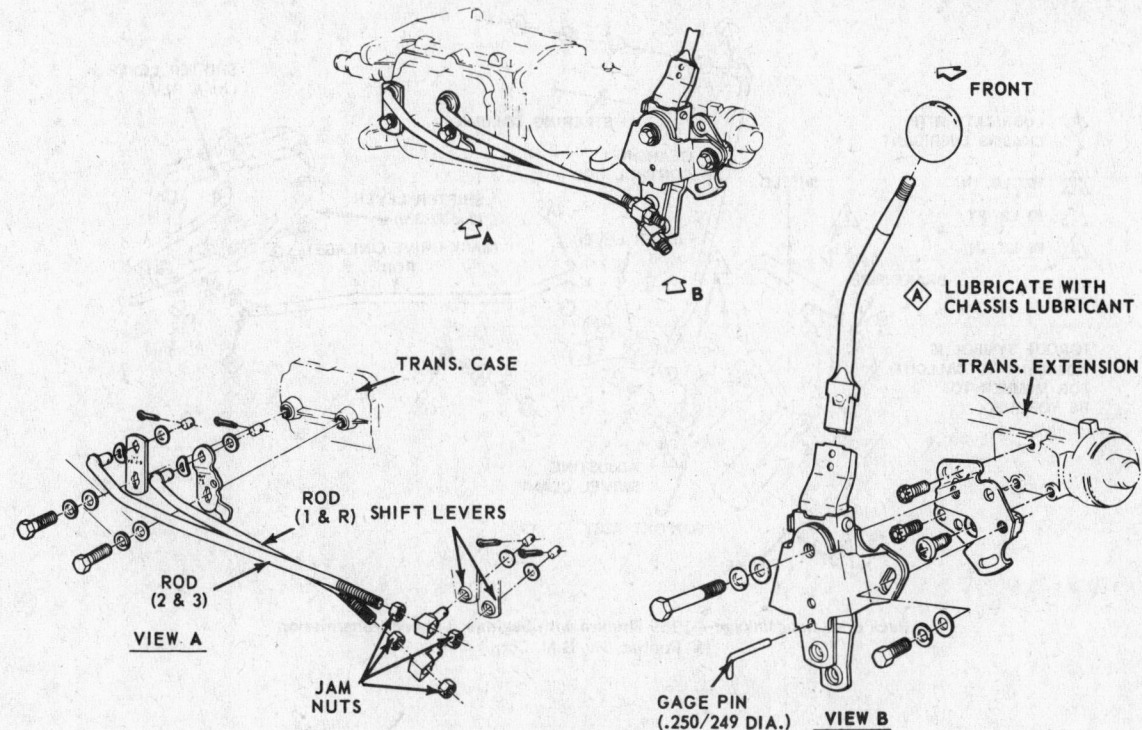

Gearshift floor linkage—1967-68 Firebird with Saginaw 3-speed transmission
(© Pontiac Div. G.M. Corp.)

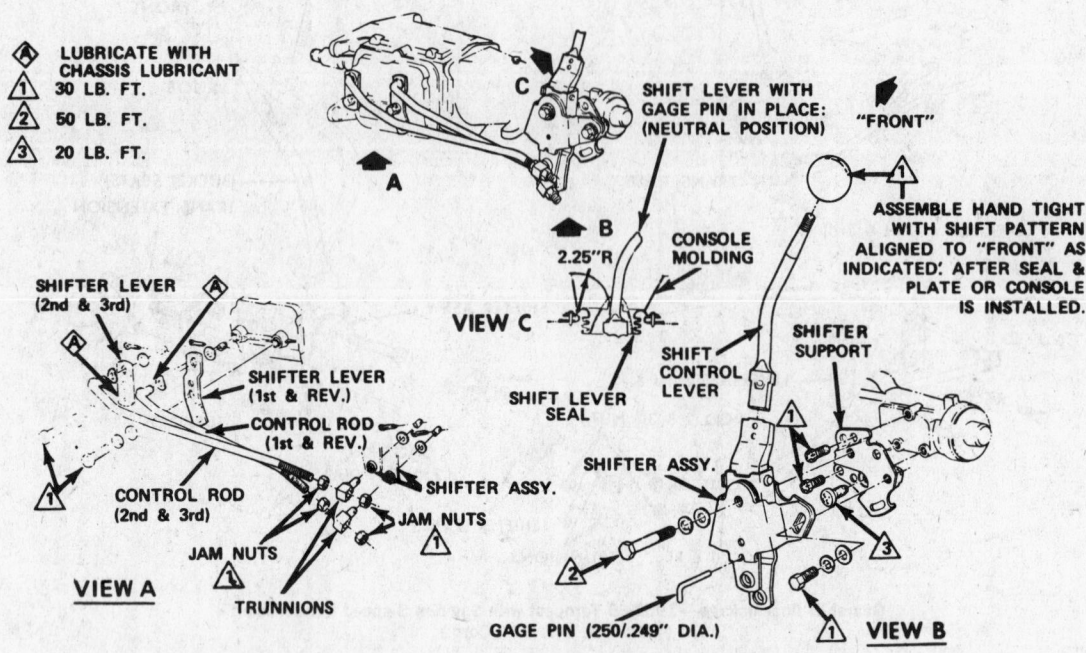

LUBRICATE WITH CHASSIS LUBRICANT
1 30 LB. FT.
2 50 LB. FT.
3 20 LB. FT.

SHIFT LEVER WITH GAGE PIN IN PLACE: (NEUTRAL POSITION)

"FRONT"

ASSEMBLE HAND TIGHT WITH SHIFT PATTERN ALIGNED TO "FRONT" AS INDICATED. AFTER SEAL & PLATE OR CONSOLE IS INSTALLED.

CONSOLE MOLDING

2.25"R

VIEW C

SHIFT LEVER SEAL

SHIFT CONTROL LEVER

SHIFTER SUPPORT

SHIFTER LEVER (2nd & 3rd)

SHIFTER LEVER (1st & REV.)

CONTROL ROD (1st & REV.)

SHIFTER ASSY.

CONTROL ROD (2nd & 3rd)

SHIFTER ASSY.

JAM NUTS

JAM NUTS

TRUNNIONS

GAGE PIN (250/.249" DIA.)

VIEW B

VIEW A

Gearshift floor linkage—1969 Firebird with Saginaw 3-speed transmission
(© Pontiac Div. G.M. Corp.)

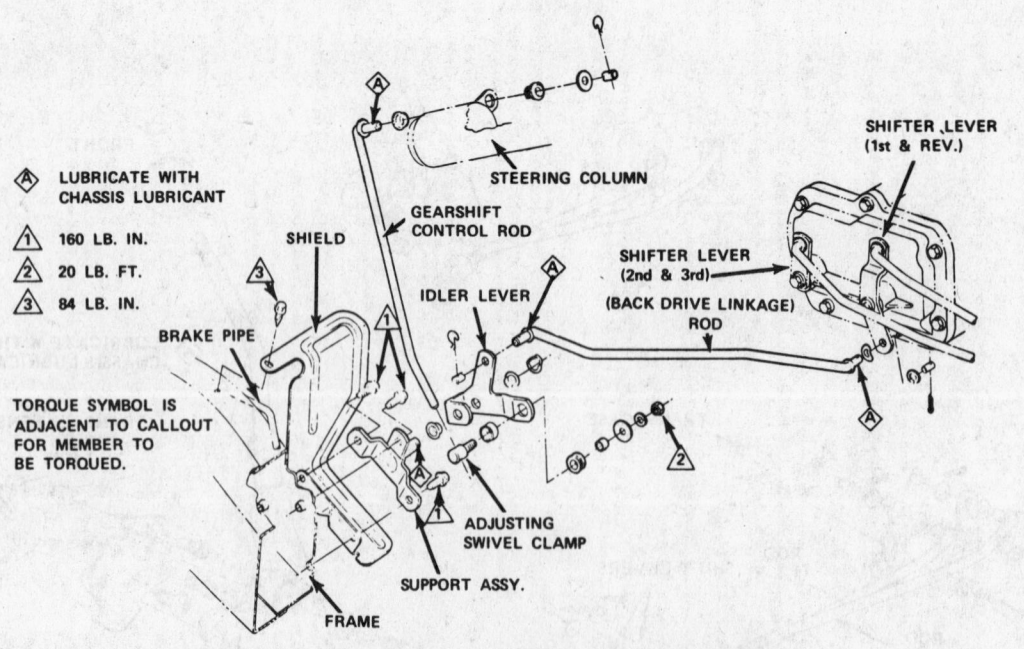

LUBRICATE WITH CHASSIS LUBRICANT
1 160 LB. IN.
2 20 LB. FT.
3 84 LB. IN.

STEERING COLUMN

SHIFTER LEVER (1st & REV.)

SHIELD

GEARSHIFT CONTROL ROD

SHIFTER LEVER (2nd & 3rd)

(BACK DRIVE LINKAGE) ROD

BRAKE PIPE

IDLER LEVER

TORQUE SYMBOL IS ADJACENT TO CALLOUT FOR MEMBER TO BE TORQUED.

ADJUSTING SWIVEL CLAMP

SUPPORT ASSY.

FRAME

Back drive floor linkage—1969 Firebird with Saginaw 3-speed transmission
(© Pontiac Div. G.M. Corp.)

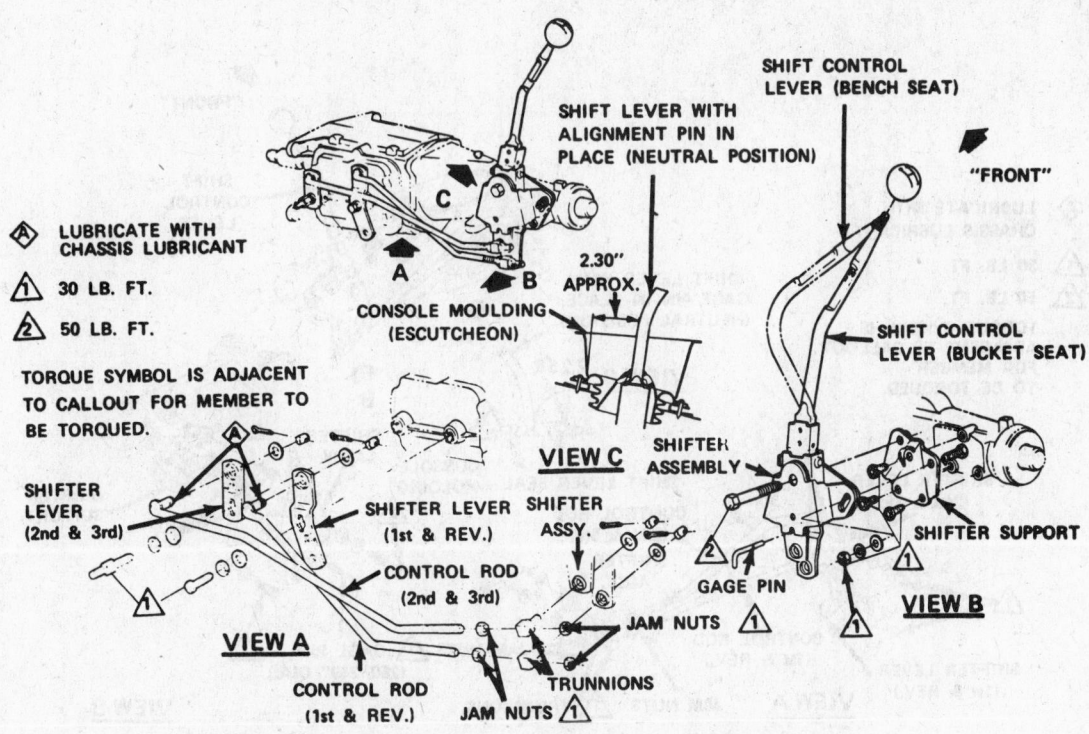

Gearshift floor linkage—1969 Tempest with Saginaw 3-speed transmission
(© Pontiac Div. G.M. Corp.)

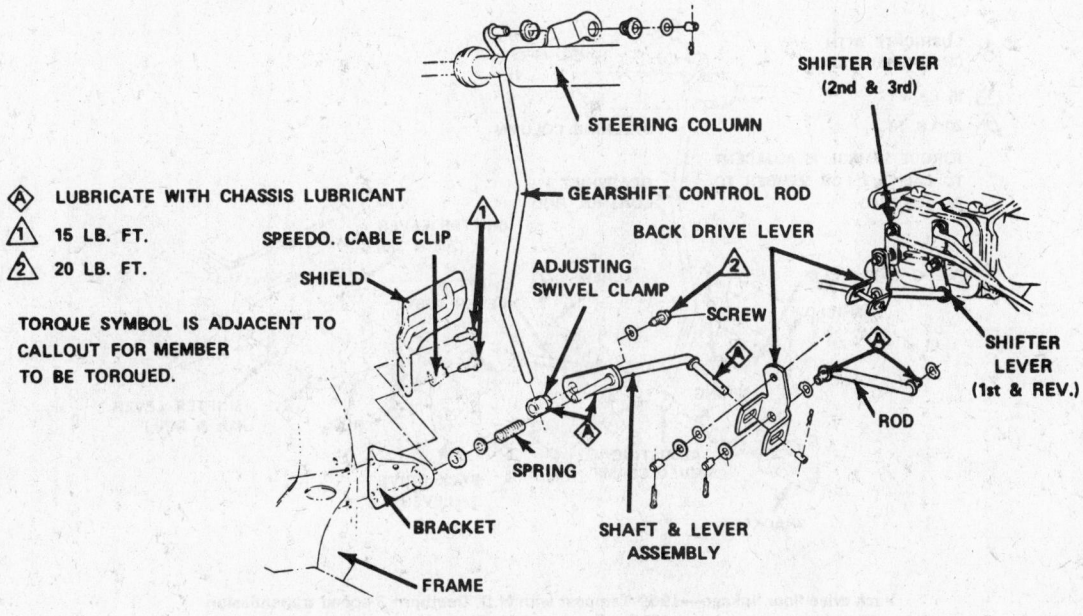

Back drive floor linkage—1969 Tempest with Saginaw 3-speed transmission
(© Pontiac Div. G.M. Corp.)

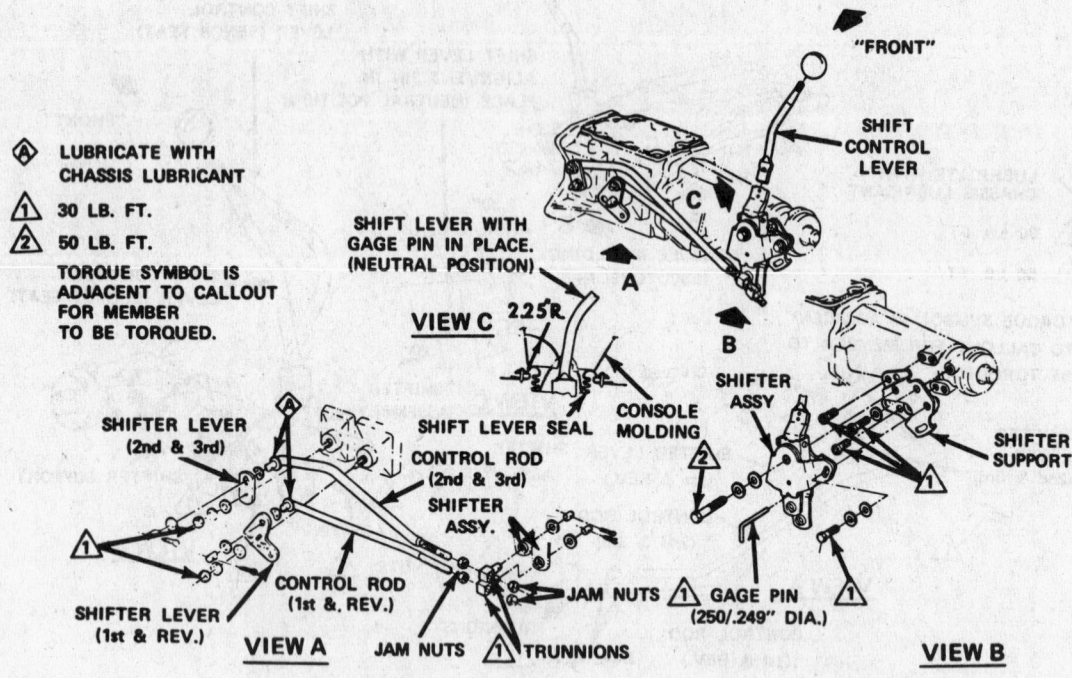

LUBRICATE WITH
CHASSIS LUBRICANT

1 30 LB. FT.

2 50 LB. FT.

TORQUE SYMBOL IS
ADJACENT TO CALLOUT
FOR MEMBER
TO BE TORQUED.

"FRONT"

SHIFT
CONTROL
LEVER

SHIFT LEVER WITH
GAGE PIN IN PLACE.
(NEUTRAL POSITION)

VIEW C   2.25R

CONSOLE
MOLDING

SHIFT LEVER SEAL

SHIFTER
ASSY

SHIFTER LEVER
(2nd & 3rd)

CONTROL ROD
(2nd & 3rd)

SHIFTER
ASSY.

SHIFTER
SUPPORT

CONTROL ROD
(1st &. REV.)

JAM NUTS

1 GAGE PIN
(250/.249" DIA.)

SHIFTER LEVER
(1st & REV.)

VIEW A   JAM NUTS   1 TRUNNIONS

VIEW B

Gearshift floor linkage—1969 Tempest with H.D. Dearborn 3-speed transmission
(© Pontiac Div. G.M. Corp.)

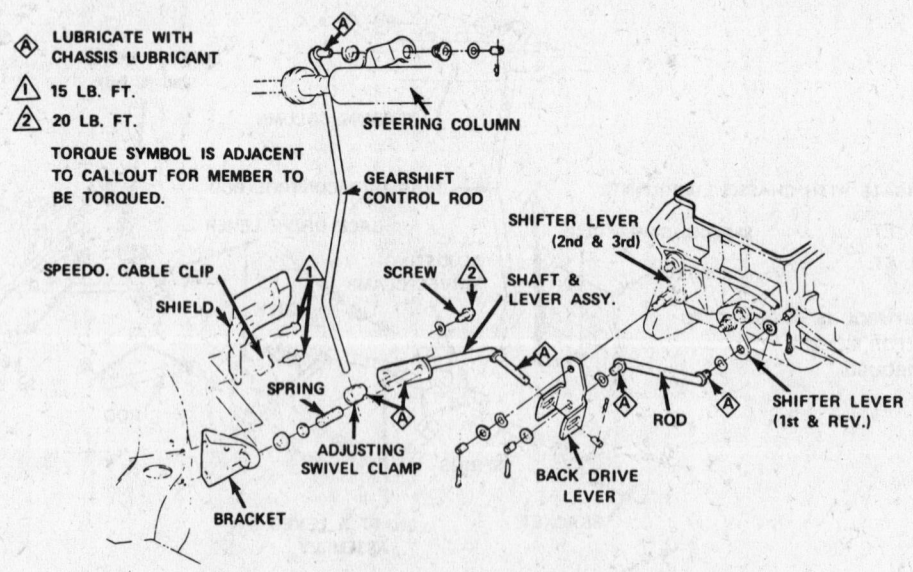

LUBRICATE WITH
CHASSIS LUBRICANT

1 15 LB. FT.

2 20 LB. FT.

TORQUE SYMBOL IS ADJACENT
TO CALLOUT FOR MEMBER TO
BE TORQUED.

STEERING COLUMN

GEARSHIFT
CONTROL ROD

SPEEDO. CABLE CLIP

SHIELD

SCREW

SHAFT &
LEVER ASSY.

SHIFTER LEVER
(2nd & 3rd)

SPRING

ADJUSTING
SWIVEL CLAMP

BACK DRIVE
LEVER

ROD

SHIFTER LEVER
(1st & REV.)

BRACKET

Back drive floor linkage—1969 Tempest with H.D. Dearborn 3-speed transmission
(© Pontiac Div. G.M. Corp.)

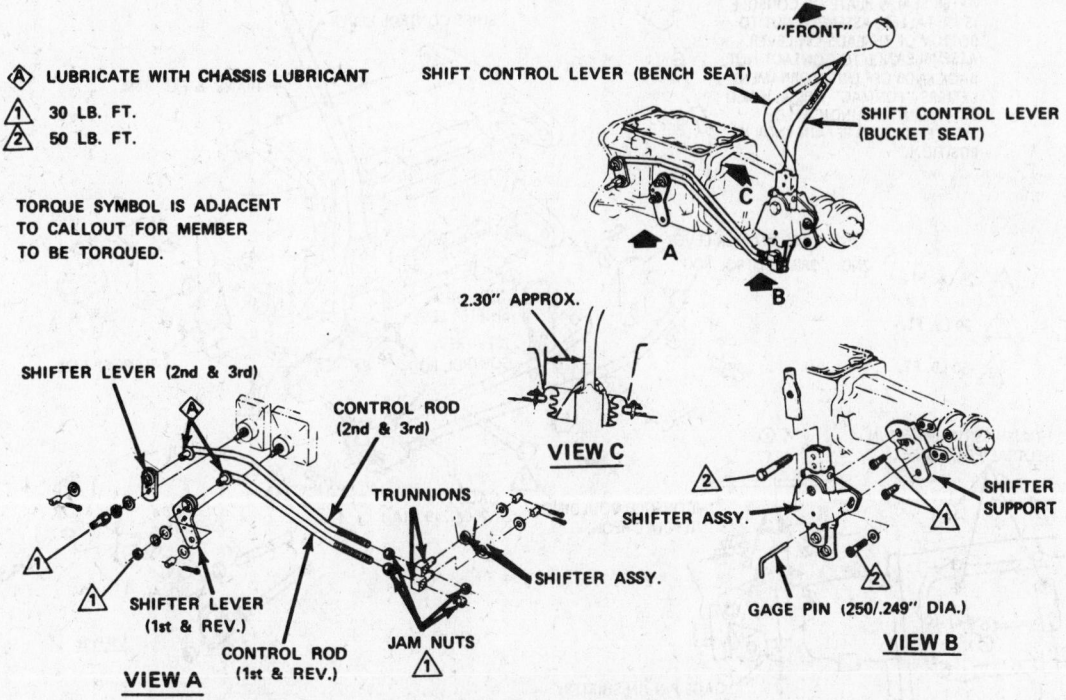

Gearshift floor linkage—1969 Firebird with H.D. Dearborn 3-speed transmission
(© Pontiac Div. G.M. Corp.)

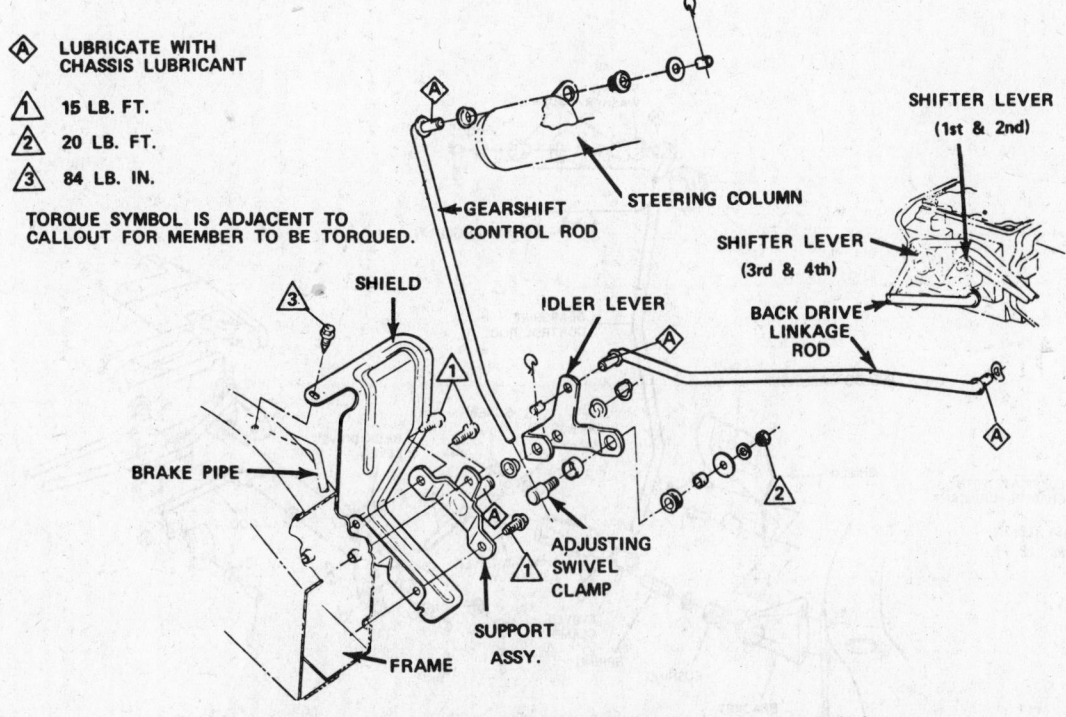

Back drive floor linkage—1969 Firebird with H.D. Dearborn 3-speed transmission
(© Pontiac Div. G.M. Corp.)

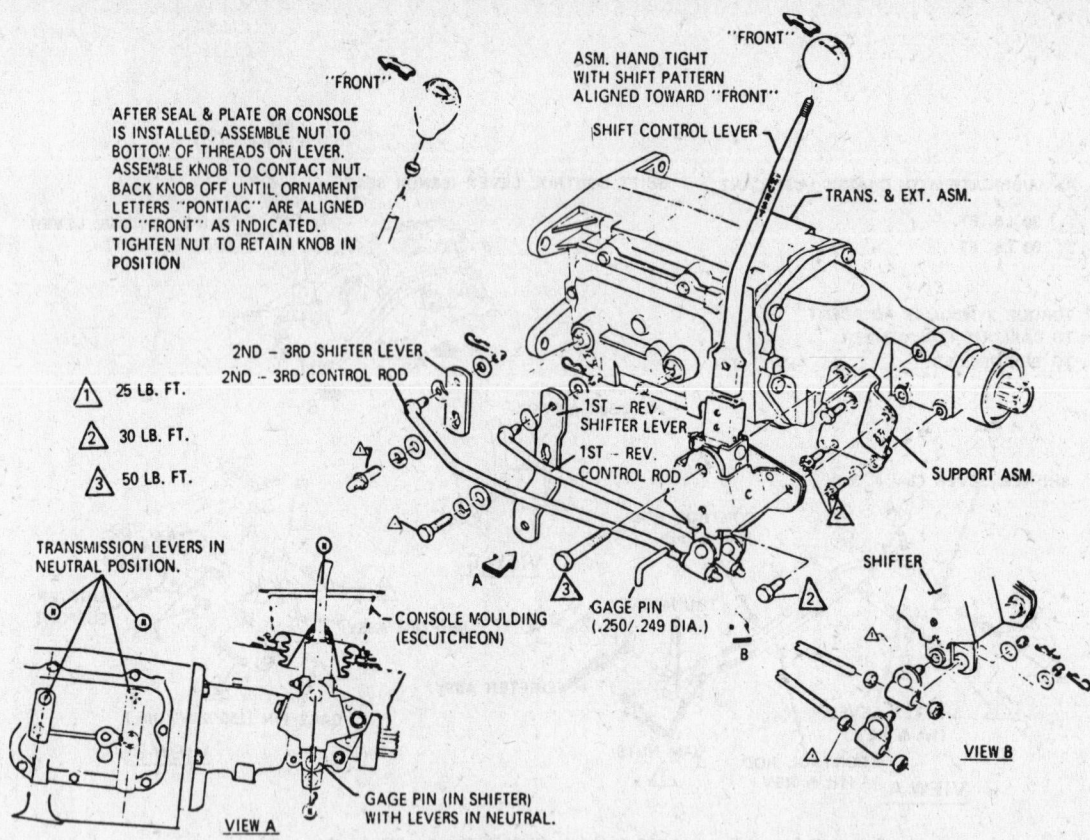

AFTER SEAL & PLATE OR CONSOLE IS INSTALLED, ASSEMBLE NUT TO BOTTOM OF THREADS ON LEVER. ASSEMBLE KNOB TO CONTACT NUT. BACK KNOB OFF UNTIL ORNAMENT LETTERS "PONTIAC" ARE ALIGNED TO "FRONT" AS INDICATED. TIGHTEN NUT TO RETAIN KNOB IN POSITION

"FRONT"

"FRONT"

ASM. HAND TIGHT WITH SHIFT PATTERN ALIGNED TOWARD "FRONT"

SHIFT CONTROL LEVER

TRANS. & EXT. ASM.

2ND – 3RD SHIFTER LEVER
2ND – 3RD CONTROL ROD

1ST – REV. SHIFTER LEVER
1ST – REV. CONTROL ROD

SUPPORT ASM.

⚠1 25 LB. FT.
⚠2 30 LB. FT.
⚠3 50 LB. FT.

TRANSMISSION LEVERS IN NEUTRAL POSITION.

SHIFTER

CONSOLE MOULDING (ESCUTCHEON)

GAGE PIN (.250/.249 DIA.)

VIEW B

GAGE PIN (IN SHIFTER) WITH LEVERS IN NEUTRAL.

VIEW A

Gearshift floor linkage—1970 with H.D. Muncie 3-speed transmission
(© Pontiac Div. G.M. Corp.)

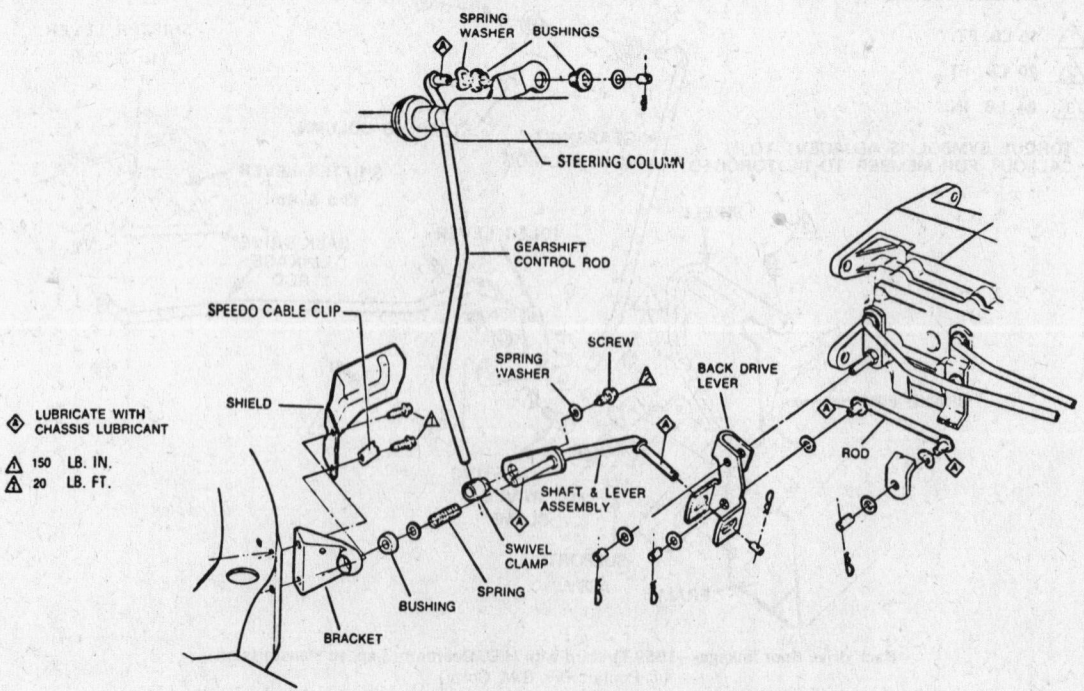

SPRING WASHER   BUSHINGS

STEERING COLUMN

GEARSHIFT CONTROL ROD

SPEEDO CABLE CLIP

SHIELD

SPRING WASHER

SCREW

BACK DRIVE LEVER

ROD

SHAFT & LEVER ASSEMBLY

◇ LUBRICATE WITH CHASSIS LUBRICANT
⚠ 150 LB. IN.
⚠ 20 LB. FT.

SWIVEL CLAMP

BUSHING   SPRING

BRACKET

Back drive floor linkage—1970 with H.D. Muncie 3-speed transmission
(© Pontiac Div. G.M. Corp.)

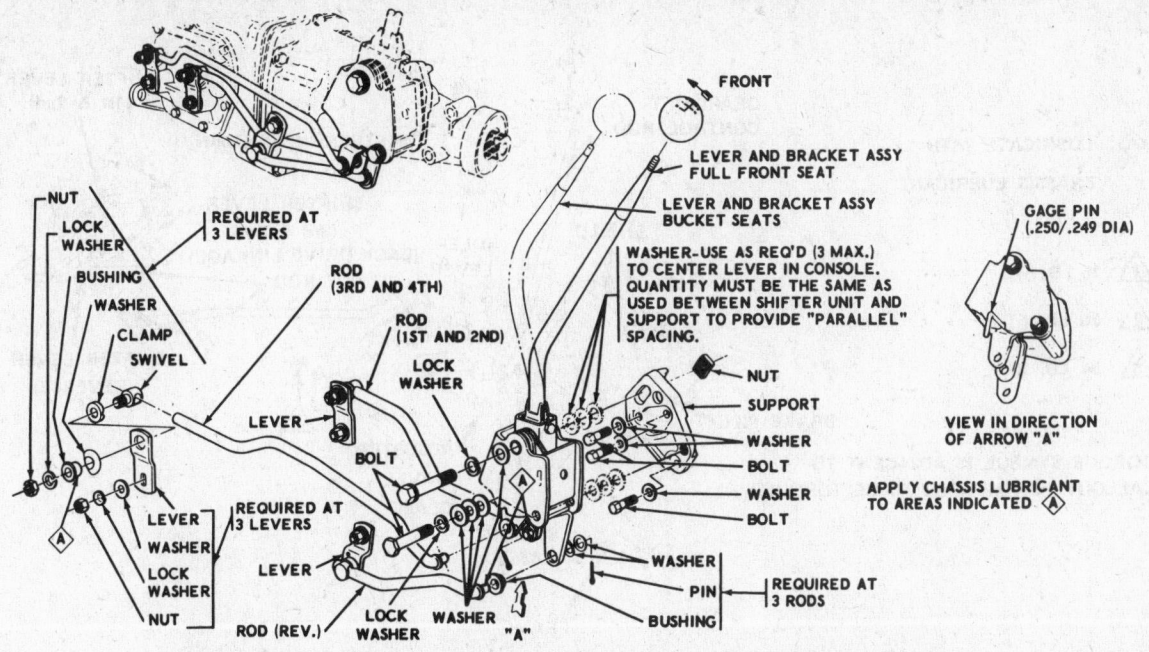

Gearshift floor linkage—typical 1964-66 4-speed transmission (© Pontiac Div. G.M. Corp.)

Four-Speed Manual Transmission

Linkage Adjustment—Floor

1964-68 All

1. Place gearshift lever in Neutral.
2. Loosen three swivel nut assemblies.
3. Insert a 1/4 in. drill rod into gauge pin hole in shifter.

A   LUBRICATE WITH CHASSIS LUBRICANT

△1 30 LB. FT.
△2 50 LB. FT.
△3 20 LB. FT.

TORQUE SYMBOL IS ADJACENT TO CALLOUT FOR MEMBER TO BE TORQUED.

4. Position transmission shift levers in Neutral.
5. Install swivel assemblies, adjusting length so that they fit into transmission shift levers without binding; tighten swivel nuts to 8-12 ft. lbs. up to 1966; 30 ft. lbs. for 1967-68.
6. Remove gauge pin and check shift pattern.

1969-70 Muncie and 1969 Saginaw Transmission

1. Place gearshift lever in Neutral.
2. Loosen adjusting swivel clamp on gearshift control rod.

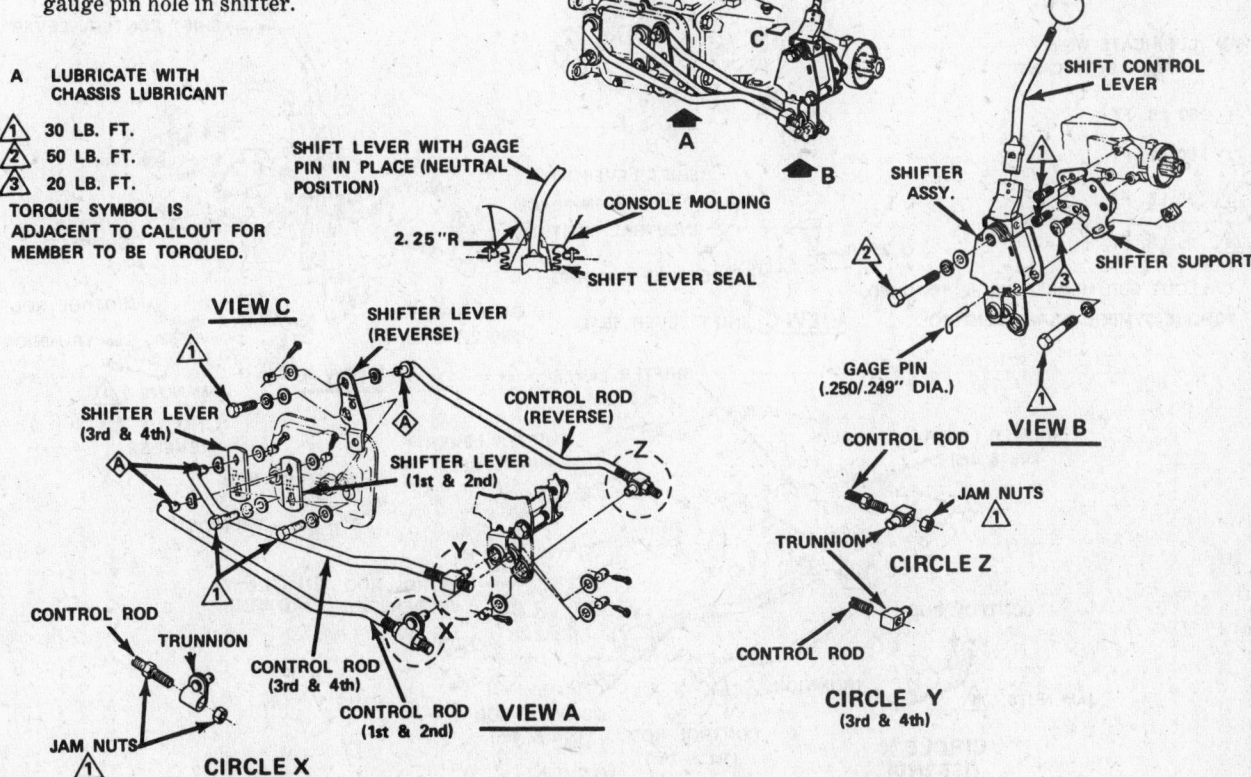

Gearshift floor linkage—typical 1967-69 Firebird with Saginaw 4-speed transmission (© Pontiac Div. G.M. Corp.)

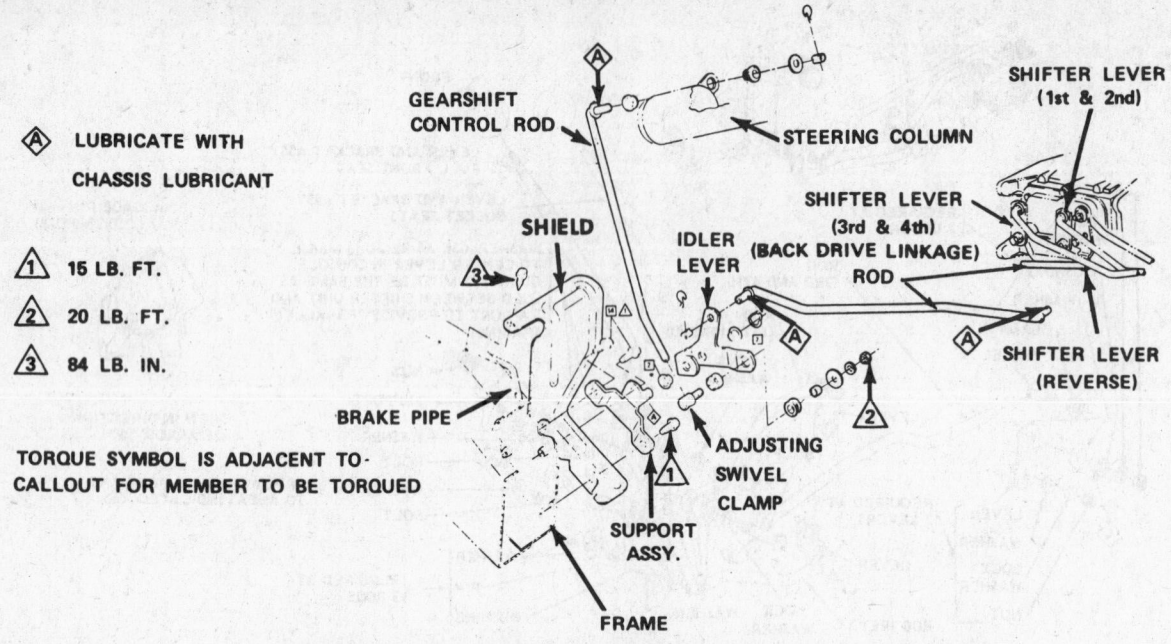

LUBRICATE WITH
CHASSIS LUBRICANT

△1 15 LB. FT.

△2 20 LB. FT.

△3 84 LB. IN.

TORQUE SYMBOL IS ADJACENT TO
CALLOUT FOR MEMBER TO BE TORQUED

Back drive floor linkage—1969 Firebird with Saginaw 4-speed transmission
(© Pontiac Div. G.M. Corp.)

3. For 1969 Saginaw, loosen trunnion locknuts (X) and (Z) on 1st-2nd and Reverse shift rods, then disconnect trunnion (Y) from lever (view A). Loosen (X), (Y), and (Z) for Muncie-equipped 1969-70 Tempest, GTO.

4. Insert a 1/4 in. drill rod into gauge pin hole in shifter.

5. If gearshift lever is not properly aligned with floor opening (view C-1969, view A-1970):

a. *Console*—loosen two shifter to support bolts and align shifter as per illustration; tighten bolts.

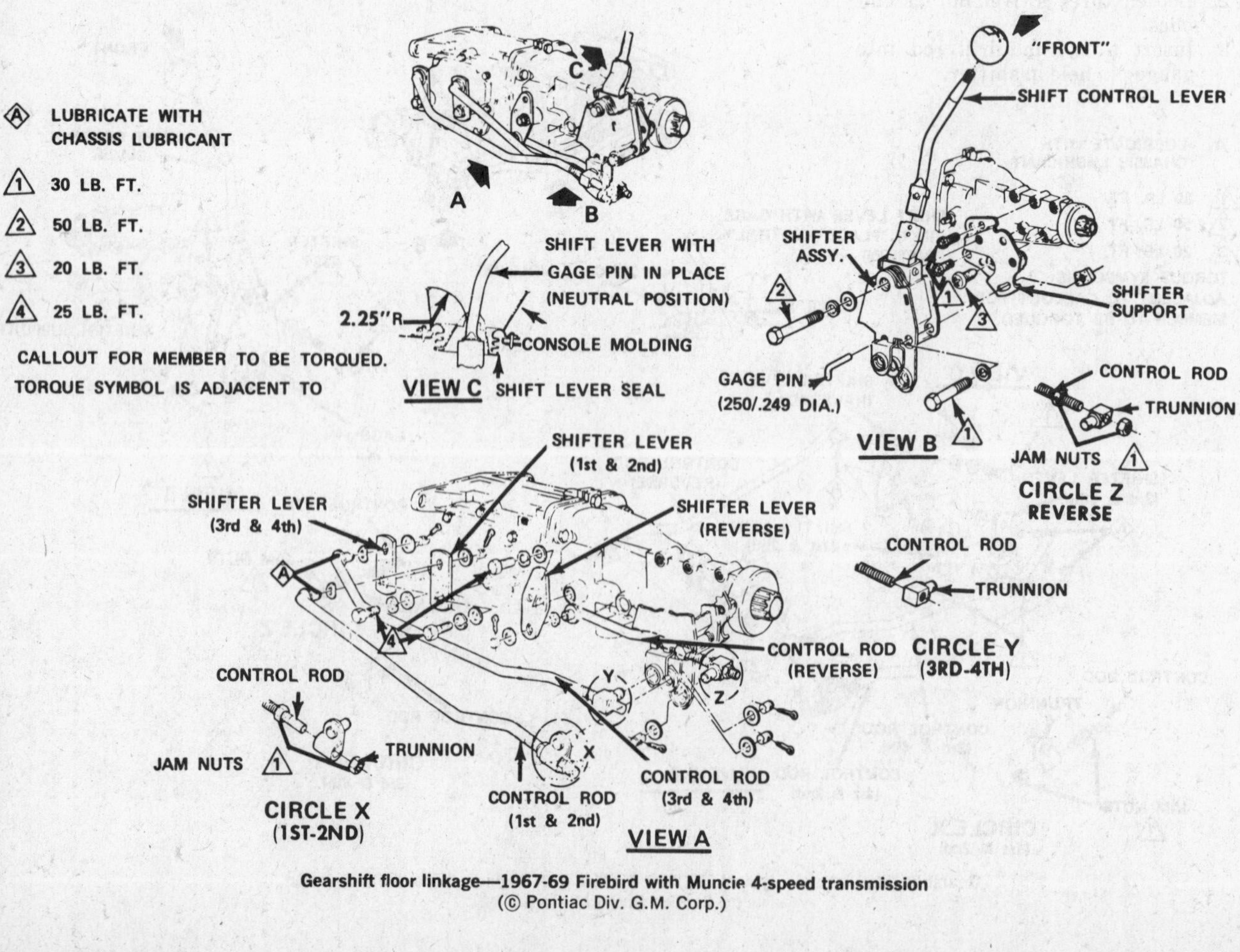

LUBRICATE WITH
CHASSIS LUBRICANT

△1 30 LB. FT.

△2 50 LB. FT.

△3 20 LB. FT.

△4 25 LB. FT.

CALLOUT FOR MEMBER TO BE TORQUED.
TORQUE SYMBOL IS ADJACENT TO

Gearshift floor linkage—1967-69 Firebird with Muncie 4-speed transmission
(© Pontiac Div. G.M. Corp.)

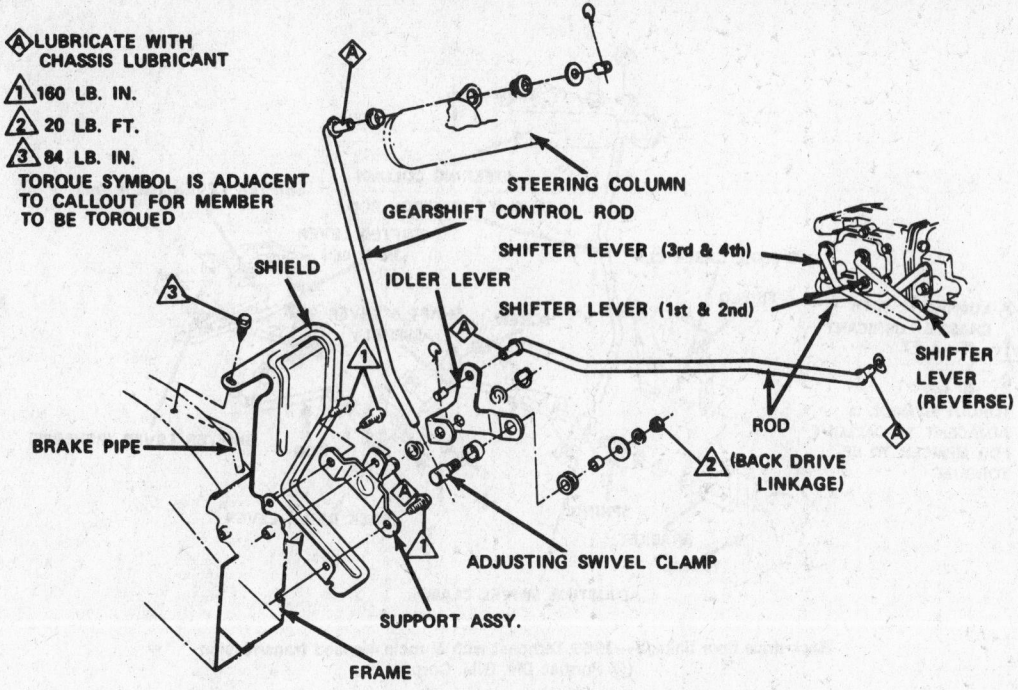

Δ LUBRICATE WITH CHASSIS LUBRICANT

⚠1 160 LB. IN.

⚠2 20 LB. FT.

⚠3 84 LB. IN.

TORQUE SYMBOL IS ADJACENT TO CALLOUT FOR MEMBER TO BE TORQUED

STEERING COLUMN

GEARSHIFT CONTROL ROD

SHIFTER LEVER (3rd & 4th)

SHIFTER LEVER (1st & 2nd)

SHIELD

IDLER LEVER

SHIFTER LEVER (REVERSE)

BRAKE PIPE

ROD

(BACK DRIVE LINKAGE)

ADJUSTING SWIVEL CLAMP

SUPPORT ASSY.

FRAME

Back drive floor linkage—1969 Firebird with Muncie 4-speed transmission
(© Pontiac Div. G.M. Corp.)

b. *Without console*—loosen two shifter to support bolts and center shifter in boot; tighten bolts.

6. Place transmission shift levers in Neutral and tighten locknuts (X) and (Z) to 30 ft. lbs. for 1969 Saginaw.

7. Align trunnion (Y) with hole in 3rd-4th shifter lever, insert trunnion and secure with washer and cotter pin for 1969 Saginaw.

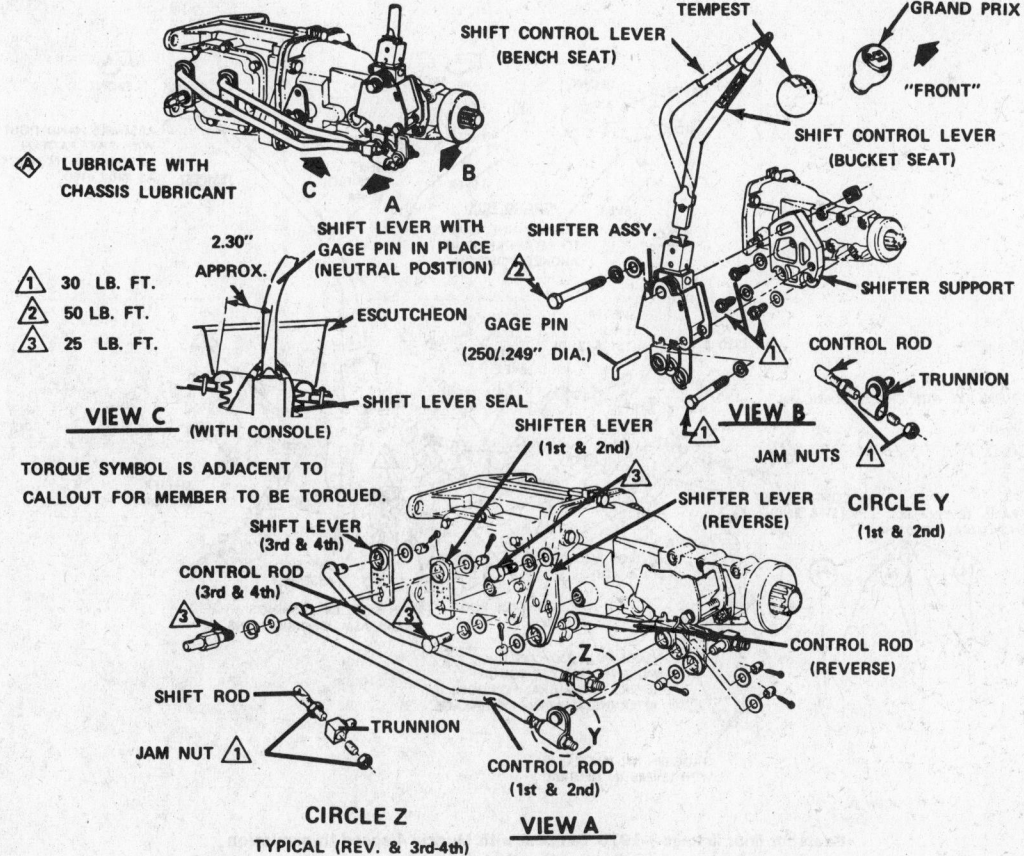

TEMPEST

SHIFT CONTROL LEVER (BENCH SEAT)

GRAND PRIX

"FRONT"

SHIFT CONTROL LEVER (BUCKET SEAT)

Δ LUBRICATE WITH CHASSIS LUBRICANT

C   A   B

SHIFTER ASSY.

SHIFTER SUPPORT

⚠1 30 LB. FT.

⚠2 50 LB. FT.

⚠3 25 LB. FT.

2.30" APPROX.

SHIFT LEVER WITH GAGE PIN IN PLACE (NEUTRAL POSITION)

ESCUTCHEON

GAGE PIN (250/.249" DIA.)

CONTROL ROD

TRUNNION

SHIFT LEVER SEAL

JAM NUTS

VIEW C (WITH CONSOLE)

VIEW B

SHIFTER LEVER (1st & 2nd)

CIRCLE Y (1st & 2nd)

TORQUE SYMBOL IS ADJACENT TO CALLOUT FOR MEMBER TO BE TORQUED.

SHIFTER LEVER (REVERSE)

SHIFT LEVER (3rd & 4th)

CONTROL ROD (3rd & 4th)

CONTROL ROD (REVERSE)

SHIFT ROD

TRUNNION

Z

JAM NUT

Y

CONTROL ROD (1st & 2nd)

CIRCLE Z
TYPICAL (REV. & 3rd-4th)

VIEW A

Gearshift floor linkage—1967-69 Tempest with Muncie 4-speed transmission
(© Pontiac Div. G.M. Corp.)

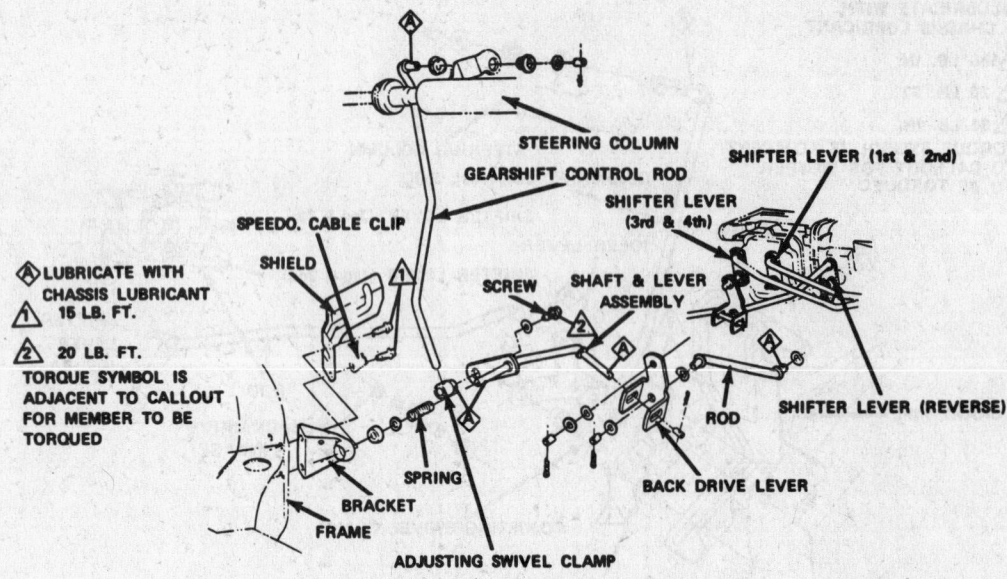

Back drive floor linkage—1969 Tempest with Muncie 4-speed transmission
(© Pontiac Div. G.M. Corp.)

Tighten (X), (Y), and (Z) to 30 ft. lbs. for 1969-70 Muncie-equipped Tempest, GTO.

8. Remove gauge pin and check shift pattern.

9. Place gearshift lever in Reverse, set steering column lower lever in Lock position and lock ignition.

10. Push up on gearshift control rod to take up lash in steering column lock mechanism, then tighten adjusting swivel clamp nut to 20 ft. lbs.

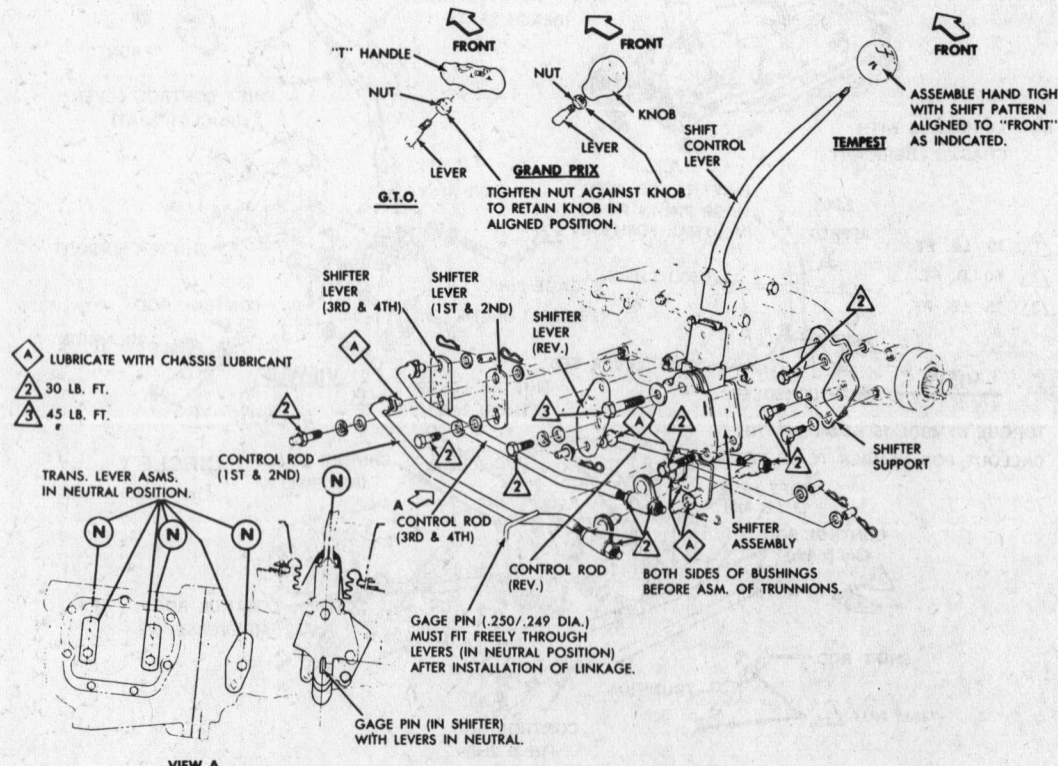

Gearshift floor linkage—1970 Tempest with Muncie 4-speed transmission
(© Pontiac Div. G.M. Corp.)

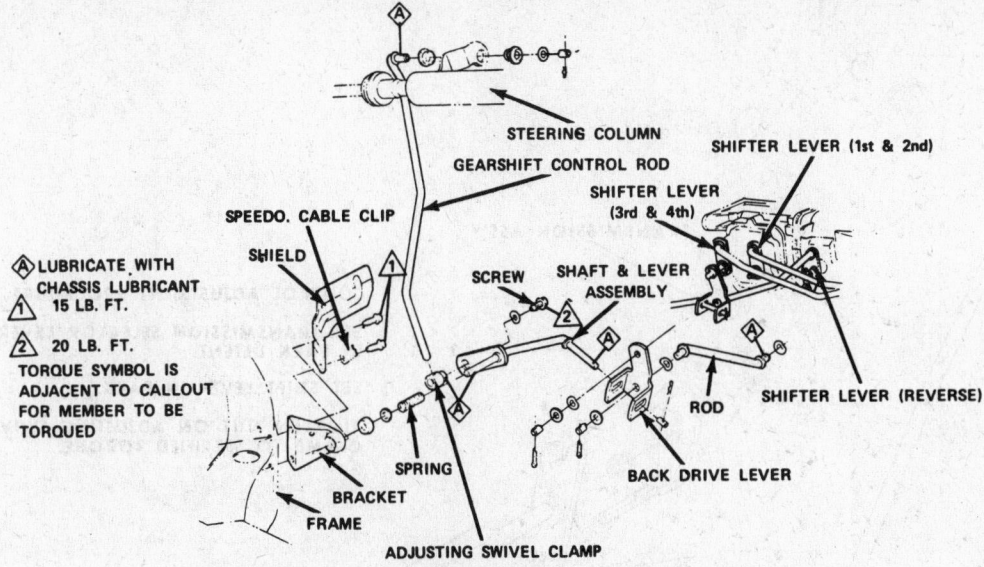

LUBRICATE WITH CHASSIS LUBRICANT

△ 15 LB. FT.

A 20 LB. FT.

TORQUE SYMBOL IS ADJACENT TO CALLOUT FOR MEMBER TO BE TORQUED

STEERING COLUMN

GEARSHIFT CONTROL ROD

SHIFTER LEVER (1st & 2nd)

SHIFTER LEVER (3rd & 4th)

SPEEDO. CABLE CLIP

SHIELD

SCREW

SHAFT & LEVER ASSEMBLY

SHIFTER LEVER (REVERSE)

ROD

BACK DRIVE LEVER

SPRING

BRACKET

FRAME

ADJUSTING SWIVEL CLAMP

Back drive floor linkage—1970 Tempest with Muncie 4-speed transmission
(© Pontiac Div. G.M. Corp.)

Automatic Transmission

Transmission R & R—1964-70

1. Raise car and provide support for front and rear of car.
2. Disconnect front exhaust pipe bolts at the exhaust manifold and at the connection of the intermediate exhaust pipe location (single exhaust only.) On dual exhaust, the exhaust pipes need not be removed, just loosened.
3. Remove driveshaft.
4. Place suitable jack under transmission and fasten transmission securely to jack.
5. Remove vacuum modulator hose from vacuum modulator, then remove T.C.S. wire.
6. Loosen cooler line bolts and separate cooler lines from transmission.
7. Remove transmission mounting pad to crossmember bolts.
8. Remove transmission crossmember support to frame rail bolts. Remove crossmember.
9. Disconnect speedometer cable.

10. Loosen shift linkage and downshift cable.
11. Disconnect and remove transmission filler pipe.
12. Support engine at oil pan.
13. Remove transmission flywheel cover pan.
14. Mark flywheel and converter pump for reassembly in same positions, and remove converter pump to flywheel bolts.
15. Remove transmission case to engine block bolts.

16. Mcve transmission rearward to provide clearance between converter pump and crankshaft.
17. Lower transmission and move to bench.
18. To install, reverse removal procedure.

Shift Linkage Adjustment
1964-68

See illustrations.

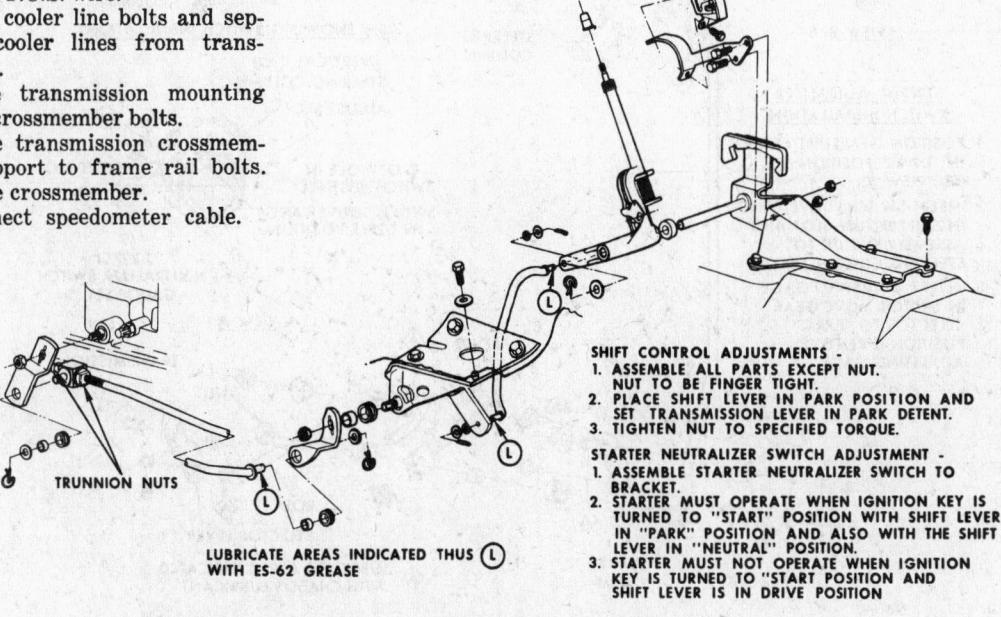

SHIFT CONTROL ADJUSTMENTS -
1. ASSEMBLE ALL PARTS EXCEPT NUT. NUT TO BE FINGER TIGHT.
2. PLACE SHIFT LEVER IN PARK POSITION AND SET TRANSMISSION LEVER IN PARK DETENT.
3. TIGHTEN NUT TO SPECIFIED TORQUE.

STARTER NEUTRALIZER SWITCH ADJUSTMENT -
1. ASSEMBLE STARTER NEUTRALIZER SWITCH TO BRACKET.
2. STARTER MUST OPERATE WHEN IGNITION KEY IS TURNED TO "START" POSITION WITH SHIFT LEVER IN "PARK" POSITION AND ALSO WITH THE SHIFT LEVER IN "NEUTRAL" POSITION.
3. STARTER MUST NOT OPERATE WHEN IGNITION KEY IS TURNED TO "START POSITION AND SHIFT LEVER IS IN DRIVE POSITION

TRUNNION NUTS

LUBRICATE AREAS INDICATED THUS (L) WITH ES-62 GREASE

Automatic console gearshift linkage—1964 Tempest (© Pontiac Div. G.M. Corp.)

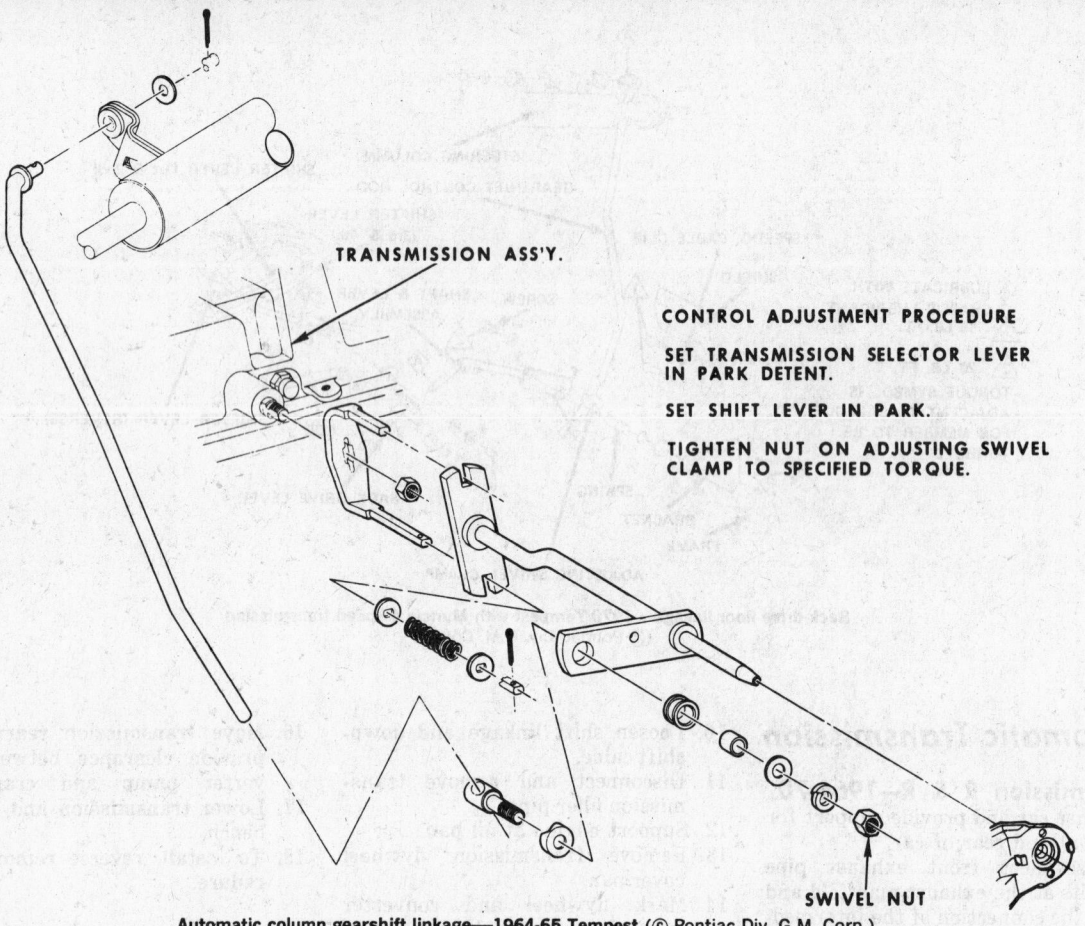

TRANSMISSION ASS'Y.

CONTROL ADJUSTMENT PROCEDURE

SET TRANSMISSION SELECTOR LEVER
IN PARK DETENT.

SET SHIFT LEVER IN PARK.

TIGHTEN NUT ON ADJUSTING SWIVEL
CLAMP TO SPECIFIED TORQUE.

SWIVEL NUT

Automatic column gearshift linkage—1964-65 Tempest (© Pontiac Div. G.M. Corp.)

TRANSMISSION

TRANSMISSION
SELECTOR LEVER

DR

VIEW B

ADJUSTING GAGE   NEUTRALIZER
SWITCH

STEERING
COLUMN

STEERING
COLUMN

VIEW SHOWING SWITCH INSTALLATION

**STARTER NEUTRALIZER
SWITCH INSTALLATION**

1. POSITION GEARSHIFT LEVER
   IN "DRIVE" POSITION.
   (SEE VIEW A)

2. INSERT SWITCH "DRIVE TANG"
   IN SHIFTER TUBE SLOT AND
   ASSEMBLY SWITCH TO
   STEERING COLUMN JACKET.

3. INSERT ADJUSTING GAGE
   IN SWITCH. MOVE GEAR
   SELECTOR TO "PARK"
   POSITION & REMOVE
   ADJUSTING GAGE.

ROD

VERTICAL ℄ OF
STEERING COLUMN

ADJUST SLOT

℄ OF HOLE IN
SWITCH ASSEMBLY

SWITCH "DRIVE TANG"
IN DRIVE POSITION

PK

RESET SLOT

DR

J-22701
NEUTRALIZER SWITCH
GAUGE SET (5)

VIEW A

LEVER

SUPPORT

NUT

TRANSMISSION

ROD

SELECTOR LEVER

FRAME

BRACKET

SWIVEL
CLAMP

LUBRICATE AREAS INDICATED
WITH CHASSIS LUBRICANT

Turbo-Hydramatic column gearshift linkage—1967-68 Firebird (© Pontiac Div. G.M. Corp.)

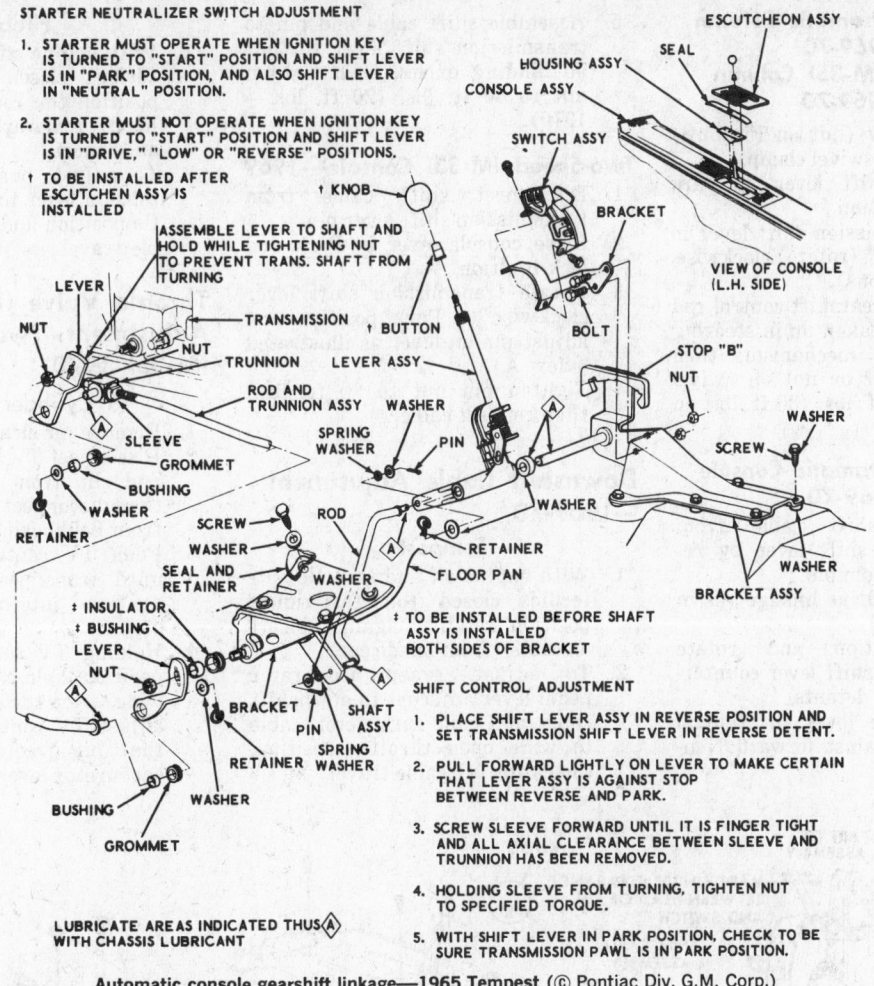

STARTER NEUTRALIZER SWITCH ADJUSTMENT

1. STARTER MUST OPERATE WHEN IGNITION KEY IS TURNED TO "START" POSITION AND SHIFT LEVER IS IN "PARK" POSITION, AND ALSO SHIFT LEVER IN "NEUTRAL" POSITION.

2. STARTER MUST NOT OPERATE WHEN IGNITION KEY IS TURNED TO "START" POSITION AND SHIFT LEVER IS IN "DRIVE," "LOW" OR "REVERSE" POSITIONS.

† TO BE INSTALLED AFTER ESCUTCHEN ASSY IS INSTALLED

ASSEMBLE LEVER TO SHAFT AND HOLD WHILE TIGHTENING NUT TO PREVENT TRANS. SHAFT FROM TURNING

SHIFT CONTROL ADJUSTMENT

1. PLACE SHIFT LEVER ASSY IN REVERSE POSITION AND SET TRANSMISSION SHIFT LEVER IN REVERSE DETENT.

2. PULL FORWARD LIGHTLY ON LEVER TO MAKE CERTAIN THAT LEVER ASSY IS AGAINST STOP BETWEEN REVERSE AND PARK.

3. SCREW SLEEVE FORWARD UNTIL IT IS FINGER TIGHT AND ALL AXIAL CLEARANCE BETWEEN SLEEVE AND TRUNNION HAS BEEN REMOVED.

4. HOLDING SLEEVE FROM TURNING, TIGHTEN NUT TO SPECIFIED TORQUE.

5. WITH SHIFT LEVER IN PARK POSITION, CHECK TO BE SURE TRANSMISSION PAWL IS IN PARK POSITION.

‡ TO BE INSTALLED BEFORE SHAFT ASSY IS INSTALLED BOTH SIDES OF BRACKET

LUBRICATE AREAS INDICATED THUS △ WITH CHASSIS LUBRICANT

Automatic console gearshift linkage—1965 Tempest (© Pontiac Div. G.M. Corp.)

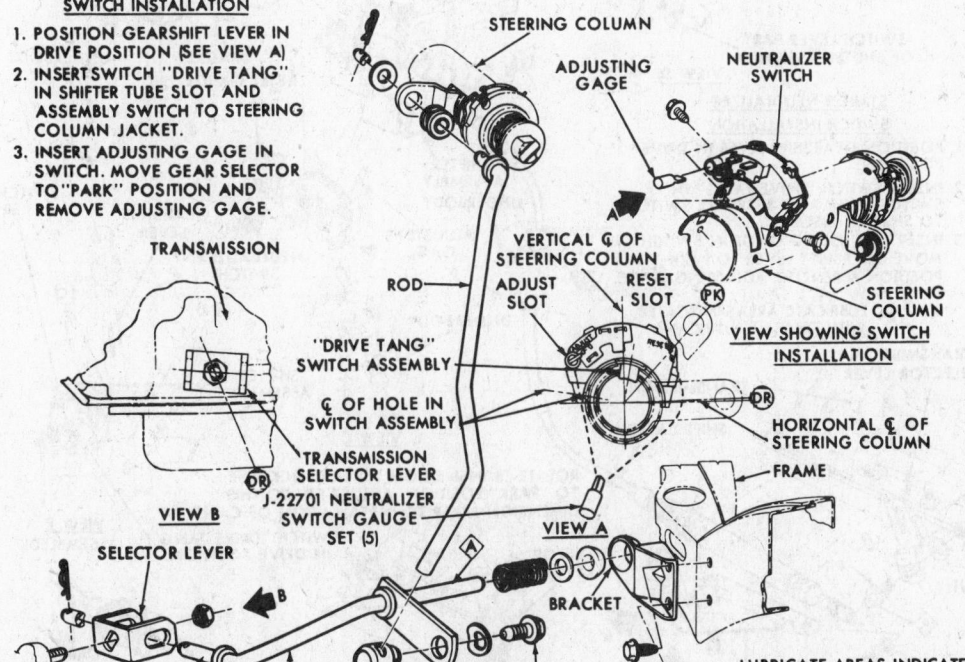

STARTER NEUTRALIZER SWITCH INSTALLATION

1. POSITION GEARSHIFT LEVER IN DRIVE POSITION (SEE VIEW A)

2. INSERT SWITCH "DRIVE TANG" IN SHIFTER TUBE SLOT AND ASSEMBLY SWITCH TO STEERING COLUMN JACKET.

3. INSERT ADJUSTING GAGE IN SWITCH. MOVE GEAR SELECTOR TO "PARK" POSITION AND REMOVE ADJUSTING GAGE.

VIEW SHOWING SWITCH INSTALLATION

VIEW B

J-22701 NEUTRALIZER SWITCH GAUGE SET (5)

VIEW A

LUBRICATE AREAS INDICATED WITH CHASSIS LUBRICANT

Turbo-Hydramatic column gearshift linkage—1967-68 GTO (© Pontiac Div. G.M. Corp.)

All Turbo-Hydramatic Column —1969-70
Two-Speed (M-35) Column —1969-70

1. Loosen screw (nut on Firebird) on adjusting swivel clamp.
2. Place gearshift lever in Park and lock ignition.
3. Place transmission shift lever in Park detent (rotate clockwise, see illustrations).
4. Push up on gearshift control rod until lash is taken up in steering column lock mechanism, then tighten screw or nut on swivel clamp to 20 ft. lbs. (30 ft. lbs. on Firebird).

All Turbo-Hydramatic Console —1969-70

1. Disconnect shift cable from transmission shift lever by removing nut from pin.
2. Adjust back drive linkage (as in Step 4, above).
3. Unlock ignition and rotate transmission shift lever counterclockwise two detents.
4. Place console lever in Neutral and move against forward Neutral stop.

5. Assemble shift cable and pin to transmission shift lever so that no binding exists, then tighten nut to 30 ft. lbs. (20 ft. lbs.—1970).

Two-Speed (M-35) Console—1969

1. Disconnect shift cable from transmission shift lever pin.
2. Place console lever in Park and lock ignition.
3. Rotate transmission shift lever clockwise to Park position and adjust pin on lever as illustrated (view A).
4. Tighten pin nut to 30 ft. lbs., then connect cable.

Downshift Cable Adjustment —1969-70

Tempest

1. With engine off and throttle butterflies closed (off fast idle), position retainer against insert on cable (from inside car).
2. To adjust, grasp accelerator pedal lever adjacent to downshift cable and pull carburetor cable to wide open throttle position. Check for full cable travel.

Firebird

1. With engine off and throttle butterflies closed (off fast idle), position the retainer (under the hood) rearward against washer and insert.
2. To adjust, push carburetor extension lever to wide open throttle position and check for full cable travel.

Throttle Valve (TV) Linkage Adjustment—Two-Speed (M-35) 1970

6 Cylinder Models

1. Remove air cleaner.
2. Disconnect TV control rod swivel and clip from carburetor lever, then disconnect TV return spring from bellhousing.
3. Push TV control rod rearward until transmission TV lever is against internal transmission stop.
4. Holding TV control rod in this position, hold carburetor lever in wide open throttle position and adjust TV control rod swivel so that pin freely enters hole in carburetor lever without binding.

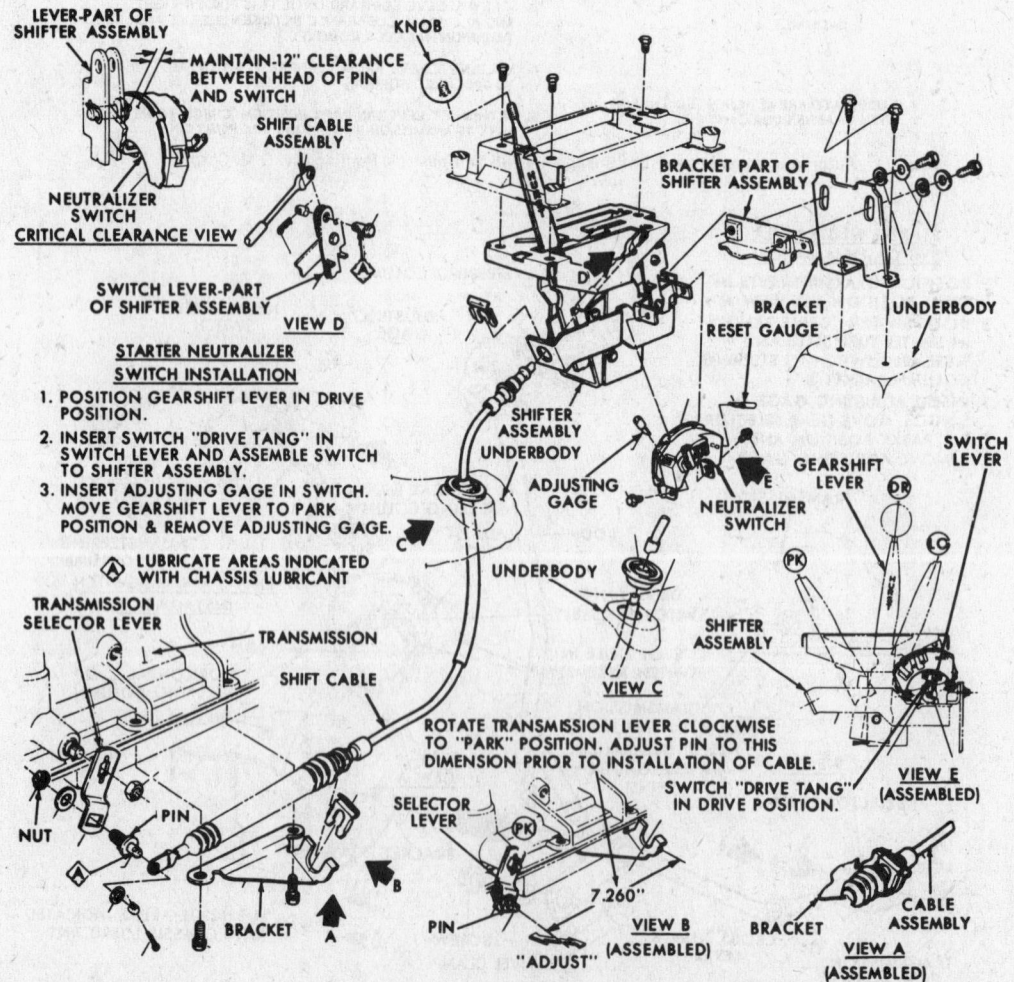

Turbo-Hydramatic console gearshift linkage—1967-68 GTO (© Pontiac Div. G.M. Corp.)

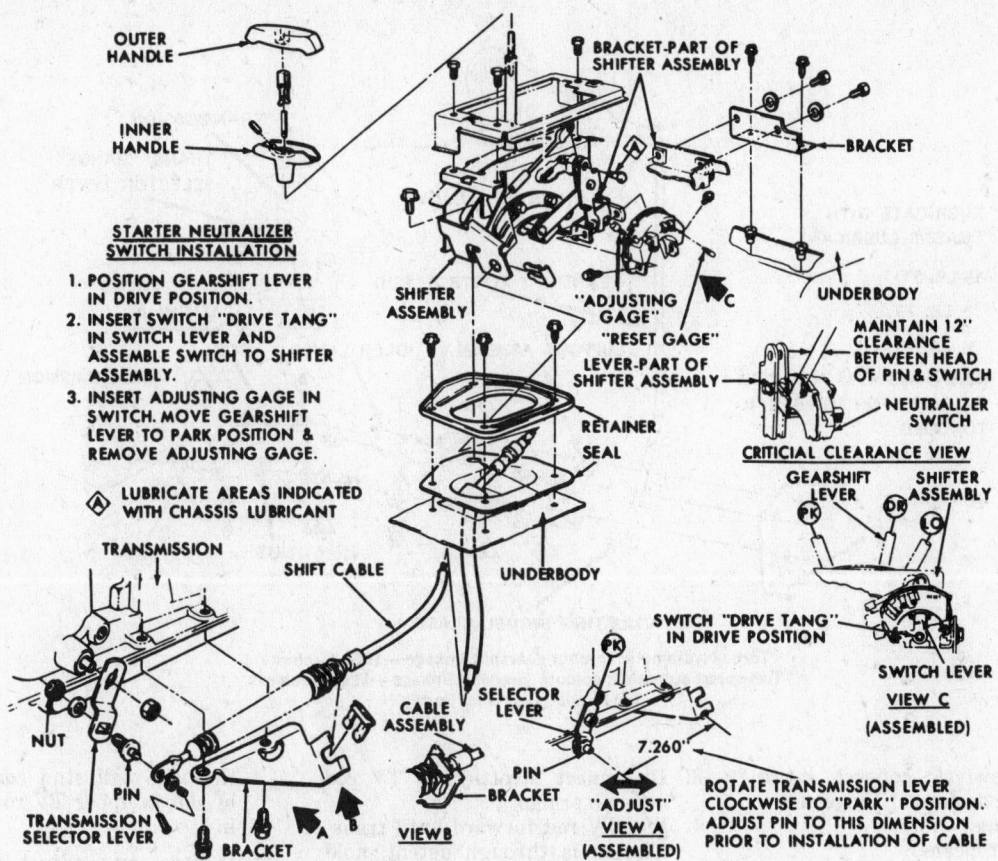

STARTER NEUTRALIZER SWITCH INSTALLATION

1. POSITION GEARSHIFT LEVER IN DRIVE POSITION.
2. INSERT SWITCH "DRIVE TANG" IN SWITCH LEVER AND ASSEMBLE SWITCH TO SHIFTER ASSEMBLY.
3. INSERT ADJUSTING GAGE IN SWITCH. MOVE GEARSHIFT LEVER TO PARK POSITION & REMOVE ADJUSTING GAGE.

Ⓐ LUBRICATE AREAS INDICATED WITH CHASSIS LUBRICANT

OUTER HANDLE

INNER HANDLE

BRACKET-PART OF SHIFTER ASSEMBLY

BRACKET

SHIFTER ASSEMBLY

"ADJUSTING GAGE"

"RESET GAGE"

LEVER-PART OF SHIFTER ASSEMBLY

UNDERBODY

MAINTAIN 12" CLEARANCE BETWEEN HEAD OF PIN & SWITCH

NEUTRALIZER SWITCH

CRITICIAL CLEARANCE VIEW

RETAINER SEAL

TRANSMISSION

SHIFT CABLE

UNDERBODY

GEARSHIFT LEVER

SHIFTER ASSEMBLY

SWITCH "DRIVE TANG" IN DRIVE POSITION

SWITCH LEVER

VIEW C (ASSEMBLED)

NUT

PIN

TRANSMISSION SELECTOR LEVER

BRACKET

CABLE ASSEMBLY

SELECTOR LEVER

PIN BRACKET

VIEW B

"ADJUST" VIEW A (ASSEMBLED)

7.260"

ROTATE TRANSMISSION LEVER CLOCKWISE TO "PARK" POSITION. ADJUST PIN TO THIS DIMENSION PRIOR TO INSTALLATION OF CABLE

Turbo-Hydramatic console gearshift linkage—1967-68 Firebird (© Pontiac Div. G.M. Corp.)

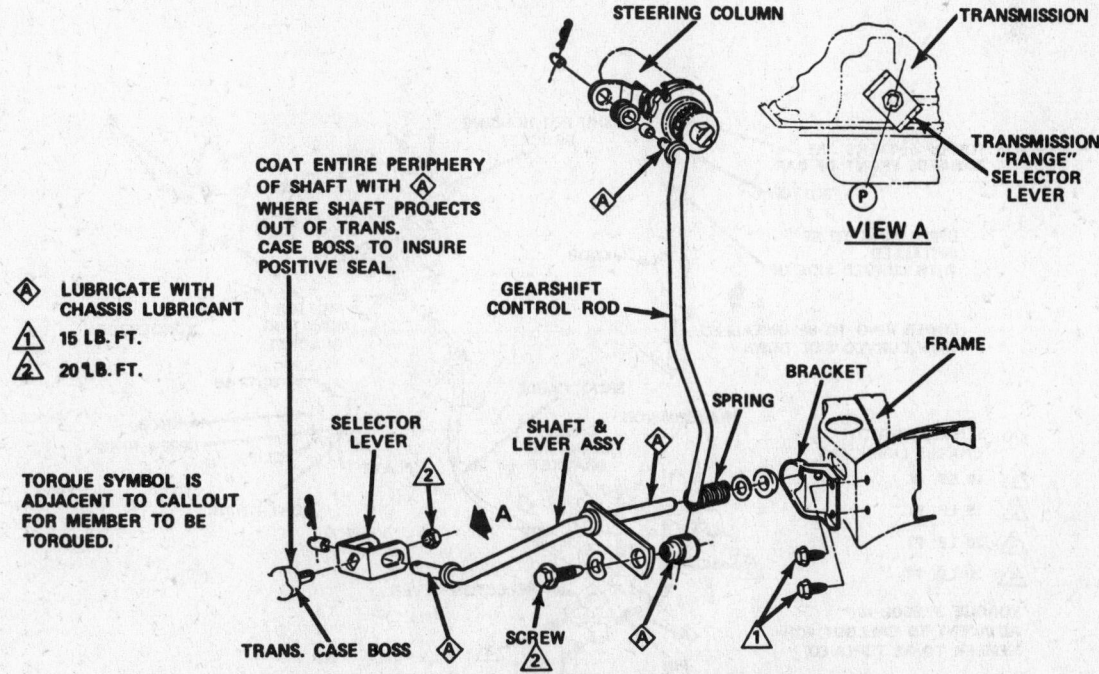

COAT ENTIRE PERIPHERY OF SHAFT WITH Ⓐ WHERE SHAFT PROJECTS OUT OF TRANS. CASE BOSS. TO INSURE POSITIVE SEAL.

Ⓐ LUBRICATE WITH CHASSIS LUBRICANT

△1 15 LB. FT.

△2 20 LB. FT.

TORQUE SYMBOL IS ADJACENT TO CALLOUT FOR MEMBER TO BE TORQUED.

STEERING COLUMN

TRANSMISSION

TRANSMISSION "RANGE" SELECTOR LEVER

VIEW A

GEARSHIFT CONTROL ROD

FRAME

BRACKET

SPRING

SELECTOR LEVER

SHAFT & LEVER ASSY

TRANS. CASE BOSS

SCREW

Turbo-Hydramatic column gearshift linkage—1969-70 Tempest and GTO
Two-speed automatic column gearshift linkage—1969-70 Tempest
(© Pontiac Div. G.M. Corp.)

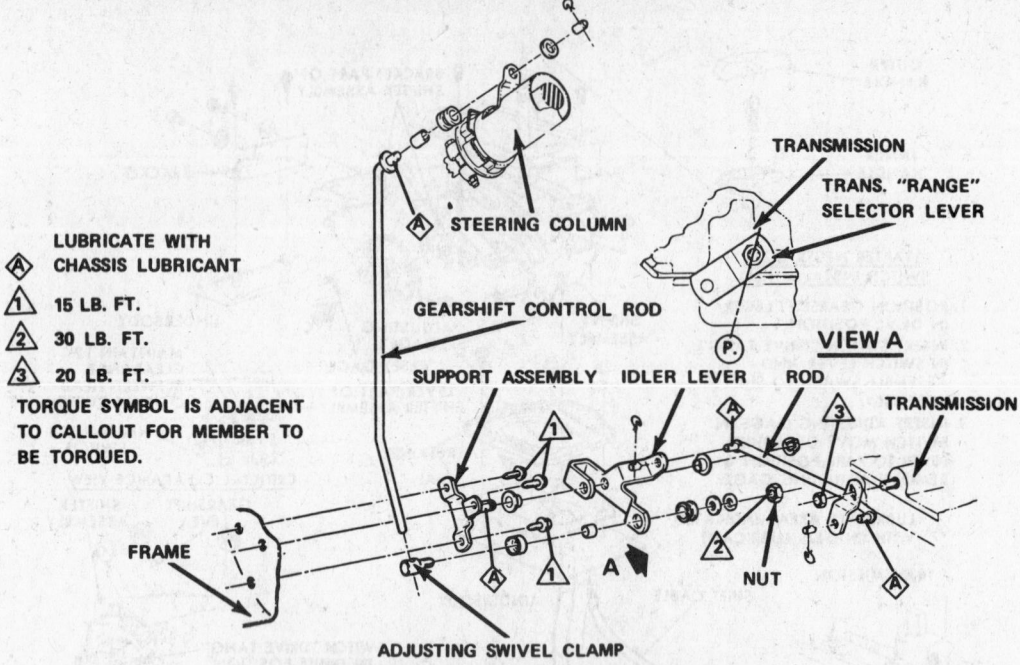

LUBRICATE WITH CHASSIS LUBRICANT

⚠ 15 LB. FT.

⚠ 30 LB. FT.

⚠ 20 LB. FT.

TORQUE SYMBOL IS ADJACENT TO CALLOUT FOR MEMBER TO BE TORQUED.

Turbo-Hydramatic column gearshift linkage—1969 Firebird
Two-speed automatic column gearshift linkage—1969 Firebird
(© Pontiac Div. G.M. Corp.)

5. Secure swivel, connect return spring and check linkage action for binding.
6. Install air cleaner.

V8 Models

1. Remove air cleaner.
2. Disconnect accelerator linkage at carburetor.

3. Disconnect throttle and TV rod return springs.
4. Pull TV rod forward until transmission is through detent, hold in this position and open carburetor butterflies to wide open position.
5. The butterflies must reach wide open position at the same time

that the ball stud contacts end of slot in upper TV rod (± 1/32 in.).

6. If necessary, adjust swivel end of upper TV rod.
7. Connect linkage and springs, then check linkage for binding.
8. Install air cleaner.

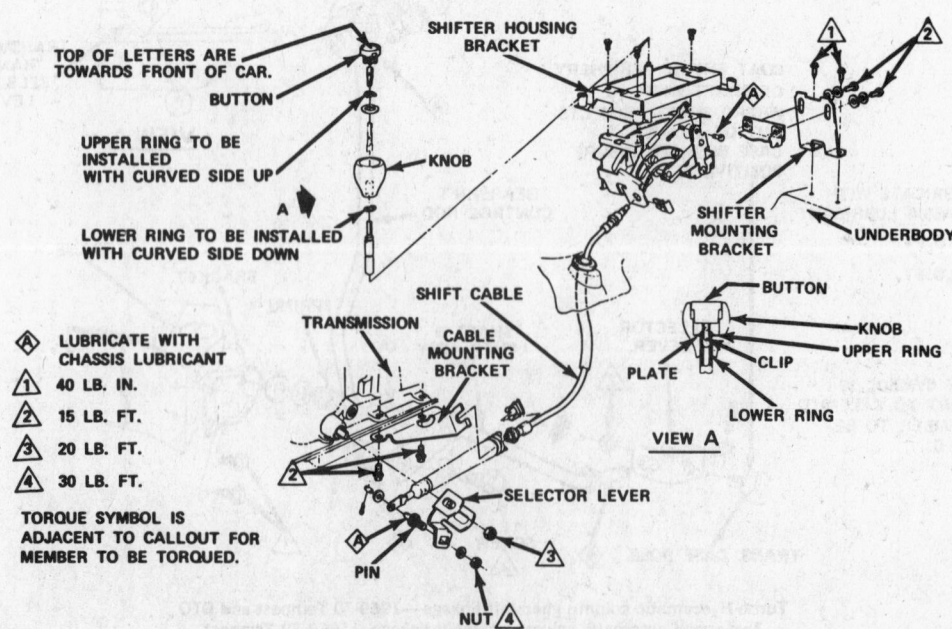

LUBRICATE WITH CHASSIS LUBRICANT

⚠ 40 LB. IN.

⚠ 15 LB. FT.

⚠ 20 LB. FT.

⚠ 30 LB. FT.

TORQUE SYMBOL IS ADJACENT TO CALLOUT FOR MEMBER TO BE TORQUED.

Turbo-Hydramatic console gearshift—1969-70 Tempest
Two-speed automatic console gearshift linkage—1969 Tempest
(© Pontiac Div. G.M. Corp.)

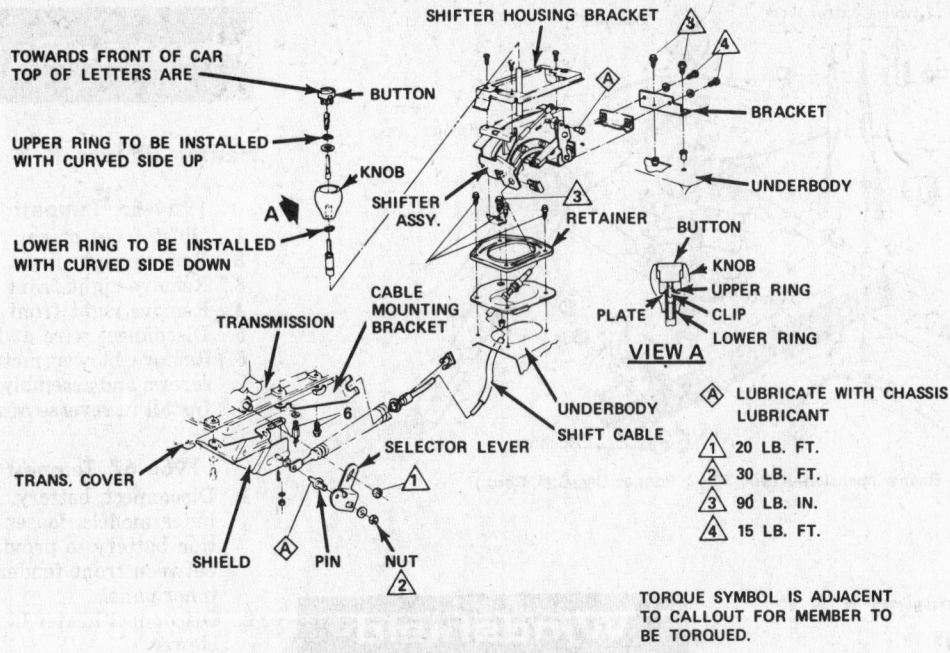

SHIFTER HOUSING BRACKET

TOWARDS FRONT OF CAR
TOP OF LETTERS ARE
— BUTTON

UPPER RING TO BE INSTALLED
WITH CURVED SIDE UP

BRACKET

KNOB
SHIFTER
ASSY.

UNDERBODY

A

RETAINER

BUTTON

LOWER RING TO BE INSTALLED
WITH CURVED SIDE DOWN

KNOB
UPPER RING

CLIP

PLATE

LOWER RING

CABLE
MOUNTING
BRACKET

TRANSMISSION

VIEW A

Ⓐ LUBRICATE WITH CHASSIS
LUBRICANT

UNDERBODY
SHIFT CABLE

SELECTOR LEVER

△1 20 LB. FT.

TRANS. COVER

△2 30 LB. FT.

△3 90 LB. IN.

SHIELD   PIN   NUT

△4 15 LB. FT.

TORQUE SYMBOL IS ADJACENT
TO CALLOUT FOR MEMBER TO
BE TORQUED.

Turbo-Hydramatic console gearshift linkage—1969 Firebird
Two-speed automatic console gearshift linkage—1969 Firebird
(© Pontiac Div. G.M. Corp.)

Rear Axles, Suspension

Troubleshooting and Adjustment

General instructions covering the troubles of the rear axle and how to repair and adjust it, together with information on installation of rear axle bearings and grease seals, are given in the Unit Repair Section.

Capacities of the drive axle are given in the Capacities tables of this section.

Rear Axle Housing R & R —Coil Springs

1. Jack up the back of the car to allow sufficient room to work, then place another jack under the rear axle housing.
2. Disconnect the driveshaft.
3. Remove both axle shafts, wire backing plates out of the way, and remove brake hose bracket and brake line.
4. Disconnect the shocks at the other end, then slowly lower the rear axle housing until springs can be removed.
5. Remove the bolt on each side which holds the torque arm to the rear axle housing, disconnect lower control arms.
6. Slide the housing assembly out.
7. To install, reverse removal procedure, tightening shock nuts to 55-75 ft. lbs., upper and lower control arm bolts to 75-100 ft. lbs. (nuts to 60-85 ft. lbs.), U-joint bolts to 14-20 ft. lbs.

NOTE: tighten nuts and bolts to specification with car on ground and suspension fully loaded. Do not exceed 20 ft. lbs. on U-joint bolts, as bearings can be distorted.

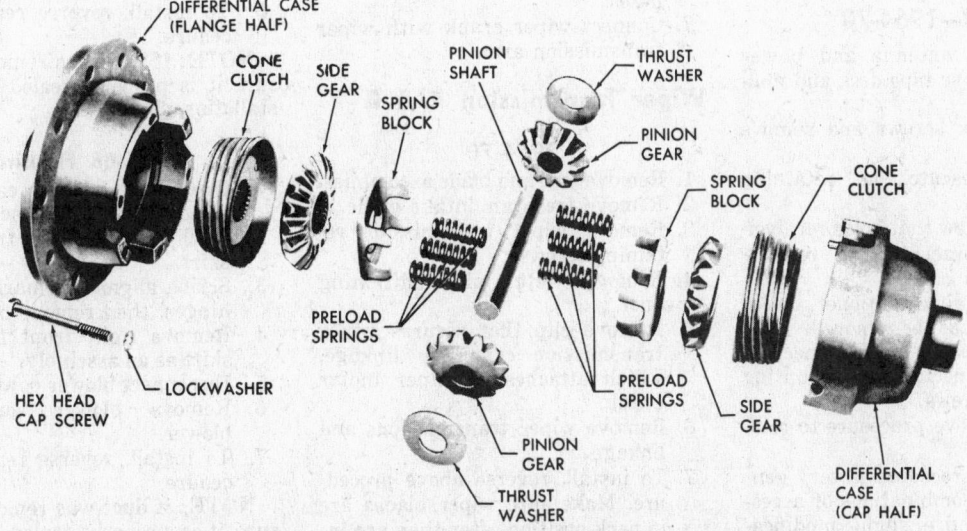

DIFFERENTIAL CASE
(FLANGE HALF)

CONE
CLUTCH

SIDE
GEAR

PINION
SHAFT

THRUST
WASHER

SPRING
BLOCK

PINION
GEAR

SPRING
BLOCK

CONE
CLUTCH

PRELOAD
SPRINGS

PRELOAD
SPRINGS

SIDE
GEAR

HEX HEAD
CAP SCREW

LOCKWASHER

PINION
GEAR

THRUST
WASHER

DIFFERENTIAL
CASE
(CAP HALF)

Safe-T-Track differential (© Pontiac Div. G.M. Corp.)

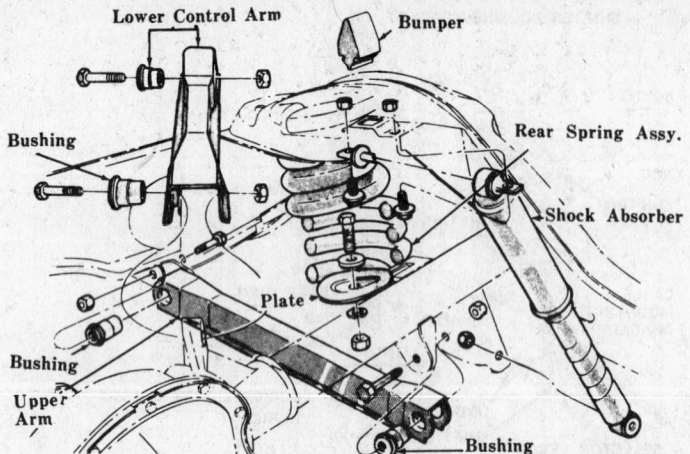

Rear suspension—1964-70 (© Pontiac Div. G.M. Corp.)

Rear Axle Housing R & R —Leaf Springs

1. Follow Steps 1-3 in *Rear Axle Housing R&R—Coil Springs.*
2. Disconnect shock absorbers.
3. Disconnect rear spring from its shackles and brackets and lower rear axle housing.
4. Remove axle housing from beneath vehicle.
5. To install, reverse removal procedure, connecting front spring eyes first. Tighten bracket nuts to 100 ft. lbs. and shackle pin nuts to 50 ft. lbs.

NOTE: tighten nuts and bolts to specification with car on ground and suspension fully loaded. Do not exceed 20 ft. lbs. on U-joint bolts, as bearings can be distorted.

Radio

Radio R & R—1964-70

1. Disconnect antenna and power leads, remove tape deck and multiplex.
2. Loosen hex screws and remove knobs.
3. Remove escutcheon retaining nuts.
4. Remove screw that holds receiver to panel bracket, then remove ash tray.
   NOTE: with air conditioner, outlet duct and bezel must be removed.
5. Remove speaker by disconnecting output connector and mounting bracket screws.
6. Reverse above procedure to reinstall.

NOTE: this procedure is very general, and some combinations of accessories may require slight modifications.

Windshield Wipers

Motor R & R

1964-70

1. Remove hoses and wire terminals that are connected to wiper unit.
2. Remove clip that secures wiper crank to wiper transmission arm.
   NOTE: this clip is under instrument panel on 1964-66 models, under leaf screen on depressed-park motors, and accessible only after firewall bolts are removed on some standard motors. On some models, the wiper arm must be removed to facilitate motor removal.
3. Remove screws that secure wiper assembly to firewall.
4. Install a gasket on the motor.
5. Position wiper assembly on firewall and secure.
6. Connect wire terminals and hoses.
7. Connect wiper crank with wiper transmission arm.

Wiper Transmission R & R

1964-70

1. Remove arm and blade assemblies.
2. Remove fresh air intake grille.
3. Remove wiper transmission retaining screws.
4. Remove left air-conditioning duct.
5. Remove clip that secures wiper transmission crank to linkage, which attaches to wiper motor crank.
6. Remove wiper transmissions and linkage.
7. To install, reverse above procedure. Make sure wiper blades are in park position after they are installed.

Heater System

Heater Blower R & R

1964-65 Tempest & GTO

1. Hoist front of car.
2. Remove right front wheel.
3. Remove right front headlamp.
4. Remove right front fender.
5. Disconnect wire at blower.
6. Remove blower motor attaching screws and assembly.
7. Install in reverse of above.

1966-67 Tempest & GTO

1. Disconnect battery. On six cylinder models, loosen and reposition battery to provide clearance between front fender and fender inner panel.
2. Disconnect heater blower wire at blower.
3. Remove blower motor to blower housing bolts.
4. Remove motor and impeller assembly from the blower housing and position with impeller facing toward engine below right hood hinge.
5. Remove blower to impeller retaining nut and washer, then separate.
6. Move motor, then impeller, separately along top of fender panel to front opening, then remove.
7. Install by reversing removal procedure.

1968-69 Tempest & GTO, 1969 Firebird

1. Remove battery and battery tray.
2. Remove inner fender skirt.
3. Remove blower power wire.
4. Remove blower retaining screws and blower.
5. To install, reverse removal procedure.

NOTE: if duct was removed, make sure it is properly sealed during installation.

1967-68 Firebird

1. Disconnect battery cables, then remove battery and battery tray.
2. Unclip heater hoses from fender skirt.
3. Scribe alignment marks at hood hinges, then remove hood.
4. Remove right front fender and skirt as an assembly.
5. Disconnect blower power wire.
6. Remove blower screws and blower.
7. To install, reverse removal procedure.

NOTE: if duct was removed, make sure it is properly sealed during installation.

1970 Tempest & GTO

1. Jack up front of car and remove right front wheel.
2. Cut access hole along stamped outline on right fender skirt, using an air chisel.
3. Disconnect blower power wire.
4. Remove blower.
5. To install, reverse removal procedure, covering access hole with a metal plate secured with sealer and sheet metal screws.

Heater Core R & R

1964-65 Tempest & GTO

1. Drain radiator.
2. Remove glove compartment.
3. Disconnect hoses at heater.
4. Disconnect control cables at core and case assembly.
5. Remove wire connector from resistor assembly at top left side of heater outlet duct by prying connector up with flat screwdriver.
6. Remove nuts that hold heater to air inlet duct, and remove heater assembly.

7. Remove core and case assembly.
8. Install in reverse of above.

1966-67 Tempest & GTO

1. Drain radiator and remove glove compartment.
2. Disconnect heater hoses at the heater.
3. Disconnect heater control cables at heater.
4. Remove front wheel.
5. Remove wire connector from resistor at top left side of heater air outlet by prying connector up with a screwdriver.
6. Cut a one inch hole in the skirt.
7. Remove six nuts from the heater to air inlet duct and remove heater.
8. Install by reversing removal procedure.
9. Patch the skirt hole.

1968 Tempest and GTO

1. Disconnect heater hoses at heater.
2. Remove glove compartment.
3. Remove five nuts which secure heater to firewall.

4. Pull case from firewall, then disconnect cables and wire connector from resistor.

1967-1969 Firebird,1969-70 Tempest & GTO

1. Drain radiator.
2. Disconnect heater hoses at air inlet assembly.

NOTE: the water pump hose goes to top heater core pipe, the other hose (from rear of right cylinder head on V8, center of block on 6) goes to the lower heater core pipe.

3. Remove nuts from core studs on firewall (under hood).
4. From inside the car, pull the heater assembly from the firewall.
5. Disconnect control cables and wires, then remove heater assembly.
6. To remove core, unhook retaining springs.
7. To install, reverse removal procedure, making sure core is properly sealed during installation.

Ford and Thunderbird

Index

514

YEAR IDENTIFICATION

FORD

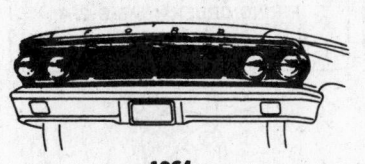

1964

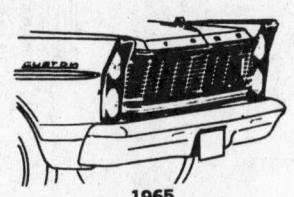

1965

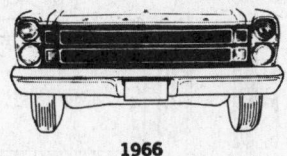

1966

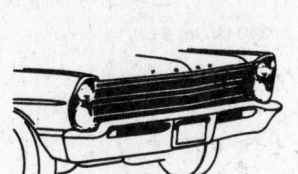

1967

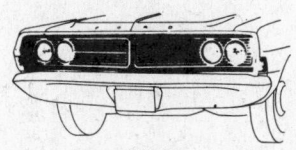

1968

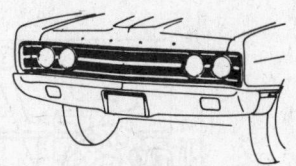

1969

1969 LTD

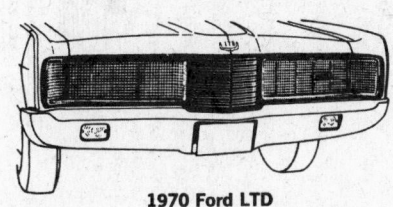

1970 Ford LTD

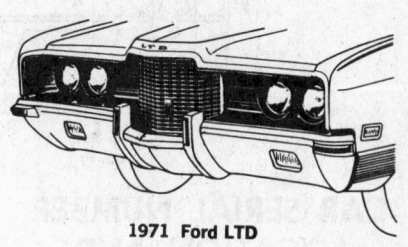

1971 Ford LTD

THUNDERBIRD

1964

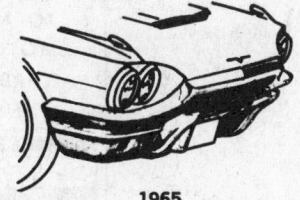

1965

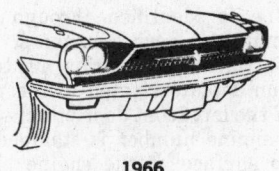

1966

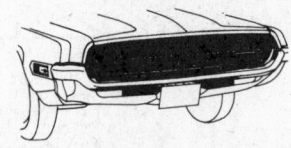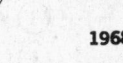

1967

1968

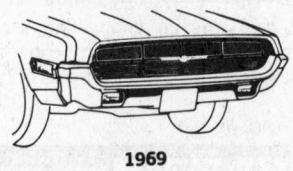

1969

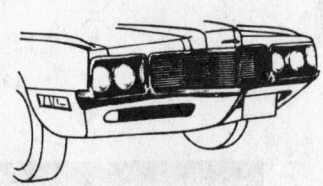

1970 Thunderbird

FIRING ORDER

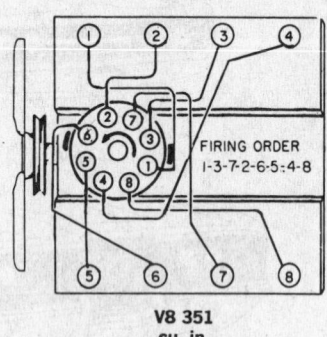

FIRING ORDER 1-5-3-6-2-4

223 cu. in. 6 cyl.

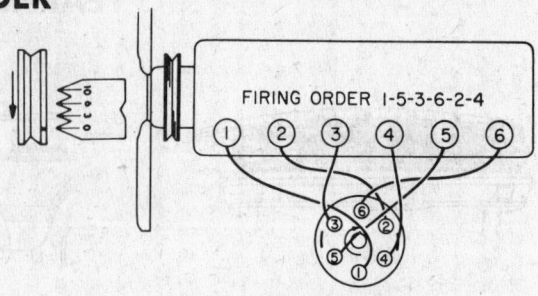

FIRING ORDER 1-5-3-6-2-4

240 cu. in. 6 cyl.

FIRING ORDER 1-3-7-2-6-5-4-8

V8 351 cu. in.

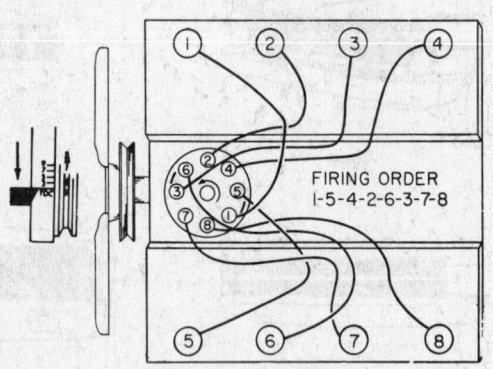

FIRING ORDER 1-5-4-2-6-3-7-8

V8 except 351 cu. in.

CAR SERIAL NUMBER LOCATION AND ENGINE IDENTIFICATION

1964-67

Engine is identified through car serial number. Serial numbers, and other pertinent information, are to be found on a plate riveted to the rear edge of the left front door pillar.

The engine number is stamped on the top surface of the engine block near the crankcase breather pipe, front left side.

The car serial number is composed of two sections, the first five units giving the year, assembly plant, model and engine type. The second section gives the consecutive production serial number. The year of pro-

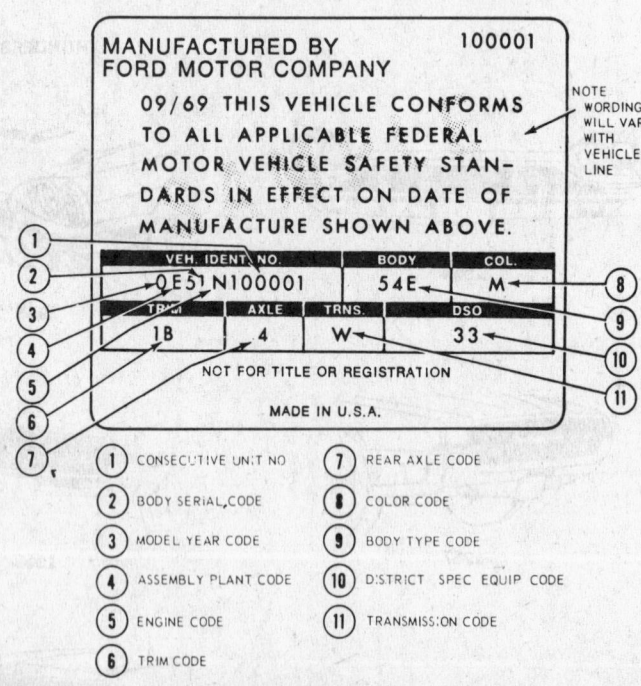

MANUFACTURED BY FORD MOTOR COMPANY

09/69 THIS VEHICLE CONFORMS TO ALL APPLICABLE FEDERAL MOTOR VEHICLE SAFETY STANDARDS IN EFFECT ON DATE OF MANUFACTURE SHOWN ABOVE.

100001

NOTE WORDING WILL VARY WITH VEHICLE LINE

VEH. IDENT. NO.	BODY	COL.
0E51N100001	54E	M

TRIM	AXLE	TRNS.	DSO
1B	4	W	33

NOT FOR TITLE OR REGISTRATION

MADE IN U.S.A.

① CONSECUTIVE UNIT NO ⑦ REAR AXLE CODE
② BODY SERIAL CODE ⑧ COLOR CODE
③ MODEL YEAR CODE ⑨ BODY TYPE CODE
④ ASSEMBLY PLANT CODE ⑩ DISTRICT SPEC EQUIP CODE
⑤ ENGINE CODE ⑪ TRANSMISSION CODE
⑥ TRIM CODE

1970-71 Vehicle Certification Label

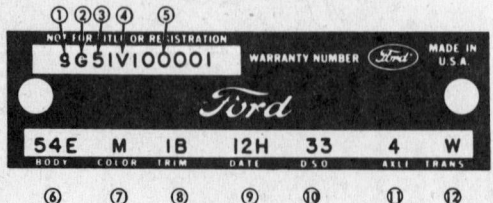

NOT FOR TITLE OR REGISTRATION 9G51V100001 WARRANTY NUMBER MADE IN U.S.A.

Ford

54E	M	1B	12H	33	4	W
BODY	COLOR	TRIM	DATE	DSO	AXLE	TRANS

1 Model year code
2 Assembly plant code
3 Body serial code
4 Engine code
5 Consecutive unit number
6 Body type code
7 Color code
8 Trim code
9 Date code
10 District—special equipment code
11 Rear axle code
12 Transmission code

9Y83N100001

Typical vehicle identification number (VIN) tab

duction is given by the first digit. (For example, "6" = 1966, "9" = 1969). The fifth digit, a letter, represents the engine identification code. (See table.)

1968-71

The serial number can be found on a plate attached to the top of the instrument panel, visible through the windshield. Information for identification remains essentially the same as for 1963-67 models.

Vehicle Certification Label 1970-71

The label is located on the rear face of the driver's door. The upper portion contains the name of the manufacturer, the month and year of manufacture, and the certification statement. The lower portion of the label is interpreted in the illustration.

Engine Identification Code

Cu. In. Displ.	Type	Year and Code						
		1964	1965	1966	1967	1968	1969	1970
289—V8	2-BBL.	C	C	C	C			
240—6	2-BBL.		V		V	V	V	V
240—6	taxi, police			B	B, E	B, E	B, E	B, E
233—6	2-BBL.	V						
302—V8	2-BBL.					F	F	F
302—V8	taxi, police						D	D
351—V8	W2-BBL. ②							H
352—V8	4-BBL.	X	X	X				
390—V8	2-BBL.			Y	Y	Y	Y	Y
390—V8	4-BBL.	Z①	Z①	Z①	Z①	Z①		
427—V8	4-BBL.	Q		Q	Q			
427—V8	2-4-BBL.	R	R	R	R			
428—V8	Police, 4-BBL.				P	P	P	P
428—V8	4-BBL.			Q①	Q①	Q①		
429—V8	2-BBL.						K	K
429—V8	4-BBL.						N①	N①

① —Used in Thunderbird also.
② —Windsor engine.

VALVE SPECIFICATIONS

YEAR AND MODEL		SEAT ANGLE (DEG.)		FACE ANGLE (DEG.)	VALVE LIFT INTAKE (IN.)	VALVE LIFT EXHAUST (IN.)	VALVE SPRING PRESSURE (VALVE OPEN) LBS. @ IN.	VALVE SPRING INSTALLED HEIGHT (IN.)	STEM TO GUIDE CLEARANCE (IN.)		STEM DIAMETER (IN.)	
		In.	Ex.						Intake	Exhaust	Intake	Exhaust
1964-65	6 Cyl.—223 Cu. In.	45	45	45	.369	.369	100 @ 1.78	1.780	.001-.0024	.0028-.0042	.3416-.3423	.3398-.3405
	6 Cyl.—240 Cu. In.	45	45	45	.376	.400	190 @ 1.33	1.700	.001-.0027	.001-.0027	.3416-.3423	.3416-.3423
	V8—289 Cu. In.; All	45	45	45	.368	.380	170 @ 1.39	1.780	.001-.0027	.002-.0037	.3416-.3423	.3406-.3413
	V8—352 Cu. In.; All	45	45	45	.408	.408	190 @ 1.42	1.820	.001-.0024	.001-.0024	.3711-.3718	.3711-.3718
	V8—390 Cu. In.; All	45	45	45	.408	.408	200 @ 1.42	1.820	.001-.0024	.001-.0024	.3711-.3718	.3721-.3728
	V8—427 Cu. In.; All	30	45	②	.500	.500	200 @ 1.42	1.820	.001-.0024	.002-.0034	.3711-.3718	.3701-.3708
1966	6 Cyl.—240 Cu. In.	45	45	45	.376	.400	190 @ 1.33	1.700	.001-.0027	.001-.0027	.3416-.3423	.3416-.3423
	V8—289 Cu. In.	45	45	45	.368	.380	170 @ 1.39	1.780	.001-.0027	.002-.0037	.3416-.3423	.3416-.3423
	V8—352 Cu. In.	45	45	45	.408	.408	190 @ 1.42	1.820	.001-.0024	.001-.0024	.3711-.3718	.3711-.3718
	V8—390 Cu. In.	45	45	45	.408	.408	220 @ 1.38	1.820	.001-.0024	.001-.0024	.3711-.3718	.3721-.3728
	V8—427 Cu. In.	30	45	②	.500	.500	270 @ 1.42	1.820	.001-.0024	.002-.0034	.3711-.3718	.3701-.3708
	V8—428 Cu. In.	45	45	45	.516	.516	270 @ 1.32	1.820	.001-.0024	.001-.0024	.3711-.3718	.3711-.3718
1967	6 Cyl.—240 Cu. In.	45	45	45	.376	.400	190 @ 1.33	1.700	.001-.0024	.001-.0027	.3416-.3423	.3416-.3423
	V8—289 Cu. In.	45	45	45	.368	.380	170 @ 1.39	1.780	.001-.0024	.002-.0037	.3416-.3423	.3416-.3423
	V8—390 Cu. In.	45	45	45	.437	.437	220 @ 1.38	1.820	.001-.0024	.001-.0024	3711-.3718	.3721-.3728
	V8—427 Cu. In.	30	45	②	.500	.500	270 @ 1.32	1.820	.001-.0024	.002-.0034	.3711-.3718	.3701-.3708
	V8—428 Cu. In.	45.	45	45	.437	.437	270 @ 1.32	1.820	.001-.0024	.001-.0024	.3711-.3718	.3711-.3718
1968-69	6 Cyl.—240 Cu. In.	45	45	45	.376	.400	190 @ 1.33	1.700	.001-.0024	.001-.0027	.3416-.3423	.3416-.3423
	V8—302 Cu. In.	45	45	45	.427	.426	180 @ 1.23	1.660	.001-.0027	.0015-.0032	.3416-.3423	.3411-.3418
	V8—390 Cu. In. 2-BBL.	45	45	45	.427	.437	220 @ 1.38	1.820	.001-.0024	.001-.0024	.3711-.3718	.3721-.3728
	V8—390 Cu. In. 4-BBL.	45	45	45	.437	.437	270 @ 1.32	1.820	.001-.0024	.001-.0024	.3711-.3718	.3721-.3728
	V8—428 Cu. In.	45	45①	①	.437	.437	270 @ 1.32	1.820	.001-.0024	.001-.0024	.3711-.3718	.3711-.3718
	V8—428 Cu. In. Interceptor	45	45①	①	.437	.437	270 @ 1.32	1.820	.001-.0024	.001-.0024	.3711-.3718	.3711-.3718
	V8—429 Cu. In.	45	45	45	.437	.437	306 @ 1.36	1.820	.001-.0027	.001-.0027	3416-.3423	.3416-.3423
1970	6 Cyl.—240 Cu. In.	45	45	45	.376	.400	190 @ 1.33	1.700	.001-.0024	.001-.0027	.3416-.3423	.3416-.3423
	V8—302 Cu. In.	45	45	45	.427	.426	180 @ 1.23	1.660	.001-.0027	.0015-.0032	.3416-.3423	.3411-.3418
	V8—351 Cu. In.	45	45	45	.418	.448	215 @ 1.34	1.790	.001-.0027	.0015-.0032	.3416-.3423	.3411-.3418
	V8—429 Cu. In.	45	45	45	.437	.437	306 @ 1.36	1.820	.001-.001	.001-.001	.3711-.3718	.3706-.3713
1971	6 Cyl.—240 Cu. In.	45	45	45	.376	.400	197 @ 1.30	1.700	.0010-.0027	.0010-.0027	.3416-.3423	.3416-.3423
	V8—302 Cu. In.	45	45	45	.368	.380	200 @ 1.31	1.690	.0010-.0027	.0015-.0032	.3416-.3423	.3411-.3418
	V8—351 Cu. In.	45	45	45	.418	.448	215 @ 1.34	1.790	.0010-.0027	.0015-.0032	.3416-.3423	.3411-.3418
	V8—390 Cu. In.	45	45	45	.427	.431	220 @ 1.38	1.820	.0010-.0027	.0015-.0032	.3711-.3718	.3706-.3713
	V8—400 Cu. In.	45	45	45	.427	.431	220 @ 1.38	1.820	.0010-.0027	.0015-.0032	.3416-.3423	.3411-.3418
	V8—429, Cu. In. 2-4-BBL.	45	45	45	.442	.486	253 @ 1.33	1.810	.0010-.0027	.0010-.0027	.3416-.3423	.3416-.3423

① —Beginning mid-year 1968, the 428 Cu. In. engine, except Police, used 30° intake valve seat angle and 29° intake valve face angle. ② —Intake 29°, exhaust 44°

TUNE-UP SPECIFICATIONS

YEAR	MODEL	SPARK PLUGS		DISTRIBUTOR		IGNITION TIMING (Deg.) ▲	CRANKING COMP. PRESSURE (Psi)	VALVES			FUEL PUMP PRESSURE (Psi)	IDLE SPEED (Rpm) ★
		Type	Gap (In.)	Point Dwell (Deg.)	Point Gap (In.)			Tappet (Hot) Clearance (In.)		Intake Opens (Deg.)		
								Intake	Exhaust			
1964-65	6 Cyl.—223 Cu. In.; M.T.	BTFG	.034	37	.025	4B	150	Zero	Zero	23B	4.5	475
	6 Cyl.—223 Cu. In.; A.T.	BTFG	.034	37	.025	10B	150	Zero	Zero	23B	4.5	475
	6 Cyl.—240 Cu. In.; M.T.	BF42	.034	37	.025	6B	150	Zero	Zero	23B	4.5	475
	6 Cyl.—240 Cu. In.; A.T.	BF42	.034	37	.025	8B	150	Zero	Zero	23B	4.5	475
	V8—289 Cu. In.; M.T.	BF42	.034	27	.015	6B	150	Zero	Zero	20B	5.0	575
	V8—289 Cu. In.; A.T.	BF42	.034	27	.015	10B	150	Zero	Zero	20B	5.0	500
	V8—352 Cu. In.; M.T.	BF42	.034	27	.015	6B	180	Zero	Zero	22B	5.0	500
	V8—352 Cu. In.; A.T.	BF42	.034	27	.015	10B	180	Zero	Zero	22B	5.0	475
	V8—390 Cu. In.; M.T.	BF42	.034	27	.015	4B	180	Zero	Zero	26B	5.0	500
	V8—390 Cu. In.; A.T.	BF42	.034	27	.015	6B	180	Zero	Zero	26B	5.0	500
	V8—427 Cu. In.; M.T.	BF32	.030	35	.020	8B	180	.025H	.025H	5A	6.0	700
1966	6 Cyl.—240 Cu. In.; M.T.	BTF42	.034	39	.025	6B⊙	150	Zero	Zero	23B	5.0	525⊙
	6 Cyl.—240 Cu. In.; A.T.	BTF42	.034	39	.025	12B⊙	150	Zero	Zero	23B	5.0	525
	V8—289 Cu. In.	BF42	.034	28	.017	6B⊙	150	Zero	Zero	20B	5.0	★
	V8—352 Cu. In.	BF42	.034	28	.017	10B	180	Zero	Zero	22B	6.0	550⊙
	V8—390 Cu. In.	BF42	.034	28	.017	10B⊙	180	Zero	Zero	26B	6.0	600⊙
	V8—427 Cu. In.	BF32	.034	23	.020	8B	180	.025H	.025H	5A	6.0	800
	V8—428 Cu. In.	BF42	.030	28	.017	10B⊙	180	Zero	Zero	16B	6.0	★
1967	6 Cyl.—240, M.T.	BF42	.034	39	.025	6B⊙	150	Zero	Zero	12B	5	525⊙
	6 Cyl.—240, A.T.	BF42	.034	39	.025	10B⊙	150	Zero	Zero	12B	5	525⊙
	V8—289, All	BF42	.034	28	.015	6B⊙	150	Zero	Zero	16B	5	525⊙
	V8—390, 2-BBL., A.T.	BF42	.034	28	.015	10B⊙	150	Zero	Zero	16B	5	500⊙
	V8—390, 4-BBL., M.T.	BF42	.034	28	.015	10B	150	Zero	Zero	16B	5	575
	V8—390, 4-BBL., A.T.	BF42	.034	28	.015	10B	150	Zero	Zero	16B	5	500
	V8—427 All	BF32	.030	23	.020	8B	180	.025H	.025H	48B	6	800
	V8—428 All	BF42	.034	28	.017	10B⊙	180	Zero	Zero	16B	6	600
1968	6 Cyl. 240	BF42	.034	39	.025	6B	150	Zero	Zero	12B	5	★
	V8—302	BF32	.034	27	.021	6B	150	Zero	Zero	15B	5	625●●
	V8—390 2-BBL.	BF32	.034	27	.021	6B	150	Zero	Zero	13B	6	625●●
	V8—390, 4-BBL.	BF32	.034	27	.021	6B	150	Zero	Zero	16B	6	625●●
	V8—390, G.T.	BF32	.034	27	.021	6B	150	Zero	Zero	16B	6	700●●
	V8—428	BF32	.034	27	.021	6B	150	Zero	Zero	16B	6	625●●
	V8—428 Interceptor	BF32	.034	27	.021	6B	150	Zero	Zero	18B	6	N.A.
	V8—429	BF42	.034	27	.021	6B	150	Zero	Zero	16B	6	550
1969	6 Cyl.—240	BF42	.034	39	.027	6B	150	Zero	Zero	6B	5	★
	V8—302, 2-BBL.	BF42	.034	27	.021	6B	150	Zero	Zero	6B	5	625●●
	V8—390, 2-BBL.	BF42	.034	27	.021	6B	150	Zero	Zero	6B	5	625●●
	V8—428, 4-BBL.	BF32	.034	29	.021	6B	180	Zero	Zero	6B	6	550
	V8—429, 2-BBL.	BF42	.034	29	.021	6B	180	Zero	Zero	6B	6	550
	V8—429, 4-BBL.	BF42	.034	29	.021	6B	180	Zero	Zero	6B	6	550
1970	6 Cyl.—240	BF42	.034	39	.027	6B	150	Zero	Zero	6B	4-6	★
	V8—302, 2-BBL.	BF42	.034	27	.021	6B	150	Zero	Zero	6B	4-6	625●●
	V8—351, 2-BBL.	BF42	.034	27	.017	6B	170	Zero	Zero	11B	4-6	600
	V8—390, 2-BBL.	BF42	.034	27	.021	6B	170	Zero	Zero	18B	5-6	650
	V8—428, 4-BBL.	BF32	.035	24-29	.021	6B	190	Zero	Zero	18B	5-6	600
	V8—429, 2-BBL.	BF42	.034	27	.021	6B	180	Zero	Zero	6B	5.5-6.5	550
	V8—429, 4-BBL.	BF42	.034	27	.021	6B	180	Zero	Zero	6B	5.5-6.5	550

TUNE-UP SPECIFICATIONS

YEAR	MODEL	SPARK PLUGS Type	SPARK PLUGS Gap (In.)	DISTRIBUTOR Point Dwell (Deg.)	DISTRIBUTOR Point Gap (In.)	IGNITION TIMING (Deg.) ▲	CRANKING COMP. PRESSURE (Psi)	VALVES Tappet (Hot) Clearance (In.) Intake	VALVES Tappet (Hot) Clearance (In.) Exhaust	Intake Opens (Deg.)	FUEL PUMP PRESSURE (Psi)	IDLE SPEED (Rpm) *
1971	6 Cyl.—240	BF42	.034	38	.027	6B	150	Zero	Zero	12B	4-6	500 ①
	V8—302, 2-BBL.	BF42	.034	27	.021	6B	150	Zero	Zero	16B	4-6	575 ②
	V8—351, 2-BBL.	BF42	.034	27	.021	6B	170	Zero	Zero	11B	4-6	575 ②
	V8—390, 2-BBL.	BF42	.034	27	.021	6B	180	Zero	Zero	13B	5.5-6.5	575 ①
	V8—400, 2-BBL.	BF42	.034	27	.021	6B	180	Zero	Zero	18B	5.5-6.5	575 ③
	V8—429, 2-BBL.	BF42	.034	27	.021	6B	180	Zero	Zero	16B	5.5-6.5	600
	V8—429, 4-BBL.	BF42	.034	27	.021	6B	180	Zero	Zero	16B	5.5-6.5	600
	V8—429, 4-BBL. ④	BF42	.034	30	.017	6B	180	Zero	Zero	16B	5.5-6.5	600

B—Before top dead center.
TDC—Top dead center.
●—With mechanical, zero lash rocker arms, see text.
★—With manual transmission in N and automatic in D. Add 50 rpm if equipped with air conditioning.
▲—With vacuum advance disconnected and plugged. NOTE: These settings are only approximate. Engine design, altitude, temperature, fuel octane rating and the condition of the individual engine are all factors which can influence timing. The limiting advance factor must, therefore, be the "knock point" of the individual engine.

*—M.T.—600
A.T.—500
●●—A.T. 550 RPM
⑥—1966 with thermactor system:
timing: 6 Cyl. & V8 289—TDC
idle speed: 6 Cyl. M.T.—650 rpm
V8 352—635, 289—600
V8 390 A.T.—500

⑥—1967 with thermactor system:
6 Cyl. M.T.—TDC A.T.—4B
V8—289 exc. Hi Per.—TDC
V8—289—Hi Per.—12B
V8—390, 428—6B
①—M.T.—775
②—M.T.—700
③—M.T. 725
④—Thunderbird

CAUTION

General adoption of anti-pollution laws has changed the design of almost all car engine production to effectively reduce crankcase emission and terminal exhaust products. It has been necessary to adopt stricter tune-up rules, especially timing and idle speed procedures. Both of these values are peculiar to the engine and to its application, rather than to the engine alone. With this in mind, car manufacturers supply idle speed data for the engine and application involved. This information is clearly displayed in the engine compartment of each vehicle.

GENERAL CHASSIS AND BRAKE SPECIFICATIONS

YEAR	MODEL	CHASSIS Overall Length (In.)	CHASSIS Tire Size	BRAKE CYLINDER BORE Master Cylinder (In.) Std.	BRAKE CYLINDER BORE Master Cylinder (In.) Power	BRAKE CYLINDER BORE Wheel Cylinder Diameter (In.) Front	BRAKE CYLINDER BORE Wheel Cylinder Diameter (In.) Rear	BRAKE DRUM Diameter (In.) Front	BRAKE DRUM Diameter (In.) Rear
1964	Pass.	209.8	7.50 x 14	1.0	.875	1³/₃₂	¹⁵/₁₆	11.03	11.03
	Sta. Wagon	209.8	8.00 x 14	1.0	.875	1³/₃₂	¹⁵/₁₆	11.03	11.03
	Thunderbird	205.4	8.00 x 15	.875	.875	1³/₃₂	¹⁵/₁₆	11.09	11.09
1965	Pass.	210.0	7.35 x 15	1.0	.875	1³/₃₂	¹⁵/₁₆	11.03	11.03
	Sta. Wagon	210.0 ▲	8.15 x 15	1.0	.875	1³/₃₂	¹⁵/₁₆	11.03	11.03
	Thunderbird	205.4	8.15 x 15	.937	.937	1¹⁵/₁₆	¹⁵/₁₆	11.87 Disc.	11.03
1966	Pass.	210.0	7.35 x 15	1.0	.875	1³/₃₂ ■	³¹/₃₂	11.03⑦	11.03
	Sta. Wagon	210.9	8.15 x 15	1.0	.875	1¹/₁₆ ■	⅞	11.03⑦	11.03
	Thunderbird	205.4	8.15 x 15	.937	.937	1¹⁵/₁₆	¹⁵/₁₆	11.87 Disc.	11.03
1967	Pass.	213.0	7.75 x 15①	1.0	.937	1³/₃₂ ■	³¹/₃₂	11.02⑦	11.03
	Sta. Wagon	213.9	8.45 x 15	1.0	.937	1³/₃₂ ■	¹⁵/₁₆	11.03⑦	11.03
	Thunderbird	206.9●	8.15 x 15	1.0	1.0	1¹⁵/₁₆	¹⁵/₁₆	11.87 Disc.	11.03
1968-70	Pass.	213.9	7.75 x 15③	1.0	1.0	1³/₃₂ ④	³¹/₃₂	11.03③	11.03
	Sta. Wagon	216.9	8.45 x 15	1.0	1.0	1³/₃₂ ④	¹⁵/₁₆	11.03③	11.03
	Thunderbird	212.5	8.15 x 15	1.0	1.0	2³/₄	¹⁵/₁₆	11.82 Disc.	11.03
1971	Pass.	216.2	F78 x 15	1.0	1.0	1⅛	¹⁵/₁₆	11.03⑤	11.03
	Sta. Wagon	219.2	H70 x 15	1.0	1.0	1⅛	¹⁵/₁₆	11.03⑤	11.03
	Thunderbird	212.5	215R x 15	1.0	1.0	2³/₄	¹³/₁₆	11.72 Disc	11.03

▲—10-passenger wagon—210.9.
■—Disc. brakes opt. 1.938
●—4 Door Landau—209.4
①—Cars with 390 and 428 engines 8.15 x 15

⑦—Cars with disc brakes—11.87
③—W/302—8.15 x 15
—W/390 & 428—8.45 x 15
④—W/Disc.—2³/₄

⑤—W/Disc.—11.72
Tire Note:—All tire information is shown on a decal affixed to the inside door of the glove compartment.

GENERAL ENGINE SPECIFICATIONS

YEAR	CU. IN. DISPLACEMENT	CARBURETOR	DEVELOPED HORSEPOWER @ RPM	DEVELOPED TORQUE @ RPM (FT. LBS.)	A.M.A. HORSEPOWER	BORE AND STROKE (IN.)	ADVERTISED COMPRESSION RATIO	VALVE LIFTER TYPE	NORMAL OIL PRESSURE (PSI)
1964	6 Cyl.—223	1-BBL.	138 @ 4200	203 @ 2200	31.54	3.625 x 3.600	8.5-1	Mech.	45
	V8—289	2-BBL.	195 @ 4400	282 @ 2400	51.20	4.000 x 2.870	9.0-1	Hyd.	45
	V8—352	4-BBL.	250 @ 4400	352 @ 2800	51.20	4.000 x 3.500	9.3-1	Hyd.	45
	V8—390	2-BBL.	266 @ 4600	378 @ 2400	52.49	4.050 x 3.781	9.4-1	Hyd.	45
	V8—390	4-BBL.	300 @ 4600	427 @ 2800	52.49	4.050 x 3.781	10.1-1	Mech.	45
	V8—427	4-BBL.	410 @ 5600	476 @ 3400	57.33	4.233 x 3.781	11.5-1	Mech.	45
	V8—427	2-4-BBL.	425 @ 6000	480 @ 3700	57.33	4.233 x 3.781	11.2-1	Mech.	45
1965	6 Cyl.—240	1-BBL.	155 @ 4200	239 @ 2200	38.40	4.000 x 3.180	8.75-1	Hyd.	45
	V8—289	2-BBL.	200 @ 4400	282 @ 2400	51.20	4.000 x 2.870	9.3-1	Hyd.	45
	V8—352	4-BBL.	250 @ 4400	352 @ 2800	51.20	4.000 x 3.500	9.3-1	Hyd.	45
	V8—390	4-BBL.	300 @ 4600	427 @ 2800	52.49	4.050 x 3.781	10.1-1	Hyd.	45
	V8—390	4-BBL.	330 @ 5000	427 @ 3200	52.49	4.050 x 3.781	10.1-1	Mech.	45
	V8—427	2-4-BBL.	425 @ 6000	480 @ 3700	57.33	4.233 x 3.781	11.1-1	Mech.	45
1966	6 Cyl.—240	1-BBL.	150 @ 4000	234 @ 2200	38.40	4.000 x 3.180	9.2-1	Hyd.	45
	V8—289	2-BBL.	206 @ 4400	282 @ 2400	51.20	4.000 x 2.870	9.3-1	Hyd.	45
	V8—352	4-BBL.	250 @ 4400	352 @ 2800	51.20	4.000 x 3.500	9.3-1	Hyd.	45
	V8—390	2-BBL.	275 @ 4400	405 @ 2600	52.49	4.050 x 3.781	9.5-1	Hyd.	45
	V8—390	4-BBL.	315 @ 4600	427 @ 2800	52.49	4.050 x 3.781	10.5-1	Hyd.	45
	V8—427	4-BBL.	410 @ 5600	476 @ 3400	57.33	4.233 x 3.781	11.1-1	Hyd.	45
	V8—427	2-4-BBL.	425 @ 6000	480 @ 3700	57.33	4.233 x 3.781	11.1-1	Hyd.	45
	V8—428	4-BBL.	345 @ 4600	462 @ 2800	54.48	4.130 x 3.980	10.5-1	Hyd.	45
1967	6 Cyl.—240	1-BBL.	155 @ 4200	239 @ 2200	38.40	4.000 x 3.180	9.2-1	Hyd.	45
	V8—289	2-BBL.	200 @ 4400	282 @ 2400	51.20	4.000 x 2.870	9.3-1	Hyd.	45
	V8—390	2-BBL.	275 @ 4400	405 @ 2600	52.49	4.050 x 3.781	9.5-1	Hyd.	45
	V8—390	4-BBL.	315 @ 4600	427 @ 2800	52.49	4.050 x 3.781	10.5-1	Hyd.	45
	V8—427	4-BBL.	410 @ 5600	476 @ 3400	57.33	4.233 x 3.781	11.1-1	Mech.	45
	V8—427	2-4-BBL.	425 @ 6000	480 @ 3700	57.33	4.233 x 3.781	11.1-1	Mech.	45
	V8—428	4-BBL.	345 @ 4600	462 @ 2800	54.58	4.130 x 3.980	10.5-1	Hyd.	45
1968	6 Cyl.—240	1-BBL.	150 @ 4000	234 @ 2200	38.40	4.000 x 3.180	9.2-1	Hyd.	45
	V8—302	2-BBL.	210 @ 4600	300 @ 2600	51.20	4.000 x 3.000	9.0-1	Hyd.	45
	V8—390	2-BBL.	265 @ 4400	309 @ 2600	52.49	4.050 x 3.781	9.5-1	Hyd.	45
	V8—390	4-BBL.	315 @ 4600	427 @ 2800	52.49	4.050 x 3.781	10.5-1	Hyd.	45
	V8—428	4-BBL.	340 @ 5400	462 @ 2800	54.58	4.130 x 3.980	10.5-1	Hyd.	45
	V8—428	4-BBL.	360 @ 5400	459 @ 3200	54.58	4.130 x 3.980	10.5-1	Hyd.	45
	V8—429	4-BBL.	360 @ 4600	480 @ 2800	60.80	4.360 x 3.590	10.5-1	Hyd.	45
1969	6 Cyl.—240	1-BBL.	150 @ 4000	234 @ 2200	38.40	4.000 x 3.180	9.2-1	Hyd.	45
	V8—302	2-BBL.	210 @ 4400	295 @ 2400	51.20	4.000 x 3.000	9.5-1	Hyd.	45
	V8—390	2-BBL.	270 @ 4400	390 @ 2600	52.49	4.050 x 3.781	9.5-1	Hyd.	45
	V8—390	4-BBL.	280 @ 4400	403 @ 2600	52.49	4.050 x 3.781	10.5-1	Hyd.	45
	V8—428	4-BBL.	360 @ 5400	459 @ 3200	54.58	4.130 x 3.980	10.5-1	Hyd.	45
	V8—429	2-BBL.	320 @ 4400	460 @ 2200	60.80	4.360 x 3.590	10.5-1	Hyd.	45
	V8—429	4-BBL.	360 @ 4600	476 @ 2800	60.80	4.360 x 3.590	11.0-1	Hyd.	45
1970	6 Cyl.—240	1-BBL.	150 @ 4000	234 @ 2200	38.40	4.000 x 3.180	9.2-1	Hyd.	45
	V8—302	2-BBL.	210 @ 4400	295 @ 2400	51.20	4.000 x 3.000	9.5-1	Hyd.	45
	V8—351	2-BBL.	250 @ 4600	355 @ 2600	51.20	4.000 x 3.500	9.5-1	Hyd.	50
	V8—390	2-BBL.	270 @ 4400	390 @ 2600	52.54	4.050 x 3.781	9.5-1	Hyd.	45
	V8—428	4-BBL.	360 @ 5400	459 @ 3200	54.58	4.130 x 3.980	10.5-1	Hyd.	45
	V8—429	2-BBL.	320 @ 4400	460 @ 2200	60.80	4.360 x 3.590	10.5-1	Hyd.	45
	V8—429	4-BBL.	360 @ 4600	480 @ 2800	60.80	4.360 x 3.590	10.5-1	Hyd.	45
1971	6 Cyl.—240	1-BBL.	150 @ 4000	234 @ 2200	38.40	4.000 x 3.180	9.2-1	Hyd.	50-60
	V8—302	2-BBL.	220 @ 4600	300 @ 2600	51.20	4.000 x 3.000	9.5-1	Hyd.	45-55
	V8—351	2-BBL.	250 @ 4600	355 @ 2600	51.20	4.000 x 3.500	9.5-1	Hyd.	35-55
	V8—390	2-BBL.	265 @ 4400	390 @ 2600	52.54	4.052 x 3.784	9.5-1	Hyd.	45-65
	V8—400	2-BBL.	N.A.	N.A.	51.20	4.000 x 4.000	9.5-1	Hyd.	50-70
	V8—429	2-BBL.	320 @ 4400	460 @ 2200	60.80	4.360 x 3.590	10.5-1	Hyd.	35-75
	V8—429	4-BBL.	360 @ 4600	480 @ 2800	60.80	4.360 x 3.590	10.5-1	Hyd.	35-75

CRANKSHAFT BEARING JOURNAL SPECIFICATIONS

YEAR	MODEL	MAIN BEARING JOURNALS (IN.)				CONNECTING ROD BEARING JOURNALS (IN.)		
		Journal Diameter	Oil Clearance	Shaft End-Play	Thrust On No.	Journal Diameter	Oil Clearance	End-Play
1964–65	6 Cyl.—223 Cu. In.	2.4984–2.4988	.0009–.0032	.004–.008	3	2.2984–2.2988	.0009–.0029	.005–.011
	6 Cyl.—240 Cu. In.	2.3982–2.3990	.0008–.0024	.004–.008	5	2.1232–2.1236	.0009–.0029	.006–.013
	V8—289 Cu. In.	2.2486–2.2490	.0007–.0030	.004–.008	3	2.1232–2.1236	.0009–.0029	.006–.016
	V8—427 Cu. In.	2.7488–2.7492	.0010–.0031	.004–.010	3	2.4384–2.4388	.0013–.0032	.014–.024
	V8—Exc. Above	2.7488–2.7492	.0006–.0027	.004–.010	3	2.4384–2.4388	.0007–.0028	.006–.024
1966	6 Cyl.—240 Cu. In.	2.3982–2.3990	.0005–.0024	.004–.008	5	2.1228–2.1236	.0006–.0022	.006–.013
	V8—289 Cu. In.	2.2482–2.2490	.0005–.0024	.004–.008	3	2.1228–2.1236	.0008–.0026	.010–.020
	V8—352 Cu. In.	2.7484–2.7492	.0005–.0025	.004–.010	3	2.4380–2.4388	.0008–.0026	.010–.020
	V8—390 Cu. In.	2.7484–2.7492	.0005–.0025	.004–.010	3	2.4380–2.4388	.0008–.0026	.010–.020
	V8—427 Cu. In.	2.7484–2.7492	.0007–.0031	.004–.010	3	2.4380–2.4388	.0013–.0032	.014–.024
	V8—428 Cu. In.	2.7484–2.7492	.0005–.0025	.004–.010	3	2.4380–2.4388	.0008–.0026	.010–.020
1967	6 Cyl.—240 Cu. In.	2.3986–2.3990	.0008–.0024	.004–.008	5	2.1228–2.1236	.0006–.0026	.006–.013
	V8—289 Cu. In.	2.2482–2.2490	.0005–.0022	.004–.008	3	2.1228–2.1236	.0007–.0028	.010–.020
	V8—390 Cu. In.	2.7484–2.7492	.0005–.0025	.004–.010	3	2.4380–2.4388	.0007–.0028	.014–.020
	V8—427 Cu. In.	2.7484–2.7492	.0007–.0031	.004–.010	3	2.4380–2.4388	.0013–.0032	.014–.020
	V8—428 Cu. In.	2.7484–2.7492	.0008–.0012	.004–.010	3	2.4380–2.4388	.0008–.0022	.014–.020
1968	6 Cyl.—240 Cu. In.	2.3986–2.3990	.0008–.0024	.004–.008	5	2.1232–2.1246	.0007–.0028	.014–.020
	V8—302 Cu. In.	2.2486–2.2490	.0005–.0024	.004–.008	3	2.1232–2.1246	.0007–.0028	.014–.020
	V8—390, 427, 428, 429 Cu. In.	2.7488–2.7492	.0008–.0012	.004–.008	3	2.4384–2.4388	.0007–.0028	.014–.020
1969	6 Cyl.—240 Cu. In.	2.3982–2.3990	.0005–.0015	.004–.008	5	2.1228–2.1236	.0008–.0015	.006–.013
	V8—302 Cu. In.	2.2482–2.2490	.0005–.0015	.004–.008	3	2.1228–2.1236	.0008–.0015	.010–.020
	V8—390 Cu. In.	2.7484–2.7492	.0013–.0025	.004–.010	3	2.4380–2.4388	.0008–.0015	.010–.020
	V8—428 Cu. In.	2.7484–2.7492	.0010–.0020	.004–.010	3	2.4380–2.4388	.0020–.0030	.010–.020
	V8—429 Cu. In.	2.9994–3.0002	.0005–.0015	.004–.008	3	2.4992–2.5000	.0008–.0015	.010–.020
1970	6 Cyl.—240 Cu. In.	2.3982–2.3990	.0005–.0015	.004–.008	5	2.1228–2.1236	.0008–.0026	.006–.013
	V8—302 Cu. In.	2.2482–2.2490	.0005–.0015	.004–.008	3	2.1228–2.1236	.0008–.0026	.010–.020
	V8—351 Cu. In.	2.994–2.3002	.0013–.0025	.004–.008	3	2.3103–2.3111	.0008–.0026	.010–.020
	V8—390 Cu. In.	2.7484–2.7492	.0005–.0025	.004–.008	3	2.4380–2.4388	.0008–.0026	.010–.020
	V8—428 Cu. In.	2.7484–2.7492	.0008–.0020	.004–.008	3	2.4380–2.4388	.0008–.0026	.010–.020
	V8—429 Cu. In.	2.9994–3.0002	.0005–.0025	.004–.008	3	2.4992–2.5000	.0008–.0026	.010–.020
1971	6 Cyl.—240 Cu. In.	2.3982–2.3990	.0005–.0022	.004–.008	5	2.1228–2.1236	.0008–.0024	.006–.013
	V8—302 Cu. In.	2.2482–2.2490	.0005–.0024	.004–.008	3	2.1228–2.1236	.0008–.0026	.010–.020
	V8—351 Cu. In.	2.9998	.0012–.0029	.004–.008	3	2.3103–2.3111	.0007–.0025	.010–.020
	V8—390 Cu. In.	2.7484–2.7492	.0005–.0025	.004–.008	3	2.4380–2.4388	.0008–.0026	.010–.020
	V8—400 Cu. In.	2.7484–2.7492	.0009–.0026	.004–.008	3	2.3103–2.3111	.0008–.0026	.010–.020
	V8—429 Cu. In.	2.9994–3.0002	.0005–.0025	.004–.008	3	2.4995–2.500	.0008–.0026	.010–.020

BATTERY AND STARTER SPECIFICATIONS

YEAR	MODEL	BATTERY				STARTERS					
		Ampere Hour Capacity	Volts	Terminal Grounded	Amps.	Lock Test		No-Load Test			Brush Spring Tension (Oz.)
						Volts	Torque	Amps.	Volts	RPM	
1964	All Models	55▲	12	Neg.	580	5.0	14.8	95	12	9500	52
1965–71	All Models	45■③	12	Neg.	670	5.0	①	70	12	9500②	40

▲—65 & 70 Ampere-Hour Available.
■—55, 70, 80 & 85 Ampere-hour available.

① —9 in.—9.0
4.5 in.—15.5.

② —428, 429—11,000.
③ —428, 429—80 ampere hour.

CAPACITIES

YEAR	MODEL	ENGINE CRANKCASE ADD 1 Qt. FOR NEW FILTER	TRANSMISSIONS Pts. TO REFILL AFTER DRAINING Manual 3-Speed	4-Speed	Automatic	DRIVE AXLE (Pts.)	GASOLINE TANK (Gals.) ▲	COOLING SYSTEM (Qts.) WITH HEATER
1964	6 Cyl.	4	3	None	18½*	5	20	16
	V8, 260, 289 Cu. In.	4	3	3	18½*	5	20	15
	V8, 352 Cu. In.	5	3	3	18½*	5	20	20
	V8, 390, 406, 427 Cu. In.	5	3	3	18½*	5	20	20
	T-Bird	5	None	None	20	5	20	20
1965	6 Cyl.	4	3	None	18½*	5	20	16
	V8, 289	4	3	4	18½*	5	20	16
	V8, 352	5	3	4	18½*	5	20	20½
	V8, 390, 427	5	3	4	18½*	5	20	20½
	T-Bird	5	None	None	20	5	20	20½
1966	6 Cyl.	4	3½	None	20	5	25	13
	V8—289	4	3½	None	20	5	25	15
	V8—352	5	3½	None	20	5	25	20½
	V8—390	4	3½	4	26*	5	25	20½
	V8—427	5	None	4	None	5	25	20½
	V8—428	4	None	4	26	5	25	20½
	T-Bird	4	None	None	26*	5	22	20½
1967	6 Cyl.	4	3½	None	20	5	25	13
	V8—289	4	3½	None	20	5	25	15
	V8—390	4	3½	4	26	5	25	20½
	V8—427	5	None	4	None	5	25	20½
	V8—428	4	None	4	26	5	25	20½
	T-Bird	4	None	None	26*	5	22	20½
1968–69	6 Cyl.	4½	3½	None	20	5	25▲	13
	V8—302	5	3½	None	22	5	25▲	15
	V8—390	5	3½	4	22	5	25▲	20½
	V8—428	5	None	4	26	5	25▲	20½
	V8—429 (T-Bird)	5	None	None	26½	5	24	20½
1970	6 Cyl.—240	4½	3½	None	22	5	24½①	14.4
	V8—302	5	3½	None	22	5	24½①	15.4
	V8—351	5	3½	None	22	5	24½①	16.5
	V8—390	5	3½	None	26	5	24½①	20.1
	V8—428	5	None	4	26	5	24½①	20.1
	V8—429	5	None	None	26	5	24½①	18.6
	V8—429 T-Bird	5	None	None	26	5	24	19.4
1971	6 Cyl.—240	5	None	None	22	5	22½①	14.1
	V8—302	5	None	None	22	5	22½①	15.4③
	V8—351	5	None	None	22	5	22½①	16.5④
	V8—390	5	None	None	26	5	22½①	20.1⑤
	V8—400	5	None	None	26	5	22½①	15.4④
	V8—429	5	None	None	26	5	22½①	18.6⑦
	V8—429 T-bird	5	None	None	26	5	22½⑦	19.4

① —Sta. wagon—22 gal.
② —With nonevaporative emission control 24 gal.
③ —With A/C—15.6.
④ —With A/C—16.8
⑤ —With A/C—20.5.

⑥ —With A/C—16.4.
⑦ —With A/C—19.0.
▲—1963-65 sta. wgn.—21 gals.
1966-70 sta. wgn.—20 gals.
*—Cruise-o-matic—20 pts.

TORQUE SPECIFICATIONS

YEAR	MODEL	CYLINDER HEAD BOLTS (FT. LBS.)	ROD BEARING BOLTS (FT. LBS.)	MAIN BEARING BOLTS (FT. LBS.)	CRANKSHAFT BALANCER BOLTS (FT. LBS.)	FLYWHEEL TO CRANKSHAFT BOLTS (FT. LBS.)	MANIFOLD (FT. LBS.) Intake	MANIFOLD (FT. LBS.) Exhaust
1964-66	6 Cyl.—223 Cu. In.	105-115	40-45	95-105	70-90	75-85	26	26
	V8—260, 289; 6 Cyl. 240 Cu. In.	65-70	19-24■	60-70	70-90▲	75-85	13 1/2	15 1/2
	V8—352, 390, 428 Cu. In.	80-90	40-45	95-105	70-90	75-85	33 1/2	15
	V8—406, 427 Cu. In.	100-110	53-58	95-105	70-90	75-85	33 1/2	15
1967-68	6 Cyl.—240 Cu. In.	70-75	40-45	60-70	130-145	75-85	25	25
	V8—289, 302 Cu. In.	65-70	19-24	60-70	70-90	75-85	21	15 1/2
	V8—390, 428, 429 Cu. In.	80-90	40-45	95-105	70-90	75-85	33 1/2	15 1/2
	V8—427 Cu. In.	100-110	53-58	95-105	70-90	75-85	33 1/2	15 1/2
1969	6 Cyl.—240 Cu. In.	70-75	40-45	60-70	130-150	75-85	25	25
	V8—302 Cu. In.	65-72	19-24	60-70	70-90	75-85	24	14
	V8—390, 428 Cu. In.	80-90	40-45	95-105	70-90	75-85	33 1/2	21
	V8—429 Cu. In.	130-140	40-45	95-105	70-90	75-85	27 1/2	30 1/2
1970	6 Cyl.—240 Cu. In.	70-75	40-45	60-70	130-150	75-85	25	25
	V8—302 Cu. In.	65-72	19-24	60-70	70-90	75-85	24	14
	V8—351 Cu. In.	95-100	40-45	95-105	70-90	75-85	23-25	18-24
	V8—390, 428 Cu. In.	80-90	**	95-105	70-90	75-85	32-35	18-24
	V8—429 Cu. In.	130-140	40-45	95-105	70-90	75-85	27 1/2	30 1/2
1971	6 Cyl.—240 Cu. In.	70-75	40-45	60-70	130-150	75-85	25	25
	V8—302 Cu. In.	65-72	19-24	60-70	70-90	75-85	24	14
	V8—351 Cu. In.	95-100	40-45	95-105	70-90	75-85	23-25	18-24
	V8—390 Cu. In.	80-90	40-45	95-105	70-90	75-85	32-35	18-24
	V8—400 Cu. In.	N.A.	40-45	95-105	70-90	75-85	N.A.	N.A.
	V8—429 Cu. In.	130-140	40-45	95-105	70-90	75-85	27 1/2	30 1/2

▲—6 Cyl. 240 Cu. In. = 130-145. ■—6 Cyl. 240 Cu. In. = 40-45. **—390—40-45; 428—53-58

CYLINDER HEAD BOLT TIGHTENING SEQUENCE

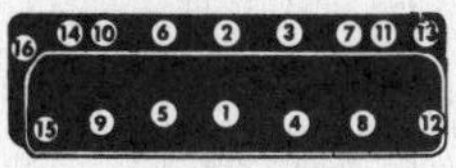

6 cyl.

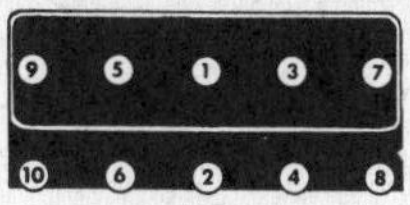

V8 (big block)

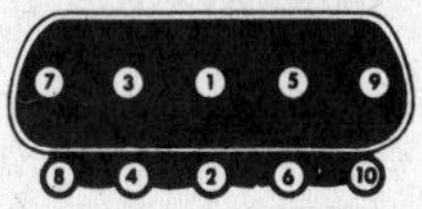

V8 260, 289, 302 and 351

WHEEL ALIGNMENT

YEAR	MODEL	CASTER Range (Deg.)	CASTER Pref. Setting (Deg.)	CAMBER Range (Deg.)	CAMBER Pref. Setting (Deg.)	TOE-IN (In.)	KING-PIN INCLINATION (Deg.)	WHEEL PIVOT RATIO Inner Wheel	WHEEL PIVOT RATIO Outer Wheel
1964	Ford	1N to 1P	0	0 to 1 1/4P	5/8P	1/8 to 1/4	7 1/8	20	17 1/2
	T-Bird	3/4N to 2 1/4N	1 1/2N	0 to 3/4P	3/8P	1/8 to 1/4	7 3/8	20	18 3/4
1965	Ford	0 to 2P	1P	1/4N to 1P	1/2P	1/32 to 7/32	7	20	17 1/2
	T-Bird	3 1/4N to 1 1/4N	1 1/2N	3/4N to 1P	1/2P	1/32 to 9/32	7	20	19 1/2
1966	Ford	1/2P to 1 1/2P	1P	1/4N to 1 1/4P	1/4P	1/8 to 1/4	7	20	17 1/2
	T-Bird	2 1/2N to 1/2P	1 1/2N	1/4N to 1 1/4P	1/2P	1/8 to 1/4	7	20	19 1/2
1967	Ford	1/2P to 1 1/2P	1P	1/4N to 1 1/4P	1/2P	1/8 to 1/4	7 3/4	20	18 1/8
	T-Bird	1/2N to 1 1/2P	1/2P	1/2P to 1 1/2P	1P	1/8 to 1/4	7 3/4	20	18 1/8
1968-69	Ford	0 to 2P	1P	1/4N to 1 1/4P	1/2P	1/16 to 5/16	7 3/4	20	18 1/8
	T-Bird	0 to 2P	1P	1/4N to 1 1/4P	1/2P	1/16 to 5/16	7 3/4	20	18 1/4
1970-71	Ford	1/2P-1 1/2P	1P	0 to 1P	1/2P	1/8 to 1/4	7 3/4	20	18 1/8
	T-Bird	1/2P-1 1/2P	1P	0 to 1P	1/2P	1/8 to 1/4	7 3/4	20	18 1/4

N—Negative.
P—Positive.

AC GENERATOR AND REGULATOR SPECIFICATIONS

YEAR	ALTERNATOR			REGULATOR						
	Part No. or Manufacturer	Field Current @ 12 V.	Output @ Generator RPM	Part No. or Manufacturer	Field Relay			Regulator		
					Air Gap (in.)	Point Gap (in.)	Volts to Close	Air Gap (in.)	Point Gap (in.)	Volts @ 75°
1964	C3SF10300A	2.9–3.1	40	C4UF10316A	.026	.018	3–4	.048	.013	14.1–14.9
	C4SF10300A	2.9–3.1	42	C4TF10316B	.026	.018	3–4	.048	.013	14.1–14.9
	C30F10300E	2.9–3.1	42							
1965	Autolite	2.8–3.3	42	Autolite	.018	N.A.	2.5–4	.052	.020	14.1–14.9
	Autolite	2.8–3.3	45							
	Autolite	2.8–3.3	55							
	Leece-Nevelle	2.8–3.3	53	Leece-Nevelle	.010	.019	1.6–2.6	.050	.019	14.1–14.9
				Leece-Nevelle	.012	.025	6.2–7.2	.050	.019	14.1–14.9
	Leece-Nevelle	2.8–3.3	60							
1966	Autolite	2.9	42	Autolite	0.014	N.A.	2.5–4	.053	.020	14.1–14.9
	Autolite	2.9	45	Autolite transistor	①	①	2.5–4	①	①	14.1–14.9
	Autolite	2.9	55							
	Leece-Nevelle	2.9	53	Leece-Nevelle 53 Amp only	.010	.019	1.6–2.6	.050	.019	14.1–14.9
	Leece-Nevelle	4.6	60	Leece-Nevelle 53 and 60 Amp	.012	.025	6.2–7.2	.050	.019	14.1–14.9
	Leece-Nevelle	2.9	60							
1967	Autolite	2.5	38	Autolite	.014	N.A.	2.5–4	N.A.	N.A.	13.9–14.9
	Autolite	2.9	42							
	Autolite	2.9	45							
	Autolite	2.9	55							
	Autolite	4.6	60	Autolite Transistor	①	①	2.5–4	①	①	13.9–14.9
	Leece-Nevelle	2.9	53	Leece-Nevelle 53 Amp only	.010	.019	1.6–2.6	.050	.019	13.9–14.9
	Leece-Nevelle	2.9	65	Leece-Nevelle 53 and 60 Amp	.012	.025	6.2–7.2	.050	.019	13.9–14.9
1968	Autolite C6AF10300C	2.9	42	Autolite	①	①	4.2–9.0	①	①	13.5–15.3
	Autolite C6AF10300G	2.9	55							
	Autolite C6TF10300F	2.9	65							
	Leece-Nevelle	2.9	65	Leece-Nevelle	0.012	0.025	6.2–7.2	0.047	0.019	13.9–14.9
1969	Autolite	2.9	42	Autolite	①	①	4.2–9.0	①	①	13.5–15.3
	Autolite	2.9	55							
	Autolite	2.9	65							
	Leece-Nevelle	2.9	65	Leece-Nevelle	0.012	0.025	6.2–7.2	0.047	0.019	13.9–14.9
1970–71	Autolite	2.9	42	Autolite	①	①	2.0–4.2	①	①	13.5–15.5
	Autolite	2.9	55							
	Autolite	2.9	65							
	Leece-Nevelle	2.9	65	Leece-Nevelle	.012	.025	6.2–7.2	.047	.019	13.9–14.9

① —Transistorized regulator—not adjustable.

GENERATOR AND REGULATOR SPECIFICATIONS

YEAR	MODEL	GENERATORS				REGULATORS		
		Field Current in Amperes At 12 Volts	Brush Spring Tension (Oz.)	Cut-Out Relay		Current & Voltage Air Gaps (In.)	Current Regulator Setting	Voltage Regulator Setting
				Air Gap (In.)	Closing Voltage			
1964	All	1.0–1.5	23	.010	12.7	.034	30	14.6–15.4

LIGHT BULBS
1965
FORD

Unit	① Candela Or Wattage	Trade Number	Unit	① Candela Or Wattage	Trade Number	Unit	① Candela Or Wattage	Trade Number
Headlight no. 1 (lower)	37.5 W.	4001	Instrument panel indicators			Illumination		
Headlight no. 2 (upper)	37.5/50 W.	4002	Hi beam	2 C.	1895	Speedometer	1.5 C.	1895
Front turn signal/parking	4/32 C.	1157-A	Oil pressure, coolant temp. hot/cold	2 C.	1895	Cluster	2 C.	1895
Rear turn signal & stop/tail	4/32 C.	1157	Alternator	2 C.	1895	Heater control panel	2 C.	1895
License plate	4 C.	1155	Turn signal	2 C.	1895	Clock	2 C.	1895
Back-up lights	32 C.	1156	Parking brake warning	2 C.	257	Radio dial	3 C.	1816
Spot light	30 W.	4405				Courtesy and/or Map (door mounted)	15 C.	1003
Dome light	15 C.	1003				Automatic transmission control	1 C.	161

①—Candela is the new International term for candlepower.

1966
FORD

Unit	① Candela Or Wattage	Trade Number	Unit	① Candela Or Wattage	Trade Number	Unit	① Candela Or Wattage	Trade Number
Standard equipment			Courtesy lamp ("C" pillar)	15 C.	1003	Spotlight 4.4" diameter	30 Watts	4405
Headlamp			Console lamp	6 C.	631	Clock	3 C.	1816
Hi-lo beam	37.5 & 50 Watts	4002	Courtesy lamp	12 C.	211	Automatic transmission	1 C.	161
Hi beam	37.5 Watts	4001	Instrument panel			Luggage compartment	15 C.	93
Front parking and turn indicator	4-32 C.	1157	Instruments	2 C.	1895	Cigar lighter	2 C.	1895
			Indicator hi beam	2 C.	1895	Engine compartment	6 C.	631
Rear lamp, stop and turn indicator (sedan)	4-32 C.	1157	Warning lights (oil & generator temperature hot/cold)	2 C.	1895	Tachometer	2 C.	1895
(sta. wag.)	4-32 C.	1157	Park brake rel. warning light	2 C.	1895	Trans. control selector indicator	1.5 C.	1445
License plate lamp	4 C.	1155	Heater control panel (fresh air)	3 C.	1895	Safety package		
Courtesy lamp (convertible)	6 C.	631	L & R turn indicator	2 C.	1895	Door lock	2 C.	1895
Dome lamp	15 C.	1003	Accessory equipment			Low fuel	2 C.	1895
Courtesy lamp (door mountings)	15 C.	1003	Back-up lamps	32 C.	1156	Door-warning open	2 C.	1895
			Radio pilot light	2 C.	1891	Emergency flasher	2 C.	1895
						Seat belt	2 C.	1895

①—Candela is the new International term for candlepower.

1967

Light Description	Ford		Thunderbird		Light Description	Ford		Thunderbird	
	Bulb No.	Candela or Wattage	Bulb No.	Candela or Wattage		Bulb No.	Candela or Wattage	Bulb No.	Candela or Wattage
Cargo lamp (station wagon)	1003	15C.	...	...	Seat belt warning	158	2C.	1891	2C.
Dome lamp	1003	15C.	...	...	Ash receptacle or cigar lighter	1895	2C.	1895	2C.
Engine compartment	631	6C.	...	...	Back-up lamps	1156	32C.	1076	32C.
Glove compartment	1895	2C.	631	6C.	Courtesy lamp	211	12C.	...	...
Headlamps hi-lo beam (outer or upper)	4002	37.5-50W	4002	37.5-50W	"C" pillar	1003	15C.	1003	15C.
hi beam (inner or lower)	4001	37.5W	4001	37.5W	Console	1816	3C.	...	...
License lamp	97	4C.	97	4C.	Convertible and inst. panel	631	6C.	...	...
Luggage compartment	93	15C.	631	6C.	Door mounting or armrest	1004	15C.	211	12C.
Map light	...	...	211	12C.	Clock and ignition key	1816	3C.	158	2C.
Park and turn signal lamp	1157-A	4-32C.	1157-A	32C.	Control nomenclature (ex heater)	1895	2C.	1891 ②	2C.
Spotlight (4.4 inch dia.)	4405	30W	4405	30W	Control, right and left	1895	2C.	①	
Taillight, stop and turn signal	1157	4-32C.	1157	32C.	Gages and speedometer	1816	3C.	1445	1.5C.
Door lock warning	1895	2C.	256	1.6C.	Heater control	158	2C.	...	...
Emergency flasher	1895	2C.	1891	2C.	Radio dial light AM	1893	1.9C.	1891	2C.
Hi-beam indicator	158	2C.	158	2C.	AM-FM	1893	1.9C.	1892	1.3C.
Low fuel warning	158	2C.	1891	2C.	Tachometer	1895	2C.	1895	2C.
Oil press., alt and temp warning	158	2C.	158	2C.	Trans. control selector indicator (console)	1445③ ,1.5C.		158	2C.
Parking brake warning	257	1.6C.	1895	2C.	Turn signal indicators (L and R)	158	2C.	53 X ④	1C.

①—1445 (1.5C.) for speedometer, 1895 (2C.) for gages.
②—1816 (3C.) for identification light.
③—161 (1C.) for automatic transmission.
④—1895 (2C.) for turn signal, 1895G (2C.) for Minnesota and Wisconsin.

1968
FORD

Unit	(1) Candela Or Wattage	Trade Number
Standard equipment		
Headlamp		
Hi-lo beam	37.5 & 50 Watts	4002
Hi-beam	37.5 Watts	4001
Front parking, side marker and turn indicator	4-32 C.	1157
Rear lamp and stop and turn indicator		
sedan	4-32 C.	1157
sta. wag.	4-32 C.	1157
License plate lamp	4 C.	97
Courtesy lamp (convertible)	6 C.	631
Dome lamp	15 C.	1003
Courtesy lamp (door mounting)	6 C.	631
Courtesy lamp ("C" pillar)	15 C.	1003
Console lamp	3 C.	1816
Cargo lamp (sta. wag.)	15 C.	1003
Instrument panel		
Instruments	2 C.	1895
Indicator hi beam	2 C.	1895
Warning lights (oil & generator temperature hot/cold)	2 C.	1895
Park brake rel. warning light	2 C.	257
Heater control panel (fresh air)	2 C.	1895
L & R turn indicator	2 C.	1895
Accessory equipment		
Back-up lamps	32 C.	1156
Radio pilot light	1.9 C.	1893
Spotlight 4.4" diameter	30 Watts	4405
Clock	3 C.	1816
Automatic transmission	2 C.	158
Luggage compartment	6 C.	631
Engine compartment	6 C.	631
Tachometer	2 C.	1895
Trans. control selector indicator	1.5 C.	1445
Safety package		
Low fuel	2 C.	1895
Door-warning open	2 C.	1895
Emergency flasher	2 C.	1895
Seat belt	2 C.	1895

(1) Candela is the new international term for candlepower.

1968
THUNDERBIRD

Light Description	Bulb No.	Candela or Wattage①
Headlights—hi & lo	4002	37.5 & 50 W
Headlights—hi beams	4001	37.5 W
Front park & turn signal	1157A	4-32 C.
Rear tail/stop/turn signal	1157	4-32 C.

①—Candela is the international term for candlepower.

②—Natural amber color.

Light Description	Bulb No.	Candela or Wattage①
Back-up light	1156	32 C.
License plate light	97	4 C.
Roof quarter light	1003	15 C.
Glove compartment	1895	2 C.
Auto. trans. select. ind. fixed & tilt column	1445	1.5 C.
Door courtesy light	212 or 212-1	6 C.
Map light	212 or 212-1	6 C.
Luggage compartment	631	6 C.
Front side marker	97NA②	4 C.
Instrument panel courtesy light	631	6 C.
Hi-beam indicators	194	2 C.
Turn signal indicators	194	2 C.
Warning—brakes & belts	194	2 C.
Ignition switch	194	2 C.
Instruments	194	2 C.
Heater controls	1895	2 C.
Control nomenclature rear vent & wipers	194	2 C.
Fog lights—clear	4415	35 W
Fog light switch	53X	1 C.
Spotlight—4.4 inch diameter	4405	30 W
Manual A/C control	1895	2 C.
ATC control	1895	2 C.
Radio pilot light	1893	1.9 C.
Park brake signal	1895	2 C.
Safety convenience package		
Low fuel warning	1891	2 C.
Door ajar warning	256	1.6 C.
Seat belt warning	1891	2 C.
Emergency flasher warning	1891	2 C.
Cigar lighter	1895	2 C.
Engine compartment	631	6 C.
Portable trunk light	1003	15 C.
Auxiliary instrument center	1895	2 C.
Tachometer	1895	2 C.
Supplemental stop light	1156	32 C.

1969
FORD

Light Description	Bulb No.	Candela or Wattage①
AM/FM radio	1891	2 C.
AM/FM/stereo radio	1892	1.3 C.
Air conditioner control	1895	2 C.
Ash tray receptacle	1445	1.5 C.
Automatic transmission selector (column mtd)	194	2 C.
Auxiliary instrument cluster charge indicator	1895	2 C.
Back-up lights	1156	32 C.
Cargo light (station wagon)	1003	15 C.

①—Candela is the international term for candlepower.

Light Description	Bulb No.	Candela or Wattage①
Clock	1895	2C.
Courtesy light (54C, 57F)	631	6C.
Courtesy light (65A)	212	6C.
Courtesy light (C pillar)	1003	15C.
Courtesy light (convertible)	631	6C.
Dome light	1003	15C.
Door ajar (convenience group)	1895	2C.
Emergency flasher	1156	32C.
Engine compartment	631	6C.
Seat belt (convenience group or instrument panel)	1895	2C.
Speed control actuator indicator	161	1C.
Spotlight 4.4 inch diameter	4405	30 W
Trans. control selector indicator (floor shift)	1445	1.5C.
Trunk light (portable)	1003	15C.
Fog lights (clear)	4415	35W
Fog light switch	53X	1C.
Front parking and turn indicator	1157	3-32C.
Front-side marker	1178-A	4C.
Glove compartment	1816	3C.
Headlight hi-beam	4001	37.5W
Headlight—hi-lo beam	4002	37.5 & 50W
Heater control panel (fresh air)	1895	2C.
Ignition switch	1895	2C.
Indicator hi-beam	194	2C.
Indicator lights (oil & alternator temperature hot/cold brakes)	194	2C.
Instruments	194	2C.
L & R turn indicator	194	2C.
License plate light	97	4C.
Low fuel (convenience group)	1895	2C.
Luggage compartment	631	6C.
Park brake release indicator light	257	1.6C.
Park brake release indicator light (convenience group)	1895	2C.
Radio pilot light	1895	2C.
Rear light and stop and turn indicator	1157	3-32C.
Rear-side marker	194	2C.

1969
THUNDERBIRD

Light Description	Bulb No.	Candela or Wattage①
ATC control	1895	2C.
Auto. trans. select. ind. fixed & tilt column	1445	1.5C.
Back-up light	1156	32C.
Brakes & belts indicator	194	2C.
Cigar lighter	1895	2C.
Control nomenclature: rear vent & wipers	194	2C.
Cornering light	97A	4C.
	or 1195	50C.

①—Candela is the international term for candlepower.

Light Description	Bulb No.	Candela or Wattage ①	Light Description	Bulb No.	Candela or Wattage ①	Light Description	Bulb No.	Candela or Wattage ①
Door ajar indicator	256	1.6C.	Headlights—hi-beams	4001	37.5 W	Parking brake signal	1895	2C.
Door courtesy light	211 or		Heater controls	1895	2C.	Radio pilot light	1893	1.9C.
	211-1	12C.	Hi-beam indicators	194	2C.	AM/FM stereo pilot light	1892	1.3C.
Emergency flasher	1891	2C.	Ignition switch	161	1C.	Rear-side marker	194	2C.
Engine compartment	631	6C.	Instrument panel courtesy light	631	6C.	Rear tail/stop/turn signal	1157	3-32C.
Fog lights—clear	4415	35 W	Instruments	194	2C.	Roof quarter light	1003	15C.
Fog light switch	53X	1C.	License plate light	97	4C.	Seat belt indicator	1891	2C.
Front park & turn signal	1157A	3-32C.	Low fuel indicator	1891	2C.	Spotlight—4.4 inch diameter	4405	30W
Front side marker	1178A	4C.	Luggage compartment	631	6C.	Supplemental stop light	1156	32C.
Glove compartment	1895	2C.	Manual A/C control	1895	2C.	Trunk light (portable)	1003	15C.
Headlights—hi & lo	4002	37.5 & 50 W	Map light	212	6C.	Turn signal indicators	168A	3C.

1970
FORD

Light Description	Candle Power or Wattage	Trade No.	Light Description	Candle Power or Wattage	Trade No.	Light Description	Candle Power or Wattage	Trade No.
Standard equipment			Door courtesy 65A	6C.	212	AM-dual channel	1.9C.	1893
Headlamps—hi and lo	37.5-50 w	4002	Door courtesy 54C, 57F	6C.	631	AM-radio	1.9C.	1893
Headlamps—hi beam	37.5 w	4001	Rear side marker lamp	2C.	194	Defogger, rear window	2C.	1895
Front park & turn signal &			Instrument panel			Electric defroster, rear window	2C.	1895
emergency flasher	3-32 C.	1157A	Cluster	2C.	1895	Headlights on	1.6C.	256
Front side marker lamp	2C.	194	Cluster lights	2C.	194	Ash tray lamp (std. on some		
Rear tail/stop/turn signal sedan	3-32C.	1157	Glove box lamp	3C.	1816	models)	2 C.	1891
Rear tail/stop/turn signal			Hi-beam-indicators	2C.	194	AM/FM radio dial light (MPX)	2C.	1891
sta. wgn.	3-32C.	1157	Turn signal indicators	2C.	194	AM/FM/MPX stereo jewel	1.3C.	1892
Rear side marker/lower bulb	12C.	105	Warning lights, (oil, alt.,			AM stereo tape	1.9C.	1893
Back-up lamps sedans	32C.	1156	temp., brakes)	2C.	194	Convenience package		
sta. wgn.	32C.	1076	Seat belt warning (R.P.O.)	2C.	1891	Low fuel	2C.	1895
License plate lamp	4C.	97	Open door warning (taxis)	2C.	1895	Door ajar	2C.	1895
Dome lamp 62, 71, 54A,			Parking brake release			Seat belt	2C.	1895
B & E, 57B	12C.	105	warning (optional)	1.6C.	257	Parking brake	2C.	1891
Courtesy lamp "C" pillars			Instruments	2C.	194	Glove compartment	3C.	1816
68, 65, 57, 54C	12C.	105	Clock or block cover	2C.	1895	Speed control actuator indicator	1C.	161
Instrument courtesy lamp			Heater control panel	2C.	1895	Luggage compartment	6C.	631
convertible	6C.	631	Optional equipment			Portable trunk lamp	15C.	1003
Cargo lamps			Spotlight	30 w	4405	Engine compartment	6C.	631
sta. wgn. models 71A/B/C,			Fog lamps, clear	35 w	4415	Floor shift-PRND21	1.5C.	1895
71E; optional, 71D/H/J	12C.	105	Fog lamp switch	1C.	53X	Auxiliary inst. cluster		
			Air conditioner control	2C.	1895	(charge indicator)	2C.	1895

1970
THUNDERBIRD

Light Description	Candle Power or Wattage	Trade No.	Light Description	Candle Power or Wattage	Trade No.	Light Description	Candle Power or Wattage	Trade No.
Standard equipment			Front side marker lamp	2C.	194	Supplemental parking lamps	6C.	90
Headlamps—hi & lo	37.5 & 50 w.	4002	Cigar lighter	1.5C.	1445	Fog lights	35 w.	4415
Headlamps—hi beams	37.5 w.	4001N	Rear side marker lamp	2C.	194	Fog lights switch	1C.	53X
Front park and turn signal	3-32C.	1157A	Instrument panel			Radio pilot light (All)	1.9C.	1893
Rear tail/stop/turn signal	3-32C.	1157	Map	6C.	212	Spotlight	30 W.	4405
License plate	4C.	97	Brake and belt warnings	2C.	194	Parking brake signal	2C.	1895
Floor console ash tray	1.5C.	1445	Glove compartment	2C.	1895	Engine compartment	6C.	631
"C" pillar	12C.	105	Instrument panel courtesy	6C.	631	Portable trunk lamp	15C.	1003
Auto. trans. quadrant	1.5C.	1445	Ash tray	1.5C.	1445	High level taillamps	32C.	1156
Door courtesy	6C.	212	Hi-beam indicator	2C.	194	Convenience check group		
Luggage compartment	6C.	631	Turn signal indicators	2C.	168A	Low fuel	2C.	1891
Glove compartment	2C.	1895	Instruments	2C.	194	Lights on	2C.	1891
Back-up lamps	32C.	1156	Control nomenclature—			Door ajar	2C.	1891
Front side marker	2C.	194A	Wiper/Washer	2C.	194	Seat belt	2C.	1891
Front cornering lamp	50C.	1196	Heater-A/C Controls	2C.	1895	Rear window electric	1.3C.	1892
			Optional equipment			Defrost warning		

FUSES AND CIRCUIT BREAKERS

1965
FORD

Circuit	Protective Device	Location
Headlights	18 Amp. circuit breaker	Incorporated in lighting switch
Instrument panel, transmission indicator and ash receptacle lights	AGA-4 fuse	Fuse panel
Turn signals and back-up lights, radio, single-speed washer motor	SFE-14 fuse	Fuse panel
dual-speed washer motor	12 Amp. circuit breaker	Incorporated in washer switch
Seat belt warning, emergency warning	SFE-20 fuse	Fuse panel
Heater blower, power antenna, door open warning	SFE-20	Fuse panel
Electric windshield wiper (single-speed)	6 Amp. circuit breaker	Incorporated in wiper switch
(dual-speed)	12 Amp. circuit breaker	
Cigar lighter	SFG-14 fuse	Fuse panel
Air conditioner (integrated system)	25 Amp. circuit breaker	On ignition switch
Spot light	SFE-7.5 fuse	Cartridge in feed wire
Overdrive	AGC-15 fuse	On overdrive relay
Clock, dome, courtesy, map, cargo, trunk, glove box	SFE-9 fuse	Fuse panel
Tail, park, license plate, and horns	15 Amp. circuit breaker	Incorporated in lighting switch
Speed control	SFE 14 Amp. fuse	Cartridge in feed wire
Transistorized ignition	AGA-2 fuse	Cartridge in feed wire
Top control motor feed, power windows and seat, back lite	20 Amp. circuit breaker	At starter relay
Air conditioner economy	AGC-15 fuse	Cartridge in feed wire

1966
FORD

Circuit	Circuit Protection		Location
	Rating	Trade No.	
Power assists			
Convertible top circuit	14 gage wire Fuse link	(C6AB-14A094-A)	In wiring near starter relay
Convertible top with power option(s)	20 Amp. C.B. replaces fuse link		On starter relay
Power windows including backlite & tailgate	20 Amp. C.B.		
Power seats (four- and six-way)	20 Amp. C.B.		
Air conditioning circuit			
Air conditioner (economy)	15 Amp.	AGC 15	Fuse cartridge in feed wire
Air conditioner (selectair)	25 Amp. C.B.		On back of ignition switch at accessory terminal
Miscellaneous circuits			
Overdrive	20 Amp. fuse	AGC 15	On overdrive relay
Transistorized ignition	2 Amp. fuse	AGA 2	Fuse cartridge in feed wire
Speed control	14 Amp. fuse	SFE 14	
Spotlight	7.5 Amp. fuse	SFE 7.5	
Automatic headlamp dimmer	4 Amp. fuse	AGA 4	
Seat belt warning and parking brake warning	No protection		
Gage circuits			
Charge indicator	No protection		
Oil pressure indicator			
Engine water—cold			
Engine water—hot			
Motors			
Windshield wiper motor	Circuit		Integral with motor
Convertible top motor	breaker		
Power window motor			
Power seat motor			

Circuit	Circuit Protection		Location
	Rating	Trade No.	
Headlamp circuit			
Headlamps	18 Amp. C.B.		
High beam indicator lamp			Integral with lighting switch.
Rear lamps (tail lights & stop lights)	15 Amp. C.B.		
Front parking lamps			
Ignition switch lamp			
License lamp			
Horns			
Dome lamp circuit			
Dome lamp	9 Amp. fuse	SFE 9	On fuse panel (dome socket).
Courtesy lamps			
Cargo lamp			
Glove compartment lamp			
Luggage compartment lamp			
Clock			
Instrument panel illumination			
	4 Amp. fuse	AGA 4	Fuse panel (instrument panel socket) This circuit is connected into the light switch rheostat.
Clock light			
Instrument cluster lights			
Ash tray light			
PRNDL—Console or column light			
Radio light			
Heater control lights			
Heater circuit			
Heater and defroster motor	20 Amp. fuse	SFE 20 or AGC 20	Fuse panel (heater socket).
Safety convenience panel lamps			
Low fuel warning			
Seat belt warning			
Door ajar warning			
Door unlock warning			
Power antenna			
Station wagon under seat heater			
Open door warning			
Cigar Lighter Circuit			
Cigar lighter	14 Amp. fuse	SFG 14	Fuse panel (cigar lighter socket).
Cigar lighter plus Emergency warning option (hang-on or convenience panel)	20 Amp. fuse (replaces 14 Amp. cigar fuse)	SFE 20	
Radio Circuit			
Radio	14 Amp. fuse	SFE 14	Fuse panel (radio socket).
Back-up lamps			
Single speed washers			
Turn signal circuit			(After fuse flasher protects circuit for turn signal).
Windshield Wipers			
Single speed wipers	6 Amp. C.B.		Integral with windshield wiper switch.
Dual speed wipers and washers	12 Amp. C.B.		

1967
FORD

Circuit	Protective Device-Fuse or Circuit Breaker (C.B.) in Amperes	Location
Headlights.	18 C.B.	Integral with light switch
High beam indicator	18 C.B.	Integral with light switch
Horns, tail- and stoplights, parking lights	15 C.B.	Integral with light switch
Ignition switch light and license light	15 C.B.	
Clock, and lamps for dome, cargo		
Courtesy, glove and luggage compartments	SFE 9	Fuse panel
Lights for clock, radio, heater control, ash tray		
Instrument cluster, transmission selector (console or column)		Fuse panel
Heater and defroster motor, power antenna, stationwagon underseat heater, and warning lamps for low fuel, seat belt, door ajar, and door lock	SFE 20 or AGC 20	Fuse panel
Cigar lighter	SFE 14	Fuse panel
Cigar lighter and emergency warning (hang-on or conv. panel)	SFE 20	Fuse panel
Radio, back-up lights, single speed washers, turn signals	SFE 14	Fuse panel (after fuse, flasher protects circuit for turn signal)
Dual speed wipers and washers (Ford)	12 C.B.	Integral with windshield wiper switch
Convertible top circuit	Fuse link ①	In wiring near starter relay
Convertible top with power option(s)	20 C.B. replaces link	On starter relay
Power windows (including backlite and tailgate)	20 C.B.	
Power seats (four- and six-way)	20 C.B.	
Air conditioner (economy)	AGC 15	Fuse cartridge in feed wire
Air conditioner (selectair)	25 C.B.	②
Overdrive	AGC 15	On overdrive relay
Speed control	AGA 5	Fuse cartridge in feed wire
Spotlight	SFE 7.5	Fuse cartridge in feed wire
Automatic headlight dimmer	AGW 4	Fuse cartridge in feed wire
Seat belt warning and parking brake warning	SFE 7.5	Fuse cartridge in feed wire
Indicators for charge, oil press., engine coolant hot and cold	No protection	
Windshield wiper motor, convertible top motor, power window motor and power seat motor	Circuit breaker	Integral with motor

① —14 gage wire (C6AB-14A094-A)
② —On back of ignition switch at accessory terminal

1967
THUNDERBIRD

Circuit	Protective Device-Fuse or Circuit Breaker (C.B.) in Amperes	Location
Electric antenna	SFE 14	Fuse panel
Low fuel level warning	7.5 SFE, AGW	Fuse panel
Speed control, seat belt warning	AGW 4	Fuse panel
Radio	SFE, AGW 7.5	Fuse panel
Windshield washer, backup lights, door open warning	SFE, AGW, 4AG 7.5	Fuse panel
Turn signal	AGC, 3AG 15	Fuse panel
Instrument panel lights, ash receptacle light	SFE 6	Fuse panel
Trans. indicator light (PRNDL), glove box light	SFE 6	Fuse panel
Interior lights, dome, courtesy, map, luggage comp, clock	SFE 14	Fuse panel
Cigar lighter	SFE 15	Fuse panel
Emergency warning indicator	SFE 20	Fuse panel
Rear light wire	5-C.B.	Rear panel-LH rear of quarter panel
Tail lights, parking lights, license light	15-C.B.	In headlight switch
Headlights	18-C.B.	In headlight switch
Stoplights	15-C.B.	Fuse panel
Heater, air conditioner	30-C.B.	Fuse panel
Electric windows, electric seats, horns	20-C.B.	Fuse panel
Windshield wiper	Hydraulically operated	No. fuse or circuit breaker

1968
FORD

Circuit	Circuit Protection		Location
	Rating	Trade No.	
Headlamp circuit			
Headlight	18 Amp. C.B.		
High beam indicator light			Integral with lighting switch.
Rear lamps (tail lights & stop lights)	15 Amp. C.B.		
Front parking and side marker light			
Ignition switch light			
License light			
Horns			
Dome lamp circuit			
Dome light	9 Amp. fuse	SFE 9	On fuse panel (dome socket).
Courtesy light			
Cargo light			
Glove compartment light			
Luggage compartment light			
Clock			
Tachometer			
Instrument panel illumination	4 Amp. fuse	SFE 4	Fuse panel (instrument panel socket). This circuit is connected into the light switch rheostat.
Clock light			
Instrument cluster lights			
Ash tray light			
PRNDL—Console or column light			
Radio light			
Heater control lights			
Heater circuit			
Heater and defroster motor	20 Amp. fuse	SFE 20 or AGC 20	Fuse panel (heater socket).
Safety convenience panel lamps			
Low fuel warning			
Seat belt warning			
Door ajar warning			
Door unlock warning			
Power antenna			
Station wagon under seat heater			
Defogger and spotlight			
Cigar lighter plus emergency warning	20 Amp. fuse	SFE 20	Fuse panel
Radio circuit			
Radio	14 Amp. fuse	SFE-14	Fuse panel (radio socket).
Back-up lamps			
Single speed washers			
Turn signal circuit			(After fuse flasher protects circuit for turn signal).
Seat belt warning, parking brake, door ajar, power window lockout circuit			
Windshield wipers			Lower center flange of instrument panel
Dual speed wipers and washers	7.5 Amp. C.B.		
Power assists			
Convertible top circuit	14 gage wire Fuse link	(C6AB-14A094-A)	In wiring near starter relay
Convertible top with power option(s)	20 Amp. C.B. replaces fuse link		On starter relay
Power windows	20 Amp. C.B.		
Including backlite & tailgate			

1968 FORD

Circuit	Circuit Protection		Location
	Rating	Trade No.	
Power seats (four-and six-way)	20 Amp. C.B.		
Air conditioning circuit			
Air conditioner (economy)	15 Amp.	AGC 15	Fuse cartridge in feed wire
Air conditioner (selectair)	25 Amp. C.B.		Center flange of instrument panel
Miscellaneous circuits			
Speed control	5 Amp. fuse	AGA 5	Fuse cartridge in feed wire
Spotlight	20 Amp. fuse	SFE 20	Fuse panel (heater socket)
Automatic headlamp dimmer	4 Amp. fuse	SFE 4 7.5	Fuse cartridge in feed wire
Parking brake warning	14 Amp. fuse	SFE 24	Fuse panel (accessory socket)
Motors			
Windshield wiper motor	Circuit		Integral with motor
Convertible top motor	breaker		
Power window and backlite motor			
Power seat motor			

1968 THUNDERBIRD

Circuit	Location	Protective Device	Circuit	Location	Protective Device	Circuit	Location	Protective Device
Dome, courtesy, map, glove compt., luggage compt. and movable steering column	Fuse panel	SFE 14①	Heater—air conditioner	Right side of dash panel	30 C.B.	Horns, power windows, power seats and power reclining seat	Fuse panel	20 C.B.
Tail, park, license, marker, and supplemental taillights	In headlight switch	15 C.B.	Instrument panel and PRNDL lights, ash tray lights	Fuse panel	SFE 6	Ammeter light	Fuse panel	SFE 14
Stop light and emergency flashers	Right side of dash panel	20 C.B.	Cigar lighter	Fuse panel	SFE 20	Dual brake warning system	Fuse panel	SFE 6
			Power radio antenna	Fuse panel	3AG 10	Load circuit	Terminal junction block and starter motor relay	Fuse link
Clock and stereo tape player	Fuse panel	SFE 15	Speed control, seat belt and door opening warning	Fuse panel	SFE 4			
Back-up lights and W/S washer	Fuse panel	SFE 7.5	Low fuel warning	Fuse panel	SFE 7.5			
Turn signal	Fuse panel	SFE 15	Headlights	In headlight switch	18 C.B.			
Radio and rear window defogger	Fuse panel	SFE 7.5						

①—With movable steering column use SFE-20

1969 FORD

Circuit	Location	Protective Device	Circuit	Location	Protective Device	Circuit	Location	Protective Device
Air conditioner feed	Center of instrument panel flange	25 C.B.	Power seat feed	On starter relay	20 C.B.	License lights / Lights on buzzer	Integral with lighting switch	15 C.B.
Air conditioner feed (dealer installed)	Cartridge in feed line at accy. terminal of fuse panel	AGC or SFE 20	Power seat motors	In motor	C.B.	Low fuel indicator	Fuse panel—left side of dash	AGC or SFE 20
Antenna—electric / Back-up lights / Cigar lighter		AGC or SFE 20	Power top feed	On starter relay	20 C.B.	Luggage compartment		SFE 14
			Power top motor / Power window motors	In motor	C.B.	Marker lights	Integral with lighting switch	15 C.B.
Clock feed		SFE 14	Power window relay feed	Fuse panel—left side of dash	AGC or SFE 20	Power window supply	On starter relay	20 C.B.
Courtesy lights / C pillar / Cargo / Dome / Door / Glove compartment / Instrument panel / Map	Fuse panel left side of dash	SFE 14	Headlights and high beam indicator	Integral with lighting switch	18 C.B.	Radio feed / Rear window defogger / Seat belt indicator	Fuse panel—left side of dash	AGC or SFE 20
			Heated back window	On starter relay	20 C.B.	Speed control	Cartridge in feed line at accy. terminal of fuse panel	AGA 5
Door ajar indicator / Emergency flasher		AGC or SFE 20	Heater—defroster motor	Fuse panel—left side of dash	SFE 14			
Open door indicator	Fuse panel—left side of dash	SFE 14 / AGC or SFE 20	Horns / Ignition switch light	Integral with lighting switch	15 C.B.	Stop lights / Taillights	Integral with lighting switch	15 C.B.
Parking brake indicator			Instrument panel lights / Clock / Gauges / Heater control / PRNDL / Radio	Fuse panel—left side of dash	SFE 4	Windshield washer	Fuse panel—left side of dash	AGC or SFE 20
Parking lights	Integral with lighting switch	15 C.B.				Windshield wiper-washer	In wiper switch	10 C.B.

1 —Fuse or circuit breaker (C.B.) in Amperes.

1969
THUNDERBIRD

Circuit	Location	Protective Device [1]
Ammeter circuit		SFE 14
Ash tray lights		SFE 6
Back-up lights		SFE 7.5
Cigar lighter (front and rear)	Fuse panel	AGC or SFE 20
Clock feed		3AG or AGC 15
Cornering lights		7AG or AGW 15
Courtesy lights Footwell Glove compartment Luggage compartment Map		8AG or AGX 20
Dual brake warning system		SFE 6
Electric seat motors	Integral part of motor	
Electric window motors		CB.
Emergency flashers	Right side of dash panel	20 CB.
Four-horn circuit	Cartridge in feed line	AGC or SFE 20
Headlights	In headlight switch	18 CB.
Headlights on buzzer circuit	Integral with lighting switch	15 CB.
Heated back window	Attached to brake pedal support	20 CB.
Heater—air conditioner	Right side of dash panel	30 CB.
Horns	Circuit breaker panel RH. dash panel	20 CB.
Indicator lights Door ajar	Fuse panel	SFE 7.5
Low fuel Seat belt (console)		

Circuit	Location	Protective Device 1
Instrument panel lights Dual brakes indicator Heater control Ignition switch PRND21 Radio dial Seat belt (instrument panel) Stereo tape player	Fuse panel	SFE 6
License Marker	In headlight switch	15 C.B.
Movable steering column	Fuse panel	8AG or AGX 20
Park	In headlight switch	15 C.B.
Power antenna	Fuse panel	3AG or AGC 10
Power seats	Circuit breaker Panel R.H.	20 C.B.
Power windows	dash	
Power window relay feed	panel	30 C.B.
Radio Rear window defogger	Fuse panel	SFE 7.5
Speed control, seat belt		7AG or AGW 15
Stereo tape player feed		3AG or AGC 15
Stoplight	Right side of dash panel	20 C.B.
Supplemental stop light circuit	Fuse panel	8 AG or AGX 20
Taillights	In headlight switch	15 C.B.
Turn signals	Fuse panel	7AG or AGW 15
Windshield washer	Fuse panel	SFE 7.5

[1] —Fuse or circuit breaker (C.B.) in amperes.

1970
FORD

Circuit	Location	Circuit Breaker (C.B.) or Fuse in Amperes
Headlights	Integral with light switch	18 C.B.
Rear tail lights Front parking lights License light Side marker Horns Headlight buzzer Stop lights	Integral with light switch	15 C.B.
Windshield wipers	In windshield wiper switch	10 C.B.
Interval windshield wipers	Upper center of instr. panel near blower switch	8.25 C.B.
Convertible top Power Windows	On starter relay	20 C.B.
Rear window defroster & seat back latch Power seats	On starter relay	20 C.B.
Air conditioner	Upper center instrument panel near blower switch	30 C.B.
Air conditioner dealer installed	Cartridge in feed line attached to accy. terminal fuse panel	20 fuse
Motors—protected by integral circuit breakers Power seat Power window Convertible top	In motor	C.B.

1970
THUNDERBIRD

Circuit	Rate	Location	Circuit	Rate	Location
Parking lights	15 AMP	In headlight switch	Power seat		
License light			Power windows	Integral	In motor assembly
Tail lights			Headlight circuit	18 AMP	In headlight switch
Marker lights			Windshield wiper	Integral	In wiper switch
Headlight buzzer			Rear window	20 AMP	On brake
Motors			defroster		pedal support

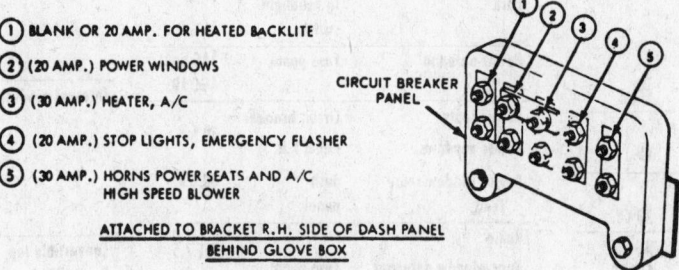

1. BLANK OR 20 AMP. FOR HEATED BACKLITE
2. (20 AMP.) POWER WINDOWS
3. (30 AMP.) HEATER, A/C
4. (20 AMP.) STOP LIGHTS, EMERGENCY FLASHER
5. (30 AMP.) HORNS POWER SEATS AND A/C HIGH SPEED BLOWER

CIRCUIT BREAKER PANEL

ATTACHED TO BRACKET R.H. SIDE OF DASH PANEL
BEHIND GLOVE BOX

Circuit breaker panel—1970 Thunderbird

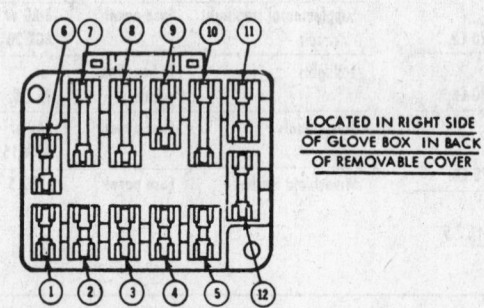

LOCATED IN RIGHT SIDE
OF GLOVE BOX IN BACK
OF REMOVABLE COVER

1. (SPARE) OR (7.5 AMP.) SAFETY CONVENIENCE LOW FUEL, DOOR AJAR, SEAT BELT (CONSOLE)
2. (15 AMP.) SEAT BELT WARNING (STD), POWER WINDOW SAFETY RELAY COIL FEED
3. (7.5 AMP.) BACK-UP LIGHTS, SPEED CONTROL
4. (15 AMP.) TURN SIGNALS & CORNERING LIGHTS
5. (7.5 AMP.) RADIO & DEFOGGER
6. (6 AMP.) DUAL BRAKE WARNING
7. (15 AMP.) STEREO TAPE RADIO
8. (20 AMP.) CIGAR LIGHTER (FRONT & REAR)
9. (20 AMP.) COURTESY LIGHTS, FOOTWELL, REAR, LUGGAGE COMPT., GLOVE BOX, MAP, SUPPLEMENTAL STOP LIGHT CIRCUIT, CLOCK FEED
10. (SPARE) OR (10 AMP.), WINDSHIELD WASHERS
11. (6 AMP.) INSTRUMENT PANEL LIGHTS, HEATER CONTROL, ASH TRAY RADIO, STEREO TAPE LIGHT, SEAT BELT LIGHT, PRND 21
12. (SPARE) OR (3 AMP.) FOR ANTI-SKID BRAKE CONTROL

Fuse panel—1970 Thunderbird

Distributor

Detailed information on distributor drive, direction of distributor rotation, cylinder numbering, firing order, point gap, point dwell, timing mark location, spark plugs, spark advance, ignition resistor location and idle speed is in the Specifications table of this section.

All 6 Cylinder

On 223 cu. in. engines the distributor is located on the right-side of the cylinder block. On 240 cu. in. engines it is on the left.

First, mark the position of the rotor and body and, to remove it, lift off

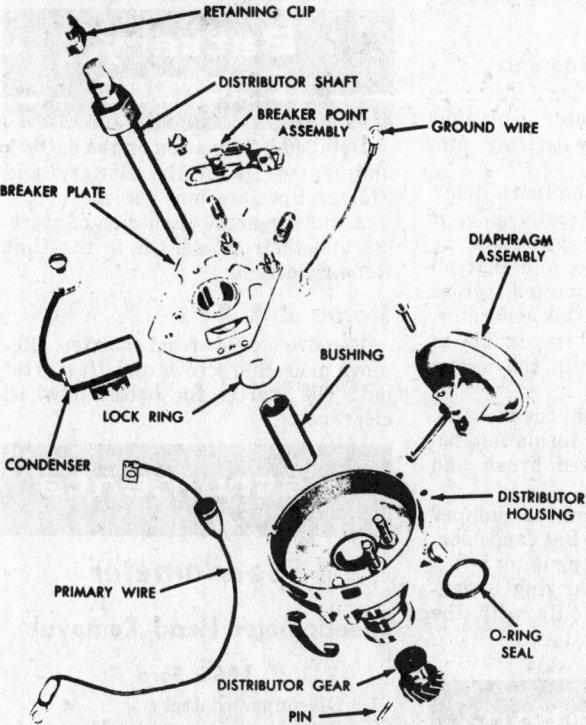

Vacuum advance distributor (© Ford Motor Co.)

the cap and wire assembly, disconnect the ignition primary wire, disconnect the vacuum line to the carburetor, remove the distributor hold-down bolt and pull the distributor out of the side of the block.

All V8s

On these models, the distributor is located in the front of the engine and is easily accessible.

First, mark the position of the rotor, also the position of the body with relation to the block. Disconnect the ignition primary wire, the vacuum lead, the distributor cap, then take out the hold-down bolt that holds the distributor down in the block and lift it up out of the block.

Do not disturb the engine after the distributor has been removed in case it should disturb the ignition timing.

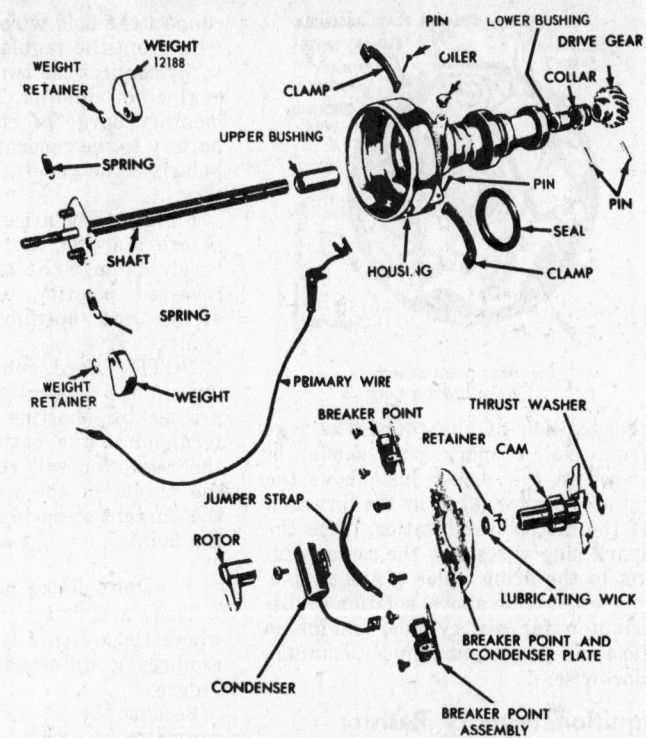

Centrifugal advance distributor (© Ford Motor Co.)

Ignition Timing

If the timing relationship has been completely fouled up, proceed to retime the ignition as follows: bring No. 1 cylinder up into the firing position. This can be checked by removing the spark plug, placing your thumb in the spark plug hole, then cranking the engine until the compression attempts to blow by your thumb. Now, slowly bring the crankshaft around until the T.D.C. mark on the crankshaft pulley lines up with the pointer. This is the approximate firing position for No. 1 cylinder.

Remove the distributor cap and mark on the outside of the distributor

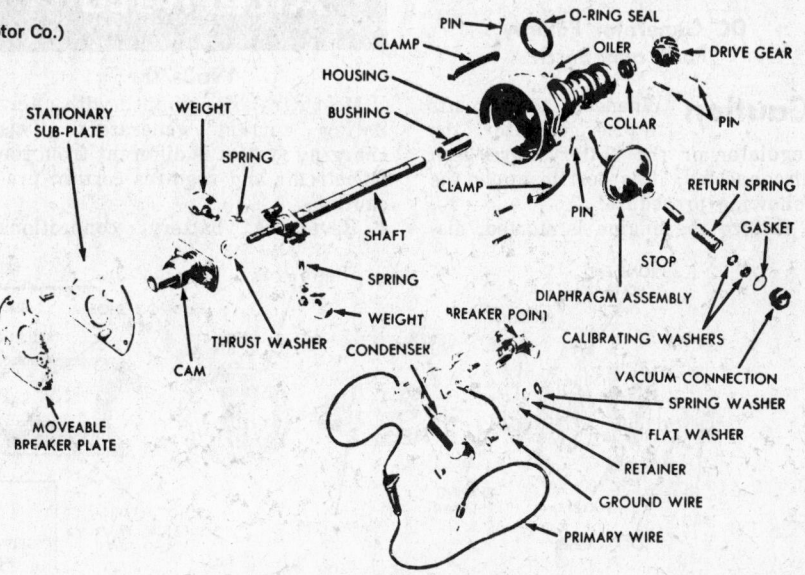

Dual advance distributor (© Ford Motor Co.)

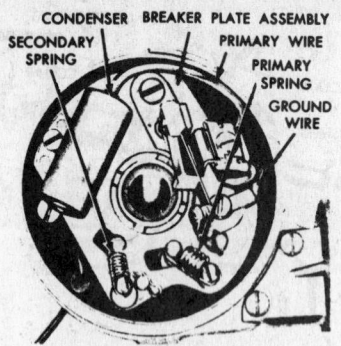

Pivotless point assembly
(© Ford Motor Co.)

the position of the rotor. The wire from No. 1 spark plug should be placed in the socket just above the rotor. Now, working in the direction of the distributor rotation, place the spark plug wires into the cap according to the firing order of the engine.

Viewed from above, rotation of distributor for six cylinder engine is clockwise; for eight cylinder, counterclockwise.

Ignition Primary Resistor

A resistance wire is used in the primary circuit of the ignition. The resistor is a wire of specific resistance running between the ignition switch and the coil. Any time difficulty is experienced with the ignition system, it is a good idea to check the resistor.

Generator, Regulator

Detailed facts on the generator and the regulator are in the Generator and Regulator Specifications table.

General information on generator and regulator repair and troubleshooting is in the Unit Repair Section.

DC Generator Polarity, Ford and Bosch

Caution Whenever the circuits to the generator, the regulator or the battery have been disconnected, it is best to apply the following procedure.

Before the engine is started, disconnect the field wire and the battery wire from the regulator and momentarily connect the two wires together, engine not running. This gives a momentary surge of current from the battery to the generator and correctly polarizes the generator with the battery.

Failure to polarize a DC generator before starting the engine may severely damage the regulator because reversed polarity causes vibration, arcing and burning of the relay points.

NOTE: Ford generators are not wired like other generators. Polarizing by shorting from the field terminal to the battery terminal on the regulator will result in burning the points of the regulator because the current runs to ground through the points.

Delco-Remy Generator

There are, however, some instances where Delco-Remy is used. This unit requires a different polarizing procedure.

Because the field of the Delco-Remy generator is grounded externally, it is important that the generator be mounted on the engine, and that all leads be properly connected before attempting to polarize this generator. If this is not done, it is possible to polarize the generator in the wrong direction.

1. Remove the brush cover band.
2. Place a piece of insulation between the insulated brush and the commutator.
3. Momentarily connect a jumper lead between the Bat. and Gen. terminals of the regulator.

This method of polarizing a generator is to be used only with the Delco-Remy DC generator.

Alternator

1963-70

Most cars are equipped with alternating current generators. This charging system is different from the DC circuit and requires certain precautions.

1. Reversing battery connections will cause damage to the one-way electrical valves, the rectifiers.
2. Booster battery connections must be made as follows: the negative terminal of the booster battery must be connected to the negative terminal of the car battery. The positive terminal of the booster battery must be connected to the positive terminal of the car battery.
3. Fast chargers should never be used as boosters to start AC circuit-equipped cars.
4. When servicing the battery with a fast charger, always disconnect car battery cables.
5. Never attempt to polarize an AC generator.

Complete alternator servicing data is in the Unit Repair Section.

Battery, Starter

Detailed information on the battery and starter is in the Battery and Starter Specifications table.

A more general discussion of starters and their troubles is in the Unit Repair Section.

Starter R & R

Remove cable from starter. Remove mounting screws and lift starter out. Tilt starter for better flywheel clearance.

Instruments

Speedometer

Speedometer Head Removal

1964 Ford

1. Disconnect battery.
2. Remove instrument cluster assembly.
3. Remove speedometer assembly retaining screws and the assembly-to-housing retaining screws
4. Remove the speedometer assembly-to-housing retaining screws and speedometer head.
5. To reinstall, reverse the removal procedure.

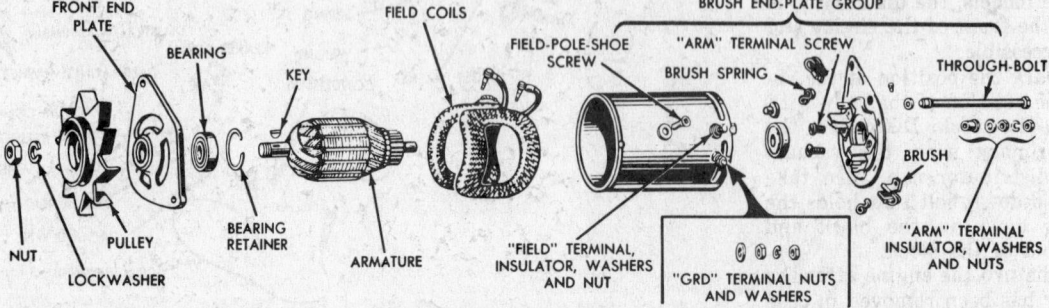

Standard generator (© Ford Motor Co.)

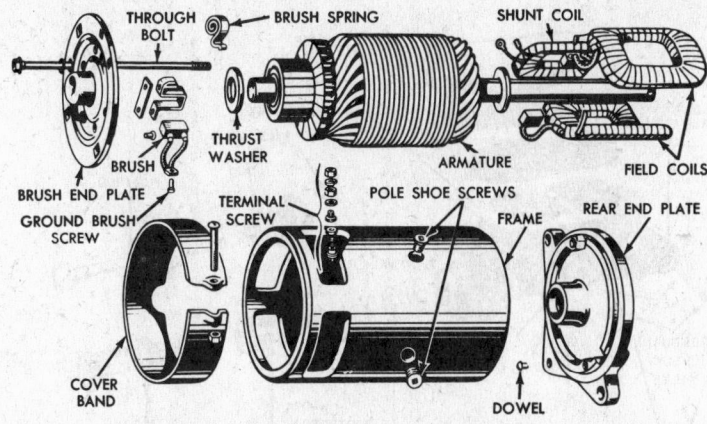

THROUGH BOLT — BRUSH SPRING — SHUNT COIL — THRUST WASHER — ARMATURE — FIELD COILS — BRUSH — BRUSH END PLATE — GROUND BRUSH SCREW — TERMINAL SCREW — POLE SHOE SCREWS — FRAME — REAR END PLATE — COVER BAND — DOWEL

Disassembled starter motor (© Ford Motor Co.)

1965-66 Ford

1. Disconnect battery, and cover steering column.
2. Remove retaining screws from instrument panel cluster pad and retainer.
3. Remove screws from upper and lower instrument panel cluster cover. Remove cover.
4. Remove radio knobs.
5. Remove friction pins retaining the instrument cluster lens and mask.
6. Disconnect the speedometer cable.
7. On 1966 models, remove the clock.
8. Remove retaining screws from the speedometer retaining bracket, and remove the bracket.
9. Remove screws retaining speedometer head, remove the standoffs from the locating posts and the speedometer head.
10. To reinstall, reverse the removal procedure.

1967 Ford

1. Disconnect battery.
2. Remove instrument panel pad.
3. Remove clock reset knob.
4. Remove four screws retaining warning light housing, disconnect fuel gauge, and remove warning light housing.
5. Remove eight friction pins retaining instrument cluster and mask. Remove lens.
6. Remove rubber spacers.
7. Remove three screws retaining speedometer assembly. Disconnect speedometer cable and remove speedometer.

1968 Ford

1. Disconnect battery.
2. Remove right and left windshield mouldings.
3. Pry moulding from right side of instrument panel pad covering the pad retaining screws.
4. Pry off two access covers located above the speedometer lens and on underside of pad.
5. Remove screws retaining instrument panel pad, and remove pad.
6. Remove radio knobs.
7. Remove button clips retaining instrument cluster mask and lens, and remove mask and lens.
8. Disconnect speedometer cable.
9. Remove screws retaining instrument panel lower pad and remove pad.
10. Remove screws from clock retainer and clock and position clock forward.
11. Remove plate under speedometer and two rubber spacers and screws retaining speedometer assembly.
12. Remove speedometer.

1969-70 Ford

1. Remove upper part of instrument panel by removing screws along lower edge, two screws in each of the defroster registers, and disconnecting the radio speaker.
2. Remove cluster opening finish panels from each side of instrument cluster.
3. Disconnect plugs to printed circuit, radio, heater and A/C fan, windshield wipers and washers, and any other electrical connection to cluster.
4. Disconnect heater and A/C control cables and speedometer cable.
5. Remove all knobs from instrument panel.
6. Remove instrument cluster trim cover.
7. Remove eight mounting screws and remove cluster.
8. Remove speedometer from cluster.

1964-70 Thunderbird

1. Disconnect the battery ground cable. Then, cover the steering column and panel where necessary to prevent paint damage.
2. Remove the radio knobs and bezel.
3. Remove headlight switch control knob and bezel nut.
4. Remove the instrument finish panel.
5. Remove the headlight switch mounting screws and push the switch toward the front of the car.
6. Remove the console panel finish moulding cap, and remove the screws retaining the left lower half of the instrument cluster housing assembly.
7. Remove the clock housing retaining screws and rotate the clock housing upward and rearward to expose the two tab screws retaining the instrument panel upper moulding.
8. Remove screws from the instrument panel upper moulding and the screws under the cluster. Pull the moulding away from the instrument panel for access to the cluster screws. Tape the tabs to prevent scratches.
9. Remove the instrument indicator cover retaining screws and remove the covers.
10. Through the indicator openings, remove the screws retaining the lower cluster to the upper cluster. Position the instrument indicator covers on the instruments to prevent damage.
11. Remove the screws retaining the speedometer cluster to the instrument panel at the top of the cluster, and pull speedometer from cluster.
12. Disconnect the cable and all wires to the speedometer head and remove the speedometer cluster.
13. Remove the speedometer housing-to-cluster mounting screws and remove housing assembly.
14. Remove the screws retaining the speedometer housing cover to the speedometer assembly and remove the speedometer.
15. To reinstall, reverse the removal procedure.

Ignition Switch

Ford and T-Bird Lock Cylinder Replacement

1. Insert key and turn to Acc. position.
2. With stiff wire in hole, depress lock pin and rotate cylinder counterclockwise, then pull out cylinder.

Switch Replacement

1. Remove cylinder as above.
2. Unscrew the bezel from the ignition switch and remove switch from panel.
3. Remove insulated plug from rear of switch.
4. Install in reverse of above.

Lighting Switch

Replacement

Ford and T-Bird

1. Disconnect battery.
2. Remove knob and shaft by pressing release knob button on switch housing and with knob in full on position.
3. Remove mounting nut from switch.
4. Remove junction block from switch.
5. Install in reverse of above.

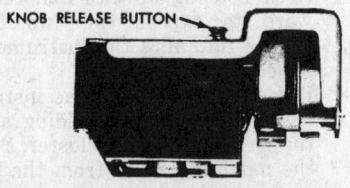

Typical headlight switch
(© Ford Motor Co.)

Brakes

Brake Information

Specific information on brake cylinder sizes is in the General Chassis and Brake Specifications table of this section.

Information on brake adjustments, band replacement, bleeding procedure, master and wheel cylinder overhaul is in the Unit Repair Section.

Information on troubleshooting and overhauling power brakes is in the Unit Repair Section.

Information on the grease seals which may need replacement is in the Unit Repair Section.

Beginning 1965 Thunderbird

Disc brakes on front are standard equipment. Information on adjustments and replacements is in Unit Repair Section.

Master Cylinder Removal

All Models

To remove the master cylinder, remove the pin that holds the master cylinder pushrod to the brake pedal. This is accessible from under the dash.

Working in the engine compartment, remove the brake lines from the back end of the master cylinder and take out the master-cylinder mounting bolts.

Beginning 1967

Since 1967, a tandem-type master cylinder is used on all models. This design divides the brake hydraulic system into two independent and hydraulically separated halves.

Details and repair procedures on this tandem system are in the Unit Repair Section.

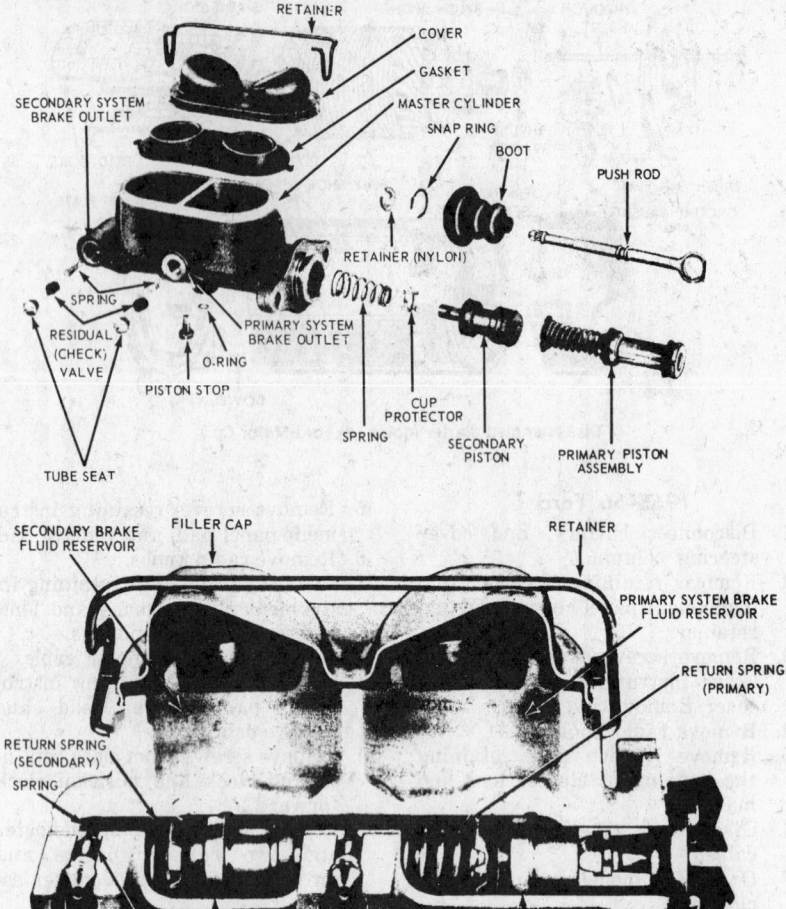

Dual type master cylinder (© Ford Motor Co.)

Beginning 1969

A single cylinder front disc brake is used on some models. See Unit Repair Section.

Brake Pedal Adjustment

Establish approximately ½ in. free pedal travel, measured at the toe board by turning the eccentric bolt, that attaches the brake pedal assembly to the master-cylinder pushrod assembly. Rotate this eccentric bolt until the play is between ¼ and ½ in.

Disc Brakes

Since 1965, disc brakes have been available on front wheels of some models. Complete service procedures are in the Unit Repair Section.

Parking Brake Lever

Hand-operated lever

Remove the brackets that hold the lever assembly up under the dash, slide the end of the cable ball joint out of its connection in the end of the lever and lift the lever off the vehicle.

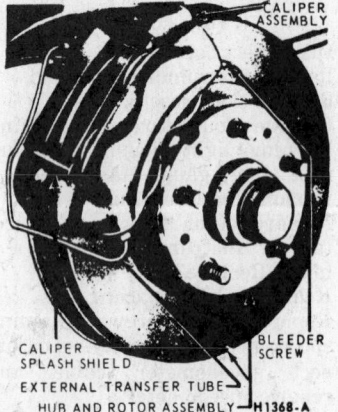

Disc brake caliper assembly
(© Ford Motor Co.)

Foot-operated Lever

The foot-operated lever is mounted under the left side of the instrument panel. Slack off on the brake cable and remove the clevis that holds the cable to the top part of the lever. Remove the brackets and lever.

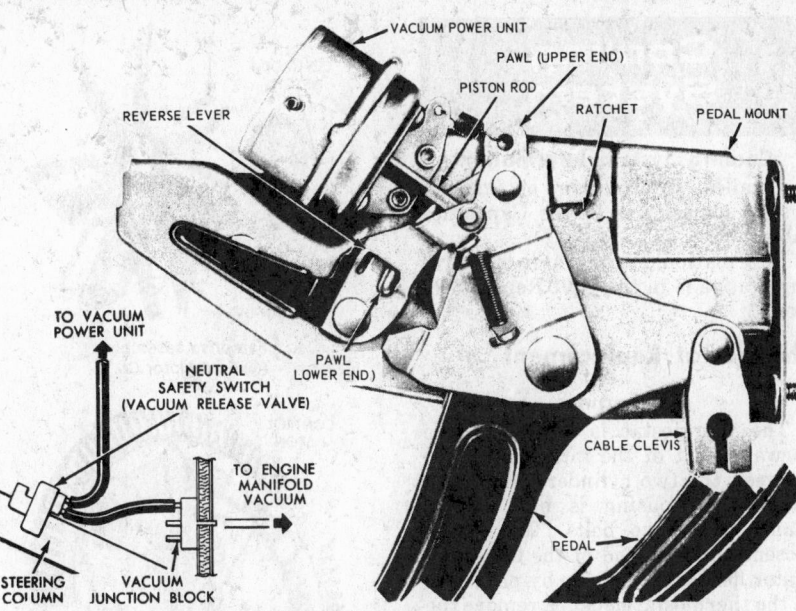

Vacuum release parking brake (© Ford Motor Co.)

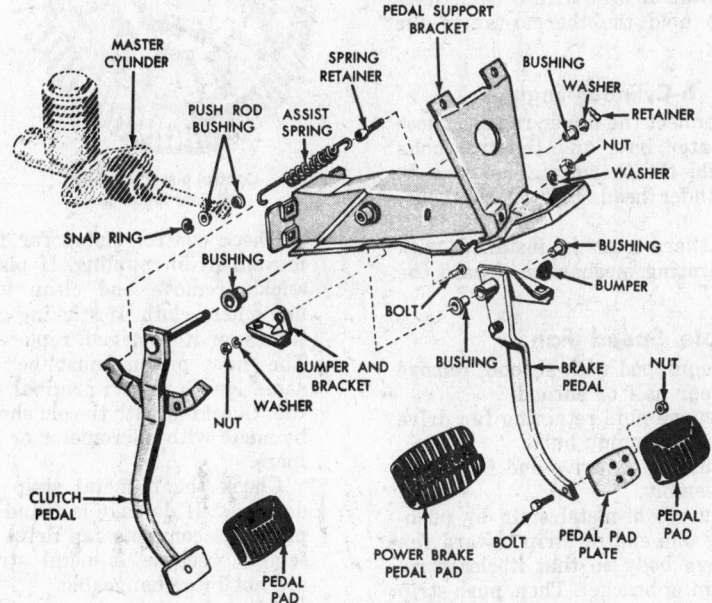

Pedal assembly and related parts (© Ford Motor Co.)

Fuel System

Fuel System Information

Data on capacity of the gas tank are in the Capacities table. Data on correct engine idle speed and fuel pump pressure are in the Tune-up Specifications table.

Information covering operation and troubles of the fuel gauge is in the Unit Repair Section.

Detailed information on the carburetor, and how to adjust it, is in the Unit Repair Section. Carter, Holley, Rochester, Ford and Stromberg carburetors are covered.

Dashpot adjustment is in the Unit Repair Section.

Fuel Pump Removal

6-Cylinder Models

The fuel pump is located low on the right front of 223 cu. in. engines, and low on the left front of 240 cu. in. engines. To remove the fuel pump, disconnect the flex line from the frame and the solid line to the carburetor, unbolt and lift off the fuel pump.

V8 Models

The fuel pump is located on the timing case cover in front of the left bank of cylinders.

Remove the input and output line from the fuel pump, unbolt and lift off the pump.

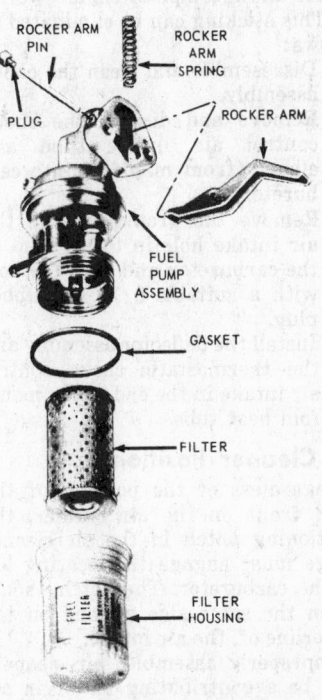

Typical Carter sealed design fuel pump
(© Ford Motor Co.)

⏻ CHILTON TIME-SAVER

Rear Brake Drum Removal

Occasional cases of rear wheel drums, frozen to the rear axle flange, require much time and effort to remove without damage.

If a rear drum resists normal efforts to remove by tapping, try the following method:

1. Drive two or three of the serrated hub bolts out of the drum and into the brake shoe area.

2. With an old screwdriver or other suitable wedge forced between the drum and axle flange through these bolt holes, tap and wedge the drum from the axle flange.

3. After the drum is removed, the bolts can be recovered and returned to their respective places in the axle flange.

Any damage to the drum can usually be corrected by a few taps with a hammer.

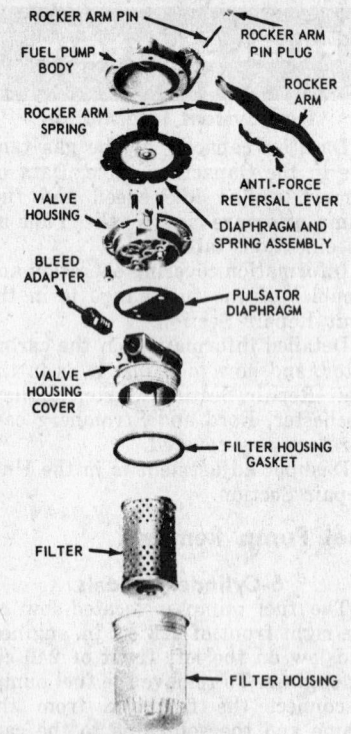

Carter fuel pump
(© Ford Motor Co.)

Sticking Choke

On some 8-cylinder cars, automatic choke malfunctioning may occur due to a sticking choke mechanism. Sticking may be due to corrosion from exhaust gases entering a split heat tube.

This sticking can also be due to excessive moisture condensation in the housing. This will cause corrosion or freezing of the choke mechanism, particularly when distances traveled do not allow complete engine warm-up. This sticking can be eliminated as follows:

1. Disassemble and clean the choke assembly.
2. Remove and discard the choke control air intake tube and elbow (from manifold into carburetor).
3. Remove the grommet from the air intake hole in the bottom of the carburetor and plug the hole with a suitable 9/16 in. rubber plug.
4. Install the deflector assembly and the thermostatic choke control air intake in the end of the manifold heat tube.

Air Cleaner Position

Regardless of the position of the word front on the air cleaner, the positioning notch in the air cleaner flange must engage the locating lug on the carburetor. The notch should be on the same side as, and on the centerline of, the air intake.

Improperly assembled air cleaners may be a contributing factor in accelerator binding, carburetor icing, and excessive fuel consumption.

Cooling System

Cooling System Information

Detailed information on cooling system capacity is in the Capacities table.

Information on the water temperature gauge is in the Unit Repair Section.

Thermostat Replacement

V8 Engines

The thermostat is located in the forward part of the intake manifold between the two cylinder heads. The thermostat housing is held to the manifold by two bolts. To replace, loosen the lower end of the upper radiator hose and the little by-pass hose at the thermostat housing, remove the thermostat housing and lift out the thermostat. A lock wire is sometimes used to hold the thermostat in the housing.

6-Cylinder Engines

Disconnect the upper radiator hose and heater hose and the two bolts that hold the thermostat housing to the cylinder head. Lift off the housing.

The thermostat is installed with the operating mechanism toward the engine.

Variable Speed Fan

1. If equipped with shroud, remove upper half of shroud.
2. Remove nuts retaining fan drive to water pump hub.
3. Remove fan drive and fan as an assembly.
4. Remove bi-metal strip by pushing one end of strip toward fan drive body so that it clears retaining bracket. Then, push strip to side so that opposite end of strip will spring out of bracket.
5. Remove control piston.

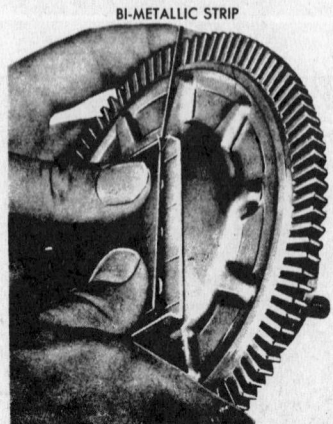

BI-METALLIC STRIP

Bi-metallic strip removal and installation
(© Ford Motor Co.)

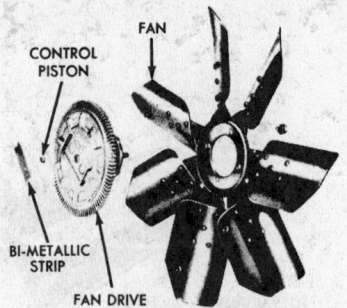

Fan drive assembly
(© Ford Motor Co.)

Control piston installed
(© Ford Motor Co.)

Check control piston for free movement in coupling. If piston sticks, remove and clean with fine emery cloth. If sticking continues, or if damaged, replace it. The new piston must be the same length as the original piston. Checking with the old should be made with micrometer or calipers.

Check the bi-metal strip for damages. If damage is found replace the complete fan drive assembly because bi-metal strips are not interchangeable.

After fan drive is assembled, clean with a clean cloth and solvent. Do not dip in any liquid.

6. Install control piston so that projection on end of piston will contact bi-metal strip.
7. Install the bi-metal strip with the identification stamp B-1 facing the fan drive.
8. Reinstall the fan.

Water Pump Removal

V8 Engines

Take off the fan belt, fan blades and fan pulley. Disconnect the water hose, unbolt and lift off the pump.

6-Cylinder Engines

Take off the fan belt, fan and pulley. Disconnect the hoses and remove the cap screws that hold the water pump assembly to the block and lift the pump.

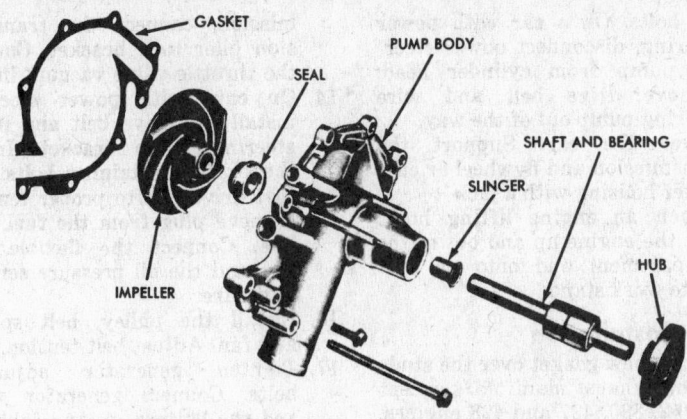

289 V8 water pump assembly (© Ford Motor Co.)

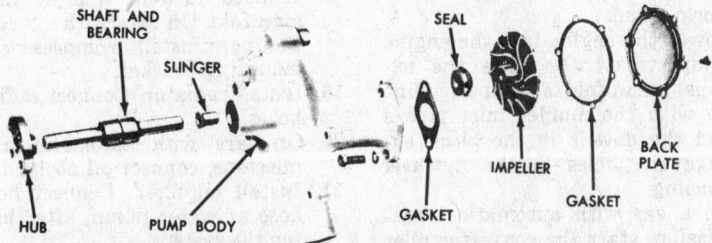

Water pump assembly—352, 390 and 428 cu. in. engines (© Ford Motor Co.)

Radiator Core Removal

Disconnect the upper and lower radiator hoses and take out the center air deflector side screws on models with air deflector. Then, take out the cap screws that secure each side of the radiator to the front fender apron support. The core can then be lifted straight up and off the vehicle.

Engine

References

Engines in full size Fords from 1964 to 1971 fall into five families. The first engine family has only one representative, the 223 cu. in. six cylinder. 1964 Fords are the last cars to use this engine. It has intake and exhaust manifolding on the left-hand side of the engine, which is very unusual for Ford sixes. The engine that replaces the 223 cu. in. six as the basic six in full size Fords is the 240 cu. in. six which is used to the present day. The intake manifolding on this six is mounted conventionally on the right-hand side and is detachable, unlike the intake manifolding on Ford sixes in smaller cars. The 289, 302, 351W V8 engines are the most popular engines in full sized Fords. The 289 and 302 are identical in exterior appearance and are notably compact, about 20 in. across. The larger displacement 351W is wider and bulkier although nearly identical in layout and conformation. All three have trapezoidal shaped valve covers. The 352, 390, 427 and 428 family of engines is recognizable by its unusual

intake manifold that extends under valve covers. The engines of this family are identical in exterior appearance. These engines are used widely in full size Fords and the 390 engine was standard equipment in Thunderbirds from 1961 to 1968, with the 428 as an option in 1966 and 1967.

The 429 engine is the first of a new series of big block Ford engines. At present, it is available in two-barrel and four-barrel versions. The engine is identifiable by its great bulk, and

by the tunnel port configuration noticeable in the shape of its intake manifold.

Valves

Detailed information on the valves, is in the Valve Specifications table.

A general discussion of valve clearance and a chart showing how to read pressure and vacuum gauges when using them to diagnose engine troubles are in the Unit Repair Section.

Valve tappet clearance for each engine is given in the Tune-up Specifications table.

Bearings

Detailed information on engine bearings is in the Crankshaft Bearing Journal Sizes table.

Engine crankcase capacities are listed in the Capacities table.

Approved torque wrench readings and head-bolt tightening sequences are covered in the Torque Specifications table.

Exhaust Emission Control

In compliance with anti-pollution laws, the Ford Motor Company has adopted a distributor and a modified carburetor with some engine changes to reduce terminal exhaust fumes to an acceptable level. This method is known as Ford's Improved Combustion System (IMCO).

The plan supersedes (in most cases) the previous method used to conform to 1966-67 California laws. The new system phases out (except with stick shift and special purpose engine applications) the thermactor, or afterburner type of exhaust treatment.

The IMCO concept utilizes broader,

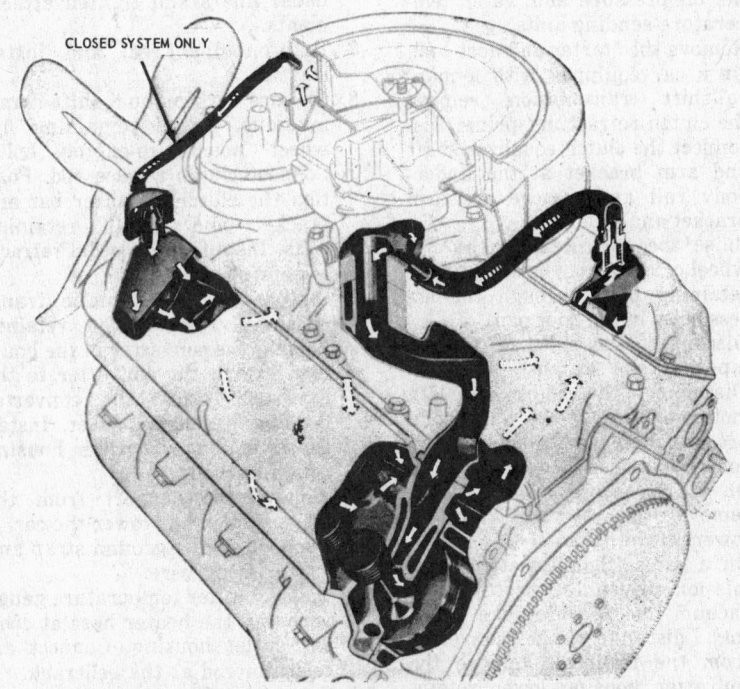

289 V8 positive crankcase ventilation system (© Ford Motor Co.)

yet more critical, distributor control through carburetor modification.

Since 1968, all car makers have posted idle speeds and other pertinent data relative to the specific engine application in a conspicuous place in the engine compartment.

For details on the IMCO system, see the Unit Repair Section.

Engine Removal

1. Scribe the hood hinge outline on the underside of the hood, disconnect the hood and remove.
2. Drain the entire cooling system and oil from engine oil pan.
3. Remove the air cleaner, disconnect the battery at the cylinder head. On automatic transmission-equipped cars, disconnect oil cooler lines at the radiator.
4. Remove upper and lower radiator hoses and radiator.
5. Remove fan, fan belt, and upper pulley.
6. Disconnect the heater hoses at the water pump and the carburetor spacer.
7. Disconnect the generator wires at the generator, the starter cable at the starter, the accelerator rod at the carburetor and on the 6-cylinder engine, the choke control cable at the carburetor.
8. On vehicles with power brakes, disconnect vacuum line at intake manifold. On cars with air conditioning, remove compressor from mounting bracket and position it out of the way. Leave refrigerant lines connected.
9. Disconnect fuel tank line at the fuel pump and plug the line.
10. Disconnect the coil primary wire at the coil. Disconnect wires at the oil pressure and water temperature-sending units.
11. Remove the starter and dust seal.
12. On a car equipped with a manual-shift transmission, remove the clutch retracting spring. Disconnect the clutch equalizer shaft and arm bracket at the underbody rail and remove the arm bracket and equalizer.
13. Raise the car. Remove the flywheel or converter housing upper retaining bolts through the access holes in the floor pan.
14. Disconnect the exhaust pipe or pipes at the exhaust manifold. Disconnect the right and left motor mount at the underbody bracket. Remove the flywheel or converter housing cover.
15. On a car with manual shift, remove the flywheel housing lower retaining bolts.
16. On a car with automatic transmission, disconnect throttle valve vacuum line at the intake manifold, disconnect the converter from the flywheel. Remove the converter housing lower retaining bolts. On a car with power steering, disconnect power steering pump from cylinder head. Remove drive belt and wire steering pump out of the way.
17. Lower the car. Support the transmission and flywheel or converter housing with a jack.
18. Attach an engine lifting hook. Lift the engine up and out of the compartment and onto an adequate work stand.

Engine Installation

1. Place a new gasket over the studs of the exhaust manifold/s except on 352, 390, 427 and 428 engines.
2. Attach engine sling and lifting device. Then lift engine from work stand.
3. Lower the engine into the engine compartment. Be sure the exhaust manifold/s properly line up with the muffler inlet pipe/s and the dowels in the block engage the holes in the flywheel housing.
   On a car with automatic transmission, start the converter pilot into the crankshaft.
   On a car with manual-shift transmission, start the transmission main drive gear into the clutch disc. If the engine hangs up after the shaft enters, rotate the crankshaft slowly (with transmission in gear) until the shaft and clutch disc splines mesh.
4. Install the flywheel or converter housing upper bolts.
5. Install engine support insulator to bracket retaining nuts. Disconnect engine lifting sling and remove lifting brackets.
6. Raise front of car. Connect exhaust line/s and tighten attachments.
7. Position dust seal and install starter.
8. On cars with manual-shift transmissions, install remaining flywheel housing-to-engine bolts. Connect clutch release rod. Position the clutch equalizer bar and bracket and install retaining bolts. Install clutch pedal retracting spring.
9. On cars with automatic transmissions, remove the retainer holding the converter in the housing. Attach the converter to the flywheel. Install the converter housing inspection cover. Install the remaining converter housing retaining bolts.
10. Remove the support from the transmission and lower the car.
11. Connect engine ground strap and coil primary wire.
12. Connect water temperature gauge wire and the heater hose at coolant outlet housing. Connect accelerator rod at the bellcrank.
13. On cars with automatic transmission, connect the transmission filler tube bracket. Connect the throttle valve vacuum line.
14. On cars with power steering, install the drive belt and power steering pump bracket. Install the bracket retaining bolts. Adjust drive belt to proper tension.
15. Remove plug from the fuel tank line. Connect the flexible fuel line and the oil pressure sending unit wire.
16. Install the pulley, belt spacer, and fan. Adjust belt tension.
17. Tighten generator adjusting bolts. Connect generator wires and the battery ground cable.
18. On vehicles with power brakes, connect vacuum line at intake manifold. On cars with air conditioning; install compressor on mounting bracket.
19. Install radiator. Connect radiator hoses.
20. On cars with automatic transmissions, connect oil cooler lines.
21. Install oil filter. Connect heater hose at water pump, after bleeding the system.
22. Bring crankcase to level with correct grade of oil. Run engine at fast idle and check for leaks. Install air cleaner and make final engine adjustments.
23. Install and adjust hood.
24. Road test car.

Replacement Engines

Both new and factory-approved rebuilt engines are available for all of these models.

Ordinarily, a factory-approved rebuilt engine does not include the manifolds, oil pan or flywheel. Short blocks also are available which include the block assembly less the timing case cover and flywheel.

Cylinder Block Core Hole Plugs

6-Cylinder Engines

There are two large cylinder-block core hole plugs located in the back of the clutch bell housing. In order to service either or both of these plugs, it is necessary to remove the clutch bell housing. This job is best accomplished by removing the engine.

Rocker Arm Removal

6-Cylinder Engines Prior to 1965

Remove the rocker chamber cover and the screws that hold the oil feed lines to the rocker shaft. Remove the feed lines by prying up with a pair of pliers out of the cylinder head.

Back off on the rocker arm adjusting screws until the pressure is taken off the valve springs, then unbolt the rocker shaft assembly.

352, 390, 427 and 428 cu. in. V8 Engines

Remove the rocker cover assembly. It may be necessary to disconnect the spark plug wires in order to get it clear. Carefully scrape off the gasket.

Crank engine until No. 1 piston is at 45° ATDC at beginning of power stroke, identified by XX on crankshaft damper. Starting on No. 4 cylinder on the right bank, loosen rocker shaft support bolts in sequence two turns at a time until free. Remove bolts and lift off rocker assembly. Repeat procedure starting with No. 5 cylinder on left bank.

If the rockers are to be disassembled, carefully mark the position of each unit as it is taken off so that it

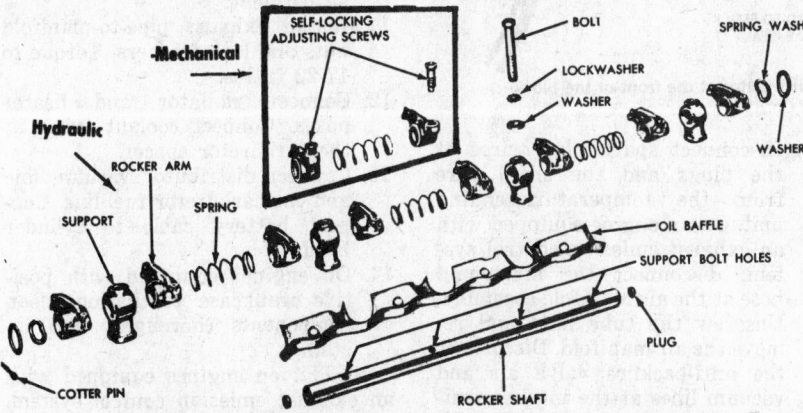

Rocker arms and shaft—V8 except 289 cu. in. (© Ford Motor Co.)

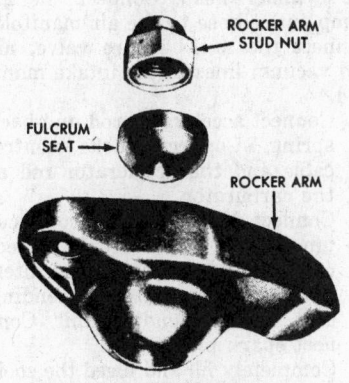

Rocker arm—1965-69 6 cyl. and 289 cu. in. V8 (© Ford Motor Co.)

can be reinstalled in the position from which is was removed. Reinstall in reverse order.

1965-69—6 Cylinder; and 289, 302, 351, 429

These engines use separate rocker arm suspension. Each rocker arm pivots individually on a fulcrum seat supported by a stud pressed into the cylinder head.

1. Remove the rocker cover assembly.
2. Remove rocker arm adjusting nuts, fulcrum seats and rocker arms.

Pushrod Removal

6-Cylinder Engines Prior to 1965

The valve pushrods may be removed from the engine without removing the rocker shaft by taking off the valve rocker cover and backing off on the rocker adjusting screw until all of the spring tension is taken off the valve.

The rocker can then be slid sideways which will permit removal of the pushrod.

Caution The rocker arms at each end of the engine cannot be shifted in this manner, and the two end pushrods cannot be removed until after the cylinder head has been taken off.

Removing pushrods (© Ford Motor Co.)

V8 Engines and Late Model

The pushrods may be lifted out through the cylinder head after the rocker shaft has been removed, or in the case of the 289, 302, 351 and 429 engines, and the 1965-67, 6 cylinder, after the rocker arm has been loosened and turned to an out-of-the-way position.

Engine Manifolds

Intake and Exhaust Manifold Removal

All 6-Cylinder Engines

Disconnect all lines to the carburetor, including gas, vacuum, throttle, and governor lines. Remove the air cleaner. Unbolt and lift off the carburetor.

Detach the exhaust pipe at the flange, remove the bolts that secure the manifold to the header block and lift off the manifold.

Intake Manifold Removal

352, 390, 427 and 428 Engines

Remove air cleaner and upper radiator hose. Disconnect coil and all lines to carburetor, including gas, vacuum and throttle lines. Remove distributor and draw distributor cap and ignition wires away from manifold. Remove rocker covers. Remove rocker shaft assembly in accordance with the procedure in rocker arm removal in this section. Withdraw pushrods from engine in sequence, and save them to replace in their original bore. Remove manifold bolts. Install lifting eyes in the rocker cover bolt holes left front and right rear of the manifold, and lift manifold off engine with hoist. The manifold may be removed by hand, but it is very heavy. Remove baffle plate beneath manifold by prying up with screwdriver.

289, 302, 351, and 429 Engines

Remove air cleaner and upper radiator hose. Disconnect all lines to carburetor. Remove coil, PCV tubing, heater hoses and all other lines from

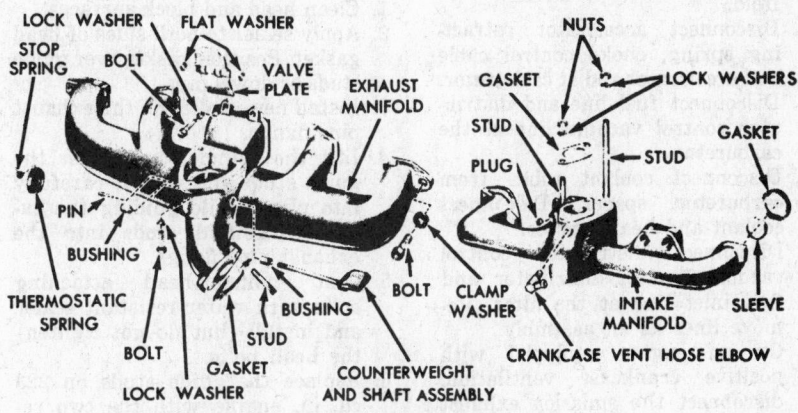

6 cyl. 223 manifold assemblies (© Ford Motor Co.)

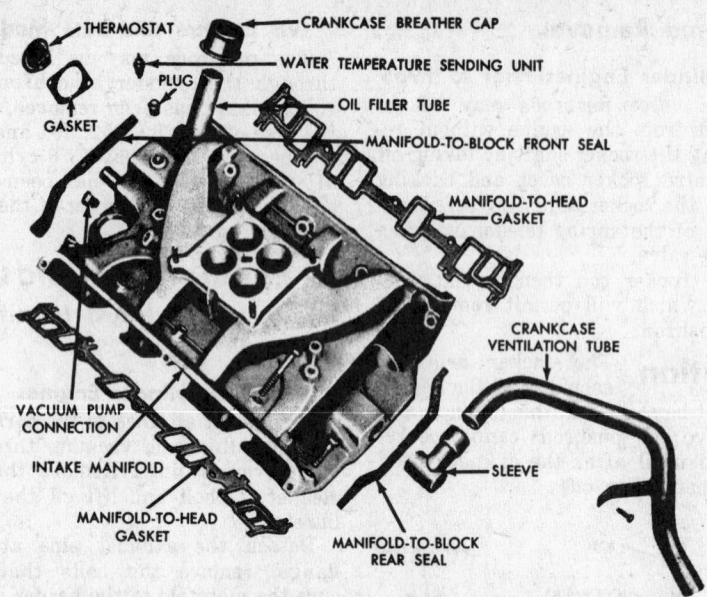

THERMOSTAT
CRANKCASE BREATHER CAP
WATER TEMPERATURE SENDING UNIT
PLUG
OIL FILLER TUBE
GASKET
MANIFOLD-TO-BLOCK FRONT SEAL
MANIFOLD-TO-HEAD GASKET
CRANKCASE VENTILATION TUBE
VACUUM PUMP CONNECTION
INTAKE MANIFOLD
SLEEVE
MANIFOLD-TO-HEAD GASKET
MANIFOLD-TO-BLOCK REAR SEAL

Intake manifold typical of V8 engines with distributor at the front of the block
(© Ford Motor Co.)

manifold. Disconnect battery ground. Remove distributor cap, spark plug wiring, and distributor. Disconnect water pump bypass hose. Remove manifold bolts. Remove manifold.

Exhaust Manifold Removal

All V8 Models

Disconnect the exhaust pipe flanges. The bolts are accessible from underneath the car. Remove the bolts that hold the exhaust manifolds to the head and lift off the manifolds.

On models equipped with a heater, many prefer to disconnect the heater when working on the right side exhaust manifold. While the manifold can be taken off with the heater in place, it is easier to take the heater off first.

Cylinder Head

6-Cylinder Head Removal

1. Drain coolant and remove air cleaner. Disconnect battery cable at cylinder head.
2. Disconnect exhaust pipe at manifold.
3. Disconnect accelerator retracting spring, choke control cable and accelerator rod at carburetor.
4. Disconnect fuel line and distributor control vacuum line at the carburetor.
5. Disconnect coolant tubes from carburetor spacer. Disconnect coolant and heater hoses.
6. Disconnect distributor control vacuum line at distributor and fuel inlet line at the filter. Remove lines as an assembly.
7. On an engine equipped with positive crankcase ventilation, disconnect the emission exhaust tube.

.8. Disconnect spark plug wires at the plugs and the small wire from the temperature-sending unit. On an engine equipped with an exhaust emission control system, disconnect the air pump hose at the air manifold assembly. Unscrew the tube nuts and remove the air manifold. Disconnect the anti-backfire valve air and vacuum lines at the intake manifold. On a car equipped with power brakes, disconnect the brake vacuum line at the intake manifold.
9. Remove rocker arm cover.
10. Remove rocker arm shaft assembly. On 240 cu. in. engines, loosen the rocker arm stud nut so that the rocker arm can be rotated to one side. Remove valve pushrods and keep them in sequence.
11. Remove one cylinder-head bolt from each end and install two 7/16 in. x 14 guide studs on 223 cu. in. engines.
12. Remove remaining cylinder-head bolts, then remove cylinder head.

6-Cylinder Head Installation

1. Clean head and block surfaces.
2. Apply sealer to both sides of head gasket. Position gasket over guide studs or dowel pins.
3. Install new gasket on the exhaust pipe flange.
4. Lift the cylinder head over the guide studs and slide it carefully into place while guiding the exhaust manifold studs into the exhaust pipe flange.
5. Coat cylinder-head attaching bolts with water-resistant sealer and install (but do not tighten) the head bolts.
6. Replace the guide studs on 223 cu. in. engine with the two remaining head bolts, then torque

the head, in proper sequence, and in three progressive steps to 75 ft. lbs.
7. Lubricate both ends of the pushrods and insert them in their original bores and sockets.
8. Lubricate valve stem tips and rocker arm pads.
9. Install valve rocker arm shaft assembly and torque attaching pedestal bolts, in progressive steps, to 30-35 ft. lbs. on 223 cu. in. engine.
   NOTE: on engines, beginning 1965, position the rocker arms and tighten the stud nuts enough to hold the pushrods in position. Adjust valve lash, as outlined in a later paragraph, valve adjustment, fulcrum type.
10. Do a preliminary, cold, valve lash adjustment.
11. Install exhaust pipe-to-manifold nuts and lock washers. Torque to 17-22 ft. lbs.
12. Connect radiator and heater hoses. Connect coolant tubes at the carburetor spacer.
13. Connect distributor vacuum line and the carburetor fuel line. Connect battery cable to cylinder head.
14. On engines equipped with positive crankcase ventilation, clean components thoroughly and install.
    NOTE: on engines equipped with an exhaust emission control system, install the air manifold assembly on the cylinder head. Connect the air pump outlet hose to the air manifold. Connect the anti-backfire valve, air and vacuum lines to the intake manifold.
15. Connect accelerator rod pull-back spring. Connect choke control cable and the accelerator rod at the carburetor.
16. Connect distributor control vacuum line at distributor. Connect carburetor fuel line at fuel filter.
17. Connect temperature - sending unit wire at sending unit. Connect spark plug wires.
18. Completely fill and bleed the cooling system.
19. Run engine for a minimum of 30 minutes at 1200 rpm to stabilize engine temperature. Then, check for coolant and oil leaks.
20. Adjust engine idle mixture and speed. Check valve lash and adjust, if necessary.
21. Install valve rocker arm cover, then the air cleaner.

V8 Heads, 352, 390, 427, 428, Removal

1. Remove intake manifold as previously described.
2. Remove any remaining accessories.
3. Remove exhaust manifolds.
4. Unbolt and remove heads.

V8 Heads, 260, 289, 302, 351 and 429 cu. in., Removal

1. Remove the intake manifold and carburetor as an assembly.
2. Remove rocker arm covers.
3. On cars equipped with air conditioning, isolate and remove the compressor.
4. If the left cylinder head is involved on a car with power steering, remove the steering pump and bracket and remove the drive belt. Tie assembly out of the way.
5. If the left cylinder head is involved on a car equipped with exhaust emission control system, disconnect the hose from the air manifold on the left cylinder head.
6. If the right head is involved, remove the alternator mounting bracket bolt and spacer, ignition coil and air cleaner inlet duct from the right cylinder head.
7. If the right cylinder head is to be removed on an engine equipped with an exhaust emission control system, remove the air pump and bracket. Disconnect the hose from the right cylinder head.
8. Disconnect the exhaust manifold/s at the exhaust pipe/s.
9. Loosen rocker arm stud nuts so that the arms can rotate to the side to clear the pushrods. Remove the pushrods. On 351 engines, remove exhaust manifold to get access to lower cylinder head bolts.
10. Remove cylinder-head bolts and lift off cylinder head.

V8 Heads Installation

Reverse above procedure, (see Valve Lash Adjustment under, Valves, in the following paragraphs).

Valve System

References

Since 1963, all Ford engines have operated at zero clearance, with the exception of some power options. These special, high out-put engines are equipped with provision for adjustment and should operate efficiently with about .025 in. lash. All other engines use hydraulic valve lifts except the 6 cylinder, 223 cu. in. engine. This model uses a zero lash mechanical adjuster built into the rocker arm.

6 Cylinder Prior to 1965, Dismantling

1. Remove cylinder head as described in previous paragraph.
2. Reach down through the pushrod area with a magnet, or other suitable extractor, and remove the tappets. Keep the tappets in order, so that they can be installed in their original location.
3. Loosen but do not remove rocker shaft support screws two turns at a time in sequence.
4. Remove cap screw and bracket from No. 4 valve rocker shaft support. Pull oil inlet line out of support, then pull line out of block with pliers.
5. Remove cap screw from No. 1 valve rocker shaft support. Remove oil outlet line and bracket.
6. Remove remaining rocker shaft support screws and lift off rocker assembly.
7. Dismantle shaft and rockers by removing pin and spring washer from each end of the shaft.
8. Slide rocker arms, springs and pedestals off the shaft. (Be sure to identify the parts for proper assembly sequence.)

6 Cylinder Prior to 1965 Assembly

1. Lubricate and assemble all rocker shaft components as illustrated.
2. If the end plugs were removed from the rocker shaft, install new plugs, cup side out, to each end of the shaft.
3. Do valve grinding job or any other service to the cylinder head, check valve spring condition, install new oil seals on valve stems, check springs and install valve assembly in place. Compress springs and install keepers.
4. Install valve tappets in their proper bores.
5. Apply cylinder-head gasket sealer to both sides of a new gasket and locate gasket on cylinder block. Install cylinder head and torque, in three progressive steps, to 75 ft. lbs.
6. Apply lubriplate, or suitable substitute, to both ends of pushrods, then insert pushrods into their respective places.
7. Install rocker assembly on head.
8. Install oil outlet line, bracket and cap screw on No. 1 rocker shaft support.
9. Install oil inlet line with new O-ring seal in position at No. 4 rocker shaft support.
10. Install rocker arm and shaft. Tighten rocker-shaft pedestal bolts in progressive steps to 30-35 ft. lbs.

Mechanical Adjustable Valve Adjustment—Primary Step

1. Make primary valve adjustment in the following manner, and continue to install rocker covers and fill cooling system.

NOTE: tappets must be adjusted while on the low radius of the cam.

2. If the distributor has not been disturbed and ignition timing is reasonably correct, proceed as follows: rotate crankshaft until the distributor rotor points to No. 1 plug wire tower of the distributor cap. Adjust valves in cylinder firing order according to rotor position.
3. If the distributor is out of time or has been removed from the engine: turn the crankshaft until No. 1 piston is at the top of its compression stroke, (intake valve of No. 6 cylinder just beginning to open), and the crankshaft damper is on T.D.C. Make three chalk marks on the crankshaft damper, 120° apart, starting with T.D.C. These marks will divide crankshaft travel into three parts, or six segments, of each engine cycle. Valve adjustment can be made in firing sequence, beginning with No. 1 on T.D.C. and progressing through the regular order of firing by advancing one chalk mark, (120 crankshaft degrees) at a time.

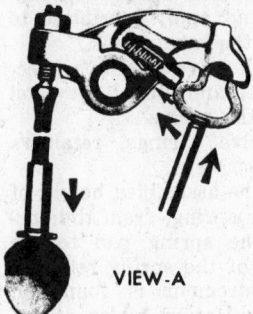 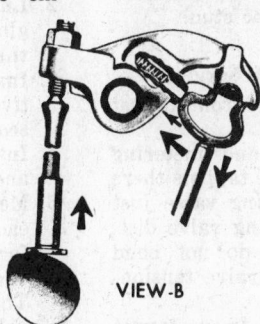 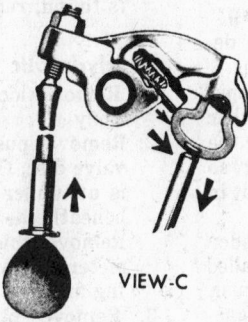 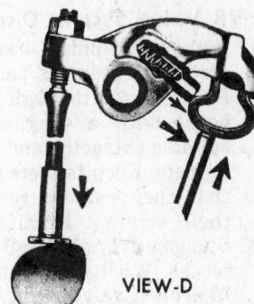

VIEW-A VIEW-B VIEW-C VIEW-D

Zero lash rocker arms (© Ford Motor Co.)

Mechanical Adjustable Valves Adjustment—Final Step

NOTE: be sure engine is at regular operating temperature by running at least thirty minutes.

1. With engine idling, check valve clearance with a feeler gauge. Adjust clearance, if necessary, to .016 in. (hot) for both intake and exhaust.

Zero Lash Mechanical Automatic Adjusters

1963-64, 223 Cu. In.

This rocker arm has an eccentric at the valve stem end. It is held against the stem by a spring loaded plunger. As wear occurs, or parts expand, due to heat, the plunger-loaded eccentric holds zero lash.

To make initial setting (engine off):

1. Set No. 1 piston at T.D.C. compression stroke.
2. Adjust intake and exhaust rockers for No. 1 cylinder by tightening screw until eccentric pushes plunger completely into bore.
3. Back out adjusting screw until adjusting mark is directly over valve stem. Torque locknut to 35 ft. lbs.
4. Repeat these steps for balance of cylinders.
5. Start engine and idle. Make recheck for location of adjusting mark. Make any minor readjustment to position the mark directly over valve stem.

Dismantling

6 Cylinder, Beginning 1965

1. Remove the coolant outlet housing and thermostat.
2. Remove temperature-sending unit.
3. Remove intake and exhaust manifolds and the carburetor assembly.
4. Remove spark plugs. Then, clean carbon from the combustion area before removing the valves.
5. Compress valve springs, remove valve locks, then release spring tension and dismantle the valve assembly.
6. Assemble by reversing the above procedure.

V8 Valve System Dismantling

1. Remove cylinder heads as described in previous paragraph.
2. Reach down through the tappet bores with a magnet or other suitable extractor and remove the tappets. Keep tappets in order, so that they can be reinstalled in their original location.
3. Support cylinder head on wooden blocks. With valves still installed to protect valve seats, wire brush the head clean, do not scratch gasket surfaces.

Valve stem seal removal
(© Ford Motor Co.)

4. Remove rocker arm adjusting nuts, fulcrum seats, and rocker arms from 289, 302, 351W and 429 engines. Remove rocker shaft assemblies from others. Compress valve springs, remove spring retainer keys, retainers and springs. Remove and discard stem seals and remove valves. On 1970 and 1971 302 cu. in. engines, remove stem caps from stems of exhaust valves.
5. Remove the exhaust manifolds.
6. Inspect condition of manifolds and heads.
7. On the right exhaust manifold, be sure that the automatic choke air inlet and outlet holes are completely open. Clean the maze screen in the passage with cleaning solvent.
8. Perform necessary valve seating and guide services.
9. Check valve springs for squareness and free length. Springs should not be more than 5/64 in. out of square, or 1/8 in. shorter than a new spring. A more accurate spring value can be obtained with a spring tester (see Valve Specifications chart).
10. On 289, 302, 351W and 429 engines, check rocker arm studs in cylinder head for damage. Also check for evidence of coolant leak at stud base which may indicate a loose stud. Service replacements are available in standard size and .015 in. oversize. If any looseness is found, replace the stud.

Hydraulic Tappet Service

1. Remove lock ring from tappet body.
2. Remove pushrod and metering valve disc. On some tappets there is an upper metering valve just beneath the metering valve disc. Remove this, but do not bend metering valve or valve tensioning finger.
3. Remove plunger from tappet body.

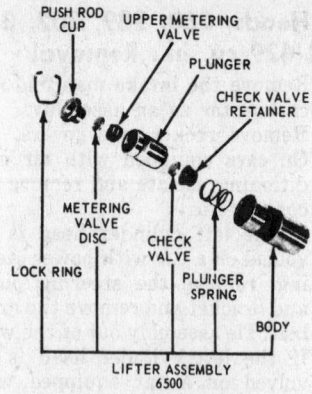

Typical hydraulic valve lifter
(© Ford Motor Co.)

4. Invert the plunger and remove the check valve retainer by carefully prying up on it with a screwdriver.
5. Remove the check valve and the spring.
6. Remove the plunger spring from the tappet body.
7. Soak all tappet components in solvent (lacquer thinner works well on these tar and varnish substances).
8. After tappet parts have been thoroughly cleaned and blown dry, place the plunger upside down on a clean surface.
9. Place the check valve over the hole in the bottom of the plunger. Place check valve spring on top of the check valve.
10. Position check valve retainer over the spring and check valve, then push the retainer down into place on the plunger.
11. Place plunger spring and plunger into the tappet body.
12. Place upper metering valve, if any, metering valve disc and pushrod cup in the plunger. Depress plunger and install the lock ring.
13. Test all hydraulic tappets (both new and used) before installing them in an engine. The factory quotes a bleed-down time of from 10 to 80 seconds, using special test oil and their approved tester.

V8 Valve System Assembly

1. Install hydraulic tappets into their respective bores.
2. Lubricate valve stems with engine oil and apply lubriplate to the tip of the valve stems.
3. Insert each valve into its respective guide, and install new oil seals on the stems.
4. Install valve springs, retainers and retainer keys.
5. Measure the assembled height of each valve spring, from the surface of the spring pad to the underside of the spring retainer. This masurement is found in valve specification tables at beginning of section. If assembled

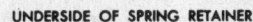

UNDERSIDE OF SPRING RETAINER

SURFACE OF SPRING PAD

Valve spring assembled height
(© Ford Motor Co.)

height is not within limits, install necessary spacer/s between the spring pad and the spring to correct the measurement.

Caution Do not install spacers, unless necessary. Use of spacers, in excess of recommendations, will result in overstressing the valve springs and overloading the camshaft lobes. This could cause spring breakage and camshaft wear.

6. On 352, 390, 428 and 427 cu. in. engines, position new sealer coated gasket on block, install head, and torque to specifications. Install intake manifold and torque to specification. Insert pushrods into their respective bores after lubricating both ends. Install and torque down rocker shafts to 40-45 ft. lbs. On other V8s, install rocker arms and fulcrum seats. On engines equipped with pedestal-mounted rocker arms, install rocker arms and fulcrum seats. Install adjusting nuts, finger tight.
Position new sealer-coated head gasket, install cylinder head, and progressively torque to specification.
Lubricate both ends of pushrods and install them into their respective tappets.
Install the exhaust manifold/s. Torque manifold bolts to specification and bend over the bolt locking tabs.

7. Connect water pump by-pass hose to the pump and the water outlets housing, tighten the clamps.
8. Install fuel pump and the fuel filter and adapter assembly.
9. Rotate crankshaft to bring No. 1 piston up to T.D.C. of the compression stroke.
10. Position distributor in the block with the rotor at No. 1 firing position and the points just opening. Install distributor hold-down clamp.
11. Install carburetor choke heat tube and the carburetor. Install distributor cap and plug wires as an

assembly. Install ignition coil, and connect high tension lead at the coil.
12. Install and connect the fuel filter to carburetor line, the fuel pump to fuel filter line, and the distributor to carburetor vacuum line.
13. Check engine oil, fill and bleed cooling system, start engine, make final adjustments and roadtest car.

Valve Lash—Hydraulic

Valve stem to rocker arm clearance is zero. However, conditions may change the original setting and adjustment may become necessary.

Clearance must be determined by measuring the clearance between the closed valve stem and the rocker arm (with tappets empty and collapsed).

Valve lash should be as follows:

1965-70	240	.082-.152
1964-66	289	.082-.152
1967-70	289, 302	.067-.167
1969-70	351W	.083-.183
1964	352, 390	.083-.183
1965-66	352, 390, 428	.050-.150
1967-70	390, 428	.100-.200
1968-70	429	.075-.175

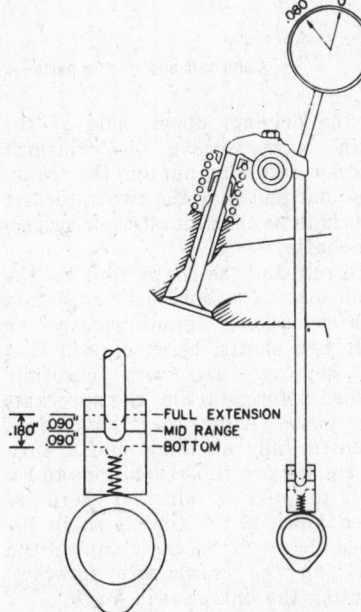

Checking tappet operational range

If the engine is not equipped with adjustable rocker arms or individual rocker pedestals (as with the 240-6, 289, 302, 351W or 429 V8s), proper lash must be obtained by using longer or shorter pushrods. These rods are available in .060 in. options, above or below standard.

Valve Springs

Valve springs should be checked for pressure at any time they are removed.

There is, however, a quick, easy way to check valve springs if they

have all been removed from the car. That is to place all the valve springs beside one another on a level flat surface, and if they are all the same height, only one of the springs needs to be checked. If the one spring comes up to specifications, it can be safely assumed that the rest of them will, also.

⏱ CHILTON TIME-SAVER

Tappet Removal

To remove and replace tappets from 352, 390, 427, and 428 engines without removing the intake manifold, first remove rocker covers and rocker assembly. Then remove pushrods from their bores. Locate tappet or tappets to be moved by shining a light through pushrod bores. Use a magnet or claw tool to seize tappet and withdraw it through pushrod bore. It may be necessary on some tappets to move them over and draw them through a larger adjoining pushrod bore, but tappets should always be replaced in their original holes.

Timing Case

Timing Gear Cover Removal

6-Cylinder Engines

The timing case cover is located to the front of the cylinder block by two dowel pins. It is held in place with hex-head screws. There are, also, two screws fed up from the oil pan into the timing case cover. The cover, too, is used as the engine front support.

To remove the timing case cover, it is recommended that the oil pan be removed. The oil pan gasket contacts the bottom of the timing case cover, making it an extremely difficult job to remove the timing case cover without destroying or damaging the gasket so that it is no longer fit for service.

Take out the engine front support bolts and support the engine on a jack. Take off the radiator core and the fan assembly, remove the vibration damper and unbolt the timing case cover.

V8 Engines

To remove the timing case cover from the front of these engines, first remove the oil pan since there is a gasket between the lower part of the timing case cover and the oil pan. It will be very difficult to prevent oil leaks if the oil pan is left in place and an attempt is made to seal between the bottom of the timing case cover and the oil pan.

Remove the radiator core, the fan, fan pulley, the water pump and the water by-pass pipe back of the water pump. On air conditioned cars, remove compressor mounting bracket and position compressor out of the way. Remove bolts attaching the condenser, and position it forward. Do not disconnect any refrigerant lines. Take off the crankshaft pulley and the fuel pump.

Take off the bolts that hold the timing case cover to the front of the block and slide the cover off the crankshaft.

It is always good practice to install a new oil seal when replacing the timing case cover.

Caution Due to the structure of the crankshaft, it is very important that the proper tools be used to draw the damper or pulley into place on the crankshaft. (Do not use a hammer; breakage of the cast crankshaft can result.)

Timing Case Oil Seal Replacement

All Models

To replace the oil seal, it is necessary to take off the timing case cover and drive the seal out with a pin punch. Clean out the recess in the cover and install a new seal using a special driving tool.

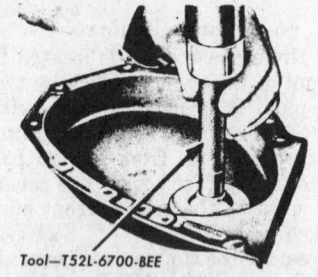

Tool—T52L-6700-BEE

Timing case oil seal installation—6 cyl.
(© Ford Motor Co.)

Coat the new seal with grease to reduce friction when installing and starting the car.

Timing Sprocket and/or Chain Replacement

Caution Do not rotate the crankshaft or camshaft unless the timing gears/sprockets are installed and in time. Damage may result by interference between the camshaft lobes and the crankshaft.

6-Cylinder Engines—1963-64

To replace the timing chain for wear or looseness, first align the timing marks so that there are 12 pins of the chain between the mark on the cam and the mark on the crankshaft sprocket. These should be measured

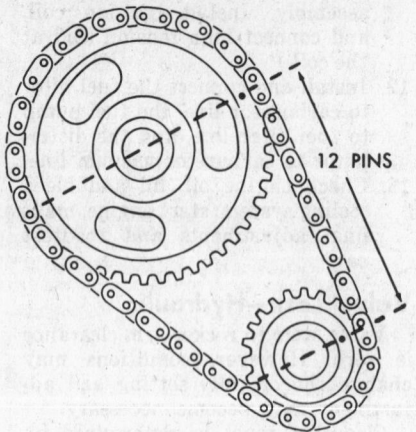

Timing mark alignment—223 6 cyl.

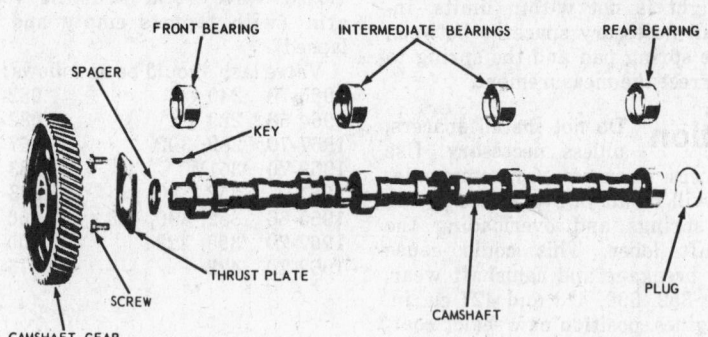

Camshaft and related parts—240 cu. in. 6 cyl. (© Ford Motor Co.)

on the driving, upper, side of the chain. Now, remove the camshaft sprocket retaining nut and the crankshaft nut and slide the two sprockets and chain as one unit off their respective shafts.

To reinstall the chain, first set the chain over both sprockets and start both sprockets simultaneously on their two shafts, being certain that the keyways are very carefully aligned before starting the sprockets into place. After they have been mounted fully on their shafts, turn the engine two full revolutions to recheck to make certain that there are seven links of the timing chain between the mark on the cam and the mark on the crankshaft sprocket, counting the link at each mark.

Check the chain for wear or looseness. Some police or taxi 223 cu. in. engines have timing gears in place of the timing chain. To remove, align timing marks by rotating engine. Remove crankshaft front oil slinger, camshaft retaining bolt and washers, and timing gears. To replace, install keys in camshaft and crankshaft. Install gears taking care that timing marks align. Replace crankshaft oil slinger and camshaft retaining bolt and washers.

6-Cylinder Engines—1965-70

1. With cover removed and push-rods and valve lifters out, remove camshaft.

2. With camshaft mounted in press, press off timing gear.
3. Remove oil slinger from crankshaft and with puller, remove crankshaft timing gear and key.
4. To reinstall, place camshaft in engine. Install spacer. Install thrust plate. Insert key in keyway.
5. Place gear on camshaft with timing gear marks in proper location for alignment. Tools T64T-6306A and T65L-6306A are recommended for this operation.
6. Install the thrust plate screws and torque to 35-45 ft. lbs.
7. Install crankshaft key in keyway and install crankshaft gear. Install oil slinger.

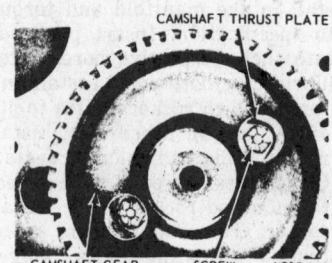

Camshaft gear removal—240 cu. in. 6 cyl. (© Ford Motor Co.)

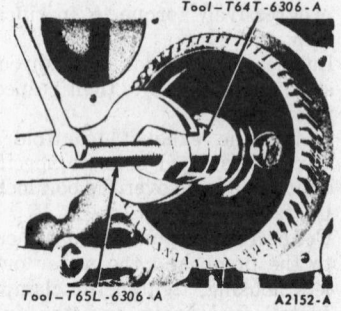

Camshaft gear installation—240 cu. in. 6 cyl. (© Ford Motor Co.)

8. Recheck timing marks, end-play and gear runout and backlash.

V8

1. Remove the timing case cover as outlined under the paragraph devoted to this subject, and remove the cam and crankshaft sprockets.

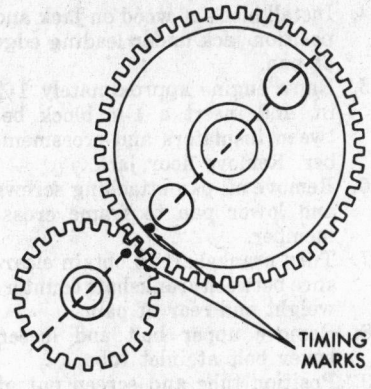

Timing mark alignment—240 6 cyl.

2. Before taking the sprockets off, however, arrange them so that the timing marks are in a direct line with each other and the center of the shafts.
3. Remove camshaft sprocket cap screw, washer, and fuel pump eccentric.
4. Remove the cam and crankshaft sprocket.
5. Arrange the new chain and/or sprockets on the bench to occupy the same relative position just explained.
6. Now, carefully start the sprockets upon the shafts with the chain in place and, after being certain that they are started straight, push them all the way onto the shafts and secure. Install camshaft sprocket cap

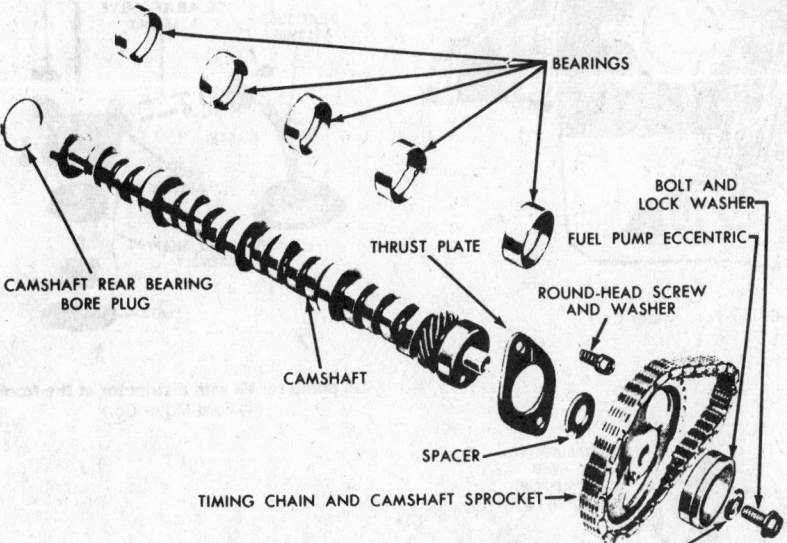

V8 camshaft and related parts (© Ford Motor Co.)

screw washer and fuel pump eccentric.
7. Turn the engine two full revolutions to make sure that the timing marks are in the correct position.

Engine Lubrication

6 Cylinder—No Oil Pressure

Some instances of no oil pressure have occurred on the 6-cylinder engine.

Design characteristics and wear bring about the failure responsible for this condition.

The gear-type oil pump is entirely within the crankcase, in line with, and driven by, the distributor gear through a hex shaft.

This shaft is pinned in the lower end of the distributor driveshaft by a roll pin. The bottom end of the hex shaft seats in the female end of the oil pump drive. It is possible for the roll pin to shear or work out of place.

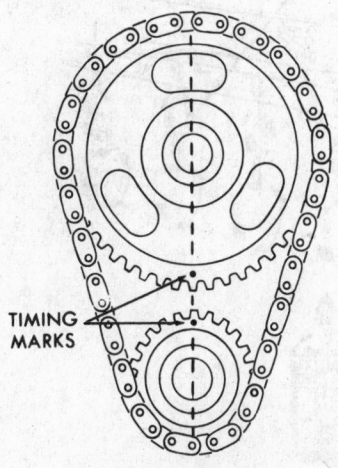

Timing mark alignment— V8 front mounted distributor

This will allow the hex driveshaft to slip down and out of the distributor drive and cause oil pump failure.

⏻ CHILTON TIME-SAVER

In the event of no-oil-pressure, remove the distributor and check the condition of this hex drive. If the above situation exists, it may be possible to recover the hex shaft with the aid of a magnet or a mechanical finger tool. If recovery is possible, the shaft can be reattached to the distributor with a new roll pin. The alternative is to drop the oil pan and remove the oil pump to retrieve the hex.

Oil Pan Removal

1964—6 Cylinder

1. Drain crankcase. Remove oil dipstick.
2. Six cylinder with standard transmission, remove clutch retracting spring.
3. Six cylinder, remove stabilizer bar-to-underbody retaining nuts and pull the bar downward.
4. V8 models, remove starter motor.
5. Remove oil pan retaining bolts and crank the engine as required to obtain clearance, then remove the oil pan.
6. Install oil pan in the reverse order of removal. Torque the 1/4 in. x 20 cap screws to 7-9 ft. lbs., and 5/16 in. x 18 cap screws to 9-11 ft. lbs.
7. Install oil dipstick. Fill crankcase to level, then run the engine and check for leaks.

1965-70—6 Cylinder

1. Drain crankcase and cooling system.
2. Disconnect upper hose at outlet elbow and lower hose at radiator. Remove radiator.
3. Disconnect flexible fuel line at fuel pump.
4. With automatic transmission, disconnect kickdown rod at bellcrank assembly. On car with standard transmission, disconnect clutch linkage.
5. Raise car on hoist.
6. Disconnect starter cable at starter. Remove retaining bolts and remove starter.
7. Remove nuts on both engine front support insulator-to-support bracket.
8. Remove bolt and insulator, rear support insulator-to-crossmember and insulator-to-transmission extension housing.
9. Raise transmission, remove support insulator, lower transmission to crossmember.

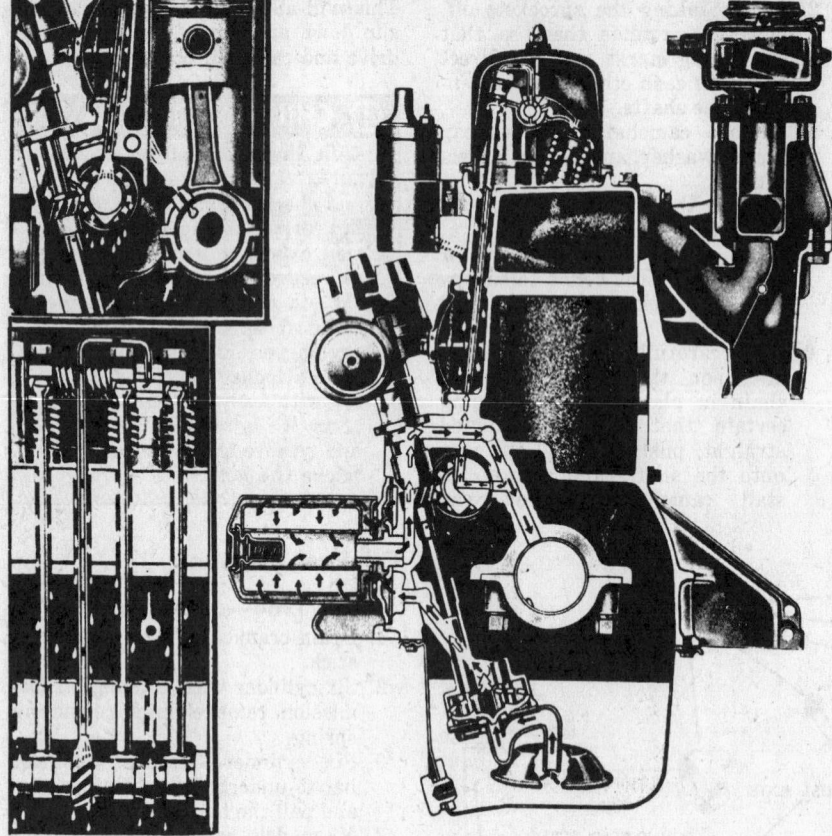

Engine lubrication—6 cyl. 1963-64 (© Ford Motor Co.)

4. Install block of wood on jack and position jack under leading edge of pan.
5. Raise engine approximately 1¼ in. and insert a 1-in. block between insulators and crossmember. Remove floor jack.
6. Remove oil pan attaching screws and lower pan to frame crossmember.
7. Turn crankshaft to obtain clearance between crankshaft counterweight and rear of pan.
8. Remove upper bolt and loosen lower bolt at inlet tube.
9. Position tube and screen out of the way and remove the pan.
10. Install in reverse of above.

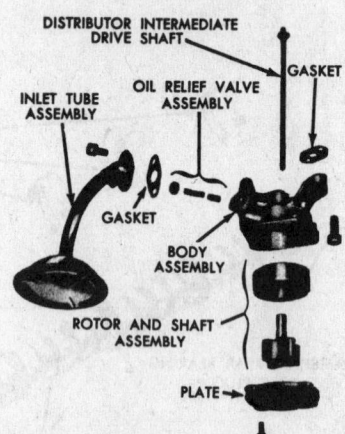

Oil pump for V8 with distributor at the front (© Ford Motor Co.)

10. Raise engine with transmission jack and place 3-in. thick wood blocks between both front support insulators and intermediate support brackets.
11. Remove oil pan retaining bolts and oil pump mounting bolts. With oil pump in pan, rotate crankshaft as needed to remove pan.
12. Install in reverse of above.

1964-70—289, 302, 351 Cu. In. V8

1. Remove oil level dipstick. Drain oil pan.
2. Disconnect stabilizer bar from lower control arms, and pull ends down.
3. Remove oil pan attaching bolts and position pan on front cross-member.
4. Remove one oil inlet tube bolt and loosen the other to position tube out of way to remove pan.
5. Turn crankshaft as required for clearance to remove pan.
6. Install in reverse of above.

1964-70—V8s Except 289, 302, 351 Cu. In.

1. Raise car and place safety stands in position. Drain oil from crankcase.
2. Disconnect stabilizer bar links and pull ends down.
3. Remove nuts and lock washers, engine front support insulator-to-intermediate support bracket.

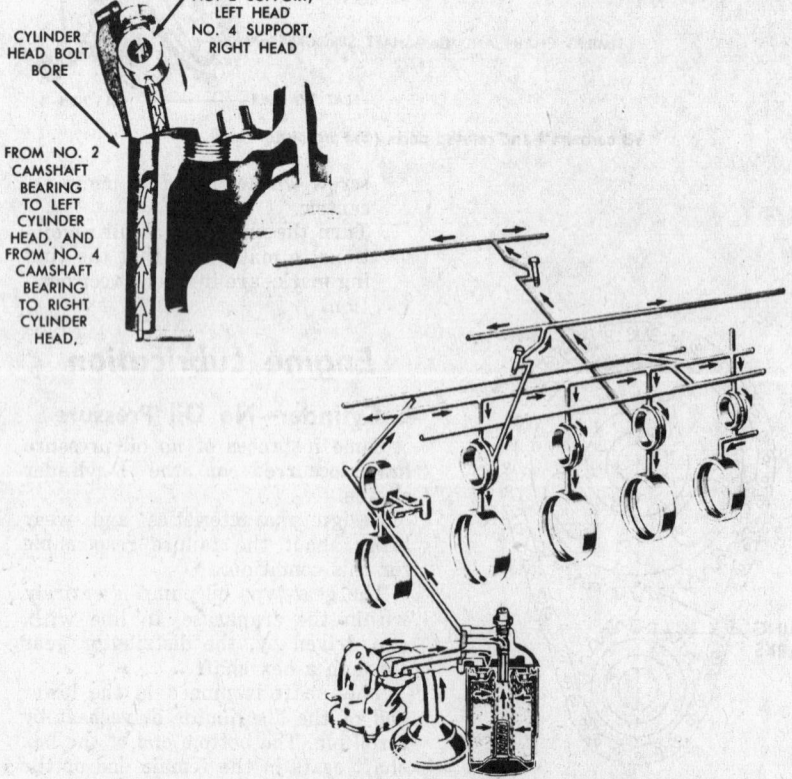

Engine lubrication—'8 (© Ford Motor Co.)

Difficulty in oil filter replacement, starts with oil filter elements being turned on too tightly. The unit may be too tight to remove by hand, and it may collapse in the grip of a tool that applies enough squeeze to grip the element hard enough to turn it.

Alternate Method of Removal

1. Raise the car on a jack or hoist and place a drip pan under the filter.
2. With a 12-14 in. slender punch drive a hole in the element from one side to the other.
   NOTE: before punching the hole, consider the angle required for the punch to act as a lever with the least interference.
3. With the drift all the way through the filter and acting as a lever, turn the unit counterclockwise enough to break it loose.
4. Final loosening and removal can now be accomplished by hand.

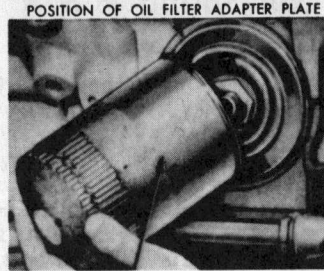

POSITION OF OIL FILTER ADAPTER PLATE

OIL FILTER ELEMENT

Check filter adaptor plate for proper position
(ⓒ Ford Motor Co.)

Oil Filter Replacement

1. Coat the gasket on the new filter with oil.
2. Place the new filter in position on the block.
3. Hand-tighten until contact is made between the filter gasket and the adapter face.
4. Tighten by further turning the filter one-half turn.
5. Run the engine at fast idle and check for oil leaks.
6. Check the oil and bring crankcase to level if necessary.

Connecting Rods and Pistons

Removal

All Models

Take off the cylinder heads and oil pan. Select the pistons in the down

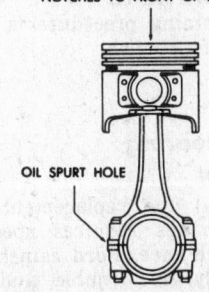

NOTCHES TO FRONT OF ENGINE

OIL SPURT HOLE

Piston and rod assembly—6 cyl.

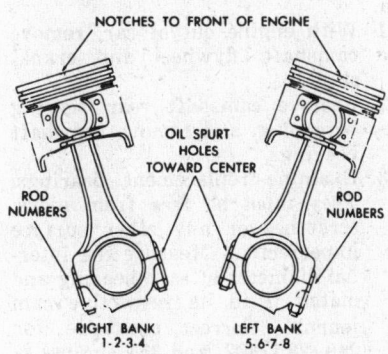

NOTCHES TO FRONT OF ENGINE

OIL SPURT HOLES TOWARD CENTER

ROD NUMBERS

ROD NUMBERS

RIGHT BANK 1-2-3-4

LEFT BANK 5-6-7-8

Piston and rod assembly—V8

position and, using a good cylinder ridge reamer, remove the ridge from the top of these cylinders. Turn the crankshaft and repeat this reaming operation until all of the cylinders have been ridge-reamed.

From underneath the car, remove the two bolts from the connecting rod cap on the rods that are in the down position. Push the connecting rod and piston assembly part way up the bore and immediately replace the cap on the bottom of the connecting rod. This is to eliminate the possibility of losing the bearing or getting the caps mixed up.

Inspection of Pistons

All Models

Carefully inspect the pistons, particularly the thrust face, for scratches or scores. Do not confuse an ordinary wear pattern with scoring. Scoring on the piston is generally accompanied by a comparable scoring on the cylinder walls, whereas the wear pattern usually leaves a perfectly smooth and, in fact, somewhat glassy cylinder wall.

Cylinders showing the slightest scoring will have to be rebored and pistons with scores will have to be resized or replaced.

Fitting Rings

Complete instructions are supplied by each ring manufacturer with each package set of rings. These instructions should be followed.

Rear Main Bearing Oil Seal—Except 6-Cylinder, 240 Engine and 1968-70 390-428 Engines

On both 6 and 8 cylinder models, a packing-type seal is used in back of

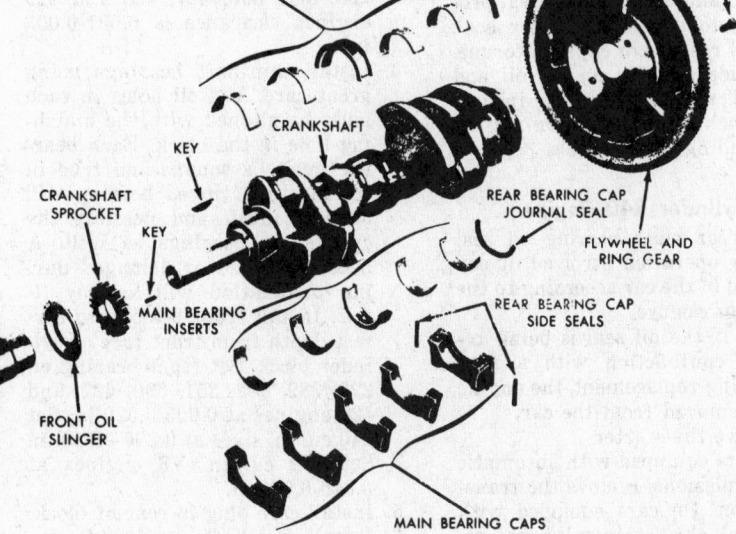

CYLINDER BLOCK JOURNAL SEAL

MAIN BEARING INSERTS

BOLT

KEY

CRANKSHAFT

CRANKSHAFT SPROCKET

KEY

MAIN BEARING INSERTS

REAR BEARING CAP JOURNAL SEAL

FLYWHEEL AND RING GEAR

REAR BEARING CAP SIDE SEALS

POWER STEERING PUMP PULLEY

DAMPER

SLEEVE

BOLT

WASHER

FRONT OIL SLINGER

MAIN BEARING CAPS

V8 crankshaft and related parts (ⓒ Ford Motor Co.)

the rear main bearing. To replace the upper half of the seal, it is necessary to remove the crankshaft. The lower half of the seal may be replaced by taking down the rear main bearing cap and inserting new packing, letting the packing protrude approximately 1/16 in. above the cap at either side. Bolt the cap into place and immediately take it down to determine if the seal has riveted over. If it has riveted over, trim just the riveted portion with a razor blade, and bolt the cap back up again. Remove the cap again to determine if the main bearing cap is seating properly in the block. The reason for permitting the oil seal packing to protrude is so that it will tend to compress the upper half of the packing more tightly into the retainer in the cylinder block. In this way it is sometimes possible to prevent leaks at the rear main bearing without replacing the upper half of the oil seal.

1968-70 390 and 428 Cu. In. V8 Engines

1. Loosen all main bearing cap bolts, lowering crankshaft slightly but not more than 1/32 in.
2. Remove rear main cap, and remove upper and lower halves of seal. On block half of seal, use seal removing tool or insert a small metal screw into end of seal with which to draw it out.
3. Clean seal grooves with solvent and dip replacement seal in clean engine oil.
4. Install upper seal half in its groove in block with lip toward front of engine by rotating it on seal journal of crankshaft until approximately 3/8 in. protrudes below parting surface.
5. Tighten other main caps and torque to specification.
6. Install lower seal half in rear main cap with lip to front and approximately 3/8 in. of seal protrudes to mate with upper seal.
7. Install rear main cap and torque.
8. Dip side seals in engine oil and install them. Tap seals in last half inch if necessary. Do not cut protruding ends of seals.

6-Cylinder 240 Engine

If the rear main bearing oil seal is the only operation involved, it can be replaced in the car according to the following procedure.

NOTE: if the oil seal is being replaced in conjunction with a rear main bearing replacement, the engine must be removed from the car.

1. Remove the starter.
2. On cars equipped with automatic transmissions, remove the transmission. On cars equipped with manual shift transmissions, remove the transmission, clutch,

flywheel and engine rear cover plate.
3. With an awl, punch holes in the main bearing oil seal, on opposite sides of the crankshaft and just above the bearing cap to cylinder block split line. Insert a sheet metal screw in each hole. With two large screwdrivers, pry the oil seal out.
4. Clean the oil recess in the cylinder block, main bearing cap and the crankshaft sealing surface.
5. Lubricate the entire oil seal. Then, install and drive the seal into its seat .005 in. below the face of the cylinder block with tool T-65L-6701-A.
6. The remaining procedure is the reverse of removal.

Camshaft Bearing Replacement

The removal and replacement of camshaft bearings requires special tools and much care. Ford camshaft bearings rarely give trouble, and so unless they are known to be bad we advise leaving the original bearings in.

1. With engine out of car, remove camshaft, flywheel and crankshaft.
2. Remove camshaft rear bearing bore plug, and remove camshaft bearings.
3. Examine replacement bearings. They must be free from nicks, scratches or any other surface imperfection. Measure the internal diameter of each bearing and match it to its respective cam journal. Correct clearance for 240, 289, 302, and 351 engines is 0.001-0.002 in. On 289, 302 and 351 engines camshaft bearings decrease in diameter from front to rear. Do not mix bearings. On 223, 352, 390, 427, 428 and 429 engines clearance is 0.001-0.003 in.
4. Install camshaft bearings using great care. The oil holes in each must be aligned with the matching hole in the block. Each bearing must be square and true in the block. A tipped bearing will destroy itself, and possibly the crankshaft bearings as well. A bearing nicked or damaged during installation will destroy itself. Install front bearing to correct depth from front face of cylinder block. Set front bearing on 223, 289, 302, 351, 390, 427, and 428 engines at 0.005-0.020 in. Set 240 cu. in. sixes at 0.200-0.350 in. Set 429 cu. in. V8 engines at 0.040-0.060 in.
5. Install core plug in rear of block.
6. Install camshaft, crankshaft, flywheel and related parts.

Front Suspension

Adjustment

General instructions covering the front suspension, and how to repair and adjust it, together with information on installation of front wheel bearings and grease seals, are in the Unit Repair Section.

Definitions of the points of steering geometry are in the Unit Repair Section. This article also covers troubleshooting front end geometry and irregular tire wear.

Figures covering the caster, camber, toe-in, kingpin inclination, and turning radius are in the Front Wheel Alignment table of this section.

Extended Lubrication

The following front suspension and steering parts will require lubrication only after long mileage periods: two upper ball joints, two lower ball joints, one idler arm joint, two cross-link joints, four tie-rod ball joints, one power steering valve joint.

These places are factory-packed with a special compound and, when replaced, only the original-type lubricant should be used.

Coil Spring R & R

Except 1963-66 T-Bird

1. Raise car and support with stands placed back of lower arms.
2. If equipped with drum type brakes, remove the wheel and brake drum as an assembly. Remove the brake backing plate attaching bolts and remove the backing plate from the spindle. Wire the assembly back out of the way.
3. If equipped with disc brakes, remove the wheel from the hub. Remove two bolts and washers that hold the caliper and brake hose bracket to the spindle. Remove the caliper from the rotor and wire it back out of the way. Then, remove the hub and rotor from the spindle.
4. Disconnect lower end of the shock absorber and push it up to the retracted position.
5. Disconnect stabilizer bar link from the lower arm.
6. Remove cotter pins from the upper and lower ball joint stud nuts.
7. Remove two bolts and nuts holding the strut to the lower arm.
8. Loosen the lower ball joint stud nut two turns. Do not remove this nut.
9. Install spreader tool T57P-3006-

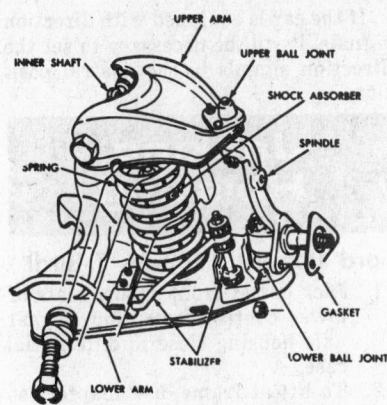

Typical front suspension—1963-64
(© Ford Motor Co.)

Typical front suspension—1965-70 (© Ford Motor Co.)

A between the upper and lower ball joint studs.

10. Expand the tool until the tool exerts considerable pressure on the studs. Tap the spindle near the lower stud with a hammer to loosen the stud in the spindle. Do not loosen the stud with tool pressure only.
11. Position floor jack under the lower arm and remove the lower ball joint stud nut.
12. Lower floor jack and remove the spring and insulator.
13. Install by reversing above procedure.

T-Bird—1963-66

1. Raise the car and position stands under the suspension lower arms.
2. Remove wheel and brake assem-

bly as described in Steps 2 and 3 of the previous paragraph.
3. Raise the car slightly in order to lower the suspension upper arm. Install spring tool T63P-5310-A. Slide the tool bearing and upper plate over the shaft screw against the shaft nut. Insert the tool assembly through the upper opening in the spring housing so that the shaft screw goes through the top of the coil spring. The tool upper plate holes must accommodate the studs in the frame spring tower.
4. From under the car, place the tool lower plate under the fourth spring coil from the bottom.

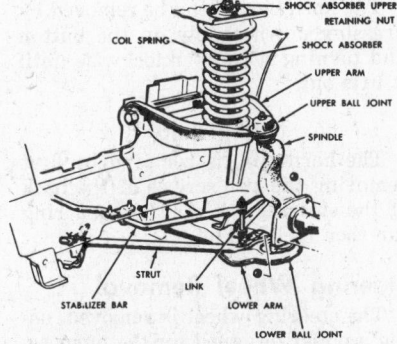

T-Bird front suspension—1963-66
(© Ford Motor Co.)

Secure the plate to the coil by installing the tool retainer in the groove in the shaft screw.
5. Insert a ½ in. square flex-handle wrench in the drive hole in the lower plate to prevent the tool with spring from turning. While holding the tool, compress the spring by turning the tool shaft nut clockwise.
6. Remove the two nuts and lock washers that attach the upper arm inner shaft to the chassis, and swing the arm out of the way. The arm pivots on the ball joint.
7. Remove the bolt that holds the clip and brake line to the chassis, then move the brake line out of the way.
8. Disconnect the stabilizer bar from the link at both left- and right-hand suspension lower arms by removing the bar-to-link attaching nuts and upper bushings. Position the bar out of the way.
9. Fully release the spring tension by turning the tool shaft nut counterclockwise. Hold the lower plate of the tool with the ½ in. square drive wrench so that the tool will not turn or snap loose during spring release.
10. Remove the spring tool, then remove the spring from the car.
11. Install by reversing above procedure.

Steering Gear

Steering Gear Adjustment
Instructions covering the adjustment of the steering gear are in the Unit Repair Section.

Manual Steering Gear Assembly Removal

1. Disconnect and tag the wire from the column underneath the dash. Remove the horn ring and take off the steering wheel.
2. Remove the turn indicator lever and the gear shift lever from the top of the steering column. If equipped with automatic transmission, remove the selector dial and cover. Throw back the floor mat and take off the floor cover plate. Disconnect the speedometer cable from the instrument panel and remove the right-hand cover plate from the cable. Disconnect the gearshift rods at the bottom of the steering column. Disconnect the neutral safety switch wire at the bottom of the clutch return-spring bracket.

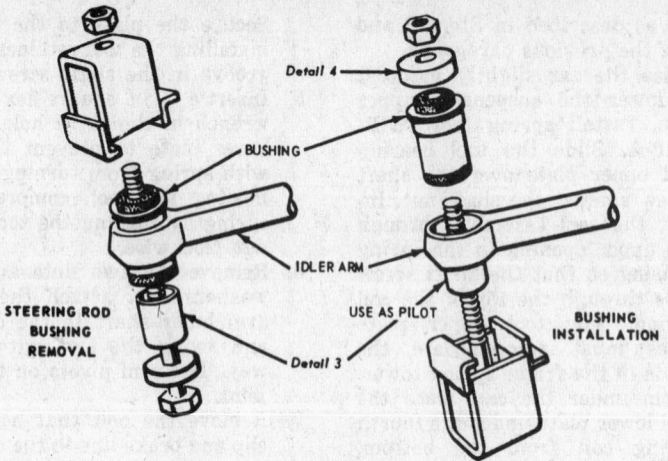

Installing idler arm bushing (© Ford Motor Co.)

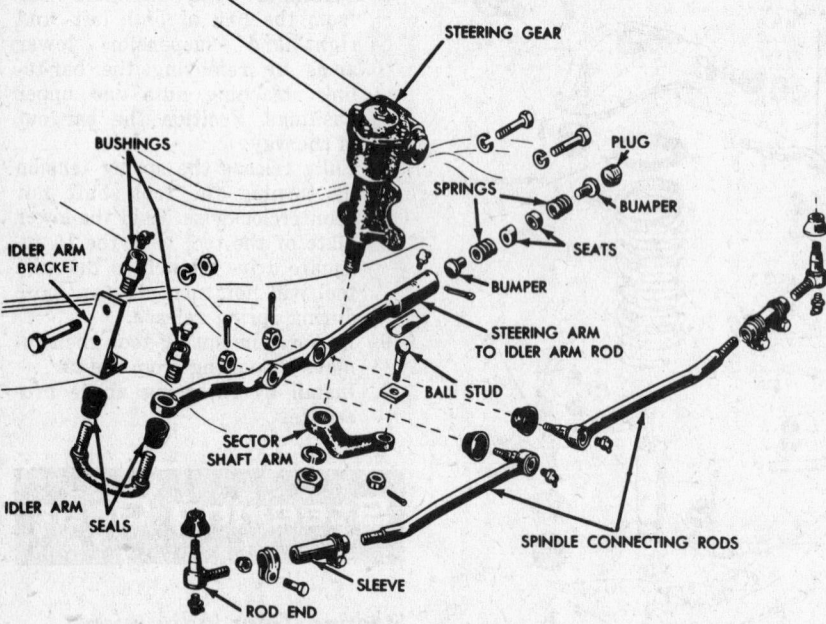

Typical steering linkage (© Ford Motor Co.)

3. On cars with single exhaust systems, disconnect the exhaust pipe for access to the pitman arm and remove the pitman arm. Remove the bolts that hold the steering mast jacket to the under side of the dash panel, then remove the bolts that hold the gear box itself to the frame. Lift the steering gear assembly out through the engine compartment on standard transmission jobs, up into the body on automatic cars.

On cars with a flexible joint in steering shift, disconnect joint and remove without disturbing upper portion.

T-Bird

Disconnect the wires from the bottom of the gear housing, loosen the lower steering column clamp and remove the screws from the upper steering column bracket. Remove the horn ring from the steering wheel and take off the upper steering column, upper steering shaft and steering wheel as an assembly from the car. Remove the pitman arm and take out the bolts that hold the steering gear housing to the frame side member in order to remove the lower steering gear assembly from the car.

Horn Button Removal

The horn button can be removed by pressing down evenly on the button and turning it counterclockwise until it lifts out.

Horn Ring

The horn ring is removed by first removing the two screws at the back of the steering wheel. The horn ring can then be lifted off.

Steering Wheel Removal

The steering wheel is removed, using a puller designed for the purpose, after the horn button and the steering wheel nut have been taken off.

If the car is equipped with direction signals, it will be necessary to set the direction signals in the neutral position.

Jacking, Hoisting

Ford (Except 1963-66 T-Bird)

1. Jack car at front spring seats of lower control arms, and at rear axle housing close to differential case.
2. To lift at frame, use adapters so that contact will be made at points shown. Adapters should support at least 12 sq. in.

T-Bird—1963-66

3. (With twin post lift) Front: place adapters to contact lower suspension arms. Rear: place forks not more than one in. from the circular welds near the differential housing.
4. (With frame contact lift) Place adapters at points shown on diagram and covering an area not less than 24 sq. in.

Clutch

Clutch Pedal Adjustment

1. The clutch release fork rod should be adjusted so that there is from 7/8 to 1 1/8 in. free-play of the clutch pedal, measured at the toe board. This is the only practical adjustment that can be made with the clutch mounted in the car.
2. Make a final check with the engine running at 3000 rpm, and transmission in neutral. Under this condition, centrifugal weights on release fingers may reduce the clearance. Readjust, if necessary, to obtain at least 1/2-in. free-play while maintaining the 3000 rpm to prevent fingers contacting release bearing. This is important.

All other adjustments require that the clutch be removed.

Clutch Removal

1. Raise car on jackstands, disconnect driveshaft from rear U-joint, and slide driveshaft off transmission output shaft. Plug rear extension with a special tool or rags to prevent loss of transmission oil.
2. Disconnect speedometer cable, transmission linkage, and all other lines to transmission.
3. Support engine with jack and disconnect transmission mount on transmission extension housing.

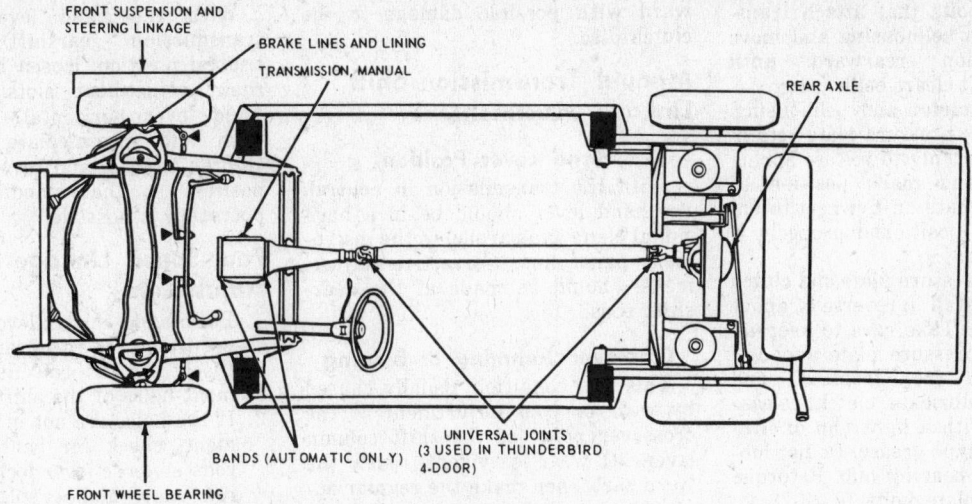

FRONT SUSPENSION AND STEERING LINKAGE

BRAKE LINES AND LINING

TRANSMISSION, MANUAL

REAR AXLE

BANDS (AUTOMATIC ONLY)

UNIVERSAL JOINTS (3 USED IN THUNDERBIRD 4-DOOR)

FRONT WHEEL BEARING

▲ STEERING LINKAGE
△ FRONT SUSPENSION
■ LIFT POINT

Ford hoist lifting positions (© Ford Motor Co.)

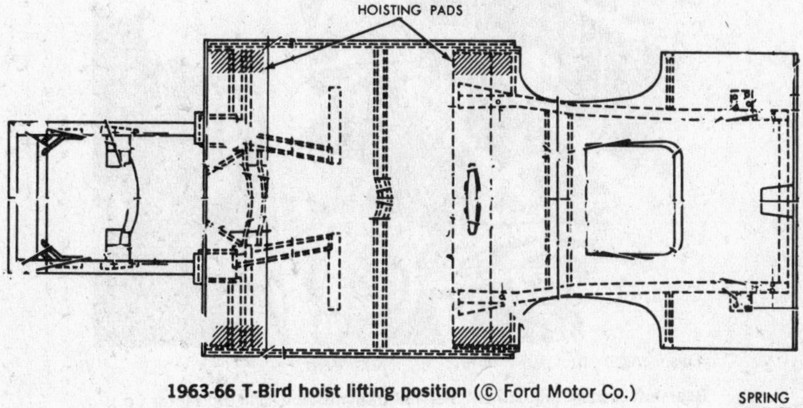

HOISTING PADS

1963-66 T-Bird hoist lifting position (© Ford Motor Co.)

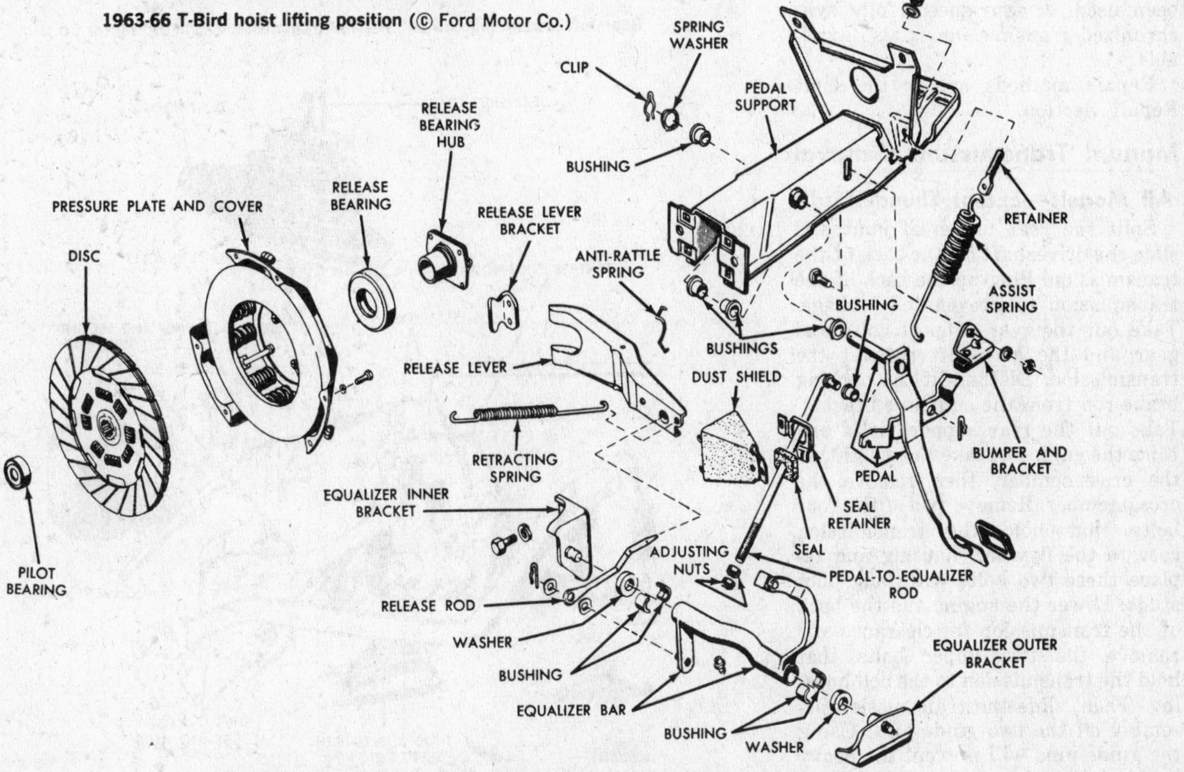

PRESSURE PLATE AND COVER

DISC

RELEASE BEARING

RELEASE BEARING HUB

RELEASE LEVER BRACKET

CLIP

SPRING WASHER

PEDAL SUPPORT

BUSHING

RETAINER

ANTI-RATTLE SPRING

ASSIST SPRING

RELEASE LEVER

BUSHING

BUSHINGS

DUST SHIELD

RETRACTING SPRING

BUMPER AND BRACKET

PEDAL

EQUALIZER INNER BRACKET

SEAL RETAINER

PILOT BEARING

ADJUSTING NUTS

SEAL

PEDAL-TO-EQUALIZER ROD

RELEASE ROD

WASHER

BUSHING

EQUALIZER OUTER BRACKET

EQUALIZER BAR

BUSHING

WASHER

Exploded view of the complete clutch and pedal assembly (© Ford Motor Co.)

4. Remove bolts that attach transmission to bellhousing and move transmission rearward until input shaft clears bellhousing.
5. Remove starter and bellhousing.
6. Loosen six pressure plate retaining bolts evenly to release spring tension, and mark position of pressure plate on flywheel to ensure it is positioned properly if reinstalled.
7. Remove pressure plate and clutch disc. Reinstall in reverse of above procedure. Take care to keep all parts of pressure plate and disc clean and free from oil and grease. Lubricate clutch release bearing with a light film of lithium-soap type grease. Do not lubricate the bearing hub. Retorque pressure plate bolts to 23-28 ft. lbs. for 1964-66 models and 12-20 ft. lbs. for 1967-70 models.

Transmissions

Transmission Information

Transmission refill capacities are in the Capacities table.

General information and exploded views, together with troubleshooting charts, are included in the Unit Repair Section.

Troubleshooting and repair of overdrive units are covered in the Unit Repair Section.

Since 1963, a fully synchronized three-speed manual transmission has been used. A four-speed, fully synchronized transmission is also available.

Repair methods are in the Unit Repair Section.

Manual Transmission Removal

All Models—Except Thunderbirds

Split the rear universal joint and slide the driveshaft off the back of the transmission. Plug up the back of the transmission to prevent oil leakage. Take out the rear support bolts and gear and the gearshift rods at the transmission. Disconnect the parking brake rod from the equalizer bracket. Take out the rear support bolts and raise the engine to take the weight off the crossmember, then remove the crossmember. Remove two of the four bolts that hold the transmission case to the flywheel housing and replace these two bolts with long pilot studs. Lower the engine and the back of the transmission for clearance and remove the two upper bolts that hold the transmission to the bell housing. Then, slide the transmission assembly off the two guide pins. Using the guide pins will prevent the heavy transmission from springing downward with possible damage to the clutch disc.

Manual Transmission Shift Linkage Ajustment

Hand Lever Position

With the transmission in neutral, the hand lever should be in a horizontal plane and parallel to the instrument panel line. Corrective adjustments should be made at the gearshift rods.

Crossover Jamming or Binding

This is a condition, usually caused by wear or poor adjustment at the crossover point of the shift column levers. If wear is evident, renew the worn part, then make the regular adjustment as follows:

With the hand lever and the transmission gearshift levers in neutral position, loosen both rods at their adjustment slots. With the hand lever horizontal, tighten the high and intermediate rod. Then tighten the low and reverse rod in a position so that smooth crossover operation is possible.

Four-Speed Linkage Adjustment

1. Place hand shifter lever in neutral position, then raise car on a hoist.
2. Insert a ¼ in. rod into the alignment holes of the shift levers.
3. If the holes are not in exact alignment, check for bent connecting rods or loose lever locknuts at the rod ends. Make replacements or repairs, then adjust as follows.

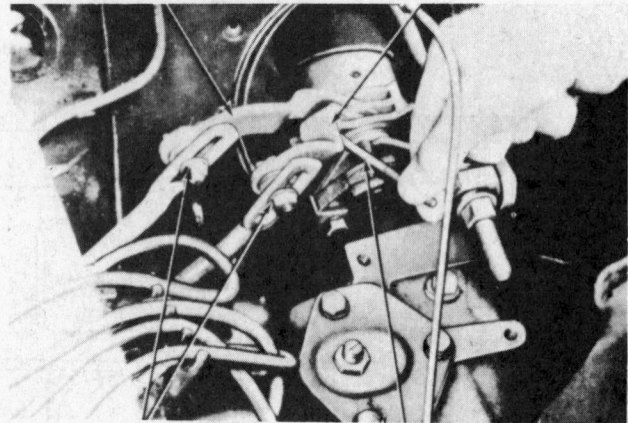

FIRST AND REVERSE LEVER   SECOND AND THIRD LEVER

GEARSHIFT ROD ADJUSTMENT NUTS   ALIGNMENT PIN

Gearshift linkage adjustment 3-speed transmission (© Ford Motor Co.)

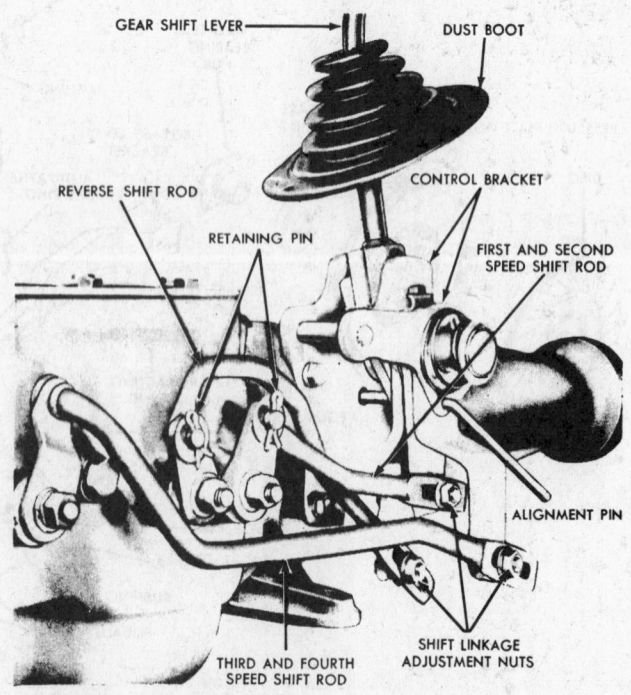

GEAR SHIFT LEVER   DUST BOOT

REVERSE SHIFT ROD   CONTROL BRACKET

RETAINING PIN   FIRST AND SECOND SPEED SHIFT ROD

ALIGNMENT PIN

THIRD AND FOURTH SPEED SHIFT ROD   SHIFT LINKAGE ADJUSTMENT NUTS

Adjusting shift linkage, 4-speed transmission (© Ford Motor Co.)

4. Loosen the three rod-to-lever retaining locknuts and move the levers until the 1/4 in. gauge rod will enter the alignment holes. Be sure that the transmission shift levers are in neutral, and the reverse shifter lever is in the neutral detent.

5. Install shift rods and torque locknuts to 12-15 ft. lbs. on 1964-65 transmissions and 18-23 ft. lbs. on 1966-70 transmissions.

6. Remove the 1/4 in. gauge rod.

7. Operate the shift levers to assure correct shifting.

8. Lower the car and road-test.

Automatic Transmission

Dashpot Adjustment 1964-67 Six Cylinder

1. Adjust throttle position to fast idle position and turn dashpot adjusting screw out until it is clear of dashpot plunger assembly.

2. Turn in screw until it contacts plunger. Then turn in screw specified number of turns against plunger.

Dashpot—1964-67 6 cylinder
(© Ford Motor Co.)

Models	Number of Turns
1964 240 six	3 1/2
1965 240 six	3 1/4—3 3/4
1966 240 six	3 1/2
1967 240 six	
without emission control	6
with emission control	2

3. Check accelerator pump setting.

Dashpot Adjustment All Others

1. With engine idle speed and mixture properly adjusted and with engine at operating temperature, loosen dashpot locknut.

2. Hold throttle in closed position and depress dashpot plunger.

Measure clearance between plunger and cam. Adjust dashpot adjusting nut to give proper clearance.

3. Tighten locknut and check setting of accelerator pump.

Manual Linkage Adjustment

1. With engine off, loosen clamp at shift lever so shift rod is free to slide.

2. Position selector lever in D1 position (large green dot) on dual range transmissions. On select shift transmission (P R N D 2 1) position lever in D position.

3. Shift lever at transmission into D1 detent position on dual range

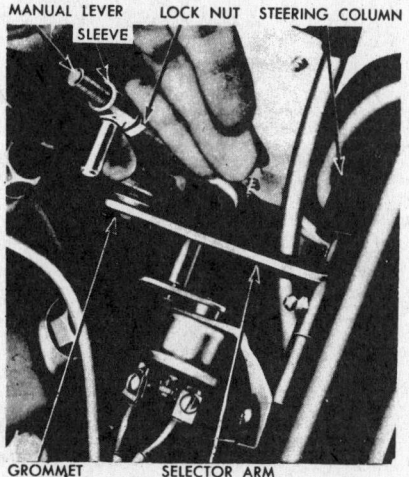

MANUAL LEVER SLEEVE   LOCK NUT   STEERING COLUMN

GROMMET   SELECTOR ARM

Adjusting manual control valve linkage
(© Ford Motor Co.)

transmissions or into D position on select shift transmissions.

NOTE: D1 position is second from rear on all dual range transmissions. D position is third from rear on all column shift select shift transmissions and 1966-68 console shift select shift transmissions. D position is fourth from rear on 1969-70 console shift select shift transmissions.

4. Tighten clamp and torque nut to 8-13 ft. lbs. on 1964-68 column shifts and 10-20 ft. lbs. on 1969-70 column shifts. Torque 1964-68 console shift nuts to 20-25 ft. lbs. and 1969-70 console shift nuts to 10-20 ft. lbs.

Neutral Start Switch Adjustment —1964-67 Except Console Shift Cruis-o-matics and All 1967 Console Shifts

1. With manual linkage properly adjusted, try to engage starter in each position on quadrant. Starter should engage only in Park and Neutral positions.

2. To adjust, loosen screws that locate switch on steering column.

3. Place shift lever in neutral detent.

4. Rotate switch until gauge pin (No. 43 drill) can be inserted in gauge pin hole 31/64 in.

5. Tighten down locating screws and check starter engagement in each position as in Step 1.

Neutral Start Switch Adjustment —Column Shift 1968-70

1. With manual linkage properly ad-

Dashpot Adjustment

Note: Not all engines have dashpots.

Model	Clearance (in.)	
	Manual	Automatic
1964 two barrel carburetors		1/16 - 3/32
1964 four barrel carburetors		1/16 - 3/32
1965 two barrel carburetors		5/64
1965 four barrel carburetors		5/16
1966 two barrel carburetors		0.060 - 0.090
1966 four barrel carburetors		0.060 - 0.090
1967 289 2V with emission control		0.110 - 0.140
1967 390 2V with emission control		0.080 - 0.110
1967 390 4V with emission control		1/8
1967 428 4V with emission control		1/8
1967 428 Police with emission control		1/8
1968 240 Six without thermactor pump		0.080
1968 240 Six with thermactor pump		0.100
1968 302 2V		0.125
1968 390 2V		0.125
1968 390 4V		0.093
1968 428 4V		0.093
1968 428 Police with thermactor pump		0.109
1969 240 six	0.080	0.080
1969 302 2V	1/8	1/8
1969 351 2V	7/64	
1969 351 4V	3/32	—
1969 390 2V	1/8	1/8
1969 390 4V	1/8	—
1969 428 Police	—	7/64
1969 429 2V	—	1/8
1969 429 4V	3/32	

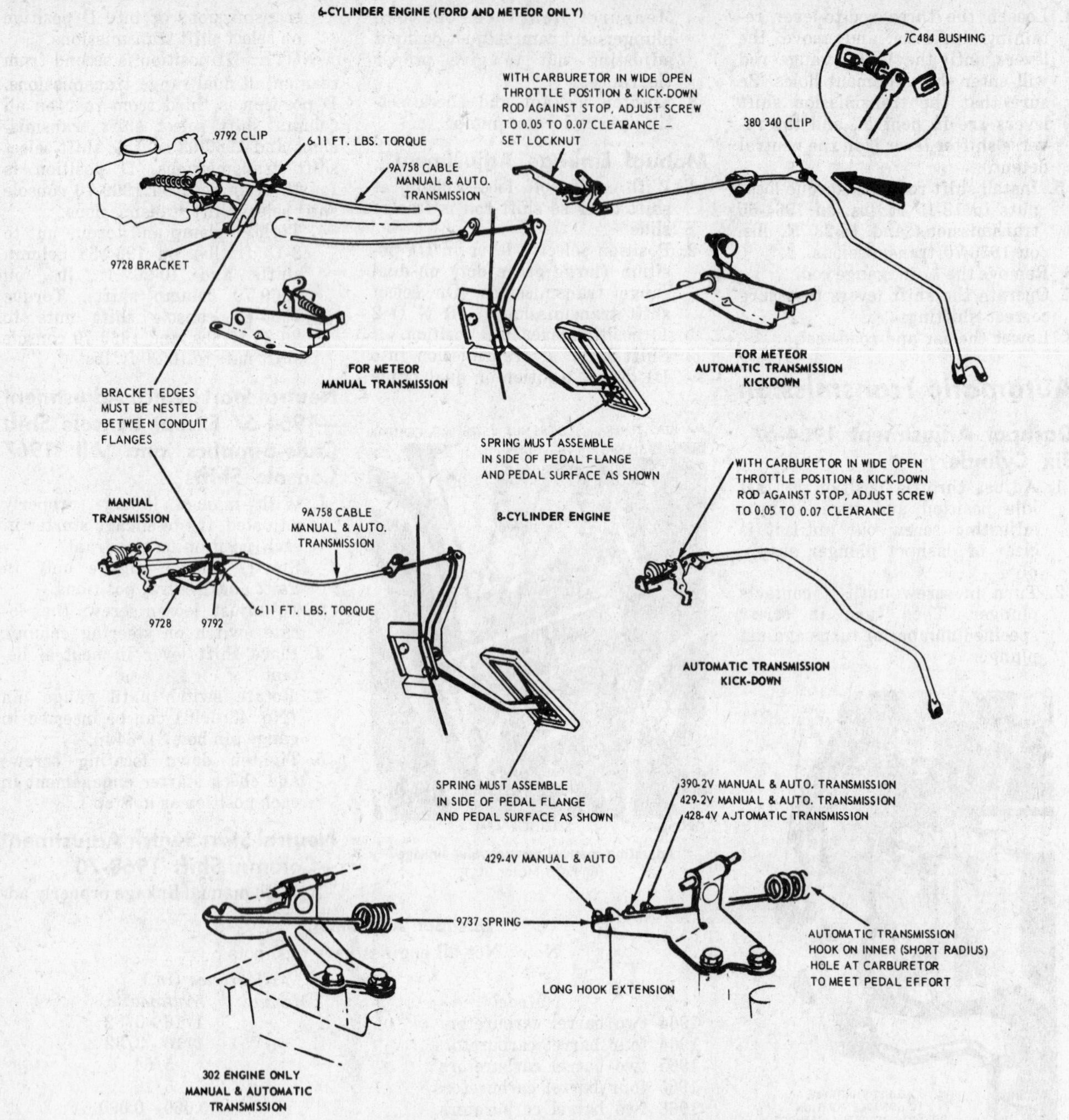

6-CYLINDER ENGINE (FORD AND METEOR ONLY)

7C484 BUSHING

WITH CARBURETOR IN WIDE OPEN
THROTTLE POSITION & KICK-DOWN
ROD AGAINST STOP, ADJUST SCREW
TO 0.05 TO 0.07 CLEARANCE –
SET LOCKNUT

380 340 CLIP

.9792 CLIP

6-11 FT. LBS. TORQUE

9A758 CABLE
MANUAL & AUTO.
TRANSMISSION

9728 BRACKET

FOR METEOR
AUTOMATIC TRANSMISSION
KICKDOWN

BRACKET EDGES
MUST BE NESTED
BETWEEN CONDUIT
FLANGES

FOR METEOR
MANUAL TRANSMISSION

SPRING MUST ASSEMBLE
IN SIDE OF PEDAL FLANGE
AND PEDAL SURFACE AS SHOWN

WITH CARBURETOR IN WIDE OPEN
THROTTLE POSITION & KICK-DOWN
ROD AGAINST STOP, ADJUST SCREW
TO 0.05 TO 0.07 CLEARANCE

MANUAL
TRANSMISSION

9A758 CABLE
MANUAL & AUTO.
TRANSMISSION

8-CYLINDER ENGINE

9728   9792

6-11 FT. LBS. TORQUE

AUTOMATIC TRANSMISSION
KICK-DOWN

SPRING MUST ASSEMBLE
IN SIDE OF PEDAL FLANGE
AND PEDAL SURFACE AS SHOWN

390-2V MANUAL & AUTO. TRANSMISSION
429-2V MANUAL & AUTO. TRANSMISSION
428-4V AUTOMATIC TRANSMISSION

429-4V MANUAL & AUTO

9737 SPRING

AUTOMATIC TRANSMISSION
HOOK ON INNER (SHORT RADIUS)
HOLE AT CARBURETOR
TO MEET PEDAL EFFORT

LONG HOOK EXTENSION

302 ENGINE ONLY
MANUAL & AUTOMATIC
TRANSMISSION

Manual linkage adjustment—1969-70 (© Ford Motor Co.)

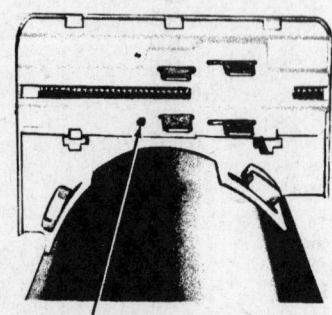

BOTTOM VIEW OF SWITCH

Gauge Pin HOLE

Neutral start switch—1964-67 Cruise-o-matic
(© Ford Motor Co.)

justed, try to engage starter in each position on quadrant. Starter should engage only in neutral or park position.

2. Place shift lever in neutral detent.

3. Disconnect start switch wires at plug connector. Disconnect vacuum hoses, if any. Remove screws securing neutral start switch to steering column and remove switch. Remove actuator lever along with Type III switches.

4. With switch wires facing up, move actuator lever fully to the left and insert gauge pin (No. 43 drill) into gauge pin hole at point A. See accompanying figure. On Type III switch, be sure gauge pin is inserted a full ½ in.

5. With pin in place, move actuator lever to right until positive stop is engaged.

6. On Type I and Type II switches, remove gauge pin and insert it at point B. On Type III switches, remove gauge pin, align two holes in switch at point A and reinstall gauge pin.

7. Reinstall switch on steering column. Be sure shift lever is engaged in neutral detent.

8. Connect switch wires and vacuum hoses and remove gauge pin.

9. Check starter engagement as in Step 1.

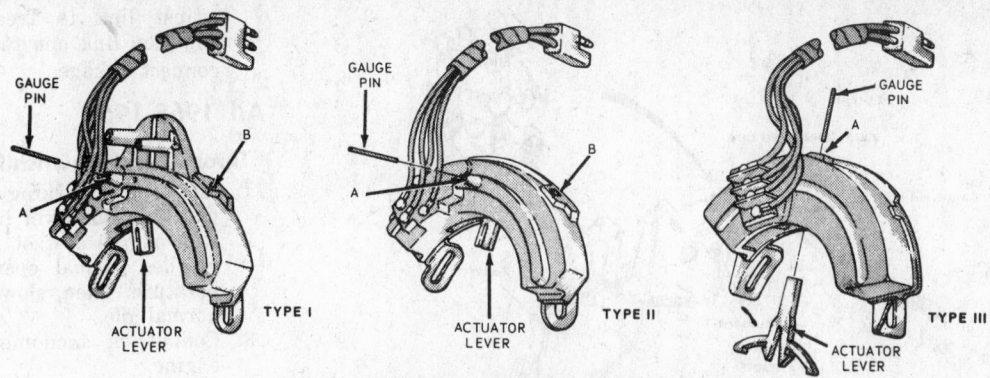

Neutral start switches—1968-70 (© Ford Motor Co.)

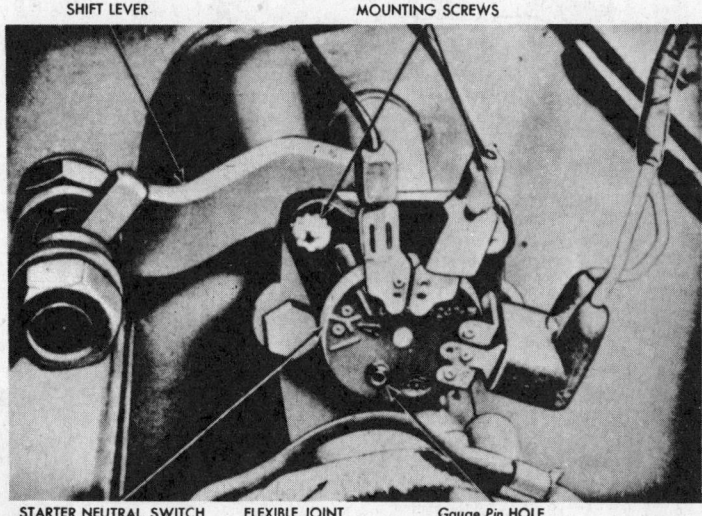

Neutral start switch—column mounted C4 and C6 1964-67 (© Ford Motor Co.)

Neutral Start Switch Adjustment —1964-66 Console Shift Cruiso-matics and 1967 Console Shift C4 and C6 Transmissions

1. With manual linkage properly adjusted, try to engage starter in each position on quadrant. Starter should engage only in Neutral or Park position.
2. Remove handle from shift lever and chrome trim panel from top of console.
3. Place lever in Neutral, remove quadrant retaining screws and indicator light, and lift quadrant from console.
4. Loosen switch attaching screws, and move shift lever back and forth until gauge pin (No. 43 drill) can be inserted into gauge pin holes.
5. Place shift lever in neutral position, and slide switch back and forward until switch actuating lever contacts shift lever.
6. Reassemble shift quadrant in console.
7. Check starter engagement as in Step 1.

Neutral Start Switch Adjustment —Console Shift 1968-70

1. With manual linkage properly

adjusted, try to engage starter at each position on quadrant. Starter should engage only in Neutral and Park positions.
2. Remove shift handle from shift lever, and console from vehicle.
3. Loosen switch attaching screws, and move shift lever back and forward until gauge pin (No. 43 drill) can be inserted fully.
4. Place shift lever firmly against neutral detent stop and slide switch back and forward until switch lever contacts shift lever.

5. Tighten switch attaching screws, and check starter engagement as in Step 1.
6. Reinstall console and shift linkage.

Throttle Linkage Adjustments

6 Cylinder 1964

1. With engine at normal operating temperature and throttle in curb idle position against stop screw, adjust accelerator linkage to secure 4 5/16 in. clearance from tip of pedal to floor pan.
2. Adjust downshift link to secure free entry of gauge pin (1/4 in. drill rod) in gauge holes.

8 Cylinder 1964

1. With engine at normal operating temperature and throttle in curb idle position against stop screw, adjust two-speed and three-speed accelerator linkage to secure 4 5/16 in., or (C-4 Dual) 3⅝ in. clearance from tip of pedal to floor pan.
2. Disconnect carburetor control link from bellcrank.
3. Install 1/4 in. gauge pin in gauge holes and lift carburetor link to normal operating position. While holding against carburetor screw

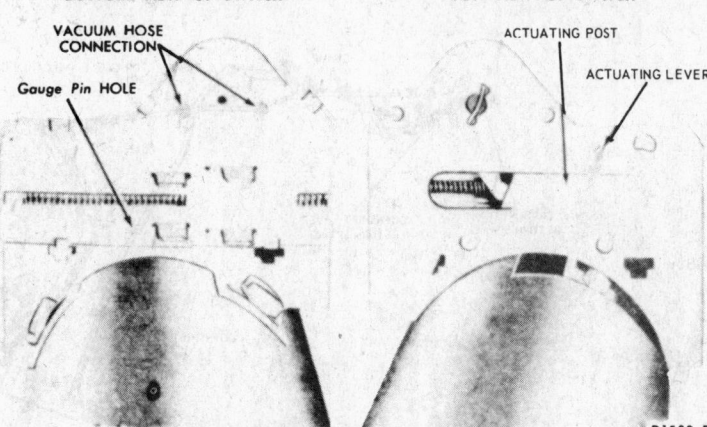

Neutral start switch—T-Bird (© Ford Motor Co.)

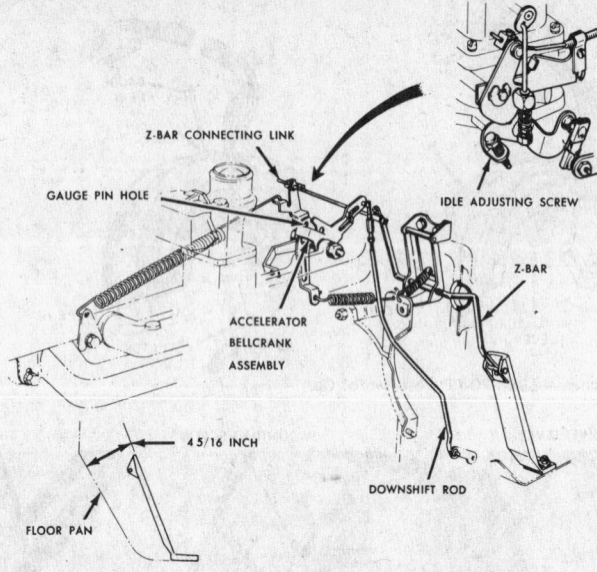

Throttle linkage—C4 dual range 6 cyl. 1964 (© Ford Motor Co.)

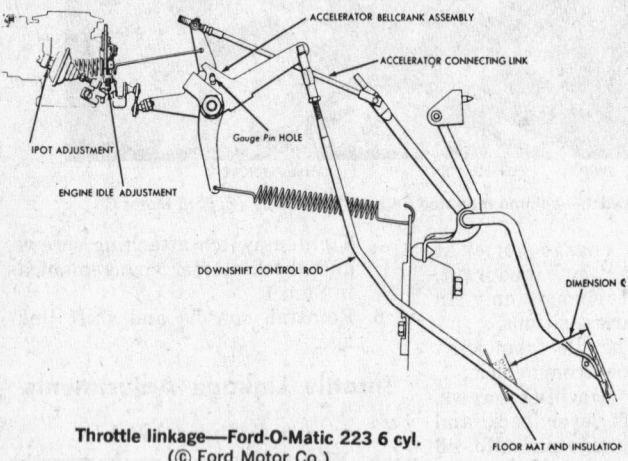

Throttle linkage—Ford-O-Matic 223 6 cyl.
(© Ford Motor Co.)

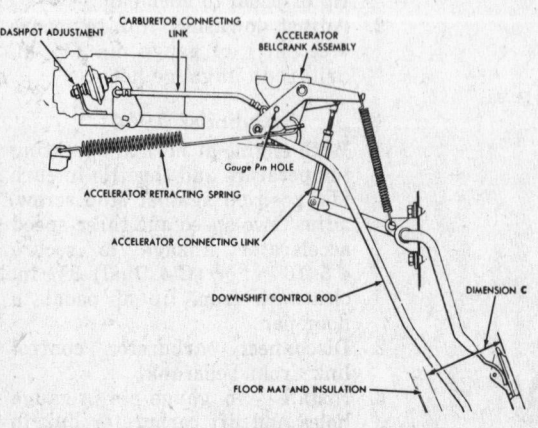

Throttle linkage—Ford-O-Matic 1964 352 and 390 V8
(© Ford Motor Co.)

adjust link to free fit. Then lengthen link one full turn. Reconnect linkage.

All 1965-1968

Throttle and Downshift Linkage

1. Apply the parking brake and place selector lever in N.
2. Run engine at fast idle until it reaches normal operating temperature. Then, slow it down to normal idle.
3. Connect a tachometer to the engine.
4. Adjust engine to specified idle speed with selector in D1 or D2. Due to the vacuum parking brake release (if so equipped), the parking brake will not hold while selector is in D1 or D2. Keep service brake applied.
5. When satisfied that idle speed is correct, stop engine and adjust dashpot clearance. Check clearance between dashpot plunger and throttle lever.
6. With engine stopped, disconnect carburetor return spring from throttle lever and loosen accelerator cable conduit attaching clamp.

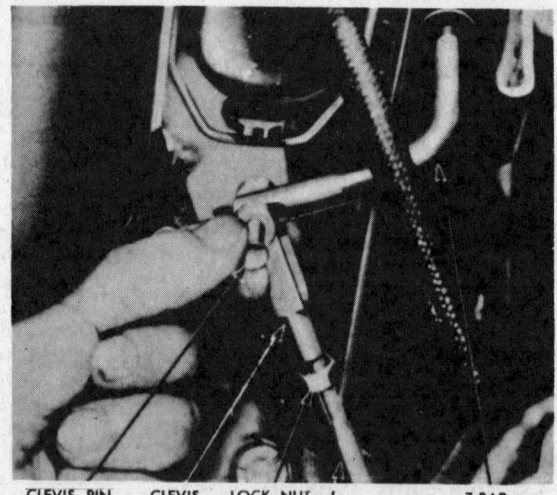

Adjusting throttle valve control rod (© Ford Motor Co.)

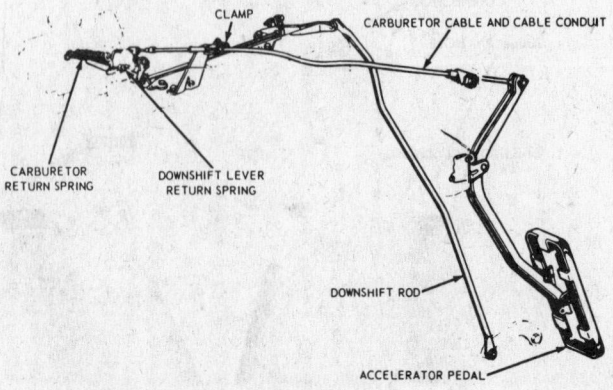

Throttle linkage—6 cyl. 1965-69 (© Ford Motor Co.)

7. With accelerator pedal to floor and throttle lever wide open, slide cable conduit to rear (to left on six cylinder engine) to remove slack from cable. Tighten clamp.
8. Disconnect downshift lever return spring and hold throttle lever wide open. Depress downshift rod to "through detent stop". Set downshift lever adjusting screw against throttle lever.
9. Connect carburetor return spring and downshift lever return spring.

Beginning 1969—All

1. Disconnect downshift lever return spring.
2. Hold throttle shaft lever wide open, and hold downshift rod against "through detent stop".
3. Adjust downshift screw to provide 0.050-0.070 in. clearance between screw and throttle shaft lever. On 240 cu. in. engine, tighten locknut.
4. Connect downshaft lever return spring.

C4 Automatic Transmission R&R

1. Raise vehicle and remove three converter cover attaching bolts and remove cover.
2. Remove two converter drain plugs, drain converter and reinstall plugs.
3. Remove drive shaft and plug of extension housing.
4. Remove vacuum line hose from vacuum unit and from attaching clip.
5. Remove extension housing to crossmember attaching bolts.
6. Remove speedometer cable.
7. Disconnect exhaust pipes from manifolds.
8. Remove parking brake cable from equalizer lever.
9. Remove fluid filler tube and drain transmission fluid. On 1964 cars, loosen eleven pan bolts and tip corner of pan to drain transmission fluid.
10. Disconnect oil cooler lines. On 223 cu. in. six remove tube retaining bracket and remove transmission with filler tube attached.
11. Remove all linkages from transmission.
12. Disconnect starter cable and remove starter.
13. Remove four torque converter to flywheel attaching nuts.
14. Position jack to support transmission and secure with safety chain. Remove four crossmember bolts and lower crossmember.
15. Remove five converter to engine bolts. Lower transmission and remove from car.
16. Reverse above procedure to install transmission.

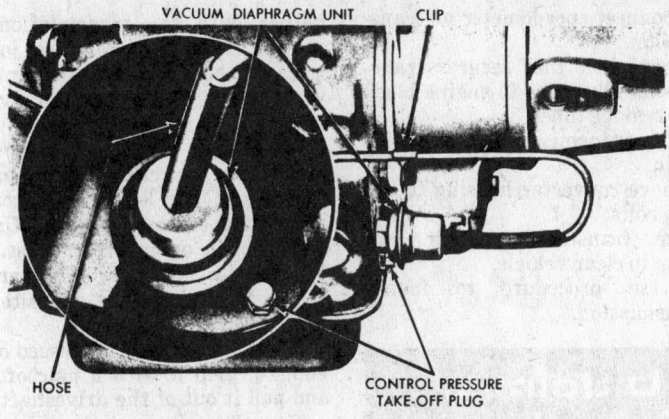

Vacuum diaphragm and control pressure connecting point (© Ford Motor Co.)

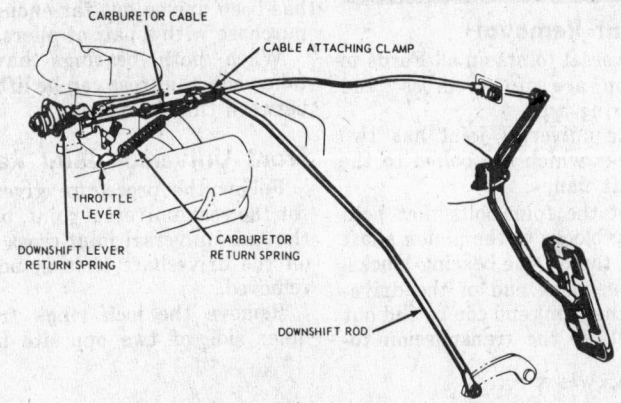

Throttle linkage—V8 1965-69 (© Ford Motor Co.)

Cruise-o-matic Automatic Transmission R&R

After 1967, Cruise-o-matics are designated as FMX or MX transmissions, depending on application.

1. Remove two upper bolts and lock washers which attach converter to engine.
2. Raise vehicle, remove cover from lower front side of converter housing, and drain fluid from transmission.
3. Remove one of converter drain plugs. Rotate converter 180° and remove other plug. Do not attempt to turn converter with wrench on converter stud nuts.
4. Disconnect filler tube from transmission case.
5. Disconnect vacuum hose from vacuum diaphragm unit and from extension housing clip.
6. Remove flywheel to converter nuts and flat washers. Install drain plugs in converter and torque to 15-28 ft. lbs. Install converter housing front plate to hold converter when removing transmission.
7. Disconnect and remove starter.
8. Disconnect oil cooler lines and remove vent tube.
9. Disconnect all linkages.
10. Disconnect speedometer and remove drive shaft.
11. Support transmission on stand, remove transmission mount bolts, and remove crossmember.
12. Support rear of engine, remove remaining engine attaching bolts, and lower transmission.
13. Reverse above procedure to install transmission.

C6 Automatic Transmission R&R

1. Raise vehicle and drain transmission and converter.
2. Remove driveshaft and plug extension housing.
3. Disconnect and remove starter.
4. Remove four converter to flywheel attaching nuts.
5. Disconnect parking brake front cable from equalizer.
6. Remove two crossmember to frame attaching bolts.
7. Remove two transmission mount bolts.
8. Disconnect all linkages.
9. Disconnect exhaust pipes from manifolds.
10. Raise transmission on transmission jack.
11. Disconnect parking brake cables from equalizer.
12. Remove crossmember and lower transmission to gain access to oil cooler lines.
13. Disconnect oil cooler lines from transmission and disconnect vacuum line from vacuum diaphragm and retaining clip.

14. Disconnect speedometer at transmission.
15. Remove bolt that secures transmission filler tube to engine block and remove tube.
16. Secure transmission to jack with chain.
17. Remove converter housing to engine bolts.
18. Move transmission back and down to clear vehicle.
19. Reverse procedure to install transmission.

U Joints, Drive Lines

Rear Joint Removal

The universal joints on all Fords in this section are of the cross- and needle-bearing-type.

The rear universal joint has two pillow blocks which are bolted to the pinion shaft flange.

Take out the four bolts that hold the bearing blocks to the pinion shaft and gently tap off the bearing blocks.

Lower the back end of the driveshaft and the front end can be slid out of the back of the transmission together with the transmission yoke portion of the front universal joint.

Carry the assembly—the front universal joint complete, the driveshaft and the rear universal joint crossover—to the bench and remove the cross from the rear universal joint by taking out the lock rings from the inner side of the bearings. Using a large punch or an arbor press, drive one of the bearings in toward the center forcing out the opposite bearing.

When the bearing is pressed out far enough, grip it with a pair of pliers and pull it out of the driveshaft yoke. Now, drive the cross in the opposite direction until the remaining bearing has been driven out far enough for a purchase with a pair of pliers.

When both bearings have been taken out, the cross can be lifted from between the two yokes.

Front Universal Joint Removal

Follow the procedure given above for the rear universal joint, but leave the rear universal joint cross in place on the driveshaft if it is not to be removed.

Remove the lock rings from the inner side of two opposite bearings

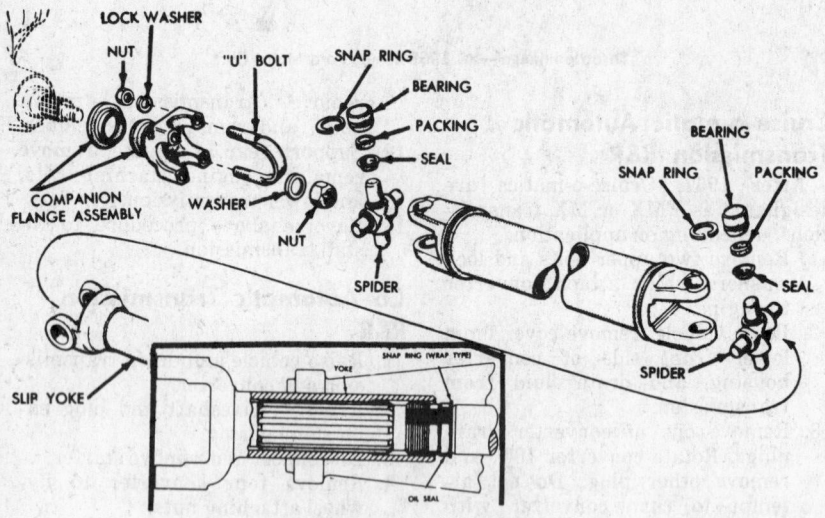

Driveshaft and universal joint (© Ford Motor Co.)

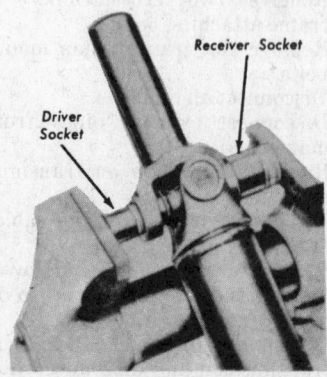

U-joint removal
(© Ford Motor Co.)

and press on the outer side of one of the bearings, forcing the crossover. This will force the bearing on the opposite side out of its yoke.

Remove the forced-out bearing and press the cross in the opposite direction to force the other bearing out.

Repeat this procedure on the third and fourth bearings.

When installing the new bearings in the universal joint yoke, it is possible to put them in with a driver of some type, but it is recommended that this work be done in an arbor press since a heavy jolt on the needle bearings can very easily misalign them.

Drive Axles

Troubleshooting and Adjustment

General instructions covering the troubles of the drive axle, and how to repair and adjust it, together with information on installation of rear axle bearings and grease seals and data on the equa-lock differential are in the Unit Repair Section.

Capacities of the drive axle are in the Capacities tables.

Rear Spring Removal

1963-64

1. To remove rear springs, disconnect the shock absorber at the spring clip plate and remove the U-bolt nuts.
2. Take out the stud at the front hanger and the two bolts at the rear shackle.
3. The spring can then be lifted out from under the car.

NOTE: when installing a new spring the short end of the spring goes toward the front of the car.

The short end of the spring is the one having the shortest distance from the spring eye to the center bolt.

4. Spring clips should be tightened to 45-50 ft. lbs. torque.

1965-69

Coil springs are used at rear.

The rear suspension is a coil-link design. Large, low-rate coil springs are mounted between rear axle pads and frame supports. Parallel lower arms extend forward of the spring seats to rubber frame anchor to accommodate driving and breaking forces. A third link is mounted between the axle and the frame to control torque reaction forces from the rear wheels.

Lateral (side sway) motion of the rear axle is controlled by a rubber bushed rear track bar, linked laterally between the axle and frame.

1. Place car on hoist and lift under rear axle housing. Place jack stands under frame side rails.
2. Disconnect track bar at the rear axle housing bracket.
3. Disconnect rear shock absorbers from the rear axle housing brackets.
4. Disconnect hose from axle housing vent.
5. Lower hoist with axle housing until coil springs are released.
6. Remove spring lower retainer with bolt, nut, washer and insulator.
7. Remove spring with large rubber insulator pads from car.
8. Install in reverse of above.

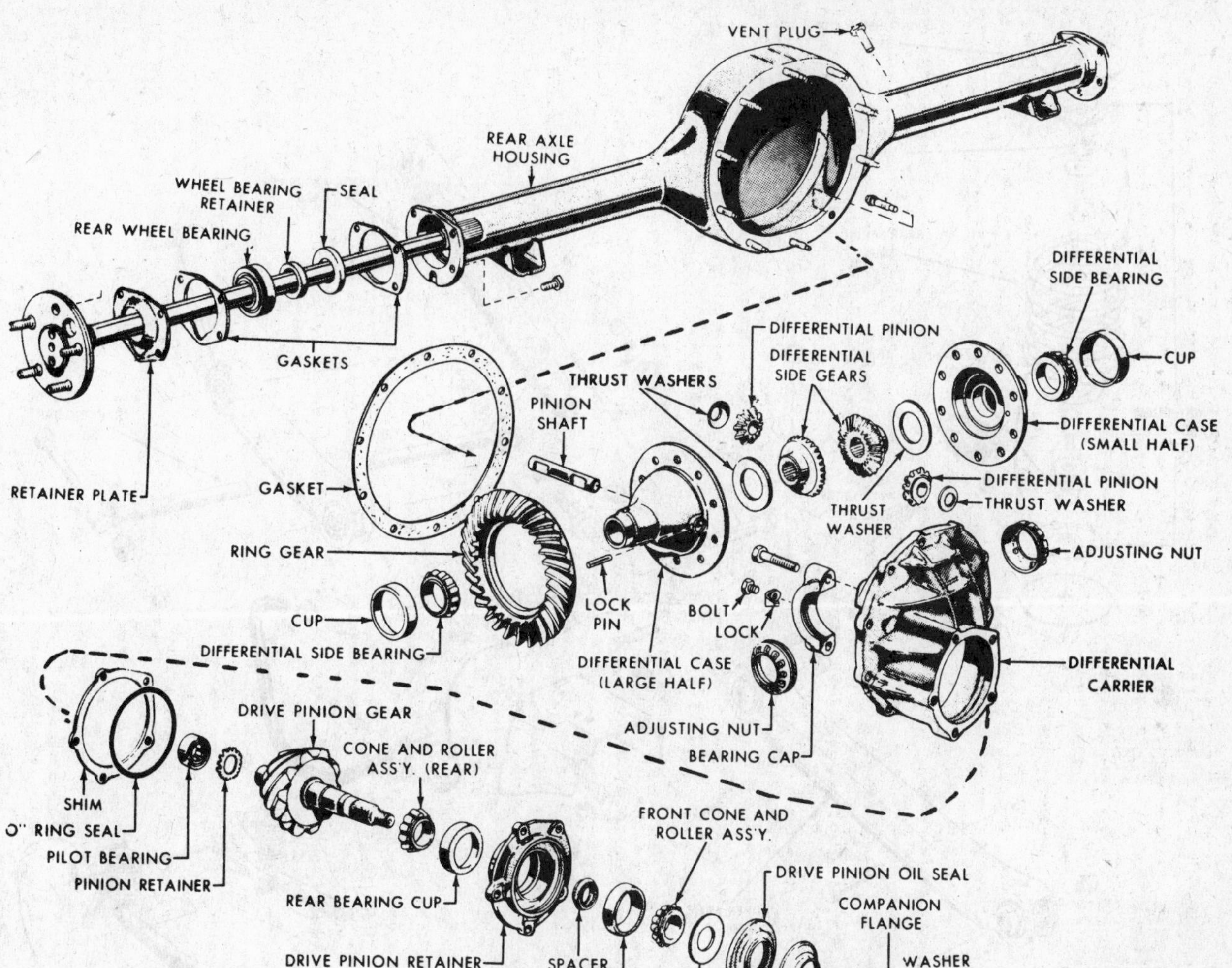

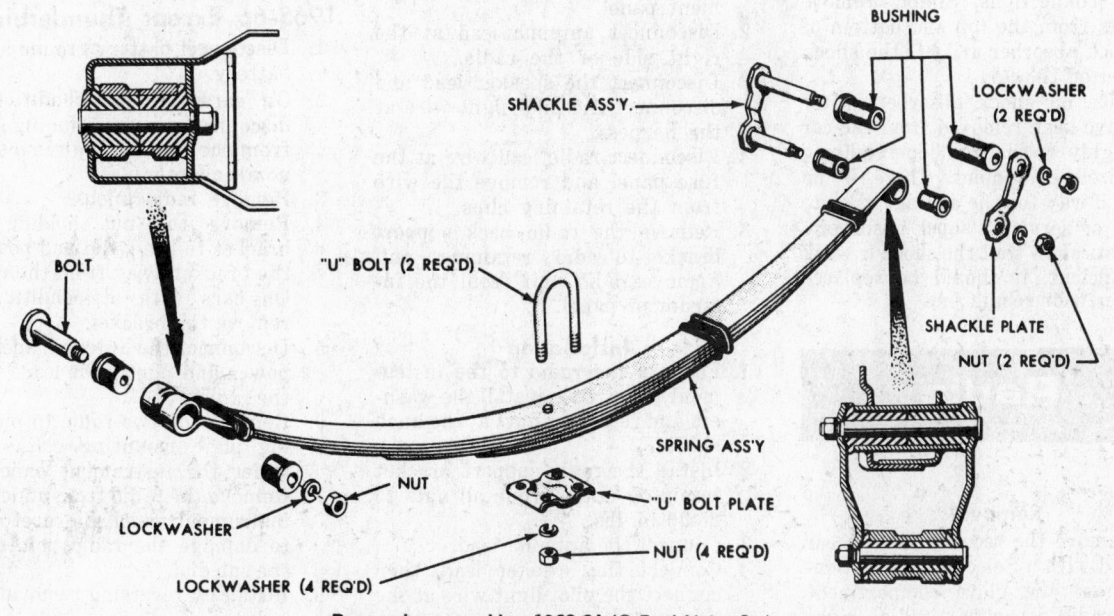

VENT PLUG

REAR AXLE HOUSING

WHEEL BEARING RETAINER

SEAL

REAR WHEEL BEARING

GASKETS

RETAINER PLATE

DIFFERENTIAL SIDE BEARING

CUP

DIFFERENTIAL PINION

DIFFERENTIAL SIDE GEARS

THRUST WASHERS

PINION SHAFT

DIFFERENTIAL CASE (SMALL HALF)

DIFFERENTIAL PINION THRUST WASHER

THRUST WASHER

GASKET

RING GEAR

ADJUSTING NUT

CUP

DIFFERENTIAL SIDE BEARING

LOCK PIN

BOLT LOCK

DIFFERENTIAL CASE (LARGE HALF)

ADJUSTING NUT

BEARING CAP

DIFFERENTIAL CARRIER

DRIVE PINION GEAR

CONE AND ROLLER ASS'Y. (REAR)

FRONT CONE AND ROLLER ASS'Y.

"O" RING SEAL

SHIM

PILOT BEARING

PINION RETAINER

REAR BEARING CUP

DRIVE PINION OIL SEAL

COMPANION FLANGE

WASHER

NUT

DRIVE PINION RETAINER

SPACER

FRONT BEARING CUP

SLINGER

DEFLECTOR

Rear axle components (© Ford Motor Co.)

SHACKLE ASS'Y.

BUSHING

LOCKWASHER (2 REQ'D)

BOLT

"U" BOLT (2 REQ'D)

SHACKLE PLATE

NUT (2 REQ'D)

NUT

LOCKWASHER

SPRING ASS'Y

"U" BOLT PLATE

LOCKWASHER (4 REQ'D)

NUT (4 REQ'D)

Rear spring assembly—1963-64 (© Ford Motor Co.)

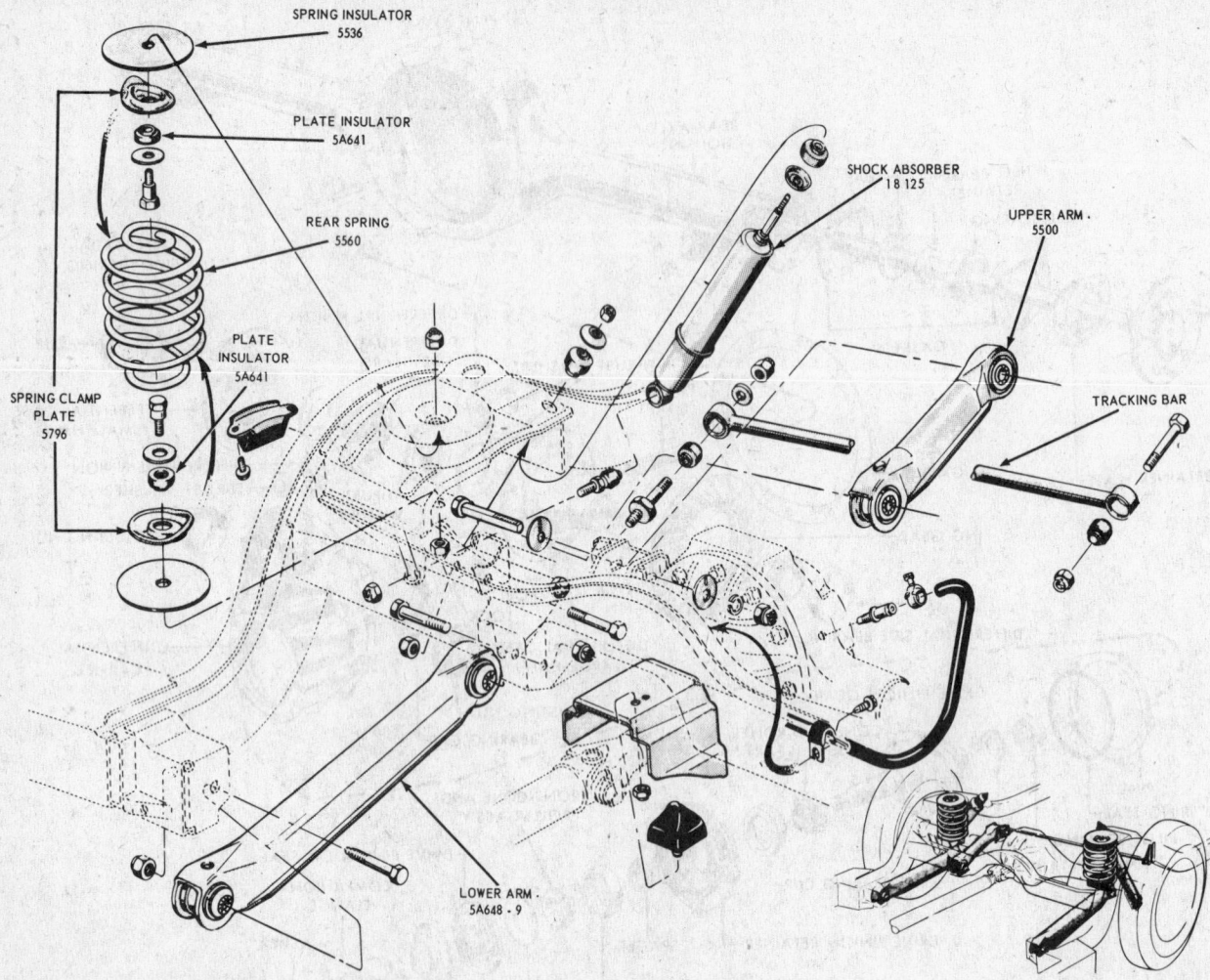

SPRING INSULATOR
5536

PLATE INSULATOR
5A641

REAR SPRING
5560

PLATE
INSULATOR
5A641

SPRING CLAMP
PLATE
5796

SHOCK ABSORBER
18 125

UPPER ARM .
5500

TRACKING BAR

LOWER ARM .
5A648 - 9

Rear suspension—beginning 1965 (© Ford Motor Co.)

Rear Shock Absorber Replacement

Rear shock absorbers on all Fords are straddle-mounted and are held to rubber bushings at both the top and bottom connections. Simply remove the nuts from the top and bottom of the shock absorber and lift the shock absorber off the car.

Service on shock absorbers after they have been removed from the car is a highly specialized job requiring specialized equipment. There is no practical way for the average shop to rebuild or service a shock absorber.

If tests show that the shock is weak or inefficient, it should be replaced with a new or rebuilt one.

Radio

1964

Removal

To remove the receiver from a car equipped with a Select-Air Conditioner, remove the glove compartment liner, radio speaker, radio right

mounting support. Remove the radio through the speaker opening.
Remove the receiver as follows:
1. Pull off radio control knobs and remove the nuts and washers holding the radio to the instrument panel.
2. Disconnect antenna lead at the right side of the radio.
3. Disconnect the speaker lead and disconnect the pilot light wire at the harness.
4. Disconnect radio lead wire at the fuse panel and remove the wire from the retaining clips.
5. Remove the radio back support bracket-to-radio retaining nut. Remove radio unit from the instrument panel.

Installation
1. Position the radio to the instrument panel, then install the washers and retaining nuts at the knob shafts.
2. Install the radio support bracket retaining nut. Torque all nuts to 25-30 in. lbs.
3. Connect the antenna lead.
4. Connect the speaker lead, then connect the pilot light wire at the harness.

5. Connect the lead wire at the fuse panel mounted on the headlight switch.
6. Install radio control knobs, then check radio performance.

1965-66 Except Thunderbird
1. Disconnect battery ground at the battery.
2. On cars with air-conditioning, disconnect air-conditioning ducts from the plenum and remove the nozzle adapter plate.
3. Remove radio knobs.
4. Remove the nut holding the bracket to the radio and position the bracket away from the radio. On cars with air-conditioning, remove the bracket.
5. Disconnect the antenna, speaker, power and pilot light leads from the radio.
6. Remove the two radio to mounting plate mounting bolts from under the instrument panel.
7. Remove the radio from under the instrument panel. Be careful not to damage the radio pointer on the sub dial.
8. Install by reversing removal procedure.

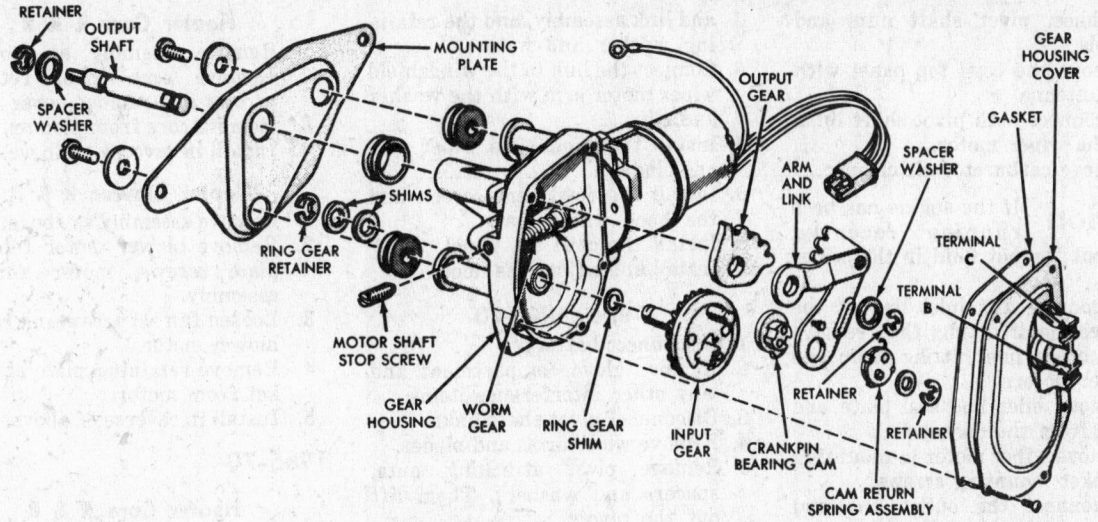

Two-speed electric windshield wiper motor (© Ford Motor Co.)

1965-66 Thunderbird

1. Pry off the right and left side console mouldings. Remove both console moulding retainers.
2. Pry off the right and left instrument panel chrome mouldings.
3. Remove the six screws and two bolts retaining the right and left side finish panels to the instrument panel. Pull the finish panels away from the instrument panel.
4. Remove the screws holding the right and left side console finish panels. Remove the finish panels.
5. Remove the two screws holding the lower end of the radio to the support brackets.
6. Remove radio knobs and the bezel mounting nuts and washers.
7. Disconnect the antenna and speaker connectors and remove the radio.
8. Install by reversing removal procedure.

1967 Except Thunderbird

1. Disconnect battery.
2. Remove cigarette lighter and radio knobs.
3. Remove instrument cluster retaining screws and instrument cluster.
4. Remove radio support nut and two bolts which retain radio in cluster.
5. Move radio out of cluster and disconnect antenna, speaker and power leads.
6. Reverse procedure to install radio.

1967-69 Thunderbird

1. Disconnect battery.
2. Remove inspection cover plate beneath steering wheel.
3. If necessary remove vacuum motor on inboard tilt swing column to provide clearance. Place shift lever in any position but Park before removing motor.

4. Remove two knobs and discs from radio.
5. Rpmove sleeve or fader control and two hex mounting nuts from radio shafts.
6. Remove radio rear support bracket.
7. Slide radio forward and down toward inspection hole.
8. Disconnect all leads and remove radio.
9. Reverse procedure to install radio.

1968 Except Thunderbird

1. Disconnect battery.
2. Remove mouldings from pillar windows.
3. Unsnap mouldings from right side of instrument panel.
4. Remove pop off access covers from cluster area.
5. Remove four screws attaching right half of pad to instrument panel.
6. Remove two screws attaching left side of pad to instrument panel above cluster.
7. Remove one screw attaching each end of instrument panel lower pad to upper pad and remove upper pad from vehicle.
8. Pull all knobs from radio control shafts.
9. Remove ten retaining buttons and remove lens and mask from instrument cluster.
10. Remove two screws from blackout cover at right of speedometer and remove cover.
11. Remove four screws attaching radio front plate to instrument cluster.
12. Pull radio out and disconnect leads.
13. Reverse procedure to install radio.

1969 Except Thunderbird

1. Remove radio knobs and wiper and washer knobs.

2. Remove lighter and pull off heater switch knobs.
3. Remove ten screws retaining instrument panel trim cover assembly and remove.
4. Remove lower rear radio support bolt.
5. Remove three nuts retaining radio in instrument panel and pull radio halfway out.
6. Disconnect all leads and remove radio.
7. Reverse procedure to install radio.

Windshield Wipers

Motor R & R

Ford, 1964-70

NOTE: if the car is equipped with air-conditioner, remove the glove compartment, liner, radio speaker, radio, and the right defroster nozzle-to-instrument panel brace. Then, proceed as with non-air-conditioned cars.

1. Remove the two mounting bolts, then lower the motor.
2. Disconnect the motor vacuum line, or the feed wire at the bullet connector.
3. Disconnect control cable at wiper motor.
4. Lift out the wiper motor.
5. Install by reversing removal procedures.

T-Bird, 1964-1970

This motor works by hydraulic pressure taken from the power steering system. During wiper operation, a part of the fluid supply is bypassed through the wiper motor by a valve on the motor.

1. Remove speaker grille and speaker.
2. Disconnect antenna at the radio.
3. Remove wiper arm and blade as-

semblies, pivot shaft nuts and bezels.

4. Remove the cowl top panel with the antenna.
5. Disconnect both pivot shaft links at the wiper motor.
6. Remove carburetor air cleaner.

Caution If the engine has been running recently, watch out for hot fluid in the wiper system.

7. Disconnect the inlet line at the wiper motor fitting, then remove the brass inlet fitting from the wiper motor.
8. Remove inlet line seal plate and seal from the dash panel.
9. Remove wiper motor-to-mounting bracket mounting screws.
10. Disconnect the outlet line and the control cable from the motor, then remove the motor.
11. If replacing the wiper motor, transfer all fittings.
12. Position motor to the bracket area, connect and adjust the control cable.
13. Start outlet line in the fitting, position the motor on its mount and install attaching screws. Tighten outlet fitting.
14. Connect the links to the wiper motor arm.
15. Position the rubber seal on the brass inlet fitting, and install the fitting into the motor.
16. Install the seal plate over the seal and inlet fitting.
17. Connect inlet line.
18. Start engine and check operation of wiper motor. Stop engine and bring power steering reservoir to level.
19. Reinstall cowl top panel, antenna and other parts which were removed at the start of the operation.
20. Install bezels, nuts and wiper arm assemblies. Install the air cleaner.

Transmission or Linkage R & R

Ford, 1964-70

1. Remove the hood, cowl panel, windshield washer nozzles and hood ground strap. On cars with a radio, swing the cowl top panel around and lay it across the air cleaner without removing the antenna.
2. Remove the cowl vent screen and water seal.
3. Remove the windshield wiper arm and blade. Disconnect the wiper link at the motor arm.
4. Remove the windshield wiper pivot shaft nut and washer. Guide the pivot shaft down into the air duct and remove the pivot and link assembly.
5. Install the windshield wiper pivot

and link assembly, and the retaining washer and nut.

6. Connect the link to the windshield wiper motor arm with the washer and clip.
7. Install the windshield wiper arm and blade.
8. Install the cowl vent screen and the hood water seal.
9. Install the cowl top panel.
10. Install and adjust the hood.

T-Bird, 1964-70

1. Disconnect battery.
2. Remove glove compartment and any other interference items.
3. Disconnect pivot shaft links.
4. Remove wiper arms and blades.
5. Remove pivot attaching nuts, spacers and washers. Then, lift out the pivots.
6. Install by reversing removal procedure.

Heater System

Through 1964

Heater R & R

1. Partially drain cooling system.
2. Disconnect fresh air inlet control cable at instrument panel.
3. Disconnect right air inlet boot at blower housing.
4. Disconnect defroster valve cable at plenum chamber.
5. Disconnect defroster hoses at outlets.
6. Disconnect wires from resistor and motor.
7. Disconnect heater hoses at firewall.
8. Remove plenum chamber retaining nuts at firewall and set assembly on floor.
9. Disconnect control valve cable at valve. Remove assembly from car.
10. Install in reverse of above.

Heater Core R & R

1. Remove assembly, as above.
2. Remove core cover retaining screws and remove cover.
3. Remove core from plenum.
4. Install in reverse of above.

Heater Blower R & R

1. Remove assembly as above.
2. Remove blower motor retaining plate screws, motor and fan assembly.
3. Loosen fan set screw and remove blower motor.
4. Remove retaining plate and gasket from motor.
5. Install in reverse of above.

1965-70

Heater Core R & R

1. Partially drain cooling system.
2. Remove heater hoses at core.
3. Remove retaining screws, core cover and seal from plenum.
4. Remove core from plenum.
5. Install in reverse of above, applying a thin coat of silicone to the pads.

T-Bird 1964-66

Heater Core R & R

1. Partially drain cooling system.
2. Disconnect hoses from heater core.
3. Remove lower instrument panel.
4. Remove right-hand trim panel from console.
5. Disconnect defroster hoses at heater.
6. Disconnect Bowden control cables from heater control head and from fresh-air door on heater.
7. Disconnect wiring.
8. Under hood, remove nuts holding heater to dash.
9. Remove rear support screw near the fresh air intake.
10. Ease heater to floor of car. Use

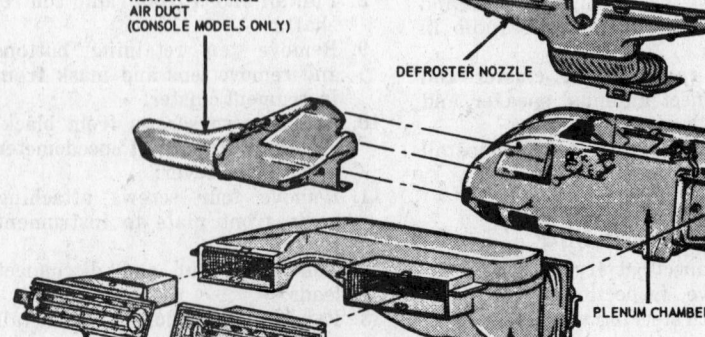

HEATER DISCHARGE AIR DUCT (CONSOLE MODELS ONLY)

DEFROSTER NOZZLE

PLENUM CHAMBER HEAT DAMP

LT.-REGISTERS-RT.

REGISTER AIR DUCT

Comfort stream heater installation—1968 Ford (© Ford Motor Co.)

care not to spill coolant on carpets.
11. Install in reverse of above.

T-Bird 1966-70

Heater Core R & R

1. Drain cooling system.
2. Disconnect heater hoses at the core.
3. Remove the moulding from the right side lower instrument panel.
4. Remove two screws that hold the moulding on the right side of the console. Remove the moulding to gain access to the left bolt holding the lower panel.

5. Remove the right lower instrument panel held by three screws and two bolts.
6. Disconnect the resistor plug. Remove seven retaining screws and the recirculating air grille.
7. Remove two screws from the front of the heater core to case.
8. Loosen seven nuts from the engine side of the dash, holding the heater assembly.
9. Pull the heater assembly away from the dash. Remove four screws, disconnect a cable and position the fresh-air duct out of the way to gain access to the core rear screws.
10. Remove the two screws from the

rear side of the core and remove the core.
11. Install by reversing removal procedure.

T-Bird Through 1970

Heater Blower R & R

The blower motor and fan assembly can be removed from the engine compartment. Disconnect the wiring. Remove mounting screws, then the blower motor and fan assembly.

Jeep

Index

NOTE: Service procedures for the 225 V6 and 350 V8 engines can be found in the Buick Special Car Section. Service procedures for the 232 OHV 6, 390 V8 and the 327 V8 can be found in the Rambler Car Section. Service procedures for the 153 OHV 4 can be found in the Chevy II Car Section.

YEAR IDENTIFICATION

MD-M38A1 Military

CJ-6 Universal Jeep

C-101 Jeepster

C-101 Jeepster Commando

C-101 Jeepster Convertible

1414 Super Wagoneer

CJ-3B Universal Jeep

CJ-5 Universal Jeep

DJ-3A Dispatcher 2WD

1413 Jeep Panel Delivery

1414 Jeep Wagoneer

4-75 Jeep 4x4 Utility Wagon

FIRING ORDER

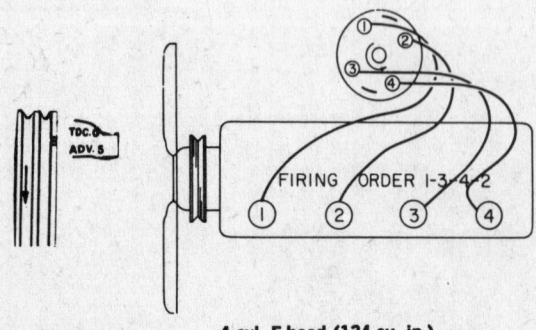
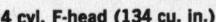
4 cyl. F-head (134 cu. in.)

FIRING ORDER 1-3-4-2

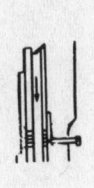
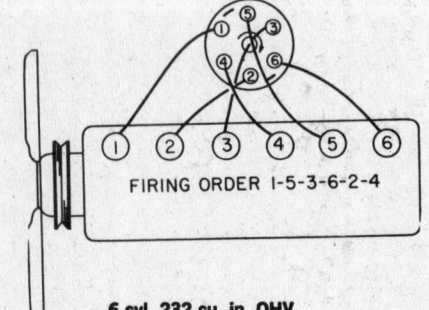
6 cyl. 232 cu. in. OHV

FIRING ORDER 1-5-3-6-2-4

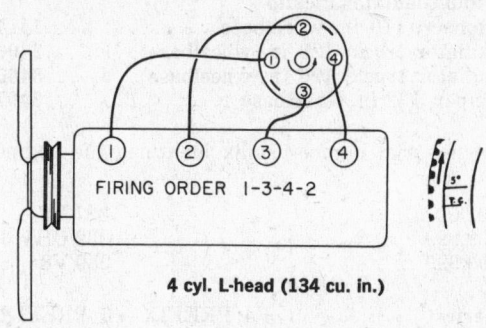

4 cyl. L-head (134 cu. in.)

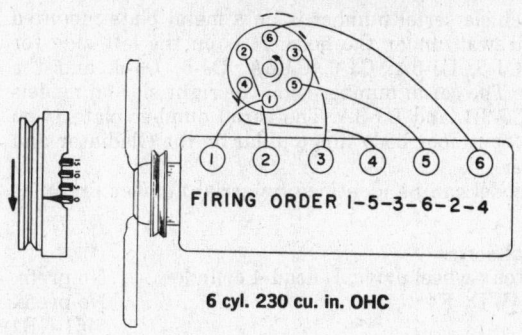

6 cyl. 230 cu. in. OHC

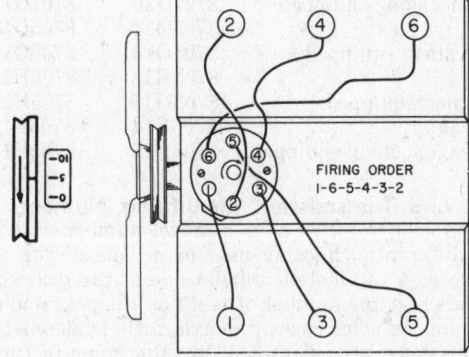

V6 225 cu. in.

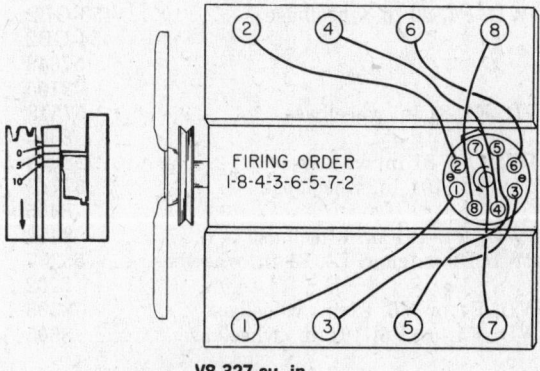

V8 327 cu. in.

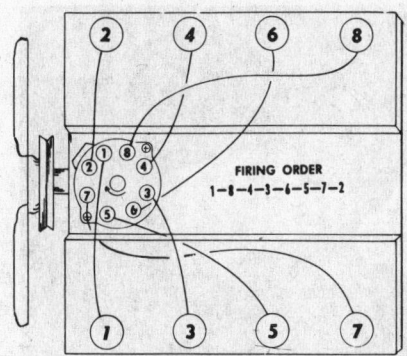

V8 350 cu. in.

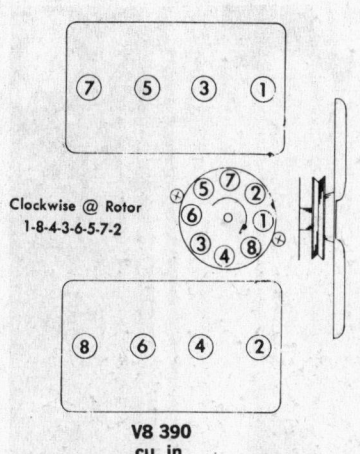

V8 390 cu. in.

ENGINE IDENTIFICATION

Code Locations

A—Front of block above water pump.
B—Lower right front corner of block behind coil.
C—Front face of left bank, below head.
D—Stamped on boss above oil filter.
E—On alternator bracket.
F—Left cylinder head between front two spark plugs.
G—Near fuel filter on left side, near lift pump on right side
H—On tag attached to right rocker cover

YEAR	MODELS	ENGINE	ENGINE CODE	LOCATION
1964	MD, CJ3B, CJ5, CJ6	4/F-head/134	4J	A
	DJ-3A	4/L-head/134	3J	A
	J162, J164	6/OHC/230	ND60C, TD60C	B
	CJ5, CJ6	Diesel/192		G
1965	MD, CJ5, CJ6	4/F-head/134	4J	A
	CJ6	V6/225	KLH	C
	J162, J164	6/OHC/230	ND60C, TD60C	B
	CJ5, CJ6	4/Diesel/192		G
1965	MD, CJ5, CJ6	4/F-head/134	4J	A
	CJ6	V6/225	KLH	C
	J162, J164	6/OHC/230	ND60C, TD60C	B
	CJ5, CJ6	4/Diesel/192		G
	J162, J164	V8/327	E	E
	CJ5, CJ6	4/Diesel/192		G
1967	MD, CJ5, CJ6, DJ5, DJ6,	4/F-head/134	4J	A
	J162, J164	V8/327	E	E
	CJ5, CJ6	4/Diesel/192		G
1968–70	CJ5, CJ6, DJ5, DJ6, C101	4/F-head/134	4J	A
	J164	V8/350	KPO	F
	J162, J164	V8/390	X	H

The vehicle serial number is on a metal plate mounted on the firewall under the hood. It is on the left side for models CJ-5, CJ-5A, CJ-6, CJ-6A, DJ-5, DJ-6, and the Jeepster. The serial number is on the right side on models CJ-2A, CJ-3B, and DJ-3A. The serial number plate is on the left front door body hinge pillar on the Gladiator and Wagoneer.

The model can be identified by serial number prefix as follows:

Universal Series

CJ-2A, four-wheel drive, L-head 4 cylinder	No prefix
CJ-3A, 4WD, F4	No prefix
	451-GB1
	452-GB1
	453-GB1
CJ-3B, 4WD, F4, 80 in. wheelbase	453-GB2
	454-GB2
	57348
	8105
CJ-5, 4WD, F4, 81 in. wheelbase	57548
	8305
CJ-5A, 4WD, F4, 81 in. wheelbase	8322
CJ-6, 4WD, F4, 101 in. wheelbase	57748
	8405
CJ-6A, 4WD, F4, 101 in. wheelbase	8422
DJ-3A, 2WD Dispatcher, L4, 81 in. wheelbase	56337
	8202
DJ-5, 2WD, F4 or V6, 81 in. wheelbase	8505
DJ-6, 2WD, F4, or V6, 101 in. wheelbase	8605

Wagoneer and Gladiator Pickup

J-100 Wagoneer, 110 in. wheelbase	1414
J-2000 Gladiator truck, 120 in. wheelbase	2406
J-3000 Gladiator truck, 126 in. wheelbase	3406
J-3000 Camper, 132 in. wheelbase	3407

The vehicle serial number suffix identifies the engine model:

50,001–199,999	327 V8
200,001–299,999	232 OHV 6
300,001–399,999	350 V8

Jeepster Series	F4 PREFIX	V6 PREFIX
Convertible	8701O14	8701O16
Convertible, smog equipped	8701O15	8701O17
Roadster	8705O14	8705O16
Roadster, smog equipped	8705O15	8705O17
Pickup	8705H14	8705H16
Pickup, smog equipped	8705H15	8705H17
Station wagon	8705F14	8705F16
Station wagon, smog equipped	8705F15	8705F17

Axle and Transmission Identifying Numbers

Axles on Jeep vehicles have a model number cast into the axle differential housing near or on one of the reinforcing webs. A tag installed under one of the gear cover screw heads lists the number of teeth on the gear and also on the pinion. In some cases, the axle ratio is also listed.

Transmissions are identified by the manufacturer's model number on a metal plate attached to the transmission case cover.

Serial number plate—CJ-2A, CJ-3B, DJ-3A
(© Kaiser Jeep Corp.)

Serial number plate—Jeepster
(© Kaiser Jeep Corp.)

Serial number plate—
CJ-5, CJ-5A, CJ-6, CJ-6A, DJ-5, DJ-6
(© Kaiser Jeep Corp.)

Serial number plate—Wagoneer and Gladiator
(© Kaiser Jeep Corp.)

GENERAL ENGINE SPECIFICATIONS

MODEL	CU. IN. DISPLACEMENT	CARBURETOR (BBL.)	DEVELOPED HORSEPOWER @ RPM	DEVELOPED TORQUE @ RPM (FT. LBS.)	A.M.A. HORSEPOWER	BORE & STROKE (IN.)	ADVERTISED COMPRESSION RATIO	VALVE LIFTER TYPE	NORMAL OIL PRESSURE (PSI)
4 Cyl.—L Head	134.2	1	60 @ 4000	105 @ 2000	15.63	3.125 x 4.375	7.0-1	Mech.	35 @ 2000 rpm
4 Cyl.—F Head	134.2	1	75 @ 4000	114 @ 2000	15.63	3.125 x 3.375	7.4-1/6.9-1/ 7.8-1	Mech.	35 @ 2000 rpm
6 Cyl.—OHC	230.5	1	140 @ 4000	210 @ 1750	26.77	3.344 x 4.315	8.5-1	Mech	50 @ 2000 rpm
6 Cyl.—OHV ①	232	1	145 @ 4300	215 @ 1600	33.75	3.75 x 3.50	8.5-1	Hyd.	50
V6 ②	225	2	160 @ 4200	235 @ 2400	15.63	3.750 x 3.400	9.0-1	Hyd.	33 @ 2400 rpm
V8 ③	327	2	250 @ 4700	340 @ 2600	51.20	4.00 x 3.25	8.7-1	Hyd.	55
V8 ④	350	2	230 @ 4400	350 @ 2400	46.2	3.80 x 3.85	9.0-1	Hyd.	37 @ 2400 rpm
V8 ⑤	390	4	325 @ 5000	420 @ 3200	55.51	4.165 x 3.574	10.0-1	Hyd.	45–85
4 Cyl.—Perkins Diesel	192.4	Rotor-distributor pump	60 @ 3000	143 @ 1350	19.6	3.5 x 5.0	16.5-1	Mech.	—

① —Use American Motors 6 Cyl.—232 cu. in. service procedures for this engine.

② —Use Buick Special V6—225 cu. in. service procedures for this engine.

③ —Use American Motors V8—327 cu. in. service procedures for this engine.

④ —Use Buick V8—350 cu. in. service procedures for this engine.

⑤ —Use American Motors V8—390 cu. in. service procedures for this engine.

TUNE-UP SPECIFICATIONS

YEAR	MODEL	SPARK PLUGS		DISTRIBUTOR		IGNITION TIMING (Deg.) ▲	CRANKING COMP. PRESSURE (Psi)	VALVES			FUEL PUMP PRESSURE (Psi)	IDLE SPEED (Rpm) ★
		Type	Gap (In.)	Point Dwell (Deg.)	Point Gap (In.)			Tappet (Hot) Clearance (In.) Intake	Exhaust	Intake Opens (Deg.)		
1964	4 Cyl.—134 Cu. In., L Head	Champ.-J8	.030	40	.020	5B	90–110	.016 Cold	.016 Cold	9B	3	600
1964-71	4 Cyl.—134 Cu. In., L Head	Champ.-J8	.030	40	.020	5B	120–130	.016 Cold	.016 Cold	9B	3	600
1964-65	6 Cyl.—230 Cu. In., OHC	Champ.-L17Y	.030	38	.020	5B	150	.008	.008	15B	4½	550
1966-71	6 Cyl.—232 Cu. In., OHV	Champ.-N-14Y	.035	32	.016	①	145	Zero	Zero	12½B	5	650-700 ②
1965-71	V6—225 cu. in.	AC-44S	.035	30	.016	5B	165	Zero	Zero	24B	5	550
1966-69	V8—327 Cu. In.	Champ.-H-14Y	.035	30	.016	③	145	Zero	Zero	12½B	4½	④
1969-70	V8—350 Cu. In.	AC-R45-TS ⑤	.030	30	.016	⑥	165	Zero	Zero	24B	5	650-700 ⑦
1971	V8—390 Cu. In.	Champ.-N-12Y	.035	30	.016	5B	145	Zero	Zero	18½B	6	650

▲—With vacuum advance disconnected. NOTE: These settings are only approximate. Engine design, altitude, temperature, fuel octane rating and the condition of the individual engine are all factors which can influence timing. The limiting advance factor must, therefore, be the "knock point" of the individual engine.

★—With manual transmission in N and automatic in D.

●—With premium fuel 8° B.

B—Before top dead center.

① —W/o emission control, 5°B on dist. model 1110340, 0° on dist. model 1110444, 0° w..emission control.

② —550 rpm w. emission control.

③ —5° B w/o emission control, 0° w emission control.

④ —650-700 rpm w/o emission control; w. emission control-550/500, standard/ automatic.

⑤ —Champion H-14Y, .033-.037 in.

⑥ —0° on dist. model 1111330, 1111474, and 1111938. 5°B on model 11116964.

⑦ —W/o emission control, 550/600, standard/automatic.

CAUTION

General adoption of anti-pollution laws has changed the design of almost all car engine production to effectively reduce crankcase emission and terminal exhaust products. It has been necessary to adopt stricter tune-up rules, especially timing and idle speed procedures. Both of these values are peculiar to the engine and to its application, rather than to the engine alone. With this in mind, car manufacturers supply idle speed data for the engine and application involved. This information is clearly displayed in the engine compartment of each vehicle.

VALVE SPECIFICATIONS

ENGINE	SEAT ANGLE (DEG.) In.	SEAT ANGLE (DEG.) Ex.	INTAKE VALVE LIFT (IN.)	EXHAUST VALVE LIFT (IN.)	VALVE SPRING PRESSURE INTAKE & EXHAUST Outer	VALVE SPRING PRESSURE INTAKE & EXHAUST Inner	STEM TO GUIDE CLEARANCE (IN.) Intake	STEM TO GUIDE CLEARANCE (IN.) Exhaust	STEM DIAMETER (IN.) In.	STEM DIAMETER (IN.) Ex.	VALVE GUIDE REMOVABLE
4 Cyl.—L Head, 134 Cu. In.	45	45	.351	.351	53 @ $2^7/_{64}$	None	.0015	.0035	.373	.371	Yes
4 Cyl.—F Head, 134 Cu. In.	46	46	.260	.351	73 @ $1^{21}/_{32}$	None	.0014	.0035	.373	.371	Yes
6 Cyl.—230 Cu. In., OHC	45	45	.375	.375	57 @ $1^1/_4$	None	.007	.0035	.340	.339	Yes
6 Cyl.—232 Cu. In., OHV	30	45	.375	.375	100 @ $1^{13}/_{16}$	None	.0020	.0020	.373	.373	No
V6—225 Cu. In.	45	45	.401	.401	①	None	.0020	.0025	.3407	.3407	No
V8—327 Cu. In.	30	45	.375	.375	88 @ $1^{13}/_{16}$	None	.0020	.0020	.372	.372	Yes
V8—350 Cu. In.	45	45	.371	.384	75 @ 1.727	None	.0025	.0025	.372	.372	No
V8—390 Cu. In.	30	45	.425	.425	94 @ $1^{13}/_{16}$	None	.0020	.0020	.372	.372	No

① —Intake—64 @ 1.727, exhaust—64 @ 1.64.

CRANKSHAFT BEARING JOURNAL SPECIFICATIONS

ENGINE	MAIN BEARING JOURNALS (IN.) Journal Diameter	MAIN BEARING JOURNALS (IN.) Oil Clearance	MAIN BEARING JOURNALS (IN.) Shaft End-Play	MAIN BEARING JOURNALS (IN.) Thrust On No.	CONNECTING ROD BEARING JOURNALS (IN.) Journal Diameter	CONNECTING ROD BEARING JOURNALS (IN.) Oil Clearance	CONNECTING ROD BEARING JOURNALS (IN.) End-Play
4 Cyl.—L Head	2.333	.0016	.005	1	1.9375	.0010	.007
4 Cyl.—F Head	2.333	.0019	.005	1	1.9375	.0014	.007
V6—225 Cu. In.	2.4995	.0009	.006	2	2.0000	.0021	.010
6 Cyl.—230 Cu. In., OHC	2.375	.0015	.005	Rear	2.0623	.0015	.009
6 Cyl.—232 Cu. In., OHV	2.4988	.0012	.005	3	2.0952	.0008	.009
V8—327 Cu. In.	2.4991	.0018	.005	1	2.2486	.0015	.010
V8—350 Cu. In.	2.9995	.0010	.006	3	2.0000	.0012	.010
V8—390 Cu. In.	2.7481 / 2.7471 rear	.002	.006	3	2.2478	.002	.012

GENERAL CHASSIS AND BRAKE SPECIFICATIONS

YEAR	MODEL	CHASSIS Wheel Size	CHASSIS Tire Size	Master Cylinder (In.) Power	BRAKE CYLINDER BORE Wheel Cylinder Diameter (In.) Front	BRAKE CYLINDER BORE Wheel Cylinder Diameter (In.) Rear	BRAKE DRUM Diameter (In.) Front	BRAKE DRUM Diameter (In.) Rear
	DJ-3A, DJ-5, DJ-6	13 x 4.00 J / 15 x 4.00 J	6.50 x 13 / 6.85 x 15	1.0	$1^1/_8$	$^{13}/_{16}$	9	9
	CJ-3B, CJ-5, CJ-6 (early)	16 x 4.50 E	6.00 x 16	1.0	1	$^3/_4$	9	9
	CJ-5A, CJ-6A, CJ-5, CJ-6 (late)	15 x 5.5 K	7.35 x 15	1.0	1	$^{13}/_{16}$	10	10
	Jeepster	15 x 5.5 K	—	—	1 / $1^1/_8$	$^{13}/_{16}$ / 1	10 / 11	10 / 11
	Wagoneer Gladiator 1414 2406 W 3406 W 2406 X 3406 X	15 x 5.5 K	— / 8.55 x 15 / 7.75 x 15	1.0	$1^1/_8$	1	11	11
	Gladiator 2406Y, 3306C, 3406Y	16 x 6.00 L / 16 x 5.50 F	7.00 x 16	1.0	$1^1/_8$	1	12	12
	Gladiator, Dual Rear Wheel 2406Z 3406Z	16 x 6.00 L / 16 x 5.50 F	7.00 x 16	$1^1/_8$	$1^1/_8$	$1^1/_4$	12	13
	Gladiator, Camper 3407Z	16 x 5.50F	7.50 x 16	1.0	$1^1/_8$	$1^1/_8$	$12^1/_8$	$12^1/_8$
	Gladiator, Dual Rear Wheel J200, J300	16 x 6.00 L / 16 x 5.50 F	7.50 x 16	1.0	$1^1/_8$	$1^1/_4$	12	13

TORQUE SPECIFICATIONS

ENGINE	CYLINDER HEAD BOLTS (FT. LBS.)	ROD BEARING BOLTS (FT. LBS.)	MAIN BEARING BOLTS (FT. LBS.)	CRANKSHAFT BALANCER BOLT (FT. LBS.)	FLYWHEEL TO CRANKSHAFT BOLTS (FT. LBS.)	MANIFOLD (FT. LBS.)	
						Intake	Exhaust
4 Cyl.—134 Cu. In., F Head, L Head	60–70	35–45	65–75	60–70	35–41	29–35	29–35
6 Cyl.—230 Cu. In., OHC	80–95	40–45	85–95	100–130	35–40	15–20	35–40
6 Cyl.—232 Cu. In., OHV	80–85	27–30	75–85	70–80	100–110	20–25	20–25
V6—225 Cu. In.	65–80	30–40	95–120	140■	50–65	25–35	15–20
V8—327 Cu. In.	58–62	46–50	▲	70–80	100–110	20–25	20–25
V8—350 Cu. In.	●	35	110	140–180	60	50	18
V8—390 Cu. In.	105–115	35–40	95–105	50–60	100–110	40–45	30–35

■ Minimum   ▲ 50–55, rear; 80–85, except rear.   ● 75, metal gasket; 80, composition.

CYLINDER HEAD BOLT TIGHTENING SEQUENCE

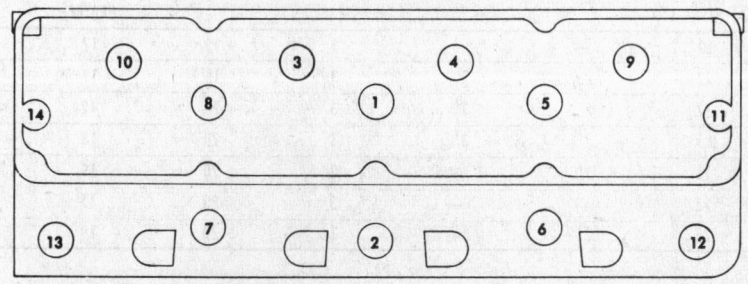

V8 390 cu. in.

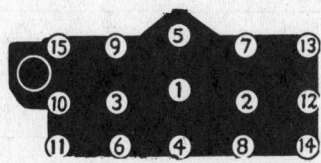

4 cyl. L-head or F-head (134 cu. in.)

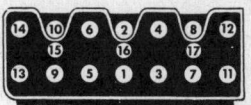

6 cyl. OHC (230 cu. in.)

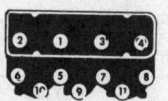

V6 225 cu. in.

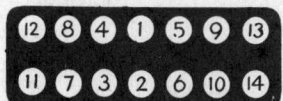

6 cyl. OHV (232 cu. in.)

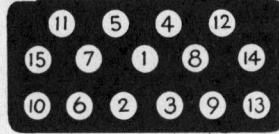

V8 327 cu. in.

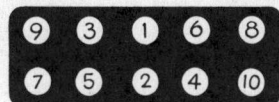

V8 350 cu. in.

GENERATOR AND REGULATOR SPECIFICATIONS

	GENERATORS—6 VOLT			REGULATORS—6 VOLT			
Make	Model No.	Output (AMPS)	Brush Spring Tension (Oz.)	Model No.	Regulated Voltage	Regulated Amperage	Cutout Relay Closing Voltage
Auto-Lite	GDZ4817	35	35–53	VRP-6003 VRP-4007	7.1–7.3	49	6.3–6.8 @ 1000 rpm
	GDZ6001	35	35–53	VBO-4601			
	GGW4801	45	35–53	VBO-4601C	7.1–7.3	49	6.3–6.8 @ 1000 rpm
	GGW7404	45	18–36				
Delco-Remy	1102811		28	1972063	6.9–7.4	42–47	5.9–6.7

	GENERATORS—12 VOLT			REGULATORS—12 VOLT			
Make	Model No.	OUTPUT (AMPS)	Brush Spring Tension (Oz.)	Model No.	Regulated Voltage	Regulated Amperage	Cutout Relay Closing Voltage
Auto-Lite	GJC-7002	30	18–36	VRX-6009	14.3–14.7	39	12.6–13.6 @ 1325 rpm
	GJP-7202	35	18–36	VBO-4201E-4A	14.3–14.7	39	12.6–13.6 @ 1325 rpm
	GJP-7402A	35	18–36				
	GJP-7401A	35	18–36				
Delco-Remy	1102096	35	28	1972029	14.2–14.4	36	11.8–13.5

WHEEL ALIGNMENT

MODEL	CASTER Pref. Setting (Deg.)	CAMBER Pref. Setting (Deg.)	TOE-IN (In.)	KING-PIN INCLINATION (Deg.)	WHEEL PIVOT RATIO Inner Wheel	Outer Wheel
(Early) CJ	3°	1°	$\frac{3}{64} - \frac{3}{32}$	$7\frac{1}{2}°$	20	20
DJ-3A	3°	1°	$\frac{3}{64} - \frac{3}{32}$	$7\frac{1}{2}°$	20	22
CJ-3B, CJ-5, CJ-6, DJ-5, DJ-6, CJ-5A, CJ-6A	3°	1°30'	$\frac{3}{64} - \frac{3}{32}$	$7\frac{1}{2}°$	20	20
Jeepster	3° Preset	1°30' Preset	$\frac{3}{64} - \frac{3}{32}$	$7\frac{1}{2}°$	—	—
Wagoneer, Gladiator	3°	1°30'	$\frac{3}{64} - \frac{3}{32}$	$7\frac{1}{2}°$	—	—

CAPACITIES

MODEL	ENGINE CRANKCASE ADD 1 Qt. FOR NEW FILTER	TRANSMISSIONS MANUAL Add 1 Pt. for Overdrive	AUTOMATIC (Pts.)	FRONT DRIVE AXLE (Pts.)	TRANSFER CASE (Pts.)	REAR DRIVE AXLE (Pts.)	GAS TANK (Gals.)	COOLING SYSTEM (Qts.) WITH HEATER
4 Cyl.—2 Whl. Drive	4	$1\frac{1}{2}$	22	—	—	2	15	12
4 Cyl.—4 Whl. Drive	4	3	22	$2\frac{1}{2}$	$3\frac{1}{2}$	3	15	12
6 Cyl.—OHC	5	$8\frac{3}{4}$	17	2	$3\frac{1}{4}$	3	20	12
V6—225	4	$2\frac{3}{4}$	N.A.	$2\frac{1}{2}$	$3\frac{1}{4}$	3	20	13
6 Cyl. OHV & V8 (327)	5	$2\frac{3}{4}$	19	$2\frac{1}{2}$	—	3	20	19
V8—350	4	$2\frac{3}{4}$	22	—	—	3	20	15
V8-390	4	$2\frac{3}{4}$	22	$2\frac{1}{2}$	$3\frac{1}{4}$	3	20	19

ALTERNATOR AND REGULATOR SPECIFICATIONS

Make	ALTERNATOR Model No.	Field Current Draw @ 10 V (Amps)	Output (Amps)	REGULATOR Model No.	Type	Regulated Voltage	Regulated Amperage
Motorola	A12NW528 A12NW526	1.2–1.7	35	R2K1	Transistorized	14.2–14.6	35
Motorola	A12NAM453 A12NAM451-S A12NW526 A12NW525 A12NW527	1.2–1.7	35	TVR-12-W1 4	Transistorized	14.2–14.6	35
American Motors	3195534-5-2	2.4–2.5	35	3195003	Transistorized	15	35
(390V8)	—	1.8–2.4	55		Transistorized	15	55

BATTERY AND STARTER SPECIFICATIONS

MODEL	BATTERY Ampere Hour Capacity	Volts	STARTERS Make	Lock Test Amps.	Volts	Torque (Ft-lbs)	No-Load Test Amps.	Volts	RPM	Brush Spring Tension (Oz.)
Universal, Dispatcher 4-Cyl.	100	6	Auto-Lite	335	2.0	6.0	65	5	4,300	42–53
			Delco-Remy	600	3.0	15	60	5	6,000	24
Universal, Dispatcher 4-Cyl.	50	12	Auto-Lite	170/280	4.0	1.5/6.2	50	10	4,400/5,300	31–47
			Delco-Remy	435	5.8	10.5	75	10.3	6,900	35
Jeepster 4-Cyl.	50	12	Prestolite	405	—	9	50	10	5,300	32–40
Universal, Diesel	—	12	Lucas M45G	—	—	—	—	—	—	—
232 6 Cyl.	50	12	Prestolite	405	4.0	9	60	10	4,200	32–40
230 6 Cyl, OHC	50	12	—	405	4.0	9	60	10	4,200	36
V6	50	12	Delco-Remy	—	—	—	75	10.6	6,200	32–40
327V8	60	12	Prestolite	405	4.0	9	60	10	4,200	32–40
350V8	60/70	12	Delco-Remy	300–360	3.5	9	65–100	10.6	3,600–5,100	35
390V8	70	12	FoMoCo	500	4.5	10	70	12	9,500	40

FUSES

Universal—6 Volt

Directional Signal	14 A
Heater	14 A

Universal—12 Volt (to 1965)

Directional Signal	9 A
Heater	9 A

Universal—12 Volt (1965 on)

Heater	15 A
Backup	9 A
Wipers	14 A
Directional Signal	14 A
4-Way Flasher	14 A

Universal—1970

Heater	9 A
Directional Signal	9 A
4-Way Flasher	14 A

Jeepster

Clock	1 A
Heater	15 A
Radio	2 A
Backup	9 A
Wipers	14 A
Directional Signal	14 A
4-Way Flasher	14 A

Jeepster—1970

Heater	15 A
Radio	2 A
Wipers	14 A
Turn Signal	9 A
4-Way Flasher	9 A
Backup	9 A

Wagoneer and Gladiator

Clock	2 A
Heater	14 A
Radio	2 A
Parking Brade	9 A
Backup	9 A
Overdrive Relay	14 A
Wipers	14 A
4 WD Warning	9 A
Directional Signal	9 A
4-Way Flasher	14 A

Wagoneer and Gladiator—1969

Heater	14 A
Radio	2 A
Wipers	14 A
Directional Signal	9 A
4-Way Flasher	9 A
Brake Warning	9 A

LIGHT BULBS

Universal—6 Volt

Headlight	5040-S or 6006
Parking Light	63
Park and Directional	1158
Stop, Tail, and Directional	1158
Headlight, Directional, Charge, and Oil Pressure Indicators	51
Instrument Light	55
Directional Flasher Type	P229D

Universal—12 Volt

Headlight	5400-S or 6012
Parking Light	67
Park and Directional	1176 or 1034
Stop, Tail, and Directional	1034
Headlight Indicator	53 or 57
Directional Indicator	53
Charge Indicator	53 or 57
Oil Pressure Indicator	57
Instrument Light	57
Directional Flasher Type	524

Jeepster

Headlight	6012
Parking and Directional	1157
License Plate Light	67 or 1155
Headlight, Directional, Charge, Oil Pressure, and Cluster Indicator Lights	158
Clock	1816
Radio	1892
Courtesy	90
Backup	1156

Jeepster and Universal—1970

Headlight	6012
Parking and Directional	1157
Marker and Reflector	194
Stop, Tail, and Rear Turn Signal	1157
License Plate Light	67 or 1155
Backup	1156

Wagoneer and Gladiator

Headlight	6012
Front Parking Light	1034 or 1157
Stop, Tail, and Directional	1034 or 1157
License Plate Light	67 or 1155
Headlight, Directional, Charge, Oil Pressure, and Cluster Indicators	57 or 158
Dome Lamp	212
Heater Control	1816
Clock	1816
Ignition Switch and Shift Selector	1445
4 WD Indicator	53
Parking Brake	257 or 57
Radio	1892
Courtesy	90
Backup	1073 or 1156
Glove Compartment	1891
4-Way Flasher	53

Wagoneer and Gladiator—1969

Headlight	6012
Stop, Tail, Turn Signal, Parking	1157
Marker and Reflector	194
License Plate Light	1155
Backup	1156

Distributor

Detailed information on distributor drive, direction of distributor rotation, cylinder numbering, firing order, point gap, point dwell, timing mark location, spark plugs, spark advance, ignition resistor location and idle speed is in the Specifications table of this section. Further information on troubleshooting, general tune-up procedures, how to replace ignition wires, how to install points and condensers, how to choose the proper spark plug and how to adjust timing is in the Unit Repair Section.

Removal

The distributor assembly on all Jeep 4 and 6 cylinder in-line engines is located on the right side of the engine.

To remove it, take off the distributor cap and wire assembly and bend out of the way. Remove the ignition primary wire from the side of the distributor and take off the vacuum lines to the carburetor. Mark the distributor housing and the engine crankcase to insure correct reinstallation. Note the position of the rotor. Remove the bolt that holds the distributor down into the block and lift it off the engine.

Installation

If the crankshaft has not been rotated, simply match the marks made on removal and reinstall the rotor as noted on removal.

If the crankshaft has been rotated, turn the crankshaft until No. 1 cylinder is at the top of the compression stroke as indicated by air being forced out from No. 1 spark plug opening. Install distributor with rotor pointing toward No. 1 spark plug wire on the distributor cap and the points just opening. Start the engine and finish the timing procedure with a timing light.

Generator, Regulator

Detailed facts on the generator and the regulator are in the Generator and Regulator Specifications table of this section.

General information on generator and regulator repair and troubleshooting is in the Unit Repair Section.

Generator Polarity

Caution Whenever the circuits to the generator, the regulator, or the battery have been disconnected it is best to apply the following procedure:

Before the engine is started, momentarily short from the Bat. to the Gen. terminals of the regulator with a screwdriver. This gives a momentary surge of current from the battery to the generator, and correctly polarizes the generator with the battery.

Failure to polarize the generator before starting the engine may severely damage the regulator, because reversed polarity causes vibration, arcing, and burning of the relay points.

AC Generator

The purpose of this unit is to satisfy the increase in electrical loads that have been imposed upon the car battery by modern conditions of traffic and driving patterns.

The AC generator is explained in the Unit Repair Section.

Caution Since the AC generator and regulator are designed for use on only one polarity system, the following precautions must be observed:

1. The polarity of the battery, generator and regulator must be matched and considered before making any electrical connections in the system.

2. When connecting a booster battery, be sure to connect the negative battery terminals together and the positive battery terminals together.

3. When connecting a charger to the battery, connect the charger positive lead to the battery positive terminal. Connect the charger negative lead to the battery negative terminal.

4. Never operate the AC generator on open circuit. Be sure that all connections in the circuit are clean and tight.

5. Do not short across or ground any of the terminals on the AC generator.

6. Do not attempt to polarize the AC generator.

7. Do not use test lamps of more than 12 volts for checking diode continuity.

8. Avoid long soldering times when replacing diodes or transistors. Prolonged heat is damaging to these units.

9. Disconnect the battery ground terminal when servicing any AC system. This will prevent the possibility of accidentally reversing polarity.

Starter R & R

Disconnect battery and starter wires. Remove attaching bolts and lift out starter.

Instruments

Speedometer

Jeep Universal, Dispatcher

To remove speedometer:
1. Disconnect battery ground cable.
2. Remove mounting screws from rear of dash.
3. Disconnect wiring and speedometer cable.
4. Remove unit from instrument panel. Fuel gauge and temperature gauge can now be removed from speedometer unit.
5. Reinstall by reversing removal procedure.

Gladiator, Wagoneer, Jeepster

To remove speedometer:
1. Disconnect battery ground cable.
2. Disconnect speedometer cable and electrical connector.
3. Compress retainer springs and slide instrument cluster forward from instrument panel.
4. Pry off glass frame clips. Remove frame and glass.
5. To remove fuel gauge, temperature gauge, or speedometer, remove locknuts and remove these units from the front.
6. Reinstall by reversing removal procedure.

Brakes

Specific information on brake lining sizes is in the General Chassis and Brake Specifications table.

Information on brake adjustments, band replacement, bleeding procedure, master and wheel cylinder overhaul is in the Unit Repair Section.

Information on troubleshooting and overhauling power brakes is in the Unit Repair Section.

Information on the grease seals which may need replacement is in the Unit Repair Section.

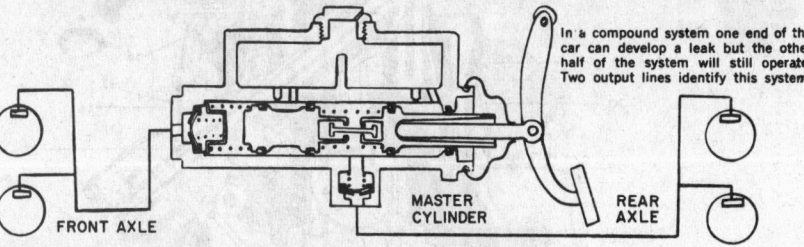

In a compound system one end of the car can develop a leak but the other half of the system will still operate. Two output lines identify this system.

Dual type master cylinder

Fuel System

Data on capacity of the gas tank is in the Capacities table. Data on correct engine idle speed and fuel pump pressure are in the Tune-Up Specifications table.

Information covering operation and troubles of the fuel gauge is in the Unit Repair Section.

Detailed information on the carburetor, and how to adjust it, is in the Unit Repair Section under the make of carburetor being used on the engine being worked on.

Fuel Pump

The fuel pump is mounted on the left side of the engine and is actuated by an eccentric on the camshaft. On the OHC 6, the fuel pump is mounted at the right front and is actuated by a pushrod leading from an eccentric at the front of the camshaft. A double action fuel pump is used on vehicles with vacuum powered windshield wipers, and a single action pump on those with electric wipers. Some fuel pumps have an external sediment bowl on the top of the pump. The screen and sediment bowl should be cleaned twice yearly.

Cooling System

Detailed information on cooling-system capacity is in the Capacities table of this section.

Information on the water temperature gauge is in the Unit Repair Section.

Caution Do not run cold water over the outside of pressurized radiators without first removing the radiator cap. When the cap is left on and the cold water hits the hot radiator, the steam in the radiator condenses very rapidly and sometimes collapses the top radiator tank. This is most likely to happen if the coolant level is below normal.

Water Pump Removal

Remove the fan blades and the fan pulley, then take out the attaching screws and lift the pump assembly from the front of the block.

It is not necessary to remove the radiator.

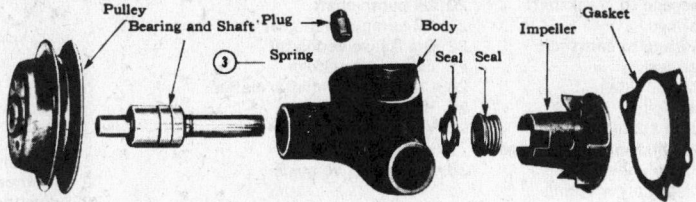

Water pump—134 cu. in. 4 cyl. (© Willys Corp.)

Engine

The Specifications table lists all the available facts about the engine.

Engine crankcase capacities are listed in the Capacities table.

Torque wrench readings and head-bolt tightening sequences are covered in the Torque Specifications table.

Information on the engine marking code will be found in the Engine Identification table at the start of this section.

For information on the 225 cu. in. V6 and the 350 cu. in. V8, refer to the Buick Special Section. For information on the 232 cu. in. OHV 6, 390 V8 and the 327 cu. in. V8, see the Rambler Section. For the 153 cu. in. OHV 4, see the Chevy II Section. Information on L-head engines given in this section can be applied, in general, to the 226 cu. in. L-head 6. Information on F-head engines can also be applied, in general, to the 150 and 161 cu. in. F-head 6 engines.

OHC 6

The Tornado 230 OHC is an overhead cam type engine. All valve and valve parts are entirely on the cylinder head, and are removable with the head as an assembly.

The camshaft has only six lobes. Each lobe operates both intake and exhaust valves through rocker arms.

A fully counterbalanced crankshaft is mounted in four main bearings with end-play controlled at the rear.

An oil pump mounted outside at the left front, supplies oil under pressure through drilled passages to the main and rod bearings. An external tube carries oil from the block to the cylinder head to lubricate the camshaft and rockers. An oil fitting in front of block spurts oil on the chain.

Bearings

Detailed information on engine bearings is in the Crankshaft Bearing Journal Sizes table.

Valves

Detailed information on the valves, the type of valve guide and the location of valve timing marks is in the Valve Specifications table.

A general discussion of valve clearance, and a chart showing how to read pressure and vacuum gauges when using them to diagnose engine troubles, is in the Unit Repair Section as well as a chart on engine troubleshooting.

Valve tappet clearance for each engine is given in the Tune-up Specifications table.

Engine Removal

F 4 and L 4

1. Drain cooling system.
2. Disconnect battery ground cable.
3. Remove air cleaner and disconnect breather hose at oil filler.
4. Disconnect choke and throttle controls.
5. Disconnect fuel line and windshield wiper hose at fuel pump.
6. Remove radiator stay bar, if so equipped.
7. Remove radiator and heater hoses.
8. Remove fan blades, fan hub, radiator, and shroud.
9. Remove starter motor.
10. Disconnect:
    a. Alternator or generator.
    b. Ignition primary wire at coil.

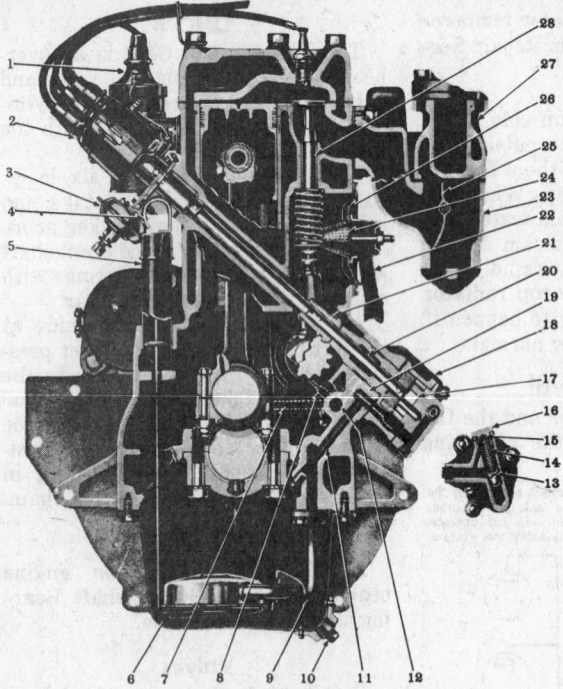

L4 engine (© Willys Corp.)

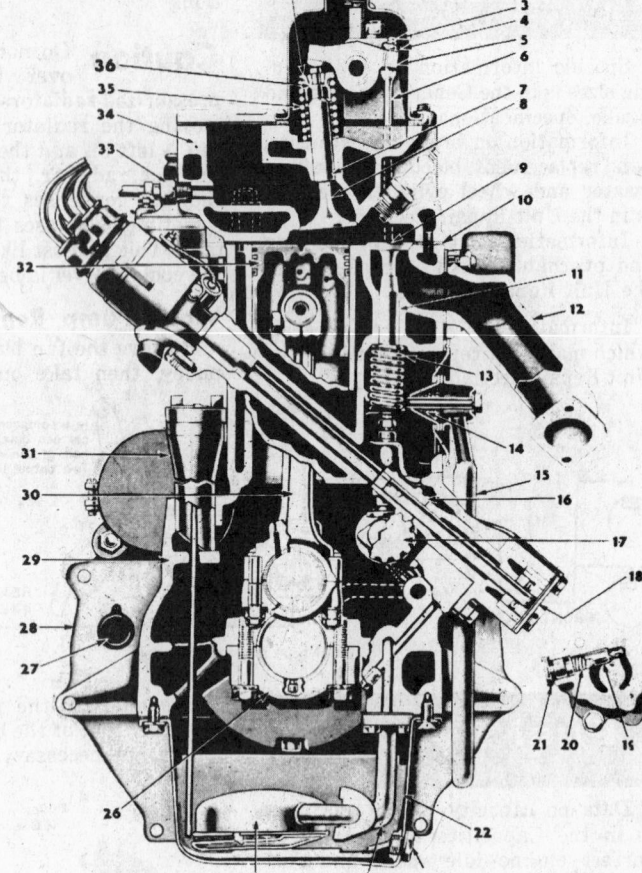

F4 134 engine (© Willys Corp.)

1 Ignition coil
2 Ignition distributor
3 Vacuum spark control
4 Oil level measuring rod
5 Oil filter tube
6 Floating oil intake
7 Oil passage to crankshaft bearings
8 Oil passage to camshaft bearings
9 Floating oil intake
10 Oil drain plug
11 Oil pump suction passage
12 Main oil distributing passage
13 Oil relief plunger
14 Oil relief plunger spring
15 Oil relief plunger spring shim
16 Oil relief plunger spring retainer
17 Oil pump pinion
18 Oil pump
19 Oil pump discharge passage
20 Oil pump shaft
21 Oil pump driven gear
22 Crankcase ventilator
23 Exhaust manifold
24 Crankcase ventilator baffle
25 Heat control valve
26 Valve spring cover
27 Intake manifold
28 Exhaust valve guide

1 Inlet valve spring retainer
2 Breather cap
3 Adjusting screw
4 Adjusting screw nut
5 Rocker arm
6 Push rod
7 Intake valve guide
8 Intake valve
9 Exhaust valve
10 Cylinder head gasket
11 Exhaust valve guide
12 Exhaust manifold
13 Exhaust valve spring
14 Ventilator baffle
15 Crankcase ventilator
16 Oil pump driven gear
17 Camshaft
18 Oil pump
19 Relief plunger
20 Relief plunger spring
21 Relief spring retainer
22 Oil pan
23 Oil pan drain plug
24 Oil float support
25 Floating oil intake
26 Crankshaft
27 Rear engine plate
28 Cylinder block
29 Connecting rod
30 Oil filler tube
31 Piston
32 Vacuum tube connection
33 Cylinder head
34 Inlet valve spring
35 Oil seal

c. Oil pressure and temperature sending units.
d. Exhaust pipe from manifold.
e. Engine ground strap.
11. Attach lifting device to engine. Unbolt and remove front engine supports.
12. Remove flywheel housing bolts.
13. Pull engine forward until clutch clears the flywheel housing. Lift engine from vehicle.
14. Install in reverse of above.

OHC Tornado 230

1. Drain engine oil.
2. Remove oil filter.
3. Drain cooling system.
4. Using clean pan under clutch slave cylinder, disconnect line at slave cylinder.
5. Remove hood from hinges.
6. Remove radiator and heater hoses.
7. Disconnect transmission oil cooler lines, if automatic transmission.
8. Remove radiator.
9. Disconnect wires from engine sending units, starter, generator and coil.
10. Remove generator, fan belt, fan and fan spacer.
11. Remove battery and battery supports.
12. Disconnect ground cable and engine ground strap at right front engine support.
13. Disconnect exhaust pipe at manifold, and bracket at flywheel housing.
14. Remove carburetor.
15. Disconnect flexible line, frame-to-engine.
16. Remove front support bolts.
17. Install suitable lifting fixture on engine, preferably at upper left rear and upper right front of engine.
18. Remove transmission.
19. Raise engine slowly and pull forward to clear. Continue to raise until clear and lift from vehicle.
20. Install in reverse of above.

V8

1. Remove hood from hinges.
2. Remove air cleaner.
3. Drain radiator and cylinder block.
4. Disconnect radiator and heater hoses.
5. If equipped with automatic

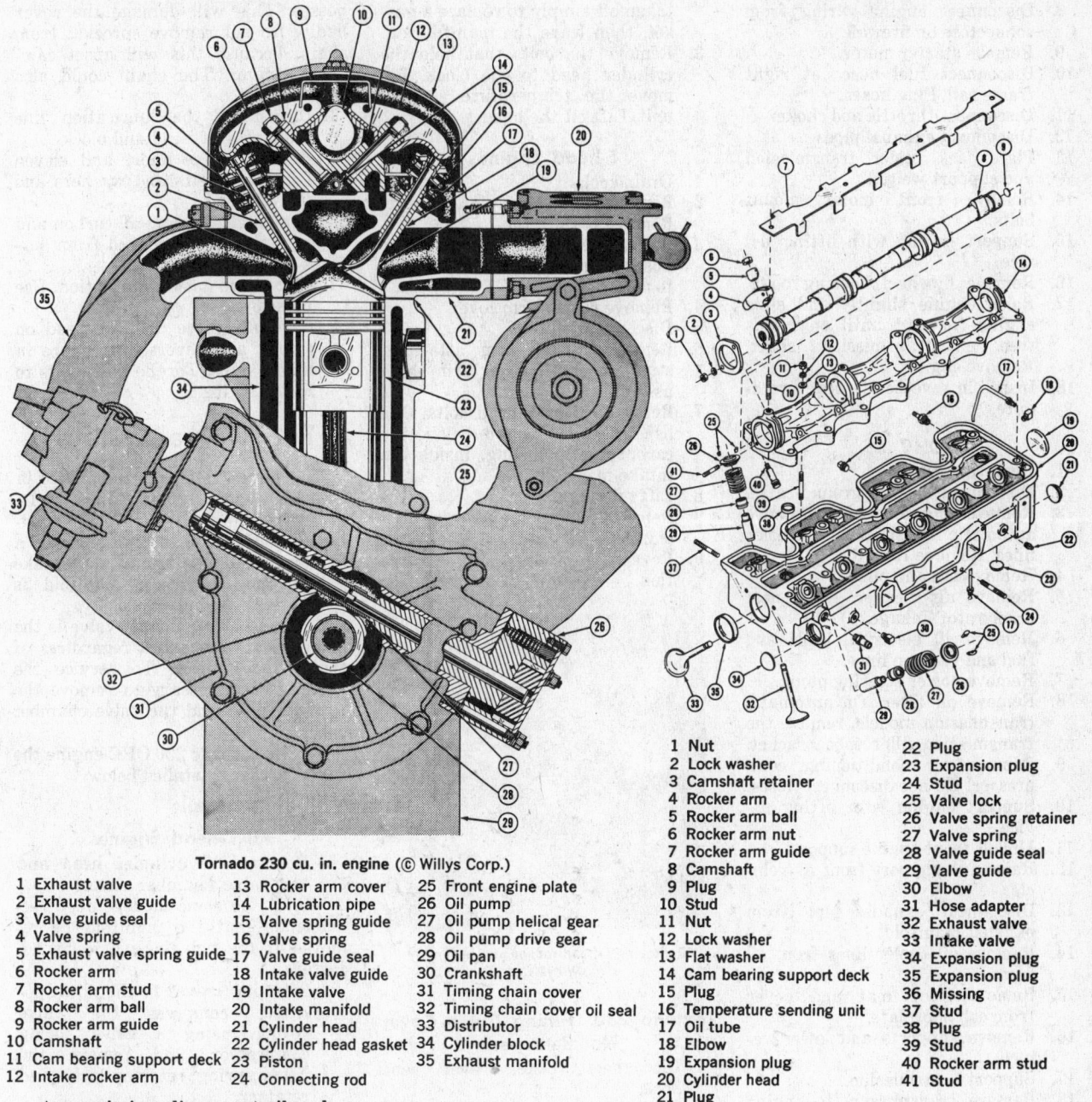

Tornado 230 cu. in. engine (© Willys Corp.)

1 Exhaust valve	13 Rocker arm cover	25 Front engine plate
2 Exhaust valve guide	14 Lubrication pipe	26 Oil pump
3 Valve guide seal	15 Valve spring guide	27 Oil pump helical gear
4 Valve spring	16 Valve spring	28 Oil pump drive gear
5 Exhaust valve spring guide	17 Valve guide seal	29 Oil pan
6 Rocker arm	18 Intake valve guide	30 Crankshaft
7 Rocker arm stud	19 Intake valve	31 Timing chain cover
8 Rocker arm ball	20 Intake manifold	32 Timing chain cover oil seal
9 Rocker arm guide	21 Cylinder head	33 Distributor
10 Camshaft	22 Cylinder head gasket	34 Cylinder block
11 Cam bearing support deck	23 Piston	35 Exhaust manifold
12 Intake rocker arm	24 Connecting rod	

1 Nut	22 Plug
2 Lock washer	23 Expansion plug
3 Camshaft retainer	24 Stud
4 Rocker arm	25 Valve lock
5 Rocker arm ball	26 Valve spring retainer
6 Rocker arm nut	27 Valve spring
7 Rocker arm guide	28 Valve guide seal
8 Camshaft	29 Valve guide
9 Plug	30 Elbow
10 Stud	31 Hose adapter
11 Nut	32 Exhaust valve
12 Lock washer	33 Intake valve
13 Flat washer	34 Expansion plug
14 Cam bearing support deck	35 Expansion plug
15 Plug	36 Missing
16 Temperature sending unit	37 Stud
17 Oil tube	38 Plug
18 Elbow	39 Stud
19 Expansion plug	40 Rocker arm stud
20 Cylinder head	41 Stud
21 Plug	

transmission, disconnect oil cooler lines at radiator. Remove radiator.

6. Remove fan, belt and hub.

7. Drain engine oil and remove filter.

8. Disconnect temperature-sender lead, pressure-sender lead; coil, starter solenoid, alternator and distributor leads.

9. Disconnect accelerator cable at carburetor throttle shaft lever and at accelerator cable support bracket.

10. Disconnect heater system vacuum valve hose at intake manifold.

11. Disconnect flexible fuel line, frame-to-crankcase at frame end. Plug end of hose.

12. Disconnect exhaust pipe at both manifolds.

13. Remove air conditioning compressor. Do not disconnect hoses.

14. If equipped with manual transmission, disconnect linkage at clutch. Disconnect clutch crossshaft support brackets at flywheel housing and frame.

15. Install suitable lifting fixture on engine.

16. Support transmission on jack.

17. Remove nuts, engine-to-front support brackets.

18. With manual transmission, remove cap screws, transmission-to-clutch housing. With automatic transmission, remove cap screws attaching transmission housing-to-flywheel housing adapter.

19. Pull engine forward and upward until free from transmission or clutch.

20. Install in reverse of above.

V6

1. Remove hood if necessary.

2. Disconnect battery ground cable.

3. Remove air cleaner.

4. Drain coolant.

5. Disconnect radiator hoses.

6. Remove radiator support bars.

7. On Universal Series, remove radiator. On Jeepster Series, disconnect headlamp wiring from block on left fender, disconnect horn wiring from horn, disconnect oil cooler lines if automatic transmission, remove front fenders, radiator, and grille as a unit.

8. Disconnect engine wiring from connectors on firewall.
9. Remove starter motor.
10. Disconnect fuel hoses at right frame rail. Plug hoses.
11. Disconnect throttle and choke.
12. Disconnect exhaust pipes.
13. Place jack under transmission and support weight.
14. Remove front motor mount bolts.
15. Support engine with lifting device.
16. Remove flywheel housing bolts.
17. Raise engine slightly and slide engine forward until engine is free of transmission shaft. Remove engine.
18. Install in reverse of above procedure.

OHV 6

1. Remove hood.
2. Disconnect battery ground cable.
3. Remove radiator hoses. Remove automatic transmission oil cooler lines. Remove radiator.
4. Remove fan and pulley.
5. Remove air cleaner. Disconnect accelerator linkage.
6. Remove all electrical leads and fuel and vacuum lines.
7. Remove power steering pump.
8. Remove oil filter. On automatic transmission models, remove the transmission filler tube bracket.
9. Remove air conditioning compressor. Do not disconnect hoses.
10. Support engine with lifting device.
11. Unbolt front engine support.
12. Raise and support front of vehicle.
13. Disconnect exhaust pipe from exhaust manifold.
14. Remove oil cooler lines from oil pan.
15. Remove floor mat and cover from cab floorplate.
16. Remove supports and lower vehicle.
17. Support transmission.
18. Remove transmission to engine adapter plate bolts on automatic transmission models. On manual transmission models, remove clutch housing to engine bolts. Remove engine.
19. Install in reverse of above.

Cylinder Head

Cylinder Head Removal

L-head Engines

1. Drain coolant. Remove the radiator hose and disconnect all fuel lines to the carburetor and fuel pump, and all lines and fittings to the carburetor. Take off the carburetor.
2. If the valves are to be worked on, the manifold should be removed. However, if the head is being

taken off simply to replace a gasket, then leave the manifold on.
3. Remove the bolts that hold the cylinder head to the block. Remove the temperature sending unit. Lift off the head.

F-head Engines

1. Drain coolant.
2. Remove upper radiator hose. Remove carburetor.
3. On early engines, remove bypass hose from front of cylinder head to water pump.
4. Remove rocker arm cover.
5. Disconnect oil line.
6. Remove rocker arm attaching stud nuts and rocker arm shaft assembly.
7. Remove cylinder head bolts. One head bolt is located below the carburetor mounting, inside the intake manifold.
8. Lift off cylinder head.
9. Remove pushrods and valve lifters.
10. Reverse procedure for installation.

Removing F-head rocker arm assembly
(© Willys Corp.)

Tornado 230 (Timing Chain Cover Not Removed)

1. Disconnect rocker cover vent hose.
2. Remove cover cap nuts, washers and cable brackets from rocker cover studs.
3. Lift off cover and gasket and four seal washers.
4. Install camshaft sprocket tool W-268 on cover studs. Insert hook in sprocket and tighten nut to relieve tension on camshaft.
5. Remove capscrew, lock and flat washer, and fuel pump eccentric from camshaft sprocket.
6. Pull sprocket forward from pilot on camshaft. With sprocket still in timing chain release tension on tool W-268 and allow sprocket to rest on bosses in timing chain cover.

Caution Do not turn engine while sprocket is removed from camshaft and resting on

bosses. This will damage the cover badly. Do not remove sprocket from chain, because this will upset camshaft timing. The chain could also be damaged.

7. Disconnect the lubrication line from rear of head and block.
8. Remove three short and eleven long head-bolts and washers and lift off head.

To reinstall: be sure all carbon and foreign matter is removed from surfaces and bolt holes.

1. Place head gasket in position. Use no sealer.
2. Carefully place cylinder head on gasket and reverse procedure in above steps. Torque head bolts to 65-70 ft. lbs.

The Valve System

Two valve systems are available in Jeep 4-cylinder engines. One is the L-head-type and the second the F-head-type. On the F-head-type engines, a cylinder head containing the intake valves and an intake manifold is placed on the standard block.

Service on the exhaust valve is the same on all Jeep cars, regardless of the cylinder head. To service the valves it is necessary to remove the cylinder head and the valve chamber cover.

On the Tornado 230 OHC engine the procedures are detailed below.

Valve Removal

All L-head Engines

1. Remove the cylinder head and the valve chamber cover.
NOTE: on some models, Jeep recommends that the manifold be removed if the valves have to be adjusted.
2. Selecting valves in the down position, compress the valve spring, using a valve spring compressor, and remove the valve spring retainer locks and retainer.
3. Release the pressure from the valve spring and pull the valve up to the top of the cylinder block. Remove the spring.
4. If the valve tends to bind on its way through the guide, push it back in again and dissolve the gum and tar from the bottom of the valve stem which prevents it from coming up through the guide.

F-head Engines

On F-head engines, the exhaust valves are removed as described above. However, the inlet valve is in the cylinder head and is removed on the bench.

Tornado 230 OHC Engines

1. Place head on blocks to provide hand clearance under it.

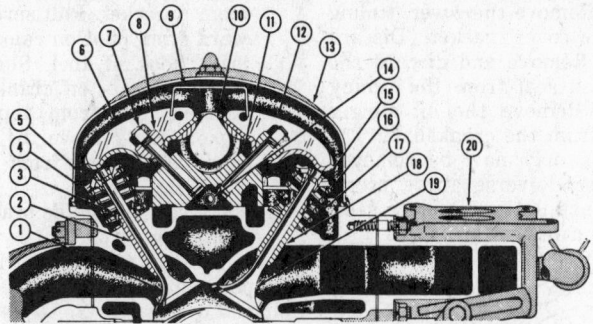

Tornado 230 cutaway valve section (© Willys Corp.)

1 Exhaust valve	11 Cam bearing support deck
2 Valve guide	12 Rocker arm
3 Valve guide seal	13 Rocker arm cover
4 Valve spring	14 Oil tube
5 Valve spring retainer	15 Valve spring retainer
6 Rocker arm	16 Valve spring
7 Rocker arm stud	17 Valve guide seal
8 Rocker arm ball	18 Valve guide
9 Rocker arm guide	19 Intake valve
10 Camshaft	20 Intake manifold

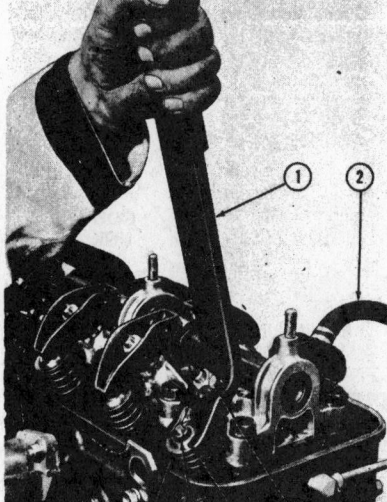

Valve spring tool—Tornado 230
(© Willys Corp.)

1 Valve spring
   compressor tool
   W-267
2 Air hose
3 Rocker arm nut
4 Valve spring
5 Valve lock
6 Valve spring
   retainer

Adjusting valve clearance—Tornado 230
(© Willys Corp.)

2. Remove valve locks and springs. (Discard the valve guide seal.)
To reinstall: always use new guide seals.
1. With head on blocks insert the valves. Apply engine oil to stems.
2. Install seals on stems and push down to seat squarely on guide.
3. Compress spring and insert retainers and valve locks.

Valve Adjustment

Tornado 230 OHC Engines

To adjust valves in this engine: if rocker arms are off, install in reverse of removal, as described in Cylinder Head Removal. Or, if arms have not been removed, proceed as follows:
1. Turn down rocker arm nut but do not tighten.
2. Align each arm on stud and install guide to hold it in place.
3. Adjust the nut to .006 in. clearance on intake valves and .008 in. on exhaust. Always be sure arm is completely off cam when making the adjustment.

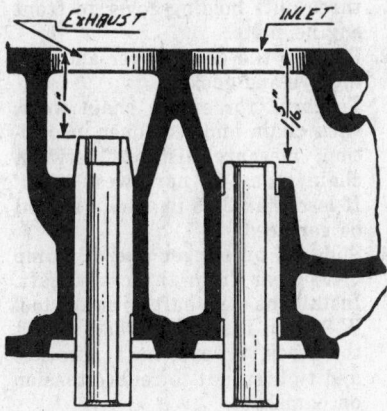

Valve guide installation dimensions—L-head
(© Willys Corp.)

Valve Guide Replacement

L-Head

The old guides are removed by driving them through the block into the valve compartment. The valve guides can also be removed by use of a puller. When replacing the guides, the top of the exhaust guide must be positioned 1 in. below the top face of the cylinder block. The intake guide must be 1-5/16 in. below the top face of the cylinder block. A reamer should be run through the new guides after they have been correctly positioned. This is done to remove any distortion caused by the driving operation.

Valve guide installation dimensions—F-head
(© Willys Corp.)

F-Head

The old guides can be driven or pulled out. The top end of the exhaust valve guide must be exactly 1 in. below the top face of the cylinder block. The intake guide should be driven in, tapered end first, into position from the bottom of the cylinder head. When properly positioned, the end of the guide is just flush with the end of the valve guide bore in the cylinder head. A reamer should be run through the new guides after installation, to remove distortion caused by the driving operation.

Valve Lifters

Valve lifters in all Jeep engines except the 230 OHC are the mushroom type, and require removal of the engine and camshaft in order to replace them.

On the L-head-type valve, the lifter is fitted with a self-locking tappet screw.

Overhead valves are adjusted at the rocker arm.

Camshaft and Bearings Inspection and Replacement

Tornado 230 OHC Engine Removal (Head Removed)

1. Lift off rocker arm guide.
2. Turn rocker arms that have no tension, parallel to camshaft.

Then rotate shaft until more arms are free and turn them parallel. Continue until all arms are off cams.

3. Remove retainer from front of camshaft bearing deck and pull shaft forward from bearings.

4. By removing three nuts with washers, bearing deck can now be lifted free.

Inspection

1. Clean camshaft thoroughly with solvent. See that all oil passages are clear. Runout of shaft must not exceed .0005 in. Check diameter of journals.
   Front .... 1.9975—1.9965 in.
   No. 2 .... 1.8725—1.8715 in.
   No. 3 .... 1.7475—1.7465 in.
   No. 4 .... 1.3725—1.3715 in.

2. Check deck for cracks and distorsions.

3. Check diameter of bearing bores.
   Front .... 1.9995—2.0005 in.
   No. 2 .... 1.8745—1.8755 in.
   No. 3 .... 1.7495—1.7505 in.
   No. 4 .... 1.3745—1.3755 in.
   Maximum running clearance .004 in.

Installation

1. Place camshaft bearing deck on cylinder head and install nuts and washers. Tighten evenly.

2. Lubricate the bearings with engine oil and slide shaft into the bearings from the front, taking care not to damage bearings.

3. Secure shaft in place with the retainer plate and nuts and washers.

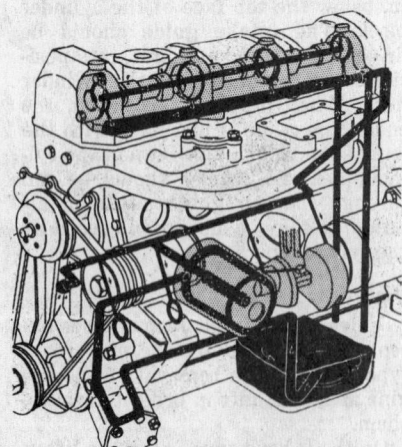

Engine lubrication system—
Tornado 230 engine
(© Willys Corp.)

Timing Case

Timing Gear Cover Removal

L-Head and F-Head

To remove the timing case cover, it is necessary to take off the radiator, the fan, and pulleys. Remove bolts, nuts, and lockwashers from timing gear cover. Remove the cover, timing pointer, and cover gasket. Discard the gasket. Remove and discard the crankshaft oil seal from the timing gear cover. Remove the oil slinger and spacer from the crankshaft. The timing gears may now be removed with a puller. Reverse above procedure for reinstallation. Replace seals and gasket on reinstallation.

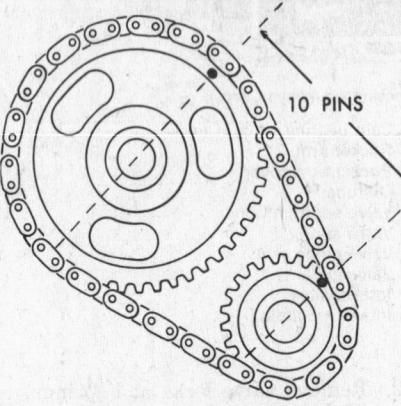

Aligning timing marks—226 6 cyl.

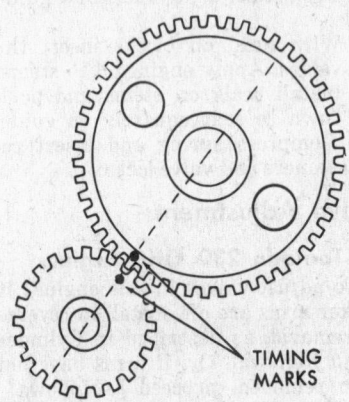

Aligning timing marks—4 cyl.

Timing Case Cover and Chain Removal

Tornado 230 OHC Engine

1. Remove hose from water port on cover. Then remove eight bolts, lock washers and flat washers, and three nuts holding cover to front engine plate.

2. Remove the chain cover and lifting eye and gaskets.

3. To check for excess chain wear, with chain and tensioner in position, measure distance between chain sides at narrowest point. If less than 3.38 in., chain should be replaced.

4. Slide off oil slinger and oil pump drive gear from the crankshaft.

5. Install the camshaft sprocket tool W-268 on the cover studs. Insert the hook in camshaft sprocket and tighten nut to relieve tension on camshaft.

6. Remove capscrew, lock and flat washer and fuel pump eccentric from sprocket. Pull sprocket forward from pilot on camshaft. Release hook of tool. Slide crankshaft sprocket off crankshaft.

7. Remove pin from top of tensioner, and remove the tensioner blade and spring from the lower mounting stud.

8. Remove chain guide bracket from front engine plate.

To reinstall: Reverse the above. Proper camshaft timing is accomplished by turning No. 1 piston to top of stroke with crankshaft sprocket key up, at twelve o'clock position, and camshaft sprocket dowel down at six o'clock position. Both No. 1 intake and exhaust valves should be closed.

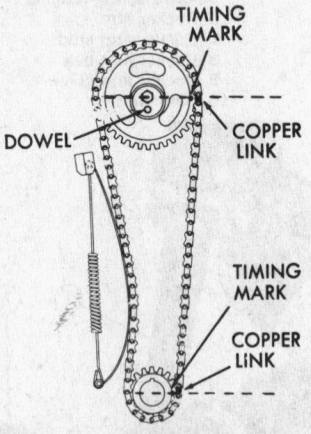

Aligning timing marks—230 OHC
(© Willys Corp.)

Front Crankshaft Oil Seal Removal

230 OHC

1. Remove vibration damper.

2. Thread housing of tool into crankshaft oil seal. Turn tool screw clockwise to force out seal.

To reinstall: be sure there are no burrs to damage new seal.

3. Coat outer edge of seal with a good sealing compound and place the seal on shaft with lip toward inside. Drive seal into place with tool.

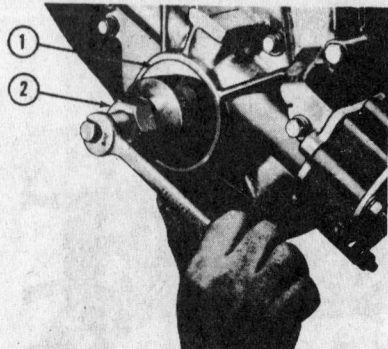

Removing front oil seal from 230 OHC
(© Willys Corp.)

1 Oil seal
2 W-270 puller

Connecting Rods and Pistons

Piston Assembly Removal

1. Remove cylinder head and oil pan.
2. Remove ridge from tops of cylinder bores, using a ridge reamer.
3. One at a time, remove connecting rod caps and push piston assemblies out of the top of the block. Number all pistons, connecting rods, and caps on removal.
4. Remove old rings with ring expander.
5. Release piston pin lock screws and force out pins.
6. Check cylinder bores for distortion, taper, and other evidence of excessive wear. Bore and hone as necessary.

Fitting Rings and Pins

When new rings are installed without reboring the cylinders, the cylinder wall glaze should be broken by honing.

New piston rings should be checked for end gap and clearance in cylinder bores and ring grooves. When fitting new rings, compression ring side clearance should be no more than .006 in. for four-cylinder engines. Any clearance greater than this indicates a need for a new piston. End gap of compression rings is checked by pushing the ring to the bottom of the bore in which it will operate by use of a piston. Compression ring end gap should be at least .007 in. for four-cylinder engines.

Piston pins should be a push-fit. If they are excessively loose, the piston must be replaced.

When assembling the rod and piston assemblies to the four-cylinder engine, the T-slot in the piston should be on the left side. The oil spurt hole should face to the right, away from the camshaft and the T-slot. On the 226 cu. in. six-cylinder engine, the oil spray hole should face the camshaft. On the 230 OHC engine, the "F" marking on the piston should be to the front and the oil spurt hole to the right.

Piston to cylinder wall clearance should be checked with a ribbon

gauge and a spring scale. The scale should register 5-10 lbs. to pull a .003 in. ribbon gauge between the piston and the cylinder wall on four-cylinder engines.

Connecting Rod Bearings

1. Remove connecting rod cap. Wipe all oil from the bearing area.
2. Place a piece of Plastigage lengthwise along the bottom center of the lower bearing shell. Install cap and bearing shell. Torque to the figure specified in the Torque Specifications Table.
3. Remove the cap and shell. The gauge material will be found flattened and adhering to either the bearing shell or the crankpin. Do not remove it.
4. Using the scale that is supplied with the Plastigage, measure the flattened material at its widest point. The number within the graduation which comes closest to the width of the gauging material indicates the bearing to crankpin clearance in thousandths of an inch.
5. The correct clearance for four-cylinder engines is .0010 to .0019 in.
6. When installing new bearings, try standard size, then each undersize bearing in turn until the correct clearance is obtained.
7. Clean off the Plastigage material, oil the bearing thoroughly, reinstall the bearing cap and shell, and torque to the specified figure.
8. After the cap has been torqued, it should be possible to move the connecting rod back and forth on the crankpin .004 to .010 in.

NOTE: when the front connecting rod bearing on a four-cylinder engine must be replaced because of scoring or burning, check the timing gear oil jet. If the jet has an aperture of .070 in., replace it with the later production jet, which has a .040 in. aperture. This jet entered production with engine number 3J-166871.

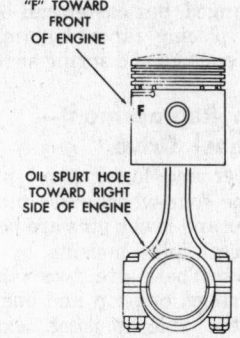

Piston to rod assembly—OHC 6 230

Piston Assembly Installation

1. Assemble piston to piston rod by pushing in and locking the pin.

2. Using a ring expander, install piston rings. Install bottom (oil) ring first, center ring second, and top ring last.
3. Coat all bearing surfaces, rings, and piston skirt with engine oil.
4. Turn down the crankpin of the cylinder being worked on.
5. Make sure that the gaps in the rings are not aligned with each other.
6. Using a ring compressor, install the piston and rod assembly into the cylinder and carefully tap down until the rod bearing is solidly seated on the crankpin.
7. Install cap and lower bearing shell. Torque to the specified figure.
8. Install all piston assemblies in the same manner. Be sure that the rods and pistons are installed in the correct manner as described above under Fitting Rings and Pins.
9. Install cylinder head and oil pan.

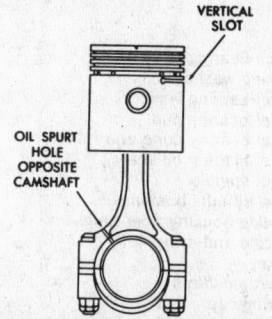

Piston to rod assembly—L4 and F4

Front Suspension

General instructions covering the front suspension, and how to repair and adjust it, together with information on installation of front wheel bearings and grease seals, are in the Unit Repair Section.

Definitions of the points of steering geometry are covered in the Unit Repair Section. This section also covers troubleshooting front end geometry and irregular tire wear.

Figures covering the caster, camber, toe-in, and kingpin inclination are in the Front Wheel Alignment table.

Tire and brake size figures are in the General Chassis and Brake Specifications table.

Live Front Axle

On four-wheel-drive models, caster is adjusted by placing shims between the spring pads on the front axle housing and the top of the spring.

This method should only be used where it is necessary to adjust both sides of the vehicle the same amount.

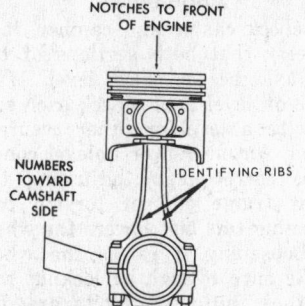

Piston to rod assembly—OHV 6 232

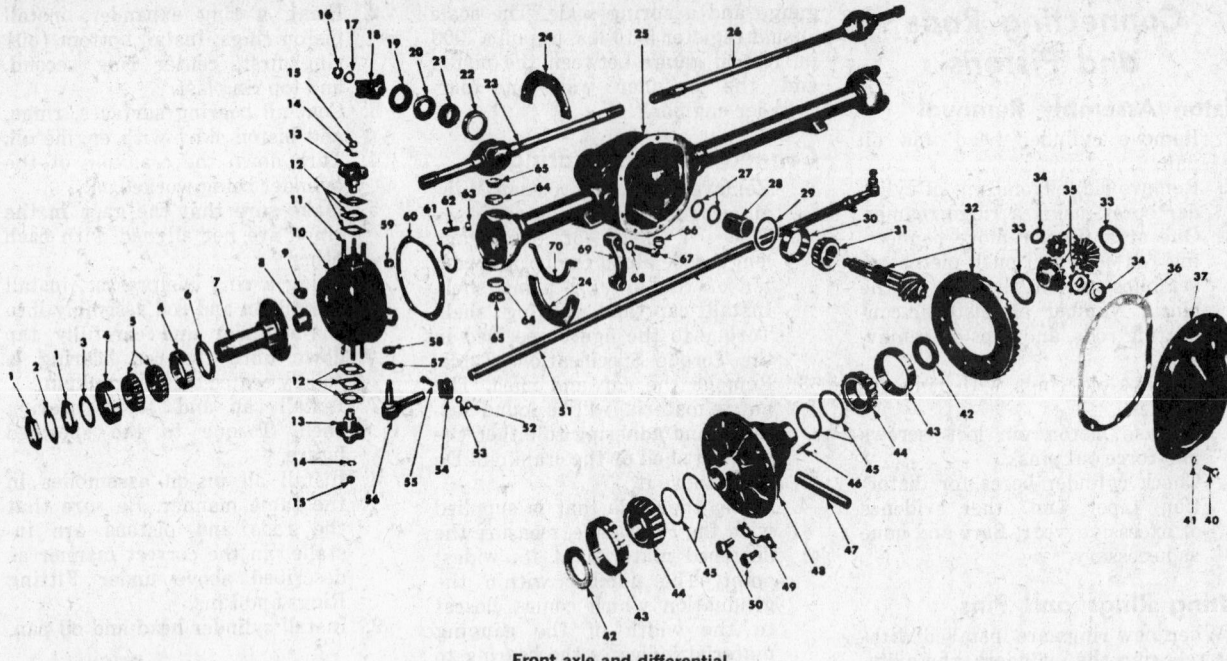

Front axle and differential
(© Willys Corp.)

1 Wheel bearing nut
2 Locking washer
3 Wheel bearing washer
4 Wheel bearing cup
5 Wheel bearing cone and rollers
6 Wheel bearing oil seal
7 Wheel spindle
8 Wheel spindle bushing
9 Knuckle housing filler plug
10 Knuckle and arm—left
11 No key
12 Adjusting shims
13 Bearing cap
14 Lockwasher
15 Bearing cap bolt
16 Pinion nut
17 Pinion washer
18 Universal joint yoke
19 Pinion shaft oil seal
20 Bearing oil slinger
21 Bearing cone and rollers
22 Bearing cup
23 Right universal joint and shaft
24 Knuckle oil seal retainer
25 Front axle housing
26 Left universal joint and shaft
27 Axle shaft oil seal
28 Axle shaft guide
29 Pinion bearing adjusting shims
30 Pinion shaft bearing cup
31 Pinion shaft bearing cone and rollers
32 Ring gear and pinion
33 Side gear thrust washer
34 Pinion mate thrust washer
35 Differential gears
36 Housing cover gasket
37 Housing cover
40, 41 Housing cover screw and lockwasher
42 Retaining ring
43 Differential bearing cup
44 Differential bearing cone and rollers
45 Differential bearing adjusting shims
46 Pinion shaft lock pin
47 Pinion shaft
48 Differential case
49 Ring gear lock strap
50 Ring gear bolts
51 Steering tie rod
52 Tie rod clamp nut
53 Lockwasher
54 Tie rod socket clamp
55 Tie rod clamp screw
56 Tie rod socket
57 Dust cover
58 Spring

59 Tie rod stud nut
60 Oil seal and backing ring
61 Universal joint thrust washer
62 Knuckle stop bolt
63 Stop bolt nut
64 King pin bearing cup
65 King pin bearing cone and rollers
66 Housing drain plug
67 Cap bolt
68 Washer
69 Differential bearing cap
70 Steering knuckle oil seal

In the event the caster angle is incorrect because of sag of the front leaf springs, it is advisable to replace the springs rather than use caster wedges. Camber can not be adjusted.

Solid Front Axle

Universal models with two wheel drive have a solid front axle. The axle may be either of tubular construction or a forged I-beam. Springs may be slung either under or over the axle. Standard caster and camber are built into the front axle. Camber cannot be changed, but caster can be adjusted by placing tapered shims between the springs and spring seats.

King-Pin Replacement— Four Wheel Drive

An upper and lower pivot pin is used in the four-wheel-drive models. Both upper and lower pins are held to the universal joint housing by four cap screws. There are two tapered roller bearings, one top and one bottom, in the axle housing, exactly centered across the neutral point of the constant velocity universal joint. If it is desired to replace the tapered roller bearings, it will be necessary to remove the universal joint outer housing. This involves the wheel and

the stationary axle. If the pivot pins, only, are to be replaced, they can be removed by taking out the four cap-screws that hold them to the universal joint housing. Shims are used under the head of the pivot pins to insure zero up- and down-play in the tapered roller bearings. To remove the pivot pins, take off the wheel, brake drum, and brake backing plate.

Independent Suspension

Wagoneer and Gladiator

Some two-wheel drive vehicles are equipped with an independent front suspension system with torsion bars, an upper support assembly and a tubular front axle. The axle is in two sections, hinged at the center. Knuckle supports are mounted between the axle ends and the upper supports by means of lower and upper ball joints.

The independent front suspension is also available on four-wheel drive vehicles. The general difference is that the lower support axle is hinged at the differential.

Independent Suspension Alignment

To check caster and camber, it is necessary that both sections of the front axle be on exact level. With vehicle on level floor, check each section with a level or other accurate method. An uneven or unlevel condition is corrected by adjusting the bolt at frame end of torsion bar. Tightening this bolt lowers the wheel end. Loosening it raises the wheel end. Be sure to tighten locknut any time these adjusting bolts are disturbed.

To adjust caster, loosen jam nuts at each end of strut tube and turn tube until specified caster is obtained, then tighten locknuts and recheck caster.

To adjust camber, loosen the control-arm-to-ball-joint bolts that are mounted in slotted holes. Move ball joints until specified camber is obtained and retighten the bolts.

Ball Joint Replacement— Wagoneer and Gladiator

Ball joints are bolted to the support arms and mounted in the knuckle support by the conventional taper with castellated nut.

remove from vehicle. The split grommet or seal at front will come out with bar.

Installation

1. With front of car raised and supported under frame with wheels and axles hanging free, insert the hex end of bar in the hex tube of upper support.
2. Keeping the bar parallel to frame, slide adjusting arm on bar, holding lower end of arm against frame.
3. Mark the relative position of the arm to the bar on the serrations for further reference. Then remove the arm.

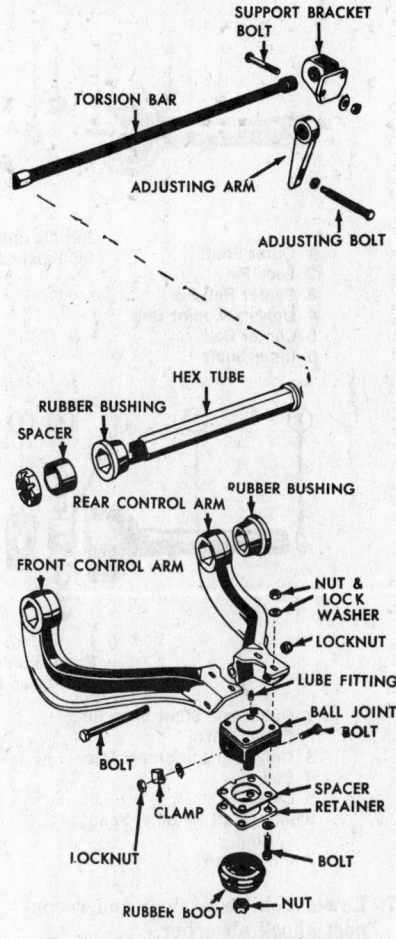

Torsion bar—Wagoneer and Gladiator
(© Willys Corp.)

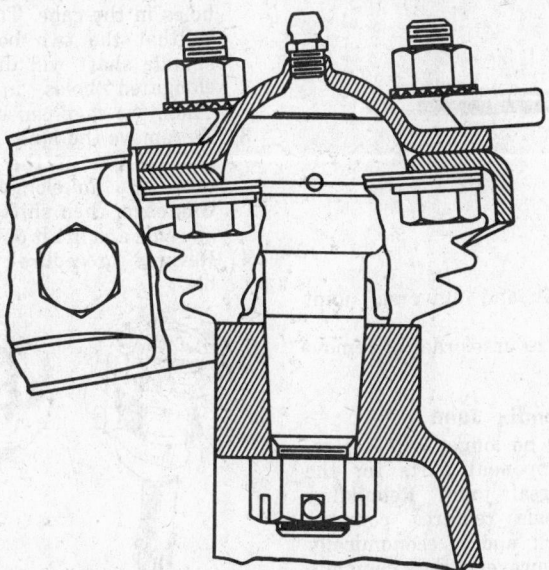

Upper ball joint—Wagoneer and Gladiator (© Willys Corp.)

Torsion Bar Replacement— Wagoneer and Gladiator

The torsion bars are marked on the front (hex) end with either L or R to indicate the side. Never alter this location.

The adjusting arm at the rear of the bar has two adjusting bolt indentations. When installing the arm on the bar, place it on the bar with the upper indentation next to frame on J-100 vehicles. On J-200 and J-300 vehicles, install the arm on the bar with the lower indentation next to frame.

Removal

1. Disconnect front shock absorber at lower end.
2. Keeping the bar parallel to frame, wheel and axle hang free.
3. Back off adjusting screw at rear of bar until inner end of screw is flush with frame at inside of side rail.
4. Remove the two bolts attaching the rear bar anchor support to the frame side rail.
5. Slide torsion bar, with adjusting arm and bracket, to the rear to

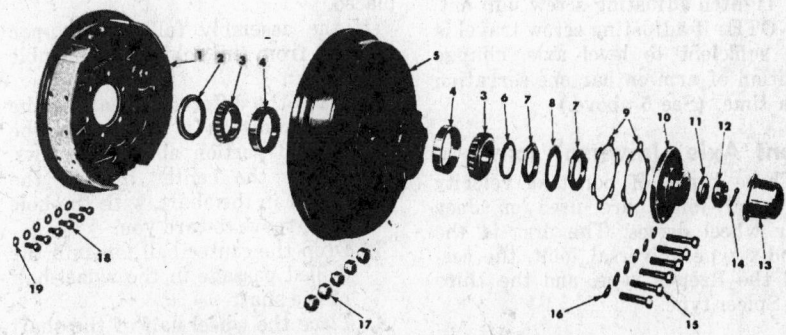

Front wheel drive attaching parts (© Willys Corp.)

1 Left front brake assembly
2 Oil seal
3 Bearing cone and roller
4 Bearing cup
5 Hub and brake drum
6 Bearing locking washer
7 Lock nut
8 Bearing nut locking washer
9 Adjusting shim
10 Driving flange
11 Axle shaft nut washer
12 Axle shaft nut
13 Hub cap
14 Cotter pin
15 Driving flange bolt
16 Lockwasher
17 Wheel nut
18 Backing plate bolt
19 Lockwasher

4. Install the sealing grommet between bar and hex tube by sliding the bar back about 1 in. Place the grommet on bar and slide into position in tube. Be sure a good seal is obtained.
5. Use a wooden wedge and pry the rear of bar away from rail so that adjusting arm and support bracket can be installed with lower end against frame and two serrations separated from the mark made in Item 3, above.
6. Remove wedge, then bolt bracket to frame. Torque nuts to 45-50 ft. lb.

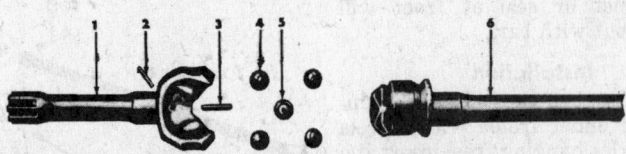

Bendix universal joint
(© Kaiser Jeep Corp.)

1 Outer Shaft
2 Lock Pin
3 Center Ball Pin
4 Universal Joint Ball
5 Center Ball
6 Inner Shaft

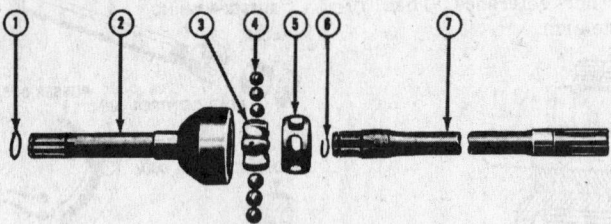

Rzeppa universal joint
(© Kaiser Jeep Corp.)

1 Outer Axle Shaft Snap Ring
2 Outer Shaft
3 Universal Joint Inner Race
4 Ball
5 Cage
6 Axle Shaft Retainer Snap Ring
7 Inner Shaft

7. Lower vehicle to floor and reconnect shock absorber.
8. Level axle with adjusting screw, as described in the paragraphs covering front alignment, previously covered.
9. Tighten adjusting screw jam nut
   NOTE: if adjusting screw travel is not sufficient to level axle, change position of arm on bar one serration at a time. (See 5 above.)

Front Axle Universal Joints

Three types of constant velocity universal joints are used on Jeep four wheel drives. The first is the Bendix type universal joint, the second the Rzeppa type, and the third the Spicer type.

Removal

1. Remove the wheel.
2. Remove hub dust cap.
3. Remove axle shaft driving flange bolts.
4. Apply and hold foot brakes. Remove axle shaft flange with a puller.
5. Release locking lip on lockwasher, remove outer nut, lockwasher, adjusting nut, and bearing lockwasher. Use a special wrench for these nuts.
6. Remove wheel hub and drum assembly with the bearings.
7. Remove hydraulic brake tube, backing plate screws, spindle,

axle shaft and universal joint assembly.
8. Install in reverse order of removal.

Bendix Joint

The factory no longer supplies replacement component parts for the Bendix universal joint. Rebuilding these universals requires complex shop equipment and is economically unfeasible. However, the complete universal joint assembly may be replaced.

If the assembly falls apart upon removal from the vehicle, reassemble as follows:
1. Place the differential half of the axle shaft in a vise, with the ground portion above the jaws.
2. Install the center ball in the socket in the shaft, with the hole and groove toward you.
3. Drop the center ball pin into the drilled passage in the wheel half of the shaft.
4. Place the wheel half of the shaft on the center ball. Then slip three balls into the raceways.
5. Turn the center ball until the groove lines up with the raceway for the remaining ball. Slip the ball into the raceway and straighten up the wheel end of the shaft.
6. Turn the center ball until the center ball pin drops into the hole in the ball.
7. Install the retainer pin and prick punch both ends to lock it in place.
8. After reassembly, grasp both ends of the shaft and twist the ends back and forth. Should excessive wear be indicated by

back lash or lost motion, the assembly should be replaced.

Rzeppa Joint

1. Dismantle the Rzeppa joint, remove the three screws (some axles have no such screws) that hold the front axle shaft to the joint itself, and pull the shaft out of the splined inner race. To take out the axle shaft retainer, remove the retainer ring on the shaft. Push down on the various points of the inner race and cage until the balls can be removed with the help of a small screwdriver.
2. There are two large elongated holes in the cage. Turn the cage so that the two bosses in the spindle shaft will drop into the elongated holes in the cage. Then, the cage can be lifted out.
3. To remove the inner race, turn it so that one of the bosses, will drop into an elongated hole in this cage, then shift the race to one side and lift it out.
4. Reverse procedure to reassemble.

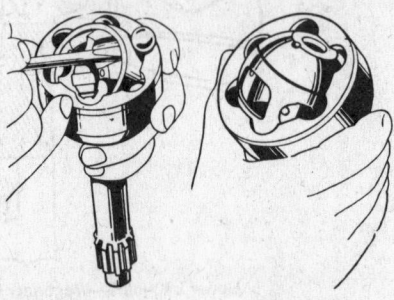

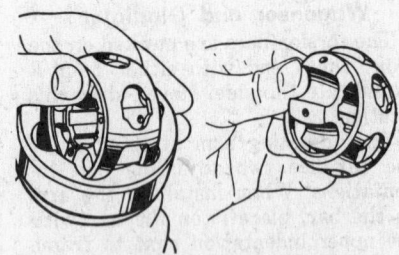

Disassembly of the Rzeppa joint
(© Willys Corp.)

Spicer Joint

The Spicer universal joint is a cross and roller unit with needle bearings. It is quite similar in design to the Spicer driveshaft universal joint. Disassembly is as follows:
1. Remove snap-rings.
2. Press on end of one bearing until opposite bearing is pushed from yoke arm.
3. Turn joint over. Press first bearing back out of arm by pressing on exposed end of journal shaft. Repeat this operation for the other two bearings, then lift out journal assembly by sliding to one side.

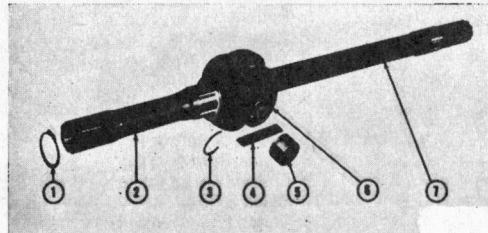

Spicer universal joint
(© Kaiser Jeep Corp.)

1 Outer Axle Shaft Snap Ring
2 Outer Shaft
3 Bearing Retainer Snap Ring
4 Bearing Rollers
5 Bearing Retainer
6 Universal Joint Journal Assembly
7 Inner Shaft

4. Wash all parts in solvent and inspect for wear. Replace all worn parts.
5. Install new gaskets on the journal assembly. Make certain that the grease channel in each journal trunnion is open.
6. Pack bearing cones one-third full of grease and install the rollers.
7. Assemble in reverse order of disassembly. If the joint binds when assembled, tap the arms lightly to relieve any pressure on the bearings at the end of the journal.

Steering Gear

Instructions covering the overhaul of the steering gear are in the Unit Repair Section.

Steering Gear Assembly Removal

Universal Series
1. Remove directional signal unit from steering column.
2. Remove steering column bracket from instrument panel.
3. Remove upper section of the floor pan.
4. Disconnect shift rods from lower end of steering column.
5. Disconnect horn wire at lower end of steering assembly.
6. Remove steering gear arm.
7. Unbolt steering gear housing from frame. Remove assembly by bringing it up through the floor pan opening.
8. Install in reverse of above.

Jeepster, Gladiator, and Wagoneer
1. Disconnect the steering gear from the steering column by removing the flexible coupling clamp screw.
2. Disconnect the drag link from the steering gear arm.

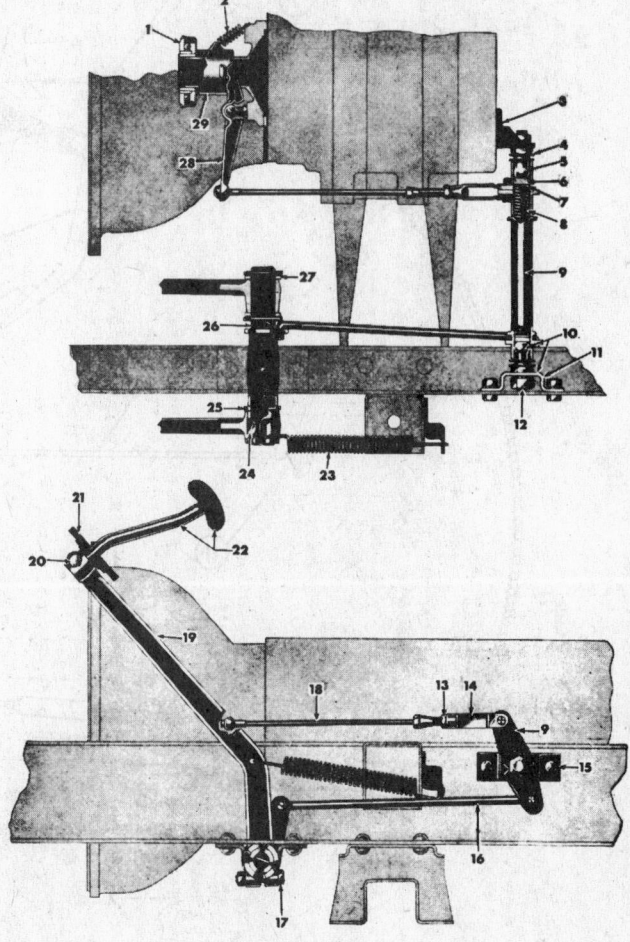

Clutch linkage—Universal series
(© Kaiser Jeep Corp.)

1 Clutch Release Bearing
2 Carrier Spring
3 Bracket
4 Dust Seal
5 Ball Stud
6 Pad
7 Retainer
8 Control Tube Spring
9 Control Lever and Tube
10 Ball Stud and Bracket
11 Frame Bracket
12 Ball Stud Nut
13 Yoke Lock Nut
14 Adjusting Yoke
15 Bolt
16 Pedal Release Rod
17 Pedal Clamp Bolt
18 Control Cable
19 Clutch Pedal
20 Screw and Lockwasher
21 Draft Pad
22 Pedal Pad and Shank
23 Retracting Spring
24 Pedal to Shaft Key
25 Washer
26 Pedal Shaft
27 Master Cylinder Tie Bar
28 Control Lever
29 Bearing Carrier

3. Remove housing bolts from frame. Remove steering gear from engine compartment.
4. Reverse procedure to install.

Power Steering
1. Disconnect hoses from pressure and return ports. Raise hoses above pump to prevent oil loss.
2. Remove flexible coupling clamp screw.
3. Remove steering gear arm.
4. Remove steering gear mounting bolts. Remove steering gear from vehicle.
5. Reverse procedure to install.

Clutch

A single dry disc clutch is used on all Jeep models. Except for adjust-ing the pedal clearance, no service is possible on the clutch assembly unless it is removed from the car.

Clutch Pedal Adjustment

Universal Series, Jeepster
Adjust the linkage at the turnbuckle on the cable from the clutch fork control lever so that the clutch pedal can be depressed 1 in. before clutch disengagement starts.

Gladiator, Wagoneer
1. Disconnect adjustable rod from clutch pedal.
2. Adjust clutch pedal stop bolt for a positive over-center action. This is done by turning the stop bolt in or out to obtain clutch pedal travel of ¼-½ in. before the over-center spring assists

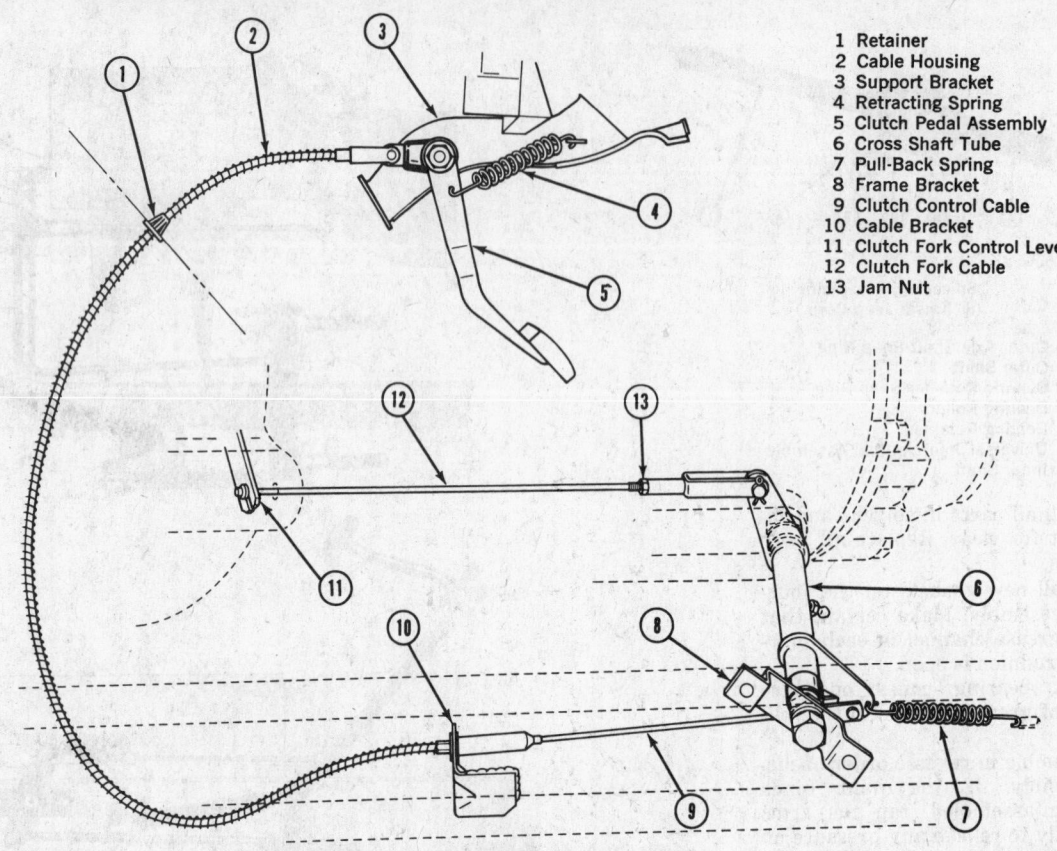

1 Retainer
2 Cable Housing
3 Support Bracket
4 Retracting Spring
5 Clutch Pedal Assembly
6 Cross Shaft Tube
7 Pull-Back Spring
8 Frame Bracket
9 Clutch Control Cable
10 Cable Bracket
11 Clutch Fork Control Lever
12 Clutch Fork Cable
13 Jam Nut

Clutch linkage—Jeepster (© Kaiser Jeep Corp.)

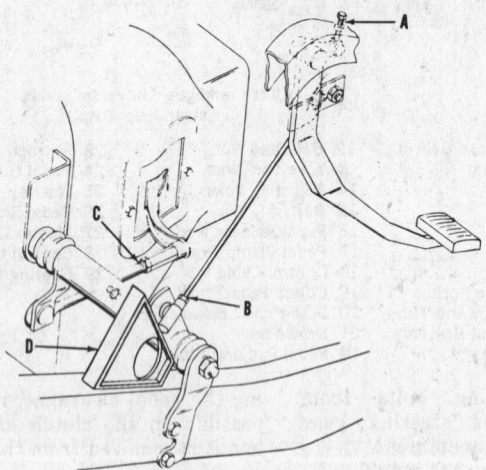

Clutch linkage—Wagoneer and Gladiator
(© Kaiser Jeep Corp.)

A Pedal Height Adjustment Point
B Cross Shaft Arm Position Adjustment
C Clutch Pedal Free Play Adjustment
D Tool W-341 or W-317

the pedal to the floorboard. The pedal need not be the same height as the brake pedal.

3. Adjust rod from brake pedal to cross-shaft so that proper angle is obtained between arm of cross-shaft and top of left frame rail. For the 350 V8, the correct angle is thirty degrees. For the 327 V8 and the 232 OHV 6, the correct angle is forty-nine de-

grees. This angle may be measured with a protractor.

4. Adjust the cross-shaft to throw-out lever link so that the clutch pedal can be depressed 1 in. before clutch engagement starts.

Clutch Assembly Removal

All Models Except 350 V8 and 225 V6

1. Remove transmission.
2. Disconnect clutch linkage.
3. Mark the clutch pressure plate and engine flywheel so the clutch

will be reassembled in the same position.

4. Remove the clutch pressure plate bracket bolts alternately, a little at a time, to prevent distortion.
5. Remove the pressure plate assembly and driven plate from the flywheel.

350 V8 and 225 V6

For clutch removal and installation on models equipped with these engines, refer to the Buick Special Section.

Clutch Assembly Installation

The clutch release bearing is lubricated during assembly and need not be lubricated at any other time.

1. Put a very small amount of light grease in the flywheel pilot bushing. Install the driven plate with short end of hub toward flywheel. Place the pressure plate assembly in position.
2. Using a clutch plate aligning arbor or a spare transmission main shaft, align the driven plate splines. Tighten the pressure plate screws evenly. Remove aligning arbor.
3. Assemble flywheel housing to engine. Make sure that the clutch release bearing carrier return spring is hooked in place. Reverse order of clutch assembly removal procedure to complete reinstallation.

Standard Transmission

Transmission refill capacities are in the Capacities table of this section.

Troubleshooting and repair of transmission, transfer case, and overdrive units are covered in the Unit Repair Section.

Transmission and Transfer Case Removal

The transmission and transfer case can be removed as a unit. These instructions apply to both three and four-speed transmissions, and also generally to two-wheel drive models which have no transfer case.

Universal Series

1. Drain transmission and transfer case.
2. Remove floor pan inspection plate.
3. Remove shift lever and shift housing, or disconnect remote control rods, depending on the model.
4. If vehicle has power take off, remove shift lever.
5. Disconnect front and rear driveshafts from transfer case. Disconnect power take off drive shaft.
6. Disconnect speedometer cable.
7. Disconnect hand brake cable.
8. Disconnect clutch release cable or rod.
9. Place jacks under engine and transmission, protecting the oil pan with a wooden block.
10. Remove rear crossmember.
11. Unbolt transmission from flywheel housing.
12. Force transmission to right to disengage clutch control lever tube ball joint.
13. Lower jacks. Slide transmission and transfer case rearward until clutch shaft clears the flywheel housing.
14. Lower transmission jack. Remove assembly from under vehicle.
15. To install reverse procedure.

Jeepster

The procedure for these models is the same as that outlined above for the Universal series, with the substitution of the following steps:

2. Remove right front seat, floor mat, and floorboard center section. Disconnect backup switch wires.
10. Remove rear crossmember. Remove transmission and transfer case stabilizer brackets.

Wagoneer and Gladiator

The procedure for these models is the same as that outlined above for the Universal series, with the substitution of the following step:

2. Remove the transmission access cover from the floor.

Shift Linkage Adjustment

Universal Series—Column Shift

1. Disconnect shift rods at transmission.
2. Put transmission levers in neutral positions.
3. Lock gearshift levers in neutral positions by putting a 1/4 in. dia. rod through the gearshift levers and housing at the bottom of the steering column.
4. Adjust length of shift rods and reconnect.
5. If shifting from first to second is difficult or transmission hangs in first gear, shorten the first-reverse rod one turn at a time until the condition is corrected.

Jeepster—Console Shift

1. Remove plug from hole in left side of console. If there is no hole, cut a 1 1/8 in. dia. hole or remove the console.
2. Lift the shift tower rubber cover and remove the plug in the shift tower. Move the selector lever to the neutral position.
3. Loosen the adjusting nuts at the transmission.
4. Insert a 5/16 in. diam. rod through the holes in the console and shift tower, and through the aligning holes in the two shift levers. Check that transmission shift levers are in neutral positions.
5. Torque adjusting nuts to 15-20 ft. lbs.
6. Remove adjusting rod, replace plugs, check shifting action. If selector lever interferes with console, relocate console.

Wagoneer and Gladiator— Column Shift

1. Put selector lever in neutral position.
2. Loosen shift rod adjusting nuts at transmission. Place shift levers in neutral positions.
3. Insert a 3/16 in. dia. rod through remote control shift levers and housing at bottom of steering column.
4. Torque adjusting nuts to 15-20 ft. lbs. Remove adjusting rod.

Transfer Case Removal

The transfer case can also be removed as a separate unit.

Jeepster

1. Drain transmission and transfer case. The automatic transmission need not be drained.
2. Remove transfer case shift lever.

3. Disconnect driveshafts.
4. Disconnect speedometer cable.
5. Disconnect brake cable (directly under transfer case).
6. Disconnect clutch cables from cross-shaft.
7. Disconnect clutch control cross-shaft ball joint at frame.
8. Remove transfer case stabilizer bracket.
9. Remove bolts securing transfer case to transmission. Slide transfer case rearward to remove.
10. Reinstall in reverse order of above steps.

Gladiator and Wagoneer

This procedure is the same as that detailed above for removal of the Jeepster transfer case with the substitution of the following steps:

5. Disconnect parking brake spring from fuel tank flange. Remove clevis pin from brake cable connecting bracket.
6. Remove exhaust pipe bracket bolts.
7. Delete Step 7.
8. Delete Step 8.

Universal Series

1. Drain transmission and transfer case.
2. Disconnect brake cable.
3. Disconnect driveshafts.
4. Disconnect speedometer cable.
5. Disconnect transfer case shift levers.
6. Remove cover plate on rear of transfer case. Remove cotter key, nut, and washer from transmission main shaft.
7. If possible, remove transfer case main drive gear from transmission mainshaft.
8. Remove transfer case mounting bracket bolts and nuts.
9. Unbolt transmission from transfer case.
10. Remove transfer case. If main drive gear has not been removed in Step 7, proceed as follows: Brace the end of the transmission mainshaft so that it cannot move in the transmission, pull the transfer case to the rear to remove the rear. Be careful that the transmission mainshaft bearing, which bears in both housings, remains in the transmission case.
11. Installation is the reverse of removal procedure.

Automatic Transmission

When automatic transmission trouble is reported, a road test and careful diagnosis are in order. Transmission Removal and Replacement and Link-

age Adjustments are covered in the following paragraphs. The automatic transmission currently used in Jeep vehicles is the General Motors Turbo Hydra-Matic. For test procedures, transmission overhaul and other detailed information, see the Unit Repair Section.

Linkage Adiustment Cable Operated Linkage

1. Disconnect control cable at transmission lever and place transmission shift lever in neutral.
2. Place control lever in neutral position.
3. Loosen two nuts at upper end of control cable housing and work housing up or down until cable exactly matches position of lever on transmission.
   Lock nuts on housing and connect cable to lever.

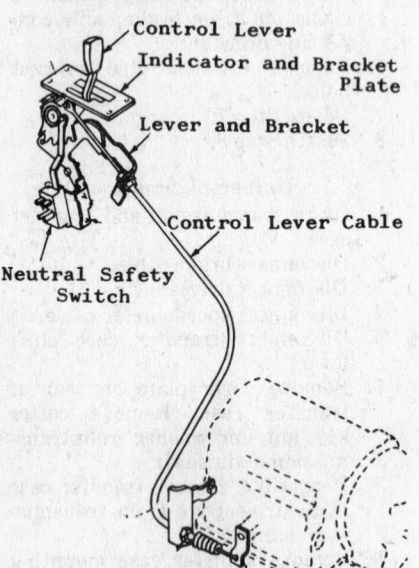

Cable operated linkage
(© Willys Corp.)

Rod Operated Linkage— Column or Console Shift

1. Remove adjusting block from transmission shift lever.
2. Make sure that transmission lever is in neutral position.
3. Place selector lever in Neutral position and hold firmly against the stop.
4. Loosen locknuts on adjusting block. Replace block on transmission lever. Tighten locknuts to 6-12 ft. lbs.
5. Check shifting operation. If console interferes with shift selector lever, reposition console.

Dashpot Adjustment

A dashpot is provided on automatic transmission Jeep vehicles to prevent the throttle from closing too rapidly and causing the engine to stall. The dashpot is located on the throttle linkage at the carburetor. Adjustments are made with the engine at normal idling speed. There should be a clearance of approximately 3/32 in. between the dashpot stem and the throttle lever. If stalling persists, this clearance may be lessened as necessary.

Neutral Safety Switch —All Models

1. Check that the shift linkage is correctly adjusted.
2. Set the handbrake and the footbrake. Put the handlever on the steering column in drive position. Hold the ignition key (or starter button) on and slowly move the handlever toward neutral until the starter cranks and the engine runs.
3. Without moving the lever further, press the accelerator to determine whether or not the transmission is really in neutral.
4. If all is correct, the engine will have started when the handlever got to the neutral position, and the transmission will not be in gear.
5. Adjust the neutral safety switch by turning it and its mounting bracket until the above conditions have been met.

Transmission Removal

1. Disconnect battery ground strap. Release parking brake. Raise and support vehicle.
2. Drain and remove transfer case. Tie driveshafts out of the way.
3. Disconnect:
   a. Electrical lead case connector.
   b. Vacuum line at modulator.
   c. Oil cooler lines.
   d. Shift linkage.
4. Remove exhaust pipe assembly from muffler and exhaust manifolds.
5. Remove hand brake cable plate from crossmember.
6. Support transmission with a jack.
7. Remove crossmember.
8. Remove converter dust shield from transmission case.
9. Remove bolts attaching torque converter to flex plate.
10. Lower transmission until jack is barely supporting it.
11. Remove transmission mounting bolts.
12. Raise transmission to its normal position, slide rearward, and lower away from vehicle. Keep rear of transmission lower than front to avoid dropping the converter.
13. Installation is the reverse of removal.

The drive of four wheel drive Universal and Jeepster models from the transfer case to the front and rear axles is through two tubular driveshafts. Each driveshaft has two cross and roller universal joints. Two wheel drive models drive through a single driveshaft with two universal joints. Each driveshaft has a splined slip joint at one end to allow for variations in length.

Jeepster V6 models have a front driveshaft made up of two shafts with three universal joints and one slip joint. The shorter of the two shafts, directly ahead of the transfer case, has a shaft support bearing.

Gladiator and Wagoneer models use several types and sizes of driveshafts, depending upon various transmission and equipment options. These are divided into two basic types. The first is that with two cross and roller joints and a slip joint. The second has a cross and roller joint at one end, and a ball and trunnion universal joint at the other end. Automatic transmission Gladiator and Wagoneer models have the slip joint end of the front driveshaft at the axle rather than at the transfer case, as on all other models.

Cross and Roller Universal Joint

Snap Ring Type Disassembly and Repair

1. Remove snap-rings.
2. Press on end of one bearing until opposite bearing is pushed from yoke arm.
3. Turn joint over. Press first bearing back out of arm by pressing on exposed end of journal shaft. Repeat this operation for the other two bearings, then lift out journal assembly by sliding to one side.
4. Wash all parts in solvent and inspect for wear. Replace all worn parts.
5. Install new gaskets on the journal assembly. Make certain that the grease channel in each journal trunnion is open.
6. Pack bearing cones one-third full of grease and install the rollers.
7. Assemble in reverse order of disassembly. If the joint binds when assembled, tap the arms lightly to relieve any pressure on the bearings at the end of the journal.

U-Bolt Type Disassembly and Repair

Remove the attaching U-bolts to

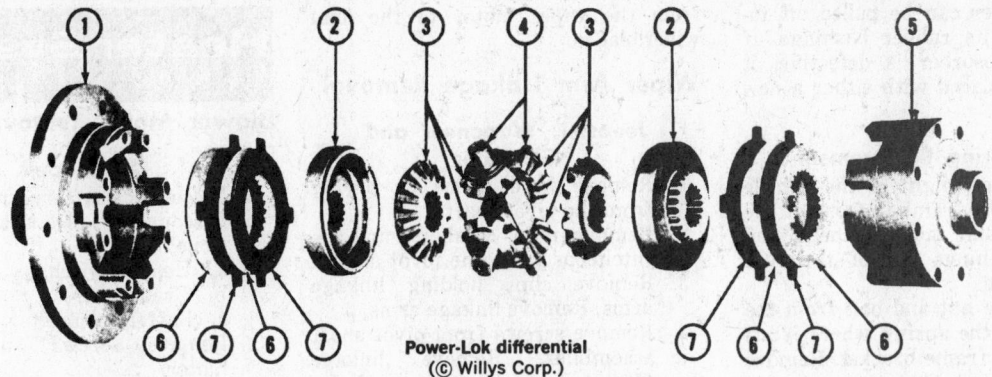

Power-Lok differential
(© Willys Corp.)

1 Differential case flange half
2 Side gear differential ring
3 Side gear and pinion mate gear set
4 Cross pinion mate shaft
5 Differential case button half
6 Differential clutch friction disc
7 Differential clutch friction plate

release one set of bearing races. Slide the driveshaft into the yoke flange to remove the races. The rest of the disassembly and repair procedure is the same as that given above for the snap ring type of cross and roller joint. The correct U-bolt torque is 15-20 ft. lbs.

Ball and Trunnion Universal Joint

Disassembly and Repair

1. Clamp shaft firmly in vise.
2. Bend grease cover lugs away from universal joint body. Remove cover and gasket.
3. Remove two clamps from dust cover. Push joint body toward driveshaft tube. Remove two each, centering buttons, spring washers, ball and roller bearings, and thrust washers, from trunnion pin.
4. Press trunnion pin from ballhead.
5. If ballhead is bent out of alignment or if trunnion pin bore is worn or damaged, replace driveshaft.

To reassemble:
6. Secure larger end of dust cover to joint body with larger of two clamps. Install smaller clamp. Fit cover over ballhead shaft.
7. Push universal joint cover toward driveshaft tube. Press trunnion pin into centered position. If trunnion pin is not centered, imbalance will result.
8. Install thrust washers, ball and roller bearings, spring washer, and centering buttons on trunnion pin. Compress centering buttons. Move joint body to hold buttons in place.
9. Insert breather between dust cover and ballhead shaft, along length of shaft. Breather must extend no more than ½ in. beyond dust cover. Tighten clamp screw to secure cover to shaft. Cut away any portion of dust cover protruding under clamps.

10. Pack raceways around ball and roller bearings with about two ounces of universal joint grease.
11. Position gasket and grease cover on body. Bend lugs of cover into notches of body. Move body back and forth to distribute grease in raceways.

Rear Axles, Suspension

General instructions covering the drive axle and how to repair and adjust it, together with information on installation of drive axle bearings and grease seals, are given in the Unit Repair Section.

Capacities of the drive axle are given in the Capacities table.

The drive axle assembly on all Jeep cars is of the shim adjusted type. The front driving axle on four-wheel drive models is serviced in exactly the same way as the rear driving axle and many of the parts are interchangeable.

Rear Axle Removal

Support the back end of the vehicle on stands on the frame in front of the rear springs. Disconnect the driveshaft at the rear universal joint companion flange. Disconnect the brake hoses and cables and shock absorbers.

Take off the U-Bolt nuts that hold the rear spring to the rear axle housing tubes. Lower the springs to the floor and pull out the rear axle assembly from under the vehicle.

Rear Springs, Shocks

Rear Suspension

Rear Shock Absorber Removal

The bottoms of rear shock absorbers are mounted to a single bolt on the U-bolt plate of the rear spring.

The tops of shock absorbers are mounted to a bracket on the frame at the top.

Remove the locknuts and washers from the top and bottom and the

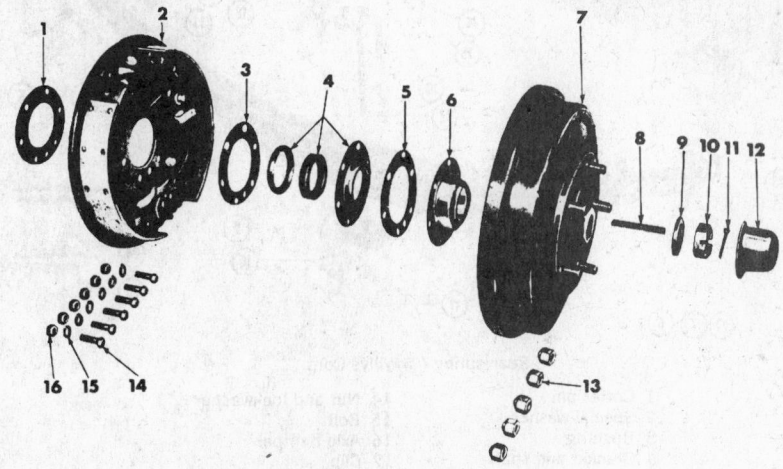

Rear wheel attaching parts (© Willys Corp.)

1 Retaining flange	9 Axle shaft washer
2 Brake assembly	10 Axle shaft nut
3 Adjusting shim	11 Cotter pin
4 Grease retainer	12 Hub cap
5 Grease shield gasket	13 Wheel nut
6 Grease shield	14 Axle flange bolt
7 Hub and drum	15 Lockwasher
8 Axle shaft key	16 Nut

shock absorber can be pulled off together with its rubber bushings. If the shock absorber is defective, it should be replaced with either a new or rebuilt one.

Rear Spring Replacement

Support the weight of the vehicle on jacks on the frame in front of the rear spring and take the bolts and threaded bushings out of the rear spring shackle.

Remove the nut and bolt from the front eye of the spring where it attaches to the frame bracket. Remove the U-bolts that hold the spring to the axle and lower the spring from under the vehicle.

Windshield Wipers

On the Universal series, the windshield wiper vacuum motors, arms, and blades are attached directly to the windshield and are completely accessible for service.

Jeepster, Gladiator and Wagoneer windshield wipers are powered by a single electric motor mounted to the underside of the cowl. A system of mechanical linkage transmits power from the wiper motor to the dual wiper blades.

Wiper Arm Linkage Removal

Jeepster, Wagoneer and Gladiator

1. Remove wiper arms by prying from the pivot shafts.
2. Remove nuts, washers, and escutcheons from the pivot shafts.
3. Remove clips holding linkage arms. Remove linkage arms.
4. Remove screws from pivot shaft assemblies. Remove linkage from under instrument panel.
5. Reverse procedure to install.

Wiper Motor Removal

Jeepster, Wagoneer and Gladiator

1. Disconnect battery ground cable.
2. Disconnect wiring at wiper switch.
3. Disconnect mechanical linkage from motor.
4. Disconnect washer pump hoses on models with integral wiper motor and washer pump assembly.
5. Remove bolts securing wiper motor. Remove wiper motor assembly.
6. Reverse procedure to install.

Heater System

Blower Motor Removal

All Models

1. Disconnect battery ground cable.
2. Disconnect electrical connections:
   a. Heater switch.
   b. Ground wire.
   c. Battery connector.
3. Remove screws and remove motor.
4. Reverse procedure to install.

Heater Core Removal

Jeepster

1. Drain coolant.
2. Remove battery and battery box.
3. Disconnect two coolant hoses and heater core drain hose.
4. Remove four locknuts from heater assembly studs. Remove two Bowden wires from studs.
5. Remove glove compartment.
6. Remove defroster hoses.
7. Disconnect heater wire harness. Remove harness from clips.
8. Remove heater assembly.
9. Mark duct halves to insure proper reassembly. Remove screws fastening duct halves together. Remove screws securing heater core to duct. Remove heater core.
10. Reverse procedure to install.

Wagoneer and Gladiator

1. Drain coolant.
2. Disconnect:
   a. Temperature control cable.
   b. Heater hoses.
   c. Heater resistor wires.
3. Remove four nuts securing core and duct to firewall. Two of the nuts are inside the vehicle.
4. Remove core and duct.
5. Mark duct halves to insure proper reassembly. Remove screws fastening duct halves together. Remove screws securing heater core to duct. Remove heater core.
6. Reverse procedure to install.

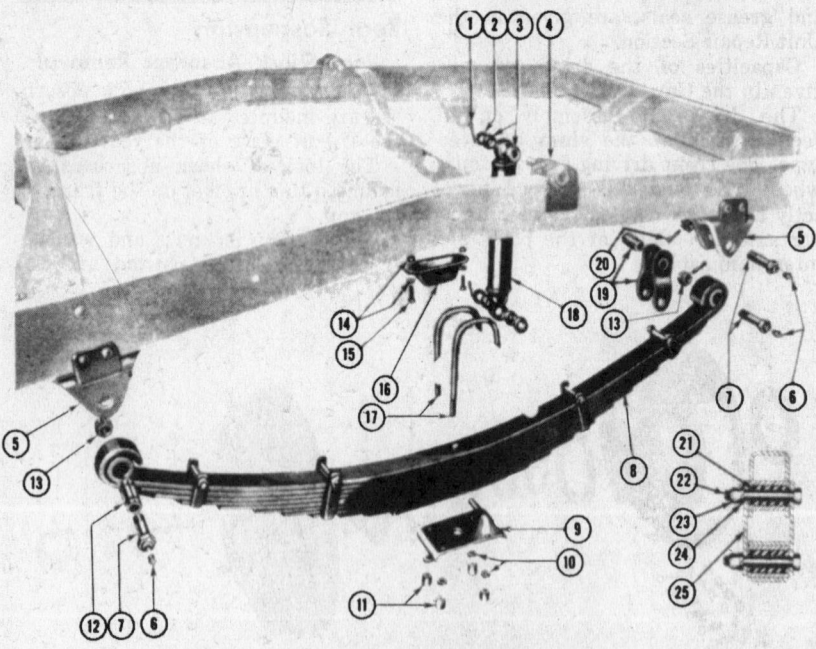

Rear spring (© Willys Corp.)

1 Cotter pin	14 Nut and lockwasher
2 Special washer	15 Bolt
3 Bushing	16 Axle bumper
4 Bracket and shaft	17 Clip
5 Rear spring hanger	18 Rear shock absorber
6 Lubrication fitting	19 Rear spring shackle
7 Bolt	20 Cotter pin
8 Rear spring	21 Silent block bushing
9 Plate and shaft	22 Bolt
10 Lockwasher	23 Nut
11 Nut	24 Lockwasher
12 Rear spring eye bushing	25 Spring shackle side plate
13 Nut	

Lincoln Continental and Mark III

Index

597

YEAR IDENTIFICATION

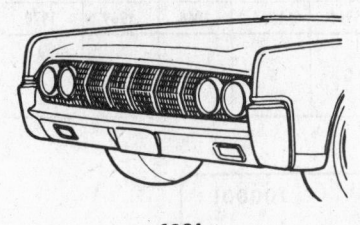

1964

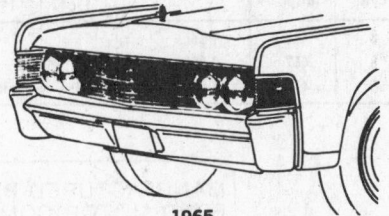

1965

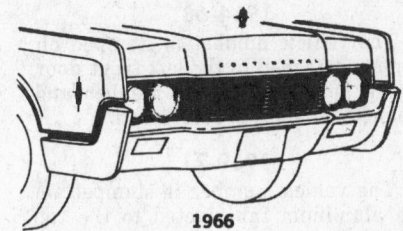

1966

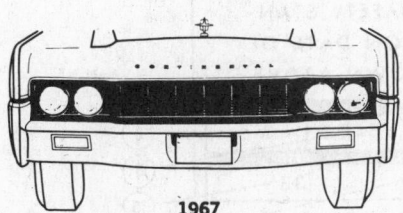

1967

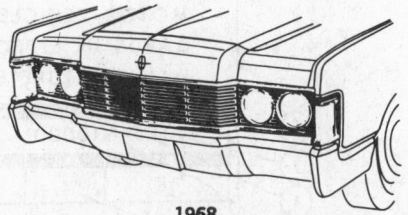

1968

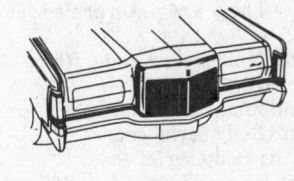

1969

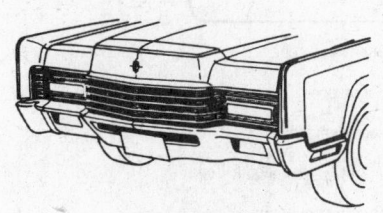

1970 Continental

1970 Mark III

1971 Lincoln Continental

1971 Continental Mark III

FIRING ORDER

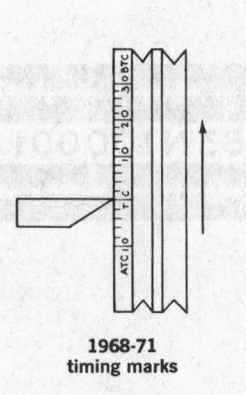

1968-71
timing marks

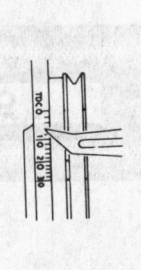

1964-67
timing marks

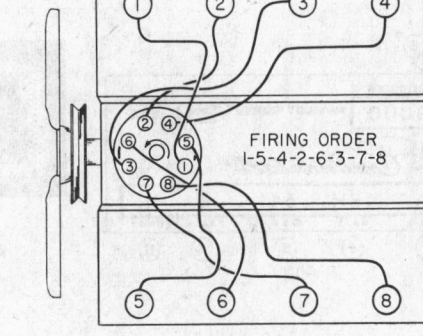

FIRING ORDER
1-5-4-2-6-3-7-8

1964-71

CAR SERIAL NUMBER LOCATION

1964-68

The vehicle number is stamped on a plate attached to the left front door hinge post. The engine number and vehicle number are identical.

1969-71

The vehicle number is stamped on an aluminum tab riveted to the top of the instrument panel, visible through the driver's side windshield. The serial number contains eleven units. They are interpreted as follows:

First: Model year code. (9 = 1969, 0 = 1970, 1 = 1971)
Second: Assembly plant code.
Third: Body serial code.
Fourth: Body serial code.
Fifth: Engine code. (A = 460 cu. in.)
Last six: Consecutive unit number.

Vehicle Certification Label

1970-71

The vehicle certification label is attached to the left-hand door jamb. This label has, on its upper portion, the name of the manufacturer, the month and year of manufacture, and the certification statement. The label also shows the vehicle identification number, which must match that on the dashboard. See the illustration for interpretation of the lower portion of the label.

Engine Identification

No. Cyls.	Cu. In. Displ.	Type	YEAR AND CODE						
			1964	1965	1966	1967	1968	1969	1970
8	460	4-BBL.						A	A
8	462	4-BBL.			G	G	G		
8	430	4-BBL.	N	N					

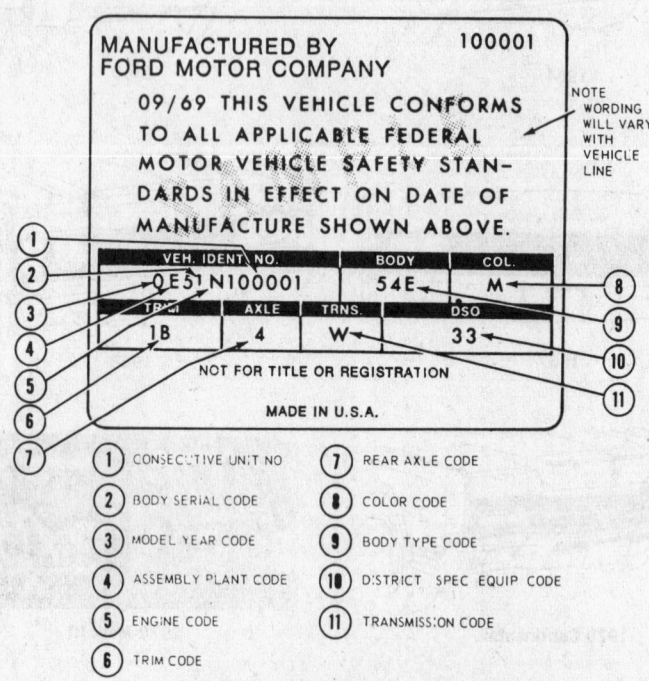

Vehicle certification label—1970-71

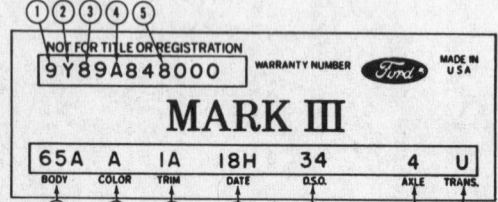

1 Model year code	7 Color code
2 Assembly plant code	8 Trim code
3 Body serial code	9 Date code
4 Engine code	10 District—special equipment code
5 Consecutive unit number	11 Rear axle code
6 Body type code	12 Transmission code

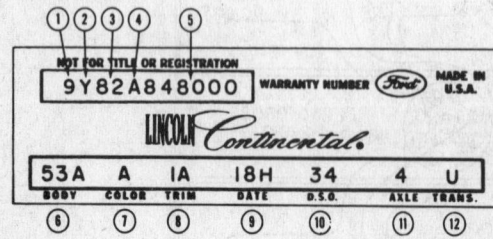

GENERAL ENGINE SPECIFICATIONS

YEAR	CU. IN. DISPLACEMENT	CARBURETOR	DEVELOPED HORSEPOWER @ RPM	DEVELOPED TORQUE @ RPM (FT. LBS.)	A.M.A. HORSEPOWER	BORE STROKE (IN.)	ADVERTIZED COMPRESSION RATIO	VALVE LIFTER TYPE	NORMAL OIL PRESSURE (PSI)
1964-65	430	4-BBL.	320 @ 4600	465 @ 2600	59.7	4.300 x 3.700	10.1-1	Hyd.	50
1966-68	462	4-BBL.	340 @ 4600	485 @ 2800	61.4	4.380 x 3.830	10.25-1	Hyd.	45
1968-69	460	4-BBL.	365 @ 4600	500 @ 2800	60.8	4.360 x 3.850	10.5-1	Hyd.	35-60 ①
1970-71	460	4-BBL.	365 @ 4600	500 @ 2800	60.8	4.360 x 3.850	10.5-1	Hyd.	35-60 ①

① —At 2,000 rpm.

TUNE-UP SPECIFICATIONS

YEAR	MODEL	SPARK PLUGS Type	SPARK PLUGS Gap (In.)	DISTRIBUTOR Point Dwell (Deg.)	DISTRIBUTOR Point Gap (In.)	IGNITION TIMING (Deg.) ▲	CRANKING COMP. PRESSURE (Psi)	VALVES Tappet (Hot) Clearance (In.) Intake	VALVES Tappet (Hot) Clearance (In.) Exhaust	Intake Opens (Deg.)	FUEL PUMP PRESSURE (Psi)	IDLE SPEED (Rpm) ★
1964-65	All	BF42	.035	26-28	.015	6B	180	Zero	Zero	22B	5-6	465
1966-67	Std. Ign.	BTF42	.034	26-31	.017	10B	180	Zero	Zero	20B	5-6	500 ⊙
	Transistor Ign.	BTF42	.034	22-24	.020	10B	180	Zero	Zero	20B	5-6	500 ⊙
1968	All	BTF42	.034	24-29	.021	6B	180	Zero	Zero	20B	5-6	550
1968-69	460 Cu. In.	BF42	.034	26-31	.017	10B	190	Zero	Zero	16B	4-5	600
1970-71	460 Cu. In.	BF42	.034	26-31	.017	10B	190	Zero	Zero	16B	4-5	600

★—With manual transmission in N and automatic in D.

▲—With vacuum advance disconnected. NOTE: These settings are only approximate. Engine design, altitude, temperature, fuel octane rating and the condition of the individual engine are all factors which can influence timing. The limiting advance factor must, therefore, be the "knock point" of the individual engine.

⊙ —500-525 with Thermactor system.

B—Before top dead center.

Caution

General adoption of anti-pollution laws has changed the design of almost all car engine production to effectively reduce crankcase emission and terminal exhaust products. It has been necessary to adopt stricter tune-up rules, especially timing and idle speed procedures. Both of these values are peculiar to the engine and to its application, rather than to the engine alone. With this in mind, car manufacturers supply idle speed data for the engine and application involved. This information is clearly displayed in the engine compartment of each vehicle.

VALVE SPECIFICATIONS

YEAR AND MODEL		SEAT ANGLE (DEG.)	FACE ANGLE (DEG.)	VALVE LIFT INTAKE (IN.)	VALVE LIFT EXHAUST (IN.)	VALVE SPRING PRESSURE (VALVE OPEN) LBS. @ IN.	VALVE SPRING INSTALLED HEIGHT (IN.)	STEM TO GUIDE CLEARANCE (IN.) Intake	STEM TO GUIDE CLEARANCE (IN.) Exhaust	STEM DIAMETER (IN.) Intake	STEM DIAMETER (IN.) Exhaust
1964-65	All	45 45	44	.408	.408	220-240 @ 1.39	1.64	.0008-.0018	.0020-.0030	.3711-.3722	.3701-.3709
1966-68	All	45 45	44	.442	.442	220-240 @ 1.39	1.65	.0008-.0025	.0010-.0027	.3710-.3717	.3708-.3715
1968-69	460 Cu. In.	45 45	44	.443	.486	240-266 @ 1.33	1.81	.0010-.0027	.0010-.0027	.3416-.3423	.3416-.3423
1970	460 Cu. In.	45 45	44	.443	.486	240-266 @ 1.33	1.81	.0010-.0027	.0010-.0027	.3416-.3423	.3416-.3423
1971	460 Cu. In.	45 45	44	.443	.486	240-266 @ 1.33	1.81	.0010-.0027	.0010-.0027	.3416-.3423	.3416-.3423

CRANKSHAFT BEARING JOURNAL SPECIFICATIONS

YEAR	MODEL	MAIN BEARING JOURNALS (IN.) Journal Diameter	MAIN BEARING JOURNALS (IN.) Oil Clearance	MAIN BEARING JOURNALS (IN.) Shaft End-Play	MAIN BEARING JOURNALS (IN.) Thrust On No.	CONNECTING ROD BEARING JOURNALS (IN.) Journal Diameter	CONNECTING ROD BEARING JOURNALS (IN.) Oil Clearance	CONNECTING ROD BEARING JOURNALS (IN.) End-Play
1964-68	All	2.8994-2.9002	.0016	.006	3	2.5992-2.6000	.0016	.006
1968-71	460 Cu. In.	2.9994-3.0002	.0005-.0015	.004-.008	3	2.4992-2.5000	.0008-.0015	.010-.020 ①

① —Both rods.

BATTERY AND STARTER SPECIFICATIONS

YEAR	MODEL	BATTERY Ampere Hour Capacity	BATTERY Volts	BATTERY Terminal Grounded	STARTERS Lock Test Amps.	STARTERS Lock Test Volts	STARTERS Lock Test Torque	STARTERS No-Load Test Amps.	STARTERS No-Load Test Volts	STARTERS No-Load Test RPM	STARTERS Brush Spring Tension (Oz.)
1964-65	All	80	12	Neg.	670	4.0	15.5	70	12.0	9,500	40
1966-71	All	85	12	Neg.	670	5.0	15.5	70	12.0	9,500	40

CAPACITIES

| YEAR | MODEL | ENGINE CRANKCASE ADD 1 Qt. FOR NEW FILTER | TRANSMISSIONS Pts. TO REFILL AFTER DRAINING | | | DRIVE AXLE (Pts.) | GASOLINE TANK (Gals.) | COOLING SYSTEM (Qts.) WITH HEATER |
| | | | Manual | | Automatic | | | |
			3-Speed	4-Speed				
1964–65	All Models	5	None	None	23	4.8	24	25
1966–67	All Models	5	None	None	26	5.5	25.5	23
1968–69	462 Cu. In.	5	None	None	27	5.0	25.5	23.5
1968–69	Mark III	4	None	None	26.5	5.0	25.5	19.3
1970–71	Mark III	4	None	None	26.5	5.0	24	19.4
1970–71	460 Cu. In.	4	None	None	27	5.0	24	19.6

TORQUE SPECIFICATIONS

| YEAR | MODEL | CYLINDER HEAD BOLTS (FT. LBS.) | ROD BEARING BOLTS (FT. LBS.) | MAIN BEARING BOLTS (FT. LBS.) | CRANKSHAFT BALANCER BOLT (FT. LBS.) | FLYWHEEL TO CRANKSHAFT BOLTS (FT. LBS.) | MANIFOLD (FT. LBS.) | |
							Intake	Exhaust
1964–68	All	125–135 ①	40–45	95–105	75–90	75–85	23–28	15–21
1968–70	460 Cu. In.	130–140 ②	40–45	95–105	70–90	75–85	25–30	28–33

① — 1966–68 462 Cu. In. 135–145.
② — In three steps: Step 1—75
Step 2—105
Step 3—130–140

WHEEL ALIGNMENT

| YEAR | MODEL | CASTER | | CAMBER | | TOE-IN (In.) | KING-PIN INCLINATION (Deg.) | WHEEL PIVOT RATIO | |
		Range (Deg.)	Pref. Setting (Deg.)▲	Range (Deg.)	Pref. Setting (Deg.)■			Inner Wheel	Outer Wheel
1964	Continental	³/₄N to 2¹/₄N	1¹/₂N	0 to ³/₄P	³/₈P	¹/₁₆ to ³/₁₆	7	20	17³/₄
1965	Continental	2¹/₂N to ¹/₂N	1¹/₂N	¹/₄P to 1¹/₂P	³/₄P	¹/₃₂ to ⁹/₃₂	7	20	17³/₄
1966–69	Continental	2¹/₂N to ¹/₂N	1¹/₂N	¹/₄P to 1¹/₂P	³/₄P	0 to ¹/₄	7	20	17³/₄
1968–69	Mark III	0 to 2P	1	¹/₄N to 1¹/₄P	¹/₂P	¹/₁₆ to ⁵/₁₆	7³/₄	20	18¹/₁₆
1966–70	Continental	2¹/₂N to ¹/₂N	1¹/₂N	¹/₄P to 1¹/₂P	³/₄P	0 to ¹/₄	7	20	17³/₄
1968–70	Mark III	0 to 2P	1	¹/₄N to 1¹/₄P	¹/₂P	¹/₁₆ to ⁵/₁₆	7³/₄	20	18¹/₁₆
1971	Continental	2¹/₂N to ¹/₂N	1¹/₂N	¹/₄P to 1¹/₂P	³/₄P	0 to ¹/₄	7	20	17³/₄
	Mark III	0–2P	1	¹/₄N to 1¹/₄P	¹/₂P	¹/₁₆ to ⁵/₁₆	7³/₄	20	18¹/₁₆

■—Not to vary more than ¹/₂ degree from one side to the other.
▲—Not to vary more than ¹/₄ degree from one side to the other.
N—Negative.
P—Positive.

GENERAL CHASSIS AND BRAKE SPECIFICATIONS

| YEAR | MODEL | CHASSIS | | BRAKE CYLINDER BORE | | | BRAKE DRUM | |
| | | Overall Length (In.) | Tire Size | Master Cylinder (In.) | Wheel Cylinder Diameter (In.) | | Diameter (In.) | |
					Front	Rear	Front	Rear
1964	All	216¹/₃	9.15 x 15	²¹/₃₂	1³/₃₂	¹⁵/₁₆	11.06	11.06
1965	All	216¹/₃	9.15 x 15	1.0	1¹⁵/₁₆	¹⁵/₁₆	11.87 Disc	11.03■
1966	All	220.9	9.15 x 15	1.0	1¹⁵/₁₆	¹⁵/₁₆	11.87 Disc	11.03■
1967–69	All Exc. Mark III	220.9 ①	9.15 x 15	1.0	1¹⁵/₁₆	¹⁵/₁₆	11.96 Disc	11.09②
1968–69	Mark III	216.1	8.55 x 15	1.0	2³/₄	¹⁵/₁₆	11.72 Disc	11.03■
1970	All Exc. Mark III	225.0	9.15 x 15	1.0	1¹⁵/₁₆	¹⁵/₁₆	11.72 Disc	11.09②
1970	Mark III	216.1	225R15	1.0	2.755	¹⁵/₁₆	11.72 Disc	11.03■

■—Refinishing Limit—11.090.
① —1969, 224.2.
② —Refinishing Limit—11.150.

AC GENERATOR AND REGULATOR SPECIFICATIONS

YEAR	MODEL	ALTERNATOR			REGULATOR						
		Field Current Draw @ 12V.	Output @ Alternator RPM		Model	Field Relay			Regulator		
			900	6500		Air Gap (In.)	Point Gap (In.)	Volts to Close	Air Gap (In.)	Point Gap (In.)	Volts at 125°
1964-65	Ford Std.	2.9-3.1	10	40	Ford	.020	.019	2½-4	.052	.020	13.8-14.6
	Ford Opt.	2.9-3.1	10	42	Ford	.020	.019	2½-4	.052	.020	13.8-14.6
	Ford Opt.	2.8-3.3	10	55	Ford	.018	.019	2-3	.052	.020	13.8-14.6
1966-67	Autolite	4.4-4.8	10	60	Autolite	.014	—	2½-4	.052	.020	13.8-14.6
1968-69	Autolite	2.9-3.1	22	55	Autolite	.018	.019	2½-4	.052	.020	13.3-15.3
1970-71	Autolite ① DOLF—10300	2.9-3.1	22	55	Autolite	N.A.	N.A.	N.A.	N.A.	N.A.	13.3-14.6
	Autolite ② DOAF—10300	2.8-3.3	28	65	Autolite	N.A.	N.A.	N.A.	N.A.	N.A.	13.8-15.3

① —Integral regulator.
② —Opt.; required with heated back window.

CYLINDER HEAD BOLT TIGHTENING SEQUENCE

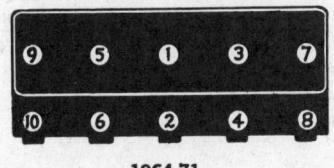

1964-71

FUSES AND CIRCUIT BREAKERS

1964

Headlamps............................. 18 Amp. CB
Glove box, interior lights .... 14 Amp. Fuse
Clock .... 2 Amp. Fuse
Cigar lighter........................... 15 Amp. Fuse
Back up lights, windshield washer, parking,
  tail, dash, stop/turn, radio .... 7.5 Amp. Fuse
Power seats............................ 30 Amp. CB
Power windows......................... 20 Amp. CB

CIRCUIT PROTECTION—1965-67 LINCOLN

Circuit	Location	Protective Device-Fuse Or Circuit Breaker (C.B.) In Amperes	Circuit	Location	Protective Device-Fuse Or Circuit Breaker (C.B.) In Amperes
Cigar lighter (front)	Fuse panel	AGC 15	Top control neutral relay	R.H. cowl side panel	10 C.B.
Cigar lighter (rear)	Fuse panel	AGC 15	Stoplights and emergency warning	R.H. cowl side panel	15 C.B.
Interior lights: dome, courtesy, map, glove box and luggage compartment	Fuse panel	SFE 14	Taillights, license light, parking lights, ash tray light	In headlight switch	15 C.B.
Electric antenna	Fuse panel	AGC 10	Headlights	In headlight switch	18 C.B.
Instrument panel lights, trans. indicator light (PRNDL)	Fuse panel	SFE 6	Upper back panel motor control	In luggage comp't.	30 C.B.
Speed control	Fuse panel	SFE 14	Deck lock control and top lock motor control	In luggage comp't.	30 C.B.
Radio	Fuse panel	SFE 7.5	Top control motor feed	Terminal junction block and starter motor relay	Fuse assy (fuse link wire) (6-in. long no. 12 gauge wire)
Back-up lights and windshield washer	Fuse panel	SFE 15			
Turn signals	Fuse panel	SFE 15	Power circuit	Terminal junction block and starter motor relay	Fuse assy (fuse link wire) (6-in. long no. 14 gauge wire)
Warning lights: open door, deck lid open, seat belt	Fuse panel	SFE 5			
Heater and air-conditioner	R.H. cowl side panel	30 C.B.	Windshield wiper	Hydraulically operated (no fuse or circuit breaker)	
Electric seats and horns	R.H. cowl side panel	30 C.B.			
Electric windows	R.H. cowl side panel	20 C.B.			

CIRCUIT PROTECTION—1968 LINCOLN

Circuit	Location	Protective Device	Circuit	Location	Protective Device
Cigar lighter (front)	Fuse panel	AGC 15	Warning lights: open door, deck lid open, seat belt	Fuse panel	SFE 7.5
Cigar lighter (rear)	Fuse panel	AGC 15			
Interior lights: dome, courtesy, map, glove box and luggage compartment and clock feed	Fuse panel	SFE 14	Heater and air-conditioner	R.H. cowl side panel	30 C.B.
			Electric seats and horns	R.H. cowl side panel	30 C.B.
			Electric windows	R.H. cowl side panel	20 C.B.
Electric antenna	Fuse panel	AGC 10	Stoplights and emergency warning	R.H. cowl side panel	10 C.B.
Instrument panel lights, trans. indicator light (PRNDL)	Fuse panel	SFE 6	Taillights, license light, parking lights, ash tray light	In headlight switch	15 C.B.
Speed control, pwr. window by-pass switch, A/C and heater relay, defogger blower	Fuse panel	SFE 14	Headlights	In headlight switch	18 C.B.
			Power circuit	Terminal junction block and starter motor relay	Fuse assy. (fuse link wire) (6-in. long no. 14 gauge wire)
Radio	Fuse panel	SFE 7.5			
Back-up lights, windshield washer, turn signals	Fuse panel	SFE 15	Windshield wiper	Hydraulically operated	(No fuse or circuit breaker)

CIRCUIT PROTECTION—1969 LINCOLN CONTINENTAL

Circuit	Location	Protective Device	Circuit	Location	Protective Device
Ash receptical light (front)	Fuse panel	SFE 14	Instrument panel lights: clock, defogger indicator, fuel gauge, heater/defroster and A/C control lights, radio dial light, speed control indicator	Fuse panel	SFE 6
Ash receptical light (rear)	In headlight switch	15 C.B.			
Back-up lights	Fuse panel	7AG or AGW 15			
Cigar lighter (front)	Fuse panel	3AG or AGC 15			
Cigar lighter (rear)	Fuse panel	3AG or AGC 15	License light	In headlight switch	15 C.B.
Clock feed	Fuse panel	SFE 14	Marker lights	In headlight switch	15 C.B.
Courtesy lights: dome, map, glove box, luggage compartment, reading lights, C pillar lights	Fuse panel	SFE 14	Power window motors	Integral part of motor	C.B.
			Radio	Fuse panel	SFE 7.5
Defogger motor	Fuse panel	SFE 14	Speed control circuit	Fuse panel	SFE 14
Electric antenna	Fuse panel	3AG or AGC 10	Stereo tape player	Fuse panel	SFE 7.5
Emergency flasher	R.H. cowl side panel	15 C.B.	Stop lights	R.H. cowl side panel	15 C.B.
Headlights	In headlight switch	18 C.B.	Taillights	In headlight switch	15 C.B.
Heater and air-conditioner	R.H. cowl side panel	30 C.B.	Transmission indicator light (PRNDL)	Fuse panel	SFE 6
Horns	R.H. cowl side panel	30 C.B.	Turn signals	Fuse panel	7 AG or AGW 15
Ignition switch light	Integral with lighting switch	15 C.B.	Windshield washer	Fuse panel	7 AG or AGW 15
Indicator lights: brake, deck lid open, engine temperature, low fuel, oil pressure, open door, seat belt	Fuse panel	SFE 7.5	Optional equipment power circuit	Terminal junction block and starter motor relay	Fuse assembly (fuse link wire) (7 inch long no. 14 gauge wire)
			Parking lights	In headlight switch	15 C.B.
			Power seats	R.H. cowl side panel	30 C.B.
			Power seats motors	Integral part of motor	C.B.
			Power windows	R.H. cowl side panel	20 C.B.
			Power window by-pass switch	Fuse panel	SFE 14

FUSES AND CIRCUIT BREAKERS NOT IN PANEL— 1970–71 LINCOLN CONTINENTAL

Circuit	Rate	Location	Circuit	Rate	Location
Automatic headlamp dimmer	4 AMP SFE-4	In-line fuse	Headlights, headlight door open warning light	18 AMP C.B.	In headlight switch
Parking lights, license light, tail lights, ash receptacle light, marker lights	15 AMP C.B.	In headlight switch	Motors, power seat, power windows	C.B.	In motor assembly
			Windshield wiper	7.5 AMP Fuse	In wiper switch

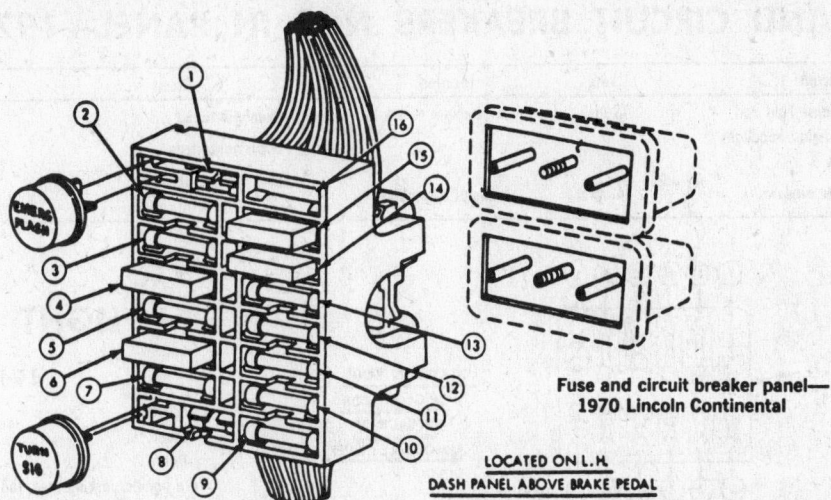

**Fuse and circuit breaker panel—
1970 Lincoln Continental**

LOCATED ON L.H.
DASH PANEL ABOVE BRAKE PEDAL

① (SPARE) OR (3 AMP. FUSE) SURE TRACK (BRAKE SKID CONTROL)

② (7.5 AMP. FUSE) WARNING LAMPS, DOOR AJAR, DECK LID OPEN, LOW FUEL, SEAT BELT, DUAL BRAKE

③ (20 AMP. FUSE) CIGAR LIGHTER (FRONT), DOOR LOCK SOLENOID

④ (20 AMP. C.B.) STOPLAMPS & EMERGENCY WARNING SYSTEM

⑤ (20 AMP. FUSE) CIGAR LIGHTER (REAR)

⑥ (30 AMP. C.B.) POWER SEAT, HORNS

⑦ (15 AMP. FUSE) COURTESY LAMPS, DOORS, "C" PILLAR READING, LUGGAGE COMPT., MAP, GLOVE BOX, & CLOCK FEED, SEAT BACK LATCH CONTROL, DOME LAMP

⑧ (6 AMP. FUSE) (INSTR. PANEL LAMPS) UPPER & LOWER CLUSTER ILLUM. CLOCK, HTR. & A/C, W/S WIPER, MAP LAMP SWITCH ILLUM., RADIO LAMP & PRNDL LAMP

⑨ (7.5 AMP. FUSE) SPEED CONTROL, DEFOGGER, W/S WASHER & SEAT BELT WARNING (STD.)

⑩ (15 AMP. FUSE) TURN SIGNAL & CORNERING LAMPS & BACK-UP LAMPS

⑪ (25 AMP. FUSE) HEATED BACK WINDOW

⑫ (7.5 AMP. FUSE) POWER WINDOW SAFETY RELAY COIL FEED

⑬ (7.5 AMP. FUSE) RADIO, POWER ANTENNA & STEREO TAPE PLAYER

⑭ (20 AMP. C.B.) POWER WINDOWS

⑮ (30 AMP. C.B.) AIR CONDITIONER, HEATER & DEFROSTER

⑯ (SPARE)

CIRCUIT PROTECTION 1968–69 MARK III

Circuit	Circuit Protection—Fuse or Circuit Breaker (C.B.)		Location
	Rating	Trade No.	
Headlight Circuit	18 Amp. C.B.		Integral with Lighting Switch
Tail Lights, Running Lights, License Plate Light, Parking Lights, Marker Lights	15 Amp. C.B.		Integral with Lighting Switch
Instrument Panel and Instrument Cluster Illumination	6 Amp.	SFE 6	Fuse Panel
Clock Feed, Courtesy Lights, Luggage Compartment Light, Glove Compartment Light, Map Light and Reading Lights	14 Amp. ①	SFE 14①	Fuse Panel
Warning Lights (except Low Fuel Warning)	7.5 Amp.	SFE 7.5	Fuse Panel
Low Fuel Warning	7.5 Amp.	SFE 7.5	Fuse Panel
Windshield Washer and Back-Up Lights	7.5 Amp.	SFE 7.5	Fuse Panel
Turn Signal	7.5 Amp.	SFE 7.5	Fuse Panel
Ammeter	14 Amp.	SFE 14	Fuse Panel
Speed Control	7.5 Amp.	SFE 7.5	Fuse Panel
Radio	7.5 Amp.	SFE 7.5	Fuse Panel
Power Antenna	10 Amp.	3AG 10	Fuse Panel
Front Cigar Lighter and Stereo	15 Amp.	SFE 15	Fuse Panel
Rear Cigar Lighter	15 Amp.	SFE 15	Fuse Panel
Power Seats, Windows and Horns	30 Amp. C.B.		Circuit Breaker Panel (Seat and window motors also protected by integral circuit breakers)
Stop Lights and Emergency Warning	10 Amp. C.B.		Circuit Breaker Panel
Heater and Air Conditioner	30 Amp. C.B.		Circuit Breaker Panel
Charging Circuit	Fusible Link		NFusible Link Block (below starter relay in engine compartment)
Headlight Dimmer	4 Amp.		Fuse Cartridge in Feed Wire
Anti-Skid Brake Control	3 Amp.	SFE3	Fuse Panel

① —Use 20 Ampere SFE 20 fuse if equipped with moveable steering column.

FUSES AND CIRCUIT BREAKERS NOT IN PANEL—1970–71 MARK III

Circuit	Rate	Location	Circuit	Rate	Location
Parking lights, license light, tail lights, marker lights, headlight buzzer, motors	15 AMP	In headlight switch	Headlight circuit	18 AMP	In headlight switch
			Windshield wiper	Integral	In wiper switch
Power seat, power windows	Integral	In motor assembly	Rear window defroster	20 AMP	On brake pedal support

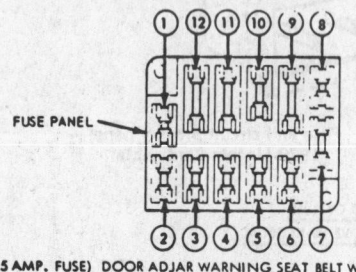

FUSE PANEL →

LOCATED IN RIGHT SIDE
OF GLOVE BOX
IN BACK OF
REMOVABLE COVER

① (7.5 AMP. FUSE) DOOR ADJAR WARNING SEAT BELT WARNING BRAKE SYS. WARNING ENGINE TEMP FUEL GAUGE

② (7.5 AMP. FUSE) LOW FUEL WARNING

③ (7.5 AMP. FUSE) POWER WINDOW SAFETY RELAY COIL FEED.

④ (7.5 AMP. FUSE) BACK-UP LAMPS SPEED CONTROL

⑤ (7.5 AMP. FUSE) TURN SIGNALS

⑥ (7.5 AMP. FUSE) WINDSHIELD WASHER REAR WINDOW DEFOGGER

⑦ (SPARE OR 3 AMP. FUSE) SURE-TRACK BRAKE SYS.

⑧ (6 AMP. FUSE) ILLUMINATION LAMPS INSTRUMENT PANEL CLUSTER RADIO PRND 21 HEATER/A.C. CONTROLS ASH TRAY

⑨ (15 AMP. FUSE) POWER ANTENNA

⑩ (14 AMP. FUSE) COURTESY LAMPS, CLOCK FEED, LUGGAGE COMPT. LAMP, GLOVE COMPT. LAMP, ENGINE COMPT. LAMP, MAP LAMP AND REAR READING LAMPS

⑪ (15 AMP. FUSE) FRONT LIGHTERS, RADIO AND STEREO TAPE

⑫ (15 AMP. FUSE) REAR LIGHTERS

Fuse panel—1970 Mark III

① BLANK OR 20 AMP. FOR HEATED BACKLITE

② (20 AMP.) POWER WINDOWS

③ (30 AMP.) HEATER, A/C

④ (10 AMP.) STOP LIGHTS, EMERGENCY FLASHER

⑤ (30 AMP.) HORNS POWER SEATS AND A/C HIGH SPEED BLOWER

Circuit breaker panel—1970 Mark III

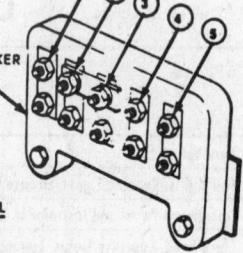

CIRCUIT BREAKER PANEL →

ATTACHED TO BRACKET R.H. SIDE OF DASH PANEL
BEHIND GLOVE BOX

LIGHT BULBS

1964

Headlamps
Inner	4001; 37.5 W
Outer	4002; 50–37.5 W
Turn signals, parking, stop, tail	1157; 32/4 CP
Back-up lamps	1141; 21 CP
Indicators, cigar lighter	1895; 2 CP
Control housing, glove compartment, parking brake, door open warning	257; 2 CP
C-Pillar	1003; 15 CP
License plate	1155; 4 CP
Luggage compartment	631; 6 CP
Engine compartment	93; 15 CP
Ash receptacle	1445; 1.5 CP
Dome lamp	1003; 15 CP
Speedometer, clock	158; 2 CP
Radio dial	1891; 2 CP

1965–67 LINCOLN CONTINENTAL

Light Description	Bulb No.	Candlepower or Wattage	Light Description	Bulb No.	Candlepower or Wattage
Engine compartment	93	15 C.P.	Ash receptacle or cigar lighter	1445	1.5 C.P.
Glove compartment	1895	2 C.P.	Back-up lamps	1156	32 C.P.
Headlamps hi-lo beam (outer or upper)	4002	37.5–50W	Courtesy lamp	631	6 C.P.
hi beam (inner or lower)	4001	37.5W	("C" pillar)	1003	15 C.P.
License lamp	97	4 C.P.	(console)	1895	2 C.P.
Luggage compartment	631	6 C.P.	(door mounting or armrest)	212	6 C.P. (fuse type)
Map light	631	6 C.P.	Clock and ignition key	158	2 C.P.
Park and turn signal lamp	1157-A	4–32 C.P.	Control nomenclature (ex. c. heater)	158	2 C.P.
Taillight, stop and turn signal	1157	32 C.P.	Control, right and left	1895	2 C.P.
Door lock warning	158	2 C.P.	Gages and speedometer	158	2 C.P.
Emergency flasher	158	2 C.P.	Heater control	158	2 C.P.
Hi-beam indicator	158	2 C.P.	Radio dial light AM	1893	1.9 C.P.
Low fuel warning	1895	2 C.P.	AM-FM	1893	1.9 C.P.
Oil press., alt and temp warning	158	2 C.P.	Trans. control selector indicator	1445	1.5 C.P.
Parking brake warning	158	2 C.P.	Turn signal indicators (L and R)	158	2 C.P.
Seat belt warning	158	2 C.P.	Trunk warning	158	2 C.P.
Utility lamp (double end) dealer	1155	4 C.P.	Cruise control indicator	158	2 C.P.
accessory	1143	32 C.P.			

1968 LINCOLN CONTINENTAL

Light Description	Bulb No.	Candela or Wattage [1]	Light Description	Bulb No.	Candela or Wattage [1]
Engine compartment	631	6C.	Courtesy light (roof)	631	6C.
Glove compartment	1895	2C.	(quarter)	1003	15C.
Headlights hi-lo beam	4002	37.5–50W	(console)	1895	2C.
hi beam	4001	37.5W	(door mounting or armrest)	212	6C. (fuse type)
License light	97	4C.	Clock and ignition key	1895	2C.
Luggage compartment	631	6C.	Control nomenclature (ex. C. heater)	1895	2C.
Map light	212	6C.	Control, right and left	1895	2C.
Park and turn signal light	1157NA[2]	4—32C.	Gauges and speedometer	194	2C.
Taillight, stop and turn signal	1157	4—32C.	Heater control	1895	2C.
Door lock warning	1895	2C.	Radio dial light AM	1893	1.9C.
Emergency flasher	1895	2C.	AM-FM	1893	1.9C.
Hi-beam indicator	1895	2C.	Trans. control selector indicator (tilt)	1445	1.5C.
Low fuel warning	1895	2C.	Turn signal indicators (L and R)	1895	2C.
Oil press., alt. and temp. warning	1895	2C.	Trunk warning	1895	2C.
Seat belt warning	1895	2C.	Cruise control indicator	1445	1.5C.
Utility light (double end) dealer	1155	4C.	Trans. control selector indicator (fixed)	194	2C.
accessory	1143	32C.	Spotlight	4405	30 W
Ash receptacle or cigar lighter	1445	1.5C.	Manual A/C control	1895	2C.
Back-up lights	1156	32C.	ATC control	1895	2C.

[1] — Candela is the international term for candlepower
[2] — Natural amber color.

1969 LINCOLN CONTINENTAL

Light Description	Bulb No.	Candela or Wattage	Light Description	Bulb No.	Candela or Wattage
Ash receptacle or cigar lighter	1445	1.5C.	Luggage compartment	631	6C.
ATC control	1895	2C.	Manual A/C control	1895	2C.
Back-up lights	1156	32C.	Map light	212	6C.
Clock and ignition key	1895	2C.	Oil press., alt. and temp indicator	1895	2C.
Courtesy light			Park and turn signal light	1157NA[1]	3–32C.
C pillar	1003	15C.	Radio dial light		
glove compartment	1895	2C.	AM	1893	1.9C.
door	212	6C.	AM/FM	1891	2C.
Cruise control indicator	1445	1.5C.	AM/FM stereo	1892	1.3C.
Door lock warning	1895	2C.	Seat belt indicator	1895	2C.
Engine compartment	631	6C.	Side marker lights (front)	1178A	4C.
Glove compartment	1895	2C.	Side marker lights (rear)	194	2C.
Headlights			Speed control actuator indicator	1445	1.5C.
hi-lo beam	4002	37.5–50W	Spotlight	4405	30W
hi beam	4001	37.5W	Taillight, stop and turn signal	1157	3–32C.
Heater control	1895	2C.	Trans. control selector indicator (fixed)	158	2C.
Hi-beam indicator	1895	2C.	Trans. control selector indicator (tilt)	1445	1.5C.
Instrument cluster illumination	1895	2C.	Trunk light (portable)	1003	15C.
License light	97	4C.	Trunk open indicator	1895	2C.
Low fuel indicator	1895	2C.	Turn signal indicators (L and R)	1895	2C.

[1] — Natural amber color.

1970–71 LINCOLN CONTINENTAL

Light Description	Trade No.	Candle Power or Wattage	Light Description	Trade No.	Candle Power or Wattage
Headlamps			L & R turn signal indicator	1895	2 c.p.
hi-lo beam	4002	37.5 & 50 watts	Hi beam indicator	1895	2 c.p.
hi-beam	4001	37.5	Heater control	1895	2 c.p.
Front parking lamps and turn signal	1157A	3–32 c.p.	Courtesy lamp	631	6 c.p.
Front side marker	194-A	2 c.p.	Illumination	1895	2 c.p.
Rear side marker	1895	2 c.p.	Speedometer	1895	2 c.p.
Rear lamp, stop and turn signal	1157	3–32 c.p.	Clock	1895	2 c.p.
Rear seat reading	105	12 c.p.	Ash tray (instr. panel and rear)	1445	1.5 c.p.
Luggage compartment	631	6 c.p.	Door lock nomenclature	1895	2 c.p.
License plate	97	4 c.p.	Low fuel warning	1895	2 c.p.
Back-up lamp	1156	32 c.p.	Auto. trans. ind. lamp	1445	1.5 c.p.
Courtesy lamp (door) 4-door	212	6 c.p. fuse type	Radio AM signal seeking	1893	1.9 c.p.
Map	212	6 c.p.	Radio AM-FM Stereo	1891	2.0 c.p.
*212-1 used in 2-door models			Radio AM-stereo tape	1893	1.9 c.p.
Instrument panel			Engine compartment	631	6 c.p.
Warning lights	1891	2 c.p.	Speed control illumination	1445	1.5 c.p.
Glove compartment	1895	2 c.p.			

1969 MARK III

Light Description	Bulb No.	Candela or Wattage	Light Description	Bulb No.	Candela or Wattage
ATC control	1895	2C.	Hi-beam indicators	194	2C.
Auto. trans. select. ind.—fixed &			Ignition switch	161	1C.
tilt column	1445	1.5C.	Instrument panel courtesy light	631	6C.
Back-up light	1156	32C.	Instruments	194	2C.
Brakes & belts indicator	194	2C.	License plate light	97	4C.
Cigar lighter	1895	2C.	Low fuel indicator	194	2C.
Control nomenclature: rear vent &			Luggage compartment	90	6C.
wipers	194	2C.	Manual A/C control	1895	2C.
Door ajar indicator	1891	2C.	Map light	212	6C.
Door courtesy light	211 or 211-1	12C.	Parking brake signal	1895	2C.
Emergency flasher	1891	2C.	Radio pilot light	1893	1.9C.
Engine compartment	631	6C.	AM/FM stereo pilot light	1892	1.3C.
Fog lights—clear	4415	35 W	Rear-side marker	97	4C.
Fog light switch	53X	1C.	Rear tail/stop/turn signal	1157	3–32C.
Front park & turn signal	1157NA	3–32C.	Roof quarter light	1003	15C.
Front side marker	1178A	4C.	Seat belt indicator	1891	2C.
Glove compartment	1895	2C.	Trunk light (portable)	1003	15C.
Headlights—hi & lo	4002	37.5 & 50 W	Trunk open indicator	1891	2C.
Headlights—hi-beams	4001	37.5 W	Turn signal indicators	168A	3C.
Heater controls	1895	2C.			

1970–71 MARK III

Light Description	Trade No.	Candle Power or Wattage	Light Description	Trade No.	Candle Power or Wattage
Headlight—hi-lo beam	4002	37.5 & 50 W	Overhead console warning lights	1891	2 c.p.
hi beam	4001	37.5 W	Courtesy lights	631	6 c.p.
Front park & turn signal	1157NA	3–32 c.p.	Hi-beam indicator	194	2 c.p.
Rear tail/stop/turn signal	1157	3–32 c.p.	Turn signal indicators	168-A	3 c.p.
Back up light	1156	32 c.p.	Warning lights, brake & low fuel	194	2 c.p.
License plate light	97	4 c.p.	Warning light, rear window defroster	1892	1.3 c.p.
Rear seat reading light	105	12 c.p.	Instrument illumination light	194	2 c.p.
Glove compartment light	1895	2 c.p.	Radio dial light	1893	1.9 c.p.
Transmission control selector indicator light	1445	1.5 c.p.	Heater/air cond./auto. temp. control light	1895	2 c.p.
Door courtesy lights	212	6 c.p.	Wiper control lights	194	2 c.p.
Map light	212	6 c.p.	Cigar lighter light	1895	2 c.p.
Front side marker light	97NA	4 c.p.	Speed control light	53X	1 c.p.
Rear side marker light	97	4 c.p.	Ash tray light	1445	1.5 c.p.
Luggage compartment light	90	6 c.p.			

Distributor

Detailed information on distributor drive, direction of distributor rotation, cylinder numbering, firing order, point gap, cam dwell, timing mark location, spark plugs, spark advance and idle speed is in the Specifications table of this section. Further information on troubleshooting, general tune-up procedures, how to replace ignition wires, how to install points and condensers, how to choose the proper spark plug, adjust timing, is in the Tune-up Specifications table.

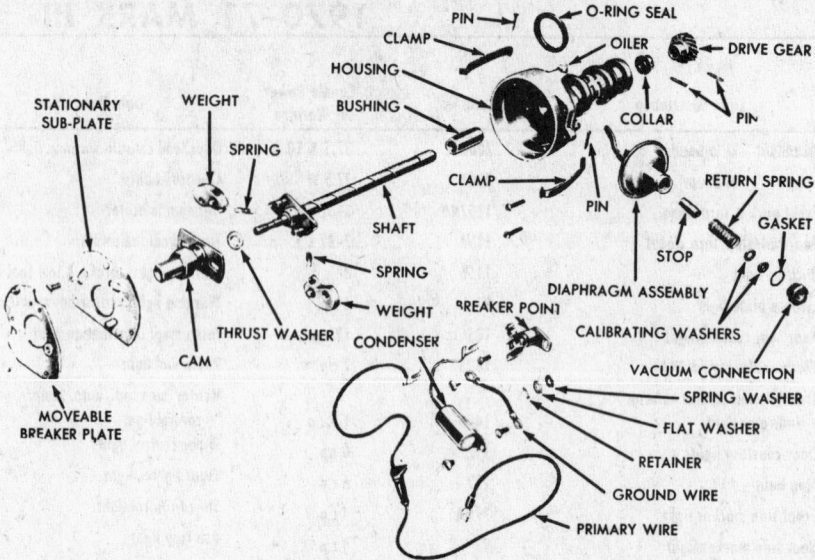

Distributor (© Ford Motor Co.)

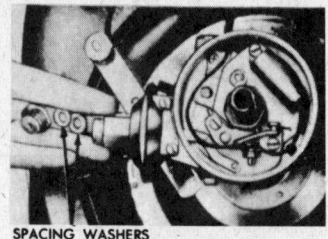

SPACING WASHERS

Vacuum advance unit
(© Ford Motor Co.)

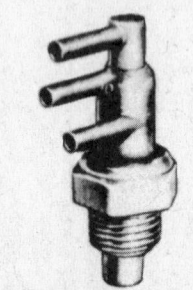

Distributor vacuum control valve
(© Ford Motor Co.)

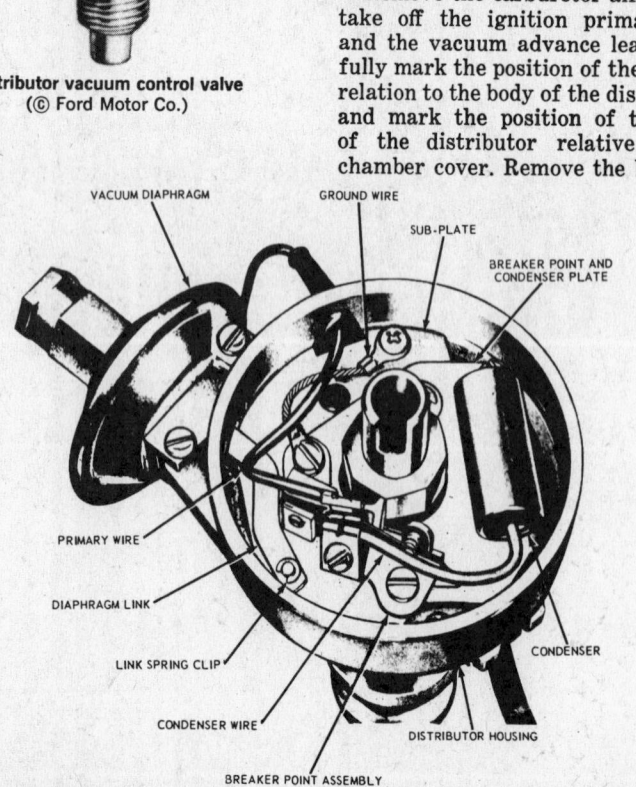

Breaker plate installed—V8 engine, single diaphragm distributor (© Ford Motor Co.)

Ignition Primary Resistor

On all Lincoln Continental and Mark III models; 12-volt systems, a resistor wire is used in the ignition, primary circuit. It is located between the ignition switch and the coil. If difficulty is experienced with the ignition, first check to make certain that the resistor wire is functioning properly.

Distributor Removal

The distributor is located at the front of the engine between the cylinder banks.

Remove the carburetor air cleaner, take off the ignition primary lead and the vacuum advance lead. Carefully mark the position of the rotor in relation to the body of the distributor, and mark the position of the body of the distributor relative to the chamber cover. Remove the bolt that

holds the distributor down and lift it out. The marks are made so that the distributor can be reinstalled without having to retime the ignition.

Ignition Timing

All

The ignition timing is marked on the vibration damper and it is recommended that a stroboscopic-type timing light be used to determine the exact point of ignition.

Ignition Retiming

If the timing relationship has been fouled up, retime the ignition as follows: bring No. 1 cylinder up into the firing position. This can be checked by removing the spark plug, placing your thumb in the spark plug hole and then cranking the engine until the compression attempts to blow by your thumb. Now, slowly bring the crankshaft around until the T.D.C. mark on the crankshaft pulley lines up with the pointer. This is the approximate firing position for No. 1 cylinder.

Remove the distributor cap and mark the position of the rotor on the outside of the distributor. The wire from No. 1 spark plug should be placed in the socket just above the rotor. Now, working in the direction of distributor rotation, place the spark plug wires into the cap according to the firing order of the engine.

Ignition Wires

The best way to replace ignition wires is to put them in, one at a time, and use the old wire as a come along for the new wire. However, if all the wires are removed and it is necessary to replace them, proceed as follows: remove the spark plug from No. 1 cylinder and crank the engine with the thumb in the cylinder hole until compression is felt to blow by the thumb.

Continue turning the engine very slowly until the ignition timing mark comes up on the vibration damper. This places No. 1 cylinder in firing position.

Remove the distributor cap and note the position of the rotor by marking its position on the cap. Place the first wire in the tower directly over the tip of the rotor and put that wire on No. 1 plug. Now, moving in a counterclockwise direction, place the wires in the cap and to the spark plugs according to the firing order of the engine, which is 1-5-4-2-6-3-7-8. Keep in mind that the cylinders are numbered:

Right front 1-2-3-4
Left front 5-6-7-8

NOTE: to reduce the probability of induced voltage cross fire in the ignition system, never run spark plug wires close to each other and parallel for more than a couple of inches. Crossing the wires at an angle is OK. Don't tape the cables together.

Spark Plug Replacement

The plugs are above the exhaust manifolds and can be reached easily.

Alternator, Regulator

Facts on the alternator and the regulator are in the Alternator and Regulator Specifications table.

General information on alternator and regulator repair and troubleshooting is in the Unit Repair Section under the heading Basic Electrical Diagnosis.

Alternator

Cars are equipped with alternating current generators. This charging system is different from the DC circuit, and requires certain precautions.

1. Reversing battery connections will cause damage to the one-way electrical valves, the rectifiers.
2. Booster battery connections must be made as follows: the negative terminal of the booster battery must be connected to the negative terminal of the car battery. The positive terminal of the booster battery must be connected to the positive terminal of the car battery.
3. Fast charges should never be used as boosters to start AC circuit equipped cars.
4. When servicing the battery with a fast charger, always disconnect car battery cables.
5. Never attempt to polarize an AC generator.

Complete alternating servicing data are in the Unit Repair Section.

Battery, Starter

A more general discussion of batteries is in the Unit Repair Section.

A more general discussion of starters and their troubles is in the Unit Repair Section.

Starter R & R

Disconnect starter cable, raise car, turn front wheels fully to the right. Remove the two bolts attaching the steering idler arm, remove starter mounting bolts. Remove starter.

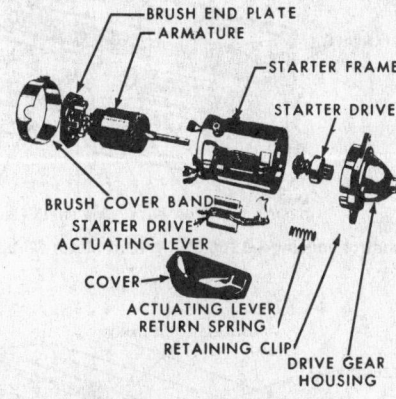

Starter motor—disassembled view
(© Ford Motor Co.)

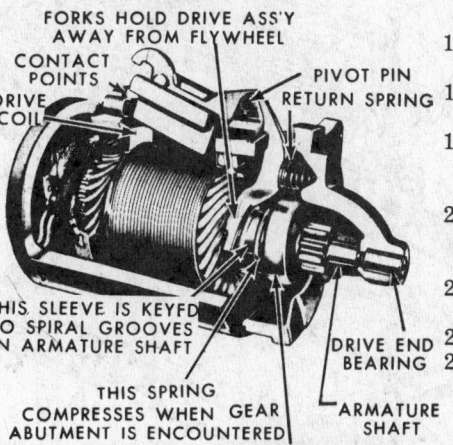

Starter drive disengaged
(© Ford Motor Co.)

Battery Location

Battery is located under the hood on the right front fender skirt.

Instruments

Speedometer and Instrument Removal

1964-65 Cluster Removal

1. Disconnect battery.
2. Remove heater, air conditioner, blower and radio control knobs.

3. Remove the screws holding instrument bezel to panel and remove bezel.
4. Remove the finish molding from upper edge of instrument panel.
5. Remove screws from trim collar at steering column and remove collar.
6. Remove screws from right and left lower panel end plates.
7. Remove screws from plate over steering column and remove plate.
8. Remove lock cylinder from ignition switch.
9. Remove screws from upper edge of lower instrument panel.
10. Remove jam nuts and disconnect speedometer trip reset and clock reset from lower edge of lower panel.
11. Remove hood lock release control handle and bracket from lower panel.
12. Remove speed control head and mounting bracket from lower panel.
13. Remove fuse block cover plate in glove compartment, and remove screws holding fuse block to lower panel.
14. Disconnect compartment courtesy light.
15. Disconnect rear deck lid lock release control handle.
16. Remove instrument switch from lower panel. Do not disconnect wires.
17. Remove headlight switch. Do not disconnect wires.
18. Remove knob from wiper control lever.
19. Remove wiper control bezel, wiper and washer control from panel.
20. Remove bolts at right and left sides of lower panel to inner cowl panel.
21. Remove ash tray and disconnect tray light and lighter leads.
22. Remove lower panel from car.
23. Reinstall by reversing the above procedure.

1966-69 Cluster Removal Except Mark III

1. Disconnect battery.
2. Remove six screws in air conditioning register casting assembly and pull register rearward until it hangs on ducts.
3. Remove three screws between cluster hood and crash pad.
4. Remove four screws in lower windshield moulding and remove moulding.
5. Remove two screws and two nuts in forward edge of crash pad.
6. Disconnect radio speaker lead and remove crash pad.
7. Remove radio knobs and bezels.
8. Remove heater, air conditioning control, and windshield wiper control knobs.
9. Remove four nuts and four re-

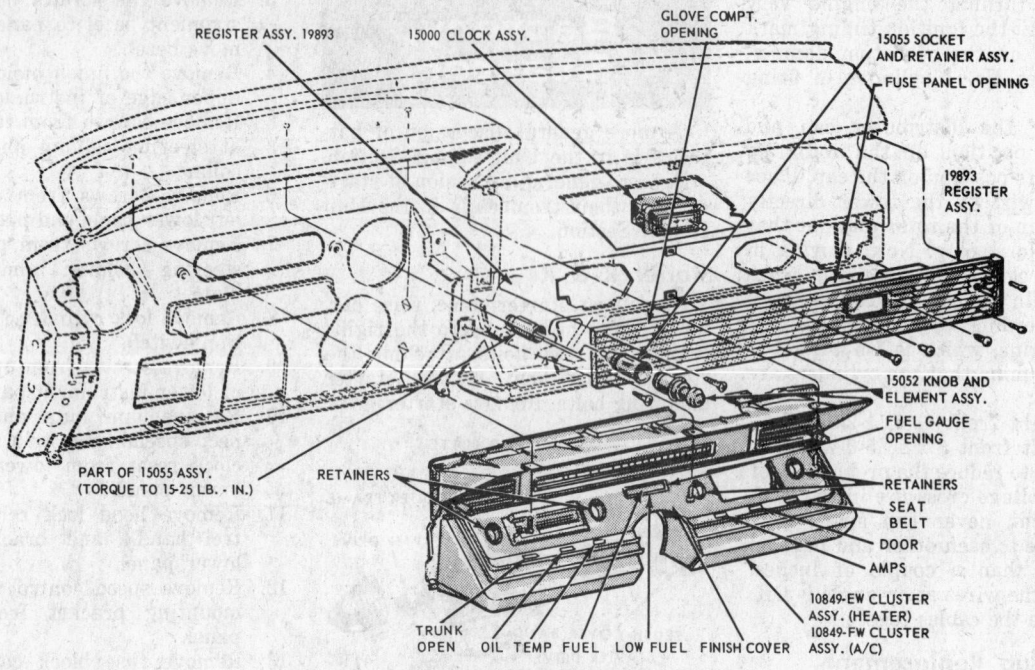

REGISTER ASSY. 19893
15000 CLOCK ASSY.
GLOVE COMPT. OPENING
15055 SOCKET AND RETAINER ASSY.
FUSE PANEL OPENING
19893 REGISTER ASSY.
15052 KNOB AND ELEMENT ASSY.
FUEL GAUGE OPENING
PART OF 15055 ASSY. (TORQUE TO 15-25 LB. - IN.)
RETAINERS
RETAINERS
SEAT BELT
DOOR
AMPS
10849-EW CLUSTER ASSY. (HEATER)
10849-FW CLUSTER ASSY. (A/C)
TRUNK OPEN   OIL TEMP FUEL   LOW FUEL   FINISH COVER

Lower control housing—1966-69 (© Ford Motor Company)

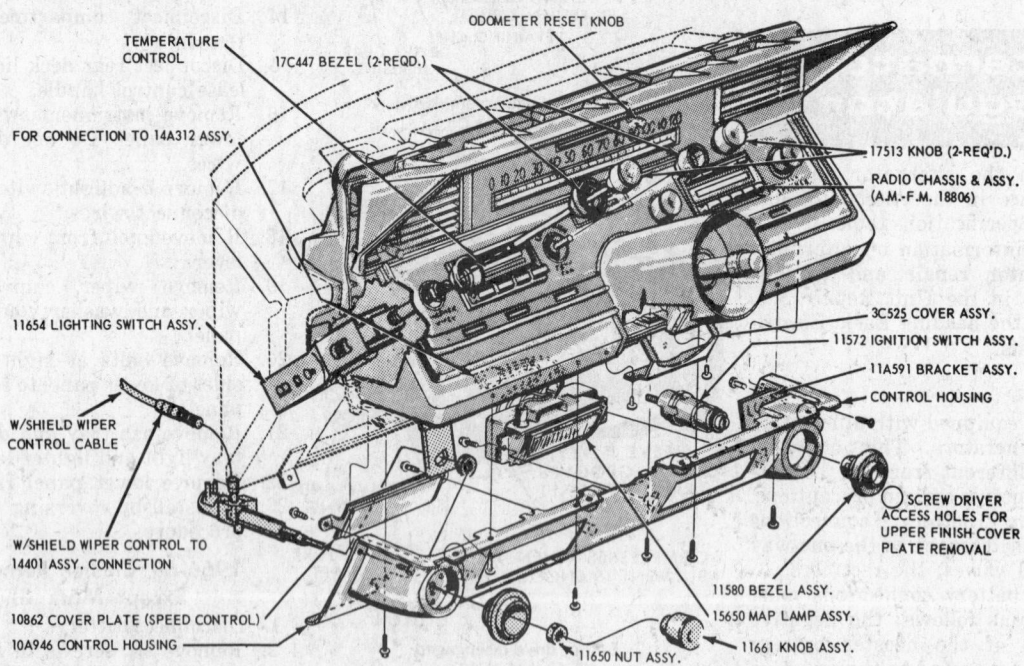

TEMPERATURE CONTROL
ODOMETER RESET KNOB
17C447 BEZEL (2-REQD.)
FOR CONNECTION TO 14A312 ASSY.
17513 KNOB (2-REQ'D.)
RADIO CHASSIS & ASSY. (A.M.-F.M. 18806)
11654 LIGHTING SWITCH ASSY.
3C525 COVER ASSY.
11572 IGNITION SWITCH ASSY.
11A591 BRACKET ASSY.
CONTROL HOUSING
W/SHIELD WIPER CONTROL CABLE
SCREW DRIVER ACCESS HOLES FOR UPPER FINISH COVER PLATE REMOVAL
W/SHIELD WIPER CONTROL TO 14401 ASSY. CONNECTION
10862 COVER PLATE (SPEED CONTROL)
10A946 CONTROL HOUSING
11580 BEZEL ASSY.
15650 MAP LAMP ASSY.
11650 NUT ASSY.
11661 KNOB ASSY.

Typical instrument panel cluster—1966-69 (© Ford Motor Company)

tainers on shafts of radio, heater, air conditioning, and windshield wiper control.

10. Remove eight screws in lower control housing and drop housing.

11. Remove four screws from underside of instrument panel attaching finish cover and remove cover.

12. Remove odometer reset knob.

13. Remove three screws retaining lower edge of upper cluster finish panel.

14. Remove four screws retaining upper edge of upper cluster finish panel.

15. Remove nine screws retaining mask and lens to cover and remove mask and two piece lens.

16. Remove eight screws retaining speedometer dial plate to cover and remove dial. Reverse procedure to install.

1970 Cluster Removal Except Mark III

1. Remove instrument panel pad. Do not remove cluster trim cover, if only instrument cluster is to be removed.

2. Reach under instrument panel and disconnect instrument cluster printed circuit plug from receptacle.

3. From passenger side of instrument panel, remove cluster to cluster housing retaining screws and swing cluster away from housing.

4. From underside of housing unhook pointer control cable from PRNDL pointer lever.

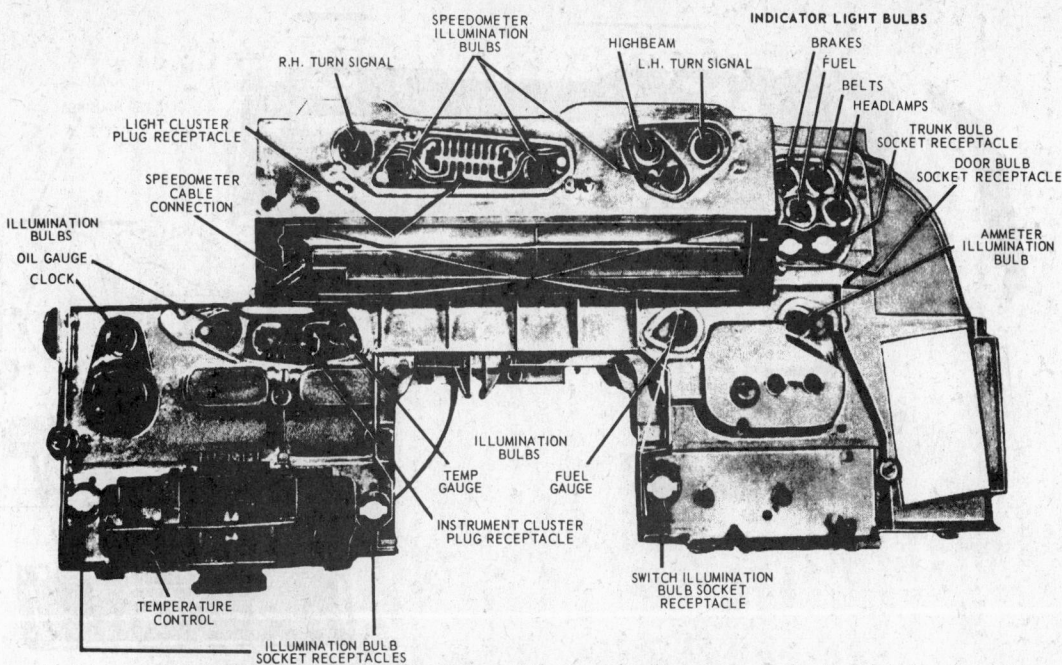

SPEEDOMETER
ILLUMINATION
BULBS

R.H. TURN SIGNAL   HIGHBEAM   INDICATOR LIGHT BULBS
                          L.H. TURN SIGNAL   BRAKES
                                              FUEL
LIGHT CLUSTER   BELTS
PLUG RECEPTACLE   HEADLAMPS
                                              TRUNK BULB
SPEEDOMETER   SOCKET RECEPTACLE
CABLE   DOOR BULB
CONNECTION   SOCKET RECEPTACLE

ILLUMINATION   AMMETER
BULBS   ILLUMINATION
OIL GAUGE   BULB
CLOCK

                          ILLUMINATION
                          BULBS
                          TEMP   FUEL
                          GAUGE   GAUGE

                          INSTRUMENT CLUSTER
                          PLUG RECEPTACLE

                          SWITCH ILLUMINATION
                          BULB SOCKET
                          RECEPTACLE
TEMPERATURE
CONTROL

ILLUMINATION BULB
SOCKET RECEPTACLES

1970 Lincoln Continental instrument cluster—rear view (© Ford Motor Company)

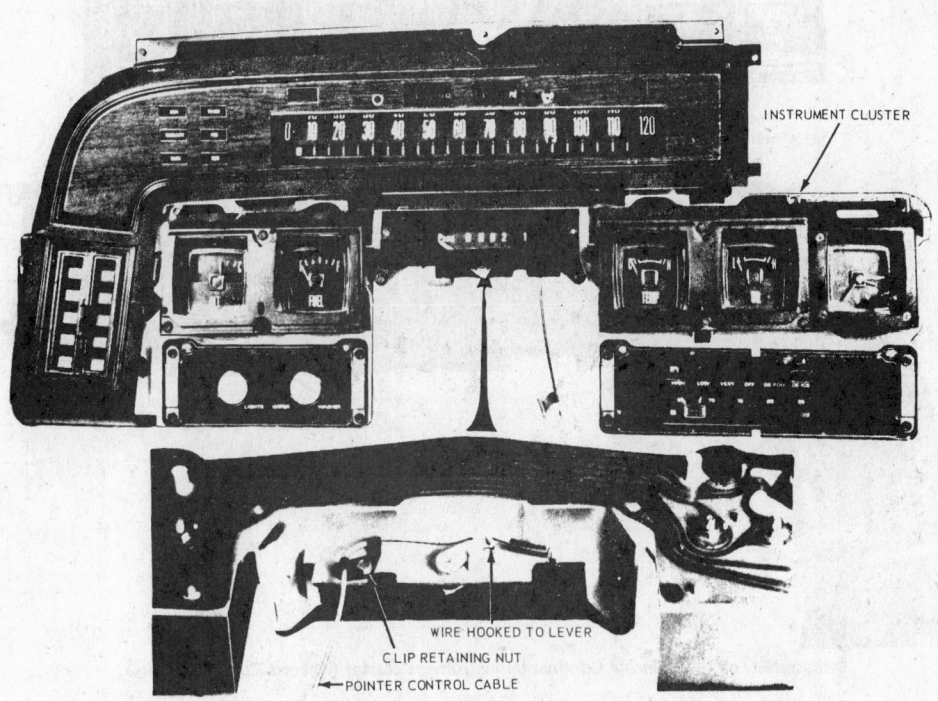

INSTRUMENT CLUSTER

WIRE HOOKED TO LEVER
CLIP RETAINING NUT
POINTER CONTROL CABLE

1970 Lincoln Continental instrument cluster—front view (© Ford Motor Company)

5. Remove cable retaining clip (one nut) from cluster and remove cluster from vehicle.

1968-70 Mark III Instrument Cluster R & R

Removal

1. Disconnect battery ground cable.
2. Remove five attaching screws from upper edge of instrument cluster pad and retainer assembly. Remove the pad and retainer from face of instrument cluster.

3. Remove clock knob. Remove eight push-type retainer buttons from the instrument cluster mask and remove the mask.
4. Remove three screws holding the speedometer to the cluster and pull the speedometer from the cluster. Disconnect the two speedometer cable-to-cluster retaining screws and clamps, release the tab of the plastic retainer and remove it from the cable. If equipped with speed control, disconnect speedometer

cable at speed-control-unit instead.
5. Remove eight screws holding the instrument cluster to the instrument panel. Remove the three screws holding the rear vent and wiper control pod. Pull the cluster and pod out of the panel.
6. Disconnect the multiple connector and the low fuel warning lights at the printed circuit. Then, remove the instrument cluster.

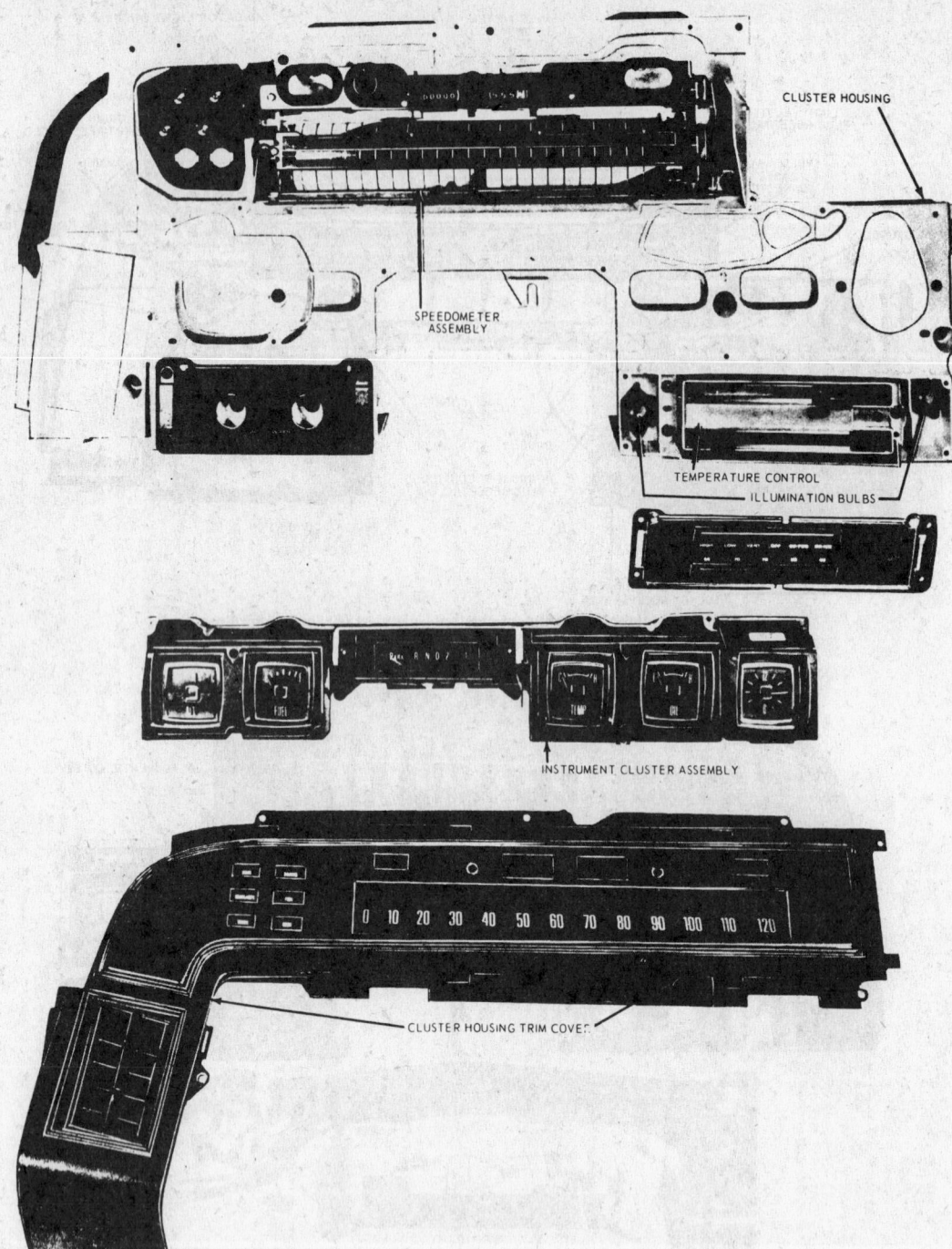

Disassembly of 1970 Lincoln Continental instrument cluster (© Ford Motor Company)

Installation

1. Connect instrument cluster multiple connector to the printed circuit. Install the low fuel warning and dual brake warning lights to the cluster.
2. Guide the speedometer cable through the cluster housing and install the plastic retainer to the cable.
3. Position the instrument cluster to the panel and insert the eight retainer screws. Insert the two speedometer cable retaining screws and clamps.
4. Position the speedometer to the cluster and insert the three retaining screws. Connect the speedometer cable to the speed-control-unit, if so equipped.
5. Position the rear vent and wiper control pod to the panel and insert the three retaining screws.
6. Position the instrument cluster mask and install the eight push-type retainer buttons. Install the clock knob.
7. Position instrument cluster pad and retainer assembly to the face of the cluster and insert the five screws holding the upper edge of the pad to the instrument panel pad.
8. Connect battery ground cable.

Speedometer R & R

1964-65

1. Disconnect battery.
2. Remove heater, air conditioner, blower and radio control knobs.
3. Remove screws holding plate over steering column. Remove plate.
4. Remove screws holding instrument panel bezel to panel and remove the bezel.

5. Disconnect speedometer reset control cable from instrument panel.
6. Remove speedometer to panel retaining screws, position speedometer out of panel and disconnect cable. Remove speedometer and reset cable as a unit.
7. Reinstall by reversing the above.

1966-69 Except Mark III

1. Disconnect battery.
2. Remove six screws from the air-conditioner register casting. Pull register rearward until it is free.
3. Remove two screws in the applique on the left side of cluster.
4. Remove two screws between cluster hood and crash pad.
5. Remove four screws in the lower windshield moulding, and the moulding.
6. Remove two screws and two nuts from the forward edge of the crash pad.
7. Disconnect radio speaker lead and remove crash pad.
8. Remove radio knobs and bezels.
9. Remove heater and air conditioner control knobs.
10. Remove four nuts and retainers on shafts of the radio, heater, air-conditioner and wiper controls.
11. Remove eight screws from the lower control housing and lower the housing.
12. Remove four screws from the underside of the instrument panel attaching the finish cover, and remove cover.
13. Remove odometer reset knob.
14. Remove three screws holding the lower edge of the upper cluster finish panel.
15. Remove four screws holding the upper edge of the upper cluster finish cover and remove cover.
16. Remove four screws holding the right-hand cluster air-conditioning register. Disconnect hose and remove it.
17. Disconnect speedometer cable at the head.
18. Remove three screws that hold the head to the cluster housing, and remove the speedometer head.

1970 Except Mark III

1. Remove retaining screws and remove cluster housing trim panel from cluster housing.
2. Reach up behind instrument panel and disconnect speedometer cable.
3. Remove speedometer to cluster housing retaining screws.
4. Lift speedometer from cluster housing.

1968-70 Mark III

To remove speedometer, follow Steps 1 through 3 under Instrument Cluster R & R.

4. Remove the three screws holding the speedometer to the cluster and pull the speedometer head from the cluster. Transfer the speedometer mounting plate to the replacement speedometer (assuming that a new unit is to be installed). Install the speedometer with the three retaining screws. Complete the installation by following Steps 6 through 8 under Instrument Cluster Installation.

Cable and Housing Renewal

1968-70 Mark III

1. Remove instrument cluster pad and mask, following Steps 2 and 3 under Instrument Cluster R & R.
2. Remove speedometer head as described above.
3. Through the speedometer head opening in the cluster, remove the speedometer housing and retainer screws, clamps, and plastic retainer from the cluster housing. Depress the knurled surface of the cable housing attachment and pull the cable housing off the retainer (collar). Pull the cable housing free from the cluster and partially feed the cable housing through the dash.
4. From the engine compartment, pull the cable housing and grommet from the dash panel.
5. Raise car and disconnect the cable from transmission.
6. Open clip on bottom surface of floor pan and remove the cable.

Ignition Switch Replacement

1964-65
Ignition Lock R & R

1. Insert key and turn to Acc. position.
2. With stiff wire in hole depress lock pin and rotate cylinder counterclockwise. Pull out cylinder.
3. Unscrew bezel from switch and remove switch from panel.
4. Remove insulated plug from rear of switch.
5. Install in reverse of above.

1966-69 Except Mark III

1. Disconnect battery.
2. Remove eight screws from lower control housing, drop the housing.
3. Remove nut holding the wiring connector at the ignition switch and remove connector.
4. Unscrew the bezel and remove the switch.

1968-69 Mark III

1. Disconnect battery.

2. Insert ignition key in switch. Turn key to accessories position. Insert a paper clip end in the hole at the edge of the lock tumbler cylinder. Lightly depress the clip end while turning the key counterclockwise past the accessory position. This will release the lock cylinder from the switch assembly. Pull out the lock cylinder with the key. If only the lock cylinder is to be replaced, proceed to Step 8.
3. Remove bezel nut with special tool T65L-700-A.
4. Lower switch assembly, and remove accessory wire retaining nut. Depress the tabs securing the multiple connector to the rear of the switch with a modified tool 18918-A. Modification of the tool consists of grinding the inside of the tool jaws to produce more clearance around the rear of the switch. Pull multiple connector from switch and remove the switch.
5. To install the switch, depress tabs on multiple connector and plug connector into the switch assembly. Be sure tabs lock into place.
6. Position switch in its retainer and install bezel nut.
7. Insert key in the switch assembly and turn the key to accessory position. Place cylinder and key in switch. Push cylinder into the switch until it is fully seated, then turn key to lock position. Check by turning key to verify lock cylinder operation.
8. Reconnect battery.

1970 All

1. Disconnect battery.
2. Remove shrouding from steering column, and detach and lower steering column from brake support bracket.
3. Disconnect switch wiring at multiple plug.
4. Remove two nuts that retain switch to steering column.
5. Detach switch plunger from switch actuator rod and remove switch.
6. To re-install switch, place both locking mechanism at top of column and switch itself in lock position for correct adjustment. To hold column in lock position, place automatic shift lever in PARK or manual shift lever in reverse, and turn to LOCK and remove key. New switches are held in lock by plastic shipping pins. To pin existing switches, pull switch plunger out as far as it will go and push back in to first detent. Insert 3/32 in. diameter wire in locking hole in top of switch.
7. Connect switch plunger to switch actuator rod.
8. Position switch on column and

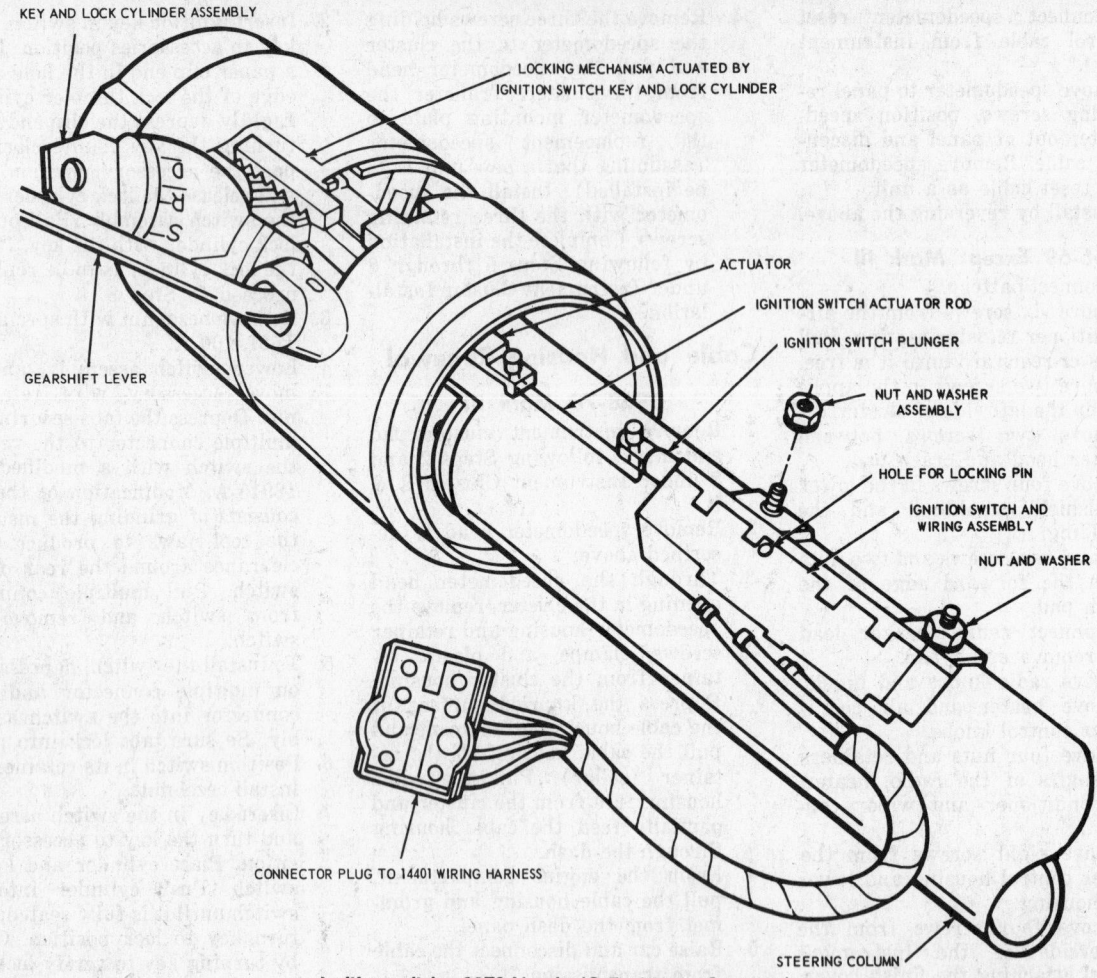

KEY AND LOCK CYLINDER ASSEMBLY

LOCKING MECHANISM ACTUATED BY
IGNITION SWITCH KEY AND LOCK CYLINDER

ACTUATOR

IGNITION SWITCH ACTUATOR ROD

IGNITION SWITCH PLUNGER

NUT AND WASHER
ASSEMBLY

HOLE FOR LOCKING PIN

IGNITION SWITCH AND
WIRING ASSEMBLY

NUT AND WASHER

GEARSHIFT LEVER

CONNECTOR PLUG TO 14401 WIRING HARNESS

STEERING COLUMN

Ignition switch—1970 Lincoln (© Ford Motor Company)

install attaching nuts. Do not tighten them.

9. Move switch up and down to locate mid position of rod lash, and then tighten nuts.
10. Remove locking pin or wire.
11. Attach steering column to brake support bracket and install shrouding.

Lighting Switch

Replacement

Except Mark III
1. Disconnect battery.
2. Remove knob and shaft by pressing release knob button on switch housing and with knob in full on position.
3. Remove moulding nut from switch.
4. Remove junction block from switch.
5. Install in reverse of above.

1968-70 Mark III
1. Disconnect battery ground cable.
2. Remove seven screws holding lower finish panel to lower side of instrument panel.
3. Remove control knob and shaft from headlight switch. This is

done by pressing the release button on the underside of the switch (with shaft pulled all the way out).
4. Remove bezel and nut from headlight switch.
5. Remove two screws from switch-to-instrument panel.
6. Remove switch from panel. Disconnect wire multiple connector and vacuum hoses from the switch.
7. Install by reversing removal procedure.

Rear Lights

1963-65
To replace tail, stop and turn signal bulbs, pull luggage compartment liner away from lamp body and remove bulbs.

Neutral Safety and Back-Up Light Switch Assembly

Switch Adjustment

1964-65
1. With manual lever properly adjusted, loosen the two switch attaching bolts.
2. With the transmission lever in

neutral, rotate the switch and insert the gauge pin A (No. 43 drill shank end) into the gauge pin holes of the switch. The gauge pin must be inserted to a full 13/64 in. into all three gauge holes of the switch.
3. Torque the switch attaching bolts to 55-75 ft. lbs. Then, remove gauge pin from the switch.
4. Check operation of the switch. The engine should start only with the selector lever in neutral and park. The back-up light should burn only with the selector in reverse.

1966-67
1. With manual linkage properly adjusted, try to engage starter in each position on quadrant. Starter should engage only in park and neutral positions.
2. To adjust loosen screws that locate switch on steering column.
3. Place shift lever in neutral detent.
4. Rotate switch until gauge pin (No. 43 drill) can be inserted into gauge pin hole 31/64 in.
5. Tighten down locating screws and check starter engagement in each position as in step one.

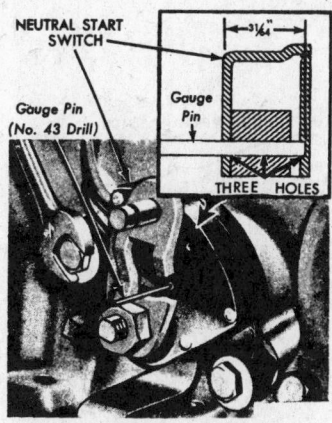

Neutral start switch—1966-67
(© Ford Motor Company)

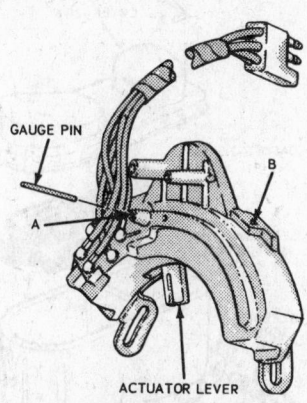

Neutral start switch—1968-70
(© Ford Motor Company)

1968-70

1. With manual linkage properly adjusted, try to engage starter in each position on quadrant. Starter should engage only in park or neutral positions.
2. Place shift lever in neutral detent.
3. Disconnect start switch wires at plug connector. Disconnect vacuum hoses if any. Remove screws securing neutral start switch to steering column and remove switch. Remove actuator lever along with Type III switches.
4. With switch wires facing up move actuator lever fully to the left and insert gauge pin (No. 43 drill) into gauge pin hole at point A. See accompanying figure. On Type III switch, be sure gauge pin is inserted a full 1/2 in.
5. With pin in place, move actuator lever to right until positive stop is engaged.
6. On Type I and Type II switches remove gauge pin and insert it at point B. On Type III switches remove gauge pin, align two holes in switch at point A and reinstall gauge pin.
7. Reinstall switch on steering column. Be sure shift lever is engaged in neutral detent.
8. Connect switch wires and vacuum hoses and remove gauge pin.
9. Check starter engagement as in Step 1.

Brakes

Information on brake adjustments, band replacement, bleeding procedure, master and wheel cylinder overhaul is in the Unit Repair Section.

Information on troubleshooting and overhauling power brakes is in the Unit Repair Section.

Information on the grease seals which may need replacement, is in the Unit Repair Section.

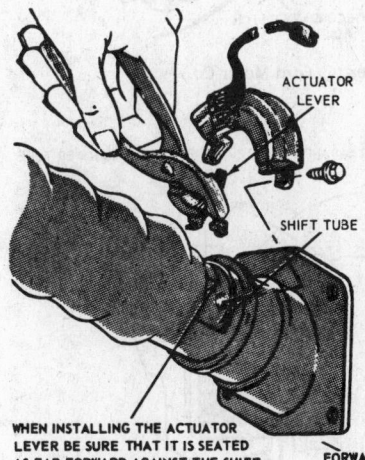
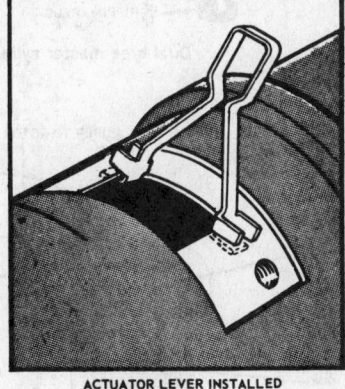

Removing or installing switch actuator—1968-70 (© Ford Motor Company)

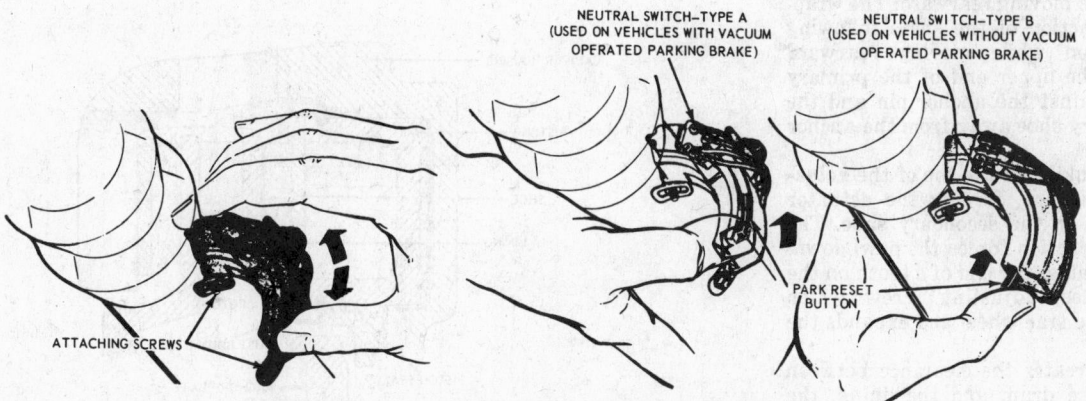

Adjusting neutral start switch—1968-70 (© Ford Motor Company)

1965-70

Front wheel disc brakes, a master cylinder with diaphragm-type seal, and new tandem diaphragm vacuum power booster are used. See Unit Repair Section.

Brake Pedal Clearance

The pedal clearance is adjusted at the brake pushrod and should allow approx. ¾ in. to 1 in. free motion of the pedal before the master cylinder piston starts to move.

Disc Brakes

Since 1965, disc brakes have been available on front wheels of some models. Complete service procedures are covered in the Unit Repair Section.

Parking Brake Cable Adjustment

1. Fully release the parking brake pedal.
2. Raise the car on a hoist.
3. Adjust the pedal cable to about 10 in., measured from the cable attachment at the crossmember to the cable adjusting nut.
4. Depress the parking brake pedal one notch from normal, released position.
5. Loosen locknut on equalizer rod and turn the forward nut inward toward the front of the car until a moderate drag is felt when turning the rear wheels.
6. Holding forward nut in position, tighten locknut. Lock the adjustment at the equalizer.
7. Release parking brake, and make sure that the brake shoes return to the fully released position.

Brake Adjustment—1963-64 All Models 1965-70 Rear Brakes Only

A self-adjuster has been added to the service brake mechanism.

General Description

The automatic adjusters operate only when the brakes are applied as the car is moving rearward. The wrap-around action of the shoes following the drum while moving rearward forces the upper end of the primary shoe against the anchor pin and the secondary shoe away from the anchor pin.

The link holds the top of the actuator stationary, forcing the actuator to pivot on the secondary shoe. The pivoting action forces the pawl downward against the end of a tooth on the star wheel adjusting screw which turns the star wheel and expands the shoes.

The greater the clearance between the brake drum and the lining, the greater the travel of the secondary

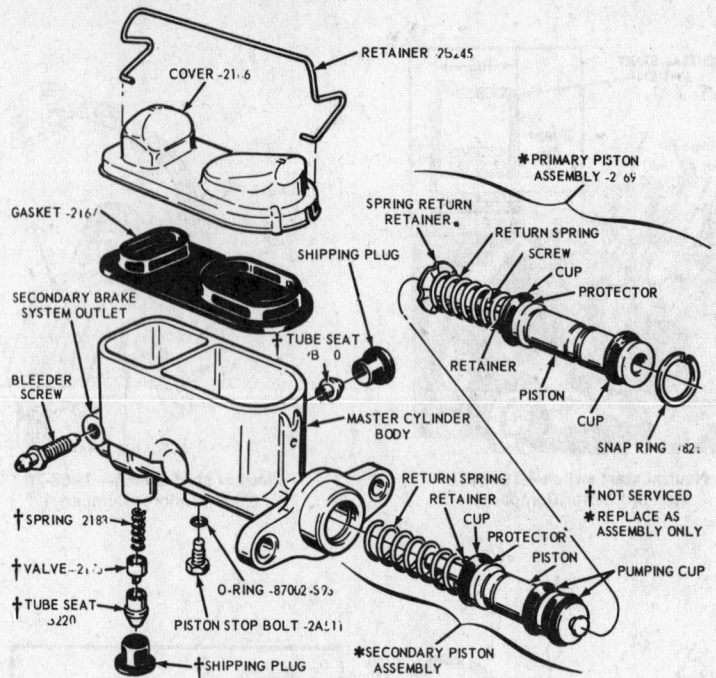

Dual type master cylinder (© Ford Motor Co.)

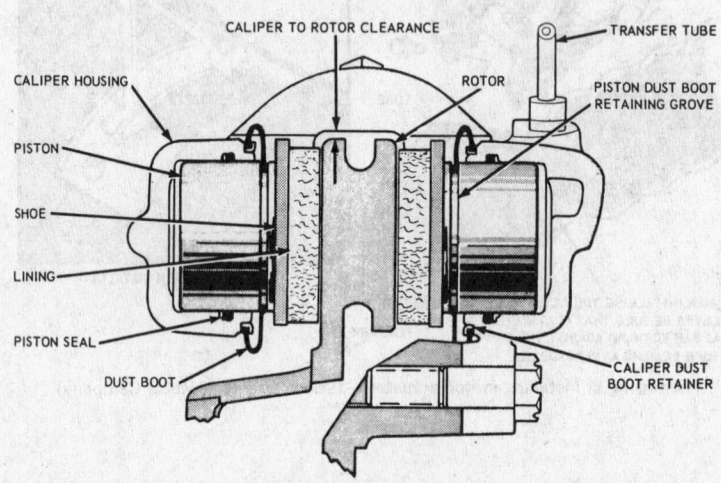

Disc brake caliper assembly—1965-69 Continental except Mark III (© Ford Motor Co.)

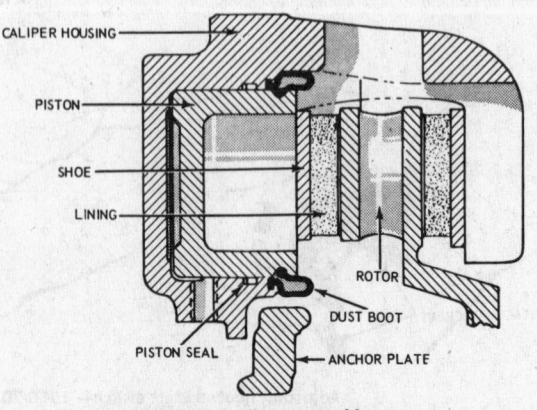

Disc brake caliper assembly—
1970 Continental and 1968-70 Mark III
(© Ford Motor Company)

shoe away from the anchor. The more the secondary shoe moves, the greater the adjustment. When the brakes are adjusted correctly, there will not be sufficient travel of the secondary shoe to permit the actuator to pivot and force the pawl to engage against the end of a tooth of the star wheel adjusting screw, and turn the wheel to expand the shoe.

When the brakes are applied as the car is moving forward, the self-adjuster does not operate because the wraparound action of the shoes forces the secondary shoe against the anchor pin.

The rear brake assembly is basically the same as the front brake, except that the conventional parking brake operating lever, spring, and parking brake strut rod are used in the rear brake. The anchor pin on all brakes can be adjusted when necessary.

Whenever removing or replacing a star wheel assembly, special care must be exercised to be certain that the correct star wheel assembly is installed on the right brake drum to enable the self-adjuster to function properly.

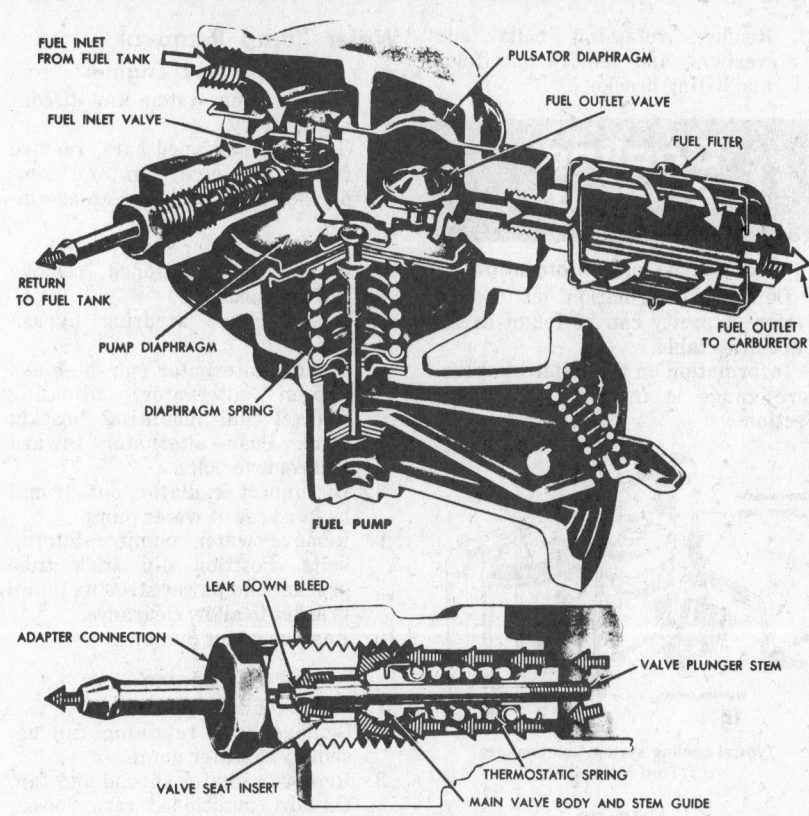

Fuel pump and thermostatic valve (© Ford Motor Co.)

Fuel System

Fuel System Information

Data on capacity of the gas tank are in the Capacities table. Data on correct engine idle speed and fuel pump pressure are in the Tune-Up Specifications table.

Information covering operation and troubles of the fuel gauge is in the Unit Repair Section.

Detailed information on the carburetor, and how to adjust it, is in the Unit Repair Section. Carter, Holley, Rochester, and Stromberg carburetors are covered.

Dashpot adjustment is in the Unit Repair Section under the same heading as that of the automatic transmission used in the car.

Fuel Pump Removal

1963-68

The fuel pump is mounted on the top portion of the engine front cover.

To remove the pump, disconnect the fuel and vapor discharge connections. The pump can then be unbolted and lifted off.

On power-steering models, the bolts are accessible from under the car.

1968-70 460 Engine

The fuel pump is mounted on the left side of the cylinder front cover.

A separate in-line fuel filter is used. The filter cannot be serviced. Renew it in case of obstruction. This pump is spring loaded in opposition to cam-

shaft eccentric lobe action, and is conventional.

The pump is Carter-built and can not be serviced.

Exhaust System

Intake Manifold Removal

Remove the carburetor air cleaner and the radiator upper hose and by-pass hose. Remove distributor. Remove all lines to the carburetor, including the throttle rods, and take off the carburetor. Remove the ignition coil and the carburetor throttle rod bracket from the manifold. Disconnect the heat-sending unit wire and the power-brake vacuum lines. Disconnect all attaching lines to the intake manifold. Unbolt and lift off the manifold.

Exhaust Manifold Removal— 430 and 462 Engines

1. Remove air cleaner and air inlet duct assembly. Block rear wheels and set parking brake. Raise front of car and install safety stands.
2. Disconnect the exhaust manifold/s at the resonator inlet pipe/s.
3. Detach the engine front support insulators from the underbody side members.

4. Place a jack under the front edge of the oil pan. Raise the front of the engine about 2 in. to make clearance for removal of the exhaust manifold/s.

Caution When raising the engine, use care to prevent forcing the engine against the automatic temperature control case (if so equipped) in the engine compartment. Position 2 in. wood blocks between the front support insulators and underbody side members. Remove the jack and let the engine rest on the wood blocks.

5. Unlock and remove the manifold lower retaining bolts. Remove safety stands and lower the car.
6. Remove the two automatic choke tubes from the right exhaust manifold.
7. If the engine is equipped with an exhaust emission control system, remove the air hose/s from the air manifold/s.
8. Unlock and remove the exhaust manifold retaining nuts, manifold/s and gasket/s.
9. Install by reversing the above.

Exhaust Manifold Removal— 460 Engine

1. Remove air cleaner and warm air duct assembly to remove right exhaust manifold.
2. Disconnect manifolds at exhaust pipe.

3. Remove retaining bolts and washers, and remove manifolds and lifting brackets.

Cooling System

Cooling System Information

Detailed information on cooling system capacity can be found in the Capacities table.

Information on the water temperature gauge is in the Unit Repair Section.

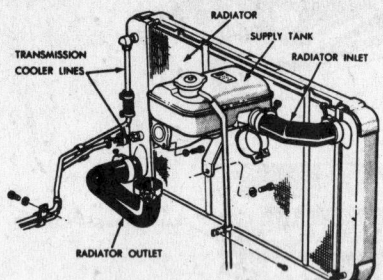

Typical cooling system components
(© Ford Motor Co.)

1963-70

The radiator is cellular-tubular cross flow. The coolant tanks provide a means of cooling the power steering fluid and the transmission fluid. A supply tank provides a means of controlling surge and permits lowering of hood line.

A power booster fan is used on air-conditioned cars.

Radiator Core Removal

On all models, the radiator is removed in practically the same way. Disconnect the radiator hoses, disconnect the wires from the radiator frame and take out the bolts that hold the radiator core to its mounting. Lift it straight up.

Water Pump Removal

430 and 462 Engines

1. Drain cooling system and disconnect battery.
2. On air conditioned cars, remove fan drive clutch, fan, and compressor drive pulley as an assembly.
3. Remove radiator supply tank.
4. If not air conditioned, remove fan and spacer.
5. Loosen clamp securing bypass hose.
6. Remove alternator splash shield.
7. Loosen alternator adjusting bracket and mounting bracket bolts. Push alternator inward and remove bolts.
8. Disconnect radiator outlet and heater hose at water pump.
9. Remove water pump retaining bolts. Position dip stick tube bracket and power steering pump bracket to allow clearance.
10. Remove water pump.

460 Engine

1. Drain cooling system.
2. Remove bolts retaining fan assembly to water pump.
3. Remove radiator shroud and fan.
4. On air conditioned cars, loosen compressor drive belt.
5. Loosen alternator mounting bolts and remove alternator drive belt.
6. Remove water pump pulley.
7. Disconnect radiator lower hose, heater hose, and bypass hose at water pump.
8. Remove water pump bolts and remove water pump.

Engine

References

The Specifications table of this section lists all the available facts about the engines.

Lincoln Engines—1964-1970

From 1958 to 1965 Lincoln used a bulky, heavy 430 cu. in. V8. Big though it was, this engine was admirably suited for use in a luxury car. The engine was characterized by a relatively short stroke, a deep Y-block construction, and most unusual cylinder heads. The head mating surface was flat with no indentation for a combustion chamber, and was mounted on the block at approximately 60° to the axis of the bore. The angle between the cylinder head and the piston formed the combustion chamber.

In 1966 Lincoln increased the bore to 4.380 in. and the stroke to 3.830 in. for 462 cu. in. This engine continued in use through 1968.

During 1968 the 460 cu. in. engine was introduced. Throughout that year it was mixed indiscriminately with the 462 engine in Lincoln production, only the new Mark III used the 460 exclusively. Because of this mixing and because the bore, 4.36 in., and stroke, 3.85 in., and displacement are close to those of the older engine, many people confuse the two. The new engine differs substantially. It has canted valves, stud mounted rocker arms, semi-hemispherical combustion chambers, tunnel ports, a block split at crankshaft centerline, an intake manifold that replaces the valley cover, and a much lighter weight. Since 1969 this engine has been used exclusively.

Valves

Methods of troubleshooting hydraulic valve lifters are in the Unit Repair Section.

Detailed information on the valves is in the Valve Specifications table.

A general discussion of valve clearance, and a chart showing how to read pressure and vacuum gauges when using them to diagnose engine troubles, is in the Unit Repair Section, and, under the same head, a chart on engine troubleshooting.

Valve tappet clearance for each engine is given in the Tune-Up Specifications table.

Bearings

Detailed information on engine bearings will be found in the Crankshaft Bearing Journal Sizes table of this section.

Exhaust Emission Control

In compliance with anti-pollution laws, the Ford Motor Company has adopted a distributor and a modified carburetor with some engine changes, to reduce terminal exhaust fumes to an acceptable level. This method is known as Ford's Improved Combustion System (IMCO).

The plan supersedes (in most cases) the previous method used to conform to 1966-67 California laws. The new system phases out (except with stick shift and special purpose engine applications) the thermactor,

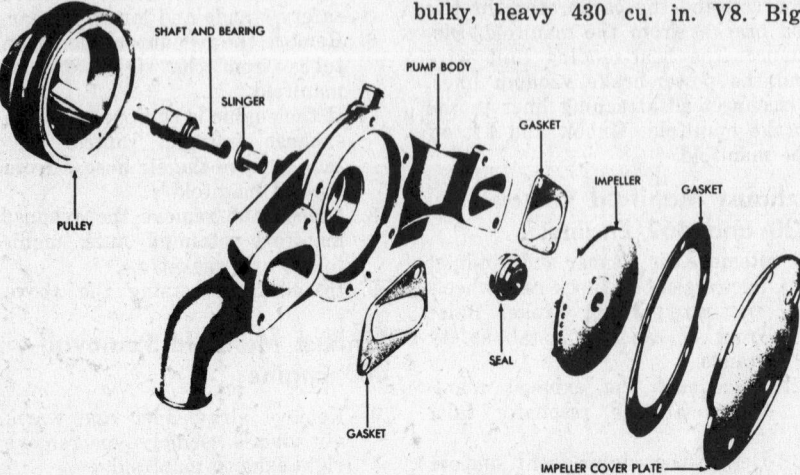

Water pump—430 and 462 V8 (© Ford Motor Co.)

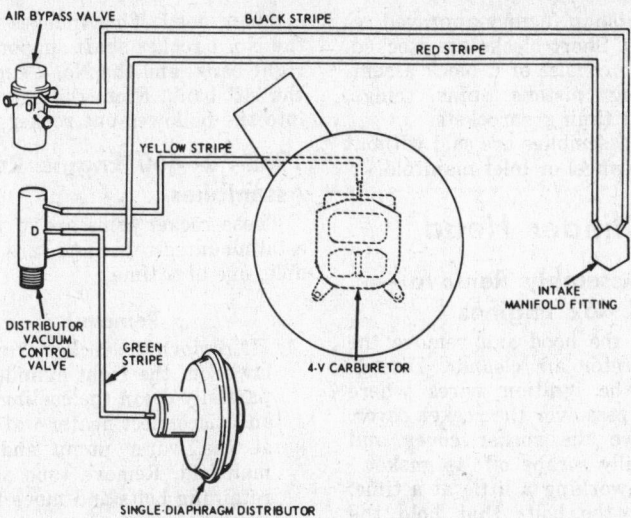

Ignition vacuum schematic with Thermactor emission with or without A/C
(© Ford Motor Co.)

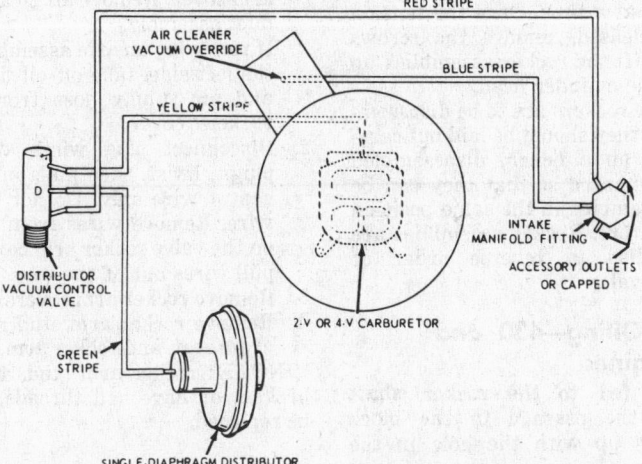

Ignition vacuum schematic with Imco emission system (© Ford Motor Co.)

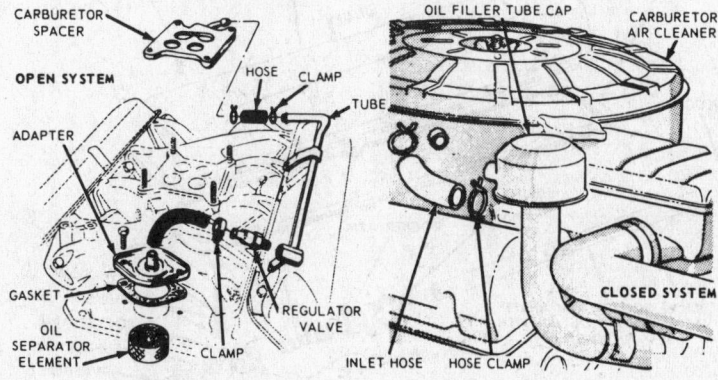

Crankcase ventilating valve—430 and 462 V8 (© Ford Motor Co.)

or afterburner type of exhaust treatment.

The IMCO concept utilizes broader, yet more critical, distributor control through carburetor modification.

Since 1968, all car makers have posted idle speeds and other pertinent data, relative to the specific engine application in a conspicuous place in the engine compartment.

For details on the IMCO system, consult the Unit Repair Section.

Engine Assembly Removal

1963-70

Engine R & R is for the engine only, without the transmission attached.

1. Raise the hood, and cover or mask all parts of the car that could be scratched during R & R procedures.
2. Set the parking brake and raise the car. Put stands beneath the underbody front crossmember.
3. Drain the engine cooling system and the engine oil pan.
4. Scribe the hinge outline on the underside of the hood. Remove hood.
5. If the engine is equipped with an exhaust emission control system, remove the crankcase vent filter from the air cleaner. Remove carburetor air cleaner and air inlet duct assembly. Disconnect the battery ground.
6. Remove both engine radiator hoses.
7. Disconnect heater hoses at intake manifold and water pump. Disconnect power brake and power booster line from the intake manifold connection and position it to one side.
8. Disconnect heater vacuum hose from the intake manifold.
9. Disconnect automatic transmission vacuum line at the intake manifold.
10. Remove transmission tube slotted bracket from the right rear exhaust manifold mounting stud.
11. Disconnect battery ground strap at cylinder block.
12. Disconnect primary wires at the coil. Disconnect wires from temperature-sending unit and the fast idle solenoid (air-conditioned cars).
13. Disconnect wire from oil pressure-sending unit. Detach wiring loom from valve rocker arm cover and position it out of the way.
14. Disconnect transmission fluid lines at the radiator. Remove transmission fluid filter from underbody side member (if car is so equipped).
15. Remove fuel hose mounting bracket from radiator. Remove heat shield from fuel pump. Disconnect hoses from fuel pump.
16. On air-conditioned cars, remove fan drive clutch to water pump pulley retaining bolts. Remove fan drive clutch, fan and compressor pulley from the car as a unit.
17. Remove fan blade and spacer assembly from water pump pulley.
18. On an air-conditioned car, remove the radiator lower mounting bolts. Remove the radiator to condenser attaching bolts and the radiator-to-radiator support bolts. Remove the radiator. Disconnect the compressor clutch wire and remove compressor support bracket. Remove the compressor high and low pressure

service valve caps, and fully seat the high and low pressure service valves.

Loosen the manifold gauge port caps slightly, and slowly bleed the refrigerant from the compressor. Tighten the port caps when all the refrigerant has been bled. Remove the high and low pressure valves and muffler assembly from the compressor. Cover all openings to prevent contamination. Position service valves and lines against the fender apron for clearance purposes. Disconnect fast idle control rod at accelerator shaft assembly.

19. Remove lower mounting bolts from radiator. Remove the radiator.
20. Disconnect the transmission and accelerator linkage at the bellcrank. Secure the linkage to the dash panel for engine clearance purposes.
21. Remove access cover from the converter housing. Remove underbody splash shield at lower front of transmission.
22. Remove resonator inlet pipes from the exhaust manifolds.
23. Remove the clamp holding the power steering pump outlet line to the mounting bracket. Disconnect power steering outlet line from the power steering pump. Drain the fluid into a clean container.
24. Remove the nuts and washers that hold the engine front support insulators to the underbody side members.
25. Remove the starter attaching bolts. Remove the starter.
26. Disconnect the generator wires. Detach the oil cooler inlet and outlet transfer line retaining clip from the cylinder block.
27. Remove the flywheel to converter retaining nuts.
28. Remove lower converter housing to cylinder block retaining bolts.
29. Install a transmission support under the transmission.
30. Remove the upper converter housing to cylinder block retaining bolts.
31. On a car equipped with air conditioning, remove compressor support bracket from cylinder head and intake manifold.
32. Attach engine lifting eyes (tool 6000-K) (on Mark III T53L-300-A) to the exhaust manifolds.
33. Install lifting sling and attach to chain hoist. With plenty of help, carefully raise and remove engine from car.
34. Install by reversing removal procedure.

Replacement Engines

Replacement engines are available for all Lincoln models, and they are also available in factory-approved rebuilt types. Short blocks are stocked, also, which consist of a block assembly having pistons, pins, rings, valves and timing sprockets.

Engine assemblies are sold without oil pan, flywheel or inlet manifold.

Cylinder Head

Rocker Assembly Removal— 430 and 462 Engines

1. Raise the hood and remove the carburetor air cleaner. Disconnect the ignition wires where they cross over the rocker cover. Remove the rocker cover and carefully scrape off its gasket.
2. Now, working a little at a time, loosen the bolts that hold the rocker brackets to the cylinder head so that the tension of the valve springs will be left off, a little at a time. Once the tension is released, remove the screws and lift the rocker assemblies up off the cylinder head.
3. If the rockers are to be disassembled, they should be laid out carefully on a bench, disassembled and marked so that they can be reassembled in the same position as before. Rocker assemblies are installed in reverse order of removal.

Rocker Oiling—430 and 462 Engines

Oil is fed to the rocker shaft through the passage in the block that lines up with the hole in the cylinder head. This indexes through the No. 1 rocker shaft support on the right bank, and the No. 2 support on the left bank. From there it goes up into the hollowed-out rocker shafts.

1968-70—460 Engine Rocker Assemblies

These rocker arms are of the pedestal-mounted-type and are removable, one at a time.

Removal

1. If removing a rocker arm assembly from the right cylinder head, partially drain the cooling system and disconnect heater water tubes at the water pump and intake manifold. Remove tube assembly retaining bolts and move tube assembly out of the way. Remove crankcase ventilation regulator valve and hose from valve rocker arm cover. Remove air cleaner and duct assembly.
   If removing an arm assembly from the left side, take off oil filler cap and air supply hose from valve rocker cover.
2. Disconnect plug wires at spark plugs. Twist, then pull, on molded cap of wire only. Do not pull the wire. Remove wires from bracket on the valve rocker arm covers and pull wires out of the way.
3. Remove rocker arm covers.
4. Remove rocker arm stud nut, fulcrum seat, and rocker arm.

NOTE: rocker arm studs that are broken, or have bad threads, should be replaced.

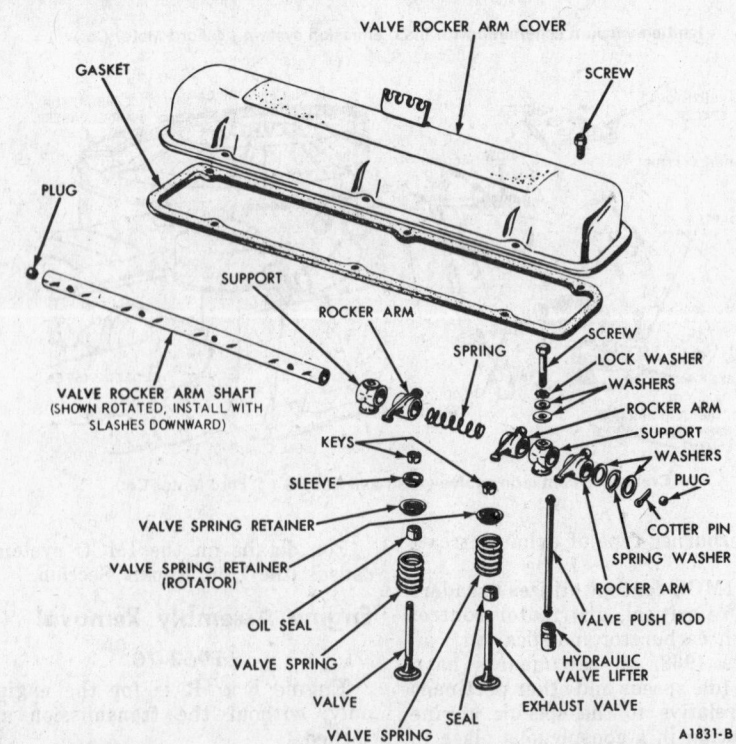

Rocker shaft and valve components—430 and 462 V8 (© Ford Motor Co.)

If the stud is broken, flush with the head, drill and use an easy-out.

When installing the new stud, lubricate the threads, then torque to 65-75 ft. lbs.

Installation

1. Apply lubriplate to top of valve stem.
2. Lubriplate fulcrum seat and socket. Install rocker arm, fulcrum seat and stud nut.
3. Adjust valve clearance according to recommendations.
4. Clean rocker arm covers and cylinder head gasket surfaces.
5. Apply oil-resistant sealer to one side of new cover gaskets. Apply cemented side of gaskets in rim of covers.
6. Position covers on cylinder heads. Install and torque cover bolts to 2½-4 ft. lbs. Two minutes later, retorque attaching bolts to same specifications.
7. Route spark plug wires in brackets on valve rocker covers. Reconnect plug wires.
8. Install heater tube assembly, if disconnected, and fill cooling system.
9. On the right valve rocker arm cover, install crankcase ventilation regulator valve and hose.
10. Install air cleaner and duct, and adjust assembly, if removed. On the left rocker arm cover, install oil filler cap and air supply hose.

Cylinder Head Removal

All Engines

Follow the procedure given for the rocker shafts and for the inlet manifolds. Disconnect the exhaust pipes at their flanges, remove the bolts that hold the cylinder heads to the cylinder blocks and lift off the heads.

Disassembly of Cylinder Heads

1. Remove cylinder heads.
2. Compress valve springs using valve spring compressor.
3. Remove valve locks or keys.
4. Release valve springs.

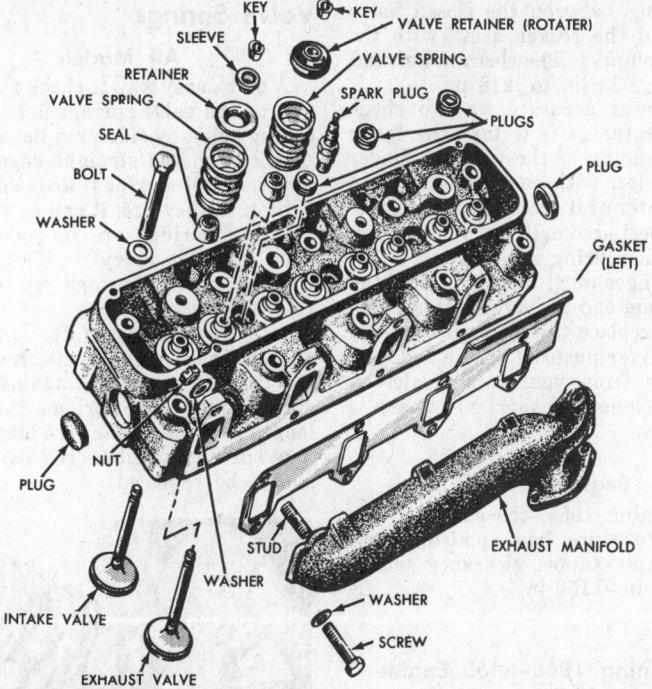

Cylinder head and parts layout—430 and 462 V8 (© Ford Motor Co.)

⏻ CHILTON TIME-SAVER

Cylinder Head Removal— 460 Engines

On 460 engines it is not necessary to remove rockers to remove head. Simply loosen rocker adjusting nuts until rocker can be turned to side and withdraw pushrods in sequence. Remove cylinder head bolts and remove heads.

Disassembly of cylinder head (© Ford Motor Company)

5. Remove valve springs, retainers, oil seals, and valves.

NOTE: if a valve does not slide out of the guide easily, check end of stem for mushrooming or heading over. If head is mushroomed, file off excess, remove and discard valve. If valve is not mushroomed, lubricate stem of valve, remove, and check for stem wear or damage.

Valve System

All Models

Caution The hydraulic valve lifters are not to be interchanged from one bore to the other. They should be carefully marked so that they will be returned to the bore from which they were removed.

None of the internal parts of the lifters are interchangeable with any other lifter.

Valve Adjustment

1963-64

Valve stem to rocker arm (operating) clearance is zero; however, conditions may change the original setting and adjustment may become necessary.

Clearance must be determined by

Cylinder head removal 460 V8 (© Ford Motor Company)

measuring between the closed valve stem and the rocker arm (with the tappets empty). The clearance should be from .078 in. to .218 in.

The most accurate way to check these clearances is to take the feeler gauge reading of the valves, cylinder by cylinder, with each piston at top dead center of its compression stroke. This is best accomplished by following the regular firing order.

If changes must be made, grind the valve stem end to increase this clearance or replace the pushrod with the next shorter pushrod. These rods are available from your local dealer in .060 in. (longer or shorter) phases.

Beginning 1965

Beginning 1965, the same adjustment procedure has applied. However, empty tappet clearance should be 0.050 in.-0.150 in.

Beginning 1968—460 Engine

Cylinder numbering has not been changed, right bank—1, 2, 3, 4. Left Bank—5, 6, 7, 8.

Valves on the right bank are, from front to rear, I-E-I-E-I-E-I-E.

Valves on the left bank are, from front to rear, E-I-E-I-E-I-E-I.

1. Remove valve rocker arm covers.
2. Disconnect the brown lead (I terminal) and the red and blue lead (S terminal) at the starter relay. Install an auxiliary starter switch between the battery and S terminals of the starter relay. Crank the engine with the ignition switch off until No. 1 cylinder is at the top of its compression stroke (distributor rotor pointing to distributor cap No. 1 plug wire tower).
3. To adjust the intake and exhaust valve clearance for No. 1 cylinder, loosen the rocker arm stud nut until there is looseness in the pushrod. Then, tighten the nut to remove all of the rod-to-rocker arm clearance. This may be determined by attempting to rotate the pushrod with the fingers. When pushrod-to-rocker arm clearance has been eliminated, tighten the stud nut exactly one additional turn. This will place the hydraulic lifter plunger in the desired operating range (about midway of its travel limits).
4. Rotate crankshaft to the next cylinder in sequence of firing order (the distributor rotor and major point opening will indicate this approximate position).
5. Repeat this valve adjustment procedure on each succeeding set of valves, following the regular firing sequence until the entire valve train has been adjusted.

Valve Springs

All Models

A quick easy way to check the condition of the valve springs is to put the springs side-by-side on a flat surface. Check with the straight edge to see whether all springs are the same height. If they are, it can be assumed that the springs are in good condition, since it is extremely unlikely that all of the springs will collapse the same amount.

If, however, any of the springs is shorter than the rest, the free length will have to be checked against a new spring. Only those springs whose free length is equal to the free length of a new spring should be reused. Others should be replaced.

Adding spacer to correct the valve spring assembled height
(© Ford Motor Co.)

VALVE GUIDE

Tool—6085-H
Reamer Kit
6085-H-1 Standard Size
6085-H-2 0.003 Oversize
6085-H-3 0.015 Oversize
6085-H-4 0.030 Oversize
or Kit T58P-6085-B

Reaming valve guides
(© Ford Motor Company)

that cleans up wear. If a large oversize is required, it is best to approach that size through stages by using a series of reamers of increasing diameters. This helps to maintain the concentricity of the guide bore with the valve seat. The correct valve guide to stem clearance is in front of this section. As an alternative, some local automotive machine shops will fit replacement guides that use standard stem valves.

⏻ CHILTON TIME-SAVER

The following is a method for replacing valve springs, oil seals or spring retainers without removing the cylinder head.

1. Entirely dismantle a spark plug and save the threaded shell.
2. To this shell, braze or weld an air chuck.
3. Remove the valve rocker cover. Remove the rocker arm from the valve to be worked on.
4. Remove the spark plug from the cylinder to be worked on.
5. Turn the crankshaft to bring the piston of this cylinder down, away from possible contact with the valve head. Sharply tap the valve retainer to loosen the valve lock.
6. Then turn the crankshaft to bring the piston in this cylinder to the Exact Top of its Compression Stroke.
7. Screw in the chuck-equipped spark plug shell.
8. Hook up an air hose to the chuck and turn on the pressure (about 200 lbs.).
9. With a strong and constant supply of air holding the valve closed, compress the valve spring and remove the lock and retainer.
10. Make the necessary replacements and reassemble.

NOTE: it is important that the operation be performed exactly as stated, in this order. The piston in the cylinder must be on exact top-center to prevent air pressure from turning the crankshaft.

Valve Guides

Lincolns use integral valve guides. Lincoln dealers offer valves with oversize stems for worn guides. To fit these, enlarge valve guide bores with valve guide reamers to an oversize

Hydraulic Value Lifter Disassembly

Disassemble lifters for cleaning only. No repairs are permitted.

1. Grasp lock ring with needle nose pliers and remove. Depress

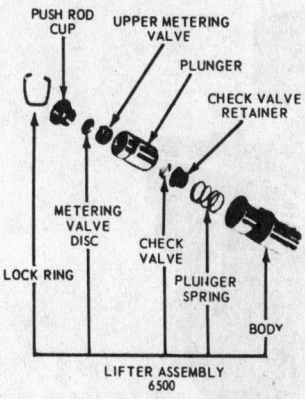

Typical hydraulic lifter
(© Ford Motor Company)

plunger to gain clearance for pliers.

2. Remove pushrod cup, metering valve disc, and upper metering valve, if any. Do not bend upper metering valve.
3. Remove plunger assembly, and plunger spring.
4. Remove spring, check valve retainer and check valve from plunger.
5. Clean thoroughly in solvent and assemble.

Timing Case Cover Chain and Sprockets

Timing Case Cover Removal
1963-68, Except 460 Engine

1. Drain entire engine cooling system.
2. Disconnect all radiator hoses and the overflow pipe from the coolant supply tank.
3. Remove the bolt securing the supply tank brace, engine ground strap and battery ground cable to the water pump. Remove supply tank thermostat and gasket. Loosen coolant by-pass hose at water pump.
4. On a car equipped with air-conditioning, loosen the compressor support bracket bolts. Remove

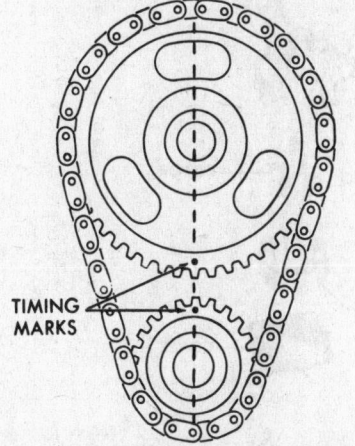

Valve timing alignment marks

TIMING MARKS

the drive belt, then, remove the fan drive clutch and fan assembly and the compressor drive pulley as a unit.
5. Loosen the generator, remove the fan and generator drive belts. Remove the fan blade assembly, spacer and mounting bolts from the water pump pulley as a unit.
6. Disconnect the dipstick tube bracket at the water pump. Remove the steering pump reservoir bracket from the water pump and loosen the remaining bracket mounting bolts to allow clearance for removal of the water pump.
7. Remove the water pump. Remove the crankshaft damper attaching bolt. Remove the damper.
8. Disconnect power steering lines at the pump and plug the lines. Remove bolts holding the power steering reservoir to the front cover.
9. Remove the crankshaft damper key and remove the power steering pump.
10. Remove the shield from the fuel pump, fuel lines from the pump; remove the pump. Remove the cup-type plug from the top of the cylinder front cover by using a long punch. Remove the fuel pump push rod.
11. Raise front of car and install safety stands.
12. Remove cylinder front cover attaching bolts and remove front cover.
13. Install by reversing above procedure.

1968-70 460 Engine

Removal

1. Drain cooling system and crankcase.
2. Remove oil pan and oil pump.
3. Remove fan blades from water pump shaft.
4. Remove radiator (fan) shroud.
5. Disconnect all radiator hoses at engine. Disconnect oil cooler lines.
6. Remove radiator.
7. Loosen alternator. Loosen air conditioner idler pulley. Remove drive belts with water pump pulley.
8. Remove air conditioner compressor (do not open compressor lines to expose the sealed air-conditioner system to atmosphere).
9. Remove crankshaft pulley attaching bolt and washer. Remove damper with tool T58P-631-B and remove Woodruff key from crankshaft.
10. Disconnect power steering pressure line at pump. Drain fluid.
11. Remove steering pump.
12. Loosen by-pass hose at water

pump. Disconnect heater hose at pump.
13. Disconnect and plug fuel inlet line at fuel pump. Disconnect fuel line at carburetor fuel pump. Remove fuel pump.
14. Remove front cover-to-block attaching bolts. Remove front cover and water pump as an assembly. Discard gasket.
15. If a new front cover is to be installed, change the water pump at this time.
16. Check timing chain deflection, at this time, by rotating crankshaft in a clockwise direction enough to take up the slack on the right hand side of the chain (as facing the open chain). Establish a reference mark on the block and measure from this point to the left side of the chain. This measurement when deflected should not exceed ½ in. If deflection is more than ½ in., replace chain and both sprockets.
17. If chain and sprockets are being removed, crank the engine until timing marks on the sprockets are at their closest related points and on a center line with both crankshaft and camshaft centers.
18. Remove camshaft sprocket capscrew, washer, and fuel pump eccentric. Slide off timing chain, sprockets and chain as an assembly.

Installation

1. Install chain and sprockets as an assembly with sprocket timing marks directly toward each other and on a centerline with the crankshaft and camshaft.
2. Install fuel pump eccentric, washer, and attaching cap screw. Torque camshaft sprocket attaching screw to 40-45 ft. lbs. Lubricate chain and sprockets with engine oil.
3. After cleaning mating surfaces, coat the areas with oil-resistant sealer and position gasket on cylinder block.
4. Install front cover aligning tool No. T68P-6019-A to the cylinder front cover so keyway in tool pilot is aligned with key in crankshaft. Position cover and pilot over crankshaft and slide cover on against cylinder block. Coat cover retaining screws with oil-resistant sealer and install screws. While holding in on alignment tool, torque attaching screws to 10-13 ft. lbs. Remove alignment tool.
5. Apply lubriplate to oil seal rubbing surface of steering pump inner hub. Apply mixture of white lead and oil to crankshaft stub in preparing damper installation. Install power steering pump.

6. Install crankshaft damper Woodruff key and press on crankshaft damper with tool No. T64T-6306-A. Do not hammer damper into place. Install damper retainer screw and washer. Torque to 75-90 ft. lbs.

7. Coat new fuel pump gasket with oil-resistant sealer and place on fuel pump. Install fuel pump. Connect fuel lines to fuel pump.

8. Install oil pump and oil pan.

9. Install air-conditioner compressor and water pump.

10. Install water pump pulley and all drive belts.

11. Position radiator to lower support, position upper support to radiator retaining bolts. Connect air coolant hoses. Connect oil cooler lines.

12. Place fan assembly inside radiator shroud and set in position on water pump hub. Install shroud to radiator screws and tighten. Insert and tighten fan attaching screws.

13. Adjust belt tension. Tighten alternator retaining bolts and compressor idler pulley.

14. Fill and bleed cooling system. Fill crankcase.

15. Run engine at fast idle and check for coolant and oil leaks. Set ignition timing.

Valve Timing Procedure and/or Timing Sprocket Replacement

Remove the radiator core, vibration damper and timing case cover, as explained in paragraphs devoted to these subjects.

Turn the crankshaft so the mark on the crankshaft and the mark on the cam sprocket are as near as possible to each other, and in line between shaft centers. Then, remove the camshaft sprocket and chain. Arrange the chain on the old (or new) sprockets so that the marks will be as near as possible to each other and in line between the shaft centers. Then, replace the camshaft sprocket with the chain fitted over it onto the camshaft and bolt in place. Turn the engine two full revolutions, then make certain that the marks on the sprockets are in line between the shaft centers.

Engine Lubrication

Oil Pan Removal

1963-68 430 and 462 Engines

1. Position No. 1 piston to 15° B.T.D.C. Remove the oil dipstick.

2. Disconnect the fan shroud and place it over the fan.

3. Set the parking brake, then raise the car. The car must be supported in a manner that will not in-

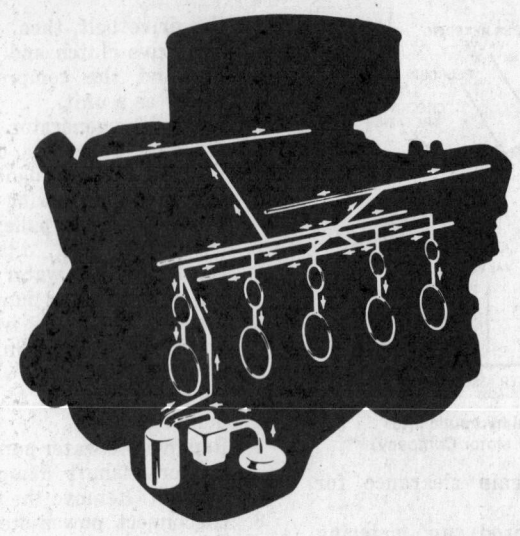

Lubrication system—430 and 462 V8 (© Ford Motor Co.)

terfere with the lowering of the pan.

4. Drain crankcase.

5. Remove generator splash shield. Disconnect engine lateral restrictor.

6. To provide clearance, remove engine front support insulator to underbody side member retaining nuts. With a block of wood on a floor jack under the front edge of the oil pan, raise the engine about an inch. Insert a 1/2 in. block of wood between the insulators and the underbody side members. Remove the floor jack.

7. Remove the end attachments of the front stabilizer bar and rotate the ends of the bar downward to position the center of the bar up away from the oil pan.

8. Remove oil pan attaching bolts. Free the oil pan from the cylinder block. Remove the two bolts that hold the oil pump pick-up tube and screen assembly to the

oil pump, and allow the tube and screen to drop into the oil pan. Remove the oil pan.

9. Install the oil pan by reversing the removal procedure.

Oil Pan Removal

1968-69 460 Engine

1. Disconnect radiator shroud.

2. Raise car on a hoist and drain crankcase.

3. Disconnect idler arm from underbody.

4. Loosen starter mounting bolts.

5. Remove cylinder block to converter housing bolts.

6. Disconnect engine front support insulators from underbody crossmember. Place floor jack under front of oil pan (block of wood between jack and oil pan). Raise engine just enough to insert 1 in. wood blocks between insulators and underbody side members. Remove floor jack.

7. Remove end attachments of front

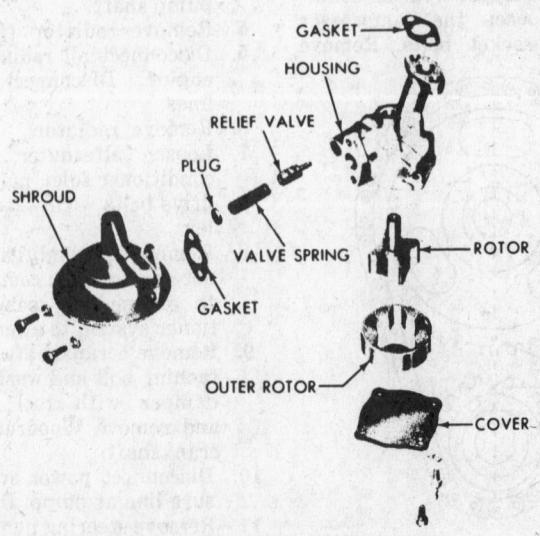

Disassembled oil pump (© Ford Motor Co.)

stabilizer bar and rotate ends of bar down to raise center of bar. Remove oil filter.

8. Remove oil pan mounting bolts and lower oil pan to underbody crossmember. Remove splash shield from right side of oil pan.

9. Disconnect pressure line at power steering pump. Remove bolts holding the pump to cylinder front cover and rotate pump to clear the oil pan. Remove the oil pan.

Oil Pan Installation

1968-70 460 Engine

1. Install oil pan in reverse order of removal and torque attaching bolts to 6-9 ft. lbs. Torque oil pump-to-cylinder block bolts to 20-25 ft. lbs.

Oil Filter Replacement

1963-70—All Models

The oil filter is located on the left side of the engine toward the front and is simply unscrewed from its fitting. Access to the filter is from under the vehicle.

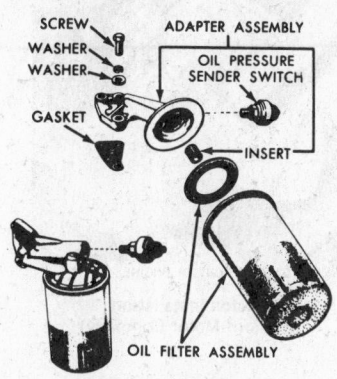

Oil filter and adapter
(© Ford Motor Co.)

Connecting Rods and Pistons

Rod and Piston Assembly Removal

On all Lincoln Continental and Mark II models, the rod and piston

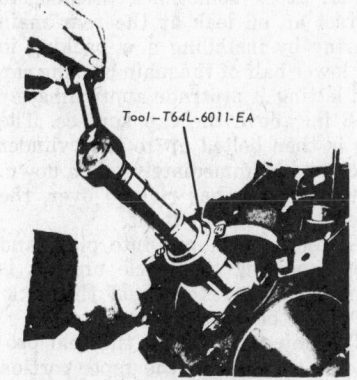

Removing cylinder ridge with ridge reamer
(© Ford Motor Company)

assemblies are removed through the top of the block.

Remove the oil pan and cylinder heads.

Start with any pistons that are down and remove the ring ridge from the top of the cylinder wall with a ring ridge reamer or a bearing scraper.

From underneath the car, take off the lower half of the connecting rod on those rods from which the cylinder ridge has been removed. Carefully mark the cap so that it can be replaced in the same position on the same rod, or install the cap on the rod immediately.

Push the upper half of the rod and piston assembly up out of the top of the block.

Repeat on the rest of the cylinders and piston assemblies.

Piston Inspection

Examine the piston, particularly the thrust surface for scores and scratches, and, if any are found, immediately examine the cylinder wall to see if there are matching scratches. If there are, the cylinder will have to be rebored or honed.

If there are scores or scratches on the piston, but none on the cylinder wall, the piston can usually be dressed down, providing of course it isn't marred enough to spoil the fit.

Piston Pins

1963-70 All Models

The piston pin is a press fit in the connecting rod. The bearing is in the piston.

If the pin can be moved in the connecting rod with less than 20 ft. lbs. torque, the press fit between the piston and the rod is insufficient, and either the pin or the connecting rod (or both) must be replaced.

The pin should be pressed in so that the connecting rod is centered between the bosses of the piston without the pin protruding at either side.

Connecting Rods

All Models

Connecting rods used in all Lincoln engines are fitted with an individual-type connecting rod bearing.

Lincoln does not recommend adjusting these rod bearings. However, it is possible to secure a good working adjustment by placing a feather or taper-type shim between the lower part of the rod bearing and the cap. As much as .004 in. excessive play may be taken up by this method.

Piston and Rod Assembly Installation (into the Engine)

1963-70 All Models

The pistons are assembled to the engine so that the V-notch at the top

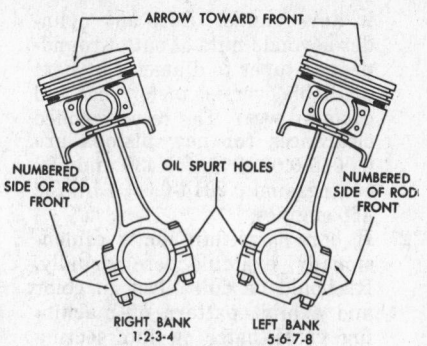

Correct relation of piston to rod—
430 and 462 V8

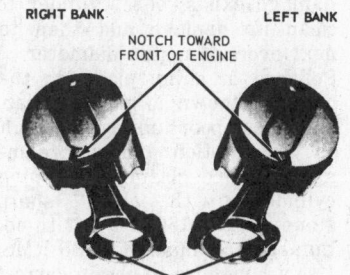

Correct piston and rod positions 460 engines
(© Ford Motor Company)

of the piston faces the front of the engine.

A double check will show that the numbers at the bottom of the connecting rods away from the camshaft on both banks.

Piston Rings

Replacement

Before replacing rings, inspect cylinder bores.

1. Using internal micrometer measure bores both across thrust faces of cylinder and parallel to axis of crankshaft at minimum of four locations equally spaced. The bore must not be out of round by more than 0.005 in. and it must not "taper" more than 0.010 in. "Taper" is the difference in wear between two bore measurements

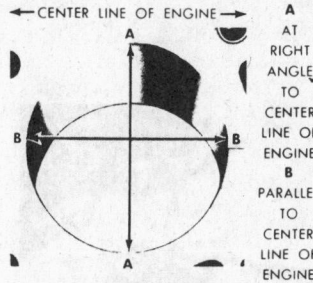

1. OUT-OF-ROUND = DIFFERENCE BETWEEN A AND B
2. TAPER = DIFFERENCE BETWEEN THE A MEASUREMENT AT TOP OF CYLINDER BORE AND THE A MEASUREMENT AT BOTTOM OF CYLINDER BORE

Cylinder bore out of roundness and taper
(© Ford Motor Company)

in any cylinder. Bore any cylinder beyond limits of out of roundness or taper to diameter of next available oversize piston that will clean up wear. The recommended clearances for new pistons are 0.0015-0.0021 in. for 430 and 462 engines and 0.0014-0.0021 in. for 460 engines.

2. If bore is within limits dimensionally, examine bore visually. It should be dull silver in color and exhibit pattern of machining cross hatching intersecting at about 45 degrees. There should be no scratches, tool marks, nicks, or other damage. If any such damage exists, bore cylinder to clean up damage and then to next oversize piston diameter. Polished or shiny places in the bore are known as glazing. Glazing causes poor lubrication, high oil consumption and ring damage. Remove glazing by honing cylinders with clean, sharp stones of No. 180-220 grit to obtain surface finish of 15-35 RMS. Use a hone also to obtain correct piston clearance and surface finish in any cylinder that has been bored.

⏻ CHILTON TIME-SAVER

This or any other machining operation should be done with the cylinder block completely disassembled. Hot tank cylinder block after honing or boring. To remove minor glazing when honing equipment is not available, run emery cloth back and forth across glazed area perpendicular to axis of bore. Scrub block and bores thoroughly with soap and water to remove all grit after using emery cloth.

NOTE: the emery cloth method should be used only as a last resort as it is a method much inferior to honing.

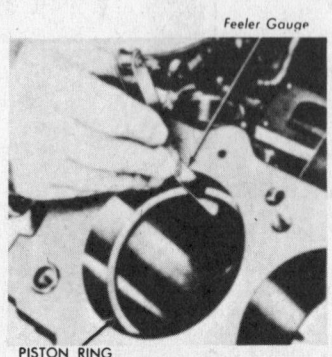

Feeler Gauge

PISTON RING

Checking ring gap
(© Ford Motor Company)

	Ring Gaps (in.)		
Year and Engine	Top Compression	Bottom Compression	Oil Control
1964-66 430 and 462	0.010-0.020	0.010-0.020	0.015-0.066
1967-68 462	0.010-0.035	0.010-0.020	0.015-0.066
1968 460	0.010-0.031	0.010-0.031	0.015-0.066
1969-70 460	0.010-0.020	0.010-0.020	0.010-0.035

3. If cylinder bore is in satisfactory condition, place each ring in bore in turn and square it in bore with head of piston. Measure ring gap. If ring gap is greater than limit, get new ring. If ring gap is less than limit, file

	Side Clearance (in.)		
Year and Engine	Top Compression	Bottom Compression	Oil Control
1964-68 430 and 462	0.0020-0.0035	0.0020-0.0035	Snug
1968-70 460	0.0020-0.0040	0.0020-0.0040	Snug

end of ring to obtain correct gap.

4. Check ring side clearance by installing rings on piston, and inserting feeler gauge of correct dimension between ring and lower land. Gauge should slide freely around ring circumfer-

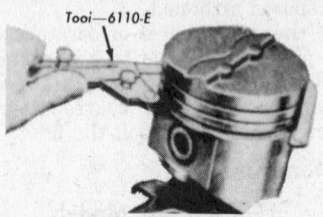

Tool—6110-E

Cleaning ring grooves
(© Ford Motor Company)

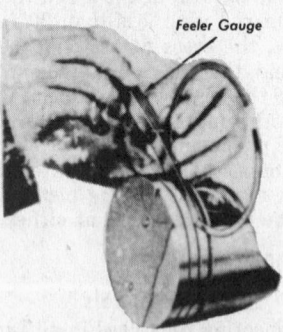

Feeler Gauge

Checking ring side clearance
(© Ford Motor Company)

ance without binding. Any wear will form a step on lower land. Replace any pistons having high steps. Before checking ring side clearance be sure ring grooves are clean and free of carbon, sludge, or grit.

5. Space ring gaps at equidistant intervals around piston circumference. Be sure to install piston in its original bore. Install short lengths of rubber tubing over connecting rod bolts to prevent damage to rod journal. Install

ring compressor over rings on piston. Lower piston rod assembly into bore until ring compressor contacts block. Using wooden handle of hammer push piston into bore while guiding rod onto journal.

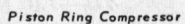

Piston Ring Compressor

ARROW TOWARD FRONT OF ENGINE

Piston installation
(© Ford Motor Company)

Rear Main Bearing Oil Seal

All Models

The rear main bearing oil seal on all Lincoln models is a packing-type seal which requires the removal of the crankshaft to replace the upper half. The lower half may be replaced, however, by removing the crankshaft rear main bearing cap and inserting a new packing into the packing retainer. It is sometimes possible to correct an oil leak at the rear main bearing by installing new packing in the lower half of the main bearing cap and letting it protrude approximately 1/16 in. above the cap surface. The cap is then bolted up to the cylinder block and is immediately taken down. If the packing has riveted over, the riveted portion is cut off.

Again, bolt the cap into place and keep repeating this cycle until it is either necessary to cut off the packing or the bearing cap finally seats.

The object of letting the seal protrude is to compress the upper portion of the rear main bearing seal to prevent a leak at that point.

Top Half, Rear Main Bearing Oil Seal Replacement

All Models

The following method has proven a distinct advantage in most cases.

1. Drain engine oil and remove oil pan.
2. Remove rear main bearing cap.
3. With a 6 in. length of 3/16 in. brazing rod, drive up on either exposed end of the top half of oil seal. When the opposite end of the seal starts to protrude, have a helper grasp it with pliers and pull, gently, while the driven end is being tapped. It is surprising how easily most of these seals can be removed by this method.

Fabric-Type Seal Installation

1. Obtain a 12 in. piece of copper wire (about the same gauge as that used in the strands of an insulated battery cable).
2. Thread one strand of this wire through the new seal, about ½ in. from the end, bend back and make secure.
3. Thoroughly saturate the new seal with engine oil.
4. Push the copper wire up through the oil seal groove until it comes down on the opposite side of the bearing.
5. Pull (with pliers) on the protruding copper wire while the crankshaft is being turned and the new seal is slowly fed into place.
   CAUTION: this snaking operation slightly reduces the diameter of the new seal and care will have to be used to keep the seal from slipping too far through the top half of the bearing.
6. When an equal amount of seal is extending from each side, cut off the copper wire close to the seal and tamp both ends of the seal up into the groove (this will tend to expand the seal again).
   NOTE: don't worry about the copper wire left in the groove. It is too soft to cause damage.
7. Replace the seal in the cap in the usual way, and replace the oil pan.

Jacking, Hoisting

Drive-On Hoist

Care should be exercised when driving the car on to a hoist because the body may contact the upright flanges on the hoist with subsequent damage. The approach ramp should be built up slightly if the angle of approach is too steep.

Rail-Type Hoist

The forks which contact the rear axle must be carefully positioned to avoid damage to the shock absorbers.

Forklift Hoist

The rear post fork, if not adjustable to width, may require special adapters to avoid damaging the rear shock absorbers.

Frame Contact Hoist

Particular care must be exercised when using a frame contact hoist. Specific areas marked on the underbody are designated as hoisting areas.

The lifting areas at the front of the vehicle are clearly designated by corrugated metal plates. These plates are bolted to the underbody midway between the front edge of the door and the rear edge of the front fender wheel opening. The front hoist pads or adapter arms, must be positioned on these corrugated plates.

The lifting areas at the rear of the vehicle are located at the edge of the underbody approximately 15 in. forward of the front edge of the rear wheel opening cover panel. The rear hoist pads, or adapter arms, must not be positioned forward of this point.

Floor Jack (Support)

Various acceptable jacking locations are available when it is necessary to raise any one portion of the vehicle. However, when jacking against sheet metal, a wood block 2 x 4 of suitable length should be placed between the jack and the sheet metal to prevent damaging or deforming the metal. Do not attempt to raise one entire side of the body by placing a jack midway between the front and rear wheels. This procedure will probably result in permanent damage.

Jacking at the Front

Each wheel can be raised independently by placing a floor jack under the spring seat pocket in the lower suspension arm.

Jacking at the Rear

Each wheel can be raised independently by placing a floor jack under the rear axle housing.

Bumper Jack

The bumper jack lifting points are similar to those specified in previous models. There is a mounting bracket at the underside edge at the front and rear bumpers for a bumper jack.

Front Suspension

References

General instructions covering the front suspension, and how to repair and adjust it, together with information on installation of front wheel bearings and grease seals, are in the Unit Repair Section.

Definitions of the points of steering geometry are in the Unit Repair Section. This Section also covers troubleshooting front-end geometry and irregular tire wear.

Beginning 1969

Regulation of the various aspects of steering geometry are basically unchanged from former models. There are, however, tools available to aid in the adjustment of caster and camber. Tool T65P-3000-D is now available from your Lincoln-Mercury dealer to help with this adjustment operation.

Figures covering the caster, camber, toe-in, kingpin inclination and turning radius are in the Front Wheel Alignment table.

Wheelbase, tread and tire size figures are in the General Chassis and Brake Specifications table.

Front Spring and Lower Arm

Removal and Replacement 1964-69 Except Mark III

1. Raise the car. Place a support under each underbody side rail to the rear of the lower arm, in the lifting pad area.
2. Remove wheel and tire assembly, then, remove the hub and drum. On disc brake-equipped cars: remove two bolts and washers that attach the caliper to the spindle. Remove the caliper from the rotor and wire it to the underbody. Remove the hub and rotor from the spindle.
3. Loosen splash shield to provide clearance at the end of the arm when it is lowered.
4. Remove the shock absorber, and disconnect the stabilizing strut from the lower arm. Disconnect the stabilizer bar from the suspension bar.
5. Remove the cotter pin and loosen the castellated nut attaching the lower ball joint to the spindle.
6. Place a box wrench over the lower end of the ball joint remover tool between the two spin-

dle pivot points. (The tool, T57P-3006-A, should seat firmly against the ends of both studs, and not against the lower stud nut.)

7. Turn the wrench until both studs are under tension, then, rap the spindle near the lower stud to loosen the stud from the spindle. Do not loosen the stud with tool pressure alone.

8. Place a jack under the outer end

MEASURE FROM UNDERSIDE OF CROSS MEMBER POCKET, ADJACENT TO FRONT SIDE OF JOUNCE BUMPER BRACKET, TO TOP OF DRAG STRUT FLAT

FRONT RIDE HEIGHT (INCHES)

NORMAL	5¾
MAXIMUM	6⅛
MINIMUM	5³⁄₁₆

POINT A

Steering Wheel

Steering Wheel Removal

All Models Except 1968-70

Remove cap and emblem assembly by simply prying it up off the steering wheel. Remove the steering wheel nut and, using a puller, pull the steering wheel off the sector shaft.

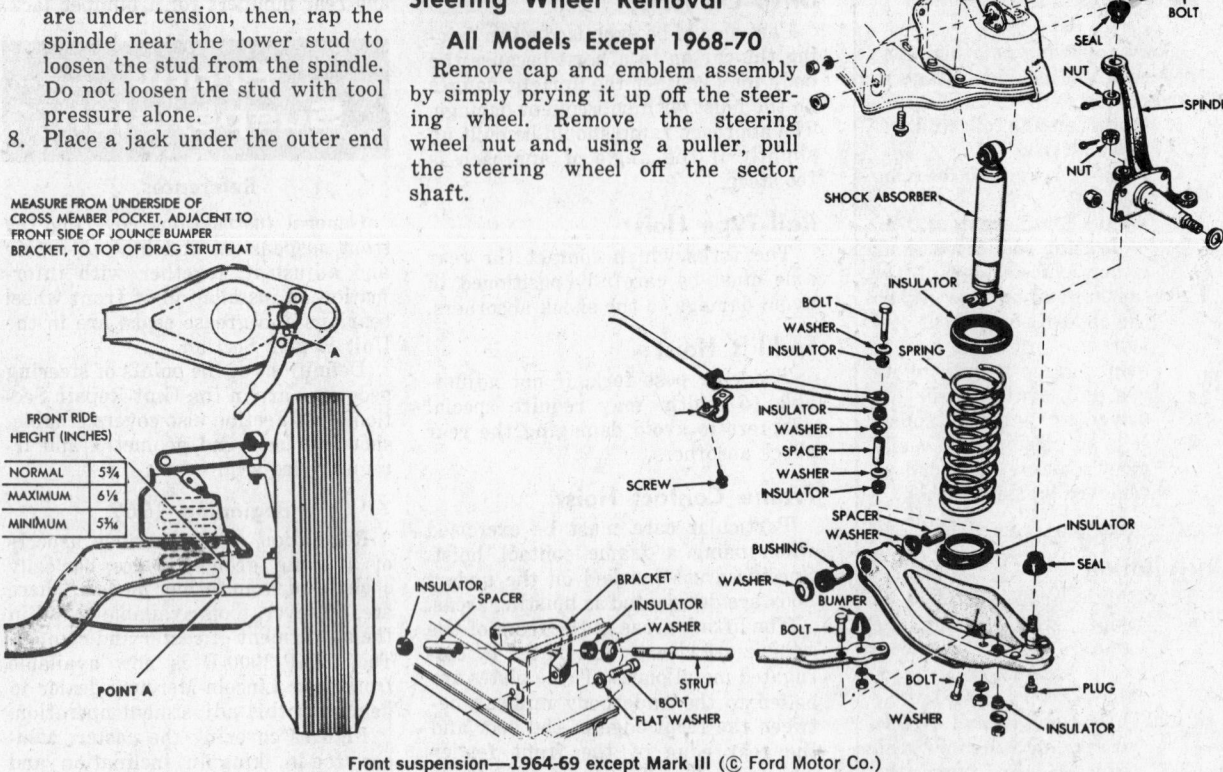

Front suspension—1964-69 except Mark III (© Ford Motor Co.)

of the lower arm and raise the arm several inches.

9. Install spring compressor tool 5310-T inside the spring with the jaws of the tool toward the center of the car.

10. Remove the nut from the ball joint stud. Lower the jack until the spindle and spring are free, and remove the spring and insulators.

11. Remove the lower arm to cross-member nut, bolt, washers, and spacer, then remove arm.

12. Install by reversing the removal procedure.

Front Spring Removal—1968-70 Mark III and 1970 Continental

1. Raise vehicle and support front end of frame with jack stands.
2. Place jack under lower arm to support it.
3. Disconnect lower end of shock absorber from lower arm.
4. Remove bolts that attach strut and rebound bumper to lower arm.
5. Disconnect lower end of sway bar stud from lower arm.
6. Remove nut and bolt that secures inner end of lower arm to cross-member.
7. Lower jack slowly to relieve spring pressure on lower arm then remove spring.

Beginning 1968-70

Working from the underside of the steering wheel spokes, remove the two screws that secure the wheel crash pad. Do not use a hammer or knock-off-type puller. Striking the puller or shaft is very likely to cause damage to the bearings or collapsible steering column.

Button, Horn Blowing Ring Removal

First, look to see if the horn ring is mounted on top of, or underneath, the steering wheel. If it is mounted underneath the steering wheel, pry up the medallion, remove the steering wheel and the horn blowing ring can then be detached from the top of the steering column.

If the horn blowing ring is mounted on top of the steering wheel, simply pry off the medallion in the center. This will give access to the wheel nut that holds the blowing ring in place.

Steering Gear

Power Steering Gears 1963-70

Troubleshooting and repair instructions covering power steering gears are in the Unit Repair Section.

Steering Mechanism Adjustment

All Models

Complete instructions on the adjustment of this and all other manual steering gears are in the Unit Repair Section.

Steering Gear Assembly Removal (From Car)

1964-65

1. Disconnect the power gear hoses. Remove the bolts that hold the flexible coupling to the steering gear flange, raise the car and remove the pitman arm.
2. Disconnect the exhaust pipe at the flange.
3. Clamp the front wheels to the extreme right position and take out the cap screws that secure the gear assembly to the under body front side member. Remove the gear.
4. The gear is reinstalled in the reverse of the procedure which removed it.

1966-1967

1. Disconnect pressure and return line from steering gear. Cap each line and plug inlets in steering gear.

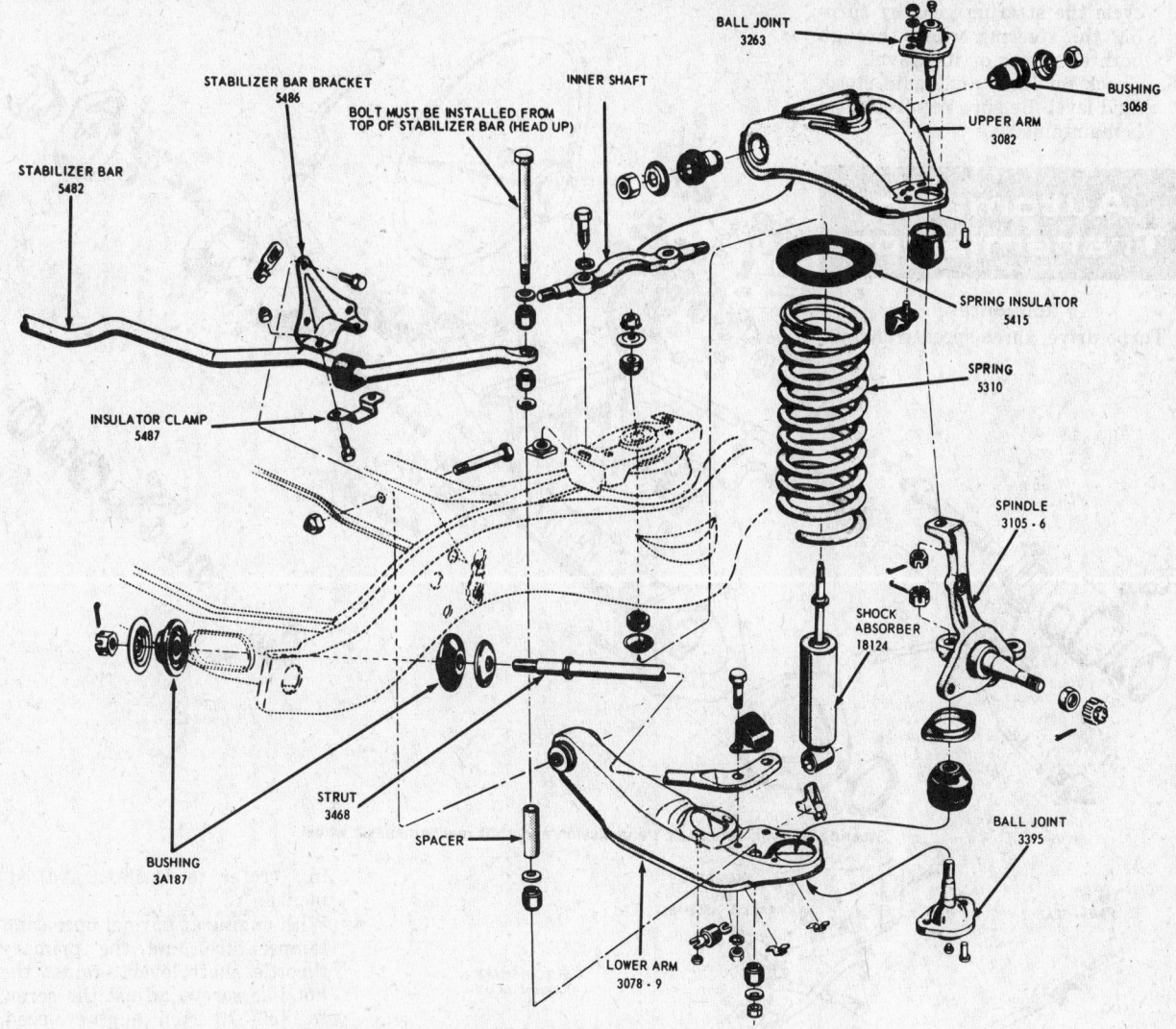

1970 Lincoln Continental and 1968-70 Mark III (© Ford Motor Company)

2. Remove ground strap from steering gear housing.
3. Remove bolt that attaches flex coupling to steering gear.
4. Remove bolt that attaches left brace to torque box. Loosen bolt that secures brace to side rail and swing brace to one side.
5. Remove pitman arm from sector shaft.
6. Remove pipe between manifold and resonator.
7. Disconnect linkage rod from equalizer shaft. Remove equalizer stud from side rail. Move equalizer shaft up and out of way. Do not lose stud or bushings.
8. Remove bolt from lower end of fender splash shield. Move splash shield to one side to gain access to steering gear.
9. Remove three steering gear attaching bolts. Support gear before completely removing last bolt.
10. Move steering gear down to free it from flex joint. Rotate it coun-

terclockwise to provide clearance between side rail and engine.
11. Remove steering gear and remove three pads from gear. Reverse procedure to install.

1968-70

1. Disconnect hydraulic lines from the steering gear. Plug lines and ports to protect from leaking and the entry of dirt.
2. Remove the two bolts that hold the flex coupling to the steering gear and to the column.
3. Raise the car and remove the sector shaft attaching nut.
4. Remove pitman arm from the sector shaft with tool T64P-3590-F.
5. Support steering gear, then remove the three steering gear attaching bolts.
6. Work the steering gear free of the flex coupling and remove it from the car.
7. If the flex coupling stayed on the input shaft, lift it off the shaft at this time.

8. Install by sliding the flex coupling into place on the steering shaft. Turn steering wheel so that the spokes are in the horizontal position.
9. Center the steering gear input shaft.
10. Slide the steering gear input shaft into the flex coupling and into place on the frame side rail. Install the three attaching bolts and torque them to specifications.
11. Be sure that the front wheels are in a straight ahead position, then install the pitman arm on the sector shaft. Install and tighten the sector shaft to pitman arm attaching nut to 150-225 ft. lbs.
12. Move the flex coupling into place on the input shaft and steering column shaft, install and tighten attaching bolts.
13. Connect and tighten the fluid pressure and the return lines to the steering gear.
14. Fill the power steering pump and

cycle the steering gear by turning the steering wheel through both extremes of its travel.

15. Check for leaks and again check fluid level. Be sure required level is maintained.

Automatic Transmission

References
Turbo-drive, three-speed transmis-

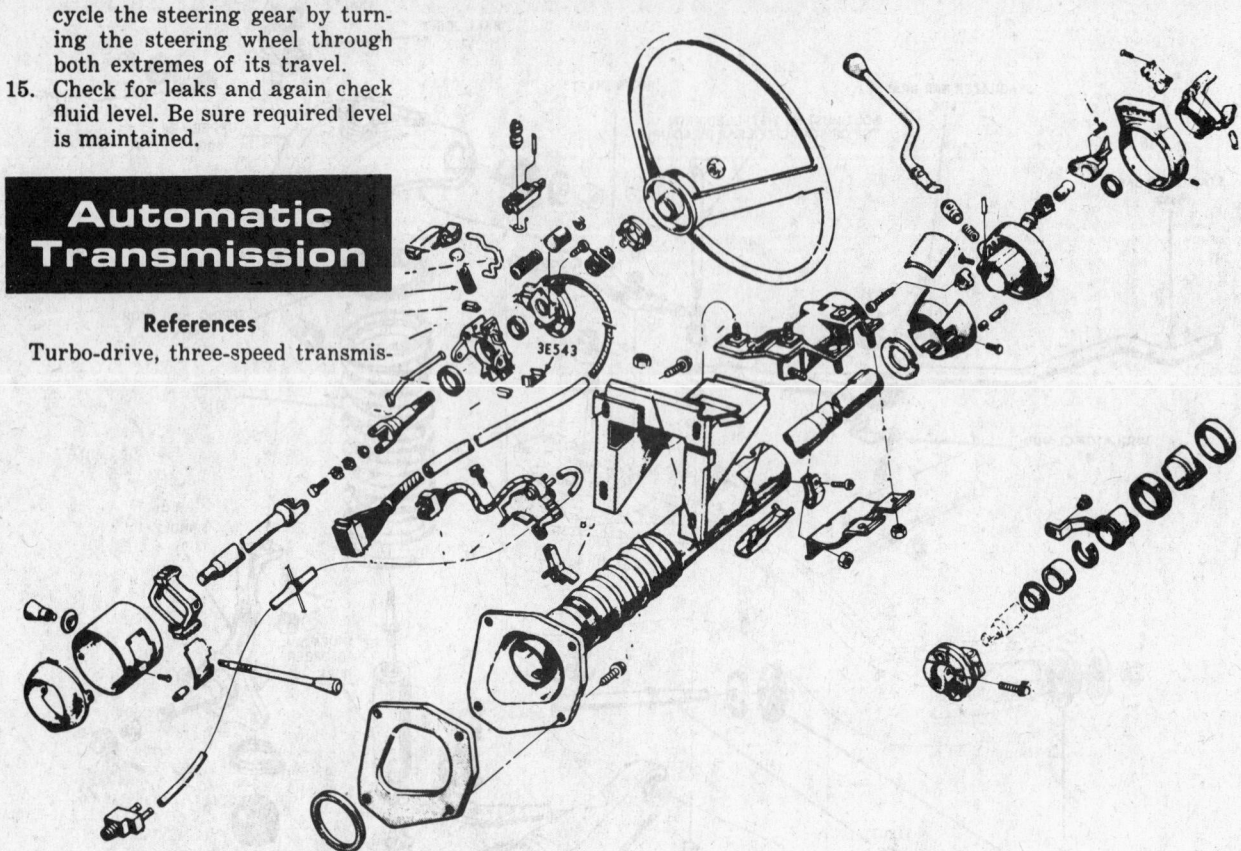

Steering column, automatic transmission and shift mechanism tilt wheel

SEAL

PLUG

PLUGS

SPINDLE ARM
CONNECTING ROD

IDLER ARM
AND
BRACKET ASSEMBLY

PLUG

SECTOR SHAFT
(PITMAN ARM)

SEALS

STEERING
CONNECTING LINK

SEAL

ADJUSTING SLEEVE

ROD END ASSEMBLY

PLUG

Disassembled view of typical steering linkage (© Ford Motor Co.)

sions were used through 1965. Since 1966, the C-6 Dual Range, three-speed transmission has been used. Detailed information on both these transmissions is in the Unit Repair Section.

NOTE: all linkage adjustments must be made in the following sequence:

1. Engine idle speed
2. Accelerator pedal height
3. Downshift rod
4. Manual linkage

Twin-Range Turbo-Drive

1963-65 Idle Speed Adjustment

1. Apply parking brake, start engine and place selector lever in D2 position.
2. Hook-up tachometer.
3. Check dashpot for bottom clearance. Hold carburetor lever against hot idle screw and depress dashpot plunger by hand. Clearance should be .060 to .90

in., (refer to Dashpot Adjustment).

4. With engine at normal operating temperature, and the primary throttle shaft lever against the hot idle screw, adjust the screw to 450-575 rpm engine speed, with selector lever in D2 position.

Carburetor Rod Adjustment

1. Disconnect the transmission throttle rod at C, and carburetor rod at A.
2. Insert ¼ in. gauge pin through alignment holes in bellcrank bracket and tab at D.
3. Hold carburetor lever against hot idle screw, adjust length of carburetor rod between A and B for free fit into carburetor lever. Lengthen the rod one turn.
4. Remove ¼ in. gauge pin from bellcrank and connect rod to carburetor lever.
5. Recheck alignment of gauge pin holes in bellcrank bracket and tab. Further adjustment of carburetor rod may be needed. The ¼ in. gauge pin holes must be in alignment to assure correct carburetor-to-transmission calibration.

Throttle Rod Adjustment

1963-65

1. Raise the hood. Connect pres-

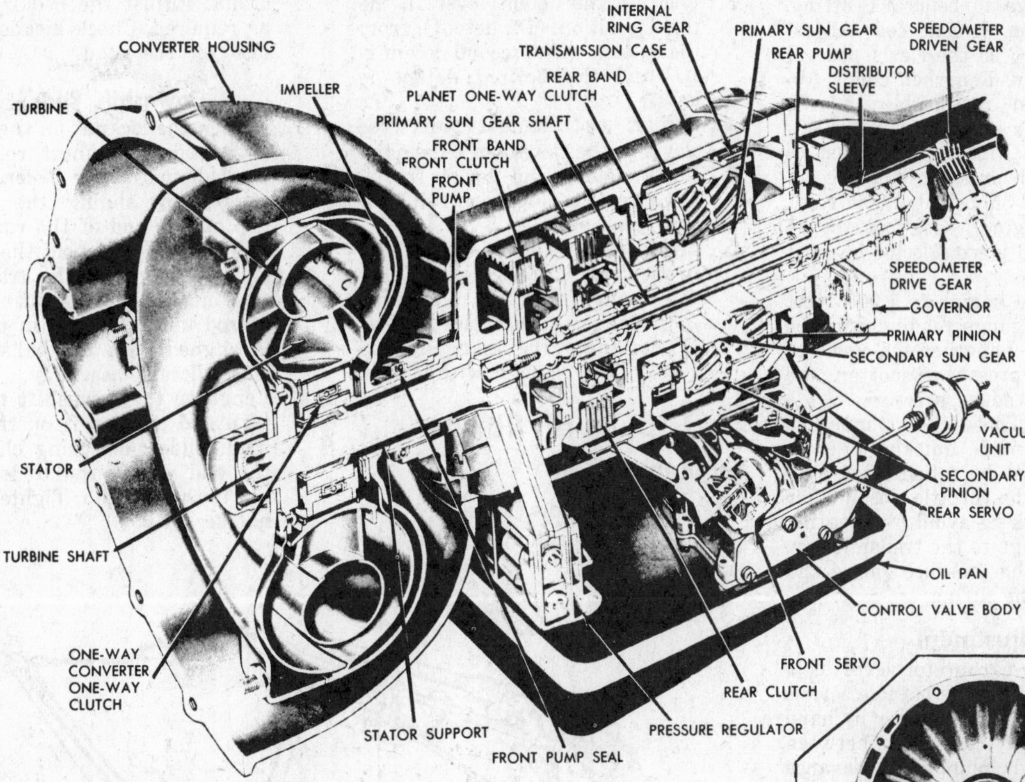

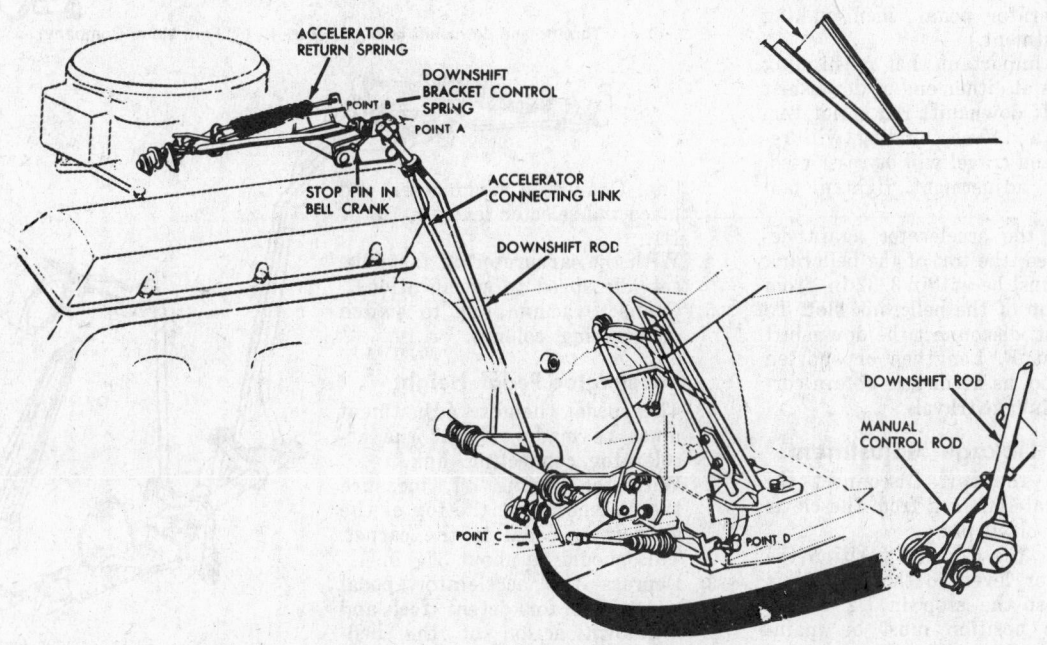

CONVERTER HOUSING

TURBINE

IMPELLER

PLANET ONE-WAY CLUTCH

PRIMARY SUN GEAR SHAFT

FRONT BAND
FRONT CLUTCH
FRONT PUMP

REAR BAND

TRANSMISSION CASE

INTERNAL RING GEAR

PRIMARY SUN GEAR
REAR PUMP

SPEEDOMETER DRIVEN GEAR

DISTRIBUTOR SLEEVE

SPEEDOMETER DRIVE GEAR

GOVERNOR

PRIMARY PINION

SECONDARY SUN GEAR

VACUUM UNIT

SECONDARY PINION

REAR SERVO

OIL PAN

CONTROL VALVE BODY

STATOR

TURBINE SHAFT

ONE-WAY CONVERTER ONE-WAY CLUTCH

STATOR SUPPORT

FRONT PUMP SEAL

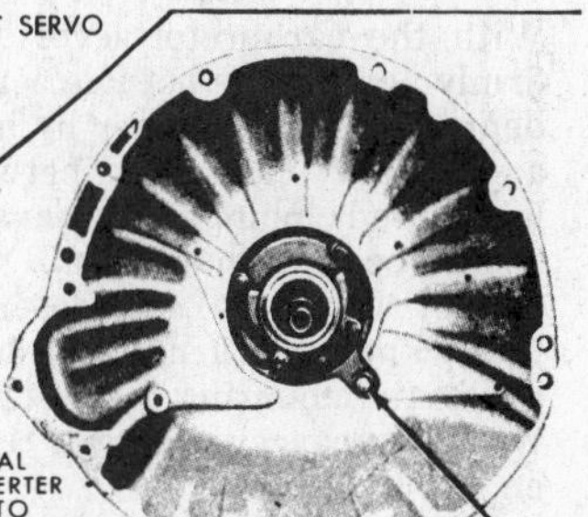

PRESSURE REGULATOR

REAR CLUTCH

FRONT SERVO

ADDITIONAL BOLTS-CONVERTER HOUSING TO TRANSMISSION CASE

BOLT

Turbo-Drive transmission, 1964-65

Additional bolts attaching converter housing to transmission case—1964-65

ACCELERATOR RETURN SPRING

DOWNSHIFT BRACKET CONTROL SPRING

POINT B

POINT A

STOP PIN IN BELL CRANK

ACCELERATOR CONNECTING LINK

DOWNSHIFT ROD

DOWNSHIFT ROD

MANUAL CONTROL ROD

POINT C

POINT D

Typical linkage adjustments (© Ford Motor Co.)

sure gauge to bellcrank fitting.
2. Pull up on throttle rod. Adjust clevis at C so the clevis pin enters freely. Lengthen clevis 3½ turns and reassemble throttle rod. (This is a preliminary setting only. Final setting must be made with pressure gauge reading of 80-85 psi at 1,000 rpm.,
3. Start engine, set foot brake firmly and place selector lever at D2.
4. Accelerate engine to 1,000 rpm. Pressure gauge reading must be 80-85 psi. Lengthen the clevis to increase pressure, shorten the clevis to reduce pressure.

Caution These checks must be made quickly. The selector must be returned to neutral and the throttle closed after each check to avoid overheating and damage to the transmission.
5. Remove the pressure gauge and seal up the fitting.

Dashpot Adjustment

1. With the carburetor lever held firmly against the hot idle screw, depress dashpot plunger by hand and check clearance between lever and plunger. Clearance should be .060 in.-.090 in.
2. Adjust clearance by loosening the locknut and turning the dashpot in its mounting bracket.
3. When clearance is correct, tighten the locknut.

Downshift Rod Adjustment

1. With ignition key off, push accelerator to full kickdown position. Check position of bellcrank slot and pin. The pin must be within 3/32 in. from the top of the bellcrank slot. (Release accelerator pedal when making adjustment.)
2. It is important that no binding exists at either end of downshift rod. If downshift rod is not vertical and free, binding will result and travel will be restricted. After adjustment, tighten jam nuts.
3. With the accelerator again depressed, the top of the bellcrank pin must be within 3/32 in. from the top of the bellcrank slot. To adjust, disconnect the downshift rod at F. Lengthen or shorten the rod, as needed, to obtain correct linkage travel.

Manual Linkage Adjustment

1. Raise the car. Disconnect the manual shift rod from the clevis at the clevis pin.
2. Lower the car and position the selector lever so the pointer is against the stop in D2 range. (This position must be maintained throughout the manual adjustment.)

3. Position the detent lever in the third position (D2 detent) from the bottom. (The second detent is D1 and the bottom detent is low.)
4. Adjust the manual shift rod clevis so the clevis pin enters the clevis detent and lever freely. Lengthen the clevis one full turn and reassemble.
5. Check the position of the pointer in each range. If the pointer does not line up with the letters in each range, adjust the letters. Do not adjust the linkage to correct pointer register with letters.

C-6 Dual Range

Beginning 1966

Engine Idle Speed
If the car is equipped with air-

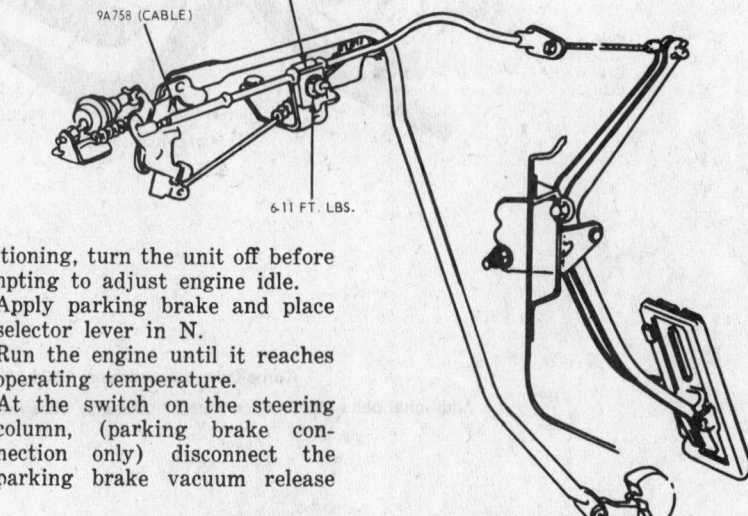

Throttle and downshift linkage—Mark III (© Ford Motor Company)

conditioning, turn the unit off before attempting to adjust engine idle.
1. Apply parking brake and place selector lever in N.
2. Run the engine until it reaches operating temperature.
3. At the switch on the steering column, (parking brake connection only) disconnect the parking brake vacuum release

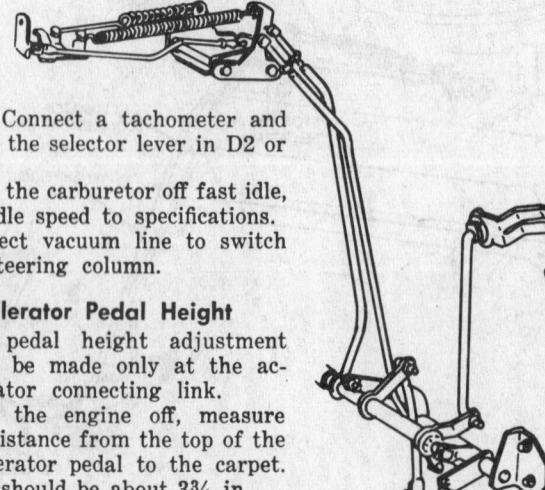

line. Connect a tachometer and place the selector lever in D2 or D1.
4. With the carburetor off fast idle, set idle speed to specifications.
5. Connect vacuum line to switch on steering column.

Accelerator Pedal Height
1. The pedal height adjustment must be made only at the accelerator connecting link.
2. With the engine off, measure the distance from the top of the accelerator pedal to the carpet. This should be about 3¾ in.
3. Depress the accelerator pedal and check for detent feel and kickdown action of the bell-

crank. Adjust the pedal height, as required. Check kickdown action on the road.

Downshift Rod
1. Loosen the locknut on the downshift rod. Disconnect rod from the ballstud on the bellcrank assembly by sliding the spring clip off the end of the rod.
2. Pull upward and hold the downshift rod against the transmission internal stop. Adjust length of rod until the hole in the rod is aligned with the ballstud on the bellcrank assembly.
3. Lengthen the downshift rod one turn and position it on the ballstud. Slide the spring clip over the end of the rod to lock the rod to the ballstud. Tighten locknut securely.

Throttle and downshift linkage—1966-70 Continental (© Ford Motor Company)

4. Be sure the bellcrank outer bracket is against the stop pin. If it is not, lengthen the downshift rod one turn. If the rod is too long, there will be no upshift.

Manual Linkage Adjustment

1. If the car is equipped with a tilt wheel steering, position the column up as far as possible. With the engine off, place the selector lever against the stop in the D1 (large dot) position. Raise the car, and remove linkage splash shield.
2. Disconnect the adjustable link from the transmission manual shift lever on the transmission.
3. Be sure the transmission shift is fully engaged in D1, the second detent from the bottom. The bottom detent is L (low).
4. Loosen locknut on adjustable link, then pull down on the link to hold the selector lever against the D1 stop. Adjust the link by turning the lower end until the hole in the link aligns with the stud on the transmission manual lever. Connect it to the transmission shift lever.
5. Check selector lever through all positions to secure correct adjustment.

Beginning 1968-69—All Models

Due to government regulations concerning exhaust emission control, on any relative or contributing operational aspect such as carburetor adjustment, idle speed, timing adjustment or any transmission functions, see the Unit Repair Section.

Transmission Removal & Replacement

1963-65 Turbo-Drive

To remove the transmission from any Lincoln or Lincoln Continental, it is necessary to remove the gear box and converter housing assembly as a unit. Two additional converter housing-to-transmission case bolts are located behind the converter assembly, and are not accessible until after the converter has been removed.

Refer to illustration for identity of underbody components.
1. Raise the hood and disconnect accelerator linkage (downshift rod) from bellcrank.
2. Remove two upper bolts and one bolt from right side of converter housing to engine block.
3. Remove one inner upper bolt attaching the starter motor to the converter housing.
4. Install remote control starter switch to the starter solenoid and put the transmission in neutral. Raise the car.

5. Disconnect the engine stabilizer bar bracket from the converter housing. (See illustration.) Remove converter housing lower plate.
6. Drain the converter.
7. Disconnect the fluid filler tube at the transmission and drain transmission.
8. Remove exhaust crossover pipe.
9. Disconnect oil cooler lines at transmission.
10. Index the rear universal joint and pinion flange to help upon reassembly. Disconnect propeller shaft at rear universal joint and remove the shaft.
11. Disconnect parking brake cables and the spring from equalizer bar. Remove the bar.
12. Remove two remaining bolts from starter motor to converter housing. Disconnect starter cable from electrical power box. Slide starter motor forward over the crossmember and secure out of the way.
13. Remove control linkage splash shield. Disconnect manual and throttle control linkage from the transmission.
14. Disconnect speedometer at transmission.
15. Remove nut and washer from engine rear mount lower stud.
16. Secure transmission jack under, and raise the transmission off the rear mount crossmember. (See illustration.)
17. Remove two attaching bolts from the rear mount to the transmission extension housing.
18. Remove four attaching bolts and nuts from the crossmember to underbody brackets. (See illustration.)
19. Remove the rear mount from the crossmember. Then, remove the crossmember by sliding it over the left exhaust pipe to the rear of the left mounting bracket.
20. Lower the transmission jack slightly, and remove the remaining three converter housing to engine block bolts.
21. Move transmission toward the rear, lower, and remove from under the car.
22. Remove converter from converter housing.
23. Remove six bolts holding the converter housing to transmission case. Remove converter housing.

NOTE: replace transmission by reversing the above procedure.

For further detail and overhaul data on transmissions, see Unit Repair Section.

Removal and Replacement
1966-69

The R & R procedure is essentially the same as for the earlier Turbo-

drive transmissions. However, factory recommendations are as follows:
1. Raise hood and disconnect starter neutral switch wires.
2. Disconnect the transmission oil filler tube from the manifold.
3. Raise the car and remove the bolts that attach the reinforcement plate at the rear of the transmission oil pan. Remove the plate.
4. With a drain pan under the transmission, loosen the transmission oil pan bolts and slowly drain and remove the pan. After all of the oil is out of the transmission, reinstall the pan, using about four bolts.
5. Remove two bolts that attach the cover to the lower end of the converter housing.
6. Remove two drain plugs from the converter housing and allow it to drain.
7. Remove four nuts that attach the converter to the drive plate.
8. Lift the filler tube from the transmission case.
9. Disconnect starter cable, then, remove the starter.
10. Disconnect fluid cooler lines from transmission.
11. Disconnect vacuum hose from the diaphragm.
12. Disconnect manual and downshift rods from the transmission.
13. Disconnect speedometer cable from extension housing.
14. Remove the three bolts that hold the manual and downshift control rod splash shield to the side rail and remove the shield.
15. Remove lower bellcrank bracket lower attaching bolt. Pivot the bracket to allow the bellcrank to hang free.
16. Pry upper bellcrank out of converter housing and allow it to hang free.
17. Disconnect the driveshaft from the real axle and remove it from the transmission.
18. Remove converter housing-to-cylinder block lower attaching bolts.
19. Loosen parking brake adjusting nut at the equalizer and remove the retracting spring. Disconnect rear brake cables and remove the equalizer.
20. Remove the two nuts that attach the engine rear mounts to the crossmember.
21. Place a transmission jack under the transmission and raise it just high enough to remove the weight from the crossmember.
22. Remove crossmember-to-frame attaching nuts and remove the crossmember.
23. Remove engine rear support-to-

extension housing attaching bolts and remove support.

24. Secure the transmission to the jack with a safety chain. Lower the transmission and remove the upper converter housing-to-cylinder block attaching bolts.
25. Move the transmission away from the cylinder block. Lower it and remove it from under the car.
26. Remove the converter and mount the transmission in a holding fixture.
27. Replace the transmission by reversing removal procedure.

NOTE: transmission and servicing procedures can be found in the Unit Repair Section.

1968-70, Mark III

The C-6 transmission is used, as in previous models, however some modifications warrant some R & R procedure changes.

Removal

1. From the engine compartment, remove the fluid filler tube bracket attaching screw that secures it to the rear of the right cylinder head. Lift tube and dipstick from the transmission.
2. Remove the starting motor upper attaching bolt using a long extension.
3. Remove the two converter housing upper attaching bolts.
4. Raise the car on a hoist.
5. Remove dust shield from the front end of the converter housing.
6. Crank engine until one of the converter drain plugs is accessible. Remove the plug.
7. Crank engine until the other plug is at the bottom. Place a drain pan to catch the fluid.
8. Remove the driveshaft.
9. Remove the side rail support brace attaching bolts and remove the brace.
10. After the fluid has drained from the converter, replace the plug.
11. Place drain pan under the transmission pan and loosen the pan attaching bolts. Allow the fluid to drain. Finally, remove all of the pan bolts except two on the same side. After the fluid has drained, install the bolts on the opposite side of the pan to hold it in place.
12. Remove converter-to-flywheel attaching nuts.
13. Disconnect downshift linkage from the transmission downshift lever.
14. By using tool T67P-7341-A, remove the selector rod from the manual shift lever.
15. Remove the two screws that attach the shift rod bellcrank

bracket to the converter housing and remove the bracket.
16. Disconnect the vacuum diaphragm hose from the upper end of the vacuum tube.
17. Disconnect the speedometer cable from the extension housing and place it out of the way.
18. Remove the starting motor-to-lower attaching bolts and place the motor to one side.
19. Disconnect the two oil cooler lines from the right side of the transmission.
20. Disconnect the muffler inlet pipes at the exhaust manifold and allow the pipes to hang.
21. Remove the vibration absorber from the extension housing.
22. Remove the two engine rear support-to-extension housing attaching bolts.
23. Place transmission jack under the transmission and raise it just enough to remove the weight from the support.
24. Remove two support attaching bolts and remove the support.
25. Lower the jack just enough to remove the weight.
26. Remove the four remaining converter housing-to-cylinder block attaching bolts and the accelerator linkage stop from the left side of the housing.
27. Carefully lower the transmission and remove it from under the car.
28. Remove the converter, then mount the transmission in a holding fixture.

Installation

1. Install in reverse of removal procedure. Torque converter drain plugs to 14-28 ft. lbs.
2. Install converter on stator support.
3. Secure transmission on a transmission jack.
4. Rotate converter so that the studs and drain plugs are aligned with their holes in the flywheel.
5. Move transmission toward the cylinder block until they contact each other. Install and torque the attaching bolts to 40-50 ft. lbs. Before tightening the center bolt on the left side, make sure that the accelerator linkage stop bracket is properly positioned so that the left upper bolt may be installed later.
6. Connect the fluid cooler lines to their fittings on the right side of the transmission.
7. Raise the transmission and install the engine rear support and tighten the bolts. Be sure that the handbrake equalizer is in position.
8. Lower transmission and remove the jack.

9. Install engine rear support-to-extension housing attaching bolts.
10. Install driveshaft.
11. Connect speedometer cable to extension housing.
12. Position starting motor on the converter housing and secure it with the two lower bolts.
13. Install the torque converter-to-flywheel attaching nuts and torque to 20-30 ft. lbs.
14. Install converter housing dust shield.
15. Secure the frame side rail support brace with the attaching bolts and lock washers.
16. Connect the downshift rod to the transmission downshift lever.
17. Connect the vacuum diaphragm hose to the upper end of the vacuum tube.
18. Using tool No. T67P-7341-A, install a new grommet in the manual lever. Then, secure the manual selector rod to the lever.
19. Position the vibration absorber to the transmission extension housing and secure it with the three attaching bolts.
20. Connect the muffler inlet pipes to the exhaust manifolds.
21. Lower the car and install the two converter housing upper bolts. Torque them to specifications.
22. Install starter motor upper bolt.
23. Place a new O-ring on the end of the fluid filler tube and insert it in the transmission case. Secure the tube to the right cylinder head wih the attaching screw and lock washer.
24. Fill the transmission to the proper level with the specified amount of fluid.
25. Adjust manual and throttle linkage as required.

1970 Lincoln Continental

1. Raise hood and remove transmission dipstick.
2. Raise vehicle on hoist and remove bolt that secures transmission filler tube to cylinder head.
3. Remove bolts that attach reinforcement plate at rear of transmission oil pan, and remove plate.
4. Remove three bolts that attach manual and downshift control rod splash shield to side rail and remove shield.
5. Place drain pan under transmission. Loosen all transmission pan attaching bolts and allow fluid to drain. Remove pan. After fluid is drained, replace pan.
6. Remove two bolts that attach cover to lower end of converter housing.
7. Remove two front support bracket bolts at converter housing.

8. Remove drain plug from converter and allow to drain.
9. Remove four nuts that attach converter to flywheel.
10. Lift fluid filler tube from transmission case.
11. Remove two idler arm bracket bolts from frame side rail and allow idler arm to hang free.
12. Disconnect starter cable, and remove starter.
13. Disconnect oil cooler lines from transmission.
14. Disconnect vacuum hose from diaphragm.
15. Loosen parking brake adjuster nut at equalizer and remove retracting spring. Disconnect rear brake cables and remove equalizer from bracket.
16. Disconnect speedometer cable from extension housing.
17. Disconnect driveshaft from transmission flange and position out of way.
18. Disconnect downshift rod from downshift lever.
19. Remove manual selector rod from selector lever.
20. Remove lower bell crank.
21. Pry upper bell crank out of converter housing.
22. Remove cooler lines from clip on cylinder head.
23. Remove converter housing to cylinder block lower attaching bolts.
24. Remove two nuts that attach engine rear mounts to crossmember.
25. Place transmission jack under transmission and raise it enough to remove weight from crossmember.
26. Remove engine rear support to extension housing attaching bolts and remove supports.
27. Remove crossmember.
28. Secure transmission to jack with safety chain.
29. Lower transmission and remove upper converter housing to cylinder block attaching bolts.
30. Move transmission away from cylinder block.
31. Lower transmission and remove from vehicle.
32. Remove converter. Reverse procedure to install.

U Joints, Drive Lines

Universal Joint and Drive Shaft Removal

1963-70 All Models

A Hotchkiss-type drive line is used with three universal joints. A single cross and needle-type at rear and a double one at front. To remove the assembly:

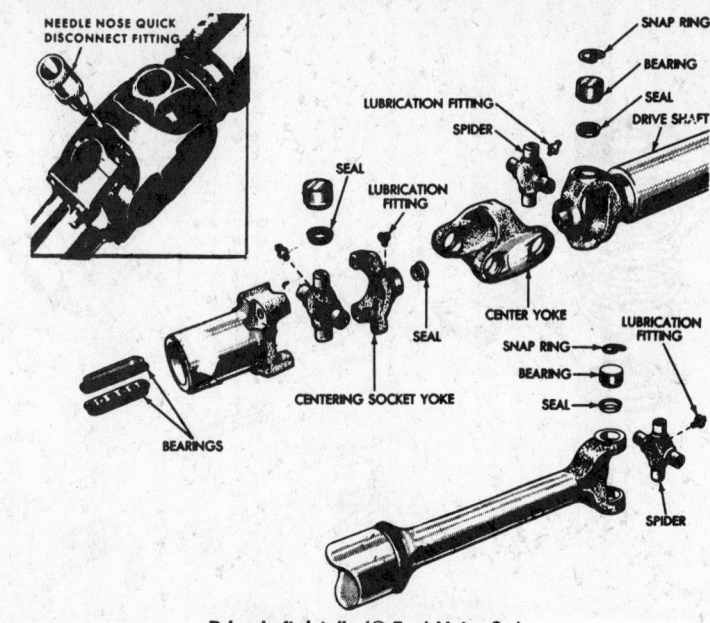

Driveshaft details (© Ford Motor Co.)

1. Mark the slip yoke, companion flange and driveshaft so reassembly will be in original position to maintain drive line balance.
2. Disconnect rear bearing U-bolts.
3. Remove cap screws holding front joint to slip yoke and slide slip yoke forward onto transmission output shaft.
4. Lower front end first.

Driveshaft Disassembly

1963-70 All Models

The rear joint is serviced in the conventional way.

The two front joints and centering yoke are similarly serviced. A special adapter is required for lubricating the centering yoke. (See illustration.)

Rear Axles, Suspension

Rear Spring Replacement

1963-69 Except Mark III

Longitudinal leaf springs are used on these models.

They are held with a single bolt at the front, and with a shackle at the back.

Take the weight of the car on a frame in front of the rear spring. Unbolt and remove the rear shackle. Disconnect the rear shock absorber and remove the four U-bolts from each side of the spring saddle where the spring is held to the rear axle housing. Lower the spring to the ground, remove the nut that holds the front pin in the frame bracket and drive the front spring pin out.

The spring can then be slid from under the car.

Replace in reverse of the procedure which removed it.

1968-70 Mark III and 1970 Continental

Coil springs are used at rear suspension.

The rear suspension is a coil-link design. Large, low-rate coil springs are mounted between rear axle pads and frame supports. Parallel lower arms extend forward of the spring seats to rubber frame anchor to accommodate driving and braking

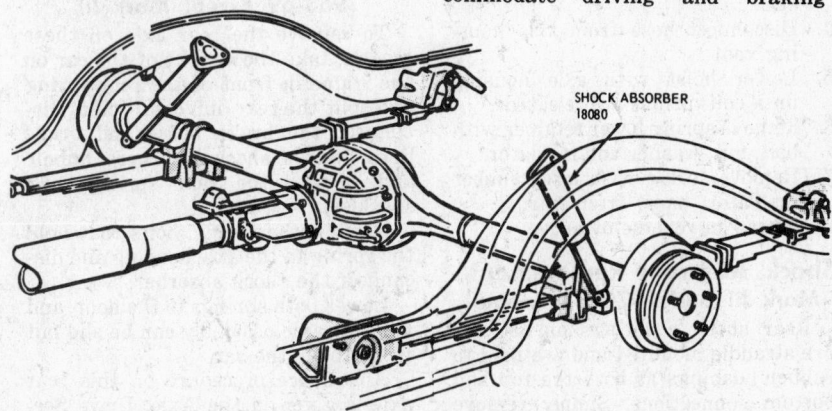

Rear suspension—1964-69 except Mark III

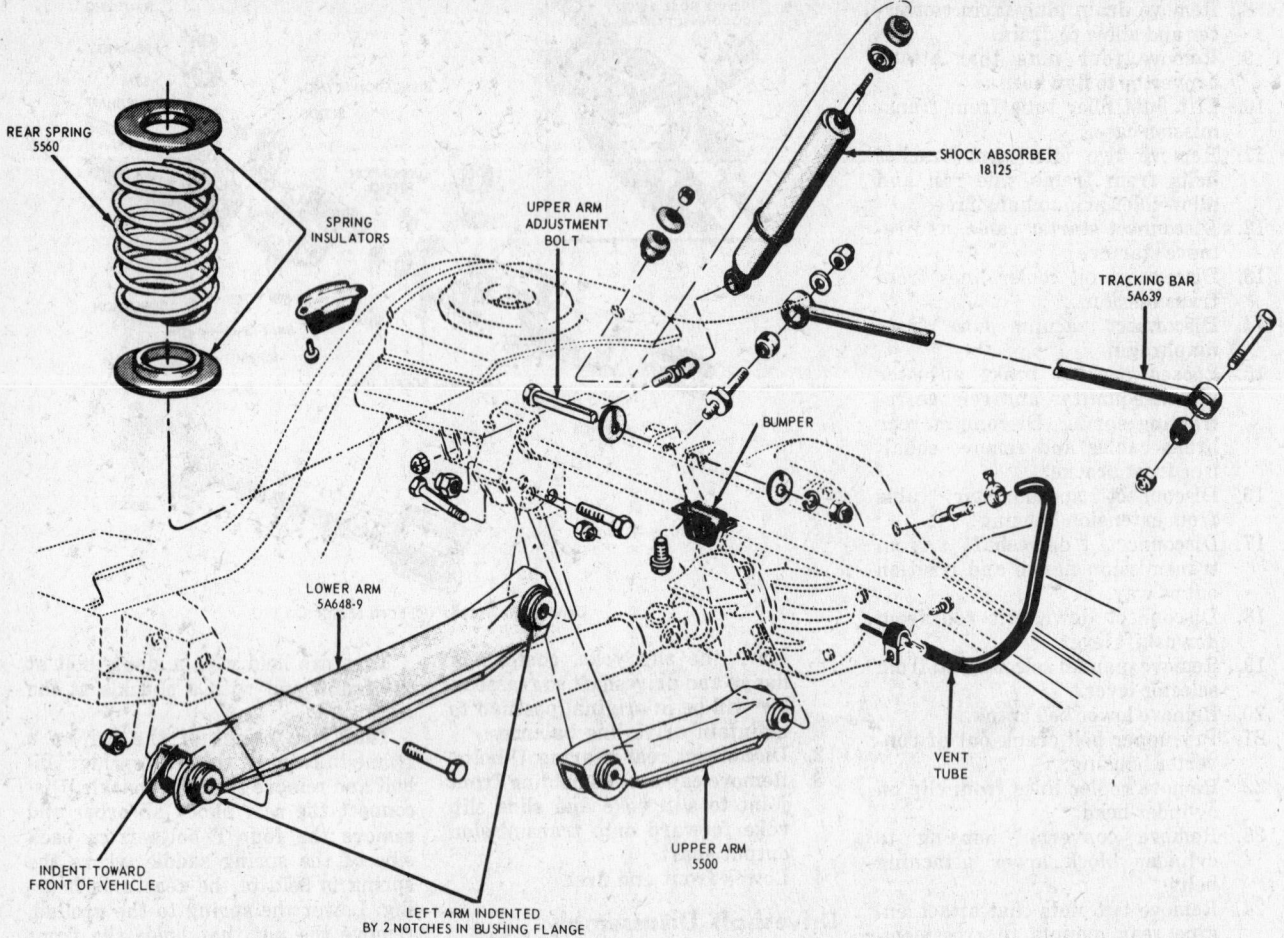

REAR SPRING
5560

SPRING INSULATORS

UPPER ARM ADJUSTMENT BOLT

SHOCK ABSORBER 18125

TRACKING BAR 5A639

BUMPER

LOWER ARM 5A648-9

VENT TUBE

INDENT TOWARD FRONT OF VEHICLE

LEFT ARM INDENTED BY 2 NOTCHES IN BUSHING FLANGE

UPPER ARM 5500

1970 Lincoln Continental and 1968-70 Mark III (© Ford Motor Company)

forces. A third link is mounted between the axle and the frame to control torque reaction forces from the rear wheels.

Lateral (side sway) motion of the rear axle is controlled by a rubber bushed rear track bar, linked laterally between the axle and frame.

1. Place car on hoist and lift under rear axle housing. Place jack stands under frame side rails.
2. Disconnect track bar at the rear axle housing bracket.
3. Disconnect rear shock absorbers from the rear axle housing brackets.
4. Disconnect hose from axle housing vent.
5. Lower hoist with axle housing until coil springs are released.
6. Remove spring lower retainer with bolt, nut, washer and insulator.
7. Remove spring with large rubber insulator pads from car.
8. Install in reverse of above.

Shock Absorber Replacement
Mark III and 1970 Continental

Rear shock absorbers on all cars are straddle mounted and are held to rubber bushings at both the top and bottom connections. Simply remove the nuts from the top and bottom of

the shock absorber and lift the shock absorber off the car.

Service on shock absorbers after they have been removed from the car is a highly specialized job, requiring specialized equipment. There is no practical way for the average shop to rebuild or service a shock absorber.

If tests show that the shock is weak or inefficient, it should be replaced with a new or rebuilt one.

Axle Assembly Removal

1963-69 Except Mark III

To remove the rear axle on these models, take the weight of the car on the frame in front of the rear spring and split the rear universal joint. Disconnect the brake hoses and brake lines, and the shock absorbers, unbolt and remove the rear spring back shackles.

Remove the four U-bolts that hold the spring to the axle housing and disconnect the shock absorber.

Lower both springs to the floor, and the rear axle assembly can be slid out from under the car.

All service procedure on this rear axle is given in the Axle, Drive Section. See index.

Mark III and 1970 Continental

The Mark III and 1970 Continental axle are not removed as an assembly.

1. Raise vehicle on hoist and remove wheels.
2. Remove rear hubs and drums. Back off brake shoes if drums do not come easily.
3. Remove nuts that secure rear wheel bearing retainer through hole in axle housing flange.
4. Pull each axle out of housing using properly designed puller. Do not damage oil seals.
5. Secure brake carrier plate to axle housing with one attaching nut.
6. Remove drive shaft.
7. Clean housing around carrier to prevent ingress of dirt.
8. Remove carrier nuts and drain housing.
9. Remove carrier.
10. Position safety stands under frame rear members, and support axle housing with floor jack or hoist.
11. Disengage brake line from clips that hold it on axle.
12. Disconnect vent tube from rear axle housing.
13. Remove brake backing plates and

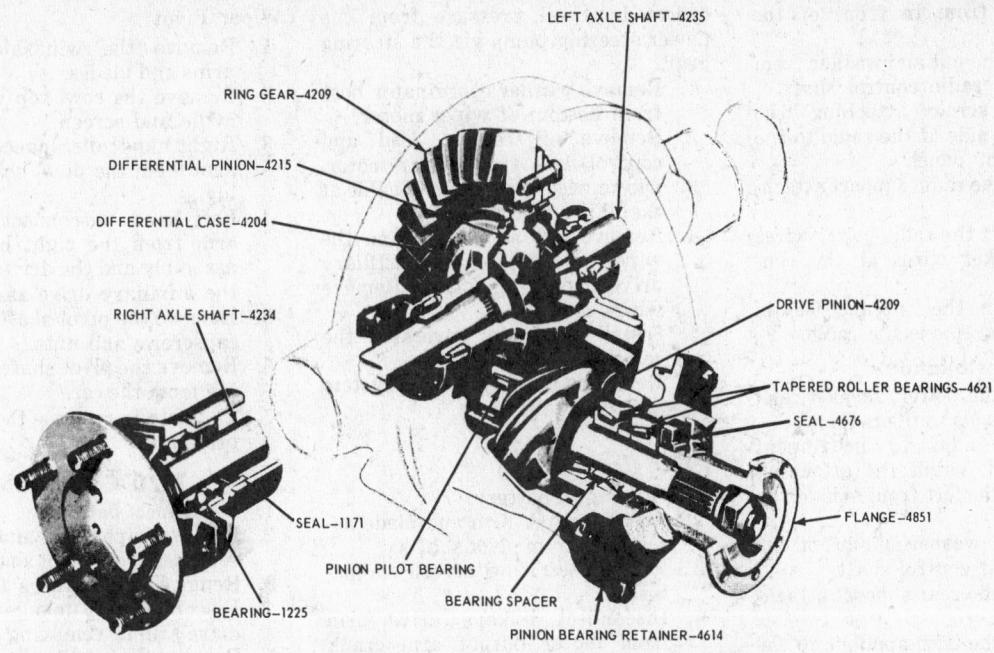

Lincoln rear axle assembly (© Ford Motor Company)

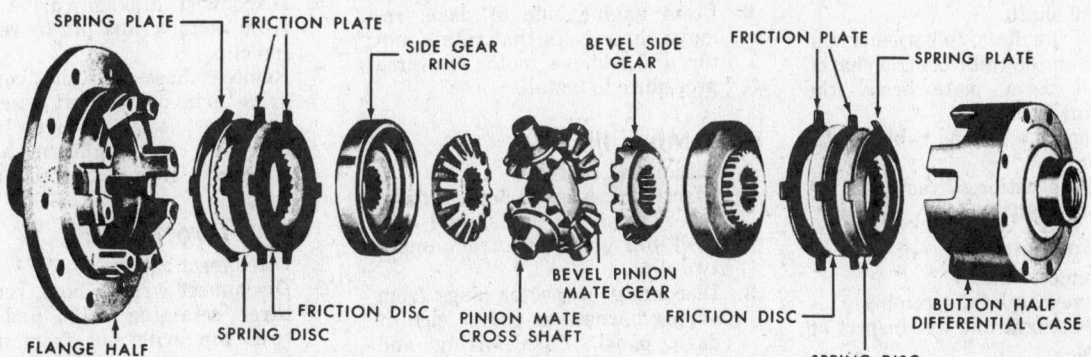

Directed power differential (© Ford Motor Co.)

secure them with wire. Do not disconnect brake line.

14. Disconnect lower studs of shocks from axle.
15. Remove retaining nut and washer and disconnect track bar from mounting stud on axle housing.
16. Lower axle housing until springs are released.
17. Lift out springs.
18. Remove nuts, washers, and pivot bolts that connect lower arms to axle. Disconnect arms from axle.
19. Remove pivot bolt, nut, lock washer, and two eccentric washers.
20. Lower axle housing and remove from vehicle. Reverse procedure to install.

Radio

Radio R & R

1963-65

1. Disconnect battery.

2. Remove control knobs and two jam nuts and lock washers retaining outer bezel.
3. Remove outer bezel and four screws and large bezel.
4. Remove two screws attaching center trim panel to support and slide panel away from windshield.
5. Disconnect feed and lamp connectors, and antennae lead in.
6. Remove three attaching screws and pull radio and support out of instrument panel.
7. Discover speaker leads.
8. Remove two jam nuts and washers and remove support.
To reinstall: reverse the above.

1966-69 Except Mark III

1. Disconnect battery.
2. Remove eight screws in the lower control housing.
3. Disconnect lead from the speaker/s.
4. Disconnect power antenna lead.
5. Disconnect lead to the foot op-

erated switch for AM-FM radios on cars so equipped.
6. Disconnect one two-way disconnect for power and pilot light.
7. Remove the two knobs and two bezels on the selector shafts. Remove the two nuts and two retainers on the selector shafts.
8. Remove the two screws holding the radio bracket to the lower reinforcement on the instrument panel.
9. Remove the two nuts and washers from the selector shafts. Disconnect the antenna lead and remove the radio.
10. To install, reverse the above.

1968-70—Mark III

Removal

1. Disconnect ground from battery.
2. Pull knobs off the control shafts.
3. Remove the cover plate located below the steering column.
4. Remove nut from the right radio control shaft.
5. Remove six screws and the trim

applique from in front of the radio.

6. Remove the nut and washer from the right radio control shaft.
7. Remove screw attaching the front left side of the radio to the instrument panel.
8. Remove the radio support attaching screw.
9. Disconnect the radio power wires and speaker wires at the connectors.
10. Disconnect the antenna lead-in cable and remove the radio.

Installation

1. Connect the power, speaker, and antenna leads to the radio.
2. Position radio to instrument panel and install the attaching screw at the left front side of the radio.
3. Install the washer and nut on the radio right control shaft.
4. Install radio rear support attaching screw.
5. Position the trim applique to the instrument panel and install the six attaching screws.
6. Install the nut on the radio right control shaft.
7. Install the discs, felt washer and knobs on the radio control shafts.
8. Install cover plate below the steering column.
9. Connect the ground cable to the battery.
10. Check operation of radio and set the push buttons.

1970 Continental

1. Disconnect battery.
2. Remove map light assembly.
3. Remove right and left inspection covers.
4. Remove lower instrument panel pad.
5. Remove glove box, open ashtray, and leave it open.
6. Remove glove box switch.
7. Through glove box opening remove two nuts retaining radio finish panel to instrument panel.
8. Remove radio knobs.
9. Remove two screws at top of finish panel. Position panel out and disconnect cigar lighter and light from right panel.
10. Through glove box opening remove nut from lower right corner of center finish panel.
11. Remove radio top support nut and three mounting screws. Pull radio out. Disconnect power leads and antenna cable. Remove radio. Reverse procedure to install.

Windshield Wipers

Motor R & R

1963-69

The wiper motor is hydraulically operated by oil pressure from the power steering pump via the steering gear.

1. Remove washer coordinator hose from bottom of wiper motor.
2. Remove oil return, feed and control lines from wiper motor.
3. Disconnect wiper control cable at the wiper motor.
4. Remove two screws holding the wiper motor to the auxiliary drive mounting plate. Remove wiper motor.
5. Install motor by reversing the removal procedure.
6. Refill the power steering system and bleed the lines.

1970 Continental

1. Disconnect battery.
2. Remove wiper arm and blade assemblies from pivot shafts.
3. Remove left cowl screen for access.
4. Disconnect linkage drive arm from motor output arm crank pin by removing retaining clip.
5. Disconnect two push on wire connectors from the motor.
6. From engine side of dash, remove three bolts that retain motor and remove motor. Reverse procedure to install.

1970 Mark III

1. Disconnect battery.
2. Disconnect washer hose, remove three retaining bolts, and pull cowl top grille out from under two clips.
3. Disconnect connector plugs from wiring harness at engine side of dash panel. Push wiring and plugs along with grommet through hole in dash.
4. Remove four motor to cowl retaining bolts. Lift motor out, and at the same time pull wiper arm and blade assembly to left for access to motor crank pin clip. Remove clip and disconnect drive link from motor crank pin. Remove three retaining bolts and separate motor from mounting plate and cover and wiring harness assembly. Reverse procedure to install.

Transmission R & R

1964-69

Auxiliary Drive

1. Remove both wiper arms and blades from the pivot assemblies.
2. Remove the ventilation grille and screen.
3. Remove both washer nozzles.
4. Disconnect the pivot shaft drive arm from the auxiliary drive.
5. Remove the four screws and the auxiliary drive assembly.
6. To install, reverse the removal procedure.

Wiper Pivot

1. Remove the windshield wiper arms and blades.
2. Remove the cowl top ventilation grille and screen.
3. Right hand: disconnect the drive arm from the pivot shaft assembly.
4. Left hand: disconnect the drive arm from the right hand pivot assembly and the drive arm from the auxiliary drive assembly.
5. Remove the pivot shaft retaining capscrews and nuts.
6. Remove the pivot shaft and housing from the car.
7. To install, reverse the removal procedure.

1970 Continental

1. Disconnect battery.
2. Remove wiper arm and blade assemblies from pivot shafts.
3. Remove cowl screens for access.
4. Disconnect left linkage arm from drive arm by removing clip.
5. Remove three bolts retaining left pivot shaft assembly through cowl opening.
6. Disconnect linkage drive arm from motor crank pin by removing clip.
7. Remove three bolts that connect drive arm pivot shaft assembly to cowl. Remove pivot shaft drivearm and rightarm as an assembly. Reverse procedure to install.

1970 Mark III

1. Disconnect battery.
2. Disconnect washer hose, remove three retaining bolts, and pull cowl top grille out from under two clips.
3. Remove wiper arm and blade assemblies.
4. To remove left pivot shaft, loosen right pivot shaft retaining bolts, and remove left pivot shaft retaining bolts. Remove connecting clip, and disconnect left pivot shaft link from right pivot shaft crank pin. Work left pivot shaft and link assembly toward right and out through cowl opening.
5. To remove right pivot shaft, remove three retaining bolts and disconnect linkage from crank pin by removing clip. Lift pivot shaft assembly from cowl. Reverse procedure to install.

Heater System

Heater Core Removal

1966-69 Continental

1. Disconnect battery, remove air cleaner, and drain coolant from system.
2. Disconnect hoses at heater core.

3. Remove harness clamp on top of evaporator-heater case.
4. Remove temperature blender door variable actuator.
5. Remove heater core cover plate retaining screws and cover plate.
6. Lift out heater core. Reverse procedure to install.

1968-70 Mark III

1. Remove hood and air cleaner and drain engine coolant.
2. Disconnect both hydraulic lines at wiper motor and position them to one side on 1968-69 Mark III.
3. Disconnect heater hoses at heater core and position hoses and water valve away from housing.
4. Disconnect vacuum supply hose on top of housing and remove oil pressure sending unit from back of engine.
5. Remove transmission dip stick and tube assembly.
6. Disconnect multiple connector leading to icing switch.
7. Remove evaporator housing front cover.
8. Remove heater core housing cover.
9. Remove heater core retaining bracket and remove heater core.

1970 Continental

1. Drain engine coolant.
2. Disconnect vacuum junction valve from dash panel and move valve and vacuum hoses away from case.
3. Disconnect speed control servo and bracket assembly if so equipped from dash panel and move it away from case.
4. Disconnect multiple connector from blower resistor and remove harness from clip on case.
5. Disconnect heater hoses from heater case and remove hose support clamp from case. Move hoses and water valve away from case.
6. Remove seven case cover to case flange attaching screws and wire harness clip.
7. Remove six cover to back plate stud nuts.
8. Remove one upper case to dash panel mounting screw.
9. Remove two case to dash panel mounting stud nuts, one on inboard mounting flange and one below case on lower flange.
10. Carefully move heater core assembly forward to clear mounting studs and lift it up and out of vehicle.
11. Remove two spring clips from core tubes on front of core cover.
12. Remove three screws from core end plate and remove plate.
13. Remove heater core and mounting gasket assembly from core cover and remove gasket from core. Reverse procedure to install.

Heater Blower Removal

1966-69 Continental

The blower motor is mounted in the right fender well.

1. Remove right fender splash shield secured with six bolts.

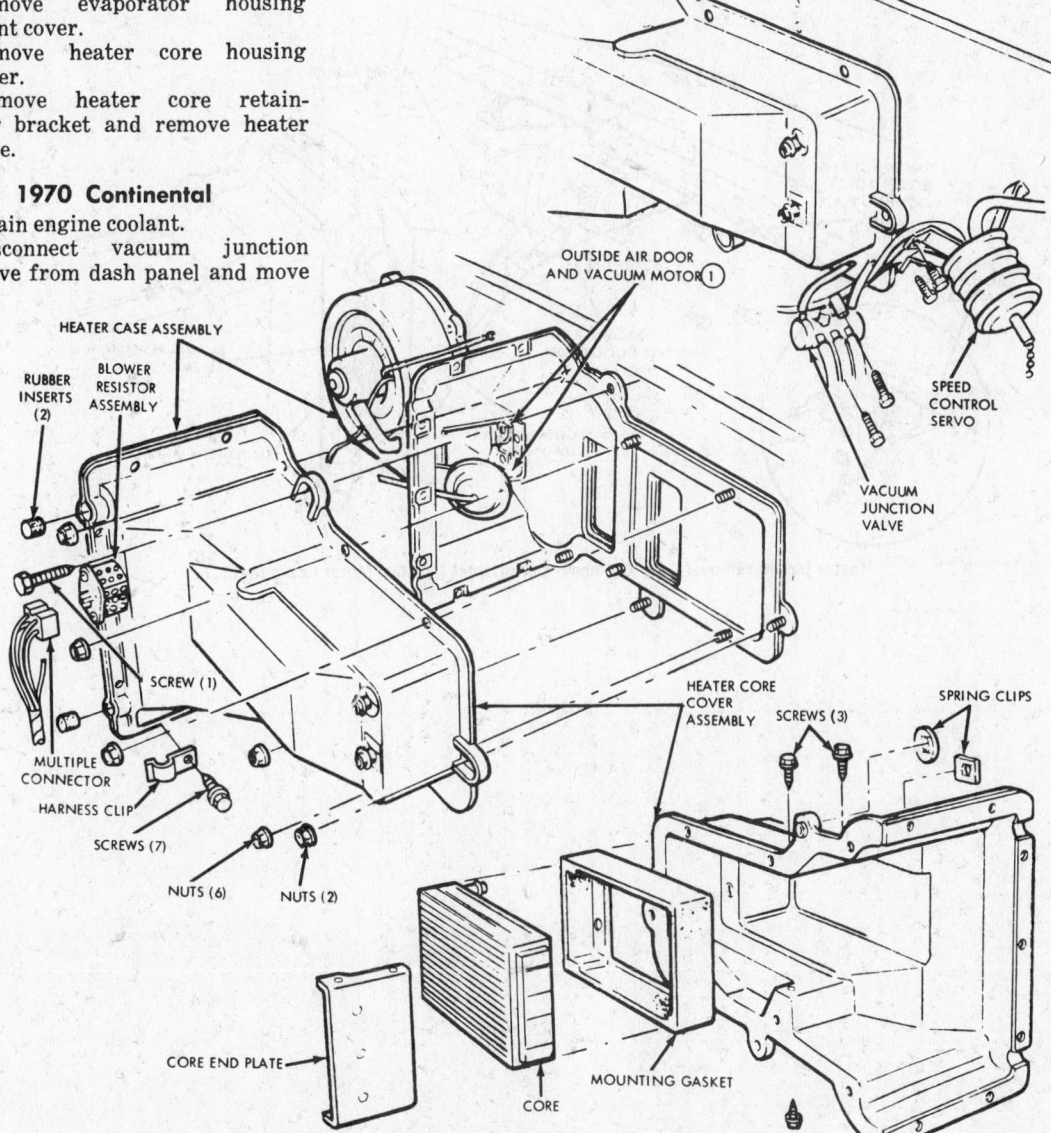

HEATER CASE ASSEMBLY
RUBBER INSERTS (2)
BLOWER RESISTOR ASSEMBLY
OUTSIDE AIR DOOR AND VACUUM MOTOR (1)
SPEED CONTROL SERVO
VACUUM JUNCTION VALVE
SCREW (1)
MULTIPLE CONNECTOR
HARNESS CLIP
SCREWS (7)
NUTS (6)
NUTS (2)
CORE END PLATE
CORE
MOUNTING GASKET
HEATER CORE COVER ASSEMBLY
SCREWS (3)
SPRING CLIPS

Heater core removal—1970 Lincoln Continental (© Ford Motor Company)

2. Remove four screws retaining blower motor to housing.
3. Disconnect electrical leads and remove motor.
4. Remove four nuts retaining motor housing to air door assembly and remove one nut through motor opening.
5. Remove housing and disassemble boot.
6. From inside car, remove right cowl trim panel and remove three screws retaining air door assembly to cowl.
7. From under fender remove screw at top.
8. Disconnect vacuum hose and remove air door assembly. Reverse procedure to install.

1968-70 Mark III

1. Remove right cowl side trim panel.

2. Remove screws retaining duct to cowl side panel and sound baffle and remove duct.
3. Disconnect lead wire to blower motor.
4. Remove one screw from motor mounting plate, rotate motor mounting plate clockwise to unlock plate from case and remove motor and wheel assembly through opening in cowl side of panel. Reverse procedure to install.

1970 Continental

1. Remove hood.
2. Remove right hood hinge and right fender inner support brace as an assembly.
3. Disconnect blower motor air cooling tube from motor.
4. Disconnect motor lead wire from

harness and ground wire from dash panel.
5. Disconnect rear section of right front fender panel apron from fender around wheel opening and remove two lower fender to cowl mounting screws.
6. Separate fender apron from fender wheel opening so that apron can be pushed downward away from blower motor.
7. Remove four blower motor plate screws. Move motor and wheel forward out of blower scroll and remove assembly through opening while applying pressure to fender apron to enlarge opening at hinge area. Reverse procedure to install.

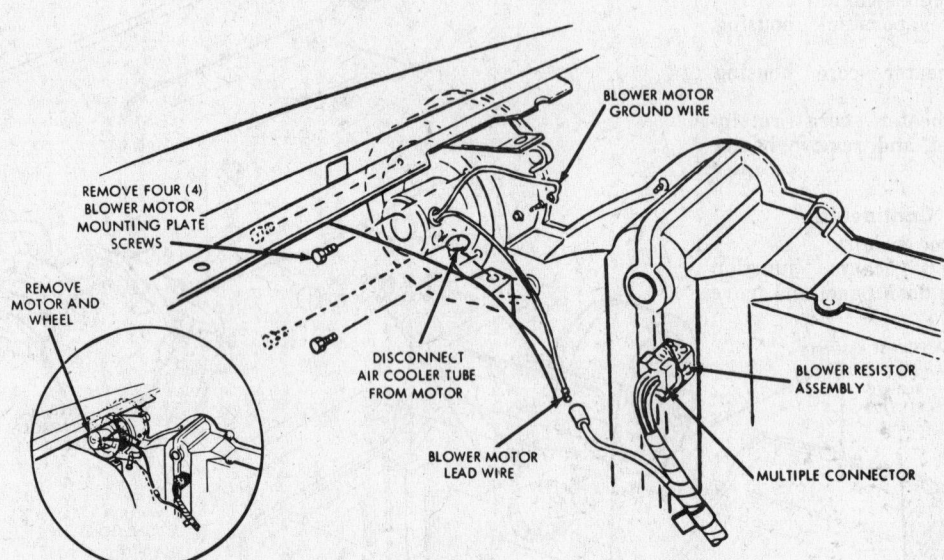

Heater blower removal—1970 Lincoln Continental (© Ford Motor Company)

Index

YEAR IDENTIFICATION

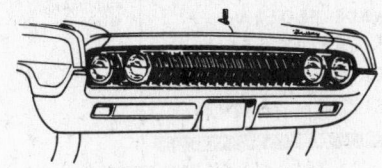
1964

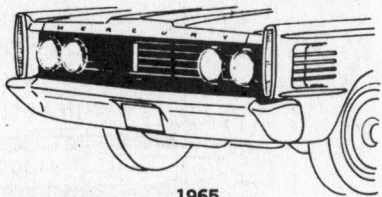
1965

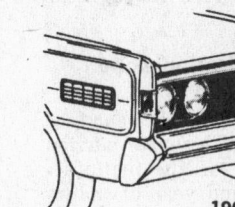
1966

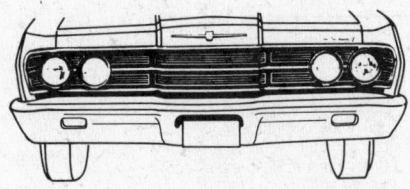
1967

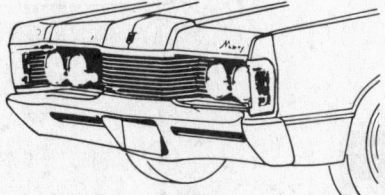
1968

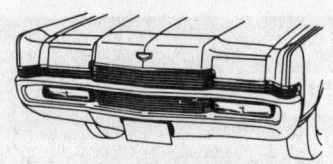
1969

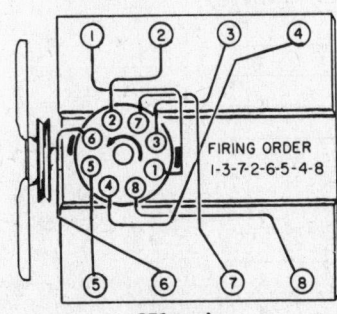
1969 (Marquis)

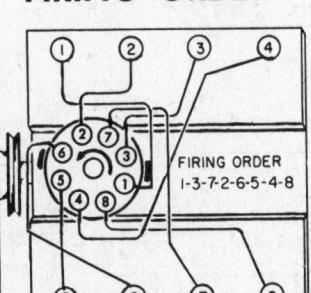
1970 Monterey

1970 Marquis Brougham

1971

FIRING ORDER

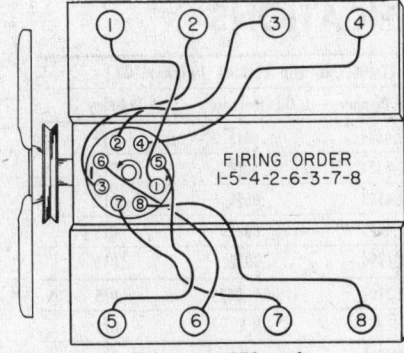

FIRING ORDER
1-3-7-2-6-5-4-8

351 cu. in.

FIRING ORDER
1-5-4-2-6-3-7-8

V8 except 351 cu. in.

CAR SERIAL NUMBER LOCATION AND ENGINE IDENTIFICATION

1964-67

Engine is identified through car serial number. Serial numbers, and other pertinent information, are to be found on a plate riveted to the rear edge of the left front door pillar.

The engine number is stamped on the top surface of the engine block near the crankcase breather pipe, front left side.

The car serial number is composed of two sections, the first five units giving the year, assembly plant, model and engine type. The second section gives the consecutive production serial number. The year of production is given by the first digit. (For example, "6" = 1966, "9" = 1969.) The fifth digit, a letter, represents the engine identification code. (See table.)

1968-71

The serial number can be found on a plate attached to the top of the instrument panel, visible through the windshield.

Vehicle Certification Label 1970-71

The label is located on the rear face of the driver's door. The upper portion contains the name of the manufacturer, the month and year of manufacture, and the certification statement. The lower portion of the label is interpreted in the illustration.

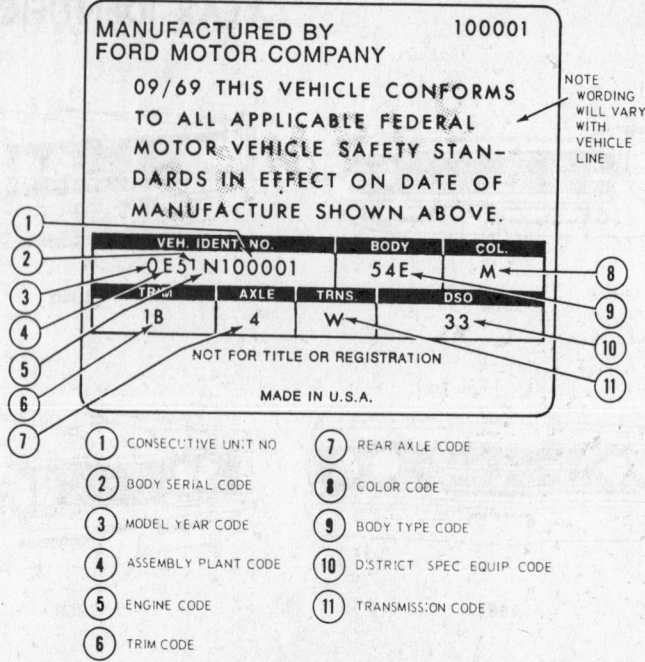

MANUFACTURED BY FORD MOTOR COMPANY   100001

09/69 THIS VEHICLE CONFORMS TO ALL APPLICABLE FEDERAL MOTOR VEHICLE SAFETY STANDARDS IN EFFECT ON DATE OF MANUFACTURE SHOWN ABOVE.

NOTE WORDING WILL VARY WITH VEHICLE LINE

VEH. IDENT. NO.	BODY	COL.
0E51N100001	54E	M

TRIM	AXLE	TRNS.	DSO
1B	4	W	33

NOT FOR TITLE OR REGISTRATION

MADE IN U.S.A.

1 CONSECUTIVE UNIT NO
2 BODY SERIAL CODE
3 MODEL YEAR CODE
4 ASSEMBLY PLANT CODE
5 ENGINE CODE
6 TRIM CODE
7 REAR AXLE CODE
8 COLOR CODE
9 BODY TYPE CODE
10 DISTRICT SPEC EQUIP CODE
11 TRANSMISSION CODE

1970-71 Vehicle Identification Label

NOT FOR TITLE OR REGISTRATION
9254Y500001   WARRANTY NUMBER   Ford   MADE IN U.S.A.

MERCURY

54C	M	3B	14H	34	6	U
BODY	COLOR	TRIM	DATE	DSO	AXLE	TRANS

1 Model year code
2 Assembly plant code
3 Body serial code
4 Engine code
5 Consecutive unit number
6 Body type code
7 Color code
8 Trim code
9 Date code
10 District—special equipment code
11 Rear axle code
12 Transmission code

Ford   9Y83N100001   Ford

Typical vehicle identification number (VIN) tab

Engine Identification Code

Mercury

Cu. In. Displ.	Type	YEAR AND CODE						
		1964	1965	1966	1967	1968	1969	1970
390-V8	2-BBL.	H	H	H	H	X	X	
390-V8	2-BBB. H.C.	Y	Y	Y	Y	Y	Y	Y
390-V8	4-BBL.	Z	Z	Z		Z		
410-V8	4-BBL.			M	M			
427-V8	4-BBL.	Q						
427-V8	2-4-BBL.	R	R	R	R			
428-V8	4-BBL. Police			P	P	P	P	P
428-V8	4-BBL.			Q	Q	Q		
429-V8	2-BBL.						K	K
429-V8	4-BBL.						N	N

CRANKSHAFT BEARING JOURNAL SPECIFICATIONS

YEAR	MODEL	MAIN BEARING JOURNALS (IN.)				CONNECTING ROD BEARING JOURNALS (IN.)		
		Journal Diameter	Oil Clearance	Shaft End-Play	Thrust On No.	Journal Diameter	Oil Clearance	End-Play
1964-65	V8—390, 406, 427 Cu. In.	2.7488	.0020	.006	3	2.4384	.0019	.019
1966-68	All	2.7488	.0015	.008	3	2.4384	.0018	.015
1969-70	V8—390 Cu. In.	2.7488	.0020	.007	3	2.4383	.0025	.015
	V8—428	2.7488	.0015	.007	3	2.4383	.0025	.015
	V8—429 Cu. In.	3.0000	.0010	.006	3	2.4996	.0012	.015
1971	V8—351 2-BBL.	2.9998	.0012-.0029	.004-.008	3	2.3107	.007-.0025	.015
	V8—400 2-BBL.	2.7488	.009-.0026	.004-.008	3	N.A.	N.A.	.015
	V8—429	2.9998	.005-.0025	.004-.008	3	2.4996	.0012	.015

GENERAL ENGINE SPECIFICATIONS

YEAR	CU. IN. DISPLACEMENT	CARBURETOR	DEVELOPED HORSEPOWER @ RPM	DEVELOPED TORQUE @ RPM (FT. LBS.)	A.M.A. HORSEPOWER	BORE AND STROKE (IN.)	ADVERTISED COMPRESSION RATIO	VALVE LIFTER TYPE	NORMAL OIL PRESSURE (PSI)
1964	V8—390, Std.	2-BBL.	250 @ 4400	378 @ 2400	52.5	4.050 x 3.781	9.0-1	Hyd.	52-62
	V8—390, Opt.	2-BBL.	266 @ 4400	378 @ 2400	52.5	4.050 x 3.781	9.4-1	Hyd.	52-62
	V8—390	4-BBL.	300 @ 4600	427 @ 2800	52.5	4.050 x 3.781	10.8-1	Hyd.	52-52
	V8—427	4-BBL.	410 @ 5600	476 @ 3400	57.4	4.230 x 3.781	11.5-1	Mech.	45-60
	V8—427	2-4-BBL.	425 @ 6000	480 @ 3700	57.4	4.230 x 3.781	11.5-1	Mech.	45-60
1965	V8—390	2-BBL.	250 @ 4400	378 @ 2400	52.5	4.050 x 3.781	9.4-1	Hyd.	52-62
	V8—390	4-BBL.	300 @ 4600	427 @ 2800	52.5	4.050 x 3.781	10.1-1	Hyd.	52-62
	V8—390	4-BBL.	330 @ 5000	427 @ 3200	52.5	4.050 x 3.781	10.1-1	Mech.	52-62
	V8—427	4-BBL.	410 @ 5600	476 @ 3400	57.4	4.230 x 3.781	11.1-1	Mech.	45-60
	V8—427	2-4-BBL.	425 @ 6000	480 @ 3700	57.4	4.230 x 3.781	11.1-1	Mech.	45-60
1966	V8—390	2-BBL.	265 @ 4400	397 @ 2600	52.5	4.050 x 3.781	9.5-1	Hyd.	35-55
	V8—390	2-BBL.	275 @ 4400	405 @ 2600	52.5	4.050 x 3.781	9.5-1	Hyd.	35-55
	V8—410	4-BBL.	330 @ 4600	444 @ 2800	52.5	4.050 x 3.980	10.5-1	Hyd.	35-55
	V8—428	4-BBL.	345 @ 4600	462 @ 2800	54.5	4.130 x 3.980	10.5-1	Hyd.	35-55
	V8—428	4-BBL.	360 @ 5400	459 @ 3200	54.5	4.130 x 3.980	10.5-1	Hyd.	35-55
1967	V8—390	2-BBL.	265 @ 4400	401 @ 2600	52.5	4.050 x 3.781	9.5-1	Hyd.	35-55
	V8—390	2-BBL.	275 @ 4400	405 @ 2600	52.5	4.050 x 3.781	9.5-1	Hyd.	35-55
	V8—410	4-BBL.	330 @ 4600	444 @ 2800	52.5	4.050 x 3.980	10.5-1	Hyd.	35-55
	V8—428	4-BBL.	345 @ 4600	462 @ 2800	54.5	4.130 x 3.980	10.5-1	Hyd.	35-55
	V8—428	4-BBL.	360 @ 5400	459 @ 3200	54.5	4.130 x 3.980	10.5-1	Hyd.	35-55
1968	V8—390	2-BBL.	265 @ 4400	390 @ 2600	52.5	4.050 x 3.781	9.5-1	Hyd.	35-55
	V8—390	2-BBL.	280 @ 4400	403 @ 2600	52.5	4.050 x 3.781	10.5-1	Hyd.	35-55
	V8—390	4-BBL.	315 @ 4600	427 @ 2800	52.5	4.050 x 3.781	10.5-1	Hyd.	35-55
	V8—428	4-BBL.	340 @ 4600	462 @ 2800	54.5	4.130 x 3.980	10.5-1	Hyd.	35-55
	V8—428	4-BBL.	360 @ 5400	459 @ 3200	54.5	4.130 x 3.980	10.5-1	Hyd.	35-55
1969	V8—390	2-BBL.	270 @ 4400	390 @ 2600	52.5	4.050 x 3.781	9.5-1	Hyd.	35-55
	V8—390	2-BBL.	280 @ 4400	403 @ 2600	52.5	4.050 x 3.781	10.5-1	Hyd.	35-55
	V8—428	4-BBL.	360 @ 5400	459 @ 3200	54.5	4.130 x 3.980	10.5-1	Hyd.	35-55
	V8—429	2-BBL.	320 @ 4400	460 @ 2200	60.8	4.360 x 3.590	10.5-1	Hyd.	35-55
	V8—429	4-BBL.	360 @ 4600	476 @ 2800	60.8	4.360 x 3.590	11.0-1	Hyd.	35-55
1970-71	V8—390	2-BBL.	265 @ 4400	390 @ 2600	52.5	4.050 x 3.781	9.5-1	Hyd.	35-55
	V8—351	2-BBL.	250 @ 4600	355 @ 2600	51.20	4.000 x 3.500	9.5-1	Hyd.	35-55
	V8—400	2-BBL.	N.A.	N.A.	51.20	4.000 x 4.000	9.5-1	Hyd.	50-70
	V8—429	2-BBL.	320 @ 4400	460 @ 2200	60.8	4.360 x 3.590	10.5-1	Hyd.	35-55
	V8—428	4-BBL.	360 @ 5400	459 @ 3200	54.5	4.130 x 3.980	10.5-1	Hyd.	35-55
	V8—429	4-BBL.	360 @ 4600	480 @ 2800	60.8	4.360 x 3.590	10.5-1	Hyd.	35-55

TORQUE SPECIFICATIONS

YEAR	MODEL	CYLINDER HEAD BOLTS (FT. LBS.)	ROD BEARING BOLTS (FT. LBS.)	MAIN BEARING BOLTS (FT. LBS.)	CRANKSHAFT BALANCER BOLT (FT. LBS.)	FLYWHEEL TO CRANKSHAFT BOLTS (FT. LBS.)	MANIFOLD (FT. LBS.) Intake	MANIFOLD (FT. LBS.) Exhaust
1964-65	V8—390 Cu. In.	80-90	40-45	95-105	70-90	75-85	32-35	12-18
	V8—406, 427 Cu. In.	100-110	53-58	95-105	70-90	75-85	15	27
1966-68	All	80-90	40-45	95-105	70-90	75-85	32-35	18-24
1969-71	V8—390 Cu. In.	80-90	40-45	95-105	70-90	75-85	32-35	18-24
	V8—428 Cu. In.	80-90	53-58	95-105	70-90	75-85	32-35	18-24
	V8—429 Cu. In.	130-140	40-45	95-105	70-90	75-85	25-30	28-33

CYLINDER HEAD BOLT TIGHTENING SEQUENCE

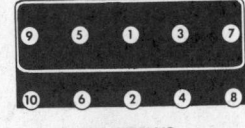

All OHV V8

TUNE-UP SPECIFICATIONS

YEAR	MODEL	SPARK PLUGS		DISTRIBUTOR		IGNITION TIMING (Deg.) ▲	CRANKING COMP. PRESSURE (Psi)	VALVES			FUEL PUMP PRESSURE (Psi)	IDLE SPEED (Rpm) ★
		Type	Gap (In.)	Point Dwell (Deg.)	Point Gap (In.)			Tappet (Hot) Clearance (In.)		Intake Opens (Deg.)		
								Intake	Exhaust			
1964	V8—390 Cu. In., Std.	BF42	.035	27	.015	6B	180	Zero	Zero	26B	5	500
	V8—390 Cu. In., Hi. Perf.	BF32	.035	27	.020	10B	190	.025	.025	28B	6	500
	V8—406 Cu. In., 4-BBL.	BF32	.035	27	.020	8B	180	.025	.025	15½A	6	675
	V8—406 Cu. In., 3-2-BBL.	BF32	.035	27	.020	8B	180	.025	.025	15½A	6	675
	V8—427 Cu. In., Std. Ign.	BF32	.035	35	.020	8B	180	.025	.025	8A	6	675
	V8—427 Cu. In., T. Ign.	BF32	.035	23	.020	8B	180	.025	.025	8A	6	675
1965	V8—390 Cu. In., 2-BBL.	BF42	.035	27	.015	6B	190	Zero	Zero	18B	5-6	575
	V8—390 Cu. In., 4-BBL., M.T.	BF42	.035	27	.020	4B	190	Zero	Zero	26B	5-6	575
	V8—390 Cu. In., 4-BBL., A.T.	BF42	.035	27	.020	6B	190	Zero	Zero	26B	5-6	500
	V8—390 Cu. In., 4-BBL., M.T.	BF32	.035	27	.020	4B	190	.025	.025	28½B	5-6	575
	V8—390 Cu. In., 4-BBL., A.T.	BF32	.035	27	.020	6B	190	.025	.025	28½B	5-6	575
	V8—427 Cu. In., 2-4-BBL.	BF32	.035	35	.020	8B	180	.025	.025	8A	5-6	800
1966-67	V8—390 Cu. In.	BF42	.035	28	.017	10B⊙	180	Zero	Zero	16B	5-6	600①
	V8—410 Cu. In.	BF42	.035	28	.017	10B⊙	180	Zero	Zero	16B	5-6	600①
	V8—428 Cu. In.	BF42	.035	28	.017	10B⊙	180	Zero	Zero	16B	5-6	600①
1968	V8—390, 2-BBL.	BF32	.035	28	.017	6B	180	Zero	Zero	13B	5-6	625②
	V8—390, 428, 4-BBL.	BF32	.035	28	.017	6B	180	Zero	Zero	16B	5-6	625②
	V8—428 Interceptor	BF32	.035	28	.017	6B	180	Zero	Zero	18B	5-6	625②
1969	V8—390, 2-BBL. (270 HP)	BF42	.035	28	.017	6B	180	Zero	Zero	13B	5-6	650②
	V8—390, 2-BBL. (280 HP)	BF42	.035	28	.021	6B	180	Zero	Zero	16B	5-6	650②
	V8—428	BF32	.035	28	.017	6B	180	Zero	Zero	18B	5-6	600
	V8—429, 2-BBL.	BF42	.035	28	.017	6B	180	Zero	Zero	16B	5-6	550
	V8—429, 4-BBL.	BF42	.035	28	.017	6B	180	Zero	Zero	16B	5-6	550
1970	V8—390, 2-BBL. (265 HP)	BF42	.035	24-29	.021	6B	180	Zero	Zero	13B	5-6	650②
	V8—428, 4-BBL.	BF32	.035	24-29	.021	6B	190	Zero	Zero	18B	5-6	600
	V8—429, 2-BBL. (320 HP)	BF42	.035	24-29	.021	6B	180	Zero	Zero	16B	5-6	550
	V8—429, 4-BBL. (360 HP)	BF42	.035	24-29	.021	6B	180	Zero	Zero	16B	5-6	550
1971	V8—351, 2-BBL.	BF42	.034	24-29	.021	6B	180	Zero	Zero	11B	5-6	700①
	V8—400, 2-BBL.	BF42	.034	25-29	.021	6B	180	Zero	Zero	18B	5-6	700③
	V8—429, 2-BBL., 4-BBL.	BF42	.035	25-29	.021	6B	180	Zero	Zero	18B	5-6	700

★—With manual transmission in N and automatic in D. Add 50 rpm when equipped with air conditioning.

▲—With vacuum advance disconnected. NOTE: These settings are only approximate. Engine design, altitude, temperature, fuel octane rating and the condition of the individual engine are all factors which can influence timing. The limiting advance factor must, therefore, be the "knock point" of the individual engine.

⊙—With Thermactor System: Timing—All—6 B.
①—With A.T.—500 rpm.
②—With A.T.—550 rpm.
③—With A.T.—575 rpm.
A.T.—Automatic transmission.
M.T.—Manual transmission.
B—Before top dead center.

Caution

General adoption of anti-pollution laws has changed the design of almost all car engine production to effectively reduce crankcase emission and terminal exhaust products. It has been necessary to adopt stricter tune-up rules, especially timing and idle speed procedures. Both of these values are peculiar to the engine and to its application, rather than to the engine alone. With this in mind, car manufacturers supply idle speed data for the engine and application involved. This information is clearly displayed in the engine compartment of each vehicle.

BATTERY AND STARTER SPECIFICATIONS

YEAR	MODEL	BATTERY				STARTERS						Brush Spring Tension (Oz.)
		Ampere Hour Capacity	Volts	Group Number	Terminal Grounded	Lock Test			No-Load Test			
						Amps.	Volts	Torque	Amps.	Volts	RPM	
1964-65	390 Cu. In., M.T.	55	12	4NF	Neg.	580	5	14.8	110	12	5,200	48
	390 Cu. In., A.T.	65	12	4NF	Neg.	580	5	14.8	110	12	5,200	48
	406, 427 Cu. In.	70	12	3MSA	Neg.	580	5	14.8	110	12	5,200	48
1966-67	V8—390 Cu. In.	55	12	4NF	Neg.	670	5	15.5	70	12	5,200	40
	V8—410, 428 Cu. In.	70	12	3MSA	Neg.	670	5	15.5	70	12	5,200	40
1968	V8—390 Cu. In.	45	12	17M1A	Neg.	670	5	15.5	70	12	5,200	40
	V8—390 Cu. In.	55	12	172B	Neg.	670	5	15.5	70	12	5,200	40
	V8—390 Cu. In.	70	12	17H3	Neg.	670	5	15.5	70	12	5,200	40
	V8—428 Cu. In.	80	12	17H3A	Neg.	670	5	15.5	70	12	5,200	40
1969	V8—390 Cu. In.	55	12	24F	Neg.	670	5	15.5	70	12	9,500	40
	V8—390 Cu. In.	70	12	27F	Neg.	670	5	15.5	70	12	9,500	40
	V8—428 Cu. In.	80	12	27HF	Neg.	700	4	15.5	70	12	11,000	40
	V8—429 Cu. In.	80	12	27HF	Neg.	700	5	15.5	70	12	10,000	40
1970	V8—390 Cu. In.	55	12	24F	Neg.	670	5	15.5	70	12	9,500	40
	V8—390 Cu. In.	70	12	27F	Neg.	670	5	15.5	70	12	9,500	40
	V8—429 Cu. In.	80	12	27HF	Neg.	700	5	15.5	70	12	10,000	40
1971	V8—351 Cu. In.	55	12	24F	Neg.	670	5	15.5	70	12	9,500	40
	V8—400 Cu. In.	80	12	24F	Neg.	670	5	15.5	70	12	9,500	40
	V8—429 2-BBL.	80	12	24HF	Neg.	700	5	15.5	70	12	9,500	40
	V8—429 4-BBL.	80	12	24HF	Neg.	700	5	15.5	70	12	10,000	40

GENERAL CHASSIS AND BRAKE SPECIFICATIONS

YEAR	MODEL	CHASSIS			BRAKE CYLINDER BORE		BRAKE DRUM	
		Overall Length (In.)	Tire Size	Master Cylinder (In.)	Wheel Cylinder Diameter (In.)		Diameter (In.)	
					Front	Rear	Front	Rear
1964	All	215.5	8.00 x 14	1.0 ①	1.094	.969	11.03	11.03
1965	Sedan	218.4	8.15 x 15	1.0 ①	1.094	.969	11.03	11.03
	Sta. Wagon	214.5	8.15 x 15	1.0 ①	1.062	.969	11.03	11.03
1966	Sedan	220.4	8.15 x 15	1.0 ①	1.094	.969	11.03	11.03
	Sta. Wagon	216.5	8.15 x 15	1.0 ①	1.062	.969	11.03	11.03
1967	Ex. Sta. Wagon	218.5	8.15 x 15	1.0 ②	1.094③	¹⁵/₁₆	11.03 ④	11.03
	Sta. Wagon	213.8	8.15 x 15	1.0 ②	1.094③	¹⁵/₁₆	11.03 ④	11.03
1968	Exc. Sta. Wagon	220.1	8.15 x 15	1.0	1.094	¹⁵/₁₆	11.03	11.03
	Sta. Wagon	215.4	8.45 x 15	1.0	1.094	¹⁵/₁₆	11.03	11.03
	All w/Disc. Brakes	221.8	8.15 x 15	1.0	2³/₄	¹⁵/₁₆	11.96 (Disc)	11.03
1969	w/Drum Brakes	221.8	⑤	1.0	1.125	¹⁵/₁₆	11.03	11.03
	w/Disc. Brakes	221.8	⑤	1.0	2³/₄	¹⁵/₁₆	11.72	11.03
1970	w/Drum Brakes	⑥	⑤	1.0	1.125	¹⁵/₁₆	11.03	11.03
	w/Disc. Brakes	⑥	⑤	1.0	2³/₄	¹⁵/₁₆	11.72	11.03
1971	Exc. Sta. Wagon	224.7	⑤	1.0	1.125⑦	¹⁵/₁₆⑧	...	11.03
	Sta. Wagon	220.5	⑤	1.0	.938⑦	¹⁵/₁₆⑧	...	11.03

① —With power brakes 0.875.
② —With power .9375.
③ —With disc brakes 1.938.
④ —With Discs 11.87.
⑤ —All tire information is shown on a decal affixed to the inside door of the glove compartment. (G78 x 15, H78 x 15 or H70 x 15)
⑥ —Monterey — 221.8, Marquis — 224.3, Sta. Wagon — 220.5.
⑦ —With disc 2.750.
⑧ —With drum 0.81.

AC GENERATOR AND REGULATOR SPECIFICATIONS

| YEAR | ALTERNATOR | | | REGULATOR | | | | | | |
| | Part No. or Manufacturer | Field Current @ 12 V. | Output @ Generator RPM | Part No. or Manufacturer | Field Relay | | | Regulator | | |
					Air Gap (in.)	Point Gap (in.)	Volts to Close	Air Gap (in.)	Point Gap (in.)	Volts @ 75°
1964	C3SF10300A	2.9–3.1	40	C4UF10316A	.026	.018	3–4	.048	.013	14.1–14.9
	C4SF10300A	2.9–3.1	42	C4TF10316B	.026	.018	3–4	.048	.013	14.1–14.9
	C30F10300E	2.9–3.1	42							
1965	Autolite	2.8–3.3	42	Autolite	.018	N.A.	2.5–4	.052	.020	14.1–14.9
	Autolite	2.8–3.3	45							
	Autolite	2.8–3.3	55							
	Leece-Nevelle	2.8–3.3	53	Leece-Nevelle	.010	.019	1.6–2.6	.050	.019	14.1–14.9
				Leece-Nevelle	.012	.025	6.2–7.2	.050	.019	14.1–14.9
	Leece-Nevelle	2.8–3.3	60							
1966	Autolite	2.9	42	Autolite	0.014	N.A.	2.5–4	.053	.020	14.1–14.9
	Autolite	2.9	45	Autolite transistor	①	①	2.5–4	①	①	14.1–14.9
	Autolite	2.9	55							
	Leece-Nevelle	2.9	53	Leece-Nevelle 53 Amp only	.010	.019	1.6–2.6	.050	.019	14.1–14.9
	Leece-Nevelle	4.6	60	Leece-Nevelle 53 and 60 Amp	.012	.025	6.2–7.2	.050	.019	14.1–14.9
	Leece-Nevelle	2.9	60							
1967	Autolite	2.5	38	Autolite	.014	N.A.	2.5–4	N.A.	N.A.	13.9–14.9
	Autolite	2.9	42							
	Autolite	2.9	45							
	Autolite	2.9	55							
	Autolite	4.6	60	Autolite Transistor	①	①	2.5–4	①	①	13.9–14.9
	Leece-Nevelle	2.9	53	Leece-Nevelle 53 Amp only	.010	.019	1.6–2.6	.050	.019	13.9–14.9
	Leece-Nevelle	2.9	65	Leece-Nevelle 53 and 60 Amp	.012	.025	6.2–7.2	.050	.019	13.9–14.9
1968	Autolite C6AF10300C	2.9	42	Autolite	①	①	4.2–9.0	①	①	13.5–15.3
	Autolite C6AF10300G	2.9	55							
	Autolite C6TF10300F	2.9	65							
	Leece-Nevelle	2.9	65	Leece-Nevelle	0.012	0.025	6.2–7.2	0.047	0.019	13.9–14.9
1969	Autolite	2.9	42	Autolite	①	①	4.2–9.0	①	①	13.5–15.3
	Autolite	2.9	55							
	Autolite	2.9	65							
	Leece-Nevelle	2.9	65	Leece-Nevelle	0.012	0.025	6.2–7.2	0.047	0.019	13.9–14.9
1970–71	Autolite	2.9	42	Autolite	①	①	2.0–4.2	①	①	13.5–15.5
	Autolite	2.9	55							
	Autolite	2.9	65							
	Leece-Nevelle	2.9	65	Leece-Nevelle	.012	.025	6.2–7.2	.047	.019	13.9–14.9

① —Transistorized regulator—not adjustable.

VALVE SPECIFICATIONS

YEAR AND MODEL		SEAT ANGLE IN. (DEG.)	SEAT ANGLE EX. (DEG.)	FACE ANGLE (DEG.)	VALVE LIFT INTAKE (IN.)	VALVE LIFT EXHAUST (IN.)	VALVE SPRING PRESSURE (VALVE OPEN) LBS. @ IN.	VALVE SPRING INSTALLED HEIGHT (IN.)	STEM TO GUIDE CLEARANCE (IN.) Intake	STEM TO GUIDE CLEARANCE (IN.) Exhaust	STEM DIAMETER (IN.) Intake	STEM DIAMETER (IN.) Exhaust
1964-65	V8—390 Cu. In.	45	45	44	.408	.408	200 @ 1.42	1.83	.001-.0024	.001-.0024	.3711-.3718	.3693-.3700
	V8—406, 427 Cu. In.	30	45	①	.500	.500	200 @ 1.42	1.83	.001-.0024	.002-.0034	.3711-.3718	.3701-.3708
1966-67	All	45	45	44	.437	.437	200 @ 1.42	1.82	.001-.0024	.002-.0034	.3711-.3718	.3693-.3700
1968	V8—390, 2-BBL.	45	45	44	.427	.430	200 @ 1.42	1.82	.001-.0024	.001-.0024	.3711-.3718	.3693-.3700
	V8—390, 428, 4-BBL.	45	45	44	.427	.437	220 @ 1.38	1.82	.001-.0024	.001-.0024	.3711-.3718	.3711-.3718
	V8—428 Interceptor	45	45	44	.480	.490	220 @ 1.38	1.82	.001-.0024	.001-.0024	.3711-.3718	.3711-.3718
1969	V8—390, Lo. Comp.	45	45	44	.427	.430	200 @ 1.42	1.82	.001-.0024	.001-.0024	.3711-.3718	.3693-.3700
	V8—390, Hi. Comp.	45	45	44	.440	.440	220 @ 1.38	1.82	.001-.0024	.001-.0024	.3711-.3718	3711-.3718
	V8—428	30	45	44	.481	.490	220 @ 1.38	1.82	.001-.0024	.001-.0024	.3711-.3718	.3711-.3718
	V8—429	45	45	44	.443	.486	220 @ 1.38	1.81	.001-.0027	.001-.0027	.3416-.3423	.3416-.3423
1970	V8—390, Lo. Comp.	45	45	44	.427	.430	200 @ 1.42	1.82	.001-.0027	.001-.0027	.3711-.3718	.3706-.3713
	V8—390, Hi. Comp.	45	45	44	.440	.440	220 @ 1.38	1.82	.001-.0027	.001-.0027	.3711-.3718	.3706-.3713
	V8—428	45	45	44	.481	.490	220 @ 1.38	1.82	.001-.0027	.001-.0027	.3711-.3718	.3711-.3718
	V8—429	45	45	44	.443	.486	220 @ 1.38	1.81	.001-.0027	.001-.0027	.3416-.3423	.3416-.3423
1971	V8—351-2 BBL.	45	45	44	.418	.448	220 @ 1.34	1.79	.0010-.0027	.0015-.0032	.3423-.3416	.3418-.3411
	V8—400-2 BBL.	45	45	44	.418	.448	220 @ 1.34	1.79	.0010-.0027	.0015-.0032	.3423-.3416	.3418-.3411
	V8—429-2 BBL.	45	45	44	.418	.448	250 @ 1.33	1.81	.0010-.0027	.001-.0032	.3423-.3416	.3418-.3411
	V8—429—4 BBL.	45	45	44	.418	.448	250 @ 1.33	1.81	.0010-.0027	.001-.0032	.3423-.3416	.3418-.3411

① —Intake 29° Exh. 44°.

WHEEL ALIGNMENT

YEAR	MODEL	CASTER Range (Deg.)	CASTER Pref. Setting (Deg.)▲	CAMBER Range (Deg.)	CAMBER Pref. Setting (Deg.)■	TOE-IN (In.)	KING-PIN INCLINATION (Deg.)	WHEEL PIVOT RATIO Inner Wheel	WHEEL PIVOT RATIO Outer Wheel
1964	All	½N to ½P	0	¼P to 1P	⅝P	⅛ to ¼	7	24¼	20
1965	All	0 to 2P	1P	¼N to 1P	½P	⅛ to ¼	7	20	17½
1966	All	½P to 1½P	1P	¼N to ¾P	¼P	⅛ to ¼	7	20	17½
1967	All	½P to 1½P	1P	¼N to 1¼P	½P	⅛ to ¼	7¾	20	18⅛
1968	All	½P to 1½P	1P	¼N to ¾P	¼P	⅛ to ¼	7½	20	18¾
1969	All	0 to 2P	1P	¼N to 1¼P	½P	1/16 to 5/16	7½	20	19½
1970-71	All	0 to 2P	1P	¼N to 1¼P	½P	1/16 to 5/16	7½	20	19½

N—Negative.
P—Positive.
▲—Preferred Caster: Within ½° one side to other.
■—Preferred Camber: Within ¼° one side to other.

CAPACITIES

YEAR	MODEL	ENGINE CRANKCASE ADD 1 Qt. FOR NEW FILTER	TRANSMISSIONS Pts. TO REFILL AFTER DRAINING Manual	TRANSMISSIONS Pts. TO REFILL AFTER DRAINING Automatic	DRIVE AXLE (Pts.)	GASOLINE TANK (Gals.)	COOLING SYSTEM (Qts.) WITH HEATER
1964-65	390 Cu. In.	4	3	20	5	20	20½
	406, 427 Cu. In.	5	3¼	None	5	20	20½
1966-67	V8—390 Cu. In.	4	3½	20	5	25②	20½
	V8—410, 428 Cu. In.	4	3½ ①	26	5	25②	20½
1968	All Models	5	3	20	5	24②	20½
1969	All Models	4	3½	25½	5	24②	20½
1970-71	All Models	4	3½	25½	5	24.5②	③

①—1966-67, 4-speed—4 pts.
②—Sta. Wagon—23.5 gals.
③—429 cu. in.—18.6; with A/C—19.0

FUSES AND CIRCUIT BREAKERS

1965

Circuit	Protective Device	Location
Headlights	18 Amp. circuit breaker	Incorporated in lighting switch
Instrument panel, transmission indicator and ash receptacle lights	AGA-5 Fuse	Fuse panel
Turn signals and back-up lights, radio, single-speed washer motor	SFE-14 Fuse	Fuse panel
dual-speed washer motor	12 Amp. circuit breaker	Incorporated in washer switch
Seat belt warning, emergency warning	SFE-20 Fuse	Fuse panel
Heater blower, power antenna, door open warning	SFE-20	Fuse panel
Electric windshield wiper (single-speed)	6 Amp. circuit breaker	Incorporated in wiper switch
(dual-speed)	12 Amp. circuit breaker	
(intermittent—Mercury)	7 Amp. circuit breaker	
Cigar lighter	SFG-14 Fuse	Fuse panel
Air conditioner (integrated system)	25 Amp. circuit breaker	On ignition switch
Spot light	SFE-7.5 Fuse	Cartridge in feed wire
Overdrive	AGC-15 Fuse	On overdrive relay
Clock, dome, courtesy, map, cargo, trunk, glove box	SFE-9 Fuse	Fuse panel
Tail, park, license plate, and horns	15 Amp. circuit breaker	Incorporated in lighting switch
Speed control	SFE 14 Amp. fuse	Cartridge in feed wire
Transistorized ignition	AGA-2 fuse	Cartridge in feed wire
Top control motor feed, power windows and seat, back lite	20 Amp. circuit breaker	At starter relay
Air conditioner economy		Cartridge in feed wire

1966

Circuit	Circuit Protection		Location
	Rating	Trade No.	
Power assists			
Convertible top circuit	14 gage wire fuse link	(C6AB-14A094-A)	In wiring near starter relay
Convertible top with power option(s)	20 Amp. C.B. replaces fuse link		On starter relay
Power windows including backlite & tailgate	20 Amp. C.B.		
Power seats (four- and six-way)	20 Amp. C.B.		
Air conditioning circuit			
Air conditioner (economy)	15 Amp.	AGC 15	Fuse cartridge in feed wire
Air conditioner (selectair)	25 Amp. C.B.		On back of ignition switch at accessory terminal
Miscellaneous circuits			
Speed control	14 Amp. fuse	SFE 14	Fuse cartridge in feed wire
Spotlight	7.5 Amp. fuse	SFE 7.5	
Automatic headlamp dimmer	4 Amp. fuse	AGA 4	
Seat belt warning and parking brake warning	No protection		
Gage circuits			
Charge indicator	No protection		
Oil pressure indicator			
Engine water—cold			
Engine water—hot			
Motors			
Windshield wiper motor	Circuit breaker		Integral with motor
Convertible top motor			
Power window motor			
Power seat motor			
Headlamp circuit			

1966

Circuit	Circuit Protection		Location
	Rating	Trade No.	
Headlamps	18 Amp. C.B.		Integral with lighting switch
High beam indicator lamp			
Rear lamps (tail lights & stop lights)	15 Amp. C.B.		
Front parking lamps			
Ignition switch lamp			
License lamp			
Horns			
Dome lamp circuit			
Dome lamp	9 Amp. fuse	SFE 9	On fuse panel (dome socket)
Courtesy lamps			
Cargo lamp			
Map lamp			
Glove compartment lamp			
Luggage compartment lamp			
Clock			
Tachometer			
Instrument panel illumination			
Mercury	5 Amp. fuse	AGA 5	Fuse panel (instrument panel socket). This circuit is connected into the light switch rheostat.
Clock light			
Instrument cluster lights			
Ash tray light			
RPNDL—console or column light			
Radio light			
Heater control lights			
Medallion lamp (no clock on Parklane)			
Heater circuit			
Heater and defroster motor	20 Amp. fuse	SFE 20 or AGC 20	Fuse panel (heater socket)
Safety convenience panel lamps:			
Low fuel warning			
Seat belt warning			
Door ajar warning			
Door unlock warning			
Cornering lamps			
Power antenna			
Station wagon under seat heater			
Open door warning			
Cigar lighter circuit			
Cigar lighter	14 Amp. fuse	SFG 14	Fuse panel (cigar lighter socket)
Cigar lighter plus emergency warning option (hang-on or convenience panel)	20 Amp. fuse (Replaces 14 Amp. cigar fuse)	SFE 20	
Radio circuit			
Radio	14 Amp. fuse	SFE 14	Fuse panel (radio socket)
Back-up lamps			
Single speed washers			
Intermittent speed washers			
Turn signal circuit			(After fuse flasher protects circuit for turn signal)
Windshield wipers			
Single speed wipers	6 Amp. C.B.		Integral with windshield wiper switch
Intermittent wipers	6 Amp. C.B. (single speed) 7 Amp. C.B. (two-speed)		

1967

Circuit	Protective Device-Fuse or Circuit Breaker (C.B.) in Amperes	Location
Intermittent wipers (Mercury)	6 C.B.	Integral with windshield wiper switch
Convertible top circuit	Fuse link ①	In wiring near starter relay
Convertible top with power option(s)	20 C.B. replaces link	On starter relay
Power windows (including backlite and tailgate)	20 C.B.	
Power seats (four- and six-way)	20 C.B.	
Air conditioner (economy)	AGC 15	Fuse cartridge in feed wire
Air conditioner (Selectair)	25 C.B.	②
Speed control	AGA 5	Fuse cartridge in feed wire
Spotlight	SFE 7.5	Fuse cartridge in feed wire
Automatic headlight dimmer	AGW 4	Fuse cartridge in feed wire
Seat belt warning and parking brake warning	SFE 7.5	Fuse cartridge in feed wire
Indicators for charge, oil press., engine coolant hot and cold	No protection	
Windshield wiper motor, convertible top motor, power window motor and power seat motor	Circuit breaker	Integral with motor
Headlights	18 C.B.	Integral with light switch
High beam indicator	18 C.B.	Integral with light switch
Horns, tail- and stoplights, parking lights	15 C.B.	Integral with light switch
Ignition switch light and license light	15 C.B.	
Clock, tachometer (Mercury) and lamps for dome, cargo	SFE 9	Fuse panel
Courtesy, glove and luggage compartments, map (Mercury)		
Lights for clock, radio, heater control, ash tray	(Mercury) AGA 5	Fuse panel
Instrument cluster, transmission selector (console or column), medallion (Mercury)	(Ford) AGA 4	
Heater and defroster motor, cornering lamps (Mercury), power antenna, stationwagon underseat heater, and warning lamps for low fuel, seat belt, door ajar, and door lock	SFE 20 or AGC 20	Fuse panel
Cigar lighter	SFE 14	Fuse panel
Cigar lighter and emergency warning (hang-on or conv. panel)	SFE 20	Fuse panel
Radio, Back-up lights, single speed washers, intermittent speed washers (Mercury), turn signals	SFE 14	Fuse panel (after use, flasher protects circuit for turn signal)

① —14 gage wire (C6AB-14A094-A)
② —On back of ignition switch at accessory terminal

1968

Circuit	Circuit Protection		Location
	Rating	Trade No.	
Headlamp circuit			
Headlight	18 Amp. C.B.		Integral with lighting switch
High beam indicator light			
Rear lamps (tail lights & stop lights)	15 Amp. C.B.		
Front parking and side marker light			
Ignition switch light			
License light			
Horns			
Dome lamp circuit			
Dome lamp	9 Amp. fuse	SFE 9	On fuse panel (dome socket)
Courtesy light			
Cargo light			
Map light			
Glove compartment light			
Luggage compartment light			
Clock			
Tachometer			
Instrument panel illumination			

1968

Circuit	Circuit Protection		Location
	Rating	Trade No.	
Mercury	5 Amp. fuse	AGA 5	Fuse panel (instrument panel socket). This circuit is connected into the light switch rheostat
Clock light			
Instrument cluster lights			
Ash tray light			
PRNDL—console or column light			
Radio light			
Heater control lights			
Heater circuit			
Heater and defroster motor	20 Amp. fuse	SFE 20 or AGC 20	Fuse panel (heater socket)
Safety convenience panel lamps:			
Low fuel warning			
Seat belt warning			
Door ajar warning			
Door unlock warning			
Power antenna			
Station wagon under seat heater			
Defogger and spot light			
Cigar lighter plus emergency warning	20 Amp. fuse	SFE 20	Fuse panel
Radio circuit			
Radio	14 Amp. fuse	SFE-14	Fuse panel (radio socket)
Back-up lamps			
Single speed washers			
Intermittent speed washers			
Turn signal circuit			(After fuse flasher protects circuit for turn signal)
Seat belt warning, parking brake, door ajar			
Windshield wipers			Integral with windshield wiper switch
Intermittent wipers (Mercury)	7.5 Amp. C.B. (two-speed)		
Power assists			
Convertible top circuit	14 gage wire fuse link	(C6AB-14A094-A)	In wiring near starter relay
Convertible top with power option(s)	20 Amp. C.B. replaces fuse link		On starter relay
Power windows including backlite & tailgate	20 Amp. C.B.		
Power seats (four- and six-way)	20 Amp. C.B.		
Air conditioning circuit			
Air conditioner (Economy)	15 Amp.	AGC 15	Fuse cartridge in feed wire
Air conditioner (Selectair)	25 Amp. C.B.		Center flange of instrument panel
Miscellaneous circuits:			
Speed control	5 Amp. fuse	AGA 5	Fuse cartridge in feed wire
Spotlight	20 Amp. fuse	SFE 20	Fuse panel (heater socket)
Automatic headlamp dimmer	4 Amp. fuse	SFE 4 7.5	Fuse cartridge in feed wire
Parking brake warning	14 Amp. fuse	SFE 24	Fuse panel (accessory socket)
Motors:			
Windshield wiper motor	Circuit breaker		In switch
Convertible top motor			
Power window and backlite motor			
Power seat motor			

1969

Circuit	Location	Protective Device ①
Air conditioner feed	Center of instrument panel flange	25 C.B.
Air conditioner, feed (dealer installed)	Cartridge in feed line at accy. terminal of fuse panel	AGC or SFE 20
Ammeter	Fuse panel—	SFE4
Antenna—Electric	Side of dash	AGC or SFE 20
Back-up lights		
Cigar lighter		
Clock feed		SFE 14
Courtesy lights		
C Pillar	Fuse panel—	SFE 14
Cargo	Left side of dash	
Dome		
Door		
Glove compartment		
Instrument panel map		
Door adjar indicator emergency flasher		AGC or SFE 20
Open door indicator	Fuse panel— left side of dash	SFE 14
Parking brake indicator		AGC or SFE 20
Parking lights	Integral with lighting switch	15 C.B.
Power seat feed	On starter relay	20 C.B.
Power seat motors	In motor	C.B.
Power top feed	On starter relay	20 C.B.
Power top motor	In motor	C.B.
Power window motors		
Power window relay feed	Fuse panel— left side of dash	AGC or SFE 20
Headlights and high beam indicator	Integral with lighting switch	18 C.B.
Heated back window	On starter relay	20 C.B.
Heater—defroster motor	Fuse panel— left side of dash	SFE 14
Horns ignition switch light	Integral with lighting switch	15 C.B.
Instrument panel lights	Fuse panel— left side of dash	SFE 4
Clock		
Gauges		
Heater control		
PRNDL		
Radio		
License lights	Integral with lighting swtich	15 C.B.
Lights on buzzer		
Low fuel indicator	Fuse panel— left side of dash	AGC or SFE 20
Luggage compartment		SFE 14

① —Fuse or circuit breaker (C.B.) in amperes.

Circuit	Location	Protective Device ①
Marker lights	Integral with lighting switch	15 C.B.
Power window supply	On starter relay	20 C.B.
Radio feed	Fuse panel— left side of dash	AGC or SFE 20
Rear window defogger		
Seat belt indicator		
Speed control	Cartridge in feed line at accy. terminal of fuse panel	AGA 5
Stop lights	Integral with lighting switch	15 C.B.
Taillights		
Windshield washer	Fuse panel— left side of dash	AGC or SFE 20
Windshield wiper-washer	In wiper switch	10 C.B.

1970

Circuit	Location	Circuit Breaker (C.B.) or Fuse in Amperes
Headlights	Integral with light switch	18 C.B.
Rear tail lights	Integral with light switch	15 C.B.
Front parking lights		
License light		
Side marker		
Horns		
Headlight buzzer		
Stop lights		
Windshield wipers	Integral with wiper switch	7 C.B.
Inteval windshield wipers	Mounted to lower center flange of instrument panel	8.25 C.B.
Convertible top	On starter relay	20 C.B.
Power windows		
Rear window defogger & seat back latch	On starter relay	20 C.B.
Power seats		
Air conditioner	Mounted to lower center flange of instrument panel	25 C.B.
Stop lights	Mounted to lower R.H. flange of instrument panel	20 C.B.
Motors—protected by integral circuit breakers	In motor	C.B.
Power seat		
Power window		
Convertible top		

LIGHT BULBS

1965

Unit	Candela* or Wattage	Trade Number
Headlight no. 1 (inner)	37.5W	4001
Headlight no. 2 (outer)	37.5/50W	4002
Front turn signal/parking	4/32C.	1157-A
Rear turn signal & stop/tail	4/32C.	1157
License plate	4C.	1155
Back-up lights	21C.	1156
Spot light	30W	4405
Dome light	15C.	1003
Instrument panel indicators		
Hi beam	2C.	158
Oil pressure gauge	2C.	158
Ammeter	2C.	158
Turn signal	2C.	1895
Parking brake warning	2C.	257
Illumination		
Speedometer	2C.	158
Cluster	2C.	158
Heater control	2C.	158
Clock	2C.	1895
Radio dial	2C.	1891
Courtesy and/or Map (door mounted)	15C.	1003
Automatic transmission control	1C.	161

*Candela is the new international term for candlepower.

1966

Unit	Candela ① or Wattage	Trade Number
Standard equipment		
Headlamps		
Hi-lo beam	37.5 & 50 Watts	4002
Hi-beam	37.5 Watts	4001
Front park and turn signal lamp	4-32C.	1157
Rear lamp and stop and turn signal (pass. car)	4-32C.	1157
Dome lamp	15C.	1003
License lamp	4C.	1155
Courtesy lamp (convertible)	6C.	631
Cargo lamp (sta. wag. 77B)	15C.	1003
Back-up lamp (sta. wag. & pass. car)	21C.	1141
Rear, stop and turn (station wagon)	4 & 32C.	1157
Courtesy lamp (door mounting)	15C.	1003
Auto. trans. selector	2C.	1895
Instrument panel		
Medallion light	1.5C.	1445
Gages, speedometer	2C.	158
Hi-beam indicator	2C.	158
L & R signal indicator	2C.	158
Glove compartment	2C.	1895

① —Candela is the new international term for candlepower

1966

Unit	Candela① or Wattage	Trade Number
Automatic transmission	1C.	161
Control nomenclature	2C.	1895
Speedometer pointer	1.5C.	1445
Rear window/top nomenclature	2C.	1895
W/S wiper nomenclature	2C.	1895
Lights nomenclature	2C.	1895
Dome lamp—swivel	15C.	1003
Dome lamp—swivel	6C.	631
Courtesy lamp ("C" pillar reverse back)	15C.	1003
Courtesy lamp ("C" pillar, fastback)	6C.	631
Cornering lamp	50C.	1195
Heater control	2C.	1895
Accessory equipment		
Warning indicator panel	2C.	1895
Warning indicator panel	1.6C.	257
Clock	3C.	1816
Engine compartment lamp	15C.	93
Radio pilot light	1.9C.	1893
Radio on-off light (AM-FM only)	1.3C.	1892
Instrument warning lamp kit	2C.	1895
Luggage compartment lamp	6C.	631
Spotlamp (4.40 diameter)	30 Watts	4405
Air conditioner	2C.	1895
Cargo lamp (station wagon)	15C.	1003
Parking brake warning	1.6C.	257
Back-up lamp	21C.	1141
Courtesy lamp (instrument panel)	6C.	631
Oil temp. warning	1.6C.	257
Cigar lighter	2C.	1895
Tachometer	2C.	1895
Trans. control selector indicator	1.5C.	1445

1967

Light Description	Bulb No.	Candela or Wattage
Cargo lamp (station wagon)	1003	15C.
Cornering lamp (Mercury)	1195	50C.
Dome lamp	1003①	15C.
Engine compartment	93	15C.
Glove compartment	1816	3C.
Headlamps		
Hi-lo beam (outer or upper)	4002	37.5-50W
Hi beam (inner or lower)	4001	37.5W
License lamp	97	4C.
Luggage compartment	631	6C.
Map light	631	6C.
Park and turn signal lamp	1157-A	4-32C.
Spotlight (4.4 inch dia.)	4405	30W
Taillight, stop and turn signal	1157	4-32C.

① —631 (6C.) for dual swivel dome lights, 1003 (15C.) for swivel dome lights.
② —1445 (1.5C.) for speedometer pointer and medallion.
③ —161 (1C.) for automatic transmission.

Light Description	Bulb No.	Candela① or Wattage
Door lock warning	1895	2C.
Emergency flasher	1895	2C.
Hi-beam indicator	158	2C.
Low fuel warning	158	2C.
Oil press., alt and temp warning	158	2C.
Parking brake warning	257	1.6C.
Seat belt warning	158	2C.
Ash receptacle or cigar lighter	1445	1.5C.
Back-up lamps	1156	32C.
Courtesy lamp	631	6C.
("C" pillar)	①	
(Console)	1816	3C.
(Convertible and inst. panel)	631	6C.
(Door mounting or armrest)	631	6C.
Clock and ignition key	1816	3C.
Control nomenclature (ex heater)	1895②	2C.
Control, right and left	158②	2C.
Gages and speedometer	158	2C.
Heater control	1895	2C.
Radio dial light		
AM	1893	1.9C.
AM-FM	1893	1.9C.
Tachometer	1895	2C.
Trans. control selector indicator (console)	1445③	1.5C.
Turn signal indicators (L and R)	158	2C.

1968

Unit	Candela① or Wattage	Trade Number
Standard equipment		
Headlamps		
Hi-lo beam	37.5 & 50 Watts	4002
Hi-beam	37.5 Watts	4001
Front park and turn signal lamp	4-32C.	1157NA
Rear lamp and stop and turn signal (pass. car)	4-32C.	1157
Dome lamp	15C.	1003
License lamp	4C.	97
Courtesy lamp (convertible)	6C.	631
Cargo lamp (sta. wag. 77B)	15C.	1003
Back-up lamp (sta. wag. & pass. car)	32C.	1156
Rear, stop and turn (sta. wag.)	4 & 32C.	1157
Courtesy lamp (door mounting)	6C.	631
Auto. trans. selector	2C.	158
Instrument panel		
Gages, speedometer	2C.	194
Hi-beam indicator	2C.	194
L & R signal indicator	2C.	194
Glove compartment	3C.	1816
Automatic transmission	2C.	158
Control nomenclature	2C.	1895
Speedometer pointer	2C.	194

① —Candela is the new international term for candle-power.

Unit	Candela① or Wattage	Trade Number
Rear window/top nomenclature	2C.	1895
W/S wiper nomenclature	2C.	1895
Lights nomenclature	2C.	1895
Front side marker	4C.	97NA
Courtesy lamp ("C" pillar)	15C.	1003
Heater control	2C.	1891
Accessory equipment		
Warning indicator panel	2C.	1891
Warning indicator panel	1.6C.	256
Clock	2C.	194
Engine compartment lamp	6C.	631
Radio pilot light	1.9C.	1893
Instrument warning lamp kit	2C.	1895
Luggage compartment lamp	6C.	631
Spotlamp (440 diameter)	30 Watts	4405
Air conditioner	2C.	1895
Cargo lamp (station wagon)	15C.	1003
Parking brake warning	2C.	194
Back-up lamp	32C.	1156
Courtesy lamp (instrument panel)	6C.	631
Tachometer	2C.	1895
Trans. control selector indicator	2C.	158
Speed control indicator	2C.	194
Map light	3C.	1816
Ash tray light	2C.	194
Fog lights (amber)	35 Watts	4415A
Fog light switch	1C.	161

1969

Light Description	Bulb No.	Candela① or Wattage
Air conditioner control	1895	2C.
Ash tray light	1445	1.5C.
Auto. trans. selector (column mounted)	1445	1.5C.
Auxiliary instrument cluster	1895	2C.
Back-up light	1156	32C.
Cargo light (station wagon)	1003	15C.
Convenience indicator panel	1891	2C.
Convenience indicator panel	256	1.6C.
Courtesy light (C pillar)	1003	15C.
Courtesy light (door mounting)	212	6C.
Dome light	1003	15C.
Engine compartment light	631	6C.
Fog lights (amber)	4415A	35W
Fog light switch	161	1C.
Front park and turn signal light (high series) ②	1157A	3-32C.
Front park, side marker and turn signal light (lo series) ②	1157NA	3-32C.
Front side marker (high series) ②	1178A	4C.
Luggage compartment light	631	6C.
Map light	1816	3C.

① —Candela is the international term for candle power.
② —High series Mercury models are: Marquis, Brougham Marauder and Marquis Colony Park. All other Mercury models are low series.

1969

Light Description	Bulb No.	Candela① or Wattage
Parking brake indicator	194	2C.
Radio		
AM/FM/MPX dial light	1891	2C.
Pilot light	1893	1.9C
Stereo jewel light	1892	1.3C.
Rear side marker	194	2C.
Gages, speedometer	194	2C.
Glove compartment	1816	3C.
Headlights—hi beam	4001	37.5W
Headlights—hi-lo beam	4002	37.5 & 50W
Heater control	1895	2C.
Indicator lights		
Alternator	194	2C.
Brakes	194	2C.
Cold	194	2C.
Door ajar	256	1.6C.
High beam	194	2C.
Hot	194	2C.
Lights on	256	1.6C.
Low fuel	1891	2C.
Oil	194	2C.
Instrument indicator light kit	1895	2C.
L & R signal indicator	194	2C.
License light	97	4C.
Rear stop light (high series) ②	1156	32C.
Rear tail, stop, and turn signal	1157	3-32C.
Rear tail light (high series) ②	1095	4C.
Rear window/top nomenclature	1895	2C.
Speed control indicator	161	1C.
Spot light (4.40 diameter)	4405	30W
Trunk light (portable)	1003	15C.

1970

Light Description	Candle Power or Wattage	Trade No.	Light Description	Candle Power or Wattage	Trade No.
Standard equipment headlamps			Hot	2 c.p.	194
Hi-lo beam	37.5 & 50W	4002	Brakes	2 c.p.	194
Hi-beam	37.5W	4001	Seat belt	2 c.p.	194
Front park and turn signal lamp	3-32 c.p.	1157	Parking brake	2 c.p.	194
Rear lamp and stop and turn			Defog	2 c.p.	194
signal (pass. car)	3-32 c.p.	1157	L & R signal indicator	2 c.p.	194
Dome lamp	12 c.p.	105	Glove compartment	3 c.p.	1816
License lamp	4 c.p.	97	Select-shift transmission		
Courtesy lamp (conv.)	6 c.p.	631	selector—column & console	2.5 c.p.	1445
Cargo lamp (sta. wag.)	12 c.p.	105	Courtesy lamp (inst. panel)	6 c.p.	631
Back-up lamp (sta. wag. & pass.			Ash tray—inst. panel	1.5 c.p.	1445
car)	32 c.p.	1156	Heater control	1.5 c.p.	1445
Rear stop and turn (sta. wag.)	3-32 c.p.	1157	Accessory equipment		
Rear running lamp (sta. wag.)	4 c.p.	1095	Clock	2 c.p.	194
Courtesy lamp (door mtd.)	6 c.p.	212	Engine compartment lamp	6 c.p.	631
Courtesy lamp ("C" pillar)	12 c.p.	105	Radio and stereo tape pilot light	2 c.p.	1893
Courtesy lamp—inst. panel	6 c.p.	631	AM/FM Multiplex and AM Pilot	2 c.p.	1893
Front & rear side marker	2 c.p.	194	Luggage compartment lamp	6 c.p.	631
Instrument panel			Spotlamp (4.40 dia.)	30W	4405
Gauges, speedometer	2 c.p.	194	Air conditioner	2 c.p.	1895
Hi-beam indicator	2 c.p.	194	Fog lamps (amber)	35W	4415A
Warning lights			Fog lamp switch (illum.)	1 c.p.	161
Oil	2 c.p.	194	Map lamp	3 c.p.	1816
Alternator	2 c.p.	194	Tachometer	2 c.p.	1895

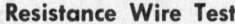

Distributor

1964-70 Models

To remove distributor, first remove the cap and mark the position of the rotor in relation to the distributor body, then, mark the position of the distributor body in relation to the block so that the unit can be reinstalled without having to retime the ignition.

Disconnect the vacuum advance line and the ignition wire, then remove the holddown bolt and pull the distributor up out of the block.

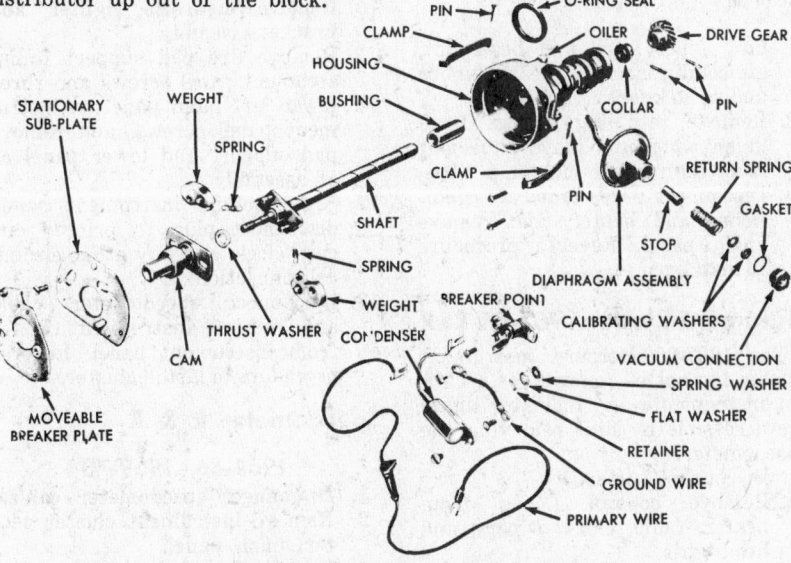

Dual advance distributor (© Ford Motor Corp.)

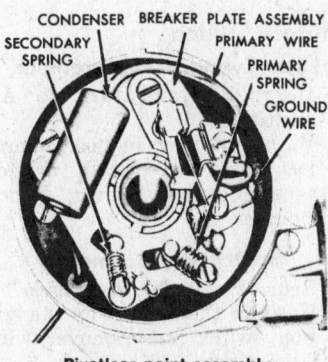

Pivotless point assembly
(© Ford Motor Co.)

Firing Order and Timing Mark Location

See beginning of this section.

Ignition Primary Resistor

A resistance is used in the primary circuit of the ignition system. This resistor is built into the primary wire from the ignition switch to the coil.

If there is difficulty with the ignition, make absolutely certain that the primary resistor is functioning properly.

Resistance Wire Test

1. Connect the red lead of a voltmeter to the ignition coil terminal of the ignition switch, and the black voltmeter lead to the coil primary terminal.
2. Install a jumper wire from the distributor terminal of the coil to a good ground on the distributor housing.
3. Turn the lights, and all the accessories, off.
4. Turn the ignition switch on. If the voltmeter reading is 6.6 volts or less, the resistance wire is satisfactory.
5. If the reading is greater than 6.6 volts, replace the resistance wire.

Resistance Wire Replacement

1. Cut the brown wire and the red wire (with a green band) from the upper disconnect at the dash panel. Cut the wires as close to the quick disconnect as possible.
2. Solder a male bullet-type terminal to the brown wire and to the red wire (with the green band). Make a single terminal of the two wires.
3. Using a female bullet terminal connector, connect the wires to one end of the service replacement resistance wire. Do not splice the resistance wire.
4. Drill a 3/4 in. hole through one of the accessory dimples in the dash panel and install a grommet.
5. Thread one end of the service replacement resistance wire through the grommet in the dash panel and connect it to the jumper wire at the ignition switch.
6. Cut off and discard (at the point where it enters the taped area

of the wiring harness) the length of defective resistance wire which is not enclosed in the taped portion of the wiring harness.

Ignition Timing

If the timing relationship has been altered, proceed to retime the ignition as follows: bring No. 1 cylinder up into the firing position. This can be checked by removing the spark plug. Then placing your thumb in the spark plug hole and cranking the engine until the compression attempts to blow by your thumb, slowly bring the crankshaft around until the TDC mark on the crankshaft pulley lines up with pointer. This is firing position for No. 1 cylinder.

Remove the distributor cap and mark, on the outside of the distributor, the position of the rotor. The wire from No. 1 spark plug should be placed in the socket just above the rotor. Now, working in the direction of distributor rotation, place the spark plug wires into the cap according to the firing order of the engine.

Generator, Regulator

Generator or Alternator Service

Most 1964 Mercurys use a DC generator, but since 1965 all Mercurys use an AC generator commonly known as an alternator.

On V8s, the generator or alternator is located on the right bank of cylinders at the front, under the side of the cylinder head.

To remove either generator or alternator, loosen the bolt that holds the tension bar on the drive belt in order to release the belt tension and slip the belt off the pulley. Remove the wires, then the generator or alternator may be detached either at the bracket or at the two swivel bolts. On these models it is easier to remove the bracket than it is to separate the generator or alternator from its swivel.

See Basic Electrical Diagnosis in Unit Repair Section for more complete information on the AC generator and regulator.

Battery, Starter

Starter System Service

See Unit Repair Section.

Starter R & R

1964-70

Disconnect battery, starter cable and cable bracket below No 4 spark

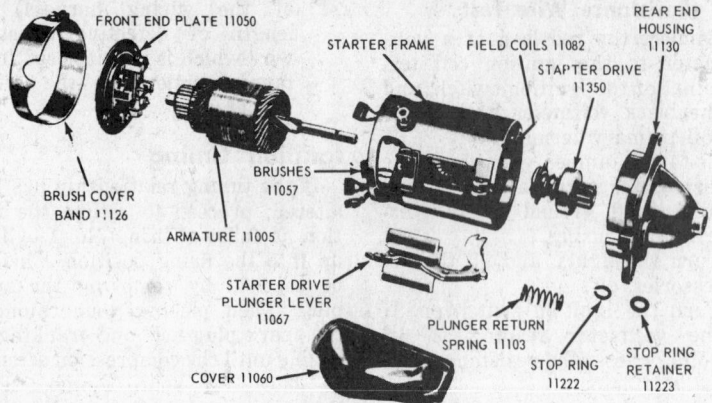

FRONT END PLATE 11050
STARTER FRAME
FIELD COILS 11082
REAR END HOUSING 11130
STAPTER DRIVE 11350
BRUSH COVER BAND 11126
BRUSHES 11057
ARMATURE 11005
STARTER DRIVE PLUNGER LEVER 11067
PLUNGER RETURN SPRING 11103
STOP RING 11222
STOP RING RETAINER 11223
COVER 11060

Starter (ⓒ Ford Motor Co.)

plug. Turn front wheels full right. Remove attaching screws. Pull starter out and up and while in vertical position, rotate unit to permit end plate lug clearance. Lower between idler and tie-rod.

To install, position starter, tighten bolts and torque to 1-20 ft. lbs.

Instruments

Since all work on an instrument cluster is done in very tight quarters, it is imperative that the battery is disconnected before an attempt is made to work under the dash in order to avoid short-circuiting any of the wires.

Perhaps the easiest way to work on any or all of the instruments is to remove the entire instrument cluster and pull it into the driver's compartment.

Instrument Cluster 1964 R&R

1. Disconnect battery.
2. Disconnect speedometer cable at head.
3. Remove nine screws retaining instrument cluster to instrument panel. Pull out cluster and place on steering column.
4. Remove two nuts retaining main wiring loom in clips at back of cluster.
5. Disconnect main wiring loom lead and ammeter lead at junction box in center of cluster. Pull ammeter lead out of loom.
6. Disconnect wires to fuel, temperature, and oil pressure gauges, constant voltage regulator, and turn signal flasher.
7. Remove lights from cluster and remove cluster. To install, reverse above procedure.

Instrument Cluster Center Finish Panel 1965-66 R&R

On 1965-66 Mercurys once finish panel is removed, all instruments are readily accessible and removable.
1. Disconnect battery.
2. Remove headlight knob and bezel

nut, wiper knob and bezel nut, air conditioner and heater knobs and radio knobs.
3. Remove retaining screws from finish panel and move panel out from instrument cluster a little.
4. Disconnect wires from air conditioner and heater, and remove finish panel. Reverse procedure to install panel.

Instrument Cluster 1967-68 R&R

The 1967-68 Mercury instrument cluster can not be removed as a unit, but instruments and indicator lights are accessible by first removing the speedometer.
1. Disconnect battery.
2. Remove control knobs from heater, radio, clock, wipers, and headlights.
3. Remove four retaining screws around speedometer bezel and remove bezel.
4. Remove screws retaining right and left finish panels and remove panels.
5. Remove four button retainers from speedometer dial cover and remove speedometer dial.
6. Remove four screws retaining speedometer, and move speedometer out a little.
7. Disconnect speedometer cable and instrument voltage regulator. Remove speedometer and housing.
8. Remove two nuts retaining speedometer to housing and remove speedometer. Reverse procedure to install speedometer.

Instrument Cluster 1969-70 R&R

1. Disconnect battery.
2. Remove wiper knob and bezel, cigarette lighter element and finish panel.
3. Remove retaining screws and remove left and right end finish panels.
4. Remove screw that attaches lower pad to upper pad at each end.
5. Remove screws that retain upper finish panel to its support at each

defroster opening. Remove six upper finish panel to pad screws, and lift off upper finish panel.
6. Remove four screws that attach upper pad and retainer assembly to pad supports and remove upper pad assembly.
7. Remove two screws that attach right half of lower pad and retainer assembly to pad supports, remove screws that attach left half of lower pad to lower left instrument panel, and remove lower pad and retainer assembly.
8. Remove windshield wiper nut.
9. Remove bracket from left end of pad support.
10. Remove cigarette lighter and bracket assembly.
11. Remove five pad support to instrument panel screws and three lower left hand panel to instrument panel screws, and remove pad support and lower panel as an assembly.
12. From behind instrument panel, disconnect plug to printed circuit, clock, and any other electrical connection.
13. Disconnect speedometer cable and remove instrument cluster from instrument panel. Reverse procedure to install cluster.

Speedometer R & R

1964-66, 1969-70

1. Disconnect speedometer cable.
2. Remove instrument cluster center finish panel.
3. Remove speedometer head retaining screws and speedometer.

Ignition Switch

Lock Cylinder Replacement

1. Insert key and turn to Acc. position.
2. With stiff wire in hole, depress lock pin and rotate cylinder counterclockwise, then, remove cylinder.

Switch Replacement

1. Remove cylinder, as above.
2. Unscrew the bezel from the ignition switch and remove switch from panel.
3. Remove insulated plug from rear of switch.
4. Install in reverse of above.

Lighting Switch

Replacement

1. Disconnect battery.
2. Remove knob and shaft by pressing release knob button on switch

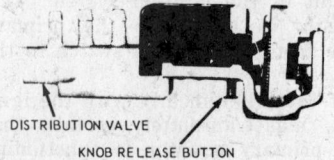

DISTRIBUTION VALVE
KNOB RELEASE BUTTON

Headlight switch with distribution valve

housing, with knob in on position.
3. Remove molding nut from switch.
4. Remove junction block from switch.
5. Install in reverse of above.

Rear Lights

All tail light, stop light and turn signal bulbs are readily accessible, either through the trunk compartment or, when so equipped, by removing the screws and lenses.

Back-Up Light Switch— Manual Transmission

The back-up light switch may be located in either of two places. On cars with column shift selector and linkage controls, the switch is at the bottom of the column.

On cars with consoles and floor shift selector, the switch is on the left side of the transmission at the shift control bracket.

Brakes

Information on brake adjustments, band replacement, bleeding procedure, master and wheel cylinder overhaul is in the Unit Repair Section.

Information on troubleshooting and overhauling power brakes can be found in the Unit Repair Section.

Information on the grease seals which may need replacement can also be found in the Unit Repair Section.

Brake Pedal Adjustment

Adjust the brake pedal so that there is approximately ½ in. freeplay of the pedal before the pushrod contacts the piston in the master cylinder. This adjustment is made at the eccentric bolt that holds the brake pushrod to the brake pedal under the dash.

Brake Adjustment

All adjustment and service on these brakes is given in the Unit Repair Section.

1964-70 Self Adjusters

The service brake has a self-adjusting brake shoe mechanism, consisting of a link, actuator, pawl, and pawl return spring. The looped end of the link is attached to the anchor pin, and the hooked end to the actuator. The actuator is held against the secondary shoe by means of the hold-down cup and spring. The pawl is connected to the actuator and held in position by the pawl return spring.

The automatic adjusters operate only when the brakes are applied as the car is moving rearward. The wraparound action of the shoes, following the drum while moving rearward,

forces the upper end of the primary shoe against the anchor pin and the secondary shoe away from the anchor pin.

The link holds the top of the actuator stationary, forcing the actuator to pivot on the secondary shoe. The pivoting action forces the pawl downward against the end of a tooth on the star-wheel adjusting screw, which turns the star-wheel and expands the shoes.

The greater the clearance between the brake drum and the lining, the greater the travel of the secondary shoe away from the anchor. The farther the secondary shoe moves, the greater the adjustment. When the brakes are adjusted correctly, there will not be sufficient travel of the secondary shoe to permit the actuator to pivot and force the pawl to engage against the end of a tooth of the star wheel adjusting screw, turning the wheel to expand the shoe.

When the brakes are applied as the car is moving forward, the self-adjuster does not operate because the wraparound action of the shoe forces the secondary shoe against the anchor pin.

The rear brake assembly is basically the same as the front brake, except that the conventional parking brake operating lever, spring, and parking brake strut rod are used in the rear brake. The anchor pin on all brakes can be adjusted when necessary.

Whenever removing or replacing a star wheel assembly, make certain that the correct star wheel assembly is installed on the right brake drum.

Disc Brakes

Since 1965, disc brakes are available on the front wheels of some mod-

els. Complete service procedures are covered in the Unit Repair Section.

Parking Brake Linkage Adjustment

Check the parking brake cables when the brakes are fully released. If the cables are loose, adjust them as follows:
1. Fully release the parking brake pedal.
2. Depress the parking brake pedal one notch from its normal released position.
3. Raise the car.

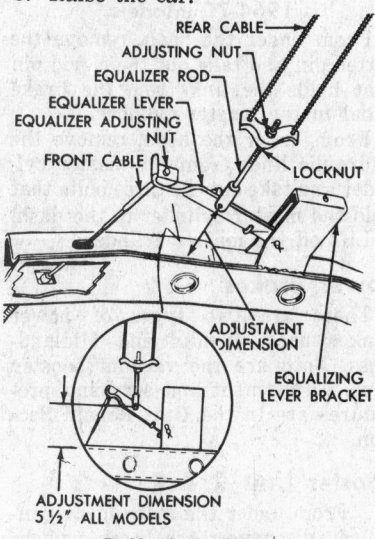

Parking brake linkage
(© Ford Motor Co.)

4. Turn the adjusting nut forward against the equalizer until a moderate drag is felt when turning the rear wheels.
5. Release the parking brake, and make sure that the brake shoes return to the fully released position.

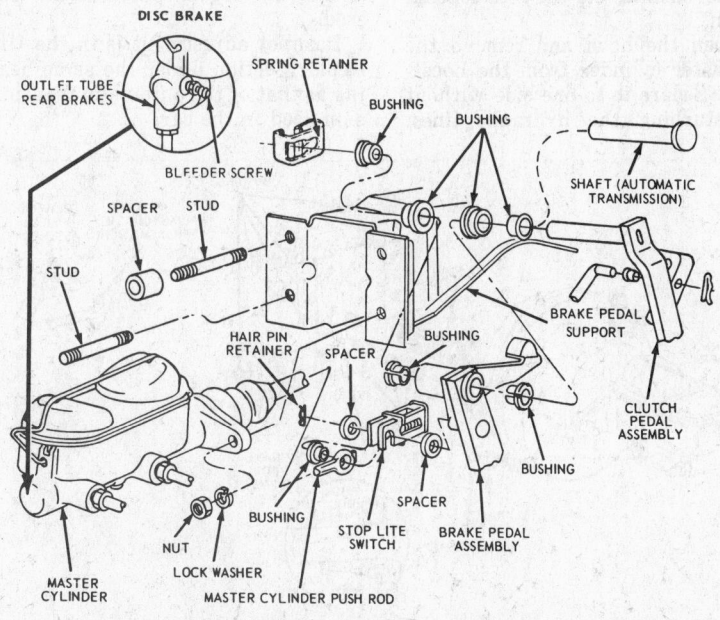

Master cylinder—1967-70 standard brake

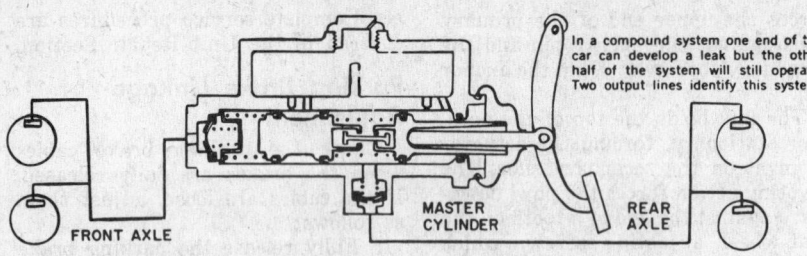

In a compound system one end of the car can develop a leak but the other half of the system will still operate. Two output lines identify this system.

FRONT AXLE   MASTER CYLINDER   REAR AXLE

Dual type master cylinder

Master Cylinder Removal

1964-70 Models

From under the dash, remove the cotter pin and take out the clevis pin that holds the link from the brake pedal to the master cylinder.

From under the hood, remove the hydraulic lines from the master cylinder and take out the three bolts that hold the master cylinder to the dash.

Lift off the master cylinder.

Power Brakes

There are two types of power brakes used, Bendix and Midland-Ross. Both are the vacuum booster type. Adjustment and servicing procedures are in the Unit Repair Section.

Booster Unit R & R

1. From under the instrument panel, disconnect the booster pushrod link from the brake pedal. To do this, disconnect the stop light wires at the connector. Remove the hairpin retainer. Slide the stop light switch from the brake pedal pin, far enough for the switch outer hole to clear the pin, then lift the switch straight upward from the pin. Slide the pushrod, nylon washers and bushing off the brake pedal pin.
2. Open the hood and remove the master cylinder from the booster. Secure it to one side without disturbing the hydraulic lines.

3. Disconnect vacuum hose from booster unit. If the car has an automatic transmission, disconnect the transmission vacuum hose.
4. Remove the four mounting brackets to dash panel attaching nuts and washers. Remove the booster assembly from the dash panel. Remove the four spacers.
5. Remove the pushrod link boot from the dash panel.
6. Install by reversing the removal procedure. See Unit Repair Section for adjustment procedures.

Fuel System

Data on capacity of the gas tank are in the Capacities table. Data on correct engine idle speed and fuel pump pressure are in the Tune-up Specifications table.

Information covering operation and troubles of the fuel gauge is in the Unit Repair Section.

Detailed information on the carburetor, and how to adjust it, is in the Unit Repair Section under the specific heading of the make of carburetor being used on the engine being worked on. Carter, Holley, Rochester and Stromberg carburetors are covered.

Dashpot adjustment is in the Unit Repair Section under the same heading as that of the automatic transmission used in the car.

Fuel Pump

1964-70 V8 Models

The fuel pump is located at the front of the engine in the center above the fan pulley.

To remove, disconnect the flex line and the copper input line and then remove the bolts that hold it to the casting and lift off the fuel pump.

Fuel Filter

1964 Models

The fuel filters for these engines are located between the fuel pump and carburetor. They use replaceable elements.

To replace the element, unscrew the lower section and remove element and gasket. Clean lower bowl. Install new element and gasket. Coat this new gasket with light engine oil. Install lower cover hand tight plus one-eighth turn. Start engine and test for leaks.

1965-69

An inline filter is used on all engines except the 427 cu. in. V8. It is one piece and can not be cleaned. Replace the unit, if necessary. The 427 cu. in. V8 uses a filter with a replacement element.

Exhaust System

Exhaust Pipe Replacement

The exhaust manifold flange bolts are accessible from underneath the car with a long extension.

Work plenty of derusting oil around the joint between the exhaust pipe and the muffler, then disconnect the flange bolts, remove the clamp that holds the exhaust pipe to the muffler and separate the pipe from the muffler. The pipe can then be threaded out from under the car. On cars with dual exhaust systems, each pipe can be removed separately. On models with single exhaust systems, the crossover is accomplished in back of the transmission ahead of the muffler.

Muffler Removal

Disconnect the exhaust pipe at the flange and squirt penetrating or derusting oil around the joint between the exhaust pipe and the muffler and between the muffler and the tail pipe. Then, remove the clamps that hold these two together. Pull the exhaust pipe forward until it comes out of the muffler, then tilt the muffler downward and pull it from the tail pipe.

Tail Pipe Replacement

Some models are fitted with a res-

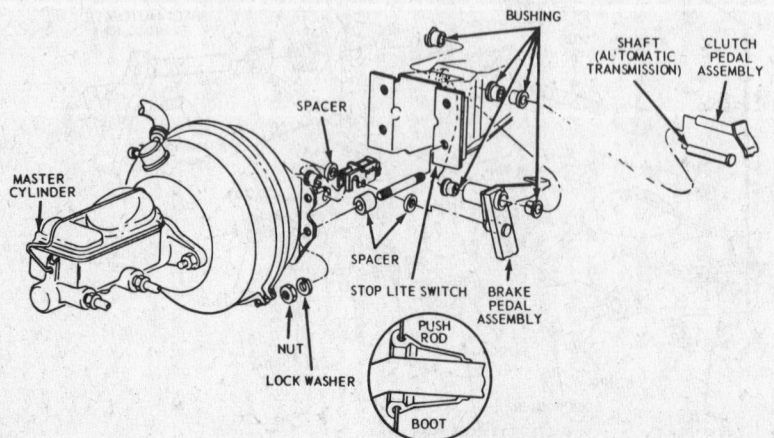

BUSHING

SHAFT (AUTOMATIC TRANSMISSION)   CLUTCH PEDAL ASSEMBLY

SPACER

MASTER CYLINDER

SPACER

STOP LITE SWITCH   BRAKE PEDAL ASSEMBLY

NUT

LOCK WASHER

PUSH ROD

BOOT

Power brake master cylinder installation

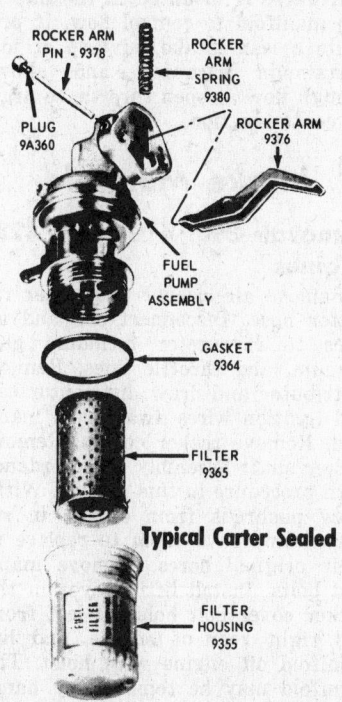

Typical Carter Sealed

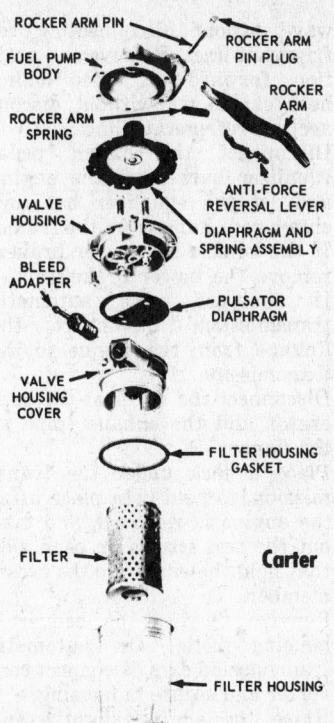

Carter

Typical fuel pumps (© Ford Motor Co.)

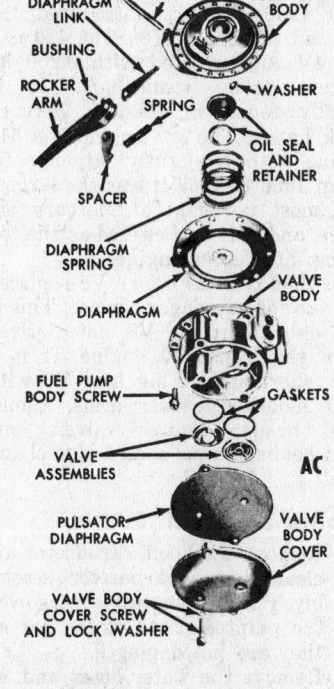

AC

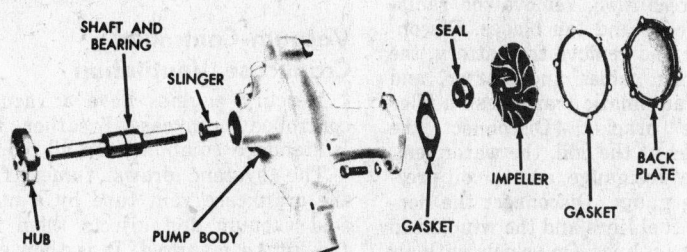

352 and 390 V8 water pump assembly

onator, which is an integral part of the tail pipe; it is not sold separately.

Remove the brackets that hold the tail pipe to the under side of the car, squirt derusting oil on the joint between the muffler and the tail pipe, remove the clamp and separate the tail pipe from the muffler, threading the entire unit toward the front of the car.

Exhaust Manifold Removal

Disconnect the exhaust pipe flanges. The bolts are accessible from underneath the car. Remove the bolts that hold the exhaust manifolds to the head and lift off the manifolds.

On models equipped with a heater, disconnect the heater when working on the right-side exhaust manifold. While the manifold can be taken off with the heater in place, it is easier to take the heater off first.

Cooling System

Thermostat Removal

1964-70 Models

The thermostat is located in the thermostat housing in the front of the intake manifold. Disconnect the upper radiator hose, at its lower end, and the by-pass hose. Unbolt and lift off the thermostat housing, then take out the thermostat.

Water Pump Removal

1964-70 Models

1. Drain cooling system. Remove supply tank (where used).

2. Disconnect lower radiator hose, heater hose and remove belt, fan, spacer and pulley.
3. Remove pump retaining bolts and lift off pump.

NOTE: on cars equipped with power steering, air conditioning or exhaust emission control systems, pump interference may be encountered. In these cases, the offending pump/s or compressor should be removed, if possible, without disconnecting lines or hoses.

Power Surge Automatic Cooling Fan Removal

1964-69 All Engines With Power Boost Fan

Remove the four screws and lock washers that hold the power boost fan assembly to the water pump hub, then remove the boost assembly.

Remove the four nuts and lock washers which attach the fan proper to the power boost fluid unit and take off the fan blade assembly. Reassemble in reverse order.

Radiator Core Removal

1964-70

Detach the top and bottom radiator hoses at the radiator, remove the bolts at the side that hold the radiator to the radiator support, and lift the radiator assembly straight up.

Engine

1964-70

The standard Mercury engine throughout this period has been the 390 cu. in. V8. This is a conventional V8 of rugged design with a deep block that extends well below crankshaft centerline. This engine in two-barrel and four barrel form is found in more Mercurys than all other engines combined.

Mercury offered a 427 cu. in. V8 in 1964 and 1965, but this engine is rarely seen in a Mercury anywhere but on the race track where the '64s were extremely successful. The engine is essentially a 390 with a larger

bore and substantially stronger crankshaft, rods, pistons, valve gear and bearing supports.

In 1966 Mercury introduced the 410 and 428 cu. in. V8s. The 410 is a 390 4V engine fitted with 0.200 in. longer stroke crankshaft. It is usually found in 1966-67 Mercury Park Lanes. The 428 engine is a 410 engine with a slightly larger bore. From 1966 to 1968 it was the largest and most powerful of Mercury engines, and it is still offered as the top police intercepter engine.

In 1969 the 429 cu. in. V8 replaced the 428 for passenger car use. This is the only Mercury V8 not derived from the basic 390 engine. It is a very short stroke, big bore V8 with stud mounted rocker arms, tunnel port heads, canted valves, and semi-hemispherical combustion chambers.

Engine Removal

1. Remove the hood, carburetor air cleaner, and carburetor assembly. Place protecting covers over the painted parts of the car so they are not damaged.
2. Remove the water hoses and, as a precaution, remove the radiator core and fan blades. Disconnect and remove the battery, the starter cables and starter, and the automatic transmission filler tube bracket. Disconnect the wires at the coil, the water temperature gauge, and the oil pressure gauge. Disconnect the flexible fuel lines and the windshield vacuum lines. On vehicle with air conditioning, remove compressor and bracket and position out of

way without disconnecting refrigerant lines. Remove and position forward air conditioning heat exchanger without disconnecting refrigerant lines.
3. Disconnect the clutch pedal equalizer bars from the engine and let the equalizer bar and clutch rods hang from the frame.
4. If the vehicle has power brakes, remove the power brake unit.
5. If equipped with automatic transmission, disconnect the linkage from the engine to the transmission.
6. Disconnect the wires at the generator and the exhaust pipe at the flanges.
7. Place a jack under the transmission, to hold it in place after the engine is removed, and take out the two screws on each side that hold the engine to the crossmember.
8. Remove flywheel or converter housing bolts. On automatic transmission cars disconnect converter and secure in housing.
9. Raise the engine slightly and slide it forward off the front of the transmission.

Vacuum-Controlled Crankcase Ventilation

Mercury engines have a vacuum-controlled crankcase breather that is standard equipment on all models.

The system draws fumes from the crankcase vent tube by a manifold vacuum and injects them into the intake manifold. It is then combined with the air-fuel mixture and added to combustion.

A valve is mounted in the line to the manifold to control flow. It prevents a lean intake mixture at idle (maximum vacuum), and allows enough flow at open throttle to draw off contamination.

Intake Manifold

Removal—390, 410, 427, 428 Engines

Remove air cleaner and upper radiator hose. Disconnect coil and all lines to carburetor including gas, vacuum, and throttle lines. Remove distributor and draw distributor cap and ignition wires away from manifold. Remove rocker covers. Remove rocker shaft assembly in accordance with procedure in this section. Withdraw pushrods from engine in sequence and save them to replace in their original bores. Remove manifold bolts. Install lifting eyes in the rocker cover bolt holes at left front and right rear of engine, and lift manifold off engine with hoist. The manifold may be removed by hand, but it is very heavy. Remove baffle plate beneath manifold by prying up with screwdriver.

Removal—429 Engine

Remove air cleaner and upper radiator hose. Disconnect all lines to carburetor. Remove coil, PCV tubing, heater hoses, and all other lines from manifold. Disconnect battery ground. Remove distributor cap, spark plug wiring, and distributor. Disconnect water pump bypass hose. Remove manifold bolts. Remove manifold.

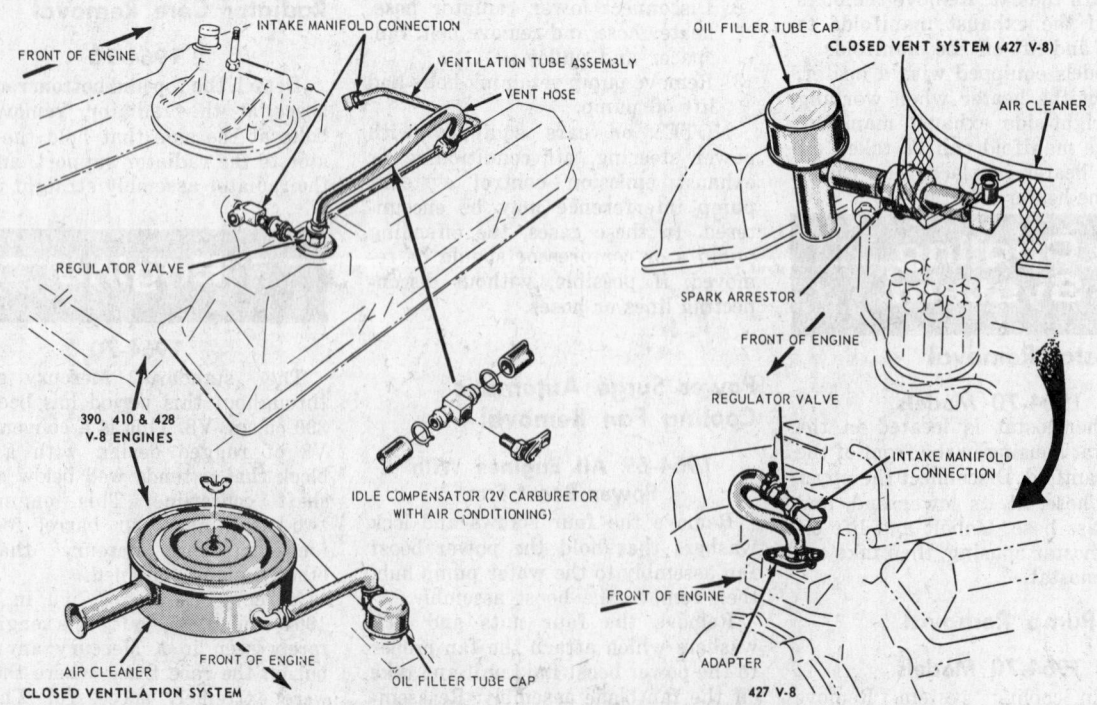

V8 positive crankcase ventilation system (© Ford Motor Co.)

Cylinder Head

Cylinder Head Removal

390, 410, 427, and 428 Engines

1. Remove intake manifold as described elsewhere in this section.
2. Remove exhaust manifolds and any remaining accessories from heads.
3. Remove head bolts breaking torque in reverse of torquing sequence.
4. Lift off heads.

429 Engines

1. Remove intake manifold as described elsewhere in this section.
2. Remove exhaust manifolds and any remaining accessories.
3. Remove rocker covers.
4. Loosen rocker adjustment until rockers can be pivoted to one side.
5. Withdraw pushrods in sequence and save to install in sequence.
6. Remove head bolts breaking torque in reverse of torquing sequence.
7. Lift off heads.

Rocker Arm and Shaft Removal

390, 410, 427, and 428 Engines

1. Remove rocker covers.
2. Crank engine until No. 1 piston is at 45 degrees ATDC at beginning of power stroke, identified by XX on crankshaft.
3. Starting on No. 4 cylinder on the right bank, loosen rocker shaft support bolts in sequence two turns at a time until free.
4. Remove bolts and lift off rocker assembly.
5. Repeat procedure starting with No. 5 cylinder on left bank.
6. Remove pushrods in sequence so that they may be installed in their original bores.

429 Engine

1. Remove rocker covers.
2. Loosen rocker arm stud nuts and turn rocker arms to side.

3. Remove pushrods in sequence so that they may be installed in their original bores.
4. Remove rocker arm stud nuts, fulcrum seats, and rocker arms.

The Valve System

All V8 Models

Due to the use of both solid and hydraulic lifters on V8 engines, it is recommended that the Tune-Up Specifications be checked for proper application.

The valves are removed by taking off the cylinder head, transferring it to a bench, and using a C-type or lever-type valve spring compressor. Compress the spring and remove the key. Release the spring and the valve can be pulled out into the combustion chamber side of the head.

If the valve has a tendency to stick as it comes through the guide, thoroughly clean the exposed portion of the stem so that all oil residue is removed, and the valve will come out readily. Sometimes a valve is found that has been headed over by the rocker. Use a small file to remove edge from end of stem and remove and discard valve.

CHILTON TIME-SAVER

Frequently valves become bent or warped, or their seats become blocked with carbon or other material. Left unattended, these situations can cause burnt valves, damaged cylinder heads and other expensive trouble. To detect leaking valves early, perform this test whenever the cylinder head is removed.

1. After removing head, replace spark plugs. Removing spark plugs before removing heads eliminates breakage.
2. Place head on bench with valves, springs, retainers, and keys installed combustion chambers up.
3. Pour enough gasoline into each combustion chamber to completely cover both valves. Watch combustion chambers for two minutes for any air bubbles that indicate leakage.

Valve Guide Replacement

All Models

The cylinder head does not have separate valve guides, instead the guide is cast integrally with the cylinder head. Mercury offers oversize replacement valves in the event the valve fits poorly. To fit these, enlarge valve guide bores with valve guide reamer to diameter of replacement valve stem plus recommended valve guide clearance. As an alternative, some local automotive machine shops will fit replacement guides that use the original valves.

Valve Springs

Any time the valve spring is removed from the engine it should be checked.

When a spring tester is not available, place all the springs on a level surface and put a new spring alongside the old ones. If all of the old springs come up to the same free length as the new spring, it can be assumed that all of the springs are good. If one or more of the springs is shorter than the new one, they should all be tested with a spring tester and the defective ones replaced.

Hydraulic Valve Lifters

Overhead Valve V8 Models

The lifters can be pulled up out of their bores after the pushrods and

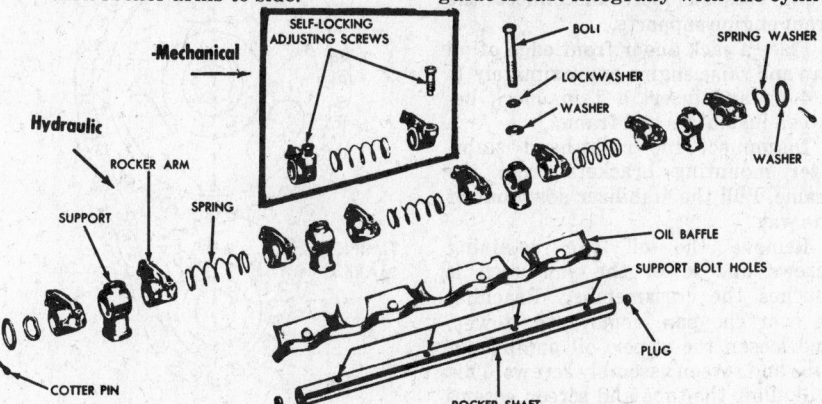

Valve rocker arm assembly—V8 (© Ford Motor Co.)

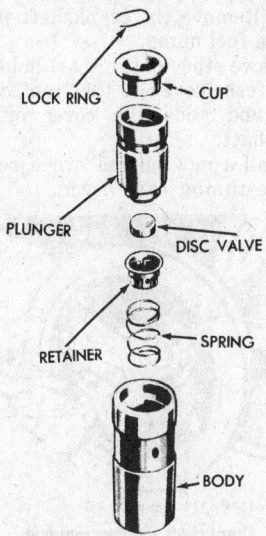

Typical hydraulic valve lifter

the intake manifold have been removed.

If there is a tendency for the lifters to stick in their bores, penetrating oil should be tried before they are forced out of their bores. Hydraulic valve lifters may be disassembled for cleaning as described in the Ford-Thunderbird section. Do not interchange parts between lifters or attempt any repairs to lifters.

Valve Lash—Hydraulic

Valve stem to rocker arm clearance is zero, however, conditions may change the original setting and adjustment becomes necessary.

Clearance must be determined by measuring the clearance between the closed valve stem and the rocker arm, (with tappets empty and collapsed).

Valve lash should be as follows:
1964 390 ....0.083-0.183 in.
1965-66 390,
410, 428 ....0.050-0.150 in.
1967-70 390,
410, 428 ....0.100-0.200 in.
1969-70 429 ....0.075-0.175 in.

If the engine is not equipped with adjustable rocker arms, proper lash must be obtained by using longer or shorter pushrods. These rods are available in .060 in. options, above or below standard.

Timing Case

Timing Chain Cover Removal

V8 Engines

To remove the timing case cover remove the oil pan because there is a gasket between the lower part of the timing case cover and the oil pan, and it will be difficult to prevent oil leaks if the oil pan is left in place and an attempt is made to seal between the bottom of the timing case cover and the oil pan.

Remove the radiator core, the fan, fan pulley, the water pump and the water by-pass pipe back of the water pump. Remove the crankshaft pulley and the fuel pump.

Remove the bolts that hold the timing case cover to the front of the block and slide the cover off the crankshaft.

Install a new oil seal when replacing the timing case cover.

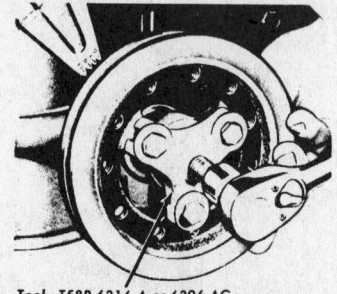

Tool—T58P-6316-A or 6306-AG
Crankshaft damper removal
(© Ford Motor Co.)

Caution Due to the structure of the crankshaft, it is very important that the proper tools be used to draw the damper or pulley into place. (Do not use a hammer. Breakage of the cast crankshaft can result.)

Timing Case Oil Seal Replacement

All Models

To replace the oil seal, remove the timing case cover and drive the seal out with a pin punch. Clean out the recess in the cover and install a new seal, using a special driving tool.

Coat the new seal with grease to reduce friction when installing and starting the car.

Timing Sprocket and/or Chain

V8

1. Remove timing case cover.
2. Remove crankshaft oil slinger.
3. Rotate engine until timing marks are aligned with each other and with centers of sprockets.
4. Remove cap screw and fuel pump eccentric.
5. Slide both sprockets and timing chain forward and remove them as an assembly.
6. To install, arrange chain and sprockets on bench until timing marks are aligned with each other.
7. Slide both sprockets and chain onto shafts as an assembly.
8. Install fuel pump eccentric and cap screw on camshaft.
9. Install oil slinger on crankshaft.
10. Install timing case cover and oil pan.

Engine Lubrication

Oil Pan Removal

1964-70 Models

Crank the engine until No. 1 cylinder is in the firing position. Drain the oil, remove the oil level dipstick and disconnect the battery. Remove the splash shield from the frame front crossmember. Disconnect the front engine supports.

Place a jack under front edge of oil pan and raise engine approximately 1-1/4 in. and insert a 1 in. block between insulators and frame.

Disconnect the right-hand stabilizer mounting brackets from the frame. Pull the stabilizer down out of the way.

Remove the oil pan retaining screws and lower the pan until it touches the crossmember. Reaching in over the pan, remove the lower, and loosen the upper, oil pump inlet tube and screen assembly screws. This will allow the tube and screen assembly to swing free, permitting the oil

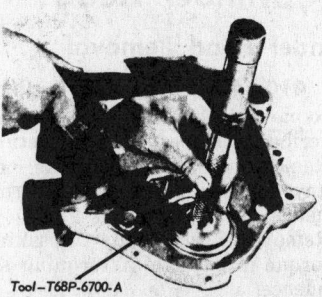

Tool—T68P-6700-A
Timing case oil seal replacement
(© Ford Motor Co.)

Timing chain removal
(© Ford Motor Co.)

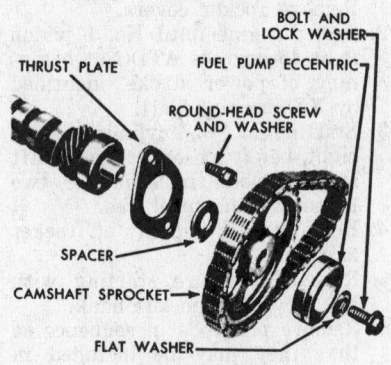

Camshaft thrust plate
(© Ford Motor Co.)

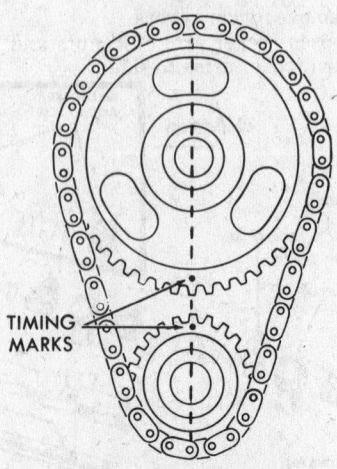

Timing gear marks, V8 with front mounted distributor

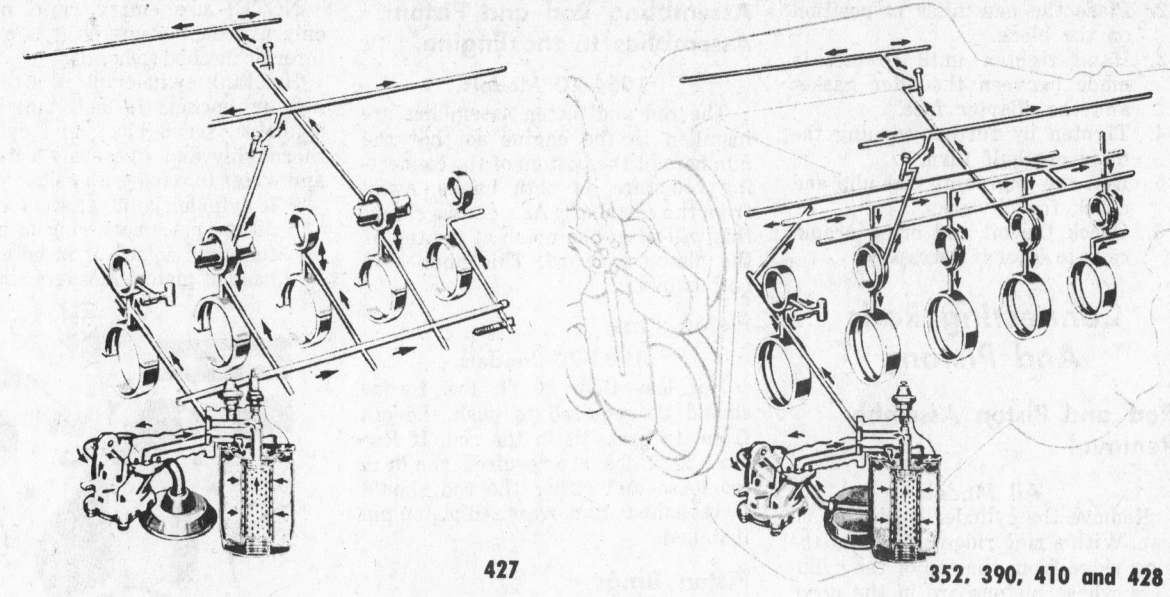

427

352, 390, 410 and 428

V8 engine lubrication (© Ford Motor Co.)

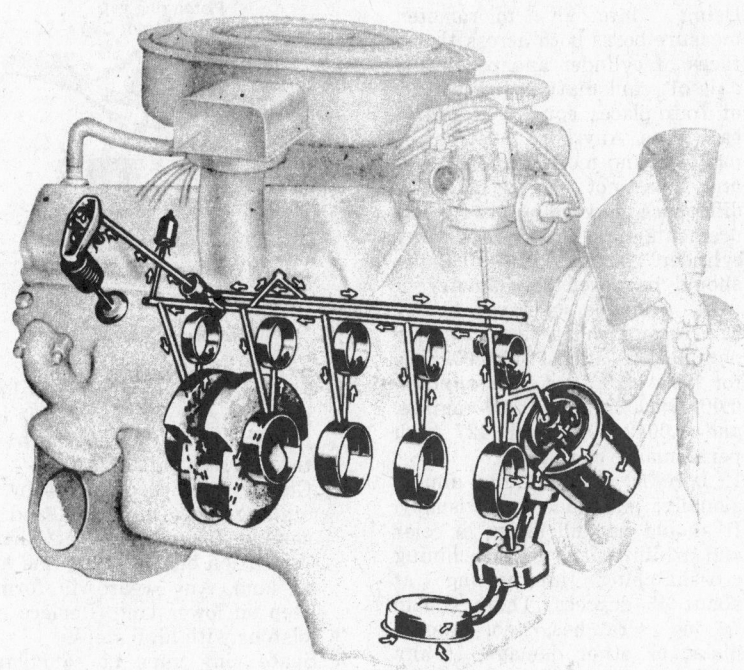

429 lubrication (© Ford Motor Co.)

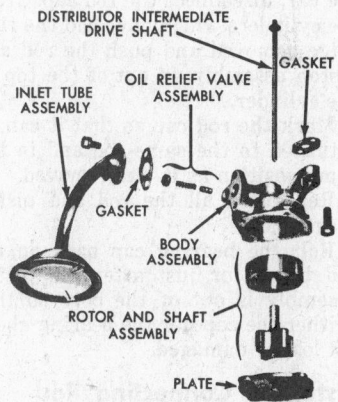

Oil pump—V8
(© Ford Motor Co.)

The oil pump flange is bolted up to the underside of the block.

Oil Filter Assembly

Full flow, disposable-type filters are used. No tools are required to change this filter.

Replacement

1. Coat the gasket on the new filter with oil.

pan baffle to clear. Or, remove both bolts and let the screen drop into the pan. Lower oil pan to floor.

The pan is reinstalled in reverse order of removal. It should have a new gasket and the block should be carefully cleaned of the old gasket.

Set the oil pan up on the cross-member and tighten the oil pump upper and lower screws.

Oil Pump Assembly

1964-70 Models

The oil pump is accessible after the oil pan has been removed. It is located in the oil pan at the front of the block.

⏻ CHILTON TIME-SAVER

Oil filter replacement trouble starts with the elements being turned on too tightly. The unit may be too tight to remove by hand, and it may collapse in the grip of a tool that applies enough pressure to grip the element hard enough to turn it.

Alternative Method of Removal

1. Raise the car on a jack or hoist and place a drip pan under the filter.

2. With a 12-14 in. slender punch drive a hole in the element from one side to the other.

   NOTE: before punching the hole, consider the angle required for the punch to act as a lever, with the least interference.

3. With the drift all the way through the filter and acting as a lever, turn the unit counterclockwise to break it loose.

4. Final loosening and removal can be accomplished by hand.

2. Place the new filter in position on the block.
3. Hand tighten until contact is made between the filter gasket and the adapter face.
4. Tighten by further turning the filter one-half turn.
5. Run the engine at fast idle and check for oil leaks.
6. Check the oil and bring crankcase to level if necessary.

Connecting Rods And Pistons

Rod and Piston Assembly Removal

All Models

Remove the cylinder heads and oil pan. With a ring ridge cutter, cut the ring ridge from the top of the cylinders whose pistons are in the down position. Working from underneath the car, disconnect the rod caps from the cylinders which have had the ring ridge removed and push the rod and piston assembly up out of the top of the cylinder.

Mark the rod cap so that it can be returned to the same rod and in the same position as it was removed.

Repeat on all the rod and piston assemblies.

Bolt the bearing cap back on the rod before or just after the piston assembly is out of the bore so that neither the cap nor the bearing shells get lost or damaged.

Piston to Connecting Rod Assembly

1964-70 Models

Set the connecting rod in a vise so that the numbers on the bottom of the rod face you. With the rod held in this position, the notch at the top of the piston will go to your left hand on the left bank of cylinders, and to your right hand on the right bank of cylinders. This will place the oil squirt hole of the rod on the side facing the camshaft when the pistons are mounted in the engine.

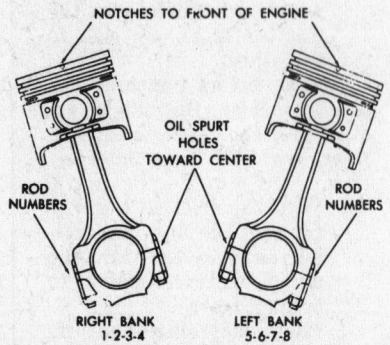

NOTCHES TO FRONT OF ENGINE

OIL SPURT HOLES TOWARD CENTER

ROD NUMBERS   ROD NUMBERS

RIGHT BANK 1-2-3-4   LEFT BANK 5-6-7-8

Piston and rod assembly—8 cyl.
(© Ford Motor Co.)

Assembling Rod and Piston Assemblies to the Engine

1964-70 Models

The rod and piston assemblies are installed in the engine so that the numbers at the bottom of the connecting rod face on both banks, away from the camshaft. As a double check, this will place the notch at the top of the piston forward. This applies to both banks.

Piston Pins

1964-70 Models

Not less than 20 ft. lbs. torque should be required to push the pin from its press fit in the rod. If less than 20 ft. lbs. are required, the fit is too loose and either the rod should be rebushed or an oversized piston pin installed.

Piston Rings

Before replacing rings, inspect the cylinder bores.

1. Using internal micrometer, measure bores both across thrust faces of cylinder and parallel to axis of crankshaft at a minimum of four places equally spaced in each bore. Any bore must not be out of round more than 0.005 in. and must not exceed 0.010 in. difference between any two measurements in a single bore. Any cylinder beyond these limits should be bored to diameter of next available oversize piston. The recommended clearances for new pistons are 0.0015-0.0023 in. for 390, 410, and 428 engines, 0.0014-0.0022 in. for 429 engines, and 0.0042-0.0066 for 427 high performance engines.
2. If bore is within limits dimensionally, examine bore visually. It should be dull silver in color and exhibit pattern of machining crosshatching intersecting at about 45 degrees. There should be no scratches, tool marks, nicks, or other damage. If any such damage exists, bore cylinder to clean up damage and then bore to next available oversize piston diameter. Polished or shiny places in the bore are known as glazing. Glazing causes poor lubrication, high oil consumption, and ring damage. Remove glazing by honing cylinders with clean sharp stones of No. 180-220 grit to obtain surface finish of 15-35 RMS. Use hone also after a cylinder has been bored to obtain correct piston clearance and cylinder surface finish. To remove minor glazing when honing equipment is not available, run emery cloth back and forth across glazed areas perpendicular to axis of bore until glazing disappears.

NOTE: use emery cloth method only as a last resort as it is a much inferior method to honing.

Hot tank cylinder block after honing or boring. If hot tank is not available, scrub block and cylinders thoroughly and rigorously with soap and water to remove all grit.

3. If cylinder is in satisfactory condition, place each ring in bore in turn and square it in bore using head of piston. Measure ring gap.

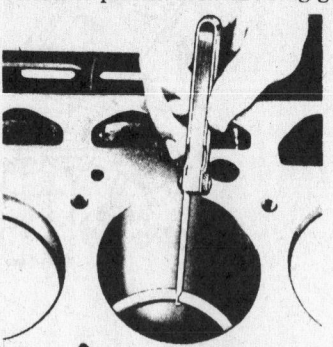

Piston ring gap
(© Ford Motor Co.)

Piston ring side clearance
(© Ford Motor Co.)

If ring gap is greater than limit get new ring. If ring gap is less than limit, file end of ring to obtain correct gap.

4. Check ring side clearance by installing rings on piston and inserting feeler gauge of correct dimension between ring and lower land. Any wear will form a step on lower land. Replace any pistons with high steps.
5. Space ring gaps at equidistant intervals around piston circum-

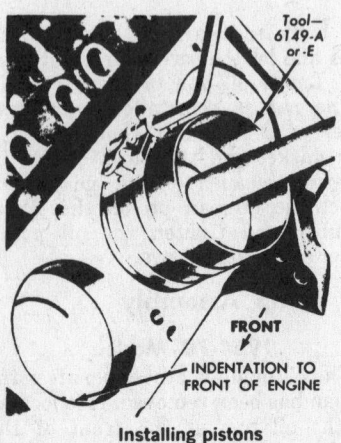

Tool— 6149-A or -E

FRONT

INDENTATION TO FRONT OF ENGINE

Installing pistons
(© Ford Motor Co.)

ference. Be sure to install each piston in its original bore. Install short lengths of rubber tubing over connecting rod bolts to prevent damage to rod journal. Install ring compressor over rings on piston. Lower piston rod assembly into bore until ring compressor contacts block. Using wooden handle of hammer push piston into bore while guiding rod onto journal.

Rear Main Bearing Oil Seal

A packing-type oil seal is used in back of the rear main bearing. The lower half of this oil seal is held in place by the flywheel front plate. The upper half is mounted in a container, which is fitted to a tongued-out portion of the cylinder block itself. To replace the rear main bearing oil seal, it is necessary to remove the engine from the car and disassemble the flywheel and flywheel front plate.

⏻ CHILTON TIME-SAVER

Top Half, Rear Main Bearing Oil Seal Replacement (Wick-Type)

The following method has proven a distinct advantage in most cases and saves many hours of labor.

1. Drain engine oil and remove oil pan.
2. Remove rear main bearing cap.
3. With a 6 in. length of 3/16 in. brazing rod drive up on either exposed end of the top half oil seal. When the opposite end of the seal starts to protrude, have a helper grasp it with pliers and pull gently while the driven end is being tapped.

Woven Fabric Type Seal Installation Except 1968-70 390 and 428

1. Obtain a 12 in. piece of copper wire (about the same gauge as that used in the strands of an insulated battery cable).
2. Thread one strand of this wire through the new seal, about ½ in. from the end, bend back and make secure.
3. Thoroughly saturate the new seal with engine oil.
4. Push the copper wire up through the oil seal groove until it comes down on the opposite side of the bearing.
5. Pull (with pliers) on the protruding copper wire while the crankshaft is being turned and the new seal is slowly fed into place.

Ring Side Clearance

Year and Engine	Top Comp.			Bottom Comp.			Oil Control
	Min.	Max.	Replace	Min.	Max.	Replace	
1964-66 390, 410, 427, 428	0.0024	0.0041	0.006	0.002	0.004	0.006	SNUG
1967-69 390, 410, 428, 429	0.002	0.004	0.006	0.002	0.004	0.006	SNUG
1967 427	0.0024	0.0041	0.006	0.002	0.004	0.006	SNUG

Ring Gap

Year and Engine	Top Comp.	Bottom Comp.	Oil Control
1964 all	0.015-0.036	0.015-0.036	0.015-0.066
1965 all	0.010-0.020	0.010-0.020	0.015-0.066
1966-68 390, 410, and 427	0.010-0.031	0.010-0.020	0.015-0.066
1966-68 428	0.010-0.020	0.010-0.020	0.015-0.066
1969 390	0.010-0.020	0.010-0.020	0.015-0.066
1969 428 and 429	0.010-0.020	0.010-0.020	0.010-0.035

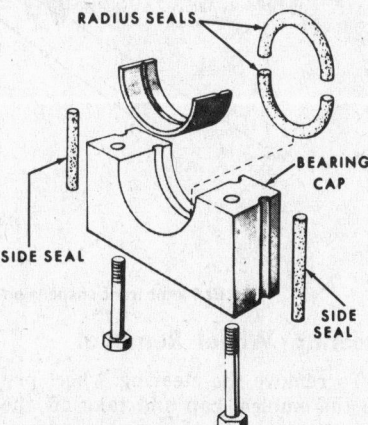

Rear main bearing cap and seals
(© Ford Motor Co.)

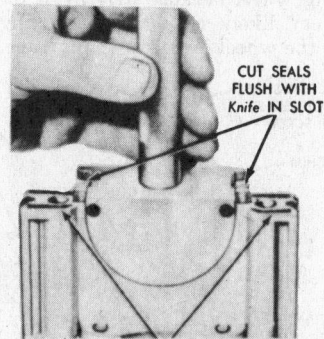

Installing crankshaft rear oil seal
(© Ford Motor Co.)

Caution This snaking operation slightly reduces the diameter of the new seal and care will have to be used to keep the seal from slipping too far through the top half of the bearing.

6. When an equal amount of seal is extending from each side, cut off the copper wire ends of the seal up into the groove (this will tend to expand the seal again).

NOTE: don't worry about the copper wire left in the groove, it is too soft to cause damage.

7. Replace the seal in the cap and replace the oil pan.

Neoprene Seal Installation 1968-70 390 and 428

1. Loosen all main bearing cap bolts, lowering crankshaft slightly but not more than 1/32 in.
2. Remove main cap and remove upper and lower halves of seal. On block half of seal, use seal removing tool or insert a small metal screw into end of seal with which to draw it out.
3. Clean seal grooves with solvent and dip replacement seal in engine oil.
4. Install upper seal half in its groove in block with lip towards front of engine by rotating it on seal journal of crankshaft until approximately 3/8 in. protrudes below parting surface.
5. Tighten other main caps and torque to specification.
6. Install lower seal half in rear main cap with lip to front and approximately 3/8 in. of seal protruding to mate with upper seal.
7. Install rear main cap and torque.
8. Dip side seals in engine oil and install them. Tap seals in last ½ in. if necessary. Do not cut protruding ends of seal.

Front Suspension

See Unit Repair Section.

1964 Cushion Link Suspension

1964 Mercurys have cushion link suspension at all four wheels.

This suspension allows the wheels to move rearward, as well as upward, as a better means to insulate shock.

In the front, the cushion link is at the front pivot of the lower arm. It is an arrangement of two pivot pins

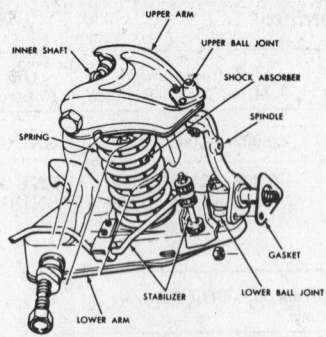

Front suspension—1963-64
(© Ford Motor Co.)

in shackle form. The upper pin is in a large elliptical rubber bushing and controls horizontal movement. The shackle is positioned in neutral, unless disturbed by road bumps.

When road shock is encountered, fore and aft, as well as vertical, movement is permitted.

At the rear, a rubber insulated tension shackle is mounted at the front of each rear spring. Again, rearward movement is allowed to cushion the shock when a road bump is encountered.

Extended Lubrication

The following front suspension and steering parts will require lubrication only after long mileage periods; two upper ball joints, two lower ball joints, one idler arm joint, two cross-link joints, four tie-rod ball joints, one power steering valve joint.

These are factory packed with a special compound and when repacked, only the original type lubricant can be used.

1965-69

A new tension strut type front suspension is used. In design, it is very similar to the Lincoln Continental type.

Steering Wheel

Horn Button or Horn Blowing Ring Removal

To remove the horn button, simply pry it up on all Mercury models.

Caution On some models, the emblem cap was held by three screws from underneath the steering wheel. Before prying up the cap check to see if there are three screws; if so, remove them and then lift off the horn button.

Horn blowing rings are mounted underneath the steering wheel and to take them off it is first necessary to remove the steering wheel.

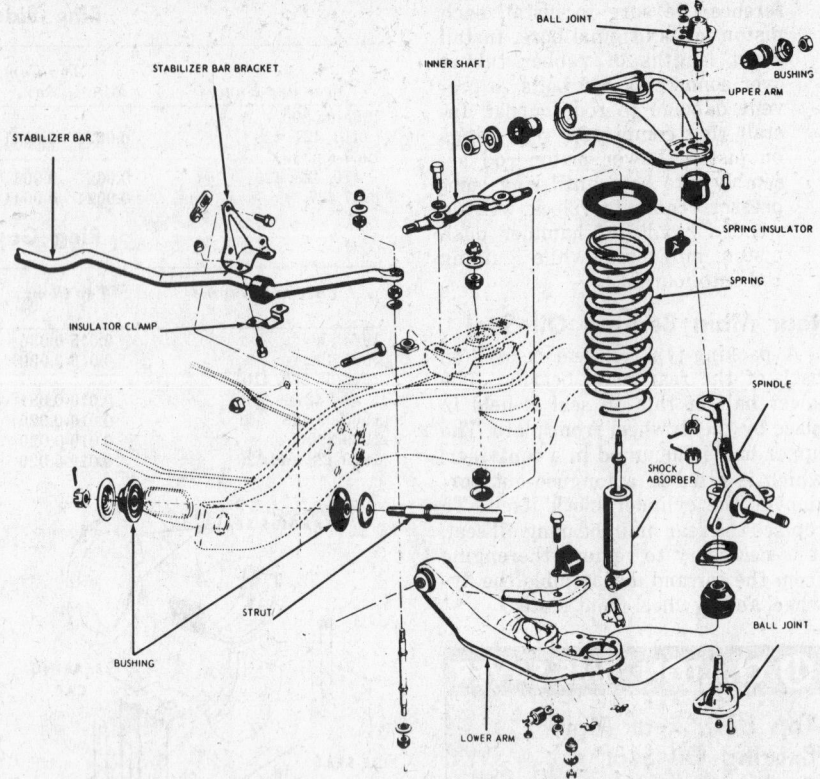

Ball joint front suspension—1965-70 (© Ford Motor Co.)

Steering Wheel Removal

To remove the steering wheel pry up the emblem cap and take off the nut that mounts the steering wheel to the steering tube. A puller should always be used when removing the steering wheel because driving it off will very likely result in damaging either the wheel or the steering tube.

Manual Steering

Steering Gear Assembly Removal

1964-70

A flexible shaft coupling is used. On these models the coupling can be separated, the pitman arm removed

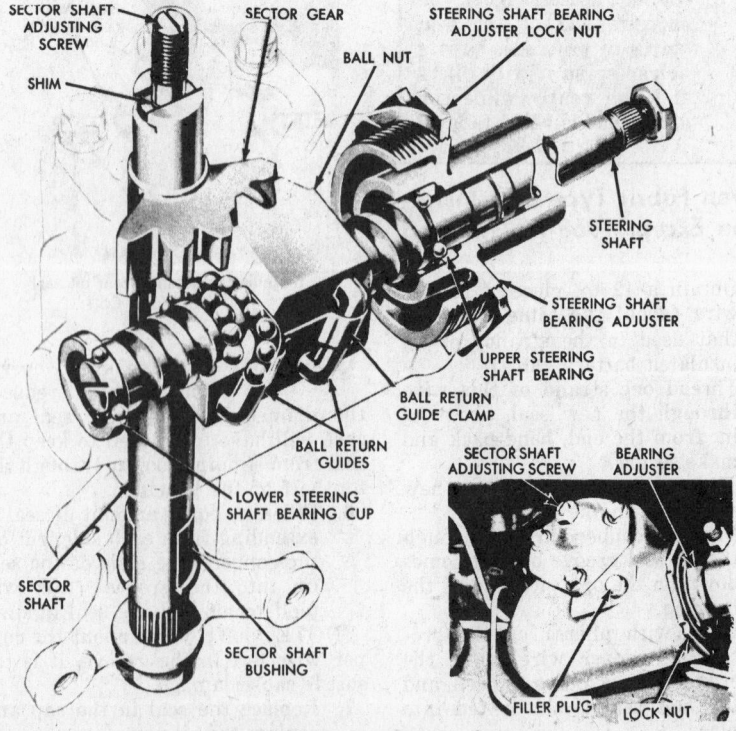

Steering gear (© Ford Motor Co.)

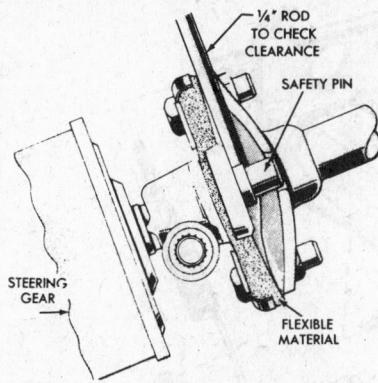

Checking clearance between steering gear and steering shaft (© Ford Motor Co.)

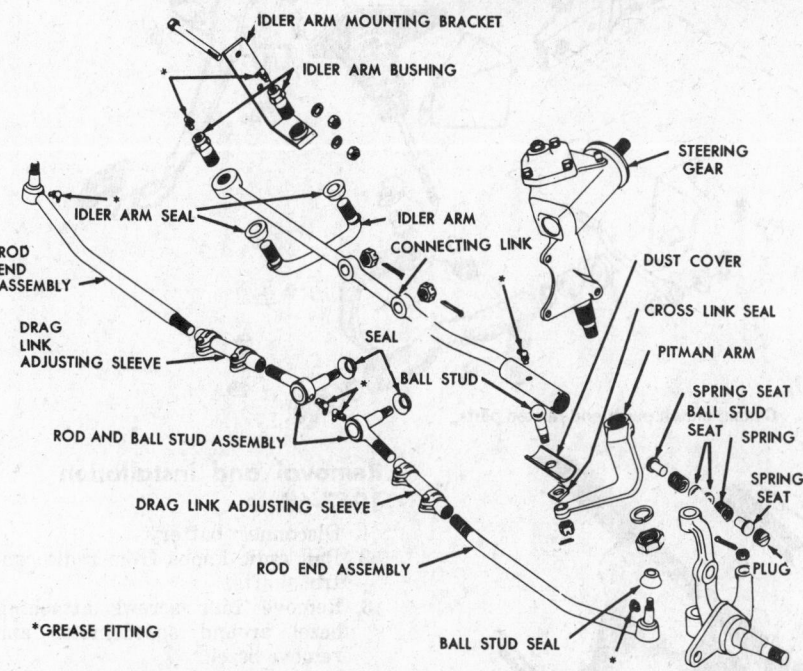

Typical steering linkage (© Ford Motor Co.)

and mounting bolts to frame removed. Then gear may be lifted out, without disturbing upper column and controls.

All Models
Complete instructions for adjusting this and all other steering gears are given in the Unit Repair Section.

Power Steering

All service on the power steering gear is given in the Unit Repair Section.

Power Steering Pump Removal

1964-70 Models
On these models, the power steering pump is belt driven directly off the front of the crankshaft and is mounted on the left cylinder head.

The power cylinder is an integral part of the gear housing.
1. Drain power steering fluid from

pump reservoir by disconnecting fluid return line at reservoir.
2. Disconnect pressure line from pump.
3. Remove three bolts from front of pump and one bolt from rear that attaches pump to mounting bracket.
4. Disconnect belt from pulley, and remove pump.
5. Reverse procedure to install pump.

Standard Clutch

The standard clutch in all Mercury models is of the single plate, dry disc type, having a coil spring pressure plate.

Clutch Assembly Removal and Replacement
See Ford-Thunderbird section.

Standard Transmission

See Ford-Thunderbird section.

Automatic Transmission

See Ford-Thunderbird section.

U Joints, Drive Lines

Universal Joint and/or Driveshaft Removal
See Ford-Thunderbird section.

Drive Axle, Suspension

Note description of cushion link under Front Suspension.

Rear Spring Replacement
See Ford-Thunderbird section.

Rear Axle Removal
See Ford-Thunderbird section.

Improved Traction Axle Identification
The following is a simple rule-of-thumb way to distinguish between the standard and the improved-type units.
1. Raise both rear wheels off the ground.
2. With the parking brake off, turn one wheel forward (by hand) and note the direction of rotation of the other wheel.
3. If the other wheel turns in the same direction as the one being turned, the rear axle is of the improved-type.
4. If the other wheel turns in the opposite direction, the axle is of standard design.

Rear Shock Absorber Replacement
The rear shock absorber is replaced by detaching it at the axle housing at the bottom and from the frame at the top. Rubber grommets are used, top and bottom, on the shock absorber.

Radio

Removal and Installation 1964
1. Disconnect the battery.
2. With air-conditioning in the car, it is necessary to remove glove compartment.
3. Disconnect antenna lead, speaker plug, dial lamp lead at bullet connector and power lead at main junction block.
4. Remove control knobs, with two nuts and washers from front of radio.
5. Remove bracket retaining nut from left side of radio and the attaching screw from the right side.
6. Lift out radio. (On air-conditioned cars, the radio is lifted out through the speaker grille opening.)

Removal and Installation 1965-66
1. Disconnect battery.
2. On cars with air conditioning, disconnect air conditioning ducts from plenum and remove nozzle

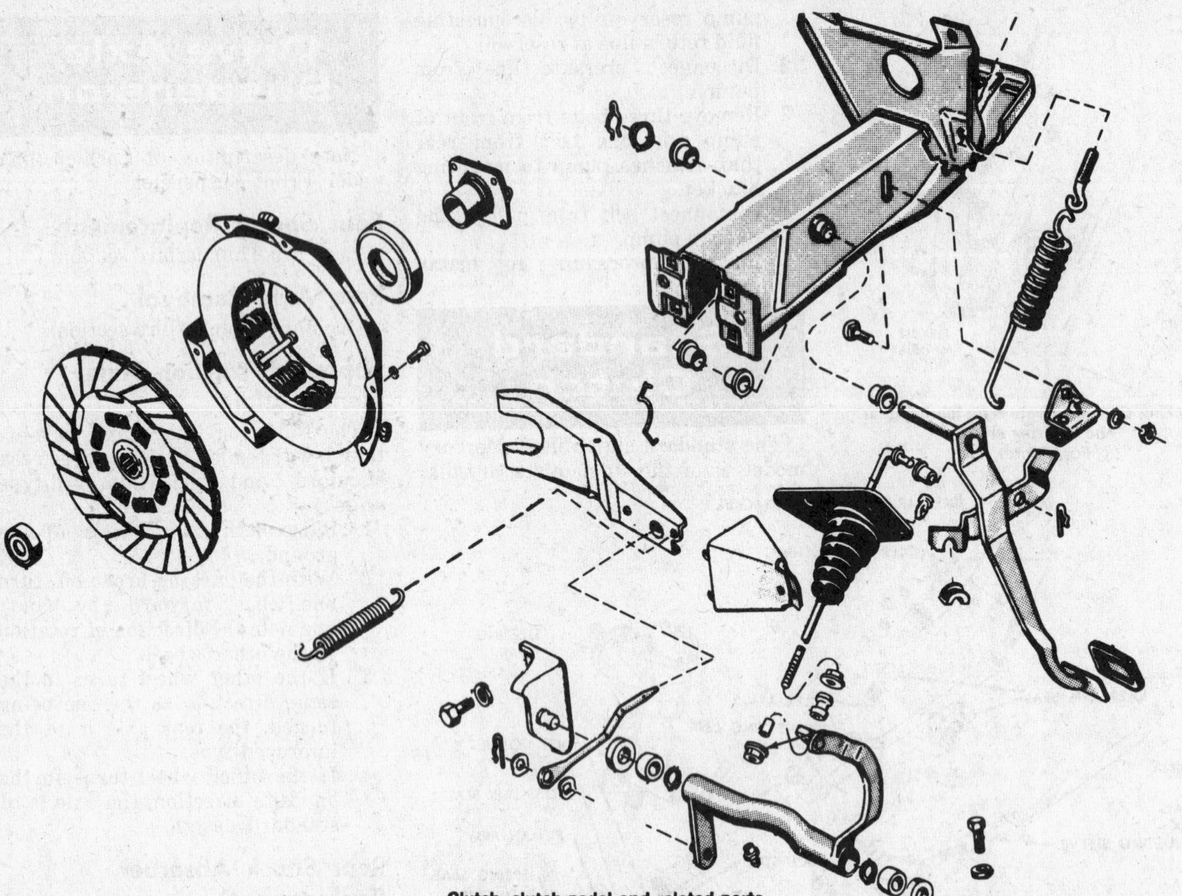

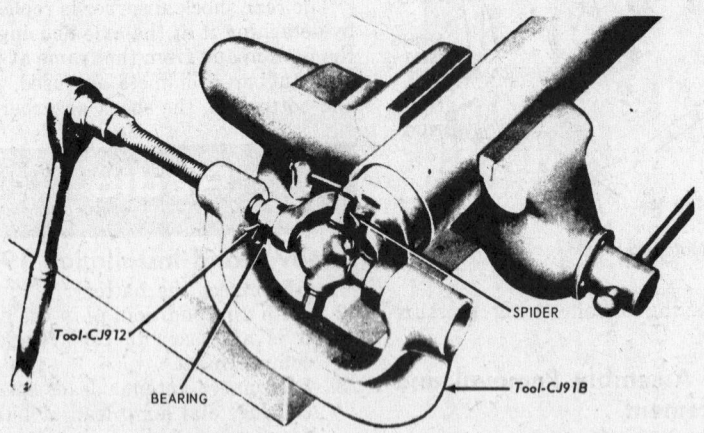

Clutch, clutch pedal and related parts

Installing universal joint bearing (© Ford Motor Co.)

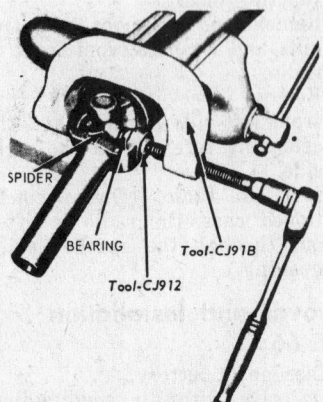

Removing universal joint bearing
(© Ford Motor Co.)

adapter plate, ash tray and slide bracket.
3. Remove radio knobs.
4. Remove nut retaining bracket to radio and move bracket away from radio. On air conditioned cars remove bracket.
5. Disconnect antenna, speaker, power, and pilot light leads from radio.
6. Remove two mounting nuts and washers.
7. Carefully remove radio from under instrument panel.
8. Remove radio. Reverse procedure to install radio.

Removal and Installation 1967-68

1. Disconnect battery.
2. Pull radio knobs from radio control shafts.
3. Remove four screws attaching bezel around speedometer and remove bezel.
4. Pull wiper control knob off shaft.
5. Remove clock and heater knobs.
6. Remove four screws attaching cluster right finish cover and remove cover.
7. Remove nut attaching rear support to radio.
8. Remove four screws attaching radio and mounting plate to instrument panel.
9. Pull radio out of panel and disconnect radio feed wires and speaker and antenna lead.
10. Remove mounting plate from front of radio. Reverse procedure to install radio.

Removal and Installation 1969-70

1. Disconnect battery.
2. Remove radio knobs and remove nut from radio shaft.
3. Remove radio rear support nut and nut retaining radio to instrument panel.
4. Lower radio and disconnect antenna, radio, power, and speaker lead wires.

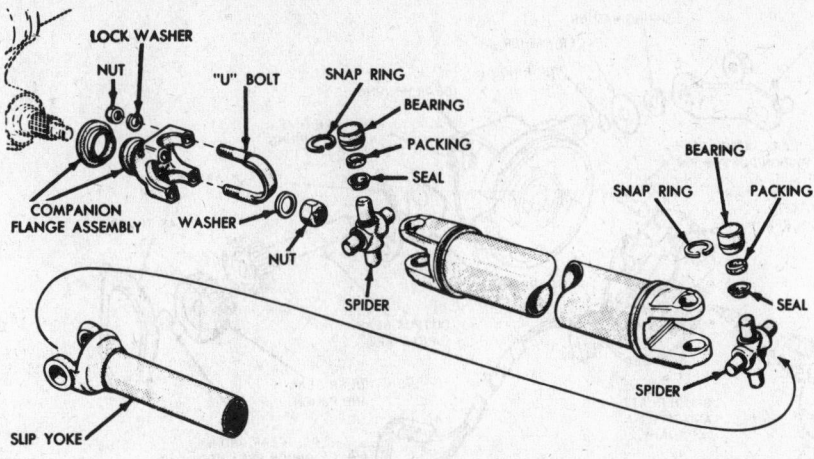

Driveshaft and universal joint assembly (© Ford Motor Co.)

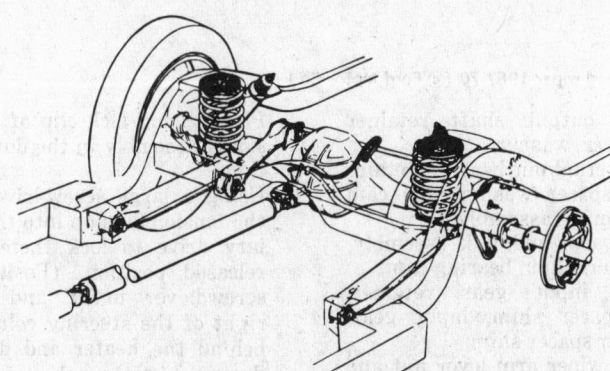

Rear suspension (© Ford Motor Co.)

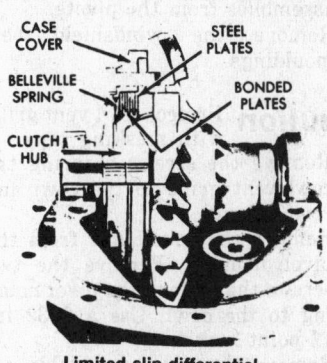

Limited slip differential
(© Ford Motor Co.)

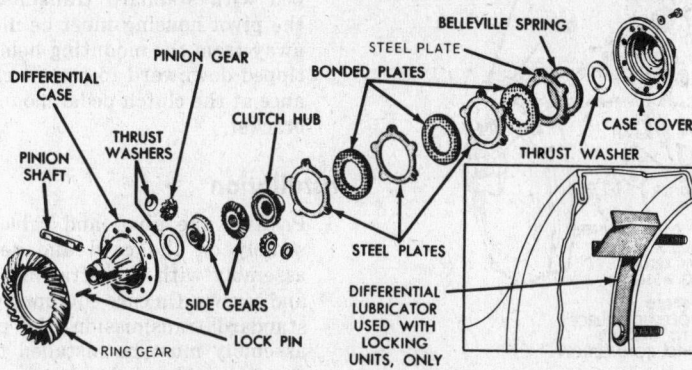

Limited slip differential assembly (© Ford Motor Co.)

5. Remove radio. Reverse procedure to install radio.

Windshield Wipers

Motor R & R

1964-66

1. Disconnect the battery.
2. On air-conditioned cars, remove the glove compartment, radio and the radio speaker.
3. Disconnect the electrical connector and the control cable from the motor.

4. Remove the two motor attaching bolts and, if necessary, remove the ten cowl grille attaching screws and the grille.
5. Remove the nut and washer holding the motor to the auxiliary drive. Remove the motor.
6. Install by reversing above procedure.

1967-68

1. Remove wiper arm and blade assemblies from pivot shafts.
2. Remove cowl intake screen.
3. Disconnect wiper links at motor output arm pin.
4. Disconnect motor harness by removing retention clip.
5. Disconnect motor harness connector and remove motor.
6. Remove bolts holding motor to bracket assembly and remove bracket assembly. Reverse procedure to install motor.

1969-70

1. Remove wiper arm and blade assemblies from pivot shafts.
2. Remove cowl top grille for access to inner side of dash panel.
3. Disconnect linkage drive arm from motor output crankpin by removing retaining pin.
4. Disconnect two push on wire connectors from motor.
5. Remove three bolts that retain motor to dash and remove motor.

Two Speed Wiper Motor Disassembly

1967-70

1. Remove gear cover retaining screws, ground terminal, and cover.
2. Remove gear and pinion retainer.
3. Remove idler gear and pinion and thrust washer.
4. Remove motor through bolts, motor housing, switch terminal insulator sleeve, and armature. Do not pound motor housing magnet assembly as ceramic magnets may be damaged.
5. Mark position of output arm with respect to output shaft for assembly. Remove output arm, retaining nut, spring washer, flat washer, output gearshaft assembly, thrust washer, parking switch lever, and parking switch lever washer.
6. Remove brushes and brush springs.
7. Remove brush plate and switch assembly, and switch contact to parking lever pin. Reverse procedure to assemble.

Two Speed Oscillating Wiper Motor Disassembly

1967-68

1. Remove gear housing cover plate and gasket.

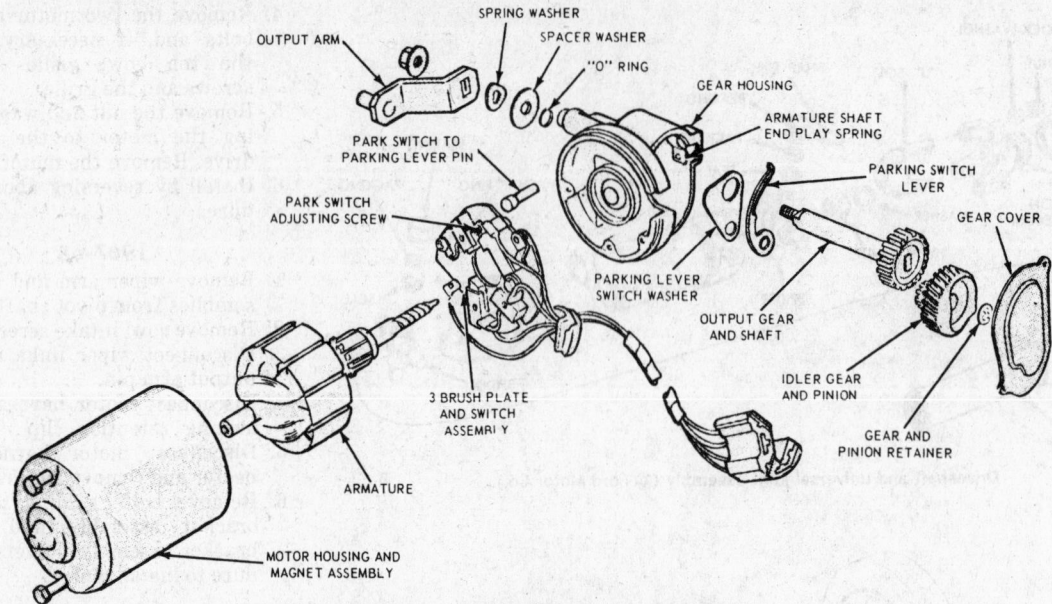

OUTPUT ARM
SPRING WASHER
SPACER WASHER
"O" RING
GEAR HOUSING
ARMATURE SHAFT END PLAY SPRING
PARKING SWITCH LEVER
GEAR COVER
PARK SWITCH TO PARKING LEVER PIN
PARK SWITCH ADJUSTING SCREW
PARKING LEVER SWITCH WASHER
OUTPUT GEAR AND SHAFT
IDLER GEAR AND PINION
GEAR AND PINION RETAINER
3 BRUSH PLATE AND SWITCH ASSEMBLY
ARMATURE
MOTOR HOUSING AND MAGNET ASSEMBLY

Disassembled two speed wiper 1967-70 (© Ford Motor Co.)

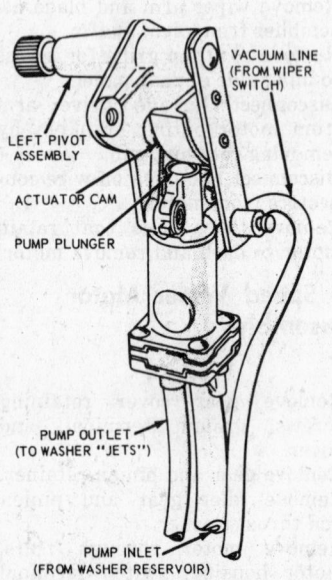

VACUUM LINE (FROM WIPER SWITCH)
LEFT PIVOT ASSEMBLY
ACTUATOR CAM
PUMP PLUNGER
PUMP OUTLET (TO WASHER "JETS")
PUMP INLET (FROM WASHER RESERVOIR)

Windshield washer pump assembly (© Ford Motor Co.)

2. Remove output shaft retainer and spacer washer.
3. Remove crankpin bearing retainer and spacer washer, and cam return spring assembly.
4. Remove arm and link assembly.
5. Remove crankpin bearing cam.
6. Remove input gear retainer, outer spacer shim, input gear, and inner spacer shim.
7. Remove wiper arm lever nut and lock washer.
8. Remove wiper arm lever, spacer, and output shaft and gear assembly.
9. Remove output shaft by tapping with fiber hammer. The worm drive gear and armature are not serviced. Reverse procedure to assemble.

Wiper Transmission R & R

1963-66

1. Disconnect battery.
2. Remove glove compartment and radio.

3. Place spring lock clip of the tensioner assembly in the down position.
4. Using a large screwdriver, push the tension lugs up into the auxiliary drive to lock them in the released position. (Position the screwdriver under and to the right of the steering column and behind the heater and defroster ducts.) Lift the cable ends from the auxiliary drive.
5. Remove the wiper arm and blade assemblies from the pivots.
6. Remove the windshield belt mouldings.

Caution Protect cowl vent grille with masking tape.
7. Remove the screws securing the cowl vent grille to the cowl and remove grille.
8. Remove the round nut from the pivot housing. Remove the two screws that attach the pivot housing to the cowl. Use a 9/32 in. 12 point socket.
9. Remove the pivot and cable assembly from the underside of the instrument panel. On cars equipped with standard transmission, the pivot housing must be moved away from the mounting hole and tipped downward to permit clearance at the clutch pedal mounting bracket.

Installation

1. Position the pivot and cable assembly on the cowl and secure assembly with the attaching nut and screws. On cars equipped with standard transmission, the pivot assembly must be installed from the right side of clutch mounting bracket.

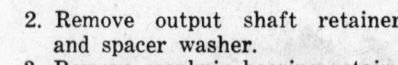

RETAINER
SPACER WASHER
OUTPUT SHAFT
SHIMS
RING GEAR RETAINER
MOTOR SHAFT STOP SCREW
WORM GEAR
RING GEAR SHIM
INPUT GEAR
GEAR HOUSING
OUTPUT GEAR
ARM AND LINK
GEAR HOUSING COVER
GASKET
SPACER WASHER
TERMINAL C
TERMINAL B
RETAINER
CRANKPIN BEARING CAM
RETAINER
CAM RETURN SPRING ASSEMBLY

TESTING CONNECTIONS
TO PARK
B TERMINAL TO VOLTAGE SOURCE
C TERMINAL TO *Blue* AND *Red*
Yellow TO *White Black* TO GROUND

TO OPERATE AT HIGH SPEED
Yellow TO VOLTAGE SOURCE
Blue TO *White Black* TO GROUND
B AND **C** TERMINALS
AND *Red* NO CONNECTION

TO OPERATE AT LOW SPEED
Red AND *Yellow* TO VOLTAGE SOURCE
Blue TO *White Black* TO GROUND
B AND **C** TERMINALS NO CONNECTION

2-speed oscillating type wiper motor—1967-68 (© Ford Motor Co.)

2. With the tension lug on the auxiliary drive locked in the compressed position, install the cables on the auxiliary drive. Then move the spring clip to the up position, and compress the prongs to release the tension lugs.

3. Position the cowl vent grille on the cowl and install the sheet metal screws that attach the grille to the cowl. Install the belt mouldings.

4. Connect the battery cable to the battery.

5. To locate wiper pivots, turn ignition key to Acc position. Turn wiper control knob to operate wiper momentarily.

6. Install the wiper arm and blade assemblies so that they point toward the right side of windshield and contact the belt moulding.

7. Check the wiper operation.

8. Install the radio and glove compartment.

1967-68

1. Remove wiper arms and blades.

2. Remove retaining screws, cowl ventilator grille, hood pad, and windshield washer nozzles.

3. Through cowl opening remove clip retaining wiper links to motor drive arm.

4. Remove pivot shaft retaining screws and remove pivot and links as an assembly from right side of cowl.

5. Remove link to pivot retaining clip.

1969-70

1. Remove wiper arms and blades from pivot shafts.

2. Remove cowl top grille.

3. Remove three screws at right pivot shaft and disconnect right link from plate assembly on inner side center of dash panel by removing clip.

4. Lift pivot and link assembly out of cowl.

5. Remove left pivot and link assembly by same procedure as right.

6. Disconnect drive arm from motor, remove four screws retaining pivot plate to dash, and remove assembly through cowl. Reverse procedure to install.

Heater System

1964

Heater R & R

1. Partially drain cooling system.

2. Disconnect fresh air inlet control cable at instrument panel.

3. Disconnect right air inlet boot at blower housing.

4. Disconnect defroster valve cable at plenum chamber.

5. Disconnect defroster hoses at outlets.

6. Disconnect wires from resistor and motor.

7. Disconnect heater hoses at firewall.

8. Remove plenum chamber retaining nuts at firewall and lay assembly on floor.

9. Disconnect control valve cable at valve. Remove assembly from car.

10. Install in reverse of above.

Heater Core R & R

1. Remove assembly as above.

2. Remove core cover retaining screws and remove cover.

3. Remove core from plenum.

4. Install in reverse of above.

Heater Blower R & R

1. Remove assembly as above.

2. Remove blower motor retaining plate screws and remove motor and fan assembly.

3. Loosen fan set screw and remove blower motor.

4. Remove retaining plate and gasket from motor.

5. Install in reverse of above.

1965-70

Heater Core R & R

1. Partially drain cooling system.

2. Remove heater hoses at core.

3. Remove retaining screws, core cover and seal from plenum.

4. Remove core from plenum.

5. Install in reverse of above, applying a thin coat of silicone to the pads.

Heater Blower R & R

The blower motor and fan assembly can be removed from the engine compartment. Disconnect the wiring. Remove mounting screws, then the blower motor and fan assembly.

Mercury Capri

Index

YEAR IDENTIFICATION

1970-71 two-door sedan

FIRING ORDER

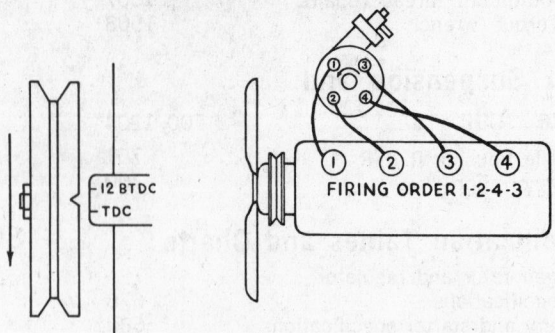

FIRING ORDER 1-2-4-3

Left illus. Timing marks; right illus. Firing order

IDENTIFICATION

The identification plate is found under the hood, riveted to the inner fender panel. Interpretation of the plate is as follows:

A type drive (1 = left-hand drive)
B type engine (L1 = 1600 cc.)
C type transmission (5 = four-speed floorshift)
D axle ratio (V = 3.89:1)
E paint code
  B = ermine white
  7 = amber gold metallic
  1 = blue mink metallic
  5 = fern green metallic
  6 = aquatic jade metallic
  H = red
F trim code
G S.V.C. reference (indicates date of manufacture when car is

shipped elsewhere for final assembly)
H vehicle serial number

Serial Number Interpretation

The serial number consists of eleven digits, both letters and numbers, arranged in five sections.
  1st digit: letter—product source (G = Germany)
  2nd digit: letter—assembly plant (A = Cologne, B = Genk)
  3rd & 4th digits: letters—body type (EC = two-door sedan)
  5th & 6th digits: letters—assembly code (The first letter denotes year of manufacture, the second the month. See chart below.)

7th to 11th digits: sequential serial number (from 00001-99999)

Assembly Code Interpretation
YEAR CODE

Month	K-1970	L-1971
Jan.	L	C
Feb.	L	K
Mar.	Y	D
Apr.	S	E
May	T	L
June	J	Y
July	U	S
Aug.	M	T
Sept.	P	J
Oct.	B	U
Nov.	R	M
Dec.	A	P
	G	

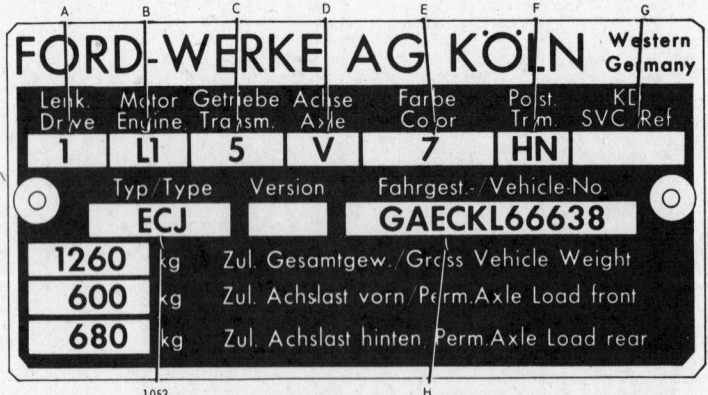

A	B	C	D	E	F	G

FORD-WERKE AG KÖLN Western Germany

Lenk. Drive	Motor Engine	Getriebe Transm.	Achse Axle	Farbe Color	Post. Trim	KD SVC Ref.
1	L1	5	V	7	HN	

Typ/Type	Version	Fahrgest.-/Vehicle-No.
ECJ		GAECKL66638

1260 kg	Zul. Gesamtgew./Gross Vehicle Weight
600 kg	Zul. Achslast vorn/Perm. Axle Load front
680 kg	Zul. Achslast hinten Perm. Axle Load rear

1053   H

Vehicle identification plate (© Ford Motor Co.)

AC GENERATOR AND REGULATOR SPECIFICATIONS

		ALTERNATOR			REGULATOR						
		Brush Spring Pressure (Oz.)	Output			Field Relay			Regulator		
YEAR	MODEL		Volts	Amps.	Model	Air Gap (In.)	Point Gap (In.)	Volts to Close	Air Gap (In.)	Point Gap (In.)	Volts @5000 RPM 2
1970-71	17ACR	7-10	14.1-14.5	35	17ACR①	—	—	—	—	—	14.1-14.5

① —Integral with alternator.

② —Alternator rpm. Alternator speed is 1.88 x engine speed.

VALVE SPECIFICATIONS

YEAR AND MODEL		SEAT ANGLE (DEG.)	INTAKE VALVE LIFT (IN.)	EXHAUST VALVE LIFT (IN.)	VALVE SPRING PRESSURE INTAKE & EXHAUST		STEM TO GUIDE CLEARANCE (IN.)		STEM DIAMETER (IN.)	VALVE GUIDE REMOVABLE
					Outer	Inner	Intake	Exhaust		
1970-71	Two-door sedan	45	.315	.319	44-49 @ 1.263"	—	.0008-.0030	.0017-.0039	②	No①

① —Sleeves available for repairs.
② —Intake—0.3095-0.3105.
   Exhaust—0.3086-0.3096.

CRANKSHAFT BEARING JOURNAL SPECIFICATIONS

YEAR AND MODEL		MAIN BEARING JOURNALS (IN.)				CONNECTING ROD BEARING JOURNALS (IN.)		
		Journal Diameter	Oil Clearance	Shaft End-Play	Thrust On No.	Journal Diameter	Oil Clearance	End-Play
1970-71	1600 cc.	①	.001-.002	.003-.011	3	1.9368-1.9376	.0004-.0024	.004-.010

① —Blue color code—2.1253-2.1257
   Red color code—2.1257-2.1261
   Green color code—2.1153-2.1157
   Yellow color code—2.1157-2.1161

TUNE-UP SPECIFICATIONS

YEAR	MODEL	SPARK PLUGS		DISTRIBUTOR		IGNITION TIMING (Deg.) ▲	CRANKING COMP. PRESSURE (Psi)	VALVES			FUEL PUMP PRESSURE (Psi)	IDLE SPEED (Rpm) ★
		Type	Gap (In.)	Point Dwell (Deg.)	Point Gap (In.)			Tappet (Hot) Clearance (In.)		Intake Opens (Deg.)		
								Intake	Exhaust			
1970-71	Two-door sedan	Autolite AG22	.025	38-40	.025	6B	165	.010②	.017②	17B	3½-5	830-870①

▲—With vacuum advance disconnected. Add 50 rpm when equipped with air conditioning. NOTE: These settings are only approximate. Engine design, altitude, temperature, fuel octane rating and the condition of the individual engine are all factors which can influence timing. The limiting advance factor must, therefore, be the "knock point" of the individual engine.

B—Before top dead center.
① —Fast idle = 1775 rpm.
② —Cold = intake .008-.010.
   exhaust .018-.020.

CAUTION

General adoption of anti-pollution laws has changed the design of almost all car engine production to effectively reduce crankcase emission and terminal exhaust products. It has been necessary to adopt stricter tune-up rules, especially timing and idle speed procedures. Both of these values are peculiar to the engine and to its application, rather than to the engine alone. With this in mind, car manufacturers supply idle speed data for the engine and application involved. This information is clearly displayed in the engine compartment of each vehicle.

GENERAL CHASSIS AND BRAKE SPECIFICATIONS

YEAR AND MODEL		CHASSIS		BRAKE CYLINDER BORE				BRAKE DRUM Diameter (In.)	
		Overall Length (In.)	Tire Size	Master Cylinder (In.)		Wheel Cylinder Diameter (In.)			
				Std.	Power	Front	Rear	Front	Rear
1970-71	Two-door sedan	167.8	165R x 13①	—	N.A.	2.125	.75	②	9.0

① —185R x 13 optional.
② —Front disc 9.625 in. diameter x .50 in. thick.

CAPACITIES

YEAR AND MODEL		ENGINE CRANKCASE (Qts.) Incl.	TRANSMISSIONS Pts. TO REFILL AFTER DRAINING			DRIVE AXLE (Pts.)	GASOLINE TANK (Gals.)	COOLING SYSTEM (Qts.) with Heater
			Manual		Automatic			
			3-Speed	4-Speed				
1970-71	Two-door sedan	4.25	—	2⅞①	—	2.4	12	7.25

① —SAE 80 E.P.

BATTERY AND STARTER SPECIFICATIONS

| YEAR | BATTERY | | | | STARTERS | | | |
| | Ampere Hour Capacity | Volts | Group Number | Terminal Grounded | No-Load Test | | | Brush Spring Tension (Oz.) |
					Amps.	Volts	RPM	
1970–71	55	12	24F①	Neg.	N.A.	N.A.	N.A.	28

① —Requires slight modification of battery tray and clamps. Drill new hole halfway between the 55 Amp./hr. and 66 Amp./hr. holes, use existing clamp with nut and bolt through new hole. Must use terminal adapter clamp with existing cables—positive terminal to rear of car.

GENERAL ENGINE SPECIFICATIONS

YEAR	CU. IN. DISPLACEMENT	CARBURETOR	DEVELOPED HORSEPOWER @ RPM	DEVELOPED TORQUE @ RPM (FT. LBS.)	A.M.A. HORSEPOWER	BORE & STROKE (IN.)	COMPRESSION RATIO	VALVE LIFTER TYPE	NORMAL OIL PRESSURE (PSI)
1970–71	97.51	Autolite, 1-BBL.	71 @ 5000	91 @ 2,800	16.1	3.188 x 3.056	8:1	Mech.	35–40

WHEEL ALIGNMENT

| YEAR AND MODEL | CASTER | | CAMBER | | TOE-IN (In.) | KING-PIN INCLINATION (Deg.) | WHEEL PIVOT RATIO | |
	Range (Deg.)	Pref. Setting (Deg.)	Range (Deg.)	Pref. Setting (Deg.)			Inner Wheel	Outer Wheel
1970–71 Two-door sedan	0°30'–1°30'①	—	0°30'–0°30'①	—	.09–.15	7°30'–8°30'①	N.A.	N.A.

① —Not adjustable.
N.A.—Not available.

TORQUE SPECIFICATIONS

| YEAR | CYLINDER HEAD BOLTS (FT. LBS.) | ROD BEARING BOLTS (FT. LBS.) | MAIN BEARING BOLTS (FT. LBS.) | CRANKSHAFT BALANCER BOLT (FT. LBS.) | FLYWHEEL TO CRANKSHAFT BOLTS (FT. LBS.) | MANIFOLD (FT. LBS.) | |
						Intake	Exhaust
1970–71	65–70	30–35	65–70	24–28	50–55	15–18	15–18

CYLINDER HEAD BOLT TIGHTENING SEQUENCE

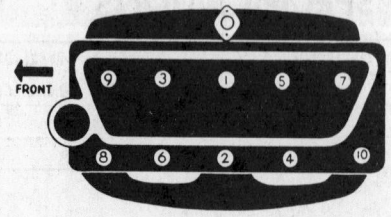

1600 cc.

LIGHT BULBS

Headlights . . . . 2x 50 watts
Side lights . . . . 2x 5 watts
Rear turn signals . . . . 2x 28 watts
License plate . . . . 2x 5 watts
Indicators . . . . 4x 2.2 watts
Side markers . . . . 4x 5 watts
Clock . . . . 1x 1.2 watts
Front turn signals . . . . 2 x 7/28 watts
Tail and stop . . . . 2 x 7/28 watts
Interior . . . . 1x 6 watts
Instrument panel . . . . 4x 2.2 watts

FUSES AND CIRCUIT BREAKERS

Back-up light . . . . 10 AMPS.
All others . . . . 8 AMPS.

Ignition System

Distributor

The Autolite distributor is mounted on the right-hand side of the engine and is gear-driven by the camshaft. It incorporates both centrifugal and vacuum advance mechanisms. The vacuum *advance* action is controlled by carburetor vacuum, while the vacuum *retard* action is controlled by intake manifold vacuum. The IMCO system, described in more detail in the Unit Repair Section, is used to control exhaust emissions.

Removal

1. Unsnap the two clips and remove the distributor cap.
2. Disconnect the vacuum lines from the distributor.
3. Matchmark the distributor housing and the engine block, then scribe another mark on the housing to indicate the rotor position.
4. Remove the bolt that holds the distributor, then carefully pull out the unit.

Disassembly

1. Remove the condenser lead and condenser, then remove wires from points.

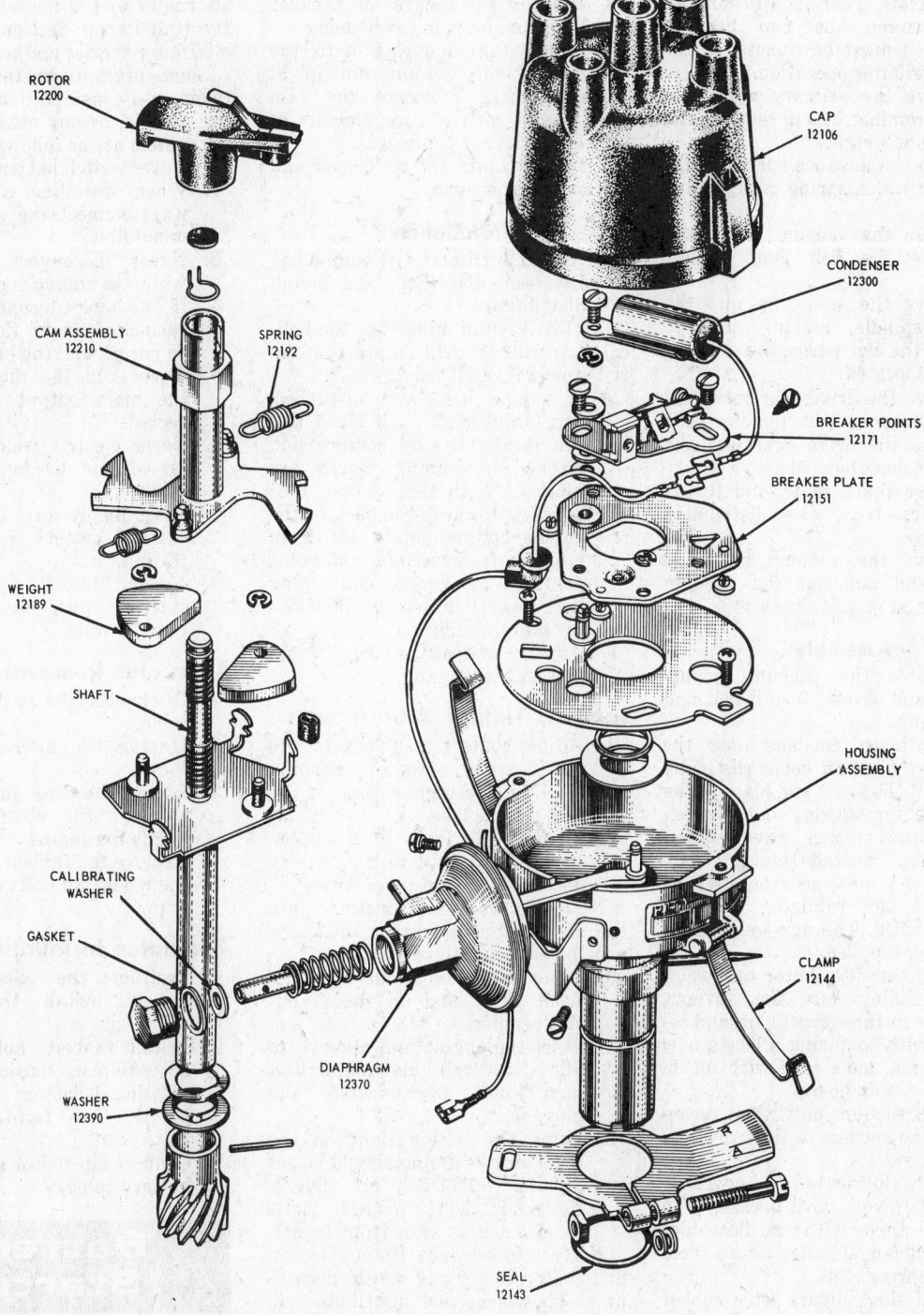

ROTOR 12200

CAM ASSEMBLY 12210

SPRING 12192

WEIGHT 12189

SHAFT

CALIBRATING WASHER

GASKET

WASHER 12390

DIAPHRAGM 12370

CAP 12106

CONDENSER 12300

BREAKER POINTS 12171

BREAKER PLATE 12151

HOUSING ASSEMBLY

CLAMP 12144

SEAL 12143

Distributor (© Ford Motor Co.)

2. Remove breaker point assembly.
3. Remove the snap-ring on the vacuum unit pivot post, then remove the two breaker plate hold-down screws and the breaker plate.
4. Remove the large snap-ring on the pivot post.
5. Remove the flat washer, the two wave washers, and the upper contact breaker plate.

NOTE: it may be necessary to move the breaker plate to disengage the hold-down pin from the slot in the lower plate. There is a grounding spring between the two breaker plates which must be reinstalled for proper distributor operation.

6. Remove the primary wire rubber grommet, then remove the governor weights.
7. Unclip the advance springs, first noting which spring goes to each post.
8. Remove the vacuum unit, then remove the felt cam spindle pad.
9. Remove the snap-ring and the cam spindle, making sure to mark the slot where the advance stop is located.
10. Remove the drive gear retaining pin, using a pin punch, then remove the drive gear and the two washers located above it.
11. Remove shaft, plate, and thrust washers from the distributor housing.
12. Remove the vacuum unit end bolt and pull out the vacuum spring, stop, and shims.

Assembly

1. Assemble the vacuum spring, stop, and shims; install bolt and seal ring.
2. Slide thrust washers onto the distributor shaft below the plate, then fit the shaft and plate to the distributor housing and slide on the thrust washer, wave washer, and gear. Install retaining pin.

NOTE: if a new gear or shaft is installed, a new retaining pin hole must be drilled. The new gear has a pilot hole as supplied.

3. Install the distributor cam spindle, making sure the advance stop is in the correct slot, and secure with snap-ring. Place a new felt wick, moistened with oil, in the cam spindle top.
4. Install vacuum unit, then reconnect the springs to their original posts.
5. Lightly lubricate the governor weight pivots with grease, then install them with the flat sides toward cam spindle; secure them with spring clips.
6. Install the primary wire rubber grommet, connect the grounding spring to the pivot post, then install upper breaker plate. The

hold-down spindle must enter the "keyhole" slot.
7. Hold this assembly in place using the two spring washers, flat washer, and large snap-ring.
8. Check the clearance between the two breaker plates, using a feeler gauge, at the point underneath the nylon bearing nearest the hold-down pin; clearance should not exceed 0.010 in.

NOTE: to reduce clearance, thread the nut further onto the hold-down screw.

9. Position and secure the breaker plate assembly in the housing.
10. Install the snap-ring onto the end of the vacuum unit pivot post, then lubricate the cam spindle with a tiny amount of Lubriplate or equivalent.
11. Install points and condenser and adjust point gap.

Installation

1. Align matchmarks, if engine has not been disturbed, and install distributor.

NOTE: keep in mind that the helical gear will tend to rotate the distributor as it is pushed down.

2. If engine has been disturbed, turn crankshaft until No. 1 piston is at TDC on compression stroke and timing marks are aligned. With the vacuum unit pointing towards the back of the engine approximately 45° from crankshaft centerline, and rotor pointing to No. 2 spark plug wire tower, insert the distributor into the engine.
3. Adjust contact breaker points and ignition timing.

Ignition Timing Adjustment

1. Adjust contact breaker points to specification, with the rubbing block on the highest part of the cam spindle lobe.
2. Connect the leads of a strobe-type timing light to the battery and to No. 1 spark plug wire.

NOTE: a small nail, inserted into the distributor cap tower, makes a good hook-up terminal.

3. Clean the timing marks on the front cover and on the crankshaft pulley.
4. Start the engine and allow it to idle at 800 rpm, distributor vacuum lines disconnected and plugged.
5. Shine the timing light on the index area—timing should be set at 12° BTDC for IMCO-equipped Capri models (left-hand mark as seen from front).

NOTE: to advance timing, loosen distributor clamp and rotate distributor opposite normal distributor rotation; to retard timing, rotate in the same direction as distributor normally rotates.

Alternator

The charging system consists of the battery, which will be covered separately, the alternator, the regulator and the wires and cables required to connect these units. The alternator used is an English Lucas unit, model 17ACR, having a 35 Amp. output. The alternator is driven by a V-belt from the engine at 1.88 times engine speed.

The regulator is integral with the alternator and is non-adjustable. See the Unit Repair Section for complete alternator service and testing.

Some precautions that should be taken into consideration when working on this, or any other, AC charging system are as follows:

1. Never switch battery polarity.
2. When installing a battery, always connect the grounded terminal first.
3. Never disconnect the battery while the engine is running.
4. If the molded connector is disconnected from the alternator, do not ground the hot wire.
5. Never run the alternator with the main output cable disconnected.
6. Never electric weld around the car without disconnecting the alternator.
7. Never apply any voltage in excess of battery voltage during testing.
8. Never "jump" a battery for starting purposes with more than 12 volts.

Alternator Removal

1. Disconnect the battery negative cable.
2. Unplug the alternator connectors.
3. Loosen the three mounting bolts and tilt the alternator in towards the engine.
4. Remove the fanbelt, then remove the mounting bolts and the alternator.

Alternator Installation

1. Position the alternator and loosely install the mounting bolts.
2. Install fanbelt, hold alternator so as to place tension on the belt (½ in. deflection at belt midpoint), then tighten mounting bolts.
3. Connect alternator plugs and the battery cable.

Battery, Starter

Battery

The battery used is a standard

12-volt, lead-acid unit having a capacity of 55 Amp./hrs. See the Unit Repair Section of this manual for a general discussion of batteries and their tests.

Starter

The Lucas starter motor used is very similar to American-made units of the same type. See the Unit Repair Section of this manual for overhaul procedures.

Removal

1. Disconnect the battery.
2. Disconnect the motor wires from the starter motor.
3. Remove the solenoid hold-down nuts, then remove the solenoid.
4. Set the parking brake, jack up the front of the car and support on axle stands.
5. Remove the two lower starter hold-down bolts, then loosen the upper bolt.
6. While supporting the starter with one hand, remove the top bolt and lower the starter.

Installation

1. Position the starter motor and install upper bolt loosely.
2. Install two lower bolts, then tighten all three bolts to 20-25 ft. lbs.
3. Install the solenoid, then reconnect starter wires. Tighten main terminal nuts to 24-26 ft. lbs.
4. Connect the battery cables.

Instruments

The standard instruments consist of two clusters and four warning lights. The left-hand cluster contains a speedometer and an odometer, while the right-hand cluster contains the fuel and temperature gauges. These two gauges are powered through a 5-volt voltage regulator mounted on the rear of the speedometer. The warning lights are for turn signals (green), high beam (blue), alternator (red), and oil pressure (amber). Light bulbs can be replaced without removing the cluster.

A bank of rocker switches is mounted in the center of the instrument panel. These switches control the windshield wipers, emergency flashers, and the tandem brake system test circuit.

Cluster Removal and Installation

1. Remove the two steering column hold-down bolts, then lower the steering column.
2. Remove the five Phillips-head screws, then pull the instrument panel forward.
3. Disconnect the speedometer

cable and wiring, remove the four remaining Phillips-head screws, and remove the cluster.
4. To service any one gauge, simply remove it from the cluster.
5. To install, reverse removal procedure.

Headlight Switch R&R

1. Disconnect battery ground cable.
2. Pull the connector from the back of the switch.
3. From behind the sub-panel, push the switch to one side and remove.
4. To install, reverse removal procedure.

Ignition Switch R&R

1. Disconnect battery ground cable.
2. Remove steering column shroud.
3. Place key on "O" position, then disconnect switch wires.
4. Remove the two screws that hold switch to steering lock, then remove switch.

Ignition Key Warning Buzzer R&R

1. Remove parcel tray.
2. Disconnect the two leads to the buzzer.
NOTE: the buzzer is located beside the flasher unit on the brake pedal support.
3. Remove buzzer retaining screw and buzzer.
4. To install, reverse removal procedure.

Steering Lock

Removal

1. Disconnect battery ground cable.
2. Remove steering column shroud, then remove screws that hold upper steering column.
3. Turn steering column to gain access to headless bolts, then disconnect leads to ignition switch and lock.
4. Drill out the headless bolts and remove steering lock.

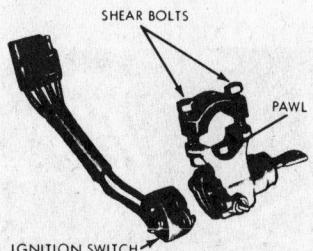

Steering lock (© Ford Motor Co.)

Installation

1. Position the lock assembly, with ignition key in place, on the steering column.
2. Withdraw lock to allow pawl to enter steering shaft.
3. Place clamp in position, install

"shear head" bolts, and tighten bolts until heads break off. Make sure pawl operates freely.
4. Connect wires to switch terminals, reposition steering column, and replace shroud. Tighten upper steering column screws.
5. Connect battery cable.

Brakes

Hydraulic brakes, with fixed-caliper, dual-piston front discs and self-adjusting rear drums, are used. The system is the standard tandem type, having two separate brake circuits. A failure in one brake circuit will cause the pressure differential valve to switch braking effort from that circuit which, in so doing, activates another switch that controls the warning light on the dashboard. The pressure differential valve is similar to American-made units, except for a special centering procedure. See the Unit Repair Section.

The brake booster is non-serviceable and must be replaced if defective. Vacuum to the booster is supplied via a check valve, and both sides of the diaphragm normally are under equal vacuum. When the brake pedal is depressed, atmospheric pressure is admitted to the rear diaphragm chamber, and the vacuum in the forward chamber pulls the diaphragm forward to apply additional braking force to the master cylinder.

The parking brake is operated through a floor-mounted lever located between the front seats. Pulling the lever transmits force through a two cable linkage to operate the rear drum brakes. A self-adjusting feature operates when there is excessive clearance between the brake shoes and drums.

Master Cylinder Removal

1. Siphon the fluid from the reservoir.
2. Disconnect brake lines from master cylinder.
3. Remove the master cylinder retaining nuts.
4. Remove the master cylinder, being careful not to damage the vacuum seal.

Master Cylinder Installation

1. Position the master cylinder and fluid seal on the pushrod.
2. While holding in this position, thread the brake line fittings a few turns.
3. Bolt the master cylinder to the vacuum booster, then tighten brake lines.
4. Fill the reservoir with an approved type brake fluid.
NOTE: FoMoCo C6AZ-19542-A is recommended.

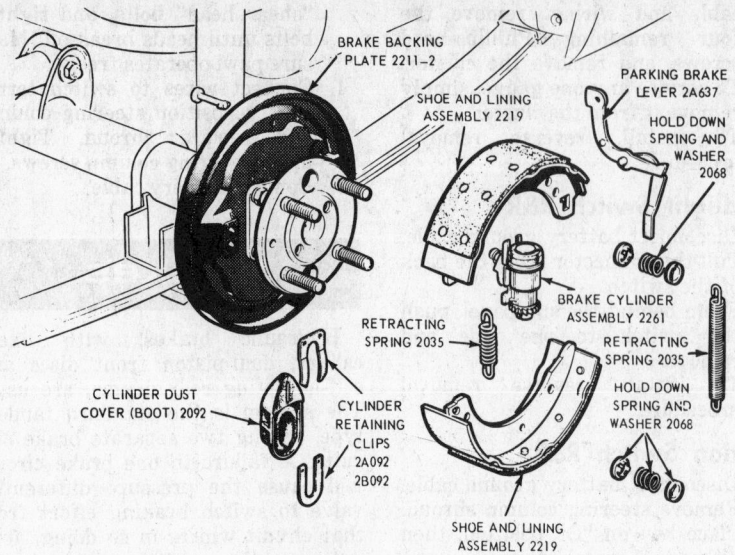

Rear drum brake (© Ford Motor Co.)

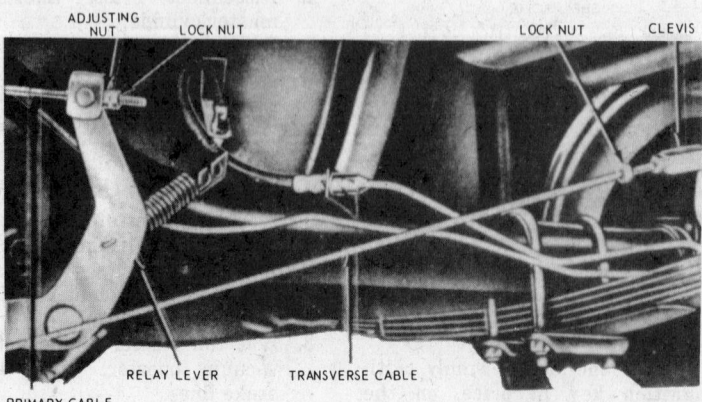

Front disc brake (© Ford Motor Co.)

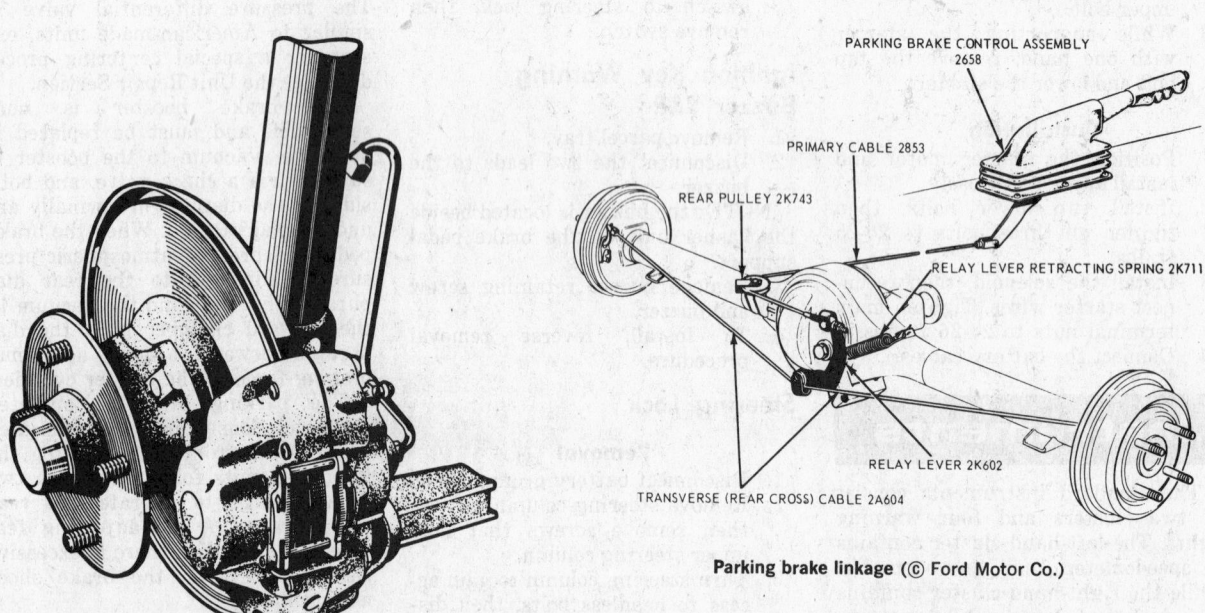

Parking brake linkage (© Ford Motor Co.)

5. Bleed the brake system, both front and rear.
NOTE: see Unit Repair Section.
6. Check brake operation.

Power Brake Booster R&R

1. Disconnect brake pushrod clevis pin from brake pedal, then remove pin.
2. Remove master cylinder retaining nuts and wire cylinder out of the way, being careful not to damage the fluid seal.
3. Remove the vacuum hose from the brake booster.
4. Remove the booster to firewall retaining screws, then remove booster and seal.
5. Remove retaining bracket and gasket from booster.
6. To install, reverse removal procedure, using new gaskets throughout where possible.

Parking Brake Adjustment

1. Jack up the rear of the car and support on axle stands.
2. Release the parking brake and make sure primary cable is free and greased.
3. Loosen the primary cable rear locknut and adjust cable length until primary cable has no slack and relay lever is just clear of the rear axle housing.
4. Loosen the locknut on the transverse cable end near the right-hand rear brake and adjust the cable so that there is no slack. The parking brake operating levers must be on their stops.
5. Tighten locknuts and lower car.
NOTE: never adjust the cables to take up for rear brake wear. If brake wear seems to be causing cable slack, pull the rear drums and check the self-adjusters.

Parking brake linkage adjustment points (© Ford Motor Co.)

Parking Brake Lever R&R

1. Jack up the front of the car and support on axle stands.
2. Chock the rear wheels and release the parking brake.
3. Remove the carpet around the brake lever.
4. Remove the spring clip and clevis pin that holds the primary cable to the lever.
5. Remove the six sheet metal screws and the brake lever boot.
6. Remove the two lever hold-down screws and the lever assembly.
7. To install, reverse removal procedure, smearing clevis pin with grease.
8. Check primary cable adjustment.

Fuel System

Carburetor

The carburetor used on the 1600 cc. engine is an Autolite single-barrel downdraft unit having idle, main, power valve, and accelerator pump systems. Cars equipped with exhaust emission controls (all those imported) have a tamper-proof slow-running volume screw that limits the rich mixture setting. All carburetors have an automatic choke. See the Unit Repair Section.

Carburetor R&R

1. Lift the hood and remove the air cleaner.
2. Disconnect fuel and vacuum lines at the carburetor.
3. Disconnect decel valve line at carburetor.
4. Disconnect accelerator linkage, drain cooling system, and disconnect automatic choke.

5. Remove carburetor hold-down nuts and washers and the carburetor.
6. To install, reverse removal procedure, using a new gasket.

Carburetor Adjustment

Accurate adjustment of the carburetor requires the use of an exhaust gas analyzer. Using the analyzer, adjust for a 13.4-13.9 air/fuel ratio at 780-820 rpm.

If an analyzer is not available, the carburetor can be roughly adjusted in the following manner:

1. Connect a tachometer between coil-to-distributor primary wire and ground. Adjust idle to 820 rpm, turning the curb idle screw.
2. Adjust the lower idle screw to obtain maximum rpm—do not break the limit stop.
3. Screw in the lower idle screw to obtain a 20-40 rpm drop from the maximum idle speed.
4. Reset the idle speed, if necessary, using the curb idle screw to obtain 870 rpm.

Air Cleaner Assembly

The air cleaner used on the Capri has a temperature-operated valve and duct assembly built-in, the failure of which will reduce the effectiveness of the emission control system.

The system is designed to provide heated air (+90° F.) to the carburetor during normal operation. A thermostatic bulb is connected through a spring-loaded linkage to a flap valve. This valve is designed to shut off the flow of hot air when underhood temperatures approach 110° F.

Testing

With the duct and valve assembly installed, the flap valve should be in the *up* position, shutting off the cold air intake, so long as ambient air temperature is below 85°F. If the valve is not in the **up** position, check for valve wear or breakage in the duct. The entire duct and valve assembly must be replaced as a unit if wear is in evidence.

To further test the unit, remove it from the car and immerse the ther-

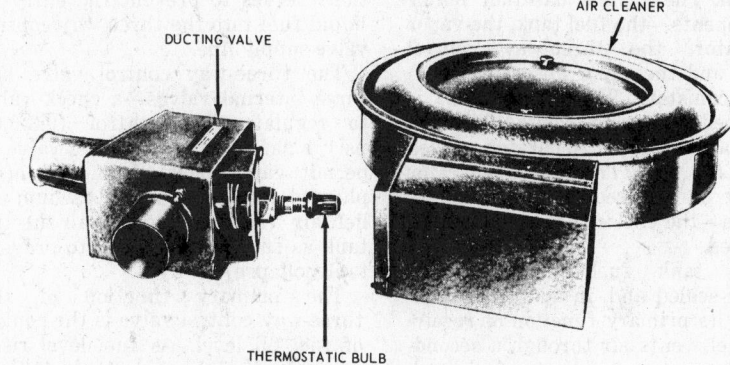

Air cleaner and duct assembly (© Ford Motor Co.)

mostatic bulb in water. Heat the water to 85°F., allow the temperature to stabilize for about five minutes, then observe valve—it should be in the **up** position. Increase water temperature to 110°F. and again wait five minutes. The flap valve should be in the **down** position to shut off the hot air. If the flap does not work as indicated, the thermostatic bulb must be replaced as a unit.

Fuel Pump

A diaphragm-type mechanical fuel pump, mounted on the right-hand side of the engine, supplies fuel to the carburetor. The fuel pump has a gauze screen and a glass sediment bowl which must be cleaned periodically.

Removal and Installation

1. Lift the hood, then disconnect the fuel lines at the pump.
   NOTE: plug the lines, as gas may siphon from tank.

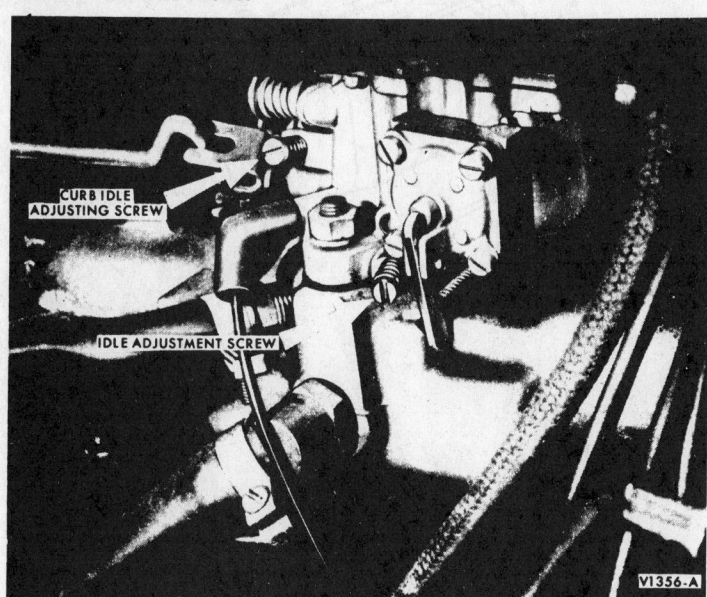

Carburetor adjustment (© Ford Motor Co.)

Installing fuel pump (© Ford Motor Co.)

2. Remove the two bolts and washers that hold the pump to the block.
3. Remove the fuel pump, carefully lifting the lever to clear the cam eccentric.
4. To install, reverse removal procedure, tightening bolts to 12-15 ft. lbs.

Evaporative Emission Control System

All cars imported into the U.S. are equipped with a fuel vapor control system. The system has four major components—the fuel tank, the vapor separator, the three-way control valve, and the vapor absorbing charcoal canister. The fuel tank is equipped with a non-vented filler cap and has the vapor separator welded to its top side. The vapor separator cannot be serviced separately if defective—the entire fuel tank must be replaced.

The tank fuel filler neck is double-sealed and, in addition to fulfilling its primary function of receiving fuel, vents air through a secondary chamber and indicates fuel level.

Charcoal canister—tighten center bolt to 15-18 ft. lbs. (© Ford Motor Co.)

The vapor separator (see illustration) serves to prevent the entry of liquid fuel into the three-way control valve supply line.

The three-way control valve has three internal valves—a check valve to regulate fuel control (0.3-0.65 psi), a safety pressure relief valve to permit vapor blow-off in case of a plugged vapor line, and a vacuum relief air valve to replace air in the tank as fuel is consumed (to prevent tank collapse).

The primary function of the three-way control valve is the control of fuel fill level. As fuel level rises, vapor is discharged into the atmosphere through the fill control tube and the space between the inner filler tube and the outer neck. When the fuel level is high enough to cover the fill control tube orifice, vapor can no longer escape and the filler tube begins to fill up. At this point, neither vapor nor fuel can flow through the system to the check valve, since at least 0.3 psi is required to open the valve and the pressure head in the filler is still insufficient. The tank level is therefore controlled by the level in the fill control tube and is maintained, when full, at approximately 90% of full tank volume to allow for thermal expansion.

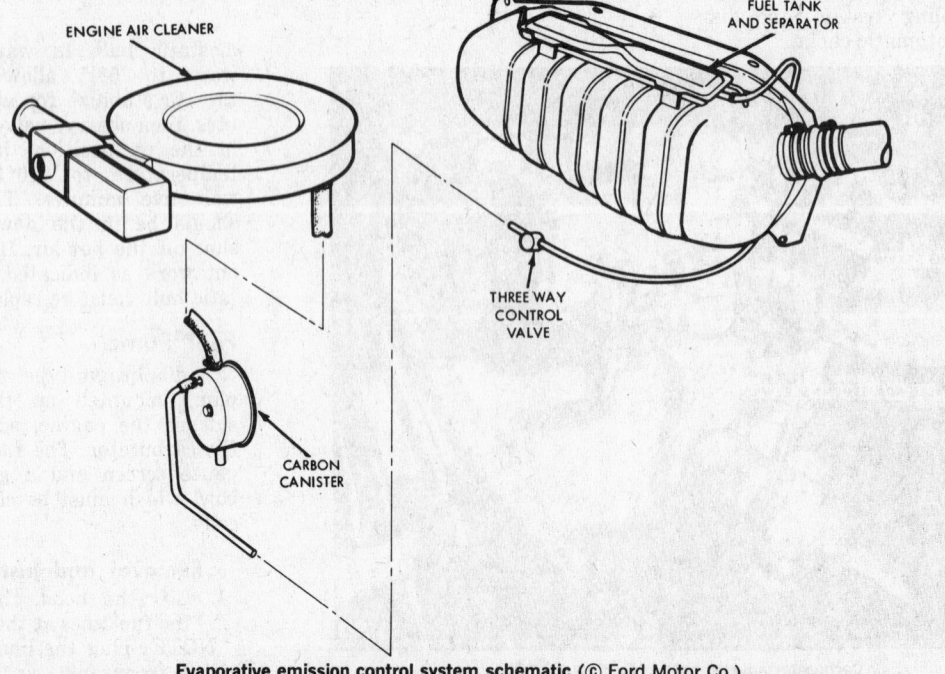

ENGINE AIR CLEANER

FUEL TANK AND SEPARATOR

THREE WAY CONTROL VALVE

CARBON CANISTER

Evaporative emission control system schematic (© Ford Motor Co.)

Three-way control valve (© Ford Motor Co.)

Exhaust System R&R—Complete

1. Jack up the car and support on axle stands.
2. Loosen the rear muffler inlet clamp.
3. Unhook rear O-rings and remove rear muffler assembly.
4. Release front pipe from bracket.
5. Remove front pipe-to-manifold clamp.
6. Unhook front O-rings and remove front assembly.
7. To install, first install front assembly, then rear assembly.
8. Tighten all bolts loosely until entire system is in place.
9. Tighten manifold clamp nuts to 15-20 ft. lbs., rear assembly clamp to 10-12 ft. lbs., and bracket to 12-15 ft. lbs.
10. Start engine and check for exhaust leaks, then lower car to floor.

When thermal expansion takes place, vapor flows when a vapor pressure of 0.3 to 0.65 psi is developed in the line from the separator unit. The vapor passes through the now open three-way control valve, through the system, to the charcoal canister in the engine compartment. Vapor is there absorbed by the activated charcoal until the car is started, whereupon the "stored" vapor is sucked through a connecting hose into the air cleaner and burned.

If the line should become restricted (between charcoal canister and check valve), the vapor pressure is permitted to rise to 0.7 to 1.5 psi, at which point the pressure relief valve portion of the three-way control valve opens to allow the vapor to pass into the atmosphere.

The vacuum relief valve portion of the three-way control valve always remains closed until a pressure differential of 0.25 psi is reached (between atmospheric pressure and vacuum in tank). When this point is reached, the valve opens to the atmosphere and air is allowed to pass into the tank, through a filter, until pressure balance is achieved.

Throttle Linkage Adjustment

1. To adjust the accelerator pedal, place a 0.015 in. feeler gauge between the return stop and the pedal arm. Adjust the return stop until the front face of the pedal is 4.84 in. from the floor.
2. After adjusting accelerator pedal, leave the feeler gauge in place and disconnect the outer throttle cable retainer from the valve cover (under the hood).
3. Allow the throttle cable to assume its natural position, then reattach the retainer. Check that linkage operates properly, after removing feeler gauge.

Exhaust System

The exhaust system consists of two parts—a front pipe, resonator and front muffler assembly; and a rear muffler and tailpipe assembly. The exhaust system is clamped together and suspended on four rubber O-rings, in addition to a bracket from the transmission that holds the front pipe.

Cooling System

The cooling system consists of the water pump, fan, thermostat, radiator, and connecting lines. Coolant is circulated from the bottom of the radiator up through the water pump and into the cylinder block and cylinder head to the thermostat. If the engine is at operating temperature (or hotter), the coolant is returned to the radiator top tank, from where it flows down through the radiator tubes to be cooled by air. If the engine is cold, the coolant flows through a bypass hose to allow the coolant in the block and head to warm up quickly.

Water Pump R&R

1. Lift the hood, then drain the cooling system.
2. Remove lower radiator hose and heater hose from water pump.
3. Loosen alternator adjusting and mounting bolts, pivot alternator toward engine, and remove V-belt.
4. Remove fan and pulley, then remove pump bolts and pump.

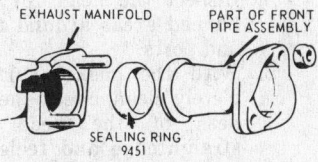

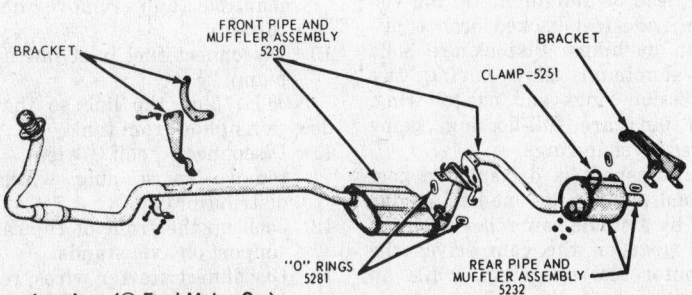

Exhaust system (© Ford Motor Co.)

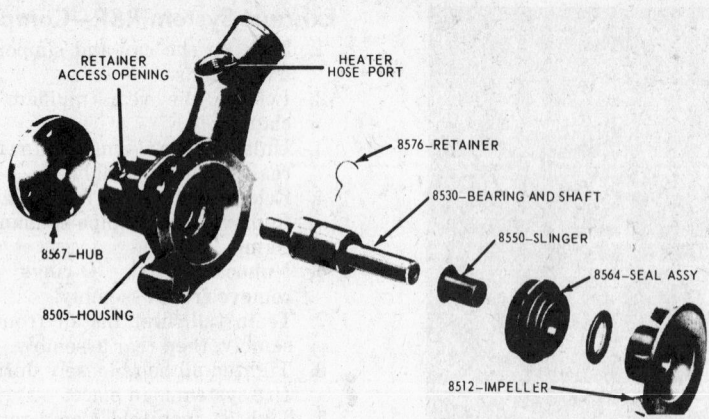

Water pump—parts are available for repairs (© Ford Motor Co.)

5. To install, reverse removal procedure, tightening alternator bolt to 15-18 ft. lbs.

NOTE: use a new gasket, with Permatex.

Radiator R&R

1. Lift the hood and drain the cooling system.

NOTE: radiator cap should be opened to allow pressure release.

2. Disconnect radiator top and bottom hoses, remove hold-down bolts, and lift out radiator.
3. Remove the overflow hose from the filler neck.
4. To install, reverse removal procedure.

NOTE: cap should be left off with engine running to allow air to bleed out of system.

Engine

The engine used in the Capri is a 1600 cc., four-cylinder, inline overhead valve unit having a cross-flow cylinder head and piston-shaped combustion chambers. The cylinder bores are machined in the cast-iron block and are cooled by full-length water jackets.

The crankshaft is made of cast iron and runs in five main bearings having steel backed copper/lead or lead/bronze inserts. End-play is controlled by half thrust washers on each side of the center main bearing.

The connecting rods are forged H-section units having steel backed copper/lead or aluminum/tin big end inserts and steel backed bronze piston pin bushings. Pistons are solid skirt aluminum alloy having two compression rings and one oil ring. Piston pins are full-floating, being retained by snap-rings.

The camshaft is driven in a conventional manner, at one-half engine speed, by a single-row roller chain. A helical gear on the cam drives the distributor and oil pump, while an eccentric operates the fuel pump.

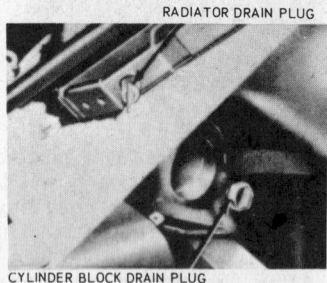

Coolant drain plugs (© Ford Motor Co.)

The cast-iron cylinder head is machined flat on the down side, having no cast-in combustion chambers. Valve guides are integral, although guide replacement is possible and sleeves are available from the dealer.

Engine R&R

1. Remove hood, after scribing matchmarks around hinges.
2. Disconnect battery cables.
3. Drain the cooling system.

NOTE: drain engine block as well as radiator.

4. Disconnect radiator hoses and remove radiator.
5. Remove air cleaner assembly.
6. Disconnect heater hoses from water pump and intake manifold.
7. Disconnect throttle linkage.
8. Disconnect oil pressure and temperature sender wires, then disconnect alternator wires.

NOTE: it is a good idea to tag these wires.

9. Disconnect exhaust pipe from manifold and remove hot air tubes.
10. Disconnect fuel inlet line at fuel pump.

NOTE: plug the line so that gas does not siphon from tank.

11. Disconnect coil wires, then remove spark plug wires and distributor cap.
12. Jack up the front of the car and support on axle stands.
13. Disconnect starter wires, remove starter and oil pan shield.

14. Remove clutch cover.
15. Lower the car to the floor.
16. Remove clutch housing-to-engine bolts, then install lifting brackets and chain hoist.
17. Disconnect the front motor mounts, while supporting engine with chain hoist.
18. Place an axle stand or wooden block under the transmission.
19. Raise the engine slightly, while pulling forward to separate the transmission drive shaft from the clutch; lift engine out of car.
20. To install, reverse removal procedure.

NOTE: don't forget to attach engine ground strap.

Cylinder Head

Removal

1. Remove the air cleaner, then disconnect the fuel line at the pump and the carburetor.
2. Drain the cooling system.
3. Disconnect spark plug wires, then disconnect heater and vacuum hoses from the intake manifold and choke housing.
4. Disconnect temperature sender wire, then disconnect exhaust pipe at manifold flange.
5. Disconnect throttle linkage and distributor vacuum line at carburetor.
6. Remove thermostat housing and thermostat.
7. Remove rocker arm cover and gasket, then remove the rocker shaft bolts, evenly, and the rocker shaft assembly.
8. Remove pushrods and place them aside in proper order for correct installation.
9. Remove cylinder head bolts, head, and gasket.

Cleaning and Inspection

1. Use a scraper and a wire brush to clean carbon deposits from head, following up with a kerosene-soaked rag.
2. Remove the valves and springs.
3. Remove spark plugs and manifolds, then carefully clean the gasket surfaces of the head and manifolds.
4. Inspect the head for cracks or burned areas around the valves and ports.
5. With the gasket surface completely clean, check the straightness of the head using a straightedge and feeler gauges. Generally, 0.001 in. per inch of head length is acceptable.

Installation

1. Place head gasket on the block.

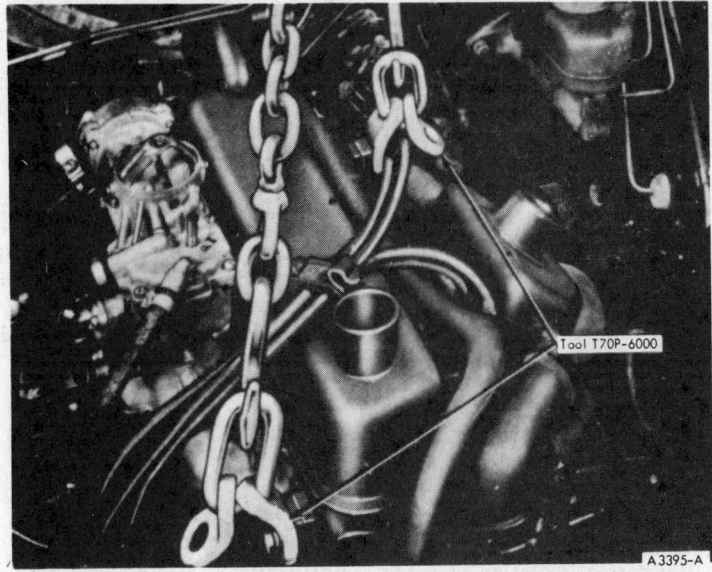

Removing/installing engine—note location of lifting brackets for proper balance
(© Ford Motor Co.)

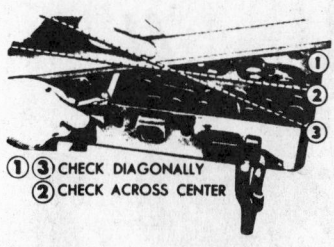

Checking cylinder head straightness
(© Ford Motor Co.)

2. Position the cylinder head and install the bolts. Tighten evenly in sequence (see illustration under **Torque Specifications** table) to proper torque.
   NOTE: manifolds may be installed prior to placing head on block.
3. Install pushrods in correct order, then place rocker arms and shaft assembly on head and locate pushrods in rocker arm screws. Tighten rocker arm bolts to 25-30 ft. lbs.
4. Adjust valve clearance, then install rocker arm cover and gasket.
5. Continue installation by reversing Steps 1-6 of removal procedure.

Exhaust Emission Control—1970-71

The IMCO (improved combustion) system is used on the 1600 cc. Capri engine. With this system, exhaust emissions are controlled by improving combustion efficiency during deceleration from the higher speed ranges.

The system, as used on this model, consists of three major components—the deceleration valve, the dual-diaphragm distributor, and a modified carburetor.

Deceleration Valve

The deceleration valve is connected to the intake manifold by a nut and tapered adapter. The valve itself is controlled by two springs, the tension of which is set at the factory by the nylon screw on the cap. Basically, it operates like a power brake vacuum servo. One side of a diaphragm is subject to atmospheric pressure, while the other side is open to intake manifold vacuum. When the valve is opened by the strong vacuum in the intake manifold, an air fuel mixture is drawn, at the same time, from the carburetor into the manifold. When this occurs, pressure in the manifold rises slightly and increases the volume of mixture into the cylinders. This greater volume results in better burning and reduced hydrocarbon emissions.

Deceleration valve (© Ford Motor Co.)

Dual-Diaphragm Distributor

The distributor has a normal set of centrifugal advance weights and vacuum diaphragm advance, with the addition of another diaphragm controlled by manifold vacuum. This second diaphragm acts to retard the spark under deceleration and idle, when manifold vacuum is greatest. While this decreases the power of the engine at these times, there is an increase in the braking effect of the engine and hydrocarbon emissions are reduced further.

Carburetor

The Autolite single-barrel carburetor is calibrated for use with the IMCO system. Major differences between this and other carburetors of the same type are the limit stop on the mixture screw and a modified carburetor float.

Valve System

The valves are mounted vertically in the cylinder head, the intake valve heads being larger than the exhaust valve heads. The exhaust valves are stellite-coated for better heat and wear resistance, while the intake valves are coated with diffused aluminum for the same reason. The factory does not recommend grinding the intake valves or lapping the intake valve seats, because the grinding operation removes the coating and shortens the life of the valve. Exhaust valves, on the other hand, may be ground if necessary.

Valve stems are phosphate-coated for better wear resistance. Valve guides are cast integral with the head, although sleeves are available if guides become worn. In addition, valves are available with 0.003 and 0.015 in. oversize stem diameters.

The valve keepers do not grip the stem, allowing the valves to rotate freely during operation.

Valve Removal

1. Remove cylinder head, as previously described.
2. Compress valve springs, using valve spring compressor.
3. Remove valve locks or keys.
4. Release valve springs.
5. Remove valve springs, retainers, oil seals, and valves. Check valve spring squareness—5 64 in. is maximum permissible tolerance.
NOTE: if a valve does not slide out of the guide easily, check end of stem for mushrooming or heading over. If head is mushroomed, file off excess material, remove and discard valve. If valve is not mushroomed, lubricate stem, remove valve and check guide for galling. Valve seat

width (minimum and desirable) is 1/16 in. for intakes, 5/64 in. for exhausts.

Removing rocker arm assembly
(© Ford Motor Co.)

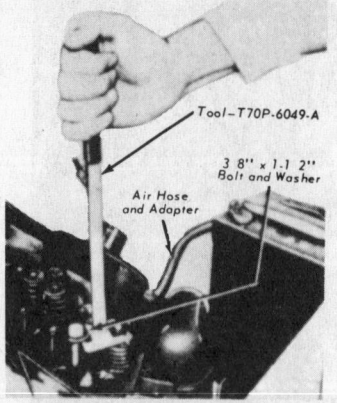

Removing valve spring with cylinder head installed
(© Ford Motor Co.)

umn are fully open. At this point, the valves in the second column can be checked and adjusted, using a 7/16 in. box wrench and feeler gauges. Clearance should be 0.008-0.010 in. for intake valves, 0.018-0.020 in. for exhaust valves when using this method.

The valves also can be adjusted with the engine running. In this case, clearance should be set to specifications found in the **Tune-up Specifications** table.

Valves Open	Valves to Adjust
1 and 6	3 and 8
2 and 4	5 and 7
3 and 8	1 and 6
5 and 7	2 and 4

Valve Clearance Adjustment

Consulting the chart, rotate the engine in the normal direction of rotation until the valves in the first col-

Timing Cover

Removal

1. Drain coolant.

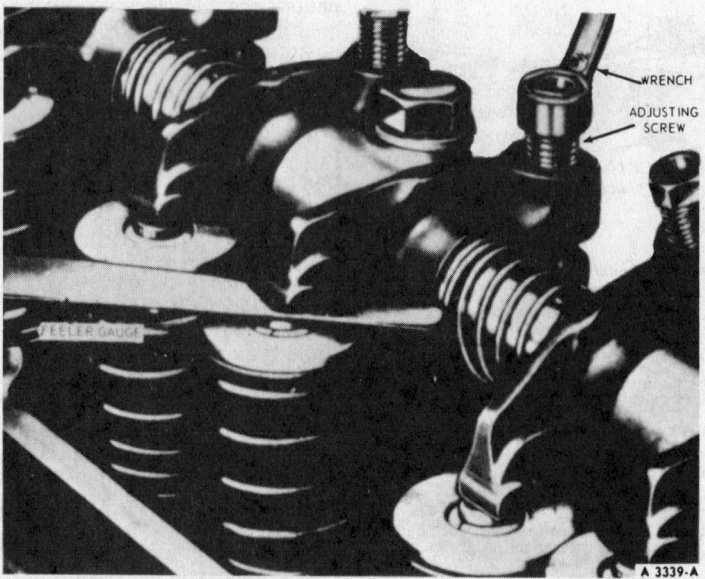

Adjusting valve clearance (© Ford Motor Co.)

CHILTON TIME-SAVER

The following is a method for replacing valve springs, oil seals or spring retainers without removing the cylinder head.

1. Entirely dismantle a spark plug and save the threaded shell.
2. To this shell, braze or weld an air chuck.
3. Remove the valve rocker cover. Remove the rocker arm from the valve to be worked on.
4. Remove the spark plug from the cylinder to be worked on.
5. Turn the crankshaft to bring the piston of this cylinder down, away from possible contact with the valve head. Sharply tap the valve retainer to loosen the valve lock.
6. Then turn the crankshaft to bring the piston in this cylinder to the Exact Top of its Compression Stroke.
7. Screw in the chuck-equipped spark plug shell.
8. Hook up an air hose to the chuck and turn on the pressure (about 200 lbs.).
9. With a strong and constant supply of air holding the valve closed, compress the valve spring and remove the lock and retainer.
10. Make the necessary replacements and reassemble.

NOTE: it is important that the operation be performed exactly as stated, in this order. The piston in the cylinder must be on exact top-center to prevent air pressure from turning the crankshaft.

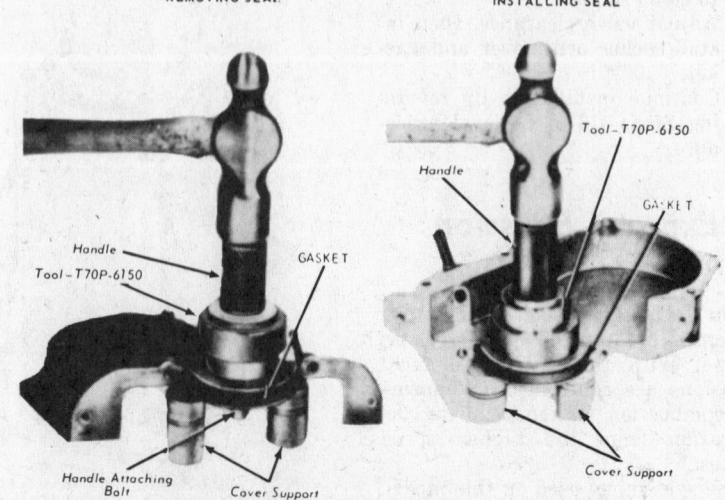

Removing and installing front cover oil seal (© Ford Motor Co.)

2. Disconnect radiator hoses at the engine, then remove radiator.
3. Remove fanbelt, fan, and water pump pulley.
4. Remove the water pump.
5. Remove the crankshaft pulley, using a puller only.
6. Remove the front cover.

NOTE: cover is secured by four oil pan bolts as well.

Installation

1. Remove the old oil seal by driving out to the rear.
2. Install new seal by driving in from the rear.
3. Position a new gasket, new cork seal, and new front portion of oil pan seal (if necessary) using Permatex, then align the cover and install the bolts. Front cover bolts are tightened to 5-7 ft. lbs., oil pan bolts to 7-9 ft. lbs.

NOTE: make sure oil seal is concentric with crankshaft.

4. Position crankshaft pulley, making sure keyway is properly aligned, then draw into place and tighten bolt to 24-28 ft. lbs.
5. Continue installation by reversing Steps 1-4 of removal procedure.

Camshaft and Timing Chain

The camshaft runs in three bearings in the block. The front and rear camshaft bearings are approximately ¾ in. wide, the front bearing having an oil hole for the rocker arm oil feed, while the center camshaft bearing is approximately ⅝ in. wide.

The camshaft is retained by an iron thrust plate bolted to the front face of the block. A single-row timing chain, with tensioner, operates the camshaft. The timing chain runs across a synthetic rubber pad on the tensioner arm. This pad wears in service so that the chain runs in two grooves. These grooves are essential

to the life of the timing chain and never should be removed.

Timing Chain R&R

1. Remove the front cover.
2. Remove the crankshaft oil slinger.
3. Remove the camshaft sprocket and disconnect the timing chain.
4. To install, position the timing chain on the sprockets so that the sprocket marks will face each other when the chain is installed.
5. Place the sprocket in position,

Timing chain tensioner (© Ford Motor Co.)

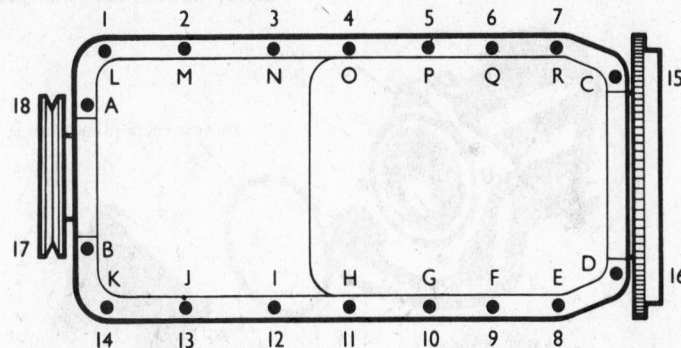
Oil pan bolt tightening sequence (© Ford Motor Co.)

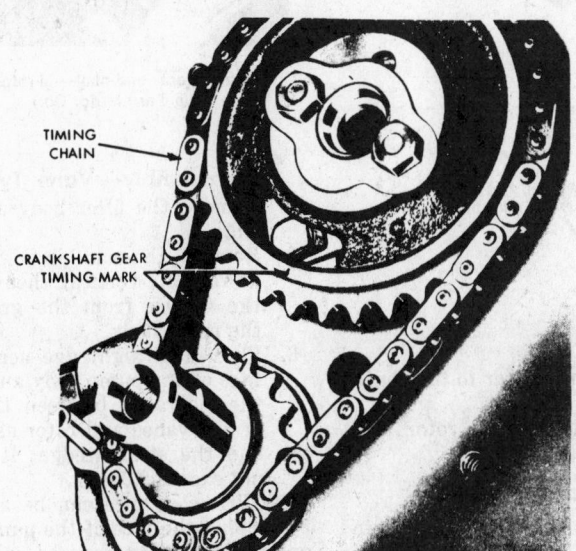
Valve timing mark alignment (© Ford Motor Co.)

with chain in place, and tighten bolts to 12-15 ft. lbs. Bend up the lock tabs.
6. Install the oil slinger and front cover.

Engine Lubrication

Oil Pan R&R

1. Drain the engine oil and disconnect the battery cables.
2. Disconnect throttle linkage at the carburetor.
3. Set the parking brake, jack up the front of the car and support on axle stands.
4. Place a jack under the transmission.
5. Remove the front motor mount bolts.
6. Remove oil pan shield.
7. Remove the starter motor.
8. Remove the oil pan and gasket, jacking up the transmission to gain clearance.
9. Remove cork packing strips, then clean block surface.
10. Remove oil pump pickup tube screen and soak it in gasoline or kerosene, then blow dry with compressed air.
11. To install, first position gaskets with Permatex, then place the cork packing strips in place with the chamfered ends in the grooves and install the oil pan and bolts. Tighten bolts to 7-9 ft. lbs., following the torque sequence illustrated (first the letters, then the numbers).
12. Continue installation by reversing Steps 1-7.

Oil Pump

The oil pump and filter assembly is bolted to the right side of the block and can be serviced with the engine installed in the car.

Two types of oil pump have been installed during production—an eccentric bi-rotor type and a sliding vane type. These pumps are readily identified by their end covers—the eccentric bi-rotor type has four recesses cast into its cover while the

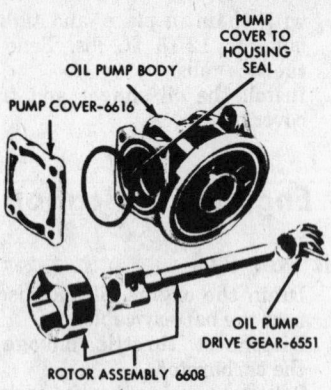

Eccentric bi-rotor oil pump
(© Ford Motor Co.)

Checking pump rotor clearance—bi-rotor type (© Ford Motor Co.)

sliding vane type has a flat cover. These two pumps are interchangeable, although their internal parts are not.

Removal and Installation

1. Lift the hood and place a drain pan under the oil pump.
2. Remove the three bolts which hold the pump and filter assembly.
   NOTE: tighten these bolts to 13-15 ft. lbs. when installing pump.
3. Remove filter from pump.
4. To install, reverse removal procedure.

4. Check clearance between outer rotor and housing; it should not exceed 0.010 in.
   NOTE: if clearance is excessive, a new rotor assembly and/or pump body must be installed.
5. Check the clearance between the rotor faces and the pump housing, using a straightedge; clearance should not exceed 0.005 in.
   NOTE: pump body face can be lapped on a glass plate, using fine valve grinding compound, to correct clearance.
6. If necessary, remove the outer rotor, drive out the pin that

shaft until the gear teeth are 2¼ in. from mounting flange. Drill a 0.125 in. hole at right angles to the shaft through the gear shoulder 1.3 in. from the mounting flange, then insert a retaining pin and peen the ends.
3. Install the outer rotor, with the chamfered face towards the pump body.
4. Install O-ring into pump body groove, then position the end plate with the machined face towards the rotors.
5. Install filter and seal ring, tightening center bolt to 12-15 ft. lbs.
   NOTE: use a new aluminum seal washer on the center bolt.

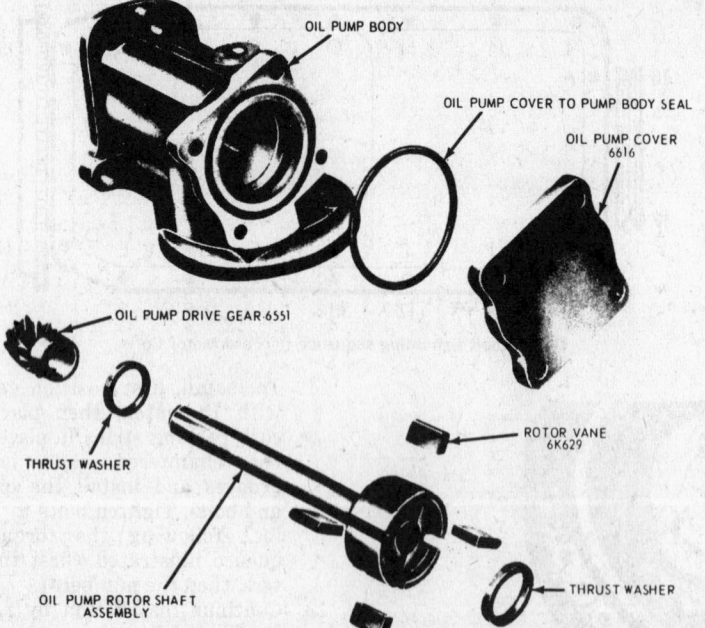

Sliding vane oil pump (© Ford Motor Co.)

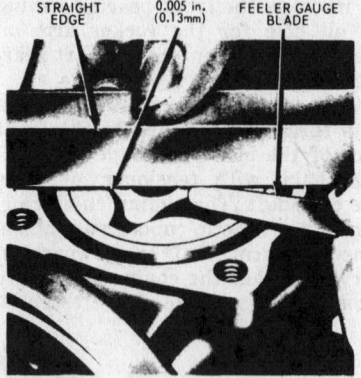

Checking rotor end-play—bi-rotor type
(© Ford Motor Co.)

Disassembly—Bi-Rotor Type

1. Remove oil filter and seal ring.
2. Remove end plate and O-ring from groove in pump body.
3. Check clearance between inner and outer rotor lobes; it should not exceed 0.006 in.
   NOTE: rotors are sold only in matched pairs.

holds the drive gear to the shaft, and pull off the gear.
7. Pull out the inner rotor and drive shaft.

Assembly—Bi-Rotor Type

1. Place inner rotor and shaft into pump body.
2. Press the drive gear onto the

Disassembly—Vane Type

1. Remove the filter body and seal ring.
2. Remove the end plate, keeping drive shaft vertical, then remove the O-ring from the groove in the pump body.
3. Place a straightedge across the face of the pump body and check the clearance between the face of the vanes and rotor assembly and the straightedge; it should not exceed 0.005 in.
   NOTE: clearance can be adjusted by lapping the face of the pump body on a glass plate covered with fine valve grinding compound.

4. Rotate the oil pump until one of the vanes is in the middle of the cam base, then check the clearance between the rotor and pump body at the closest point; clearance should not exceed 0.005 in.
NOTE: clearance in excess of this figure indicates a worn pump body, which means that the entire pump must be replaced.
5. With the rotor in the same position, center the locating ring and check the clearance between the opposite vane edge and the pump body; clearance should not exceed 0.010 in.
NOTE: if clearance is excessive, vanes must be replaced.
6. Check vane-to-locating groove clearance; clearance should be less than 0.005 in.
NOTE: if clearance is excessive, install new vanes and recheck; if clearance is still excessive, the rotor grooves are worn and a new rotor and shaft assembly must be installed.
7. If necessary, remove the vanes and outer locating ring, drive

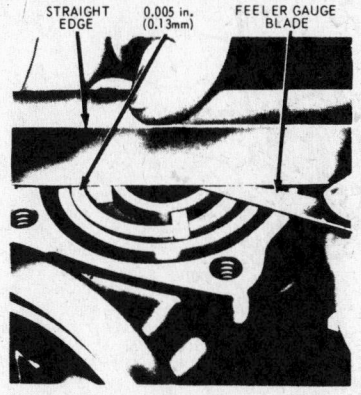

Checking pump rotor end-play—vane type
(© Ford Motor Co.)

out the gear retaining pin, and pull off the gear. Remove drive shaft and rotor arm assembly, and the inner ring, from the pump housing.

Assembly—Vane Type

1. Place the vane locating inner ring in the pump housing, then install drive shaft and rotor assembly.
2. Press the drive gear onto the shaft until the far end of the gear teeth are 2¼ in. from the mounting flange. Drill a 0.125 in. hole, at right angles to the shaft, through the gear shoulder 1.3 in. from the mounting flange. Insert a retaining pin and peen the ends.
3. Install the vane locating outer ring, then locate the vanes in the rotor grooves with the curved edges outwards.

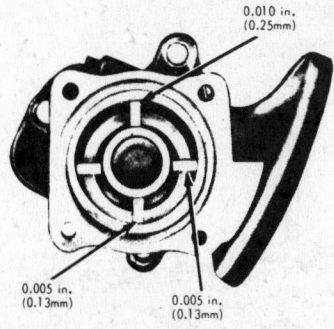

Checking vane and rotor clearances—vane type
(© Ford Motor Co.)

4. Install an O-ring in the groove of the pump body, then install the end plate with the machined face towards the rotor.
5. Install new filter and seal, install a new aluminum washer on the center bolt, and tighten the center bolt to 12-15 ft. lbs.

Pistons, Connecting Rods, and Main Bearings

Piston Rings

Engine disassembly is entirely conventional. Before replacing rings, inspect the cylinder bores. The cylinder bores are machined directly into the block. Cylinder liners (cast-iron, dry-type) are available, however, in 0.020 in. oversize (O.D.). When liners are installed, connecting rod clearance slots must be cut in their bases and the liners honed to proper size.

1. Using internal micrometer measure bores both across thrust faces of cylinder and parallel to axis of crankshaft at minimum of four locations equally spaced. The bore must not be out of round by more than 0.005 in. and it must not "taper" more than 0.010 in. "Taper" is the difference in wear between two bore measurements in any cylinder. Bore any cylinder beyond limits of out of roundness or taper to diameter of next available oversize piston that will clean up wear. Recommended clearance for new pistons (in bore) is 0.0013-0.0019 in. Clearance between block face and piston top at TDC should be 0.085-0.103 in.
2. If bore is within limits dimensionally, examine bore visually. It should be dull silver in color and exhibit a pattern of machining cross hatching intersecting at about 30 degrees. There should be no scratches, tool marks, nicks, or other damage.

If any such damage exists, bore cylinder to clean up damage and then to next oversize. Pistons are available in 0.0025, 0.015, and 0.030 in. oversizes. Identification of pistons is found in the chart.

Piston Marking Code

Piston Size	Std.	0.0025 o/s	0.015 o/s	0.030 o/s
Code	THA TRA	TJA	TKA	TLA

Polished or shiny places in the bore are known as glazing. Glazing causes poor lubrication, high oil consumption and ring damage. Remove glazing by honing cylinders with clean, sharp stones of No. 180-220 grit to obtain a surface finish of 15-35 RMS. Use a hone also to obtain correct piston clearance and surface finish in any cylinder that has been bored.

⏱ CHILTON TIME-SAVER

This or any other machining operation should be done with the cylinder block completely disassembled. Hot tank cylinder block after honing or boring. To remove minor glazing when honing equipment is not available, run emery cloth back and forth across glazed area perpendicular to axis of bore. Scrub block and bores thoroughly with soap and water to remove all grit after using emery cloth.
NOTE: the emery cloth method should be used only as a last resort as it is a method much inferior to honing.

3. If cylinder bore is in satisfactory condition, place each ring in bore in turn and square it in bore with head of piston. Measure ring gap; it should be 0.009-0.014 in. If ring gap is greater than limit, get new ring. If ring gap is less than limit, file end of ring to obtain correct gap.
4. Check ring side clearance by installing rings on piston, and inserting feeler gauge of correct dimension between ring and lower land. Clearance for compression rings is 0.0016-0.0036 in., for oil rings 0.0018-0.0038 in. Gauge should slide freely around ring circumference without binding. Any wear will form a step on lower land. Replace any pistons having high steps. Before checking ring side clearance be sure ring grooves are clean and free of carbon, sludge, or grit.

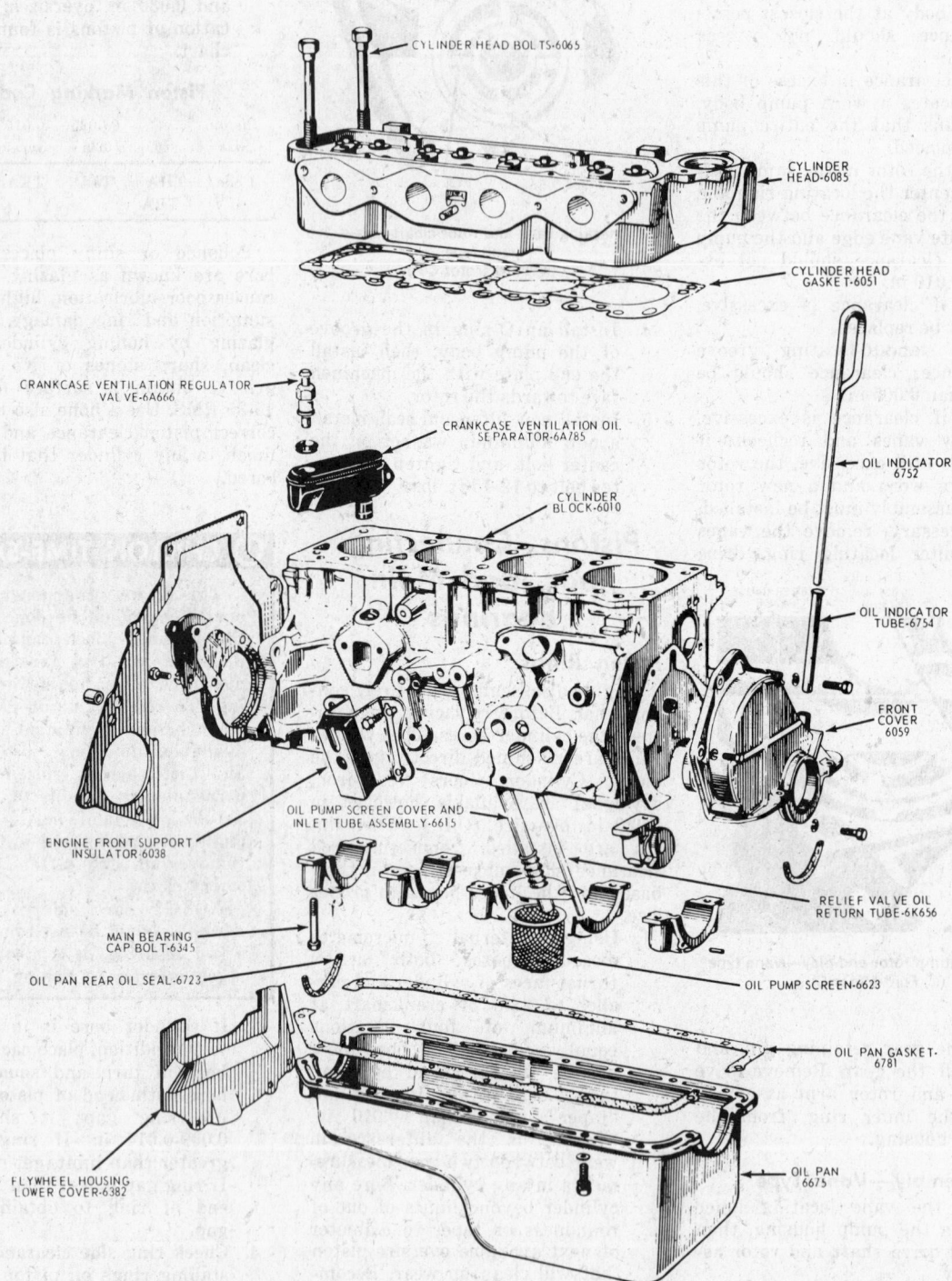

CYLINDER HEAD BOLTS-6065

CYLINDER HEAD-6085

CYLINDER HEAD GASKET-6051

CRANKCASE VENTILATION REGULATOR VALVE-6A666

CRANKCASE VENTILATION OIL SEPARATOR-6A785

OIL INDICATOR 6752

CYLINDER BLOCK-6010

OIL INDICATOR TUBE-6754

FRONT COVER 6059

ENGINE FRONT SUPPORT INSULATOR-6038

OIL PUMP SCREEN COVER AND INLET TUBE ASSEMBLY-6615

RELIEF VALVE OIL RETURN TUBE-6K656

MAIN BEARING CAP BOLT-6345

OIL PAN REAR OIL SEAL-6723

OIL PUMP SCREEN-6623

OIL PAN GASKET-6781

FLYWHEEL HOUSING LOWER COVER-6382

OIL PAN 6675

Engine block and cylinder head, showing related parts (© Ford Motor Co.)

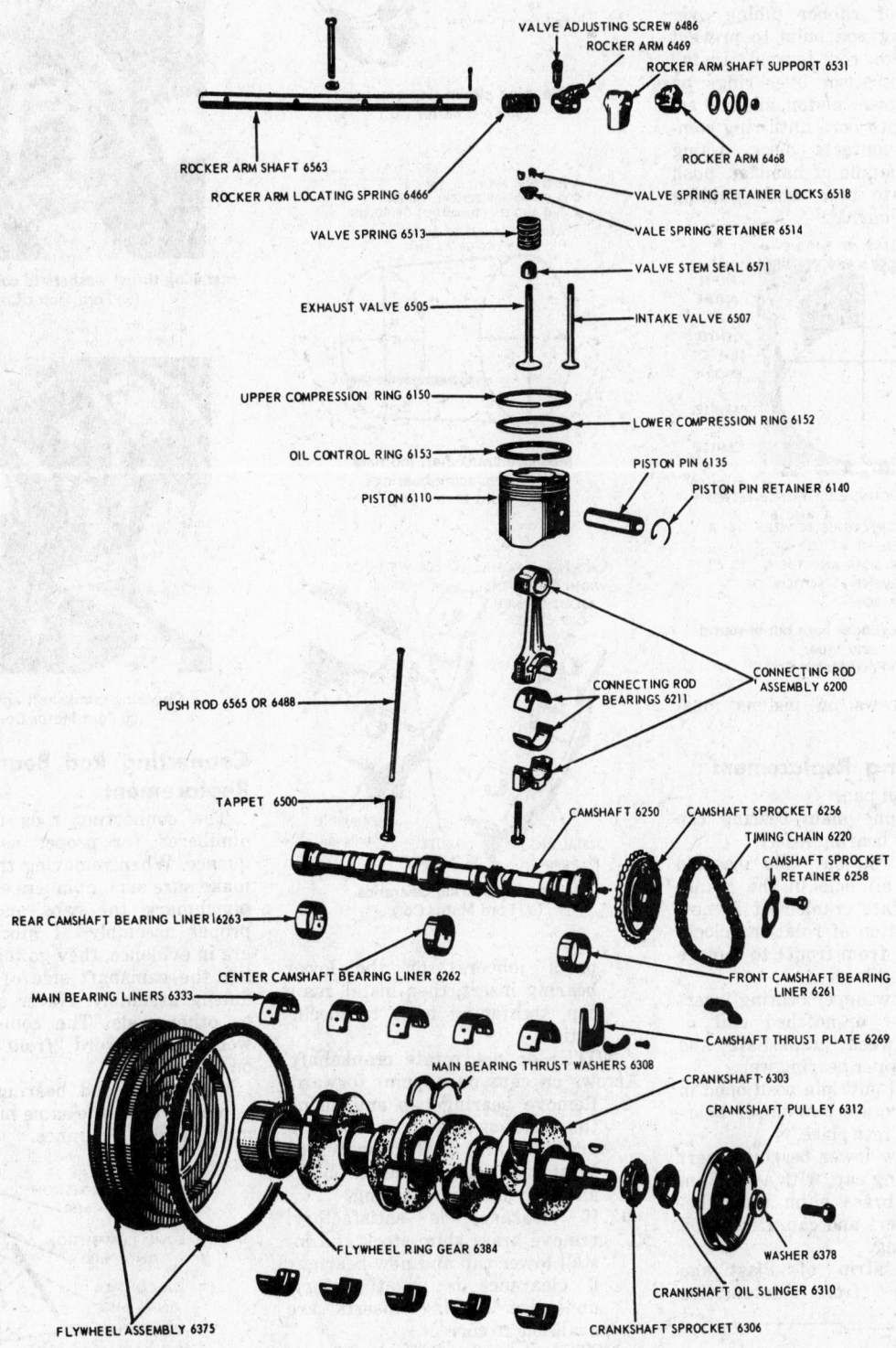

VALVE ADJUSTING SCREW 6486
ROCKER ARM 6469
ROCKER ARM SHAFT SUPPORT 6531
ROCKER ARM 6468
ROCKER ARM SHAFT 6563
ROCKER ARM LOCATING SPRING 6466
VALVE SPRING RETAINER LOCKS 6518
VALE SPRING RETAINER 6514
VALVE SPRING 6513
VALVE STEM SEAL 6571
EXHAUST VALVE 6505
INTAKE VALVE 6507
UPPER COMPRESSION RING 6150
LOWER COMPRESSION RING 6152
OIL CONTROL RING 6153
PISTON PIN 6135
PISTON 6110
PISTON PIN RETAINER 6140
CONNECTING ROD ASSEMBLY 6200
CONNECTING ROD BEARINGS 6211
PUSH ROD 6565 OR 6488
CAMSHAFT 6250
CAMSHAFT SPROCKET 6256
TIMING CHAIN 6220
CAMSHAFT SPROCKET RETAINER 6258
TAPPET 6500
REAR CAMSHAFT BEARING LINER 6263
CENTER CAMSHAFT BEARING LINER 6262
FRONT CAMSHAFT BEARING LINER 6261
MAIN BEARING LINERS 6333
CAMSHAFT THRUST PLATE 6269
MAIN BEARING THRUST WASHERS 6308
CRANKSHAFT 6303
CRANKSHAFT PULLEY 6312
FLYWHEEL RING GEAR 6384
WASHER 6378
CRANKSHAFT OIL SLINGER 6310
FLYWHEEL ASSEMBLY 6375
CRANKSHAFT SPROCKET 6306

Engine internal components (© Ford Motor Co.)

5. Space ring gaps at equidistant intervals around piston circumference. Be sure to install piston in its original bore. Install short lengths of rubber tubing over connecting rod bolts to prevent damage to rod journal. Install ring compressor over rings on piston. Lower piston, and rod assembly into bore until ring compressor contacts block. Using wooden handle of hammer, push piston into bore while guiding rod onto journal.

←— CENTER LINE OF ENGINE —→

A AT RIGHT ANGLE TO CENTER LINE OF ENGINE
B PARALLEL TO CENTER LINE OF ENGINE

1. OUT-OF-ROUND = DIFFERENCE BETWEEN **A** AND **B**
2. TAPER = DIFFERENCE BETWEEN THE **A** MEASUREMENT AT TOP OF CYLINDER BORE AND THE **A** MEASUREMENT AT BOTTOM OF CYLINDER BORE

Measuring cylinder bore out-of-round and taper
(© Ford Motor Co.)

NOTE: arrows on pistons must point forward.

Main Bearing Replacement

1. Remove oil pan.
2. Remove one main bearing cap and lower bearing insert.
3. With a "roll out" pin inserted into the oil hole in the crankshaft, rotate crankshaft in normal direction of rotation (clockwise seen from front) to remove top bearing insert.
4. Oil the new upper bearing insert and place un-notched end of shell between crankshaft and notched upper bearing web.
5. With "roll out" pin positioned in oil hole, rotate new upper bearing insert into place.
6. Install new lower bearing insert into bearing cap, with a 0.002 in. strip of brass shim stock between insert and cap. Do not oil this bearing.
7. Place a strip of Plastigage (available from automotive

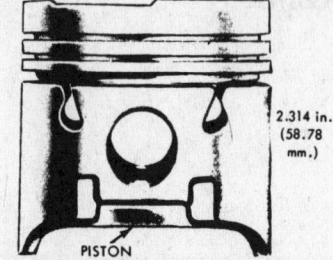

2.314 in. (58.78 mm.)

PISTON

Piston gauging point (© Ford Motor Co.)

Cleaning piston ring grooves
(© Ford Motor Co.)

Ring Groove Cleaner

A VS B ══ VERTICAL TAPER
C VS D ══ HORIZONTAL TAPER
A VS C AND B VS D ══ OUT-OF-ROUND
CHECK FOR OUT-OF-ROUND AT EACH END OF JOURNAL

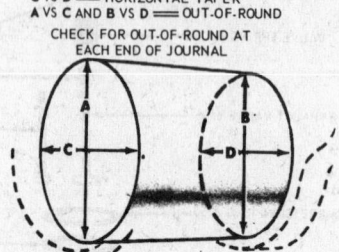

Measure crankshaft journals before replacing bearings
(© Ford Motor Co.)

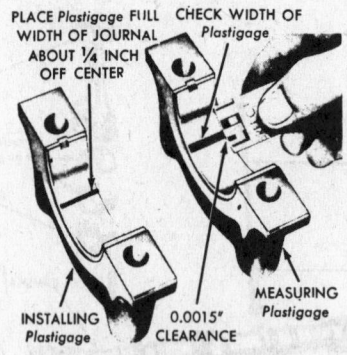

PLACE *Plastigage* FULL WIDTH OF JOURNAL ABOUT ¼ INCH OFF CENTER

CHECK WIDTH OF *Plastigage*

INSTALLING *Plastigage*

0.0015" CLEARANCE

MEASURING *Plastigage*

Measuring bearing clearance
(© Ford Motor Co.)

parts jobbers) on the lower bearing insert, then install main cap, tightening bolts to specification.

NOTE: do not rotate crankshaft. Arrows on caps must point forward.

8. Remove bearing cap and, using the scale on the envelope, measure the width of the compressed Plastigage; this is the oil clearance for this main bearing.
9. If clearance is satisfactory, remove brass shim stock and install lower cap and new bearing; if clearance is unsatisfactory, undersize bearing inserts are available to correct.

NOTE: never file main caps to adjust clearance.

10. Replace the remaining main bearings in the engine by following Steps 2-9, then install oil pan.

NOTE: don't forget to check crankshaft end-play—it may be necessary to replace thrust washers on center main bearing to correct for this.

CRANKSHAFT THRUST WASHERS

A3401-A

Installing thrust washers to correct end-play
(© Ford Motor Co.)

0.003-0.011 in (0.08-0.28mm) Feeler Gauge

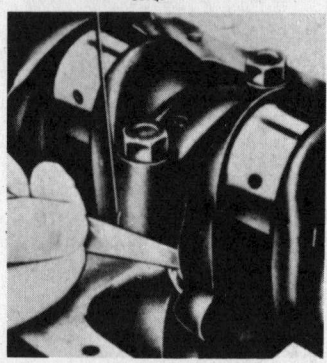

Checking crankshaft end-play
(© Ford Motor Co.)

Connecting Rod Bearing Replacement

The connecting rods usually are numbered for proper assembly sequence. When removing the rod caps, make sure such numbers exist. If not, punchmark the caps and rods for proper assembly. If stock numbers are in evidence, they go together, facing the camshaft side of the block, during assembly. Never switch caps to other rods. The connecting rod web has the word "front" embossed on it.

Connecting rod bearing clearance is measured in the same manner as is main bearing clearance.

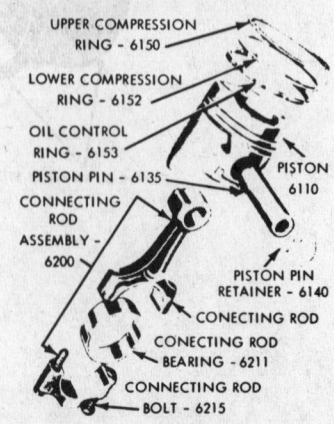

UPPER COMPRESSION RING - 6150
LOWER COMPRESSION RING - 6152
OIL CONTROL RING - 6153
PISTON PIN - 6135
CONNECTING ROD ASSEMBLY - 6200
PISTON - 6110
PISTON PIN RETAINER - 6140
CONECTING ROD
CONECTING ROD BEARING - 6211
CONNECTING ROD BOLT - 6215

Connecting rod and piston assembly
(© Ford Motor Co.)

Front Suspension

The independent front suspension, of the McPherson strut type, utilizes variable rate coil springs and hydraulic shock absorbers. Lateral front wheel movement is controlled by a track control arm and front to rear movement by a stabilizer bar.

Downward movement is limited by rebound stops inside the integral suspension unit, while upward movement is restricted by the compression limit of the coil springs. In practice, a rubber bumper is hit before the coil spring is completely compressed.

The front suspension geometry, except toe-in, is not adjustable. Any tolerance deviations require disassembly and repair of the affected part/s.

Front Suspension Removal

1. Lift the hood, then jack up the car and support on chassis stands.
2. Remove the front wheel.
3. Disconnect the brake line from the suspension unit.
4. Place a jack under the control arm and jack up the suspension.
5. Remove the cotter pin and the castle nut that holds the tierod end to the steering arm, then separate the joint and remove the jack.
6. Remove the cotter pin and the castle nut that holds the tie-rod arm to the suspension, then disconnect the control arm.
7. Remove the three bolts that hold the suspension to the fender well (from under the hood), then remove the assembly, with brake caliper, from the car.

Front Suspension Disassembly

1. Install spring compressors on the coil spring.
2. Compress the coil spring slightly, then unscrew the piston rod nut and remove the retainer.
3. Remove the top mount and the upper spring seat, coil spring, and rubber bumper.
4. Using a large pipe wrench, remove the bumper platform.
5. Remove the O-ring from upper guide assembly.
6. Check the top edge of the piston rod (machined area); it must be free of burrs and nicks.
NOTE: remove any burrs with a fine oil stone. If this is not done, the bushing will be damaged as it is removed.
7. Lift the piston rod until the gland and bushing assembly is free of the outer casing, then slide the gland assembly from the rod.
8. Empty the fluid into a container. Pull the piston rod—complete with piston, cylinder, and compression valve—from the outer casing.

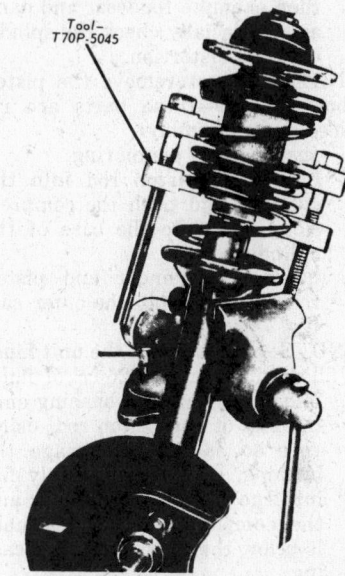

Tool—T70P-5045

Compressing coil spring—unit out of car (© Ford Motor Co.)

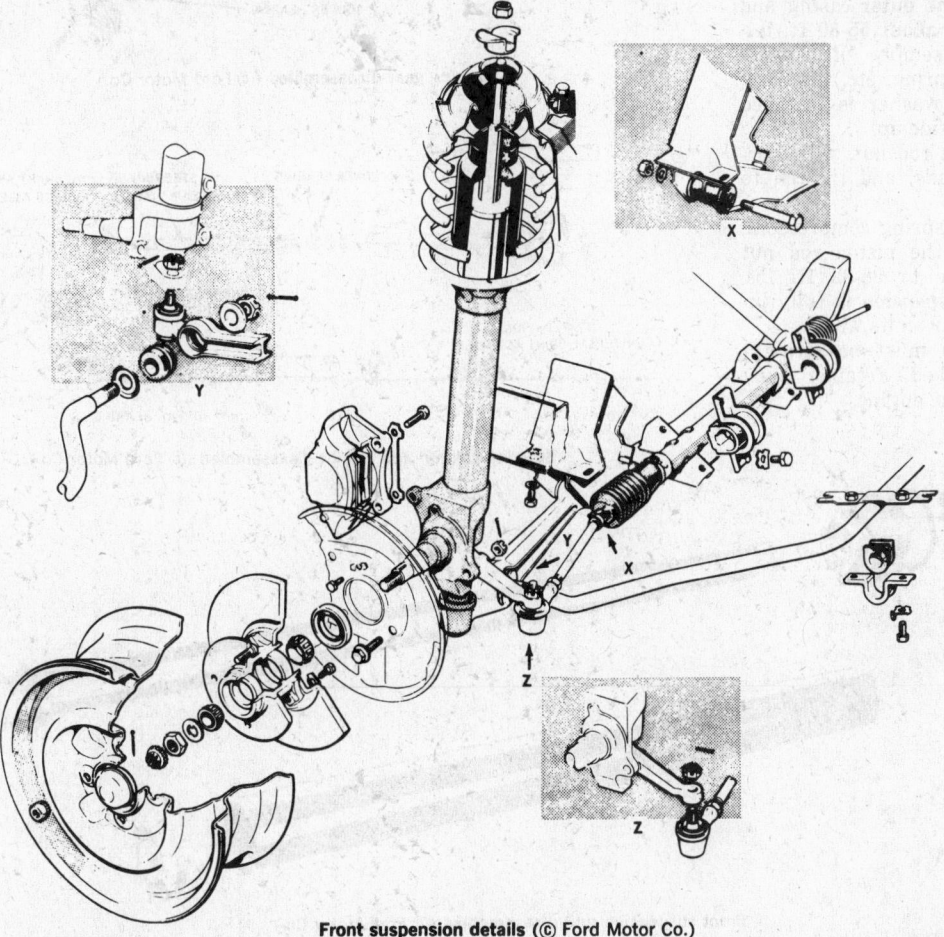

Front suspension details (© Ford Motor Co.)

9. Remove the piston rod from the cylinder by pushing the compression valve out of the base while pushing the rod downwards.
10. Remove the ring from the piston.

Front Suspension Assembly

1. Wash all parts with solvent, then examine for wear and damage. Especially check the spindle body for distortion.
   NOTE: do not remove the piston from the rod—these parts are replaced as an assembly.
2. Install a new piston ring.
3. Insert the piston rod into the cylinder and push the compression valve into the base of the cylinder.
4. Insert the cylinder and piston rod assembly into the outer casing.
   NOTE: at this point, the unit must be filled with exactly 326 cc. of fluid.
5. Slide the gland and bushing onto the end of the piston rod, using care so as not to damage the bushing, until the assembly fits into the end of the cylinder and the complete internal assembly is below the top of the outer casing.
6. Locate a rubber O-ring on top of the gland and bushing assembly.
7. Screw the bumper platform onto the top of the outer casing and tighten it to about 55-60 ft. lbs.
8. Continue assembly of components (coil spring, etc.), making sure dished washer is installed with convex side up.
9. Install piston rod nut, with Loctite on threads, and tighten to 5-10 ft. lbs.
10. Remove coil spring compressors, then loosen the piston rod nut and retorque to 28-32 ft. lbs. when the suspension is fully installed and car on its wheels.
    NOTE: wheels must be straight ahead and "cranked" retainer must face in towards the engine.

Front Suspension Installation

1. Lift the suspension into position and secure using the three bolts through the wheel well, tightening bolts to 15-18 ft. lbs.
2. Reconnect the control arm ball stud to the suspension unit, tighten nut to 30-35 ft. lbs., and install a new cotter pin.
3. Reconnect tie-rod end to steering arm and tighten castle nut to 18-22 ft. lbs., then install a new cotter pin.
4. Reconnect brake line to bracket, then bleed brakes.
5. Replace wheel and lower the car.

Steering Gear

The rack and pinion steering gear is rubber-mounted to the front crossmember. Movement of the steering wheel is transmitted through a universal joint and flexible coupling to the pinion, which in turn moves the rack laterally to turn the wheels. Steering is approximately 3½ turns lock to lock, with an overall steering ratio of 17.7:1.

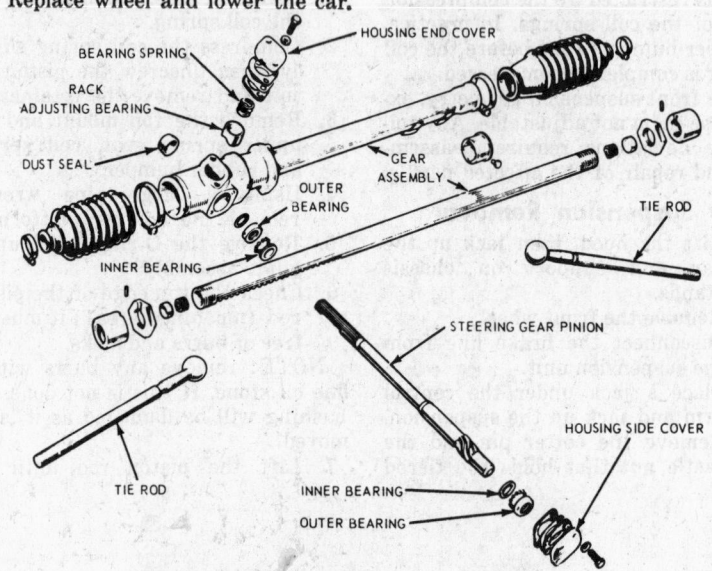

Steering gear disassembled (© Ford Motor Co.)

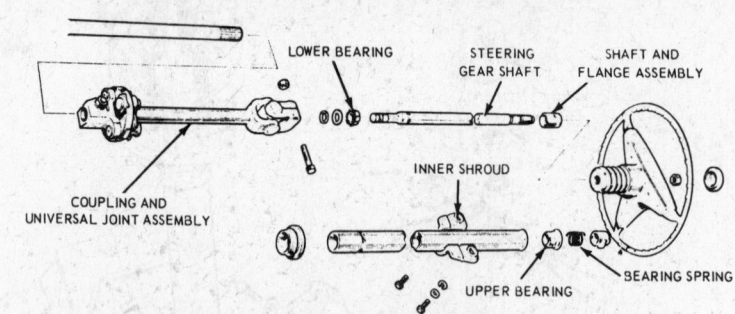

Steering column and U-joint disassembled (© Ford Motor Co.)

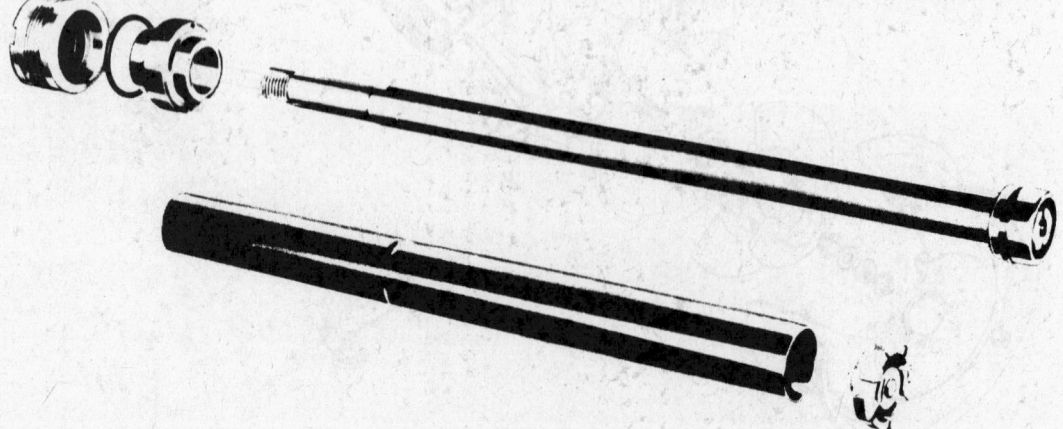

Front suspension unit disassembled (© Ford Motor Co.)

There are two possible adjustments to the steering gear—rack damper adjustment and pinion bearing preload adjustment. Both of these adjustments are made by varying the thickness of the shim pack under the steering box cover plate. It is necessary to have the gear removed from the car for these two adjustments.

The tie-rods are adjustable to provide a toe-in adjustment range; caster and camber angles are built-in and are not adjustable.

One caution that should be followed: when the car is jacked up with the front wheels off the ground, never rapidly turn wheels from lock to lock. If this is done, pressure within the steering may blow off the bellows.

Steering Gear R&R

1. Place wheels in straight-ahead position.
2. Jack up the front of the car and support on axle stands.
3. Remove the nut and bolt that secure the flexible coupling to the pinion splines.
4. Bend back the lock tabs and remove the bolts that hold the gear box to the crossmember.
5. Remove the cotter pins and castle nuts that hold tie-rod ends to steering arms, then separate the ball studs from the tapered fittings.
6. Remove the steering gear from the car.
   NOTE: it is necessary to turn one wheel against the stop to give enough clearance between the steering gear and the sway bar.
7. Remove the tie-rod ends and locknuts, counting the number of turns required to remove each one.

8. To install, reverse removal procedure tightening gear box-to-crossmember bolts to 15-18 ft. lbs., tie-rod ends to 18-22 ft. lbs., and flexible coupling to 12-15 ft. lbs.

Steering Wheel R&R

1. Align front wheels in straight-ahead position.
2. Pry out the steering wheel emblem to expose the nut.
3. Remove the steering shaft nut and pull the wheel off the shaft, using a universal steering wheel puller.
4. To install, reverse removal procedure, tightening nut to 20-25 ft. lbs.

Clutch

The clutch used is a 7.5 in. diameter unit having a diaphragm-spring pressure plate. The clutch is cable-operated and operates in a conventional manner.

Adjustment

1. Remove release lever boot and lubricate the ball end of the cable with MS_2 grease. Replace boot.
2. Loosen locknut and, with the clutch pedal pulled back to its stop on the pedal bracket, adjust the nut to give 0.138-0.144 in. clearance between the nut and the shoulder of the fitting.
   NOTE: this should result in $1/2$-$3/4$ in. free-play at clutch pedal.
3. Tighten the locknut, then grease the pedal end of the cable.

Clutch, Clutch Housing, and Transmission R&R

1. Lift the hood and disconnect the battery cables.
2. Disconnect throttle linkage at the carburetor.
3. Loosen gearshift knob locknut, then remove knob and locknut.
4. Remove the seven screws that hold the console to the floor, then remove the console.
5. Bend up the lock tab and unscrew the plastic nut, then remove gearshift lever.
6. Jack up the car at all corners and support on axle stands.
7. Matchmark the rear U-joint where it connects to the pinion flange, then remove the four bolts and separate the U-joint.
8. Remove the two bolts that hold the center driveshaft bearing carrier, then slide the driveshaft to the rear to pull the front splined yoke from the transmission extension housing.
   NOTE: plug the opening in the transmission with rags to prevent oil loss.

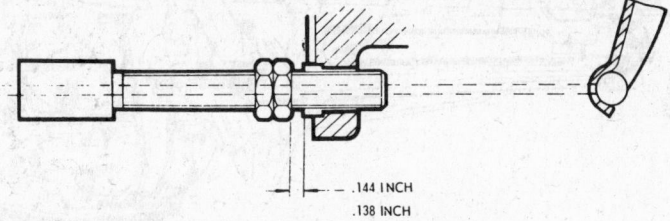

.144 INCH
.138 INCH

Clutch adjustment (© Ford Motor Co.)

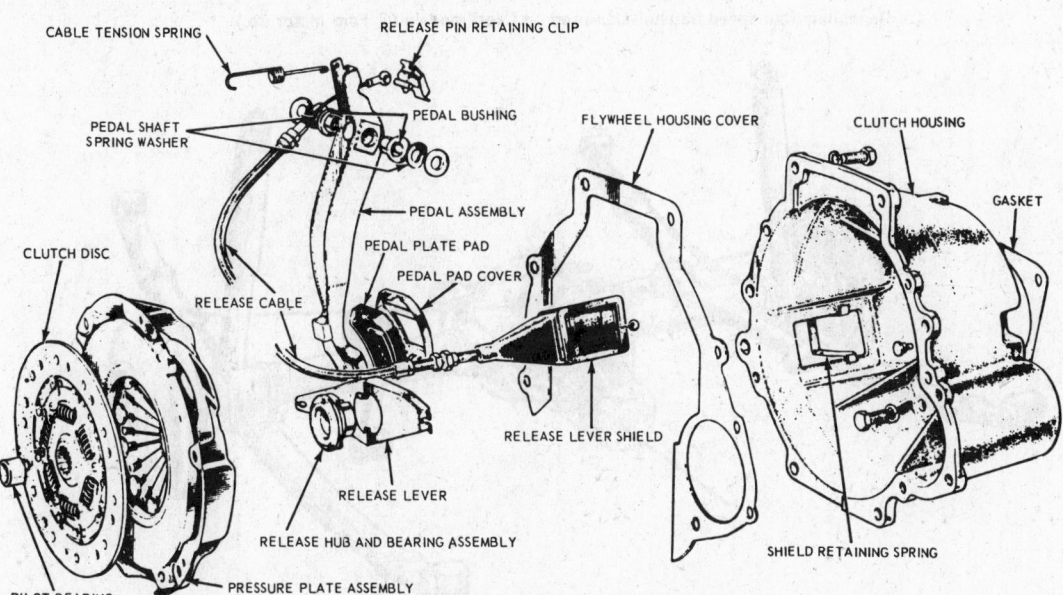

CABLE TENSION SPRING
RELEASE PIN RETAINING CLIP
PEDAL SHAFT SPRING WASHER
PEDAL BUSHING
FLYWHEEL HOUSING COVER
CLUTCH HOUSING
PEDAL ASSEMBLY
GASKET
CLUTCH DISC
PEDAL PLATE PAD
PEDAL PAD COVER
RELEASE CABLE
RELEASE LEVER SHIELD
RELEASE HUB AND BEARING ASSEMBLY
RELEASE LEVER
SHIELD RETAINING SPRING
PILOT BEARING
PRESSURE PLATE ASSEMBLY

Clutch components (© Ford Motor Co.)

9. Remove the snap-ring that holds the speedometer cable, then pull the cable from the transmission and wire it out of the way.
10. Disconnect the exhaust pipe from the manifold.
11. Disconnect the clutch cable from the release lever, then remove the starter motor bolts and swing the starter motor away from the working area (still connected to its wires).
12. Remove the bolts that hold the clutch housing to the engine block.

NOTE: the top bolt holds a ground strap that must be replaced.

13. Remove the clutch inspection cover, then place a screw jack, insulated with a block of wood, under the rear of the engine.
14. Remove the four bolts that hold the transmission crossmembers to the body, then slide the transmission to the rear (while supporting its weight) and detach it from the engine.
15. Loosen the six pressure plate retaining bolts, evenly, then remove the pressure plate and clutch disc.

16. If pilot bearing is to be removed, use a slide hammer. New bearing must be carefully tapped into place so that the bearing is seated 0.156-0.175 in. below crankshaft flange level.
17. Place new clutch disc in position on flywheel, after coating splines with MS_2 grease, and center it using a dummy pilot shaft or old transmission mainshaft. Place pressure plate on flywheel and install bolts, tightening them evenly all around to 12-15 ft. lbs.
18. Remove pilot shaft, then install clutch housing by reversing Steps 1-14.

NOTE: the two lower clutch housing bolts are longer than the others. Tighten all clutch housing bolts to 40-45 ft. lbs.

Standard Transmission

The Dagenham four-speed transmission is identical to that used on the 1970-71 Cortina; all forward speeds are synchronized. See the Unit Repair Section for complete service procedures.

Transmission R&R

See *Clutch, Clutch Housing, and Transmission R&R* in this car section.

Rear Suspension

The rear suspension consists of conventional three-leaf, longitudinal, semi-elliptic springs, located asymmetrically to the rear axle carrier and fastened by U-bolt clamps. Two radius arms, one on each side, prevent excessive lateral axle movement, while standard telescopic shock absorbers control vertical rebound. The shock absorbers are staggered (front of axle on right-hand side, rear of axle on left-hand side) to provide additional control of axle tramp and wheel hop during acceleration and deceleration.

The drive axle is the Salisbury type—see the Unit Repair Section.

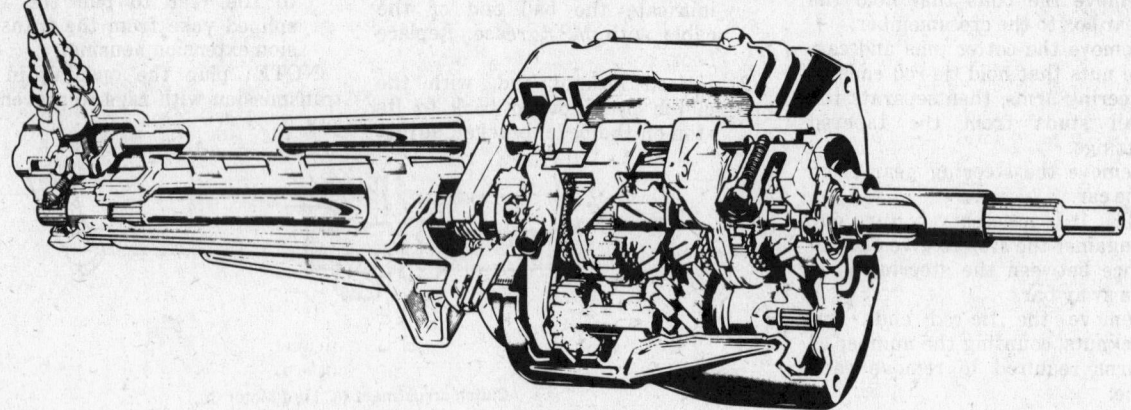

Dagenham four-speed transmission used on Capri models (© Ford Motor Co.)

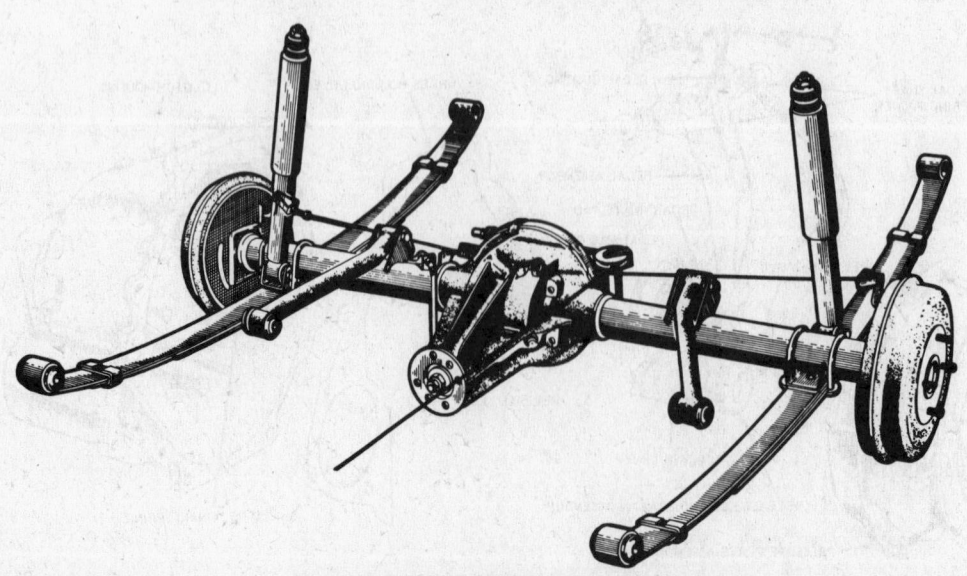

Rear suspension assembled (© Ford Motor Co.)

Rear Axle Housing R&R

1. Jack up the rear of the car, support on chassis stands, and remove the wheels.
2. Release the parking brake and disconnect the cable at the lever on the rear of the axle housing.
3. Disconnect the brake line at the junction block on the axle.
4. Matchmark the rear U-joint and pinion flange, then remove the four bolts and washers.
5. Jack up the center section slightly, then disconnect the lower shock absorber bolts.
6. Disconnect radius arms at rear axle.
7. Remove U-bolt nuts, U-bolts, and spring plates, then remove axle housing through wheel opening.
8. To install, reverse removal procedure.

Windshield Wipers

Motor R&R

1. Disconnect battery ground cable.
2. Remove wiper arms and the two nuts that hold the wiper spindles.
3. Remove front parcel tray.
4. Disconnect defroster vent hose, then disconnect the two control cables.
5. Remove the two screws that hold the motor to its bracket, then disconnect wires to motor.
6. Remove wiper motor assembly.
7. To install, reverse removal procedure.

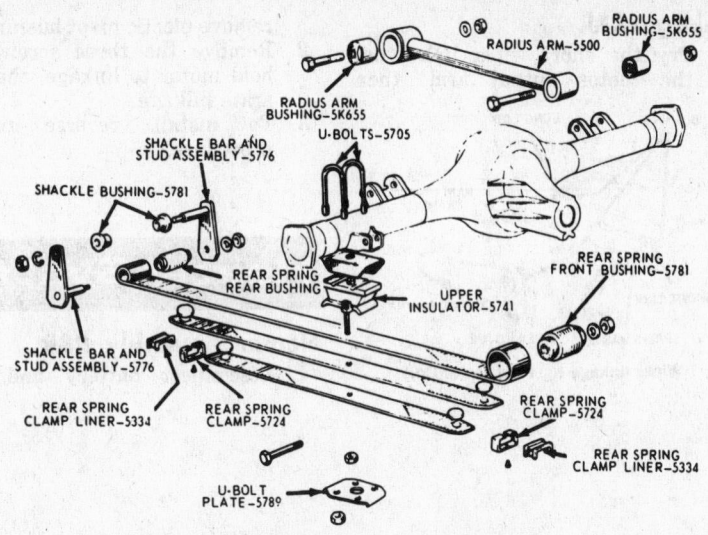

Rear suspension disassembled (© Ford Motor Co.)

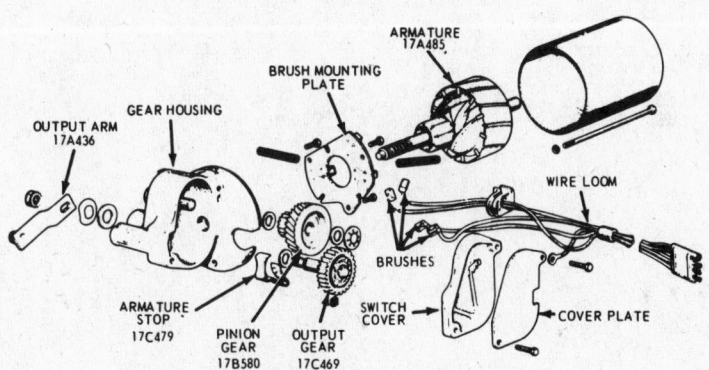

Wiper motor disassembled (© Ford Motor Co.)

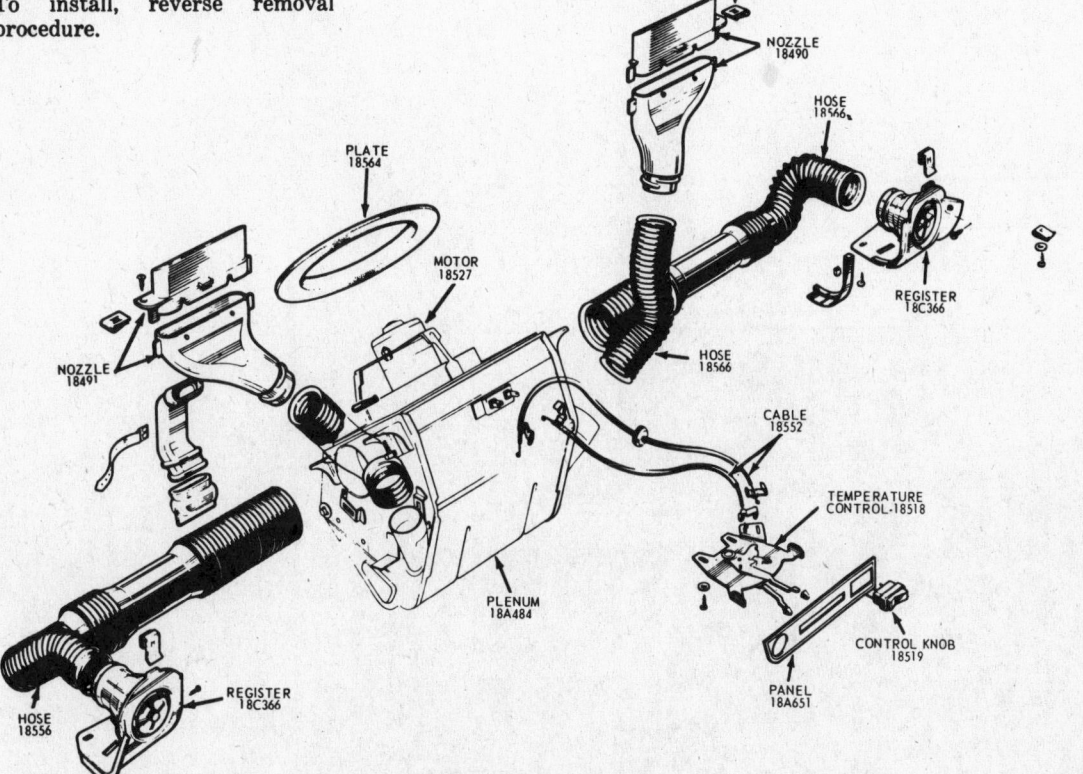

Heater system (© Ford Motor Co.)

Linkage R&R

1. Pry the short wiper link from the motor output arm, then

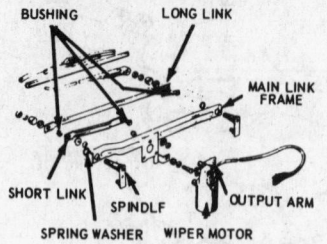

BUSHING   LONG LINK

MAIN LINK FRAME

SHORT LINK   SPINDLF   OUTPUT ARM

SPRING WASHER   WIPER MOTOR

Wiper linkage (© Ford Motor Co.)

remove plastic pivot bushing.
2. Remove the three screws that hold motor to linkage, then separate linkage.
3. To install, reverse removal procedure.

Heater System

Heater Assembly R&R

1. Disconnect battery and drain cooling system.
2. Remove parcel tray, then, from inside engine compartment, disconnect heater hoses from panel.
3. Remove two screws, then heater hose plate and gasket, from panel.
4. Remove heater control cables, then disconnect heater motor wires.
5. Remove ducts from the heater, then remove four bolts and heater.
6. To install, reverse removal procedure.

Index

YEAR IDENTIFICATION

OLDS

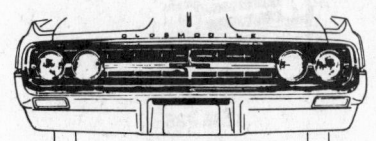

1964 "98"

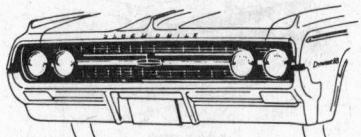

1964 Jetstar 1

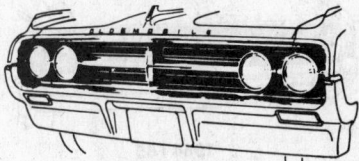

1964 Starfire "88"

1965 Starfire "88"

1965 "98"

1965 Jetstar

1966 Dynamic "88"

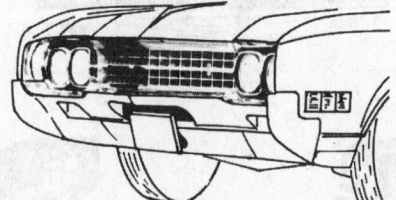

1966 "98"

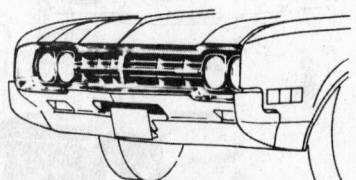

1966 Starfire "88"

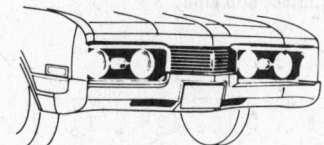

1967 Delta "88"

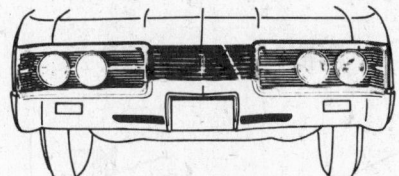

1967 Delmont

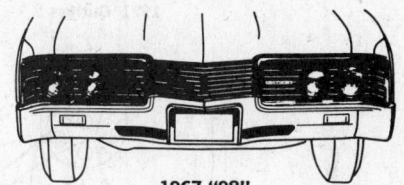

1967 "98"

1968 "88"-"98"

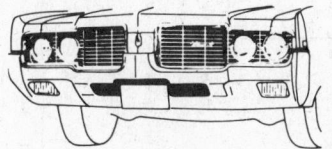

1969 "88"

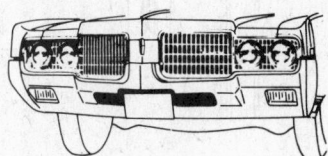

1969 "98"

1970 98 Series

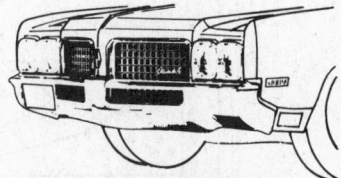

1970 Delta 88

1971 Delta 88

1971 98

YEAR IDENTIFICATION

F-85

1964 F85

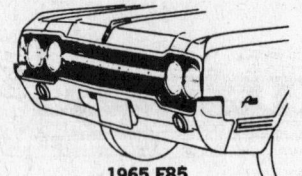

1965 F85

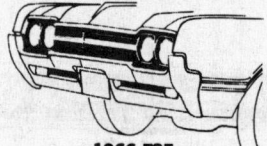

1966 F85

1967 F85

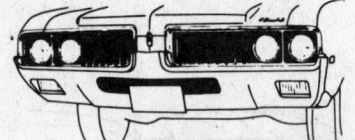

1967 F85

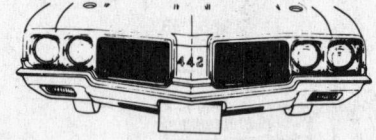

1968 F85

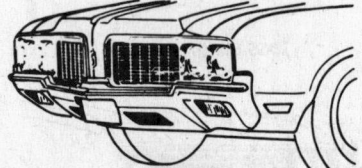

1968 "442"

1969 F85

1970 4-4-2

1970 F85

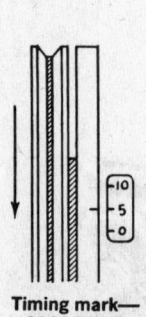

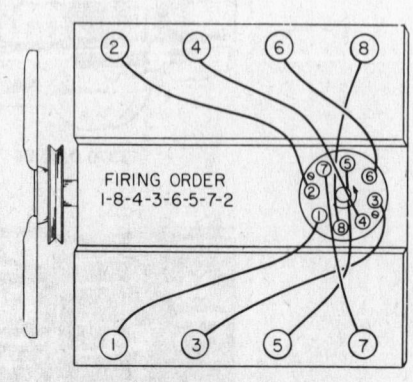

1971 Cutlass S

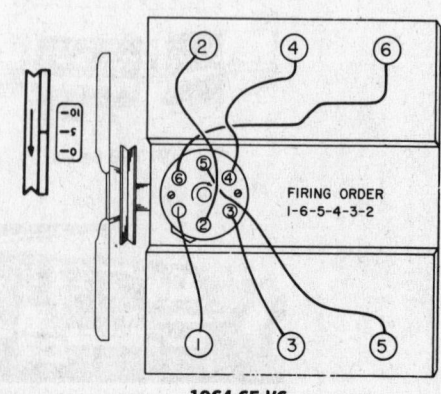

1971 Cutlass Supreme

FIRING ORDER

FIRING ORDER 1-5-3-6-2-4

1966-70 OHV 6 cyl.

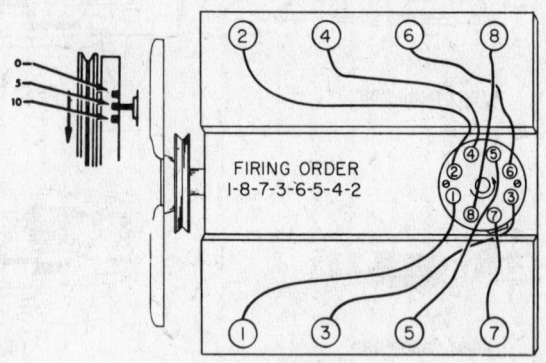

FIRING ORDER 1-8-7-3-6-5-4-2

1964 394 V8

Timing mark—
1964-70 V8

FIRING ORDER 1-8-4-3-6-5-7-2

1965-70 330, 350, 400, 425, 455 V8

FIRING ORDER 1-6-5-4-3-2

1964-65 V6

ENGINE IDENTIFICATION

YEAR	ENGINE	PRODUCTION CODE	CODE LOCATION
1964	394 V8 2-BBL.	H	A
	330 V8	T	B
1965	425 V8 2-BBL.	M	B
	425 V8 4-BBL.	N	B
	330 V8	U	B
1966	425 V8 2-BBL.	M	B
	425 V8 4-BBL.	N	B
	330 V8	V, X	B
1967	425 V8 2-BBL.	P	B
	425 V8 4-BBL.	R	B
	330 V8	V, X	B
1968–69	350 V8 2-BBL.	TB, TD, TL	C
	350 V8 4-BBL.	TN	C
	455 V8 2-BBL.	UA, UB	C
	455 V8 4-BBL.	UN, UO	C
	455 V8 2-BBL. (low comp.)	UC, UD, UJ	C
1970–71	350 V8 2-BBL.	QA, QI, QJ, TC, TD, TL	C
	350 V8 4-BBL. 10.25 CR	QN, QP, QV	C
	350 V8 4-BBL. 10.5 CR	QD, QX	C
	455 V8 2-BBL. 9.0 CR	UC, UD, UJ	C
	455 V8 2-BBL. 10.25 CR	TX, TY	C
	455 V8 4-BBL., 10.25 CR	TP, TQ, TU, TV, TW, UN, UO	C
	455 V8 4-BBL. 10.5 CR	TS, TT	C
	455 V8 4-BBL. 10.25 CR W 33	UL	C

Code location: A—Top of center exhaust port, left cylinder head.
B—Stamped on machined pad at front of right cylinder head.
C—Tape attached directly to front of oil fill tube.

CAR SERIAL NUMBER LOCATION

1964-67
Located on left front door hinge pillar.

1968-70
Left side of instrument panel, visible through windshield.

1964
First digit (6 or 8) indicates V6 or V8
Second digit indicates series

1965-71
First digit—Oldsmobile Division
Second digit indicates series
Third digit indicates engine
   Odd no. = 1965—V6,
   1966-71—L6
   Even no. = V8
Fourth and fifth digits indicate body type
Sixth digit indicates year
Seventh digit (letter) indicates plant

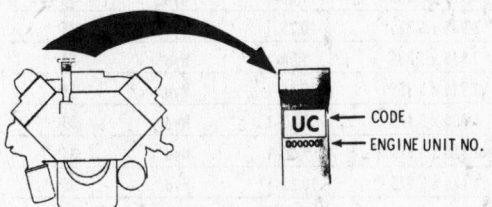

V8 engine identification code location— 1968-71

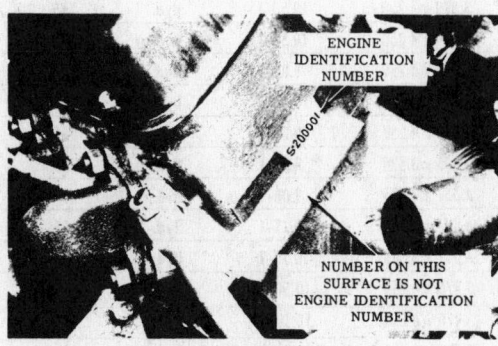

V8 engine identification code location— 1964-67

F-85 ENGINE IDENTIFICATION

YEAR	ENGINE	PRODUCTION CODE	CODE LOCATION
1964	225 V6	KH	C
	330 V8	T	D
1965	225 V6	LH	C
	330 V8	U	D
	440 V8	W	D
1966	250 L6 1-BBL.	F	B
	250 L6	T	B
	330 V8	V, X	D
	400 V8	V	D
1967	250 L6 1-BBL.	F	B
	330 V8	V, X	D
	400 V8	V	D
1968–69	250 L6 1-BBL.	VA, VB, VE, VF	B
	350 V8 2-BBL.	QA, QB, QI, QJ	E
	350 V8 4-BBL. ①	QN, QP, QV, QX	E
	400 V8 2-BBL.	QL	E
	400 V8 4-BBL.	QR, QS, QW	E
	400 V8 4-BBL. (W-30)	QT, QU	E
1970–71	250 L6 1-BBL.	VB, VF	B
	350 V8 2-BBL. 9.0 CR	QA, QI, OJ, TC, TD, TL	C
	350 V8 4-BBL. 10.25 CR	QN, QP, QU	C
	350 V8 4-BBL. 10.5 CR	QD, QX	C

Code location: A—Front of right cylinder head.
B—Right side of engine block, directly behind distributor.
C—Right cylinder head-to-block deck face.
D—Stamped on machined pad at front of right cylinder head.
E—Tape attached directly to front of oil fill tube.
①—QX = W31 engine.

GENERAL ENGINE SPECIFICATIONS

YEAR	CU. IN. DISPLACEMENT	CARBURETOR	DEVELOPED HORSEPOWER @ RPM	DEVELOPED TORQUE @ RPM (FT. LBS.)	A.M.A. HORSEPOWER	BORE AND STROKE (IN.)	ADVERTISED COMPRESSION RATIO	VALVE LIFTER TYPE	NORMAL OIL PRESSURE (PSI)
1964	V6—225	1-BBL.	155 @ 4400	225 @ 2400	33.8	3.750 x 3.400	9.00-1	Hyd.	35
	V8—330	2-BBL.	210 @ 4400	325 @ 2400	49.6	3.938 x 3.385	9.00-1	Hyd.	35
	V8—330	4-BBL.	260 @ 4800	345 @ 3200	49.6	3.938 x 3.385	10.25-1	Hyd.	35
	V8—330	2-BBL.	225 @ 4600	335 @ 2400	49.6	3.938 x 3.385	10.25-1	Hyd.	35
	V8—330	4-BBL.	260 @ 4800	345 @ 3200	49.6	3.938 x 3.385	10.25-1	Hyd.	35
	V8—394	2-BBL.	280 @ 4400	430 @ 2400	54.0	4.125 x 3.688	10.25-1	Hyd.	35
	V8—394	4-BBL.	330 @ 4600	440 @ 2800	54.0	4.125 x 3.688	10.25-1	Hyd.	35
	V8—394	4-BBL.	345 @ 4800	440 @ 3200	54.0	4.125 x 3.688	10.50-1	Hyd.	35
1965	V6—225	1-BBL.	155 @ 4400	225 @ 2400	33.8	3.750 x 3.400	9.00-1	Hyd.	35
	V8—330	2-BBL.	250 @ 4800	335 @ 2800	49.6	3.938 x 3.385	9.00-1	Hyd.	35
	V8—330	4-BBL	315 @ 5200	360 @ 3600	49.6	3.938 x 3.385	10.25-1	Hyd.	35
	V8—425	2-BBL.	280 @ 4400	425 @ 2400	54.0	4.125 x 3.975	9.00-1	Hyd.	35
	V8—425	2-BBL.	310 @ 4400	450 @ 2400	54.0	4.125 x 3.975	10.25-1	Hyd.	35
	V8—425	4-BBL.	360 @ 4800	470 @ 2800	54.0	4.125 x 3.975	10.25-1	Hyd.	35
	V8—425	4-BBL.	370 @ 4800	470 @ 3200	54.0	4.125 x 3.975	10.50-1	Hyd.	35
1966	L6—250	1-BBL.	155 @ 4200	240 @ 2000	36.0	3.875 x 3.530	8.50-1	Hyd.	35
	V8—330	2-BBL.	250 @ 4800	335 @ 2800	49.6	3.938 x 3.385	9.00-1	Hyd.	35
	V8—330	4-BBL.	320 @ 5200	360 @ 3600	49.6	3.938 x 3.385	10.25-1	Hyd.	35
	V8—400	4-BBL.	350 @ 5000	440 @ 3600	51.2	4.000 x 3.975	10.50-1	Hyd.	35
	V8—425	2-BBL.	310 @ 4400	450 @ 2400	54.0	4.125 x 3.975	10.25-1	Hyd.	35
	V8—425	4-BBL.	365 @ 4800	470 @ 3200	54.0	4.125 x 3.975	10.25-1	Hyd.	35
	V8—425	4-BBL.	375 @ 4800	470 @ 3200	54.0	4.125 x 3.975	10.50-1	Hyd.	35
1967	L6—250	1-BBL.	155 @ 4200	240 @ 2000	36.0	3.875 x 3.530	8.50-1	Hyd.	35
	V8—330	2-BBL.	250 @ 4800	335 @ 2800	49.6	3.938 x 3.385	9.00-1	Hyd.	35
	V8—330	2-BBL.	260 @ 4800	355 @ 2800	49.6	3.938 x 3.385	10.25-1	Hyd.	35
	V8—330	4-BBL.	310 @ 5200	340 @ 3600	49.6	3.938 x 3.385	9.00-1	Hyd.	35
	V8—330	4-BBL.	320 @ 5200	360 @ 3600	49.6	3.938 x 3.385	10.25-1	Hyd.	35
	V8—400	4-BBL.	350 @ 5000	440 @ 3600	51.2	4.000 x 3.975	10.50-1	Hyd.	35
	V8—425	2-BBL.	300 @ 4400	430 @ 2400	54.0	4.125 x 3.975	9.00-1	Hyd.	35
	V8—425	2-BBL.	310 @ 4400	450 @ 2400	54.0	4.125 x 3.975	10.25-1	Hyd.	35
	V8—425	4-BBL.	365 @ 4800	470 @ 3200	54.0	4.125 x 3.975	10.25-1	Hyd.	35
	V8—425	4-BBL.	375 @ 4800	470 @ 3200	54.0	4.125 x 3.975	10.50-1	Hyd.	35
1968	L6—250	1-BBL.	155 @ 4200	240 @ 2000	36.0	3.875 x 3.530	8.50-1	Hyd.	35
	V8—350	2-BBL.	250 @ 4400	355 @ 2600	52.7	4.057 x 3.385	9.00-1	Hyd.	35
	V8—350	4-BBL.	300 @ 4800	390 @ 3600	52.7	4.057 x 3.385	10.25-1	Hyd.	35
	V8—350	4-BBL.	310 @ 4800	390 @ 3200	52.7	4.057 x 3.385	10.25-1	Hyd.	35
	V8—400	2-BBL.	290 @ 4600	425 @ 2400	47.9	3.870 x 4.250	9.00-1	Hyd.	45
	V8—400	4-BBL.	350 @ 4800	440 @ 3200	47.9	3.870 x 4.250	10.50-1	Hyd.	45
	V8—400	4-BBL.	360 @ 4800	440 @ 3600	47.9	3.870 x 4.250	10.50-1	Hyd.	45
	V8—455	2-BBL.	310 @ 4200	490 @ 2400	54.5	4.125 x 4.250	9.00-1	Hyd.	35
	V8—455	4-BBL.	320 @ 4200	500 @ 2400	54.5	4.125 x 4.250	10.25-1	Hyd.	35
	V8—455	4-BBL.	365 @ 4600	510 @ 3000	54.5	4.125 x 4.250	10.25-1	Hyd.	35
1969	L6—250	1-BBL.	155 @ 4200	240 @ 2000	36.04	3.875 x 3.530	8.50-1	Hyd.	35
	V8—350	2-BBL.	250 @ 4400	355 @ 2600	52.7	4.057 x 3.385	9.00-1	Hyd.	35
	V8—350	4-BBL.	310 @ 4800	390 @ 3200	52.7	4.057 x 3.385	10.25-1	Hyd.	35
	V8—350 (W31)	4-BBL.	325 @ 5400	360 @ 3600	52.7	4.057 x 3.385	10.50-1	Hyd.	35
	V8—400	4-BBL.	325 @ 4600	440 @ 3000	47.9	3.870 x 4.250	10.50-1	Hyd.	45
	V8—400	4-BBL.	350 @ 4800	440 @ 3200	47.9	3.870 x 4.250	10.50-1	Hyd.	45
	V8—400	4-BBL.	360 @ 5400	440 @ 3600	47.9	3.870 x 4.250	10.50-1	Hyd.	45
	V8—455	2-BBL.	310 @ 4200	490 @ 2400	54.5	4.125 x 4.250	9.00-1	Hyd.	35
	V8—455	4-BBL.	365 @ 4600	510 @ 3000	54.5	4.125 x 4.250	10.25-1	Hyd.	35
	V8—455	4-BBL.	390 @ 5000	500 @ 3200	54.5	4.125 x 4.250	10.25-1	Hyd.	35

GENERAL ENGINE SPECIFICATIONS, continued

YEAR	CU. IN. DISPLACEMENT	CARBURETOR	DEVELOPED HORSEPOWER @ RPM	DEVELOPED TORQUE @ RPM (FT. LBS.)	A.M.A. HORSEPOWER	BORE AND STROKE (IN.)	ADVERTISED COMPRESSION RATIO	VALVE LIFTER TYPE	NORMAL OIL PRESSURE (PSI)
1970	L6—250	1-BBL.	155 @ 4200	240 @ 2000	36.04	3.875 x 3.530	8.50-1	Hyd.	35
	V8—350 (L65)	2-BBL.	250 @ 4400	355 @ 2600	52.7	4.057 x 3.385	9.00-1	Hyd.	35
	V8—350 (L74)	4-BBL.	310 @ 4800	390 @ 3200	52.7	4.057 x 3.385	10.25-1	Hyd.	35
	V8—350 (W31)	4-BBL.	325 @ 5400	360 @ 3600	52.7	4.057 x 3.385	10.50-1	Hyd.	35
	V8—455 (L33)	2-BBL.	320 @ 4200	500 @ 2400	54.5	4.125 x 4.250	10.25-1	Hyd.	35
	V8—455 (L31)	4-BBL.	365 @ 4600	510 @ 3000	54.5	4.125 x 4.250	10.25-1	Hyd.	35
	V8—455 (W33)	4-BBL.	390 @ 5000	500 @ 3200	54.5	4.125 x 4.250	10.25-1	Hyd.	35
	V8—455 (W30)	4-BBL.	370 @ 5400	500 @ 3600	54.5	4.125 x 4.250	10.50-1	Hyd.	35
	V8—455 (L30)	2-BBL.	310 @ 4200	490 @ 2400	54.5	4.125 x 4.250	9.00-1	Hyd.	35
1971	L6—250	1-BBL.	155 @ 4200	240 @ 2000	36.4	3.875 x 3.530	8.5-1	Hyd.	35
	V8—350 (L65)	2-BBL.	250 @ 4400	355 @ 2600	52.7	4.057 x 3.385	9.00-1	Hyd.	35
	V8—350 (L74)	4-BBL.	250 @ 4400	390 @ 3200	52.7	4.057 x 3.385	—	Hyd.	35
	V8—350 (W31)	4-BBL.	—	360 @ 3600	52.7	4.057 x 3.385	—	Hyd.	35
	V8—455 (L33)	2-BBL.	—	500 @ 2400	54.5	4.125 x 4.250	—	Hyd.	35
	V8—455 (L31)	4-BBL.	—	510 @ 3000	54.5	4.125 x 4.250	—	Hyd.	35
	V8—455 (W33)	4-BBL.	—	500 @ 3200	54.5	4.125 x 4.250	—	Hyd.	35
	V8—455 (W30)	4-BBL.	—	500 @ 3600	54.5	4.125 x 4.250	—	Hyd.	35
	V8—455 (L30)	2-BBL.	—	490 @ 2400	54.5	4.125 x 4.250	—	Hyd.	35

VALVE SPECIFICATIONS

YEAR AND MODEL	SEAT ANGLE (DEG.)	FACE ANGLE (DEG.)	VALVE LIFT INTAKE (IN.)	VALVE LIFT EXHAUST (IN.)	VALVE SPRING PRESSURE (VALVE OPEN) LBS. @ IN.	VALVE SPRING INSTALLED HEIGHT (IN.)	STEM TO GUIDE CLEARANCE (IN.) INTAKE	STEM TO GUIDE CLEARANCE (IN.) EXHAUST	STEM DIAMETER (IN.) INTAKE	STEM DIAMETER (IN.) EXHAUST
1964 V6—225	45	45	.391	.401	168 @ 1.26	1.64	③	③	.3407-.3412	.3402-.3407 ④
V8—330	45	45	.401	.401	193-207 @ 1.20	1.60	.0010-.0027	.0015-.0032	.3425-.3432	.3420-.3427
V8—394, 2-BBL.	45	45	.435	.435	180 @ 1.43	1.84	.0010-.0025	.0015-.0030	.3427-.3432	.3422-.3427
V8—394, 4-BBL.	45	45	.444	.435	180 @ 1.43	1.84	.0010-.0025	.0015-.0030	.3427-.3432	.3422-.3427
1965 V6—225	45	45	.391	.401	168 @ 1.26	1.64	③	③	.3407-.3412	.3402-.3407 ④
V8—330	45	45	.401	.401	180-194 @ 1.27	1.67	.0010-.0027	.0015-.0032	.3425-.3432	.3420-.3427
V8—425, 400	①	②	.429	.429	180-194 @ 1.27	1.67	.0010-.0027	.0015-.0032	.3425-.3432	.3420-.3427
1966-67 L6—250	45	45	.388	.388	180-192 @ 1.27	1.66	.0010-.0027	.0010-.0027	.3410-.3417	.3410-.3417
V8—330	45	46	.401	.401	180-194 @ 1.27	1.67	.0010-.0027	.0015-.0032	.3425-.3432	.3420-.3427
V8—400	①	②	.431	.433	180-194 @ 1.27	1.67	.0010-.0027	.0015-.0032	.3425-.3432	.3420-.3427
V8—425	①	②	.429	.429	180-194 @ 1.27	1.67	.0010-.0027	.0015-.0032	.3425-.3432	.3420-.3427
1968-69 L6—250	45	45	.388	.388	180-192 @ 1.27	1.66	.0010-.0027	.0010-.0027	.3410-.3417	.3410-.3417
V8—350	45	46	.435	.435	180-194 @ 1.27	1.67	.0010-.0027	.0015-.0032	.3425-.3432	.3420-.3427
V8—400, 2-BBL.	①	②	.430	.432	180-194 @ 1.27	1.67	.0010-.0027	.0015-.0032	.3425-.3432	.3420-.3427
V8—400, 4-BBL.	①	②	.472	.472	180-194 @ 1.27	1.67	.0010-.0027	.0015-.0032	.3425-.3432	.3420-.3427
V8—455	45	46	.435	.435	180-194 @ 1.27	1.67	.0010-.0027	.0015-.0032	.3425-.3432	.3420-.3427
1970 L6—250	45	45	.388	.388	180-192 @ 1.27	1.66	.0010-.0027	.0010-.0027	.3410-.3417	.3410-.3417
V8—350	①	②	.400	.400	180-194 @ 1.27	1.67	.0010-.0027	.0015-.0032	.3425-.3432	.3420-.3427
V8—455	45	46	.435	.435	180-194 @ 1.27	1.67	.0010-.0027	.0015-.0032	.3425-.3432	.3420-.3427
V8—455 (W33, W30)	①	②	.472	.472	180-194 @ 1.27	1.67	.0010-.0027	.0015-.0032	.3425-.3432	.3420-.3427
1971 L6—250	45	45	.388	.388	180-192 @ 1.27	1.66	.0010-.0027	.0010-.0027	.3410-.3417	.3410-.3417
V8—350	①	②	.400	.400	180-194 @ 1.27	1.67	.0010-.0027	.0015-.0032	.3425-.3432	.3420-.3427
V8—455	45	46	.435	.435	180-194 @ 1.27	1.67	.0010-.0027	.0015-.0032	.3425-.3432	.3420-.3427
V8—455 (W33, W30)	①	②	.472	.472	180-194 @ 1.27	1.67	.0010-.0027	.0015-.0032	.3425-.3432	.3420-.3427

① —Intake 30°, Exhaust 45°
② —Intake 30°, Exhaust 46°
③ —Guide tapers from top to bottom with bigger dimension at top.
  Intake .001-.003
  Exhaust .0015-.0035
④ —Guide tapers from top to bottom with bigger dimension at top.

TUNE-UP SPECIFICATIONS

YEAR	MODEL	SPARK PLUGS		DISTRIBUTOR		IGNITION TIMING (Deg.) ▲	CRANKING COMP. PRESSURE (Psi)	VALVES Tappet (Hot) Clearance (In.)		Intake Opens (Deg.)	FUEL PUMP PRESSURE (Psi)	IDLE SPEED (Rpm) *
		Type	Gap (In.)	Point Dwell (Deg.)	Point Gap (In.)			Intake	Exhaust			
1964	F-85, V6—All	44S	.035	30	.016	5B	160	Zero	Zero	24B	$4\frac{1}{2}$–$5\frac{3}{4}$	550
	F-85, V8—2-BBL.	45S	.035	30	.016	$7\frac{1}{2}$B	180	Zero	Zero	22B	$7\frac{1}{2}$–$8\frac{1}{2}$	550
	F-85, V8—4-BBL.	44S	.035	30	.016	$7\frac{1}{2}$B	180	Zero	Zero	22B	$7\frac{1}{2}$–$8\frac{1}{2}$	550
	Jetstar 88, 2-BBL.	45	.030	30	.016	$7\frac{1}{2}$B	180	Zero	Zero	24B	$5\frac{1}{2}$	550
	Jetstar 88, 4-BBL.	44	.030	30	.016	$7\frac{1}{2}$B	180	Zero	Zero	24B	$5\frac{1}{2}$	550
	Dynamic 88, M.T.	45	.030	30	.016	$2\frac{1}{2}$B	185	Zero	Zero	14B	$5\frac{1}{2}$	550
	Dynamic 88, A.T.	45	.030	30	.016	5B	185	Zero	Zero	14B	$5\frac{1}{2}$	550
	All Others, M.T.	44	.030	30	.016	$2\frac{1}{2}$B	185	Zero	Zero	11B	$5\frac{1}{2}$	550
	All Others, A.T.	44	.030	30	.016	5B	185	Zero	Zero	11B	$5\frac{1}{2}$	550
1965	F-85, V6—All	44S	.035	30	.016	5B	160	Zero	Zero	24B	$4\frac{1}{2}$–$5\frac{3}{4}$	550
	F-85, V8—2-BBL.	45S	.035	30	.016	$7\frac{1}{2}$B	180	Zero	Zero	22B	$7\frac{1}{2}$–$8\frac{1}{2}$	550
	F-85, V8—4-BBL.	44S	.035	30	.016	$7\frac{1}{2}$B	180	Zero	Zero	22B	$7\frac{1}{2}$–$8\frac{1}{2}$	550
	Jetstar 88, 2-BBL.	45S	.030	30	.016	$7\frac{1}{2}$B	180	Zero	Zero	24B	$5\frac{1}{2}$	550
	Jetstar 88, 4-BBL.	44S	.030	30	.016	$7\frac{1}{2}$B	180	Zero	Zero	24B	$5\frac{1}{2}$	550
	Dynamic 88, 2-BBL.	45S	.030	30	.016	$7\frac{1}{2}$B	170	Zero	Zero	21B	$5\frac{1}{2}$	850
	Series 98	44S	.030	30	.016	5B	180	Zero	Zero	21B	$5\frac{1}{2}$	850
	Jetstar-1, Starfire	44S	.030	30	.016	5B	180	Zero	Zero	21B	$5\frac{1}{2}$	850
1966	F-85, OHV 6, 1-BBL.	46N	.035	32	.019	5B	130	Zero	Zero	62B	$4\frac{1}{2}$●	500
	F-85, V8, 2-BBL.	45S	.030	30	.016	$7\frac{1}{2}$B	180	Zero	Zero	22B	$7\frac{1}{2}$–$8\frac{1}{2}$	550
	F-85, V8, 4-BBL.	44S	.030	30	.016	$7\frac{1}{2}$B	180	Zero	Zero	22B	$7\frac{1}{2}$–$8\frac{1}{2}$	550
	F-85, V8, 4-4-2, 4-BBL.	44S	.030	30	.016	$7\frac{1}{2}$B	190	Zero	Zero	21B	6	600
	Jetstar 88, 2-BBL.	44S	.030	30	.016	$7\frac{1}{2}$B	180	Zero	Zero	24B	$7\frac{3}{4}$–9	550
	Jetstar 88, 4-BBL.	44S	.030	30	.016	$7\frac{1}{2}$B	180	Zero	Zero	24B	$7\frac{3}{4}$–9	550
	Dynamic 88, 2-BBL.	44S	.030	30	.016	$7\frac{1}{2}$B	170	Zero	Zero	21B	$7\frac{3}{4}$–9	500
	Dynamic 88, 4-BBL.	44S	.030	30	.016	$7\frac{1}{2}$B	180	Zero	Zero	21B	$7\frac{3}{4}$–9	550
	Delta 88, 2-BBL.	44S	.030	30	.016	$7\frac{1}{2}$B	170	Zero	Zero	21B	$7\frac{3}{4}$–9	600
	Delta 88, 4-BBL.	44S	.030	30	.016	$7\frac{1}{2}$B	180	Zero	Zero	21B	$7\frac{3}{4}$–9	600
	Starfire, 4-BBL.	44S	.030	30	.016	$7\frac{1}{2}$B	180	Zero	Zero	21B	$7\frac{3}{4}$–9	600
1967	L6—250, 1-BBL. (155 HP)	46N	.035	32	.016	5B	130	Zero	Zero	62B	$4\frac{1}{2}$●	500
	V8—330, 2-BBL. (250 HP)	45S	.030	30	.016	$7\frac{1}{2}$B	180	Zero	Zero	12B ①	8	550
	V8—330, 2-BBL. (260 HP)	44S	.030	30	.016	$7\frac{1}{2}$B	180	Zero	Zero	21B	8	550
	V8—330, 4-BBL. (310 HP)	44S	.030	30	.016	$7\frac{1}{2}$B	180	Zero	Zero	21B	8	550
	V8—330, 4-BBL. (320 HP)	44S	.030	30	.016	$7\frac{1}{2}$B	180	Zero	Zero	21B	8	550
	V8—400, 4-4-2 (350 HP)	44S	.030	30	.016	$7\frac{1}{2}$B	180	Zero	Zero	30B ②	7	600
	V8—425, 2-BBL. (300 HP)	44S	.030	30	.016	5B	160	Zero	Zero	21B	8	550
	V8—425, 2-BBL. (310 HP)	44S	.030	30	.016	$7\frac{1}{2}$B	180	Zero	Zero	21B	8	550
	V8—425, 4-BBL. (365 HP)	44S	.030	30	.016	$7\frac{1}{2}$B	180	Zero	Zero	21B	8	550
	V8—425, 4-BBL. (375 HP)	44S	.030	30	.016	$7\frac{1}{2}$B	190	Zero	Zero	21B	8	550
1968	L6—250, M.T. (155 HP)	46N	.035	32	.016	4B	130	Zero	Zero	16B	$3\frac{1}{2}$–$4\frac{1}{2}$	650
	L6—250, A.T. (155 HP)	46N	.035	32	.016	6B	130	Zero	Zero	16B	$3\frac{1}{2}$–$4\frac{1}{2}$	550
	V8—350, 2-BBL. (250 HP)	45S	.030	30	.016	5B	160	Zero	Zero	16B	$5\frac{1}{2}$–7	650■
	V8—350, 4-BBL. (300 HP)	44S	.030	30	.016	$7\frac{1}{2}$B	180	Zero	Zero	16B	$5\frac{1}{2}$–7	650■
	V8—350, 4-BBL. (310 HP)	44S	.030	30	.016	$7\frac{1}{2}$B	180	Zero	Zero	16B	$5\frac{1}{2}$–7	650■
	V8—400, 2-BBL. (290 HP)	44S	.030	30	.016	TDC	180	Zero	Zero	21B	$5\frac{1}{2}$–7	725
	V8—400, 4-BBL. (350 HP)	44S	.030	30	.016	TDC	180	Zero	Zero	30B	$5\frac{1}{2}$–7	725
	V8—400, 4-BBL. (360 HP)	44S	.030	30	.016	TDC	180	Zero	Zero	30B	$5\frac{1}{2}$–7	725
	V8—455, 2-BBL. (310 HP)	45S	.030	30	.016	10B	180	Zero	Zero	20B	$5\frac{1}{2}$–7	550
	V8—455, 2-BBL. (320 HP)	45S	.030	30	.016	10B	180	Zero	Zero	20B	$5\frac{1}{2}$–7	550
	V8—455, 4-BBL. (365 HP)	45S	.030	30	.016	10B	190	Zero	Zero	20B	$5\frac{1}{2}$–7	550
1969	L6—250, M.T. (155 HP)	46N	.030	35	.016	TDC	130	Zero	Zero	16B	$3\frac{1}{2}$–4	700
	L6—250, A.T. (155 HP)	46N	.030	35	.016	4B	130	Zero	Zero	16B	$3\frac{1}{2}$–4	500
	V8—350, 2-BBL. (250 HP)	45S	.030	30	.016	6B	180	Zero	Zero	16B	$5\frac{1}{2}$–7	850
	V8—350, 4-BBL. (310 HP)	44S	.030	30	.016	8B	180	Zero	Zero	16B	$5\frac{1}{2}$–7	850
	V8—350, 4-BBL. (325 HP)	44S	.030	30	.016	12B	180	Zero	Zero	16B	$5\frac{1}{2}$–7	850

TUNE-UP SPECIFICATIONS, continued

YEAR	MODEL	SPARK PLUGS		DISTRIBUTOR		IGNITION TIMING (Deg.) ▲	CRANKING COMP. PRESSURE (Psi)	VALVES			FUEL PUMP PRESSURE (Psi)	IDLE SPEED (Rpm) ★
		Type	Gap (In.)	Point Dwell (Deg.)	Point Gap (In.)			Tappet (Hot) Clearance (In.)		Intake Opens (Deg.)		
								Intake	Exhaust			
	V8—400, 4-BBL. (325 HP)	44S	.030	30	.016	8B ③	180	Zero	Zero	56B	5-8	850
	V8—400, 4-BBL. (350 HP)	44S	.030	30	.016	8B ③	180	Zero	Zero	56B	5-8	850
	V8—400, 4-BBL. (360 HP)	44S	.030	30	.016	8B ③	180	Zero	Zero	56B	5-8	850
	V8—455, 2-BBL. (310 HP)	45S	.030	30	.016	6B	180	Zero	Zero	20B	5-8	850
	V8—455, 4-BBL. (365 HP)	44S	.030	30	.016	8B	180	Zero	Zero	20B	5-8	850
	V8—455, 4-BBL. (390 HP)	44S	.030	30	.016	10B	190	Zero	Zero	20B	5-8	850
1970	L6—250, M.T. (155 HP)	R46T	.035	31-34	.019	TDC	130	Zero	Zero	16B	4-5	750
	L6—250, A.T. (155 HP)	R46T	.035	31-34	.019	4B	130	Zero	Zero	16B	4-5	600
	V8—350, 2-BBL. (250 HP)	R46S	.030	30	.016	10B ④	180	Zero	Zero	16B	5½-6½	750 ⑤
	V8—350, 4-BBL. (310 HP)	R45S	.030	30	.016	10B	180	Zero	Zero	16B	5½-6½	650 ⑤
	V8—350, 4-BBL. (325 HP)	R45S	.030	30	.016	14B	180	Zero	Zero	40B	5½-6½	750 ⑥
	V8—455, 2-BBL. (310 HP)	R46S	.030	30	.016	8B	180	Zero	Zero	20B	5-6½	675 ⑤
	V8—455, 4-BBL. (365 HP)	R45S	.030	30	.016	8B	180	Zero	Zero	20B	5-6½	575
	V8—455, 4-BBL. (390 HP)	R44S	.030	30	.016	12B	190	Zero	Zero	24B	5-6½	600
1971	L6—250 M.T. (155 HP)	R46T	.035	31-34	.019	**	—	Zero	Zero	16B	4-5	750
	L6—250 A.T. (155 HP)	R46T	.035	31-34	.019	**	—	Zero	Zero	16B	4-5	650
	V8—350, 2-BBL. (250 HP)	R46S	.030	30	.016	**	—	Zero	Zero	16B	5½-6½	750 ⑤
	V8—350, 4-BBL. (310 HP)	R45S	.030	30	.016	**	—	Zero	Zero	16B	5½-6½	650 ⑤
	V8—350, 4-BBL. (325 HP)	R45S	.030	30	.016	**	—	Zero	Zero	40B	5½-6½	750 ⑥
	V8—455, 2-BBL. (310 HP)	R46S	.030	30	.016	**	—	Zero	Zero	20B	5½-6½	675 ⑤
	V8—455, 4-BBL. (365 HP)	R45S	.030	30	.016	**	—	Zero	Zero	20B	5½-6½	575
	V8—455, 4-BBL. (390 HP)	R44S	.030	30	.016	**	—	Zero	Zero	24B	5½-6½	600

★—With manual transmission in N and automatic in D. If equipped with air conditioning add 50 rpm.

▲—With vacuum advance disconnected. NOTE: These settings are only approximate. Engine design, altitude, temperature, fuel octane rating and the condition of the individual engine are all factors which can influence timing. The limiting advance factor must, therefore, be the "knock point" of the individual engine.

●—If equipped with exhaust emission control—5½-7.

■—550 rpm with automatic transmission.

①—Vista Cruiser = 21B.

②—Auto. transmission = 21B.

③—Std. transmission = 2°B.

B—Before top dead center.

TDC—Top dead center.

A.T.—Automatic transmission.

M.T.—Manual transmission.

④—10B for intermediate cars with 350 2-BBL. (250 HP). 8B for full size cars with 350 2-BBL. (250 HP).

⑤—575 rpm for A.T.

⑥—625 rpm for A.T.

**—See engine decal.

Caution

General adoption of anti-pollution laws has changed the design of almost all car engine production to effectively reduce crankcase emission and terminal exhaust products. It has been necessary to adopt stricter tune-up rules, especially timing and idle speed procedures. Both of these values are peculiar to the engine and to its application, rather than to the engine alone. With this in mind, car manufacturers supply idle speed data for the engine and application involved. This information is clearly displayed in the engine compartment of each vehicle.

BATTERY AND STARTER SPECIFICATIONS

YEAR	MODEL	BATTERY			STARTERS						Brush Spring Tension (Oz.)
		Ampere Hour Capacity	Volts	Terminal Grounded	Lock Test			No-Load Test			
					Amps.	Volts	Torque	Amps.	Volts	RPM	
1964		61	12	Neg.	Not Recommended			62	10.6	7,800	35
		61	12	Neg.	Not Recommended			80	10.6	7,800	35
		62▲	12	Neg.	Not Recommended			80	10.6	4,350	35
		62▲	12	Neg.	Not Recommended			80	10.6	4,350	35
		62▲	12	Neg.	Not Recommended			87	10.6	5,000	35
1965-71		44	12	Neg.	Not Recommended			85	10.6	5,000	35
		61	12	Neg.	Not Recommended			80	10.6	4,350	35
		70■	12	Neg.	Not Recommended			62	10.6	7,800	35

▲—With high compression engine 70 Amp.

■—With high compression engine 73 or 75 Amp.

CRANKSHAFT BEARING JOURNAL SPECIFICATIONS

YEAR	MODEL	MAIN BEARING JOURNALS (IN.)				CONNECTING ROD BEARING JOURNALS (IN.)		
		Journal Diameter	Oil Clearance	Shaft End-Play	Thrust On No.	Journal Diameter	Oil Clearance	End-Play
1964-66	V6—225	2.4995	.0012	.006	2	2.000	.0013	.010
	V8—330	2.4995	.0012	.006	3	2.000	.0013	.007
	V8—394, 425	3.0000	.0020	.006	5▲	2.500	.0015	.007
1968-70	L6—250	2.2988	.0003-.0029	.002-.006	7	1.999-2.000	.0007-.0027	.0085-.0135
	V8—350	2.4995	.0005-.0021①	.004-.008	3	2.1238-2.1248	.0004-.0033	.002-.013
	V8—455	3.0000	.0005-.0021②	.004-.008	3	2.4988-2.4998	.0004-.0033	.002-.013
1971	L6—250	2.2988	.0003-.0029	.002-.006	7	1.999-2.000	.0007-.0027	.0085-.0135
	V8—350	2.4995	.0005-.0021	.004-.008	3	2.1238-2.1248	.0004-.0033	.002-.013
	V8—455	3.0000	.0005-.0021	.004-.008	3	2.4988-2.4998	.0004-.0033	.002-.013

▲—1965-66, Thrust On No. 3. ★—No. 5, .0027. ①—No. 5, .0015-.0031 ②—No. 5, .0020-.0034

TORQUE SPECIFICATIONS

YEAR	MODEL	CYLINDER HEAD BOLTS (FT. LBS.)	ROD BEARING BOLTS (FT. LBS.)	MAIN BEARING BOLTS (FT. LBS.)	CRANKSHAFT BALANCER BOLT (FT. LBS.)	FLYWHEEL TO CRANKSHAFT BOLTS (FT. LBS.)	MANIFOLD (FT. LBS.)	
							Intake	Exhaust
1964-69	6 Cyl. OHV	90-95	35-45	60-70	Press Fit	55-65	25-35	20-25
	F-85①	50-55	30-35	50-60■	150	85-95	25-30	18-24
	Exc. F-85	60-80	38-48	80-120▲	100●	85-95	28	22
1970-71	L6—250	95	35	65	Press fit	60	②	②
	V8—All	80	42	120③	160 min	④	23-35	20-25

▲—No. 1 thru 4—80-120. No. 5—120-160. ②—Outer clamp—20 ft. lbs., others 30 ft. lbs.
●—1965—100. ③—V8—350—80 on No. 1, 2, 3 and 4, 120 on No. 5
■—No. 5—65-70. V8—455—125 on all
①—1964-69—refer to 6 cyl. OHV and V8. ④—A.T. 60 ft. lbs.; M.T. 90 ft. lbs.

CYLINDER HEAD BOLT TIGHTENING SEQUENCE

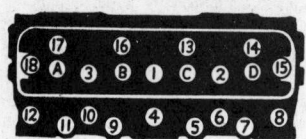

1964 394 V8

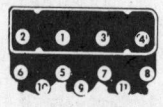

1964-65 V6

1964-70 330, 350, 400, 425, 455 V8

1966-70 OHV 6 cyl.

AC GENERATOR AND REGULATOR SPECIFICATIONS

YEAR	Part No.	ALTERNATOR			REGULATOR						
			Amp Output @ Engine RPM			Field Relay			Regulator		
		Field Current Draw @ 12V.	775 RPM	1950 RPM	Part No.	Air Gap (In.)	Point Gap (In.)	Volts to Close	Air Gap (In.)	Point Gap (In.)	Volts at 125°
1964	1100624	1.9-2.2	20-27	32-39	1119507	.015	.030	6.3-8.3	.060	.014	13.5-14.4
	1100656	1.9-2.3	20-27	32-39	1119515	.015	.030	6.3-8.3	.060	.014	13.5-14.4
1965-66	1100705	2.2-2.6	20-27	37	1119515	.015	.030	6.3-8.3	.060	.014	13.5-14.4
	1100696	2.2-2.6	20-27	42	1119515	.015	.030	6.3-8.3	.060	.014	13.5-14.4
	1100704	2.2-2.6	24	37	1119515	.015	.030	6.3-8.3	.060	.014	13.5-14.4
	1100699	2.2-2.6	26	42	1119515	.015	.030	6.3-8.3	.060	.014	13.5-14.4
	1100686	2.2-2.6	33	55	1119515	.015	.030	6.3-8.3	.060	.014	13.5-14.4
1967-70	1100767	2.2-2.6	20-27	37	1119515	.015	.030	6.3-8.3	.060	.014	13.5-14.4
	1100734	2.2-2.6	20-27	42	1119515	.015	.030	6.3-8.3	.060	.014	13.5-14.4
1971	1100769	2.2-2.6	20-26	37	1119515	.015	.030	6.3-8.3	.060	.014	13.5-14.4
	1100736	2.2-2.6	20-27	42	1119515	.015	.030	6.3-8.3	.060	.014	13.5-14.4

GENERAL CHASSIS AND BRAKE SPECIFICATIONS

| YEAR | MODEL | CHASSIS | | | BRAKE CYLINDER BORE | | BRAKE DRUM | |
| | | Overall Length (In.) | Tire Size | Master Cylinder (In.) | Wheel Cylinder Diameter (In.) | | Diameter (In.) | |
					Front	Rear	Front	Rear
1964	F-85, V6	203.0	6.50 x 14	1.0	1.0	⅞	9½	9½
	F-85, V8	203.0	7.00 x 14	1.0	1.0	⅞	9½	9½
	Jetstar 88	215.3	7.50 x 14	1.0	1⅛	1.0	9½	9½
	98	222.3	8.50 x 14	1.0	1⅛	1.0	11	11
	All others	215.3	8.00 x 14	1.0	1⅛	1.0	11	11
1965	F-85	204.4	6.95 x 14 ⑩	1.0	1.0	⅞	9½	9½
	Deluxe, Cutlass	204.4	7.35 x 14	1.0	1.0	⅞	9½	9½
	Jetstar 88	216.9	7.75 x 14	1.0	1⅛	1.0	9½	9½
	98	222.9	8.55 x 14	1.0	1⅛	1.0	11	11
	All others	216.9	8.25 x 14	1.0	1⅛	1.0	11	11
1966	F-85, Exc. 442	204.0	6.95 x 14 ⑩	1.0	1⅛	⅞	9½	9½
	F-85, 442	204.0	7.75 x 14	1.0	1⅛	⅞	9½	9½
	Jetstar 88	217.0	7.75 x 14	1.0	1⅛	1.0	11	11
	Dynamic & Delta 88	223.0	8.25 x 14	1.0	1⅛	1.0	11	11
1967	F-85, Exc. 442	204.0	7.75 x 14	1.0①	1¹/₁₆	⅞	9½⑦	9½
	442	204.0	7.75 x 14	1.0①	1¹/₁₆④	1.0⑥	9½⑦	9½
	88 Series	217.0	8.55 x 14	1.0①	1⅛③	1.0	11⑧	11
	98 Series	223.0	8.85 x 14	1.0①②	1⅛③⑤	1.0	11①	11
	Vista Cruiser	209.5	8.25 x 14	1.0①	1¹/₁₆④	¹⁵/₁₆	9½⑦	9½
1968	F-85	205.6	7.75 x 14	1.0	1⅛	¹⁵/₁₆	9½	9½
	442	201.6	F70 x 14	1.0	1⅛	¹⁵/₁₆	9½	9½
	88 Series	217.8	8.55 x 14	1.0	1³/₁₆	1.0	11	11
	98 Series	223.7	8.85 x 14	1.0	1³/₁₆	1.0	11	11
	Vista Cruiser	217.5	8.25 x 14	1.0	1.06	1.0	9½	9½
1969	F-85	201.9⑨	7.75 x 14	1.0	1⅛⑪	¹⁵/₁₆	9½⑫	9½
	Cutlass Sta. Wag.	212.6	7.75 x 14	1.0	1⅛⑪	¹⁵/₁₆	9½⑫	9½
	442	201.9	F.70 x 14	1.0	1⅛⑪	¹⁵/₁₆	9½⑫	9½
	88 Series	218.6	8.55 x 15	1.0	1³/₁₆⑪	1.0	11⑬	11
	98 Series	224.4	8.85 x 15	1.0	1³/₁₆⑪	1.0	11⑬	11
	Vista Cruiser	217.6	8.25 x 14	1.0	1⅛⑪	¹⁵/₁₆	9½⑫	9½
1970-71	F-85	203.2⑭	F78 x 14	1.0	1⅛⑪	⅞	9½⑫	9½
	Cutlass Sta. Wag.	213.2	G78 x 14	1.0	1⅛⑪	¹⁵/₁₆	9½⑫	9½
	442	203.2	G70 x 14	1.0	1⅛⑪	⅞	9½⑫	9½
	88 Series	219.1	H78 x 15	1.0	1³/₁₆⑪	¹⁵/₁₆	11⑬	11
	98 Series	225.2	J78 x 15	1.0	2¹⁵/₁₆	¹⁵/₁₆	11⑬	11
	Vista Cruiser	218.2	H78 x 14	1.0	1⅛⑪	1.0	9½⑫	9½

①—Disc Brakes = 1.125.
②—Metallic Lining = ⅞.
③—Disc Brakes = 1¹⁵/₁₆.
④—Disc Brakes = 2¹/₁₆.
⑤—Metallic Lining = 1³/₁₆.
⑥—Disc Brakes Front, Rear = ¹³/₁₆.
⑦—Disc Diameter = 11.
⑧—Disc Diameter = 11⅞.
⑨—Sedans = 205.9.
⑩—Vista Cruiser = 7.75 x 14.
⑪—With Disc Brake = 2¹⁵/₁₆.
⑫—Disc Diameter = 10.88.
⑬—Disc Diameter = 11.80.
⑭—203.2—Coupe and Conv.
202.2—Pillar Sedan and Holiday Sedan.

WHEEL ALIGNMENT

| YEAR | MODEL | CASTER | | CAMBER | | TOE-IN (In.) | KING-PIN INCLINATION (Deg.) | WHEEL PIVOT RATIO | |
		Range (Deg.)	Pref. Setting (Deg.)	Range (Deg.)	Pref. Setting (Deg.)			Inner Wheel	Outer Wheel
1964	F-85	½N to ½N	1¼N	¼N to ½P	¼P	0 to ⅛	7½	23	20
	Exc. F-85	1N to 0	½N	¼N to ½P	¼P	⅛ to ³/₁₆	10	23	20
1965-67	F-85	½N to 2N	1½N	¼N to ½P	¼P	⅛ to ³/₁₆	7½-8	23	20
	Exc. F-85	½N to 1½N	½N	¼N to ½P	¼P	⅛ to ³/₁₆	10	23	20
1968-69	Exc. 88 & 98	½N to 2N	1¼N①	¼N to ½P	⅛P	⅛ to ¼	9	20	18¾
	88 & 98 Series	½N to 1½N	1¼N①	¼N to ½P	⅛P	⅛ to ³/₁₆	11	20	18½
1970-71	Exc. 88 & 98	½N to 2N	1¼N①	¼N to ½P	⅛P	⅛ to ¼	9	20	18¾
	88 & 98 Series	½N to 1½N	1¼N①	¼N to ½P	⅛P	⅛ to ³/₁₆	11	20	18½

①—Power Steering. ¾N.   P—Positive   N—Negative.

CAPACITIES

| YEAR | MODEL | ENGINE CRANKCASE ADD 1 Qt. FOR NEW FILTER | TRANSMISSIONS Pts. TO REFILL AFTER DRAINING | | | DRIVE AXLE (Pts.) | GASOLINE TANK (Gals.) | COOLING SYSTEM (Qts.) WITH HEATER |
| | | | Manual | | Automatic | | | |
			3-Speed	4-Speed				
1964	F-85, V6	4	2	$2\frac{1}{2}$	21	$2\frac{1}{2}$	20	11
	F-85, V8	4	2	$2\frac{1}{2}$	21	$2\frac{1}{2}$	20	17
	Jetstar 88	4	$2\frac{1}{2}$	$2\frac{1}{2}$	21	$5\frac{1}{2}$	21	17
	All others	4	$2\frac{1}{2}$	None	21	$5\frac{1}{2}$	21	$20\frac{1}{4}$
1965	F-85, V6	4	2	$2\frac{1}{2}$	18	$2\frac{1}{2}$	20	11
	F-85, V8	4	2	$2\frac{1}{2}$	18	$2\frac{1}{2}$	20	17
	Jetstar 88	4	2	$2\frac{1}{4}$	18	3	25	17
	All others	4	3	$2\frac{1}{4}$	18	$4\frac{3}{4}$	25	18
1966	F-85, L6	4	$3\frac{1}{2}$	None	18	$3\frac{3}{4}$	20	$11\frac{3}{4}$
	F-85, V8	4	$3\frac{1}{2}$	$2\frac{1}{4}$	18	$3\frac{3}{4}$	20●	$15\frac{1}{4}$
	442, V8	4	5	$2\frac{1}{4}$	18	$3\frac{3}{4}$	20	$16\frac{1}{4}$
	Jetstar 88	4	5	$2\frac{1}{4}$	18	$3\frac{3}{4}$	25	17
	All others	4	5	$2\frac{1}{4}$	18	$5\frac{1}{2}$	25	$17\frac{1}{2}$
1967	F-85, L6	4	$3\frac{1}{2}$	None	19	$3\frac{1}{2}$	20	$11\frac{3}{4}$
	F-85, V8	4	$3\frac{1}{2}$	$2\frac{1}{4}$	19	$3\frac{1}{2}$	20	$15\frac{1}{4}$
	442 (400 Eng.)	4	5	$2\frac{1}{4}$	23	3	20	$16\frac{1}{4}$
	Series with 425 Eng.	4	5	$2\frac{1}{4}$	23	$5\frac{1}{4}$	25	$17\frac{1}{2}$
	Vista Cruiser (330 Eng.)	4	$3\frac{1}{2}$	$2\frac{1}{4}$	19	$3\frac{1}{2}$	24	$15\frac{1}{4}$
	Delmont, All	4	5	$2\frac{1}{4}$	19■	3	25	$16\frac{1}{2}$
1968-69	L6	4	$3\frac{1}{2}$	None	19	$3\frac{3}{4}$	20	$12\frac{1}{2}$
	V8—350	4	$3\frac{1}{2}$	$2\frac{1}{4}$	19	$3\frac{3}{4}$	20	$15\frac{1}{4}$
	V8—400 (442)	4	5	$2\frac{1}{4}$	23	$3\frac{3}{4}$	20	$16\frac{1}{4}$
	V8—455	4	$3\frac{1}{2}$	None	23	$5\frac{1}{4}$	25	$17\frac{1}{2}$
1970-71	L6	4	$3\frac{1}{2}$	None	20	$3\frac{3}{4}$	20	$12\frac{1}{2}$
	V8—350	4	$3\frac{1}{2}$	$2\frac{1}{4}$	20	$3\frac{3}{4}$	20①	$15\frac{1}{4}$
	V8—455	4	$4\frac{1}{2}$	None	20	$5\frac{1}{4}$	25	$17\frac{1}{2}$

● —1967 Vista Cruiser = 24 Gal.
■ —When Equipped with 425 Engine = 23.
① —88 Series with V8 350 equipped with 25 gal.

LIGHT BULBS

1964-70

Headlamps (low beam) .... 4002; 50/37.5 W
  (high beam) .... 4001; 37.5 W
Stop, tail, park, turn signals 1157
Dome... 1004
License plate lamp............................. 97
Courtesy, arm rest, map lamp 90
Ignition switch, ash tray 1445
Heater control, map light, tach. 1895
Warning and indicators 158

LIGHT BULBS—F-85

1964-70

Headlamps (outer) .... 4002; 50/37.5 W
Headlamps (inner) .... 4001; 37.5 W
Turn signals, stop, tail, park .... 1157; 32/4 CP
License............................... 97; 4 CP
Indicator and warning lights .... 158; 2 CP
Ash tray, shift indicator, cruise control ...... 1445; 1 CP
Heater control, tachometer, console comp.,
  parking brake warning .... 1895; 2 CP
Dome............................... 211; 12 CP
Courtesy lamp........................... 90; 6 CP

FUSES AND CIRCUIT BREAKERS— OLDSMOBILE AND F-85

1964-65

Headlamps .... 20 AMP C.B.
Power seats and windows .... 20 AMP C.B.
Air conditioners, power top, defogger.......... AGC-25
Ash tray, instrument panel AGA-3
Back-up, cornering, fuel gauge, glove box,
  license, parking brake, radio, tail, trunk,
  underhood SFE-9
Cigar, clock, courtesy, cruise control, dome,
  heater, stop, windshield wiper SFE-20

1966

Same as 1965 except for the following:
Air conditioner, power seats, power top AGC-30
License, tail, trunk AGW-15

1967

Same as 1966 except for the following:
Cornering AGW-15

1968

Same as 1967 except for the following:
Back-up, cornering, hazard, side marker, tail,
  turn signals SFE-20
Air conditioner, dome, cigar, courtesy, heater,
  defogger, trunk, underhood AGC-25
Instrument panel............................. SFE-4
Radio, transmission control AGC-10

1969

Same as 1968 except for the following:
Back-up, transmission control SAE-9
Cornering, license, side marker, tail............ AGW-15
Cruise control, underhood SFE-20
Parking brake............................. AGC-10

1970

Same as 1969 except for the following:
Courtesy, license, tail SFE-20

Distributor

References

Detailed information on distributor drive, direction of distributor rotation, cylinder numbering, firing order, point gap, point dwell, timing mark location, spark plugs, spark advance, and idle speed is in the Specifications table. Further information on troubleshooting, general tune-up procedures, and adjusting timing is in the Unit Repair Section.

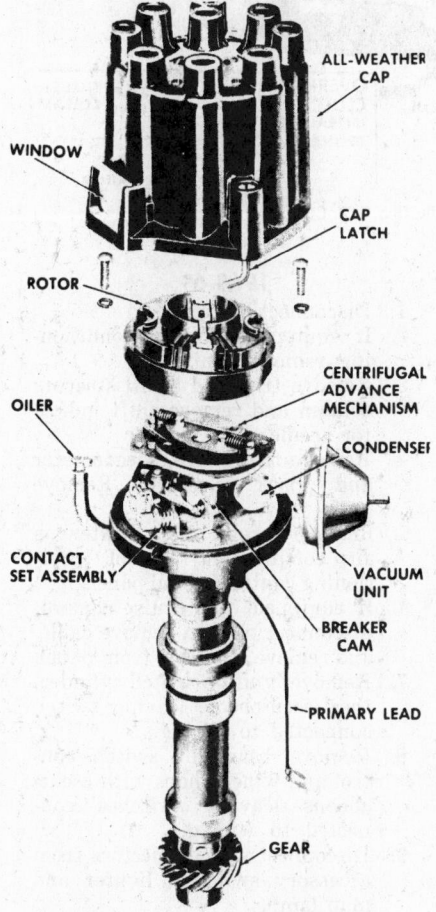

Distributor details, externally adjusted type
(© Oldsmobile Div. G.M. Corp.)

Distributor Removal

V6 and V8

1. Remove distributor cap, primary wire and vacuum line at the distributor.
2. Scribe a mark on the distributor body, locating the position of the rotor, and scribe another mark on the distributor body and engine block, showing the position of the body in the block.
3. Remove the hold-down screw and lift the distributor out of the block.

NOTE: V6 engine design requires special attention to spark plug wire placement in distributor cap towers. Due to distributor cam shape, it is important that the original distributor (plug wire) locations be maintained. This is an aspect peculiar to this 90°, V6 engine.

Distributor Installation

1. If engine has been disturbed, rotate the crankshaft to bring the piston of No. 1 cylinder to the top of its compression stroke.
2. Position the distributor in the block with the rotor at No. 1 firing position. Make sure the oil pump intermediate driveshaft is properly seated in the oil pump.
3. Install the distributor lock but do not tighten.
4. Rotate the distributor body clockwise until the breaker points are starting to open. Tighten the retaining screw.
5. Connect the primary wire and the vacuum line to the distributor, then install distributor cap.
6. Start the engine and check the timing with a timing light.

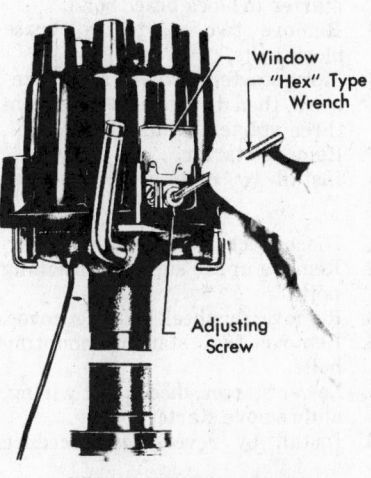

Adjusting distributor points
(© Oldsmobile Div. G.M. Corp.)

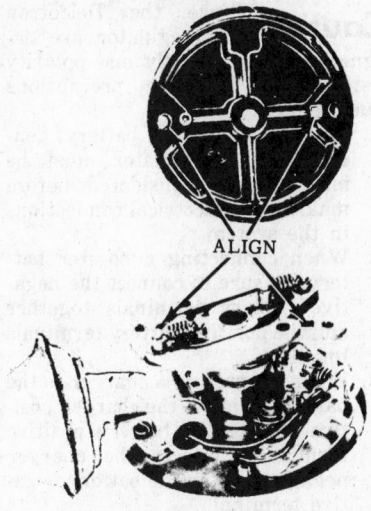

Rotor installation
(© Oldsmobile Div. G.M. Corp.)

OHV 6

The distributor assembly is mounted on the right side of the block and is driven directly from the camshaft.

To remove the distributor, first detach the vacuum lines to the vacuum advance unit and lift off the distributor cap.

The distributor body is held to the block by a single cap screw that holds the octane selector plate down against the block. Remove the retaining screw and lift the distributor out of the block.

AC Generator (Delcotron)

1964-70

An alternating current generator is used. This unit is Delco-Remy Delcotron. The purpose of this unit is to satisfy the increase in electrical loads that have been imposed upon the car battery by conditions of traffic and driving patterns.

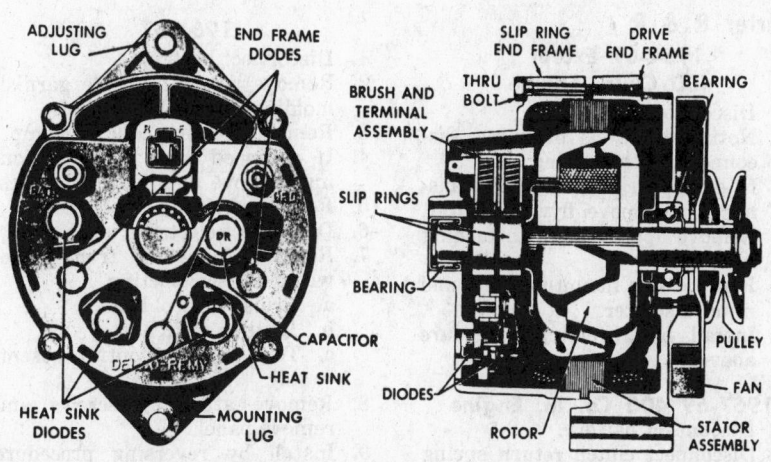

Delcotron (© Oldsmobile Div. G.M. Corp.)

Caution Since the Delcotron and regulator are designed for use on only one polarity system, the following precautions must be observed:

1. The polarity of the battery, generator, and regulator must be matched and considered before making any electrical connections in the system.
2. When connecting a booster battery, be sure to connect the negative battery terminals together and the positive battery terminals together.
3. When connecting a charger to the battery, connect the charger positive lead to the battery positive terminal. Connect the charger negative lead to the battery negative terminal.
4. Never operate the Delcotron on open circuit. Be sure that all connections in the circuit are clean and tight.
5. Do not short across or ground any of the terminals on the Delcotron regulator.
6. Do not attempt to polarize the Delcotron.
7. Do not use test lamps of more than 12 volts for checking diode continuity.
8. Avoid long soldering times when replacing diodes or transistors. Prolonged heat is damaging to these units.
9. Disconnect the battery ground terminal when servicing any A.C. system. This will prevent the possibility of accidental reversing of polarity.

Battery, Starter

Specifications on the battery and starter are in the Battery and Starter Specifications table.

A more general discussion of starters and their troubles is in the Unit Repair Section.

Starter R & R

1964-69 Except 400 Cu.In. Engine

1. Disconnect battery.
2. Noting positions of wires, disconnect starter wiring.
3. If equipped with manual transmission, remove flywheel cover.
4. Remove upper support attaching bolt.
5. Remove two mounting bolts and remove starter.
6. Install by reversing procedure above.

1967-69 400 Cu. In. Engine

1. Disconnect battery.
2. Disconnect clutch return spring at clutch release yoke.

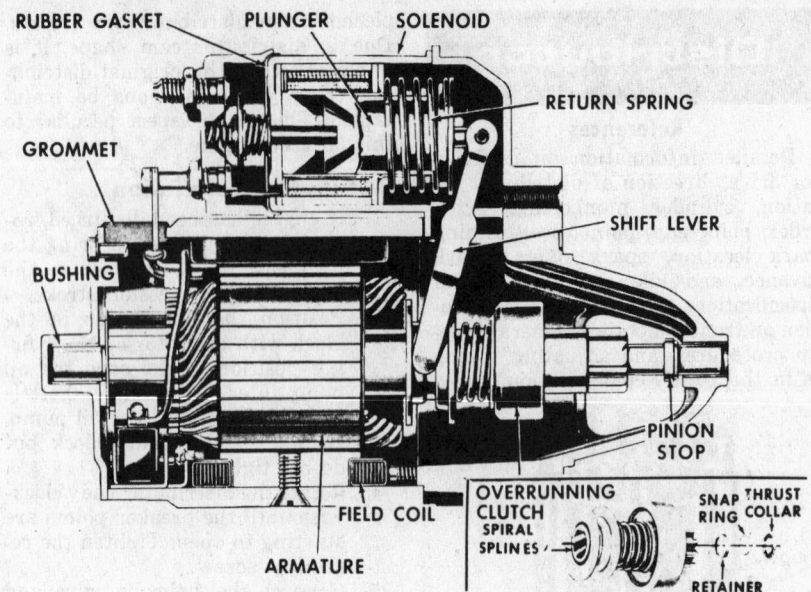

Starter (© Oldsmobile Div. G.M. Corp.)

3. Disconnect exhaust pipe from left manifold.
4. Loosen upper and remove lower starter to block brace bolts.
5. Remove two starter to brace block bolts.
6. Move starter forward and downward, then disconnect wires from three starter terminals.
7. Remove starter.
8. Install by reversing procedure.

1970

1. Disconnect battery and hoist car.
2. Remove upper support attaching bolt.
3. Remove flywheel housing cover.
4. Remove two starter mounting bolts.
5. Lower starter, disconnect wiring, and remove starter.
6. Install by reversing procedure.

Instruments

1964-67

1. Disconnect battery.
2. Remove windshield side garnish moldings on 1964-66 models.
3. Remove steering column clamp.
4. If equipped with air conditioning, remove manifold and hoses.
5. Remove shift indicator needle.
6. Disconnect electrical leads.
7. Remove following from panel without disconnecting.
   a. Cruise control head.
   b. Ignition switch.
   c. Temperature control assembly.
8. Remove attaching screws and remove panel.
9. Install by reversing procedure above.

1968-69

1. Disconnect battery.
2. If equipped with air conditioning, remove manifold.
3. Remove trim pad below steering column and remove shift indicator needle.
4. Remove fuel gauge, speedometer and clock clusters. Remove radio.
5. Remove screws from heater or air conditioning control, positioning control behind panel.
6. If equipped with cruise control, disconnect wiring, remove cable, and remove assembly from panel.
7. Remove ignition switch cylinder, then escutcheon, leaving switch connected to wiring.
8. Remove headlight switch control and wiper knobs with escutcheons leaving switches connected to wiring.
9. Disconnect wire connectors from accessory switches, lighter and map lamp.
10. Remove ash tray.
11. If equipped with air conditioning, remove left outlet housing by removing the knob and screw.
12. Remove panel attaching screws.
13. Install by reversing procedure above.

1970

1. Disconnect battery.
2. Remove headlight switch knob and escutcheon, leaving switch behind panel with wiring attached.
3. Remove heater or air conditioning control, positioning behind panel. Do not disconnect wiring, hoses or cables.
4. Remove glove compartment and remove trim cap.

5. Remove shift indicator needle, if so equipped.
6. Remove attaching screw and bolts.
7. Disconnect wire connectors.
8. Remove screws attaching main wiring harness.
9. Disconnect speedometer cable.
10. Remove nuts from steering column bracket and lower column until wheel rests on seat.
11. Remove screws from panel to column support.
12. Remove attaching screws above cluster.
13. Remove panel assembly by moving toward right side of car.
14. Install by reversing procedure above.

Instrument Panel Removal— F-85

1964-67

1. Disconnect battery.
2. Disconnect speedometer cable, wiring connectors and clip.
3. If equipped with air conditioning, disconnect cables, vacuum hoses, duct and manifold.
4. Remove screws that secure steering column and nuts that secure column clamp and lower column.
5. Remove attaching nuts and remove panel with controls attached.
6. Install by reversing procedure above.

1968-69

1. Disconnect battery.
2. Remove glove compartment as illustrated.
3. If equipped with air conditioning, remove right and left ducts and manifold.
4. Remove screws from heater or air conditioning control. Do not disconnect hoses or cables.
5. Disconnect antenna lead-in and speaker connector at radio.
6. Disconnect wiring connectors.
7. Remove escutcheon from ignition switch, positioning switch behind dash.
8. Disconnect speedometer cable.
9. Remove bolt attaching wiring harness clamp.
10. Remove trim cap below steering column.
11. Remove bolts and nuts attaching steering column, lowering column until wheel rests on seat.
12. Remove panel.
13. Install by reversing procedure above.

1970

Instruments can be removed after instrument trim panel is removed. Trim panel is removed by removing attaching screws and pulling top of panel rearward, then up.

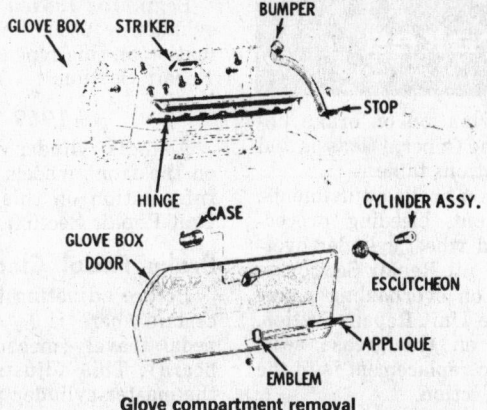

Glove compartment removal
(© Oldsmobile Div. G.M. Corp.)

Printed Circuit Removal

1. Remove all cluster lamp sockets, and circuit attaching screws.
2. Remove hex nuts from fuel gauge mounting studs.
3. Lift printed circuit from cluster.

Printed Circuit Installation

When installing, reverse removal procedure. Do not overtighten attaching screws or nuts. This could cause cracking and opening the circuit.

Fuel Gauge Removal and Installation

The fuel gauge can be removed without removing the instrument cluster as follows:
1. Disconnect positive battery cable.
2. Remove cluster cover.
3. Disconnect fuel gauge wiring connector and the light socket from the printed circuit.
4. Remove the two attaching nuts and withdraw fuel gauge from front of the cluster.
5. Install in reverse of above. Use extreme care to prevent damage.

Ignition Switch Replacement

1964-67

1. Disconnect battery.
2. Place switch in off position and insert wire in small hole in cylinder face. While pushing in on wire turn cylinder counterclockwise and pull cylinder from case.
3. Remove nut from passenger side of dash.
4. Pull switch from under dash and remove wiring connector.
5. To remove theft resistant connector, where used, switch must be out from under dash. With screwdriver, depress tangs and separate the connector.
6. Install in reverse order of above.

1969-70

The ignition switch is mounted on the right side of the steering column and includes a lock for both the selector lever and the steering gear.

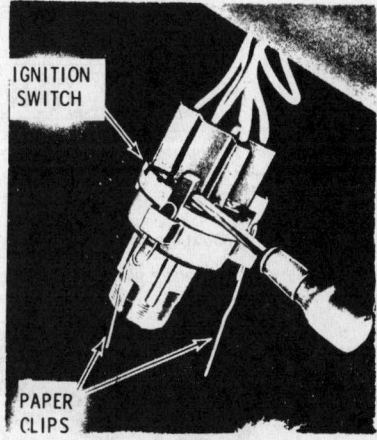

Ignition switch
(© Oldsmobile Div. G.M. Corp.)

Lighting Switch Replacement

1. Disconnect battery.
2. Pull knob out to on position.
3. Reach under instrument panel, depress the switch shaft retainer, and remove knob and shaft assembly.
4. Remove retaining ferrule nut.
5. Remove switch from instrument panel.
6. Disconnect the multi-plug connector from the switch.
7. Install in reverse of above. (In checking lights before installation, switch must be grounded to test dome lights).

WITH KNOB PULLED TO "HEADLIGHT ON" POSITION, DEPRESS PULL ROD RELEASE BUTTON ON TOP OF SWITCH AND PULL KNOB AND ROD FROM SWITCH

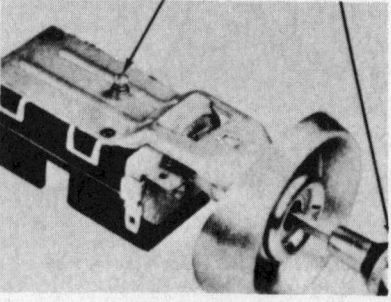

Headlight switch
(© Oldsmobile Div. G.M. Corp.)

Brakes

Specific information on brake lining sizes is in the General Chassis and Brake Specifications table.

Information on brake adjustments, band replacement, bleeding procedure, master and wheel cylinder overhaul is in the Unit Repair Section.

Information on overhauling power brakes is in the Unit Repair Section.

Information on the grease seals which may need replacement is in the Unit Repair Section.

Beginning 1967, a dual-type master cylinder is used. For detailed information on this type cylinder, see Unit Repair Section.

1969-70

A single-cylinder disc brake is used on the front wheels of some models. Information on this brake is in the Unit Repair Section.

Brake Pedal Clearance

Before adjusting the brakes, make certain there is ½ in.—¾ in. free-pedal travel (measured at the toe-board). This adjustment is made at the master-cylinder pushrod.

Parking Brake Cable Adjustment

1. Release parking brake. Check for proper pedal clearance.
2. Adjust rear cables by tightening equalizer adjusting nut to obtain heavy drag at rear brakes.
3. Loosen equalizer adjusting nut seven full turns. Tighten locknut.

Power Cylinder Removal

From inside the car, detach the brake pushrod from the brake pedal. From underneath the car, detach the vacuum hose at the vacuum cylinder and disconnect the hydraulic line from the front of the slave cylinder.

Remove the nuts that hold the vacuum unit up to the toe-board and let the vacuum unit come out through the bottom of the car.

It is tight, but the unit will come down.

On models using Moraine power brakes, be sure to record and use the proper number of shims found between master and vacuum cylinders.

Fuel System

Data on capacity of the gas tank are in the Capacities table. Data on correct engine idle speed and fuel pump pressure are in the Tune-up Specifications table.

Information covering operation and troubles of the fuel gauge is in the Unit Repair Section.

Detailed information on the carburetor and how to adjust it is in the Unit Repair Section. Carter, Holley, Rochester and Stromberg carburetors are covered.

Dashpot adjustment is in the Unit Repair Section under the same heading as that of the automatic transmission used in the car, and in the carburetor section.

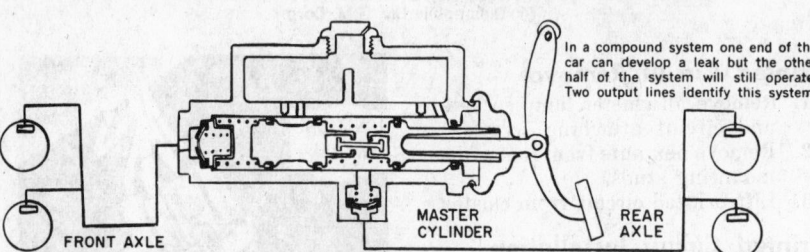

In a compound system one end of the car can develop a leak but the other half of the system will still operate. Two output lines identify this system.

Dual type master cylinder (© Oldsmobile Div. G.M. Corp.)

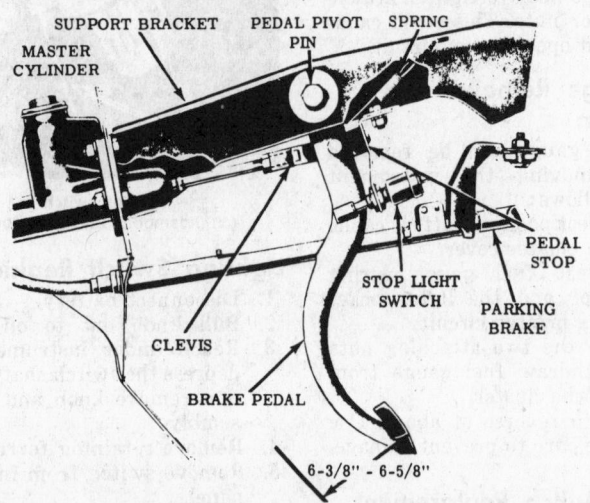

Brake pedal and support—F-85 (© Oldsmobile Div. G.M. Corp.)

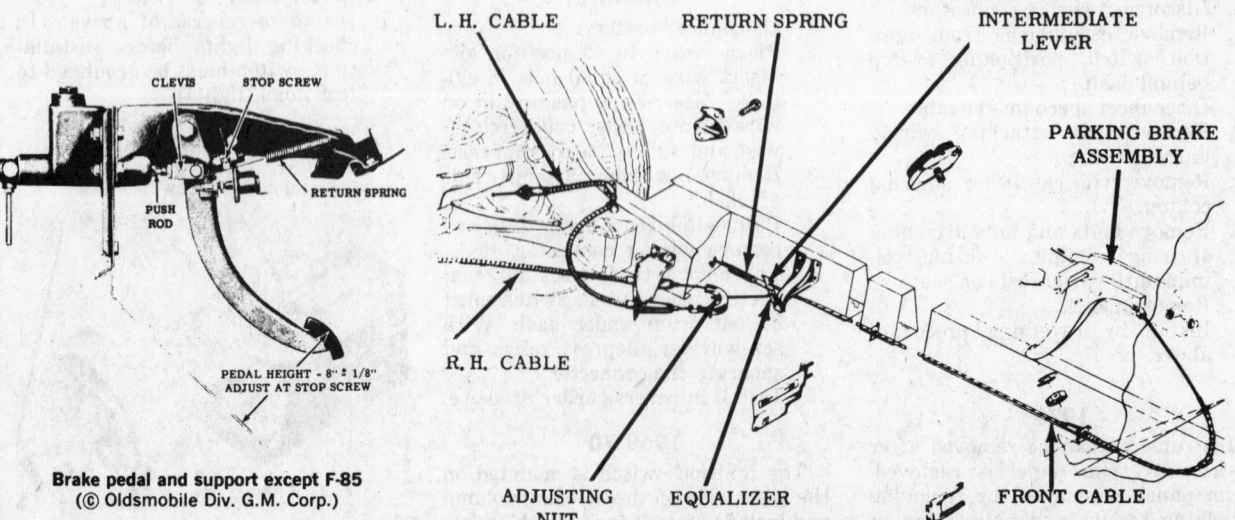

Brake pedal and support except F-85 (© Oldsmobile Div. G.M. Corp.)

Parking brake linkage (© Oldsmobile Div. G.M. Corp.)

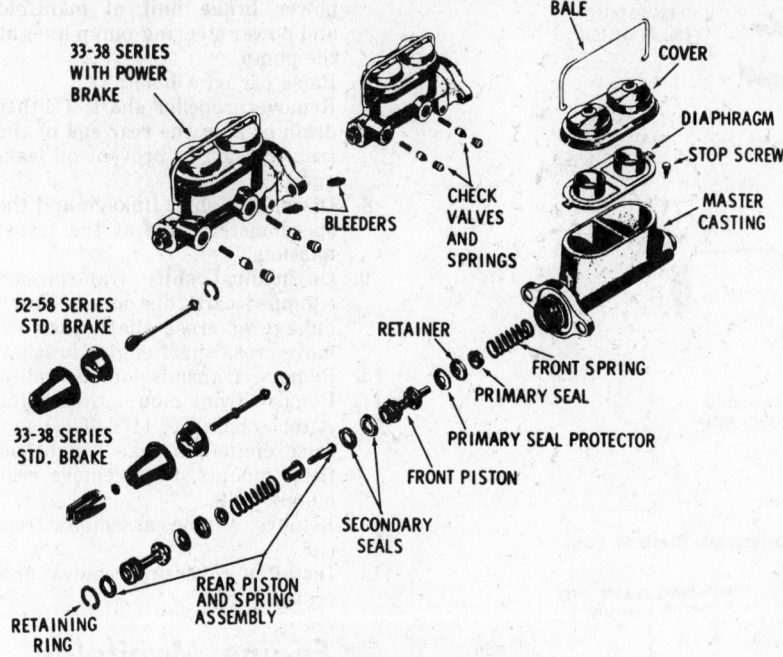

Master cylinder (Moraine)
(© Oldsmobile Div. G.M. Corp.)

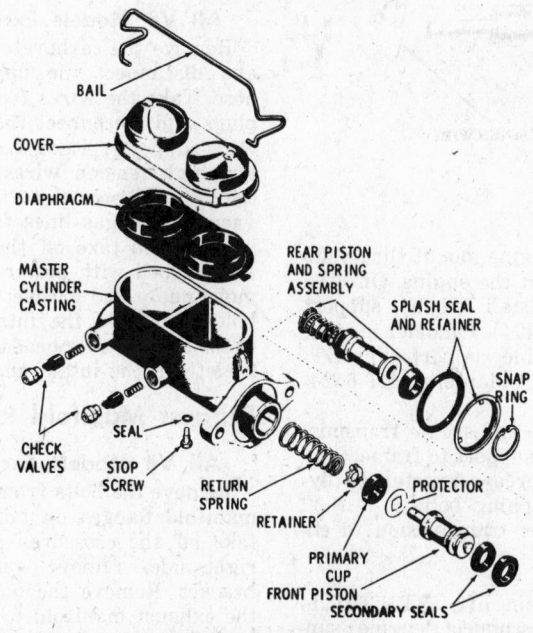

Master cylinder (Bendix)
(© Oldsmobile Div. G.M. Corp.)

TIP FUEL PUMP AND
POSITION ARM UNDER
FUEL PUMP ECCENTRIC

Installing fuel pump
(© Oldsmobile Div. G.M. Corp.)

Fuel Pump Removal

Disconnect the fuel and vacuum lines, take out the two mounting bolts that hold it to the timing case cover and lift off the pump.

The back bolt can be reached with a universal socket and short extension.

Cooling System

Detailed information on cooling system capacity is in the Capacities table.

Information on the water temperature gauge is in the Unit Repair Section.

Thermostatic Vacuum Switch

If equipped with exhaust emission control and air-conditioning, a thermostatic vacuum switch is used to increase ignition timing advance at idle when higher coolant temperatures are encountered. This change from part to full vacuum advance at engine idle, tends to improve cooling at idle.

If a heating condition exists at idle speeds only, disconnect the distributor vacuum hose at the thermostatic switch. When engine temperature is above 230° F., there should be no vacuum at this port at idle. If engine temperature is above 230° F. and vacuum is present at port D, check distributor advance.

Radiator Core Removal

NOTE: on models with an oil cooler, the oil cooler lines will have to be disconnected and capped.

On models with the horns attached to the core brackets, remove the horns.

The radiator core can be removed from all Oldsmobile models without taking off the water pump. Remove the radiator upper baffle plate and radiator hoses, headlight wires, hood cables, etc., and unbolt the core assembly from its support and raise it up out of the car. It may be necessary to rotate the fan blades in order to keep them out of the way.

Water Pump Removal

Drain the cooling system and slack off the generator and power steering belts, remove the fan and fan pulley, remove bolts that hold the water pump to front of engine and lift off pump.

Only one side of the gasket should be coated with gasket compound when reinstalling.

It is a good idea to dip the bolts in sealer before inserting them in the pump.

Engine

Valve Information

Detailed information on valves and valve guides is in the Valve Specifications table.

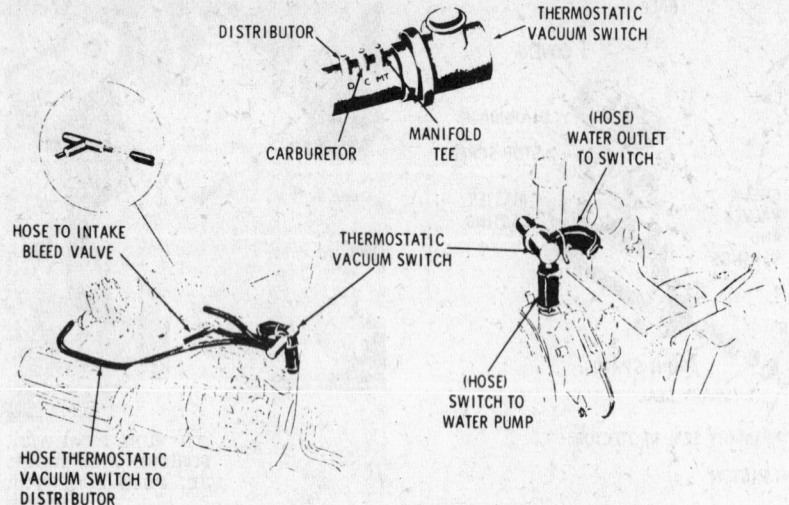

Thermostatic vacuum switch K-19 (© Oldsmobile Div. G.M. Corp.)

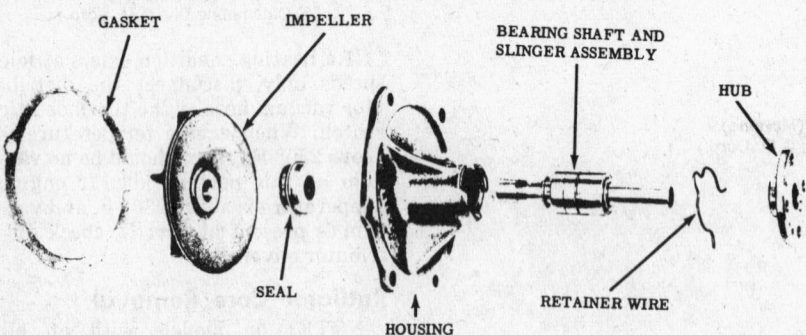

Water pump (© Oldsmobile Div. G.M. Corp.)

Bearing Information

Detailed information on engine bearings will be found in the Crankshaft Bearing Journal Sizes table.

Engine Information

Engine crankcase capacities are listed in the Capacities Table.

Approved torque wrench readings and head bolt tightening sequences are covered in the Torque Specifications table.

Information on the engine marking code is in the Model Year Identification table.

Engine Removal

1964-70 V6 and V8

1. Drain cooling system, remove hood, and disconnect battery.
2. Disconnect radiator hoses, heater hoses, vacuum and power steering hoses, starter cable at junction block, engine to ground strap, fuel hose from fuel line, wiring and accelerator linkage.
3. Remove fan and fan pulley, coil and upper radiator support.
4. Raise car on hoist.
5. Disconnect exhaust pipes at manifold.
6. Remove torque converter cover and install a converter holding tool.
7. Remove engine mount thru bolts and support the engine. On F-85 models, install engine support bar (BT-6424). On older models, install engine support bar (BT-3016). Use adapters (BT-6424-2).
8. Lift engine and secure transmission chain support to frame.
9. Remove three converter to flywheel attaching bolts.
10. Remove six transmission to engine bolts.
11. Lower the car.
12. Secure chain lift (BT-6606) to engine, disconnect engine support bar and remove engine.

Engine Removal

1965-70 OHV 6 Engine

1. Drain cooling system and the engine oil.
2. Remove carburetor air cleaner, disconnect battery, remove hood.
3. Remove radiator shroud, fan blade and pulley.
4. Disconnect wires at starter solenoid, generator, temperature switch, oil pressure switch and at the coil.
5. Disconnect accelerator linkage at the manifold, exhaust pipe at manifold, fuel pump (from tank) at fuel pump, vacuum line to

power brake unit at manifold and power steering pump lines at the pump.
6. Raise car on a hoist.
7. Remove propeller shaft. (Either drain or plug the rear end of the transmission to prevent oil leakage.)
8. Disconnect shift linkage and the speedometer cable at the transmission.
9. On manual shift transmission equipped cars, disconnect clutch linkage at cross shaft, then remove cross shaft engine bracket.
10. Remove transmission assembly.
11. Remove front mount thru bolts.
12. Attach chain lift (BT-6606) and raise engine to take weight off front mounts, then remove rear mount bolts.
13. Remove engine assembly from car.
14. Install by reversing removal procedure.

Engine Manifolds

Intake Manifold Removal

All V8 Models Except F-85

Remove the carburetor air cleaner and disconnect the upper radiator hose. Take the wires from the spark plugs and disconnect the spark plug wire supports. Take distributor cap and high tension wires off the distributor. Disconnect the throttle, vacuum and gas lines from the carburetor and take off the carburetor. If equipped with power steering, remove pump and bracket. Remove the bolts that hold the intake manifold to the two cylinder heads. The coil can be left on the intake manifold.

Exhaust Manifold Removal

All V8 Models Except F-85

Remove the bolts from the exhaust manifold flanges on both sides and take off the crossover pipe. On the right side, remove generator and bracket. Remove the bolts that hold the exhaust manifold to the cylinder head and lift off the exhaust manifolds.

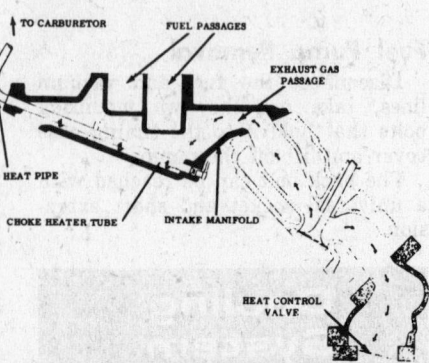

Exhaust flow (heat control closed)
(© Oldsmobile Div. G.M. Corp.)

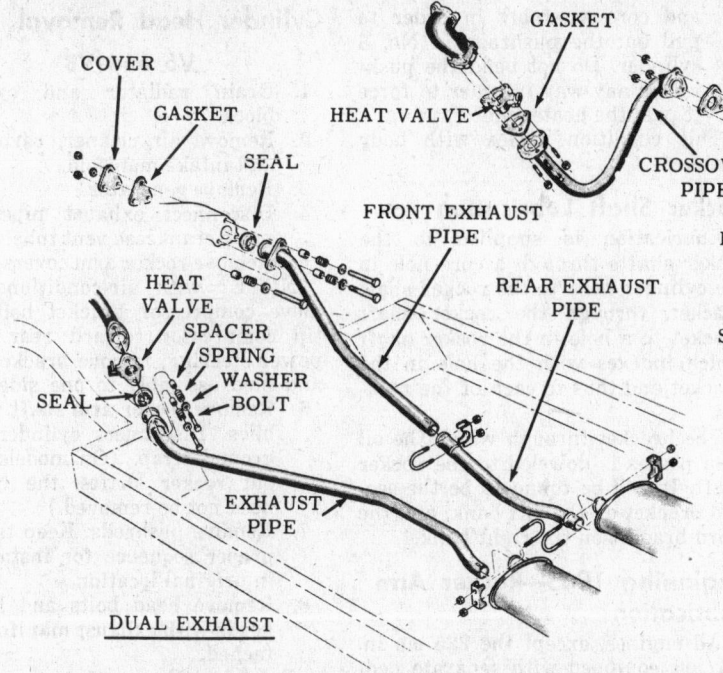

Typical exhaust system except F-85 (© Oldsmobile Div. G.M. Corp.)

OHV 6 Models

This engine uses a combined intake and exhaust manifold, equipped with thermostatic heat-riser valve.

To remove the manifold assembly disconnect the exhaust pipe flange and remove all connections to the carburetor. Take off the vacuum lines on the manifold and also at the carburetor.

Remove the carburetor and unbolt the manifold.

Before reinstalling the manifold, thoroughly clean out the ports to prevent turbulence particularly in the intake manifold.

Intake Manifold Removal

F-85 V6 & V8

1. Drain cooling system.
2. Remove carburetor air cleaner. Disconnect throttle return from the carburetor. Disconnect and remove the coil.
3. Disconnect temperature indicator wire from sending unit.
4. Disconnect accelerator and transmission linkage at carburetor. Disconnect throttle return spring.
5. Slide front thermostat by-pass hose clamp back on the hose. Disconnect upper radiator hose at outlet.
6. Disconnect heater hose at the temperature control valve inlet. Force the end of the hose down to permit coolant to drain from intake manifold.
7. Remove 12 manifold-to-head attaching bolts.
8. Remove intake manifold and carburetor as an assembly by sliding rearward to disengage the thermostat by-pass hose from the water pump. Remove intake manifold gasket sound absorber.
9. Be sure there is no coolant present. Then remove intake manifold gasket clamps and remove the gasket. Remove rubber gasket seal.
10. Install by reversing above plan. Torque to 25-30 ft. lb.

Exhaust Manifold Removal

F-85 V6 & V8

(Cylinder heads off)
1. Remove exhaust manifold-to-exhaust pipe attaching bolts.
2. On the right side, remove generator rear attaching bolt.
3. Unlock and remove exhaust manifold-to-cylinder-head bolts. Remove the manifold.
4. Install by reversing above. Torque to 18-24 ft. lbs.

Exhaust Emission Control

The air injection system (K-19) is used on all General Motors cars (with a few design exceptions).

The General Motors Corporation has adopted a special system of terminal exhaust treatment. This plan supersedes (in most cases) the method used to conform to 1966-67 California laws. The new system cancels out (except with stick shift and special purpose engine applications) the use of the A.I.R. method previously used.

The new concept, combustion control system (C.C.S.) utilizes engine modification, with emphasis on carburetor and distributor changes. Any of the methods of terminal exhaust treatment requires close and frequent attention to tune-up factors of engine maintenance.

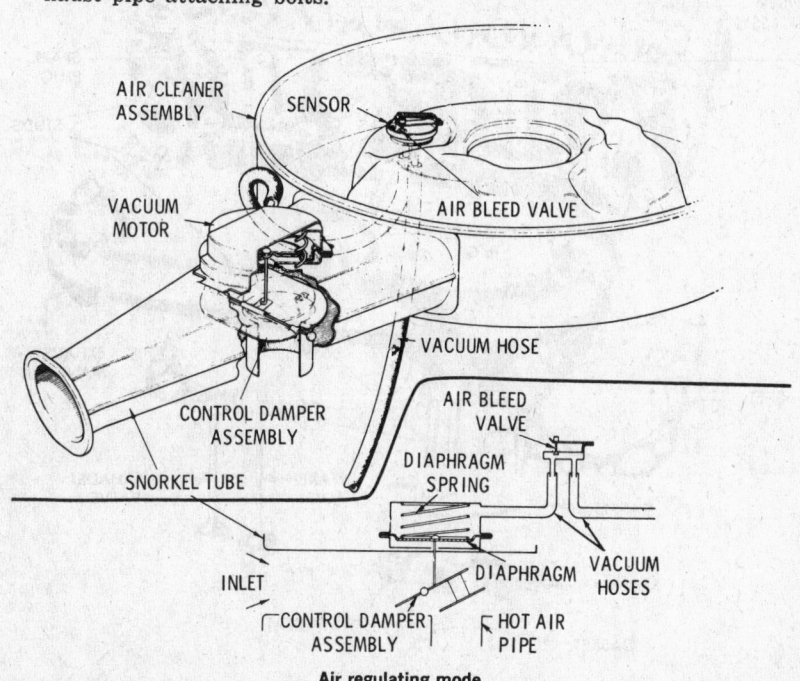

Air regulating mode

All car manufacturers post idle speeds and other pertinent data relative to the specific engine-car application in a conspicuous place in the engine compartment.

For details, consult the Unit Repair Section.

Cylinder Head

Rocker Shaft Removal

1964 V8 and All V6 Models

1. Remove the bolts that hold the rocker cover assembly to the cylinder head and lift off the rocker cover.
2. A little at a time, loosen the bolts that hold the rocker brackets to the cylinder head, and lift off the rocker assemblies.
3. If the pushrods are to be pulled up through the cylinder head, be careful to break the seal created by the oil at the lifter plunger in order to prevent pulling the lifter plunger up with the push rod.
4. On some models it may be necessary to loosen the heater motor and core assembly in order to pull out the pushrods in No. 8 cylinder. Do not bend the push-rod in any way in order to force it past the heater motor.

This condition varies with body styles.

Rocker Shaft Lubrication

Lubrication is supplied to the rocker shafts through a core hole in the cylinder head to the rocker shaft bracket, through the rocker shaft bracket to a hole in the rocker shaft which indexes with the hole in the bracket and thus to each of the rockers.

The bracket through which the oil then passes is doweled to the rocker shaft. It will be found to be the second bracket on the left bank, and the third bracket on the right bank.

Beginning 1965—Rocker Arm Lubrication

All engines, except the 225 cu. in. V6, are equipped with separate pedestal-mounted valve rocker levers. With this design, oil is fed up the pushrod to the rocker arms.

Cylinder Head Removal

V6 and V8

1. Drain radiator and cylinder block.
2. Remove air cleaner, carburetor and intake manifold.
3. Remove generator.
4. Disconnect exhaust pipes and move crankcase vent tube.
5. Remove rocker arm covers.

NOTE: with air-conditioner, remove compressor bracket bolts and tip compressor toward rear. With power steering, remove bracket bolts and hold assembly to one side.

6. Remove rocker arm shaft assemblies. Disconnect cylinder head ground strap. (On models without rocker shafts, the rockers need not be removed.)
7. Remove pushrods. Keep them in proper sequence for installation in original location.
8. Remove head bolts and lift off heads with exhaust manifolds attached.

To reinstall: reverse the above. Oldsmobile recommends use of a sealer on both sides of head gaskets and on head bolts. Two guide studs will allow easier location of heads.

OHV 6 Engine

1. Drain cooling system and remove manifold assembly and valve mechanism.
2. Remove fuel and vacuum line from retaining clip and disconnect wires from temperature sending units.
3. Disconnect upper radiator hose and battery ground strip.
4. Remove coil.
5. Remove cylinder head bolts, then head and gasket.

NOTE: place cylinder head on two blocks to prevent damage.

Valve System

On all models, work on the cylinder head on the bench, compress the spring with a C-type or lever-type valve spring compressor, remove the key and release the valve spring, which will permit the valve to be pulled through the head.

Valve Guide Replacement

1964 V8 Models

The old valve guide is driven out through the top of the cylinder head, and the new one is installed by driving from the top of the cylinder head down toward the firing chamber.

After the new guides have been driven into place immediately try a valve in the guide to make sure it was not warped or twisted in the driving process. If it is, it will be necessary to ream it to the correct size.

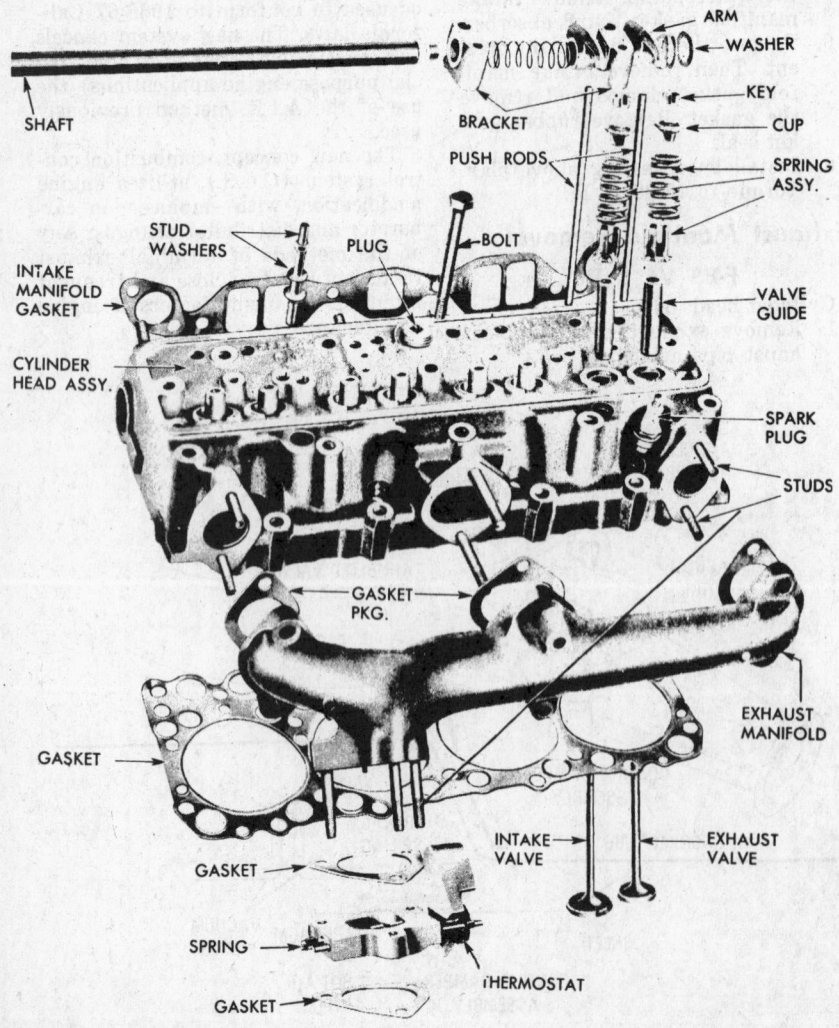

SHAFT

ARM
WASHER
KEY
CUP
SPRING ASSY.

BRACKET
PUSH RODS

STUD—WASHERS
PLUG
BOLT

INTAKE MANIFOLD GASKET

CYLINDER HEAD ASSY.

VALVE GUIDE

SPARK PLUG

STUDS

GASKET PKG.

EXHAUST MANIFOLD

GASKET

GASKET

INTAKE VALVE

EXHAUST VALVE

SPRING

THERMOSTAT

GASKET

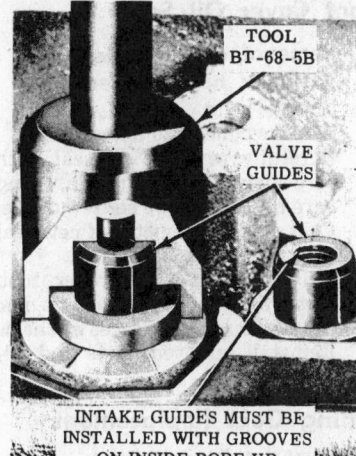

INTAKE GUIDES MUST BE
INSTALLED WITH GROOVES
ON INSIDE BORE UP

Installing valve guides—1963-64
(© Oldsmobile Div. G.M. Corp.)

Hydraulic Valve Lifter Adjustment

Pedestal Type

1. Crank the engine until No. 1 piston is at the top of its compression stroke, (distributor rotor pointing to No. 1 distributor cap tower).
2. Both valves of No. 1 cylinder may be adjusted by backing off on the pedestal nuts until there is play in the pushrods, (intake and exhaust) of that cylinder. Now tighten these two pedestal nuts to remove all pushrods to rocker arm clearance. This may be noted by rotating the pushrods with the fingers while the nuts are being tightened. When the pushrod does not easily rotate with the fingers, all of the play has been removed.
3. The adjusting nut should then be tightened one full turn. This

should place the hydraulic plunger in the center of its travel.
4. Rotate the crankshaft to bring the next cylinder, in sequence of firing order, to the top of its compression stroke. Adjust both valves of that cylinder.
5. Continue to adjust valve lash throughout the remaining cylinders, in the order of firing.

Valve Springs

All Models

Perhaps the easiest and quickest way to check a valve spring is to lay the spring on a flat, level surface and compare its free length with that of a new valve spring. If a new valve spring is not available, lay all the springs alongside each other to see if they are all the same length. If they are, all of the springs are in good condition, because it is unlikely that all of them would collapse exactly the same amount.

If they vary in length, they must be checked either on a spring compression type checker or tested for free length against a new valve spring.

Hydraulic Valve Lifters

Hydraulic valve lifters are used on all Oldsmobile engines. Oil under metered pressure is supplied to these valves through core holes in the cylinder block. The lifter operates normally at zero clearance. Lifters can be taken out of the engine without removing the cylinder head after the rocker assembly and intake manifold have been removed. Simply pull out the pushrods and lift out the lifter assemblies. Lifter assemblies are not interchangeable one bore to the other nor are the internal parts of the lifter interchangeable with each other.

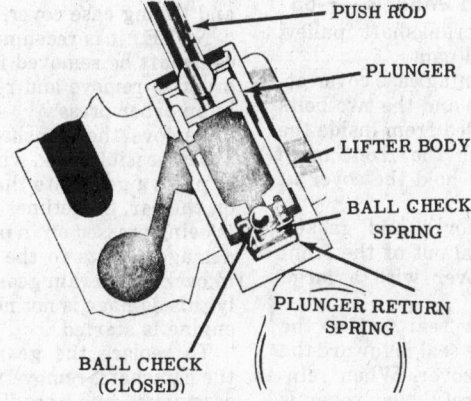

Hydraulic valve lifter
(© Oldsmobile Div. G.M. Corp.)

Timing Case

V8 Engines

The timing case cover and the water pump housing are a one-piece casting.

1. Drain the cooling system and disconnect the radiator and heater hoses, take off the radiator core and remove the fan blades and pulley.
2. Remove the vibration damper.
3. Place a jack under the engine, take a light load on the jack and remove the two bolts that attach the front of the engine to the frame.
4. Remove the oil pan (see Engine Lubrication section). Crank the engine until No. 3 or No. 7 cylinder is in the firing position. This will be when the distributor rotor points to the wire leading to No. 3 or No. 7 spark plug.
5. Take off the fuel pump. Now remove the front cover attaching bolts and lift the cover assembly off the front of the engine.

Valve Timing Procedure

V8 and V6

On all Oldsmobile models except the L6, a chain is used to drive the camshaft. The construction is such that the chain can be worn even bad-

ly without seriously affecting the valve timing. If the chain is worn badly enough to cause the timing to jump or it becomes necessary to replace either the chain or the sprockets or both, proceed as follows:

1. Remove the timing case cover and take off the camshaft gear.
NOTE: the fuel pump operating cam is bolted to the front of the camshaft sprocket and the sprocket is located to the camshaft by means of a dowel.
2. Remove the timing chain and the camshaft sprocket and if the crankshaft sprocket is to be replaced remove it also at this time.

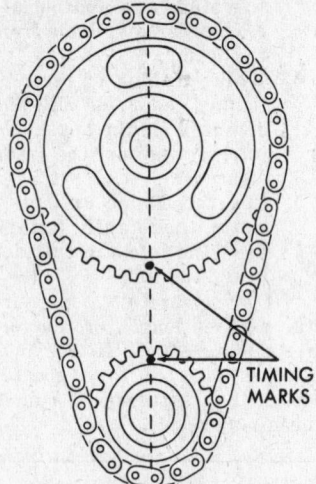

Timing marks—except inline 6

3. Reinstall the crankshaft sprocket being careful to start it with the keyway in perfect alignment since it is rather difficult to correct for misalignment after the gear has been started on the shaft. Turn the timing mark on the crankshaft gear until it points directly toward the center of the camshaft. Mount the timing chain over the camshaft gear and start the camshaft gear up on to its shaft with the timing marks as close as possible to each other and in line between the shaft centers. Rotate the camshaft to align the shaft with the new gear.

A dowel pin is used for alignment. Secure the camshaft gear and check to see that the mark on the crankshaft sprocket and the mark on the camshaft sprocket are as described above. Valves timed in this manner are correct regardless of which piston is at top center. It may be necessary, however, to retime the ignition since there is a possibility it will be 180° out of position.

V6 Engines

The cover includes part of the water pump, oil pump and filter mounting and mounting for fuel pump.

Front Cover Removal

1. Drain cooling system.
2. Disconnect heater hose, by-pass hose and both radiator hoses. Disconnect oil pressure switch wire.
3. Remove crankshaft pulley, fan and fan pulley, and all belts.
4. Remove distributor cap, vacuum hose, generator and mounting bracket.
5. Remove fuel pump hoses, fuel pump and two front oil pan bolts.
6. Remove nine cover-to-block attaching bolts and remove cover.
NOTE: whenever the front cover is removed, it is necessary to remove the oil pump cover and to completely pack the pump gear housing with petrolatum. This is to prime the oil pump and insure immediate oil pressure to engine parts.

Front Cover Installation

1. Install new cover gasket with a good sealing compound.
2. Install the cover.
3. Oil attaching bolts and install. Torque evenly to 30-35 ft. lbs.
4. Apply special seal lubricant on pulley seal surface.
5. Install pulley, pulley bolt and pull the pulley into place. Torque to 140-160 ft. lbs.
6. Connect oil pressure switch.
7. Install generator mounting bracket and adjust the link.
8. Install fuel pump, using a new gasket and sealer.
9. Oil fuel pump bolts and torque to 20-25 ft. lbs.
10. Connect fuel lines.
11. Install distributor (see distributor paragraph in this section.)
12. Connect vacuum advance hose and primary wire.
13. Install distributor cap and wires.
14. Connect all of the hoses.
15. Install fan pulley, fan and four bolts, torqueing to 20 ft. lbs.
16. Install and adjust belts.
17. Refill and bleed cooling system.

1966-70, OHV 6 Models—F-85

1. Remove the crankshaft pulley. Remove the oil pan.
2. Remove the timing case cover attaching screws and the two bolts that are installed from inside the engine through the front main bearing cap to hold the cover at the bottom.
3. Remove the cover and gasket. Pry the old seal out of the front side of the cover with a large screwdriver.
4. Install the new seal so that the open end of the seal is toward the inside of the cover. When reinstalling, be careful that cover is positioned to hold seal concentric to the shaft.
5. Tighten the screws and the two bolts inside the engine to 6-7½ ft. lbs.

Front Cover Oil Seal

Removal and Installation (Cover not removed)

1. Remove belts.
2. Remove pulley and hubs.
3. Carefully pry out old seal with screwdriver or thin punch, using care not to damage shaft surface.
4. Coat outside diameter of new seal with proper sealer.
5. Drive in seal with proper tool, using care not to distort it nor damage mating surfaces or shaft.
6. Reinstall removed parts and adjust belts.

Timing Gear Replacement

1966-70, OHV 6 Engines—F-85

Timing gears are arranged so that (unless deliberately disturbed) the valve timing will remain as set at the factory. Unless the gears are badly worn or seriously damaged, the valve timing will remain constant within reasonable limits.

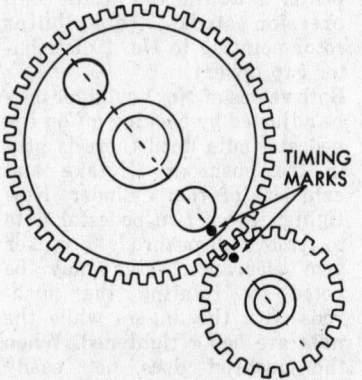

Timing marks—inline 6

If it becomes necessary to replace the timing gears due to wear or damage, remove the radiator core, disconnect the front motor mounts and jack up the front of the engine. Remove the fan belt, fan pulley, oil pan and timing case cover.

NOTE: it is recommended that the camshaft be removed from the car in order to remove and replace the gear in an arbor press.

Remove the camshaft in order to avoid possible risk when attempting to press a gear onto the shaft in place on the car. Sometimes when the gear is being pressed on in place on the car, damage results to the thrust washer in back of the cam gear. Unfortunately this damage is not noticed until the engine is started.

To replace the gear by removing the camshaft, remove the rocker arm assemblies and the distributor, take out all of the pushrods and take out all of the lifters. The camshaft may then be pulled out toward the front of the engine. It will be necessary to retime the ignition.

Runout of the timing gear should not exceed .004 in. Backlash between the two gears should not be less than .004 in. nor more than .006 in. End-play of the thrust plate should be .001-.005 in.

Engine Lubrication

Oil Pan Removal And Installation

1. Raise car and support on stands.
2. Drain engine oil.
3. Disconnect exhaust pipe at crossover, where used.
4. If standard transmission equipped, loosen clutch equalizer-to frame attaching bolts and on some models, remove starter.
5. Remove steering idler arm bracket-to-suspension crossmember attaching bolts.
6. Support engine with a padded jack under the oil pan.
7. Remove bolts and nuts attaching engine mounts to mount brackets.
8. Raise engine.
9. Remove flywheel housing bolts. Then remove housing.
10. Remove oil pan bolts and lower the oil pan enough to remove oil pump pipe and screen-to-cylinder block attaching bolts.
11. Rotate crankshaft to provide maximum clearance at the front end of oil pan. Move the front of the pan to the right and lower the pan through opening between crossmember and steering linkage intermediate shaft.
12. Install by reversing removal procedure. Torque to 6-15 ft. lbs.

Oil Pump

Models with Pump on Front Cover

The oil pump is located in the engine front cover where it is connected by a drilled passage in the cylinder crankcase to an oil screen housing and stand pipe assembly.

Oil is drawn into the pump through the screen and pipe. Oil is discharged from the pump to the oil pump cover assembly. The cover assembly consists of an oil pressure relief valve, an oil filter by-pass valve and a nipple for installation of an oil filter. The oil pressure relief valve limits oil pressure to a maximum of 33 psi. The oil filter by-pass valve opens if the filter becomes clogged to the extent that 4½ to 5 psi. difference exists between the filter inlet and exhaust. This is a safeguard for oil passage to the main engine oil galleries in case of filter stoppage. See engine front cover.

Rear Main Bearing Oil Seal

Except OHV 6 Engine—1966-70

A packing-type rear main bearing oil seal is used in all models of Oldsmobile. To replace the upper half of this seal, remove the engine and crankshaft. The lower half of the seal may be replaced, however, by removing the rear main bearing cap, taking out the old packing and installing new packing, permitting it to protrude slightly from the bearing cap. Bolt the bearing cap up into place and immediately remove it to determine if the extended packing has riveted over, preventing the cap from seating properly. If it has, trim off the riveted over portion only and rebolt the cap. Repeat this operation until the cap seats firmly without riveting over the protruding portion of the oil seal. The reason this is done is so the lower oil seal will have a tendency to compress the upper oil seal, forcing it to a tighter fit around the upper half of the crankshaft. In this way, it is frequently possible to prevent an oil leak in the upper half of the rear main bearing packing without actually replacing the packing.

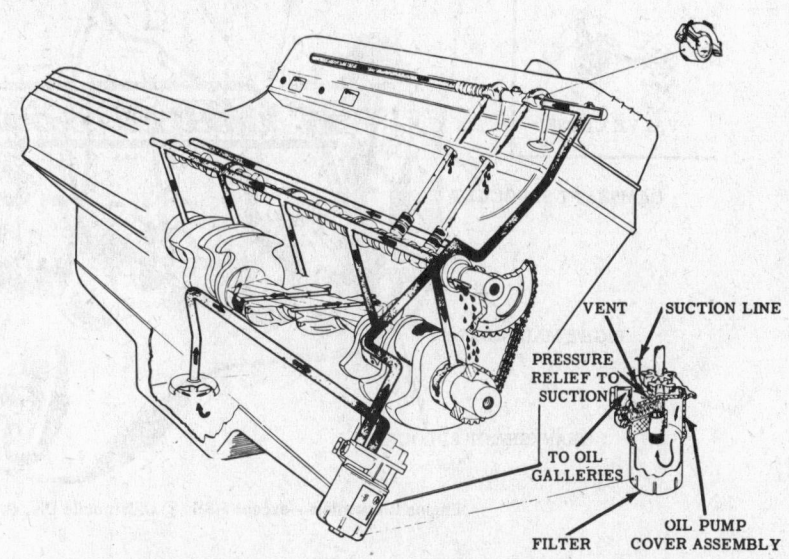

VENT   SUCTION LINE

PRESSURE RELIEF TO SUCTION

TO OIL GALLERIES

FILTER   OIL PUMP COVER ASSEMBLY

Engine oil flow—F-85 V8 and V6 (© Oldsmobile Div. G.M. Corp.)

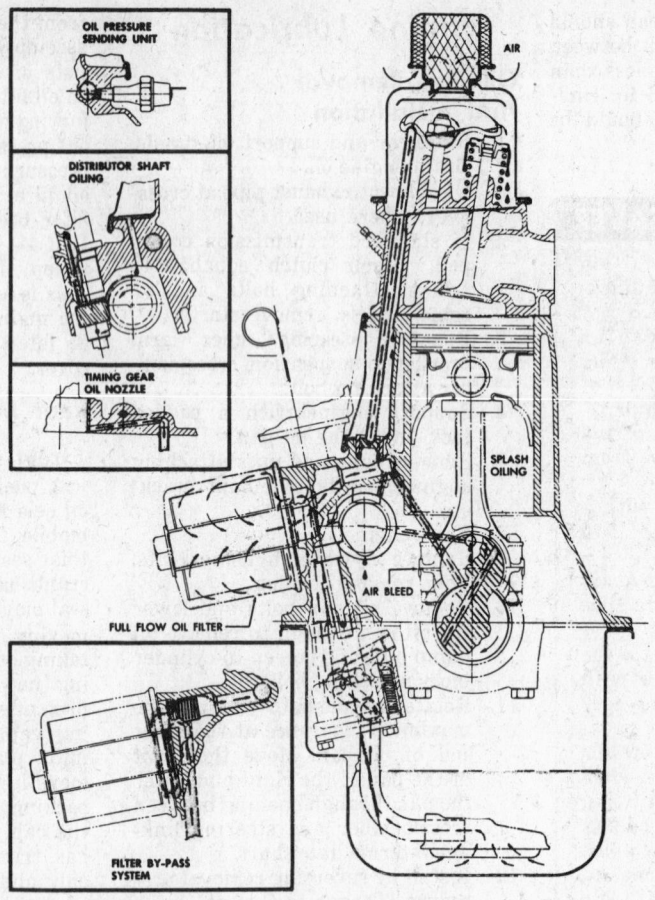

Engine oil flow—inline 6 (© Oldsmobile Div. G.M. Corp.)

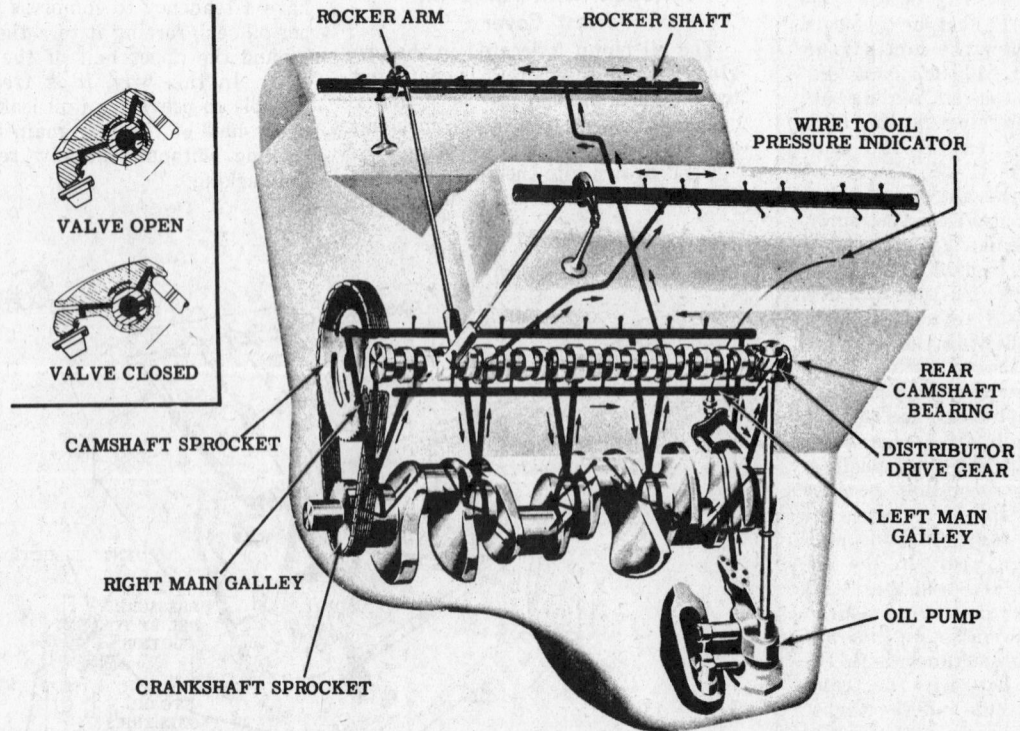

Engine lubrication—except F-85 (© Oldsmobile Div. G.M. Corp.)

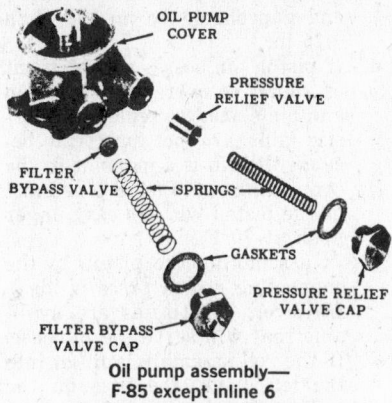

OIL PUMP COVER

PRESSURE RELIEF VALVE

FILTER BYPASS VALVE → SPRINGS

GASKETS

PRESSURE RELIEF VALVE CAP

FILTER BYPASS VALVE CAP

Oil pump assembly—F-85 except inline 6
(© Oldsmobile Div. G.M. Corp.)

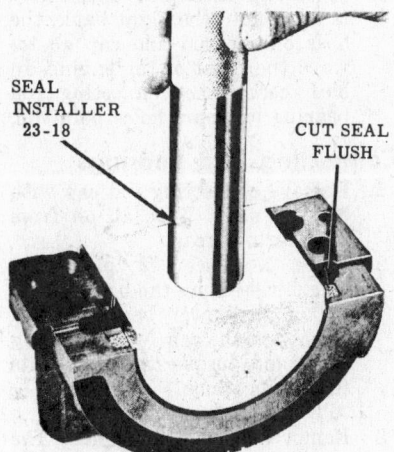

SEAL INSTALLER 23-18

CUT SEAL FLUSH

Installing oil seal
(© Oldsmobile Div. G.M. Corp.)

OHV 6 Engine—1966-70

The rear main bearing oil seal is of moulded design and can be replaced (both halves) without removal of the crankshaft.

NOTE: always replace both halves as a unit. Install with the lip facing toward front of the engine.

1. With oil pan removed, remove the rear main bearing cap.
2. Remove oil seal from the groove by lifting the end tab, then clean seal groove.
3. Lubricate the lip and O.D. of a new seal with engine oil. Keep oil off the parting line surface.
4. Insert seal into cap and roll into place with fingers. Use light pressure on the seal to prevent cutting the O.D. of the seal with the sharp edges of the groove. Be sure the tabs of the seal are properly located in the cross grooves.
5. To remove upper half of seal, use a small hammer to tap a brass pin punch on one end of seal until it protrudes far enough to be removed with pliers.
6. Lubricate the lip and O.D. of a new seal with engine oil. Keep oil off parting line surface. Gradually push with a hammer handle, while turning crankshaft, until seal is rolled into place. Be

Top Half, Rear Main Bearing Oil Seal Replacement

The following method has proven a distinct advantage in most cases.

1. Drain engine oil and remove oil pan.
2. Remove rear main bearing cap.
3. With a 6 in. length of 3/16 in. brazing rod, drive up on either exposed end of the top half oil seal. When the opposite end of the seal starts to protrude, have a helper grasp it with pliers and pull gently while the driven end is being tapped. It is surprising how easily most of these seals can be removed by this method.

New Woven Fabric-Type Seal Installation

1. Obtain a 12 in. piece of copper wire (about the same gauge as that used in the strands of an insulated battery cable).
2. Thread one strand of this wire through the new seal, about ½ in. from the end, bend back and make secure.
3. Thoroughly saturate the new seal with engine oil.
4. Push the copper wire up through the oil seal groove until it comes down on the opposite side of the bearing.
5. Pull (with pliers) on the protruding copper wire while the crankshaft is being turned and the new seal is slowly fed into place.
CAUTION: this snaking operation slightly reduces the diameter of the new seal and care will have to be used to keep the seal from slipping too far through the top half of the bearing.
6. When an equal amount of seal is extending from each side, cut off the copper wire close to the seal and tamp both ends of the seal up into the groove. This will tend to expand the seal again.
NOTE: don't worry about the copper wire left in the groove, it is too soft to cause damage.
7. Replace the seal in the cap in the usual way and replace the oil pan.

careful that seal bead on O.D. of seal is not cut.
7. Install rear main bearing cap (with new seal) and torque to specifications. Be sure cross seal tabs are in place and properly seated.

Connecting Rods And Pistons

Piston Assembly Removal

1. Remove cylinder head.
2. Remove oil pan.
3. Examine cylinder bores for top ridge. If ridge exists, remove it before taking pistons out.
4. Number all the pistons, connecting rods and caps. Starting at

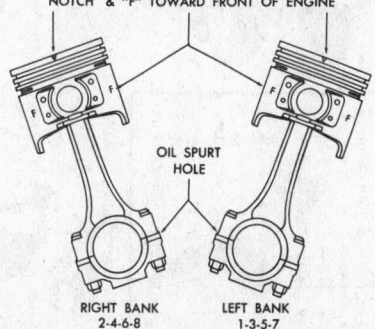

NOTCH & "F" TOWARD FRONT OF ENGINE

OIL SPURT HOLE

RIGHT BANK 2-4-6-8   LEFT BANK 1-3-5-7

Piston and rod assembly—V8 except 215

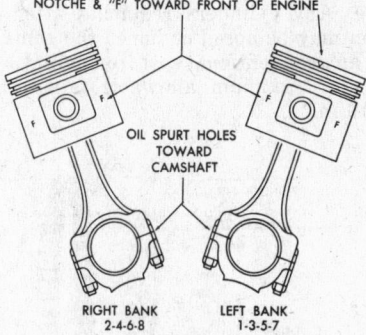

NOTCHE & "F" TOWARD FRONT OF ENGINE

OIL SPURT HOLES TOWARD CAMSHAFT

RIGHT BANK 2-4-6-8   LEFT BANK 1-3-5-7

Piston and rod assembly—V8 215

the front, the right bank is numbered 2-4-6-8. The left bank is numbered, 1-3-5-7.

With the V6 engine, the right bank is numbered 2-4-6; the left bank, 1-3-5, from the front. The OHV inline 6 engine has an order of 1-2-3-4-5-6.

5. With No. 4 crankpin straight down, remove cap and bearing shell from No. 1 connecting rod. Install connecting rod bolt guides to hold upper half of the bearing shell in place.
6. Push piston and rod assembly up out of the cylinder. Then remove bolt guides and reinstall cap and bearing shell on the rod.
7. Remove the remaining rod and

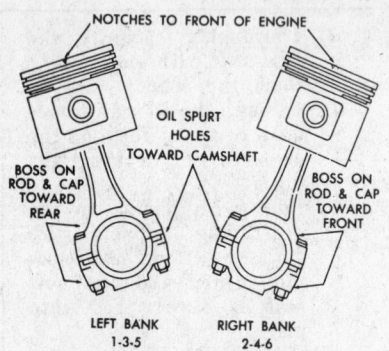

Piston and rod assembly—V6 225

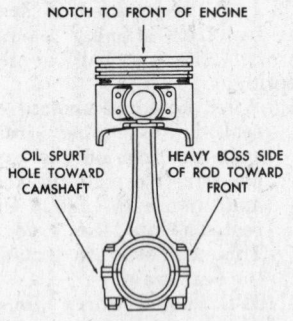

Piston and rod assembly—inline 6

Fitting Rings and Pins

1. When new rings are installed without reboring the cylinders, cylinder wall glaze should be broken. This can be done by using the finest grade stones in a cylinder hone.
2. New piston rings must be checked for clearance in piston grooves and for cap in cylinder bores.
3. When fitting new rings to new pistons the side clearance for compression rings should be .002 to .004 in. for V8 and .003 to .005 in. for 6 cylinder engines. Side clearance of the oil ring should be .0001 to .0051 for V8 engines through 1967, .0035 to .0095 for 6 cylinder engines through 1967, .001 to .005 for later 6 cylinder and 350 cu. in. V8, .001 to .010 for 400 cu. in. V8 and .002 to .008 for 455 cu. in. V8.
4. Check end gap of compression rings by placing them in the bore in which they will operate. Then push them to the bottom of the bore with a piston. Now measure the end gap in each ring. The end gap should be no less than .015 in.
5. If piston pin bosses are worn out of round or oversize, the piston and pin should be replaced. Oversize pins are not practical because the pin is a press fit in the connecting rod. Piston pins must fit the piston with an easy finger push at 70°F.
6. In assembling the piston to the connecting rod, a press is ideal. However, substitutes are available that will serve the purpose.
7. If the rod assembly is to go into the left bank, the boss on the rod and cap go toward the rear of the engine. If the rod assembly is to go into the right bank, the boss on the rod and cap go toward the front of the engine. In both cases, the connecting rod bearing oil spurt holes point up.

Connecting Rod Bearings

1. Remove connecting rod cap with bearing shell. Wipe all oil from the bearing area.
2. Place a piece of plastigage lengthwise along the bottom center of the lower bearing shell. Then install cap with bearing shell and torque the bolt nuts to specifications.
   NOTE: do not turn crankshaft
3. Remove the cap and shell. The gauge material will be found flattened and adhering to either the bearing shell or the crankpin. Do not remove it.
4. Using the scale that comes with the gauge, measure the flattened gauge material at its widest point. The number within the

piston assemblies in the same manner.

8. Carefully remove oil rings with piston ring expander.
9. Carefully press out the old pin.
   NOTE: check the cylinder bores for out-of-round, taper or other damage. Any cylinders requiring attention may be bored or honed the same as any conventional cast iron cylinder block. Maximum allowable taper is .010 in.

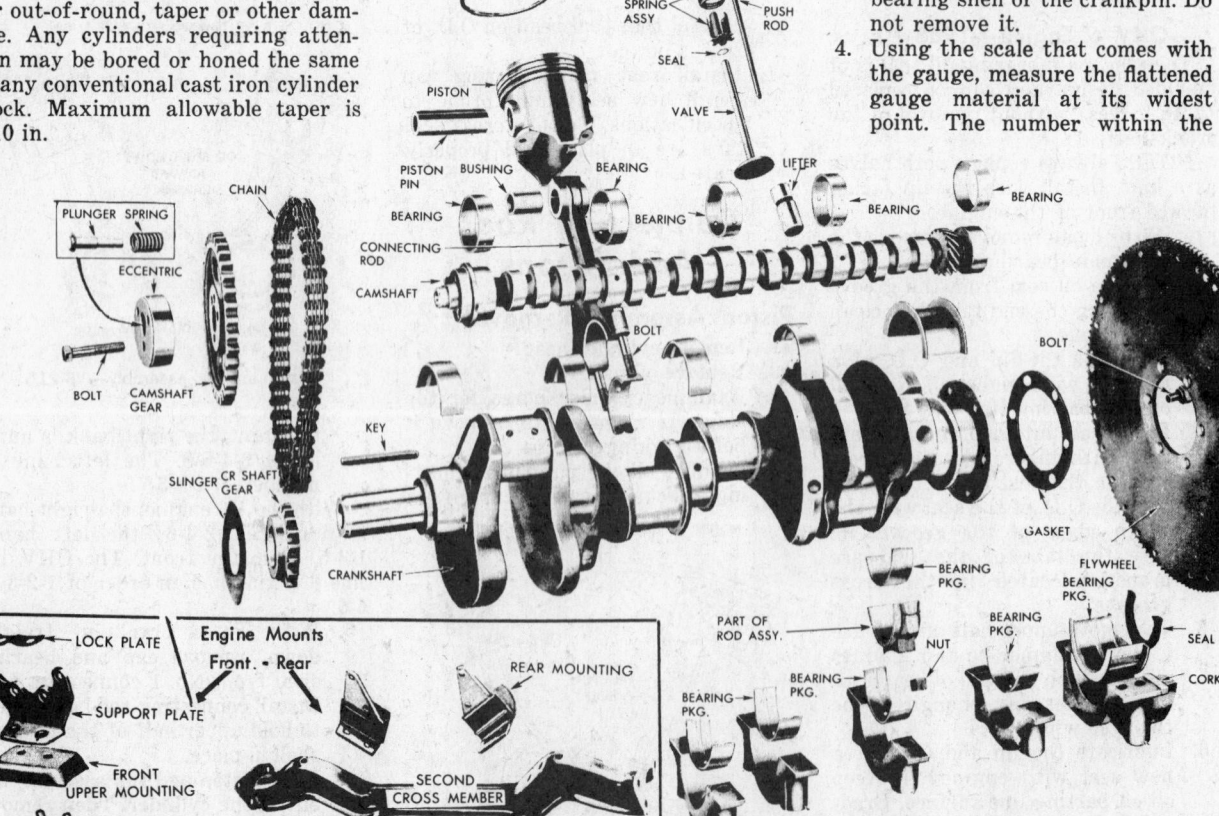

Internal layout of Rocket V8 engine (© Oldsmobile Div. G.M. Corp.)

graduation which comes closest to the width of the gauging material indicates the bearing to crankpin clearance in thousandths of an inch.

5. Desired clearance for a new bearing is given in the specifications. If the bearing has been in service, it is wise to install a new bearing if clearance exceeds .0035 in.

6. If a new bearing is required, try a standard, then each undersize bearing in turn until one is found that is within specifications.

7. With the proper bearing selected, clean off the gauging material, reinstall the bearing cap and torque.

8. After the bearing cap has been torqued, it should be possible to move the connecting rod back and forth on the crankpin, the extent of end clearance.

Piston Assembly Installation

1. Carefully assemble the piston to the connecting rod (press in the pin).

2. Remove piston and rod from the press. Rock the piston on the pin to be sure pin or piston boss was not damaged during the pressing operation.

3. Install ring expander in lower ring groove. Position the ends of the expander above the piston pin where groove is not slotted. The ends of the expander must butt together.

4. Install oil ring rails over expander with gaps up on same side of piston as oil spurt hole in connecting rod.

5. Install compression rings, (with a ring expander) in top and center groove.

6. Coat all bearing surfaces, rings and piston skirt with engine oil.

7. Position the crankpin of the cylinder being worked on, down.

8. Remove connecting rod bearing cap and with upper bearing shell correctly seated in the rod, install connecting rod bolt guides.

9. Make sure the gaps in the two oil ring rails are up toward the center of the engine. Make sure the gaps of the compression rings are not in line with each other or the oil ring rails. Be sure the ends of the oil ring spacer-expander are butted and not overlapped.

10. With a good ring compressor, install the piston and rod assembly into the cylinder bore and carefully tap down until the rod bearing is solidly seated on the crankpin.

11. Remove the connecting rod bolt guides and install cap and lower bearing shell. Torque to 30-35 ft. lbs.

12. Install other piston and rod assemblies in the same manner. When the assemblies are all installed, the oil spurt holes will be up. The rib on the edge of the rod cap will be on the same side as the conical boss on the connecting rod web. These marks will be toward the other connecting rod on the same crankpin.

13. Accumulated end clearance between rod bearings on any crankpin should be .006 in.—.014 in.

14. Install oil screen and oil pan.

15. Install cylinder heads.

NOTE: before starting a new or reconditioned engine, it is advisable to pack the oil pump with petroleum jelly to insure pump priming for immediate lubrication. See Engine Front Cover Installation in this section.

Front Suspension

References

General instructions covering the front suspension and how to repair and adjust it, together with information on installation of front wheel bearings and grease seals, are given in the Unit Repair Section.

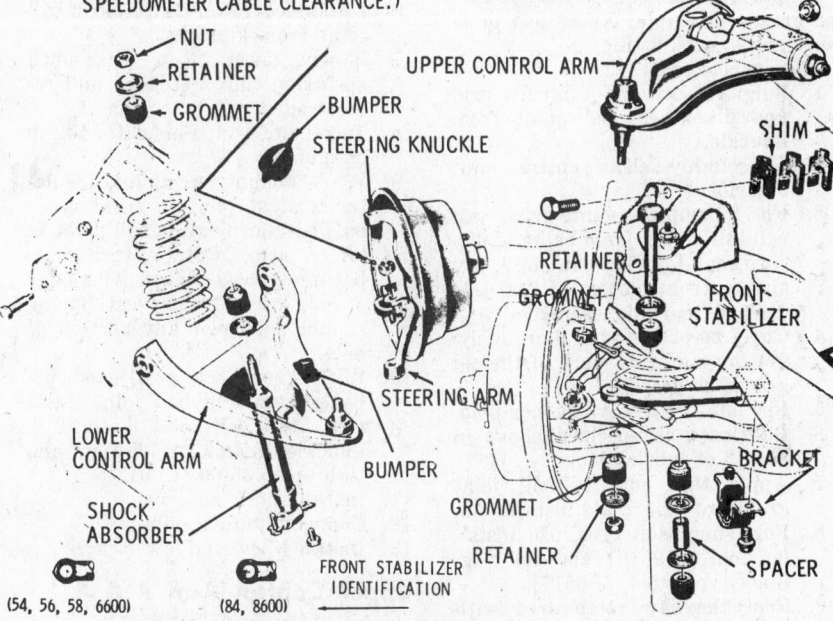

1965-70 front suspension—except F-85 (© Oldsmobile Div. G.M. Corp.)

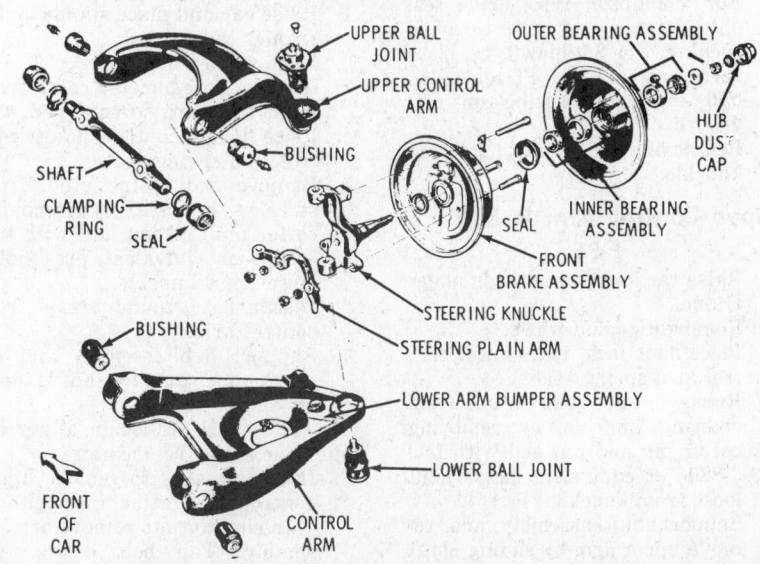

1964-70 front suspension—F-85 (© Oldsmobile Div. G.M. Corp.)

Definitions of the points of steering geometry are covered in the Unit Repair Section.

Overall length, brake cylinder and tire size figures are in the General Chassis and Brake Specifications table.

Checking Ball Joint Wear

Check the illustration showing the proper procedure in the Unit Repair Section.

A new type ball joint is used. Service is unlikely unless noise develops, then lubrication may be required. To lubricate:

1. Place car on hoist so that there is free access to joint.
2. Thoroughly clean plug area, then discard the rubber plug.
3. Use hand-operated ball-type gun, using ball joint grease only. Install new plug.

Important: do not overfill; .01 lb. of grease is ample. Never use pressure gun or zerk fitting.

To replace seal:

1. Support lower control arm and disconnect ball joint from knuckle.
2. Thoroughly clean entire joint exterior.
3. For Thompson joint, drive seal retaining ring from ball and discard seal. For Saginaw joint, pry spring from bottom of seal and discard seal and spring.
4. Clean pivot and stud thoroughly, getting rid of as much of the old grease as possible.
5. Pry out and discard rubber plug.
6. Lubricate as specified above in Step 3 of lubrication.
7. Apply thin coat of ball joint grease to seal to aid installation.
8. For Thompson type, use installer tool (J-9516) and for Saginaw type, tool (J-9517).
9. Coat the saw tooth area with ball joint grease.
10. For Thompson type, drive seal onto joint, being sure to avoid cocking. For Saginaw type, place spring around tool. Place seal on ball joint and spring on seal groove.
11. Reassemble ball joint to steering knuckle.

Upper Control Arm R & R
F-85

1. Raise car and place stands under frame.
2. Remove tire and wheel.
3. Place floor jack under lower control arm spring seat.
4. Remove ball joint stud from steering knuckle, by removing cotter pin and nut and with tool J-8806 or equivalent, press joint loose from knuckle.
5. Support hub assembly and remove upper arm by sliding shaft off end of bolts.

NOTE: mark or locate alignment shims for easier reassembly.

6. Attach arm assembly to frame using original shims. Torque to 65 ft. lbs.
7. Install ball joint. Torque nut to 40 ft. lbs. (minimum).
8. Install hub, drum and wheel assembly and lower car to floor.

Lower Control Arm and/or Front Spring R & R
All Models

1. Raise front of car and support by stands under frame.
2. Remove tire and wheel.
3. Disconnect stabilizer link from lower arm.
4. Remove shock absorber.
5. Place floor jack under lower arm, between spring seat and ball joint. Tool BT-6505 may be used to compress spring.
6. Disconnect lower control arm ball joint from knuckle.
7. Slowly lower floor jack until spring is fully extended and remove spring.
8. To reinstall, tape insulator to top of spring.
9. While holding spring and insulator against pilot in front cross bar, tilt spring so it will pivot in lower arm. Rotate spring so bottom coil will index with edge of hole in arm spring seat. Spring should not cover any portion of hole.
10. With floor jack positioned between seat and ball joint, raise arm until ball joint is tight in knuckle. Install ball joint nut and tighten to 35-60 ft. lbs.
11. Install shock absorber.
12. Connect stabilizer link.
13. Install wheel and lower car.

Upper Control Arm R & R
Except F-85

1. Raise car and place stands under frame.
2. Remove tire and wheel.
3. Remove speedometer cable from knuckle where so equipped.
4. Place floor jack under lower control arm spring seat.
5. Remove ball joint stud from steering knuckle, by removing cotter pin and nut and with tool J-8806 or equivalent, press joint loose from knuckle.
6. Disconnect ground strap from control arm.
7. Support hub assembly and remove upper arm by sliding shaft off end of bolts.

NOTE: mark or locate alignment shims for easier reassembly.

8. It is necessary to remove upper control arm attaching bolts to gain clearance to remove arm assembly. Tap bolt down with brass drift. Pry bolt up with box wrench. Using a suitable pry bar and block of wood, pry bolts from frame.
9. Remove control arm from car.
10. To reinstall, position bolts loosely in frame and install pivot shaft on bolts.
11. Install lock washers and nuts and with brass drift, drive attaching bolts into frame.
12. Install alignment shims, placing them in position from which they were removed. Torque nuts to 70 ft. lbs.
13. Connect ball joint and torque to 40 ft. lbs.
14. Attach ground strap.
15. Install speedometer cable and wheel and tire and lower car to floor.

Jacking, Hoisting

When supporting car on floor jack or floor stand, the car should be supported at the suspension points only. Under no condition should the car be supported at extreme ends of frame or side rail.

When using frame contact lift, position contact pads to lift the side rails at points shown on diagram.

Never use bumper jack other than that provided with car.

Steering Wheel

Pry up the center medallion, then remove the nut that holds the steering wheel in place. This also holds the horn blowing ring contacts.

Deluxe horn ring caps are retained to the horn ring hub by a screw which is accessible from underneath the cap.

Steering Wheel Removal

On all Oldsmobile models, it is necessary to use a puller to take off the steering wheel.

Manual Steering
References

Instructions covering the overhaul of the steering gear will be found in the Unit Repair Section.

Removal

1. Remove flex coupling flange attaching nuts and lockwashers.
2. Raise front of car and support under outer ends of lower control arms.
3. Remove pitman shaft nut and pull pitman arm from shaft with suitable puller.
4. Remove gear to frame mounting bolts. Clear speedometer cable

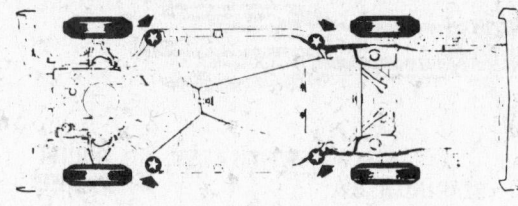

LIFT POINTS

Hoist contact points (© Oldsmobile Div. G.M. Corp.)

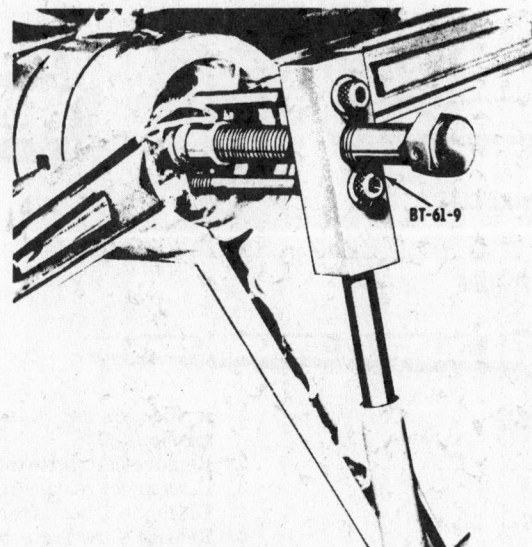

BT-61-9

Typical steering wheel removal
(© Oldsmobile Div. G.M. Corp.)

and any linkage parts from interference and remove gear assembly from car.

5. Reinstall in reverse of the above. Torque gear to frame nuts to 60-80 ft. lbs. Torque pitman arm nut to 100-120 ft. lbs.

Power Steering

References
Troubleshooting and repair instructions covering power steering gears are given in the Unit Repair Section.

Power Steering Gear Removal
Disconnect the power steering hoses and remove the four bolts that hold the coupling to the power gear. Remove the nut from the cross-shaft and pull off the pitman arm.

From underneath the vehicle, remove the bolts that hold the power gear assembly to the frame and let the power gear come down through the bottom.

Power Steering Pump Removal
Remove and cap the two hoses that run through the pump, loosen the clamp bolts so that the pump can be slid along its adjusting slot and take the belt off. Remove the three bolts that hold the pump bracket to the cylinder heads and lift off the pump.

Pitman Shaft Seal Replacement (with power steering gear in place in car)
1. Disconnect pitman arm from pitman shaft. Clean end of pitman

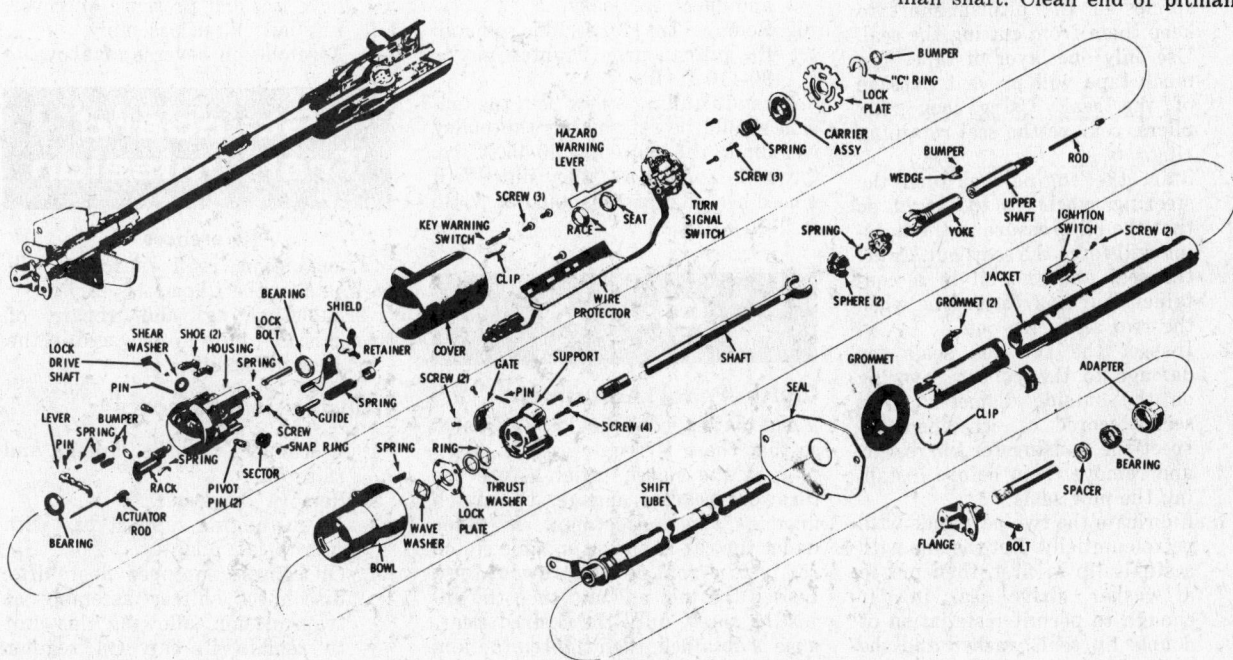

Tilt and Travel steering column (© Oldsmobile Div. G.M. Corp.)

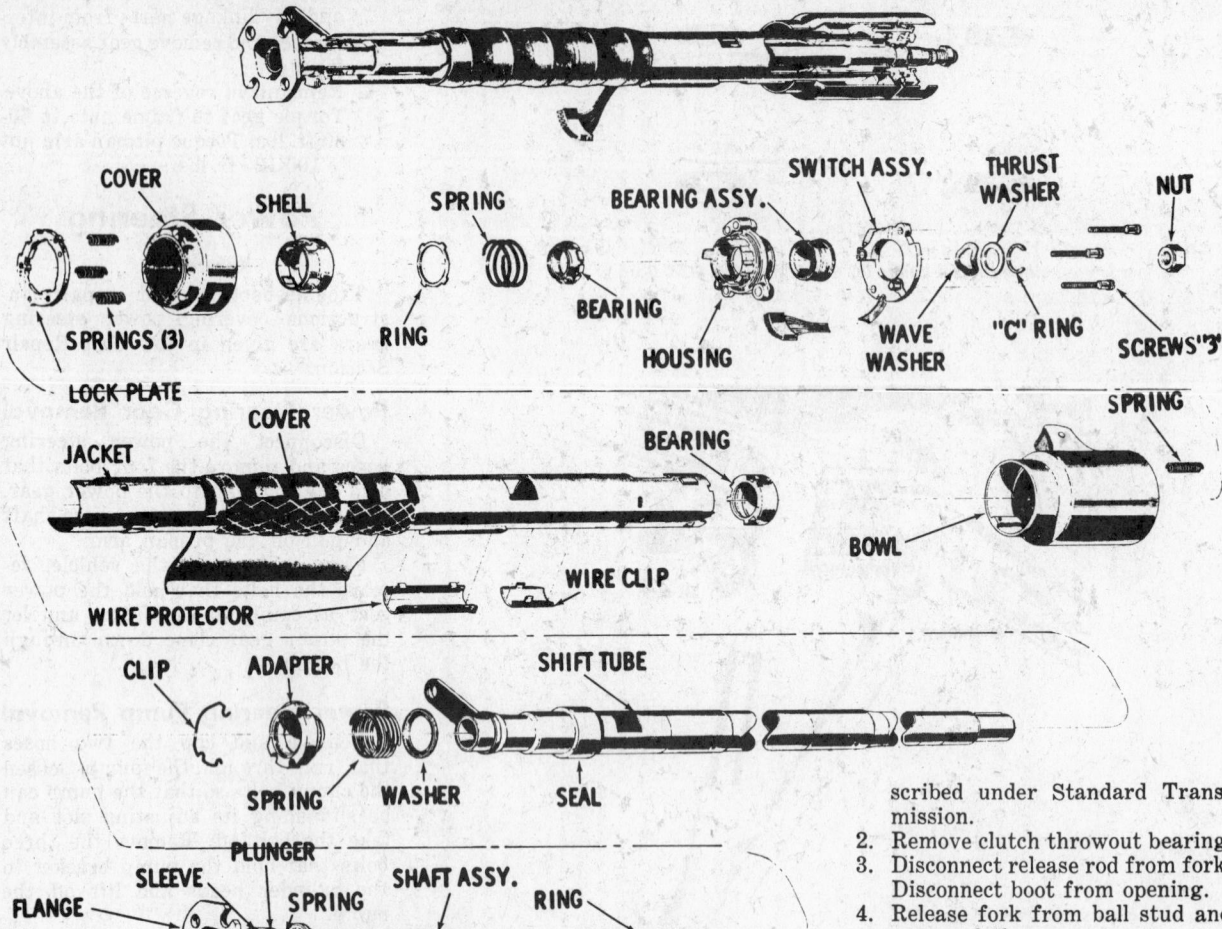

Telescoping steering column (© Oldsmobile Div. G.M. Corp.)

shaft and housing. Tape the splines of the pitman shaft to keep them from cutting the seal. Use only one layer of tape. Too much tape will prevent passage of the seal. Using lock ring pliers, remove the seal retaining ring.

2. Start the engine and turn the steering wheel to the right so that the oil pressure in the housing will force the seals out. Catch the seal and the oil in a container. Turn off the engine when the two seals are out.

3. Inspect the two old seals for damage to the rubber covering on the outside diameter. If it seems scored or scratched, inspect the housing for burrs, etc. and remove them before installing the new seals.

4. Lubricate the two new seals with petroleum jelly. Put the one with a single lip in first, then put in a washer, drive seal in far enough to permit installation of double lip seal, washer and the seal retaining ring. The first seal is not supposed to bottom in its counterbore.

5. Fill reservoir to proper level,

start engine, turn wheel to right, and check for leaks.

6. Remove the tape and reinstall the pitman arm. Tighten nut to 90-110 ft. lbs.

When installing a new belt, the tension should be set so that the pulley will slip in the belt when 40-45 ft. lbs. torque is applied to pulley nut. With a used belt the torque should be 30-35 ft. lbs.

Clutch

Clutch Pedal Adjustment

The clutch pedal should be adjusted so that there is from 1/2-3/4 in. freeplay at the clutch pedal before the throwout bearing engages the clutch fingers. This adjustment is made under the car at the adjustable clutch rod just in front of the throwout fork. Loosen the jam nut and turn the adjusting screw until the desired clearance is obtained, then tighten the jam nut.

Clutch Removal

1. Remove transmission as de-

scribed under Standard Transmission.

2. Remove clutch throwout bearing.

3. Disconnect release rod from fork. Disconnect boot from opening.

4. Release fork from ball stud and remove fork.

5. Mark clutch cover and flywheel for reassembly indexing.

6. Remove clutch cover attaching bolts. Be sure to loosen equally and gradually around cover to relieve unequal pressure on cover.

7. Lift out clutch assembly.

8. Assemble in reverse of above.

Standard Transmission

References

Transmission refill capacities will be found in the Capacities table.

Troubleshooting and repair of transmission units are covered in the Unit Repair Section.

Transmission Removal

1. Disconnect throttle linkage and raise car.

2. Remove driveshaft.

3. Install engine support bar with appropriate adapter.

4. On console equipped floorshifts, disconnect shifter assembly at transmission, allowing this unit to remain in car. On regular floorshifts, remove floor pan seal. Insert a feeler gauge, remove shift lever and remove shifter with transmission.

NOTE: this information does not pertain to column shifts.

5. Disconnect parking brake cables and remove cross support bar.
6. Disconnect speedometer cable.
7. Remove transmission upper and lower bolts.
8. Slide transmission rearward and remove.
9. Install by reversing procedure above.

Shift Linkage Adjustment

Three-Speed

1. Loosen swivels and move both transmission levers until transmission is in neutral. Detents in cover must both be engaged in neutral for proper adjustment. (Check by rotating propeller shaft with rear wheels raised.)
2. Move selector lever to neutral position with first and reverse lever aligned with second and high lever.
3. With this proper alignment, center the movement of the rods in the swivels and tighten the swivels.
4. Move selector lever through all positions to check the adjustments.

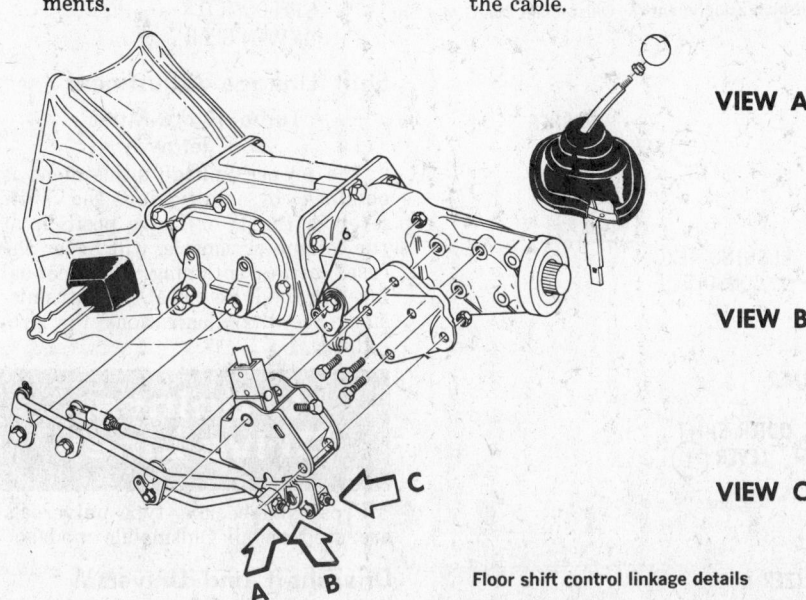

Four-Speed

1. Place transmission levers in neutral.
2. Using a 5/16 in. rod, align levers in neutral position.
3. Adjust swivels to obtain a free pin fit at the levers.

Automatic Transmission

Transmission Removal
Turbo Hydra-Matic 400

1. Remove flywheel cover.

2. Remove torque converter attaching bolts.
   NOTE: mark flywheel and converter so they can be installed in same position.
3. Install engine support bar.
4. Disconnect solenoid wires and manual shift linkage at side of transmission.
5. Disconnect oil cooler lines, vacuum modulator line and filler pipe.
6. Disconnect parking brake cable.
7. Remove driveshaft assembly.
8. Disconnect exhaust pipe bracket at rear of crossmember.
9. Position jack under transmission.
10. Remove crossmember.
11. Remove transmission to block attaching bolts.
12. Move transmission away from engine, using tool (J-21364) to hold converter in place.
13. Lower transmission and remove from car.
14. Install by reversing procedure above.

Turbo Hydra-Matic 350

1. Inside the car, slide the clip on the downshift cable to the end of the cable.

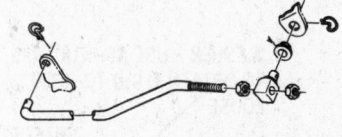

VIEW A

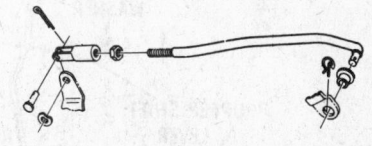

VIEW B

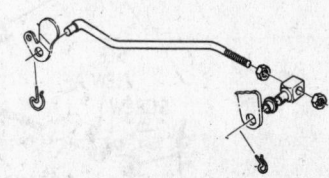

VIEW C

Floor shift control linkage details

2. Remove transmission oil level dipstick.
3. Hoist car.
4. Remove bolt and remove downshift cable from link, plugging hole.
5. Remove transmission cooler lines.
6. Remove flywheel cover pan and three bolts holding the converter to the flywheel.
7. Disconnect modulator line.
8. Remove speedometer clip and driven gear, plugging hole.
9. Remove cotter pin and manual shift linkage.
10. Remove driveshaft.

11. Remove transmission mount support bolts at transmission mount and frame.
12. Raise transmission with a jack and remove support.
13. Lower transmission slightly and remove transmission to engine bolts.
14. Remove transmission by lowering and moving rearward, using a special tool to hold converter.
15. Install by reversing procedure above.

Jetaway

1. Remove transmission filler pipe.
2. Raise car on hoist.
3. Disconnect transmission control
4. Disconnect manual rod from transmission lever.
5. Remove propeller shaft assembly.
6. Remove flywheel dust cover.
7. Install engine support bar.
8. Remove bolts at rear transmission mount.
9. Remove attaching brackets and remove cross support bar.
10. Disconnect oil cooler lines and cap immediately.
11. Disconnect modulator line.
12. Disconnect speedometer cable.

13. Remove three flywheel to converter attaching bolts.
14. Support transmission with unit lift and remove transmission to flywheel housing bolts.
15. Remove transmission by moving rearward and lowering from car, using a special tool to hold converter in place.
16. Install by reversing procedure above.

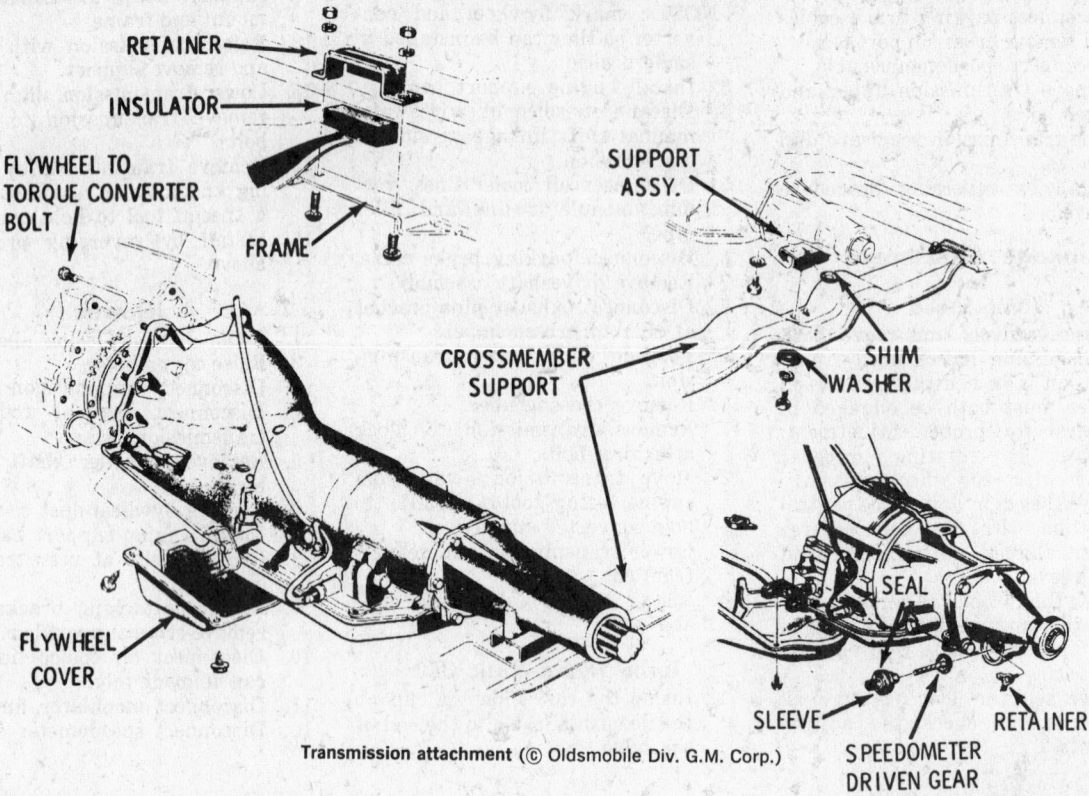

Transmission attachment (© Oldsmobile Div. G.M. Corp.)

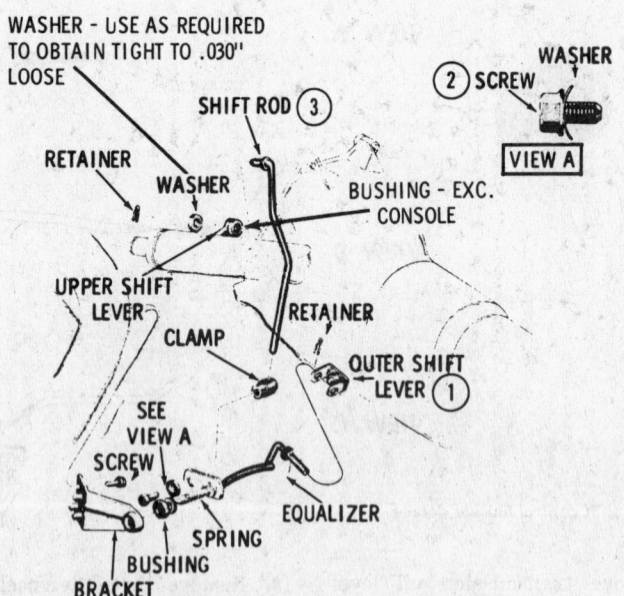

Column shift linkage adjustment for Turbo Hydra-matic transmission
(© Oldsmobile Div. G.M. Corp.)

SHIFT ROD ADJUSTMENT

1. Set transmission outer lever in drive position.
2. Hold upper shift lever against drive position stop in upper steering column (do not raise lever).
3. Tighten screw in clamp on lower end of shift rod to specified torque.
4. Check Operation:
   A. With key in "run" position and transmission in "reverse" be sure that key cannot be removed and that steering wheel is not locked.
   B. With key in "lock" position and shift lever in "park," be sure that key can be removed, that steering wheel is locked, and that the transmission remains in park when the steering column is locked.

Shift Linkage Adjustment
Turbo Hydra-Matic and Jetaway

The proper linkage adjustment is obtained by positioning the shift lever in the D or Drive position at the column or console, whichever the case, and positioning the manual lever in the D detent at the transmission. See the illustrations for procedure.

U Joints, Drive Lines

Cross and bearing-type universals are used on all Oldsmobile models.

Driveshaft and Universal Joint Removal

1. Remove the four bolts that hold the rear universal filler blocks to the pinion shaft flange and pry the universal joint off the pinion flange, lowering the back end of the shaft to the floor.
2. It is customary to tape the bearing blocks to the universal joints so that they don't get lost or dirty.
3. The front end of the shaft can then be slid off the back of the transmission shaft and carried to the bench.
4. The bearings are held into the yokes by two lock plates, one on each side.

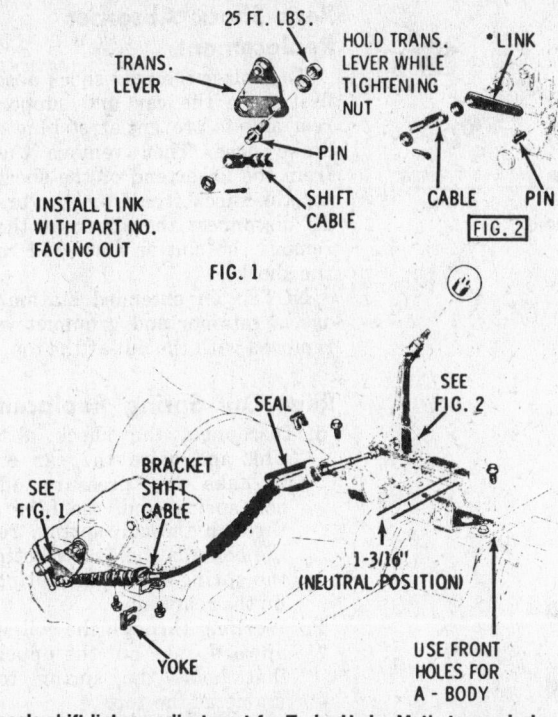

25 FT. LBS.

TRANS. LEVER

HOLD TRANS. LEVER WHILE TIGHTENING NUT

•LINK

PIN

SHIFT CABLE

CABLE   PIN

FIG. 2

INSTALL LINK WITH PART NO. FACING OUT

FIG. 1

SHIFT CABLE ADJUSTMENT

1. Loosen shift rod clamp screw, loosen pin in transmission manual lever.
2. Place shift lever in "P" position, place transmission manual lever in "P" position and ignition key in lock position.
3. Pull shift rod lightly against lock stop and tighten clamp screw.
4. Move pin in manual transmission lever to give "free pin" fit and tighten attaching nut.
5. Check Operation:
   A. Move shift handle into each gear position and see that transmission manual lever is also in detent position.
   B. With key in "run" position and transmission in "reverse," be sure that key cannot be removed and that steering wheel is not locked.
   C. With key in "lock" position and transmission in "park," be sure that key can be removed and that steering wheel is locked.

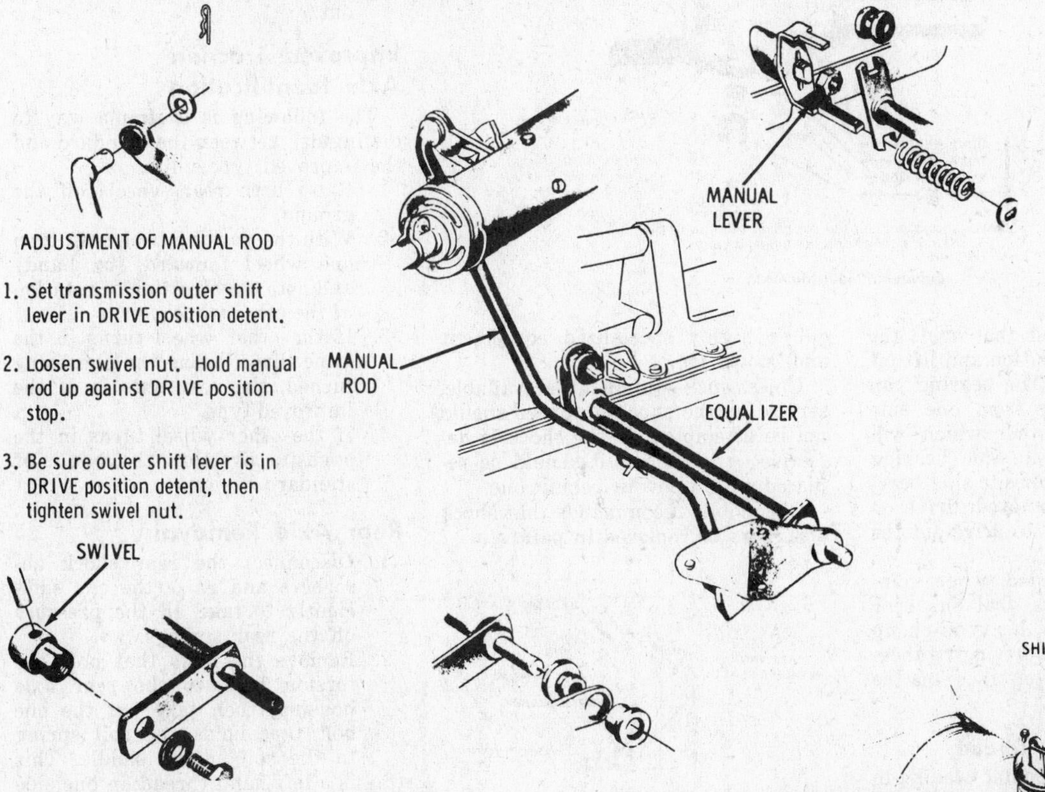

SEAL

SEE FIG. 2

BRACKET SHIFT CABLE

SEE FIG. 1

1-3/16" (NEUTRAL POSITION)

YOKE

USE FRONT HOLES FOR A - BODY

Console shift linkage adjustment for Turbo Hydra-Matic transmission
(© Oldsmobile Div. G.M. Corp.)

ADJUSTMENT OF MANUAL ROD

1. Set transmission outer shift lever in DRIVE position detent.

2. Loosen swivel nut. Hold manual rod up against DRIVE position stop.

3. Be sure outer shift lever is in DRIVE position detent, then tighten swivel nut.

SWIVEL

MANUAL LEVER

MANUAL ROD

EQUALIZER

Column shift linkage adjustment for Jetaway transmission (© Oldsmobile Div. G.M. Corp.)

MANUAL LEVER - POSITION ON TRANS WITH WORD "OUT" UPRIGHT AND OUTBOARD.

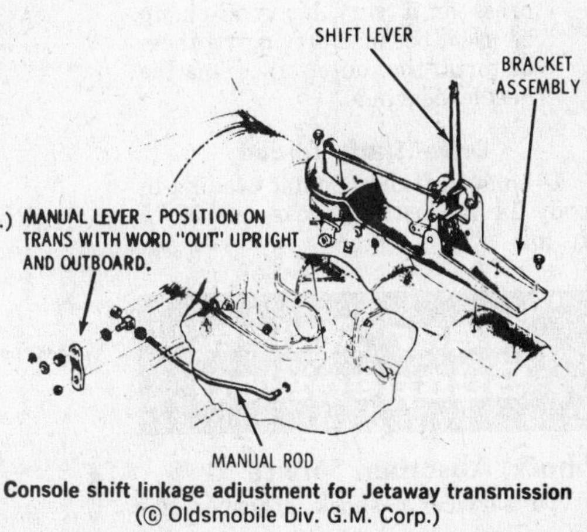

SHIFT LEVER

BRACKET ASSEMBLY

MANUAL ROD

Console shift linkage adjustment for Jetaway transmission
(© Oldsmobile Div. G.M. Corp.)

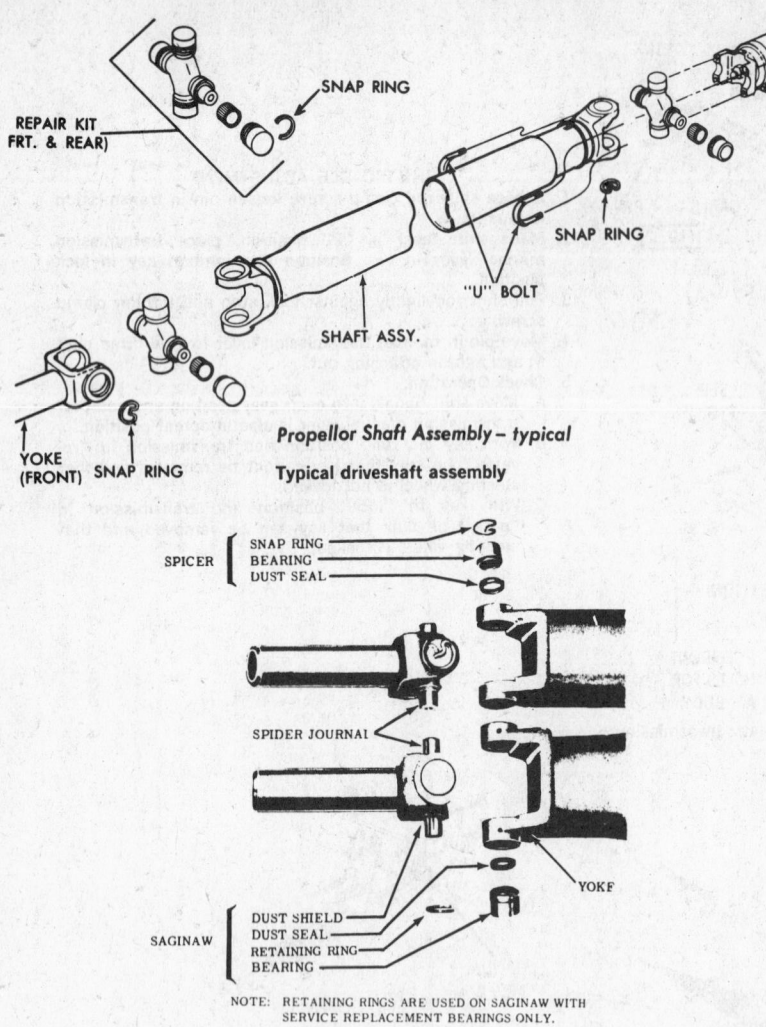

Propellor Shaft Assembly – typical

Typical driveshaft assembly

NOTE: RETAINING RINGS ARE USED ON SAGINAW WITH
SERVICE REPLACEMENT BEARINGS ONLY.

Driveshaft identification

5. Take out the nut that holds the lock plate in position and lift off the lock plate. The bearing can then be driven from one side across to the other which will drive the opposite side bearing out. Once the opposite side bearing has been removed, drive on the cross, itself, to drive out the first bearing.

6. It is recommended, when reinstalling bearings, that an arbor press or a very heavy C-clamp be used, because driving on them distorts the outer race of the needle bearings.

Drive Shaft Torque

U-bolts 16 ft. lbs. Center bearing to body 14 ft. lbs. Slip yoke nut 50-75 ft. lbs.

Rear Axles, Suspension

Shock Absorber Service

To service a shock absorber re-
quires highly specialized equipment and knowledge.

Unless such equipment is available, service on the shock absorber should not be attempted. If the shock is defective or inefficient it should be replaced with a new or rebuilt one.

Oldsmobile recommends that shock absorbers be replaced in pairs.

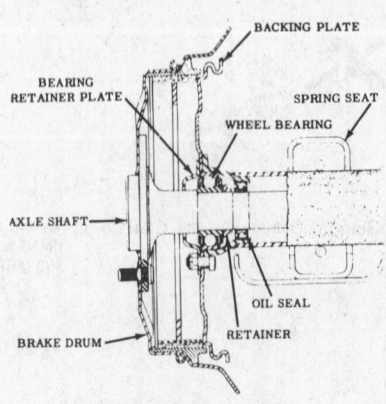

Axle shaft and related parts
(© Oldsmobile Div. G.M. Corp.)

Rear Shock Absorber Replacement

To replace the rear shock absorber, first raise the car and support the rear axle to prevent stretching of the brake hose. Then remove the nut from the lower end of the shock and tap the shock free from the bracket. To disconnect the shock at the top, remove the bolt or bolts and remove the shock.

NOTE: on extended station wagons, a retainer and grommet will be removed with the nut at the top.

Rear Coil Spring Replacement

1. Disconnect the shock absorber link and raise the car enough to take the pressure off the coil spring and, reaching down through the coil spring, remove the bolt that holds the bottom of the spring to the insulating pad on the rear axle.
2. Working through the coil spring upward, take out the upper bolt that holds the spring to the frame at the top.
3. If the car has been raised sufficiently, the spring can be lifted out.

Improved Traction Axle Identification

The following is a simple way to distinguish between the standard and the improved type units.

1. Raise both rear wheels off the ground.
2. With the parking brake off, turn one wheel forward (by hand) and note the direction of rotation of the other wheel.
3. If the other wheel turns in the same direction as the one being turned, the rear axle is of the improved type.
4. If the other wheel turns in the opposite direction, the axle is of standard design.

Rear Axle Removal

1. Disconnect the rear shock absorbers and raise the car sufficiently to take all the pressure off the rear springs.
2. Remove the bolts that hold the torsion bars to the rear axle housing, then take out the one bolt that holds the coil spring to the coil spring saddle. This is a left-hand thread on one side and a right-hand thread on the other.
3. Disconnect the brake lines and cables at the rear axle, split the rear universal joint, disconnect the sway bar and roll the rear axle assembly from under the car.

Automatic Level Control

The system consists of a vacuum-

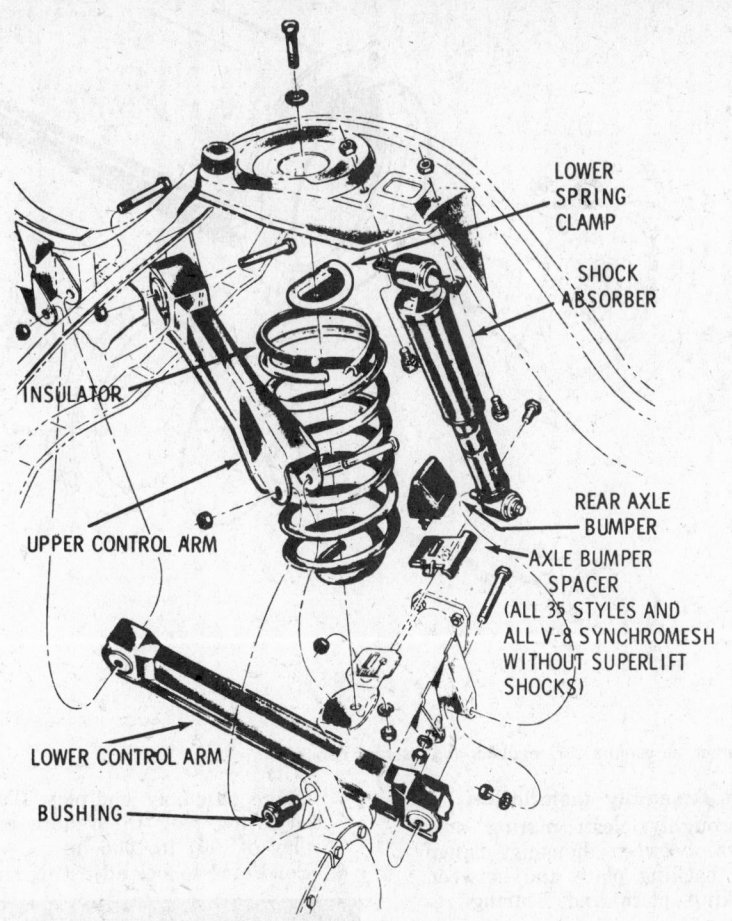

1965-70 rear suspension (© Oldsmobile Div. G.M. Corp.)

sure to these shock absorbers, lifting vehicle to normal position. As load is reduced the valve releases air and lowers vehicle to the previous normal level.

The valve is connected by a link to the right rear upper control link. A deflection of at least ½ in. is required to make it operative.

A delay mechanism is built into the valve housing. This requires 4 to 15 seconds to cause valve to operate. It prevents operation during normal road motion.

Pressure at the shock absorber units is kept equal by means of the line connecting the two units, with only one unit connected directly to the control valve. This keeps approximately 8 to 15 psi on shock absorber units at all times. The pressure is released at the control valve and the equalizing pressure is maintained through a check valve at the release fitting.

The compressor is located at the engine compartment. It is operated by vacuum surge through a line connected just forward of the carburetor insulator connection. Air at atmospheric pressure is taken into the compressor through a line connected to the air cleaner. The compressed air from the compressor is supplied to a reservoir and then to the control valve.

Any service work on this system or other parts of the vehicle that may

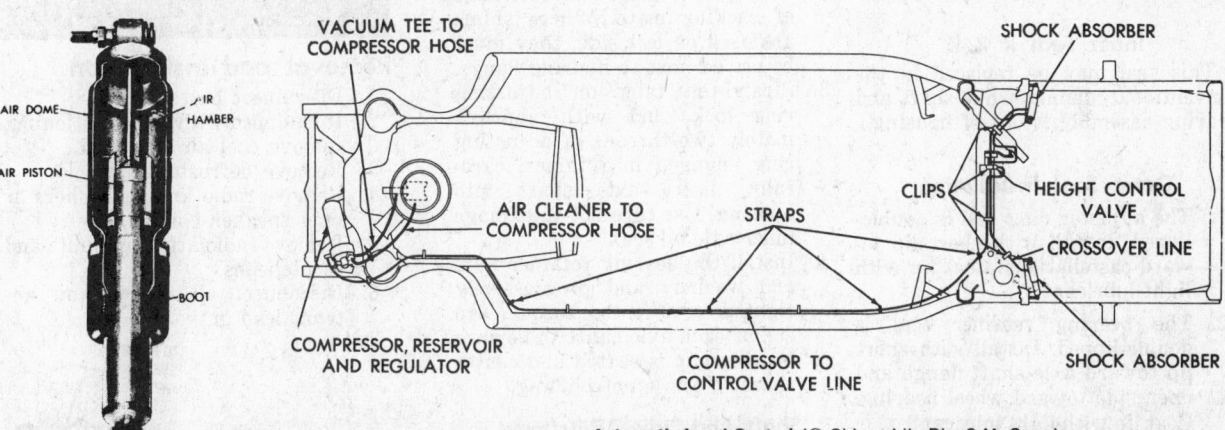

Automatic Level Control (© Oldsmobile Div. G.M. Corp.)

Superlift shock absorber
(© Oldsmobile Div. G.M. Corp.)

operated air compressor, connected to a control valve mounted at rear suspension crossmember. The valve is then connected to Superlift rear shock absorbers.

The Superlift shock absorber is essentially a conventional shock absorber enclosed in an air chamber. A pliable nylon-reinforced neoprene boot seals the air dome to air piston. It will extend or retract under the pressure controlled by the valve.

As load is added to the vehicle, the control valve admits air under pres-

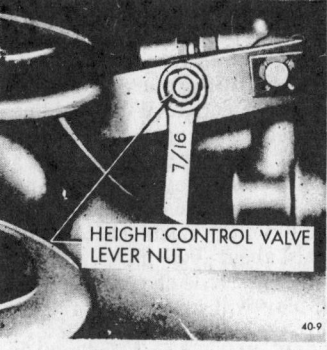

Height control valve lever nut
(© Oldsmobile Div. G.M. Corp.)

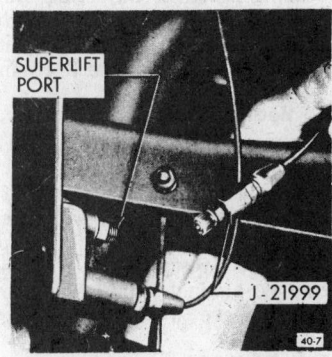

Superlift air supply line removal
(© Oldsmobile Div. G.M. Corp.)

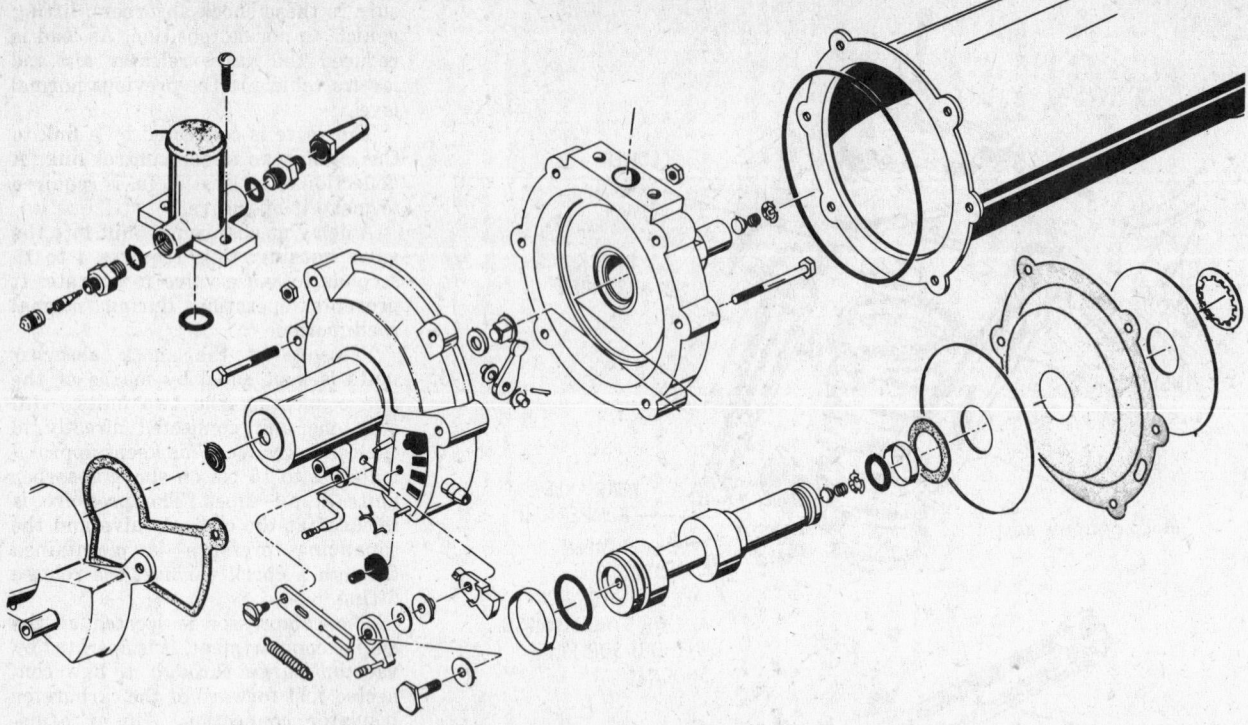

Superlift shock absorber, air compressor, regulator and reservoir details

deflate the system will require reinflation to approximately 140 psi.

All lines are ⅛ in. diameter flexible black tubing. In working on this system, use care not to kink this tubing. Keep tubing away from the exhaust system.

Inner Seal R & R

This seal may be replaced in the conventional manner while shaft and bearing assembly is out of housing.

Outer Seal Installation

1. The adjuster ring seal is double-lipped. Install with short lip toward castellations. Coat lip with light lubricant.
2. The bearing retainer seal is double-lipped. Install with short lip toward axle shaft flange and open side toward wheel bearing. Coat lip with light lubricant.
3. With proper sized adapter. Press bearing seal into adjusting ring or retainer. From face of seal to edge of bearing retainer or adjusting ring should be 3/16 in.

Bearing Installation

1. Thoroughly and carefully pack the bearing with lubricant.
2. Assemble parts in order on shaft in following sequence—adjuster ring with seal, retainer, cup, cone and pressing sleeve. Press tightly against shaft shoulder. Tool, sleeve J-4704-1, is suggested.
3. Install clinch ring with chamfer toward bearing cone.

Shaft Assembly Installation

1. Thoroughly clean mating surfaces between housing flange and backing plate and between backing plate and bearing retainer.
2. Install new gasket on each side of backing plate. Where shims are used on left side, they must be placed next to housing flange.
3. Straighten tang on adjusting ring lock, and with approximately two threads of adjusting ring engaged in retainer, carefully insert axle shaft into housing. Use care not to damage inner axle oil seal.
4. Install the bearing retainer nuts and washers and progressively tighten to 40 ft. lbs. torque. Tap end of each axle shaft to be sure bearing cup is seated and center thrust block is not binding.

Shaft End-Play Adjustment

1. Coat adjuster ring threads with waterproof sealer.
2. Tighten adjuster ring as far as possible, by hand.
3. Insert screwdriver, or other suitable tool, through access hole in axle shaft flange. While holding axle shaft flange to prevent turning, tighten adjusting ring until all end-play has been removed. Then back off ring two castellations. Each castellation equals about .005 in.
4. Tap on left axle shaft so that all end-play will be at the left axle shaft bearing.
5. Mount dial gauge at backing plate and measure shaft and flange assembly end-play. Turn adjusting ring to produce end-play of .001 in.-.006 in.
6. Bend tang to lock adjusting ring.

Radio

Removal and Installation

1. Disconnect battery.
2. If equipped with air conditioning, remove cool air manifold.
3. Remove defroster manifold.
4. Remove radio knobs, washers or rear speaker control.
5. Remove radio attaching nuts and escutcheons.
6. Disconnect all wiring and antenna lead-in.

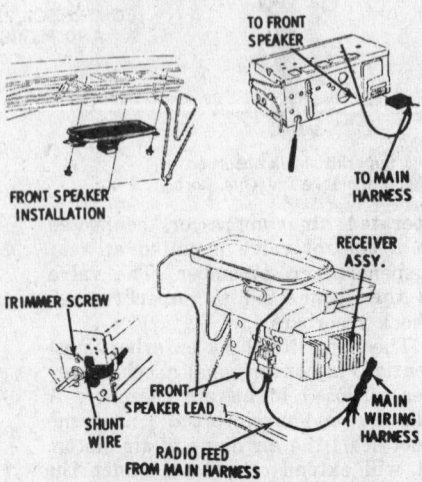

Radio and speaker installation
(52 through 86 series)
© Oldsmobile Div. G.M. Corp.)

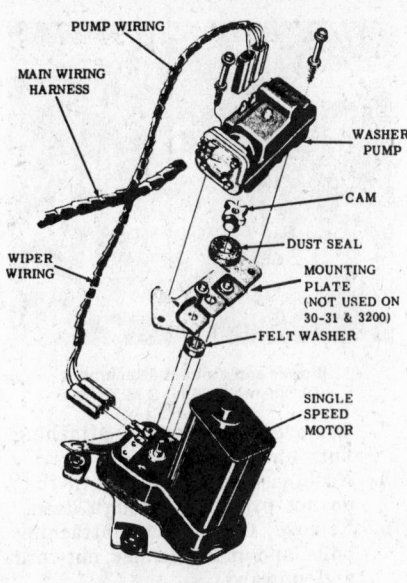

Windshield washer mounting

PUMP WIRING
MAIN WIRING HARNESS
WASHER PUMP
CAM
DUST SEAL
WIPER WIRING
MOUNTING PLATE (NOT USED ON 30-31 & 3200)
FELT WASHER
SINGLE SPEED MOTOR

Wiper Transmission R & R

All Models Except F-85

1. Remove cowl vent grille. Remove baffle.
2. Detach transmission drive linkage from wiper motor assembly.
3. Remove transmission attaching screws and remove the transmission.
4. Remove transmission and linkage arms from the cowl.
5. To install, reverse removal procedures.

NOTE: The linkage for the right transmission overlaps linkage for the left transmission at the wiper motor crank arm.

Motor R & R

F-85

The car may be equipped with a single or a two-speed windshield wiper. The motor is mounted on the engine side of the firewall. The motor drives two wiper transmissions through a crank and connecting links that are located under the instrument panel, inside the car.

1. Remove retaining clip, then disconnect drive link from crank arm.
2. Disconnect harness connector from motor terminals.
3. If car has windshield washers, remove hoses from washer pump.
4. Remove three motor attaching screws, then lift motor assembly from firewall.
5. Install by reversing removal procedure.

Wiper Transmission R & R

F-85 Removal

1. Remove wiper arms from wiper transmission.
2. Tape fender edge adjacent to cowl air inlet grill.
3. Raise hood, remove six cowl vent-to-cowl attaching screws, and remove cowl vent grill and screen.
4. Remove glove compartment, then remove defroster hoses.
5. Disconnect wiper motor to left hand transmission drive link by removing Tru-Arc snap-ring retainer and clip.
6. Remove three transmission attaching bolts from both transmissions and allow transmission assemblies to drop from cowl.
7. Remove Tru-Arc snap-ring retainer from the long drive link and remove transmission from car.

F-85 Installation

1. Install by connecting drive link to transmission under the instrument panel and retain with Tru-Arc snap-ring retainer.

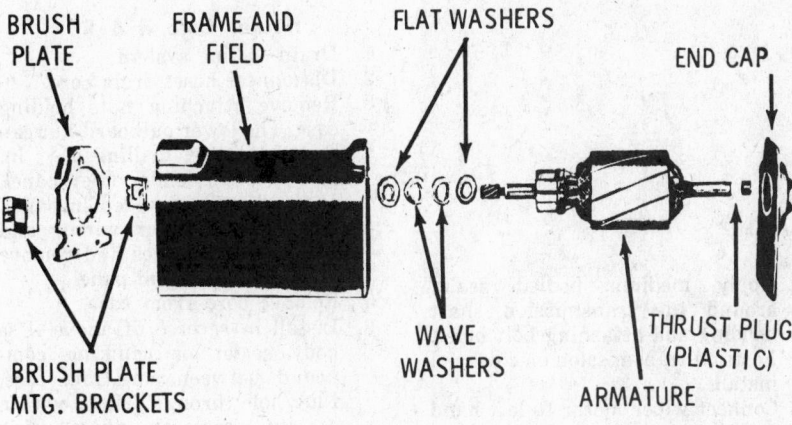

Wiper motor

BRUSH PLATE
FRAME AND FIELD
FLAT WASHERS
END CAP
BRUSH PLATE MTG. BRACKETS
WAVE WASHERS
ARMATURE
THRUST PLUG (PLASTIC)

7. Remove lower radio support bracket attaching screw.
8. Remove radio from rear of instrument panel.
9. Install by reversing removal procedure.

Windshield Wipers

Motor R & R

All Models Except F-85

1. Disconnect drive link from crank arm under the instrument panel by removing the retaining clip.
2. Disconnect harness connector from motor terminals and disconnect washer hoses, if so equipped.
3. Remove three attaching screws, then lift motor assembly from cowl, guiding the crank arm out of the hole in the cowl panel.
4. Install by reversing removal procedures.

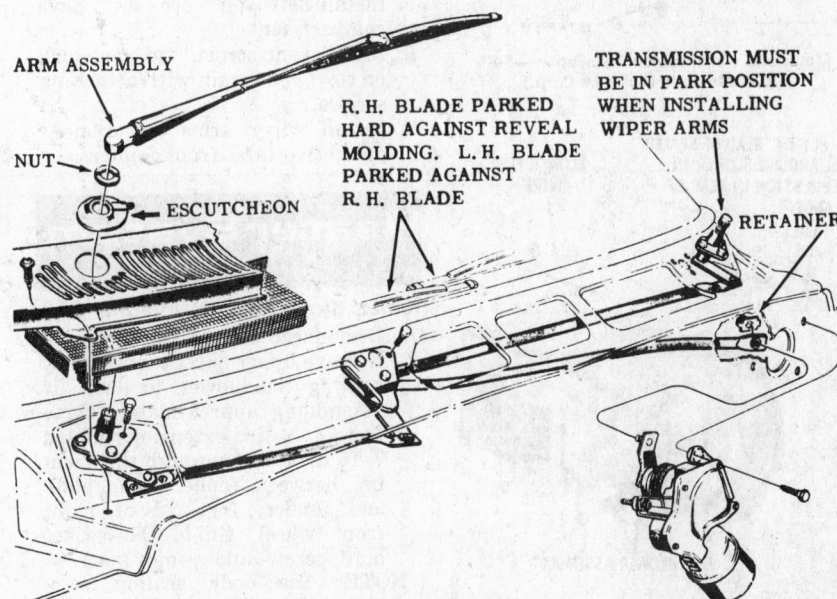

ARM ASSEMBLY
NUT
ESCUTCHEON
R. H. BLADE PARKED HARD AGAINST REVEAL MOLDING. L. H. BLADE PARKED AGAINST R. H. BLADE
TRANSMISSION MUST BE IN PARK POSITION WHEN INSTALLING WIPER ARMS
RETAINER

2-speed overlap wiper—except F-85 (© Oldsmobile Div. G.M. Corp.)

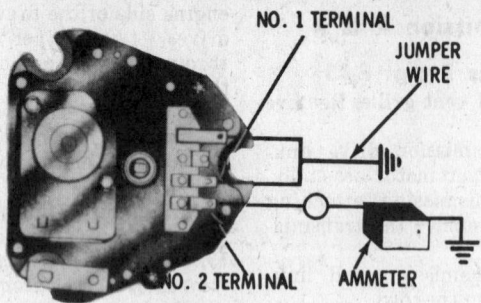

Wiper connections

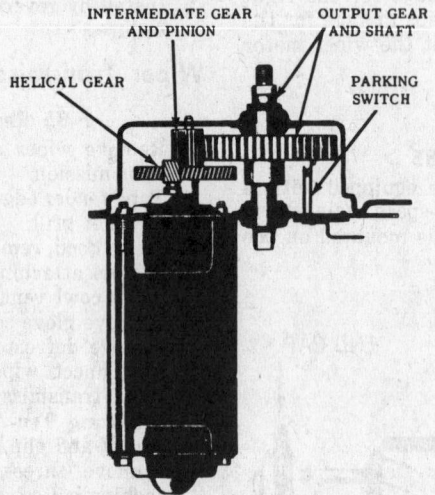

Wiper motor gear train

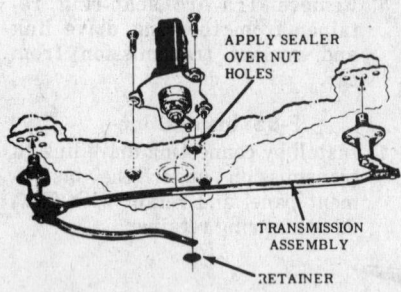

Motor and transmission installation—F-85
(© Oldsmobile Div. G.M. Corp.)

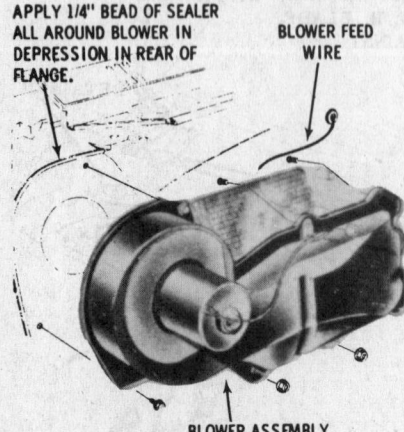

APPLY 1/4" BEAD OF SEALER ALL AROUND BLOWER IN DEPRESSION IN REAR OF FLANGE.

BLOWER FEED WIRE

BLOWER ASSEMBLY

Blower installation
(© Oldsmobile Div. G.M. Corp.)

2. Apply medium bodied sealer around the transmission shaft housing and attaching bolt holes.
3. Position transmission on cowl and install.
4. Connect wiper motor to left hand transmission drive link. Retain link at the transmission with a Tru-Arc snap-ring and at the motor with the spring clip.
5. Install defroster hoses and glove compartment.
6. Install vent screen and vent grill on cowl and retain with attaching screws.
7. Install wiper arms and remove protective tape from fenders.

Heater System

Heater Blower (With Inlet) R & R
1. Disconnect blower feed wire.
2. Remove upper sheet metal screw holding the blower to dash by assembling approximately three feet of 3/8 in. extensions and a 7/16 in. socket through the opening between fender filler plate and fender, forward of right front wheel. Guide the socket onto screw and remove screw. NOTE: use body sealing compound to hold screw in socket while removing and installing.

Blower and air inlet attachment
(© Oldsmobile Div. G.M. Corp.)

3. Remove remaining attaching nuts and screws.
4. Push case studs back until they do not protrude through dash.
5. Remove fender rear attaching bolts and move fender outward and upward.
6. Rotate blower and inlet assembly and remove from car.
7. Install in reverse of above.

Heater Core R & R
1. Drain cooling system.
2. Disconnect hoses from core.
3. Remove attaching nuts holding case. The lower outboard nut can be removed by drilling a 3/4 in. hole through fender filler panel, at the dimple provided in panel.
4. Disconnect resistor wiring and three control cables and remove case assembly from panel.
5. Remove core from case.
6. Install in reverse of above. Use body sealer or caulking compound between core and case. Plug hole through fender with a 3/4 in. accessory plug, using sealer.

Heater Blower R & R
1. Remove right front wheel. Disconnect blower motor wiring.
2. Remove screws and nuts fastening inlet assembly to dash. NOTE: the lower outboard nut can be removed by drilling a 3/4 in. hole through the fender filler panel, at dimple provided.
3. Disconnect inlet assembly from studs and remove from car.
4. Remove blower to inlet attaching screws. Remove fan attaching nut and lock washer.
5. Install in reverse of above.

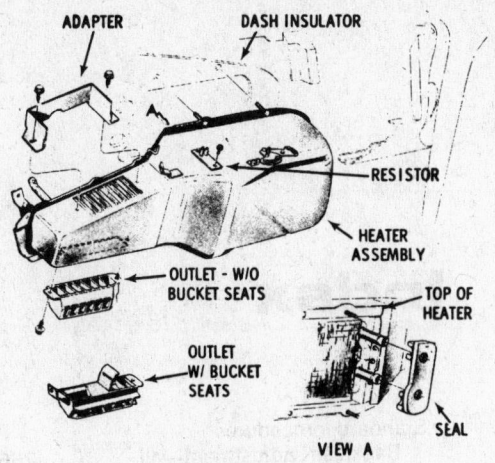

Heater assembly—F-85
(© Oldsmobile Div. G.M. Corp.)

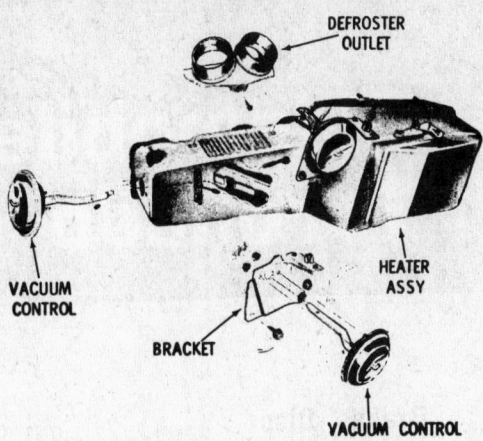

Heater assembly—except F-85
(© Oldsmobile Div. G.M. Corp.)

Oldsmobile Toronado

Index

NOTE: Information not given in this section will be found under that given for the same year models and same year and displacement engines in the Oldsmobile section.

YEAR IDENTIFICATION

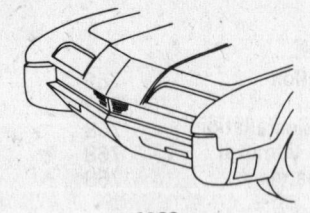
1966

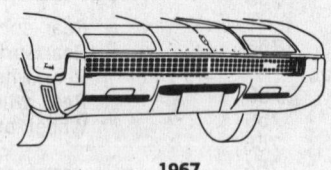
1967

1968

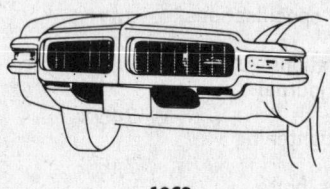
1969

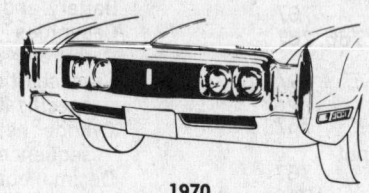
1970

1971

FIRING ORDER

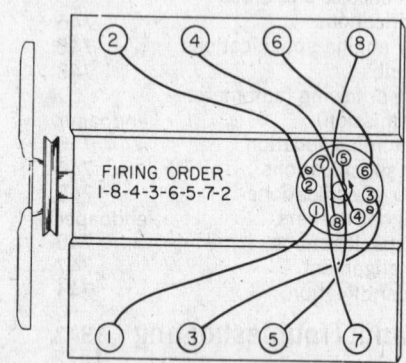

FIRING ORDER
1-8-4-3-6-5-7-2

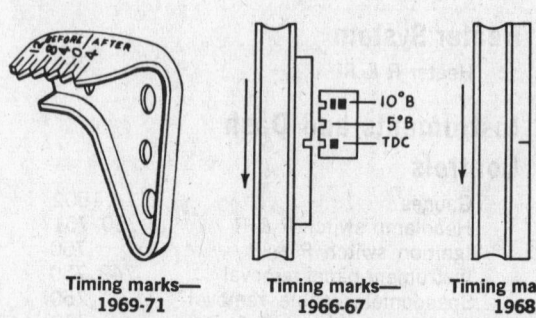

Timing marks—
1969-71

Timing marks—
1966-67

Timing marks—
1968

CAR SERIAL NUMBER LOCATION AND ENGINE IDENTIFICATION

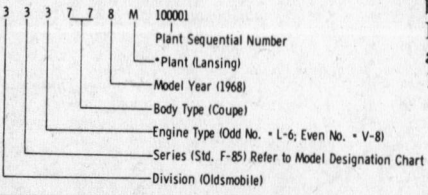

3 3 3 7 7 8 M 100001
Plant Sequential Number
*Plant (Lansing)
Model Year (1968)
Body Type (Coupe)
Engine Type (Odd No. • L-6; Even No. • V-8)
Series (Std. F-85) Refer to Model Designation Chart
Division (Oldsmobile)

*Series	Letter	Plant
3100 through 4800	M	Lansing, Michigan
	Z	Fremont, California
	G	Framingham, Mass. (Early Production)
	E	Linden, New Jersey
5400 through 6600	M	Lansing, Michigan
	X	Kansas City, Kansas
	E	Linden, New Jersey
	C	Southgate, California
	D	Atlanta, Georgia
8400 through 9400	M	Lansing, Michigan
3100 through 4400	(Number) 1	Oshawa, Canada

Vehicle identification number interpretation

Vehicle Identification Number

1966-67

The plate is located on the left front door pillar. Interpretation of the serial number is seen in the illustration. All Toronado models, Series 39687, are built in the Lansing plant in these years. Starting serial numbers for the standard Toronado and Deluxe Toronado are 394877M600001 and 396877M600002 respectively (for 1967 models).

1968-71

The identification plate is located on the dashboard, and can be seen through the left-hand side of the windshield. Interpretation of the number is the same as for 1966-67 models. The Toronado series number is 9400.

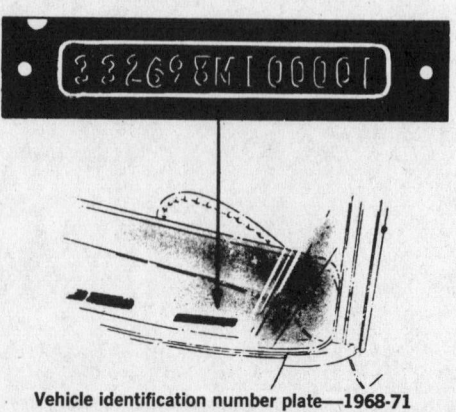

Vehicle identification number plate—1968-71

Engine Identification

1966-67

The engine identification number is stamped on a pad on the front of the right cylinder head. The code suffix letter is found in the chart. All engines are 425 cu. in. displacement.

1968-71

The engine identification number is found on a tape attached to the oil breather tube. The code letters are found in the chart. All engines are 455 cu. in. displacement.

Toronado Engine Identification Code

V8—425—Stamped on machine pad at front of right cylinder head.

No. Cyls.	Cu. In. Displ.	YEAR AND CODE				
		1966	1967	1968	1969	1970
8	425	T	T			
8	455			US, UV, UT, UW	US, UW, UT, UV	(S, US, UT-Std.) (O, UW, UV—W34)

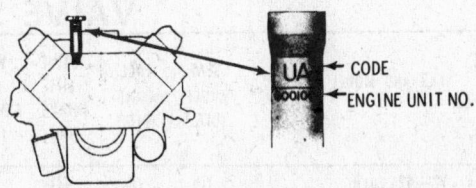

Engine code and unit number tag—1968-71

GENERAL ENGINE SPECIFICATIONS

YEAR	CU. IN. DISPLACEMENT	CARBURETOR	DEVELOPED HORSEPOWER @ RPM	DEVELOPED TORQUE @ RPM (FT. LBS.)	A.M.A. HORSEPOWER	BORE & STROKE (IN.)	ADVERTIZED COMPRESSION RATIO	VALVE LIFTER TYPE	NORMAL OIL PRESSURE (PSI)
1966-67	V8—425 Cu. In.	4-BBL.	385 @ 4800	475 @ 3200	54.0	4.125 x 3.975	10.50-1	Hyd.	35①
1968	V8—455 Cu. In.	4-BBL.	375 @ 4600	510 @ 3000	54.5	4.126 x 4.250	10.25-1	Hyd.	35①
	V8—455 Cu. In.	4-BBL.	400 @ 4800	500 @ 3200	54.5	4.126 x 4.250	10.25-1	Hyd.	35①
1969	V8—455 Cu. In.	4-BBL.	365 @ 4600	510 @ 3000	54.5	4.126 x 4.250	10.25-1	Hyd.	35①
	V8—455 Cu. In.	4-BBL.	390 @ 5000	500 @ 3200	54.5	4.126 x 4.250	10.25-1	Hyd.	35①
1970-71	V8—455 Cu. In.	4-BBL.	375 @ 4600	510 @ 3000	54.5	4.126 x 4.250	10.25-1②	Hyd.	35①
	V8—455 Cu. In. (W34)	4-BBL.	400 @ 5000	500 @ 3200	54.5	4.126 x 4.250	10.25-1②	Hyd.	35①

① —At 1500 rpm.
② —1971 slightly lower.

TUNE-UP SPECIFICATIONS

YEAR	MODEL AND CU. IN. DISPLACEMENT	SPARK PLUGS		DISTRIBUTOR		IGNITION TIMING (Deg.) ▲	CRANKING COMP. PRESSURE (Psi)	VALVES			FUEL PUMP PRESSURE (Psi)	IDLE SPEED (Rpm) ★
		Type	Gap (In.)	Point Dwell (Deg.)	Point Gap (In.)			Tappet (Hot) Clearance (In.) Intake	Exhaust	Intake Opens (Deg.)		
1966	425 Cu. In.	44S	.030	30°	.016	7½B	180	Zero	Zero	21B	7½	850
1967	425 Cu. In.	44S	.030	30°	.016	7½B	180	Zero	Zero	21B	7½	850
1968	455 Cu. In.	44S	.030	30°	.016●	7½B	180	Zero	Zero	20B	7	850
1969	455—4-BBL. (365 HP)	44S	.030	30°	.016	8B	180	Zero	Zero	20B	7	850
	455—4-BBL. (390 HP)	44S	.030	30°	.016	10B	190	Zero	Zero	20B	7	850
1970-71	455—4-BBL. (375 HP)	R45S	.030	30°	.016	8B①	180	Zero	Zero	22B	5½-7	600
	455—4-BBL. (400 HP)	R44S	.030	30°	.016	12B①	190	Zero	Zero	24B	5½-7	600

★—With automatic in D.
▲—With vacuum advance disconnected. NOTE: These settings are only approximate. Engine design, altitude, temperature, fuel octane rating and the condition of the individual engine are all factors which can influence timing. The limiting advance factor must, therefore, be the "knock point" of the individual engine.
●—W/transistor ign.—.045 in.
B—Before top dead center.
①—1971—See engine decal.

Caution

General adoption of anti-pollution laws has changed the design of almost all car engine production to effectively reduce crankcase emission and terminal exhaust products. It has been necessary to adopt stricter tune-up rules, especially timing and idle speed procedures. Both of these values are peculiar to the engine and to its application, rather than to the engine alone. With this in mind, car manufacturers supply idle speed data for the engine and application involved. This information is clearly displayed in the engine compartment of each vehicle.

TORQUE SPECIFICATIONS

YEAR	MODEL	CYLINDER HEAD BOLTS (FT. LBS.)	ROD BEARING BOLTS (FT. LBS.)	MAIN BEARING BOLTS (FT. LBS.)	CRANKSHAFT BALANCER BOLT (FT. LBS.)	DRIVEPLATE TO CRANKSHAFT BOLTS (FT. LBS.)	MANIFOLD (FT. LBS.)	
							Intake	Exhaust
1966–67	V8—425 Cu. In.	80	42	80–120▲	50	60	35	25
1968–71	V8—455 Cu. In.	80	42	120	160	60	35	25

▲—Nos. 1 thru 4—80. No. 5—120.

VALVE SPECIFICATIONS

YEAR AND MODEL	SEAT ANGLE (DEG.)	FACE ANGLE (DEG.)	VALVE LIFT INTAKE (IN.)	VALVE LIFT EXHAUST (IN.)	VALVE SPRING PRESSURE (VALVE OPEN) LBS. @ IN.	VALVE SPRING INSTALLED HEIGHT (IN.)	STEM TO GUIDE CLEARANCE (IN.)		STEM DIAMETER (IN.)	
							Intake	Exhaust	Intake	Exhaust
1966–67 V8—425 Cu. In.	①	②	.431	.433	180–194 @ 1.27	1.670	.0010–.0027	.0015–.0032	.3432–.3425	.3427–.3420
1968–69 V8—455 Cu. In.	①	②	.435	.435	180–194 @ 1.27	1.670	.0010–.0027	.0015–.0032	.3432–.3425	.3427–.3420
1970 V8—455 Cu. In.	③	③	.435	.435	180–194 @ 1.27	1.670	.0010–.0027	.0015–.0032	.3432–.3425	.3427–.3420
V8—455 Cu. In. (W34)	④	⑤	.472	.472	180–194 @ 1.27	1.670	.0010–.0027	.0015–.0032	.3432–.3425	.3427–.3420
1971 V8—455 Cu. In.	③	③	.435	.435	180–194 @ 1.27	1.670	.0010–.0027	.0015–.0032	.3432–.3425	.3427–.3420
V8—455 Cu. In. (W34)	④	⑤	.472	.472	180–194 @ 1.27	1.670	.0010–.0027	.0015–.0032	.3432–.3425	.3427–.3420

①—Intake 30°, Exh. 45°.
②—Intake 30°, Exh. 45°.
③—Intake & Exh. Seat 45°, Intake & Exh. Face 46°.
④—Intake 30°, Exh. 45°.
⑤—Intake 30°, Exh. 46°

CRANKSHAFT BEARING JOURNAL SPECIFICATIONS

YEAR	MODEL	MAIN BEARING JOURNALS (IN.)				CONNECTING ROD BEARING JOURNALS (IN.)		
		Journal Diameter	Oil Clearance	Shaft End-Play	Thrust On No.	Journal Diameter	Oil Clearance	End-Play
1966–67	V8—425 Cu. In.	3.0000	②	.004–.008	3	2.4988–2.5003	.0004–.0033	.002–.013①
1968–71	V8—455 Cu. In.	3.0000	②	.004–.008	3	2.4988–2.4998	.0004–.0033	.002–.013①

①—Both rods.
②—Nos. 1-4—.0021–.005.
　No. 5—.002–.0034.

CYLINDER HEAD BOLT TIGHTENING SEQUENCE

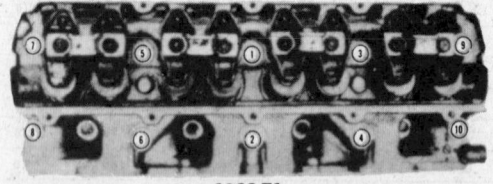

1966-71

GENERAL CHASSIS AND BRAKE SPECIFICATIONS

YEAR	MODEL	CHASSIS			BRAKE CYLINDER BORE			BRAKE DRUM	
		Overall Length (In.)	Tire Size	Master Cylinder (In.)	Wheel Cylinder Diameter (In.)			Diameter (In.)	
					Front	Rear		Front	Rear
1966	Toronado F.W.D.	211.0	8.85 x 15▲	7/8	1 1/8	7/8		11	11
1967	Toronado F.W.D.	211.0	8.85 x 15▲	1.0	1 1/8 ■	7/8		11①	11
1968	Toronado F.W.D.	211.4	8.85 x 15	1.0	1 1/8 ■	7/8		11①	11
1969	Toronado F.W.D.	214.8	8.85 x 15	1.0②	1 1/8 ③	7/8		11①	11
1970–71	Toronado F.W.D.	214.3	J78 x 15	1 1/8	2 15/16	15/16		10.88 disc	11

▲—Opt.—9.15 x 15
■—Disc = 2.06 in.
①—Disc = 11 19/64 in.
②—Disc = 1 1/16 in.
③—Disc = 2 15/16 in.

AC GENERATOR AND REGULATOR SPECIFICATIONS

YEAR	PART NO.	ALTERNATOR				REGULATOR						
		Field Current Draw @ 12V.	Output @ Generator RPM		Part No.	Field Relay			Regulator			
			775	1950		Air Gap (In.)	Point Gap (In.)	Volts to Close	Air Gap (In.)	Point Gap (In.)	Volts at 125°
1966	1100842	2.2-2.6	26	42	1119515	.015	.030	6.3-8.3	.060	014	13.5-14.4
	1100694	2.2-2.6	33	55	1119515	.015	.030	6.3-8.3	.060	.014	13.5-14.4
1967	1100842	2.2-2.6	26	42	1119515	.015	.030	6.3-8.3	.060	.014	13.5-14.4
1968-69	1100842	2.2-2.6	26	42	1119515	.015	.030	6.3-8.3	.060	.014	13.5-14.4
	1100694	2.2-2.6	33	55	1119515	.015	.030	6.3-8.3	.060	.014	13.5-14.4
1970	1100842	2.2-2.6	26	42	1119515	.015	.030	6.3-8.3	.060	.014	13.5-14.4
	1100694	2.2-2.6	33	55	1119515	.015	.030	6.3-8.3	.060	.014	13.5-14.4
1971	1100845	2.2-2.6	26	42	1119515	.015	.030	6.3-8.3	.060	.014	13.5-14.4
	1100697	2.2-2.6	33	55	1119515	.015	.030	6.3-8.3	.060	.014	13.5-14.4

BATTERY AND STARTER SPECIFICATIONS

YEAR	MODEL	BATTERY			STARTERS						Brush Spring Tension (Oz.)
		Ampere Hour Capacity	Volts	Terminal Grounded	Lock Test			No-Load Test			
					Amps.	Volts	Torque	Amps.	Volts	RPM	
1966	425 cu. in.	73	12	Neg.	Not Recommended			70-105	10.6	3,800	35
1967	425 cu. in.	73	12	Neg.	Not Recommended			70-105	10.6	3,800	35
1968-71	455 cu. in.	75	12	Neg.	Not Recommended			70-105	10.6	3,800	35

CAPACITIES

YEAR	MODEL	ENGINE CRANKCASE ADD 1. Qt. FOR NEW FILTER	TRANSMISSIONS Pts. TO REFILL AFTER DRAINING			DRIVE AXLE (Pts.)	GASOLINE TANK (Gals.)	COOLING SYSTEM (Qts.) WITH HEATER
			Manual		Automatic			
			3-Speed	4-Speed				
1966-69	Toronado F.W.D.	5	None	None	④	①	24	17.5②
1970-71	Toronado F.W.D.	5	None	None	④	③	24	18①

① —18.5 with A/C.
② —16.5 without heater; 18.0 with A/C.
③ —1966-68—4.5; 1969-71—4.0.
④ —1966—24; 1967—24 dry, 5½ refill; 1968—24 dry, 8 refill; 1969—24 dry,
    8 refill; 1970-71—24 dry, 8 refill.

WHEEL ALIGNMENT

YEAR	MODEL	CASTER		CAMBER		TOE-IN (In.)	KING-PIN INCLINATION (Deg.)	WHEEL PIVOT RATIO	
		Range (Deg.)	Pref. Setting (Deg.)	Range (Deg.)	Pref. Setting (Deg.)			Inner Wheel	Outer Wheel
1966-67	All	1½N to 2½N	2N	¼N to ½P	⅛P	0 to ¹⁄₁₆	11	20	18¼
1968-69	All	1½N to 2½N	2N	¼N to ½P	⅛P	0 to ¹⁄₁₆	11	20	18¼
1970-71	All	1¾N to 2¾N	2¼N	¼N to ½P	⅛P	0 to ¹⁄₁₆	11	20	18¼

N—Negative.
P—Positive.

FUSES AND CIRCUIT BREAKERS

1966–69

Radio, back-up lamps, fuel gauge, oil pressure
  indicator lamp, parking brake lamp, transmission
  control, temperature gauge lamp, cornering
  lamp, glove compartment lamp, underhood
  lamp .... SFE-9 Amp. Fuse
License lamp, tail lamps, trunk lamp... AGW-15 Amp. Fuse
Ash tray lamp, clock lamp, cruise control lamp
  heater, ventilator and comfortron
  lights .... AGA-3A Amp. Fuse
Heater .... SFE-20 Amp. Fuse
Heater w/A.C. .... ACE-30 Amp. Fuse
Heater w/rear window defroster ...... ACE-30 Amp. Fuse
Cruise control relay, power antenna,
  power windows .... SFE-20 Amp. Fuse
Power seat relay .... AGC-30 Amp. Fuse
Windshield wiper, stop lamps .... SFE-20 Amp. Fuse
Cigar lighter, clock, courtesy lamps
  rear quarter lamps, rear seat lamps. AGC-30 Amp. Fuse
Headlamps .... CB
Air conditioner .... CB
Electric seat and/or window motors .... CB

1970–71

Application	Name of Fuse Circuit on Fuse Block	Fuse Type and Amps
Heater, air conditioning	Htr.-A/C	AGC-25
Stop lamps, hazard warning lamps	Stop-Haz.	SFE-20
Trunk lamp, trunk release Lighter Clock Courtesy lamps Rear quarter lamps Rear seat lighter Glove box lamp Headlamp "off" delay (night watch)	Clk. Ltr.-Ctsy.	AGC-25
Rear lamp Underhood lamp Cornering lamp License lamp Side marker lamps	Tail	AGW-15

Radio	Radio	SAE-9
Power windows Cruise control Mirror map lamp	Pwr.-Relay	SFE-20
Windshield wiper	Wiper	AGC-25
Instrument panel lamp	Inst. Lps.	SAE-4
Transmission control	Gauges-Trans.	SAE-9
Back-up lamp Park brake lamp Turn signal flasher	Back-Up	SAE-9

LIGHT BULBS

1966–69

Speedometer indicator, high beam indicator
  fuel, temperature, ammeter, turn signal
  oil pressure warning, parking brake
  warning................................ 194; 2 CP
Heater, air conditioning.................. 1895; 2 CP
Radio dial .... 1893; 2 CP
Ignition switch, ash tray, cruise control....... 1445; 1 CP
Electric clock .... 1816; 2.5 CP
FM-AM Indicator........................ 2182D; .3 CP
Underhood, trunk........................ 631; 6 CP
Courtesy .... 90; 6 CP
Rear quarter lamps, door lights, glove
  compartment .... 212; 6 CP
Headlamp—upper beam .... L-4001; 37½ W
Headlamp—upper & lower beam .. L-4002; 37½ & 55 W
License................................. 97; 4 CP
Parking & turn signal .... 1157A; 4 & 32 CP
Tail & stop .... 1157; 4 & 32 CP
Back-up .... 1156; 32 CP
Cornering lamp......................... 1195; 50 CP

1970–71

Ident. No.	Quan.	C.P.	Usage
194	3	2	Speedometer, odometer and shift indicator
194	1	2	High beam indicator
194	2	2	Fuel, temperature, ammeter and oil gauge
194	2	2	Turn signal indicator
194	1	2	Brake warning light
194	1	2	Generator warning
194	1	2	Eng. temp.—stop engine warning
168	1	2.5	Oil pressure warning
1895	1	2	Heater, ventilation & air cond. control
1895	1	2	Console compartment
1445	1	1	HMT shift indicator—console only
257	1	2	Engine temperature indicator —hot
1445	1	1	Ash tray light
1445	1	1	Cruise control "on" light
1445	1	1	Rear window defogger "on" light
1445	4	1	Nameplate lights
1445	1	1	Auxiliary high beam indicator
1445	2	1	Electric clock
1893	1	2	Radio dial
563	1	4	Mirror map lamp
631	1	6	Underhood
631	1	6	Trunk
90	2	6	Courtesy
68	2	3	Front seat back
97	1	4	Courtesy light—console
212	4	6	Rear quarter lamps
212	4	6	Door lights
212	1	6	Glove box

EXTERNAL LAMPS

97	1	4	License
1157NA	2	3 & 32	Parking & turn signal
1157	4	3 & 32	Tail and stop
194	2	2	Side marker—front
194	2	2	Side marker—rear
1157	2	32	Back up
1195	2	50	Cornering lamp
L-4001	2	37½ Watt	Headlamp—upper beam
L-4002	2	37½ & 55 Watt	Headlamp—upper & lower beam

Distributor

Detailed information on distributor drive, direction of distributor rotation, cylinder numbering, firing order, point gap, point dwell, timing mark location, spark plugs, spark advance, and idle speed are in the specifications tables of this section. Further information on troubleshooting, general tune-up procedures, how to replace ignition wires, adjust timing, etc. is in the Unit Repair Section.

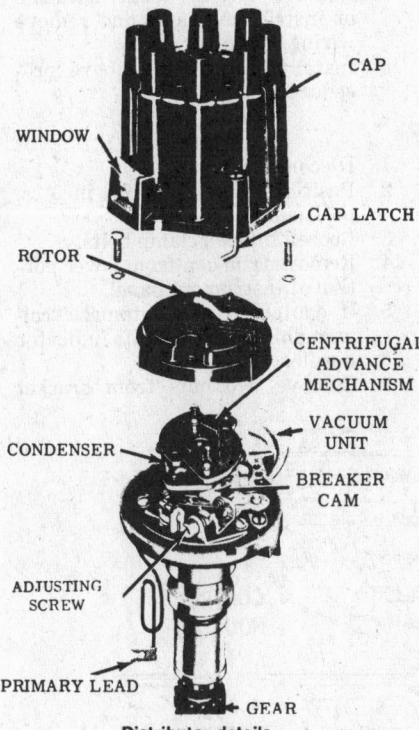

Distributor details
(© Oldsmobile Div. G.M. Corp.)

Labels: CAP, WINDOW, ROTOR, CAP LATCH, CENTRIFUGAL ADVANCE MECHANISM, VACUUM UNIT, CONDENSER, BREAKER CAM, ADJUSTING SCREW, PRIMARY LEAD, GEAR

Distributor Removal

1966-70

Remove the air cleaner, distributor cap and wire assembly. Mark the position of the rotor in relation to the body and the body in relation to the block. Do not move the engine after the distributor is removed. Disconnect the vacuum line at the distributor and remove the primary wire. Take out the bolt that holds the distributor to the block and lift off the distributor.

These models are fitted with an externally adjusted distributor having a window in the distributor cap. Final adjustment of the breaker points is done with the engine running.

The breaker points are set roughly to the correct gap, the distributor installed on the vehicle and the engine started and run at fast idle.

An electric dwell meter can be attached to the distributor.

Raise the window in the cap and, reaching in, with the engine running, turn the adjusting screw until a point dwell of 30° is obtained.

If a dwell meter is not available, turn the screw inward until the engine slows down, then outward until the engine slows down, counting the number of turns between the two positions. Set the points midway between these two positions.

Generator

Removal

All

The generator is located on top of the right bank of cylinders.

Slacken off on the tensioner and remove the fan belt, disconnect the wires to the generator, let the generator rock down and remove the swivel bolts.

The generator may now be lifted off the car.

Delcotron (AC Generator)

1966-70

An alternating current generator is used. This unit is the Delco-Remy, Delcotron. The purpose of this unit is to satisfy the increase in electrical loads that have been imposed upon the car battery by modern conditions of traffic and driving patterns.

The Delcotron is covered in the Unit Repair section of this manual under Basic Electrical Diagnosis.

Caution Since the Delcotron and regulator are designed for use on a one polarity system, the following precautions must be observed:

1. The polarity of the battery, generator and regulator must be considered and matched before making any electrical connections to the system.
2. When connecting a booster battery, be sure to connect the negative battery terminals together and the positive battery terminals together.
3. When connecting a charger to the battery, connect the charger positive lead to the battery positive terminal. Connect the charger negative lead to the battery negative terminal.
4. Never operate the Delcotron on open circuit. Be sure that all connections in the circuit are clean and tight.
5. Do not short across or ground any of the terminals on the Delcotron regulator.
6. Do not attempt to polarize the Delcotron.
7. Do not use test lamps of more than 12 volts for checking diode continuity.

8. Avoid long soldering times when replacing diodes or transistors. Prolonged heat is damaging to these units.
9. Disconnect the battery ground terminal when servicing any AC system. This will prevent the possibility of accidental reversal of polarity.

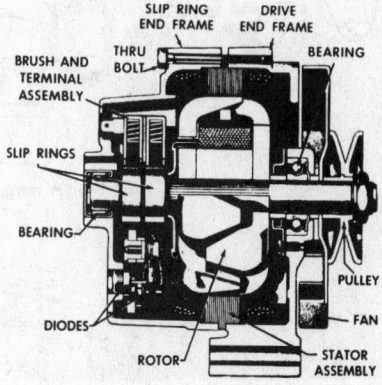

Delcotron alternator
(© Oldsmobile Div. G.M. Corp.)

Labels: SLIP RING END FRAME, DRIVE END FRAME, BRUSH AND TERMINAL ASSEMBLY, THRU BOLT, BEARING, SLIP RINGS, BEARING, DIODES, ROTOR, PULLEY, FAN, STATOR ASSEMBLY

Battery, Starter

Detailed information on the battery and starter is in the Battery and Starter Specifications table of this section.

A more general discussion of starters and their repair procedures is in the Unit Repair Section under the heading: Basic Electrical Diagnosis.

Starter R & R

Disconnect battery at junction block and solenoid wire from harness. Raise car. Remove mounting bolts and lift out starter. Cable will slide through sleeve.

Instruments

Dash Instrument Panel

Removal

1. Disconnect battery.
2. If equipped with Cruise Control, disconnect cable at the regulator.
3. Remove screws from right-hand and left-hand lower trim panels.
4. Remove transmission indicator needle.
5. Remove nuts and clamps from column to upper reinforcement. Column may rest on front seat cushion.
6. Loosen bolts at the side, approximately 1/4 in.
7. Speedometer cable and radio lead-in should be disconnected from behind control panel.
8. Remove screws at the top of the control panel.

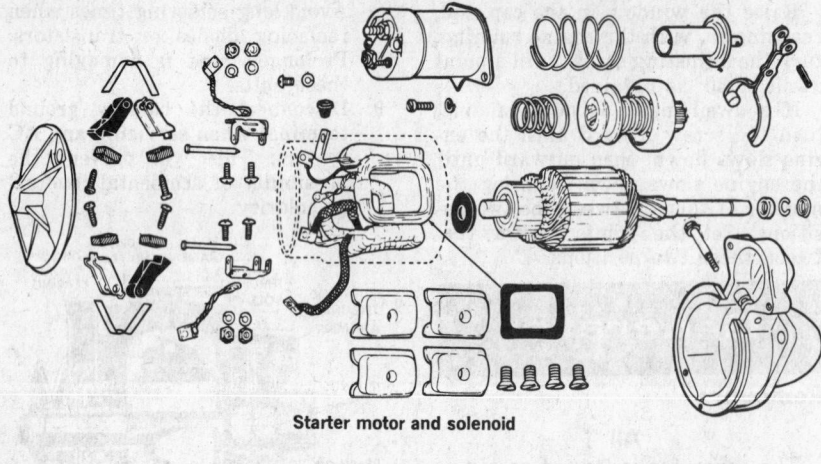

Starter motor and solenoid

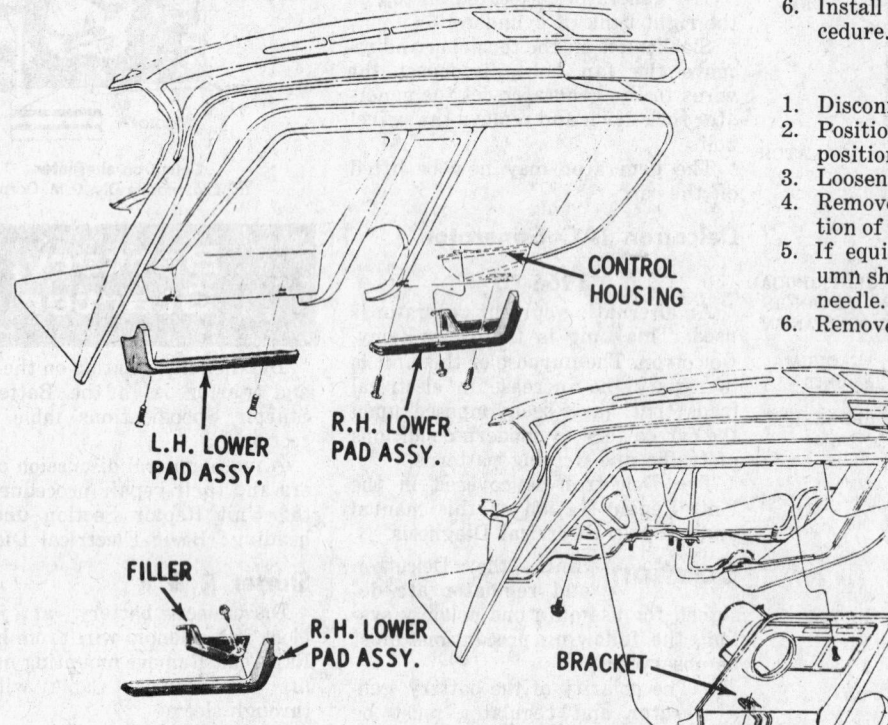

CONTROL HOUSING

L.H. LOWER PAD ASSY.

R.H. LOWER PAD ASSY.

FILLER

R.H. LOWER PAD ASSY.

CONTROL HOUSING

BRACKET

Control panel attachment (© Oldsmobile Div. G.M. Corp.)

5. Remove four attaching bolts, cover to speedometer head.
6. Install by reversing the above procedure.

Ignition Switch R & R

1966-67

1. Remove control panel.
2. Turn switch to Acc. position.
3. Insert a paper clip into the small hole in the front side of the switch and depress, while turning the key counterclockwise. The lock should pop out.
4. Remove escutcheon.
5. Remove switch from backside of instrument panel and remove wiring connector.
6. Install by reversing above procedure.

1968-70

1. Disconnect battery.
2. Position lock assembly in Acc. position.
3. Loosen toe-pan clamp bolts.
4. Remove trim cap from lower portion of instrument panel.
5. If equipped with automatic column shift, remove shift indicator needle.
6. Remove two nuts from bracket

9. If equipped with air-conditioner, disconnect outlet hose.
   NOTE: all components of the control panel can be removed at this time for any service required.
10. Disconnect main wiring harness from components.
11. Remove bolts that were originally loosened ¼ in. in Step 6.
12. Remove control panel.
13. Install by reversing the removal procedure.

Speedometer Cable

Removal—1966-67

1. Remove right-hand lower trim panel.
2. Disconnect cable at speedometer.
3. Disconnect cable at transmission.
4. Remove cable.

5. Reinstall by reversing the above procedure.

1968-70

1. Remove instrument panel.
2. Disconnect cable at speedometer.
3. Remove cable.
4. Install by reversing removal procedure.

Speedometer Head R & R

Removal

1. Remove control panel, refer to Dash Instrument Panel Removal.
2. Disconnect printed circuits. (There are three connectors).
3. Remove radio.
4. Remove eight cover-to-cluster attaching screws.

assembly and lower column until steering wheel rests on seat cushion.
7. Remove two switch attaching screws and lift switch off actuator rod.
8. Disconnect wiring from ignition switch.
9. Install by reversing removal procedure.

Headlamp Switch R & R

1. Remove lower left-hand trim panel.
2. Remove knob by first pulling the knob out to the headlight position, then depressing the spring-loaded button on switch body. Then, pull knob out of switch assembly.
3. Remove escutcheon nut.

4. Remove headlamp switch from rear of control panel.
5. Disconnect wiring and vacuum hoses.

Brakes

Brake Information

See the General Chassis and Brake Specifications table of this section.

Information on brake adjustments, lining replacement, bleeding procedure, master and wheel cylinder overhaul is in the Unit Repair Section.

Information on troubleshooting and overhauling power brakes can be found in the Unit Repair Section, also.

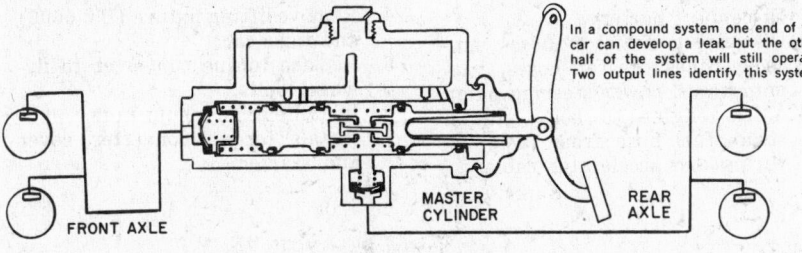

In a compound system one end of the car can develop a leak but the other half of the system will still operate. Two output lines identify this system.

Dual type master cylinder

1969-70

Since 1969, a single cylinder front wheel disc brake is used. For information on this brake, consult the Unit Repair Section of this manual.

Brake Pedal Clearance

Before adjusting the brakes, make certain there is ½ in.—¾ in. free pedal travel (measured at the toeboard). This adjustment is made at the master cylinder pushrod.

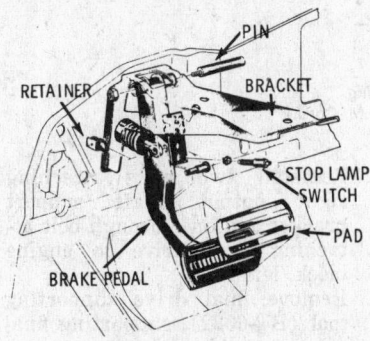

Brake pedal assembly
(© Oldsmobile Div. G.M. Corp.)

Parking Brake Cable Adjustment

The parking brake is controlled by a separate foot pedal. Adjust as follows:

1. Release parking brake.
2. Do minor service brake adjustment.
3. Check for proper length of intermediate cable by inserting a ¼

in. gauge, or drill, into the gap between the relay lever and the end of the slot in the frame. Adjust this cable at front clevis to obtain this ¼ in. fit.

4. Loosen brake equalizer nut until relay lever will move forward enough to admit a 1 3/16 in. gauge between the relay lever and the rear of the slot in the frame.
NOTE: be sure that the equalizer nut has been backed off enough to slacken the cable when the 1 3/16 in. gauge is in place.
5. Tighten the equalizer nut until the rear wheels are nearly locked up.
6. Remove 1 3/16 in. gauge from slot.

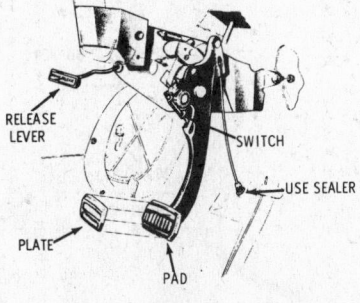

Parking brake assembly
(© Oldsmobile Div. G.M. Corp.)

NOTE: If instructions have been followed and the linkage and cables are free, the parking brake pedal should travel about 2¼ in. to set the brakes solidly.

Power Cylinder Removal

1. From inside the car, detach the brake pushrod from the brake pedal.
2. From underneath the car, detach the vacuum hose at the vacuum cylinder and disconnect the hydraulic line from the front of the slave cylinder.
3. Remove the four nuts that hold the vacuum unit up to the toeboard and let the vacuum unit come out through the bottom of the car.

Fuel System

Data on capacity of the gas tank are in the Capacities table. Data on correct engine idle speed and fuel pump pressure are in the Tune-up Specifications table. Both tables are in this section.

Information covering operation and troubles of the fuel gauge is in the Unit Repair Section.

Detailed information on the carburetor, and how to adjust it, is in the Unit Repair Section under the broad heading: Carburetors, and the specific heading of the make of carburetor being used on the engine being worked on.

Dashpot adjustment is in the Unit Repair Section under the same heading as that of the automatic transmission used in the car, and in the carburetor section.

Fuel Pump Removal

The fuel pump is located in front of the right bank of cylinders.

Disconnect the fuel and vacuum lines, take out the two mounting bolts that hold it to the timing case cover and lift off the pump.

The back bolt can be reached with a universal socket and short extension.

1969-70

A new fuel pump features a large fuel reservoir, with a filter element within the pump housing.

Cooling System

Cooling System Information

Detailed information on cooling system capacity is in the Capacities table of this section.

Information on the water temperature gauge is in the Unit Repair Section.

Thermostatic Vacuum Switch

If equipped with exhaust emission control and air-conditioning, a thermostatic vacuum switch is used to increase ignition timing advance at idle when higher coolant temperatures are encountered. This change, from part to full vacuum advance at engine idle, tends to improve cooling at idle.

If a heating condition exists at idle speeds only, disconnect the distributor vacuum hose at the thermostatic switch. When engine temperature is above 230° F., there should be no vacuum at this port at idle. If engine temperature is above 230° F., and vacuum is present at port D, check distributor advance.

Radiator Core Removal

The radiator core can be removed from all Toronado models without taking off the water pump. Remove the radiator upper baffle plate and radiator hoses, headlight wires, hood, cables, etc. Then, unbolt the core assembly from its support and take it out of the car. It may be necessary to rotate the fan blades.

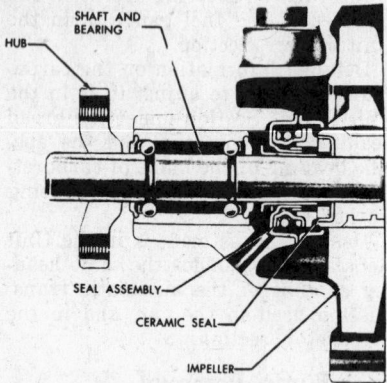

Water pump cross-section
(© Oldsmobile Div. G.M. Corp.)

Water Pump Removal

Drain the cooling system and slack off the generator and power steering belts. Remove the fan and fan pulley, then remove bolts that hold the water pump to front of engine and lift off pump.

Only one side of the gasket should be coated with gasket compound when reinstalling.

It is a good idea to dip the bolts in sealer before inserting them into the pump.

Thermostats

On all engines, the thermostat is located under the water outlet elbow in the water manifold at the front of the block.

Engine

References

In the specifications tables of this section are listed all the available facts about the engines.

Any other information not given here may be found under that given for the 425 and 455 cu. in. displacement engines in the Oldsmobile section.

Valve Information

Detailed information on the valves and the type of valve guides, is in the Valve Specifications table of this section.

Bearing Information

Detailed information on engine bearings is in the Crankshaft Bearing Journal Specifications table of this section.

Engine Reassembly Information

Engine crankcase capacities are listed in the Capacities Table of this section.

Approved torque wrench readings and head bolt tightening sequences are covered in the Torque Specifications Table of this section.

Information on the engine marking code is in the Model Year Identifications table of this section.

Engine R & R

Removal

1. Drain radiator.
2. Remove hood, marking hinge for reassembly.
3. Disconnect battery.
4. Disconnect radiator hoses and cooler lines, heater hoses, vacuum hoses, power steering pump hoses, engine to body ground strap, fuel hose from fuel line, wiring and accelerator cable.
5. Remove coil, throttle control switch bracket, radiator support and radiator
6. Raise the car.
7. Disconnect exhaust pipes at manifold.
8. Disconnect wires and remove starter.
9. Remove torque converter cover and remove three bolts securing the converter to flywheel.
10. Attach a final drive supporting tool (BT-6322) to support final drive assembly.
11. Remove two attaching bolts from right output shaft support bracket and one through bolt attaching final drive to engine block on the left side.
12. Remove engine mount to crossmember nuts.
13. Lower the car.
14. Support engine by using a lifting fixture (BT-6606).
15. Remove six bolts, transmission to engine.
16. Using suitable lifting device, lift engine from car.

Caution If car is to be moved, install converter holding tool (J-21654).

Installation

1. Attach lift fixture (BT-6606) and lower engine into position.
2. Locate engine dowels into transmission and position mount studs into front crossmember.
3. Reinstall six bolts, transmission to engine.
4. Remove lifting fixture (BT-6606) and raise car.
5. Replace torque converter to flywheel bolts.
6. Install engine mount nuts.
7. Install torque converter cover and starter.

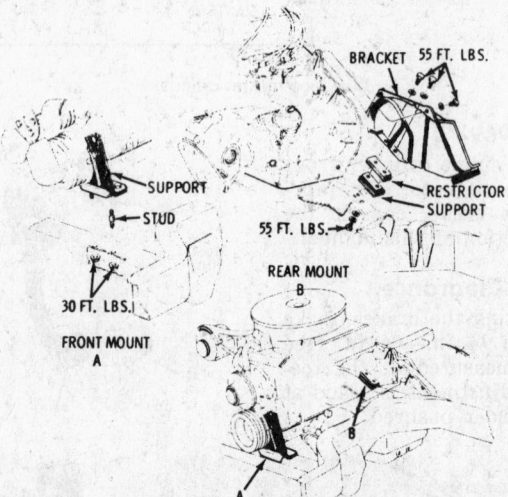

Engine mounting
(© Oldsmobile Div. G.M. Corp.)

8. Reinstall two bolts attaching right output shaft support bracket and one through bolt attaching final drive to engine block, left side.
9. Remove final drive supporting tool (BT-6322), supporting final drive assembly.
10. Connect exhaust pipes.
11. Lower the car.
12. Install coil, throttle control switch bracket, radiator and radiator support.
13. Reconnect accelerator cable, wiring, fuel hoses, engine to body ground strap, power steering pump hoses, vacuum lines, heater hoses, cooler lines and radiator hoses.
14. Install and align hood.
15. Refill the cooling system.
16. Connect battery.

Engine Lubrication

Engine Oil Pan Removal
1966-67

1. Remove engine assembly, as previously outlined.
2. Remove dipstick.
3. Drain oil.
4. Remove mount from front cover.
5. Remove oil pan attaching bolts and remove oil pan.

Engine Oil Pan Installation
1966-67

1. Apply a good sealer to both sides of pan gaskets and install on block.
2. Install front and rear seal.
3. Wipe lubricant on seal area and install pan. Torque 5/16 in. bolts to 15 ft. lbs. and ¼ in. bolts to 10 ft. lbs.
4. Reinstall mount to front cover.
5. Reinstall engine and fill crankcase as explained in charts.

Engine Oil Pan Removal
1968-70

1. Disconnect battery and remove dipstick.
2. Remove upper radiator support screws and fan shroud screws.
3. Hoist car and drain oil.
4. Disconnect engine mounts and jack front of engine up as far as possible.
5. Remove crossover pipe and starter.
6. Remove oil pan attaching bolts and remove oil pan.

Timing Case

Timing case services or services to the front cover, such as timing chain or sprockets, require the removal of the engine.

Timing Cover Removal

1. Disconnect radiator and by-pass hoses from water pump.
2. Remove cover to block attaching bolts and remove cover, timing pointer and water pump assembly.

Timing Cover Installation

1. Install new cover gasket. Apply a sealer to the gasket around water holes and place on the block.
2. Install front cover and torque to specification.
3. Reinstall engine.

Valve Timing Procedure

NOTE: camshaft end-play on these engines is controlled by a block and spring-loaded plunger.

On all Toronado models, it is necessary to remove the engine to service

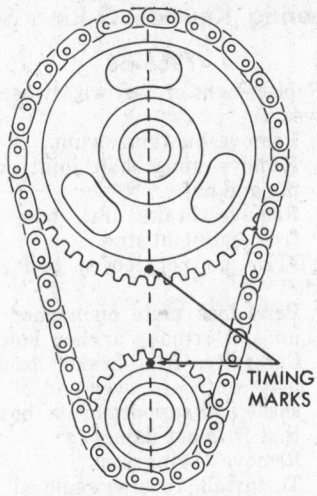

Valve timing alignment marks

the timing chain area. The construction is such that the chain can be worn very badly without seriously affecting the valve timing. If the chain is worn badly enough to cause timing to jump, or it becomes necessary to replace either the chain or the sprockets, proceed as follows:

1. Remove the engine.
2. Remove the timing case cover and the camshaft gear.

NOTE: the fuel pump operating cam is bolted to the front of the camshaft sprocket and the sprocket is located to the camshaft by means of a dowel.

3. Remove the timing chain and the camshaft sprocket and, if the crankshaft sprocket is to be replaced, remove it also at this time.
4. Reinstall the crankshaft sprocket, being careful to start it with the keyway in perfect alignment since it is rather difficult to correct for misalignment after the gear has been started on the shaft.
5. Turn the timing mark on the crankshaft gear until it points directly toward the center of the camshaft.
6. Mount the timing chain over the camshaft gear and start the camshaft gear up onto its shaft, with the timing marks as close as possible to each other and in line between the shaft centers.
7. Rotate the camshaft enough to align the shaft with the new gear. A dowel pin is used for alignment.
8. Secure the camshaft gear and check to see that the mark on the crank sprocket and the mark on the cam sprocket are as close as possible to each other, in line between shaft centers. Valves timed in this manner are correct, regardless of which piston is at top center. It may be necessary, however, to retime the ignition,

since there is a possibility it will be 180° out of position.
9. Reinstall the engine.

Connecting Rods And Pistons

Piston and Rod Assembly Removal

NOTE: it is necessary to remove the engine to service the bearing area, such as pistons, rods, engine bearings, etc.

1. Remove engine. Then, remove the oil pan and the cylinder heads.
2. Select the pistons in the down position and, using a ridge reamer, cut out the ring ridge from the top of the cylinder. If a ridge reamer is not available, a good bearing scraper will cut the ridge.
3. From underneath the engine, remove the rod bearing cap from the rod and piston assemblies where the ring ridge has been cut and push the rod and piston assembly up through the top of the block and out. Immediately replace the bearing cap on the bottom of the rod so that neither the bearing nor the bearing nuts become lost or damaged. Repeat on the rest of the cylinders.

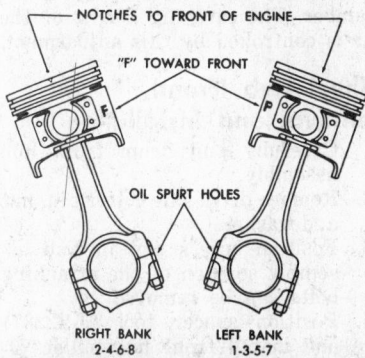

Piston and rod assembly

Steering Wheel

Steering Wheel and Horn Ring Removal and Installation

1. Turn emblem counterclockwise to release.
2. Remove steering wheel nut and washer.
3. Using a puller, remove the wheel from the steering shaft.
4. Install by reversing removal procedure.

Steering Column Removal And Installation

1. Disconnect battery, and remove steering wheel.

2. Remove right and left lower trim panels.
3. Remove screws holding the metal cover and sleeve to the floor panel.
4. Disconnect all wiring connectors from column.
5. Disconnect upper centershaft from joint assembly.
6. Disconnect shift lever from lower end of column.
7. Disconnect transmission indicator needle from shifter tube.
8. Remove clamp holding column to lower instrument panel.
9. Remove column assembly with upper center shaft still in column. It is not necessary to disturb clamp and spring at lower end of column.
10. Install by reversing removal procedure.

Front Suspension

References

The front suspension consists of control arms, stabilizer bar, shock absorbers and a right and left torsion bar. Torsion bars are used in place of conventional coil springs. The front end of the torsion bar is attached to the lower control arm. The rear of torsion bar is mounted into an adjustable arm at the torsion bar crossmember. The carrying height of the car is controlled by this adjustment.

Wheel Hub (Front)
Removal and Installation

1. Carefully pull drum from hub assembly.
2. Remove drive axle cotter pin, nut and washer.
3. Position access slot in hub assembly so each of the attaching bolts can be removed.
4. Position spacer tool (J-22237) and install front hub puller (J-21579) and slide hammer (J-2619).
5. Remove hub assembly.
6. To install, reverse removal procedure.

NOTE: O.D. of bearing must be lubricated with E.P. chassis lubricant. Use care when installing hub assembly over drive axle splines.

Steering Knuckle O-Ring Seal

1966-68

Replacement of seal with wheel removed.
1. Remove hub and drum.
2. Remove upper ball joint cotter pin and nut.
3. Remove brake line hose clip from ball joint stud.
NOTE: do not loosen ball joint stud.
4. Bend lock plate on anchor bolt up and remove anchor bolt.
5. Carefully lift brake backing plate outboard over end of axle shaft and support brake hose so that it is not damaged.
6. Remove O-ring seal.
7. To install, reverse removal procedure.

Knuckle Seals—1969-70

Right-Hand Seal Removal

1. Hoist car under lower control arms.
2. Remove drive axle cotter pin, nut and washer.
3. Remove oil filter element.
4. Remove inner C. V. joint attaching bolts and push joint outward and rearward to disengage from output shaft.
5. Remove output shaft.
6. Remove drive axle assembly.
7. Pry seal from knuckle.

Left-Hand Seal Removal

1. Hoist car under lower control arms.
2. Remove wheel and drum.
3. Remove drive axle cotter pin, nut and washer.
4. Remove hub assembly.
5. Remove tie-rod end cotter pin and nut.
6. Using hammer and brass drift, drive on knuckle until tie-rod end stud is free.
7. Place support block between drive axle and lower control arm.
8. Remove upper control arm ball joint cotter pin and nut. Remove brake hose clip from ball joint stud.
9. Using hammer and brass drift, drive on knuckle until ball joint stud is free.
10. Install spacer (J-22292-3) between lower ball joint seal and knuckle.
11. Remove lower ball joint from knuckle.
12. Remove knuckle.
13. Pry seal from knuckle.

Steering Knuckle Removal

1. Raise car and support lower control arm with floor stands.
2. Remove wheel and drum.
3. Remove hub assembly.
4. Follow Steps 2 through 6 on knuckle O-ring removal.
5. Place (J-22193) support block under drive axle to protect C. V. joint seal.
6. Using a brass drift and hammer, loosen upper ball joint stud.
7. Remove cotter pin and nut from tie rod end.
8. Using brass drift and hammer, remove tie rod end from knuckle.
9. Remove cotter pin and nut from lower ball joint.
10. Carefully place spacer (J-22292-3) between ball joint seal and knuckle.
11. Using ball joint puller (J-22292-1-2), remove lower ball joint from knuckle.
12. Remove knuckle.
13. Knuckle seal can be pried from the knuckle at this time.

Steering Knuckle Installation

1. Using seal installer (J-22234), install seal into knuckle.
2. Install lower ball joint stud into knuckle and attach nut. Do not torque.

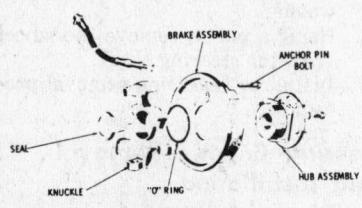

Front hub assembly
(© G.M. Corp.)

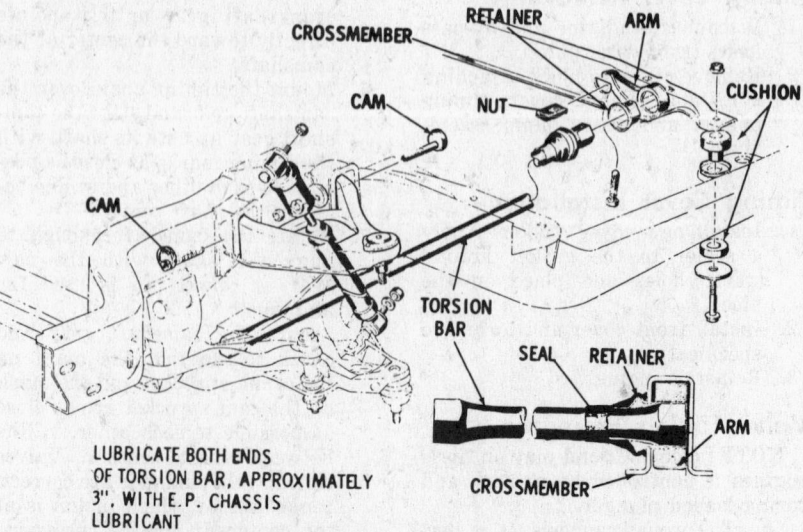

LUBRICATE BOTH ENDS OF TORSION BAR, APPROXIMATELY 3", WITH E.P. CHASSIS LUBRICANT

Front suspension torsion bar (© Oldsmoblie Div. G.M. Corp.)

3. Install tie rod and stud into knuckle and attach nut. Do not torque.
4. Install upper ball joint stud into knuckle and attach nut. Do not torque.
5. Install backing plate onto knuckle with anchor bolt and lock plate. Do not torque.
6. Remove upper ball joint attaching nut and install brake line hose clip.
7. Torque ball joints nuts to a minimum of 40 ft. lbs. Never back off to install cotter pins.

NOTE: cotter pin on upper ball must be bent up only, to prevent interference with C. V. joint seal.

8. Torque tie rod end to 45 ft. lbs. and install cotter pin.
9. Torque anchor bolt to 135 ft. lbs. and bend lock plate onto flat of bolt head.
10. Install hub assembly.
11. Install drum and wheel.
12. Remove floor stand and lower car.
13. Be sure to check camber, caster and toe-in, and adjust if necessary.

Torsion Bar Removal

1966-68

1. Hoist car and place floor stands under front frame horns. Lower front hoist.
2. Slide seal at rear of torsion bar forward.
3. Before using torsion bar remover and installer (BT-6601), remove the four nuts and center screw, then place tool over crossmember support. Align pin of tool in hole in crossmember. Install two nuts on tool and center screw, then install two remaining bolts to lock center screw. Turn center screw until seated in dimple of torsion adjusting arm.
4. Using a socket on the torsion bar adjusting bolt, turn counterclockwise; counting the number of turns necessary to remove.

NOTE: number of turns to remove the adjusting bolt will be used when installing.

5. Remove adjusting bolt and nut.
6. Turn center screw of tool BT-6601 until torsion bar is completely relaxed.
7. Place a block of wood 6 x 6 x 8 in. on hoist and raise under lower control arm until drive axle is horizontal.
8. Remove stabilizer bolt, nut, grommets, retainers and spacer. Discard the bolt.
9. Place a daub of paint on the bottom side of the torsion bar.
10. Slide torsion bar forward until it bottoms in lower control arm. Adjusting arm will drop out.

NOTE: do not mark, scratch, or in

Removing torsion bar
(© Oldsmobile Div. G.M. Corp.)

any way damage the surface of the torsion bar. Replacement will be necessary if such conditions exist.

11. Remove bolt from crossmember on the side you are removing torsion bar.
12. Remove center bolt assembly from tool BT-6601.
13. Raise crossmember and twist rearward until torsion bar clears member.
14. Slowly raise lower control arm until maximum height is obtained. Care must be exercised so the floor stands are not dislodged.
15. Raise center crossmember until contact is made with floor.
16. Pull rearward on torsion bar with hands only, until torsion bar is out of lower control arm. It may be necessary to use air, blowing into nut of lower control arm, to relieve the vacuum caused by grease.

1969-70

1. Hoist car and support at lift points.
2. Place torsion bar remover and installer (BT-6601) so that center screw is seated in dimple of torsion adjusting arm.
3. Remove torsion bar adjusting bolt, counting number of turns necessary.
4. Turn center screw of tool BT-6601 until torsion bar is completely relaxed.
5. Disconnect stabilizer link.
6. Disconnect shock absorber from lower control arm.
7. Remove bolts from lower control arm to frame.
8. Pry lower control arm from frame and move forward until torsion bar and adjusting arm can be removed.

Torsion Bar Installation

1966-68

1. Lubricate both ends of torsion bar for approximately 3 in. with extreme pressure chassis lubricant.
2. With daub of paint, in the same location as when removed, insert torsion bar into lower control bar nut and push forward until bar bottoms.
3. Pry crossmember back, then align torsion bar with hole in crossmember.
4. Lower crossmember and install center bolt into tool BT-6601.
5. Lower front lower control arm until drive axle is horizontal.
6. Install torsion bar arm and pull torsion bar rearward until fully seated in arm.
7. Install crossmember bolt through rubber mounting and torque to 40 ft. lbs.
8. Lower front lower control arm and remove 6 x 6 x 8 in. block.
9. Install new bolt into stabilizer bar to lower control arm. Torque to 14 ft. lbs., cut off bottom of bolt so that 1/4 in. is remaining below nut.
10. Using tool BT-6601, tighten up torsion bar arm to install lock plate under arm and through crossmember.
11. Lubricate threads of torsion bar adjuster bolt with extreme pressure chassis lubricant and turn nut the same number of turns used to remove.
12. Remove tool BT-6601.
13. Raise front hoist and remove floor stands from front frame horns. Lower car to floor.
14. Check carrying height. Correct by turning torsion bar adjusting bolt in or out.

1969-70

1. Lubricate both ends of torsion bar for approximately 3 in. with extreme pressure chassis lubricant.
2. Position adjusting arm into crossmember, insert torson bar into adjusting arm and lower control arm, then position lower control arm into frame brackets and install nuts and bolts loosely.
3. Connect shock to lower control arm, tightening nut to 80 ft. lbs.
4. Connect stabilizer bar to lower control arm. Torque nut to 15 ft. lbs. and cut off bolt ¼ in. below nut.
5. Place torsion bar remover and installer BT-6601 over crossmember and tighten center screw.
6. Raise hoist under lower control arms.
7. Torque lower control arm bushing nuts to 90 ft. lbs.
8. Check carrying height and adjust if necessary.

Upper Control Arm Removal

NOTE: the upper control arm is serviced as an assembly, less bushings.
1. Hoist car and remove wheel.
2. Remove upper shock attaching bolt.
3. Remove cotter pin and nut on upper ball joint.
4. Disconnect brake hose clamp from ball joint stud.
5. Using hammer and a drift, drive on spindle until upper ball joint stud is disengaged.
6. Remove upper control arm cam assemblies and remove control arm from car by guiding shock absorber through access hole in arm.

Upper Control Arm Installation

1. Guide upper control arm over shock absorber and install bushing ends into frame horns.
2. Install cam assemblies.
NOTE: front cam is mounted up, rear cam is mounted down.
3. Install ball joint stud into knuckle.
4. Install brake hose clip onto ball joint stud.
5. Install ball joint nut. Torque to 40 ft. lbs. and insert cotter pin, crimp.
NOTE: cotter pin must be crimped toward upper control arm to prevent interference with outer C. V. joint seal.
6. Install upper shock, attaching bolt and nut. Torque to 75 ft. lbs.
7. Install wheel.
8. Lower hoist.
9. Check camber, caster and toe-in, and adjust if necessary.

Upper Control Arm Bushing Removal (On the Car)

NOTE: the upper control arm bushings can be removed and installed on or off the car.
1. Hoist car and remove wheel.
2. Disconnect upper shock absorber attaching bolt.
3. Remove cam assemblies from control arms.
4. Move control arms out of frame horns and attach bushing removal tools.

Upper Control Arm Bushing Installation (On the Car)

1. Install tools and press bushings into control arm.
2. Move control arm into frame horns and install cam assemblies. Front cam is installed with the bolt in the lower position. Rear cam is installed with the bolt in the upper position.
3. Connect upper shock attaching bolt. Torque to 75 ft. lbs.
4. Replace wheel and lower car.
5. Align front wheels.

Lower Control Arm Removal —Right Side 1966-68

1. Remove drive axle assembly.
2. Hoist car and place floor stands under frame horns.
3. Position tool BT-6601 on torsion bar. Refer to Torsion Bar Removal, Steps 2 through 6. Place daub of paint on bottom side of torsion bar, when completely relaxed, as an installation aid.
4. Remove wheel.
5. Remove shock absorber.
6. Disconnect stabilizer bar from lower control arm. Discard the bolt.
7. Place tool J-22193-1 between stabilizer bar and tie rod.
8. Remove cotter pin and nut from lower ball joint stud.
9. Install tool J-22292-3, adapter under ball joint seal, then install tool J-22292-1-2 and remove ball joint stud from spindle.
10. Remove lower control arm bushing bolts and, with the aid of a helper, remove lower control arm and torsion bar, as an assembly. Torsion bar arm will drop out at this time.
11. Carefully slide torsion bar from lower control arm and store in a safe clean place.

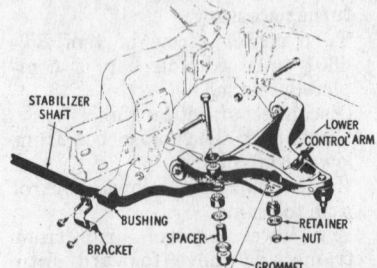

Lower control arm and related components
(© Oldsmobile Div. G.M. Corp.)

Lower Control Arm Installation —Right Side 1966-68

1. Lubricate both ends of torsion bar for approximately 3 in., with extreme pressure chassis lubricant.
2. Install torsion bar into nut of lower control arm.
3. With the aid of a helper, lift torsion bar and lower control arm assembly up until torsion bar will engage arm in crossmember and bushing ends of lower control arm engage frame horn.
NOTE: make sure daub of paint on torsion bar is in the same position as when removed.
4. Install lower control arm bushing bolts and nuts. Do not torque.
5. Install lower control arm ball joint stud into knuckle. Install and torque to 40 ft. lbs.
6. Remove tool J-22193-1 from between stabilizer bar and tie rod.
7. Using a new bolt, connect stabilizer bar to lower control arm. Refer to illustration for correct sequence of grommets, retainers and spacer. Torque to 14 ft. lbs. Cut bolt off ¼ in. below nut.
8. Install shock absorber. Torque upper bolts to 75 ft. lbs., lower nut to 75 ft. lbs.
9. Install wheel.
10. Position tool BT-6601 over torsion bar crossmember and tighten center bolt until nut can be positioned through crossmember. Lubricate adjuster bolt with E. P. chassis lubricant and turn in the same number of turns required to remove.
11. Raise hoist under lower control arms and remove floor stands.
12. Install right-hand drive axle assembly.
13. Torque lower control arm bushing bolts to 75 ft. lbs.
14. Check caster, camber and toe-in.

Lower Control Arm Removal —Left Side 1966-68

1. Remove drive axle assembly. Refer to Drive Axle Assembly Removal—Left Hand.
2. Place floor stands under frame horns.
3. Lower hoist slowly until floor stands are seated and hoist is still under lower control arm.
4. Remove shock absorber and bolt from stabilizer bar to lower control arm. Discard the bolt.
5. Install tool BT-6601 and completely relax the torsion bar. For attachment, refer to Torsion Bar Removal Steps 2 through 6.
6. Lower front hoist to floor.
7. With the aid of a helper, remove lower control arm attaching bolts to frame and carefully lower control arm and torsion bar as an

assembly. Torsion bar arm will drop out at this time.

8. Carefully slide torsion bar from lower control arm and store in a safe clean place.

Lower Control Arm Installation —Left Side 1966-68

1. Lubricate both ends of torsion bar with extreme pressure chassis lubricant for approximately 3 in.
2. Install torsion bar into nut of lower control arm.
3. With the aid of a helper, lift torsion bar and lower control arm assembly up until torsion bar will engage arm in crossmember and bushing ends of lower control arm engage frame horn.

NOTE: make sure daub of paint on torsion bar is in the same position as when removed.

4. Install lower control arm bushing bolts and nuts. Do not torque.
5. Position tool BT-6601 over torsion bar crossmember and tighten center bolt until lower control arm is in a horizontal plane.
6. Install drive axle. Refer to Drive Axle Installation—Left Side, Steps 1 through 9.
7. Install shock absorber. Torque upper bolt to 75 ft. lbs., lower nut to 75 ft. lbs.
8. Using a new bolt, connect stabilizer bar to lower control arm. Refer to illustration for correct sequence of grommets, retainers and spacer. Torque to 14 ft. lbs. Cut bolt off ¼ in. below nut.
9. Continue to tighten center bolt of tool BT-6601 until nut can be positioned through crossmember. Lubricate adjuster bolt with E. P. chassis lubricant and turn in the same amount of turns required to remove.
10. Install drum and wheel.
11. Remove floor stands and lower hoist.
12. Torque lower control arm bushing bolt to 75 ft. lbs.
13. Check caster, camber and toe-in.

Lower Control Arm R & R —1969-70

1. Hoist car and support at lift points. Remove wheel assembly.
2. Place torsion bar remover and installer BT-6601 over crossmember so that center screw is seated in dimple of torsion adjusting arm.
3. Remove torsion bar adjusting bolt and nut, counting the number of turns necessary.

NOTE: this number of turns will be used when installing, to obtain initial carrying height.

4. Turn center screw of tool BT-6601 until torsion bar is completely relaxed.
5. Disconnect shock absorber and

stabilizer link from lower control arm.
6. Remove drive axle nut.
7. Remove cotter pin and nut from lower ball joint stud.
8. Remove ball joint stud from knuckle, using ball joint puller (J-22292-1-2).
9. Push in on drive axle and pull outward on knuckle to gain clearance, then remove lower control arm from knuckle and torsion bar.
10. Install by reversing removal procedure. Check and adjust carrying height if necessary.

Lower Control Arm Bushing Removal

1. Hoist car on a two post lift.
2. Disconnect upper shock attaching bolt.
3. Remove stabilizer link bolt and discard the bolt.
4. Place floor stands under frame horns. Lower front lift to floor.
5. Install tool BT-6601.
6. Remove lower control arm bushing bolts and lower control arm until free of frame horns.
7. Install bolts through rear bushing and press out bushing.

NOTE: Because of the torsion bar nut attachment to the lower control arm, it will be necessary to use a hardened ½—20 nut, to remove the front bushing.

Lower Control Arm Bushing Removal

1. Install tools and press rear bushing into lower control arm.

NOTE: Because of the torsion bar nut attachment to the lower control arm, it will be necessary to use a hardened ½–20 nut, to remove the front bushing.

2. Raise lower control arm into frame horns and install bushing bolts and nuts. Do not torque.
3. Using tool BT-6601, turn center bolt into dimple of torsion bar arm until adjusting nut can be inserted through center frame support. Turn adjusting bolt clockwise the same number of turns needed to remove. Remove tool BT-6601.
4. Raise front lift under lower control arms, and remove floor stands.
5. Connect upper shock attaching bolt and nut. Torque to 75 ft. lbs.
6. Using a new bolt, attach stabilizer link bolt to lower control arm. Torque to 14 ft. lbs.
7. Lower car and torque lower control arm bushing bolts to 75 ft. lbs.

Ball Joint Vertical Check

1. Raise the car and position floor stands under the left and right lower control arm, as near as

possible to each lower ball joint. Car must be stable and should not rock on floor stands.
2. Position dial indicator to register vertical movement at wheel hub.
3. Place a pry bar between the lower control arm and the outer race, and pry down on the bar. Care must be used so that the drive axle seal is not damaged. The vertical reading must not exceed .125 in.

Ball Joint Horizontal Check

1. Place car on floor stands, as outlined in Step 1 in the Vertical Check.
2. Position dial indicator at the rim of the wheel, to indicate side play.
3. Grasp wheel with the hands, top and bottom, and push in on the bottom of the tire while pulling out at the top. Read gauge, then reverse the push-pull procedure. Horizontal deflection on the gauge should not exceed .125 in. at the wheel rim. This procedure checks both the upper and lower ball joints.

Lower Control Arm Ball Joint Removal

1. Remove knuckle.
2. Using hack saw, saw the three rivet heads off.
3. Using a 7/32 in. bit, drill side rivets 3/16 in. deep.
4. Using hammer and punch, drive on center rivet of joint, until joint is out of the control arm.

Lower Control Arm Ball Joint Installation

1. Install service ball joint into control arm and torque bolts and nut.
2. Reverse knuckle removal.

Lower Control Arm Ball Joint Seal Removal

The lower ball joint seal can be installed with lower control arm either in or out of the car.
1. Remove steering knuckle. Refer to Knuckle Removal, Steps 1 through 12.
2. Using hammer and chisel, drive seal from ball joint.
3. Wipe grease from ball joint and stud.

Lower Control Arm Ball Joint Seal Installation

1. Position new seal over ball joint stud.
2. Lubricate jaws of seal installer J-5504 and carefully slide jaw between seal and retainer.
3. Tap lightly with hammer on center bolt of tool J-5504 until retainer is fully seated.
4. Install knuckle.

5. Lubricate the ball joint fitting until grease appears from seal.

Stabilizer Bar Removal

1. Remove link bolts, nuts, grommets, spacers and retainers from lower control arm. Discard bolts.
2. Remove two bolts which attach dust shield to frame, both sides.
3. Remove bracket to frame attaching bolts and remove stabilizer bar from front of car.

Stabilizer Bar Installation

Reverse removal procedure.
NOTE: new link bolts are torqued to 14 ft. lbs., then cut off ¼ in. from nut.

Torsion Bar Crossmember Removal

1. Raise car on a two post hoist and place floor stand under front frame horns. Lower front hoist.
2. Disconnect parking brake cable and equalizer, and clip at torsion bar crossmember. Pull cable through the crossmember.
3. Remove torsion bars.
4. Remove tool BT-6601.
5. Raise front hoist under lower control arms until drive axles are about level.
6. Slide both torsion bars forward until they bottom in lower control arm nut.
7. Remove left intermediate exhaust pipe.
8. Remove bolts from torsion bar crossmember to frame.
9. Move torsion bar crossmember rearward until torsion bars are free and adjuster arms, bolts and nuts can be removed.
10. Daub paint on lower portion of each torsion bar to help in cor-

rect positioning when reinstalling.
11. Raise torsion bar crossmember and slide torsion bar rearward until removed.
NOTE: torsion bars are not interchangeable. Keep separate.
12. Raise torsion bar crossmember, remove rubber cushions from frame horns.
13. Move torsion bar crossmember to the right side until member clears frame, then remove.

Torsion Bar Crossmember Installation

1. Insert torsion bar crossmember above frame on right side of car and position left side over frame horn.
2. Raise crossmember so that rubber cushions and restrictors can be installed.
3. Lubricate both ends of torsion bar with extreme pressure chassis lubricant for about 3 in.
4. Insert torsion bars into nut on lower control arm and slide toward front of car until torsion bars bottom out.
NOTE: be sure daub of paint on torsion bar is in the same location as when it was removed.
5. Raise crossmember up and toward rear of car. Raise torsion bars until they enter hole provided for them in crossmember.
6. Install adjuster arms into crossmember and slide torsion bars toward the rear of car until fully seated against rear edge of crossmember.
7. Install rebound cushions, bolts, washers and nuts through frame and crossmember. Torque to 40 ft. lbs.

8. Install left intermediate exhaust pipe.
9. Lower front hoist from under control arms.
NOTE: care must be used that floor stands are fully seated on frame horns.
10. Position tool BT-6601 over crossmember and tighten center bolt until nut can be inserted through crossmember.
11. Lubricate adjuster bolt with extreme pressure chassis lubricant and turn the same number of turns required to remove it.
12. Remove tool BT-6601.
13. Install seals over retainers at rear of torsion bars.
14. Insert parking brake cable through crossmember. Install clip at crossmember.
15. Connect parking brake cable to equalizer.
16. Hoist car and remove floor stands.
17. Lower car.
18. Check parking brake cable adjustment. Adjust as necessary.

Front End Alignment

Carrying height is controlled by the adjustment setting of the torsion bar adjusting bolt. Clockwise rotation of the bolt increases the front height. It is very important that this height be considered and made correct before further steering geometry is established. Car must be on a level surface, gas tank full or a compensating weight added. Front seat must be all the way to the rear and tires inflated to 24 psi in the front and 22 psi in the rear. All doors must be closed and no passengers or additional weight should be in the car or trunk.

CAR ON LEVEL SURFACE
FUEL TANK FULL
TRUNK EMPTY
FRONT SEAT REARWARD
DOORS CLOSED
TIRES AT CORRECT PRESSURE

TO ADJUST FRONT CARRYING HEIGHT
RAISE CAR AT FRONT CROSSMEMBER
TO RELIEVE STRAIN ON ADJUSTING
BOLT. LUBRICATE ADJUSTING BOLT
BEFORE ATTEMPTING TO CHANGE
CARRYING HEIGHT.

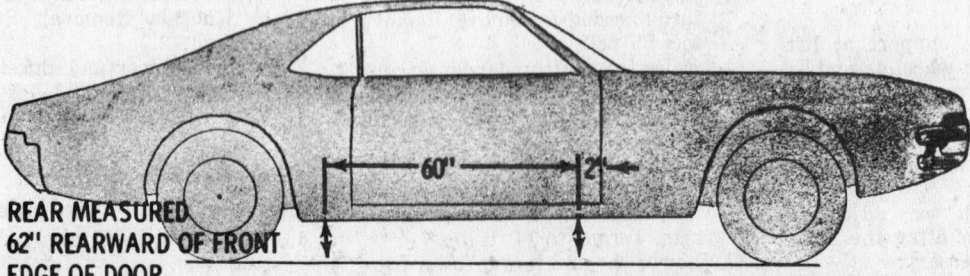

REAR MEASURED
62" REARWARD OF FRONT
EDGE OF DOOR
SPEC. 8" + 1/2" OR -1" ROCKER PANEL TO FLOOR

FRONT TO REAR - WITHIN 3/4"
SIDE TO SIDE - WITHIN 3/4"

FRONT MEASURED 2" REARWARD OF
FRONT EDGE OF DOOR
SPEC. 8" + 1/2" OR - 1" MEASURED FROM
ROCKER PANEL TO FLOOR.

Adjusting carrying height (© Oldsmobile Div. G.M. Corp.)

1. Check rocker panel to ground dimension, as illustrated. Front reading to ground should be 8 in., rear reading to ground should be 8¼ in. Front to rear should be within 1 in. Side to side should be within ⅝ in. If not within tolerance, adjust torsion bar.
2. Align car on wheel alignment equipment.
3. Raise front end and check wheel runout. Set and center the runout, then lower the car.
4. Loosen nuts on inboard side of upper control arm cam bolts.
5. Check camber and adjust if necessary with rear cam bolt. Camber should be ¼° negative to ½° positive. Camber reading of the right and left wheel should be within ½° of each other.
6. Take the caster reading. Caster should be 2½° negative to 1½° negative. If necessary to adjust caster, turn wheel to straight ahead position. Use camber reading scale for making this adjustment.
   A. Turn rear bolt so camber reading is ¼° more than original setting, for every 1° of caster change needed for correct reading. Turn to plus side of camber if caster is negative and to negative camber if caster is positive.
   B. Turn the front cam bolts so camber will return to the original proper setting that was made on the camber adjustment, which was plus ⅛°.
   C. Recheck caster reading.
   NOTE: if a problem exists, and you should run out of cam in the attempt to gain correct reading.
   A. Turn front cam bolt so high part of cam is pointing up.
   B. Turn rear cam bolt so high part of cam is pointing down.
   This is a location to start from and a correct setting should be obtainable with the above procedure.
   NOTE: torque upper control arm cam nuts to 95 ft. lbs.; hold head of bolt securely. Any movement of the cam will affect final setting and you will have to recheck caster and camber adjustment.
7. Toe-in adjustment is 0—1/16 in.
   A. Center steering wheel, raise car and check wheel runout.
   B. Loosen tie rod end nuts, and adjust to proper setting.
   C. Tighten tie rod end nuts. Torque nuts 20 ft. lbs. Position tie rod clamps so openings of clamps are facing down. This is a very necessary setting. Interference and possible trouble with front end linkage could occur if clamps snag anything while turning.

Drive Axles

Drive axles are a complete flexible assembly and consist of an axle shaft and an inner and outer constant velocity joint. Right axle shaft has a torsional damper mounted in the center. The inner constant velocity joint has complete flexibility, plus inward and outward movement. The outer constant velocity joint has complete flexibility only.

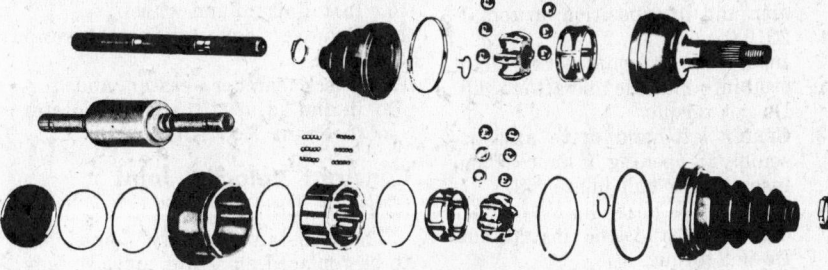

Drive axle assembly

Beginning 1967

Beginning 1967, Tri-pot inboard drive joints replace the ball spline Rzeppa joints of 1966 production. These joints consist of a three-pronged trunnion spider, needle bearing mounted balls, a universal housing, axle shaft, and rubber boot. This joint combines constant velocity universal action with axial slip motion.

Drive Axle Removal —Right Side

1. Hoist car under lower control arms.
2. Remove drive axle, cotter pin, nut and washer.
3. Remove oil filter.
4. Remove inner constant velocity joint attaching bolts.
5. Push inner constant velocity joint outward enough to disengage the right-hand final drive output shaft, then move rearward.
6. Remove right-hand output shaft bracket bolts to engine and final drive.
7. Remove right-hand output shaft.

Caution Care must be exercised so that constant velocity joints do not turn to full extremes, and that seals are not damaged against shock absorber or stabilizer bar.

Drive Axle Installation —Right Side

1. Carefully place righthand drive axle assembly into lower control arm and enter outer race splines into knuckle.
2. Lubricate final drive output shaft

seal, with special seal lubricant part No. 1050169.
3. Install right-hand output shaft into final drive and attach the support bolts to engine and brakes. Torque the bolts to 50 ft. lbs.
4. Move right-hand drive axle assembly toward front of car and align with right-hand output shaft. Install attaching bolts and torque to 65 ft. lbs.
5. Install oil filter.
6. Install washer and nut on drive axle. Torque to 60 ft. lbs. and insert cotter pin.
7. Remove floor stands and lower hoist.
8. Check engine oil. Add if necessary.

Drive Axle Removal —Left Side

1. Hoist car under lower control arms.
2. Remove wheel and drum.
3. Remove drive axle cotter pin, nut and washer.
4. Position access slot in hub assembly so that each of the attaching bolts can be removed. It will be necessary to push aside adjuster lever to remove one of the bolts.
5. Position spacers and tool J-22237. Install J-21579 and slide hammer J-2619 with adapter J-2619-1.
6. Remove hub assembly. It will again be necessary to push aside adjuster lever for clearance for hub assembly.
7. Remove tie-rod end cotter pin and nut.
8. Using hammer and brass drift, drive on knuckle until tie-rod end stud is free.
9. Remove bolts from drive axle assembly and left output shaft. Insert tool J-22193.
10. Remove upper control arm ball joint cotter pin and nut.
11. Using hammer and brass drift, drive on knuckle until upper ball joint stud is free.
12. Install J-22292-3 carefully between ball joint seal and knuckle.
13. Using tool J-22292-1-2, remove lower ball joint from knuckle. Care must be exercised so that

ball joint does not damage drive axle seal.

14. Remove knuckle and support, so that brake hose is not damaged.
15. Carefully guide drive axle assembly outboard.

NOTE: care must be exercised so that constant velocity joints do not turn to full extremes and that seals are not damaged against shock absorber or stabilizer bar.

Drive Axle Installation —Left Side

1. Carefully guide left-hand drive axle assembly onto lower control arm and into position on tool J-22193.
2. Insert lower control ball joint stud into knuckle and attach nut. Do not torque.
3. Center left-hand drive axle assembly in opening of knuckle and insert upper ball joint stud.
4. Place brake hose clip over upper ball joint stud and install nut. Do not torque.
5. Insert tie-rod end stud into knuckle and attach nut. Torque to 45 ft. lbs. Install cotter pin and crimp.
6. Lubricate hub assembly bearing

O.D. with E.P. grease and install. Torque to 65 ft. lbs.

7. Align inner constant velocity joint with output shaft and install attaching bolts. Torque to 65 ft. lbs.
8. Torque upper and lower ball joint stud nuts to 40 ft. lbs. Install cotter pins and crimp.

NOTE: upper ball joint cotter pin must be crimped toward upper control arm to prevent interference with outer constant velocity joint seal.

9. Install drive axle washer and nut. Torque to 60 ft. lbs. Install cotter pin and crimp.
10. Install drum and wheel.
11. Remove floor stands and lower hoist.
12. Check camber, caster and toe-in and adjust if necessary. Refer to Front End Alignment.

Constant Velocity Joint (Out of Car)

The constant velocity joints are to be replaced as a unit and are only disassembled for repacking and replacement of seals.

Outer C. V. Joint Disassembly

1. Insert axle assembly into vise.

Hold by the mid-portion of the axle shaft.

2. Remove inner and outer seal clamps.
3. Slide seal down axle shaft to gain access to C. V. joint.
4. Using tool KMO-0630, spread retaining ring until C. V. joint can be removed from axle spline.
5. Remove retaining ring.
6. Slide seal from axle shaft.
7. Remove grease from constant velocity joint.
8. Holding constant velocity joint with one hand, tilt cage and inner race so that one ball can be removed. Continue until all six balls are removed.
9. Turn cage 90° and, with large slot in cage aligned with land in inner race, lift out.
10. With cage and inner race assembly, turn inner race 90° in line with large hole in cage. Lift land on inner race up through large hole in cage and turn up and out to separate parts.

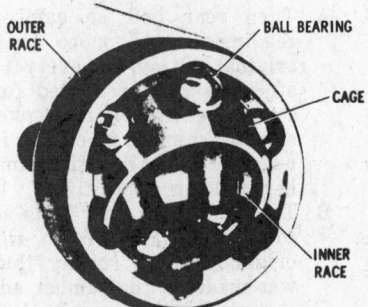

Removing balls from outer race
(© Oldsmobile Div. G.M. Corp.)

Outer C. V. Joint Assembly

1. Insert land of inner race into large hole in cage and pivot to install into cage.
2. Align inner race and pivot inner race 90° to align in outer race.
3. Insert balls into outer race one at a time until all six balls are installed. Inner race and cage will have to be tilted so that each ball can be inserted.
4. Pack constant velocity joint full of lubricant, part No. 1050530.
5. Pack inside of seal with the same lubricant, until folds of seal are full.
6. Place small keystone clamp on axle shaft.
7. Install seal onto axle shaft.
8. Install retaining ring into inner race.
9. Insert axle shaft into splines of outer constant velocity joint until retaining ring secures shaft.
10. Position seal in slot of outer race.
11. Install large keystone clamp over seal and secure.
12. Install small keystone clamp over seal and secure.

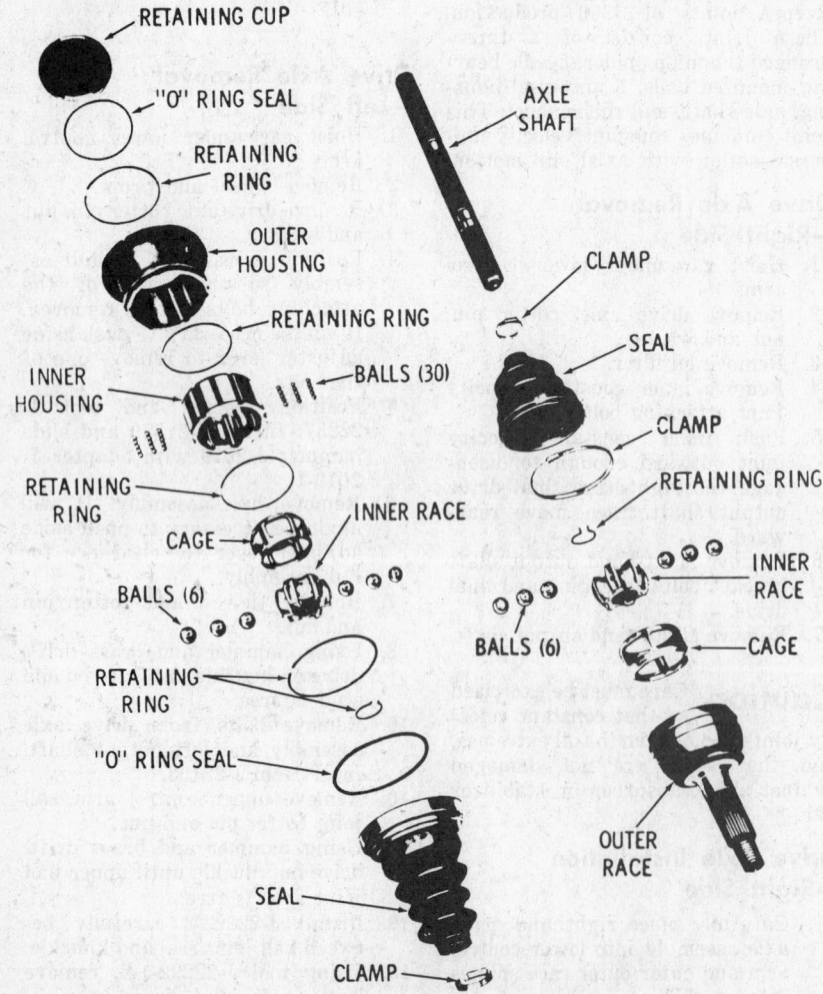

Left-hand drive axle assembly (© Oldsmobile Div. G.M. Corp.)

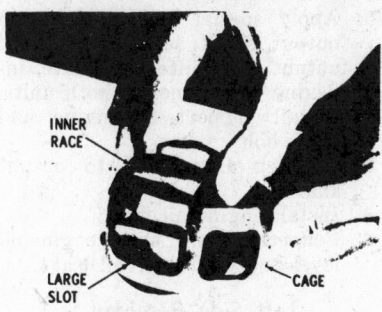

Removing inner race
(© Oldsmobile Div. G.M. Corp.)

1966 Inner C. V. Joint Disassembly

1. Insert axle assembly into vise. Clamp on mid-portion of axle shaft.
2. Remove small seal clamp.
3. Remove large end of seal from C. V. joint by prying out peened spots and driving off C. V. joint with hammer and chisel.
4. Slide seal down shaft until C. V. joint is disassembled. Remove O-ring from outer housing.
5. Wipe all grease from C. V. joint.
6. Remove axle shaft retaining ring.
7. Remove retaining ring from interior of ball spline outer housing.
8. Remove ball spline outer housing from ball spline inner housing being careful not to drop or lose balls, five in a line with six rows, total 30.
9. Follow disassembly of outer C. V. joint from Steps 4 through 9 for inner C. V. joint.
10. Remove grease from inside of ball spline outer housing.
11. Using wood block, carefully drive along outer edges from inside of ball spline outer housing and remove retaining cup.
12. Remove O-ring from interior of ball spline outer housing.
13. Remove retaining ring from interior of ball spline outer housing.
14. Remove O-ring from exterior of ball spline outer housing.
15. Remove two retaining rings from ball spline inner housing.

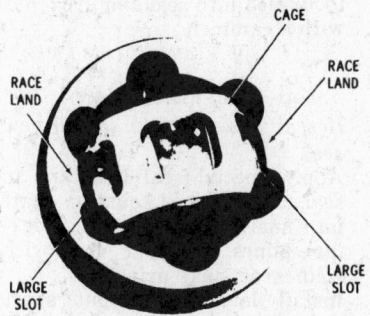

Positioning cage
(© Oldsmobile Div. G.M. Corp.)

1966 Inner C. V. Joint Assembly

1. Install two retaining rings onto inner spline housing.
2. Install retaining ring into groove of outer spline housing.
3. Install O-ring into inner groove of outer spline housing. Lubricate O-ring with special seal lubricant, part No. 1050169.
4. Install cover into outer spline housing using tools J-22247-1 and J-22247-2.
5. Insert land of inner race into large hole in case and pivot to install in cage.
6. Align inner race and cage and install into inner spline housing.
7. Insert balls into inner spline housing, one at a time, until all six balls are in. Inner race and cage will have to be tilted so that each ball can be inserted.
8. Install inner spline housing into outer spline housing. Raise inner spline housing and install five balls into each spline. (Six splines.)
9. Lower inner spline housing into outer spline housing and install retaining ring into groove at top of outer spline housing.
10. Install retaining ring into inner race.
11. Pack C. V. joint full of lubricant, part No. 1050530.
12. Place small keystone clamp on axle shaft.
13. Pack inside of seal with lubricant, part No. 1050530, until folds of seal are full.
14. Install seal onto axle shaft.
15. Install O-ring onto outer spline housing.
16. Insert axle shaft into splines of inner race until retaining ring secures shaft.
17. Insert drive axle assembly into a press and peen seal in six places, equally spaced.
18. Remove drive axle from press.
19. Position seal in groove of axle shaft until inner constant velocity joint is fully extended.

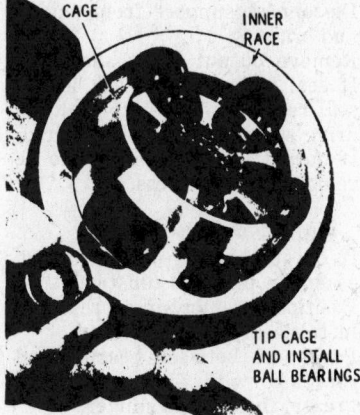

Installing balls into outer race
(© Oldsmobile Div. G.M. Corp.)

20. Install keystone clamp and secure.

Drixe Axle Disassembly

1. Remove drive axle assembly.
2. Remove outer C. V. joint seal clamps.
3. Remove inner C. V. joint seal by prying out peened spots and driving off seal.
4. Slide seals inboard on shaft.
5. Using tool KMO-0630, spread retaining rings until both C. V. joints can be removed from axle shaft.
6. Remove seals from axle.

Drive Axle Assembly

1. Pack seals with lubricant, part No. 1050530.
2. Place outer C. V. joint seal on axle, with keystone clamps in position on seal.
3. Insert axle into outer C. V. joint until retaining ring locks axle into position.
4. Position seal and clamps and secure keystone clamps.
5. Place inner C. V. joint seal on axle with keystone clamp in position on seal.
6. Insert axle into inner C. V. joint until retaining ring locks axle into position.
7. Place axle assembly into press using tools J-22247-1 and J-22247-2. Press seal into position and peen to secure.
8. Secure small keystone clamp with seal in position.
9. Install drive axle assembly.

Differential

Planetary-Type

1966-67

This type differential replaces the conventional spider and beveled axle drive pinions with a planetary gear set to distribute torque to the respective drive axles.

Engine torque is transmitted from the power train, to the main drive pinion and ring gear, to the differential housing. Torque is then applied through the planetary and sun gear mechanism to both drive axles at variable speed requirements.

While the car is moving straight ahead, the planetary gears are fixed and rotate with the differential case and ring gear, as a unit. However, when turning, the planetary gears revolve upon their individual axes with differential action, allowing the drive axles to rotate at different speeds.

This unit is not a controlled or limited slip differential.

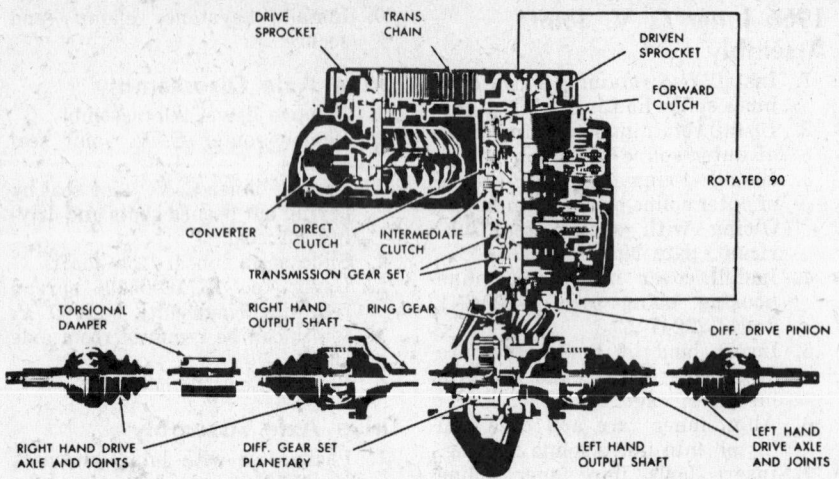

Power train cross-section—1966-67 (© Oldsmobile Div. G.M. Corp.)

5. Apply special seal lubricant to output shaft seal, then install output shaft into final drive, indexing the splines of both units.
6. Install support to engine and brace bolts.
7. Connect drive axle to output shaft.
8. Install engine oil filter.
9. Connect battery, check engine oil level and check for oil leaks.

Left Side Removal

The left-hand output shaft can normally be removed only after removing the final drive assembly from the car. However, if the left-hand axle assembly has been removed for any reason, the output shaft and seal can be removed as follows:

1. Remove right-hand output shaft

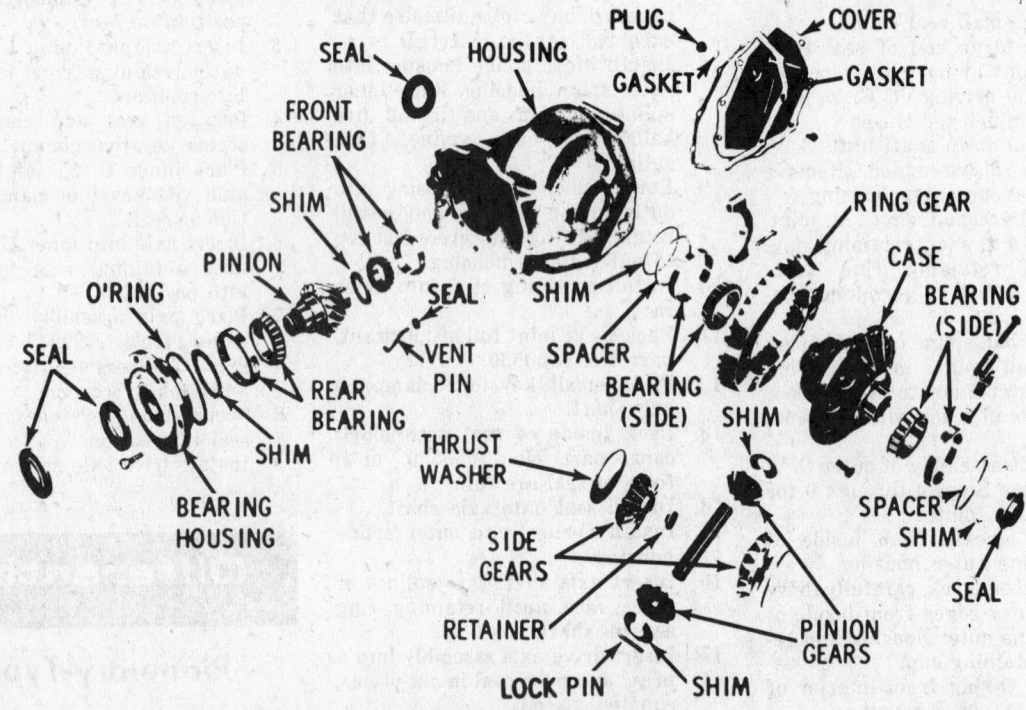

Final drive components—1968-70 (© Oldsmobile Div. G.M. Corp.)

Bevel Gear-Type

1968-70

Since 1968, a bevel gear-type differential is used on all front-wheel drive models. This design supersedes the original planetary gear-type final drive. While unit removal and installation procedures are typical, the assembly is not interchangeable with the 1966-67 design.

Overhauling the differential assembly is not encouraged. However, reconditioning procedures are given in later paragraphs of this final drive coverage. Differences in procedure will be clearly indicated.

Output Shafts and Seals

Right Side Removal

1. Disconnect battery.

2. Hoist car.
3. Remove engine oil filter.
4. Disconnect right-hand drive axle.
5. Disconnect support from engine and brace.
6. Remove output shaft assembly.
7. If seal is to be removed, install seal remover J-943 into seal and drive seal out with a hammer.
8. If output shaft bearing is to be removed, use a press.

Right Side Installation

1. If output bearing was removed, assemble parts as illustrated.
2. Position assembly in a press and install bearing until seated.
3. Pack area between bearing and retainer with wheel bearing grease, then install slinger.
4. If seal was removed, it can now be installed.

assembly, as in previous paragraphs.
2. Using a 9/16 in. socket, remove left-hand output shaft retaining bolt and left-hand shaft.
3. Install left-hand output shaft tool J-943 into seal and drive out with a hammer.

Left Side Installation

1. If seal was removed, install new seal.
2. Apply special seal lubricant to seal, then insert output shaft into final drive assembly, indexing splines of output shaft with splines of final drive.
3. Install left-hand output shaft retaining bolt and torque to 45 ft. lbs.
4. Install right-hand output shaft.

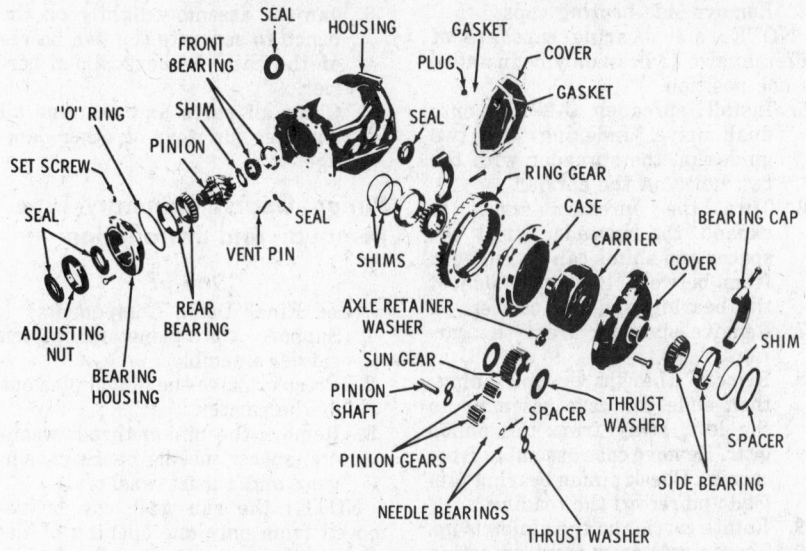

Final drive component—1966-67 (© Oldsmobile Div. G.M. Corp.)

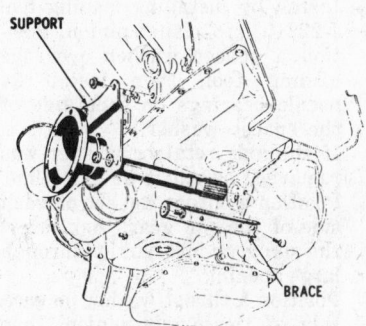

Right-hand output shaft
(© Oldsmobile Div. G.M. Corp.)

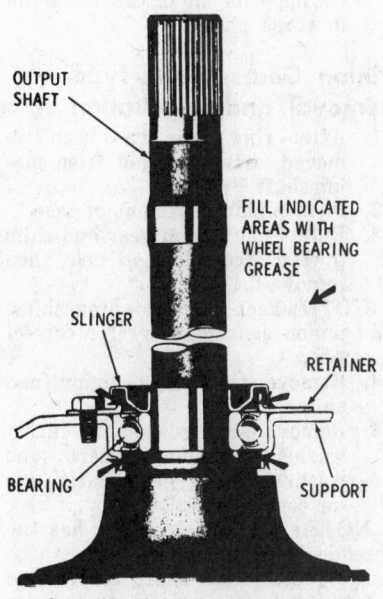

Assembly of right-hand output shaft
(© Oldsmobile Div. G.M. Corp.)

Final Drive

Removal

1. Disconnect battery.
2. See illustration. Remove bolts A, B, and C. Nut D must be removed with a special wrench, such as MAC S-147.
   NOTE: it may be necessary to remove the transmission filler tube to gain clearance.
3. Hoist the car. If a two post hoist is used, the car must be supported with floor stands at the front frame rails and the front post lowered.
4. Disconnect right and left drive axles from the output shafts.
5. Remove engine oil filter.
6. Disconnect brace from final drive, then disconnect right-hand output shaft assembly from engine.

7. Remove output shaft assembly from final drive.
8. See illustration. Remove bolt X and loosen bolts Y and Z.
9. Remove final drive cover and allow lubricant to drain.
10. Position transmission lift with adapter for final drive. Install an anchor bolt through final drive housing and lift pad.
11. See illustration. Remove bolts E, F, and G, and nut from H.
12. Move transmission lift toward front of car to disengage final drive splines from transmission.
13. Lower transmission lift and remove final drive from lift.
14. Using a 9/16 in. socket, remove the left output shaft retainer bolt, then pull output shaft from final drive.
15. Remove transmission to final drive gasket.

Installation

1. Apply special seal lubricant to both output shaft seals.
2. Install the left output shaft into the final drive. Retain with bolt and torque to 45 ft. lbs.
3. Position final drive on transmission lift and install an anchor bolt through the housing and lift pad.
4. Apply a thin film of special seal lubricant on the transmission side of the new final drive to transmission gasket. Then position gasket on the transmission.
5. Raise the transmission lift. Align the two bolt studs D and H on the transmission with their mating holes in the final drive. Move final drive until it mates with the transmission.

Caution
It may be necessary to rotate the left output shaft to align the splines on the final drive with the splines of the transmission output shaft.

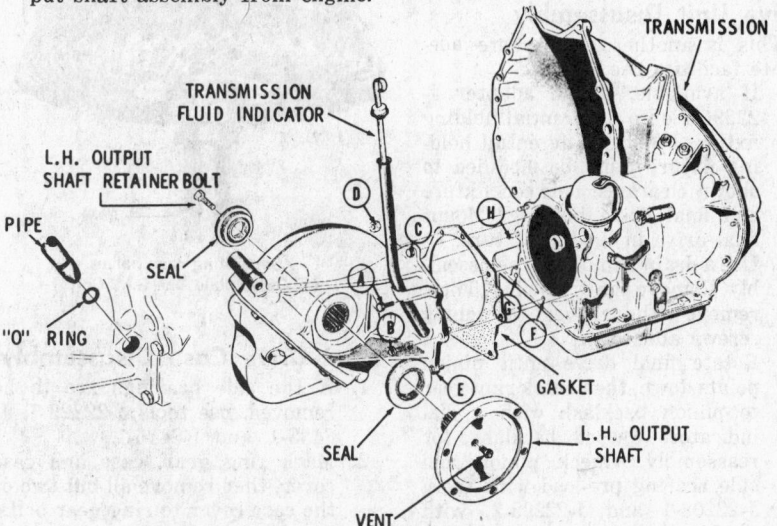

Transmission attachment bolts (© Oldsmobile Div. G.M. Corp.)

6. Install bolts E, F, and G and nut H finger tight.
7. Install bolt X and torque to 75 ft. lbs. Tighten and torque bolts Y and Z to 50 ft. lbs.
8. Loosen and remove lift from final drive.
9. Position a new cover gasket on the final drive, then install cover. Torque cover bolts to 30 ft. lbs.
10. Install right output shaft into final drive, indexing splines of output shaft with splines of final drive. Install mounting bracket and brace bolts and tighten.
11. Connect drive axles to output shafts.
12. Install oil filter.
13. Raise hoist, remove stands and lower car.
14. If filler tube was removed, attach a new O-ring and install filler tube.
15. Install bolts A, B, and C and nut D. Torque all final drive to transmission bolts to 25 ft. lbs. Torque nuts to about 25 ft. lbs.
16. Connect battery.
17. Fill final drive with four and one-half pints of lubricant, part No. 1050015.
18. Check engine oil level. Start engine and check transmission fluid level.
19. Check for any oil leaks.

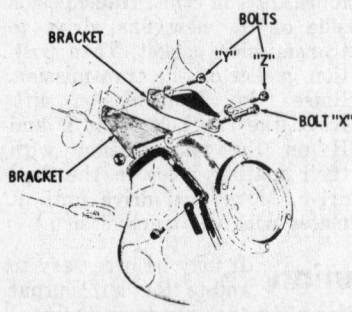

Disconnecting final drive from engine
(© Oldsmobile Div. G.M. Corp.)

Drive Unit Disassembly

This is another area where adequate facilities are a must.

1. If available, install adapter J-22296-1 onto differential holding fixture J-3289. Differential holding fixture must be modified to obtain clearance between fixture and final drive housing. Mount final drive in holding fixture.
2. Use a drain pan under the assembly. Remove the drain plug. Then, remove the cover attaching screws and cover.
3. Rotate final drive until pinion points down, then check ring gear to pinion backlash with a dial indicator. Record backlash for reassembly. Check pinion and side bearing pre-load with tools, J-22208-1 and J-22208-2, with the help of a torque wrench. Record pre-load reading.

4. Remove side bearing caps.
NOTE: side bearing caps are of different size and can only be installed in one position.
5. Install spreader J-22196 onto final drive, indexing the two guides on the spreader with the two holes on the carrier.
6. Turn the spreader screw to expand the spreader until the spacer and shims can be removed from between the small side of the bearing and the carrier.
7. Remove spreader from the carrier.
8. Remove the spacer and shims, then slide the case assembly to the left, away from the pinion gear. Remove case assembly from carrier. Check pinion bearing pre-load and record the reading.
9. Rotate carrier so the pinion is up.
10. Loosen set screw from adjusting nut.
11. Remove bearing housing bolts. Remove the drive pinion housing and remove the adjusting nut and housing from drive pinion. Remove rubber seal from bearing housing.
12. Remove rubber seal and vent wire from carrier.
13. With slide hammer J-2619 and tool J-22201, remove pinion front outer race.
14. Remove the output shaft oil seals.
15. Remove the two oil seals from the adjusting nut.
16. If necessary to remove pinion rear outer race, now is the time.

Pinion Bearing Removal

1. Remove the pinion front bearing and selective shim. Bearing can be removed with a press.
2. Remove the pinion rear bearing.

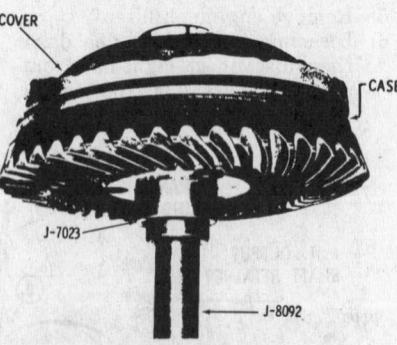

Separating case halves
(© Oldsmobile Div. G.M. Corp.)

Final Drive Case Disassembly

1. If the side bearings are to be removed, use tools J-22229-1, J-8433-1, and J-8416-1.
2. Mark ring gear, case and case cover, then remove all but two of the case cover to ring gear bolts. Loosen, but do not remove, two of the bolts 180° apart.

3. Jar the assembly lightly on the bench to separate the two halves of the case. Remove pinion carrier.
4. Clean all parts and examine all surfaces for wear or other damage.

Pinion Gears, Planetary-Type Removal and Installation

1966-67

(See Final Drive Components)
1. Support the planetary pinion carrier assembly.
2. Press or drive the pinion pins out of the carrier.
3. Remove the pinion thrust washers, spacer, needle bearings, sun gear and thrust washers.
NOTE: the sun gear can be removed from only one opening of the carrier. This opening can be identified by the thinner wall at the carrier opening.
4. After removing the sun gear, the left axle retainer washer can be removed from the carrier.
5. Install by installing loading tool J-22210 into planet pinion. Position a spacer washer over the loading tool, then install 24 needle bearings on each side of the spacer washer.
6. If the axle retainer washer was removed, install it at this time.
7. Position a thrust washer on each side of the sun gear, then insert the sun gear into carrier through large opening.
8. Position a thrust washer on each side of the planet pinion, then insert planet pinion into carrier.
9. Using a deep socket as a receiver, press pinion pin into carrier, until it bottoms.
10. Place a large punch in a vise, to be used as an anvil, and stake the opposite end of the pinion pin in three places.

Pinion Gears, Bevel-Type Removal and Installation

1. After ring gear has been removed, drive lock pin from pinion shaft.
2. Push pinion shaft out of case.
3. Rotate one pinion gear and shim toward access hole in case, then remove.
NOTE: keep corresponding shims and pinion gear together for correct assembly.
4. Remove the other pinion and shim.
5. Remove side gears and thrust washers, keeping gears and washers in proper relationship for correct installation.
NOTE: the left-side gear has the threaded retainer that secures the (short) left output shaft. If threaded retainer is to be removed, use a brass drift to prevent trouble.

6. Upon assembling pinion and side gears into the case, lubricate components with a quality extreme pressure lubricant.

7. Place side gear thrust washers over the side gear hubs and install side gears into case. Gear with threaded retainer belongs in left side of case.

8. Position one pinion (without shims) between side gears, then rotate gears until pinion is directly opposite from loading opening in case. Place other pinion between side gears so that the pinion shaft holes are in line; then rotate gears to make sure holes in pinions line up with holes in case.

9. If holes line up, rotate pinions back to loading opening just enough to permit insertion of the pinion gear shims.

10. Install pinion shaft. Drive pinion shaft retaining lock pin into position.

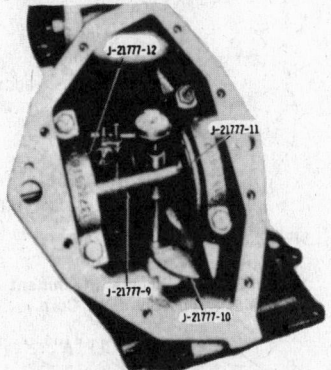

Checking pinion depth
(© Oldsmobile Div. G.M. Corp.)

Checking Pinion Depth

1. Install pinion front outer race. Drive race in until it bottoms.

2. Lubricate front bearing with final drive lubricant and install into front outer race.

3. Position tool J-21777-10 on the front bearing. Install tool J-21579 onto final drive housing and retain with two bolts. Thread screw J-21777-13 into J-21579 until tip of screw engages tool J-21777-10. Torque tool J-21777-13 to 20 in. lbs. to pre-load the bearing.

4. Remove dial indicator post from tool J-21777-9 and install discs J-21777-11 and J-21777-12. Reinstall dial indicator post.

5. Place the gauging discs in the side bearing bores and install the side bearing caps.

6. Position the dial indicator on the mounting post of the gauge shaft with the contact button touching the indicator pad. Set dial indicator to zero, then depress the dial indicator until the needle rotates three-quarters of a turn clockwise. Tighten dial indicator.

7. Position the gauge shaft assembly in the carrier so that the dial indicator contact rod is directly over the gauging area of the gauge block, and the discs are seated fully in the side bearing bores.

8. Position gauge shaft so that the indicator rod contacts the gauging area. Rotate gauge rod back and forth until the indicator reads the greatest deflection. At the point of greatest deflection, set the indicator to zero. Repeat the rocking action to verify the zero setting.

9. After zero setting is obtained, rotate gauge shaft until the indicator rod does not touch the gauging area. Read the pinion depth directly from the dial indicator.

10. Select the correct pinion shim to be used during assembly on the following basis:
    A. If a service pinion is being used, or a production pinion with no marking, the correct shim will have a thickness equal to the indicator gauge reading found in Step 9.
    B. If a production pinion is being used and it is marked + or —, the correct shim will be determined as follows:
    If the pinion is marked +, the shim thickness indicated by the dial indicator on the pinion setting gauge must be increased by the amount etched on the pinion. If the pinion is marked —, the shim thickness indicated on the dial must be decreased by the amount etched on the pinion.

11. Remove pinion depth checking tools and front bearing from carrier.

Final Drive Case Assembly

1. Install the pinion carrier into the case.

2. With the case and cover alignment marks indexed, insert four ring gear attaching bolts through case and cover. Align mark on ring gear with alignment marks on case and cover, then install ring gear onto case. Tighten the six attaching bolts alternately. Torque bolts to 85 ft. lbs.

3. If side bearings were removed, they can be installed now. Drive bearing on until it bottoms.

4. Install pinion rear bearing.

5. Position correct shim on drive pinion and install the drive pinion front bearing with tool J-21022 and a press.

6. Lubricate pinion bearings and install pinion into carrier.

7. Install seals into adjusting nut.

8. Install O-ring and vent pin onto face of carrier. Torque attaching nuts to 35 ft. lbs.

9. Install seal protector J-22236 over drive pinion, then install the adjusting nut over the seal protector and thread into the housing.

10. Assemble tools, as illustrated, and adjust pinion bearing preload. The preload is 2-10 in. lbs. for new bearings, and 2-3 in. lbs. for used bearings. Adjust new bearing preload to 4 in. lbs. while rotating the pinion and checking preload. Adjust until preload remains constant. When correct preload is obtained, tighten the set screw. Record preload reading as it will be used when making side bearing preload adjustment. Leave the tools on pinion for side bearing preload adjustment.

Side Bearing Preload Adjustment

Differential side bearing preload is adjusted by means of shims located between the side bearings and the carrier. One spacer is used on the right side only. Shims are used on both sides and come in thickness increments of .002 in. from .036 to .070 in. By changing the thickness of both side shims equally, ring gear and pinion backlash will not change.

1. Lubricate the side bearings with final drive lubricant.

2. Place differential in position in the carrier.

3. If the original ring gear and pinion are being used, subtract the reading obtained in Step 8 from the reading obtained in Step 3 of the Final Drive Disassembly procedure. This determines the original side bearing preload and will aid in determining whether thicker or thinner shims are needed to bring the side bearing preload to specifications.

4. Install original shim onto left side and spacer onto the right side.

5. Install the carrier spreader and apply just enough tension to allow the shim to be installed between the spacer and the carrier.

Checking backlash
(© Oldsmobile Div. G.M. Corp.)

6. Release tension on the spreader, install side bearing caps, then check preload. Preload should be 15-20 in. lbs. for new bearings and 5-7 in. lbs. for old bearings over the pinion bearing preload obtained in Step 11, Final Drive Case Assembly.

7. If pre-load is not within specifications, select thicker or thinner shims to bring preload within limits.

Backlash Adjustment

1. Rotate differential case a few times to seat bearings, then mount dial indicator in order to read movement at the outer edge of one of the ring gear teeth.

2. Check backlash at three points around the ring gear. Lash must not vary more than .002 in.

3. Backlash at the minimum point should be .006-.008 in. for all new gears. If original ring gear and pinion were installed, backlash should be set at the same reading obtained in Step 3 of Final Drive Disassembly procedure, if reading was within specifications.

4. If backlash was not within limits, correct by increasing thickness of one differential shim and decreasing thickness of other side shim the same amount. This will not disturb differential side bearing preload. For each .001 in. change in backlash desired, transfer .002 in. shim thickness. To decrease backlash .001 in., decrease thickness of right shim .002 in. and increase thickness of left shim .002 in. To increase backlash .002 in. increase thickness of right shim .004 in. and decrease thickness of left shim .004 in.

5. When backlash is correct, remove spreader. Install bearing caps and bolts. Torque to 50 ft. lbs.

6. Install new output shaft seals.

7. Install new gasket onto housing. Install cover, torque cover attaching bolts to 30 ft. lbs. Fill final drive to correct level.

Automatic Transmission

References

The Turbo-Hydramatic transmission used on the Toronado is a fully automatic transmission used for front-wheel drive applications. It consists primarily of a three element hydraulic torque converter, dual sprocket and link assembly, compound planetary gear set, three multiple disc clutches, a sprag clutch, a roller clutch, two band assemblies and hydraulic control system. For major repair operations on this transmis-

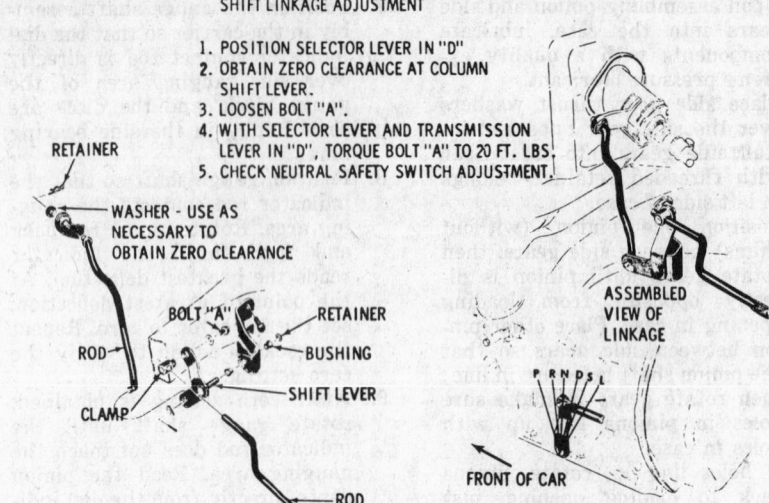

SHIFT LINKAGE ADJUSTMENT

1. POSITION SELECTOR LEVER IN "D"
2. OBTAIN ZERO CLEARANCE AT COLUMN SHIFT LEVER.
3. LOOSEN BOLT "A".
4. WITH SELECTOR LEVER AND TRANSMISSION LEVER IN "D", TORQUE BOLT "A" TO 20 FT. LBS.
5. CHECK NEUTRAL SAFETY SWITCH ADJUSTMENT.

Shift linkage adjustments (© Oldsmobile Div. G.M. Corp.)

sion, consult the Unit Repair Section of this manual.

Linkage Adjustments

See illustration for manual linkage adjustment.

Automatic Transmission Removal

1. Disconnect battery.
2. Disconnect oil cooler lines at transmission and speedometer cable at governor.
3. Install engine support bar.
4. Remove nut D and bolts A, B and C as in illustration. A special wrench, such as MAC S-147, must be used to remove nut D on 1966-67 models.
5. Remove bolts A, B, C and D as in illustrations.
6. Remove flywheel cover plate bolt A. (See illustration.)
7. Raise the car on a hoist.
8. Disconnect starter wiring, then remove starter.
9. Remove bolts B, C and D from flywheel cover plate.
10. Remove flywheel to converter bolt E. Rotate flywheel until all bolts are removed.
11. Disconnect vacuum modulator line and stator wiring.
12. Install transmission lift.

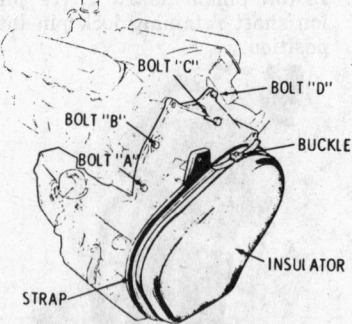

Transmission to engine attachment
(© Oldsmobile Div. G.M. Corp.)

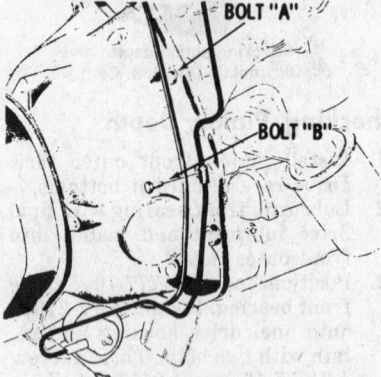

Transmission to engine attachment
(© Oldsmobile Div. G.M. Corp.)

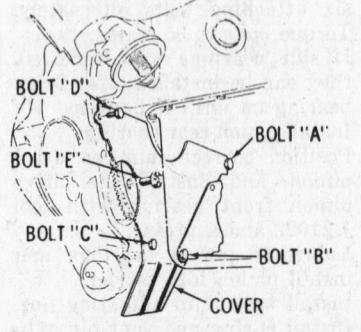

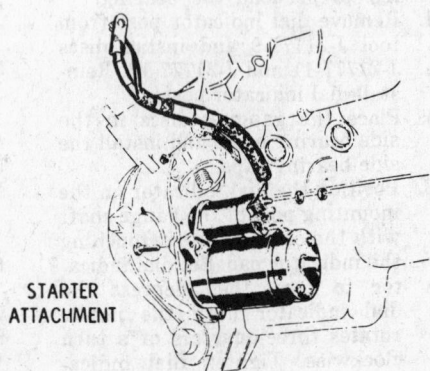

STARTER ATTACHMENT

Transmission to converter attachment (© Oldsmobile Div. G.M. Corp.)

13. Remove shift linkage.
14. Remove bolts E, F, G and nut from H. (See illustration.)
NOTE: when the last three transmission to final drive bolts are removed, a quantity of oil will be lost.
15. Remove bolts A and B. (See illustration.)
16. Remove the two upper engine mount brackets to transmission bolts A and B. (See illustration.)
17. Remove the four brackets to engine mount bolts.
18. Slide transmission rearward and down. Engine mount bracket will follow transmission down. Install converter holding tool J-21654.
19. After transmission is removed from car, the link assembly cover insulator can be removed or installed.

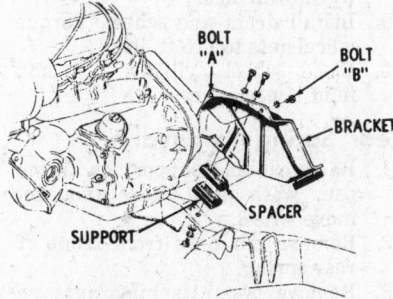

Engine mount attachment
(© Oldsmobile Div. G.M. Corp.)

Automatic Transmission Installation

When installing the transmission, the motor mount bracket must be positioned loosely on the link assembly cover until the transmission is in place. Then reverse removal procedure. Torque the bolts as follows:
Engine to torque converter housing —30 ft. lbs.
Engine bracket to transmission—55 ft. lbs.
Engine bracket to rubber mount—55 ft. lbs.
Oil cooler lines to transmission—30 ft. lbs.
Final drive to transmission nuts and bolts—25 ft. lbs.
Torque converter to flywheel bolts—30 ft. lbs.
Flywheel housing cover—5 ft. lbs.
Starter to transmission—30 ft. lbs.

After the transmission is installed, check transmission oil level. Refer to capacity chart at the beginning of this car section.

Drive and Driven Sprockets for the Transmission Drive

If it should be necessary to replace either the drive sprocket, chain, or driven sprocket, the three unit combination must be replaced as a set.

They are matched and are not to be serviced separately.

Removal

1. Remove cover housing attaching bolts.
2. Remove cover housing and gasket. Discard the gasket.
3. Install J-4646 snap-ring pliers into sprocket bearing retaining snap-rings located under the drive and driven sprockets, then remove snap-rings from retaining grooves in support housings.
NOTE: do not remove snap-rings from beneath the sprockets. Leave them in a loose position between the sprockets and the bearing assemblies.
4. Remove drive and driven sprockets, link assemblies, bearings and shaft simultaneously by alternately pulling upward on the drive and driven sprockets until the bearings are out of the drive and driven support housings.
NOTE: it may be necessary to pry up on the sprockets. Use care.

Caution Do not pry on the guide links or the aluminum case. Pry only on the sprockets.

5. Remove link assembly from drive and driven sprockets.
6. Remove two hook-type oil seal rings from turbine shaft.
7. Inspect drive and driven sprocket bearing assemblies for rough or defective bearings.
NOTE: do not remove bearing assemblies from drive or driven sprockets unless they need replacement.
8. If removal of bearing assembly from drive and/or driven sprockets is necessary, proceed as follows:
   A. Remove sprocket to bearing assembly retaining snap-ring using tool J-5589, (snap-ring pliers).
   B. Mount sprocket with turbine or input shaft placed in hole in work bench on two 2 x 4 x 10 in. pieces of wood.
   C. With a hammer and brass rod, drive the inner race alternately through each of the access openings until the bearing assembly is removed from the sprocket hub. Drive the sprocket, then turbine shaft and link assembly.

Inspection

1. Inspect drive sprocket teeth for nicks, burrs, scoring, gauling and excessive wear.
2. Inspect drive sprocket to ball bearing retaining snap-ring for damage.
3. Inspect drive sprocket ball bearing inner race mounting surface for damage.
4. Inspect turbine shaft for open lubrication passages. Run a tag

wire through the passages to make sure they are open.
5. Inspect spline for damage.
6. Inspect the ground bushing journals for damage.
7. Inspect the two hook-type oil seal grooves for damage or excessive wear.
8. Inspect the turbine shaft for cracks or distortion.
9. Inspect the link assembly for damage or loose links.
NOTE: take particular notice of the guide links. They are the wide outside links on each side of the link assembly.

Installation

Install by reversing removal procedures.

Driven Sprocket at Input Shaft Inspection

1. Inspect driven sprocket teeth for nicks, burrs, scoring, galling and excessive wear.
2. Inspect sprocket to ball bearing retaining snap-ring for damage.
3. Inspect ball bearing inner race mounting surface for damage.
4. Inspect input shaft for open lubrication holes. Run a tag wire through the holes to make sure they are open.
5. Inspect spline for damage.
6. Inspect ground bushing journals for damage.

Sprocket Bearing Installation

1. Turn sprocket so that turbine or input shaft is pointing upward.
2. Install new sprocket bearing as follows:
   A. Install snap-ring, letter side down to shaft.
   B. Assemble bearing assembly on turbine or input shaft.
   C. Using drift (J-6133-A), drive the bearing assembly onto the hub of the sprocket until it is resting on the bearing seat of the sprocket.
   D. Install sprocket to bearing assembly retaining snap-ring into groove sprocket hub.
3. Install two hook-type oil seal rings onto turbine shaft.

Checking front unit end-play
(© Oldsmobile Div. G.M. Corp.)

Front Unit End-Play Check

1. Install front unit end-play checking tool, (J-22241), into driven sprocket housing so that the urethane on the tool can engage the splines and the forward clutch housing. Let the tool bottom on the main-shaft, then withdraw it approximately $1/16$-$1/8$ in.
2. Remove two of the 5/16—18 bolts from the driven support housing.
3. Install 5/16—18 threaded slide hammer bolt with jam nut into one bolt hole in driven support housing.

NOTE: do not thread slide hammer bolt deep enough to interfere with forward clutch travel.

4. Mount dial indicator on rod and index indicator to register with the forward clutch drum which can be reached through second bolt removed from driven support housing.
5. Push end-play tool down to remove slack.
6. Push and hold output flange outward. Place a screwdriver in case opening at parking pawl area and push upward on output carrier.
7. Place another screwdriver between the metal lip of the end-play tool and the drive sprocket housing. Now, push upward on the metal lip of the end-play tool and read the resulting end-play. This should be .003-.024 in. The selective washer controlling this end-play is the phenolic thrust washer located between the driven support housing and the forward clutch housing. If more or less washer thickness is required to bring the end-play within specifications, select the proper washer from the chart.

THICKNESS	COLOR
.060 - .064	Yellow
.071 - .075	Blue
.082 - .086	Red
.093 - .097	Brown
.104 - .108	Green
.115 - .119	Black
.126 - .130	Purple

Front unit end-play selective washer thickness

Rear Suspension

Wheel Bearing Adjustment

For the rear wheel tapered roller bearings to be correctly adjusted, the following precautions should be taken:
1. The cones must be a slip fit on the spindle.
2. Inside of cones should be lubri-cated to make sure the cone creeps on the spindle.
3. Spindle nut must be a free-running fit on the threads.
4. Adjustment of rear wheel bearings should be made by continuously revolving the wheel while torquing the nut as follows:
  A. Torque adjusting nut to 25-30 ft. lbs. to seat all components thoroughly.
  B. Back off nut one-half turn, then retighten finger tight.
  C. If unable to insert cotter pin at this position, back off to nearest castellation.

Rear Wheel Spindle Removal

1. Support rear of car on stands.
2. Remove wheel and drum, then the hub assembly.
3. Disconnect brake line fitting at wheel cylinder.
4. Remove four spindle attaching bolts, and tie backing plate up, out of the way.
5. Place jack under axle, then remove four attaching bolts from center spring clamp.
6. Remove rubber insulator from spring.
7. Lower jack until spindle is clear of spring and spindle is accessible for removal.
8. Drive spindle out of axle with a hammer.

Rear Wheel Spindle Installation

1. Start new spindle, with keyway up, into axle and install the four bolts (backing plate to spindle). Tighten bolts progressively, one turn each, until spindle is fully seated.
2. Remove the bolts.
3. Position insulators on the spring. Tape if necessary to secure.
4. Raise rear axle until spring aligning pin locates into axle.
5. Replace four attaching bolts into center spring clamp. Torque to 30 ft. lbs.
6. Install backing plate onto spindle. Torque to 35 ft. lbs.
7. Connect brake line to wheel cylinder.
8. Install hub assembly and adjust wheel bearings.
9. Install drum and wheel. Torque wheel nuts to 115 ft. lbs.
10. Bleed wheel cylinder and add fluid if necessary.

Rear Spring Removal

1. Raise car and support on frame pad. With jack under axle, remove wheel.
2. Remove nut only from front of rear spring.
3. Remove two attaching nuts on rear shackle (outer). Remove rear shackle (outer).

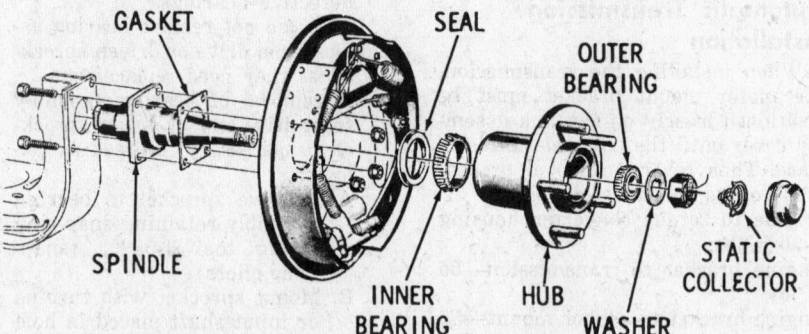

Rear hub assembly (© Oldsmobile Div. G.M. Corp.)

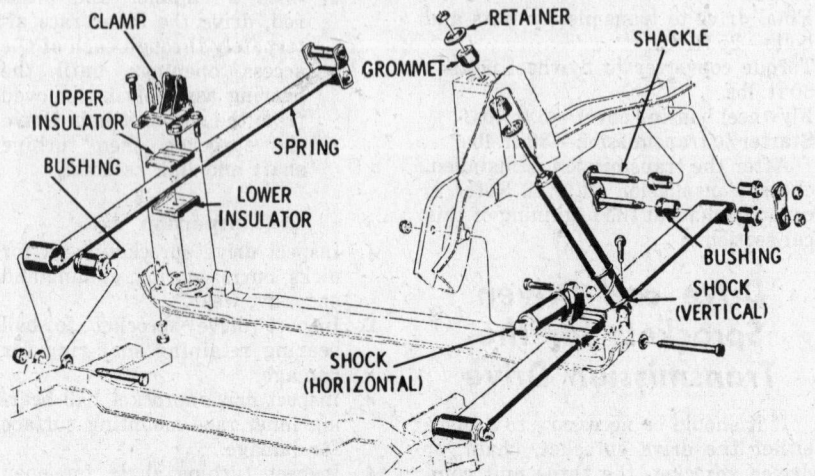

Rear suspension (© Oldsmobile Div. G.M. Corp.)

4. Remove four attaching bolts on center clamp assembly.
5. Lift center clamp up, shock will hold it in position.
6. On 1966-67 models, remove resonator bracket attaching bolts to frame and allow resonator to hang loose.
7. Lower jack until axle is free from spring.
8. Remove shackle assembly from spring and body.
9. Remove bolt from front of rear spring and remove spring.
10. If spring bushing is worn, remove and replace it.

Rear Spring Installation

To install, reverse removal procedure.

A. Front bushing is a press fit. Replacement will require an arbor press.
B. Torque resonator bracket attaching bolts to 14 ft. lbs.
C. Torque four spring center clamp assembly bolts to 30 ft. lbs.
D. Install wheel, torque to 115 ft. lbs.
E. Remove all supports and, with car on the ground, torque rear shackle bolts to 40 ft. lbs. and front spring bolt to 75-80 ft. lbs.

Rear Axle Removal

1. Lift car and support at both rear frame pads.
2. Remove wheels, drum and hub assemblies.
3. Disconnect brake lines and parking brake cable.
4. Remove backing plate attaching bolts and let backing plates rest on floor.
5. With jack under center of axle, remove eight attaching bolts from center spring clamp assemblies.
6. Disconnect brake hose.
7. Lower jack and remove axle.

Radio

R & R

1. Disconnect battery.
2. Remove both lower cluster panels.
3. Remove steering column attaching nuts and lower bracket.
4. Remove shift indicator needle.
5. Disconnect speedometer cable.
6. Remove attaching nuts from cluster lower brackets, leaving brackets attached to the instrument panel.
7. Remove two upper instrument panel screws and lay cluster assembly on steering column.
8. Remove radio knobs, washers or rear seat speaker control.
9. Remove radio attaching nuts and escutcheons.
10. Disconnect all wiring and antenna lead-in.
11. Remove lower radio support bracket attaching nut.
12. Remove radio from instrument panel.
13. Install by reversing the removal procedure.

Windshield Wipers

The windshield wiper system consists of the wiper motor and transmission assembly. It is very similar to that used on other Oldsmobile models.

Heater System

R & R

1. Disconnect blower feed and resistor wiring.
2. Disconnect vacuum hoses from the air inlet and forced vent diaphragms.
3. Disconnect temperature cable from temperature door lever.
4. Disconnect heater hoses. Keep open ends of hoses above engine coolant level to prevent loss of coolant.
5. Remove heater assembly attaching screws.
6. Remove heater assembly from the cowl.
7. If heater core is to be removed, it can be removed at this time.
8. To install, reverse the removal procedure. Be sure to apply sealer to the mounting face of the heater assembly.

Pontiac

Index

YEAR IDENTIFICATION

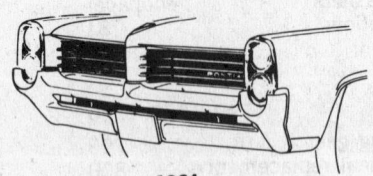

1964

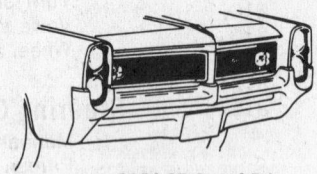

1964-65 Grand Prix

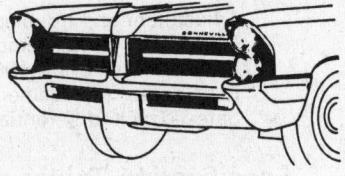

1965

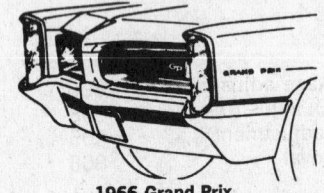

1966 Grand Prix

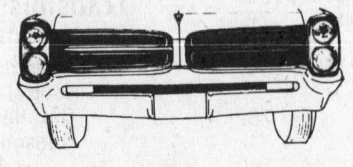

1966

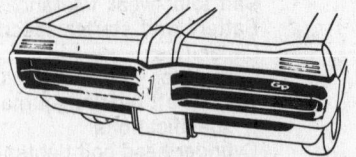

1967 Grand Prix

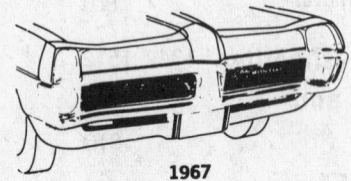

1967

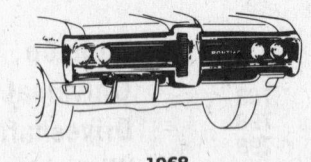

1968

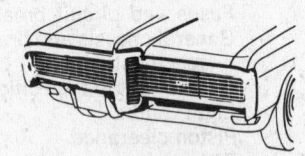

1968 Grand Prix

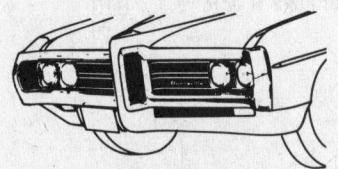

1969

1969 Grand Prix

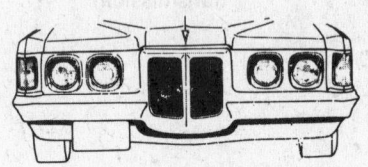

1970 Grand Prix

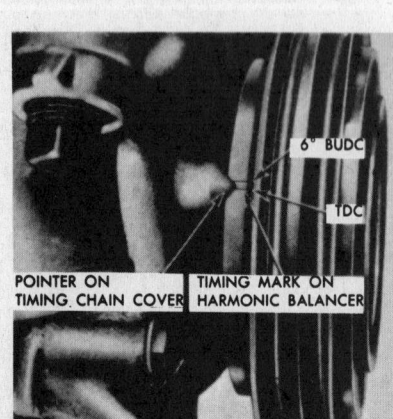

1970 Catalina

1971 Pontiac

1971 Grand Prix

FIRING ORDER

Ignition timing marks—1964-65
(© Pontiac Div., G.M. Corp.)

Ignition timing marks—1966-71
(© Pontiac Div., G.M. Corp.)

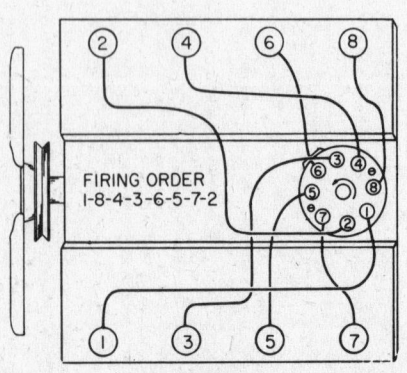

FIRING ORDER
1-8-4-3-6-5-7-2

1964-70 V8

CAR SERIAL NUMBER LOCATION

1964-67

Car serial number is found on the left front door pillar. This number is identical with engine number.
First digit: Engine
Second digit: Model series
Third digit: Year manufactured
Fourth digit: Assembly plant

1968-70

Car serial number is located on the upper left-hand side of the instrument panel, visible through the windshield.

Engine Identification

1964-70

The engine production number is found on the front of the right bank of cylinders, stamped into the block.

The engine production code is stamped on a machined pad just beneath this number. Refer to the charts for engine code interpretation.
First digit: Pontiac Division.
Second and third digit: Model series.
Fourth and fifth digit: Body style.
Sixth digit: Year manufactured.
Seventh digit: Plant.

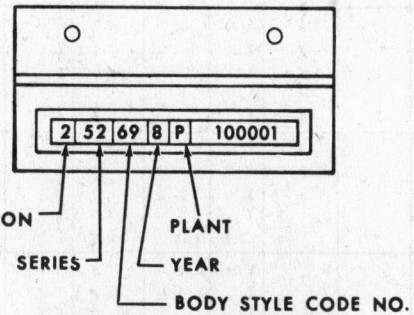

Example of plate shows the first 1968 Catalina (52) series 4-door sedan (69) style built at Pontiac plant (P.)

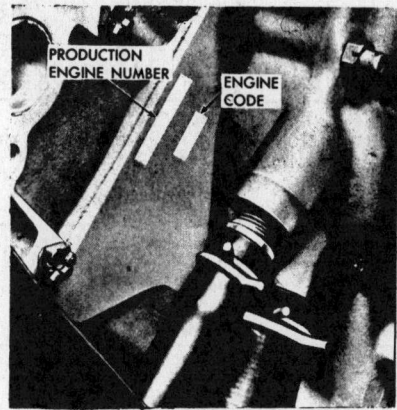

Engine serial number location

Engine Production Codes 1964–67

No. Cyls.	Cu. In. Displ.	Type	1964	1965	1966	1967
8	389	3 Spd., M.T., 2 Bbl.	01A			
8	389	3 Spd., M.T., 2 Bbl.	02B			
8	389	A.T., 2 Bbl.	04L			
8	389	A.T., 2 Bbl.	05L			
8	389	A.T., 2 Bbl.	08R			
8	389	A.T., 2 Bbl.	09R			
8	389	3 Spd., M.T., 2 Bbl.		WA	WA	
8	400	M.T., 2 Bbl.				WA
8	400	M.T., 2 Bbl. H.D.C.				WB
8	389	3 Spd., M.T., 4 Bbl.		WD		
8	400	M.T., 4 Bbl., Exh. Em.				WD
8	389	3 Spd., M.T., 4 Bbl.			WE	
8	400	M.T., 4 Bbl.				WE
8	389	3 Spd., M.T., 3-2 Bbl.	WF			
8	400	A.T., 2 Bbl., AC				WF
8	421	3 Spd., M.T., 4 Bbl.	WG	WG		
8	428	M.T., 4 Bbl., 2 + 2				WG
8	421	3 Spd., M.T., 3-2 Bbl.	WH	WH		
8	421	3 Spd., M.T., 3-2 Bbl.	WJ	WJ		
8	428	M.T., 4 Bbl.			WS	
8	421	3 Spd., M.T., 4 Bbl.		WK		
8	428	A.T., 4 Bbl., Exh. Em.				WL
8	400	M.T., 2 Bbl., Exh. Em.				WM
8	400	M.T., 2 Bbl., Exh. Em.				WN
8	389	3 Spd., M.T., 2 Bbl.	XA			
8	400	M.T., 2 Bbl.			XA	
8	389	A.T., 2 Bbl.	XB			
8	400	A.T., 2 Bbl.			XB	
8	389	A.T., 4 Bbl.	XC			
8	400	A.T., 4 Bbl.			XC	
8	428	A.T., 4 Bbl., Exh. Em., 2 + 2			XD	
8	400	A.T., 4 Bbl.			XH	
8	400	A.T., 4 Bbl., AC			XJ	
8	428	M.T., 4 Bbl., 2 + 2, Exh. Em.			XK	
8	400	M.T., 4 Bbl., Exh. Em.			XY	
8	400	M.T., 4 Bbl.			XZ	
8	400	A.T., 2 Bbl.			YA	
8	400	A.T., 2 Bbl., Exh. Em.			YB	
8	400	A.T., 2 Bbl.			YC	
8	400	A.T., 2 Bbl., AC			YD	
8	400	A.T., 4 Bbl., H.D.C.			YE	
8	400	A.T., 4 Bbl., AC, H.D.C.			YF	
8	389	A.T., 4 Bbl., AC		YF	YF	
8	421	A.T., 4 Bbl.		YH	YH	
8	400	A.T., 4 Bbl.				YH
8	421	A.T., 3-2 Bbl.		YJ	YJ	
8	421	3 Spd., M.T., 3-2 Bbl.		YK		
8	421	A.T., 3-2 Bbl.			YK	
8	428	A.T., 4-Bbl. 2 + 2				YK
8	400	A.T., 4 Bbl., Exh. Em.				YL
8	421	A.T., 4 Bbl.			YT	
8	389	A.T., 2 Bbl.			YU	
8	400	A.T., 2 Bbl., Exh. Em.				YU
8	389	A.T., 2 Bbl., AC			YV	
8	400	A.T., 2 Bbl., AC				YV
8	389	A.T., 4 Bbl.			YW	
8	400	A.T., 4 Bbl., Exh. Em., H.D.C.				YW
8	389	A.T., 4 Bbl., AC			YX	
8	400	A.T., 4 Bbl., Exh. Em., AC, H.D.C.				YX
8	428	M.T., 4 Bbl., Exh. Em., 2 + 2				YV
8	389	3 Spd., M.T., 2 Bbl.	10A			
8	389	A.T., 2 Bbl.	11H			
8	389	A.T., 2 Bbl.	12H			
8	389	A.T., 2 Bbl.	17M			
8	389	A.T., 2 Bbl., AC	18M			
8	389	3 Spd., M.T., 4 Bbl.	22B			
8	389	3 Spd., M.T., 4 Bbl.	23B			
8	389	A.T., 4 Bbl.	25K			
8	389	A.T., 4 Bbl., AC	26K			
8	389	A.T., 4 Bbl.	27P			
8	389	A.T., 4 Bbl., AC	28P			
8	389	A.T., 4 Bbl.	29N			
8	389	3 Spd., 3-2 Bbl.	32B			
8	389	A.T., 3-2 Bbl.	33-6			
8	389	A.T., 3-2 Bbl.	34N			
8	421	3 Spd., M.T., 4 Bbl.	35B			
8	421	A.T., 4 Bbl.	38S			
8	421	A.T., 4 Bbl.	43N			
8	421	3 Spd., M.T., 3-2 Bbl.	44B			
8	421	3 Spd., M.T., 3-2 Bbl.	45B			
8	421	A.T., 3-2 Bbl.	46G			
8	421	A.T., 3-2 Bbl.	47S			
8	421	A.T., 3-2 Bbl.	48N			
8	421	A.T., 3-2 Bbl.	50Q			

AC—Air conditioned.   A.T.—Automatic transmission.   M.T.—Manual transmission.   2 Bbl.—Two barrel carburetor.   4 Bbl.—Four barrel carburetor.

1964 ENGINE CODES AND SPECIAL EQUIPMENT

Horsepower	Engine Code	389 Cu. In.	421 Cu. In.	Type Trans.	Usage	23	W51 OPT 29	26	2835	2839,2847,2867 / 2840,2850,2890	Taxi Police 10.75:1	10.5:1	8.6:1	7.9:1	45 Spark Plug	44 Spark Plug	H.D. Starter	11110 54 Dist.	11110 52 Dist.	2BBL 11/16 S.M.	2BBL 11/16 H.M.	2BBL 11/16 H.M. A/C	2BBL 7/16 S.M.	2BBL 7/16 H.M.	2BBL 7/16 H.M. A/C	4BBL S.M.	4BBL H.M.	Triple 2BBL S.M.	Triple 2BBL H.M.	Cam 524009	Cam 518111 S.M.	Cam 519640	Cam 529472	Cam 9770543 421 H.O.	Single Spring	Std. Two Springs	H.D. Two Springs	Special Lifter Assembly	4 Bolt Brg Caps / Spec. Exh. Manifolds	High Output Fuel Pump	H.D. Clutch	H.D. Trans.
215	01A	XX		SM	Standard	X						X			X								X							X						X						
215	01A	XX					X					X			X								X							X						X						X
215	02B	XX			1—R.P.O.	X						X			X								X							X						X					X	X
215	02B	X										X			X								X							X						X					X	X
215	02B	X			Std. Police						X	X			X								X							X						X					X	X
215	02B	X			Std. Taxi							X			X								X							X						X					X	X
215	02B	X			Trail Prov.	X		X				X			X		X						X							X						X				X	X	X
239	03A	X			Export	X							X	X	X								X									X			X							
239	03A	X						X					X	X	X								X									X			X							
306	23B	X			Spec. Police						X	X			X	X							X					X			X					X		X	X	X	X
306	23B	X			Standard		X		X	X		X			X		X	X					X					X						X	X		X	X	X
306	23B	X			1—R.P.O.	X	X		X			X			X		X	X					X					X						X	X		X	X	X
235	22B	X			Standard					X			X		X								X					X					X					X	X	X
350	32B	X			1—R.P.O.	X X X		X X X			X			X		X X	X									X						X	X		X	X	X
320	35B	X			421	X X X		X X X			X			X		X X	X								X						X	X		X	X	X
350	44B	X			421	X X X		X X X			X			X		X X													X	X	X	X X	X X X
370	45B	X			421 H.O.	X X X		X X X			X			X		X X	X											X	X	X	X	X X X
283	10A	X			1—R.P.O.	X						X			X	X X	X	X						X	X	X	X X X
283	10A	X			1—R.P.O.			X				X			X	X X	X							X	X	X	X X X
283	10A	X			Standard	X X		X				X			X	X X	X							X	X	X	X X X
350	49N	X		315 HM	421			X X X			X			X		X X											X	X X	X
370	50Q	X			421 H.O.			X X X			X			X	X	X X	X					X			X	X X X X	X
350	34N	X			1—R.P.O.			X X X			X			X		X X									X	X	X	X
230	08R	X			2—425E			X X X		X X	X			X	X			X							X		
230	09R	X			2—425E A/C			X X X		X	X			X	X				X					X		
303	27P	X			Standard			X X			X			X	X X				X			X	X	X	
303	27P	X			1—R.P.O.	X					X			X	X X				X			X	X		
303	27P	X			Standard				X		X			X	X X				X			X	X	X
303	28P	X			Air Cond.			X X X			X			X	X X				X			X	X	X
303	28P	X							X		X			X	X X				X			X	X	X
306	29N	X			1—R.P.O.			X	X	X	X			X	X	X X				X		X	X	X	X
276	30P	X			Export			X X					X	X				X			X	X	X
283	17M	X			Standard	X					X		X	X X	X	X				X	X	
283	17M	X								X X	X		X	X X	X	X				X	X	
283	18M	X			Air Cond.						X		X	X X	X		X			X	X	X
283	18M	X				X X					X	X	X X	X			X			X	X	X
257	19M	X			Export	X			X		X	X	X			X			X	X	
320	43N	X			421	X X X					X	X	X X		X		X			X	X	X	X

Transmission Code Stamped at Engine Assy. Plant

3-Speed SM Transmission
- A 3 Speed Synchro-mesh
- B 3 Speed H.D. Synchro-mesh

375 RHM Trans. (Roto)
- G PC
- H P
- K PB
- L PE

315 HM Trans. (Super)
- M PS
- N PAS
- P PBS
- R PES

6 PG
S PAH

1—Regular production option.
2—425 is torque rating.

4-Speed SM Transmission
- 2 26 & 28 with 421 & 421-HO Engs. (exc. 2835 or 39:20 R/Axle)
- 3 26 & 28 421 & 421-HO Engs. (with 39:10 Ratio Axle exc.2835)

Q PCS

- C 23, 2835 & 29 (exc. 39:10 Ratio Axle, 421 & 421-HO Engs.)
- D 26 & 28 (exc. 39:10 Ratio Axle, 2835 or 421 & 421-HO Engs.)
- E 23, 2835 & 29 (with 39:10 Ratio Axle exc. 421 & 421-HO Engs.)
- F 26 & 28 (with 39:10 Ratio Axle exc. 2835 or 421 & 421-HO Engs.)
- T 23, 2835 & 29 with 421 & 421-HO Engs. (exc. 39:10 Ratio Axle)
- U 23, 2835 & 29 with 421 & 421-HO Engs. (with 39:10 Ratio Axle)

1964 ENGINE CODES AND SPECIAL EQUIPMENT

Horsepower	Engine Code	389 Cu. In.	421 Cu. In.	Type Trans.	Usage	23	W51 OPT.	29	26	2835	2839,2847,2867	2840,2850,2890	Taxi Police	10.75:1	10.5:1	8.6:1	7.9:1	45 Spark Plug	44 Spark Plug	H.D. Starter	11110 54 Dist.	11110 52 Dist.	2 BBL S.M. (11/16 Bore)	2 BBL H.M.	2 BBL H.M. A/C	2 BBL S.M. (7/16 Bore)	2 BBL H.M.	2 BBL H.M. A/C	4 BBL Carb. S.M.	4 BBL H.M.	Triple 2 BBL	524009	518111 S.M.	519640	529472	9770543 421 H.O.	Single Spring	Std. Two Springs	H.D. Two Springs	Special Lifter Assembly	4 Bolt Brg. Caps (3 Centers)	Special Exhaust Manifolds	High Output Fuel Pump	H.D. Clutch	H.D. Trans.
230	04L	X		375	2—425E	X	X	X								X		X	X											X				X					X							
230	05L	X		RHM	2—425 A/C	X	X	X								X		X	X												X			X					X							X
370	46G		X		421 H.O.	X	X	X					X							X	X	X										X					X			X	X	X	X	X	X	X
350	47S		X		421	X	X	X					X							X	X	X									X	X	X						X		X	X		X	X	X
303	25K	X			1—R.P.O.	X	X								X			X		X	X								X					X					X							X
303	25K	X			Standard					X					X			X		X	X								X					X					X							X
303	26K	X			Air Cond.	X	X								X			X		X	X								X					X					X							X
303	26K	X													X			X		X	X								X					X					X							X
267	11H	X			Standard	X	X								X			X		X	X	X												X					X							
267	12H	X			Air Cond.	X	X								X			X		X	X			X										X					X							X
240	13H	X			Export	X	X	X				X					X	X				X												X					X							
320	38S		X		421	X	X	X							X			X		X	X								X			X	X						X					X		X
350	336	X			1—R.P.O.	X	X	X					X					X		X	X										X	X	X						X		X					X

1967 ENGINE CODES AND SPECIAL EQUIPMENT

HORSEPOWER	ENGINE CODE	400	428	MANUAL	AUTOMATIC	2-BBL	4-BBL	QUADRAJET	7.9	8.6	10.5	10.75	9779254(U)	9779066(N)	9779067(P)	977068(S)	1111250	1111254	1111237	1111253	1111242	1111183	1111252	1111261	1111243	1111244	1111245	1111251	1111180	1111255	STD. TWO	H.D. TWO	H.D. SPECIAL	AIR COND.	A.I.R. SYSTEM	C.C.S.	SPECIAL EXHAUST MANIFOLDS
265	WA	X		X		X				X			X								X										X						
265	WB	X		X		X				X			X								X										X						
333	WD	X		X			X				X			X			X														X				X		
333	WE	X		X			X				X			X						X											0	X						
260	XB	X			X	X			X				X								X										X						
293	XC	X			X	X			X						X						X										X						
350	XH	X			X		X				X				X										X		X				0	X					
350	XJ	X			X		X				X				X										X		X				0	X		X			
350	XY	X		X			X				X			X										X			X				X					X	
350	XZ	X		X			X				X			X											X		X				0	X					
265	YA	X			X	X			X				X								X										X						
265	YB	X			X	X			X				X											X			X				X						X
290	YC	X			X	X				X			X									X									X		0				
290	YD	X			X	X				X			X									X									X		0	X			
325	YE	X			X		X			X						X						X									X		0				
325	YF	X			X		X			X						X						X									X		0	X			
360	WG		X	X	X		X			X			X										X										X	0				
376	WJ		X	X	X		X					X		X●									X										X●	0		X		X
376	Y3	X			X		X					X		X●									X										X●	0				X
360	Y2	X			X		X			X			X										X										X●	0				
376	XK	X		X			X					X			X		X										●							X	X		X
360	YH	X			X		X			X			X										X										X	0				
376	YK	X			X		X					X		X●									X										X●	0				X
360	YY		X	X			X			X			X								X												X			X		

●—With 60 PSI oil pump spring
0—Optional

1968 ENGINE CODES AND SPECIAL EQUIPMENT

DISPLACEMENT	HORSEPOWER	ENGINE CODE	TRANSMISSION		CARBURETOR		COMPRESSION RATIO			CAMSHAFT				STANDARD DISTRIBUTOR							VALVE SPRINGS		SPECIAL EXHAUST MANIFOLDS
			MANUAL	AUTOMATIC	2-BBL	QUADRAJET	8.6	10.5	10.75	9777254	9779066	9779067	9779068	1111272	1111270	1111449	1111300	1111448	1111435	1111450	STD. TWO	H.D. TWO (SPEC.)	
400 Cu. In. Engine	290	WA	X		X			X			X							X			X		
	290	WB	X		X			X			X							X			X		
	265	YA		X	X		X			X				X							X		
	290	YC		X	X			X				X		X							X		
	340	YE		X		X		X			X						X				X		
	350	XZ	X			X		X				X						X			X		
	350	XH		X		X		X			X							X			X		
428 Cu. In. Engine	375	WG	X			X	X						X						X		X	X	
	390	WJ	X			X			X		X					●						X	X
	375	YH		X		X	X				X								X		X		
	390	YK		X		X			X		X				●						X		X

●—With 60 psi oil pump spring. All cars use CCS.

1969 ENGINE CODES AND SPECIAL EQUIPMENT

400	428	HORSEPOWER	ENGINE CODE	TRANSMISSION		CARBURETOR		COMPRESSION RATIO			CAMSHAFT				DISTRIBUTOR						VALVE SPRINGS		CYLINDER HEAD	
				MANUAL	AUTOMATIC	2-BBL (2GV)	4-BBL (4MV)	8.6:1	10.5:1	10.75:1	9777254	9779066	9779067	9779068	1111253	1111940	1111946 (e)	1111952 (e)	1111960	1111959	H.D.—DUAL	H.D. SPEC.—DUAL	SMALL VALVE	LARGE VALVE
X		290	(a) WA WD	X		X			X		X						X				X		X	
X		265	(a) YA YB		X	X		X			X					X					X		X	
X		290	(a) WB WE	X		X			X		X						X				X		X	
X		290	(a) YC YD		X		X		X		X				X						X		X	
	X	360	WG	X			X		X				X						X		X			X
	X	360	(a) YL YH		X		X		X			X									X	X	X	
	X	390	WJ	X			X			X				X					X		X			X
	X	360	(b) XK		X		X		X				X								X	X		X
	X	360	(a) XE (c) XJ		X		X		X				X								X	X	X	
	X	390	YK		X		X			X			X							X	X			X
X		350	(d) WX	X			X		X				X						X		X			X
X		350	(d) XH		X		X		X			X			X						X			X
	X	370	(d) WF	X			X		X				X					X			X			X
	X	370	(d) XF		X		X		X			X									X	X		X
	X	390	(d) WL	X			X			X				X					X		X			X
	X	390	(d) XG		X		X			X			X			X					X			X
X		265	(d) YF		X	X		X			X					X					X		X	
X		340	XZ		X		X		X		X				X						X		X	

(a) Early production (small valve) engines with 30° intake valve seat angle. Later production (small valve) engines use 45° intake valve seat. NOTE: All large valve engines use 30° intake valve seat.

(b) Police Freeway Enforcer.
(c) Police Highway Patrol.
(d) Gran Prix.
(e) Uses special hardened drive gear. (With 60 psi oil pump.)

1970 ENGINE CODES AND SPECIAL EQUIPMENT

HORSEPOWER	ENGINE CODE	USAGE PONTIAC	USAGE GRAND PRIX	DISPL 350	DISPL 400	DISPL 455	TRANS MANUAL	TRANS AUTOMATIC	CARB 2 BBL (2GV)	CARB 4 BBL (4MV)	COMP 8.8:1	COMP 10.0:1	COMP 10.25:1	CAM 9777254(U)	CAM 9779066(N)	CAM 9779067(P)	CAM 9779068(S)	DIST 1112008	DIST 1112007	DIST 1111148 S	DIST 1111176 S	DIST 1111105	DIST 1112012 S	VALVE SPRINGS STD-DUAL	CYL HEAD SMALL VALVE	CYL HEAD LARGE VALVE
225	W7	X		X			X		X		X			X				X						X	X	
255	X7	X		X				X	X		X			X				X						X	X	
290	WE	X			X		X		X			X			X			X						X	X	
350	WX		X		X		X			X			X			X					X			X		X
350	XH		X		X			X	X				X	X						X				X		X
330	XZ	X			X			X	X			X			X				X					X	X	
265	YB	X	X		X			X	X		X						X		X					X	X	
290	YD	X			X			X	X			X			X				X					X	X	
370	WG	X	X			X	X		X				X				X					X		X	X	X
370	XF	X	X			X		X	X				X			X							X	X	X	X
360	YH	X				X		X	X		X					X						X		X	X	

S—Uses hardened drive gear for use with 60 psi oil pump and high tension distributor points.

GENERAL CHASSIS AND BRAKE SPECIFICATIONS

YEAR	MODEL	CHASSIS Overall Length (In.)	CHASSIS Tire Size	BRAKE CYLINDER BORE Master Cylinder (In.) Power	BRAKE CYLINDER BORE Wheel Cylinder Diameter (In.) Front	BRAKE CYLINDER BORE Wheel Cylinder Diameter (In.) Rear	BRAKE DRUM Diameter (In.) Front	BRAKE DRUM Diameter (In.) Rear
1964	Catalina, Grand Prix	220.0	8.50 x 14	1.0	$1^3/_{16}$	1.0	11	11
	Bonneville, Star Chief	213.0	8.00 x 14	1.0	$1^3/_{16}$	1.0	11	11
	Safari Wagon	213.8	8.50 x 14	1.0	$1^3/_{16}$	1.0	11	11
1965	Catalina, Grand Prix	214.6	8.25 x 14	1.0	$1^3/_{16}$	$^{15}/_{16}$	11	11
	Bonneville, Star Chief	221.7	8.25 x 14	1.0	$1^3/_{16}$	$^{15}/_{16}$	11	11
	Station Wagon	217.9	8.55 x 14	1.0	$1^3/_{16}$	$^{15}/_{16}$	11	11
1966	Catalina, Grand Prix	214.8	8.25 x 14	.875	$1^3/_{16}$	$^{15}/_{16}$	11	11
	Bonneville, Star Chief	221.8	8.25 x 14	.875	$1^3/_{16}$	$^{15}/_{16}$	11	11
	Station Wagon	218.1	8.55 x 14	.875	$1^3/_{16}$	$^{15}/_{16}$	11	11
1967	Catalina	215.6	8.25 x 14	1.0	$1^3/_{16}$	$^{15}/_{16}$	11	11
	Executive, Bonneville	222.6	8.55 x 14	1.0	$1^3/_{16}$	$^{15}/_{16}$	11	11
	Grand Prix	215.6	8.55 x 14	1.0	$1^3/_{16}$	$^{15}/_{16}$	11.8	11
	Station Wagon	218.4	8.55 x 14	1.0	$1^3/_{16}$	$^{15}/_{16}$	11	11
1968	Catalina	216.5	8.55 x 14*	1.0	$1^1/_8$	$^{15}/_{16}$	11	11
	Executive, Bonneville	223.5	8.55 x 14	1.0	$1^1/_8$	$^{15}/_{16}$	11	11
	Grand Prix	216.3	8.55 x 14	1.0	$1^1/_8$	$^{15}/_{16}$	11	11
	Station Wagon	218.4	8.55 x 14	1.0	$1^1/_8$	$^{15}/_{16}$	11	11
	Models w/Disc Brakes	—	—	1.125	$2^1/_{16}$	$^{15}/_{16}$	11.8 (Disc)	11
1969	Catalina	217.5	8.55 x 15	1.0	$1^1/_8$	$^{15}/_{16}$	11	11
	Executive	223.5	8.55 x 15	1.0	$1^1/_8$	$^{15}/_{16}$	11	11
	Bonneville	224.0	8.55 x 15	1.0	$1^1/_8$	$^{15}/_{16}$	11	11
	Grand Prix	210.2	6.78 x 14	1.0	$1^1/_8$	$^{15}/_{16}$	11	11
	Station Wag.	220.5	8.85 x 15	1.0	$1^1/_8$	$^{15}/_{16}$	11	11
	Models w/Disc Brakes	—	8.85 x 15	1.125	$2^{15}/_{16}$	$^{15}/_{16}$	11.8 (Disc)	11
1970-71	Catalina	217.9	H78 x 15	1.0	$1^1/_8$	$^{15}/_{16}$	11	11
	Executive	223.9	H78 x 15	1.0	$1^1/_8$	$^{15}/_{16}$	11	11
	Bonneville	224.6	H78 x 15	1.0	$1^1/_8$	$^{15}/_{16}$	11	11
	Grand Prix	210.2	G78 x 14	1.125	$2^{15}/_{16}$	$^7/_8$	10.9 (disc)	9.5
	Station Wag.	220.9	L78 x 15	1.0	$1^1/_8$	$^{15}/_{16}$	11	11
	Models w/Disc Brakes	—	H78 x 15	1.125	$2^{15}/_{16}$	$^{15}/_{16}$	11.7 (Disc)	11

*—Sedan—8.25 x 14.

TUNE-UP SPECIFICATIONS

YEAR	MODEL	SPARK PLUGS		DISTRIBUTOR		IGNITION TIMING (Deg.) ▲	CRANKING COMP. PRESSURE (Psi)	VALVES		Intake Opens (Deg.)	FUEL PUMP PRESSURE (Psi)	IDLE SPEED (Rpm) ★
		Type	Gap (In.)	Point Dwell (Deg.)	Point Gap (In.)			Tappet (Hot) Clearance (In.)				
								Intake	Exhaust			
1964–66	V8—Std. Eng., M.T.	45S	.035	30	.016	6B	145	Zero	Zero	22B	5–6½	500
	V8—Std. Eng., A.T.	45S	.035	30	.016	6B	160	Zero	Zero	30B	5–6½	500
	V8—389 (425E)	45S	.035	30	.016	6B	160	Zero	Zero	14B	5–6½	500
	V8—389 (425A)	45S	.035	30	.016	6B	160	Zero	Zero	29B	5–6½	500
	V8—421 Cu. In., Std.	45S●	.035	30	.016	6B	160	Zero	Zero	23B	5–6½	500
	V8—421 Cu. In., Opt.	45S●	.035	30	.016	6B	160	Zero	Zero	31B	5–6½	600
1967	V8—400, 2-BBL.	45S	.035	30	.016	6B⊙	160	Zero	Zero	22B	5–6½	600■
	V8—400, 4-BBL., M.T.	45S	.035	30	.016	6B⊙	160	Zero	Zero	23B	5–6½	600■
	V8—400, 4-BBL., A.T.	45S	.035	30	.016	6B⊙	160	Zero	Zero	30B	5–6½	500■
	V8—428, 4-BBL.	44S	.035	30	.016	6B⊙	160	Zero	Zero	23B	5–6½	700
	V8—428, H.O.	44S	.035	30	.016	6B⊙	160	Zero	Zero	31B	5–6½	700
1968	V8—400, 2-BBL. (L. Comp.)	45S	.035	30	.016	9B	160	Zero	Zero	22B	5–6½	800☆
	V8—400, 2-BBL., M.T.	45S	.035	30	.016	9B	160	Zero	Zero	30B	5–6½	800
	V8—400, 2-BBL., A.T.	45S	.035	30	.016	9B	160	Zero	Zero	22B	5–6½	600
	V8—400, 4-BBL., M.T.	45S	.035	30	.016	9B	160	Zero	Zero	23B	5–6½	850
	V8—400, 4-BBL., A.T.	45S	.035	30	.016	9B	160	Zero	Zero	30B	5–6½	600
	V8—428, 4-BBL.	44S	.035	30	.016	9B	160	Zero	Zero	23B	5–6½	850☆
	V8—428, H.O., M.T.	44S	.035	30	.016	9B	160	Zero	Zero	31B	5–6½	850
	V8—428, H.O., A.T.	44S	.035	30	.016	9B	160	Zero	Zero	23B	5–6½	650
1969	V8—400, 2-BBL. (Low C.)	45S	.035	30	.016	9B	160	Zero	Zero	22B	5–6½	800☆
	V8—400, 2-BBL., M.T.	45S	.035	30	.016	9B	160	Zero	Zero	30B	5–6½	800
	V8—400, 2-BBL., A.T.	45S	.035	30	.016	9B	160	Zero	Zero	22B	5–6½	600
	V8—400, 4-BBL., M.T.	45S	.035	30	.016	9B	160	Zero	Zero	23B	5–6½	850
	V8—400, 4-BBL., A.T.	45S	.035	30	.016	9B	160	Zero	Zero	30B	5–6½	600
	V8—428, M.T.	45S	.035	30	.016	9B	160	Zero	Zero	23B	5–6½	600
	V8—428, A.T.	44S	.035	30	.016	9B	160	Zero	Zero	23B	5–6½	600
	V8—428, H.O., M.T.	44S	.035	30	.016	9B	180	Zero	Zero	31B	5–6½	850
	V8—428, H.O., A.T.	44S	.035	30	.016	9B	180	Zero	Zero	31B	5–6½	850

▲—With vacuum advance disconnected. NOTE: These settings are only approximate. Engine design, altitude, temperature, fuel octane rating and the condition of the individual engine are all factors which can influence timing. The limiting advance factor must, therefore, be the "knock point" of the individual engine.

★—With manual transmission in N and automatic in D.

●—1966, V8—421; 44S.

■—100 rpm higher on exhaust emission cars.

⊙—Engines equipped with emission control systems—initial timing the same as standard.

☆—V8—400, 2-BBL. (Low Comp.)—M.T.—800 rpm; A.T.—600 rpm.
V8—428, 4-BBL., M.T.—850 rpm; A.T.—650 rpm.

B—Before top dead center.

①—800—M.T.
650—A.T.

②—M.T.—31B

③—With A/C—6½–8

④—M.T.—950

**—See engine decal.

CAUTION

General adoption of anti-pollution laws has changed the design of almost all car engine production to effectively reduce crankcase emission and terminal exhaust products. It has been necessary to adopt stricter tune-up rules, especially timing and idle speed procedures. Both of these values are peculiar to the engine and to its application, rather than to the engine alone. With this in mind, car manufacturers supply idle speed data for the engine and application involved. This information is clearly displayed in the engine compartment of each vehicle.

TUNE-UP SPECIFICATIONS, continued

YEAR	MODEL	SPARK PLUGS Type	SPARK PLUGS Gap (In.)	DISTRIBUTOR Point Dwell (Deg.)	DISTRIBUTOR Point Gap (In.)	IGNITION TIMING (Deg.) ▲	CRANKING COMP. PRESSURE (Psi)	VALVES Tappet (Hot) Clearance (In.) Intake	VALVES Tappet (Hot) Clearance (In.) Exhaust	VALVES Intake Opens (Deg.)	FUEL PUMP PRESSURE (Psi)	IDLE SPEED (Rpm) ★
1970	V8—350, 2-BBL., 255 HP	R46S	.035	30	.016	9B	160	Zero	Zero	22B	5-6½	800 (4)
	V8—400, 2-BBL. (Low C.)	R46S	.035	30	.016	9B	160	Zero	Zero	22B	5-6½	800☆
	V8—400, 2-BBL., M.T.	R46S	.035	30	.016	9B	160	Zero	Zero	30B	5-6½	800
	V8—400, 2-BBL., A.T.	R45S	.035	30	.016	9B	160	Zero	Zero	22B	5-6½	650
	V8—400, 4-BBL., M.T.	R46S	.035	30	.016	9B	160	Zero	Zero	23B	5-6½	850
	V8—400, 4-BBL., A.T.	R45S	.035	30	.016	9B	160	Zero	Zero	30B	5-6½	650
	V8—455, 4-BBL., 360 HP	R44S	.030	30	.016	8B	180	Zero	Zero	20B	5-6½	650 (4)
	V8—455, 4-BBL., 370 HP	R44S	.030	30	.016	10B	190	Zero	Zero	20B	5-6½	650 (4)
1971	V8—350, 2-BBL., 255 HP	R46S	.035	30	.016	9B	160	Zero	Zero	22B	5-6½	800 (1)
	V8—400, 2-BBL., (Low C.)	R46S	.035	30	.016	**	160	Zero	Zero	22B	5-6½	800 (1)
	V8—400, 2-BBL., M.T.	R46S	.035	30	.016	**	160	Zero	Zero	30B	5-6½	800
	V8—400, 2-BBL., A.T.	R46S	.035	30	.016	**	160	Zero	Zero	22B	5-6½	650
	V8—400, 4-BBL., M.T.	R46S	.035	30	.016	**	160	Zero	Zero	23B	5-6½	950
	V8—400, 4-BBL., A.T.	R45S	.035	30	.016	**	160	Zero	Zero	30B	5-6½	650
	V8—455, 4-BBL., 360 HP	R44S	.035	30	.016	**	N.A.	Zero	Zero	23B (2)	5-6½ (3)	650
	V8—455, 4-BBL., 370 HP	R44S	.035	30	.016	**	N.A.	Zero	Zero	23B (2)	5-6½ (3)	650 (4)

DELCOTRON AND AC REGULATOR SPECIFICATIONS

YEAR	ALTERNATOR PART NUMBER	ALTERNATOR Field Current Draw @ 12V.	ALTERNATOR Output @ 12V. and Engine RPM 5,000	REGULATOR Part No.	Field Relay Air Gap (In.)	Field Relay Point Gap (In.)	Field Relay Volts to Close	Regulator Air Gap (In.)	Regulator Point Gap (In.)	Regulator Volts at 125°
1964	1100678	1.9-2.3	40	1119511	.015	.030	2.3-3.7	.057	.015	13.5-14.4
	1100634	1.9-2.3	40	1119511	.015	.030	2.3-3.7	.057	.015	13.5-14.4
	1100682	2.2-2.6	50	1116366	.015	.030	2.3-3.7	.057	.015	13.5-14.4
	1100621	1.9-2.3	50	1119511	.015	.030	2.3-3.7	.057	.015	13.5-14.4
	1100674	2.6-2.9	57	9000590	.015	.030	2.3-3.7	.057	.015	13.7-14.3
1965-66	1100699	1.9-2.3	42	1119511	.015	.030	2.3-3.7	.057	.015	13.5-14.4
	1100700	1.9-2.3	55	1119511	.015	.030	2.3-3.7	.057	.015	13.5-14.4
	1100692	1.9-2.3	55	1119511	.015	.030	2.3-3.7	.057	.015	13.5-14.4
	1100702	N.A.	60	1119511	.015	.030	2.3-3.7	.057	.015	13.5-14.4
	1100758	N.A.	66	1119511	.015	.030	2.3-3.7	.057	.015	13.5-14.4
1967-69	1100699	1.9-2.3	42	1119511	.015	.030	2.3-3.7	.057	.015	13.5-14.4
	1100700	1.9-2.3	55	1119511	.015	.030	2.3-3.7	.057	.015	13.5-14.4
	1100800	1.9-2.3	55	1119511	.015	.030	2.3-3.7	.057	.015	13.5-14.4
	1100801	1.9-2.3	42	1119511	.015	.030	2.3-3.7	.057	.015	13.5-14.4
1970-71	1100704	2.2-2.6	37	1119515	.015	.030	2.3-3.7	.057	.015	13.5-14.4
	1100700	2.2-2.6	55	1119515	.015	.030	2.3-3.7	.057	.015	13.5-14.4
	1100895	2.2-2.6	61	1119515	.015	.030	2.3-3.7	.057	.015	13.5-14.4
	1117765	4.1-4.6	62	1116368	TRANSISTORIZED REGULATOR					

WHEEL ALIGNMENT

YEAR	MODEL	CASTER Range (Deg.)	CASTER Pref. Setting (Deg.)	CAMBER Range (Deg.)	CAMBER Pref. Setting (Deg.)	TOE-IN (In.)	KING-PIN INCLINATION (Deg.)	WHEEL PIVOT RATIO Inner Wheel	WHEEL PIVOT RATIO Outer Wheel
1964	All Series	2N-1N	1½N	¼N-¾P	¼P	0-1/16	4½	20	18
1965-71	All Series	2N-1N	1½N	¼N-¾P	¼P	0-1/8	8½	20	18

N—Negative.
P—Positive.

GENERAL ENGINE SPECIFICATIONS

YEAR	CU. IN. DISPLACEMENT	CARBURETOR	DEVELOPED HORSEPOWER @ RPM	DEVELOPED TORQUE @ RPM (FT. LBS.)	A.M.A. HORSEPOWER	BORE & STROKE (IN.)	ADVERTISED COMPRESSION RATIO	VALVE LIFTER TYPE	NORMAL OIL PRESSURE (PSI)
1964	389	2-BBL.	235 @ 4000	386 @ 2000	52.8	4.063 x 3.750	8.6-1	Hyd.	30-40
	389	2-BBL.	283 @ 4400	418 @ 2800	52.8	4.063 x 3.750	10.5-1	Hyd.	30-40
	389	4-BBL.	303 @ 4600	430 @ 2800	52.8	4.063 x 3.750	10.5-1	Hyd.	30-40
	389	3-2-BBL.	330 @ 4600	430 @ 3200	52.8	4.063 x 3.750	10.75-1	Hyd.	30-40
	421	4-BBL.	320 @ 4000	455 @ 2800	53.6	4.097 x 4.000	10.75-1	Hyd.	30-40
	421	3-2-BBL.	370 @ 5200	460 @ 3800	53.6	4.097 x 4.000	10.75-1	Hyd.	30-40
1965-66	389	2-BBL.	256 @ 4600	388 @ 2400	52.8	4.063 x 3.750	8.6-1	Hyd.	30-40
	389	4-BBL.	333 @ 5000	429 @ 3200	52.8	4.063 x 3.750	10.5-1	Hyd.	30-40
	389	2-BBL.	290 @ 4600	418 @ 2400	52.8	4.063 x 3.750	10.5-1	Hyd.	30-40
	389	4-BBL.	325 @ 4800	429 @ 2800	52.8	4.063 x 3.750	10.5-1	Hyd.	30-40
	389	3-2-BBL.	338 @ 4800	433 @ 3600	52.8	4.063 x 3.750	10.75-1	Hyd.	30-40
	421	4-BBL.	338 @ 4600	459 @ 2800	53.6	4.097 x 4.000	10.5-1	Hyd.	30-40
	421	3-2-BBL.	356 @ 4800	459 @ 3200	53.6	4.097 x 4.000	10.75-1	Hyd.	30-40
	421	3-2-BBL.	376 @ 5000	461 @ 3600	53.6	4.097 x 4.000	10.75-1	Hyd.	30-40
1967	400	4-BBL.	265 @ 4600	397 @ 2400	54.3	4.121 x 4.000	8.6-1	Hyd.	30-40
	400	2-BBL.	290 @ 4600	428 @ 2500	54.3	4.121 x 3.750	10.5-1	Hyd.	30-40
	400	4-BBL.	333 @ 5000	445 @ 3000	54.3	4.121 x 3.750	10.5-1	Hyd.	30-40
	400	4-BBL.	350 @ 5000	440 @ 3200	54.3	4.121 x 3.750	10.5-1	Hyd.	30-40
	400	4-BBL.	325 @ 4800	445 @ 2900	54.3	4.121 x 3.750	10.5-1	Hyd.	30-40
	400	4-BBL.	350 @ 4800	440 @ 3000	54.3	4.121 x 3.750	10.5-1	Hyd.	30-40
	428	4-BBL.	360 @ 4600	472 @ 3200	54.3	4.121 x 4.000	10.5-1	Hyd.	30-40
	428	4-BBL.	376 @ 5100	462 @ 3400	54.3	4.121 x 4.000	10.75-1	Hyd.	30-40
1968	400	2-BBL.	265 @ 4600	397 @ 2400	54.3	4.121 x 3.750	8.6-1	Hyd.	30-40
	400	2-BBL.	290 @ 4600	428 @ 2500	54.3	4.121 x 3.750	10.5-1	Hyd.	30-40
	400	4-BBL.	340 @ 4800	445 @ 2900	54.3	4.121 x 3.750	10.5-1	Hyd.	30-40
	400	4-BBL.	350 @ 5000	445 @ 3000	54.3	4.121 x 3.750	10.5-1	Hyd.	30-40
	428	4-BBL.	375 @ 4800	472 @ 3200	54.3	4.121 x 4.000	10.5-1	Hyd.	30-40
	428	4-BBL.	390 @ 5200	465 @ 3400	54.3	4.121 x 4.000	10.75-1	Hyd.	30-40
1969	400	2-BBL.	265 @ 4600	397 @ 2400	54.3	4.121 x 3.750	8.6-1	Hyd.	30-40
	400	2-BBL.	290 @ 4600	428 @ 2500	54.3	4.121 x 3.750	10.5-1	Hyd.	30-40
	400	4-BBL.	340 @ 4800	445 @ 2900	54.3	4.121 x 3.750	10.5-1	Hyd.	30-40
	400	4-BBL.	350 @ 5000	445 @ 3000	54.3	4.121 x 3.750	10.5-1	Hyd.	30-40
	428	4-BBL.	360 @ 4600	472 @ 3200	54.3	4.121 x 4.000	10.5-1	Hyd.	30-40
	428	4-BBL.	370 @ 4800	472 @ 3200	54.3	4.121 x 4.000	10.5-1	Hyd.	30-40
	428	4-BBL.	390 @ 5200	465 @ 3400	54.3	4.121 x 4.000	10.75-1	Hyd.	30-40
1970	350	2-BBL.	255 @ 4600	355 @ 2800	48.0	3.88 x 3.75	8.8-1	Hyd.	35
	400	2-BBL.	265 @ 4600	397 @ 2400	54.3	4.121 x 3.750	8.8-1	Hyd.	30-40
	400	2-BBL.	290 @ 4600	428 @ 2500	54.3	4.121 x 3.750	10.5-1	Hyd.	30-40
	400	4-BBL.	340 @ 4800	445 @ 2900	54.3	4.121 x 3.750	10.5-1	Hyd.	30-40
	400	4-BBL.	350 @ 5000	445 @ 3000	54.3	4.121 x 3.750	10.5-1	Hyd.	30-40
	455	4-BBL.	365 @ 4300	510 @ 2700	54.5	4.15 x 4.21	10.0-1	Hyd.	35
	455	4-BBL.	390 @ 4600	500 @ 3100	54.5	4.15 x 4.21	10.25-1	Hyd.	35
1971	350	2-BBL.	255 @ 4600	355 @ 2800	48.0	3.88 x 3.75	8.8-1	Hyd.	35
	400	2-BBL.	290 @ 4600	397 @ 2400	54.3	4.125 x 3.250	10.0-1	Hyd.	30-40
	400	2-BBL.	330 @ 4800	428 @ 2500	54.3	4.125 x 3.250	10.0-1	Hyd.	30-40
	400	4-BBL.	N.A.	445 @ 2900	54.3	4.125 x 3.250	N.A.	Hyd.	30-40
	400	4-BBL.	N.A.	445 @ 3000	54.3	4.125 x 3.250	N.A.	Hyd.	30-40
	455	4-BBL.	N.A.	500 @ 2700	54.5	4.15 x 4.21	N.A.	Hyd.	35
	455	4-BBL.	N.A.	500 @ 3100	54.5	4.15 x 4.21	N.A.	Hyd.	35

CYLINDER HEAD BOLT TIGHTENING SEQUENCE

VALVE SPECIFICATIONS

YEAR AND MODEL	SEAT ANGLE (DEG.)	FACE ANGLE (DEG.)	VALVE LIFT INTAKE (IN.)	VALVE LIFT EXHAUST (IN.)	VALVE SPRING PRESSURE (VALVE OPEN) LBS. @ IN.	VALVE SPRING INSTALLED HEIGHT (IN.)	STEM TO GUIDE CLEARANCE (IN.) Intake	Exhaust	STEM DIAMETER (IN.) Intake	Exhaust
1964-65 V8—389 (425A) A.T.	30	45	(4)	.405	.405	109 @ 1.15	1.53	.0021-.0038	.0026-.0043	.3407-.3414	.3402-.3409
V8—389 (425E) A.T.	30	45	(4)	.370	.370	109 @ 1.15	1.53	.0021-.0038	.0026-.0043	.3407-.3414	.3402-.3409
V8—389 M.T., All	30	45	(4)	.330	.330	109 @ 1.15	1.53	.0021-.0038	.0026-.0043	.3407-.3414	.3402-.3409
V8—389, 3-2-BBL., 421	30	45	(4)	.405	.405	114 @ 1.12	1.53	.0021-.0038	.0026-.0043	.3407-.3414	.3402-.3409
1966 V8—389 (Low Comp.)	30	45	(4)	.370	.406	(9)	1.59	.0021-.0038	.0026-.0043	.3407-.3414	.3402-.3409
V8—389 (High Comp.)	30	45	(4)	.406	.409	115 @ 1.18	1.59	.0021-.0038	.0026-.0043	.3407-.3414	.3402-.3409
V8—421	30	45	(4)	.406	.408	(10)	1.59	.0021-.0038	.0026-.0043	.3407-.3414	.3402-.3409
V8—421 H.O.	30	45	(4)	.409	.409	133 @ 1.18	1.59	.0021-.0038	.0026-.0043	.3407-.3414	.3402-.3409
1967 V8—400, 2-BBL.	30	45	(4)	.375	.410	(5)	1.59	.0016-.0033	.0021-.0038	.3407-.3414	.3402-.3409
V8—400, 4-BBL.	30	45	(4)	.407	.411	133 @ 1.17	1.59	.0016-.0033	.0021-.0038	.3407-.3414	.3402-.3409
V8—400, Grand Prix	30	45	(4)	.410	.413	133 @ 1.17	1.59	.0016-.0033	.0021-.0038	.3407-.3414	.3402-.3409
V8—428, 4-BBL.	30	45	(4)	.410	.413	133 @ 1.17	1.59	.0016-.0033	.0021-.0038	.3407-.3414	.3402-.3409
V8—428, H.O.	30	45	(4)	.414	.413	133 @ 1.17	1.59	.0016-.0033	.0021-.0038	.3407-.3414	.3402-.3409
1968 V8—400, 2-BBL.*	30	45	(4)	.376	.412	(5)	1.59	.0016-.0033	.0021-.0038	.3412-.3419	.3407-.3414
V8—400, 2-BBL., M.T.**	30	45	(4)	.410	.414	(6)	1.59	.0016-.0033	.0021-.0038	.3412-.3419	.3407-.3414
V8—400, 2-BBL., A.T.**	30	45	(4)	.376	.412	(5)	1.59	.0016-.0033	.0021-.0038	.3412-.3419	.3407-.3414
V8—400, 4-BBL., M.T.	30	45	(4)	.410	.413	(6)	1.59	.0016-.0033	.0021-.0038	.3412-.3419	.3407-.3414
V8—400, 4-BBL., A.T.	30	45	(4)	.410	.414	(6)	1.59	.0016-.0033	.0021-.0038	.3412-.3419	.3407-.3414
V8—428, 4-BBL.	30	45	(4)	.410	.413	133 @ 1.17	1.59	.0016-.0033	.0021-.0038	.3412-.3419	.3407-.3414
V8—428, H.O., M.T.	30	45	(4)	.414	.414	133 @ 1.17	1.59	.0016-.0033	.0021-.0038	.3412-.3419	.3407-.3414
V8—428, H.O., A.T.	30	45	(4)	.410	.413	133 @ 1.17	1.59	.0016-.0033	.0021-.0038	.3412-.3419	.3407-.3414
V8—350, 2-BBL.	30	45	(4)	.375	.410	127 @ 1.20	1.59	.0016-.0033	.0021-.0038	.3412-.3419	.3407-.3414
V8—350, 4-BBL.	30	45	(4)	.407	.411	133 @ 1.17	1.59	.0016-.0033	.0021-.0038	.3412-.3419	.3407-.3414
1969 V8—350, 2-BBL.	45	45	(4)	.375	.410	127 @ 1.20	1.58	.0016-.0033	.0021-.0038	.3407-.3414	.3407-.3414
V8—400, 2-BBL.*	45	45	44	.376	.412	(5)	1.58	.0016-.0033	.0021-.0038	.3407-.3414	.3407-.3414
V8—400, 2-BBL.**	45	45	44	.376	.412	(5)	1.58	.0016-.0033	.0021-.0038	.3407-.3414	.3407-.3414
V8—400, 4-BBL.	30	45	(4)	.410	.414	(6)	1.58	.0016-.0033	.0021-.0038	.3407-.3414	.3407-.3414
V8—400, 350 (1)	30	45	(4)	.410	.414	(7)	1.58	.0016-.0033	.0021-.0038	.3407-.3414	.3407-.3414
V8—428, 360 HP	30	45	(4)	.410	.413	133 @ 1.17	1.58	.0016-.0033	.0021-.0038	.3407-.3414	.3407-.3414
V8—428, 370 HP	30	45	(4)	.410	.413	133 @ 1.17	1.58	.0016-.0033	.0021-.0038	.3407-.3414	.3407-.3414
V8—428, H.O.	30	45	(4)	.410 (11)	.413 (11)	133 @ 1.17	1.58	.0016-.0033	.0021-.0038	.3407-.3414	.3407-.3414
1970-71 V8—350, 2-BBL.	45	45	44	.376	.412	(5)	1.58	.0016-.0033	.0021-.0038	.3412-.3419	.3407-.3414
V8—400, 2-BBL.*	45	45	44	.376	.412	(5)	1.58	.0016-.0033	.0021-.0038	.3412-.3419	.3407-.3414
V8—400, 2-BBL., M.T.**	45	45	44	.410	.414	(6)	1.58	.0016-.0033	.0021-.0038	.3412-.3419	.3407-.3414
V8—400, 2-BBL., A.T.**	45	45	44	.376	.412	(5)	1.58	.0016-.0033	.0021-.0038	.3412-.3419	.3407-.3414
V8—400, 4-BBL., M.T. (1)	30	45	(4)	.410	.413	(6)	1.58	.0016-.0033	.0021-.0038	.3412-.3419	.3407-.3414
V8—400, 4-BBL., A.T.	45	45	44	.410	.414	(6)	1.58	.0016-.0033	.0021-.0038	.3412-.3419	.3407-.3414
V8—455, 4-BBL., M.T.	30	45	(4)	.414	.413	(8)	1.58	.0016-.0033	.0021-.0038	.3412-.3419	.3407-.3414
V8—455, 4-BBL., A.T.	45	45	44	.410	.413	(7)	1.56	.0016-.0033	.0021-.0038	.3412-.3419	.3407-.3414
V8—455, 4-BBL., A.T. (2)	30	45	(4)	.410	.413	(7)	1.56	.0016-.0033	.0021-.0038	.3412-.3419	.3407-.3414
V8—400, 350 HP, A.T. (1)	30	45	(4)	.410	.414	(7)	1.56	.0016-.0033	.0021-.0038	.3412-.3419	.3407-.3414

① —Available only in Grand Prix (350 HP)
② —H.O. engine—370 HP (available only with A.T.)
③ —Intake 30°, Exhaust 45°
④ —Intake 29°, Exhaust 44°
⑤ —Intake 127.5 @ 1.21, Exhaust 133.7 @ 1.17
⑥ —Intake 133.4 @ 1.17, Exhaust 131.1 @ 1.16
⑦ —Intake 137.0 @ 1.15, Exhaust 137.7 @ 1.15
⑧ —Intake 134.1 @ 1.17, Exhaust 133.9 @ 1.17
⑨ —Intake 110 @ 1.21, Exhaust 115 @ 1.18
⑩ —Intake 131 @ 1.18, Exhaust 133 @ 1.18
⑪ —M.T. .414, .414
* —Low compression
** —High compression
M.T.—Manual transmission
A.T.—Automatic transmission

TORQUE SPECIFICATIONS

YEAR	MODEL	CYLINDER HEAD BOLTS (FT. LBS.)	ROD BEARING BOLTS (FT. LBS.)	MAIN BEARING BOLTS (FT. LBS.)	CRANKSHAFT BALANCER BOLT (FT. LBS.)	FLYWHEEL TO CRANKSHAFT BOLTS (FT. LBS.)	MANIFOLD (FT. LBS.) Intake	Exhaust
1964-71	All Series	80-95	35-45	85-95▲	160	90-95	40	30

▲—Rear Main, 120.

BATTERY AND STARTER SPECIFICATIONS

YEAR	MODEL	BATTERY				STARTERS						Brush Spring Tension (Oz.)
		Ampere Hour Capacity	Volts	Group Number	Terminal Grounded	Lock Test			No-Load Test			
						Amps.	Volts	Torque	Amps.	Volts	RPM	
1964-66	with Synchromesh	53	12	2SM	Neg.	Not Recommended			100	10.6	3,600	35
	with Hydramatic	61	12	29M	Neg.	Not Recommended			120	10.6	4,700	35
	Hvy. Duty Opt. ▲	70	12	25M	Neg.	Not Recommended			Not Recommended			35
1967-71	with Synchromesh	53	12	2SM	Neg.	Not Recommended			100	10.6	4,350	35
	with Hydramatic	61	12	2SM	Neg.	Not Recommended			120	10.6	4,350	35
	Hvy. Duty Opt.	70	12	25T	Neg.	Not Recommended			Not Recommended			

▲—1966

CAPACITIES

YEAR	MODEL	ENGINE CRANKCASE ADD 1 Qt. FOR NEW FILTER	TRANSMISSIONS Pts. TO REFILL AFTER DRAINING			DRIVE AXLE (Pts.)	GASOLINE TANK (Gals.)	COOLING SYSTEM (Qts.) WITH HEATER
			Manual		Automatic			
			3-Speed	4-Speed				
1964	All Exc. Safari	4	2¾	2½	18½	5½	25	19½
	Safari	4	2¾	None	18½	5½	19	19½
1965-66	All Exc. Safari	5	2¾	2½	22½ ▲	4½	26½	20
	Safari	5	2¾	None	22½ ▲	4½	24	20
1967		5	2¾	2½	19	4½	26½	18†
1968-71	Exc. Sta. Wag.	5	3½	2½	20	4½ ①	26½ ②	18†
	Sta. Wag.	5	3½	2½	20	4½	24	18†

▲—1966—19 Pts.
†—Grand Prix—18.6 Qts., 428 Cu. In. Opt. Eng. 17.2 Qts.

① —3 pts. with 8.125 in. ring gear 5 pts. with 8.875 in. ring gear.
② —21½ gals. for Grand Prix.

CRANKSHAFT BEARING JOURNAL SPECIFICATIONS

YEAR	MODEL	MAIN BEARING JOURNALS (IN.)				CONNECTING ROD BEARING JOURNALS (IN.)		
		Journal Diameter	Oil Clearance	Shaft End-Play	Thrust On No.	Journal Diameter	Oil Clearance	End-Play
1964-66	V8—OHV, Exc. 421 Cu. In.	3.000	.0018▲	.006	4	2.250	.0015	.009
	V8—421 Cu. In.	3.250	.0018▲	.007	4	2.250	.0015	.009
1967-69	V8—400 Cu. In.	3.000	.0018	.006	4	2.250	.0015	.009
	V8—428 Cu. In.	3.250	.0018	.006	4	2.250	.0015	.009
1970	V8—350 Cu. In.	3.000	.0002-.0017	.0035-.0085	4	2.250	.0005-.0025	.012-.017 ①
	V8—400 Cu. In.	3.000	.0002-.0017	.0035-.0085	4	2.250	.0005-.0025	.012-.017
	V8—455 Cu. In.	3.250	.0005-.0021	.0035-.0085	4	2.250	.0005-.0026	.012-.017
1971	V8—350 Cu. In.	3.000	.0002-.0017	.0035-.0085	4	2.250	.0005-.0025	.012-.017
	V8—400 Cu. In.	3.000	.0002-.0017	.0035-.0085	4	2.250	.0005-.0025	.012-.017
	V8—455 Cu. Inc.	3.250	.0005-.0021	.0035-.0085	4	2.250	.0005-.0026	.012-.017

▲—Bearing No. 1—.0015.
① —Total for two rods.

FUSES AND CIRCUIT BREAKERS

1964–65

Air conditioner controls, clock lamp,
  instrument panel lamps .... 4 Amp. Fuse
Air conditioner power & blower .... 30 Amp. Fuse
Back-up lamps, cruise control, windshield
  washer & wiper .... 25 Amp. Fuse
Cigar lighter, clock power, directional signal,
  stop lamp, dome lamp, license lamp,
  power antenna, tail lamps .... 14 Amp. Fuse
Heater blower .... 20 Amp. Fuse
Radio power .... 2.5 Amp. Fuse

1966

Same as 1964–65 except the following:
  Power antenna .... 20 Amp. Fuse
  Radio power .... 9 Amp. Fuse

1967–69

Same as 1964–65 except the following:
  Directional signal & radio power .... 10 Amp. Fuse
  Dome lamp, hazard warning, heater blower,
  power antenna & stop lamps .... 25 Amp. Fuse

1970–71

	Pont.	G.P.
Air cond. controls lamp (custom)	SFE-4	SFE-4
Air cond. power & blower mtr. (custom)	AGC-30	AGC-30
Ash tray lamp	SFE-4	SFE-4
Back up lamp	SFE-20	SFE-20
Cigar lighter	SFE-20	SFE-20
Cigar lighter lamp	SFE-4	SFE-4
Clock lamp	SFE-4	SFE-4
Clock Power	SFE-20	SFE-20
Console compartment lamp	—	SFE-20
Cornering lamps	SFE-20	SFE-20
Deck lid release	SFE-20	SFE-20
Directional signal & stop lamp	SFE-20	SFE-20
Dome, rear qtr. or rear courtesy lamp	SFE-20	SFE-20
Electrocruise & low fuel lamp	AGC-10	AGC-10
Hazard flasher	SFE-20	SFE-20
Heater cont. lamp	SFE-4	SFE-4
Heater blower motor	AGC-25	AGC-25
Instrument lamps	SFE-4	SFE-4
Instrument panel compartment lamps	SFE-20	SFE-20
Instrument panel courtesy lamp	SFE-20	SFE-20
License lamp	SFE-20	SFE-20
Lugg. compt. lamp	SFE-20	SFE-20
P.B. warning lamp	AGC-10	AGC-10
Power seat circuit breaker	40	40
Power tail gate wind. circuit breaker	40	—
Power side window circuit breaker	40	40
Radio dial lamp	SFE-4	SFE-4
Radio power	AGC-10	AGC-10
Rear wind. defogger	AGC-25	AGC-25
Safeguard speedo	AGC-25	AGC-25
Shift indicator lamp (column & console)	SFE-4	SFE-4
Side marker lamps frt. & rear	SFE-20	SFE-20
Tachometer lamp	—	SFE-4
Tail lamp	SFE-20	SFE-20
Under hood or utility	SFE-20	SFE-20
W/S Washer pump	AGC-25	AGC-25
W/S wiper or washer motor	AGC-25	AGC-25

LIGHT BULBS

1964–69

Ash tray .... 1445; 1 CP
Back-up .... 1156; 32 CP
Clock, heater controls, radio dial ..... 1895; 2 CP
Cornering lamp .... 1195; 50 CP
Dome lamp .... 211; 12 CP
Headlamp (low beam) .... 4002; 50/37.5 CP
  (high beam) .... 4001; 37.5 CP
License .... 67; 4 CP
Luggage compt. .... 1003; 15 CP
Tachometer .... 194; 2 CP
Underhood .... 93; 15 CP

1970–71

	Pont.	G.P.
Air cond. controls	1895	1895
Ash tray	1445	1445
Back-up lamp	1156	1157
Brake warning	194	194
Clock	1895	1895
Cigar lighter	1445	1445
Console compt.	—	1893
Cornering lamps	1195	1195
Courtesy lamp	550	89
Courtesy (sta. wag. rear compt.)	90	—
Dir. signal ind.	194	194
Dome lamp	211-1	211-1
Head lamp (lower) or (inner)	Type 1	Type 1
Head lamp (upper) or (outer)	Type 2	Type 2
Head lamp beam ind.	194	194
Heater cont.	1895	1895
Inst. panel compt.	1895	1895
License lamp	67	67
Low fuel warning	159	159
Luggage compt.	89	89
Oil pressure	194	194
Park & dir. signal	1157 NA	1157 NA
Radio dial	1895	1895
Reading lamp	1004	1004
Rr. seat arm rest (conv.)	68	—
Side marker-front	194 A	1157NA
Side marker-rear	1895 (3)	168
Side marker-rear	194 (4)	—
Shift ind. (exc. console)	1895	1893 (5)
Shift ind. (console)	—	1445
Speedo. ill. (exc. aux. ga.)	194	194
Tail stop & signal	1157	1157
Tachometer (hood)	—	194
Tachometer (i.p. mtg.)	—	1895
Temp. ga. (exc. aux. ga.)	194	194
Under hood or utility	1003	1003

Distributor

Detailed information on distributor drive, direction of distributor rotation, cylinder numbering, firing order, point gap, cam dwell, timing mark location, spark plugs, spark advance, ignition resistor location, and idle speed will be found in this section.

Distributor Removal

The distributor is located in the back of the block. To remove, disconnect the air cleaner and the distributor cap and wire assembly, mark the position of the rotor relative to the distributor body and the body relative to the block for easy replacement. Do not move the engine after removing the distributor. Detach the vacuum lines and the ignition primary wire from the distributor and remove the bolt which holds the distributor down to the block. The unit can then be lifted up. On V8 models, the drive gear is located on the bottom of the distributor. When replacing it, it will be necessary to index the groove in the bottom of the distributor shaft with the oil pump driveshaft.

Distributor Operation

An external adjustment distributor is used. The cap has a window for adjusting dwell time (cam angle) with the cap in place.

Adjustment of dwell is made on the car while the engine is operating or while the distributor is being checked on a distributor tester. The centrifugal advance parts have been relocated above the breaker plate and cam. This plan permits the cam and breaker lever to be located closer to the upper bearing for greater stability.

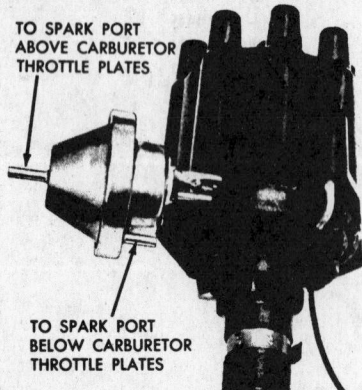

Vacuum advance valve
(© Pontiac Div. G.M. Corp.)

The breaker plate is one piece, and rotates on the outer diameter of the upper bearing. The plate is held in position by a retainer clip in the upper shaft bushing. The molded rotor serves as a cover for the cen-

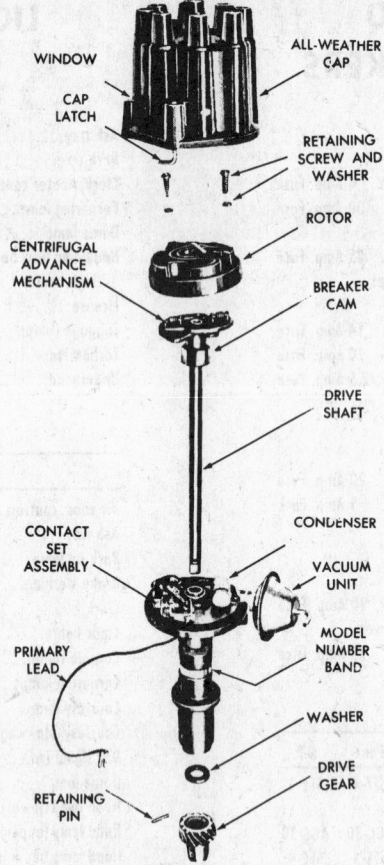

V8 distributor
(© Pontiac Div. G.M. Corp.)

trifugal advance mechanism. The vacuum control unit is mounted under the movable breaker plate and is fastened to the distributor housing.

The point set has the breaker lever spring tension and point alignment pre-set and is serviced as an assembly. Only the dwell angle requires adjustment after replacement.

Under part throttle operation, manifold vacuum is enough to actuate the vacuum control diaphragm. This

Distributor vacuum control valve
(© Pontiac Div. G.M. Corp.)

causes the movable plate to advance the spark and aid fuel economy. During acceleration, or on a heavy pull, the vacuum is insufficient to move the plate. The plate is spring-loaded, through the vacuum control diaphragm, and remains in the retarded position.

The centrifugal advance is conventional and operates through two spring-loaded weights.

On 1970 Pontiacs, the radios are more sensitive to ignition interference due to the antenna in the windshield. To combat this, all distributors have a Radio Frequency Interference Shield covering the circuit breaker plate assembly. Shield must be removed to install points or condenser, but dwell angle may be set through an opening in the shield.

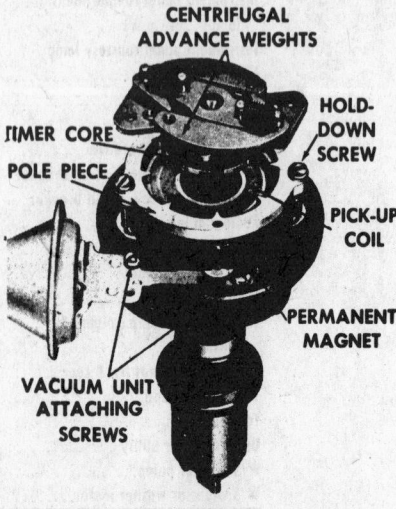

Installing Radio Interference Shield
(© Pontiac Div. G.M. Corp.)

Transistor Ignition

This system consists of a special distributor, an ignition pulse amplifier, and a special ignition coil. The distributor is simlar in external appearance to the standard V8 distributor, but the internal construction bears little resemblance to the contact-point unit. An iron timer core replaces the breaker cam. This

Transistor ignition distributor
(© Pontiac Div. G.M. Corp.)

eight-lobed timer rotates inside a magnetic pick-up assembly, which replaces the contact points and condenser. The magnetic pick-up assembly consists of a ceramic permanent magnet, a steel pole piece, and a pick-up coil.

The magnetic pick-up assembly is mounted over the distributor shaft bearing, and is rotated by the vacuum advance unit to provide automatic spark advance. Centrifugal advance is provided by the rotating timer core, which is attached to normal advance weights. Troubleshooting is found in Unit Repair.

Removal

1. Disconnect pick-up coil connector.
2. Remove distributor cap.
3. Crank engine so that rotor points to No. 1 cylinder plug tower and timing mark on crankshaft pulley are indexed with pointer.
4. Remove distributor vacuum line.
5. Remove distributor hold-down bolt and clamp, then remove distributor.

Disassembly

1. Remove rotor.
2. Remove centrifugal weight springs.
3. Remove weights.
4. Remove roll pin, drive gear and washer.
5. Remove drive shaft.
6. Remove weight support and timer core from drive shaft.
7. Remove magnetic core assembly.

AC Generator (Delcotron)

1964-70

The Delco-Remy Delcotron is used. The purpose of this unit is to satisfy the increase in electrical loads that have been imposed upon the car battery by modern conditions of traffic and driving patterns.

The Delcotron is covered in the Unit Repair Section.

Since the Delcotron and regulator are designed for use on only one polarity system, the following precautions must be observed:

1. The polarity of the battery, generator and regulator must be matched and considered before making any electrical connections to the system.
2. When connecting a booster battery, be sure to connect the negative battery terminals respectively, and the positive battery terminals respectively.
3. When connecting a charger to the battery, connect the charger positive lead to the battery positive terminal. Connect the charger

negative lead to the battery negative terminal.
4. Never operate the Delcotron on open circuit. Be sure that all connections in the circuit are clean and tight.
5. Do not short across or ground any of the terminals on the Delcotron regulator.
6. Do not attempt to polarize the Delcotron.
7. Do not use test lamps of more than 12 volts for checking diode continuity.
8. Avoid long soldering times when replacing diodes or transistors. Prolonged heat is damaging to these units.
9. Disconnect the battery ground terminal when servicing any AC system. This will prevent the possibility of accidental reversal of polarity.

C.S.I. AC Generator

See Basic Electrical Diagnosis Section in Unit Repair Section for service and test procedures.

Battery, Starter

Detailed information on the battery and starter will be found in the Battery and Starter Specifications table of this section.

A more detailed discussion of starters and their troubles can be found in the Unit Repair Section under the heading Basic Electrical Diagnosis.

Starter Removal

1. Disconnect starter motor cable from battery.
2. Remove junction box cover and disconnect solenoid wires from block. It is well to mark wire connection locations to facilitate installation.
3. Remove battery cable from clip on junction block.
4. Raise front of car and support on stands.
5. Pull cable and wire loom down to hang free.
6. Remove mounting screws and starter motor with cable and solenoid wires.
7. Remove wires from starter.
8. To reinstall reverse the above, first installing the wires to solenoid.

Instruments

Speedometer Head Removal

1964

1. Disconnect Safeguard cable, on cars so equipped, then pull

straight out.
2. Remove speedometer cluster assembly.
3. Remove face plate and lens by unsnapping.
4. Remove three screws at back of cluster.
5. Remove group carefully, noting wire routing.
6. Remove screws from numeral plate and remove speedometer head.
7. Reverse above to replace instrument.

1967

1. Remove instrument cluster.
2. Remove cluster face plate and lens by unsnapping face plate from housing.
3. Remove screws on back of cluster at speedometer cable fitting area.
4. Remove speedometer and instruments carefully. If car is equipped with Safeguard speedometer, disconnect ground wire at clip terminal and also wire retaining clip. Carefully note routing of wires before removing ground wire.
5. Remove two screws retaining numeral plate and remove speedometer head assembly.
6. Reverse procedure to install.

Cluster R & R

1964

1. Disconnect battery.
2. On air-conditioned cars, the center duct must be removed.
3. Disconnect speedometer cable.
4. Using a deep socket, remove four attaching nuts.
5. Disconnect illumination board on back side of cluster by removing screws.

NOTE: if car is equipped with Electro-cruise, the connectors and indicator lamps will have to be disconnected.

6. Push cluster toward front of car and remove cluster assembly.
7. Install by reversing removal procedure.

1967-68

1. Disconnect battery.
2. Remove instrument panel pad. If car is equipped with front speaker replace speaker wire.
3. Remove two bolts on each end of instrument panel trim plate.
4. On air conditioned cars it is necessary to remove lower duct assembly.
5. Remove bolts connecting instrument panel trim plate and lower instrument panel.
6. If car is equipped with automatic transmission, remove column cover and remove transmission indicator.
7. On Grand Prix models, it may be

necessary to disconnect vacuum lines from headlight switch.
8. On models with upper level ventilation system, remove pipes connecting to nozzle assembly.
9. If car is equipped with radio, remove bolt from radio to radio support brace.
10. If car is equipped with Safeguard speedometer, buzzer assembly must be detached.
11. Disconnect feeder of fiber optic system from cigar lighter.
12. If car is equipped with multiplex, disconnect mutiplex unit.
13. Disconnect speedometer.
14. Pull instrument panel plate out. Make sure all wires and routing clips are loose enough to allow panel to move out.
15. Disconnect wires attaching instrument cluster.
16. Remove four nuts that retain instrument cluster to instrument panel trim panel.
17. Remove cluster.
18. Replace by reversing procedure.

1969 Pontiac Except Grand Prix
1. Disconnect battery.
2. Remove lower instrument panel cover.
3. Remove radio.
4. Remove heater or air conditioning control panel.
5. Disconnect speedometer, printed circuit connector, cigar lighter and instrument panel harness from rear of cluster.
6. Remove four cluster retaining screws.
7. Position ground straps so that cluster may be pulled from studs and removed toward center of car.
8. Reverse procedure to install.

1969 Grand Prix
1. Disconnect battery.
2. Remove glove compartment.
3. Disconnect speedometer cable and wire connectors at headlamp switch, windshield wiper, turn signal, ignition switch, printed circuit, and heater or air conditioning control panel.
4. Remove lower column trim and disconnect air control at heater case.
5. Remove three screws at instrument nacelles.
6. Remove two instrument panel to column support.
7. Remove one screw at each outboard lower end of panel.
8. Remove screws retaining right and left side of instrument panel to body. To do this, remove upper vent nozzles by inserting two thin blade screwdrivers on right and left side of nozzle to disengage metal retaining clip. Pull nozzle towards rear of car to remove.

9. Remove upper instrument panel cover plate. Studs are pushed into retaining clips.
10. Remove two upper panel retaining screws and upper speaker support bracket screw.
11. Remove console to instrument panel retaining screws.
12. Loosen toe plate screws.
13. Remove column to lower instrument panel retaining nuts and drop column.
14. Pull entire instrument panel rearward on left far enough to gain access to cluster.
15. Remove ground strap retaining screws, and two instrument panel harness conduit retaining screws.
16. Remove instrument panel cluster retaining screws and remove cluster.
17. Reverse procedure to install.

1970 Pontiac Except Grand Prix
1. Disconnect battery and remove lower air conditioning duct.
2. Remove lower instrument panel trim at steering column.
3. Remove radio, radio support and ground plate.
4. Remove automatic transmission shift indicator, if so equipped.
5. Disconnect speedometer cable and heater cables at heater case.
6. Lower steering column.
7. Remove instrument panel trim plate screws and place trim plate on steering column.
8. Disconnect printed circuit.
9. Remove wire harness retaining screws and clock.
10. Remove cluster retaining nuts and cluster.
11. Reverse procedure to install.

1970 Grand Prix
1. Disconnect battery and remove lower air conditioning duct, if so equipped.
2. Remove pillar post mouldings and piller plate at windshield.
3. Remove lower instrument panel trim at steering column.
4. Disconnect speedometer cable and radio antenna lead.
5. Lower steering column.
6. Remove upper air outlet vents and instrument panel attaching screws at steering column, console, ends and upper center.
7. Remove lower defroster duct screw at heater case.
8. Move instrument pad outward on steering column.
9. Disconnect printed circuit.
10. Remove wire harness retaining screws and cluster ground screw.
11. Remove cluster mounting screws and remove cluster. Reverse procedure to install.

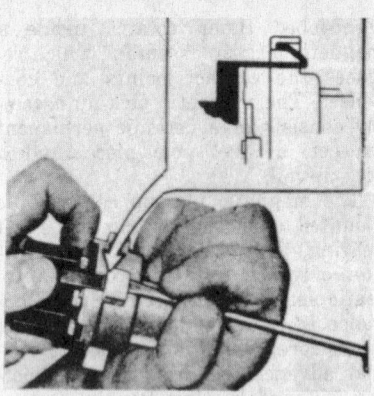

Unlocking ignition switch connector
(© Pontiac Div. G.M. Corp.)

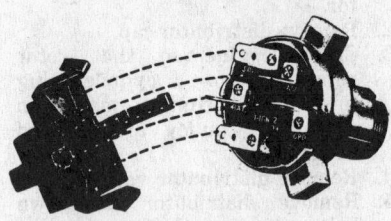

Separated ignition switch and connector
(© Pontiac Div. G.M. Corp.)

Ignition Switch Replacement

1964
1. Disconnect battery.
2. Remove lock cylinder by placing in off position and insert wire in small hole in cylinder face. While pushing in on wire, continue to turn cylinder counterclockwise and pull cylinder from case.
3. Remove nut from face of switch. (Spanner tool J-7607 will assist.)
4. Pull switch from dash and remove wiring connector. (See illustration).
5. Install in reverse of above.

1967-68
1. Remove ignition lock.
2. Remove bezel retaining switch to dash. Take care not to destroy fiber optic light.
3. Remove wire connector from ignition switch.
4. Remove switch.
5. Reverse procedure to install.

1969-70
1. Disconnect battery.
2. Loosen toe pan screws on steering column.
3. Remove column to instrument panel attaching nuts.
4. Lower column and disconnect switch wire connectors.
5. Remove switch attaching screws and remove switch.
6. To replace move key lock to OFF- LOCK position.

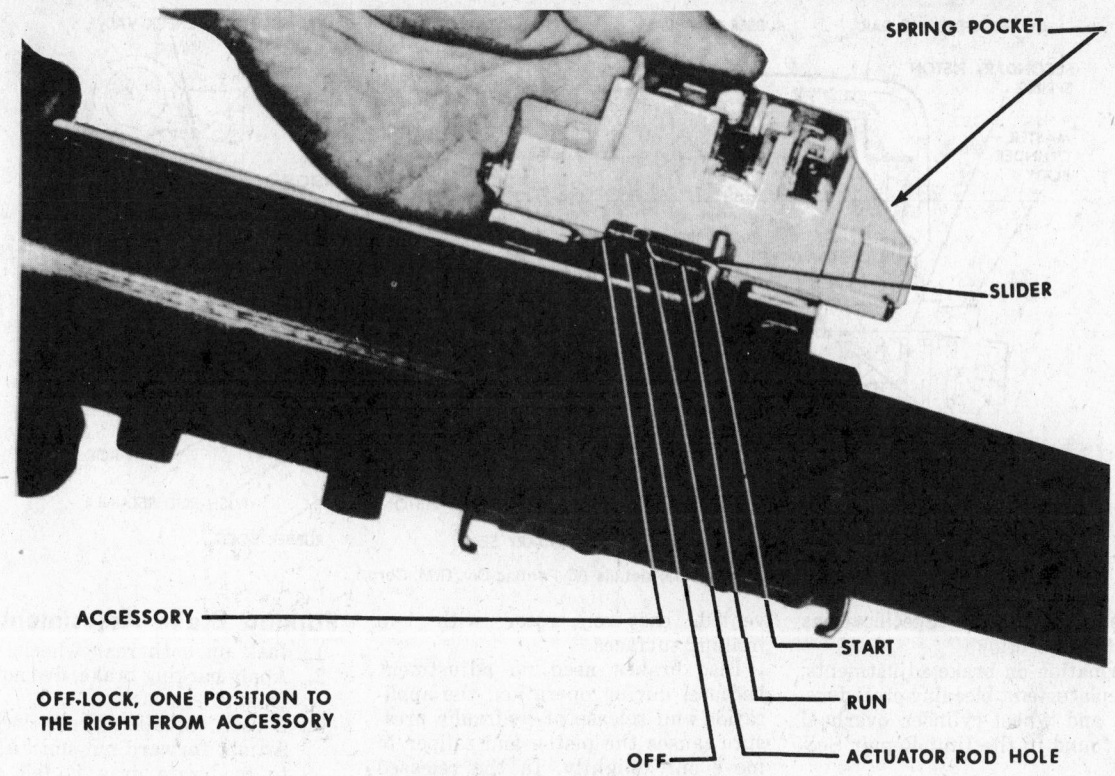

SPRING POCKET

SLIDER

ACCESSORY

START

RUN

ACTUATOR ROD HOLE

OFF

OFF-LOCK, ONE POSITION TO THE RIGHT FROM ACCESSORY

Installing ignition switch (© Pontiac Div. G.M. Corp.)

7. Move actuator rod hole in switch to OFF-LOCK position.
8. Install switch with rod in hole.
9. Position and reassemble steering column in reverse of disassembly procedure.

Lighting Switch Replacement

1964

1. Disconnect battery.
2. Pull knob out to ON position.
3. Reach under instrument panel and depress the switch shaft retainer (see illustration), and remove knob and shaft assembly.
4. Remove retaining ferrule nut.
5. Remove switch from instrument panel.
6. Disconnect multi-plug connector from switch.
7. Install in reverse of above. (In checking lights before installation, switch must be grounded to test dome lights on some models.)

1967

1. Disconnect battery.
2. Pull switch knob to ON position, push latch button on side of switch to ON position and pull out switch knob.
3. Unscrew bushing from switch and remove switch.
4. Remove push on connector with leads from light switch and connect to new switch.
5. On Grand Prix models remove vacuum connectors and connect new switch.
6. Position new switch in instru-

ment panel, and start bushing through ferrule into switch. Tighten bushing securely.
7. Insert knob into switch until end of knob engages catch.
8. Connect battery.

1968-70

1. Depress button on switch and remove knob and shaft.
2. Remove retaining nut.
3. Remove wire connector from

switch and remove switch.
4. On vacuum operated headlight models, remove vacuum connector.
5. Reverse procedure to install.

Brakes

Specific information on brake cylinder sizes can be found in the General

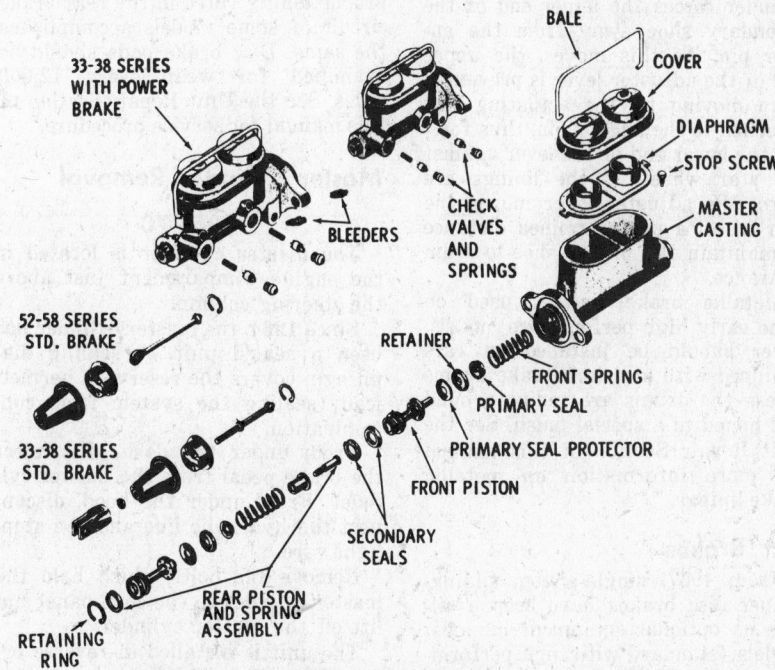

BALE

COVER

DIAPHRAGM

STOP SCREW

MASTER CASTING

33-38 SERIES WITH POWER BRAKE

BLEEDERS

CHECK VALVES AND SPRINGS

52-58 SERIES STD. BRAKE

33-38 SERIES STD. BRAKE

RETAINER

FRONT SPRING

PRIMARY SEAL

PRIMARY SEAL PROTECTOR

FRONT PISTON

SECONDARY SEALS

RETAINING RING

REAR PISTON AND SPRING ASSEMBLY

Dual type master cylinder (Delco) (© Pontiac Div. G.M. Corp.)

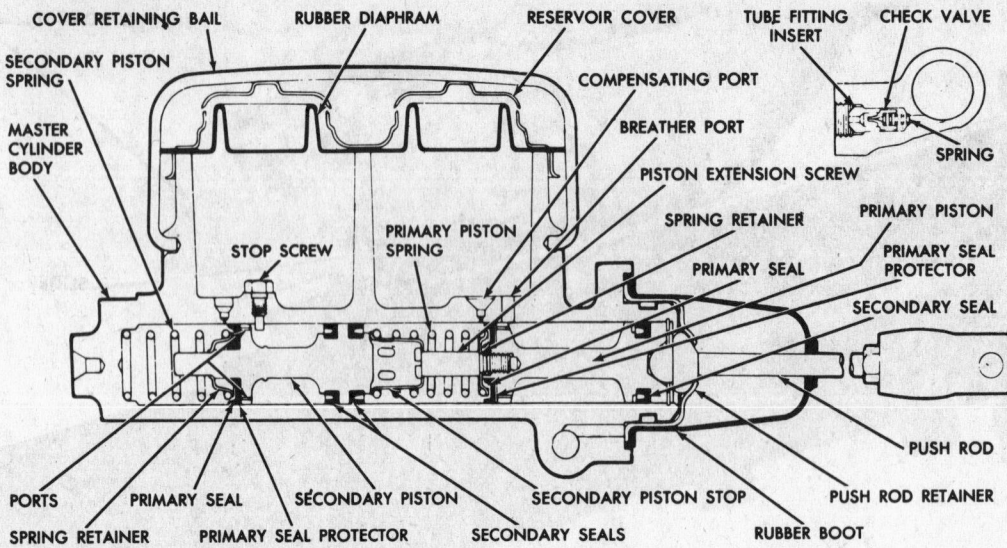

Master cylinder details (© Pontiac Div. G.M. Corp.)

Chassis and Brake Specifications table of this section.

Information on brake adjustments, band replacement, bleeding procedure, master and wheel cylinder overhaul can be found in the Unit Repair Section.

Information on troubleshooting and overhauling power brakes can be found in the Unit Repair Section.

Information on the grease seals which may need replacement can also be found in the Unit Repair Section.

Standard brakes are of the duo-servo, self-adjusting type. The self-adjusting feature operates only when the brakes are applied with car moving in reverse. When the brakes are applied, friction between the primary shoe and the drum causes the primary shoe to bear against the anchor pin. Hydraulic pressure in the wheel cylinder forces the upper end of the secondary shoe away from the anchor pin. As this moves, the upper end of the adjuster lever is prevented from moving by the actuating link attached to the anchor pin, thus forcing the lower end of the lever against the star wheel. If the linings are worn, the adjuster lever moves the star wheel a predetermined distance to maintain the proper shoe-to-drum clearance.

Metallic brake linings, used on some early high-performance models, never should be installed on cars equipped with standard brake drums unless the drums are radius ground and honed to a special finish. See the Unit Repair Section of this manual for more information on metallic brake linings.

Disc Brakes

From 1967, single-piston, sliding-caliper disc brakes have been available as optional equipment on most models (standard with high performance packages). These brakes have a vented, cast-iron rotor with two braking surfaces.

Disc brakes need no adjustment because, during operation, the application and release of hydraulic pressure causes the piston and caliper to move only slightly. In the released position, the pads do not move very far from the rotor; thus, as pads wear down, the piston simply moves farther out of the caliper bore and the caliper repositions itself on its mounting bolts to maintain proper pad-to-rotor clearance.

A metering valve in the front brake circuit prevents the discs from operating until about 75 psi exists in the system. This enables the rear drum brakes to operate in synchronization with the front discs and reduces the possibility of unequal brake application and premature lock-up. A proportioning valve in the rear brake circuit of some models accomplishes the same. Disc brake pads should be examined for wear every 12,000 miles. See the Unit Repair Section of this manual for service procedures.

Master Cylinder Removal

1964-70

The master cylinder is located in the engine compartment just above the steering column.

Since 1964, the master cylinder has been a sealed unit. A sealing diaphragm covers the reservoir, hermetically sealing the system from contamination.

From under the dash, disconnect the brake pedal from the master cylinder. From under the hood, disconnect the hydraulic line and the stoplight wire.

Remove the bolts which hold the master cylinder to the cowl panel and lift off the master cylinder.

The unit is installed in reverse order of removal.

Parking Brake Adjustment

1. Jack up both rear wheels.
2. Apply parking brake, five notches from full release.
3. Loosen equalizer rear locknut. Adjust forward nut until a light to moderate drag is felt when rear wheels are rotated.
4. Tighten the locknut.
5. Fully release the parking brake and rotate rear wheels; no drag should be felt.

Parking Brake Lever Removal

1964-70

A foot-operated parking brake is used on these models.

Remove the clevis which holds the cable to the foot-brake lever, then disconnect the bolts that hold the lever assembly to the bracket. Let the assembly come down sufficiently to get the bolts which hold the cable conduit to the bracket assembly.

Disconnect the release lever at the dash and move the entire assembly out from under the vehicle.

Parking Brake Cable

All Models

Follow the instructions given for hand- or foot-lever to disconnect the cable from the end of brake lever, then carefully thread the cable out over its pulleys and disconnect it at the handbrake end.

It is always an excellent idea to tie a string or cord to the cable so that the route of the old cable can be followed when replacing it with a new one.

Fuel System

Data on capacity of the gas tank will be found in the Capacities table. Data on correct engine idle speed and

fuel pump pressure will be found in the Tune-up Specifications table. Both the above tables can be found at the start of their sections.

Since 1966, a disposable-type fuel pump has been used. When a fuel pump failure is noted, a pump assembly replacement is in order.

Information covering operation and troubles of the fuel gauge will be found in the Unit Repair Section.

Detailed information on the carburetor and how to adjust it will be found in the Unit Repair Section under the specific heading of the make of carburetor being used on the engine being worked on. Carter, Holley, Rochester and Stromberg carburetors are covered.

Dashpot adjustments can be found in the Unit Repair Section under the same heading as that of the automatic transmission used in the car, as well as under the make and model carburetor.

Fuel Pump Removal

1964-70

1. From the left front side of the engine, disconnect the input and output line from the fuel pump.
2. Remove the bolts which hold the fuel pump to the timing case cover and lift off the pump.

NOTE: on models equipped with power steering it is possible, but somewhat difficult, to reach the

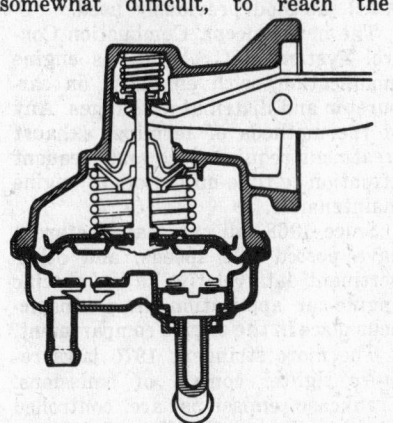

Fuel pump—1966-70
(© Pontiac Div. G.M. Corp.)

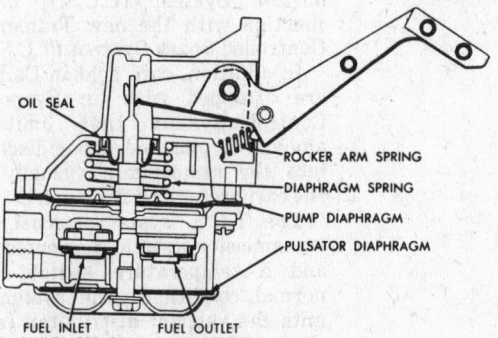

OIL SEAL
ROCKER ARM SPRING
DIAPHRAGM SPRING
PUMP DIAPHRAGM
PULSATOR DIAPHRAGM
FUEL INLET VALVE ASSEMBLY
FUEL OUTLET VALVE ASSEMBLY

Fuel pump—1964-65 (© Pontiac Div. G.M. Corp.)

mounting bolts with the steering pump in place. It may pay to slack off on the power steering pump, remove its mounting bolts and, with it still connected to its lines, lift it up out of the way.

Exhaust System

Exhaust Manifold Removal

Tab locks are used on front and rear pairs of bolts on each exhaust manifold. When removing bolts, straighten tabs from beneath car using long handled screw driver. When installing tab locks, bend tabs against sides of bolt not over top of bolt.

Left-Side Manifold

1. If the car is equipped with power steering, disconnect the power steering pump but leave it attached to its hoses and pull it up out of the way.
2. From underneath the vehicle, disconnect the exhaust crossover pipe flange.
3. If the car is equipped with power brakes, the rear bolts of the manifold are difficult to reach but they can be removed with a box wrench.
4. Remove the bolts that hold the manifold to the left cylinder head and take off the manifold.

Right-Side Manifold

From underneath the vehicle, disconnect the upper flange from the right manifold. This is the upper flange where the cross manifold, exhaust pipe and right manifold join.

From underneath the vehicle, remove the bolts that hold the manifold to the head on the back two flanges. The front flange can be removed from the top of the car with a box wrench.

Intake Manifold Removal

All Models

1. Drain coolant from petcocks on

radiator and on each side of block.

NOTE: most of the coolant can be drained from block through radiator drain by raising rear end of car approximately 15-18 in. off floor.

2. Remove air cleaner.
3. Remove water outlet fitting bolts and position fitting out of way, leaving radiator hose attached.
4. Disconnect heater hose from fitting.
5. Disconnect wire from thermogauge unit.
6. Remove spark plug wire brackets from manifold.
7. On cars equipped with power brakes, remove power brake vacuum pipe from carburetor.
8. Disconnect distributor to carburetor vacuum hoses.
9. Disconnect fuel line connecting carburetor and fuel pump.
10. Disconnect crank case vent hose from intake manifold.
11. Disconnect throttle rod from carburetor.
12. Remove screws retaining throttle control bracket assembly.
13. Remove intake manifold retaining bolts and nuts, and remove manifold and gaskets. Make sure that O-ring seal between intake manifold and timing chain cover is retained and installed during assembly.
14. Reverse procedure to install. Use plastic gasket retainers in figure to prevent manifold gaskets from slipping out of place.

Exhaust Crossover Pipe

The flange bolts that hold the exhaust crossover pipe to the manifolds are accessible from underneath the car.

Engine

References

The Specifications tables of this section list all the available facts about the engines.

Engine crankcase capacities are listed in the Capacities table.

Approved torque wrench readings and head bolt tightening sequences are covered in the Torque Specifications table, and information on the engine marking code will be found in

PLASTIC GASKET RETAINERS
GASKET

Plastic manifold retainers
(© Pontiac Div. G.M. Corp.)

the Model Year Identification table.

1964-70

Only one basic V8 engine has powered Pontiacs since 1955. It is a conventional V8 with wedged shaped combustion chambers, and a very compact block for its displacement. When it was first introduced, its stud mounted rocker arms and manifold-valley cover unit were very unusual, but today these features have been so widely adopted as to be quite conventional.

At the start of the period, the engine was used in 389 and 421 cu. in. displacements. The 389 version had a 4 1/6 in. bore and a 3¾ in. stroke. The 421 had a 4 3/32 in. bore and a 4 in. stroke as well as ¼ in. larger main bearing journals. Both engines have used four bolt main caps for their most powerful versions.

In 1967, Pontiac increased the bore of both V8s to 4.120 in., increasing the 389 to 400 cu. in. and the 421 to 428 cu. in. At the same time, Pontiac introduced a new cylinder head with larger ports and valves and larger

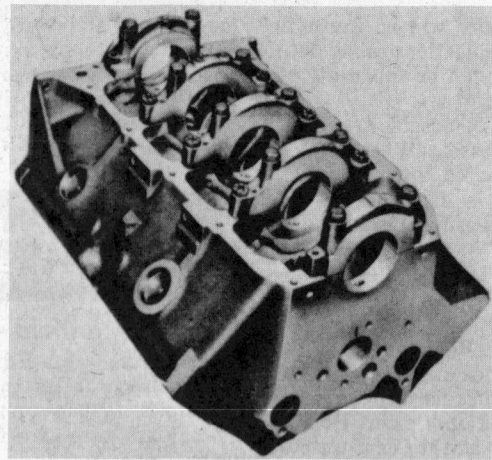

Heavy duty block with four bolt main caps
(© Pontiac Div. G.M. Corp.)

combustion chambers than used previously.

In 1970 Pontiac increased the displacement of its largest engine again by bringing the bore and stroke of its 428 engine out to 4.15 in. by 4.21 in. for 455 cu. in., and they introduced a small bore 350 cu. in. version of their

400 V8 for their base Catalina engine.

Exhaust Emission Control Systems

In compliance with anti-pollution laws involving all of the continental United States, the General Motors Corporation has elected to adopt a special system of terminal exhaust treatment. This plan supersedes (in most cases) the method used to conform to 1966-67 California laws. The new system cancels out (except with stick shift and special purpose engine applications) the use of the A.I.R. method previously used.

The new concept, Combustion Control System (C.C.S.) utilizes engine modification, with emphasis on carburetor and distributor changes. Any of the methods of terminal exhaust treatment require close and frequent attention to tune-up factors of engine maintenance.

Since 1968, all car manufacturers have posted idle speeds, and other pertinent data relative to the specific engine-car application, in a conspicuous place in the engine compartment.

The more stringent 1970 laws require tighter control of emissions. Crankcase emissions are controlled by the Closed Positive Crankcase Ventilation System, and exhaust emissions by the engine Controlled Combustion System (C.C.S.), in conjunction with the new Transmission Controlled Spark System (T.C.S.).

In addition, cars sold in California are equipped with an Evaporation Control System that limits the amount of gasoline vapor discharged into the atmosphere (usually from the carburetor and fuel tank).

The T.C.S. system consists of a transmission switch, a solenoid valve, and a temperature switch. Under normal conditions, the system permits the vacuum distributor (spark) advance to operate only in high gear (both manual and automatic transmissions). When the engine tempera-

Typical Pontiac V8 (© Pontiac Div. G.M. Corp.)

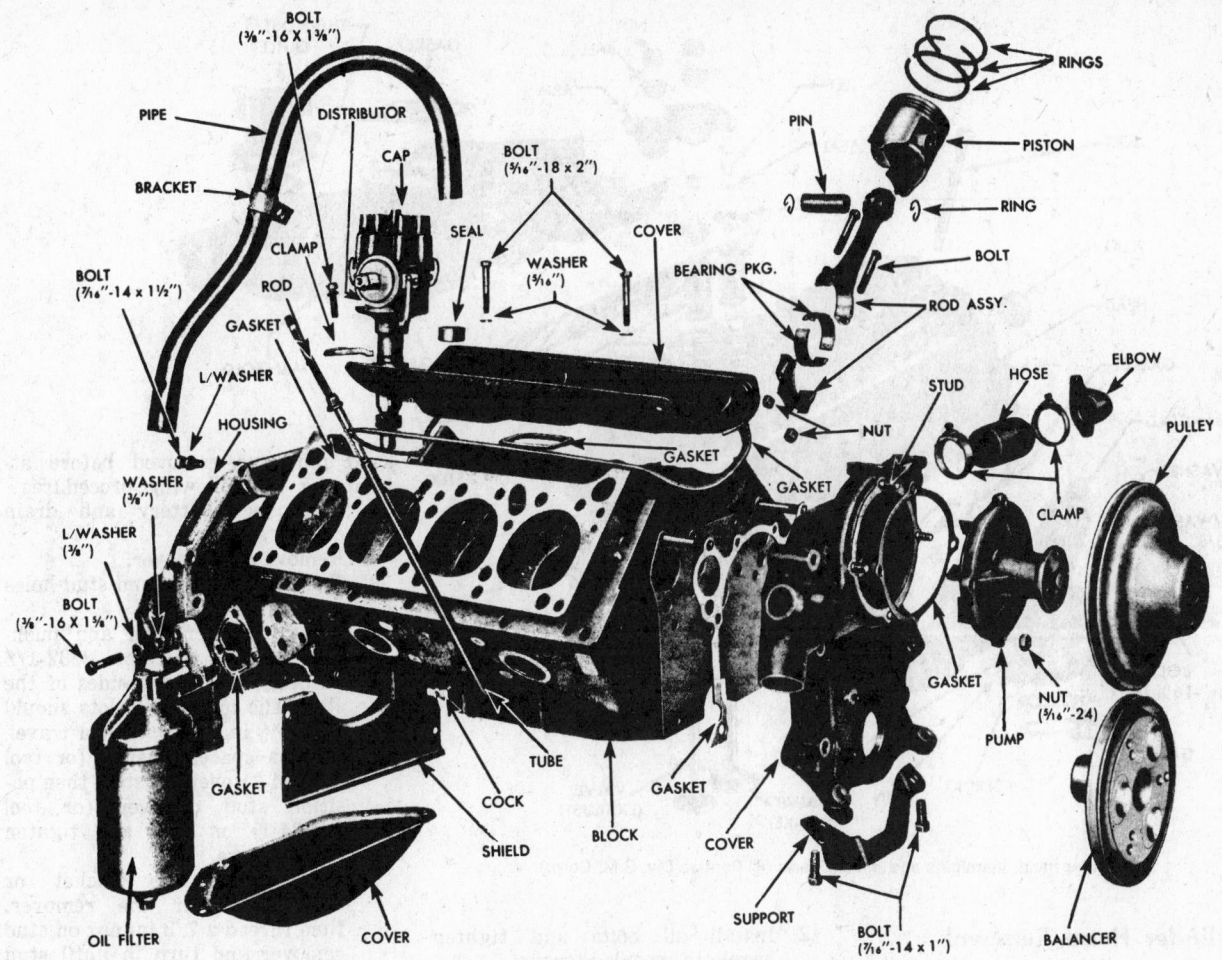

Cylinder block and related external parts (© Pontiac Div. G.M. Corp.)

ture is below 85° F., or above 220° F., however, the system allows the vacuum advance to operate normally.

For details, consult the Unit Repair Section.

Engine Removal

The engines on all Pontiac models are removed in practically the same manner. On cars equipped with automatic transmissions, the engine and transmission assembly are removed as a unit.

1. Drain the water, the engine oil and the transmission oil.
2. Remove the hood, battery, air cleaner, radiator core, power steering pump and carburetor. As a safety measure, remove the generator and starter.
3. Disconnect oil pressure, heat indicator and vacuum lines. Remove the engine side aprons and crankcase ventilator outlet pipe.
4. On cars with power steering, remove pump drive belt and pump with bracket. Do not disconnect hoses. Locate pump, with hoses out of area, to prevent damage as engine is lifted.
5. Disconnect the exhaust pipes from the flanges.
6. Remove all gearshift linkage

from the transmission and disconnect the clutch linkage at the clutch fork.
7. On models with clutch control countershaft, remove the bracket from the flywheel housing.
8. Split the rear universal joint and slide the driveshaft off the back of the transmission.
9. Place the lifting device on the engine and take a slight load on the device.
10. Remove the screws holding the front and rear insulators to their crossmembers. On models with automatic transmissions, place a jack under the transmission, take a load on the jack and remove the bolts that hold the rear support crossmember to the frame, then take down the crossmember.

Caution Remember that cars with automatics have a heavy transmission which will tend to tilt the engine towards the back. The transmission and jack should be left in place so that, as the engine is threaded forward, the transmission jack will roll, still carrying the

weight of the transmission upward until it is lifted clear by the lifting device.

Harmonic Balancer Removal

Drain cooling system. Remove the radiator assembly. Remove fan belt and position fan with wide angle to clear balancer. The harmonic balancer can then be removed from the crankshaft by the use of a special puller.

Cylinder Head

Rocker Arm Removal

1. Remove the four bolts that hold the rocker cover to the cylinder head and lift off the rocker cover.

NOTE: it is a good idea to disconnect the spark plug wires and get them out of the way, although this is not absolutely necessary.

2. On the left side rocker cover, the job is somewhat simplified by taking off the air cleaner.
3. Each rocker is held to a pressed-in stud in the cylinder head and is removed separately.
4. When reinstalling, simply run the rocker pivot nut down until it bottoms, then tighten it to 15-25 ft. lbs.

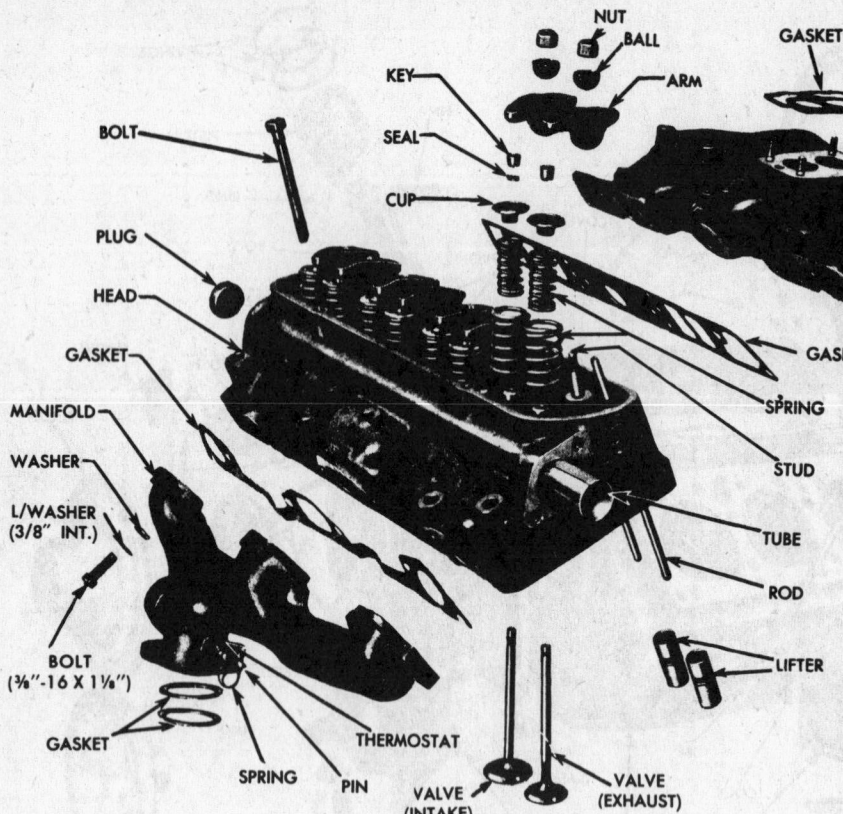

Cylinder head, manifolds and related parts (© Pontiac Div. G.M. Corp.)

Cylinder Head Removal

1. Remove intake manifold, valley cover, and rocker arm cover.
2. Loosen all rocker arm retaining nuts and pivot rockers off pushrods.
3. Remove pushrods and place in order.
4. Remove exhaust pipe flange bolts.
5. Remove battery ground strap and engine ground strap on left head; engine ground strap and automatic transmission oil filler tube bracket on right head.
6. Remove cylinder head bolts and head, with exhaust manifold attached.

NOTE: left head must be maneuvered to clear power steering and power brake units.

NOTE: on air-conditioned Firebird models, the right motor mount-to-frame bolt must be removed and the engine jacked up about 2 in. to gain access to the right rear rocker arm cover bolt and cylinder head bolt.

Cylinder Head Installation

1. Check head surface for straightness, then place a new head gasket on block.

NOTE: bolts are of three different lengths on all V8s. When bolts are properly installed, they will project an equal distance from head.

2. Install all bolts and tighten evenly to specified torque.
3. Install pushrods in original positions.
4. Position rocker arms over pushrods and tighten ball retaining nuts to 20 ft. lbs. (except Ram Air IV engines; see special procedure).
5. Replace rocker arm cover.
6. Replace valley cover.
7. Replace ground straps, oil filler tube bracket, intake manifold.
8. Install exhaust pipe flange nuts.

NOTE: most left and right cylinder heads are interchangeable within a single year. Large and small valve heads should not be used on the same engine, nor should 1964 389 cu. in. heads (standard) be interchanged with 1964 GTO heads.

Stud Removal

Pressed-in Studs

Caution This procedure can be used *only* on engines with pressed-in rocker studs. GTO and some special high performance engines have screwed-in rocker studs which are easily identified by their hex head lower portion. Another common stud-securing procedure on standard engines is "pinning" pressed-in studs by drilling through the stud boss and stud and inserting an interference-fit roll pin. Make sure any

such pins are removed before attempting the following procedure.

1. Disconnect battery and drain cooling system.
2. Remove rocker cover.
3. Pack oily rags around stud holes and engine openings.
4. Remove rocker arm and pushrod, then file two slots 3/32-1/8 in. deep on opposite sides of the stud. The top of the slots should be 1/4-3/8 in. below thread travel.
5. Place a spacer washer (or tool J-8934-3) over the stud, then position stud remover (or tool J-8934-1) on stud and tighten Allen screws.
6. Place a spacer (socket or J-8934-2) over the remover, then thread a 7/8 in. nut on stud remover and turn in until stud pulls from head.
7. If an oversize stud is to be used (0.003 in. oversize studs are available), ream stud hole to the proper size, then clean chips from area.
8. To install, coat press-pit area of stud with axle lube, then press or hammer into place. If possible use Pontiac Tool J-233 42. If tool is not available chill stud in dry ice to facilitate installation. Take care as dry ice makes studs very brittle.

Screwed-In Rocker Studs

1. Remove rocker cover.
2. Remove rocker arm and nut.
3. Remove stud, using a deep socket.
4. Install new stud, tightening to 50 ft. lbs.

Valve System

Cylinder Head Disassembly

1. Remove cylinder heads, as previously described.
2. Compress valve springs, using valve spring compressor.
3. Remove valve locks or keys.
4. Release valve springs.
5. Remove valve springs, retainers, oil seals, and valves.

NOTE: if a valve does not slide out of the guide easily, check end of stem for mushrooming or heading over. If head is mushroomed, file off excess material, remove and discard valve. If valve is not mushroomed, lubricate stem, remove valve and check guide for galling.

Valve Guides

Pontiac engines have integral valve guides. Pontiac offers valves with oversize stems for worn guides (0.001, 0.003 and 0.005 in. being available for most engines). To fit these, enlarge valve guide bores with valve guide reamers to an oversize that cleans up wear. If a large oversize is required, it is best to approach that size in stages by using a series of reamers of increasing diameter. This helps to maintain the concentricity of the guide bores with the valve seats. The correct valve stem to guide clearance is given in the Valve Specifications table at the beginning of this section.

As an alternate procedure, some local automotive machine shops fit replacement guides that use standard stem valves.

Hydraulic Valve Lifter Disassembly

Disassemble lifters for cleaning only; no repairs are permitted.
1. Grasp lock ring with needle nose pliers and remove. (Depress plunger to gain clearance.)
2. Remove pushrod cup, metering valve disc, and upper metering disc (if any). Do not bend metering disc.
3. Remove plunger assembly and plunger spring.
4. Remove spring, check valve retainer and check valve from plunger.
5. Clean all parts in solvent (lacquer thinner) and reassemble.
NOTE: internal parts are *not* interchangeable between lifters.

Valve Springs

All Models

In order to check on the condition of the valve springs, lay all of the springs on a flat surface and carefully measure across the top of the intake springs with a straightedge to see that all are the same height. If all are the same height, it may safely be assumed that they are all in good condition because it is very unlikely that they will all collapse an equal amount. Do the same with exhaust valve springs.

If one or more are found to be a different height from the rest of the springs, it is a good idea to get one new spring and carefully measure all the old springs against the new one. Those which come up to the same

height as the new spring may be considered to be in good condition. Those which do not should be replaced.

Where regular spring testing equipment is not available, this is generally considered to be a good, safe way to check the condition of the valve springs.

⏻ CHILTON TIME-SAVER

Intake or Exhaust Valve Stem Oil Seal or Spring Replacement (Without Removing Cylinder Head)

1. Remove rocker arm cover and any other interfering item.
2. Remove the spark plug from the affected cylinder.
3. Remove distributor cap and crank the engine to the firing position of the affected cylinder.
4. Remove the rocker arm, lubricate the rocker arm stud and attach spring compressor to the stud and compress the valve spring.
5. Some type of valve holder must be used through the spark plug hole.

NOTE: it will require the use of special tools, peculiar to the make of car being serviced. In the case of Pontiac:

Compressor stud J-6384-2
Compressor stud nut ..J-6348-2
Compressor J-6384-1
Valve Holder J-5961-2

With some study of our transverse engine cross-section illustration and a view of the cylinder head and rocker levers with the above tools attached, the resourceful mechanic can improvise in the event of lack of tools.

Valve Adjustment

The valve train is designed to operate through hydraulic action, at zero clearance. To maintain the built-in mechanical limits necessary for correct hydraulic valve train per-

formance, certain procedures must be followed every time an operation is performed on any part of the valve train (valves, pushrods, rockers, rocker studs, etc.).

If valves have been ground or renewed, check and correct the length of the valve stem, using valve train gauge J-8928 as follows:
1. Position rocker arm on pedestal stud and hold in place with stud installer J-8927.
2. Slip valve into place and hold it against valve seat.
3. While holding valve and rocker arm in this position, insert valve train gauge J-8928 through pushrod hole and seat snugly in pushrod seat of rocker arm.
4. With all parts seated, the step end of the gauge should be at least flush with the gasket face of the cylinder head, but should not project past the step on the gauge.
5. If step projects too far, the valve stem is too long and should be shortened by grinding.

Caution

Overheating the valve stem tip by careless grinding can soften the tempered tip, resulting in premature wear.
6. Remove gauging tools and install rocker arms, pushrods, and rocker arm retaining ball nuts.
7. Torque the nuts.
NOTE: it is very important that the rocker arm ball nuts be torqued to 15-25 ft. lbs.

Valve spring compressed
(© Pontiac Div. G.M. Corp.)

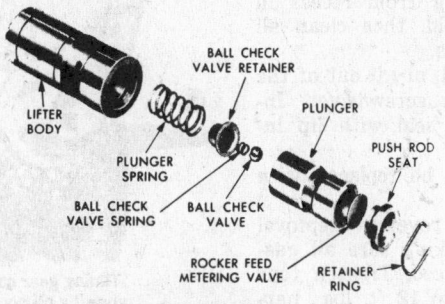

Hydraulic lifter
(© Pontiac Div. G.M. Corp.)

Hydraulic Valve Lifter Removal

Remove the rocker cover and the intake manifold, then take off the pushrod cover. Loosen the rocker arm ball nut and lift the rocker arm off the pushrod. The pushrods can be pulled up through the cylinder head and the lifters can then be pulled up out of their bores. The lifters must be returned to the bores from which they were taken.

Caution

The hydraulic lifter is a complete assembly, matchmated at the factory, and the bore of one lifter positively cannot be used in the body of another. These parts should not be mixed.

NOTE: the rear-most pushrod on the left bank cannot be lifted out on cars equipped with a defroster unit. However, the pushrod can be lifted up far enough to permit removal of the hydraulic lifter.

Hydraulic Lifter Initial Adjustment

Tighten down on the rocker adjusting nut until it bottoms, then tighten it to 15-25 ft. lbs.

There is a shoulder on the mounting stud which is precision-machined to give the correct operating position when this operation is performed.

Timing Case

Timing Case Cover Removal
Seal Replacement

1. Drain radiator and cylinder block.
2. Loosen alternator adjusting bolts.
3. Remove fan, fan pulley, and accessary drive belts.
4. Disconnect radiator hoses.
5. Remove fuel pump.
6. Remove harmonic balancer bolt and washer.
7. Remove harmonic balancer.
   NOTE: do not pry on rubber-mounted balancers.
8. Remove front four oil pan to timing cover bolts.
9. Remove timing cover bolts and nuts and cover to intake manifold bolt.
10. Pull cover forward and remove.
11. Remove O-ring from recess in intake manifold, then clean all gasket surfaces.
12. To replace seal, pry it out of the cover using a screwdriver. Install the new seal with lip inwards.
    NOTE: seal can be replaced with cover installed.
13. To install, reverse removal procedure, making sure all gaskets are replaced. Tighten four oil pan bolts to 12 ft. lbs., harmonic balancer bolt to 160 ft.

lbs., and fan pulley bolts to 20 ft. lbs.

Timing Chain and Sprocket Removal

1. Remove the radiator core, water pump and the vibration damper.
2. Support the front of the engine with a jack protected by a wood block and remove the front engine support retaining bolts. Take off the engine support.
3. Remove the oil pan front screws and the timing cover bar. Now, unbolt and remove the timing case cover.
4. Turn the crank and camshaft (if the chain is broken) until the two timing marks are in line between the shaft centers.
5. Remove the nut from the front of the camshaft sprocket.
6. Using a puller, draw the sprocket off the front of the camshaft.

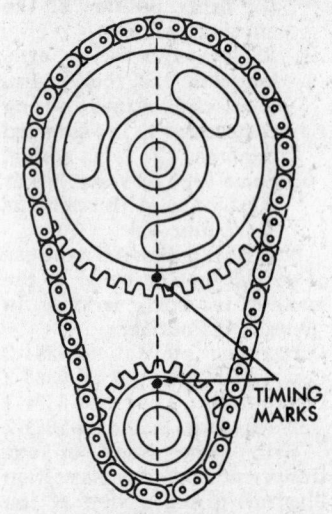

TIMING MARKS

Valve timing marks

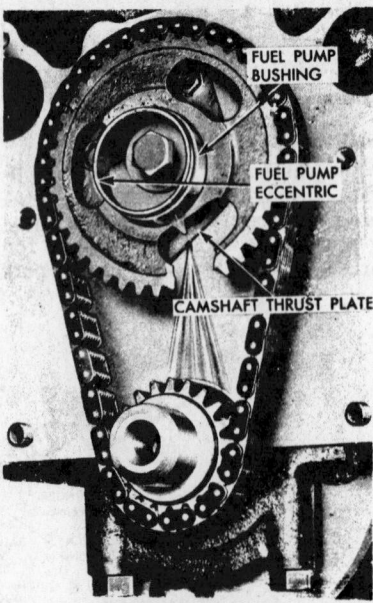

FUEL PUMP BUSHING

FUEL PUMP ECCENTRIC

CAMSHAFT THRUST PLATE

Timing gear assembly, showing oil squirt holes
(© Pontiac Div. G.M. Corp.)

7. Arrange the new chain on the sprocket and set the sprocket up over the camshaft by looping the chain over the crank sprocket. That way, when the cam sprocket engages its key, the timing marks are nearest each other and in-line between the shaft centers.
8. Secure the camshaft sprocket in this position.
NOTE: when reassembling the timing case cover, extra care should be taken to make sure that the oil seal between the bottom of the timing case cover and the front of the oil pan is still a good one. Plenty of gasket cement should be used, at this point, to prevent oil leaks.

1964-70 V8

1. Drain cooling system and remove air cleaner.
2. Disconnect all water hoses, vacuum lines and spark plug wires.
3. Disconnect accelerator linkage, temperature gauge wire, and fuel lines.
4. Remove hood latch brace.
5. Remove PCV hose, then remove rocker covers.
   NOTE: on air-conditioned models, remove alternator and bracket.
6. Remove distributor, then remove intake manifold.
7. Remove valley cover.
8. Loosen rocker arm nuts and pivot rockers out of the way.
9. Remove pushrods and lifters (keep them in proper order).
10. Remove harmonic balancer, fuel pump, and four oil pan to timing cover bolts.
11. Remove timing cover and gasket, then remove fuel pump eccentric and bushing.
12. Align timing marks, then remove timing chain and sprockets.
13. Remove camshaft thrust plate.
14. Remove camshaft by pulling straight forward, being careful not to damage cam bearings in the process.
    NOTE: it may be necessary to jack up the engine slightly to gain clearance, especially if motor mounts are worn.
15. Install new camshaft, with lobes and journals coated with heavy (SAE 50-60) oil, into the engine, being careful not to damage cam bearings.
    NOTE: most specialty cams come with a special "break-in" lubricant for the lobes and journals; if such lubricant is available, use it instead of heavy oil.
16. Install camshaft thrust plate and tighten bolts to 20 ft. lbs.
17. To install, reverse Steps 1-12, tightening sprocket bolts to 40 ft. lbs., timing cover bolts and nuts to 30 ft. lbs., oil pan bolts to 12 ft. lbs., and harmonic balancer bolt to 160 ft. lbs.

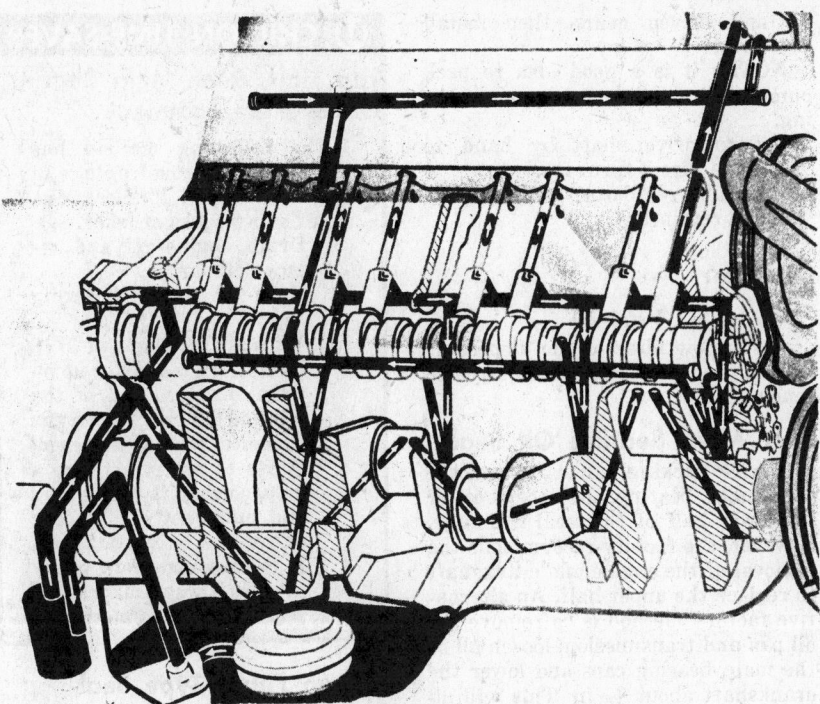

Engine lubrication (© Pontiac Div. G.M. Corp.)

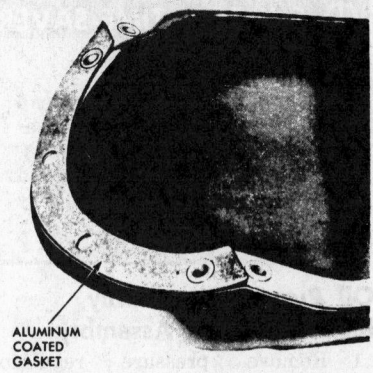

ALUMINUM
COATED
GASKET

Front oil pan gasket overlapping side gaskets
(© Pontiac Div. G.M. Corp.)

12. Remove oil pan bolts and baffle tray, if so equipped.
13. Support front of engine by chain hoist, hooking under front of intake manifold and raising as far as possible.
14. Lower oil pan as far as possible.
15. Remove front main bearing cap.
16. Turn crank shaft until No. 1 throw is up.
17. Hold crossover pipe and tie rod down and clear oil pan.
18. After cleaning and installing new gaskets, reverse the above.

NOTE: use plastic gasket retainers to hold pan gaskets in place while installing pan.

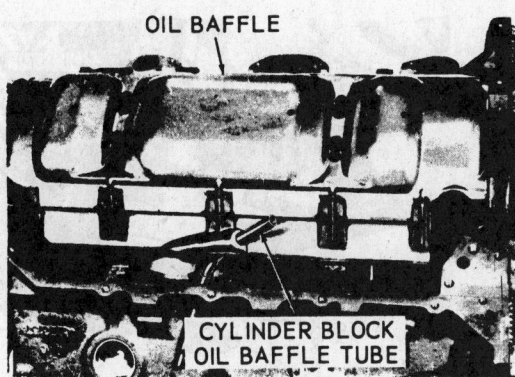

OIL BAFFLE

CYLINDER BLOCK
OIL BAFFLE TUBE

Oil baffle plate for high performance engines
(© Pontiac Div. G.M. Corp.)

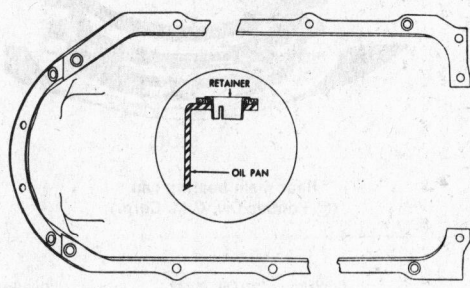

RETAINER

OIL PAN

Oil pan gasket retainers
(© Pontiac Div. G.M. Corp.)

Engine Lubrication

Oil Pan Removal

1964-70

1. Remove hood and air cleaner.
2. Drain oil pan and radiator.
3. Disconnect battery.
4. Disconnect radiator hose.
5. Remove fan guard and upper radiator support.
6. Remove coil mounting bolts to suspend coil out of way.
7. Remove exhaust pipe and crossover pipe bolts from manifolds.
8. Remove idler arm mounting bolts and lower arm and tie rods.
9. Remove front engine mount bolts.
10. Remove starter mounting bolts and lower starter.
11. Remove lower clutch housing cover.

CEMENT
GROOVE 1" to 1¼"
(BOTH SIDES)

Rear oil pan gasket positioned in bearing cap
(© Pontiac Div. G.M. Corp.)

Oil Pump Disassembly, Inspection and Assembly

1. Remove pressure regulator spring.
2. Remove cover hold-down bolts and cover.
3. Remove driven gear and drive gear, then remove drive shaft.
4. Clean all parts in solvent, especially pump pickup screen.

CAUTION: do not remove or loosen oil pickup tube or screen.

5. Inspect regulator spring for distortion, wear, or cracks.
6. Inspect regulator ball for damage.
7. Inspect pump components for wear.
8. To assemble, first install drive and driven gears, then install cover.

NOTE: it is a good idea to pack pump with Vaseline for proper priming.

9. Turn drive shaft by hand to make sure it turns freely.
10. Install regulator ball, spring, and retainer.

Oil Filter

1964-70

A full-flow filter of the disposable type is used. No tools are required to renew.

Rear Main Bearing Oil Seal

A wick packing is used to seal the rear main bearing. Replacement of the lower half of this seal is simple, however, the factory recommends the removal of the engine and crankshaft to replace the upper half. An alternative factory method is to remove the oil pan and transmission, loosen all of the main bearing caps and lower the crankshaft about 3/8 in. This will allow the removal of the old upper half rear bearing seal and the installation of a new one.

Woven Fabric-Type Seal Installation

1. Obtain a 12 in. piece of copper wire (about the same gauge as

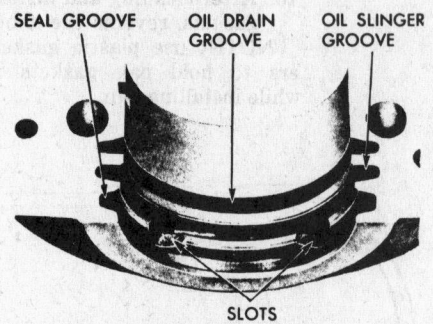

Rear main bearing cap
(© Pontiac Div. G.M. Corp.)

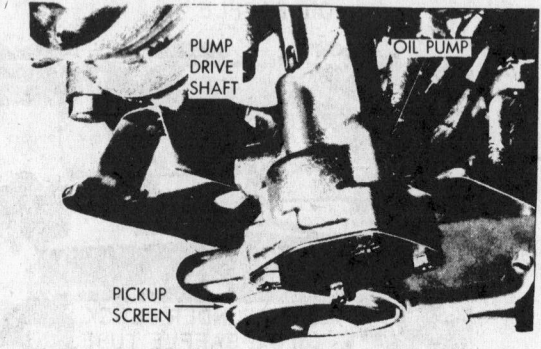

Oil pump and pump drive shaft
(© Pontiac Div. G.M. Corp.)

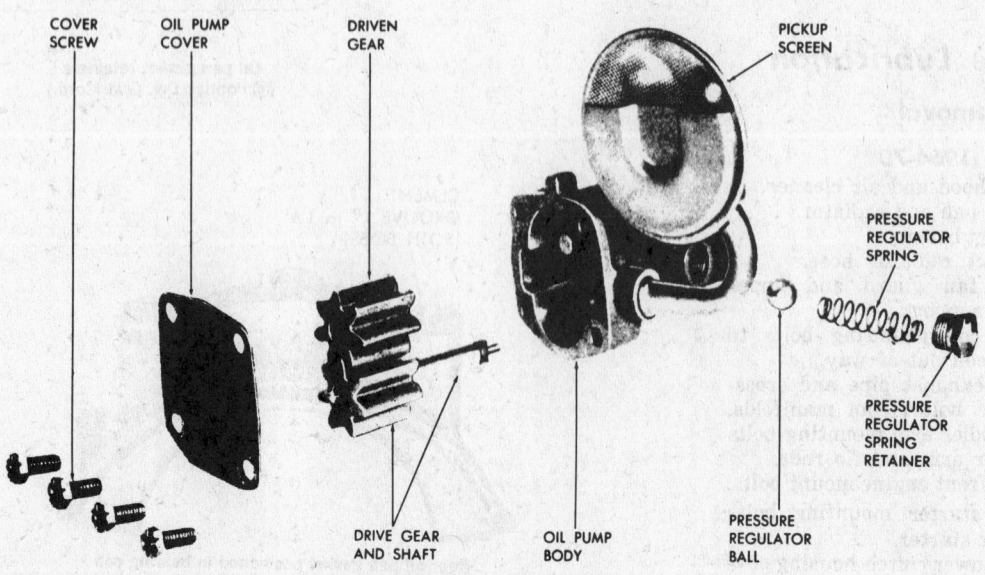

Oil pump—disassembled

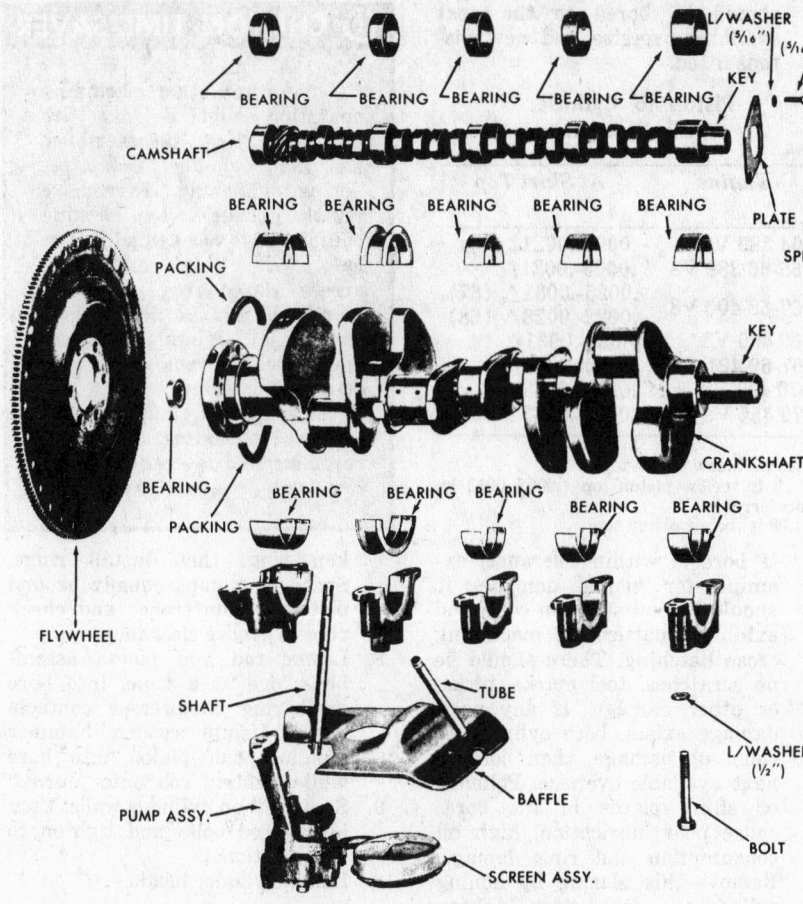

Camshaft, crankshaft and related parts (© Pontiac Div. G.M. Corp.)

that used in the strands of an insulated battery cable).

2. Thread one strand of this wire through the new seal, about ½ in. from the end, bend back and make secure.

3. Thoroughly saturate the new seal with engine oil.

4. Push the copper wire up through the oil seal groove until it comes down on the opposite side of the bearing.

5. Pull (with pliers) on the protruding copper wire while the crankshaft is being turned and the new seal is slowly fed into place.

Caution This snaking operation slightly reduces the diameter of the new seal and care will have to be used to keep the seal from slipping too far through the top half of the bearing.

6. When an equal amount of seal is extending from each side, cut off the copper wire close to the seal and tamp both ends of the seal up into the groove (this will tend to expand the seal again).

NOTE: don't worry about the copper wire left in the groove, it is too soft to cause damage.

7. Replace the seal in the cap in the usual way and replace the oil pan.

Crankshaft, Piston and Rod Assemblies

Slipper skirt tin-plated aluminum pistons with steel struts are used.

All three rings are located above the wrist pin. The letter F and the depression in the edge of the piston go to the front of the engine in all cases.

Main Bearing Replacement

1. Remove oil pan and, on V8 engines, oil baffle plate.

2. Remove one main bearing cap and lower bearing insert.

3. With a "roll out" pin, or a cotter pin bent as illustrated, inserted into the oil hole in the crankshaft, rotate crankshaft in normal direction of rotation (clockwise seen from front) to remove top bearing insert.

4. Oil the new upper bearing insert and place un-notched end of shell between crankshaft and notched upper bearing web.

5. With "roll out" pin positioned in oil hole, rotate new upper bearing insert into place.

6. Install new lower bearing insert into bearing cap, with a 0.002 in. strip of brass shim stock between insert and cap. Do not oil this bearing.

7. Place a strip of Plastigage (available from automotive parts jobbers) on the lower bearing insert, then install main cap, tightening bolts to specification.

NOTE: do not rotate crankshaft.

8. Remove bearing cap and, using the scale on the envelope, measure the width of the compressed Plastigage; this is the oil clearance for this main bearing.

9. If clearance is satisfactory, remove brass shim stock and install lower cap and new bearing; if clearance is unsatisfactory, undersize bearing inserts are available to correct.

NOTE: never file main caps to adjust clearance.

10. Replace the remaining main bearings in the engine by following Steps 2-9, then install oil pan.

NOTE: it may be necessary to remove oil pump to gain clearance.

Connecting Rod Bearing Replacement

1. Remove oil pan and, on V8 engines, oil baffle.

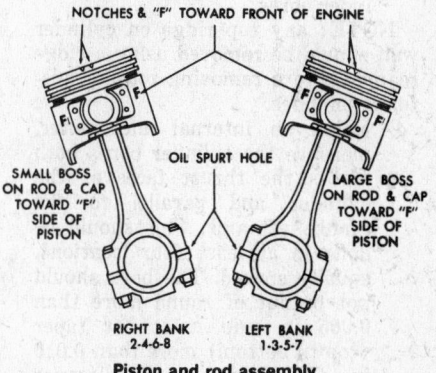

Piston and rod assembly

2. Remove oil pump.
3. Rotate crankshaft to bring bearing caps, in turn, into position for removal.
4. Remove bearing cap.
5. Push sections of rubber tubing over connecting rod bolts (to protect crankpin), then push rod and piston assembly up far enough to remove upper bearing insert.
6. Remove lower bearing shell from cap.
7. Measure crankpin for taper and wear, using a micrometer.

NOTE: some two-barrel, V8 engines have 0.010 in. undersize crankpins. These are identified by a .010 U.S. stamp on front of No. 1, and rear of No. 8, counterweight.

8. Check connecting rod bearing oil clearance, using Plastigage, in the same manner as for main bearings, then install new bearing inserts and tighten rod bolts to specified torque.

Rod and Piston Assembly Removal

1. Remove the head and oil pan and selecting pistons which are in the down position, cut the cylinder ring ridge from the top of the cylinder, using a good ring ridge reamer.

NOTE: if no reamer is available, the ridge can be taken off with a good sharp bearing scraper.

2. From underneath the car, mark the connecting rod cap and the rod itself on the camshaft side so that they can be replaced in the same cylinder and in the same position.
3. Remove the bolts and push the piston assemblies up out of the top of the block. Immediately replace the rod bearing cap on the rod so that the bearing shells do not get lost or the threads in the rod get damaged.

Piston Ring Replacement

1. Remove rod and piston assemblies.
2. Remove cylinder head/s.
3. Push connecting rod and piston assemblies up and out of the cylinder bores.

NOTE: any top ridge on cylinder wall should be removed using a ridge reamer before removing rod and piston assemblies.

4. Using an internal micrometer, measure the cylinder bores both across the thrust faces of the cylinder and parallel to the crankshaft axis. This should be done in at least four locations, equally spaced. The bore should not be out of round more than 0.005 in. and must not taper (top to bottom) more than 0.010 in. Cylinders out of tolerance

should be bored to the next available oversize and new pistons fitted.

Piston to Cylinder Wall Clearance

Engine	At Skirt Top
1964 389 V8	.0005-.0021‡
1965-66 389 V8	.0005-.0021‡
1967-68 400 V8	.0025-.0031△ (67)
	.0022-.0028△ (68)
1969 400 V8	.0025-.0031△
1967-69 421 V8	.0030-.0036△
1970 400 V8	.0025-.0033△
1970 455 V8	.0025-.0033◎

△ 1.11 in. below piston top.
‡ 1.18 in. below piston top (.0007-.0013 in. preferred).
◎ 1.08 in. below piston top

5. If bore is within tolerance, examine for visible damage. It should be dull silver in color and exhibit a pattern of machining cross hatching. There should be no scratches, tool marks, nicks, or other damage. If any such damage exists, bore cylinder to clean up damage, then bore to next available oversize. Polished or shiny places in the bores cause poor lubrication, high oil consumption and ring damage. Remove this glazing by honing cylinders with clean, sharp stones of No. 180-220 grit to obtain a surface finish of 15-35 RMS. Use a hone to correct piston clearance and surface finish any cylinder that has been bored.
6. If cylinder bore is in satisfactory condition, place each ring in bore, in turn, and square it in bore using a piston top. Measure the ring end gap with feeler gauges. If gap is greater than that specified, get new ring; if gap is less than that specified, file end of ring to bring within tolerance.
7. Clean ring grooves, using a commercial groove cleaner or a bro-

ken ring, then install rings. Space ring gaps equally around piston circumference and check ring to groove clearance.

8. Lower rod and piston assemblies, one at a time, into bore until ring compressor contacts block. Using wooden hammer handle, push piston into bore while guiding rod onto journal.
9. Squirt oil on cylinder walls, then install rod bolts and tighten to specification.
10. Install cylinder head.

Front Suspension

References

General instructions covering the front suspension, and how to repair and adjust it, together with information on installation of front wheel bearings and grease seals, are given in the Unit Repair Section.

Definitions of the points of steering geometry are covered in the Unit Repair Section.

Piston Ring Gaps

Year and Engine	Top Compression	Bottom Compression	Oil Control
1964-66 389 V8	0.016-0.026 in.	0.013-0.025 in.	0.015-0.055 in.
1964-66 421 V8	0.016-0.026 in.	0.013-0.025 in.	0.015-0.055 in.
1967-69 All V8s	0.010-0.030 in.	0.010-0.030 in.	0.015-0.055 in.
1970 400	0.010-0.030 in.	0.010-0.030 in.	0.015-0.055 in.
1970 455	0.010-0.030 in.	0.010-0.030 in.	0.015-0.055 in.

Ring Side Clearance

Year and Engine	Top Compression	Bottom Compression	Oil Control
1964-66 V8s	0.0015-0.0030 in.	0.0015-0.0035 in.	0.0005-0.0055 in.
1967-70 V8s	0.0015-0.0050 in.	0.0015-0.0050 in.	0.0015-0.0050 in.

Specifications covering the caster, camber, toe-in, kingpin inclination, and turning radius can be found in the Wheel Alignment table in this section.

Overall length and tire size figures can be found in the General Chassis and Brake Specifications table in this section.

Front Spring R & R

1964-68

1. Raise front end of car; support so that lower control arm hangs free.
2. Remove wheel and brake drum.
3. Remove brake backing plate from steering knuckle. (Do not disconnect brake line.)
4. Disconnect stabilizer from lower control arm.
5. Remove shock absorber, and install spring compressor J-7592-

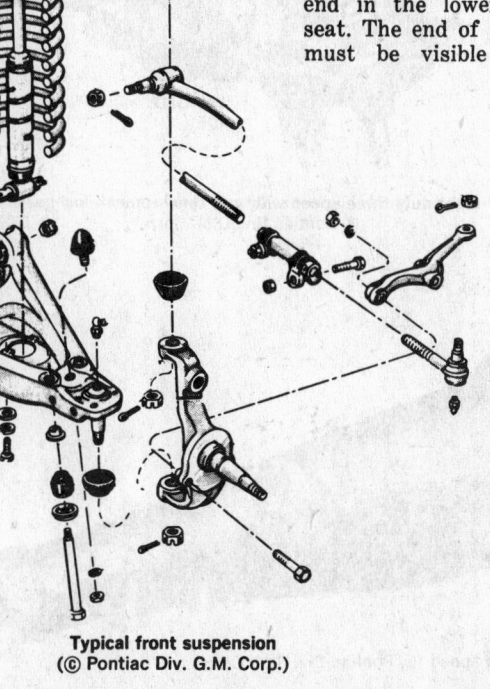

Typical front suspension
(© Pontiac Div. G.M. Corp.)

01 in the following manner:
A. Install one J-7592-7 cast plate in spring with boss down; be sure the angled center hole of the plate is aligned with axis of spring. Rotate the plate upward into highest position in coil.
B. Install another J-7592-7 plate under the third coil from the bottom with the boss up and center hole aligned with axis of spring. This plate should be slanted in the same direction and parallel with the upper plate.
C. Install long bolt up through both plates with thread end down. Install J-7592-4 retainer (cup up) and J-7592-2 locking clip through opening at upper shock bracket to hold bolt to upper plate.
D. Place J-7592-6 (ball up) thrust bearing, and J-7592-3 nut (threads down) on the bolt and screw up snug.
6. While holding upper end of rod, turn nut at lower end to compress spring.
7. Support lower control arm with a jack, then disconnect lower ball stud from steering knuckle with J-6627.
8. Support upper control arm and steering knuckle assembly by inserting a wood block between upper control arm and frame.
9. Carefully lower jack, allowing outer end of lower control arm to swing down until spring is free. Remove the spring.
10. Before replacing spring, assemble compressor J-7592 on the new spring.
11. Install spring by placing one end in the frame seat and the other end in the lower control arm seat. The end of the spring coil must be visible through the

drain hole in the lower control arm spring seat.
12. Install spring by reversing removal procedure.

1969-70

1. Jack up car and support on jack stands at frame side rails.
2. Remove shock absorber.
3. Disconnect stabilizer bar at lower control arm.
4. Support lower control arm with a hydraulic floor jack, then remove the two inner control arm to front crossmember bolts.
5. Carefully lower the control arm, allowing the spring to relax.
6. Reach in and remove spring.
   NOTE: this is probably the best all-around procedure and it can be used, with slight modification, for all models from 1964.

Jacking, Hoisting

Jack car at front spring seats of lower control arms and, at rear, at axle housing.

When using frame lift, use side rails at points shown on diagram. Be sure that adapters are properly supporting these designated areas.

Steering Gear

Manual Steering

References

Instructions covering the overhaul of the steering gear will be found in the Unit Repair Section.

Manual Steering Gear Removal and Installation

1964-70

1. Disconnect pitman arm.
2. Scribe position of steering shaft on worm shaft flange and disconnect lower flange from shaft.
3. Remove three gear-to-frame bolts and lift out gear.
To replace:
1. Align scribe marks at shaft and flange.

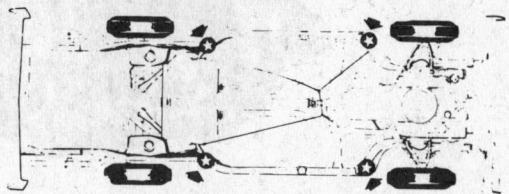

Hoist contact lifting points

2. Position gear assembly. Install three gear-to-frame bolts and tighten to 70-90 ft. lbs.
3. Install pitman arm. Tighten nut to 110-140 ft. lbs.
4. Install two flange nuts and lock washers and tighten to 10-20 ft. lbs.

NOTE: in connecting and aligning shaft and jacket, avoid metal to metal contacts to counteract noise transmission to driver.

Pitman Shaft Seal Replacement (With Steering Gear in Place in Car)

1. Disconnect pitman arm from pitman shaft. Clean end of pitman shaft and housing.
2. Tape the splines of the pitman shaft to keep them from cutting the seal. Use only one layer of tape. Too much tape will prevent passage of the seal.
3. Using lock ring pliers, remove the seal retaining ring.
4. Start the engine and turn the steering wheel to the right so that the oil pressure in the housing will force the seals out.
5. Catch the seal and the oil in a container. Turn off the engine when the two seals are out.
6. Inspect the two old seals for damage to the rubber covering on the outside diameter. If it seems scored or scratched, inspect the housing for burrs, etc., and remove them before installing the new seals.
7. Lubricate the two new seals with petroleum jelly. Install the one with single lip first, then, a washer. Drive seal in far enough to permit installation of double lip seal, washer and the seal retaining ring. The first seal is not supposed to bottom in its counterbore.
8. Fill reservoir to proper level, start engine, turn wheel to right and check for leaks.
9. Remove the tape and reinstall the pitman arm. Tighten nut to 90-110 ft. lbs.

Power Steering

References

Troubleshooting and repair instructions covering power steering gears are given in the Unit Repair Section.

Power Steering Pump Removal

1. Disconnect and cover the two hoses at the back of the pump. Remove the bolt from the tensioner bracket and take off the drive belt.
2. Tilt the pump in toward the engine and remove the bolts that hold its bracket to the cylinder head.

Power Gear Removal
1964-70

1. Scribe alignment mark on shaft and worm shaft flange, then remove two nuts and washers.
2. Disconnect oil lines from valve body.
3. Remove pitman arm.
4. Remove gear to frame bolts and lift out gear.
5. Reverse above procedure to replace. When installing mounting bolts, tighten finger-tight, only, until proper alignment is obtained.

Clutch

Clutch Pedal Adjustment

1. Disconnect return spring.
2. With pedal against stop, loosen locknut to allow adjusting rod to be turned out of swivel until the throwout bearing contacts the release fingers in the pressure plate.
3. Turn adjusting rod into swivel or pushrod 3½ turns; tighten locknut to 8-12 ft. lbs. for 1964-68, 30 ft. lbs. for 1969-70.
4. Install return spring and check pedal lash; it should be approximately 1 in.

Clutch Removal

Remove transmission, making sure that its weight is not allowed to rest on the hub of the clutch disc.

On all models, unhook the clutch pedal, pull back spring, and take out the clutch fork ball support, the clutch fork and the clutch throwout bearing. Be sure to mark the flywheel and clutch cover so that assembly can be made in the same relative position to preserve the clutch balance. A little at a time, loosen the bolts holding clutch to flywheel and remove clutch.

FLANGED SIDE

Light duty three speed with non synchromesh low gear
(© Pontiac Div. G.M. Corp.)

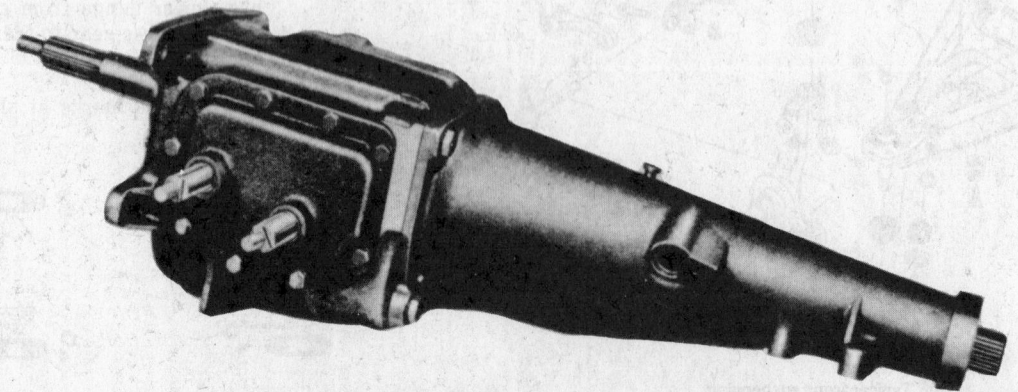

Heavy duty Borg Warner three speed (© Pontiac Div. G.M. Corp.)

Transmissions

Standard Transmission

Transmission Information

Transmission refill capacities will be found in the Capacities table of this section.

Step-by-step repair procedures are covered in the Unit Repair Section.

A general overdrive repair section is also included in the Unit Repair Section.

1964-70 full sized Pontiacs have used five different manual transmissions: 1. a light duty three speed with synchromesh on second and third only recognizable by its five bolt side cover, 2. a heavy duty Borg Warner three speed with synchromesh on second and third only, a nine bolt side cover, and a very long extension shaft, 3. a fully synchromesh, Dearborn, top cover, three speed built by Ford, 4. Borg Warner fully synchromesh four speed with a nine bolt side cover, and 5. a Muncie fully synchromesh four speed with a seven bolt side cover.

Transmission Removal

Light Duty Three Speed and Borg Warner Three Speed

1. Drain transmission if it is equipped with a drain plug.
2. Remove U-bolt nuts, lock plates, and U-bolts from rear axle drive pinion flange.
3. Use a suitable rubber band to hold bearing on to journals if tie wire has been removed to prevent loss of needle bearings when rear joint is disconnected.
4. Remove complete drive line assembly by sliding rearward to disengage yoke from splines on transmission main shaft.
5. Disconnect speedometer cable from speedometer driven gear.
6. Disconnect shift rods and remove lever and cross shaft assembly.

7. Support rear of engine with a floor jack.
8. Remove two transmission brackets to cross member retaining nuts.
9. Remove upper transmission to clutch housing bolts and insert transmission aligning studs.

NOTE: aligning studs are necessary since they support the transmission and prevent distortion of the clutch driven plate hub when the lower transmission bolts are removed.

10. Remove lower transmission to clutch housing bolts, tilt rear of extension upward to engage bracket studs from cross member support and withdraw transmission from clutch housing.
11. Tilt front downward and remove. Reverse procedure to install.

Dearborn Three Speed

1. Disconnect speedometer cable.
2. Disconnect shift control rods from transmission.
3. Remove propeller shaft.

4. Support rear of engine and remove transmission mount.
5. Remove four crossmember bolts and slide member rearward.
6. Remove two upper transmission to flywheel housing bolts and insert guide pins.
7. Remove two lower transmission to flywheel housing attaching bolts.
8. Slide transmission straight back on guide pins until main drive gear splines are free of splines in clutch friction plate.
9. Remove transmission.

Borg Warner Four Speed

1. Remove drain plug and drain transmission.
2. Remove six metal boot retainer attaching screws and slide boot over shift lever.
3. Disconnect speedometer cable from speedometer.
4. Disconnect back up light leads from back up light switch.
5. Disconnect shift control rods from shifter levers.
6. Remove U-bolt nuts, lock plates,

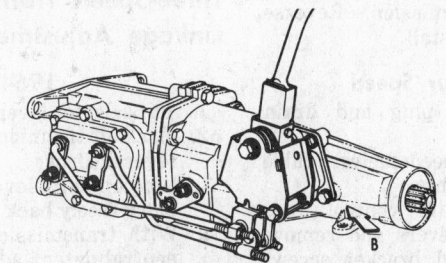

Muncie four speed transmission
(© Pontiac Div. G.M. Corp.)

Dearborn three speed fully synchromesh transmission (© Ford Motor Co.)

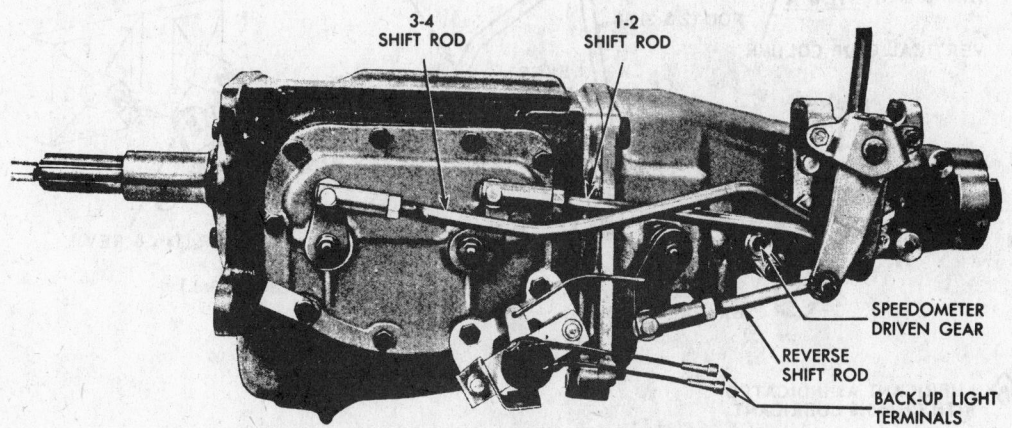

Borg Warner four speed transmission (© Pontiac Div. G.M. Corp.)

and U-bolts from rear axle drive pinion flange.

7. Use suitable rubber band to hold bearing on to journals if wire has been removed to prevent loss of needle bearings.

8. Remove complete drive line assembly by sliding rearward to disengage yoke from splines on main shaft.

9. Support rear of engine and remove two transmission extension bracket retaining nuts.

10. Remove two top transmission to clutch housing bolts and insert two transmission aligning studs.

11. Remove two lower transmission to clutch housing bolts.

12. Tilt rear of extension housing upward to disengage bracket studs from crossmember support and withdraw transmission from clutch housing.

NOTE: on long wheelbase Pontiacs (Star Chiefs and Bonnevilles), remove crossmember before removing transmission because of the additional length of the extension housing.

13. Remove transmission. Reverse procedure to install.

Muncie Four Speed

1. Remove drain plug and drain transmission.

2. Disconnect speedometer cable and back up lights.

3. Disconnect shift control rods from shifter levers and remove two levers and bracket screws and remove shift lever and bracket.

4. Remove U-bolt nuts, lock plates, and U-bolts from rear axle drive pinion flange.

5. Use rubber band to hold bearing onto journals if wire has been removed to prevent loss of needle bearings.

6. Remove complete drive shaft assembly by sliding rearward to disengage yoke from splines on transmission main shaft.

7. Support rear of engine and remove two transmission extension insulator to cross member support retaining bolts.

8. Remove two top transmission to clutch housing bolts and insert two transmission aligning studs.

9. Remove two lower transmission to clutch housing bolts.

10. Tilt rear of extension upward to disengage bracket studs from cross member support and withdraw transmission from clutch housing.

11. Remove transmission. Reverse procedure to install.

Three-Speed Transmission Linkage Adjustment

1964

1. Put selector lever in neutral.
2. Back off trunnion adjusting nuts several turns.
3. Line up shift levers so that they move freely back and forth.
4. With transmission levers in full neutral detent adjust second and third shifter rod trunnion at cross shaft lever and first and

reverse trunnion at transmission lever. Tighten trunnion nuts to 60-120 in. lbs.

5. Move selector lever to first gear and check key at first and reverse shift lever. Key should just clear lower side of opening in steering column.

6. Move selector lever to third gear and check key at second and third shift lever. Key should just clear lower side of opening in steering column.

7. Apply wheel bearing grease to all gear shift linkage joints.

8. Check complete shift pattern movement with engine off, start engine and perform shift pattern.

1967-69 Column Shift

1. Align upper and lower gear shift levers on steering column assembly in neutral position by inserting gauge pin in hole as shown in figure.

2. Loosen clamp screws at transmission gear shift control rods.

3. Position levers on transmission in neutral.

4. Torque clamp screws to 20 ft. lbs. and check shift pattern.

1967-69 Floor or Console Shift

1. Put selector lever in neutral.

2. Loosen trunnion jam nuts on transmission gear shift control rods.

3. Place transmission lever and bracket assembly in neutral and install gauge pin as shown in figure.

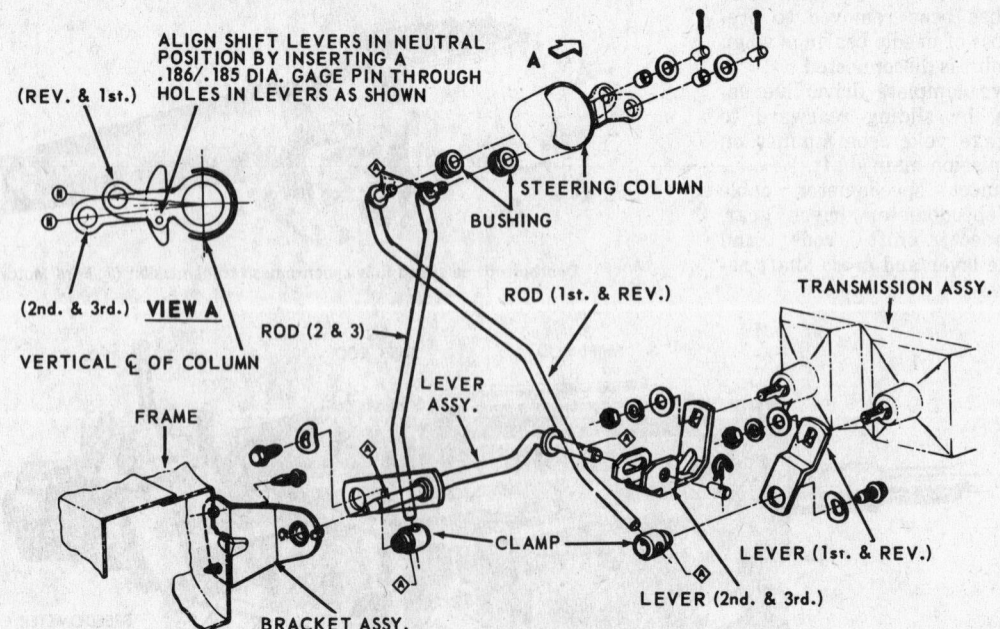

ALIGN SHIFT LEVERS IN NEUTRAL POSITION BY INSERTING A .186/.185 DIA. GAGE PIN THROUGH HOLES IN LEVERS AS SHOWN

(REV. & 1st.)

(2nd. & 3rd.) **VIEW A**

VERTICAL ₵ OF COLUMN

STEERING COLUMN

BUSHING

ROD (1st. & REV.)

ROD (2 & 3)

TRANSMISSION ASSY.

LEVER ASSY.

FRAME

CLAMP

LEVER (1st. & REV.)

LEVER (2nd. & 3rd.)

BRACKET ASSY.

⬧ LUBRICANT AS INDICATED WITH CHASSIS LUBRICANT

Standard shift controls—1967-69 (© Pontiac Div. G.M. Corp.)

4. Put levers on transmission in neutral.
5. Torque trunnion jam nuts to 30 ft. lbs.
6. Remove gauge pin and check complete shift pattern for freeness of operation.

1970

1970 standard shift Pontiacs incorporate a transmission controlled sparked switch (TCS) which prevents the distributor vacuum advance from operating when the transmission is in the lower gears.
1. Set gear shift lever in reverse and lock ignition.
2. Loosen swivel clamp screws C at first and reverse lever and D at cross shaft assembly.
3. Position second and third lever in reverse position.
4. Tighten swivel clamp screw C to 20 ft. lbs., unlock steering column and shift to neutral.
5. Align lower control levers E and F in neutral position and insert 0.185-0.186 in. diameter gauge pin through hole in lower control levers.
6. Tighten swivel clamp screw D to 20 ft. lbs., remove gauge pin and check shift pattern.
7. Shift transmission into high gear and adjust TCS switch so its plunger is fully depressed against second and third lever.

Four-Speed Linkage Adjustment

Borg Warner 4 Speed

The four-speed transmission gearshift linkage uses three shift rods and levers. Adjustment can be made with the aid of a simple gauge block,

J-9574. An alternative method is to have an assistant hold the manual shifter lever firmly in the neutral position.
1. Remove three screws that hold the chrome ring to the floor pan, then remove ring.
2. Remove three screws that hold the boot and retainer to the floor pan, then slide the boot up the shift rod.
3. Place transmission in neutral and install gauge block (or have assistant hold the manual shift lever in neutral.)

SHIFT CONTROL ADJUSTMENT

1. POSITION LEVERS ON TRANSMISSION IN NEUTRAL WITH GAGE PIN IN PLACE IN LEVER AND BRACKET ASSEMBLY (AS SHOWN IN VIEW "A") ASSEMBLE SHIFT CONTROL RODS TO LEVER AND BRACKET ASSEMBLY LEVERS AND TIGHTEN NUTS ON SWIVEL AS INDICATED.

4. Remove the cotter pin, anti-rattle washer, and clevis pin at each of the three shift levers.
5. On each shift rod, adjust the threaded clevis to permit free entry of the clevis pin into the hole in the transmission lever. This adjustment is quite critical.
6. Lubricate shift rod clevis pin and connect clevises to shift levers.
7. Remove gauge block (if used), and check for freedom and ease of shifting. If any one of the shifts is not smooth, one of the clevises may require one-half

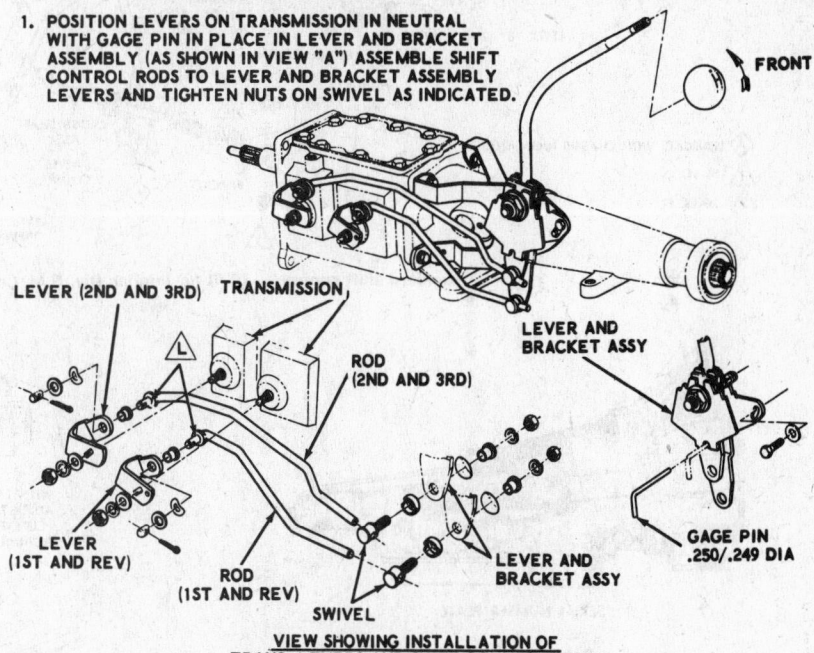

VIEW SHOWING INSTALLATION OF TRANS. LEVERS AND SHIFT CONTROLS RODS

3-speed gearshift linkage—1965-70 (© Pontiac Div. G.M. Corp.)

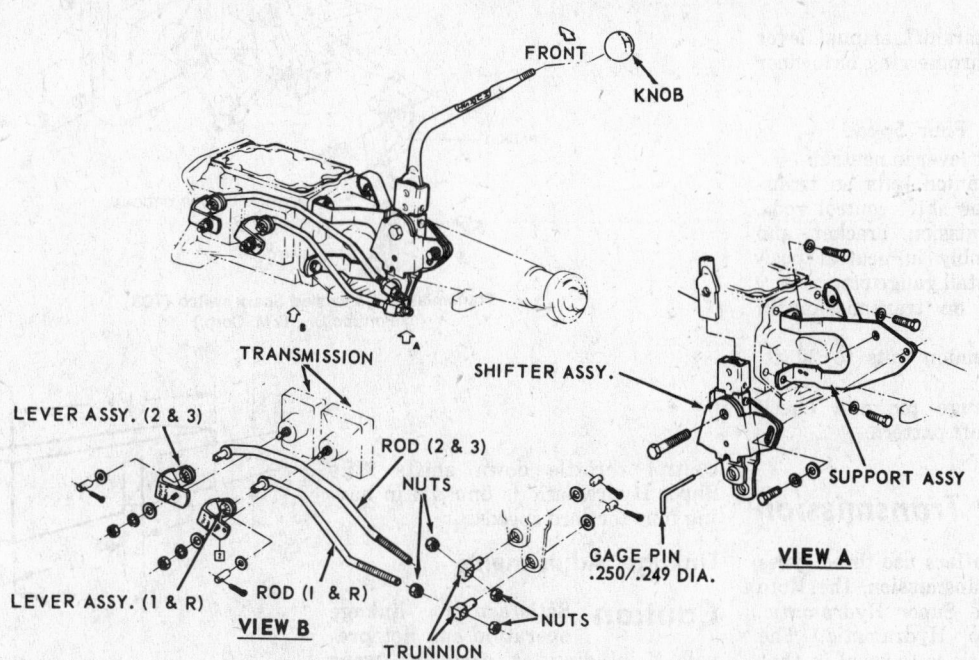

Three speed floor shift controls—1967-69 (© Pontiac Div. G.M. Corp.)

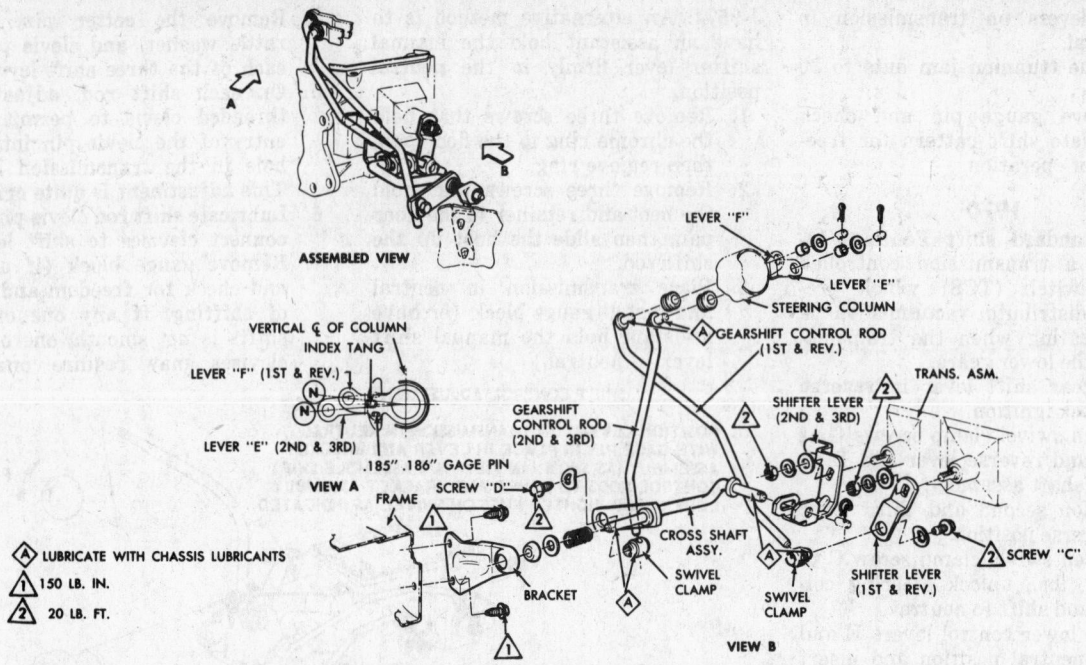

ASSEMBLED VIEW

Ⓐ LUBRICATE WITH CHASSIS LUBRICANTS
① 150 LB. IN.
② 20 LB. FT.

VIEW A

VIEW B

Standard shift controls—1970 (© Pontiac Div. G.M. Corp.)

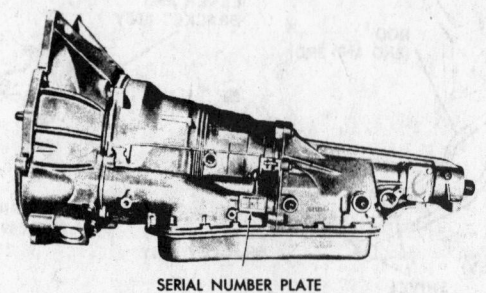

SERIAL NUMBER PLATE

Roto Hydramatic
(© Pontiac Div. G.M. Corp.)

turn.

8. Position gearshift manual lever boot and chrome ring onto floor pan.

Muncie Four Speed

1. Put selector lever in neutral.
2. Loosen trunnion nuts on transmission gear shift control rods.
3. Put transmission bracket and lever assembly in neutral position and install gauge pin.
4. Put levers on transmission in neutral.
5. Torque trunnion nuts to 30 ft. lbs.
6. Remove gauge pin and check complete shift pattern.

Automatic Transmission

Full sized Pontiacs use three types of automatic transmission, the Roto Hydramatic, the Super Hydramatic, and the Turbo Hydramatic. The Turbo Hydramatic is unusual in that it uses a non-adjustable solenoid to control throttle down shifts. The Super Hydramatic is unusual in having four forward speeds.

Linkage Adjustments

Caution Satisfactory linkage operation can not prevail if binding or excessive wear exists.

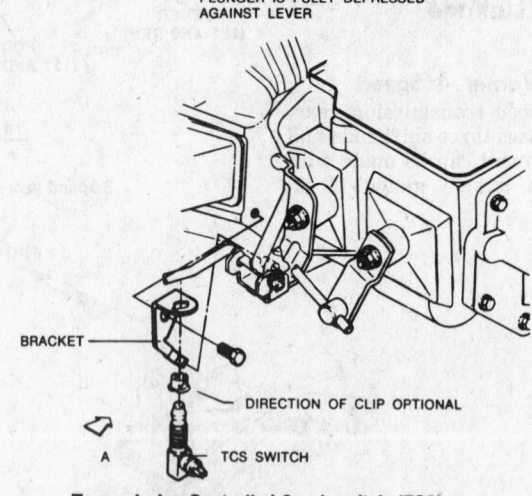

VIEW A

WITH LEVER IN 3RD POSITION ADJUST SWITCH SO THAT THE PLUNGER IS FULLY DEPRESSED AGAINST LEVER

BRACKET

DIRECTION OF CLIP OPTIONAL

TCS SWITCH

Transmission Controlled Spark switch (TCS)
(© Pontiac Div. G.M. Corp.)

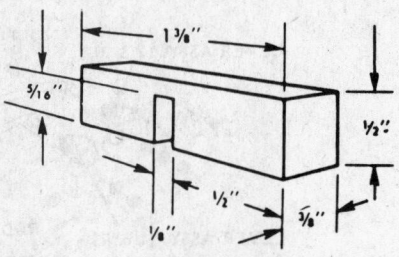

4-speed gauge block
(© G.M. Corp.)

Throttle Linkage Adjustment

Roto Hydramatic and Super Hydramatic

1. Remove carburetor air cleaner.
2. Loosen locknuts at top of transmission throttle control rod trunnion.
3. With engine at normal operating temperature and selector in drive, adjust idle to 480-500 rpm, (540-560 with air conditioning).
4. Stop the engine and install proper diameter pin through holes in throttle control lever and bracket.

   NOTE: four-barrel units have a throttle return check. Before installing the pin on these models, it is advisable to remove the throttle return check to prevent interference with linkage adjustment.
5. With throttle valves fully closed, loosen locknut and adjust length of transmission throttle control rod to carburetor so that the gauge pin is free in the hole. Then, tighten the locknut.
6. Push throttle rod to transmission, (TV rod) downward until the outer throttle lever reaches the end of its travel.

Caution Make sure that, when the lever is in this position, the upper locknut is not interfering with the trunnion.

7. While holding the throttle rod to transmission in this position, tighten upper and lower trunnion locknuts finger-tight. Shorten throttle control rod to transmission by backing off the lower trunnion nut four and one-half turns and tightening upper nut

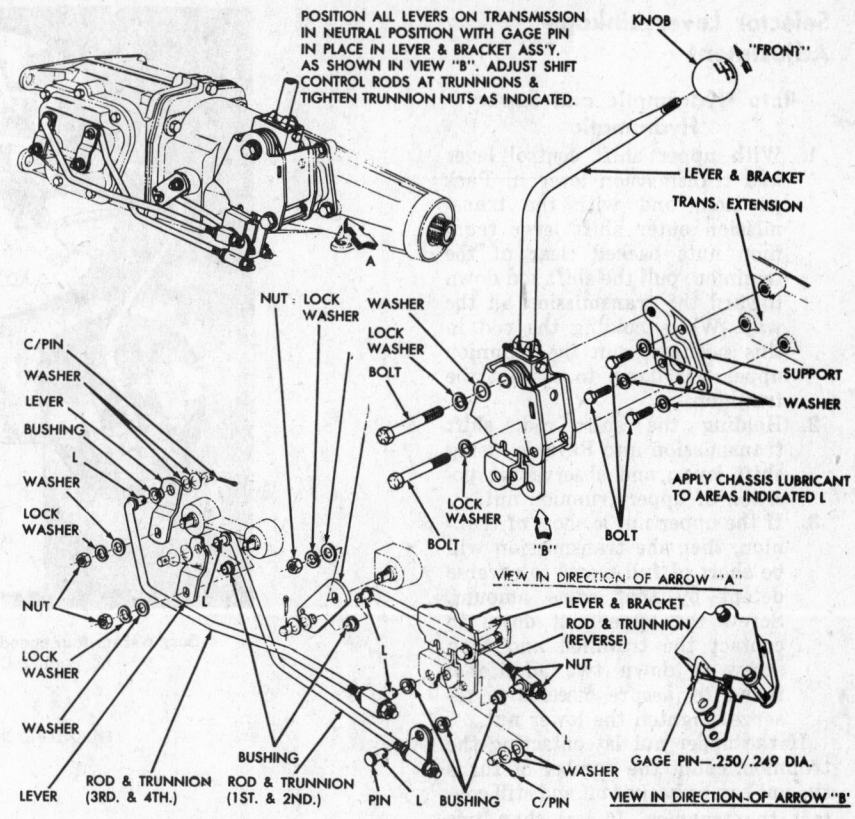

POSITION ALL LEVERS ON TRANSMISSION IN NEUTRAL POSITION WITH GAGE PIN IN PLACE IN LEVER & BRACKET ASS'Y. AS SHOWN IN VIEW "B". ADJUST SHIFT CONTROL RODS AT TRUNNIONS & TIGHTEN TRUNNION NUTS AS INDICATED.

4-speed gearshift linkage—1965 (© Pontiac Div. G.M. Corp.)

securely. Remove gauge pin.
8. Loosen locknut on carburetor throttle rod. Adjust carburetor throttle rod to obtain 4 29/64 in. from roller end of pedal rod to body toe board or 3¾ in. to carpet.
9. Tighten locknut on carburetor throttle rod.
10. Reinstall the air cleaner.
11. To complete throttle linkage adjustment, road test the car. Modify the adjustment, as required, by shortening or lengthening the throttle control rod to transmission, (TV rod) one-half turn at a time to obtain the best shift pattern.

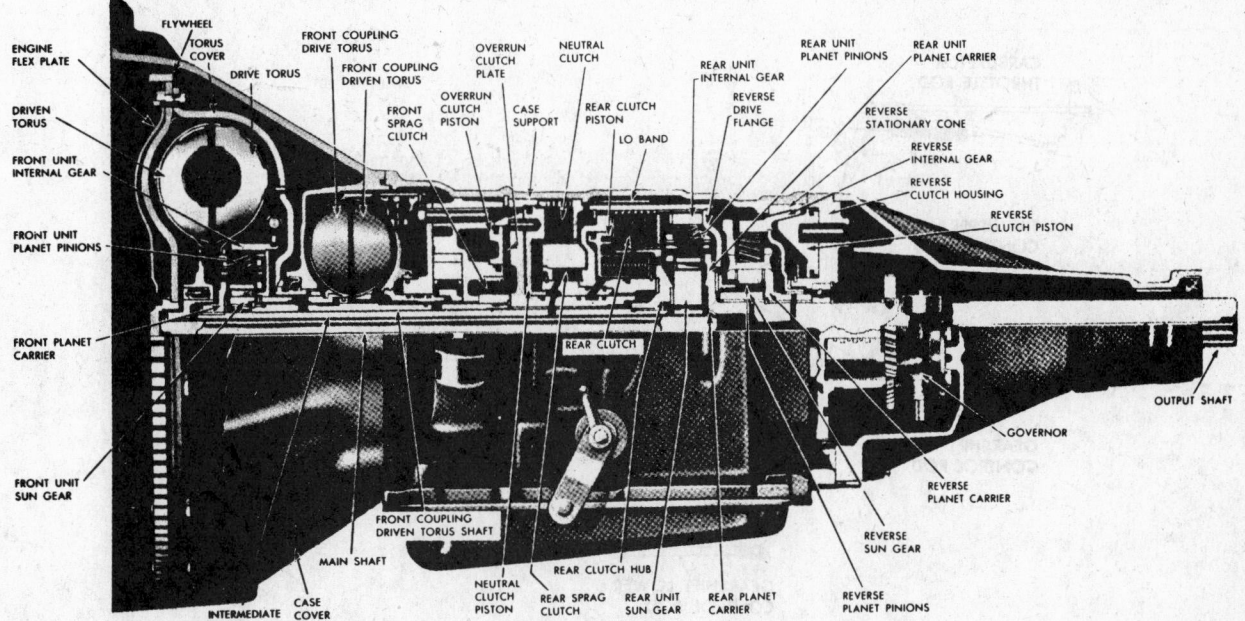

Super Hydramatic (© Pontiac Div. G.M. Corp.)

Selector Lever Linkage Adjustment

Roto Hydramatic and Super Hydramatic

1. With upper shift control lever and transmission lever in Park position, and with the transmission outer shift lever trunnion nuts backed clear of the trunnion, pull the shift rod down toward the transmission all the way. While holding the rod in this position, run the trunnion upper nut down to contact the trunnion.

2. Holding the shift rod, shift transmission into Reverse, using shift lever, and observe the position of upper trunnion nut.

3. If the upper nut is short of trunnion, then the transmission will be short of full travel to reverse detent by that same amount. Screw the upper nut down to contact the trunnion and then screw it down two additional turns to assure necessary reserve. Tighten the lower nut.

   If the upper nut is contacting the trunnion, count the number of turns the nut can be backed off and still contact the trunnion. If less than two turns, turn nut down two turns against the trunnion from the contact position and lock the lower nut. If more than two turns, turn upper nut down against the trunnion from the contact position to the original or starting position and lock the lower nut.

4. After completing above adjustment, check transmission parking lock with car on ramp or grade to insure positive lock.

5. The selector indicator must not be off index register more than

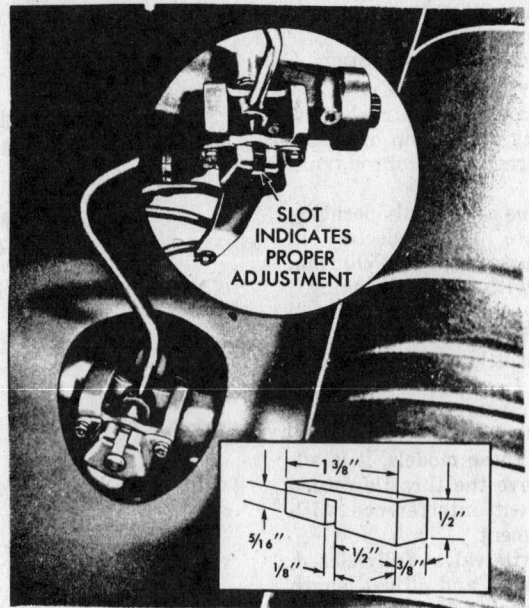

Borg Warner four speed shift linkage adjustment
(© Pontiac Div. G.M. Corp.)

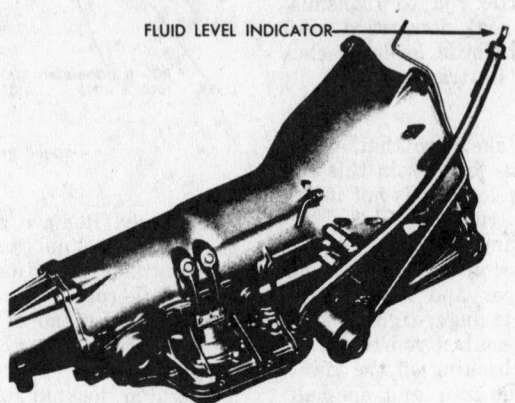

FLUID LEVEL INDICATOR

Turbo Hydramatic
(© Pontiac Div. G.M. Corp.)

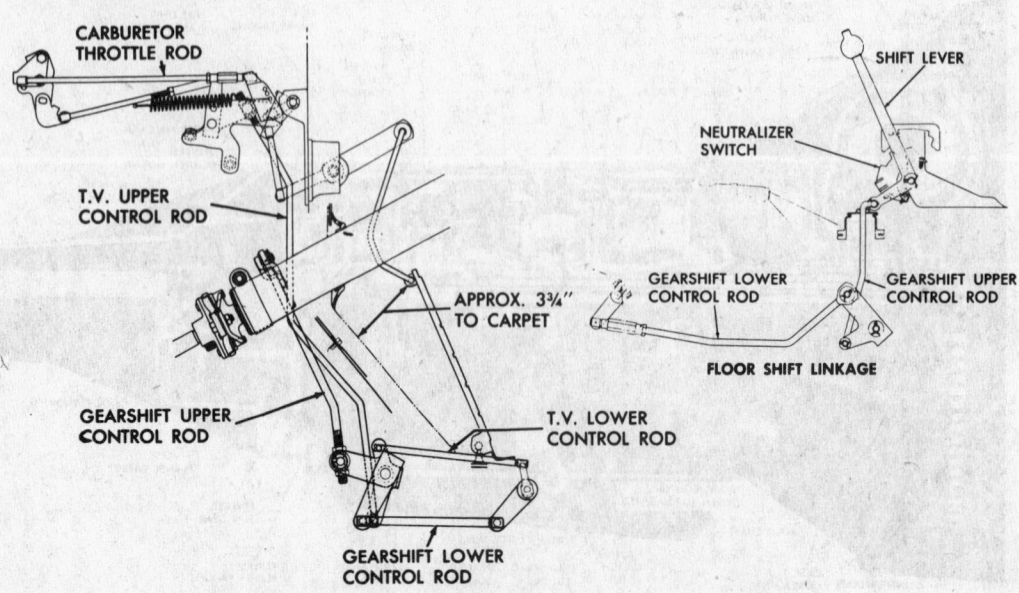

CARBURETOR THROTTLE ROD

T.V. UPPER CONTROL ROD

APPROX. 3¾" TO CARPET

GEARSHIFT UPPER CONTROL ROD

T.V. LOWER CONTROL ROD

GEARSHIFT LOWER CONTROL ROD

SHIFT LEVER

NEUTRALIZER SWITCH

GEARSHIFT LOWER CONTROL ROD

GEARSHIFT UPPER CONTROL ROD

FLOOR SHIFT LINKAGE

Automatic shift linkage adjustment—1964 (© Pontiac Div. G.M. Corp.)

1/16 in. after adjustment is complete.

1967-68 Turbo Hydramatic
See illustrations.

All Turbo-Hydramatic Column— 1969-70

1. Loosen screw (nut on Firebird) on adjusting swivel clamp.
2. Place gearshift lever in Park and lock ignition.
3. Place transmission shift lever in Park detent (rotate clockwise, see illustrations).
4. Push up on gearshift control rod until lash is taken up in steering column lock mechanism, then tighten screw or nut on swivel clamp to 20 ft. lbs.

All Turbo-Hydramatic Console— 1969-70

1. Disconnect shift cable from transmission shift lever by removing nut from pin.
2. Adjust back drive linkage (as in Step 4, above).
3. Unlock ignition and rotate transmission shift lever counterclockwise two detents.
4. Place console lever in Neutral and move against forward Neutral stop.
5. Assemble shift cable and pin to transmission shift lever so that

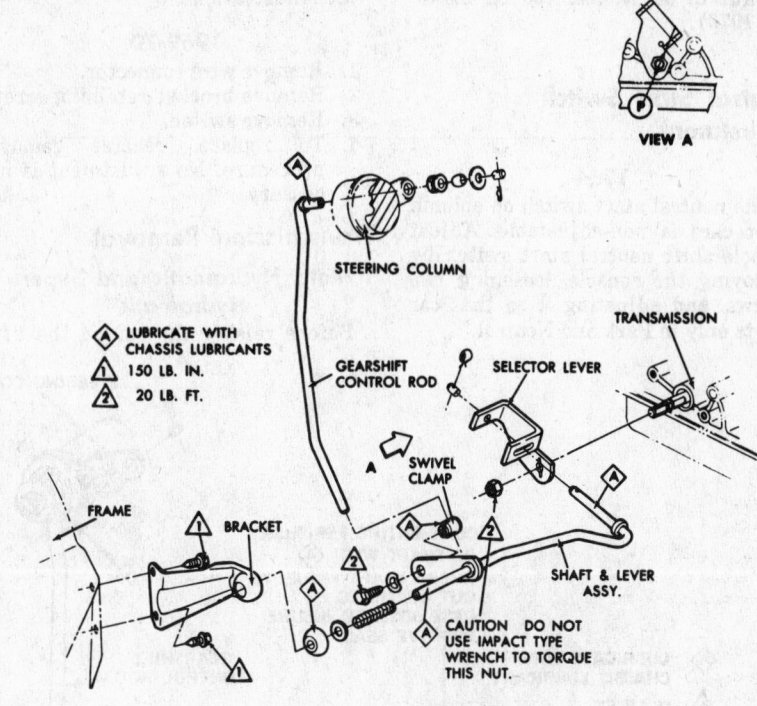

VIEW A

LUBRICATE WITH CHASSIS LUBRICANTS
150 LB. IN.
20 LB. FT.

STEERING COLUMN

TRANSMISSION

GEARSHIFT CONTROL ROD

SELECTOR LEVER

SWIVEL CLAMP

FRAME

BRACKET

SHAFT & LEVER ASSY.

CAUTION DO NOT USE IMPACT TYPE WRENCH TO TORQUE THIS NUT.

Automatic shift linkage adjustment—1969-70 column shift (© Pontiac Div. G.M. Corp.)

CONTROL ADJUSTMENT PROCEDURE
SET TRANSMISSION SELECTOR LEVER IN DRIVE DETENT. (SEE VIEW B)

SET SHIFT LEVER IN DRIVE. (SEE VIEW A)

PULL DOWN ON SHIFT ROD AND PUSH UP ON END OF LEVER ASSEMBLY TO TAKE UP SLACK IN SYSTEM.

TIGHTEN SCREW ON ADJUSTING SWIVEL CLAMP TO SPECIFIED TORQUE.

STARTER NEUTRALIZER SWITCH INSTALLATION
1. POSITION GEARSHIFT LEVER IN DRIVE POSITION. (SEE VIEW "A")

2. INSERT SWITCH "DRIVE TANG" IN SHIFTER TUBE SLOT AND ASSEMBLE SWITCH TO STEERING COLUMN JACKET USING TWO (2) #8-32 x 7/32" WASHER HEAD TAPPING SCREW.

ADJUSTMENT
1. INSERT NOMINAL ADJUSTING GAGE IN SWITCH. MOVE GEAR SELECTOR TO PARK POSITION AND REMOVE ADJUSTING GAGE.

2. THE STARTER SHOULD OPERATE ONLY WHEN IGNITION KEY IS TURNED TO "START" POSITION WITH SHIFT LEVER IN "PARK" POSITION AND IN "NEUTRAL" SHIFT POSITION. HOWEVER, IF STARTER DOES OPERATE IN OTHER POSITIONS USE THE FOLLOWING TESTS.

SWITCH ADJUSTED	STARTS IN				RESET* AND RETEST USING THIS GAGE
	PK.	REV.	N	DR.	
USING NOMINAL GAGE	X	X	X		#1 + GAGE
USING #1 + GAGE	X	X	X		#2 + GAGE
USING NOMINAL GAGE			X	X	#1 - GAGE
USING #1 - GAGE	X		X	X	#2 - GAGE

*Insert blade of nominal adjusting gage in reset slot.
Move gearshift lever slowly to "LOW" position.

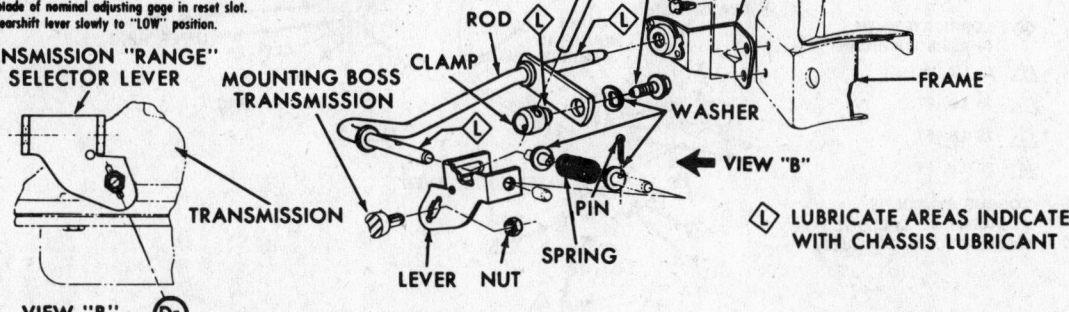

PIN
WASHER
BUSHING
STEERING COLUMN
SLEEVE
ROD

SWITCH ASM.
ADJUSTING GAGE
VIEW "A"
RIVET
STEERING COLUMN

VIEW SHOWING SWITCH INSTALLATION

VERTICAL ℄ OF STEERING COLUMN
HORIZONTAL ℄ OF STEERING COLUMN
℄ OF HOLE IN SWITCH ASS'Y.
SWITCH "DRIVE TANG" IN DRIVE POSITION
Pk
Dr
VIEW "A"

TRANSMISSION "RANGE" SELECTOR LEVER
MOUNTING BOSS TRANSMISSION

SCREW BRACKET ASM.
ROD
CLAMP
ROD
WASHER
FRAME
VIEW "B"
PIN
SPRING
LEVER NUT

TRANSMISSION

LUBRICATE AREAS INDICATED WITH CHASSIS LUBRICANT

VIEW "B" Dr.

Automatic shift linkage adjustment—1967-68 column shift (© Pontiac Div. G.M. Corp.)

no binding exists, then tighten nut to 30 ft. lbs. (20 ft. lbs.— 1970).

Neutral Start Switch Adjustment

1964

The neutral start switch on column shift cars is non-adjustable. Adjust console shift neutral start switch by removing the console, loosening two screws, and adjusting it so that car starts only in Park and Neutral.

1967-68

See illustrations.

1969-70

1. Remove wire connector.
2. Remove bracket retaining screw.
3. Remove switch.
4. To replace, reverse removal procedure. No adjustment is necessary.

Transmission Removal

Roto Hydramatic and Super Hydramatic

Before raising the car on the lift, remove one cable (either one) from the battery and release the emergency brake.

1. Remove the filler tube and drain the transmission. Push the filler tube up toward its upper bracket out of the way.
2. Disconnect propeller shaft from transmission:
   A. Remove U-bolt nuts, lock plates, and U-bolts from rear axle drive pinion flange.
   B. Use a suitable rubber band or tape to hold bearings onto U-joint journals if tie-wire is broken.

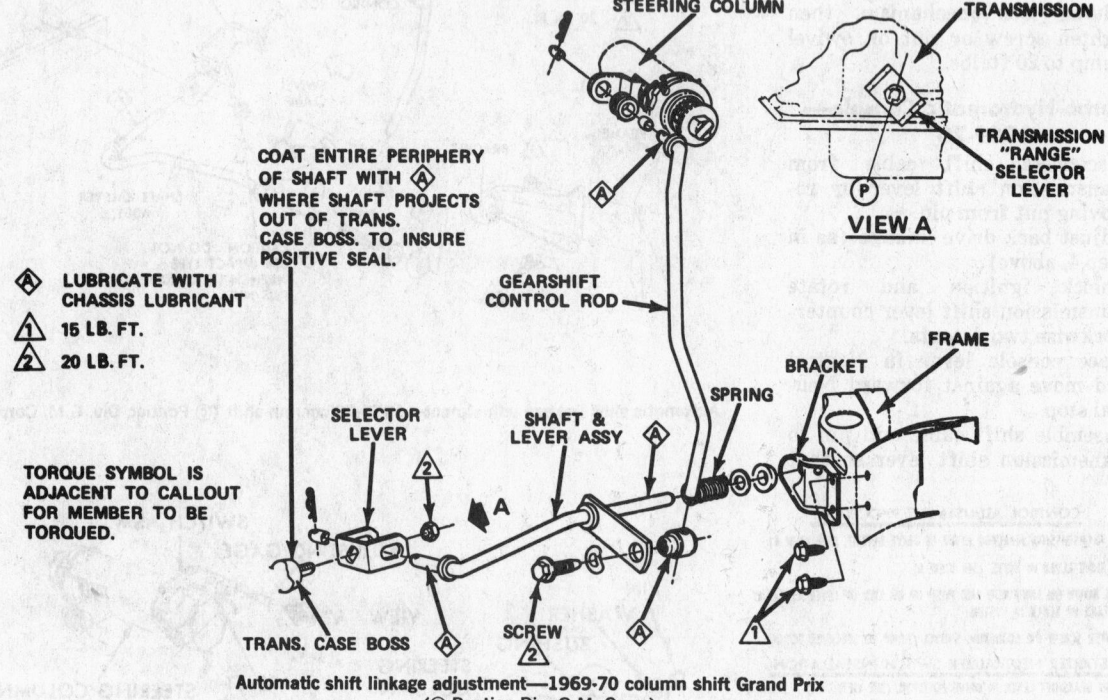

Automatic shift linkage adjustment—1969-70 column shift Grand Prix
(© Pontiac Div. G.M. Corp.)

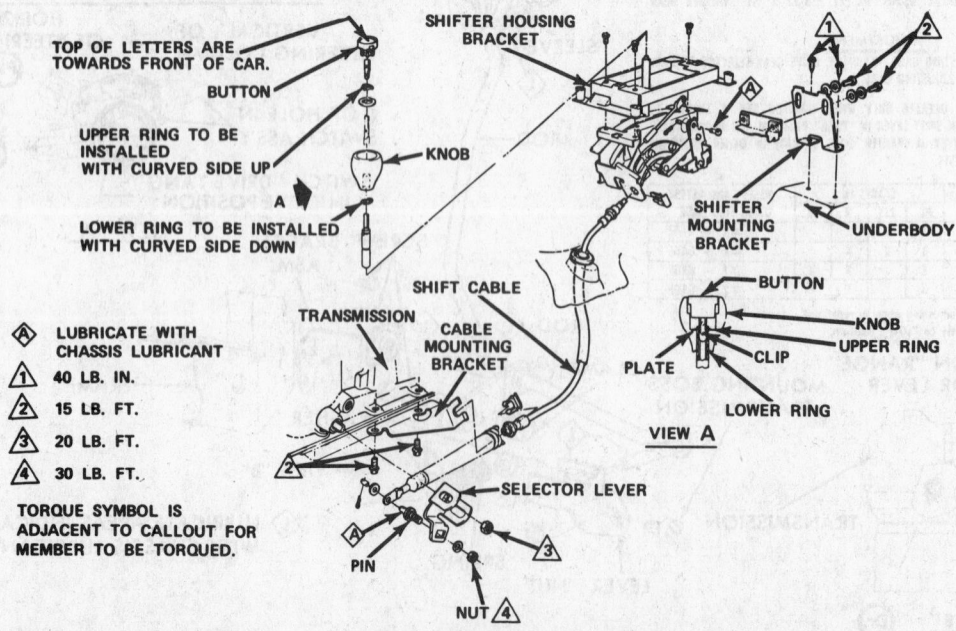

Automatic shift linkage adjustment—1969-70 console shift Grand Prix
(© Pontiac Div. G.M. Corp.)

C. Slide propeller shaft rearwards from transmission output shaft.

3. Disconnect speedometer cable from speedometer-driven gear.

4. Remove gearshift control lower rod.

5. Remove lower end of gearshift control upper rod by removing E-ring.

6. Remove the two cross-shaft bracket-to-frame attaching bolts, then remove the bracket, cross-shaft lever, and bushing from car.

7. Remove lower end of throttle control transmission rod (engine to transmission idler lever.)

8. Remove idler lever to outer throttle lever control rod.

9. Remove throttle control idler lever.

10. Remove parking brake return spring and brake cable guide hook from frame crossmember.

11. Remove oil cooler lines.

12. Loosen exhaust pipe to manifold bolts about ¼ in.

13. Remove both starter cables.

14. Remove the starter and shield by removing the two attaching bolts.

15. Remove bottom cover from bottom of case cover (three attach-ing bolts).

16. Remove the four bolts which hold the flywheel front cover plate to the transmission case cover.

17. Place special automatic transmission jack under transmission and raise it enough to support the transmission.

18. Remove two rear mount support-to-frame crossmember nuts and raise transmission so studs clear the crossmember.

19. Remove the two bolts at each end of the frame crossmember and remove crossmember.

20. Lower the transmission until the jack is barely supporting it.

21. Remove breather pipe clip bolt and remove pipe from transmission.

22. Using a long wrench extension with a U-joint, remove the remaining six transmission case cover-to-engine attaching bolts.

23. Raise transmission to its normal position, slide rearward from the engine and flywheel, and lower it away from the car.

24. Remove rear mount support from rear mount by removing a nut from each insulator.

25. Remove four rear mount to rear bearing retainer attaching screws.

26. Install by reversing above procedure.

Turbo Hydramatic

1. Remove drive shaft.

2. Disconnect speedometer cable, electrical lead to case connector, vacuum line at modulator, and oil cooler pipes.

3. Disconnect shift control linkage.

4. Support transmission with jack.

5. Disconnect rear mount from transmission and frame connector.

6. Remove two bolts at each end of frame crossmember and remove crossmember.

7. Remove converter duct shield.

8. Remove converter to flex plate bolts.

9. Loosen exhaust pipe to manifold bolts approximately ¼ in. and lower transmission until jack is barely supporting it.

10. Remove transmission to engine mounting bolts.

11. Remove transmission to its normal position, slide it rearward from engine, and remove it from car. When lowering transmission keep rear end lower than front to avoid dropping converter.

12. Reverse procedure to install.

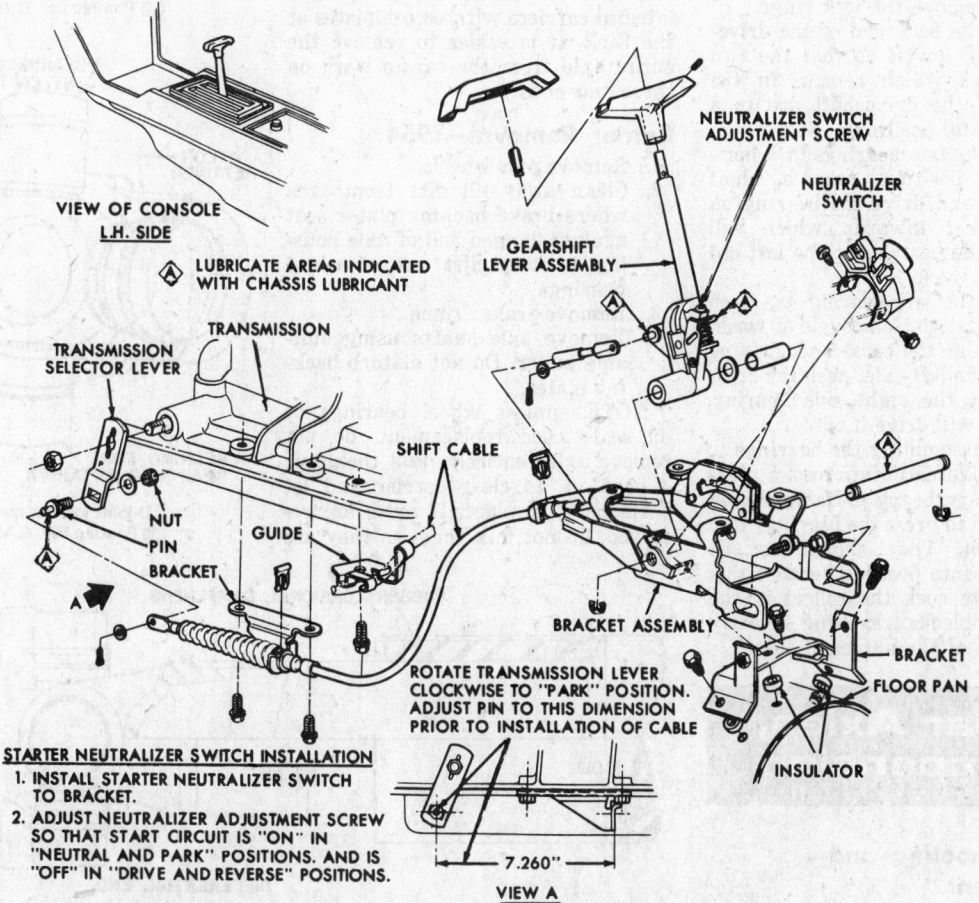

VIEW OF CONSOLE L.H. SIDE

◇ LUBRICATE AREAS INDICATED WITH CHASSIS LUBRICANT

TRANSMISSION SELECTOR LEVER

TRANSMISSION

NEUTRALIZER SWITCH ADJUSTMENT SCREW

NEUTRALIZER SWITCH

GEARSHIFT LEVER ASSEMBLY

SHIFT CABLE

NUT PIN BRACKET GUIDE

A

BRACKET ASSEMBLY

BRACKET

FLOOR PAN

INSULATOR

ROTATE TRANSMISSION LEVER CLOCKWISE TO "PARK" POSITION. ADJUST PIN TO THIS DIMENSION PRIOR TO INSTALLATION OF CABLE

STARTER NEUTRALIZER SWITCH INSTALLATION

1. INSTALL STARTER NEUTRALIZER SWITCH TO BRACKET.

2. ADJUST NEUTRALIZER ADJUSTMENT SCREW SO THAT START CIRCUIT IS "ON" IN "NEUTRAL AND PARK" POSITIONS. AND IS "OFF" IN "DRIVE AND REVERSE" POSITIONS.

← 7.260" →

VIEW A

Neutral start switch adjustment—1967-68 console shift (© Pontiac Div. G.M. Corp.)

U Joints, Drive Lines

1964-70

Two basic designs are used; one is a typical solid shaft with two joints. The other incorporates five rubber torsional dampeners. The accompanying diagram shows the rubber dampeners inserted between the solid and tubular sections.

Cross- and bearing-type universals are used on all Pontiacs.

There are two types used, however. One type held with a C-shaped lock ring; the other held with a lock plate.

1. From under the car, remove the four bolts which hold the two rear universal joint pillow blocks to the rear axle pinion shaft flange.
2. Tap the pillow blocks until they come off the flange.
3. Lower the rear end of the driveshaft and slide the front end off the splines of the transmission, together with the front universal joint.
4. Take the entire assembly to the bench. If lock-type bearings are used, remove the screws which hold the lock plates over the bearing. If lock ring-types are used, remove the lock rings.
5. Place the back end of the driveshaft in a vise so that the two bearings which remain in the end of the driveshaft are in a horizontal position.
6. With the two bearings in a horizontal position, take a blunt punch and drive the bearing on the right inwards, which will force the bearing on the left out of the yoke.
7. When the left bearing has been taken out, pack a couple of washers under the cross and, driving from the left side, push the cross against the right side bearing, which will drive it out.
8. On reassembling the bearings in the universal joint cross, a press or a very heavy C-clamp should be used to press the bearings into position. They should not be driven into position because this tends to cock the rollers in the pillow blocks, resulting in early failure of the universal joint.

Rear Axles, Suspension

Troubleshooting and Adjustment

A four link pivoted control arm suspension system is used.

Removable differential carrier assembly
(© Pontiac Div. G.M. Corp.)

General instructions covering the rear axle, and how to repair and adjust it, together with information on installation of rear axle bearings and grease seals, are given in the Unit Repair Section.

Capacities of the rear axle are given in the Capacities tables of this section.

1964 rear axles have removable carrier assemblies which allows most rear end work to be done without removing the axle housing from the car. 1965 and later rear axles have integral carriers with access plates at the back. It is easier to remove the entire axle from the car to work on these rear ends.

Carrier Removal—1964

1. Remove rear wheels.
2. Clean away all dirt from area where brake backing plates seat against flanged end of axle housing to keep dirt out of wheel bearings.
3. Remove brake drums.
4. Remove axle shafts using suitable puller. Do not disturb backing plates.

NOTE: unless wheel bearings or oil seals need replacement, do not remove axles entirely. Pull them out 4 or 5 in. to clear carrier and let them hang in place. If axles are removed, do not mix them as they are

Installing snap-ring retainer
(© Pontiac Div. G.M. Corp.)

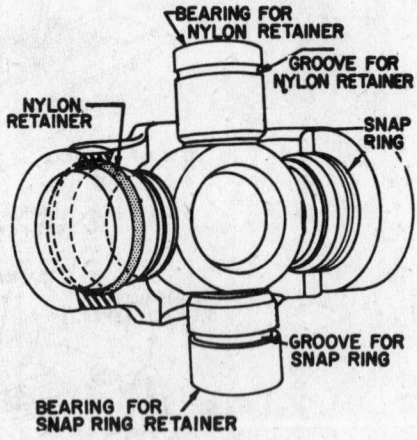

U-joint locking methods
(© Pontiac Div. G.M. Corp.)

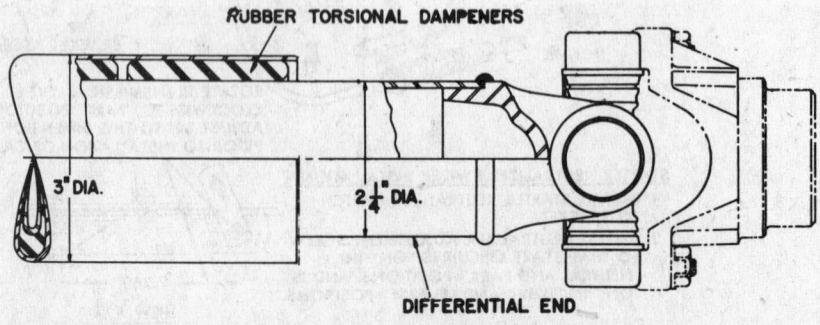

Drive shaft with rubber torsional dampeners (© Pontiac Div. G.M. Corp.)

different lengths and left hand axle has studs with left hand threads.

5. Turn down tabs on lock plates and remove U-bolts that hold rear U-joint to companion flange. Use a rubber band or tape to hold bearings on journals.

6. Clean differential bolt flange to keep dirt out of differential.

7. Drain oil by loosening all differential carrier nuts and pulling carrier out about 1/8 in.

8. Allow oil to drain thoroughly, remove nuts, and remove carrier assembly. Reverse procedure to install.

Rear Axle Removal—1965-70

1. Raise rear of car high enough to permit working underneath. Place floor jack under center of axle housing so it just starts to raise rear axle assembly. Place car stands solidly under frame members on both sides.

2. Disconnect rear universal joint from companion flange by removing two U-bolts. Use rubber band or tape to hold bearings onto journal. Support drive shaft out of the way.

3. Remove wheels, and brake drums. Both right and left wheel studs have right hand threads.

4. Remove nuts holding retaining plates and brake backing plates.

5. Remove axles with suitable puller.

6. Support brake backing plates out of way.

7. Disconnect rear brake hose bracket by removing top cover bolt. Remove brake line from housing by bending back tabs.

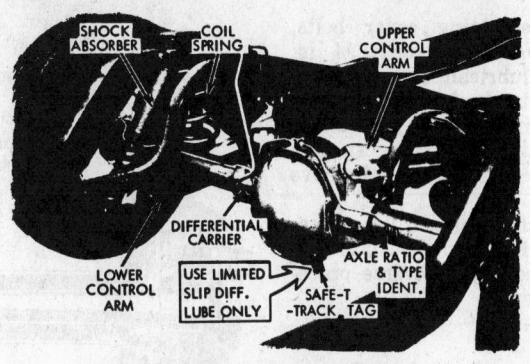

Typical rear suspension—1965-70
(© Pontiac Div. G.M. Corp.)

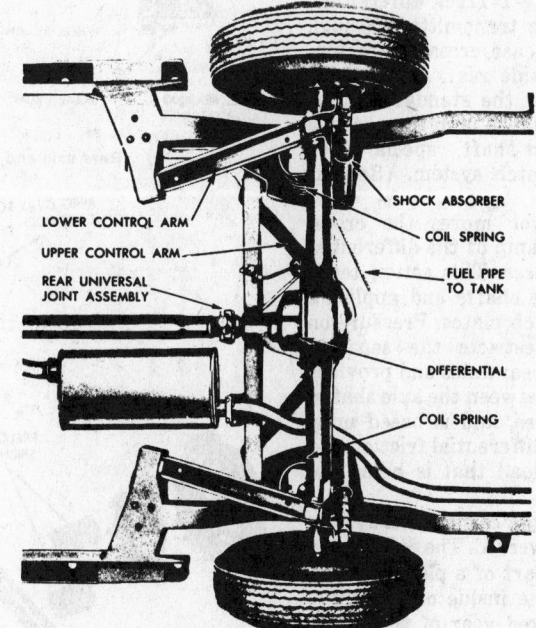

Typical rear suspension—1964
(© Pontiac Div. G.M. Corp.)

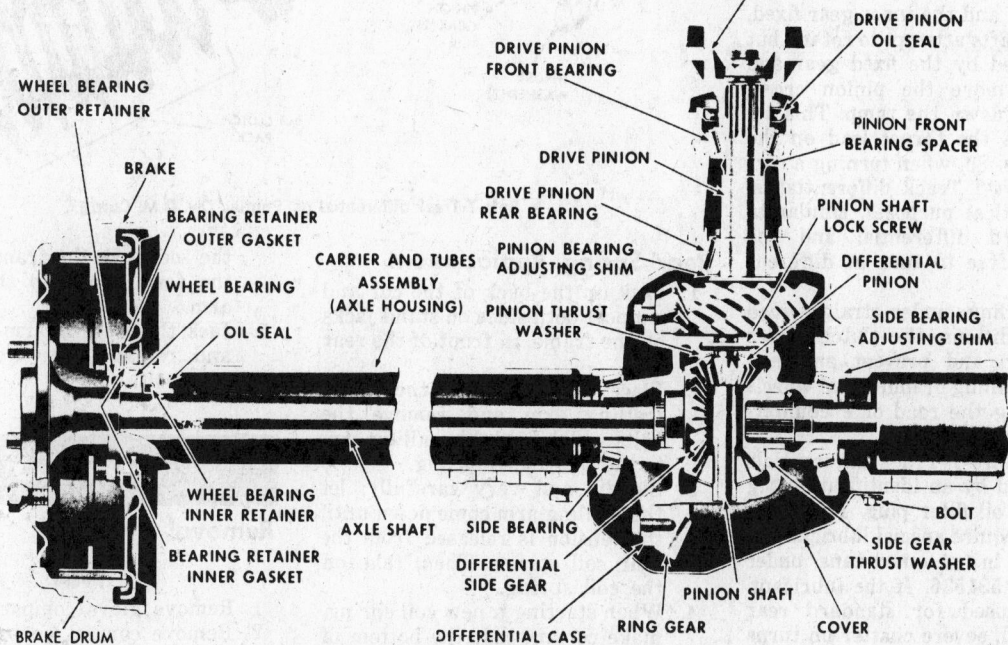

Rear axle and differential—1965-70 (© Pontiac Div. G.M. Corp.)

8. Loosen remaining cover bolts, break cover loose about 1/8 in. and allow lubricant to drain.
9. Disconnect shock absorbers at axle housing. Lower jack under axle housing until rear springs can be removed. Remove springs.
10. Disconnect upper and lower control arms from axle housing.
11. Remove housing. Reverse procedure to install.

Safe-T-Track Differential

With the Safe-T-Track differential, driving force is transmitted through the differential case, cross-shafts, pinion gears and side gears in the same manner as with the standard differential. The variance is in the use of a two-piece cross-shaft, special side gears and a clutch system. (See exploded view.)

Applied power moves the cross-shafts up the ramp of the differential case cam surfaces. This action tends to separate the shafts and applies a load to the clutch plates. Pressure on these plates restricts the separate turning of the rear axles and provides a torque ratio between the axle shafts. This ratio varies, and is based upon the amount of differential friction and the degree of load that is being applied.

When turning a corner, this system is somewhat reversed. The differential gears become part of a planetary set. The gear on the inside of the curve becomes the fixed gear of the planetary train. The outer gear of this set overruns, as does the outside wheel on the curve, having a further distance to travel. With the outer gear overrunning and the inner gear fixed, the cross-shafts attempt to rotate, but are restricted by the fixed gear and they must move the pinion cross-shafts back down the ramp. This action relieves the thrust load on the clutch plates. So, when turning a corner, the Safe-T-Track differential is, for all practical purposes, similar to the standard differential and the wheels are free to turn at different speeds.

While pulling, under straight-road driving conditions, the clutch plates are engaged and prevent any momentary spinning of individual wheels when leaving the road or encountering poor traction areas.

NOTE: Safe-T-Track differentials are identified by an identification tag next to the oil filter plug. These differentials require special lubricant. It is available, in one quart cans, under Pontiac No. 531536. If the lubricant previously used for standard rear axles is used, severe chatter on turns will result.

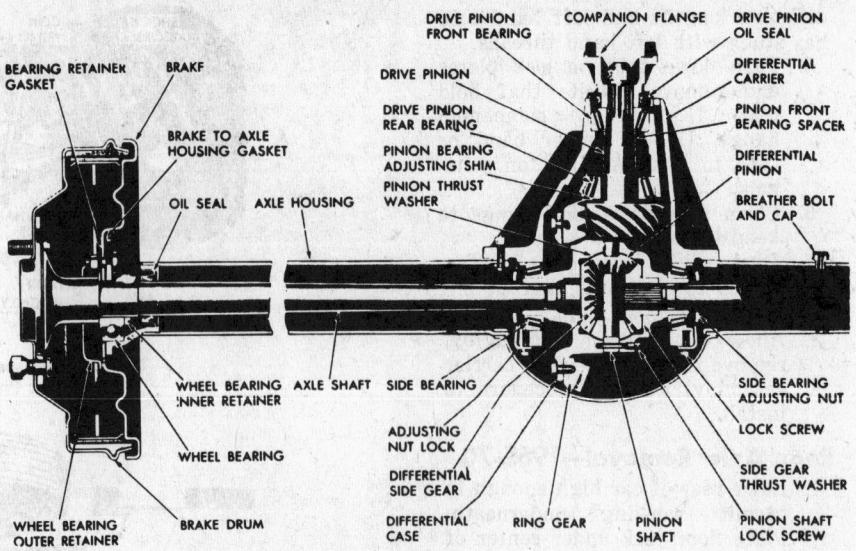

Rear axle and differential—1964 (© Pontiac Div. G.M. Corp.)

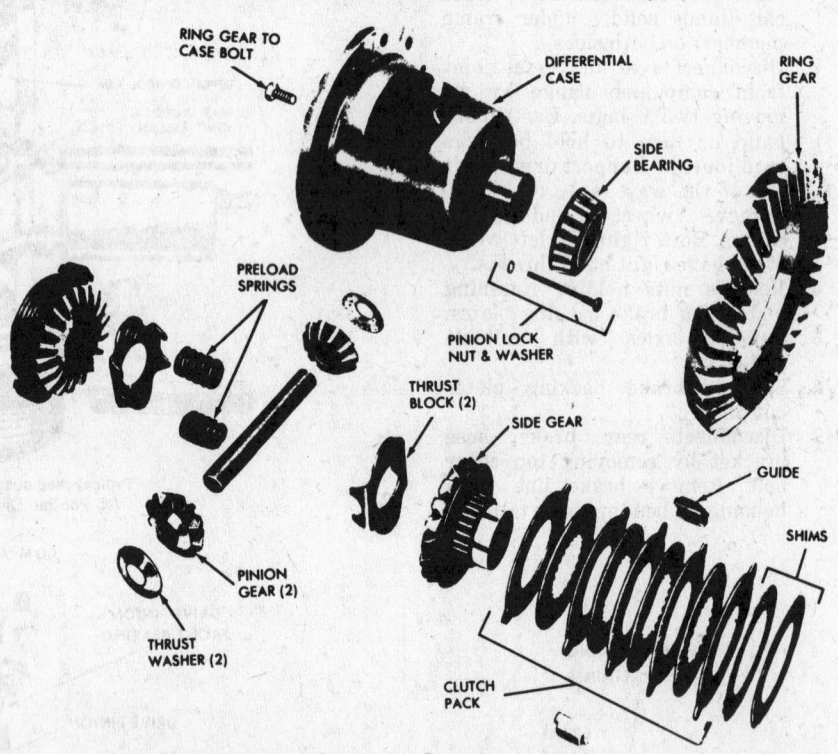

Safe-T-Track differential (© Pontiac Div. G.M. Corp.)

Rear Spring Replacement

1. Jack up the back of the car and support both sides on stand jacks on the frame, in front of the rear axle.
2. Place a jack under the lower trailing arm and remove the bolts which hold the trailing arm to the rear axle housing.
3. Slowly, and very carefully, let the trailing arm come down until the tension is released from the rear coil spring. Then, take off the coil spring.
4. When starting a new coil spring, make certain that the bottom of the coil is properly inserted into the socket in the frame and into the form plate on the trailing arm.
5. Jack the trailing arm into place and reinstall the trailing arm rear bolt.

Radio

Removal

1964

1. Remove glove compartment.
2. Remove control knobs and large nuts.

3. Remove nut and washer from bracket at left side of radio.
4. Remove fuse block connector, speaker and antennae leads.
5. Remove through compartment door.
6. Reverse above procedure to replace. If speaker is to be removed, while radio is out, remove four nuts holding to panel and lift out.

1967-68

1. Remove stereo tape player, if so equipped.
2. Remove knobs, springs, nuts, and bezels from control bushings.
3. If car is air conditioned, remove three Phillips cross head screws holding bottom air conditioning air duct and remove duct.
4. Disconnect stereo multiplex plug from radio, if so equipped.
5. Remove antenna lead in and speaker connector.
6. Remove hex head screw holding right side of radio to brace.
7. Disconnect dial light socket and lower radio to floor.
8. Remove multiplex adapter if so equipped.
9. Reverse procedure to install.

1969-70

1. Disconnect battery.
2. Remove lower air conditioning duct if equipped.
3. Remove two radio control knobs and hex nuts.
4. Remove ash tray and bracket.
5. Remove upper air conditioning duct, if so equipped.
6. Disconnect all radio connections.
7. Remove screws holding radio brace to lower edge of instrument panel and remove radio.
8. Reverse procedure to install.

Windshield Wipers

Motor R & R

1. Disconnect electrical and hose connections at wiper.
2. Disconnect wiper crank from wiper transmission linkage.
3. Remove wiper motor mounting screws, then remove the motor from the firewall.
4. Install by reversing removal procedure.

Transmission and Linkage R & R

1. Remove arm and blade assemblies.
2. Remove fresh air intake grille.
3. Remove wiper transmission retaining screws.
4. Remove retainer holding linkage which connects with wiper motor crank.

5. Remove wiper transmission and linkage.
6. Install by reversing removal procedure. Be sure wiper blades are in park position after they are installed.

Heater System

1964-65

Heater Blower R & R

1. Drain radiator.
2. Raise front of car, and remove right front wheel.
3. Remove right headlamp assembly.
4. Disconnect right front fender skirt, and move downward and toward rear of car.
5. Disconnect wires at blower.
6. Disconnect vacuum hose at air inlet duct diaphragm.
7. Disconnect heater inlet and outlet water hose at heater.
8. Remove nuts securing air duct and remove assembly.
9. Install in reverse of above.

Heater Core R & R

1. Drain radiator.
2. Remove glove compartment.
3. Disconnect hoses at heater.
4. Disconnect temperature control cable at top of core and air outlet duct.
5. Disconnect vacuum hose from defroster air valve diaphragm.
6. Remove wire connector from resistor assembly at top left side of heater air outlet duct by prying connector up with flat-bladed screwdriver.
7. Remove nuts attaching heater to air duct and remove heater.
8. Remove core and case assembly.
9. Remove heater core.
10. Install in reverse of above.

1966-67

Heater Blower, Impeller and Inlet Duct R & R

1. Disconnect wires at connector to blower.
2. Disconnect water hoses and plug openings.
3. Disconnect vacuum hose at air inlet duct diaphragm.
4. Remove nuts and screws that hold air inlet and remove assembly.
5. Remove large motor retaining ring from motor.
6. Remove motor and impeller assembly.
7. Install in reverse of above.

Heater Core R & R

1. Drain radiator.
2. Disconnect inlet and outlet hoses at heater.

3. Disconnect temperature control cable at top of heater core and case.
4. Disconnect vacuum hose from defroster and air inlet diaphragms.
5. Remove wire connector from resistor assembly at top of air outlet duct by prying up with flat blade screwdriver.
6. Remove nuts and screws securing heater to air inlet duct assembly.
7. Remove heater core and case assembly.
8. Remove heater core.
9. Install in reverse of above.
10. Adjust temperature control cable.

1968

Heater Blower Motor, Impeller and Duct R&R

1. Remove hood hinge to fender retaining bolts.
2. Prop hood and rest hinge on plenum.
3. Remove blower motor or duct retaining screws as desired.
4. Remove motor electrical lead.
5. Remove motor or duct as desired.
6. Reverse procedure to install.

Heater Core R & R

1. Drain radiator.
2. Disconnect inlet and outlet hoses at heater.
3. Disconnect temperature control cable at top of heater core and case.
4. Disconnect vacuum hose from defroster and air inlet diaphragms.
5. Remove wire connector from resistor assembly at top of air outlet duct by prying up with flat blade screwdriver.
6. Remove nuts and screws securing heater to air inlet duct assembly.
7. Remove heater core and case assembly.
8. Remove heater core.
9. Install in reverse of above.
10. Adjust temperature control cable.

1969-70

Blower Motor and Impeller

1. Jack up front of car and remove right front wheel.
2. Cut access hole along stamped outline on right fender skirt, using an air chisel.
3. Disconnect blower power wire.
4. Remove blower.
5. To install, reverse removal procedure, covering access hole with a metal plate secured with sealer and sheet metal screws.

Heater Core R & R

1. Drain radiator.

2. Disconnect heater hoses at air inlet assembly.

NOTE: the water pump hose goes to top heater core pipe, the other hose (from rear of right cylinder head on V8, center of block on 6) goes to the lower heater core pipe.

3. Remove nuts from core studs on firewall (under hood).
4. From inside the car, pull the heater assembly from the firewall.
5. Disconnect control cables and wires, then remove heater assembly.
6. To remove core, unhook retaining springs.
7. To install, reverse removal procedure, making sure core is properly sealed during installation.

YEAR IDENTIFICATION

SERIES 10 AND 20 CLASSIC AND REBEL

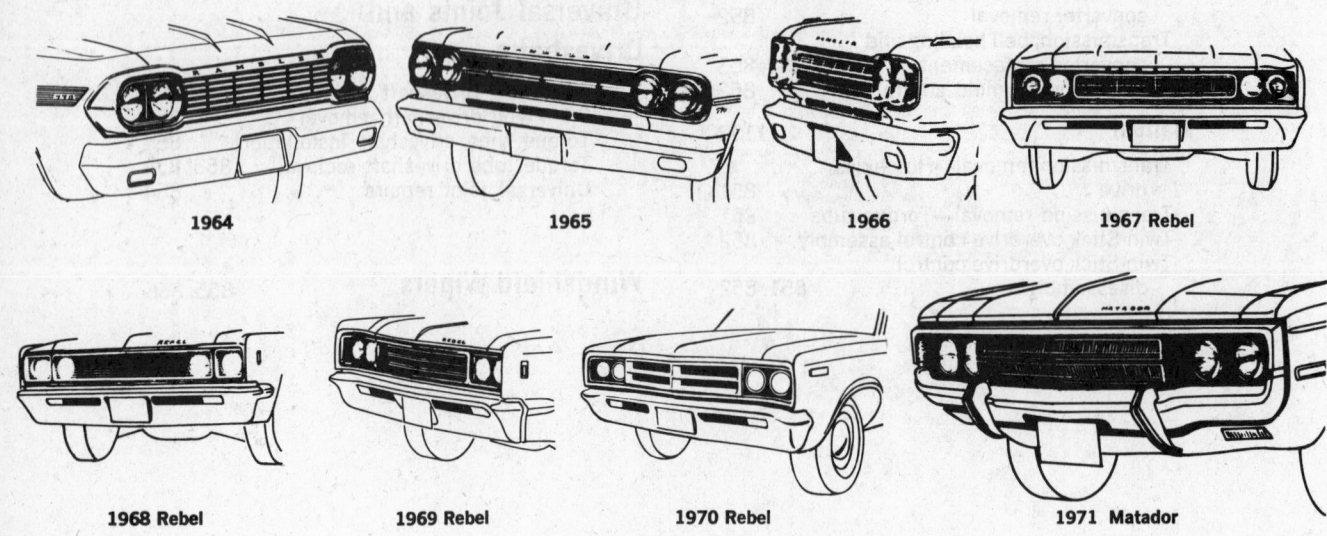

| 1964 | 1965 | 1966 | 1967 Rebel |

| 1968 Rebel | 1969 Rebel | 1970 Rebel | 1971 Matador |

SERIES 80, AMBASSADOR

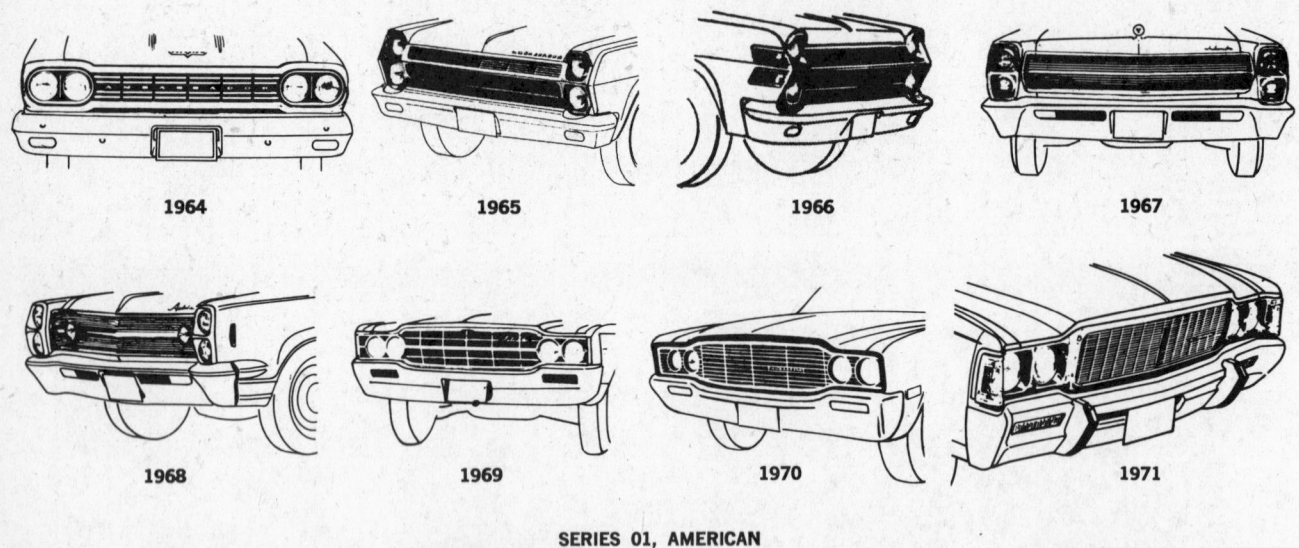

| 1964 | 1965 | 1966 | 1967 |

| 1968 | 1969 | 1970 | 1971 |

SERIES 01, AMERICAN

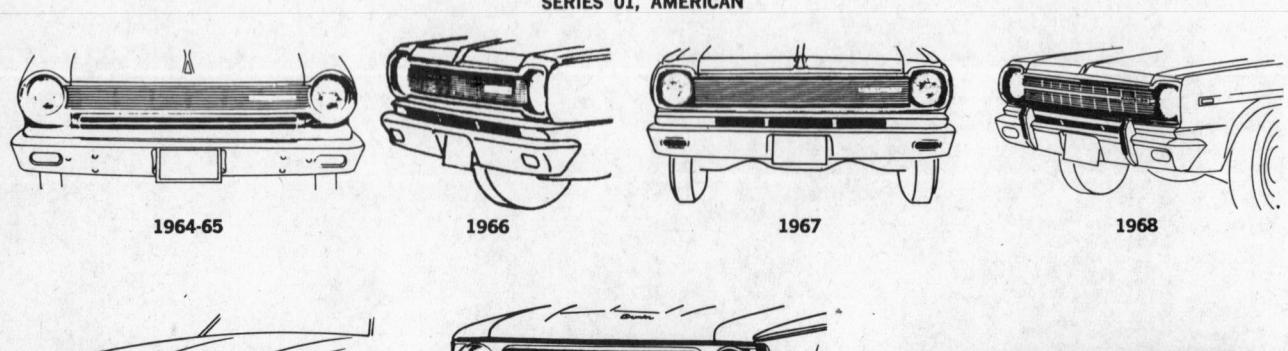

| 1964-65 | 1966 | 1967 | 1968 |

1970-71 Hornet

1970 Gremlin

1971 Gremlin

YEAR IDENTIFICATION

SERIES 50, MARLIN

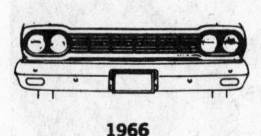

1966

1967

SERIES 70, JAVELIN

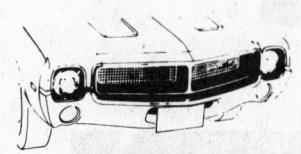

1968

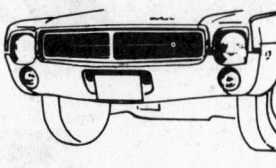

1969

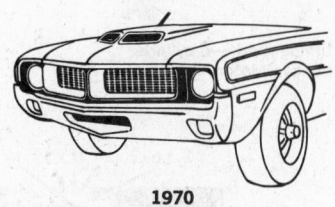

1970

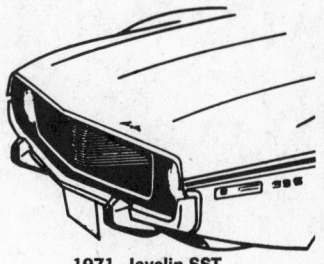

1971 Javelin SST

SERIES 30, AMX

1968-69 AMX

1970 AMX

FIRING ORDER

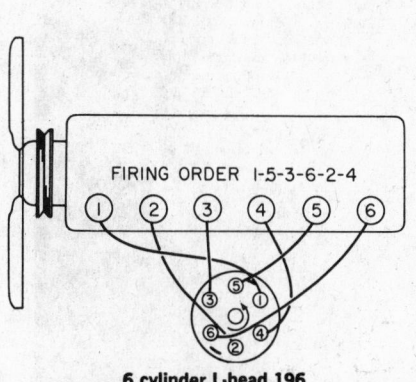

FIRING ORDER 1-5-3-6-2-4

6 cylinder L-head 196

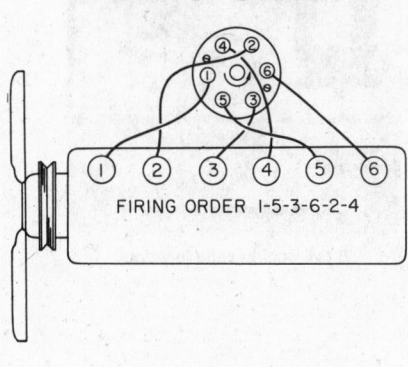

FIRING ORDER 1-5-3-6-2-4

6 cylinder OHV 232—1964

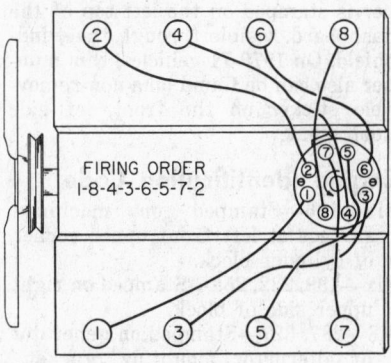

FIRING ORDER 1-8-4-3-6-5-7-2

V8 287 and 327—1964-66

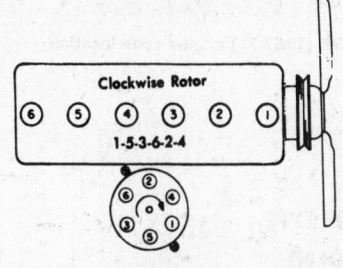

Clockwise Rotor
1-5-3-6-2-4

6 cylinder
OHV 199,
232 and 258—
1965-71

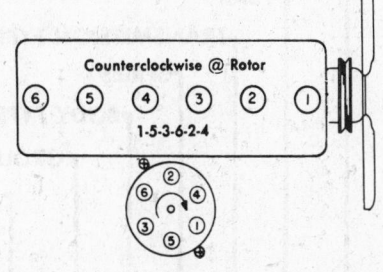

Counterclockwise @ Rotor
1-5-3-6-2-4

6 cylinder
OHV 196

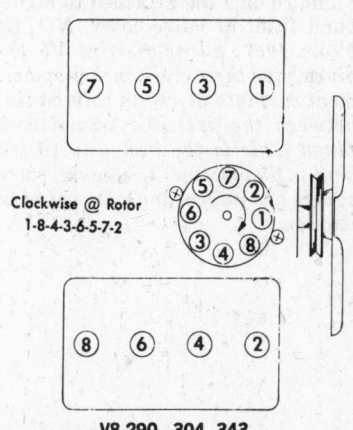

Clockwise @ Rotor
1-8-4-3-6-5-7-2

V8 290, 304, 343,
360, 390, and 401—1967-71

FIRING ORDER

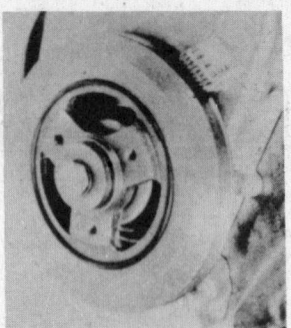

Ignition timing mark—6 cylinder
(© American Motors Corp.)

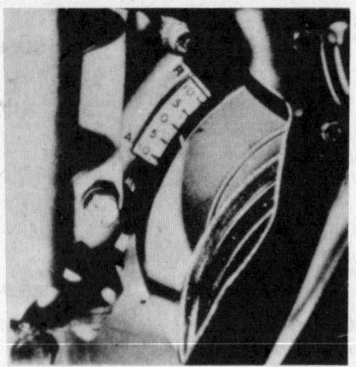

Timing marks— 290, 304,
343, 360, 390, and 401 V8

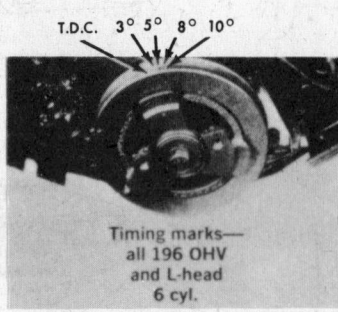

T.D.C. 3° 5° 8° 10°

Timing marks—
all 196 OHV
and L-head
6 cyl.

T.D.C. 5° 10°

Timing marks—
287 and 327 V8

CAR SERIAL NUMBER LOCATION

1964-68

The car serial number is found on a plate attached to the right front wheelwell, under the hood.

1969-71

The thirteen digit car serial number is stamped on the left top of the dashboard, visible through the windshield. On 1970-71 vehicles, this number also can be found on a non-removable sticker on the front, left-side door pillar.

Engine Identification Code

Six—196—Stamped on machined surface at left front upper corner of cylinder block.

Six—199, 232, 258—Stamped on right upper side of block.

V8—287, 327—Stamped on generator or alternator mounting bracket.

V8—290, 304, 343, 360, 390, 401—Stamped on a tag attached to right-hand front of valve cover. *NOTE: From 1967, all new series V8 engines have their cubic inch displacement cast into block, on both banks, between the first and second core plugs. This is the best way to tell engine displacement, because valve covers are interchangeable between engines.*

809L27

6 cyl. engine code location

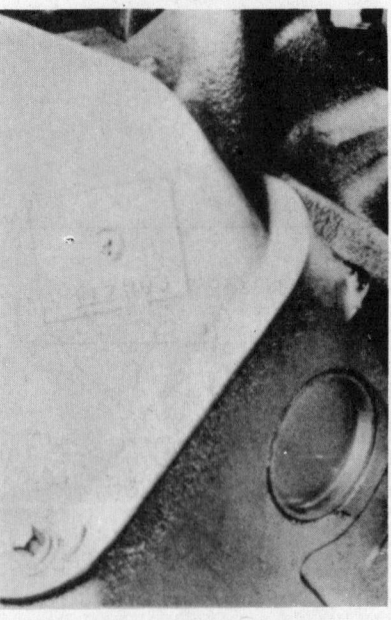

V8 (1967-71) engine code location

VEHICLE IDENTIFICATION

COMPANY
YEAR
TRANSMISSION TYPE
SERIES
BODY TYPE
GROUP
ENGINE TYPE
PLANT
SEQUENTIAL SERIAL NO.

A 9 S 0 5 0 A 1 0 0 0 0 1

Engine Identification Code

No. Cyls.	Cu. In. Displ.	Type	1964	1965	1966	1967	1968	1969	No. Cyls.	Cu. In. Displ.	Type	1964	1965	1966	1967	1968	1969	1970
6	199	American 1 BBL.			A	A		J	6	199	American		Q			J	J	
6	196	Ser. 01, 10	B						8	327	Ambassador 4 BBL.			Q				
6	232	American 2 BBL.			B	B		L	8	343	Ambassador 4 BBL.				Q	Z	Z	
8	290	American 2 BBL.				C		H	8	343	Ambassador 2 BBL.				R	S	S	
8	290	American 4 BBL.				D		N	6	232	Ambassador		S			L	L	
8	287	Ambassador		E					6	232	Marlin			S	S			
6	199	Classic 1 BBL.			E			J	8	287	Marlin		T					
6	232	American 1 BBL.				E		L	6	232	Marlin 2 BBL.				T			
6	232	Classic 1 BBL.			F			J	8	327	Classic		U					
6	232	Rebel 1 BBL.				F		L	8	327	Marlin			U				
6	196	Ser. 01, 10	G						8	290	Marlin 2 BBL.				U			
6	232	Classic 2 BBL.		•	G			L	6	232	Marlin 1 BBL.		V					
6	232	Rebel 2 BBL.				G		L	8	343	Marlin 2 BBL.			V				
8	327	Ambassador	H	H					6	232	American		W			L	L	
8	287	Classic 2 BBL.			H				8	327	Marlin 2 BBL.			W				
8	290	Rebel 2 BBL.				H		H	8	343	Marlin 4 BBL.				W	Z		
6	199	Classic			J			J	8	287	Classic	Z	Z					
8	327	Classic 4 BBL.			J				6	232	Marlin		2					
8	343	Rebel 2 BBL.				J		S	8	287	Marlin		3					
8	327	Classic 4 BBL.			K				8	327	Marlin		4					
8	343	Rebel 4 BBL.				K		Z	8	390	4-BBL.						W	X
6	232	Classic		L				L	8	360	4-BBL.							P
6	232	Ambassador 2 BBL.			M	M		L	8	360	2-BBL.							N
8	287	Ambassador 2 BBL.			N				8	304	2-BBL.							H
8	290	Ambassador 2 BBL.				N		H	6	232	2-BBL.						L	G
6	196	American		P					6	232	1-BBL.						L	E
8	327	Ambassador 2 BBL.			P				6	199	1-BBL.						J	A
6	232	Ambassador 2 BBL.					P	L	8	390	4-BBL. (Machine)							Y

AC GENERATOR AND REGULATOR SPECIFICATIONS

YEAR	Part No.	ALTERNATOR Field Current Draw @ 12V.	Output @ 15 Volts	Part No.	REGULATOR — Field Relay Air Gap (In.)	Point Gap (In.)	Volts to Close	Regulator Air Gap (In.)	Point Gap (In.)	Volts at 125°
1964–65	A-12NAM451	2.0–2.5	33	TUR-12NAM6	(Not adjustable, sealed at factory)					
1966	A-12NAM453	2.0–2.6	35	R2AM1	(Not adjustable, sealed at factory)					
	A-12NAM552	1.8–2.4	40	R2AM1	(Not adjustable, sealed at factory)					
1967–68	ALE6305(6)	2.3–2.4	40	R2AM1	(Not adjustable, sealed at factory)					
	ALK6310	2.4–2.5	35	VSC-62436	(Not adjustable, sealed at factory)					
	A12NAM453	2.0–2.6	35	R2AM1	(Not adjustable, sealed at factory)					
	ALK6309	2.4–2.5	35	VSC-62436	(Not adjustable, sealed at factory)					
	A12NAM552	1.8–2.4	40	R2AM1	(Not adjustable, sealed at factory)					
	A12NAM455	2.0–2.6	35	R2AM1	(Not adjustable, sealed at factory)					
	ALK6311(0)	2.4–2.5	35	VSC-6234L	(Not adjustable, sealed at factory)					
	A12NAM553	1.8–2.4	40	R2AM1	(Not adjustable, sealed at factory)					
1969–70	3195534(5)	2.4–2.5	35*	3195003	(Not adjustable, sealed at factory)					
	A12AM456(7)	2.0–2.6	35	R2AM4	(Not adjustable, sealed at factory)					
	A12NAM606(7)	1.8–2.4	55	R2AM4	(Not adjustable, sealed at factory)					
1971	ALK6312	2.4–2.5	35*	VSC-62437	(Not adjustable, sealed at factory)					
	A12NAM460	2.0–2.6	35	R2AM4	(Not adjustable, sealed at factory)					
	A12NAM555	1.8–2.4	55	R2AM4	(Not adjustable, sealed at factory)					

*—At 14.2 volts

TUNE-UP SPECIFICATIONS

YEAR	MODEL	SPARK PLUGS		DISTRIBUTOR		IGNITION TIMING (Deg.) ▲	CRANKING COMP. PRESSURE (Psi)	VALVES			FUEL PUMP PRESSURE (Psi)	IDLE SPEED (Rpm) ★
		Type	Gap (In.)	Point Dwell (Deg.)	Point Gap (In.)			Tappet (Hot) Clearance (In.)		Intake Opens (Deg.)		
								Intake	Exhaust			
1964–65	Ambassador—V8, A.T.	H18Y	.035	30	.020	5B	145	Zero	Zero	12½B	4½	475
	American—L-Hd. 6	H18Y	.035	32	.016	3B	130	.016C	.018C	10B	4½	550
	American—OHV 6, M.T.	H18Y	.035	32	.016	8B	145	.012	.016	12½B	4½	550
	American—OHV 6, A.T.	H18Y	.035	32	.016	10B	145	.012	.016	12½B	4½	500
	Classic (199 Cu. In.)	N14Y	.035	32	.016	5B	145	Zero	Zero	12½B	4½	550
	Classic—OHV 6, Iron Hd.	H18Y	.035	32	.016	5B	145	.012	.016	12½B	4½	550
	Classic—OHV 6, Alum. Hd.	H18Y	.035	32	.016	5B	145	Zero	Zero	12½B	4½	550
	Amer. & Classic (232, 1-BBL.)	N14Y	.035	32	.016	5B	145	Zero	Zero	12½B	4½	550
	Amer. & Classic (232, 2-BBL.)	N14Y	.035	32	.016	8B	145	Zero	Zero	12½B	4⅛	550
	Ambassador—V8, M.T.	H18Y	.035	30	.016	TDC	145	Zero	Zero	12½B	4½	550
	Ambassador—V8, A.T.	H18Y	.035	30	.016	5B	145	Zero	Zero	12½B	4½	475
	Classic—V8, M.T.	H18Y	.035	30	.016	TDC	145	Zero	Zero	12½B	4½	550
	Classic—V8, A.T.	H18Y	.035	30	.016	5B	145	Zero	Zero	12½B	4½	475
1966	American—OHV 6	N14Y	.035	32	.016	5B●	145	Zero	Zero	12½B	4½	550
	All—232—OHV 6	N14Y	.035	32	.016	5B●	145	Zero	Zero	12½B	4½	550
	All—287—V8	H14Y	.035	30	.016	5B●	145	Zero	Zero	12½B	4½	550
	All—327—V8	H14Y	.035	30	.016	5B	145	Zero	Zero	12½B	4½	550
1967	All—199—OHV 6 ⑥	N14Y	.035	32①	.016②	10B③	145	Zero	Zero	12½B	4½	525
	All—232—OHV 6 ⑥	N14Y	.035	32①	.016②	5B④	145	Zero	Zero	12½B	4½	600
	290—(200 HP)⑥	N12Y	.035	30	.016	3B	145	Zero	Zero	18½B	5½	600
	290—(225 HP)⑥	N12Y	.035	30	.016	TDC■	145	Zero	Zero	18½B	5½	600
	343—(235 HP)⑥	N12Y	.035	30	.016	TDC■	145	Zero	Zero	18½B	5½	600
	343—(280 HP)⑥	N12Y	.035	30	.016	TDC■	145	Zero	Zero	18½B	5½	600
1968–69	6 Cyl.—199, M.T.	N14Y	.035	32	.016	TDC	145	Zero	Zero	12½B	5	600
	6 Cyl.—199, A.T.	N14Y	.035	32	.016	5B	145	Zero	Zero	12½B	5	525
	6 Cyl.—232 (Rogue) A.T.	N14Y	.035	32	.016	5B	145	Zero	Zero	12½B	5	525
	6 Cyl.—232, A.T.	N14Y	.035	32	.016	TDC	145	Zero	Zero	12½B	5	525
	6 Cyl.—232, M.T.	N14Y	.035	32	.016	TDC	145	Zero	Zero	12½B	5	600
	V8—290, M.T.	N12Y	.035	30	.016	TDC■	145	Zero	Zero	18½B⑤	6	650
	V8—290, A.T.	N12Y	.035	30	.016	TDC■	145	Zero	Zero	18½B③	6	550
	V8—343, 390, M.T.	N12Y	.035	30	.016	TDC■	145	Zero	Zero	18½B③	6	650
	V8—343, 390, A.T.	N12Y	.035	30	.016	TDC■	145	Zero	Zero	18½B③	6	550
1970	6 Cyl.—199, M.T.	N14Y	.035	31–34	.016	3B	145	Zero	Zero	12½B	4–5½	⑦
	6 Cyl.—199, A.T.	N14Y	.035	31–34	.016	3B	145	Zero	Zero	12½B	4–5½	⑦
	6 Cyl.—232 (01 Series) A.T.	N14Y	.035	31–34	.016	3B	145	Zero	Zero	12½B	4–5½	⑦
	6 Cyl.—232, A.T.	N14Y	.035	31–34	.016	3B	145	Zero	Zero	12½B	4–5½	⑦
	6 Cyl.—232, M.T.	N14Y	.035	31–34	.016	3B	145	Zero	Zero	12½B	4–5½	⑦
	V8—304, M.T.	N12Y	.035	29–31	.016	5B	145	Zero	Zero	18½B	4–5½	⑥
	V8—360, 2-V.	N12Y	.035	29–31	.016	5B	145	Zero	Zero	18½B	4–5½	①
	V8—360, 390, M.T., A.T.⑨	N12Y	.035	29–31	.016	5B■	145	Zero	Zero	18½B⑤	4–5½	①
	V8—390, 4-V ⑩	N12Y	.035	29–31	.016	TDC■	145	Zero	Zero	18½B⑤	4–5½	①
	V8—390 ⑥	N10Y	.035	29–31	.016	TDC■	145	Zero	Zero	46B	4–5½	600
1971	6 Cyl.—232	N14Y	.035	31–34	.016	**	145	Zero	Zero	12½B	4–5½	⑦
	6 Cyl.—258	N14Y	.035	31–34	.016	**	145	Zero	Zero	12½B	4–5½	⑦
	V8—304, 360	N12Y	.035	29–31	.016	**	145	Zero	Zero	18½B	4–5½	①
	V8—401	N12Y	.035	29–31	.016	**	145	Zero	Zero	N.A.	4–5½	⑧

TUNE-UP SPECIFICATIONS, continued

▲—With vacuum advance disconnected. Add 50 rpm when equipped with air conditioning. NOTE: These settings are only approximate. Engine design, altitude, temperature, fuel octane rating and the condition of the individual engine are all factors which can influence timing. The limiting advance factor must, therefore, be the "knock point" of the individual engine.

★—With manual transmission in N and automatic in D.

●—With premium fuel 10B.

■—Regular fuel not recommended.

B—Before top dead center.

M.T.—Manual transmission.

A.T.—Automatic transmission.

⊙—Cars with exhaust emission control = timing, TDC.

①—Prestolite = 39°.

②—Prestolite = .019 in.

③—w/Air Guard—3 B.

④—Premium fuel = 8B.

⑤—w/H.L. Cam 46°B.

⑥—Rebel Machine.

⑦—For six cylinder engines—600 rpm M.T.; 550 rpm A.T. with air cond. off, parking brake applied, and A.T. in D1.

⑧—For V8 engines—650 rpm M.T.; 600 rpm A.T. with air cond. off, parking brake applied, and A.T. in D1.

⑨—Prior to 390 engine 209x26 (Type 1 distributor).

⑩—After 390 engine 209x26 (Type 2 distributor).

**—See engine decal.

CAUTION

General adoption of anti-pollution laws has changed the design of almost all car engine production to effectively reduce crankcase emission and terminal exhaust products. It has been necessary to adopt stricter tune-up rules, especially timing and idle speed procedures. Both of these values are peculiar to the engine and to its application, rather than to the engine alone. With this in mind, car manufacturers supply idle speed data for the engine and application involved. This information is clearly displayed in the engine compartment of each vehicle.

VALVE SPECIFICATIONS

YEAR AND MODEL	SEAT ANGLE (DEG.)	FACE ANGLE (DEG.)	VALVE LIFT INTAKE (IN.)	VALVE LIFT EXHAUST (IN.)	VALVE SPRING PRESSURE (VALVE OPEN) LBS. @ IN.	VALVE SPRING INSTALLED HEIGHT (IN.)	STEM TO GUIDE CLEARANCE (IN.)		STEM DIAMETER (IN.)	
							Intake	Exhaust	Intake	Exhaust
1964 American, 6 Cyl., L-hd.	45	44	.324	.322	79 @ 1⁷/₁₆	1³/₄	.0018–.0033	.0018–.0033	.3407–.3412	.3407–.3412
American & Classic, 6 Cyl. Iron	45	44	.371	.367	120 @ 1⁷/₁₆	1¹³/₁₆	.0015–.0035	.002–.004	.3411–.3417	.3405–.3415
Classic, 6 Cyl., OHV—Alum.	45	44	.406	.406	155 @ 1⁷/₁₆	1¹³/₁₆	.0015–.0035	.002–.004	.3411–.3417	.3405–.3415
Classic & Ambassador V8, OHV	①	②	.375	.375	155 @ 1⁷/₁₆	1¹³/₁₆	.001–.003	.001–.003	.3715–.3725	.3715–.3725
1965–66 6 Cyl.—L-hd 195.6* Cu. In.	45	44	.324	.322	79 @ 1⁷/₁₆	1³/₄	.0025–.004	.0025–.004	.3407–.3412	.3407–.3412
6 Cyl.—OHV 199 Cu. In.	①	②	.375	.375	155 @ 1⁷/₁₆	1¹³/₁₆	.001–.003	.001–.003	.3715–.3725	.3715–.3725
6 Cyl.—OHV 232 Cu. In.	①	②	.375	.375	155 @ 1⁷/₁₆	1¹³/₁₆	.001–.003	.001–.003	.3715–.3725	.3715–.3725
V8—287 Cu. In.	①	②	.375	.375	155 @ 1⁷/₁₆	1¹³/₁₆	.001–.003	.001–.003	.3718–.3725	.3718–.3725
V8—327 Cu. In.	①	②	.375	.375	155 @ 1⁷/₁₆	1¹³/₁₆	.001–.003	.001–.003	.3718–.3725	.3718–.3725
1967–68 6 Cyl.—OHV 199 Cu. In.	①	②	.254†	.254†	195 @ 1⁷/₁₆	1¹³/₁₆	.001–.003	.001–.003	.3715–.3725	.3715–.3725
6 Cyl.—OHV 232 Cu. In.	①	②	.254†	.254†	195 @ 1⁷/₁₆	1¹³/₁₆	.001–.003	.001–.003	.3715–.3725	.3715–.3725
V8—290 Cu. In.	①	②	.265†	.265†	194 @ 1¹³/₃₂	1¹³/₁₆	.001–.003	.001–.003	.3715–.3725	.3715–.3725
V8—343 Cu. In.	①	②	.265†	.265†	194 @ 1¹³/₃₂	1¹³/₁₆	.001–.003	.001–.003	.3715–.3725	.3715–.3725
V8—390 Cu. In.	①	②	.265†	.265†	250 @ 1²¹/₆₄	1¹³/₁₆	.001–.003	.001–.003	.3715–.3725	.3715–.3725
1969 6 Cyl.—OHV (199)	③	②	.254†	.254†	195 @ 1⁷/₁₆	1¹³/₁₆	.001–.003	.001–.003	.3715–.3725	.3715–.3725
6 Cyl.—OHV (232)	③	②	.254†	.254†	195 @ 1⁷/₁₆	1¹³/₁₆	.001–.003	.001–.003	.3715–.3725	.3715–.3725
V8—290 Cu. In.	①	④	.265†	.265†	200 @ 1²⁵/₆₄	1¹³/₁₆	.001–.003	.001–.003	.3715–.3725	.3715–.3725
V8—343 Cu. In.	①	④	.265†	.265†	200 @ 1²⁵/₆₄	1¹³/₁₆	.001–.003	.001–.003	.3715–.3725	.3715–.3725
V8—390 Cu. In.	①	④	.265†	.265†	200 @ 1²⁵/₆₄	1¹³/₁₆	.001–.003	.001–.003	.3715–.3725	.3715–.3725
1970 6 Cyl.—199, 232 Cu. In.	①	②	.254†	.254†	195 @ 1⁷/₁₆	1¹³/₁₆	.001–.003	.001–.003	.3715–.3725	.3715–.3725
V8—304, 360 Cu. In.	①	④	.265†	.265†	195 @ 1⁷/₁₆	1¹³/₁₆	.001–.003	.001–.003	.3715–.3725	.3715–.3725
V8—390 Cu. In.	①	④	.265†	.265†	200 @ 1²⁵/₆₄	1¹³/₁₆	.001–.003	.001–.003	.3715–.3725	.3715–.3725
V8—390 Cu. In. ⑤	①	④	.287†	.287†	189 @ 1²³/₆₄	1¹³/₁₆	.001–.003	.001–.003	.3715–.3725	.3715–.3725
1971 6 Cyl.—232 Cu. In.	①	②	.254†	.254†	195 @ 1⁷/₁₆	1¹³/₁₆	.001–.003	.001–.003	.3715–.3725	.3715–.3725
6 Cyl.—258 Cu. In.	①	②	.254†	.254†	195 @ 1⁷/₁₆	1¹³/₁₆	.001–.003	.001–.003	.3715–.3725	.3715–.3725
V8—304, 360 Cu. In.	①	④	.265†	.265†	195 @ 1⁷/₁₆	1¹³/₁₆	.001–.003	.001–.003	.3715–.3725	.3715–.3725
V8—401 Cu. In.	①	④	...	...	189 @ 1²³/₆₄	1¹³/₁₆	.001–.003	.001–.003	.3715–.3725	.3715–.3725

*—Not used in 1966.

①—Intake 30°, Exhaust 45°.

②—Intake 29°, Exhaust 44°.

③—Intake 30°, Exhaust 44°.

④—Intake 29°, Exhaust 44½°.

⑤—Rebel Machine.

†—Cam lobe lift.

GENERAL ENGINE SPECIFICATIONS

YEAR	CU. IN. DISPLACEMENT	CARBURETOR	DEVELOPED HORSEPOWER @ RPM	DEVELOPED TORQUE @ RPM (FT. LBS.)	A.M.A. HORSEPOWER	BORE & STROKE (IN.)	ADVERTIZED COMPRESSION RATIO	VALVE LIFTER TYPE	NORMAL OIL PRESSURE (PSI)
1964–65	195.6	1-BBL.	90 @ 3800	160 @ 1600	23.4	3.125 x 4.250	8.0-1	Mech.	50–58
	195.6	1-BBL.	125 @ 4200	180 @ 1600	23.4	3.125 x 4.250	8.7-1	Mech.	50–58
	195.6	2-BBL.	138 @ 4500	185 @ 1800	23.4	3.125 x 4.250	8.7-1	Mech.	60–65
	195.6	1-BBL.	127 @ 4200	180 @ 1600	23.4	3.125 x 4.250	8.7-1	Mech.	60–65
	195.6	2-BBL.	138 @ 4200	185 @ 1800	23.4	3.125 x 4.250	8.7-1	Mech.	60–65
	199	1-BBL.	128 @ 4400	182 @ 1600	33.7	3.750 x 3.000	8.5-1	Hyd.	40–50
	232	1-BBL.	145 @ 4300	215 @ 1600	33.7	3.750 x 3.500	8.5-1	Hyd.	40–50
	232	2-BBL.	155 @ 4400	222 @ 1600	33.7	3.750 x 3.500	8.5-1	Hyd.	40–50
	287	2-BBL.	198 @ 4700	280 @ 2600	45.0	3.750 x 3.250	8.7-1	Hyd.	55–60
	327	2-BBL.	250 @ 4700	340 @ 2600	51.2	4.000 x 3.250	8.7-1	Hyd.	55–60
	327	4-BBL.	270 @ 4700	360 @ 2600	51.2	4.000 x 3.250	9.7-1	Hyd.	55–60
1966	199	1-BBL.	128 @ 4400	182 @ 1600	33.7	3.750 x 3.000	8.5-1	Hyd.	55–60
	232	1-BBL.	145 @ 4300	215 @ 1600	33.7	3.750 x 3.500	8.5-1	Hyd.	55–60
	232	2-BBL.	155 @ 4400	222 @ 1600	33.7	3.750 x 3.500	8.5-1	Hyd.	55–60
	287	2-BBL.	198 @ 4700	280 @ 2600	45.0	3.750 x 3.250	8.7-1	Hyd.	55–60
	327	2-BBL.	250 @ 4700	340 @ 2600	51.2	4.000 x 3.250	9.7-1	Hyd.	55–60
	327	4-BBL.	270 @ 4700	360 @ 2600	51.2	4.000 x 3.250	9.7-1	Hyd.	55–60
1967–69	199	1-BBL.	128 @ 4400	182 @ 1600	33.7	3.750 x 3.000	8.5-1	Hyd.	55–60
	232	1-BBL.	145 @ 4300	215 @ 1600	33.7	3.750 x 3.500	8.5-1	Hyd.	55–60
	232	2-BBL.	155 @ 4400	222 @ 1600	33.7	3.750 x 3.500	8.5-1	Hyd.	55–60
	290	2-BBL.	200 @ 4600	285 @ 2800	45.0	3.750 x 3.280	9.0-1	Hyd.	55–60
	290	4-BBL.	225 @ 4700	300 @ 3200	45.0	3.750 x 3.280	10.0-1	Hyd.	55–60
	343	2-BBL.	235 @ 4400	345 @ 2600	53.3	4.080 x 3.280	9.0-1	Hyd.	55–60
	343	4-BBL.	280 @ 4800	365 @ 3000	53.3	4.080 x 3.280	10.2-1	Hyd.	55–60
	390	4-BBL.	315 @ 4600	425 @ 3200	55.5	4.165 x 3.574	10.2-1	Hyd.	55–60
1970	199	1-BBL.	128 @ 4400	182 @ 1600	33.7	3.750 x 3.000	8.5-1	Hyd.	55–60
	232	1-BBL.	145 @ 4300	215 @ 1600	33.7	3.750 x 3.500	8.5-1	Hyd.	①
	232	2-BBL.	155 @ 4400	222 @ 1600	33.7	3.750 x 3.500	8.5-1	Hyd.	①
	304	2-BBL.	210 @ 4400	305 @ 2800	45.0	3.750 x 3.440	9.0-1	Hyd.	①
	360	2-BBL.	245 @ 4400	365 @ 2400	53.3	4.080 x 3.440	9.0-1	Hyd.	①
	360	4-BBL.	290 @ 4800	395 @ 3200	53.3	4.080 x 3.440	10.0-1	Hyd.	①
	390	4-BBL.	325 @ 5000	420 @ 3200	55.5	4.165 x 3.574	10.0-1	Hyd.	①
	390*	4-BBL.	340 @ 5100	430 @ 3600	55.5	4.165 x 3.574	10.1-1	Hyd.	①
1971	232	1-BBL.	145 @ 4300	215 @ 1600	33.7	3.750 x 3.500	8.0-1②	Hyd.	①
	258	1-BBL.	160 @ 4000	245 @ 1600	33.7	3.750 x 3.900	8.5-1②	Hyd.	①
	304	2-BBL.	210 @ 4400	305 @ 2800	45.0	3.750 x 3.440	8.4-1②	Hyd.	①
	360	2-BBL.	245 @ 4400	365 @ 2400	53.3	4.080 x 3.440	8.5-1②	Hyd.	①
	360	4-BBL.	290 @ 4800	395 @ 3200	53.3	4.080 x 3.440	8.5-1②	Hyd.	①
	401	4-BBL.	330 @ 5200	430 @ 3300	55.9	4.170 x 3.680	9.5-1③	Hyd.	①

*—Rebel Machine.   ① —13 psi @ idle; 45-85 psi at 40 mph and above.   ② —For unleaded fuel.   ③ —Premium fuel.

CRANKSHAFT BEARING JOURNAL SPECIFICATIONS

YEAR	MODEL	MAIN BEARING JOURNALS (IN.)				CONNECTING ROD BEARING JOURNALS (IN.)		
		Journal Diameter	Oil Clearance	Shaft End-Play	Thrust On No.	Journal Diameter	Oil Clearance	End-Play
1964	All 6 Cyl.	2.4794	.0012	.005	1	2.0952	.0012	.010
	All V8	2.4988	.0015	.005	1	2.2486	.0017	.010
1965–66	6 Cyl., L-Hd.	2.4794	.0012	.005	1	2.0952	.0012	.010
	6 Cyl., OHV	2.4987	.0012	.005	3	2.0952	.0012	.010
	V8, All	2.4988	.0015	.005	1	2.2486	.0017	.010
1967–71	6 Cyl. OHV	2.4986-2.5001	.001-.002	.0015-.007	3	2.0934-2.0955	.001-.002	.008-.010
	V8	①	.001-.002②	.003-.008	3	③	.001-.002	.009-.015

① —Nos. 1-4—2.7474-2.7489; No. 5—2.7464-2.7479.
② —Rear main—.002-.003.
③ —V8 290, 343, 304 and 360—2.7464-2.7479; V8 390—2.2471-2.2492.

CAPACITIES

YEAR	MODEL	ENGINE CRANKCASE ADD 1 Qt. FOR NEW FILTER	TRANSMISSIONS Pts. TO REFILL AFTER DRAINING Manual 3-Speed	4-Speed	Automatic	DRIVE AXLE (Pts.)	GASOLINE TANK (Gals.)	COOLING SYSTEM (Qts.) WITH HEATER
1964	195.6—6 Cyl.	4	1½ ①	—	18	3	16	12③
	232—6 Cyl.	4	1½ ①	—	18	3	19③	10½
	287—V8	4	2¼	—	22	4	19③	19
	327—V8	4	4④	—	22	4	19③	19
1965	195.6—6 Cyl.	4	1½ ①	—	18	3	16	12⑥
	199—6 Cyl.	4	1½ ①	—	18	3	16	12⑥
	232—6 Cyl.	4	2¼ ②	3½	18	4	19⑤	10½
	287—6 Cyl.	4	3½ ⑦	3½	22	4	19③	19
	327—V8	4	4⑧	3½	22	4	19⑤	19
1966	199—6 Cyl.	4	1½ ①	—	18	3	16	10½
	232—6 Cyl.	4	2¼ ②	—	18	4	19③	10½
	287, 290—V8	4	2¼ ②	3½	22	4	19③	19
	327—V8	4	4⑦	3½	22	4	19③	19
1967	199—6 Cyl	4	1½ ①	—	18	3	16	10½
	232—6 Cyl.	4	1½ ①	—	18	4	21½⑨	10½
	290—V8	4	2½ ⑩	3½	18	4	21½⑨	14
	343—V8	4	—	3½	20	4	21½⑨	13
1968–69	199—6 Cyl.	4	1½ ①	—	18 ⑭	3	16	10½
	232—6 Cyl.	4	2½ ①	—	18 ⑭	3 ⑪	⑫	10½
	290—V8	4	3 ⑬	3½ ⑮	18 ⑭	4	⑫	14
	343. 390—V8	4	—	3½ ⑮	20	4	⑫	13
1970–71	199—6 Cyl.	4	1½	—	19	3	⑯	10½
	232, 258—6 Cyl.	4	1½	—	19	3	⑯	10½
	304—V8	4	3	—	19	4	⑯	14
	360,390,401—V8	4	—	2½	20	4	⑯	13

① Overdrive—2¾.
② Two-seat sta. wag.—19.
   Three-seat sta. wag.—17.
③ L-hd.—12, OHV—11, Alum.—10½ .
④ Overdrive—4.
⑤ Three-seat sta. wag.—17.
⑥ OHV—11.
⑦ Overdrive—3½.

⑧ Overdrive—4.
⑨ Three-seat sta. wag.—19.
⑩ Overdrive—3¾.
⑪ 01, 70—3; 10, 80—4.
⑫ 30, 70—19; 10, 80—21½
⑬ Overdrive 3.1
⑭ 1969—19.
⑮ 1969—2½.

⑯ —1970-71 Gremlin—21. 1970 Rebel and Ambassador sedans—21½; all others— 19.
1971 Ambassador sedans—19.5, Ambassador wagon—17.0.
1971 Matador sedans—19.5, Matador wagon—17.0.
1970-71 Hornet—17.0.
1971 Javelin—17.0.

BATTERY AND STARTER SPECIFICATIONS

YEAR	MODEL	BATTERY Ampere Hour Capacity	Volts	Group Number	Terminal Grounded	STARTERS Lock Test Amps.	Volts	Torque	No-Load Test Amps.	Volts	RPM	Brush Spring Tension (Oz.)
1964–66	6 Cyl., All	50	12	2-SM	Neg.	285	4.0	6.5	62	10.6	7,850	35
	V8, Classic & Ambassador	60	12	2-SMH	Neg.	290	4.0	9.0	82	10.6	4,350	48
1967–69	6 Cyl.	50	12	2-CM50	Neg.	290	4.3	6.5	63	10.6	7,850	35
	V8	60	12	2-SM60	Neg.	500	4.5	9.0	70	12.0	9,500	40
	V8	70	12	2-SH70	Neg.	500	4.5	10.0	70	12.0	9,500	40
1970–71	6 Cyl.	50	12	2-CM50	Neg.	500	4.3	6.5	65	12.0	9,250	40
	V8	60	12	2-SM60	Neg.	500	4.5	9.0	65	12.0	9,250	40
	V8 (opt.)	70	12	2-SH70	Neg.	500	4.5	10.0	65	12.0	9,250	40

GENERAL CHASSIS AND BRAKE SPECIFICATIONS

YEAR	MODEL	CHASSIS		BRAKE CYLINDER BORE			BRAKE DRUM	
		Overall Length (In.)	Tire Size	Master Cylinder (In.)	Wheel Cylinder Diameter (In.)		Diameter (In.)	
					Front	Rear	Front	Rear
1964	American, All	177.3	6.00 x 14	1.0	1.0	13/16	9	9
	Rambler, Classics, 6 Cyl.	190.0	6.50 x 14	1.0	1 1/8	15/16	9	9
	Classic & Ambassador, V8	190.0	7.50 x 14	1.0	1 1/8	7/8	10	10
1965	American (Series 01)	177.25	6.45 x 14	1.0	1 1/8	15/16	9	9
	Classic, 6 Cyl. (Series 10)	195.0	6.95 x 14	1.0	1 1/8★	15/16●	9	9
	Classic, V8 (Series 10)	195.0	7.75 x 14	1.0	1 1/8★	7/8†	10	10
	Ambassador, 6 Cyl. (Series 80)	200.0	7.35 x 14	1.0	1 3/16★	15/16●	9	9
	Ambassador, V8 (Series 80)	200.0	7.75 x 14	1.0	1 3/16★	15/16●	10	10
1966	American (Series 01)	181.0	6.45 x 14	1.0	1 1/8★	15/16	9	9
	Marlin (Series 50)	195.0	7.35 x 14▲	1.0	1 1/8★	15/16●	10	10
	Classic, 6 Cyl. (Series 10)	195.0	6.95 x 14	1.0	1 1/8★	15/16●	9	9
	Classic, 8 Cyl. (Series 10)	195.0	7.75 x 14	1.0	1 3/16★	15/16●	10	10
	Ambassador, 6 Cyl. (Series 80)	200.0	7.35 x 14	1.0	1 3/16★	15/16●	9	9
	Ambassador, 8 Cyl. (Series 80)	200.0	7.75 x 14	1.0	1 3/16★	15/16●	10	10
	Ambassador, Sta. Wag. (Series 80)	199.0	7.75 x 14	1.0	1 1/8	15/16●	10	10
1967-69	American, Rambler (Series 01)	181.0	6.45 x 14■	1.0	1 1/8②③	15/16④	9②⑦	9①
	American (Wagon) (Series 01)	181.0	6.95 x 14★★	1.0	1 1/8②③	15/16④	9②⑦	9①
	Rebel (Series 10)	197.0	7.35 x 14☆	1.0	1 1/8②③	15/16④	9②⑦	9①
	Rebel (Wagon) (Series 10)	198.0	7.75 x 14☆☆	1.0	1 3/32②③	15/16④	10⑦	10
	Marlin, 6 Cyl. (Series 50)	201.5	7.35 x 14☆	1.0	1 1/8③	15/16④	10⑦	10
	Marlin, V8 (Series 50)	201.5	7.75 x 14☆☆	1.0	1 3/16③	15/16④	10⑦	10
	Ambassador (Series 80)	202.5⑨	7.35 x 14☆	1.0	1 1/8②③	15/16④	10⑦	10
	Ambassador (Wagon) (Series 80)	203.0⑩	8.25 x 14	1.0	1 3/16③	15/16	10⑦	10
	Javelin (Series 70)	189.2	6.95 x 14①	1.0	1 3/16③	15/16④	10⑦	10
	AMX (Series 30)	177.2	7.35 x 14⑭	1.0	2.0	7/8	⑦	10
1970	Hornet (Series 01)	179.3	B78 x 14⑪	1.0	1 1/8②③	15/16④	9②⑦	9①
	Gremlin	162.3	6.00 x 13⑱	1.0	1 1/8	7/8	9	9
	Rebel (Series 10)	199.0	E78 x 14⑫	1.0	1 1/8②③	15/16③	⑳⑦	⑩
	Rebel (Wagon) (Series 10)	198.0	G78 x 14⑬	1.0	1 3/32②③	15/16③	10⑦	10
	Ambassador (Series 80)	208.0	F78 x 14⑯	1.0	1 1/8②③	15/16④	10⑦	10
	Ambassador (Wagon) (Series 80)	207.0	H78 x 14⑰	1.0	1 3/16③	15/16	10⑦	10
	Javelin (Series 70)	191.0	C78 x 14⑮	1.0	1 3/16③	15/16⑤	10⑦	10
	AMX (Series 30)	179.0	E78 x 14⑭	1.0	2.0	7/8	⑦	10
1971	Hornet 2-dr. sedan (Series 01)	179.3	6.45 x 14㉑	—	—	—	9-6 Cyl. 10-V8 ⑲	9-6 Cyl. 10-V8
	Hornet 4-dr. sedan (Series 01)	179.3	6.95 x 14㉑	—	—	—	9-6 Cyl. 10-V8 ⑲	9-6 Cyl. 10-V8
	Hornet Sportabout S/W	179.3	6.95 x 14㉑	—	—	—	9-6 Cyl. 10-V8 ⑲	9-6 Cyl. 10-V8
	Hornet SC/360	179.3	D70 x 14	—	—	—	9-6 Cyl. 10-V8 ⑲	9-6 Cyl. 10-V8
	Matador 2-dr. h.t. (Series 10)	206.0	E78 x 14⑫	—	—	—	10 ⑲	10
	Matador 4-dr sedan (Series 10)	206.0	E78 x 14⑫	—	—	—	10 ⑲	10
	Matador (Wagon)(Series 10)	205.0	G78 x 14⑬	—	—	—	10 ⑲	10
	Ambassador 2-dr (Series 80)	210.8	F78 x 14⑯	—	—	—	10 ⑲	10
	Ambassador 4-dr (Series 80)	210.8	F78 x 14⑯	—	—	—	10 ⑲	10
	Ambassador (Wagon)(Series 80)	209.7	H78 x 14	—	—	—	10 ⑲	10
	Javelin-AMX (Series 70)	191.8	C78 x 14⑮	—	—	—	9-6 Cyl. 10-V8 ⑲	9-6 Cyl. 10-V8
	Gremlin 2-pass. sedan	161.3	6.00 x 13⑳	—	—	—	9	9
	Gremlin 4-pass. sedan	161.3	6.00 x 13⑳	—	—	—	9	9

GENERAL CHASSIS AND BRAKE SPECIFICATIONS continued

★—With disc brakes—2 in.
●—With disc brakes, sedan, H.T., conv.—1 in., sta. wag.—1⅛ in.
†—With disc brakes, sedan, H.T., conv.—¹⁵⁄₁₆ in., sta. wag.—1 in.
▲—Opt.—7.75 x 14.
■—6.95 x 14 & 6.85 x 15 optional.
★★—6.85 x 15 optional.
☆—7.75 x 14 & 7.35 x 15 optional.
✩—8.25 x 14 & 7.75 x 15 optional.
①—10 in. with V8 engine.
②—V8—1⁷⁄₁₆ in.
③—Disc brakes 2 in.
④—V8—²⁹⁄₃₂ in.
⑤—6 Cyl. disc brakes 1 in.
⑥—Disc brakes 1⅛ in.
⑦—Disc diam. 11⁷⁄₁₆ in.
⑧—V8—7.35 x 14.
⑨—1969—206.5.
⑩—1969—207.

⑪—V8—C78 x 14 std., D78 x 14 and D70 x 14 optional.
⑫—F78 x 14 optional; Machine—E60 x 15 std.
⑬—H78 x 14 optional.
⑭—E70 x 14 and F70 x 14 optional; E60 x 15 available special order for 1970–71
⑮—6 cyl. size std.; 6 cyl. optional tires: D78 x 14, E78 x 14, E70 x 14. V8 size std.—D78 x 14, with E78 x 14, E70 x 14, F70 x 14 and E60 x 15 optional.
⑯—G78 x 14 and H78 x 14 optional.
⑰—Recommended size; std. size is F78 x 14
⑱—6.45 x 13 and B78 x 14 optional.
⑲—Ventilated disc brakes optional.
⑳—6.45 x 13, B78 x 14, D70 x 14 optional.
㉑—6.95 x 14, B78 x 14, C78 x 14 optional—6 cyl. C78 x 14, D78 x 14, D70 x 14 optional—V8

S/W—Sedan/wagon configuration.

TORQUE SPECIFICATIONS

YEAR	MODEL	CYLINDER HEAD BOLTS (FT. LBS.)	ROD BEARING BOLTS (FT. LBS.)	MAIN BEARING BOLTS (FT. LBS.)	CRANKSHAFT BALANCER BOLT (FT. LBS.)	FLYWHEEL TO CRANKSHAFT BOLTS (FT. LBS.)	MANIFOLD (FT. LBS.)	
							Intake	Exhaust
1964–66	6 Cyl. L-Hd. Ex. 199 & 232	60	30	70	80	105	20	25
	6 Cyl. Aluminum	50	30	58	80	105	★	☆
	6 Cyl. OHV (199 & 232)	85	30	80	80	105	20	25
	All V8	60	50	65▲	80	105	20	25
1967–71	6 Cyl.	85	30	80	80	105	20	25
	All V8	100	30	100	55	110	42	33

▲—No. 5 (rear main)—55 ft. lbs.
★—Cover screw, ⁵⁄₁₆ - 15 ft. lbs.
    ¼ - 11 ft. lbs.

☆—End nuts = 10 ft. lbs.
☆—Center flange = 25 ft. lbs.

CYLINDER-HEAD BOLT TIGHTENING SEQUENCE

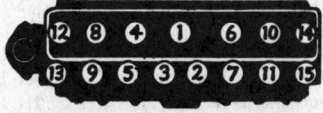

196 6 cylinder cast-iron OHV

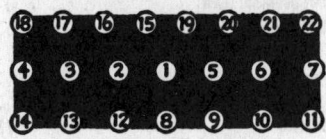

196 6 cylinder L-head

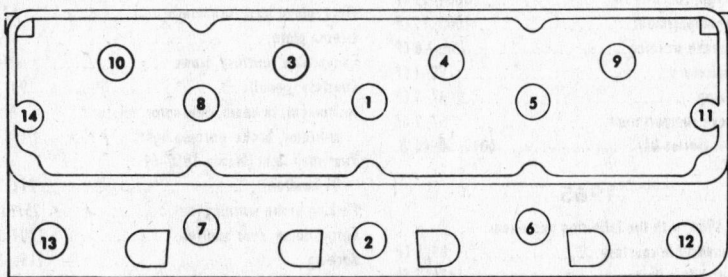

290, 304, 343, 360, 390, and 401 V8

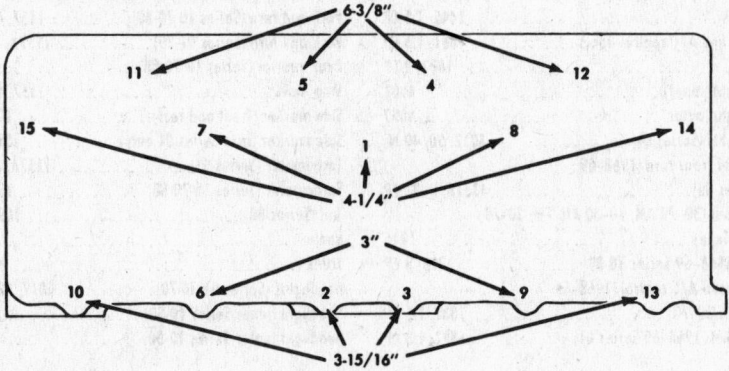

287 and 327 V8

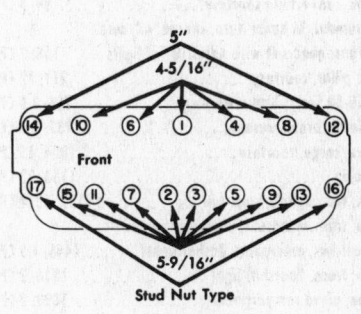

196 6 cylinder alum. OHV

WHEEL ALIGNMENT

YEAR	MODEL	CASTER		CAMBER		TOE-IN (In.)	KING-PIN INCLINATION (Deg.)	WHEEL PIVOT RATIO	
		Range (Deg.)	Pref. Setting (Deg.)	Range (Deg.)	Pref. Setting (Deg.)			Inner Wheel	Outer Wheel
1964	Series 01, W/PS	1½-2P	2P	¼N-¼P	0	1/16-3/16	8	25	22
	Series 01, WO/PS	¼N-¼P	0	¼N-¼P	0	1/16-3/16	8	25	22
	Series 10, 80, W/PS	1½N-1P	1P	¼N-¼P	0	1/16-3/16	6½	25	22
	Series 10, 80, WO/PS	¼N-¼P	0	¼N-¼P	0	1/16-3/16	6½	25	22
1965-66	Series 01, W/PS	¾P-1½P	1½P	¼N-¼P	0	1/16-3/16	6½	25	22
	Series 01, WO/PS	0-½P	¼P	¼N-¼P	0	1/16-3/16	6½	25	22
	Series 10, 80, W/PS	¾P-1½P	1½P	¼N-¼P	0	1/16-3/16	6⅛	25	22
	Series 10, 80, WO/PS	0-½P	¼P	¼N-¼P	0	1/16-3/16	6⅛	25	22
1967-68	American, Javelin, AMX	½N to ½P▲	0	⅜N-⅜P	0	1/16-3/16	6½	25	25
	Rebel	½N to ½P▲	0	⅜N-⅜P	0	1/16-3/16	6½	25	25
	Marlin	½N to ½P▲	0	⅜N-⅜P	0	1/16-3/16	6⅛	25	25
	Ambassador	½N to ½P▲	0	⅜N-⅜P	0	1/16-3/16	6⅛	25	25
1969	Series 01, 30, 70	½N to ½P▲	0	⅜N-⅜P	0	1/16-3/16	6½	25	22
	Series 10, 80	0 to 1N	½N	⅜N-⅜P	0	1/16-3/16	6½	25	22
1970-71	Manual Steering	½P to 1½P	1P	⅜N-⅜P	0	1/16-3/16	7¾	25	22
	Power Steering	½P to 1½P	1P	⅜N-⅜P	0	1/16-3/16	7¾	25	22

▲—With power steering ½P to 1½P.
N—Negative.
P—Positive.
PS—Power steering.

LIGHT BULBS

1964

Headlamps (low beam).............. 4002; 50/37.5 W.
Headlamps (high beam)................ 4001; 37.5 W.
Radio, transmission selector .... 1893; 2 CP
Panel switches, overdrive signal........... 1445; 1.5 CP
Turn signals, park, stop, tail .... 1157; 32/4 CP
Back-up .... 1156; 32 CP
Dome, cargo, courtesy.................. 1004; 15 CP
Luggage compartment .... 1003; 15 CP
Parking brake warning.................. 256; 1.6 CP
Gauge indicators....................... 158; 2 CP
License lamp........................... 67; 3 CP
Clock, glove compartment .... 57; 2 CP
Headlights (Series 01) .... 6012; 50/40 W.

1965

Same as 1964, with the following exceptions:
Trunk, convertible courtesy............ 89; 6 CP
Auto. floorshift indicator .... 1816; 2 CP

1966

Clock .... 57; 2 CP
License............................ 67; 3 CP
Trunk, convertible courtesy............ 89; 6 CP
Instrument, hi beam, turn, charge, oil, auto.
  trans. quadrant with Adjust-O-Tilt lights ..... 158; 2 CP
Rear pillar, courtesy................ 211; 12 CP
01-50-80 Series brake warning .... 256; 1.6 CP
10 Series brake warning .... 257; 1.6 CP
Dome, cargo, courtesy................ 1004; 15 CP
Back-up............................ 1156; 32 CP
Park, turn signals, stop, tail .... 1157; 32/4 CP
Auto. trans. quadrant, instrument panel
  switches, emergency flasher lights....... 1445; 1.5 CP
Auto. trans. floorshift light .... 1816; 3 CP
Radio, glove compartment .... 1893; 2 CP

1967-69

Compass light .... 53; 1 CP
Clock, glove box, tachometer .... 57; 2 CP
License plate .... 67; 4 CP
Convertible courtesy, trunk................ 89; 6 CP
Courtesy (panel)..................... 94; 15 CP
Instrument, hi beam, alternator, oil, turn
  indicator, brake warning lights .... 158; 2 CP
Rear shelf light (Marlin-1968-69
  01 hardtop)..................... 211; 12 CP
Parking brake warning (on)............... 257; 1.6 CP
Cargo, dome, rear quarter .... 1004; 15 CP
Back-up............................ 1156; 32 CP
Park, turn signals, stop, tail .... 1157; 4/32 CP
Control lights, cigarette lighter, ignition
  switch, light switch, wiper switch, gear
  selector quadrant, A/C thermostat
  lights .... 1445; 1.5 CP
Heater and A/C control lights .... 1881; 1.5 CP
Radio .... 1893; 2 CP
Headlight (inner) 4001
Headlight (outer) 4002
Headlights (Series 01) .... 6012; 50/40 W.
Park and front turn (1968-69
  Series 30)..................... 1157A; 3-32 CP
Radio dial (30-70 AM; 10-80 AM/FM; 30-70
  AM/Tape)......................... 1815
Clock (1968-69 Series 10-80) .... 1816; 3 CP
Heater and A/C control (1968-69
  Series 30-70) .... 1881; 1.5 CP
Radio dial (1968-69 Series 01) .... 1892; 1.3 CP

Headlight (inner) 4001
Headlight (outer) 4002
Headlights (Series 01) .... 6012; 50/40 W.
Radio (Series 01)..................... 1892; 2 CP

Radio dial (1968-69 Series 30-70 AM/FM) .... 1893; 2 CP
Tachometer (1968-69 Series 01-10-80) .... 1895; 2 CP
Headlights (1968-69 Series 01-30-70) .... 6012; 50/40 W.

1970-71

Back-up................................ 1156; 32 CP
Cargo (Series 10) 561
Clock (Series 01-30) .... 1816; 3 CP
Clock (Series 10).................... 57; 2 CP
Control lights........................ 1445; 1.5 CP
Console (Series 30) .... 211; 12 CP
Courtesy (Series 01-10-80) .... 94/561; 15 CP
Courtesy (Series 30-70) .... 211; 12 CP
Dome (Series 01-10-80).................. 562
Gearshift quadrant (LAW) .... 1816; 3 CP
Gearshift quadrant (WAW-WFS)............. 1893; 2 CP
Glove box (Series 01)................ 57; 2 CP
Glove box (Series 10-30-70-80)........... 1891; 2 CP
Indicators and instrument lights............. 158; 2 CP
License plate 2286V
Low fuel warning .... 257; 1.6 CP
Map light (Series 30-70)................ 563; 4 CP
Parking brake warning.................. 257; 1.6 CP
Park and turn (Series 10-70-80) .... 1157; 4/32 CP
Park and turn (Series 01-30) .... 1157A; 3-32 CP
Rear quarter (Series 10-70-80)................. 562
Stop, tail .... 1157; 4/32 CP
Side marker (front and rear)................ 194; 2 CP
Side marker (rear Series 01 only) .... 1895; 2 CP
Tachometer (Series 01)................ 1157A; 3-32 CP
Tachometer (Series 10-70-80)............. 158; 2 CP
Tail (Series 80) .... 1095; 4 CP
Radio 1815
Trunk .... 89; 6 CP
Headlights (Series 01-30-70) .... 6012; 50/40 W.
Headlight (inner Series 10-80) 4001
Headlight (outer Series 10-80) 4002

GENERATOR AND REGULATOR SPECIFICATIONS

YEAR	MODEL	GENERATORS				REGULATORS		
		Field Current in Amperes At 12 Volts	Brush Spring Tension (Oz.)	Cut-Out Relay Air Gap (In.)	Closing Voltage	Current & Voltage Air Gaps (In.)	Current Regulator Setting	Voltage Regulator Setting
1964	All 6 Cyl.	1.5	28	.020	12.7	.060	25	14.3
	Series 20, V8, OHV	1.65	28	.020	12.7	.060	30	14.3
	Series 80, V8, OHV	1.65	28	.020	12.7	.075	30	14.3

FUSES AND CIRCUIT BREAKERS

1964

Dome, courtesy, cargo, trunk, glove box, clock .... 9 AMP. fuse
Park, tail, instrument, license, clock light, trans. light .... 6 AMP. fuse
Gauges, alternator and oil lights, park warn, E-stick .... 4 AMP. fuse
Heater motor, twin-stick relay, V8 auto. trans.. .... 20 AMP. fuse
Air cond., A/C clutch .... 30 AMP. fuse
Stop, turn, back-up lights, flasher........... 9 AMP. fuse
Headlights...................... 20 AMP. CB (on switch)
Wipers........................ 6 AMP. CB (on switch)
Power windows..... 30 AMP. CB; 4–15 AMP. CB (in doors)
Power tail gate, window ..... 2–30 AMP. CB (Series 01–3–25 AMP. CB)
Overdrive relay .... 15 AMP. fuse (on relay)
Radio................................. 5 AMP. (inline)

1965

Same as 1964 with exception of Park-tail.... 9 AMP. fuse

1966

Dome, courtesy, cargo, trunk, glove box, clock .... 9 AMP. fuse
Park, tail, instrument, license, clock and trans. lights .... 9 AMP. fuse
Gauges, alternator and oil lights, brake warn......................... 4 AMP. fuse
Heater motor, air cond. motor, A/C clutch, overdrive........ 20 AMP. (30 AMP. with A/C)
Stop, turn, back-up, flasher................. 9 AMP. fuse
Radio.......................... 5 AMP. (inline)
Headlights...................... 20 AMP. CB (on switch)
Wipers........................ 5 AMP. CB (on switch)
Power windows...... 30 AMP. CB; 1–15 AMP. CB for each door; 2–25 AMP. CB for quarter windows.

Power tail gate window .... 3–25 AMP. CB
Convertible top .... 30 AMP. CB (under dash)

1967

Dome, courtesy, cargo, clock, glove box, trunk................................ 9 AMP. fuse
Tail, park, instrument, light switch, wiper switch, heater control, ignition switch, cigarette lighter, clock, license, trans. light, cruise command, A/C thermostat, radio, tachometer, ash tray lights................................. 9 AMP. fuse
Stop, turn, emergency flasher........... 9 AMP. fuse
Turn signals, radio, vibra-tone, cruise command feed, spotlight, back-up lights .... 9 AMP. fuse
Gauges, alternator and oil lights, brake warn, park lights............................. 4 AMP. fuse
Heater motor, A/C motor, A/C clutch, overdrive, auto. trans. .... 20 AMP. fuse (30 AMP. with A/C)
Headlights...................... 20 AMP. CB (on switch)
Convertible top .... 30 AMP. CB (under dash)
Power windows............... 25 AMP. CB (under dash)
Wipers .... 6 AMP. CB (in switch)
Power tail gate window .... 3–25 AMP. CB

1968–71

Dome, cargo, courtesy, clock, glove box, trunk lights........................... 9 AMP. fuse
Park, tail, instrument, light switch, wiper switch, heater control, cigarette lighter, clock, license, auto. trans., cruise command, A/C thermostat, radio, tachometer, ash tray lights................................ 9 AMP. fuse
Stop, turn ind., flasher................. 14 AMP. fuse
Turn signals, back-up, accessories .... 14 AMP. fuse
Indicators, alternator and oil lights, brake warn, parking brake lights .... 4 AMP. fuse
Heater motor, A/C motor, A/C clutch, overdrive, auto. trans. .... 20 AMP. fuse (30 AMP. with A/C)

Headlights .... 20 AMP. CB (in switch)
Power windows............... 20 AMP. CB (under dash)
Wipers .... 6 AMP. CB (in switch)
Power tailgate window................. 2–20 AMP. CB

1969–71 Fusible Links

Location	Color	Protects
Battery terminal of starter relay to main wire harness	Red	Complete wiring
Battery terminal of horn relay to main wire harness	Pink	Horn circuit
Accessory terminal of ignition switch to wire harness	Brown	Electric tailgate instrument panel switch, cigarette lighter, all accessories from fuse panel, and electric windshield wiper
Ignition terminal of ignition switch to wire harness	Yellow	Alternator exciter voltage, fuel temperature indicator, park brake warning light.

There are two additional fusible links which protect optional equipment when the car is so equipped:

Location	Color	Protects
Battery terminal of starter relay to air conditioning blower motor relay (10-80 Series)	Red	Blower motor circuit (high speed)
Battery terminal of starter relay to junction block for "Rally Pac" ammeter (30-70 Series)	Black	"Rally Pac" wiring

Distributor

Detailed information on cylinder numbering, firing order, point gap, point dwell, timing, spark plugs, and idle speed will be found in the Specifications tables. Rambler uses both Auto-Lite and Delco-Remy Distributors.

Distributor Removal

6 Cylinder Models

The distributor is mounted on the side of the engine. Remove the distributor cap and mark the position of the rotor relative to the distributor body, then mark the distributor body relative to the block. Remove the distributor hold down screw, disconnect the ignition primary wire and the vacuum advance tube. Lift the distributor out of the block.

Ignition timing marks—287/327 V8
(© American Motors Corp.)

V8 Models

1. Remove the distributor cap, mark the position of the rotor relative to the distributor body and mark the body relative to the block. Re-

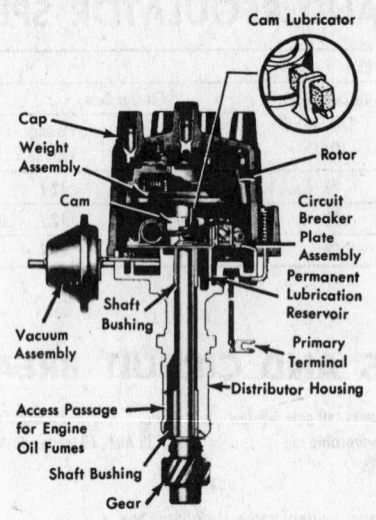

Distributor—all V8
(© American Motors Corp.)

move the carburetor air cleaner, the distributor primary wire and the distributor vacuum tube.
2. Remove the holddown bolt and take the distributor up out of the block.

The rotor and body are marked so that they can be returned to the position from which they were removed. Do not turn the engine after the distributor has been taken off.

Ignition Resistor

The ballast resistor is a white porcelain unit located on the dash panel above and behind the engine.

Late-model engines are equipped with ignition ballast resistance wire in the ignition, primary circuit.

This wire replaces the dash mounted-resistance-unit.

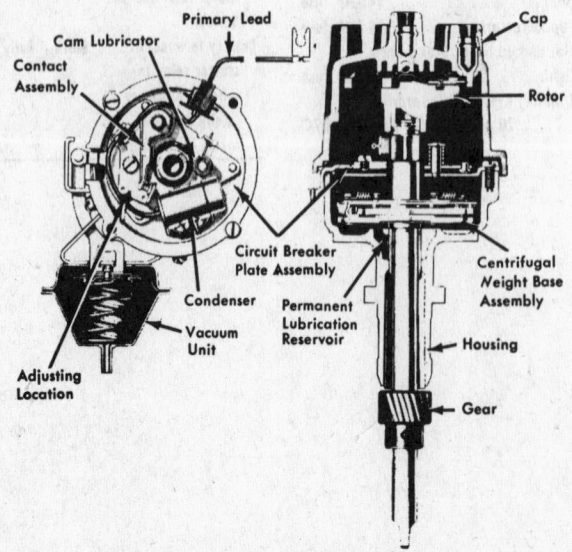

Distributor—199/232 6 cylinder
(© American Motors Corp.)

Generator, Regulator

Detailed facts on the alternator and the regulator can be found in the Alternator and Regulator Specifications table.

General information on generator and regulator repair and troubleshooting can be found in the Unit Repair section.

Regulator Removal

Disconnect plug to the regulator. Remove the metal screws which hold the regulator to the sheet metal and lift off the regulator.

AC Generator

Caution Since the alternator and regulator are designed for use on only one polarity system, the following precautions must be observed:

1. The polarity of the battery, generator and regulator must be matched and considered before making any electrical connections in the system.
2. When connecting a booster battery, be sure to connect the negative battery terminals respectively and the positive battery terminals respectively.
3. When connecting a charger to the battery, connect the charger positive lead to the battery positive terminal. Connect the charger negative lead to the battery negative terminal.
4. Never operate the alternator on open circuit. Be sure that all connections in the circuit are clean and tight.
5. Do not short across or ground any of the terminals on the alternator regulator.
6. Do not attempt to polarize the alternator.
7. Do not use test lamps of more than 12 volts for checking diode continuity.
8. Avoid long soldering times when replacing diodes or transistors. Prolonged heat is damaging to these units.
9. Disconnect the battery ground terminal when servicing any AC system. This will prevent the possibility of accidental reversing of polarity.

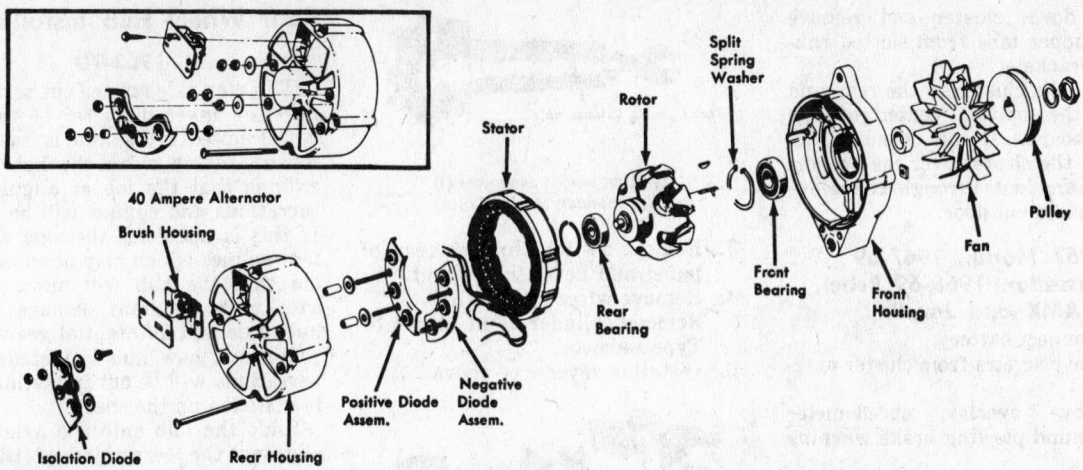

40 Ampere Alternator

Brush Housing

Isolation Diode

Rear Housing

Positive Diode Assem.

Negative Diode Assem.

Stator

Rear Bearing

Rotor

Split Spring Washer

Front Bearing

Front Housing

Fan

Pulley

35 Ampere alternator (© American Motors Corp.)

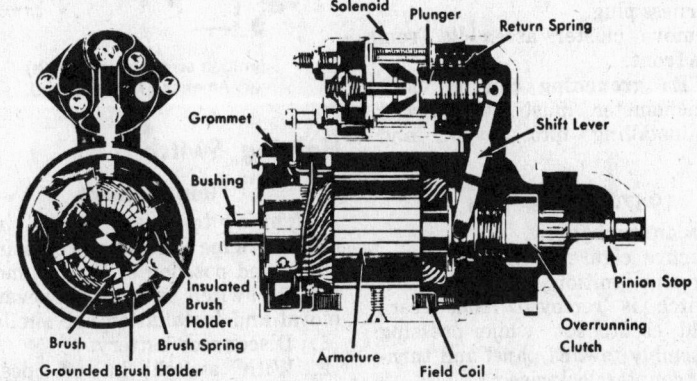

Solenoid

Plunger

Return Spring

Shift Lever

Grommet

Bushing

Insulated Brush Holder

Brush

Grounded Brush Holder

Brush Spring

Armature

Field Coil

Pinion Stop

Overrunning Clutch

Sectional view of enclosed shift lever type starter motor (© American Motors Corp.)

Battery, Starter

Detailed information on the battery and starter will be found in the Battery and Starter Specifications table.

A more general discussion of batteries will be found in the Unit Repair Section.

A more general discussion of starters and their troubles can be found in the Unit Repair section.

Starter Removal

6 Cylinder

Remove the oil filler pipe, disconnect the battery lead from the starter and the solenoid lead from the starter. From underneath the vehicle, remove the bolts which hold the starter to the bell housing and lift off the starter.

V8

Disconnect the battery wire and the solenoid wire at the starter. From underneath the vehicle, remove the bolts which hold the starter to the flywheel housing and lift off the starter.

Instruments

Speedometer and/or Instrument Removal

1964 Classic and Ambassador

1. Disconnect battery.
2. Remove plug from rear of panel; disconnect speedometer cable.
3. Remove lamp wires from connectors
4. Remove three screws that hold hood to panel; lift hood and disengage flange.
5. Remove two screws that hold cluster in place, lift up to disengage tabs, tilt unit to right and pull out.

NOTE: printed circuit must be replaced, if defective, since repairs are not feasible.

1964-65 American

1. Disconnect battery.
2. Remove ignition switch, headlight switch, fan temperature control knob and defroster control knobs.
3. Disconnect speedometer cable, harness plug and flasher harness plug.
4. Remove air and defroster control assembly and blower switch.
5. Remove the two screws on top of cluster, then pull unit through opening.

1965-66 Classic

1. Disconnect battery.
2. Remove flasher harness, printed circuit connectors and speedometer cable.
3. Remove the two 5/16 in. hex screws that hold harness.
4. Remove cluster opening molding.
5. Remove the four screws that hold instrument panel bezel to panel.
6. Remove four screws that hold cluster to panel, then remove cluster and bezel.

NOTE: remove air discharge grill on air-conditioned cars.

1965-66 Ambassador and 1966 Marlin

1. Follow Steps 1-4, above.
2. Remove windshield wiper control.
3. Remove two screws that hold bezel to panel.
4. Remove cluster and bezel from front.

NOTE: models with automatic transmission and tilt wheel; remove bracket bolts from column to allow passage of cluster.

1966 American

1. Disconnect battery.
2. Remove ash tray and bracket, radio and overlay panel.
3. Remove Weather Eye console, bulb and socket.
4. If car is air conditioned, remove air discharge grill and glove compartment.

NOTE: remove center air discharge bezel and vent to reach center screw.

5. Disconnect printed circuit plug and speedometer cable; remove harness clips.
6. Detach headlight switch, ignition switch and windshield wiper control and wire them out of the way without detaching their connections.

7. Pull down cluster and remove two upper tabs from slotted rubber brackets.
8. Work the cluster to the rear and out the bottom center of the dashboard. If air conditioned, work the cluster over the evaporator and out through the glove compartment door.

1967 Marlin, 1967-69 Ambassador, 1968-69 Rebel, AMX and Javelin

1. Disconnect battery.
2. Remove screws from cluster overlay.
3. Remove overlay, speedometer cable and parking brake warning light.
4. Remove screws that secure cluster to instrument panel; remove harness plug.
5. Remove cluster assembly from the front.

NOTE: grounding clip on clock and tachometer must be replaced when installing into Javelin and AMX.

1967-68 American

1. Disconnect battery.
2. Remove cigarette lighter assembly and ignition switch. Ignition switch is removed from rear; hold escutcheon while pressing assembly toward panel and turning counterclockwise.
3. Remove wiper control knob, nut, headlight switch knob, shaft assembly and nut.
4. Remove flasher unit, then disconnect speedometer cable.
5. Remove the two screws at the top of cluster overlay; slide cluster away from panel slightly for access to cluster plug.
6. Remove plug and pull cluster out.

Ignition Switch

Replacement

Two types of ignition switch and mountings are used on Ramblers and Americans over different years. Removal and installation of each type follows:

Type I

1. Disconnect battery.
2. Disconnect switch wires.
3. Remove screws at rear of panel and remove switch toward rear of panel.
4. With switch removed, turn key to Acc. position and insert wire (paper clip) in small hole in housing. Depress retainer while turning and pulling out cylinder.
5. Install in reverse of above.

Type II

1. Disconnect battery.
2. Depress switch and turn counterclockwise.

Ignition switch assembly (I)
(© American Motors Corp.)

3. Remove switch through rear of panel and bevel from front.
4. Remove wires from switch.
5. Remove cylinder as in Step 4 of Type 1 above.
6. Install in reverse of above.

Ignition switch assembly (II)
(© American Motors Corp.)

Lighting Switch

Replacement

Light switches are typical in all models. Some variation occurs in the shape and position of the nut mounting the switch to dash. However, removal and installation are similar.

1. Disconnect battery.
2. With switch in off position press the release button and remove the knob and shaft.
3. Remove screws attaching switch and bracket to panel.
4. Reverse for installation, positioning switch so that the shaft is lined up properly before tightening the bracket screws.

Brakes

Specific information on brake cylinder sizes can be found in the General Chassis and Brake Specifications table.

Information on brake adjustments, band replacement, bleeding procedure, master and wheel cylinder overhaul can be found in the Unit Repair Section.

Information on troubleshooting and overhauling power brakes can be found in the Unit Repair Section.

Information on the grease seals which may need replacement can be found in the Unit Repair Section.

Rear Wheel Hub Installation

1963-70

The rear axle splines cut serrations into the inner diameter of the rear wheel hub. If the hub is to be removed, match mark the hub to the axle so that the job of aligning the serrations and splines will be easier. If this is not done, the axle will cut new splines which may be so near the old that the hub will move on the axle with resultant damage to the hub, axle and differential gears.

When a new hub is installed, the serrations will be cut in the hub as it is installed on the shaft.

Slide the hub onto the axle shaft, aligning the serrations of the hub with those of the shaft. Now, install the nut and tighten the hub onto the shaft until the face of the hub is 3/16 in from the outer taper of the shaft. Nut must be torqued to 250 ft. lbs.

Master Cylinder Power Unit Removal

Remove the clevis pin from the power-unit operating rod. Disconnect the vacuum line and the hydraulic lines from the power unit, remove the stop light wires, remove the mounting bolts and lift off the power cylinder.

Parking Brake Lever Removal

Foot-operated Lever

The foot-operated lever is mounted under the left side of the instrument panel. Slack off on the brake cable and remove the clevis which holds the cable to the top part of the lever. Remove the brackets which hold the lever assembly to the side of the body and lift off the lever.

Parking Brake Cable Replacement

All Models

1. Disconnect the lower end of the cable at the cross-shaft or equalizer, disconnect it at the handbrake end.
2. Remove the brackets which retain it to the body and firewall and thread it out of the vehicle.

When a new cable is to be installed, it is always a good idea to tie the new one to the end of the old one so that it will thread through in the same route as the old one. This, sometimes, will require the service of a helper to guide it.

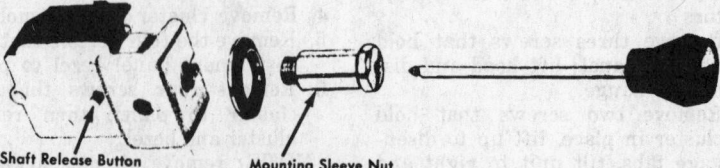

Shaft Release Button   Mounting Sleeve Nut
Light switch assembly (© American Motors Corp.)

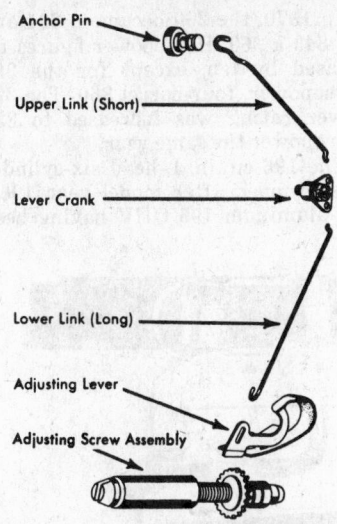

Automatic adjuster brake parts—Wagner
(© American Motors Corp.)

Automatic Adjusters

1964-69

All models are available with automatically - adjusted service brakes. This is the Bendix Duo-Servo or the Wagner Compound brake. It automatically adjusts for lining wear, as the case requires. This continuous adjustment maintains a constant pedal height and is a decided safety and economic factor throughout the entire lining life.

The adjuster is operated by the movement of the secondary shoe during reverse brake application. The automatic mechanism is attached to, and works through the standard-type star wheel adjuster. Therefore, care must be used during a reline or major brake job to assure freedom of adjuster parts movement. Care must also be used to eliminate the mixing of right with left-hand parts.

1964

The 10 series (six cylinder Rambler) includes an automatic adjuster. This adjuster is designed to operate

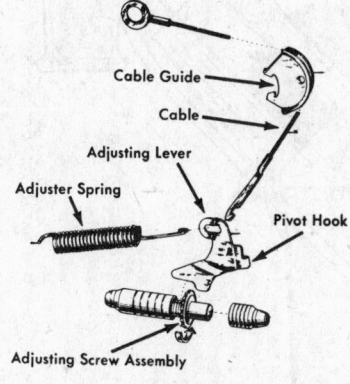

Automatic adjuster brake parts—Bendix
(© American Motors Corp.)

on the primary shoe instead of the conventional secondary type. Incorporating the automatic adjuster into the primary shoe causes the brake adjustment to be made upon forward brake application. The reverse, is the case with most other designs of brake automatic adjusters.

Dual Master Cylinder

1964-70

A dual master cylinder is used. The front reservoir supplies rear brakes, and the rear one supplies the front brakes. This allows one pair of brakes to operate, should there be a failure of the opposite pair. If lines have been disconnected, be sure to reinstall in proper place (front to rear cylinder, and rear to front.)

With pressure bleeder, the Bendix system can be bled from front reservoir by covering rear reservoir with solid cap. The Moraine type must be bled separately, front and rear.

Without pressure bleeder, keep both reservoirs nearly full.

Fuel System

Information

Data on capacity of the gas tank will be found in the Capacities table. Data on correct engine idle speed and fuel pump pressure will be found in the Tune-Up Specifications table.

Information covering operation and troubles of the fuel gauge will be found in the Unit Repair Section.

Detailed information on the carburetor and how to adjust it will be found in the Unit Repair Section.

Fuel Pump Removal

Disconnect both gas lines from the fuel pump, disconnect the vacuum line, if it is a vacuum pump. Remove the two bolts which hold it to the block and lift off the pump.

Fuel Pump—Carter

1964-70

A fuel-vacuum pump is used on Classic and Ambassador models.

The outward appearance of this pump is traditional and looks not unlike Carter's earlier double-action pumps. However, the vacuum pumping mechanism is quite different.

This is a piston-type vacuum pump, capable of much greater volume than some older models. It has been adapted to satisfy the needs of vacuum wiper motors on cars equipped with blades large and heavy enough to cover modern wrap around windshields.

An electric wiper is optional, allowing the use of a single-unit fuel pump.

Exhaust System

Exhaust Pipe Removal

All Models With Single Exhaust System

1. Disconnect the exhaust pipe at the flange on the manifold. Squirt plenty of penetrating oil between the muffler and the exhaust pipe, or, if it is frozen too solidly, heat it.
2. Remove the clamps which hold the exhaust pipe to the muffler and slide the exhaust pipe out the front of the muffler. It can be threaded out from underneath the car.
3. Reinstall the new exhaust pipe so that the pipe extends past the slots in the muffler connections.

Muffler Installation

1. Squirt a good penetrating or derusting oil at the joint between the exhaust pipe and the muffler and the joint between the muffler and the tail pipe.
2. Let this oil work in. Remove the clamps which hold the muffler to the exhaust pipe and the muffler to the tail pipe. Loosen the tail pipe hanger bracket so that tail pipe can be slid backwards slightly, and can be cleared from the muffler.
3. Let the tail pipe clear, bend the muffler downwards and slide it off the back of the exhaust pipe.
4. Reinstall in reverse order and tighten all of the clamps which were loosened to permit removal of the muffler.

Tail Pipe Installation

1. Squirt plenty of penetrating oil where the tail pipe joins the muffler and remove the clamp.
2. Remove the hangers which hold the tail pipe up to the underbody and slide the front of the tail

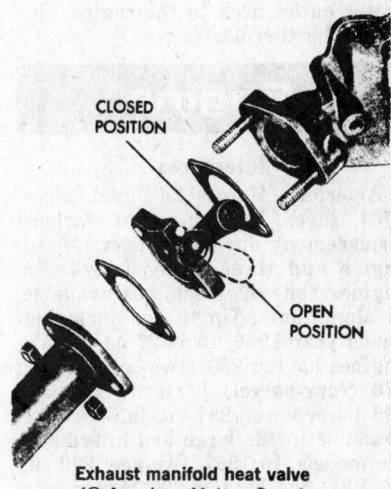

CLOSED POSITION

OPEN POSITION

Exhaust manifold heat valve
(© American Motors Corp.)

pipe out of the back of the muffler.

3. When the new tail pipe is installed, slide it far enough into the muffler to cover the slots in the muffler joint.

4. Arrange the muffler so that it hangs on each of its hangers approximately equally. Don't leave one hanger carrying all the weight so that it can rattle on the other one.

5. Equality can be accomplished easily by twisting the muffler slightly in order that pressures are reasonably equal on all hangers. This will prevent rattles.

Cooling System

Detailed information on cooling system capacity can be found in the Capacities table.

Information on the water temperature gauge can be found in the Unit Repair Section.

Radiator Core Removal

Raise the hood, drain the radiator, remove the upper and lower radiator hose. Take out the bolts which hold the radiator to its cradle and, if the car is fitted with an oil cooler on the transmission, disconnect the oil cooler lines and lift the core up and out.

Water Pump Removal

Slack off and remove the fan belt, take out the bolts which hold the fan pulley to the hub and remove the fan blades and hub assembly. The water pump can then be unbolted from the manifold and lifted off.

Thermostat Removal

The thermostat is located in the water outlet housing at the top of the cylinder head, or on V8 models in front of the manifold.

Disconnect the upper radiator hose and remove the bolts which hold the water outlet neck to the engine. Remove the thermostat.

Engine

References

American Motors has used, since 1964, seven V8 engines of various displacement, one six-cylinder L-head engine and three OHV six-cylinder engines (one of which was available in aluminum). Up to and including model year 1966, two 327 cu. in. V8 engines having 250 (two-barrel) and 270 (four-barrel) horsepower and a 198 horsepower 287 cu. in. V8 were available in the large and intermediate models. In 1967, two new 290 cu. in. V8s, one having 200 horsepower

and the other 225 (depending on carburetion), replaced these engines. In the same year, two 343 cu. in. engines having 235 and 280 horsepower were available. In 1969, a 390 cu. in. V8, based on the 343, was introduced. This engine had a forged steel crankshaft and was rated at 315 horsepower.

In 1970, the 290 became a 304 and the 343 a 360. Horsepower figures increased by ten, except for the 300 horsepower, four-barrel 360. The 390 power rating was increased to 335 horsepower the same year.

The 196 cu. in. L-head six-cylinder was dropped after model year 1965, the aluminum 196 OHV having been

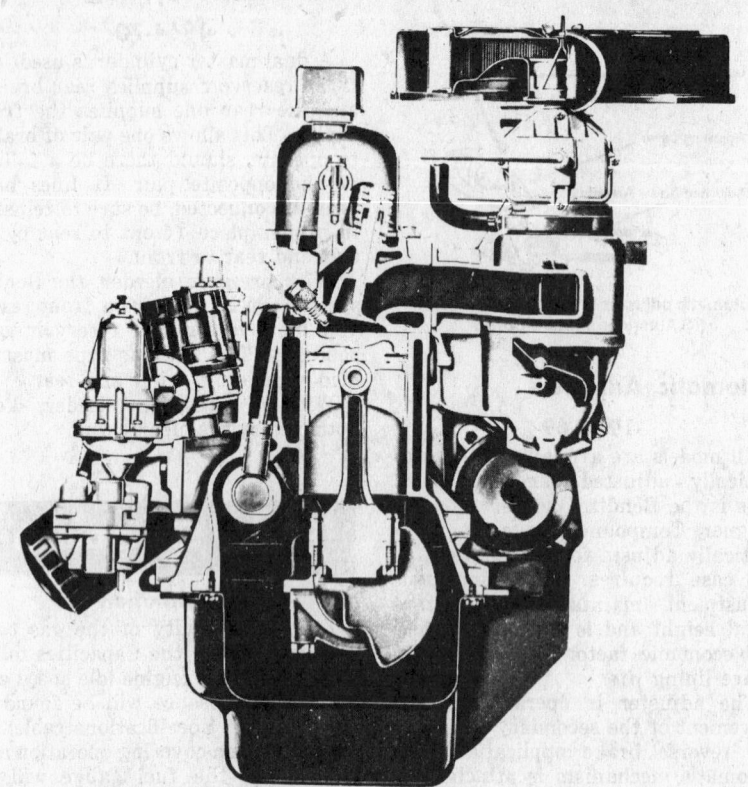

6 cylinder OHV 199/232 engine (© American Motors Corp.)

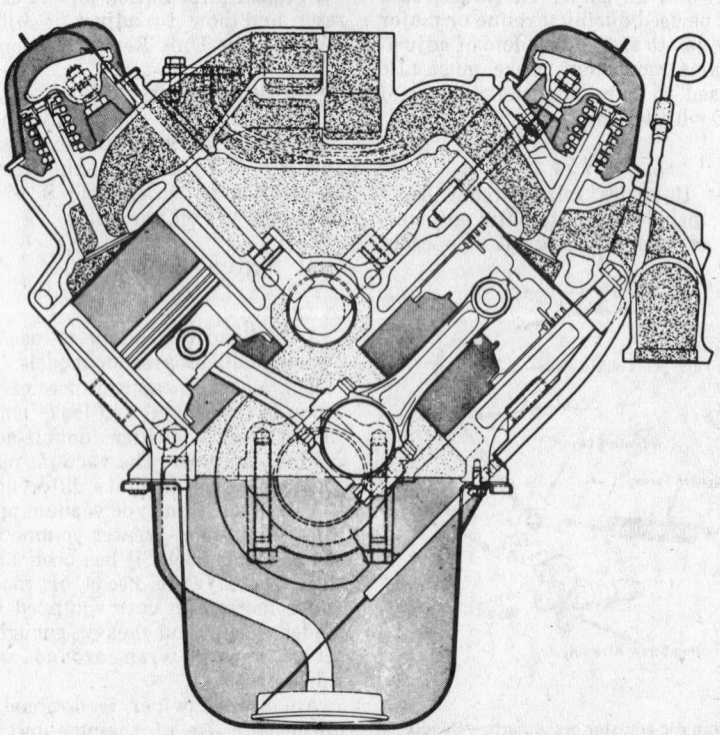

V8 OHV 290/304/343/360/390 engine (© American Motors Corp.)

dropped the previous year. The cast iron 196 OHV was used until 1965; it was replaced the next year by a more modern seven-main bearing OHV six of 199 cu. in. and two 232 cu. in. sixes of 145 and 155 horsepower (depending on whether a single- or two-barrel carburetor was used). These last engines are used up to the present, the 199 and the small 232 being available in the 1970 ½ Gremlin.

The means of determining which engines have hydraulic valve lifters and which do not will be found in the Tune-up table.

Engine crankcase capacities are listed in the Capacities table of this section.

Approved torque wrench readings and head bolt tightening sequences are covered in the Torque Specifications table.

Information on the engine marking code will be found in the Engine Identification table of this section.

6 cylinder, OHV 196 cast iron engine
(© American Motors Corp.)

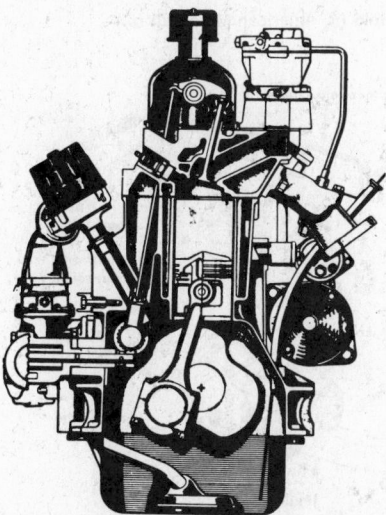

6 cylinder OHV 196 aluminum engine
(© American Motors Corp.)

Exhaust Emission Control

In compliance with anti-pollution laws involving all of the continental United States, the American Motors Corporation has elected to adopt a special system of terminal exhaust treatment. This plan supersedes (in most cases) their method previously used to conform to 1966-67 California law.

The new system phases out (except in cases of special purpose engines and applications) the Air Guard (afterburner) type, previously used in California in 1966-67.

American Motor's new approach, the MOD system, utilizes engine modification, with emphasis on carburetor and distributor changes. The new concept employs broader, yet more critical, distributor control. However, any terminal exhaust treatment renders engine adjustments more critical. Therefore, it means that close and frequent attention to tune-up maintenance details are necessary.

Since 1968, all car manufacturers have posted idle speeds, and other

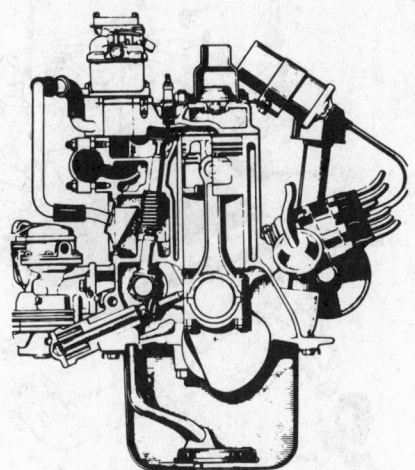

6 cylinder 196 L-head engine
(© American Motors Corp.)

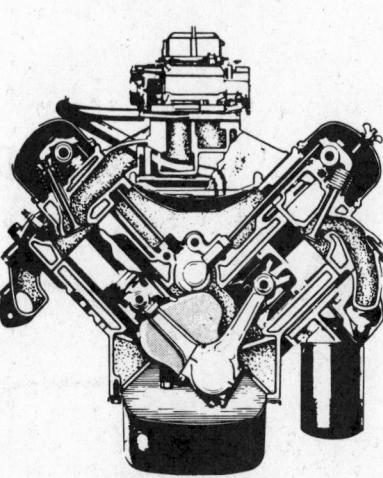

V8 287/327 engine
(© American Motors Corp.)

pertinent data relative to the specific engine-car application, in a conspicuous place in the engine compartment.

For details, consult the Unit Repair section.

Valves

Detailed information on the valves can be found in the Valve Specifications table.

A general discussion of valve clearance, and a chart showing how to read pressure and vacuum gauges when using them to diagnose engine troubles, will be found in the Troubleshooting Section. A chart on engine troubleshooting will be found, also.

Valve tappet clearance for each engine is given in the Tune-up Specifications Table of this section.

Bearings

Detailed information on engine bearings will be found in the Crankshaft Bearing Journal Sizes table.

Engine Removal

All L-head and OHV 6-cylinder 1964-70, 1964 Ambassador and Classic 287/327 V8

1. Remove hood; mark hinge position for easy assembly.
2. Remove battery.
3. Drain engine oil and cooling system.
4. On automatic transmission equipped cars: drain transmission oil and remove cooler lines (if so equipped).
5. Remove power steering pump, smog pump and air conditioner compressor (if so equipped). CAUTION: do *not* disconnect air conditioner lines.
6. Disconnect all hoses, tubes and wiring connecting engine and radiator to body and chassis. Remove radiator and starter motor for extra clearance in some models.
7. Disconnect exhaust pipe at manifold on six-cylinder; remove manifolds from cylinder heads on V8 287/327.
8. Attach chain hoist to engine, then remove flywheel housing bolts and mud pan.
9. On six-cylinder engines, support flywheel housing with screw jack or wood block.
10. If equipped with automatic transmission, remove flywheel drive plate to converter bolts.
11. On L-head and OHV 199/232 engines, support transmission and torque tube and remove rear engine crossmember.
12. On 199/232 Classic and Ambassador, disconnect torque tube from extension housing.
13. On 232 Ambassador from 1965, disconnect rear U-joint and pull driveshaft from transmission.

14. Disconnect front motor mounts at engine, then lift engine forward and out through hood opening.

1965-66 Classic, Ambassador and Marlin 287/327 V8

1. Follow Steps 1-6, inclusive, from previous procedure, then continue as follows:
2. Disconnect exhaust pipes from manifold.
3. Support engine with chain hoist.
4. Remove rear engine crossmember.
5. Disconnect brake tube bracket (rear) that is secured to body pan.
6. Disconnect rear shock absorbers at lower brackets, then disconnect torque tube at transmission extension housing.
7. Disconnect parking brake equalizer and brake cable housing at torque tube bracket. Using a chain and body jack, pull the rear axle back far enough to remove front U-joint from transmission.
8. Disconnect speedometer cable and shift rods from transmission.
9. Unbolt front motor mounts, then pull engine forward and upward through hood opening.

1967-70 Rebel, Ambassador, Marlin, American, Javelin, AMX 290/343/390 V8

1. Follow Steps 1-6, inclusive, from six-cylinder procedure, then continue as follows:
2. Disconnect exhaust pipes from manifolds.
3. Support engine with chain hoist.
4. Remove engine rear crossmember.
5. Disconnect speedometer cable at transmission.
6. Disconnect automatic transmission shift linkage or standard transmission shift rods at transmission. If equipped with American Motors four-speed shifter, remove boot from floor pan and pull the two bolts that hold shift lever to shift mechanism; the mechanism can remain attached. If equipped with Hurst linkage, the entire shifter should be removed from transmission for adequate clearance.
7. Disconnect front motor mounts and pull engine forward and upward, while supporting driveshaft as slip joint is removed from extension housing.

NOTE: if desirable, remove entire driveshaft beforehand by splitting rear U-joint.

Engine Manifolds

Intake Manifold

All 196 L-head and OHV 6 Cylinder

On these models, the intake manifold is cast integrally with the cylinder head. To remove top cover on OHV, remove linkage, lines and bolts.

287/327 V8

1. Disconnect the water outlet tube and remove the distributor.
2. Take off the air cleaner.
3. Disconnect all throttle lines across the cylinder head.
4. Disconnect the vacuum lines, the coil and ignition primary leads.

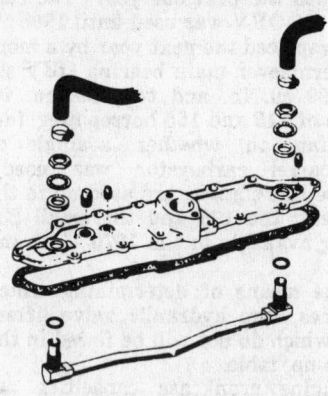

OHV 196 intake manifold (© American Motors Corp.)

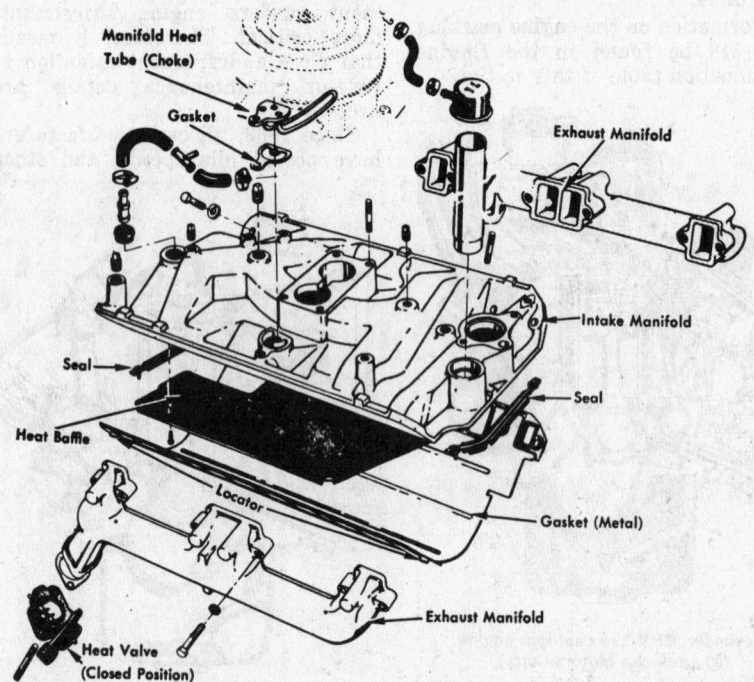

OHV 290/343 two-barrel intake manifold (© American Motors Corp.)

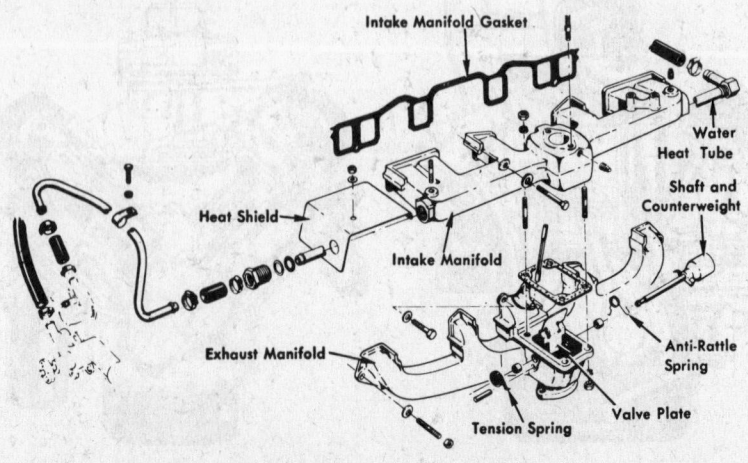

OHV 199/232 6 cylinder intake and exhaust manifold (© American Motors Corp.)

5. Remove the bolts which hold the intake manifold to both cylinder heads and lift off the manifold.

199/232 6 Cylinder

The intake manifold is mounted on the left-hand side of the engine and bolted to the cylinder head. A gasket is used between the intake manifold and the head, none is required for the exhaust manifold. The manifold on the Rogue 232 incorporates an internal water tube to supply carburetor heat; it is secured with a tapered fitting.

1. Remove air cleaner, carburetor and linkage.
2. On Rogue 232 engines, remove coolant hoses from intake manifold and plug the ends.
3. Disconnect the exhaust pipe at the manifold flange.
4. Remove manifold hold-down bolts and separate the intake and exhaust manifold from the head as a unit. Always use a new gasket when installing manifolds.

NOTE: manifold bolts and nuts are tightened to 25 ft. lbs.

290/343/390 V8

The cast iron manifold completely encloses and seals the tappet valley between the cylinder heads. The manifold contains water passages, a crankcase vent passage, exhaust crossover and induction passages. A one-piece metal gasket seals the intake manifold to cylinder head joint and also serves as an oil baffle. The left-hand carburetor bores supply cylinders No. 1, 7, 4 and 6; the right-hand bores cylinders No. 3, 5, 2 and 8.

1. Drain the radiator.
2. Disconnect throttle linkage, hoses, wiring and vacuum lines.
3. Remove manifold hold-down bolts and lift manifold straight up.

NOTE: when installing, coat gasket with sealer. Tighten bolts to 45 ft. lbs. evenly.

Exhaust Manifold Removal

All 196 L-head 6 Cylinder

The exhaust manifold on these models is actually the exhaust pipe held at the side of the cylinder head. Remove the caps which hold the pipe to the cylinder head, split the pipe at the flange and remove it from the vehicle. Tighten bolts to 10-15 ft. lbs.

V8 Models

Two exhaust manifolds are used, one on each bank. Disconnect the exhaust pipe at the flange, remove the bolts which hold the manifold to the cylinder head and lift the manifold off.

OHV 196 6 Cylinder

The cast iron exhaust manifold is removed in the following manner;
1. Disconnect exhaust pipe at flange.
2. Loosen the manifold bolts and break manifold away from engine.
3. Remove all bolts and manifold. When installing, tighten end flange bolts to 8-10 ft. lbs., center bolts to 20-25 ft. lbs.

OHV 199/232 6 cylinder

Exhaust manifold is removed along with *intake* manifold; see above.

Cylinder Head

Rocker Assembly Removal

OHV 196 6 Cylinder
Aluminum and cast iron block

The rocker arms and valve assembly are lubricated high pressure oil from an opening in the head through the front rocker support. The rocker shaft is one-piece construction, the rocker arms being retained by a snap-ring, thrust washers and a wave washer on each end of the shaft. The rocker shaft is installed with the oil holes down.

1. Remove the carburetor air cleaner and take off the nuts which hold the rocker cover to the cylinder head. Carefully remove the gasket from both the hood and the cover.
2. Working a few turns at a time, remove the bolts which hold the rocker assembly brackets to the cylinder head until the valve spring tension has been released.
3. Continue to remove the bolts until the rocker assembly is clear.

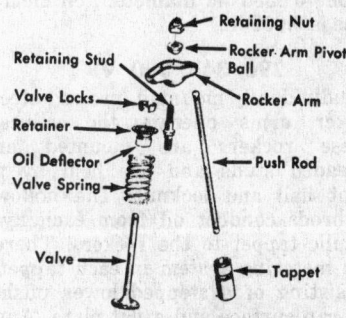

OHV 290/304/343/360/390
rocker arm assembly
(© American Motors Corp.)

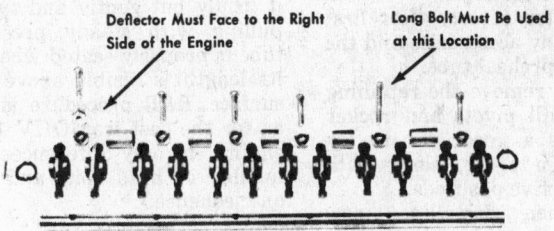

OHV 199/232 6 cylinder rocker arm assembly
(© American Motors Corp.)

OHV 196 6 cylinder rocker arm assembly (© American Motors Corp.)

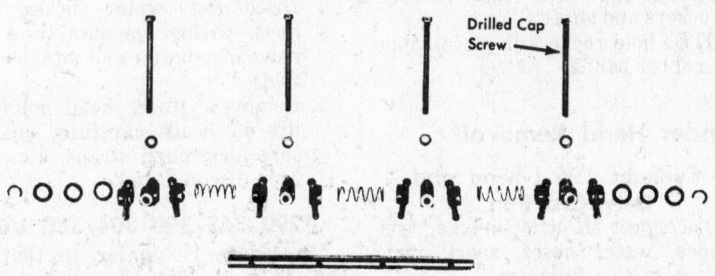

OHV 287/327 V8 rocker arm assembly (© American Motors Corp.)

287/327 V8

The rocker shaft assembly is secured to the head with four cap bolts. The hollow rocker shafts serve as oil galleries for the rocker arms, push rod ends and valve stems. Oil pressure is supplied by the valve tappet main oil gallery and enters the rocker shafts via passages in the block and heads through a special drilled bolt at the rear of each shaft. Although the shaft assemblies are interchangeable between banks, two different rocker arms are used to actuate intake and exhaust valves. Rocker shafts are 0.8580-0.8585 in. diameter; oil clearance is 0.002 in.

290/343/390 V8

Individually mounted, pressed steel rocker arms operate the valves. These rockers are mounted on threaded studs and are held by a pivot ball and locknut. The hollow pushrods conduct oil from each hydraulic tappet to the rockers. There is a metering system in each tappet, consisting of a stepped lower pushrod cap surface and a flat plate. Any loss of lubrication to the rockers usually can be traced to failure of this part, or to a blocked pushrod oil passage. The pushrods rub against the cylinder head during operation and serve to maintain the correct rocker to valve stem angle.

1. Remove valve covers, after first removing any accessories and the air cleaner preheat tube.
2. Loosen and remove the retaining locknuts, ball pivots and rocker arms. It is a good idea to lay them out in order, along with their respective pushrods.
   NOTE: when installing new threaded studs, make sure hex nut is fully seated and tightened to 65-70 ft. lbs. Retaining locknuts are tightened to 20-25 ft. lbs.

OHV 199/232 6 Cylinder

The rocker shaft assembly is secured to the cylinder head with six cap bolts. Oil pressure for rocker lubrication is supplied via No. 3 camshaft bearing from the main oil gallery to No. 5 rocker support. Rocker shaft is 0.8575-0.8585 in. diameter; oil clearance is 0.003-0.005 in.

1. Remove valve cover.
2. Unbolt cap bolts and remove rockers and shaft.
   NOTE: hold rockers in place using large rubber bands.

Cylinder Head Removal

6 Cylinder 196 L-head and OHV cast iron

1. Disconnect throttle linkage, fuel lines, water hoses, spark plug wires and vacuum line.
2. Loosen all nuts and remove; note position of battery strap.

3. Slide a heavy scraper around the head-to-block joint to free gasket. If necessary, reconnect battery ground and crank engine (coil wire out) to allow compression to free head. CAUTION: Do not hammer on head.
4. Before replacing head, wire brush studs and make sure they are tight. Clean top surface of block and lower head surface; check head for straightness. If head is out of true 0.006 in. over its length (or 0.001 in. every 1 in.) it must be resurfaced on a milling machine.
5. When installing, use sealer on both sides of gasket.

6 Cylinder 196 OHV Aluminum

The cylinder head has variable wedge type combustion chambers in conjunction with contoured top pistons. Hollow guide pins at right front and right rear corners correctly locate the head. Care must be exercised during installation to prevent damage to these pins, especially to the right front one. This pin serves as the oil delivery tube for the rocker arms and is sealed by O-rings at both ends. These O-rings must be replaced each time the head is removed.

To remove the delivery tube, grasp it firmly but gently and twist, while pulling with steady pressure. The tube is properly seated when one-half its length is visible above the block surface. R&R procedure is the same as for the cast iron OHV 196 cu. in. engine, the only differences being the number of head bolts and their torque sequence.

287/327 V8

The cylinder heads have two holes to assist head location. Maximum out of true is 0.006 in. for the entire length of head; 0.001 in every 1 in. Make sure the rear rocker arm bolts are properly installed, otherwise no oil will get to the rockers.

1. Remove oil filler tube, rocker covers, power steering pump, alternator, exhaust manifolds and air conditioner. Swing air conditioner out of the way without disconnecting its hoses.
2. Remove rockers and pushrods.
3. Disconnect water hoses, fuel lines, wiring, vacuum lines; remove distributor and intake manifold.
4. Remove cylinder head bolts and lift off heads carefully, making sure all ground straps, etc. have been disconnected.

290/343/390/304/360 V8

Procedure is similar to that for 287/327 cu. in. V8, except that rocker lubrication is accomplished through the pushrods and the distrib-

utor does not have to be removed. Cylinder head out of true is 0.006 in. for the full head length, 0.002 in. every 6 in. and 0.001 in. every 1 in. Torque sequence is different as well; see illustrations under Torque Specifications table.

NOTE: second bolt from front, bottom row, *must* have threads coated with sealer to prevent coolant leakage.

Engine Lubrication

Oil Pan Removal

NOTE: it is far easier to remove the engine in most cases.

1963-67 All 6 Cylinder Models

1. Support weight of front of car, and remove front springs.
2. Support front of engine from above the engine compartment, then remove the engine support-cushion attaching nuts at the mounting bracket.
3. Disconnect idler arm.
4. If equipped, remove front stabilizer bar.
5. Remove attaching bolts from front crossmember and pry crossmember down to gain clearance. To assist in holding crossmember down, install wood spacer blocks between the crossmember and body side sill.
6. Remove flywheel dust cover.
7. Drain oil and turn crankshaft so No. 1 piston is on upstroke, and remove pan.
8. To install, reverse the removal procedure.

1968-70 All 6 Cylinder Models

1. Disconnect front cushions from engine bracket and remove right bracket from engine.
2. Disconnect ground strap.
3. Remove valve cover and air cleaner.
4. If equipped, remove fan shroud.
5. Raise engine as far as possible.
6. Disconnect idler arm from side sill.
7. If equipped, disconnect stabilizer bar.
8. Loosen strut rod bolts at lower control arms.
9. Remove bolts retaining crossmember, and, with weight of car on wheels, pry down crossmember and insert wooden blocks to hold it down.
10. Drain oil and remove pan.
11. To install, reverse removal procedure.

1963-67 V8 Engines

1. Remove front springs. (See Front Suspension.) Support weight of front of car.
2. Support front of engine from above, and take away support cushion at engine.

3. Remove idler arm and piston rod from bracket on power steering models.
4. Remove sway bar.
5. Remove front suspension cross member from side sills.
6. Pry crossmember down to get clearance for pan. Use of wooden blocks between crossmember and side sills will aid.
7. Remove flywheel dust cover.
8. Drain oil and remove pan.

All 1966 Ambassador V8 Engines

1. Drain oil.
2. Remove flywheel dust cover.
3. Disconnect sway bar at rear mounts.
4. Remove oil pan bolts and oil pan.

All 1968-70 V8 Engines

1. Turn crankshaft until mark on damper is 180° from cover marks.
2. Disconnect engine cushion mounts from crossmember; remove fan shroud, if so equipped.
3. Disconnect ground strap; disconnect cushion mount brackets from block on American.
4. Remove starter motor.
5. Remove idler arm, except on 1969-70 Ambassador.
6. Ambassador sway bar: disconnect at side sills. American, Javelin, AMX, Rebel sway bar: loosen links at lower control arms as far as possible.
7. Disconnect shock absorbers at lower control arms on all 1969-70 except Ambassador.
8. Attach chain hoist and raise engine as far as possible.
9. Remove cushion mounts and brackets from American.
10. Loosen strut rod bolts at lower control arms; remove crossmember side sill bolts.
11. With car weight on front wheels, pry crossmember down far enough for clearance. Use wood blocks for support between crossmember and side sills.
12. Remove oil pan bolts and oil pan.

Rear Main-Bearing Oil Seal

6 Cylinder Engines

Several different kinds of rear main-bearing oil seals are used, such as wood, rubber, felt and packing. Wooden plugs are used to seal the sides of the rear main-bearing caps. Later models use synthetic rubber key strips overlapping the bearing cap and seal the sides as well as the mating surfaces. As an actual oil seal, a slinger, an integral part of the crankshaft, is used to throw the oil into the trough so that it cannot get on the clutch.

To remove and replace the neoprene seal used on 199/232 engines:
1. Remove oil pan, as previously de-

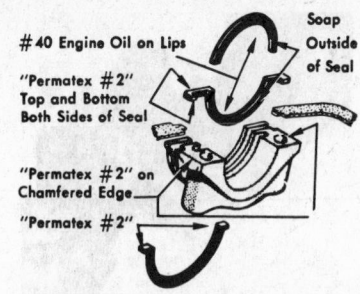

#40 Engine Oil on Lips
Soap Outside of Seal
"Permatex #2" Top and Bottom Both Sides of Seal
"Permatex #2" on Chamfered Edge
"Permatex #2"

Rear main bearing oil seal—199/232 6 cylinder and all 1967-71 V8
(© American Motors Corp.)

scribed.
2. Scrape clean all gasket surfaces, then remove rear main cap.
3. Discard lower portion of seal; drive out upper portion, using a brass drift, until it can be grasped with pliers.
4. Clean main cap, then *loosen* all remaining main cap bolts.
5. Lightly oil all surfaces, then coat the block-side surface of the new upper seal with soap and the seal lip with SAE 40 engine oil.
6. Install upper seal portion with the lip facing the front.
7. Coat the cap and block-side seal surface with Permatex No. 2.
8. Coat the back surface of new lower seal with soap, the lip with SAE 40 engine oil. Install lower seal firmly into main cap.
9. Coat both chamfered edges of rear main cap with Permatex No. 2, install bearing inserts (if removed) and tighten cap bolts to 75-80 ft. lbs.
10. Cement oil pan gasket to block; coat gasket tongues with Permatex No. 2 where they fit into rear main cap, as well as front neoprene seal.
11. Coat rear pan seal with soap and place into proper recess, then install oil pan bolts (¼ in.—5-8 ft. lbs.; 5/16 in.—10-12 ft. lbs).

287/327 V8 Engines

On these models, a packing-type rear main-bearing seal is used. The upper half can be replaced only after the crankshaft has been removed. The lower half can be replaced any time the rear main-bearing cap is taken down. To replace the lower half, take down the oil pan and remove the rear main-bearing cap. Remove the oil packing and set the new packing into the cap so that it protrudes a little at both ends. Temporarily, bolt it up into place. This will probably cause the upper portion of the oil seal to compress and rivet over just a little bit. Trim off the riveted portion and again insert the cap into position. The reason this is done is that, sometimes, the compression from the new oil seal will cause the upper seal to

come down tighter against the crankshaft and prevent leaks, even in the upper half which has not been replaced.

290/343/390/304/360 V8 Engines

A neoprene seal, consisting of two pieces, is used. Procedure is identical to that used for the 199/232 6 cylinder, except that main cap bolts are tightened to 95-105 ft. lbs.

Oil Pump Service

Lubrication, on all American Motors engines, is full pressure except to the wrist pins. Pressure is supplied by a gear-type pump.

L-head 196 and OHV 196 6 Cylinder

The oil pump is mounted on the outside of the cylinder block on the right-hand side. Pumps used on E-stick models are similar, but provide greater oil capacity. A valve body in the pump cover controls oil supply to an actuator, which engages the E-stick clutch throwout lever

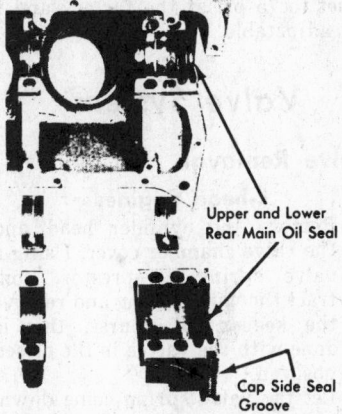

Upper and Lower Rear Main Oil Seal

Cap Side Seal Groove

Rear main bearing oil seal, 287/327 V8
(© American Motors Corp.)

when in Drive position. A pressure relief valve in the valve body opens at 10 psi to make sure engine lubrication needs are met before oil is supplied to clutch.

The engine oil pressure relief valve is found in the cylinder block where oil enters the distribution channel. The valve is not adjustable and is set to 50-58 psi at the factory (60-65 for aluminum 196 OHV). Oil pump bolts are tightened to 20-25 ft. lbs. except on E-stick units; these bolts must be tightened to 13-15 ft. lbs. only.

NOTE: always make sure connection between oil tube and block is tight.

199/232 6 Cylinder and 287/327 V8

The oil pump is driven by the distributor drive shaft. Oil pump R&R does not, however, affect distributor timing because the drive gear re-

mains in mesh with the camshaft gear.

With a straightedge across the pump body and gears, clearance should be 0.000-0.004 in. (gears should project above body). Do not disturb location of tube in pump body if possible. Now, measure clearance between gears and wall of gear cavity opposite point of gear mesh; should be 0.0005-0.0025 in. for 199/232, 0.008 in. for 287/327 V8. The oil pressure relief valve is set at the factory to 60 psi and is not adjustable.

290/343/390/304/360 V8

The oil pump is located in, and as part of, the timing cover. The pump is driven by the distributor drive shaft. Oil pump R&R does not, however, affect distributor timing.

Remove pump cover and place a straightedge across pump body and gears. Clearance should be 0.0025-0.0065 in. (gears projecting above body). Measure clearance between gears and wall of gear cavity opposite point of gear mesh; should be 0.002-0.004 in. The oil pressure relief valve is set to 75 psi at the factory and is not adjustable.

Valve System

Valve Removal

L-head Engines

1. Remove the cylinder head and the valve chamber cover. Using a valve spring compressor, contract the valve spring and remove the keeper. Of course, this is done with the valves in the closed position.
2. Let the valve spring come down, and pull the valve up out of the top of the cylinder.
3. Thoroughly clean up the valve and examine the seat for pits and scratches. If any are found, the

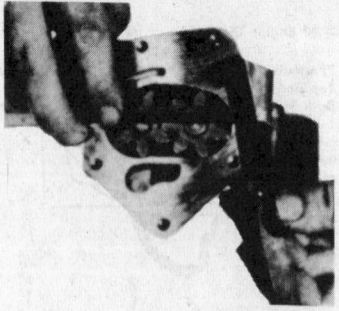

Checking oil pump gear end clearance
(© American Motors Corp.)

Checking oil pump gear to body clearance
(© American Motors Corp.)

valve face should be ground on a facing machine. When refaced, examine carefully to make sure that too much metal was not removed from the valves, since a sharp corner on the head of the valve will not last long in an engine.

4. There should be plenty of metal above the seat.
5. Examine the valve stem for wear and, if necessary, mike the valve stem in two or three places to check the amount of wear.

Overhead Valve Engines

Remove the cylinder head and carry it to a bench. Compress the

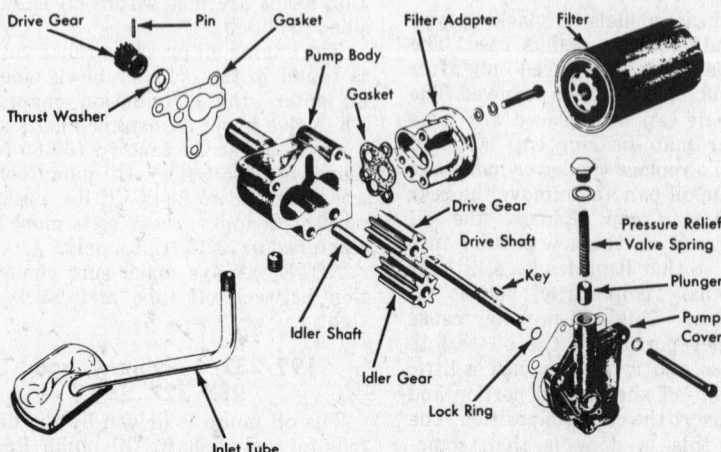

Drive Gear — Pin — Gasket — Filter Adapter — Filter
Pump Body
Gasket
Thrust Washer
Drive Gear
Drive Shaft
Pressure Relief
Valve Spring
Key
Plunger
Idler Shaft
Pump Cover
Idler Gear
Lock Ring
Inlet Tube

OHV 196 6 cylinder oil pump assembly (© American Motors Corp.)

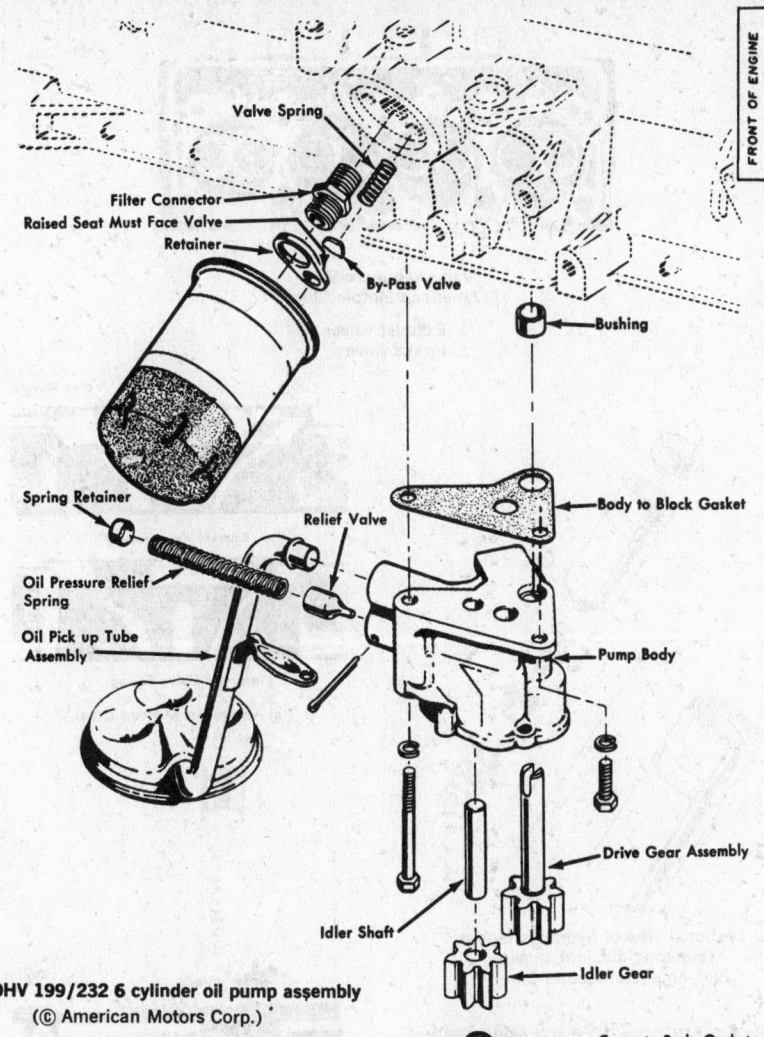

OHV 199/232 6 cylinder oil pump assembly
(© American Motors Corp.)

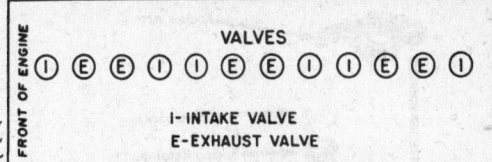

Valve sequence—6-cylinder L-head and OHV 196
(© American Motors Corp.)

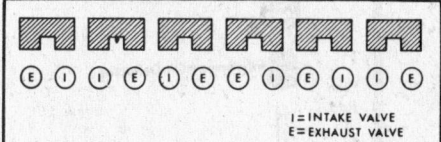

Valve sequence—OHV 6 199/232, bottom view
© American Motors Corp.

necessary to ream the guide so that the valve will fit properly. Any time a new valve guide is installed the valve should be reseated to make certain that the seat is concentric with the new guide.

Cast-iron OHV 196 6 Cylinder

With cylinder head off, remove the valves and carefully measure the distance from the top of the cylinder head to the top of the valve guide. Then, drive the guide out into the combustion chamber.

A new guide is driven in from the top towards the combustion chamber the distance noted before the old guide was removed. NOTE: 23/32 in. above head.

Sometimes, driving the guide disturbs it somewhat, making it neces-

valve spring and remove the keeper. Release the pressure from the spring and push the valve out.

Valve Spring Check

Valve spring pressure specifications are given in the General Engine Specification table.

However, a quick check can be made by placing all of the intake springs next to one another on a straight edge. If they all come up to the same height it can be presumed, with a fair degree of accuracy, that all the valve springs are in good condition since it is unlikely that all of them would collapse the same amount.

Valve Guide Removal

L-head 196 6 Cylinder

Remove the valves and, using a draw-type puller, draw the valve guide up through the top of the bore. Before the guide is pulled, the distance from the top of the cylinder head to the top of the guide should be measured and noted so that a new guide can be driven in just that amount. Sometimes, driving a new guide into the bore disturbs the top of the guide somewhat, making it

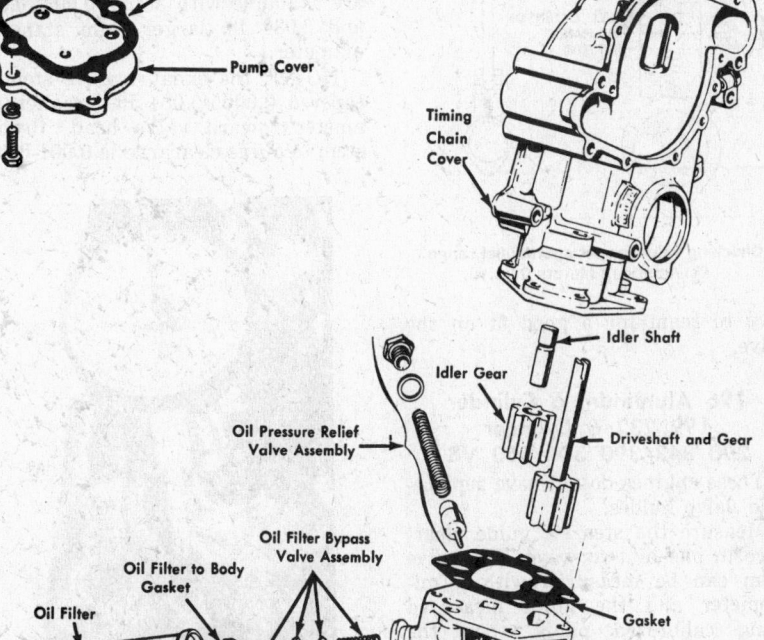

OHV 290/304/343/360/390 V8 oil pump assembly (© American Motors Corp.)

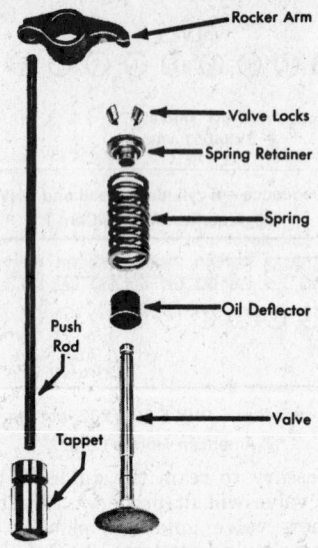

Valve assembly sequence—OHV 6 cylinder
(© American Motors Corp.)

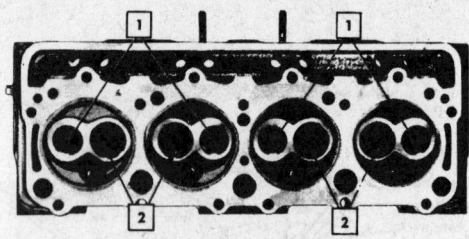

Valve sequence V8
(© American Motors Corp.)

1 Exhaust valves
2 Intake valves

Correct Valve Facing

Incorrect Valve Facing
(© American Motors Corp.)

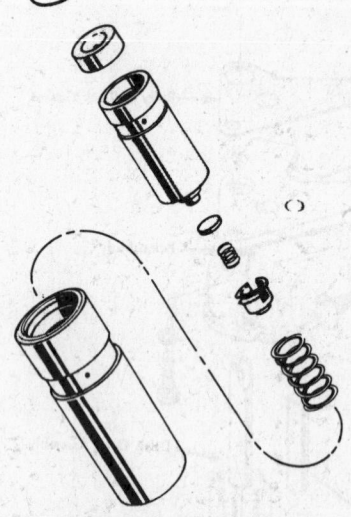

Sectional view of hydraulic tappet
(metering disc not shown)
(© American Motors Corp.)

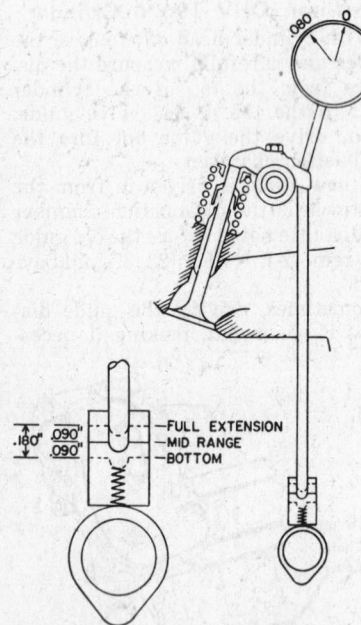

Checking valve tappet operational range
(© American Motors Corp.)

Removing valve spring keepers
(© American Motors Corp.)

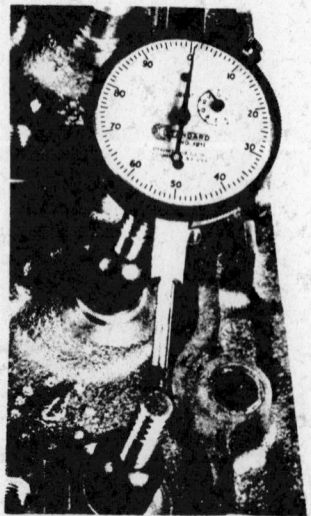

Installation of dial indicator for checking valve guide clearance
(© American Motors Corp.)

proper oversize. Three oversize valves are available with stems 0.003, 0.015 and 0.030 in. larger than standard diameter.

NOTE: the exhaust valve stem is tapered 0.0005-0.001 in., smaller diameter toward valve head. Proper stem to guide clearance is 0.001-0.003 in. for both intake and exhaust valves.

No provision is made for adjustment of hydraulic tappet travel. Tappets having various thicknesses of push rod seats are available and can be installed to get a standard center travel position under normal operating conditions.

287/327 V8

The valve stem to guide clearance is maintained through replacement of valve guides. The valve guides are an interference fit in their bores in the cylinder head, and can be replaced by driving out the old guides and driving in the new. The new guides are driven to a depth to permit ¾ in. + or — 1/64 in. to remain exposed above the cylinder head.

sary to ream for a good fit on the valve.

196 Aluminum 6 Cylinder
199/232 6 Cylinder
290/343/390/304/360 V8

These engines do not have replaceable valve guides.

Measure the stem to guide clearance in one of two ways. The valve stem can be measured with a micrometer and the guide measured using calibrated pilots, then the difference computed. The best way is to install the valve into the guide, without spring, and measure the lateral movement using a dial indicator. If stem to guide clearance is excessive, guides must be reamed to the

Tool J-21753, when driven until bottomed, establishes the required ¾ in. guide height.

Valve stem to guide clearance is .001-.003 in. for intake and exhaust.

Timing Case

Vibration Damper Removal

All Models

Remove the radiator core and the water pump. Remove the nut from the center of the pulley and, using a puller, extract the pulley from the front of the crankshaft.

Timing Case Cover R&R 199/232 OHV 6 Cylinder

The timing chain cover has a seal and oil slinger to prevent oil leakage past the crankshaft pulley hub.

1. Remove all V-belts, fan blades and pulley.
2. Remove vibration damper.
3. Remove oil pan to cover bolts and cover to block bolts.
4. Raise cover and pull oil pan front seal up far enough to extract the tabs from the holes in cover.
5. Remove cover gasket from block; cut off seal tab flush with front face of block.
6. Clean all mating surfaces and remove oil seal.
7. Install new front oil seal, using proper size arbor.
8. Install new neoprene front oil pan seal, cutting off protruding tab to match original.
9. Position cover on block and install bolts. Tighten cover bolts to

Timing chain cover assembly—199/232 6 cylinder
(© American Motors Corp.)

4-6 ft. lbs.; four lower bolts to 10-12 ft. lbs.
10. Install vibration damper, tightening bolt to 50-60 ft. lbs.

NOTE: front oil seal can be installed with cover in place only if proper tool or duplicate is available.

290/343/390/304/360 V8

The die-cast timing cover incorporates an oil seal at the vibration damper hub. This seal must be installed from the rear; therefore the cover must be removed from engine in every case to replace front seal.

1. Drain coolant and remove hoses from cover.
2. Remove distributor, fuel pump, alternator drive belt, accessory drive belts, fan and hub assembly.
3. Remove the vibration damper bolt, then pull off the damper.
4. Remove air conditioner compressor and power steering pump, if so equipped, and swing them out of the way *without* disconnecting hoses.
5. Remove the two front oil pan bolts from beneath the car, then remove the eight 9/16 in. hex head cover bolts.
6. Remove cover from block, then clean all parts and mating surfaces and remove oil seal.
7. Coat new seal lips with Vaseline and seal surface with sealer, then drive seal into cover bore until it seats against the outer cover face. Use a proper size arbor for this job.
8. Remove lower dowel pin from cylinder block; this must be replaced when cover is in position but before bolts are installed.
9. Cut the oil pan gasket flush with the block on both sides of the oil pan.
10. Cut corresponding pieces of gasket from another oil pan gasket and cement them to cover. Install neoprene oil pan front seal into cover and align cork gasket tabs with pan seal.
11. Apply Permatex No. 2 to gaskets, then position cover. Install oil pan bolts and tighten evenly until cover lines up with upper

dowel pin.
12. Install lower dowel pin, then cover to block bolts; tighten to 20-30 ft. lbs.
13. Install all removed pieces and adjust ignition timing.

Valve Timing

Camshaft (valve) timing is determined by the relationship between the camshaft sprocket and the crankshaft gear. A jumped timing chain can cause any number of problems (backfiring, spitting back, poor performance), not the least of which is a non-starting engine.

Valve Timing Check—all 6 Cylinder

To check valve timing, remove valve cover and spark plugs, then rotate crankshaft until No. 6 piston is at TDC on compression stroke. Compression stroke can be determined by holding a finger in the No. 6 spark plug hole while turning the

Installation of dial indicator for checking valve timing
(© American Motors Corp.)

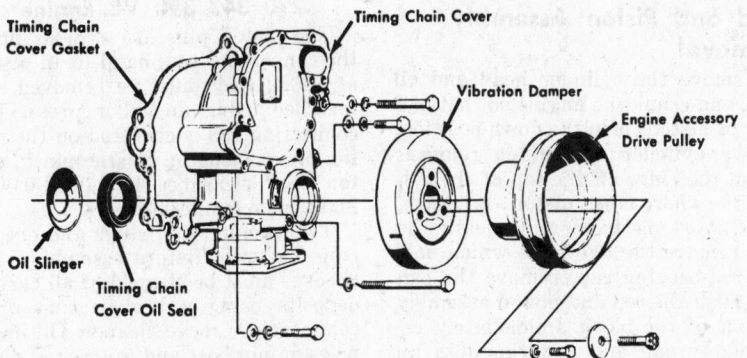

Timing chain cover assembly—290/304/343/360/390 V8 (© American Motors Corp.)

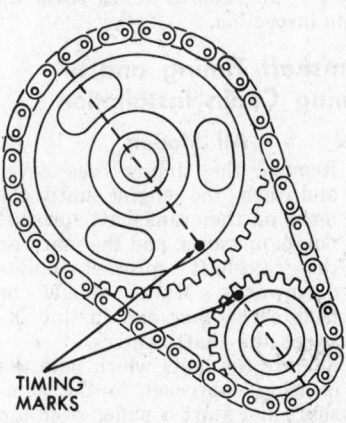

TIMING MARKS

Six-cylinder timing chain and sprockets

engine by hand—finger will be forced out by compression pressure. Set the valves of No. 1 cylinder to 0.003 in. clearance and rock the crankshaft back and forth. The exhaust valve should open before the TDC mark on the pulley lines up with the pointer—measure the actual distance. The intake valve should open the same distance *past* the pointer—if it varies more than ½ in., remove the timing cover and inspect the chain.

Valve Timing Check—all V8

To check the valve timing, remove the rocker covers and spark plugs, then rotate the crankshaft until No.

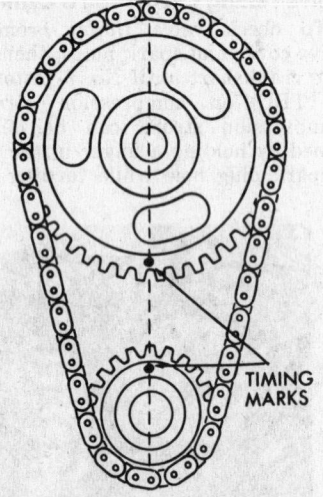

V8 timing chain and sprockets

6 piston is at TDC on compression stroke. This places No. 1 piston on TDC exhaust overlap. Rotate the crankshaft counterclockwise 90° (¼ turn) and install a dial indicator on No. 1 intake rocker pushrod end. Crank the engine in the normal direction of rotation until the pushrod moves, as indicated on the dial by a ± 0.020 in. pointer movement. In this position, the timing pointer should align with TDC mark on 290/343/390/304/360 engines; about 25/32 in. before TDC mark on 287/327 engines. If this varies more than about ¼ in., remove timing cover for chain inspection.

Camshaft Timing and/or Timing Chain Installation

All Models

1. Remove the timing case cover and turn the engine until the mark on the crankshaft sprocket points upwards, and the mark on the camshaft sprocket points downwards. Marks should be near each other and in line between the shaft centers.
2. Remove the bolts which hold the camshaft sprocket to the camshaft and start a puller over the crank gear.

3. Pull the crank gear off the front of the crankshaft.
4. On the bench, arrange the new chain over the sprockets so that the marks are nearest each other and in line between their own centers, then carry this to the engine. Start the crank gear up on its key and arrange the camshaft so that, when the three bolts line up, the marks are between shaft centers.
   NOTE: on 6 cylinder engines, locate the marked cam tooth at one o'clock position; mark on crank should be approximately at point of mesh. Count the number of links between the marks; should be 19 pins or 9½ links for OHV and L-head 196 cast iron engine, 18 pins or 9 links for OHV 196 aluminum engine, 15 pins or 7½ links for OHV 199/232 engine.

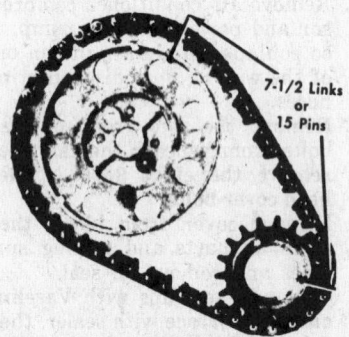

Correct timing chain installation— 199/232 6 cylinder
(© American Motors Corp.)

On all V8 engines, locate the marked cam tooth on a horizontal line at three or nine o'clock position; mark on crank at point of mesh. Count the number of links between marks; should be 20 pins or 10 links.

5. Secure the cam gear to the camshaft and force the crankshaft gear all the way on the shaft.
6. Reassemble the front of the engine.

Pistons, Connecting Rods and Main Bearings

Rod and Piston Assembly Removal

Remove the cylinder head and oil pan, and crank the engine so that one pair of pistons is in the down position. Use a cylinder ring ridge remover to cut the ridge off the top of the cylinders where the piston is down. Now, working from underneath the car, remove the two bolts which hold the rod-bearing cap, remove the cap and push the rod and piston assembly up out of the block. Immediately replace the cap and run the nuts up finger-tight so that the cap will not

get turned in, or the bearing will not be lost.
NOTE: use rubber hose on rod bolts to protect journals.

Rod and Piston Assembly Installation

On engines using split-skirt pistons, the slit in the skirt must be installed opposite the oil squirt hole in the connecting rod. Solid skirt pistons are assembled so that the boss, or dimple, (and, in some instances, the letter F) at the top of the piston is on the same side of the connecting rod as the boss. This will be found on the connecting rod channel about halfway up the rod.

The piston and rod assemblies are united to the engine from the top and the dimple, or dot, on the top of the piston goes toward the front. On those engines having split-skirt pistons, the slit in the skirt of the piston goes to the left side of the engine.

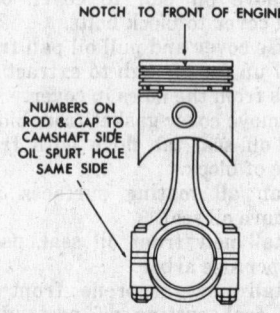

Piston and rod assembly 6-cylinder engine

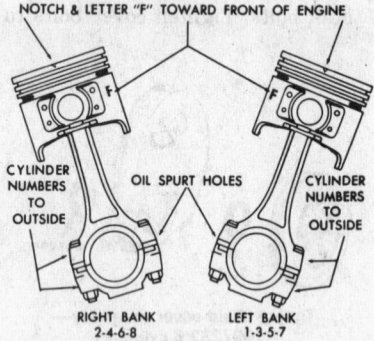

Piston and rod assembly—V8 engines

290/343/390 V8 Engine

The piston pins are a press fit in the connecting rod, hand fit in piston at 68°F; they must be removed and installed using an arbor press. The connecting rod is centered on the piston pin ± 0.030 in.; piston pin to piston bore clearance is 0.0003-0.0005 in. at room temperature.

Two compression rings and one oil ring are used. Before assembly, ring grooves must be cleaned of all carbon deposits using a broken ring or a commercial groove cleaner. Oil drain holes on pin boss and in grooves must be cleaned at the same time.

Side clearance between land and piston ring should be 0.002-0.004 in. for No. 1 and No. 2 rings, 0.000-0.005 in. for No. 3 oil control ring.

Ring end gap is measured at the bottom of the cylinder bore near the end of ring travel area; push ring down using an inverted piston and measure gap with feeler gauge. Compression rings—0.010-0.020 in.; oil ring—0.015-0.055 in. (rail gap).

Prior to installing piston and connecting rod assembly into 290/343, arrange ring gaps with No. 1 180° from No. 2 and No. 3 at least 90° from No. 2. Each rail gap on oil control ring should be 30° apart.

On 390 engine, arrange No. 1 gap 180° from No. 2; oil control expander spacer tangs must be installed in drilled holes above piston pin and oil control rails must be 180° apart.

Piston to cylinder bore clearance should be 0.001-0.0018 in. for 290 engine; 0.0012-0.002 in. for 343 engine; 0.0010-0.0018 in. for 390 engine.

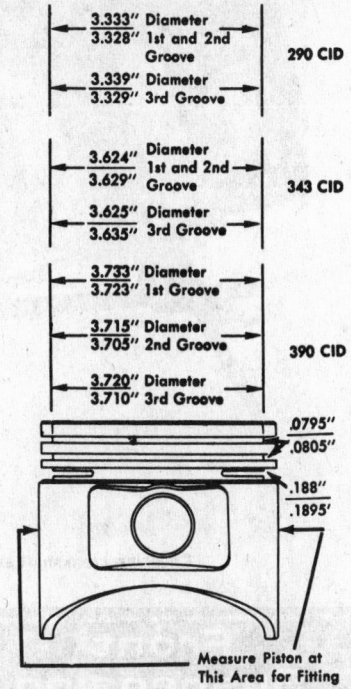

290/343/390 V8 piston dimensions
(© American Motors Corp.)

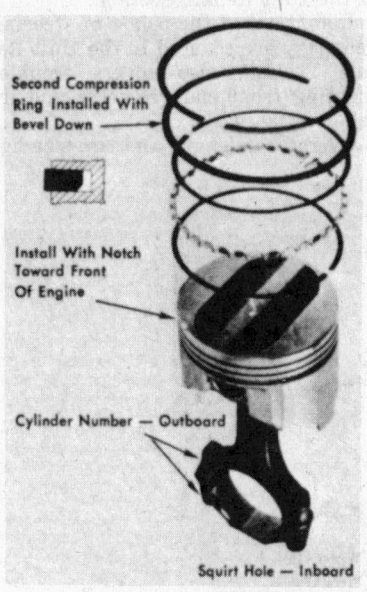

390 V8 piston assembly
(© American Motors Corp.)

Measuring ring side clearance
(© American Motors Corp.)

287/327 V8 Engine

Piston pins are a press fit in connecting rod, hand fit in piston at 68°F. Before assembly, clean ring grooves using broken ring or commercial groove cleaner. Piston to cylinder bore clearance at top land

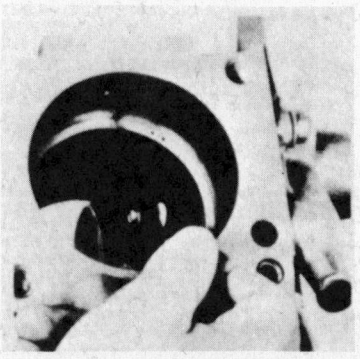

Measuring ring end gap
(© American Motors Corp.)

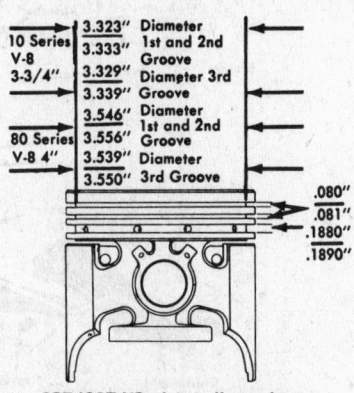

287/327 V8 piston dimensions
(© American Motors Corp.)

should be 0.028-0.032 in.; at skirt top 0.0009-0.0025 in.; at skirt bottom 0.0009-0.0015 in.

Piston ring groove clearance should be 0.002-0.006 in. for compression rings, 0.0001-0.004 in. for oil ring. Ring end gap, with ring near bottom of cylinder, should be 0.010-0.020 in. for compression rings, 0.015-0.055 in. for oil ring rail.

Piston rings should be arranged with the end gaps 120° apart, with no gap being placed over the pin boss.

199/232 6 Cylinder Engine

Pistons of the 199 engine can be identified by their flat head and two notches; 232 pistons have a concave head with one notch. Specified piston to cylinder bore clearance is 0.0005-0.0013 in.

Ring side clearance on No. 1 and No. 2 ring should be 0.0015-0.0035 in.; 0.0000-0.005 in. on oil ring. End gap on compression rings should be 0.010-0.020; 0.015-0.055 in. on oil control rail. Arrange rings in the same manner as on 290 V8 engine.

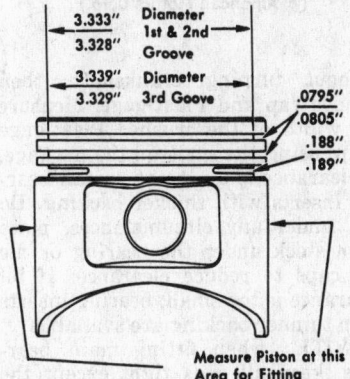

199/232 6 cylinder piston dimensions
(© American Motors Corp.)

196 L-head and OHV 6 Cylinder Engines

Piston ring side clearance in grooves should be 0.002-0.004 in. for compression rings, 0.0001-0.004 in. for oil ring. Piston to cylinder bore clear-

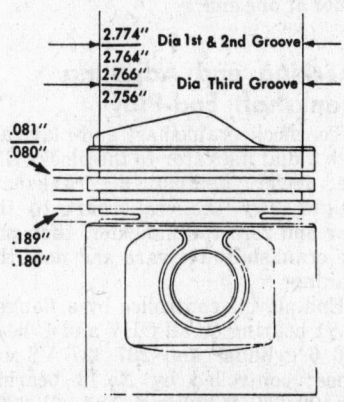

196 OHV and L-head 6 cylinder piston dimensions
(© American Motors Corp.)

ance 0.0006-0.0012 in. at skirt bottom, 0.017-0.019 in. at top land. Piston ring gap should be 0.010-0.020 in. for compression rings, 0.015-0.055 in. for oil ring.

Arrange piston rings with gaps 120° apart.

Fitting Connecting Rod and Main Bearings

To determine the amount of bearing clearance, place a piece of Plastigage, available at automotive wholesalers, in the cap (on the dry bearing) and install the cap. Tighten cap bolts to proper torque specification,

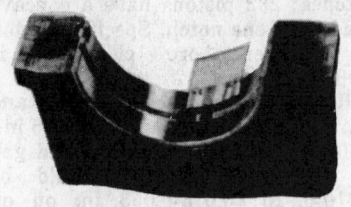

Checking bearing clearance with Plastigage
(© American Motors Corp.)

without turning crankshaft, then remove cap and Plastigage. Measure the width of the crushed Plastigage strip, using the scale on the package. If clearance is too great, install bearing inserts with thicker backing. Do not, under any circumstances, place shim stock under the bearing or file the caps to reduce clearance. If oil clearance is too small, bearing inserts with thinner backing are available.

NOTE: when fitting main bearings, keep all caps tight except the one being checked. Support weight of crank with a screw jack for best results.

Lacking a micrometer, Plastigage can be used to check journal taper. Place strip of Plastigage lengthwise along the bearing insert (aligned with long axis of engine) and tighten cap to proper torque. If the journal is tapered, the Plastigage will be wider at one end.

Checking and Adjusting Crankshaft End-Play

To check crankshaft end-play, attach a dial indicator to the block with the feeler against a crankshaft weight. Pry the crankshaft to the rear and zero the indicator, then pry the crankshaft forward and note the reading.

End-play is controlled by a flanged No. 1 bearing on all OHV and L-head 196 6 cylinder and 287/327 V8 engines; controlled by No. 3 bearing on 199/232 6 cylinder and all 290/343/390/304/360 V8 engines from 1967.

Checking crankshaft end-play (© American Motors Corp.)

Front Suspension

References

General instructions covering the front suspension and how to repair and adjust it, together with information on installation of front wheel bearings and grease seals, are given in the Unit Repair Section.

Definitions of the points of steering geometry are covered in the Unit Repair Section, also covers troubleshooting front end geometry and irregular tire wear.

Overall car length and tire size fig-

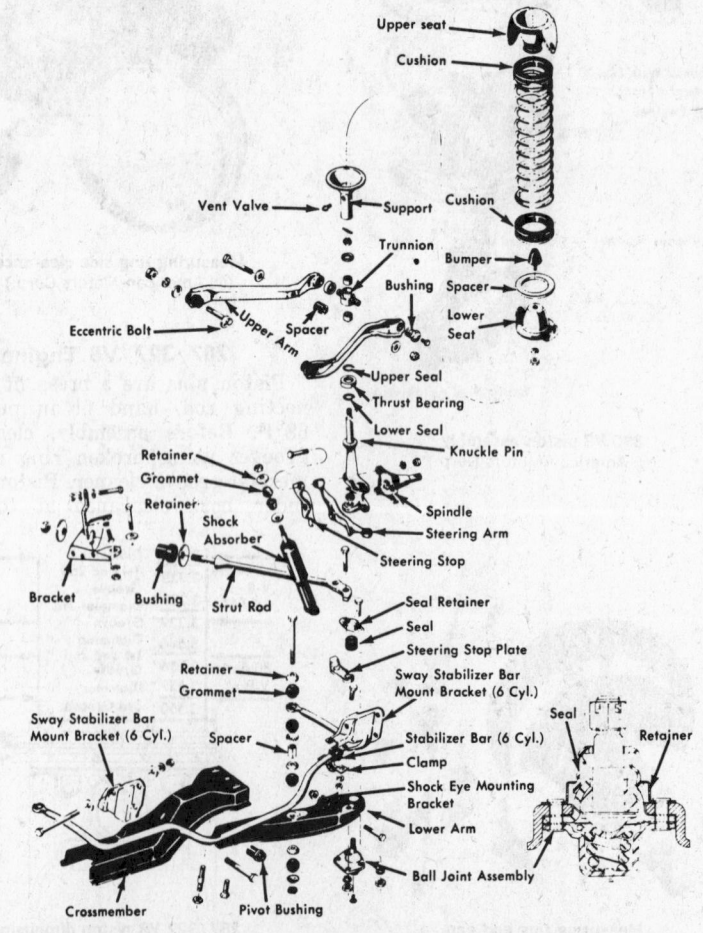

Typical front suspension, except American (© American Motors Corp.)

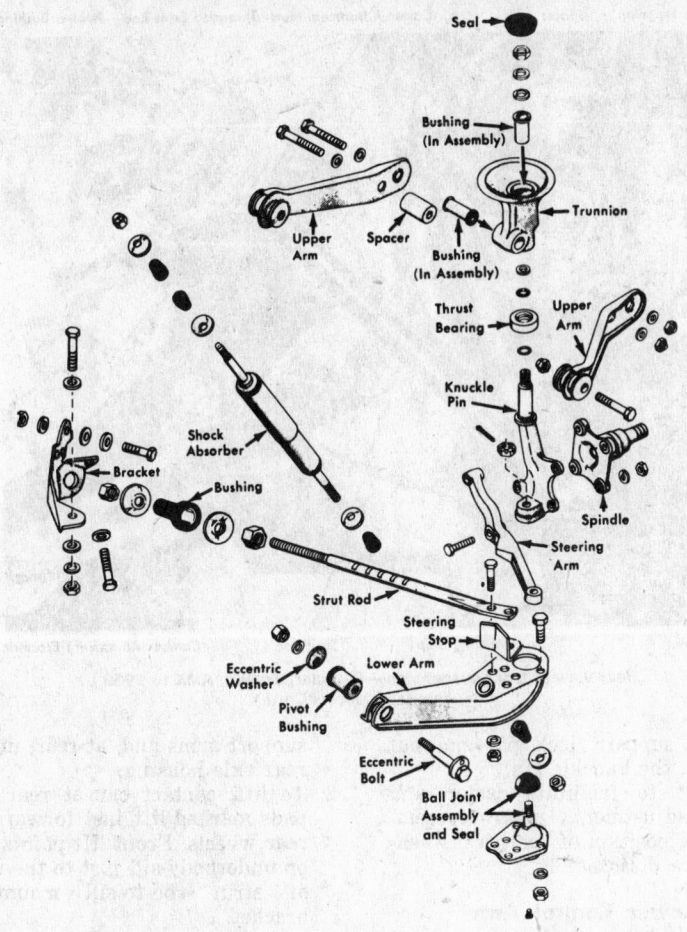

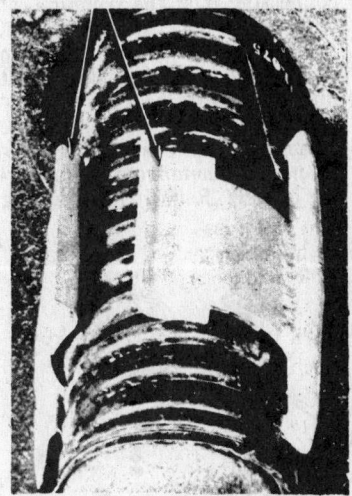

Front coil spring removal tools in place
(arrows)
(© American Motors Corp.)

ures can be found in the General Chassis and Brake Specifications table of this section.

Front Spring Removal

Raise rear of car at diagonal corner from front spring to be removed. Install hooks in holes provided in spring seats while compressed. Release load and lower rear corner. This allows spring removal from car.

To install new spring, compress it by means of hydraulic press or jack with seats in place. Be sure holes

Typical front suspension—American (© American Motors Corp.)

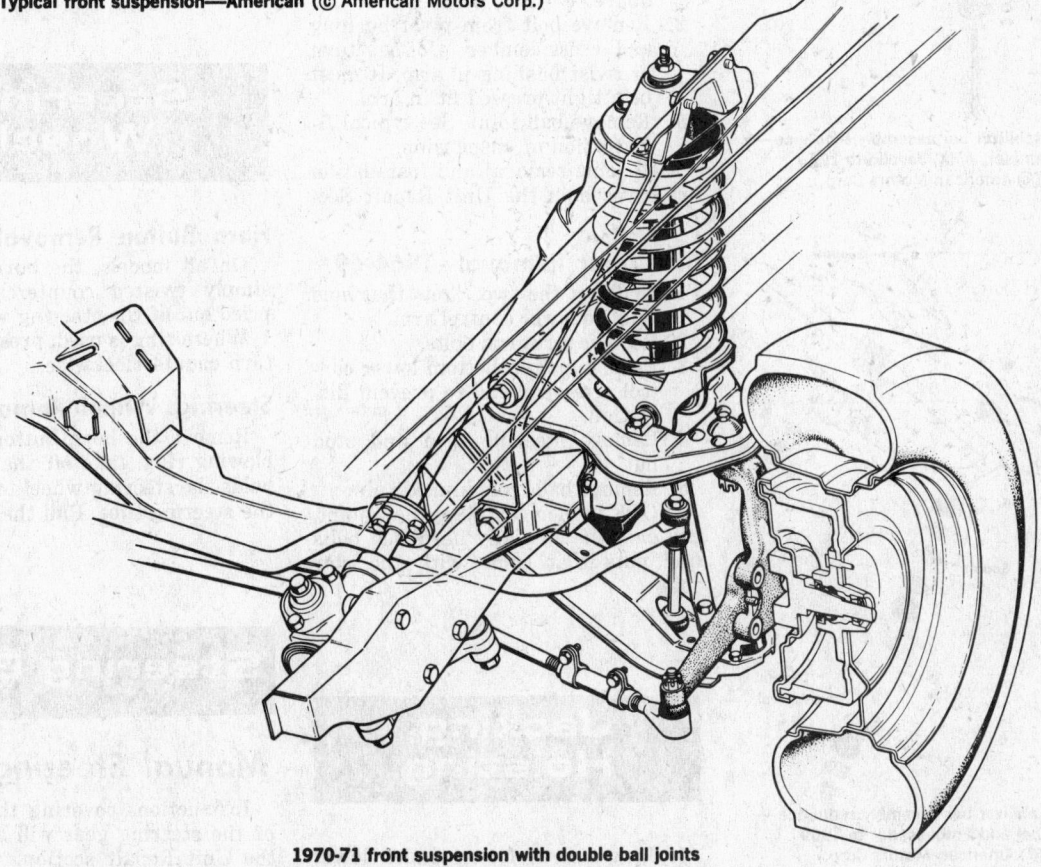

1970-71 front suspension with double ball joints

in seats are aligned. Install hooks to hold spring in compression while placing in position. Reversing the above removal procedures will control position to release hooks when spring is in place.

Control Arm Removal 1964-69

Upper Control Arm

1. Remove front spring as described above.
2. Remove front or rear arm by disconnecting it at the trunnion, at wheelhouse panel mounting bolt, and the control arm spacer. Upon reassembly, torque spacer bolt nut to 80-90 ft. lbs.
3. Both front and rear arms may be removed as an assembly by disconnecting them from the mounting bracket. Remove lower spring

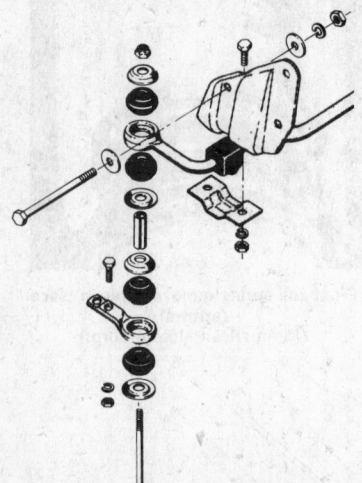

Front stabilizer bar assembly sequence— Rambler, AMX, Javelin to 1969
(© American Motors Corp.)

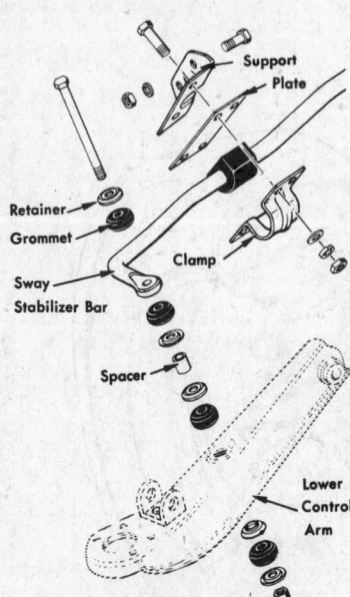

Front stabilizer bar assembly sequence— Rebel and Ambassador to 1969
(© American Motors Corp.)

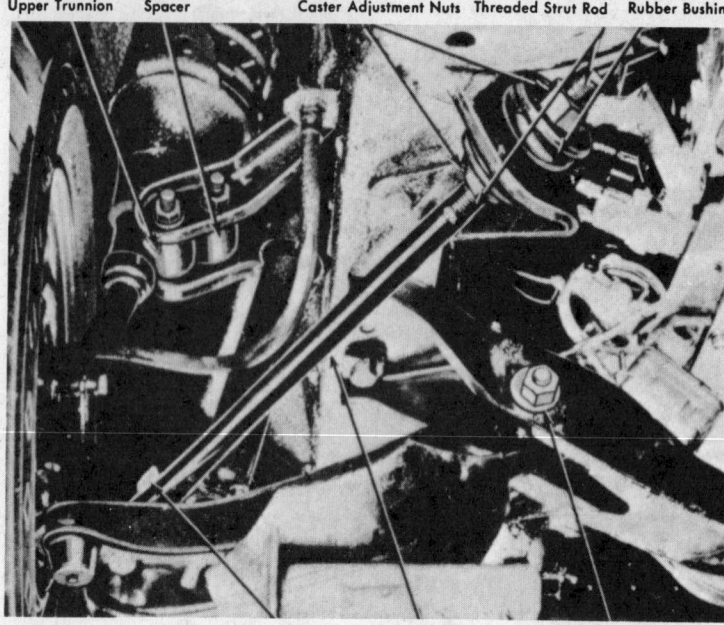

Upper Trunnion   Spacer   Caster Adjustment Nuts   Threaded Strut Rod   Rubber Bushing

Ball Joint   Steering Stop   Strut Rod   Camber Adjustment Eccentric

Rear view of front suspension—Rambler, Javelin, AMX to 1969
(© American Motors Corp.)

seat support, lock pin and nut from the knuckle pin.

NOTE: to facilitate caster and camber adjustments upon reassembly, mark position of eccentric washers before disassembly.

Lower Control Arm

1. Remove front spring as described above.
2. Remove bolt from pivot bushing and crossmember. Do not turn or twist bushing in arm. It must be a tight-pressed fit in arm.
3. Remove ball joint. See typical Illustration of suspension.

Ball joint removal and installation are covered in the Unit Repair Section.

Ball Joint Removal—1964-69

1. Drill out the two rivets that hold joint to lower control arm.
2. Remove strut rod bolts.
3. Place punch in bottom lower control arm open end to prevent distortion.
4. Remove steering arm and stud nut.
5. Remove ball from knuckle pin.

NOTE: when installing new joint, use *only* the specially hardened bolts and nuts that come with the new joint.

Jacking, Hoisting

1. Jack car, at front, under lower

support arms and, at rear, under rear axle housing.
2. To lift, contact car at rear lift pads marked lift just forward of rear wheels. Front lift points are on underbody sill just to the rear of strut rod-to-sill mounting bracket.

Steering Wheel

Horn Button Removal

On all models, the horn button is simply twisted counterclockwise or pried out of the steering wheel.

Where ring is used, press down and turn counterclockwise.

Steering Wheel Removal

Remove the horn button, or horn-blowing ring, take off the nut which holds the steering wheel to the top of the steering tube. Pull the wheel.

Steering Gear

Manual Steering Gear

Instructions covering the overhaul of the steering gear will be found in the Unit Repair section.

Power Steering Gears

Troubleshooting and repair instructions covering power steering gears are given in the Unit Repair Section.

Steering Gear Removal

All Models 1963-70

1. Disconnect battery.
2. Raise and support front of car.
3. Remove horn button, or ring, and steering wheel.
4. Remove jacket support plate screws. Loosen tube bracket to instrument panel mounting bolts.
5. Remove pitman arm.
6. Remove steering gear mounting screws, and lower gear to move out bottom.

Power Gear Pump Removal

1. Disconnect the pressure hoses from the pump, slack off on the adjusting bracket and remove the bracket bolts which hold the pump to the side of the engine. Lift off the pump.
2. On Eton pumps, to separate the oil reservoir from the pump body, remove the cover, the element, the hollow stud, the two attaching screws, the element baffle and the reinforcement plate. On Vicars pumps, take the cover off the reservoir and remove the two cap screws which hold the reservoir to the pump.

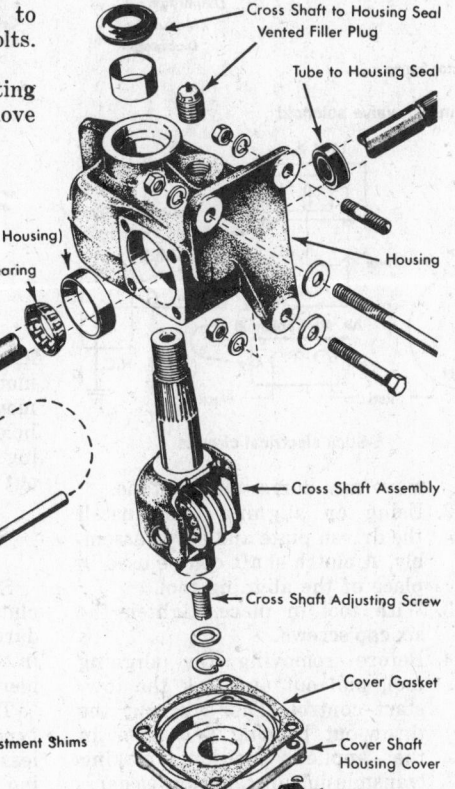

Typical steering gear—American (© American Motors Corp.)

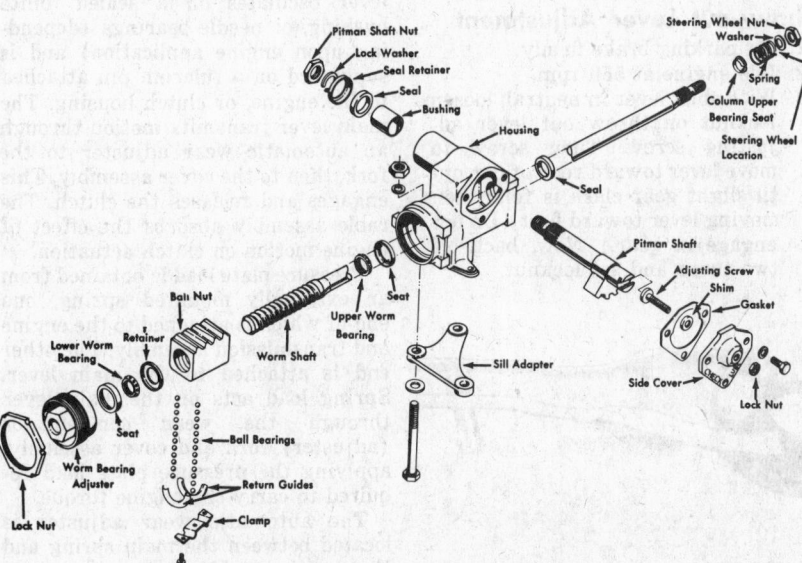

Typical steering gear, except American (© American Motors Corp.)

Clutch

Pedal Clearance Adjustment

Adjust the free-play of the clutch pedal to 1/2-3/4 in. up to 1967, 7/8-1 1/8 in. thereafter. This is done by changing the length of the link between the throw out lever and the clutch lever.

NOTE: adjust clutch pedal to floor clearance to 6 1/2 in. for models with 199 cu. in. engine. On 232 and all V8 models, adjust clutch pedal height by inserting a 5/16 in. x 4 1/2 in. long pin into the alignment holes in the pedal bracket. Adjust pedal height so that pin slides freely.

Clutch Assembly Removal

Remove the transmission and disconnect the clutch linkage, unbolt the clutch inspection pan. Release the pedal rod and return spring, which will permit removing the throw-out shaft so that the release bearing can be removed through the transmission opening. On all models, release the clutch mounting bolts evenly and remove the assembly from the flywheel. Reverse procedure to replace, using a pilot shaft to align the clutch disc with the pilot bushing.

E-Stick Clutch

1963-64

An optional clutch, engaged by engine oil pressure, is used. This pressure is controlled through a valve body so arranged that engine vacuum raises or lowers the pressure as driving conditions demand. In turn, the position of the gearshift lever controls a solenoid which changes applied manifold vacuum. This, through the valve body, regulates the required oil pressure.

With engine off and gearshift lever in neutral, a retracting spring releases pressure on clutch fingers, throw-out bearing and driven plate.

With engine idling and gearshift lever in neutral, the solenoid prevents vacuum from applying any pressure.

With engine idling, and gearshift lever in any driving position, the solenoid allows a controlled pressure (approximately 7 psi) through the valve body to servo piston. This piston, in turn, applies slight pressure to the throw-out lever, throw-out bearing, clutch cover fingers, cover plate and driven plate. The slight pressure takes up drive line slack. It is also enough pressure to hold clutch engagement for engine braking.

As engine is accelerated, controlled

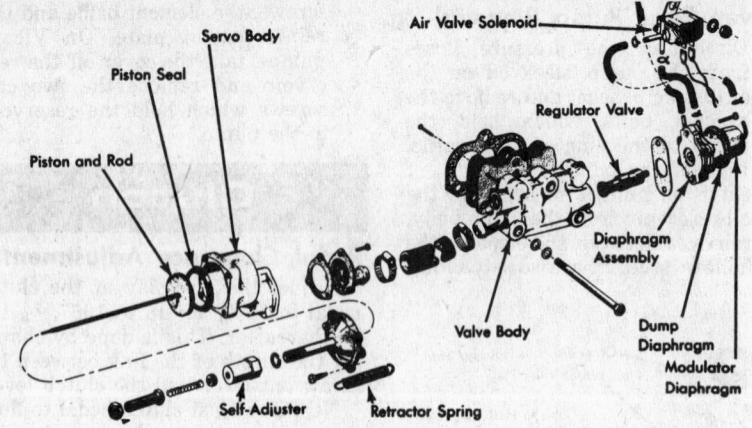

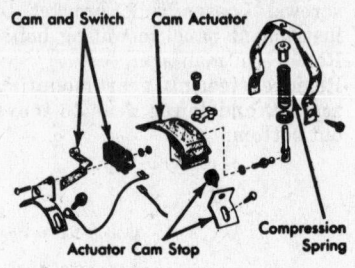

E-Stick servo, control valve and air valve solenoid

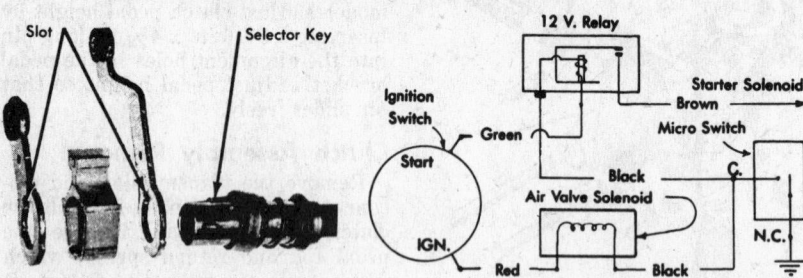

Indexing key with slot in lever

E-Stick electrical circuits

Actuator cam and cam switch

vacuum and controlled oil pressure increase clutch pressures so that there is positive clutch engagement.

Releasing of accelerator increases vacuum. This now reduces oil pressure to the point of clutch release for gear shifting. When gearshift is returned to neutral, a cam switch actuates the solenoid and controls the oil pressure through the valve body.

A tow start control handle is mounted at the steering column. This is connected to the throw-out lever by a cable, which, when pulled out, locks the throw-out lever in engaged position. This can be used for tow-starting the engine or in-gear parking.

The throw-out bearing is a ball type, permanently lubricated. It rotates at all times when the engine is running and the car is in gear. In servicing, care must be used to avoid dirt. This bearing should not be washed, as gasoline or other solvents will destroy the permanent lubricant.

E-stick Clutch Removal

1. Remove driveshaft, transmission and associated cables and linkages, as with conventional clutch.
2. Mark clutch cover, pressure plate and flywheel before removing, so the relative position can be resumed in reassembly.
3. Remove six cover cap screws and lift out clutch.

E-stick Clutch Installation

1. Check freeness of driven plate on transmission clutch shaft and re-

move any burrs with a stone.
2. Using an aligning tool, install the driven plate and cover assembly. A clutch shaft can be used in place of the aligning tool.
3. With tool in place, tighten the six cap screws.
4. Before removing the aligning tool, pull out and lock the tow-start control. This will put the throw-out bearing and lever in the applied position, making transmission installation easier.
5. Complete installation of transmission, driveshaft and associated cables and linkages.

Throw-out Lever Adjustment

1. Set parking brake firmly.
2. Idle engine at 550 rpm.
3. With shift lever in neutral, loosen locknut on throw out lever adjusting screw. Turn screw to move lever toward rear of car until slight gear clash is felt when moving lever toward first (Do not engage in gear.) Now, back off two turns and set locknut.

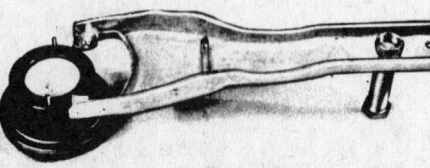

E-Stick clutch throwout lever and bearing
(© American Motors Corp.)

4. Loosen locknut on tow-start cable. Adjust hexagon adjuster until pivot clears throwout lever by 1/8 in. Tighten locknut.

Actuator Cam Adjustment

This switch is mounted on column below instrument panel, and must be set properly to release clutch when shifting gears. Move gearshift lever back and forth between second and high. The switch will emit a slight click when closing the circuit to re-

lease the clutch. With the switch mounting screws slightly loosened, move the switch by means of slotted holes just enough to equalize the spacing of the clicks when moving lever out of either gear.

S-type Clutch

Since 1966, a new self-adjusting clutch system has been used as standard equipment for 6-cylinder models in the Ambassador, Marlin and Rambler Classic series.

The new clutch is known as the S type. It is smoother in action, needs less pedal pressure, and is self-adjusting throughout its service life.

Linkage design utilizes an enclosed cable to transmit motion between the clutch pedal and the main lever. The lever oscillates on a sealed, oilite bushing, or needle bearings (depending upon engine application) and is supported on a fulcrum pin attached to the engine, or clutch housing. The main lever transmits motion through an automatic wear adjuster to the fork, then to the cover assembly. This engages and releases the clutch. The cable assembly absorbs the effect of engine motion on clutch actuation.

Pressure plate load is obtained from an externally mounted spring, one end of which is attached to the engine and transmission assembly. The other end is attached to the main lever. Spring load acts on the main lever through the wear compensator (adjuster) fork and cover assembly, applying the pressure plate load required to carry the engine torque.

The automatic wear adjuster is located between the main spring and the cover assembly.

The adjuster senses and maintains

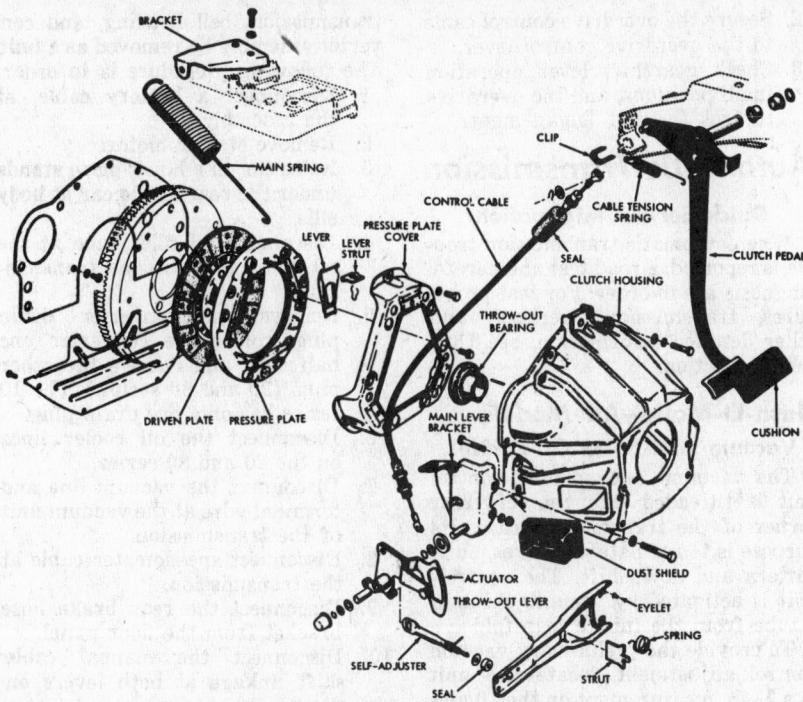

S-type clutch and linkage assembly (© American Motors Corp.)

a specific angular travel of the main lever. As the clutch facing wears, the travel of the main lever increases causing the actuating arm on the lever to contact a spring. The spring acts as a one-way clutch on a threaded push rod. The actuating arm, pushing on the spring when the clutch is released, rotates the rod and increases the overall length of the adjuster. The actuating arm also repositions the spring on the rod when the clutch is engaged. This sequence of events reduces the travel of the main lever to the desired arc and continues throughout clutch life. An annular contact ball bearing is used to carry the load between fork and cover assembly levers. This bearing is under load at all times.

NOTE: crankshaft end-play has an effect on light torque chatter. Therefore, if this type of chatter does exist, it is recommended that crankshaft end play be held to .006 in. maximum. The light bearing loads (of the type S system) at initial engagement are insufficient to prevent crankshaft end flutter until about 20% of engagement is reached.

Clutch R & R

In general, removal and installation procedures are the same as with traditional clutches.

Transmissions

Standard Transmission

General Information

Transmission refill capacities will be found in the Capacities table of this section.

Troubleshooting and repair of overdrive units are covered in the Unit Repair Section.

Synchromesh Transmission Removal

Models with Hotchkiss Drive

1. Split the rear universal joint and slide the driveshaft off the back of the transmission. (See Universal Joints and Drive Lines.)
2. Remove shift mechanism linkage to the transmission, and disconnect the clutch linkage.
3. Disconnect the overdrive mechanism (if so equipped) and remove the rear mounts.

NOTE: on V8 models with dual exhaust, exhaust pipes must be disconnected from manifolds so that rear crossmember can be removed. On Javelin and AMX models having Hurst shifter, entire shifter must be removed so that transmission can slide back far enough for removal.

4. Take out the transmission support crossmember, remove the two studs which hold the transmission to the bell housing and replace these two studs with two long pilot studs.
5. Take out the two bottom studs and slide the transmission assembly along the pilot studs and out of the car.

Models with Torque Tubes

While it is possible to remove the transmission with the torque tube in the car by prying the torque tube back as far as it will go, it is probably easier to take out with the rear axle removed. (See Universal Joints and Drive Lines.)

1. Disconnect the brake lines and brake cables. Jack the car sufficiently high so that the operator can work under it.
2. Remove the bolts that hold the front flange at the torque tube to the back of the transmission, detach the rear end of the springs and the shock absorbers, remove the U-bolts which hold the rear axle assembly to the leaf spring and roll the rear axle assembly back a couple of inches to clear the transmission.
3. Disconnect the shift levers at the transmission speedometer cable, and, if equipped with overdrive, the overdrive wiring.
4. Support the back of the engine and remove the bolts which hold the back of the transmission to the crossmember. Take out the crossmember.
5. Remove the bolts which hold the transmission to the bell housing and replace two of the bolts with long guide studs. Slide the transmission out along the guide studs and down and out of the car.

Twin-Stick Overdrive Control Disassembly

1. Place the assembly in a vise.
2. Remove the four screws that retain the gearshift pivots to the gearshift lever pivot bracket.
3. Remove the four cap screws that secure the pivot bracket to the gear shift lever housing. After removing the gearshift knob, the pivot bracket can be removed.
4. Remove the two cap screws from the retainer that holds the selec-

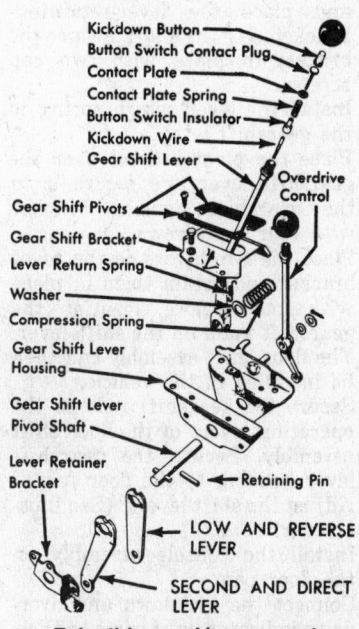

Twin-stick assembly components
(© American Motors Corp.)

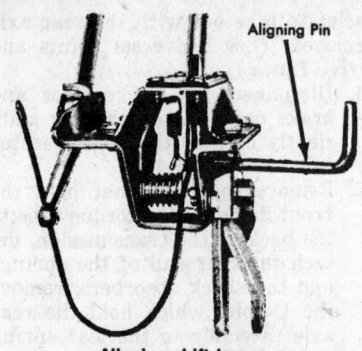

Aligning shift levers
(© American Motors Corp.)

tor levers on the gearshift lever pivot shaft. Then the levers can be removed.

5. Remove the lever return spring from the gearshift lever.
6. Use a 3/16 in. punch to remove the pin that secures the gearshift lever to the gearshift lever pivot shaft and remove the shaft.

NOTE: the gearshift lever pivot shaft and the selector pin are serviced as an assembly.

Twin-Stick Overdrive Control Assembly

1. With the gearshift lever housing held in a vise, install the gear shift lever, washer, and compression spring in the housing.
2. Install the gearshift lever pivot shaft through the housing, and the gearshift lever, washer, compression spring in the housing.
3. Install the pin in the gearshift lever and through the pivot shaft. Use a long-fiber grease as a lubricant for the pivot shaft.
4. Position the low and reverse selector lever on the pivot shaft. Then, place the second and high-selector lever on the pivot shaft and place the lever-retaining-bracket over the shaft. Secure the bracket in place with two cap screws.
5. Install the lever return spring to the gearshift lever.
6. Place the pivot bracket over the gearshift lever and secure it to the gearshift housing bracket with four cap screws.
7. Place the two pivots on the pivot bracket and retain them in place with four screws. Position the gearshift knob on the shift lever. The floor shift assembly can then be installed in the vehicle.
8. Secure the gearshift rods to the operating levers of the floor shift assembly. Secure the gearshift lever housing to the floor pan.
9. Adjust the shift levers. (See illustration)
10. Install the console assembly on the floor pan.
11. Connect the kickdown and overdrive indicator light wires to their respective connectors.

12. Secure the overdrive control cable to the overdrive control lever.
13. Check gearshift lever operation in all positions, and the overdrive control for full engagement.

Automatic Transmission

Quick Service Information

When automatic transmission trouble is reported, a road test and careful diagnosis are in order. For test procedures, transmission overhaul, and other detailed information, see Unit Repair Section.

Flash-O-Matic—All Models

Vacuum and Solenoid Control

The vacuum- and solenoid-control unit is threaded into the left rear corner of the transmission case. Its purpose is to regulate pressures, shift pattern and downshift. The control-unit is activated by vacuum through a tube from the intake manifold.

To provide the preliminary vacuum control adjustment, locate the unit to a 3/8 in. measurement on the 10 and 20 series, 23/64 in. on the 80 series. This measurement is between the back face of the transmission case and the front edge of the control unit.

1. Connect a vacuum gauge to a fitting with a T connection between the control unit and the vacuum line tube.
2. Remove the 1/8 in. pipe plug located at the left front lower side of the transmission case. Install a pressure gauge line connector at this location connected to a pressure test gauge. Start the engine and put the selector in D1.
3. Apply the parking and service brakes and accelerate the engine until 10.5 in. of vacuum is obtained on the vacuum gauge for the 10 series car, 12.1 in. of vacuum for the 20 series car and 13.8 in. of vacuum for the 80 series car.
4. At this time, check the reading on the pressure gauge. The correct pressure should be 100 psi ± 3 psi for the 10 series and 85 psi ± 3 psi for the 20 and 80 series.
5. If the correct reading is not obtained, it will be necessary to adjust the vacuum-control-unit. Rotating the vacuum unit clockwise increases the pressure, and counterclockwise decreases the pressure.

NOTE: do not operate the engine over ten seconds at any one time when performing the above test.

Transmission, Bell Housing And Converter Removal

Flash-O-Matic

Involvement may require that the

transmission, bell housing, and converter assembly be removed as a unit. The following procedure is in order:

1. Disconnect a battery cable, at the assembly.
2. Remove starter motor.
3. Raise car on a hoist, place stands under the rear of the car at body sills.
4. Disconnect oil filler tube at the oil pan and drain the transmission.
5. Remove one converter drain plug, rotate the converter one half turn and remove the other plug (20 and 80 series.) The 10 series has only one drain plug.
6. Disconnect the oil cooler lines on the 20 and 80 series.
7. Disconnect the vacuum line and terminal wire at the vacuum unit of the transmission.
8. Disconnect speedometer cable at the transmission.
9. Disconnect the rear brake hose bracket from the floor panel.
10. Disconnect the manual cable shift linkage at both levers on the transmission.
11. Remove shift cable bracket from transmission.
12. Disconnect rear shock absorbers at the rear axle.
13. Disconnect the torque tube, propeller shaft, and parking brake cable.
14. Lower rear axle and move rearward to separate torque tube and propeller shaft from transmission.
15. Attach a transmission jack to the underside of the transmission and apply a slight lifting effort.
16. Remove the crossmember to side sill bolts.
17. Lower the engine and bell-housing assembly. Block the engine on both sides from top ledges of the oil pan to the side sill crossmember. These blocks should be 2 x 2 x 5 in.
18. Remove the bell housing-to-engine bolts and bell housing lower mud pan.
19. Remove six cap screws that hold the converter to the drive plate.
20. With the transmission jack holding the transmission in alignment, pull the unit to the rear to disengage the housing and converter from the engine.
21. Lower the complete assembly and remove from the car.

NOTE: the converter cannot be taken apart, and is only serviced as a unit.

Transmission, Bell Housing And Converter Replacement

1. With the bell housing, converter and rear crossmember attached to the transmission, locate and secure the transmission to the transmission jack.

2. Raise the transmission to an approximate installation height.
3. Align the bell housing with the engine. Install the two lower engine-to-bell housing bolts in the engine block. This is to guide the bell housing dowels into place.
4. Move the entire assembly forward, guiding the converter pilot into the crankshaft.
5. Install the plain washers, long washers and nuts on the two lower bolts and install the three upper bell housing-to-engine bolts. Tighten securely.
6. Raise the transmission and install the transmission-to-side sill bolts. Then remove the two 2 x 2 x 5 in. blocks from engine and side sill crossmember.
7. Remove transmission jack.
8. Connect the exhaust pipe (20 and 80 series with single exhaust system), torque tube, propeller shaft, and shock absorbers.
9. Connect the manual cable bracket and shift linkage, rear brake hose bracket and speedometer cable.
10. Connect the vacuum line and terminal wire to the vacuum unit
11. Connect the drive plate to the torque converter and tighten the six attaching screws to 23-28 ft. lbs.
12. Install the bell housing lower mud pan.
13. Connect the filler tube to the oil pan.
14. Connect the oil cover lines, (20 and 80 series.)
15. Lower the car from the hoist, replace the starter and connect the battery.
16. Refill transmission to prescribed level.
17. Road-test car for shift pattern, manual cable linkage adjustment and leaks.

U Joints, Drive Lines

Open-type Drive Line Removal

An open tubular shaft is used with a slip joint at front. The rear joint uses a coupling to the rear axle pinion shaft. Both joints are the cross- and trunnion-type.

1. To remove assembly, raise and place stands under rear of body.
2. Disconnect parking brake cable.
3. Disconnect rear shock absorbers.
4. Disconnect rear brake line at body bracket.
5. Disconnect rear spring U-bolts.
6. Remove coupling nut (see illustration).
7. Shift rear axle assembly to rear, allowing front joint to slide from transmission shaft and separate at rear coupling.

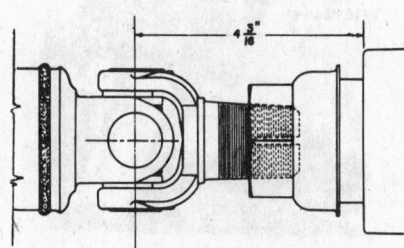

American, series 01, coupling location
(© American Motors Corp.)

Open-Type Drive Line Installation

1. Place the shaft and coupling on the rear axle pinion shaft until the center of the joint yoke bearing is 4 3/16 in. from the front face of the pinion shaft housing (see illustration).
2. Tighten coupling nut to 300 ft. lbs.

3. Slide the front slip joint onto the transmission output shaft.
4. Reposition rear axle assembly.
5. Reinstall U-bolts.
6. Reconnect brake line, parking brake cable and shock absorbers.
7. Bleed brakes.
8. Lower car to floor.

Torque Tube-Type Drive Line Removal

Two types of drive line are used. Both are torque tube drives, one with a solid shaft and one with a tubular shaft (see illustration).

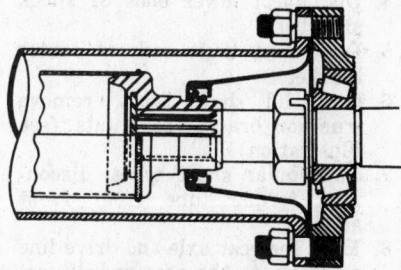

Tubular shaft in torque tube, series 10, 20 and 80
(© American Motors Corp.)

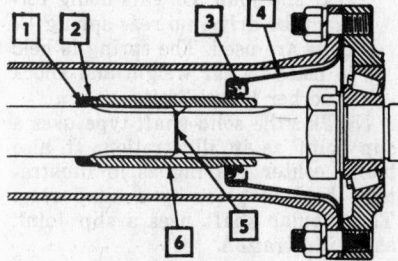

Series 10, 20 and 80 solid shaft in torque tube
(© American Motors Corp.)

1 Propeller shaft coupling coil seal
2 Propeller shaft coupling oil seal retainer
3 Torque tube rear oil seal
4 Torque tube rear oil seal retainer
5 Spacer
6 Propeller shaft coupling

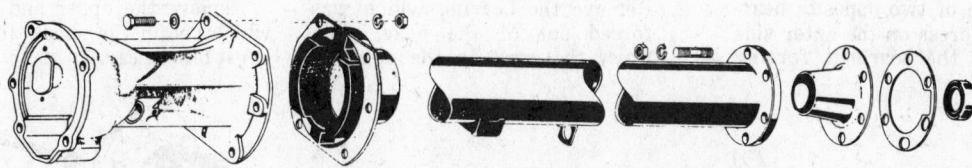

Torque tube, all except American (© American Motors Corp.)

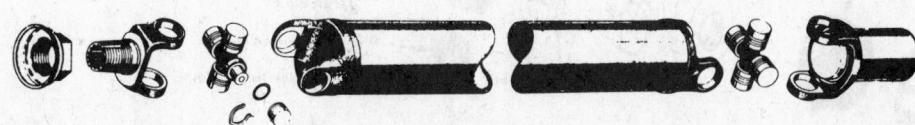

Driveshaft, American (© American Motors Corp.)

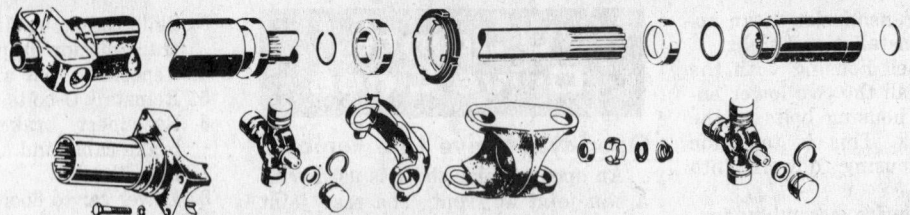

Driveshaft, 1964-66 V8 (© American Motors Corp.)

1. To remove assembly, raise and place stands under rear of body.
2. Disconnect parking brake cable.
3. Disconnect rear stabilizer bar.
4. Disconnect lower ends of shock absorbers.
5. Disconnect truss rods at center bracket.
6. On solid shaft types, remove trunnion bracket rear nuts (see illustration).
7. On tubular shaft types, disconnect torque tube from front adapter.
8. Move the rear axle and drive line assembly to the rear and disconnect the torque tube from the rear axle housing.
9. Move the tube forward to release rear slip joint. On cars using torque tube drive, no rear spring U-bolts are used. The spring is held in place by car weight and shock absorber travel limits.

   NOTE: the solid shaft-type uses a slip joint as in illustration. It also has a center bearing as in illustration, held in place by a snap ring. The tubular shaft uses a slip joint, as in illustration.

Torque Tube-Type Drive Line Installation

On both torque tube-types, reverse the removal procedures above. Use care, when inserting at slip joints, not to damage oil seals.

Universal Joint Repairs

1. Remove the lock rings from the inner side of two opposite bearings and press on the outer side of one of the bearings, forcing

Removing end bearing from the yoke

Installing end bearing

Rear universal joint exploded view

the crossover. This will force the bearing on the opposite side out of its yoke.
2. Remove the bearing which was forced out of the yoke, then press the cross in the opposite

direction to force the other bearing out.
3. Repeat this procedure on the third and fourth bearing.
4. When installing the new bearings in the universal joint yoke, it is possible to put them in with a driver of some type, but it is recommended that this work be done in an arbor press because a heavy jolt on the needle bearings can very easily misalign them, and greatly shorten their life.

Drive Axle, Suspension

Rear Suspension

Rear Spring Removal

Jack up the car and place stand jacks at the frame in front of the rear axle. Disconnect the lower end of the shock absorber and the torsion bar.

Disconnect the spring at its front hanger and rear shackle and remove the U-bolts which hold the spring to the axle housing. Remove the spring. On cars using torque tube drive, no rear spring U-bolts are used. The spring is held in place by car weight and shock absorber travel limits.

Shock Absorber Removal

Remove the upper and lower bolts which retain the shock absorber and lift it off the car.

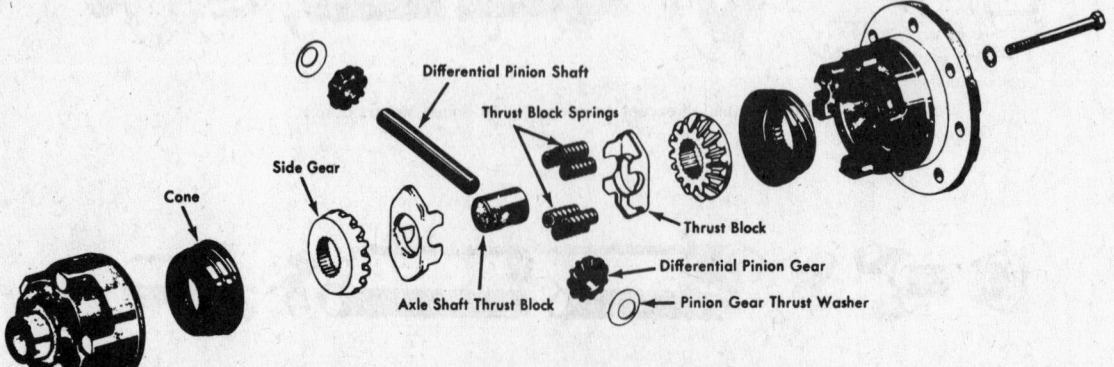

Twin-Grip differential assembly used with 199-232 cu. in. engines
(© American Motors Corp.)

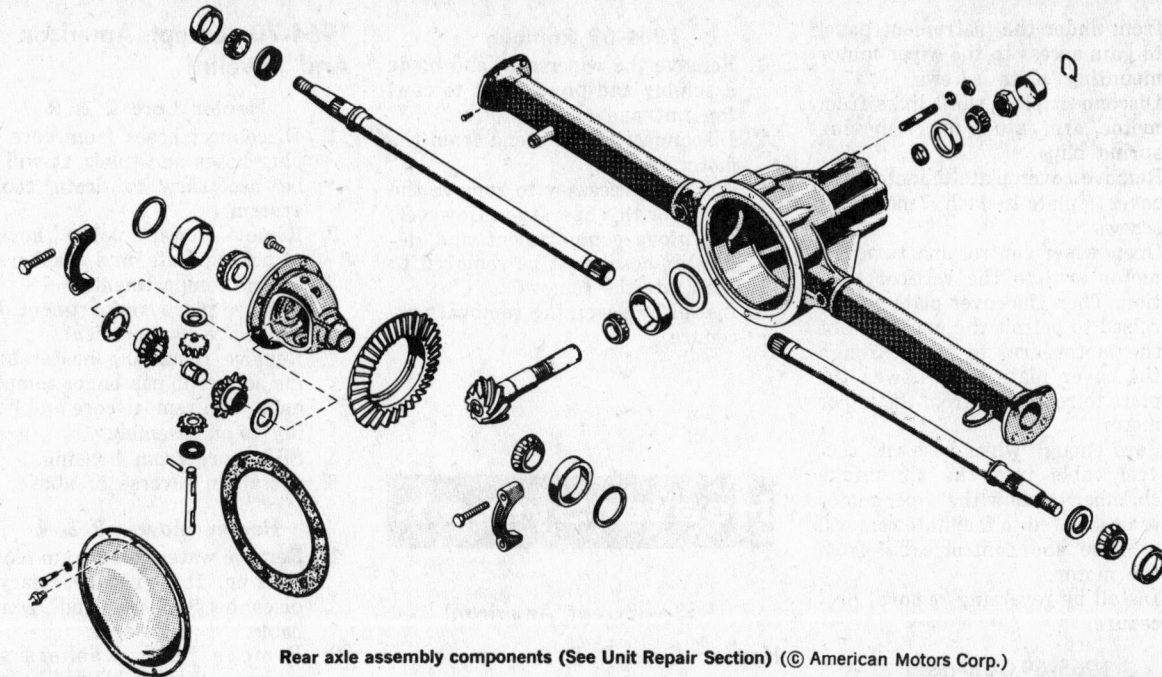

Rear axle assembly components (See Unit Repair Section) (© American Motors Corp.)

Drive Axle

Troubleshooting and Adjustment

General instructions covering the troubles of the rear axle, and how to repair and adjust it, together with information on installation of rear axle bearings and grease seals, are given in the Unit Repair Section.

Capacities of the rear axle are given in the Capacities tables.

Torque Tube Rear Axle Removal

While it is possible to remove the rear axle assembly leaving the torque tube under the car, this practice is not recommended.

1. Disconnect the brake tubes and brake lines. Disconnect the shock absorber and sway bar.
2. Remove the bolts which hold the front of the torque tube to the back of the transmission. Jack up the car and support the weight of the car on the frame in front of the rear springs.
3. Remove the bolts which hold the rear spring to the rear shackle. Remove the four U-bolts which hold the rear axle housing assembly to the rear spring, let the rear axle assembly come down with the springs and roll it out from under the vehicle. If the vehicle is jacked high enough, it is possible to roll the rear axle out on its own wheels. On cars using torque tube drive, no rear spring U-bolts are used. The spring is held in place by car weight and shock absorber travel limits.

Coil Spring with Open Drive-Shaft Rear Axle Removal

1. Raise rear of car high enough to permit working under the car. Place a floor jack under center of axle housing so it just starts to raise rear axle assembly. Place car stands solidly under body members on both sides.
2. Mark rear universal joint and pinion flange for proper indexing at the time of installation, then, disconnect rear universal joint at pinion flange. Wire the propeller shaft back out of the way.
3. Disconnect parking brake cables at the sheave. Remove cable connector and two clips. Slide cable back until free of body.
4. Disconnect rear brake hose at floor pan.
5. Disconnect shock absorbers at axle housing. Lower jack under housing until rear springs can be removed.
6. Disconnect upper control arms at axle.
7. Disconnect lower control arms at axle housing and roll rear axle assembly out from under the car.

Windshield Wipers

Motor R & R

1964 American

The motor is mounted to the motor access hole cover plate with two machine screws. The plate is attached to the dash panel under the instrument panel.

1. Remove radio and/or bezel, if so equipped.
2. Remove air conditioning unit

Proper position of wiper control cable (© American Motors Corp.)

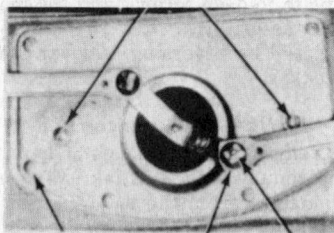

Wiper motor mounting (© American Motors Corp.)

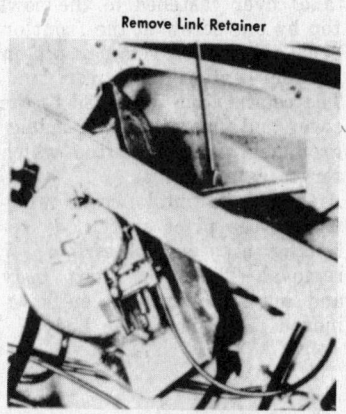

Removing link to wiper motor retainer (© American Motors Corp.)

from under the instrument panel to gain access to the wiper motor mounting plate screws.

3. Disconnect pivot shaft links from motor arm studs by removing spring clips.
4. Remove cover plate-to-motor and cover plate-to-dash mounting screws.
5. Open wiper control and turn the motor arm to the vertical position. Then the cover plate can be raised to permit the upper end of the motor arm to pass through the cover plate hole. Lower the plate to remove it from the wiper motor.
6. Feed enough wiper hose and control cable into the air intake chamber to allow the wiper motor to be lowered to facilitate removal of hose and control cable from the motor.
7. Install by reversing removal procedure.

1965-69 American

The wiper motor is mounted on the engine side of the firewall and is easily accessible from under the hood.

1963-70 Classic Rebel and Ambassador

1. Remove wiper arms and blades and the air intake cover.
2. Slide the link-to-motor retainer clip off of the motor arm stud. Remove the link from the motor.
3. Disconnect control cable and vacuum hose from the motor.
4. Remove the motor and mounting plate-to-dash screws, and the motor assembly.
5. Install by reversing removal procedure.

1964-69 American

To remove the pivot shaft body and link assembly after the link has been disconnected from the motor:
1. Remove pivot shaft spacer mounting nut, spacer and gaskets.
2. Disconnect washer hose.
3. Remove cowl ventilator air intake cover fastened to the cowl top by one screw at the front on each side and a retainer pin in the rear center of the cover.
4. The retainer pin is welded to the cover and inserted into a rubber grommet in the cowl top which serves as a retainer.
5. After the air intake cover is removed, the pivot shaft body retaining nuts are accessible for removal. The pivot shaft body and link assembly can then be removed from inside the body.

1964-69 Rambler

1. Remove the wiper arm and blade assembly and pivot shaft to cowl top nut and spacer.
2. Disconnect the link end from the motor arm.
3. It is not necessary to remove the radio for this operation. However, the glove compartment and defroster hose must be removed to facilitate the removal.
4. Install by reversing removal procedure.

Heater System

1963—(Except American)

Heater Core R & R

1. Disconnect hoses in engine compartment. Plug hoses and tubes with corks.
2. Remove two lower blower housing attaching nuts and washers in engine compartment.
3. Remove heater housing screws in passenger compartment.
4. Remove heater housing and core as an assembly.
5. Slide heater core from housing.
6. Install in reverse of above.

Blower R & R

1. Remove water valve from blower housing.
2. Remove nuts, washers and screws attaching blower housing-to-dash in engine compartment.
3. Remove housing, then fan and motor.
4. Install in reverse of above.

American—1963

Heater Core R & R

1. Disconnect blower wires.
2. Disconnect hoses. Plug hoses and core tubes. It will not be necessary to drain cooling system.
3. Remove core housing and blower housing.
4. Remove screws and slide core from housing.
5. Install in reverse of above.

Heater Blower R & R

1. Remove core housing from dash panel.
2. Motor and fan can now be removed.

1964-70 (Except American And Javelin)

Heater Core R & R

1. Disconnect hoses from core and plug hoses and tubes. It will not be necessary to drain cooling system.
2. Remove lower blower housing attaching nuts and washers in engine compartment.
3. Remove glove compartment door and glove compartment.
4. Remove remaining heater housing screws in passenger compartment, and remove core and housing as an assembly.
5. Slide core from housing.
6. Install in reverse of above.

Heater Blower R & R

1. Remove water valve from blower housing. It is not necessary to disconnect hoses and control cable.
2. Remove nuts, washers and screws attaching blower housing to dash panel in engine compartment.
3. Remove motor and fan, then separate fan from motor.
4. Install in reverse of above.

American & Javelin—1964-70

Heater Core, Defroster and Blower Housing R & R

1. Drain 1½ qts. of coolant from system.
2. Disconnect hoses from heater core tubes in engine compartment. Install corks in hoses and tubes.
3. Disconnect blower motor wires.
4. Remove housing attaching nuts at blower motor opening in dash.
5. Remove glove compartment door and glove compartment.
6. Disconnect air and defroster cables from damper levers.
7. Remove assembly.
8. Install in reverse of above.

Heater Core R & R

1. Perform the first seven steps above to remove the assembly.
2. Remove core from housing.

Heater Blower R & R

1. Remove assembly, as described in seven steps listed above.
2. Remove blower scroll cover to which blower is assembled, and remove blower from motor shaft for access to motor attaching nuts.
3. Install in reverse of above.

Valiant, Dart, Barracuda, Challenger

Index

YEAR IDENTIFICATION

VALIANT

1964

1965

1966

1967

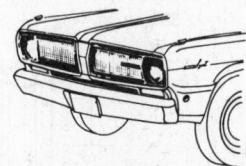
1968

1969

1970 Duster

1970 Plymouth Cuda

1971 Cuda

DART

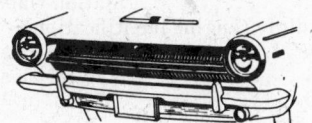
1964

1965

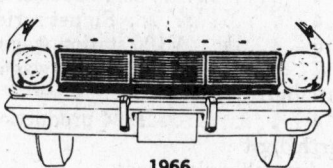
1966

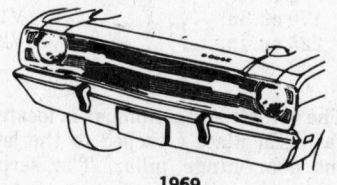
1967

1968

1969

1970 Dart

1971 Challenger

1971 Demon

FIRING ORDER

FIRING ORDER
1-8-4-3-6-5-7-2

273, 318, 340 cu. in.

FIRING ORDER 1-5-3-6-2-4

170, 225 cu. in.

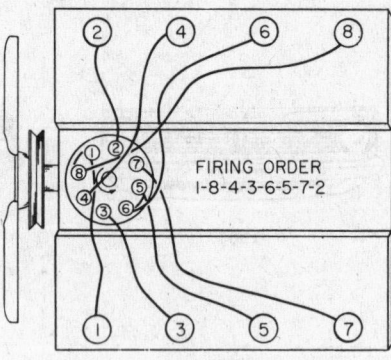

FIRING ORDER
1-8-4-3-6-5-7-2

383 cu. in.

CAR SERIAL NUMBER LOCATION AND ENGINE IDENTIFICATION

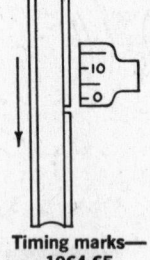

Timing marks—
1964-65
6 cyl.

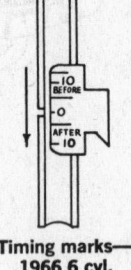

Timing marks—
1966 6 cyl.

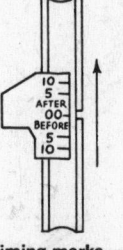

Timing marks—
1967-71
6 cyl.

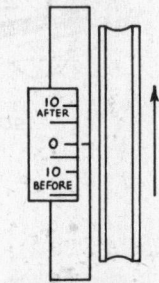

Timing marks—
1966-71 V8

1964

The vehicle number is located on a metal plate attached to the left front door hinge pillar. The serial number contains ten digits, interpreted as follows:

First digit
  1 . . . . Valiant
  7 . . . . . . Dart 170, 270 (six cyl.)
  8 . . . . Dart GT (V8)
Second digit
  1 . . . . V100 series
  3 . . . . V200 series
  4 . . . . Signet series
  5 . . . . V100 station wagon
  7 . . . . V200 station wagon
Third digit
  4 . . . . 1964 production
Fourth digit
  Assembly plant code
Fifth to tenth digits
  Series production number

Engine Production Code
  170 cu. in. V17
  225 cu. in. V22

1965

The vehicle serial number is located on a metal plate attached to the left front door hinge pillar. The serial number contains ten digits, interpreted as follows:

First digit
  V . . . . Valiant
  2 . . . . Dart (six cyl.)
  L . . . . Dart (V8)
Second digit
  1 . . . . V100 series
  3 . . . . V200 series
  4 . . . . Signet series
  5 . . . . V100 station wagon
  7 . . . . V200 station wagon
Third digit
  5 . . . . 1965 production
Fourth digit
  Assembly plant code
Fifth to tenth digits
  Series production number

1966

The vehicle serial number is located on a metal plate attached to the left front door hinge pillar. The serial number now consists of thirteen digits, interpreted as follows:
First digit
  B . . . . Barracuda
  V . . . . Valiant
  L . . . . Dart
Second digit (price range)
  E . . . . Economy

  L . . . . Low
  M . . . . Medium
Third and fourth digits
  21 . . . . 2-dr. sedan
  23 . . . . 2-dr. hardtop
  27 . . . . Convt.
  41 . . . . 4-dr. sedan
  43 . . . . 4-dr. sedan
  45 . . . . Station wagon
Fifth digit (engine identification)
  A . . . . 170 cu. in. six
  B . . . . 225 cu. in. six
  D . . . . 273 cu. in. V8
Sixth digit
  6 . . . . 1966 production
Seventh digit
  Assembly plant code
Eighth to thirteenth digits
  Series production number

1967

The vehicle serial number is located on a metal plate attached to the left front door hinge pillar. The serial number contains thirteen digits, interpreted as follows:
First digit
  L . . . . Dart
  V . . . . Valiant

Engine code location—273, 318, 340

Engine code location—383, 440

Second digit (price range)
E ....Economy
L ....Low
M ....Medium
H ....High
Third and fourth digits
21 ....2-dr. sedan
23 ....2-dr. hardtop
27 ....Convt.
41 ....4-dr. sedan
43 ....4-dr. sedan
45 ......6-pass. station wagon
46 ......9-pass. station wagon
Fifth digit (engine identification)
A ....170 cu. in. six
B ....225 cu. in. six
D ....273 cu. in. V8
Sixth digit
7 ....1967 production
Seventh digit
Assembly plant code
Eighth to thirteenth digits
Series production number

1968

The vehicle number is located on a plate on the left side of the instrument panel, visible through the windshield. The serial number is interpreted as follows:
First digit
V ....Valiant
B ....Barracuda
L ....Dart
Second digit (price range)
E ....Economy
L ....Low
M ....Medium
H ....High
S ....Special
K ....Police
T ....Taxi
O ....Superstock
Third and fourth digits
Same as 1967, except the following: 29 ....2-dr. sports hdtp.
Fifth digit (engine identification)
A ....170 cu. in. six
B ....225 cu. in. six
C ....Spec. Ord. 6

D ....273 cu. in. V8
E .....273 cu. in. High Perf. V8
F ....318 cu. in. V8
G ....383 cu. in. V8
H ...383 cu. in. High Perf. V8
N ....340 cu. in. V8
P ....340 cu. in. High Perf. V8
Sixth digit
8 ....1968 production
Seventh digit
Assembly plant code
Eighth to thirteenth digits
Series production number

1969

The vehicle number is located on a plate on the left side of the instrument panel, visible through the windshield. The serial number is interpreted as follows:
First digit
V ....Valiant
B ....Barracuda
L ....Dart
Second digit (price range)
Same as 1968, with the following exception:
X ....Fast Top
Third and fourth digits
Same as 1967-68
Fifth digit (engine identification)
Same as 1968, omitting "E" and "P"
Sixth digit
9 ....1969 production
Seventh digit
Assembly plant code
Eighth to thirteenth digits
Series production number

1970

The vehicle number is located on a plate on the left side of the instrument panel, visible through the windshield. The serial number is interpreted as follows:
First digit
V ....Valiant
B ....Barracuda

L ....Dart
J ....Challenger
Second digit (price range)
E ....Economy
L ....Low
M ....Medium
H ....High
P ....Premium
K ....Police
T ....Taxi
S ....Special
O ....Super Stock
Third and fourth digits
23 ....Two door hardtop
27 ....Convertible
29 ....Two door special hardtop
41 ....Four door sedan
Fifth digit (engine identification)
B ....198 cu. in. six
C ....225 cu. in. six
E ....Special six
G ....318 cu. in. V8
H ....340 cu. in. V8
L ....383 cu. in. V8
N ....383 high perf. V8
R ......426 cu. in. Hemi V8
T ....440 cu. in. V8
U ....440 high perf. V8
V ....440 six pack V8
Z ....Special V8
Sixth digit (model year)
Seventh digit (assembly plant)
Eighth to thirteenth digits
(series production number)

Engine Number Location

170, 198 and 225 cu. in. six engines ... stamped on joint face of block, next to No. 1 cylinder.
273, 318 and 340 cu. in. V8 engines ... stamped on front of block, just below left cylinder head.
383, 426 and 440 cu. in. V8 engine ... stamped on cylinder block pan rail, at left rear corner below starter opening.

BATTERY AND STARTER SPECIFICATIONS

YEAR	MODEL	BATTERY			STARTERS						
		Ampere Hour Capacity	Volts	Terminal Grounded	Lock Test			No-Load Test			Brush Spring Tension (Oz.)
					Amps.	Volts	Torque	Amps.	Volts	RPM	
1964-67	170	38	12	Neg.	380	4	20.0	90	11.0	2,950	32-48
	225	48	12	Neg.	380	4	24.0	85	11.0	1,950	32-48
	V8—273	48	12	Neg.	450	4	24.0	90	11.0	2,400	32-48
1968-69	6—170	38	12	Neg.	380	4	20.0	90	11.0	2,950	32-36
	6—225, 273, 318, 340	48	12	Neg.	425	4	24.0	90	11.0	2,300	32-36
	V8—383	59	12	Neg.	425	4	24.0	90	11.0	2,300	32-36
1970-71	6 & V8—198, 225, 318, 340	46	12	Neg.	425	4	—	90	11.0	2,300	32-36
	V8—360	59	12	Neg.	425	4	24.0	90	11.0	2,300	32-36
	V8—383	59	12	Neg.	425	4	24.0	90	11.0	2,300	32-36
	V8—426, 440	70	12	Neg.	425	4	24.0	90	11.0	2,300	32-36

TUNE-UP SPECIFICATIONS

YEAR	MODEL	SPARK PLUGS		DISTRIBUTOR		IGNITION TIMING (Deg.) ▲	CRANKING COMP. PRESSURE (Psi)	VALVES		Intake Opens (Deg.)	FUEL PUMP PRESSURE (Psi)	IDLE SPEED (Rpm) *
		Type	Gap (In.)	Point Dwell (Deg.)	Point Gap (In.)			Tappet (Hot) Clearance (In.)				
								Intake	Exhaust			
1964-65	All Exc. V8—273 Cu. In.	N14Y	.035	40–45	.020	2½B	145	.010	.020	8B	3½–5	550
	V8—273 Cu. In.; A.T.	N14Y	.035	28–33	.017	10B	145	.013	.021	14B	5–7	550
	V8—273 Cu. In.; M.T.	N14Y	.035	28–33	.017	5B	145	.013	.021	14B	5–7	550
1966	170 Cu. In.	N14Y	.035	40–45	.020	5B	125	.010	.020	8B	3½	550
	170 Cu. In.	N14Y	.035	40–45	.020	5A⊙	125	.010	.020	8B	3½	650
	225 Cu. In.	N14Y	.035	40–45	.020	5B	125	.010	.020	10B	3½	550
	225 Cu. In.	N14Y	.035	40–45	.020	5A⊙	125	.010	.020	10B	3½	650
	273 Cu. In. M.T.	N14Y	.035	28–32	.017	5B	135	.013	.021	14B	5–7	600
	273 Cu. In. A.T.	N14Y	.035	28–32	.017	10B	135	.013	.021	14B	5–7	600
	273 Cu. In. M.T.	N14Y	.035	28–32	.017	5A⊙	135	.013	.021	14B	5–7	650
	273 Cu. In. A.T.	N14Y	.035	28–32	.017	5A⊙	135	.013	.021	14B	5–7	650
	273 Cu. In. 4-BBL.	N9Y	.035	27–31*	.017	10B	148	.013	.021	14B	5–7	650
	273 Cu. In. 4-BBL.	N9Y	.035	27–31*	.017	5A⊙	148	.013	.021	14B	5–7	700
1967	170 Cu. In.	N14Y	.035	40–45	.020	5B	125	.010	.020	10B	3½–5	550
	170 Cu. In.	N14Y	.035	40–45	.020	5A⊙	125	.010	.020	10B	3½–5	700
	225 Cu. In.	N14Y	.035	40–45	.020	5B	125	.010	.020	10B	3½–5	550
	225 Cu. In.	N14Y	.035	40–45	.020	TDC⊙	125	.010	.020	10B	3½–5	650
	273 Cu. In. M.T.	N14Y	.035	28–32	.017	5B	135	.013	.021	14B	5–7	650
	273 Cu. In. A.T.	N14Y	.035	28–32	.017	10B	135	.013	.021	14B	5–7	650
	273 Cu. In. 2-BBL.	N14Y	.035	28–32	.017	5A⊙	135	.013	.021	14B	5–7	650
	273 Cu. In. 4-BBL.	N10Y	.035	27–31*	.017	10B	148	.013	.021	14B	5–7	650
	273 Cu. In. 4-BBL.	N10Y	.035	27–31*	.017	5A⊙	148	.013	.021	14B	5–7	700
1968	170 Cu. In., M.T.	N14Y	.035	40–45	.020	5A	125	.010	.020	10B	3½–5	700
	170 Cu. In., A.T.	N14Y	.035	40–45	.020	2½A	125	.010	.020	10B	3½–5	700
	225 Cu. In.	N14Y	.035	40–45	.020	TDC	125	.010	.020	10B	3½–5	650
	273 Cu. In., M.T.	N14Y	.035	28–33	.017	5A	135	Zero	Zero	10B	5–7	700
	273 Cu. In., A.T.	N14Y	.035	28–33	.017	2½A	135	Zero	Zero	10B	5–7	650
	318 Cu. In., M.T.	N14Y	.035	28–33	.017	5A	140	Zero	Zero	10B	5–7	650
	318 Cu. In., A.T.	N14Y	.035	28–33	.017	2½A	140	Zero	Zero	10B	5–7	600
	340 Cu. In., M.T.	N9Y	.035	27–32●	.017	TDC	140	Zero	Zero	26B	5–7	700
	340 Cu. In., A.T.	N9Y	.035	27–32●	.017	5B	140	Zero	Zero	22B	5–7	650
	383 Cu. In., M.T.	J11Y	.035	28–33	.017	TDC	150	Zero	Zero	18B	3½–5	650
	383 Cu. In., A.T.	J11Y	.035	28–33	.017	5B	150	Zero	Zero	18B	3½–5	600
1969	170 Cu. In., 1-BBL. M.T.	N14Y	.035	42–47	.020	5A	125	.010	.020	10B	3½–5	750
	170 Cu. In., 1-BBL. A.T.	N14Y	.035	42–47	.020	2½A	125	.010	.020	10B	3½–5	750
	225 Cu. In., 1-BBL. M.T.	N14Y	.035	42–47	.020	TDC	125	.010	.020	10B	3½–5	700
	225 Cu. In., 1-BBL. A.T.	N14Y	.035	42–47	.020	TDC	125	.010	.020	10B	3½–5	650
	273 Cu. In., 2-BBL. M.T.	N14Y	.035	30–35	.017	2½A	135	Zero	Zero	10B	5–7	700
	273 Cu. In., 2-BBL. A.T.	N14Y	.035	30–35	.017	2½A	135	Zero	Zero	10B	5–7	650
	318 Cu. In., 2-BBL. M.T.	N14Y	.035	30–35	.017	TDC	140	Zero	Zero	10B	5–7	700
	318 Cu. In., 2-BBL. A.T.	N14Y	.035	30–35	.017	TDC	140	Zero	Zero	10B	5–7	650
	340 Cu. In., 4-BBL. M.T.	N9Y	.035	27–32●	.017	TDC	150	Zero	Zero	22B	5–7	750
	340 Cu. In., 4-BBL. A.T.	N9Y	.035	30–35	.017	5B	150	Zero	Zero	22B	5–7	700
	383 Cu. In., 4-BBL. M.T. H.P.	J11Y	.035	30–35	.017	TDC	140	Zero	Zero	18B	3½–5	700
	383 Cu. In., 4-BBL. A.T. H.P.	J11Y	.035	30–35	.017	5B	140	Zero	Zero	18B	3½–5	650
1970	198 Cu. In., 1-BBL.	N14Y	.035	42–47	.020	⑦	125	.010	.020	10B	3½–5	750
	225 Cu. In., 1-BBL. M.T.	N14Y	.035	42–47	.020	TDC	125	.010	.020	10B	3½–5	700
	225 Cu. In., 1-BBL. A.T.	N14Y	.035	42–47	.020	TDC	125	.010	.020	10B	3½–5	650
	318 Cu. In., 2-BBL. M.T.	N14Y	.035	30–35	.017	TDC	140	Zero	Zero	10B	5–7	750
	318 Cu. In., 2-BBL. A.T.	N14Y	.035	30–35	.017	TDC	140	Zero	Zero	10B	5–7	700
	340 Cu. In., 4-BBL. M.T.	N9Y	.035	27–32●	.017	5B	150	Zero	Zero	22B	5–7	900
	340 Cu. In., 4-BBL. A.T.	N9Y	.035	30–35	.017	5B	150	Zero	Zero	22B	5–7	750
	383 Cu. In., 4-BBL. M.T. H.P.	J11Y	.035	28–32	.017	TDC	140	Zero	Zero	21B	3½–5	750
	383 Cu. In., 4-BBL. A.T. H.P.	J11Y	.035	28–32	.017	2½B	140	Zero	Zero	18B	3½–5	650

TUNE-UP SPECIFICATIONS, continued

YEAR	MODEL	SPARK PLUGS Type	Gap (In.)	DISTRIBUTOR Point Dwell (Deg.)	Point Gap (In.)	IGNITION TIMING (Deg.) ▲	CRANKING COMP. PRESSURE (Psi)	VALVES Tappet (Hot) Clearance (In.) Intake	Exhaust	Intake Opens (Deg.)	FUEL PUMP PRESSURE (Psi)	IDLE SPEED (Rpm) ★
	383 Cu. In., 2-BBL., M.T.	J14Y	.035	28-32	.014-.019	TDC	140	Zero	Zero	18B	3½-5	750
	383 Cu. In., 2-BBL. A.T.	J14Y	.035	28-32	.014-.019	2½B	140	Zero	Zero	18B	3½-5	650
	426 Cu. In., Hemi	N10Y	.035	●	.014-.019	TDC	150	Zero	Zero	36B	7-8½	900
	440 Cu. In., H.P., M.T.	J11Y	.035	28-32	.014-.019	TDC	150	Zero	Zero	21B	3½-5	900
	440 Cu. In., H.P., A.T.	J11Y	.035	28-32	.014-.019	2½B	150	Zero	Zero	21B	3½-5	800
	440 Cu. In., 3-2-BBL.	J11Y	.035	●	.014-.019	5B	150	Zero	Zero	21B	3½-5	900①
1971	198 Cu. In., 1-BBL.	N114Y	.035	42-47	.020	**	100-125	.010	.020	10B	3½-5	750
	225 Cu. In., 1-BBL. M.T.	N114Y	.035	42-47	.020	**	100-125	.010	.020	10B	3½-5	700
	225 Cu. In., 1-BBL. A.T.	N114Y	.035	42-47	.020	**	100-125	.010	.020	10B	3½-5	650
	318 Cu. In., 2-BBL. M.T.	N114Y	.035	30-35	.017	**	100-140	Zero	Zero	10B	5-7	750
	318 Cu. In., 2-BBL. A.T.	N114Y	.035	30-35	.017	**	100-140	Zero	Zero	10B	5-7	700
	340 Cu. In., 4-BBL. M.T.	N9Y	.035	●	.014-.019	**	110-150	Zero	Zero	22B	5-7	900
	340 Cu. In., 4-BBL. A.T.	N9Y	.035	30-34	.014-.019	**	110-150	Zero	Zero	22B	5-7	900
	340 Cu. In., 3-2-BBL.	N.A.	.035	●	.014-.019	**	110-150	Zero	Zero	N.A.	5-7	**
	360 Cu. In.	N.A.	.035	30-34	.014-.019	**	100-140	Zero	Zero	16B	5-7	**
	383 Cu. In., 4-BBL. M.T. H.P.	J11Y	.035	28-32	.017	**	110-150	Zero	Zero	21B	3½-5	700
	383 Cu. In., 4-BBL. A.T. H.P.	J11Y	.035	28-32	.017	**	110-150	Zero	Zero	21B	3½-5	650
	383 Cu. In., 2-BBL. M.T.	J14Y	.035	28-32	.014-.019	**	100-140	Zero	Zero	18B	3½-5	750
	383 Cu. In., 2-BBL. A.T.	J14Y	.035	28-32	.014-.019	**	100-140	Zero	Zero	18B	3½-5	650
	426 Cu. In., Hemi	N10Y	.035	●	.014-.019	**	110-150	Zero	Zero	36B	7-8½	900
	440 Cu. In. STD.	N.A.	.035	28-32	.016-.021	**	110-150	Zero	Zero	18B	3½-5	**
	440 Cu. In., M.T. H.P.	J11Y	.035	28-32	.016-.021	**	110-150	Zero	Zero	21B	3½-5	900
	440 Cu. In., M.T. H.P.	J11Y	.035	28-32	.016-.021	**	110-150	Zero	Zero	21B	3½-5	800
	440 Cu. In., 3-2-BBL.	J11Y	.035	●	.014-.019	**	110-150	Zero	Zero	21B	3½-5	900

▲—With vacuum advance disconnected and plugged. NOTE: These settings are only approximate. Engine design, altitude, temperature, fuel octane rating and the condition of the individual engine are all factors which can influence timing. The limiting advance factor must, therefore, be the "knock point" of the individual engine.

★—With manual transmission in 'N' and automatic in 'D', air conditioning on— if so equipped.

*—Both sets 36°-40°.
⊙—Engines equipped with C.A.P. system.
●—Dual Points, total—37°-42°.
A—After top dead center.
B—Before top dead center.
A.T.—Automatic transmission.

M.T.—Manual transmission.
T.D.C.—Top dead center.
H.P.—High performance.
①—With electric solenoid throttle stop connected.
②—M.T.—2½B; A.T.—TDC.
**—See engine decal.

Caution

General adoption of anti-pollution laws has changed the design of almost all car engine production to effectively reduce crankcase emission and terminal exhaust products. It has been necessary to adopt stricter tune-up rules, especially timing and idle speed procedures. Both of these values are peculiar to the engine and to its application, rather than to the engine alone. With this in mind, car manufacturers supply idle speed data for the engine and application involved. This information is clearly displayed in the engine compartment of each vehicle.

CRANKSHAFT BEARING JOURNAL SPECIFICATIONS

YEAR	MODEL	MAIN BEARING JOURNALS (IN.) Journal Diameter	No. of Main Bearings	Oil Clearance	Shaft End-Play	Thrust On No.	CONNECTING ROD BEARING JOURNALS (IN.) Journal Diameter	Oil Clearance	End-Play
1964-71	6 Cyl.	2.750	4	.0005-.0015	.002-.007	3	2.187	.0005-.0025	.006-.012
	V8—273, 318, 340, 360	2.500	5	.0005-.0015	.002-.007	3	2.125	.0005-.0025	.006-.012
	V8—383	2.625	5	.0005-.0015	.002-.007	3	2.380	.0007-.0032	.009-.017
	V8—440	2.750	5	.0005-.0015	.002-.007	3	2.380	.0007-.0032	.009-.017
	V8—426 Hemi	2.750	5	.0015-.0025	.002-.007	3	2.380	.0010-.0035	.009-.017

GENERAL CHASSIS AND BRAKE SPECIFICATIONS

YEAR	MODEL	CHASSIS Overall Length (In.)	CHASSIS Tire Size	BRAKE CYLINDER BORE Master Cylinder (In.)	BRAKE CYLINDER BORE Wheel Cylinder Diameter (In.) Front	BRAKE CYLINDER BORE Wheel Cylinder Diameter (In.) Rear	BRAKE DRUM Diameter (In.) Front	BRAKE DRUM Diameter (In.) Rear
1964	Valiant	184.0	6.50 x 13	1	1	13/16	9	9
	Dart Sedan	196.3	6.50 x 14	1	1	13/16	9	9
	Dart St. Wag.	190.2	7.00 x 14	1	1	13/16	9	9
1965	Barracuda 6 Cyl.	188.2	6.50 x 13	1	1	13/16	9	9
	Barracuda V8	188.2	7.00 x 13	1	1	13/16	9	9
	Valiant Pass.	188.2	6.50 x 13①	1	1②	13/16	9③	9③
	Valiant Sta. Wag.	188.8	6.50 x 13①	1	1②	13/16	9③	9③
	Dart Pass.	196.4	6.50 x 13①	1	1②	13/16	9③	9③
	Dart Sta. Wag.	190.2	6.50 x 13①	1	1②	13/16	9③	9③
1966	6 Cyl.—All exc. Sta. Wag.	188.3	6.50 x 13	1	1	13/16	9③	9③
	6 Cyl.—Station Wagon	189.0	7.00 x 13	1	1	13/16	9③	9③
	V8—All exc. Sta. Wag.	188.3	7.00 x 13	1	1 1/8	15/16	10	10
	V8—Station Wagon	189.0	7.00 x 13	1	1 1/8	15/16	10	10
1967	6 Cyl. exc. Conv.	195.4④	6.50 x 13	1	1	15/16	9	9
	Conv.	195.4④	7.00 x 13	1	1	15/16	9	9
	V8, All	195.4④	7.00 x 13	1	1 1/8	15/16	10	10
	6 & V8—Opt. Disc Brakes	195.4④	6.95 x 14	1	1 5/8	15/16	10 25/32	10
	Barracuda 6 Cyl.	192.8	6.95 x 14	1	1	29/32	9	9
	Barracuda V8	192.8	6.95 x 14	1	1 1/8⑤	29/32	10⑥	10
1968	Valiant 6 Cyl.	188.4	6.50 x 13⑧	1	1	13/16	9	9
	Valiant V8	188.4	7.00 x 13	1	1 1/8	15/16	10	10
	Valiant (Disc.)	188.4	6.95 x 14	1⑩	⑪	15/16	⑨	10
	Barracuda 6 Cyl.	192.8	6.95 x 14	1	1	13/16	9	9
	Barracuda V8	192.8	6.95 x 14	1	1 1/8	15/16	10	10
	Barracuda (Disc.)	192.8	D70 x 14	1⑩	⑪	15/16	⑨	10
	Dart 6 Cyl.	195.4	6.50 x 13⑧	1	1	13/16	9	9
	Dart V8	195.4	⑦	1	1 1/8	15/16	10	10
	Dart (Disc.)	195.4	6.95 x 14	1	⑪	15/16	⑨	10
1969	Valiant 6 Cyl.	188.4	6.50 x 13⑧	1	1	13/16	9	9
	Valiant V8	188.4	7.00 x 13	1	1 1/8	15/16	10	10
	Valiant Disc	188.4	6.95 x 14	⑫	⑬	15/16	⑭	10
	Barracuda 6 Cyl.	192.8	6.95 x 14	1	1	13/16	9	9
	Barracuda V8	192.8	6.95 x 14	1	1 1/8	15/16	10	10
	Barracuda Disc	192.8	6.95 x 14	⑫	⑬	15/16	⑭	10
	Dart 6 Cyl.	195.4	6.50 x 13⑧	1	1	13/16	9	9
	Dart V8	195.4	⑦	1	1 1/8	15/16	10	10
	Dart (Disc.)	195.4	6.95 x 14	⑫	⑬	15/16	⑭	10
1970–71	Valiant and Duster 6 Cyl.	188.4	⑮	1	1	13/16	9	9
	Valiant and Duster V8	188.4	⑯	1	1 3/16	15/16	10	10
	Barracuda 6 Cyl.	186.7	E78 x 14	1	1	13/16	9	9
	Barracuda V8	186.7	⑰	1	1 3/16	15/16	10	10
	Dart 6 Cyl.	196.2	D78 x 14	1	1	13/16	10	9
	Dart V8	196.2	⑱	1	1 3/16	15/16	10	10
	Challenger	191.3	⑲	1	1 3/16	15/16	10	10
	All Models (Disc.)	—	—	⑫	⑬		⑭	—

① —All V8 models—7.00 x 13.
② —All V8 models—1 1/8
③ —Optional and V8 models—10
④ —Valiant—overall length 188.4
⑤ —With disc brakes—front—1 41/64
rear—15/16
⑥ —With disc brakes—11 7/8
⑦ —273, 318, 7.00 x 383 E70 x 14
⑧ —225 Cu. I.—7.00 x 13
⑨ —Disc od. Kelsey Hayes—11.04
Bendix—11.19
Budd—11.88
⑩ —Budd 1 1/8
⑪ —Kelsey Hayes 1.638
Bendix—2
Budd—2.375
⑫ —Kelsey Hayes—1
Bendix—1 7/8
Kelsey Hayes (Floating)—1-1 1/8
⑬ —Kelsey Hayes—1.636
Bendix—2
Kelsey Hayes (Floating)—2 3/4
⑭ —Kelsey Hayes—11.04
Bendix—11.19
Kelsey Hayes (Floating)—11.75
⑮ —Coupe—B78 x 14; Sedan—C78 x 14; w/A.C.—C78 x 14.
⑯ —318 V8—C78 x 14; 318 V8 w/A.C.—D78 x 14; 340 V8—E70 x 14.
⑰ —318 V8—E78 x 14; 340 V8—E60 x 15; 383, 440 V8—F70 x 14; 426 Hemi—F60 x 15.
⑱ —318 V8—D78 x 14; 340 V8—E70 x 14.
⑲ —225 6 Cyl., 318 V8—E78 x 14; 340 V8—E60 x 15; 383, 440 V8—F70 x 14; 426 Hemi—F60 x 15.

GENERAL ENGINE SPECIFICATIONS

YEAR	CU. IN. DISPLACEMENT	CARBURETOR	DEVELOPED HORSEPOWER @ RPM	DEVELOPED TORQUE @ RPM (FT. LBS.)	A.M.A. HORSEPOWER	BORE & STROKE (IN.)	ADVERTIZED COMPRESSION RATIO	VALVE LIFTER TYPE	NORMAL OIL PRESSURE (PSI)
1964–65	170	1-BBL.	101 @ 4400	155 @ 2400	27.8	3.406 x 3.125	8.2-1*	Mech.	45–65
	225	1-BBL.	145 @ 4000	215 @ 2800	27.8	3.406 x 4.125	8.2-1‡	Mech.	45–65
	273	2-BBL.	180 @ 4200	260 @ 1600	42.0	3.625 x 3.310	8.8-1	Mech.	45–65
	273	4-BBL.	235 @ 5200	280 @ 4000	42.0	3.625 x 3.310	10.5-1	Mech.	45–65
1966	170	1-BBL.	101 @ 4400	155 @ 2400	27.8	3.406 x 3.125	8.5-1	Mech.	45–65
	225	1-BBL.	145 @ 4000	215 @ 2400	27.8	3.406 x 4.125	8.4-1	Mech.	45–65
	273	2-BBL.	180 @ 4200	260 @ 1600	42.0	3.625 x 3.310	8.8-1	Mech.	45–65
	273	4-BBL.	235 @ 5200	280 @ 4000	42.0	3.625 x 3.310	10.5-1	Mech.	45–65
1967	170	1-BBL.	115 @ 4400	155 @ 2400	27.7	3.406 x 3.125	8.5-1	Mech.	45–65
	225	1-BBL.	145 @ 4000	215 @ 2400	27.7	3.406 x 4.125	8.4-1	Mech.	45–65
	273	2-BBL.	180 @ 4200	260 @ 1600	42.2	3.625 x 3.310	8.8-1	Mech.	45–65
	273	4-BBL.	235 @ 5200	280 @ 4000	42.2	3.625 x 3.310	10.5-1	Mech.	45–65
1968	170	1-BBL.	115 @ 4400	155 @ 2400	27.7	3.406 x 3.125	8.5-1	Mech.	45–60
	225	1-BBL.	145 @ 4000	215 @ 2400	27.7	3.406 x 4.125	8.4-1	Mech.	45–60
	273	2-BBL.	190 @ 4400	260 @ 2000	42.2	3.625 x 3.310	9.0-1	Hyd.	45–65
	318	2-BBL.	230 @ 4400	340 @ 2400	48.9	3.910 x 3.310	9.2-1	Hyd.	45–65
	340	4-BBL.	275 @ 5000	340 @ 3200	52.2	4.040 x 3.310	10.5-1	Hyd.	45–65
	383	4-BBL.	300 @ 4400	400 @ 2400	57.8	4.250 x 3.375	10.0-1	Hyd.	45–65
1969	170	1-BBL.	115 @ 4400	155 @ 2400	27.7	3.406 x 3.125	8.5-1	Mech.	45–60
	225	1-BBL.	145 @ 4000	215 @ 2400	27.7	3.406 x 4.125	8.4-1	Mech.	45–60
	273	2-BBL.	190 @ 4400	260 @ 2000	42.2	3.625 x 3.310	9.0-1	Hyd.	45–65
	318	2-BBL.	230 @ 4400	340 @ 2400	48.9	3.910 x 3.310	9.2-1	Hyd.	45–65
	340	4-BBL.	275 @ 5000	340 @ 3200	52.2	4.040 x 3.310	10.5-1	Hyd.	45–65
	383	4-BBL.	330 @ 5200	410 @ 3600	57.8	4.250 x 3.375	10.0-1	Hyd.	45–65
1970	198	1-BBL.	125 @ 4500	180 @ 2000	27.7	3.406 x 3.640	8.4-1	Mech.	45–60
	225	1-BBL.	145 @ 4000	215 @ 2400	27.7	3.406 x 4.125	8.4-1	Mech.	45–60
	318	2-BBL.	230 @ 4400	320 @ 2000	48.9	3.910 x 3.310	8.8-1	Hyd.	45–65
	340	4-BBL.	275 @ 5000	340 @ 3200	52.2	4.040 x 3.310	10.5-1	Hyd.	45–65
	383	4-BBL.	330 @ 5000	425 @ 3200	57.8	4.250 x 3.375	9.5-1	Hyd.	45–65
	383	2-BBL.	290 @ 4400	390 @ 2800	57.8	4.250 x 3.375	8.7-1	Hyd.	45–65
	383 H.P.	4-BBL.	335 @ 5200	425 @ 3400	57.8	4.250 x 3.375	10.5-1	Hyd.	45–65
	426 Hemi	2-4-BBL.	425 @ 5000	490 @ 4000	57.8	4.250 x 3.750	10.25-1	Hyd.	45–65
	440	4-BBL.	375 @ 4600	480 @ 3200	59.7	4.320 x 3.750	9.7-1	Hyd.	45–65
	440 Six Pack	3-2-BBL.	390 @ 4700	490 @ 3200	59.7	4.320 x 3.750	10.50-1	Hyd.	45–65
1971	198	1-BBL.	N.A.	N.A.	27.7	3.406 x 3.640	N.A.	Mechg	45–60
	225	1-BBL.	N.A.	N.A.	27.7	3.406 x 4.125	N.A.	Mech.	45–60
	318	2-BBL.	N.A.	N.A.	48.9	3.910 x 3.310	N.A.	Hyd.	45–65
	340	4-BBL.	N.A.	N.A.	52.2	4.040 x 3.310	N.A.	Hyd.	45–65
	340	3-2-BBL.	N.A.	N.A.	52.2	4.040 x 3.310	N.A.	Hyd.	45–65
	360	N.A.	N.A.	N.A.	51.0	4.000 x 3.580	N.A.	Hyd.	45–65
	383	2-BBL.	N.A.	N.A.	57.8	4.250 x 3.375	N.A.	Hyd.	45–65
	383 H.P.	4-BBL.	N.A.	N.A.	57.8	4.250 x 3.375	N.A.	Hyd.	45–65
	426 Hemi	2-4-BBL.	N.A.	N.A.	57.8	4.250 x 3.750	N.A.	Hyd.	45–65
	440 Std.	4-BBL.	N.A.	N.A.	59.7	4.320 x 3.750	N.A.	Hyd.	45–65
	440 H.P.	4-BBL.	N.A.	N.A.	59.7	4.320 x 3.750	N.A.	Hyd.	45–65
	440 Six Pack	3-2-BBL.	N.A.	N.A.	59.7	4.320 x 3.750	N.A.	Hyd.	45–65

*—1964 Dart 8.5-1.
‡—1964 Dart 8.4-1.
H.P.—High performance.

VALVE SPECIFICATIONS

YEAR AND MODEL		SEAT ANGLE (DEG.)	FACE ANGLE (DEG.)	VALVE LIFT INTAKE (IN.)	VALVE LIFT EXHAUST (IN.)	VALVE SPRING PRESSURE (VALVE OPEN) LBS. @ IN.	VALVE SPRING INSTALLED HEIGHT (IN.)	STEM TO GUIDE CLEARANCE (IN.)		STEM DIAMETER (IN.)	
								Intake	Exhaust	Intake	Exhaust
1964	6 Cyl.	45	①	.375	.360	177 @ 1⁵⁄₁₆	1¹¹⁄₁₆	.001–.003	.002–.004	.372–.373	.371–.372
	V-8	45	45	.395	.405	145 @ 1⁵⁄₁₆	1¹¹⁄₁₆	.001–.003	.002–.004	.372–.373	.371–.372
1965–66	170	45	①	.371	.364	145 @ 1⁵⁄₁₆	1¹¹⁄₁₆	.001–.003	.002–.004	.372–.373	.371–.372
	225	45	①	.394	.390	145 @ 1⁵⁄₁₆	1¹¹⁄₁₆	.001–.003	.002–.004	.372–.373	.371–.372
	273, 2-BBL.	45	45	.395	.405	145 @ 1⁵⁄₁₆	1¹¹⁄₁₆	.001–.003	.002–.004	.372–.373	.371–.372
	273, 4-BBL.	45	45	.415	.425	177 @ 1⁵⁄₁₆	1¹¹⁄₁₆	.001–.003	.002–.004	.372–.373	.371–.372
1967	6 Cyl.	45	②	.395	.395	145 @ 1⁵⁄₁₆	1¹¹⁄₁₆	.001–.003	.002–.004	.372–.373	.371–.372
	273, 2-BBL.	45	45	.395	.405	145 @ 1⁵⁄₁₆	1¹¹⁄₁₆	.001–.003	.002–.004	.372–.373	.371–.372
	273, 4-BBL.	45	45	.415	.425	177 @ 1⁵⁄₁₆	1¹¹⁄₁₆	.001–.003	.002–.004	.372–.373	.371–.372
1968	6 Cyl.	45	①	.395	.395	145 @ 1⁵⁄₁₆	1¹¹⁄₁₆	.001–.003	.002–.004	.372–.373	.371–.372
	273, 318	45	①	.373	.399	177 @ 1⁵⁄₁₆	1¹¹⁄₁₆	.001–.003	.002–.004	.372–.373	.371–.372
	340	45	①	③	④	242 @ 1⁷⁄₃₂	1¹¹⁄₁₆	.001–.003	.002–.004	.372–.373	.371–.372
	383, 2-BBL.	45	45	.425	.435	200 @ 1⁷⁄₁₆	1⁵⁵⁄₆₄	.001–.003	.002–.004	.372–.373	.371–.372
	383, 4-BBL.	45	45	.425	.435	230 @ 1¹³⁄₃₂	1⁵⁵⁄₆₄	.001–.003	.002–.004	.372–.373	.371–.372
1969	6 Cyl.	45	①	.395	.395	145 @ 1⁵⁄₁₆	1¹¹⁄₁₆	.001–.003	.002–.004	.372–.373	.371–.372
	273, 318	45	①	.373	.399	177 @ 1⁵⁄₁₆	1¹¹⁄₁₆	.001–.003	.002–.004	.372–.373	.371–.372
	340	45	①	.429	.444	242 @ 1⁷⁄₃₂	1¹¹⁄₁₆	.001–.003	.002–.004	.372–.373	.371–.372
	383, 2-BBL.	45	45	.425	.435	200 @ 1⁷⁄₁₆	1⁵⁵⁄₆₄	.001–.003	.002–.004	.372–.373	.371–.372
	383, 4-BBL.	45	45	.450	.458	246 @ 1²³⁄₆₄	1⁵⁵⁄₆₄	.001–.003	.002–.004	.372–.373	.371–.372
1970	6 Cyl.	45	①	.395	.395	145 @ 1⁵⁄₁₆	1¹¹⁄₁₆	.001–.003	.002–.004	.372–.373	.371–.372
	318	45	①	.373	.399	177 @ 1⁵⁄₁₆	1¹¹⁄₁₆	.001–.003	.002–.004	.372–.373	.371–.372
	340	45	①	.429	.444	242 @ 1⁷⁄₃₂	1¹¹⁄₁₆	.001–.003	.002–.004	.372–.373	.371–.372
	383, 2-BBL.	45	47	.425	.435	200 @ 1⁷⁄₁₆	1⁵⁵⁄₆₄	.001–.003	.002–.004	.372–.373	.371–.372
	383, 4-BBL.	45	47	.450	.458	234 @ 1⁷⁄₁₀	1⁵⁵⁄₆₄	.001–.003	.002–.004	.372–.373	.371–.372
	383, Hi Perf.	45	47	.450	.458	234 @ 1⁷⁄₁₀	1⁵⁵⁄₆₄	.001–.003	.002–.004	.372–.373	.371–.372
	426	45	45	.490	.481	310 @ 1³⁄₈	1⁵⁵⁄₆₄	.002–.004	.003–.005	.308	.309
	440	45	47	.425	.435	234 @ 1⁷⁄₁₀	1⁵⁵⁄₆₄	.001–.003	.002–.004	.372–.373	.371–.372
	440, Hi Perf.	45	47	.450	.458	234 @ 1⁷⁄₁₀	1⁵⁵⁄₆₄	.001–.003	.002–.004	.372–.373	.371–.372
	440, 3-2-BBL.	45	47	.450	.458	234 @ 1⁷⁄₁₀	1⁵⁵⁄₆₄	.001–.003	.002–.004	.372–.373	.371–.372
1971	6 Cyl.	45	①	.406	.414	144 @ 1⁵⁄₁₆	1¹¹⁄₁₆	.001–.003	.002–.004	.372–.373	.371–.372
	318	45	①	.373	.399	177 @ 1⁵⁄₁₆	1¹¹⁄₁₆	.001–.003	.002–.004	.372–.373	.371–.372
	340, 4-BBL./6-BBL.	45	①	.429	.444	238 @ 1⁵⁄₁₆	1¹¹⁄₁₆	.001–.003	.002–.004	.372–.373	.371–.372
	360	45	①	.410	.412	177 @ 1⁵⁄₁₆	1¹¹⁄₁₆	.001–.003	.002–.004	.372–.373	.371–.372
	383, 2-BBL.	45	45	.425	.435	200 @ 1⁷⁄₁₆	1⁵⁵⁄₆₄	.001–.003	.002–.004	.372–.373	.371–.372
	383 Hi Perf.	45	45	.450	.458	246 @ 1²³⁄₆₄	1⁵⁵⁄₆₄	.001–.003	.002–.004	.372–.373	.371–.372
	426 Hemi	45	45	.490	.481	310 @ 1³⁄₈	1⁵⁵⁄₆₄	.002–.004	.003–.005	.310	.309
	440 Std.	45	45	.425	.435	200 @ 1⁷⁄₁₆	1⁵⁵⁄₆₄	.001–.003	.002–.004	.372–.373	.371–.372
	440 Hi Perf.	45	45	.450	.458	246 @ 1²³⁄₆₄	1⁵⁵⁄₆₄	.001–.003	.002–.004	.372–.373	.371–.372
	440, 3-2-BBL.	45	45	.450	.458	246 @ 1²³⁄₆₄	1⁵⁵⁄₆₄	.001–.003	.002–.004	.372–.373	.371–.372

① —Intake 45; Exhaust 43.
② —Intake 45; Exhaust 47.
③ —Manual trans. .444; Auto trans. .429
④ —Manual trans. .453; Auto trans. .444

TORQUE SPECIFICATIONS

YEAR	MODEL	CYLINDER HEAD BOLTS (FT. LBS.)	ROD BEARING BOLTS (FT. LBS.)	MAIN BEARING BOLTS (FT. LBS.)	CRANKSHAFT BALANCER BOLT (FT. LBS.)	FLYWHEEL TO CRANKSHAFT BOLTS (FT. LBS.)	MANIFOLD (FT. LBS.)	
							Intake	Exhaust
1964–71	6 Cyl.—198, 170, 225	65	45	85	Press Fit	55	10▲	10
	V8—273, 318, 340, 360	85*	45	85①	135	55	35	30
	V8—383, 440	70	45	85	135	55	35	30
	V8—426 Hemi	70–75②	75	100	135	70	③	24

▲—Intake to Exhaust Bolts—20. ★—340—95.

① —340—95.
② —Bolts and studs same torque.
③ —See text.

CAPACITIES

| YEAR | MODEL | ENGINE CRANKCASE ADD 1 Qt. FOR NEW FILTER | TRANSMISSIONS Pts. TO REFILL AFTER DRAINING | | | DRIVE AXLE (Pts.) ② | GASOLINE TANK (Gals.) | COOLING SYSTEM (Qts.) WITH HEATER (No A/C) |
| | | | Manual | | Automatic | | | |
			3-Speed	4-Speed				
1964-65	Valiant—6 Cyl.	4	6	N.A.	17	③	18	12
	Dart—6 Cyl.	4	6	7	17	③	18	12
	V8—All	4	6	7	17	③	18	18
1966	6 Cyl. 170	4	6½	N.A.	16	③	18	12
	6 Cyl. 225	4	6½	N.A.	16	③	18	13
	V8—273	4	6½	8	16	③	18	18
1967	Exc. 170	4	6½	N.A.	16	③	18	12
	6 Cyl. 225	4	6½	N.A.	16	③	18	13
	V8—273	4	6½	8	16	③	18	19
1968	6 Cyl.	4	6½	N.A.	15½	③	18	●
	V8—273	4	6	8	15½	③	18	19
	V8—318	4	6	8	15½	③	18	18
	V8—340, 383	4	N.A.	9	18½	③	18	17
1969	6 Cyl. 170	4	6½	N.A.	15½	③	18	12
	6 Cyl. 225	4	6½	N.A.	15½	③	18	13
	V8—273	4	6½	7	15½	③	18	17
	V8—318	4	6½	7	15½	③	18	16
	V8—340	4	N.A.	7	18½	③	18	16
	V8—383	4	N.A.	7	18½	③	18	16
1970-71	6 Cyl. 225, 198, 170	4	① 6½ ⑦	N.A.	17	③	18 ④	13
	V8—318	4	4¾ ①	7½	16	③	18 ④	16
	V8—340, 360	4	① 4¾	7½	16	③	18 ④	16 ⑤
	V8—383	4	N.A.	7½	19 ⑥	③	18 ④	16
	V8—426	6	N.A.	7½	16.8	5½	18 ④	15½
	V8—440	6	N.A.	7½	19.0	5½	18 ④	15½

①—All Synchromesh 3-spd., 4.75 pts. Dexron A.T. Fluid.   ●—170—12, 225—13.   ④—Barracuda, 19.
②—Use only limited slip lube for limited slip drive units.   ⑤—Valiant and Dart, 15.
③—7¼ in. axle, 2.0; 8¼ in. axle, 4.4; 8¾ in. axle, 4.4; 9¾ in. axle, 5.5.   ⑥—4-BBL. Hi Perf., 16.
⑦—Barracuda and Challenger, 4¾.

Axle	Filler location	Cover fastening
7¼ in.	Cover	9 bolts
8¼ in.	Carrier, right side	10 bolts
8¾ in.	Carrier, right side	Welded
9¾ in.	Cover	10 bolts

CYLINDER HEAD BOLT TIGHTENING SEQUENCE

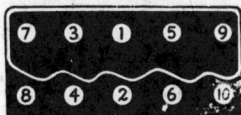

273, 318, 340 cu. in. V8

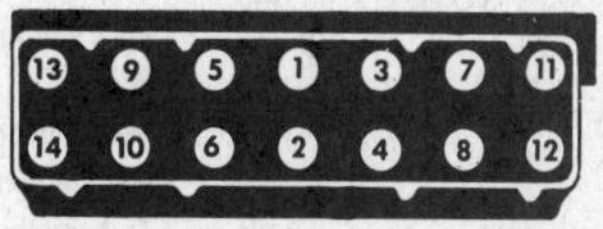

6 cyl.

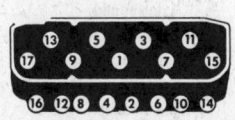

383, 440 cu. in. V8

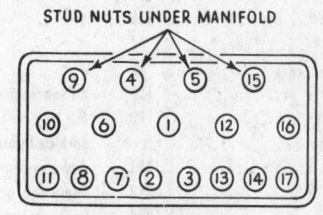

STUD NUTS UNDER MANIFOLD

426 cu. in. Hemi V8

LIGHT BULBS
DODGE

1964

	Dart
Sealed beam—single 2 filament	6012
Tail, stop & turn signal	1034
Park & turn signal	1034
Back up lamps	1073
License lamp	67
Trunk and/or under hood	1004
Glove compartment	1891
Radio	53X
Trans. control push buttons	53X
Handbrake indicator	57
Dome lamp	1004
Map lamp	90
Turn signal indicator	57
High beam indicator	57
Oil pressure warning light	57
Instrument cluster illumination	57

1965

	Dart
Sealed beam—single 2 filament	6012
Tail, stop & turn signal	*1034
Tail light	*1095
Park & turn signal	1034
Back up lights	1073
License light	67
Trunk and/or under hood	1004
Glove compartment	1891
Radio	1893
Gear shift indicator	1445
Handbrake indicator	57
Dome lamp	1004
Map lamp	90
Turn signal indicator	57
High beam indicator	57
Oil pressure warning light	57
Instrument cluster illumination	57

*67 on Dart Station Wagon.

1966

	Dart
Back-up lights	1073
Clock	57
Dome lights	1004
Door and/or pocket	90
Emergency flasher	57
Gear selector indicator	1445
Gear selector with console	57
Glove compartment	1891
Handbrake indicator	57
Heater and/or A.C. Control	**
High beam indicator	158
Instrument cluster illumination	158
License light	67
Map light	90
Oil pressure indicator	158
Panel and/or ridge light	90
Park and turn signal	1034
Radio	1893
Sealed sealed beam—single "7"	6012
Tail stop and turn signal	1034
Trunk and/or under hood light	1004

Turn signal indicator ... 158
**Not lighted

1967

	Dart
Back-up lights	1141
Brake system warning light	57
Dome and/or "C" pillar light	1004
Door, pocket panel and/or reading light	90
Fender mounted turn signal indicator	1893
Gear selector indicator	1445
Gear selector with console	57
Glove compartment	1891
Handbrake indicator	57
High beam indicator	158
Instrument cluster illumination	158
License light	67
Map light	90
Oil pressure indicator	158
Park and turn signal	1034
Portable reading light	90
Radio	1893
Sealed beam—single "7"	6012
Tachometer	1816
Tail, stop and turn signal	1034
Trunk and/or under hood light	1004
Turn signal indicator (panel)	158

1968

	Dart
Air conditioning indicator	1445
Ash tray	1445
Back-up lights	1073
Brake system warning light	57
Courtesy lamp	89
Dome and/or "C" pillar light	1004
Door, pocket panel and/or reading light	90
Fender mounted turn signal indicator	330
Gear selector indicator	1445
Gear selector with console	57
Glove compartment	1891
High beam indicator	158
Instrument cluster illumination	158
License light	67
Map light	
Oil pressure indicator	158
Park and turn signal	1034A
Radio	1816
Sealed beam—single "7"	6012
Side marker	1895
Tachometer	1816
Tail, stop and turn signal	1034
Trunk and/or under hood light	1004
Turn signal indicator (panel)	158

1969

	Dart
Air conditioning indicator	**1445
Ash tray	**1445
Back-up lights	1156 (2)
Brake system warning light	57
Dome and/or "C" pillar light	1004
Door, pocket panel and/or reading light	90
Fender mounted turn signal indicator	330 (2)
Gear selector indicator	**1445
Gear selector with console	** 57
Glove compartment	1891
High beam indicator	158
Ignition lamp	1445
Instrument cluster illumination	** 158 (4)
License light	67
Map light	89
Oil pressure indicator	158
Park and turn signal	1157 A(2)
Radio	**1816
Reverse 4-speed transmission indicator	53
Sealed beam—single "7"	6012
Tail, stop and turn signal	1157 (2)
Trunk and/or under hood light	1004
Turn signal indicator (panel)	158 (2)

***Headlamp rheostat dimming.

1970–71

	Challenger (Rallye)	Challenger	Dart
Air conditioner control	**1815	**1815	**1445
Ash tray	**1445	**1445	**1445
Back-up lights	1156 (2)	1156 (2)	1156 (2)
Clock	*	*	*
Cornering lights	1445	1445	
Dome lamp	550 or 1004	550 or 1004	1004
Door ajar indicator	1892	1892	
Pocket panel lamp	90	90	
Fender mounted turn signal indicator	330 (2)	330 (2)	330 (2)
Gear selector indicator (column)	** 161	** 161	** 161
Gear selector with console	** 57	** 57	** 57
Glove compartment	1891	1891	1891
Headlamp switch rheostat valve	15.5 Ohms	15.5 Ohms	28 Ohms
High beam indicator	57	57	158
Instrument cluster and speedometer illumination	**1893	**1816 (3)	** 158 (4)
Ignition lamp	1445	1445	1445
License light	67	67	67
Low fuel indicator	1892	1892	

LIGHT BULBS

1970–71 DODGE

	Challenger (Rallye)	Challenger	Dart
Map and courtesy lamp	89	89	89
Oil indicator	57	57	158
Park and turn signal	1157 (2)	1157 (2)	1157 (2)
Radio—AM and tape ..	**1816	**1816	**1816
Radio—AM-FM stereo	**1815	**1815	
Reverse 4-speed transmission indicator	53	53	53
Sealed beam—hi-beam (No. 1)	4001	4001	
Sealed beam—hi-lo beam (No. 2)	4002	4002	
Sealed beam—single "7"			6012
Seat belts indicator ..	1892	53	53
Stereo indicator ..	1445	1445	
Switch lighting	*	*	
Tachometer ..	*	*	
Tail, stop and turn signal	1157 (4)	1157 (4)	1157 (2)
Trunk and/or under hood lamp	1004		1004
Turn signal indicator (panel)....	57 (2)	1004	

*Included in instrument cluster lighting.
**Headlamp rheostat dimming.

ALTERNATOR AND AC REGULAR SPECIFICATIONS

YEAR	ALTERNATOR Part Number	ALTERNATOR Field Current Draw @ 12V.	ALTERNATOR Output @ Engine RPM 1250 (AMP.)	REGULATOR Part Number	REGULATOR Field Relay Air Gap (In.)	REGULATOR Field Relay Point Gap (In.)	REGULATOR Field Relay Volts to Close	REGULATOR Regulator Air Gap (In.)	REGULATOR Regulator Point Gap (In.)	REGULATOR Regulator Volts at 140°
1964–67	2098835●	2.38–2.75	30	2098300	.050	.014	13.8	.050	.015	13.2–14.2
	2098830●	2.38–2.75	35	2098300	.050	.014	13.8	.050	.015	13.2–14.2
	2098850	2.38–2.75	46	2098300	.050	.014	13.8	.050	.015	13.2–14.2
	2444599●	2.38–2.75	46	2098300	.050	.014	13.8	.050	.015	13.2–14.2
1968–69	2642538●	2.38–2.75	30	2098300	.050	.014	13.8	.050	.015	13.2–14.2
	2642537	2.38–2.75	37	2098300	.050	.014	13.8	.050	.015	13.2–14.2
1970–71	3438171	2.38–2.75	30	3438150	.050	.014	13.8	.050	.015	13.3–14.0
	3438172	2.38–2.75	37	3438150	.050	.014	13.8	.050	.015	13.3–14.0
	3438176	2.38–2.75	37	3438150	.050	.014	13.8	.050	.015	13.3–14.0

●Replaced by 2098850

WHEEL ALIGNMENT

YEAR	MODEL	FRONT END HEIGHT (In.)	CASTER Range (Deg.)	CASTER Pref. Setting (Deg.)	CAMBER Range (Deg.)	CAMBER Pref. Setting (Deg.)	TOE-IN (In.)	KING-PIN INCLINATION (Deg.)	WHEEL PIVOT RATIO Inner Wheel	WHEEL PIVOT RATIO Outer Wheel
1964–65	Manual	$1\frac{3}{4} \pm \frac{1}{8}$	1N to 0	$\frac{1}{2}$N	▲	▲	$\frac{1}{8}$	$7\frac{1}{2}$	20	17.6
	Power	$1\frac{3}{4} \pm \frac{1}{8}$	$\frac{1}{4}$P to $1\frac{1}{4}$P	$\frac{3}{4}$P	▲	▲	$\frac{1}{8}$	$7\frac{1}{2}$	20	17.6
1966–68	Manual	$2 \pm \frac{1}{8}$ ●	1N to 0	$\frac{1}{2}$N	+	+	$\frac{3}{32}$–$\frac{5}{32}$	$7\frac{1}{2}$	20	17.6
	Power	$2 \pm \frac{1}{8}$ ●	$\frac{1}{4}$P to $1\frac{1}{4}$P	$\frac{3}{4}$P	+	+	$\frac{3}{32}$–$\frac{5}{32}$	$7\frac{1}{2}$	20	17.6
1969	Manual	$2 \pm \frac{1}{8}$ ●	1N to 0*	$\frac{1}{2}$N	+	+	$\frac{3}{32}$–$\frac{5}{32}$ ⊙	$7\frac{1}{2}$	20	17.6
	Power	$2 \pm \frac{1}{8}$ ●	$\frac{1}{4}$P to $1\frac{1}{4}$P*	$\frac{3}{4}$P	+	+	$\frac{3}{32}$–$\frac{5}{32}$ ⊙	$7\frac{1}{2}$	20	17.6
1970–71	Valiant Manual	$2\frac{1}{8} \pm \frac{1}{8}$	$\frac{1}{2}$P $\pm \frac{1}{2}$	$\frac{1}{2}$N	①	②	$\frac{3}{32}$–$\frac{5}{32}$	$7\frac{1}{2}$	20	17.5
	Valiant Power	$2\frac{1}{8} \pm \frac{1}{8}$	$\frac{3}{4}$P $\pm \frac{1}{2}$	$\frac{3}{4}$P	①	②	$\frac{3}{32}$–$\frac{5}{32}$	$7\frac{1}{2}$	20	17.5
	Dart Manual	$2\frac{1}{8} \pm \frac{3}{8}$	0 to 1N	$\frac{1}{2}$N	①	②	$\frac{3}{32}$–$\frac{5}{32}$	$7\frac{1}{2}$	20	17.6
	Dart Power	$2\frac{1}{8} \pm \frac{1}{8}$	$\frac{1}{4}$P to $1\frac{1}{4}$P	$\frac{3}{4}$P	①	②	$\frac{3}{32}$–$\frac{5}{32}$	$7\frac{1}{2}$	20	17.6
	Barracuda Manual	$1\frac{3}{16} \pm \frac{1}{8}$	0 to 1N	$\frac{1}{2}$N	①	②	$\frac{1}{32}$–$\frac{7}{32}$	$7\frac{1}{2}$	20	17.5
	Barracuda Power	$1\frac{3}{16} \pm \frac{1}{8}$	$\frac{1}{4}$P to $1\frac{1}{4}$P	$\frac{3}{4}$P	①	②	$\frac{1}{32}$–$\frac{7}{32}$	$7\frac{1}{2}$	20	17.5
	Challenger Manual	$1\frac{3}{16} \pm \frac{1}{8}$	0 to 1N	$\frac{1}{2}$N	①	②	$\frac{3}{32}$–$\frac{5}{32}$	$7\frac{1}{2}$	20	17.8
	Challenger Power	$1\frac{3}{16} \pm \frac{1}{8}$	$\frac{1}{4}$P to $1\frac{1}{4}$P	$\frac{3}{4}$P	①	②	$\frac{3}{32}$–$\frac{5}{32}$	$7\frac{1}{2}$	20	17.8

① —Left wheel: $\frac{1}{2}$P $\pm \frac{1}{4}$; Right wheel: $\frac{1}{4}$P $\pm \frac{1}{4}$.
② —Left wheel: $\frac{1}{2}$P; Right wheel $\frac{1}{4}$P.
▲—Right side, $\frac{1}{8}$N to $\frac{3}{8}$P; left side, $\frac{1}{8}$P to $\frac{5}{8}$P.
+—Right side, 0 to $\frac{1}{2}$P, $\frac{1}{4}$P preferred; left side, $\frac{1}{4}$P to $\frac{3}{4}$P, $\frac{1}{2}$P preferred.
N—Negative.
P—Positive.
●—Barracuda—$1\frac{3}{8} \pm \frac{1}{8}$.
*—Dart: manual—$\frac{1}{16}$P to $1\frac{1}{16}$N; power—$\frac{3}{16}$P to $1\frac{5}{16}$P.
⊙—Dart $\frac{1}{16}$–$\frac{3}{16}$.

LIGHT BULBS

Plymouth

1964

	Valiant
Single beam 2 filament	6012
Tail, stop & turn signal	1034
Park & turn signal	1034
Back-up lamps	1073
License lamp	67
Trunk and/or under hood lamp	1004
Glove compartment	1891
Radio	53X
Transmission control push buttons	53X
Handbrake indicator	57
Dome lamp	1004
Map lamp	90
Turn signal indicator	57
High beam indicator	57
Oil pressure indicator	57
Instrument cluster illumination	57

1965-66

	Valiant, Barracuda
Single beam 2 filament	6012
Tail, stop & turn signal	1034
Park & turn signal	1034
Back-up lamps	1073
License lamp	67
Trunk and/or under hood lamp	1004
Glove compartment	1891
Radio	1893
Transmission gear shift control	1445
Handbrake indicator	57
Dome lamp	1004
Map lamp	90
Turn signal indicator	158
High beam indicator	158
Oil pressure indicator	158
Instrument and speedometer cluster illumination	158
Emergency flasher	57
Gear selector with console	53X

1967

	Valiant
Brake system warning light	57
Dome and/or "C" pillar light	1004
Door, pocket panel and/or reading light	90
Fender mounted turn signal indicator	330
Gear selector indicator	1445
Gear selector with console	57
Glove compartment	1891
Handbrake indicator	57
High beam indicator	158
Instrument cluster illumination	158

License light	67
Map light	90
Oil pressure indicator	57
Park and turn signal	1034A
Radio	1893
Sealed beam—single "7"	6012
Tail, stop and turn signal	1034
Trunk and/or under hood light	1004
Turn signal indicator (panel)	158

1968

	Barracuda	Valiant
Air conditioning indicator	1445*	1445*
Ash receiver	1445*	1445*
Back-up lights	1073	1141 (Sta. Wag.)
Brake system warning light	158	57
Courtesy lamp	89	89
Dome and/or "C" pillar light	211-1	1004
Door, pocket panel and/or reading light	90	90
Fender mounted turn signal indicator	330	330H
Gear selector indicator	1445	1445
Gear selector with console	57	57
Glove compartment	1891	1891
High beam indicator	158	158
Instrument cluster illumination	158	158
License light	67	67
Oil pressure indicator		158
Park and turn signal	1034A	1034
Radio	1816	1816
Sealed beam—single "7"	6012	6012
Side marker	1895	1895
Tachometer	*	1816
Tail, stop and turn signal	1034	1034
Trunk and/or under hood light	1004	1004
Turn signal indicator (panel)	158	158

*Included in instrument cluster.

1969

	Barracuda	Valiant
Air conditioner control and auto-temp	**1445	**1445
Ash receiver	**1445	**1445
Back-up lights	1156 (2)	1156 (2)
Brake system warning indicator	158	158
"C"-pillar light	211-1 (2)	—
Courtesy lamp	89	
Dome lamp	1004	1004
Pocket panel lamp	90	NA
Fender mounted turn signal indicator	330 (2)	330 (2)
Gear selector indicator	**1445	**1445
Gear selector with console	** 57	
Glove compartment	1891	1891

High beam indicator	158	158
Instrument cluster and speedometer illumination	** 158 (4)	** 158 (3)
Ignition lamp	1445	1445
License light	67	67
Map and courtesy lamp	89	89
Oil pressure indicator	—	158
Park and turn signal	1157A (2)	1157A (2)
Radio	**1816	**1816
Sealed beam—single "7"	6012	6012
Tachometer	*	—
Tail, stop and turn signal	1157 (2)	1157 (2)
Trunk and/or under hood lamp	1004	1004
Turn signal indicator (panel)	158 (2)	158 (2)
Reverse 4-speed transmission indicator	53	53

NA—Not available.

*—Included in instrument cluster lighting.

**—Headlamp rheostat dimming.

1970-71

	Barracuda	Valiant
Air conditioner control and auto-temp	1815	1445
Ash receiver	1445	1445
Brake system warning indicator	57	158
Courtesy lamp	1445	—
Dome lamp	50 or 1004	1004
Pocket panel lamp	90	—
Fender mounted turn signal indicator	330	330
Gear selector indicator	161	161
Gear selector with console	57	57
Glove compartment	1891	1891
High beam indicator	57	158
Instrument cluster and speedometer illumination	1816	158
Ignition lamp	1445	1445
License light	67	67
Map and courtesy lamp	89	89
Oil pressure indicator	57	158
Park and turn signal	1157	1157
Radio (FM)	1815	1815
Radio with tape	1816	1816
Sealed beam—single "7"	6012	6012
Stereo indicator	1445	—
Tail, stop and turn signal	1157	1157
Trunk and/or under hood lamp	1004	1004
Turn signal indicator (panel)	57	158
Reverse 4-speed transmission indicator	53	53
Door open indicator	1892	—
Low fuel indicator	1892	—
Seat belt indicator	1892	53
Side marker	1895	1895

FUSES AND CIRCUIT BREAKERS

PLYMOUTH

1964
FUSES

Radio	3 AG/AGC; 7.5 AMP
Heater or air conditioning	3 AG/AGC; 20 AMP
Accessories	3 AG/AGC; 15 AMP
Cigar lighter	3 AG/AGC; 20 AMP
Tail, stop, dome	3 AG/AGC; 20 AMP
Instrument lamps	3 AG/AGC; 2 AMP

CIRCUIT BREAKERS

Windshield wiper—variable speed (back of wiper switch)	6 AMP
Windshield wiper—single speed (integral with wiper switch)	5 AMP
Lighting system (integral with headlamp switch)	15 AMP
Top lift (behind left front kick panel)	30 AMP

1965–66
FUSES

Radio	5 AMP
Heater or air conditioning	20 AMP
Accessories	20 AMP
Cigar lighter	20 F MP
Tail, stop, dome	20 AMP
Instrument lamps	*

*AV-1 and AV-2—2 ampere; AR-1 and AR-2—3 ampere; AP-1 and AP-2—4 ampere.

CIRCUIT BREAKERS

Windshield wiper—variable speed (integral with wiper switch)	*
Windshield wiper—single speed (integral with wiper switch)	5 AMP
Lighting system (integral with headlamp switch)	15 AMP
Power windows, power seats, top lift (behind left front kick panel)	30 AMP
Electric door locks (behind left front kick panel)	15 AMP

*AV-1, AV-2, AR-1 and AR-2—6 ampere; AP-1 and AP-2—7½ ampere.

1967
FUSES

Accessories	20 AMP
Cigar lighter (front)	20 AMP
Emergency flasher	20 AMP
Heater or air conditioner	20 AMP
Instrument lights	2 AMP
	—Valiant
Radio	5 AMP
Tail, stop, dome	20 AMP

CIRCUIT BREAKERS

	Valiant
Convertible top (*integral with top lift switch)	30 AMP
Headlights (integral with headlight switch)	15 AMP
Windshield wipers (integral with wiper switch)	6 AMP

1968
FUSES

	Barracuda	Valiant
Accessories	20	20
Cigar lighter (front) and dome light	20	20
Heater or air conditioner	20	20
Instrument lights	2	2
Radio and back-up lamps	5	5
Tail, stop and emergency flasher	20	20

CIRCUIT BREAKERS

	Barracuda	Valiant (Amps.)
Convertible top (integral with top lift switch)	30	30
Headlights (integral with headlight switch)	15	15
Windshield wipers (integral with wiper switch)	6	6

*Behind left front cowl trim panel in Fury and V.I.P. models.

1969–71
FUSES

	Barracuda	Valiant (Amps.)
Accessory	20	20
Heater and air conditioner	20	20
Instrument lamps	3	3
Radio and back-up lamps	7.5*	7.5*
Stop and dome lamps	20	20
Tail lamps and cigar lighter	20	20

*—1970-71—20

CIRCUIT BREAKERS

	Barracuda	Valiant (Amps.)
Convertible (integral with switch)	30	—
Headlights (integral with headlamp switch)	15	15
Windshield wipers (integral with wiper switch)	6	6

Dodge

1964
FUSES

Radio	3 AG/AGC; 7.5 AMP
Heater or air cond.	3 AG/AGC; 20 AMP
Accessories	3 AG/AGC; 15 AMP
Cigar lighter	3 AG/AGC; 20 AMP
Tail, stop, dome	3 AG/AGC; 20 AMP
Instrument lamps	3 AG/AGC; 2 AMP

CIRCUIT BREAKERS

Windshield wiper—variable speed (back of wiper switch)	6 AMP
Windshield wiper—single speed (integral with wiper switch)	5 AMP
Lighting system (integral with headlamp switch)	15 AMP
Top lift (behind left front kick panel)	30 AMP

1965
FUSES

Radio	3 AG/AGC; 5 AMP
Heater or air cond.	3 AG/AGC; 20 AMP
Accessories	3 AG/AGC; 20 AMP
Cigar lighter	3 AG/AGC; 20 AMP
Tail, stop, dome	3 AG/AGC; 20 AMP
Instrument lights	3 AG/AGC; 2 AMP Dart

CIRCUIT BREAKERS

Windshield wiper—variable speed (integral with wiper switch)	6 AMP
Windshield wiper—single speed (integral with wiper switch)	5 AMP
Lighting system (integral with headlight switch)	15 AMP Dart
Top lift (behind left front kick panel)	30 AMP

1966–67
FUSES

Accessories	3 AG/AGC; 20 AMP
Cigar lighter	3 AG/AGC; 20 AMP
Heater or air cond.	3 AG/AGC; 20 AMP
Instrument lights	3 AG/AGC; 2 AMP Dart
Radio	3 AG/AGC; 5 AMP
Tail, stop, dome	3 AG/AGC; 20 AMP

CIRCUIT BREAKERS

Lighting system (integral with headlight switch)—Dart	15 AMP
Top lift (behind left front kick panel)	30 AMP
Windshield wiper (integral with wiper switch) Single speed	5 AMP
Variable speed	6 AMP

1968
FUSES

	Dart
Accessories	20 AMP
Cigar lighter (front) and dome light	20 AMP
Heater or air conditioner	20 AMP
Instrument lights	2 AMP
Radio and back-up lamps	5 AMP
Tail, stop and emergency flasher	20 AMP

CIRCUIT BREAKERS

	Dart
Convertible top (integral with top lift switch)	30 AMP
Headlights (integral with headlight switch)	15 AMP
Windshield wipers (integral with wiper switch)	6 AMP

1969
FUSES

	Dart
Accessory, tail lamps and cigar lighter	20 AMP
Emergency, stop and dome	20 AMP
Heater and air conditioner	20 AMP
Instrument lamps	3 AMP
Radio and back-up lamps	7.5 AMP

CIRCUIT BREAKERS

	Dart
Convertible top (integral with top lift switch)	30 AMP
Headlights (integral with headlight switch)	15 AMP
Windshield wipers (integral with wiper switch)	6 AMP

1970–71
FUSES

	Challenger	Dart (Amps.)
Accessory	20	20
Emergency flasher	20	—
Heater and air conditioner (blower motor)	20	20
Instrument lamps	3	3
Radio and back-up lamps	20	20
Stop and dome lamps	20	20
Tail lamps and cigar lighter	20	20

CIRCUIT BREAKERS

	Challenger	Dart (Amps.)
Convertible (on fuse block)	30	—
Headlamps (integral with headlamp switch)	20	15
Windshield wipers (integral with wiper switch) 2-speed motor	6.0	6.0
3-speed motor (optional)	7.5*	7.5*
Variable speed motor (optional)	7.5*	7.5*

Distributor

Removal

1. Remove distributor cap.
2. Disconnect primary wire.
3. Disconnect vacuum line.
4. Rotate engine crankshaft until distributor rotor is pointing at cylinder block. Scribe mark on block at this point to indicate position of rotor.
5. Remove hold down bolt and remove distributor.

Installation

If engine has not been disturbed, install distributor so that rotor aligns with scribe mark on block. If engine has been disturbed, proceed as follows.

1. Rotate crankshaft until mark on inner edge of crankshaft pulley is in line with the O (TDC) mark on timing chain case cover.
2. With distributor gasket in position, hold distributor over the mounting pad.
3. Turn the rotor to point forward, corresponding to four o'clock.
4. Install distributor so that, with distributor fully seated on the engine, the gear has spiraled to bring rotor to a five o'clock position.
5. Turn the housing until the ignition points are separating and rotor is under No. 1 cap tower.
6. Install hold-down bolt.
7. Adjust timing to specifications, using a timing light.

Exhaust Emission Control

In compliance with anti-pollution laws involving all of the continental United States, Chrysler Motors Corporation has continued to use its C.A.P. method of terminal exhaust emission control.

The origin of this design was to satisfy the requirements of California smog control for 1966-67. Based upon its success, the C.A.P. system will be used on almost all cars produced by the Chrysler Corporation in all of its divisions.

The C.A.P. concept utilizes broader, yet more critical, distributor control through carburetor modification, and by using a second vacuum-operated distributor control in the distributor timing regulator mechanism.

Any of the methods of terminal exhaust treatment requires close and frequent attention to tune-up factors of engine performance.

Since 1968, all car manufacturers have posted idle speeds and other pertinent data relative to the specific engine-car application in a conspicuous place in the engine compartment.

For details, consult the Unit Repair Section.

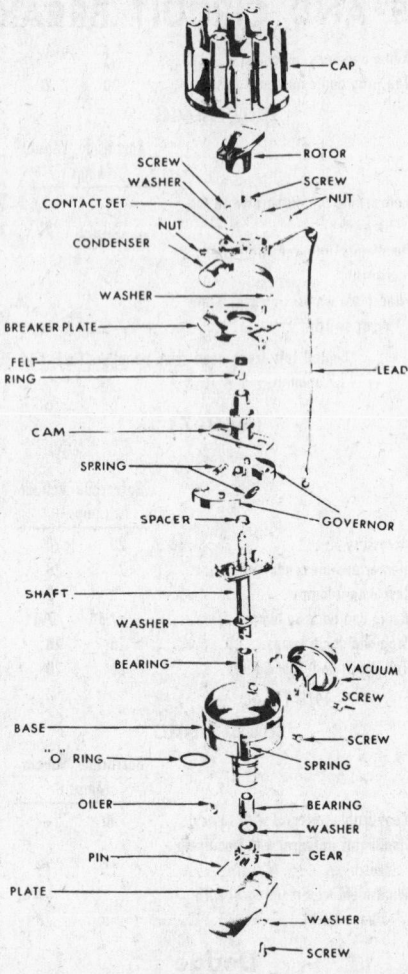

6 cylinder distributor
(© Chrysler Corp.)

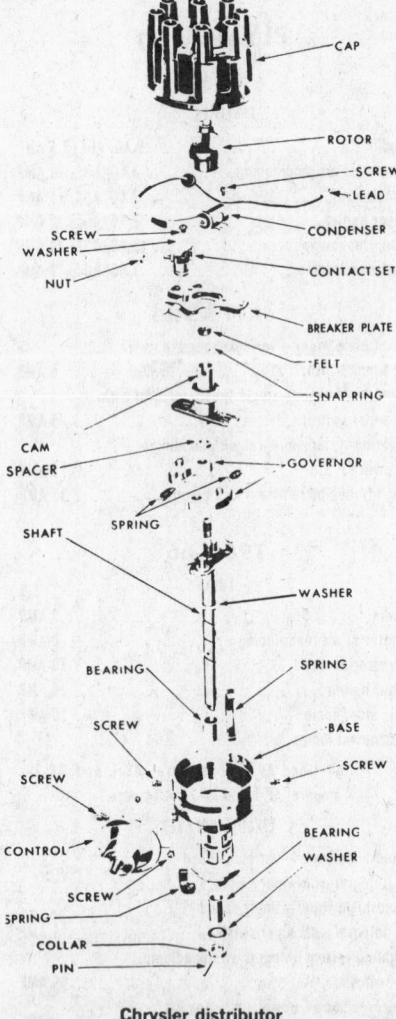

Chrysler distributor
(© Chrysler Corp.)

Alternator, Regulator

References

Details on alternator and regulator can be found in specifications table of this section.

General information on alternator and regulator repair and troubleshooting can be found in the Unit Repair Section.

Removal

To remove alternator:
1. Disconnect battery ground cable.
2. Disconnect BAT and FLD leads from alternator.
3. Remove alternator by removing two mounting bolts and belt tensioner bracket bolt.
4. To reinstall, reverse above.

NOTE: never attempt to polarize an alternator, nor short the regulator.

Starter

References

More detailed information on starters will be found in the Unit Repair Section.

Removal

1. Disconnect negative cable at battery.
2. Disconnect starter cable at starter.
3. Remove starter attaching bolts and remove starter and cylinder block seal from beneath the engine.

Installation

Reverse above procedure.

Instruments

References

Thermal type fuel and temperature

gauges are used on Valiant and Dart cars.

Both gauges receive a constant voltage supply from a voltage regulator built into the temperature gauge. They are sensitive only to changes in fuel level or temperature

The terminals on the temperature gauge housing the constant voltage regulator, internally, are marked as follows:

A: the output terminal for the controlled voltage from the constant voltage regulator.

I: the 12-volt input voltage terminal to the voltage regulator.

S: the terminal for the connection to the sending unit.

The fuel gauge will have only the controlled voltage terminal A and the terminal S for the connection to the tank sending unit.

The oil pressure indicator lamp is connected between the oil pressure sending unit in the oil pump body and the ignition terminal of the ignition switch. When oil pressure exceeds 8-12 psi, the contacts open and the light goes out.

More detailed information on instruments may be found in the Unit Repair Section.

Cluster R & R

1964-65 Valiant, Dart, and Barracuda

1. Disconnect battery.
2. Cover steering column jacket tube between steering wheel and instrument panel with masking tape to avoid scratching finish.
3. Loosen steering column clamp screws several turns to lower steering column slightly.
4. Reach up under instrument panel and disconnect speedometer cable at speedometer.
5. Remove four screws visible on face of cluster.
6. Tilt cluster downward and to right and disconnect lead wires to ammeter, ignition switch multi-connector, and printed circuit multi-connector.
7. Remove cluster. Reverse procedure to install.

1966 Valiant and Barracuda

1. Disconnect battery. Disconnect speedometer at speedometer head.
2. Remove steering column clamp, and lower the support plate at the firewall.
3. Protect top of steering column jacket against scratches.
   NOTE: on the Barracuda, remove ignition switch bezel with tool C-3824.
4. Remove four cluster mounting screws and pull cluster out far enough to disconnect printed cir-

cuit board and ignition switch multiple connectors.
5. Disconnect wires and remove cluster to work area for service.
6. Install by reversing removal procedure.

1966 Dart

1. Disconnect battery, then tape steering column jacket to protect against scratches.
2. If car is air conditioned, remove left spot cooler duct, hose and fuse block.
3. Disconnect speedometer cable at speedometer head.
4. Remove steering column support bracket and lower the column support plate at firewall.
5. Remove radio control knobs, attaching nuts, ash tray and housing, and the cigar lighter.
6. From under the dash, remove mounting nut next to heater switch.
7. Remove headlight switch knob and bezel by depressing release button on headlight switch and pulling out on headlight switch knob. Use tool C-3824 to remove bezel. Do not remove switch from panel.
8. Loosen set screw in wiper switch knob and remove switch bezel with tool C-3824. Do not remove switch from panel.
9. Remove four instrument cluster retaining screws and pull cluster out far enough to disconnect printed circuit board and ignition switch multiple connectors. Disconnect two ammeter wires, then remove cluster to work area.
10. Install by reversing removal procedure.

1967-70 Valiant and Dart and 1967-69 Barracuda

1. Disconnect battery.
2. From under instrument panel, disconnect speedometer cable from speedometer.
3. Loosen steering column floor plate attaching screws, remove steering column upper clamp, and allow steering column to rest in lowered position.
4. Tape top of steering column to prevent damage to finish.
5. Remove six mounting screws from cluster in Valiants and Barracudas, three in underside of cluster bezel and three in lower edge. Remove six screws from cluster in Darts, four from underside of upper lip and two from lower lip of cluster.
6. Reach up behind and above cluster and bend three harness clips out of the way.
7. Roll upper edge of cluster out far enough to disconnect ammeter wires, emergency flasher,

windshield wiper, and headlight switch connectors.
8. Remove cluster. Reverse procedure to install.

1970 Barracuda

1. Disconnect battery.
2. Tape steering column to prevent damage to finish.
3. Reaching up from under panel, disconnect multiple connector from left printed circuit board.
4. Remove four cluster retaining screws from underside of crash pad and four retaining screws from lower front face of cluster.
5. Remove clock reset cable.
6. Disconnect speedometer cable at speedometer.
7. Pull cluster out far enough to reach behind and disconnect right printed circuit board multiple connector, ammeter wires, vacuum gauge hose or tachometer wire, and emergency flasher switch conductor.
8. Loosen air conditioner or heater knobs and remove from slide control. Remove air conditioner or heater control mounting screws and move control out of way.
9. Depress headlight switch knob release button on bottom side of switch and pull knob and shaft out of switch.
10. Pull windshield wiper knob from shaft. Remove wiper switch bezel nut and allow switch to remain connected to wire harness.
11. Roll cluster out from panel opening face down and to the right. Reverse procedure to install.

1970 Challenger

1. Disconnect battery.
2. Remove six lamp panel mounting screws.
3. Carefully slide lamp panel out and lay on top of instrument panel. It is not necessary to disconnect wire harness.
4. Remove four switch bezel mounting screws and let bezel hang loose. It is not necessary to disconnect switches.
5. Remove steering column "Duffy" plate mounting screws and two column support clamp retaining nuts. Allow column to rest on seat.
6. From under instrument panel disconnect speedometer cable from speedometer.
7. Remove six cluster bezel mountings screws.
8. Angle bezel out in such a manner as to clear clock reset button.
9. Reach between bezel and panel and disconnect stereo control wiring harness, if so equipped. On Rallye cluster, remove clock reset button and odometer reset

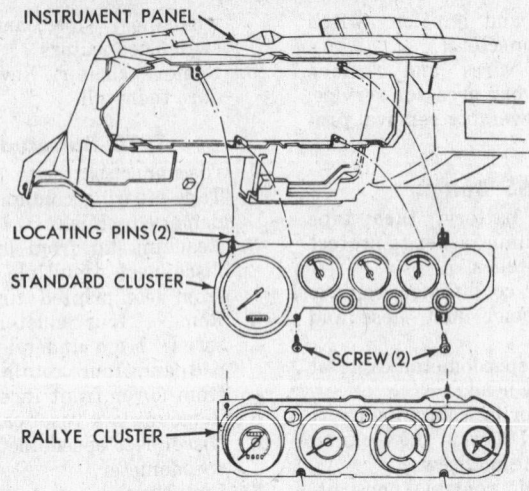

Challenger cluster removal (© Chrysler Corp.)

button from cluster bezel to remove clock.

10. Remove four cluster to panel mounting screws.

11. Disconnect all wiring harnesses and remove cluster. Reverse procedure to install.

Speedometer Removal

1. Remove instrument cluster.
2. Remove clock reset button, if so equipped.
3. Place cluster assembly on padded surface with cluster face down.
4. Remove screws attaching bezel assembly to clusterbase and remove bezel.
5. Remove attaching nuts or screws and remove speedometer.

Brakes

References

Self-energizing hydraulic brakes are used.

The service brakes are operated by a suspended foot pedal. The master cylinder is mounted on the firewall for easy access. Power brakes are optional and the power unit is located between the brake pedal and the master cylinder. The rear wheel brakes also serve as mechanically-operated parking brakes. The parking brake is applied by a foot-operated lever and cables to the rear wheel shoes.

Since 1964, self-adjusting brakes have been used.

Specific information on brake cylinder sizes can be found in the General Chassis and Brake Specifications table in this section.

Information on brake adjustments, band replacement, bleeding procedure, master and wheel cylinder overhaul can be found in the Unit Repair Section.

Since 1965, options have included disc brakes on the front of some models. Detailed information on disc brakes may be found in the Unit Repair Section.

Information on troubleshooting and overhauling power brakes can be found in the Unit Repair Section.

Information on the grease seals which may need replacement can be found in the Unit Repair Section.

Fuel System

Carburetor

Both Carter and Holley carburetors are used. For details, see Unit Repair Section.

The illustration shows the cable-type linkage used on some models.

Linkage Adjustment
1964-66 Valiant, Dart, and Barracuda with Six Cylinders and Automatic Transmission

1. Apply thin film of multi-purpose grease on outside surface of carburetor lever isolator (1), torque shaft plastic bushing (2), torque shaft ball stud (3), and bellcrank pin (10).
2. Disconnect transmission intermediate rod ball socket (5) from bellcrank ball end.
3. Disconnect choke at carburetor (4) or block choke valve in full open position. Open throttle slightly to release fast idle cam, then return throttle to curb idle.
4. Loosen locknut (8) in transmis-

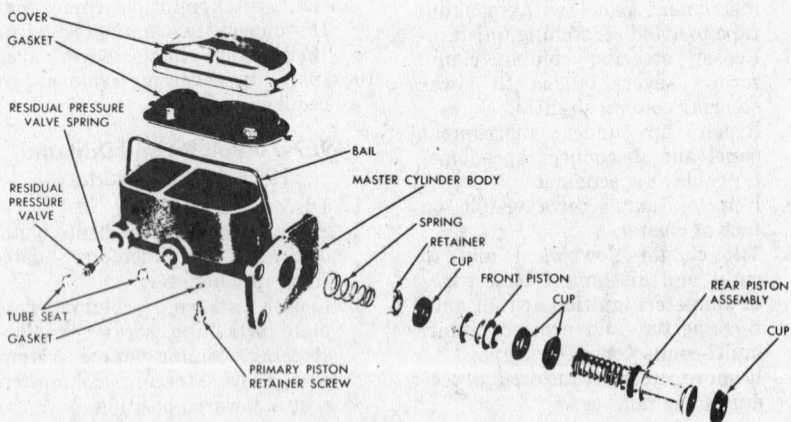

Dual type master cylinder (disc brakes) (© Chrysler Corp.)

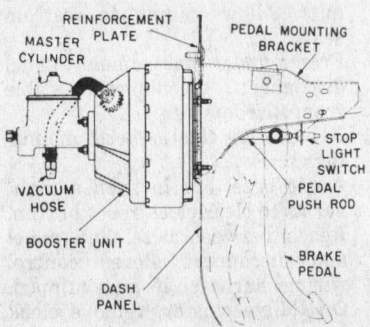

Vacuum-suspended power brake unit
(© Chrysler Corp.)

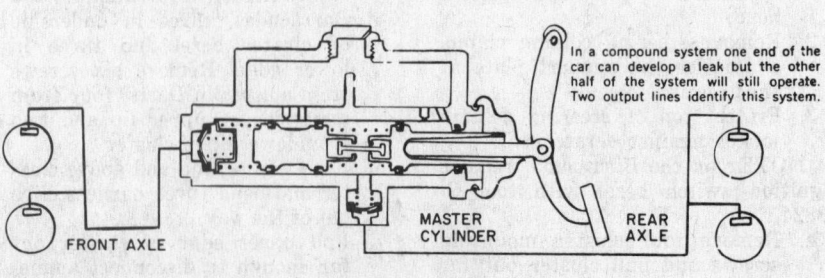

In a compound system one end of the car can develop a leak but the other half of the system will still operate. Two output lines identify this system.

Dual type master cylinder (© Chrysler Corp.)

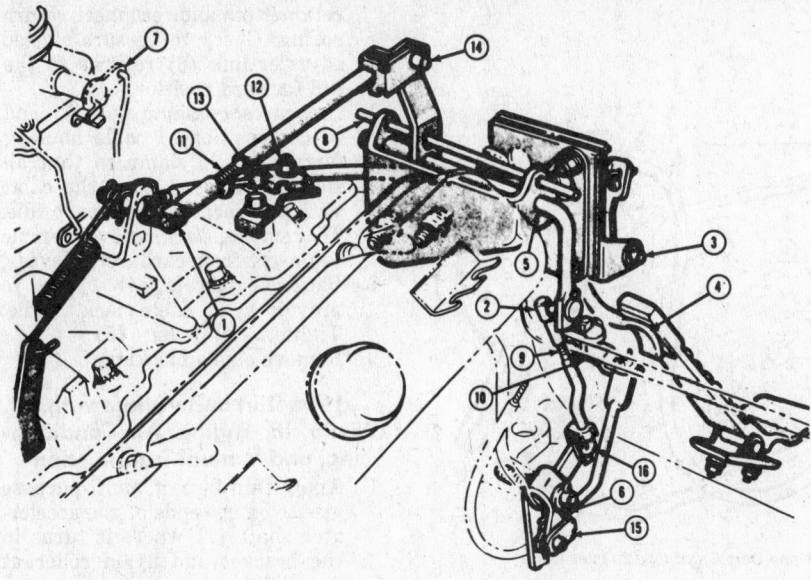

Throttle linkage adjustment—1965-66 Dart and Valiant (© Chrysler Corp.)

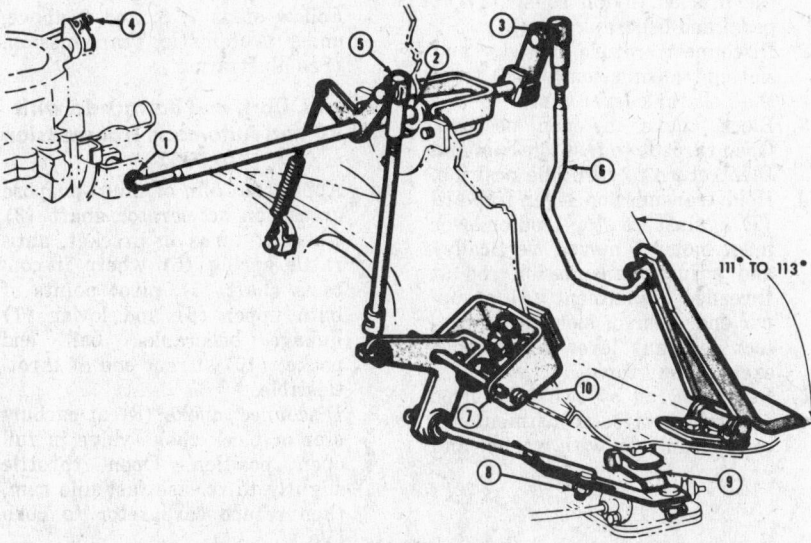

Throttle linkage adjustment—1964-66 Dart, Barracuda and Valiant 170 and 225 cu. in. (© Chrysler Corp.)

sion rod, insert 3/16 in. diameter rod (7) approximately four in. long in holes provided in transmission rod bellcrank bracket and lever assembly.

5. Move transmission lever (9) forward against stop and tighten transmission rod lock nut (8).

6. Disconnect top end of accelerator pedal rod (6). Adjust length of rod to provide pedal angle of 111 to 113 degrees. To increase pedal angle, increase length of rod by means of screw adjustment. Reinstall top end of rod.

7. Remove 3/16 in. diameter rod (7) from transmission rod bracket and lever assembly. Adjust length of transmission bellcrank to torque shaft rod with screw adjustment at top end.

The correct rod length will allow ball socket to line up with ball end, when rod is held upward against transmission stop.

8. Install ball socket on torque shaft level ball end (5).

9. Connect choke rod (4) or remove blocking fixture.

1964-66 Valiant, Dart, and Barracuda with Six Cylinders and Manual Shift

1. Apply thin film of multi-purpose grease on outside of carburetor lever isolator (1), torque shaft plastic bushing (2), and torque shaft ball stud (3).

2. Disconnect choke at carburetor (4), or block choke valve in full open position. Open throttle slightly to release fast idle cam, then return throttle to curb idle.

3. Disconnect top end of accelerator pedal rod (6). Adjust length of rod by means of screw adjustment to obtain pedal angle of 111 to 113 degrees. Reinstall top end of rod (6). Connect choke rod (4) or remove blocking fixture.

V8 Valiant with 273 Cu. In. Engine and Automatic Transmission (See Illustration)— 1965-66

1. Apply thin coat of grease to shaft (3), bottom of pedal (4) and pivots (5-6), and transmission linkage areas (14, 15, 16).

2. Disconnect transmission intermediate rod ball socket (2).

3. Block choke in open position and be sure throttle is at full curb idle.

4. Insert 3/6 x 6 in. rod in gauge holes (8) at engine-mounted bell-crank. Adjust the rod (9) at upper end. The socket must line up with ball with rod held against stop (10).

5. Assemble the socket (2) and remove the rod (8).

6. Hold throttle rod (11) forward against transmission stop (10) and adjust at threads (11) to gain proper line up with ball (1).

7. Lengthen carburetor rod four turns by turning ball socket (1) counterclockwise.

8. Assemble ball socket and reconnect return spring.

9. Loosen clamp (12) and adjust position of housing ferrule (13) so that all slack is removed from the cable with carburetor at curb idle.

10. Back off ferrule (13) 1/4 in. to provide 1/4 in. cable slack at idle. Tighten clamp nut (12).

11. Remove block from choke.

V8 Valiant with 273 Cu. In. Engine and Manual Transmission—1965-66

Follow procedures from above automatic transmission, Steps 1, 3, 9, 10 and 11.

Dart and Valiant with 170 or 225 Cu. In. Engine and Automatic Transmission (See Illustration)—1967

1. Apply thin coat of grease to points (1, 2, 3 and 10).

2. Disconnect intermediate rod at ball socket (5).

3. Block choke in open position and be sure that throttle is against curb idle stop.

4. Loosen locknut (8) and insert a 3/16 in. rod into gauge holes (1) provided in bellcrank bracket.

5. Move transmission lever (3) forward against stop and tighten clamp nut (8).

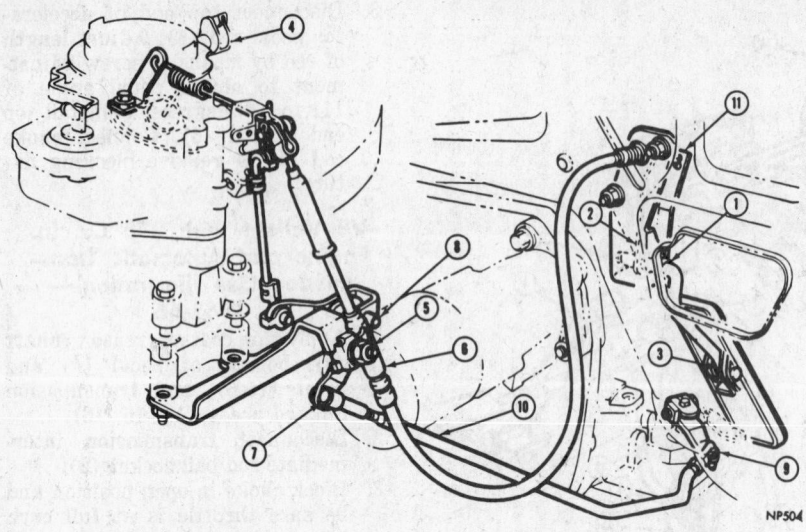

Throttle linkage adjustment—1967 Valiant and Dart 6 cyl. (© Chrysler Corp.)

6. Disconnect top of rod (6) and adjust to provide a pedal angle of 111—113°. Reconnect rod (6).
7. Remove the 3/16 in. rod (7) and adjust at ball joint (5) to create a neat connection.
8. Install the ball socket (5).
9. Remove choke block.

Dart and Valiant with 170 or 225 Cu. In. Engine and Automatic Transmission—1967

Follow procedures from above automatic transmission, Steps 1, 3, 6 and 9.

1966 Dart and Valiant with 225 Cu. In. Engine, Air Conditioning, and Automatic Transmission

1. Apply thin film of multi-purpose grease on the ends of the accelerator shaft (1) where it turns in the bracket, nylon roller (8) at pedal and bellcrank pin (2).
2. Disconnect return spring and slotted transmission rod (6) from the bellcrank lever pin.
3. Block choke in open position. Open throttle to fast idle cam and then return to curb idle position.
4. Hold transmission lever forward (7) against its stop (rod or lever must not be moved vertically) and adjust transmission rod at threaded adjustment (6) at upper end. Rear of slot should contact bellcrank lever pin without exerting any force.
5. Lengthen rod by two full turns.
6. Assemble slotted adjustment (6) to bellcrank pin with washer and

retainer pin and reconnect return spring. Check to be sure slotted adjuster link (6) returns to the full forward position.
7. Loosen cable clamp nut (5) and adjust position of cable housing ferrule (4) in clamp so that all slack is removed from the cable with the carburetor at curb idle. To remove slack, move ferrule (4) away from carburetor lever.
8. Back off ferrule (4) ¼ in. to provide ¼ in. cable slack at idle. Tighten clamp nut (5).
9. Remove choke blocking.

1966 Dart and Valiant with 225 Cu. In. Engine, Air Conditioning, and Manual Transmission

1. Apply thin film of multi-purpose grease on the ends of the accelerator shaft (1) where it turns in the bracket and nylon roller at pedal (8).
2. Block choke in open position and operate throttle to be sure throttle returns to curb idle.
3. Follow steps 7, 8, and 9 above, under Automatic Transmission. (See illustration.)

Valiant, Dart, and Barracuda with 273 V8 and Automatic Transmission —1967

1. Apply thin film of multi-purpose grease on accelerator shaft (3) where it turns in bracket, anti-rattle spring (5) where it contacts shaft (4), pivot points of both upper (6) and lower (7) linkage bellcranks, ball end pocket (15) at rear end of throttle cable.
2. Disconnect choke (8) at carburetor or block choke valve in full open position. Open throttle slightly to release fast idle cam, then return carburetor to curb idle.
3. Hold or wire transmission lever (11) firmly forward against its

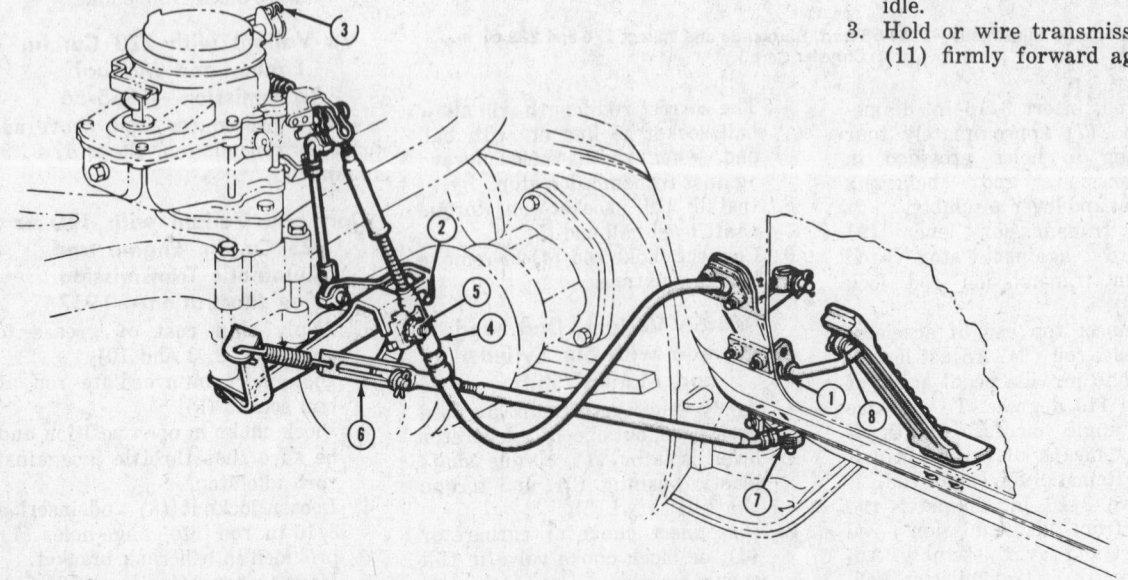

Linkage adjustment—1966 225 six cylinder with air conditioning (© Chrysler Corp.)

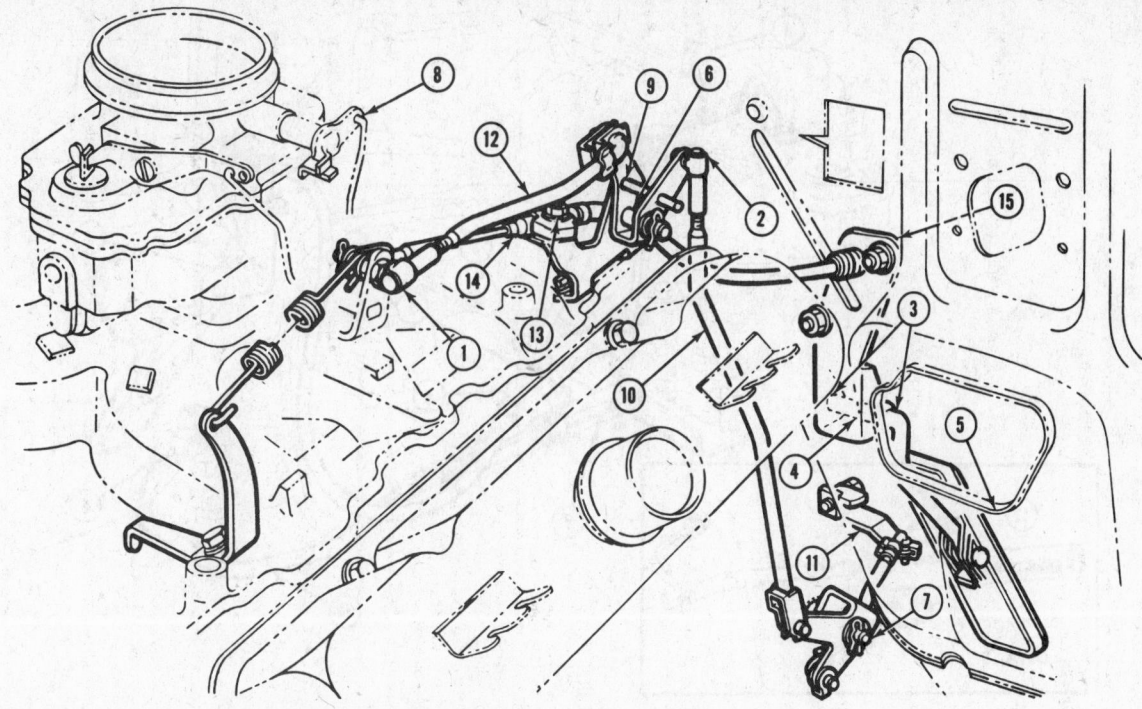

Linkage adjustment—1967 273 V8 (© Chrysler Corp.)

stop while performing adjustments in next four steps. It is important that lever remains firmly against stop during these steps to ensure correct adjustment.

4. With 3/16 in. diameter rod (9) placed in holes provided in upper bellcrank (6) and lever, adjust length of intermediate transmission rod (10) by means of threaded adjustment at upper end with slight downward effort on rod.

5. Assemble ball socket (2) to ball end and remove 3/16 in. rod (9) from upper bellcrank and lever.

6. Adjust length of carburetor rod (12) by pushing rearward on rod with slight effort and turning threaded adjustment (1) so that ball socket lines up with ball end on carburetor lever.

7. Assemble ball socket (1) to ball end, then remove wire securing transmission lever.

8. Loosen clamp nut (13), adjust position of cable housing ferrule (14) in clamp so that all slack is removed from cable with carburetor at curb idle. To remove slack from cable, move ferrule (14) in clamp in direction away from carburetor lever.

9. Back off ferrule (14) ¼ in. This provides ¼ in. free play between front edge of accelerator shaft lever and dash bracket. Tighten cable to clamp nut to 45 in. lbs.

10. Connect choke rod (8) or remove blocking fixture.

Valiant, Dart, and Barracuda with 273 V8 and Manual Transmission —1967

Follow steps 1, 2, 8, 9, and 10 in automatic transmission procedure.

Valiant, Dart, Barracuda, and Challenger with Six Cylinders and Automatic Transmission—1968-70

1. Appy thin film of multi-purpose grease on accelerator shaft (1) where it turns in bracket, anti-rattle spring (8), bellcrank pivot pin (9), and ball end and pocket (11) at rear end of throttle cable.

2. Disconnect choke (4) at carburetor or block choke valve in full open position. Open throttle to release fast idle cam, then return.

3. Hold transmission rod (10) forward with slight pressure so transmission lever (9) is against stop. Hold slotted adjusting link (7) forward against pin. Tighten adjusting link lock screw (5) to 95 in. lbs.

4. Assemble slotted adjuster link (7) to carburetor lever pin and install washer and retaining pin. Assemble transmission linkage return spring in place.

5. Check transmission linkage freedom of operation by moving slotted adjuster link (7) to full rearward position, then allow it to return slowly, making sure it returns to full forward position.

6. Loosen cable clamp nut (5), ad-

just position of cable housing ferrule (6) in clamp so that all slack is removed from cable with carburetor at curb idle. To remove slack from cable, move ferrule (6) in clamp in direction away from carburetor lever.

7. Back off ferrule (6) ¼ in. This provides ¼ in. free play between dash mounted accelerator lever and bracket. Tighten cable clamp nut (5) to 45 in. lbs.

8. Connect choke rod (4) or remove blocking fixture.

Valiant, Dart, and Barracuda with Six Cylinder and Manual Transmission—1968

Follow steps 1, 2, 6, 7, and 8 in automatic transmission procedure.

1. Apply thin film of multi-purpose grease on accelerator shaft (3) where it turns in bracket, pivot points of both upper (6) and lower (7) linkage bellcranks, and ball end and pocket (14) at rear end of throttle cable.

2. Disconnect choke (8) at carburetor or block choke valve in full open position. Open throttle slightly to release fast idle cam, then return carburetor to curb idle.

3. With 3/16 in. diameter rod (9) placed in holes provided in upper bellcrank and lever, adjust length of intermediate transmission rod (10) by means of threaded adjustment at upper end. Ball socket must line up with ball end when rod is held

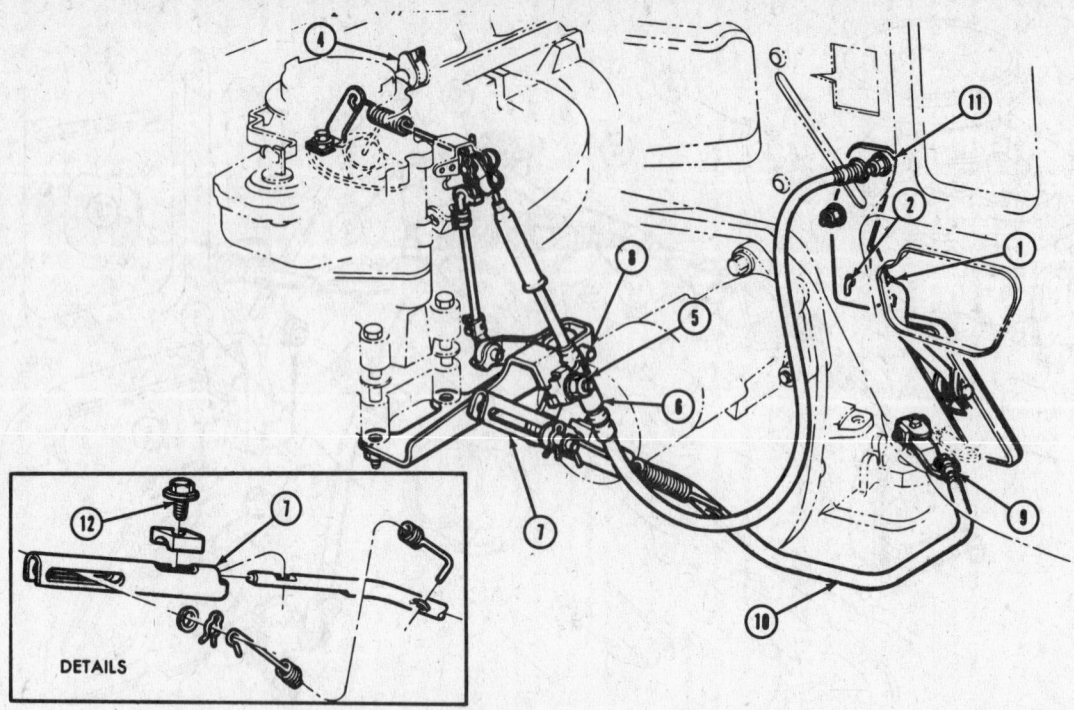

Linkage adjustment—1968-70 six cylinders (© Chrysler Corp.)

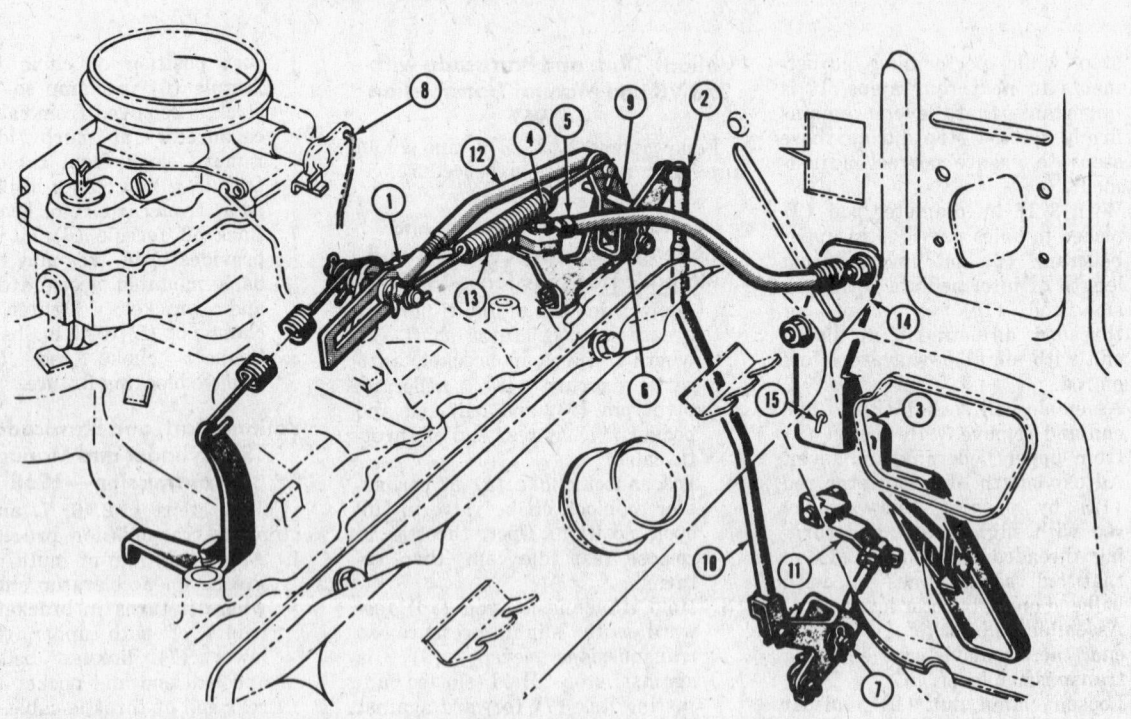

Linkage adjustment—1968-70 273, 318, and 340 V8s (© Chrysler Corp.)

upward and transmission lever is forward against its stop.

4. Assemble ball socket (2) to ball end and remove 3/16 in. rod (9) from upper bellcrank and lever.

5. Disconnect return spring (13), then adjust length of carburetor rod (12) by pulling forward on rod so transmission lever is against stop and turning threaded adjuster link. Turn link so rear end of slot just contacts carburetor lever pin when link is placed in its normal operating position against lever pin nut. Lengthen rod by two full turns of link.

6. Assemble slotted adjustment (1) to carburetor lever pin and install washer and retainer pin.

Assemble transmission linkage return spring (13) in place.

7. Check transmission linkage freedom of operation, by moving slotted adjuster link (1) to full rearward position, then allow it to return slowly, making sure it returns to full forward position.

8. Loosen cable clamp nut (4). Adjust position of cable housing

ferrule (5) in clamp so that all slack is removed from cable with carburetor at idle. To remove slack from cable, move ferrule (5) in clamp in direction away from carburetor lever.

9. Back off ferrule (5) ¼ in. This provides ¼ in. free play between front edge of accelerator shaft lever and dash bracket. Tighten cable clamp nut (4) to 45 in. lbs.

10. Connect choke (8) or remove blocking fixture.

Valiant, Dart, and Barracuda with 273 and 340 V8s and Manual Transmission—1968

Follow steps 1, 2, 8, 9, and 10 in automatic transmission procedure.

Valiant, Dart, Barracuda and Challenger with All Engines and Manual Transmission Except 426 Hemi—1969-70

1. Apply thin film of multi-purpose grease on accelerator shaft where it turns in the bracket, ball end and pocket at rear end of throttle cable.

2. Disconnect choke at carburetor or block choke valve in full open position. Open throttle slightly to release fast idle cam, then return carburetor to curb idle.

3. Loosen cable clamp nut (1) adjust position of cable housing ferrule (2) in the clamp so that all slack is removed from the cable with carburetor at idle. To remove slack from cable, move ferrule (2) in clamp away from carburetor lever.

4. Back off ferrule (2) ¼ in. To provide ¼ in. cable slack at idle. Tighten cable clamp nut to 45 in. lbs.

5. Connect choke rod or remove blocking fixture.

426 Hemi

See Dodge-Plymouth section for Hemi linkages.

Fuel Pump

To test this single action pump, the following steps should be taken:

Pressure Test

1. Insert a T-fitting between the fuel pump and the carburetor, as illustrated.

2. Connect a six in. hose between the T-fitting and the pressure gauge.

3. Connect a tachometer to the engine. Start the engine and vent the pump for a few seconds. (If this is not done, the pump will not operate at full capacity and a false reading will result.)

4. Run the engine at 500 rpm. The reading should be 3½—5 psi and remain constant or return to zero very slowly when the engine is stopped. An instant drop to zero indicates a leaky outlet valve. If the pressure is too low, a weak diaphragm main-spring or improper assembly of the diaphragm may be the cause. If the pressure is too high, the main-spring is too strong.

Vacuum Test

The vacuum test should be made with the fuel line disconnected from the carburetor.

1. Disconnect the inlet line to the fuel pump.

2. Connect a vacuum gauge to the pump inlet fitting.

3. Crank the engine with the starter. There should be a noticeable vacuum present, not alternated with blowback.

4. If blowback is present, the inlet valve, or pump valve body, is at fault.

Volume Test

The pump should supply one quart of fuel in one minute or less at 500 rpm.

If the fuel pump fails any of the above tests, the pump should be rebuilt or renewed.

Fuel Tank Sending Unit

The fuel tank on both the conventional and Suburban models is located at the rear of the body, under the floor.

The fuel gauge (tank unit) and standpipe are one unit. The filter on the end of the standpipe is replaceable. It is quite helpful in preventing water and other foreign matter from entering the fuel pump and carburetor.

When installing a fuel gauge (tank unit), be sure to push the filter down on the pipe until it is seated.

Exhaust System

Manifold

6 Cylinder Removal

1. Remove air cleaner.

2. Remove vacuum control tube at carburetor and distributor.

3. Remove fuel line and carburetor.

4. Remove three bolts that hold the flange.

5. Remove nuts and washers holding the intake and exhaust manifolds to the cylinder head.

6. Remove the assembly from the head.

7. Remove three bolts holding the intake and exhaust manifolds together.

8. Clean manifold mating and at-taching surfaces with a straight edge and feeler gauge. All mating surfaces should be flat and plane within .008 in.

6 Cylinder Installation

1. Place a new gasket between intake and exhaust manifolds and install three attaching bolts, loosely.

2. With a new gasket in place, position the complete manifold combination on the cylinder head.

3. Install conical washers (cupped side away from the nut) and nuts. Torque alternately to a final 10 ft. lbs.

4. Now, torque the three intake-to-exhaust manifold nuts to 15 ft. lbs.

5. Connect the exhaust pipe to the manifold flange and torque these two bolts to 30 ft. lbs.

6. Install carburetor and connect line, vacuum line and throttle linkage.

7. Install air cleaner. Start engine and check for intake and exhaust leaks.

V8 Removal and Installation

1. Remove exhaust pipe-to-manifold bolts and, on the left side, remove the attaching brace bolt.

2. Remove manifold - to - cylinder head bolts and remove manifold.

3. Reinstall by reversing the above. Be sure mating surfaces are clean, in complete alignment, and that heat control valve is free.

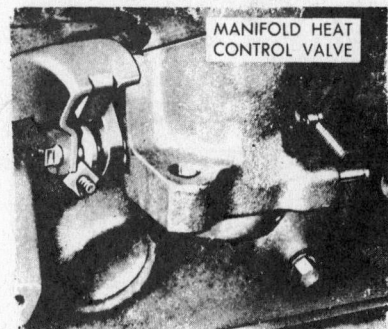

Manifold heat control valve—318 cu. in.
(© Chrysler Corp.)

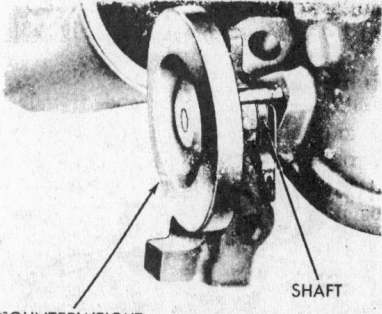

Manifold heat control valve—340 cu. in.
(© Chrysler Corp.)

Muffler and Tail Pipe

The exhaust pipe and muffler, as produced at assembly, is a one-piece unit. It is handled in the customary manner. Replacement of muffler can be done only by cutting off the exhaust pipe just ahead of the muffler. Connect the new muffler to the exhaust pipe and clamp in place with the present saddle bracket and U-bolt clamp.

Cooling System

The 180° F. thermostat is located in the aluminum housing above the water pump. A 160° F. thermostat is available for use with an alcohol-type solution. The radiator pressure cap is calibrated to maintain 14 psi.

Cooling system capacities can be found in the capacities table of this section.

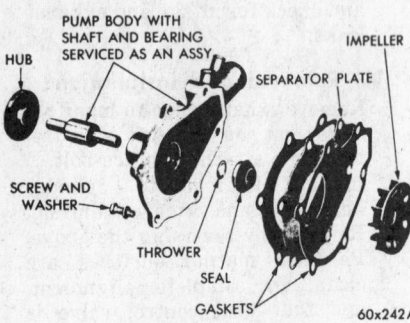

Water pump—6 cyl.
(© Chrysler Corp.)

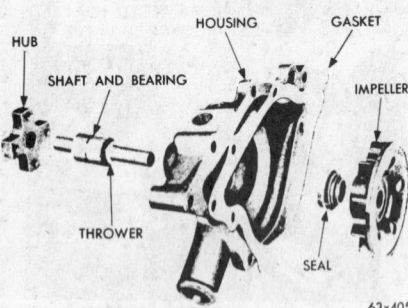

Water pump—273 and 318
(© Chrysler Corp.)

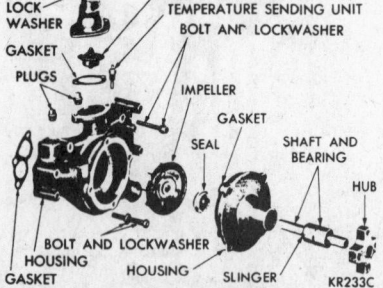

Water pump—383
(© Chrysler Corp.)

Water Pump Removal

1. Drain the cooling system.
2. Loosen the drive belt adjusting strap at the alternator and move alternator toward the engine.
3. Remove the fan, spacer, pulley and belt.
   NOTE: if a fluid drive is used in place of spacer, do not set drive unit with shaft pointing downwards or silicone fluid will drain into fan drive bearing and ruin grease.
4. Remove the pump inlet hose and the heater hose.
5. Remove clamp from the by-pass hose, where used.
6. Remove water pump body-to-housing bolts and push the pump body down, off the by-pass hose where used.

Water Pump Installation

1. Place by-pass hose clamp on the hose, then lift the pump up into the hose.
2. Attach pump body to the housing, using the two long bolts above and below the inlet hose. Tighten bolts to 30 ft. lbs.
3. Attach the inlet and heater hose.
4. Install drive belt, pulley, spacer and fan.
5. Adjust drive belt.
6. Close radiator drain cock, and fill and bleed cooling system.

Engine

References

Most Darts and Valiants are equipped with the 170 or 225 cu. in. slant six cylinder engines. These engines are very conventional in design, except that the cylinder blocks are canted 30 degrees to the right. Many parts interchange between the two and the main difference between them is that the 225 engine has a one in. longer stroke and correspondingly higher block. Most recently a 198 cu. in. six has replaced the 170 cu. in. six. It is essentially a 170 six with a stroke half way between a 170 engine and a 225 engine.

In 1964 Dart and Valiant introduced a 273 cu. in. V8 which was a redesigned and substantially lightened version of the Dodge and Plymouth 318 V8. The engine is conservative in design and very rugged. It uses a rocker shaft to carry the rockers instead of the currently more popular but more fragile stud mounted rockers. It has had throughout its production, 1964-69, the smallest bore of any contemporary American V8, and until 1968 continued to use mechanical tappets.

In 1968 Dart and Valiant added a 318 and a 340 V8 to their line up. The 318 had been redesigned to the point where it was merely a 273 with a 5/16 in. larger bore. The 340 V8 had still a larger bore in the basic 273 block, but also added heads with big valves and well formed parts which make it possibly the hottest medium V8 available.

The 383 and 440 wedge V8s and the 426 Hemi V8 are basically big car power plants and are covered in detail in the Dodge and Plymouth section.

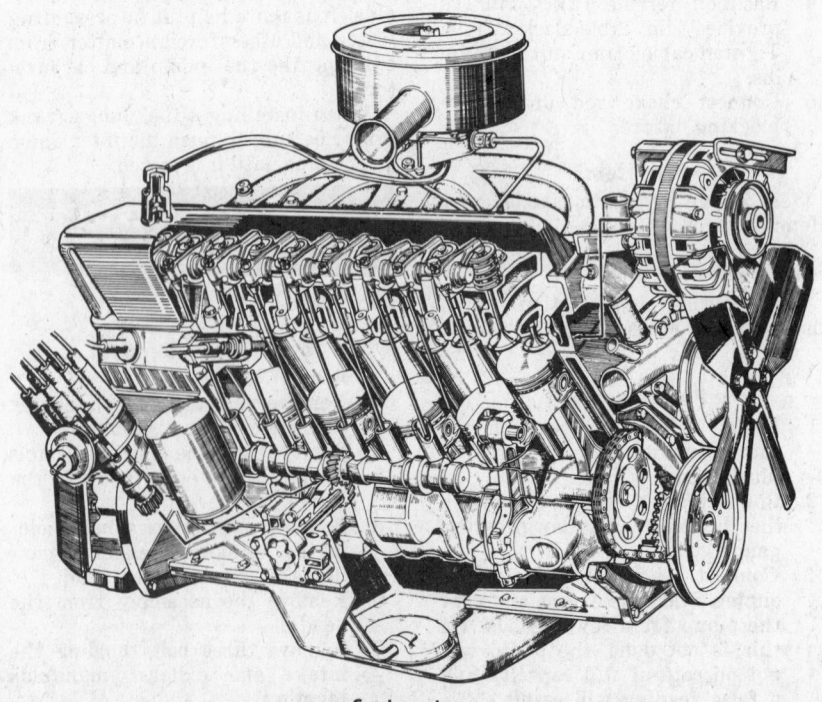

6 cyl. engine
(© Chrysler Corp.)

6 Cylinder Engine Removal And Replacement

1. Scribe the hood hinge outlines on the underside of the hood, then remove the hood.
2. Drain the cooling system, remove the battery and carburetor air cleaner.
3. Remove radiator and heater hoses, then the radiator.
4. Remove the outlet vent pipe from the cylinder head cover.
5. Disconnect fuel lines, linkage and wiring to the engine.
6. Disconnect exhaust pipe at exhaust manifold.
7. Raise car on hoist.
8. If equipped with automatic transmission, drain the converter and the transmission. Remove the oil cooler lines, filler tube and push button cable.
9. Remove the clutch torque shaft, brake cables and rods.
10. Remove the speedometer cable and gear shift rods.
11. Disconnect propeller shaft and tie out of the way.
12. Install an engine support fixture to the rear of the engine.
13. Remove the engine rear support crossmember.
14. Remove transmission mounting bolts from clutch housing.
15. Remove the transmission.
16. Lower the car.
17. Position engine lifting fixture onto the engine, and attach chain hoist to the fixture eyebolt.
18. Remove the engine support fixture.
19. Remove the engine front mounting bolts.
20. Lift the engine out of the engine compartment and lower it onto a substantial work stand.
21. To install the engine, reverse the above procedure and torque the front engine mounting bolts to 85 ft. lbs.; the rear ones to 35 ft. lbs.

V8 Engine Removal and Replacement

1. Scribe hinge outlines on hood, then remove hood.
2. Remove battery and drain cooling system.
3. Remove all hoses, fan shroud and oil cooler lines (where used.) Remove radiator.
4. Disconnect fuel lines and all wires attached to engine.
5. Remove air cleaner and carburetor.
6. Attach lifting fixture, tool C-3466, or equivalent, to carburetor flange studs.
7. Attach rear engine support, tool C-3487, and remove rear engine support.
8. Disconnect exhaust pipes at manifolds.
9. Disconnect propeller shaft, wires, cables, linkage and cooler lines and remove transmission.
10. Remove engine front mount nuts. Attach crane or other suitable lifting tool to manifold fixture and lift out engine.
11. To replace, reverse above procedure.

Cylinder Head

The cylinder head is chrome alloy cast iron and is attached to the cast iron cylinder block by 14, 7/16 x 14 bolts.

6 Cylinder Removal

1. Drain the cooling system.
2. Remove carburetor air cleaner and fuel lines.
3. Disconnect accelerator linkage.
4. Remove the vacuum line from carburetor to distributor.
5. Carefully disconnect spark plug wires by pulling straight, in line with plug.
6. Disconnect heater hose and clamp holding the by-pass hose.
7. Disconnect the heat indicator-sending-unit wire.
8. Disconnect exhaust pipe at the exhaust manifold flange.
9. Remove the intake and exhaust manifold and carburetor as an assembly.
10. Remove the outlet vent tube and cylinder head cover.
11. Remove the rocker arms and shaft.
12. Remove the pushrods and place them in order.
13. Remove the head bolts and lift off the cylinder head.
14. Place cylinder head on bench and remove the spark plugs and tubes.

6 Cylinder Installation

1. Clean carbon from the combustion area. Clean all gasket surfaces of both head and cylinder block. Install spark plugs (the aluminum plug shields act as satisfactory gasket material between spark plug body and cylinder head.)
2. If there is any cause to suspect leakage, check all surfaces with a straightedge. If out of flatness exceeds 0.00075 times the span length in any direction, replace head or machine head gasket surface. For example, on a 12 in. span the maximum allowable out of flat is 12 x 0.00075 or 0.009 in.
3. Apply a reliable sealer to the new gasket and install the gasket and cylinder head.
4. Install the 14 cylinder head bolts. Starting at the top center, tighten all cylinder head bolts to 65 ft. lbs.
5. Inspect all push rods for bends or wear. Replace if necessary.
6. Insert the pushrods, small ends down into the tappets.
7. Install rocker arms and shaft assembly with flat on the end of the shaft on top and pointing toward the front of the engine. This is necessary to provide lubrication to the rocker assemblies. Torque the attaching bolts to 30 ft. lbs.
8. Loosen the three bolts that connect the intake and exhaust manifolds. (This is necessary to obtain proper alignment.)
9. Position intake and exhaust manifold and carburetor assembly onto the cylinder head. Put the cup side of the conical washers against the manifolds, install the attaching nuts and torque to 10 ft. lbs.
10. Retighten the three intake-to-exhaust manifold bolts to 15 ft. lbs.
11. Connect the heater hose and by-pass hose clamp.
12. Connect the heat indicator sending-unit wire, the accelerator linkage and the spark plug wires.
13. Install carburetor-to-distributor vacuum line.
14. Connect exhaust pipe to the exhaust manifold.
15. Install the fuel line and carburetor air cleaner.
16. Refill the cooling system.
17. Start the engine and let run until operating temperatures have been reached.
18. Adjust valve tappet clearance to .010 in. (intake) and .020 in. (ex-

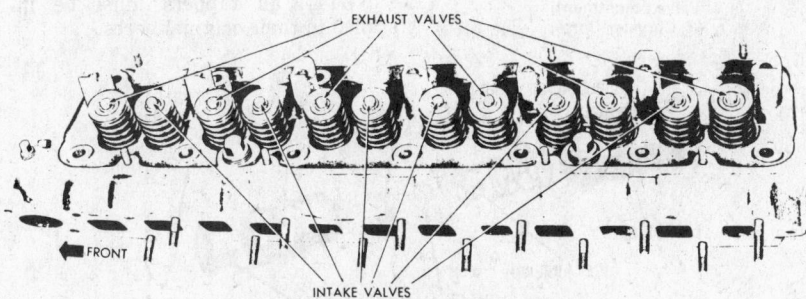

EXHAUST VALVES

FRONT

INTAKE VALVES

Cylinder head, showing valve sequence—
170 and 225 cu. in.
(© Chrysler Corp.)

haust.) The adjusting screw in the push rod end of the rocker arm should have a minimum of 3 ft. lbs. (36 in. lbs.) tension as it is turned. If less, replace the adjusting screw and the rocker arm.

19. Place the new cylinder head cover gasket in position and install cylinder head cover. Torque attaching nuts to 40 in. lbs. (3 1/3 ft. lbs.).
20. Install outlet vent tube.

273, 318, and 340 V8 Head Removal

1. Drain cooling system and disconnect battery.
2. Remove alternator, air cleaner and fuel line.
3. Disconnect accelerator linkage.
4. Remove vacuum hose between carburetor and distributor.
5. Remove distributor cap and wires. If removing heads in vehicle, remove plugs to prevent breaking them.
6. Disconnect coil wires, temperature sending wire, heater hoses, and bypass hose.
7. Remove closed ventilation system (PCV), evaporative control system if so equipped, and valve covers.
8. Remove intake manifold, ignition coil, and carburetor as an assembly.
9. Remove exhaust manifolds.
10. Remove rocker arm and shaft assemblies. Remove pushrods and identify to ensure installation in original location.
11. Remove 10 head bolts from each cylinder head and lift off heads.
12. Clean all surfaces.
13. Inspect all surfaces with straight edge if there is any reason to suspect leakage. If out of flatness exceeds 0.00075 times span length in any direction, replace head or machine mating surface. For example, if span length is 12 in., maximum out of flatness is 12 x 0.00075 or 0.009 in.
14. Reverse procedure to install.

383 and 440 Wedge V8s and 426 Hemi V8

See Dodge-Plymouth Section.

Disassembly of Cylinder Heads

1. Remove cylinder heads.
2. Compress valve springs using valve spring compressor.
3. Remove valve locks or keys.
4. Release valve springs.
5. Remove valve springs, retainers, oil seals, and valves.

NOTE: if a valve does not slide out of the guide easily, check end of

stem for mushrooming or heading over. If head is mushroomed, file off excess, remove and discard valve. If valve is not mushroomed, lubricate stem of valve, remove, and check for stem wear or damage.

Valve Train

Dodge and Plymouth don't use separate valve guides. They do sell 0.005, 0.015, and 0.030 in. oversize valves. To use these, ream the worn guides to the smallest oversize that will clean up wear. Always start with

Valve adjustment
(© Chrysler Corp.)

the smallest reamer and proceed by steps to the larger reamers as this maintains the concentricity of the guide with the valve seat.

As an alternative procedure, some local automotive machine shops bore out the stock guides and replace them with bronze or cast iron guides which use stock valve stem sizes.

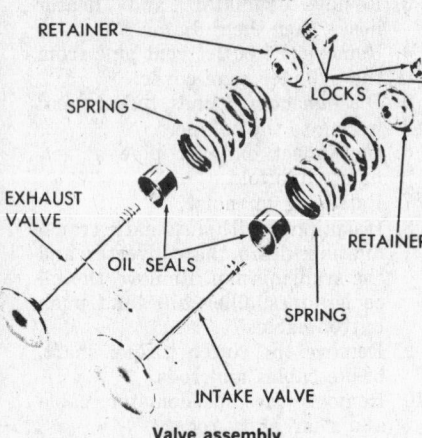

Valve assembly
(© Chrysler Corp.)

Valves

Inspect valve stems for wear. Valve stems worn more than 0.002 in. must be replaced. Inspect valves for burning or warping. Replace damaged valves. Install valves in chuck of a valve grinder or a drill, rotate valve and watch for head oscillation that indicates a bent valve. Replace all bent valves.

Hydraulic Tappets

Removal Except 426 Hemi

NOTE; only 1967-70 V8 engines have hydraulic lifters.

Tappets may be removed without removing manifolds or cylinder heads as follows:

1. Remove rocker covers and rocker shaft assembly.
2. Remove pushrods and identify to ensure installation in original bore.
3. Slide magnetic or claw tool through opening in cylinder head and seat tool firmly in head of tappet.
4. Pull tappet out of bore with twisting motion.

NOTE: all tappets must be installed in their original bores.

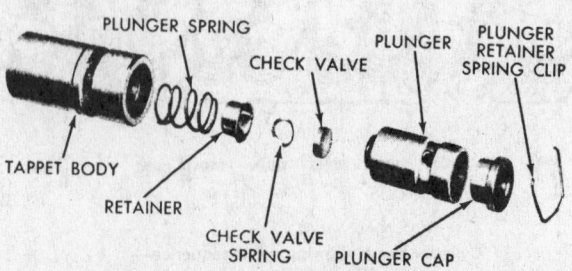

Disassembly of hydraulic tappet (© Chrysler Corp.)

Disassembly Except 426 Hemi

The 426 Hemi has used mechanical tappets until the 1970 model. The hydraulic tappets in the 1970 426 Hemi are special high performance, high speed tappets and should not be disassembled.

1. Pry out plunger retainer spring clip.
2. Clean out varnish deposits from inside tappet body above plunger.
3. Invert tappet body and remove plunger cap, plunger, flat check valve, check valve spring, check valve retainer, and plunger spring.
4. Clean tappet parts in solvent that will remove all varnish and carbon.

NOTE: tappet parts are not interchangeable between tappets. Do not mix parts.

5. If tappet plunger shows scoring or wear, or check valve is pitted, or plunger valve seats show any condition which would prevent valve from seating properly, replace tappet.

Timing Case

Removal—Six Cylinder

1. Drain cooling system.
2. Remove radiator and fan.
3. With puller, remove vibration damper.
4. Loosen engine oil pan bolts to allow clearance, and remove timing case cover and gasket.
5. Slide crankshaft oil slinger off the front of crankshaft.
6. Remove the camshaft sprocket bolt.
7. Remove the timing chain with camshaft sprocket.

Installation—Six Cylinder

1. Turn crankshaft to line up the timing mark on the crankshaft sprocket with the centerline of camshaft (without the chain.)
2. Remove the camshaft sprocket and reinstall with chain.

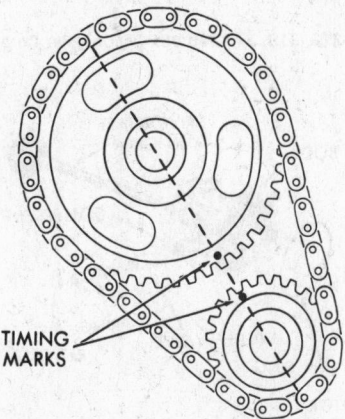

Alignment of timing marks—6 cyl.

3. Torque camshaft sprocket to 35 ft. lbs.
4. Replace oil slinger.
5. Reinstall timing case cover with new gasket and torque to 15 ft. lbs. Retighten engine oil pan to 17 ft. lbs.
6. Replace vibration damper.
7. Replace radiator and hoses.
8. Refill and bleed cooling system.

Removal—273, 318, 340 V8

1. Drain cooling system and remove radiator, fan belt and water pump assembly.
2. Remove pulley from vibration damper and bolt and washer securing vibration damper on crankshaft.
3. Pull vibration damper from end of crankshaft using suitable puller.
4. Remove fuel lines and fuel pump.

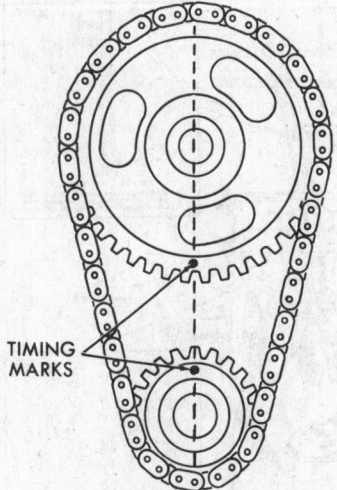

TIMING MARKS

Alignment of timing marks—V8

5. Loosen oil pan bolts and remove front bolt at each side.
6. Remove timing cover and gasket taking care not to damage oil pan gasket.
7. Slide crankshaft oil slinger from end of crank.
8. Remove camshaft sprocket attaching cupwasher, and fuel pump eccentric and remove timing chain with sprockets.
9. Reverse procedure to install. Tighten timing cover to 30 ft. lbs. and pan bolts to 15 ft. lbs.

Removal—383 and 440 Wedge and 426 Hemi

See Dodge-Plymouth section.

Engine Lubrication

Engine Oil Pan Removal

6 Cylinder 170 Cu. In. Models

1. Raise the vehicle on hoist.
2. Drain oil pan.

3. Remove ball joints from steering linkage center link.
4. Remove dust shield.
5. Remove motor mount stud nuts.
6. Lower vehicle. Remove horns and horn brackets. Disconnect battery.
7. Attach an engine life plate and raise engine 1½—2 in.
8. Again, raise vehicle on hoist. Remove pan bolts and pan.

Engine Oil Pan Removal

6 Cylinder 225 Cu. In. Models

1. Raise vehicle on hoist.
2. Drain oil pan.
3. Remove ball joints from steering linkage center link.
4. Remove dust shield.
5. Remove oil pan bolts, rotate crankshaft to clear counterweights and remove pan.

Engine Oil Pan Removal

V8 Models

1. Disconnect battery. Remove dipstick.
2. Raise vehicle on hoist. Drain oil. Remove engine to converter left housing brace.
3. Remove ball joints from center steering link.
4. Remove crossover pipe from manifolds. On Dart, it can hang. On Valiant, remove completely.
5. Remove oil pan bolts and pan.

Engine Oil Pan Installation

Replace oil pan in reverse of removal instructions and torque attaching bolts to 16¾ ft. lbs. Torque engine front mounting bolts to 85 ft. lbs. on 170 cu. in. models.

Oil Pump

Six Cylinder Removal

1. Drain radiator, disconnect upper and lower hoses, and remove fan shroud.
2. Raise vehicle on hoist, support front of engine with jack stand placed under right front corner of oil pan, and remove engine mount bolts. Do not support engine at crankshaft pulley or vibration damper.
3. Raise engine approximately 1½ to 2 in.
4. Remove oil filter, oil pump attaching bolts, and pump assembly.

273, 318, and 340 Engines

1. Remove oil pan.
2. Remove oil pump from rear main bearing cap.

Disassembly

Six Cylinder

1. Remove pump cover and seal ring.

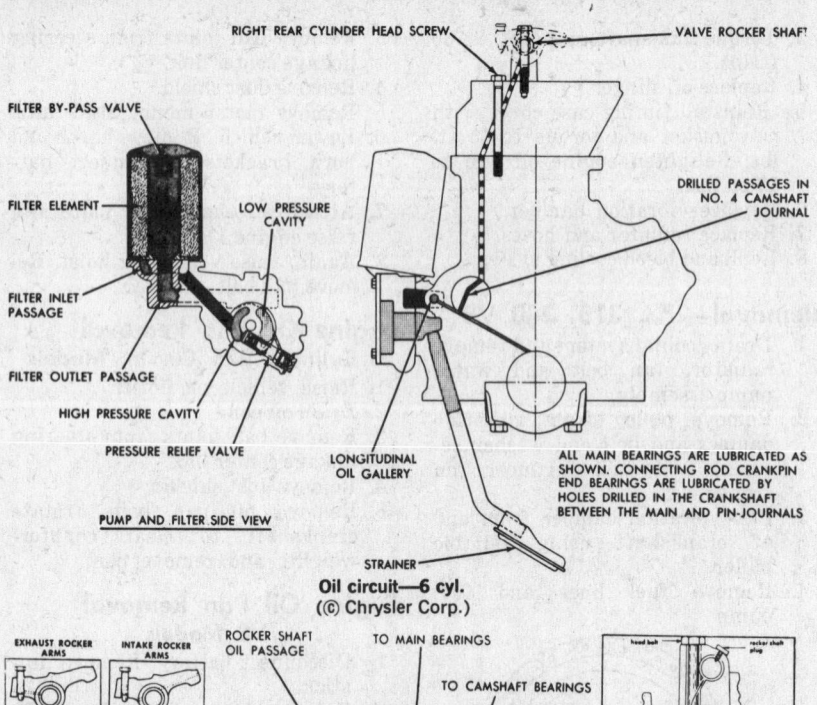

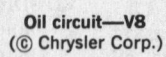

Oil circuit—6 cyl.
(© Chrysler Corp.)

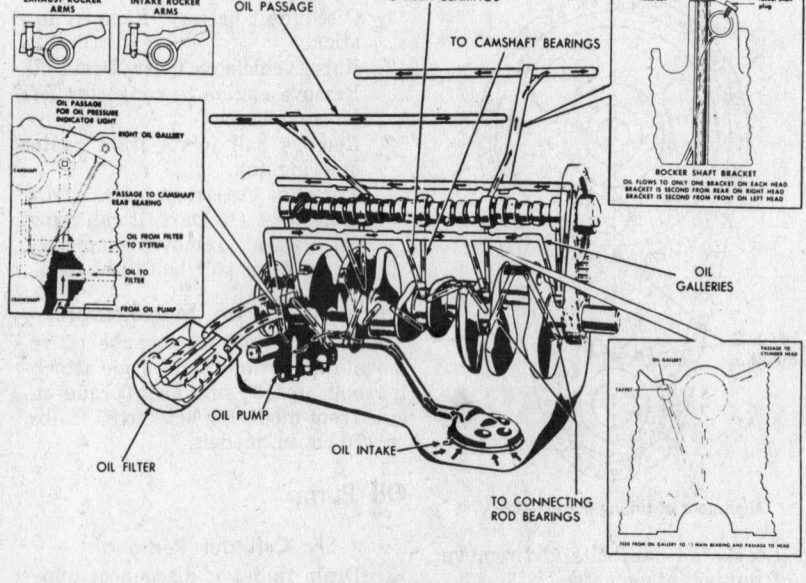

Oil circuit—V8
(© Chrysler Corp.)

2. Press off drive gear. Support gear to keep load off aluminum body.
3. Remove pump rotor and shaft and lift out outer pump rotor.
4. Remove oil pressure relief valve plug and lift out spring and plunger.

273, 318 and 340

1. Remove cotter pin from relief valve, drill 1/8 in. hole into relief valve cap, and insert a self-threading sheet metal screw into cap.
2. Clamp sheet metal screw in vise, and supporting oil pump body, tap body with a soft headed hammer until cap comes out. Discard cap and remove spring and relief valve.
3. Remove oil pump cover bolts and lockwashers, and lift off cover.
4. Discard oil seal ring.
5. Remove pump rotor and shaft and remove outer rotor.

Inspection—All Engines

1. Clean all parts thoroughly in solvent.
2. Inspect mating face of oil pump cover. Mating face should be smooth with no scratches or grooving. Replace if scratched or grooved.
3. Lay straight edge across oil pump cover face. If 0.0015 in. feeler gauge can be inserted be-

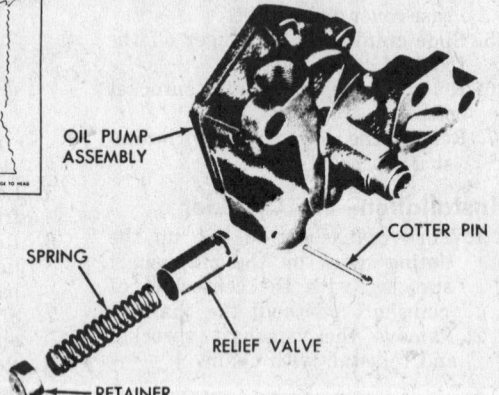

Relief valve—273, 318, and 340 V8s (© Chrysler Corp.)

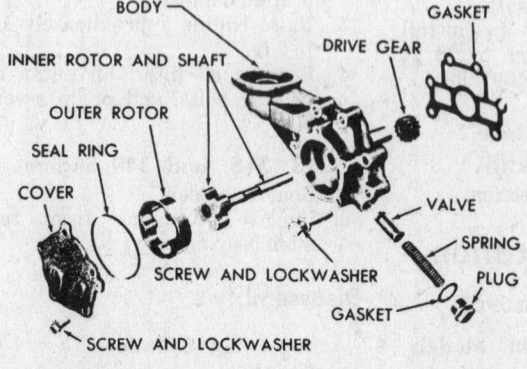

Disassembly of six cylinder oil pump (© Chrysler Corp.)

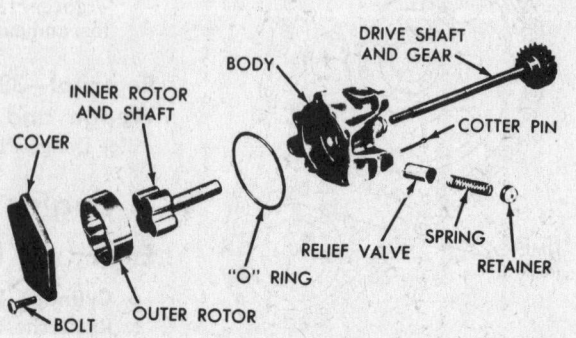

Exploded view of oil pump—V8 (© Chrysler Corp.)

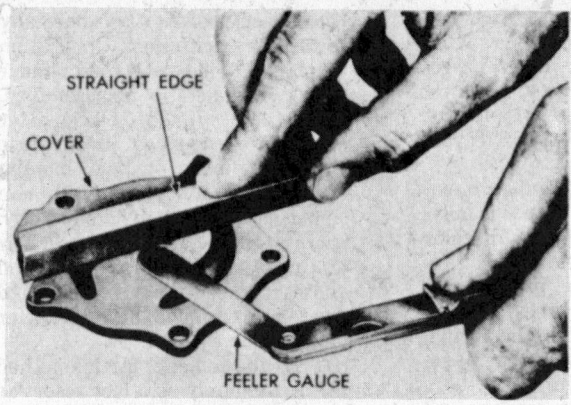

Measuring oil pump cover flatness (© Chrysler Corp.)

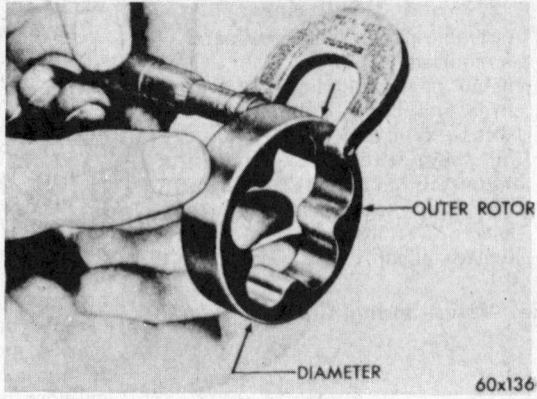

Measuring oil pump outer rotor (© Chrysler Corp.)

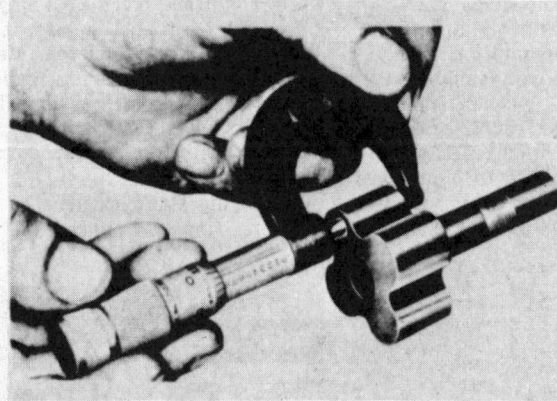

Measuring oil pump inner rotor (© Chrysler Corp.)

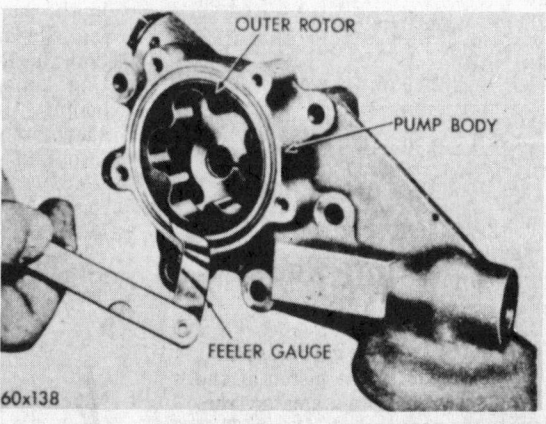

Measuring outer rotor to pump body clearance (© Chrysler Corp.)

Measuring clearance over rotors (© Chrysler Corp.)

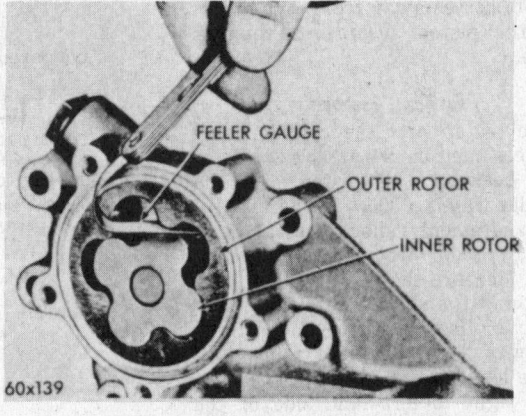

Measuring clearance between rotors (© Chrysler Corp.)

tween cover and straight edge, replace cover.

4. Measure outer rotor. If rotor length is less than 0.649 in. for six cylinders, 0.825 in. for 273 and 318 V8s, or 0.943 in. for larger V8s, replace rotor. If rotor diameter is less than 2.469 in. for either sixes or eights, replace rotor.
5. Measure inner rotor. If rotor length is less than 0.649 in. for six cylinders, 0.825 in. for 273 and 318 V8s, or 0.942 in. for larger V8s, replace rotor.
6. Install outer rotor in pump body

and holding rotor to one side measure clearance between rotor and body. If clearance is more than 0.014 in. replace oil pump body.

7. Install inner rotor into pump body and place straight edge across body between bolt holes. If feeler gauge greater than 0.004 in. can be inserted between rotors and body, replace body.
8. Measure clearance between tips of inner and outer rotors where they are opposed. If clearance exceeds 0.010 in., replace inner and outer rotors.

Assembly and Installation

Six Cylinder

1. Assemble pump using new parts as required.
2. Install new seal rings between cover and body. Tighten cover bolts to 95 in. lbs.
3. Install pump on engine. Tighten attaching bolts to 200 in. lbs.
4. Lower engine and replace engine mount bolts.
5. Replace fan shroud, connect upper and lower hoses, and fill radiator.

273 and 318 Engines

1. Assemble pump using new parts as required.
2. Install new seal rings between cover and body. Tighten cover bolts to 95 in. lbs.
3. Fill pump with oil to prime it and install pump and strainer on rear main bearing cap. Tighten bolts to 35 ft. lbs.
4. Replace oil pan.

Relief Valve Spring Chart

Color	Length	Free-loaded length	Compression (lbs.)
Gray (light)	2.19 in.	1.60 in.	11.85-12.85
Red (Std.)	2.29 in.	1.60 in.	14.85-15.85
Brown (heavy)	2.39 in.	1.60 in.	17.90-18.90

Pistons and Connecting Rods

References

The pistons are cam ground so that the diameter across the piston at the pin boss is less than its diameter from major to minor thrust faces. This allows for expansion under normal operating conditions. Expansion forces the pin bosses away from each other, and the piston assumes a rounder shape.

Fitting Pistons

The piston and the cylinder wall surfaces must be dry and clean. Clearance between the piston and the cylinder wall is .0005—.0015 in.

Measurements should be taken at normal room temperature, about 70°F. Measure the piston diameter at the top of the skirt, 90° to the piston pin axis. Measure the cylinder walls half way down the bore, transverse to the length of the engine (measure from right to left-hand side of the bore.)

Fitting Rings

Before replacing rings, inspect cylinder bores.

1. Using internal micrometer measure bores both across thrust faces of cylinder and parallel to axis of crankshaft at minimum of four locations equally spaced. The bore must not be out of round by more than 0.005 in. and it must not "taper" more than 0.101 in. "Taper" is the difference in wear between two bore measurements in any cylinder. Bore any cylinder beyond limits of out of roundness or taper to diameter of next available oversize piston that will clean up wear. The recommended clearances for new pistons are 0.0005-0.0015 in. for six cylinders and 273, 318 and 340 V8s.

2. If bore is within limits dimensionally, examine bore visually. It should be dull silver in color and exhibit pattern of machining cross hatching intersecting at about 60 degrees. There should be no scratches, tool marks, nicks, or other damage. If any such damage exists, bore cylinder to clean up damage and then to next oversize piston diameter. Polished or shiny places in the bore are known as glazing. Glazing causes poor lubrication, high oil consumption and ring damage. Remove glazing by honing cylinders with clean, sharp stones of No. 180-220 grit to obtain surface finish of 15-35 RMS. Use a hone also to obtain correct piston clearance and surface finish in any cylinder that has been bored. This or any other machining operation should be done with the cylinder block completely disassembled. Hot tank cylinder block after honing or boring. To remove minor glazing when honing equipment is not available, run emery cloth back and forth across glazed area perpendicular to axis of bore. Scrub block and bores thoroughly with soap and water to remove all grit after using emery cloth.

NOTE: the emery cloth method should be used only as a last resort as it is a method much inferior to honing.

3. If cylinder bore is in satisfactory condition, place each ring in bore in turn and square it in bore with head of piston. Measure ring gap. If ring gap is greater than limit, get new ring. If ring gap is less than limit, file end of ring to obtain correct gap.

4. Check ring side clearance by installing rings on piston, and in-

←CENTER LINE OF ENGINE→

A AT RIGHT ANGLE TO CENTER LINE OF ENGINE
B PARALLEL TO CENTER LINE OF ENGINE

1. OUT-OF-ROUND = DIFFERENCE BETWEEN A AND B
2. TAPER = DIFFERENCE BETWEEN THE A MEASUREMENT AT TOP OF CYLINDER BORE AND THE A MEASUREMENT AT BOTTOM OF CYLINDER BORE

Cylinder bore taper and out of roundness (© Ford Motor Co.)

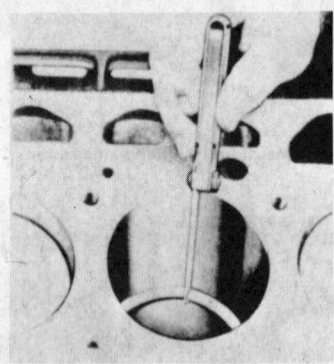

Measuring piston ring gap (© Ford Motor Co.)

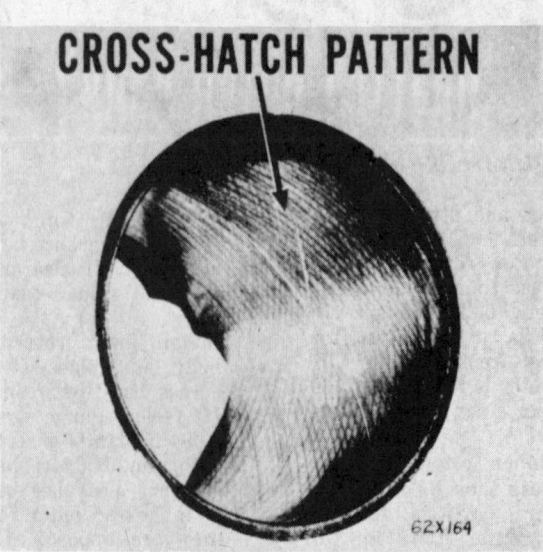

CROSS-HATCH PATTERN

62X164

Cylinder bore surface finish (© Chrysler Corp.)

Ring Gap

Year and Engine	Top Compression	Bottom Compression	Oil Control
1964-65 Six	0.010-0.020 in.	0.010-0.020 in.	0.015-0.062 in.
1966-70 Six	0.010-0.020 in.	0.010-0.020 in.	0.015-0.055 in.
1965-70 273, 318, and 340 V8s	0.010-0.020 in.	0.010-0.020 in.	0.015-0.055 in.

Ring Side Clearance

Year and Engine	Top Compression	Bottom Compression	Oil Control
1964-65 Six	0.0015-0.0040 in.	0.0015-0.0040 in.	0.009 in. max.
1964-65 273 V8	0.0015-0.0030 in.	0.0015-0.0040 in.	0.009 in. max.
1966-70 Six and 273, 318 and 340 V8s	0.0015-0.0040 in.	0.0015-0.0040 in.	0.0002-0.0050 in.

serting feeler gauge of correct dimension between ring and lower land. Gauge should slide freely around ring circumference without binding. Any wear will form a step on lower land. Replace any pistons having high steps. Before checking ring side clearance be sure ring grooves are clean and free of carbon, sludge, or grit.

Cleaning ring grooves
(© Ford Motor Co.)

Measuring piston ring side clearance
(© Ford Motor Co.)

5. Space ring gaps at equidistant intervals around piston circumference. Be sure to install piston in its original bore. Install short lengths of rubber tubing over connecting rod bolts to prevent

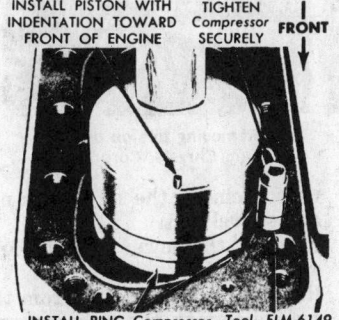

INSTALL PISTON WITH INDENTATION TOWARD FRONT OF ENGINE — TIGHTEN Compressor SECURELY — FRONT

INSTALL RING Compressor Tool—FLM-6149 WITH RETAINER TOWARD SKIRT

Installing piston (© Ford Motor Co.)

damage to rod journal. Install ring compressor over rings on piston. Lower piston rod assembly into bore until ring compressor contacts block. Using wooden handle of hammer push piston into bore while guiding rod onto journal.

Rod Bearings and Piston Assemblies

Rod bearings are conventional insert-type lead-base babbitt on steel shells. They are to be fitted to .0005-.0015 in. clearance.

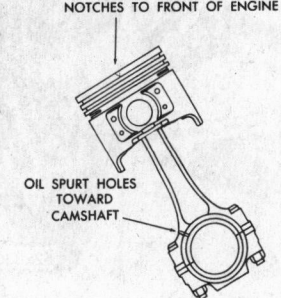

NOTCHES TO FRONT OF ENGINE

OIL SPURT HOLES TOWARD CAMSHAFT

Piston and rod assembly—6 cyl.
(© Chrysler Corp.)

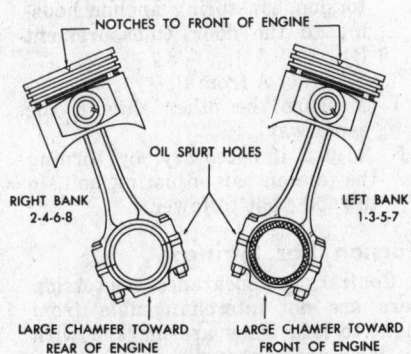

NOTCHES TO FRONT OF ENGINE

OIL SPURT HOLES

RIGHT BANK 2-4-6-8 — LEFT BANK 1-3-5-7

LARGE CHAMFER TOWARD REAR OF ENGINE — LARGE CHAMFER TOWARD FRONT OF ENGINE

Piston and rod assembly—V8
(© Chrysler Corp.)

1. Position the compression ring gaps opposite each other and 90° from the oil control ring gap.
2. Have the oil ring expander gap toward the right (camshaft or distributor side of the engine.) Turn the oil ring gap toward the left (manifold) side of the engine.

3. Immerse the piston head and rings in a container of clean engine oil. Compress the rings, taking care not to disturb their position, and insert the piston-and-rod-assembly into its respective cylinder bore. The notch on top of the piston must point toward the front of the engine. The split hole, in the rod bearing end of the connecting rod, must point toward the left (manifold side) of the engine for proper cylinder wall and piston lubrication.
4. Protect the connecting rod crankpin journal and the connecting rod bolts with a couple of turns of masking tape applied to the threaded end of the rod bearing bolts. This is a good move in the absence of special tools. The angle of this engine requires some type of guide to lead the rod bearing squarely onto the crankshaft journal.
5. Turn the crankshaft so that the respective connecting rod journal is on bottom dead center and on centerline of its cylinder bore.
6. Tap the piston down into the cylinder bore, using a hammer handle. At the same time, guide the connecting rod bearing into position on the crankshaft journal.
7. Lubricate the rod cap, install the attaching nuts and torque to 45 ft. lbs.

Front Suspension

References

Definitions of the points of steering geometry are covered in the Unit Repair Section.

Figures covering the caster, camber, toe-in, kingpin inclination, and turning radius can be found in the Wheel Alignment table of this section.

Tire size figures can be found in the General Chassis and Brake Specifications table.

Torsion-Aire front suspension is similar in principle to that used on other cars of the Chrysler family. The torsion bars are used in the front, only, and are connected to the lower control arms and anchored to the body under the front floor.

Ball joints are used at the upper and lower ends of the steering knuckles. Cam-type caster and camber adjustment methods are employed. Direct-acting Oriflow shock absorbers are attached between the lower control arms, and stamped steel towers welded into the body-frame unit.

Radius rods are used on the lower control arms and extend forward to attach to the K-shaped engine support member. The upper control arm

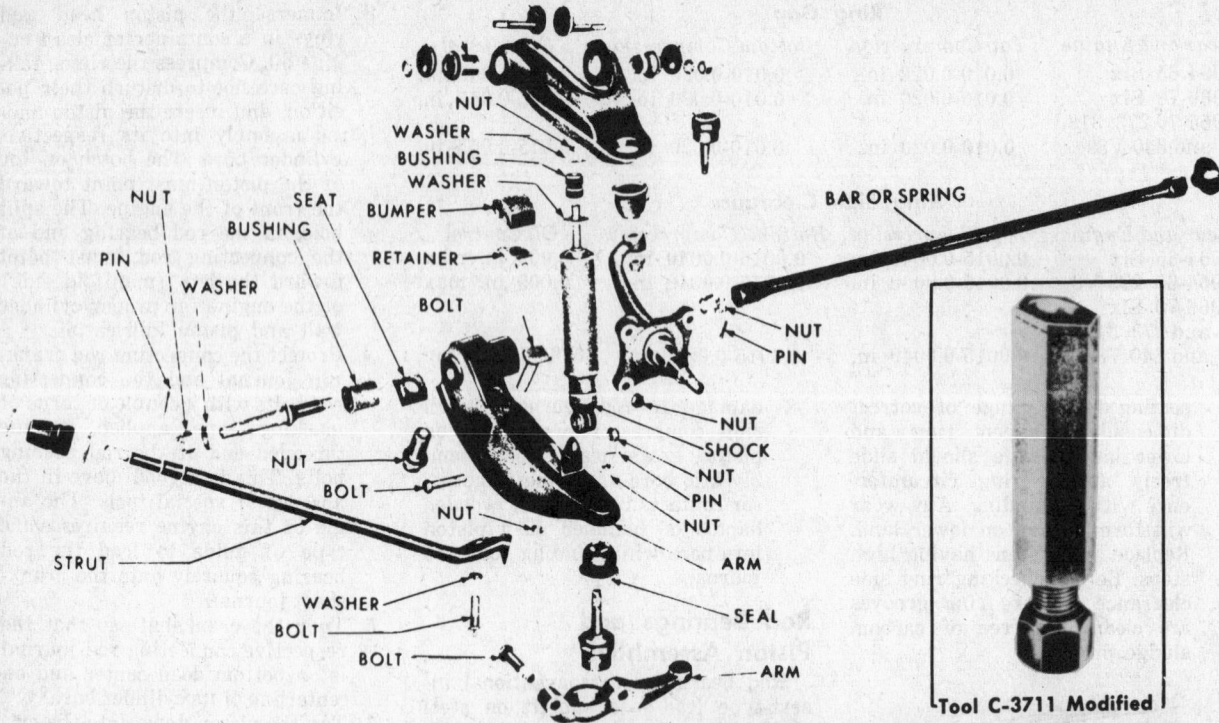

NUT
WASHER
BUSHING
WASHER
BUMPER
RETAINER
BOLT

NUT
SEAT
BUSHING

NUT
PIN
WASHER

BAR OR SPRING

NUT
PIN

NUT
SHOCK
NUT
PIN
NUT

NUT
BOLT
NUT

STRUT

WASHER
BOLT
BOLT

ARM

SEAL

ARM

Torsion bar front suspension (© Chrysler Corp.)

Tool C-3711 Modified

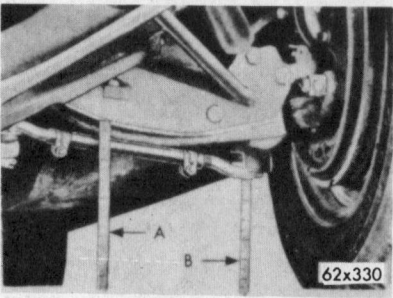

Front suspension height
(© Chrysler Corp.)

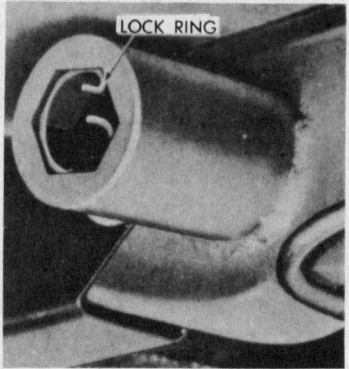

LOCK RING

Torsion bar rear anchor lock ring
(© Chrysler Corp.)

TORSION BAR

TOOL

Removing torsion bar
(© Chrysler Corp.)

is angled downward, toward the rear, to reduce nose dive during brake application.

Peculiar to this suspension is a compression-type lower ball joint and a torsion bar adjustment in the lower control arm.

The torsion bar adjustment permits rotating the bar relative to the lower control arm to obtain correct front-end height settings. This adjustment is located at the front end of the bar, and permits easy adjustment during wheel alignment operations.

Lubrication

Balloon-type and semi-permanently lubricated steering linkage and front suspension ball joints are used. Relubrication at these points is required at about 36,000 miles, or three year periods (whichever comes first.) However, the balloon seals should be inspected for leaks, or other damage, two or three times a year.

When lubricating these points, use only multi-mileage long-life chassis grease. Remove the threaded plug from each joint to be lubricated, and temporarily install lube fittings. Inject lubricant, while feeling the seal with the fingers. Stop just before the seal starts to balloon. Remove lube fittings and reinstall plugs.

Front Height Adjustment (Without Gauge)

1. Jounce the car and measure from the lower ball joint to the floor (measurement A).
2. Measure from the control arm

torsion bar spring anchor housing to the floor (measurement B).
3. Subtract A from B.
4. Measure the other side in the same way.
5. Adjust, if necessary, by turning the torsion bar adjusting bolt, in to raise; out to lower.

Torsion Bar Springs

Contrary to appearance, the torsion bars are not interchangeable from right to left. They are marked with an R or an L, according to their location.

Removal

1. Lift the car, by the body, high enough to free the front suspension of all load. If the car is to be raised with jacks, place jack under center of K-member and raise until suspension is free of all load.
2. Release load from torsion bar by backing off anchor adjusting

nuts. Remove the adjusting nut and swivel bolt.
3. Remove the lower control arm strut.
4. Remove the lock spring from the rear of torsion bar rear anchor.
5. Install tool C-3728, or other suitable clamp, and remove torsion

bar rearward by striking the clamping tool with a hammer.

NOTE: do not apply heat to the front or rear anchors. Do not scratch or otherwise mar the skin of the torsion bar during removal or installation.

6. Remove the clamping tool and slide the rear anchor balloon seal off the front end of the bar.
7. Remove torsion bar by sliding the bar rearward and out through the rear anchor.

Installation

1. Clean the hex openings of both front and rear anchors, also clean the male ends of the torsion bar.
2. Feed the torsion bar through the rear anchor.
3. Slide the balloon-type seal over the torsion bar, with the large cupped side of the seal facing the rear.
4. Coat both ends of the torsion bar with multi-purpose grease.
5. When starting the bar into the anchor in the lower control arm, position the adjusting arm about 60° below the horizontal plane. This will permit windup for future adjustment.
6. Position the lock ring into the rear anchor, then move torsion bar rearward until the bar contacts the lock ring.
7. Position swivel bolt on the control arm and hold in place while installing the adjusting nut and seat. Tighten the adjustment about ten turns before lowering car to the floor.
8. Pack the annular opening in the rear anchor with multi-purpose grease. Slide the rear anchor balloon type seal into position over the rear anchor until the lip of the seal fits in the groove.
9. Install lower control arm strut.
10. Lower car to the floor and adjust front suspension height.

Jacking, Hoisting

Jack car at front control arms and at rear under axle housing.

To lift at frame use adapters, so that contact will be made at points shown. Lifting pads must extend beyond sides of supporting structure.

Steering Gear

Instructions for the adjustment of the steering gear will be found in the Unit Repair Section.

Correct Vehicle Height

Model and Year	Height
1964-66 Valiant and Dart and 1965-66 Barracuda with standard suspension	2 in. ± 1/8 in.
1964-66 Valiant and Dart and 1965-66 Barracuda with heavy duty suspension	2-3/8 in. ± 1/8 in.
1967-70 Valiant and Dart and 1967 Barracuda	2-1/8 in. ± 1/8 in.
1968-69 Barracuda	1-3/8 in. ± 1/8 in.
1970 Barracuda and Challenger	1-3/16 in. ± 1/8 in.

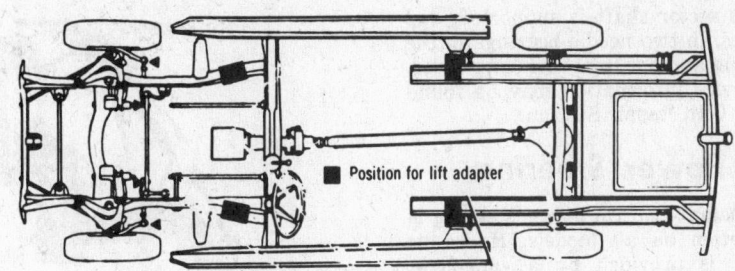

Positioning lift adapter (© Chrysler Corp.)

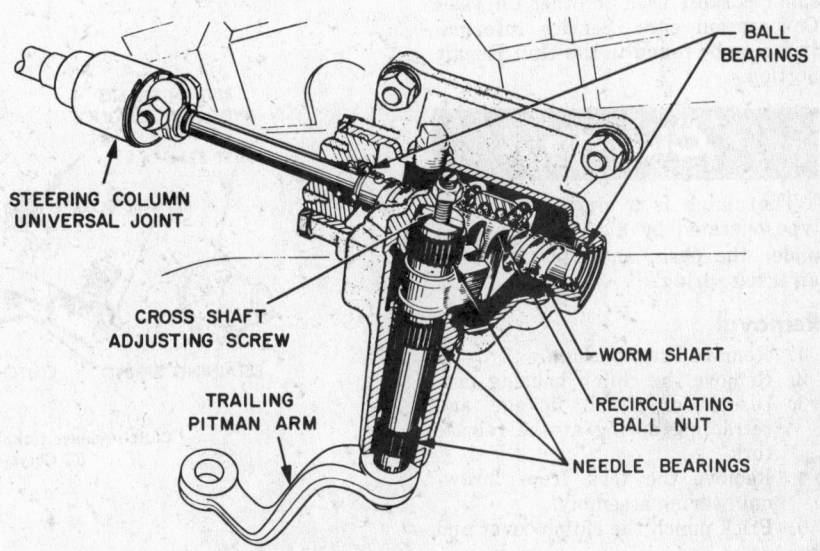

Manual steering gearbox (© Chrysler Corp.)

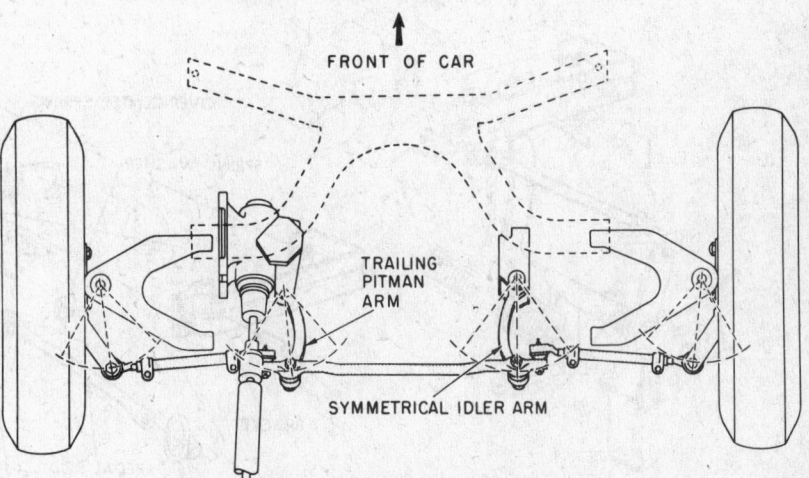

ALL ARMS AND LINKS SWING THROUGH SIMILAR ARCS
MINIMIZING EFFORT THROUGHOUT THE ENTIRE STEERING RANGE

Steering linkage arrangement (© Chrysler Corp.)

Manual Steering

A worm and recirculating ball type steering gear is used with the manual steering system.

The worm shaft is supported at each end by ball-type thrust bearings.

The sector shaft includes an integral sector gear which meshes with helical grooves on the worm shaft ball nut.

The sector shaft is supported, and rotates, in two needle bearings in the housing and one in the housing cover.

Service information may be found in the Unit Repair Section.

Power Steering

Constant-Control power steering is an option on all models. Hydraulic power is provided by a vane-type, belt-driven pump. A double-groove pump pulley is used. The power steering gear and pump are essentially the same as those used on other Chrysler Corporation cars. Service information may be found in the Unit Repair Section.

Clutch

The clutch is a single dry plate-type operated by a pedal suspended under the dash, and equipped with an assist spring.

Removal

1. Remove Transmission.
2. Remove the clutch housing pan.
3. Disconnect clutch linkage and retracting spring at the release fork.
4. Remove the fork from throwout bearing assembly.
5. Prick punch the clutch cover and flywheel to assure correct relocation when installing.
6. Remove the six clutch-to-flywheel bolts and remove the clutch cover and driven plate.

Installation

Install the clutch in reverse order and adjust clutch linkage.

Linkage Adjustment

Adjust the length of the release fork rod until there is 5/32 in. free-play measured at the outer end of the clutch fork. This will produce the 1 in. free-play required at the pedal.

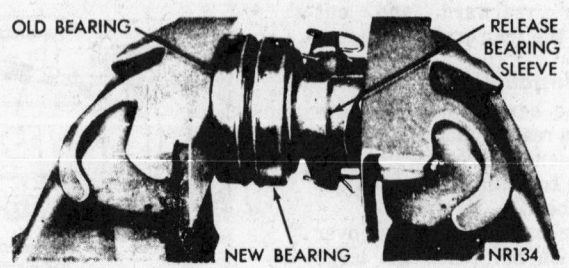

Replacing clutch release bearing
(© Chrysler Corp.)

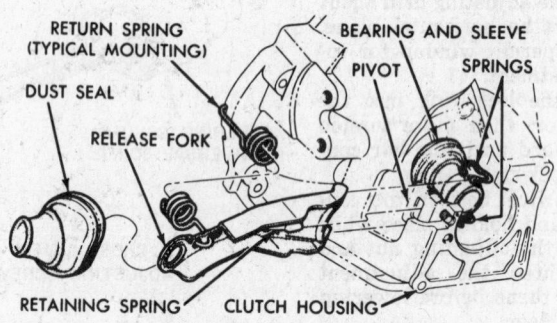

Clutch release fork, bearing and sleeve
(© Chrysler Corp.)

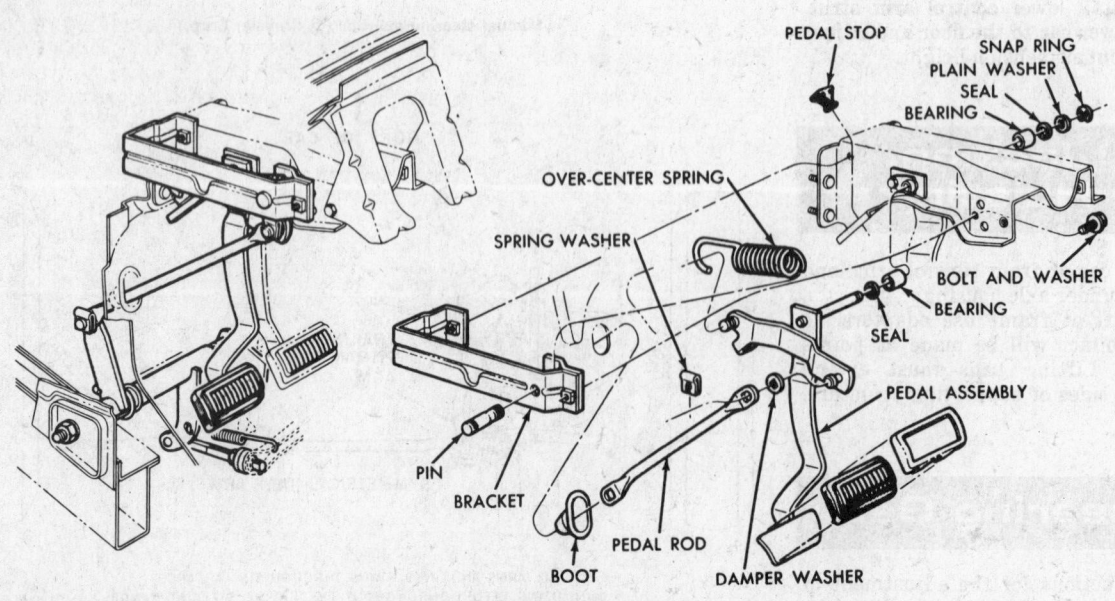

Typical clutch pedal and linkage (© Chrysler Corp.)

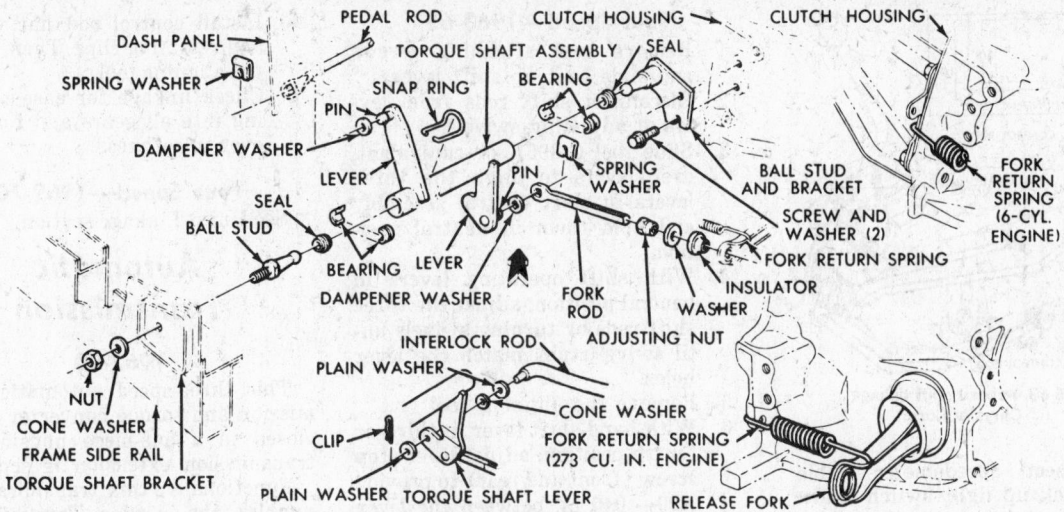

Typical clutch linkage (© Chrysler Corp.)

Standard Transmission

For details on reconditioning, see Unit Repair Section.

Dodge and Plymouth have used four synchromesh manual transmissions in recent years. Six cylinders use a three speed, top cover transmission with synchromesh on second and third. V8s up to 1970 use a similar three speed, top cover transmission with synchromesh on second and third. 1970 V8s use a new fully synchromesh, side cover three speed. All Dodge and Plymouth four speed cars use a fully synchromesh, side cover four speed.

Removal

Top Cover Three Speeds

1. Drain Transmission.

2. Disconnect drive shaft at rear universal joint. Carefully pull shaft yoke out of transmission.
3. Disconnect speedometer cable and back up light switch.
4. Install engine support fixture or jack up engine about one in. and block in place.
5. Disconnect transmission extension housing from center crossmember.
6. Support transmission with jack and remove crossmember. Remove bolts that attach transmission to clutch housing.
7. Slide transmission rearward until pinion shaft clears clutch disc before lowering transmission.
8. Lower transmission and remove.

Fully Synchromesh, Side Cover Three Speed

1. Remove shift rods from transmission levers.
2. Drain transmission fluid.
3. Disconnect drive shaft at rear universal joint. Mark both parts for reassembly.
4. Carefully pull yoke out of transmission extension.
5. Disconnect speedometer and back up lights.
6. Remove part of exhaust if it blocks transmission.
7. Raise engine slightly and block in place.
8. Support transmission with jack, and remove crossmember.
9. Remove transmission to clutch housing bolts.
10. Slide transmission to rear until drive pinion shaft clears clutch disc, lower transmission, and remove from vehicle.

Four Speed

1. Raise vehicle on a hoist and drain transmission.
2. Disconnect all shift controls from transmission levers. Remove three bolts securing shift unit to extension housing.
3. Disconnect propellor shaft at rear universal joint. Carefully pull yoke out of transmission extension.

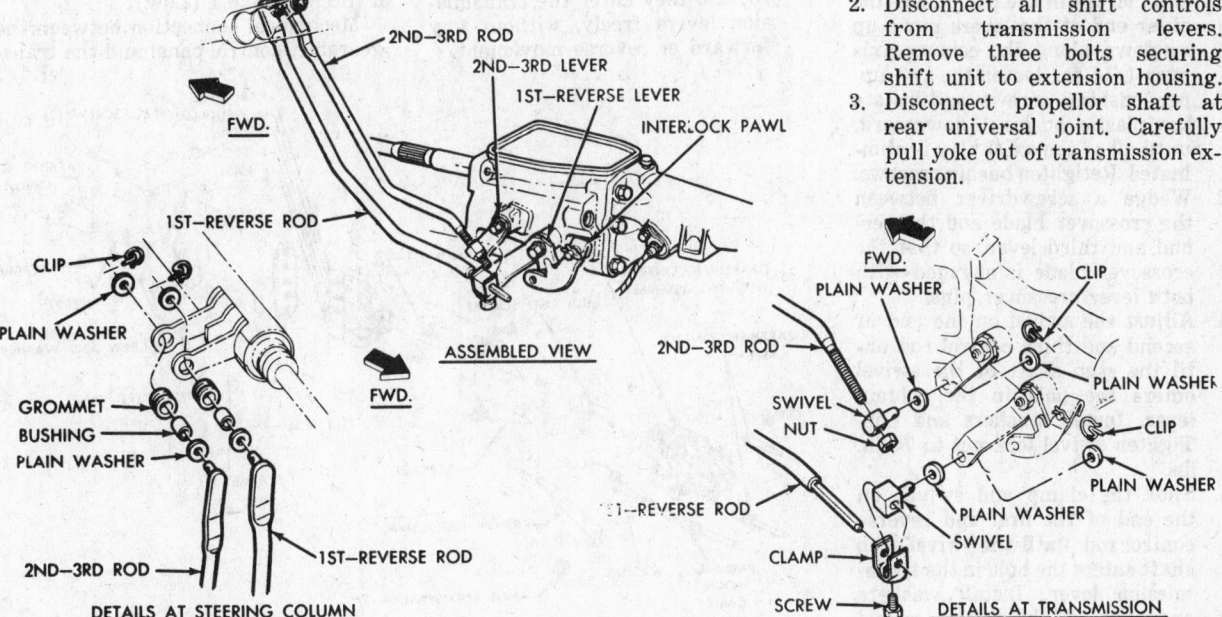

ASSEMBLED VIEW

DETAILS AT STEERING COLUMN

DETAILS AT TRANSMISSION

3-speed shift linkage (© Chrysler Corp.)

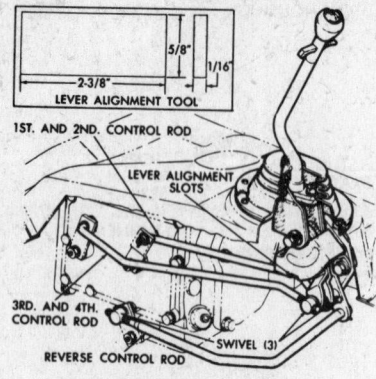

1966-68 4-speed shift linkage
(© Chrysler Corp.)

4. Disconnect speedometer cable and back up light switch leads.
5. Disconnect left exhaust pipe or dual exhausts. Disconnect parking brake cable.
6. Raise engine slightly and block in place.
7. Disconnect transmission extension from crossmember.
8. Remove crossmember.
9. Support transmission with jack. Remove clutch housing to transmission bolts.
10. Slide transmission to rear until drive pinion shaft clears clutch disc.
11. Lower transmission and remove from vehicle.

Linkage Adjustment

1963-70

With the second and third control rod disconnected from the lever on the column, and first and reverse control rod disconnected at the transmission, levers in neutral:

1. Check for axial freedom of the shift levers in the column. If the outer end of the levers move up or down along the column axis over 1/16 in., loosen the two upper bushing screws and rotate the plastic bushing, downward, until all of the axial play is eliminated. Retighten bushing screws.
2. Wedge a screwdriver between the crossover blade and the second and third lever, so that the crossover blade is engaged with both lever crossover pins.
3. Adjust the swivel on the end of second and third control rod until the stub shaft of the swivel enters the hole in the column lever. Install washers and clip. Tighten swivel lock nut to 70 in. lbs.
4. Slide the clamp and swivel, on the end of the first and reverse control rod, until the swivel stub shaft enters the hole in the transmission lever. Install washers and clip. Tighten the swivel clamp bolt to 100 in. lbs.

Four-Speed—1963-65

1. Remove boot attaching screws and slide boot up shift lever.
2. Disconnect shift rods from levers at adjusting swivels.
3. Slide tool C-3951, or equivalent, over levers to align the three levers in shift control assembly and hold them in neutral position.
4. With shift operating levers in neutral position, adjust the three shift rods by turning swivels until swivel stubs match the lever holes.
5. Remove the aligning tool.
6. With hand-shift lever in third or fourth position, adjust lever stop screw (front and rear) to provide .020—.040 in. between the lever and the stops.

Caution Because there is no reverse gear interlock, it is very important that the transmission linkage adjustments be correctly performed in order to prevent the possibility of engaging two gears at once.

For cars with cable-type throttle control, and all 1965 models, see notes under fuel system.

Four Speed—1966-68

1. Make up a lever aligning tool from 1/16 in. thick metal, as illustrated.
2. With transmission in neutral, disconnect all control rods from the transmission levers.
3. Insert lever aligning tool through the slots in the levers, making sure it is through all the levers and against the back plate.
4. Now that all levers are locked in neutral, adjust length of control rods so they enter the transmission levers freely, without any forward or reverse movement.

5. Install control rod flat washers and retaining clips. Then, remove the aligning tool.
6. Check linkage for ease of shifting into all gears and for crossover smoothness.

Four Speed—1969-70
See Hurst Linkage section.

Automatic Transmission

1963-70

This three-speed automatic transmission and torque converter are enclosed in a one-piece housing. The transmission extension is separate.

Functionally, this transmission resembles the earlier Torqueflite. It uses the same gear selection and performs about the same. In design detail, however, it is entirely different. It is more compact and the internal parts are engineered to accommodate the torque requirements of the 6 cylinder and V8 engines.

Within the transmission extension, there is a parking pawl operated by a sliding lever. The parking pawl locks the transmission output shaft.

General information, troubleshooting and repairs will be found in the Unit Repair Section.

Towing Instructions

If the car is to be towed for any reason, lift the rear end or remove the driveshaft.

Push Button Gearshift Control Unit

The transmission is operated from a dash panel unit. The unit consists of five push buttons, identified by R (Reverse), N (Neutral), D (Drive), 2 (Second) and 1 (Low).

Mechanical connection between the gearshift control panel and the trans-

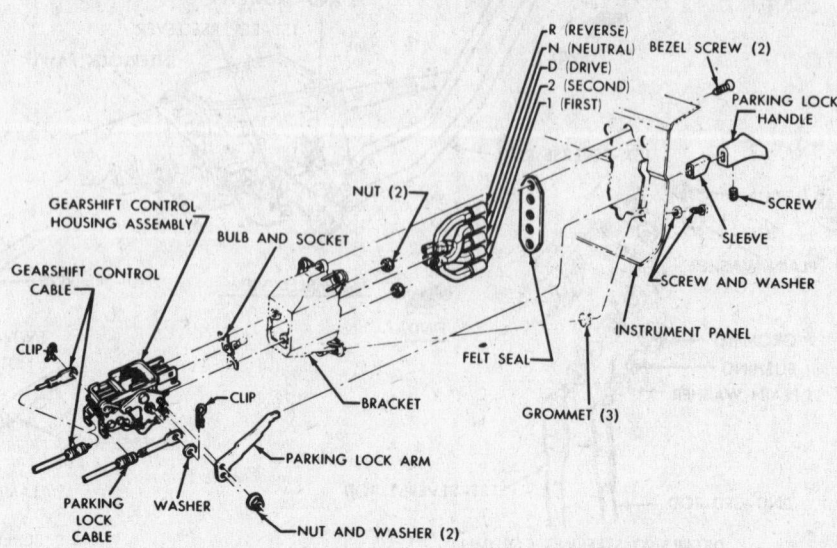

Removing and installing gear selector (© Chrysler Corp.)

mission manual control valve is obtained through the use of a single push-pull cable. One end of the wire is connected to the cable actuator in the control panel. The other end is attached to the manual control valve lever in the transmission.

Should the R button be pushed in while the car speed is above 10 mph, the manual control lever will go to the Neutral position. When the car speed drops below 10 mph, it will be necessary to re-engage the R button to attain Reverse.

A back-up light switch (when so equipped) is included in the gearshift control panel. It is operated by the R (Reverse) button.

Removal

1. Disconnect one battery cable.
2. Disconnect back-up light switch connectors (if so equipped) at push button control, at the rear of the panel.

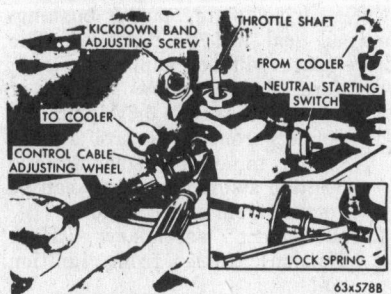

Removing gearshift control cable
(© Chrysler Corp.)

3. Remove screws from push button face plate and remove the buttons by pulling them off their slides. Remove lamp bulb.
4. Remove control housing stud nuts (they are now accessible) and remove control cable from the rear of the panel.
5. Remove hairpin lock holding the control cable to the actuator and the screws that hold the cable bracket to the control housing.

Installation

1. Insert end of cable in the actuator and reassemble hairpin clip. Install cable.
2. Carefully guide the unit into position from the rear of the instrument panel and install the attaching stud nuts from the front side of instrument panel.
3. Install lamp bulb in push button control and reinstall push buttons onto actuator slides. Replace the face plate.
4. Connect back-up and push button lamp wires.

Push Button Lamp Replacement

The push button lamp bulb can be replaced by taking the face plate off

and removing one or two of the center push buttons.

Control Cable (Transmission End) Removal

1. Lift car on a hoist and drain about two quarts of transmission fluid.
2. Depress 1 (Low) push button to position cable for removal from transmission.
3. Disconnect wire from neutral starting switch and remove switch.
4. Remove push button control cable-to-transmission adjusting wheel lock screw.
5. Insert screwdriver through neutral starting switch opening. Push screwdriver gently against upper projecting portion of the cable lock-spring, and pull outward on the cable to remove cable from adapter and transmission case.

Control Cable Installation

1. Have a helper engage the R button and hold it firmly engaged for the duration of the cable attachment operation.
2. Back the adjusting wheel off on the cable housing (counterclockwise) until two or three threads are exposed behind the wheel on the guide.
3. Lubricate the cable with transmission fluid. Insert cable in transmission case and push inward, making sure the lockspring engages the cable. Adjust cable as outlined in Gearshift Control Cable Adjustment.
4. Refill transmission with automatic transmission fluid (type A, suffix A) to proper level.

Caution

While in the process of making adjustments and tests, do not stall-test the torque converter. For safety, and to prevent internal damage to the transmission, a wide-open throttle stall test should be avoided.

Lubrication

Refer to the Capacities table.

Gearshift Control Cable Adjustment

1. Raise car on hoist. Have a helper hold the R button firmly depressed.
2. Remove control cable adjustment wheel lock screw at the left side of transmission.
3. Back the adjustment wheel off on the cable guide until only two or three threads show behind the wheel.

NOTE: be sure the adjustment wheel turns freely on the guide. Lub-

ricate the cable guide threads with transmission fluid.

4. Hold the control cable guide centered in the hole of the transmission case and apply only enough inward force (2-3 lbs.) to bottom the assembly at the reverse detent. While holding the cable bottomed, rotate the adjustment wheel clockwise until it contacts the case.
5. Turn the wheel clockwise; just enough to make the next adjustment hole in the wheel line up with the screw hole in the case.
6. Counting this hole as number one, keep turning the wheel clockwise until the fifth hole lines up with the screw hole in the case.
7. Install lock screw and torque to 30-50 in. lbs.

Neutral Starting Switch

To properly adjust and test:
1. With the correct cable adjustment certain, depress the N button.
2. Raise the car and drain about two quarts of transmission fluid.
3. Unscrew the neutral starting switch from the transmission case. Check the alignment of the switch operating lever relative to the switch opening in the case.
4. Place cupped washer and O-ring over threads of the switch, then screw switch into transmission case a few turns.
5. Hook up a test lamp; one lead to battery current and the other lead to the switch terminal. Screw the switch into the transmission case until the lamp lights, now, seat the switch by turning it an additional one-third to one-half.

NOTE: the switch must be tight enough to prevent an oil leak. If it isn't, add a thin washer and readjust the switch.

6. Remove test lamp and reconnect the regular wire to the switch.
7. Add fluid to bring transmission to the prescribed level.

1965-70 Steering Column Gearshift

The steering column gear shift sequence is (left to right): P (Park), R (Reverse), N (Neutral), D (Drive), 2 (Second), and 1 (Low). Two cables connect the column with the transmission as in the push button-type shift.

Adjustments at the transmission end are the same as in the push button-type. To adjust the sprag lever and detent plate, hold the lever firmly in the 1 (Low) position.

1. Rotate the sprag lever pivot clockwise until the slot in the sprag lever is tight against the

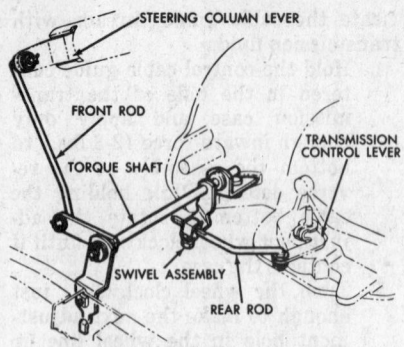

**Automatic column shift linkage—
except 1970 Barracuda and Challenger**
(© Chrysler Corp.)

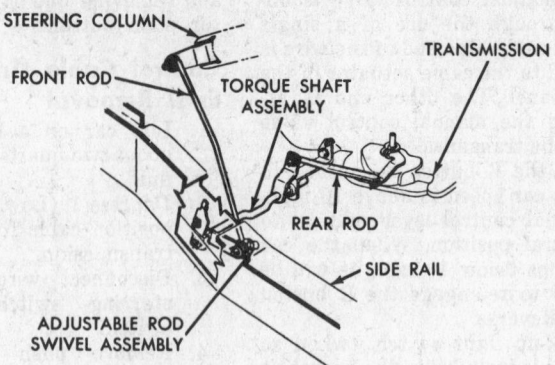

**Column shift linkage—
1970 Barracuda and Challenger**
(© Chrysler Corp.)

pin, then tighten the three pivot screws.

2. With the pawl spring in place, adjust the detent plate to align the end detent with the pin on the detent pawl. Tighten the plate screws.

3. Move the lever in and out of 1 (Low) several times and inspect the adjustment.

4. Install the cable bracket and tighten the retaining screws, if it has been disturbed.

Console Gearshift Removal

1. Disconnect battery
2. Remove gearshift handle knob set screw with Allen wrench. Unscrew knob from cable, then remove spring and handle.
3. Remove two screws from front

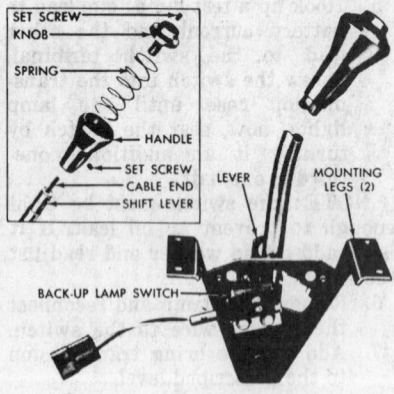

Automatic console shift linkage
(© Chrysler Corp.)

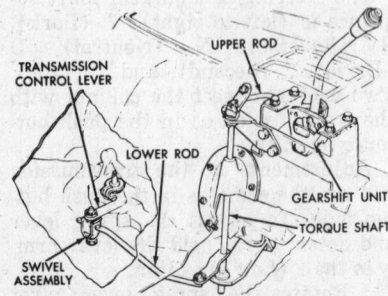

Automatic console shift linkage
(© Chrysler Corp.)

corners of console. Raise console lid and remove two screws from bottom of tray.

4. Raise console enough to disconnect dial lamp and other electrical connections.

5. Lift off console assembly.
6. Disconnect back-up lamp switch wires.
7. Disconnect upper rod from gearshift unit.
8. Remove four gearshift-to-floor bracket bolts and remove the unit.

Console Gearshift Installation

1. Install gearshift unit in its bracket, install and tighten four retaining bolts.
2. Attach upper rod to the unit. Then, connect back-up lamp switch wires.
3. Lower console down over shift lever. Insert dial lamp into its housing, then make all electrical connections.
4. Position console over floor brackets. Install and tighten retaining screws, then install upper finish plate.
5. Install gearshift handle and spring. Secure the handle to lever with set screw.
6. With lever in neutral, install knob by threading it onto cable end until serrated surface on knob is about 1/32 in. above top of handle. Secure knob with set screw.

Console Gearshift Linkage Adjustment

1. With gearshift selector lever in park position, loosen bolt in lower rod adjusting lever.
2. Move transmission control lever all the way to the rear (park detent.)
3. With control lever in transmission in park position detent, and selector lever in park position, tighten the bolt in the lower rod adjusting lever securely.
4. Reconnect the battery.

Torqueflite Removal and Installation

Remove transmission and torque converter as an assembly or the converter drive plate, pump bushing, and oil seal will be damaged. The drive plate will not support a load; therefore, do not allow weight of transmission to rest on drive plate.

1. Connect remote control starter switch to starter solenoid and position switch so engine can be rotated from under vehicle.
2. Disconnect secondary (High Tension) cable from ignition coil.
3. Remove cover plate from in front of converter to provide access to converter drain plug and mounting bolts.
4. Rotate engine with remote control starter switch to bring drain plug to six o'clock position. Drain torque converter and transmission.
5. Mark converter and drive plate to aid in reassembly. There is an offset hole in crankshaft flange bolt circle, inner and outer circle of holes in drive plate, and four tapped holes in front face of coverter so these parts will be installed in original position.
6. Rotate engine with remote control switch to locate two converter to drive plate bolts at five and seven o'clock positions. Remove bolts, rotate engine and remove two more bolts. Do not rotate converter by prying as this will distort drive plate. Do not engage starter if drive plate is not attached to converter with at least one bolt or if transmission to block bolts have been loosened.
7. Disconnect battery.
8. Remove starter.
9. Disconnect wire from neutral start switch.
10. Disconnect gear shift rod from transmission lever. Remove gearshift torque shaft from trans-

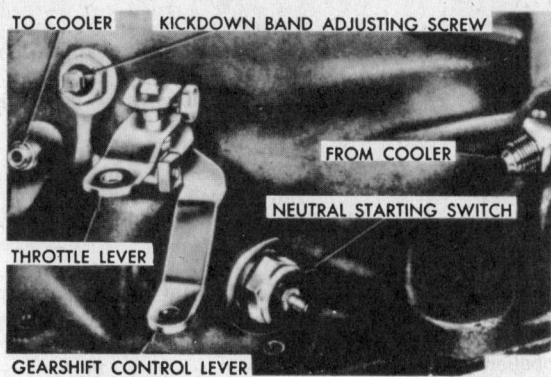

Torqueflite external controls (© Chrysler Corp.)

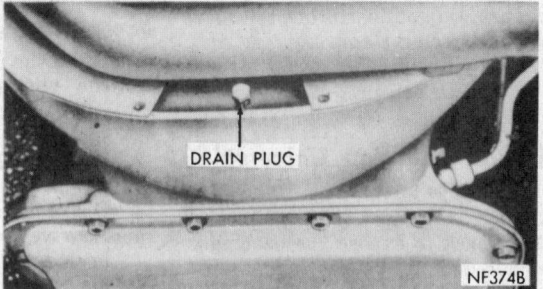

Converter drain plug (© Chrysler Corp.)

mission housing and left side rail. On console shifts, remove two bolts securing gearshift torque shaft lower bracket to extension housing. Swing bracket out of way for transmission removal. Disconnect gearshift rod from transmission lever.

11. Disconnect throttle rod from bellcrank at left side of transmission bell housing.
12. Disconnect oil cooler lines at transmission and remove oil filler tube. Disconnect speedometer cable.
13. Disconnect drive shaft at rear universal joint. Carefully pull shaft assembly out of extension housing.
14. Remove transmission mount to extension housing bolts.
15. Raise engine slightly and block in place.
16. Remove crossmember attaching bolts and remove crossmember.
17. Support transmission with jack.
18. Attach a small "C" clamp to edge of converter housing to hold converter in place during removal of transmission.
19. Remove converter housing retaining bolts. Carefully work transmission to rear of engine dowels and disengage converter hub from end of crankshaft.
20. Lower and remove transmission. Remove "C" clamp and remove converter. Reverse procedure to install.

U Joints, Drive Lines

Six cylinder cars and 1964-65 eight cylinder cars with manual transmission use a drive shaft with a ball-and-trunnion-type front universal joint and a cross-and-roller-type rear universal joint. 1964-65 Darts and Valiants with eight cylinder engines and automatic transmission and all subsequent models use cross-and-roller-type universal joints both front and rear.

Drive Shaft R & R

1. Remove both clamps from the pinion yoke and remove the bearings.
2. Remove the front universal joint bolts at the transmission flange.

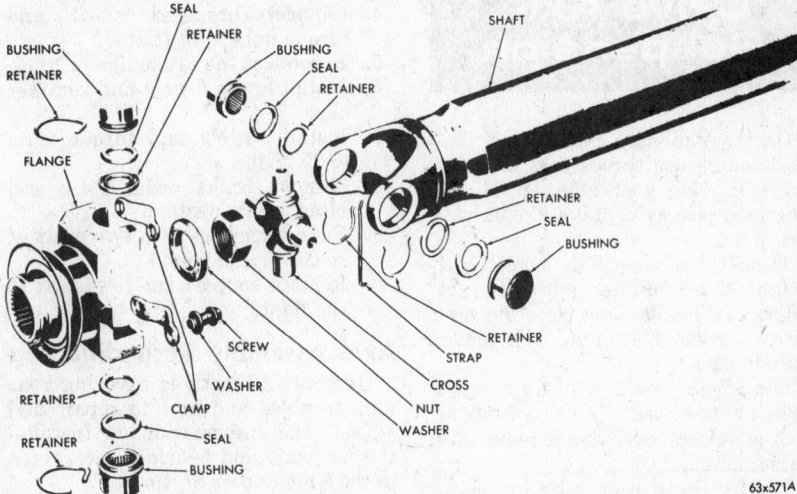

Rear cross and roller universal joint (© Chrysler Corp.)

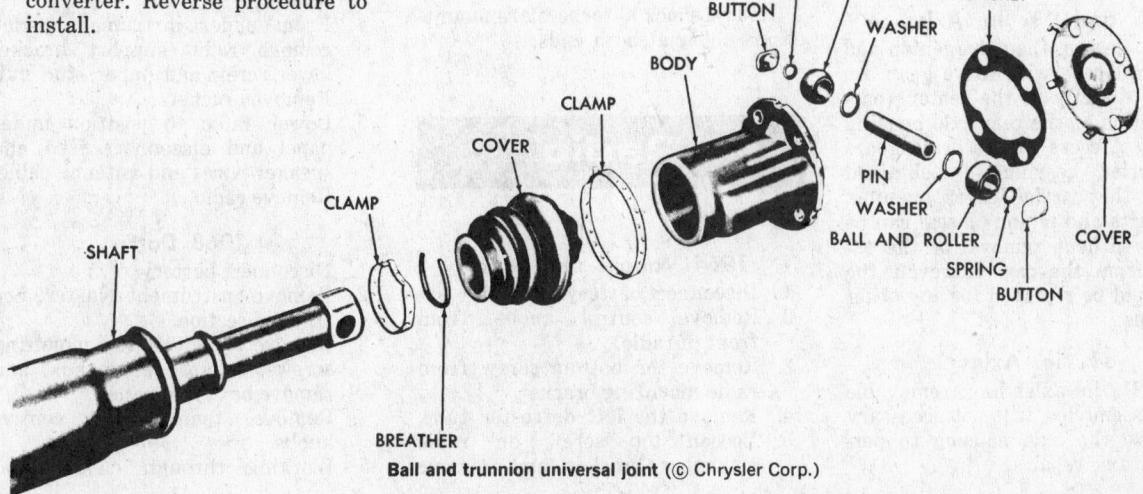

Ball and trunnion universal joint (© Chrysler Corp.)

NOTE: to install, reverse above procedure. Do not overtorque the rear universal bearing clamp nuts, as distortion may result. The correct torque is 14 ft. lbs.

Ball and Trunnion Universal Joint Replacement

Remove the boot and slide the housing back to expose the bearings. The ball and needle bearing assemblies can now be removed. If the pin, or housing, is worn, a press should be used on the pin. The pin must be a very tight fit in the propeller shaft.

Cross and Roller Universal Joint Replacement

Remove the retainers and push one roller and bushing assembly out by pressing the opposite side in. The remaining one can be pushed out by using the cross to press it.

Drive Axle, Suspension

Darts, Valiants, Barracudas, and Challengers use three types of axles, a 7¼ in. ring gear axle, an 8¾ in. ring gear size axle, and a 9¾ in. ring gear axle.

The 7¼ in. axle is a small lightweight axle for use with six cylinders. It has a one piece housing with a removable nine bolt cover plate in back.

The 8¾ in. axle is a larger axle for eight cylinder cars. It has a housing with a welded cover and a removable carrier assembly.

The 9¾ in. axle is a husky heavy duty axle for large V8s and high performance applications. It has a one piece housing and a 10 bolt cover plate.

7¼ and 9¾ In. Axles

Due to design, the drive pinion and differential case with drive gear are mounted directly to the center (carrier) section of the rear axle housing assembly. Access to the drive gears and carrier bearings is obtained through the carrier cover opening. Axle shafts and pinion oil seal can be changed without removal of the assembly from the car. However, the unit should be removed for any other operations.

8¾ In. Axles

Since 8¾ in. axles have removable carrier assemblies, it is not necessary to remove the axle housing to perform any axle repairs.

One Piece Axle Housing Assembly Removal

1. Raise the rear of the car to have the wheels clear the floor. Support the body in front of rear springs.
2. Block the brake pedal up.
3. Remove rear wheels.
4. Disconnect hydraulic flex line.
5. Remove the cotter pin and unhook the parking brake rear cable rod from the intermediate arm.
6. Disconnect the propeller shaft at the rear end.
7. Remove rear spring U-bolts and shock absorbers.
8. Remove axle assembly from the car.

Axle Housing Installation

1. Position the rear axle housing spring seats over the spring center bolts.
2. Install U-bolts and shock absorbers. Torque U-bolts to 45 ft. lbs.
3. Connect parking brake cable at intermediate arm.
4. Connect propeller shaft and torque nuts to 14 ft. lbs.
5. Reconnect the brake line or lines.
6. Install brake drums and retainer clips.
7. Install wheels and torque nuts to 55 ft. lbs.
8. Remove brake pedal block and bleed brake system.
9. Fill differential with two pints of specified compound.
10. Remove support and lower car to the floor.

Axle Assembly Reconditioning

General instructions covering rear axle troubles and how to repair and adjust, and information on installation of seals and bearings, are given in the Unit Repair Section.

Rear Springs

Rear springs are leaf semi-elliptic-type, using rubber bushings in the eye of each end of the main leaf.

Oriflow shock absorbers are mounted in rubber at both ends.

Radio

Removal

1964 Valiant and Dart

1. Disconnect battery.
2. Remove control knobs from front of radio.
3. Remove the bottom screw from radio mounting bracket.
4. Remove the left defroster tube.
5. Loosen top screw on radio mounting bracket and remove bracket.
6. Disconnect speaker and antenna leads.
7. Remove mounting nuts from front of radio.
8. Remove radio bezel.
9. Lower radio and disconnect radio power feed cable.
10. Remove radio from under instrument panel.

1965-67 Valiant and Barracuda and 1965 Dart

1. Disconnect battery.
2. Remove control knobs from front of radio.
3. From under instrument panel disconnect radio feed wire at connector.
4. Remove bottom screw from radio mounting bracket.
5. Remove left defroster tube.
6. Loosen top screw on radio mounting bracket and remove bracket.
7. Disconnect antenna and speaker leads.
8. Remove mounting nuts from front of radio and remove radio bezel.
9. Remove radio from below instrument panel.

1966 Dart

1. Disconnect battery.
2. If air conditioned, remove spot cooler tubes and spot cooler.
3. Disconnect antenna, speaker, and power supply leads.
4. Remove radio control knobs and mounting nuts.
5. Remove radio support bracket. Rotate front edge of radio down and remove from under panel.

1967 Dart and 1968 Valiant and Barracuda

1. Disconnect battery.
2. Remove radio control knobs and mounting nuts.
3. On air conditioned vehicles remove two outlet duct retaining bolts and remove duct. Remove right defroster hose and hose bracket.
4. From under instrument panel, remove radio support bracket lower screw and upper stud nut. Remove bracket.
5. Lower radio to position under panel and disconnect feed and speaker wires and antenna cable. Remove radio.

1968 Dart

1. Disconnect battery.
2. Remove instrument cluster. See front of section.
3. Remove six glove box mounting screws, collapse glove box, and remove box from panel.
4. Remove temperature control knobs.
5. Working through cluster and

glove box openings, remove two heater or air conditioning mounting stud nuts and move controls out of way.

6. Remove center bezel seven mounting screws and remove center bezel.
7. Remove radio mounting bracket.
8. Disconnect speaker and antenna leads.
9. Remove ashtray by removing four mounting screws.
10. Remove two radio mounting screws.
11. Remove radio from under instrument panel.

1969 Valiant, Dart, Barracuda, and Challenger

1. Disconnect battery.
2. From under panel disconnect speaker and wiring leads at radio.
3. Remove two radio mounting nuts from radio mounting bracket.
4. Move radio down and out from under instrument panel.

1970 Valiant, Dart, Barracuda, and Challenger

1. Disconnect battery.
2. From under panel, disconnect speaker and wiring leads at radio.
3. Remove channel selector shaft and knobs if so equipped.
4. Remove two radio mounting screws from under panel and loosen radio to lower support bracket mounting nut.
5. Move radio rearward, down, and out from under instrument panel.

Windshield Wipers

Motor Removal

1964-66 Valiant, Dart, and Barracuda

1. Disconnect the wiper link at the wiper motor. For variable speed motors, note position of follower cam.
2. Disconnect the wiper motor lead wires at the wiper motor.
3. Remove the three nuts attaching the wiper motor and wiper motor bracket assembly to the cowl panel. Pull the motor bracket assembly down from the bracket mounting studs and out from underneath the instrument panel.
4. Install by reversing removal procedure.

1967-70 Valiant and Dart and 1967-69 Barracuda

1. Disconnect battery.
2. Disconnect wiper motor wiring harness.
3. Remove three wiper motor mounting nuts. On vehicles with air conditioning it is easier to remove crank arm nut and crank arm from under instrument panel first and omit steps 4 and 5.
4. Work motor off mounting studs far enough to gain access to crank arm mounting nuts.
CAUTION: do not force or pry motor from mounting studs as drive link can be easily distorted.
5. Using ½ in. open end wrench,

remove motor crank arm nut. Carefully pry arm off shaft.
6. Remove wiper motor.

1970 Barracuda and Challenger

1. Disconnect battery.
2. Carefully remove wiper arm and blade assemblies.
3. Remove left cowl screen.
4. Remove drive crank arm retaining nut and drive crank. Disconnect wiring to motor.
5. Remove three wiper motor mounting nuts and remove motor.

Linkage Removal

1964-66 Valiant, Dart, and Barracuda Single Speed

1. From under instrument panel remove retaining clip and core washer from pivot arm pins.
2. Remove link and pin from motor crank.
3. Withdraw assembly from under instrument panel and remove drive link and bushing.

1964-66 Valiant, Dart, and Barracuda Variable Speed

1. Remove connecting link retaining clips and washers from pivot pins.
2. Remove connecting link.
3. Remove drive link retainer clip and washers from left pivot pin.
4. Remove retainer and washer from motor drive crank pin.
5. Remove drive link.
6. Remove cam, release, parking spring, and washer for examination, lubrication, or replacement.

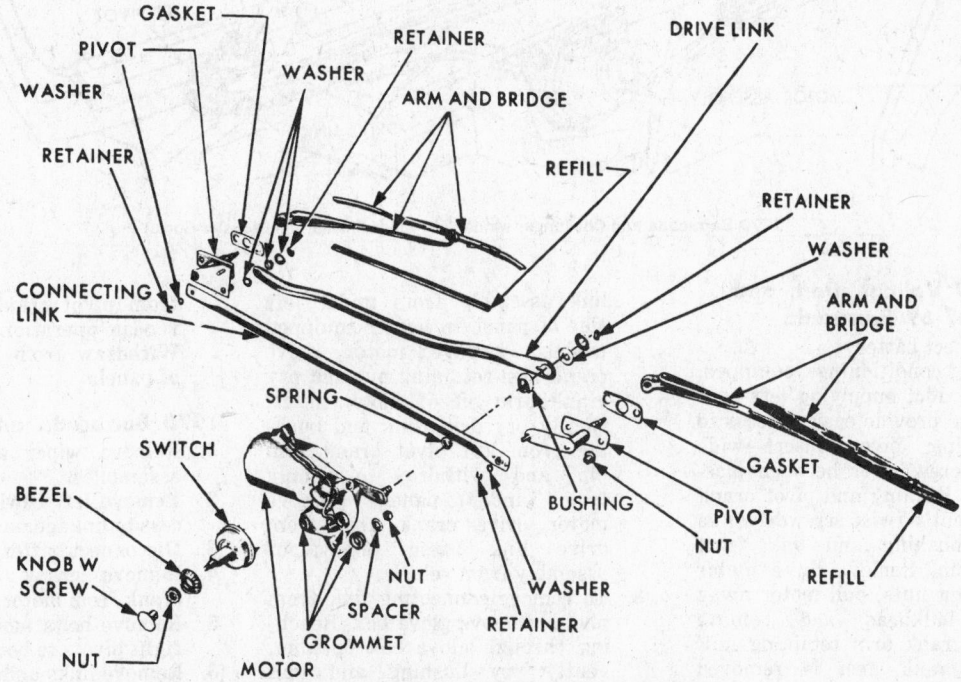

1964-66 single speed windshield wiper (© Chrysler Corp.)

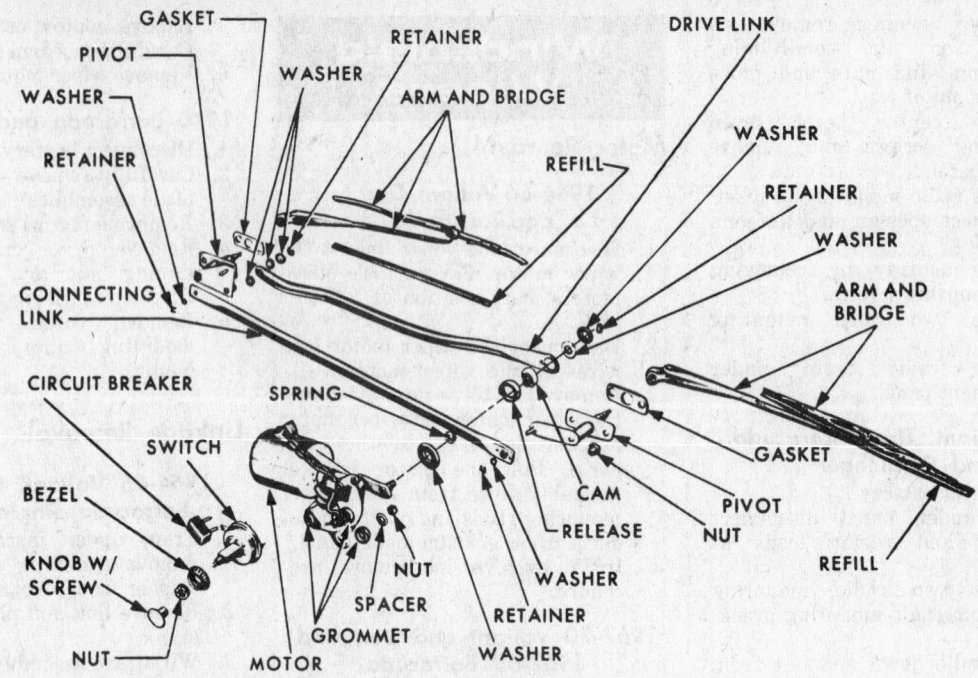

1964-66 variable speed windshield wiper (© Chrysler Corp.)

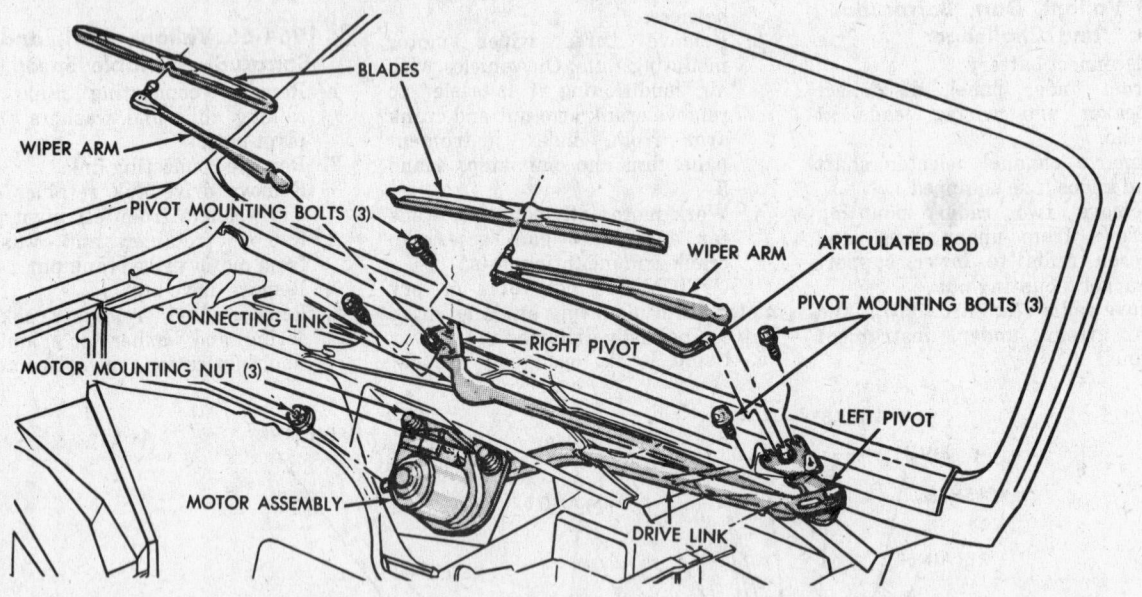

1970 Barracuda and Challenger windshield wiper linkage (© Chrysler Corp.)

1967-70 Valiant, Dart, and 1967-69 Barracuda

1. Disconnect battery.
2. If air conditioning equipped, remove duct supplying left spot cooler to provide easier access to left wiper pivot. Insert wide blade screwdriver between plastic link bushing and pivot crank arm. Gently twist screwdriver to force bushing and link from pivot pin. Remove three motor mounting nuts, pull motor away from bulkhead and remove motor crank arm retaining nut. After crank arm is removed from motor shaft, remove drive link assembly from under left side of panel. In heater equipped models, remove motor drive crank arm retaining nut and pry crank arm off of motor shaft. Gently pry drive link and bushing from left pivot crank arm pin and withdraw assembly from under panel. Remove motor drive crank arm from drive link after removal of assembly from vehicle.
3. To remove connecting link from pivots, remove glove box. Reaching through glove box opening, gently pry bushing and link from right pivot pin. Lift link from pivot crank arm pin and repeat operation at left pivot. Withdraw from under left side of panel.

1970 Barracuda and Challenger

1. Remove wiper arm and blade assemblies.
2. Remove left cowl screen for access to linkage.
3. Disconnect battery.
4. Remove crank arm nut and crank from motor shaft.
5. Remove bolts mounting left and right pivots to body.
6. Remove links and pivots through cowl top opening.

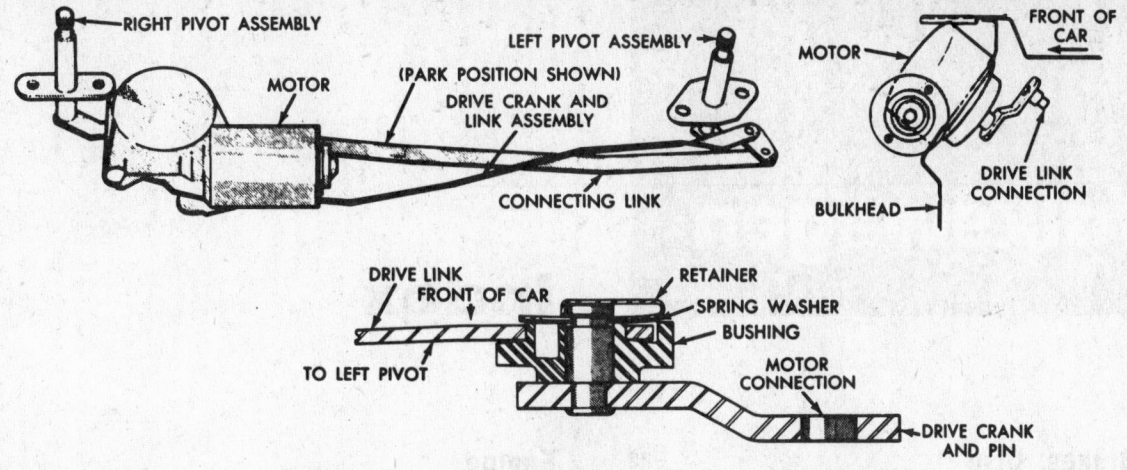

1967-70 Valiant and Dart and 1967-69 Barracuda windshield wiper linkage
(© Chrysler Corp.)

Heater System

Heater Blower R & R

1964-65 Valiant and Dart and 1965 Barracuda

1. Disconnect battery.
2. Disconnect heater ground wire.
3. Loosen fresh air intake duct clamp at blower end and remove duct from blower assembly.
4. Remove screw attaching blower to plenum.
5. Remove heater blower assembly from heater housing.
6. From inside heater housing, disconnect blower assembly wires.

Heater Core R & R

1964-65 Valiant and Dart and 1965 Barracuda

1. Drain coolant as necessary.
2. Disconnect heater hoses at core and temperature control valve.
3. Remove screws that hold temperature control valve cover to heater housing.
4. Carefully withdraw control valve from opening in heater housing and disconnect valve control cable.
5. Disconnect hose connecting water valve to core.
6. Remove water valve and valve cover.
7. Remove screws that attach core housing to cowl panel and remove housing.
8. Insert screwdriver between cowl and heater housing, and carefully pry heater housing from cowl.
9. Remove screws that secure core to housing and remove core.
10. Install in reverse of above.

1966-70 Valiant and Dart and 1966-69 Barracuda

To service heater blower or heater core, it is necessary to remove heater assembly from vehicle.

Heater Removal

1. Drain radiator.
2. Disconnect heater hoses from heater and remove heater hoses to dash retainer plate.
3. Remove heater motor seal retainer plate from dash panel.
4. Disconnect heater-defroster and temperature control cables from heater assembly.
5. Remove heater motor resistor wire from resistor.
6. Remove defroster tubes from heater assembly.
7. Disconnect heater housing support rod from fresh air duct.
8. Remove heater assembly.

Heater Disassembly

1. Remove seal from around heater motor mounting studs.
2. Remove spring clips holding spacers and heater motor to heater housing.
3. Remove fan from heater motor.
4. Remove mounting support plate from heater motor.
5. Remove heater motor resistor assembly from heater housing.
6. Remove fresh air door seal from either inner or outer heater housing half only.
7. Remove retainer clips attaching heater housing halves together.
8. Separate heater housing halves.
9. Remove screw attaching seal retainer and seal around heater core tubes.
10. Remove heater core tube support clamp.
11. Remove screws attaching heater core to heater housing and remove core.
12. Remove seal retainer and seal from heater core.

1970 Challenger and Barracuda

Heater Removal

1. Disconnect battery.
2. Drain coolant.
3. Disconnect heater hoses from core tubes at dash panel. Plug core tubes to prevent spilling coolant on interior of car.
4. Remove three mounting nuts from studs around blower motor and remove flange and air seal.
5. Unplug antenna from radio and place wire to one side.
6. Remove screw from housing to plenum support rod on right side of housing above fresh air opening.
7. Disconnect three air door cables.
8. Disconnect wires from blower motor resistor.
9. Tip unit down and out from under instrument panel.

Blower Motor Removal

1. Remove heater assembly from car.
2. Disconnect blower motor lead from resistor block and ground wire from mounting plate.
3. Remove six sheet metal screws and six retaining clips holding blower motor assembly from housing.
4. Remove blower wheel from motor shaft.
5. Remove two retaining nuts and separate motor from mounting plate.

Volkswagen

Index

BASIC BODY TYPES

Convertible, Type 1

Fastback sedan, Type 3

Type 1

Squareback Sedan, Type 3

Micro Bus, Type 2

Sedan (Beetle), Type 1

YEAR IDENTIFICATION

1949
(© Volkswagen)

1950
(© Volkswagen)

1951
(© Volkswagen)

1952
(© Volkswagen)

1953
(© Volkswagen)

1954
(© Volkswagen)

1955
(© Volkswagen)

1956
(© Volkswagen)

1957
(© Volkswagen)

1958
(© Volkswagen)

1959
(© Volkswagen)

1960
(© Volkswagen)

YEAR IDENTIFICATION

1961
(© Volkswagen)

1962
(© Volkswagen)

1963
(© Volkswagen)

1964
(© Volkswagen)

1965
(© Volkswagen)

1966
(© Volkswagen)

1967
(© Volkswagen)

1968
(© Volkswagen)

1969
(© Volkswagen)

1970
(© Volkswagen)

FIRING ORDER

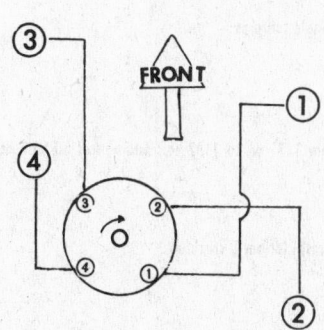

CAR SERIAL NUMBER

The only positive way to identify the year of manufacture is to compare the chassis number with the table below. Chassis numbers are found in two locations:

(1) Under the rear seat, stamped on the frame tunnel.
(2) Behind the spare tire in the trunk.

Table of Volkswagen Types

Model Number (LHD)	Description
111	VW 1300A sedan
115	VW 1300A sedan with folding sunroof
113	VW 1500 sedan
117	VW 1500 sedan with steel sunroof
141	VW 1500 Karmann Ghia Convertible
143	VW 1500 Karmann Ghia Coupe
151	VW 1500 Convt. (4-seater)
211-215	Delivery Van
221-225	Micro Bus
231-237	Kombi
241	Deluxe Micro Bus (9-seater)
251	Deluxe Micro Bus (7-seater)
261-267	Pick-up
271-273	Ambulance
281-285	Micro Bus (7-seater)
311	Fastback sedan (1600TL)
313	Fastback sedan with steel sunroof
315	1600A sedan
317	1600A sedan with steel sunroof
343	1600L Karmann Ghia Coupe
345	1600L Karmann Ghia Coupe with steel sunroof
361	1600L Squareback sedan
363	1600L Squareback sedan with steel sunroof
365	1600A Squareback sedan
367	1600A Squareback sedan with steel sunroof

CAPACITIES

YEAR AND MODEL	ENGINE CRANKCASE (Pts.)	REAR AXLE TRANSMISSION CASE (Pts.)	GASOLINE TANK (Gals.)	AIR CLEANER (Pts.)
Type 1,3 4-Speed	5.3	6.3	10.5	.53
Type 2	5.3	7.4	15.8	.95

TYPE 1 IDENTIFICATION

Year	Chassis Numbers	Major Body Changes	Major Mechanical Changes
1970		Louvers in rear hood; modified lights and reflectors; ignition lock buzzer warning.	1600 CC 57 h.p. engine; vapor emission control system (California only).
1969	119,000,001–	Electric rear window defroster; ignition lock combined with steering lock. Locking gas flap.	Double-jointed rear axle standard.
1968	118,000,001–118,1016,098	Raised bumpers, front and rear; vertical bumper guards eliminated. Built-in headrest in front seats; extensive padding in front compartment and dashboard. Seat belts in rear standard. External gas filler.	Exhaust emission control. Collapsible steering column. Automatic stick shift optional.
1967	117,000,001–118,000,000	Back-up lights. Retractable seat belts. Armrest for driver. Locking buttons on doors. Parking light built into turn signals. Narrower chrome trim. Volkswagen nameplate on engine lid. Two-speed windshield wipers. Headlights now vertical in indented fenders.	Dual brake system. Increased horsepower, from 50 to 53. (SAE) Larger engine, from 1300 cc to 1500 cc. 12-volt electrical system. Number of fuses increased from 8 to 10. More powerful starter motor. Equalizer spring rear axle.
1966	116,000,0001–116,1,021,298	Number "1300" on engine lid. Flat hub caps; ventilated wheel discs. Four-way flasher system. Dimmer switch on turn signals. Defroster outlet in center of dash. Front seat backs equipped with safety locks.	Increased engine size, from 1200 cc. to 1300 cc. Increased engine output, from 40 hp. to 50 hp. (SAE)
1965	115,000,0001–115,979,200	Larger windows, narrower window and door posts. Heater control levers now mounted on tunnel, formerly a twisting knob. Push-button handle on engine lid. Back of rear seat convertible to a flat platform.	No major changes.
1964	5,677,119–6,502,399	Steel sliding sunroof; crank operated. Wider license plate light. Non-porous leatherette upholstery replaced by porous vinyl material.	No major changes.
1963	4,846,836–5,677,118	Sunroof equipped with folding handle. Fresh air heating system. Nylon window guides. Introduction of leatherette headliner; formerly "mouse fuzz." Wolfsburg hood crest eliminated from front hood.	No major changes.
1962	4,010,995–4,846,835	Spring-loaded hood. Addition of seat belt mounting points. Gasoline gauge on dashboard; formerly only a reserve fuel tap. Size of tail lights increased. Sliding covers for front floor heating outlets. Windshield washer added; compressed air type.	Worm and roller steering; formerly worm and sector. Tie rod ends permanently lubricated.
1961	3,192,507–4,010,994	Flatter gasoline tank. Increased front luggage space. Windshield washer; pump-type. Key slot in doors now vertical; formerly horizontal. Starter switch now non-repeat.	Increased engine output, from 36 hp. to 40 hp. (SAE). Automatic choke. Push-on electrical connectors. First gear now synchromesh; all forward speeds now synchromesh.
1960	2,528,668–3,192,506	"Dished" steering wheel. Push-button door handles; formerly lever-type. Foot rest for passenger. Padded sunvisor.	Front anti-sway bar added. Generator output increased to 180 watts, formerly 160. Steering damper added.
1959	2,007,616–2,528,667	No major changes.	Stronger clutch springs. Fan belt improved. Frame given additional reinforcement.
1958	1,600,440–2,007,615	Larger rear window and windshield. Front turn signal lights moved to top of fenders. Radio grill moved to far left of dashboard.	Wider brake drums and shoes.
1957	1,246,619–1,600,439	Doors fitted with adjustable striker plates. Front heater outlets moved rearward, to within five inches of door. Tubeless tires used; formerly tube-type.	No major changes.
1956	929,746–1,246,618	Tail light housings raised two inches. Steering wheel spoke moved lower and off-center. Heater knob moved forward. Adjustable front seat backs; formerly non-adjustable. Increased front luggage space.	Dual tail pipe; formerly single tail pipe.
1955	722,935–929,745	Flashing turn signal lights replace "semaphore"-type flappers. Indicators mounted near outside bottom of front fenders.	No major changes.
1954	575,415–722,934	Starter switch combined with ignition switch; formerly a separate button on dashboard. Interior courtesy light added.	Increased engine size, from 1131 cc. to 1192 cc. Addition of oil-bath air cleaner.
1953	428,157–575,414	Oval rear window replaces two-piece split rear window. Vent window handles now provided with a lock button.	No major changes.
1952	313,830–428,156	Vent windows added. Body vent flaps eliminated. Window crank geared down from 10½ to 3½ turns. Door added to glove compartment. Turn signal control to steering wheel; formerly on dashboard. 5.60 x 15 tires. Formerly 5.00 x 16.	Top three gears synchromesh; formerly crashbox.
1951	220,472–313,829	Vent flaps in front quarter-panel of body. Wolfsburg crest above front hood handle.	No major changes.
1950	138,555–220,471	Ash tray added to dashboard.	Hydraulic brakes; formerly mechanical.
1949	91,922–138,4554	Pull release for front hood; formerly locking handle.	Solex carburetor now standard equipment.

BATTERY AND STARTER

	BATTERY						STARTER		
Amp. Hr. Capacity	Volts	SAE Group Number	Terminal Grounded	Lock Test			Amps.	No-Load Test Volts	RPM
				Amps	Volts				
66 or 77	6	7L1	Neg.	500	3.5		80	5.5	5400
(1967–70) 44	12	—	Neg.	250–285	—		38–45	12	7150

TUNE-UP SPECIFICATIONS

MODEL	SPARK PLUGS Type	Gap (In.)	DISTRIBUTOR Gap (In.)	Dwell (Deg.)	INITIAL TIMING ▲ (Deg.)	VALVE CLEARANCE (Cold) (In.)★	IDLE SPEED (Hot) RPM
25 HP	L-10	.026	.016	50	5B	.004	550
36 HP	L-10S	.026	.016	50	7½B	.004	550
40 HP	L-10S	.026	.016	50	*	**	550
1300 SERIES	L-10S	.026	.016	50	7½B	.004	550
1500 SERIES	L-85	.026	.016	50	①	***	550
1600 SERIES	L-87Y	.026	.016	50	7½B	***	550

*—7½°B up to 8/66 models, 10°B from engine no. DO 095 049.

**—Up to chassis no. 9,299,999, .008 intake; .012 exhaust. Chassis no. 9,300,000 up, .004 intake and exhaust or .008 intake and .012 exhaust. Refer to decal on engine fan housing for positive clearance figures.

***—1500/1600 models, single carburetor versions: up to no. 672,748—.008 intake; .012 exhaust. From no. 672,749—.004 intake and exhaust. 1500/1600 models, dual carburetor versions: up to no. 672,297—.008 intake; .012 exhaust. From no. 672,298—.004 intake and exhaust.

B—Before top dead center.

★—If any conflict exists between the figures given here and the decal on an individual engine, use the figure given on the decal. Most Volkswagen automobiles have had this decal since 1964. Also, most rebuilt engines, regardless of year, have this decal.

▲—NOTE: These settings are only approximate. Engine design, altitude, temperature, fuel octane rating and the condition of the individual engine are all factors which can influence timing. The limiting advance factor must, therefore, be the "knock point" of the individual engine.

①—10°B; 0° for exhaust emission control.

GENERAL ENGINE SPECIFICATIONS

MODEL OR HORSEPOWER	DISPLACEMENT (CC./CU. IN.)	BORE & STROKE (IN.)	HORSEPOWER (SAE)	TORQUE (FT. LBS.)	COMPRESSION RATIO	OIL PRESSURE (PSI)	CRANKING COMPRESSION (PSI)
25 HP Type 1	1131/69.02	2.953 x 2.520	25 @ 3300	51 @ 2000	5.8-1	7.0 #	85-105
36 HP Type 1 (to 7/60)	1192/72.74	3.0315 x 2.520	36 @ 3700	60 @ 2400	6.6-1	7.0 #	100-120
40 HP Type 1	1192/72.74	3.0315 x 2.520	40 @ 3900	60.7 @ 2000	7.1-1	7.0 #	100-129
42 HP Type 1 (from 8/60)	1192/72.74	3.0315 x 2.520	41.5 @ 3900	65 @ 2400	7.0-1	7.0 #	100-128
1300 Type 1 (from 8/65)	1285/78.3	3.0315 x 2.720	50 @ 4600	69 @ 2600	7.3-1	7.0 #	107-135
1500 Type 1 (from 8/66)	1493/91.10	3.2677 x 2.7165	53 @ 4200	78 @ 2600	7.5-1	7.0 #	114-142
1500 Type 2 (from 7/65)	1493/91.10	3.2677 x 2.7165	51 @ 4000	74 @ 2600	7.8-1	7.0 #	121-142
1500 Type 2 (from 8/65)	1493/91.10	3.2677 x 2.7165	53 @ 4200	78 @ 2600	7.5-1	7.0 #	114-142
1500 Type 3 (to 7/65)	1493/91.10	3.2677 x 2.7165	54 @ 4200	84 @ 2800	7.8-1	7.0 #	121-142
1500 Type 3 (from 8/65)	1493/91.10	3.2677 x 2.7165	54 @ 4200	84 @ 2800	7.5-1	7.0 #	114-142
1500 S Type 3 (to 7/65)	1493/91.10	3.2677 x 2.7165	66 @ 4800	84 @ 3000	8.5-1	7.0 #	135-164
1600 Type 3 (from 8/65)	1584/96.66	3.360 x 2.7165	65 @ 4600	87 @ 2800	7.7-1	7.0 #	114-142
1600 Type 1	1584/96.66	3.360 x 2.7165	57 @ 4400	—	7.5-1	7.0#	114-142

#—28 PSI @ 2500 rpm.

GENERATOR AND REGULATOR

GENERATOR Bosch Part No.	Brush Spring Pressure (oz.)	Field Resistance	REGULATOR Bosch Part No.	Cut-Out Relay Volts To Close	Reverse Current	Maximum Current (Amps.)	Voltage Regulator Setting
LJ/REF160/6/2500L19	16-21	1.20-1.32	RS/TA160/6A3	5.5-6.3	2.0-5.5	40	7.4-8.1
LJ/REF160/6/2500L17	16-21	1.20-1.32	RS/TAA160/6/1	6.4-6.8	2.0-5.5	40	7.4-8.1
LJ/REF160/6/2500L21	16-21	1.20-1.32	RS/TAA160/6/1	6.4-6.8	2.0-5.5	40	7.4-8.1
RED130/6/2600AL16	16-21	1.20-1.32	RS/TA130/150/6A1	5.5-6.3	2.0-7.5	33	7.4-8.1
LJ/REF160/6/2500L4	16-21	1.20-1.32	RS/TA160/6A3	5.5-6.3	2.0-5.5	40	7.4-8.1
LJ/REG/180/6/2500L2	16-21	1.20-1.32	RS/TA780/6/1	6.2-6.8	2.0-5.5	45	—
1967—L1/GEG450M12/3700FL	—	—	RS/VA450/M12/A5	12.4-13.1	3.0-9.0	30	12.8-13.6

When a unit is being overhauled, any parts that have either reached or are close to the wear limit should be renewed.

Engine; 1.2 liter (40 bhp), 1.3, 1.5, 1.6

	Tolerance (new) mm. (in.)	Wear Limit
Cylinder seating depth in cylinder head:		
1.2, 1.3	13.7–13.8 (.539–.543)	—
1.5, 1.6	13.75–13.85 (.541–.545)	—
Cylinder, out of round:		
all	.01 (.0004)	—
Clearance between piston and cylinder:		
1.2, 1.3	.04–.05 (.0016–.002)	.20 (.008)
1.5, 1.6	.04–.06 (.0016–.0024)	.20 (.008)
Side clearance, upper compression ring:		
1.2, 1.3	.07–.09 (.0027–.0035)	.12 (.005)
1.5, 1.6	.08–.11 (.0031–.0043)	.12 (.005)
Side clearance, lower compression ring:		
all	.05–.07 (.002–.0027)	.10 (.004)
Side clearance, oil scraper ring:		
all	.03–.05 (.001–.002)	.10 (.004)
Gap, both compression rings:		
all	.30–.45 (.012–.018)	.90 (.035)
Gap, oil scraper ring:	.25–.40 (.010–.016)	.95 (.037)
Maximum weight difference between pistons in one engine:		
all	5 grams	10 grams (when repairing)
Maximum weight difference of connecting rods in one engine:		
all	5 grams	10 grams (when repairing)
Clearance between piston pin and connecting rod bushing:		
all	.01–.02 (.0004–.0008)	.04 (.0016)
Connecting rod/connecting rod journal, radial clearance:		
1.2, 1.3	.02–.08 (.0008–.003)	15 (.006)
1.5, 1.6	.02–.07 (.0008–.0027)	.15 (.006)
Connecting rod/connecting rod journal, side clearance:		
all	.10–.40 (.004–.016)	.70 (.028)

	Tolerance (new) mm. (in.)	Wear Limit
Crankshaft, main journal radial clearance, all engines (taking into account preload of housing):		
Bearing #1, #3	.04–.10 (.0016–.004)	.18 (.007)
Bearing #2 (from 8/65)	.03–.09 (.001–.0035)	.17 (.0067)
Bearing #4	.05–.10 (.002–.004)	.19 (.0075)
Crankshaft run-out at #2 and 4 main journals (#1 and 3 journal on V-blocks):		
1.2, 1.3	—	.03 (.001)
1.5, 1.6	—	.02 (.0008)
Crankshaft/main journal 1, side clearance:		
all	.07–.13 (.0027–.005)	.15 (.006)
Crankshaft/main journal 2, side clearance:		
all	.06–.13 (.0024–.005)	.15 (.006)
Crankshaft, maximum out-of-balance:		
all	8 cmg (approx. 11 in. oz.)	—
Main bearing journals, maximum out of round:		
all	—	.03 (.001)
Connecting rod journal, maximum out-of-round:		
all engines	—	.03 (.001)
Crankcase bore for crankshaft bearings:		
#1, #2, & #3	65.00–65.02 (2.559–2.5598)	65.03 (2.5601)
#4	50.00–50.03 (1.9685–1.9696)	50.04 (1.9700)
Diameter, oil seal seat	90.00–90.05 (3.5433–3.5452)	
Camshaft, diameter of crankcase bearing bores, from August, 1962:		
all	27.50–27.52 (1.082–1.083)	
Crankcase bores for camshaft, diameter:		
all	25.02–25.04 (.9850–.9857)	
Camshaft/camshaft bearings, radial clearance (taking housing preload into account):		
all	.02–.05 (.0008–.002)	.12 (.0047)

	Tolerance (new) mm. (in.)	Wear Limit
Camshaft thrust bearing, axial clearance:		
all	.06–.11 (.0024–.0043)	.14 (.0055)
Camshaft center bearing, run-out between centers:		
all	.02 (.0008)	.04 (.0016)
Camshaft gear backlash:	.00–.05 (.000–.002)	
Flywheel, lateral run-out measured at center of friction surface:		
all	max. .30 (.012)	
Flywheel, maximum out-of-balance:		
all	5 cmg (approx. 7 in. oz.)	
Shoulder for oil seal, outside diameter:		
all	69.9–70.1 (2.7519–2.7598)	69.4 (2.7322)
Turning down tooth width:		
all	maximum 2.0 (.08)	
Valve stem, intake diameter:		
all	7.95–7.94 (.3130–.3126)	7.90 (.3110)
Valve stem, exhaust diameter:		
all	7.92–7.91 (.3118–.3114)	7.87 (.3098)
Maximum out-of-round of valve stem:		
all	.01 (.0004)	
Valve head diameter, intake		
1.2	31.5 (1.239)	
1.3	33.0 (1.299)	
1.5 and 1.6	35.5 (1.396)	
Valve head diameter, exhaust:		
1.2 and 1.3	30.0 (1.181)	
1.5 and 1.6	32.0 (1.259)	
Valve guide intake inside diameter:		
all	8.00–8.02 (.3150–.3157)	8.06 (.3173)
Valve guide, exhaust, inside diameter:		
all	8.00–8.02 (.3150–.3157)	8.06 (.3173)
Rock of valve guide/valve stem, intake:		
all	.21–.23 (.008–.009)	.80 (.031)

	Tolerance new mm. in.	Wear Limit
Rock of valve guide/valve stem, exhaust: all	.28–.32 (.011–.012)	.80 (.031)
Valve seat width, intake: all	1.3–1.6 (.05–.06)	
Valve seat width, exhaust: all	1.6–2.0 (.06–.08)	
Valve seat/head, run-out: all	.01 (.0004)	.02 (.0008)
Valve springs, loaded lengths and loads: all	31 mm / 33.4 mm / 34.3 mm	126 + or − 8.8 lbs. / 96.4 + or − 6.6 lbs. / 102 + or − 5 lbs.
Valve rocker arms, inside diameter: all	18.00–18.02 (.708–.709)	18.04 (.710)
Valve rocker shaft, diameter: all	17.98–17.97 (.7077–.7073)	17.95 (.7066)
Cam follower bore in crankcase diameter: all	19.00–19.02 (.7480–.7485)	19.05 (.7509)
Cam follower, diameter: all	18.98–18.96 (.7471–.7463)	18.93 (.7452)
Compression, maximum permissible difference between cylinders in a single engine: all	14 psi	
Oil pump, axial clearance between gears/housing, measured without preload and with gasket: all	.07–.18 (.0027–.007)	.20 (.008)
Axial clearance between gears/housing, without gasket: all		.10 (.004)
Oil pump, backlash of gears: all	.03–.08 (.001–.003)	.15 (.006)
Pressure at which oil pressure relief valve opens: all	.15–.45 atu (2.1–6.4 psi)	

	Tolerance (new) mm. in.	Wear Limit
Oil pressure, with SAE 30 oil and an oil temperature of 70°C. (158°F.) all engines: at 550 rpm	minimum .5 atu (7 psi)	
at 2500 rpm	minimum 2.0 atu (28 psi)	
Loaded length of 23.6 mm. oil pressure relief spring, required load: all	7.75 kg (17 lbs.)	
Clutch pressure plate, run-out: all	maximum .10 (.004)	
Clutch release plate, run-out: all	maximum .30 (.012) .60 (.024) for diaphragm spring clutch	
Clutch complete, unbalance: all	maximum 15 cmg (.21 in. oz.)	
Clutch plate, run-out: all	.8 (.32)	
Free play at clutch pedal: all	10–20 (.4–.8)	
Automatic clutch (Type 1, Saxomat), shift lever contact gap:	.25 (.010)	

Rear Axle and Transmission Full Synchromesh

	Tolerance (new)	Wear Limit
1st and 4th speed gearwheels, axial clearance: all	.10–.25 (.004–.010) (Keep to lower limit, .004)	
Synchronizer rings/gear wheels, clearance between end faces of toothed rings: all	1.10 (.043)	.60 (.024)

	Tolerance (new) mm. (in.)	Wear Limit
Shift fork/operating sleeve axial clearance for 1st/2nd, 3rd/4th speeds: all	.10–.30 (.004–.012)	
Drive shaft, front (surface for 3rd gear needle bearing, run-out): all	maximum .02 (.0008)	
Rear axle shaft; flange/fulcrum plates/differential side gears (4 parts) clearance all	.04–.24 (.0015–.009)	.25 (.010)
Rear axle shaft; flange/differential side gear clearance, measured across the convex faces: all	.03–.10 (.001–.004)	.20 (.008)
Rear axle shafts; measured between centers, measured at bearing seat, run-out: all	maximum .05 (.002)	
Preload of transmission case halves and final drive covers on differential ball bearings: all	.14 (.005)	
Plastic packing/transmission case/axle tube/final drive cover clearance: all	.25–.35 (.010–.014)	.40 (.016)
Run-out of reduction gear shaft (Type 2):	maximum .01 (.0004)	
Inside diameter, gearshift housing bushes: all	15.05–15.03 (.592–.591)	15.25 (.600)
Inner shift lever, diameter: all	15.00–14.96 (.590–.588)	14.75 (.580)
Starter bush, inside diameter: all	12.55–12.57 (.493–.494)	12.65 (.497)
Starter shaft/bush radial clearance: all	.09–.14 (.0035–.005)	.25 (.010)

TORQUE SPECIFICATIONS

Engine

Description	Class, Thread	Torque ft. lbs.
Nuts, crankcase halves: (except 1.2 liter, 36 bhp engine: 22)	5D M 12 x 1.5	25
Crankcase screws, nuts:	5S M 8	14
Cylinder head nuts: (except 1.2 liter, (36 bhp: 26–27)	5D M 10	23
Rocker shaft nuts:	8G M 8	14–18
Flywheel gland nut:	8G M 28 x 1.5	217
Connecting rod bolts:	10K M 9 x 1	32 ± 3.6
Special nut for fan:	5S M 12 x 1.5	40–47
Nut on generator pulley: (for 65 bhp engine: 32)	5S M 12 x 1.5	40–47
Bolt, crankshaft pulley:	9S 20 K	for 1600 (29–36) (94–108)
Spark plugs:	M 14 x 1.25	22–29
Oil drain plug: (for 36 bhp engine: 22–29)	5S M 14 x 1.5	25

Description	Class, Thread	Torque ft. lbs.
Clutch to flywheel:	8G M8 x 1.5	18
Insert for spark plug, 1.2 liter engine:	5S M 18 x 1.5	50–54

Transmission and Rear Axle (Fully Synchronized)

Description	Class, Thread	Torque ft. lbs.
Double ball bearing:	C 35N M 35 x 1.5	87
Double taper roller bearing:	C 35N M 35 x 1.5	144
Retaining screws, pinion bearing:	10K M 10 x 1.5	36
Pinion nut:	CK 45K M 22 x 1.5	43*
Drive shaft nut:	CK 45K/C 35 M 22 x 1.5	43*
Reverse lever guide screw:	8G M 7 x 1	14
Selector fork screws:	C45 KN M 8 x 1.25	14

Description	Class, Thread	Torque ft. lbs.
Ring gear screws:	10K M 10 x 1.5	43
Final drive cover nuts:	8G M 8 x 1.25	22
Axle tube retainer nuts:	6G M 8 x 1.25	14
Oil drain plug and oil filler plug:	Muk 7 M24 x 1.5	14
Rear axle shaft nut (type 1 and type 3):	C45KN M 24 x 1.5	217
Nut on driven shaft (type 2 from 8/63):	6S M 30 x 1.5	108
Nut on rear axle driven shaft (type 2 from chassis number 1144303):	C45KN M 30 x 1.5	217 (250 max.)
Transmission carrier on frame:	8G M 18 x 1.5	166
Spring plate/ reduction gear screws (type 2):	10K M 12 x 1.5	72–87

*—Tighten first to 86 ft. lbs. and then back off, and finally tighten to 43 ft. lbs.

CYLINDER-HEAD BOLT TIGHTENING SEQUENCE

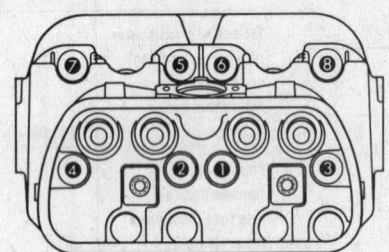

Tighten nuts to 1 mkg (7 ft. lbs.) first in the order shown

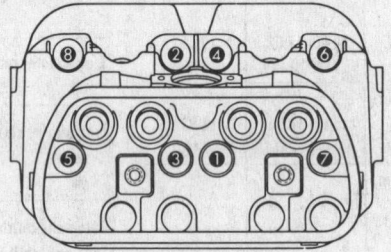

Fully tighten nuts to specified torque in this order

GENERAL CHASSIS AND WHEEL ALIGNMENT SPECIFICATIONS
(Unloaded)

Model	REAR TRACK (mm.)	TIRE SIZE WHEEL SIZE	CASTER (Deg.)	CAMBER (Deg.)	TOE-IN (mm.)	WHEEL PIVOT RATIO	
						Inner	Outer
1200 Type 1 (to July '65)	1300 mm.	5.60 x 15/4J15	2° ± 15'	0°40' ± 30'Δ	2–4.5 mm.	34°	28°
1200 Type 1 (After July '65)	1300 mm.	5.60 x 15/4J15	2° ± 15'	0°30' ± 15'Δ	2–4.5 mm.	34°	28°
1300 Type 1	1300 mm.	5.60 x 15/4J15	2° ± 15'	0°30' ± 15'Δ	2–4.5 mm.	34°	28°
1500 Type 1 (swing axle)	1358 mm.	5.60 x 15/4J15	2° ± 15'	0°30' ± 15'	2–4.5 mm.	34°	28°
Type 2—1968 Double-link axle	1423 mm.	7.00 x 14/5JK14	3° ± 40'	40' ± 15'	—	—	—
Type 3 (Swing axle)	—	6.00 x 15/4½J15	—	1°20' ± 10'Δ	3.6–6 mm.	—	—

Δ—Rear wheel chamber 3° ± 30'

CRANKSHAFT BEARING JOURNAL SPECIFICATIONS

MODEL OR HORSEPOWER	MAIN BEARING JOURNALS (IN.)			CONNECTING ROD JOURNALS (IN.)
	No. 1–3	No. 4	End-Play	
25 HP	1.9685	1.5748	.003–.005	1.9685
36 HP	1.9685	1.5748	.003–.005	1.9685
40 HP	1.9685	1.5748	.003–.005	1.9685
1200, 1300 Series	2.1654	1.5748	—	2.1654
1500, 1600 Series	2.1654	1.5748	—	2.1653

NOTE: See Tolerances and Wear Limits Chart.

VALVE SPECIFICATIONS

MODEL OR HORSEPOWER	SEAT ANGLE (DEG.)	SPRING PRESSURE (Lbs. @ Inches of Length)	INTAKE OPENS (DEG.)	VALVE STEM (IN.)			
				Diameter		Clearance	
				Intake	Exhaust	Intake	Exhaust
25 HP	45°	74 @ 1.10	2.5°B	.2740	.2736	.0019	.0023
36 HP	45°	74 @ 1.10	2.5°B	.2740	.2736	.0019	.0023
40 HP	45°	74 @ 1.10	2.5°B	.2740	.2736	.0019	.0023
1200, 1300 SERIES	45°	74 @ 1.10	4.0°B	.3128	.3116	.0022	.0032
1500, 1600 SERIES	45°	74 @ 1.10	1.0°B	.3126–.3130	.3114–.3118	—	—

B—Before top dead center.

LIGHT BULBS

VW 1960 (6 Volt)

Headlamps	6006
Parking	63
Stop & tail	1154
Turn signals—front	1129
Turn signals—rear	1154
License plate	63

VW 1961–66 (6 Volt)

Same as 1960, except

Rear turn signals	1129
License plate	63

1967–70 (12 Volt)

Headlamp	6012
Stop & tail	67
Front park, turn signal, tail and stop lights	1034
Rear turn signal—backup	1073
License lamp	89

Tune-Up

Tune-up and diagnosis operations are performed in the same way as on any other four cycle automotive engine. See charts on preceding pages for specifications.

Compression

1. Blow foreign matter from the plug wells. Then loosen all spark plugs one turn.
2. Start engine and accelerate to blow out disturbed carbon.
3. Stop engine and remove plug wires and spark plugs.
4. Block choke and throttle in wide-open position.
5. Hook up starter remote control cable and insert compression gauge into plug port.
6. Crank engine through about four compression strokes and record highest reading.
7. Do likewise with the remaining cylinders. Variation between cylinders should not exceed 20 psi.
NOTE: see General Engine Specifications Table for proper reading for each engine. Only if compression pressures are within limits should tune-up be continued.

Spark Plugs

Use a good spark plug tester, if available. A visual check should disclose any worn electrodes, glazed, broken or blistered porcelains and heavy oil or carbon deposits. Clean or replace spark plugs as required (see tune-up chart for correct type), install new plug gaskets and torque to 20-25 ft. lbs.

Ignition System

1. Replace brittle or otherwise damaged spark plug wires.
2. Tighten all ignition system connections.
3. Remove distributor cap, clean cap and inspect for cracks, carbon tracks, and burned or corroded terminals. Replace cap, if necessary.
4. Clean rotor and inspect for damage.
5. Check distributor governor advance action (if distributor is so equipped) by twisting the rotor in a clockwise direction as far as possible. Release the rotor to see if the spring tension is sufficient to return it to the retarded position. In the case of sluggish or partial return, the distributor must be disassembled for corrective measures.
6. Check vacuum spark control mechanism (if distributor is so equipped) by pushing the breaker plate connecting lever against the diaphragm spring, then releasing it to see if the spring returns it to full retard. Correct any interference or binding, if present.
7. Examine points, clean or replace as necessary.
8. Adjust distributor points.

Battery and Cables

1. Visually inspect battery case, cables and carrier for any condition which would interfere with good service. Make corrections.
2. Measure the specific gravity of the electrolyte in each cell. If it is below 1.230 (at about 80°F), recharge the battery or further check for a drain in the charging circuit.
3. If a high rate discharge tester is used to check the battery, cells in satisfactory condition should give a reading of 2.2 volts each. A reading of 2.0 volts, or less, indicates a battery in need of attention.

Generator and Blower Drive Belt

1. Inspect condition of belt.
2. Check and adjust, if necessary, for correct tension. If necessary to adjust drive belt, add or subtract spacer washers from the two halves of the generator drive pulley. Removing spacers from between the two halves increases the effective pulley diameter, increasing belt tension. Adding spacers between the pulley halves decreases belt tension. When all spacer washers have been removed, the belt must be replaced.
NOTE: belt has proper tension, when 1/2-3/4 in. deflection can be produced when slight thumb pressure is applied to the belt, halfway between the two pulleys.
On type 1 and 2 models, the generator belt drives the cooling fan; these vehicles must never be operated with a broken or slipping belt. On type 3 vehicles, the belt drives only the generator. These vehicles may be operated for short distances without the belt, but only under emergency conditions. Both types use the same method of adjustment.

Valve Adjustment

Remove valve covers and proceed as follows:
1. Rotate crankshaft until No. 1 cylinder is up on compression stroke and the distributor rotor points to No. 1 plug wire pick-up. Both intake and exhaust valves in No. 1 cylinder are now closed. The valve tappets are on the low point of No. 1 cylinder camshaft lobe.
2. Check and adjust valve lash with a feeler gauge. (See specifications chart for clearance).
Since 1964, a sticker on the fan housing has specified clearances.
3. Lock adjusting screw and recheck clearance.
4. Rotate engine 1/2 turn counterclockwise (reverse of running rotation), using a spark plug wrench on the generator pulley nut. Set valves on No. 2 cylinder.
5. Rotate another 1/2 turn counterclockwise; set valves on No. 3 cylinder on the other side of the car.
6. Rotate 1/2 turn counterclockwise once more; set valves on No. 4 cylinder.
7. Replace valve covers with new gaskets. Replace distributor cap.
It is also possible to adjust valves in normal firing order sequence, 1-4-3-2, rotating the engine clockwise. It is helpful to mark the fan pulley directly opposite the 0° mark in order to determine precisely when the engine is turned 1/2 revolution.

Fuel System

Fuel Tank R & R

The type 1 and 3 fuel tank is located at the front of the car, under the hood, with a fuel control valve, filter and a fuel screen inside the tank. On type 2 vehicles, the tank is at the rear, ahead of the engine.
To clean the outside filter, access is possible from under the car, or through an inspection hole in the body. The tank must be removed for internal screen cleaning.
1. Remove spare wheel, jack and tools.
2. Remove trunk floor covering.
3. Lift sender unit cover off.
4. Disconnect fuel gauge cable.
5. Pull fuel hose off the line in frame and seal it. The hose and clip remain on tank pipe.
6. Remove fuel tank breather pipe.
7. Remove four tank retaining screws and lift out the tank.
8. Install by reversing removal procedures.

Fuel Pump R & R and Adjustment

1. Disconnect fuel line and hose from pump.
2. Remove mounting nuts from flange.
3. Remove pump.
4. Remove pump pushrod, intermediate flange and gaskets.
5. When installing, fit intermediate flange onto the crankcase with two new gaskets and insert pushrod (rounded end down.) The complete stroke, from high lobe of cam to low radius should be about 0.160 in. This can be obtained by selecting the proper number of flange gaskets.

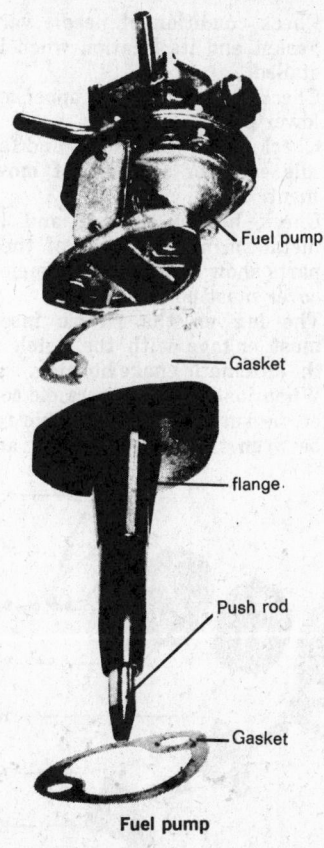

Fuel pump

- Fuel pump
- Gasket
- flange
- Push rod
- Gasket

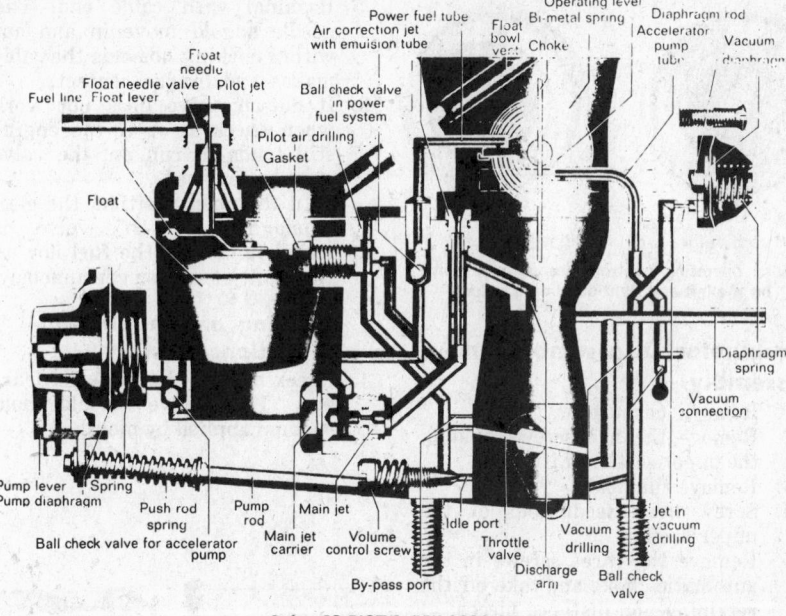

Solex 28 PICT-1 (VW 1200 A)

Carburetor Removal

NOTE: Solex carburetors are used on all models. Prior to 1961, a model 28PCI, with manual choke, was used. Since 1961, a model 28PICT or a 30PICT, with automatic choke, is used. Make sure screens are clean and all connections tight.

1. Remove pre-heater pipe from air cleaner intake.
2. Pull crankcase breather pipe from air cleaner. Disconnect thermostatic control wire from air cleaner.
3. Loosen clamp on cleaner and remove.
4. Remove fuel and vacuum hoses from carburetor.
5. Pull automatic choke and electro-magnetic pilot jet cables off.
6. Detach accelerator cable from throttle valve lever and remove spring, spring plate and cable pin.
7. Remove two nuts and take off carburetor.

Carburetor Installation

1. Fit a new gasket on intake manifold flange.
2. Install carburetor and tighten nuts evenly.
3. Secure accelerator cable so that there is about 0.040 in. play between throttle valve lever and the stop on the carburetor body when the accelerator pedal is fully depressed.
4. Start engine, warm up, then adjust idle.

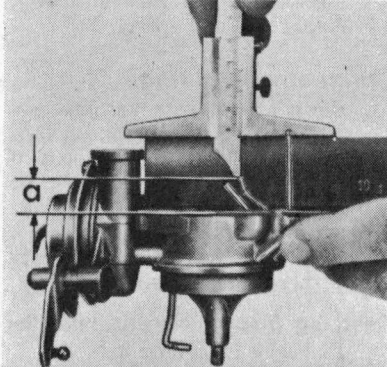

The height of the injector tube opening from the upper part of the 32 PDSIT-2/3 carburetor should be 12 mm. for engines up to No. T 0244544, 9.0 mm. thereafter.

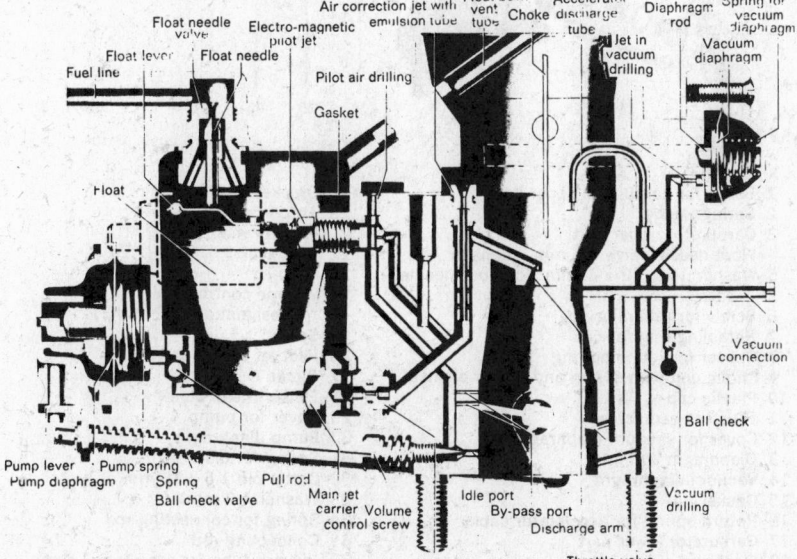

Solex 30 PICT-1 (VW 1300)

Automatic choke terminal

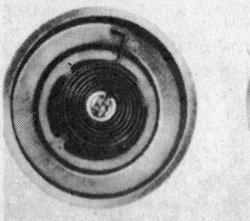 

Left carburetor   Right carburetor

Choke operating controls are not the same on the left and right-hand carburetors.

Carburetor Disassembly and Assembly

1. Remove carburetor.
2. Remove the five screws holding the upper section. Lift it off.
3. Remove the float.
4. Screw float needle out of the upper section.
5. Remove the three screws in the automatic choke and take off the retaining and distance pieces.
6. Remove the ceramic plate, bimetal spring heater element and plastic cap.
7. Screw out the air correction jet with the emulsion tube and the electromagnetic pilot jet.
8. Screw out the main jet and the volume control screw.
9. Remove pump lever cotter pin from the connector rod.
10. Remove four pump cover retaining screws and take off cover, diaphragm and spring.

Carburetor Cleaning

1. Clean all parts in gasoline, except the automatic choke cover.
2. Remove pilot jet from the cut-off valve, using two wrenches. Do not grip the valve in a vise, damage may result.
3. Clean jets, valves and drillings with compressed air.

Checking Electromagnetic Pilot Jet

1. Pull cable from terminal.
2. Turn ignition on and touch terminal with cable end. The needle should move in and out with a clicking noise as the cable makes and breaks contact.
3. If cut-off valve does not work when checked, or if the engine still tends to run on, the valve must be replaced.

In the open position, the electromagnetic cut-off valve is switched off and the fuel flow to the pilot jet is open continuously.

Checking and Assembling Upper Section

1. Check float needle valve for leakage. This valve should hold vacuum applied by mouth.
2. Check condition of needle valve gasket and its location when installed.
3. Check gasket between upper and lower parts.
4. Check choke valve shaft and fast idle cam for freedom of movement.
5. Check heater element and bimetal spring. If either of these parts show damage, the complete cover must be replaced.
6. The lug on the plastic insert must engage with the notch in the automatic choke housing.
7. When installing the ceramic cover, be sure that the ceramic rod between the heater element and

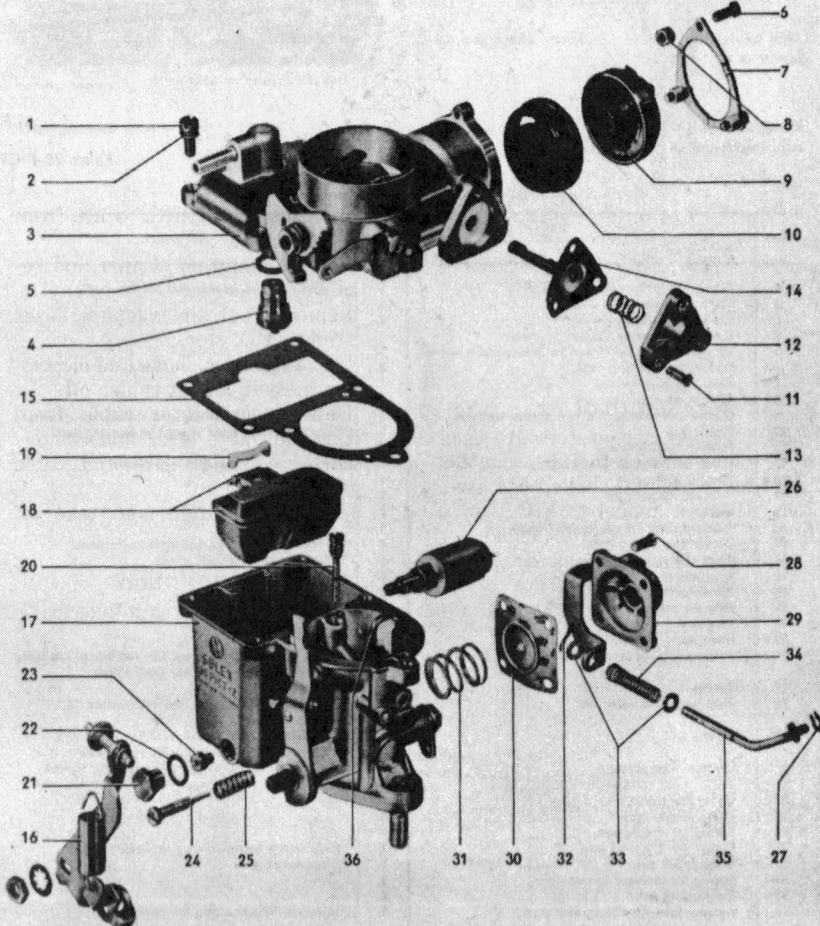

Main jet is removed after unscrewing plug (1). Volume control screw (2) determines low-speed mixture and is removable. Volume control screw should not be turned in too tightly, as its seat could be damaged.

1 Screw for carburetor upper part
2 Spring washer
3 Carburetor upper part
4 Float needle valve 1.5 mm. diameter
5 Washer 15 x 12 x 1 mm. for float needle valve
6 Screw for retaining ring
7 Retaining ring for cap
8 Spacer for retaining ring
9 Choke unit with spring and heater element
10 Plastic cap
11 Fillister head screw
12 Cover for vacuum diaphragm
13 Diaphragm spring
14 Vacuum diaphragm
15 Gasket
16 Return spring for accelerator cable
17 Carburetor lower part
18 Float and pin

19 Bracket for float pin
20 Air correction jet
21 Plug for main jet
22 Plug seal
23 Main jet
24 Volume control screw (designation 1, 2 and 3)
25 Spring
26 Pilot jet cut-off valve "A"
27 Circlip
28 Fillister head screw
29 Cover for pump
30 Pump diaphragm
31 Spring for diaphragm
32 Cotter pin 1.5 x 15 mm.
33 Washer 4.2 mm.
34 Spring for connecting rod
35 Connecting rod
36 Injector tube for accelerator pump

Exploded view, Solex PICT-2 carburetor, standard equipment on 1968-69 1500 Type 1

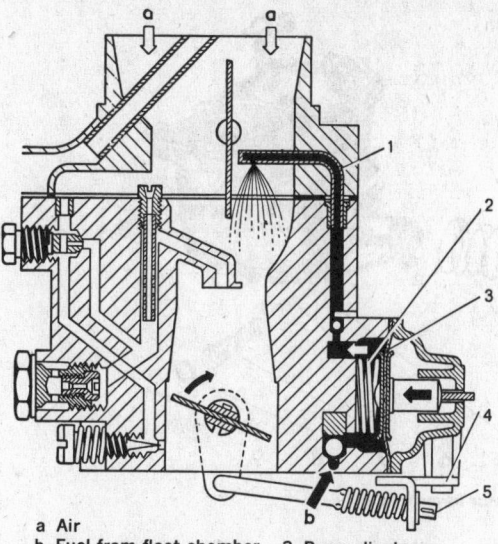

a Air
b Fuel from float chamber
1 Injection tube
2 Pump spring
3 Pump diaphragm
4 Pump lever
5 Connecting rod

Typical Solex carburetor, showing internal parts

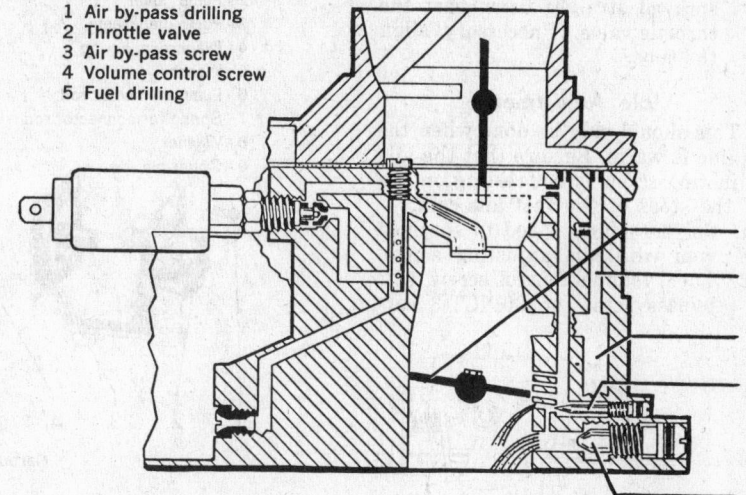

1 Air by-pass drilling
2 Throttle valve
3 Air by-pass screw
4 Volume control screw
5 Fuel drilling

Solex 30 PICT-3 carburetor used in 1970 Type 1. The volume control screw is factory set. Idling adjustments are made with the air bypass screw. (© Volkswagen)

the bi-metallic spring is properly located.

8. The operating lever must engage the hooked end of the bi-metallic spring when the cover is installed.

9. Install outer cap and retaining ring with three screws and distance pieces.

10. Turn the cap until the mark on the ceramic cover is in line with the center marker on the automatic choke housing. Do not overtorque the screws.

11. Lightly lubricate the fast idle cam on the choke valve shaft.

Checking and Assembling Lower Section

1. Check pump diaphragm for leakage and replace, if necessary.

2. When tightening screws of pump cover, the pump lever should be pressed away from the float chamber so that the diaphragm is secure in the pressure stroke position.

3. Place cir-clip on pump rod so there is 0.012-0.020 in. clearance between clip and carburetor body. Insert cotter pin into inner hole.

4. Test float for leakage by immersion in hot water. If bubbles appear, the float needs replacement.

5. Check throttle shaft clearance. Excessive shaft or body wear permits an air leak at this point, upsetting idle operation and causing hard starting. Replace the shaft and re-bush the housing, if necessary.

6. Check volume control screw. The tapered portion must not be grooved, bent or pressure marked.

7. Install the float.

8. Check position of accelerator pump injector tube in the carburetor bore. When the throttle

1 Gasket	14 Venturi	27 Float
2 Fuel pipe	15 Relay lever	28 Main jet
3 Float pin	16 Vacuum connection	29 Volume control screw
4 Float needle valve	17 Bi-metal coil	30 Connecting rod
5 Float needle	18 Intermediate lever	31 Idling mixture port
6 Pilot jet	19 Fast idle cam	32 By-pass port
7 Pilot air bleed drilling	20 Stop lever	33 Idle adjustment screw
8 Air correction jet	21 Pump lever	34 Throttle valve
9 Vent passage for float chamber	22 Pump diaphragm	35 Vacuum drilling
10 Emulsion tube with ventilation jet	23 Connecting rod spring	36 Discharge arm
11 Power fuel pipe	24 Diaphragm spring	37 Vacuum piston
12 Choke valve	25 Ball pressure valve	38 Piston rod
13 Injector tube accelerator pump	26 Ball suction valve	39 Operating rod

The Solex 32 PDSIT-2 used on the dual-carb 1600 Type 3 engine. This is the left-hand carburetor, which contains a double vacuum drilling for the advance mechanism.

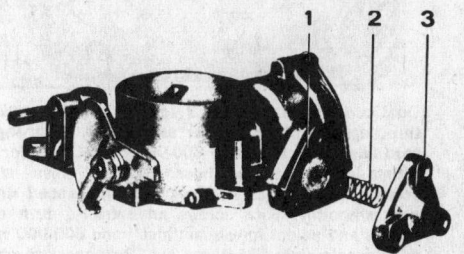

1 Vacuum diaphragm
2 Diaphragm spring
3 Cover for vacuum diaphragm

Carburetor top assembly

is opened, the fuel jet must be sprayed straight down past the throttle valve. If necessary, align the tube.

Idle Adjustment

This should only be done when the engine is warm. Be sure that the idle adjusting screw is not resting on one of the steps of the fast idle cam.

1. Set hot idling speed to specified rpm with idle adjusting screw.
2. Turn volume control screw (air bypass screw on 30PICT-3 car-

1 - Screw
2 - Pump cover
3 - Pump diaphragm
4 - Diaphragm spring
5 - Washer
6 - Pump connector rod
7 - Spring for connector rod
8 - Washer
9 - Cotter pin

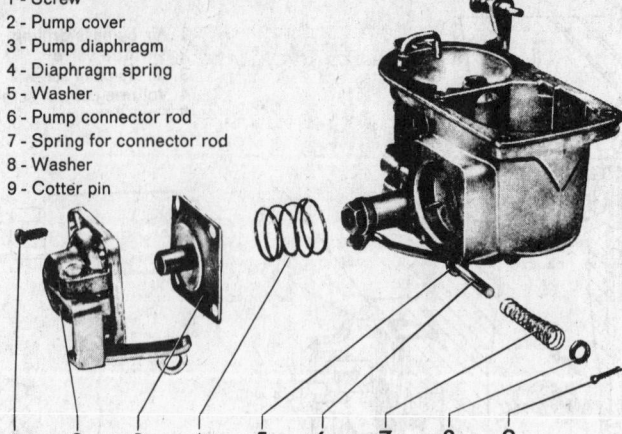

Carburetor lower assembly

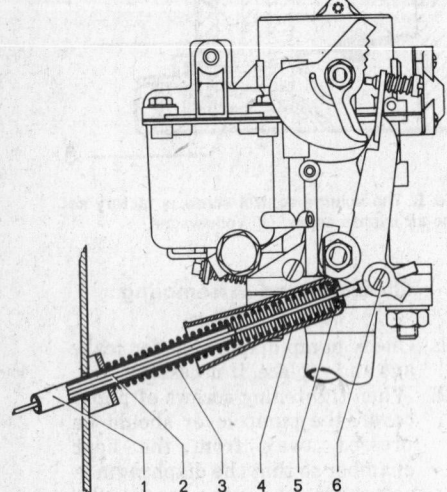

1 Accelerator cable guide tube
2 Accelerator cable
3 Accelerator cable spring
4 Spring sleeve
5 Spring seat
6 Accelerator cable swivel pin

Throttle linkage details—1100, 1200, 1300

buretor) in slowly until the engine speed begins to drop, then turn the screw out until the engine runs smoothly. Next, turn the volume control screw about one-quarter turn clockwise.
3. Engine should not stall when throttle is snapped shut from wide open to idle.

Accelerator Cable Removal and Installation

1. Disconnect accelerator cable from throttle valve lever.
2. Compress the spring at carburetor and remove spring seat. Remove sleeve and spring.
3. Detach bolt from accelerator pedal and disconnect cable from bolt.

4. Pull accelerator cable from its guide tube in the fan housing (toward the front.)
5. Pull plastic hose from cable.
6. Remove rubber boot at end of cable guide tubes in the frame.
7. Pull the cable towards front, out of the guide tube.
8. Install accelerator cable with universal grease as a lubricant.
9. Make sure that the rubber boot and the plastic hose are correctly seated to keep water from entering the guide tubes.
10. Special care must be used when attaching the accelerator cable to the throttle valve lever to avoid breakage of the cable at full throttle. Open the throttle valve so that there is clearance of about 0.040 in. between the throttle lever and the stop at the carburetor body.
11. Fully depress accelerator pedal and connect cable to throttle valve.

Throttle Regulator

The exhaust emission control device used on type 1 and 2 vehicles, 1968-70, is the throttle valve regulator. This device holds the throttle open slightly on deceleration to prevent an excessively rich mixture.

Removal

1. Disconnect vacuum hose.
2. Remove three screws.
3. Remove retainer ring. Take out regulator. On 1970 models, the throttle regulator consists of two parts, connected by a hose. The operating part is mounted at the carburetor, and the control part is located on the left sidewall of the engine compartment.
4. Unhook control rod at lever.
5. Reverse procedure to install.

Adjustment

1. Engine must be at operating temperature, with automatic choke fully open.

Dual carburetors on Type 3 1500 and 1600 engines are balanced by removing right-hand throttle rod and air cleaner assemblies. With engine at normal temperature, adjust each carb idle speed to obtain 800-900 rpm. Check for equal air flow by listening at carb throat with a length of rubber hose. An equal "hiss" should be heard. Adjust volume (mixture) screws, one at a time, clockwise until speed drops, then counterclockwise until engine runs smoothly. Turn screws an extra ¼ turn counterclockwise, then go to idle speed screws and adjust for equal "hiss" and 800-900 rpm. Recheck mixture screws; they should be approximately 1½ turns out. Reconnect throttle rod, increase speed to 1400 rpm (to check high-speed balance), and listen at each carb throat for equal hiss. If not equal, rod must be changed in length to obtain proper synchronization. When synchronized, install air cleaners, keeping in mind that the outer wingnuts must be tightened before center fastener. If center wingnut is tightened first, the tops of the carburetors could be moved enough to upset balance.

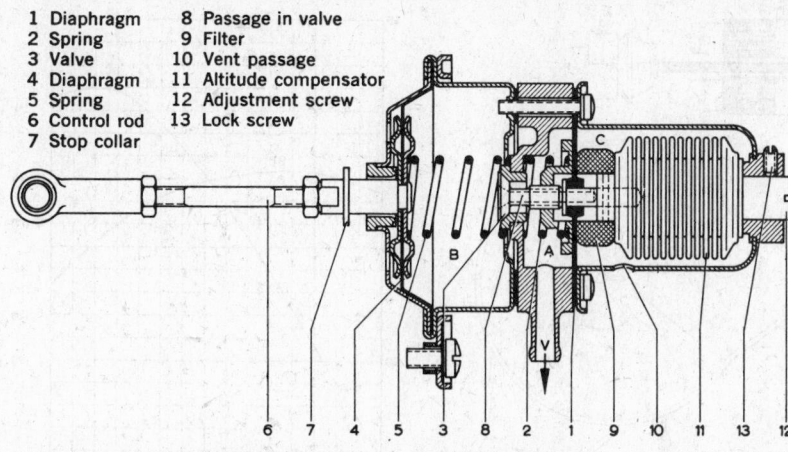

1 Diaphragm   8 Passage in valve
2 Spring   9 Filter
3 Valve   10 Vent passage
4 Diaphragm   11 Altitude compensator
5 Spring   12 Adjustment screw
6 Control rod 13 Lock screw
7 Stop collar

Throttle regulator, 1968-69 (© Volkswagen)

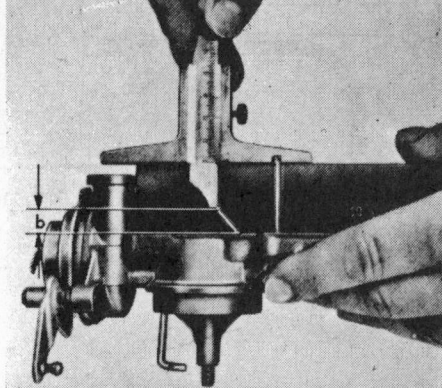

Height of power fuel tube opening (b) from upper part of carburetor should be 15 mm. on Type 3 1600 carburetors equipped with power fuel system.

2. Start engine. Turn regulator adjusting screw clockwise until control rod just starts to move throttle valve lever. The stop collar on the control rod will be against the regulator body. Engine speed should be 1700-1800 rpm.
3. If speed is too high, shorten control rod.
4. After adjustment, tighten lock nuts on control rod.
5. Turn regulator adjusting screw counterclockwise until an idle speed of 850 rpm is obtained.
6. Increase engine speed to 3000 rpm, then release throttle valve lever. Engine should take 3-4 seconds to return to idle.

Incorrect throttle regulator adjustment may cause erratic idle, excessively high idle speed, and backfiring on deceleration.

VW 1.6 Liter Fuel Injection

Since 1968, type 3 engines have been equipped with an electronic fuel injection system. This system is unique in design, as the accompanying illustrations show, and should not be confused with some American designs now in use.

The manufacturer claims more efficient combustion and a marked reduction in the amount of carbon-monoxide and unburned hydrocarbons in terminal exhaust, this reduction being accomplished by a constant monitoring system of sensing devices and combustion controls.

The control unit is the brain of the system. It controls the quantity of fuel to be injected relative to engine speed, intake manifold pressure and

Measuring throttle valve gap of 32 PDSIT-2/3. Gap should be .60-.65 mm. (.024-.026 in.).

engine temperature. When the ignition is switched on, current is supplied to the control unit by means of a main relay. This unit also controls the fuel pump, which is supplied with

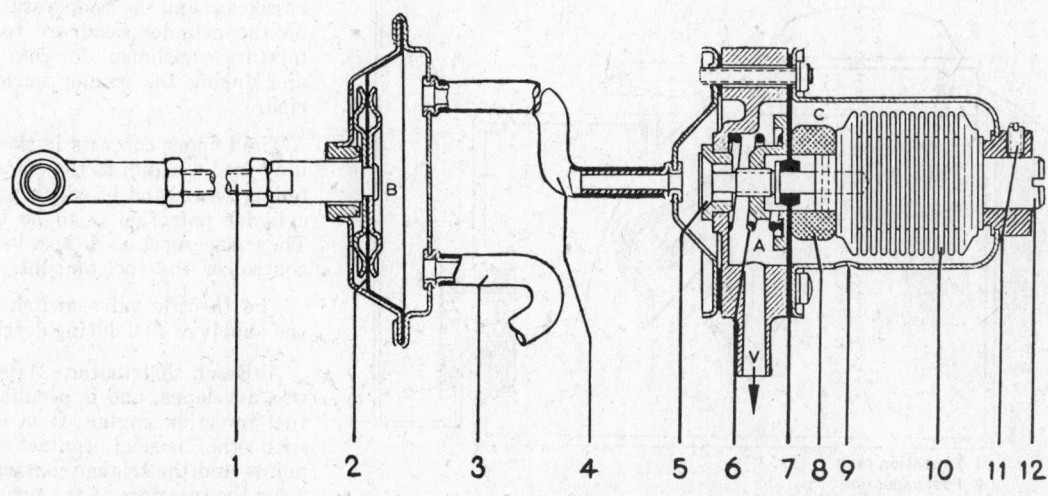

1 Pull rod
2 Operating diaphragm
3 Hose to the vacuum drilling in the carburetor
4 Hose between operating and control part
5 Valve
6 Spring
7 Control diaphragm
8 Plastic foam filter
9 Drilling
10 Altitude corrector
11 Lock screw
12 Adjusting screw

Throttle valve regulator, 1970 (© Volkswagen)

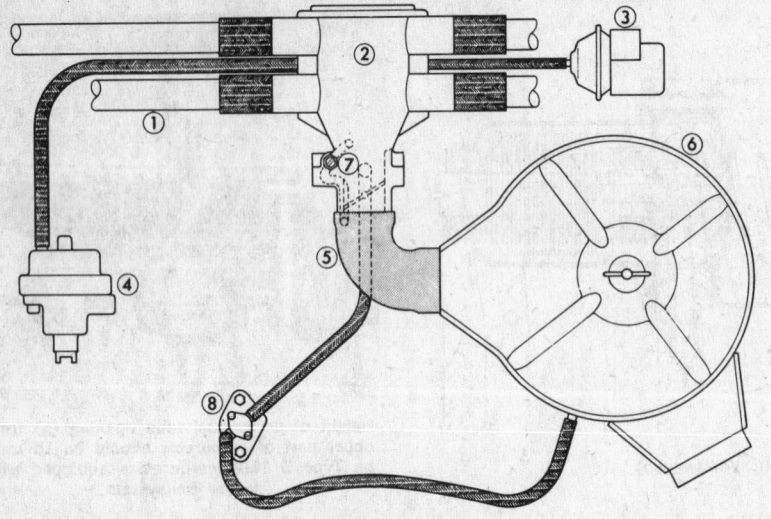

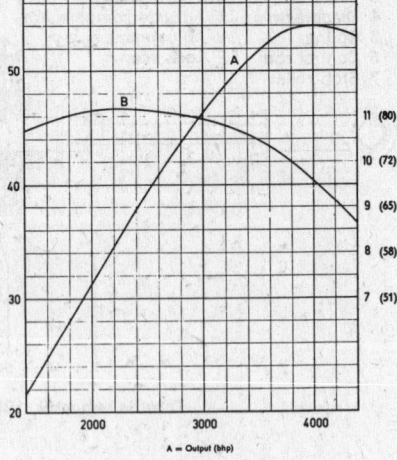

1 Intake pipes	3 Pressure switch	5 Idle air circuit	7 Idle air screw
2 Intake air distributor	4 Pressure sensor	6 Air cleaner	8 Auxiliary air regulator

Air system of the fuel-injected engine

A = Output (bhp)
B = Torque in mkg (lb. ft.)

**Horsepower and torque curves,
fuel-injected 1600 engine**

current via the pump relay only when the engine is running.

The control unit opens the injector valves electrically, in pairs, to cylinders 1 and 4 and to cylinders 2 and

3. The amount of fuel injected depends upon the length of time the injectors are kept open, and is metered according to engine demands. This is made possible by the fact that the in-

jectors are under constant fuel pressure.

The duration of injection is computed by the control unit. The data which is processed in the electronic control unit is fed to it by the various sensing devices fitted on the engine. The control unit is mounted in the left air duct and connected by a special cable harness to the sender units. These are as follows:

The pressure sensor, which controls the basic fuel quantity according to engine loading, is influenced by intake manifold pressure.

The pressure switch controls the fuel enrichment for full load, and is operated by the pressure differential between intake manifold and the surrounding (outside) air pressure.

The temperature sensor in the crankcase and the temperature sensor on the cylinder head are to control mixture enrichment for cold starting and during the engine warm-up period.

The trigger contacts in the distributor feed a signal to the control unit, telling when, and in which particular cylinder pair fuel is to be injected. They also serve as a speed-sensitive control of the fuel amount.

The throttle valve switch cuts off the supply of fuel during deceleration.

A Bosch distributor, 311905205M, was developed, and is peculiar to the fuel injection engine. It is equipped with the normal contact breaker points and the trigger contacts 1 and 2 for the injectors of the two pairs of cylinders.

Ignition timing for the fuel injection engine is to be set at 0° (T.D.C.)

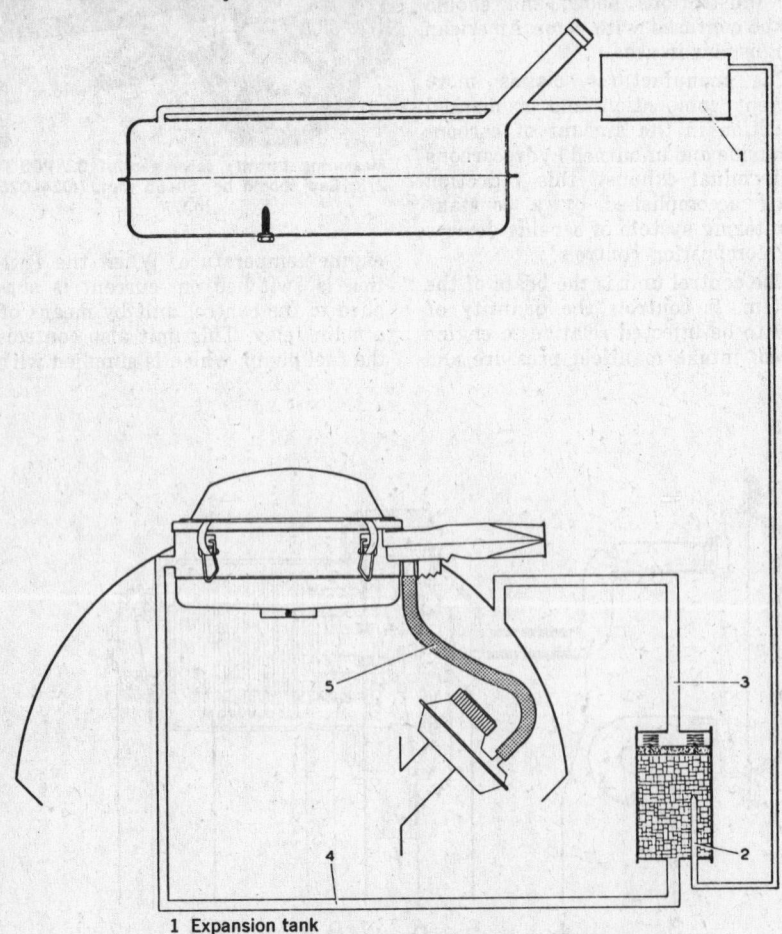

1 Expansion tank
2 Fuel vapor line
3 Pressure hose (from cooling air blower to filter container)
4 Suction hose (from filter container to oil bath air cleaner)
5 Crankcase ventilation

Activated charcoal filter system used on all VW vehicles sold in California, 1970
(© Volkswagen)

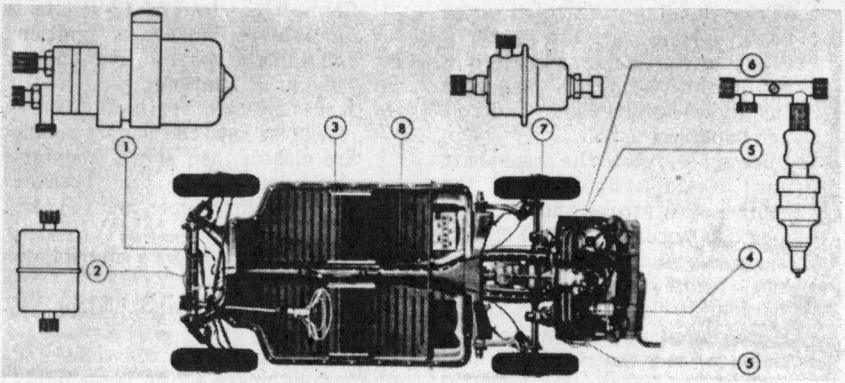

1 Electric fuel pump
2 Filter
3 Pressure line
4 Ring main
5 Electromagnetic fuel injectors
6 Distributor pipes
7 Pressure regulator
8 Return line

Fuel system of the fuel-injected engine

CONTROL UNIT

PRESSURE SENSOR

MAIN RELAY

INJECTOR VALVES

TRIGGER CONTACTS

PUMP RELAY

TEMPERATURE SENSOR #1
TEMPERATURE SENSOR #2

THROTTLE VALVE SWITCH

PRESSURE SWITCH

FUEL PUMP

INJECTOR VALVES

Fuel injection system schematic diagram

Distributor

Removal

1. Remove distributor cap, primary wire and vacuum line at distributor.
2. Turn engine until contact end of distributor rotor is aligned with notch in distributor rim denoting No. 1 firing position. Scribe a mark on the distributor body and engine block, showing the position of the body in the block.
3. Remove the distributor hold-down screw and lift the distributor up and out of the engine.

Installation

1. If the crankshaft has been rotated, turn the engine until the piston of No. 1 cylinder is at the top of its compression stroke.

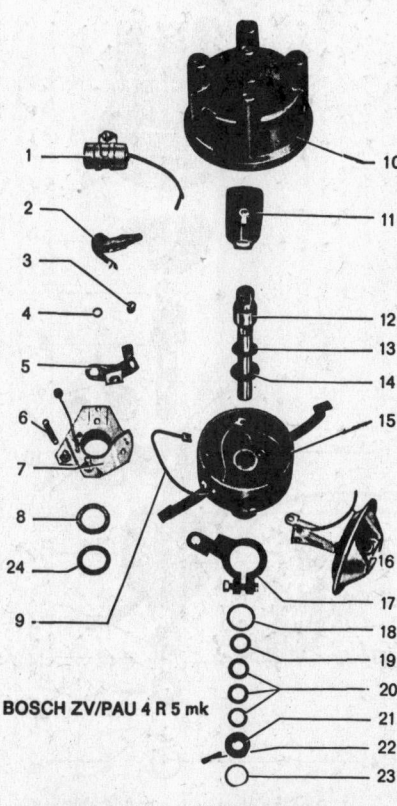

BOSCH ZV/PAU 4 R 5 mk

1 Condenser	11 Rotor
2 Contact breaker arm	12 Distributor shaft
3 Securing screw with flat and spring washers	13 Steel washer
	14 Fiber washer
	15 Distributor housing
4 Insulating washer	16 Vacuum advance
5 Contact breaker point	17 Clip
	18 Sealing ring
7 Breaker plate with ground cable	19 Fiber washer
	20 Shim
8 Plastic washer	21 Driving dog
	22 Pin
9 Low tension cable	23 Locking spring
10 Distributor cap	24 Shim

Bosch distributor (© Volkswagen)

Exhaust emission controlled engines are timed with a strobe light, as opposed to static timing method which has been used in the past.

2. Position the distributor to the block so that the vacuum control unit is in its normal position.

3. Position the rotor to point toward notch in rim at five o'clock position. Install distributor on the engine and push down to engage distributor drive.

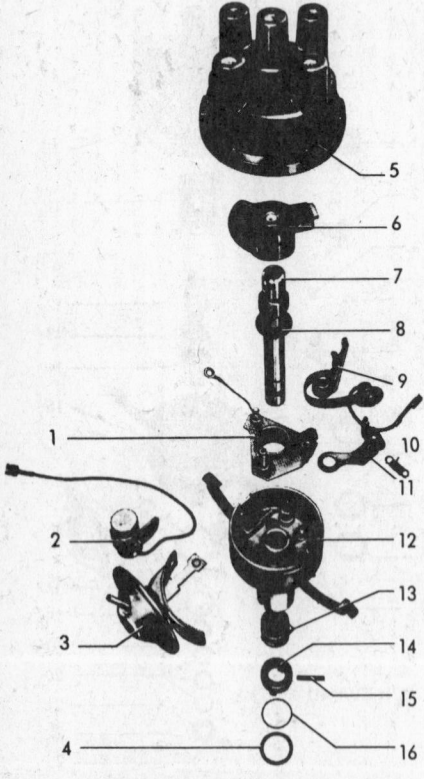

1 Breaker plate with ground cable
2 Condenser
3 Vacuum advance unit
4 Sealing ring
5 Distributor cap
6 Rotor
7 Distributor shaft
8 Fiber washer
9 Contact breaker arm with spring
10 Return spring
11 Contact breaker point
12 Distributor housing
13 Steel washers
14 Driving dog
15 Pin
16 Locking ring

VW distributor

4. While holding the distributor down in place, jog the starter a few times to make sure the distributor driveshaft is engaged. Install hold-down clamp and bolt and snug up the bolt.
Once again, rotate the crankshaft until No. 1 cylinder is at the compression stroke and the harmonic balancer mark is on 0°.

5. Turn distributor body slightly until points just open. Tighten distributor clamp bolt.

6. Place distributor cap in position and see that the rotor lines up with the notch and the terminal for No. 1 spark plug.

7. Install cap, distributor primary wire and check plug wires in the cap.

8. Start engine and set timing according to the tune-up chart. Reconnect vacuum hose to vacuum control assembly.

When No. 1 cylinder of the Type 3 1600 is at its firing point, the slot of the distributor drive shaft must form an angle of approximately 60°, as shown, with the smaller segment of the slot pointed towards the oil cooler. On exhaust emission controlled Type 1 engines, the narrow part of the slot must be toward the pulley.

Ignition Timing (cold engine)

Always set breaker point gap to the proper specification before setting ignition timing. Depending on the model year, there may be one, two, or three timing notches in the crankshaft pulley. These correspond, from left to right, to 0° B.T.D.C., 7.5° B.T.D.C., and 10° B.T.D.C. Exhaust emission controlled engines generally have only the 0° mark.

1. Crank the engine until the timing mark on the crankshaft pulley lines up approximately with the vertical crankcase joint or timing pointer, and the distributor rotor arm is in the position for firing No. 1 cylinder. (See mark on rim of distributor base).

NOTE: do not set timing on No. 3 cylinder firing position or timing will be retarded about 3°.

2. Loosen clamp screw of distributor retainer.

3. Connect one lead of 12-volt test lamp to terminal 1 of ignition coil and the other to ground.

4. Switch on ignition.

5. Rotate the distributor body clockwise until the contact points are closed, then slowly counterclockwise until the breaker points begin to open and the test lamp lights.

6. Tighten clamp screw distributor retainer.

7. Reinstall rotor and distributor cap.

The ignition is correctly timed for all four cylinders if the lamp lights when the right-hand mark on the pulley is exactly in its highest, or lowest, position (in line with the crankcase joint) while slowly cranking the engine.

NOTE: the adjustment of the ignition firing point on a cold engine must always be done with a 12-volt test lamp. A stroboscopic lamp should not be used, as it will alter the entire setting range. However, it is recommended that exhaust emission controlled engines using the fully vacuum advanced distributor be timed by using a stroboscopic light for best results. If, in exceptional cases, it is necessary to adjust the timing with a warm engine, the following points should be noted:

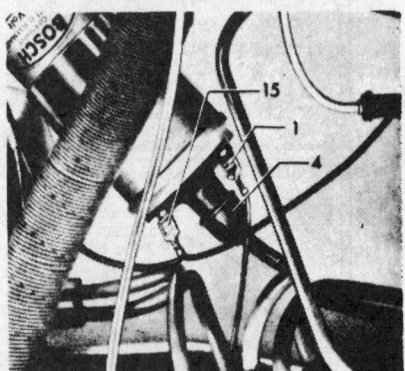

1 To distributor breaker points
4 High-voltage lead to distributor cap
15 To ignition/starter switch

Connections at the coil

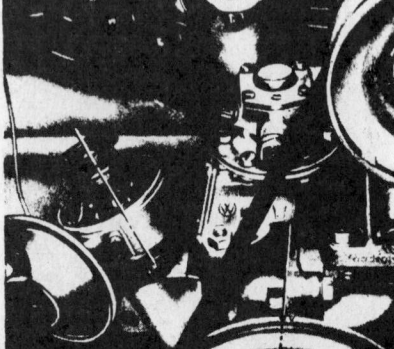

Relationship of rotor to timing marks at No. 1 cylinder firing position

Adjusting Ignition Timing With Engine Warm

Accurate control of timing setting is almost impossible when engine temperature is above 122°F. When engine is warm but not exceeding 122°F, the timing may be set but must be advanced about 2.5° beyond the normal setting. This setting should be rechecked with a cold engine at the first opportunity.

Generator, Regulator

The generator was of the 6-volt, DC-type through 1966. A 12-volt unit has been used from that date. It is equipped with ball bearings that are factory packed with high melting point grease. Lubrication, in general, is necessary only during overhaul. Never use ordinary grease. Brush wear should be checked if trouble is experienced or when doing repairs on the generator.

Service operations in no way differ from conventional procedures as covered in the Unit Repair Section. A voltmeter should be used to test the generator, as follows:

1. With a fully charged battery in the car, disconnect the lead cable from right-hand (outside) terminal of the regulator. Connect the positive lead of a voltmeter to this same terminal of the regulator. Then, ground the negative voltmeter lead.
2. Start the engine and increase rpm until voltage is at a maximum. Voltage should gradually increase to 7.4-8.1 volts.
3. Switch off engine. Needle should jump back to Zero just before engine stops.

Generator R & R

1. Disconnect battery ground cable.
2. Disconnect cables from regulator.
3. Remove air cleaner and carburetor.
4. Remove fan belt.
5. Remove generator hold-down strap.
6. Remove cooling air thermostat. The screws can be removed with a socket inserted through the holes in the lower air deflector plate.
7. Detach warm air hoses from fan housing.

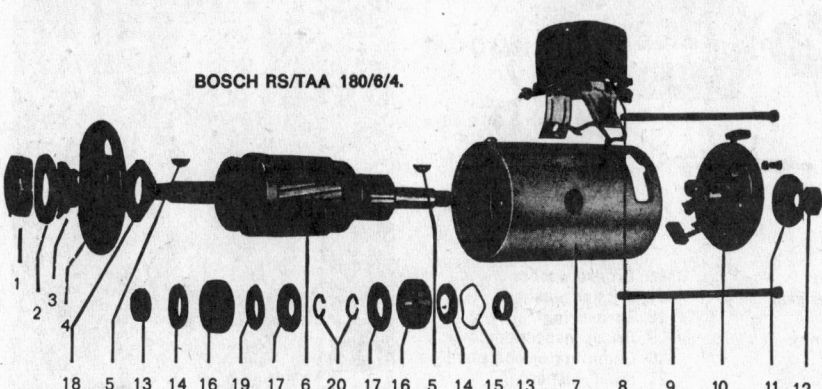

BOSCH RS/TAA 180/6/4.

1 Fan nut	7 Housing and field assembly	11 Spacer washer	16 Ball bearing
2 Carrier plate	8 Regulator	12 Pulley nut	17 Oil slinger
3 Fan hub	9 Housing screws	13 Spacer ring	18 Flange
4 End plate	10 Brush holder end plate	14 Oil slinger	19 Cover washer
5 Woodruff key		15 Spring ring	20 Circlip
6 Armature			

Bosch Generator (© Volkswagen)

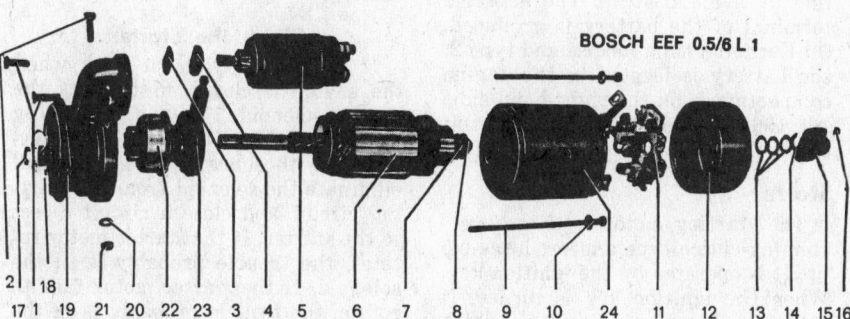

BOSCH EEF 0.5/6 L 1

1 Securing screws	7 Steel washer	13 Shims	19 Intermediate bracket
2 Bearing bolt	8 Synthetic washer	14 Lockwasher	20 Spring washer
3 Support washer	9 Housing screws	15 End cap	21 Nut
4 Rubber seal	10 Washer	16 Screws	22 Drive pinion
5 Solenoid	11 Brush holder	17 Circlip	23 Shift lever
6 Armature	12 End plate	18 Stop ring	24 Housing

Bosch generator (© Volkswagen)

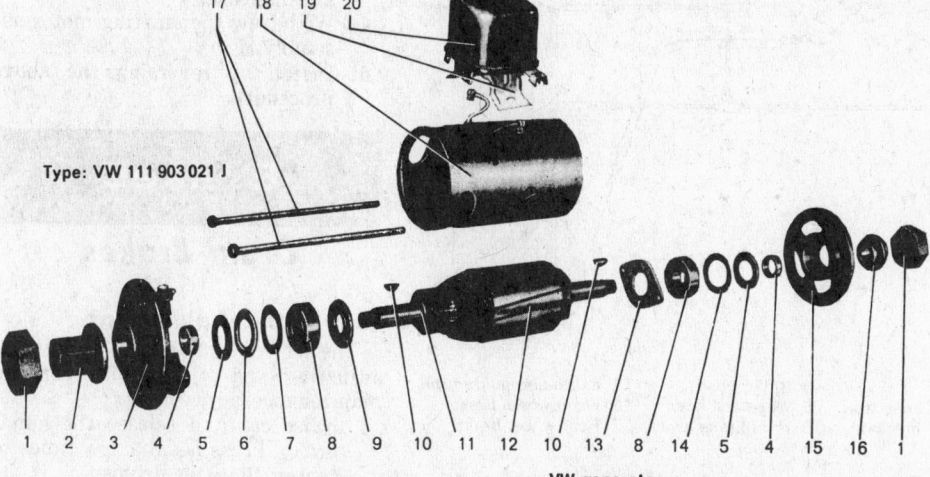

Type: VW 111 903 021 J

1 Nut
2 Pulley hub
3 Brush holder end plate
4 Spacer ring
5 Felt washer
6 Retainer
7 Thrust ring
8 Ball bearing
9 Washer
10 Key
11 Spacer
12 Armature
13 Bearing retainer
14 Thrust ring
15 End plate
16 Fan hub
17 Housing screws
18 Housing and field assembly
19 Slotted screw
20 Regulator

VW generator

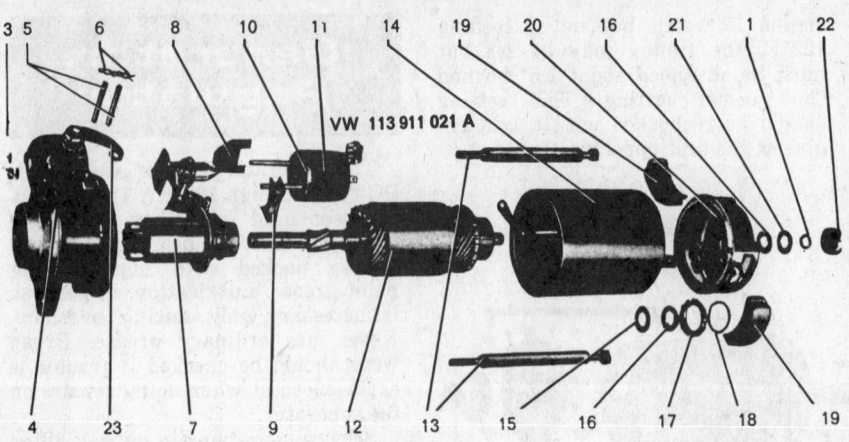

VW 113 911 021 A

VW starter

1	Circlip	8	Insulating plate	16	Bronze washer
2	Cup washer	9	Moulded rubber seal	17	Friction washer
3	Nuts and lockwashers	10	Insulating disc	18	Thrust ring
4	Intermediate bracket	11	Solenoid housing	19	Brush inspection cover
5	Pivot pins	12	Armature	20	Commutator end plate
6	Spring clips	13	Housing screws	21	Steel washer
7	Drive pinion with linkage and solenoid core	14	Housing and field assembly	22	Cap
		15	Steel washer	23	Connecting strip

VW sedans now incorporate a new carburetor preheating system, whereby a cold engine receives all intake air from cylinder head area.

8. Remove two fan housing screws and lift fan housing.
9. Remove four screws in fan housing cover and take off generator and fan.
10. Install by reversing above procedure.

Battery, Starter

Battery

The 6-volt battery consists of three cells and has a capacity of 66 or 77 amp hours at a 20 hr. discharge rate.

The battery is located on the right-hand side of the frame under the rear seat on the type 1 sedan and VW convertible, and is securely held in position by a metal strap. The negative terminal of the battery is grounded. On Karmann Ghia models and type 2, the battery is located in the engine compartment, on the right-hand side. All 1967-70 models have a 12 volt battery.

Starter

The starting motor is of the over-running-clutch-type and produces 0.5 hp. It is operated by the ignition key. When the ignition key is turned, it actuates a solenoid switch, which connects the motor to the battery causing the pinion to engage. For overhaul procedures, see Unit Repair Section.

Check the Starter

If the starter will not work when the switch is closed, first, check the starter solenoid. This can be done by bridging the two heavy solenoid terminals with a heavy wire. This will eliminate the solenoid from the starting circuit and close a circuit direct to the starter. If the starter motor rotates, the trouble probably is in the solenoid. If the starter motor fails to rotate, the trouble probably is in the starter motor. The solenoid can be removed without removing the starter.

Starter R & R

1. Disconnect negative battery lead.
2. Disconnect wiring from solenoid.
3. Remove two retaining nuts which attach starter to transmission case.
4. Withdraw the starting motor assembly.
5. Install by reversing the above procedure.

Brakes

Drum Brakes

Foot Brake Adjustment

The service (foot brake) is not self-adjusting and will require periodic maintenance adjustments.

1. Raise car and release the handbrake. Press pedal a few times to center shoes in drums.

1	Brake pedal	5	Brake line	8	Brake hose	11	Cable and guide tube
2	Master cylinder	6	Three-day connection	9	Wheel cylinder	12	Front wheel brake
3	Fluid reservoir	7	Brake hose bracket	10	Handbrake lever	13	Rear wheel brake
4	Stoplight switch						

Single circuit hydraulic brake system, vehicles with drum brakes front and rear

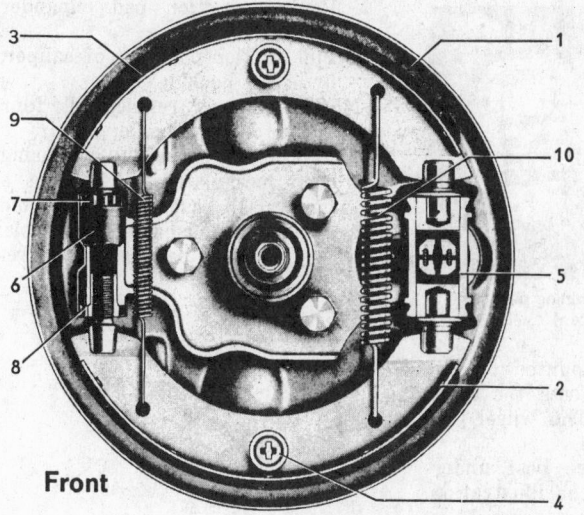

Front

Drum brakes (© Volkswagen)

1 Brake back plate
2 Secondary (trailing) brake shoe
3 Primary (leading) brake shoe
4 Hold-down spring and spring seal
5 Wheel cylinder
6 Anchor block
7 Adjusting nut
8 Adjusting nut
9 Return spring
10 Return spring

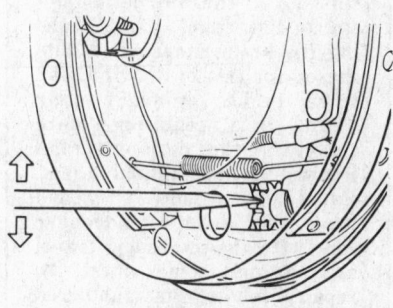

Adjusting brakes—drum type

2. Turn the wheel to be adjusted, forward until the hole in the brake drum is in line with one of the brake shoe star wheel adjusting nuts. This is not necessary on brakes with adjusting holes in the backing plates.
3. Insert a brake shoe adjusting spoon, or screwdriver, through the hole and turn the adjusting nut. Alternately turn the wheel and tighten the star wheel adjuster until a light drag is felt when turning the road wheel. Back off adjustment three or four teeth to allow wheel to rotate freely.
4. Repeat this procedure on the other adjusting star wheel. Note the opposite turning direction of the two star wheel adjusters.
5. Repeat this operation on the other three wheels.
6. Check hydraulic system reservoir for fluid. Then, road-test vehicle.

Parking Brake Adjustment

Unless adjustments are of major proportions, or parts replacement needed, the parking brake may be adjusted inside the vehicle.
1. Raise both rear wheels.
2. Slide off rubber ring, then fold back the parking brake lever rubber boot until cable adjusting nuts are accessible.
3. Back off locknuts, then tighten adjusting nuts to a point where the rear wheels will still turn freely when the hand-brake is in off position.
4. Pull the hand-lever up two notches, then check that both rear wheels have the same value of brake hold. Application to the fourth notch should lock the wheels to hand turning.
5. Secure the locknuts and re-position hand-lever rubber boot.

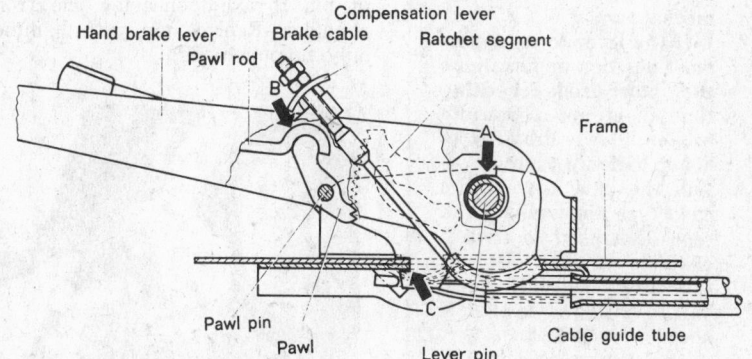

Handbrake lever and associated parts

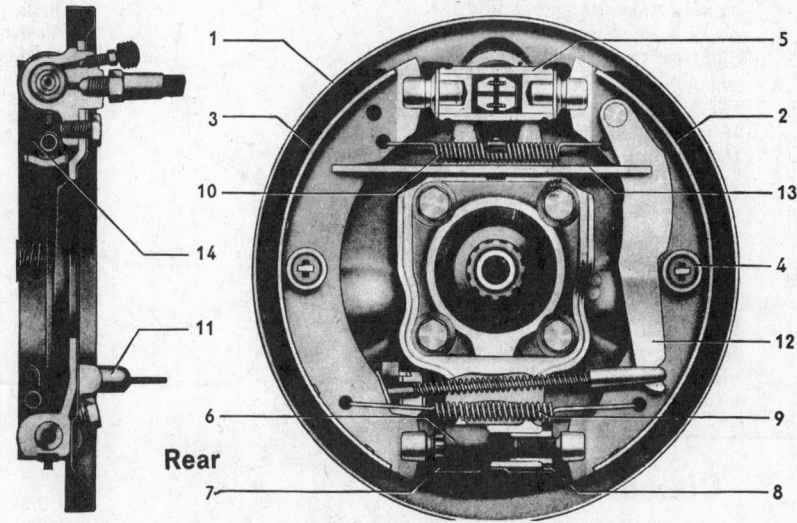

Rear

Drum brakes (© Volkswagen)

1 Brake back plate
2 Secondary (trailing) brake shoe
3 Primary (leading) brake shoe
4 Hold-down spring and spring seat
5 Wheel cylinder
6 Anchor block
7 Adjusting nut
8 Adjusting nut
9 Return spring
10 Return spring
11 Brake cable sleeve
12 Brake shoe lever
13 Operating link
14 Clip

⏻ CHILTON TIME-SAVER

On parking brake levers of this type, the ratchet sometimes disengages itself, making the brake inoperative. The reason for this is that the rear wheel brakes gradually wear down, thus requiring more travel in the parking brake mechanism to tighten them. Eventually, a point is reached where the ratchet can no longer cope with the increasing travel and the unit comes apart. To repair, follow the numbered steps, but remember, a lasting repair can only be accomplished by adjusting the rear wheel brakes, thus bringing the system back to normal operating tolerance.

1. Completely remove the four nuts which secure and lock the cables to the mechanism.
2. Lift the lever to its uppermost position and examine the pawl rod, checking that it is not physically broken. Slowly drop rod to about halfway position.
3. Pull the pawl rod off the pawl (see illustration) and reposition pawl so that it contacts ratchet (B). At the same time push down on the pawl rod so that it hooks over the pawl.
4. Slowly lower lever to full down position, then carefully check for proper operation by pushing button and slowly pulling up on lever a short distance. The ratchet mechanism should make a clicking sound, indicating that the pawl rod is successfully hooked over the pawl.
5. If the lever comes free, the operation must be repeated. A study of the illustration will help greatly, as it indicates the correct positioning of the components
6. When the operation of the handbrake lever is satisfactory, place two of the nuts on the cable ends and adjust.

Disc Brakes

The disc brake used on the front of some models consists of three basic parts: splash shield, brake disc and caliper. The operation, in principle, is similar to the disc brakes described in the Unit Repair Section.

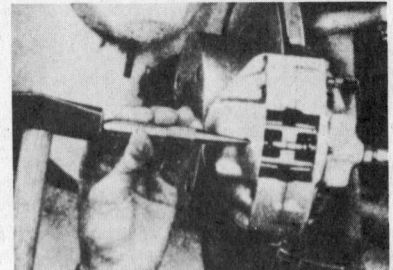
Removing pad retaining pin
(© Volkswagen)

A splash shield mounted on the steering knuckle protects the inner surface of the disc. The wheel protects the outer surface.

The components are best understood by a study of the illustration.

Friction Pads R & R

1. Remove wheel.
2. Drive out pad retaining pin(s).
NOTE: use a punch the exact size of pin to avoid shearing the front shoulder. Replace pin with hammer only, no punch.

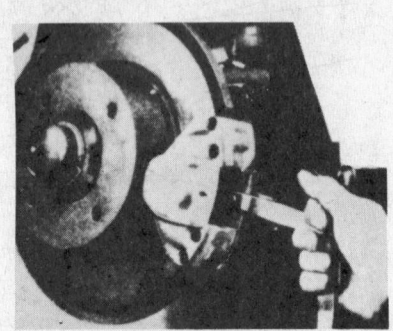
Removing friction pad
(© Volkswagen)

3. Remove friction pad expander spring.
4. Pull friction pads out of caliper, by use of hook.
NOTE: always replace all four pads, never a single pad or a pair.
5. Thoroughly clean all areas using good methylated spirits, not a mineral oil based solvent. Avoid damage from sharp-edged tools. Any damaged parts should be replaced.

Disc brake pads should be checked for wear every 6,000 miles. Pads less than 2.0 mm. (a) should be replaced.

6. Push pistons back into their cylinders. This may cause the master cylinder to overflow. Insert the friction pads into caliper. Be sure they move freely.
7. Install a new friction pad expander spring.
8. Insert the retaining pin into caliper. Use new pin(s) if any corrosion is present.
9. Pump pedal several times to seat all parts. Then, check fluid and road-test.

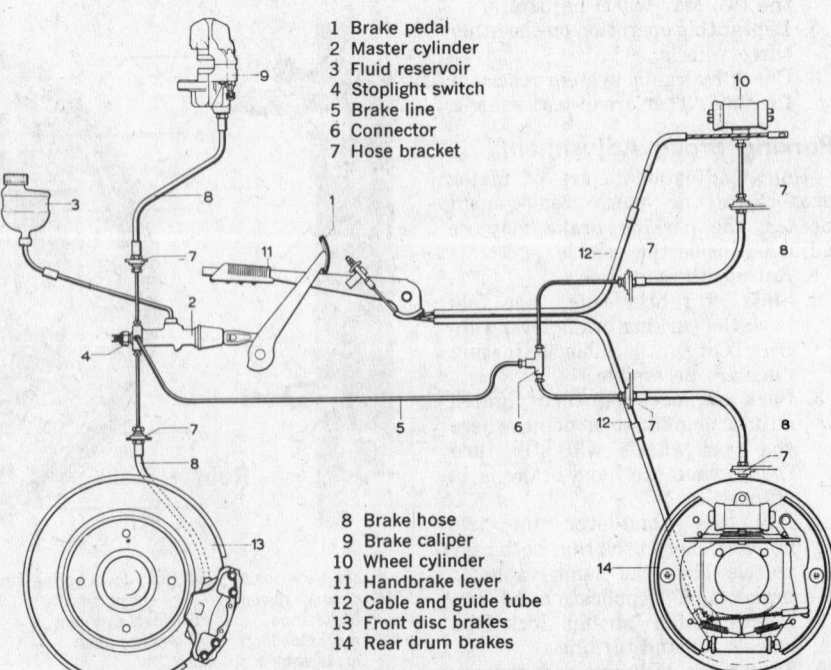

1 Brake pedal
2 Master cylinder
3 Fluid reservoir
4 Stoplight switch
5 Brake line
6 Connector
7 Hose bracket

8 Brake hose
9 Brake caliper
10 Wheel cylinder
11 Handbrake lever
12 Cable and guide tube
13 Front disc brakes
14 Rear drum brakes

Hydraulic brake system, 1600 vehicles with disc/drum brakes front and rear

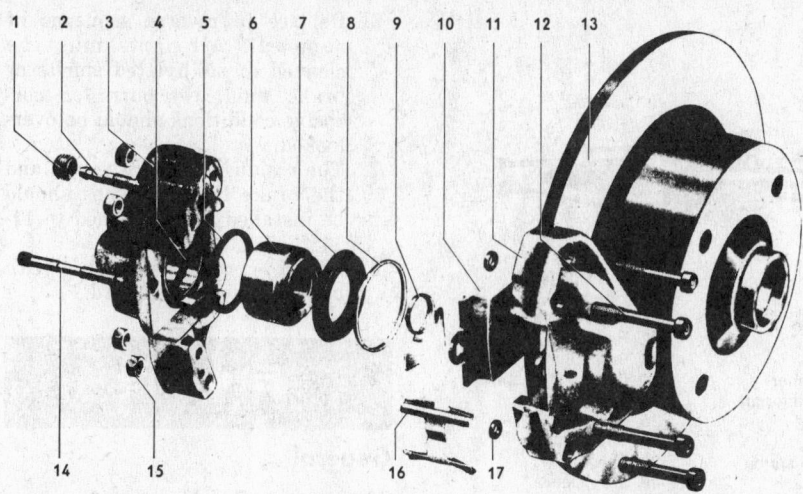

1 2 3 4 5 6 7 8 9 10 11 12 13

14   15   16   17

1 Bleeder valve dust cap	7 Rubber boot	13 Brake disc
2 Bleeder valve	8 Spring ring	14 Friction pad retaining pin
3 Brake caliper inner housing	9 Piston retaining plate	15 Nut
4 Groove for rubber seal	10 Friction pad	16 Spreader spring
5 Rubber seal	11 Brake caliper outer housing	17 Fluid channel "O" ring
6 Brake caliper piston	12 Caliper housing securing bolt	

Disc brake parts (© Volkswagen)

Brake Caliper R & R

1. Remove front wheel.
2. Remove brake hose, cap, and bleeder valve dust cap.
3. Bend back mounting bolt lock plate.
4. Remove caliper attaching bolts.
5. Remove caliper.
6. In reinstalling, clean all mating surfaces and steering knuckle.
7. Install mounting bolts and torque to 43 ft. lbs. The bolts and locking plates should be renewed.
8. Bleed the brake system. Be sure to replace dust caps on bleeder valves.
9. Road-test.

Brake Caliper Repair

1. Remove friction pads.
2. Remove caliper.
3. Mount caliper in vise using vise clamps.
4. Remove piston retaining plates.
5. With screwdriver, pry out rubber boot spring ring. Do not damage boot.
6. With plastic rod, remove boot.
7. Remove one piston with air pressure, holding the second piston with retaining pliers.
8. Remove rubber seal with a plastic rod.
9. To reassemble, first clean with methylated spirits or brake fluid.
10. Replace any parts showing wear, corrosion or physical damage. A damaged cylinder requires the replacement of a complete caliper.
11. Install new rubber boot and spring ring.
12. Install retaining plate.
13. Now, follow above procedures for the second piston.

Piston retaining pliers are needed to push both pistons back

Brake Disc R & R

1. Remove wheel.
2. Remove caliper from knuckle and hang on the tie rod with wire hook.

3. Remove wheel bearing clamp nut and remove disc.
4. In replacing, check splash shield for damage and replace if needed.
5. Reinstall disc and adjust bearing.

Dual Master Cylinder

Steps applying to the former single cylinder-type are not covered here; only steps that apply to the dual-type are used.

Fluid Reservoir

A dual-section reservoir is used so that loss of fluid in one portion will not cause failure of the entire system.

Pushrod

To obtain proper master cylinder action, the pushrod must be set to obtain pedal free movement of 0.20-0.28 in.

If a new rod is to be used, it must be an exact predetermined length. That is, from end of ball to center of pin hole—5.433 ±.019 in. Pushrod length is set at the factory.

Master Cylinder Repair

1. Remove boot.
2. Remove stop screw.
3. Remove spring stop-ring.
4. Remove internal parts.
5. Remove residual pressure valves and light switches.

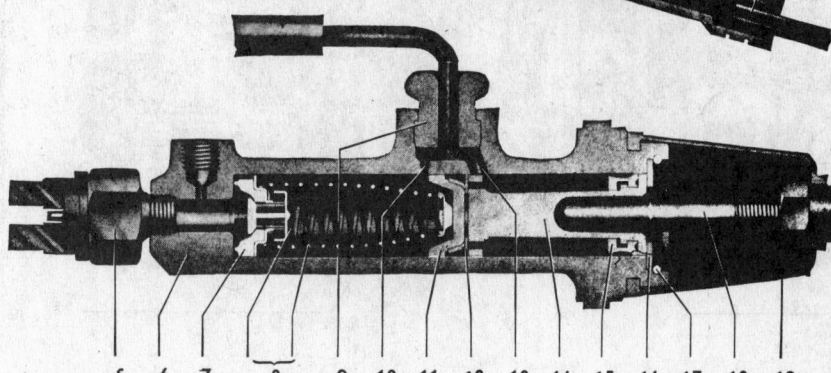

5 6 7   8   9 10 11 12 13 14 15 16 17 18 19

1 Cover	6 Master cylinder body	11 Main cup	16 Piston stop plate
2 Fluid reservoir	7 Check valve	12 Piston washer	17 Lock ring
3 Bracket	8 Piston return spring	13 Intake port	18 Piston push rod
4 Line	9 Rubber plug	14 Piston	19 Boot
5 Stop light switch	10 By-pass port	15 Secondary cup	

Single circuit master cylinder (© Volkswagen)

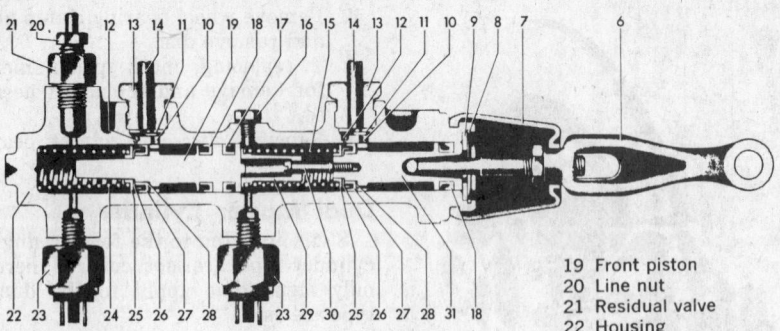

6. Replace in reverse sequence of removal. All parts must be cleaned in methylated spirits or brake fluid. No burrs or corrosive conditions should be overlooked.

7. The residual pressure valves and the brake light switches should be installed and tightened to 11-14 ft. lbs.

8. Install protective cap with breather hole downward.

1 Cap	10 Feed port	19 Front piston
2 Reservoir	11 Sealing washer	20 Line nut
3 Feed line	12 Compensating port	21 Residual valve
4 Line seal	13 Sealing plug	22 Housing
5 Line nut	14 Feed line	23 Brake light switch
6 Push rod	15 Rear piston spring	24 Front piston spring
7 Boot	16 Stop screw	25 Spring plate
8 Spring ring	17 Seal	26 Support ring
9 Stop ring	18 Cup-secondary	27 Cup-primary
		28 Cup washer
		29 Stop sleeve
		30 Limiting screw
		31 Rear piston

Dual master cylinder (© Volkswagen)

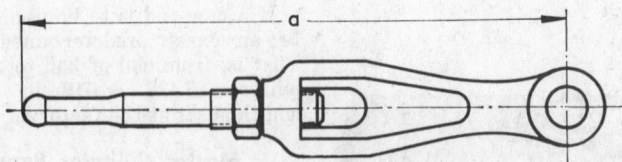

Master cylinder pushrod length (© Volkswagen)

Engine

General

Crankcase

The crankcase is a light metal alloy die casting. The crankcase halves are machined in pairs to very close tolerances. In consequence, replacements must be made in pairs.

Crankshaft

The crankshaft is heat-treated at all bearing points. The shaft is supported by four special light-metal bearings. No. 2 bearing (as seen from the clutch) is of the split type. No. 1 bearing is lead-coated and takes crankshaft thrust. The flywheel,

1 Fan housing	7 Oil pressure switch	13 Oil filter and breather	19 Push rod	25 Generator (dynamo)
2 Ignition coil	8 Valve	14 Pre-heating pipe	20 Heat exchanger	26 Flywheel
3 Oil cooler	9 Cylinder	15 Connecting rod	21 Thermostat	27 Crankshaft
4 Intake manifold	10 Piston	16 Spark plug	22 Cam follower	28 Camshaft
5 Ignition distributor	11 Oil pressure relief valve	17 Cylinder head	23 Throttle ring	29 Oil pump
6 Fuel pump	12 Fan	18 Rocker arm	24 Carburetor	30 Oil strainer

Engine assembly (© Volkswagen)

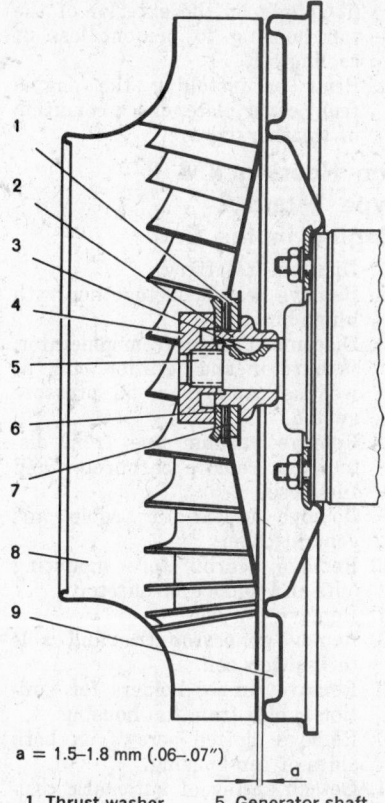

a = 1.5–1.8 mm (.06–.07")

1 Thrust washer   5 Generator shaft
2 Fan hub   6 Lock washer
3 Woodruff key   7 Spacer washers
4 Special nut   8 Fan
   9 Fan cover

Cooling blower

with starter gear, is held by a gland nut and additionally held to the crankshaft by four dowel pins. The timing gear and distributor drive gear are held in place by Woodruff keys. The type 1 and 2 fan drive pulley is bolted to the crankshaft. On the type 3 engine, the cooling fan is attached directly to the crankshaft. An oil seal is fitted to the clutch side of the crankshaft and an oil slinger and oil return thread to the other side.

Connecting Rods

The crankshaft ends of the four connecting rods contain replaceable lead-bronze bearings. Piston pin ends are provided with bronze bushings. Piston pins are full-floating.

Pistons

Pistons are of light-metal alloy and are equipped with three rings, the bottom one being an oil ring.

Cylinders

The four cylinders are cast of special material, and are interchangeable. Cylinders are provided with integral cooling fins.

Cylinder Heads

Each pair of cylinders has one mutual detachable cylinder head, a light-alloy casting. The cylinder heads are also provided with fins for cooling and they have shrunk-in valve seat inserts and valve guides. No gasket is used between cylinder and cylinder head. However, copper-asbestos gaskets, between the flanges of cylinder and cylinder head, prevent leakage of combustion gases.

Valve Train

The camshaft is supported by three bearings in the crankcase. It is driven, through helical gears, by the crankshaft. The camshaft gear is of light metal. Valves are operated by cam lobes via cam followers, pushrods and rocker arms. Each cam operates, in turn, one of the valves of two opposed cylinders. Exhaust valves are plated with chrome-nickel steel.

Cooling System

The type 1 and 2 engine is cooled by a fan attached to the extended generator shaft. It is driven by the crankshaft pulley through an adjustable V-belt at about twice engine rpm. The type 3 engine has a fan attached to the end of the crankshaft. The fan draws air through an opening in the fan housing. The air then passes through the fins of cylinders and cylinder heads. Air flow is directed by deflector plates, some in the fan housing and others covering the cylinders. The throttle ring at the air intake opening of the fan housing is thermostatically controlled, which insures quickly attained and steadily maintained operating temperatures.

Engine Lubrication

Pressure-fed lubrication includes a special oil cooler. The gear-type oil pump is located on the distributor side of the engine and is driven by the camshaft. Oil is drawn from the lowest point of the crankshaft and forced into the oil passages via the oil cooler. Part of the oil is fed through the crankshaft and main bearings to the rod bearings. Another amount lubricates the camshaft and bearings. The remainder is fed through hollow pushrods to the rocker arms and valve stems. Cylinder walls, pistons and piston pins are lubricated by splash. Oil then returns to the crankcase, to be filtered by a gauze strainer before again entering circulation.

The oil cooler on the crankcase is positioned in the ducted air flow. It is so located that oil forced by the pump must pass through it before reaching lubrication points. In cold weather, an oil pressure relief valve makes possible direct engine lubrication without the colder, thicker oil passing through the oil cooler.

Oil Pressure Warning Light

A pressure warning light switch is fitted to the oil pressure circuit on the oil pipe between the oil pump and cooler. This switch opens an electric

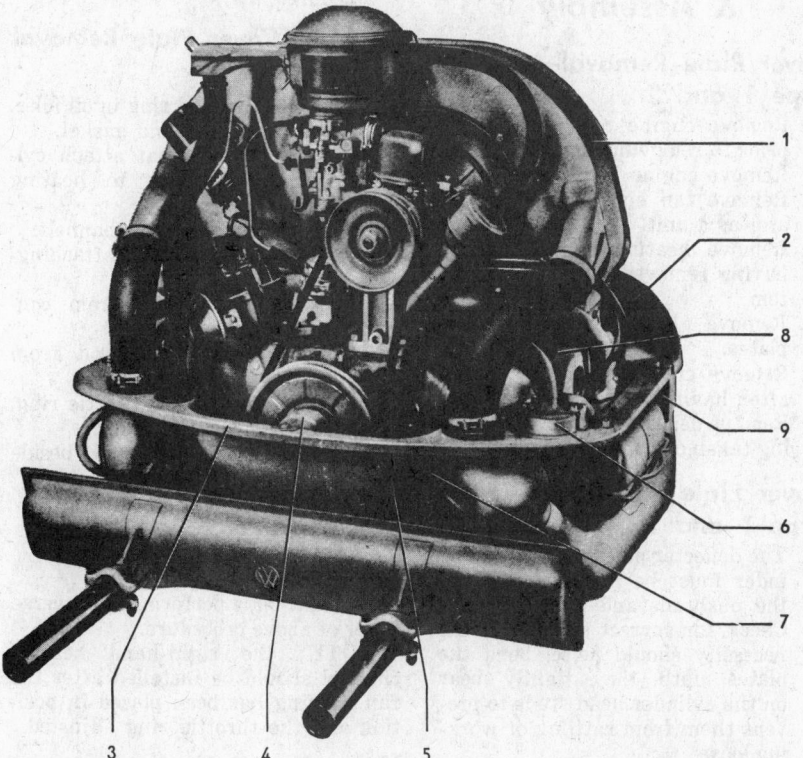

1 Fan housing
2 Front engine cover plate
3 Rear engine cover plate
4 Fan pulley cover
5 Fan pulley lower plate
6 Pre-heater pipe sealing plate
7 Air deflector plate
8 Cylinder cover plate
9 Screening plate

Engine external parts

contact at pressures from 2.1-6.3 psi. The warning lamp lights up when switching on the ignition, and when oil pressure is too low.

ENGINE R & R

Removal

This operation requires that the vehicle clear the floor by about 3 ft.

1. Disconnect battery ground strap. Disconnect generator cables. Disconnect regulator wires on Type 1 vehicles with regulator mounted on generator.
2. Remove air cleaner. Remove rear engine cover plate on type 1 models.
3. Rotate type 1 distributor to clear rear cover plate. This is not necessary on 1967-70 models.
4. Disconnect throttle cable from carburetor(s). Disconnect wires from automatic choke, coil, electromagnetic cut-off jet, and oil pressure sending unit.
5. Disconnect and plug fuel hose.
6. On type 3, remove oil dipstick, rubber boot between oil filler and body, cooling air intake bellows, warm air hose, and rear engine support.
7. Raise rear of car about three feet.
8. Remove hoses between engine and heat exchangers. Disconnect heater flap cables. Unscrew two lower engine mounting nuts.
9. Place a suitable jack under engine.
10. Support engine. Have an assistant hold two upper engine mounting bolts. Remove nuts.
11. Roll engine and jack backwards slightly so that clutch clears main drive shaft.
12. Lower engine.

Installation

This is a reversal of the preceding operations, but the following points should be observed:

1. Install engine only with engine rear cover plate removed.
2. Loosen mounting screw on the distributor. Turn distributor.
3. Check central position of clutch plate.
4. Check clutch release bearing and clutch release plate for wear and cracks. Renew, if necessary.
5. Examine needle bearing in flywheel gland nut for wear and pack it with (0.35 oz.) universal grease.
6. Lubricate starter, shaft bushing, starter drive pinion, and flywheel gear ring with multi-purpose grease.
7. Lubricate main drive shaft splines and pilot with transmission oil. Apply with a clean dry cloth.
8. Thoroughly clean transmission case and engine flange.

9. When replacing engine, care must be taken to prevent damage to gland nut needle bearing and clutch release bearing. Avoid bending the main driveshaft. To facilitate entry of main drive shaft into clutch plate and gland nut needle bearing, rotate engine at V-belt (engage a gear to steady main drive shaft).

To avoid interference with the function of the automatic clutch, take care to route the connecting hoses so that they are not kinked or jammed when installing the engine. This applies, particularly, to the small, diameter pipe from the control valve to the carburetor venturi, which will only work properly if routed in the original production manner.

10. When installing engine, first insert the lower engine mounting bolts into their corresponding holes in transmission case flange. Press engine firmly against flange until it is contacting properly all around. Slightly tighten the upper mounting bolt nuts; then, the lower ones. After that, fully tighten in the same order.
11. Adjust accelerator cable correctly.
12. Adjust ignition timing.
13. Install type 3 rear engine support with enough washers to raise engine about 2-3 mm.

Engine Disassembly & Assembly

Cover Plate Removal— Type 1 and 2

1. Remove engine rear cover plate prior to removing engine.
2. Remove engine front cover plate.
3. Remove fan housing and generator as a unit.
4. Remove heating channels after having removed the exhaust system.
5. Remove both cylinder cover plates.
6. Remove crankshaft pulley cover after having removed the pulley.
7. Remove deflector plates after having taken off valve pushrod tubes.

Cover Plate Installation— Type 1 and 2

1. The deflector plates below the cylinder must be installed prior to the pushrods and pushrod tubes Check for correct seating. If the necessity should arise, bend the plates until they tightly bear on the cylinder head studs to prevent them from rattling or working loose.
2. When replacing cylinder cover plates, attention should be paid to condition and sealing of spark plug rubber caps.
3. The cylinder cover plates should

fit snugly on the exterior of the fan housing to prevent loss of cooling air.

4. Prior to installing the engine front cover plate, check condition of weatherstrip.

Fan Housing R & R— Type 1 and 2 (Engine in the Car)

1. Disconnect battery.
2. Remove rear hood together with hinge bracket.
3. Disconnect cables from generator, carburetor and ignition coil, as well as cable from oil pressure switch.
4. Remove vacuum line from distributor, remove carburetor and fuel hose.
5. Remove accelerator cable and conduit tube.
6. Remove carburetor mounting nuts and remove carburetor.
7. Remove fan belt.
8. Remove generator strap and cable to ignition coil.
9. Remove rubber holders for ignition cables from fan housing.
10. Remove slotted screws on both sides of fan housing.
11. Detach spring of automatic cooling air control and remove throttle-ring screws.
12. Lift off fan housing and generator as a unit.
13. Install in reverse of above procedure.

Cylinder Cover Plate Removal Type 1 and 2

1. Remove threaded ring in oil filler. Remove oil filler and gasket.
2. Remove screws that attach cylinder cover plates to heating channels.
3. Remove both heating channels.
4. Remove thermostat attaching-screw.
5. Unscrew thermostat from connector rod.
6. Disconnect connector rod from operating lever.
7. Remove nut of the throttle ring, then withdraw the shaft.
8. Remove cover plate screws beside both ends of intake manifold.
9. Remove both cover plates.

Cylinder Cover Plate Installation—Type 1 and 2

Installation is performed in reverse order of above procedure.

NOTE: the right-hand heating channel should be installed after the fan housing has been placed in position and the throttle ring adjusted.

Valve Rocker Mechanism Removal

1. Remove valve cover.

NOTE: if valve covers are dirty, clean thoroughly before removing.

Removing rocker shaft
(© Volkswagen)

Any dirt falling on the rockers or valve springs can cause considerable damage and eventual engine failure.
2. Remove rocker arm shaft retaining nuts.
3. Remove rocker arm shaft and rocker arms.
4. Remove the stud seals.

Rocker arm support location
(© Volkswagen)

Valve Rocker Mechanism Installation

Install by reversing the removal procedure, plus;
1. Install new stud seals. Ball ends of pushrods must be centered in sockets of rocker arms. Torque nuts to 14 ft. lbs.
2. To help valves rotate during operation, rocker arm adjusting screws should contact the tip of the valve stem slightly off center. (To the right.)
3. Adjust valves.
4. Reinstall valve cover with new gasket.

Disassembly and Assembly

1. Remove spring clips from rocker arm shaft.
2. Remove washers, rocker arms, and bearing supports.
   Recently, the sealing between cylinder head and valve cover was modified.
   The sealing surface on the cylinder head, below the intake man-

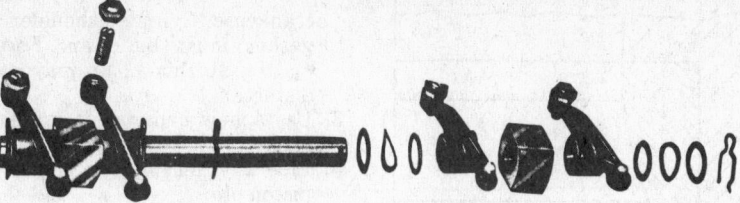

Rocker arm shaft assembly

ifold, is now straight, and the valve cover has been modified to suit.
The older version valve cover and cork gasket must *not* be used for engines with modified cylinder heads. If these parts are used, sealing near the intake port will be defective.
Note the following points:
1. Check rocker arm shaft for wear.
2. Examine seats and ball sockets of rocker arms and valve adjusting screws for wear.
3. Loosen adjusting screws prior to installing rocker arms.

Cylinder Head

Removal

1. Remove cylinder-head nuts.
2. Lift off cylinder head.

Installation

During installation, the following points should be noted.
1. Check for cracks or any other damage to combustion chambers and exhaust ports. Cracked heads must be replaced.
2. Check studs for security in the crankcase. If necessary, use Helicoil, or its equivalent, as a repair method.
3. There is no gasket used between the upper edge of the cylinder and the corresponding contact surface of the cylinder head.
4. Renew gasket between shoulder of cylinder and cylinder head. The slotted side of gasket must be toward the cylinder head.
5. Make sure that the oil seals at the ends of the pushrod tubes are correctly seated.
6. Install cylinder head nut washers.

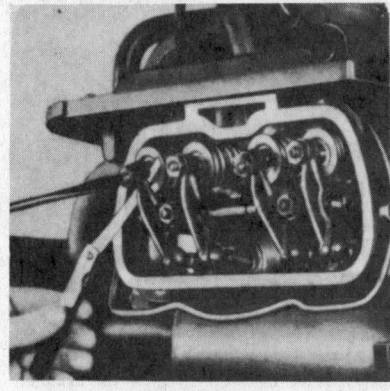

Valve adjustment
(© Volkswagen)

7. Coat cylinder head nuts with graphite paste, then screw them down finger tight. Tighten, in correct sequence, to 7 ft. lbs. torque.
8. Then torque, in correct sequence, to 22-23 ft. lbs. (27 ft. lbs. for 1959 and earlier.)

Valve Servicing

The valves are removed from their respective locations, serviced and reinstalled in the same way as their counterparts in American manufacture. However, the following items will bear attention.

Replacement of valve guides is not possible without tools and other facilities capable of chilling and shrinking the guides into place. Deposits and other buildup may be removed by reaming or broaching.

Damaged valve seat inserts may be reconditioned with a 45° seat cutter. Do not chamfer the 15° outer edge to exceed the outer diameter of the insert. Inserts cannot be replaced without special facilities capable of chilling and shrinking the insert into place.

To rework the valve seat, cut the seat to 45°. Stop cutting as soon as an even and concentric seat has been

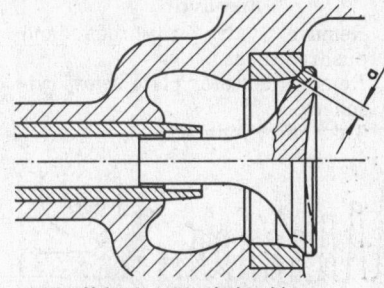

**Valve to seat relationship—
seat width = .05-.06 in. (intake)
.06-.08 in. (exhaust)**

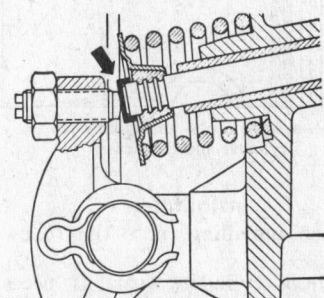

Valves with damaged stem ends can be made usable by fitting small caps on the stems. The cap is placed on the stem before fitting the rocker arms.

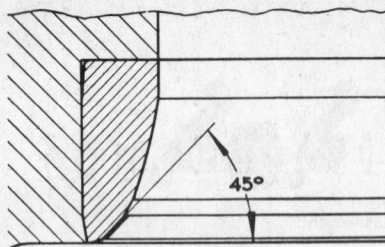

Valve seat angle = 45°

achieved. Then, slightly chamfer the lower valve seat with a 75° cutter. Next, cut the 15° seat chamfer on the upper seat edge until the correct seat width is obtained.

After grinding the valve, or lapping the valve face and seat, blue test the sealing surface as a precaution against gas leaks.

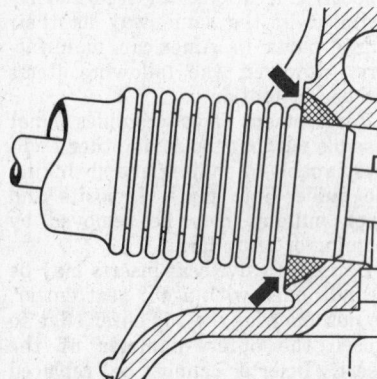

When installing cylinder head, make sure that the oil seals at the ends of the push rod tube are properly seated

Cylinders R & R

Removal

1. Remove valve pushrods and pushrod tubes.
2. Remove deflector plate below cylinders.
3. Take off cylinders.

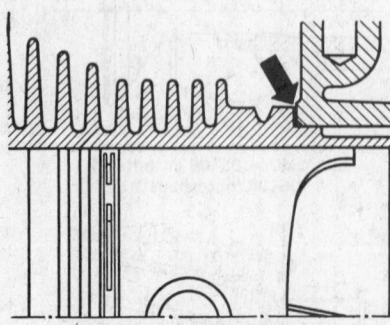

Always use a new gasket between cylinder and crankcase

Installation

When installing, note the following:

1. Check cylinder wear. If necessary, replace with another matched cylinder and piston assembly of the same size.
2. Check cylinder seating surface on

crankcase. Cylinder shoulder and gasket must be clean. Foreign matter at this point may cause distortion.
3. Use a new cylinder to crankcase gasket.
4. Liberally oil piston, rings, and piston pin.
5. Compress rings with compressing tool. Be sure ring gaps are adequate and staggered on the piston. The oil ring must be inserted into the cylinder so that its gap is positioned up when the pistons are in their horizontal position in the engine.
6. Lubricate cylinder wall. Crankcase studs must not contact cylinder cooling fins.
7. Install deflector plates. Make sure they are correctly seated.

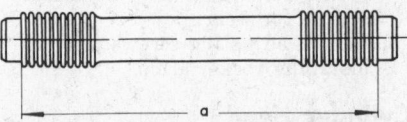

Pushrod tube length (a) should be 190-191 mm. for 1300, 1500 and 1600 engines; 180.5-181.5 mm. for the 40 H.P. 1200 engine

8. Install pushrod tubes with new seals and pushrods. Insert the tubes so that the seam is facing upwards. Before installing used tubes, they should be stretched to assure a proper length for sealing.

Removing cylinder
(© Volkswagen)

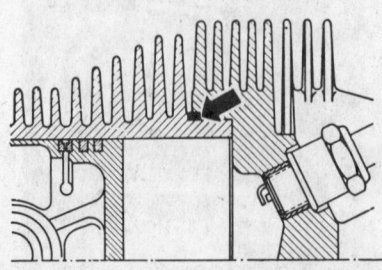

Renew gasket between shoulder of cylinder and cylinder head. The slotted side of the gasket must be toward the cylinder head.

Relation of piston and cylinder
(© Volkswagen)

Pistons R & R

Removal

1. Remove cylinder.
2. Mark the piston, and position.
3. Remove piston pin circlip.
4. Heat piston to 176°F.
5. Remove piston pin with a pilot drift.
6. If piston rings are to be removed, use a piston ring removing tool.

Installation

Assembling of piston and rod components is not unlike most engines of American manufacture. However, note the following:

1. Carefully clean the piston and ring grooves (don't scratch or otherwise damage ring grooves.)
2. Check for piston wear. If replacement is necessary, replace by one of a corresponding size. Difference in weight of pistons in any one engine must not exceed 10 grams.
3. Use piston rings of correct size. Compression ring gaps should be 0.012 in.-0.018 in. with a maximum of 0.035 in.
Oil scraper ring gap should be 0.010 in.-0.016 in. with a maximum of 0.035 in.
When reinstalling the cylinder, make sure the ring gap is at the top and that each piston ring gap is offset by 120°.

Check piston ring side clearance with a feeler gauge. Side clearance should be as follows:
No. 1 ring (compression) 0.0026 in.-0.0036 in. (1200, 1300). .0031-.0043 in. (1500, 1600).
No. 2 ring (compression) 0.0018 in.-0.0028 in. (All)
No. 3 ring (oil scraper) 0.0010 in.-0.0020 in. (All)

Make sure that the compression rings are installed right-side-up. (The marking "top" or "oben" toward the top of the piston.)

Insert the piston pin circlip, which faces toward the flywheel. Because piston pin holes are offset, make sure

An explanation of the marking of the pistons

A Arrow (indented or stamped on) which must point toward the flywheel when piston is installed.
B Details of piston pin bore size indented or stamped on (s = black, w = white).
C Paint spot indicating matching size (blue, pink, green).
D The letter near the arrow corresponds to the index of the part number of the piston concerned. It serves as an identification mark.
E Details of weight grading (+ or −) indented or stamped on.
F Paint spot indicating weight grading (brown = −weight, grey = +weight).
G Details of piston size in mm.

that the arrow (or word "vorn") points toward the flywheel. This offset is to help accommodate thrust loads that amplify, and lead to, objectionable piston slap.

Check and fit piston pin. The pin may be found to be a light finger-push fit in the piston, even when the piston is cold. This condition is quite normal, even to the extent of the pin sliding out of the piston of its own weight. For pistons with a hole diameter in excess of 0.7874 in., oversize pins are available.

Clearance between piston pin and connecting rod bushing should be .0004-.0008 in. If clearance is near

the wear limit of 0.0016 in., renew the piston pin and the rod bushing. It is not advisable to install an oversize pin in this case.

Heat the piston, in oil, to about 176° F in all cases where the pin is not a light finger-push fit in the cold piston.

Insert the other circlip. It is important that the circlips fit in their grooves perfectly.

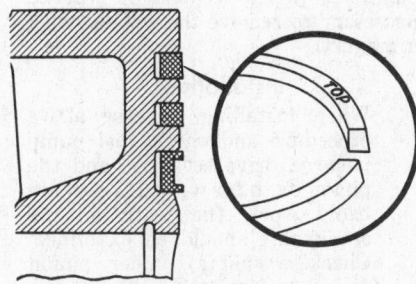

Piston rings must be installed as shown

Piston to Cylinder Clearance

The fitting clearance between piston and cylinder should be 0.0014-0.0021 in., with a wear limit of 0.008 in. Maximum out-of-round limit is 0.0004 in.

Besides wear, oil consumption is a good indication of whether or not a new cylinder and piston should be installed. If oil consumption is more than 1 qt./600 miles, the engine probably needs reconditioning.

Oil Strainer R & R

Removal
1. Remove retaining nuts from strainer bottom plate.
2. Remove the bottom plate.
3. Remove strainer and gaskets.

Installation
Install by reversing above procedure, plus the following precautions:
1. Be sure that the suction pipe is correctly seated in the strainer.

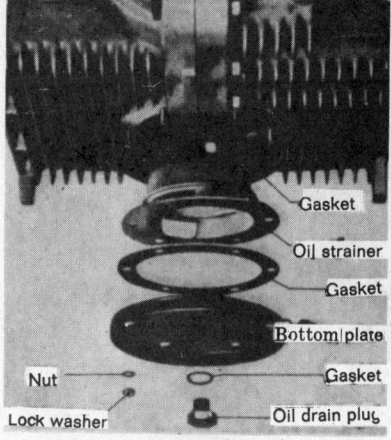

Oil strainer
(© Volkswagen)

2. Make sure that all of the sealing surfaces are straight and clean. To avoid distorting the bottom plate, especially if using thicker gaskets, do not over-torque.
3. Always check the oil pressure relief valve when any discrepancies are found in oil circulation or pressure, or when the oil cooler is leaky. If the plunger sticks, in closed position, there is a possibility of causing a leak in the oil cooler. If the plunger sticks in open position, oil will flow directly back to the sump.

Oil Cooler R & R
1. Remove oil cooler retaining nuts.
2. Remove oil cooler and gaskets.
3. Install by reversing above procedure, plus the following precautions:
   A. Check studs and bracket on cooler for tightness. Cooler should withstand 85 psi pressure.
   B. If cooler leaks, check oil pressure relief valve.
   C. Ribs of oil cooler should not touch one another. The partition sheet should not be loose.
   D. Use new gaskets.

Oil Pump R & R
1. Remove nuts on oil pump cover, then remove the cover and gasket.
2. Remove gears.
3. Pull the oil pump body with extractor.
4. Check all pump parts for wear. Excess wear will cause low oil pressure.
   Gear backlash should be, 0.0012-0.0032 in. Gear end-play should not exceed 0.004 in.
5. Check idler gear pin for tightness. If necessary, peen it securely in position, or replace pump body.
6. Install pump body with new gasket.
7. Insert oil pump pilot shaft in-

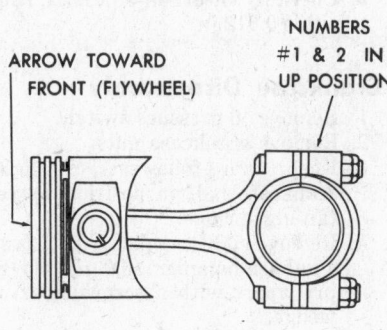
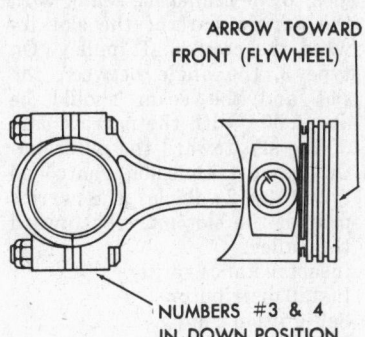

WHILE ASSEMBLING RODS TO CRANKSHAFT, (RODS IN VERTICAL POSITION) ALL ROD NUMBERS ARE ON RIGHT SIDE, VIEWED FROM FLYWHEEL END.

Piston and rod installation (© Volkswagen)

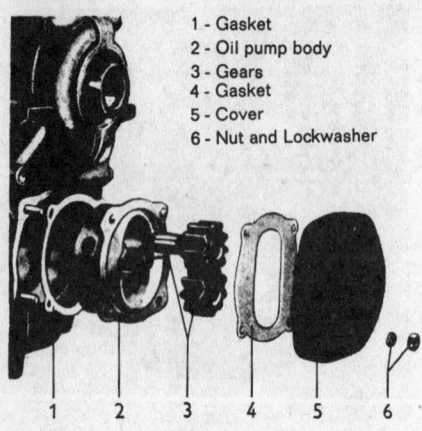

1 - Gasket
2 - Oil pump body
3 - Gears
4 - Gasket
5 - Cover
6 - Nut and Lockwasher

1　2　3　4　5　6

Oil pump
(© Volkswagen)

stead of oil pump drive shaft into body.

8. Turn camshaft 360°. This should center the oil pump drive shaft with the slot in the camshaft.
9. Mark the pump body so that the correct fit of the oil pump can be checked after the cover has been installed.
10. Remove oil pump pilot. Install gears.
11. Place a straight edge across pump body and gears, then use a feeler gauge between straight edge and gears. Clearance should not exceed 0.004 in. with gasket removed.
12. Use new gasket without sealant, and install pump cover.

NOTE: with engine installed in type 1 and 2, the oil pump can be removed after the engine cover plate, crankshaft pulley and cover plate below the pulley have been removed. Type 3 oil pump removal requires engine removal.

Distributor Drive Pinion R & R

Removal

1. Loosen distributor clamp bolt, and lift out distributor.

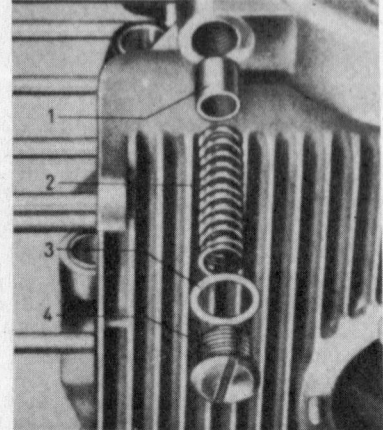

1 Plunger　3 Gasket
2 Spring　4 Plug

Exploded view, oil pressure relief valve. If oil cooler is leaky, this may be the cause.

2. Remove fuel pump and intermediate flange, gaskets, and fuel pump pushrod.
3. Remove distance spring on distributor drive pinion.
4. Pull drive pinion out with extractor.
5. Remove washer(s) under drive shaft.

Caution (Don't drop washer into the crankcase.) If engine is in the vehicle, it will be necessary to remove the washer with a magnet.

Installation

1. When installing, reverse above procedure and check fuel pump pushrod drive eccentric and the pinion teeth for wear. If teeth are badly worn, the teeth on the crankshaft should be examined. Check washer(s) under pinion for wear. Renew, if worn.
2. Position No. 1 cylinder to firing point, then insert distributor drive gear. On type 1 and 2 vehicles, the offset slot in the top of the pinion must be at right an-

Distance spring

Drive pinion

Washer

Distributor drive
(© Volkswagen)

gles to the crankcase seam, with the narrow part of the slot toward the crankshaft pulley. On type 3, the angle between the slot and the seam should be about 60°, with the narrow part of the slot toward the oil cooler. On exhaust emission controlled type 1 and 2 vehicles, the narrow part of the slot must be toward the pulley.
3. Insert distance spring.
4. Install distributor.
5. Set ignition timing.
6. Install fuel pump.

NOTE: in recent production, the thrust surface for the distributor drive shaft in the left crankcase half was machined 0.196 in. deeper. At the

same time, the distributor drive shaft was modified. The shaft now has a thrust shoulder with a flat machined on one side.

When this modified distributor drive is used, the drive can only be removed and installed when No. 1 cylinder is at firing point. The flat on the distributor drive is then toward the distributor drive gear. The modified drive cannot be installed in the former crankcase.

Flywheel R & R

The flywheel is attached to the crankshaft with a gland nut, and is located by four dowels. There is a metal or paper gasket between flywheel and crankshaft. An oil seal is recessed in the crankcase casting at No. 1 main. A needle bearing, which supports the main drive shaft, is located in the gland nut.

Removal

1. Remove clutch pressure plate.
2. Remove clutch driven plate.
3. Loosen gland nut (using 36mm special wrench and flywheel retainer). Remove guide plate of special wrench.
4. Remove gland nut.
5. Withdraw flywheel.

Installation

Install by reversing removal procedure, plus the following.
1. Adjust crankshaft end-play.
2. Check needle bearing in gland nut for wear. Lubricate needle bearing with about 1/3 oz. universal joint grease.
3. Renew flywheel gasket.

NOTE: to minimize engine imbalance, the crankshaft, flywheel, and clutch are marked at their heaviest points. Upon assembly, be sure that the marks on these units are offset by 120°. If but two of these parts are marked, the marks should be offset by 180°.
4. Tighten gland nut to 217 ft. lbs.
5. Check flywheel run-out. Max. run-out is 0.012 in.

Crankcase Disassembly

1. Remove oil pressure switch.
2. Remove crankcase nuts.
3. Keep cam followers of right crankcase half in position by retaining springs.
4. Remove right crankcase half (use a rubber hammer). Do not try to pry apart with a screwdriver or wedge.
5. Remove crankshaft oil seal.
6. Remove camshaft end-plug.
7. Take out crankshaft and camshaft.
8. Remove cam followers.
9. Remove bearing shells and oil pressure relief valve.

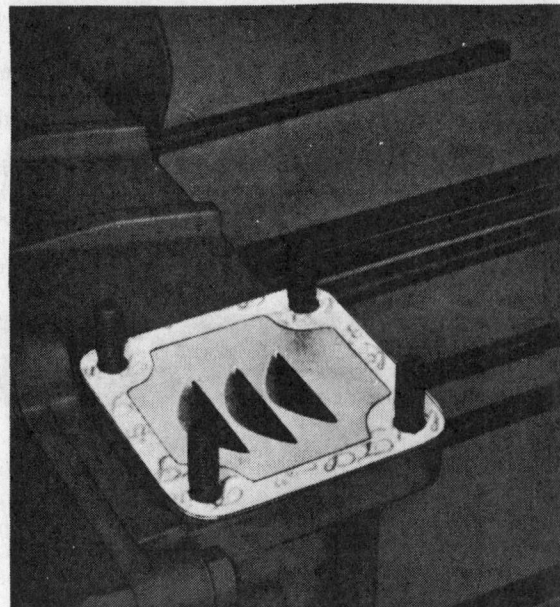

1968-70 Type 1 vehicles employ a new
baffle plate which prevents excessive oil
being drawn out through the crankcase breather

Crankcase Assembly

In addition to reversing disassembly procedure, check the following.
1. Check crankcase for damage and cracks.
2. Clean crankcase thoroughly, especially mating and sealing surfaces. Flush and blow out all ducts and passages.
3. Check oil suction pipe for leaks.
4. Install cam followers.
5. Check and install oil pressure switch.
6. Note timing marks on the gears, and be sure crankshaft oil slinger is correctly installed.
7. Install crankshaft, with bearings well lubricated.
8. Install camshaft.
9. Install camshaft end plug, using sealing compound. Install thrust washers and crankshaft oil seal. The oil seal must rest squarely on the bottom of its recess in the crankcase.
10. Spread a thin film of sealing compound on the crankcase joining faces. Use care so that no sealing compound enters the oil passages of crankshaft or camshaft bearings.
11. Keep cam followers of right crankcase half in place by using retaining springs.
12. Join crankcase halves and evenly torque the 12mm nuts to 24-26 ft. lbs. Torque the 8mm nuts to 13-15 ft. lbs.
NOTE: first tighten the 8mm nut which is beside the 12mm stud of crankshaft bearing No. 1. Only then should the 12mm nuts be tightened fully.

Crankshaft End-Play

Check

Crankshaft end-play should be read with a dial indicator and should be 0.0028 in.-0.0047 in., the wear limit being 0.006 in. With engine installed in the vehicle, the indicator may be mounted to read shaft end-play at the pulley. If convenient, an indicator may be mounted on the flywheel end and end-play read directly from the flywheel.

Adjustment

1. Force the installed crankshaft towards the flywheel side of the engine (with flywheel removed) so that it contacts the inner thrust surface of No. 1 bearing.
2. Insert dial gauge in flywheel seat so that it contacts the crankshaft, and measure distance from crankshaft face to outer face of main bearing No. 1.
3. Place a gauge on the flywheel flange and measure depth of crankshaft seat.
4. The thickness of the shims to be used is decided by the difference in both readings (taking into account the gasket).
Shims for this purpose are available in various thicknesses. Desirable end-play is obtained by adding or subtracting shims at the outer end of the main bearing. Never use more than one gasket.

Checking crankshaft end-play
(© Volkswagen)

Crankshaft and Connecting Rods R & R

Removal

1. Open the crankcase.
2. Remove camshaft.
3. Remove crankshaft.

Installation

1. The crankcase must not have sharp edges at points of junction.
2. Check dowel pins for tightness.
3. Place one half of No. 2 crankshaft bearing in the crankcase.
4. Slide on crankshaft bearing No. 1 so that the dowel pin hole is toward the flywheel.
5. Install crankshaft. Make sure that the dowel pins are correctly seated in the crankshaft bearings.
6. Note the marks on the timing gears when installing camshaft.

Connecting Rods R & R

Removal

1. Remove crankshaft and clamp in position.
2. Remove connecting rod clamping bolts and connecting rods and caps.

Installation

1. Check weight of connecting rods. The difference in weight of the connecting rods in one engine must not be in excess of 10

Removal and installation of
crankcase sections
(© Volkswagen)

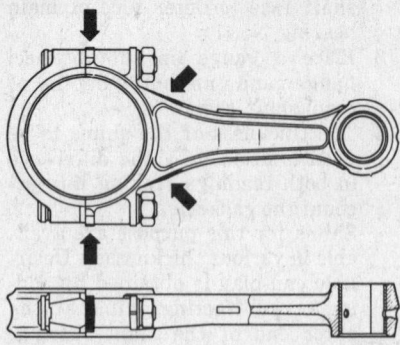

Maximum weight difference between connecting rods in one engine is 10 grams. Metal can be removed from the arrowed portions of the connecting rod.

grams to maintain proper engine balance. If necessary, metal should be removed from the heavier connecting rods at the points indicated on the drawing.
2. Inspect piston pin bushing. With a new bushing, the correct clearance is indicated by a light finger-push fit of the pin at room temperature.
3. Check and, if necessary, correct connecting rod alignment.
4. Reinsert connecting rod bearing shells after all parts have been thoroughly cleaned, and assemble connecting rods on crankshaft. The identification numbers stamped on connecting rods and bearing caps must be on one side.
5. Tighten connecting rod bolts to a torque of 36 ft. lbs. A slight pretension between the bearing halves, which is likely to occur when tightening connecting rod bolts, can be eliminated by light hammer taps. The connecting rods, lubricated with engine oil prior to assembly, must slide onto the crank pin by their own weight. The connecting rod bushings must not be scraped, reamed or filed during assembly.
6. Secure connecting rod bolts in place, using chisel.

Muffler R & R

(Engine Installed)

Removal
1. Raise car at rear and support on horses.
2. Remove rear cover plate.
3. Remove four nuts at flanges of pre-heating pipe.
4. Loosen clamps on tail pipes and remove pipes.
5. Loosen clamps at front exhaust pipes and heater boxes.
6. Take four nuts off muffler flanges.
7. Draw muffler back and remove from below. Remove gasket from cylinder head flanges, muffler and preheating pipe.

Installation
Install in reverse order of removal, plus the following precautions:
1. Check muffler and exhaust pipes for cracks and damage. If necessary, the pipes can be straightened.
   The welded joint of the muffler and tail pipe is particularly susceptible to damage by impacts. Leaks may result in exhaust fumes entering the engine compartment and interior of the car when the heater is turned on.
   Always replace all bent or out-of-round tail pipes. If the cartridges are no longer serviceable, the tail pipes have to be replaced.
2. Use new gaskets.
3. There should be a perfect seal at

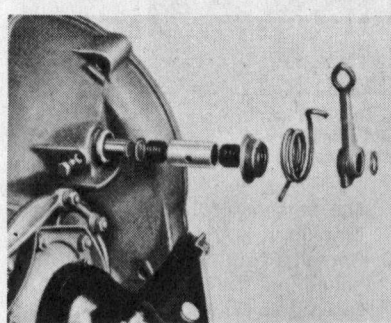

Exploded view, clutch release shaft. Beginning with the 1966 models, a plastic bushing is used.

connection to front exhaust pipes.
4. Push tail pipes into the exhaust pipes and make sure there is a perfect seal at the connection. The tail pipes should protrude approximately 7.5 in.
5. With the engine in place, the tail pipes must not touch the lower edge of the body. If necessary, remove tail pipes and heat exhaust pipes prior to bending them.

Clutch

Operation
Pedal movement is transmitted through a cable to the release mechanism. As pressure is applied to the pedal, the release levers are moved inward by the release bearing. The pressure plate is moved away from the driven plate, disengaging the clutch.

Adjustment
Adjustment is limited to clutch pedal free-play of 0.4-0.8 in. An adjustment of the pressure plate is only necessary when the clutch is disassembled for parts replacement. The clutch release bearing is a carbon thrust ring or ball type bearing and requires no maintenance.

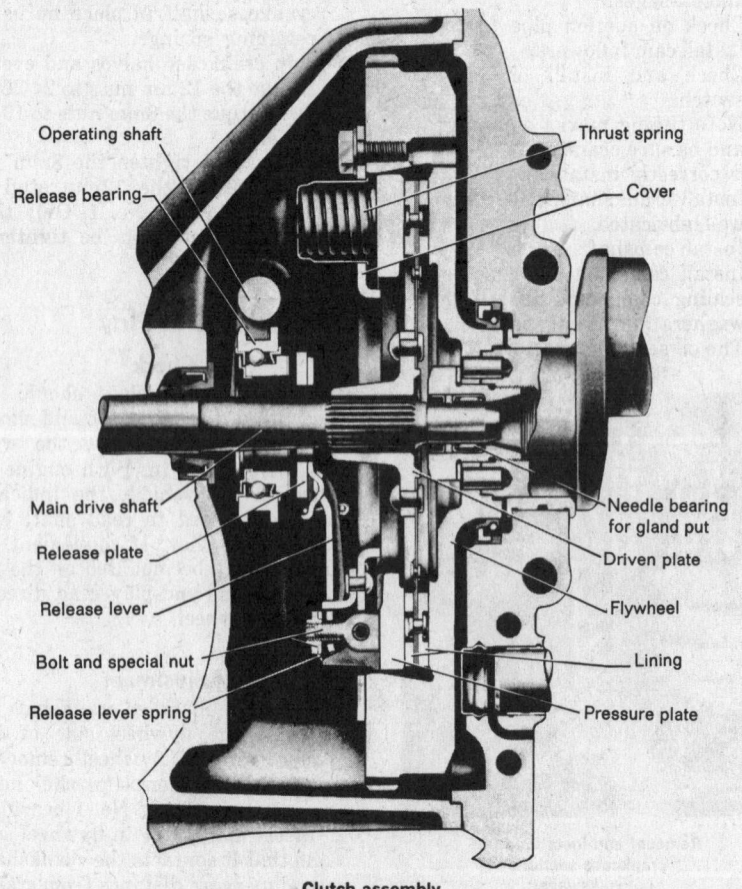

Operating shaft
Thrust spring
Release bearing
Cover
Main drive shaft
Needle bearing for gland put
Release plate
Driven plate
Release lever
Flywheel
Bolt and special nut
Lining
Release lever spring
Pressure plate

Clutch assembly

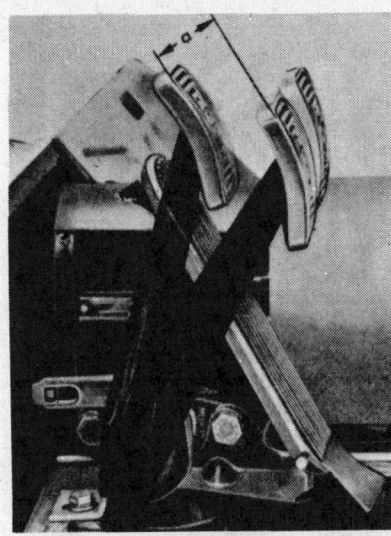

Free-play at the clutch pedal should be 10-20 mm. (.4-.8 in.)

Clutch Removal

1. Remove engine.
2. Remove clutch to flywheel attaching bolts by gradually, and alternately, backing the bolts out of the flywheel.
3. Take off clutch cover.
4. Lift out clutch-driven (lined) plate.

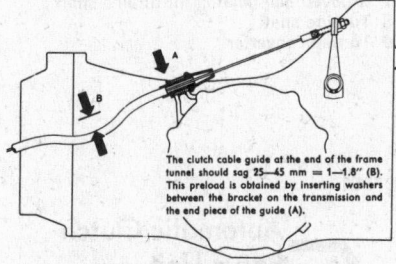

The clutch cable guide at the end of the frame tunnel should sag 25—45 mm = 1—1.8" (B). This preload is obtained by inserting washers between the bracket on the transmission and the end piece of the guide (A).

Clutch cable location and adjustment (see text)

Clutch Installation

Upon installation, reverse above procedure, plus the following:

1. Inspect and resurface pressure plate if worn in excess of 0.008 in. The friction surface should be polished.
2. Inspect driven plate and renew it if any doubt of reliability exists.
3. Examine release levers and springs.
4. Inspect release plate for damage or heat discoloration.
5. Check release bearing or carbon thrust ring. Renew complete release bearing, if necessary.
6. Inspect bearing points of clutch operating shaft in transmission case for wear.
    NOTE: a ball - type release bearing of different construction is available. Because of greater durability, it should be used in-stead of the carbon type.
7. Lubricate needle bearing in flywheel gland nut with about 10 gr. of universal grease.
8. Reinstall driven (lined) plate, using a pilot mandrel to insure correct centering alignment.
9. Evenly, and alternately, tighten clutch-to-flywheel attaching bolts.
10. Check proper distance and parallelism between clutch cover contact face at flywheel and the clutch release plate. Use a clutch adjustment gauge.

Pedal Adjustment

Measured at the clutch pedal, the free play should be 0.4-0.8 in. Clearance may be adjusted at the adjusting nut on the cable end.

1. Release locknut on the threaded cable end.
2. Adjust clutch clearance by turning the adjusting nut until proper pedal clearance is obtained.
3. Secure the adjustment with the locknut.

NOTE: late models have a self locking wing nut for adjustment.

Clutch Cable R & R

1. Remove left rear wheel.
2. Disconnect cable from operating lever on transmission.
3. Pull rubber boot off guide tube and cable.
4. Disconnect master cylinder piston pushrod and unbolt pedal assembly.
5. Unhook cable from pedal cluster.
6. Pull cable forward through hole uncovered when pedal cluster was removed.
7. Install in reverse of above. Lubricate thoroughly with chassis lubricant. Insert enough washers between case bracket and cable guide to cause guide to sag slightly at transmission.

VW Automatic Stickshift— Type 1

Since 1968, Volkswagen has offered an Automatic Clutch Control Three Speed Transmission. This unit is called the automatic stick shift.

It consists of a three speed gear box connected to the engine through a hydrodynamic torque converter. Between the converter and gearbox is a vacuum-operated clutch, which automatically separates the power flow from the torque converter while in the process of changing gear ratios.

While the torque converter components are illustrated here, the picture is for familiarization purposes only. The unit cannot be serviced. It is a welded unit, and must be replaced as a complete assembly.

The power flow passes from the engine via converter, clutch and gearbox to the final drive, which, as with the conventional gearbox, is located in the center of the transmission housing.

The converter functions as a conventional clutch for starting and stopping. The shift clutch serves only for engaging and changing the speed ranges. Friction-wise, it is very lightly loaded.

There is an independent oil supply for the converter provided by an engine driven pump and a reservoir. The converter oil pump, driven off the engine oil pump, draws fluid from the reservoir and drives it around a circuit leading through the converter and back to the reservoir.

This circuit also furnishes cooling for the converter fluid.

Operation

The control valve is activated by a very light touch to the top of the shift selector knob which, in turn, is

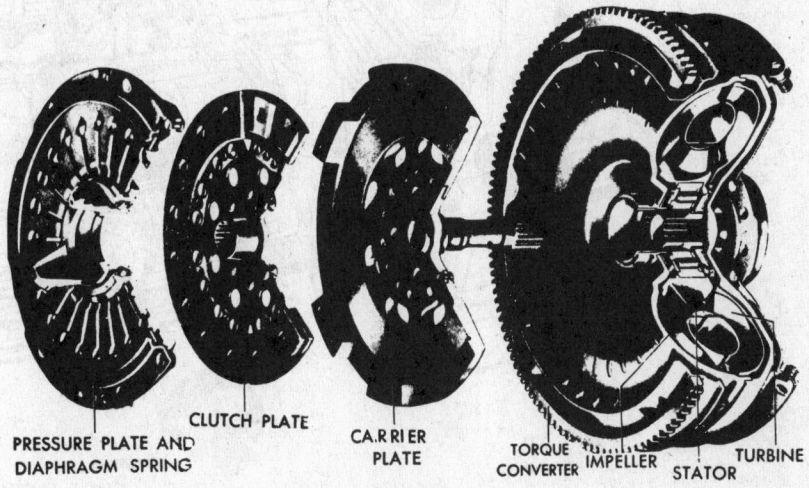

PRESSURE PLATE AND DIAPHRAGM SPRING   CLUTCH PLATE   CARRIER PLATE   TORQUE CONVERTER   IMPELLER   STATOR   TURBINE

Converter components

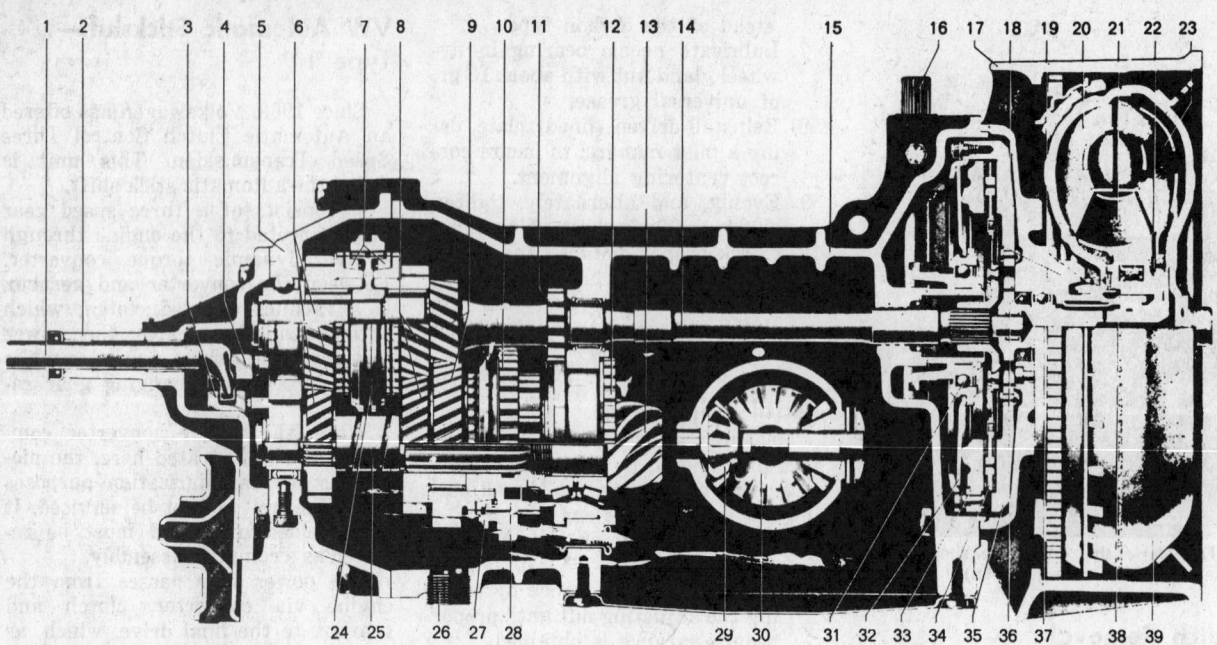

1 Inner transmission shift lever	12 Tension nut for driving pinion	26 Magnetic oil drain plug
2 Gear shift housing	13 Drive pinion	27 Synchronizing ring, 1st driving range
3 Selector shaft for 1st and driving range	14 Drive shaft	28 Clutch gear, 1st and reverse driving range
4 Selector shaft for 2nd and 3rd driving range	15 Transmission housing	29 Differential pinion
5 Gear carrier	16 Release shaft for separator clutch	30 Differential side gear
6 Gear train, 3rd driving range	17 Converter housing	31 Release bearing for separator clutch
7 Synchronizing rings, 3rd and 4th driving range	18 Support tube for converter freewheel	32 Cup spring
8 Gear train, 2nd driving range	19 Shaft sealing ring for torque converter	33 Pressure plate
9 Gear train, 1st driving range	20 Impeller	34 Drive plate
10 Operating sleeve, 1st and reverse driving range	21 Stator	35 Clutch carrier plate
11 Drive, driving range	22 Freewheel	36 Shaft sealing ring for converter housing
	23 Turbine	37 Grooved ball bearing for turbine shaft
	24 Operating sleeve, 2nd and 3rd driving range	38 Turbine shaft
	25 Axial spring, 2nd and 3rd driving range	39 Torque converter

Power transmission components—automatic stick shift

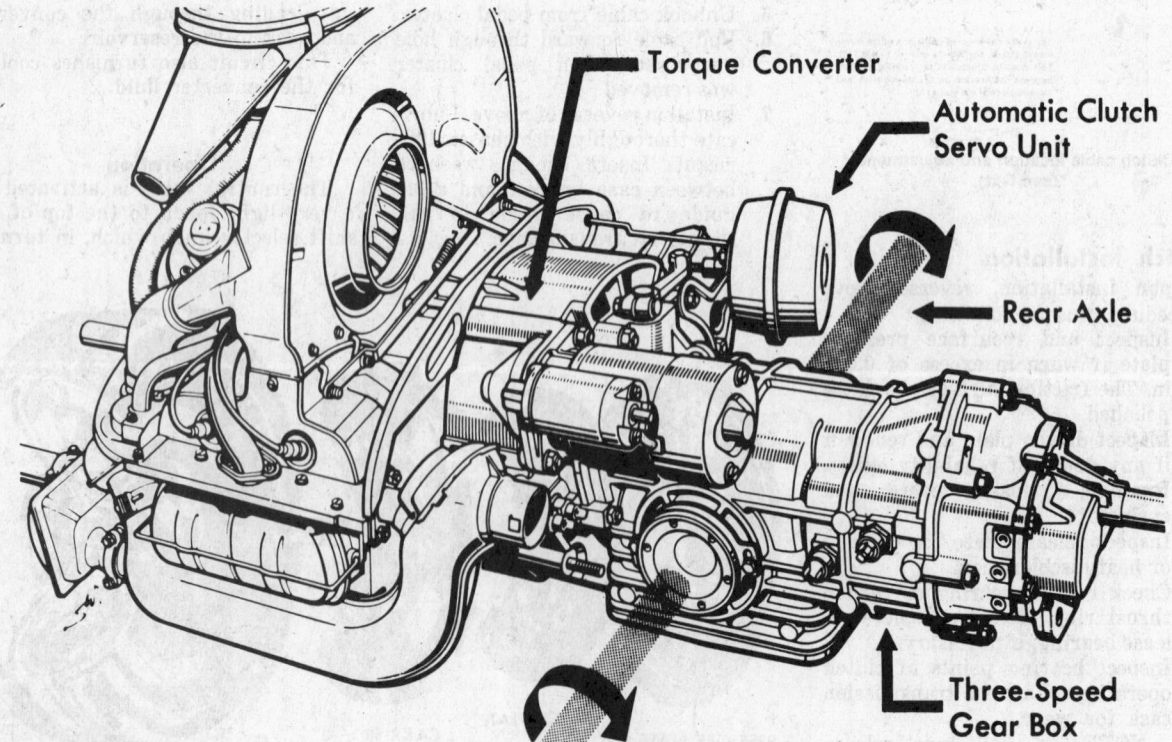

Torque Converter

Automatic Clutch Servo Unit

Rear Axle

Three-Speed Gear Box

Rear Axle

Automatic stick shift

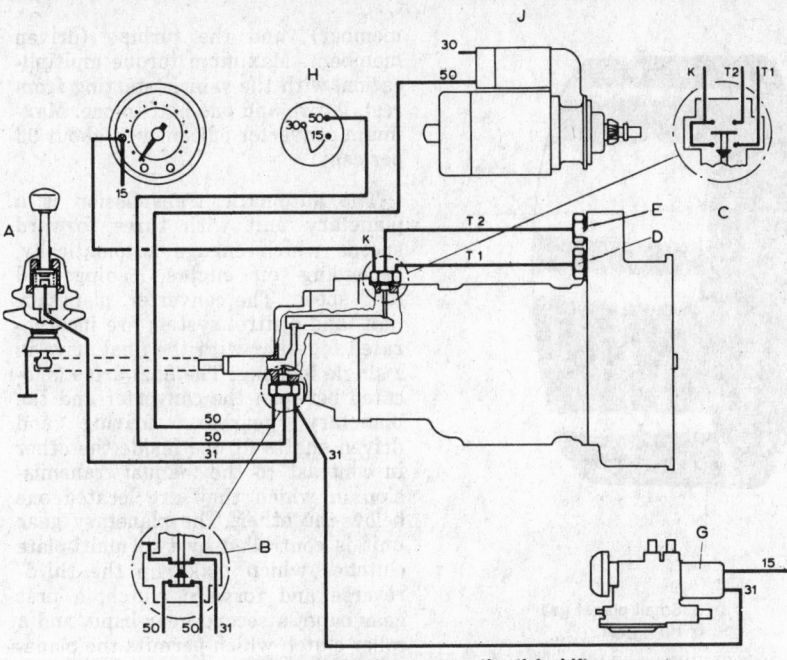

A Selector lever with switch
B Starter locking switch with bridging switch
C Selector switch
D Temperature switch for 3rd speed range
E Temperature switch for 2nd speed range
F Warning light
G Electro-magnet for control valve
H Ignition switch
J Starter F

Electrical circuits—automatic stick shift

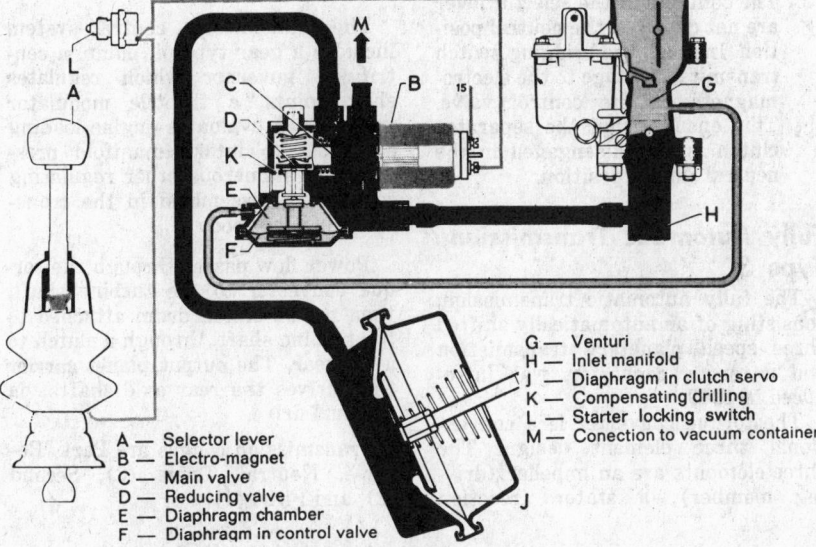

A — Selector lever
B — Electro-magnet
C — Main valve
D — Reducing valve
E — Diaphragm chamber
F — Diaphragm in control valve

G — Venturi
H — Inlet manifold
J — Diaphragm in clutch servo
K — Compensating drilling
L — Starter locking switch
M — Conection to vacuum container

Oil circuits—automatic stick shift

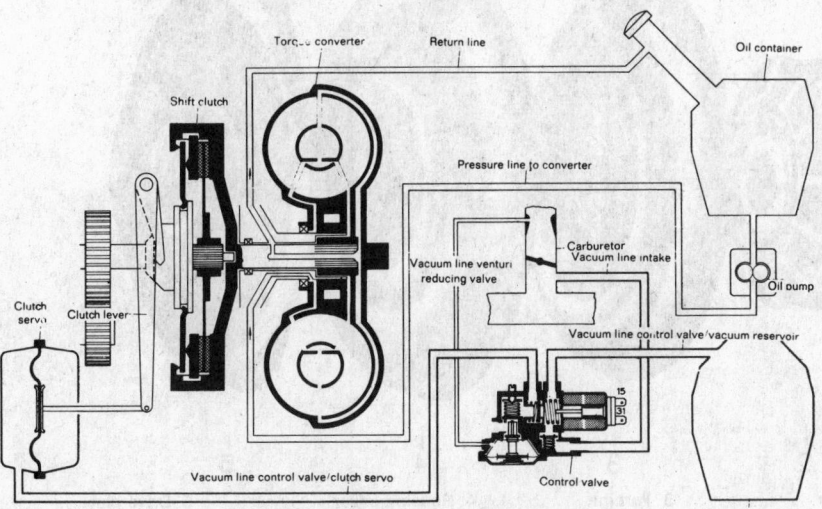

Vacuum circuits—automatic stick shift

connected to an electro-magnet. It has two functions.

At the beginning of the selection process, it has to conduct the vacuum promptly from the intake manifold to the clutch servo, so that the shift clutch disengages at once, and thus interrupts the power flow between converter and transmission. At the end of the selection process, it must, according to driving conditions, automatically ensure that the shift clutch engages at the proper speed. It may neither slip nor engage too harshly. The control valve can be adjusted for this purpose.

As soon as the selector lever is moved to the engaged position, the two contacts in the lever close the circuit. The electro-magnet is then under voltage and operates the main valve. By this means the clutch servo is connected to the engine intake manifold, and at the same time the connection to the atmosphere is closed. In the vacuum space of the servo system, a vacuum is built up, the diaphragm of the clutch servo is moved by the difference with atmospheric pressure and the shift clutch is disengaged via its linkage. The power flow to the gearbox is interrupted and the required speed range can be engaged. The process of declutching, from movement of the selector lever up to full separation of the clutch, lasts about 1/10 sec. The automatic can, therefore, declutch faster than would be possible by means of a foot-operated clutch pedal.

When the selector lever is released after changing the speed range, the switch interrupts the current flow to the electro-magnet, which then returns to its rest position and closes the main valve. The vacuum is reduced by the reducing valve and the shift clutch re-engages.

Clutch engagement takes place, quickly or slowly, according to engine loading. The clutch will engage suddenly, for example, at full throttle, and can transform the full drive moment into acceleration of the car. Or, this can be effected slowly and gently if the braking force of the engine is to be used on overrun. In the part-load range, too, the duration of clutch re-engagement depends on the throttle opening, and thus the depression in the carburetor venturi. This results in smooth, pleasant driving under all conditions.

1 Small sungear   3 Large sungear   5 Small planet gear
2 Planet carrier   4 Large sungear   6 Ring gear

Automatic transmission planetary gear unit (© Volkswagen)

Vanes on the outside of the converter housing aid in cooling. However, in the case of abnormal prolonged loading (lugging a trailer over mountain roads in second or third speed), converter heat may exceed maximum permissible temperature. This condition will cause a red warning light to function in the speedometer.

To illustrate this warning light circuit, the starter locking switch must also be considered. This, combined with a bridging switch, is operated by the inner transmission shift lever. It performs two functions.

1. With a speed range engaged, the electrical connection to the starter is interrupted. The engine, therefore, can only be started in neutral.

2. The contacts in the selector lever are not closed in the neutral position. Instead, the bridging switch transmits a voltage to the electromagnets of the control valve. This ensures that the separator clutch is also disengaged in the neutral shifter position.

Fully Automatic Transmission Type 3

The fully automatic transmission, consisting of an automatically shifted three speed planetary transmission and a torque converter, was introduced in 1969.

The torque converter is a conventional three element design. The three elements are an impeller (driving member), a stator (reaction member), and the turbine (driven member). Maximum torque multiplication, with the vehicle starting from rest, is two and one-half to one. Maximum converter efficiency is about 96 per cent.

The automatic transmission is a planetary unit with three forward speeds which engage automatically, depending on engine loading and road speed. The converter, planetary unit, and control system are incorporated together with the final drive in a single housing. The final drive is located between the converter and the planetary gearbox. Driving and driven shafts fit one inside the other in contrast to the manual transmission in which they are located one below the other. The planetary gear unit is controlled by two multi-plate clutches which make up the third-reverse and forward clutch, a first gear band, a second gear band, and a roller clutch which permits the planetary ring gear to rotate only in the direction of drive.

The transmission control system includes a gear type oil pump, a centrifugal governor which regulates shift points, a throttle modulator valve which evaluates engine loading according to intake manifold pressure, and numerous other regulating components assembled in the transmission valve body.

Power flow passes through the torque converter to the turbine shaft, then to the clutch drum attached to the turbine shaft, through a clutch to a sungear. The output planet carrier then drives the rear axle shafts via the final drive.

Transmission ranges are Park, Reverse, Neutral, Drive (3), Second (2), and First (1).

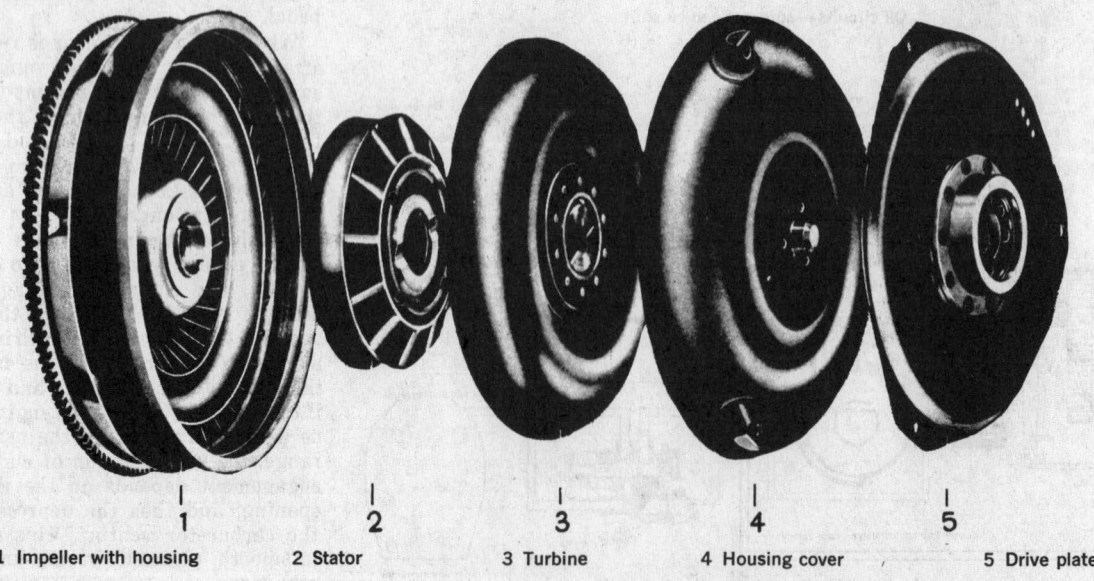

1 Impeller with housing   2 Stator   3 Turbine   4 Housing cover   5 Drive plate

Automatic transmission torque converter (© Volkswagen)

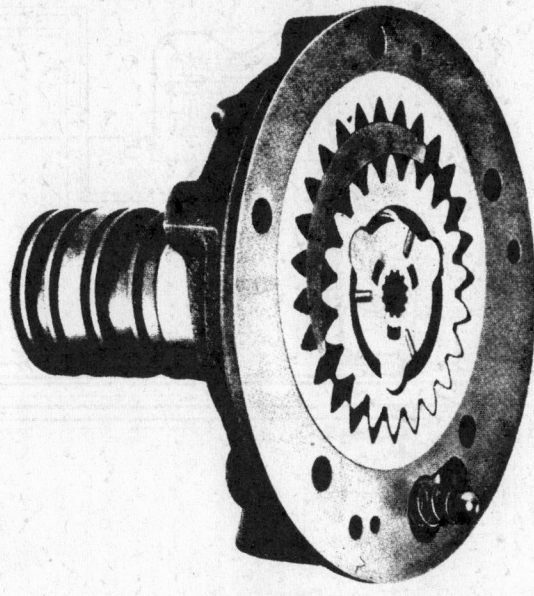

Automatic transmission oil pump (© Volkswagen)

1 Manual valve
2 Solenoid for kickdown valve
3 Oil strainer
4 Transfer plate
5 Valve body
6 Vacuum unit for primary throttle modulator valve

Automatic transmission valve body (© Volkswagen)

Ratios in Planetary Gearbox
1st — 2.65   3rd — 1.0
2nd — 1.59   Reverse — 1.8
Final drive ratio — 3.67

Capacities

Torque converter and planetary gearbox: approx. 6 liters (12.6 US pts./10.6 Imp. pts.) of ATF (Automatic Transmission Fluid) (refill with approx. 3 liters (6.3 US pts./5.3 Imp. pts.)

Use Factory Approved ATF Only

Final drive: approx. 1 liter (2.1 US pts./1.8 Imp. pts.) of SAE 90 hypoid oil

Transmission, Rear Axle

References

The transmission case is cast of light alloy. 1960 and earlier models have a two piece case. If replacements are needed, both halves of the case must be replaced, as they are cast and machined in pairs.

Gear selection is four speeds forward, and one reverse. It is of synchromesh design, with later models (from 1960) synchronized in all four forward gears.

Unlike cars of American manufacture, there is no direct (1:1) drive.

Transmission work requires the removal of the entire power train, engine, clutch, transmission and differential.

Gear Selection

A gear shifting rod is contained in the frame tunnel and links the gear selector lever with the transmission. The synchronizing mechanism consists of clutch gear, synchronizer shifting plates, synchronizer stopring and operating sleeve.

Transaxle R & R

The transmission, being an integral part of the power train, requires engine removal to facilitate transmission or differential services of any kind.

With the engine removed from the car:

1. Remove the rear wheels.

2. Disconnect brake lines at rear wheels, plug the lines.
3. Disconnect parking brake cables from push bar at the frame and withdraw cables from conduit tubes.
4. Remove bolts at rear axle shaft bearing.
5. Disconnect clutch release cable from operating shaft lever and pull it from the guide plate.
6. Disconnect wires from starter motor.
7. From the access hole under the rear seat, disconnect the shift rod in back of the coupling.
8. Remove nuts from mounting studs at front of transmission.
9. Remove lower shock absorber mounting bolts. Mark position of rear torsion bar radius arm with relation to rear axle bearing housing.

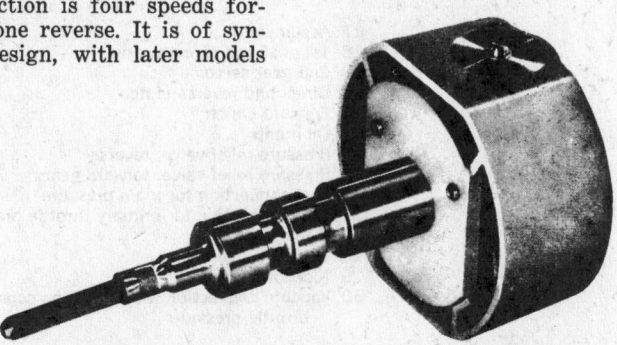

Automatic transmission centrifugal governor (© Volkswagen)

Volkswagen's new collapsible safety steering column

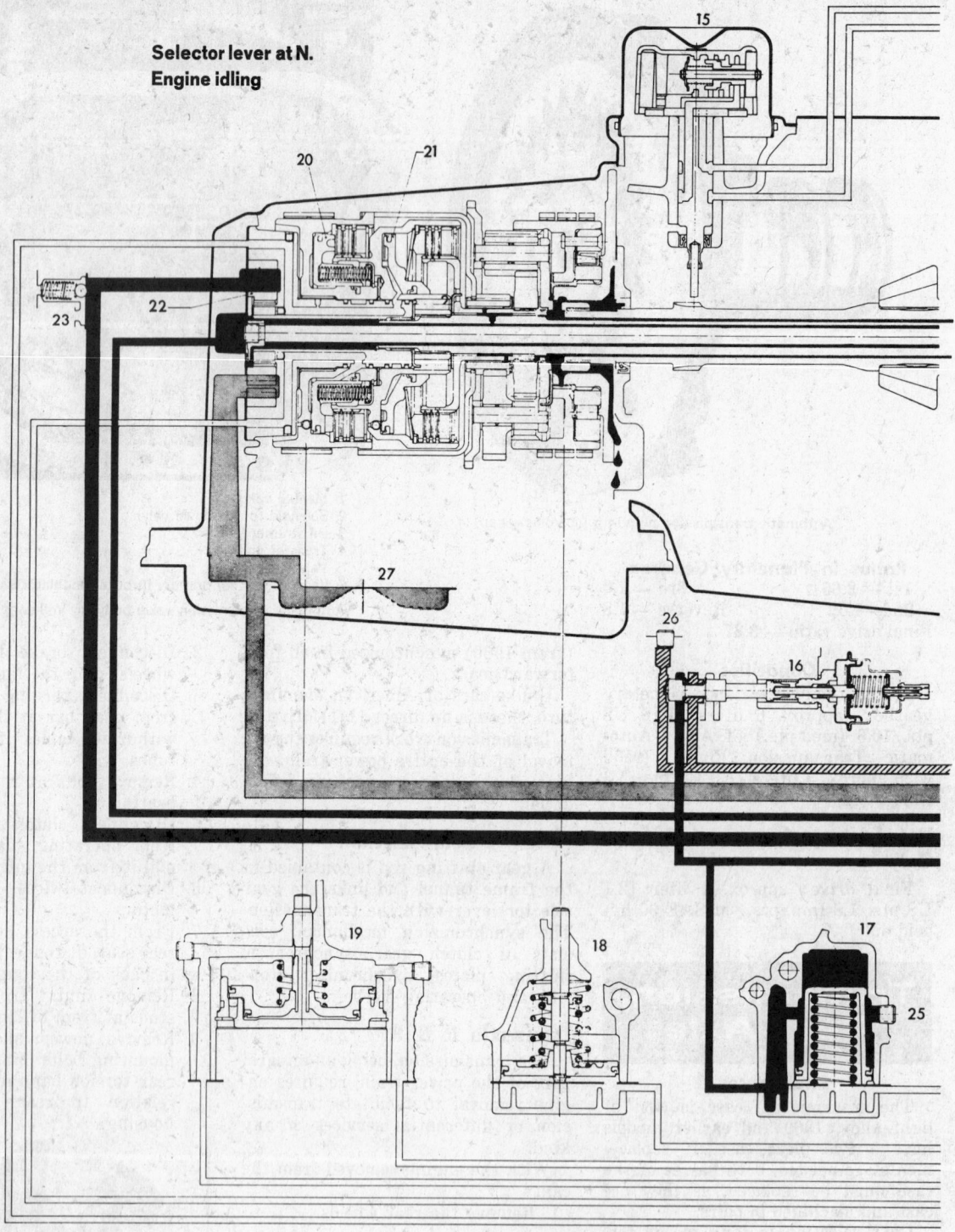

**Selector lever at N.
Engine idling**

1 Main valve	17 Accumulator
2 Two stage pressure valve	18 1st gear and reverse servo
3 Pressure limiting valve	19 2nd gear servo
4 Manual valve	20 Direct and reverse clutch
5 Converter valve	21 Forward clutch
6 Secondary throttle pressure valve with 1st gear plug	22 Oil pump
7 Kickdown valve	23 Pressure relief valve, reverse
8 Shift valve 1—2	24 Pressure relief valve, forward gears
9 Shift valve 2—3	25 Test connection for main pressure
10 3—2 control valve	26 Test connection for primary throttle pressure
11 1st gear ball valve	27 Oil strainer
12 Ball valve for 1st gear and reverse	28 Torque converter
13 Ball valve for drive and reverse	29 Kickdown switch
14 2—3 valve	30 Vacuum connection with adjusting screw for primary throttle pressure
15 Governor	
16 Primary throttle pressure valve	

Automatic transmission fluid flow diagram (© Volkswagen)

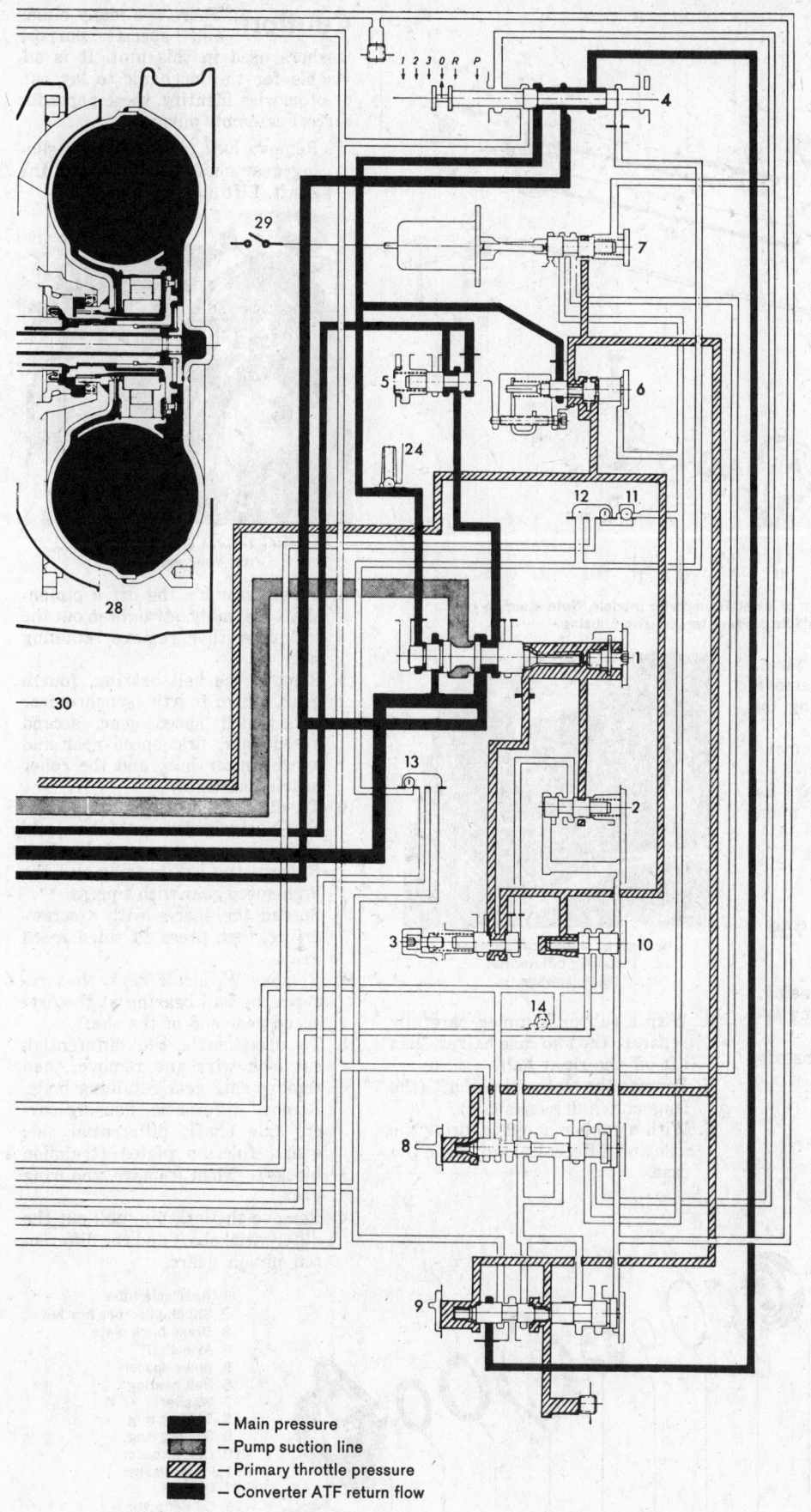

1 2 3 0 R P

- ▬ Main pressure
- ▨ Pump suction line
- ▥ Primary throttle pressure
- ▬ Converter ATF return flow

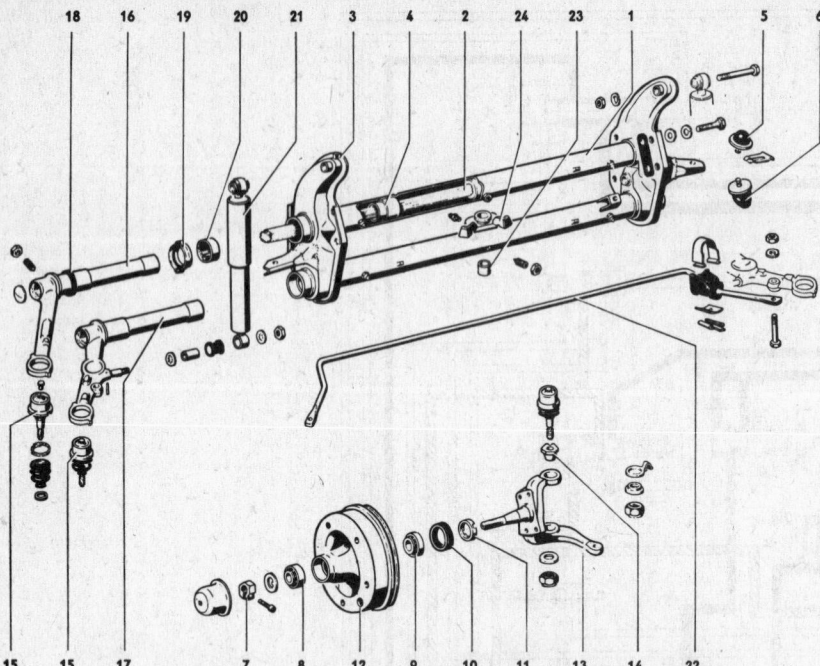

Exploded view, front suspension of latest Transporter models. Note steering gear swing lever in middle of lower torsion arm housing.

10. Disconnect starter motor wires.
11. Disconnect ground strap. Remove nuts from compensator spring, on models so equipped.
12. Support unit with jack. Remove two 27 mm. attaching bolts.
13. Withdraw transaxle to rear. Be careful not to damage the main drive shaft.
14. Reverse procedure to install.

Split-Type Transaxle Reconditioning Through 1959

Disassembly

1. Remove the gear selector housing.
2. Remove clutch release bearing assembly.
3. Remove bolts holding transaxle halves together.

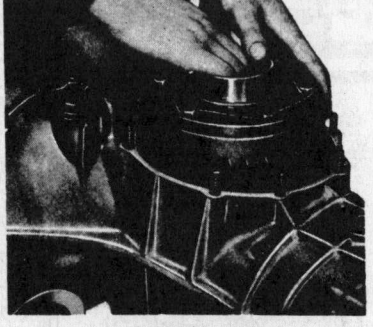

Be sure of shim location when installing differential
(© Volkswagen)

4. With a rubber hammer, carefully separate the two case halves, then lift off the right half.
5. Remove the main drive shaft (the long clutch-driven shaft).
6. With a rubber hammer, drive the axle and differential unit from the case.

Caution There are many shims and special purpose washers used in this unit. It is advisable for the mechanic to lay out, or otherwise identify, these parts for correct assembly purposes.

7. Remove lock pin at reverse sliding gear shaft and drive out the shaft. Lift out reverse gear.

Removing reverse drive gear snap-ring
(© Volkswagen)

8. To disassemble the drive pinion-shaft assembly, straighten out the lockplate, then remove retaining nut.
9. Remove the ball bearing, fourth gear, third-fourth synchronizer hub, third speed gear, second speed gear, first speed gear and synchronizer hub, and the roller bearing assembly.
10. To disassemble the main drive-shaft, straighten out the lockplate, then remove retaining nut. Remove the ball bearing and the high-speed gear with a press.
11. Spread the spacer with a screwdriver, then press off third speed gear.
12. Remove Woodruff keys, then remove the ball bearing at the first speed gear end of the shaft.
13. To disassemble the differential, cut lock wire and remove, then remove ring gear retaining bolts. Remove differential housing cover, axle shaft, differential side gears, fulcrum plates (trunnion blocks). Then remove the ring gear.
14. Remove the lock pin, pull out the differential shaft and the differential pinion gears.

1 Rear axle tube
2 Shock absorber bracket
3 Brake back plate
4 Axle shaft
5 Inner spacer
6 Ball bearing
7 Washer
8 Sealing ring
9 Sealing ring
10 Outer spacer
11 Paper gasket
12 Oil seal
13 Oil deflector
14 Cover
15 Cover retaining screw

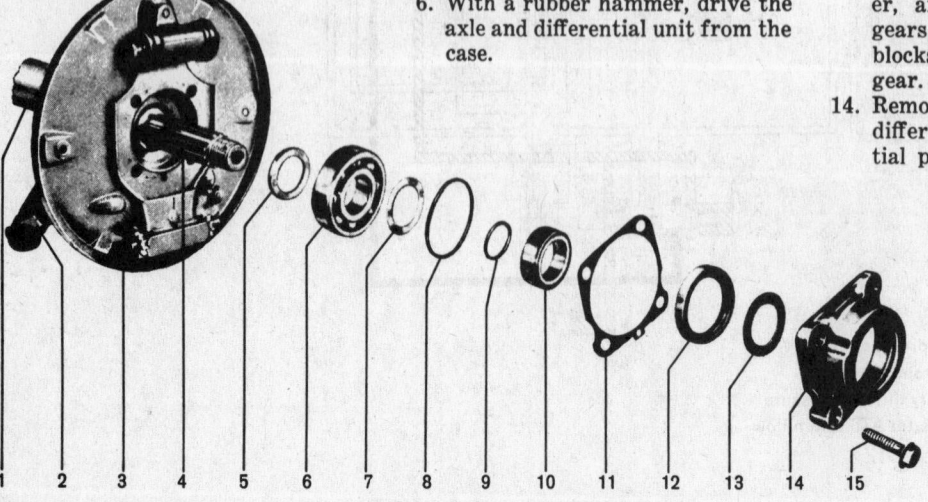

Exploded view, rear wheel bearing and rear axle assembly

The front end of the 1968-70 Transporter. Arrows point to the five grease points—four for the torsion bars, one in the center for the steering.

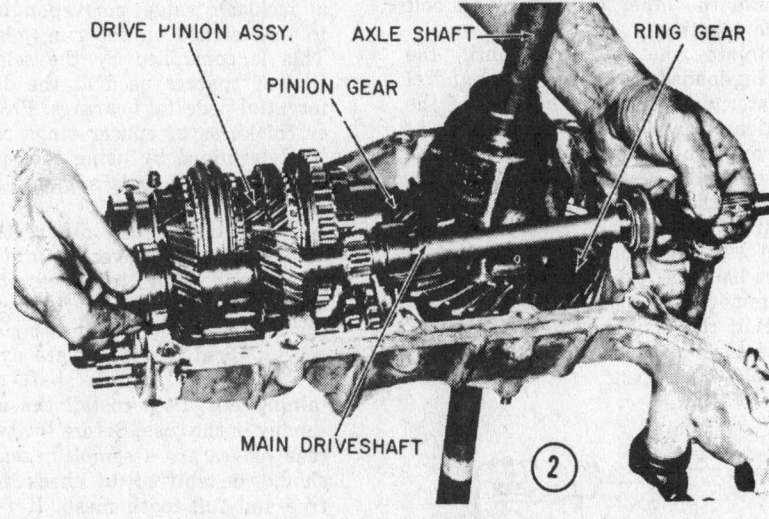

DRIVE PINION ASSY.   AXLE SHAFT   RING GEAR

PINION GEAR

MAIN DRIVESHAFT

Transmission—power train—differential carrier (split type) (© Volkswagen)

Assembly

1. To assemble the main drive shaft, install the two Woodruff keys, then position third speed gear on the drive shaft. Separate the spacer tube with a screwdriver and locate it over the Woodruff keys. Install fourth speed gear, then, with an arbor press, push the gears into position until third gear is tightly seated against second gear.
2. Install both bearings, the lockring, lock plate, and the retaining nut.
3. Torque the locknut to 30-35 ft. lbs., then bend the lockplate to secure the nut.
4. To assemble the drive pinion shaft, install the roller bearing inner race, by expanding it in an oil bath of 190°F, then sliding it over the shaft and into position.
5. Install the roller bearing. Install the end-play shim that was originally used. However, some modification, described later, may be necessary.
6. Slide the second speed clutch gear over the splines.
7. Install synchronizer and second speed gear.
8. Slide bushing into second speed gear, then install third speed gear and synchronizer ring over the bushing.
9. Assemble synchronizer hub for third and fourth speed gears by sliding the operating sleeve onto the clutch gear, at the same time pulling the three plates into position and securing with the two snap-rings. The ring ends must be located in one sector between two shifter plates.
10. Slide the synchronizer unit onto the drive pinion shaft and rotate the synchronizer ring until the shifting notches engage the slots. The top of the synchronizer hub should be even with the top end of the shaft splines by ± 0.002 in.
11. Place fourth speed synchronizer ring against the hub, then slide fourth speed gear bushing into position, followed by fourth speed gear.
12. The thrust washer may now be installed, and end-play of the gear train taken. End-play should be from 0.004-0.010 in. Shims are available in increments of 0.004 in., 0.008 in. and 0.012 in. selection.
13. Install correct amount of shims (three or four shims are usually necessary to obtain proper cone clearance and travel.)
14. Install the double-row ball bearing.
15. Position and hold the drive pinion assembly so that the locknut on the end of the shaft may be tor-

Be sure to torque main drive shaft nut and
drive pinion nut
(© Volkswagen)

qued to 80-87 ft. lbs. Do not lock
the nut at this time.

16. Complete assembly of first gear
by installing the three coil
springs and shifter plates. Be
sure the plates are positioned
with their ends under the snap-
ring.

17. Press all shifter plates down and
slide first gear over the hub.

18. To assemble the differential, insert each axle shaft into its corresponding side gear and fulcrum
plates (trunnion blocks.) With a
feeler gauge, this play (between
axle and fulcrum plates) should
be from 0.002-0.009 in. Oversize
fulcrum plates are available.

19. Install axle shaft and side gear
assembly, differential gears,
shaft, and the lock pin. Stake the
pin to secure it.

20. Install the other axle, side gear,
and housing assembly.

21. Install bolts and torque them to
43-45 ft. lbs. Safety-wire these
bolts as a security measure.

22. At this point, special equipment
is required to accurately relate
the ring gear with the pinion.

NOTE: in the absence of special
differential gauges and measuring
equipment, the red-lead tooth pat-
terning method has been used, however, the practice is not advisable.
Following is a brief description of
the procedure when using special
tools supplied by Volkswagen.

According to design, the face of the
drive pinion is positioned 59.22 mm.
from the centerline of the ring gear.
This measurement must be flexible
to the extent of manufacturing tolerances. The deviation from standard is
stamped on the edge of the ring gear.
There is, also, a matching number
etched on the edge of the ring gear
and on the end of the pinion gear.
This number identifies this set as being matched. Backlash value is also
etched on the edge of the ring gear.

To establish the location of the pinion in the case, install and hold the
pinion into the case as far and firmly
as possible. Install the mandrel of the
gauging tool in one half of the case.
Install the other case half and bolt
them together.

Rotate the mandrel until the
spring-loaded plunger of the mandrel
is at right angles to the face of the
pinion. Now, release the plunger to
permit contact with the face of the
pinion. Lock the plunger in this position, then rotate the mandrel so
that the plunger is away from contact with the pinion. Separate the two
case halves so that the mandrel may
be removed for measurement.

Half the diameter of the mandrel,
plus the length of the extended pin,
should correspond with the value

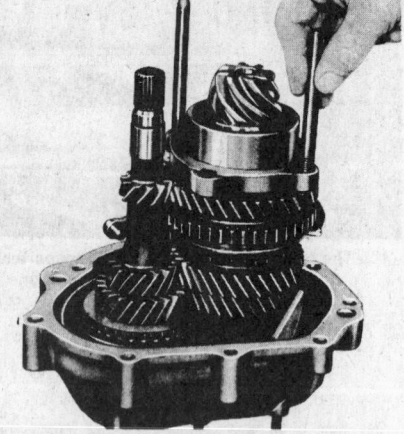

Use guide screws when installing transmission
(© Volkswagen)

etched on the edge of the ring gear.
If these values do not agree, the pin-
ion-positioning shim installed in previous paragraph No. 12 must be
changed.

23. Carrier bearings must carry a
preload sufficient to spring the
case 0.005-0.007 in., while maintaining a good tooth pattern and
a backlash value corresponding
to that etched on the ring gear.
This is controlled by the selection of spacers used at the differential pedestal bearings. Proper thickness of spacer rings can
be determined by using the special measuring tools supplied by
VW.

24. After installing the differential
assembly into its recess in the
left-hand housing, install reverse
sliding gear and the shaft. Insert
the lock pin through the bearing
recess. Bend the lock plate over
the main drive pinion shaft retaining nut, then install the assembly in the case. Before the two
case halves are assembled, gears
should be shifted to check for
free and full tooth mesh. If full
mesh is not made, loosen the positioning setscrew and move the
shifter fork enough to centralize
the gear. Reverse shaft lockscrew
is accessible from inside the case,
the other two are reached by removing threaded plugs from outside the case.

25. Coat the mating surfaces of the
housing halves with sealer. Assemble the two halves and torque
the retaining bolts to 15 ft. lbs.

26. The double row rear pinion bearing must be preloaded by 0.001-
0.004 in. This is done by inserting shims between the bearing
and gear shift housing at the rear
of the transmission. To do this,
take a depth gauge reading of the
bearing recess in the gear shift
housing. Then, measure the
amount that the bearing race extends beyond the surface of the

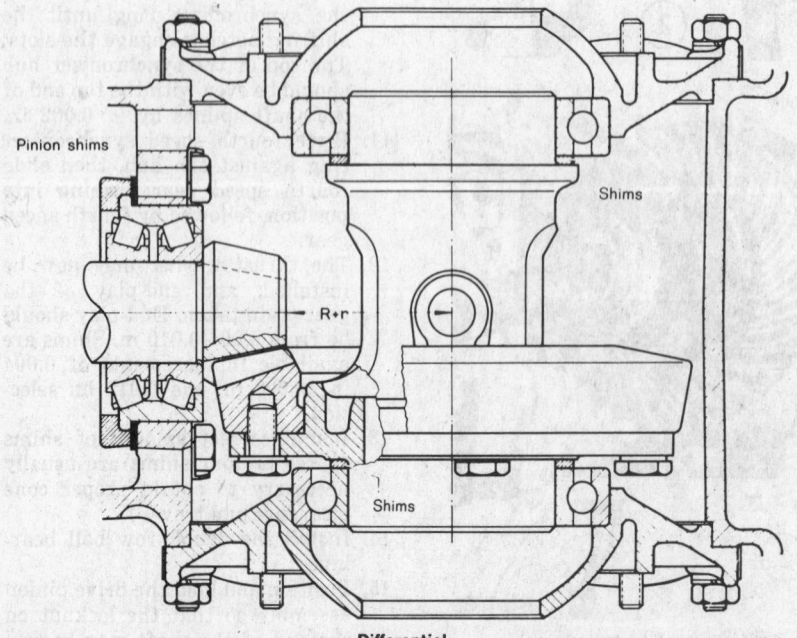

Pinion shims

R+r

Shims

Shims

Differential

case. Paper shims are available to make up this aforementioned preload of 0.001-0.004 in.

27. Repeat this measuring and shimming procedure at the mainshaft bearing and add enough shims to produce a 0.001-0.004 in. preload.
28. Install the gearshift cover and test-shift the unit.
29. Install trans-axle by reversing the removal procedure.

Tunnel-Type Transaxle Reconditioning— Beginning 1960

Disassembly

1. Remove gear selector housing.
2. Pry off lock plates which secure drive pinion and driveshaft nuts.
3. Lock transmission by engaging both reverse and high gears.
4. Remove drive pinion and main driveshaft nuts. Then remove and discard lock plates.
5. Remove gear carrier stud nuts. Remove ground strap and accelerator cable retainer.
6. Position the assembly so that the left-hand final drive cover faces up.
7. Remove stud nuts from left-hand final drive cover, then remove the cover.
8. Attach tool VW 297 to the right-

Pressing out differential (tool VW 297)
(© Volkswagen)

hand final drive cover studs, then press out the differential.

NOTE: when removing the differential, make note of the thicknesses and positioning of the differential shims to simplify reassembly.

9. Loosen retaining ring that secures reverse gear on the main driveshaft. Then, slide reverse gear toward the rear and screw main driveshaft apart.
10. Remove reverse gear and retaining ring and withdraw the rear main shaft toward the rear. Don't damage the oil seal.
11. Remove tool VW 297 and right-hand final drive cover.
12. Release lock tabs and remove attaching screws, then remove

drive pinion ball bearing retainer.
13. Press transmission out of case with tool VW 296 or suitable alternate. Note thickness of pinion shims to simplify reassembly.
14. Spread and remove snap-ring, then pull off reverse drive gear.
15. Remove Woodruff key and withdraw reverse gear shaft and thrust washer from transmission case.
16. Remove security screw from reverse gear shaft needle bearing spacer sleeve.
17. With a suitable drift, drive out the reverse gear shaft needle bearings and spacer sleeve.
18. Remove security screw from the

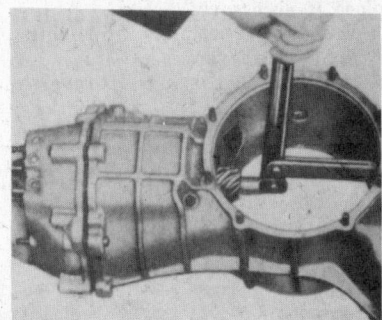

Removing transmission from case (tool VW 296)
(© Volkswagen)

Double-jointed rear suspension, 1969-70 Type 3 (© Volkswagen)

needle bearing of the main drive-shaft.

19. With a suitable drift, drive out the main driveshaft needle bearing.
20. Press out ball bearings from both final drive covers.
21. Remove clutch release bearing and operating shaft.

Assembly

NOTE: clean and inspect the case and all components for wear, damage or any indication of malfunction, and replace as necessary. The starting motor armature, brushes and bushings should be inspected and dealt with, accordingly. Check clutch operating shaft and bushings for wear, replace if necessary.

1. Press ball bearings into both, final drive covers.
2. Insert needle bearings for reverse gear shaft and spacer sleeve, then secure.
3. Install main driveshaft needle bearing with a suitable drift, then secure.
4. Install reverse shaft, Woodruff key thrust washer and gear, then secure with snap-ring.
5. Locate drive pinion shims over bearing, then turn two 4 in. guide studs into bearing retainer to assure retainer and shim alignment during transmission-to-case assembly.
6. Push reverse selector and sliding gear onto reverse lever and engage reverse gear.
7. With a new carrier gasket in place, carefully insert the transmission into the case (the 4 in. guide screws will help at this time).
8. Remove guide screws, then install transmission-to-case attaching screws and lock plates. Torque to 36 ft. lbs.
9. Lubricate lip of oil seal, then install rear half of main drive-shaft. Screw both halves of the driveshaft together. Back them off until the splines of the reverse gear are in line, then install reverse gear snap-ring.
10. With new gasket in place, install right-hand, final drive cover. Torque to 18 ft. lbs.
11. With shims properly inserted, install differential in case.
12. Now, install gear carrier nuts and torque to 14 ft. lbs.
13. Lock transmission by engaging both reverse and high gears at the same time.
14. Torque the main driveshaft nut to 87 ft. lbs. Loosen the nut, then retighten to a final 43 ft. lbs. tor-

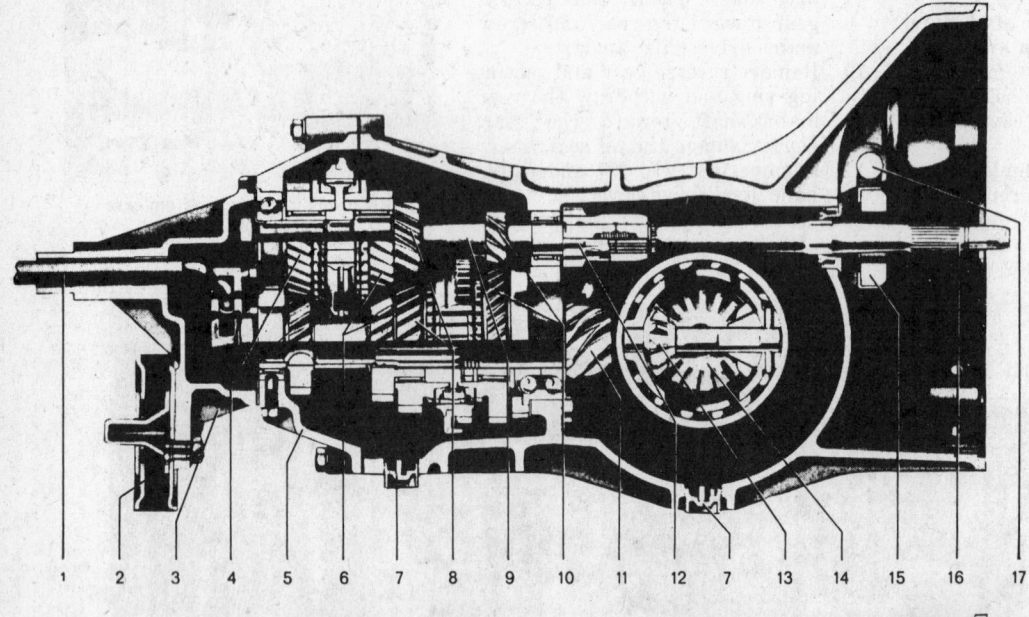

1	Transmission shift lever
2	Bonded rubber mounting
3	Gear shift housing
4	4th speed
5	Gear carrier
6	3rd speed
7	Oil drain plugs
8	2nd speed
9	Main drive shaft, front
10	1st speed
11	Drive pinion
12	Reverse gear
13	Differential pinion
14	Differential side gear
15	Clutch release bearing
16	Main drive shaft, rear
17	Clutch operating shaft
18	Reverse sliding gear
19	Oil filler plug
20	Reverse shaft
21	Reverse drive gear
22	Ring gear
23	Rear axle shaft
24	Fulcrum plate
25	Differential housing

Transmission—power train—differential carrier (tunnel type) (© Volkswagen)

que. Secure with lock plate.
15. Torque the drive pinion nut to 43 ft. lbs. and secure with lock plate.

Caution
When installing gearshift housing, make sure that the three selector shafts are in neutral.

Gear shift housing mounted (tool VW 294)
(© Volkswagen)

Gear Carrier Disassembly

1. Remove reverse selector fork, including reverse sliding gear, from reverse lever.
2. Remove and note thickness of drive pinion ball bearing shims.
3. Clamp gear carrier in vise equipped with soft jaws.
4. Loosen shifting fork lock screws, then remove first-second shifting fork.
5. Withdraw the shifting fork shaft from third-fourth shifting fork.
6. For security, place a strong rubber band around first-second operating sleeve and main shaft.
7. Mount the assembly in a press, case end up, and press on the main shaft, to remove transmission from gear carrier.
8. Remove screw which holds drive pinion needle bearing, then press out bearing.
9. Press out main driveshaft bearing.
10. Clamp gear carrier in vise with soft jaws, then remove reverse lever guide screw.
11. Withdraw reverse gear selector

Gear carrier mounted (tool VW 294)
(© Volkswagen)

shaft and remove reverse lever guide.
12. Withdraw first-second selector shaft and remove reverse lever from support.
13. Withdraw third-fourth selector shaft.
14. Remove plungers and detent balls. Then, hook out detent springs with a small screwdriver.

Gear Carrier Assembly

NOTE: check all components for damage and wear. Replace parts, if necessary. Free length of detent springs should be 1.0 in. The wear, or fatigue, limit is 0.9 in. A force of 33-44 ft. lbs. applied to the ends of the selector shafts should be required to unseat the detent balls when shifting.
1. Insert detent springs through the selector shaft holes.
NOTE: due to design, the springs for first-second and reverse gear detents can be more easily installed by inserting them into the top halves first.
2. Install reverse selector shaft including reverse lever and reverse lever guide.
3. Install selector shafts for first-second and for third-fourth gears. Don't forget the two interlocking plungers. This is a safety feature guarding against shifting into two gears at one time.
4. Check drive needle bearing, and main driveshaft ball bearing for condition. Replace, if necessary. Secure drive pinion needle bearing in the gear carrier.
5. Position gear carrier on a suitable support and press main driveshaft ball bearing into position.
6. Check selector forks for wear. Fork-to-operating sleeve clearance should be 0.004-0.012 in. Clearance greater than this warrants parts replacement.
7. Position the selector fork for third-fourth gears. Then, press transmission into the gear carrier. While pressing, take care that the third-fourth selector fork does not become jammed. Also, with a heavy rubber band, secure first-second gears to the main driveshaft.
8. Install first-second selector fork.
9. Attach reverse gear fork with reverse sliding gear onto selector lever.
10. Adjust gear selector forks.

Selector Fork Adjustment

NOTE: at this point, special tools are required to make the adjustments.
1. Place transmission, drive pinion shims and gasket for gear carrier on test tool VW 294 and secure gear carrier with four screws.
2. Tighten drive-pinion bearing-retainer, with two screws located

Note location of heavy rubber band. This procedure is used when pressing the gear assembly out of/into the gear carrier
(© Volkswagen)

diagonally, to 36 ft. lbs. torque.
3. Push crank of the test tool onto splines of main driveshaft so that the main shaft is locked by the crank. Engage first and second gears.
4. With a torque wrench, tighten the main driveshaft nut to 87 ft. lbs. Then, loosen the nut and retighten to 43 ft. lbs. and lock it.
5. Tighten drive pinion nut to 43 ft. lbs. and lock it.
6. Attach gearshift housing and shifting handle. By attaching the gearshift housing, correct seating of the main drive shaft bearing in its recess in the gear carrier is guaranteed.
7. Position first-second and third-fourth selector forks so that they move freely in the operating sleeve while in neutral, and when various gears are engaged.
8. Position reverse gear selector fork so that reverse sliding gear is centered between the operating sleeve and second gear of the main driveshaft with second gear engaged. The fork must also engage properly in reverse gear of the drive pinion when reverse gear is engaged.
9. Using a T-handle, torque wrench and socket, tighten the selector fork locking screws to 18 ft. lbs. Tighten the reverse lever guide screw to 14 ft. lbs. torque.
10. Remove gearshift housing and transmission from tool VW 294.

Main Driveshaft Oil Seal Removal

1. Remove the engine.
2. Remove clutch release bearing.
3. Carefully remove the faulty oil seal from the transmission case.

Main Driveshaft Installation

1. Lightly coat the outer edge of the new seal with sealing com-

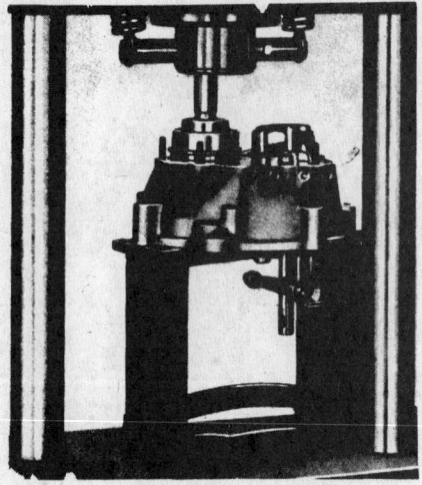

Pressing main drive shaft ball bearing into
position in gear carrier
(© Volkswagen)

pound. Lubricate the main drive-shaft and the seal lip.
2. Slide oil seal onto the main driveshaft, then drive it into position with a suitable driving sleeve or pipe.

Drive Pinion Reconditioning

Disassembly

1. Press out inner race of needle bearing and fourth gear, then remove Woodruff from the shaft.
2. Remove spacer sleeve, concave washer shims and the concave washer.
3. Remove second and third gears with needle cage and second gear synchronizer stopring.
4. Remove clutch gear for first and second gears, including springs, shifting plates and operating sleeve. Disassemble parts.

Inspection

1. Inspect all components for wear or other damage; replace as necessary. If pinion and ring gear require replacement, a matched set is in order (note matching number on both gears.)
2. If drive pinion or ball bearing is replaced, the drive pinion and ring gear must be adjusted.
3. Whenever a damaged gear is replaced, the mating gear should also be replaced. Worn or otherwise damaged first and second speed gears require a replacement of the front main driveshaft.
4. Clean and check all synchronizer components for wear. Clearance between the synchronizer stop ring face and the clutch teeth of the corresponding gear should be about 0.043 in. If the wear limit of 0.024 in. has been reached, the stop ring should be replaced.
5. If a gear will not engage, even though the clutch is fully released, the probable cause is wear in the slots of the stop ring. Replace worn parts.

Assembly

1. In preparation, the inner races of the ball bearing and the needle bearing inner race for first gear should be heated in an oil bath to about 194°F.
2. Slide one inner ball bearing race onto the drive pinion.
3. Slide the ball bearing onto the drive pinion. Then, slide the

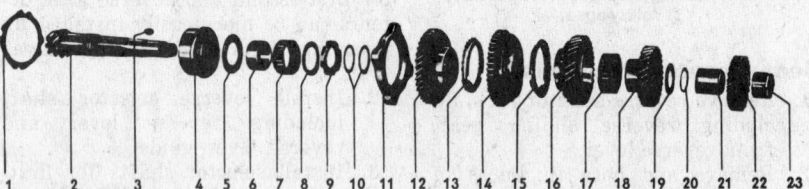

1 Shim	13 Synchronizer stop ring (1st gear)
2 Drive pinion	14 Clutch gear for 1st and 2nd gears, and reverse gear
3 Woodruff key for 4th gear	15 Synchronizer stop ring (2nd gear)
4 Ball bearing	16 2nd gear
5 Thrust washer for 1st gear	17 Needle cage (2nd gear)
6 Needle bearing inner race (1st gear)	18 3rd gear
7 Needle cage (1st gear)	19 Concave washer
8 Thrust washer for needle bearing (1st gear)	20 Shims for concave washer
9 Round nut	21 Spacer sleeve
10 Shims, end play 1st gear	22 4th gear
11 Ball bearing retainer	23 Inner race, needle bearing in gear carrier
12 1st gear	

Drive pinion component (© Volkswagen)

The double-jointed rear axles of the 1968-70 Transporter

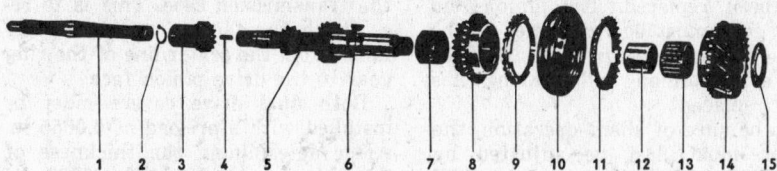

1 Main drive shaft rear half
2 Snap ring for reverse gear
3 Reverse gear on drive shaft
4 Stud
5 Main drive shaft front half
6 Woodruff key for clutch gear
7 Needle cage (3rd gear)
8 3rd gear
9 Synchronizer stop ring (3rd gear)
10 Clutch gear (3rd and 4th speeds)
11 Synchronizer stop ring (4th gear)
12 Needle bearing inner race (4th gear)
13 Needle cage (4th gear)
14 4th gear
15 Thrust washer (4th gear)

Main drive shaft components (© Volkswagen)

second inner race on so that the bearing parts numbers are exactly opposite each other.

4. Slide first gear thrust washer and needle bearing inner race onto the drive pinion.

5. With components mounted in a press, push all parts into correct position.

6. Tighten round nut to 90 ft. lbs. torque.

7. Install shims for first gear. After the clutch gear for first and second gears has been installed, end-play of 0.004-0.010 in. should be checked between thrust washer and first gear. Correct, if necessary. Shims of various thicknesses are available.

8. Position first speed stop ring on the cone surface of the gear. (First and second synchronizer stop rings are not interchangeable.)

9. Assemble the synchronizer unit for first and second gears. Slide operating sleeve on the clutch gear so that its shifting plate slots are in line with the slots in the clutch gear. With the shifting plates in position, install the two snap-rings, offset to each other.

10. Slide the synchronizer assembly onto the drive pinion. The longer hub should be toward the face of the drive pinion splines. Turn first speed stop ring until shifting plates engage with the slots.

11. Adjust the concave washer to produce the prescribed spring travel of 0.007 ± 0.0004 in.

12. The fourth speed gear and needle bearing inner race should be heated in an oil bath of 194°F. before being pressed into position.

13. Insert Woodruff key for fourth gear into drive pinion.

14. Slide fourth gear onto the drive pinion, wide shoulder facing the spacer sleeve.

15. Press fourth speed gear and needle bearing inner race fully into position.

Main Driveshaft Reconditioning

Disassembly

1. Remove thrust washer, fourth gear, needle bearing cage and stop ring.

2. Remove fourth speed needle bearing inner race, clutch gear for third and fourth speed and third gear.

3. Remove needle bearing cage for third gear.

4. Strip down synchronizing unit for third and fourth gears.

Inspection

1. Clean and inspect all parts for wear or other damage.

2. Place front main driveshaft between two centers and, with a dial indicator, check for runout at the contact surface of third gear needle bearing. Runout must not exceed 0.0006 in.
NOTE: if excessive runout warrants replacement of the front main driveshaft, the gear wheels for first and second speeds on the drive pinion must also be replaced at the same time.

3. Check clearance between the stop ring face and the clutch teeth of the corresponding gear with a feeler gauge. Normal clearance is 0.043 in. If a wear limit of 0.024 in. has been reached, stop rings need replacement.

4. If a gear resists engagement, even though the clutch is fully released, it may be due to misalignment of the teeth of the stop ring with the splines of the operating sleeve. This is caused by wear in the slots of the stop ring.

5. Check fourth gear thrust washer for wear, replace, if necessary.

Assembly

1. Assemble synchronizing unit for third and fourth gears. To hold lash between the clutch gear and operating sleeve to a minimum, the sleeve and clutch gear are matched and etched for identification. Position the shifting plates and install the two snaprings, offset to each other. Be sure that the ring ends engage behind the shifting plates.

2. Insert clutch gear Woodruff key into the main driveshaft. Then, place third gear synchronizer stop ring on the cone of the gear.

3. Press clutch gear for third and fourth gears into position. The identifying figure "4" on the clutch gear must be towards fourth gear. Third gear is lifted slightly and turned until the stop ring engages in the shifting plates.

4. Press fourth gear needle bearing inner race into position.

Differential Reconditioning

Disassembly

1. Put differential in holding fixture.

2. Cut and remove safety lock wire and ring gear attaching screws.

3. Lift off ring gear.

4. After driving out the lock pin, push out the differential pinion shaft, then remove differential pinions.

Assembly

1. Check differential pinion concave thrust surfaces in the differential housing. If scored or worn, replace differential housing.

2. Install differential pinion gears and shaft, then install the pinion shaft lock pin and peen it into place.

3. Examine ring gear for wear or damage. If necessary to replace, ring and drive pinion must be replaced as a matched pair.
NOTE: replacement of drive pinion and ring gear, or the differential housing, requires readjustment of the transmission.

4. Install and tighten ring gear attaching screws to 43 ft. lbs.

5. Insert ring gear attaching screw safety wire, to place a clockwise force on the attaching screws. Twist ends of safety wire and cut off.

Pinion and Ring Gear Adjustment

Quiet operation with minimum wear of final drive is directly dependent upon pinion and ring gear relationship. For this reason, drive pinion and ring gears are produced in matched pairs and are so identified. Silent operation is obtained by adjusting the drive pinion endwise, with the ring gear lifted enough out of the fully engaged position (without backlash) to insure backlash being within the prescribed tolerance of 0.0067—0.0098 in. Any tolerance difference

from standard is measured and marked on the pinion face.

Normally, it is necessary to readjust the ring gear and drive pinion only if parts which directly affect the adjustment have had to be replaced. It is satisfactory to readjust the ring gear if the differential housing, final drive cover or a differential bearing has been replaced. The pinion and ring gear must be readjusted if the transmission case, the gear itself, or the drive pinion ball bearing has been replaced.

To be sure of silent operation, the pinion must first be adjusted by installing shims between the ball bearing and the contact surface at the transmission case. This is to re-establish the factory setting of distance from the center line of the ring gear to the drive pinion face.

Both final drive covers must be installed with a preload of 0.0055 in. After determining the thickness of the shims, a preload of 0.0028 in. must be considered on both sides.

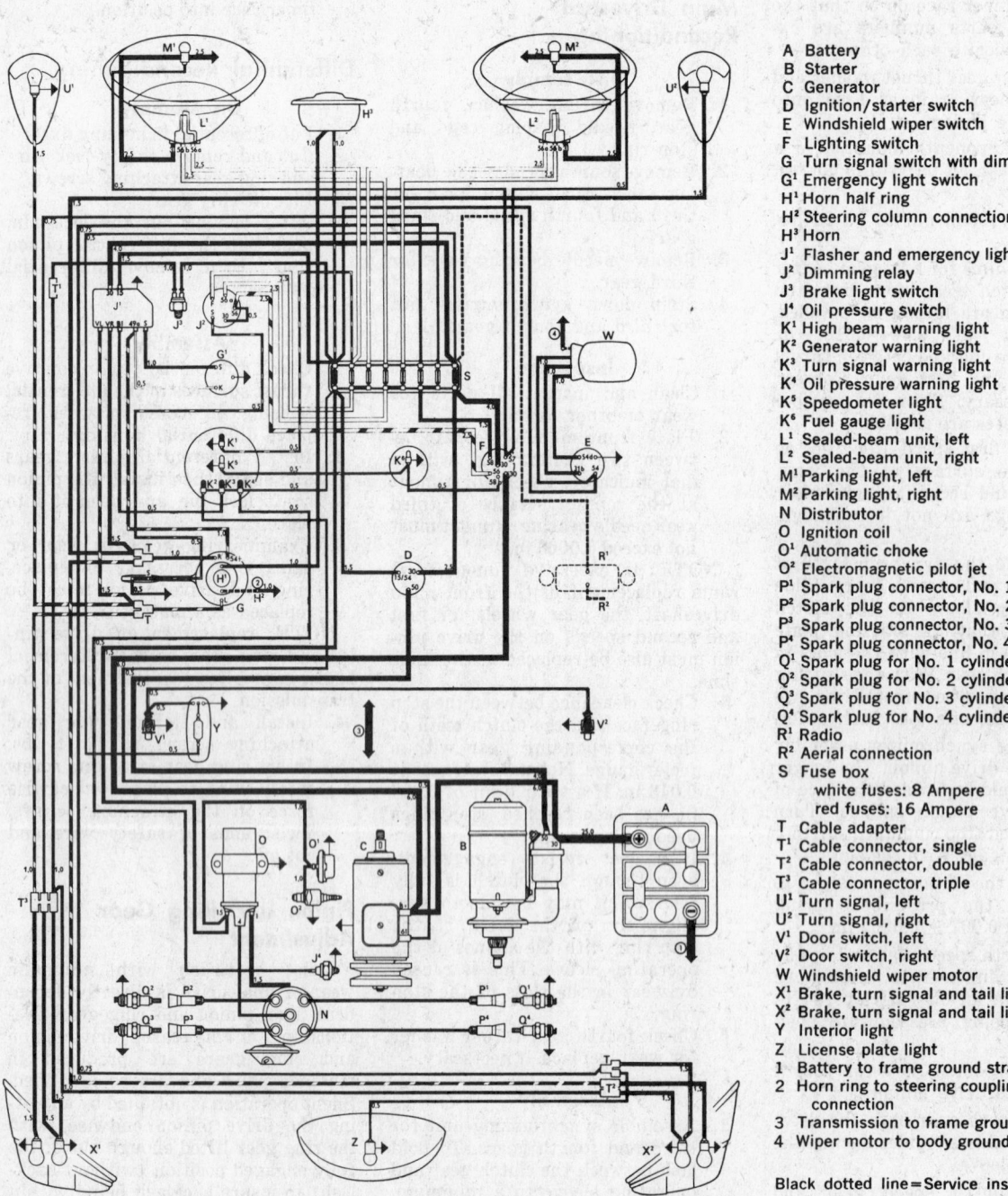

A Battery
B Starter
C Generator
D Ignition/starter switch
E Windshield wiper switch
F Lighting switch
G Turn signal switch with dimmer switch
G^1 Emergency light switch
H^1 Horn half ring
H^2 Steering column connection
H^3 Horn
J^1 Flasher and emergency light relay
J^2 Dimming relay
J^3 Brake light switch
J^4 Oil pressure switch
K^1 High beam warning light
K^2 Generator warning light
K^3 Turn signal warning light
K^4 Oil pressure warning light
K^5 Speedometer light
K^6 Fuel gauge light
L^1 Sealed-beam unit, left
L^2 Sealed-beam unit, right
M^1 Parking light, left
M^2 Parking light, right
N Distributor
O Ignition coil
O^1 Automatic choke
O^2 Electromagnetic pilot jet
P^1 Spark plug connector, No. 1 cylinder
P^2 Spark plug connector, No. 2 cylinder
P^3 Spark plug connector, No. 3 cylinder
P^4 Spark plug connector, No. 4 cylinder
Q^1 Spark plug for No. 1 cylinder
Q^2 Spark plug for No. 2 cylinder
Q^3 Spark plug for No. 3 cylinder
Q^4 Spark plug for No. 4 cylinder
R^1 Radio
R^2 Aerial connection
S Fuse box
   white fuses: 8 Ampere
   red fuses: 16 Ampere
T Cable adapter
T^1 Cable connector, single
T^2 Cable connector, double
T^3 Cable connector, triple
U^1 Turn signal, left
U^2 Turn signal, right
V^1 Door switch, left
V^2 Door switch, right
W Windshield wiper motor
X^1 Brake, turn signal and tail lights, left
X^2 Brake, turn signal and tail lights, right
Y Interior light
Z License plate light
1 Battery to frame ground strap
2 Horn ring to steering coupling ground connection
3 Transmission to frame ground strap
4 Wiper motor to body ground strap

Black dotted line = Service installation

Wiring diagram, VW 1300 Type 1 (from August, 1965)

Details of new tensioning ring for pinion, 1969-70 Type 2. The flats on the outer race of the double tapered roller bearing rest against mating flats in the transmission housing and prevent turning of the race when the tensioning ring is adjusted.
(© Volkswagen)

Threaded bearing rings replacing side covers on 1969-70 Type 2. These permit easier adjustment of the rear axle drive and are secured by a lock plate as shown.
(© Volkswagen)

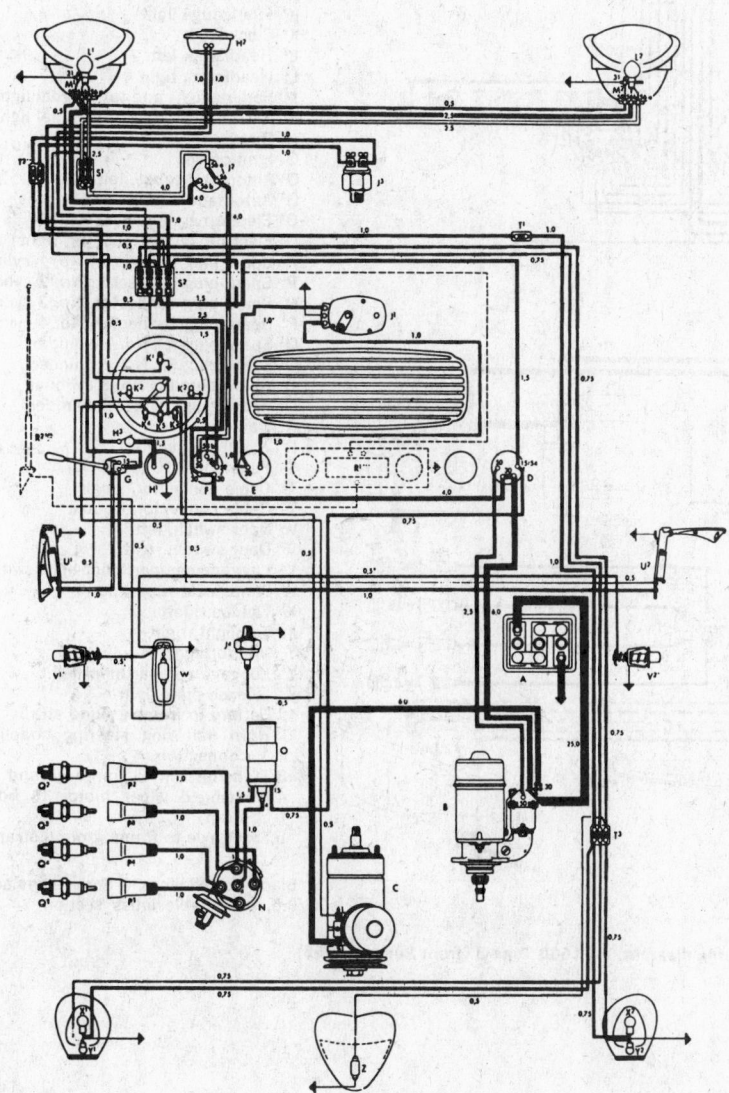

A	Battery
B	Starter
C	Generator
D	Ignition/starter switch
E	Windshield wiper switch
F	Lighting switch
G	Turn signal switch
H^1	Horn button
H^2	Horn
H^3	Slip ring switch
I	Not given
J^1	Wiper motor
J^2	Dimming switch
J^3	Stoplight switch
J^4	Oil pressure sending unit
K^1	High beam indicator
K^2	Speedometer light
K^3	Oil pressure warning light
K^4	Generator warning light
K^5	Turn signal indicator
L^1	Headlamp (left)
L^2	Headlamp (right)
M^1	Parking lamp (left)
M^2	Parking lamp (right)
N	Distributor
O	Ignition coil
P^1-P^4	Spark plug caps
Q^1-Q^4	Spark plugs
R^1	Radio
R^2	Antenna
S^1	Fuse box
S^2	Fuse box
T^1	Single terminal block
T^2	Double terminal block
T^3	Triple terminal block
U^1	Turn signal indicator
U^2	Turn signal indicator
V^1	Courtesy light switch
V^2	Courtesy light switch
W	Dome light
X^1	Left rear stoplight
X^2	Right rear stoplight
Y^1	Left tail light
Y^2	Right tail light
Z	License plate light

Wiring diagram, VW Type 1 Sedan (1954-57)

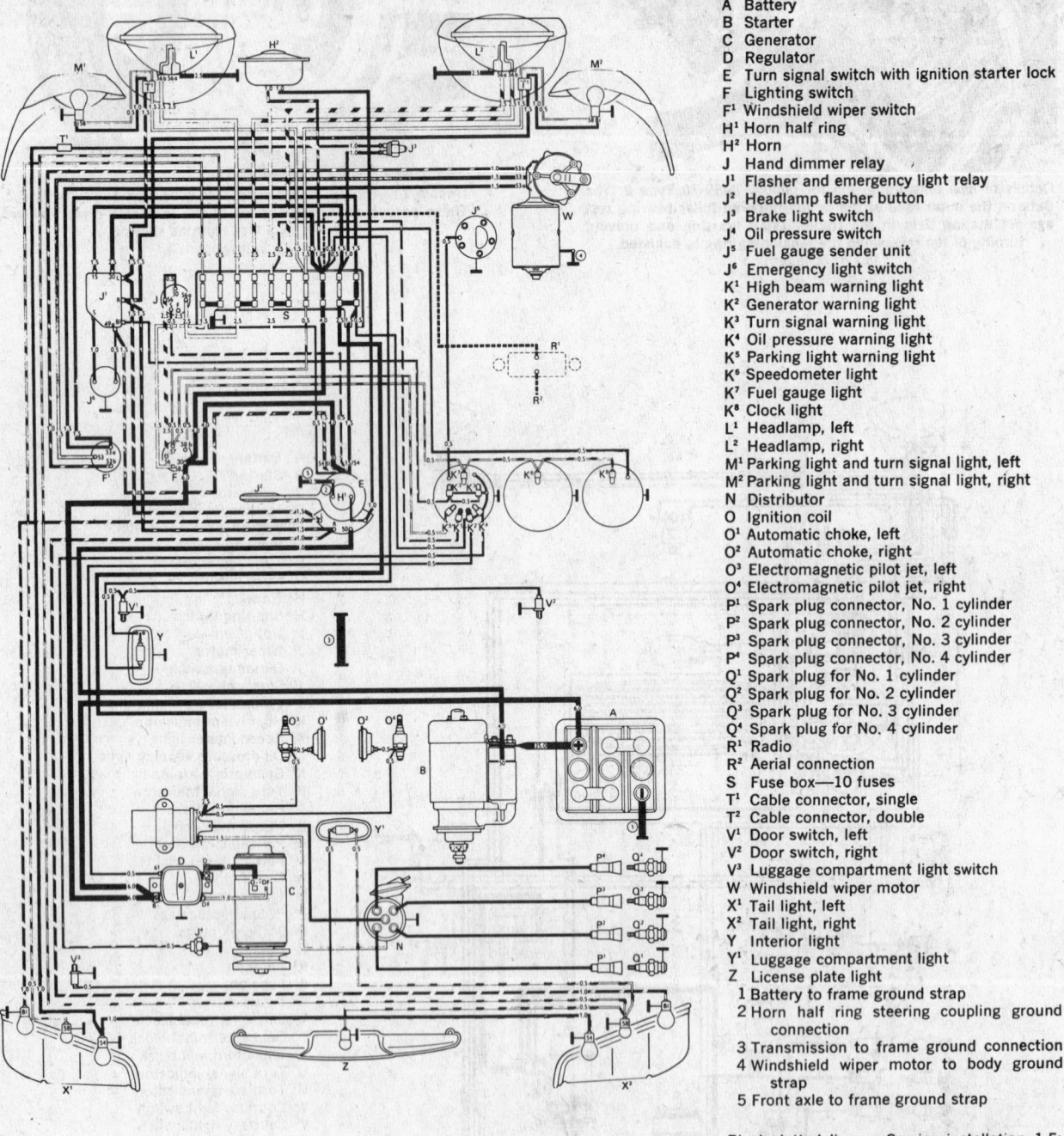

A   Battery
B   Starter
C   Generator
D   Regulator
E   Turn signal switch with ignition starter lock
F   Lighting switch
F¹ Windshield wiper switch
H¹ Horn half ring
H² Horn
J   Hand dimmer relay
J¹ Flasher and emergency light relay
J² Headlamp flasher button
J³ Brake light switch
J⁴ Oil pressure switch
J⁵ Fuel gauge sender unit
J⁶ Emergency light switch
K¹ High beam warning light
K² Generator warning light
K³ Turn signal warning light
K⁴ Oil pressure warning light
K⁵ Parking light warning light
K⁶ Speedometer light
K⁷ Fuel gauge light
K⁸ Clock light
L¹ Headlamp, left
L² Headlamp, right
M¹ Parking light and turn signal light, left
M² Parking light and turn signal light, right
N   Distributor
O   Ignition coil
O¹ Automatic choke, left
O² Automatic choke, right
O³ Electromagnetic pilot jet, left
O⁴ Electromagnetic pilot jet, right
P¹ Spark plug connector, No. 1 cylinder
P² Spark plug connector, No. 2 cylinder
P³ Spark plug connector, No. 3 cylinder
P⁴ Spark plug connector, No. 4 cylinder
Q¹ Spark plug for No. 1 cylinder
Q² Spark plug for No. 2 cylinder
Q³ Spark plug for No. 3 cylinder
Q⁴ Spark plug for No. 4 cylinder
R¹ Radio
R² Aerial connection
S   Fuse box—10 fuses
T¹ Cable connector, single
T² Cable connector, double
V¹ Door switch, left
V² Door switch, right
V³ Luggage compartment light switch
W   Windshield wiper motor
X¹ Tail light, left
X² Tail light, right
Y   Interior light
Y¹ Luggage compartment light
Z   License plate light
1   Battery to frame ground strap
2   Horn half ring steering coupling ground
    connection
3   Transmission to frame ground connection
4   Windshield wiper motor to body ground
    strap
5   Front axle to frame ground strap

Black dotted lines = Service installation 1.5;
0.5, etc.: Cable cross section

Wiring diagram, VW 1600 Type 3 (from August, 1965)

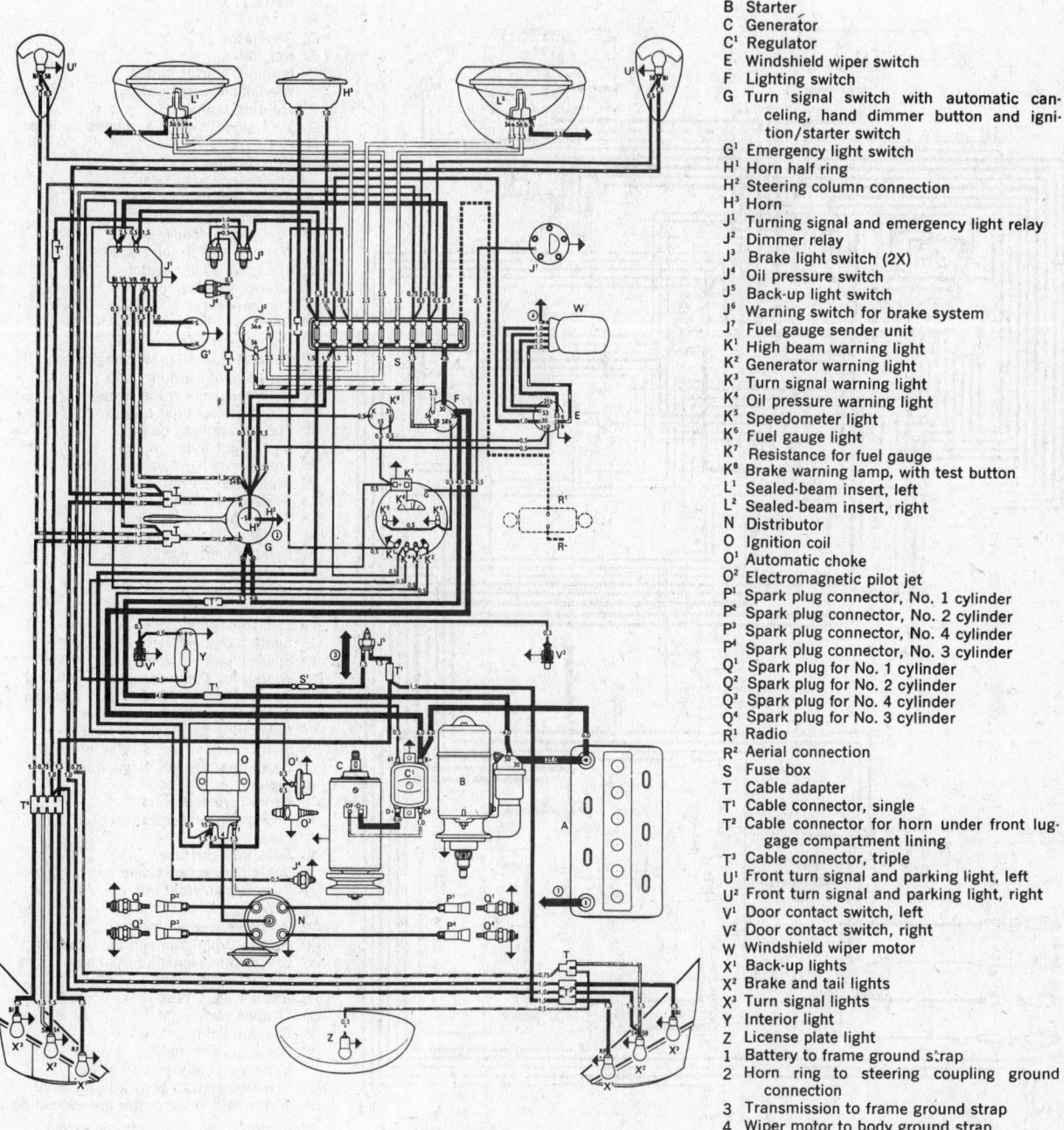

A Battery
B Starter
C Generator
C¹ Regulator
E Windshield wiper switch
F Lighting switch
G Turn signal switch with automatic can-
    celing, hand dimmer button and igni-
    tion/starter switch
G¹ Emergency light switch
H¹ Horn half ring
H² Steering column connection
H³ Horn
J¹ Turning signal and emergency light relay
J² Dimmer relay
J³ Brake light switch (2X)
J⁴ Oil pressure switch
J⁵ Back-up light switch
J⁶ Warning switch for brake system
J⁷ Fuel gauge sender unit
K¹ High beam warning light
K² Generator warning light
K³ Turn signal warning light
K⁴ Oil pressure warning light
K⁵ Speedometer light
K⁶ Fuel gauge light
K⁷ Resistance for fuel gauge
K⁸ Brake warning lamp, with test button
L¹ Sealed-beam insert, left
L² Sealed-beam insert, right
N Distributor
O Ignition coil
O¹ Automatic choke
O² Electromagnetic pilot jet
P¹ Spark plug connector, No. 1 cylinder
P² Spark plug connector, No. 2 cylinder
P³ Spark plug connector, No. 4 cylinder
P⁴ Spark plug connector, No. 3 cylinder
Q¹ Spark plug for No. 1 cylinder
Q² Spark plug for No. 2 cylinder
Q³ Spark plug for No. 4 cylinder
Q⁴ Spark plug for No. 3 cylinder
R¹ Radio
R² Aerial connection
S Fuse box
T Cable adapter
T¹ Cable connector, single
T² Cable connector for horn under front lug-
    gage compartment lining
T³ Cable connector, triple
U¹ Front turn signal and parking light, left
U² Front turn signal and parking light, right
V¹ Door contact switch, left
V² Door contact switch, right
W Windshield wiper motor
X¹ Back-up lights
X² Brake and tail lights
X³ Turn signal lights
Y Interior light
Z License plate light
1 Battery to frame ground strap
2 Horn ring to steering coupling ground
    connection
3 Transmission to frame ground strap
4 Wiper motor to body ground strap

Black dotted line = Optional extras or service
installation

Wiring diagram, VW 1500 Type 1 sedan and convertible (1968)

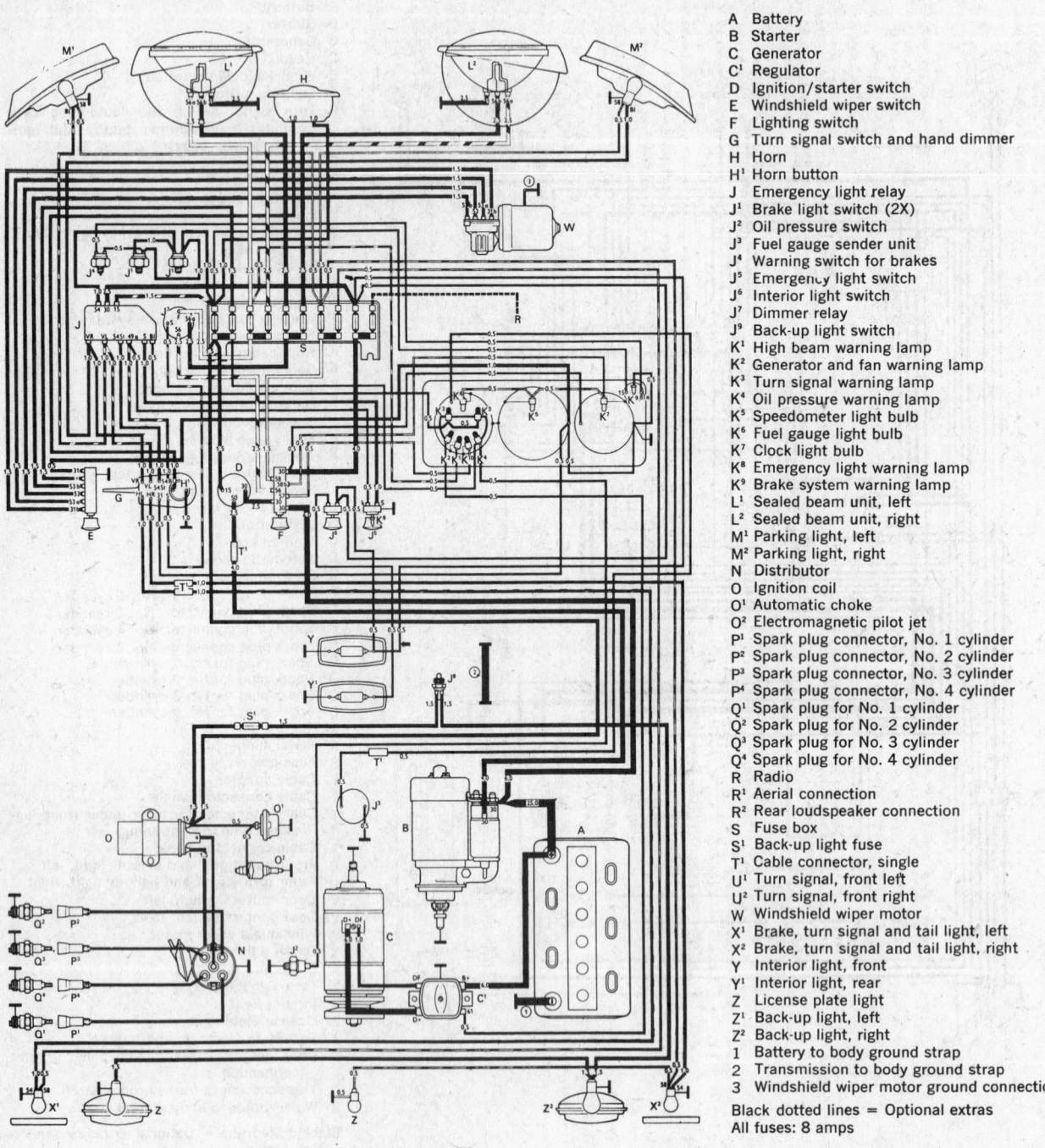

A   Battery
B   Starter
C   Generator
C^1 Regulator
D   Ignition/starter switch
E   Windshield wiper switch
F   Lighting switch
G   Turn signal switch and hand dimmer
H   Horn
H^1 Horn button
J   Emergency light relay
J^1 Brake light switch (2X)
J^2 Oil pressure switch
J^3 Fuel gauge sender unit
J^4 Warning switch for brakes
J^5 Emergency light switch
J^6 Interior light switch
J^7 Dimmer relay
J^9 Back-up light switch
K^1 High beam warning lamp
K^2 Generator and fan warning lamp
K^3 Turn signal warning lamp
K^4 Oil pressure warning lamp
K^5 Speedometer light bulb
K^6 Fuel gauge light bulb
K^7 Clock light bulb
K^8 Emergency light warning lamp
K^9 Brake system warning lamp
L^1 Sealed beam unit, left
L^2 Sealed beam unit, right
M^1 Parking light, left
M^2 Parking light, right
N   Distributor
O   Ignition coil
O^1 Automatic choke
O^2 Electromagnetic pilot jet
P^1 Spark plug connector, No. 1 cylinder
P^2 Spark plug connector, No. 2 cylinder
P^3 Spark plug connector, No. 3 cylinder
P^4 Spark plug connector, No. 4 cylinder
Q^1 Spark plug for No. 1 cylinder
Q^2 Spark plug for No. 2 cylinder
Q^3 Spark plug for No. 3 cylinder
Q^4 Spark plug for No. 4 cylinder
R   Radio
R^1 Aerial connection
R^2 Rear loudspeaker connection
S   Fuse box
S^1 Back-up light fuse
T^1 Cable connector, single
U^1 Turn signal, front left
U^2 Turn signal, front right
W   Windshield wiper motor
X^1 Brake, turn signal and tail light, left
X^2 Brake, turn signal and tail light, right
Y   Interior light, front
Y^1 Interior light, rear
Z   License plate light
Z^1 Back-up light, left
Z^2 Back-up light, right
1   Battery to body ground strap
2   Transmission to body ground strap
3   Windshield wiper motor ground connection

Black dotted lines = Optional extras
All fuses: 8 amps

Wiring diagram, VW Transporter, Type 2 (from August, 1967)

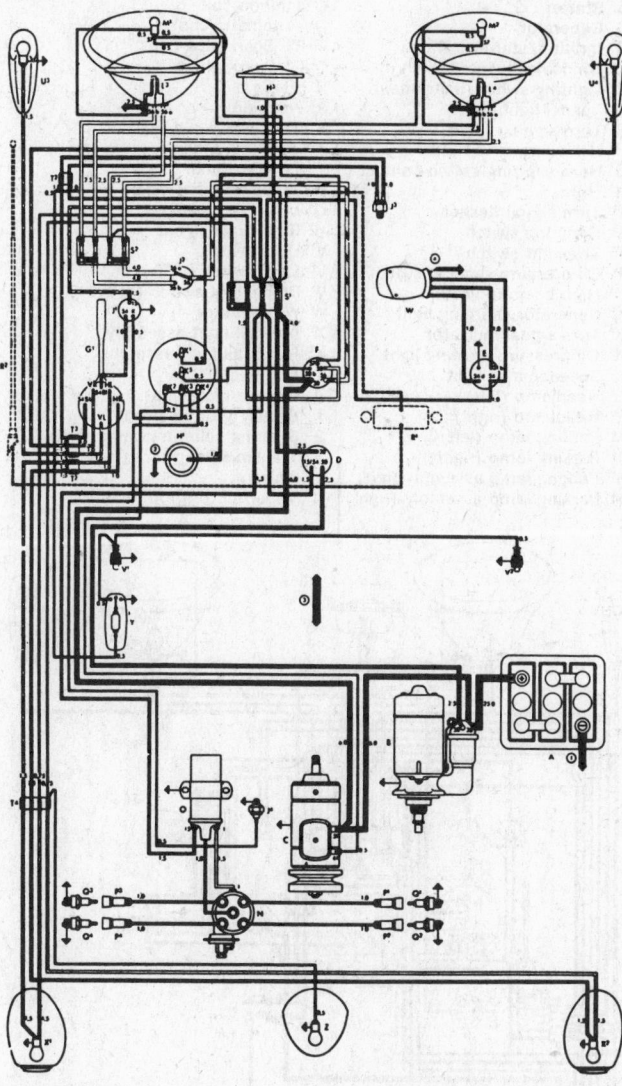

A   Battery
B   Starter
C   Generator
D   Ignition/starter switch
E   Windshield wiper switch
F   Instrument panel lighting switch
G^1 Turn signal switch
H^1 Horn ring
H^3 Horn
J^1 Turn signal relay
J^2 Dimming switch
J^3 Stoplight switch
J^4 Oil pressure sending switch
K^1 High beam indicator light
K^2 Generator warning light
K^3 Turn signal indicator light
K^4 Oil pressure warning light
K^5 Speedometer light
L^3 Headlamp (left)
L^4 Headlamp (right)
M^1 Left parking lamp
M^2 Right parking lamp
N   Distributor

O   Ignition coil
P^1-P^4 Spark plug caps
Q^1-Q^4 Spark plugs
R^1 Radio
R^2 Antenna
S^1 Fuse box
S^2 Fuse box
T^2 Double terminal block
T^3 Triple terminal block
T^4 Quadruple terminal block
U^3 Left turn signal
U^4 Right turn signal
V^1 Door switch
V^2 Door switch
W   Windshield wiper motor
X^1 Left tail light assembly
X^2 Right tail light assembly
Y   Dome light
Z   License plate light
1   Battery ground strap
2   Steering column ground strap
3   Transmission to frame ground strap
4   Wiper motor ground strap

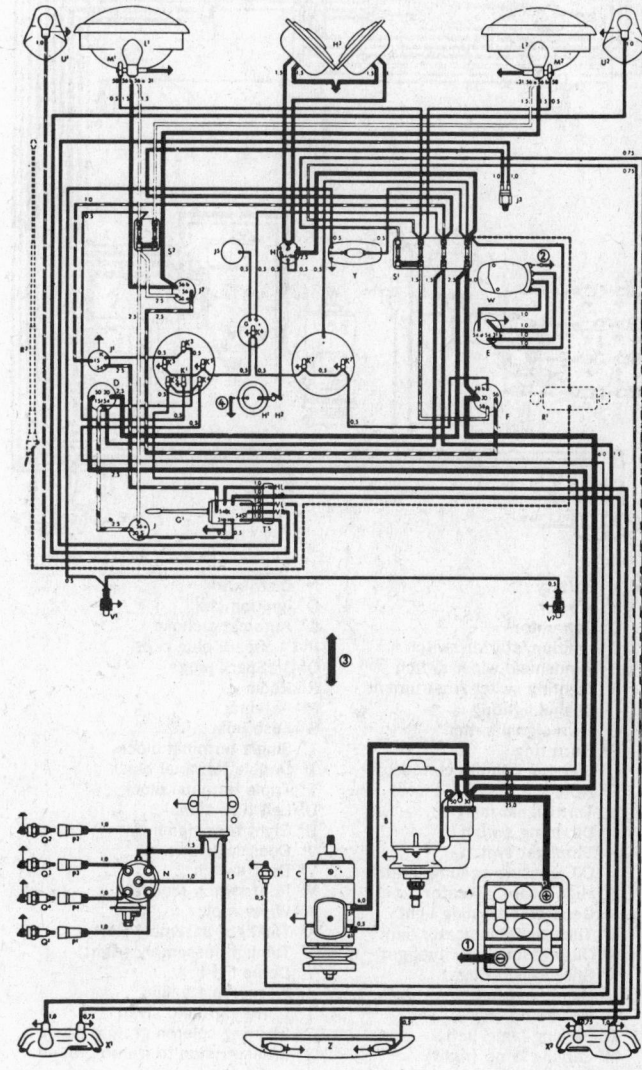

A   Battery
B   Starter
C   Generator
D   Ignition/starter switch
E   Windshield wiper switch
F   Instrument panel lighting switch
G^1 Turn signal switch
H^1 Horn ring
H^3 Horn
J^1 Turn signal relay
J^2 Dimming switch
J^3 Stoplight switch
J^4 Oil pressure sending unit
K^1 High beam indicator
K^2 Generator warning light
K^3 Turn signal indicator
K^4 Oil pressure warning light
K^5 Speedometer light
L^3 Headlamp (left)
L^4 Headlamp (right)
M^1 Parking lamp (left)
M^2 Parking lamp (right)
N   Distributor

O   Ignition coil
P^1-P^4 Spark plug caps
Q^1-Q^4 Spark plugs
R^1 Radio
R^2 Antenna
S^1 Fuse box
S^2 Fuse box
T^2 Double terminal connector
T^3 Triple terminal connector
T^4 Quadruple terminal connector
U^3 Turn signal indicator
U^4 Turn signal indicator
V^1 Left door switch
V^2 Right door switch
W   Windshield wiper motor
X^1 Left tail light assembly
X^2 Right tail light assembly
Y   Dome light
Z   License plate light
1   Battery ground strap
2   Steering column ground strap
3   Transmission to frame ground strap
4   Wiper motor ground strap

Wiring diagram, VW Type 1 sedan (1958-60)

Wiring diagram, VW Karmann Ghia, Type 1 (1958-60)

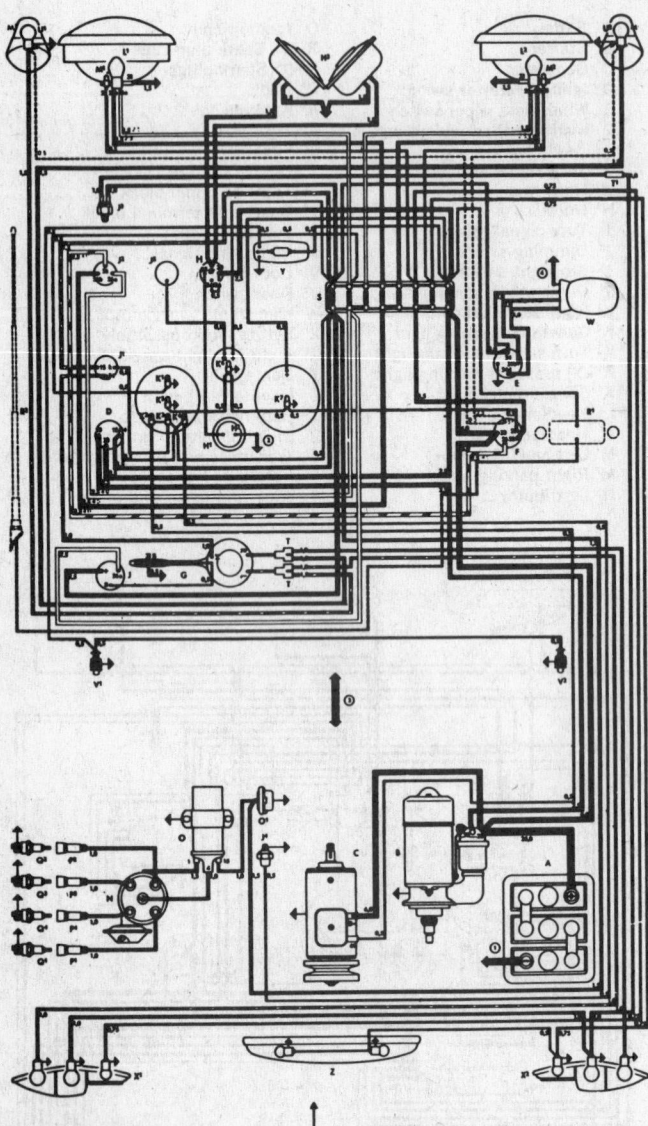

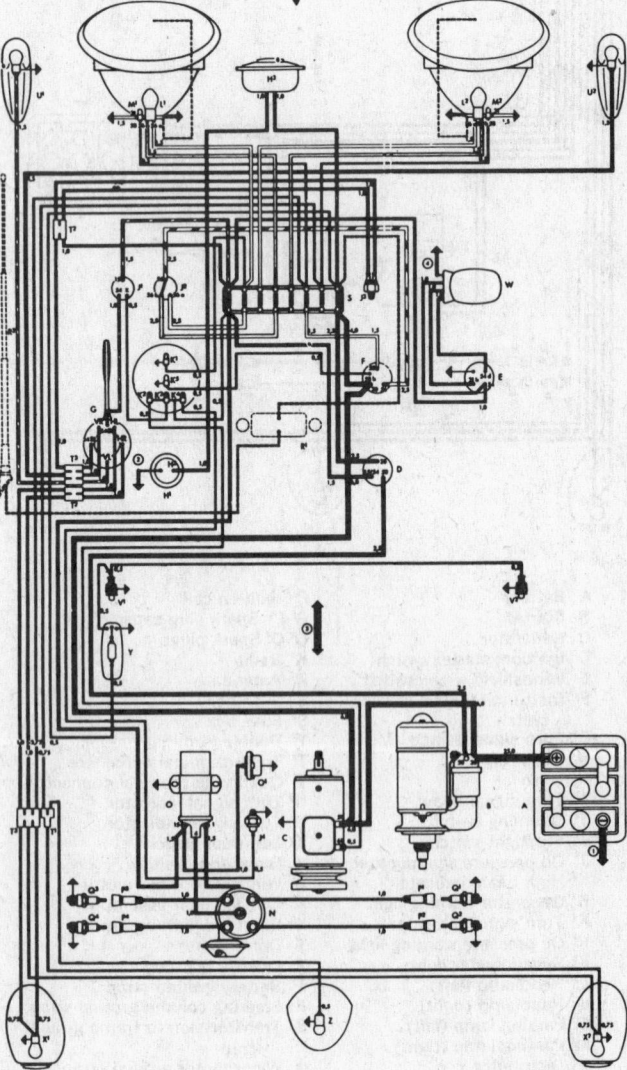

A   Battery
B   Starter
C   Generator
D   Ignition/starter switch
E   Windshield wiper switch
F   Lighting switch/instrument
    panel lighting
G   Turn signal switch
H¹ Horn ring
H² Horn slip ring switch contact
H³ Horn
J¹ Turn signal flasher
J² Dimming switch
J³ Stoplight switch
J⁴ Oil pressure sending switch
K¹ High beam indicator
K² Generator charging light
K³ Turn signal indicator
K⁴ Oil pressure warning light
K⁵ Speedometer light
L¹ Headlamp (left)
L² Headlamp (right)
M¹ Parking lamp (left)
M² Parking lamp (right)
M³ Parking lamp assembly (left)
M⁴ Parking lamp assembly (right)

N   Distributor
O   Ignition coil
O¹ Automatic choke
P¹-P⁴ Spark plug caps
Q¹-Q⁴ Spark plugs
R¹ Radio
R² Antenna
S   Fuse box
T¹ Terminal block
T² Dual terminal block
T³ Triple terminal block
U¹ Left turn signal
U² Right turn signal
V¹ Door switch
V² Door switch
V³ Dome light switch (convt.)
W   Wiper motor
X¹ Left tail light assembly
X² Right tail light assembly
Y   Dome light
Z   License plate light
1   Battery ground strap
2   Steering column ground strap
3   Transmission to frame ground
    strap
4   Wiper motor ground strap

A   Battery
B   Starter
C   Generator
D   Ignition/starter switch
E   Windshield wiper switch
F   Lighting switch/instrument
    panel lighting
G   Turn signal switch
H¹ Horn ring
H² Steering column connector
H³ Horn
J¹ Turn signal relay
J² Dimming switch
J³ Stoplight switch
J⁴ Oil pressure sending switch
K¹ High beam indicator
K² Generator warning light
K³ Turn signal indicator light
K⁴ Oil pressure warning light
K⁵ Speedometer light
L¹ Headlamp (left)
L² Headlamp (right)
M¹ Parking lamp (left)
M² Parking lamp (right)
M³ Parking lamp assembly (left)
M⁴ Parking lamp assembly (right)

N   Distributor
O   Ignition coil
O¹ Automatic choke
P¹-P⁴ Spark plug caps
Q¹-Q⁴ Spark plugs
R¹ Radio
R² Antenna
S   Fuse box
T¹ Single terminal block
T² Double terminal block
T³ Triple terminal block
U¹ Left turn signal
U² Right turn signal
V¹ Door switch
V² Door switch
V³ Roof switch (convt.)
W   Wiper motor
X¹ Tail light assembly (left)
X² Tail light assembly (right)
Y   Dome light
Z   License plate light
1   Battery ground strap
2   Steering column ground strap
3   Transmission to frame ground
    strap
4   Wiper motor ground strap

Wiring diagram, VW Karmann Ghia, Type 1 (1961-62)   Wiring diagram, VW Type 1 sedan (1961-62)

Unit Repair Section

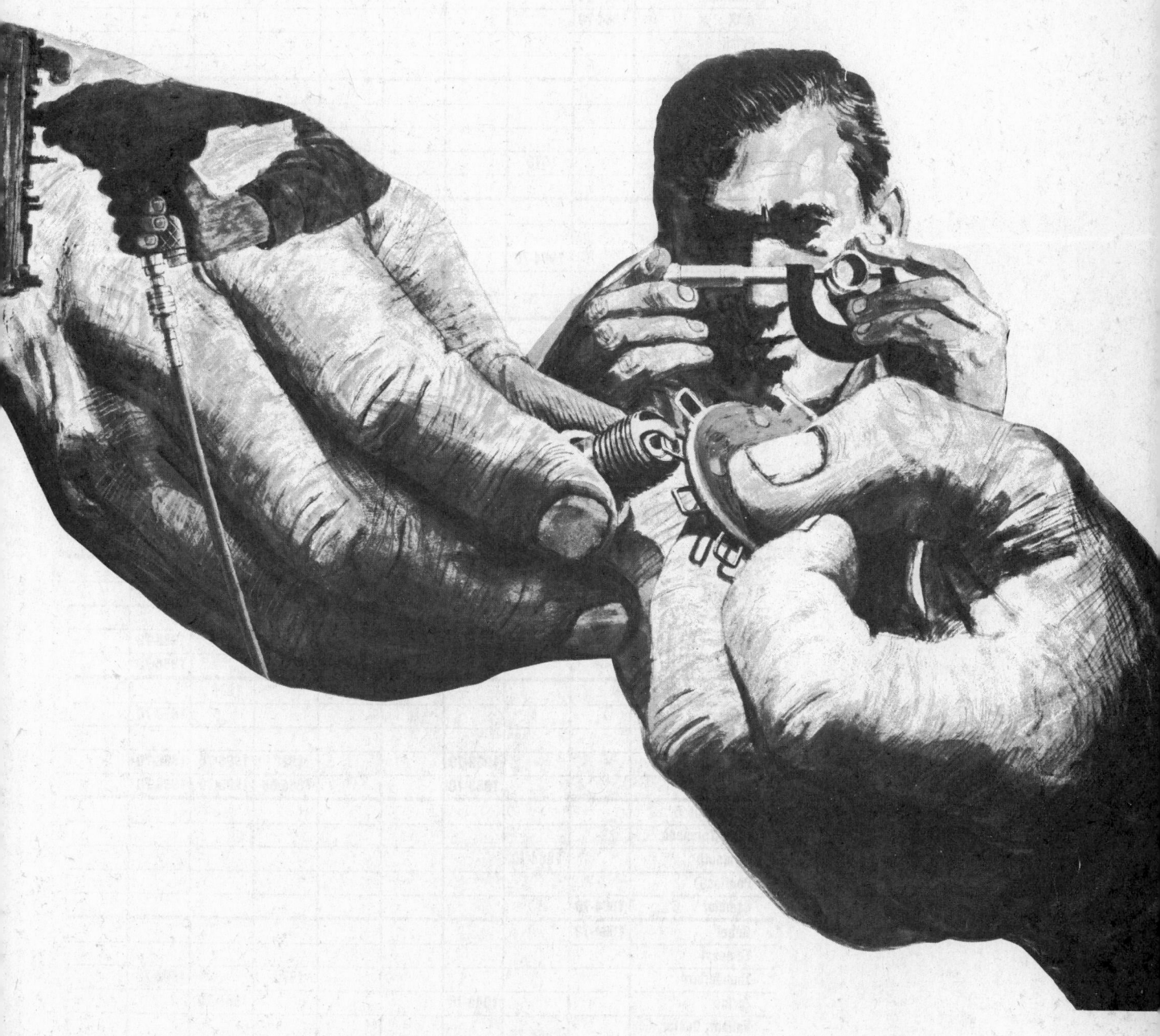

Model	American Motors — Flash-O-Matic Shift-Command	Chrysler Corp. — Torque-flite	Ford Motor Company						
			FMX Select-Shift (3-speed)	Multi-drive	Twin Range Turbo-Drive	Cruise-O-Matic (3-speed)	C4 (3-speed), C4-S	C6 (3-speed)	Merc-O-Matic (3-speed)
Page Number	958	980	1002	1002	1002	1002	1023	1056	1002
Ambassador	1964-70								
American	1964-70								
AMX	1964-70								
Buick									
Buick Special									
Cadillac									
Cad. Eldorado									
Camaro									
Challenger		1970							
Chevelle									
Chevrolet									
Chevy II, Nova									
Chrysler		1964-70							
Comet				1964-65		1968	1964-68	1966-67	1966-67
Corvette									
Cougar			1969-70				1967-70	1966-70	1967-68
Dart		1964-70							
Dodge		1964-70							
Fairlane			1969-70			1964-68	1965-70	1966-70	
Falcon						1964-68	1964-70		
Firebird									
Ford			1969-70			1964-68	1964-70	1966-70	
F-85, Cutlass									
Gremlin	1970								
GTO									
Hornet	1970								
Imperial		1964-70							
Javelin	1968-70								
Lincoln Cont.					1964-65			1966-70	
Mark III								1968-70	
Maverick							1970		
Mercury								1966-70	
Monte Carlo									
Montego			1969-70			1968	1968-70	1966-70	
Mustang			1969-70			1965-68	1965-70	1966-70	
Oldsmobile①									
Olds. Toronado									
Plymouth		1964-70							
Pontiac①									
Rambler	1964-70								
Rebel	1964-70								
Tempest									
Thunderbird						1970		1966-70	
Torino			1969-70				1968-70		
Valiant, Duster, Barracuda		1964-70							
Willys Jeep									

① Dual-coupling Hydra-Matic and Roto-Hydra-Matic were used on some 1964-65 models. These transmissions not covered in this manual.

Identification Chart

Power-glide	Type C Torque Drive	Turbo-Hydra-matic M-40	Jet-away	2-Speed Automatic	Super Turbine 300	Super Turbine 400	Turbo-Hydra-matic 350	Turbo-Hydra-matic 400	Kaiser Jeep Hydra-matic 400	Model
1066	1078	1112	1080	1080	1080	1112	1092	1112	1112	Page Number
										Ambassador
										American
										AMX
					1964-69	1964-70				Buick
					1964-69		1969-70	1967-70		Buick Special
								1964-70		Cadillac
								1967-70		Cad. Eldorado
1967-69	1969-70						1969-70	1968-70		Camaro
										Challenger
1964-70							1969-70	1966-70		Chevelle
1964-70							1969-70	1966-70		Chevrolet
1964-70	1969-70						1969-70	1969-70		Chevy II, Nova
										Chrysler
										Comet
1964-67								1968-70		Corvette
										Cougar
										Dart
										Dodge
										Fairlane
										Falcon
1970		1968-70		1967-69			1969-70			Firebird
										Ford
			1964-69				1969-70	1966-70		F-85, Cutlass
										Gremlin
		1967-70		1964-66						GTO
										Hornet
										Imperial
										Javelin
										Lincoln Cont.
										Mark III
										Maverick
										Mercury
1970							1970	1970		Monte Carlo
										Montego
										Mustang
			1964-69					1965-70		Oldsmobile①
								1966-70		Olds. Toronado
										Plymouth
		1966-70								Pontiac①
										Rambler
										Rebel
1970		1969-70		1964-69			1969-70			Tempest
										Thunderbird
										Torino
										Valiant, Duster, Barracuda
									1965-70	Willys Jeep

957

Automatic Transmissions
American Motors Corporation
Flash-O-Matic—Shift-Command

Application

All American Motors products, 1964-70

Diagnosis

This diagnosis guide covers the most common symptoms and is an aid to careful diagnosis. The items to check are listed in the sequence to be followed for quickest results. Thus, follow the checks in the order given for the particular transmission type.

Flash-O-Matic

CONDITIONS		Corrections		
		Group 1	Group 2	Group 3
Engage-ments	Harsh	BCD	LS	cf
	Delayed forward	ABCD	KLS	a
	Delayed reverse	ACDF	KLS	a
	None	ABCJ	KLS	aklmo
	No forward D-1	AC	KLS	abi
	No forward D-2	ABC	KLS	ab
	No reverse	ABCFE	OKS	aeh
	No neutral	C	LS	c
Upshifts	No 1-2	ACE	MKLNS	ay
	No 2-3	ACE	LNS	aety
	Shift points too high	ABCE	LS	a
	Shift points too low	ABE	LS	a
Upshift Quality	1-2 Delayed followed close by			
	2-3 shift	ABCE	MKLNS	abg
	2-3 slips	ABCE	KLNS	aegt
	1-2 harsh	ABCE	MLS	b
	2-3 harsh	ABCE	MLS	f
	1-2 ties up	AF	LS	fj
	2-3 ties up	AC	MN	
Downshifts	No 2-1 in D-1	ABCE	LS	iy
	No 2-1 in L-range	ABCE	LOS	hy
	No 3-2	ABE	LNS	gy
	Shift points too high	ABCE	LS	a
	Shift points too low	ABCE	LS	a
Forced Downshifts	2-1 slips	ABC	S	bi
	3-2 slips	ABC	MKLNS	aegt
	3-1 shifts above—mph.	CE	KLNS	ag
	2-1 harsh			abi
	3-2 harsh	ABC	MLNS	ef
Reverse	Slips or chatters	ABCF	OKLS	aceht
	Tie up	A	MNLS	ac

Line Pressure	Low idle pressure	ACDE	KLS	am
	High idle pressure	AB	LS	
	Low stall pressure	ABE	KLS	amy
	High stall pressure	AB	LS	
Stall Speed	Too low (200 rpm or more)	H		o
	Too high D-1	ABCJ	KLS	uvabiko
	Reverse too high	ABCFJ	KLOS	uvheko
Others	No push starts	A		n
	Poor acceleration	H		yo
	Noisy in neutral		S	fpdo
	Noisy in park		S	pdo
	Noisy in all gears		S	pro
	Noisy in 1st & 2nd gear only			prw
	Park brake does not hold	C	Q	q
	Oil out breather	AGE	KL	anx
	Oil out fill tube	AGI	K	anx
	Ties up in low, 1st gear		MKLNS	fa
	Ties up in D-1, 1st gear		MKLNS	fa
	Ties up in D-1 or D-2, 2nd gear	FC	LOS	faj
	Ties up in D-1 or D-2, 3rd gear	FC	MKLNOS	taj

Correction Code Key

Group 1
On The Car Without Draining or Removing Oil Pan

A. Check oil level
B. Check oil pressure (gauges)
C. Manual linkage adjustment
D. Engine idle speed
E. Governor inspection
F. Rear band adjustment
G. Check dip stick length
H. Engine tune-up
I. Breather restricted
J. Broken propeller shaft or axle shaft

Group 2
On The Car After Draining and Removing Oil Pan

K. Oil tubes missing or damaged
L. Valve body attaching bolts loose or missing
M. Front band adjustment
N. Front servo, remove, disassemble and inspect
O. Rear servo, remove, disassemble and inspect
P. Tube or servo seal rings missing, broken, leaking
Q. Parking linkage inspection
R. Pressure regulator (V8)
S. 1. Remove control valve assembly and inspect for loose screws.
   2. Replace control valve if no defects in Step 1.
   NOTE: Road test car after Step 1 if defect is corrected.
   Road test car after control valve replacement.

Group 3
Bench Overhaul

a. Sealing rings missing or broken
b. Front clutch slipping, worn plates or faulty parts
c. Front clutch seized or distorted plates
d. Front clutch hub thrust washer missing (detectable in N, P, R only)
e. Rear clutch slipping, worn or faulty parts
f. Rear clutch seized or distorted plates
g. Front band worn or broken
h. Rear band worn or broken
i. One-way (sprag) clutch slipping or incorrectly installed
j. One-way (sprag) clutch seized

k. Broken input shaft
l. Front pump drive tangs or converter hub broken
m. Front pump worn
n. Rear pump
o. Converter
p. Front pump
q. Parking linkage
r. Planetary assembly
s. Fluid distributor sleeve in output shaft (V8)
t. Rear clutch piston ball check leaks
u. Broken output shaft
v. Broken gears
w. Forward sun gear thrust washer missing
x. Breather baffle missing
y. Output shaft plug missing (6 cyl.)

Shift-Command

Engage-ments	Harsh	cf		Forced Downshifts	2-1 Slips	blz
	Delayed forward	az			3-2 Slips	aegt
	Delayed reverse	a			3-1 Shifts above——mph.	ag
	None	aklmo			2-1 Harsh	abi
	No forward D	abiz			3-2 Harsh	ef
	No forward 2	abz		Reverse	Slips or chatters	aceht
	No reverse	aeh			Tie up	ac
	No neutral	c		Line Pressure	Low idle pressure	am
Upshifts	No 1-2	ay			Low stall pressure	amy
	No 2-3	aety		Stall Speed	Too low (200 rpm or more)	o
	Shift points too high	a			Too high D	uvabikoz
	Shift points too low	a			Reverse too high	uvheko
Upshift Quality	1-2 Delayed followed close by 2-3 shift	abg		Others	Poor acceleration	yo
	2-3 Slips	aegt			Noisy in Neutral	fpdo
	1-2 Harsh	b			Noisy in Park	pdo
	2-3 Harsh	f			Noisy in all gears	pro
	1-2 Ties up	fj			Noisy in 1st & 2nd gear only	prw
Downshifts	No 2-1 in D	iy			Park brake does not hold	q
	No 2-1 in 1	hy			Oil out breather	axx-1
	No 3-2	gy			Oil out fill tube	axx-1
	Shift points too high	a			Ties up in 1, 1st gear	fa
	Shift points too low	a			Ties up in D, 1st gear	fa
					Ties up in 2nd gear	faj
					Ties up in 3rd gear	faj
					Chatters——D, 2 or 1	abz

Correction Code Key

a. Sealing rings missing, leaking or broken
b. Front clutch slipping, worn plates or faulty parts
c. Front clutch seized or distorted plates
d. Front clutch hub thrust washer missing (detectable in N, P, R only)
e. Rear clutch slipping, worn or faulty parts
f. Rear clutch seized or distorted plates

g. Front band worn or broken
h. Rear band worn or broken
i. One-way (sprag) clutch slipping or incorrectly installed
j. One-way (sprag) clutch-seized
k. Broken input shaft
l. Pump drive tangs or converter hub broken
m. Pump worn
o. Converter
p. Pump
q. Parking linkage
r. Planetary assembly

s. Fluid distributor sleeve in output shaft (V8)
t. Rear clutch piston ball check leaks
u. Broken output shaft
v. Broken gears
w. Forward sun gear thrust washer missing
x. Breather baffle missing
x-1 Fluid aerated or overfull
y. Output shaft plug missing (6 cyl.)
z. Front clutch piston check valve leaks

Identification

The transmission identification tag is located on the left-hand side of the transmission. This tag shows the Borg-Warner model number, the serial number, and the American Motors part number.

General Information

The transmission consists of a hydraulically-controlled planetary gear box with torque converter. The torque converter is a sealed unit and cannot be serviced separately.

Model identification tags
(© American Motors)

In general, all transmission types are similar in design. There are variations, however, in extension housing configuration due to the use of both torque tube and open drivelines, and some light duty units (such as late models 36, 37 and 40) have aluminum cases. Cast iron cases are used on models 11 and 12 for heavy duty applications.

The Shift-Command transmission is basically the same as the Flash-O-Matic, except that the shift control mechanism is modified so that the transmission can be held in any one gear without automatic upshifting.

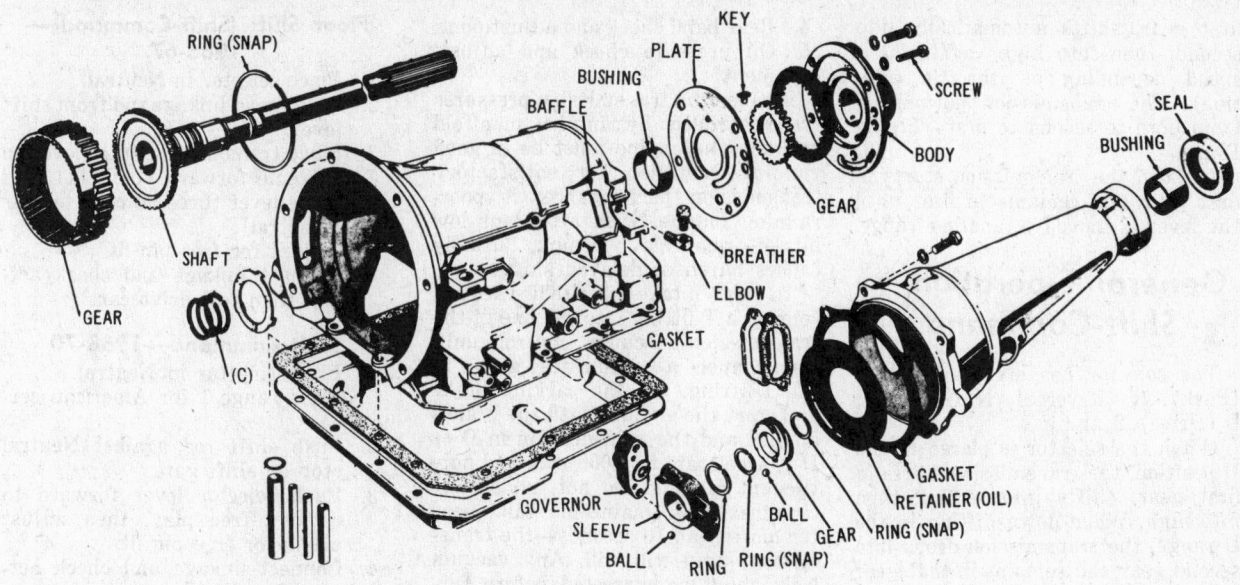

Typical Flash-O-Matic transmission (© American Motors)

General Operation— Flash-O-Matic

The selector has six positions—P (Park), R (Reverse), N (Neutral), D-2 (Drive), D-1 (Drive), and L (Low).

All normal driving is done in D-2 or D-1. In D-2, the transmission starts in second gear, then automatically shifts into high (at a speed depending on throttle position). Also in D-2, the transmission downshifts from high to second.

In D-1, the transmission starts in

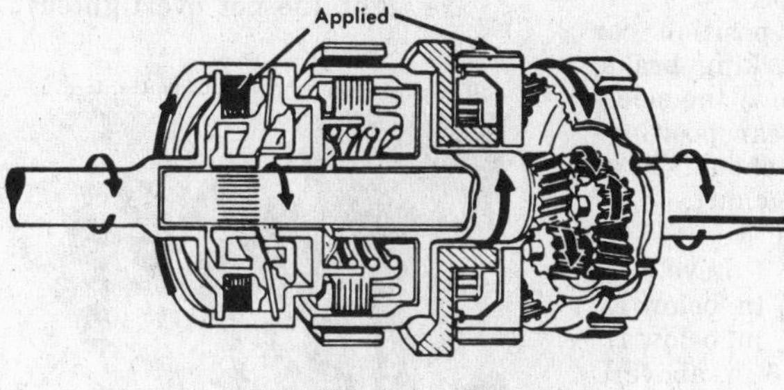

Power flow in low (L-range)
(© American Motors)

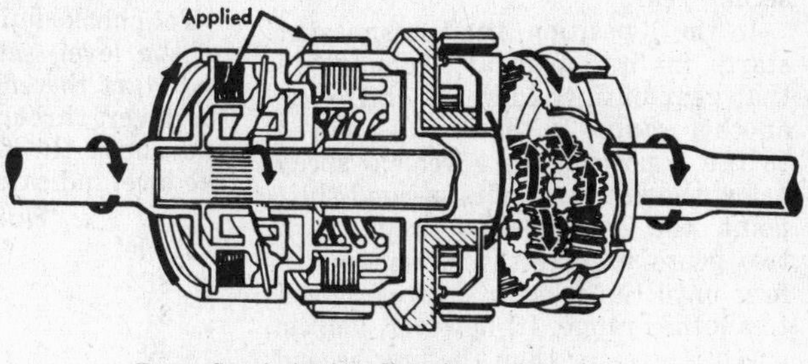

Power flow in intermediate (second gear)
(© American Motors)

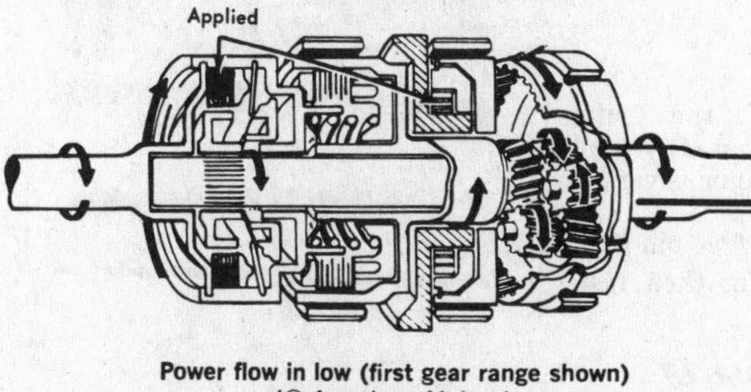

Power flow in low (first gear range shown)
(© American Motors)

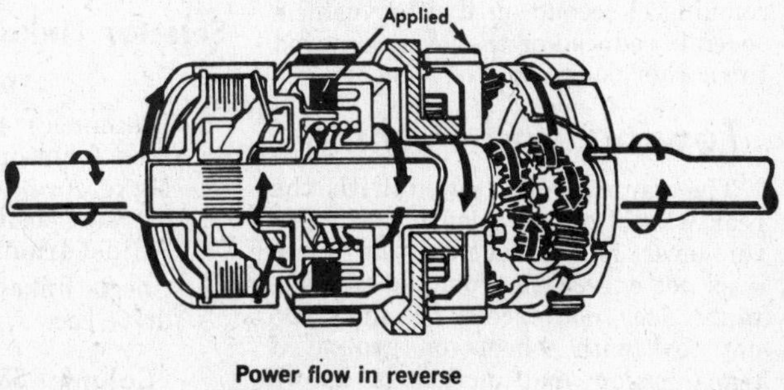

Power flow in reverse
(© American Motors)

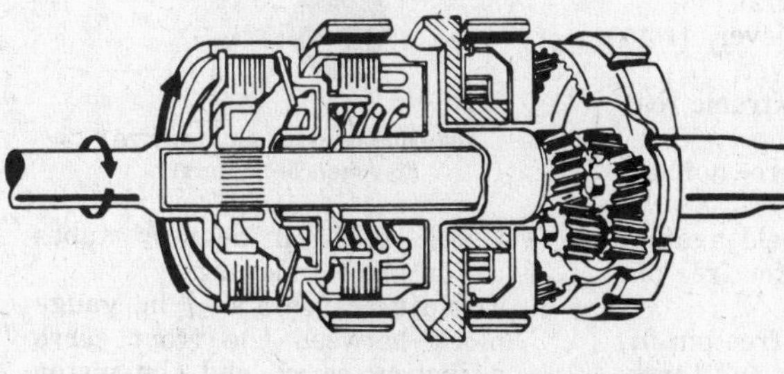

Power flow in neutral
(© American Motors)

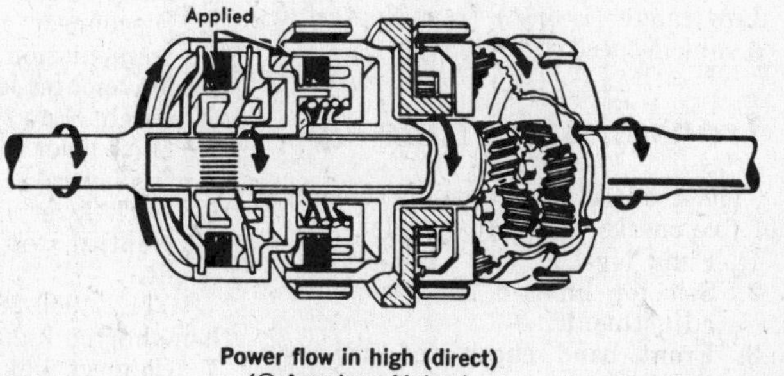

Power flow in high (direct)
(© American Motors)

first gear, shifts automatically into second, then into high (again, at a speed depending on throttle position). The transmission downshifts from high to second to first when in D-1.

In Low, the transmission starts in first gear and remains in first until the lever is moved to another range.

General Operation— Shift-Command

The selector has six positions—P (Park), R (Reverse), N (Neutral), D (Drive), 2, and 1.

When the selector is placed in the D position, the transmission starts in first gear, shifts into second, then into high. When downshifting in the D range, the transmission drops into second gear and remains in that gear until either the full-throttle automatic upshift takes place or the accelerator pedal is released.

In the 2 position, the transmission starts in second gear and remains in that gear until the lever is shifted to another range. If the lever is moved to the 2 position from D, at a speed below the second-third shift point, the transmission drops down into second gear.

In the 1 position, the transmission starts in first gear and remains in that gear until the lever is shifted to another range. If the lever is moved to the 1 position at a vehicle speed below the automatic first-second shift point, the transmission drops down two gears into first, and remains in first until the selector lever is moved to another range. If, however, vehicle speed is greater than the first-second shift point, the transmission drops down only one gear, into second, and remains in second until either vehicle speed is reduced or the lever is moved to another position.

Towing and Pushing

The car may be towed with the rear wheels on the road if the selector lever is in Neutral, tow speed does not exceed 35 mph, and tow distance does not exceed 50 miles. Do not tow with wheels on ground if transmission malfunction is known or suspected.

To push-start the car, shift into Low range (L or 1) from Neutral at a vehicle speed of 15-20 mph.

Transmission Tune-up

The transmission tune-up consists of five checks and adjustments:
1. Fluid level.
2. Selector linkage inspection and adjustment.
3. Front band check and adjustment.
4. Rear band check and adjustment.
5. Oil pressure check and adjustment.

Because the transmission pressures are controlled by intake manifold vacuum, the engine must be in good running condition for satisfactory transmission performance. A poor-running engine, with attendant low intake manifold vacuum, usually causes harsh or delayed shifting.

To test intake manifold vacuum, connect a T-fitting into the line at the transmission vacuum control unit, then connect a vacuum test gauge to the T-fitting. Set the parking brake and start the engine. With the brakes applied and the transmission in D or D-1, accelerate to 1,000 rpm and note vacuum reading. Do not, under any circumstances, maintain stall speed for more than 10 seconds—the transmission fluid will boil. Any vacuum leaks must be corrected before further transmission testing can continue.

Fluid Level

The "full" (F) mark on the dipstick is calibrated for normal operating temperature. If the transmission is filled to the 'F' mark when cool, the fluid may froth and be forced out the vent or filler tube.

To check fluid cold, position car on the level, set the parking brake, and start the engine. Move the selector lever through all gear positions and check the fluid—it should be at the level indicated in the chart.

Fluid Level

Model	Level
36	1/4 in. below L
37	1/4 in. below L
40	5/16 in. above L
11	L
12	L

Selector Linkage

1964-65

1. Disconnect gearshift rod trunnion from operating shaft lever.
2. Make sure transmission lever is in rear position.
3. Adjust trunnion for free pin fit.
4. Check linkage action, then test drive car.

Column Shift—1966-67

1. Turn off ignition switch.
2. Place selector in Neutral.
3. Disconnect manual lever from transmission outer lever.
4. Move outer lever to extreme rear notch (Low).
5. Move lever forward three notches to Neutral.
6. With lever linkage held against Neutral stop, adjust for free pin fit.
NOTE: 1966 287/327—free pin fit, then shorten linkage three full turns.
7. Connect linkage and check action, then test drive car.

Floor Shift (Shift-Command)— 1966-67

1. Place selector in Neutral.
2. Disconnect linkage rod from shift lever.
3. Move transmission outer lever to extreme forward notch (1).
4. Move lever three notches to rear (Neutral).
5. Adjust for free pin fit.
6. Connect linkage and check action, then test drive car.

Shift-Command—1968-70

1. Place selector in Neutral.
NOTE: range 1 for American series.
2. Push shift rod against Neutral stop on shift gate.
3. Push selector lever forward to remove free play, then adjust clevis for free pin fit.
4. Connect linkage and check action, then test drive car.

Front Band

1. Drain transmission fluid and remove oil pan.
2. Check for debris, "scorched" smelling fluid, and loose parts.
3. Check pick up screen for clogging, then check that all valve body cap screws and servo bolts are tight (do not overtighten).

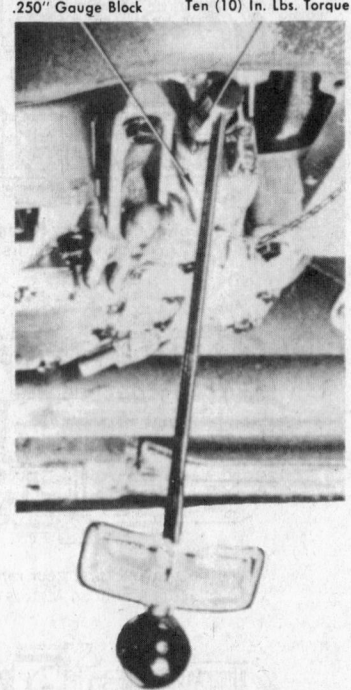

.250" Gauge Block   Ten (10) In. Lbs. Torque

Adjusting front servo—1967-70
(© American Motors)

4. Check that oil delivery tubes are in place and snug.
5. To adjust, insert a 1/4 in. gauge block between the front servo adjusting screw and the piston rod, then tighten adjusting screw to 9 in. lbs. for 1964-67 trans-

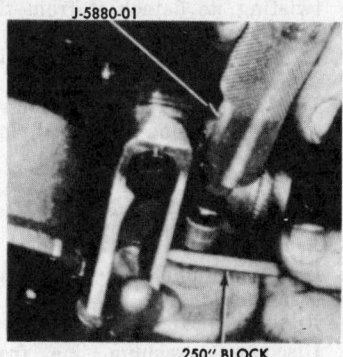

.250" BLOCK

Adjusting front servo—1964-66 V8, 1967 343 V8
(ⓒ American Motors)

J-9639
.250" Block

Adjusting front servo—1964-66 6 cylinder
(ⓒ American Motors)

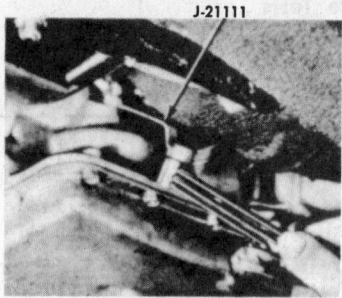

Adjusting rear band—1964-66 6 cylinder
(ⓒ American Motors)

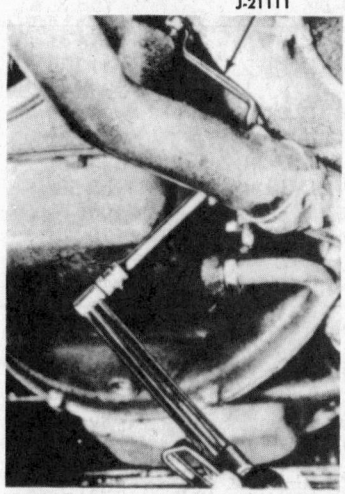

J-21111

Adjusting rear band—1964-66 V8
(ⓒ American Motors)

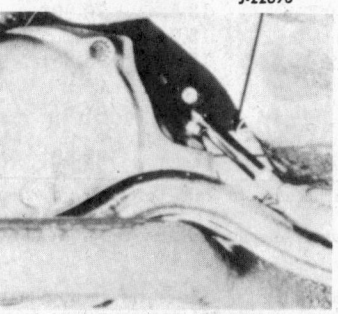

J-22698

Adjusting rear band—1967-70
(ⓒ American Motors)

missions, 10 in. lbs. for 1968-70 transmissions.

Caution the adjusting screw on M11 and M12 transmissions has a left-hand thread.

6. For 1969-70 only: inspect the adjuster wire for proper clearance —one screw thread must be exposed between the wire and the servo actuating lever.
7. Clean and install the oil pan, using a new gasket, then install proper quantity of approved fluid.

Rear Band

1. Place a hydraulic jack under the transmission, then remove the four crossmember fasteners and the crossmember.
NOTE: not necessary to remove crossmember on most Ambassador models. On AMX and Javelin equipped with power steering, lift the hood before lowering transmission to keep the power steering fluid reservoir wingnut from hitting the hood.
2. Lower the transmission, then loosen the rear band adjusting screw locknut and tighten the adjusting screw to 9 ft., lbs. for

1964, 10 ft. lbs. for 1965 and later, transmissions.
3. Back off adjusting screw ¾ turn for M36, M37, and M40 transmissions, 1¼ turns for M11 and M12 transmissions. Tighten locknut to 28 ft. lbs.
NOTE: 1966 287/327 transmission —1½ turns, then tighten locknut to 35-40 ft. lbs.
4. Raise transmission and replace crossmember.

Control Pressure

Remove the pipe plug on the left-hand side of the transmission case, then install a pressure gauge (0-400 psi). Connect a T-fitting and vacuum gauge into the line to the vacuum

control unit, chock the wheels, set the parking brake, and bring the engine to normal operating temperature.

Place selector lever in Reverse and, with brakes locked, accelerate engine to obtain vacuum in chart (or 1,000 rpm). Make sure the pressure is the same as that listed for the appropriate vacuum and rpm. Do not exceed 10 seconds at stall speed.

If pressure is too low, remove the vacuum control unit line and turn the adjusting screw clockwise. If pressure is too high, turn adjusting screw counterclockwise. For M36, M37, and M40 transmission, one turn of the adjusting screw equals approximately 18 psi pressure change; for M11 and M12 transmissions, 10 psi pressure change.

If correct pressure cannot be obtained, check the vacuum control unit. The pushrod on the M11 and M12 transmissions should be 3.439 in. long (with altitude correction type control unit). The pushrod on the M36, M37, and M40 transmissions should be 4.116 ± .005 in. long (diaphragm type control unit).

If pressure is O.K. in Reverse, check the pressure in the forward ranges. Do not, however, try to adjust the pressure in the forward ranges. If pressure is low in all forward ranges, but correct in Reverse, remove and inspect the governor for sticking or loose screws. If pressure is correct in all ranges, but the transmission does not upshift automatically when in Drive, check that the governor is not stuck closed.

Service—Transmission in Car

Oil Pan and Screen R&R— 1964-70 All

1. Jack up the car and disconnect the oil filler tube at the transmission, then drain the transmission oil.
2. Remove oil pan screws and lockwashers, then remove oil pan and gasket.
3. Remove Alnico magnet from the head of the rear servo hold-down bolt, then remove inlet oil screen.
NOTE: do not use air pressure, or any solvents other than ATF, to clean.
4. To install, reverse removal procedure, tightening oil pan screws to 15 ft. lbs.

Downshift Solenoid R&R— 1967-70

1. Drain oil and remove oil pan, then disconnect solenoid wire from transmission case terminal.
2. Push in on the solenoid, while

Control Pressure Tests

Year	Engine	Engine RPM	Selector Range	Manifold Vacuum (in. Hg.)	Pressure (PSI)
1964-65	6/195 OHV	1,000	D-1		100±10▲
	6/195 L-head	1,000	D-1		108±10▲
	6/all 195	500	D-1		60±5 ▲
	6/199 and 232	1,000	D-1		100±10▲
	6/199 and 232	500	R,L,D		60±5 ▲
	V8/287	500	N	11.8	85±3†
	V8/327	500	N	13.8	85±3†
1966	6/199 and 232	1,000	D-1		100±10▲
	6/199 and 232	500	R,L,D		60±5 ▲
	V8/287	1,000	D-1		70±5 ▲
	V8/327	1,000	D-1		75±3 ▲
1967	6/199 and 232	1,000	R		95‡
	6/199 and 232	1,000	D-1,D-2,L		90-100‡
	V8/290	1,000	R		95‡
	V8/290	1,000	D-1,D-2,L		90-100‡
	6/199 and 232	500	R		55-68‡
	V8/290	500	R,D-1,D-2,L		55-68‡
	6/199 and 232	500	D-1,D-2,L		55-68‡
	V8/343	1,000	R		100‡
	V8/343	1,000	D-1,D-2,L		70-80‡
	V8/343	500	R		57-67‡
	V8/343	500	D-1,D-2,L		42-52‡
1968-70	6/199	◎	R	9.0	95‡
	6/199	◎	1,2,D	9.0	90-100‡
	6/199	idle	D,R,1,2		55-68‡
	6/232	◎	R	13.5	95‡
	6/232	◎	1,2,D	13.5	90-100‡
	6/232	idle	D,R,1,2		55-68‡
	V8/290, 304 2-v	◎	1,2,D	13.5	90-100‡
	V8/290, 304 2-v	◎	R	13.5	95‡
	V8/290, 304 2-v	idle	D,R,1,2		55-68‡
	V8/290 4-v, 304 4-v,343	◎	R	13.5	100‡
	V8/290 4-v, 304 4-v, 360, 343	◎	1,2,D	13.5	70-80‡
	V8/390 4-v	◎	R	15.0	120‡
	V8/390 4-v	◎	1,2,D	15.0	75-85‡
	V8/290 4-v, 304 4-v, 343, 360, 390	Idle	R		57-67‡
	V8/290 4-v, 304 4-v, 343, 360, 390	Idle	1,2,D		42-52‡

▲ To adjust, lengthen throttle valve cable carburetor to raise pressure, shorten cable to lower pressure.
† To adjust loosen locknut and screw in vacuum solenoid unit to raise pressure, unscrew unit to lower pressure.
‡ To adjust, see text.
◎ At whatever speed vacuum in column 5 is attained. In some cases, the vacuum may be low due to altitude or a new engine. Here, an external vacuum source must be used. To utilize this vacuum source, disconnect engine vacuum line, connect external source to control unit, and adjust source to specified amount. Then, test transmission pressure.

Front servo components—1968-70 M36, M37, M40 (© American Motors)

Front servo—1964-70 light-duty transmission
(© American Motors)

twisting, to detach it from the control valve.
CAUTION: do not lose downshift valve spring.
3. To install, reverse removal procedure, using a new O-ring.

Control Valve R&R—1967-70 M36, M37, and M40

1. Disconnect filler tube and drain transmission oil.
2. Remove oil pan.
3. Disconnect vacuum line from control unit.
4. Remove vacuum control unit and pushrod.
5. Remove rear servo oil tube, rear clutch oil tube, front servo release tube, and front servo apply tube.
6. Disconnect solenoid wire at terminal, then remove the three control valve-to-case attaching screws.
7. Remove control valve assembly.
8. To install, begin with the pump pressure and converter "in" and "out" tubes, then install a new O-ring on the pump suction tube and install the tube into the pump.
9. Engage the manual lever link in the detent lever and align the control valve with the oil tubes in the pump.
CAUTION: care must be exercised during this operation, because a bent pump suction tube will cause the oil to foam.
10. Install the control valve-to-case cap screws, finger-tight, while holding the valve tightly against the case. Torque cap screws to 75 in. lbs.
11. Continue installation by reversing Steps 1-6.

Control Valve R&R—1967-70 M11 and M12

1. Follow Steps 1-4 of M36, M37, and M40 procedure.
2. Loosen front servo adjusting screw until servo arm hits servo body.

NOTE: adjusting screw has a left-hand thread. Hold the bent tang end of the adjuster wire in a counter-clockwise direction while adjusting screw is being turned clockwise.

3. Loosen the front servo attaching bolts.
4. Remove valve body hold-down bolts—three at front, three at rear.
5. Hold servo about ¼ in. away from case, then disengage front servo pressure tubes by moving valve body away from servo.
6. Remove valve body.
7. To install, reverse removal procedure.

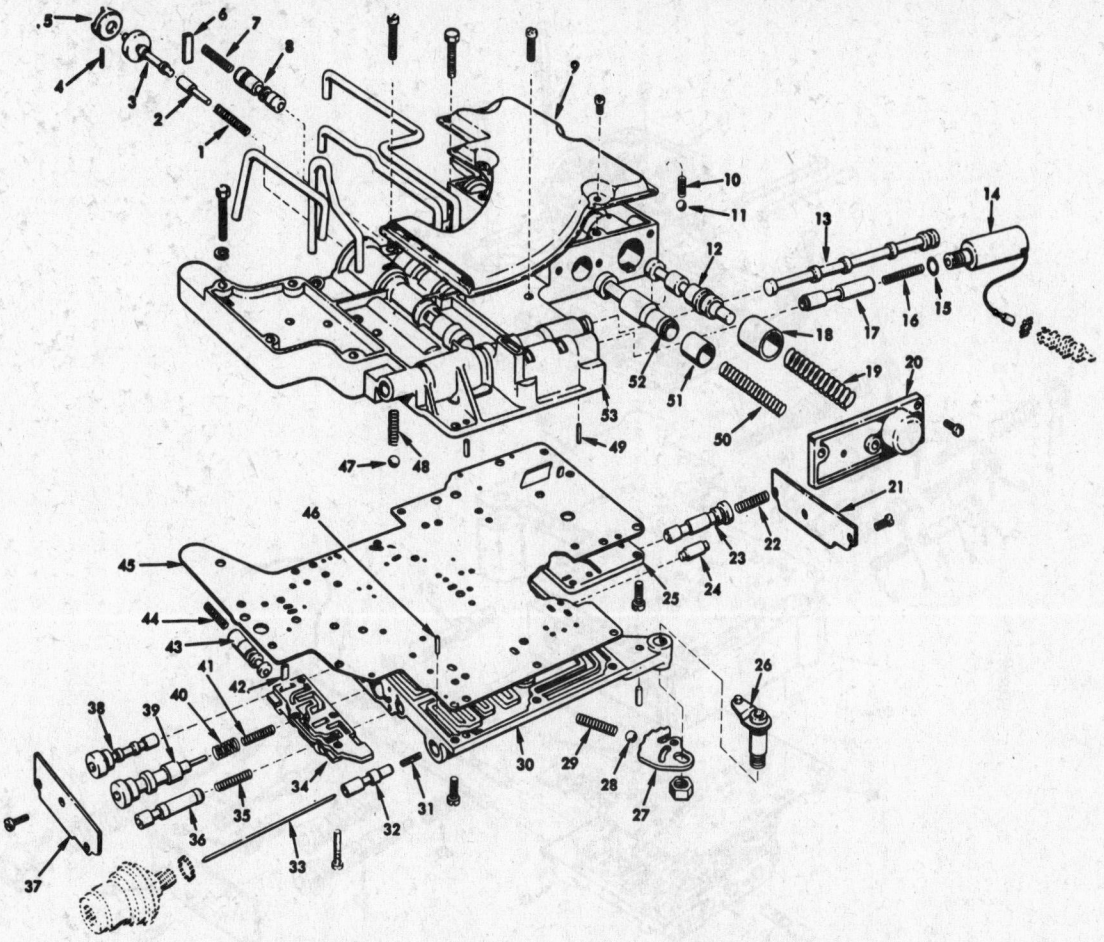

Control valve assembly sequence, models 36, 37, and 40—1967-69 American Motors
(© American Motors)

1 Modulator valve spring
2 Modulator valve plug
3 Modulator valve
4 Dowel pin
5 Modulator valve spacer
6 Servo orifice control valve stop
7 Servo orifice control valve spring
8 Servo orifice control valve
9 Oil screen
10 Converter out spring*
11 Converter out ball*
12 Primary regulator valve
13 Manual valve
14 Downshift solenoid
15 "O" ring
16 Downshift valve return spring
17 Downshift valve
18 Primary regulator valve sleeve
19 Primary regulator valve spring
20 Lower valve body end plate
21 Front end plate, upper body
22 1-2 shift valve spring
23 1-2 shift valve plunger
24 2-3 shift valve plunger
25 Oil tube plate
26 Manual control valve lever

27 Detent ball
29 Detent spring
30 Upper valve body
31 Throttle valve spring
32 Throttle valve
33 Throttle valve control rod
34 Front clutch orifice control valve body
35 Range control valve spring
36 Range control valve
37 Read end plate, upper body
38 1-2 shift valve
39 2-3 shift valve
40 2-3 shift valve outer spring
41 2-3 shift valve inner spring
42 Front clutch orifice control valve stop
43 Front clutch orifice control valve
44 Front clutch orifice control valve spring
45 Separator plate
46 Throttle valve retaining pin
47 2-3 shift valve circuit ball
48 2-3 shift valve circuit spring
49 Solenoid retaining pin
50 Secondary regulator valve spring
51 Secondary regulator valve sleeve
52 Secondary regulator valve
53 Lower valve body
* Required for air cooled transmissions

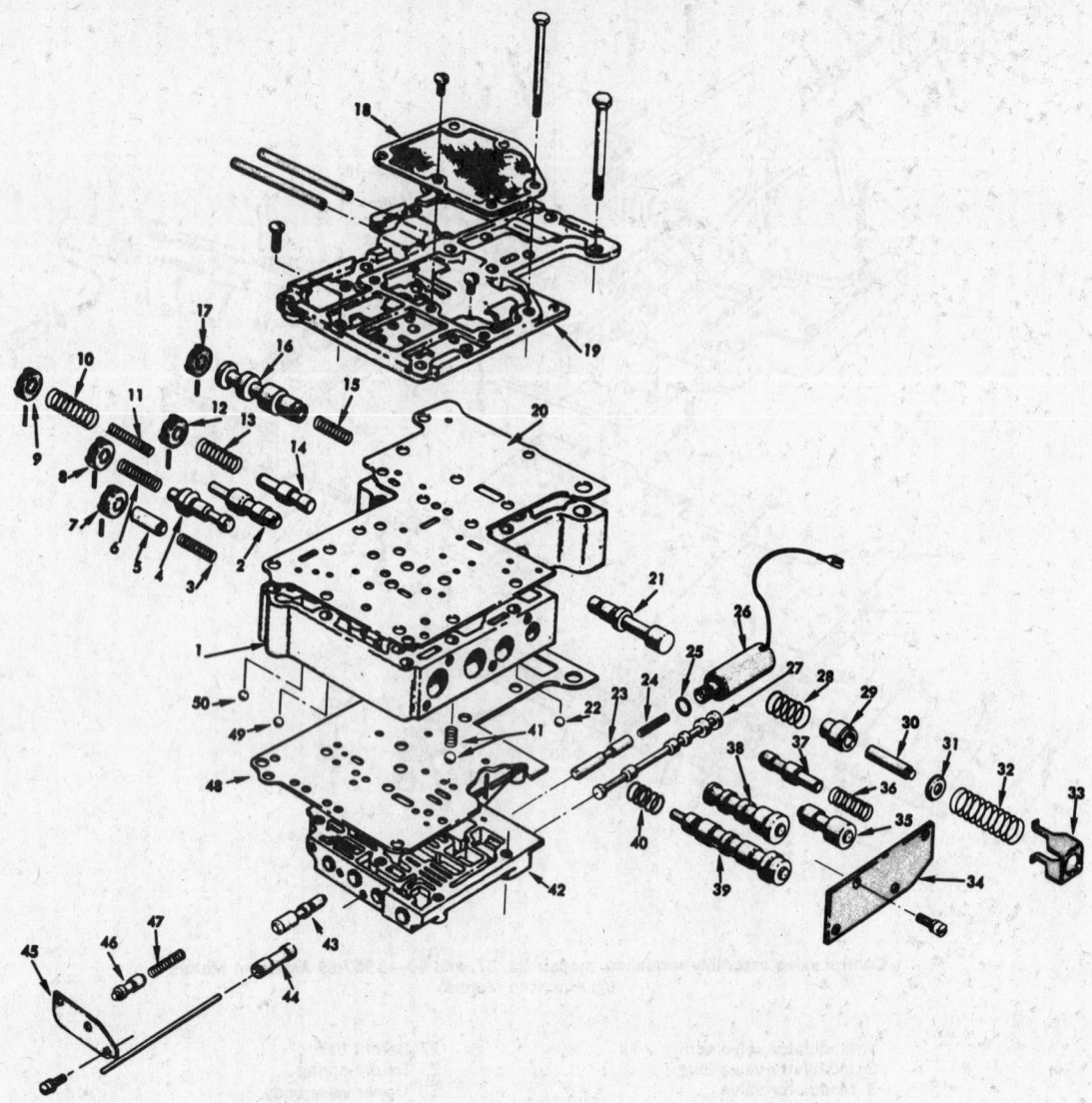

Valve body assembly sequence, models 11 and 12, 1968-69 American Motors
(© American Motors)

1 Main valve body
2 Servo regulator valve
3 2-3 shift valve spring
4 1-2 shift valve
5 2-3 shift valve
6 1-2 shift valve spring
7 2-3 shift valve plug
8 1-2 shift valve plug
9 Servo regulator valve plug
10 Servo regulator valve spring
11 Servo regulator valve inner
     spring M-12 only
12 Governor modulator valve plug
13 Governor modulator valve spring
14 Governor modulator valve
15 Secondary regulator valve spring
16 Secondary regulator valve
17 Secondary regulator valve plug
18 Screen
19 Valve body cover
20 Lower separator plate
21 Primary regulator valve
22 Orifice control directional
     check ball
23 Downshift valve
24 Downshift valve spring
25 Solenoid "O" ring

26 Downshift solenoid
27 Manual valve
28 Primary regulator valve spring
     (inner)
29 Primary regulator valve plug
30 Primary regulator valve spacer
31 Primary regulator spring seat
32 Primary regulator spring
33 Primary regulator spring retainer
34 Valve body end plate
35 Cutback valve
36 Throttle modulator valve spring
37 Throttle modulator valve
38 1-2 shift valve
39 2-3 shift valve
40 2-3 shift valve spring
41 2-3 circuit check ball and spring
42 Upper valve body
43 Servo regulator timer valve
44 Throttle valve
45 Upper valve body end plate
46 Orifice control valve
47 Orifice control valve spring
48 Upper separator plate
49 Rear servo release check ball
50 1-2 directional check ball

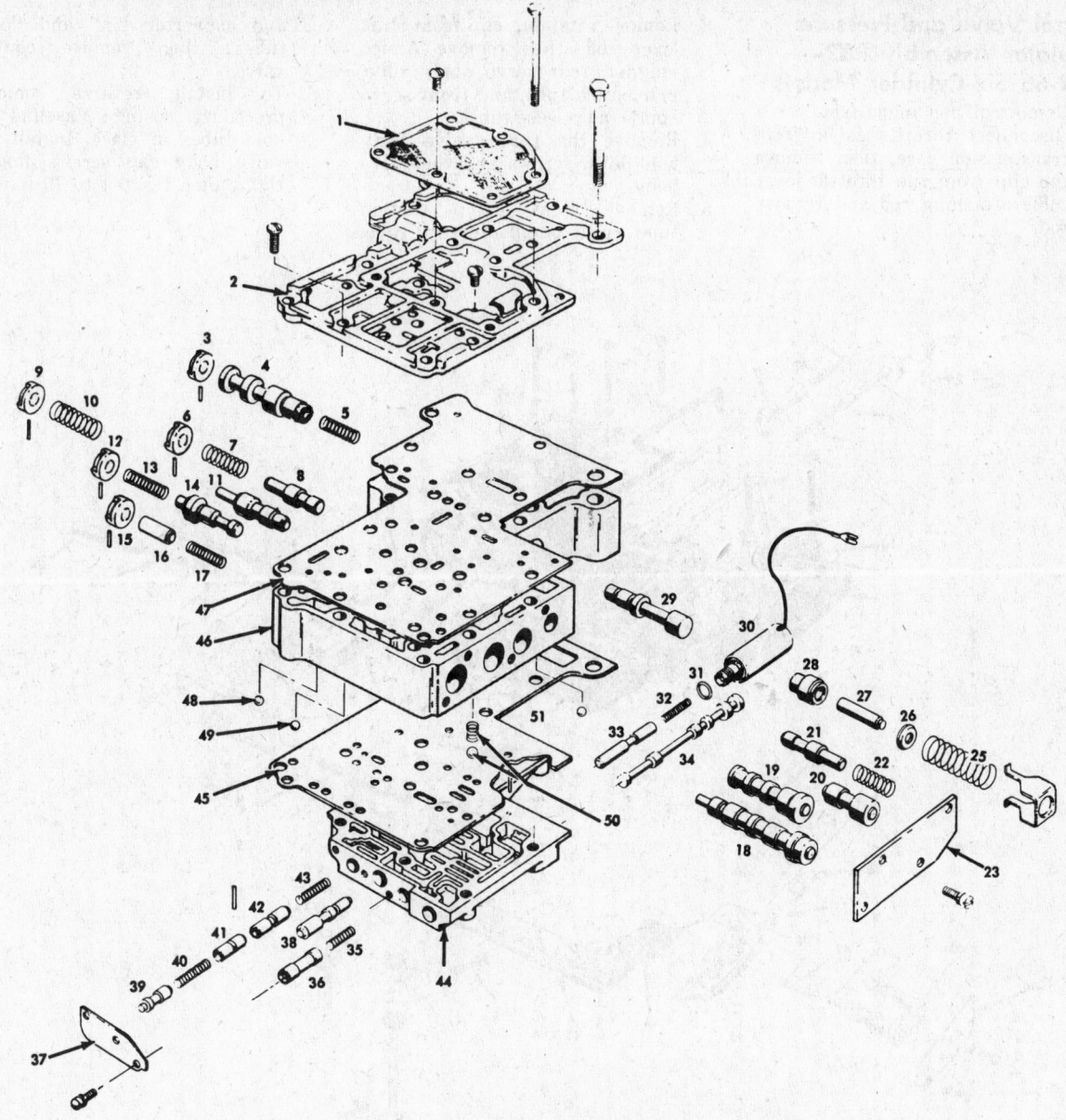

Control valve assembly sequence—1967 343 cu. in. V8 (© American Motors)

1 Screen	17 2–3 shift valve spring	34 Manual valve
2 Valve body cover	18 2–3 shift valve	35 Throttle valve spring
3 Secondary regulator valve, plug	19 1–2 shift valve	36 Throttle valve
4 Secondary regulator valve	20 Cutback valve	37 Upper valve body end plate
5 Secondary regulator valve spring	21 Throttle modulator valve	38 Servo regulator timer valve
6 Governor modulator valve plug	22 Throttle modulator valve spring	39 Orifice control valve
7 Governor modulator valve spring	23 Valve body end plate	40 Orifice control valve spring
8 Governor modulator valve	24 Primary regulator spring retainer	41 Valve spacer
9 Servo regulator valve plug	25 Primary regulator spring	42 3–2 control valve
10 Servo regulator valve spring	26 Primary regulator spring seat	43 3–2 control valve spring
11 Servo regulator valve	27 Primary regulator valve spacer	44 Upper valve body
12 1–2 shift valve plug	28 Primary regulator valve plug	45 Upper separator plate
13 1–2 shift valve spring	29 Primary regulator valve	46 Main valve body
14 1–2 shift valve	30 Kickdown solenoid	47 Lower separator plate
15 2–3 shift valve plug	31 Solenoid "O" ring	48 1–2 directional check ball
16 2–3 shift valve	32 Kickdown valve spring	49 Rear servo release check ball
	33 Kickdown valve	50 2–3 circuit check ball
		51 Orifice control directional check ball

Control Valve and Pressure Regulator Assembly R&R— 1964-66 Six-Cylinder Models

1. Remove oil pan and gasket.
2. Disconnect throttle cable from transmission case, then remove the clip from the throttle lever cable attaching rod and remove rod.
3. Remove retaining clip from shift lever rod, then remove Alnico magnet, rear servo apply tube, rear clutch tube, and front servo apply and release tubes.
4. Remove the three valve body hold-down screws and the valve body.
5. Remove front pump suction tube, front pump pressure tube, and converter "in" and "out" tubes; then remove control valve.
6. To install, reverse removal procedure, using Vaseline to hold tubes in place. Install the valve body cap screws finger-tight, then tighten to 75 in. lbs.

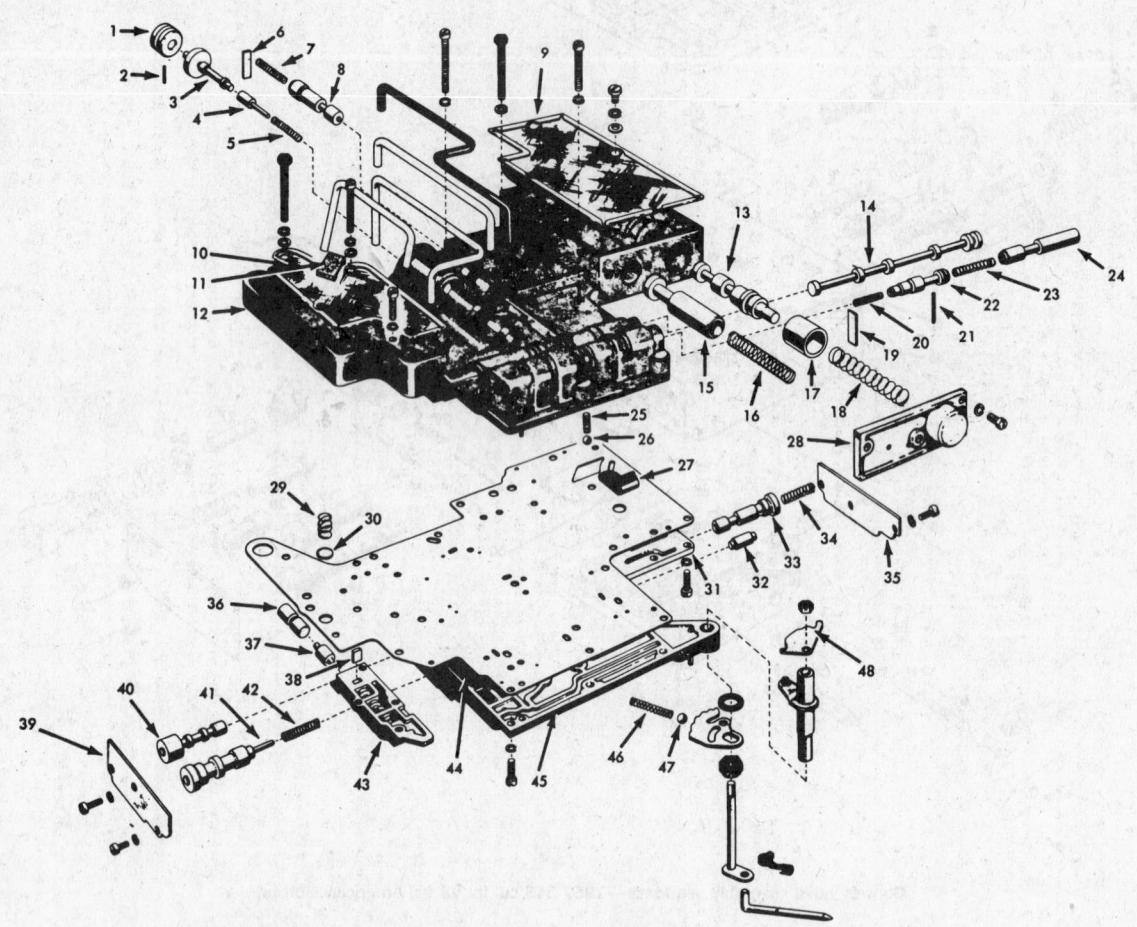

Control valve assembly sequence—1964 6 cylinder (© American Motors)

1 Modulator valve spacer	25 Converter out check spring
2 Dowel pin	26 Converter out check ball
3 Modulator valve	27 Front pump check valve
4 Modulator valve plug	28 Lower valve body end
5 Modulator valve spring	29 Rear pump check valve spring
6 Servo orifice control valve stop	30 Rear pump check valve
7 Servo orifice control valve spring	31 Oil tube plate
8 Servo orifice control valve	32 2–3 shift valve plunger
9 Front oil screen	33 1–2 shift valve plunger
10 Rear oil screen	34 1–2 shift valve spring
11 Magnet	35 Front end plate
12 Lower valve body	36 D-1 and D-2 control spacer
13 Primary regulator valve	37 D-1 and D-2 control valve
14 Manual valve	38 D-1 and D-2 control spacer stop
15 Secondary regulator valve	39 Rear end plate
16 Secondary regulator valve spring	40 1–2 shift valve
17 Primary regulator valve sleeve	41 2–3 shift valve
18 Primary regulator valve spring	42 2–3 shift valve spring
19 Throttle valve stop	43 D-1 and D-2 control valve body
20 Throttle valve return spring	44 Separator Plate
21 Throttle valve dowel pin	45 Upper valve body
22 Throttle valve	46 Detent spring
23 Throttle valve spring	47 Detent ball
24 Downshift valve	48 Throttle control cam

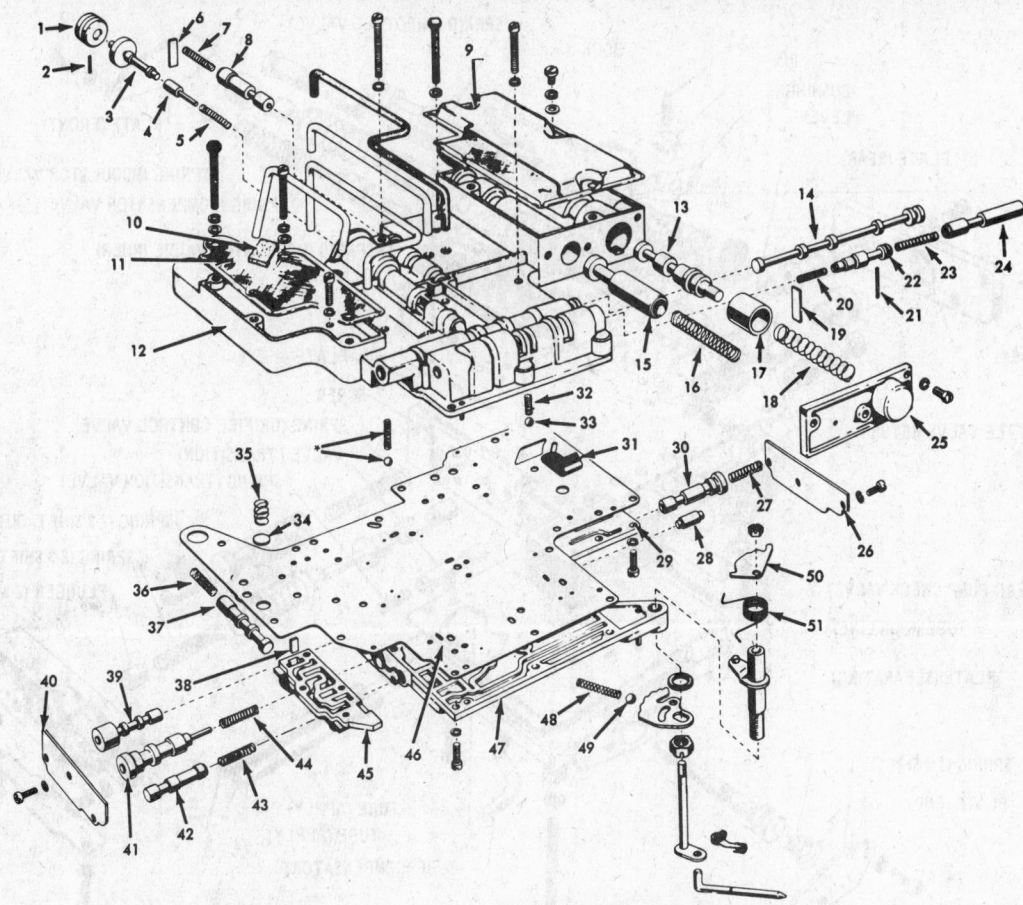

Control valve assembly sequence—1965-66 6 cylinder (© American Motors)

1 Modulator valve spacer
2 Dowel pin
3 Modulator valve
4 Modulator valve plug
5 Modulator valve spring
6 Servo orifice control valve stop
7 Servo orifice control valve spring
8 Servo orifice control valve
9 Front oil screen
10 Magnet
11 Rear oil screen
12 Lower valve body
13 Primary regulator valve
14 Manual valve
15 Secondary regulator valve
16 Secondary regulator valve spring
17 Primary regulator valve sleeve
18 Primary regulator valve spring
19 Throttle valve stop
20 Throttle valve return spring
21 Throttle valve dowel pin
22 Throttle valve
23 Throttle valve spring
24 Downshift valve
25 Lower valve body end
26 Front end plate
27 1–2 shift valve spring

28 2–3 shift valve plunger
29 Oil tube plate
30 1–2 shift valve plunger
31 Front pump check valve
32 Converter out check spring
33 Converter out check ball
34 Rear pump check valve
35 Rear pump check valve spring
36 Front clutch orifice control valve spring
37 Front clutch orifice control valve
38 Front clutch orifice control valve stop
39 1–2 shift valve
40 Rear end plate
41 2–3 shift valve
42 Range control valve
43 Range control valve spring
44 2–3 shift valve spring
45 Range control valve body
46 Separator plate
47 Upper valve body
48 Detent spring
49 Detent ball
50 Throttle control cam
51 Throttle control cam return spring
*2–3 shift valve circuit ball and spring
shift-command only

Control Valve R&R—1964-66 V8 Models

1. Remove the oil pan, gasket, and screen.
2. Remove the large control pressure tube and the small compensator pressure tube from the pressure regulator and control valve assembly.
3. Loosen front servo cap screw, then depress downshift valve and move the cam past the downshift valve.
4. Disconnect the throttle control rod from the cam, then remove the three control valve-to-case cap screws.
5. Disengage the front servo pressure tubes from the control valve assembly, then remove the assembly from the transmission.
6. To install, **reverse removal procedure**, tightening front servo cap screw to 30-35 ft. lbs., and control valve cap screws to 8-10 ft. lbs.

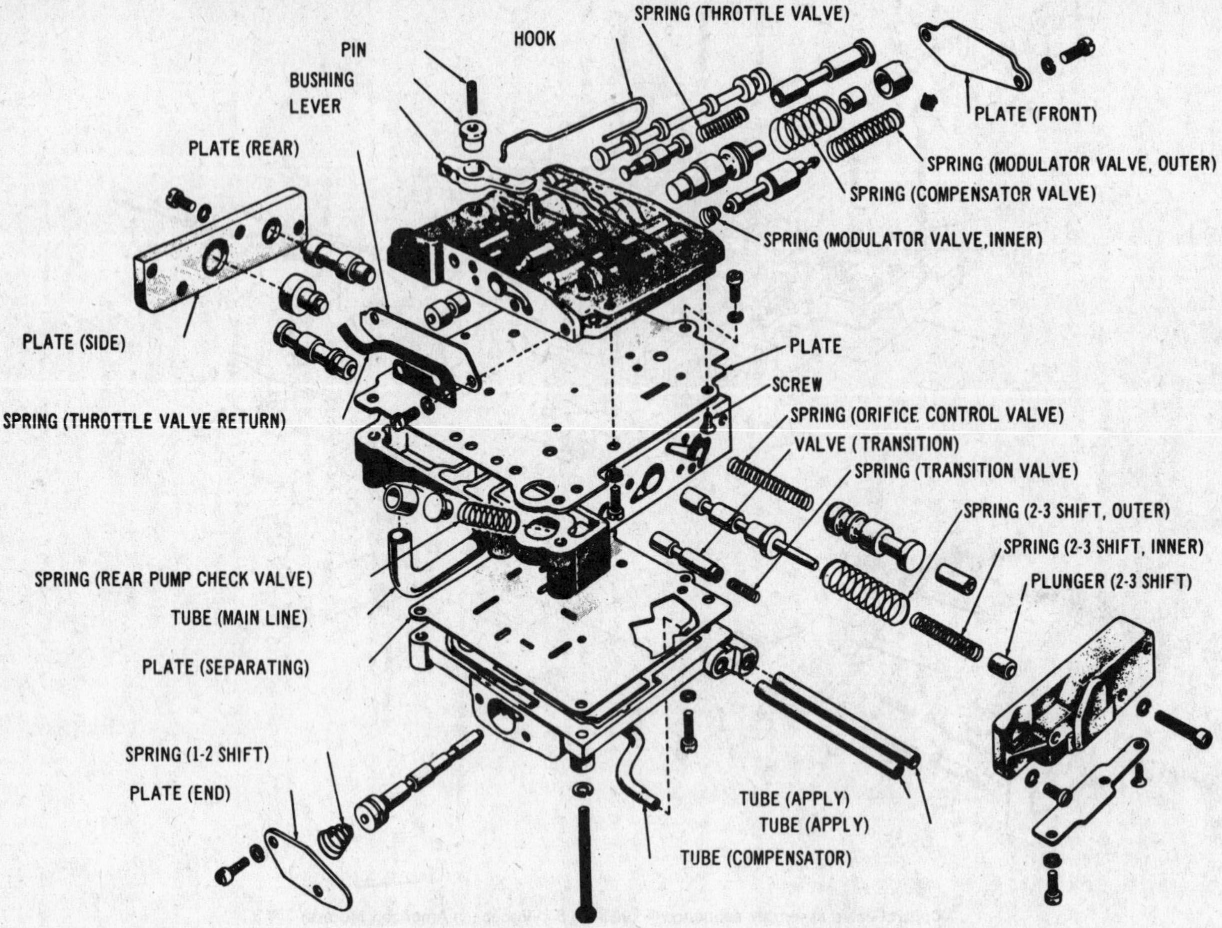

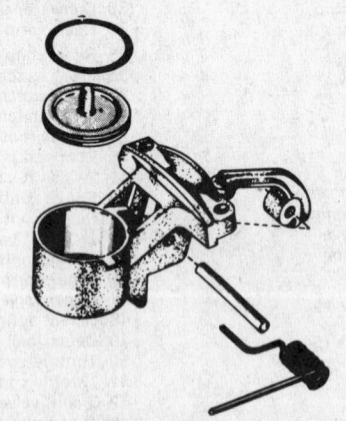

PIN
BUSHING
LEVER
HOOK
SPRING (THROTTLE VALVE)
PLATE (FRONT)
PLATE (REAR)
SPRING (MODULATOR VALVE, OUTER)
SPRING (COMPENSATOR VALVE)
SPRING (MODULATOR VALVE, INNER)
PLATE (SIDE)
PLATE
SCREW
SPRING (THROTTLE VALVE RETURN)
SPRING (ORIFICE CONTROL VALVE)
VALVE (TRANSITION)
SPRING (TRANSITION VALVE)
SPRING (2-3 SHIFT, OUTER)
SPRING (REAR PUMP CHECK VALVE)
SPRING (2-3 SHIFT, INNER)
TUBE (MAIN LINE)
PLUNGER (2-3 SHIFT)
PLATE (SEPARATING)
SPRING (1-2 SHIFT)
PLATE (END)
TUBE (APPLY)
TUBE (APPLY)
TUBE (COMPENSATOR)

Valve assembly—1964 with Shift-Command (© American Motors)

Rear servo—1964-70 light-duty transmission
(© American Motors)

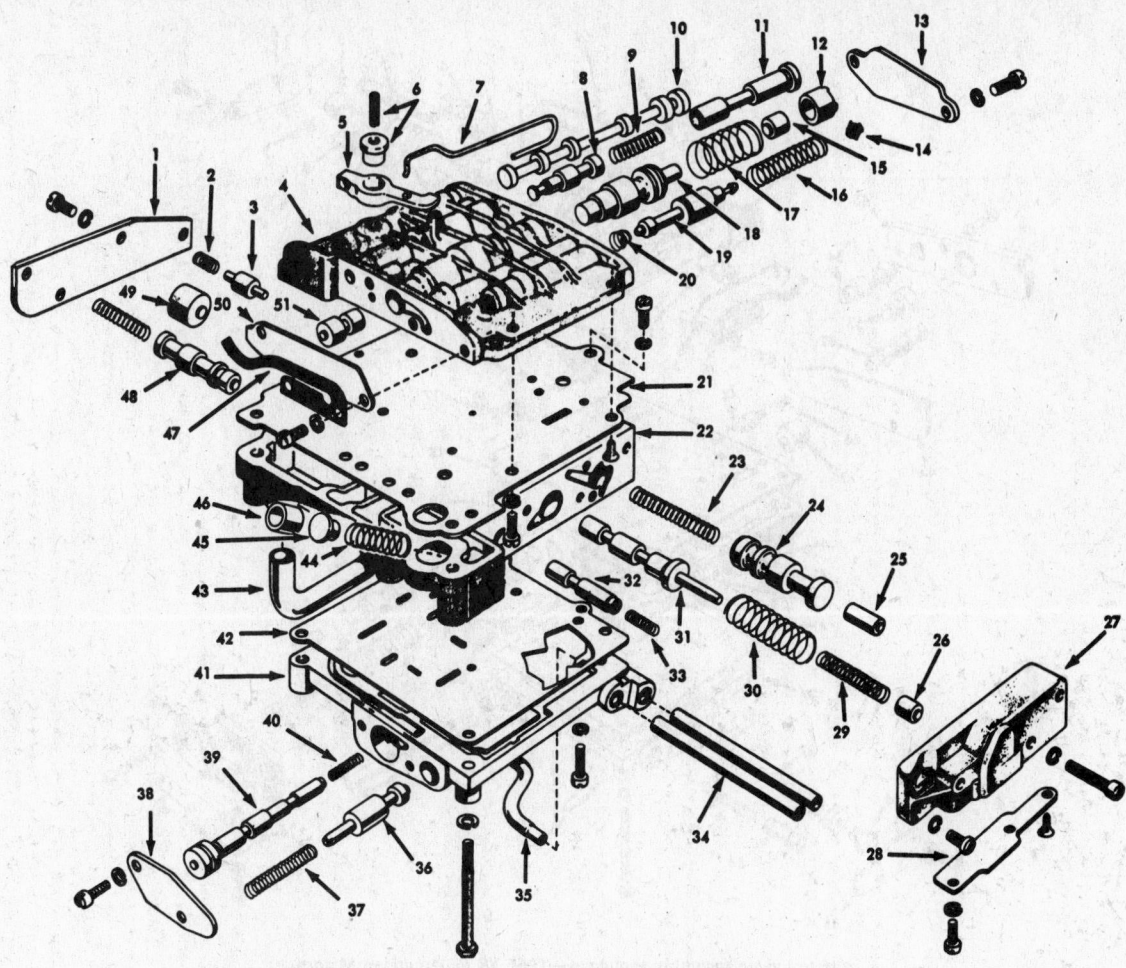

Control valve assembly sequence—1964 V8 (© American Motors)

1 Side plate
2 Range control valve spring
3 Range control valve
4 Upper body
5 Vacuum control lever
6 Retainer and pin
7 Vacuum control lever hook
8 Throttle valve
9 Throttle valve spring
10 Manual valve
11 Downshift valve
12 Compensator valve sleeve
13 Front plate
14 Modulator valve outer spring retainer
15 Compensator valve plug
16 Modulator valve spring, outer
17 Compensator valve spring
18 Compensator valve
19 Modulator valve
20 Modulator valve spring, inner
21 Body plate
22 Lower body
23 Orifice control valve spring
24 Orifice control valve
25 Orifice control valve plug
26 2–3 shift valve plug

27 End body
28 End body plate
29 2–3 shift valve spring, inner
30 2–3 shift valve spring, outer
31 2–3 shift valve
32 Transition valve
33 Transition valve spring
34 Front servo tubes
35 Compensator tube
36 Governor safety valve
37 Governor safety valve spring
38 Cover end plate
39 1–2 shift valve
40 1–2 shift valve spring
41 Cover—lower body
42 Separator plate
43 Control pressure tube
44 Rear pump check valve spring
45 Rear pump check valve
46 Rear pump check valve sleeve
47 Throttle valve return spring
48 Reverse plug
49 Governor plug
50 Rear plate
51 Compensator cut-off valve

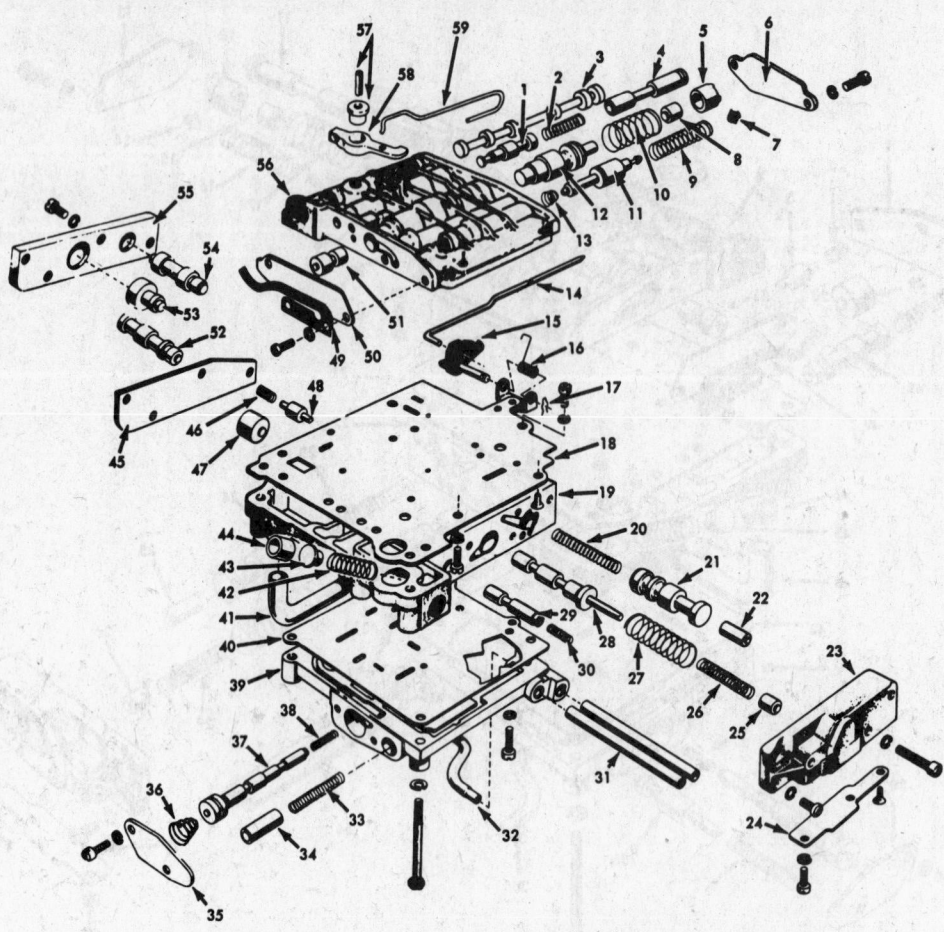

Control valve assembly sequence—1965 V8 (© American Motors)

1 Throttle valve	#33 Front servo release orifice valve spring
2 Throttle valve spring	#34 Front servo release orifice valve
3 Manual valve	35 Cover end plate
4 Downshift valve	%36 1–2 shift valve spring
5 Compensator valve sleeve	37 1–2 shift valve
6 Front plate	#38 1–2 shift valve spring
7 Modulator valve outer spring retainer	39 Cover—lower body
8 Compensator valve plug	40 Separator plate
9 Modulator valve spring, outer	41 Control pressure tube
10 Compensator valve spring	42 Rear pump check valve spring
11 Modulator valve	43 Rear pump check valve
12 Compensator valve	44 Rear pump check valve sleeve
13 Modulator valve spring, inner	#45 Side plate
**14 Throttle rod	#46 Range control valve spring
**15 Cam	#47 Governor plug
**16 Cam return spring	#48 Range control valve
**17 Retainer clip	49 Throttle valve return spring
18 Body plate	50 Rear plate
19 Lower body	51 Compensator cut-off valve
20 Orifice control valve spring	%52 Body plug
21 Orifice control valve	%53 Governor plug
22 Orifice control valve plug	%54 Range control valve
23 End body	%55 Side plate
24 End body plate	56 Upper body
25 2–3 shift valve plug	*57 Retainer and pin
26 2–3 shift valve spring, inner	*58 Vacuum control lever
27 2–3 shift valve spring, outer	*59 Vacuum control lever hook
28 2–3 shift valve	*Items used only with vacuum control
29 Transition valve	**Items used only with throttle cable control
30 Transition valve spring	#Items used only with column shift
31 Front servo tubes	%Items used with Shift-Command
32 Compensator tube	

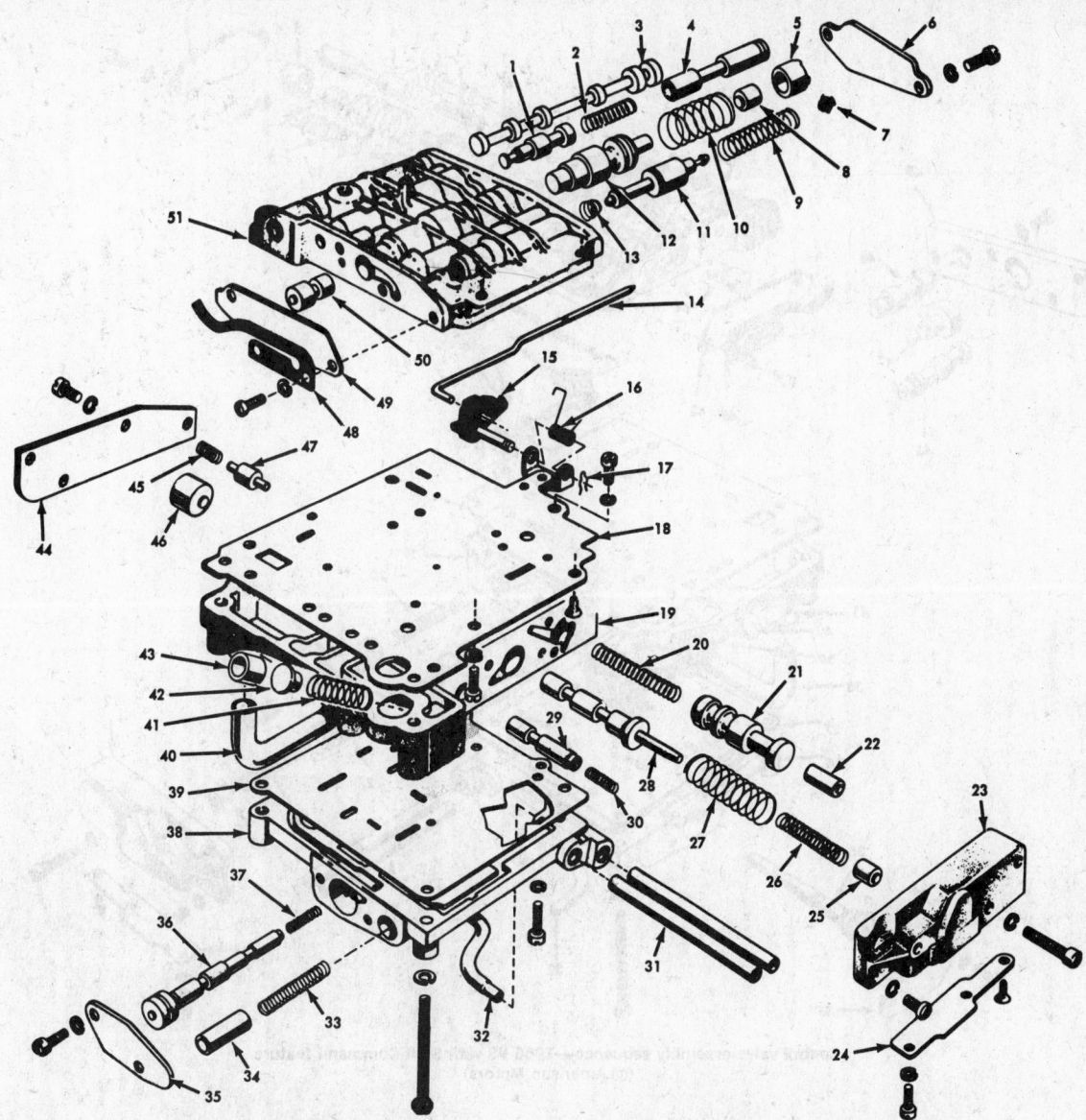

Control valve assembly sequence—1966 V8 with column shift (© American Motors)

1 Throttle valve	27 2–3 shift valve spring, outer
2 Throttle valve spring	28 2–3 shift valve
3 Manual valve	29 Transition valve
4 Downshift valve	30 Transition valve spring
5 Compensator valve sleeve	31 Front servo tubes
6 Front plate	32 Compensator tube
7 Modulator valve outer spring retainer	33 Front servo release orifice valve spring
8 Compensator valve plug	34 Front servo orifice valve
9 Modulator valve spring, outer	35 Cover end plate
10 Compensator valve spring	36 1–2 shift valve
11 Modulator valve	37 1–2 shift valve spring
12 Compensator valve	38 Cover—lower body
13 Modulator valve spring, inner	39 Separator plate
14 Throttle rod	40 Control pressure tube
15 Cam	41 Rear pump check valve spring
16 Cam return spring	42 Rear pump check valve
17 Retainer clip	43 Rear pump check valve sleeve
18 Body plate	44 Side plate
19 Lower body	45 Range control valve spring
20 Orifice control valve spring	46 Governor plug
21 Orifice control valve	47 Range control valve
22 Orifice control valve plug	48 Throttle valve return spring
23 End body	49 Rear plate
24 End body plate	50 Compensator cut-off valve
25 2–3 shift valve plug	51 Upper body
26 2–3 shift valve spring, inner	

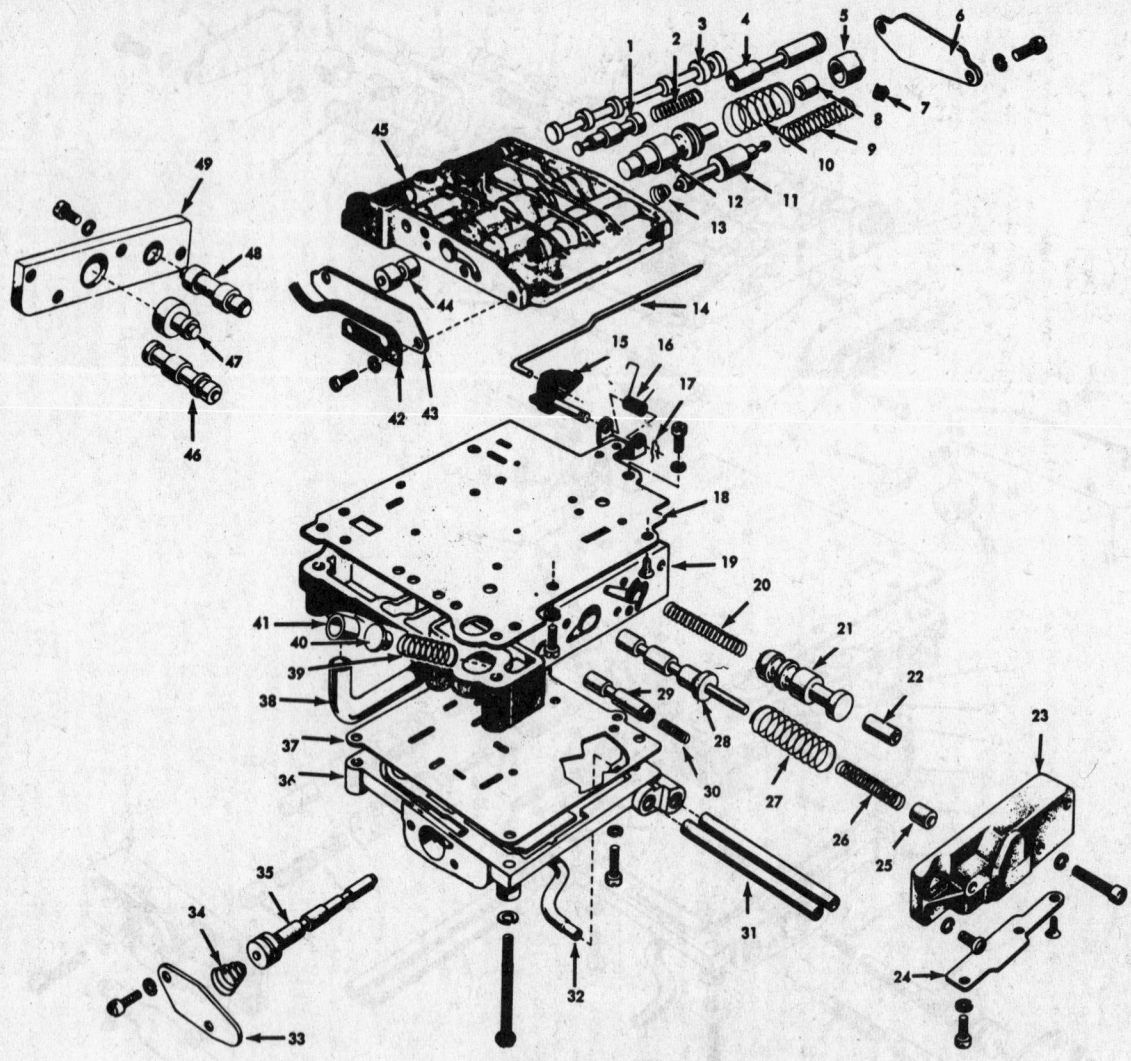

Control valve assembly sequence—1966 V8 with Shift-Command feature
(© American Motors)

1 Throttle valve	26 2–3 shift valve spring, inner
2 Throttle valve spring	27 2–3 shift valve spring, outer
3 Manual valve	28 2–3 shift valve
4 Downshift valve	29 Transition valve
5 Compensator valve sleeve	30 Transition valve spring
6 Front plate	31 Front servo tubes
7 Modulator valve outer spring retainer	32 Compensator tube
8 Compensator valve plug	33 Cover end plate
9 Modulator valve spring, outer	34 Shift valve spring
10 Compensator valve spring	35 1–2 shift valve
11 Modulator valve	36 Cover—lower body
12 Compensator valve	37 Separator plate
13 Modulator valve spring, inner	38 Control pressure tube
14 Throttle rod	39 Rear pump check valve spring
15 Cam	40 Rear pump check valve
16 Cam return spring	41 Rear pump check valve sleeve
17 Retainer clip	42 Throttle valve return spring
18 Body plate	43 Rear plate
19 Lower body	44 Compensator cut-off valve
20 Orifice control valve spring	45 Upper body
21 Orifice control valve	46 Body plug
22 Orifice control valve plug	47 Governor plug
23 End body	48 Range control valve
24 End body plate	49 Side plate
25 2–3 shift valve plug	

Pressure Regulator Valve R&R —1964-66 V8 Models

1. Remove the oil pan, gasket, and screen.
2. Remove the lubrication tube from the pressure regulator valve assembly and the rear pump.
3. Remove the large control pressure tube and the small compensator tube from the pressure regulator valve and control valve assembly.
4. Hold in the spring retainer, then disengage the tabs and slowly release the pressure—the springs and pilots then will come out.
5. Pull the converter pressure regulator valve (the small one) out about ½ in. and drain the converter fluid into a pan.
6. Remove the pressure regulator screws and the regulator.
7. To install, reverse removal procedure, tightening screws to 17-22 ft. lbs.

Control Shaft Inner Levers or Manual Control Shaft Seals R&R—1964-70

1. Disconnect filler tube and drain transmission oil.
2. Remove oil pan and gasket.
   NOTE: and throttle control cable, if so equipped.
3. Remove control valve assembly.
4. Remove the shift rod from the outer manual lever.

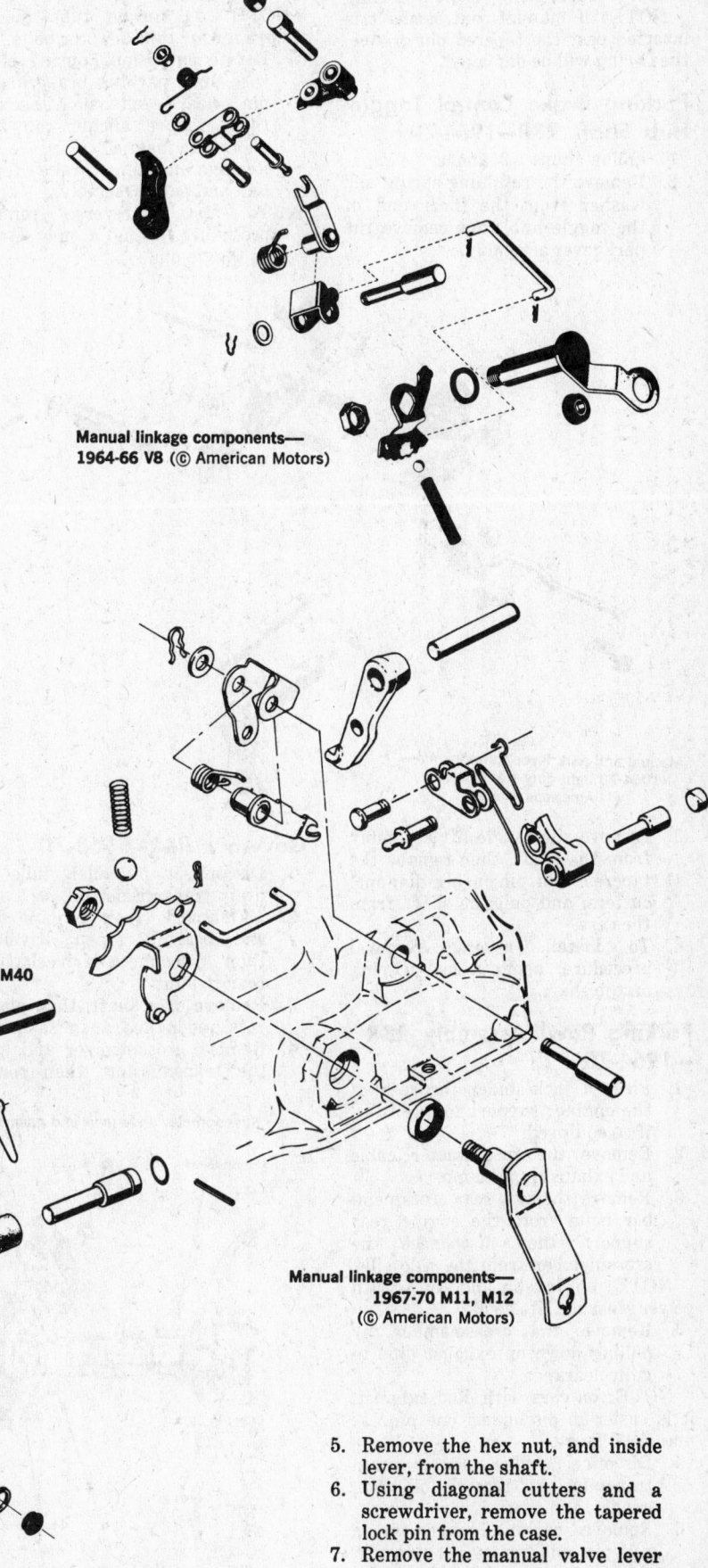

Manual linkage components— 1964-66 V8 (© American Motors)

Manual linkage and park lever—1967-70 M36, M37, M40 (© American Motors)

Manual linkage components— 1967-70 M11, M12 (© American Motors)

5. Remove the hex nut, and inside lever, from the shaft.
6. Using diagonal cutters and a screwdriver, remove the tapered lock pin from the case.
7. Remove the manual valve lever shaft from the case.

8. To install, reverse removal procedure, using a new O-ring.

NOTE: if manual control shaft is inserted past the tapered pin groove, the O-ring will be damaged.

Parking Brake Control Toggle Hub Shaft R&R—1964-70

1. Follow Steps 1-3, above.
2. Remove the retaining spring and washer from the inner end of the toggle hub, then remove the park lever assembly.

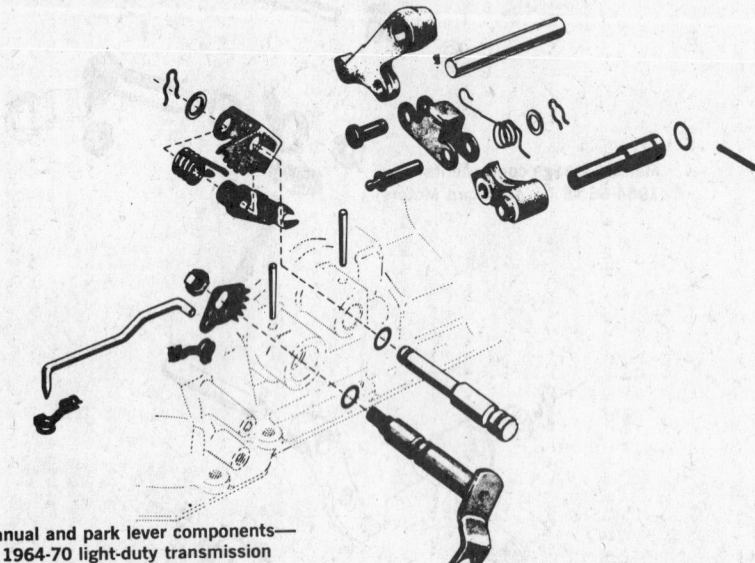

Manual and park lever components—1964-70 light-duty transmission
(© American Motors)

3. Remove the toggle lift assembly from the shaft, then remove the tapered lock pin, using diagonal cutters, and pull the shaft from the case.
4. To install, reverse removal procedure, using a new O-ring on the shaft.

Parking Pawl Assembly R&R —1964-70

1. Place a jack under the rear of the engine; support torque tube, if so equipped.
2. Remove the speedometer cable and exhaust pipe clamp.
3. Remove the two rear crossmember bolts from the engine rear support, then disconnect the crossmember from the side sills.

NOTE: on Javelin and AMX with power steering, lift hood.

4. Remove the crossmember by pulling down on exhaust pipe to gain clearance.

NOTE: on cars with dual exhaust, it is easier to disconnect one pipe at manifold flange.

5. Disconnect filler tube, drain transmission; remove oil pan, gasket, and control valve.
6. Remove parking brake toggle roll pin and remove toggle pin, then remove rear extension

housing-to-case cap screws and lockwashers.

NOTE: on torque tube models, remove torque tube adapter bolts.

7. Rotate extension housing clockwise until parking brake anchor pin clears extension housing, then remove anchor pin with pliers or a magnet.
8. Remove parking brake toggle link and pawl assembly.
9. To install, reverse removal procedure, using a new O-ring on toggle pin.

Governor R&R—1968-70

1. Disconnect oil filler tube and drain transmission oil.
2. Matchmark rear U-joint yoke and bearing (open driveline), then disconnect driveshaft at rear U-joint.
3. Remove driveshaft, then place a jack under the transmission.
4. Remove crossmember and lower the transmission, then remove

the speedometer cable and gear from the extension housing.

5. Remove the extension housing and the governor body retaining bolts, then remove governor from sleeve.
6. Remove the retainer from the stem of the governor valve weight, then remove the spring, weight and valve from the governor body and inspect the weight stem for grooves. Grooved stems can be polished with crocus cloth.
7. Remove the side plate from the governor body and clean the body and oil passages.

NOTE: during assembly, tighten side plate bolts to 25 in. lbs. only—plate goes to rear of transmission.

8. To install, reverse removal procedure, tightening counterweight screws to 75 in. lbs. (for 1967-70) or 50-60 in. lbs. (for 1964-66) to prevent valve body distortion.

Governor R&R—1964-67

1. Remove inspection hole cover plate from extension housing.

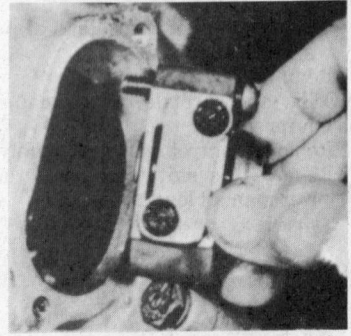

Governor removal
(© American Motors)

2. Rotate shaft until governor lines up with hole, then remove governor.
3. Follow Steps 6-8, above.

Speedometer drive gear and governor—1964-66 6 cylinder, 1967-70 M36, M37, M40
(© American Motors)

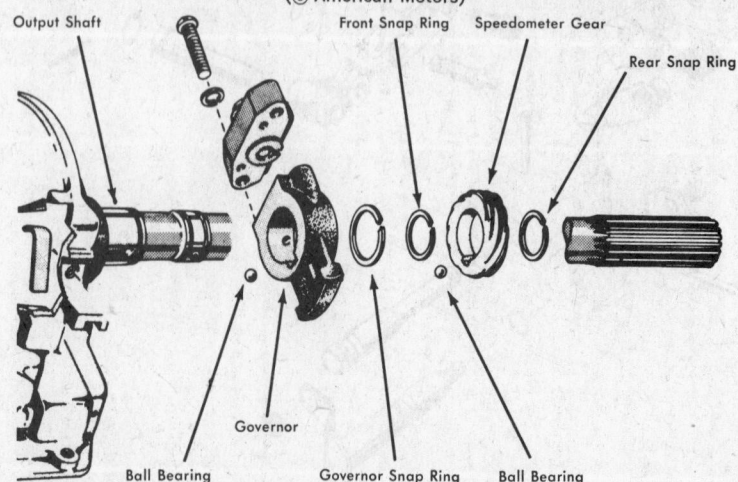

Output Shaft — Front Snap Ring — Speedometer Gear — Rear Snap Ring — Governor — Ball Bearing — Governor Snap Ring — Ball Bearing

Extension Housing Rear Oil Seal R&R—1964-70

1. Jack up the car and remove the driveshaft and/or torque tube from the rear of the extension housing.
2. Remove oil seal, using a slide hammer.
3. Position new seal in housing, with spring towards front.
4. Drive seal into place, using a large socket, then reverse Steps 1-2.

Service—Transmission Out of Car

See Car Section for assembly removal and installation.

End-Play Check

This procedure should be made before any of the main units are disturbed in the transmission.

Checking transmission end-play
(© American Motors)

1. Remove one front pump retaining screw and mount dial indicator so that contact rests against input shaft.
2. With front clutch assembly pried toward rear of case, set dial indicator to zero. Now, pry between parking gear and case to move units toward front of case. Note dial indicator reading. Limits should be .010-.029 in. (.015 in. preferred) for V8; .009-.032 for six-cylinder.

Sub-Assembly Removal

Front Pump—1964-70

1. Remove oil pressure tube, suction tube, and O-ring, then remove converter "in" and "out" tubes from pump.
2. Remove the oil seal from the pump body.
3. Remove pump cap screws, pump, and selective thrust washer.

Front Clutch, Rear Clutch, and Primary Sun Gear Assemblies—1964-70

1. Remove front and rear clutch assemblies and sun gear assembly by inserting a screwdriver be-

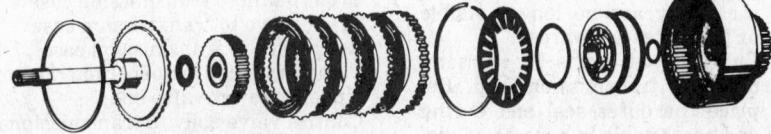

Front clutch—1964-70 typical
(© American Motors)

tween rear clutch drum and center support. Keep parts together as a unit.
2. Remove front band through front of case. (Remove from

Front sun gear and shaft—1964-69 typical
(1970 has thrust washer in place of Torrington bearing)
(© American Motors)

bottom of case on 1964-66 V8. 1967 343 V8, and 1968-70 290 4-v, 343, and 390 V8 models.)

Center Support and Planet Carrier Assemblies—1964-70

1. Remove two cap screws from sides of case, then remove center support, sprag clutch, and planet carrier.
2. Remove rear band through front of case. (Remove through rear of case on 1964-66 V8, 1967 343 V8, and 1968-70 290 4-v, 343, and 390 V8 models.)

NOTE: bands are interchangeable on 1964 6-cylinder models. Bands do not interchange thereafter and can be identified on 1965-66 6-cylinder models by the red lining on front band, brown lining on rear band. From 1967-70, the bands are not interchangeable and can be identified by the one-piece construction of the front band and three-piece construction of the rear band.

Rear Extension Housing and Output Shaft—1964-70

1. Remove extension housing cap screws, then remove extension housing and gasket.

2. Remove the rear output shaft snap-ring, then remove speedometer drive gear, ball bearing, and front snap-ring.
3. Remove governor snap-ring, governor assembly, and ball bearing from output shaft.

4. Remove the six large screws from the rear case adapter, then remove adapter.
5. Remove the three oil rings from the output shaft, then remove output shaft and rear thrust washer through the front end of the case.

Rear Pump—1964-66

1. Remove the six large Fillister head screws and the small Fillister head screw from the rear pump body.
2. Remove rear pump body and driven gear.
3. Remove the three oil rings from the output shaft, then the rear pump drive gear, drive gear key, and rear pump body plate.
4. Remove the output shaft and rear thrust washer through the front end of the case.

Component Reconditioning

Front Clutch Disassembly and Assembly—1964-66 V8; 1967 343 V8; 1968-70 290 4-v, 304 4-v, 343, 360, 390 V8

1. Remove cover, snap-ring, then input shaft and fiber washer from hub.
2. Remove the four driven and three drive plates, pressure plate, and hub from the drum.
3. Using spring compressor on release spring, remove snap-ring and release spring.

NOTE: front cover snap-ring and release spring snap-ring are not the same—the thicker snap-ring goes against the release spring.

Output shaft assembly sequence—1964-70 heavy duty transmission (© American Motors)

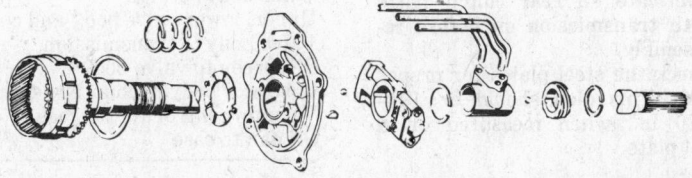

4. Force the piston from drum bore by applying air pressure to the clutch apply hole.
5. The outer seal and inner O-ring may now be removed.
6. Check all parts and replace as needed. The lined plates should be replaced when the grooves in the lining are no longer visible or the plates are warped.
7. Lubricate all parts in reassembly with transmission fluid. Replace the outer seal and O-ring and reassemble in reverse of disassembly procedure.

Front Clutch Disassembly and Assembly—1964-66 6 Cyl.; 1967 6 Cyl. and 290 V8; 1968-70 6 Cyl., 290 2-v V8

1. With front and rear clutch and planet carrier on bench, lift front clutch and input shaft from rear clutch.
2. Remove input shaft snap-ring, then input shaft.
3. Remove hub and fiber thrust washer, drive and driven plates and pressure plate from front clutch cylinder.
4. Remove snap-ring from return spring, then remove spring.
5. Using air pressure at oil apply hole, blow annular piston from cylinder.
6. Inspect all parts and replace seal and O-ring, together with any other required parts.
7. Reverse above sequences in re-installation.

Rear Clutch Disassembly—All

1. Remove seal rings and thrust washers from sun gear shaft and lift from shaft. Remove rear clutch drum.
2. Remove sun gear from planet carrier.
3. Remove snap-ring from rear drum and remove drive, driven and pressure plates.
4. Using spring compressor on release spring, remove snap-ring and release spring.
5. Apply air pressure and remove piston from drum.
6. Remove piston seal and O-ring.
7. Inspect all parts and replace as required. Always renew oil seal and O-ring.

Rear Clutch Assembly—1964-66 V8; 1967 343 V8; 1968-70 290 4-v, 304 4-v, 343, 360, 390 V8

1. Lubricate all rear clutch parts with transmission oil before reassembly.
2. Check the steel plates for proper dish. This dish should be .010-.020 in., when measured on a flat plate.

Torque Specifications

V8 up to and incl. 1966

	Foot-Pounds	Inch-Pounds
Converter to drive plate stud nuts	23-28	
Transmission case to converter housing	40-45	
Front pump to transmission case	17-22	
Front servo to transmission case	30-35	
Rear servo to transmission case	40-45	
Planetary center support to transmission case	20-25	
Upper to lower valve body	4-6	
Control valve body to transmission case	8-10	
Pressure regulator to transmission case	17-22	
Extension to transmission case	28-38	
Oil pan to transmission case	10-20	
Case assembly—gauge hole plug	10-15	
Rear servo adjusting screw locknut	35-40	
Front servo adjusting screw locknut	20-25	
Detent lever attaching nut	35-40	
Front pump cover screws		25-35
Rear pump cover (1/4—20)		80-90
Rear pump cover (10—24)		25-35
Governor inspection cover		50-60
Transmission vent assembly	7-10	
Governor valve body to counterweight		50-60
Governor valve body cover		20-30
Pressure regulator cover screws		20-30
Control valve body plug	10-14	
Control valve lower body plug	7-15	

M36, M37, M40 and 6 Cyl. up to and incl. 1966

	Foot-Pounds
Converter to drive plate	30-35
Transmission case to converter housing	17-22
Rear extension to case	30-40
Oil pan to case	10-20
Front servo to case	9-12
Rear servo to case	20
Pump adapter to front pump housing	17-22
Pump adapter to case	10-18
Rear pump to case	6.2*
Rear oil screen	6.2
Governor valve body to counter weight	6.2
Governor valve body cover to governor	2-3
Case line pressure plug	6-10
Rear case adapter to case	6.2★
Center support to case	20
Valve body to case (1/4—20)	6.2
Valve body screws (10—24)	2-3

* Up to and including 1966 only.
★ After 1967.

1967-70 M11 and M12

	Foot-Pounds	Inch-Pounds
Converter to flex plate	35	
Converter housing to engine	28	
Transmission to converter housing	55	
Case line pressure plug	15	
Front pump assembly to pump body	20	
Front pump assembly to transmission case	20	
Manual control lever to shaft	45	
Center support to case	25	
Front servo adjusting screw locknut	20	
Front servo to case	35	
Extension housing to case	35	
Valve body screws		30
Upper, lower valve body and cover bolts	10	
Valve body to transmission	10	
Oil screen to valve body		30
Governor body to counterweight		75
Vacuum control unit to case	15	
Oil pan to case	15	

Rear clutch components—1964-70 (© American Motors)

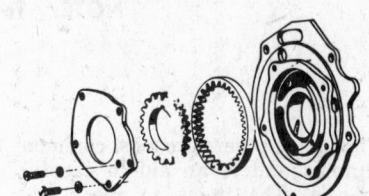

Sprag clutch and planet carrier—
typical 1964-70
(© American Motors)

3. Install piston. Press spring and retainer into position and install snap-ring.

4. Install the drive and driven plates and the pressure plate, then install the snap-ring. (Note: the dished plates are installed with the dished side up and the pressure plate with the smooth side down.)

5. Follow the assembly procedure in reverse of the removal procedure.

NOTE: side clearance between grooves and new cast-iron rings must be less than 0.005 in. to prevent pressure loss.

Rear Clutch Assembly 1964-70 6 Cyl.; 1967 290 V8; 1968-70 290 2-v V8

1. Lubricate all parts with transmission oil before reassembly.

2. Check the steel plates for proper dish. This dish should be .010-.015 in., when measured on a flat surface.

3. Be sure that the 30 needle bearings are properly installed into the drum. Use petrolatum to hold in place during assembly.

4. Place small O-ring on cylinder hub and seal ring on the annular piston, then place in piston installing tool.

5. Place tool in rear clutch drum, then drive into drum.

6. Install return spring, retainer and snap-ring into drum and compress spring. Then, install snap-ring.

7. Install drive and driven plates and pressure plate. (Note: the dished plates are installed with the dished side up.)

8. Install pressure plate and snap-ring.

9. Place sun gear shaft in planet carrier, with one Torrington bearing on top side of sun gear, and install clutch assembly onto shaft.

10. Install two oil seal rings onto sun gear shaft.

11. Install steel and bronze thrust washers, using petrolatum to hold them in place.

12. Install front clutch cylinder onto shaft, aligning it with plates in rear drum.

13. Install front clutch hub and fiber thrust washer.

14. Install front clutch pressure plate, with smooth side up, then install drive and driven plates.

15. Install input shaft and snap-ring.

Sprag Clutch—1964-70 All

1. A sprag clutch is incorporated in the pinion carrier. Twenty-four sprags are installed into the clutch assembly.

2. Fit the outer race into the pinion carrier assembly.

NOTE: planet carrier should free-wheel in counterclockwise rotation, lock-up clockwise, when viewed from rear.

Front and Rear Pumps

Normal disassembly and inspections are required for scratches, scores and excessive wear. See illustrations for reassembly sequences.

Rear Pump—1964-66 V8
(© American Motors)

Front pump components—1964-70 (© American Motors)

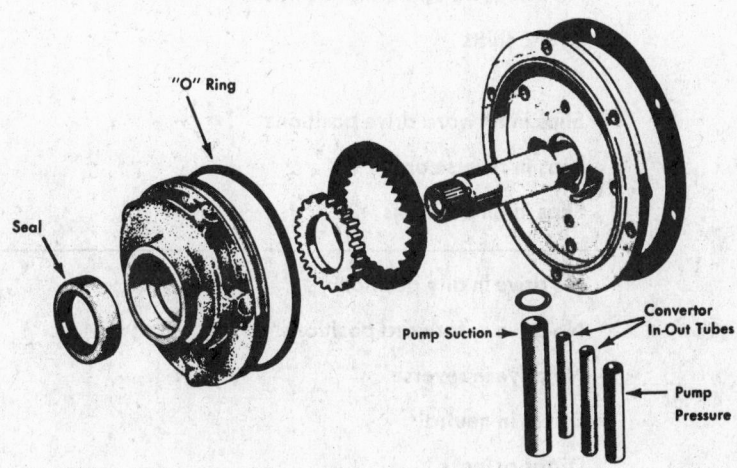

"O" Ring

Seal

Pump Suction

Convertor In-Out Tubes

Pump Pressure

Rear pump components—1964-66 (1965-66 has no plate) (© American Motors)

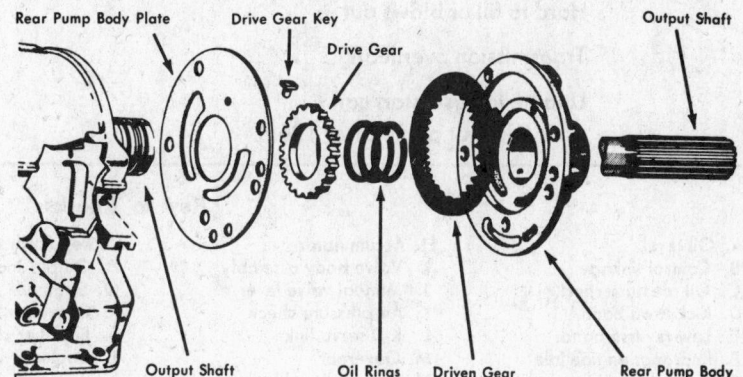

Rear Pump Body Plate

Drive Gear Key

Drive Gear

Output Shaft

Output Shaft

Oil Rings

Driven Gear

Rear Pump Body

Chrysler Corporation

Torqueflite

Application
Chrysler, 1964–70
Challenger 1970

Dart, 1964–70
Dodge, 1964–70
Barracuda 1964–70

Valiant, 1964–70
Plymouth, 1964–70

NOTE: Torqueflite Transmissions have no rear pump starting with 1966 Models.

Diagnosis

This guide covers the most common symptoms and is an aid to careful diagnosis. The items to check are listed in sequence to be followed for quickest results. Follow the checks in the order given for the particular transmission type.

TROUBLE SYMPTOMS	Items to Check In Car	Out of Car
Harsh N to D or N to R shift	CDEFGIJ	ab
Delayed Shift—N to D	ACIHKJ	ca
Runaway on upshift and 3-2 kickdown	ABCDLIHK	b
Harsh upshift and 3-2 kickdown	BCDLIHJ	b
No upshift	ABCDLIHKM	b
No kickdown on normal downshift	ABNCDLIHKM	d
Erratic shifts	ABNCFOPQ IKMJ	c
Slips in forward drive positions	ACIHK	abd
Slips in reverse only	CEGIK	
Slips in all positions	ACOIK	e
No drive in any position	ACOQIKJ	ca
No drive in forward positions	CDOLIHK	abd
No drive in reverse	CEGKMI	b
Drives in neutral	NIJ	a
Drags or locks	DELG	abfd
Noises	AORPMIJ	a
Hard to fill or blows out	AORSTQI	
Transmission overheats	ADEORTMI	abe
Unable to push-start car (No rear pump after 1965)	ACEGI	g

Key to Checks

A. Oil level
B. Control linkage
C. Oil pressure check
D. Kickdown band
E. Low-reverse band
F. Improper engine idle
G. Servo linkage

H. Accumulator
I. Valve body assembly
J. Manual valve lever
K. Air pressure check
L. K-D servo link
M. Governor
N. Gear shift cable

O. Regulator valve and/or spring
P. Output shaft bushing
Q. Strainer
R. Converter control valve
S. Breather clogged
T. Cooler or lines

a. Front-kickdown clutch
b. Rear clutch
c. Front pump and/or sleeve
d. Overrunning clutch
e. Converter
f. Planetary
g. Rear pump (not after 1965)

Torqueflite Specifications

	A-904	*A-727*
Type—three speed with torque converter	Yes	Yes
Torque converter dia. std.	10¾ in.	11¾ in.
1968-70 High Perf.		10¾ in.
Converter oil capacity		
1964	18 pts.	20 pts.
1965	17 pts.	19½ pts.
1966-67	16 pts.	18½ pts.
1968-69 std.	15½ pts.	18½ pts.
1968-69 High Perf.		15½ pts.
1970 std.	17 pts.	19 pts.
1970 High Perf.		16½ pts.
Cooling method	Water	Water
Lubrication pump	Rotor type	Rotor type
Clutches (No. of Plates and Discs)		
Front plates 1964 170 six	3	
Front discs 1964 170 six	3	
Rear plates 1964 170 six	2	
Rear discs 1964 170 six	3	
Front plates 1964 225 six	4	4
Front discs 1964 225 six	4	4
Rear plates 1964 225 six	2	2
Rear discs 1964 225 six	3	3
Front plates 1964 all V8s		4
Front discs 1964 all V8s		4
Rear plates 1964 318 V8		2
Rear discs 1964 318 V8		3
Rear plates 1964 large V8s		3
Rear discs 1964 large V8s		3
Front plates 1965-70 170/225 six	3	
Front discs 1965-70 170/225 six	3	
Rear plates 1965-70 170/225 six	2	
Rear discs 1965-70 170/225 six	3	
Front plates 1965-69 273 V8	4	
Front discs 1965-69 273/V8	4	
Rear plates 1965-69 273/V8	3	
Rear discs 1965-69 273/V8	4	
Front plates 1965 police and taxi six		4
Front discs 1965 police and taxi six		4
Rear plates 1965 police and taxi six		2
Rear discs 1965 police and taxi six		3
Front plates 1965 318 V8		4
Front discs 1965 318 V8		4
Rear plates 1965 318 V8		2
Rear discs 1965 318 V8		3
Front plates 1965-66 large V8s		4
Front discs 1965-66 large V8s		4
Rear plates 1965-66 large V8s		3
Rear discs 1965-66 large V8		4
Front plates 1966 318 and police and taxi six		4
Front discs 1966 318 and police and taxi six		4
Rear plates 1966 318 and police and taxi six		3
Rear discs 1966 318 and police and taxi six		4
Front plates 1967-70 318 and police and taxi six		3
Front discs 1967-70 318 and police and taxi six		3
Rear plates 1967-70 318 and police and taxi six		3
Rear discs 1967-70 318 and police and taxi six		4
Front plates 1968-69 318 V8	4	
Front discs 1968-69 318 V8	4	
Rear plates 1968-69 318 V8	3	
Rear discs 1968-69 318 V8	4	
Front plates 1967-70 340, 383, 440 V8		4
Front discs 1967-70 340. 383, 440 V8		4
Rear plates 1967-70 340, 383, 440 V8		3
Rear discs 1967-70 340, 383, 440 V8		4
Front plates 1967-70 426 Hemi		5
Front discs 1967-70 426 Hemi		5
Rear plates 1967-70 426 Hemi		3
Rear discs 1967-70 426 Hemi		4
Pump Clearances		
End clearance 1964-65 front pump	0.001-0.0025 in.	0.001-0.0025 in.
End clearance 1965-65 rear pump	0.0015-0.003 in.	0.0015-0.003 in.
End clearance 1966	0.001-0.0025 in.	0.001-0.0025 in.
NOTE: after 1965, Torqueflites do not have rear pumps		
End clearance 1967-70	0.0015-0.003 in.	0.0015-0.003 in.
Outer rotor to case bore 1969-70	0.004-0.008 in.	0.004-0.008 in.
Outer rotor tip to inner rotor tip measured diametrically 1969-70	0.005-0.010 in.	0.005-0.010 in.
Planetary assembly end play 1969-70	0.006-0.033 in.	0.010-0.037 in.
Drive train end play		
1964	0.023-0.068 in.	0.030-0.069 in.
1965	0.023-0.089 in.	0.036-0.084 in.
1966-67	0.022-0.091 in.	0.036-0.084 in.
1968	0.030-0.089 in.	0.036-0.084 in.
1969-70	0.030-0.089 in.	0.037-0.084 in.
Clutch Plate Clearance		
1964-65 Front 3 disc	0.042-0.087 in.	
1964-65 front 4 disc	0.056-0.104 in.	
1964-65 rear 3 disc	0.018-0.036 in.	0.022-0.042 in.
1965 rear 4 disc	0.024-0.049 in.	0.026-0.054 in.
1964-66 front		0.024-0.123 in.
1966-70 front 3 disc	0.042-0.087 in.	
1966-70 front 4 disc	0.056-0.104 in.	
1966 rear 3 disc	0.026-0.055 in.	
1966 rear 4 disc	0.032-0.055 in.	
1966-67 rear		0.037-0.060 in.
1967-70 rear	0.032-0.055 in.	
1967-70 front 3 disc		0.036-0.086 in.
1967-70 front 4 disc		0.024-0.125 in.
1968-70 front, 4 disc, high perf.		0.066-0.123 in.
1967 front, 5 disc, Hemi		0.029-0.151 in.
1968-70 front, 5 disc, Hemi		0.022-0.079 in.
1968-70 rear		0.025-0.045 in.
Snap Rings, Front and Rear Clutches		
Rear (selective)	0.060-0.062 in.	0.060-0.062 in.
	0.068-0.070 in.	0.074-0.076 in.
	0.076-0.78 in.	0.088-0.090 in.
Output shaft (forward end) 1965-70	0.040-0.044 in.	0.048-0.052 in.
	0.048-0.052 in.	0.055-0.059 in.
	0.059-0.065 in.	0.062-0.066 in.
Output shaft bearing 1964		0.086-0.088 in.
Output shaft bearing 1965	0.061-0.063 in.	
Thrust Washers		
Output shaft to input shaft (selective)	0.052-0.054 in. (natural) 0.068-0.070 in. (red) 0.083-0.085 in. (black)	
Reaction shaft support to front clutch retainer (selective)		0.061-0.063 in. (green) 0.084-0.086 in. (red) 0.102-0.104 in. (yellow)

	A-904	A-727
Front planetary gear to driving shell	0.060-0.062 in.	
Driving shell (steel) (2)	0.034-0.036 in.	
Rear planetary gear to driving shell	0.060-0.062 in.	
Front annulus gear support	0.121-0.125 in.	
Front clutch to rear clutch	0.043-0.045 in.	
Front clutch to reaction shaft support	0.043-0.045 in.	
Driving shell thrust plate		0.034-0.036 in.
Rear planetary gear to driving shell		0.062-0.064 in.
Output shaft to input shaft		0.062-0.064 in.
Front planetary gear to annulus gear		0.062-0.064 in.
Front annulus gear to driving shell		0.062-0.064 in.
Front clutch piston retainer to rear		
Clutch piston retainer		0.061-0.063 in.
Rear planetary gear to annulus gear 1967-70		0.034-0.036 in.

Transmission Applications

Model	Year and Engines
A 904	1964 170 and 225 sixes
A 904 G	1965-70 170, 198, and 225 sixes
A 904 LA	1965-68 273 V8 and 1969-70 318 V8
A 904 A	1969 273 V8
A 727	1964 all V8s and police and taxi six
A 727 A	1965-70 318 V8, 1968-70 340 V8, and 1965-68 225 police and taxi six
A 727 RG	1969-70 225 police and taxi six
A 727 B	1965-70 all B block engines, 361, 383, 413, 426, 426 Hemi, and 440 V8s

General

The Torqueflite transmission combines a torque converter with a fully automatic three-speed gear system. The converter housing and transmission case are combined into one aluminum casting. The transmission consists of two multiple-disc clutches, an overrunning clutch, two servos and bands, and two planetary gear sets. This provides three forward speeds and one reverse.

The torque converter is driven by the crankshaft, through a bolted-on, flexible driving plate. Cooling of the converter-transmission assembly is controlled by circulating the transmission fluid through a cooler core, located in the radiator lower tank. The torque converter assembly is a sealed unit, impractical to service in the field, except for cleaning.

For information relative to applica-

A-904 type Torqueflite transmission (© Chrysler Corp.)

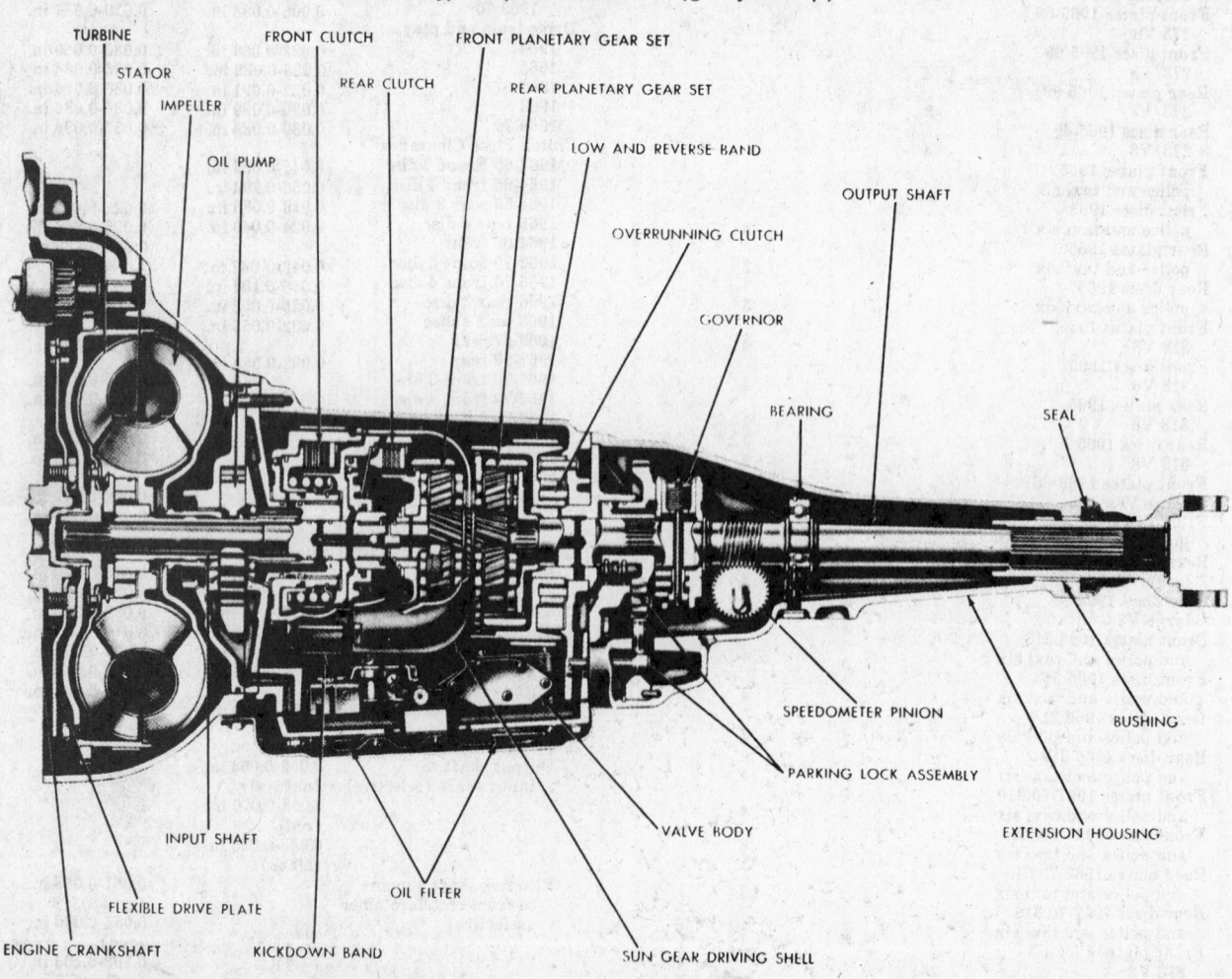

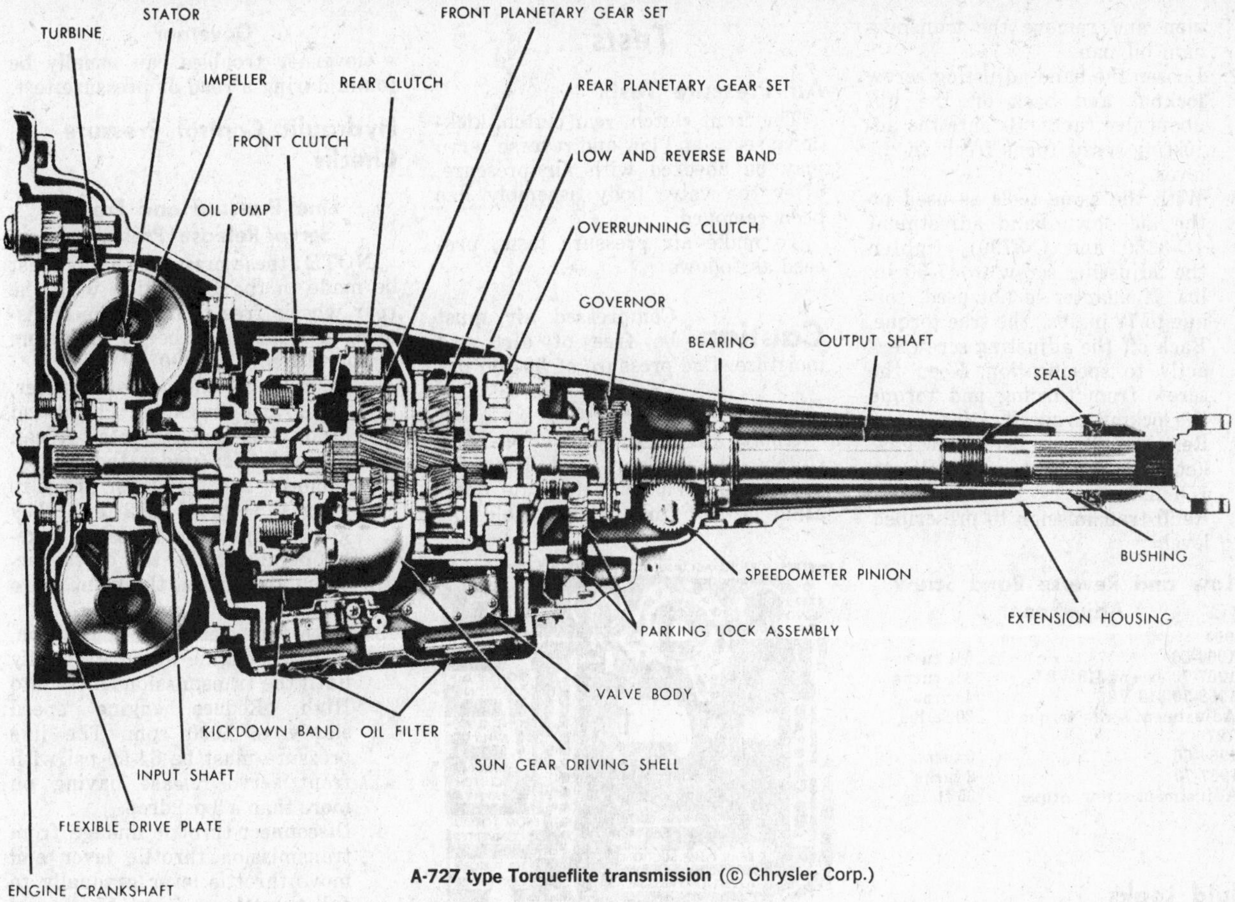

TURBINE

STATOR

IMPELLER

FRONT CLUTCH

OIL PUMP

FRONT CLUTCH

REAR CLUTCH

FRONT PLANETARY GEAR SET

REAR PLANETARY GEAR SET

LOW AND REVERSE BAND

OVERRUNNING CLUTCH

GOVERNOR

BEARING

OUTPUT SHAFT

SEALS

BUSHING

SPEEDOMETER PINION

EXTENSION HOUSING

PARKING LOCK ASSEMBLY

VALVE BODY

KICKDOWN BAND   OIL FILTER

SUN GEAR DRIVING SHELL

INPUT SHAFT

FLEXIBLE DRIVE PLATE

ENGINE CRANKSHAFT

A-727 type Torqueflite transmission (© Chrysler Corp.)

tion, such as linkage adjustments, and unit removal and replacement, refer to the car section.

Starting in 1966, the rear pump has been eliminated from the Torqueflite transmission assembly.

Operating differences are essentially, that the governor receives oil from the manual control valve rather than from the rear pump, then distributes it to the shift valves, according to the pressures supplied to the governor.

Band Adjustments

Kickdown Band

The kickdown band adjusting screw is located on the left-hand side of the transmission case near the throttle lever shaft.

1. Loosen the locknut and back off about five turns. Be sure the adjusting screw is free in the case.
2. Using wrench, tool C-3380 with adapter C-3790, or similar tools, torque the adjusting screw to 47-50 in. lbs. If adapter is not used, tighten adjusting screw to 72 in. lbs. which is the true torque.
3. Back off the adjusting screw, exactly to specification. Keep the screw from turning, and torque the locknut to specification.

Kickdown Band Adjustments

A-904

1964-70 170 six	2⅝ turns
1964-70 198 and 225 six	2 turns
1965-69 273 V8	2 turns
1968-70 318 V8	2 turns

A-727

1964-70 sixes and V8 except Hemi	2 turns
1968-70 Hemi V8	1½ turns

A-904 kickdown adjusting screw torque	25 ft. lbs.
A-727 kickdown adjusting screw torque	29 ft. lbs.

Low and Reverse Band

Access to the low and reverse band requires oil pan removal.

1. Raise the car, drain transmis-

Bottom view of transmission—pan removed (© Chrysler Corp.)

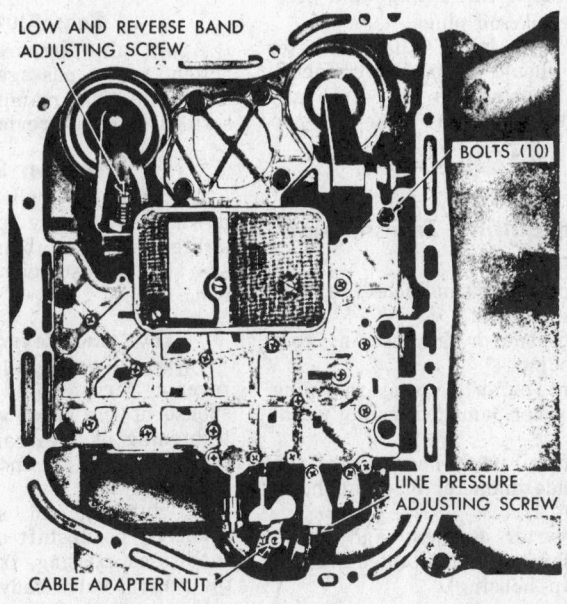

LOW AND REVERSE BAND ADJUSTING SCREW

BOLTS (10)

LINE PRESSURE ADJUSTING SCREW

CABLE ADAPTER NUT

sion and remove the transmission oil pan.

2. Loosen the band adjusting screw locknut and back off the nut about five turns. Be sure the adjusting screw turns freely in the lever.

3. With the same tools as used on the kickdown band adjustment (C-3380 and C-3790), tighten the adjusting screw to 47-50 in. lbs. If adapter is not used, torque to 72 in. lbs., the true torque.

4. Back off the adjusting screw exactly to specification. Keep the screw from turning and torque the locknut to specification.

5. Reinstall oil pan, using new gasket, and torque the pan bolts to 150 in. lbs.

6. Refill transmission to prescribed level.

Low and Reverse Band Screw Adjustment

A-904
1964-66	5¼ turns
1967-70 except 318 V8	3¼ turns
1968-70 318 V8	4 turns
Adjustment screw torque	20 ft. lbs.

A-727
1964-66	3 turns
1967-70	2 turns
Adjustment screw torque	35 ft. lbs.

Fluid Leaks

Some leaks that can normally be corrected without transmission removal are:

1. Transmission output shaft oil seal.
2. Extension housing gasket.
3. Speedometer pinion seal and cable seal.
4. Oil filter tube seal.
5. Oil pan gasket and drain plug.
6. Gearshift control cable seal.
7. Throttle shaft seal.
8. Neutral starting switch seal.
9. Oil cooler line fittings and pressure take-off plugs.

Oil found inside the converter housing should be positively identified as transmission oil before any major transmission work is performed.

Leaks Requiring Transmission Removal

1. Fractures or sand holes in transmission case.
2. Sand hole or fracture in front oil pump.
3. Front pump housing retaining screws or damaged sealing washers.
4. Front oil pump housing seal (on outside diameter of pump housing).
5. Converter assembly and impeller shaft oil seal (located in front pump housing).

Tests

Air Pressure Tests

The front clutch, rear clutch, kickdown servo and low and reverse servo may be checked with air pressure, after the valve body assembly has been removed.

To make air pressure tests, proceed as follows:

Caution Compressed air must be free of dirt and moisture. Use pressure of 30-100 psi.

Front Clutch

Apply air pressure to the front clutch apply passage and listen for a dull thud. This will indicate operation of the front clutch. Hold the

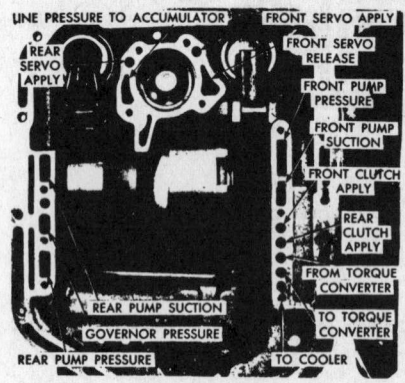

Air pressure checks
(© Chrysler Corp.)

air pressure at this point for a few seconds and check for excessive oil leaks.

NOTE: if a dull thud cannot be heard in the clutch, place finger tips on clutch housing and again apply air pressure. Movement of piston can be felt as clutch is applied.

Rear Clutch

Apply air pressure to the rear clutch apply passage and proceed in an identical manner as that described in the previous paragraph.

Kickdown Servo

Air pressure applied to the kickdown servo apply passage should tighten the front band. Spring tension should be sufficient to release the band.

Low and Reverse Servo

Direct air pressure into the low and reverse servo apply passage. Response of the servo will result in a tightening of the rear band. Spring tension should be enough to release the band.

If clutches and servos operate properly, no upshift or erratic shift conditions existing, trouble exists in the control valve body assembly.

Governor

Governor troubles can usually be found during a road or pressure test.

Hydraulic Control Pressure Checks

Line Pressure and Front Servo Release Pressure

NOTE: these pressure checks must be made in the D position with the rear wheels free to turn. The transmission fluid must be at operating temperature (150°-200° F).

1. Install an engine tachometer, then, raise the car on a hoist and locate the tachometer so it can be read from under the car.

2. Connect two 0-100 psi pressure gauges (tool C-3292 or other good gauges) to pressure take-off points at the top of the accumulator and at the front servo release.

3. With the selector in D position, increase engine speed gradually until the transmission shifts into High. Reduce engine speed slowly to 1000 rpm. The line pressure must be 54-60 psi with front servo release having no more than a 3 psi drop.

4. Disconnect throttle linkage from transmission throttle lever and move throttle lever gradually to full throttle position. Line pressure must rise to maximum of 90-96 psi just before or at kickdown into low gear. Front servo release pressure must follow line pressure up to kickdown point and should not be more than 3 psi below line pressure. If pressure is not 54-60 psi at 1000 rpm, adjust line pressure.

If line pressure is not as above, adjust the pressure as outlined under the heading: Hydraulic Control Pressure Adjustments—"Line Pressure."

If front servo release pressures are less than specified, and line pressures are within limits, there is excessive leakage in the front clutch and/or front servo circuits.•

Lubrication Pressures

A lubrication pressure check should be made when line pressure and front servo release pressures are checked.

1. Install a T fitting between the cooler return line fitting and the fitting hole in the transmission case at the rear left side of the transmission. Connect a 0-100 psi pressure gauge to the T-fitting.

2. At 1000 engine rpm, with throttle closed and transmission in High, lubrication pressure should be 5-15 psi. Lubrication pressure will approximately double as throttle is opened to maximum line pressure.

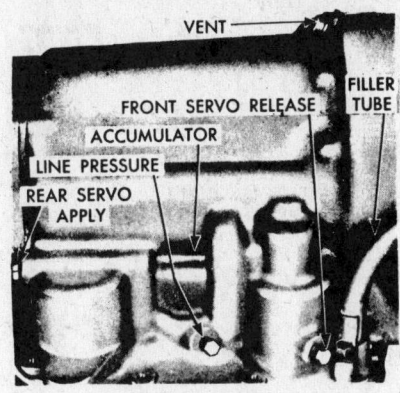

Pressure test locations (right side of case)
(© Chrysler Corp.)

Rear Servo Apply Pressure

1. Connect a 0-300 psi pressure gauge, tool C-3293 or its equivalent, to the apply pressure take-off point at the rear servo.
2. With the control in the R position, and the engine running at 1600 rpm, the reverse servo apply pressure should be 230-280 psi.

Governor Pressure

1. Connect a 0-100 psi gauge (same as the one used for line pressure and front servo release pressure) to the governor pressure take-off point. This location is at the lower left rear corner of the extension mounting flange.
2. Governor pressure should fall within limits in chart.

Pressure should change smoothly with car speeds.

If governor pressures are incorrect at the prescribed speeds, the governor valve and/or weights are probably sticking.

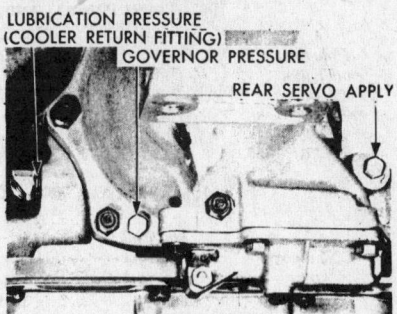

Pressure test locations (rear of case)
(© Chrysler Corp.)

Governor Pressure Chart

Axle Ratio to Vehicle Speed

SIX CYLINDERS
1964 VALIANT AND DART

170		225			Pressure Limits (PSI)
3.23	3.55	2.93	3.23	3.55	
16-17	14-15	17-19	15-17	14-15	15
29-39	27-35	29-38	26-34	24-31	45
59-66	54-60	60-69	55-64	50-57	75

FULL-SIZED DODGES AND PLYMOUTHS

				Pressure Limits (PSI)
9.93	3.23	3.31	3.55	
18-20	16-18	16-18	15-17	15
30-40	27-36	25-35	25-33	45
63-73	57-67	56-65	53-61	75

1965-66 ALL MODELS

6.50x13 Tires		6.95x14 Tires		7.35x14 Tires			Pressure Limits (PSI)
3.23	3.55	2.93	3.55	2.93	3.23	3.55	
16-17	14-16	17-19	14-16	17-19	15-17	14-16	15
36-45	33-41	35-46	30-37	36-47	33-42	30-38	50
59-67	53-61	61-70	50-58	62-71	56-65	51-59	75

1967

Valiant and Dart		Belvedere, Fury, and Coronet		Pressure Limits (PSI)
2.76	2.93	2.93	3.23	
17-20	16-19	17-19	15-17	15
44-52	36-44	37-45	33-41	50
70-77	61-67	62-69	56-62	75

1968-70

170			198 and 225			Pressure Limits (PSI)
2.76	2.93	3.23	2.76	2.93	3.23	
18-20	17-19	15-17	18-20	17-20	16-18	15
44-57	41-48	37-48	38-46	36-46	34-41	50
69-76	65-72	59-65	64-70	60-70	57-62	75

EIGHT CYLINDERS
1964 BELVEDERE, FURY AND CORONET 318, 361, 383 V8s

Standard		High Perf.		Pressure Limits (PSI)
2.76	3.23	2.93	3.23	
19-21	16-18	17-20	16-18	15
41-51	35-44	38-48	35-43	50
69-76	59-65	65-72	59-65	75

1964 DODGE 880, CHRYSLER, AND IMPERIAL

Newport and 880 8.00x14 Tires	New Yorker and 880 8.50x14 Tires	300	Imperial	Pressure Limits (PSI)
2.76	2.76	3.23	2.93	
20-22	20-23	17-19	20-23	15
44-53	45-54	37-45	44-53	50
73-80	75-82	62-69	74-81	75

1965-66 BELVEDERE, FURY, CORONET, AND DART

7.00x13		7.35x14		7.75x14		Pressure Limits (PSI)
2.93	3.23	2.93	3.23	2.93	3.23	
18-19	16-17	17-19	15-17	18-20	16-18	15
40-51	37-46	36-47	33-42	38-47	35-42	50
66-75	60-68	62-71	56-65	65-72	59-65	75

1965-66 POLARA, MONACO, AND 880

8.25x14	8.55x14	8.25x14	8.55x14	Pressure Limits (PSI)
2.76	2.76	3.23	3.23	
19-22	20-22	16-19	17-19	15
45-51	45-52	35-43	36-44	50
70-78	71-79	60-66	61-88	75

1965-66 CHRYSLER AND IMPERIAL

Newport	300	New Yorker	Imperial	Pressure Limits (PSI)
2.76	3.23	3.23	2.93	
19-22	17-19	17-19	19-22	15
42-51	36-44	36-44	43-51	50
70-78	61-68	61-68	71-78	75

1967 DART, VALIANT, AND BARRACUDA

2.93	3.23	Pressure Limits (PSI)
17-19	15-17	15
42-50	43-49	50
67-74	64-67	75

1967 CORONET, CHARGER, AND BELVEDERE

2.94	3.23	Pressure Limits (PSI)
17-19	15-18	15
43-51	39-46	50
69-76	63-69	75

1967 FURY, POLARA, AND MONACO

Polara and Fury 2.94	Fury 3.23	Polara and Monaco 2.76	Pressure Limits (PSI)
18-20	16-18	19-22	15
44-52	40-48	48-57	50
70-76	64-71	76-84	75

1967 CHRYSLER AND IMPERIAL

Newport 2.76	300 and New Yorker 2.76	Imperial 2.94	Pressure Limits (PSI)
19-22	19-22	19-22	15
48-57	49-58	48-57	50
76-84	78-86	77-85	75

1968-69 BELVEDERE, FURY, CORONET, CHARGER, VALIANT, DART AND BARRACUDA

273 and 318 V8s			340, 383, and 440 V8s			Pressure Limits (PSI)
2.76	2.93	3.23	2.76	3.23	3.55	
18-20	17-19	15-17	19-21	14-16	15-17	15
45-53	42-50	38-44	46-55	39-46	44-50	50
71-78	67-74	60-66	74-82	58-65	64-71	75

1968-69 POLARA AND MONACO

2.76	2.94	3.23	Pressure Limits (PSI)
19-22	18-20	16-18	15
48-57	44-52	40-48	50
76-84	71-78	64-71	75

1968-70 CHRYSLER AND IMPERIAL

Chrysler 2.76	3.23	Imperial 2.94	Pressure Limits (PSI)
20-22	17-19	19-22	15
48-57	41-49	48-57	50
77-85	66-73	77-85	75

1970

318, 383, and 440 V8s			Pressure Limits (PSI)
2.76	3.23	3.55	
19-21	15-17	14-17	15
46-55	44-50	39-51	50
74-82	64-71	58-71	75

HIGH PERFORMANCE V8s
1964 DODGE, PLYMOUTH, AND CHRYSLER

426 Wedge		Chrysler 300K	Pressure Limits (PSI)
3.23	3.91	3.23	
15-18	12-15	17-20	15
56-63	46-52	50-56	50
82-90	68-74	73-80	75

1965 DODGE AND PLYMOUTH 426 WEDGE

3.23	3.91	Pressure Limit (PSI)
16-18	13-15	15
45-51	37-42	50
67-73	55-60	75

1967 DODGE AND PLYMOUTH

426 Hemi 3.23	Pressure Limit (PSI)
19-22	15
54-61	50
79-87	75

1968-70 DODGE AND PLYMOUTH

426 Hemi 3.23	383 High Perf., 440 High Perf. and 440 Six Pack 3.23	Pressure Limit (PSI)
20-23	16-19	15
55-64	46-52	50
82-90	68-73	75

Hydraulic Control Pressure Adjustments

Line Pressure

An incorrect throttle pressure setting will cause incorrect line pressure even though line pressure adjustment is correct. Always inspect and correct throttle pressure adjustment before adjusting line pressure.

NOTE: before adjusting line pressure, measure distance between manual valve (valve in 1-low position) and line pressure adjusting screw. This measurement must be 1⅞ in. Correct by loosening spring retainer screws and repositioning spring retainer. The regulator valve may cock and hang up in its bore if spring retainer is out of position.

If line pressure is not correct remove valve body assembly to adjust. The correct adjustment is 1-5/16 in. measured from valve body to inner edge of adjusting nut. Vary adjustment slightly to obtain specified line pressure.

One complete turn of the adjusting screw (Allen head) changes closed throttle line pressure about 1.66 psi. Turning the screw counterclockwise increases pressure, clockwise decreases pressure.

Throttle Pressure

Because throttle pressures cannot be checked, exact adjustments should

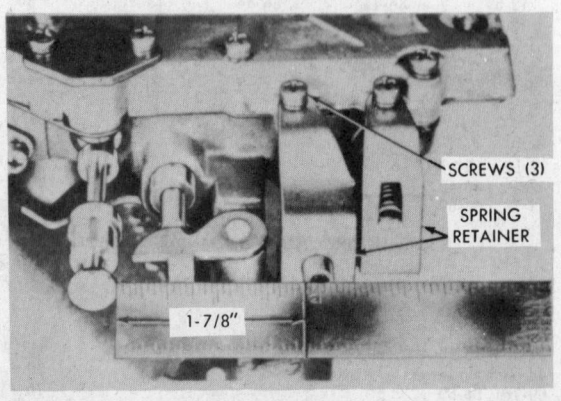

Measuring distance between manual valve and line pressure adjusting screw (© Chrysler Corp.)

SCREWS (3)
SPRING RETAINER
1-7/8"

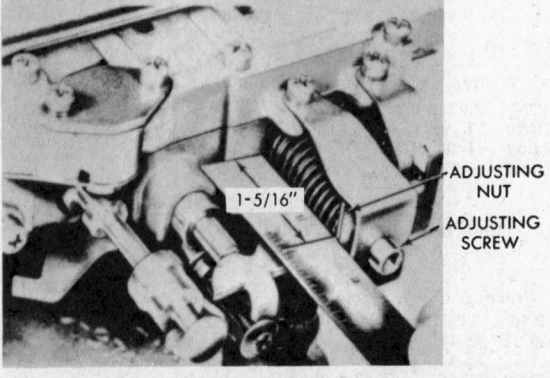

Line pressure adjustment (© Chrysler Corp.)

1-5/16"
ADJUSTING NUT
ADJUSTING SCREW

be checked and made correct whenever the valve body is disturbed.

1. Remove the valve body assembly, as outlined in a succeeding coverage entitled, Valve Body Assembly and Accumulator Piston.
2. Loosen throttle lever stop screw locknut and back off the screw about five turns.
3. Insert gauge pin of tool C-3763 between the throttle lever cam and the kickdown valve.
4. Push on the tool and compress the kickdown valve against its spring, so that the throttle valve is completely bottomed inside the valve body.
5. As the spring is being compressed, finger tighten the throttle lever stop screw against the throttle lever tang, with the lever cam touching the tool and the throttle valve bottomed. (Be sure the adjustment is made with the spring fully compressed and the valve bottomed in the valve body.)
6. Remove the tool and secure the stop screw locknut.

Service Operations in Car

Some sub-assemblies can be removed for repairs without removing the transmission from the car. Detailed reconditioning of sub-assemblies is covered further in the text.

Speedometer Pinion

Removal and Installation 1964-65

1. Remove screw and retainer which holds the cable to the extension housing. Carefully work the pinion and sleeve out of the housing.

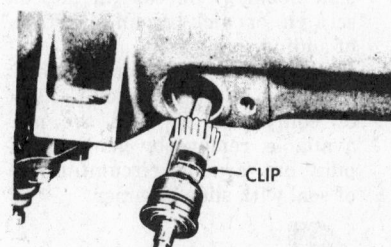

Removing or installing speedometer pinion (© Chrysler Corp.)

2. Replace the pinion and/or oil seal by prying the clip off the pinion and sliding the pinion and seal assembly from the cable.
3. If transmission fluid is found in the cable housing, replace the seal inside the pinion bore, then, slide the pinion over the end of the cable and secure it with the spring clip.
4. To install, push the pinion and sleeve into the extension housing as far as possible, then install the retainer screw. Tighten to 150 in. lbs.

Removal and Installation 1966-70

Rear axle gear ratio and tire size determine pinion gear size.

1. Remove bolt and retainer securing speedometer pinion adapter in extension housing.
2. With cable housing connected, carefully work adapter and pinion out of extension housing.
3. If transmission fluid is found in cable housing, replace seal in adapter. Start seal and retainer ring in adapter, then push them into adapter with Tool C4004 or equivalent until tool bottoms.

CAUTION: before installing pinion and adapter assembly make sure adapter flange and mating area on extension housing are perfectly clean. Dirt or sand will cause misalignment and speedometer pinion gear noise.

4. Note number of gear teeth and install pinion gear into adapter.
5. Rotate pinion gear and adapter assembly so that number on adapter corresponding to number of teeth on gear is in six o'clock position as assembly is installed.
6. Install retainer and bolt with retainer tangs in adapter positioning slots. Tap adapter firmly into extension housing and tighten retainer bolt to 100 in. lbs.

Output Shaft Oil Seal

Replacement—1964-65 A-904

1. Disconnect drive shaft at transmission flange.
2. Hold transmission flange with Tool C-3281 or equivalent and remove retaining nut and

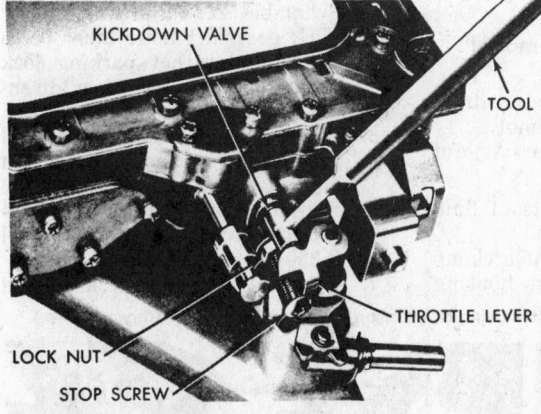

Throttle pressure adjustment (© Chrysler Corp.)

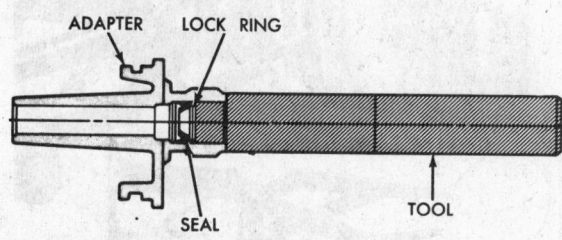

Installing speedometer pinion seal (© Chrysler Corp.)

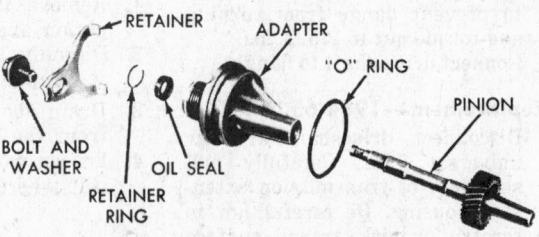

Speedometer drive (© Chrysler Corp.)

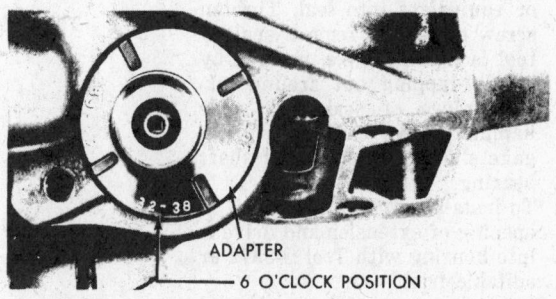

Speedometer pinion and adapter installed (© Chrysler Corp.)

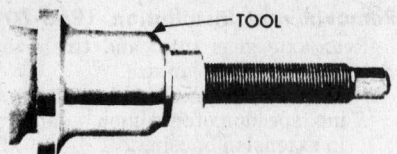

Removing output shaft oil seal
(© Chrysler Corp.)

washer. Slide flange off output shaft.
3. Screw taper threaded end of Tool C-748 or equivalent into seal, and tighten screw of tool to remove seal. If tool is not available, remove by gently tapping

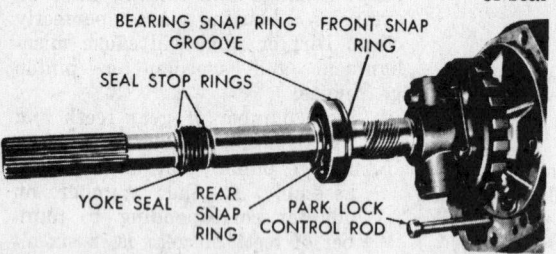

Output shaft bearing and yoke seal (© Chrysler Corp.)

out around circumference of seal with slide hammer. Take care not to engage slide hammer on tail shaft bearing.
4. To install new seal, place seal in opening of extension housing with lip of seal facing inward. Drive seal into housing with Tool C-3837 or suitable drift.
5. Install transmission output shaft flange. Install washer with three projections toward flange and nut with its convoluted surface contacting washer. Hold flange with Tool C-3281 or equivalent to prevent flange from turning and torque nut to 175 ft. lbs.
6. Connect drive shaft to flange.

Replacement—1964-65 A-727
1. Disconnect driveshaft at rear universal joint. Carefully pull shaft out of transmission extension housing. Be careful not to scratch or nick ground surface on sliding spline yoke.
2. Cut boot end off seal, then screw taper threaded end of Tool C-748 or equivalent into seal. Tighten screw of tool to remove seal. If tool is not available, remove by gently tapping out around circumference of seal with slide hammer. Take care not to engage slide hammer or tail shaft bearing.
3. To install new seal, place seal in opening of extension and drive it into housing with Tool C-3972 or suitable drift.
4. Carefully guide front universal joint into extension housing and

onto output shaft splines. Connect driveshaft to rear axle pinion shaft yoke.

Replacement—1966-70 All
1. Mark parts for reassembly. Disconnect driveshaft at rear universal joint. Carefully pull shaft yoke out of transmission extension housing. Be careful not to scratch or nick ground surface of sliding spline yoke.
2. Remove extension housing yoke seal with Tool C-3994 or C-3985 or equivalent. If tools are not available, remove by gently tapping out around circumference of seal with slide hammer.
3. To install new seal, place seal in opening of extension housing and drive it into housing with Tool C-3995 or C-3972 or suitable drift.
4. Carefully guide front universal joint yoke into extension housing and onto the mainshaft splines. Align marks made at removal and connect driveshaft to pinion shaft yoke.

Extension Housing Removal 1964-65
1. Remove the speedometer drive pinion and sleeve assembly.
2. Disconnect front universal joint at companion flange.
3. Drain about two quarts of fluid from the transmission.
4. Loosen parking lock cable clamp bolt where cable enters housing

cover. Tap end of clamp bolt lightly to release its hold on the cable. Remove housing cover lower plug. Insert screwdriver through hole. While exerting pressure against projecting portion of cable lock-spring, withdraw lock cable.
5. Remove two bolts securing extension housing to crossmember insulator.
6. Raise transmission slightly with service jack to clear crossmember. Remove crossmember attaching bolts and remove crossmember, insulator, and spring assembly.
7. Remove extension housing to transmission bolts. tap housing lightly with soft mallet to break it loose from transmission, and remove housing.

1966-70
1. Mark parts for reassembly. Disconnect driveshaft at rear universal joint. Carefully pull shaft out of extension housing.
2. Remove speedometer pinion and adapter assembly. Drain approximately two quarts of fluid from transmission.
3. Remove bolts securing extension housing to crossmember. Raise transmission slightly with service jack and remove center crossmember and support assembly.
4. Remove extension housing to transmission bolts. On console shifts. remove two bolts securing gearshift torque shaft lower bracket to extension housing. Swing bracket out of way.

NOTE: gearshift lever must be in 1-low position so that parking lock control rod can be engaged or disengaged with parking lock sprag.
5. Remove two screws, plate, and gasket from bottom of extension housing mounting pad.
6. Spread large snap ring from output shaft bearing with Tool C-3301 or equivalent.
7. With snap ring spread as far as

Removing or installing extension housing (© Chrysler Corp.)

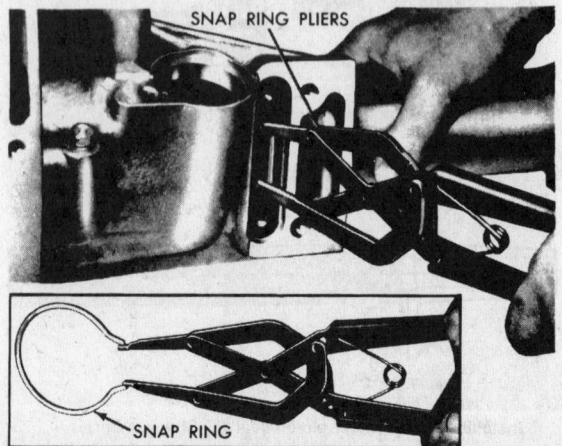

possible tap extension housing gently off output shaft bearing.

8. Carefully pull extension housing rearward to bring parking lock control rod knob past parking sprag and remove housing.

Governor

Removal

1. Remove extension housing.
NOTE: remove output shaft support bearing if so equipped.
2. With a screwdriver, carefully pry the snapring from the weight end of governor valve shaft. Slide the valve and shaft assembly out of the governor housing.

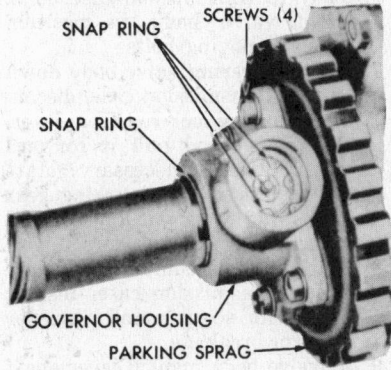

Governor shaft and weight snap-rings
(© Chrysler Corp.)

3. Remove the large snap-ring from the weight end of the governor housing and lift out the governor weight assembly.
4. Remove snap-ring from inside governor weight, remove inner weight and spring from the outer weight.
5. Remove the snap-ring from behind the governor housing, then slide the governor housing and support assembly from the output shaft. If necessary, remove the four screws and separate the governor housing from the support.

Cleaning and Inspection

The primary cause of governor operating trouble is sticking of the valve or weights. This is brought about by dirt or rough surfaces. Thoroughly clean and blow dry all of the governor parts, crocus cloth any burrs or rough bearing surfaces and clean again. If all moving parts are clean and operating freely, the governor may be reassembled.

Installation

1. Assemble the governor housing to the support, then finger tighten the screws. Be sure the oil passage of the governor housing

aligns with the passage in the support.
2. Align the master spline of the support with the master spline on the output shaft, slide the assembly into place. Install the snap-ring behind the governor housing. Torque bolts to 100 in. lbs.
3. Assemble the governor weights and spring, fasten with snap-ring inside of large governor weight. Place the weight assembly in the governor housing shaft retaining snap-ring. Install output shaft support bearing if so equipped.
4. Place the governor valve on the valve shaft, insert the assembly into the housing and through the governor weights. Install the shaft retaining snap-ring. Install output shaft support bearing if so equipped.
5. Install the extension housing. Connect the propeller shaft.

Rear Oil Pump, Through 1965

Removal

1. Remove extension housing.
2. Remove governor and support.
3. Unscrew the rear oil pump cover retaining bolts and remove the pump cover.
4. Draw a line across the face of the inner and outer pump rotors (with dye) so that they may be reinstalled in the same position.

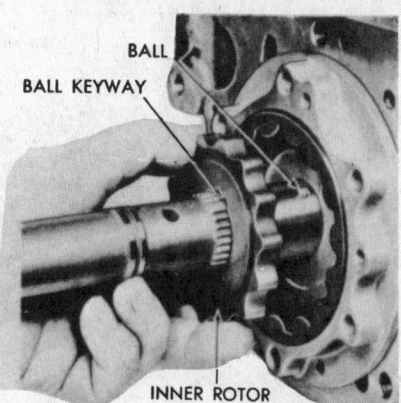

Removing or installing rear oil pump inner rotor—1963-65
(© Chrysler Corp.)

5. Slide off the inner rotor, (don't drop the small driving ball). Remove the outer rotor from the pump body.
NOTE: if replacement of the pump body is necessary, the transmission must be dismantled to allow driving the pump body out (rearward) with a wood block.

Inspection

Clean and blow dry all pump parts and examine contacting surfaces for evidence of wear, burrs or other dam-

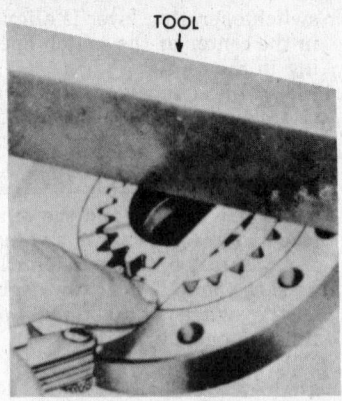

Checking oil pump rotor clearance
(© Chrysler Corp.)

age. With the parts cleaned and reassembled, place a straightedge across the face of the rotors and pump body. With a feeler gauge, check clearance between the straightedge and the face of the rotors. Clearance limits are at front of section.

Installation

1. Place the outer rotor in the pump body.
2. Rotate the output shaft so that the inner rotor driving ball (or ball key) pocket is up. Drop the ball into the pocket and slide the inner rotor onto the output shaft, in alignment with the ball.
3. Position the outer rotor to align its dye mark with the mark on the inner rotor.
4. Install the pump cover with the attaching bolts turned in a few threads. Slide the aligning fixture, tool C-3762 all the way in until it bottoms against the rotor.
5. Install the governor and support.
6. Install extension housing, brake assembly and connect the propeller shaft.
7. Connect the parking brake cable (drum-type).

Neutral Starting Switch

Removal

1. Drain about two quarts of transmission fluid.
2. Disconnect wire from switch and unscrew from transmission case.

Installation and Test

1. With the proper control cable adjustment assured, and the N selector depressed, make sure the

Neutral starting switch
(© Chrysler Corp.)

switch operating lever is aligned in the center of the switch opening in the case.

2. Place the cupped washer and O ring over the threads of the switch. Screw the switch into the transmission case a few turns.

3. Connect one lead of a test lamp to battery current and the other lead to the switch terminal. Screw into the transmission case until the lamp lights. Now, tighten the switch an additional one-third to one-half turn.

NOTE: the switch must be tight enough to prevent oil leaks. If not, add a thin washer and readjust the switch.

4. Remove test lamp and reattach the regular wire to the switch.

5. Bring transmission fluid to correct level.

Valve Body Assembly and Accumulator Piston

Removal 1964-65

1. Loosen pan bolts and drain all of the fluid from the transmission.
2. Remove the oil pan and gasket.
3. Loosen the clamp bolt and lift the throttle lever, washer and

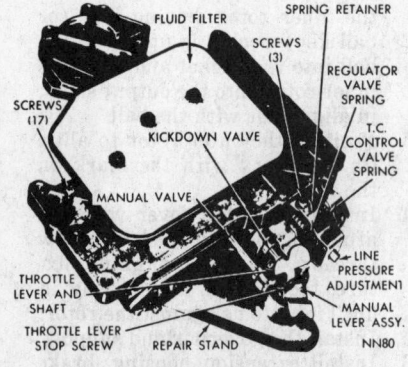

Regulator and converter control valves—
assembled view
(© Chrysler Corp.)

seal from the transmission throttle lever shaft.

4. Shift the manual control into 1 (Low) position to expose the nut holding the cable adapter to the manual lever. Remove the nut

Parking lock components (© Chrysler Corp.)

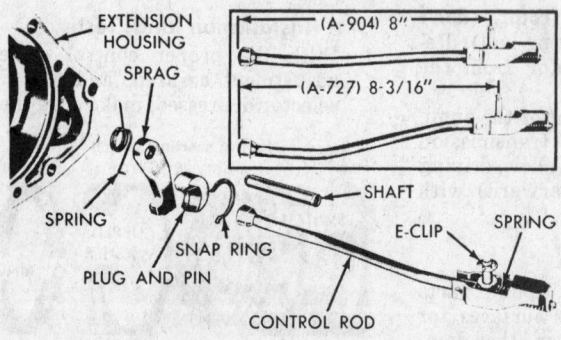

and disengage the adapter from the lever.

5. With a drain pan under the transmission, remove the ten hex-head valve body to transmission case bolts. Hold the valve body in position while removing bolts.

6. Lower the valve body assembly. Be careful not to cock the throttle lever shaft in the case hole or lose the accumulator spring.

7. Insert tool C-434 inside the accumulator piston and remove the piston from the transmission case. Inspect the piston assembly for scoring, broken rings and wear. Replace as required.

NOTE: servicing the valve body is outlined later in the text.

Installation 1964-65

1. All mating surfaces must be clean and free of burrs.
2. Install accumulator piston into the transmission case.
3. Position accumulator spring on the valve body.
4. Position the valve body assembly in place in the transmission and start all the retaining bolts.
5. Snug the bolts down evenly then torque to 100 in. lbs.
6. Connect the control cable adapter to the manual lever and install retaining nut.
7. Install seal, flat washer and throttle lever onto the throttle shaft. Tighten the clamp bolt.

8. Install clean oil pan with new gasket.
9. Add fluid to correct level.

Removal 1966-70

1. Raise vehicle on hoist and loosen oil pan bolts, tap pan to break it loose, allowing fluid to drain.
2. Remove pan and gasket.
3. Disconnect throttle and gearshift linkage from levers on transmission. Loosen clamp bolts and remove levers.
4. Remove E clip securing parking lock rod to valve body manual lever.
5. Remove backup light and neutral start switch.
6. Place drain pan under transmission, remove ten hex head valve body to transmission case bolts. Hold valve body in position while removing bolts.
7. While lowering valve body down out of transmission case, disconnect parking lock rod from lever. To remove rod pull it forward out of case. If necessary rotate driveshaft to align parking gear and sprag to permit knob on end of control rod to pass sprag.
8. Withdraw accumulator piston from transmission case. Inspect piston for scoring, and rings for wear or breakage.
9. If valve body manual lever shaft seal requires replacement, drive it out of case with punch.
10. Drive new seal into case with 15/16 in. socket and hammer.

Parking lock control rod (© Chrysler Corp.)

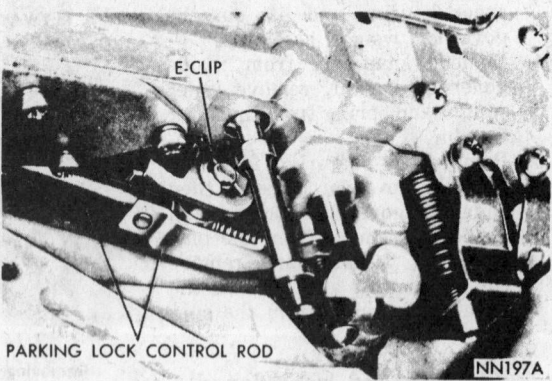

Installing valve body manual lever shaft oil seal
(© Chrysler Corp.)

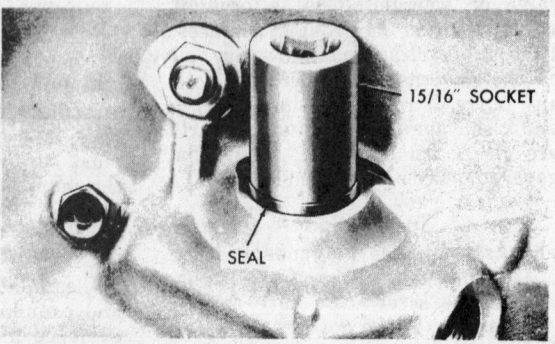

Installation 1966-70

1. If parking lock rod was removed, insert it through opening in rear of case with knob positioned against plug and sprag. Move front end of rod toward center of transmission while exerting rearward pressure on rod to force it past sprag. Rotate driveshaft if necessary.
2. Install accumulator piston in transmission case.
3. Place accumulator spring on valve body.
4. Place valve body manual lever in low position. Lift valve body into its approximate position, connect parking lock rod to manual lever and secure with E clip. Place valve body in case, install retaining bolts finger tight.
5. With neutral start switch installed, place manual lever in neutral position. Shift valve body if necessary to center neutral finger over neutral switch plunger. Snug bolts down evenly. Torque to 100 in. lbs.
6. Install gearshift lever and tighten clamp bolt. Check lever shaft for binding by moving lever through all detents. If lever binds, loosen valve body bolts and re-align.
7. Make sure throttle shaft seal is in place, install flat washer and lever and tighten clamp bolt. Connect throttle and gearshift linkage and adjust as required.
8. Install oil pan using new gasket. Add transmission fluid to proper level.

Detailed Unit Reconditioning

The following reconditioning data covers the removal, disassembly, inspection, repair, assembly and installation procedures for each sub-assembly in detail.

NOTE: in the event that any part has failed in the transmission, the converter should be thoroughly flushed to insure the removal of fine particles that may cause damage to the reconditioned transmission.

Oil Pan Removal

1. Secure transmission in a repair stand.
2. Unscrew attaching bolts and remove oil pan and gasket.

Valve Body Removal

1964-65

1. Unscrew nut and remove control cable adapter from valve body manual lever.
2. Remove (ten) hex-head valve body-to-transmission case bolts. (Hold the valve body in position while removing bolts.)
3. Lift the valve body out of the transmission case, don't cock the throttle lever shaft.

1966-70

1. Loosen clamp bolts and remove throttle and gearshift levers from transmission.
2. Remove backup light and neutral start switch.
3. Remove ten hex head valve body to transmission bolts. Remove E clip securing parking lock rod to valve body manual lever.
4. While lifting valve body upward out of transmission case, disconnect parking lock rod from lever.

Accumulator Piston and Spring

Removal

1. Lift the spring from the accumulator piston and withdraw the piston from the case.

Checking Drive Train End-Play

1. Attach a dial indicator to the extension housing and seat the plunger on the end of the output shaft.

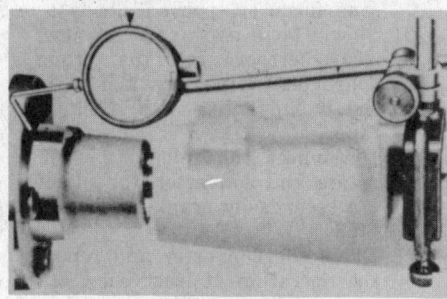

Checking drive end-play
(© Chrysler Corp.)

2. Pry the output shaft out and tap it in to register the extreme shaft end-play.
3. Record this reading for possible future use. Correct end play is found at front of section.

Extension Housing Removal

1. See Service Operations in Car.

Governor and Support Removal

1. Remove the snap-ring from the weight end of the governor valve shaft. Slide the valve and shaft assembly from the governor housing.
2. Remove the snap-ring from behind the governor housing, then slide the governor housing and support from the output shaft.

Rear Oil Pump Through 1965

Removal

1. Unscrew the rear oil pump cover retaining bolts and remove the cover.
2. Mark the face of the inner and outer pump rotor (with dye) so they may be reassembled in the same relationship.
3. The inner rotor is keyed to the output shaft by a small ball. Remove outer rotor from pump body.

NOTE: if the rear oil pump body is to be replaced, drive it rearward, out of the case (with a wood block), after the transmission case has been stripped.

Front Oil Pump and Reaction Shaft Support Removal

1. Remove front pump housing retaining bolts.
2. Attach a tool to the pump housing flange, using the eleven and four o'clock hole locations.

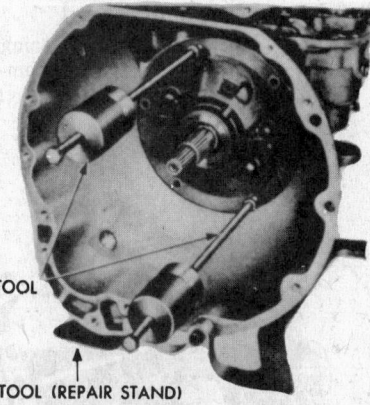

Removing or installing front oil pump and reaction shaft support assembly
(© Chrysler Corp.)

3. Bump outward, evenly, with the tool to withdraw oil pump and reaction shaft support assembly from the case.

Front Band and Front Clutch

Removal

1. Loosen the front band adjuster, remove the head strut and slide the band from the case.
2. Slide the front clutch assembly from the case.

Input Shaft and Rear Clutch

Removal

Grasp the input shaft and slide the shaft and rear clutch assembly out of the case.

NOTE: don't lose the thrust washer located between the rear end of the input shaft and the front end of the output shaft.

Planetary Gear Assemblies, Sun Gear, Driving Shell, Low and Reverse Drum Removal

While hand-supporting the output shaft and driving shell, care-

fully slide the assembly forward and out of the case.

Rear Band and Low-Reverse Drum Removal

Remove low-reverse drum, loosen rear band adjuster, remove band strut and link, and remove band from case. On 1968-70 A-904LA transmissions with double wrap band, loosen band adjusting screw and remove band and low-reverse drum.

Overrunning Clutch Removal

1. Notice the established position of the overrunning clutch rollers and springs before disassembly.
2. Slide out the clutch hub and remove rollers and springs.
3. Remove low and reverse drum thrust washer from inside the overrunning clutch case on 1964-65 transmissions.

Kickdown Servo Removal

1. Compress kickdown servo spring using engine valve spring compressor. Then remove snap ring.

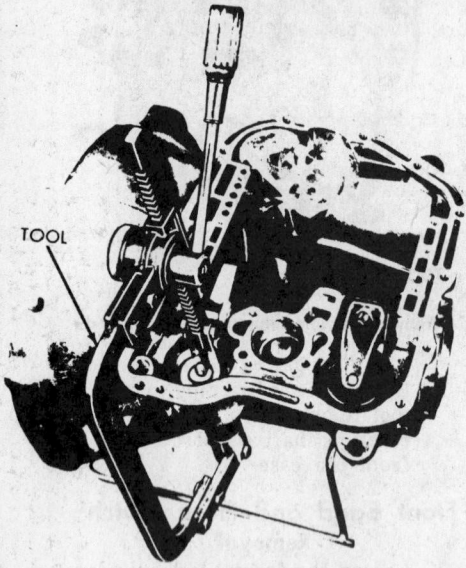

TOOL

Compressing kickdown servo spring
(© Chrysler Corp.)

2. Remove the rod guide, spring and piston rod from the case. Don't damage the piston rod or guide during removal.
3. Withdraw piston from the transmission case.

Low and Reverse Servo Removal

1. Using a suitable tool, depress the piston spring retainer and remove the snap ring.
2. Remove the spring retainer, spring, servo piston and plug assembly from the case.

Flushing the Torque Converter

1. Connect the converter assembly to the crankshaft mounting plate. Remove one drain plug and empty the converter of transmission fluid.
2. Insert a screwdriver into the converter and turn the stator hub (large splined hub) counterclockwise until one of the 1/8 x 3/8 in. slots of this assembly is visible, at the top. A second opening, directly below, provides an ample opening for the kerosene flush (if poured slowly).
3. Slowly pour two quarts of new, clean kerosene into the torque converter. Wipe dry and close the hub opening with masking tape.
4. Disconnect the coil wire, then rotate the converter for about ten seconds by cranking the engine.
5. Drain the converter and repeat the operation at least once, but as many times as is required to thoroughly flush all of the flakes and dirt from the unit.
6. Now, with both plugs removed, rotate the converter several times; this should further remove pocketed solvent and dirt. Replace the plugs and remove

the converter assembly from its mounting plate.

Before removing any of the transmission sub-assemblies, thoroughly clean the exterior of the unit, preferably by steam. When disassembling, each part should be washed in a suitable solvent, and either set aside to drain or dried with compressed air. Do not wipe with shop towels. All of the transmission parts require extremely careful handling to avoid nicks and other damage to the accurately machined surfaces.

Sub-Assembly Reconditioning

The following procedures cover the disassembly, inspection, repair and assembly of each sub-assembly as removed from the transmission.

The use of crocus cloth is permissible but not encouraged as extreme care must be used to avoid rounding off sharp edges of valves. The edge portion of valve body and valves is very important to proper functioning.

NOTE: use all new seals and gaskets, and coat each part with automatic transmission fluid, type A, suffix A, during assembly.

Valve Body

Disassembly

NOTE: this area is extremely critical, and sensitive to distortion. Never clamp any portion of the valve body or transfer plate in a vise. Clean with new solvent and dry with compressed air. Start all valves into their respective chambers with a twisting motion, seeing that they are well lubricated with automatic transmission fluid.

1. With the valve body on a clean repair stand, remove three attaching screws and the oil screen.
2. Hold the spring retainer bracket against spring tension, remove the three bracket retaining screws.

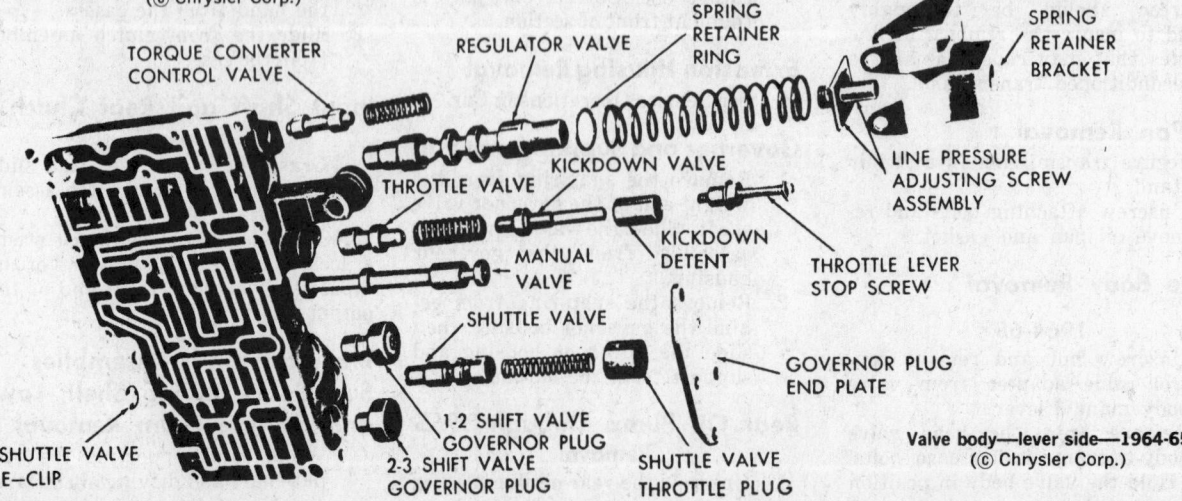

TORQUE CONVERTER CONTROL VALVE

REGULATOR VALVE

SPRING RETAINER RING

SPRING RETAINER BRACKET

THROTTLE VALVE

KICKDOWN VALVE

LINE PRESSURE ADJUSTING SCREW ASSEMBLY

MANUAL VALVE

KICKDOWN DETENT

THROTTLE LEVER STOP SCREW

SHUTTLE VALVE

GOVERNOR PLUG END PLATE

SHUTTLE VALVE E-CLIP

1-2 SHIFT VALVE GOVERNOR PLUG

2-3 SHIFT VALVE GOVERNOR PLUG

SHUTTLE VALVE THROTTLE PLUG

Valve body—lever side—1964-65
(© Chrysler Corp.)

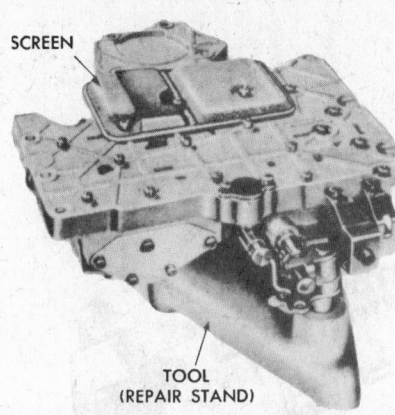

SCREEN

TOOL
(REPAIR STAND)

Removing or installing oil screen
(© Chrysler Corp.)

3. Remove spring bracket, torque converter valve spring, and the regulator valve spring with line pressure adjusting screw assembly.

NOTE: do not alter the setting of the line pressure adjusting screw and nut. The nut has an interference thread and does not turn easily on the screw.

4. Slide the regulator valve and spring retainer ring from the valve body. Slide torque converter control valve from the valve body.

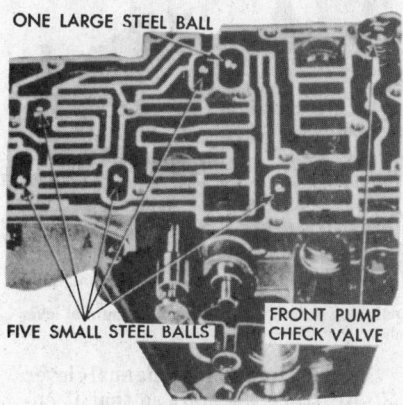

ONE LARGE STEEL BALL

FIVE SMALL STEEL BALLS

FRONT PUMP CHECK VALVE

Front pump check valve and steel ball locations—1964-68
(© Chrysler Corp.)

5. Remove the (14) transfer plate retaining screws. Lift the transfer plate and steel plate assembly off the valve body.

6. Invert the transfer plate assembly and remove the stiffener plate. Remove the remaining screws securing the steel plate to the transfer plate, and lift off the steel plate.

Remove rear pump check valve and spring on 1964-65 transmissions.

7. On 1964-65 transmissions, remove reverse blocker valve cover and lift out spring and valve.

8. On 1964-68 transmissions, note location of six steel balls in valve body. On 1969-70, note location of seven steel balls. (one of them is larger than the others and is in the larger chamber). Remove the steel balls, front pump check valve and spring.

9. On 1964-65 transmissions, invert valve body and lay it on clean paper. Remove E-clip from the throttle lever shaft. While holding manual lever detent ball and

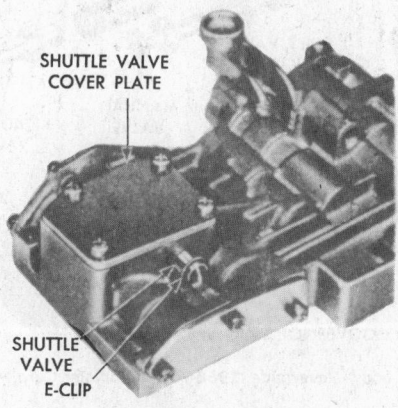

SHUTTLE VALVE COVER PLATE

SHUTTLE VALVE E-CLIP

Shuttle valve cover, shuttle valve and retaining E-clip
(© Chrysler Corp.)

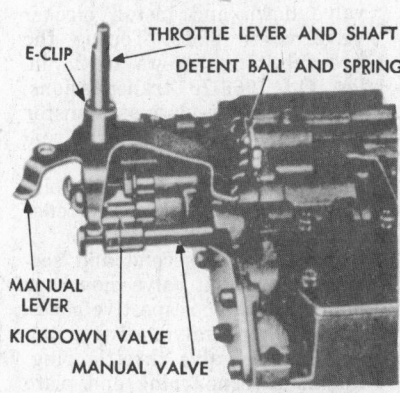

E-CLIP

THROTTLE LEVER AND SHAFT

DETENT BALL AND SPRING

MANUAL LEVER

KICKDOWN VALVE

MANUAL VALVE

Manual lever, detent ball and spring, throttle lever and shaft, manual valve, and kickdown valve
(© Chrysler Corp.)

spring in their bore with tool C-3765, or similar tool, slide manual lever from the throttle shaft. Remove detent ball and spring.

10. Remove manual valve from valve body.

11. Remove throttle lever and shaft from body.

12. Remove shuttle valve cover plate. Remove E-clip from the exposed end of shuttle valve.

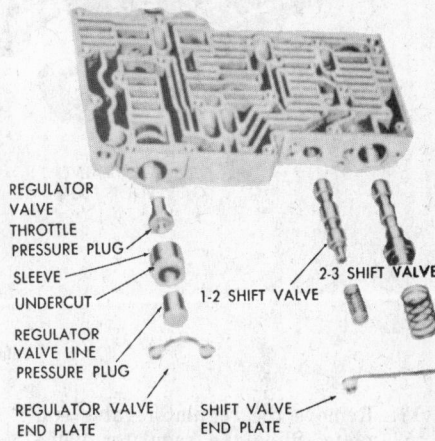

REGULATOR VALVE THROTTLE PRESSURE PLUG

SLEEVE

UNDERCUT

REGULATOR VALVE LINE PRESSURE PLUG

REGULATOR VALVE END PLATE

2-3 SHIFT VALVE

1-2 SHIFT VALVE

SHIFT VALVE END PLATE

Valve body—shift valve side
(© Chrysler Corp.)

13. Remove throttle lever stop screw, being careful not to disturb the setting.

14. Remove kickdown detent, kickdown valve, throttle valve spring and throttle valve.

15. Remove governor plug and end plate. Tip up the valve body to allow the shuttle valve throttle plug, spring, shuttle valve and the shift valve governor plugs to slide out.

NOTE: the first-second valve plug has a longer stem.

16. Remove the shift valve end plate and slide out the two springs and valves.

Steel ball location, 1969-70 (© Chrysler Corp.)

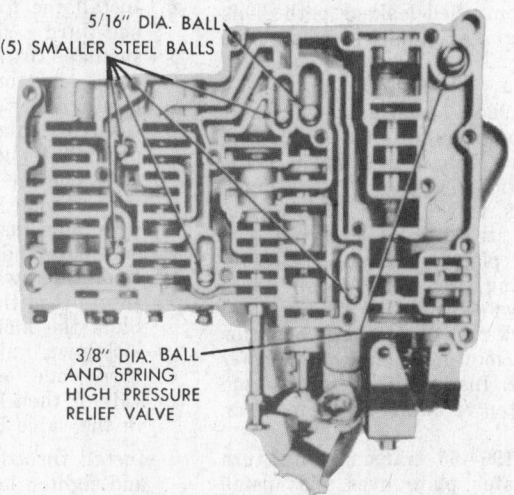

5/16" DIA. BALL

(5) SMALLER STEEL BALLS

3/8" DIA. BALL AND SPRING HIGH PRESSURE RELIEF VALVE

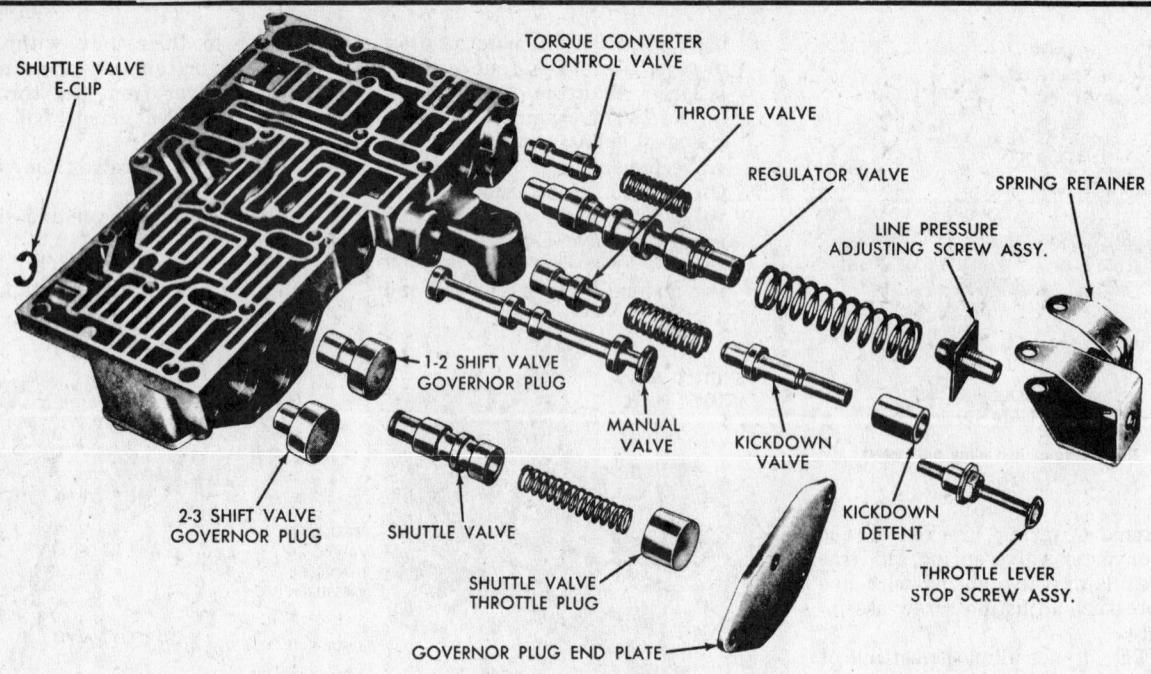

SHUTTLE VALVE
E-CLIP

TORQUE CONVERTER
CONTROL VALVE

THROTTLE VALVE

REGULATOR VALVE

SPRING RETAINER

LINE PRESSURE
ADJUSTING SCREW ASSY.

1-2 SHIFT VALVE
GOVERNOR PLUG

MANUAL
VALVE

KICKDOWN
VALVE

KICKDOWN
DETENT

2-3 SHIFT VALVE
GOVERNOR PLUG

SHUTTLE VALVE

THROTTLE LEVER
STOP SCREW ASSY.

SHUTTLE VALVE
THROTTLE PLUG

GOVERNOR PLUG END PLATE

Valve body—lever side, 1966-70 (© Chrysler Corp.)

17. Remove the regulator valve end plate. Slide the regulator valve line pressure plug, sleeve, and the regulator valve throttle pressure plug from the body.

Cleaning and Inspection

Inspect all components for scores, loose or bent levers, burrs and warping. Don't straighten bent levers; renew them. Loose levers may be silver soldered at the shaft. Burrs and minor nicks may be carefully removed with crocus cloth. Check for valve body warping or distortion with a surface plate (plate glass will do) and a feeler gauge. Do not attempt to service a distorted plate or valve body, this is a very critical area. Check all springs for distortion or fatigue. Check valves for scores and freedom of movement in the bores, they should fall of their own weight, in and out of the bore. The front and rear pump check valves are provided with a controlled leakage path. This keeps the rear pump primed.

Assembly

1. On 1964-65 transmissions, insert the rear pump check valve and spring into the transfer plate. Position the steel plate on pump check valve in its bore with a thin steel scale and install four steel plate to transfer plate retaining screws. Torque these screws evenly to 28 in. lbs. Check rear pump check valve for free movement in the transfer plate. Install stiffener plate and tighten retaining screw to 28 in. lbs.

2. On 1964-65 transmissions, turn transfer plate over and install

reverse blocker valve spring and valve. Rotate value until it seats through the steel plate. Hold the valve down and install blocker valve cover plate. Torque the two retaining screws to 28 in. lbs. On 1966-70 transmissions, place separator plate on transfer plate. Install stiffener plate and retaining screws exactly as shown. Make sure all bolt holes are aligned. Tighten stiffener plate screws to 28 in. lbs.

3. Insert the first-second and second-third shift valve governor plugs into their respective bores. Install shuttle valve, valve spring and shuttle valve throttle plug. Install governor plug end plate and torque the four retaining screws to 28 in. lbs.

4. Install E-clip onto end of shuttle valve. Install shuttle valve cover plate and torque the four retaining screws to 28 in. lbs.

5. Install the first-second and second-third shift valves and springs. Install shift valve and plate and torque the three retaining screws to 28 in. lbs.

6. Insert regulator valve throttle pressure plug, sleeve (with the undercut on the sleeve toward the end plate), and the line pressure plug. Install regulator valve end plate and torque the two retaining screws to 28 in. lbs.

7. Insert throttle valve and spring. Slide the kickdown detent onto kickdown valve (with counterbore side of detent toward valve), then insert the assembly in the valve body.

8. Install throttle lever stop screw and tighten locknut finger tight.

9. Insert manual valve into the valve body.

10. Install throttle lever and shaft on the valve body. Insert detent spring and ball into its bore in the valve body. Depress ball and

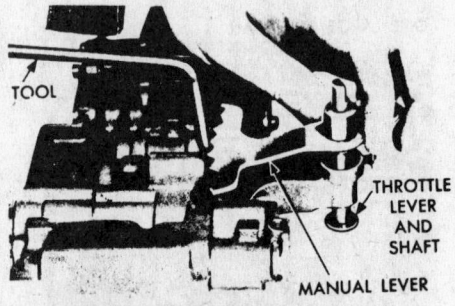

TOOL

THROTTLE
LEVER
AND
SHAFT

MANUAL LEVER

Installing detent ball, spring and manual lever (© Chrysler Corp.)

spring and slide manual lever over throttle shaft so that it engages manual valve and detent ball. Install the retaining E-clip onto the throttle shaft.

11. Position the valve body assembly on the bench or holding stand.

12. Place six steel balls in the valve body chambers (with large ball in the large chamber). Place the front pump check valve and spring in the valve body. On 1969-70 transmissions, install spring and ⅜ in. high pressure relief valve ball.

13. Position transfer plate assembly on the valve body. Hold front pump check valve in its bore with a thin steel scale. Install the (14) retaining screws, starting at the center and working outward. Torque screws to 28 in. lbs.

14. Install the torque converter valve, regulator valve and spring retainer ring.
15. Place the torque converter valve spring and the regulator valve spring over the ends of their respective valves. Place line pressure adjusting screw assembly on the end of the regulator valve spring, with the long dimension of the nut at right angles to the valve body.
16. Install spring retainer bracket (make sure the converter valve spring is engaged on the tang in the bracket). Torque the three bracket retaining screws to 28 in. lbs.
17. Install oil strainer and torque the three retaining screws to 28 in. lbs.

Important: after reconditioning the valve body, adjust the throttle and line pressure as outlined in the car section of this manual. However, if line pressure was satisfactory before disassembly, do not change this adjustment.

Accumulator Piston and Spring

Inspect both seal rings for wear and freedom in the piston grooves. Check the piston for scores, burrs, nicks and wear. Check the piston bore for corresponding damage and check piston spring for distortion and fatigue. Replace parts as required.

Governor

Disassembly
1. Remove the large snap-ring from the weight end of governor housing and lift out the governor weight assembly.

2. Remove the snap-ring from inside the governor weight, remove the inner weight and spring from the outer weight.
NOTE: thoroughly clean all parts in a suitable and clean solvent. Check for damage and free movement before assembly.
3. If lugs on support gear are damaged, remove four bolts and separate support from governor body.

Assembly
1. If support was separated from governor body, assemble and tighten bolts finger tight.
2. Assemble the governor weights and spring, then secure with snap-ring inside large governor weight.
3. Place the weight assembly in the governor housing and install snap-ring.

Rear Oil Pump, through 1965, Inspection

Clean and inspect oil pump body and cover for wear, gouging or any other type of damage. Inspect rotors for scoring or pitting. With rotors cleaned and assembled in the pump body, apply a straightedge across the face of rotors and pump body. With a feeler gauge, check clearance between straightedge and face of rotors. Clearance limits are at front of section.

Front Pump and Reaction Shaft Support

Disassembly
The illustration shows the front oil pump and reaction shaft support disassembled.

1. Remove bolts from rear side of reaction shaft support and lift support from the pump.
2. Dye-spot the face of the inner and outer rotors so they may be reinstalled in their original relationship, then remove the rotors.
3. Remove the rubber seal ring from front pump body flange.
4. Drive out the oil seal with blunt punch.

Inspection
Clean and inspect interlocking seal rings on the reaction shaft support for wear or broken interlocks, be sure they turn freely in their grooves. Check all machined surfaces of pump body and reaction shaft support for scuff marks and burrs. Inspect pump rotors for scores and pits. With rotors clean and installed into the pump body, apply a straightedge across the face of the rotors and pump body. With a feeler gauge, check straightedge to rotor face clearance. Limits are in front of section.

Assembly
1. Place reaction shaft support in assembling tool C-3759 and place it on the bench, with the support hub resting on the bench. Screw two pilot studs, tool C-3283, or satisfactory substitutes, into threaded holes of reaction shaft support flange.
2. Assemble rotors with dye marks aligned, place rotors in center of the support. The two driving lugs inside rotor must be next to the face of the reaction shaft support.
3. Lower pump body over pilot studs, insert tool C-3756 or substitute through pump body and

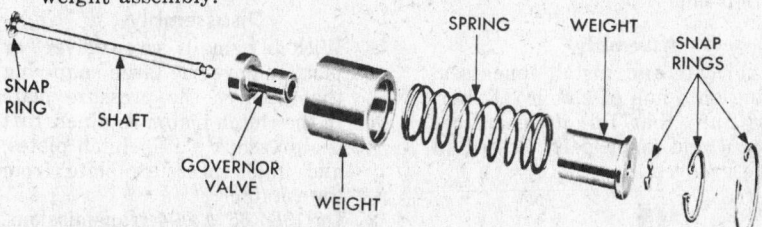

Governor valve, weights, spring and shaft (© Chrysler Corp.)

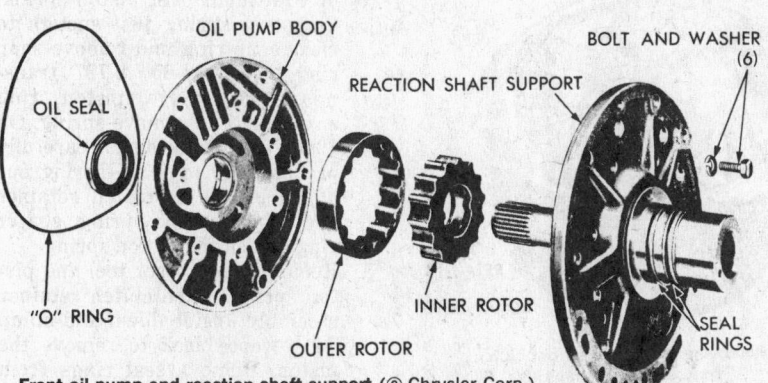

Front oil pump and reaction shaft support (© Chrysler Corp.)

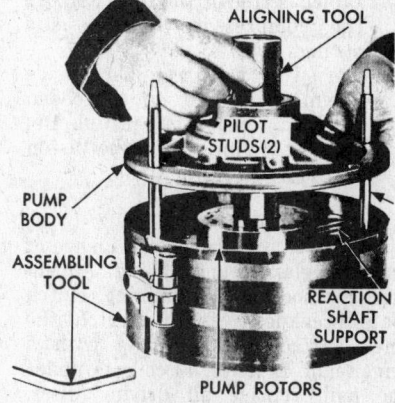

Assembling front pump and reaction shaft support (© Chrysler Corp.)

engage pump inner rotor. Turn the rotors, with the tool, to enter them into pump body. With the pump body firmly against the reaction shaft support, tighten ring squeezer or clamping tool securely.

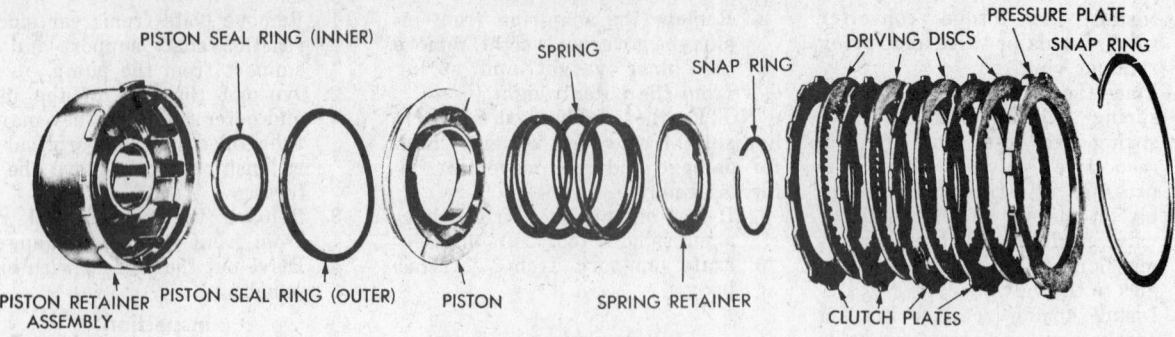

Front clutch assembly—A-904
(© Chrysler Corp.)

4. Invert the front pump and re-action shaft support assembly, with the clamping tool intact. Install support to pump body bolts. Remove clamping tool, pilot studs and rotor aligning studs.
5. Insert new oil seal into opening of front oil pump housing (with lip of seal facing inward). Drive seal into housing.

Front Clutch
Disassembly
Exploded view of front clutch assembly is illustrated.
1. With screwdriver or pick, remove large snap-ring, which holds the pressure plate in the clutch piston retainer. Lift pressure plate and clutch plates out of the retainer.
2. Install compressor, tool C-3575 for A-904 transmissions, or C-3863 for A-727 transmissions, or similar tool, over piston spring retainer (see illustration). Compress spring and remove snap-ring, then, slowly release tool until the spring retainer is free of the hub. Remove the compressor, retainer and spring.
3. Turn the clutch retainer upside down and bump on a wooden block to remove the piston. Remove seal rings from the piston and clutch retainer hub.

Inspection
Inspect clutch discs for evidence of burning, glazing and flaking. A general method of determining clutch plate breakdown is to scratch the lined surface of the plate with a finger nail. If material collects under the nail, replace all driving discs. Check driving splines for wear or burrs. Inspect steel plates and pressure plate surfaces for discoloration, scuffing or damaged driving lugs. Replace if necessary.
Check steel plate lug grooves in clutch retainer for smooth surfaces. Plate travel must be free. Inspect band contacting surface of clutch retainer, being sure the ball moves freely. Check seal ring surfaces in clutch

retainer for scratches or nicks, light annular scratches will not interfere with the sealing of neoprene rings.

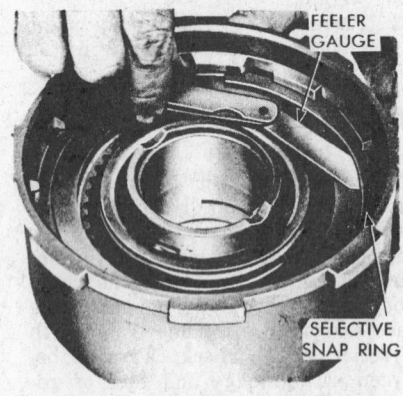

Checking front clutch plate clearance
(© Chrysler Corp.)

Inspect inside bore of piston for score marks. If light marks exist, polish with crocus cloth. Check seal ring grooves for nicks and burrs. Inspect neoprene seal rings for deterioration, wear and hardness. Check piston spring, retainer, and snap-ring for distortion and fatigue.

Assembly
1. Lubricate and install inner seal ring onto hub of clutch retainer. Be sure that lip of seal faces down and is properly seated in the groove.

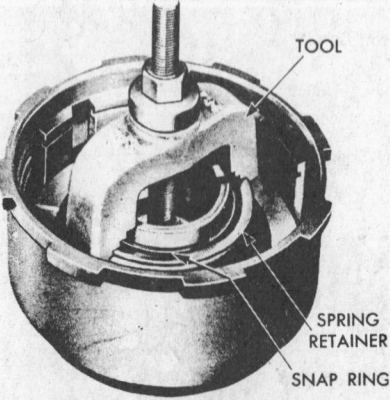

Removing or installing front clutch spring retainer snap-ring
(© Chrysler Corp.)

2. Lubricate and install outer seal ring onto clutch piston, with lip of seal toward the bottom of the clutch retainer. Place piston assembly in retainer and, with a twisting motion, seat the piston in the bottom of the retainer.
3. Place spring on the piston hub and position spring retainer and snap-ring on spring. Compress spring with tool, or suitable ring compressor, and seat snap-ring in the hub groove. Remove compressor.
4. Lubricate all clutch plates, then, install a steel plate, followed by a lined plate, until all plates are installed. Install the pressure plate and snap-ring. Be sure the snap-ring is correctly seated.
5. With front clutch assembled, insert a feeler gauge between the pressure plate and snap-ring. The clearance should be to specification. If not, install a snap-ring of proper thickness.

Rear Clutch
Disassembly
1. With a small screwdriver or pick, remove the large snap-ring that secures the pressure plate in the clutch piston retainer. Lift the pressure plate, clutch plates, and inner pressure plate from the retainer.
2. On 1964-65 A-904 transmissions, install compressor Tool C-3760 or equivalent over piston spring. Compress spring just enough to clear snap ring and remove snap ring. On 1964-65 A-727 transmissions, remove piston ring snap ring and remove spring. On 1966-70 transmissions, carefully pry one end of wave spring out of its groove in clutch retainer and remove wave spring, spacer ring and clutch piston spring.
3. Remove compressor tool and piston spring. Turn clutch retainer assembly upside down and bump on a wood block to remove the piston. Remove seal rings from the piston.

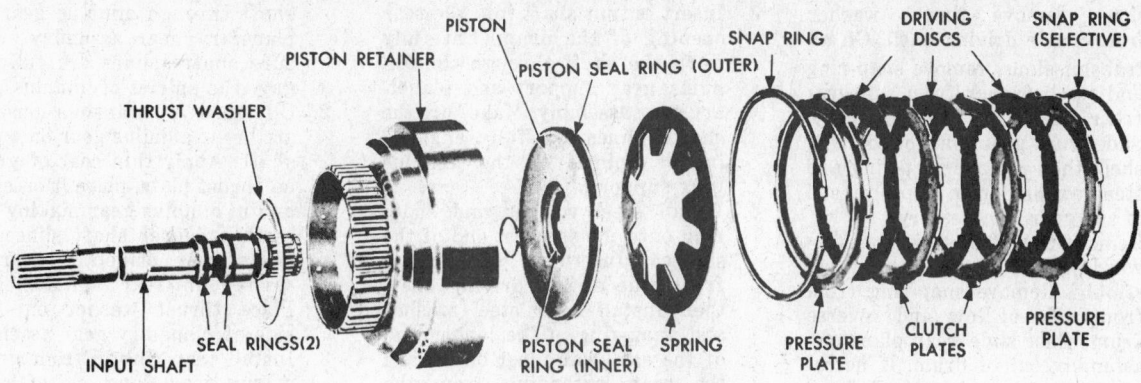

Rear clutch assembly—A-904
(© Chrysler Corp.)

Inspection

Inspect driving discs for indication of damage; handle as previously outlined under front clutch inspection.

Assembly

1. Lubricate, then install inner and outer seal rings onto the clutch piston. Be sure the seal lips face toward the head of the clutch retainer and seals are properly seated in the piston grooves.
2. Place piston assembly in retainer and, with a twisting motion, seat piston in bottom of retainer.
3. On 1964-65 A-904 transmissions, place spring over piston with outer edge of spring positioned below snap ring groove. Install compressor Tool C-3760 or equivalent over spring. Compress spring just enough to install snap ring. Remove tool. On 1964-65 A-727 transmissions, place spring over piston with outer edge of spring positioned below snap ring groove. Start one end of snap ring in groove. Make sure spring is exactly centered on piston. Progressively tap snap ring into groove. Be sure snap ring is fully seated in

groove. On 1966-70 transmissions, place clutch piston spring and spacer ring on top of piston in clutch retainer. Make sure spring and spacer ring are placed in retainer recess. Start one end of wave spring in retainer groove. Progressively push or tap spring into place making sure it is fully seated in groove.
4. Install inner pressure plate into clutch retainer, with raised portion of plate resting on the spring.
5. Lubricate all clutch plates, then install one lined plate, followed by a steel plate, until all plates are installed. Install outer pressure plate and snap-ring.
6. With rear clutch completely assembled, insert a feeler gauge between the pressure plate and snap-ring. The clearance should be to specification. If not, install snap-ring of proper thickness to obtain the required clearance.

NOTE: rear clutch plate clearance is very important to obtaining satisfactory clutch performance. Clearance is influenced by the use of various thickness outer snap-rings.

Planetary Gear Train
Disassembly

Refer to illustrations for assembly and disassembly of these units.

1. Remove thrust washer from forward end of output shaft.
2. Remove snap-ring from forward end of output shaft, then, slide front planetary assembly from the shaft.
3. On A-904 transmissions, remove snap ring and thrust washer from forward hub of planetary gear assembly. Slide front annulus gear and support off planetary gear set. Remove thrust washer from rear side of planetary gear set. If necessary, remove snap ring from front of annulus gear to separate support from annulus gear. On A-727 transmissions, slide front annulus gear off planetary gear set. Remove thrust washer from rear side of planetary gear set.
4. Slide the sun gear, driving shell, and rear planetary assembly, with low and reverse drum, from the output shaft.
5. Remove sun gear and driving shell from the rear planetary assembly. On A-727 transmis-

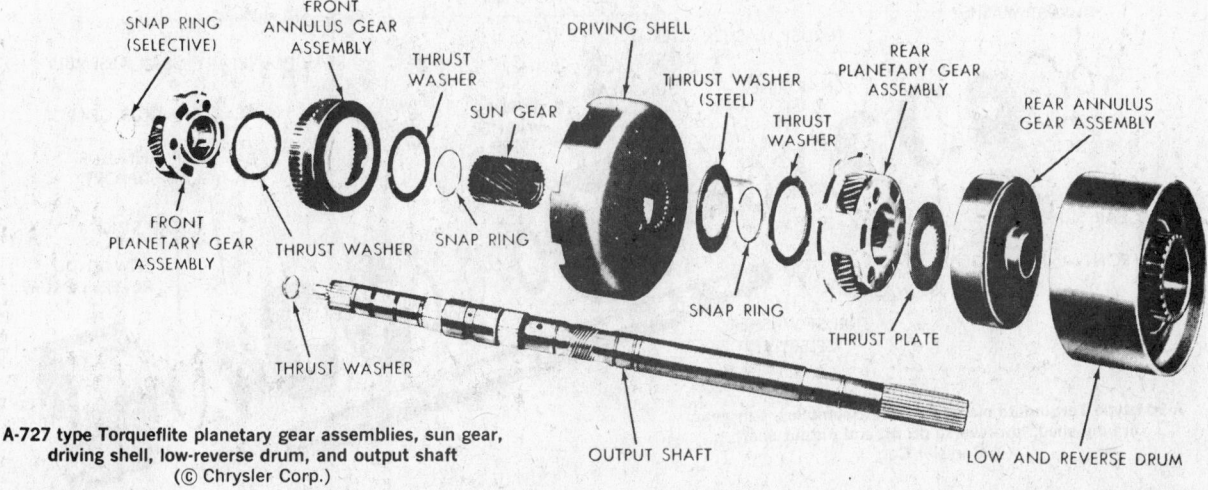

A-727 type Torqueflite planetary gear assemblies, sun gear, driving shell, low-reverse drum, and output shaft
(© Chrysler Corp.)

sions, remove thrust washer from inside driving shell. On all transmissions, remove snap-ring and steel washer from sun gear (rear side of driving shell). Slide sun gear out of driving shell, then remove snap-ring and steel washer from opposite end of sun gear, if necessary.

6. Remove thrust washer from forward side of rear planetary assembly. Remove snap-ring from front side of low and reverse drum, then slide rear planetary assembly out of drum. If necessary, remove snap-ring from rear of annulus gear in order to separate the support from the annulus gear.

Inspection

Inspect output shaft bearing surfaces for burrs or other damage. Light scratches or burrs may be polished out with crocus cloth or a fine stone. Check speedometer drive gear for damage, and make sure all oil passages are clear.

Check bushings in the sun gear for wear or scores. Replace sun gear assembly if bushings show wear or other damage. Inspect all thrust washers for wear and scores. Replace if necessary. Check lock rings for distortion and fatigue. Inspect annulus gear and driving gear teeth for damage. Inspect planetary gear carrier for cracks and the pinions for broken or worn gear teeth.

Assembly—A-904

1. Locate the rear annulus gear support in the annulus gear and install snap-ring.
2. Position the rear planetary gear assembly in the rear annulus gear, then slide the assembly into low and reverse drum. Put thrust washer on the front side of the planetary gear assembly.

3. Insert output shaft into the rear opening of the drum. Carefully work the shaft through the annulus gear support and planetary gear assembly. Make sure the shaft splines are fully engaged in the splines of the annulus gear support.
4. Install steel washer and snap-ring onto the shortest end of the sun gear. Insert sun gear through front side of the driving shell, then install rear steel washer and snap-ring. (The longer end of the sun gear must be toward the rear, extending from the driving shell).
5. Carefully slide the driving shell and sun gear assembly onto the output shaft, engaging the sun gear teeth with the planetary pinion teeth.
6. Place front annulus gear support in annulus gear and install snap-ring.
7. Position front planetary gear assembly in front annulus gear, place thrust washer over planetary gear assembly hub and install snap-ring. Position thrust washer on rear side of planetary gear assembly.
8. Carefully work the front planetary and annulus gear assembly onto the output shaft, meshing the planetary pinions with the sun gear teeth.
9. With all components properly assembled, install snap-ring onto the front end of the output shaft.

Assembly—A-727

1. On 1964-65 transmissions, place rear planetary gear assembly in rear annulus gear. Place thrust washer on front side of planetary gear assembly. Insert output shaft in rear opening of rear annulus gear. Carefully work

shaft through annulus gear and planetary gear assembly. Make sure shaft splines are fully engaged in splines of annulus gear.
2. On 1966-70 transmissions, install rear annulus gear on output shaft. Apply thin coat of grease on thrust plate, place it on shaft, and in annulus gear making sure teeth are over shaft splines. Position rear planetary gear assembly in rear annulus gear. Place thrust washer on front side of planetary gear assembly.
3. Install snap ring in front groove of sun gear (long end of gear). Insert sun gear through front side of driving shell. Install rear steel washer and snap ring.
4. Carefully slide driving shell and sun gear assembly on output shaft, engaging sun gear teeth with rear planetary pinion teeth. Place thrust washer inside front driving shell.
5. Place thrust washer on rear hub of front planetary gear set. Slide assembly into front annulus gear.
6. Carefully work front planetary and annulus gear assembly on output shaft, meshing planetary pinions with sun gear teeth.
7. With all components properly positioned, install selective snap ring on front end of output shaft. Measure end play of assembly. Adjust end play with selective snap rings.

Overrunning Clutch Inspection

Inspect clutch rollers for smooth round surfaces, they must be free of flat spots, chipped edges and flaking. Inspect roller contacting surfaces on both cam and race for pock marks and roller wear-marks. Check springs for distortion and fatigue and inspect low and reverse drum thrust. On 1966-70 A-727 transmissions, inspect cam set screw for tightness.

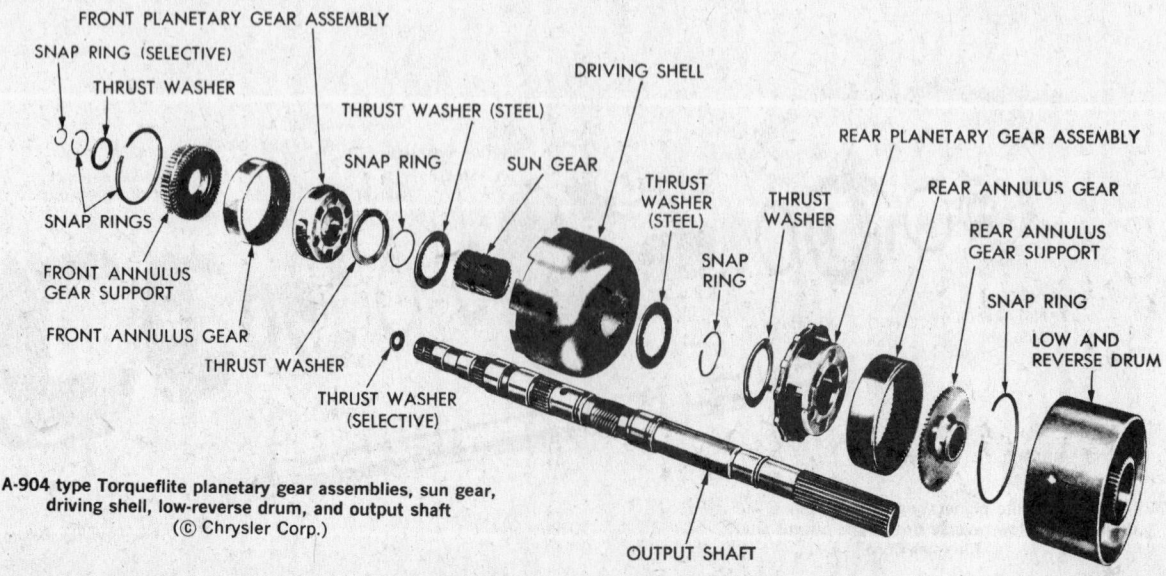

A-904 type Torqueflite planetary gear assemblies, sun gear, driving shell, low-reverse drum, and output shaft
(© Chrysler Corp.)

Kickdown Servo and Band
Inspection

See illustration for an exploded view of the kickdown servo.

Inspect piston and guide seal rings for wear, and be sure of their free-

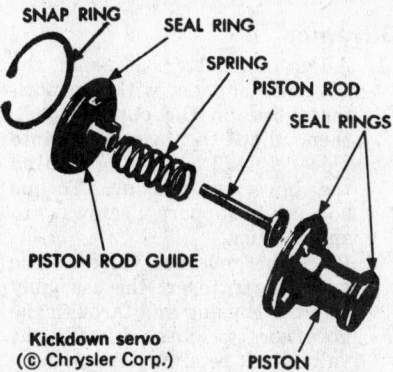

Kickdown servo
(© Chrysler Corp.)

dom in grooves. It is not necessary to remove seal rings, unless circumstances warrant. Inspect piston for scores, burrs or other damage. Check fit of guide on piston rod. Check piston for distortion and fatigue. Inspect band lining for wear and fit of lining material to the metal band. This lining is grooved; if grooves are not still visible at the ends or any part of the band, replace the band. Inspect band for distortion or cracked ends.

Low and Reverse Servo and Band
Disassembly

The illustration shows an exploded view of the low and reverse servo.

Remove snap-ring from piston and remove the piston plug and spring.

Inspection

Inspect neoprene seal ring for damage, rot, or hardness. Check piston

ends or at any part of the band, replace the band. Inspect the band for distortion or cracked ends.

Assembly

Lubricate and insert the piston plug and spring into the piston, and secure with the snap-ring.

Sub-Assemblies Installation

The following assembly procedures include the installation of sub-assemblies into the transmission case and adjustment of the drive train end-play. Do not use force to assemble any of the mating parts. Always use new gaskets during the assembly operations.

NOTE: use only automatic transmission fluid, type A, suffix A, or fluid of equivalent chemical structure, to lubricate automatic transmission parts during, or after, assembly.

Rear Oil Pump Body
1964-65

The following procedures should be followed closely when installing a new rear pump body, or reinstalling the original pump body, in order to prevent pump body distortion.

1. Cut a piece of .002-.003 in. thick wrapping paper, slightly smaller than the outside diameter of the outer rotor, to use as a shim during installation.
2. Chill the pump body to approximately 0°F. in a deep freeze, or with dry ice.
3. Quickly place pump body in the case, and install inner and outer rotors. Smear a daub of grease on the face of the rotors, center the paper shim on face of rotors, then install pump cover and tighten retaining bolts firmly.

overrunning clutch housing, then, place the clutch hub (race) on the thrust washer.
2. Install springs and rollers, as shown in illustration.

1966-70

With transmission case in upright position, insert clutch hub inside cam. Install overrunning clutch rollers and springs as shown in figure.

Low and Reverse Servo and Band

1. Carefully work servo piston assembly into the case with a twisting motion. Place spring, retainer and snap-ring over the piston.
2. Using the screw portion of tool C-3322, or suitable substitute, depress the spring and install the snap-ring.
3. Position rear band in the case, install the short strut, then connect the long lever and strut to the band. Screw in band adjuster just enough to hold struts in place. Install low-reverse drum. On 1966-70 A-727 transmissions, be sure long link and anchor assembly is installed to provide running clearance for low-reverse drum.

Low-Reverse Band A-904-LA 1968-70 318 V8 Only

This transmission has a double wrap band supported at two points by a band reaction pin in case and acted upon at one point by a servo lever adjusting screw.

1. Push band reaction pin with new O-ring into case flush with gasket surface.
2. Place band in case resting two lugs against band reaction pin.
3. Install low-reverse drum into overrunning clutch and band.

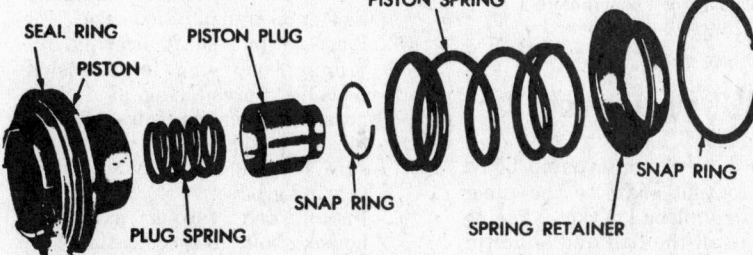

Low and reverse servo
(© Chrysler Corp.)

and piston plug for nicks, burrs, scores and wear. The piston plug must operate freely in the piston. Check the piston bore in the case for scores or other damage. Examine springs for distortion and fatigue.

Check band lining for wear and the fit of the lining to the metal band. This lining has a grooved surface; if the grooves are worn away at the

4. After the pump body has warmed to room temperature, remove the pump cover, paper shim and rotors.

Overrunning Clutch
1964-65

1. With the transmission case positioned upright, place low and reverse drum thrust washer in

Overrunning clutch
(© Chrysler Corp.)

4. Install operating lever with pivot pin flush in case and adjusting screw touching center lug on band.

Kickdown Servo

1. Carefully insert servo piston into case bore. Install piston rod, two springs and guide. Depress guide and install snap-ring.

NOTE: A-904 transmissions and maximum performance A-727 transmissions use only one small spring.

Planetary Gear Assemblies, Sun Gear, Driving Shell, Low and Reverse Drum

1. While supporting the assembly in the case, insert output shaft through the overrunning clutch hub. Carefully work the assembly rearward, engaging the drum splines with splines of the overrunning clutch hub.

Caution Be careful not to damage the ground surfaces of the output shaft during installation.

2. Apply a daub of grease to the selective thrust washer and install washer on the front end of the output shaft.

NOTE: if the drive train end-play was not within specifications when checked (Checking Drive Train End-Play), replace the thrust washer with one of proper thickness.

Input Shaft and Rear Clutch

1. Turn transmission in an upright position, with the output shaft pointing downward.
2. Align the rear clutch plate inner splines, then lower the input shaft and clutch assembly into position in the case.
3. Carefully work the clutch assembly, in a circular motion, to engage the clutch splines with the splines of the kickdown annulus gear support.
4. Daub one side of the fiber thrust washer with heavy grease, then position washer in the recess on the front face of the rear clutch retainer.

Front Clutch

1. Align the front clutch plate inner splines, then lower the clutch assembly into position in the case.
2. Carefully work the clutch assembly, in a circular motion, to engage clutch splines with splines of the rear clutch piston retainer. Be sure the front clutch driving lugs are fully engaged in the slots of the driving shell.

Front Band

1. Slide the band over the front clutch assembly.
2. Install band strut, screw in the adjuster just enough to hold the band in place.

Front Oil Pump and Reaction Shaft Support

1. Screw (two) pilot studs into front pump opening in the case.
2. Place a new rubber seal ring in groove on outer flange of pump. Be sure the seal ring is not twisted.
3. Install the assembly into the case, tap lightly with a soft mallet if necessary. Install four bolts, remove pilot studs, install remaining bolts and pull down evenly, then, torque the bolts to specification.
4. Rotate the pump rotors until the two small holes in the handle of the tool are vertical. This is to locate the inner rotor so the converter impeller shaft will engage the inner rotor lugs during installation.

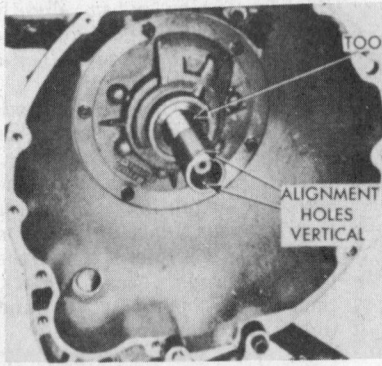

Aligning front pump rotors
(© Chrysler Corp.)

Rear Oil Pump Through 1965

1. Place outer rotor in pump body.
2. Turn output shaft so the inner rotor driving ball pocket is facing up. Install the ball and slide the inner rotor onto the output shaft, in alignment with the ball.
3. Position the outer rotor so the dye marks will be aligned, then push the inner rotor into mesh

with the outer rotor. (see illustration).

4. Install the pump cover, with the retaining bolts threaded a few turns. Slide aligning sleeve, tool C-3762, all the way in, until it bottoms against the rotors.

Governor

1. Align the master spline of the governor support with the master spline on the output shaft, then slide the assembly into place. Install snap-ring behind the governor housing. Torque housing-to-support screws to specification.
2. Place the governor valve on the valve shaft, insert the assembly into the housing and through the governor weights. Install the valve shaft retaining snap-ring.

Extension Housing

1964-65

1. Using new gasket, slide extension housing into place. Install retaining bolts and washers, and tighten bolts to 25 ft. lbs.
2. Install transmission output shaft flange. Install washer with its three projections toward flange and install nut with its convoluted surface contacting washer. Hold flange so that it does not move and torque nut to 175 ft. lbs. Torque reading must be taken as nut passes over hump.

1966-70

1. Install snap ring in front groove on output shaft.
2. Install bearing on shaft with its outer race ring groove toward front. Press or tap bearing tight against front snap ring.
3. Install rear snap ring.
4. Place new extension housing gasket on transmission case.
5. Place output shaft bearing retaining snap ring in extension housing. Spread ring as far as possible, then carefully tap extension housing into place. Make sure snap ring is fully seated in bearing groove.
6. Install and torque extension housing bolts to specification.
7. Install gasket, plate, and two screws on bottom of extension housing mounting pad.
8. Install speedometer pinion and adapter assembly.

Torque Reference

	Foot-Pounds	Inch-Pounds
Kickdown band adjusting screw locknut (eight cylinder cars)	29	—
Kickdown band adjusting screw locknut (six cylinder cars)	—	—
Kickdown lever shaft plug	25	150
Reverse band adjusting screw locknut (eight cylinder cars)	35	—
Reverse band adjusting locknut (six cylinder cars)	20	—
Cooler line fitting 1964-68	—	75
1969-70		110
Control cable adjusting wheel bolt		
Converter drive plate to crankshaft bolt	—	40
Converter drive plate to torque converter bolt	55	—
Extension housing to transmission case bolt	—	270
Extension housing to insulator mounting bolt	24	—
1964-67	35	—
1968-70	40	—
Extension housing—crossmember to frame bolt	75	—
Front oil pump housing to transmission case bolt	—	150
Governor body to parking sprag bolt	—	100
Neutral starter switch	25-30	—
Oil pan bolt	—	150
Output shaft flange nut 1964-65 only	175	—
Overrunning clutch cam set screw	—	40
Reaction shaft support to front oil pump bolt		
1964-67 all and 1968-69 A-727	—	150
1968-69 A-904	—	125
1970 all	—	160
Rear oil pump cover bolt 1964-65 only	—	140
Transmission to engine bolt	25-30	—
Valve body screw 1964-67	—	28
1968-70	—	35
Valve body to transmission case bolt	—	100

3-Speed Automatic (FMX)

Application

The FMX 3-speed automatic transmission presented here has been used in Ford Motor Company cars under various titles, such as, Cruise-O-Matic, Merc-O-Matic, Select Shift, Multi-Drive, and Twin Range Turbo-Drive. The basic transmission has remained the same since the beginning, except for small modifications. The various models of this transmission and the cars in which they are used are listed below:

FMX SELECT SHIFT: Ford—1969-70; Meteor—1969-70; Cougar—1969-70; Fairlane—1969-70; Montego—1969-70; Mustang—1969-70.

CRUISE-O-MATIC: Ford—1964-68; Fairlane—1964-68; Falcon—1964 (Fordomatic), 1965-68; Mustang—1965-68; Comet—1968; Montego—1968; Cyclone—1968; Maverick—1970; Thunderbird—1970.

MERC-O-MATIC: Mercury Cyclone & GT—1967; Mercury Comet 202—1967; Mercury Capri—1967; Mercury Caliente—1967; Mercury Cougar—1967-68; Mercury Comet—1964-65 (Multi-Drive) and 1966.

TWIN RANGE TURBO-DRIVE: Lincoln Continental—1964-65.

Diagnosis

This diagnosis guide covers the most common symptoms and is an aid to careful diagnosis. The items to check are listed in the sequence to be followed for quickest results. Thus, follow the checks in the order given for the particular transmission type.

TROUBLE SYMPTOMS	Items to Check	
	In Car	Out of Car
Rough initial engagement in D1 or D2	KBWFEG	
1-2 or 2-3 shift points incorrect or erratic	ABCDWEL	
Rough 2-3 shifts	BGFEJ	
Engine overspeeds on 2-3 shift	BGEF	m
No 1-2 or 2-3 upshift	DECJG	bcf
No 3-1 shift	KBE	
No forced downshift	LWE	
Runaway engine on forced downshift	GFEJB	c
Rough 3-2 or 3-1 shift at closed throttle	KBE	
Creeps excessively	KZ	
Slips or chatters in first gear, D1	ABWFE	acfi
Slips or chatters in second gear	ABGWFEJ	ac
Slips or chatters in R	AHWFEIB	bcf
No drive in D1	CE	i
No drive in D2	ERC	acf
No drive in L	CER	acf
No drive in R	HIERC	bef
No drive in any selector position	ACWFER	cd
Lockup in D1	CIJ	bgc
Lockup in D2	CHI	bgci
Lockup in L	GJE	bjc
Lockup in R	GJ	agc
Parking lock binds or does not hold	C	g
Transmission overheats	OFG	l
Engine will not push-start	ACFE	ec
Maximum speed too low, poor acceleration	Y	l
Transmission noisy in N and P	F	ad
Noisy transmission during coast 30-20 mph with engine stopped		e
Transmission noisy in any drive position	F	hbad
Fluid leaks	MNOPQSTUX	jkl

Key to Checks

A. Fluid level
B. Vacuum diaphragm unit or tubes
C. Manual linkage
D. Governor
E. Valve body
F. Pressure regulator
G. Front band
H. Rear band
I. Rear servo
J. Front servo
K. Engine idle speed
L. Downshift linkage
M. Converter drain plug
N. Oil pan, filler tube and/or seals
O. Oil cooler and/or connections
P. Manual or throttle shaft seals
Q. Pipe plug, side of case
R. Perform air pressure checks
S. Extension housing-to-case gasket or washer
T. Center support bolt lock washer
U. Extension housing rear oil seal
W. Make control pressure check
X. Speedometer drive gear adaptor seal
Y. Engine performance
Z. Vehicle brakes
a. Front clutch
b. Rear clutch
c. Hydraulic system leakage
d. Front pump
e. Rear pump
f. Fluid distributor sleeve—output shaft
g. Parking linkage
h. Planetary assembly
i. Planetary oneway clutch
j. Engine rear oil seal
k. Front oil pump seal
l. Front pump-to-case seal or gasket
m. Rear clutch piston air bleed valve

General Information

This section provides procedures for testing, inspecting, adjusting, and repairing the FMX 3-speed automatic transmission. Where there are differences in procedures or specifications for various model changes, these differences will be outlined.

The Ford FMX 3-speed automatic transmission (see illustration) is a three-speed unit that provides automatic upshifts and downshifts through three forward gear ratios and also provides manual selection of first and second gears. The transmission consists of a torque converter, planetary gear assembly, two multiple disc clutches, and a hydraulic control system.

The FMX transmission cools its transmission fluid through a cooler core in the radiator lower tank when a steel converter is used. If an aluminum converter is used, the transmission fluid is air-cooled.

Towing

If a disabled car is to be towed a short distance, it may be towed safely by placing the selector lever in the Neutral position and releasing the parking brake. The driver of the towed car should be careful to maintain a safe distance behind the towing vehicle. The towing speed should not exceed 30 mph.

If the disabled car is to be towed a long distance, or the transmission cannot be put in the Neutral gear, or if there is no driver for the towed car, raise the car so the rear wheels clear the ground and tow the car backwards.

Starting the Engine without the Starter

The transmission may be used to start an engine when the starter cannot be used by moving the disabled car at a road speed of about 25 mph while in Neutral and then placing the transmission in Low gear while the ignition is on. The resultant turning over of the engine will start the engine and the car may move under its own power. The disabled car may be towed or pushed but extreme care must be taken not to lose control of the car's motion and cause a collision. When the engine starts, the car will move forward quickly and must be slowed enough to avoid a collision. If the disabled car is being pushed, the rear car must drop back before the transmission is engaged to avoid hitting the front car.

Transmission Checks

Before performing any of the tests and adjustments given in this part of this section, the following preliminary checks should be done:

Transmission Fluid Level Check

1. Position the car on a level place and firmly apply the parking brake.

   NOTE: on cars equipped with a vacuum brake release, disconnect the release line and plug the end of the line. Otherwise, the parking brake will not hold the car in any drive gear of the transmission.

2. Start the engine and run at normal idle speed. If the transmission fluid is cold, the engine should idle at a fast idle speed (about 1200 rpm) until the fluid is warm. When the transmission fluid reaches its normal operating temperature, slow the engine to normal idle speed.

3. Shift the transmission selector lever through all drive positions briefly and then put it in the Park position. Do not shut off the engine during the fluid level check.

4. Locate the transmission fluid dipstick and clean all dirt and grease from the cap before removing it. On some cars the dipstick is located in the right rear corner of the engine compartment. On other cars, the dipstick is located under the front floormat to the right of center.

5. Pull the dipstick out of the filler tube, wipe it clean, and reinsert it into the filler tube until it is fully seated.

6. Pull the dipstick out again and see where the fluid level is on the dipstick. The correct level is between the ADD and FULL marks on the dipstick. If necessary, add enough fluid through the filler tube to raise the fluid to the correct level. *Do not overfill the transmission.* Replace the dipstick in the filler tube.

7. If a vacuum brake release line was disconnected, reconnect it and test for proper operation.

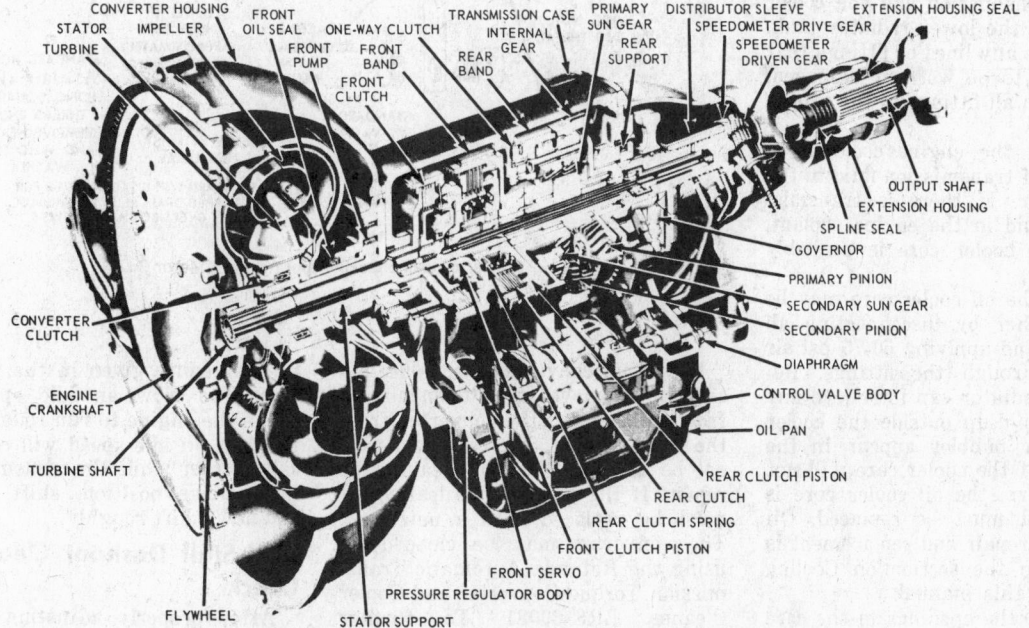

Typical FMX automatic tranmission. Three-speed Ford Cruise-O-Matic, Mercury Merc-O-Matic, Mercury Multi-Drive, Lincoln Continental Twin Range Turbo Drive
(© Ford Motor Co.)

Fluid Aeration Check

If the transmission is overfilled with too much fluid, the fluid will be whipped up into a foamy or aerated condition. This condition will cause low control pressure and the fluid may be forced out of the vent or through ruptured seals.

Check the transmission fluid level for low level conditions that may indicate fluid leaks or, at the very least, poor operation.

Transmission Fluid Leakage Checks

Make the following checks if a leakage is suspected from the transmission case:

1. Clean all dirt and grease from the transmission case.
2. Inspect the speedometer cable connection at the extension housing of the transmission. If fluid is leaking here, disconnect the cable and replace the rubber seal.
3. Inspect the oil pan gasket and attaching bolts for leaks. Tighten any bolts that appear loose to the proper torque (10-13 ft. lbs.). Recheck for signs of leakage. If necessary, remove the pan attaching bolts and old pan gasket and install new gasket and reinstall the pan and its attaching bolts.
4. Check filler tube connection at the transmission for signs of leakage. If tube is leaking, tighten the connection to stop the leak. If necessary, disconnect the filler tube, replace the O-ring, and reinstall the filler tube.
5. Inspect all fluid lines between the transmission and the cooler core in the lower radiator tank. Replace any lines or fittings that appear to be worn or damaged. Tighten all fittings to the proper torque.
6. Inspect the engine coolant for signs of transmission fluid in the radiator. If there is transmission fluid in the engine coolant, the oil cooler core is probably leaking.

NOTE: the oil cooler core may be tested further by disconnecting all lines to it and applying 50-75 psi air pressure through the fittings. Remove the radiator cap to relieve any pressure build-up outside the cooler core. If air bubbles appear in the coolant or if the cooler core will not hold pressure, the oil cooler core is leaking and must be replaced. Oil cooler core repair and replacement is discussed in the section on Cooling Systems in this manual.

7. Inspect the openings in the case where the downshift control lever shaft and the manual lever shaft are located for leaks. If ne-

cessary, replace the defective seal.
8. Inspect all plugs or cable connections in the transmission for signs of leakage. Tighten any loose plugs or connectors to the proper torque according to the specifications.
9. Remove the lower cover from the front of the bellhousing and inspect the converter drainplugs for signs of leakage. If there is a leak around the drainplugs, loosen the plug and coat the threads with a sealing compound and tighten the plug to the proper torque.

NOTE: fluid leaks from around the converter drainplug may be caused by engine oil leaking past the rear main bearing or from the oil gallery plugs. To determine the exact cause of the leak before beginning repair procedures, an oil-soluble aniline or fluorescent dye may be added to the transmission fluid to find the source of the leak and whether the transmission is leaking. If a fluorescent dye is used, a black light must be used to detect the dye.

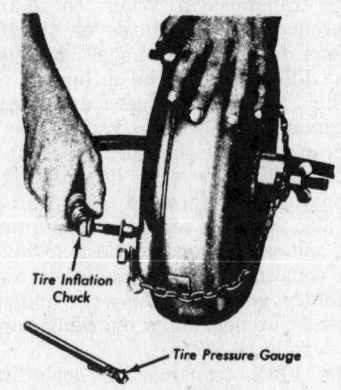

Converter leak tool installation (© Ford Motor Co.)

Engine Idle Speed Check

Check the idle speed of the engine and adjust it at the carburetor using

verter with 20 psi air pressure. Then, place the converter in a tank of water and watch for air bubbles. If no air bubbles are seen, the converter is not leaking.

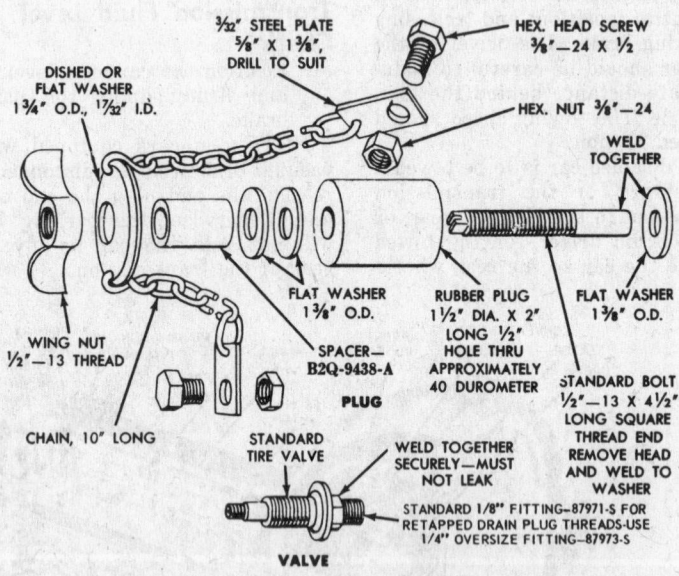

Converter leak checking tool (© Ford Motor Co.)

If further converter checks are necessary, remove the transmission from the car and the converter from the transmission. The converter cannot be disassembled for cleaning or repair. If the converter is leaking, it must be replaced with a new unit. The converter may be cleaned by using the Rotunda Automatic Transmission Torque Converter and Cooler Cleaner LRE-60081. To further check the converter for leaks, assemble and install the converter leak checking tool shown and fill the con-

the procedure given in the Car Section. Too slow an idle speed will cause the engine to run roughly; and too fast an idle speed will cause the car to creep while the transmission is in a Drive position, shift harshly, and downshift roughly.

Anti-Stall Dashpot Clearance Check

After properly adjusting the engine idle speed, check the clearance of the anti-stall dashpot using the procedure given in the Car Section.

Manual Linkage Checks

Correct manual linkage is necessary for the proper operation of the manual valve which helps control fluid pressure to various transmission components. Improperly adjusted manual linkage may cause fluid leaks if not corrected. See the section on linkage adjustments in the Car Section.

Control Pressure Check for Automatic Transmissions

When the vacuum diaphragm unit operates properly and the downshift linkage is adjusted correctly, all transmission shifts (automatic and kickdown) should occur within the specified road speed limits. If these shifts do not occur within the limits or if the transmission slips during a shift point, perform the following procedure to locate the problem:

1. Connect the Automatic Transmission Tester (see illustration) as follows:

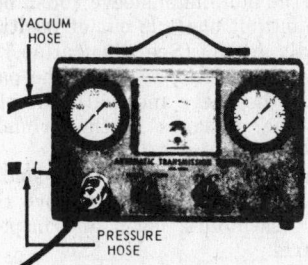

Rotunda ARE-2905 automatic transmission tester
(© Ford Motor Co.)

a. Tachometer cable to engine.
b. Vacuum gauge hose to the transmission vacuum diaphragm unit (see illustration).

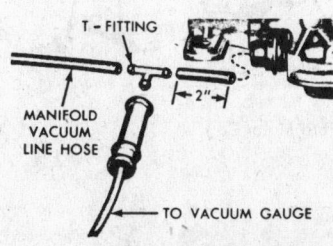

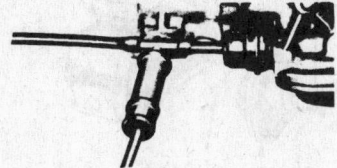

Typical vacuum test line connections
(© Ford Motor Co.)

c. Pressure gauge to the control pressure outlet on the transmission (see illustration).

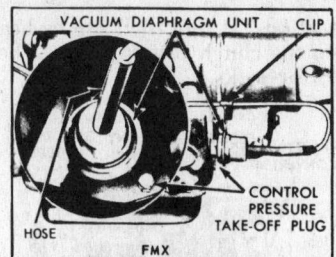

Typical control pressure connecting points
(© Ford Motor Co.)

2. Apply the parking brake and start the engine. On a car equipped with a vacuum brake release, disconnect the vacuum line or use the service brakes since the parking brake will release automatically when the transmission is put in any Drive position.

3. Check engine idle speed and throttle and downshift linkage for correct operation. Check the transmission diaphragm unit for leaks.

Vacuum Diaphragm Unit Check

Non-Altitude Compensating Type

1. Remove the vacuum diaphragm unit from the transmission after disconnecting the vacuum hose. See illustration.

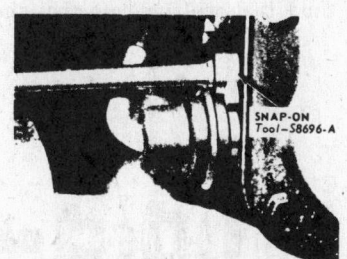

Removing or installing vacuum diaphragm
(© Ford Motor Co.)

2. Adjust a distributor tester equipped with a vacuum pump until the vacuum gauge shows 18 inches with the vacuum hose blocked. See illustration.

Testing transmission vacuum unit for leakage
(© Ford Motor Co.)

3. Connect the vacuum hose to the vacuum diaphragm unit and note the reading on the vacuum gauge. If the reading is 18 inches of vacuum, the vacuum diaphragm unit is good. While removing the vacuum hose from the vacuum diaphragm unit, hold a finger over the end of the control rod. As the vacuum is released, the internal spring of the vacuum diaphragm unit will push the control rod out.

Altitude-Compensating Type

The vacuum diaphragm unit may be checked for damaged or ruptured bellows by doing the procedure below:

1. Remove the diaphragm and the throttle valve rod from the transmission.

2. Insert a rod into the diaphragm unit until it is seated in the hole. Make a reference mark on the rod where it enters the diaphragm (see illustration).

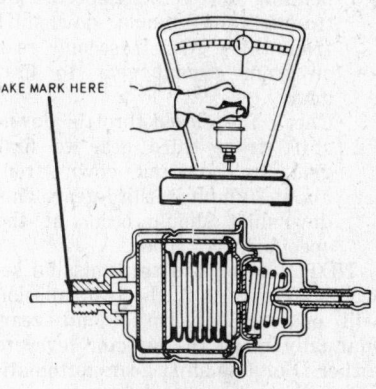

BELLOWS INTACT

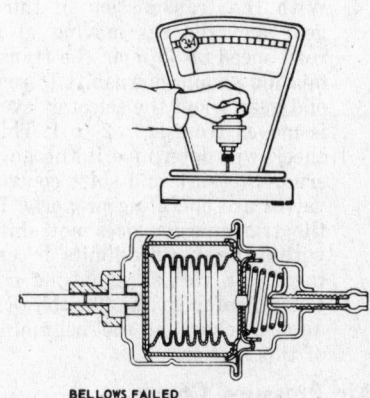

BELLOWS FAILED

Checking vacuum unit bellows-altitude compensating type
(© Ford Motor Co.)

3. Place the diaphragm unit on a weighing scale with the end of the rod resting on the weighing pan and gradually press down on the diaphragm unit.

4. Note the force (in pounds) at which the reference mark on the rod moves into the diaphragm. If the reference mark is still vis-

ible at 12 pounds of pressure on the scale, the diaphragm bellows is good. But if the reference mark on the rod disappears before four pounds of force, the diaphragm bellows are damaged and the diaphragm unit must be replaced.

Shift Point Checks for Automatic Transmissions

To determine if the transmission is shifting at the proper road speeds, do the following procedure:

1. Check the minimum throttle up-shifts by placing the transmission selector lever in the Drive position and noting the road speeds at which the transmission shifts from first gear to second gear to third gear. All shifts should occur within the specified limits.

2. While driving in third gear, depress the accelerator pedal past the detent (to the floor). Depending on vehicle speed, the transmission should downshift from third gear to second gear or from second gear to first gear.

3. Check the closed-throttle downshift from third gear to first gear by coasting down from about 30 mph in third gear. This downshift should occur at the specified road speed.

NOTE: when the transmission selector lever is at 2, the transmission will operate only in second gear. Manually move the selector lever to either D or 1 to shift gears automatically.

4. With the transmission in third gear and the car moving at a road speed of 35 mph, the transmission should downshift to second gear when the selector lever is moved from D to 2 to 1. This check will determine if the governor pressure and shift control valves are operating properly. If the transmission does not shift within the specified limits or certain gear ratios cannot be obtained, refer to the Trouble Diagnosis chart at the beginning of this section.

Air Pressure Checks

If the car will not move in one or more ranges, or, if it shifts erratically, the items at fault can be determined by using air pressure at the indicated passages.

Drain the transmission and remove the oil pan and the control valve assembly.

NOTE: oil will spray profusely during this operation.

Front Clutch

Apply sufficient air pressure to the front clutch input passage. (See il-

lustration.) A dull thud can be heard when the clutch piston moves. Check also, for leaks.

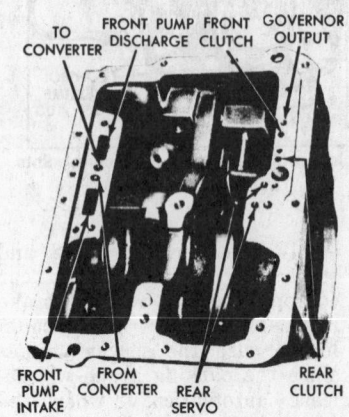

FRONT PUMP INTAKE   FROM CONVERTER   REAR SERVO   REAR CLUTCH

TO CONVERTER   FRONT PUMP DISCHARGE   FRONT CLUTCH   GOVERNOR OUTPUT

Case fluid hole identification—FMX automatic (© Ford Motor Co.)

Governor

Remove the governor inspection cover from the extension housing. Apply air to the front clutch input passage. (See illustration). Listen for a sharp click and watch to see if the governor valve snaps inward as it should.

Rear Clutch

Apply air to the rear clutch passage (See illustration) and listen for the dull thud that will indicate that the

rear clutch piston has moved. Listen also for leaks.

Front Servo

Apply air pressure to the front servo apply tube (See illustration.) and note if front band tightens. Shift the air to the front servo release tube, which is next to the apply tube, and watch band release.

Rear Servo

Apply air pressure to the rear servo apply passage. The rear band should tighten around the drum.

Conclusions

If the operation of the servos and clutch is normal with air pressure, the no-drive condition is due to the control valve and pressure regulator valve assemblies, which should be disassembled, cleaned and inspected.

If operation of the clutches is not normal; that is, if both clutches apply from one passage or if one fails to move, the aluminum sleeve (bushing) in the output shaft is out of position or badly worn. (See illustration.)

Use air pressure to check the passages in the sleeve and shaft, and also check the passages in the primary sun gear shaft.

If the passages in the two shafts and the sleeve are clean, remove the clutch assemblies, clean and inspect the parts.

Erratic operation can also be caused by loose valve body screws. When re-

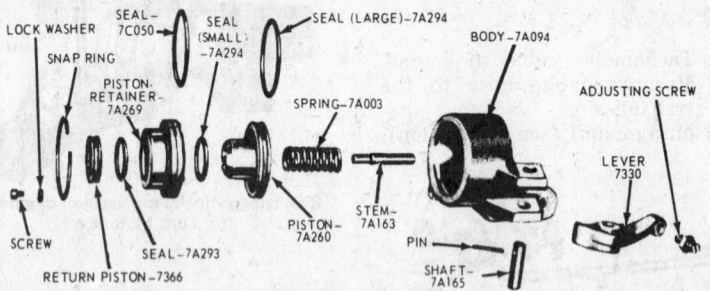

Front servo disassembled (© Ford Motor Co.)

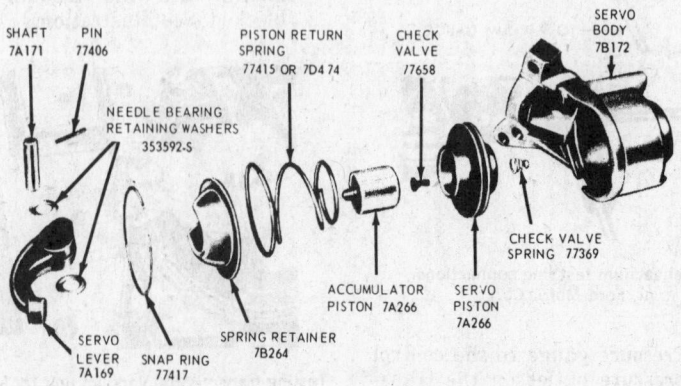

Rear servo disassembled (© Ford Motor Co.)

installing the valve body be careful to tighten: the pressure regulator valve to case bolts to 17-22 ft. lbs., the pressure regulator valve cover screws to 20-30 in. lbs., the control valve body screws to 20-30 in. lbs., the 1/4-20 capscrew (lower to upper valve body) to 4-6 ft. lbs., and the control valve body to case bolts to 8-10 ft. lbs.

In-Vehicle Adjustments and Repairs

The adjustments and repairs presented in this part of the section on FMX transmissions may be done without removing the entire transmission from the car. Some of these procedures will require the use of special tools and instruments. A list of special tools is given at the end of the section.

Transmission Fluid Drain and Refill

Normal maintenance and lubrication requirements do not include periodic changes of transmission fluid. Only when it is necessary to remove the pan for major repairs or adjustments will it be necessary to replace the transmission fluid. At this time the converter, oil cooler core, and cooler lines should be thoroughly flushed out to remove any dirt or deposits that might clog these units later.

When filling a completely dry (no fluid) transmission and converter, install five quarts of transmission fluid and then start the engine. Shift the selector lever through all positions briefly and set at Park position. Check the fluid level and add enough fluid to raise the level to between the ADD and FULL marks on the dipstick. *Do not overfill the transmission.*

The procedure for a partial drain and refill of the transmission fluid is given below:

1. Raise the car on a hoist or jack stands.
2. Place a drain pan under the transmission pan.
3. Loosen the pan attaching bolts to allow the fluid to drain.
NOTE: on FMX Transmissions (also called Cruise-O-Matic and Merc-O-Matic) used in 1968 and earlier models of Ford, the transmission fluid is drained by disconnecting the fluid filler tube from the transmission fluid pan.
4. When the fluid has stopped draining to level of the pan flange, remove the pan bolts starting at the rear and along both sides of the pan, allowing the pan to drop and drain gradually.

5. When all the transmission fluid has drained, remove the pan and the fluid filter and clean them. Discard the old pan gasket.
6. After completing the transmission repairs or adjustments, install the fluid filter screen, a new pan gasket, and the pan on the transmission. Tighten the pan attaching bolts to the proper torque (10-13 ft. lbs.).
7. Install three quarts of transmission fluid through the filler tube. If the filler tube was removed to drain the transmission, install the filler tube using a new O-ring.
8. Start and run the engine for a few minutes at low idle speed and then at the fast idle speed (about 1200 rpm) until the normal operating temperature is reached. *Do not race the engine.*
9. Move the selector lever through all positions and place it at the Park position. Check the fluid level, and add fluid till the level is between the ADD and FULL marks on the dipstick. *Do not overfill the transmission.*

Band Adjustments

Front Band Adjustment

When it is necessary to adjust the front band of the transmission, perform the following procedure:

1. Drain the transmission fluid and remove the oil pan, fluid filter screen, and clip. The same transmission fluid may be reused if it is filtered through a 100-mesh screen before being installed. Only transmission fluid in good condition should be used.
2. Clean the pan and filter screen and remove the old gasket.
3. Loosen the front servo adjusting screw locknut. See illustration.

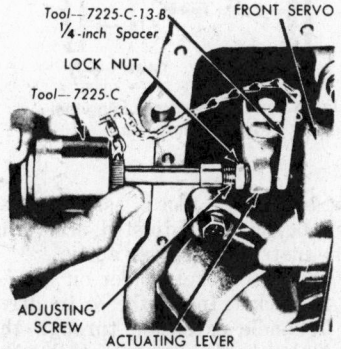

Adjusting front band—Typical
(© Ford Motor Co.)

NOTE: special band adjusting wrenches are recommended to do this operation correctly and quickly. See the list of special tools.

4. Pull back the actuating rod and insert a 1/4 in. spacer bar between the adjusting screw and the servo piston stem. Tighten the adjusting screw to 10 in. lbs. torque. Remove the spacer bar and tighten the adjusting screw an additional 3/4 turn. Hold the adjusting screw fast and tighten the locknut securely (20-25 ft. lbs.).
NOTE: on 1964-65 Lincoln transmissions, the procedure is slightly different. The locknut should be loosened and the adjusting screw backed off. Retighten the adjusting screw to 10 ft. lbs. torque, then back it off exactly three full turns. Tighten the locknut securely. *Severe damage to the transmission may result if the adjusting screw is not backed off exactly three full turns.*
5. Install the transmission fluid filter screen and clip. Install the pan with a new pan gasket.
6. Refill the transmission to the FULL mark on the Dipstick. Start the engine, run for a few minutes, shift the selector lever through all positions, and place it in Park. Recheck the fluid level again and add fluid to proper level if necessary.

Alternate Front Band Adjustment

An alternate method of adjusting the front band is presented here. The method is the same as the one given above except for the tools that are used. The procedure is as follows:

1. Drain the transmission and remove and clean the pan and fluid filter. Discard the old gasket.
2. Loosen the front servo adjusting screw locknut two full turns with a 9/16 in. wrench. Check that the adjusting screw rotates freely.
3. Pull the actuating lever back and insert the 1/4 in. spacer bar between the servo piston stem and the adjusting screw. Install the socket drive handle on the 9/16 in. socket.
4. Insert the T-handle extension through the socket handle and put the screwdriver socket on the T-handle extension.
5. Put the assembled adjusting tool on the adjusting screw so that the screwdriver socket is on the screw and the 9/16 in. socket is on the adjusting screw locknut.
6. Put a torque wrench on the T-handle extension and tighten the adjusting screw to 10 in. lbs. torque.
7. Remove the 1/4 in. spacer bar and tighten the adjusting screw an additional 3/4 turn. Hold the adjusting screw steady and tighten the locknut to 20-25 ft. lbs. torque.

8. Install the fluid filter screen, clip, and pan with a new gasket on the transmission. Refill the transmission as given earlier.

Rear Band Adjustments

The rear band of the FMX transmission may be adjusted by any of the methods given below. On most cars the basic external band adjustment is satisfactory. The internal band adjustment procedure may be done when the external adjustment procedure cannot be done correctly. On cars with a console floor shift, the entire console and shift lever and linkage will have to be removed to gain access to the rear band external adjusting screw.

Rear Band External Adjustment

The procedure for adjusting the rear band externally is as follows:

1. Locate the external rear band adjusting screw on the transmission case, clean all dirt from the threads, and coat the threads with light oil.
   NOTE: the adjusting screw is located on the upper right side of the transmission case. Access is often through a hole in the front floor to the right of center under the carpet.
2. Loosen the locknut on the rear band external adjusting screw. On 1970 Cougar and Mustang cars, use the special tool illustrated to loosen the locknut.

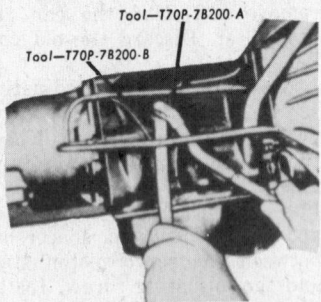

Adjusting rear band externally— 1970 Mustang and Cougar
(© Ford Motor Co.)

Adjusting rear band externally— 1970 Ford, Meteor, Fairlane, and Montego
(© Ford Motor Co.)

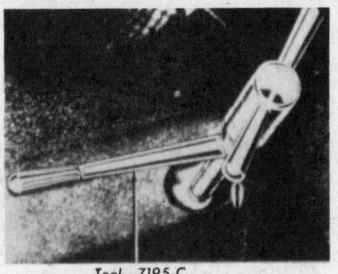

Adjusting rear band externally— 1964 to 1969 Ford and Meteor
(© Ford Motor Co.)

3. Using the special preset torque wrench shown, tighten the adjusting screw until the handle clicks at 10 ft. lbs. torque. If the adjusting screw is tighter than 10 ft. lbs. torque, loosen the adjusting screw and retighten to the proper torque.
4. Back off the adjusting screw 1½ turns. Hold the adjusting screw steady while tightening the locknut to the proper torque (35-40 ft. lbs.). *Severe damage may result if adjusting screw is not backed off exactly 1½ turns.*

Alternate Rear Band External Adjustment

The alternate method of adjusting the rear band is basically the same as the procedure given above. The difference between the two methods is that the alternate procedure uses the band adjusting tool shown in the

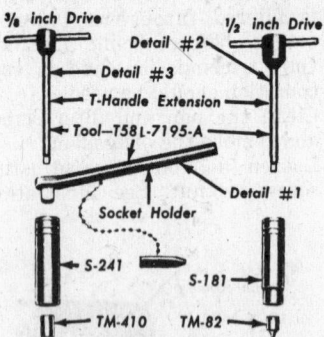

Front and rear band adjusting tools
(© Ford Motor Co.)

illustration. The tool is assembled as given below and adjusted according to the instructions given above.

1. Put the socket holder on the ¾ in. socket. Insert the ⅜ in. drive T-handle extension through the socket holder and socket. Put the 5/16 in., 8-point socket on the extension. Place a torque wrench on the T-handle extension.
2. Place the assembled adjusting tool on the adjusting screw so that it engages the adjusting screw and locknut and do the procedure given above.

3. After adjusting the rear band correctly, hold the adjusting screw steady and tighten the locknut securely.

Rear Band Internal Adjustment

The rear band is adjusted internally by doing the following procedure:

1. Drain the transmission fluid. If it is to be reused, pour the fluid through a 100-mesh screen as it drains from the transmission. Reuse the transmission fluid only if it is in good condition.
2. Remove and clean the pan, fluid filter, and clip. Discard the old pan gasket.
3. Loosen the rear servo adjusting locknut.
4. Pull the adjusting screw end of the actuating lever away from the servo body and insert the spacer tool (see illustration) be-

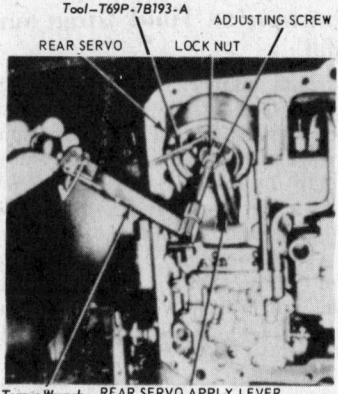

Adjusting rear band internally— 1969 Fairlane, Montego, Mustang, and Cougar
(© Ford Motor Co.)

tween the servo accumulator piston and the adjusting screw. *Be sure the flat surfaces of the tool are placed squarely between the adjusting screw and the accumulator piston. Tool must not touch servo piston and the handle must not touch the servo piston spring retainer.*
5. Using a torque wrench with an Allen head socket, tighten the adjusting screw to 24 in. lbs. torque.
6. Back off the adjusting screw exactly 1½ turns. Hold adjusting screw steady and tighten the locknut securely. Remove the spacer tool.
7. Install the fluid filter, clip, and pan with a new gasket.
8. Fill the transmission with the correct amount of fluid.

Transmission Component Removal and Installation

The various components of the FMX transmission that may be re-

moved while the transmission is in the car are given in this part of the FMX transmission section. Installation is often the reverse of the removal instructions except for adjustment and alignment. Repair of the individual components is given in the repair section.

Governor Assembly Removal and Installation

The governor assembly may be removed from transmissions built before 1968 through an extension housing access or inspection plate. On 1968 and later models of the FMX transmission, the extension housing must be removed with all attaching parts to remove the governor assembly from the output shaft. It may be necessary to remove the entire transmission from the car to do this removal procedure.

The removal procedure for the governor assembly is as follows:

1. Remove the governor inspection plate from the right side of the

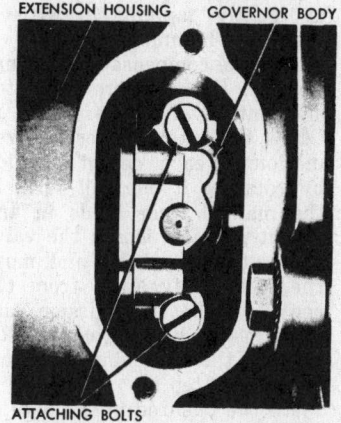

Removing governor inspection plate on extension housing—1964 to 1967
(© Ford Motor Co.)

extension housing (see Illustration), and rotate the driveshaft until the governor appears in the opening.

2. Remove the two screws holding the governor body to the counter weight sleeve. Be careful not to drop the screws or the valve into the extension housing.

3. Disassemble the governor by removing the two screws holding the side plate. Inspect the governor valve for scoring and free movement in its bore. Be sure the spring, sleeve, and plug are not damaged. Blow valve body passages clear of obstructions.

4. Reassemble the governor assembly (see illustration), aligning the passages in the sleeve and body. Tighten the valve body cover screws with 20-30 in. lbs. torque.

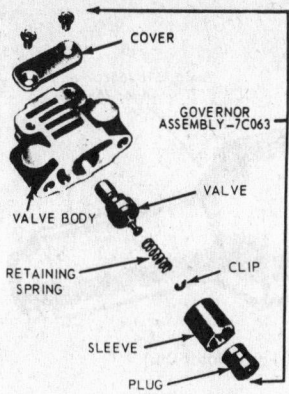

Governor disassembled
(© Ford Motor Co.)

5. Install the governor assembly on the counterweight sleeve, tightening the mounting screws with 50-60 in. lbs. torque. Replace the governor inspection plate on the extension housing.

Extension Housing Removal and Installation

1. Raise the car high enough for easy access to the extension housing.

2. Drain the transmission.

3. Disconnect the driveshaft from the rear axle and slide the front yoke from the extension housing.

4. Disconnect the speedometer cable from the extension housing.

5. Remove the rear engine support nuts. Place a transmission jack under the transmission and raise it enough to clear the crossmember.

6. Remove the bolts and nuts securing the crossmember to the side rails of the frame. Move the crossmember out of the way.

7. Remove the attaching bolts holding the engine rear support to the extension housing and remove the rear support from the extension housing.

8. Remove all the extension housing attaching bolts, slide the housing off the output shaft, and discard the gasket.

9. Installation is the reverse of removal procedures. Tighten the

housing attaching bolts to 30-40 ft. lbs. torque. Check that the output shaft rotates freely by hand. If it binds or feels tight, check the needle bearing and race for correct position.

Extension Housing Bushing and Rear Seal Removal and Installation

1. Disconnect the driveshaft from the extension housing.

2. Carefully remove the rear seal from the housing using the special tools shown in the illustration.

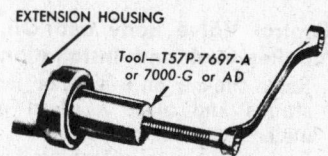

Removing extension housing bushing
(© Ford Motor Co.)

3. Remove the extension housing bushing using the bushing remover tool (see illustration). *Be careful not to damage the spline seal.*

4. Install new bushing in the extension housing using the special tool (see illustration).

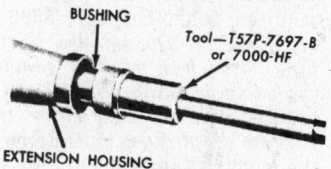

Installing extension housing bushing
(© Ford Motor Co.)

5. Inspect the universal joint yoke sealing surface for scoring or gouges. Replace the yoke if damaged.

6. Inspect the housing counterbore for burrs and, if necessary, smooth with crocus cloth.

7. Using the special tool shown, install the new rear seal into the end of the extension housing. Check that the seal is firmly seated.

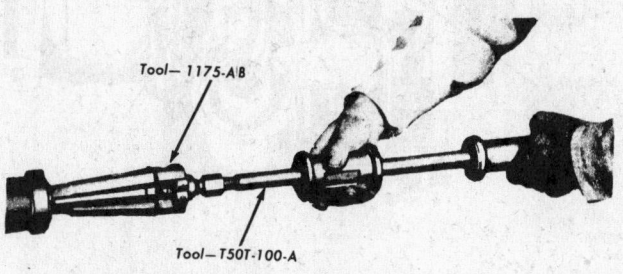

Removing extension housing seal (© Ford Motor Co.)

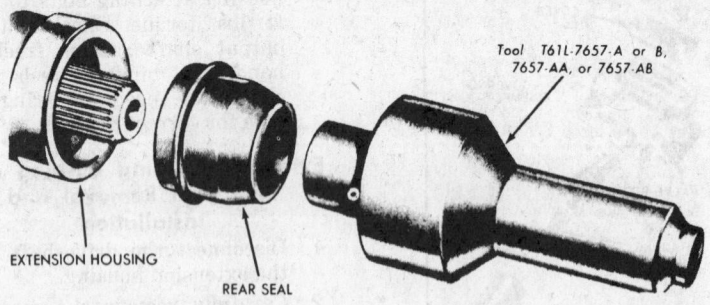

EXTENSION HOUSING

REAR SEAL

Tool T61L-7657-A or B, 7657-AA, or 7657-AB

Installing extension housing seal (© Ford Motor Co.)

Control Valve Body and Oil Pan Removal and Installation

1. Raise the car on a hoist or jackstands and place a drain pan under the transmission.
2. Loosen the pan attaching bolts so that the transmission fluid drains down to the level of the pan flange. Then, starting at the rear of the pan and working forward along the sides, remove the pan attaching bolts until all the fluid has drained.
3. Remove the oil pan, fluid filter screen, and clip and clean them thoroughly. Discard the old pan gasket.
4. Remove the vacuum diaphragm assembly using the special tool required (Snap-On tool S8696-A or FCO-24). *Do not use pliers, pipe wrenches, etc. to remove the vacuum diaphragm unit. Do not let any solvents enter the vacuum diaphragm unit.* Remove the push rod and the fluid screen and its retaining clip.

5. Remove the small compensator pressure tube. See illustration.
6. Disconnect the main pressure oil tube by carefully loosening the end connected to the control valve body first and then removing the tube from the pressure regulator unit.
   CAUTION: be sure to remove the tube in this manner. Otherwise, the tube could be kinked or bent causing improper fluid pressures and possible damage to the transmission.
7. Loosen the front servo attaching bolts about three turns.
8. Remove the three control valve body attaching bolts and carefully lower the valve body, sliding it off the front servo tubes. *Do not damage the valve body or the tubes.*
9. Disassemble the control valve body into the various parts as shown in the illustration in the major repair section.

10. When installing the control valve body, align the front servo tubes with the holes in the valve body. Shift the manual lever to the 1 detent and place the inner downshift lever between the downshift lever stop and the downshift valve. Be sure the manual lever engages the actuating pin in the manual detent lever.
11. Loosely install the control valve body attaching bolts and move the control valve body toward the center of the transmission case until there is a clearance of 0.050 in. between the manual valve and the actuating pin on the manual detent lever.
12. Tighten the attaching bolts to 8-10 ft. lbs. torque. Ensure that the rear fluid filter retaining clip is installed under the valve body.
13. Install the main pressure oil tube, connecting the end to the pressure regulator unit first and then connecting the other end to the main control valve assembly by gently tapping it with a soft-faced hammer.
14. Install the compensator pressure tube on the pressure regulator and control valve body.
15. Check the manual lever for free motion in each detent position by rotating it one full turn. If the manual lever binds in any detent position, loosen the valve body attaching bolts and move the valve body away from the center of the transmission case until the binding is relieved. Retighten the attaching bolts according to step 12.
16. Place the pushrod in the bore of the vacuum diaphragm unit and insert the other end of the pushrod into the threaded opening in the case. Install the vacuum diaphragm unit and tighten it to 20-30 ft. lbs. torque.
17. Tighten the front servo attaching bolts to 30-35 ft. lbs. torque.
18. Adjust the front band.
19. Install the fluid filter and its retaining clip.
20. Install the oil pan with a new pan gasket, tightening the pan attaching bolts to 10-13 ft. lbs. torque.
21. Adjust the rear band. If necessary to adjust the rear band internally, do the adjustment immediately after the front band adjustment.
22. Fill transmission with fluid. Start and run engine for a few minutes and check the fluid level after shifting the transmission through all positions. *Do not overfill the transmission.*
23. Check the adjustment of the transmission control linkage.

COMPENSATOR PRESSURE TUBE

MAIN PRESSURE OIL TUBE

VACUUM DIAPHRAGM

FRONT SERVO FLUID TUBES

PRESSURE REGULATOR

REAR SCREEN RETAINING CLIP

FRONT SERVO

REAR SERVO

Typical hydraulic control system (© Ford Motor Co.)

Pressure Regulator Removal and Installation

The pressure regulator unit may be removed from the transmission while it is in the car.

1. Drain the transmission of fluid and remove the pan, fluid filter screen, and its retaining clip. Discard the used pan gasket.
2. Remove the compensator pressure tube from between the control valve body and the pressure regulator. See illustration for location of components.
3. Remove the main pressure oil tube by gently prying off the end connected to the control valve first and then disconnect the other end from the pressure regulator. *Be sure to remove the tube in this order to prevent kinking or bending it.*
4. Loosen the spring retainer clip and carefully release the spring tension on the pressure springs. Remove the valve springs, retainer and valve stop, and the valves from the pressure regulator body.
5. Remove the pressure regulator attaching bolts and washers and take the regulator body out of the transmission case.
6. After cleaning, inspection, and reassembly, install the pressure regulator unit in the transmission by reversing the procedures in steps 1 through 5.

Front Servo Removal and Installation

1. Drain the transmission fluid from the transmission case and remove the pan, fluid filter screen, and its retaining clip.
2. Remove the vacuum diaphragm unit.
3. Loosen the three control valve body attaching bolts.
4. Remove the front servo attaching bolts, hold the band strut steady with the fingers, and remove the front servo unit.
5. After repairing the front servo unit, install it by first positioning the front band forward in the transmission case with the end of the band facing downward. Be sure the front servo anchor pin is placed in the case web. Align the large end of the servo strut with the servo actuating lever, and align the small end with the band end.
6. Rotate the band, strut, and servo to align the anchor end of the band with the anchor in the case. Push the servo unit onto the control valve body tubes.
7. Install the attaching bolts and tighten them to 30-35 ft. lbs. torque.
8. Tighten the control valve body attaching bolts to 8-10 ft. lbs.

torque. Check the clearance (0.050 in.) between the manual valve and the manual lever actuating pin.
9. Adjust the front band.
10. Install the vacuum diaphragm unit and its pushrod.
11. Install the fluid filter screen, its retaining clip, and the pan with a new pan gasket.
12. Fill the transmission with fluid.
13. Adjust the downshift and manual shift linkage.

Rear Servo Removal and Installation

1. Drain the transmission fluid from the transmission, and remove the pan, fluid filter screen, and its retaining clip.
2. Remove the vacuum diaphragm unit (see Vacuum Diaphragm Removal procedure).
3. Remove the control valve body and the two front servo tubes.
4. Remove the rear servo attaching bolts, hold the actuating and anchor struts with the fingers, and remove the rear servo unit.
5. Before installing the rear servo unit, position the servo anchor strut on the servo band and rotate the band to engage the strut.
6. While holding the servo anchor strut in place with the fingers, position the actuating lever strut and install the rear servo unit in place.
7. Loosely install the rear servo attaching bolts, with the longer bolt in the inner bolt hole.
8. Move the rear servo unit towards the center of the transmission case against the attaching bolts. While holding the servo in this position, tighten the attaching bolts to 40-45 ft. lbs. torque.
9. Install the two front servo tubes and the control valve body. Check for proper clearance (0.050 in.) between the manual valve and the manual actuating pin.
10. Adjust the rear band using the procedures given in the Rear Band Adjustments.
11. Install the fluid filter screen, its retaining clip, and the oil pan with a new gasket. Fill the transmission with fluid.

Parking Pawl Removal and Installation

The parking pawl assembly may be removed from the transmission while it is in the car.

1. Raise the car and drain the fluid from the transmission. *Do not remove the pan.*
2. Place the special engine support bar (T65E-6000J) under the converter housing.

3. Remove the driveshaft from the rear axle and slide the front yoke out of the extension housing.
4. Remove the two nuts securing the engine rear support to the crossmember.
5. Place a transmission jack under the transmission and raise the engine and transmission enough to take the weight off the crossmember.
6. Remove the bolts and nuts securing the crossmember to the frame side rails and slide the crossmember out of the way. Lower the jack slowly until the engine and transmission are resting on the special engine support bar.
7. Remove the two rear support-to-extension housing bolts and remove the rear support.
8. Disconnect the speedometer cable from the extension housing.
9. Remove the pan, fluid filter screen, and its retaining clip.
10. Loosen the rear band adjusting screw locknut and tighten the adjusting screw to 24 in. lbs. torque.

NOTE: the rear band will tighten around the planet carrier and hold the planet carrier and clutch assemblies in position during repair operations on the parking pawl.

11. Remove the small compensator pressure tube from the pressure regulator and control valve body.
12. Remove the main pressure oil tube from between the main control valve body and the pressure regulator. Be sure to disconnect the end of the tube connected to the control valve body first to avoid kinking or bending the tube.
13. Disconnect the vacuum diaphragm unit after removing the vacuum hose.
14. Loosen the front servo attaching bolts.
15. Remove the three control valve body attaching bolts and lower the valve body while carefully pulling it off the front servo tubes.
16. Remove the rear servo attaching bolts and the rear servo unit.
17. Remove the extension housing attaching bolts and housing.
18. Remove the output shaft and rear support assembly.
19. Remove the parking pawl pin from the case with a magnet.
20. Working from inside the case, tap the shoulder of the toggle lever pin to move the retaining plug part way out of the case. Remove the plug with pliers.
21. Loosen and remove the toggle lever pin by alternately sliding the toggle lever from side to side

and pushing outward at the same time until it is loose. Lift the pawl and toggle lever from the case as one assembly.

22. After replacing worn or damaged parts, reinstall the new parking pawl and link assembly with the toggle lever pin and plug.
23. Secure the pawl to the case with the pawl pin.
24. Install the rear support and output shaft after putting a new gasket on the rear support and the thrust washer is in place. Make sure the pressure tubes are positioned properly in the case.
25. Position a new gasket on the extension housing and secure the extension housing to the transmission case by tightening the attaching bolts.
26. Install the rear servo and struts.
27. Install the main control valve body assembly as given in the procedure on *Oil Pan and Control Valve Body Removal and Installation*.
28. Tighten the front servo attaching bolts to 30-35 ft. lbs. torque.
29. Install the main pressure oil tube and the small compensator pressure tube to the control valve body and the pressure regulator.
30. Adjust the front band.
31. Adjust the rear band.
32. Install the vacuum control rod, the vacuum diaphragm unit, and attach the vacuum hose.
33. Install the fluid filter screen, retaining clip, and pan with a new gasket.
34. Connect the speedometer cable to the extension housing.
35. Secure the engine rear support to the extension housing with the two attaching bolts.
36. Place transmission jack under the transmission and raise the transmission and engine high enough to position and secure the rear support crossmember to the frame side rails. Tighten the attaching bolts and nuts to the specified torque.
37. Lower the transmission jack till the transmission and engine are resting on the rear support crossmember. Install and tighten to specified torque the nuts securing the rear support to the crossmember. Remove the transmission jack and the engine support bar.
38. Adjust the transmission control linkage.
39. Install the driveshaft and lower the car.
40. Fill the transmission with correct amount of fluid.

Transmission Overhaul Procedures

The transmission overhaul procedures presented here are the checks and repairs that must be done with the transmission out of the car. Disassembly of each transmission subassembly is illustrated by exploded views of the subassembly showing how the individual parts fit together. Reassembly of the subassembly is often the reverse of the disassembly procedure except for alignment, special tolerances, etc.

Procedures for removing the transmission from the car and reinstalling it back in the car are given in the Car Section.

During all repairs to the transmission subassemblies, the following instructions must be followed:

1. Be sure that no dirt or grease gets in the transmission. All parts must be clean. *Remember—a little dirt can disable a transmission completely if it gets in a fluid passage.*
2. Handle all transmission parts carefully to avoid burring or nicking bearing or mating surfaces.
3. Lubricate all internal parts of the transmission with clean transmission fluid before assembling. *Do not use any other lubricants except on gaskets or thrust washers which may be coated lightly with vaseline to ease assembly.*
4. Always use new gaskets when assembling the parts of the transmission.
5. Tighten all bolts and screws to the recommended torque limits using a torque wrench.

Transmission Disassembly

1. Thoroughly clean the outside of the transmission to prevent dirt or grease from getting inside the mechanism. *Do this before removing any subassembly.*
2. Place the transmission in the transmission holder. See illustration.
3. Remove the transmission oil pan, gasket, and fluid filter retaining clip.
4. Lift the fluid filter screen off the forward tube, and then off the rear tube.
5. Remove the spring seat from the pressure regulator. *Maintain constant pressure on the spring seat and release slowly to prevent spring distortion and personal injury.*
6. Remove the pressure regulator springs and pilots but do not remove the valves yet.
7. Loosen but do not remove the pressure regulator attaching bolts.
8. Remove the small compensator pressure tube from the pressure regulator and the control valve body.
9. Remove the main pressure oil tube from the pressure regulator and the main control valve body assembly. Gently pry off the end connected to the main control valve body first and then remove the tube from the pressure regulator. *Failure to do this may kink or bend the tube causing damage to the transmission later.*
10. Loosen the front and rear servo band adjusting screws five turns. Loosen the front servo attaching bolts three turns.
11. Remove the vacuum diaphragm unit and pushrod.
12. Remove the control valve body attaching bolts and align the levers to allow removal of the valve body. Lift the valve body up and pull it off the servo tubes. Place the valve body on a clean surface.
13. Remove the pressure regulator from the case. Keep the control

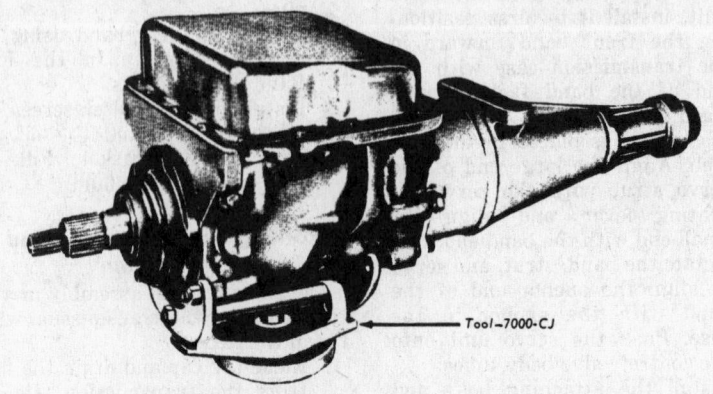

Tool-7000-CJ

Transmission mounted in holding fixture (© Ford Motor Co.)

pressure valve and the converter pressure valve in the regulator body to avoid damaging the valves.

14. Remove the front servo apply, accumulator, and release tubes by twisting and pulling at the same time. Remove the front servo attaching bolts. While holding the front servo strut with the fingers, lift the front servo from the case.

15. Remove the rear servo attaching bolts. While holding the actuating and anchor struts with the fingers, lift the rear servo from the case.

Transmission End-Play Check

The transmission end play check is done as given below:

1. Remove one of the front pump attaching bolts and mount the dial indicator support tool in the hole. Mount a dial indicator on the support so that a contact rests on the end of the turbine shaft. See illustration.

Case and Extension Housing Parts Removal

1. Remove the remaining front pump attaching bolts and the front pump assembly from the case. If necessary, tap the cap screw bosses to loosen the front pump from the case.

2. Remove the five case-to-extension housing attaching bolts. These bolts also hold the rear pump in the case. While holding the rear pump in position remove the housing carefully.

3. Remove the output shaft assembly (see illustration). To aid in removing the assembly, insert a screwdriver between the output shaft ring gear and the pinion carrier and move the assembly rearward. Be careful not to bend the pressure tubes between the rear support or distributor sleeve and the case as they are removed.

NOTE: on output shaft assemblies used in 1964 to 1966 transmissions, the speedometer drive gear is remov-

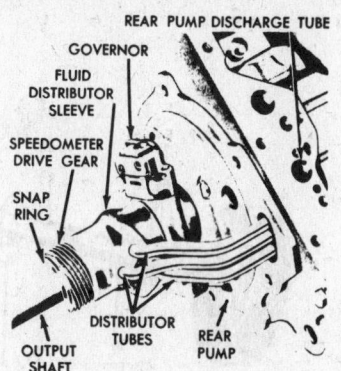

Rear pump and output shaft installed—1964-67 FMX transmission
(© Ford Motor Co.)

sleeve both rearward and forward on the shaft, and then, tapping the drive gear loose from the output shaft with a hammer and a small drift. The speedometer drive gear drive ball should fall out of its place in the output shaft, but it may need to be picked out. See the accompanying illustrations.

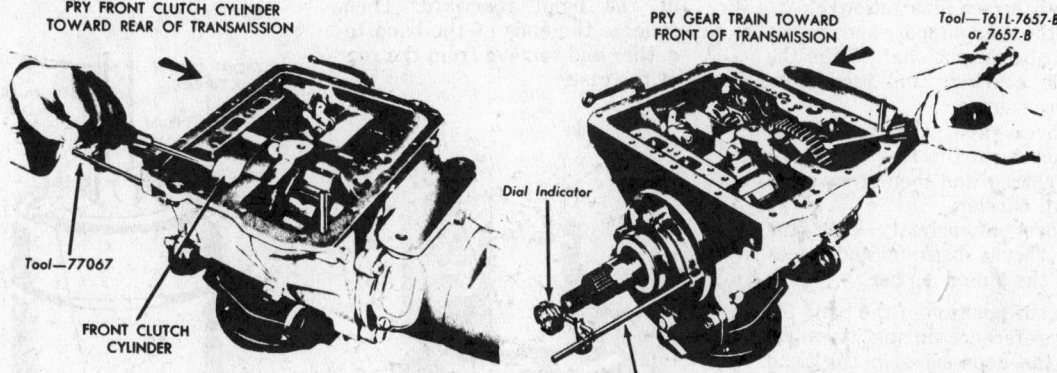

Transmission end play check (© Ford Motor Co.)

2. Install the extension housing seal replacer on the output shaft to provide support for the shaft.

3. Using a screwdriver, move the front clutch cylinder to the rear of the transmission case as far as possible. Set the dial indicator to zero while holding a slight pressure on the screwdriver.

4. Remove the screwdriver. Insert it between the large internal gear and the case and move the front clutch cylinder to the front of the case.

5. Record the indicator reading for later use during transmission reassembly. The end-play reading should be between 0.010-0.029 in. If the reading is not within these limits, a new selective thrust washer must be used when reassembling the transmission.

6. Remove the dial indicator, its support tool, and the extension housing seal replacer.

able. The nylon speedometer drive gear may be replaced by removing the snap-ring securing it on the shaft, prying the oil distributor

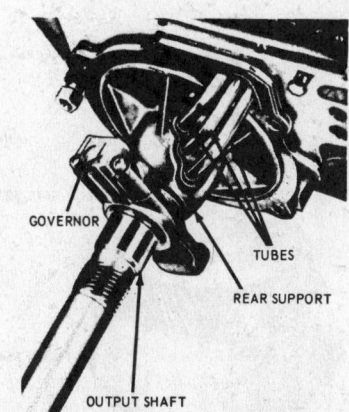

Rear support and output shaft installed—1968-70 FMX transmission
(© Ford Motor Co.)

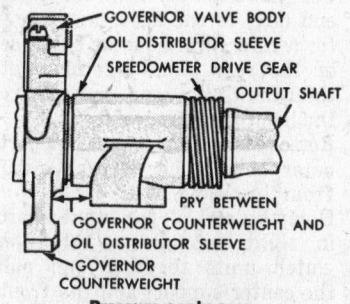

Pressure apply area for removing speedometer gear
(© Ford Motor Co.)

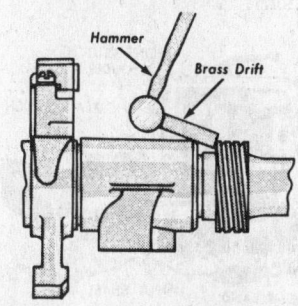

Removing speedometer drive gear—1964-66 Cruise-O-Matic (FMX) transmission
(© Ford Motor Co.)

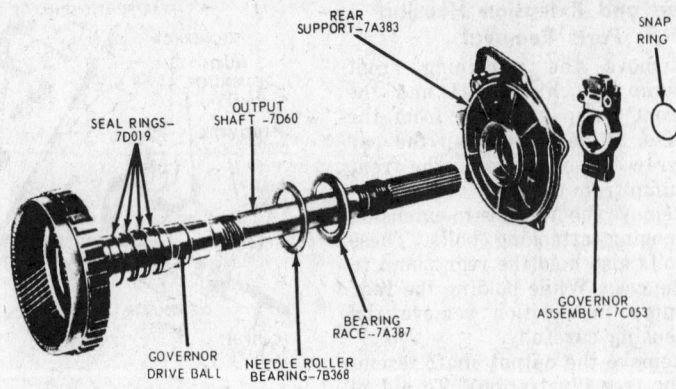

Output shaft disassembled (© Ford Motor Co.)

4. Remove the four seal rings from the output shaft using the fingers to prevent breaking them.

5. Remove the governor snap-ring from the output shaft. With a soft-faced hammer, tap the governor assembly off the output shaft. Remove the governor drive ball (see illustration).

6. Pull the rear support and gasket from the output shaft. Take the needle bearings and race from the rear support.

7. Remove the selective thrust washer from the rear of the pinion carrier and then remove the pinion carrier.

8. Remove the primary sun gear rear thrust bearing and races from the pinion carrier.

9. Note the position of the band for later reference during assembly. Use the depression in the band end next to the adjusting screw for a reference. Squeeze the ends of the rear band together, tilt the band to the rear, and remove the band from the case.

10. Remove the two center support outer bolts (one on each side) from the case.

11. Hold the end of the input shaft in tight enough to hold the clutch units together and pull the center support and the front and rear clutch assemblies from the case as one unit (see illustration).

12. Install the clutch assemblies in the bench fixture as shown. See illustration.

13. Remove the thrust washer from the front of the input shaft.

14. To remove the front band, position the band ends between the case webbing and tilt the bottom of the band rearward. Then, squeeze the ends of the band together and remove from the rear of the case.

Input shaft and clutch in holding fixture
(© Ford Motor Co.)

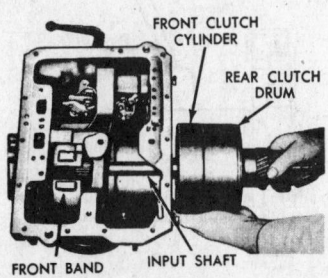

Removing or installing input shaft and clutch
(© Ford Motor Co.)

15. Lift the front clutch assembly from the primary sun gear shaft.

16. Remove the bronze and the steel thrust washers from the rear clutch assembly. Wire the thrust washers together to assure correct installation.

17. Remove the front clutch seal rings from the primary sun gear shaft.

18. Lift the rear clutch assembly from the primary sun gear shaft.

19. Remove the rear clutch seal rings from the primary sun gear shaft. *Do not break the seal rings.*

20. Remove the primary sun gear shaft front thrust washer.

Rear Brake Drum Support Bushing Removal and Installation

If the rear brake drum support bushing is to be replaced, press the bushing from the support as shown in the illustration. Then, press a new bushing into the brake drum support with the tool shown.

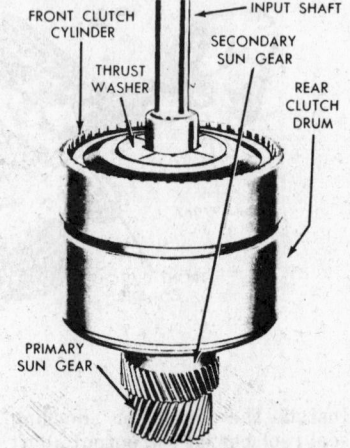

Installing output shaft bushing.
Tool used is T64L-7003-A7 (see arrow).
(© Ford Motor Co.)

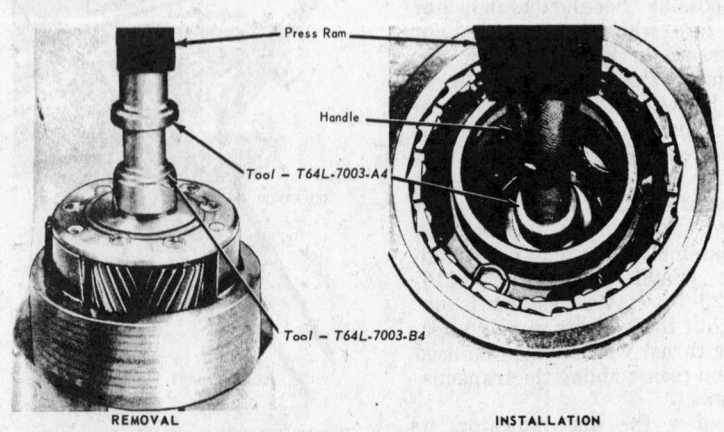

Replacing rear brake drum support bushing (© Ford Motor Co.)

Output Shaft Bushing Removal and Installation

1. Remove the worn bushing by using a cape chisel to break through the bushing wall.
2. Pry the loose ends of the bushing up with an awl and remove the bushing.
3. Install a new bushing on the output shaft hub by pressing it on with the installation tool as shown in the illustration.

Primary Sun Gear Shaft Seal Ring Replacement

1. Place the primary sun gear shaft in the clutch bench fixture.
2. Check the fit of the seal rings in their respective bores. If equipped with cast iron seal rings, there should be a clearance of 0.002-0.009 in. between the ends. If equipped with teflon seals that are worn or damaged, cut the seals from the shaft with a sharp knife. Do not score the ring grooves.
3. Replace the teflon seals with cast iron seal rings, and check for free movement in the groove.

Rear Clutch Disassembly and Assembly

The rear clutch may be removed and installed by doing the following procedure:

1. Remove the clutch pressure plate snap-ring and the pressure plate from the drum. Also, remove the waved cushion spring and the composition and steel plates.
NOTE: in 1967-70 transmissions, composition plates replaced bronze plates and a waved cushion spring was added.
2. Compress the spring with the tool shown and remove the snap-ring with snap-ring pliers.
3. Carefully guide the spring retainer while releasing spring pressure so that the retainer does not lock in the snap-ring grooves.

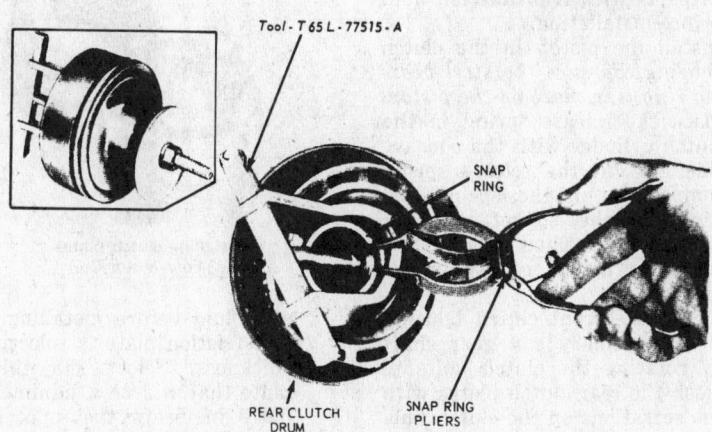

Tool - T 65 L - 77515 - A

SNAP RING

REAR CLUTCH DRUM

SNAP RING PLIERS

Removing rear clutch spring snap ring (© Ford Motor Co.)

4. Position the primary sun gear shaft in the rear clutch. Place an air hose nozzle in one of the holes in the shaft and block the other hole. Force the clutch piston out of the clutch drum with air pressure. *Hold one hand over the piston to prevent damaging it.*
5. Remove the inner and outer seal rings from the clutch piston.
6. Remove the rear clutch sun gear bushing if it is worn or damaged by cutting it out with a cape chisel. See illustration. When the bushing is cut through, pry the ends up with an awl and remove the bushing.
7. Install a new rear clutch sun gear bushing using the tool shown in the illustration.

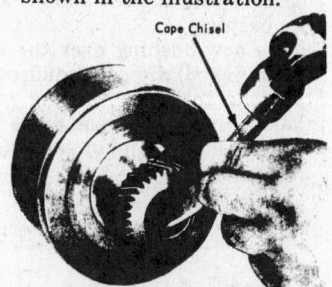

Cape Chisel

Removing rear clutch sun gear bushing (© Ford Motor Co.)

Tool - T64L-7003-A3 OR Tool - T64L-7003-A4

Handle

Installing rear clutch sun gear bushing (© Ford Motor Co.)

8. Install new inner and outer seal rings on the piston.
9. When installing the piston in the clutch drum, lubricate the piston seals and the tools with clean transmission fluid.
10. Push the small fixture down over the cylinder hub.
11. Insert the piston into the large fixture with the seal toward the thin-walled end. See illustration.

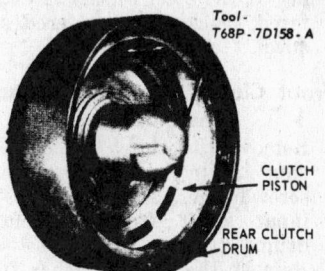

Tool - T68P - 7D158 - A

CLUTCH PISTON

REAR CLUTCH DRUM

Installing rear clutch piston (© Ford Motor Co.)

12. Hold the piston and large fixture and insert as one unit into the cylinder. Push down over the small fixture until the large tool stops against the shoulder in the cylinder; then push the piston down, out of the tool, until it bottoms in the cylinder. Remove the tools.
13. Install the clutch release spring and position the retainer on the spring.
14. Install the spring compressor retainer and install the snap-ring. *While compressing the spring, guide the retainer to avoid catching on the snap-ring groove. Be sure the snap-ring is fully seated in the groove.*
15. Install the composition and steel clutch plates. If new plates are installed, soak them in clean transmission fluid for 15 minutes before assembling them.
16. Install a steel clutch plate and the waved cushion spring. Then, install steel and composition friction plates alternately starting with a steel plate.
17. Install the clutch pressure plate with the bearing surface down. Then install the snap-ring, being sure the snap ring is fully seated in its groove.
18. Check the free pack clearance between the pressure plate and the snap-ring with a feeler gauge. The clearance should be 0.030-0.055 in. If the clearance is not within this range, install a selective snap-ring with one of the following thicknesses: 0.060-0.064, 0.074-0.078, 0.088-0.092, 0.102-0.106 in. Insert the correct snap-ring and recheck the clearance.
19. Install the thrust washer on the primary sun gear shaft. Lubri-

cate all parts with transmission fluid and install the two center seal rings.

20. Install the rear clutch on the primary sun gear shaft. *Be sure all loose needle bearings are in the hub correctly.* Install the two seal rings in the front grooves.

21. Install the steel and the bronze thrust washers on the front of the secondary sun gear assembly. If the steel washer is chamfered, place the chamfered side down.

Front Clutch Disassembly and Assembly

1. Remove the clutch cover snapping ring by prying it up with a screwdriver, and remove the input shaft from the clutch drum.

2. Remove the thrust washer from the thrust surface of the clutch hub. Lift the clutch hub straight up to remove it from the clutch drum.

3. Remove the composition and the steel clutch plates and the pressure plate from the clutch drum.

4. Place the front clutch spring compressor on the release spring, position the clutch drum on the bed of an arbor press, and compress the release spring with the arbor press until the release spring snap-ring can be removed (see illustration).

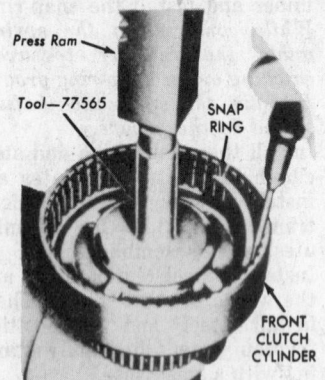

Removing or installing front clutch snap ring
(© Ford Motor Co.)

5. Remove the clutch release spring from the clutch drum.

6. Using the special air nozzle shown in the illustration, place the air nozzle against the clutch apply hole in the front clutch housing and force the piston out of the housing.

7. Remove the piston inner seal from the clutch housing and remove the piston outer seal from the groove in the piston.

8. Remove the input shaft bushing if it is worn or damaged. Use a cape chisel and cut along the bushing seam until the chisel

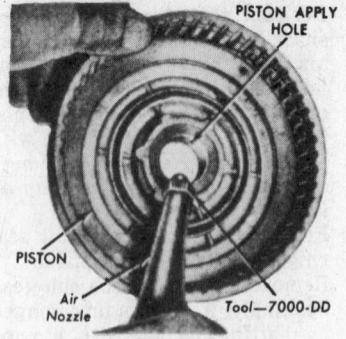

Removing front clutch piston
(© Ford Motor Co.)

breaks through the bushing wall. Pry up the loose ends of the bushing with an awl and remove the bushing.

9. Put a new bushing over the end of the installation tool and position the tool and bushing over the bushing hole. Then, press the bushing into the input shaft (see illustration).

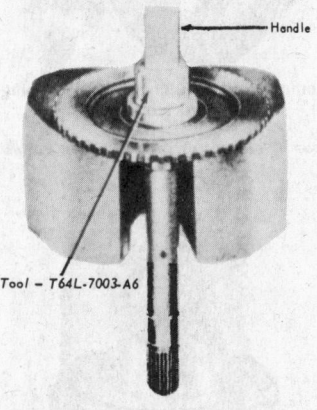

Installing input shaft bushing
(© Ford Motor Co.)

10. Install a new piston inner seal ring in the clutch cylinder and a new piston outer seal in the groove in the piston. Lubricate all parts with transmission fluid before installation.

11. Install the piston in the clutch housing. *Be sure the steel bearing ring is in place on the piston.*

12. Place the release spring in the clutch cylinder with the concave side up. Put the release spring compressor on the spring and compress the spring with an arbor press. Then, install the snap-ring in the groove in the piston.

13. Install the front clutch housing on the primary sun gear shaft by rotating the clutch units to mesh the rear clutch plates with the serrations on the clutch hub. *Do not break the seal rings.*

14. Install the clutch hub in the clutch cylinder with the deep

counterbore down. Install the thrust washer on the clutch hub. See illustration.

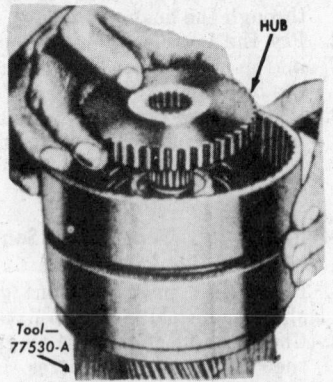

Installing clutch hub
(© Ford Motor Co.)

15. Install the pressure plate in the clutch cylinder with the bearing surface up (see illustration). Install the composition and the

Installing pressure plate
(© Ford Motor Co.)

steel clutch plates alternately, starting with a composition plate (see illustration). Lubricate all the plates with transmis-

Installing clutch plates
(© Ford Motor Co.)

sion fluid before installing. The last friction plate is selective in thickness. Select the thickest plate that will be a minimum of 0.010 in. below the input shaft shoulder in the cylinder. All other plates should be the thinnest available.

16. Install the turbine shaft in the clutch cylinder, and then install the snap-ring in the groove in the piston.

17. Install the thrust washer on the turbine shaft.

Front Pump Disassembly and Assembly

The front pump is disassembled and reassembled according to the procedure below:

1. Remove the stator support attaching screws and remove the stator support. Mark the top surface of the pump driven gear with Prussian blue to assure correct assembly. *Do not scratch the pump gears.*

2. Remove the drive and driven gears from the pump body (see illustration). Inspect the pump

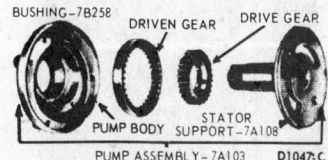

Front pump disassembled
(© Ford Motor Co.)

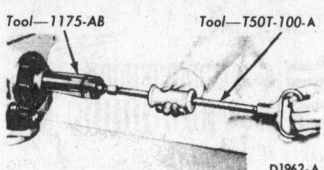

Removing front pump seal
(© Ford Motor Co.)

6. Clean and inspect the counterbore, smoothing any rough spots with crocus cloth.

7. Remove the pump body from the transmission case.

8. Coat the new seal with a good sealing compound. Then place the seal in the pump body and drive the seal into the pump body with the tool shown in the illustration. The tool may be reworked to install the latest type of seal if necessary.

9. Place the pump driven gear in the pump body with the mark facing down. Install the drive gear in the pump body with the chamfered side of the flats facing down.

10. Install the stator support and attaching screws. Check the pump gears for free and easy rotation.

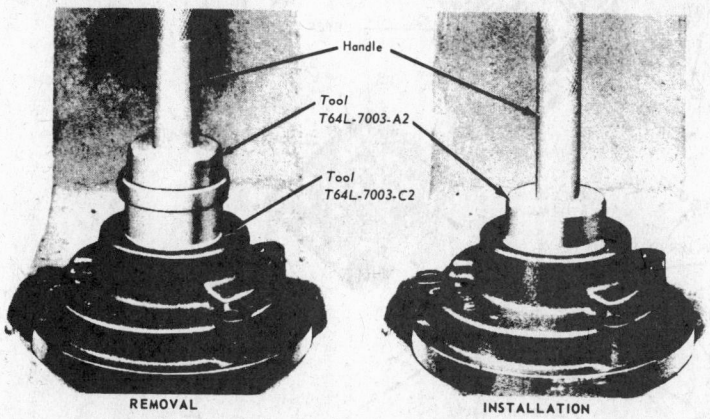

Replacing front pump housing bushing (© Ford Motor Co.)

body housing, gear pockets and crescent for scoring and burrs.

3. If the pump housing bushing needs replacing, press it from the front housing as shown in the illustration. Press a new bushing into the pump housing with the tools shown.

4. If any parts other than the stator support or bushings are defective, replace the pump as a unit. Minor burrs and scores may be removed with crocus cloth. The stator support is serviced separately.

5. Bolt the front pump to the transmission case with capscrews. Install the oil seal remover and pull the front seal from the pump body. See illustration.

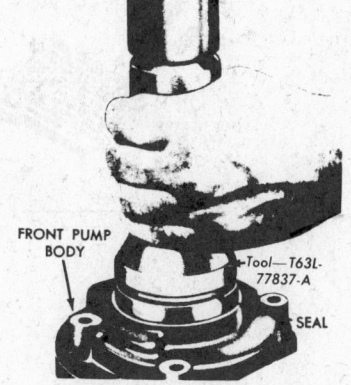

Installing front pump seal
(© Ford Motor Co.)

Rear Pump Disassembly and Assembly

The rear pump that is found in 1964 to 1967 transmissions may be disassembled and reassembled using the procedure given below:

1. When the rear pump has been removed from the output shaft, remove the screws and lockwashers holding the pump cover to the pump body (see illustration). Remove the pump cover.

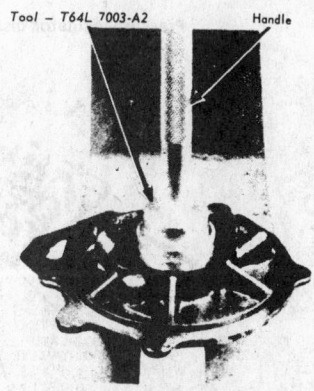

Installing rear support housing seal
(© Ford Motor Co.)

2. Mark the top faces of the pump drive and the pump driven gears with Prussian blue for realignment during assembly. *Do not scratch or gouge the pump gears.*

3. Remove the drive and driven gears from the pump body.

4. After cleaning and inspection, place the pump drive gear in the body with the mark facing up. Place the pump driven gear in the body with its mark facing up and in line with the drive gear mark. Install the pump cover, attaching screws and lockwashers. Tighten the screws to proper torque.

5. Check the pump for free rotation of the gears.

Rear Support Bushing Removal and Installation

1. Remove the three pressure tubes from the support housing.

2. Remove the rear support bushing if worn or damaged by cutting it with a cape chisel along the bushing seal until it breaks. Pry up the loose ends and remove the bushing.

3. Press a new bushing into the support housing with the tool shown in the illustration.

4. Install the pressure tubes.

Pressure Regulator Disassembly and Assembly

1. Remove the valves from the regulator body.

2. Remove the regulator body cover

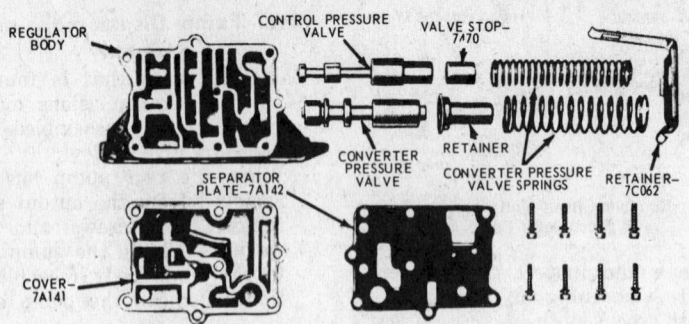

REGULATOR BODY

CONTROL PRESSURE VALVE

VALVE STOP- 7A70

CONVERTER PRESSURE VALVE

RETAINER

CONVERTER PRESSURE VALVE SPRINGS

RETAINER- 7C062

SEPARATOR PLATE-7A142

COVER- 7A141

Pressure regulator disassembled (© Ford Motor Co.)

attaching screws, and remove the cover (see illustration).

3. Remove the separator plate.
4. Clean all parts in solvent and dry with dry compressed air.
5. Inspect the regulator body and cover mating surfaces for burrs. Check the fluid passages for obstructions. Inspect the control pressure and converter pressure valves and their bores for burrs and scoring. Remove all burrs carefully with crocus cloth. Be sure the valves move freely in their bores when dry. Inspect the valve springs for distortion.
6. Reassemble the pressure regulator by placing the separator plate and the regulator cover on the regulator body and installing the attaching screws. Tighten the screws to 20-30 in. lbs. torque. Insert the valves in the regulator body.

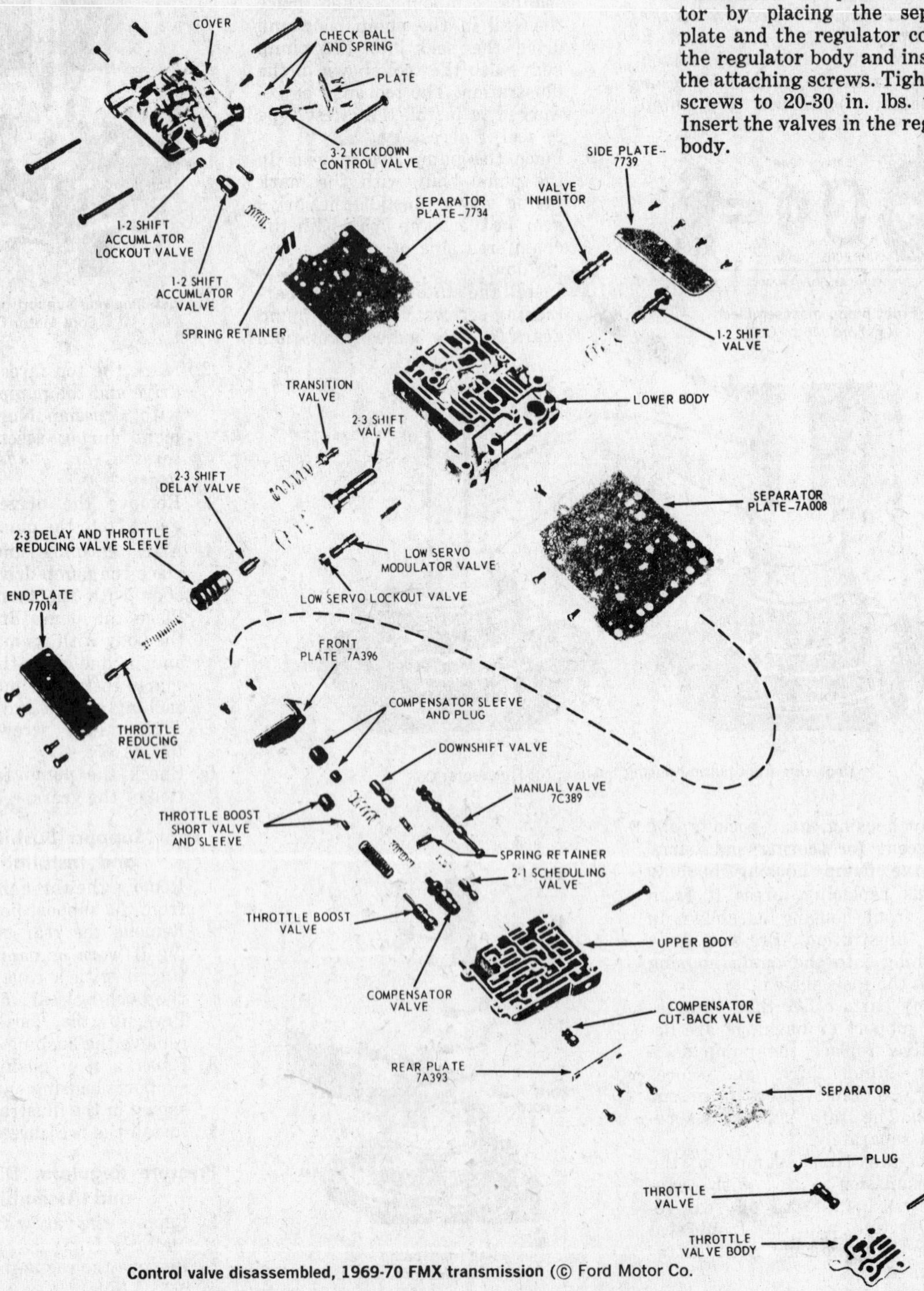

COVER

CHECK BALL AND SPRING

PLATE

3-2 KICKDOWN CONTROL VALVE

1-2 SHIFT ACCUMLATOR LOCKOUT VALVE

1-2 SHIFT ACCUMLATOR VALVE

SPRING RETAINER

SEPARATOR PLATE-7734

VALVE INHIBITOR

SIDE PLATE- 7739

1-2 SHIFT VALVE

TRANSITION VALVE

2-3 SHIFT VALVE

LOWER BODY

2-3 SHIFT DELAY VALVE

2-3 DELAY AND THROTTLE REDUCING VALVE SLEEVE

LOW SERVO MODULATOR VALVE

LOW SERVO LOCKOUT VALVE

SEPARATOR PLATE-7A008

END PLATE 77014

FRONT PLATE 7A396

THROTTLE REDUCING VALVE

COMPENSATOR SLEEVE AND PLUG

DOWNSHIFT VALVE

MANUAL VALVE 7C389

THROTTLE BOOST SHORT VALVE AND SLEEVE

SPRING RETAINER

2-1 SCHEDULING VALVE

THROTTLE BOOST VALVE

UPPER BODY

COMPENSATOR VALVE

COMPENSATOR CUT-BACK VALVE

REAR PLATE 7A393

SEPARATOR

PLUG

THROTTLE VALVE

THROTTLE VALVE BODY

Control valve disassembled, 1969-70 FMX transmission (© Ford Motor Co.

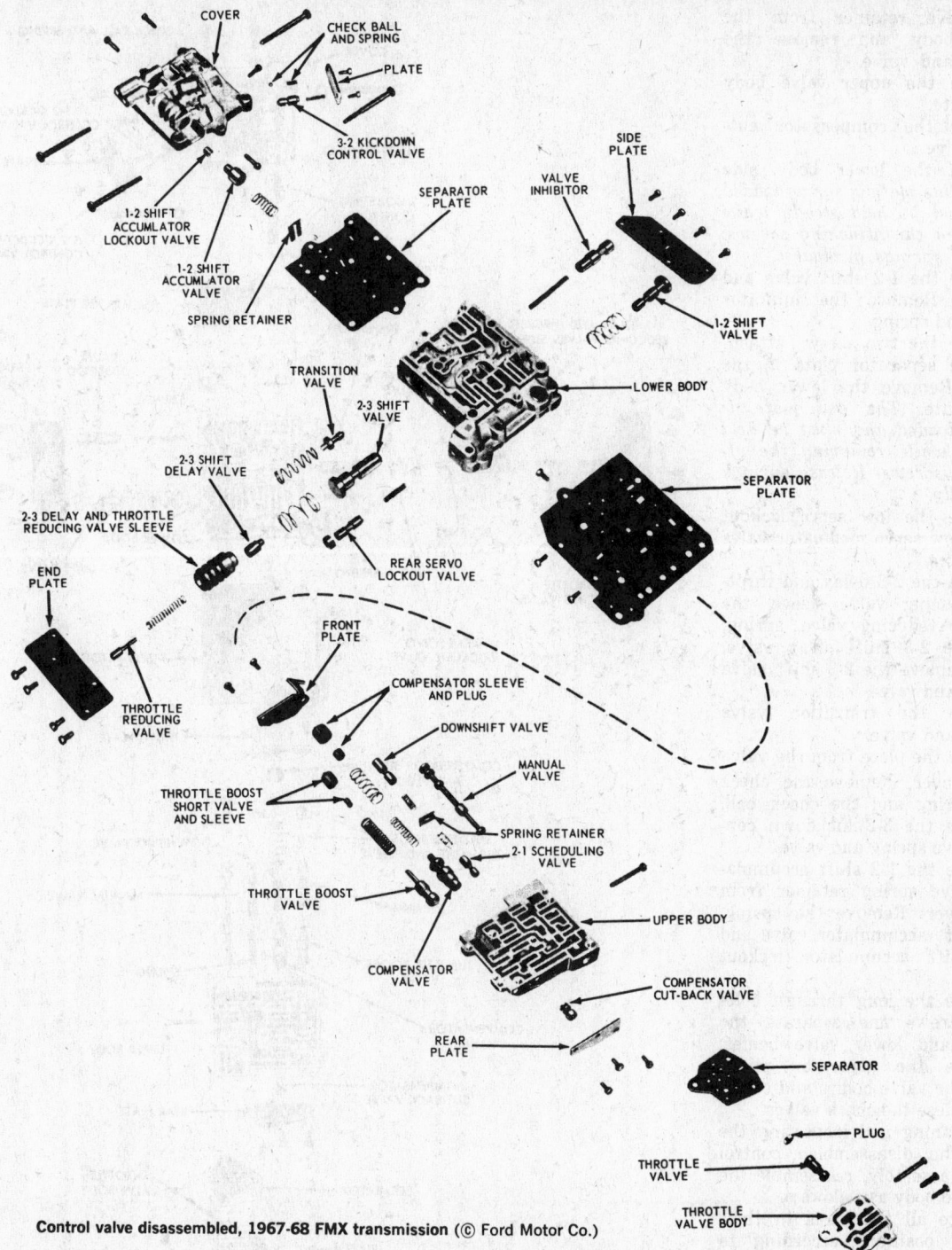

Control valve disassembled, 1967-68 FMX transmission (© Ford Motor Co.)

Control Valve Body Disassembly and Assembly

The control valve body is disassembled according to the procedure given below. The reassembly is basically the reverse of the disassembly procedure but careful reference should be made to the accompanying illustration showing the entire control valve body disassembled. During the disassembly of the control valve body assembly, be careful not to damage the valves and keep everything clean. Place the valve body on a clean surface while disassembling it. *Do not separate the upper and lower valve bodies and cover until after the valves have been removed.*

1. Remove the manual valve (see illustration).
2. Remove the throttle valve body and the separator plate. Do not lose the check valve when removing the separator plate from the valve body. Remove the throttle valve and plug.
3. Remove one attaching screw from the separator plate holding it to the lower valve body.

Remove the upper body front plate. *Since the plate is spring-loaded, hold the plate steady while removing the attaching screws.*

4. Remove the compensator sleeve and plug, and remove the compensator valve springs and the compensator valve.
5. Remove the throttle boost short valve and sleeve. Remove the throttle boost valve spring and valve.
6. Remove the downshift valve and spring. Remove the 2-1 schedul-

ing valve retainer from the valve body and remove the spring and valve.

7. Remove the upper valve body rear plate.

8. Remove the compensator cut-back valve.

9. Remove the lower body side plate. *This plate is spring-loaded and must be held steady while removing the attaching screws. Release springs gradually.*

10. Remove the 1-2 shift valve and spring. Remove the inhibitor valve and spring.

11. Remove the two screws attaching the separator plate to the cover. Remove the lower body end plate. *The end plate is spring-loaded and must be held steady while removing the attaching screws. Release springs gradually.*

12. Remove the low servo lockout valve, low servo modulator valve and spring.

13. Remove the 2-3 delay and throttle reducing valve sleeve, the throttle reducing valve, spring, and the 2-3 shift delay valve. Also, remove the 2-3 shift valve spring and valve.

14. Remove the transition valve spring and valve.

15. Remove the plate from the valve body cover. Remove the check ball spring and the check ball. Remove the 3-2 kickdown control valve spring and valve.

16. Remove the 1-2 shift accumulator valve spring retainer from the cover. Remove the spring, 1-2 shift accumulator valve and 1-2 shift accumulator lockout valve.

17. Remove the long through bolts and screws and separate the upper and lower valve bodies. Remove the separator plates from the valve bodies and cover. Do not lose the check valves.

After cleaning and inspecting the parts of the disassembled control valve body assembly, reassemble the control valve body as follows:

1. Arrange all the parts in their correct positions according to the illustration. When inserting the valves and plugs, rotate them in their bores to avoid damaging the soft body castings.

2. Place the check valve in the upper body (see illustration) and then position the separator plate on the body.

3. Position the lower body on the upper body and loosely install the attaching screws.

4. Put the check valve in the cover and position the cover and separator plate on the lower body. Install the four long through bolts.

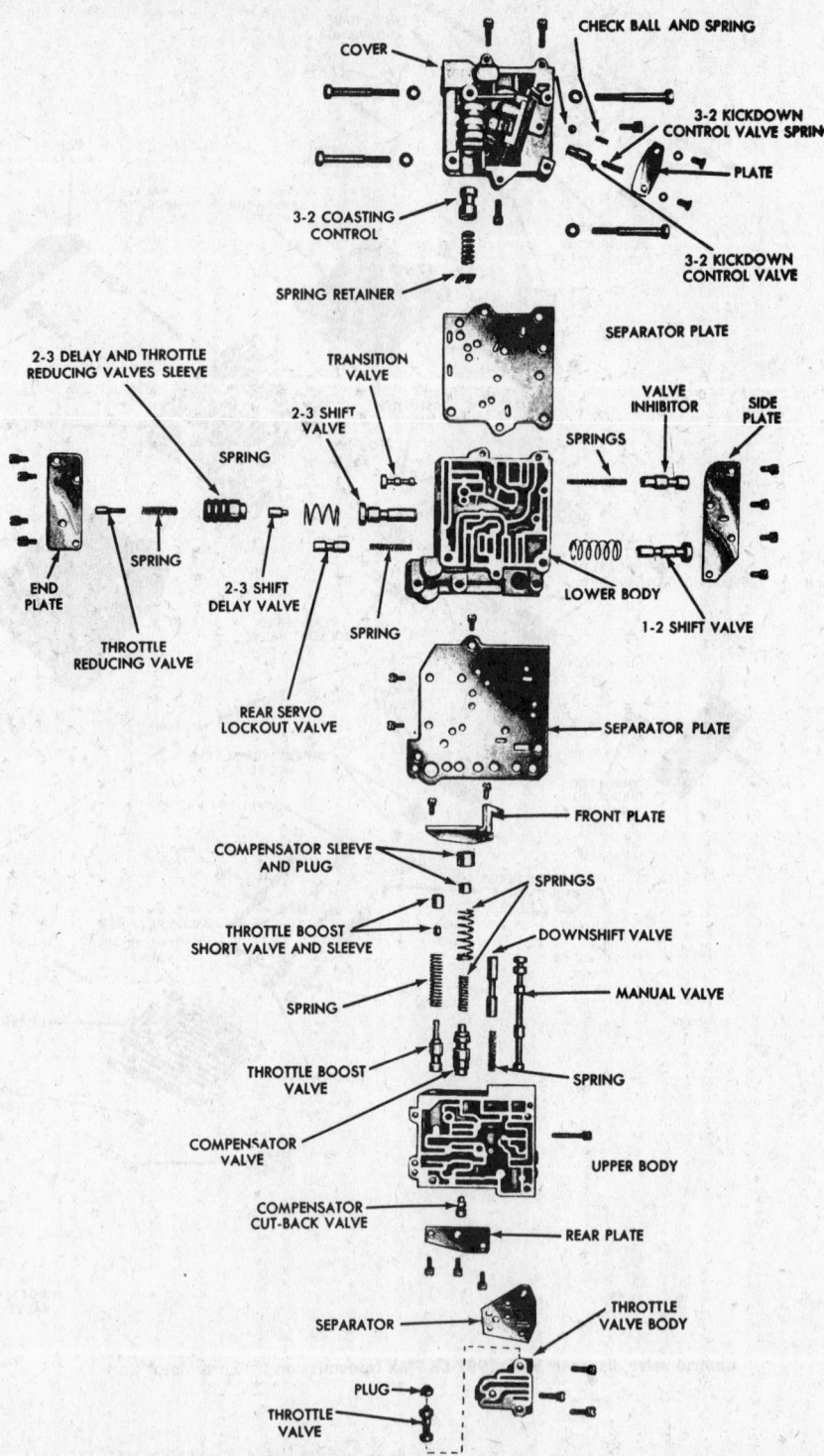

Control valve disassembled, 1964-66 FMX transmission (© Ford Motor Co.)

5. Align the separator with the upper and lower valve body attaching bolt holes and install the attaching bolts, tightening them to 48-72 in. lbs. torque. *Excessive tightening of the bolts may distort the valve bodies, causing valves or plugs to stick.*

6. Install the 3-2 kickdown control valve and spring and the check ball and spring in the cover. Install the plate.

7. Insert the 1-2 shift accumulator lockout valve, the 1-2 accumulator valve, and the spring in the cover. Install the valve spring retainer.

8. Install the transition valve and spring in the lower body. Also install the 2-3 shift valve and

spring, the 2-3 shift delay valve, and the spring and throttle reducing valve in the sleeve. Slide the valve and sleeve into the lower body.

9. Install the low servo lockout valve spring, the low servo modulator, and low servo lockout valves. Install the lower body end plate.

10. Install the inhibitor valve spring and valve in the lower body. Install the 1-2 shift valve spring and valve. Install the lower body side plate.

11. Install the compensator cut-back valve in the upper body and install the upper body rear plate.

12. Install the 2-1 scheduling valve, spring, and spring retainer in the body. Install the downshift valve spring and valve.

13. Install the throttle boost valve and spring and the throttle boost short valve and sleeve.

14. Install the compensator valve, inner and outer compensator springs and the compensator sleeve and plug.

15. Position the front plate correctly and press it against the upper body while installing the attaching screws.

16. Install the throttle valve, plug and check valve in the throttle valve body. Position the separator on the upper body and install the throttle valve body with the three attaching screws.

17. Install the four screws attaching the cover to the lower body, the two screws attaching the separator plate to the upper body, and one screw attaching the separator plate to the lower body. Tighten the cover and body screws to the proper torque.

Front Servo Disassembly and Assembly

The front servo is disassembled and reassembled according to the procedure given below. Refer to the disassembled view of the front servo to identify the parts.

1. Remove the servo piston retainer snap-ring. Since the servo piston is spring-loaded, hold the piston steady while removing the snap-ring and release the spring pressure gradually.

2. Remove the servo piston retainer, servo piston, and the return piston from the servo body. It may be necessary to tap the piston stem lightly with a soft-faced hammer to separate the piston retainer from the servo body.

3. On PHD models, remove the screw and washer from the end of the piston stem and separate the piston retainer, return piston, and the servo piston.

4. Remove all seal rings and the spring from the servo body.

5. Inspect the servo body for cracks and the piston bore and the servo piston stem for scores. Check all fluid passages for obstructions.

6. Check the actuating lever for free movement and wear. If it is necessary to replace the actuating lever shaft, remove the retaining pin and push the shaft out of the bracket. If the shaft is not held in place by a retaining pin, it is held in place by serrations on one side and it is a press-fit. To remove this type of shaft, press on the end opposite the serrations. Inspect the adjusting screw threads and the threads in the lever.

7. Check the servo spring and servo band strut for distortion.

8. Inspect the servo band lining for excessive wear and bonding to the metal. Replace the lining if it is worn so much that the grooves are not clearly seen. Inspect the band ends for cracks and check the band for distortion.

9. Lubricate all parts of the front servo by soaking in transmission fluid to ease assembly.

10. Install the inner and outer seal rings on the piston retainer. Install new O-rings on the return piston and the servo piston. See the illustration and accompanying table for the servo components used in each model of the FMX transmission.

11. On PHD models, tap the piston stem into the servo piston. Insert the servo piston in the pis-

ton retainer. Tap the return piston onto the piston stem and into the piston retainer. Be sure the dished side of the return piston is toward the servo piston. Fasten the return piston to the stem with a screw and plain washer.

12. Place the servo piston release spring in the servo body. Install the servo piston, retainer, and return piston in the servo body as an assembly. Compress this assembly into the body and secure it with the snap-ring. *Be sure the snap-ring is fully seated in the groove.*

13. Install the adjusting screw and locknut in the actuating lever if they were removed.

Rear Servo Disassembly and Assembly

1. Remove the servo actuating lever shaft retaining pin with a $\frac{1}{8}$-inch punch. Remove the shaft and actuating lever needle bearings and thrust washers. See illustration.

2. Press down on the servo spring retainer and remove the snap ring, releasing the pressure on the retainer slowly to prevent the spring from flying out.

3. Remove the retainer and the servo spring.

4. Force the servo piston out of the body with air pressure, holding one hand over the piston to prevent damage.

5. Remove the accumulator piston from the servo piston.

6. On transmissions from 1964 to 1967 a piston seal ring must be removed. FMX transmissions in

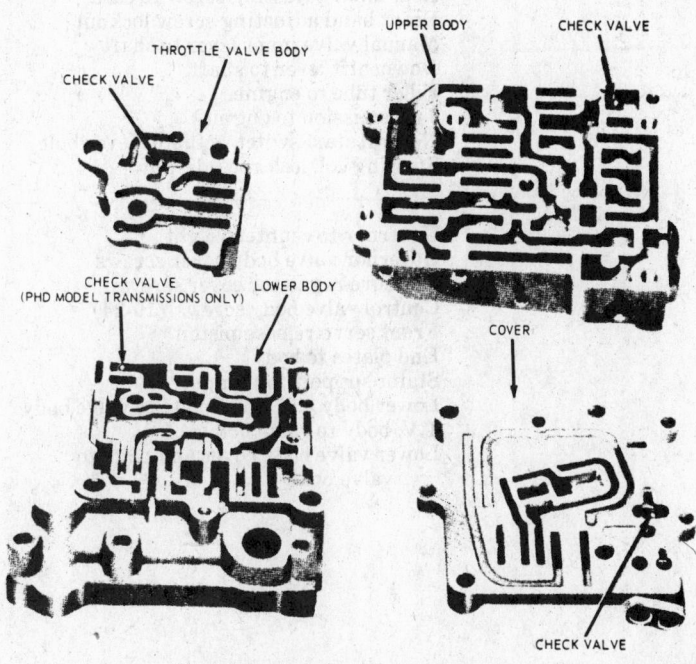

Check valve locations (© Ford Motor Co.)

1968 to 1970 do not have this seal ring.

7. On those transmissions that have a seal ring, install a new seal ring in the servo piston.

8. Install the accumulator piston in the servo piston. Install the servo piston in the servo body. Install the servo spring with the small coiled end against the servo piston. *Be sure to lubricate all parts with transmission fluid to aid assembly.*

9. Install the spring retainer, compressing the spring with a C-clamp. Then, install the snap-ring in the groove.

10. Install the needle bearings in the actuating lever. Then, install the actuating lever and thrust washers with the socket in the lever bearing on the piston stem. Install the actuating lever shaft, aligning the retaining pin holes, and install the pin.

11. Check the actuating lever for free movement.

Torque Limits

When assembling the various components of the FMX automatic transmission, the bolts and nuts must be tightened within a specified range to prevent damage to the bolts, nuts or the transmission conponents. The torque limits for the FMX and C6 automatic transmissions are listed below in the accompanying tables.

Torque Limits for FMX Automatic Transmission

	Ft. Lbs.
Converter to flywheel	23-28
Converter hsg. to trans. case	40-50
Front pump to trans. case	17-22
Front servo to trans. case	30-35
Rear servo to trans. case	40-45
Upper valve body to lower valve body	4-6
Oil pan to case	10-13
Converter cover to converter hsg.	12-16
Regulator to case	17-22
Planetary support to trans. case	20-25
Control valve body to trans. case	8-10
Diaphragm assy. to case	20-30
Cooler return check valve	9-12
Extension assy. to trans. case	30-40
Pressure gauge tap	7-15
Converter drain plug	15-28
Rear band adjusting screw to case	35-40
Front band adjusting screw locknut	20-25
Manual valve inner lever to shaft	20-30
Downshift lever to shaft	17-20
Filler tube to engine	20-25
Transmission to engine	40-50
Neutral start switch actuator lever bolt	6-10
Steering col. lock rod adj. nut	10-20
	In. Lbs.
Governor to counterweight	50-60
Governor valve body cover screws	20-30
Pressure regulator cover screws	20-30
Control valve body screws (10-24)	20-30
Front servo release piston	20-30
End plates to body	20-30
Stator support to pump	25-35
Lower body and cover plate to valve body	20-30
T.V. body to valve body	20-30
Lower valve body cover and plate to valve body	48-72

3-Speed Automatic (C4 and C4S)

Application

C4

Comet, 1964-68
Cougar, 1967-70
Fairlane, 1965-70

Falcon, 1964-70
Ford, 1964-70
Maverick, 1970
Montego, 1968-70

Mustang, 1965-70

C4S

Maverick, 1970

Diagnosis

This diagnosis guide covers the most common symptoms and is an aid to careful diagnosis. The items to check are listed in the sequence to be followed for quickest results. Thus, follow the checks in the order given for the particular transmission type.

TROUBLE SYMPTOMS	Items to Check	
	In Car	Out of Car
Rough initial engagement in D1 or D2	KBUFEG	a
1-2 or 2-3 shift points incorrect or erratic	ABLCDUER	
Rough 1-2 shifts	BJGUEF	
Rough 2-3 shifts	BJUFGER	bl
Dragged out 1-2 shift	ABJUGEFR	c
Engine overspeeds on 2-3 shift	CABJUEFG	bl
No 1-2 or 2-3 upshift	CLBDUEGJ	bc
No 3-1 shift in D1 or 3-2 shift in D2	DE	
No forced downshift	LEB	
Runaway engine on forced downshift	UJGFEB	c
Rough 3-2 or 3-1 shift at closed throttle	KBJEF	
Shifts 1-3 in D1 and D2	GJBEDR	
No engine braking in first gear—manual low	CHIEDR	
Creeps excessively	KW	
Slips or chatters in first gear, D1	ABUFE	acg
Slips or chatters in second gear	ABJGUFER	ac
Slips or chatters in R	ABHUIFER	bcl
No drive in D1	ACUER	g
No drive in D2	ACUJER	cg
No drive in L	ACUEIR	cg
No drive in R	ACHUIER	bcl
No drive in any selector position	ACUFER	cd
Lockup in D1		bec
Lockup in D2	HI	becg
Lockup in L	GJ	bec
Lockup in R	GJ	aec
Parking lock binds or does not hold	C	e
Transmission overheats	OFBU	i
Maximum speed too low, poor acceleration	VW	i
Transmission noisy in N and P	AF	df
Transmission noisy in any drive position	AF	fadg
Fluid leaks	AMNOPQSTBIJX	hik
Car moves forward in N	C	a

Key to Checks

A. Fluid level
B. Vacuum diaphragm unit or tube
C. Manual linkage
D. Governor
E. Valve body
F. Pressure regulator
G. Intermediate band
H. Reverse band
I. Reverse servo
J. Intermediate servo
K. Engine idle speed
L. Downshift linkage—inner lever position
M. Converter drain plug

N. Oil pan and/or filler tube gaskets/seals
O. Oil cooler and/or connections
P. Manual or downshift lever shaft seal
Q. Pipe plug, side of case
R. Perform air pressure checks
S. Extension housing-to-case gasket or washers
T. Extension housing rear oil seal
U. Make control pressure test
V. Engine performance
W. Vehicle brakes
X. Speedometer driven gear adaptor seal

a. Forward clutch
b. Reverse—high clutch
c. Hydraulic system leakage
d. Front pump
e. Parking brake linkage
f. Planetary assembly
g. Planetary one-way clutch
h. Engine rear oil seal
i. Front oil pump seal
j. Converter oneway clutch
k. Front pump-to-case seal or gasket
l. Reverse—high clutch piston air bleed valve

General Information

This section provides instructions for testing, inspecting, adjusting, and repairing the Ford C4 automatic and C4S semi-automatic transmissions. Where there are differences in procedures or specifications, these differences will be given.

The Ford C4 automatic transmission (see illustration) is a three speed unit that provides automatic upshifts and downshifts through three forward gear ratios and also provides manual selection of first and second gears.

The Ford C4S semi-automatic transmission is a manually operated power shift transmission which does not require a clutch pedal. The transmission is similar to the C4 automatic transmission except for differences in the control valve body and there is no vacuum diaphragm, throttle rod, governor, and the inner and outer downshift lever assemblies.

Both transmissions consist of a torque converter, planetary gear train, two multiple disc clutches, a one-way clutch, and a hydraulic control system. The only adjustments necessary on these transmissions during normal maintenance are band adjustments on the intermediate and low-reverse bands. The transmission fluid is cooled through an oil cooler core in the radiator lower tank when a steel converter is used. If an aluminum converter is used, the transmission fluid is air-cooled.

Towing

If a disabled car is to be towed a short distance, it may be towed safely by placing the selector lever in the Neutral position and releasing the parking brake. The driver of the towed car should be careful to maintain a safe stopping distance behind the towing vehicle. The towing speed should not exceed 30 mph.

If the disabled car must be towed a long distance, or the transmission cannot be put into Neutral gear, or there is no driver for the disabled car, raise the rear of the car until the wheels clear the ground and tow the car backwards.

Starting the Engine Without the Starter

The transmission may be used to start an engine when the starter cannot be used by moving the disabled car at a road speed of about 25 mph while in neutral gear and then placing the transmission in Low gear with the ignition switch on. The transmission will turn the engine over fast enough to start and the car will move under its own power. Extreme care must be taken to prevent the car from rushing forward so fast that it hits another car or obstacle. Also, if the disabled car is being pushed, the rear car must drop back before the start is attempted to avoid hitting the front car when it stops moving forward momentarily.

Transmission Checks

Before doing any of the tests and adjustments given in this section, the following preliminary checks should be done.

Transmission Fluid Level Check

1. Position the car on a level place and firmly apply the parking brake.
   NOTE: on cars equipped with a vacuum brake release, disconnect the release line and plug the end of the line. If this is not done, the parking brake will not hold the car when the transmission is in any drive gear.
2. Start the engine and run at normal idle speed. If the transmission fluid is cold, the engine should idle at a fast idle speed (about 1200 rpm) until the fluid is warm. When the transmission fluid reaches its normal operating temperature, slow the engine to normal idle speed.
3. Shift the transmission through all drive positions briefly and then put it in the Park position. Do not shut off the engine during the fluid level check.
4. Locate the transmission fluid dipstick and clean all dirt and grease from the cap before removing it. On some cars, the dipstick is located in the right rear corner of the engine compartment. On other cars, the dipstick is located under the front floormat to the right of center.
5. Pull the dipstick out of the filler

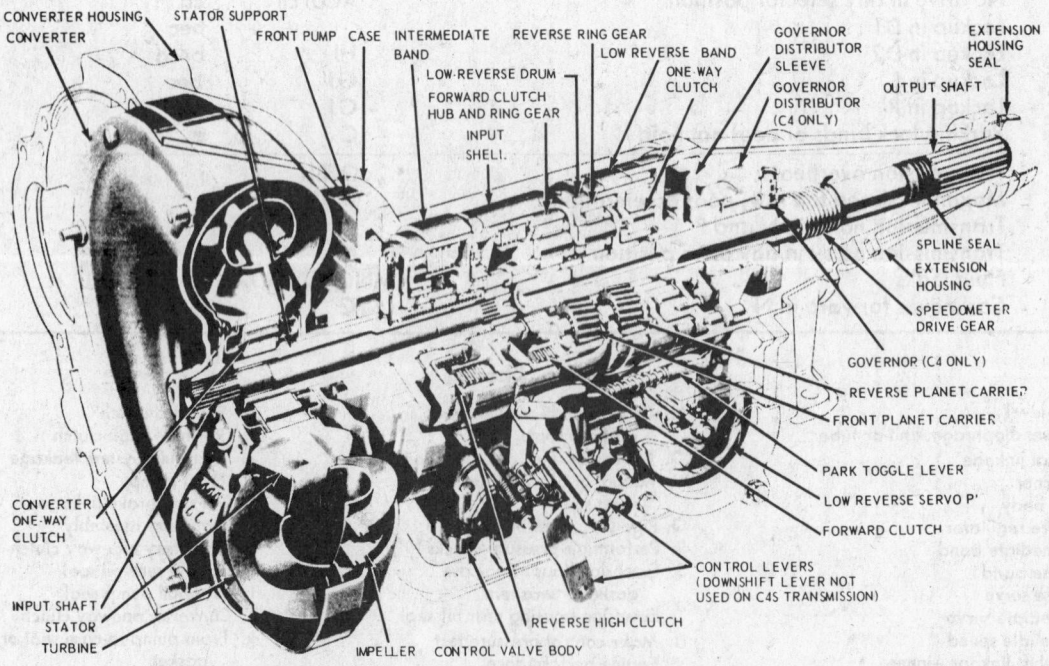

C4 Automatic and C4S semi-automatic transmission (© Ford Motor Co.)

tube, wipe it clean, and reinsert it into the filler tube until it is fully seated.

6. Pull the dipstick out again and see where the fluid level is on the dipstick. The correct level is between the ADD and FULL marks on the dipstick. If necessary, add enough fluid through the filler tube to raise the fluid level to the correct level. *Do not overfill the transmission.* Replace the dipstick in the filler tube.

7. If a vacuum brake release line was disconnected, reconnect it and test for proper operation.

Fluid Aeration Check

If the transmission is overfilled with too much fluid, the fluid will be whipped up into a foamy or aerated condition. This condition will cause low control pressure and the fluid may be forced out of the vent or through ruptured seals.

Check the transmission fluid level for low level conditions that may indicate leaks or cause poor operation.

Transmission Fluid Leakage Checks

Make the following checks if a leakage is suspected from the transmission case:

1. Clean all dirt and grease from the transmission case.

2. Inspect the speedometer cable connection at the extension housing of the transmission. If fluid is leaking here, disconnect the cable and replace the rubber seal.

3. Inspect the oil pan gasket and attaching bolts for leaks. Tighten any bolts that appear loose to the proper torque (10-13 ft-lbs) Recheck for signs of leakage. If necessary, remove the pan attaching bolts and old pan gasket and install new gasket and reinstall the pan and its attaching bolts.

4. Check filler tube connection at the transmission for signs of leakage. If tube is leaking, tighten the connection to stop the leak. If necessary, disconnect the filler tube, replace the O-ring, and reinstall the filler tube.

5. Inspect all fluid lines between the transmission and the cooler core in the lower radiator tank. Replace any lines or fittings that appear to be worn or damaged. Tighten all fittings to the proper torque.

6. Inspect the engine coolant for signs of transmission fluid in the radiator. If there is transmission fluid in the engine coolant, the oil cooler core is probably leaking.

The oil cooler core may be test-

ed further by disconnecting all lines to it and applying 50-75 psi air pressure through the fittings. Remove the radiator cap to relieve any pressure buildup outside the cooler core. If air bubbles appear in the coolant or if the cooler core will not hold pressure, the oil cooler core is leaking and must be replaced. Oil cooler core repair and replacement is discussed in the section on Cooling Systems in this manual.

7. Inspect the openings in the case where the downshift control lever shaft and the manual lever shaft are located for leaks. If necessary, replace the defective seal.

8. Inspect all plugs or cable connections in the transmission for signs of leakage. Tighten any loose plugs or connectors to the proper torque according to the specifications

9. Remove the lower cover from the front of the bellhousing and inspect the converter drainplugs for signs of leakage. If there is a leak around the drainplugs, loosen the plug and coat the threads with a sealing compound and tighten the plug to the proper torque.

NOTE: fluid leaks from around the converter drainplug may be caused by engine oil leaking past the rear main bearing or from the oil gallery plugs. To determine the exact cause of the leak before beginning repair procedures, an oil-soluble aniline or fluorescent dye may be added to the transmission fluid to find the source of the leak and whether the transmission is leaking. If a fluorescent dye is used, a black light must be used to detect the dye.

If further converter checks are necessary, remove the transmission from the car and the converter from the transmission. The converter cannot be disassembled for cleaning or repair. If the converter is leaking, it must be replaced with a new unit. The converter may be cleaned by using the Rotunda Automatic Transmission Torque Converter and Cooler Cleaner LRE-60081. To further

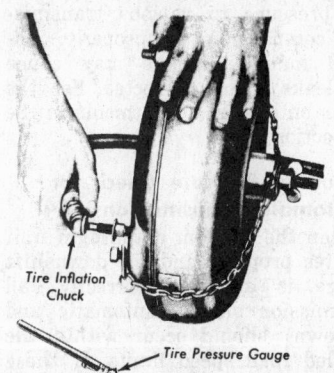

Converter leak checking tool installation
(© Ford Motor Co.)

check the converter for leaks, assemble and install the converter leak checking tool shown and fill the converter with 20 psi air pressure. Then, place the converter in a tank of water and watch for air bubbles. If no air bubbles are seen, the converter is not leaking.

Engine Idle Speed Check

Check the idle speed of the engine and adjust it at the carburetor using the procedure given in the Fuel System section. Too slow an idle speed will cause the engine to run roughly; and too fast an idle speed will cause the car to creep while the transmission is in a Drive position, shift harshly, and downshift roughly.

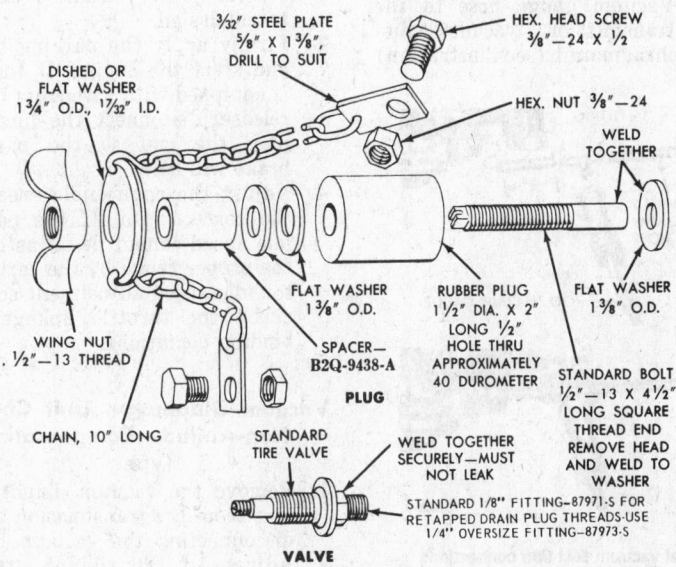

Converter leak checking tool (© Ford Motor Co.)

Anti-Stall Dashpot Clearance Check

After properly adjusting the engine idle speed, check the clearance of the anti-stall dashpot using the procedure given in the Car Section.

Manual Linkage Checks

Correct manual linkage is necessary for the proper operation of the manual valve which helps control fluid pressure to various transmission components. Improperly adjusted manual linkage may cause fluid leaks if not corrected. See the section on linkage adjustments in the Car Section.

Control Pressure Check for Automatic Transmissions (C4)

When the vacuum diaphragm unit operates properly and the downshift linkage is adjusted correctly, all transmission shifts (automatic and kickdown) should occur within the specified road speed limits. If these shifts do not occur within the limits or if the transmission slips during a shift point, perform the following procedure to locate the problem:

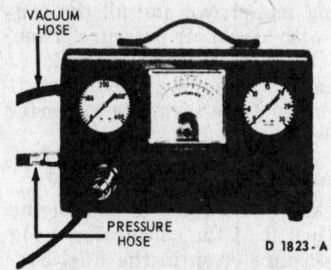

Rotunda ARE-2905 automatic transmission tester (© Ford Motor Co.)

1. Connect the Automatic Transmission Tester (see illustration) as follows:
   a. Tachometer cable to engine
   b. Vacuum gauge hose to the transmission vacuum diaphragm unit (see illustration)

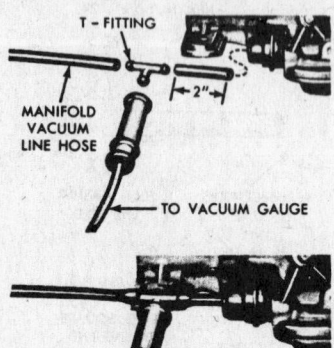

Typical vacuum test line connections (© Ford Motor Co.)

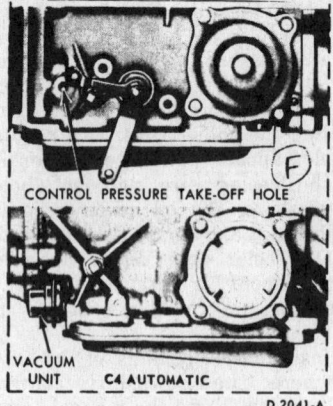

Typical vacuum diaphragm and control pressure connecting points (© Ford Motor Co.)

2. Apply the parking brake and start the engine. On a car equipped with a vacuum brake release, disconnect the vacuum line or use the service brakes since the parking brake will release automatically when the transmission is put in any Drive position.
3. Check engine idle speed and throttle and downshift linkage for correct operation. Check the transmission diaphragm unit for leaks.

Control Pressure Check for Semi-Automatic Transmissions (C4S)

If the shifts do not occur within the proper road speeds or the transmission slips during a shift point, do the following procedure to find the possible trouble:
1. Connect the tachometer of the automatic transmission tester to the engine.
2. Attach the pressure gauge to the control pressure outlet at the transmission.
3. Firmly apply the parking brake and start the engine. If the car is equipped with a vacuum brake release, disconnect the line and plug the end so the parking brake will hold.
4. Adjust the engine idle speed to the correct rpm. If the engine idle speed cannot be adjusted to the proper range by the carburetor idle speed adjustment screw, check the throttle linkage for binding conditions.

Vacuum Diaphragm Unit Check —Non-Altitude Compensating Type

1. Remove the vacuum diaphragm unit from the transmission after disconnecting the vacuum hose.
2. Adjust a distributor tester equipped with a vacuum pump

until the vacuum gauge shows 18 inches with the vacuum hose blocked. See illustration.

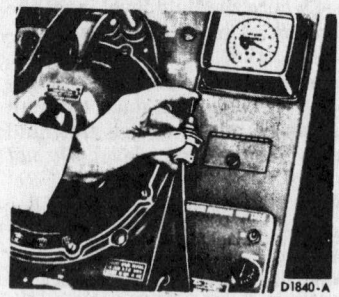

Testing transmission vacuum diaphragm for leakage (© Ford Motor Co.)

3. Connect the vacuum hose to the vacuum diaphragm unit and note the reading on the vacuum gauge. If the reading is 18 inches of vacuum, the vacuum diaphragm unit is good. While removing the vacuum hose from the vacuum diaphragm unit, hold a finger over the end of the control rod. As the vacuum is released, the internal spring of the vacuum diaphragm unit will push the control rod out.

Vacuum Diaphragm Unit Check —Altitude-Compensating Type

The vacuum diaphragm unit may be checked for damaged or ruptured bellows by doing the procedure below:
1. Remove the diaphragm and the throttle valve rod from the transmission.
2. Insert a rod into the diaphragm unit until it is seated in the hole. Make a reference mark on the rod where it enters the diaphragm (see illustration).
3. Place the diaphragm unit on a weighing scale with the end of the rod resting on the weighing pan and gradually press down on the diaphragm unit.
4. Note the force (in pounds) at which the reference mark on the rod moves into the diaphragm. If the reference mark is still visible at 12 pounds of pressure on the scale, the diaphragm bellows is good. But if the reference mark on the rod disappears before four pounds of force, the diaphragm bellows are damaged and the diaphragm unit must be replaced.

Shift Point Checks for Automatic Transmissions

To determine if the transmission is shifting at the proper road speeds, do the following procedure:
1. Check the minimum throttle up-

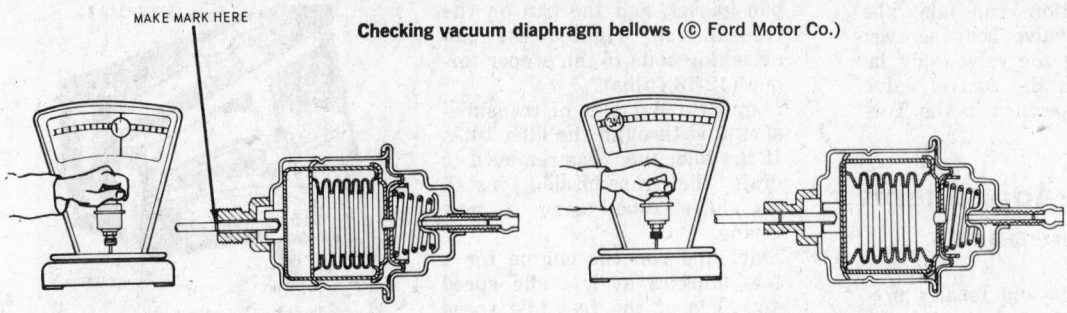

Checking vacuum diaphragm bellows (© Ford Motor Co.)

BELLOWS INTACT   BELLOWS FAILED

shifts by placing the transmission selector lever in the Drive position and noting the road speeds at which the transmission shifts from first gear to second gear to third gear. All shifts should occur within the specified limits.

2. While driving in third gear, depress the accelerator pedal past the detent (to the floor). Depending on vehicle speed, the transmission should downshift from third gear to second gear or from second gear to first gear.

3. Check the closed-throttle downshift from third gear to first gear by coasting down from about 30 mph in third gear. This downshift should occur at the specified road speed.

NOTE: when the transmission selector lever is at 2, the transmission will operate only in second gear. Manually move the selector lever to either D or 1 to shift gears automatically.

4. With the transmission in third gear and the car moving at a road speed of 25 mph, the transmission should downshift to second gear when the selector lever is moved from D to 2 to 1. This check will determine if the governor pressure and shift control valves are operating properly. If the transmission does not shift within the specified limits or certain gear ratios cannot be obtained, refer to the Trouble Diagnosis chart at the beginning of this section.

Air Pressure Checks

If the car will not move in one or more ranges, or, if it shifts erratically, the items at fault can be determined by using air pressure at the indicated passages.

Drain the transmission and remove the oil pan and the control valve assembly.

NOTE: oil will spray profusely during this operation.

Front Clutch

Apply sufficient air pressure to the front clutch input passage. (See il-

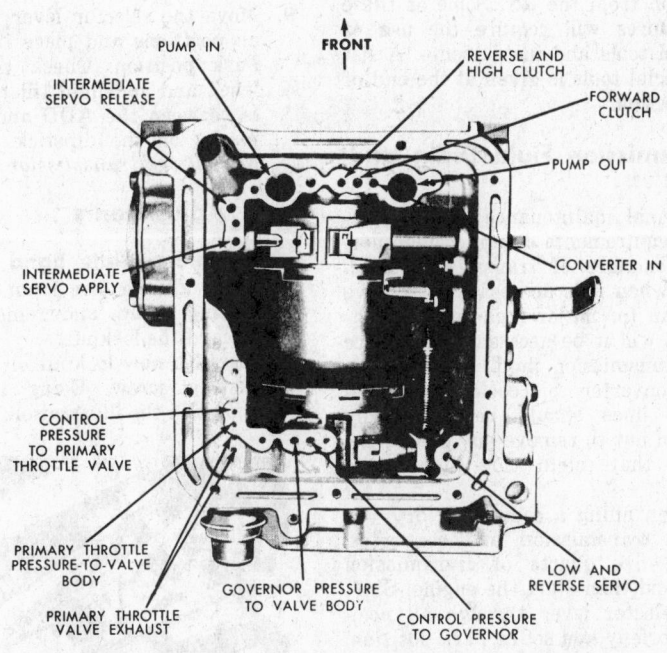

C4 automatic transmission fluid passage identification (© Ford Motor Co.)

lustration.) A dull thud can be heard when the clutch piston moves. Check also, for leaks.

Governor

Remove the governor inspection cover from the extension housing. Apply air to the front clutch input passage. (See illustration). Listen for a sharp click and watch to see if the governor valve snaps inward as it should.

Rear Clutch

Apply air to the rear clutch passage (See illustration) and listen for the dull thud that will indicate that the rear clutch piston has moved. Listen also for leaks.

Front Servo

Apply air pressure to the front servo apply tube (See illustration.) and note if front band tightens. Shift the air to the front servo release tube, which is next to the apply tube, and watch band release.

Rear Servo

Apply air pressure to the rear servo apply passage. The rear band should tighten around the drum.

Conclusions

If the operation of the servos and clutches is normal with air pressure, the no-drive condition is due to the control valve and pressure regulator valve assemblies, which should be disassembled, cleaned and inspected.

If operation of the clutches is not normal; that is, if both clutches apply from one passage or if one fails to move, the aluminum sleeve (bushing) in the output shaft is out of position or badly worn. (See illustration.)

Use air pressure to check the passages in the sleeve and shaft, and also check the passages in the primary sun gear shaft.

If the passages in the two shafts and the sleeve are clean, remove the clutch assemblies, clean and inspect the parts.

Erratic operation can also be caused by loose valve body screws. When reinstalling the valve body be careful to tighten the control valve body screws as specified in the Torque Limits table.

In-Vehicle Adjustments and Repairs

The adjustments and repairs presented in this part of the section on FMX transmissions may be done without removing the entire transmission from the car. Some of these procedures will require the use of special tools and instruments. A list of special tools is given at the end of the section.

Transmission Fluid Drain and Refill

Normal maintenance and lubrication requirements do not include periodic changes of transmission fluid. Only when it is necessary to remove the pan for major repairs or adjustments will it be necessary to replace the transmission fluid. At this time the converter, oil cooler core, and cooler lines should be thoroughly flushed out to remove any dirt or deposits that might clog these units later.

When filling a completely dry (no fluid) transmission and converter, install five quarts of transmission fluid and then start the engine. Shift the selector lever through all positions briefly and set at Park position. Check the fluid level and add enough fluid to raise the level to between the ADD and FULL marks on the dipstick. *Do not overfill the transmission.*

The procedure for a partial drain and refill of the transmission fluid is given below:

1. Raise the car on a hoist or jack stands.
2. Place a drain pan under the transmission pan.
3. Loosen the pan attaching bolts to allow the fluid to drain.
NOTE: on some models of the C4 and C4S transmissions, the transmission fluid is drained by disconnecting the fluid filler tube from the transmission fluid pan.
4. When the fluid has stopped draining to level of the pan flange, remove the pan bolts starting at the rear and along both sides of the pan, allowing the pan to drop and drain gradually.
5. When all the transmission fluid has drained, remove the pan and the fluid filter and clean them. Discard the old pan gasket.
6. After completing the transmission repairs or adjustments, install the fluid filter screen, a new

pan gasket, and the pan on the transmission. Tighten the pan attaching bolts to the proper torque (12-16 ft-lbs).
7. Install three quarts of transmission fluid through the filler tube. If the filler tube was removed to drain the transmission, install the filler tube using a new O-ring.
8. Start and run the engine for a few minutes at low idle speed and then at the fast idle speed (about 1200 rpm) until the normal operating temperature is reached. *Do not race the engine.*
9. Move the selector lever through all positions and place it at the Park position. Check the fluid level, and add fluid till the level is between the ADD and FULL marks on the dipstick. *Do not overfill the transmission.*

Band Adjustments

Intermediate Band

1. Clean all the dirt from the adjusting screw and remove and discard the locknut.
2. Install a new locknut on the adjusting screw. Using the tool shown in the illustration, tighten

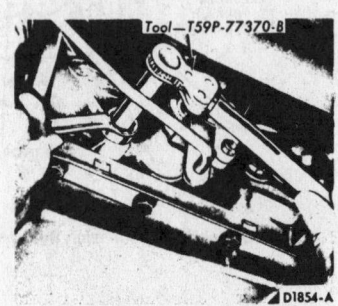

Adjusting intermediate band
(© Ford Motor Co.)

the adjusting screw until the wrench clicks and breaks at 10 ft-lbs. torque.
3. Back off the adjusting screw *Exactly 1¾ turns.*
4. Hold the adjusting screw steady and tighten the locknut to the proper torque.

Low-Reverse Band

1. Clean all dirt from around the band adjusting screw, and remove and discard the locknut.
2. Install a new locknut on the adjusting screw. Using the tool shown in the illustration, tighten the adjusting screw until the wrench clicks and breaks at 10 ft-lbs torque.
3. Back off the adjusting screw *Exactly 3 full turns.*
4. Hold the adjusting screw steady and tighten the locknut to the proper torque.

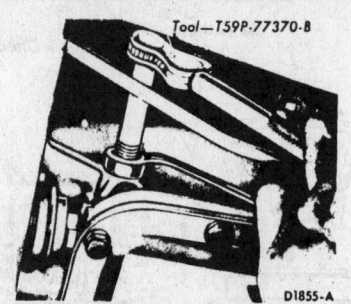

Adjusting low-reverse band
(© Ford Motor Co.)

Transmission Component Removal and Installation

The components that may be removed from and installed in the transmission while it is in the car are discussed in this part of the C4 and C4S transmission section. Only procedures for removing and installing these components are given here. Disassembly and repair of the units is given in a later part.

To avoid repetition, the following tasks should be done before proceeding with the removal and installation of components.

1. Raise the car so the transmission is accessible from under the car.
2. Drain the fluid from the transmission. Some models are drained by removing the filler tube from the transmission oil pan. Others are drained by removing the oil pan attaching bolts gradually from the rear of the pan. If the same fluid is to be reused, filter it through a 100 mesh screen. Reuse the fluid only if it is in good condition.
3. Remove the oil pan attaching bolts, the oil pan, and the old gasket. Discard gasket.
4. Be sure to have a good transmission jack available and a holding device for the transmission if it is removed from the car later.

Oil Pan and Control Valve Removal and Installation—All Cars Except Falcon

1. Do all the preliminary operations given at the beginning of this section.
2. On 1970 model C4 transmissions only, shift the transmission to Park position and remove the two bolts holding the manual detent spring to the control valve body and case.
3. Remove all the valve body-to-case attaching bolts. Hold the manual valve in place and remove the valve body from the case. If the manual valve is not held in place, it could be bent or damaged.
4. Refer to the Component Disassembly and repair section for

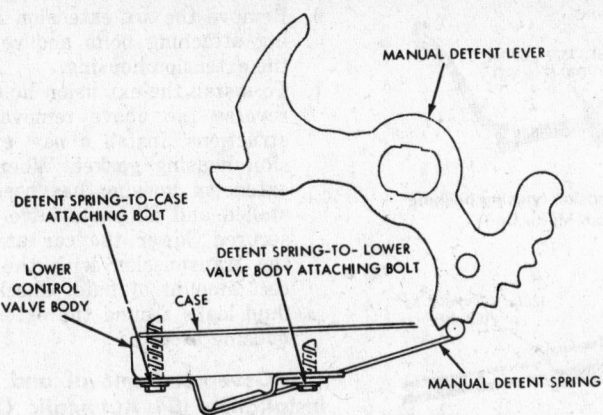

Control valve body with detent spring installed (© Ford Motor Co.)

control valve body repair procedures.

5. Thoroughly clean the old gasket material from the case and remove the nylon shipping plug from the oil filler tube hole. This nylon plug is installed before shipment and should be discarded when the transmission oil pan is removed.

6. Be sure the transmission is in the Park position (manual detent lever is in P detent position). Install the valve body in the case. Position the inner downshift lever between the downshift lever stop and the downshift valve. Be sure the two lands on the end of the manual valve engage the actuating pin on the manual detent lever.

7. Install seven valve body attaching bolts but do not tighten them.

8. Place the detent spring on the lower valve body and install the spring-to-case bolt finger tight.

9. While holding the detent spring roller in the center of the manual detent lever, install the detent spring-to-lower valve body bolt and tighten it to 80-120 in-lbs. torque.

10. Tighten the remainder of the control valve body attaching bolts to 80-120 in-lbs. torque.

11. Put a new gasket on the oil pan, install pan in place, and install and tighten all the pan attaching bolts to the proper torque.

12. If the filler tube was removed, reinstall it and tighten securely. If necessary, replace the oil seal around the filler tube to prevent leakage.

13. Lower the car and fill the transmission with enough fluid to bring the level up to the FULL mark on the dipstick. Check for fluid leaks at this time.

Oil Pan and Control Valve Body Removal and Installation —Falcon

The removal procedure for the oil pan and control valve body from the Falcon is similar to the procedure given above for the oil pan and control valve body removal from other cars except for some differences that are given here.

1. Do all operations to remove the oil pan and drain the fluid from the transmission as usual. It is not necessary to put the transmission in Park position.

2. Remove the valve body attaching bolts. Remove the valve body from the case and the transmission inner control levers.

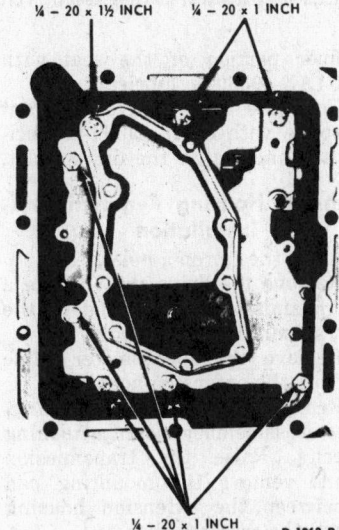

¼ – 20 x 1½ INCH   ¼ – 20 x 1 INCH

¼ – 20 x 1 INCH   D 1863-B

Control valve body attaching bolts
(© Ford Motor Co.)

3. Refer to the Component Disassembly and Repair section for control valve body repair operations.

4. Thoroughly clean and remove all traces of old gasket material and dirt from the case. Remove and discard the nylon shipping plug from the filler tube if present. Install the valve body to the case, engaging the transmission inner control levers with the valve body manual and downshift valves.

5. Install the valve body attaching bolts, tightening them to the proper torque. Operate the external manual and downshift levers to check for proper travel of the valve body manual and downshift valves.

6. Place a new gasket on the oil pan and install the oil pan and the attaching bolts. Tighten the bolts to the proper torque.

7. Lower the car and fill the transmission with enough fluid to raise the fluid level to the FULL mark on the dipstick. Check for leaks.

Intermediate Servo Removal and Installation

1. Raise the car and remove the four servo cover attaching bolts (right-hand side of case). Remove the cover and identification tag (*do not lose tag*).

NOTE: to gain access to the servo on a Maverick, the crossmember must be removed.

2. Remove the gasket, piston, and piston return spring.

3. Install the piston return spring in the case. Place a new gasket on the cover. Install the piston and cover in the transmission case, using two 5/16-18 x 1¼ bolts 180 degrees apart to align the cover against the case.

4. Install the transmission identification tag and two attaching bolts. Remove the two 1¼ bolts and install the other two cover attaching bolts. Tighten all cover attaching bolts to the proper torque.

5. On a Maverick, position the crossmember and install the attaching bolts, tightening them to the proper torque.

6. Adjust the intermediate band. Lower the car and fill the transmission with enough fluid to raise the fluid level to the FULL mark on the dipstick.

7. If the intermediate band cannot be adjusted correctly, remove the oil pan and control valve body and see if the struts are installed correctly. Adjust the struts and reinstall the control valve body and oil pan with a new gasket. Refill the transmission with fluid.

Low-Reverse Servo Piston Removal and Installation

1. Raise the car on a hoist.

2. Loosen the reverse band adjusting screw locknut and tighten the adjusting screw to 10 ft-lbs torque. This operation will hold the band strut against the case and prevent it from falling when the reverse servo piston is removed.

3. Remove the four servo cover at-

taching bolts and remove the servo cover and seal from the case.

4. Remove the servo piston from the case. *The piston and piston seal are bonded together and must be replaced together.*

5. Install the servo piston assembly into the case. Place a new cover seal on the cover and position them by installing two 5/16—18 bolts, 1¼ in. long, at 180 degrees apart on the case. Install two cover attaching bolts with the identification tag.

6. Remove the two positioning bolts and install the other two cover bolts. Tighten all the cover attaching bolts to the proper torque.

7. Adjust the low-reverse band. Lower the car and fill the transmission with enough fluid to raise the fluid level to the FULL mark on the dipstick.

8. If the low-reverse band cannot be adjusted properly, the transmission must be drained and the oil pan and valve body removed. Check the alignment of the band struts. Reinstall the valve body and the oil pan with a new gasket, and refill the transmission with fluid.

Extension Housing Bushing and Rear Seal Removal and Installation

1. Disconnect the drive shaft from the transmission.

2. If only the rear seal needs replacing, carefully remove it with a tapered chisel or use the tools shown in the illustration. Remove the bushing as shown. Be careful not to damage the spline seal with the bushing remover.

3. Install the new bushing, using the special tool shown.

4. Before installing a new rear seal, inspect the sealing surface of the universal joint yoke for scores. If the universal joint yoke is scored, replace the yoke.

5. Inspect the housing counterbore for burrs and remove them with crocus cloth if necessary.

6. Install the new rear seal into the housing, using the tool shown in the illustration. The seal should be firmly seated in the housing. Coat the inside diameter of the

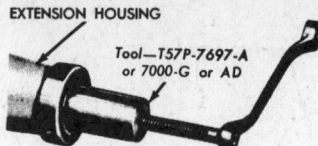

Removing extension housing bushing
(© Ford Motor Co.)

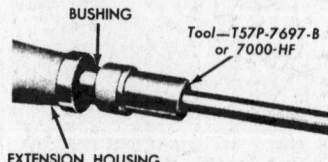

Installing extension housing bushing
(© Ford Motor Co.)

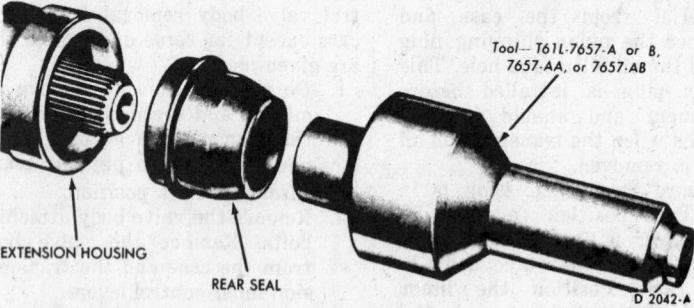

Installing extension housing seal (© Ford Motor Co.)

fiber portion of the seal with C1AZ-19590-B lubricant.

7. Coat the front universal joint spline with C1AZ-19590-B lubricant and install the drive shaft.

Extension Housing Removal and Installation

1. Raise the car on a hoist.

2. Remove the drive shaft. Place a transmission jack under the transmission for support.

3. Remove the speedometer cable from the extension housing.

4. Remove the extension housing-to-crossmember mount attaching bolts. Raise the transmission and remove the mounting pad between the extension housing and the crossmember.

5. Loosen the extension housing attaching bolts to drain the transmission fluid.

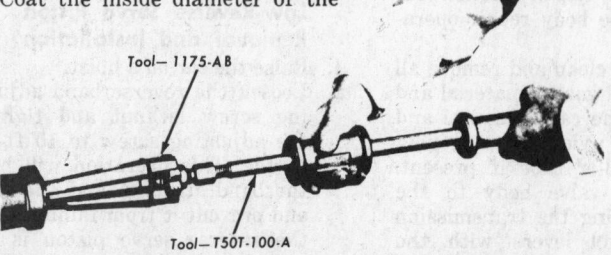

Removing extension housing seal (© Ford Motor Co.)

6. Remove the six extension housing attaching bolts and remove the extension housing.

7. To install the extension housing, reverse the above removal instructions. Install a new extension housing gasket. When the extension housing has been installed and all parts have been secured, lower the car and fill the transmission with the correct amount of fluid. Check for fluid leaks around the extension housing area.

Governor Removal and Installation (C4 Automatic Only)

1. After removing the extension housing according to the instructions above, remove the governor housing-to-governor distrib-

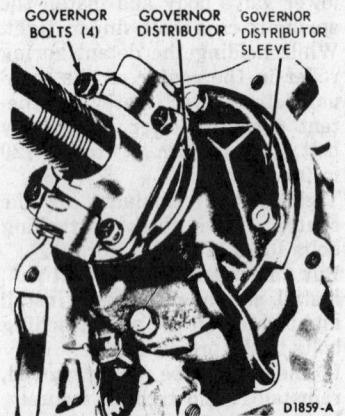

utor attaching bolts. Remove the governor housing from the distributor.

2. Refer to the Component Disassembly and Repair section for instructions on repairing the governor assembly.

Governor assembly installed
(© Ford Motor Co.)

3. Install the governor housing on the governor distributor and tighten the attaching bolts to the proper torque.

4. Install the extension housing with a new gasket according to the instructions above.

5. When the extension housing has been installed and all bolts have been tightened to the proper tor-

que, lower the car and fill the transmission with fluid to the proper level. Check around the extension housing area for leaks.

Transmission Overhaul Procedures

The transmission overhaul procedures presented here are the checks and repairs that must be done with the transmission out of the car. Disassembly of each transmission subassembly is illustrated by exploded views of the subassembly showing how the individual parts fit together. Reassembly of the subassembly is often the reverse of the disassembly procedure except for alignment, special tolerances, etc.

Procedures for removing the transmission from the car and reinstalling it back in the car are given in the Car Section.

During the transmission disassembly and reassembly operations, ten thrust washers that are installed between the subassemblies of the gear train must be removed and reinstalled correctly. Since it is very important that these thrust washers be installed correctly, they are shown in their positions and they are numbered for further identification. The No. 1 thrust washer is located at the front pump, and the No. 10 thrust washer is located at the parking pawl ring gear.

During all repairs to the transmission subassemblies, the following instructions must be followed:

1. Be sure that no dirt or grease gets in the transmission. All parts must be clean. *Remember—a little dirt can disable a transmission completely if it gets in a fluid passage.*
2. Handle all transmission parts carefully to avoid burring or nicking bearing or mating surfaces.
3. Lubricate all internal parts of the transmission with clean transmission fluid before assembling. *Do not use any other lubricants except on gaskets or thrust washers which may be coated lightly with vaseline to ease assembly.*
4. Always use new gaskets when assembling the parts of the transmission.
5. Tighten all bolts and screws to the recommended torque limits using a torque wrench.

Transmission Disassembly

Disassemble the transmission by following the procedures below:
1. Thoroughly clean the outside of the transmission to prevent dirt or grease from getting inside the

mechanism. *Do this before removing any subassembly.*
2. Place the transmission in the transmission holder. See illustration.

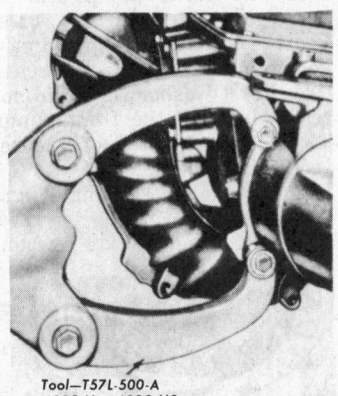

Tool—T57L-500-A
or 6005-M or 6005-MS   D1862-A

Transmission mounted in holding fixture
(© Ford Motor Co.)

3. Remove the converter from the transmission front pump and converter housing.
4. On a C4 automatic transmission, remove the transmission vacuum unit with the tool shown in the

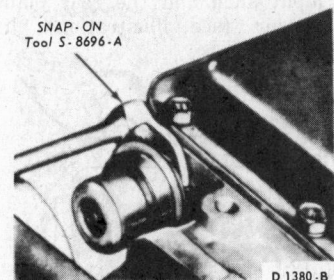

SNAP-ON
Tool S-8696-A

D 1380-B

Removing vacuum diaphragm unit
(© Ford Motor Co.)

illustration. Remove the vacuum unit gasket and the control rod.
5. On a C4 automatic transmission, remove the primary throttle valve from the opening at the rear of the case.

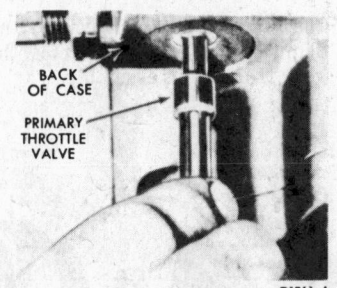

BACK OF CASE

PRIMARY THROTTLE VALVE

D1861-A

Removing or installing primary throttle valve
(© Ford Motor Co.)

6. Remove the transmission pan attaching bolts, oil pan, and gasket.

7. Remove the control valve body attaching bolts and then lift the control valve body from the case.

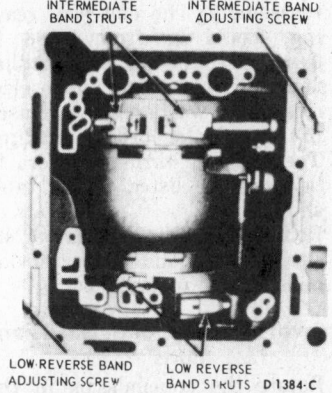

INTERMEDIATE BAND STRUTS   INTERMEDIATE BAND ADJUSTING SCREW

LOW-REVERSE BAND ADJUSTING SCREW   LOW REVERSE BAND STRUTS   D 1384-C

Band adjusting screws and struts
(© Ford Motor Co.)

8. Loosen the intermediate band adjusting screw and remove the intermediate band struts from the case. Loosen the low-reverse band adjusting screw and remove the low-reverse band struts.

Transmission End-Play Check

1. Remove one of the converter housing attaching bolts and mount the dial indicator support tool in the hole. Mount a dial indicator on the support so that a contact rests on the end of the input shaft. See illustration.

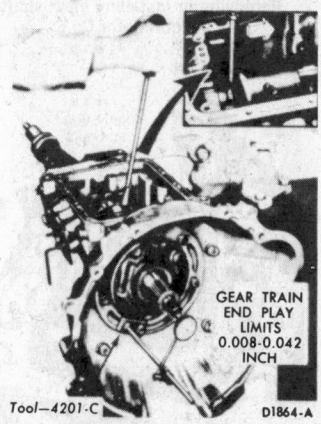

GEAR TRAIN END PLAY LIMITS 0.008-0.042 INCH

Tool—4201-C   D1864-A

Checking transmission end-play
(© Ford Motor Co.)

2. Install the extension housing seal replacer tool on the output shaft to provide support and alignment for the shaft.
3. Using a screwdriver, move the input shaft and the gear train to the rear of the case as far as possible. Set the dial indicator at zero while holding a slight pressure on the screwdriver.
4. Remove the screwdriver and insert it behind the input shell. Move the input shell and the

front part of the gear train forward.

5. Record the dial reading for later reference during transmission reassembly. The end play reading should be from 0.008 to 0.042 in. If the end play reading is not within this range, selective thrust washers must be used to obtain the proper reading. The selective thrust washers to be used are listed in the table shown.

6. Remove the dial indicator, its support bar, and the extension housing seal replacer tool.

Removal of Case and Extension Parts

1. Rotate the transmission in the holding fixture until it is in a vertical position with the converter housing up.

2. Remove the five converter housing attaching bolts and remove the converter housing from the transmission case.

Removing or installing input shaft
(© Ford Motor Co.)

Removing front pump (© Ford Motor Co.)

3. Remove the seven front pump attaching bolts. Remove the front pump by inserting a screwdriver behind the input shell and pushing it forward until the front pump seal is above the edge of the case. Remove the front pump and gasket from the case. If the selective thrust washer No. 1 did not come out with the front pump, lift it from the top of the reverse-high clutch.

4. Remove the intermediate and low-reverse adjusting screws from the case. Rotate the in-

Position of intermediate band for removal or installation
(© Ford Motor Co.)

termediate band to align the band ends with the clearance hole in the case. Remove the intermediate band from the case.

5. Using a screwdriver between the input shell and the rear planet carrier (see illustration), lift

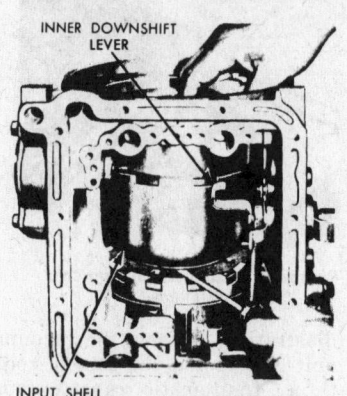

Lifting input shell and gear train
(© Ford Motor Co.)

the input shell upward and remove the forward part of the gear train as an assembly.

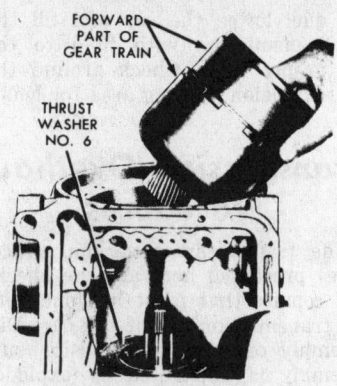

Removing or installing forward part of gear train
(© Ford Motor Co.)

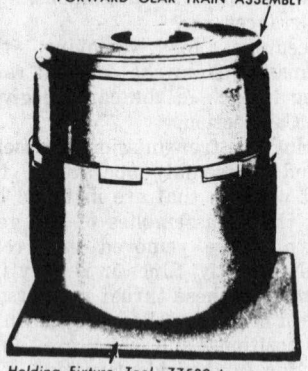

Forward part of gear train positioned in holding fixture
(© Ford Motor Co.)

6. Place the forward part of the gear train in the holding fixture shown.

7. With the gear train in the holding fixture, remove the reverse-high clutch and drum from the forward clutch. If thrust washer No. 2 did not come out with the front pump, remove the thrust washer from the forward clutch cylinder. If a selective spacer was used, remove the spacer. Remove the forward clutch from the forward clutch hub and ring gear.

8. If the thrust washer No. 3 did not come out with the forward clutch, remove the thrust washer from the forward clutch hub and

Forward part of gear train disassembled (© Ford Motor Co.)

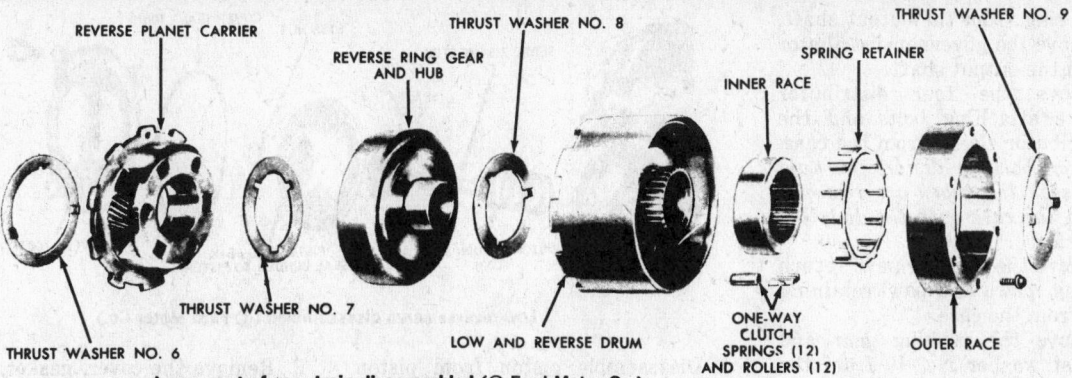

REVERSE PLANET CARRIER · THRUST WASHER NO. 8 · THRUST WASHER NO. 9 · REVERSE RING GEAR AND HUB · SPRING RETAINER · INNER RACE · THRUST WASHER NO. 7 · THRUST WASHER NO. 6 · LOW AND REVERSE DRUM · ONE-WAY CLUTCH SPRINGS (12) AND ROLLERS (12) · OUTER RACE

Lower part of gear train disassembled (© Ford Motor Co.)

lift the forward clutch hub and ring gear from the front planet carrier.

9. Remove thrust washer No. 4 and the front planet carrier from the input shell.

10. Remove the input shell, sun gear and thrust washer No. 5 from the holding fixture.

11. From inside the transmission case, remove thrust washer No. 6 from the top of the reverse planet carrier. Remove the reverse planet carrier and thrust washer No. 7 from the reverse ring gear and hub.

12. Move the output shaft forward and, with the tool shown in the illustration, remove the reverse

reverse drum from the one-way clutch inner race. Remove the one-way clutch inner race by rotating the race clockwise as it is removed.

14. Remove the 12 one-way clutch rollers, springs and the spring retainer from the outer race. *Do not lose or damage any of the 12 springs or rollers. The outer*

race of the one-way clutch cannot be removed from the case until the extension housing, output shaft and governor distributor sleeve are removed.

15. Remove the transmission from the holding fixture. Place the transmission on the bench in a vertical position with the extension housing up. Remove the four extension housing attaching bolts, the extension housing, and the gasket from the case.

16. Pull outward on the output shaft and remove the output shaft and governor distributor assembly (if so equipped) from the governor distributor sleeve.

17. On a C4 automatic transmission, remove the governor distributor

REVERSE RING GEAR HUB RETAINING RING

D 2044-A

Removing or installing reverse ring gear hub retaining ring (© Ford Motor Co.)

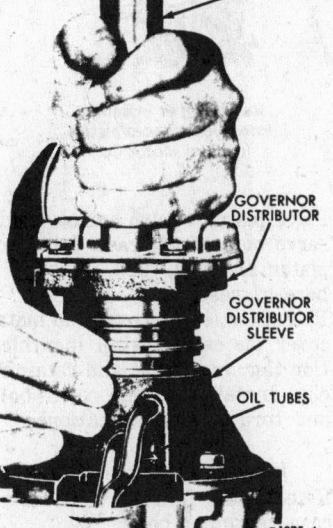

OUTPUT SHAFT · GOVERNOR DISTRIBUTOR · GOVERNOR DISTRIBUTOR SLEEVE · OIL TUBES

D1875-A

Removing or installing output shaft and governor distributor (© Ford Motor Co.)

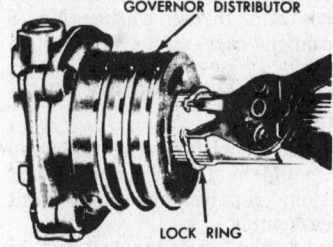

GOVERNOR DISTRIBUTOR · LOCK RING

Removing or installing governor distributor snap-ring (© Ford Motor Co.)

ring gear hub-to-output shaft retaining ring. Remove the reverse ring gear and hub from the output shaft. Remove the thrust washer No. 8 from the low and reverse drum.

13. Remove the low-reverse band from the case. Remove the low-

LOW-REVERSE BAND

Removing or installing low-reverse band (© Ford Motor Co.)

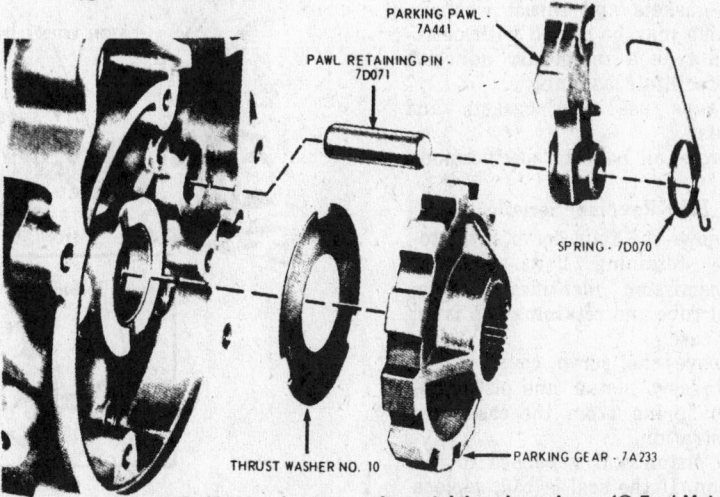

PARKING PAWL - 7A441 · PAWL RETAINING PIN - 7D071 · SPRING - 7D070 · THRUST WASHER NO. 10 · PARKING GEAR - 7A233

Parking pawl, return spring, retaining pin, and gear (© Ford Motor Co.)

lock ring from the output shaft. Remove the governor distributor from the output shaft.

18. Remove the four distributor sleeve attaching bolts and the distributor sleeve from the case. *Do not bend or distort the fluid tubes as the tubes are removed from the case with the distributor sleeve.*

19. Remove the parking pawl return spring, pawl, and pawl retaining pin from the case.

20. Remove the parking gear and thrust washer No. 10 from the case.

21. Remove the six one-way clutch outer race attaching bolts with the tool shown. As the bolts are removed, hold the outer race

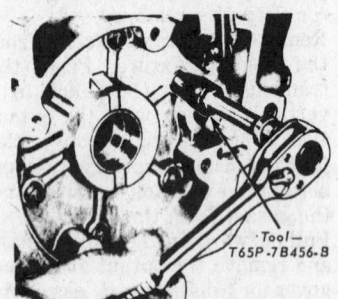

Removing one-way clutch outer race attaching bolts
(© Ford Motor Co.)

that is located inside the case in position. Then, remove the outer race and thrust washer No. 9 from the case.

Component Disassembly and Assembly

To avoid repetition, the following rules should be followed during examination, repairs or replacement and reassembly:

1. Avoid dirt and damage due to careless handling.
2. Lubricate all internal parts, before assembly, with transmission fluid. Use no other fluids, except on gaskets and thrust washers, which may be coated with petrolatum as a temporary adhesive to facilitate assembly.
3. Always use new gaskets and seals.
4. Torque all bolts to specifications.

Low-Reverse Servo

1. Remove the four servo cover-to-case retaining bolts. Remove transmission identification tag, vent tube and retaining clip from the case.
2. Remove the servo cover, cover seal, servo piston, and piston return spring from the case. See illustration.
3. The piston seal is bonded to the piston. If the seal is bad, replace piston and seal assembly.

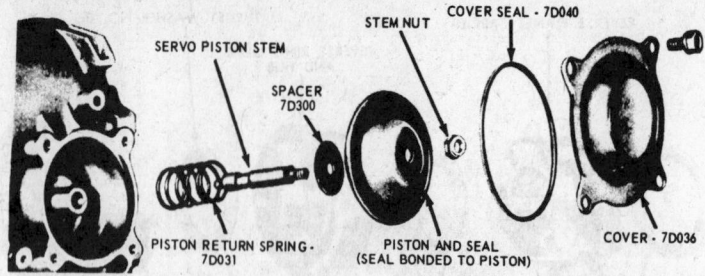

Low-reverse servo disassembled (© Ford Motor Co.)

4. Disassemble piston from piston rod.
5. Install (See illustration) new servo piston and torque the nut to specifications.

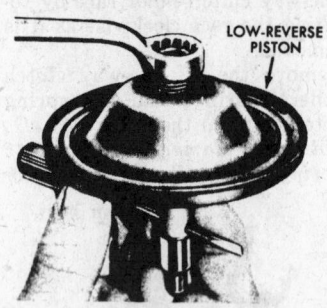

Removing or installing low-reverse servo piston
(© Ford Motor Co.)

6. Place piston return spring in the servo bore of the case. Lubricate piston and seal, then, install it in bore of case.
7. Place new seal on cover and install cover on case. Install identification tag, vent tube, and retaining clip. Install the four cover bolts and torque to specifications.

Intermediate Servo

1. Remove the four servo cover-to-case retaining bolts.

2. Remove the cover, gasket, piston and piston return spring. See illustration.
3. On a transmission used with six cylinder engines, remove the intermediate servo piston from the cover by forcing air pressure into the pressure hole in the cover.
4. Remove seal rings from the piston and cover.
5. Install new seal on the cover and servo piston. Lubricate and install the piston into the cover.
6. Install the piston return spring into the servo bore of the case.
7. With a new gasket in place, position the piston and cover assembly in the case.

NOTE: use two 5/16-18 x 1¼ in. bolts, 180° apart, to position the cover on the case. Install two cover attaching bolts. Remove the two pilot bolts, then, install the regular bolts and torque all of the retaining bolts to specifications.

Downshift and Manual Linkage

1. Downshift outer and inner levers have been removed from the case during removal and disassembly procedures.

From inside the case, remove the upper retaining ring and the flat washer from the manual lever link. Remove the upper end of the

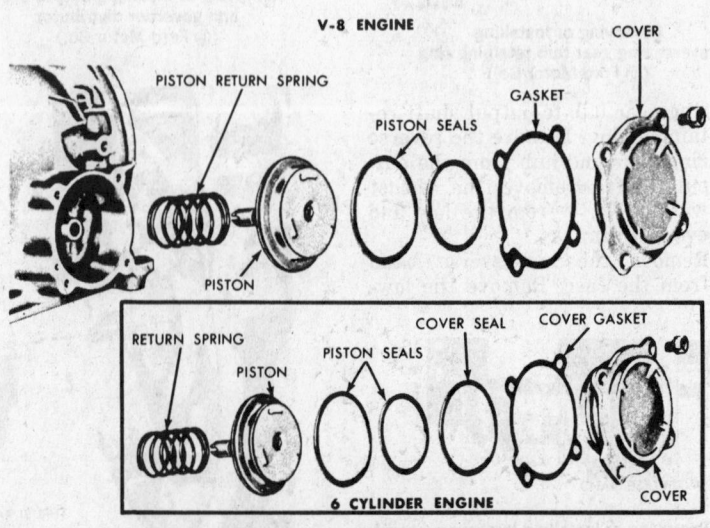

Intermediate servo disassembled (© Ford Motor Co.)

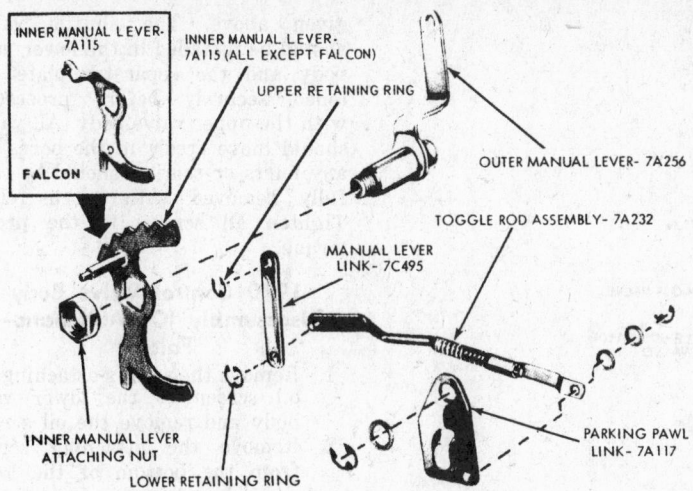

Transmission case internal linkage (© Ford Motor Co.)

lever link from the case retaining pin.

2. From the back of the transmission case, remove the upper retaining ring and flat washer from

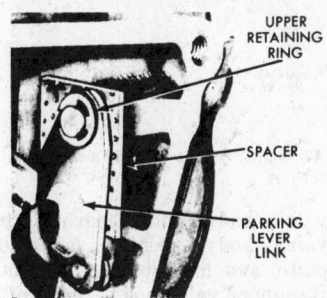

Parking pawl link (© Ford Motor Co.)

the parking pawl link. Remove the pawl link and spacer from the case retaining pin.

3. From the back of the transmission case, remove the parking pawl link, toggle rod, and the manual lever link as an assembly.

Removing or installing parking pawl link
(© Ford Motor Co.)

4. Remove the rear parking pawl link lower retaining ring, flat washer and link from the toggle rod.
5. Remove the manual lever link lower retaining ring, flat washer, and link from the toggle rod.
6. Install manual and parking pawl links, flat washers and retaining rings to the toggle rods.
7. Remove the inner manual lever retaining nut and lever. Remove

the outer manual lever from the case. Remove manual lever seal.

Removing manual lever seal
(© Ford Motor Co.)

8. Install new seal onto manual lever.
9. Install outer manual lever into the case. Install inner manual lever and retaining nut. Torque nut to specifications.
10. From back of the transmission case, install parking toggle rod and link into the case. Install parking pawl link spacer onto the case retaining pin. Dimpled side of the spacer should be facing the center of the transmission case.
11. Install the parking pawl link onto the case retaining pin. Install flat washer and retaining ring.
12. Position inner manual lever behind the manual lever link, with the cam on the lever contacting the lower link pin.

13. Install the upper end of the manual lever link onto the case retaining pin. Install flat washer and retaining ring.
14. Operate the manual lever and check for correct linkage operation.

Control Valve Body Disassembly and Assembly

Disassembly and assembly procedures are given for all models of the C4 and C4S automatic transmission control valve bodies. Where there are differences in construction, they are given. Refer to the illustrations of the different control valve bodies that show how they fit together while doing these procedures.

1970 Control Valve Body Disassembly (C4 Automatic— All Except Falcon)

1. Remove the screws attaching the oil screen to the valve body and remove the oil screen. Be careful not to lose the throttle pressure limit valve and spring when separating the oil screen from the valve body.
2. Remove the attaching screws from the lower valve body and separate the upper and lower valve bodies, the gasket, separator plate and the hold-down plate. Be careful not to lose the upper valve body shuttle valve and check valve when separating the upper and lower valve bodies.
3. Slide the manual valve out of the body.
4. Pry the low servo modulator valve retainer from the body and remove the retainer plug, spring, and valve from the valve body. While working in the same bore, pry the retainer, spring, and the downshift valve from the valve body (see illustration).
5. Depress the throttle booster plug and remove the retaining pin. Then, remove the plug, valve, and spring from the valve body.
6. Remove the cover over the cut-

Upper and lower control valve bodies disassembled— all 1970 C4 transmissions except Falcon (© Ford Motor Co.)

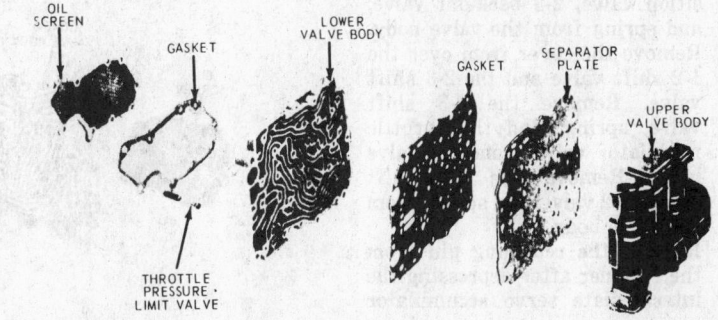

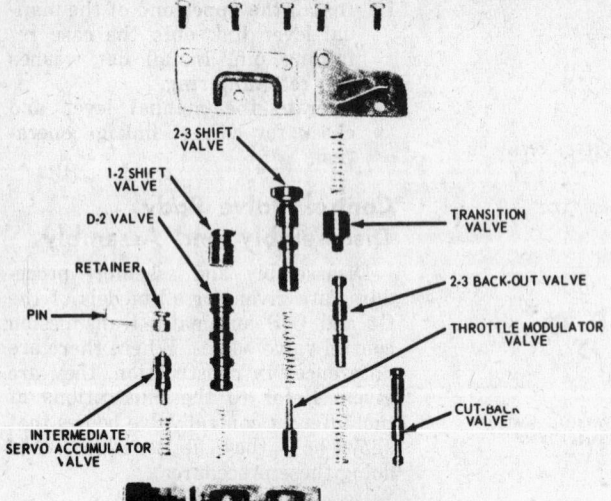

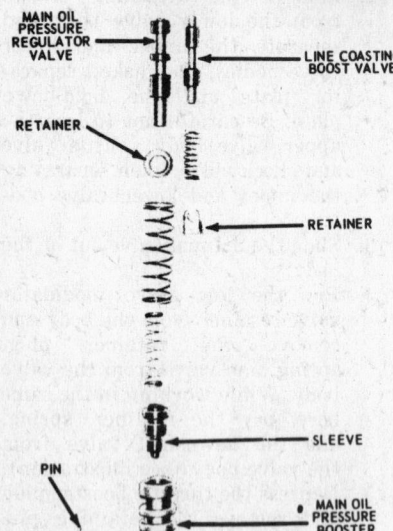

given above. The shuttle valves should be installed in the lower valve body and the separator plate fastened securely before proceeding with the upper valve body. All valves should move freely in the bores and any burrs or scoring should be carefully removed with crocus cloth. Tighten all screws to the proper torque.

1970 Control Valve Body Disassembly (C4 Automatic—Falcon)

1. Remove the screws attaching the oil screen to the lower valve body and remove the oil screen.
2. Remove the attaching screws from the bottom of the lower

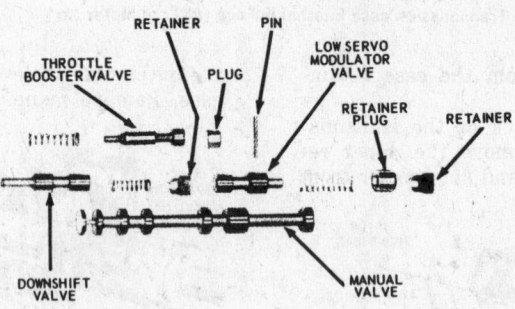

Upper control valve body disassembled—1970 C4 transmissions except Falcon
(© Ford Motor Co.)

back valve and the transition valve from the valve body.

7. Remove the cut-back valve and the transition valve spring, transition valve, 2-3 back-out valve, and spring from the valve body.
8. Remove the cover from over the 1-2 shift valve and the 2-3 shift valve. Remove the 2-3 shift valve, spring, and the throttle modulator valve from the valve body. Remove the 1-2 shift valve, D-2 valve, and spring from the valve body.
9. Remove the retaining pin from the retainer after depressing the intermediate servo accumulator

valve. Remove the retainer, intermediate servo accumulator valve, and spring from the valve body
10. Depress the main oil pressure booster valve and remove the retaining pin. Remove the main oil pressure booster valve, sleeve, springs, retainer, and main oil pressure regulator valve from the valve body.
11. Remove the line coasting boost valve retaining clip, the spring, and line coasting boost valve from the valve body.

1970 Control Valve Body Assembly (C4 Automatic—All Except Falcon)

Assembly procedures are the reverse of the disassembly procedures

valve body and separate the valve body, gasket, separator plate, and hold-down plate from the upper valve body. Be careful not to lose the upper valve body rubber ball shuttle valve and spring when separating the upper and lower valve bodies.
3. Depress the manual valve detent spring and remove the retaining pin. Then, remove the spring and manual valve detent plug from the valve body (see illustration).
4. Slide the manual valve from the valve body.
5. Remove the retainer clip, spring, and the downshift valve from the valve body.
6. Remove the stop pin after depressing the throttle booster valve and plug. Then, remove

Upper and lower control valve bodies disassembled—all 1970 Falcon transmissions
(© Ford Motor Co.)

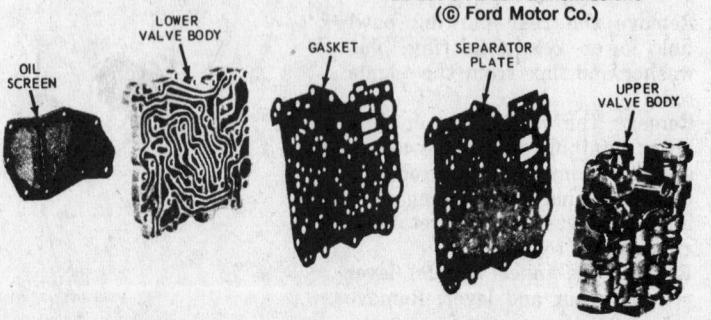

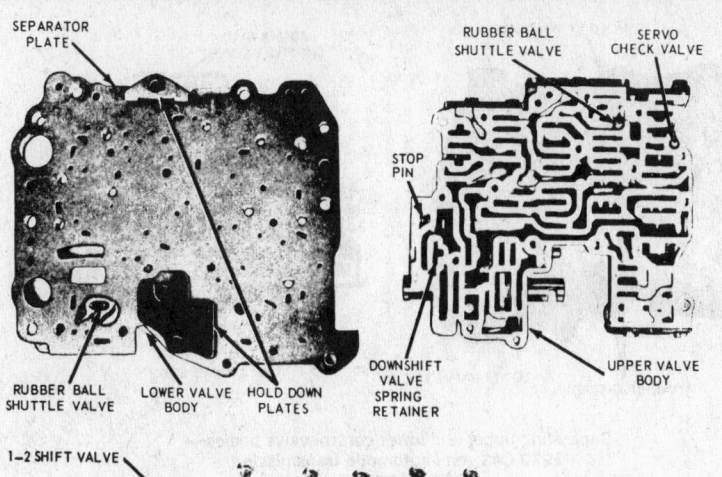

SEPARATOR PLATE

RUBBER BALL SHUTTLE VALVE

SERVO CHECK VALVE

STOP PIN

RUBBER BALL SHUTTLE VALVE   LOWER VALVE BODY   HOLD DOWN PLATES   DOWNSHIFT VALVE SPRING RETAINER   UPPER VALVE BODY

Separating upper and lower control valve bodies— all 1970 Falcon transmissions (© Ford Motor Co.)

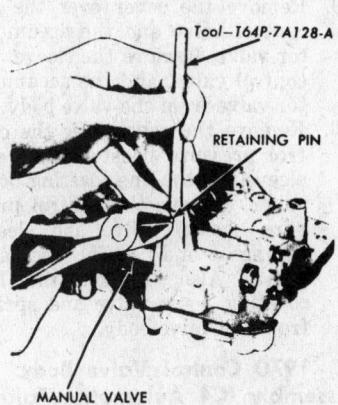

Tool—T64P-7A128-A

RETAINING PIN

MANUAL VALVE

Removing manual valve— 1970 Falcon C4 transmissions (© Ford Motor Co.)

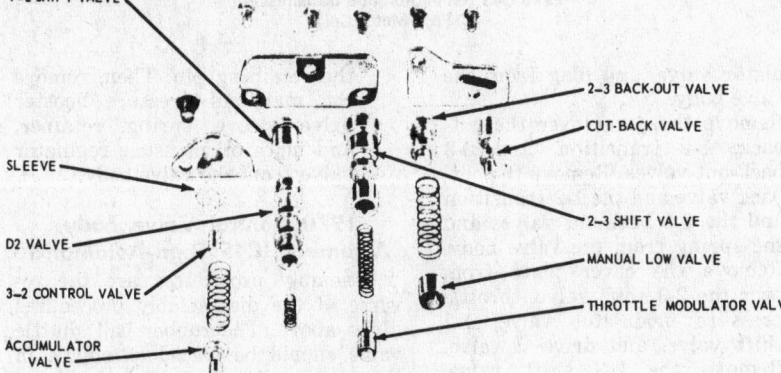

1-2 SHIFT VALVE

2-3 BACK-OUT VALVE

CUT-BACK VALVE

SLEEVE

D2 VALVE

2-3 SHIFT VALVE

3-2 CONTROL VALVE

MANUAL LOW VALVE

THROTTLE MODULATOR VALVE

ACCUMULATOR VALVE

the plug, throttle booster valve, and spring from the valve body.

7. Remove the cover over the cut-back valve and the 2-3 back-out and manual low valves. Remove the cut-back valve and the 2-3 back-out valve, spring, and the manual low valve from the valve body.

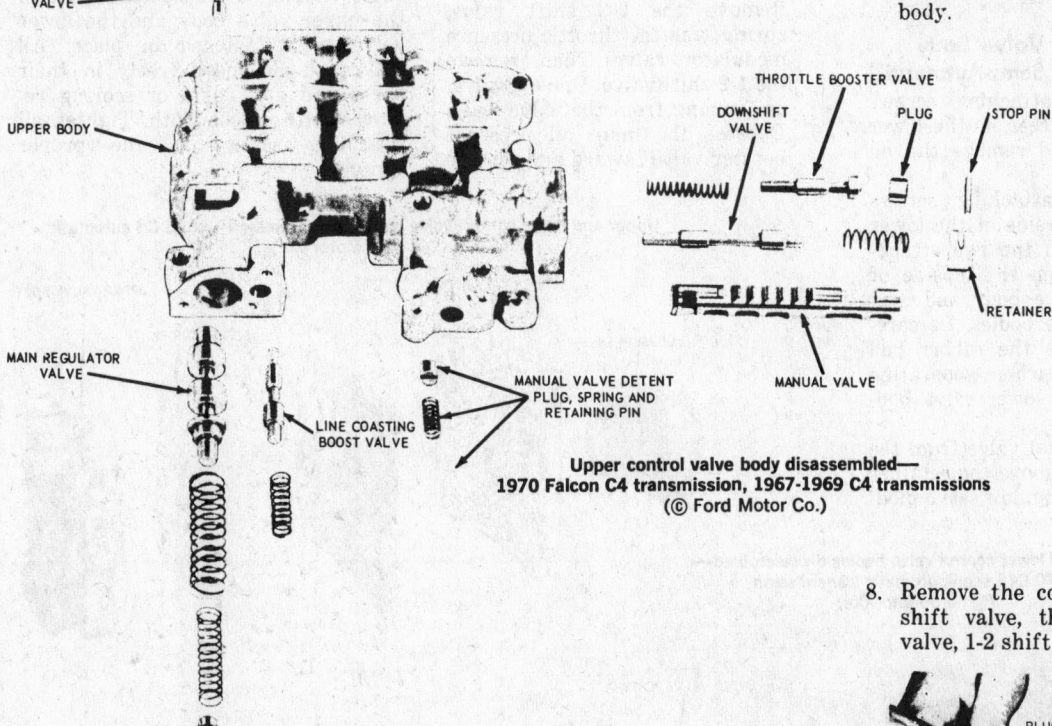

UPPER BODY

THROTTLE BOOSTER VALVE

DOWNSHIFT VALVE

PLUG

STOP PIN

RETAINER

MANUAL VALVE

MAIN REGULATOR VALVE

LINE COASTING BOOST VALVE

MANUAL VALVE DETENT PLUG, SPRING AND RETAINING PIN

Upper control valve body disassembled— 1970 Falcon C4 transmission, 1967-1969 C4 transmissions (© Ford Motor Co.)

CONTROL PRESSURE BOOSTER VALVE AND SLEEVE

8. Remove the cover over the 2-3 shift valve, throttle modulator valve, 1-2 shift valve, and the D2

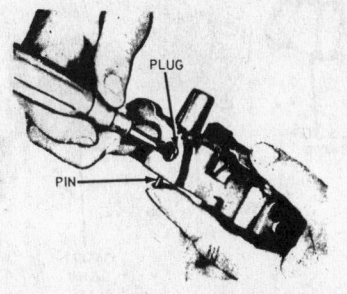

PLUG

PIN

Removing or installing throttle booster valve —1970 Falcon C4 transmission (© Ford Motor Co.)

valve and remove these valves and their springs from the valve body.

9. Remove the cover over the 3-2 control valve and the accumulator valve. Remove the sleeve, 3-2 control valve, and the accumulator valve from the valve body.

10. Remove the cover over the control pressure booster valve and sleeve and the line coasting boost valve. Remove the control pressure booster valve and sleeve, springs, and main regulator valve. Also, remove the line coasting boost valve and spring from the valve body.

1970 Control Valve Body Assembly (C4 Automatic—Falcon)

Assembly procedures are the reverse of the disassembly procedures given above. The rubber ball shuttle valve should be placed in the lower valve body as shown and the separator plate, hold-down plates, and attaching screws installed on the lower valve body (see illustration). All valves should move freely in their bores and any burrs or scoring should be removed with crocus cloth. Tighten all attaching screws to the proper torque.

1970 Control Valve Body Disassembly (C4S Semi-Automatic)

1. Remove the attaching screws from the oil screen on the lower valve body and remove the oil screen.

2. Remove the attaching screws from the underside of the lower valve body and the two attaching screws from the topside of the upper valve body and separate the valve bodies. Be careful not to lose the rubber ball shuttle valve when separating the upper and lower valve bodies.

3. Slide the manual valve from the valve body. Remove the retainer clip, plug, spring, low servo modulator valve, and plug from the valve body.

4. Remove the cover over the cutback, 1-2 transition, and 2-3 back-out valves. Remove the cutback valve and the 1-2 transition and the 2-3 back-out valves and the spring from the valve body.

5. Remove the cover plate from over the 2-3 shift valve, throttle pressure modulator valve, 1-2 shift valve, and drive 2 valve. Remove the 2-3 shift valve, spring, and the throttle pressure modulator valve. Then, remove the 1-2 shift valve, Drive 2 valve, and spring from the valve body.

6. Depress the main oil pressure booster valve inward and remove

the retaining pin. Then, remove the main oil pressure booster valve, sleeve, spring, retainer, and main oil pressure regulator valve from the valve body.

1970 Control Valve Body Assembly (C4S Semi-Automatic)

Assembly procedures are the reverse of the disassembly procedures given above. The rubber ball shuttle valve should be placed as shown in the upper valve body and the lower valve body fastened in place. All valves should move freely in their bores and any burrs or scoring removed with crocus cloth. Tighten all attaching screws to the proper torque.

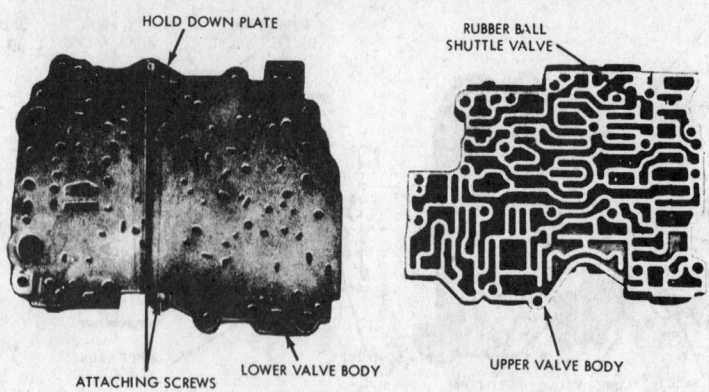

Separating upper and lower control valve bodies—1970 C4S semi-automatic transmission
(© Ford Motor Co.)

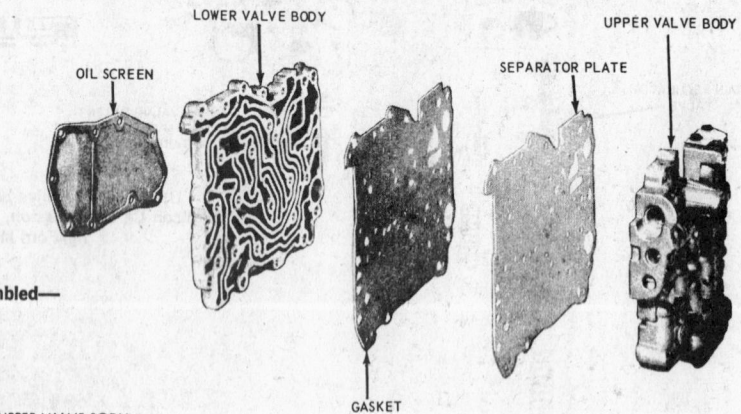

Upper and lower control valve bodies disassembled—1966 C4 automatic
(© Ford Motor Co.)

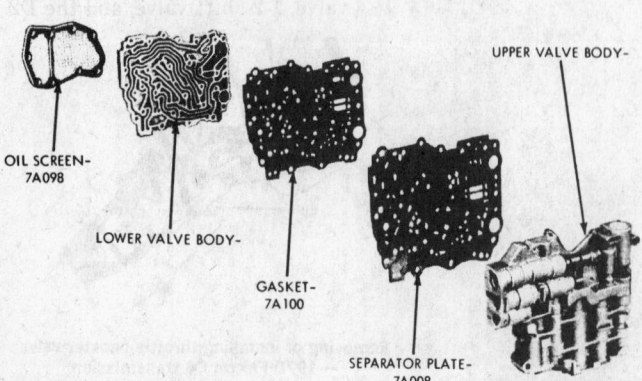

Upper and lower control valve bodies disassembled—1970 C4S semi-automatic transmission
(© Ford Motor Co.)

1967-69 Control Valve Body Disassembly and Assembly (C4 Automatic)

The procedures for the disassembly and assembly of the 1967-1969 control valve bodies are the same as those given for the 1970 control valve body (C4 Automatic—Falcon) The same control valve body is used for transmissions made in those years.

1965-66 Control Valve Body Disassembly and Assembly (C4 Automatic)

The disassembly and assembly procedures for the 1965-66 control valve bodies are similar to those given for the 1970 control valve body (C4 automatic). There are some differences in construction between these control valve bodies which may be seen in the illustrations. The assembly instructions are the reverse of the disassembly instructions.

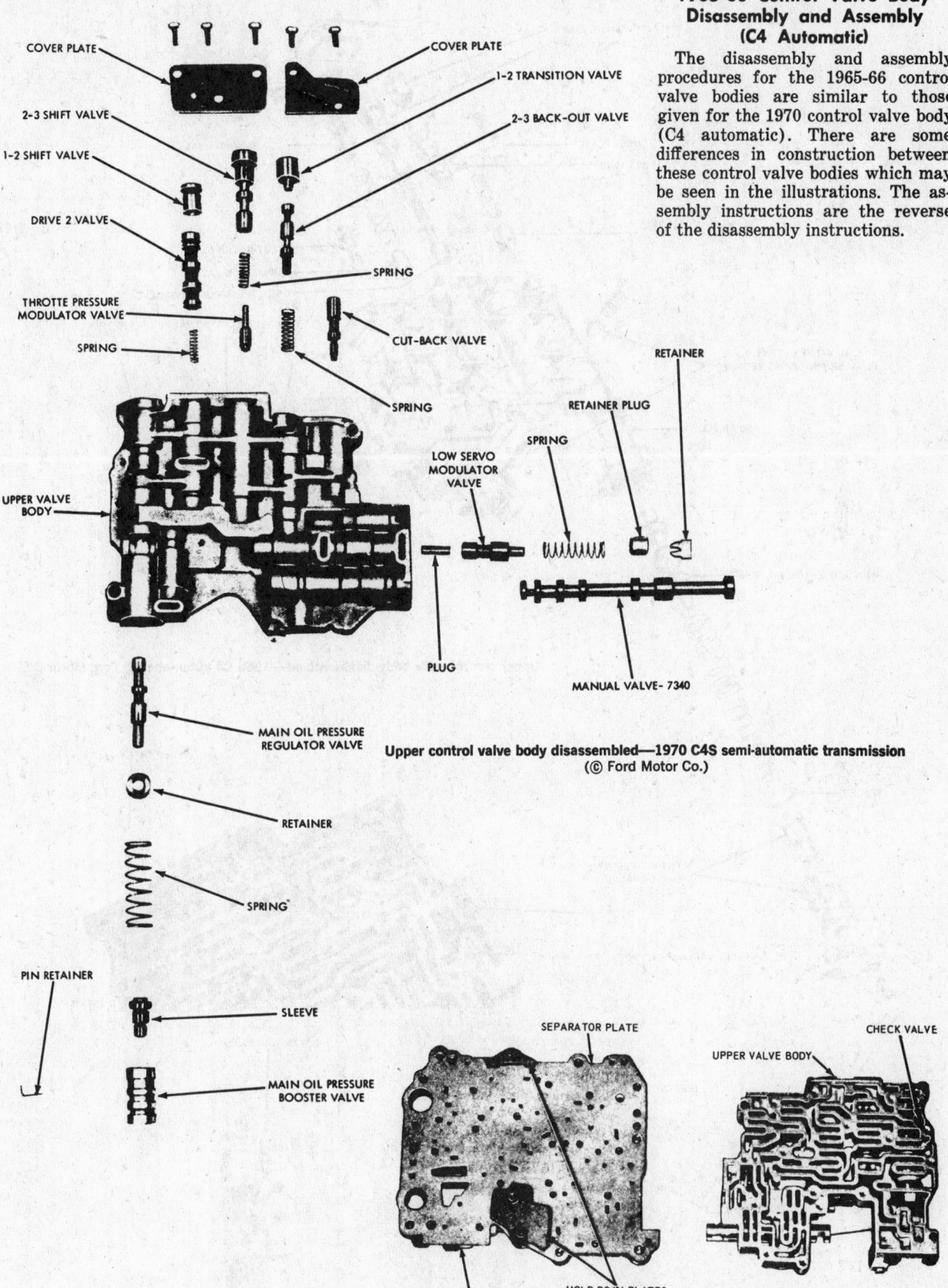

Upper control valve body disassembled—1970 C4S semi-automatic transmission
(© Ford Motor Co.)

Separating upper and lower control valve bodies—
1965 and 1966 C4 automatic
(© Ford Motor Co.)

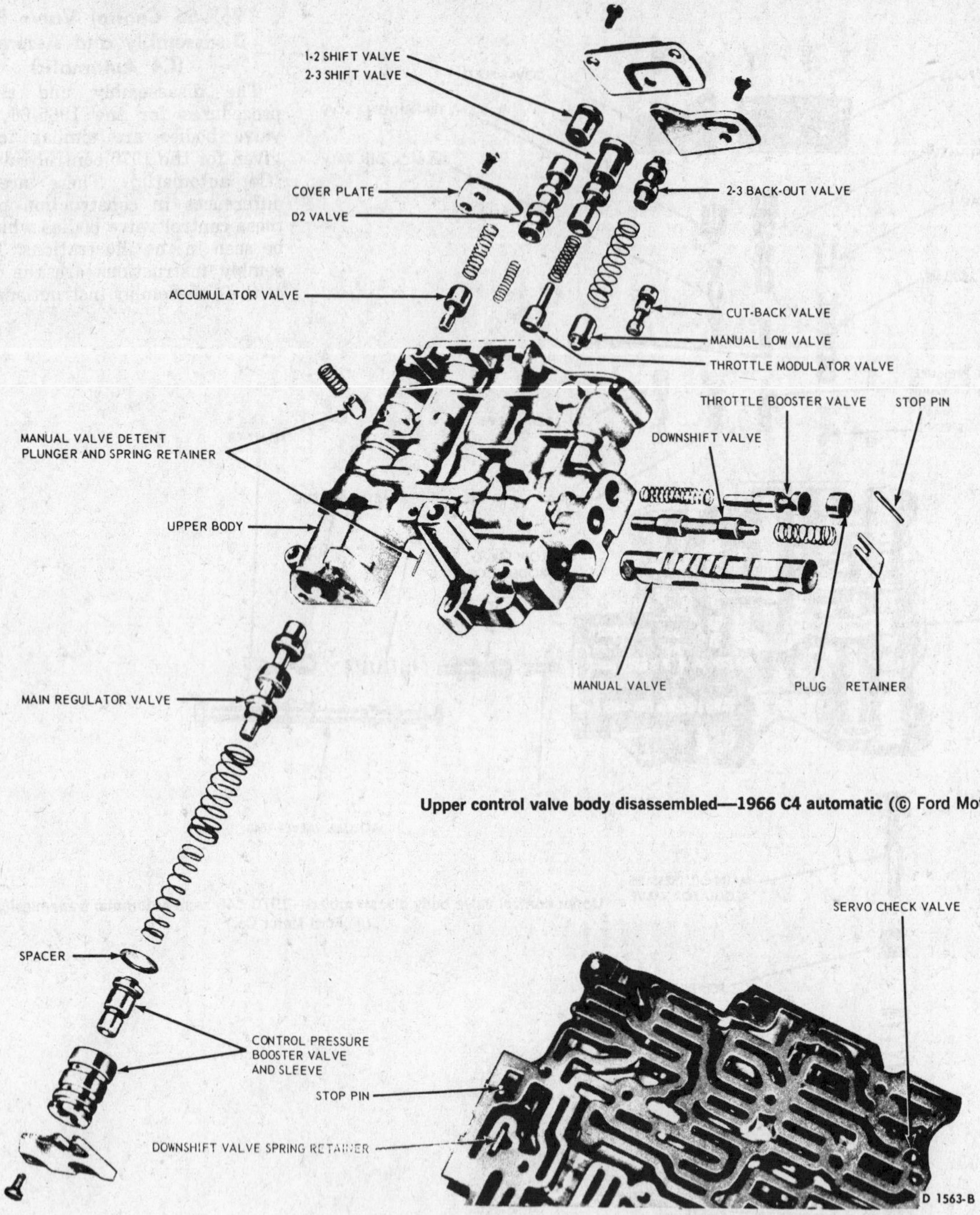

1-2 SHIFT VALVE

2-3 SHIFT VALVE

COVER PLATE

D2 VALVE

2-3 BACK-OUT VALVE

ACCUMULATOR VALVE

CUT-BACK VALVE

MANUAL LOW VALVE

THROTTLE MODULATOR VALVE

THROTTLE BOOSTER VALVE   STOP PIN

MANUAL VALVE DETENT
PLUNGER AND SPRING RETAINER

DOWNSHIFT VALVE

UPPER BODY

MAIN REGULATOR VALVE

MANUAL VALVE   PLUG   RETAINER

Upper control valve body disassembled—1966 C4 automatic (© Ford Motor Co.)

SERVO CHECK VALVE

SPACER

CONTROL PRESSURE
BOOSTER VALVE
AND SLEEVE

STOP PIN

DOWNSHIFT VALVE SPRING RETAINER

D 1563-B

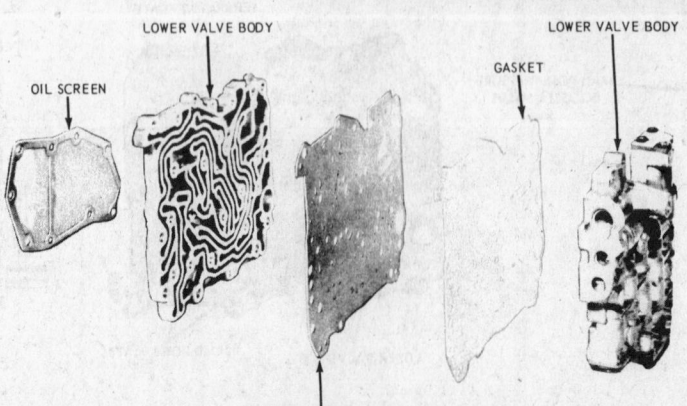

LOWER VALVE BODY

LOWER VALVE BODY

OIL SCREEN

GASKET

SEPARATOR PLATE

Upper and lower control valve bodies disassembled—1965 C4 automatic
(© Ford Motor Co.)

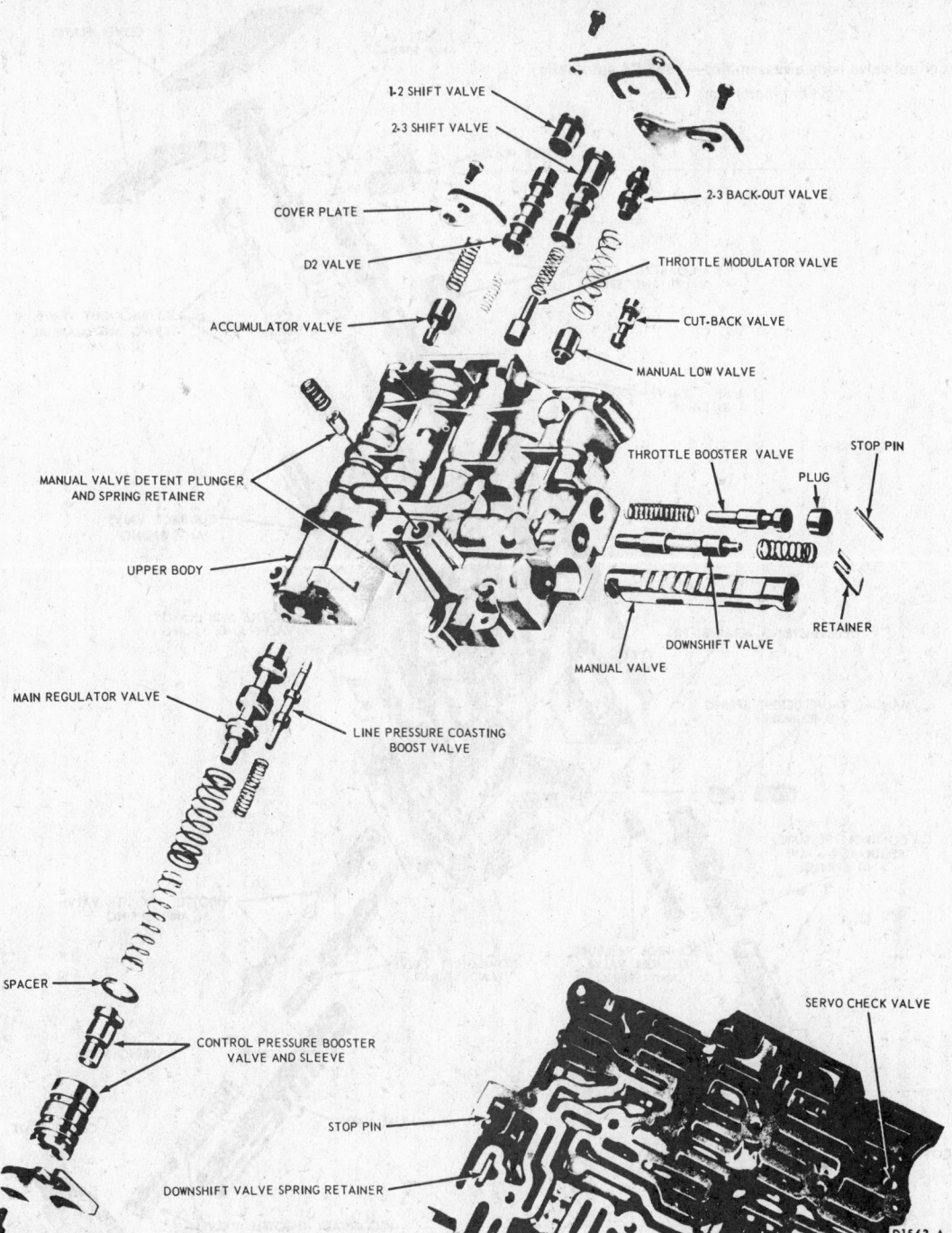

Upper control valve body disassembled—1965 C4 automatic (© Ford Motor Co.)

1964 Control Valve Body Disassembly and Assembly (C4 Automatic)

The 1964 control valve body is disassembled according to instructions given for the 1970 control valve body (C4 automatic-Falcon) and as shown in the disassembled views of the control valve body. The assembly instructions are the reverse of the disassembly instructions. Install the valves and springs as shown and be sure they move freely in their bores. Any burrs or scoring should be removed with crocus cloth. Tighten all screws to the proper torque.

Front Pump Disassembly

1. Remove the four seal rings from the stator support and, in 1964-67 models, the O-ring seal from the pump housing.
2. Remove the five bolts that secure the stator support to the front pump housing. Remove stator support from pump housing.
3. Replace the stator bushings if they are worn or damaged. Use a cape chisel to cut the bushing through. Then, pry up the loose ends of the bushing with an awl and remove the bushing. Press a new bushing in the stator support.
4. Replace the bushing in the pump housing with the tools shown,

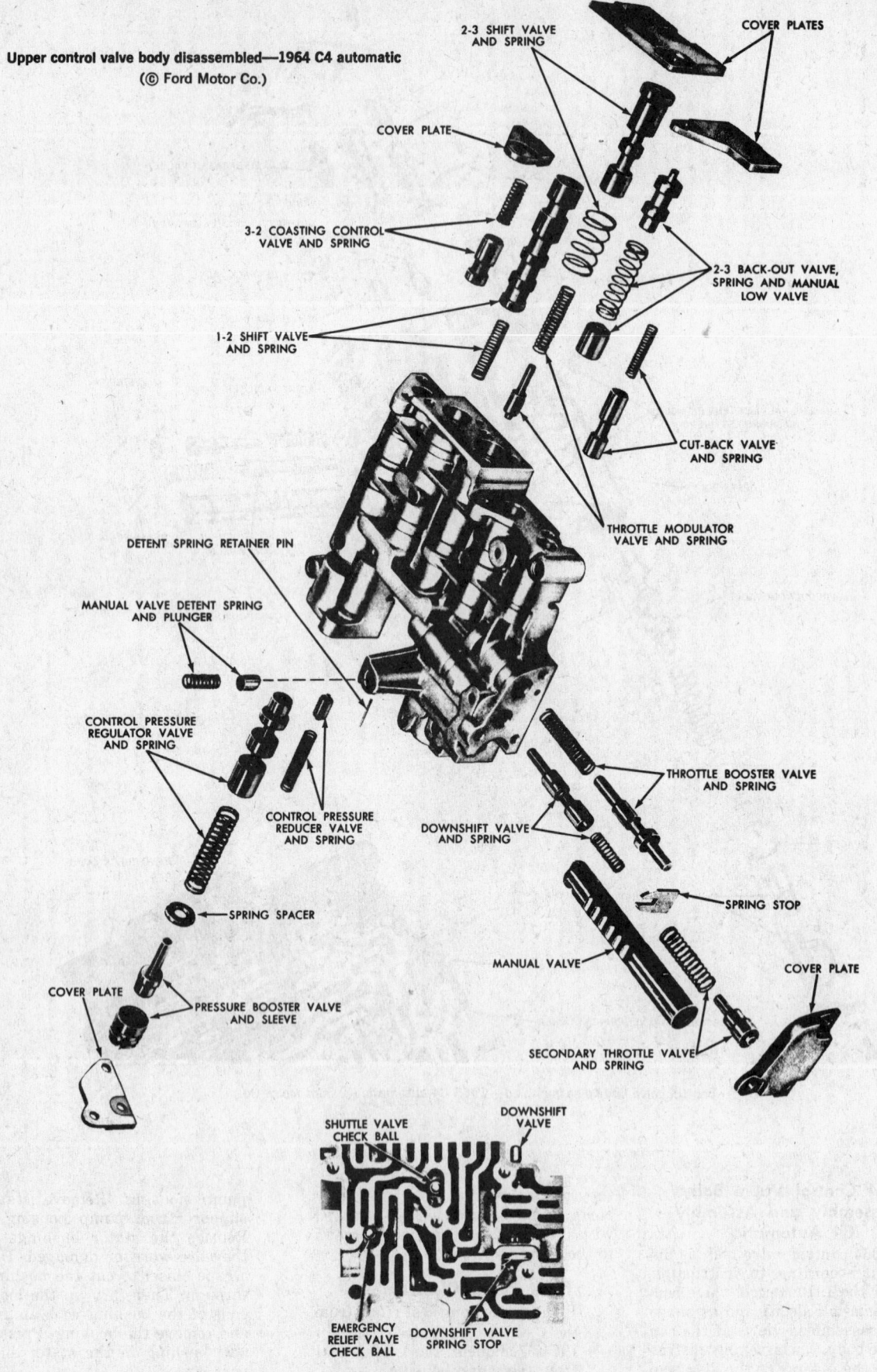

Upper control valve body disassembled—1964 C4 automatic
(© Ford Motor Co.)

2-3 SHIFT VALVE AND SPRING

COVER PLATES

COVER PLATE

3-2 COASTING CONTROL VALVE AND SPRING

2-3 BACK-OUT VALVE, SPRING AND MANUAL LOW VALVE

1-2 SHIFT VALVE AND SPRING

CUT-BACK VALVE AND SPRING

THROTTLE MODULATOR VALVE AND SPRING

DETENT SPRING RETAINER PIN

MANUAL VALVE DETENT SPRING AND PLUNGER

CONTROL PRESSURE REGULATOR VALVE AND SPRING

CONTROL PRESSURE REDUCER VALVE AND SPRING

THROTTLE BOOSTER VALVE AND SPRING

DOWNSHIFT VALVE AND SPRING

SPRING STOP

SPRING SPACER

MANUAL VALVE

COVER PLATE

COVER PLATE

PRESSURE BOOSTER VALVE AND SLEEVE

SECONDARY THROTTLE VALVE AND SPRING

SHUTTLE VALVE CHECK BALL

DOWNSHIFT VALVE

EMERGENCY RELIEF VALVE CHECK BALL

DOWNSHIFT VALVE SPRING STOP

Upper control valve body check ball
and spring location—1964 C4 automatic
(© Ford Motor Co.)

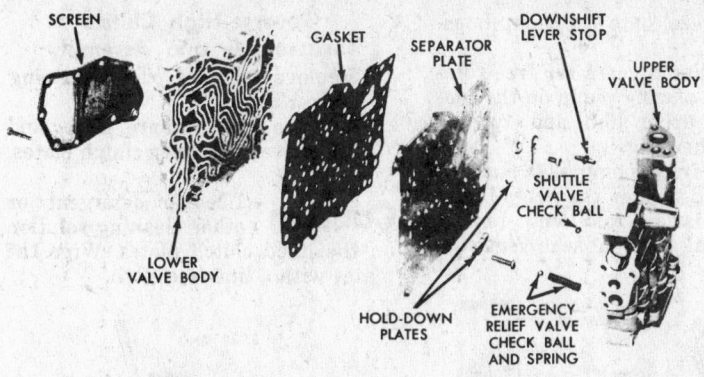

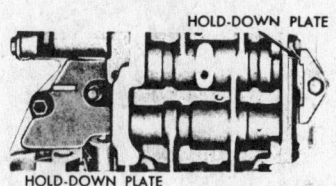

Hold down plate location—1964 C4 automatic
(© Ford Motor Co.)

Upper and lower control valve bodies separated—1964 C4 automatic
(© Ford Motor Co.)

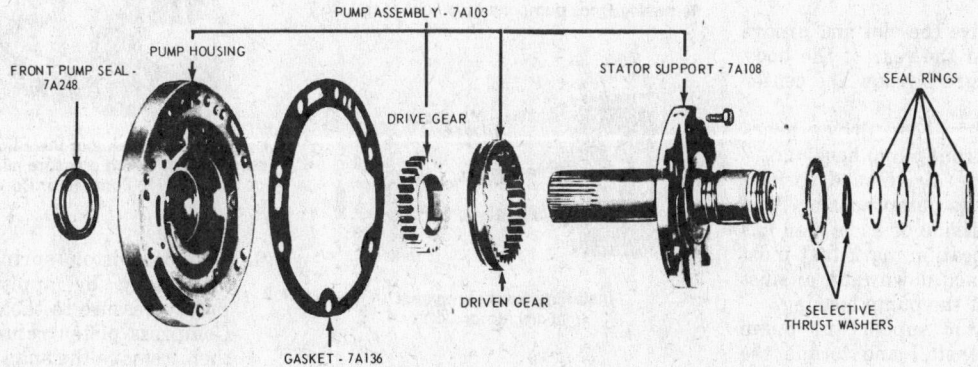

Front pump and stator support disassembled—1968-1970 C4 automatic
(© Ford Motor Co.)

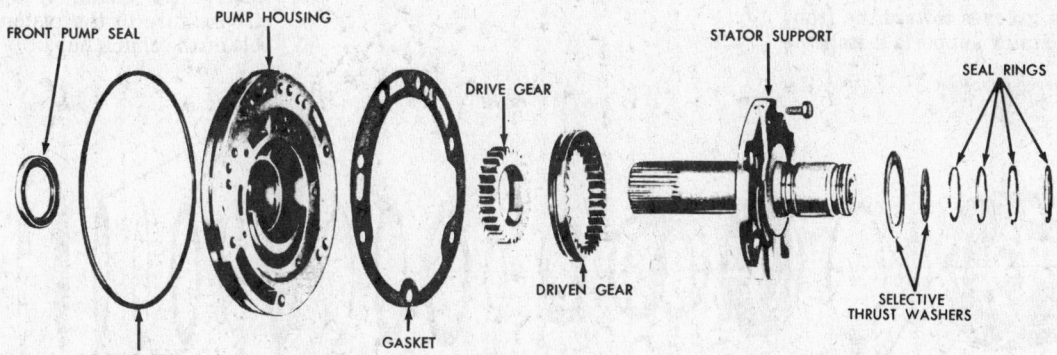

Front pump and stator support disassembled—1964-1967 C4 automatic
(© Ford Motor Co.)

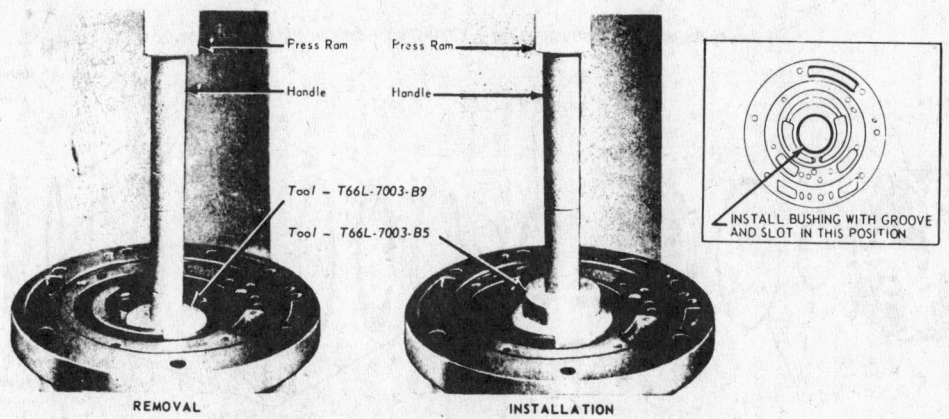

Replacing front pump housing bushing (© Ford Motor Co.)

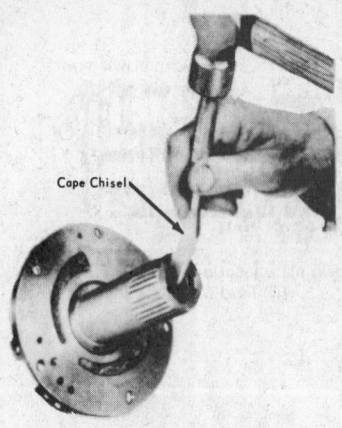

Removing stator support bushing
(© Ford Motor Co.)

making sure the slot and groove are toward the rear of the body and 60 degrees below the center line.

5. Remove drive and driven gears from the front pump housing.
6. Install the drive and driven gears in the pump housing. The chamfered side of each gear has an identification mark that must be positioned downward, against the face of the pump housing.
7. Place stator support in pump housing. Install and torque the five retaining bolts.
8. Install four seal rings onto the stator support. The two large oil rings are assembled first, in the oil ring grooves toward the front of the stator support. Install the

O-ring seal onto the pump housing.
9. Check pump gears for free rotation by placing pump on the converter drive hub and turning pump housing.
10. If the front pump seal must be replaced, mount the pump in the transmission case and remove the seal with a seal removing tool.

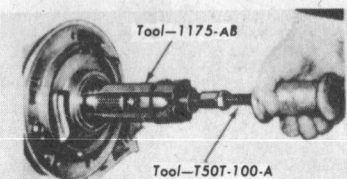

Removing front pump seal (© Ford Motor Co.)

Installing front pump seal
(© Ford Motor Co.)

Reverse-High Clutch Disassembly and Assembly

1. Remove pressure plate retaining snap-ring.
2. Remove the pressure plate, and the drive and driven clutch plates.

Caution Use no detergent or other cleaning solution on the lined clutch plates. Wipe the plates with a lint-free cloth.

Removing or installing reverse-high clutch pressure plate snap-ring
(© Ford Motor Co.)

3. Remove piston spring retainer snap ring by applying arbor press pressure to the clutch hub. Compress piston return springs, then remove the snap ring.
4. Back off press pressure, remove the retainer and ten piston return springs.
5. Remove the piston by applying air pressure to the piston apply hole of the clutch hub.

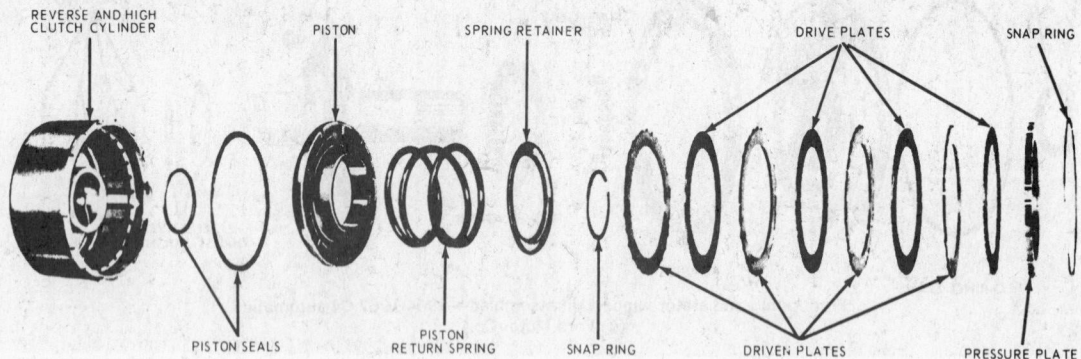

Reverse-high clutch disassembled—1970 C4 automatic (© Ford Motor Co.)

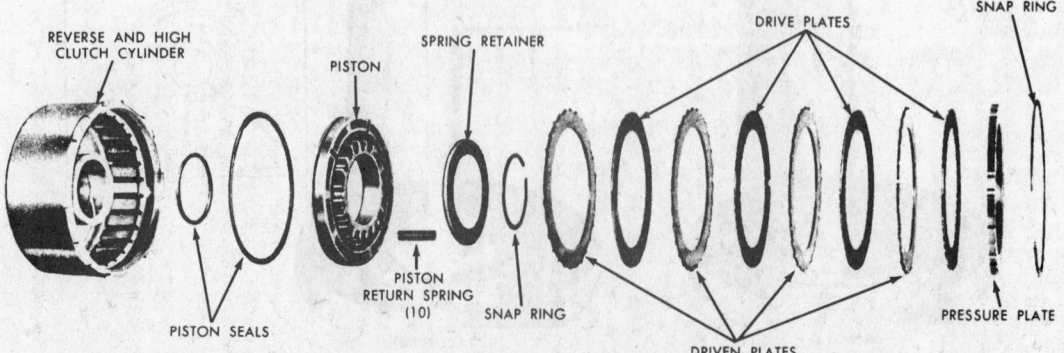

Reverse-high clutch disassembled—1964-1969 C4 automatic (© Ford Motor Co.)

**Removing or installing reverse-high clutch piston spring snap-ring—
1970 C4 automatic**
(© Ford Motor Co.)

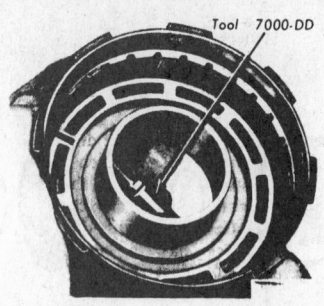

Removing reverse-high clutch piston
(© Ford Motor Co.)

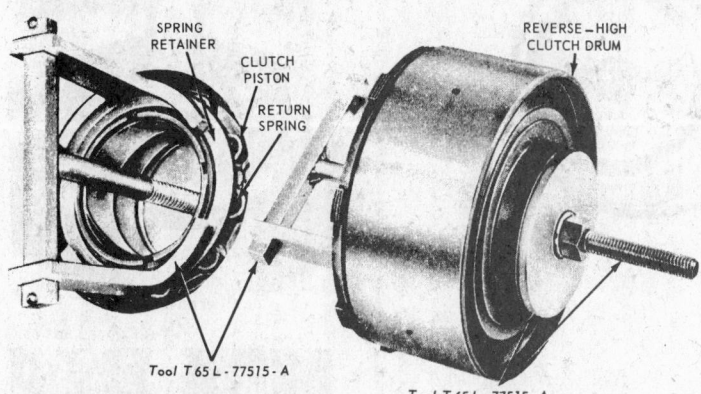

**Removing or installing reverse-high clutch piston spring snap-ring—
1964-69 C4 automatic**
(© Ford Motor Co.)

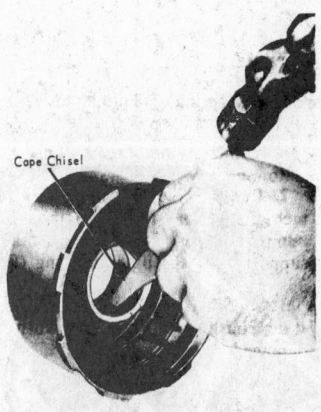

Removing reverse-high clutch bushing
(© Ford Motor Co.)

6. Remove piston outer seal from the piston and the piston inner seal from the clutch drum.

7. Remove the drum bushing if it is worn or damaged. Use a cape chisel to cut a shallow groove ¾ in. long in the bushing seam until it is broken. Pry up the loose ends of the bushing with an awl and remove the bushing. *To prevent leakage at the stator support O-rings, do not nick or damage the hub surface with the chisel.*

8. Position the drum and a new bushing in an arbor press and press the bushing into the drum with the tool shown in the illustration.

9. Install a new inner seal into the clutch drum and a new outer seal onto the clutch piston. Lubricate and install the piston into the clutch drum.

10. Locate the ten piston return springs on the piston. Place retainer on top of the springs. Compress assembly with an arbor press, and install the retainer snap-ring.

11. Soak new composition plates in transmission fluid before installation. Install clutch plates, alternately, starting with a steel plate, then a composition plate. The last plate installed is the pressure plate with the internally-chamfered side up.

12. Install the pressure plate retaining snap-ring.

Forward Clutch Disassembly and Assembly

1. Remove the clutch pressure plate retaining snap ring.

2. Remove the pressure plate, drive and driven plates from the hub.

3. Remove the disc spring retaining snap-ring.

4. Apply air pressure to the clutch piston pressure hole to remove the piston from the hub.

5. Remove piston outer seal and the inner seal from the clutch hub.

6. Install new piston seals onto the clutch piston and drum.

7. Lubricate and insert the piston into the clutch hub. Install the disc spring and retaining snap-ring.

8. Install the lower pressure plate, with the flat side up and the radiused side downward. Install one composition clutch plate and alternately install the drive and driven plates. The last plate to be installed will be the upper pressure plate, with the internally-chamfered side up.

9. Install pressure plate retaining snap-ring.

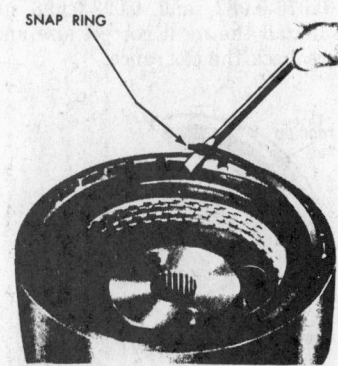

**Removing or installing
forward clutch pressure plate snap-ring**
(© Ford Motor Co.)

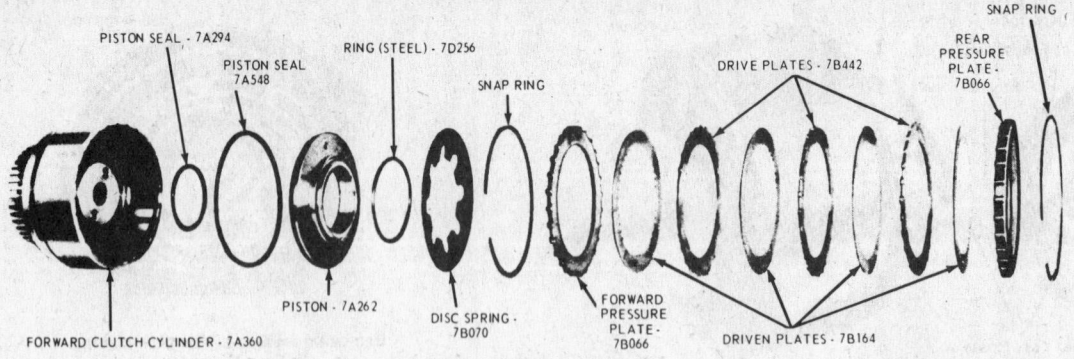

Forward clutch disassembled (© Ford Motor Co.)

Removing disc spring (© Ford Motor Co.)

REMOVAL

INSTALLATION

Replacing forward clutch hub bushing
(© Ford Motor Co.)

10. With a feeler gauge, check clearance between the snap ring and the pressure plate. Downward pressure on the plate should be used when making this check. Clearance should be 0.025-0.050 in.

11. If clearance is not within specifications, selective snap-rings are available in the following thicknesses: 0.050-0.054, 0.064-0.068, 0.078-0.082, and 0.092-0.096 in. Install the most correct size and recheck the clearance.

Forward Clutch Hub and Ring Gear Disassembly and Assembly

1. Remove forward clutch hub retaining snap-ring.
2. Separate clutch hub from ring gear.
3. Press the bushing from the clutch hub.
4. Install a new bushing into the clutch hub as shown.
5. Install clutch hub into ring gear.
6. Install hub retaining snap-ring.

Forward Planet Carrier Disassembly and Assembly

1. With a small punch inserted into the end of the planet gear retaining pin hole, force the shaft retaining pins outward. With side cutters, remove retaining pins.
2. Remove three planet gear pins, three gears, thrust washers and spacers.
3. Remove outer race, thrust bearing, and the inner race.
4. Install the inner race, thrust bearing and outer race.
5. Install 21 bearing rollers into each planet gear. Hold rollers in place with petrolatum. Position gears, spacers, and washers in the carrier.

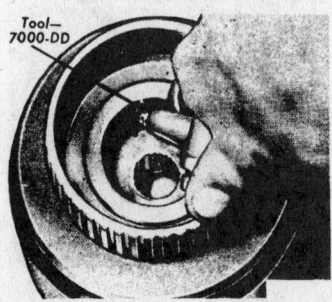

Removing forward clutch piston
(© Ford Motor Co.)

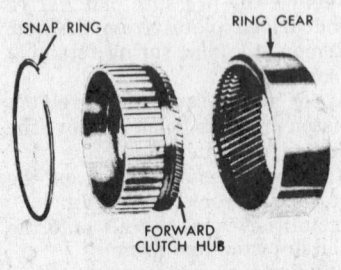

Forward clutch hub and ring gear disassembled
(© Ford Motor Co.)

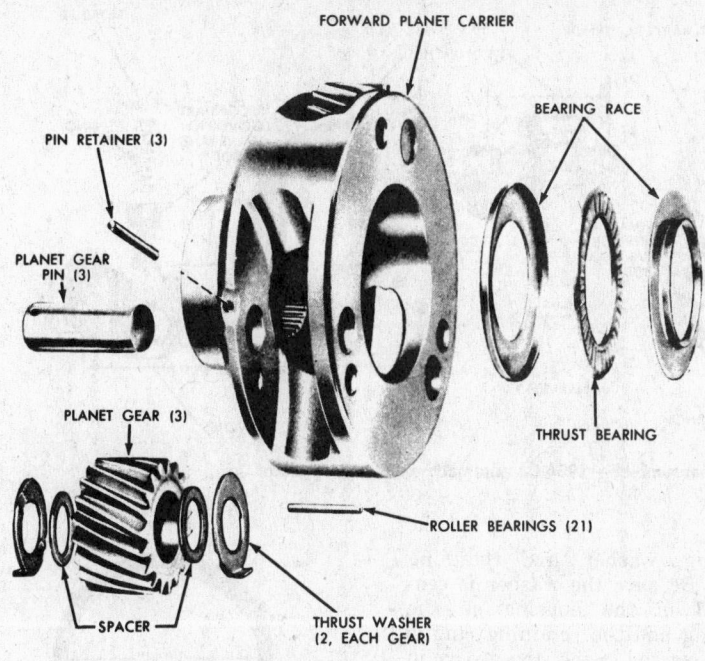

Forward planet carrier disassembled (© Ford Motor Co.)

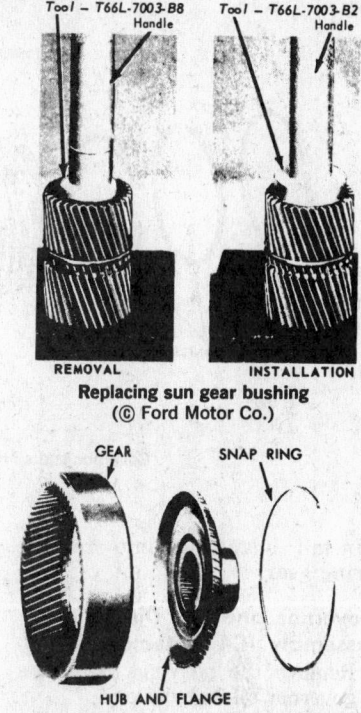

Replacing sun gear bushing
(© Ford Motor Co.)

Reverse ring gear and hub disassembled
(© Ford Motor Co.)

6. Install gear pins and align retaining pin holes. Install retaining pins.

7. When only the thrust bearing and races have to be removed, remove only two of the planet gear pins. Move the gears outward far enough to allow removal of the thrust bearing and races. This will prevent the bearing rollers from falling out. Reassemble the thrust bearing and races, two gears and pins.

Input Shell and Sun Gear Disassembly and Assembly

1. Remove external snap-ring from sun gear.
2. Remove thrust washer No. 5 from input shell and sun gear.
3. From inside the shell, remove the sun gear. Remove internal snap-ring from sun gear.
4. If the sun gear bushings are to be replaced, use the tool shown in the illustration and press both bushings through the gear.
5. Press a new bushing into each end of the sun gear.
6. Install internal snap-ring onto sun gear. Install sun gear into the input shell.
7. Install thrust washer No. 5 onto sun gear and input shell.
8. Install external snap-ring onto sun gear.

Reverse Ring Gear and Hub Disassembly and Assembly

1. Remove hub retaining snap-ring from reverse ring gear.
2. Remove the hub from reverse ring gear.
3. Install hub into reverse ring gear.

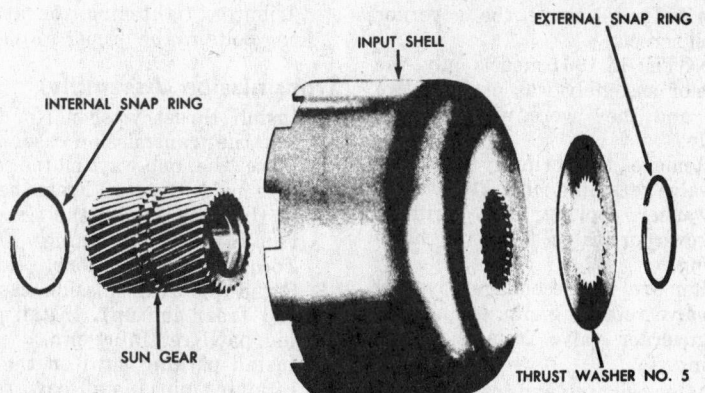

Input shell and sun gear disassembled (© Ford Motor Co.)

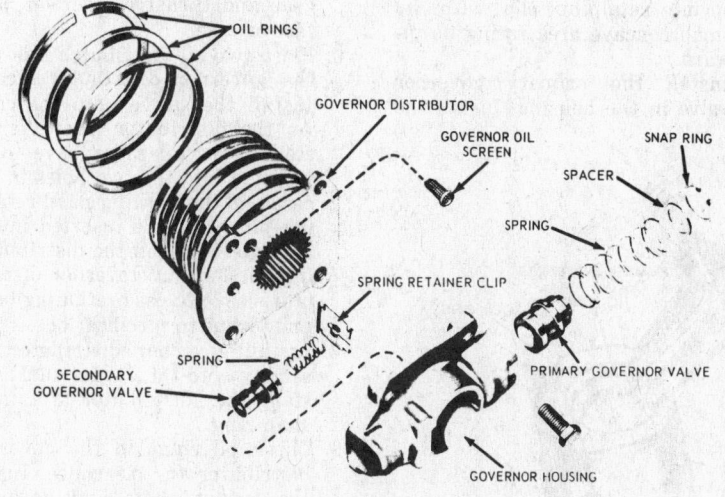

Governor and oil distributor disassembled—1965-70 C4 automatic
(© Ford Motor Co.)

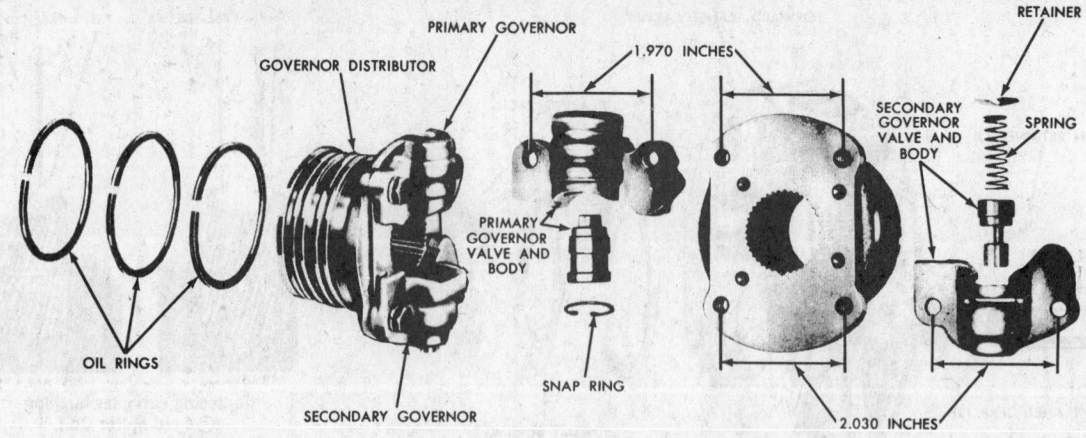

Governor and oil distributor disassembled—1964 C4 automatic only (© Ford Motor Co.)

4. Install snap-ring into reverse ring gear.

Governor and Oil Distributor Disassembly (C4 Automatic Only)

1. Remove the oil rings from the governor oil distributor.
2. Remove the governor housing attaching bolts and remove the governor assembly from the distributor. Remove the governor oil screen.
   NOTE: in 1964 models only, the governor assembly was made in two parts and they were removed separately.
3. Remove the primary governor valve retaining ring. Remove the washer, spring, and primary governor valve from the housing.
4. Remove the secondary governor valve retaining clip, spring, and governor valve from the housing.
5. After cleaning and inspecting all governor parts, install the secondary governor valve in its housing. Install the spring and spring retaining clip with its small concave area facing downward.
6. Install the primary governor valve in the housing. Install the

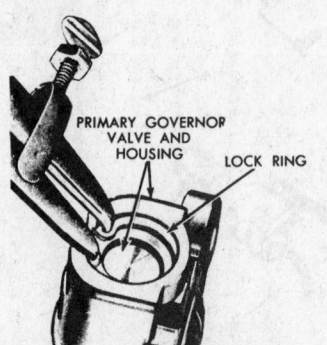

Removing or installing retaining ring
(© Ford Motor Co.)

spring, washer, and retaining clip. Be sure the washer is centered in the housing on the spring and the retaining ring is fully seated in the ring groove in the housing.
7. Install the oil rings on the governor distributor. Install the governor oil screen and mount the governor assembly on the distributor, tightening the attaching bolts to the proper torque.

Transmission Assembly

1. Install thrust washer No. 9 inside the transmission case.
2. Place the one-way clutch outer race inside the case. From the rear of the case, install the six outer race-to-case retaining bolts. Torque to specifications.
3. Stand the transmission case on end (rear end up). Install parking pawl retaining pin.
4. Install parking pawl on the case retaining pin. Install pawl return spring.
5. Install thrust washer No. 10 onto the parking pawl gear. Place gear and thrust washer on back face of case.
6. Place two oil distributor tubes in the governor distributor sleeve. Install the sleeve onto the case. As the distributor sleeve is installed, the oil tubes have to be inserted into the two holes in the case and the parking pawl retaining pin has to be inserted in the alignment hole in the distributor.
7. Install the four governor distributor sleeve-to-case retaining bolts and torque to specifications.
8. Install governor distributor assembly onto the output shaft. Install the distributor retaining snap-ring.
9. Check oil rings in the governor distributor for clearance. Install the output shaft and governor distributor into the distributor sleeve.

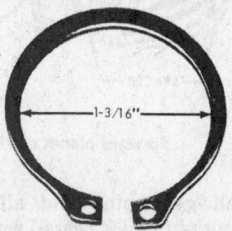

RETAINS REVERSE RING GEAR AND
HUB TO OUTPUT SHAFT

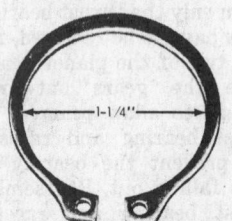

RETAINS GOVERNOR DISTRIBUTOR
TO OUTPUT SHAFT

Governor and reverse ring hub retaining snap-ring identification
(© Ford Motor Co.)

10. With a new gasket in place on the extension housing, install the extension housing, vacuum tube clip and the extension housing to case retaining bolts. Torque bolts to specifications.
11. Rotate transmission case so that front end is up, making sure that thrust washer No. 9 is in position at the bottom of the case.
12. On the bench, install the 12 one-way clutch springs onto the spring retainer.
13. Place the one-way clutch spring retainer, with springs installed, into the outer race, located inside the transmission.
14. Check the position of each spring on the spring retainer. Install the inner race inside the spring retainer and 12 springs.
15. Starting at the back of the one-way clutch outer race, install the 12 clutch rollers. Each spring will be partially compressed as the

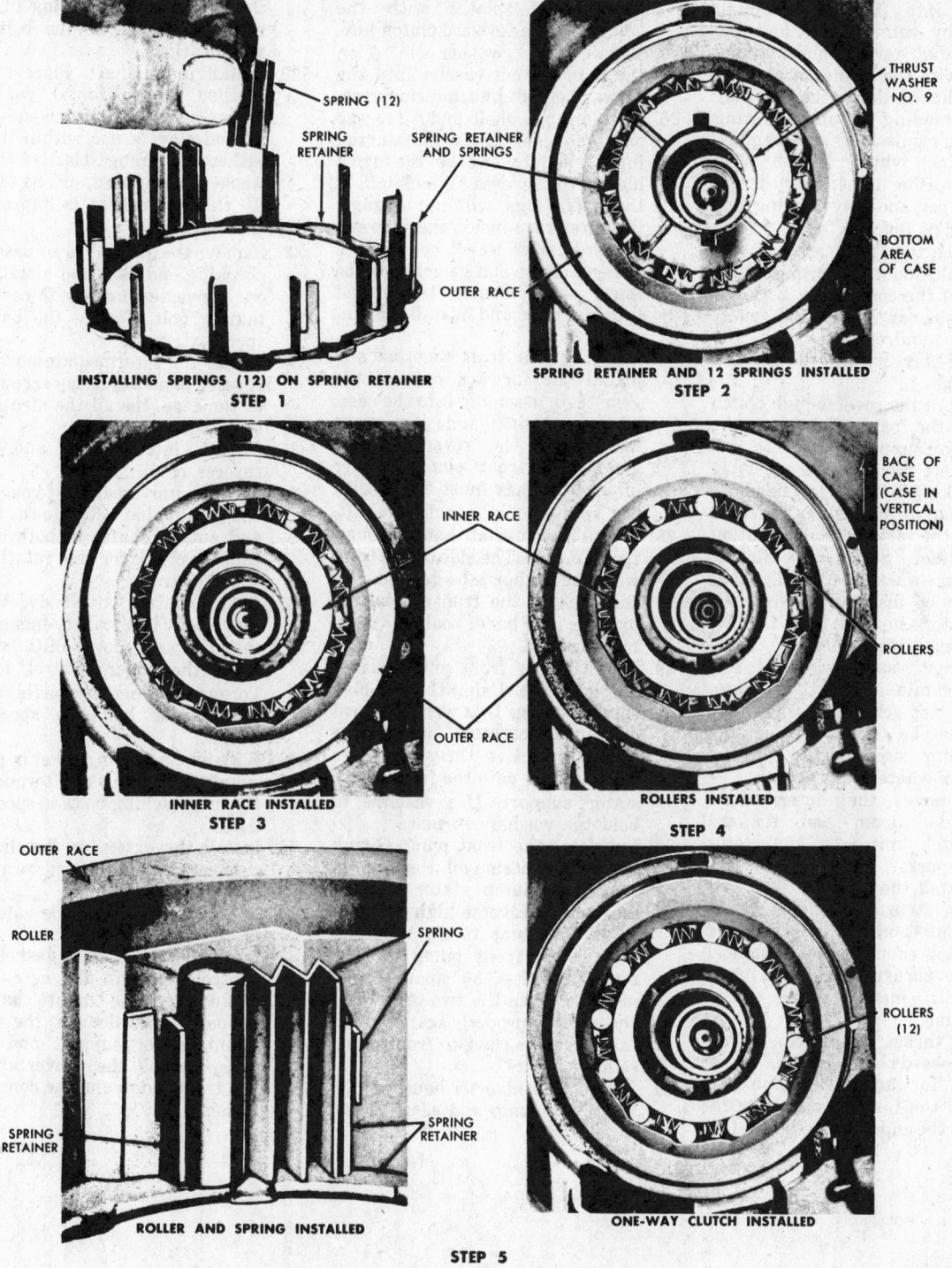

INSTALLING SPRINGS (12) ON SPRING RETAINER
STEP 1

SPRING RETAINER AND 12 SPRINGS INSTALLED
STEP 2

INNER RACE INSTALLED
STEP 3

ROLLERS INSTALLED
STEP 4

ROLLER AND SPRING INSTALLED

ONE-WAY CLUTCH INSTALLED

STEP 5

One-way clutch assembly and installation (© Ford Motor Co.)

roller is installed between the outer and inner race.

16. After the clutch has been assembled, rotate the inner race clockwise to center the rollers and springs. Install low and reverse drum. The splines of the drum must engage the splines of the one-way clutch inner race. Check the clutch operation by rotating the low and reverse drum. The drum should rotate clockwise but

should not rotate counterclockwise.

17. Install thrust washer No. 8 on top of the low and reverse drum. Install the low and reverse band into the case, with the end of the band for the small strut toward the low-reverse servo.

18. Install the reverse ring gear and hub onto the output shaft.

19. Move the output shaft forward and install the reverse ring gear

hub-to-output shaft retaining ring.

20. Place thrust washers Nos. 6 and 7 on the reverse planet carrier. Install planet carrier into the reverse ring gear and engage the tabs of the carrier with the slots in the low-reverse drum.

21. From inside the transmission case, install the inner downshift lever.

22. On the bench, install the forward

clutch into the reverse-high clutch by rotating the units to mesh the reverse-high clutch plates with the splines of the forward clutch. Using the end-play check reading obtained during the transmission disassembly, determine which No. 2 steel thrust washer is necessary to get the proper end-play reading and proceed as follows:

a. Place the stator support vertically on the bench and install the correct No. 2 thrust washer or washer and saucer as required to bring the end-play within the correct range.

b. Install the reverse-high clutch and the forward clutch on the stator support.

c. Invert the complete assembly making sure the intermediate brake drum bushing is seated on the forward clutch mating surface. Select the thickest fiber washer (No. 1) that can be inserted between the stator support and the intermediate brake drum thrust surfaces and still maintain a slight clearance. Do not select a washer that must be forced between the stator support and the intermediate brake drum.

d. Remove the intermediate brake drum and forward clutch unit from the stator support.

e. Install the selected No. 1 and No. 2 thrust washers on the front pump stator support using enough vaseline to hold the thrust washers in place while installing the front pump.

23. Install thrust washer No. 3 onto the forward clutch.

24. Install forward clutch hub and ring in the forward clutch by rotating the units to mesh the forward clutch plates with the splines on the forward clutch hub.

25. Install thrust washer No. 4 on the front planet carrier into the forward clutch hub and ring gear.

26. Install input shell and sun gear onto the gear train. Rotate the input shell to engage the drive lugs of the reverse-high clutch. If the drive lugs will not engage, the outer race inside the forward planet carrier is not centered in the end of the sun gear inside the input shell. Center the thrust bearing race and install the input shell.

27. Hold the gear train together and install the forward part of the gear train assembly into the case. The input shell sun gear must mesh with the reverse pinion gears. The front planet carrier internal splines must mesh with the splines of the output shaft.

28. Install intermediate band through front of case. The side of the band with the anchor tabs goes toward the back of the transmission. If using a new band, soak in transmission fluid.

29. Install a new front pump gasket onto the case. Using the end-play check readings that were obtained during disassembly, install the correct selective thrust washers Nos. 1 and 2 onto the front pump stator support. Use vaseline to hold the washers in place.

30. Lubricate the front pump O-ring with transmission oil, then install the front pump stator support into the reverse-high clutch. Align the pump to the case and install two front pump to case retaining bolts, 180° apart. Alternately tighten the two bolts until the pump is properly seated in the case. Remove the two front pump retaining bolts.

31. Install the converter housing onto the front pump and case. Install the six converter housing to case retaining bolts. Torque bolts to specifications.

32. Install input shaft. Place transmission in horizontal position, then check transmission end play. If end-play is not within limits, either the wrong selective thrust washers were used, or one of the 10 thrust washers is improperly positioned.

33. Remove the dial indicator used for checking end play and install the one converter housing to case retaining bolt. Torque the bolt to specifications.

34. Install the intermediate and low-reverse band adjusting screws into the case. Install the struts for each band.

35. Adjust intermediate and low-reverse band.

36. Install a universal joint yoke onto the output shaft. Rotate the input and output shafts in both directions to check for free rotation of the gear train.

37. Install control valve body. As the valve body is installed, engage the manual and down-shift valves with the inner control levers. Torque the eight control valve body-to-case bolts to specifications.

38. With a new oil pan gasket in place, install the oil pan and torque the eleven attaching bolts to specifications.

39. Install the extension housing-to-case retaining bolts and torque to specifications.

40. Install primary throttle valve into the transmission case.

41. Install vacuum unit, gasket and control rod into the case.

42. Make sure the input shaft is properly installed in the front pump, stator support and gear train. Install the converter into the front pump and the converter housing.

Specifications

Control Pressure at Zero Output Shaft Speed—1964

Engine Speed	Throttle	Manifold Vac.—Ins. Hg.	Selector Lever Position	Control Pressure (P.S.I.)
Idle	Closed	*Above 18	P, N, D1, D2, L	55-62
			R	55-96
As required	As required	17.5-16.5	D1, D2, L	Pressure starts to increase
As required	As required	10	D1, D2, L	95-110
As required	As required	3	D1, D2, L	138-148
			R	213-227

*Vehicles checked at high altitudes

Control Pressure at Zero Output Shaft Speed—1965, 1966, 1967

Engine Speed	Throttle	Manifold Vac. Ins. Hg.	Selector Lever Position	Control Pressure (P.S.I.)
Idle	Closed	*Above 18	P, N, D1, D2, L	55-62
			R	55-100
As required	As required	17 Approx.	D1, D2, L	Pressure starts to increase
As required	As required	10	D1, D2, L	96-105
As required	As required	3	D1, D2, L	138-148
			R	215-227

*Vehicles checked at high altitudes

Control Pressure at Zero Output Shaft Speed

C4 Transmission—1968, 1969

Engine Speed	Throttle	Manifold Vac. Ins. Hg	Range	P.S.I.
Idle	Closed	Above 18①	P, N, D, 2, 1	55-61
				55-102
			R	55-182
As required	As required	10	D, 2, 1	93-101
As required	Wide open	Below 1.0	D, 2, 1	142-150
			R	254-268

Control Pressure at Zero Output Shaft Speed—1970

Engine Speed	Throttle	Manifold Vac. Ins. Hg.	Range	P.S.I.
Idle	Closed	① Above 18	P. N. D.	52-69
			2, 1	80-110
			R	100-180
As required	As required	10	D, 2, 1	96-110
As required	As required	3.0	D, 2, 1	135-155
			R	220-250

At altitudes above sea level it may not be possible to obtain 18″ of vacuum at idle. For idle vacuums of less than 18″ refer to following table to determine idle speed pressure specification in forward driving ranges (D1, D2, or L).

Engine Vacuum	Control Pressure (PSI)
17	55-65
16	55-71
15	55-77
14	55-83
13	55-89
12	55-95
11	55-101

At altitudes above sea level it may not be possible to obtain 18″ of vacuum at idle. For idle vacuums of less than 18″ refer to following table to determine idle speed pressure specification in forward driving ranges (D1, D2, or L).

Engine Vacuum	Control Pressure (PSI)
17	55-62
16	55-68
15	55-74
14	55-80
13	55-87
12	55-93
11	55-99

① At altitudes above sea level, it may not be possible to obtain 18 inches of engine vacuum at idle. For idle vacuums of less than 18 inches, refer to the following table to determine idle speed pressure specifications in D driving range.

Engine Vacuum	Line Pressure
17 inches	55-66
16 inches	55-71
15 inches	55-76
14 inches	55-81
13 inches	55-86
12 inches	55-91
11 inches	55-96

① At altitudes above sea level, it may not be possible to obtain 18 inches of engine vacuum at idle. For idle vacuums of less than 18 inches, refer to the following table to determine idle speed pressure specification in D driving range.

Engine Vacuum	Line Pressure
17 inches	52-74
16 inches	52-78
15 inches	52-84
14 inches	52-90
13 inches	52-95
12 inches	52-100
11 inches	52-106

Control Pressure—C4S Semi-Automatic—1970

Engine Speed	Throttle	Range	P.S.I.
Idle	Closed	P, N, H1, 2, 1	70-130
		R	70-260
1400 rpm	As required	H1, 2, 1	110-130
1400 rpm	As required	R	210-260

Checks and Adjustments—1964

Operation	Specification
Transmission end play check	0.008-0.042 inch *Selective thrust washers available
Turbine and stator end play check	0.060 inch (maximum)
Intermediate band adjustment	Adjust screw to 10-foot-pounds torque, and back off 1½ turns
Low-reverse band adjustment	Adjust screw to 10-foot-pounds torque, and back off 3 turns
Forward clutch pressure plate to snap ring clearance	0.022 to 0.042 inch Selective snap ring thicknesses 0.092-0.088 0.078-0.074 0.064-0.060

*Selective Thrust Washers (No. 1 and 2 used in pairs)

Thrust Washer No. 1			Thrust Washer No. 2
Composition Thrust Washer	Color of Washer	No. Stamped on Washer	Metal Thrust Washer
.108-.104	Blue	5	.109-.107
.091-.087	Yellow	4	.092-.090
.074-.070	Black	3	.075-.073
.057-.053	Tan	2	.058-.056
.042-.038	Green	1	.043-.041

Checks and Adjustments—1965

Operation	Specification
Transmission end play check	0.008-0.042 inch Selective thrust washers available
Turbine and stator end play check	0.060 inch (maximum)
Intermediate band adjustment	Adjust screw to 10-foot-pounds torque, and back off 1¾ turns
Low-reverse band adjustment	Adjust screw to 10-foot-pounds torque, and back off 3 turns
Forward clutch pressure plate to snap ring clearance	0.022 to 0.042 inch Selective snap ring thicknesses 0.092-0.088 0.078-0.074 0.064-0.060
Reverse-high clutch pressure plate to snap ring clearance	0.060-0.080 inch Selective snap ring thicknesses 0.102-0.106 0.092-0.088 0.078-0.074 0.064-0.060

Selective Thrust Washers (No. 1 and 2)

Thrust Washer No. 1			Thrust Washer No. 2
Composition Thrust Washer	Color of Washer	No. Stamped on Washer	Metal Thrust Washer
.138-.142	Purple		
.121-.125	Red		
.108-.104	Blue	5	.109-.107
.091-.087	Yellow	4	.092-.090
.074-.070	Black	3	.075-.073
.057-.053	Tan	2	.058-.056
.042-.038	Green	1	.043-.041

Checks and Adjustments—1966, 67

Operation	Specification
Transmission end play	0.008-0.042 inch Selective thrust washers available
Turbine and stator end play check	0.060 inch (maximum)
Intermediate band adjustment	Adjust screw to 10 ft-lbs torque, then back off 1¾ turns and tighten locknut to spec.
Low-reverse band adjustment	Adjust screw to 10 ft-lbs torque, then back off 3 turns and tighten locknut to spec.

Forward clutch pressure plate-to-snap ring clearance	No. of Composition Plates in Clutch	Clearance Specification
	4	0.020-0.036 in.
	5	0.026-0.042 in.

Selective snap ring thicknesses
0.102-0.106
0.088-0.092
0.074-0.078
0.060-0.064

Reverse-high clutch pressure plate-to-snap ring clearance	0.050-0.066 inch Selective snap ring thicknesses 0.102-0.106 0.088-0.092 0.074-0.078 0.060-0.064

Selective Thrust Washers (No. 1 and 2)

Thrust Washer No. 1			Thrust Washer No. 2
Composition Thrust Washer	Color of Washer	No. Stamped on Washer	Metal Thrust Washer
.138-.142	Purple		
.121-.125	Red		
.108-.104	Blue	5	.109-.107
.091-.087	Yellow	4	.092-.090
.074-.070	Black	3	.075-.073
.057-.053	Tan	2	.058-.056
.042-.038	Green	1	.043-.041

Checks and Adjustments C4 Transmission—1968

Operation	Specification
Transmission end play	0.008-0.042 inch (selective thrust washers available)
Turbine and stator end play	New or rebuilt 0.023 max. Used 0.040 max.
Intermediate band adjustment	Adjust screw to 10 ft-lbs torque, then back off 1¾ turns.
Low-reverse band Adjustment	Adjust screw to 10 ft-lbs torque, then back off 3 turns.
Selective snap ring thickness	.102-.106, .088-.092, .074-.078, .060-.064

Selective Thrust Washers

Nylon Thrust Washer W/Tangs	Color of Washer	No. Stamped On Washer	Metal Thrust Washer
.053-.0575	Red		
.070-.074	Green	5	.109-.107
.087-.091	Natural	4	.092-.090
.104-.108	Black	3	.075-.073
.121-.125	Yellow	2	.058-.056
.138-.142	Blue	1	.043-.041

Checks and Adjustments C4 Transmission—1969

Operation	Specification
Transmission end play	0.008-0.042 inch (selective thrust washers available)
Turbine and stator end play	Model PEB—New or rebuilt 0.044 max., used 0.060 max.
	Model PEE, PEA new or rebuilt 0.023 max., used 0.040 max.
Intermediate band adjustment	Remove and discard lock nut. Adjust screw to 10 ft-lbs torque, then back off 1¾ turns. Install new lock nut and torque to specification.
Low-reverse band adjustment	Remove and discard lock nut. Adjust screw to 10 ft-lbs torque, then back off 3 turns. Install new locknut and torque to specification.
Selective snap ring thickness	.102-.106, .088-.092, .074-.078, .060-.064

Selective Thrust Washers

Thrust Washer No. 1		Thrust Washer No. 2	
Nylon Thrust Washer W/Tangs	Color of Washer	No. Stamped On Washer	Metal Thrust Washer
.053-.0575	Red	1	.041-.043
.070-.074	Green	2	.056-.058
.087-.091	Natural (White)	3	.073-.075
.104-.108	Black	4	.090-.092
.121-.125	Yellow	5	.107-.109

Checks and Adjustments C4—1970

Operation	Specification
Transmission end play	0.008-0.042 inch (selective thrust washers available)
Turbine and stator end play	Model PEB, PEG— new or rebuilt 0.044 max. Used 0.060 max.
	Model PEE, PEA, PEF—New or rebuilt 0.023 max. Used 0.040 max.
Intermediate band adjustment	Remove and discard lock nut. Adjust screw to 10 ft-lbs torque, then back off 1¾ turns. Install new lock nut and torque to specification.
Low-reverse band adjustment	Remove and discard lock nut. Adjust screw to 10 ft-lbs torque, then back off 3 turns. Install new lock nut and torque to specification.

Control Valve Body Spring Identification—1964

Spring	Total Coils	Free Length	O.D.	Wire Diameter	Lbs. Load	Length
Main oil pressure reducer valve	19	1.42	.281	.035	3.012	.887
Cut back control valve	19	1.00	.200	.020	1.05	.59
Manual valve detent	10	.74	.300	.047	8.0	.601
3-2 coasting control valve	10	.83	.285	.023	1.0	.476
Line pressure relief	17	1.15	.198	.041	7.0	.80
Secondary throttle valve	14	1.581	.350	.026	1.728	.565
2-3 backout control valve	11½	1.26	.45	.032	1.47	.580
Main oil pressure regulator valve	16	1.92	.48	.049	9.09	.876
Throttle pressure modulator valve	19	1.29	.29	.029	2.15	.620
Control 1-2 shift valve	14	1.58	.26	.026	1.7	.70
Throttle downshift valve	12	.79	.24	.023	1.15	.48
2-3 shift valve	6½	1.18	.53	.031	1.38	.39
Throttle pressure booster valve	16¾	1.589	.311	.035	4.80	.84

Selective snap ring thickness	.050-.054, .064-.068, .078-.082, .092-.096

Selective Thrust Washers

Thrust washer No. 1		Thrust Washer No. 2	
Nylon Thrust Washer W/Tangs	Color of Washer	No. Stamped On Washer	Metal Thrust Washer
.053-.0575	Red	1	.041-.043
.070-.074	Green	2	.056-.058
.087-.091	Natural (White)	3	.073-.075
.104-.108	Black	Spacer	.032-.036①
.121-.125	Yellow		

① This is a selective spacer. The spacer must be installed next to the stator support to obtain correct end play.

Control Valve Body Spring Identification—1965

Spring	Total Coils	Free Length	O.D.	Wire Diameter	Lbs. Load	Length
Manual valve detent	9	.74	.295	.045	7.5	.601
Line pressure coasting boost	15	1.12	.287	.032	3.5	.620
2-3 backout control valve	11½	1.26	.45	.032	1.47	.580
Main oil pressure regulator valve	9.75	1.86	.615	.047	6.2	.208
Throttle pressure modulator valve	15	1.513	.292	.0286	3.573	.620
Control 1-2 shift valve	13	.95	.230	.019	1.0	.450
Throttle downshift valve	9	1.042	.360	.031	3.0	.476
Throttle pressure booster valve	15	.458	.243	.036	5.25	.84
Line pressure boost (inner)	12.5	2.055	.450	.032	2.98	.564
Intermediate band accumulator valve	12.5	1.107	.325	.025	1.10	.551

Control Valve Body Spring Identification— 1966, 1967

Spring	Total Coils	Free Length	O.D.	Wire Diameter	Lbs. Load	Length
Manual valve detent	9	.74	.295	.045	7.5	.601
2-3 backout control valve	10	1.515	.450	.026	1.353	.580
Main oil pressure regulator valve	9.75	1.86	.615	.047	6.200	.608
Throttle pressure modulator	15	1.513	.292	.286	3.675	.620
Control 1-2 shift valve	13	.950	.230	.019	1.00	.450
Throttle downshift valve	9	1.042	.360	.031	3.0	.476
Throttle pressure booster valve	15	1.421	.326	.036	5.250	.730
Control pressure booster valve	23.5	.965	.290	.032	1.00	.715
Intermediate band accumulator valve	12.5	1.107	.325	.025	1.100	.551

Control Valve Body Spring Identification
C4 Transmission—1968

Spring	Total Coils	Free Length	O.D.	Wire Dia.	Lbs. Load	Length	Color Code
Manual valve detent	9	0.74	0.295	0.045	7.5	0.601	None
2-3 backout control valve	10	1.515	0.450	0.026	1.353	0.580	White
Main oil press. reg. valve	12	2.53	0.615	0.047	7.24	0.716	Pink
Throttle press. modulator	15	1.513	0.292	0.028	3.675	0.620	Yellow
Drive 2 valve	10	0.735	0.230	0.019	0.80	0.450	Gray
Throttle downshift valve	9	0.962	0.380	0.034	3.44	0.440	Dk. Green
Throttle press. booster valve	15	1.39	0.249 I.D.	0.036	5.250	0.730	Purple
Control press. booster valve	12.8	1.66	0.350	0.028	1.63	0.696	None
Inter. band accumulator valve	11	1.38	0.375 I.D.	0.024	1.00	0.400	White
Line coasting boost valve	10	1.03	0.346	0.034	4.42	0.464	White

Control Valve Body Spring Identification
C4 Transmission—1969

Spring	Total Coils	Free Length	O.D.	Wire Dia.	Lbs. Load	Length	Color Code
Manual valve detent all except PEC-EI	10	.764	.295	.44	6.5	.601	White
Model PEC-EI	9	.74	.295	.045	7.5	.601	None
2-3 backout control valve	10	1.515	.450	.026	1.353	.580	White
Main oil press. reg. valve	12	2.53	.615	.047	7.24	.716	Pink
Throttle press. modulator	15	1.513	.292	.028	3.675	.620	Yellow
Drive 2 valve	10	.735	.230	.019	.80	.450	Gray
Throttle downshift valve	9	.962	.380	.034	3.44	.440	Dk. Green
Throttle press. booster valve	15	1.39	.249 I.D.	.036	5.250	.730	Purple
Control press. booster valve	12.8	1.66	.350 I.D.	.028	1.63	.696	None
Inter. band accumulator valve ①	11	1.38	.375 I.D.	.024	1.0	.400	Purple
Inter. band accumulator valve ②	10	1.293	.375 I.D.	.0258	1.25	.400	None
Line press coasting boost valve	10	1.03	.346	.034	4.42	.464	White

① Used on models PEA-A2, MI, NI,
② Used on models PEB-B2, C2, PEE

Control Valve Body Spring Identification
C4 Automatic—1970

Spring	Total Coils	Free Length	O.D.	Wire Dia.	Lbs. Load	Length	Color Code
Manual valve detent		Leaf Type			7.25	.542	None
2-3 backout valve	13	1.345	.345	.0258	1.45	.620	Gray
1-2 transition valve:	12	1.150	.330	.023	.95	.480	Dk. Green
Throttle downshift valve	10	0.816	.280	.0301	3.00	.500	None
Low servo modulator valve	12	1.270	.380	.0268	1.54	.553	Orange
Throttle pressure booster valve all except PEE-AC1, AH1, M1, V1, PEF	15	1.109	.281	.0332	4.50	.620	None
model PEE-AC1, AH1, M1, V1, PEF	15	1.39	.285	.036	5.25	.730	Purple
Throttle pressure limit valve	14	1.192	.295	.0379	6.25	.770	Brown
Throttle pressure modulator valve	15	1.513	.292	.0286	3.575	.620	Yellow
Line pressure coast boost valve	10	1.023	.340	.0332	4.10	.494	Dk. Blue
Drive 2 valve	13	.950	.230	.019	1.00	.450	Violet
Int. servo accumulator valve Model PEE-AC1, AH1, M1, V1, AG1	9	.680	.300	.0244	1.00	.390	Orange
model PEA-A3, PEE-AD1, AE1, AF1, PEB, PEF	8	.680	.300	.0258	1.50	.390	Lt. Green
model PEA-M2, N2	9	.845	.300	.258	2.00	.390	None
Main oil pressure reg., valve inner	11	1.40	.407	.0286	1.49	.739	None
Main oil pressure reg. valve outer	9	1.667	.668	.0507	7.30	.586	Pink

Control Valve Body Spring Identification
C4S Semi-Automatic—1970

Spring	Total	Free Length	O.D.	Wire Dia.	Length	Lbs. Load
2-3 backout control valve	10	1.031	.356	.034	4.44	.464
Main oil press reg. valve	9	1.667	.671	.0507	7.30	.586
Throttle press modulator	15	1.24	.297	.034	4.50	.740
Drive 2 valve	9.3	.594	.230	.030	3.0	.450
Low servo modulator valve	12	1.412	.380	.028	1.89	.553

Torque Limits—C4 and C4S Automatic Transmissions—Ft. Lbs.

Item	1964	1965	1966	1967	1968	1969	1970
				Ft-lbs			
Converter to flywheel	20-30	20-30	20-30	20-30	20-30	23-28	23-28
Converter hsg. to trans.	28-40	28-40	28-40	28-40	28-40	28-40	28-40
Front pump to trans. case	28-40	28-40	28-40	28-40	28-40	28-40	28-40
Outer race to case	13-20	13-20	13-20	13-20	13-20	13-20	13-20
Oil pan to case	12-16	12-16	12-16	12-16	12-16	12-16	12-16
Rear servo cover to case	12-20	12-20	12-20	12-20	12-20	12-20	12-20
Stator support to pump	12-20	12-20	12-20	12-20	12-20	12-20	12-20
Converter cover to converter hsg.	10-13	10-13	10-13	10-13	12-16	12-16	12-16
Intermediate servo cover to case	12-20	12-20	12-20	12-20	16-22	16-22	16-22
Extension hsg. to case	28-40	28-40	28-40	28-40	28-40	28-40	28-40
Converter drain plug	20-30	20-30	20-30	20-30	20-30	20-30	20-30
Pressure gauge tap	9-15	9-15	9-15	9-15	9-15	9-15	9-15
Manual control lever nut	35-45	35-45	30-40	30-40	30-40	8-12	8-12
Downshift lever to case	12-20	12-20	12-16	12-16	12-16	12-16	12-16
Filler tube to engine	20-25	20-25	20-25	20-25	20-25	20-25	20-25
Filler tube to pan						32-42	32-42
Diaphragm assy to case	15-23	15-23	15-23	15-23	15-23	15-23	15-23
Distributor sleeve to case	12-20	12-20	12-20	12-20	12-20	12-20	12-20
Reverse servo piston to rod	12-20	12-20	12-20	12-20	*	*	*
Transmission to engine			40-50	40-50	40-50	40-50	40-50
Transmission to engine: Falcon and Mustang						23-33	23-33
Cooler bracket & oil pan to case			12-16	12-16			
Band adjust stop to case	35-45	35-45	35-40	35-40	35-45	35-45	35-45
Yoke to output shaft					60-80	60-80	

* Tighten to 10 ft-lbs and back off ⅝ turn.

Torque Limits—C4 and C4S Automatic Transmissions—In. Lbs.

Item	1964	1965	1966	1967	1968	1969	1970
End plate to valve body screw	20-35	20-35	20-35	20-35	20-35	20-35	20-35
Lower to upper valve body bolts	20-35	20-35	40-55	40-55	40-50	40-55	40-55
Screen to valve body screws	20-35	20-35	40-55	40-55	40-55	40-55	40-55
Neutral switch to case screws	30-45	30-45	55-75	55-75		55-75 *	55-75 *
Neutral switch to column						20	20
Screen & valve body to case bolts	80-120	80-120	80-120	80-120	80-120	80-120	80-120
Screen to valve body bolts	80-120	80-120	80-120	80-120	80-120	80-120	80-120
Governor body to distributor (collector) body bolts	80-120	80-120	80-120	80-120	80-120	80-120	80-120
Cooler line fittings	80-120	80-120	80-120	80-120	80-120	80-120	80-120
Reinforcement plate to body					40-55	40-55	40-55
Accumulator plate to body					80-120	50-90	50-90
Inner downshift lever stop					40-55	40-55	40-55

* Mustang and Cougar only

3-Speed Automatic (C6)

Application
Comet, 1966-67
Cougar, 1966-70
Fairlane, 1966-70

Ford & T-Bird, 1966-70
Lincoln, 1966-70
Mercury, 1966-70

Montego, 1966-70
Mustang, 1966-70

Diagnosis

This diagnosis guide covers the most common symptoms and is an aid to careful diagnosis. The items to check are listed in the sequence to be followed for quickest results. Thus, follow the checks in the order given for the particular transmission type.

TROUBLE SYMPTOMS	Items to Check	
	In Car	Out of Car
No drive in D, 2 and 1	CWER	ac
Rough initial engagement in D or 2	KBWFE	a
1-2 or 2-3 shift points incorrect or erratic	ABLCDWER	
Rough 1-2 upshifts	BJGWEF	
Rough 2-3 shifts	BJWFGER	br
Dragged out 1-2 shift	ABJWGEFR	c
Engine overspeeds on 2-3 shift	CABJWEFG	br
No 1-2 or 2-3 shift	CLBDWEGJ	bc
No 3-1 shift in D	DE	
No forced downshifts	LEB	
Runaway engine on forced 3-2 downshift	WJGFEB	c
Rough 3-2 or 3-1 shift at closed throttle	KBJEF	
Shifts 1-3 in D	GJBEDR	
No engine braking in first gear—1 range	CHEDR	
Creeps excessively	K	
Slips or chatters in first gear, D	ABWFE	aci
Slips or chatters in second gear	ABJGWFER	ac
Slips or chatters in R	ABHWFER	bcr
No drive in D only	CWE	i
No drive in 2 only	ACWJER	c
No drive in 1 only	ACWER	c
No drive in R only	ACHWER	bcr
No drive in any selector lever position	ACWFER	cd
Lockup in D only		gc
Lockup in 2 only	H	bgci
Lockup in 1 only		gc
Lockup in R only		agc
Parking lock binds or does not hold	C	g
Transmission overheats	OFBW	ns
Maximum speed too low, poor acceleration	YZ	n
Transmission noisy in N and P	AF	d
Transmission noisy in first, second, third or reverse gear	AF	hadi
Fluid leak	AMNOPQS UXBJ	jmp
Car moves forward in N	C	a

Key to Checks

A. Fluid level
B. Vacuum diaphragm unit or tubes restricted—leaking—adjustment
C. Manual linkage
D. Governor
E. Valve body
F. Pressure regulator
G. Intermediate band
H. Low-reverse clutch
J. Intermediate servo
K. Engine idle speed
L. Downshift linkage—including inner level position
M. Converter drain plugs

N. Oil pan gasket, filler tube or seal
O. Oil cooler and connections
P. Manual or downshift lever shaft seal
Q. ⅛ inch pipe plugs in case
R. Perform air pressure check
S. Extension housing-to-case gasket
U. Extension housing rear oil seal
W. Perform control pressure check
X. Speedometer driven gear adapter seal
Y. Engine performance
Z. Vehicle brakes
a. Forward clutch

b. Reverse-high clutch
c. Leakage in hydraulic system
d. Front pump
g. Parking linkage
h. Planetary assembly
i. Planetary one-way clutch
j. Engine rear oil seal
m. Front pump oil seal
n. Converter one-way clutch
p. Front pump to case gasket or seal
r. Reverse-high clutch piston air bleed valve
s. Converter pressure check valves

General Information

The C6 automatic transmission is a three-speed unit that provides automatic upshifts and downshifts through the three forward gear ratios and also provides manual selection of first and second gears. The converter housing and the fixed splines which engage the splined OD of the low-reverse clutch steel plates are both cast integrally into the case.

Only one band (intermediate) is used in the C6 transmission. This band, along with the forward clutch, is used to obtain intermediate gear. Only the adjustment of the intermediate band can be done without major disassembly.

The C6 automatic transmission is very similar to the C4 automatic transmission, and most of the maintenance and overhaul procedures given for the C4 transmission will apply to the C6 transmission with some modifications of certain components. One important difference between the C6 and the C4 transmissions is that the C6 transmission uses a low-reverse clutch in place of the low-reverse band used in the C4 transmission. Otherwise, the gear trains are the same, as are the clutch combinations. The hydraulic control systems are very similar, except for minor differences in design. All components which are different from the C4 components are illustrated and the procedures to repair them are given.

There is no drain plug in the transmission pan. To drain the transmission oil, remove all of the oil pan attaching bolts, except the two at the front; loosen these, and let the pan tilt and drain.

Intermediate Band Adjustment

1. Raise the car on a hoist or place on jack stands.
2. Clean threads of the intermediate band adjusting screw.
3. Loosen adjustment screw locknut.
4. Tighten band adjustment until torque wrench T59P-77370-B overruns itself. Then, back off the screw one complete turn. Tighten the adjusting screw locknut to 35-45 ft. lbs.

TOOL - T 59 P - 77370 - B

Adjusting intermediate band
(© Ford Motor Co.)

Transmission Disassembly

1. Thoroughly clean outside of transmission to prevent possible contamination during reassembly.
2. If one is available, secure the unit in a repair stand. Drain the oil.

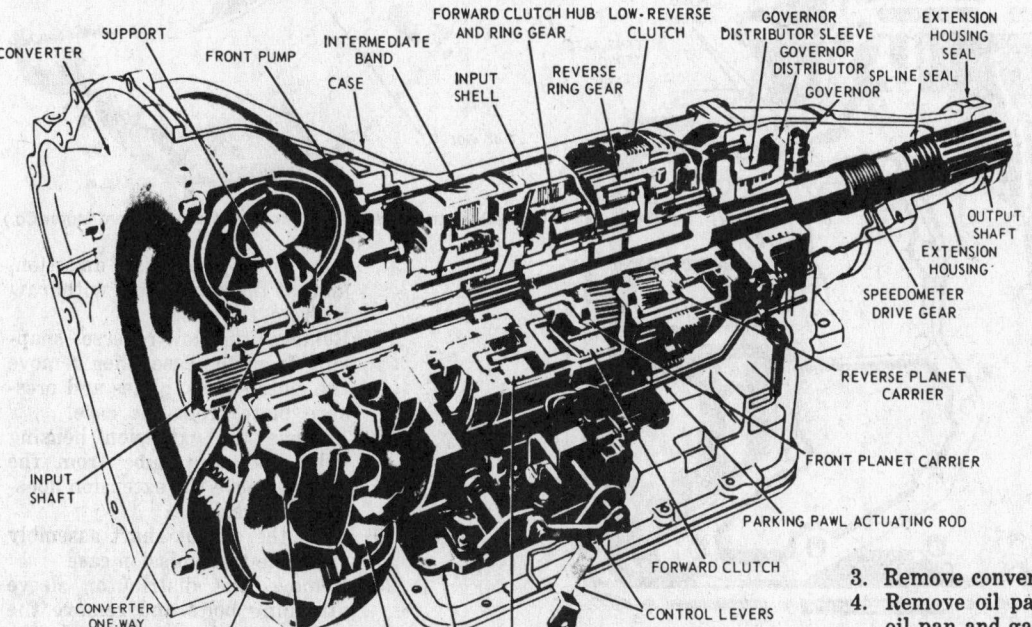

CONVERTER
SUPPORT
FRONT PUMP
INTERMEDIATE BAND
CASE
INPUT SHELL
FORWARD CLUTCH HUB AND RING GEAR
REVERSE RING GEAR
LOW-REVERSE CLUTCH
GOVERNOR DISTRIBUTOR SLEEVE
GOVERNOR DISTRIBUTOR
GOVERNOR
SPLINE SEAL
GOVERNOR
EXTENSION HOUSING SEAL
OUTPUT SHAFT
EXTENSION HOUSING
SPEEDOMETER DRIVE GEAR
REVERSE PLANET CARRIER
FRONT PLANET CARRIER
PARKING PAWL ACTUATING ROD
INPUT SHAFT
CONVERTER ONE-WAY CLUTCH
TURBINE
STATOR
IMPELLER
REVERSE-HIGH CLUTCH
CONTROL VALVE BODY
FORWARD CLUTCH
CONTROL LEVERS

C6 automatic transmission (© Ford Motor Co.)

3. Remove converter from the unit.
4. Remove oil pan attaching bolts, oil pan and gasket.

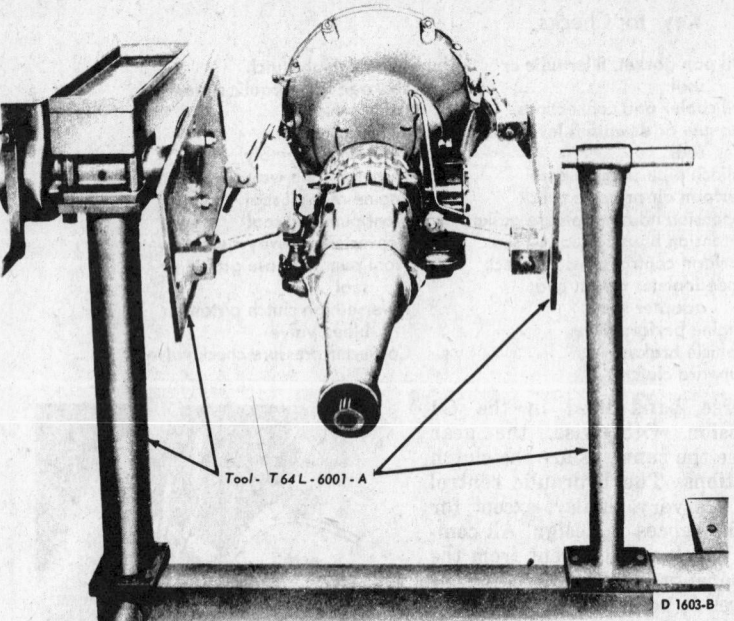

Tool - T 64 L - 6001 - A

D 1603-B

Transmission mounted in holding fixture (© Ford Motor Co.)

5. Remove valve body attaching bolts and the valve body.
6. Attach a dial indicator to the front pump. Install tool T61L-7657-B into the extension housing to center the shaft.
7. Pry the gear train to the rear of the case and, at the same time, press the input shaft inward until it bottoms. Set the dial indicator to read zero.
8. Pry the gear train forward and note the amount of gear train end-play. Record the end-play to simplify assembly. Remove the dial indicator from the pump and the tool from the extension housing.

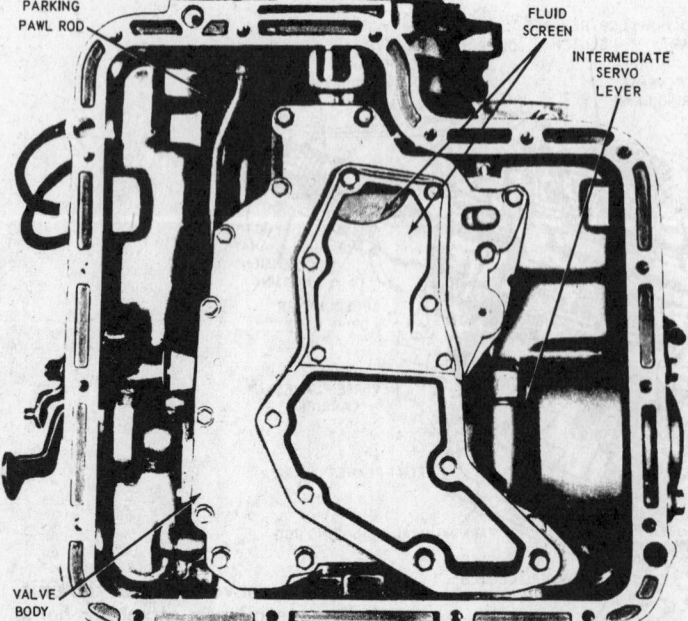

PARKING PAWL ROD

FLUID SCREEN

INTERMEDIATE SERVO LEVER

VALVE BODY

D 1604-A

Transmission with pan removed (© Ford Motor Co.)

9. Slip the input shaft out of the front pump. Remove the vacuum diaphragm, rod and primary throttle valve from the case.
10. Remove the front pump attaching bolts. Pry the gear train forward to remove the pump.
11. Loosen the band adjusting screw and remove the two struts.
12. Rotate the band 90°, counterclockwise, to align the ends with the slot in the case. Slide the band off the reverse-high clutch drum.
13. Remove the forward part of the gear train as an assembly.
14. Remove the large snap-ring that holds the reverse planet carrier in the low-reverse clutch hub. Lift the planet carrier from the drum.
15. Remove the snap-ring that holds the reverse ring gear and hub on the output shaft. Slide the ring gear and hub from the shaft.
16. Rotate the low-reverse clutch

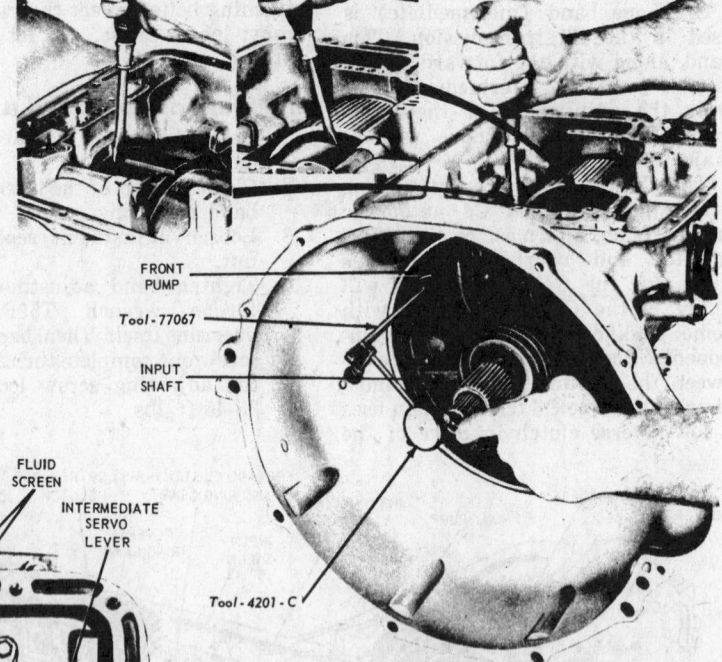

FRONT PUMP

Tool - 77067

INPUT SHAFT

Tool - 4201 - C

Checking transmission gear train end-play (© Ford Motor Co.)

hub, in a clockwise direction, and, at the same time, withdraw it from the case.
17. Remove the low-reverse snap-ring from the case, then remove the clutch discs, plates and pressure plate from the case.
18. Remove the extension housing bolts and vent tube from the case. Remove the extension housing and gasket.
19. Slide the output shaft assembly from the transmission case.
20. Remove the distributor sleeve attaching bolts and remove the sleeve parking gear and the thrust washer.

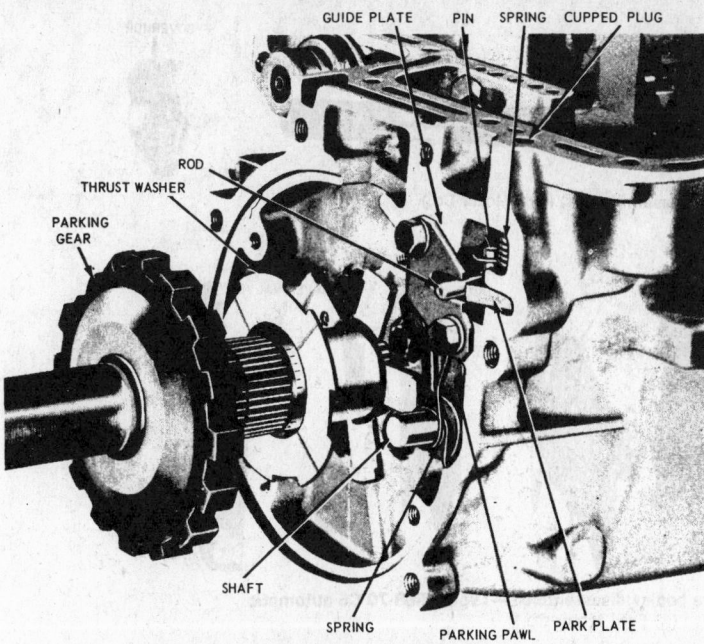

Parking pawl mechanism (© Ford Motor Co.)

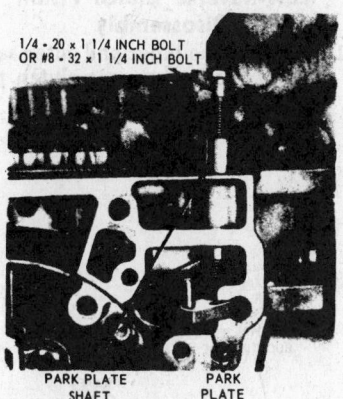

1/4 - 20 x 1 1/4 INCH BOLT
OR #8 - 32 x 1 1/4 INCH BOLT

PARK PLATE SHAFT   PARK PLATE

Removing park plate shaft
(© Ford Motor Co.)

21. Compress the low-reverse clutch piston release spring with tool T-65P-77515-A. Remove the snap-ring. Remove the tool and spring retainer.
22. Remove the one-way clutch inner race attaching bolts from the rear of the case. Remove the inner race from the inside of the case.
23. Remove the low-reverse clutch piston from the case.

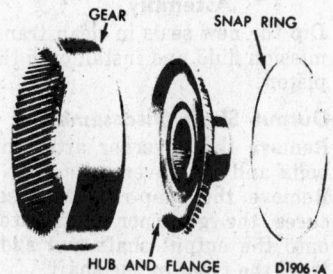

GEAR   SNAP RING

HUB AND FLANGE   D1906-A

Output shaft hub and ring gear disassembled
(© Ford Motor Co.)

Component Disassembly and Assembly

NOTE: For component services, see Component Disassembly and Assembly, in the (C-4 dual range) 3-speed aluminum transmission, immediately preceding this coverage. The exceptions are the following, in the area of the low-reverse clutch.

Output Shaft Hub and Ring Gear Disassembly

1. Remove the hub retaining snap-ring from the ring gear.
2. Lift hub from ring gear.

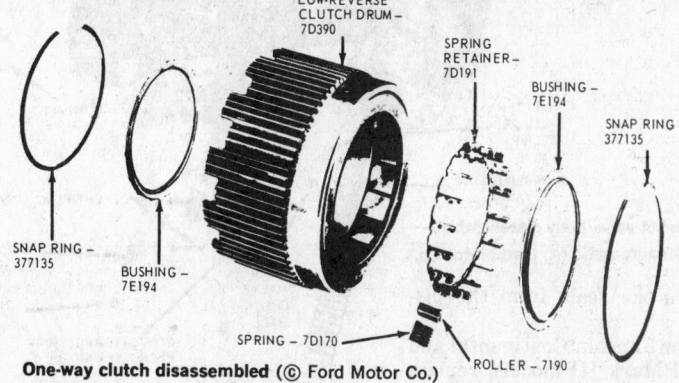

LOW-REVERSE CLUTCH DRUM – 7D390

SPRING RETAINER – 7D191

BUSHING – 7E194

SNAP RING 377135

SNAP RING – 377135

BUSHING – 7E194

SPRING – 7D170

ROLLER – 7190

One-way clutch disassembled (© Ford Motor Co.)

Output Shaft Hub and Ring Gear Assembly

1. Position the hub in the ring gear.
2. Secure the hub with the retaining snap-ring. Make sure the snap-ring is fully engaged with the groove.

One-Way Clutch Disassembly

1. Remove the snap-ring and rear bushing from the rear of the low-reverse clutch hub.

2. Remove the springs and rollers from the spring retainer and lift the spring retainer from the hub.
3. Remove the remaining bushing and snap-ring from the hub.

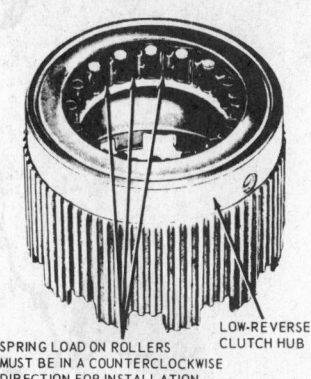

SPRING LOAD ON ROLLERS MUST BE IN A COUNTERCLOCKWISE DIRECTION FOR INSTALLATION

LOW-REVERSE CLUTCH HUB

Installing one-way clutch (© Ford Motor Co.)

One-Way Clutch Assembly

1. Install a snap-ring in the forward snap-ring groove of the low-reverse clutch hub.
2. Place the low-reverse clutch hub on the bench with the forward end down as shown.
3. Place the forward clutch bushing against the snap-ring with the flat side up. Install the one-way clutch spring retainer on top of the bushing. Be sure to install the retainer in the hub so that the springs load the rollers in a counterclockwise direction when looking down at the unit.
4. Install a spring and roller into each of the spring retainer compartments by slightly compressing each spring and placing the roller between the spring and the spring retainer.
5. Install the rear bushing on top of the retainer with the flat side down.
6. Install the remaining snap-ring at the rear of the low-reverse clutch hub to secure the assembly.

Servo Disassembly

1. Apply air pressure to the port in the servo cover in order to remove the piston and stem.

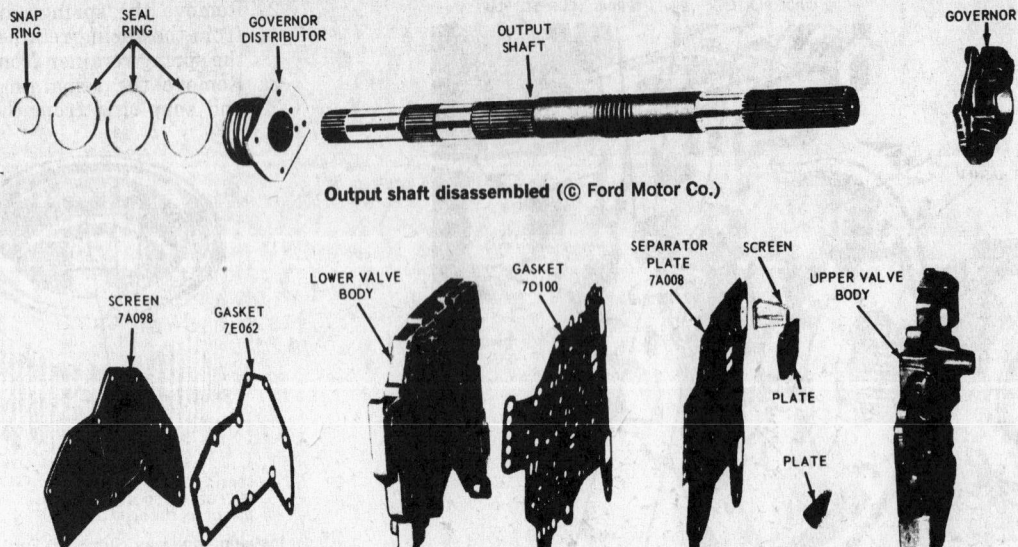

Output shaft disassembled (© Ford Motor Co.)

Upper and lower control valve bodies disassembled—1966, 1968-70 C6 automatic
(© Ford Motor Co.)

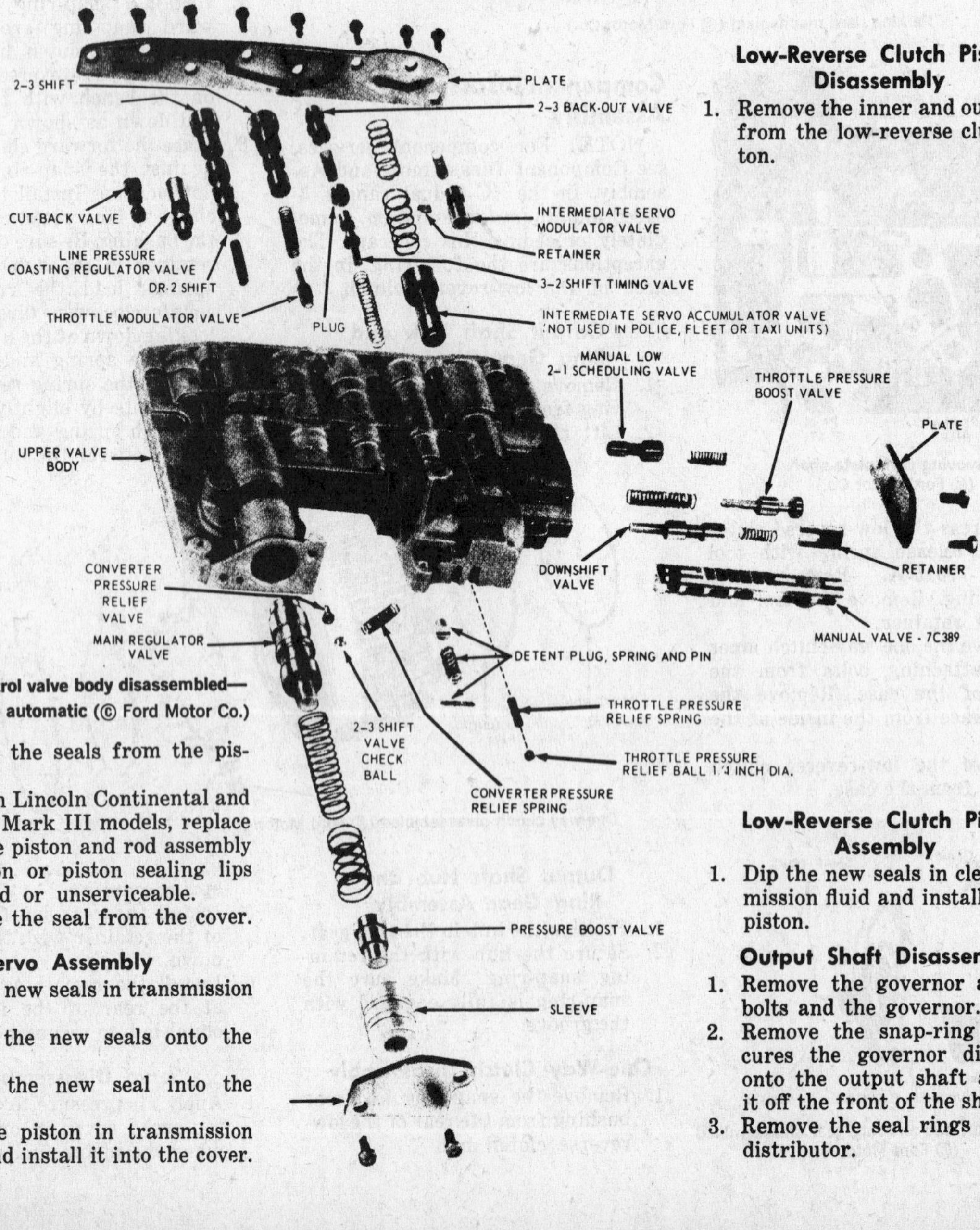

Upper control valve body disassembled—
1968-70 C6 automatic (© Ford Motor Co.)

2. Remove the seals from the piston.

NOTE: on Lincoln Continental and Continental Mark III models, replace the complete piston and rod assembly if the piston or piston sealing lips are damaged or unserviceable.

3. Remove the seal from the cover.

Servo Assembly

1. Dip the new seals in transmission fluid.
2. Install the new seals onto the piston.
3. Install the new seal into the cover.
4. Dip the piston in transmission fluid and install it into the cover.

Low-Reverse Clutch Piston Disassembly

1. Remove the inner and outer seals from the low-reverse clutch piston.

Low-Reverse Clutch Piston Assembly

1. Dip the new seals in clean transmission fluid and install onto the piston.

Output Shaft Disassembly

1. Remove the governor attaching bolts and the governor.
2. Remove the snap-ring that secures the governor distributor onto the output shaft and slide it off the front of the shaft.
3. Remove the seal rings from the distributor.

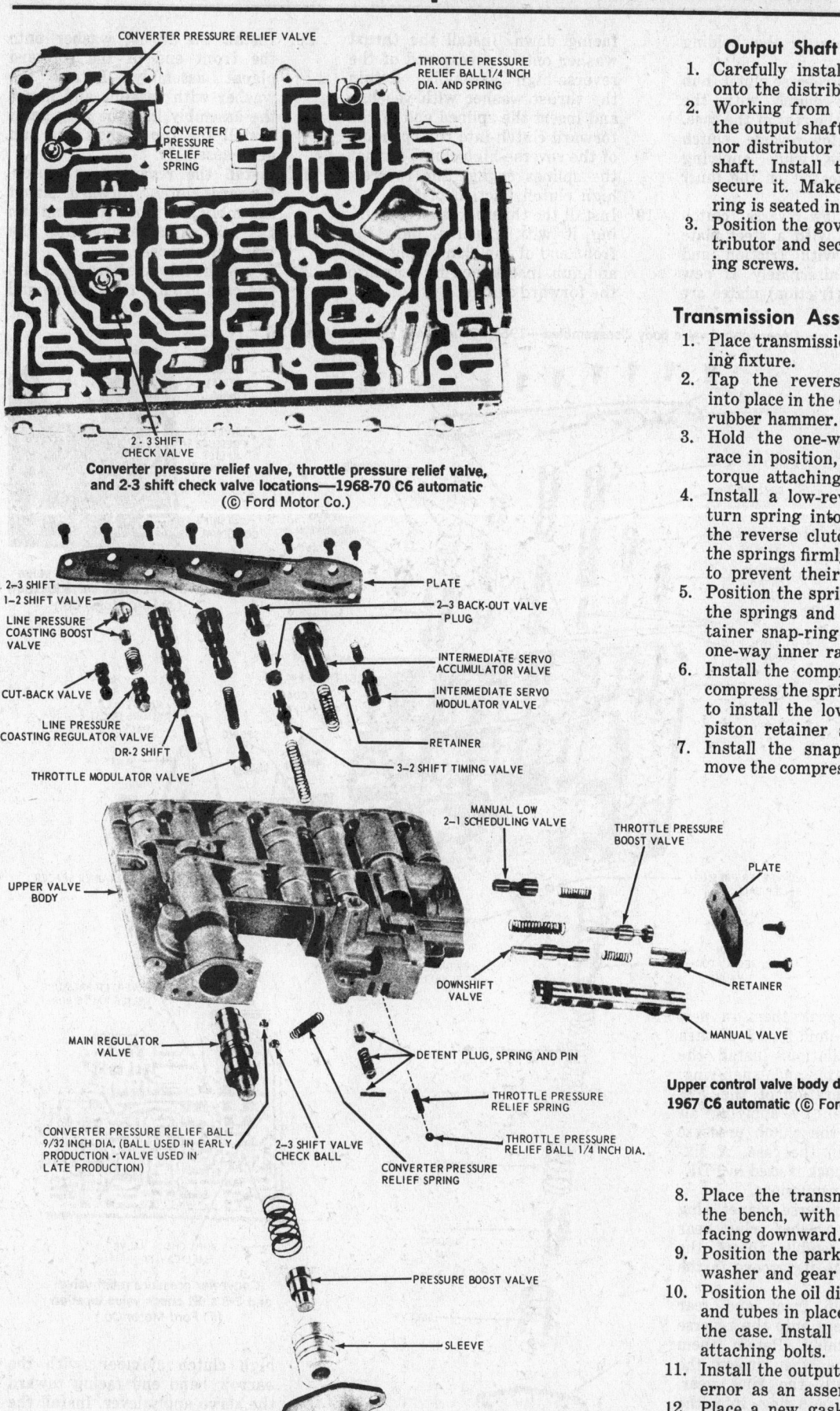

Converter pressure relief valve, throttle pressure relief valve, and 2-3 shift check valve locations—1968-70 C6 automatic (© Ford Motor Co.)

Upper control valve body disassembled—1967 C6 automatic (© Ford Motor Co.)

Output Shaft Assembly

1. Carefully install new seal rings onto the distributor.
2. Working from the front end of the output shaft, slide the governor distributor into place on the shaft. Install the snap-ring to secure it. Make sure the snap-ring is seated in the groove.
3. Position the governor on the distributor and secure with attaching screws.

Transmission Assembly

1. Place transmission case in a holding fixture.
2. Tap the reverse clutch piston into place in the case with a clean rubber hammer.
3. Hold the one-way clutch inner race in position, then install and torque attaching bolts.
4. Install a low-reverse clutch return spring into each pocket of the reverse clutch piston. Press the springs firmly into the piston to prevent their falling out.
5. Position the spring retainer over the springs and position the retainer snap-ring in place on the one-way inner race.
6. Install the compressing tool and compress the springs just enough to install the low-reverse clutch piston retainer snap-ring.
7. Install the snap-ring, then remove the compressing tool.

8. Place the transmission case on the bench, with the front end facing downward.
9. Position the parking gear thrust washer and gear on the case.
10. Position the oil distributor sleeve and tubes in place on the rear of the case. Install and torque the attaching bolts.
11. Install the output shaft and governor as an assembly.
12. Place a new gasket on the rear of the transmission case. Position the extension housing on the case and install attaching bolts.

13. Place the case in the holding fixture.
14. Align the low-reverse clutch hub and one-way clutch with the inner race at the rear of the case. Rotate the low-reverse clutch hub clockwise, while applying pressure to seat it on the inner race.
15. Install the low-reverse clutch plates. Start with a steel plate and follow with friction and steel plates, alternately. If new composition (friction) plates are

facing down. Install the thrust washer onto the rear end of the reverse-high assembly. Retain the thrust washer with vaseline and insert the splined end of the forward clutch into the open end of the reverse-high clutch so that the splines engage the reverse-high clutch friction plates.
19. Install the thrust washer, retaining it with vaseline, onto the front end of the planet ring gear and hub. Insert the ring gear into the forward clutch.

20. Install the thrust washer onto the front end of the forward planet assembly. Retain the washer with vaseline and insert the assembly into the ring gear. Install the input shell and sun gear assembly.
21. Install the reverse-high clutch, forward clutch, forward planet assembly and input shell and sun gear, as an assembly, into the transmission case.
22. Insert the intermediate band into the case around the reverse and

Upper control valve body disassembled—1966 C6 automatic (© Ford Motor Co.)

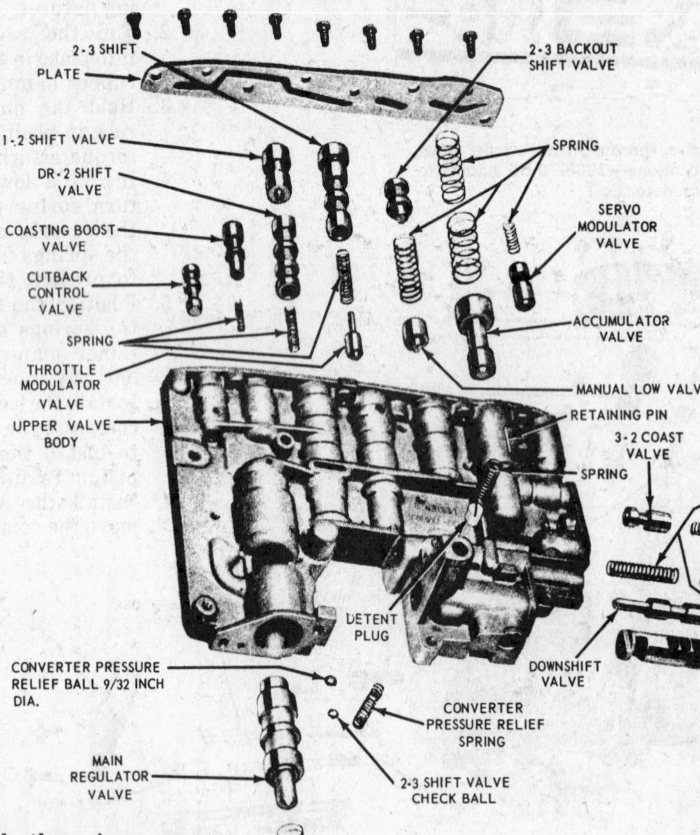

2-3 SHIFT PLATE

2-3 BACKOUT SHIFT VALVE

1-2 SHIFT VALVE

DR-2 SHIFT VALVE

COASTING BOOST VALVE

CUTBACK CONTROL VALVE

SPRING

THROTTLE MODULATOR VALVE

UPPER VALVE BODY

SPRING

SERVO MODULATOR VALVE

ACCUMULATOR VALVE

MANUAL LOW VALVE

RETAINING PIN

3-2 COAST VALVE

SPRING

SPRING

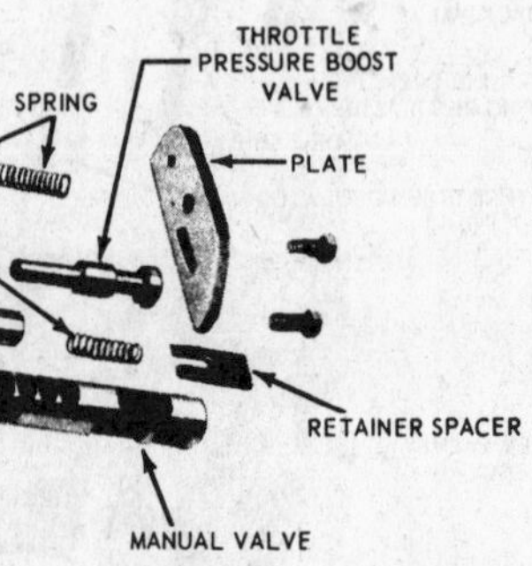

THROTTLE PRESSURE BOOST VALVE

PLATE

DETENT PLUG

DOWNSHIFT VALVE

RETAINER SPACER

MANUAL VALVE

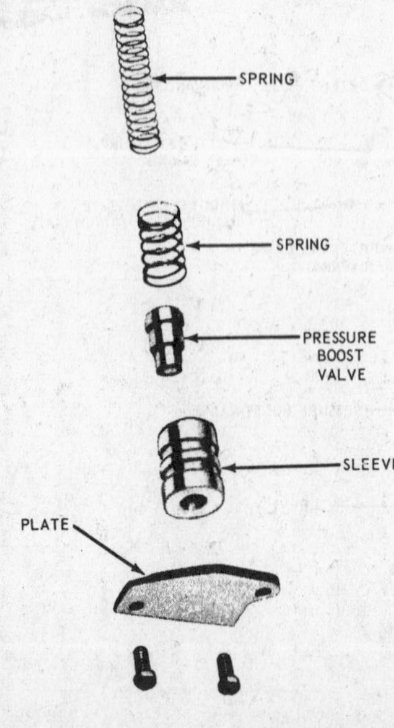

CONVERTER PRESSURE RELIEF BALL 9/32 INCH DIA.

MAIN REGULATOR VALVE

CONVERTER PRESSURE RELIEF SPRING

2-3 SHIFT VALVE CHECK BALL

SPRING

SPRING

PRESSURE BOOST VALVE

SLEEVE

PLATE

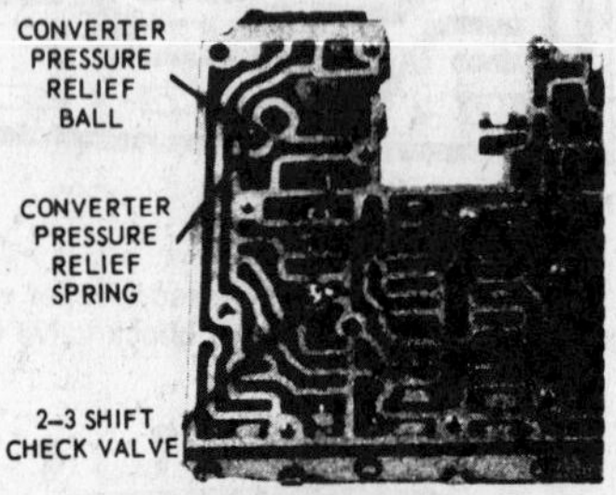

CONVERTER PRESSURE RELIEF BALL

CONVERTER PRESSURE RELIEF SPRING

2-3 SHIFT CHECK VALVE

Converter pressure relief valve and 2-3 shift check valve location (© Ford Motor Co.)

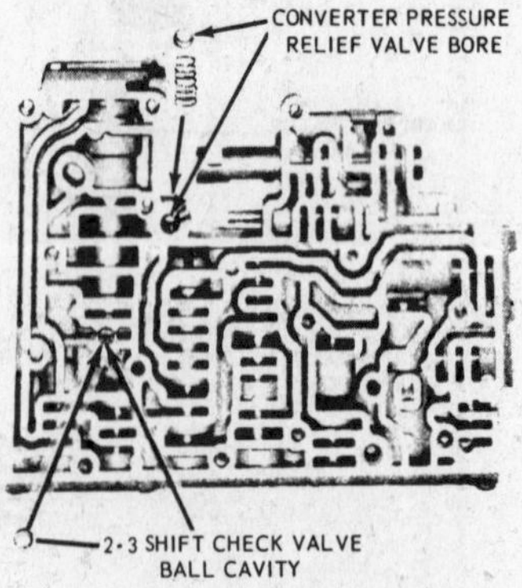

CONVERTER PRESSURE RELIEF VALVE BORE

2-3 SHIFT CHECK VALVE BALL CAVITY

Converter pressure relief valve and 2-3 shift check valve location (© Ford Motor Co.)

being used, soak them in new transmission fluid for 15 minutes before installation. Install the pressure plate and snap-ring. Test the operation of the low-reverse clutch by applying air pressure at the clutch pressure apply hole in the case. A six-plate clutch pack is used in PDE-A model transmissions.
16. Install the reverse planet ring gear thrust washer, ring gear and hub assembly. Insert the snap-ring into its groove in the output shaft.
17. Assemble the front and rear thrust washers onto the reverse planet assembly. Retain them with vaseline, then insert the assembly into the ring gear. Install the snap-ring into the ring gear.
18. Place the reverse-high clutch on the bench, with the front end

high clutch cylinder, with the narrow band end facing toward the servo apply lever. Install the struts and tighten the band adjusting screw just enough to retain the band.

23. Place a selective thickness bronze thrust washer on the rear shoulder of the stator support and retain it with vaseline. Lay a new gasket on the rear mounting face of the pump and position it on the case, being careful not to damage the O-ring. Install six of the seven mounting bolts.
24. Adjust the intermediate band as previously stated in Intermediate Band Adjustment. Then, install the input shaft.
25. Install the tool (4201-C dial indicator) at the seventh mounting bolt and check the transmission end-play. Remove the tool, then install the remaining bolt.
26. Install the control valve into the case, making sure that the levers engage the valves properly.
27. Install the primary throttle valve, rod, and the vacuum diaphragm into the case.
28. Install a new oil pan gasket, and the oil pan.
29. Install the converter assembly.
30. Install the transmission into the car.

Approximate Refill Capacities

Lincoln Continental—1968-70	13 qt.
All other models—1968-70	12¾ qt.
All models—1966-67	13.3 qt.

Checks and Adjustments—1966-67

Operation	Specification
Transmission end play	0.008-0.044 inch selective thrust washers available
Turbine and stator end play	0.060 in.
Intermediate band adjustment	Adjust screw to 10 ft-lbs torque, then back off one full turn and tighten lock nut to specification.
Forward clutch pressure plate-to-snap ring clearance	0.048-0.061 in.
Selective snap ring thicknesses	0.065-0.069 in. 0.083-0.087 in. 0.074-0.078 in.
Reverse-high clutch pressure plate-to-snap ring clearance	Transmission Models
	PDD—B, C, K, T, U, S and W PDD—D, L, F, H and N
	0.022-0.036 Inch 0.027-0.043 Inch
Selective snap ring clearances	0.065-0.069 in. 0.083-0.087 in. 0.074-0.078 in.

Checks and Adjustments—1968-70

Operation	Specification
Transmission end play	0.008-0.044 (selective thrust washers available)
Turbine and stator end play	New or rebuilt 0.021 in. max. Used 0.030 in. max. ①
Intermediate band adjustment	Remove and discard lock nut. Adjust screw to 10 ft-lbs torque, then back off 1 turn, install new lock nut and tighten lock nut to specification.
Forward clutch pressure plate-to-snap ring clearance	0.031-0.044
Selective snap ring thicknesses	0.056-0.060 in., 0.065-0.069 in., 0.074-0.078 in., 0.083-0.087 in., 0.092-0.096 in.
Reverse-high clutch pressure plate-to-snap ring clearance	Transmission Models
	PGA, PJA 0.022-0.036 in.
	PGB-AF2, F3, G3, PJB, PJC-A, B, E, F, PJD 0.027-0.043 in.
Selective snap ring thicknesses	0.065-0.069 in., 0.074-0.078 in, 0.083-0.087 in.

① To check end play, exert force on checking tool to compress turbine to cover thrust washer wear plate. Set indicator at zero.

Selective Thrust Washers—1966-70

Identification No.	Thrust Washer Thickness—Inch	Identification No.	Thrust Washer Thickness—Inch
1	0.056-0.058	4	0.103-0.105
2	0.073-0.075	5	0.118-0.120
3	0.088-0.090		

Control Pressure at Zero Governor RPM—1968-70

Engine speed		Idle			As required		As required		
Throttle		Closed			As required		As required		
Manifold vacuum (inches Hg)		Above 18 ①			10		Below 1.0		
Range		Control Pressure (psi) P, N, D, 2, 1	R	TV Pressure (psi)	Control Pressure (psi) D, 2, 1	TV Pressure (psi)	Control Pressure (psi) D, 2, 1	R	TV Pressure (psi)
	Barometric Pressure in Inches HG	Nominal Altitude (Feet)							
psi @ Barometric Pressure ②	29.5	Sea Level	56-62	71-86	7-10	100-115	40-44	160-190	240-300	77-84
psi @ barometric pressure ③	28.5	1000	49-59	65-80	4-7	99-114	37-41	158-176	233-290	74-80
	27.5	2000	49-56	60-75	2-5	96-111	35-39	156-174	228-284	72-78
	26.5	3000	49-56	56-71	0-3	91-106	32-36	151-169	222-277	69-75
	25.5	4000	49-56	56-65	0	88-103	30-34	146-164	215-269	66-72
	24.5	5000	49-56	56-65	0	84-98	27-31	143-161	211-264	64-70
	23.5	6000	49-56	56-65	0	80-95	25-29	138-156	204-256	61-67

Manifold Vacuum	Barometric Pressure at 29.5 Inches ②		Barometric Pressure at 24.5 Inches ④	
	T.V.	Cont.	T.V.	Cont.
17	11-14	56-69	0-1	49-56
16	15-18	56-75	2-5	49-56
15	20-22	56-84	7-9	49-61
14	23-26	56-92	10-13	56-67
13	28-31	56-98	15-18	56-75
12	32-35	56-105	19-22	56-84
11	36-40	56-111	23-27	56-92

① It may not be possible to obtain 18 inches of engine vacuum at idle. For idle vacuums of less than 18 inches the following table provides idle speed pressure specifications in D range:

② These specifications (with altitude compensating diaphragm) apply at observed barometric pressure of 29.5 inches (nominal sea level)

③ Specifications for barometric pressures of less than 29.5 inches.

④ At barometric pressures between 29.5 inches and 24.5 inches idle, pressures should fall between the values shown.

Control Pressure at Zero Output Shaft Speed— Altitude Compensating Type Vacuum Diaphragm—1966-67

Engine speed		Idle			As required		As required		As required		
Throttle		Closed			As required		s required		Open through detent		
Manifold vacuum (inches Hg)		Above 17			15		10		Below 1.0		
Range		Control Pressure (psi) P, N, D1, D2, L	R	TV Pressure (psi)	Control Pressure (psi) D1, D2, L	TV Pressure (psi)	Control Pressure (psi) D1, D2, L	TV pressure (psi)	Control Pressure (psi) D1, D2, L	R	TV Pressure (psi)
	Barometric pressure in inches HG	Nominal altitude (feet)									
psi @ barometric pressure	29.5	Sea Level	51-66	72-108	0-13	70-78	20-22	98-109	40-44	157-172	230-252	80-84
	28.5	1000	51-59	72-104	0-11	67-75	18-20	94-105	38-41	149-163	220-242	77-81
	27.5	2000	51-59	72-99	0-8	63-71	15-17	91-101	36-38	145-159	215-236	74-78
	26.5	3000	51-59	72-99	0-5	59-67	12-14	87-97	33-36	142-157	211-231	72-76
	25.5	4000	51-59	72-92	0-3	57-64	10-12	85-94	31-33	139-154	207-227	70-74
	24.5	5000	51-59	72-83	0	51-60	7-9	81-90	28-31	136-151	202-222	67-72
	23.5	6000	51-59	72-83	0	51-60	5-7	78-87	26-28	132-147	198-217	64-68

Torque Limits—1968-70

	Ft-Lbs
Converter to flywheel	20-30
Front pump to trans. case	12-20
Overrunning clutch race to case	18-25
Oil pan to case	12-16
Stator support to pump	12-16
Converter cover to converter hsg.	12-16
Guide plate to case	12-16
Intermediate servo cover to case	10-14
Diaphragm assy. to case	15-23
Distributor sleeve to case	12-16
Extension assy. to trans. case	25-30
Pressure gauge tap	9-15
Band adj. screw locknut to case	35-45
Cooler tube connector lock	25-35
Converter drain plug	14-28
Manual valve inner lever to shaft	30-40①
Downshift lever to shaft	12-16
Filler tube to engine	20-25
Transmission to engine	40-50
Steering col. lock rod adj. nut	10-20
Neutral start switch actuator lever bolt	6-10

	In-Lbs
End plates to body	20-30
Inner downshift lever stop	20-30
Reinforcement plate to body	20-30
Screen and lower to upper valve body	50-60②
Neutral switch to case	55-75
Neutral switch-to column	20
Control assy. to case	90-125③
Gov. body to collector body	80-120④
Oil tube connector	80-145

① 1969 torque limits— 8-12 ft.-lbs.
② 1968 torque limits— 40-50 ft.-lbs.
③ 1968 torque limits—100-120 ft.-lbs.
④ 1968 torque limits—100-120 ft.-lbs.

Torque Limits—1966-67

	Ft-Lbs
Pressure gauge tap	9-15
Servo cover to case	10-14
Parking rod guide plate to case	12-16
Outer downshift lever to shaft	12-16
Distributor sleeve to case	12-16
Support assy. to pump body	12-16
Oil pump to case	12-16
Oil pan to case	12-16
Conv. hsg. lower cover to trans.	12-16
Converter drain plugs	14-28
Diaphragm assy. to case	15-23
One-way clutch race to case	18-25
Flywheel to converter	20-30
Extension hsg. to case	25-30
Connector assy. to case	25-35
Manual lever to shaft	30-40
Band adj stop to case	35-45
Transmission to engine	40-50

	In-Lbs
Plate to control assy.	20-30
Lower to upper valve body	40-50
Upper to lower valve body	40-50
Screen & lwr v.b. to upper v.b.	40-50
Neutral switch assy. to case	55-75
Control assy. to case	100-120
Governor to collector	100-120

General Motors Corporation

Powerglide

Application

Camaro, 1967-69
Chevelle, 1964-70
Chevrolet, 1964-70

Chevy II, 1964-68
Corvette, 1964-67
Firebird, 1970

Monte Carlo, 1970
Nova, 1969-70
Tempest, 1970

Diagnosis

This diagnosis guide covers the most common symptoms and is an aid to careful diagnosis. The items to check are listed in the sequence to be followed for quickest results. Thus, follow the checks in the order given for the particular transmission type.

TROUBLE SYMPTOMS	Items to Check	
	In Car	Out of Car
Car will not move in any selector position	ABC	cab
Engine speed flares, as slipping clutch	ABDFE	def
Engine speed flares on upshift	ADBG	ghi
Transmission will not upshift	HIJ	klm
Harsh upshift	JDGK	
Harsh deceleration downshift	DLMNG	
No downshift	MHI	
Clutch failure, burnt plates	DAHO	hi
Excessive creep in drive	JL	
Car creeps in neutral	J	hd
No drive in reverse	J	jng
Improper shift points	JIH	l
Unable to push-start car		m
Oil leaks	PQG	oa
Oil forced out at filler tube	AQ	p

Key to Checks

A. Oil level
B. Oil screen
C. Pressure regulator valve
D. Band adjustment
E. Servo seal
F. Servo blocked
G. Vacuum modulator or line
H. Governor
I. Throttle valve

J. Throttle linkage
K. Hydraulic modulator valve
L. Too high idle speed
M. Valves malfunctioning
N. Make pressure tests
O. Driving too fast in low
P. Oil leaks at external points
Q. Oil cooler or lines

a. Front pump
b. Input shaft
c. Front pump priming valve
d. Low band
e. Low band linkage
f. Converter stator
g. Clutch feed blocked
h. High clutch

i. Front clutch relief valve
j. Reverse clutch relief valve
k. Low clutch valve stuck
l. Rear pump priming valve
m. Rear pump or drive
n. Low clutch
o. Front pump attaching bolts
p. Pump circuit leakage

General Description

The Powerglide transmission is a two speed unit with a one piece aluminum case and an aluminum case extension.

Driving ranges are low, high and reverse, with a throttle controlled downshift to low range available for sudden acceleration.

The oil pump is the gear type; the pump housing is used as the front bulkhead of the transmission. The torque converter is a three element welded unit bolted to the engine flywheel; it drives through a two speed planetary gearset. Low range uses a band clutch; drive and reverse ranges use disc clutches. The valve body assembly is bolted to the bottom of the transmission case and is accessible for service when the oil pan is removed. The vacuum modulator is located on the left rear face of the transmission case; the modulator valve bore is in the upper part of the valve body assembly. The governor is mounted to the output shaft, inside the case extension.

Transmission removal and installation, shift linkage adjustment, low band adjustment in the car, neutral safety switches, and transmission downshift linkages are covered in the car section.

Fluid Change Schedule

For 1964-66 models, it is not necessary to change the automatic transmission fluid except when the transmission is removed. For 1967 models, the manufacturer recommends draining the transmission sump every 12,000 miles; for 1968-70 models, every 24,000 miles. 1½ quarts of fluid should be added to Camaro and Chevy II models; 2 quarts to all other models. The fluid level should then be rechecked. This should be done with the engine idling, the selector lever in Neutral, and the transmission at operating temperature. Total transmission capacities are shown in the Capacities chart, in the applicable Car Section.

Transmission Disassembly

Extension, Governor and Rear Oil Pump—1964-66 Models

1. Place transmission in a holding fixture, if possible.

2. Remove converter holding tool, then lift off the converter.
3. If replacement is necessary, remove speedometer driven gear. Loosen cap screw and retainer clip and remove gear from extension.
4. Remove transmission extension by removing five attaching bolts. Note seal ring on rear pump body.
5. Remove speedometer drive gear from output shaft.
6. Remove C-clip from governor shaft of the weight side of governor, then remove the shaft and governor valve from the opposite side of the governor assembly and the two belleville springs.
7. Loosen the governor drive screw and slide the governor over the end of the output shaft.
8. Remove four bolts holding the rear oil pump to the transmission case and remove the pump body, drain back baffle, extension seal ring, drive and driven gears.
9. Remove oil pump drive pin. (This is of extreme importance.)
10. Remove the rear pump wear plate.

1 Reverse ring gear
2 Reverse clutch pack snap ring
3 Reverse clutch pressure plate
4 Reverse clutch reaction plates
5 Reverse clutch drive plates
6 Reverse clutch cushion spring
7 Reverse clutch piston return spring retainer snap ring
8 Reverse clutch piston return spring retainer

9 Reverse clutch piston return springs
10 Reverse clutch piston inner seal
11 Reverse clutch piston
12 Reverse clutch piston outer seal
13 Transmission case
13A Transmission case screen
14 Servo piston return spring
15 Servo piston rod
16 Servo piston apply spring seat
17 Servo piston apply spring

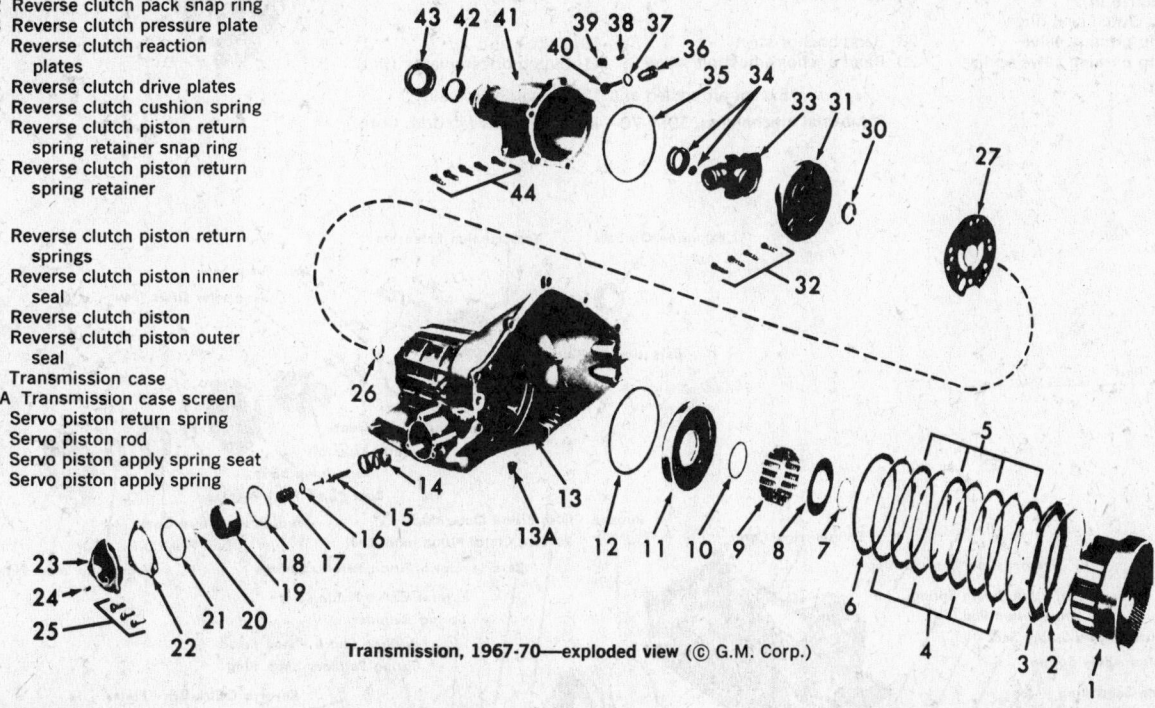

Transmission, 1967-70—exploded view (© G.M. Corp.)

18 Servo piston seal ring
19 Servo piston
20 Servo piston rod spring retainer
21 Servo cover seal
22 Servo cover gasket
23 Servo cover
24 Servo cover plug
25 Servo cover bolts
26 Transmission case bushing
27 Gasket
30 Governor support bushing
31 Governor support
32 Governor support to case attaching bolts

33 Governor assembly
34 Speedometer drive gear and clip
35 Seal
36 Speedometer shaft fitting
37 Speedometer shaft fitting oil seal
38 Lock plate attaching screw
39 Lock plate
40 Speedometer driven gear
41 Transmission extension
42 Extension bushing
43 Extension oil seal
44 Extension to case attaching screws

Extension, Governor and Governor Support—1967-69 Models

1. Proceed with Steps 1-7 from the procedure for 1964-66 models given above.
2. Remove four bolts holding the governor support to the transmission case and remove the support body, gasket, and extension seal ring.
3. Go on to Step 11.

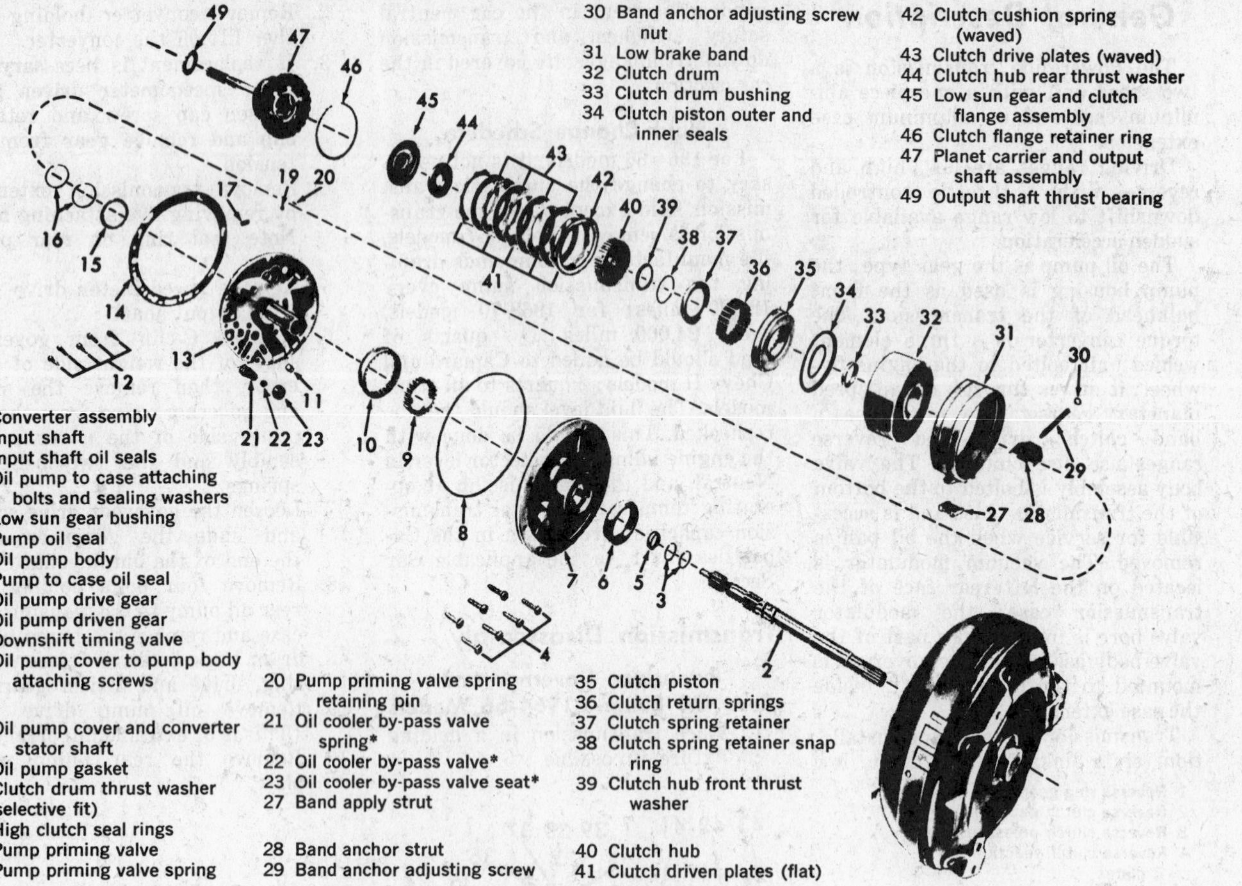

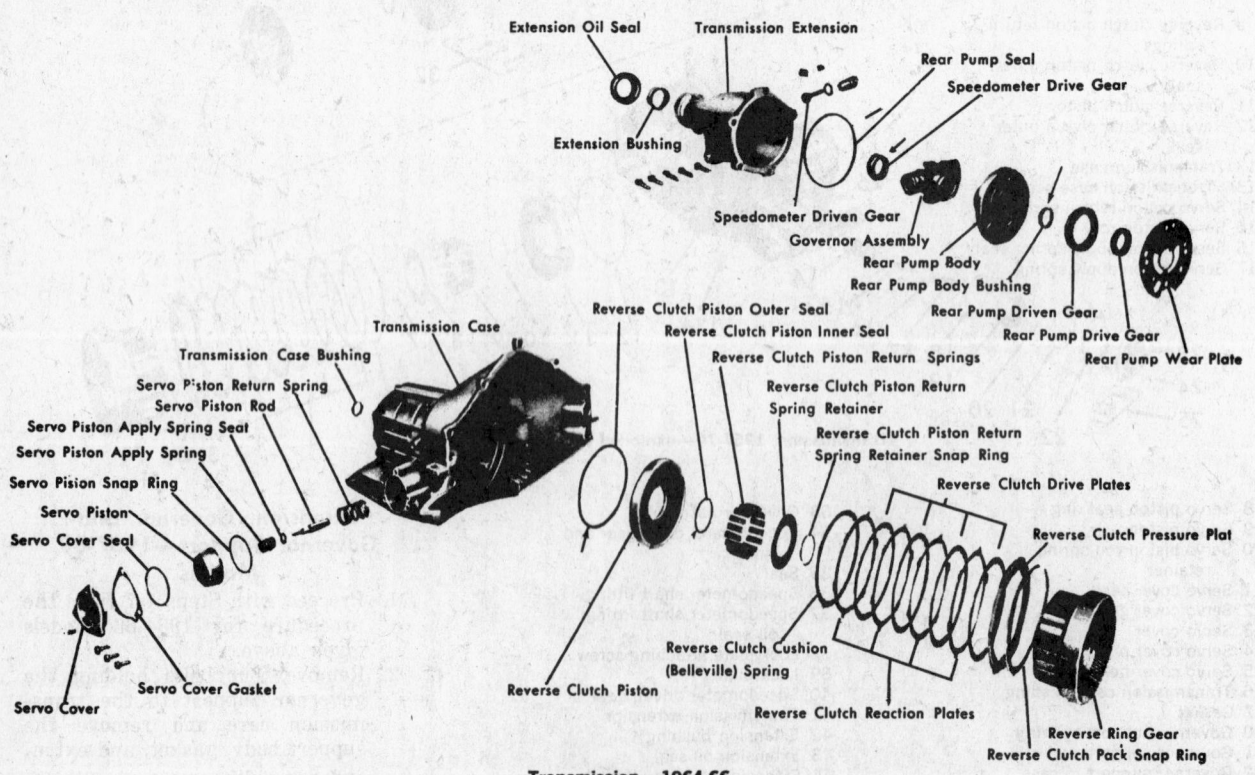

30 Band anchor adjusting screw
    nut
31 Low brake band
32 Clutch drum
33 Clutch drum bushing
34 Clutch piston outer and
    inner seals

42 Clutch cushion spring
    (waved)
43 Clutch drive plates (waved)
44 Clutch hub rear thrust washer
45 Low sun gear and clutch
    flange assembly
46 Clutch flange retainer ring
47 Planet carrier and output
    shaft assembly
49 Output shaft thrust bearing

1 Converter assembly
2 Input shaft
3 Input shaft oil seals
4 Oil pump to case attaching
    bolts and sealing washers
5 Low sun gear bushing
6 Pump oil seal
7 Oil pump body
8 Pump to case oil seal
9 Oil pump drive gear
10 Oil pump driven gear
11 Downshift timing valve
12 Oil pump cover to pump body
    attaching screws
13 Oil pump cover and converter
    stator shaft
14 Oil pump gasket
15 Clutch drum thrust washer
    (selective fit)
16 High clutch seal rings
17 Pump priming valve
18 Pump priming valve spring

20 Pump priming valve spring
    retaining pin
21 Oil cooler by-pass valve
    spring*
22 Oil cooler by-pass valve*
23 Oil cooler by-pass valve seat*
27 Band apply strut

28 Band anchor strut
29 Band anchor adjusting screw

35 Clutch piston
36 Clutch return springs
37 Clutch spring retainer
38 Clutch spring retainer snap
    ring
39 Clutch hub front thrust
    washer

40 Clutch hub
41 Clutch driven plates (flat)

*Except air cooled and 11" converter models.
Internal mechanism, 1967-70—exploded view (© G.M. Corp.)

Extension Oil Seal   Transmission Extension

Rear Pump Seal
Speedometer Drive Gear

Extension Bushing

Speedometer Driven Gear
Governor Assembly
Rear Pump Body
Rear Pump Body Bushing
Rear Pump Driven Gear
Rear Pump Drive Gear
Rear Pump Wear Plate

Transmission Case

Transmission Case Bushing
Servo Piston Return Spring
Servo Piston Rod
Servo Piston Apply Spring Seat
Servo Piston Apply Spring
Servo Pision Snap Ring
Servo Piston
Servo Cover Seal

Servo Cover

Servo Cover Gasket

Reverse Clutch Piston Outer Seal
Reverse Clutch Piston Inner Seal
Reverse Clutch Piston Return Springs
Reverse Clutch Piston Return
Spring Retainer
Reverse Clutch Piston Return
Spring Retainer Snap Ring
Reverse Clutch Drive Plates
Reverse Clutch Pressure Plat

Reverse Clutch Cushion
(Belleville) Spring
Reverse Clutch Piston
Reverse Clutch Reaction Plates
Reverse Ring Gear
Reverse Clutch Pack Snap Ring

Transmission—1964-66
(© G.M. Corp.)

Removing governor valve and shaft
(© G.M. Corp.)

REAR PUMP
DRIVE PIN

Removing rear oil pump drive pin
(© G.M. Corp.)

Transmission Internal Components

11. Rotate holding fixture, or turn the transmission, until the front end is pointing up. Then remove the seven front oil pump bolts. (The bolt holes are of unequal spacing to prevent incorrect location upon installation.)

12. Remove the front oil pump and stator shaft assembly and the selective fit thrust washer using an inertia puller or substitute.

13. Release tension on the low band adjustment, then with transmission horizontal, grasp the transmission input shaft and carefully work it and the clutch drum out of the case. Be careful not to lose

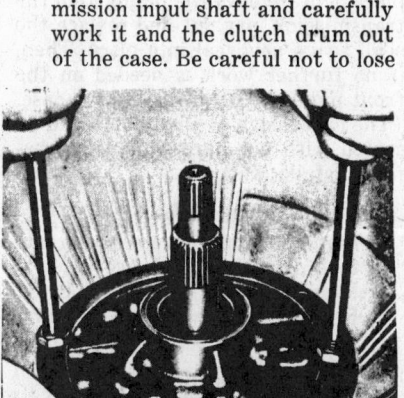

Removing front pump
(© G.M. Corp.)

LOW SUN GEAR
THRUST WASHER

LOW SUN GEAR
BUSHING

Removing clutch drum and input shaft
(© G.M. Corp.)

the low sun gear bushing from the input shaft. The low sun gear thrust washer will probably remain in the planet carrier.

14. The low brake band and struts may now be removed.

15. Remove the planet carrier and the output shaft thrust caged bearing from the front of the transmission.

16. Remove reverse ring gear if it did not come out with the planet carrier.

17. With a large screwdriver, remove the reverse clutch pack retainer ring and lift out the reverse clutch plates and the cushion spring.

18. Install reverse piston spring compressor through rear bore of the case, with the flat plate on the rear face of the case, and turn down wing nut to compress

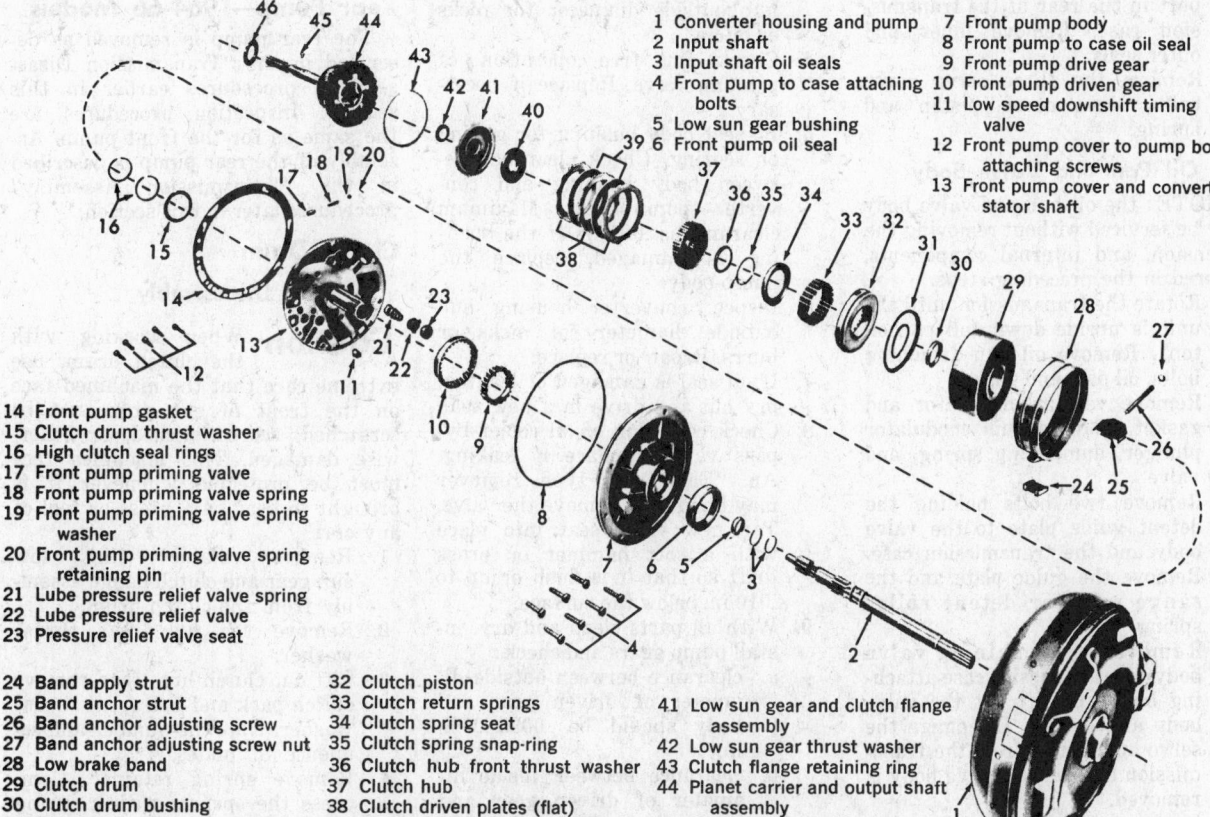

1 Converter housing and pump	7 Front pump body
2 Input shaft	8 Front pump to case oil seal
3 Input shaft oil seals	9 Front pump drive gear
4 Front pump to case attaching bolts	10 Front pump driven gear
5 Low sun gear bushing	11 Low speed downshift timing valve
6 Front pump oil seal	12 Front pump cover to pump body attaching screws
	13 Front pump cover and converter stator shaft

14 Front pump gasket
15 Clutch drum thrust washer
16 High clutch seal rings
17 Front pump priming valve
18 Front pump priming valve spring
19 Front pump priming valve spring washer
20 Front pump priming valve spring retaining pin
21 Lube pressure relief valve spring
22 Lube pressure relief valve
23 Pressure relief valve seat

24 Band apply strut	32 Clutch piston	41 Low sun gear and clutch flange assembly
25 Band anchor strut	33 Clutch return springs	42 Low sun gear thrust washer
26 Band anchor adjusting screw	34 Clutch spring seat	43 Clutch flange retaining ring
27 Band anchor adjusting screw nut	35 Clutch spring snap-ring	44 Planet carrier and output shaft assembly
28 Low brake band	36 Clutch hub front thrust washer	45 Rear pump drive pin
29 Clutch drum	37 Clutch hub	46 Output shaft thrust bearing
30 Clutch drum bushing	38 Clutch driven plates (flat)	
31 Clutch piston outer and inner seals	39 Clutch drive plates (waved) (2)	
	40 Clutch hub rear thrust washer	

Internal mechanism—1964-66
(© G.M. Corp.)

Applying air pressure to remove rear piston
(© G.M. Corp.)

the rear piston spring retainer and springs. Then remove the snap ring. A spring compressor may be made up from a suitable length bolt and large flat washers.

19. Remove the compression tool, the reverse piston spring retainer, and the 17 piston return springs.

20. Remove the rear piston by applying air pressure to the reverse port in the rear of the transmission case. Remove inner and outer seals.

21. Remove the three servo cover bolts, servo cover, piston and spring.

Oil Pan and Valve Body

NOTE: the oil pan and valve body may be serviced without removing the extension, and internal components, covered in the preceding steps.

22. Rotate the transmission until the unit is upside down (oil pan on top). Remove oil pan attaching bolts, oil pan, and gasket.

23. Remove vacuum modulator and gasket, and vacuum modulator plunger, dampening spring, and valve.

24. Remove two bolts holding the detent guide plate to the valve body and the transmission case. Remove the guide plate and the range selector detent roller spring.

25. Remove the remaining valve body-to-transmission case attaching bolts and lift out the valve body and gasket. Disengage the servo apply tube from the transmission case as the valve body is removed.

26. If necessary, the throttle valve, shift and parking actuator levers and the parking pawl and bracket may be removed.

Unit Assembly Overhaul

Converter and Stator

The converter is a welded assembly and no internal repairs are possible. Check the seams for stress or breaks and replace the converter if necessary.

Front Pump

Seal Replacement

If the front pump seal requires replacement, remove the pump from the transmission, pry out and replace the seal. Drive new seal into place. Then, if no further work is needed on the front pump, reinstall it into the case. (The outer edge of the seal should be coated with non-hardening sealer before installation.)

Disassembly

1. Remove pump cover-to-body attaching bolts and the cover.
2. Remove pump gears from body.
3. Remove rubber seal from pump body.

Inspection

1. Wash all parts in solvent. Blow out all oil passages.
2. Inspect pump gears for nicks or damage.
3. Inspect body and cover faces for nicks or scoring. Inspect cover hub outside diameter for nicks or burrs.
4. Check for free operation of priming valve. Replace if necessary.
5. Inspect body bushing for galling or scoring. Check clearance between body bushing and converter pump hub. Maximum clearance is .005 in. If the bushing is damaged, replace the pump body.
6. Inspect converter housing hub outside diameter for nicks or burrs. Repair or replace.
7. If oil seal is damaged or leaking, pry out and drive in a new seal.
8. Check condition of oil cooler bypass valve. Replace if leaking. An "Easy-Out" type remover may be used to remove the valve. Tap new valve seat into place with a soft hammer or brass drift so that it is flush or up to .010 in. below the surface.
9. With all parts clean and dry, install pump gears and check:
   a. clearance between outside diameter of driven gear and body should be .0035-.0065 in.
   b. clearance between inside diameter of driven gear and crescent should be .003-.009 in.
   c. gear end clearance should be .0005-.0015 in.

Assembly

1. Remove the input shaft, clutch drum, low band and struts as outlined under "Transmission Disassembly."
2. Install downshift timing valve, conical end out, into place in the pump cover to a height of 17/32 in. measured from shoulder of valve assembly to face of pump cover.
3. Oil the drive and driven gears and install them into the pump body.
4. Set pump cover in place over the body and loosely install two attaching bolts.
5. Place pump assembly, less the rubber seal ring, upside down into the pump bore of the case. Install remaining attaching bolts and torque to 20 ft. lbs.
6. Remove pump assembly from case bore. Replace the clutch drum and input shaft, low band and struts as described under "Transmission Assembly."
7. Renew rubber seal ring in its groove in the pump body and install the pump assembly in place in the case bore, using a new gasket. Be sure that the selective fit thrust washer is in place.
8. Install attaching bolts. (Use new bolt sealing washers if necessary.)

Rear Pump—1964-66 Models

The rear pump is removed as described in the "Transmission Disassembly" procedures earlier in this section. Inspection procedures are the same as for the front pump. Assembly of the rear pump is described in the "Transmission Assembly" procedures later in this section.

Clutch Drum

Disassembly

Caution When working with the clutch drum, use extreme care that the machined face on the front of the drum not be scratched, scored, nicked, or otherwise damaged. This machined face must be protected whenever it is brought to bear on a press or tool of any sort.

1. Remove retainer ring, low sun gear and clutch flange assembly from the clutch drum.
2. Remove the hub rear thrust washer.
3. Lift out clutch hub, then remove clutch pack and hub front thrust washer. Note the number and sequence of plates.
4. Remove spring retainer. Compress the springs with a spring compressor or an arbor press enough to permit removal of the retainer snap-ring. Then, releasing pressure on the springs,

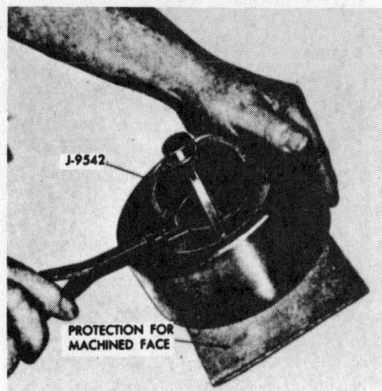

Removing clutch spring retainer snap-ring
(© G.M. Corp.)

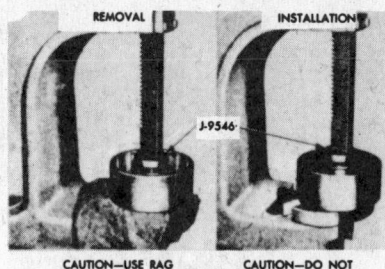

Removing and installing clutch drum bushing
(© G.M. Corp.)

1969-70 Models	L-4 & 230L-6 Pass. Cars	307 V-8; 230 L-6 Truck; 250-L-6 Exc. Taxi & Hvy. Duty Chassis 292 L-6 Truck	250L-6 Taxi & Hvy. Duty Chassis; 350 V-8 Truck	350V-8
Drive Plate	3	4	5	5
Driven Plate	4	5	6	6
Cushion Spring	1	1	None	1

TORQUE DRIVE

1969-70 Models	L-4, L-6
Drive Plate	5
Driven Plate	6
Cushion Spring	1

350 and 396 V-8 Selective Driven Plate Chart

Plate Stack Height (Less Selective Plate)	Plate Part Number	Color Code	Plate Thickness
.903-.872	3883903	Orange	.060±.0025
.872-.798	3883904	Blue	.090±.0025

remove retainer and the 24 springs.

5. Lift up on the piston with a twisting motion to remove it from the drum, then remove inner and outer seals.

Inspection

1. Wash all parts in solvent, blow out all passages, and air dry. Do not use rags to dry parts.
2. Check drum bushing for scoring or excessive we r.
3. Check steel ball relief valve in clutch drum. Be sure that it is free to move and that orifice in front face of drum is open. If ball is loose enough to come out, or not loose enough to rattle, replace drum. Do not attempt replacement or restaking of ball.
4. Check fit of low sun gear and clutch flange assembly in clutch drum slots. There should be no appreciable radial play.
5. Check low sun gear for nicks or burrs. Check gear bore for wear.
6. Check clutch plates for burring, wear, pitting, or metal pick-up. Faced plates should be a free fit over clutch hub; steel plates should be a free fit in clutch drum slots.
7. Check condition of clutch hub splines and mating splines of clutch faced plates.
8. Check clutch pistons for cracks or distortion.

Clutch Drum Bushing Replacement

If replacing drum bushing, carefully press out the old bushing. Then press (don't hammer) the new bushing into place from the machined face side of the drum. Press only far enough to bring the bushing flush with the clutch drum. Do not force the tool against the clutch drum machined face.

Assembly

1. Install new piston inner seal into hub of clutch drum with seal lip toward front of transmission.
2. Install new piston seal into clutch piston. Seal lips must be pointed toward the clutch drum, (front of transmission). Lubricate the seals and install piston into clutch drum with a twisting motion.
3. Place 24 springs in position on the piston, then place the retainer on the springs.
4. Depress the retainer plate and springs far enough to permit installation of the spring retainer snap-ring into its groove on the clutch drum hub.
5. Install the hub front washer with its lip toward the clutch drum, then install the clutch hub.
6. Install cushion spring if used. Install the steel reaction plates and drive (faced) plates alternately, beginning with a steel reaction plate.

NOTE: the number and sequence of plates varies with the power and torque requirements of the car model involved. On some models, the first driven plate is a selective fit. See the clutch assemblies chart for details.

Clutch Assemblies

Up to 1968 Models	L-4 & 230L-6 Pass. Cars.	307 V-8; Pass. Car; 250L-6 Pass. Cars (exc. Taxi & Hvy. Duty Chassis) 327V-8 Pass. Cars	250L-6 Taxi & Hvy. Duty Chassis; 396V-8 Pass. Cars.	350V-8 396V-8
Drive Plate	3	4	5	5
Driven Plate	4	5	6	6
Cushion Spring	1	1	None	1

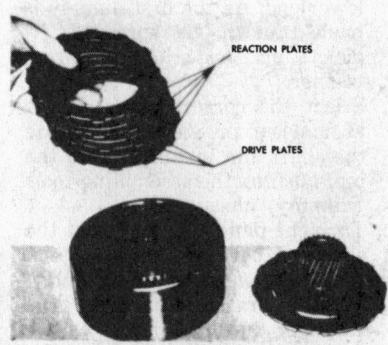

Installing clutch plates
(© G.M. Corp.)

7. Install the rear hub thrust washer with its flange toward the low sun gear, then install the low sun gear and flange assembly and secure with retaining ring. When installed, the openings in the retainer ring should be adjacent to one of the lands of the clutch drum.
8. Check assembly by turning the clutch hub. If free, assembly is OK.

Low Band

Due to band design and transmission characteristics, this band should require very little attention. However, while the transmission is disassembled, the band should be thoroughly cleaned, then replaced if any trace of scoring, burning, cracks, or excessive wear or damage is found.

Planet Assembly and Input Shaft

Inspection

1. Wash planet carrier and input shaft in cleaning solvent, blow out all passages, and air dry. Do not use rags to dry parts.
2. Inspect planet pinions for nicks or other tooth damage.
3. Check end clearance of planet gears. The clearance should be .006-.030 in.
4. Check input sun gear for tooth damage. Check thrust washer for damage.
5. Inspect output shaft bearing surface and input pilot bushing for nicks or scoring.
6. Inspect input shaft splines for nicks or damage. Check fit in

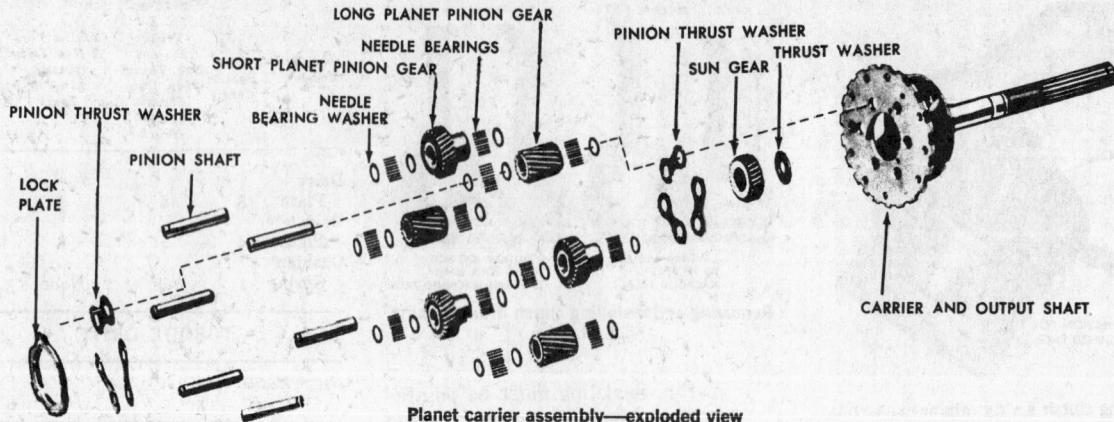

LONG PLANET PINION GEAR
NEEDLE BEARINGS
SHORT PLANET PINION GEAR
PINION THRUST WASHER
SUN GEAR
THRUST WASHER
PINION THRUST WASHER
NEEDLE BEARING WASHER
PINION SHAFT
LOCK PLATE
CARRIER AND OUTPUT SHAFT

Planet carrier assembly—exploded view
(© G.M. Corp.)

clutch hub, input sun gear, and turbine hub.

7. Check oil seal rings for damage; rings must be free in input shaft ring grooves. Remove rings and insert in stator support bore. Check to see that hooked ring ends have clearance. Replace rings on shaft.

Repairs

NOTE: some large planet carrier assemblies have the pinion shafts flared at each end for retention in the carrier. No overhaul of this type of carrier assembly should be attempted. If inspection shows excessive wear or damage, replace the entire carrier assembly.

1. Place the planet carrier assembly in a padded vise so that the front (parking lock gear end) of the assembly is up.
2. Using a prick punch, mark each pinion shaft and the carrier assembly so that, when reassembling, each shaft will be returned to its original location.
3. Remove pinion shaft lock plate screws and rotate plate counterclockwise far enough to remove it.
4. Starting with a short planet pinion, drive the lower end of the pinion shaft up until the shaft is above the press fit area of the output shaft flange. Feed a dummy shaft into the short planet pinion from the lower end, pushing the planet pinion shaft ahead of it until the tool is centered in the pinion and the pinion shaft is removed.
5. Remove short planet pinion.
6. Remove dummy shaft, needle and bearing spacers from short pinion.
NOTE: twenty needle bearings are used in each end of each gear and are separated by a bearing spacer in the center.
7. By following Steps 4, 5, and 6, remove the adjacent long planet pinion that was paired, by thrust

washers, to the short pinion now removed.
8. Remove upper and lower thrust washers.
9. Remove and disassemble remaining planet pinions, in pairs, as above.
10. Remove low sun gear needle thrust bearing, input sun gear, and thrust washer.
11. Wash all parts in solvent and blow dry, then inspect.
12. Inspect input shaft bushing in base of output shaft. If damaged, it may be removed by using a slide hammer. New bearing can be installed by using pilot end of input shaft as press tool.
13. Using dummy shaft, assemble needle bearings and spacer (20 rollers in each end) in one of the long planet pinions. Use petroleum jelly to aid in holding the rollers in position.
14. Position long planet gear, with dummy shaft centered in the pinion and with thrust washers at each end, in the planet carrier. Oil grooves on thrust washers must be toward the gears.
NOTE: long pinions are located opposite the closed portions of the carrier and short pinions are located in the openings.
15. Feed a second dummy shaft in from the top, picking up the upper thrust washer and the pinion and pushing the already installed dummy shaft out the lower end. As the first dummy is pushed down, be sure that it picks up the lower thrust washer.
16. Select the correct pinion shaft, as marked in Step 2, lubricate the shaft and install it from the top, pushing the assembling tools (dummy) ahead of it.
17. Turn the pinion shaft so that the slot or groove at the upper end faces the center of the assembly.
18. With a brass drift, drive the shaft in until the lower end is flush with the lower face of the planet carrier.

19. Following the same procedure as outlined in Steps 13 through 18, assemble and install a short planet pinion into the planet carrier adjacent to the long pinion now installed.
NOTE: the thrust washers, already installed with the long planet pinion, also serve for this short planet pinion, because the two pinions are paired together on one set of thrust washers.
20. Install the input sun gear thrust washer, input sun gear, and low sun gear needle thrust bearing.
21. Assemble and install the remaining planet pinions, in pairs, as previously explained.
22. Check end clearance of planet gears. This clearance should be .006-.030 in.
23. Place the shaft lock plate in position. Then, with the extended portions of the lock plate aligned with slots in the planet pinion shafts, rotate the lock plate clockwise until the three attaching screw holes are accessible.
24. Install lock plate attaching screws and torque to 2½ ft. lbs.

Governor

The governor assembly is a factory balanced unit. If body replacement is needed, the two sections must be replaced as a unit.

Disassembly

NOTE: the governor valve and shaft were removed in Step 6 of "Transmission Disassembly" procedures.

1. Remove the outer weight by sliding toward center of body.
2. Remove smaller inner weight retaining snap-ring and remove inner weight and spring.
3. Remove the four body assembly bolts and separate the body, hub and gasket. Remove the two seal rings.

Inspection

1. Clean all parts in solvent and air dry.

2. Check all parts. Replace all bent, scored, or otherwise damaged parts. Body and hub must be replaced as a unit.

Assembly

1. Reassemble governor weights and install into body bore. Replace seal rings on hub.
2. Slide hub into place on output shaft and lock into place with drive screw. Install gasket and governor body over output shaft, install governor shaft, line up properly with output shaft and install body attaching bolts. Torque bolts to 6-8 ft. lbs.
3. Check governor weight for free fit in body after the four attaching bolts are torqued. If the weight sticks or binds, loosen the bolts and retorque.

Valve Body

Removal

Remove valve body, as described under "Transmission Disassembly". If performing the operation on the car, the vacuum modulator valve, oil pan and gasket, guide detent plate and range selector detent roller spring must be removed in order to remove the valve from the transmission.

Disassembly

1. Remove manual valve, suction screen and gasket.
2. Remove cover bolts, then remove lower valve body and transfer plate from upper valve body. Discard gaskets.
3. Remove the front and rear pump check valves and springs from 1964-66 models.

4. From the upper valve body, remove the throttle valve and detent valve and the downshift timing valve as follows:
   A. Throttle Valve and Detent Valve—Remove the retaining pin by wedging a thin screwdriver between its head and the valve body, then removing the detent valve assembly and throttle spring. Tilt valve body to allow the throttle valve to fall out. If necessary, remove the C clip and disassemble the detent valve assembly.

   NOTE: do not change adjustment of hex nut on the detent valve assembly. This is a factory setting and should not normally be changed. However, some adjustment is possible if desired. See "Throttle Valve Adjustment," in later text.

   B. Downshift Timing Valve — Drive out the roll pin, remove valve spring and downshift timing valve.
5. From the lower valve body, remove the low-drive shift valve and the pressure regulator valve as follows:
   A. Low-Drive Shift Valve—Remove the snap-ring and tilt valve body to remove low-drive regulator valve sleeve and valve assembly, valve spring seat, valve springs and the shifter valve.
   B. Pressure Regulator Valve—Remove the snap-ring, then tilt valve body to remove the hydraulic modulator valve sleeve and valve, pressure regulator valve spring seat, spring, damper valve, spring seat and valve.

Inspection

1. Clean all parts in solvent; air dry. Use no rags.
2. Check all valves and valve bores for burrs or other deformities which could cause valve hang-up.

Assembly

NOTE: see the valve body illustration for identification of upper and lower valve body gaskets for 1964-66 models. For 1967-70 models, the gaskets are identical.

1. Replace valve components in proper bores, reversing the steps of disassembly outlined above.
2. Place front and rear pump check valves and springs into place in upper valve body on 1964-66 models only. Install the gasket and transfer plate.
3. Install lower valve body and gasket and install attaching bolts. Torque to 15 ft. lbs.
4. Install valve body onto transmission, as outlined under "Transmission, Assembly" in later text.

HUB OIL SEAL RINGS   HUB DRIVE SCREW   SNAP RING'S

HUB   GASKET   INNER WEIGHT   SPRING   OUTER WEIGHT

DAMPING SPRING BODY

DAMPING SPRING

SHAFT

VALVE

SNAP RING

BODY TO HUB SCREWS AND LOCK WASHERS

Governor—exploded view
(© G.M. Corp.)

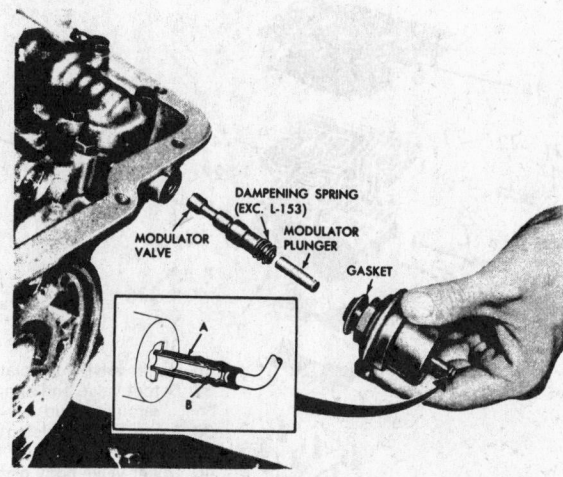

DAMPENING SPRING (EXC. L-153)

MODULATOR VALVE   MODULATOR PLUNGER   GASKET

A   B

Vacuum modulator, dampening spring, plunger, and valve
(© G.M. Corp.)

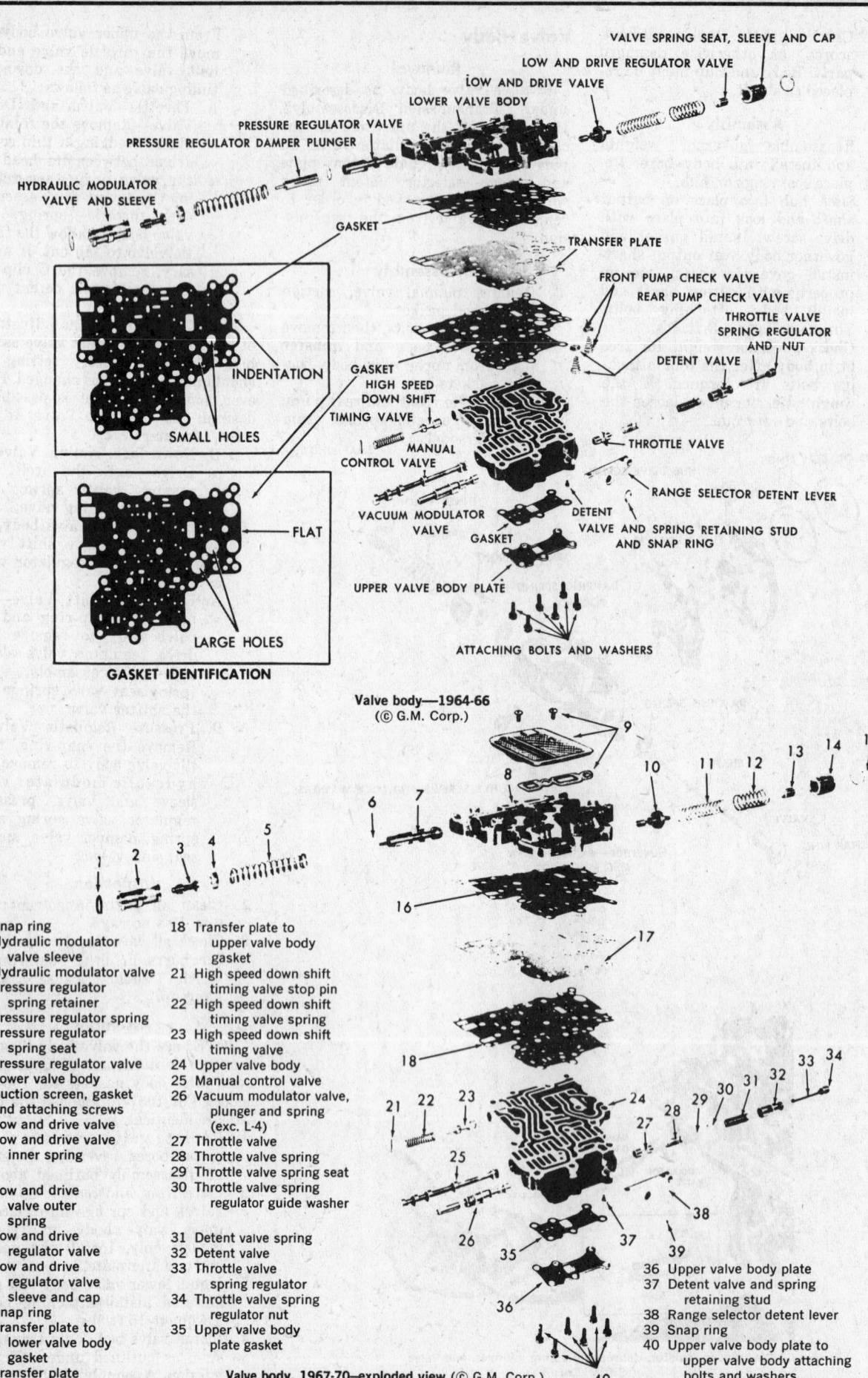

VALVE SPRING SEAT, SLEEVE AND CAP
LOW AND DRIVE REGULATOR VALVE
LOW AND DRIVE VALVE
LOWER VALVE BODY
PRESSURE REGULATOR VALVE
PRESSURE REGULATOR DAMPER PLUNGER
HYDRAULIC MODULATOR VALVE AND SLEEVE
GASKET
TRANSFER PLATE
FRONT PUMP CHECK VALVE
REAR PUMP CHECK VALVE
THROTTLE VALVE SPRING REGULATOR AND NUT
DETENT VALVE
INDENTATION
SMALL HOLES
GASKET
HIGH SPEED DOWN SHIFT TIMING VALVE
MANUAL CONTROL VALVE
THROTTLE VALVE
RANGE SELECTOR DETENT LEVER
FLAT
VACUUM MODULATOR VALVE
GASKET
DETENT
VALVE AND SPRING RETAINING STUD AND SNAP RING
LARGE HOLES
UPPER VALVE BODY PLATE
GASKET IDENTIFICATION
ATTACHING BOLTS AND WASHERS

Valve body—1964-66
(© G.M. Corp.)

1 Snap ring
2 Hydraulic modulator valve sleeve
3 Hydraulic modulator valve
4 Pressure regulator spring retainer
5 Pressure regulator spring
6 Pressure regulator spring seat
7 Pressure regulator valve
8 Lower valve body
9 Suction screen, gasket and attaching screws
10 Low and drive valve
11 Low and drive valve inner spring
12 Low and drive valve outer spring
13 Low and drive regulator valve
14 Low and drive regulator valve sleeve and cap
15 Snap ring
16 Transfer plate to lower valve body gasket
17 Transfer plate

18 Transfer plate to upper valve body gasket
21 High speed down shift timing valve stop pin
22 High speed down shift timing valve spring
23 High speed down shift timing valve
24 Upper valve body
25 Manual control valve
26 Vacuum modulator valve, plunger and spring (exc. L-4)
27 Throttle valve
28 Throttle valve spring
29 Throttle valve spring seat
30 Throttle valve spring regulator guide washer
31 Detent valve spring
32 Detent valve
33 Throttle valve spring regulator
34 Throttle valve spring regulator nut
35 Upper valve body plate gasket

36 Upper valve body plate
37 Detent valve and spring retaining stud
38 Range selector detent lever
39 Snap ring
40 Upper valve body plate to upper valve body attaching bolts and washers

Valve body, 1967-70—exploded view (© G.M. Corp.)

Vacuum Modulator

The vacuum modulator is mounted on the left rear of the transmission and can be serviced from beneath the car.

Removal

1. Remove vacuum line at the modulator.
2. Unscrew the modulator from the transmission with a thin 1 in. tappet-type wrench.
3. Remove vacuum modulator valve.

Inspection

1. Check the vacuum modulator plunger and valve for nicks and burrs. If such damage cannot be repaired with a stone, replace the part.
2. Check the vacuum modulator for leakage with a vacuum source. If the modulator leaks, replace the assembly.

Installation

Reverse removal procedure.

Transmission Assembly

NOTE: if removed, assemble manual linkage to case, as described in Steps 1-7.

1. Install parking lock pawl and shaft and insert a new E-ring retainer.
2. Install parking lock pawl pullback spring over its boss at rear of pawl. The short leg of the spring should locate in the hole in the pawl.
3. Install parking pawl reaction bracket with its two bolts.
4. Fit the actuator assembly between the parking pawl and the bracket.
5. Insert outer shift lever into the case. Pick up inner shift lever and parking lock assembly. Tighten Allen-head lock.
6. Insert outer throttle valve lever and shaft, special washer, and O-ring into case and pick up inner throttle valve lever. Tighten Allen-head lock.
7. Thread low band adjusting screw into case.

NOTE: to prevent possible binding between throttle lever and range selector controls, allow .010-.020 clearance between inner throttle valve lever and inner shift lever.

Transmission Internal Components

8. Install inner and outer rear piston seals onto reverse piston and, (with lubrication) install piston into the case.
9. With transmission case facing up, install the 17 reverse piston springs and their retainer ring.
10. Install spring compressing tool. Compress the return springs, allowing the retaining ring snap-

Installing reverse piston
(© G.M. Corp.)

ring to be installed. Remove the compressor.
11. Install the cushion spring.
12. Lubricate and install reverse clutch pack, beginning with a reaction spacer plate and alternating with the faced plates until all plates are installed.

NOTE: the number and sequence of plates varies with the power and torque requirements of the car model involved.

The notched lug on each reaction plate is installed in the groove at the seven o'clock position in the case. Then, install the thick pressure plate which has a dimple in one lug to align with the same slot in the case as the notched lugs on the other reaction plates.

13. Install clutch plate retainer ring.
14. Turn rear of transmission case down.
15. Align the internal lands and grooves of the reverse clutch pack faced plates, then engage the reverse ring gear with these plates. This engagement must be made by feel while turning the ring gear.
16. Place output shaft thrust bearing over the output shaft and install the planetary carrier and output shaft into the transmission case.
17. Move transmission to horizontal position.

Installing gearset
(© G.M. Corp.)

18. The two input shaft seal rings should be in place on the shaft. Install clutch drum (machined face first) onto the input shaft and install the low sun gear bushing against shoulder.
19. Install clutch drum and input shaft assembly into case, aligning thrust washer on input shaft and indexing low sun gear with the short pinions on the planet carrier.
20. Remove rubber seal ring from the front pump body and install front pump, gasket and selective fit thrust washer into case. Install pump-to-case bolts.
21. To check for correct thickness of the selective fit thrust washer, move transmission so that output shaft points down and proceed as follows:

Checking end-play for proper thrust washer selection
(© G.M. Corp.)

A. Mount a dial indicator so that the indicator plunger is resting on the end of the input shaft. Zero the indicator.
B. Push up on the output shaft and watch the total dial movement.
C. The indicator should read .028-.059 in. If reading is not within specifications, remove front pump, change to a thicker or thinner selective thrust washer. Repeat above checking procedure.

NOTE: washers are available in thicknesses of .061, .078, .092 in. and .106 in.

22. Install servo piston, piston ring, and spring into the servo bore. Then, using a new gasket and O-ring, install the servo cover.
23. Remove front pump and selective fit washer from the case, and install the low brake band, anchor and apply struts into the case. Tighten the low band adjusting screw enough to prevent struts from falling out of case.
24. Place the seal ring in the groove around front pump body and the two seal rings on the pump cover extension. Install the pump, gasket and thrust washer into the case. Install all pump bolts. Torque bolts to 15 ft. lbs.

Extension, Governor and Rear Oil Pump—1964-66 Models

25. Turn transmission so that output shaft points upward. Install rear pump wear plate, drive pin, and drive gear, indexing gear to drive pin.
26. Install rear pump body and driven gear, drain back baffle and pump-to-case attaching bolts.
27. Install governor over output shaft. Install governor shaft and valve, two Belleville washers, and retaining C-clips. Center shaft in output shaft bore and tighten governor drive screw.

28. Install speedometer drive gear onto output shaft.
29. Place extension seal ring over rear pump body and install transmission extension and five retaining bolts.
30. If removed, replace speedometer driven gear.

Extension, Governor and Governor Support—1967-70 Models

25. Turn transmission so that output shaft points upward.
26. Install governor support and gasket, drain back baffle, and support to case attaching bolts.

27. Install governor over output shaft. Install governor shaft and valve, urethane washer, and retaining C clips. Center shaft in output shaft bore and tighten governor hub drive screw.
28. Install speedometer gear to output shaft.
29. Place extension seal ring over governor support. Install transmission extension and five retaining bolts.
30. Replace speedometer driven gear.

Checking drive gear to pump body clearance
(© G.M. Corp.)

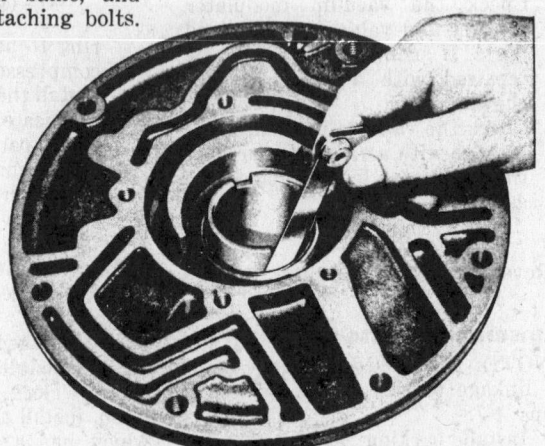

Checking pump body bushing to converter pump hub clearance
(© G.M. Corp.)

Checking gear end-play
(© G.M. Corp.)

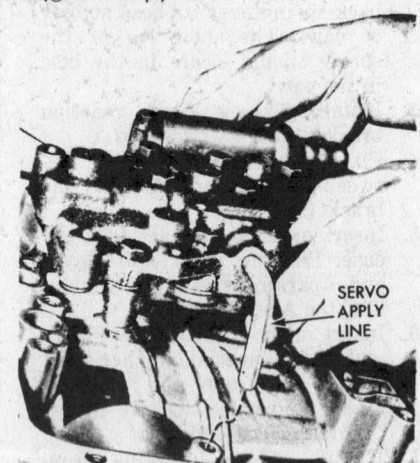

SERVO APPLY LINE

Installing valve body
(© G.M. Corp.)

Checking driven gear to crescent clearance
(© G.M. Corp.)

Installing detent guide plate
(© G.M. Corp.)

Oil Pan and Valve Body

31. With transmission upside down, the manual linkage and the selector lever detent roller installed, install the valve body with a new gasket. (Carefully guide the servo apply line into its boss in the case as the valve body is set in place.) Position the manual valve actuating lever fully forward to more easily pick up the manual valve. Install six mounting bolts and the range selector detent roller spring. On 1967-70 models, install new gasket and suction screen to valve body.
32. Install the guide plate. Install attaching bolts.
33. Install vacuum modulator valve, the vacuum modulator and the gasket.
34. Install oil pan, using a new gasket, then the pan attaching bolts.
35. Install the converter and safety holding strap.

Low Band Adjustment

Tighten the low servo adjusting screw to 40 in. lbs. for 1964-66 models; 70 in. lbs. for 1967-70 models.

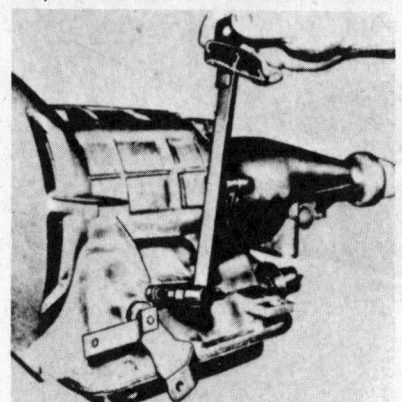

Low band adjustment
(© G.M. Corp.)

The input and output shaft must be rotated simultaneously to properly center the low band on the clutch drum. Then, back off four complete turns for a band which has been in use for 6,000 miles or more, or three turns for one in use less than 6,000 miles, and tighten the locknut.

Caution The amount of back-off is very critical. Back off exactly three or four turns.

Throttle Valve TV Adjustment

No provision is made for checking TV pressures. However, if operation of the transmission is such that some adjustment of the TV is indicated, pressures may be raised or lowered by adjusting the position of the jam nut on the throttle valve assembly. To raise TV pressure 3 psi, back off the jam nut one full turn.

Conversely, tightening the jam nut one full turn, lowers TV pressure 3 psi. A difference of 3 psi in TV pressure will cause a change of about 2-

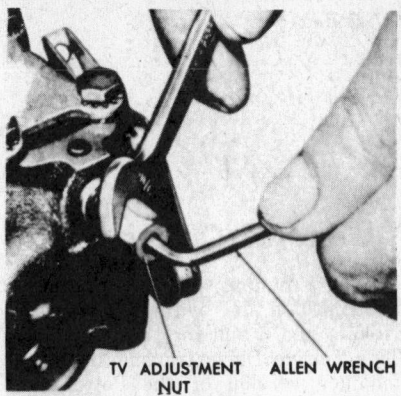

TV ADJUSTMENT NUT   ALLEN WRENCH
TV adjustment
(© G.M. Corp.)

3 mph in the wide open throttle upshift point. The end of the TV adjusting screw has an Allen head so the screw may be held stationary while the jam nut is locked.
NOTE: use care in changing this adjustment, as no pressure tap is provided to check TV pressure.

Torque References

	Foot-Pounds	
	1964-66 Models	1967-70 Models
Transmission case to engine	25-30	35
Oil pan to case	7-10	7-10
Extension to case	20-30	20-30
Servo cover to case	15-20	20
Front pump to case	13-17	13-17
Front pump cover to body attaching bolts	15-20	20
Pinion shaft lock plate attaching screws	2-3	2-3
Governor body to hub	6-8	6-8
Governor hub drive screw	6-8	6-8
Rear pump or governor support to case	8-11	8-11
Valve body to case	8-11	15
Suction screen attaching screws	2-3	2-3
Upper valve body plate	3-5	5
Lower to upper valve body	13-15	15
Inner control lever allen head locks	6-8	2½
Low band adjusting locknut	13-17	13-17
Converter to engine	15-20	35

Torque Drive

Application
Camaro with 6 Cylinder Engine, 1969-70
Nova with 4 or 6 Cylinder Engine, 1969-70

General Description

A new manual shift Torque Drive transmission is available for the Nova 4 and 6 and for the Camaro 6. This Torque Drive transmission is a modified version of the Powerglide transmission, with the automatic shifting aspects removed.

The new transmission has five con-

trol positions: Park, Reverse, Neutral, High, and First. Forward motion is accomplished by placing the selector in first position. After gradual acceleration to about 20 mph, the selector should be manually shifted into high. Operation of the transmission in high range at car speeds less than 20 mph is not recommended,

because this may lead to overheating of the transmission oil.

Operation of the car in first at speeds above 55 mph will cause undue strain on engine components, due to the high engine rpm associated with these speeds.

The selector lever may be moved freely between High and First, but

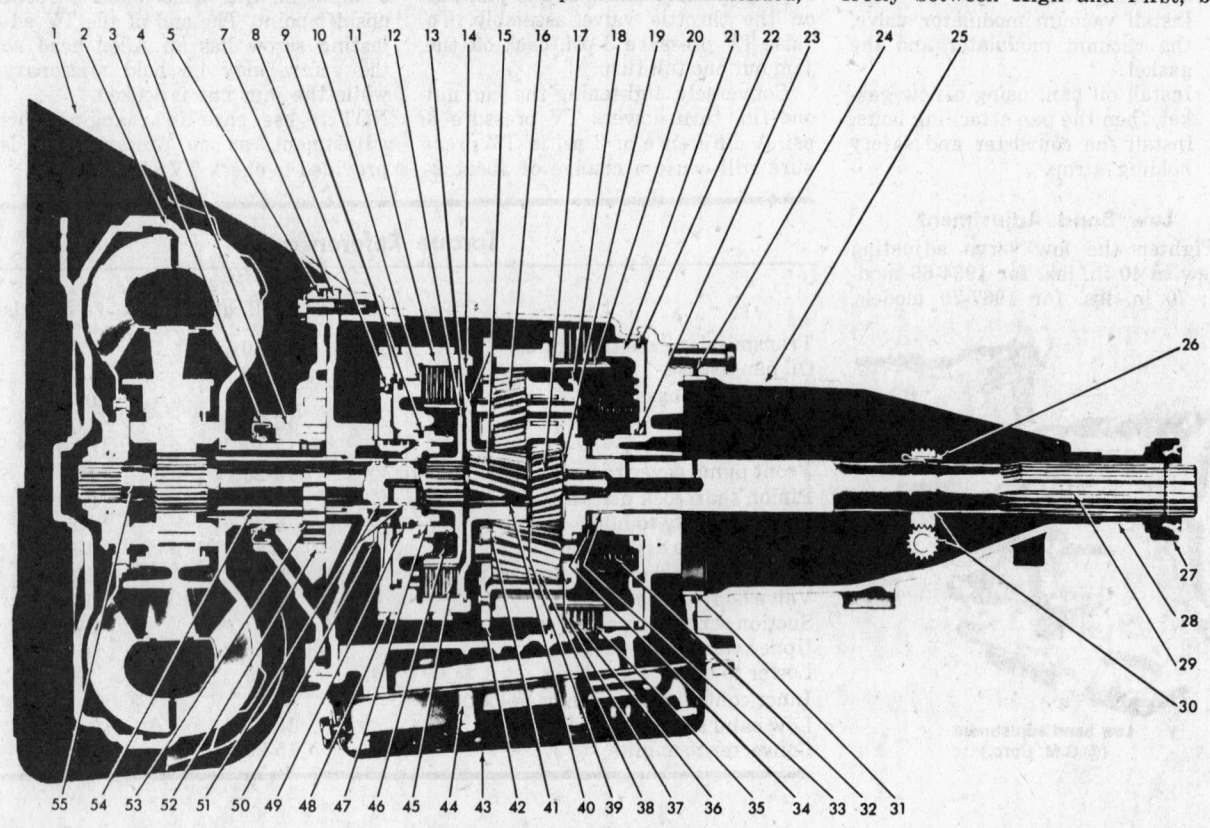

Torque drive transmission—sectional view

1 Transmission case
2 Welded converter
3 Oil pump seal assembly
4 Oil pump body
5 Oil pump body square ring seal
6 Not given
7 Oil pump cover
8 Clutch relief valve ball
9 Clutch piston inner and outer seal
10 Clutch piston
11 Clutch drum
12 Clutch hub
13 Clutch hub thrust washer
14 Clutch flange retainer ring
15 Low sun gear and clutch flange assembly
16 Planet short pinion
17 Planet input sun gear
18 Planet carrier
19 Planet input sun gear thrust washer
20 Ring gear

21 Reverse piston
22 Reverse piston outer seal
23 Reverse piston inner seal
24 Extension seal ring
25 Extension
26 Speedometer driven gear
27 Extension rear oil seal
28 Extension rear bushing
29 Output shaft
30 Speedometer drive and driven gear
31 Pilot ring
32 Reverse piston return springs, retainer and retainer ring
33 Transmission rear case bushing
34 Output shaft thrust bearing
35 Reverse clutch cushion spring (waved)
36 Reverse clutch pack
37 Pinion thrust washer
38 Planet long pinion

39 Low sun gear needle thrust bearing
40 Low sun gear bushing (splined)
41 Pinion thrust washer
42 Parking lock gear
43 Transmission oil pan
44 Valve body
45 High clutch pack
46 Clutch piston return spring, retainer and retainer ring
47 Clutch drum bushing
48 Low brake band
49 High clutch seal rings
50 Clutch drum thrust washer (selective)
51 Turbine shaft seal rings
52 Oil pump driven gear
53 Oil pump drive gear
54 Stator shaft
55 Input shaft

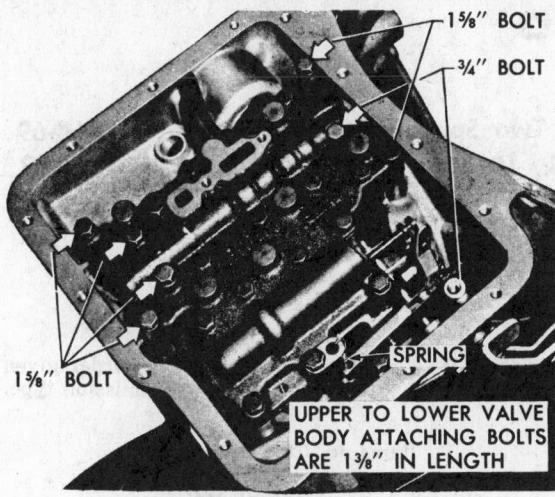

1⅝" BOLT
¾" BOLT
1⅝" BOLT
SPRING
UPPER TO LOWER VALVE BODY ATTACHING BOLTS ARE 1⅜" IN LENGTH

Valve body removal

J—9542

Removing rear piston spring retainer snap-ring

Suggested pinion shaft markings

Checking planet gear end-play

must be raised in order to shift from High to Neutral or Reverse. The lever must also be raised in order to shift into, or out of, Park.

Service Procedures

The Torque Drive transmission is a Powerglide transmission with the automatic shifting provisions omitted. Lubrication, maintenance and service information is the same as that for Powerglide. Some internal differences are as follows:

Transmission Case

The Torque Drive transmission does not use a vacuum modulator or governor. Therefore, the Powerglide case has been modified to remove the vacuum modulator tapped hole and the governor feed passage. Machining of the case, in the governor support area, has also been eliminated, since the governor and governor support are not used. Due to the elimination of the governor support, which also

performed the function of extension-to-case alignment, a pilot ring has been added to perform this function.

Park Lock and Range Selector Lever

The Torque Drive transmission does not use a throttle valve, therefore, the park lock and range selector lever is different from that used in the Powerglide, in that it does not incorporate the throttle valve lever hole.

Low Servo

The Torque Drive transmission uses higher rated return and apply springs and a longer piston rod to achieve the shifting characteristics needed for manual shift application.

Front Pump

The front pump has a new down-shift timing valve, stator shaft, tand pump cover. The stator shaft and cover incorporate larger orifices for

lubrication and converter feed. The pump body is the same as Powerglide.

Input Shaft

The input shaft is new only to the extent that it has larger feed orifices.

Carrier Assembly

The Torque Drive carrier is different from Powerglide in that the output shaft has a lube orifice extending to the former Powerglide rear pump drive pin hole. Oil is agitated at this point to lubricate the extension housing bushing and the speedometer gears.

Valve Body

All valves except the pressure regulator, hydraulic modulator, and manual shift have been eliminated from the valve body. The valve body has the raised letters T. D. in the center of the casting of the upper and lower portions.

Type 300

Application

Jetaway—Oldsmobile, 1964-69

Olds F-85, 1964-69

Super Turbine 300—Buick, 1964-69

Buick Special, 1964-69

Two Speed Automatic—Tempest, 1964-69

Firebird, 1967-69

Diagnosis

This diagnosis guide covers the most common symptoms and is an aid to careful diagnosis. The items to check are listed in the sequence to be followed for quickest results. Thus, follow the checks in the order given for the particular transmission type.

TROUBLE SYMPTOMS*	In Car	Out of Car
No drive in any selector position	ABCDE	a
Erratic operation and slippage—light acceleration	FAG	
Slippage or flare coasting to stop or cornering	HFIGJ	
Sluggish standing start	K	b
No reverse	L	cd
Slips in any range	FAGLA	
Harsh N to drive shift at idle	GH	
No upshift	MGNL	
Long shift time—not positive engagement	GL	e
Engine flares on upshift	L	fae
Late upshift	GLMNOP	
Erratic up- or down-shifts	NFAGQ	
No wide open throttle downshift	MQP	
Engine flares on wide open throttle downshift	IBGLPA	
Delayed engagement in manual low	G	
No stator action	KQP	gh
Oil surges out breather	AR	ij
Drive clutch plates burned—low band and reverse clutch ok	L	fdae
Drive clutch and reverse clutch burned, low band burned	GLA	

* Items referring to Stator Control refer to Variable Pitch Converter Models only.

Key to Checks

A. Oil level
B. Make pressure check
C. Linkage adjustment (external)
D. Linkage adjustment (internal)
E. Pressure regulator
F. Clogged filter or suction pipe leaks
G. Modulator and/or line

H. Engine idle
I. Low band adjustment
J. Low servo
K. Stator switch and/or valve
L. Valves, body and/or leaks
M. Detent solenoid or valve
N. Governor

O. Governor roll pin
P. Valve body bolt torque
Q. Crossed solenoid wires
R. Cooler or line leaks
a. Front pump
b. Stator valve or solenoid
c. Reverse clutch and/or seals

d. Reverse clutch feed
e. Front clutch and/or seals
f. Front clutch check ball
g. Reactor shaft bushing
h. Input shaft seals
i. Front pump line leaks
j. Front pump cover leaks

General Information

The illustration is typical. While design and repair procedures are similar, components vary with car application and should not be considered interchangeable.

This transmission consists of a torque converter and a two-speed planetary transmission.

A variable pitch stator is used in some 1964-65 models. These models use a digerent valve body design.

In either case, the torque converter is not serviceable, and must be replaced as a unit.

Minor Maintenance

Fluid Checking

The transmission must be at normal operating temperature before the fluid level is checked. Normal operating temperature is reached after about fifteen miles of highway driving. Check the fluid level with the vehicle level, the engine idling, and the transmission in Park. The fluid level on the dipstick should be at the full mark.

Capacity

See the Capacities table of the car section for quantity of fluid required to refill transmission after overhaul. The amount given is for both transmission and torque converter. To avoid overfilling and resultant damage, check the fluid level after adding about two-thirds of the required amount. To check fluid level after overhaul:

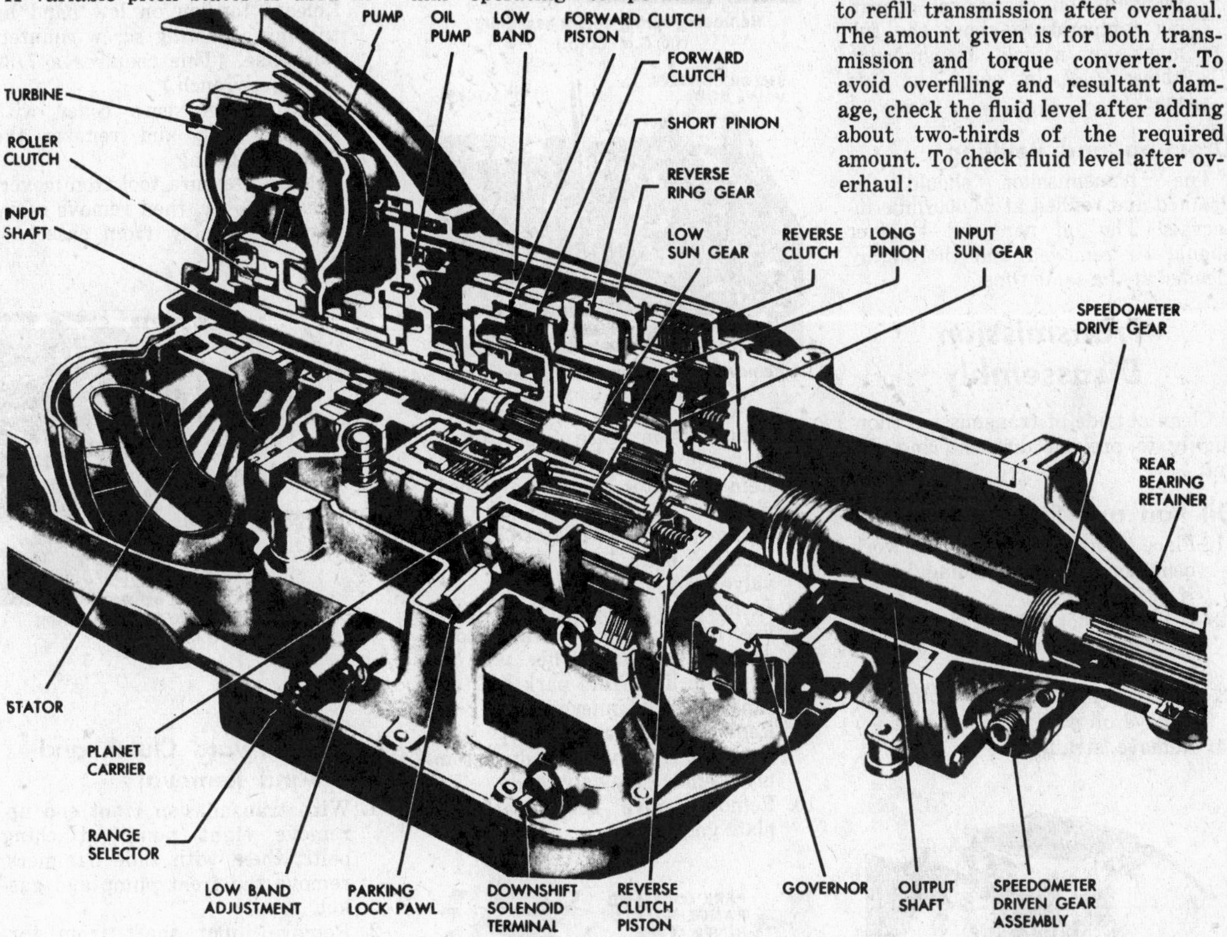

Cross-section of Type 300 transmission
(© G.M. Corp.)

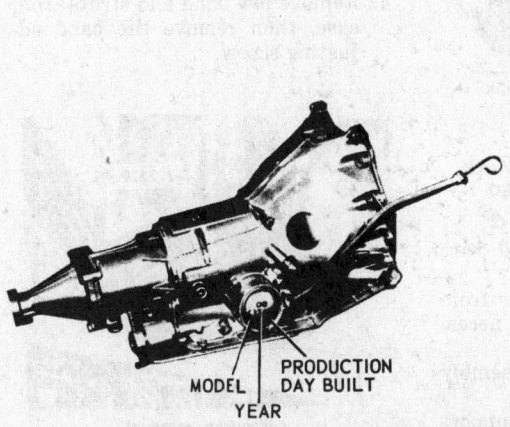

Location of serial and model number (© G.M. Corp.)

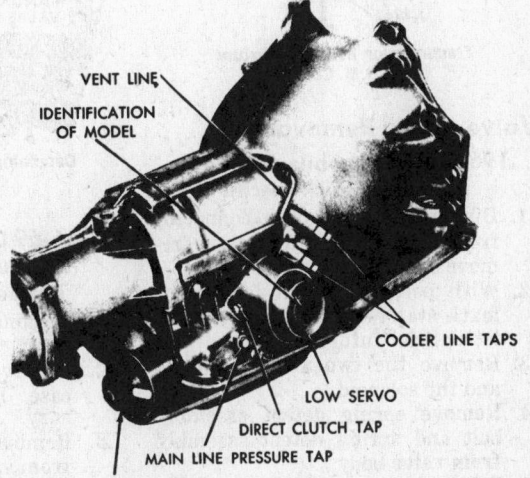

External features of Type 300 (© G.M. Corp.)

1. Start engine, operate shift lever through all ranges slowly. Do not race engine.
2. Place shift lever in Park with engine idling.
3. Fluid level should be no higher than 1/2-3/4 in. below Add mark on dipstick. Correct level is determined by the relation of the fluid level to the mark on the dipstick, rather than by the amount added. Do not fill a transmission at normal operating temperature above the full mark, or a cold transmission above 1/2-3/4 in. below the Add mark.

Draining and Refilling

The transmission should be drained and refilled at 24,000 mile intervals. The oil pan and strainer should be removed and thoroughly cleaned at the same time.

Transmission Disassembly

Clean outside of transmission thoroughly to prevent dirt entering the unit.

Oil Pan and Strainer Removal

1. Place transmission on clean work bench or in special holding fixture if available.
2. If oil has not been drained, drain oil and pull converter from the case.
3. Remove 14 attaching bolts and remove oil pan.
4. Remove strainer.

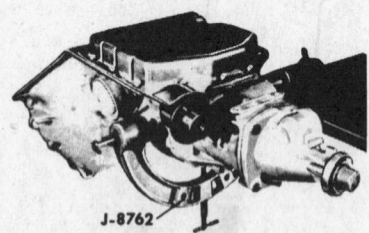

Transmission in holding fixture
(© G.M. Corp.)

Valve Body Removal

1964-65 Oldsmobile and F-85, Buick and Buick Special

1. Disconnect solenoid connector from solenoid switch, then remove switch from case.
2. With permanent marking agent, mark stator control solenoid with letter S for future reference.
3. Remove the two attaching bolts and the solenoid.
4. Remove spring detent assembly bolt and spring detent assembly from valve body.
5. Remove seven bolts holding the stator control valve body to the

Removing spring detent assembly
(© G.M. Corp.)

Removing stator control valve body bolts
(© G.M. Corp.)

transmission case and remove stator control valve body.
6. Remove stator control valve body plate.
7. Remove 11 valve body-to-case attaching bolts, (Do not remove valve body).
8. Remove manual control valve link by rotating the valve body, counterclockwise. This will detach the link from park lock and range selector inner valve.
9. Remove manual control valve and link from valve body assembly. Remove the valve body.
10. Remove valve body plate and plate gasket.

Detaching manual control valve link
(© G.M. Corp.)

1966-69 Oldsmobile, F-85, Buick, and Buick Special, 1964-69 Tempest, 1967-69 Firebird

1. Disconnect detent solenoid wire from case connector.
2. Remove case connector from case. Replace O ring as necessary.
3. Remove spring detent assembly from valve body.
4. Remove oil channel support plate.

5. Remove eleven body to case bolts. Do not remove valve body.
6. Remove manual control valve link from inner park lock and range selector lever.
7. Remove manual control valve and link from valve body assembly. Remove valve body.
8. Remove valve body plate and gasket.

Low Servo Piston Removal

1. Release tension on low band by turning adjusting screw counterclockwise. (This requires a 7/32 in Allen wrench.)
2. Press in on servo cover with suitable tool, and remove the cover snap-ring.
3. Release pressure tool from cover, remove cover, then remove servo piston assembly from case.

Removing low servo cover snap-ring
(© G.M. Corp.)

Pump, Forward Clutch and Low Band Removal

1. With transmission front end up, remove eight pump attaching bolts, then, with slide hammers, remove the front pump and gasket.
2. Remove input shaft from forward clutch drum.
3. Remove forward clutch assembly by lifting straight out of case. (Be sure low band has first been released.)
4. Remove low band and struts from case, then remove the band adjusting screw.

Oil pump removal
(© G.M. Corp.)

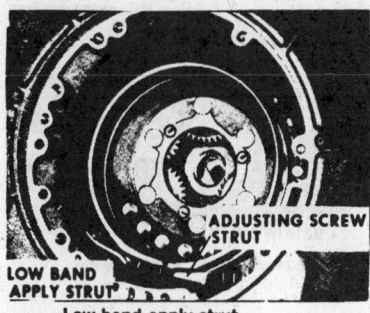

Low band apply strut
(© G.M. Corp.)

Removing low band adjusting screw
(© G.M. Corp.)

Rear Bearing Retainer and Component Removal

1. With transmission horizontal, remove speedometer driven gear sleeve retainer with ½ in. wrench. Remove sleeve and gear.
2. Remove four attaching bolts and remove rear bearing retainer and retainer to case oil seal.
3. Pry out retainer rear oil seal.

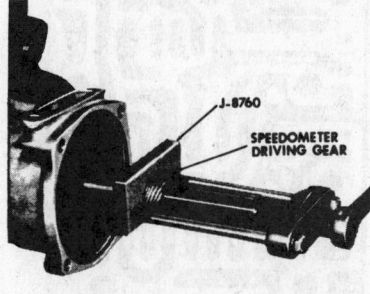

Removing speedometer driven gear sleeve retainer
(© G.M. Corp.)

Pulling off speedometer drive gear
(© G.M. Corp.)

4. Remove rear bearing retainer bushing by collapsing the bushing with a screwdriver.
5. On 1964-65 models, with transmission in Park, pull speedometer drive gear with suitable puller. On 1966-69 models, depress the retainer clip and slide the drive gear off the output shaft.

Governor and Vacuum Modulator Removal

1. Remove three attaching bolts, then remove governor cover and gasket.
2. With a twisting motion, slide the governor assembly out of its bore.
3. With a ½ in. socket, remove vacuum modulator retainer bolt. Then, remove the retainer and the vacuum modulator valve assembly.

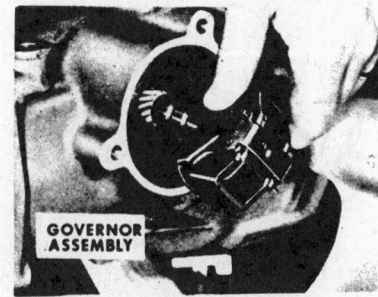

Removing governor assembly
(© G.M. Corp.)

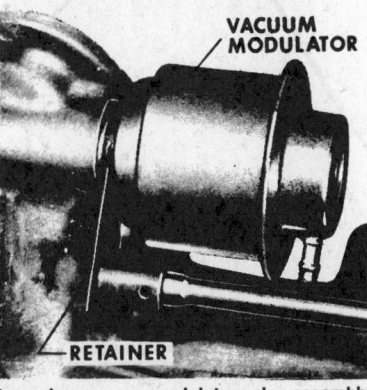

Removing vacuum modulator valve assembly
(© G.M. Corp.)

Planetary Gear Set, Reverse Clutch and Parking Lock Mechanism Removal

1. Carefully remove planet carrier assembly.
2. Lift reverse ring gear from case.
3. Remove needle thrust bearing and two races from rear of planet carrier.
4. With transmission in vertical position, remove reverse clutch pack snap-ring, then lift reverse clutch pressure plate from transmission case.
5. Remove reverse clutch pack, then the cushion spring, if installed.
6. To remove reverse piston, install

spring compressor onto reverse piston eturn seat Install flat plate over threaded shaft at rear of case. Apply pressure to compressor, then remove piston snap-ring.
7. Remove compression tool and lift off piston seat and 17 return springs.
8. With transmission in horizontal position, remove reverse clutch piston with compressed air applied to valve body.
9. Remove two parking lock bracket bolts with ½ in. socket, then remove the lock bracket.
10. Remove range selector shaft retainer.
11. With a 9/16 in. wrench, loosen the nut that holds the outer range selector lever to inner park lock and range selector lever.
12. Slide outer range selector lever out of case. Remove nut, inner park lock, and range selector lever.

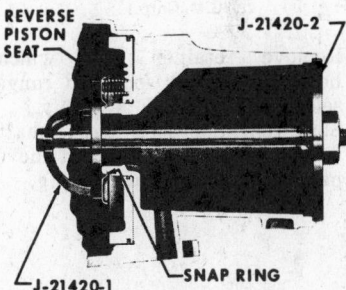

Details of spring compressor
(© G.M. Corp.)

Location of reverse piston return springs
(© G.M. Corp.)

Reverse piston air passage
(© G.M. Corp.)

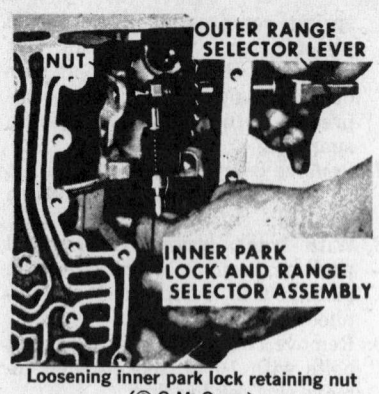

Loosening inner park lock retaining nut
(© G.M. Corp.)

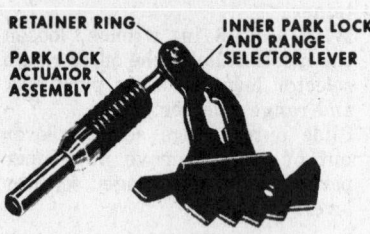

Park lock actuator assembly
(© G.M. Corp.)

Component Disassembly and Assembly

Valve Body Disassembly

1. Remove two bolts which hold the stator, where installed, and detent solenoid valves. Remove solenoid valve, gasket, spring, stator and detent valve. (Note cutout notch on solenoid valve gasket on models with stator solenoid.)
2. Depress shift control valve sleeve and remove retaining pin. Remove shift control valve sleeve, shift control valve, spring, and shift valve.
3. Depress modulator limit spring. Invert valve body so retaining

13. Remove retainer ring which holds inner park lock and range selector to park lock assembly.
14. Slide parking lock pawl shaft from parking lock pawl. Remove parking lock pawl and spring.

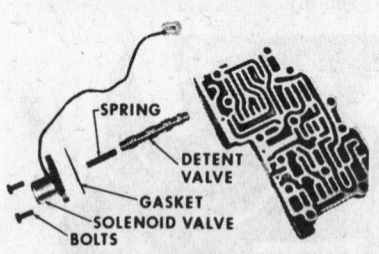

Removing solenoid valve
(© G.M. Corp.)

Removing shift control valve
(© G.M. Corp.)

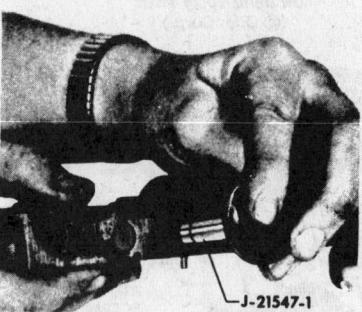

Depressing modulator limit spring
(© G.M. Corp.)

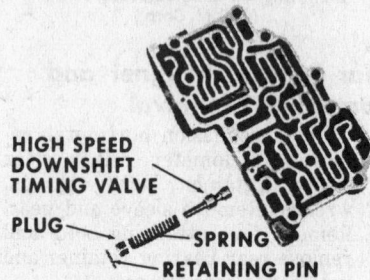

Removing high speed downshift timing valve
(© G.M. Corp.)

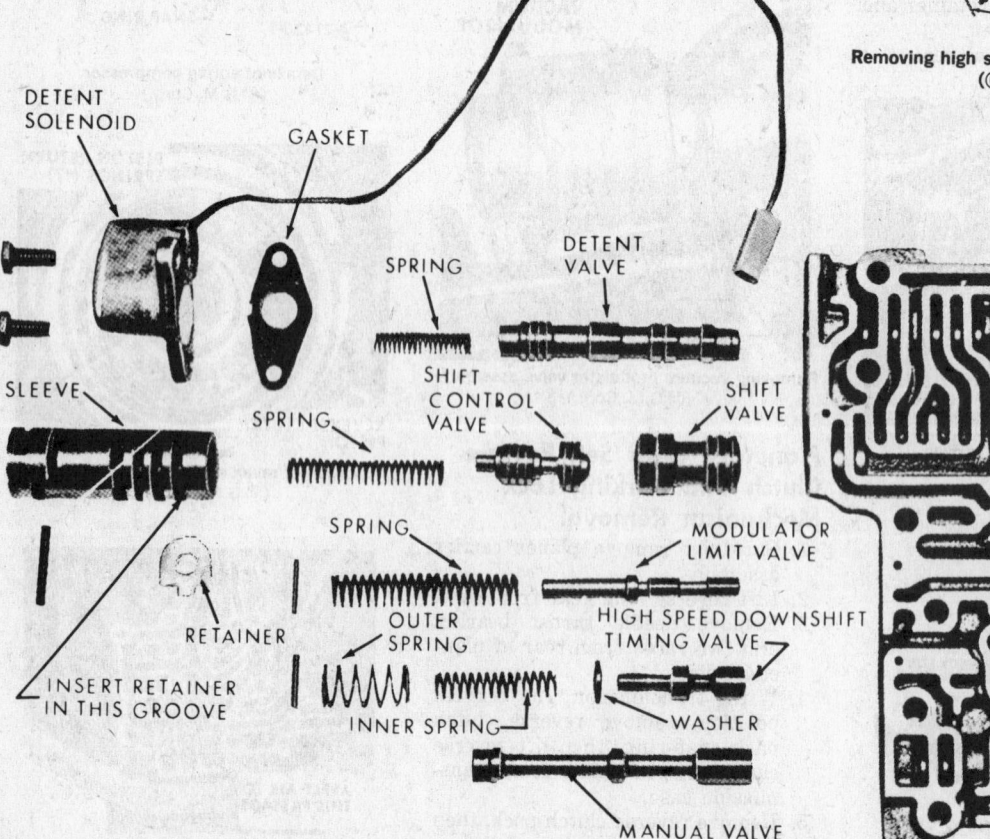

Valve assemblies, 1966-69 models (© G.M. Corp)

pin will fall free. Remove spring and valve from body.

4. Depress high speed downshift timing valve plug and remove pin by inverting valve body. Remove valve assembly. There may be one or two springs in the assembly. Some assemblies have no plug.

5. Clean all valves and valve body in solvent and blow dry with air. Inspect and test each valve in its bore. All valves must move freely of their own weight.

Valve Body Assembly

1. Install high speed downshift timing valve assembly. Install retaining pin.

2. Install modulator limit valve and spring into bore of valve body. Compress spring and install retaining pin.

3. Install shift valve assembly. Install retaining pin.

4. Install detent valve and spring. Install gasket to solenoid, with notch facing bottom of valve body. Install solenoid to valve body with two 1/4 in. bolts.

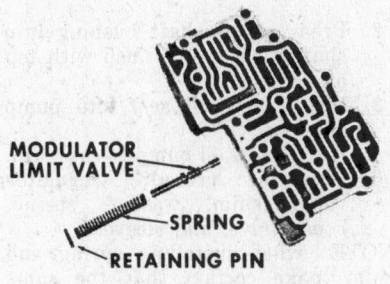

Installing modulator limit valve
(© G.M. Corp.)

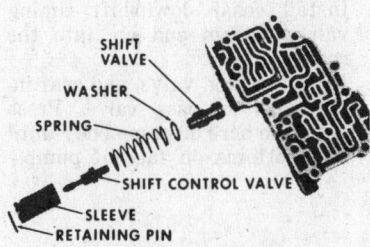

Installing shift valve assemblies
(© G.M. Corp.)

Stator Control Valve Body Disassembly

1964-65 Buick and Buick Special, Oldsmobile and F-85

1. Compress stator control valve plug. Invert valve body to release retaining pin. Remove plug, spring and valve from body.

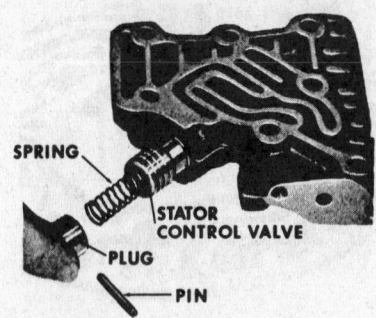

Removing stator control valve
(© G.M. Corp.)

Stator Control Valve Body Assembly

1964-65 Buick and Buick Special, Oldsmobile and F-85

1. Install stator control valve, spring, and plug into bore of valve body. Compress plug and install retaining pin.

Low Servo Disassembly

1964-65

1. Remove low servo piston seal.

2. Compress low servo piston. Exercise extreme caution in compress-

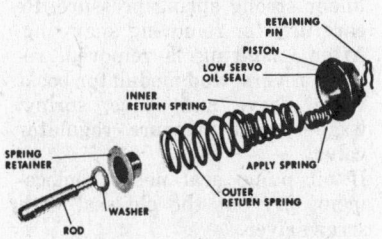

Low servo piston assembly, 1964-65
(© G.M. Corp.)

Compressing low servo piston
(© G.M. Corp.)

ing this piston so as to avoid damage.

3. Use pressure application in an arbor press.

4. Compress low servo enough to relieve pressure on retaining ring. Remove retaining ring and gradually relieve tension on the servo assembly.

5. With tension relieved, remove piston low servo, inner apply spring, the outer return springs, spring retainer, washer, and piston apply rod.

1966-69

1. Remove secondary piston seal.

2. Compress primary piston in arbor press and remove retaining snap ring.

3. Remove primary piston, cushion spring, return spring, apply pin, and washer.

4. Inspect primary piston lip seal. If nicked, torn, or worn, replace piston.

Low Servo Assembly

1964-65

1. Assemble inner and outer return springs into the servo. Install spring retainer.

2. Install this assembly into the ram press with compression tools.

3. Install piston apply rod and washer through hole in press plate, compress assembly, then install retainer pin.

4. Install low servo piston seal.

1966-69

1. Install washer into secondary piston assembly.

2. Install cushion and return springs into secondary piston assembly.

3. Install primary piston. Compress piston. Install new retaining snap ring.

Oil Pump Disassembly

1. Remove the two hook-type oil sealing rings (metal) from the pump hub.

2. Remove pump cover to forward clutch drum thrust washer.

3. Remove oil pump to case seal and discard.

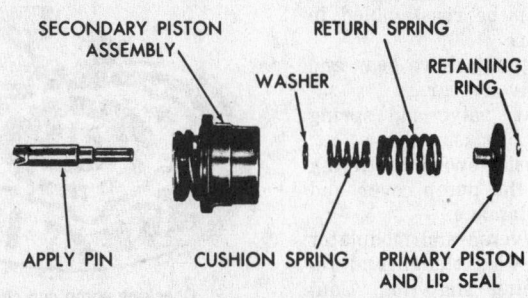

Low servo piston assembly, 1966-69 (© G.M. Corp)

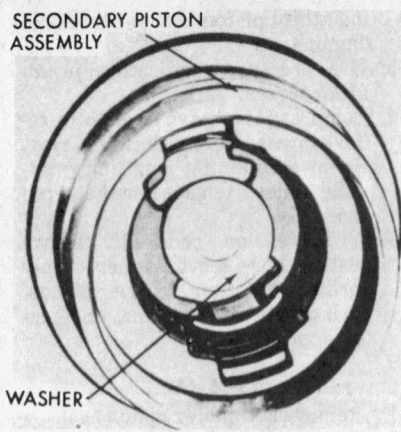

SECONDARY PISTON ASSEMBLY

WASHER

Installing washer into secondary piston
(© G.M. Corp)

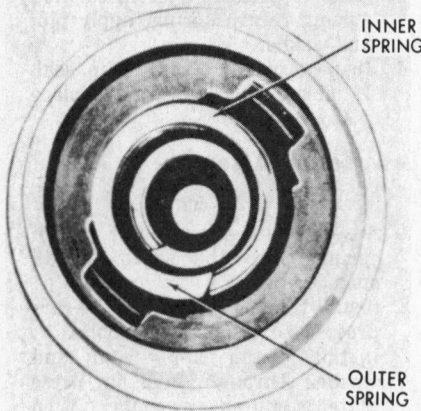

INNER SPRING

OUTER SPRING

Installing cushion and return springs into secondary piston
(© G.M. Corp.)

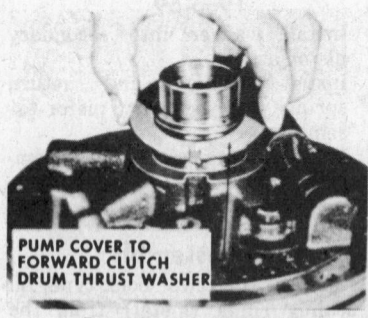

PUMP COVER TO FORWARD CLUTCH DRUM THRUST WASHER

Removing thrust washer
(© G.M. Corp.)

4. Remove five pump cover bolts with a ½ in. socket. Remove pump cover.
5. Mark, but do not scar, gear faces, so gears can be reassembled in same manner.
6. Remove oil pump drive gear and oil pump driven gear.
7. Remove seat, valve and spring from cooler by-pass valve.
8. Remove coast downshift timing valve from the pump cover and inspect for damage.
9. Compress reverse and modulator boost valve with thumb and remove retaining snap-ring. Caution, the boost valve sleeve is

J-2619
J-21361

Coast downshift timing valve
(© G.M. Corp.)

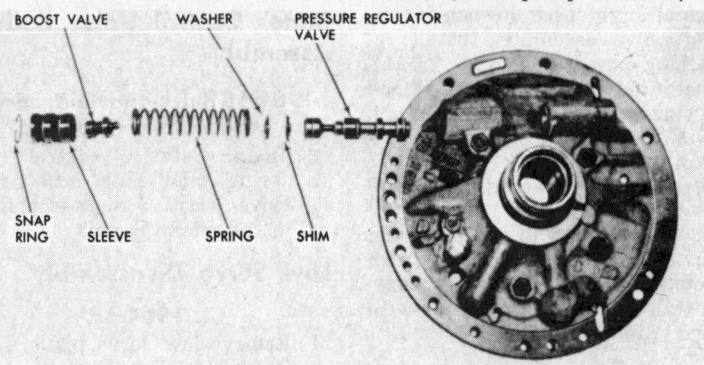

BOOST VALVE   WASHER   PRESSURE REGULATOR VALVE

SNAP RING   SLEEVE   SPRING   SHIM

Pressure regulator valve (© G.M. Corp.)

under strong spring pressure. Be careful after removing snap-ring.
10. After snap-ring is removed, remove reverse and modulator boost valve sleeve and valve, spring, washer, and pressure regulator valve.
11. If oil pump seal needs replacement, pry out the old seal with screwdriver.
12. If oil pump bushing needs replacement, press bushing out of pump body. If stator shaft bushing needs replacement, drive out bushing.

Oil Pump Inspection

1. Wash all parts in solvent. Air dry.
2. Inspect pump gears for nicks or damage.
3. Inspect pump body for nicks or damage.
4. Install pump gears, aligning marks previously made on gears.
5. Install pump body on converter hub. Check gear end clearance in

housing using a straight edge and a feeler gauge or a dial indicator. Allowable end clearance is .0005-.0015 in. for 1964 Buick and 1964-68 Pontiac; .0008-.0018 in. for 1964-66 Oldsmobile; and .0005-.0035 in. for 1965-69 Buick, 1967-69 Oldsmobile, and 1969 Pontiac.

Oil Pump Assembly

1. If oil pump bushing is being replaced, press the new bushing into the pump body until it is flush or up to .010 in. below opposite face (front pump seal site).

2. Press stator shaft bushing into shaft until tool is flush with top of shaft.
3. Install new oil seal into pump housing.
4. Install new oil pump-to-case seal.
5. Assemble pressure regulator valve, shim, washer, spring, boost valve, and sleeve.
NOTE: when installing spring and shim make certain that the same color spring and the original number of shims are installed.
6. Compress boost valve sleeve with thumb, then install retaining snap-ring.
7. Install coast downshift timing valve, bottom end up, into the cover.
8. Install spring, valve, and seat into cooler by-pass valve. Press seat into bore of pump body until tool bottoms on face of pump.

Checking pump end clearance
(© G.M. Corp.)

SNAP RING PLIERS

Installing boost valve snap-ring
(© G.M. Corp.)

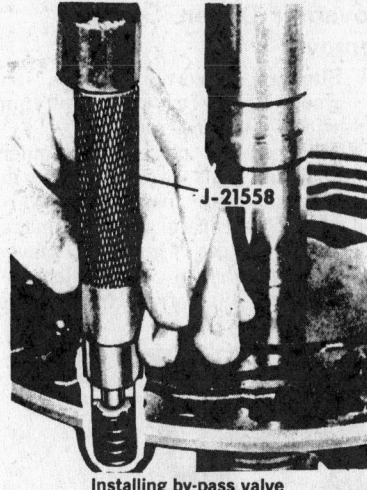

Installing by-pass valve
(© G.M. Corp.)

Use of clamp to obtain pump alignment
(© G.M. Corp.)

9. Install pump cover to pump body. Install five retaining bolts, do not tighten. Tighten clamping band around pump to get proper alignment.

Forward Clutch Disassembly

1. Remove low sun gear and flange assembly retaining snap-ring.
2. Remove low sun gear and flange.
3. Remove clutch hub rear thrust washer.
4. Lift forward clutch hub from pack and remove clutch hub front thrust washer.
5. Remove the clutch pack.
6. Compress the spring retainer. Remove the snap-ring. Then, remove the compression tool and

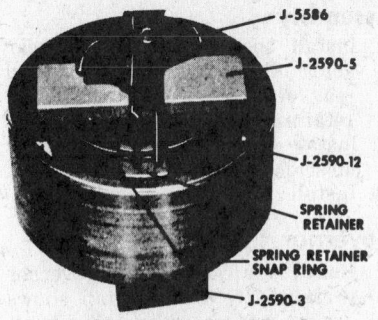

Compressing forward clutch unit spring retainer
(© G.M. Corp.)

Removing spring retainer
(© G.M. Corp.)

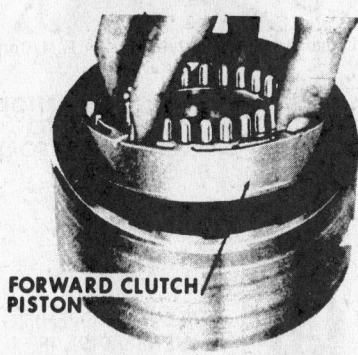

Removing forward clutch piston
(© G.M. Corp.)

lift out the spring retainer and 24 coil springs.
7. Lift out the piston, with a twisting motion, and remove piston seal.
8. Remove piston inner seal.
9. If clutch drum bushing is to be replaced, press bushing from clutch drum.
10. If low sun gear and flange bushings are to be replaced, press out the bushing.

Forward Clutch Unit Inspection

1. Wash all parts in solvent. Air dry.
2. Check steel ball in forward clutch drum. Make certain that ball is free to move in hole. Check that orifice leading to front of clutch drum is open.
3. Check clutch lined plates and steel driven plates.

Forward Clutch Assembly

1. Install bushing in forward clutch drum. Press bushing into bore until tool bottoms on the hub.
2. Press new bushing into low sun gear until tool is flush with top of low sun gear.
3. Lubricate with transmission fluid then install new forward clutch piston inner seal with seal lip pointing downward.
4. Lubricate with transmission oil and install new forward clutch piston outer seal into clutch piston, seal lip down.
5. Install forward clutch piston into clutch drum by using a loop of wire to start lip of seal into drum bore.
6. Reassemble return springs, retainer and snap-ring.
7. With spring retainer in place, compress spring retainer with the same tools used to disassemble the unit, then install the snap-ring. Remove compression tool.

Removing clutch piston inner oil seal
(© G.M. Corp.)

Installing forward clutch piston into clutch drum
(© G.M. Corp.)

Installing clutch hub
(© G.M. Corp.)

Removing low sun gear and flange assembly retaining snap-ring
(© G.M. Corp.)

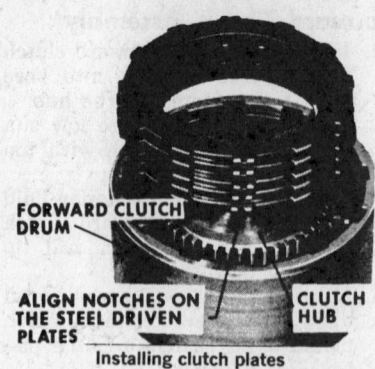

FORWARD CLUTCH DRUM

ALIGN NOTCHES ON THE STEEL DRIVEN PLATES

CLUTCH HUB

Installing clutch plates
(© G.M. Corp.)

LOW SUN GEAR AND FLANGE ASSEMBLY

CLUTCH HUB REAR THRUST WASHER

Installing clutch hub rear thrust washer
(© G.M. Corp.)

8. Install clutch hub front thrust washer to clutch hub (retain with petroleum jelly), aligning tangs in clutch hub with grooves in thrust washer. Install clutch hub.

9. Align notches on steel driven plates. Install steel plates and lined drive plates alternately, beginning with a steel plate.

10. Install clutch hub rear thrust washer with its flange toward low sun gear and flange assembly.

11. Install low sun gear and flange assembly.

12. Install snap-ring. Position so gap in ring is located between slots in drum.

Speedometer Driven Gear Disassembly

1. Remove speedometer driven gear and oil seal (large O-ring).

2. Remove speedometer driven gear shaft oil seal (small seal).

Speedometer Driven Gear Assembly

1. Install speedometer driven gear shaft oil seal, lip of seal toward rear of gear sleeve. Install seal retaining ring.

2. Install speedometer driven gear oil seal.

3. Install speedometer driven gear.

Governor Inspection

1. Check for secondary governor weight tab wear. If tab shows wear, replace governor assembly.

2. Check governor feed port opening. If opening is less than .019 in., replace governor assembly.

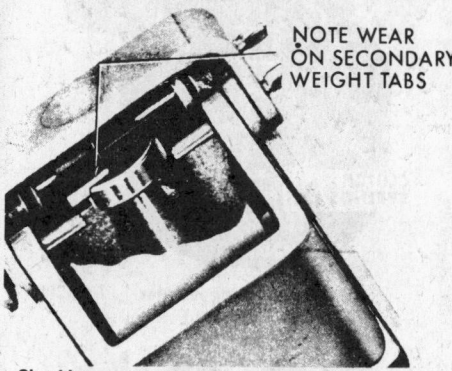

NOTE WEAR ON SECONDARY WEIGHT TABS

Checking governor weight tab wear (© G.M. Corp.)

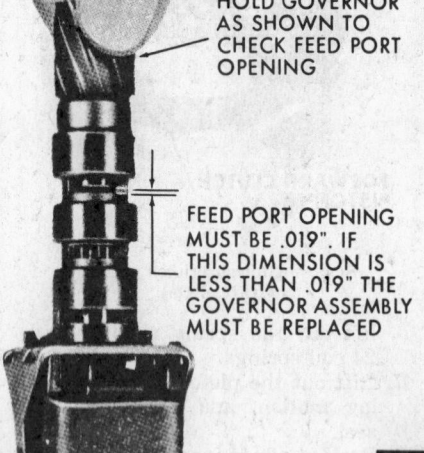

HOLD GOVERNOR AS SHOWN TO CHECK FEED PORT OPENING

FEED PORT OPENING MUST BE .019". IF THIS DIMENSION IS LESS THAN .019" THE GOVERNOR ASSEMBLY MUST BE REPLACED

Checking governor feed port dimension
(© G.M. Corp.)

Governor Driven Gear Removal

1. Support governor sleeve on wood block. Drive out roll pin with 1/8 in. drill rod.

NOTE: do not place block under nylon gear. This would result in breaking the gear inside the governor sleeve. Be careful not to damage machined surfaces of governor sleeve.

2. Remove driven gear. Remove chips or burrs from inside governor sleeve.

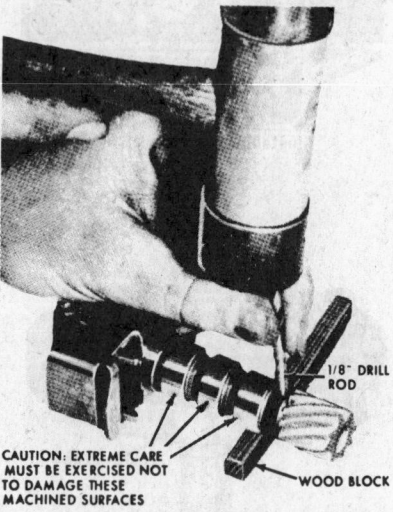

1/8" DRILL ROD

CAUTION: EXTREME CARE MUST BE EXERCISED NOT TO DAMAGE THESE MACHINED SURFACES

WOOD BLOCK

Driving out governor gear pin (© G.M. Corp.)

Governor Driven Gear Installation

1. Place shim supplied in replacement gear kit between second and third lands of governor sleeve. Support governor in press by shim. Do not support or hammer on rear of governor.

2. Press new gear into sleeve until seated against sleeve.

3. Drill out existing hole in governor sleeve to 1/8 in. Drill half-way through from each end.

4. Support end of governor sleeve (not gear) on wooden block. In-

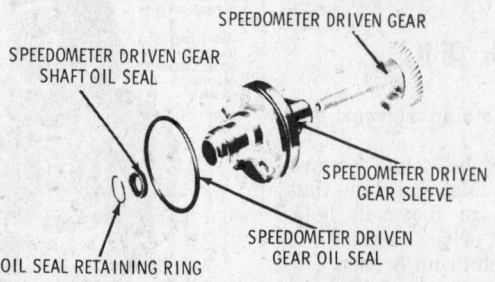

SPEEDOMETER DRIVEN GEAR

SPEEDOMETER DRIVEN GEAR SHAFT OIL SEAL

SPEEDOMETER DRIVEN GEAR SLEEVE

SPEEDOMETER DRIVEN GEAR OIL SEAL

OIL SEAL RETAINING RING

Speedometer driven gear (© G.M. Corp.)

MAKE CERTAIN GEAR IS SEATED AGAINST SLEEVE

SHIM SUPPORTS

SHIM

NO ATTEMPT SHOULD BE MADE TO SUPPORT GOVERNOR OTHER THAN SHOWN, OR VALVE WILL BE DAMAGED

Pressing on governor gear (© G.M. Corp.)

stall new roll pin, staking both ends.

5. Check for burrs on sleeve. Check that valve is free in its bore.

Planet Carrier Disassembly

1. Scribe a mark on each pinion shaft and on the planet carrier assembly in order to facilitate re-assembly.
2. Remove three pinion shaft lock plate screws and lock washers.
3. Slightly rotate, then remove lock plate.
4. Start with the short planet pinion, insert brass drift into front of carrier.
5. Remove pinion shaft and pinion gear from carrier.
6. Remove the other two short planet pinions in the same way.
7. Remove brass drift, needle bearings, and two thrust washers from each of the short planet pinion gears.

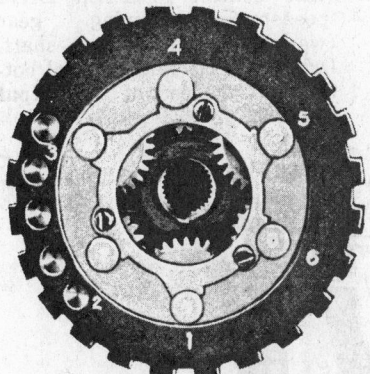

Marking pinion shafts and planet carrier
(© G.M. Corp.)

PLANET PINION SHAFT LOCK PLATE SCREW & LOCKWASHER ASSEMBLY

Removing pinion shaft lock plate screws
(© G.M. Corp.)

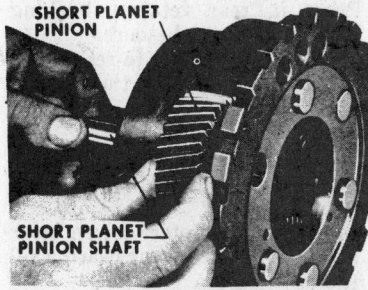

SHORT PLANET PINION

SHORT PLANET PINION SHAFT

Removing a short planet pinion and shaft
(© G.M. Corp.)

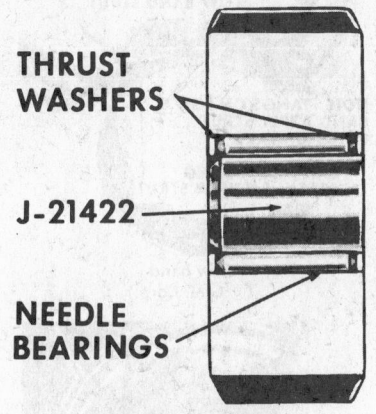

THRUST WASHERS

J-21422

NEEDLE BEARINGS

Short pinion gear
(© G.M. Corp.)

LOW SUN GEAR NEEDLE THRUST BEARING

Removing low sun gear needle thrust bearing
(© G.M. Corp.)

8. Remove low sun gear needle thrust bearing.
9. Remove input sun gear and sun gear thrust washer.
10. Insert brass drift through long planet pinion, then remove the long planet pinion.
11. Remove front planet pinion thrust washer and long planet pinion gear.

FRONT PLANET PINION THRUST WASHER

LONG PLANET PINION GEAR

Removing front planet pinion thrust washer
(© G.M. Corp.)

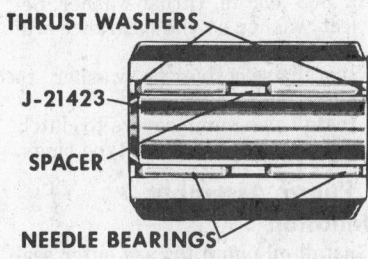

THRUST WASHERS

J-21423

SPACER

NEEDLE BEARINGS

Long pinion gear
(© G.M. Corp.)

12. Remove brass drift, needle bearings, spacer, and two thrust washers from long planet pinion.
13. Repeat Steps 10-12 to remove remaining long planet pinions.
14. If cleaning and inspection reveals worn or damaged parts, rebuild the unit.

Planet Carrier Assembly

Assemble by reversing the above disassembling procedure.

NOTE: the front thrust washers, installed with the long planet pinions, also act as thrust washers for the short pinions because they are paired together.

Transmission Assembly

Range Selector and Parking Lock Actuator Installation

1. Hold locking pawl and spring in the case with parking lock pawl shaft.
2. Install outer shift lever shaft seal, lip pointing inward.
3. Lubricate, then insert outer range selector lever and shaft assembly into the case.
4. Assemble parking lock actuator assembly to inner parking lock and range selector.
5. Install inner park lock and range selector onto outer range selector lever. Install nut onto range selector lever. Long end of selector lever goes toward bottom of transmission.
6. Slide outer range selector lever into case and tighten the attaching nut.
7. Install range selector shaft retainer.
8. Install parking lock bracket to transmission case.

Reverse Clutch Installation

1. Lubricate with transmission fluid, then install outer and inner piston seals.
2. With transmission case up-ended, install the reverse clutch piston into the case.
3. Install 17 piston return springs.
4. Position piston spring seat on the return springs and lay the retainer ring in place.
5. Compress return spring seat far enough to seat the snap-ring. Seat the snap-ring, then remove compressor.
6. Install reverse clutch cushion spring, if used, with the dish down.
7. Align notches on the steel driven plates. Install steel plates and lined drive plates alternately, starting with a steel plate. The notched lugs on the driven plates should all be aligned and in the five o'clock groove in the case.

NOTE: steel plates are dished and should all face in the same direction.

8. Install reverse clutch pressure plate, with identification mark installed in the five o'clock groove.
9. Install clutch pack snap-ring.
10. For 1964-65 models only: insert feeler gauge between any reaction plate and adjacent faced plate.

When the dimension between the reaction plate and its adjacent plate is within .058-.021 in. use the reaction plate with one identification mark. If the dimension is from .095-.058 in. use a plate with two marks. If the dimension is .133-.095 in. use a plate with three identification marks.

Planetary Gear Set Installation

1. Assemble thrust bearing race with lip, needle bearing, and a second plain thrust bearing race to the rear face of the planetary gear set. Retain with grease.
2. Install reverse ring gear into the case.
3. Install planetary gear set into the case.

Installing thrust bearing races
(© G.M. Corp.)

Low Servo and Band, and Forward Clutch Installation

1. Install low servo piston assembly into case.
2. Install servo cover oil seal, then install the cover.
3. Compress the cover and install retainer snap-ring.
4. With transmission case up-ended, install band adjusting screw into case.
5. Install low band. Install apply strut and band adjusting screw strut. Tighten band adjustment enough to prevent struts from falling out.
6. Install forward clutch assembly by turning slightly to engage low sun gear with planet pinions. Install pump into case. Install pump to case bolts.
7. Check forward clutch to oil pump clearance by installing a shim selector gauge, then check end-

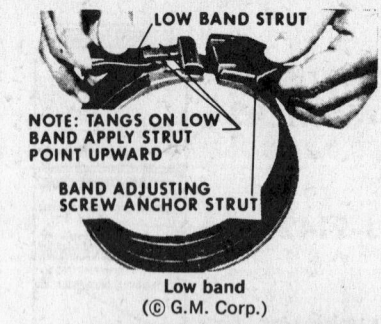

NOTE: TANGS ON LOW BAND APPLY STRUT POINT UPWARD

BAND ADJUSTING SCREW ANCHOR STRUT

Low band
(© G.M. Corp.)

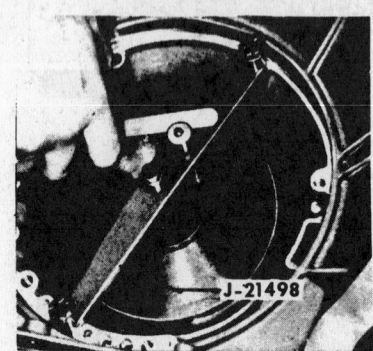

Checking oil pump end-play with feeler gauge
(© G.M. Corp.)

play with a feeler gauge. The measurement taken with the gauge determines the thickness of the thrust washer to be used. End-play can also be measured with a feeler gauge.

If it is not possible to insert a feeler gauge, use a .099-.095 in. thrust washer. If a .001-.020 in. feeler is acceptable, use a .081-.077 in. thrust washer. If a .021-.040 in. feeler is acceptable, use a .063-.059 in. thrust washer. Select washer so the clearance will be .022-.054 in.

8. Install selective fit washer to pump cover hub.
9. Install two pump covers to clutch drum oil sealing hook-type rings.

Oil Pump Assembly Installation

1. Install oil pump to case outer seal then a new pump gasket and guide pins.

2. Install input shaft oil rings, then install the shaft into the oil pump.
3. Apply transmission fluid to edge of pump, and install pump into the case.
4. Remove pump guide pins, then install eight retaining bolts with new O-rings or washer type seals under their heads. Torque pump bolts to 16-24 ft. lbs.

Low Band Adjustment

1. Tighten adjusting screw to 40 in. lbs. (about 3½ ft. lbs.). Now back off adjustment four turns, then lock with locknut. Install adjusting screw cap.

Speedometer Drive Gear Installation

1964-65

1. With transmission in horizontal position, install speedometer driving gear.
2. Place control in Park. Drive speedometer driving gear (worm) onto the output shaft. Drive gear on until the tool bottoms on the end of the output shaft.

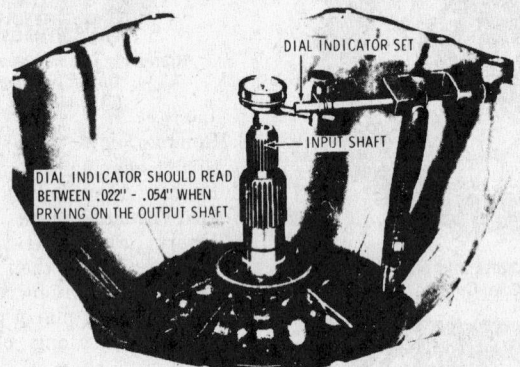

DIAL INDICATOR SET

INPUT SHAFT

DIAL INDICATOR SHOULD READ BETWEEN .022" - .054" WHEN PRYING ON THE OUTPUT SHAFT

Checking oil pump end-play with a dial indicator
(© G.M. Corp.)

1966-69

1. Place retainer in hole in output shaft.
2. Align slot in speedometer drive gear with retainer clip. Slide gear onto shaft.

Rear Bearing Retainer Bushing, Oil Seal, Bearing Retainer and Speedometer Driven Gear Installation

1. Press in rear retainer bushing until flush with the bore.
2. Install output shaft-to-rear bearing retainer oil seal, then install the retainer-to-transmission case oil seal.
3. Place rear bearing retainer against the case and install four retaining bolts.
4. Install speedometer driven gear assembly into the rear bearing retainer.
5. Install driven gear sleeve retainer.

Valve Body Installation

1964-65

1. With transmission upside down, install valve body to plate gasket, then the valve body plate.
2. Install manual control valve and link into valve body assembly.
3. Install manual control valve link into park lock and range selector inner lever.
4. Install eleven valve body-to-case attaching bolts. Torque to 8-11 ft. lbs.
5. Install stator control valve plate.
6. Install stator control solenoid and gasket to stator control valve body.
7. Install spring detent assembly. Torque bolt to 8-12 ft. lbs.
8. Install solenoid switch, then install the solenoid connector.
9. Install oil strainer pipe to case seal onto the pipe, lubricate the seal, then install the pipe into the transmission case.
10. Install oil strainer.
11. Install strainer attaching bolt or clip.
12. With a new gasket on the oil pan, install pan and attach pan to case. Torque all fourteen bolts to 10-12 ft. lbs.

1966-69

This valve body installation procedure is the same as that given above for 1964-65 models, with the substitution of the following steps:

5. Install oil channel support plate. Torque bolts to 8-12 ft. lbs.
6. Delete Step 6.
12. With a new gasket on the oil pan, install pan and attach pan to case. Torque fourteen bolts to 10-16 ft. lbs.

Governor and Vacuum Modulator Installation

1. Slide the governor into its bore in the case. Turn the governor assembly so teeth on governor gear engage teeth on output shaft.
2. Install governor gasket and cov-

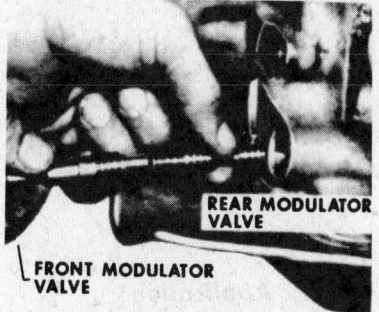

Installing modulator valve
(© G.M. Corp.)

er to case. Torque bolts to 8-12 ft. lbs.
3. Slide rear modulator valve into front modulator valve, then install into bore in case.
4. Install case-to-modulator oil seal, then install modulator into the case.
5. Install vacuum modulator retainer so the retainer tang points toward the vacuum modulator. Torque bolts to 8-12 ft. lbs.

Checking the Converter

NOTE: the converter is a welded unit, therefore, it is not serviceable. If the converter is bad, renew the unit.

Making converter end-play check
(© G.M. Corp.)

1. Install converter pressure check fixture and tighten.
2. Fill converter with 80 psi air.
3. Submerge in water and check for bubbles.
4. Then, check for end-play by installing converter end-play checking fixture and tightening the brass nut.
5. Tighten the hex nut.
6. Install dial indicator and set at zero.
7. Loosen hex nut. When nut is fully loosened, the reading obtained on the dial indicator will be converter end-play. If clearance is .050 in. or more, or the oil has the appearance of containing aluminum particles (aluminum paint), replace the converter.

Torque References

	Foot-Pounds
Case to cylinder block	30-40
Converter cover pan to transmission case	8-12
Cooler pipes to case	25-35
Low band adjusting screw nut	20-30
Pump cover to body	16-24
Transmission case bolt	8-12
Valve body to case	8-12
Solenoid valve to valve body	8-12
Vacuum modulator to case	8-12
Pump assembly to case	16-24
Rear bearing retainer to case	25-35
Oil pan to case	10-16
Governor cover to case	8-12
Speedometer sleeve retainer to bearing retainer	5-10

Type 350

Application

Buick Special, 1969-70
Chevrolet, Chevelle,
Nova, Camaro, 1969-70

Pontiac Tempest
and Firebird, 1969-70

Diagnosis

This diagnosis guide is a list of the most common troubles and their causes. The items to check are listed in the sequence to be followed.

TROUBLE SYMPTOMS	Items to Check	
	In Car	Out of Car
Slips in all ranges	ACDFGMO	N
Drive slips—no 1st gear	ACDFGMO	NQV
No 1-2 upshift	BEFGH	SU
1-2 upshift, early or late	BEFH	
Slips, 1-2 upshift	ACFGHL	NSU
Harsh 1-2 upshift	BCG	
No 2-3 upshift	FI	R
2-3 upshift, early/late	BCEFIK	
Slips, 2-3 upshift	ACFGI	NR
Harsh 2-3 upshift	BCGIL	
No full throttle downshift	K	
2-3 upshift, wide open throttle only	BK	
L₁ gear—no engine braking	GHJM	PQU
Car drives in neutral	K	Q
Slips in reverse	ACDFGHMO	RT
1-2, 2-3 shift noisy	A	RSY
Noisy in all ranges	ADFO	NXYW
Spews oil out of breather	AD	

Key to Checks

A. Low oil level/water in oil
B. Vacuum leak
C. Modulator and/or valve
D. Strainer and/or gasket leak
E. Governor valve/screen
F. Valve body gasket/plate
G. Pressure regulator and/or boost valve
H. 1-2 shift valve

I. 2-3 shift valve
J. Manual low control valve
K. Detent valve and linkage
L. 2-3 accumulator
M. Manual valve linkage
N. Pump gears
O. Gasket screen-pressure
P. Band—intermediate overrun roller clutch

Q. Forward clutch assembly
R. Direct clutch assembly
S. Intermediate clutch assembly
T. Low & reverse clutch assembly
U. Intermediate roller clutch
V. Low & reverse roller clutch
W. Parking pawl/linkage
X. Converter assembly
Y. Gear set and bearings

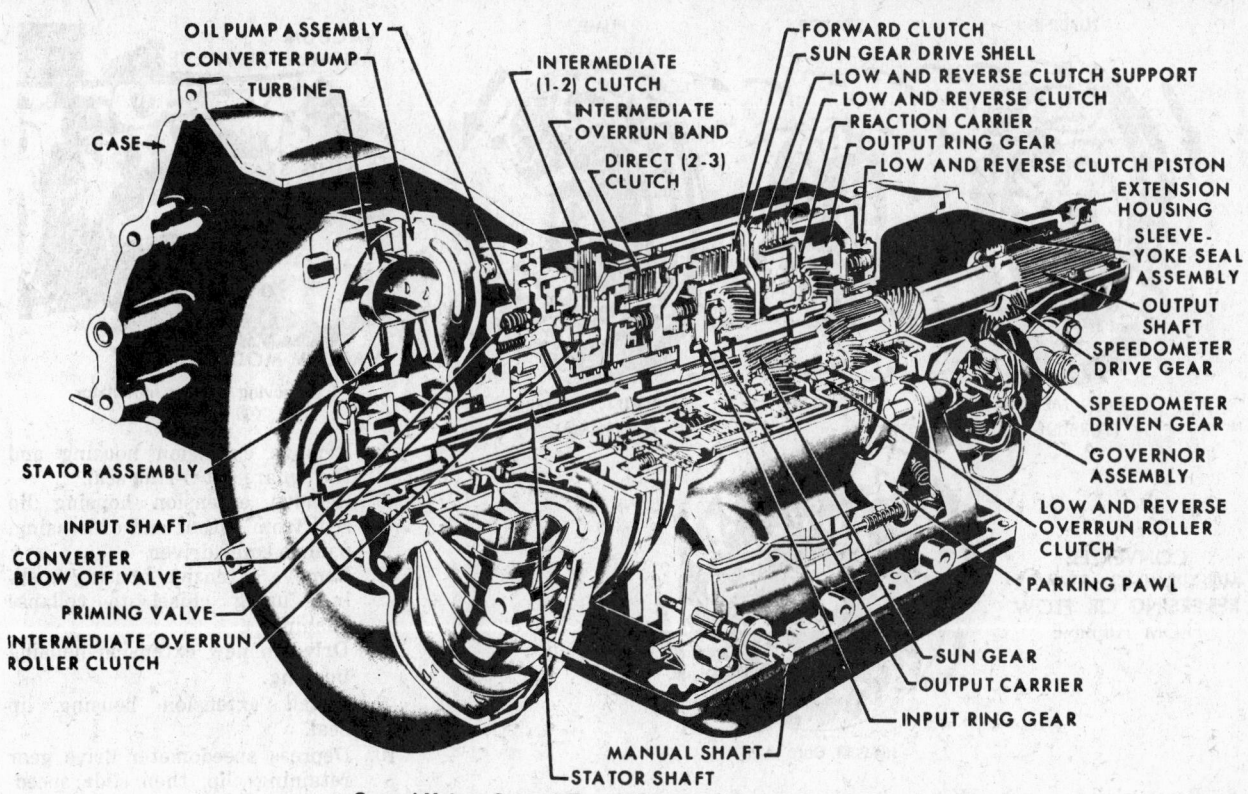

General Motors, 3-speed, Type 350 transmission (© G.M. Corp.)

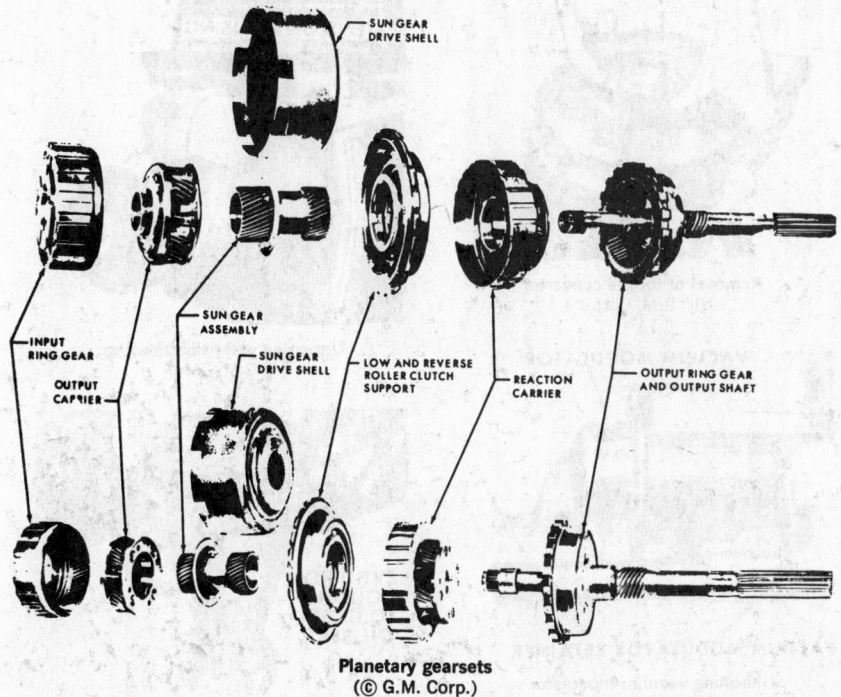

Planetary gearsets
(© G.M. Corp.)

The illustration is typical. While design and repair procedures are similar, R & R methods and components may vary with car application. Therefore, some differences may exist.

General Information

The "350" three-speed transmission is a fully automatic unit, consisting of a three-element hydraulic torque converter and two planetary gear sets. Four multiple-disc clutches, two roller clutches, and an intermediate overrun band provide the friction elements needed to obtain the desired function of the two planetary gear sets.

The torque converter couples the engine to the planetary gears through oil and hydraulically provides added torque multiplication when required. The two planetary gear sets give three forward ratios and one reverse. The torque converter consists of three basic elements:

1. Converter pump—driving member
2. Turbine—driven member
3. Stator—reaction member

Changing of the gear ratios is fully automatic and relative to car speed and engine torque input. Car speed and engine torque signals are constantly fed to the transmission to provide the proper gear ratio for maximum efficiency and performance.

The shift quadrant has six positions marked as follows:

P, R, N, D, L_2, L_1

Shift positions of the type 350 and type 400 are identical.

Capacity and General Specifications

Oil Capacity (dry)	20 Pints
Cooling	Water
Oil Filter Type	Bottom Suction Screen
Clutches	4
Roller Clutches	2
Band (non adjustable)	1

Draining and Refilling

The transmission fluid should be drained and refilled at 24,000 mile intervals. To do this:

1. Drain transmission oil.
2. Remove oil pan attaching screws.

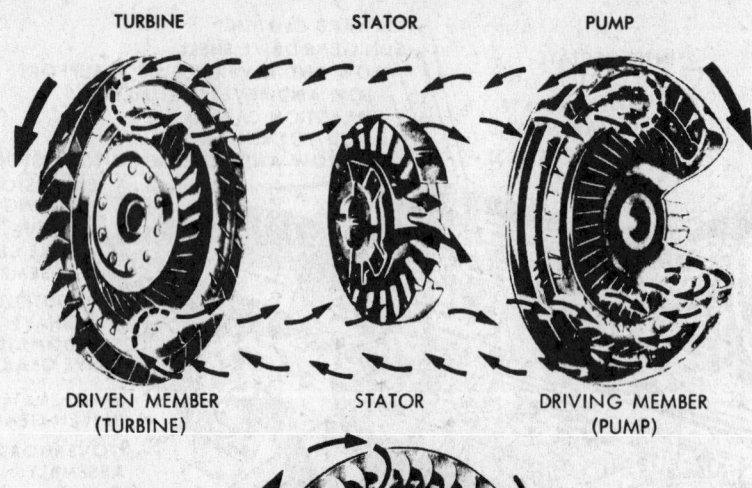

TURBINE   STATOR   PUMP

DRIVEN MEMBER (TURBINE)   STATOR   DRIVING MEMBER (PUMP)

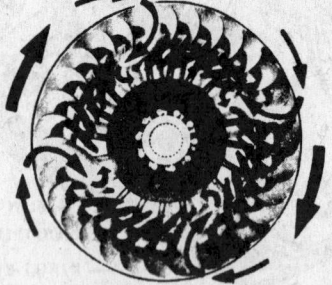

CONVERTER MULTIPLYING, STATOR REVERSING OIL FLOW FROM TURBINE

(© G.M. Corp.)

3. Remove oil pan and gasket.
4. Remove oil strainer screws and strainer.
5. Thoroughly clean oil strainer and oil pan in solvent. Blow dry with air.
6. Renew oil strainer gasket and oil pan gasket.
7. Install oil strainer.
8. Install oil pan. Torque attaching screws to 10-12 ft. lbs.
9. Add about 5 pints of transmission fluid.
10. Start engine. Do not race engine. Move selector lever slowly through each range. Check fluid level with transmission in Park.
11. With transmission cold, add enough fluid to bring level to ¼ in. below the Add mark. With the transmission hot, add enough fluid to bring level to Full mark.
12. In case of transmission overhaul, proceed as in steps 9-11. Start with 5 pints of fluid if converter has not been replaced; 8 pints if it has been replaced. Total capacity is about 20 pints, but correct level is determined by the mark on the dipstick rather than by amount added.

Transmission Disassembly

Clean ouside of transmission thoroughly to prevent dirt from entering the unit.
1. With transmission in a holding fixture, lift off torque converter assembly.

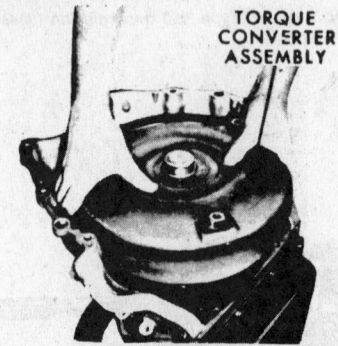

TORQUE CONVERTER ASSEMBLY

Removal of torque converter
(© G.M. Corp.)

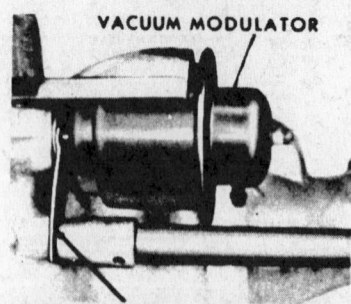

VACUUM MODULATOR

VACUUM MODULATOR RETAINER

Unbolting vacuum modulator
(© G.M. Corp.)

2. Remove vacuum modulator assembly attaching bolt and retainer.
3. Remove vacuum modulator assembly, O-ring seal and modulator valve from the case.
4. Remove four extension housing-to-case attaching bolts. On reassembly, torque attaching bolts to 35 ft. lbs.

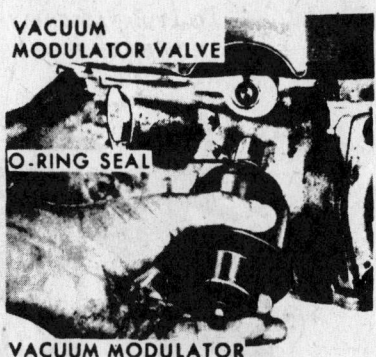

VACUUM MODULATOR VALVE

O-RING SEAL

VACUUM MODULATOR

Removing vacuum modulator
(© G.M. Corp.)

5. Remove extension housing and the square cut O-ring seal.
6. Remove extension housing lip seal from output end of housing, using a screwdriver.
7. Remove extension housing bushing, using chisel to collapse bushing.
8. Drive in new extension housing bushing.
9. Install extension housing lip seal.
10. Depress speedometer drive gear retaining clip, then slide speedometer drive gear off output shaft.

EXTENSION HOUSING

Unbolting extension housing
(© G.M. Corp.)

EXTINSION HOUSING TO CASE OIL SEAL

Removing extension housing O-ring
(© G.M. Corp.)

11. Remove governor cover retainer wire with a screwdriver.
12. Remove governor cover and O-ring seal from case, then remove O-ring seal from governor cover.
13. Remove governor assembly from case.

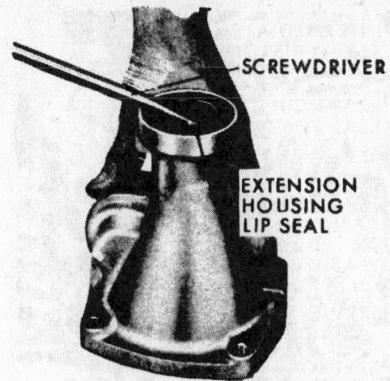

Removing extension housing lip seal
(© G.M. Corp.)

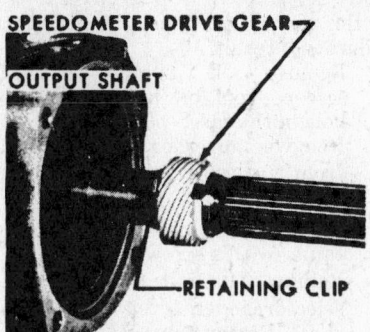

Speedometer drive gear
(© G.M. Corp.)

Removing oil pan
(© G.M. Corp.)

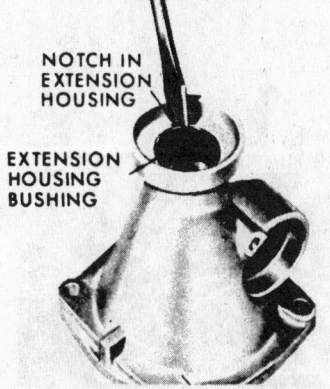

Removing extension housing bushing
(© G.M. Corp.)

Releasing governor retainer wire
(© G.M. Corp.)

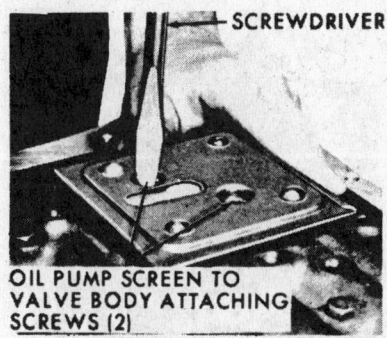

Detaching oil screen
(© G.M. Corp.)

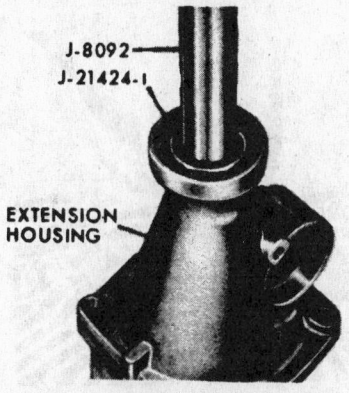

Driving in extension housing bushing
(© G.M. Corp.)

Removing governor cover
(© G.M. Corp.)

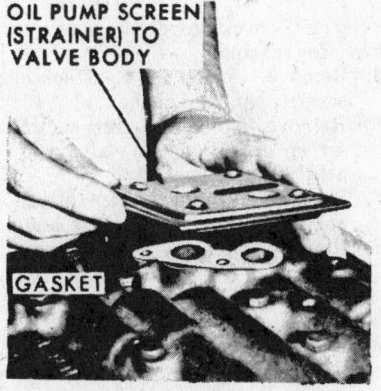

Oil screen gasket
(© G.M. Corp.)

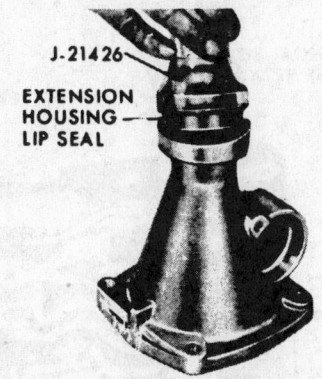

Installing new extension housing lip seal
(© G.M. Corp.)

Removing governor assembly
(© G.M. Corp.)

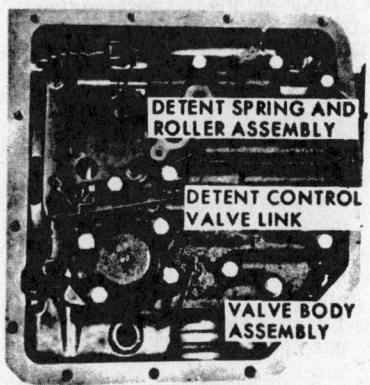

Detent roller and spring
(© G.M. Corp.)

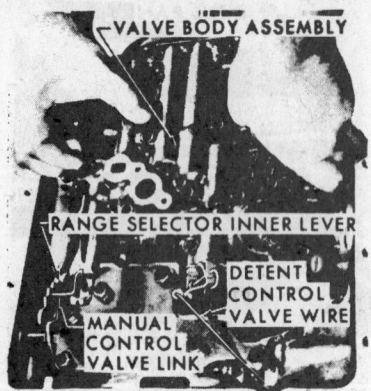

Manual control valve link
(© G.M. Corp.)

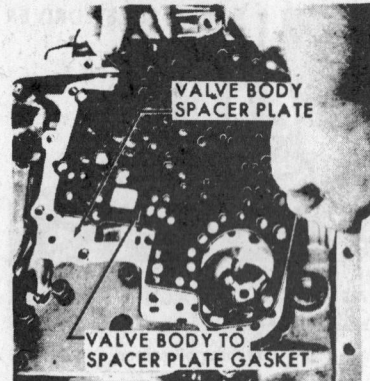

Removing valve body to spacer plate gasket
(© G.M. Corp.)

NOTE: check governor bore and sleeve for scoring.

14. Remove oil pan attaching screws, pan, and gasket.
15. Remove two oil pump suction screen (strainer) to valve body attaching screws.
16. Remove oil pump screen (strainer) and gasket from valve body.
17. Remove detent roller and spring assembly from valve body. Remove valve body-to-case attaching bolts.
18. Remove manual control valve link from range selector inner lever. Remove valve body. Remove detent control valve link from detent actuating lever.

NOTE: refer to later text for valve body disassembly and passage identification.

NOTE: at this time, when handling valve body assembly, do not touch sleeves, because retaining pins may fall into the transmission.

19. Remove valve body-to-spacer plate gasket.
20. Remove spacer support plate gasket. Remove spacer support plate.
21. Remove valve body spacer plate and plate-to-case gasket.
22. Remove four check balls from passages, as illustrated, in case face.

NOTE: if, during reassembly, any of the balls are omitted or installed in the wrong locations, transmission failure will result.

23. Remove oil pump pressure screen from oil pump pressure hole in the case.
24. Remove governor feed screen from governor feed passage in case.
25. Remove manual shaft to case retainer with a screwdriver.
26. Remove jam nut holding range selector inner lever to manual shaft. Remove manual shaft.
27. Disconnect parking pawl actuating rod from range selector inner lever. Remove both from case.
28. Remove manual shaft to case lip oil seal, using a screwdriver.
29. Remove parking lock bracket.

Removing spacer support plate
(© G.M. Corp.)

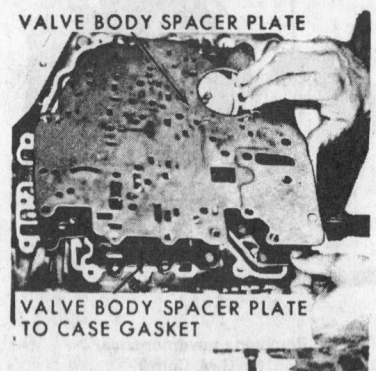

Removing valve body spacer plate and gasket
(© G.M. Corp.)

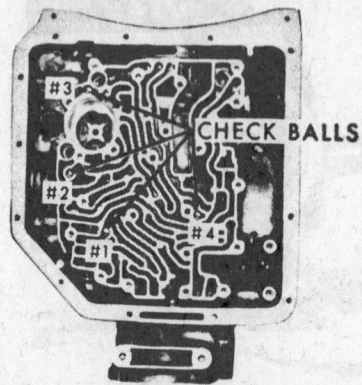

Check ball locations
(© G.M. Corp.)

Intermediate servo
(© G.M. Corp.)

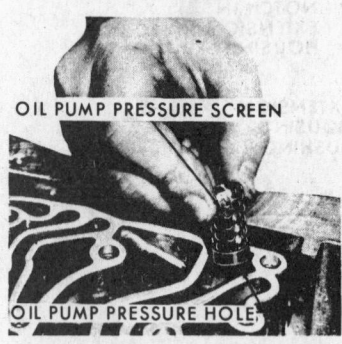

Removing oil pump pressure screen
(© G.M. Corp.)

Removing governor screen
(© G.M. Corp.)

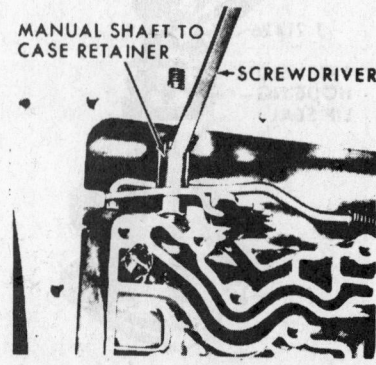

Removing manual shaft to case retainer
(© G.M. Corp.)

RANGE SELECTOR INNER
LEVER TO MANUAL SHAFT
NUT TORQUE TO 30 LB.FT

Loosening range selector inner lever jam nut
(© G.M. Corp.)

RANGE SELECTOR
INNER LEVER

MANUAL SHAFT

PARKING PAWL
ACTUATING ROD

Removing range selector inner lever
(© G.M. Corp.)

PUMP ASSEMBLY
TO CASE ATTACHING
BOLTS (8)

Oil pump
(© G.M. Corp.)

SCREWDRIVER

MANUAL SHAFT TO
CASE LIP SEAL

Removing manual shaft to case lip seal
(© G.M. Corp.)

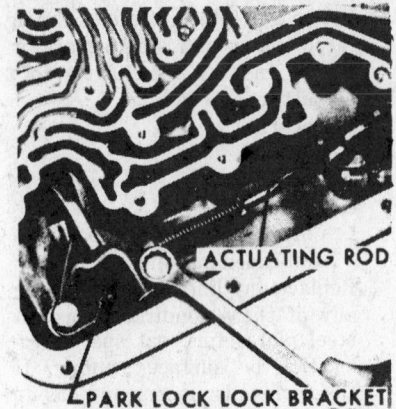

ACTUATING ROD

PARK LOCK LOCK BRACKET

Unbolting parking lock bracket
(© G.M. Corp.)

PARKING PAWL

PARKING PAWL
DISENGAGING SPRING

Disconnecting parking pawl spring
(© G.M. Corp.)

PARK PAWL

PARK PAWL
SHAFT

PARKING PAWL SHAFT
RETAINING PLUG

Parking pawl shaft details
(© G.M. Corp.)

30. Disconnect and remove parking pawl disengaging spring.
31. Remove parking pawl shaft retaining plug stake marks. Cock parking pawl on shaft. Using drift and hammer, tap on drift to force the parking pawl shaft from the case. Remove the parking pawl.
32. Remove intermediate servo piston and metal oil seal ring. Remove washer, spring seat, and apply pin.
33. Using band, apply pin selection tool J-23071, and straight edge,

apply firm downward pressure on selection pin. If the tool is below the straight edge surface, use a long apply pin. The long pin is identified by a groove on the band lug end. If the tool is above the straight edge surface, the short pin should be used. The short pin has no identifying marks.
34. Remove eight pump attaching bolts with washer-type seals. Discard seals.
35. Install two threaded slide hammers into threaded holes in pump body. Tighten jam nuts and remove pump assembly from the case. For disassembly and reassembly of oil pump, refer to paragraphs later in the text.
36. Remove pump assembly to case gasket and discard.
37. Remove intermediate clutch cushion spring.
Refer to exploded view of intermediate clutch.

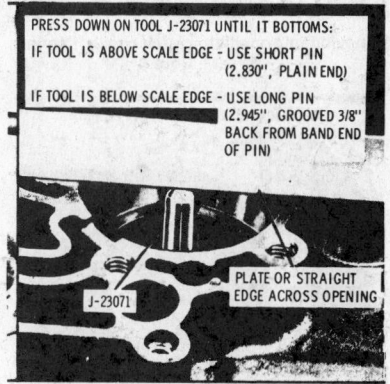

PRESS DOWN ON TOOL J-23071 UNTIL IT BOTTOMS:
IF TOOL IS ABOVE SCALE EDGE - USE SHORT PIN
(2.830", PLAIN END)
IF TOOL IS BELOW SCALE EDGE - USE LONG PIN
(2.945", GROOVED 3/8"
BACK FROM BAND END
OF PIN)

PLATE OR STRAIGHT
EDGE ACROSS OPENING

J-23071

Checking band apply pin length
(© G.M. Corp.)

SLIDE
HAMMERS
J-7004

TIGHTEN JAM NUTS

Removing oil pump
(© G.M. Corp.)

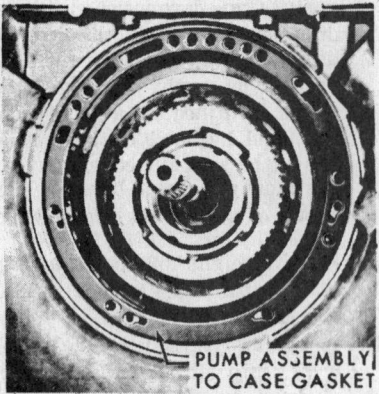

Pump to case gasket
(© G.M. Corp.)

Removing intermediate clutch cushion spring
(© G.M. Corp.)

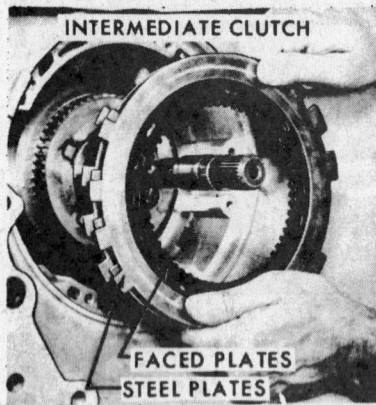

Removing intermediate clutch plates
(© G.M. Corp.)

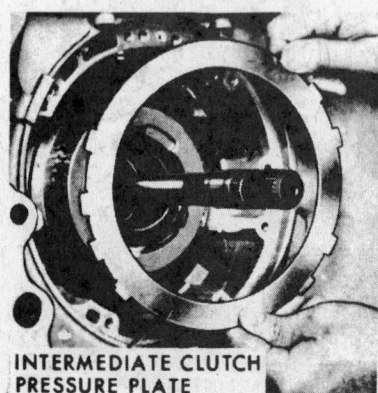

Removing intermediate clutch pressure plate
(© G.M. Corp.)

38. Remove intermediate clutch faced plates and steel separator plates. Inspect lined plates for:
   a. pitting and flaking.
   b. wear.
   c. glazing.
   d. cracking.
   e. charring.
   f. chips or metal particles imbedded in lining.
   Replace any lined plates showing any of these conditions. Inspect steel plates for heat spot discoloration or surface scuffing. If the surface is smooth and has an even color smear, the steel plates may be re-used.
39. Remove intermediate clutch pressure plate.
40. Remove intermediate overrun brake band.
41. Remove direct and forward clutch assemblies.
42. Remove forward clutch housing-to-input ring gear—front thrust washer.
   NOTE: washer has three tangs.
43. Remove output carrier-to-output shaft snap-ring.
44. Remove input ring gear.
45. Remove input ring gear to out-put carrier thrust washer.
46. Remove output carrier assembly.

Intermediate overrun brake band
(© G.M. Corp.)

Removing forward and direct clutch assemblies
(© G.M. Corp.)

Removing forward clutch housing to input ring gear front thrust washer
(© G.M. Corp.)

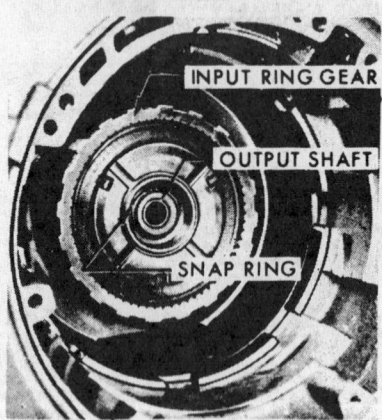

Output carrier to output shaft snap ring
(© G.M. Corp.)

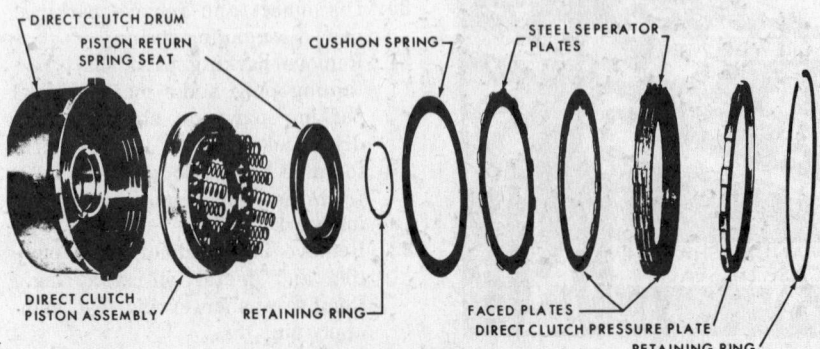

Direct clutch assembly
(© G.M. Corp.)

47. Remove sun gear drive shell assembly.
48. Remove low and reverse roller clutch support to case retaining ring.
49. Remove low and reverse roller clutch support to case retaining ring.
50. Remove low and reverse roller clutch, support assembly, and retaining spring.
51. Remove low and reverse clutch faced plates and steel separator plates.
52. Remove reaction carrier assembly from output ring gear and shaft assembly.
53. Remove output ring gear and shaft assembly from case.
54. Remove reaction carrier to output ring gear tanged thrust washer.
55. Remove output ring gear to output shaft snap-ring. Remove output ring gear from output shaft.

NOTE: do not over-stress snapring upon reassembly. Use new snapring.

56. Remove output ring gear-to-case needle bearing assembly.

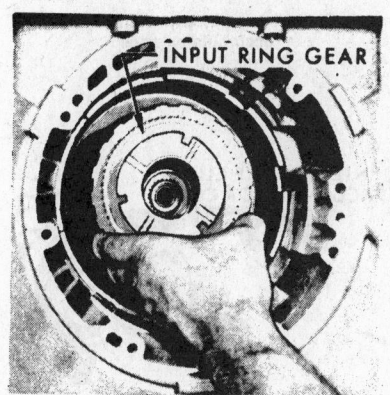

Removing input ring gear
(© G.M. Corp.)

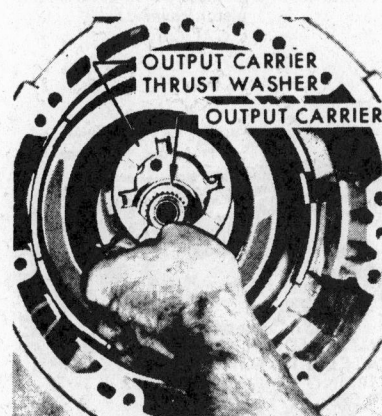

Removing input ring gear to output carrier thrust washer
(© G.M. Corp.)

Removing output carrier assembly
(© G.M. Corp.)

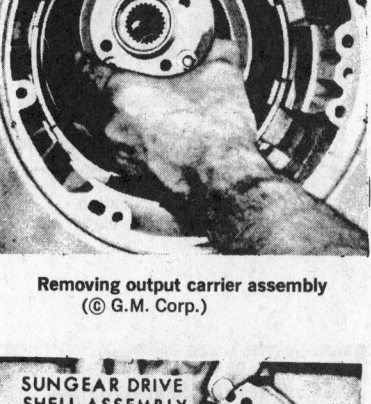

Removing sun gear drive shell assembly
(© G.M. Corp.)

Removing low and reverse roller clutch support to case retaining ring
(© G.M. Corp.)

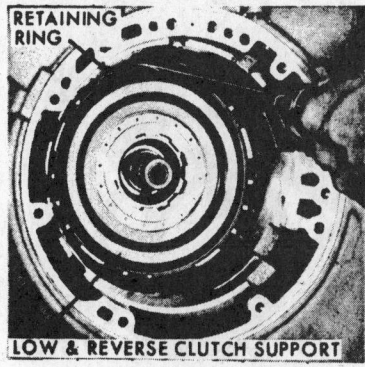

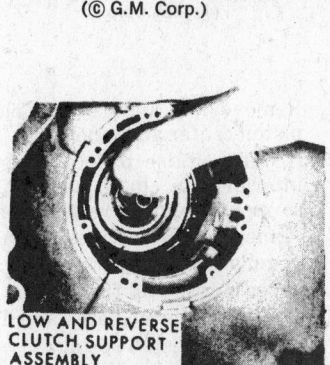

Removing low and reverse clutch support assembly
(© G.M. Corp.)

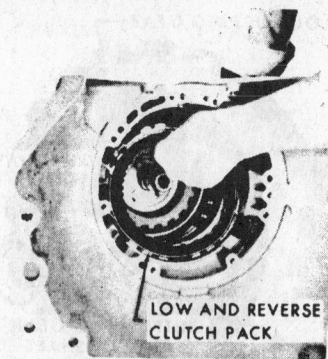

Removing low and reverse clutch pack
(© G.M. Corp.)

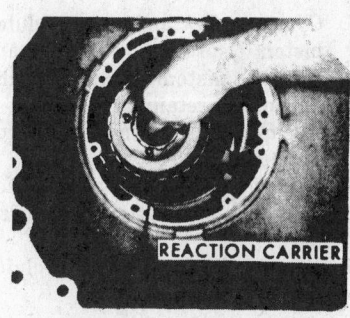

Removing reaction carrier assembly
(© G.M. Corp.)

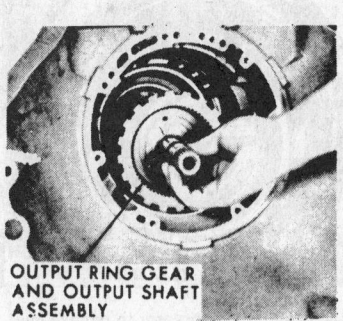

Removing output ring gear and shaft assembly
(© G.M. Corp.)

Removing reaction carrier to output ring gear thrust washer
(© G.M. Corp.)

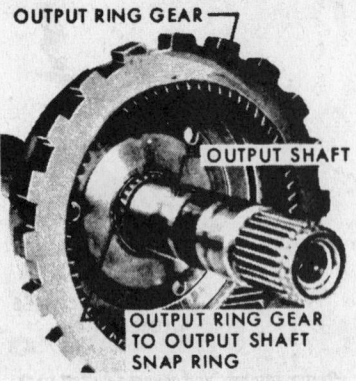

Output ring gear to output shaft snap-ring
(© G.M. Corp.)

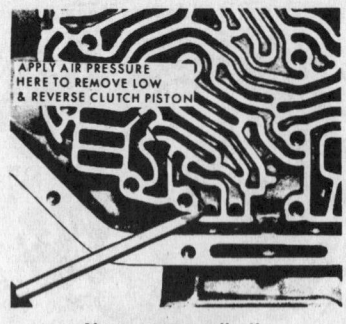

Air pressure application
to remove low and reverse clutch piston
(© G.M. Corp.)

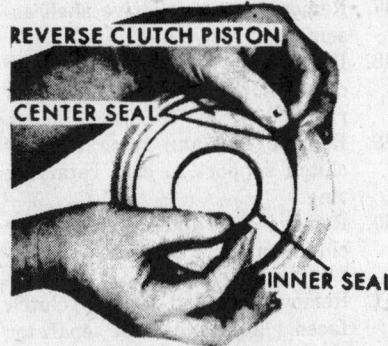

Removing low and reverse clutch piston
inner and center seals
(© G.M. Corp.)

57. Compress low and reverse clutch piston spring retainer and remove piston retaining ring and spring retainer.
58. Remove 17 piston return coil springs from piston.
59. Remove low and reverse clutch piston by application of air pressure.
60. Remove low and reverse clutch piston outer seal.

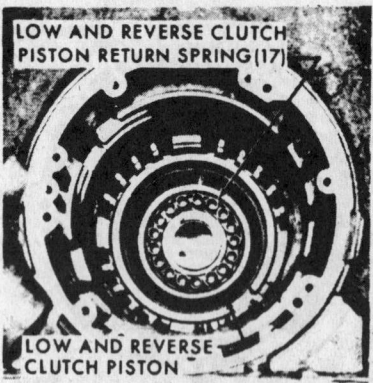

Location of low and reverse piston
return springs
(© G.M. Corp.)

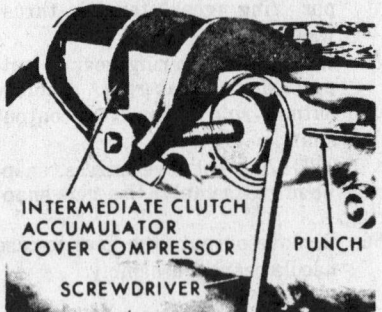

Compressing intermediate clutch
accumulator cover
(© G.M. Corp.)

Using spring compressor to remove low
and reverse clutch piston retaining ring
(© G.M. Corp.)

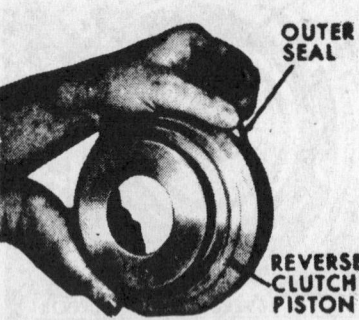

Removing low and reverse clutch piston
outer seal
(© G.M. Corp.)

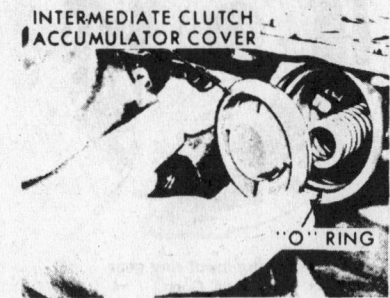

Removing intermediate clutch accumulator
piston cover
(© G.M. Corp.)

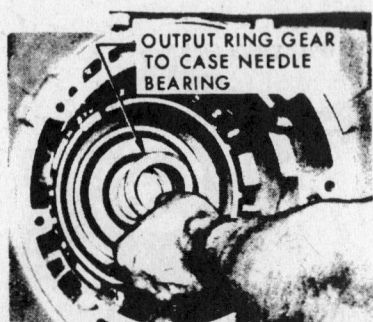

Removing output ring gear to case
needle bearing
(© G.M. Corp.)

61. Remove low and reverse clutch piston center and inner seal.
62. Install suitable tool to compress intermediate clutch accumulator cover and remove retaining ring.
63. Remove intermediate clutch accumulator piston cover. Remove cover O-ring seal from case.
64. Remove intermediate clutch accumulator piston spring.
65. Remove intermediate clutch accumulator piston assembly. Remove inner and outer hook-type oil seal rings if required.

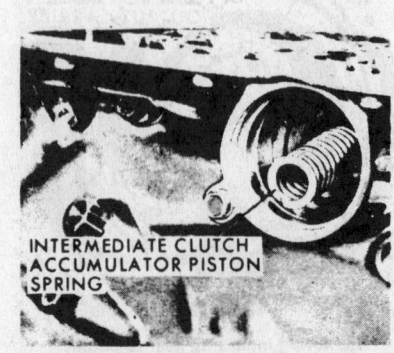

Location of accumulator piston spring
(© G.M. Corp.)

INTERMEDIATE CLUTCH ACCUMULATOR PISTON ASSEMBLY

OUTER SEAL RING

(© G.M. Corp.)

Component Disassembly

Valve Body

Disassembly

1. Position valve body assembly with cored face up and direct clutch accumulator piston pocket located as illustrated in valve body assembly illustration.
2. Remove manual valve from lower left-hand bore, A.
3. From lower right-hand bore, B, remove the pressure regulator valve train retaining pin, boost valve sleeve, intermediate boost valve, reverse and modulator

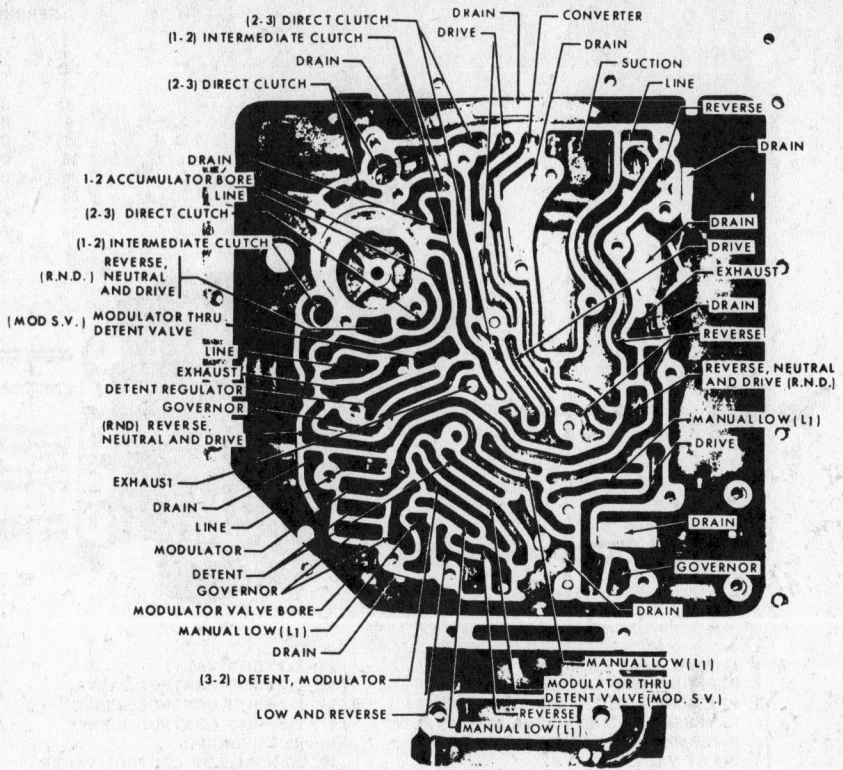

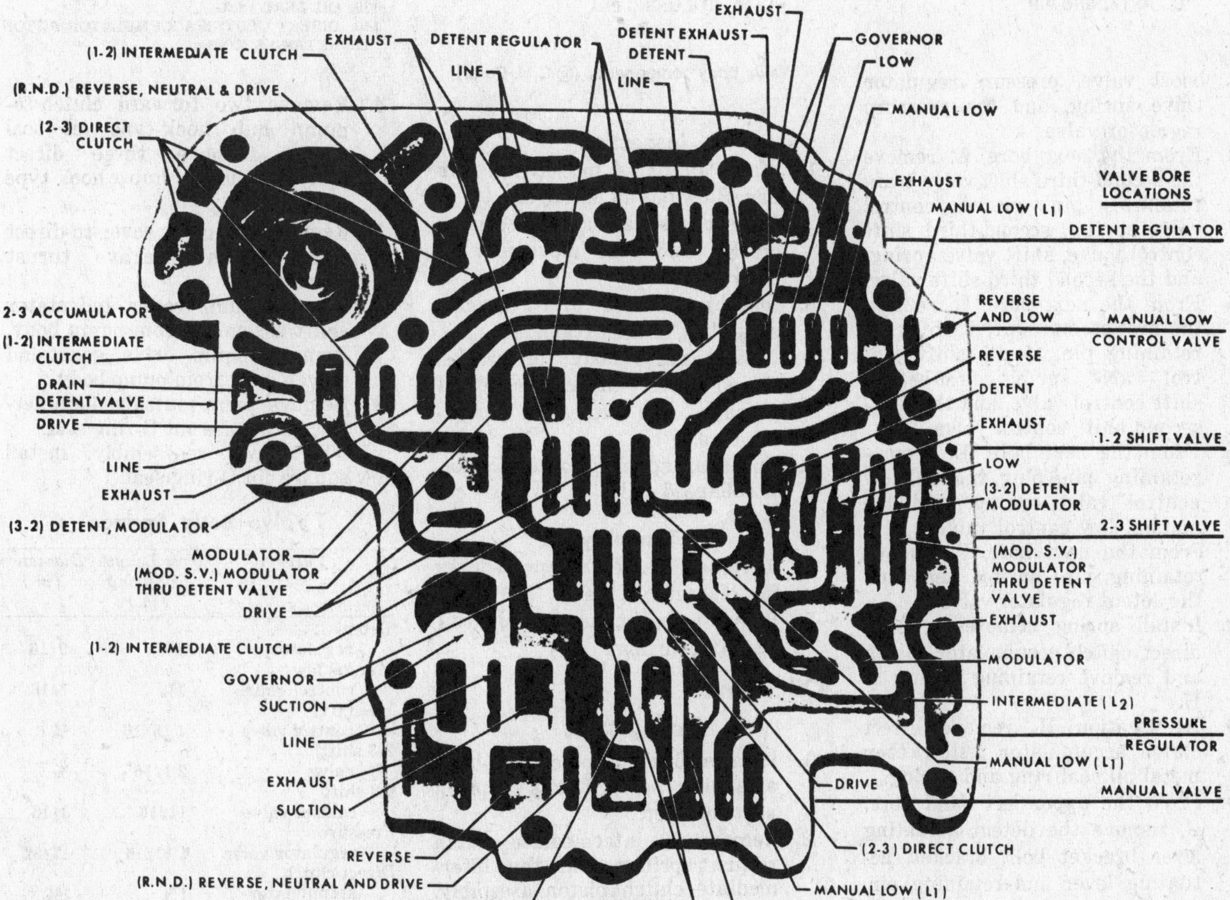

Identification of case face oil channels (© G.M. Corp.)

Identification of valve body oil channels (© G.M. Corp.)

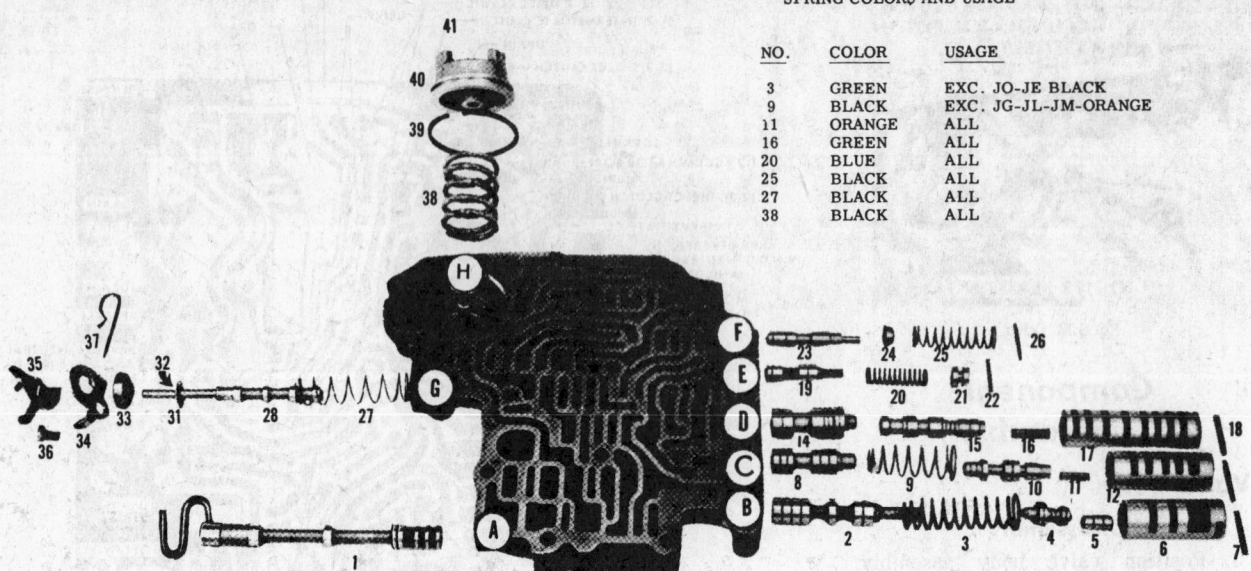

SPRING COLORS AND USAGE

NO.	COLOR	USAGE
3	GREEN	EXC. JO-JE BLACK
9	BLACK	EXC. JG-JL-JM-ORANGE
11	ORANGE	ALL
16	GREEN	ALL
20	BLUE	ALL
25	BLACK	ALL
27	BLACK	ALL
38	BLACK	ALL

Valve body components (© G.M. Corp.)

A 1. MANUAL VALVE AND LINK ASSEMBLY
B 2. PRESSURE REGULATOR VALVE
3. PRESSURE REGULATOR VALVE SPRING
4. REVERSE AND MODULATOR BOOST VALVE
5. INTERMEDIATE BOOST VALVE
6. BOOST VALVE SLEEVE
7. RETAINING PIN
C 8. 2-3 SHIFT VALVE
9. 2-3 SHIFT VALVE SPRING
10. 2-3 SHIFT CONTROL VALVE
11. 2-3 SHIFT CONTROL VALVE SPRING
12. 2-3 SHIFT CONTROL VALVE SLEEVE
13. RETAINING PIN

14. 1-2 SHIFT VALVE
15. 1-2 SHIFT CONTROL VALVE
D 16. 1-2 SHIFT CONTROL SPRING
17. 1-2 SHIFT CONTROL SLEEVE
18. RETAINING PIN
19. MANUAL LOW CONTROL VALVE
E 20. MANUAL LOW CONTROL SPRING
21. PLUG
22. RETAINING PIN
23. DETENT REGULATOR VALVE
F 24. DETENT REGULATOR SPRING SEAT
25. DETENT REGULATOR SPRING
26. RETAINING PIN

27. DETENT VALVE OUTER SPRING
28. DETENT VALVE

31. DETENT VALVE OUTER SPRING SEAT
G 32. DETENT VALVE SPRING RETAINER
33. DETENT VALVE STOP
34. DETENT VALVE BRACKET
35. DETENT VALVE ACTUATING LEVER
36. RETAINING BOLT
37. RETAINING PIN
38. DIRECT CLUTCH ACCUMULATOR SPRING
H 39. OIL SEAL RING
40. DIRECT CLUTCH ACCUMULATOR PISTON
41. RETAINER RING

boost valve, pressure regulator valve spring, and the pressure regulator valve.

4. From the next bore, C, remove the second-third shift valve train retaining pin, sleeve, control valve spring, second-third shift control valve, shift valve spring, and the second-third shift valve.

5. From the next bore, D, remove the first-second shift valve train retaining pin, sleeve, shift control valve spring, first-second shift control valve, and the first-second shift control valve.

6. From the next bore, E, remove retaining pin, plug, manual low control valve spring, and the manual low control valve.

7. From the next bore, F, remove retaining pin, spring, seat and the detent regulator valve.

8. Install spring compressor onto direct clutch accumulator piston and remove retaining E-ring at H.

9. At location H, remove direct clutch accumulator piston, then metal oil seal ring and spring.

10. From the upper left hand bore, G, remove the detent actuating lever bracket bolt, bracket, actuating lever and retaining pin, stop, spring retainer, seat, outer spring, and the detent valve.

Inspection

1. Wash all parts in solvent. Air dry. Blow out all passages.
2. Inspect all valves for scoring, cracks, and free movement in their bores.
3. Inspect sleeves for cracks, scratches, or distortion.
4. Inspect valve body for cracks, scored bores, interconnected oil passages, and flatness of mounting face.
5. Check all springs for distortion or collapsed coils.

Assembly

1. Reverse disassembly procedures for assembly. Refer to Valve Body Springs chart for identification of springs.

Oil Pump

Disassembly

1. Remove five pump cover-to-body attaching bolts. Remove spring seat retainer.
2. Remove 30 intermediate clutch return springs and the intermediate clutch piston assembly.
3. Remove intermediate clutch piston inner and outer seals.

4. Remove two forward clutch-to-pump hub hook-type oil seal rings. Remove three direct clutch-to-pump hub hook-type oil rings.
5. Remove pump cover-to-direct clutch drum selective thrust washer.
6. Remove pump cover and stator shaft assembly from pump body.
7. Remove pump drive gear and driven gear from pump body.
8. Remove pump outside diameter-to-case square cut O-ring seal.
NOTE: upon reassembly, install new square cut O-ring seal.

Valve Body Springs

Valve	Free Length of Spring (in.)	Diameter (in.)
Detent regulator	1⅞	9/16
Manual low control valve	1½	7/16
1-2 shift control valve	1 15/16	¼
2-3 shift valve	2 1/16	⅞
2-3 shift control valve	11/16	3/16
Pressure regulator valve	1 11/16	17/32
Direct clutch accumulator	1¾	1½
Detent valve	1⅞	¾

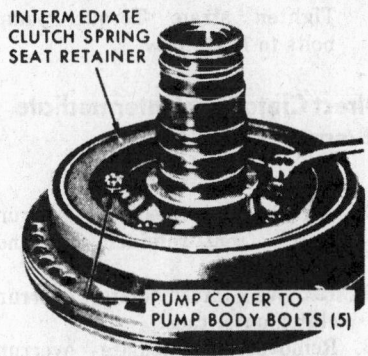

Removing oil pump cover bolts
(© G.M. Corp.)

INTERMEDIATE CLUTCH SPRING SEAT RETAINER

PUMP COVER TO PUMP BODY BOLTS (5)

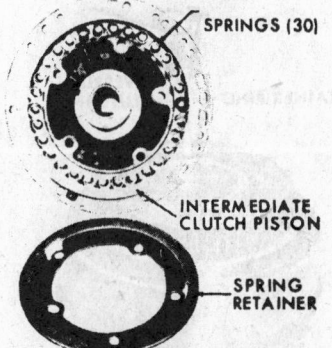

Details of intermediate clutch return springs
(© G.M. Corp.)

SPRINGS (30)

INTERMEDIATE CLUTCH PISTON

SPRING RETAINER

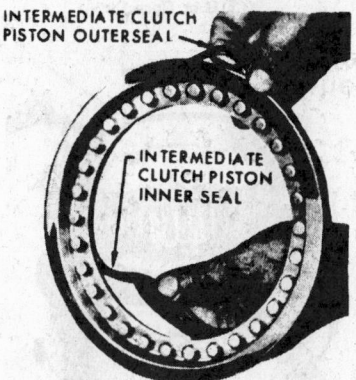

Removal of intermediate clutch piston seal
(© G.M. Corp.)

INTERMEDIATE CLUTCH PISTON OUTER SEAL

INTERMEDIATE CLUTCH PISTON INNER SEAL

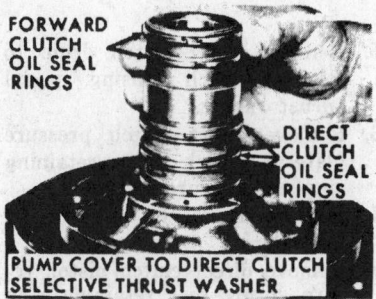

Removal of oil seal rings and selective thrust washer
(© G.M. Corp.)

FORWARD CLUTCH OIL SEAL RINGS

DIRECT CLUTCH OIL SEAL RINGS

PUMP COVER TO DIRECT CLUTCH SELECTIVE THRUST WASHER

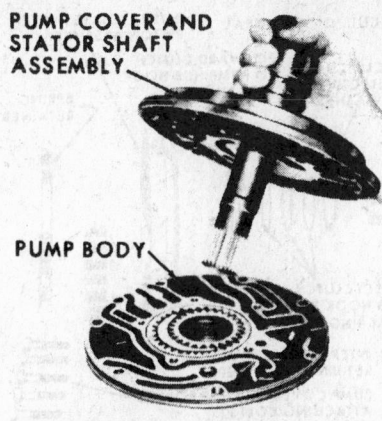

Removing pump cover and stator shaft
(© G.M. Corp.)

PUMP COVER AND STATOR SHAFT ASSEMBLY

PUMP BODY

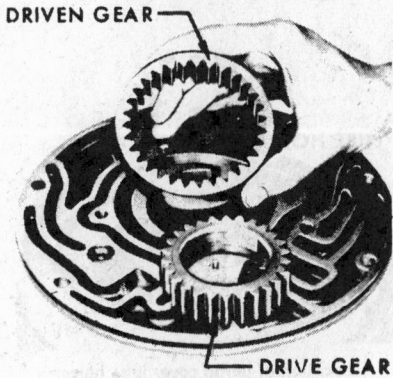

Removal of pump gears
(© G.M. Corp.)

DRIVEN GEAR

DRIVE GEAR

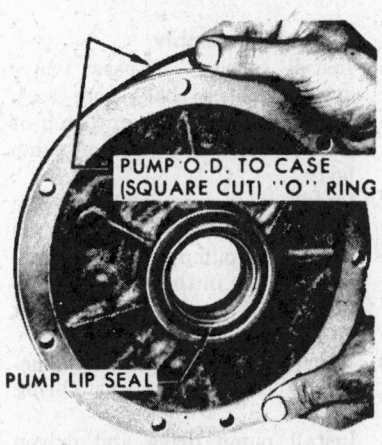

Removal of pump to case O-ring
(© G.M. Corp.)

PUMP O.D. TO CASE (SQUARE CUT) "O" RING

PUMP LIP SEAL

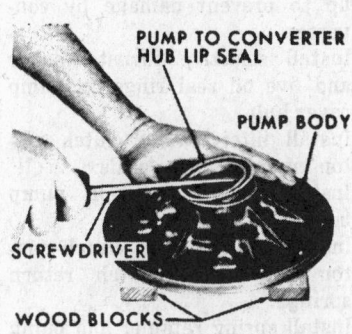

Prying out pump to converter hub lip seal
(© G.M. Corp.)

PUMP TO CONVERTER HUB LIP SEAL

PUMP BODY

SCREWDRIVER

WOOD BLOCKS

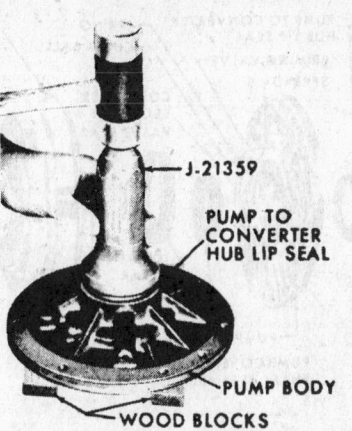

Driving in lip seal
(© G.M. Corp.)

J-21359

PUMP TO CONVERTER HUB LIP SEAL

PUMP BODY

WOOD BLOCKS

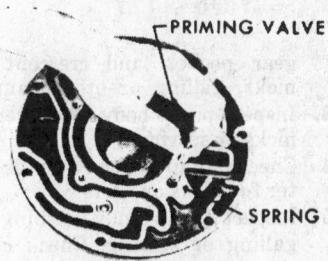

Removal of priming valve
(© G.M. Corp.)

PRIMING VALVE

SPRING

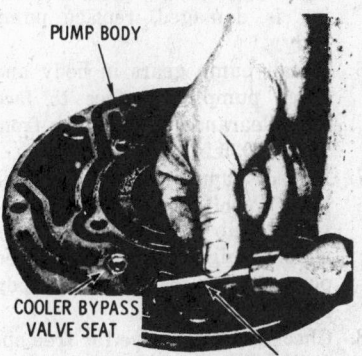

Removing cooler by-pass valve seat
(© G.M. Corp.)

PUMP BODY

COOLER BYPASS VALVE SEAT

5/16" DIA. ROD

9. Pry out pump body to converter hub lip seal, using a screwdriver. Place pump on wood blocks to prevent damage to surface finish.

10. Install pump-to-converter hub lip seal, using seal driver. Examine after installation to be sure the sealing surface is not damaged.

11. Remove oil pump priming valve and spring.

12. Remove cooler by-pass valve seat. Pack cooler by-pass passage with grease. Insert 5/16 in. dia. rod and tap with hammer to lift seat. Remove check ball and spring.

Inspection

1. Wash all parts in solvent. Air dry. Blow out all passages.

2. Inspect drive and driven gears,

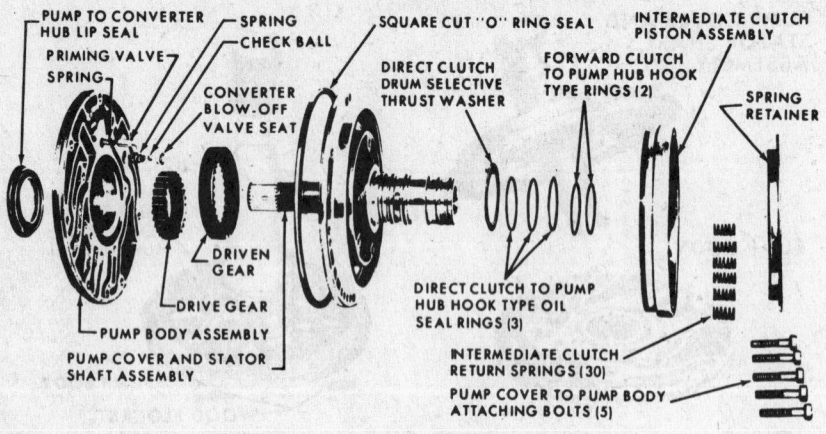

Components—oil pump assembly (© G.M. Corp.)

gear pocket, and crescent for nicks, galling or other damage.

3. Inspect pump body and cover for nicks or scoring.

4. Check pump cover outer diameter for nicks or burrs.

5. Inspect pump body bushing for galling or scoring. Check clearance between pump body bushing and converter hub. It must be no more than .005 in. If bushing is damaged, replace pump body.

6. Install pump gears in body and check pump body face to face gear clearance. It should be from .0008-.0035 in.

7. Inspect pump body to converter hub lip oil seal. Inspect converter hub for nicks or burrs which might have damaged pump lip oil seal or pump body bushing.

8. Check priming valve for free operation. Replace if necessary.

9. Check condition of cooler bypass valve. Replace valve if it leaks excessively.

10. Check all springs for distortion or collapsed coils.

11. Check oil passages in pump body and in pump cover.

12. Inspect three pump cover stator shaft bushings for galling or scoring. If they are damaged, remove using a slide hammer. Drive the new bushings into place. The front stator shaft bushing must be .250 in. below the front face of the pump body. The center bushing should be 1-1/32 in. below the face of the pump cover hub. The rear bushing should be flush or up to .010 in. below the face of the pump cover hub.

13. Check three pump cover and hub lubrication holes to make certain they are not restricted.

Location of pump cover lube holes
(© G.M. Corp.)

Assembly

1. Install cooler by-pass valve spring, check ball and seat. Press seat into bore until top of seat is flush with face of pump body.

2. Install oil pump priming valve and spring. The priming valve is used on all pump bodies having a reamed hole in the priming valve area and on all replacement pump assemblies.

3. Install new pump outside diameter to case square cut O-ring seal.

4. Install pump drive and driven gears. If drive gear has offset tangs, assemble with tangs face up to prevent damage by converter.

5. Install selective thrust washer and five oil seal rings on pump cover hub.

6. Install intermediate clutch piston inner and outer seals.

7. Install pump cover on pump body.

8. Install intermediate clutch piston and thirty clutch return springs.

9. Install spring retainer and pump cover bolts. Position aligning strap over pump body and cover.

Tighten strap. Torque pump bolts to 18 ft. lbs.

Direct Clutch and Intermediate Overrun Roller Clutch

Disassembly

1. Remove intermediate overrun clutch front retainer ring and retainer.

2. Remove intermediate overrun clutch outer race.

3. Remove intermediate overrun roller clutch assembly.

4. Remove intermediate overrun roller clutch cam.

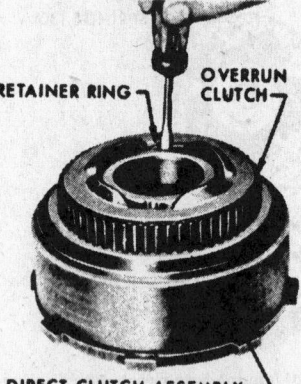

Removing intermediate overrun clutch front retainer ring
(© G.M. Corp.)

Removing intermediate overrun clutch outer race
(© G.M. Corp.)

5. Remove direct clutch drum to forward clutch housing special thrust washer.

6. Remove direct clutch pressure plate to clutch drum retaining ring and pressure plate.

7. Remove direct clutch lined and steel plates.

8. Remove direct clutch piston return spring seat retaining ring and spring seat.

9. Remove 17 clutch return coil springs and piston.

Removing intermediate overrun roller clutch assembly
(© G.M. Corp.)

INTERMEDIATE OVERRUN ROLLER CLUTCH ASSEMBLY

DIRECT CLUTCH DRUM

DIRECT CLUTCH HOUSING TO FORWARD CLUTCH HOUSING THRUST WASHER

Removing thrust washer
(© G.M. Corp.)

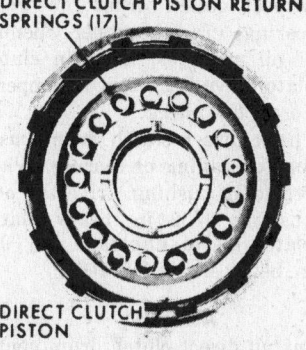

DIRECT CLUTCH PISTON RETURN SPRINGS (17)

DIRECT CLUTCH PISTON

Location of direct clutch return springs
(© G.M. Corp.)

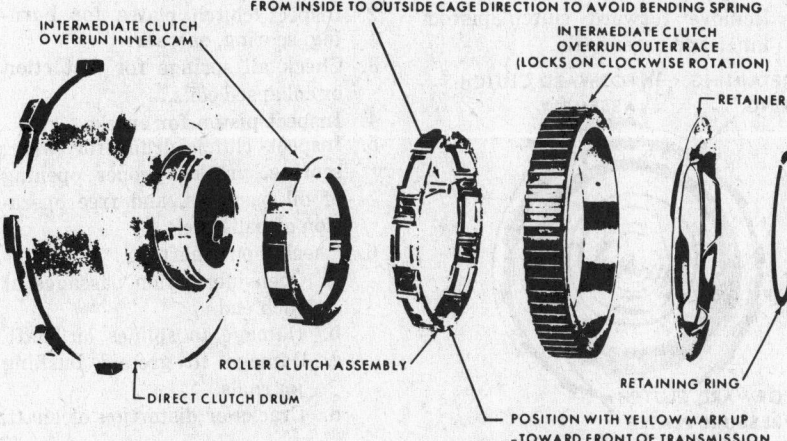

CAUTION: IF ROLLER FALLS OUT DURING ASSEMBLY OPERATION—REINSTALL ROLLER FROM INSIDE TO OUTSIDE CAGE DIRECTION TO AVOID BENDING SPRING

INTERMEDIATE CLUTCH OVERRUN INNER CAM

INTERMEDIATE CLUTCH OVERRUN OUTER RACE (LOCKS ON CLOCKWISE ROTATION)

RETAINER

ROLLER CLUTCH ASSEMBLY

DIRECT CLUTCH DRUM

RETAINING RING

POSITION WITH YELLOW MARK UP —TOWARD FRONT OF TRANSMISSION

Intermediate overrunning roller clutch assembly (© G.M. Corp.)

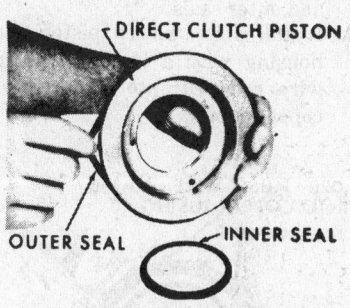

DIRECT CLUTCH PISTON

OUTER SEAL   INNER SEAL

Removing direct clutch piston seals
(© G.M. Corp.)

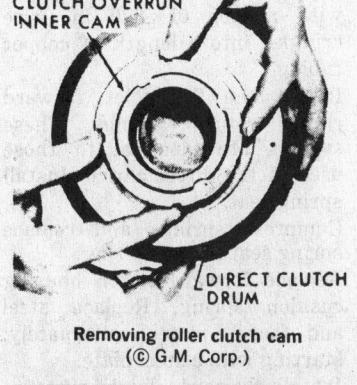

INTERMEDIATE CLUTCH OVERRUN INNER CAM

DIRECT CLUTCH DRUM

Removing roller clutch cam
(© G.M. Corp.)

RETAINING RING
PRESSURE PLATE
DIRECT CLUTCH DRUM

Removing direct clutch retaining ring
(© G.M. Corp.)

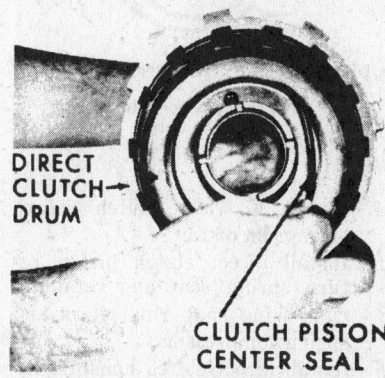

DIRECT CLUTCH DRUM

CLUTCH PISTON CENTER SEAL

Removing direct clutch drum center seal
(© G.M. Corp.)

10. Remove direct clutch piston inner and outer seals.
11. Remove direct clutch piston center seal.

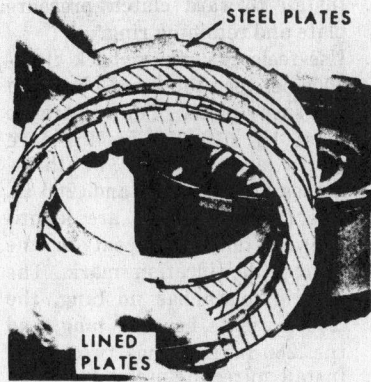

STEEL PLATES

LINED PLATES

Direct clutch plates (© G.M. Corp.)

DIRECT CLUTCH HOUSING   J-2590-3
                               J-2590-5

RETAINING RING

SNAP RING PLIERS

SPRING RETAINER

Removing spring seat retaining ring
(© G.M. Corp.)

Inspection

1. Wash all parts in solvent. Blow out all passages. Air dry.
2. Inspect clutch plates for burning, scoring, or wear.
3. Check all springs for collapsed coils or distortion.
4. Inspect piston for cracks and free operation of ball check. Ball should be loose enough to rattle but not to fall out.
5. Check overrun clutch inner cam and outer race for scratches, wear, or indentations.
6. Inspect overrun roller clutch assembly rollers for wear. Check springs for distortion.

7. Inspect clutch drum for wear, scoring, cracks, proper opening of oil passages, wear on clutch plate drive lugs, and free operation of ball check.

8. Check direct clutch drum bushing for galling or scoring. When replacing bushing, drive the new bushing 9/32 in. below clutch plate side of hub face and .010 in. below slot in hub face.

Assembly

1. Install direct clutch drum center seal.
2. Install direct clutch piston inner and outer seals.
3. Install direct clutch piston into housing with a loop of .020 in. wire crimped into a length of copper tubing.

.020" MUSIC WIRE CRIMPED INTO COPPER TUBING

PISTON

DIRECT CLUTCH HOUSING

Installing clutch piston into housing
(© G.M. Corp.)

4. Install 17 violet clutch return springs on piston.
5. Install direct clutch piston return spring seat and retaining ring using snap ring pliers and spring compressor.
6. Install direct clutch housing. Install steel and faced plates alternately, beginning with a steel plate.
7. Install direct clutch pressure plate in clutch drum. Install retaining ring.
8. Install intermediate overrun roller clutch inner cam on hub of direct clutch drum.
9. Replace intermediate overrun roller clutch assembly.
10. Install intermediate overrun clutch outer race. The outer race must free-wheel in the counterclockwise direction only.
11. Replace intermediate overrun clutch retainer and retainer ring. If the retainer ring is dished, install the ring so that it compresses the retainer.
12. Install direct clutch drum to forward clutch housing needle thrust washer on hub of roller clutch inner race.

Forward Clutch

Disassembly

1. Remove forward clutch drum to pressure plate retaining ring.
2. Remove forward clutch pressure plate.
3. Remove forward clutch housing faced plates, steel plates and cushion spring.
4. Compress springs. Remove forward clutch piston return spring seat retaining ring and spring seat.
5. Remove 21 clutch return springs (violet in color).
6. Remove forward clutch piston assembly.
7. Remove forward clutch piston inner and outer seals.

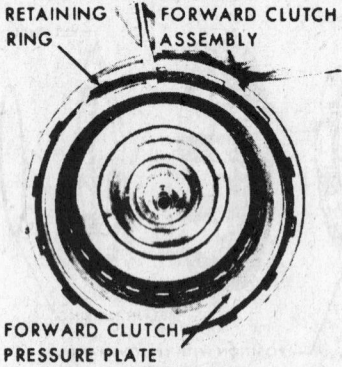

RETAINING RING

FORWARD CLUTCH ASSEMBLY

FORWARD CLUTCH PRESSURE PLATE

Removing forward clutch pressure plate retaining ring
(© G.M. Corp.)

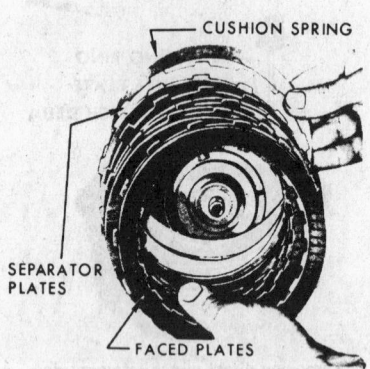

CUSHION SPRING

SEPARATOR PLATES

FACED PLATES

Forward clutch plates (© G.M. Corp.)

FORWARD CLUTCH PISTON SPRING SEAT

RETAINING RING

Forward clutch piston retaining ring and spring seat
(© G.M. Corp.)

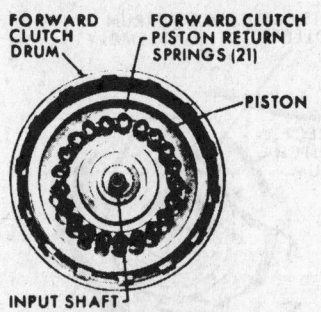

FORWARD CLUTCH DRUM

FORWARD CLUTCH PISTON RETURN SPRINGS (21)

PISTON

INPUT SHAFT

Forward clutch springs
(© G.M. Corp.)

Inspection

1. Wash all parts in solvent. Blow out all passages. Air dry.
2. Inspect clutch plates for burning, scoring, or wear.
3. Check all springs for distortion or collapsed coils.
4. Inspect piston for cracks.
5. Inspect clutch drum for wear, scoring, cracks, proper opening of oil passages, and free operation of ball check.
6. Check input shaft for:
   a. Open lubrication passages at each end.
   b. Damage to splines or shaft.
   c. Damage to ground bushing journals.
   d. Cracks or distortion of shaft.

Assembly

1. Install forward clutch piston inner and outer seals.
2. Install forward clutch piston with a loop of .020 in. wire crimped into a length of copper tubing.
3. Install the 21 violet forward clutch return springs. These springs are identical to those used in the direct clutch. Install spring seat.
4. Compress springs and replace spring seat retaining ring.
5. Replace forward clutch housing cushion spring. Replace steel and faced plates alternately, starting with a steel plate.
6. Install forward clutch pressure plate and retaining ring.
7. Use feeler gauge to check clearance between top faced plate and pressure plate. The clearance must be .010-.080 in. Pressure plates are available in thicknesses of .200, .230, and .260 in. The pressure plates are identified by tangs adjacent to the source identification mark. The .200 in. plate has no tang, the .230 in. plate has one tang, and the .260 in. plate has two tangs.
8. Install direct clutch drum on input shaft and align faced

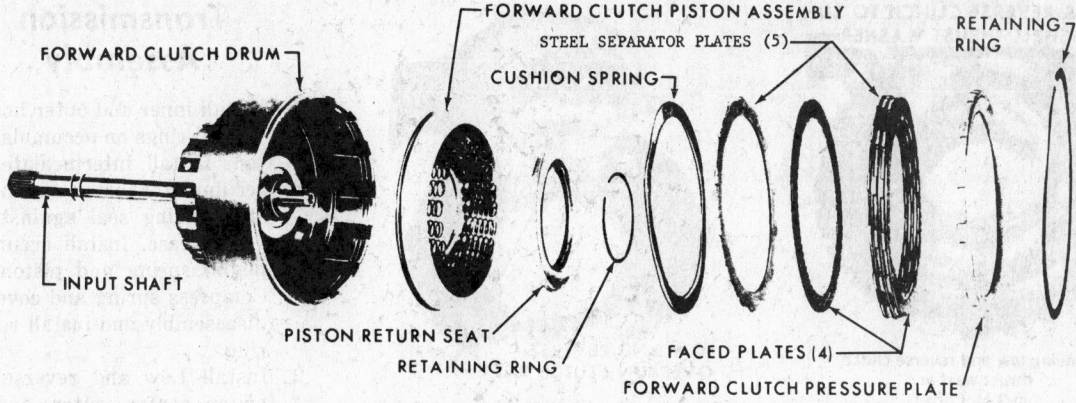

Forward clutch assembly (© G.M. Corp.)

Labels: FORWARD CLUTCH PISTON ASSEMBLY, STEEL SEPARATOR PLATES (5), CUSHION SPRING, RETAINING RING, FORWARD CLUTCH DRUM, INPUT SHAFT, PISTON RETURN SEAT, RETAINING RING, FACED PLATES (4), FORWARD CLUTCH PRESSURE PLATE

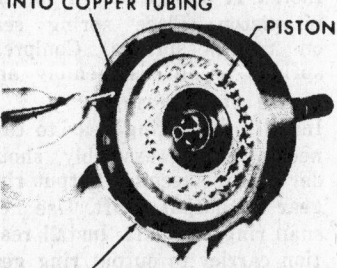

Installing forward clutch piston
(© G.M. Corp.)

Labels: .020" MUSIC WIRE CRIMPED INTO COPPER TUBING, PISTON, FORWARD CLUTCH HOUSING

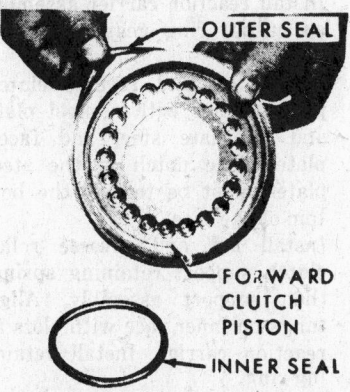

Installing forward clutch piston seals
(© G.M. Corp.)

Labels: OUTER SEAL, FORWARD CLUTCH PISTON, INNER SEAL

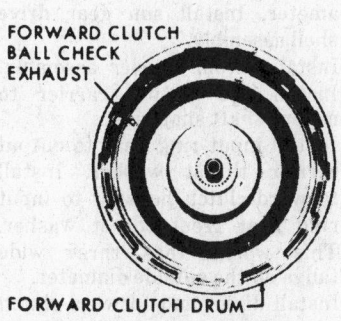

Forward clutch ball check
(© G.M. Corp.)

Labels: FORWARD CLUTCH BALL CHECK EXHAUST, FORWARD CLUTCH DRUM

plates with splines on forward clutch housing.

Sun Gear and Sun Gear Drive Shell

Disassembly
1. Remove sun gear to sun gear drive shell rear retaining ring.
2. Remove sun gear to drive shell flat rear thrust washer.

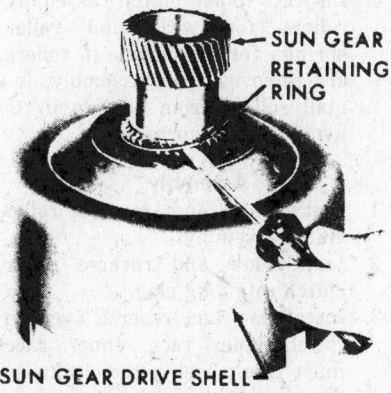

Removing sun gear retaining ring
(© G.M. Corp.)

Labels: SUN GEAR RETAINING RING, SUN GEAR DRIVE SHELL

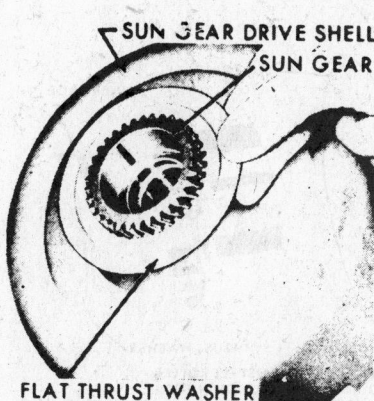

Removing sun gear rear thrust washer
(© G.M. Corp.)

Labels: SUN GEAR DRIVE SHELL, SUN GEAR, FLAT THRUST WASHER

3. Remove sun gear and front retaining ring from drive shell.
4. Remove front retaining ring from sun gear.

Inspection
1. Wash all parts in solvent. Air dry.
2. Inspect sun gear and sun gear drive shell for wear or damage.
3. Inspect sun gear bushings for galling or scoring. Drive out damaged bushings. Install new bushings flush to .010 in. below surface of counterbore.

Assembly
1. Install new front retaining ring on sun gear. Be careful not to overstress this ring.
2. Install sun gear and retaining ring in drive shell.

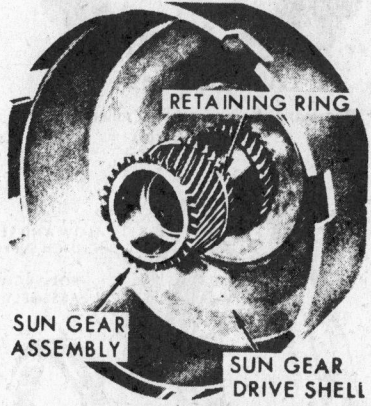

Sun gear front retaining ring
(© G.M. Corp.)

Labels: RETAINING RING, SUN GEAR ASSEMBLY, SUN GEAR DRIVE SHELL

3. Install sun gear to drive shell flat rear thrust washer.
4. Install new sun gear to sun gear drive shell retaining ring.

Low and Reverse Roller Clutch Support

Disassembly
1. Remove low and reverse clutch to sun gear shell thrust washer.

LOW & REVERSE CLUTCH TO SUN GEAR SHELL THRUST WASHER

Removing low and reverse clutch thrust washer
(© G.M. Corp.)

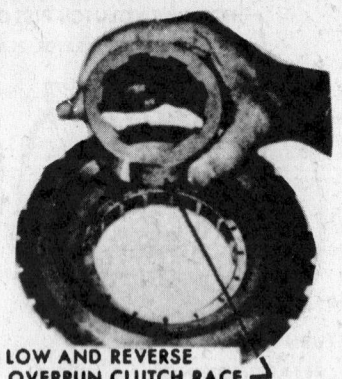

LOW AND REVERSE OVERRUN CLUTCH RACE

Removing low and reverse overrun clutch inner race
(© G.M. Corp.)

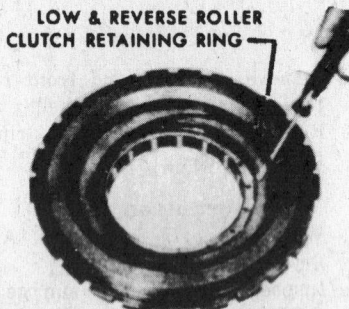

LOW & REVERSE ROLLER CLUTCH RETAINING RING

Removing low and reverse roller clutch retaining ring
(© G.M. Corp.)

LOW & REVERSE ROLLER CLUTCH ASSEMBLY

Removing low and reverse roller clutch
(© G.M. Corp.)

2. Remove low and reverse overrun clutch inner race.
3. Remove low and reverse roller clutch retaining ring.
4. Remove low and reverse roller clutch assembly.

Inspection

1. Wash all parts in solvent. Air dry.
2. Inspect roller clutch inner and outer races for scratches, wear, or indentations.
3. Inspect roller clutch assembly rollers for wear and roller springs for distortion. If rollers are removed from assembly, install rollers from outside in to avoid bending springs.

Assembly

1. Install low and reverse roller clutch assembly.
2. Install low and reverse roller clutch retaining ring.
3. Install low and reverse overrun clutch inner race. Inner race must free-wheel in the clockwise direction only.

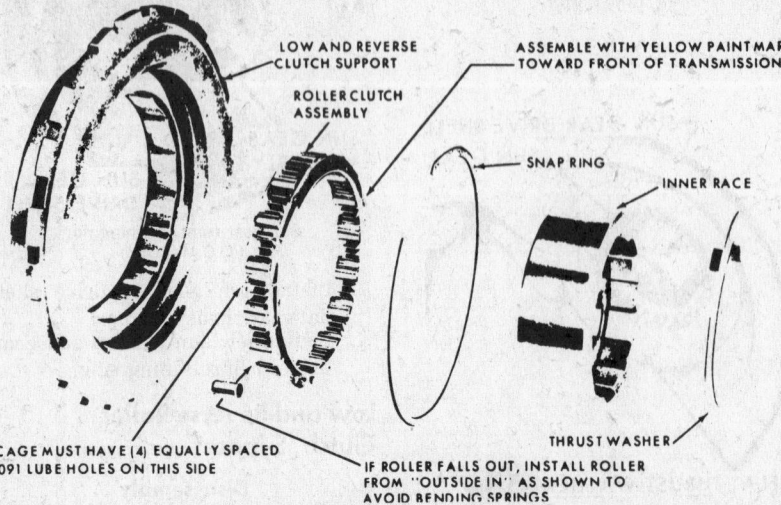

LOW AND REVERSE CLUTCH SUPPORT

ROLLER CLUTCH ASSEMBLY

ASSEMBLE WITH YELLOW PAINT MARK TOWARD FRONT OF TRANSMISSION

SNAP RING

INNER RACE

THRUST WASHER

CAGE MUST HAVE (4) EQUALLY SPACED .091 LUBE HOLES ON THIS SIDE

IF ROLLER FALLS OUT, INSTALL ROLLER FROM "OUTSIDE IN" AS SHOWN TO AVOID BENDING SPRINGS

Low and reverse roller clutch assembly
(© G.M. Corp.)

Transmission Assembly

1. Install inner and outer hook type oil seal rings on accumulator piston. Install intermediate clutch accumulator piston assembly. Install O-ring seal against shoulder in case. Install accumulator piston spring and piston cover. Compress spring and cover as in disassembly and install retaining ring.
2. Install Low and reverse clutch piston center, outer, and inner seal on piston. Install piston assembly. Notch in low and reverse clutch piston must be installed adjacent to parking pawl. Install 17 orange return springs in piston. Place spring seat on return springs. Compress springs as in disassembly and install retaining ring.
3. Install output ring gear to case needle bearing assembly, shoulder down. Assemble output ring gear on output shaft. Use new snap ring on shaft. Install reaction carrier to output ring gear tanged thrust washer. This washer has three narrow tangs on the outside diameter. Install output ring rear and shaft assembly in case.
4. Install reaction carrier assembly on output ring gear and shaft assembly.
5. Install low and reverse clutch plates. Start with a steel plate and alternate steel and faced plates. The notch in the steel plates must be toward the bottom of the case.
6. Install low and reverse roller clutch support retaining spring, then support assembly. Align tangs on inner race with slots in reaction carrier. Install retaining ring.
7. Install sun gear to roller clutch thrust washer on the roller clutch inner race. This washer has four tangs on the inside diameter. Install sun gear drive shell assembly.
8. Install output carrier assembly. Install new output carrier to output shaft snap ring.
9. Install input ring gear to output carrier thrust washer. Install forward clutch housing to input ring gear front thrust washer. This washer has three wide tangs on the outside diameter.
10. Install direct and forward clutch assemblies in case. Align plates with splines and insert until

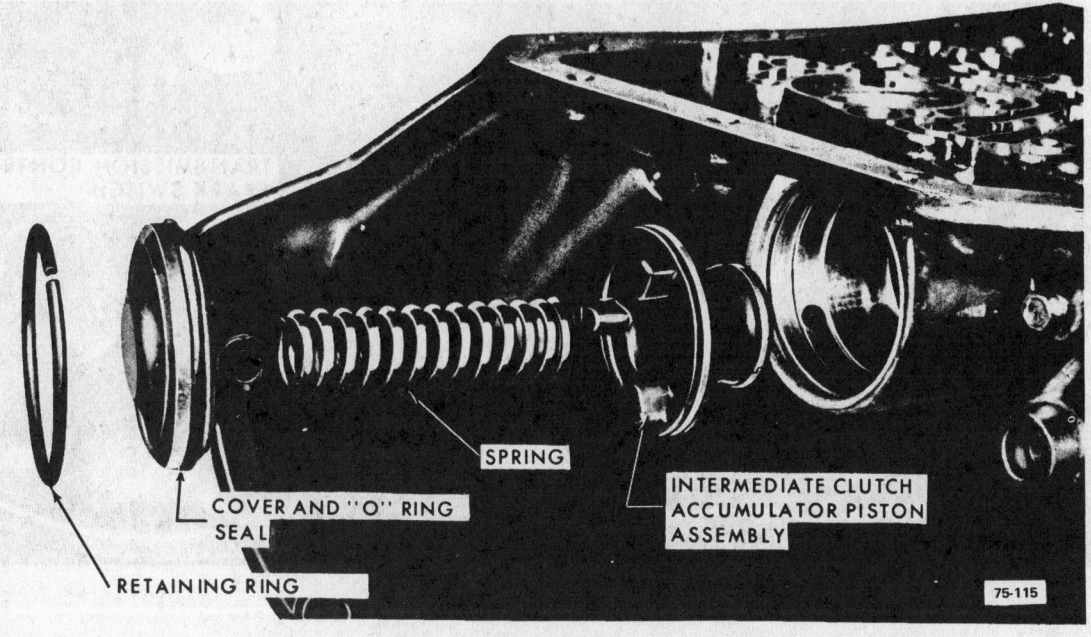

SPRING

COVER AND "O" RING SEAL

INTERMEDIATE CLUTCH ACCUMULATOR PISTON ASSEMBLY

RETAINING RING

75-115

Intermediate clutch accumulator (© G.M. Corp.)

tangs are 1/8 in. below notches in drive shell.

11. Install intermediate overrun brake band, intermediate clutch pressure plate, and steel and faced clutch plates. Insert a steel plate first, then alternate faced and steel plates. Install intermediate clutch cushion spring.

12. Lubricate case bore, then install new pump to case gasket in bore. Install the proper selective thrust washer to obtain .033-.064 in. shaft end play with pump installed. The selective thrust washer has one tang on the outside diameter.

13. Install pump into case. Drive in until seated. Install new washer type seals on pump attaching bolts. Torque to 15-25 ft. lbs. If the input shaft cannot be rotated clockwise as the pump is being bolted into place, the direct clutch and forward clutch housing has not been properly installed to index the faced plates with their respective parts. This condition must be corrected before the pump is bolted into place.

14. Invert transmission and install intermediate servo spring, spring seat, apply pin, washer, and piston assembly.

15. Install parking pawl tooth toward the inside of case. Install parking pawl shaft into case through disengaging spring. Install disengaging spring on parking pawl and slide shaft through pawl. Drive in pawl shaft retainer plug using a 3/8 in. drift. Drive plug flush to .010 in. below face of case. Stake plug in three places. Install park lock bracket. Torque bolts to 29 ft. lbs.

16. Install range selector inner lever to parking pawl actuating rod. Install actuating rod under the park lock bracket and parking pawl.

17. Drive in new manual shaft lip seal using a 7/8 in. drift. Insert manual shaft through case and range selector inner lever. Torque manual shaft jam nut to 30 ft. lbs. Install manual shaft retainer.

18. Replace governor screens in case. Replace oil pump pressure screen in oil pump pressure hole in case. Ring end of screen goes in first. Replace four check balls in correct passages in case face. Replace valve body spacer plate to case gasket, spacer plate, spacer support plate, and seven bolts. Do not tighten bolts yet. Install valve body to spacer plate gasket.

19. Install detent control valve link in detent actuating lever. Install valve body while installing manual control valve link in range selector inner lever. When handling valve body, be careful not to depress sleeves as retaining pins may fall out into the transmission.

20. Torque valve body attaching bolts to 13 ft. lbs. Install detent roller and spring to valve body.

21. Replace oil pump screen and gasket to valve body. Replace oil pump screen attaching screws.

22. Install pan and gasket. Torque bolts to 12 ft. lbs.

23. Check governor bore and sleeve for scoring. Install governor assembly into case. Install new O-ring on governor cover and install governor cover. Use a brass drift on the flange of the cover, but do not hammer on the cover. Replace governor cover retainer.

24. Place retainer clip on output shaft. Slide speedometer drive gear over output shaft until secured by clip.

25. Position yoke seal on output shaft. Tap seal into place with a suitable tool.

26. Install extension housing lip seal if not replaced on disassembly. Install extension housing square cut O-ring seal. Replace extension housing. Torque bolts to 35 ft. lbs.

27. Lubricate new modulator O-ring, then install on modulator. Replace vacuum modulator valve in case, bolting down retainer.

28. Install torque converter to transmission, engaging lugs, and install a holding tool to prevent converter from dropping off.

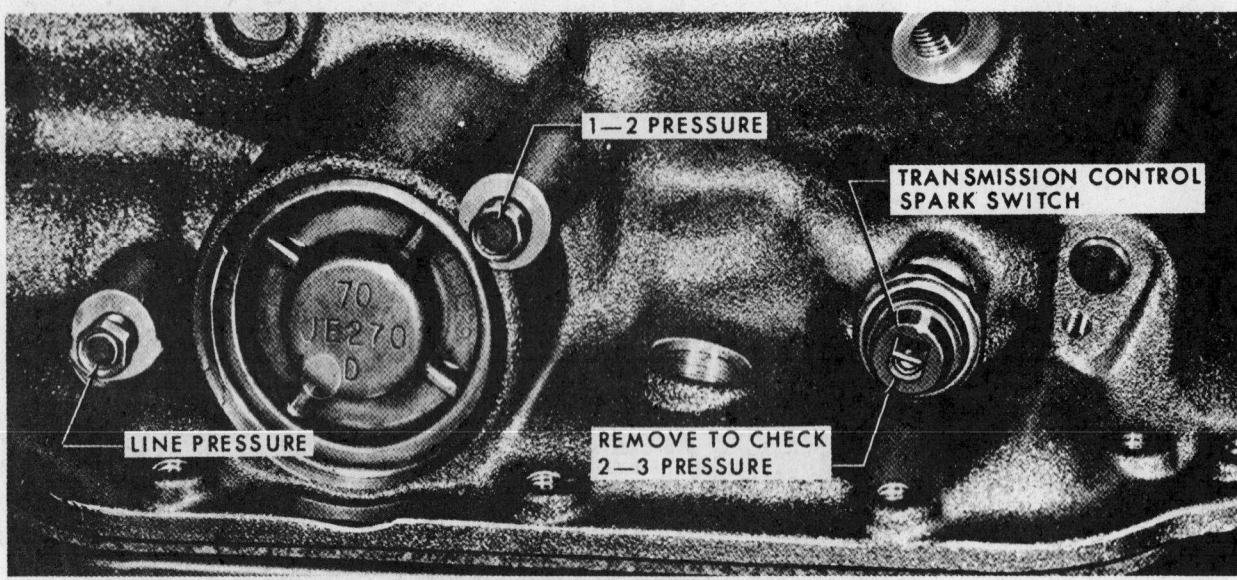

Side view of transmission (© G.M. Corp.)

Torque Specifications

Oil pan to transmission case	13 ft. lbs.
Pump assembly to transmission case	20 ft. lbs.
Vacuum modulator retainer to case	12 ft. lbs.
Valve body assembly to case	13 ft. lbs.
Oil channel support plate to case	13 ft. lbs.
Pump body to pump cover	15 ft. lbs.
Parking lock bracket to case	29 ft. lbs.
Extension housing to case	35 ft. lbs.
Inside shift nut	30 ft. lbs.
External test plugs to case	8 ft. lbs.
Cooler fitting to case	30 ft. lbs.
Oil pickup screen to valve body	36 in. lbs.
Detent valve actuating lever bracket to valve body	48 in. lbs.

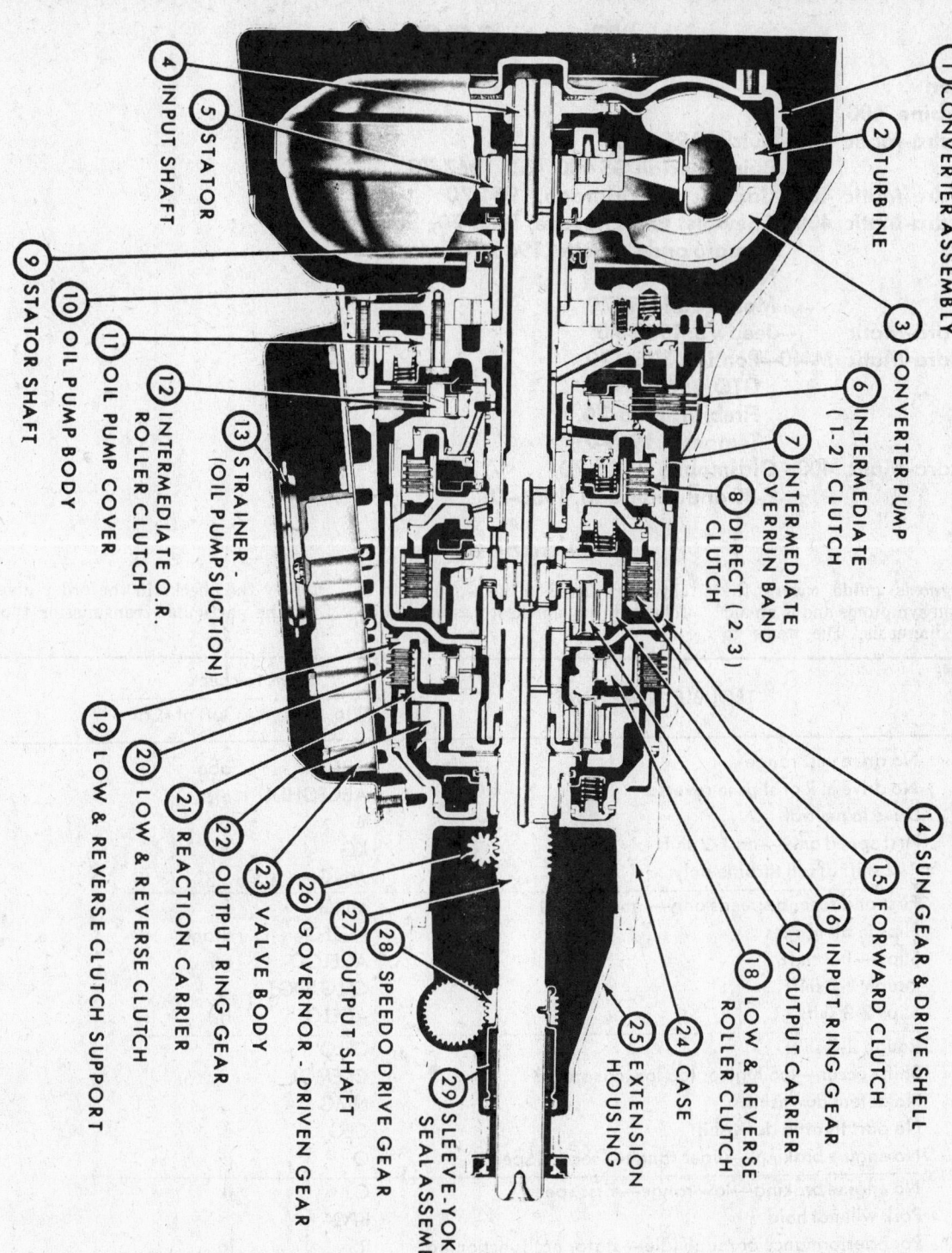

GENERAL MOTORS TYPE 350 TRANSMISSION

1. CONVERTER ASSEMBLY
2. TURBINE
3. CONVERTER PUMP
4. INPUT SHAFT
5. STATOR
6. INTERMEDIATE (1-2) CLUTCH
7. INTERMEDIATE OVERRUN BAND
8. DIRECT (2-3) CLUTCH
9. STATOR SHAFT
10. OIL PUMP BODY
11. OIL PUMP COVER
12. INTERMEDIATE O.R. ROLLER CLUTCH
13. STRAINER - (OIL PUMP SUCTION)
14. SUN GEAR & DRIVE SHELL
15. FORWARD CLUTCH
16. INPUT RING GEAR
17. OUTPUT CARRIER
18. LOW & REVERSE ROLLER CLUTCH
19. LOW & REVERSE CLUTCH SUPPORT
20. LOW & REVERSE CLUTCH
21. REACTION CARRIER
22. OUTPUT RING GEAR
23. VALVE BODY
24. CASE
25. EXTENSION HOUSING
26. GOVERNOR - DRIVEN GEAR
27. OUTPUT SHAFT
28. SPEEDO DRIVE GEAR
29. SLEEVE-YOKE SEAL ASSEMBLY

Type 400

Application
Super Turbine 400,
Turbo Hydra-Matic 400—Buick, 1964-70,
Buick Special, GS 400, 455, 1967-70
Turbo Hydra-Matic 400—Cadillac and Eldorado, 1964-70
Turbo Hydra-Matic 400—Chevrolet and Chevelle, 1966-70,
Camaro and Corvette, 1968-70,
Nova, 1969-70,
Monte Carlo, 1970
Turbo Hydra-Matic   —Jeep V8, 1965-70
Turbo Hydra-Matic M-40—Pontiac, 1966-70,
GTO, 1967-70,
Firebird, 1968-70,
Tempest, 1969-70
Turbo Hydra-Matic 400—Oldsmobile, 1965-70,
F-85 and Toronado, 1966-70

Diagnosis

This diagnosis guide covers the most common symptoms and is an aid to careful diagnosis. The items to check are listed in the sequence to be followed for quickest results. Thus, follow the checks in the order given for the particular transmission type.

TROUBLE SYMPTOMS	Items to check	
	In Car	Out of Car
No drive in D range	ABCD	abc
No drive in R or slips in reverse	ABCEGHIJL	efa
Drive in neutral	B	a
First speed only—no 1-2 shift	KG	g
1-2 shift at full throttle only	MNG	
First and second speeds only—no 2-3 shift	MNG	f
Slips in all ranges	ACEFG	dfaħc
Slips—1-2 shift	ACELGO	gh
Rough 1-2 shift	CELGJOGJ	hi
Slips 2-3 shift	ACELG	fh
Rough 2-3 shift	CELO	
Shifts occur—too high or too low car speed	CKEMGL	
No detent downshifts	NMG	
No part throttle downshift	CEG	
No engine braking—super range—second speed	O	i
No engine braking—low range—first speed	GJ	ik
Park will not hold	BPQ	
Poor performance or rough idle—stator not functioning	RS	lo
Noisy transmission	TF	dafgn

Key to Checks

A. Oil level
B. Manual linkage (external)
C. Check oil pressure
D. Manual control disconnected inside
E. Modulator and/or lines
F. Clogged strainer or intake leaks
G. Valves, body and/or leaks
H. Reverse feed passages

I. Valve check balls
J. Rear servo and accumulator
K. Governor and/or feed line seals
L. Pump regulator and boost valve
M. Detent solenoid
N. Detent switch
O. Front servo and accumulator
P. Internal linkage

Q. Parking pawl and/or link
R. Stator switch
S. Valve body-stator section
T. Cooler or lines
a. Front clutch
b. Clutch feed seals and gaskets
c. Low sprags
d. Front pump
e. Rear band
f. Direct clutch

g. Intermediate clutch
h. Pump-to-case gasket
i. Intermediate check valve ball in case
j. Front band
k. Rear band
l. Turbine shaft
m. Converter assembly
n. Planetary assembly

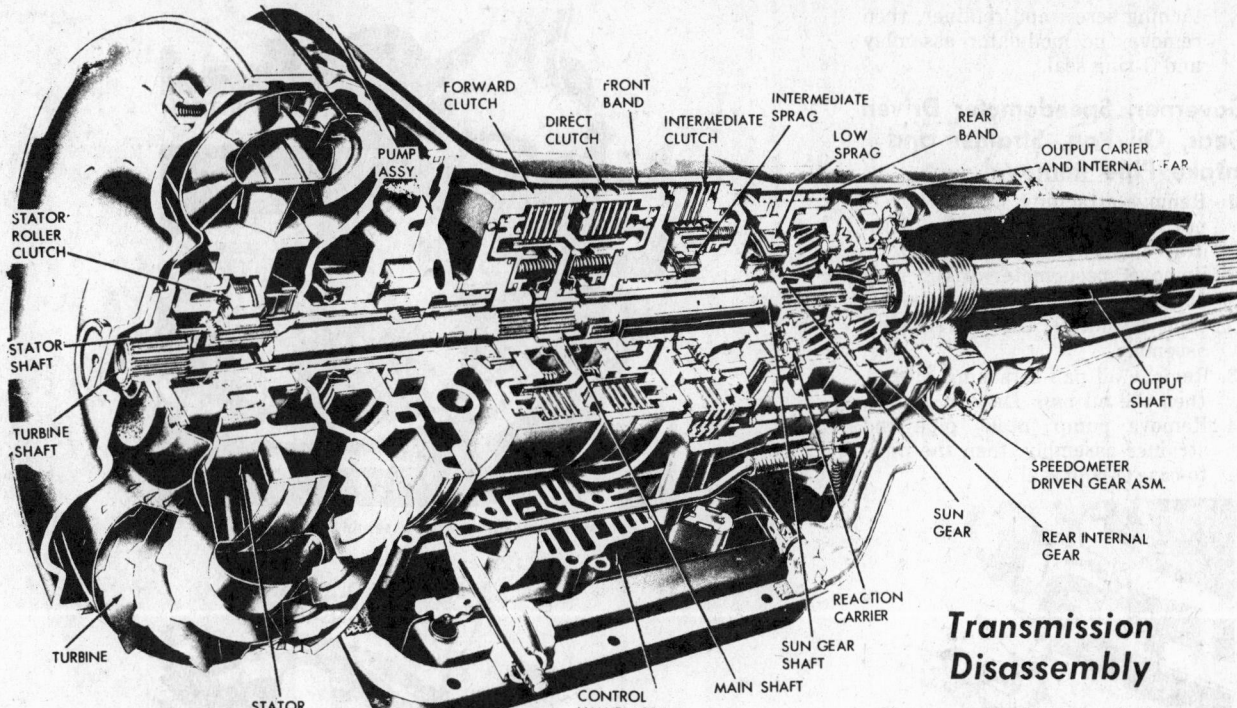

General Motors, Type 400 transmission
(© G.M. Corp.)

The illustration is typical. While design and repair procedures are similar, components vary with car application and should not be considered interchangeable.

General Information

The General Motors type 400 transmission is a fully automatic unit consisting of a three element torque converter and a three speed and reverse planetary gear set. Three multiple disc clutches, one sprag, one roller clutch, and two bands control the planetary gear set operation.

The torque converter consists of a driving member, a driven member, and a stator assembly. The stator has a roller clutch which allows it to turn in one direction only. As the speed of the driving member increases, the roller clutch allows the stator to free-wheel on its shaft. This causes the torque converter to cease its torque multiplication function and to operate as a fluid coupling with a drive ratio of approximately one to one.

Fluid Specifications

Oil capacity, transmission
　and converter
　　Approx. 22 pints
Capacity between marks
　on dip stick
　　1 pint
Type of oil
　Automatic transmission fluid
　　Type A
Drain and refill
　24,000 miles

Draining and Refilling

The transmission fluid should be drained and refilled at 24,000 mile intervals. To do this:
1. Drain transmission oil.
2. Remove oil pan attaching screws.
3. Remove oil pan and gasket.
4. Remove strainer or filter.
5. Thoroughly clean oil strainer or filter and oil pan in solvent. Blow dry with air.
6. Renew oil strainer gasket and oil pan gasket.
7. Install oil strainer or filter.
8. Install oil pan. Torque attaching screws to 10-12 ft. lbs.
9. Add about five pints of transmission fluid.
10. Start engne. Do not race engine. Move selector lever slowly through each range. Check fluid level.
11. With transmission cold, add enough fluid to bring level to $\frac{1}{4}$ in. below the Add mark. With the transmission at normal temperature, add enough fluid to bring the level to the Full mark. About fifteen miles of highway driving is required to bring the transmission to normal temperature.
12. In case of transmission overhaul, proceed as in Steps 9-11. Start with five pints of fluid if converter has not been replaced; eight pints if it has been replaced. Total capacity is about 22 pints, but correct level is determined by the mark on the dipstick rather than by amount of fluid added.

Transmission Disassembly

Clean outside of the unit thoroughly to prevent dirt from entering the unit.
1. With transmission in a work cradle or on a clean bench, lift the converter straight off the transmission input shaft.
2. With transmission bottom-up, remove modulator assembly at-

Transmission in holding fixture
(© G.M. Corp.)

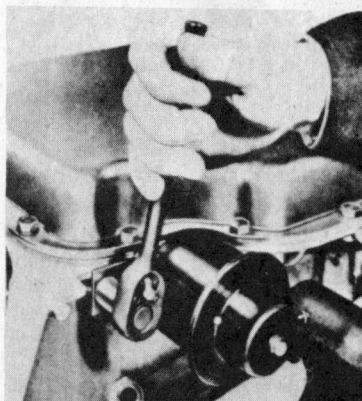

Removing modulator retaining bolt
(© G.M. Corp.)

taching screw and retainer, then remove the modulator assembly and O-ring seal.

Governor, Speedometer Driven Gear, Oil Pan, Strainer and Intake Pipe Removal

1. Remove attaching screws, governor cover and gasket, then withdraw governor from the case.
2. Remove speedometer driven gear attaching screw and retainer, then withdraw the driven gear assembly.
3. Remove oil pan attaching screws, then the oil pan. Discard gasket.
4. Remove pump intake pipe and strainer assembly, then the pipe-to-case O-ring seal.

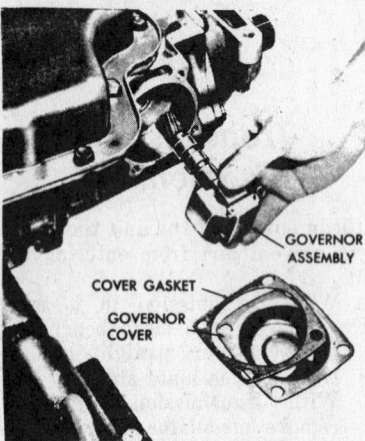

Removing governor
(© G.M. Corp.)

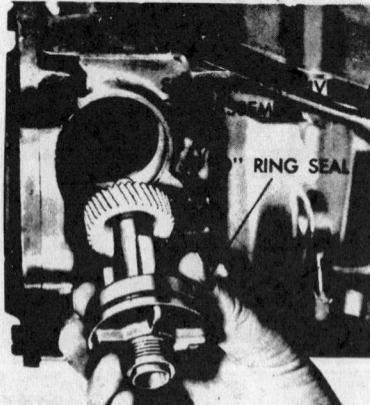

Removing speedometer driven gear
(© G.M. Corp.)

Control Valve Assembly, Governor Pipes and Detent Spring Assembly Removal

1. Disconnect the lead wire from the pressure switch assembly. Remove the switch assembly from the valve body, if necessary. This switch is installed on 1970 models only.
2. Remove control valve body attaching screws and detent roller and spring assembly.

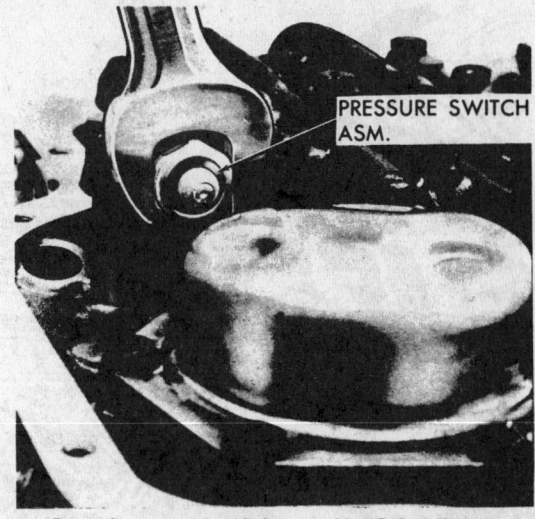

Removing pressure switch assembly (© G.M. Corp.)

Removing detent roller and spring assembly (© G.M. Corp.)

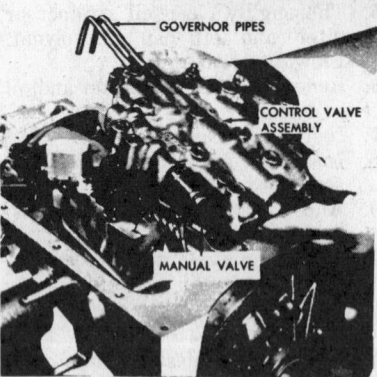

Removing control valve assembly
(© G.M. Corp.)

NOTE: do not remove solenoid attaching screws.

Remove control valve body and governor pipes.

3. Remove the governor pipes from valve body. Then remove valve body-to-spacer gasket.

Rear Servo, Solenoid, Valve Body Spacer, Front Servo, Manual Detent and Parking Linkage Removal

1. Remove rear servo cover attaching screws, the cover and gasket,

then the rear servo assembly from the case.
2. Remove servo accumulator springs.
3. Disconnect the solenoid leads from connector terminal. Withdraw connector and O-ring seal.
4. Remove solenoid attaching screws, solenoid and gasket.
5. Remove valve body spacer plate and gasket.
6. Remove six check balls from cored passages in transmission case.

Removing rear servo from case
(© G.M. Corp.)

Removing case connector (© G.M. Corp.)

7. Remove the front servo assembly.
8. Loosen the jam nut which holds the detent lever to the manual shaft. Then remove detent lever from manual shaft and remove manual shaft.
9. Remove parking actuator rod and detent lever assembly. Then remove detent lever E-ring and the detent lever.
10. Remove attaching screws and park bracket; then the parking pawl return spring.
11. Remove parking pawl shaft retainer, then the parking pawl shaft, O-ring and pawl.

Removing parking pawl shaft shaft and O-ring (© G.M. Corp.)

Rear Oil Seal and Extension Housing Removal

1. Pry rear oil seal from extension housing.
2. Remove housing attaching bolts, then remove extension housing and housing-to-case oil seal.

Oil Pump, Forward Clutch and Gear Unit Removal

1. Pry front seal from the pump. Then remove pump attaching bolts.
2. With slide hammers attached, remove pump from transmission case. Discard pump-to-case seal ring. Discard pump-to-case gasket.

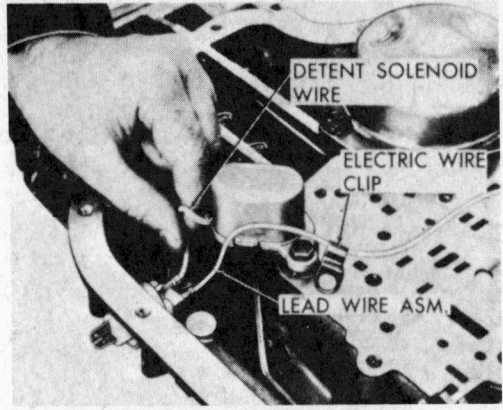

Disconnecting solenoid leads (© G.M. Corp.)

Location of check balls (© G.M. Corp.)

Removing pump body with slide hammers (© G.M. Corp.)

3. Remove turbine shaft from transmission.
4. Remove forward clutch assembly. Be sure that the bronze thrust washer came out with the clutch housing assembly.
5. Remove the direct clutch assembly. Remove front band and sun gear shaft.
6. Remove the case center support-to-case bolt and center support locating screw. 1969-70 models have no locating screw.
7. Remove intermediate clutch backing plate-to-case snap-ring. Then remove the backing plate, three composition, and three steel clutch plates.

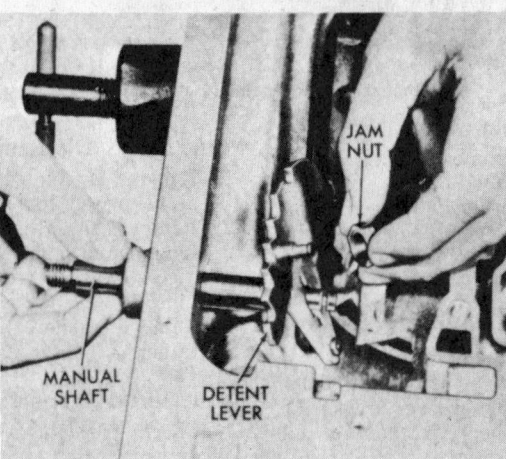

Removing jam nut and manual shaft (© G.M. Corp.)

Removing center support bolt (© G.M. Corp.)

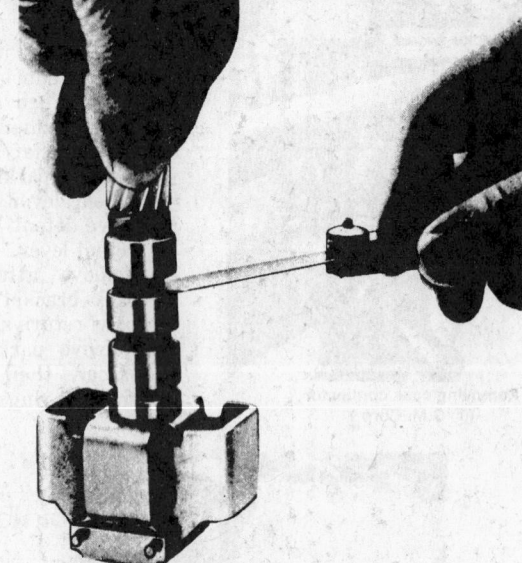

Checking governor feed port opening (© G.M. Corp.)

Removing gear assembly
(© G.M. Corp.)

8. Remove the center support-to-case retaining snap-ring.
9. Remove the entire gear unit assembly.
10. Remove the output shaft-to-case thrust washer from the rear of the output shaft, or from inside the case.
11. Remove rear unit selective washer from transmission case.
12. Remove rear band assembly.
13. Remove support to case spacer from inside case.

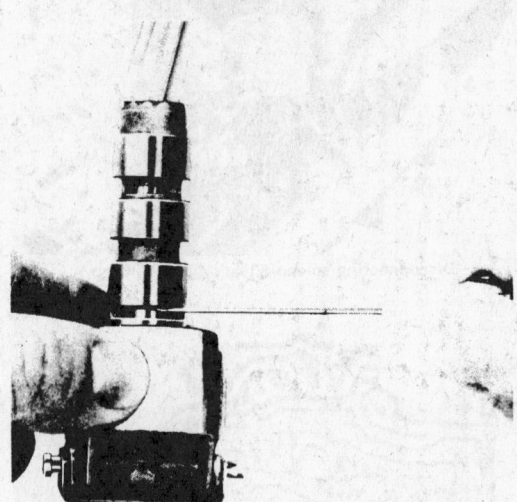

Checking governor exhaust port opening (© G.M. Corp.)

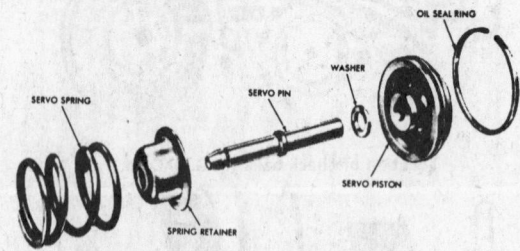

Front servo components (© G.M. Corp.)

Component Disassembly, Inspection, and Assembly

Governor

All components of the governor, except the driven gear, are a select fit and so calibrated. Therefore, service this unit as an assembly.

Clean and inspect all parts for wear or other damage. Check valve opening at feed port with a feeler gauge, holding the governor with the weights extended completely outward. Check valve opening at exhaust port, holding governor with weights completely inward. If either opening is less than .020 in., replace governor assembly.

If a new governor drive gear is installed, a new pin hole must be drilled 90 degrees from the original hole.

Front Servo Inspection

1. Inspect servo pin for damage.
2. Inspect servo piston for damaged oil ring groove, cracks, or porosity. Check freedom of oil seal ring in groove.
3. Check fit of servo pin in piston.

Rear Servo Disassembly

1. Remove E-ring which holds the servo piston to band apply pin.
2. Remove servo piston and seal from band apply pin. Remove second washer from band apply pin.
3. Remove washer, spring, and retainer.

Rear Servo Inspection

1. Check freedom of accumulator rings in piston.
2. Check fit of band apply pin in

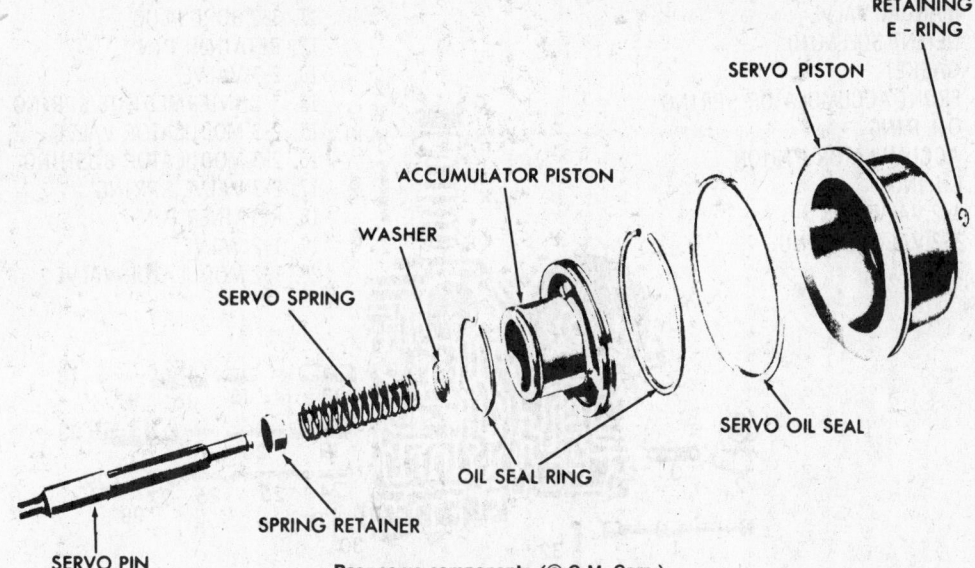

Rear servo components (© G.M. Corp.)

servo piston. Inspect pin for scores or cracks.

3. Inspect accumulator and servo pistons for cracks or porosity.

Rear Servo Assembly

1. Install spring retainer, spring, and washer on band apply pin.
2. Install band apply pin, retainer, spring, and washer into bore of servo piston and secure with E-ring.
3. Install oil seal on servo piston. Install outer and inner oil seal rings on accumulator piston and assembly into bore of servo piston.

Control Valve Body Disassembly

When disassembling control valve body, be careful to identify springs so that they can be replaced in their proper locations.

1. Position valve assembly with cored face up and accumulator pocket on bottom.
2. Remove manual valve from upper bore.
3. Compress accumulator piston spring and remove E-ring retainer. Remove accumulator piston and spring.
4. Press out retaining pin from upper right bore. Remove 1-2 modulator bushing, 1-2 regulator valve and spring, 1-2 detent valve, and 1-2 shift valve. 1-2 regulator valve and spring may be inside of 1-2 modulator bushing.
5. Press out retaining pin from center right bore. Remove 2-3 modulator bushing, 2-3 shift valve spring, 2-3 modulator valve, 3-2 intermediate spring, and 2-3 shift valve. 2-3 modulator valve will be inside of 2-3 modulator bushing.
6. Press out retaining pin from lower right bore. Hold hand over bore, as plug may pop out. Remove plug, 3-2 valve spring, spacer, and 3-2 valve.
7. Holding hand over bore, press out retainer pin from upper left bore. Remove bore plug, detent valve, detent regulator valve, spacer, and detent regulator valve spring.
8. Pry out grooved retainer ring from lower left bore with long nose pliers. Remove bore plug, 1-2 accumulator bushing, 1-2 accumulator valve, secondary spring, primary 1-2 accumulator valve, and spring.
9. Remove governor oil feed screen from oil feed hole in valve body.

Control Valve Body Inspection

1. Wash control valve body, valves and other parts in solvent. Do not allow valves to bump together. Air dry parts and blow out all passages.

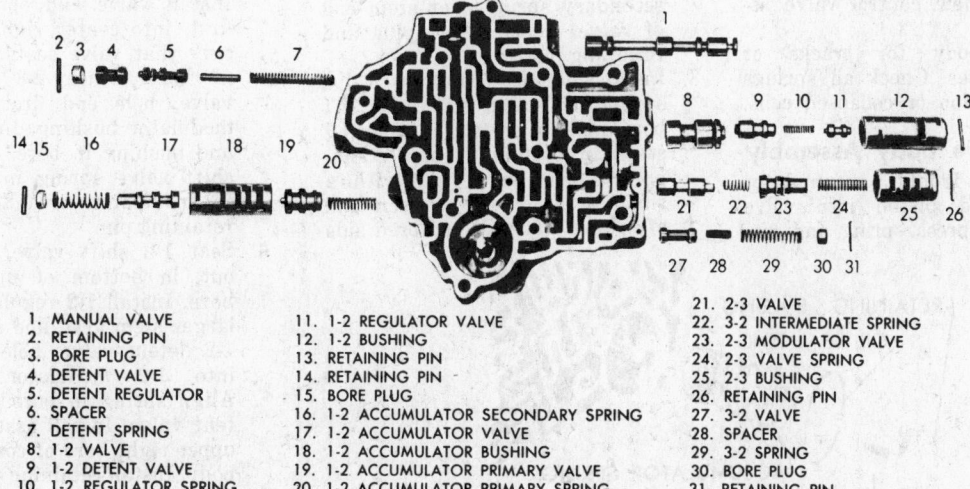

1. MANUAL VALVE	11. 1-2 REGULATOR VALVE
2. RETAINING PIN	12. 1-2 BUSHING
3. BORE PLUG	13. RETAINING PIN
4. DETENT VALVE	14. RETAINING PIN
5. DETENT REGULATOR	15. BORE PLUG
6. SPACER	16. 1-2 ACCUMULATOR SECONDARY SPRING
7. DETENT SPRING	17. 1-2 ACCUMULATOR VALVE
8. 1-2 VALVE	18. 1-2 ACCUMULATOR BUSHING
9. 1-2 DETENT VALVE	19. 1-2 ACCUMULATOR PRIMARY VALVE
10. 1-2 REGULATOR SPRING	20. 1-2 ACCUMULATOR PRIMARY SPRING

21. 2-3 VALVE	
22. 3-2 INTERMEDIATE SPRING	
23. 2-3 MODULATOR VALVE	
24. 2-3 VALVE SPRING	
25. 2-3 BUSHING	
26. RETAINING PIN	
27. 3-2 VALVE	
28. SPACER	
29. 3-2 SPRING	
30. BORE PLUG	
31. RETAINING PIN	

Control valve body components (© G.M. Corp.)

1. MANUAL VALVE
2. DETENT SOLENOID
3. GASKET
4. FRONT ACCUMULATOR SPRING
5. OIL RING
6. ACCUMULATOR PISTON
7. E-RING
8. 3-2 VALVE PIN
9. 3-2 VALVE SPRING
10. 3-2 VALVE

11. 3-2 BORE PLUG
12. RETAINER PIN
13. 2-3 VALVE
14. 3-2 INTERMEDIATE SPRING
15. 2-3 MODULATOR VALVE
16. 2-3 MODULATOR BUSHING
17. 2-3 VALVE SPRING
18. RETAINER PIN
19. 1-2 VALVE
20. 1-2 MODULATOR VALVE

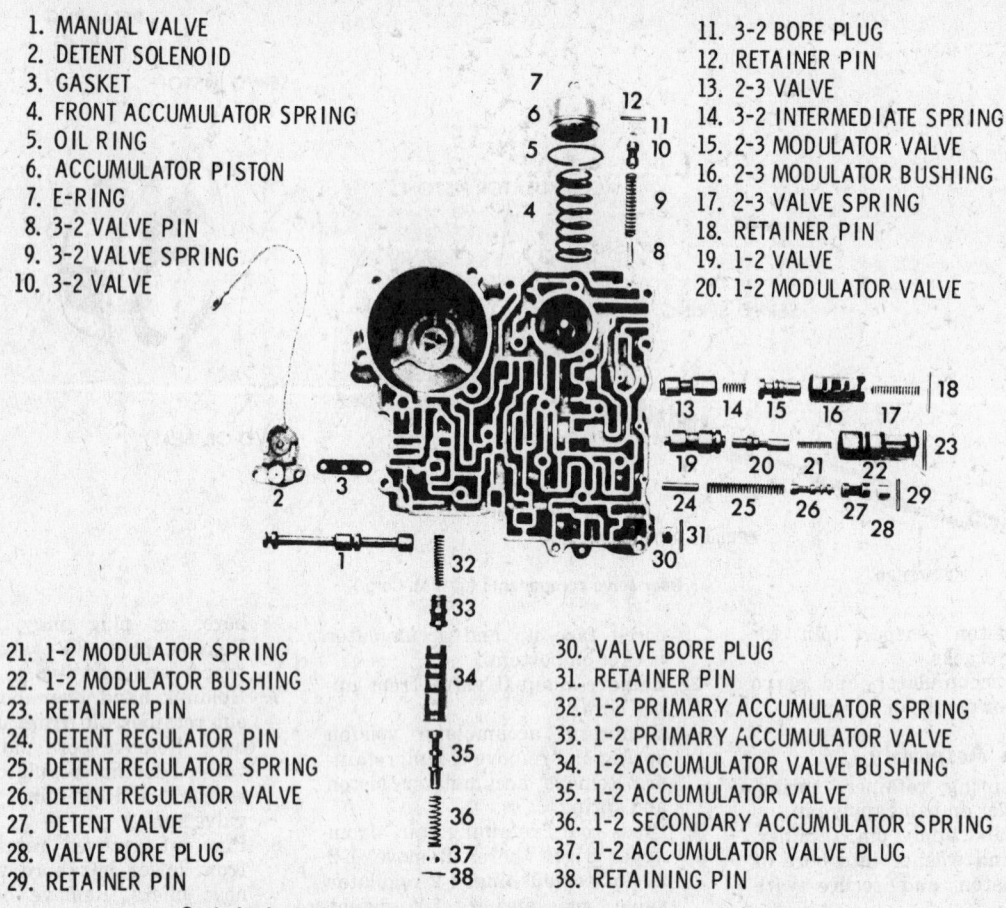

21. 1-2 MODULATOR SPRING
22. 1-2 MODULATOR BUSHING
23. RETAINER PIN
24. DETENT REGULATOR PIN
25. DETENT REGULATOR SPRING
26. DETENT REGULATOR VALVE
27. DETENT VALVE
28. VALVE BORE PLUG
29. RETAINER PIN

30. VALVE BORE PLUG
31. RETAINER PIN
32. 1-2 PRIMARY ACCUMULATOR SPRING
33. 1-2 PRIMARY ACCUMULATOR VALVE
34. 1-2 ACCUMULATOR VALVE BUSHING
35. 1-2 ACCUMULATOR VALVE
36. 1-2 SECONDARY ACCUMULATOR SPRING
37. 1-2 ACCUMULATOR VALVE PLUG
38. RETAINING PIN

Control valve body components, front wheel drive (Toronado) (© G.M. Corp.)

2. Inspect all valves and bushings carefully. Burrs may be removed with a fine stone or fine crocus cloth and light oil. Be careful not to round off shoulders of valves.
3. Test all valves and bushings for free movement in their bores. All valves should fall freely of their own weight.
4. The manual valve is the only valve that can be serviced separately. If any of the other valves are defective or damaged, install a new control valve assembly.
5. Inspect body for cracks or scored bores. Check all springs for distortion or collapsed coils.

Control Valve Body Assembly

1. Replace front accumulator spring and piston into valve body. Compress spring and piston, assuring that piston pin is correctly aligned with hole in piston and that oil seal ring does not catch on lip of bore when installing piston. Secure piston and spring with E-ring retainer.
2. Install 1-2 accumulator primary spring, 1-2 primary valve, and 1-2 accumulator bushing into lower left bore. Place 1-2 accumulator secondary valve, stem end out, into the 1-2 accumulator bushing. Place 1-2 accumulator secondary spring over stem end of valve. Replace bore plug and retaining pin.
3. Install detent regulator valve spring and spacer into upper left bore, making certain that spring seats in bottom of bore. Compress spring and hold with a small screwdriver between end of spring and wall on cored side of valve body. Insert detent regulator valve, stem end out, and detent valve, small land first. Insert bore plug, press inward, remove screwdriver, and install retaining pin.
4. Insert 3-2 valve in bottom right bore. Place spacer inside 3-2 valve spring and insert spacer and spring in bore. Install bore plug and retaining pin.
5. Install 3-2 intermediate spring on stem end of 2-3 shift valve. Install valve and spring, valve first, into center right bore. Be sure that valve seats in bottom of bore. Place 2-3 modulator valve, hole end first, into 2-3 modulator bushing. Install valve and bushing in bore. Install 2-3 shift valve spring into hole in 2-3 modulator valve. Secure with retaining pin.
6. Seat 1-2 shift valve, stem end out, in bottom of upper right bore. Install 1-2 regulator valve, larger stem first, and spring and 1-2 detent valve, hole end first, into 1-2 modulator bushing. Align spring in bore of 1-2 detent valve. Install assembly into upper right bore of control valve body. Install retaining pin.
7. Replace governor oil feed screen assembly in governor oil feed hole.

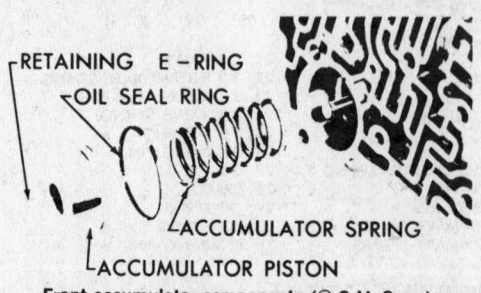

RETAINING E-RING
OIL SEAL RING
ACCUMULATOR SPRING
ACCUMULATOR PISTON

Front accumulator components (© G.M. Corp.)

8. Install manual valve with detent pin groove to the right.

Oil Pump Disassembly

1. Place the pump over a hole in the bench, shaft down, cover up.
2. Compress regulator boost valve bushing against the pressure regulator spring and remove the snap-ring.
3. Remove the boost valve bushing

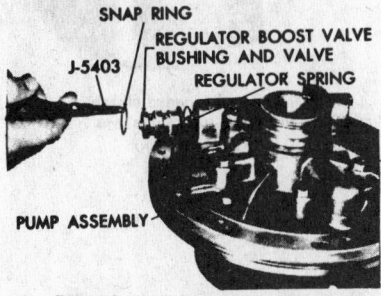

Removing regulator valve snap-ring
(© G.M. Corp.)

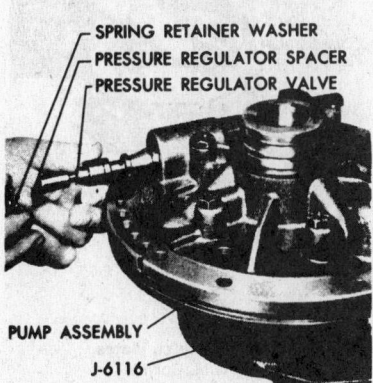

Removing regulator valve and spring
(© G.M. Corp.)

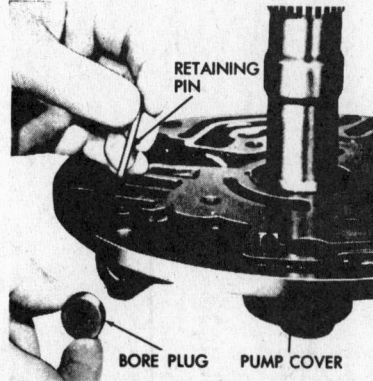

Removing retaining pin and bore plug
(© G.M. Corp.)

and valve, then the spring.

4. Remove valve spring retainer and spacer/s, if present, and regulator valve.
5. Remove pump cover to body attaching bolts, then remove the cover.
6. Remove the retaining pin and bore plug from the pressure regulator bore.
7. Remove the two hook-type oil rings from the pump cover.

Removing hook-type oil rings
(© G.M. Corp.)

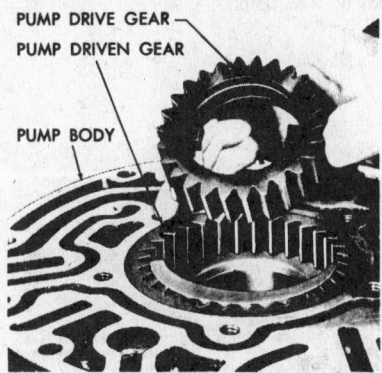

Marking drive and driven gears for assembly
(© G.M. Corp.)

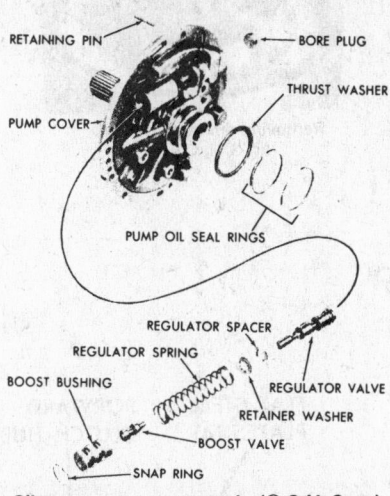

Oil pump cover components (© G.M. Corp.)

8. Remove pump to forward clutch housing selective washer.
9. Mark drive and driven gears for reassembly, then remove the gears.

Oil Pump Inspection

1. Inspect drive gear, driven gear, gear pocket, and crescent for scoring, galling, or other damage.
2. Replace pump gears in pump and check pump body face to face gear clearance. The clearance should be .0008-.0035 in.
3. Check pump body face for scoring or nicks. Check oil passages

Pump body to gear face clearance
(© G.M. Corp.)

in pump body for roughness or obstructions. Check condition of cover bolt attaching threads. Check for flatness of pump body face. Check pump body bushing for scores or nicks. If a new bushing is installed, drive it in flush to .010 in. below the gear pocket face.

4. Replace pump attaching bolt seals if necessary.
5. Check pump cover face for flatness. Check for scoring or chips in pressure regulator bore. Check that all passages are unobstructed. Check for scoring or damage at pump gear face. Check that breather hole in pump cover is open.
6. Check condition of stator shaft splines and bushings.
7. Check oil ring grooves for damage or wear. Check selective thrust washer for wear.
8. Make sure that pressure regulator valve and boost valve operate freely.

Oil Pump Assembly

1. Install drive and driven gears into the pump body, alignment marks up and in proper index, (drive gear with drive tangs up).
2. Install pressure regulator spring spacer/s if required, retainer and spring into the pressure regulator bore.
3. Install pressure regulator valve from the opposite end of the bore, stem end first.
4. Install boost valve into bushing, stem end out, and install both parts into the pump cover by compressing the bushing against the spring. Install the snap-ring.
5. Install the regulator valve bore plug and retaining pin into opposite end of bore.
6. Install the front unit selective thrust washer over the pump cover delivery sleeve.
7. Install two hook-type oil seal rings.
8. Assemble pump cover to pump body with attaching bolts. (Leave bolts one turn loose at this time.)
9. Place pump aligning strap over

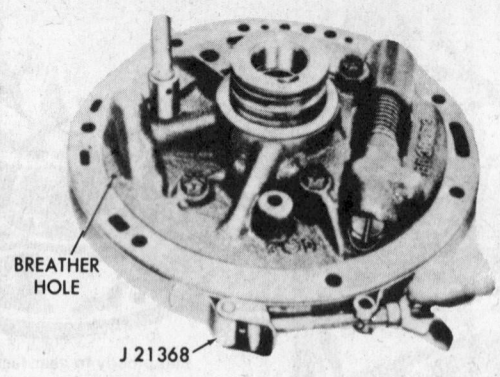

BREATHER HOLE

J 21368

Aligning pump cover and body (© G.M. Corp.)

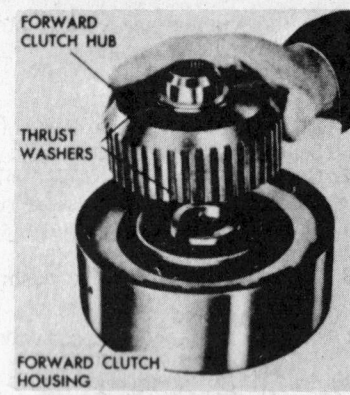

FORWARD CLUTCH HUB

THRUST WASHERS

FORWARD CLUTCH HOUSING

Removing forward clutch hub and thrust washers (© G.M. Corp.)

pump body and cover, then tighten tool.

10. Install pump cover bolts. Remove clamp and install pump-to-case O-ring seal and gasket.

Forward Clutch Disassembly

1. Remove forward clutch housing-to-direct clutch hub snap-ring.
2. Remove direct clutch hub.
3. Remove forward clutch hub and thrust washers.
4. Remove five composition and five steel clutch plates. Press out the turbine shaft.
5. Compress the spring retainer and remove the snap-ring.
6. Remove the compressor, snap-ring, spring retainer and 16 release springs, then lift out the piston.
7. Remove inner and outer piston seals.
8. Remove center piston seal from the forward clutch housing.

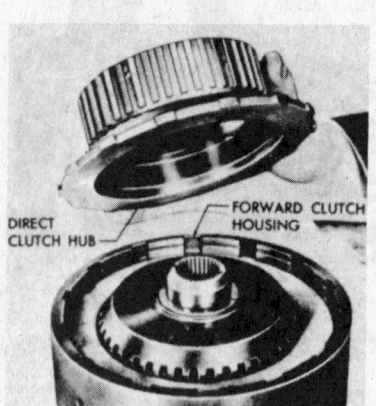

DIRECT CLUTCH HUB

FORWARD CLUTCH HOUSING

Removing direct clutch hub (© G.M. Corp.)

COMPOSITION PLATES

STEEL PLATES

FORWARD CLUTCH HOUSING

Removing clutch plates (© G.M. Corp.)

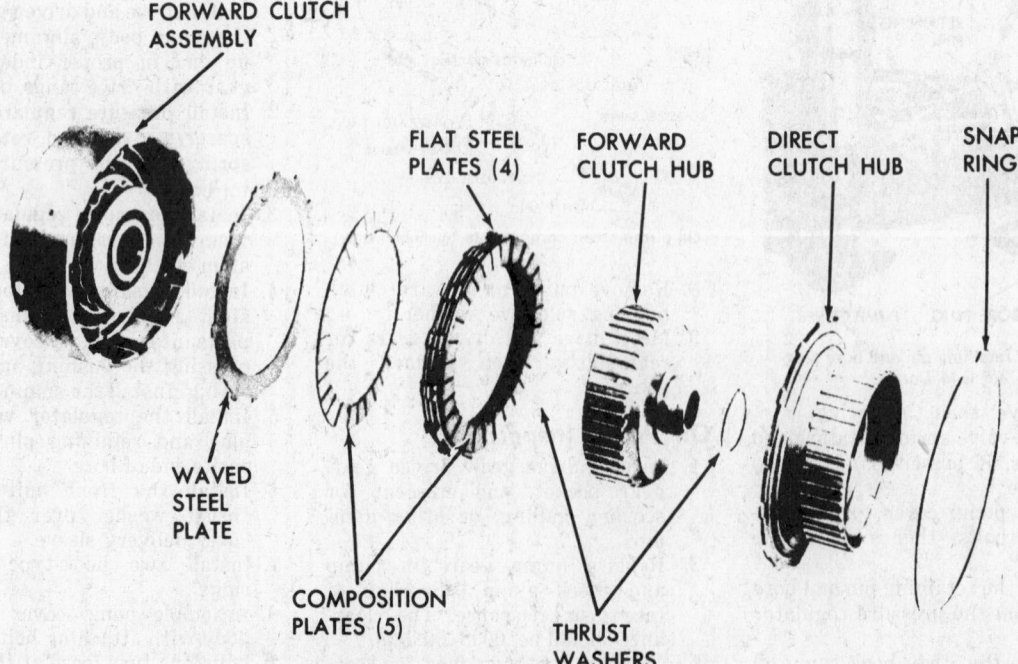

FORWARD CLUTCH ASSEMBLY

FLAT STEEL PLATES (4)

FORWARD CLUTCH HUB

DIRECT CLUTCH HUB

SNAP RING

WAVED STEEL PLATE

COMPOSITION PLATES (5)

THRUST WASHERS

Forward clutch assembly components (© G.M. Corp.)

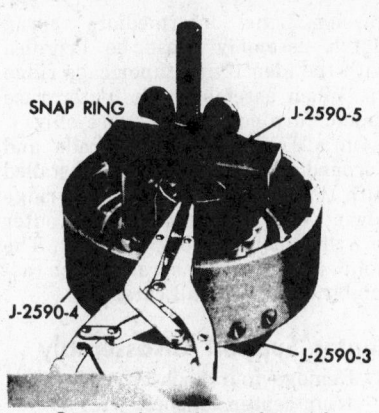

Compressing spring retainer and
removing snap-rings
(© G.M. Corp.)

Removing clutch release springs
(© G.M. Corp.)

Forward Clutch Inspection

1. Inspect clutch plates for burning, scoring, or wear.
2. Inspect springs for collapsed coils or distortion.
3. Check clutch hubs for worn splines, thrust faces, and open lubrication holes.
4. Check piston for cracks.
5. Check clutch housing for wear, scoring, open oil passages, and free operation of ball check.
6. Inspect turbine shaft for:
   a. Open lubrication passages at each end.
   b. Spline damage.
   c. Damage to ground bushing journals.
   d. Cracks or distortion of shaft.

Forward Clutch Assembly

1. Place new inner and outer oil seals on clutch piston, lips away from spring pockets.
2. Place a new center seal on the clutch housing, lip faces up.
3. Place a seal protector (thimble) tool into clutch drum and install the piston.
4. Install 16 clutch release springs into pockets in the piston.
5. Lay the spring retainer and the snap-ring on the springs.
6. Compress springs using compres-

sor, and install snap-ring. Press in turbine shaft.
7. Install the direct clutch hub washers; retain with petroleum jelly.
8. Place forward clutch hub into forward clutch housing.
9. Lubricate with transmission oil and install the clutch pack, five composition, and five steel plates, starting with steel and alternating steel and composition.
10. Install clutch hub and retaining snap-ring.
11. Place forward clutch housing on pump delivery sleeve and air-check clutch operation.

Direct Clutch and Intermediate Sprag Disassembly

1. Remove intermediate clutch retainer snap-ring, then the retainer.
2. Remove clutch outer race, bushings and sprag assembly.
3. Invert the unit and remove backing plate to clutch housing snap-ring.
4. Remove direct clutch backing plate, five composition, and five steel plates.
5. Using clutch compressor tool compress the spring retainer and remove the snap-ring.

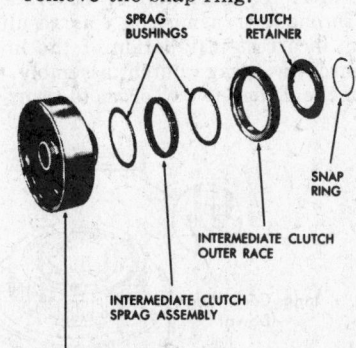

Sprag assembly
(© G.M. Corp.)

Air checking forward clutch (© G.M. Corp.)

6. Remove retainer and 16 piston release springs.
7. Remove direct clutch piston, then remove the outer and inner seals from the piston.
8. Remove center piston seal from the direct clutch housing.

Direct Clutch and Intermediate Sprag Inspection

1. Check for popped or loose sprags.
2. Check sprag bushings for distortion or wear.
3. Inspect inner and outer races for scratches or wear.
4. Check clutch housing for cracks,

Removing sprag outer race
(© G.M. Corp.)

wear, proper opening of oil passages, and wear on clutch plate drive lugs.
5. Check clutch plates for wear or burning.
6. Check backing plate for scratches or damage.
7. Check clutch piston for cracks and free operation of ball check.

Direct Clutch and Intermediate Sprag Assembly

1. Install a new clutch piston seal onto the piston, lips facing away

Backing plate to clutch housing snap-ring
(© G.M. Corp.)

Checking operation of direct clutch with air
(© G.M. Corp.)

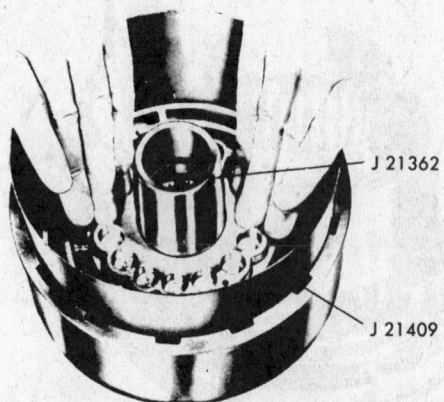

Installing direct clutch piston (© G.M. Corp.)

from spring pockets. Apply transmission fluid to oil seals.

2. Install a new outer piston seal, and a new center seal onto the clutch housing, lip of seal facing up.

3. Place seal protectors over hub and clutch housing, then install clutch piston.

4. Install 16 springs into piston, place retainer and snap-ring on retainer.

5. With clutch compressor, compress the clutch and install snap-ring.

6. Install five composition and five steel clutch plates, starting with steel and alternating composition and steel.

7. Install clutch backing plate, then install the backing plate snap-ring.

8. Invert the unit and install one sprag bushing, cup side up, over the inner race.

9. Install sprag assembly into outer race.

10. With ridge on inner cage facing down, start sprag and outer race over inner race with clockwise turning motion.

NOTE: outer race should not turn counterclockwise.

11. Install sprag bushing over sprag cup side down.

12. Install clutch retainer and snap-ring.

13. Place direct clutch assembly over center support and air check operation of direct clutch.

NOTE: it is normal for air applied to reverse passage to escape from direct clutch passage. Air applied to direct clutch passage should move direct clutch.

Eldorado and Toronado Applications

NOTE: due to front-wheel drive application, the transmission output shaft turns in the opposite direction from that required by a conventional, rear-wheel drive vehicle.

Failure of the direct clutch to function after reconditioning an Eldorado or Toronado transmission may result from incorrect installation of the intermediate sprag clutch assembly.

In the Eldorado or Toronado trans-

mission, the intermediate sprag clutch assembly must be installed with the identifying inner cage ridge up. When assembled, the outer race should turn counterclockwise only.

On all cars, except Eldorado and Toronado, this assembly is installed with the identifying inner cage ridge down. When assembled, the outer race should turn clockwise only. The same intermediate sprag clutch assembly is used on all cars.

Center Support Disassembly

1. Remove four hook-type oil rings from center support.

2. Using clutch fingers, compress the spring retainer and remove the snap-ring.

3. Remove spring retainer and 12 clutch release springs.

4. Remove intermediate clutch piston.

5. Remove inner piston seal.

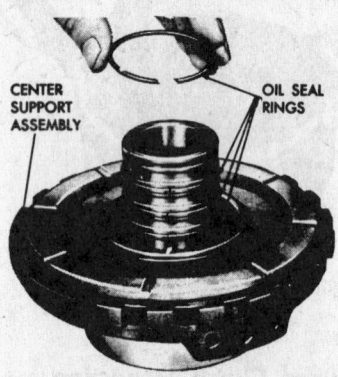

Center support oil ring removal
(© G.M. Corp.)

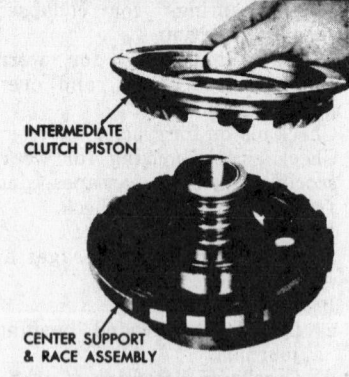

Removing intermediate clutch piston
(© G.M. Corp.)

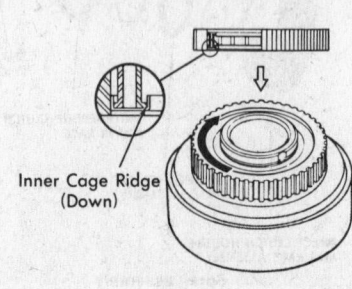

REAR-WHEEL DRIVE   FRONT-WHEEL DRIVE

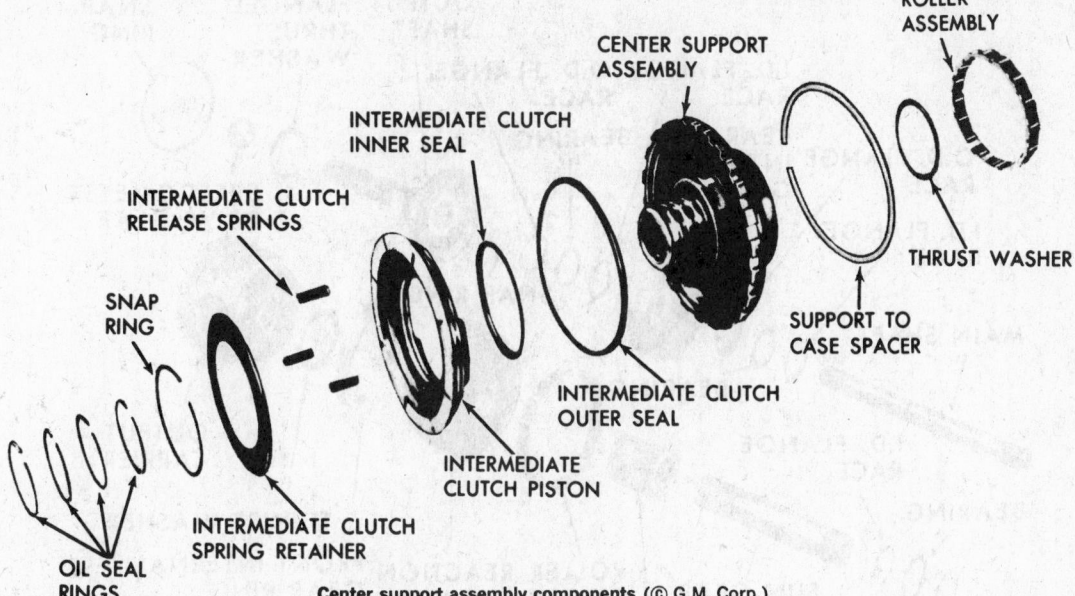

Center support assembly components (© G.M. Corp.)

NOTE: do not remove the three screws holding the roller clutch inner race to the center support.

6. Remove outer piston seal.

Center Support Inspection

1. Inspect roller clutch inner race for scratches or indentations. Check that lubrication hole is open.
2. Check bushing for scoring, wear, or galling. If bushing is replaced, drive new bushing into bore until it is flush to .010 in. below the top of the oil delivery sleeve.
3. Check oil seal rings and ring grooves in the center support tower for damage.
4. Make air pressure check of oil passages to be sure they are not interconnected.
5. Inspect piston sealing surfaces for scratches. Inspect piston seal grooves for damage. Check piston for cracks or porosity.
6. Check release springs for distortion or collapsed coils.
7. Check support to case spacer for burrs or raised edges. Repair with a stone or fine sand paper.

Center Support Assembly

1. Install new inner and outer seals on the piston, lip on inner seal facing away from the spring pocket.
2. Install inner spring protector tool onto the center support hub; lubricate the seal and install the piston.
3. Install 12 release springs into the piston.
4. Place spring retainer and snapring over the springs.
5. Using the clutch spring compressor, compress the springs and install the snap-ring.

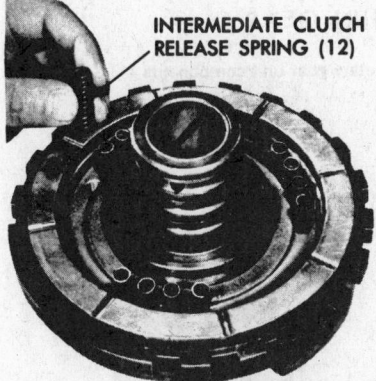

Installing release springs
(© G.M. Corp.)

6. Install four hook-type oil rings.
7. Air check the operation of intermediate clutch piston.

Torque Converter Inspection

The torque converter is a welded assembly and must be serviced as a unit. If converter output shaft has more than .050 in. end-play, renew the unit.

Check for leaks as follows:

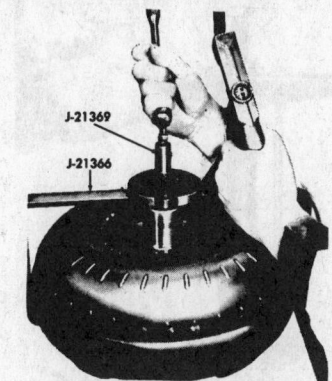

Air testing torque converter
(© G.M. Corp.)

A. Install converter leak test fixture and tighten.
B. Fill converter with air, 80 psi.
C. Submerge in water and check for bubbles.

Planetary Gear Unit Disassembly

1. Remove center support assembly.
2. Remove center support to reaction carrier thrust washer.
3. Remove center support to sun gear races and thrust bearing. One race may already have been removed with the center support.
4. Remove reaction carrier and roller clutch assembly.
5. Remove front internal gear ring from output carrier assembly.
6. Remove sun gear.
7. Take off reaction carrier to output carrier thrust washer.
8. Turn carrier assembly over. Remove output shaft to output carrier snap ring.
9. Output shaft may now be removed.
10. Measure to determine speedometer drive gear location with relation to the end of the shaft for reassembly. Remove nylon speedometer drive gear by depressing retaining clip and sliding gear off output shaft. Remove steel speedometer drive gear with a suitable puller.
11. Remove output shaft to rear internal gear thrust bearing and two races.
12. Remove rear internal gear and mainshaft. Remove rear internal gear to sun gear thrust bearing and two races. Remove rear internal gear to mainshaft snap ring. Remove mainshaft.

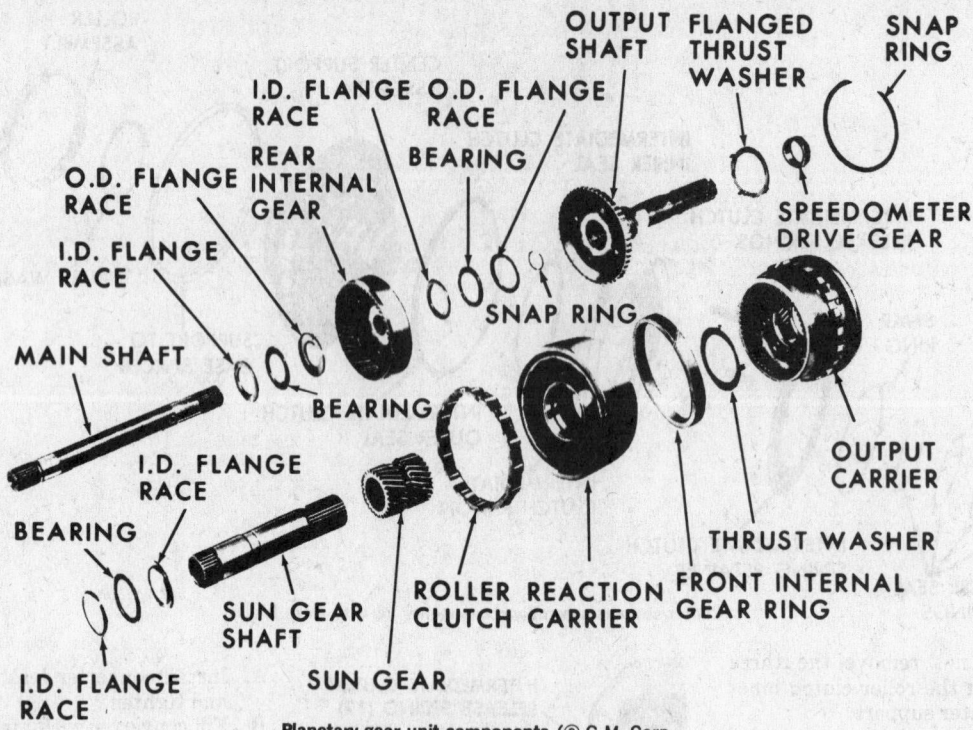

Planetary gear unit components (© G.M. Corp.,

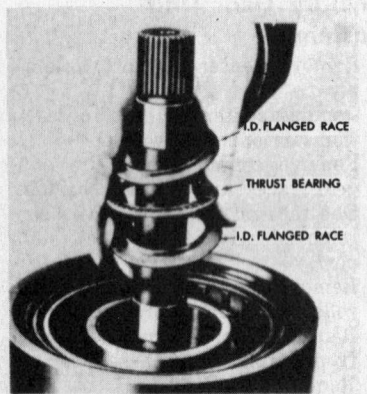

Removing center support to sun gear races
(© G.M. Corp.)

Removing reaction to output carrier
thrust washer
(© G.M. Corp.)

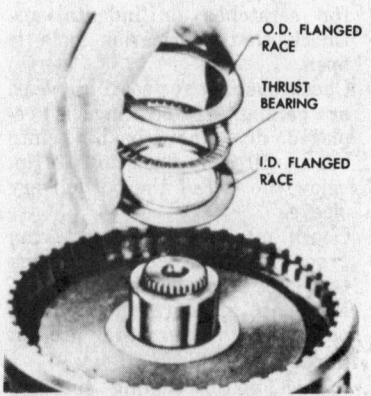

Removing output shaft to internal gear races
(© G.M. Corp.)

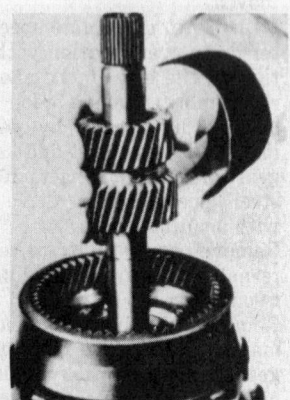

Removing sun gear
(© G.M. Corp.)

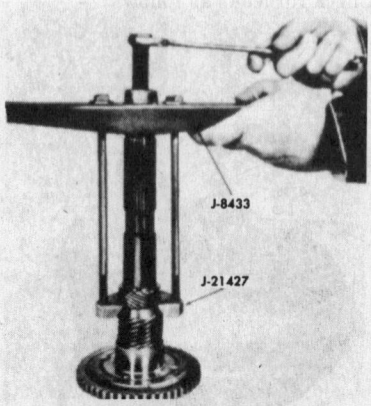

Removing speedometer drive gear
(© G.M. Corp.)

Inspection of Reaction Carrier, Roller Clutch, and Output Carrier Assembly

1. Insert band surface reaction carrier for burring or scoring.
2. Check roller clutch outer race for scoring or wear. Check thrust washer for wear.
3. Check bushing for damage. If bushing is damaged, replace reaction carrier.
4. Check reaction carrier pinions for damage, rough bearings, or excessive tilt. Check pinion end play. It should be .009-.024 in. Pinions may be replaced if necessary.
5. Check roller clutch for damaged members. Check roller clutch cage for damage.

Checking reaction carrier pinion end-play (© G.M. Corp.)

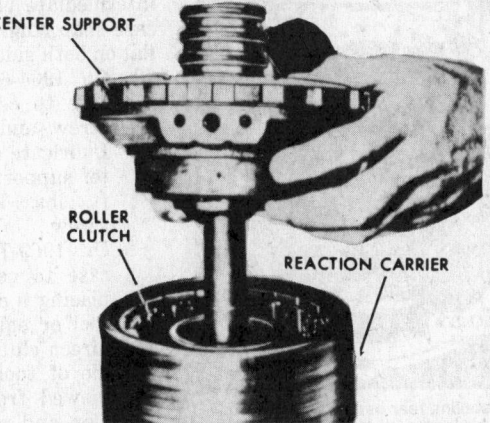

Installing center support assembly (© G.M. Corp.)

Checking output carrier pinion end-play (© G.M. Corp.)

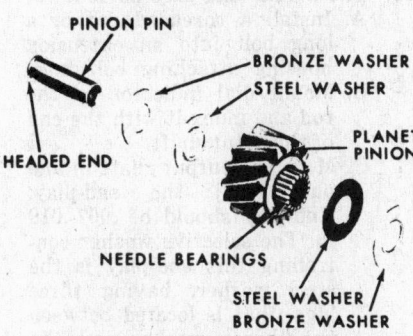

Planet pinion components (© G.M. Corp.)

6. Inspect front internal gear (output carrier) for damaged teeth. Inspect output carrier pinions for damage, rough bearings, and excessive tilt. Check pinion end play. It should be .009-.024 in.
7. Inspect parking pawl lugs for cracks or damage. Inspect output locating splines for damage.
8. Check front internal gear ring for flaking.

Planetary Gear Unit Assembly

1. Install rear internal gear onto end of mainshaft, then install the snap-ring.
2. Install sun gear to internal gear thrust races and bearings against inner face of rear internal gear as follows:
   A. Place large race against the internal gear, with flange facing up.
   B. Place thrust bearing against race.
   C. Place small race against the

bearing, with inner flange facing the bearing, or down.
3. Install the output carrier over the mainshaft so that the pinions mesh with the rear internal gear.
4. Place the above portion of the assembly through a hole in the bench so that the mainshaft hangs downward.
5. Install the rear internal gear to output shaft thrust races and bearings as follows:
   A. Small diameter race against internal gear, center flange facing up.
   B. Bearing onto the race.
   C. Second race onto the bearing, outer flange cupped over the bearing.
6. Install output shaft into the output carrier assembly.
7. Install output shaft to output carrier snap-ring.
8. Turn assembly over and support it so that the output shaft hangs downward.
9. Install the reaction carrier to output carrier thrust washer, tabs facing down and in their pockets.
10. Install sun gear, splines, chamfer down. Install gear ring over output carrier.
11. Install sun gear shaft, then the reaction carrier.
12. Install the center support to sun gear thrust races and bearing as follows:

Installing center support to sun gear races (© G.M. Corp.)

A. Install the large race, center flange up over the sun gear shaft.
B. Install thrust bearing.
C. Install the second race, center flange up.
13. Install roller clutch into reaction carrier outer race. Install the center support to reaction carrier thrust washer into the recess in the support. Retain with petroleum jelly.
14. Install center support into reaction carrier and roller clutch assembly. With reaction carrier held, center support should only turn counterclockwise.
15. Install a gear unit assembly holding tool to hold units in place. Install output shaft to case thrust washer tabs in pockets and retain with petroleum jelly.

Transmission Assembly

Parking Mechanism, Rear Band and the Complete Gear Assembly Installation

1. Install O-ring seal onto parking pawl shaft, then install parking pawl, tooth toward inside of case.
2. Install the pawl shaft retaining clip and the return spring,

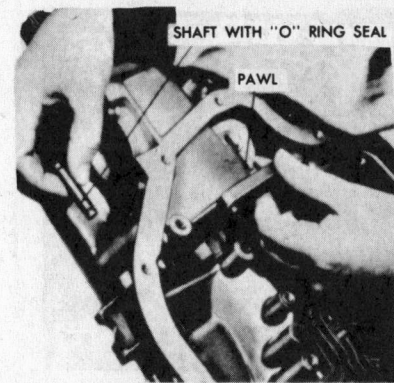

Installing parking pawl (© G.M. Corp.)

Installing rear band assembly
(© G.M. Corp.)

FRONT BAND

REAR BAND

(© G.M. Corp.)

square-end hooked on the pawl.
3. Install parking brake bracket guides over pawl, using two attaching bolts.
4. Install rear band assembly so that the two lugs index with the two anchor pins. Install support to case spacer with ring gap adjacent to band anchor pin.
5. Install rear selective washer into slots provided inside rear of transmission case. Dip washer in transmission fluid.
6. Install the complete gear unit assembly into the case.
7. Lubricate and install center support to case snap-ring. Install bevel side up.
NOTE: the support to case spacer is .040 in. thick and is flat on both sides. The center support to case snap-ring has one side beveled. The

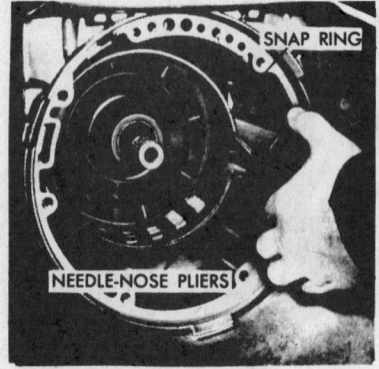

Installing snap-ring
(© G.M. Corp.)

intermediate clutch backing plate to case snap ring is .093 in. thick and is flat on both sides.
8. On 1964-68 models, install the case to center support locating screw and torque to 5 ft. lbs. Lubricate and install case to center support bolt and torque to 22 ft. lbs. Remove the locating screw.
On 1969-70 models, install the case to center support bolt by placing a center support locating tool or equivalent tool into case direct clutch passage with handle of tool pointing to right as viewed from front of transmission and parallel to bell housing mounting face. Apply pressure downward on tool handle to hold the center support firmly against the case splines. Torque the case to center support bolt to 22 ft. lbs.
9. Install three steel and three composition clutch plates. Alternate the plates, starting with steel.
10. Install the backing plate, ridge up, then the snap-ring. Locate

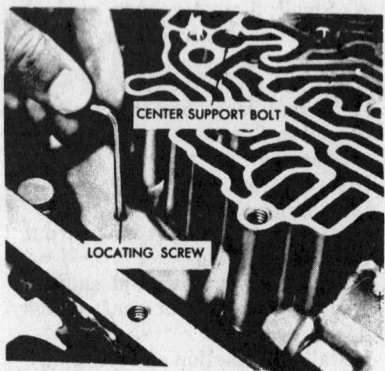

Installing center support locating screw, 1964-68 models
(© G.M. Corp.)

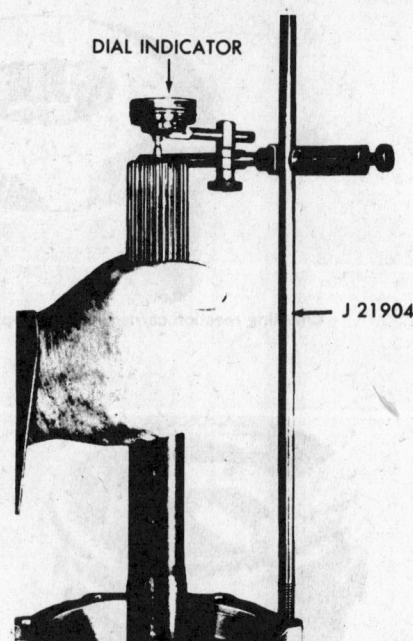

DIAL INDICATOR

J 21904

Checking rear end-play (© G.M. Corp.)

snap-ring gap opposite band anchor pin.
11. Check rear end-play as follows:
  A. Install a threaded rod or a long bolt into an extension housing attaching bolt hole.
  B. Mount dial indicator on the rod and index it with the end of the outut shaft.
  C. Move the output shaft in and out. Read the end-play. End-play should be .007-.019 in. The selective washer controlling this end-play is the steel washer, having three lugs, that is located between the thrust washer and the rear face of the transmission case.
If a different washer thickness is required to obtain proper end-play,

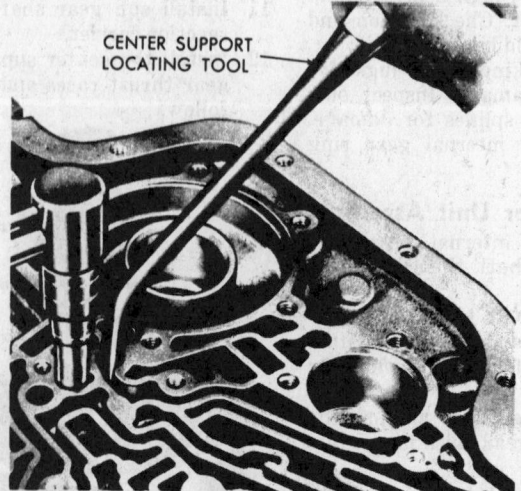

CENTER SUPPORT
LOCATING TOOL

Installing center support bolt, 1969-70 models
(© G.M. Corp.)

it can be selected from the following chart.

1964-67

Thickness	Notches
.078-.082 in.	None
.086-.090 in.	1 tab side
.094-.098 in.	2 tabs side
.102-.106 in.	1 tab O.D.
.110-.114 in.	2 tabs O.D.
.118-.122 in.	3 tabs O.D.

1968-70

Thickness	Notches and/or Numeral	
.074—.078 in.	None	1
.082—.086 in.	1 Tab Side	2
.090—.094 in.	2 Tabs Side	3
.098—.102 in.	1 Tab O.D.	4
.106—.110 in.	2 Tabs O.D.	5
.114—.118 in.	3 Tabs O.D.	6

Front Band, Direct Clutch and Forward Clutch Installation

1. Install front band, with band anchor hole placed over the band anchor pin, and apply lug facing the servo hole.
2. Install the direct clutch and intermediate sprag assembly. (Removal of direct clutch plates may help.)
3. Install forward clutch hub to direct housing bronze thrust washer onto the forward clutch hub. Retain with petroleum jelly.
4. Install forward clutch assembly, indexing the direct clutch hub so the end of the mainshaft will be flush with the end of the forward clutch hub. Use turbine shaft as a tool.
5. Install turbine shaft; end with the short spline goes into the forward clutch housing.
6. Install pump-to-case gasket onto the case face and install the front pump assembly and all but one attaching bolt and seal. Torque to 18 ft. lbs.
NOTE: if turbine shaft cannot be rotated as pump is being pulled into place, forward or direct clutch housing has not been properly installed to index with all clutch plates. This condition must be corrected before pump is fully pulled into place.
7. Drive in a new front seal.
8. Check front unit end-play as follows:
   A. Install one rod of slide hammer, with 5/16-18 in. thread, into the empty pump assembly attaching bolt hole.
   B. Mount dial indicator on the rod and adjust the indicator probe to contact end of turbine shaft.
   C. Hold output shaft forward while pushing turbine shaft rearward to its stop.

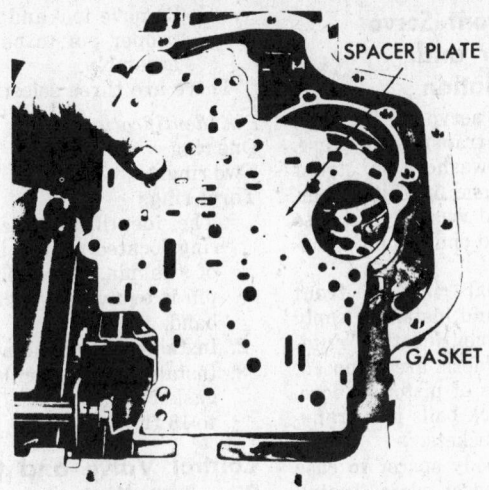

Installing gasket and spacer plate (© G.M. Corp.)

   D. Set dial indicator to zero.
   E. Pull turbine shaft forward.
   F. Read end-play, as registered on dial. The reading should be .003-.024 in.
The selective washer controlling this end-play is located between the pump cover and the forward clutch housing. If more, or less, washer thickness is required to bring end-play within specifications, make selection from the thickness-color chart.

Thickness	Color
.060-.064 in.	Yellow
.071-.075 in.	Blue
.082-.086 in.	Red
.093-.097 in.	Brown
.104-.108 in.	Green
.115-.119 in.	Black
.126-.130 in.	Purple

9. Remove dial indicator and install the remaining front pump attaching bolt and seal. Torque to 18 ft. lbs.

Rear Extension Housing Assembly Installation

1. Install extension housing-to-case O-ring seal onto extension housing.
2. Attach extension housing to transmission case. Torque to 22 ft. lbs.
3. Drive in a new extension housing rear seal.

Manual Linkage Installation

1. Install new manual shift shaft seal into the case.
2. Insert actuator rod into the manual detent lever from the side opposite the pin.
3. Install actuator rod plunger under the parking bracket and over the pawl.
4. Install manual lever and shaft through the case and detent lever, then lock with hex nut on the manual shift shaft. (Be sure detent retaining nut is tight.) Install retaining pin.

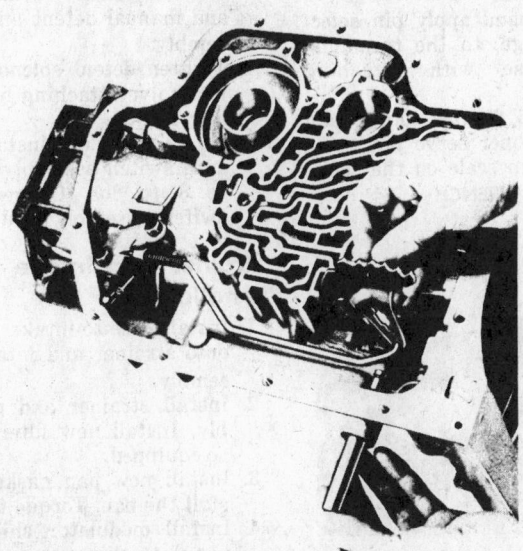

Positioning actuating rod plunger (© G.M. Corp.)

Check Balls, Front Servo Gaskets, Spacer and Solenoid Installation

1. Install front servo spring and retainer into transmission case.
2. Install flat washer on front servo pin, on end opposite taper. Install pin and washer into case so that tapered end of pin is contacting band.
3. Install oil seal ring on front servo piston, and install on apply pin so that identification numbers on shoulders are exposed. Check freeness of piston in bore.
4. Install six check balls into transmission case pockets.
5. Install valve body spacer to case gasket and spacer plate. Install detent solenoid and gasket, with connector facing outer edge of case. Do not tighten bolts at this time.
6. Install O-ring seal on electrical connector. Lubricate and install electrical connector with locator tab in notch on side of case. Connect detent wire and lead wire to electrical connector. Be sure to install electrical wire clip.

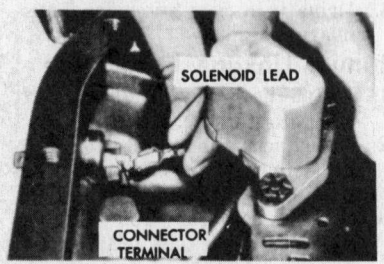

Installing solenoid
(© G.M. Corp.)

Rear Servo Installation

1. Before installing the rear servo assembly, check the band apply pin, using rear band apply fixture as follows:
   A. Attach band apply pin selection gauge, to the transmission case with attaching screws.
   B. Apply 25 ft. lbs. torque and select proper servo pin to be used from scale on the tool.

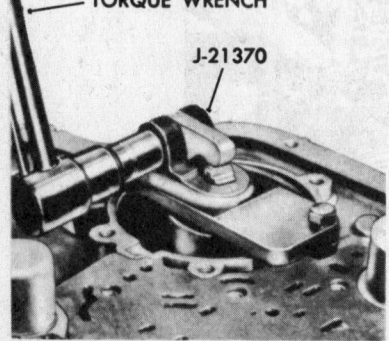

Checking band apply pin
(© G.M. Corp.)

C. Remove tool and make note of proper pin to be used during assembly.

There are three selective pins:

Pin Identification	Pin Size
One ring	Short
Two rings	Medium
Three rings	Long

The identification consists of a ring located on the band lug end of the pin. Selecting the proper pin is equivalent to adjusting the band.

2. Install rear accumulator spring.
3. Install servo assembly, then the gasket and cover. Torque bolts to 18 ft. lbs.

Control Valve and Governor Pipe Installation

1. Install control valve-to-spacer gasket, then install governor pipes into the control valve body assembly.
2. Install two guide pins, then install the control valve body and governor pipe assembly into the transmission.

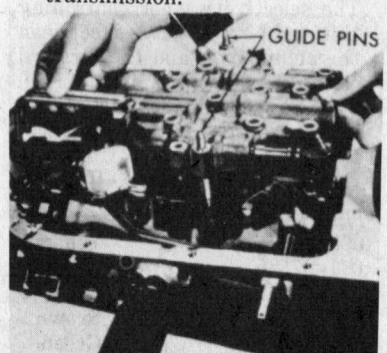

Installing control valve assembly
(© G.M. Corp.)

NOTE: be sure the manual valve is properly indexed with the pin on the manual detent lever.

3. Remove guide pins and install valve assembly attaching bolts and manual detent and roller assembly.
4. Tighten detent solenoid and control valve attaching bolts to 8 ft. lbs. torque.
5. On 1970 models, install the pressure switch assembly and torque to 8 ft. lbs. Connect pressure switch assembly lead wire.

Strainer and Intake Pipe Installation

1. Install case-to-intake pipe O-ring onto strainer and intake pipe assembly.
2. Install strainer and pipe assembly. Install new filter on models so equipped.
3. Install new pan gasket, then install the pan. Torque to 12 ft. lbs.
4. Install modulator shield and all pan attaching screws. Torque pan attaching screws.

Modulator Assembly Installation

1. Install modulator valve into the case, stem end out.
2. Install O-ring seal onto the vacuum modulator, then install assembly into the case.
3. Install modulator retainer and attaching bolt. Torque to 18 ft. lbs.

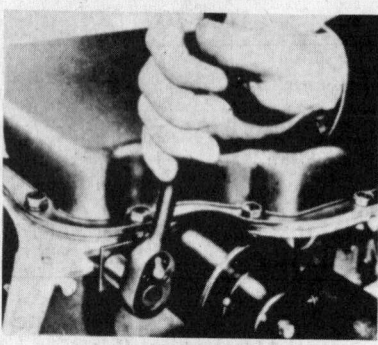

Installing modulator retainer
(© G.M. Corp.)

Governor and Speedometer Driven Gear Installation

1. Install governor assembly.
2. Attach governor cover and gasket with four bolts. Torque to 18 ft. lbs.
3. Install speedometer driven gear assembly. Install retainer and attaching bolt.

Installing speedometer driven gear retainer and bolt
(© G.M. Corp.)

Converter Installation

1. Place transmission in cradle or portable jack.
2. Install converter assembly to pump assembly, making certain that the converter hub drive slots are fully engaged with the pump drive gear tangs and that the converter is installed all the way toward the rear of the transmission.

Torque Specifications

Foot-Pounds

Pump cover bolts	18
Parking pawl bracket bolts	18
Center support bolt	22
Pump-to-case attaching bolts	18
Extension-to-case attaching bolts	22
Rear servo cover bolts	18
Detent solenoid bolts	8
Control valve body bolts	8
Bottom pan attaching screws	12
Modulator retainer bolt	18
Governor cover bolts	18
Manual lever-to-manual shaft nut	20
Linkage swivel clamp screw	20
Transmission-to-engine mounting bolts	40
Rear mount-to-transmission bolts	40
Oil cooler line	16
Filter retainer bolt	10
Pressure switch assembly	8

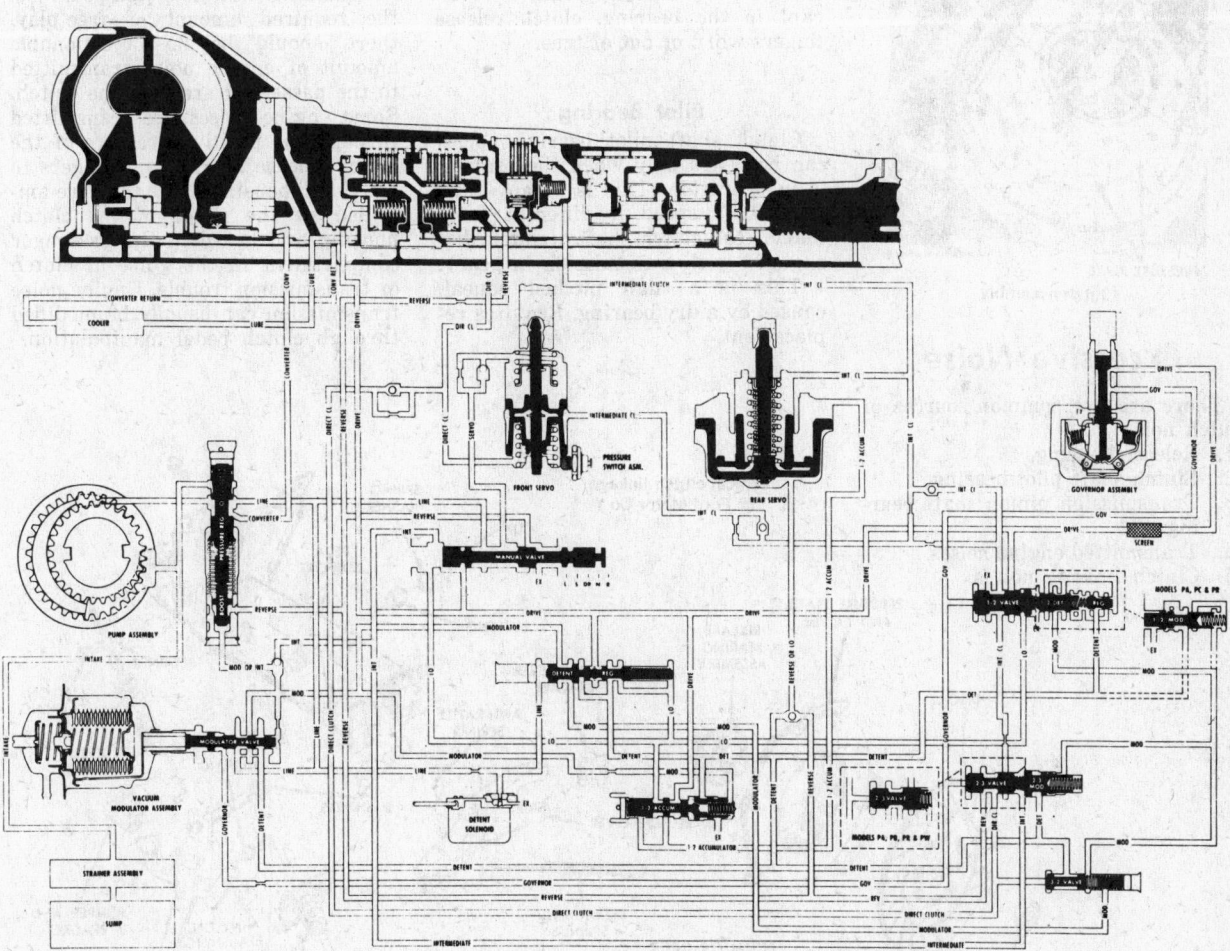

Oil flow, Neutral—engine idling (© G.M. Corp.)

Clutch Problems
Problems and Solutions

When diagnosing problems in any area, there is no substitute for careful examination and experience. The following are some symptoms that may accompany clutch troubles.
1. Excessive noise.
2. Clutch chatter or grab.
3. Clutch slip.
4. Clutch drag or failure to release.
5. Pedal pulsation.
6. Low clutch facing life.
7. Gear lock up or hard shifting.
8. Hard pedal.

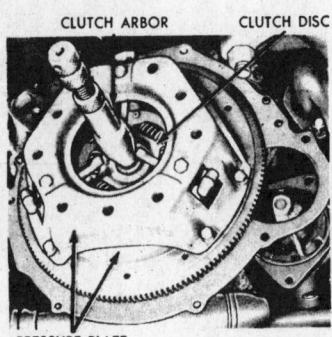

Clutch assembly

Excessive Noise

There are five common sources of clutch noise:
1. Release bearing.
2. Clutch shaft pilot bearing.
3. Transmission pinion shaft bearing.
4. Transmitted engine noises.
5. Clutch linkage noises.

Release Bearing

Release bearing noises vary with the degree of bearing failure. A dry or damaged bearing usually makes a shrill or scraping sound when depressing the clutch pedal to the point of release finger-to-bearing contact. This means that the noise should be audible at the lower end of clutch pedal free-play. Continued use of a car, with the release bearing in this condition, is damaging to the clutch release fingers.

Usual cause of release bearing failure is overwork—caused by riding the clutch. Other causes are not enough pedal free-play, lack of lubricant in the bearing, clutch release fingers worn or out of true.

Pilot Bearing

Clutch shaft pilot bearing noises can be heard only when the bearing is in operation. This is at any time crankshaft speed is different from that of the clutch shaft, (clutch disengaged with transmission in gear).

This is a high pitched squeal, caused by a dry bearing. Requires replacement.

Transmission Pinion Shaft Bearing

A rough, or otherwise damaged, transmission pinion (input) shaft bearing noise can be heard only when the clutch is engaged, with transmission in any shift position. The noise is usually quite noticeable with the gears in neutral. This noise should diminish and completely disappear as the transmission pinion gear slows down and stops after clutch release. This noise is easily distinguished from release bearing noise because of the opposite conditions of encounter.

Transmitted Engine Noises

Assuming that the clutch pedal has the required amount of free-play, there should be no objectionable amount of engine noise transmitted to the passenger area via the clutch. Some engine noises are transmitted through the positive pressure of the clutch release bearing and fingers to the clutch housing. Here they are amplified by the shape of the clutch housing and heard in the passenger compartment in the guise of clutch or transmission trouble. Engine noise transmission can usually be modified through clutch pedal manipulation.

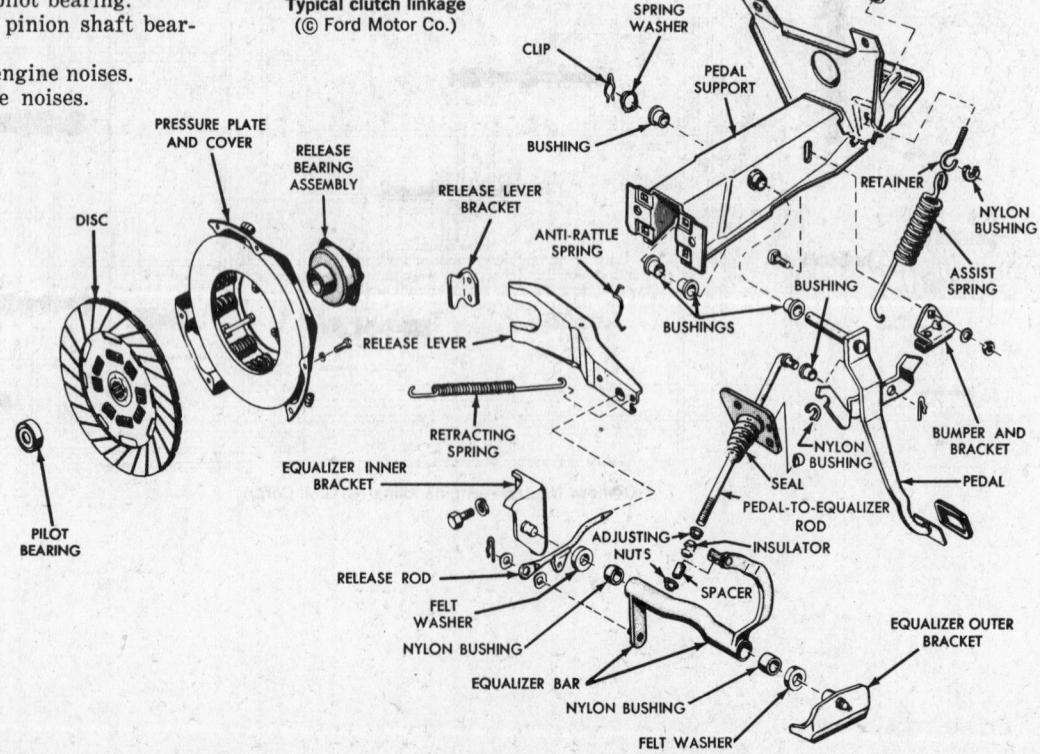

Typical clutch linkage
(© Ford Motor Co.)

Clutch Linkage Noise

Clutch linkage noise is usually a clicking or snapping sound that can be heard or felt in the pedal itself when moving it completely up or down. Locating the cause of trouble and correcting it is a matter of repositioning and lubrication. The trouble may be in the clutch assist spring, the retract spring, the release bearing lever, or even at the release bearing.

Clutch Chatter or Grab

Usually the cause of clutch chatter or grab can be located within the clutch assembly. To correct the trouble will require the removal of the clutch. However, symptoms resembling clutch trouble may be misleading and originate in other areas.

In order to isolate the cause of the problem, we suggest that the following items be checked in this order.

1. Be sure that the clutch linkage is in adjustment and not binding. If necessary, lubricate, align and adjust linkage.
2. Check for worn or loose engine or transmission mounts. If necessary, tighten or replace mounts.
3. Check for wear, looseness or misalignment of universal joints. Check attaching bolts on clutch pressure plate, transmission and clutch housing. Tighten, align or replace as necessary.
4. Check freedom of movement of the clutch release bearing on its sleeve. Free up or replace as necessary.
5. Check for oil or grease on the flywheel, friction disc or pressure plate.
6. Check for trueness of the friction disc, and that the disc hub is not binding on the splines of the transmission input shaft (clutch shaft).
7. Be sure that the disc or the pressure plate is not broken.
8. Examine clutch pressure plate and cover plate assembly for cracks or heat discoloration.

Clutch Slip

Clutch slippage is usually most noticeable when pulling away, and during acceleration from a standing start. A severe, but positive, test for slippage is to start the engine, set the parking brake and apply the service brakes; shift the transmission into high gear and release the clutch pedal while accelerating the engine. A clutch in good condition should hold

and stall the engine. If the clutch slips, the cause may be one or more of the following:

1. Improper linkage adjustment (not enough free-play).
2. Broken or disconnected parts.
3. Clutch linkage or lever mechanism binding or broken, not allowing full pressure plate application.
4. Friction disc oil-saturated or excessively worn.
5. Pressure plate worn, springs weak from temper loss or failure (damaging heat will usually cause parts to appear blue).

Clutch Drag or Failure to Release

There are many reasons for clutch drag (spin) or failure to release. The following conditions, therefore, apply to unmodified versions of standard vehicles. Changing the driven plate mass (replacing the standard driven plate with a heavy duty unit), changing transmission oil viscosity, etc. may influence clutch spin-time. Three seconds is a good, typical, spin-time for the standard transmission and clutch, driven under normal conditions, in average temperate zone climates.

The friction disc and some of the transmission gears spin briefly after clutch disengagement, so normal clutch action should not be confused with a dragging clutch.

Clutch drag, failure to release or abnormal spin-time may be caused by one or more of the following:

1. Improper clutch linkage adjustment.
2. Clutch plate hub binding on the transmission input (pinion) shaft.
3. A warped or bent friction disc or pressure plate; or loose friction material on the driven disc.
4. The transmission input shaft may be binding or sticking in the pilot bearing.
5. Misalignment of transmission to the engine.
6. Transmission lubricant low or not heavy enough.

Pedal Pulsation

This condition can be felt by applying light foot pressure to the clutch pedal with the engine idling. It may be caused by any of the following:

1. Bent or uneven clutch release finger adjustment.
2. Excessive flywheel runout due to,

bent wheel or crankshaft flange; or the flywheel may not be properly seated on the crankshaft flange.
3. Release bearing cocked on transmission bearing retainer.
4. Poor alignment of transmission with the engine.

Low Clutch Facing Life

This sort of complaint warrants a close study of the operator's driving habits. Poor clutch facing wear may be caused by any of the following:

1. Riding the clutch.
2. Drag strip type operation.
3. Continuous overloading, or the hauling of heavy trailers or other equipment.
4. Holding the car from drifting backward on a grade; by slipping the clutch instead of using the brakes.
5. Improper pedal linkage adjustment (free-play and pedal height).
6. Rough surface on flywheel or pressure plate.
7. Presence of oil or water on clutch facing.
8. Weak pressure plate springs, causing clutch creep or slip.

Gear Lock Up or Hard Shifting

This trouble is so closely related to Clutch Drag Or Failure To Release that diagnosis should be conducted in the same way as given under that heading. If, after checking the items listed and finding that the transmission still locks up or is hard to shift, the trouble probably lies in the transmission cover or shifter assembly, or in the transmission proper. In that case, transmission work is needed.

Hard Pedal

A stiff clutch pedal or a clutch release that requires abnormal pedal pressure may result from one or more of the following:

1. Dry and binding clutch linkage and levers.
2. Linkage out of alignment.
3. Improper (heavy) retracting spring.
4. Dry or binding release bearing sleeve or transmission bearing retainer.
5. Assist spring missing or improperly adjusted.
6. Wrong type clutch assembly (heavy duty) being used.

Manual Transmissions

Application Index

Type Numbers Refer to Sections in Text

Make	Model	3-Spd. Type	4-Spd. Type
American Motors	1964–70 199–232 Six	24	
	1964–67 232, 287, 290	23	
	1966–67 Six HD		
	1964, 1966 327 V8	22	
	1968–70 232, 290, 304, 360	20	
	1966, 287, 327		25
	1967–69 290, 343		
	1967–70 390		
	1970 360		
Barracuda	1964–70		4
	1966–70 Six	3	
	1966–69	2	
	1970	1	
Buick	Le Sabre, 1970 Wildcat, 1966–70	6	
	Le Sabre, 1966–68	15	
	Le Sabre, 1964–65	13	
	Le Sabre Wagon, 1964 Wildcat, 1964–65	22	25
Buick Special	1966–70	15	
	1967–70 GS		18
	1965–69 GS, 1966–69 V8	6	
	1964–65 except Skyroof Wagon and GS	13	
	1964–65		25
	1966 GS		
Camaro	1967–70	15	18, 19
	1969–70	14	
	1967–68	21	
Capri	1970		8
Challenger	1970 Six	3	
	1970	1	4
Charger	1967–70 Six	3	
	1966–70		4
	1967–69	2	
	1970	1	
Chevelle	1966–70	15	19
	1969–70	14	
	1964–70		18
	1966 (396), 1967–68	21	
	1964–65	13	
Chevrolet	1969–70	14	
	1966–70	15	19
	1964–70		18
	1965–67 (396, 427), 1968	21	
	1964–65	13	
Chevy II	1969–70	11	
	1966–70	12	19
	1964–70		18
	1968	21	
	1964–65	13	
Chrysler	1970	1	
	1964–69	2	
	1964–67		4
Comet	1964–70 V8		9
	1964–66 Six	5	7
	1964–66 V8, 1967 All	6	
	1964 V8		25
Corvair	1966–69	11	17
	1960–65	10	16
Corvette	1967–70		19
	1964–70		18
	1966–69	15	
	1966–68	21	
	1964–65	13	
Cougar	1967–70	6	9
Dart	1970	1	
	1964–70 Six	3	
	1964–70		4
	1964–69	2	
Dodge	1970	1	
	1964–70		4
	1964–70, Six	3	
	1964–69	2	
Fairlane	1965–70 V8		9
	1964–66 V8, 1967–70 All	6	
	1964–66 Six	5	7
	1964 V8		25
Falcon	1967–70 All; 64–66 V8	6	
	1965–70 V8		9
	1964–67 Six	5	
	1964–66 Six		7
	1964 V8		20
Firebird	1970	14	
	1967–70	15	18
	1968–69 V8	6	
	1967–68 Six		19
	1967 V8	21	
Ford	1964–70	6	
	1965–70		9
	1964 V8		25
Jeep	Gladiator and Wagoneer	27	
	Gladiator and Wagoneer (327 V8)	22	
	Gladiator and Wagoneer (232, 327)		26
	Universal (4 Cyl.)		28
	Dispatcher 2-Wheel Drive (4 Cyl.)	24	
	Universal (4 Cyl.), Universal and Jeepster (V6), Jeepster (4 Cyl.)	23	
	Gladiator and Wagoneer (232), Universal (V6), Jeepster, All 350 V8	20	
Mercury	1965–70		9
	1964–70	6	
	1964 V8		25
Monte Carlo	1970		18
Montego	1968–70	6	9

1132

Make	Model	3-Spd. Type	4-Spd. Type
Mustang	1965–70		9
	1965–66 V8, 1967–70 All	6	
	1965–66 (6 Cyl.)	5	7
	1965 V8		25
Oldsmobile	1966–70	6	
	1964, except Jetstar 88	12	
	Jetstar 88, 1964–65	13	
	1965		18
Oldsmobile F-85	1970, F-85 and 4-4-2	14	
	1966–70	15	
	1964–70		18
	1965–69 V8	6	
	1964–65	13	
Plymouth	1970	1	
	1964–70		4
	1964–70 Six	3	

Make	Model	3-Spd. Type	4-Spd. Type
	1964–69	2	
Pontiac	1970 GP	14	
	1968–70		18
	1966–70	6	
	1964	13	
	1964 HD	21	
Tempest	1970, Tempest and GTO	14	
	1966–70	15	
	1964–70		18
	1965–69 V8	6	
	1967–68 Six		19
	1964–65	13	
Valiant	1970	1	
	1964–70		4
	1964–70 Six	3	
	1966–69	2	

Section Page Numbers

Section	Page	Section	Page	Section	Page
1	1134	6	1147	11	1161
2	1138	7	1149	12	1163
3	1140	8	1152	13	1165
4	1142	9	1155	14	1167
5	1145	10	1159	15	1169

Section	Page	Section	Page	Section	Page
16	1171	21	1180	26	1193
17	1174	22	1182	27	1193
18	1174	23	1184	28	1193
19	1176	24	1186		
20	1177	25	1190		

Diagnosis chart—1134

Manual Transmission Index

Manual Transmissions

Diagnosis

Jumping out of High Gear
1. Misalignment of transmission case or clutch housing.
2. Worn pilot bearing in crankshaft.
3. Bent transmission shaft.
4. Worn high speed sliding gear.
5. Worn teeth in clutch shaft.
6. Insufficient spring tension on shifter rail plunger.
7. Bent or loose shifter fork.
8. End-play in clutch shaft.
9. Gears not engaging completely.
10. Loose or worn bearings on clutch shaft or mainshaft.

Sticking in High Gear
1. Clutch not releasing fully.
2. Burred or battered teeth on clutch shaft.
3. Burred or battered transmission main-shaft.
4. Frozen synchronizing clutch.
5. Stuck shifter rail plunger.
6. Gearshift lever twisting and binding shifter rail.
7. Battered teeth on high speed sliding gear or on sleeve.
8. Lack of lubrication.
9. Improper lubrication.
10. Corroded transmission parts.
11. Defective mainshaft pilot bearing.

Jumping out of Second Gear
1. Insufficient spring tension on shifter rail plunger.
2. Bent or loose shifter fork.
3. Gears not engaging completely.
4. End-play in transmission mainshaft.
5. Loose transmission gear bearing.
6. Defective mainshaft pilot bearing.
7. Bent transmission shaft.
8. Worn teeth on second speed sliding gear or sleeve.
9. Loose or worn bearings on transmission mainshaft.
10. End-play in countershaft.

Sticking in Second Gear
1. Clutch not releasing fully.
2. Burred or battered teeth on sliding sleeve.
3. Burred or battered transmission main-shaft.
4. Frozen synchronizing clutch.
5. Stuck shifter rail plunger.
6. Gearshift lever twisting and binding shifter rail.
7. Lack of lubrication.
8. Second speed transmission gear bearings locked will give same effect as gears stuck in second.
9. Improper lubrication.
10. Corroded transmission parts.

Jumping out of Low Gear
1. Gears not engaging completely.
2. Bent or loose shifter fork.
3. End-play in transmission mainshaft.
4. End-play in countershaft.
5. Loose or worn bearings on transmission mainshaft.
6. Loose or worn bearings in countershaft.
7. Defective mainshaft pilot bearing.

Sticking in Low Gear
1. Clutch not releasing fully.
2. Burred or battered transmission main-shaft.
3. Stuck shifter rail plunger.
4. Gearshift lever twisting and binding shifter rail.
5. Lack of lubrication.
6. Improper lubrication.
7. Corroded transmission parts.

Jumping out of Reverse Gear
1. Insufficient spring tension on shifter rail plunger.
2. Bent or loose shifter fork.
3. Badly worn gear teeth.
4. Gears not engaging completely.

5. End-play in transmission mainshaft.
6. Idler gear bushings loose or worn.
7. Loose or worn bearings on transmission mainshaft.
8. Defective mainshaft pilot bearing.

Sticking in Reverse Gear
1. Clutch not releasing fully.
2. Burred or battered transmission main-shaft.
3. Stuck shifter rail plunger.
4. Gearshift lever twisting and binding shifter rail.
5. Lack of lubrication.
6. Improper lubrication.
7. Corroded transmission parts.

Failure of Gears to Synchronize
1. Binding pilot bearing on mainshaft, will synchronize in high gear only.
2. Clutch not releasing fully.
3. Detent springs weak or broken.
4. Weak or broken springs under balls in sliding gear sleeve.
5. Binding bearing on clutch shaft.
6. Binding countershaft.
7. Binding pilot bearing in crankshaft.
8. Badly worn gear teeth.
9. Scored or worn cones.
10. Improper lubrication.
11. Constant mesh gear not turning freely on transmission mainshaft. Will synchronize in that gear only.

Gears Spinning When Shifting into Gear from Neutral
1. Clutch not releasing fully.
2. In some cases an extremely light lubricant in transmission will cause gears to continue to spin for a short time after clutch is released.
3. Binding pilot bearing in crankshaft.

Type-1
A-230 Fully Synchronized Chrysler 3-Speed

Application

Barracuda, 1970
Challenger, 1970
Charger, 1970

Chrysler, 1970
Dart, 1970
Dodge, 1970

Plymouth, 1970
Valiant, 1970

Disassembly
Shift Housing and Mechanism
1. Shift to second gear.
2. Unbolt and remove side cover with shift mechanism.
   If shaft O-ring seals need replacement:

3. Pull shift forks out of shafts.
4. Remove nuts and operating levers from shafts.
5. Deburr shafts. Remove shafts.

Drive Pinion Retainer and Extension Housing
1. Unbolt pinion bearing retainer from front of transmission case. Remove retainer and gasket. Pry off retainer oil seal.
   For clearance:
2. With a brass drift, tap drive pinion as far forward as possible. Rotate cut away part of second gear next to countershaft gear.

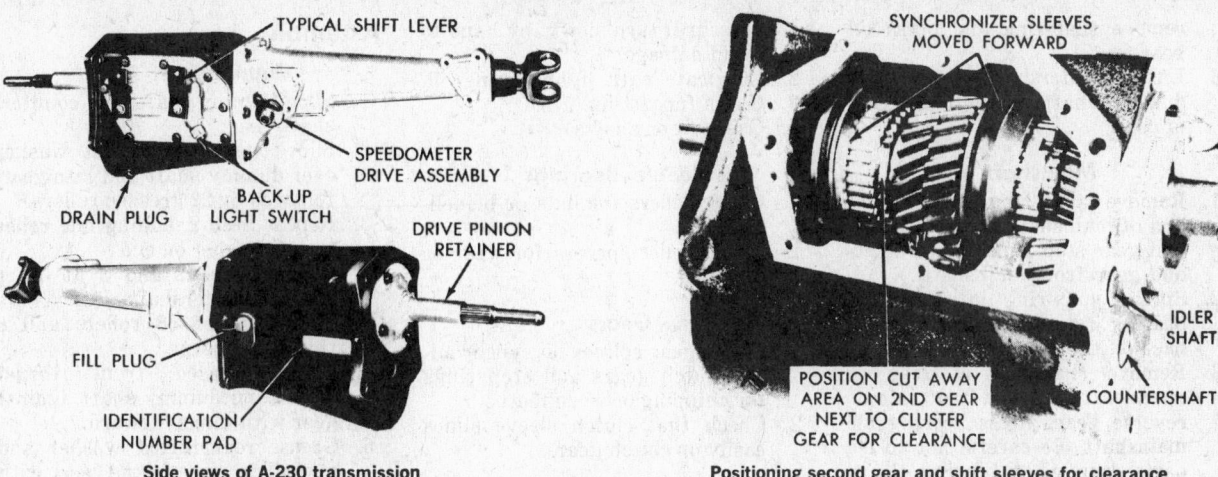

Side views of A-230 transmission
(© Chrysler Corp.)

TYPICAL SHIFT LEVER
SPEEDOMETER DRIVE ASSEMBLY
BACK-UP LIGHT SWITCH
DRAIN PLUG
DRIVE PINION RETAINER
FILL PLUG
IDENTIFICATION NUMBER PAD

SYNCHRONIZER SLEEVES MOVED FORWARD
IDLER SHAFT
COUNTERSHAFT
POSITION CUT AWAY AREA ON 2ND GEAR NEXT TO CLUSTER GEAR FOR CLEARANCE

Positioning second gear and shift sleeves for clearance
(© Chrysler Corp.)

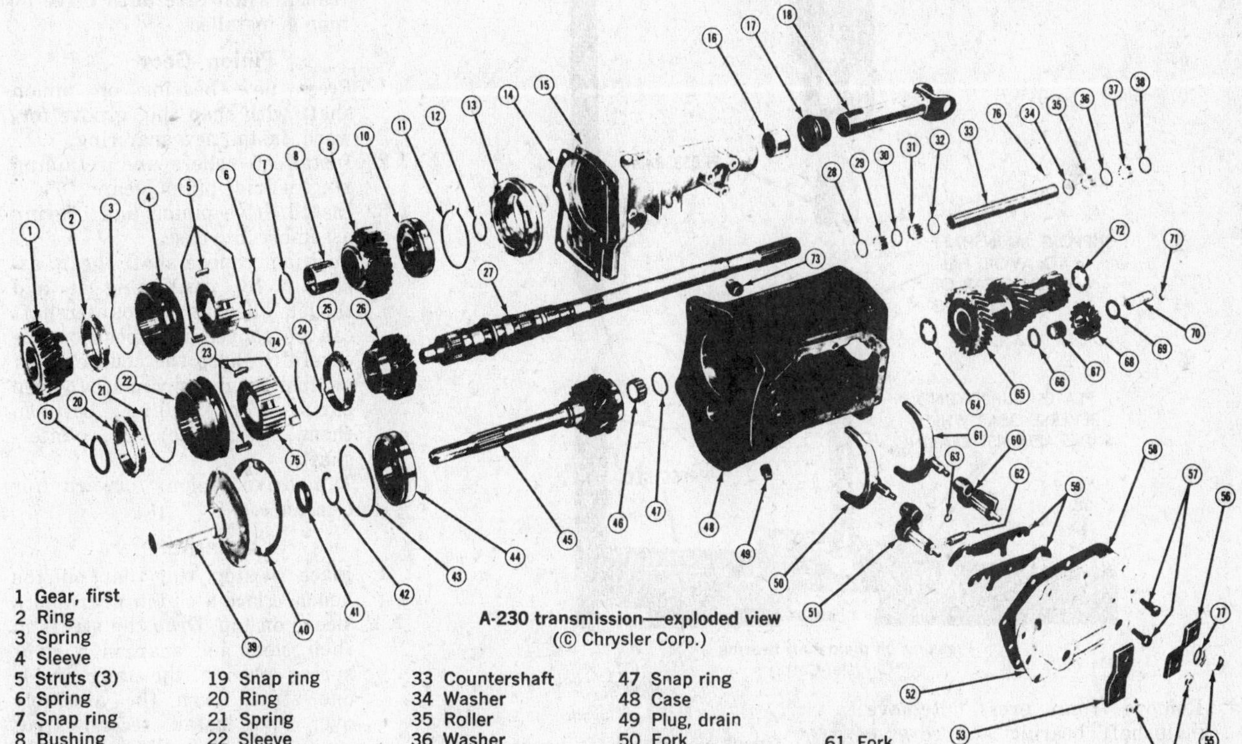

A-230 transmission—exploded view
(© Chrysler Corp.)

1 Gear, first	19 Snap ring	33 Countershaft	47 Snap ring
2 Ring	20 Ring	34 Washer	48 Case
3 Spring	21 Spring	35 Roller	49 Plug, drain
4 Sleeve	22 Sleeve	36 Washer	50 Fork
5 Struts (3)	23 Struts (3)	37 Roller	51 Lever
6 Spring	24 Spring	38 Washer	52 Housing
7 Snap ring	25 Ring	39 Retainer	53 Lever
8 Bushing	26 Gear, second	40 Gasket	54 Nut, locking
9 Gear, reverse	27 Shaft, output	41 Seal	55 Switch
10 Bearing	28 Washer	42 Snap ring	56 Lever
11 Snap ring	29 Roller	43 Snap ring	57 Bolt
12 Snap ring	30 Washer	44 Bearing	58 Gasket
13 Retainer	31 Roller	45 Pinion, drive	59 Lever, interlock
14 Gasket	32 Washer	46 Roller	60 Lever
15 Extension			
16 Bushing			
17 Seal			
18 Yoke			

61 Fork	71 Key
62 Spring	72 Washer
63 Snap ring	73 Plug, filler
64 Washer	74 Gear, clutch
65 Gear, countershaft	75 Gear, clutch
66 Washer	76 Key
67 Roller	77 Gasket
68 Gear, idler	
69 Washer	
70 Shaft	

Shift second-third synchronizer sleeve forward.

3. Remove speedometer pinion adapter retainer. Work adapter and pinion out of extension housing.

4. Unbolt extension housing. Break housing loose with plastic hammer and carefully remove.

Idler Gear and Mainshaft

1. Insert dummy shaft in case to push reverse idler shaft and key out of case.

2. Remove dummy shaft and idler gear together to prevent losing rollers.

3. Remove both tanged idler gear thrust washers.

4. Remove mainshaft assembly through rear of case.

Countershaft Gear and Drive Pinion

1. Using a mallet and dummy shaft, tap the countershaft rearward enough to remove key. Drive countershaft out of case,

maintaining contact between countershaft and dummy shaft so that washers will not drop out.

2. Lower countershaft gear to bottom of case.

3. Remove snap-ring from pinion bearing outer race (outside front of case).

4. Drive pinion shaft into case with plastic hammer. Remove assembly through rear of case.

5. If bearing is to be replaced,

remove snap-ring and press off bearing.

6. Lift countershaft gear and dummy shaft out through rear of case.

Mainshaft

1. Remove snap-ring from front end of mainshaft along with second gear stop ring. Remove second gear from mainshaft.
2. Spread snap-ring in mainshaft bearing retainer. Slide retainer back off the bearing race.
3. Remove snap-ring at rear of mainshaft. Support front side of reverse gear. Press bearing off mainshaft. Be careful not to let parts drop when bearing clears shaft.

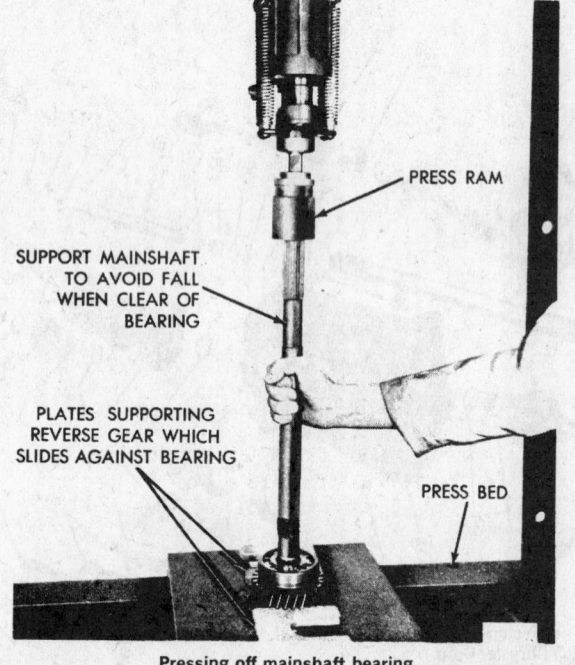

SUPPORT MAINSHAFT TO AVOID FALL WHEN CLEAR OF BEARING

PRESS RAM

PLATES SUPPORTING REVERSE GEAR WHICH SLIDES AGAINST BEARING

PRESS BED

Pressing off mainshaft bearing
(© Chrysler Corp.)

4. Remove from press. Remove mainshaft bearing and reverse gear from shaft.
5. Remove snap-ring from rear of shaft. Slide first-reverse synchronizer assembly off splines and remove rearward. Remove stop ring and first gear rearward.

Inspection

1. Clean all parts with solvent.
2. Dry with compressed air.

Case

1. Check for cracks, stripped threads, and burrs or nicks on machined surfaces. Dress off any burrs with a fine file. Stripped threads may be repaired by use of Helicoil inserts.

Ball Bearings

1. Do not spin bearings with air

pressure; turn slowly by hand to avoid damage.
2. Lubricate with light engine oil.
3. Check for pitting.
4. Check fit on shafts.

Needle Bearings

1. Check rollers for flats or brinelling.
2. Check roller spacers for wear or galling.

Gears

1. Check gear splines on synchronizer clutch gears and stop rings for chipping or worn teeth.
2. Check that clutch sleeve slides easily on clutch gear.

3. Check countershaft gear and all gear teeth for chipping, broken teeth, or excessive wear. Stone off small nicks or burrs.
4. If oil seal contact area on drive pinion shaft is pitted, rusted, or scratched, replace the pinion.

Synchronizer Stop Rings

1. Check for cracks or wear.
2. Check new rings for good fit on gear cones with minimum wobble.

Mainshaft

1. Check mainshaft gear and bearing mating surfaces for galling or excessive wear.
2. Check snaprings for burred edges. Remove burrs with a fine file.
3. Check synchronizer clutch gear splines on shaft for burrs.

Assembly

Countershaft Gear

1. Slide dummy shaft into countershaft gear.
2. Slide one roller thrust washer over dummy shaft and into gear, followed by 22 greased rollers.
3. Repeat Step 2, adding one roller thrust washer on end.
4. Repeat steps 2 and 3 at other end of countershaft gear. There is a total of 88 rollers and 6 thrust washers.
5. Place greased front thrust washer on dummy shaft against gear with tangs forward.
6. Grease rear thrust washer and stick it in place in the case, with tangs rearward. Place countershaft gear assembly in bottom of transmission case until drive pinion is installed.

Pinion Gear

1. Press new bearing on pinion shaft with snap ring groove forward. Install new snap ring.
2. Install 15 rollers and retaining ring in drive pinion gear.
3. Install drive pinion and bearing assembly into case.
4. Position countershaft gear assembly by positioning it and thrust washers so countershaft can be tapped into position. Be careful to keep the countershaft against the dummy shaft to keep parts from falling between them. Install key in countershaft.
5. Tap drive pinion forward for clearance.

Mainshaft

1. Place a stop ring flat on the bench. Place a clutch gear and a sleeve on top. Drop the struts in their slots and snap in a strut spring placing the tang inside one strut. Turn the assembly over and install second strut spring, tang in a different strut.
2. Slide first gear and stop ring over rear of mainshaft and against thrust flange between first and second gears on shaft.
3. Slide first-reverse synchronizer assembly over rear of mainshaft, indexing hub slots to first gear stop ring lugs.
4. Install first-reverse synchronizer clutch gear snap-ring on mainshaft.
5. Slide reverse gear and mainshaft bearing in place. Press bearing on shaft, supporting inner race of bearing. Be sure snap-ring groove on outer race is forward.
6. Install bearing retaining snap-ring on mainshaft. Spread snap-ring in retainer groove and slide it over the bearing. Seat ring in groove.

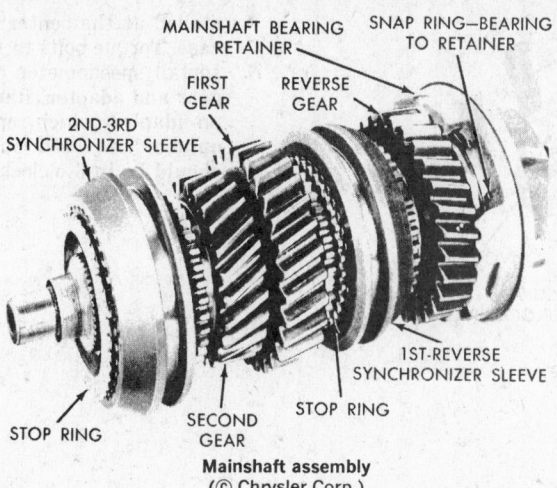

Mainshaft assembly
(© Chrysler Corp.)

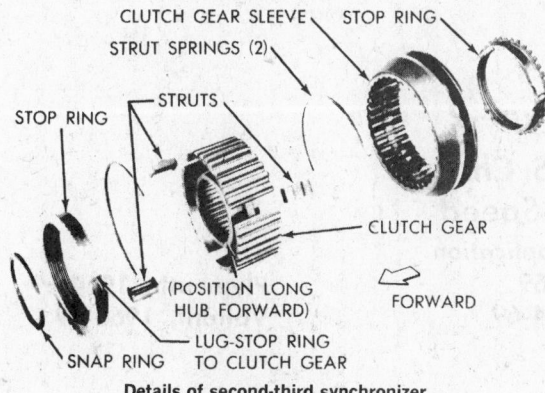

Details of second-third synchronizer
(© Chrysler Corp.)

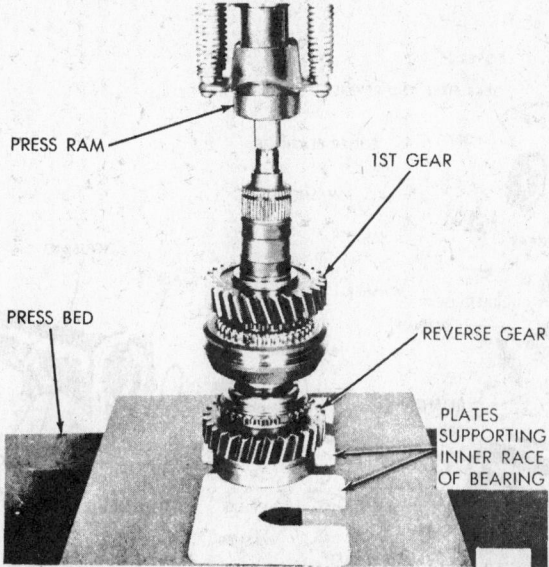

Pressing on mainshaft bearing
(© Chrysler Corp.)

11. Insert mainshaft assembly into case. Tilt assembly to clear cluster gears and insert pilot rollers in drive pinion gear. If assembly is correct, the bearing retainer will bottom to the case without force. If not, check for a misplaced strut, pinion roller, or stop ring.

Reverse Idler Gear

1. Place dummy shaft into idler gear. Insert 22 greased rollers.
2. Position reverse idler thrust washers in case with grease.
3. Position idler gear and dummy shaft in case. Install idler shaft and key.

Extension Housing

1. Remove extension housing yoke seal with special remover, or equivalent tool. Drive bushing out from inside housing.
2. Align oil hole in bushing with oil slot in housing. Drive bushing into place. Drive new seal into housing with special installation tool, or equivalent tool.
3. Install extension housing and gasket to hold mainshaft and bearing retainer in place.

Drive Pinion Bearing Retainer

1. Install outer snap-ring on drive pinion bearing. Tap assembly back until snap-ring contacts case.
2. Using seal installer tool or equivalent, install a new seal in retainer bore.
3. Position main drive pinion bearing retainer and gasket on front of case. Coat threads with sealing compound, install bolts, torque to 30 ft. lbs.

Gearshift Mechanism and Housing

1. If removed, place two interlock levers on pivot pin with spring hangers offset toward each other, so that spring installs in a straight line. Place E-clip on pivot pin.
2. Grease and install new O-ring seals on both shift shafts. Grease housing bores. Push each shaft into its bore.
3. Install spring on interlock lever hangers.
4. Rotate each shift shaft fork bore to straight up position. Install shift forks through bores and under both interlock levers.
5. Position second-third synchronizer sleeve to rear, in second gear position. Position first-reverse synchronizer sleeve to middle of travel, in neutral position. Place shift forks in the same positions.
6. Install gasket and gearshift mechanism. The bolt with the extra long shoulder must be in-

7. Place second gear over front of mainshaft with thrust surface against flange.
8. Install stop ring and second-third synchronizer assembly against second gear. Install second-third synchronizer clutch gear snap-ring on shaft.

9. Move second-third synchronizer sleeve forward as far as possible. Install front stop ring, coated with grease to hold it in position, inside sleeve with lugs indexed to struts.
10. Rotate cut out on second gear toward countershaft gear for clearance.

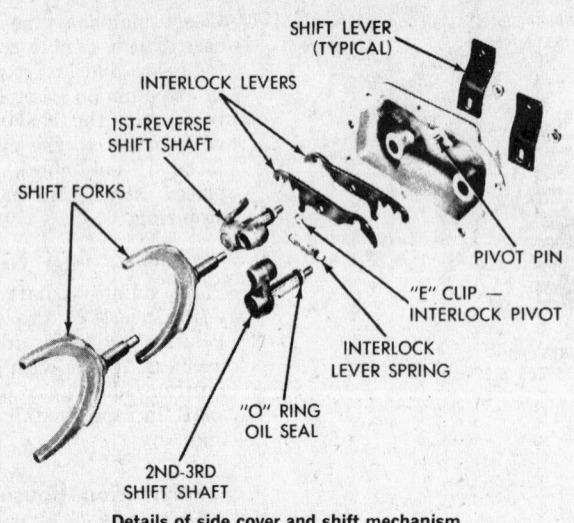

SHIFT LEVER (TYPICAL)

INTERLOCK LEVERS

1ST-REVERSE SHIFT SHAFT

SHIFT FORKS

PIVOT PIN

"E" CLIP — INTERLOCK PIVOT

INTERLOCK LEVER SPRING

"O" RING OIL SEAL

2ND-3RD SHIFT SHAFT

Details of side cover and shift mechanism
(© Chrysler Corp.)

stalled at the center rear of the case. Torque bolts to 15 ft. lbs.

7. Install speedometer drive pinion gear and adapter. Range number on adapter, which represents the number of teeth on the gear, should be in 6 o'clock position.

Type-2
A-745 Chrysler
3-Speed
Application

Barracuda, 1966-69   Dart, 1964-69   Plymouth, 1964-69
Charger, 1967-69   Dodge, 1964-69   Valiant, 1966-69
Chrysler, 1964-69

RING

STOP RING

SYNCHRONIZER UNIT   CLUTCH GEAP SLEEVE

STOP RING

RING

COVER BOLT

GEAR - SECOND   COVER

SPRING   GEAR-FIRST AND REVERSE

CLUTCH GEAR   FORK

SNAP RING   ROLLER BEARINGS

WASHER

WASHER

COUNTERSHAFT GEAR   WASHER

SPACER   IDLER GEAR

FORK   WASHER

ROLLER BEARINGS   HOUSING   SEAL

WASHER   GASKET

WASHER   BEARING

ROLLER BEARINGS   LEVER

WASHER   DETENT BALL

PIN   MAINSHAFT   BEARING

LEVER   SPRING   SNAP RINGS

SLEEVE   KEY   NUT

WASHER   DETENT BALL   LOCKWASHER

GASKET   KEY   BOLT AND WASHER

ROLLER BEARINGS   PLUG   PLUG

WASHER   WASHER

RETAINER   IDLER SHAFT

SEAL   COUNTERSHAFT

OIL SLINGER   TAPERED PIN

BEARING   CASE   LEVER

DRIVE PINION   PLUG   NUT

GASKET   SNAP RING   LEVER

SNAP RING   ROLLER BEARINGS   SEALS   WASHER

BOLT AND WASHER   TAPERED PIN   BUSHING NUT

GROMMET   WASHER   LOCKWASHER BUSHING

LOCKWASHER   GROMMET

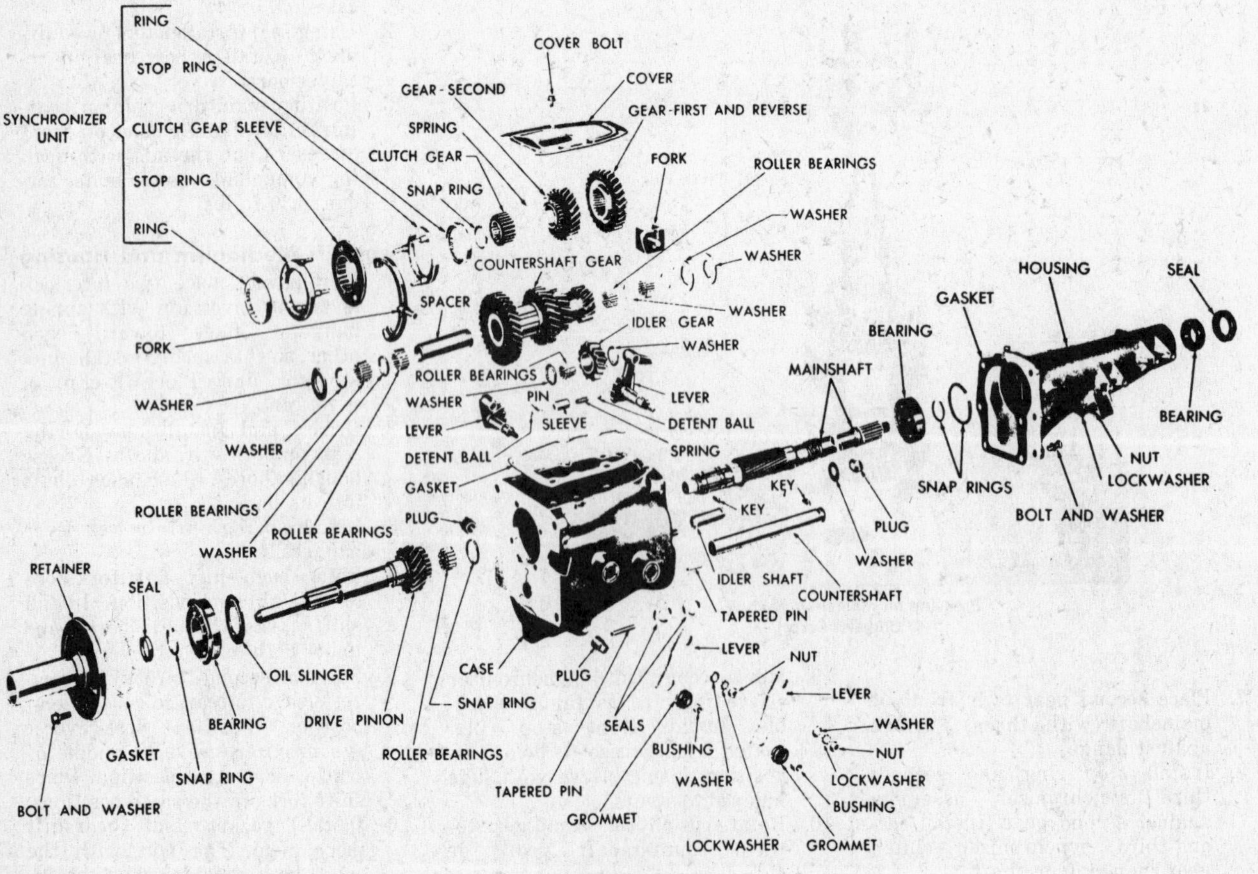

A-745 Chrysler

Disassembly

1. Remove output flange nut, then the drum and flange assembly, if so equipped. Remove parking brake assembly, if so equipped.
2. Remove case cover. Measure synchronizer float with feeler gauges. This measurement is taken between the end of a synchronizer pin and the opposite synchronizer outer ring. A measurement from .050-.090 in. is acceptable for 1954-68 models. The measurement should be .060-.117 in. for 1969 models.
3. Remove five bolts and one nut attaching the extension housing to the transmission case.
4. Remove the extension housing.
5. Remove the mainshaft rear bearing, if it did not come off with the extension housing.
6. Remove transmission case cover and gasket.
7. Remove four bolts from the drive pinion bearing retainer, then remove the retainer.
8. When removing the drive pinion and bearing assembly from the transmission case, slide the front inner stop ring from the short splines on the pinion as the assembly is being removed from the case.
9. Remove the snap-ring that holds the main drive pinion bearing onto the shaft.
10. Press bearing off pinion shaft and remove oil slinger.
11. Remove mainshaft pilot bearing snap-ring from the cavity of the pinion gear.
12. Remove the 15 pilot roller bearings.
13. Remove seal from pinion retainer.
14. Remove mainshaft rear bearing snapring from groove in mainshaft rear bearing bore in the case.
15. Slide the mainshaft and rear bearing assembly to the rear, until the rear bearing is out of the case.
16. Remove synchronizer assembly from case.
17. Remove second and third-speed shift fork.
18. Remove synchronizer clutch gear snap-ring.
19. Remove synchronizer clutch gear, second-speed gear, and first and reverse sliding gear from the mainshaft.
20. Withdraw mainshaft and bearing out through the rear of the case.
21. Remove the synchronizer clutch gear, second-speed gear, low and reverse sliding gear, and low and reverse shift fork from the case.
22. With a dummy shaft, drive the countershaft toward the rear of the case until the small key can

be removed from the countershaft.
23. Drive the countershaft the remaining way out of the case.
24. Lift the cluster gear, the thrust washers and the dummy shaft assembly out of the case.
25. Dismantle the cluster gear, (88 rollers, four spacer rings and the center spacer from the cluster).
26. With a blunt drift, drive the reverse idler shaft toward the rear of the case far enough to remove the key from the shaft.
27. Completely remove the shaft from the case, then remove the idler gear.
28. Remove the thrust washers and 22 rollers.
29. With a small punch, remove low and reverse gear lever shaft tapered lock pin by driving it toward the top of the transmission case.
30. Remove the second and third gear lever shaft in the same manner.
31. Remove the lever shafts from the transmission case. Don't lose the spring-loaded detent balls.
32. Remove the interlock sleeve, spring, pin and detent balls.
33. Remove both lever shaft seals and discard same.

Assembly

1. Place oil slinger on the main drive pinion with the offset outer portion next to the drive pinion teeth.
2. Place the main drive pinion bearing on the pinion shaft with the outer snap-ring away from the pinion gear.
3. Press the bearing into position so it is seated firmly against the oil slinger and pinion gear.
4. Install the bearing retaining snap-ring on the pinion shaft. Be sure the snap-ring is seated in its groove.
5. Coat the 15 pilot bearing rollers with heavy grease and install them in the cavity at the rear of the main drive pinion.
6. Install the snap-ring.
7. Place the bearing spacer in the center of the bore in the cluster gear and use the dummy shaft to assist in assembling the roller bearings.
8. Install a row of 22 rollers next to one end of the spacer, using heavy grease to hold them.
9. Place one of the four bearing spacer rings next to the row of rollers, and install another row of 22 rollers next to the spacer ring.
10. Install another spacer ring at the outside end of the second row of bearing rollers.
11. At the opposite end of the cluster gear bore, install the remaining spacer rings and bearing rollers in the same sequence as listed in Steps 8, 9, and 10.

12. With a small amount of grease, install the front thrust washer on the dummy shaft at the front end of the cluster gear, with the tabs outward.
13. Install the tabbed rear thrust washer onto the dummy shaft against the rear of the cluster gear with the tabs positioned in the grooves provided in the cluster gear.
14. Install the remaining rear thrust washer plate onto the rear of the gear and dummy shaft with the step in the washer facing upward, as viewed from the rear.
15. Align tabs of the front thrust washer vertically to index with notches in the transmission case and with the step in the rear thrust washer positioned upward. Insert the cluster gear and dummy shaft in the transmission case.
16. Using the countershaft, drive the dummy shaft forward, out of the case. Countershaft end play should be .0045-.028 in.
17. Position a dummy shaft in the reverse idler gear and, using heavy grease, install the 22 roller bearings into the gear.
18. Place the thrust washers at each end of the reverse idler gear, then position the assembly in the transmission case with the chamfered end of the gear teeth toward the front.
19. Insert the reverse idler shaft into the bore at the rear of the case, with the keyway to the rear, pushing the dummy shaft toward and out of the front of the transmission.
20. With the keyway in proper alignment, insert the key and continue driving the shaft forward until the key seats in the recess.
21. Install two new lever shaft seals in the transmission case.
22. Lubricate and install second and third-speed lever shaft in the bores of the case.
23. Install the second and third speed lever shaft lock pin in the hole in the case. Drive it in firmly, in a downward direction.
24. Place interlock parts in the case in the following order: ball, sleeve, spring, pin and ball.
25. Enter low and reverse lever shaft in the case bore, depress the detent ball against spring tension and push the lever shaft firmly into position, in order to prevent the ball from escaping.
26. Install low and reverse lever shaft lock pin in the case by driving it downward.
27. Place low and reverse fork in the lever shaft, with the offset toward the rear.
28. While holding the low and reverse sliding gear in position in the

fork, with the hub extension to the rear, insert the mainshaft with the rear bearing through the rear of the case and into the sliding gear.

29. Place synchronizer stop ring spring, then the rear stop ring, on the synchronizer splines of the second-speed gear. Install the second-speed gear onto the mainshaft. Synchronizer shims must be added if synchronizer float is more than the maximum in Step 2, disassembly. If float was less than minimum, the six pins must be shortened.

30. Install the synchronizer clutch gear on the mainshaft with the shoulder to the front.

31. Select the thickest synchronizer clutch gear snap-ring that can be used, and install it in the mainshaft groove.

32. Check clearance between clutch gear and second-speed gear. Clearance should be .004-.014 in.

33. Hold the synchronizer clutch gear sleeve and two outer rings together with pins properly entered into the holes in the clutch gear sleeve and with the clutch gear sleeve engaged in the groove of the second- and third-speed shift fork, position the fork in the second and third-speed lever shaft.

34. While holding the synchronizer parts and fork in position, slide the mainshaft forward, entering the synchronizer clutch gear into the clutch gear sleeve and at the same time entering the mainshaft rear bearing in the case bore.

35. While still holding the synchronizer parts in position, tap the mainshaft forward until the rear bearing bottoms in the case bore.

36. Install the mainshaft rear bearing snap-ring into place in the case bore.

37. Install a new seal in the drive pinion retainer.

38. Place the synchronizer front inner ring in position in the front outer ring, and enter the main drive pinion through the case bore.

39. Engage the splines on the rear of the pinion with the inner stop ring, and tap the drive pinion into the case until the outer snap-ring on the pinion bearing is against the transmission case.

40. Place the drive pinion bearing retainer over the pinion shaft and against the transmission case. While holding the retainer against the transmission case, measure the clearance between the retainer and case.

41. Select a gasket .003-.005 in. thicker than the clearance found.

42. Install and tighten the front bearing retainer bolts to 30 ft. lbs. torque.

43. Install a new seal in the extension housing.

44. Install extension housing. Torque the attaching bolts and nuts to 50 ft. lbs.

45. Install the parking brake assembly, on vehicles so equipped.

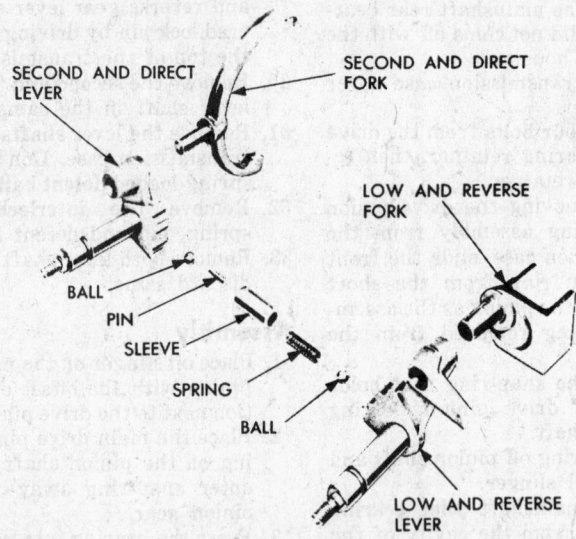

Shift forks and shafts

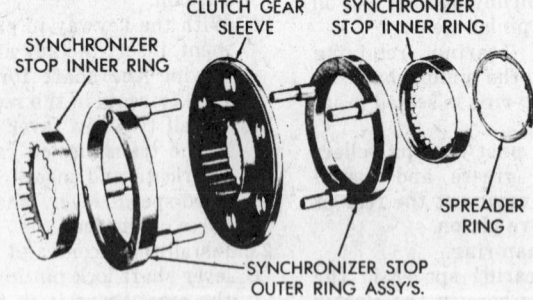

Synchromesh assembly

46. Install the parking brake drum (if so equipped) and flange assembly. Install the washer and nut and torque to 175 ft. lbs.

47. Install the drain plug in the transmission case.

48. Install the gearshift operating levers, and torque to 12 ft. lbs.

49. Install the back-up light switch.

50. Install the speedometer cable and drive gear. Bring transmission to lubricant level.

Type-3
A-903 Chrysler
3-Speed

Application

Charger (6 Cyl.), 1967-70
Challenger (6 Cyl.), 1970
Barracuda (6 Cyl.), 1966-70

Dart (6 Cyl.), 1964-70
Dodge (6 Cyl.), 1964-70

Plymouth (6 Cyl.), 1964-70
Valiant (6 Cyl.), 1964-70

Disassembly

1. Remove output shaft yoke.
2. Remove the bolts that attach he extension housing to the transmission case. Remove the housing.
3. Remove extension housing oil seal.
4. Remove the transmission case cover. Measure synchronizer float with feeler gauges. This measurement is taken between the end of a synchronizer pin and the opposite synchronizer

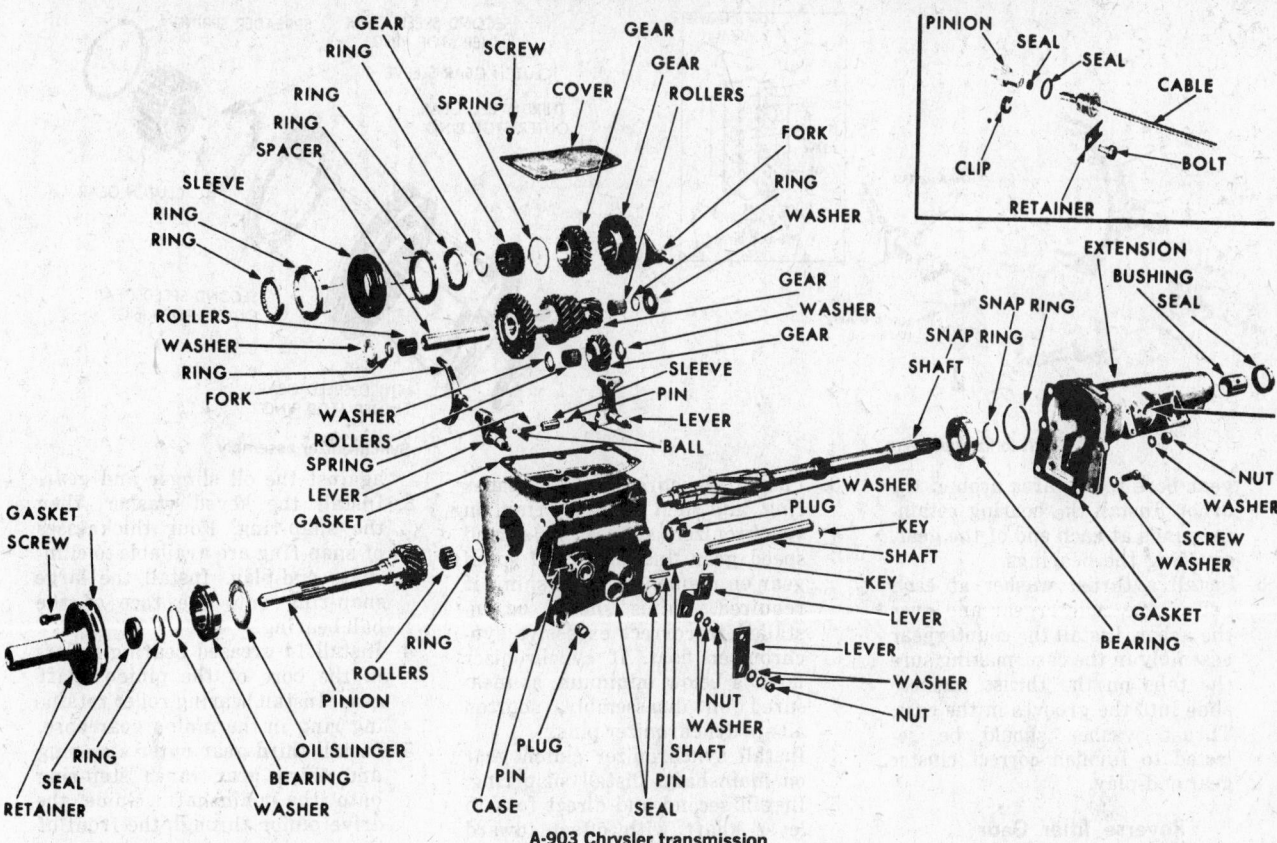

A-903 Chrysler transmission

outer ring. A measurement from .050-.090 in. is acceptable for 1964-68 models. This measurement should be .060-.117 in. for 1969 models.

5. Remove the attaching bolts and remove the main drive pinion bearing retainer. Then grasp the pinion shaft and pull the assembly out of the case.

Caution Be careful not to bind the inner synchronizer ring on the drive pinion clutch teeth.

6. Remove the snap-ring that locks the main drive pinion bearing onto the pinion shaft. Remove the bearing washer, press the shaft out of the bearing and remove the oil slinger.
7. Remove the snap-ring from the pilot bearing in the end of the drive pinion and remove the 14 rollers.
8. With the transmission in reverse, remove the outer center bearing snap-ring, then partially remove the mainshaft.
9. Cock the mainshaft, then remove the clutch sleeve, the outer synchronizer rings, the front inner ring and the second-third shift box.
10. Remove clutch gear retaining snap ring and slide the clutch gear off the end of the mainshaft.
11. Slide the second-speed gear, stop

ring and synchronizer spring off the mainshaft.
12. Remove the low and reverse sliding gear and shift fork, as the mainshaft is completely withdrawn from the case.
13. Check cluster gear end-play. End-play should be .005-.022 in. This measurement will determine thrust washer value at reassembly.
14. Drive the countershaft rearward, removing key, and out of the case.
15. Lift the gear cluster and thrust washers out of the case. Remove the needle bearings, (22 each end) and spacer from the cluster.
16. Drive the reverse idler shaft toward the rear and out of the case. Remove key.
17. Lift the reverse idler gear, thrust washers and 22 needle bearings out of the case.
18. Remove gearshift operating levers from their respective shafts.
19. Drive out tapered retaining pin from either of the two lever shafts, then withdraw the shaft from inside the transmission case. (The detent balls are spring-loaded, as the shaft is being withdrawn, the balls will fall to the bottom of the case.)
20. Remove the interlock sleeve, spring, pin and both balls from the case. Drive out the remaining tapered pin, then slide the

lever shaft out of the transmission.
21. Remove the lever shaft seals and discard them.

Assembly

1. Install two new shift lever shaft seals in the case.
2. Carefully insert low and reverse lever shaft into the rear of the case, through the seal and into position. Lock with a tapered pin. Turn lever until the center detent is in line with the interlock bore.
3. Slide the interlock sleeve in its bore in the case, followed by one of the interlock balls. Then, install interlock spring and pin.
4. Place the remaining interlock ball on top of the interlock spring.
5. Depress the interlock ball and at the same time install the second and high lever shaft into the fully seated position, with the center detent aligned with the detent ball. Secure the shaft with the remaining tapered pin.
6. Install the operating levers and secure to the shafts with nuts. Torque the nuts to 18 ft. lbs.

Countershaft (Cluster) Gear

1. Slide the dummy shaft and tubular spacer into the bore of the countergear.
2. Grease and install 22 bearing rollers into each end of the counter-

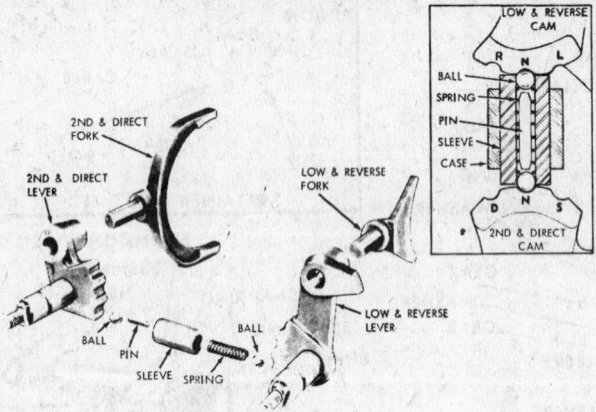

Shift forks and levers

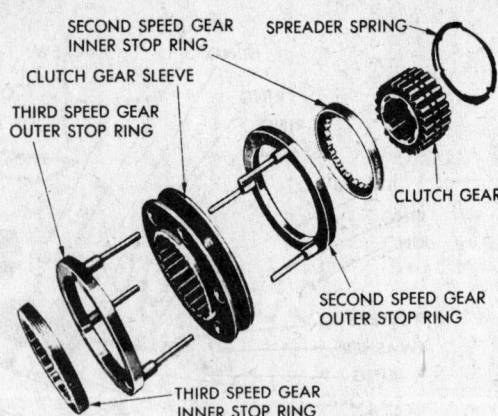

Synchronizer assembly

gear bore in the area around the arbor. Install the bearing retaining rings at each end of the gear, covering the bearings.
3. Install a thrust washer at each end of the countergear and over the arbor. Install the countergear assembly in the case, making sure the tabs on the thrust washers slide into the grooves in the case. Thrust washer should be selected to furnish correct cluster gear end-play.

Reverse Idler Gear

1. Coat the bore of the reverse idler gear with grease, then slide dummy shaft into the bore, then install 22 bearing rollers in the bore and around the dummy shaft.
2. Install a new thrust washer at each end of the gear and over the arbor.
3. With the beveled end of the teeth forward, slide the gear into position in the case. Install the reverse idler shaft in its bore in the rear of the case. Install Woodruff key and align with the keyway in the case.
4. Align the idler gear with the shaft, then drive the shaft into the case and gear until the key seats in recess.

Mainshaft

1. Install rear bearing on mainshaft and install selective fit snap ring.
2. Hold low and reverse sliding gear in position with shift fork. Insert mainshaft with rear bearing through rear of case and into the sliding gear.

3. Place synchronizer spreader ring, and then rear stop ring, on synchronizer splines of second speed gear. Install second speed gear on mainshaft, with shims if required. Shims should be installed to correct excessive synchronizer float. If synchronizer float is below minimum, as measured on disassembly, shorten all six synchronizer pins.
4. Install synchronizer clutch gear on mainshaft. Install snap ring.
5. Install second and direct fork in lever shaft with offset toward rear of transmission. Hold synchronizer clutch gear sleeve and two outer rings together, with pins in holes in clutch gear sleeve. Engage second and direct fork with clutch gear sleeve.
6. While holding synchronizer parts and fork in position, slide mainshaft forward, starting synchronizer clutch gear into clutch gear sleeve and mainshaft rear bearing into the case bore. Synchronizer parts must be correctly positioned before mainshaft is positioned.
7. While holding synchronizer parts in position, tap mainshaft forward until rear bearing bottoms in the case bore.
8. Install mainshaft rear bearing selective fit snap ring into groove in case bore.

Drive Pinion (Clutch Shaft)

1. Slide the oil slinger over the pinion shaft and down against the gear.
2. Slide the bearing over the pinion shaft (ring groove away from the gear), then press to a firm seat

against the oil slinger and gear.
3. Install the keyed washer, then the snap-ring. Four thicknesses of snap-ring are available to eliminate end-play. Install the large snap-ring onto the race of the ball bearing.
4. Install 14 greased bearing rollers in the bore of the pinion shaft gear. Install bearing roller retaining ring in the pinion gear bore.
5. Install third gear outer stop-ring and third gear inner stop-ring onto the mainshaft. Guide the drive pinion through the front of the case and engage the inner stop-ring with the clutch teeth, then seat the bearing so the large snap-ring is hard against the case.
6. Install a new seal in the pinion bearing retainer.
7. Install the gasket on the retainer and install with attaching bolts torqued to 30 ft. lbs.

Extension Housing

1. Install a new rear mainshaft bushing, and a new oil seal.
2. Protect the oil seal with thimble-type seal protector, and with gasket attached, slide the extension housing over the mainshaft and down against the case. Attach with bolts torqued to 50 ft. lbs.
3. Install flange assembly and secure with new washer and nut. Torque the nut to 140 ft. lbs.
4. Grease the cover gasket, and install gasket on cover. Torque attaching bolts to 12 ft. lbs.
5. Install drain plug and back-up light switch (if so equipped) and tighten securely. Refill transmission to proper level.

Type-4
A-833 Chrysler
4-Speed

Application

Barracuda, 1964-70
Challenger, 1970
Charger, 1966-70

Chrysler, 1964-67
Dart, 1964-70

Dodge, 1964-70
Plymouth, 1964-70
Valiant, 1964-70

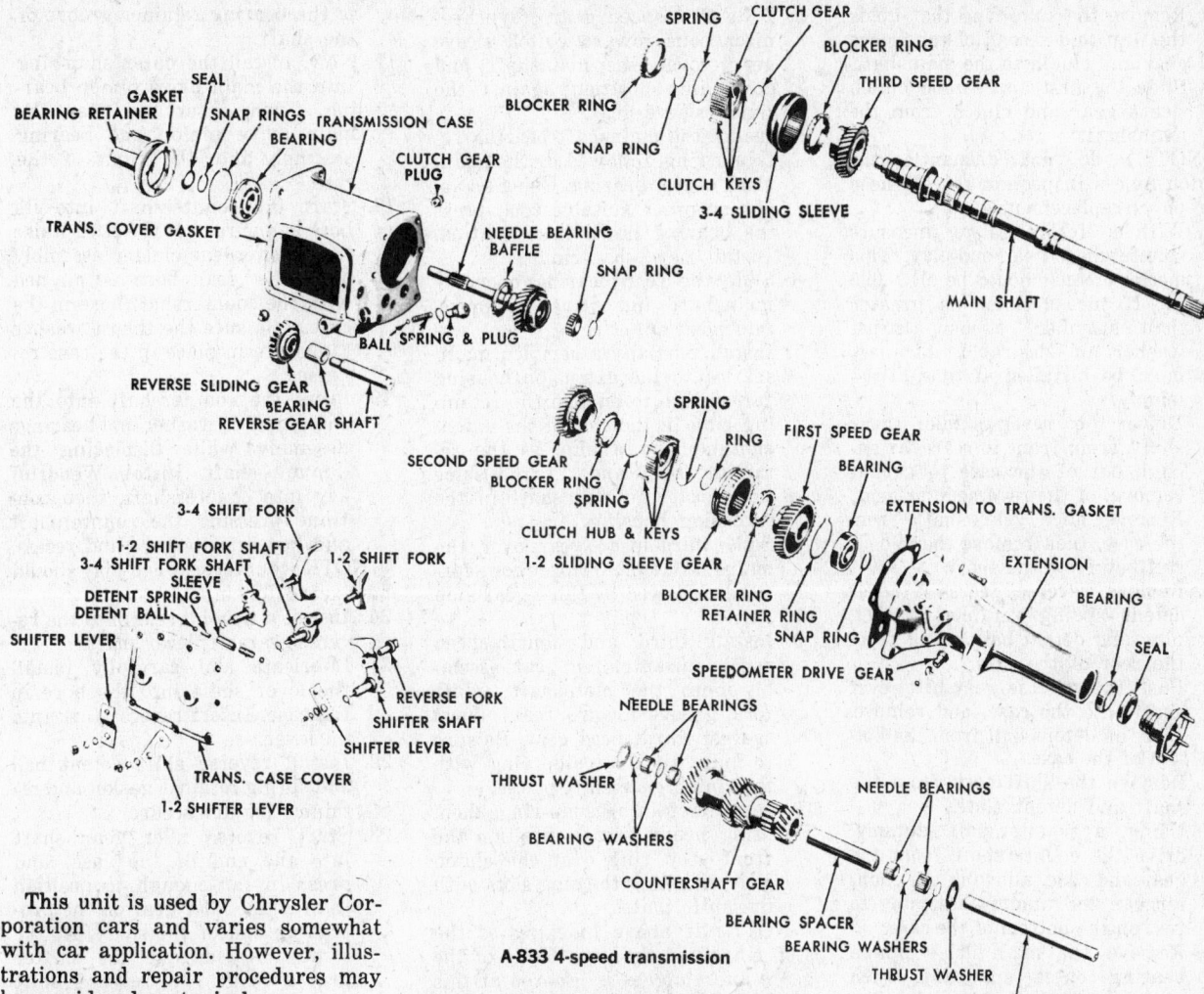

A-833 4-speed transmission

This unit is used by Chrysler Corporation cars and varies somewhat with car application. However, illustrations and repair procedures may be considered as typical.

Disassembly

1. If available, mount transmission in a repair stand.
2. Disconnect gearshift control rods from the shift control levers and the transmission operating levers.
3. Remove the two gearshift control housing mounting bolts.
4. Remove gearshift control housing from the transmission extension housing or mounting bracket (if so equipped).
5. Remove the gearshift control housing mounting bracket bolts, then, remove the bracket (if so equipped).
6. Remove back-up light switch (if so equipped).
7. Remove output companion flange nut and washer, then pull the flange from the mainshaft (output shaft).
8. Remove gearshift housing-to-transmission case attaching bolts.
9. With all levers in the neutral detent position, pull housing out and away from the case.

NOTE: if first and second, or third and fourth shift forks remain in engagement with the synchronizer sleeves, work the sleeves and remove forks from the case.

10. Remove nuts, lock washers and flat washers that hold first-second, and third-fourth-speed shift operating levers to the shafts.
11. Disengage shift levers from the flats on the shafts and remove levers.
12. Remove gearshift lever shafts out of the housing, allowing detent balls to fall free. Remove seals and discard.
13. Slide interlock sleeve, interlock pin and spring from the housing.
14. Remove main drive pinion bearing retainer attaching bolts, then slide retainer and gasket from the main drive shaft. Remove the pinion oil seal.
15. Remove the attaching bolts that hold the tailshaft extension housing to the transmission case.
16. Slide the third-fourth synchronizer sleeve slightly forward, then, using a soft hammer, tap rearward on the extension housing. Slide housing and mainshaft assembly out and away from the case.
17. Remove the snap-ring that holds the third-fourth synchronizer clutch gear and sleeve. Then,

slide third-fourth synchronizer assembly from the end of the mainshaft.
18. Slide third speed gear and stop ring from the mainshaft.

NOTE: do not separate third-fourth-speed synchronizer clutch gear, sleeve, shift plates or spring unless replacement is required.

19. With long nose pliers, compress the snap-ring that retains the mainshaft center bearing in the extension housing.
20. With snap-ring compressed, pull the mainshaft assembly and bearing out of the extension housing.
21. Remove and discard extension housing rear oil seal.
22. Remove rear bearing from the mainshaft by inserting steel plates on the front side of first-speed gear, then, with an arbor press, force the rear bearing from the mainshaft.
23. Remove the snap-ring that holds the mainshaft bearing onto the shaft.
24. Remove mainshaft bearing, retainer ring, first-speed gear, and first-speed stop ring.

25. Remove the snap-ring that holds the first and second clutch sleeve gear and clutch to the mainshaft.
26. Slide the first and second clutch sleeve gear and clutch from the mainshaft.

NOTE: do not dismantle the clutch unless inspection reveals need for parts replacement.

27. With a feeler gauge, measure countershaft gear end-play. This measurement should be .015-.025 in. If measurement is greater than specified, a new thrust washer of desirable thickness must be installed at time of assembly.
28. Drive the reverse idler gear shaft, from front to rear, far enough out of the case to permit removal of the reverse idler gear.
29. Remove idler gear shaft from the case, then remove the Woodruff key from the shaft.
30. Remove reverse gearshift lever detent spring retainer, gasket, plug and detent ball spring from the rear of the case.
31. Push the reverse gearshift lever shaft into the case, and remove. Lift the detent ball from the bottom of the case.
32. Remove the shift fork from the shaft and detent plate.
33. Using a countershaft dummy, drive the countershaft from the gear and case, allowing the countergear and dummy assembly to rest on the bottom of the case.
34. Remove the main drive pinion bearing outer snap-ring, then with a soft hammer, drive the main drive pinion into the case and remove.
35. Remove the main drive pinion bearing outer snap-ring, then, with an arbor press, remove the bearing from the main drive pinion. Remove the oil slinger.
36. Lift the countergear cluster from the bottom of the case.
37. Remove the countergear dummy shaft, 76 bearing rollers, thrust washers and tubular spacer from the center of the countergear.

Assembly

1. Slide the second-speed gear over the mainshaft (synchronizer cone toward rear) and down into position against the shoulder on the shaft.
2. Slide first and second clutch sleeve gear assembly (including second gear stop ring) over the mainshaft. Be sure shift fork groove is toward the front and down into position against second-speed gear, (stop ring must be indexed with the shift plates). Install a new snap-ring to secure.
3. Slide low gear stop ring over the shaft and down into position and index with the shift plates.

4. Slide first-speed gear, (synchronizer cone toward clutch sleeve gear) over the mainshaft and down into position against the clutch sleeve gear.
5. Install the mainshaft bearing retainer ring, followed by the mainshaft center bearing. Using an arbor or other suitable tool, press the bearing down into position. Install new snap-ring.
6. Slide the rear bearing over the mainshaft and drive, or press, into position.
7. Install partially assembled mainshaft into the extension housing far enough to engage the retaining ring in the slot in the extension housing. Compress the retaining ring and, at the same time, seat the mainshaft in the extension housing.
8. Slide third-speed gear over the mainshaft, synchronizer cone forward, followed by third gear stop ring.
9. Install third and fourth-speed synchronizer clutch gear assembly onto the mainshaft (shift fork groove toward rear) down against third-speed gear. Be sure to index the rear stop ring with the clutch gear shift plates.
10. Install retaining snap-ring, then, using heavy grease, position the front stop ring over the clutch gear, indexing the ring slots with the shift plates.

NOTE: if above indexing of the stop rings and the positioning of the gears and clutches is ignored at this point, damage will most likely result when mating the extension housing to the transmission case.

11. Grease the bore of the countergear at each end, then install the roller bearing tubular spacer (centered). Insert the countergear dummy shaft.
12. Grease each bearing roller, then install 19 bearing rollers at each end of the gear. Now, install a flat spacer onto each end of the dummy shaft and into the gear, followed by 19 more bearing rollers and a spacer ring into each end of the countergear.
13. Grease the tanged thrust washers and install them, one over each end of the dummy shaft, with the tangs toward the case (away from the gear).
14. Lay the countergear assembly into the bottom of the case.
15. To install the main drive pinion, slide the bearing oil slinger over the main drive pinion shaft, then, press the main drive pinion bearing on the pinion shaft. (Be sure the outer snap-ring groove is toward the front). Seat bearing all the way, against shoulder on gear.
16. Install a new inner snap-ring in-

to the bearing retainer groove of the shaft.
17. Now, install the outer snap-ring into the main drive pinion bearing. Then, insert and tap the main drive pinion and bearing assembly into the front of the case.
18. Start the countershaft into its bore at the rear of the case. Raise the countergear cluster assembly until the gear bore is aligned with the countershaft bore in the case. (Be sure the thrust washer tangs are in place in the case recesses.)
19. Press the countershaft into the countergear, washer and bearings assembly while displacing the dummy shaft. Install Woodruff key into countershaft, then continue pressing the countershaft and key into its bore and recess.

NOTE: countergear end-play should not exceed .029 in.

20. Install a new oil seal onto the reverse gearshift lever shaft.
21. Lubricate and carefully install the lever shaft into the bore in the case. Insert reverse fork into the lever.
22. Install reverse shift detent ball and spring retainer gasket and retainer. Tighten securely.
23. Start reverse idler gear shaft into the end of the case, and press in far enough to position the reverse idler gear on the protruding end of the shaft. At the same time, engage the shifter groove with the reverse shift fork.
24. With reverse idler gear properly positioned, install Woodruff key into the sliding gear shaft, then finish seating the shaft and key flush with the end of the case.
25. Grease, then position a new gasket on the end of the extension housing.
26. Center reverse sliding gear on its shaft, then carefully insert the mainshaft assembly into the case. (Be sure of the indexing of third and fourth-speed stop rings and shifter plates.)
27. Move third and fourth-speed clutch sleeve slightly toward the front, and, at the same time, align the end of the mainshaft with the main drive pinion. Push in on the extension housing assembly until it is entirely seated against the rear of the case.
28. Install extension-to-case attaching bolts and torque to 50 ft. lbs.
29. Install back-up light switch (if so equipped).
30. Move reverse sliding gear ahead to neutral position.
31. Slide interlock sleeve into position in the gearshift housing. Lubricate and slide a new seal

over a shifter shaft and down into its groove.

32. Install the gearshift lever shaft into position in the housing, then install the gearshift operating lever onto the flats of the shaft, (lever pointing up). Install flat washer, lock washer and nut. Tighten securely.

33. Place a detent ball in the sleeve, followed by the poppet spring and interlock pin.

34. Lubricate and slide a new seal over the other shifter shaft and down into its groove.

35. As with the first gearshift lever shaft, push the shaft into position in the housing, then install the operating lever onto the flats of the shaft (lever pointing up). Install flat washer, lock washer and nut and tighten securely.

36. Place remaining detent ball on the poppet spring, compress the ball and spring with a small screwdriver, then, push the shafts in until seated. Turn the shafts until the balls drop into the neutral position detent.

37. Place transmission on its side, gearshift cover opening up.

38. Install a shift fork onto each synchronizer sleeve collar, and, with both sleeves in neutral position, install the shift housing and new gasket.

39. Install attaching bolts and tighten to 12 ft. lbs. (The center bolt on each side of the cover is a pilot bolt and should be installed first.)

40. Lubricate and install a new oil seal in the main drive pinion retainer bore, then install the retainer and gasket. Install attaching bolts, torqued to 15-20 ft. lbs.

41. Install gearshift control and rod assembly on the extension housing, then, secure rods with washers and clips.

42. Install output companion flange, washer and nut. Torque to 175 ft. lbs.

Type-5
2.77 Ford 3-Speed

Application

Fairlane (6 Cyl.), 1964-66
Falcon (6 Cyl.), 1964-67

Comet (6 Cyl.), 1964-66
Mustang (6 Cyl.), 1965-66

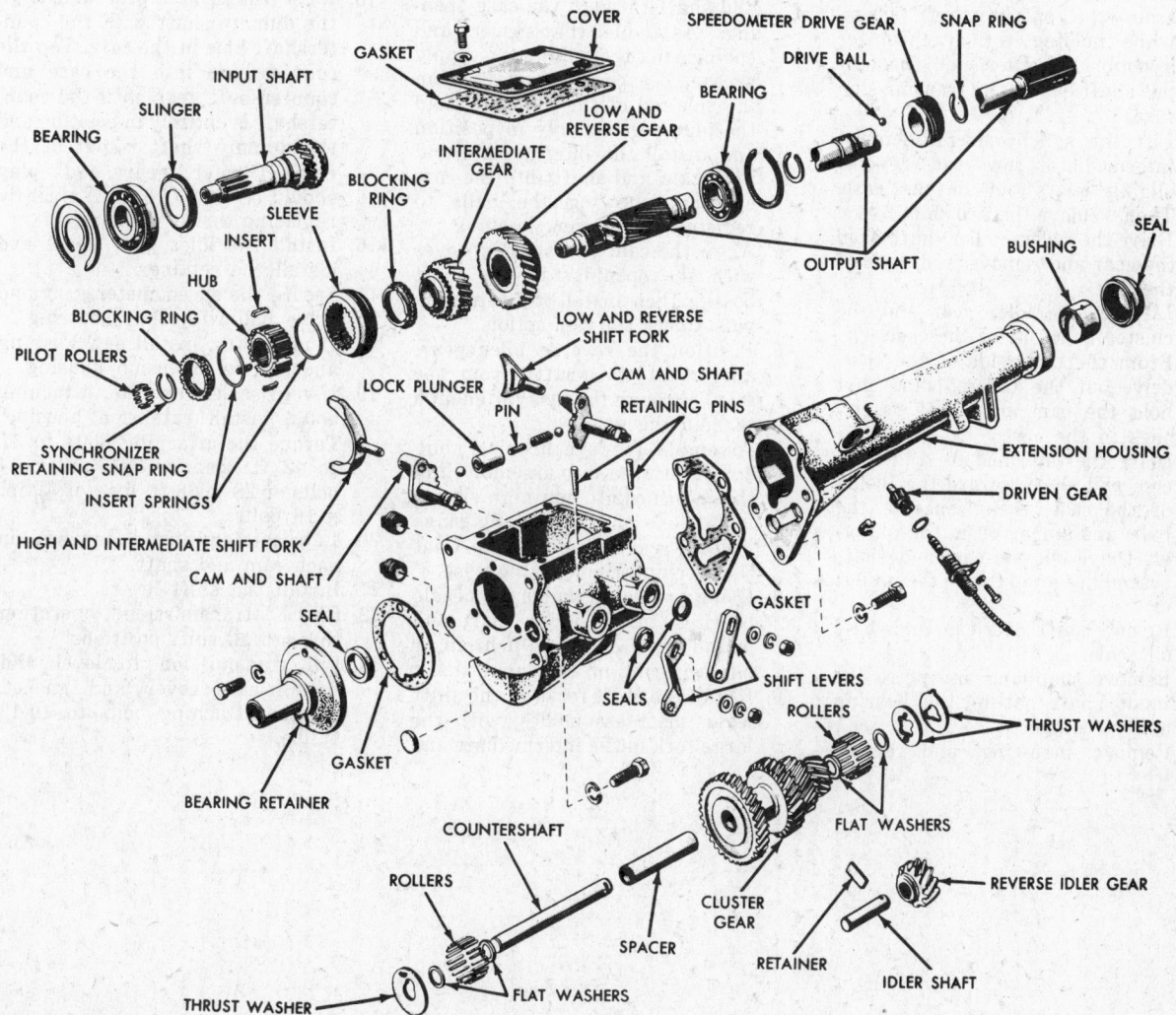

2.77 Ford 3-speed transmission
(© Ford Motor Co.)

Disassembly

1. Drain the unit and remove transmission cover and gasket.
2. Remove outer shift levers and the interlock linkage.
3. Remove extension housing attaching bolts and remove the engine rear support bracket and the extension housing and gasket. Tap the end of the output shaft with a soft hammer and withdraw extension housing.
4. Remove speedometer drive gear snap-ring. Then, remove the drive gear and drive ball from the output shaft.
5. Remove idler and countershaft retainer.
6. With a dummy shaft, drive the shaft out of the cluster gear and case. Leave the dummy shaft in the cluster, at rest, in the bottom of the case.
7. Remove input shaft bearing retainer and gasket.
8. Remove input shaft assembly and front synchronizer blocking ring from the case.
9. Remove synchronizer snap-ring from the output shaft. Then, while holding the synchronizer assembly together, pull the output shaft out of the transmission case.
10. Lift the synchronizer assembly, intermediate, low and reverse sliding gears out of the case. Then, remove the two shift forks.
11. Drive the reverse idler shaft from the gear and remove from rear of the case.
12. Lift reverse idler gear and the cluster gear out of the case.
13. From the underside of the case, drive out the tapered pins that hold the cam and shaft assemblies in the case.
14. Drive the intermediate and high cam and shaft toward the inside of the case, then, separate the balls and spring from the plunger. Drive out the cam and shaft assemblies and remove the plunger.
15. Remove shift lever cam and shaft oil seals.
16. Remove snap-ring and press the input shaft out of the bearing and oil slinger.
17. Remove snap-ring, and remove the bearing from the output shaft.
18. Dismantle the components, such as synchronizer, countershaft cluster, etc., then clean and inspect parts.

Assembly

1. Press the input shaft bearing and oil slinger onto the input shaft, then install the snap-ring onto the shaft.
2. Press the output shaft bearing onto the output shaft, install snap-ring on the shaft.
3. Insert the tubular spacer and dummy shaft into the cluster gear assembly. Position one flat washer at each end of the spacer. Apply grease to the needle bearings and assemble the needle rollers around the dummy shaft at each end of the gear. Apply grease to the other two flat washers and thrust washers and assemble at each end of the cluster gear. Position the gear cluster assembly in the bottom of the transmission case.
4. Install reverse and low shift cam and shaft through the case opening. Assemble the spacer and spring in the plunger. Apply grease to each ball and position in each end of the plunger. Hold the plunger assembly in position and install the intermediate and high cam and shaft into the case opening, allowing the balls to register in the cam detents.
5. Align the cam and shaft grooves with the openings in the shaft bosses, then install the retaining pins. Check the cam action.
6. Position the reverse idler gear, and insert the shaft (from the rear) through the case far enough to hold the gear.
7. Assemble the synchronizer hub and insert the hub assembly into the intermediate and high sleeve. Install one of the blocking rings into the rear side of the hub. Coat the blocking rings with grease.
8. Using grease, assemble the needle bearings in the input shaft and install the front synchronizer blocking ring on the input shaft.
9. Install the shift forks in the shift lever shaft assemblies, with the large fork in the intermediate and high shaft assembly. The web of the low and reverse fork must be to the rear of the shaft center.
10. Start the output shaft through the rear opening of the transmission case. Place the low and reverse gear on the shaft, followed by the intermediate gear. Tilt the output shaft enough to allow the rear shift fork to engage the sliding gear groove.
11. With the longer hub forward, slide the synchronizer assembly onto the output shaft and engage the synchronizer sleeve in the intermediate and high shift fork.
12. Install the synchronizer hub snap ring.
13. Position the input shaft and front synchronizer blocking ring.
14. Place a new gasket on the input shaft bearing retainer. Install bearing retainer oil seal, then install the bearing retainer. Line up the drain groove in the retainer with the oil hole in the case, then with sealer on the retainer attaching screws, torque them to 12-15 ft. lbs.
15. Raise the cluster gear and align the dummy shaft with the countershaft hole in the case. Tap the countershaft into the case and countershaft gear until the countershaft is entirely in position and the dummy shaft is out of the case. Cluster gear end play should be .004-.018 in. Adjust by replacing thrust washers.
16. Install the idler gear shaft and install the retainer.
17. Secure the speedometer gear and drive ball with the snap-ring.
18. If necessary, install new bushing and seal in extension housing.
19. Using sealer on the attaching bolts, install extension housing. Torque the attaching bolts to 37 to 42 ft. lbs. for 7/16 in. x 14 bolts or 28 to 38 ft. lbs. for 3/8 in. x 16 bolts.
20. Lubricate and install new seal on each cam and shaft.
21. Install the shift levers.
22. Check transmission operation through all shift positions.
23. Fill transmission to level and install case cover and gasket. Torque attaching bolts to 10-13 ft. lbs.

Type-6
3.03 Fully Synchronized
Ford 3-Speed

Application

Buick LeSabre, 1970
Buick Wildcat, 1966-70
Buick Special, 1965-69 (GS),
   1966-69 (V8)
Comet, 1964-66 (V8),
   1967 (All)
Cougar, 1967-70
Fairlane, 1964-66 (V8),
   1967-70 (All)

Falcon, 1964-66 (V8),
   1967-70 (All)
Firebird (V8), 1968-69
Ford, 1964-70
Maverick, 1970
Mercury, 1964-70
Meteor, 1969-70
Montego, 1968-70

Mustang, 1965-66 (V8),
   1967-70 (All)
Oldsmobile, 1966-70
Olds F-85 (V8), 1965-69
Tempest (V8), 1965-69
Pontiac, 1966-70
Pontiac GP, 1969

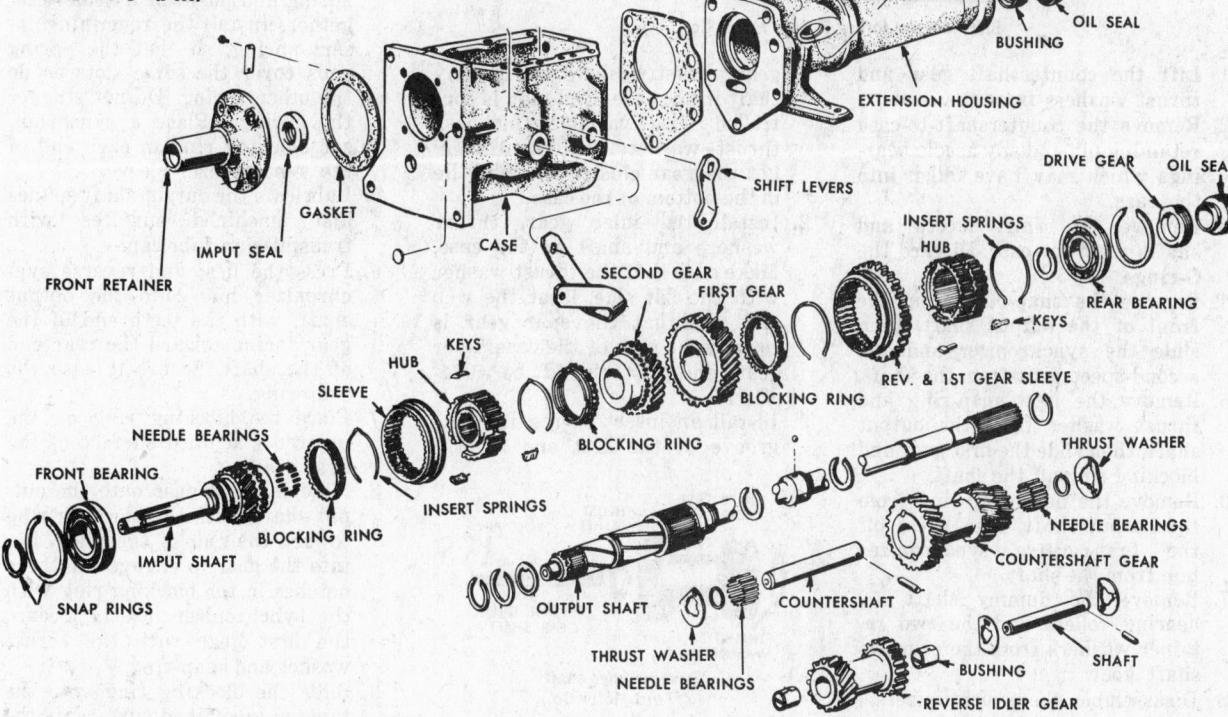

3.03 Ford 3-speed transmission (© Ford Motor Co.)

Disassembly

1. Drain the lubricant, then remove the cover bolts and the case cover.
2. Remove the five attaching screws, then remove the extension housing from the transmission case.
3. Remove the four attaching screws, then remove the front bearing retainer from the case.
4. Remove the filler plug. Working through the filler plug hole, drive the roll pin out of the case and countershaft with a small punch.
5. With a dummy shaft, push the countershaft out of the rear of the case until the countershaft cluster gear can be lowered to the bottom of the case. Remove the countershaft from the rear of the case.
6. Remove the snap-ring. Lift the input gear and shaft from the front of the case. Press the shaft out of the bearing.

7. Remove the snap-ring that holds the speedometer gear onto the shaft. Slide the speedometer gear off the output shaft. Remove the speedometer gear lock ball.
8. Remove the snap-ring that holds the output shaft bearing on the shaft. With a puller, remove the bearing from both the case and shaft.
9. Place both shaft levers in the neutral position.
10. Remove the set screw that holds the detent springs and plugs in the case. Remove a detent spring and plug from the case.
11. Remove the set screw that holds the first and reverse shift fork to the shift rail. Slide first and reverse shift rail out through the rear of the case.
12. Rotate the first and reverse shift fork upward, then lift it from the case.

13. Remove the set screw that holds the second and third shift fork to the shift rail. Rotate the shift rail 90°.
14. With a magnet, lift the interlock plug from the case.
15. Tap on the inner end of the second and third shift rail to remove the expansion plug from the front of the case. Remove the shift rail.
16. Remove second and third detent plug and spring from the detent bore.
17. Rotate the second and third shift fork upward, then lift it from the case.
18. Lift the output shaft out through the top of the case.
19. Working through the front bearing opening, drive the reverse idler shaft out through the rear of the case.
20. Lift the reverse idler gear and two thrust washers from the case.

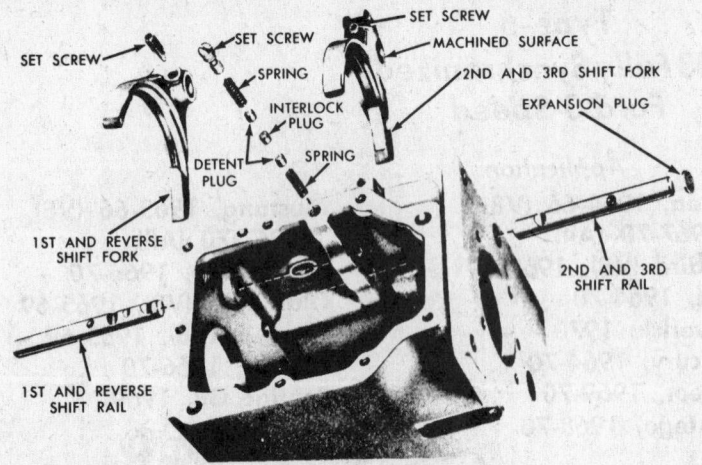

Shift rail and forks (© Ford Motor Co.)

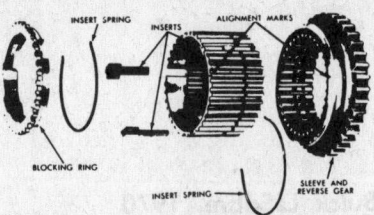

First and reverse synchronizer
(© Ford Motor Co.)

21. Lift the countershaft gear and thrust washers from the case.
22. Remove the countershaft-to-case retaining pin and any needle bearings which may have fallen into the case.
23. Remove the shift levers and shafts from the case. Discard the O-rings.
24. Remove the snap-ring from the front of the output shaft, then slide the synchronizer and the second-speed gear from the shaft.
25. Remove the next snap-ring and thrust washer from the output shaft, then slide the first gear and blocking ring off the shaft.
26. Remove the next snap-ring from the output shaft, then press off the first-reverse synchronizer hub from the shaft.
27. Remove the dummy shaft, 50 bearing rollers and the two retainer washers from the countershaft gear.
28. Disassemble the synchronizers.

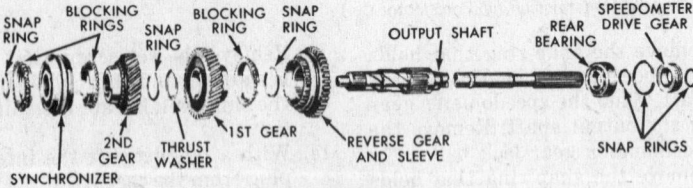

Output shaft
(© Ford Motor Co.)

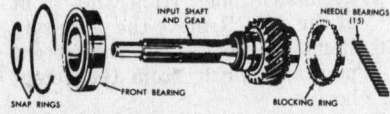

Input shaft and bearing
(© Ford Motor Co.)

Assembly

1. Coat the bore in each end of the countershaft gear with grease. Hold the dummy shaft in the gear and install 25 bearing rollers and a retainer washer in each end of the gear. Install the countershaft gear, thrust washers and dummy shaft in the case. End-play is controlled with variable thickness thrust washers to .004-.018 in. Let the gear cluster assembly lie in the bottom of the case.
2. Install the idler gear, thrust washers and shaft in the case. Make sure that the thrust washer with the flat side, is at the web end and that the spur gear is toward the rear of the case. Idler gear end-play should be .004-.018 in.
3. Install an insert spring into the groove of the first and reverse

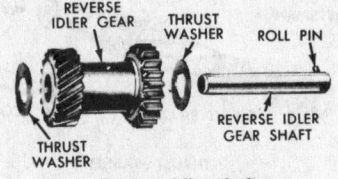

Reverse idler shaft
(© Ford Motor Co.)

synchronizer hub. Be sure that the spring covers all insert grooves. Start the hub in the sleeve, being sure the alignment marks are properly indexed. Position the three inserts in the hub and be sure the small end is over the spring and that the shoulder is on the inside of the hub. Slide the sleeve and reverse gear onto the hub until the detent is engaged. Install the other insert spring in the front of the hub to hold the inserts against it.
4. Install one insert spring into a

groove of the second-third synchronizer hub. With the alignment marks on the hub and sleeve aligned, start the hub into the sleeve. Place the three inserts on top of the retaining spring and push the assembly together. Install the remaining insert spring, so that the spring ends cover the same slots as do the other spring. Do not stagger the springs. Place a synchronizer blocking ring in each end of the synchronizer sleeve.
5. Lubricate the output shaft splines and machined surfaces with transmission lubricant.
6. Press the first and reverse synchronizer hub onto the output shaft, with the teeth end of the gear facing toward the rear end of the shaft. Secure it with the snap-ring.
7. Place the blocking ring on the tapered machined surface of the first gear.
8. Slide the first gear onto the output shaft, with the blocking ring toward the rear of the shaft. Rotate the gear to engage the three notches in the blocking ring with the synchronizer inserts. Secure the first gear with the thrust washer and snap-ring.
9. Slide the blocking ring onto the tapered, machined surface of the second gear. Slide the second gear, with blocking ring and the second and third gear synchronizer, onto the mainshaft. The tapered machined surface of the second gear must be toward the front of the shaft. Secure the synchronizer with a snap-ring.
10. Install new O-rings onto the two shift lever shafts. Lubricate the shafts with transmission fluid and install them into the case. Secure each shift lever onto its shaft.
11. Coat the bore of the input shaft with a light coat of grease. Install the 15 bearing rollers into the bore.
12. Position the output shaft assembly in the case.
13. Place a detent plug spring and a plug in the case. Place a second and third-speed shift fork in the synchronizer groove. Rotate the fork into position and install the second and third-speed shift rail.

Move the rail inward until the detent plug engages the forward notch (second). Secure the fork to the shaft with a set screw. Move the synchronizer to the neutral position.

14. Install the interlock plug in the case.

15. Place first and reverse shift fork in the groove of the first and reverse synchronizer. Rotate the fork into position and install the first and reverse shift rail. Move the rail inward until the center notch is aligned with the detent bore. Secure the fork to the shaft with a set screw. Install the remaining detent plug and spring. Secure the detent spring with the slotted head set screw. Tighten set screw until the head is flush with the case.

16. Install a new expansion plug in the case front.

17. Install the input shaft and gear in the front of the case.

18. Place front bearing retainer (with new gasket in place) on the case with the oil return groove at the bottom. Torque attaching screws to 30 ft. lbs.

19. Install the large snap-ring on the rear bearing. Place the bearing on the output shaft, with the snapring end toward the rear of the shaft. Press bearing into place and secure with a snap-ring.

20. Hold the speedometer drive gear lock ball in the detent and slide the speedometer gear into place. Secure the gear with a snap-ring.

21. Lift the countershaft gear cluster up into place, and, by entering the countershaft at the rear of the case, push the dummy shaft out of the gear and transmission case. Before the countershaft is completely in place, align the roll pin

hole in the shaft with the hole in the case.

22. Working through the filler hole, install a roll pin into the case and countershaft.

23. Install filler and drain plugs in the case.

24. Coat a new extension housing gasket with sealer and install it on the case.

25. Apply sealer to attaching screws and secure extension housing to the case by torquing the screws to 42 to 50 ft. lbs.

26. With transmission in gear, pour lubricant over the entire gear train while rotating the input or output shaft.

27. Install the transmission cover, with a new sealer-coated gasket in place, and torque the nine attaching screws to 14-19 ft. lbs.

28. Check operation of transmission in all of the gear positions.

Type-7

Comet (6 Cyl.), 1964-66
Falcon, 1964-66

Application

Fairlane (6 Cyl.), 1964-66
Mustang (6 Cyl.), 1965-66

Dagenham transmission
(© Ford Motor Co.)

Disassembly

1. Disconnect the three shift rods from the shift levers.
2. Remove the three bolts that hold the shift selector assembly to the extension housing.
3. Insert a reworked screwdriver through the clutch release lever and unhook the retainer that holds the clutch release lever to the retainer bracket.
4. Remove the four bolts that hold the flywheel housing to the transmission.
5. Remove the eight attaching bolts and the shift cover from the transmission.
6. Remove the three bolts and the input bearing retainer from the front of the transmission.
7. Remove the four bolts that hold the extension housing and output bearing adapter to the transmission case. Remove the extension housing.
8. Working from the front of the case, drive the countershaft toward the rear, until it is clear of the front wall of the case. Using a dummy shaft, push the countershaft out until the dummy shaft and gear cluster drop out of position.
9. Remove the output shaft assembly from the rear of the case.
10. Remove the input gear and bearing from the front of the case.
11. Lift the countershaft gear assembly out through the cover opening. Note that the smallest diameter thrust washer is between the rear of the countershaft gear and the case and that the larger diameter steel and bronze washers are at the front.
12. Thread a 5/16 in. x 24 bolt into the rear end of the reverse idler gear shaft and pull the shaft out with a puller. Remove the idler gear.
13. Remove the snap-ring from the input shaft and pull the input shaft bearing.
14. Straighten the output shaft locknut tab and remove the nut and lock.
15. Remove the speedometer gear, drive ball, and gear spacer.
16. With the output shaft in a press, remove the bearing, adapter, first gear, first and second synchronizer assembly and second gear from the shaft.
17. Remove the snap-ring from the front of the third-fourth synchronizer on the output shaft. Press the third gear and the third-fourth synchronizer from the shaft. Before disassembling the synchronizers, scribe an alignment mark across the hub and sleeve. These marks will furn-

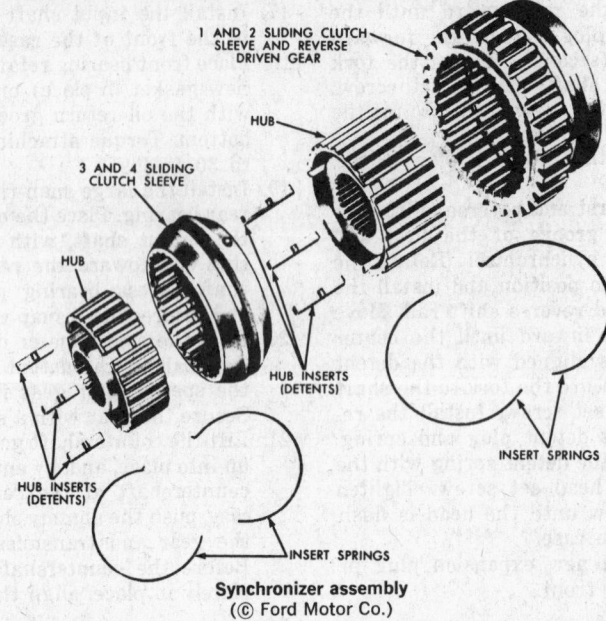

Synchronizer assembly
(© Ford Motor Co.)

ish identification at the time of assembly. Remove the front and rear insert springs from both synchronizer assemblies. Slide the sleeves off the hubs. Remove hub inserts.

18. Remove snap-ring from the end of the selector shaft.
19. Remove the flat washer and spring.
20. Remove two bolts, then pull the retainer, selector levers and bracket from the shaft.
21. Drive the short selector lever pin from the shaft.
22. Drive the long trunnion pin from the shaft and remove the trunnion and shaft.
23. Remove the shifter levers from the cam and shafts.
24. Remove the roll pin from the upper fork shaft and remove the shaft and forks.
25. Remove the reverse cam and shaft.
26. Rotate the reverse fork and shaft assembly to disengage the detent ball, then remove the fork and shaft.
27. Remove the first-second and third-fourth shift cam assemblies.
28. Push the interlock sleeve, spring and remaining ball out of the cover.
29. Remove first-second and third-fourth to reverse interlock pins from the reverse fork and shaft bosses in the cover.

Assembly

1. Place the first-second and third-fourth-to-reverse interlock pins in the holes in the reverse fork and shaft bosses.
2. Install the third-fourth shift cam into the cover.
3. Install the parts of the first-second to third-fourth cam interlock, sleeve, ball, spring and another ball, in that order.
4. Hold the third-fourth ball in neutral and the first-second ball depressed while the first-second cam is installed in the cover.
5. Install the first-second and the third-fourth levers, washers, and nuts.
6. Check clearance between the interlock, detent sleeve and the first-second and third-fourth shift cams in, and between, all shift positions. Sleeve to cam clearance must be 0.0005 to 0.010 in. Various length service sleeves are available.
7. With the first-second and third-fourth shift cams in neutral and the third-fourth-to-reverse interlock pin resting on the cams, install the reverse shaft detent spring and ball and the reverse fork and shaft.
8. Install the reverse shift cam through the cover and into the aligned fork and shaft. Install the reverse cam operating lever.
9. Position the first-second and third-fourth forks onto the shift cams and install the fork shaft. Align the shaft hole with the one in the cover and install the lock pin.
10. Check all positions for freedom of operation and detent action.
11. In assembling the transmission, place the long detent inserts into the slots in the first-second synchronizer hub and slide the combination sleeve and reverse gear over it, being sure that the scribe marks are aligned. Snap the insert springs into place. The tab on each spring must set into the under-side of an insert.

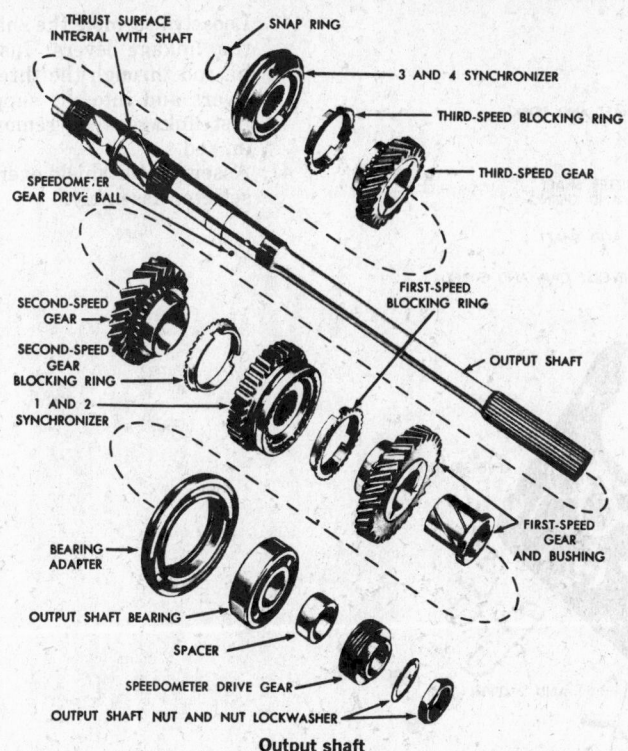

Output shaft
(© Ford Motor Co.)

12. Position the short inserts into the slots in the third-fourth synchronizer hub and slide the clutch sleeve over it, being sure that the scribe marks are aligned. Install the insert springs in the same way as with the first-second synchronizer.

13. Place the second gear on the rear of the output shaft with the clutch teeth and tapered synchronizer end toward the rear. Install a blocking ring with the clutch teeth toward the front.

14. Install the first and second synchronizer and reverse gear assembly on the rear of the output shaft with the shift fork groove to the rear.

15. Install the first-speed blocking ring with the clutch teeth to the rear and the slots engaging the synchronizer inserts.

16. Slide the first gear and sleeve onto the output shaft, taper and clutch teeth to the front, and the sleeve shoulder to the rear.

17. Assemble the output shaft ball bearing into the recess in the bearing adapter. Position the adapter and bearing on the rear of the output shaft, with the adapter forward. Hold the first gear and sleeve (bushing) forward and place the assembly in a press with the tool applied to the bearing inner race. Press the bearing on until it is firmly against the first gear sleeve.

18. Place the spacer, speedometer gear drive ball, speedometer gear,

lock and nut on the output shaft. Torque nut to 80 ft. lbs., then loosen. Torque the nut to 20-25 ft. lbs. and bend over the flat lock washer.

19. Set the third-speed gear on the front of the output shaft, with the clutch teeth toward the front. Place a blocking ring on the rear.

20. Install the third-fourth synchronizer with the wide thrust surface of the hub toward the rear. Align the blocking ring slots with the synchronizer inserts.

21. Install the snap-ring into its groove on the front of the output shaft.

22. Place a dummy shaft in the countershaft gear cluster. Starting at either end, drop a steel washer over the shaft and into the gear. Grease each bearing roller and install 22 of them into the gear. Lay another steel washer on the ends of the needles and the thrust washers.

23. Repeat the above operation for the other end of the gear.

24. Position the cluster gear in the case with the two thrust washers at the front. Allow the gear assembly to lie in the bottom of the case. Check gear end play. It must be .004-.018 in. Correct by replacing thrust washers.

25. Press the input bearing onto the input shaft, with the outer race ring groove toward the front.

26. Install snap-rings onto the bearing outer race and the gear shaft.

27. Using grease, install 17 bearing

rollers into the bore of the input gear.

28. Install input gear assembly. Place the fourth gear blocking ring on the rear of the input gear, with the clutch teeth forward.

29. Enter the output shaft assembly through the rear of the case and guide the output shaft front pilot into the input gear bore and bearings. Be sure that the fourth gear synchronizer blocking ring slots index with the inserts on the third-fourth synchronizer assembly.

30. Raise the countershaft gear until the countershaft can be inserted from the rear of the case into the gear and bearings. Push on the shaft until it contacts the front of the case. Position the flat on the rear of the countershaft on a horizontal plane so it will align with the slot in the extension housing; tap the shaft into place.

31. Install the reverse idler gear with the fork groove toward the rear and the idler shaft flat horizontal and parallel with the countershaft flat.

32. Position a new gasket to the rear of the case using non-drying-sealer.

33. Install extension housing. Align the dowel and be sure that the housing is seated squarely on the case before tightening the attaching bolts. If two long and two short bolts are used, install the two long bolts in the upper right and lower left holes.

34. Using sealer, install new gasket on the input shaft bearing retainer. With drain slot facing downward, install the bearing retainer. Seal and torque the bolts to 12-15 ft. lb.

35. Place the first-second and third-fourth synchronizer in neutral and the reverse idler gear in reverse position. Set reverse shift lever in reverse position. Install a new shift cover gasket onto the case with sealer. Install the shift cover. With sealer-coated threads, torque the shift cover attaching bolts to 12-15 ft. lbs.

36. Install the flywheel housing and with seal coated threads, torque the attaching bolts to 40-45 ft. lbs.

37. Install clutch release bearing onto the clutch release lever. Position the release lever through the housing from inside and clip the lever retainer onto its hook.

38. Install the shift selector assembly on the extension housing.

39. Insert the shift rods into the cam levers and secure them with the spring washers and fasteners. Tighten the reverse cam lever nut.

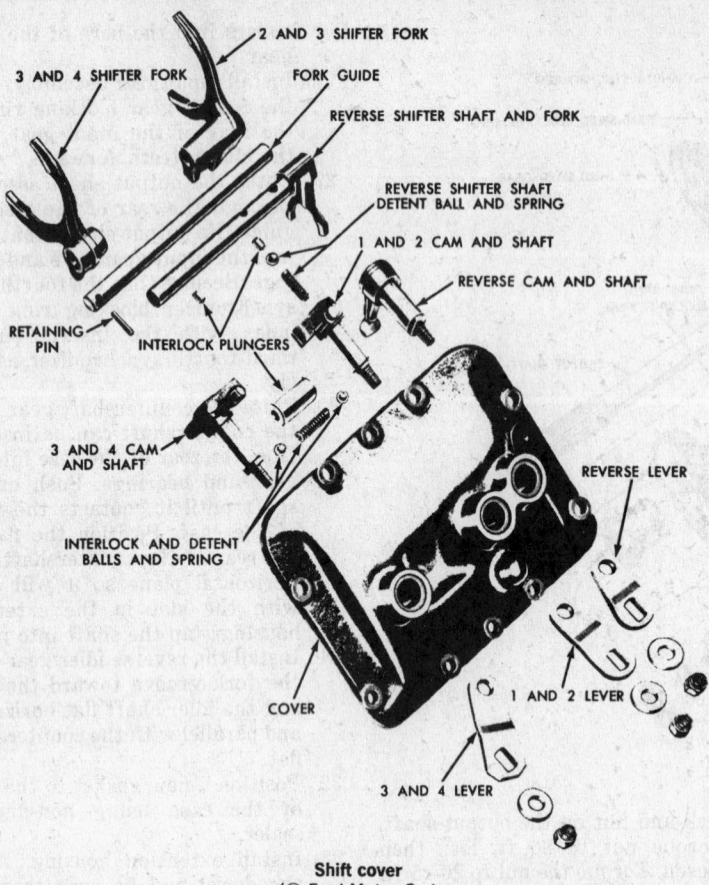

3 AND 4 SHIFTER FORK
2 AND 3 SHIFTER FORK
FORK GUIDE
REVERSE SHIFTER SHAFT AND FORK
REVERSE SHIFTER SHAFT DETENT BALL AND SPRING
1 AND 2 CAM AND SHAFT
REVERSE CAM AND SHAFT
RETAINING PIN
INTERLOCK PLUNGERS
3 AND 4 CAM AND SHAFT
INTERLOCK AND DETENT BALLS AND SPRING
REVERSE LEVER
1 AND 2 LEVER
COVER
3 AND 4 LEVER

Shift cover
(© Ford Motor Co.)

40. Loosely assemble the shift rods to the linkage levers. Insert a 1/4 in. rod through the three linkage levers and into the support. Adjust linkage, then remove the 1/4 in. rod.
41. Assemble the shield over the shift selector assembly.

Type-8
Ford (Capri) 4-Speed
Application
Capri, 1970

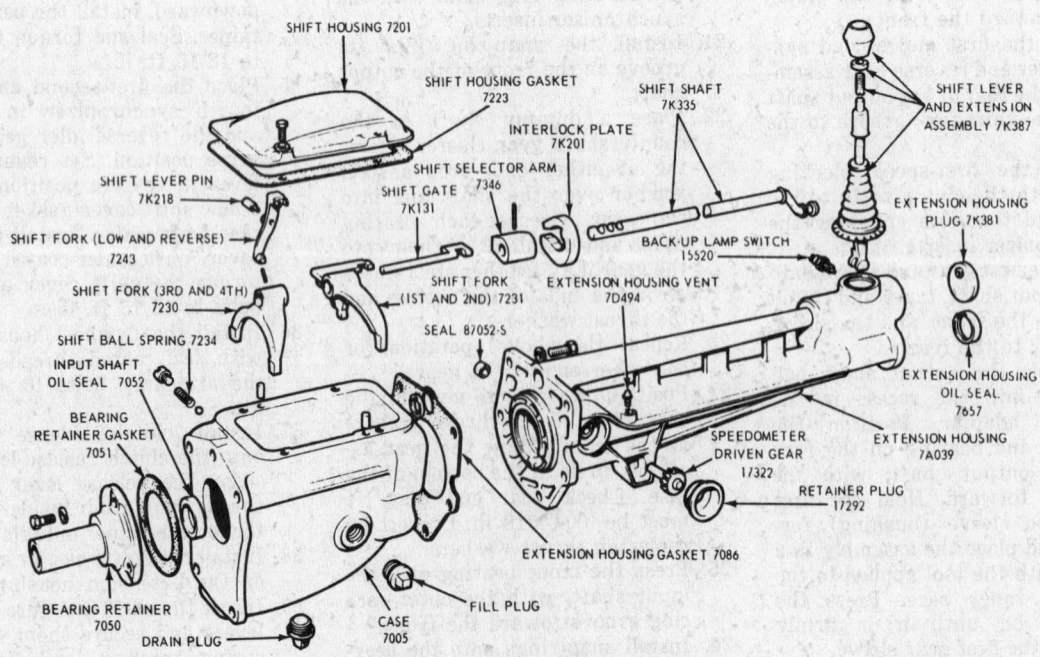

SHIFT HOUSING 7201
SHIFT HOUSING GASKET 7223
INTERLOCK PLATE 7K201
SHIFT SHAFT 7K335
SHIFT LEVER AND EXTENSION ASSEMBLY 7K387
SHIFT SELECTOR ARM
SHIFT GATE 7K131
SHIFT LEVER PIN 7K218
EXTENSION HOUSING PLUG 7K381
SHIFT FORK (LOW AND REVERSE) 7243
BACK-UP LAMP SWITCH 15520
SHIFT FORK (3RD AND 4TH) 7230
SHIFT FORK (1ST AND 2ND) 7231
EXTENSION HOUSING VENT 7D494
SHIFT BALL SPRING 7234
SEAL 87052-S
INPUT SHAFT OIL SEAL 7052
EXTENSION HOUSING OIL SEAL 7657
BEARING RETAINER GASKET 7051
EXTENSION HOUSING 7A039
SPEEDOMETER DRIVEN GEAR 17322
RETAINER PLUG 17292
BEARING RETAINER 7050
FILL PLUG
DRAIN PLUG
CASE 7005
EXTENSION HOUSING GASKET 7086

Capri transmission external parts (© Ford Motor Co.)

Transmission Disassembly

1. Remove four bolts and top cover plate.
2. Pry plug from rear of extension housing.
3. Remove plunger screw from right side of case.
4. Working through the top cover opening, use a punch to remove the pin securing the shift selector arm to the shift shaft.
5. Pull the shift shaft rearward, being careful not to drop the shift selector arm and the interlock plate.
6. Move the first-second and third-fourth gear synchronizer hubs toward the input shaft bearing.
7. If necessary, remove the shift shaft plunger spring from the case. The plunger screw was removed in Step 3.

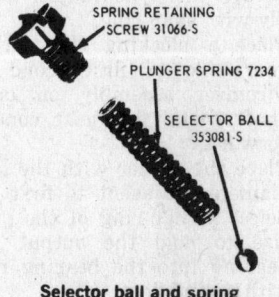

SPRING RETAINING SCREW 31066-S

PLUNGER SPRING 7234

SELECTOR BALL 353081-S

Selector ball and spring
(© Ford Motor Co.)

8. Remove the pin from the third-fourth shift fork. Remove the fork.
9. Unbolt extension housing from case. With a plastic hammer, tap the extension housing slightly rearward. Rotate the housing until the countershaft lines up with the notch in the housing flange.
10. Tap the countershaft rearward with a brass drift until it is just clear of the front of the case. Push the countershaft out with a dummy shaft. Lower the cluster gear to the bottom of the case.
11. Remove extension housing and output shaft assembly. The third-fourth synchronizer sleeve must be pushed forward for clearance.
12. Unbolt front bearing retainer from case. Remove retainer and gasket.
13. Remove input shaft oil seal.
14. Remove the snap ring around the input shaft bearing. Tap the input shaft gear and bearing assembly out of the transmission with a brass drift. Remove the needle roller bearing from the recess in the end of the input shaft gear.
15. Remove the cluster gear, two thrust washers, and the dummy shaft from the case. Remove 20

needle rollers and a retaining washer from each end of the cluster gear.
16. Assemble a nut, a flat washer, and a sleeve on a 5/16 in. x 24 UNF threaded bolt. Screw the bolt into the reverse idler shaft and tighten to pull out the shaft.
17. Remove the low-reverse shift fork from the lever pin inside the case. Do not remove the pin.

Component Disassembly

Third-Fourth Synchronizer

1. Remove fourth gear blocking ring from input shaft gear side of assembly.
2. Remove synchronizer hub snapring from forward end of output shaft and discard.
3. Support third gear. Press the output shaft out of the third-fourth gear synchronizer and third gear. Be careful not to drop the output shaft.
4. Pull the sleeve off the hub. Remove the inserts and springs.
5. Check all parts for wear. Synchronizer hub and sleeve should be replaced if worn or damaged.

First-Second Synchronizer

1. Remove plug in extension housing. Remove speedometer driven gear.

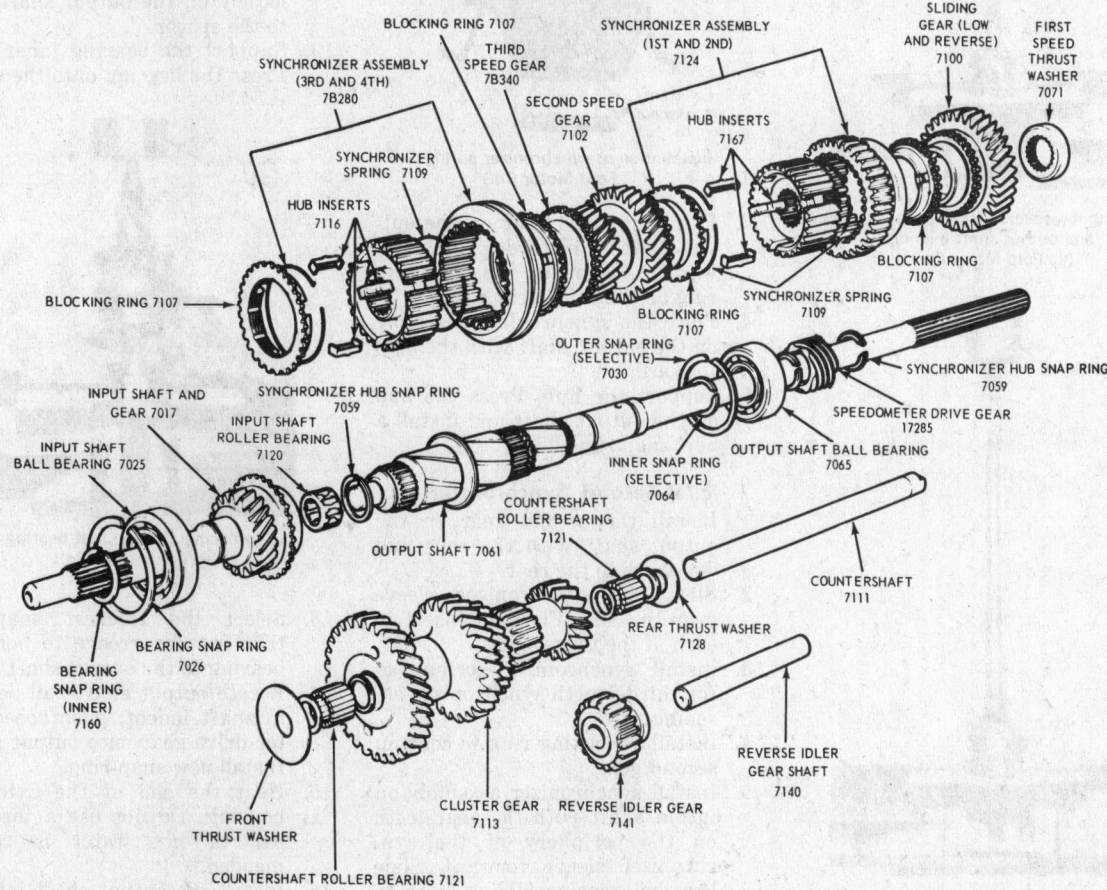

Transmission internal parts (© Ford Motor Co.)

2. Remove snap-ring holding output shaft bearing to extension housing. With a plastic hammer, tap output shaft assembly out of housing.
3. Remove snap-ring holding speedometer drive gear. Pull off gear. Remove snap-ring holding output shaft bearing.
4. Support low and reverse sliding gear. Press low and reverse sliding gear, spacer, and output shaft bearing from the output shaft.
5. Remove snap-ring holding first-second synchronizer assembly to output shaft.
6. Support second gear. Press second gear and first-second synchronizer assembly from output shaft.

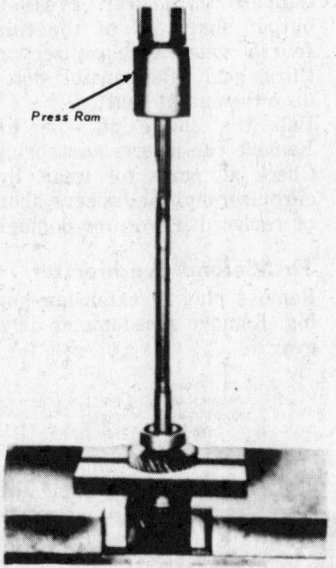

Removing low-reverse sliding gear, spacer, and output shaft bearing
(© Ford Motor Co.)

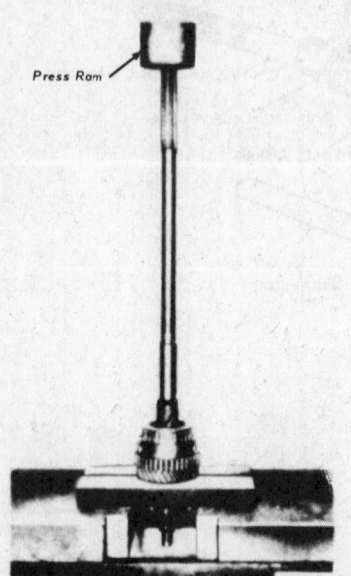

Removing first-second synchronizer assembly
(© Ford Motor Co.)

7. Dismantle synchronizer assembly. Replace synchronizer hub or sleeve if worn or damaged. The output shaft bearing must be replaced.

Input Shaft and Gear

1. Remove and discard input shaft snap-ring.
2. Press off input shaft bearing.

Component Assembly

Third-Fourth Synchronizer

1. Slide gear over hub. Locate an insert in each slot.
2. Install a synchronizer spring inside the sleeve beneath the inserts; The spring tang should fit into an insert. Install the other spring on the opposite side, fitting the tang into the same insert. When viewed from the edge, the springs should run in opposite directions.

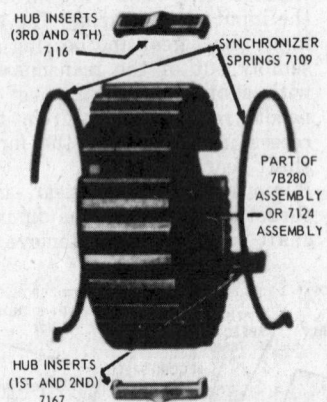

HUB INSERTS (3RD AND 4TH) 7116
SYNCHRONIZER SPRINGS 7109
PART OF 7B280 ASSEMBLY OR 7124 ASSEMBLY
HUB INSERTS (1ST AND 2ND) 7167

Installation of synchronizer springs
(© Ford Motor Co.)

3. Place the third gear on the output shaft with the dog teeth forward. Assemble the blocking ring on the third gear cone.
4. Place the synchronizer assembly on the output shaft with the boss forward.
5. Support the hub. Press the hub on the output shaft and install a new snap-ring.

First-Second Synchronizer

1. Install the second gear on the output shaft with the cone and dog teeth to the rear.
2. Slide the synchronizer sleeve over the hub. Place an insert in each of the three slots.
3. Install synchronizer springs as for third-fourth synchronizer assembly.
4. Install a blocking ring to cone on second gear.
5. Install synchronizer assembly on output shaft with the gear teeth on the periphery of the synchronizer sleeve forward. Slide low and reverse sliding gear to the rear of the synchronizer hub.

Replacing first-second synchronizer assembly
(© Ford Motor Co.)

6. Support the sliding gear. Press synchronizer assembly onto output shaft as far as possible.
7. Secure the synchronizer assembly with snap-ring.
8. Place a blocking ring on first gear side of first-second synchronizer assembly on output shaft. Install first gear, cone side forward.
9. Place the spacer with the larger diameter adjacent to first gear.
10. Select a snap-ring of the proper size to hold the output shaft bearing into the bearing recess with no end float.
11. Position the selected snap-ring loosely on the output shaft next to the spacer.
12. Support the bearing inner race. Press the bearing onto the shaft.

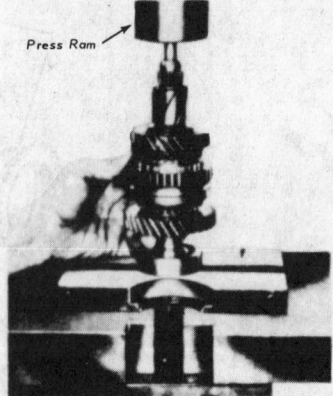

Replacing output shaft bearing
(© Ford Motor Co.)

13. Select the thickest snap-ring that fits the groove to hold the bearing to the output shaft.
14. Locate output shaft ball bearing in shaft indent, push speedometer drive gear onto output shaft. Install new snap-ring.
15. Heat the end of the extension housing. Do not use a torch. A pan of hot water is recommended.
16. Install the output shaft into the extension housing. Install the

snap-ring securing the output shaft bearing to the housing.

17. Replace the speedometer driven gear. Install a new plug, using sealer.

Input Shaft and Gear

1. Support the input shaft bearing inner race. Press the bearing onto the shaft.
2. Install the snap-ring securing the bearing to the input shaft.

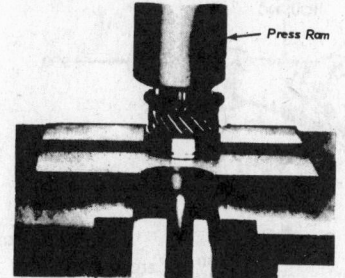

← Press Ram

Replacing input shaft bearing
(© Ford Motor Co.)

Transmission Assembly

1. Slide the low-reverse lever onto the lever pin inside the case.
2. Push the idler shaft into the case. Place the reverse idler gear on the shaft. Locate the low-reverse lever in the gear groove. Tap the reverse idler shaft into position with a soft hammer.
3. Slide a dummy shaft into the cluster gear. Push a retainer washer into the gear bore. Grease and install 20 needle rollers and the second retaining washer. Install the washers and rollers at the other end of the gear. Grease and install the thrust washers with their convex side into the gear recess.
4. Place the cluster gear in the bottom of the case. Position the thrust washers with the flat upward.
5. Place the input shaft and gear in the case. Using a brass drift, tap bearing outer race into place. Be careful not to damage the dog teeth on the input shaft gear with the cluster gear. Install the bearing snap-ring.
6. Place the input shaft needle bearing in the input shaft gear recess.
7. Drive a new oil seal into the input shaft retainer. Cover the input shaft splines. Install a new gasket on the transmission front face. Check that the retainer oil groove is lined up with the oil passage in the case. Coat the bolts with sealer and install them with lock-washers.
8. Locate the fourth gear blocking ring on the input shaft gear cone.
9. Install a new oil seal in the shift shaft aperture. Drive the seal in with a socket.
10. Install a new sealer coated gasket to the extension housing.
11. Pull the third-fourth synchronizer sleeve forward. Slide the extension housing and output shaft into position. Align the cutaway on the extension housing with the countershaft aperture in the rear face of the case.
12. Using loops of cord, lift the cluster gear into mesh with the output and input shaft gears. Take care not to drop the countershaft thrust washers.
13. Tap the countershaft into place, driving out the dummy shaft, ensuring that the lug on the rear of the countershaft fits into the recess on the extension housing flange.
14. Push the extension housing onto the transmission case. Apply sealer to bolts. Torque to 30-35 ft. lbs.
15. Replace both shift forks. Secure third-fourth fork with a new pin.
16. Position shift forks to synchronizer sleeves. Move synchronizer hubs into neutral positions.
17. Grease shift shaft oil seal in rear of case. Slide shift shaft through extension housing. Position shift selector arm and interlock plate so that interlock plate locates in cutouts in shift forks. Pass the shift shaft through the shift selector arm and forks until the pin holes are aligned.
18. Replace the plunger ball and spring. Replace the retaining screw, using sealer.
19. Install the pin through the shift selector arm and shift shaft.
20. Apply sealer to plug. Tap plug into rear of extension housing.
21. Install top cover and gasket.
22. Refill transmission with 2.8 pints SAE 80 oil.

Type-9
Ford 4-Speed

Application

Comet V8, 1964-70
Cougar, 1967-70
Fairlane V8, 1965-70

Falcon V8, 1965-70
Ford, 1965-70
Mercury, 1965-70

Montego, 1968-70
Mustang V8, 1965-70

Disassembly

1. Remove retaining clips and flat washers from the shift rods at the levers.
2. Remove shift linkage control bracket attaching screws and remove shift linkage and control bracket.
3. Remove cover attaching screws. Then lift cover and gasket from the case.
4. Remove extension housing attaching screws. Then, remove extension housing and gasket.
5. Remove input shaft bearing retainer attaching screws. Then, slide retainer from the input shaft.
6. Working a dummy shaft in from the front of the case, drive the countershaft out the rear of the case. Let the countergear assembly lie in the bottom of the case.
7. Locate first - second - speed gear shift lever in neutral. Locate third fourth-speed gear shift lever in third-speed position.
8. Remove the lock bolt that holds the third-fourth-speed shift rail detent spring and plug the left side of the case. Remove spring and plug with a magnet.
9. Remove the detent mechanism set screw from top of case. Then, remove the detent spring and plug with a small magnet.
10. Remove attaching screw from the third-fourth-speed shift fork. Tap lightly on the inner end of the shift rail to remove the expansion plug from front of case. Then, withdraw the third-fourth-speed shift rail from the front. (Do not lose the interlock pin from rail.)
11. Remove attaching screw from the first-second-speed shift fork. Slide the first-second shift rail from the rear of case.
12. Remove the interlock and detent plugs from the top of the case with a magnet.
13. Remove the snap-ring or disengage retainer that holds the speedometer drive gear to the output shaft. Slide the gear from the shaft, then remove speedometer gear drive ball.
14. Remove the snap-ring used to hold the output shaft bearing to the shaft. Remove output shaft bearing.
15. Remove the input shaft bearing and blocking ring from the front of the case.

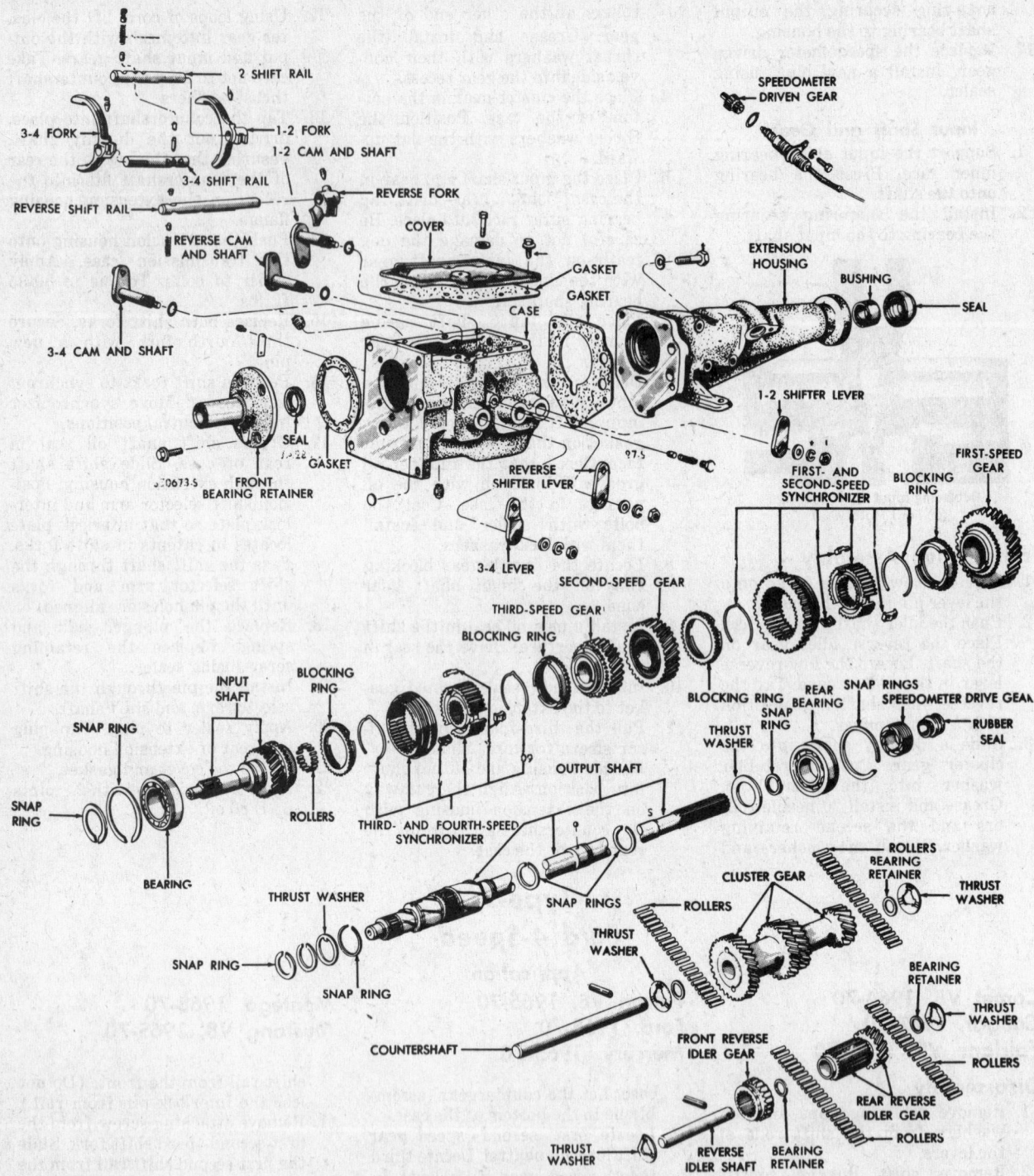

2 SHIFT RAIL
3-4 FORK
1-2 FORK
1-2 CAM AND SHAFT
3-4 SHIFT RAIL
REVERSE SHIFT RAIL
REVERSE FORK
REVERSE CAM AND SHAFT
COVER
GASKET
GASKET
CASE
SPEEDOMETER DRIVEN GEAR
EXTENSION HOUSING
BUSHING
SEAL
3-4 CAM AND SHAFT
1-2 SHIFTER LEVER
SEAL
GASKET
FRONT BEARING RETAINER
REVERSE SHIFTER LEVER
3-4 LEVER
FIRST- AND SECOND-SPEED SYNCHRONIZER
BLOCKING RING
FIRST-SPEED GEAR
SECOND-SPEED GEAR
THIRD-SPEED GEAR
BLOCKING RING
BLOCKING RING
REAR BEARING
SNAP RINGS
SPEEDOMETER DRIVE GEAR
INPUT SHAFT
BLOCKING RING
SNAP RING
THRUST WASHER
SNAP RING
RUBBER SEAL
SNAP RING
ROLLERS
OUTPUT SHAFT
BEARING
THIRD- AND FOURTH-SPEED SYNCHRONIZER
THRUST WASHER
SNAP RINGS
ROLLERS
ROLLERS
BEARING RETAINER
THRUST WASHER
CLUSTER GEAR
SNAP RING
THRUST WASHER
COUNTERSHAFT
THRUST WASHER
BEARING RETAINER
THRUST WASHER
ROLLERS
REAR REVERSE IDLER GEAR
ROLLERS
FRONT REVERSE IDLER GEAR
THRUST WASHER
REVERSE IDLER SHAFT
BEARING RETAINER

Ford 4-speed transmission (© Ford Motor Co.)

16. Move output shaft to the right side of case. Then, maneuver the forks to permit lifting them from the case.

17. Support first-speed gear to prevent it sliding from the shaft, then lift output shaft from the case.

18. Remove reverse gear shift fork attaching screw. Rotate the reverse shift rail 90°, then, slide the shift rail out the rear of the case. Lift out the reverse shift fork.

19. Remove the reverse detent plug and spring from the case with a magnet.

20. Using a dummy shaft, remove the reverse idler shaft from the case.

21. Lift reverse idler gear and thrust washers from the case. Be careful not to drop the bearing rollers or the dummy from the gear.

22. Lift the countergear, thrust washers, rollers and dummy shaft assembly from the case.

23. Remove the snap-ring from the front of the output shaft Then,

slide the third-fourth synchronizer blocking ring and the third-speed gear from the shaft.

24. Remove the next snap-ring and the second-speed gear thrust washer from the shaft. Slide the second-speed gear and the blocking ring from the shaft.

25. Remove the snap-ring, then slide the first-second synchronizer, blocking ring and the first-speed gear from the shaft.

26. Remove the thrust washer from rear of the shaft.

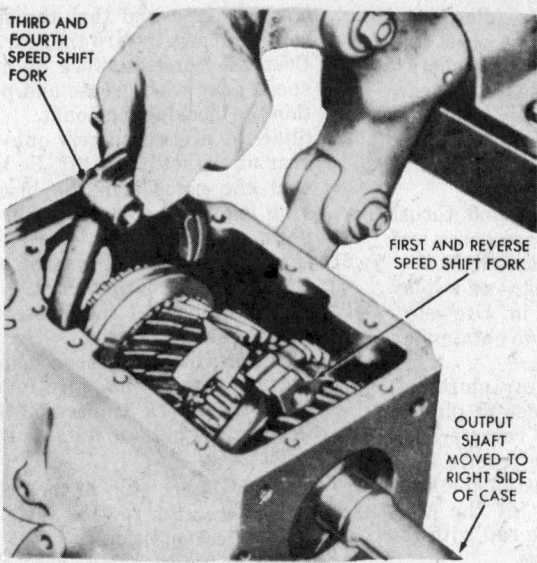

THIRD AND FOURTH SPEED SHIFT FORK

FIRST AND REVERSE SPEED SHIFT FORK

OUTPUT SHAFT MOVED TO RIGHT SIDE OF CASE

Removing shift forks (© Ford Motor Co.)

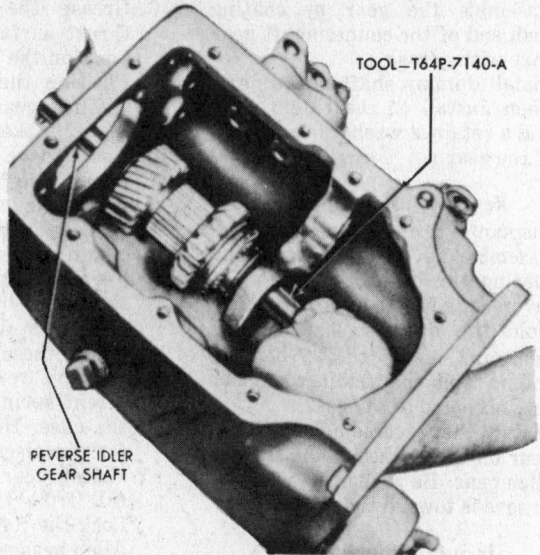

TOOL—T64P-7140-A

PEVERSE IDLER GEAR SHAFT

Removing reverse idler shaft (© Ford Motor Co.)

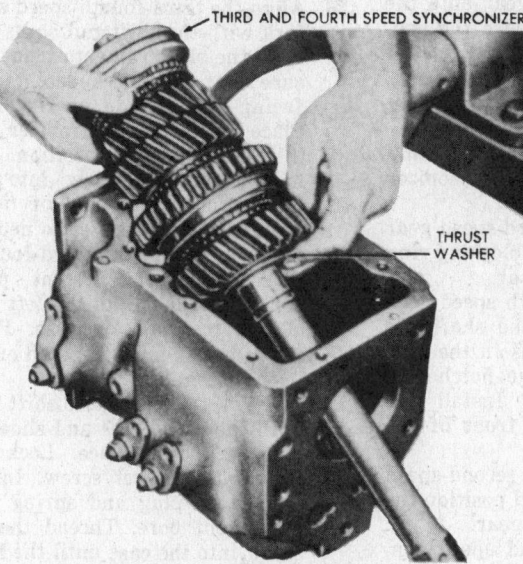

THIRD AND FOURTH SPEED SYNCHRONIZER

THRUST WASHER

Removing output shaft (© Ford Motor Co.)

TOOL T52T-6500-DJD

ROTATE DETENT 90°

REVERSE SHIFT RAIL

Rotating the reverse shift rail (© Ford Motor Co.)

Unit Repairs

Cam and Shaft Seals

1. Remove attaching nut and washers from each shift lever, then remove the three levers.
2. Remove the three cams and shafts from inside the case.
3. Replace the old O-rings with new ones that have been well-lubricated.
4. Slide each cam and shaft into its respective bore in the transmission.
5. Install the levers and secure them with their respective washers and nuts.

Input Shaft Bearing

1. Remove the snap-ring that holds the bearing to the shaft.
2. Press the shaft gear from the bearing.
3. Press a new bearing onto the input shaft.
4. Secure the bearing with a snap-ring.

Synchronizers

1. Push the synchronizer hub from each synchronizer sleeve.
2. Separate the inserts and springs from the hubs. Do not mix parts of the first-second with parts of third-fourth synchronizers.
3. To assemble, position the hub in the sleeve. Be sure the alignment marks are properly indexed.
4. Place the three inserts into place on the hub. Install the insert springs so that the irregular surface (hump) is seated in one of the inserts. Do not stagger the springs.

Countershaft Gear

1. Dismantle the countershaft gear assembly.

2. Assemble the gear by coating each end of the countershaft gear bore with grease.
3. Install dummy shaft in the gear. Then install 21 bearing rollers and a retainer washer in each end of the gear.

Reverse Idler Gear

1. Dismantle reverse idler gear.
2. Assemble reverse idler gear by coating the bore in each end of reverse idler gear with grease.
3. Hold the dummy shaft in the gear and install the 22 bearing rollers and the retainer washer into each end of the gear.
4. Install the reverse idler sliding gear on the splines of the reverse idler gear. Be sure the shift fork groove is toward the front.

Input Shaft Seal

1. Remove the seal from the input shaft bearing retainer.
2. Coat the sealing surface of a new seal with lubricant, then press the new seal into the input shaft bearing retainer.

Assembly

1. Grease the countershaft gear thrust surfaces in the case. Then, position a thrust washer at each end of the case.
2. Position the countershaft gear, dummy shaft, and roller bearings in the case.
3. Align the gear bore and thrust washers with the bores in the case. Install the countershaft.
4. With the case in a horizontal position, countershaft gear end-play should be from .004-.018 in. Use thrust washers to obtain play within these limits.
5. After establishing correct end-play, place the dummy shaft in the countershaft gear and allow the gear assembly to remain on the bottom of the case.

6. Grease the reverse idler gear thrust surfaces in the case, and position the two thrust washers.
7. Position the reverse idler gear, sliding gear, dummy, etc. in place. Make sure that the shift fork groove in the sliding gear is toward the front.
8. Align the gear bore and thrust washers with the case bores and install the reverse idler shaft.
9. Reverse idler gear - end - play should be .004-.018 in. Use selective thrust washers to obtain play within these limits.
10. Position reverse gear shift rail detent spring and detent plug in the case. Hold the reverse shift fork in place on the reverse idler sliding gear and install the shift rail from the rear of the case. Lock the fork to the rail with the Allen head set screws.
11. Install the first-second synchronizer onto the output shaft.
12. Slide second-speed gear onto the front of the shaft with the synchronizer coned surface toward the rear.
13. Install the second - speed gear thrust washer and snap-ring.
14. Slide the third-speed gear onto the shaft with the synchronizer coned surface front.
15. Coat the cone of third-speed gear with grease. Place a blocking ring on the third-speed gear.
16. Slide the third-fourth speed gear synchronizer onto the shaft. Be sure that the inserts in the synchronizer engage the notches in the blocking ring. Install the snap-ring onto the front of the output shaft.
17. Coat the cone of second-speed gear with grease and position the blocking ring on the gear.
18. Slide the first-second speed synchronizer onto the rear of the output shaft. Be sure that the inserts engage the notches in the

blocking ring and that the shift fork groove is toward the rear.
19. Coat the coned surface of first-speed gear with grease and position the blocking ring on it.
20. Slide the first-speed gear onto the rear of the output shaft. Be sure that the notches in the blocker ring engage the synchronizer inserts.
21. Install heavy thrust washer onto the rear of the output shaft.
22. Lower the output shaft assembly into the case.
23. Position the first-second speed shift fork and the third-fourth-speed shift fork in place on their respective gears. Rotate them into place.
24. Place a detent plug in the detent bore. Place the reverse shift rail into neutral position.
25. Coat the third-fourth-speed shift rail interlock pin with grease, then position it in the shift rail.
26. Align the third-fourth-speed shift fork with the shift rail bores and slide the shift rail into place. Be sure that the three detents are facing the outside of the case. Place the front synchronizer into third-fourth-speed position and install the set screw into the third - fourth - speed shift fork. Move the synchronizer to neutral position. Install the third-fourth-speed shift rail detent plug, spring and bolt into the left side of the transmission case. Place the interlock plug (tapered ends) in the detent bore.
27. Align first-second-speed shift fork with the case bores and slide the shift rail into place. Lock the fork with the set screw. Install the detent plug and spring into the detent bore. Thread the set screw into the case until the head is flush with the case.
28. Coat the input gear bore with a small amount of grease. Then install the 15 bearing rollers.
29. Place the input shaft gear in the case. Be sure that the output shaft pilot enters the roller bearing of the input shaft gear.
30. With a new gasket on the input bearing retainer, dip attaching bolts in sealer, install bolts and torque to 30-36 ft. lbs.
31. Install the output shaft bearing, then install the snap-ring to hold the bearing.
32. Position the speedometer gear drive ball in the output shaft and slide the speedometer drive gear into place. Secure gear with snap-ring.
33. Align the countershaft gear bore and thrust washers with the bore in the case. Install the countershaft.
34. With a new gasket in place, install and secure the extension

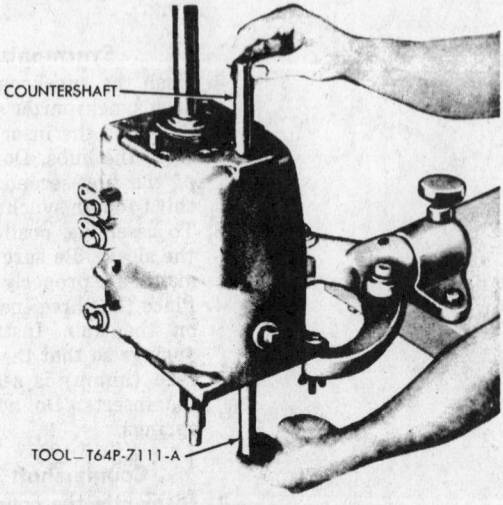

COUNTERSHAFT

TOOL—T64P-7111-A

Installing countershaft (© Ford Motor Co.)

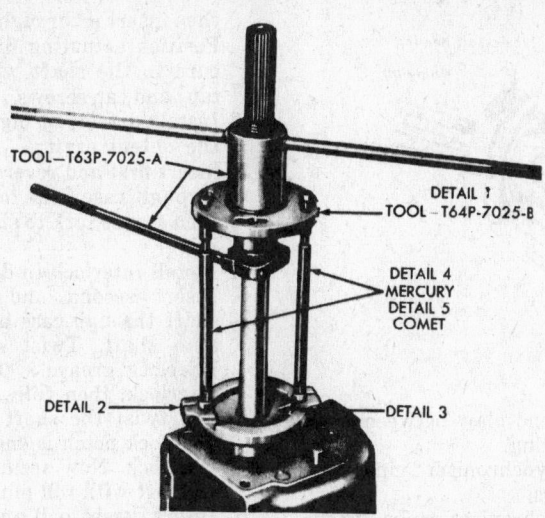

Installing output shaft bearing (© Ford Motor Co.)

TOOL—T63P-7025-A

DETAIL 1
TOOL—T64P-7025-B

DETAIL 4
MERCURY
DETAIL 5
COMET

DETAIL 2

DETAIL 3

housing. Dip the extension housing screws in sealer, then torque screws to 42-50 ft. lbs.

35. Install the filler plug (torque 10-20 ft. lbs.) and the drain plug (torque 20-30 ft. lbs.), the drain plug is magnetic.

36. Pour in four pints of mild E.P. gear oil over the entire gear train while rotating the input shaft.

37. Place each shift fork in all positions to make sure they function properly.

38. With a new cover gasket in place, install the cover. Dip attaching screws in sealer, then torque screws to 14-19 ft. lbs.

39. Coat the third-fourth speed shift rail plug bore with sealer. Install a new plug.

40. Secure each shift rod to its respective lever with a spring washer, flat washer and retaining pin.

41. Position the shift linkage control bracket to the extension housing. Install and torque the attaching screws to 12-15 ft. lbs.

Type-10
Corvair 3-Speed

Application
Corvair, 1960-65

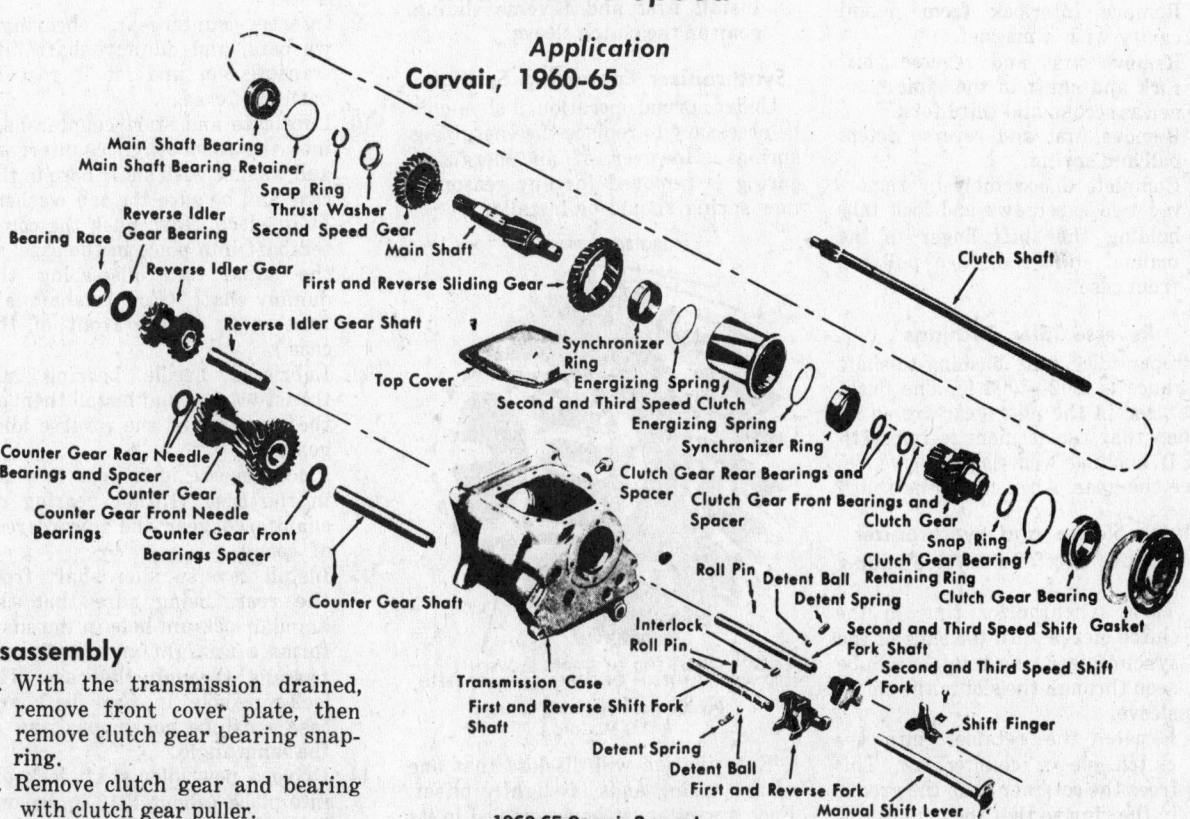

Main Shaft Bearing
Main Shaft Bearing Retainer
Snap Ring
Thrust Washer
Second Speed Gear
Main Shaft
First and Reverse Sliding Gear
Reverse Idler
Gear Bearing
Bearing Race
Reverse Idler Gear
Reverse Idler Gear Shaft
Top Cover
Synchronizer Ring
Energizing Spring
Second and Third Speed Clutch
Energizing Spring
Synchronizer Ring
Clutch Gear Rear Bearings and Spacer
Clutch Gear Front Bearings and Spacer
Clutch Gear
Clutch Shaft
Snap Ring
Clutch Gear Bearing Retaining Ring
Clutch Gear Bearing
Gasket
Counter Gear Rear Needle Bearings and Spacer
Counter Gear
Counter Gear Front Needle Bearings
Counter Gear Front Bearings Spacer
Counter Gear Shaft
Roll Pin
Detent Ball
Detent Spring
Interlock
Roll Pin
Transmission Case
First and Reverse Shift Fork Shaft
Detent Spring
Detent Ball
First and Reverse Fork
Manual Shift Lever
Second and Third Speed Shift Fork Shaft
Second and Third Speed Shift Fork
Shift Finger

1960-65 Corvair 3-speed transmission (© G.M. Corp.)

Disassembly

1. With the transmission drained, remove front cover plate, then remove clutch gear bearing snap-ring.
2. Remove clutch gear and bearing with clutch gear puller.
3. Remove transmission top cover and gasket.
4. Remove snap-ring from mainshaft groove. Snap-ring is just behind the mainshaft bearing.
5. Drive, or press, the mainshaft from the transmission.
6. Remove thrust washer, second-

speed gear, sliding gear and second- and third-speed clutch from the case by lifting out through top cover hole.

7. To remove the mainshaft bearing from the case, fully expand its

snap-ring and tap bearing out by driving on the outer race.

8. Remove countergear by driving out the countershaft with a dummy shaft from the front of the case.

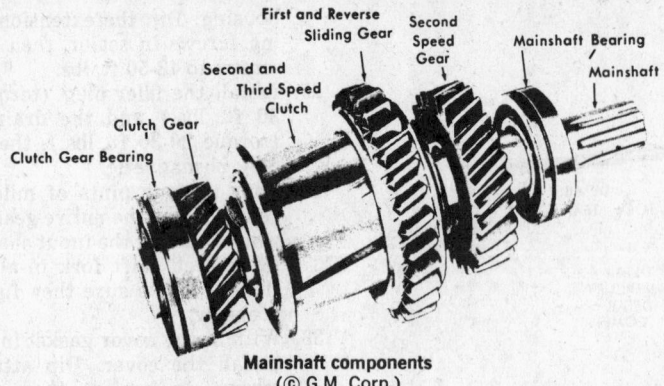

Mainshaft components
(© G.M. Corp.)

9. Drive reverse idler shaft lock pin into the shaft. Then, drive reverse idler shaft out of the case, toward the front. The expansion plug at the front of the case will be driven out by the reverse idler shaft. Remove the caged needle bearing and the thrust washer used at the rear of the reverse idler.
10. Remove detent cover and remove second and third detent ball and spring.
11. Drive out roll pin holding second and third shift fork to shaft, then tap out shaft with a drift to remove both fork and shaft.
12. Remove interlock from detent cavity with a magnet.
13. Remove first and reverse shift fork and shaft in the same manner as second and third fork.
14. Remove first and reverse detent ball and spring.
15. Complete disassembly by removing two capscrews and lock tabs holding the shift finger to the manual shift rod, then pull rod from case.

Reverse Idler Bushings

Proper idler gear bushing-to-shaft clearance is .002—.004 in. The bushings used in the idler gear are so installed that replacement is impractical. If bushing wear is excessive, replace the gear. Also, check the shaft.

Clutch Sleeve and Synchronizer

1. Remove first and reverse sliding gear.
2. Turn synchronizer ring in the clutch sleeve until the ends of the synchronizer ring retainer can be seen through the slot in the clutch sleeve.
3. Expand the retainer into the clutch sleeve counterbore. This frees the retainer from the groove in the ring so that the ring can be slipped out.
4. Check synchronizer cones for wear or looseness in clutch sleeve. If cones are damaged, it will be necessary to replace the clutch sleeve assembly and both synchronizer rings. Clutch sleeve should be replaced if there is more than

.030 in. end-play between cone and snap-ring.
5. Inspect synchronizer rings for smoothness.
6. Place synchronizer rings in the cones and check to see that the rings do not rock. Excessive rocking indicates a poor fit and justifies renewal.
7. Install the synchronizer ring retainers in the counterbores in the ends of the clutch sleeves.
8. Expand each retainer in the counterbore, lubricate the synchronizer rings with light oil and install in clutch sleeve.
9. Install first and reverse sliding gear on the clutch sleeve.

Synchronizer Energizing Springs

Under normal operation, it shouldn't be necessary to replace the energizing springs. However, if an energizing spring is removed for any reason, a new spring should be installed.

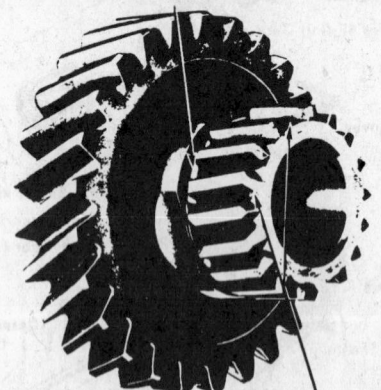

ENERGIZING SPRING

LOCATE OFFSET END OF SPRING BETWEEN 3RD AND 4TH TEETH OF EITHER BANK OF TEETH

Position of energizing spring
(© G.M. Corp.)

Examination will disclose that one of the spring ends is slightly offset. Each spring must be assembled in its groove in the clutch gear and the second-speed gear with the offset, or locking end, between the third and fourth teeth of the two banks of teeth on these gears.

Transmission Assembly

1. Lubricate manual control shaft,

then insert it through seal in case. Position actuating finger and secure to the shaft with two lock tabs and cap screws.
2. Install detent spring and ball in the detent cavity.
3. Insert first and reverse fork shaft through case, slip fork on shaft, then secure fork to shaft with roll pin.
4. Install interlock in detent cavity.
5. Insert second and third fork shaft through case and slip fork onto shaft. Twist shaft so its interlock groove is 90° from the interlock, then fully insert shaft and twist the shaft so that the interlock notch is engaged by the interlock. Now secure shift fork to shaft with roll pin.
6. Insert detent ball and spring for second and third shift fork, then install detent cavity cover and gasket.
7. With the aid of cup grease, install the 25 rollers into each end of the countergear. Then insert countergear dummy shaft.
8. Grease countergear thrust washers and place one of each on each end of countergear. Be sure tabs on washers fit into notches in case.
9. Insert countergear, bearings, washers, and dummy shaft into transmission and let it rest in bottom of case.
10. Lubricate and start countershaft into the case. Align countergear and washer with shaft bore in the case and be sure the tab washers fit in place. Now, push the countershaft into place in the case, at the same time displacing the dummy shaft. (Countershaft is a friction fit in the front of the case.)
11. Lubricate needle bearing and thrust washer and install them on the rear end of the reverse idler gear.
12. Place reverse idler gear in place in the case (thrust bearing on chamfered gear end toward rear of case.)
13. Install reverse idler shaft from the rear, being sure that the angular lock-pin hole in the shaft forms a straight and continuous passage through the case. The lock-pin hole in the shaft and case must be continuous and at the same angle.
14. Drive a new idler shaft lock pin into place (about 1/16 in. beyond being flush with the outer surface of the case).
15. Clean out expansion plug counterbored seat at the front end of the idler shaft bore. Insert a new sealer-coated expansion plug into the counterbore and expand into place with a sharp blow.

16. Install mainshaft rear bearing into case by tapping on the outer race of the bearing until bearing seats against retainer ring in case. Then, expand retainer and tap bearing until retainer ring seats in bearing groove.
17. Install sliding gear on second and third-speed clutch.
18. Install second-speed gear in second- and third-speed clutch.
19. Place the assembled second- and third-speed clutch, sliding gear and second-speed gear in case.
20. From the front, insert the mainshaft through the bores of the second- and third-speed clutch and second-speed gear. Then, in-

stall the thrust washer onto the mainshaft with its oil grooves toward the gear.
21. Tap the front of the mainshaft, until the ring groove is accessible behind the rear bearing, and install the snap-ring. Final end clearance between snap-ring and bearing inner race should be no more than .004 in. Variable thickness thrust washers are available, ranging from .086 to .097 in.
22. With some cup grease in the mainshaft pilot hole of the clutch gear, install the roller bearings and small spacer. Then, install the large spacer and remaining roller bearings.

23. Align the synchronizer lands with the clutch gear blank teeth. Install smaller snap-ring into clutch gear.
24. Tap the clutch gear bearing into the case, then open the retainer ring and fully seat the bearing.
25. Install front cover plate and gasket and secure with four screws.
26. Install and secure top cover and gasket.
27. Install expansion plug in the case opening at the rear of the manual shift shaft.

Type-11
Corvair Fully Synchronized 3-Speed

Application
Corvair, 1966-69

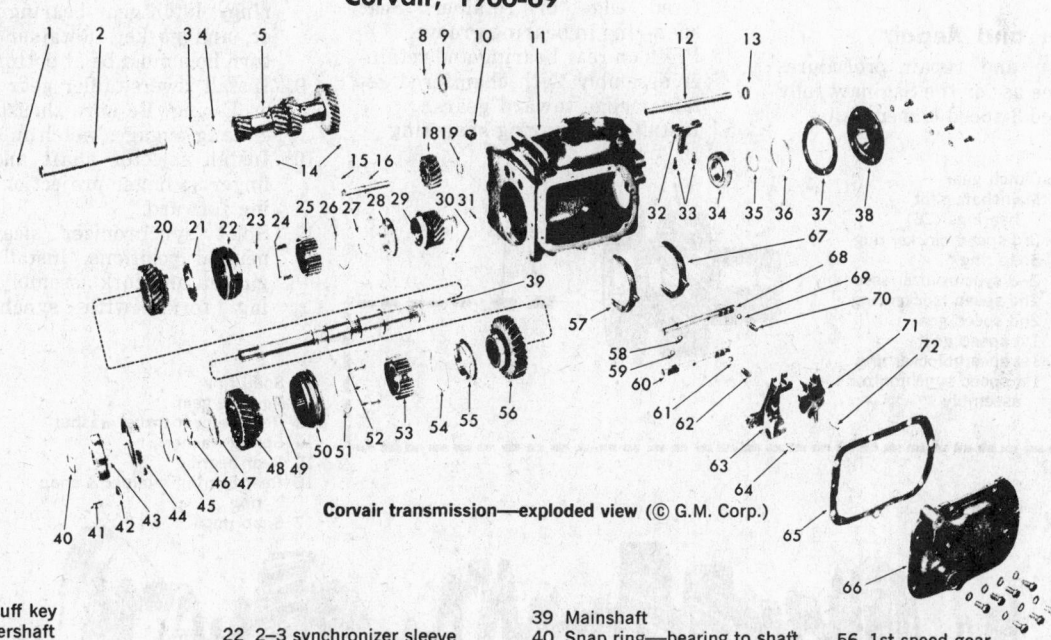

Corvair transmission—exploded view (© G.M. Corp.)

1 Woodruff key	22 2–3 synchronizer sleeve	39 Mainshaft
2 Countershaft	23 Synchronizer key retainer	40 Snap ring—bearing to shaft
3 Thrust washer	24 Synchronizer keys	41 Retainer bolt
4 Needle washer	25 Synchronizer hub	42 Retainer strap
5 Countergear	26 Synchronizer key retainer	43 Mainshaft rear bearing
8 Needle washer	27 Snap ring	44 Bearing retainer
9 Thrust washer	28 3rd speed blocker ring	45 Snap ring—bearing to retainer
10 Filler plug	29 Clutch gear	46 Spring washer
11 Case	30 Pilot bearings	47 Thrust washer
12 Shift selector shaft	31 Snap ring	48 Reverse gear
13 Seal	32 Shifter finger	49 Snap ring
14 Needle bearings	33 Screws & L. Washers	50 1st & reverse synch. sleeve
15 Woodruff key	34 Clutch gear bearing	51 Synchronizer key retainer
16 Reverse idler shaft	35 Snap ring—bearing to gear	52 Synchronizer keys
18 Reverse idler gear	36 Snap ring—bearing to case	53 Synchronizer hub
19 Retainer "E" ring	37 Gasket	54 Synchronizer key retainer
20 2nd speed gear	38 Clutch gear retainer	55 1st speed blocker ring
21 2nd speed blocker ring		56 1st speed gear

56 1st speed gear	
57 1st & reverse shift fork	
58 Interlock ball	
59 Interlock ball	
60 1st & reverse shift rail	
61 Detent ball	
62 Detent spring	
63 1st & reverse shifter head	
64 Roll pin	
65 Gasket	
66 Side Cover	
67 2nd & 3rd shift fork	
68 2nd & 3rd shift rail	
69 Detent ball	
70 Detent spring	
71 2nd & 3rd shifter head	
72 Roll pin	

Transmission Disassembly

1. Remove side cover, gasket, and shift forks.
2. Remove clutch gear bearing cover, gasket, and bearing snap-ring.

3. Remove clutch gear bearing by tapping from inside case.
4. Remove reverse idler gear retainer E-ring.
5. Remove rear bearing retainer strap.

6. Remove rear bearing retainer, mainshaft, pilot bearings, and clutch gear together from rear of case.
7. Drive out reverse idler shaft and key. Remove gear.

8. Drive countershaft and key out with a dummy shaft. Remove countergear and thrust washers.
9. Remove screws holding shift finger to selector shaft. Remove shaft from front. Pry out and replace selector shaft seal.

Mainshaft Disassembly

1. Remove second-third speed sliding clutch hub snap-ring. Remove clutch assembly, second speed blocker ring, and second gear from front of shaft.
2. Remove rear bearing snap-ring.
3. Support reverse gear. Press on rear of shaft to remove reverse gear, thrust washer, spring washer, rear bearing, and bearing retainer assembly.
4. Spread snap-ring. Press rear bearing from retainer.
5. Remove first-reverse sliding clutch hub snap-ring. Remove clutch assembly, first speed blocker ring, and first gear from rear of shaft.

Inspection and Repair

Inspection and repair procedures are the same as for the Saginaw fully synchronized 3-speed transmission.

1 Clutch gear
2 Mainshaft pilot bearings (76)
3 3rd speed blocker ring
4 Snap ring
5 2–3 synchronizer assembly
6 2nd speed blocker ring
7 2nd speed gear
8 1st speed gear
9 1st speed blocker ring
10 1st speed synchronizer assembly

Press onto splines. Be sure blocker ring notches align with synchronizer assembly keys.
    d. snap-ring. Both synchronizer snap-rings are identical.
2. Turn rear of shaft up. Install:
    a. first gear with clutching teeth up.
    b. blocker ring with clutching teeth down.
    c. first-reverse synchronizer assembly with fork slot down. Press onto splines. Be sure blocker ring notches align with synchronizer assembly keys.
    d. snap-ring.
    e. reverse gear with clutching teeth down.
    f. steel reverse gear thrust washer.
    g. spring washer.
3. Expand rear bearing retainer snap-ring and assemble rear bearing into retainer with snap-ring groove toward chamfered edge of retainer. Seat snap-ring in bearing groove.
4. Press on rear bearing and retainer assembly with chamfered retainer edge toward gears.
5. Install rear bearing snap-ring.

Install countershaft and key. Align thrust washer tangs with case notches.
3. Install reverse idler gear, shaft, and key. Do not install E-ring yet.
4. Install two rows of mainshaft pilot bearings into clutch gear cavity. Install third speed blocker ring onto clutch gear clutching surface with teeth toward gear.
5. Place clutch gear, pilot bearings, and third speed blocker ring assembly over front of mainshaft. Do not install bearing yet. Be sure blocker ring notches align with second-third synchronizer assembly keys.
6. Install clutch gear and mainshaft together into case from rear. Install rear bearing retainer strap.
7. Install front bearing outer snap-ring to bearing. Place bearing into case bore.
8. Install clutch gear hub snap-ring, clutch gear bearing retainer, and gasket. Retainer oil return hole must be at bottom.
9. Install reverse idler gear retainer E-ring. Be sure thrust washer tang engages notch in case.
10. Install selector shaft and shift finger so finger projection is facing forward.
11. Shift synchronizer sleeves to neutral positions. Install cover, gasket, and fork assembly, aligning forks with synchronizer

11 Snap ring
12 Reverse gear
13 Reverse gear thrust washer
14 Spring washer
15 Rear bearing
16 Rear bearing retainer & snap ring
17 Snap ring

Clutch gear and mainshaft assembly (© G.M. Corp.)

Mainshaft Assembly

1. Turn front of shaft up. Install:
    a. second gear with clutching teeth up.
    b. blocker ring with clutching teeth down. All three blocker rings are identical.
    c. second-third synchronizer assembly with fork slot down.

Transmission Assembly

1. Using a dummy shaft, load 27 roller bearings and a thrust washer at each end of countergear. Hold in place with grease.
2. Place countergear assembly through case rear opening with a tanged thrust washer at each end. Turn tang away from gear.

sleeve grooves and selector finger with shifter heads.

Torque Specifications

Clutch gear
    retainer .... 20 ft. lbs.
Side cover .... 20 ft. lbs.
Shift finger .... 10 ft. lbs.
Rear bearing
    retainer strap .... 10 ft. lbs.

Type-12
FSC Selector Type
3-Speed
Application
Oldsmobile, Except Jetstar 88, 1964

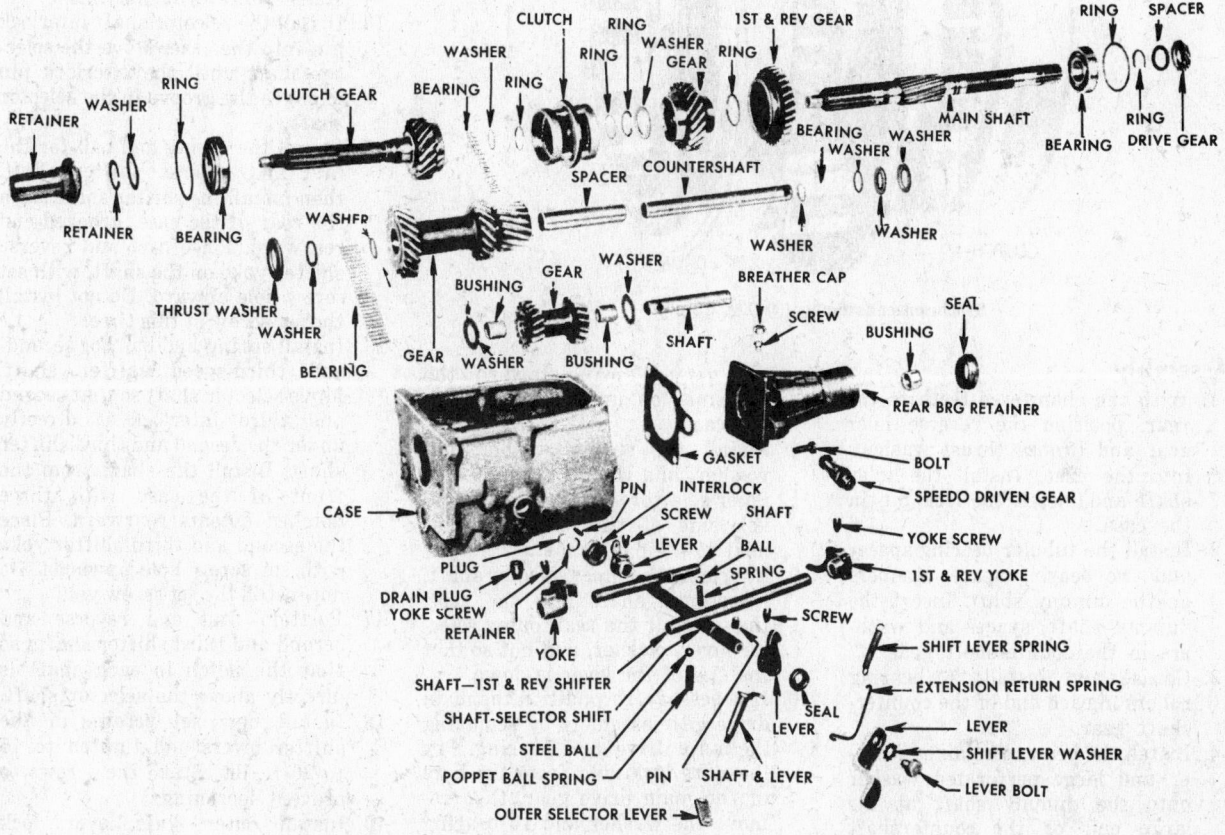

FSC selector type 3-speed transmission (© Oldsmobile Div., G.M. Corp.)

Disassembly

1. Remove the toggle spring, spring extension and spring clip from the outer shift lever. Remove the transmission cover and gasket.
2. Remove the rear bearing retainer and gasket.
3. Remove the two set screws from the two shifter yokes.
4. Pull the mainshaft rearward until the rear bearing clears the case.
5. Remove the synchronizing clutch from the mainshaft.
6. Remove snap-ring holding the second-speed gear onto the mainshaft.
7. Remove the keyed thrust washer, the second-speed gear, and the rear thrust washer from the mainshaft.
8. Remove the low and reverse gear retaining ring and slide the gear off the mainshaft.
9. Pull the mainshaft from the rear of the case.
10. Loosen the outer shift lever bolt. Position the lever so that the inner shift levers are vertical, then remove the outer shift lever.
11. Remove the set screws from the inner shift levers.
12. Pull the selector shaft away from the second and third-speed shifter shaft and remove the interlock retainer.
13. Drive the selector shaft out through the right side of the case. The Welch plug will be driven out by the shaft. Do not allow the levers of the interlock to drop into the case.
14. Push the first and reverse shifter shaft out through the rear of the case. Remove the first and reverse shifter yoke, ball and spring.
15. Push the second and third shifter shaft out through the front of the transmission case, don't let the poppet ball and spring fly out. Remove the second and third shifter yoke, ball and spring.
16. Remove the first and reverse interlock pin from the case near the selector shaft seals.
17. Drive the countershaft lock pin into the shaft.
18. Remove the retaining ring from the main drive gear bearing outer race and tap the drive gear and bearing assembly toward the rear of the case. Remove the main drive gear assembly from the case.
19. Drive the countershaft out through the rear of the case, using a dummy shaft.
20. Remove the countershaft gear, needle bearings, washers and dummy shaft from the case as an assembly.
21. Remove the transmission outer selector lever nut, washer, lever and seal, then remove the inner selector shaft and lever assembly.
22. Drive the reverse gear shaft lock pin into the shaft.
23. Drive the reverse idler shaft out of the rear of the case.
24. Remove the reverse idler gear and thrust washers from the case.
25. Remove the lock pin from the reverse idler gear shaft.

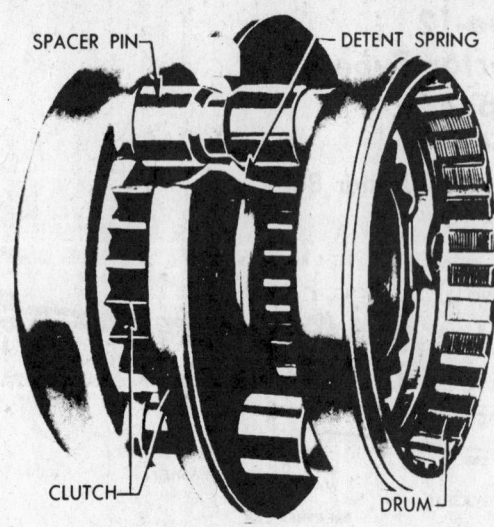

SPACER PIN — DETENT SPRING — CLUTCH — DRUM

Synchronizer assembly (© G.M. Corp.)

Assembly

1. With the chamfered teeth to the rear, position the reverse idler gear and bronze thrust washers into the case. Install the idler shaft and install the lock pin in the case.
2. Install the tubular bearing spacer and two bearing spacer washers on the dummy shaft. Insert the dummy shaft, spacer and washers in the countershaft gear.
3. Grease and install 26 bearing rollers in each end of the countershaft gear.
4. Install the bearing retainer washer and large perforated washer onto the dummy shaft at the large end of the countershaft gear.
5. Install the steel thrust washer on the dummy shaft at the small end of the countershaft gear, indexing the four tangs with the four slots in the gear. Then install the bronze and steel thrust washer onto the tool.
6. Hold the countershaft gear assembly in alignment with the countershaft holes in the case. Push the countershaft into place, from the rear of the case, while displacing the dummy shaft. Before the countershaft is entirely in place, align the lock pin hole in the countershaft with the hole in the case. Then, finish driving the shaft into the case with a brass drift. Lock the shaft into the case.
7. To replace the selector shaft seals, pry the old seals from the case. Coat the lip of the new seals with lubricant, then coat the outer diameter of the seals with sealer. Install the seal, without the garter spring in the case, by using a socket. When it bottoms in the case install the seal, with

the garter spring toward the transmission, until it is flush with the case.

8. Install the spring washer, flat washer, and the oil seal onto the inner selector shaft, with the convex side of the spring washer next to the flat washer.
9. Lubricate the inner selector shaft, insert the shaft into the case, then install the seal, outer selector lever, washer, and nut so that the end of the lever is down.
10. If necessary, recondition the main drive gear assembly by removing the main drive gear bearing. Pry the wire lock ring from the bore of the main drive gear, then remove the washer and 14 bearing rollers. To assemble the main drive gear assembly, reverse the above procedure. Grease the bearing rollers before installing them into the end of the main drive gear. When installing the ball bearing onto the main drive gear, position the shielded side of the bearing toward the gear teeth.
11. Install the main drive gear into the front of the case. Install bearing retainer ring.
12. Install a new Welch plug, coated with sealer, into the side of the case opposite the selector shaft seals.
13. Install the selector shaft and shifter levers as follows: lubricate the seals and insert the selector shaft until it just protrudes beyond the inside of the case. Engage first and reverse shifter lever with the inner selector lever in the case, then depress the inner selector lever, while sliding the selector shaft through the first and reverse shifter lever. The flat ground surface of the shifter lever must face left side of case.

Install the second and third shifter lever onto the selector shaft, installing the flat ground surface of the lever toward right side of case. Place the second and third speed interlock on the selector shaft. Install a new interlock retainer onto the shaft. Do not install set screws at this time.

14. Install the selector shaft interlock pin into the case. Move the selector shaft until the interlock pin engages the groove in the selector shaft.
15. Install the spring and ball for the first and reverse shifter shaft, then install the shifter shaft from the rear of the case, grooved-end rearward. Place first and reverse shifter yoke on the shaft, with set screw hole upward. Do not install the set screw at this time.
16. Install spring and ball for second- and third-speed shifter shaft. Move selector shaft so that second and third interlock is directly under the second and third shifter shaft. Install the shaft from the front of the case with three notched detents rearward. Place the second and third shifter yoke with set screw hole upwards. Do not install the set screw yet.
17. Position first and reverse and second and third shifter shafts so that the notch in each shaft is directly above the selector shaft.
18. Install new set screws in the shifter levers and tighten to 15 to 20 ft. lbs. Stake the screws to prevent loosening.
19. Install outer shift lever, lock washer and bolt on the selector shaft.
20. After servicing any of the items that make up the transmission mainshaft assembly, insert the shaft through the bore in the rear of the case. Then slide the first and reverse gear onto the shaft, with the flat side of the gear rearward. Install first and reverse gear retaining ring into groove in spline.
21. Line up the wire spacer ring in the groove on the mainshaft with the machined thrust washer keyway groove on the second speed gear bearing surface.
    NOTE: there are two grooves machined full length of the second-speed gear bearing surface of the main-shaft. The shallow groove is for lubrication purposes only and should not be obstructed. The deep groove is to match the tangs of the two second-speed gear thrust washers.
22. Install the second-speed gear thrust washer, indexing the tang with the proper groove in the mainshaft.

23. Place second-speed gear on the mainshaft with the clutch surface forward, then install outer thrust washer and retainer with a new retaining ring.

24. Install the synchronizing drum onto the mainshaft with the counterbored end of the gear toward the second-speed gear. Engage synchronizing drum with the second- and third-speed shifter yoke and index the first and reverse shifter yoke. Then, tap the mainshaft forward until it pilots in the main drive gear and the rear bearing seats in the case.

25. Install the rear bearing retaining ring.

26. Install new set screws in the shifter yokes and torque to 15-20 ft. lbs. Stake the screws to prevent loosening.

27. If the mainshaft rear bearing retainer seal needs replacement, pry out the old one. Apply a light coating of sealer to the outer diameter of a new seal and drive it into the bearing retainer.

28. Lubricate the rear bearing retainer bushing and oil seal, and, with a new gasket in place, install the bearing retainer and torque the retaining screws to 28-33 ft. lbs.

29. With a new gasket in place, insta the transmission cover and torque the attaching screws to 10-12 ft. lbs.

30. Install the toggle spring, spring clip and the outer shift lever.

Type-13
Muncie 3-Speed

Application

Buick Le Sabre, 1964-65
Buick Special, 1964-65
  (Except Skyroof Wagon
  and G.S.)
Chevrolet, 1964-65

Chevelle, 1964-65
Chevy II, 1964-65
Corvette, 1964-65
Oldsmobile Jetstar 88,
  1964-65

Olds F-85, 1964-65
Pontiac, 1964
Tempest, 1964-65

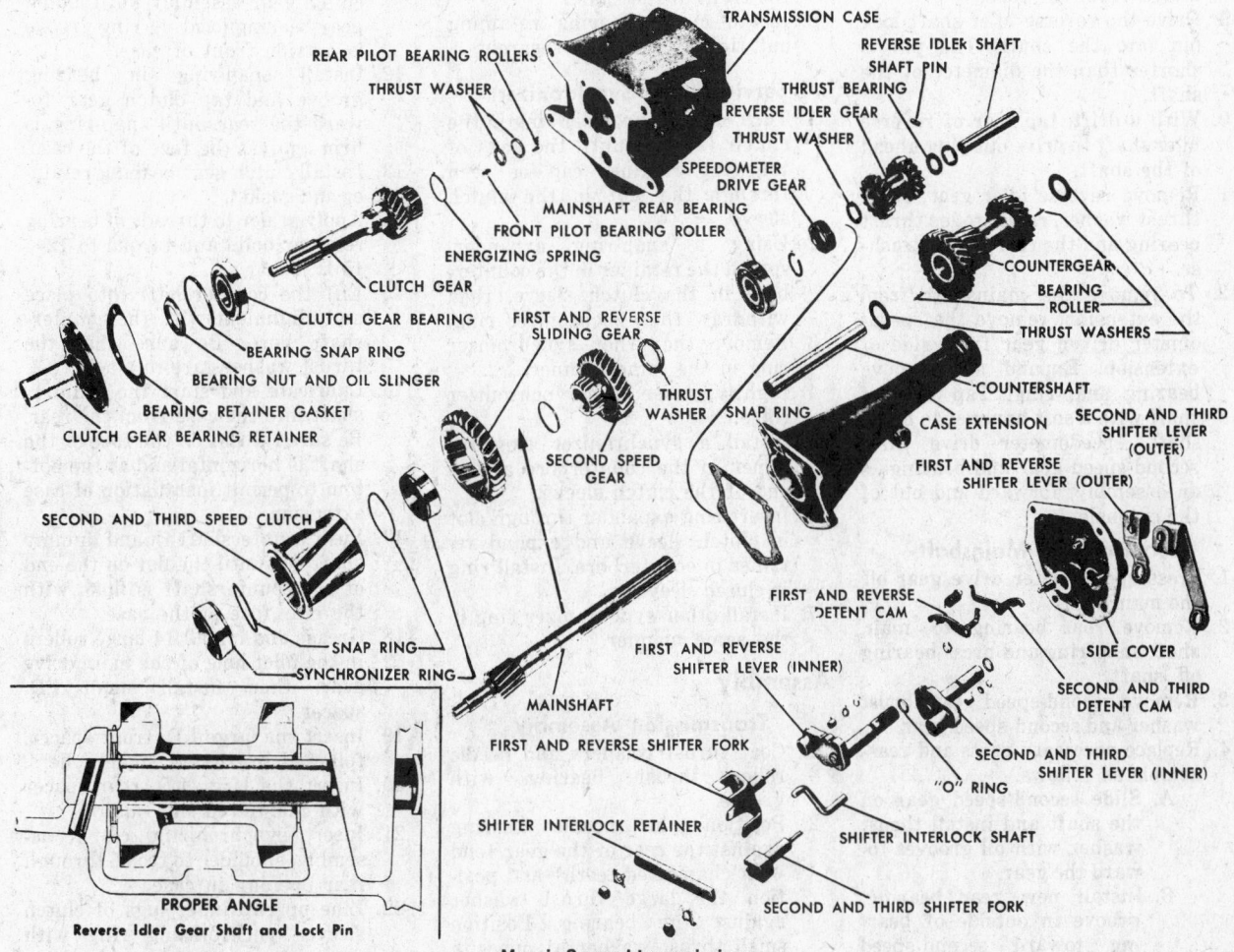

PROPER ANGLE
Reverse Idler Gear Shaft and Lock Pin

Muncie 3-speed transmission
(© G.M. Corp.)

Disassembly

1. Remove bolts from the transmission side cover and remove the cover and gasket.
2. Remove rear extension attaching bolts and pull extension and mainshaft assembly out of the transmission case, leaving second- and third-speed clutch and first and reverse sliding gear in the case.
3. Slide first and reverse gear from synchronizing clutch sleeve, then remove them separately through side opening in transmission case.
4. Remove 24 rear and 14 larger front pilot bearing rollers from inside the main drive (clutch) gear.
5. Remove four clutch gear bearing retainer bolts and washers and remove the retainer.
6. Remove the countershaft by driving it from front to rear, using a dummy shaft. Lower the countergear to the bottom of the case.
7. Remove the clutch gear bearing snap-ring. Tap end of shaft to move gear and bearing assembly into case and remove through side of case.
8. Remove the counter gear with dummy, 50 rollers, and washers intact, from the case.
9. Drive the reverse idler shaft lock pin into the shaft. The pin is shorter than the diameter of the shaft.
10. With a drift, tap rear of reverse idler shaft to drive out plug ahead of the shaft.
11. Remove reverse idler gear, front thrust washer, radial roller thrust bearing and the rear thrust washer.
12. To remove the mainshaft from the extension, remove the speedometer driven gear from side of extension. Expand and remove bearing snap-ring. Tap rear of shaft with a soft hammer to bring shaft, speedometer drive gear, second-speed gear and bearing as an assembly, forward and out of the extension.

Servicing the Mainshaft

1. Press speedometer drive gear off the main shaft.
2. Remove rear bearing - to - main shaft snap-ring and press bearing off shaft.
3. Remove second-speed gear thrust washer and second-speed gear.
4. Replace necessary parts and reassemble as follows.
   A. Slide second-speed gear on the shaft and install thrust washer with oil grooves toward the gear.
   B. Install new rear bearing, groove in outside of bearing toward second-speed gear.
   C. Select one of four available snap-rings so that end-play of bearing on shaft does not exceed .004 in.
   D. Start the speedometer drive gear on shaft with chamfered inside diameter of gear toward the bearing. Press the gear on the shaft until forward face of gear is 53/64 in. from rear face of bearing.

Servicing the Clutch Gear

1. With clutch gear in a soft-jawed vise, remove bearing retaining nut and oil slinger. The nut and oil slinger is a one piece steel casting with a left-hand thread and is staked in place on the clutch gear shaft.
2. To remove the bearing from the clutch gear, up-end the front of the transmission case on the bench. Install the gear and bearing in the front of the case with the snap-ring on the bearing.
3. Press the shaft out of the bearing and into the case.
4. Tap the bearing out of the case.
5. Press clutch gear bearing onto the clutch gear. Be sure the bearing locating ring groove is toward the front of the shaft.
6. Install clutch bearing retaining nut, tighten and stake securely.

Servicing the Synchronizer

1. Turn one synchronizer ring in the clutch (sleeve) until the ends of the ring retainer can be seen through the slot in the clutch sleeve.
2. Using a snap-ring expander, spread the retainer in the counterbore in the clutch sleeve, then withdraw the synchronizer ring.
3. Remove the other synchronizer ring in the same manner.
4. Lightly grease both synchronizer rings.
5. Install a synchronizer ring retainer in the counterbore at one end of the clutch sleeve.
6. Insert ring expander through slot in clutch sleeve and expand retainer in counterbore. Install ring in clutch sleeve.
7. Install other synchronizer ring in the same manner.

Assembly

Transmission Assembly

1. Coat thrust washers and needle roller thrust bearing with grease.
2. Position the needle bearing against the rear of the gear (end with chamfered teeth) and position the large thrust washer against the bearing. Position small thrust washer at opposite end of gear.
3. From the rear of the case, install idler shaft, aligning lock pin hole in shaft with hole in case.
4. Position reverse idler gear in the case so the radial roller thrust bearing is toward the rear of the case and the gear is lined up with the shaft. Tap shaft from rear until lock pin holes are lined up.
5. Coat new idler shaft pin with sealer and drive it in about 1/16 in. Peen the hole slightly.
6. Install new idler shaft expansion plug in front of case.
7. Grease both ends of the countergear bore and insert the dummy shaft.
8. Install 25 bearing rollers at each end of countergear.
9. Grease the bearing thrust washers and countergear thrust washers and place bearing thrust washers, followed by countergear thrust washers, at both ends of the countergear, making sure that the tangs on the countergear thrust washers face out.
10. Let the gear, dummy shaft and washer assembly lie in the bottom of the case.
11. From inside the case push clutch gear assembly through opening in front face of case and tap clutch gear assembly until clutch gear bearing locating ring groove is outside front of case.
12. Install snap-ring in bearing groove and tap clutch gear toward the rear until snap-ring is firm against the face of the case.
13. Install clutch gear bearing retainer and gasket.
14. Apply sealer to threads of bearing retainer bolts and torque to 12—15 ft. lbs.
15. Lift the countershaft into place and alignment with the countershaft bore. Be sure that the thrust washers are in place.
16. Lubricate and start the countershaft into the case from the rear. Be sure the flat on the face of the shaft is horizontal and at the bottom to permit installation of case extension.
17. Push countershaft in and dummy shaft out until the flat on the end of the countershaft is flush with the rear face of the case.
18. Grease and install 14 large rollers in the pilot hole of the main drive gear, then install small I.D. spacer.
19. Insert the large I.D. front spacer, followed by the 24 small rollers.
20. Insert the large I.D. rear spacer with chamfered side out.
21. Insert synchronizing clutch assembly, shoulder to front, through rear opening in case.
22. Line up two inner lugs of clutch sleeve synchronizing ring with two wide grooves in main drive

gear and slide clutch assembly onto main drive gear.

23. Through side opening in the case, insert first and reverse sliding gear, with wide inner bevel and small round depression toward the rear. Pilot the first and reverse sliding gear onto the clutch.

24. Insert mainshaft assembly in the transmission case extension, and by spreading the mainshaft bearing snap-ring, tap front end of mainshaft until snap-ring seats in mainshaft bearing groove.

25. Affix new extension housing gasket to transmission case.

26. Align clutch splines on the mainshaft with clutch splines on second-speed gear so that they will receive the two inner lugs of the synchronizer ring of second and third-speed clutch. Mark for identification.

27. Insert mainshaft assembly through opening at rear of transmission case. Be sure the two inner lugs of the synchronizing ring engage the previously marked grooves of mainshaft and second speed gear.

28. Rotate extension housing to line up with case, insert five extension housing bolts with sealer on them. Leave them finger-tight.

29. Set transmission assembly top side up. Maneuver second and third speed clutch until extension housing fits flush against transmission case. Torque attaching bolts to 40 to 45 ft. lbs.

30. Install speedometer driven gear, and side cover assembly.

Type-14
Muncie Fully Synchronized 3-Speed

Application

Camaro, 1969-70
Chevelle, 1969-70
Chevrolet, 1969-70

Chevy II, 1969-70
Firebird, 1970
Olds F-85, 4-4-2, 1970

Pontiac Grand Prix, 1970
Tempest, GTO, 1970

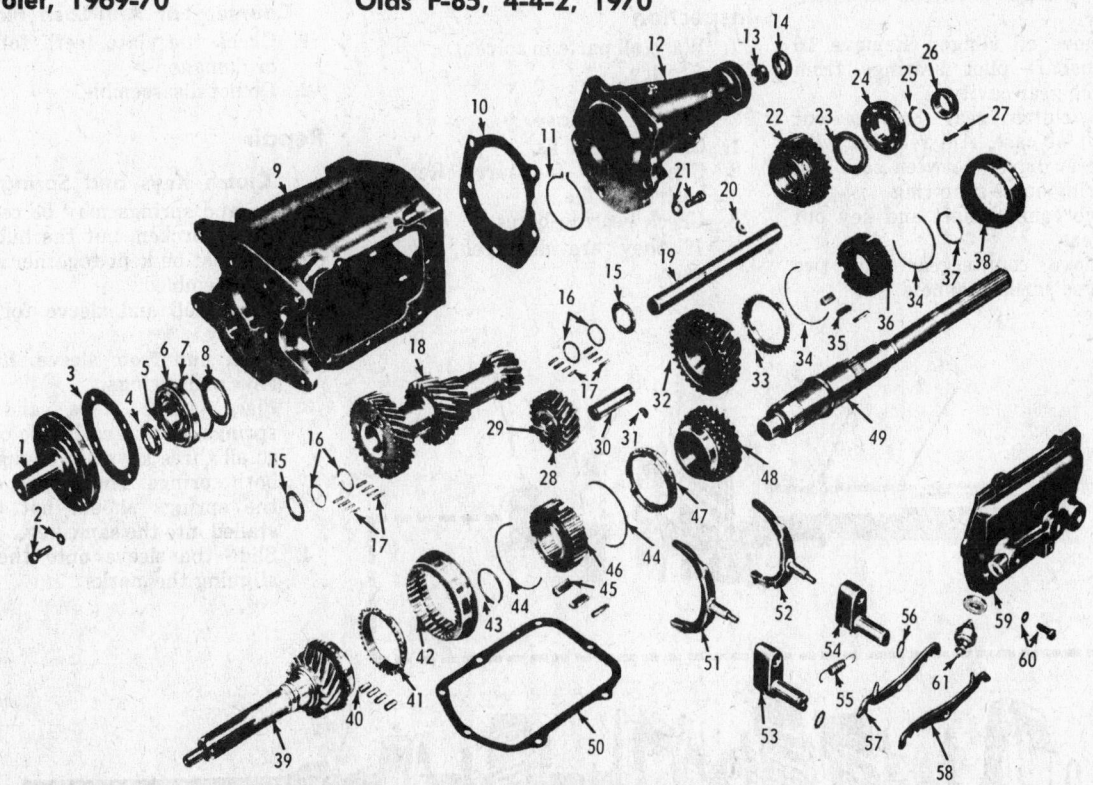

Muncie transmission—exploded view (© G.M. Corp.)

1 Bearing retainer	17 Needle bearings	35 Synchronizer keys	49 Mainshaft
2 Bolt and lock washer	18 Countergear	36 1st and reverse synchronizer hub assembly	50 Gasket
3 Gasket	19 Countershaft	37 Snap ring	51 2nd and 3rd shifter fork
4 Oil seal	20 Woodruff key	38 1st and reverse synchronizer collar	52 1st and reverse shifter fork
5 Snap ring (bearing-to-main drive gear)	21 Bolt (extension-to-case)	39 Main drive gear	53 2-3 shifter shaft assembly
6 Main drive gear bearing	22 Reverse gear	40 Pilot bearings	54 1st and reverse shifter shaft assembly
7 Snap ring bearing	23 Thrust washer	41 3rd speed blocker ring	55 Spring
8 Oil slinger	24 Rear bearing	42 2nd and 3rd synchronizer collar	56 O-ring seal
9 Case	25 Snap ring	43 Snap ring	57 1st and reverse detent cam
10 Gasket	26 Speedometer drive gear	44 Synchronizer key spring	58 2nd and 3rd detent cam
11 Snap ring (rear bearing-to-extension)	27 Retainer clip	45 Synchronizer keys	59 Side cover
12 Extension	28 Reverse idler gear	46 2nd and 3rd synchronizer hub	60 Bolt and lock washer
13 Extension bushing	29 Reverse idler bushing	47 2nd speed blocker ring	
14 Oil seal	30 Reverse idler shaft	48 2nd speed gear	
15 Thrust washer	31 Woodruff key		
16 Bearing washer	32 1st speed gear		
	33 1st speed blocker ring		
	34 Synchronizer key spring		

Transmission Disassembly

1. Remove side cover and shift forks.
2. Unbolt extension. Rotate extension to line up groove in extension flange with reverse idler shaft. Drive reverse idler shaft and key out of case with a brass drift.
3. Move second-third synchronizer sleeve forward. Remove extension housing and mainshaft assembly.
4. Remove reverse idler gear from case.
5. Remove third speed blocker ring from clutch gear.
6. Expand snap-ring which retains mainshaft rear bearing. Tap gently on end of mainshaft to remove extension.
7. Remove clutch gear bearing retainer and gasket.
8. Remove snap-ring. Remove clutch gear from inside case by gently tapping on end of clutch gear.
9. Remove oil slinger. Remove 16 mainshaft pilot bearings from clutch gear cavity.
10. Slip clutch gear bearing out front of case. Aid removal with a screwdriver between case and bearing outer snap-ring.
11. Drive countershaft and key out to rear.
12. Remove countergear and two tanged thrust washers.

Mainshaft Disassembly

1. Depress speedometer drive gear retaining clip. Slide off gear. Some speedometer drive gears, made of metal, must be pulled off.
2. Remove rear bearing snap-ring.
3. Support reverse gear, Press on rear of mainshaft to remove reverse gear, thrust washer, and rear bearing. Be careful not to cock the bearing on the shaft.
4. Remove first and reverse sliding clutch hub snap-ring.
5. Support first gear. Press on rear of mainshaft to remove clutch assembly, blocker ring, and first gear.
6. Remove second and third speed sliding clutch hub snap-ring.
7. Support second gear. Press on front of mainshaft to remove clutch assembly, second speed blocker ring, and second gear from shaft.

Inspection

1. Wash all parts in solvent.
2. Air dry.

Case

1. Check for cracks.
2. Check faces for burrs. Remove with a fine file.
3. Check bearing bores for damage. If they are damaged, replace case.

Front and Rear Bearings

1. Do not spin bearings with air pressure; turn them slowly by hand.
2. Lubricate bearings with light oil. Turn slowly to check for roughness.

Bearing Rollers

1. Check for wear; replace if worn.
2. Check countershaft and reverse idler shaft.
3. Replace all worn washers.

Gears

1. Check for wear, chips, or cracks.
2. If reverse gear bushing is worn or damaged, replace entire gear.
3. Check to see that both clutch sleeves slide freely on their hubs.

Reverse Idler Gear Bushing

This bushing may not be serviced separately. If the bushing requires replacement, replace the gear.

Countergear Anti-Lash Plate

1. Check the plate teeth for wear or damage.
2. Do not disassemble.

Repair

Clutch Keys and Springs

Keys and springs may be replaced if worn or broken, but the hubs and sleeves must be kept together as originally assembled.

1. Mark hub and sleeve for reassembly.
2. Push hub from sleeve. Remove keys and springs.
3. Place three keys and two springs, one on each side of hub, so all three keys are engaged by both springs. The tanged end of the springs should not be installed into the same key.
4. Slide the sleeve onto the hub, aligning the marks.

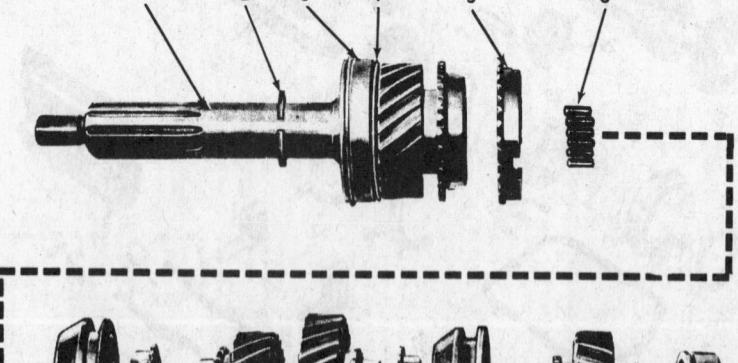

Clutch gear and mainshaft assembly (© G.M. Corp.)

1 Clutch gear	7 Retaining clip	13 Reverse gear	18 Shoulder (part of mainshaft)
2 Snap ring	8 Mainshaft	14 Snap ring	19 2nd speed gear
3 Clutch gear bearing	9 Speedo drive gear	15 1st speed synchronizer	20 2nd speed blocker ring
4 Oil slinger	10 Snap ring	assembly	21 2-3 synchronizer assembly
5 3rd speed blocker ring	11 Rear bearing	16 1st speed blocker ring	22 Snap ring
6 Mainshaft pilot bearings (16)	12 Reverse gear thrust washer	17 1st speed gear	

Extension Oil Seal and Bushing

1. Remove seal.
2. Using bushing remover and installer, or other suitable tool, drive bushing into extension housing.
3. Drive new bushing in from rear. Lubricate inside of bushing and seal. Install new oil seal with extension seal installer or suitable tool.

Clutch Bearing Retainer Oil Seal

1. Pry old seal out.
2. Install new seal using seal installer or suitable tool. Seat seal in bore.

Mainshaft Assembly

1. Turn front of mainshaft up.
2. Install second gear with clutching teeth up; the rear face of the gear butts against the flange on the mainshaft.
3. Install a blocking ring with clutching teeth downward. All three blocking rings are the same.
4. Install second and third synchronizer assembly with fork slot down. Press it onto mainshaft splines. Both synchronizer assemblies are identical but are assembled differently. The second-third speed hub and sleeve is assembled with the sleeve fork slot toward the thrust face of the hub; the first-reverse hub and sleeve, with the fork slot opposite the thrust face. Be sure that the blocker ring notches align with the synchronizer assembly keys.
5. Install synchronizer snap-ring.

Both synchronizer snap-rings are the same.
6. Turn rear of shaft up.
7. Install first gear with clutching teeth upward; the front face of the gear butts against the flange on the mainshaft.
8. Install a blocker ring with clutching teeth down.
9. Install first and reverse synchronizer assembly with fork slot down. Press it onto mainshaft splines. Be sure blocker ring notches align with synchronizer assembly keys and both synchronizer sleeves face front of mainshaft.
10. Install snap-ring.
11. Install reverse gear with clutching teeth down.
12. Install steel reverse gear thrust washer with flats aligned.
13. Press rear ball bearing onto shaft with snap-ring slot down.
14. Install snap-ring.
15. Install speedometer drive gear and retaining clip.

Transmission Assembly

1. Place a row of 29 roller bearings, a bearing washer, a second row of 29 bearings, and a second bearing washer at each end of the countergear. Hold in place with grease.
2. Place countergear assembly through rear case opening with a tanged thrust washer, tang away from gear, at each end. Install countershaft and key from rear of case. Be sure that thrust washer tangs are aligned with notches in case.
3. Place reverse idler gear in case.

Do not install reverse idler shaft yet.
4. Expand snap-ring in extension. Assemble extension over mainshaft and onto rear bearing. Seat snap-ring.
5. Load 16 mainshaft pilot bearings into clutch gear cavity. Assemble third speed blocker ring onto clutch gear clutching surface with teeth toward gear.
6. Place clutch gear assembly, without front bearing, over front of mainshaft. Make sure that blocker ring notches align with keys in second-third synchronizer assembly.
7. Stick gasket onto extension housing with grease. Assemble clutch gear, mainshaft, and extension to case together. Make sure that clutch gear teeth engage teeth of countergear antilash plate.
8. Rotate extension housing. Install reverse idler shaft and key.
9. Torque extension bolts to 45 ft. lbs.
10. Install oil slinger with inner lip facing forward. Install front bearing outer snap-ring to bearing. Slide bearing into case bore.
11. Install snap-ring to clutch gear stem. Install bearing retainer and gasket. Torque bolts to 20 ft. lbs. Retainer oil return hole must be at 6 o'clock.
12. Shift both synchronizer sleeves to neutral positions. Install side cover, aligning shifter forks with synchronizer sleeve grooves.
13. Torque side cover bolts to 20 ft. lbs.

Type-15
Saginaw Fully Synchronized 3-Speed
Application

Buick LeSabre, 1966-68
Buick Special, 1966-70
Camaro, 1967-70
Chevelle, 1966-70

Chevrolet, 1966-70
Chevy II, 1966-70
Corvette, 1966-69

Firebird, 1967-70
Olds F-85, 1966-70
Tempest, 1966-70

Transmission Disassembly

1. Remove side cover assembly and shift forks.
2. Remove clutch gear bearing retainer.
3. Remove clutch gear bearing to gear stem snap-ring. Pull clutch gear outward until a screwdriver can be inserted between bearing and case. Remove clutch gear bearing.
4. Remove speedometer driven gear and extension bolts.
5. Remove reverse idler shaft

snap-ring. Slide reverse idler gear forward on shaft.
6. Remove mainshaft and extension assembly.
7. Remove clutch gear and third speed blocker ring from inside case. Remove 14 roller bearings from clutch gear.
8. Expand the snap-ring which retains the mainshaft rear bearing. Remove the extension.
9. Using a dummy shaft, drive the countershaft and key out the rear of the case. Remove the gear, two tanged thrust washers,

and dummy shaft. Remove bearing washer and 27 roller bearings from each end of countergear.
10. Use a long drift to drive the reverse idler shaft and key through the rear of the case.
11. Remove reverse idler gear and tanged steel thrust washer.

Mainshaft Disassembly

1. Remove second and third speed sliding clutch hub snap-ring from mainshaft. Remove clutch assembly, second speed blocker

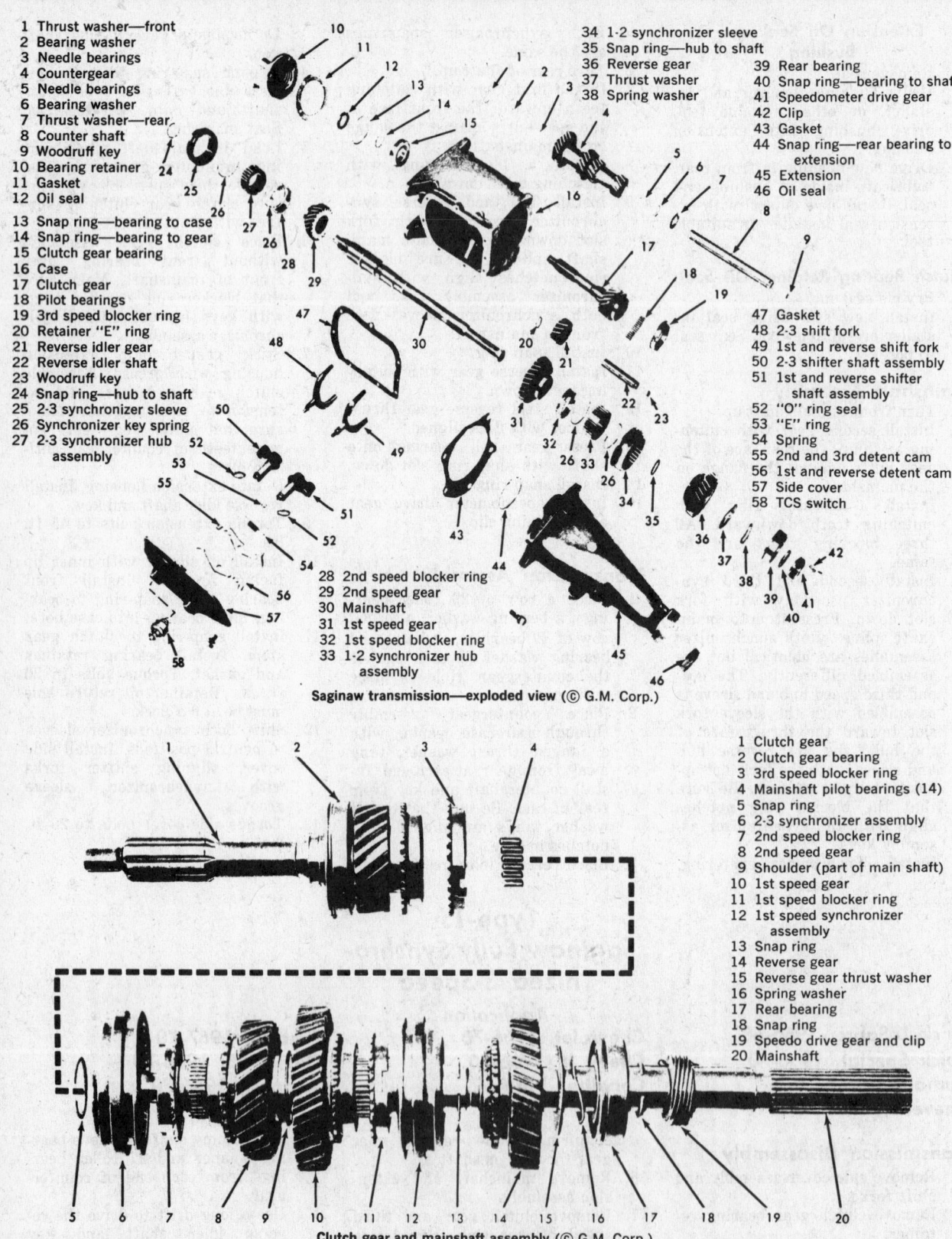

1 Thrust washer—front
2 Bearing washer
3 Needle bearings
4 Countergear
5 Needle bearings
6 Bearing washer
7 Thrust washer—rear
8 Counter shaft
9 Woodruff key
10 Bearing retainer
11 Gasket
12 Oil seal
13 Snap ring—bearing to case
14 Snap ring—bearing to gear
15 Clutch gear bearing
16 Case
17 Clutch gear
18 Pilot bearings
19 3rd speed blocker ring
20 Retainer "E" ring
21 Reverse idler gear
22 Reverse idler shaft
23 Woodruff key
24 Snap ring—hub to shaft
25 2-3 synchronizer sleeve
26 Synchronizer key spring
27 2-3 synchronizer hub
   assembly

28 2nd speed blocker ring
29 2nd speed gear
30 Mainshaft
31 1st speed gear
32 1st speed blocker ring
33 1-2 synchronizer hub
   assembly

34 1-2 synchronizer sleeve
35 Snap ring—hub to shaft
36 Reverse gear
37 Thrust washer
38 Spring washer

39 Rear bearing
40 Snap ring—bearing to shaft
41 Speedometer drive gear
42 Clip
43 Gasket
44 Snap ring—rear bearing to
   extension
45 Extension
46 Oil seal

47 Gasket
48 2-3 shift fork
49 1st and reverse shift fork
50 2-3 shifter shaft assembly
51 1st and reverse shifter
   shaft assembly
52 "O" ring seal
53 "E" ring
54 Spring
55 2nd and 3rd detent cam
56 1st and reverse detent cam
57 Side cover
58 TCS switch

Saginaw transmission—exploded view (© G.M. Corp.)

1 Clutch gear
2 Clutch gear bearing
3 3rd speed blocker ring
4 Mainshaft pilot bearings (14)
5 Snap ring
6 2-3 synchronizer assembly
7 2nd speed blocker ring
8 2nd speed gear
9 Shoulder (part of main shaft)
10 1st speed gear
11 1st speed blocker ring
12 1st speed synchronizer
   assembly
13 Snap ring
14 Reverse gear
15 Reverse gear thrust washer
16 Spring washer
17 Rear bearing
18 Snap ring
19 Speedo drive gear and clip
20 Mainshaft

Clutch gear and mainshaft assembly (© G.M. Corp.)

ring, and second gear from front of mainshaft.

2. Depress speedometer drive gear retaining clip. Remove gear. Some units have a metal speedometer drive gear which must be pulled off,

3. Remove rear bearing snap-ring.

4. Support reverse gear. Press on rear of mainshaft. Remove reverse gear, thrust washer, spring washer, rear bearing, and snap-ring. When pressing off the rear bearing, be careful not to cock the bearing on the shaft.

5. Remove first and reverse sliding

clutch hub snap-ring. Remove clutch assembly, first speed blocker ring, and first gear.

Inspection

1. Wash all parts in solvent.
2. Air dry.

Case

1. Check for cracks.
2. Check faces for burrs. Remove with a fine file.
3. Check bearing bores for damage. If they are damaged, replace case.

Front and Rear Bearings

1. Do not spin bearings with air pressure; turn them slowly by hand.
2. Lubricate bearings with light oil. Turn slowly to check for roughness.

Bearing Rollers

1. Check for wear; replace if worn.
2. Check countershaft and reverse idler shaft for wear or damage.
3. Replace all worn washers.

Gears

1. Check for wear, chips, or cracks.
2. If reverse gear bushing is worn or damaged, replace entire gear.
3. Check that both clutch sleeves slide freely on their hubs.

Reverse Idler Gear Bushing

This bushing may not be serviced separately. If the bushing requires replacement, replace the gear.

Countergear Anti-Lash Plate

1. Check the plate teeth for wear or damage.
2. Do not disassemble unit.

Repair

Clutch Keys and Springs

Keys and springs may be replaced if worn or broken, but the hubs and sleeves are matched pairs and must be kept together.

1. Mark hub and sleeve for reassembly.
2. Push hub from sleeve. Remove keys and springs.
3. Place three keys and two springs, one on each side of hub, in position, so all three keys are engaged by both springs. The tanged end of the springs should not be installed into the same key.
4. Slide the sleeve onto the hub, aligning the marks.
NOTE: a groove around the outside of the synchronizer hub marks the end that must be opposite the fork slot in the sleeve when assembled.

Extension Oil Seal and Bushing

1. Remove seal.
2. Using bushing remover and installer tool, or other suitable tool, drive bushing into extension housing.
3. Drive new bushing in from the rear. Lubricate inside of bushing and seal. Install new oil seal with extension seal installer tool or other suitable tool.

Clutch Bearing Retainer Oil Seal

1. Pry old seal out.
2. Install new seal using seal installer or suitable tool. Seat seal in bore.

Mainshaft Assembly

1. Turn front of mainshaft up.
2. Install second gear with clutching teeth up; the rear face of the gear butts against the flange on the mainshaft.
3. Install a blocker ring with clutching teeth down. All three blocker rings are the same.
4. Install second and third speed synchronizer assembly with fork slot down. Press it onto mainshaft splines. Both synchronizer assemblies are the same. Be sure that blocker ring notches align with synchronizer assembly keys.
5. Install synchronizer snap-ring. Both synchronizer snap-rings are the same.
6. Turn rear of shaft up.
7. Install first gear with clutching teeth up; the front face of the gear butts against the flange on the mainshaft.
8. Install a blocker ring with clutching teeth down.
9. Install first and reverse synchronizer assembly with fork slot down. Press it onto mainshaft splines. Be sure blocker ring notches align with synchronizer assembly keys.
10. Install snap-ring.
11. Install reverse gear with clutching teeth down.
12. Install steel reverse gear thrust washer and spring washer.
13. Press rear ball bearing onto shaft with snap-ring slot down.
14. Install snap-ring.
15. Install speedometer drive gear and retaining clip. Press on metal speedometer drive gear.

Type-16 Corvair 4-Speed

Application
Corvair, 1960-65

Transmission Assembly

1. Using dummy shaft, load a row of 27 roller bearings and a thrust washer at each end of countergear. Hold in place with grease.
2. Place countergear assembly into case through rear. Place a tanged thrust washer, tang away from gear, at each end. Install countershaft and key, making sure that tangs align with notches in case.
3. Install reverse idler gear thrust washer, gear, and shaft with key from rear of case. Be sure thrust washer is between gear and rear of case with tang toward notch in case.
4. Expand snap-ring in extension. Assemble extension over rear of mainshaft and onto rear bearing. Seat snap-ring in rear bearing groove.
5. Install 14 mainshaft pilot bearings into clutch gear cavity. Assemble third speed blocker ring onto clutch gear clutching surface with teeth toward gear.
6. Place clutch gear, pilot bearings, and third speed blocker ring assembly over front of mainshaft assembly. Be sure blocker rings align with keys in second-third synchronizer assembly.
7. Stick extension gasket to case with grease. Install clutch gear, mainshaft, and extension together. Be sure clutch gear engages teeth of countergear anti-lash plate. Torque extension bolts to 45 ft. lbs.
8. Place bearing over stem of clutch gear and into front case bore. Install front bearing to clutch gear snap-ring.
9. Install clutch gear bearing retainer and gasket. The retainer oil return hole must be at the bottom. Torque retainer bolts to 10 ft. lbs.
10. Install reverse idler gear shaft E-ring.
11. Shift synchronizer sleeves to neutral positions. Install cover, gasket, and forks, aligning forks with synchronizer sleeve grooves. Torque side cover bolts to 10 ft. lbs.
12. Install speedometer driven gear.

Disassembly

1. Remove the front and side covers from the case.
2. Remove plug, detent spring, and ball from third-fourth detent

channel at left rear of case.
3. Drive roll pin from third-fourth shift fork, then remove third-

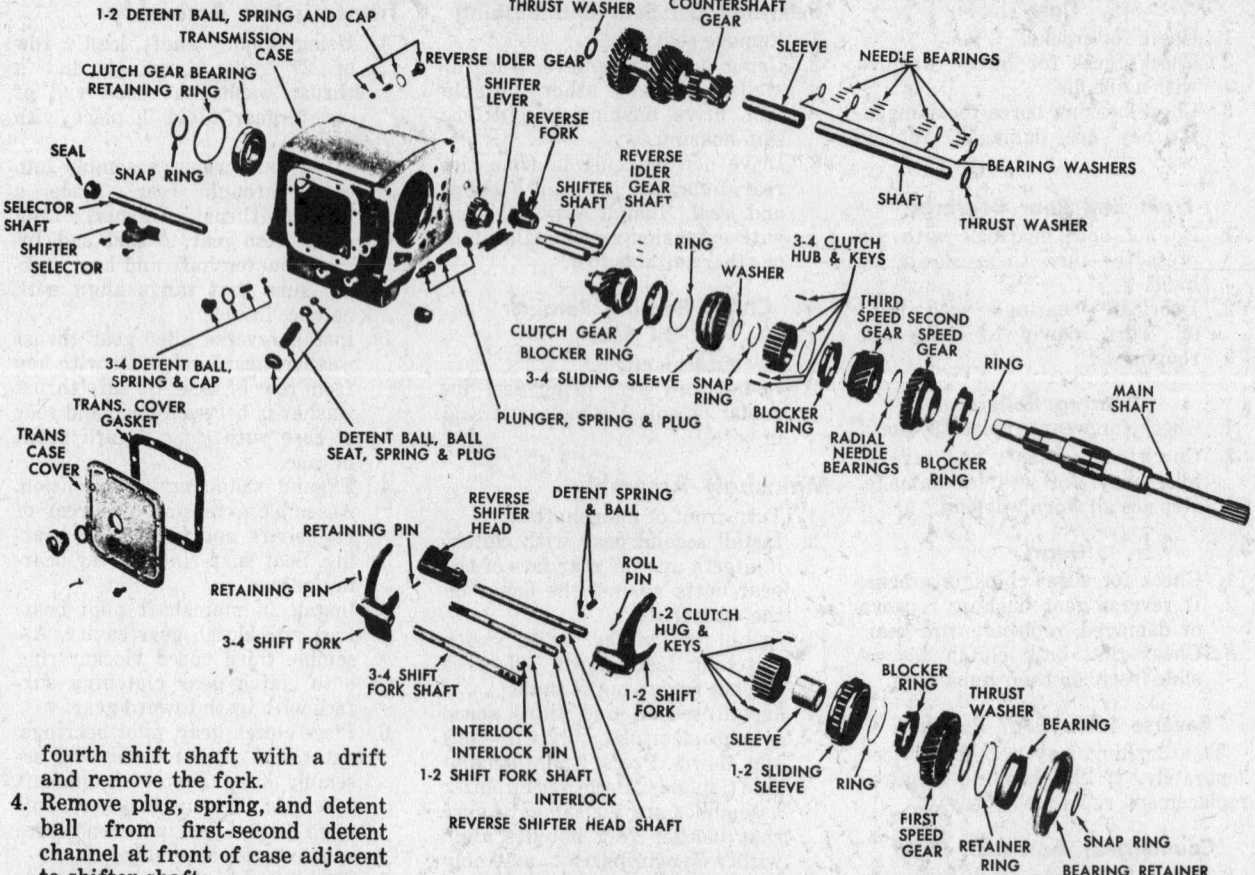

Corvair 4-speed transmission
(© G.M. Corp.)

fourth shift shaft with a drift and remove the fork.

4. Remove plug, spring, and detent ball from first-second detent channel at front of case adjacent to shifter shaft.

5. Move first-second shift fork into second gear, then remove roll pin holding fork to shaft. Remove shift fork and shaft by tapping the shaft out of the case in either direction.

6. Remove the snap-rings at the clutch gear, between the clutch gear and bearing, and between the clutch gear bearing and case.

7. Remove the clutch gear bearing with a puller.

8. Remove two snap-rings holding the rear mainshaft bearing retainer in place.

9. Remove mainshaft with assembled gears and bearing retainer.

10. Drive roll pin from reverse shifter head, then remove shifter head and shaft from the case in either direction.

11. Drive out reverse idler gear and reverse shifter fork shafts from the case by driving rearward. Remove idler gear and reverse shifter fork from the case.

12. Remove reverse shifter lever by lifting off its pin in the case.

13. With a dummy shaft, drive the countershaft rearward until the countergear and thrust washers are free. Lift the countergear assembly out of the case.

14. Remove two countergear thrust washers from the case. Then, with a magnet, remove the two interlocks and the detent ball and

spring remaining in the third-fourth detent channel.

15. Remove shift finger from selector shaft by first flattening the lock tabs securing the two retaining screws, then removing the shifter shaft plug in rear of case. Tap shaft rearward and out of case.

16. With a suitable tool, remove reverse inhibitor plug, spring, ball seat, and ball. Remove the other plug, spring, and reverse inhibitor plunger.

17. Disassemble the mainshaft and countergear assemblies. Wash and inspect all parts.

Component Assembly

Synchronizers

1. Place the two springs in position (one on each side of the hub), so a tanged-end of each spring falls into the same keyway in the hub.

2. Holding the keys in position, align etched marks in hub and sleeve, then slide hub into sleeve.

Countergear

1. Install the dummy shaft and the tubular bearing spacer into the bore of the countergear.

2. Grease the countergear bore and install 23 bearing rollers and a retaining washer.

3. Install another row of 23 rollers and a retaining washer.

4. Treat the other end of the countergear in the same manner, then set countergear assembly aside.

Mainshaft Rear Bearing

If inspection reveals the need for a new bearing, place the bearing and retainer in a press. Expand retainer ring and press out bearing. Install a new bearing by reversing the above procedure.

Mainshaft

1. Install first-second synchronizer hub onto mainshaft with groove downward, then place first gear sleeve on mainshaft. Press both first gear sleeve and synchronizer hub onto mainshaft until they bottom.

2. Install blocker ring in rear of first-second synchronizer, being sure the notches in blocker ring engage keys in synchronizer unit. NOTE: the blocker rings of the first-second synchronizer have a slightly longer hub than those of the third-fourth synchronizer. Then slide first-speed gear and its thrust washer onto the mainshaft.

3. Press rear bearing and retainer assembly onto the mainshaft. Secure with a selective snap-ring. Maximum end-play between rear

face of rear bearing and snap-ring will be .005 in.

4. Invert the mainshaft, then install the second blocker ring (long hub) in front side of first-second synchronizer.

5. Install second-speed gear with clutching teeth toward first-second synchronizer, then place radial needle thrust bearing on second-speed gear.

6. Install third-speed gear, clutching teeth upward, onto mainshaft and seat it against the radial needle bearing.

7. Place third-fourth blocker ring on cone surface of third-speed gear, then press third-fourth synchronizer unit onto blocker ring, with extended neck toward third gear. Be sure notches in blocker ring engage clutch keys in synchronizer unit. Install second blocker ring onto third-fourth synchronizer unit. Install snap-ring.

8. If clutch gear roller bearings have become displaced, load 34 needle bearings into innermost diameter and 38 needle bearings into outermost diameter. Hold needles in place with petroleum jelly.

9. Slide clutch gear onto mainshaft
   NOTE: place the clutch gear on a bench with its pilot bore up, needle bearing retainer in place, and insert the mainshaft into the clutch gear. Set assembled mainshaft aside for later installation into transmission.

Transmission Assembly

1. Install reverse inhibitor plug, then the spring and reverse inhibitor plunger.

2. Install reverse inhibitor detent ball, ball seat, spring, and plug.

3. Coat selector shaft with grease, then insert through seal from the inside of the case.

4. Attach shift finger to selector shaft with two bolts and lock tabs. Then install drain plug in case at rear of selector shaft.

5. With tab facing out, place thrust washer on large end of countergear and place countergear (large end forward) in bottom of case. Raise countergear into bore alignment, then push the dummy shaft part way out of the gear and into the case front hole just enough to hold the countergear in place. Be sure the thrust washer tab engages notch in case.

6. With the tab facing out, insert rear thrust washer, insert countergear shaft at rear of case and press the dummy shaft entirely out of the gear and the front of the case with the countershaft.

Countershaft should be flush with the rear face of the case.

7. Place reverse shift lever on pin in case with tapered end away from the reverse inhibitor.

8. Place the reverse idler gear shift fork in the case with its pin toward the front. Engage the fork pin with the reverse shifter lever, then insert the shift fork shaft.

9. With the reverse idler gear shift fork fully rearward, engage the reverse idler gear and the shift fork. Then, align the lock-pin hole or key groove in the idler gear shaft with the hole in the rear face of the case and slide the shaft fully into the case. Install pin or key in the shaft.

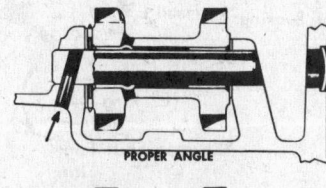

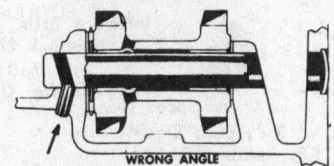

Reverse idler, showing locking pin installation (© G.M. Corp.)

10. Insert a detent spring and ball in the third-fourth detent channel, checking that the spring goes fully to the bottom of the channel.

11. Lay out the shift fork shafts to avoid confusion. With interlock notches aligned, the reverse shaft can be identified by the pin hole centered in the shaft. The first-second shaft has two interlock notches connected by a drilled hole for the interlock pin. The third-fourth shaft roll pin hole is closest to the end of the shaft opposite the detent notched end.

12. Depress the detent ball and spring in the third-fourth detent channel slightly and insert the reverse shifter head shaft partially into the case to compress the detent. Then, engage the pin of the reverse shifter head with the yoke of the reverse shift lever. Check that the shaft pin hole is aligned with the pin hole in the shifter head and push the shaft through until the pin holes in the head and shaft align. Secure head to shaft with roll pin.

13. Install the mainshaft assembly into the case by shifting the synchronizers into second and fourth simultaneously to provide clearance to pass the countergear. Insert the mainshaft through the rear bearing bore into the case.

Align rear bearing retainer snap-ring, with snap-ring cuts out, in rear face of case, then tap rear bearing retainer into case until flush with the rear face.

14. With the clutch gear in place, the large snap-ring in the bearing groove, tap bearing into the front of the case and over the clutch hub until large snap-ring seats against front of case.
    NOTE: hold rear bearing retainer against case while tapping the front bearing, to prevent retainer from working out of case.

15. Retain clutch gear in bearing with selective snap-ring. With proper snap-ring installed, maximum end-play between the bearing and snap-ring will be .005 in.

16. Install small snap-ring in clutch gear.

17. Prior to installing the first-second shift fork, shift first-second synchronizer and third-fourth synchronizer to neutral. Then, install one interlock into the third-fourth detent channel.

18. With the interlock pin in the hole in the interlock end, push the first-second shift fork shaft partially into the case. The interlock end of the shaft goes to the rear of the case. Engage first-second shift fork, identified by the through gate at the shift location, with the first-second synchronizer. Align pin holes in the shaft and fork. Secure the fork to shaft with a roll pin.

19. Install detent ball, spring, gasket, and cap in first-second detent channel. Cap used at this location has the longer shank.

20. Drop remaining interlock into third-fourth detent channel, then push third-fourth shift fork shaft partially through hole in front of case. Engage third-fourth shift fork in third-fourth synchronizer, align pin hole in shift fork and shaft. Secure shift fork shaft with roll pin.

21. Install remaining detent ball, spring, nylon washer, and cap in third-fourth detent channel at left rear of third-fourth case.

22. Before installing front bearing retainer and side cover, test operation of shift forks by actuating the shift selector lever with a small pin punch inserted into the hole in the shifter shaft. Install front bearing retainer, using a new gasket. Then, secure with seven bolts torqued to 15-20 ft. lbs. Complete assembly of transmission by installing side cover, with new gasket, and securing with eight bolts torqued to 3-4 ft. lbs.

Type-17
Corvair 4-Speed
Application
Corvair, 1966-69

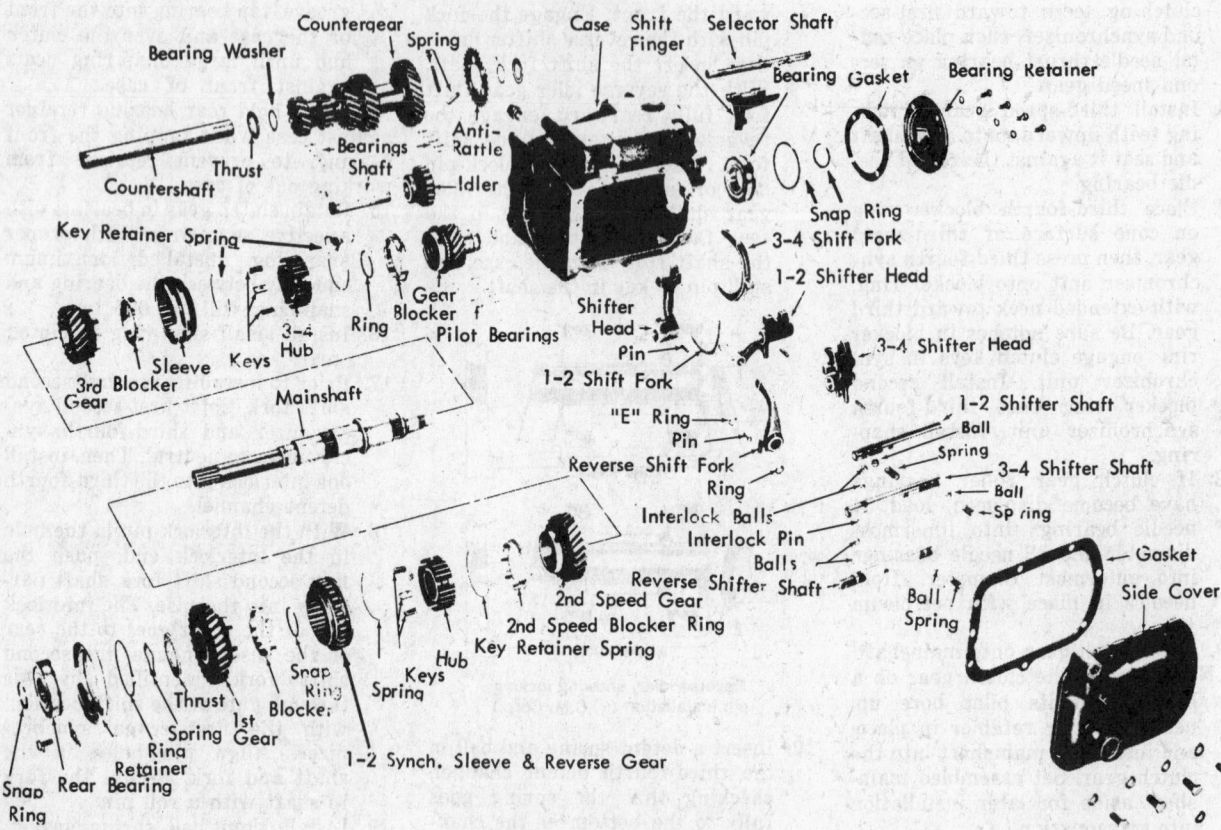

1966-69 Corvair 4-speed transmission

This transmission is very similar to the 1966-69 three-speed transmission. Many parts are interchangeable. Service procedures, as covered in that section, may be used with the following exceptions:

Mainshaft Assembly

The synchronizer assembly at the front of the mainshaft is used for the third-fourth shift rather than the second-third shift. The synchronizer at the rear of the mainshaft is used for the first-second shift rather than the first-reverse shift.

Gear teeth in the first-second synchronizer sleeve distinguish it from the third-fourth sleeve.

All parts except the gears and the first-second synchronizer sleeve in the four-speed transmission mainshaft assembly are also used in the three-speed transmission assembly. Starting from the front, the gears on the mainshaft are third, second and first. A fourth blocker ring is used between the first-second synchronizer assembly and first gear on the four-speed transmission.

Reverse Idler Gear Parts

A sliding reverse idler gear and shaft retained by a Woodruff key is used. A snap-ring stop is used to limit forward movement of the reverse idler gear.

Cover Assembly

The cover assembly is on the left side of transmission. The controls, except the selector shaft and finger, are located in the cover.

Type-18
Muncie 4-Speed
Application

Buick Special (GS), 1967-70
Camaro, 1967-70
Chevelle, 1964-70
Chevrolet, 1964-70
Chevy II, 1964-70

Corvette, 1964-70
Firebird, 1967-70
Monte Carlo, 1970
Oldsmobile, 1965

Olds F-85, 1964-70
Pontiac, 1968-70
Pontiac (GP), 1968-70
Tempest, 1964-70

Disassembly

1. Remove side cover and shift controls after draining.
2. Remove bolts and bolt lock strips from front bearing retainer and remove retainer and gasket.
3. Lock up transmission by shifting into two gears and remove main drive gear retaining nut.

4. Return gears to neutral and remove lock pin from reverse shifter lever boss and pull shaft out about 1/8 in. This will disengage reverse shift fork from reverse gear.

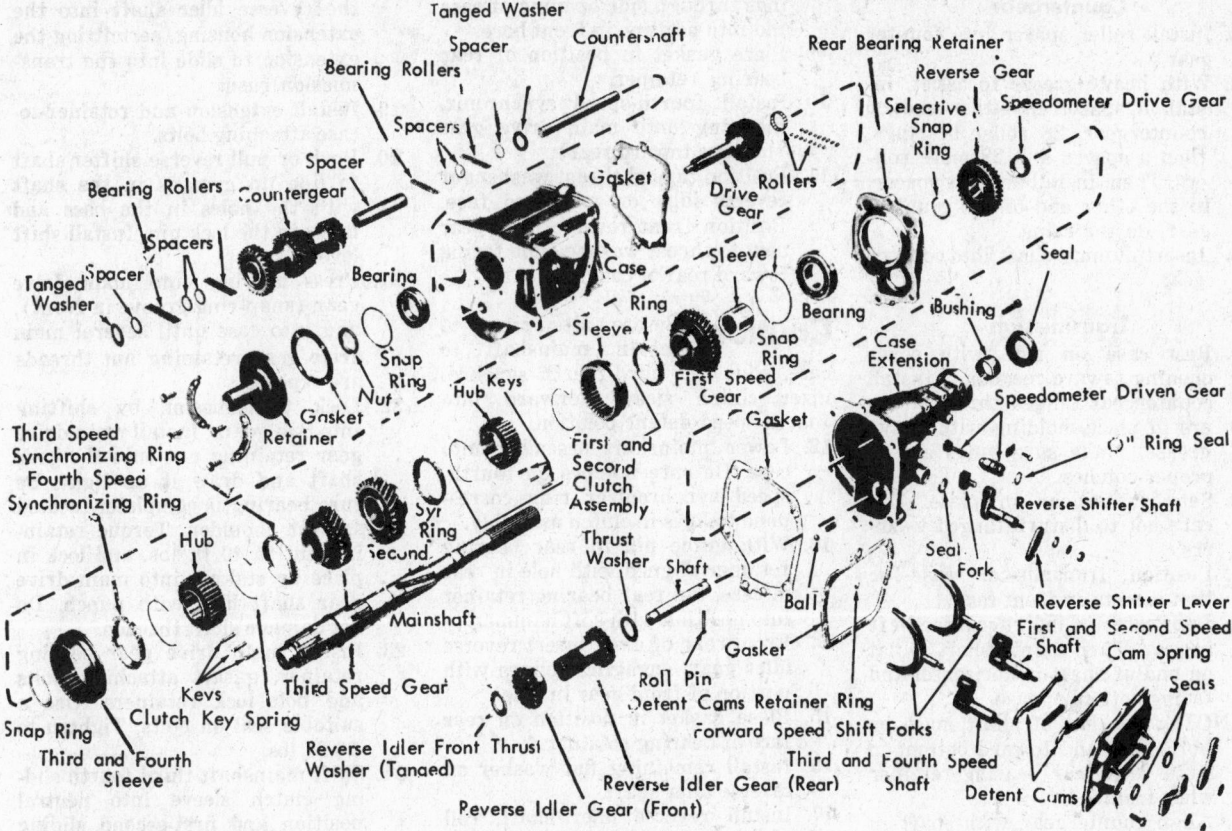

Muncie 4-speed transmission

5. Remove extension case attaching bolts. Tap extension with soft hammer toward rear. When idler shaft is out as far as it will go, move extension to left so reverse fork clears gear and remove extension and gasket.

6. Remove reverse idler gear, flat washer, shaft and roll spring pin.

7. Remove speedometer and reverse gears.

NOTE: slide third-fourth synchronizer clutch sleeve to fourth-speed gear position (forward) before trying to remove mainshaft assembly from case.

8. Remove rear bearing retainer and mainshaft assembly from case by tapping bearing retainer with soft hammer.

9. Unload bearing rollers from main drive gear and remove fourth-speed synchronizer blocking ring.

10. Lift front half of reverse idler gear with tanged thrust washer from case.

11. Press main drive gear down from bearing.

12. Tap front bearing and snap-ring from case.

13. From front of case, press out countershaft. Then, remove the countershaft gear and both tanged washers.

14. Remove the rollers (112) six .050 in. spacers and roller spacer from counter gear.

15. Remove mainshaft front snap-ring and slide third and fourth-speed clutch and third-speed gear and synchronizer ring from front of mainshaft.

16. Spread rear bearing retainer snap-ring and press mainshaft out of retainer.

17. Remove mainshaft snap-ring. Support second-speed gear and press on rear of mainshaft to remove rear bearing, first-speed gear and sleeve, first-speed synchronizing ring, first-second-speed synchronizer clutch, second-speed synchronizer ring and second-speed gear.

After thoroughly cleaning case and all parts, make thorough inspection and replace required parts. In checking bearings do not spin at high speeds, but rather clean and rotate by hand to detect roughness and unevenness. Spinning can damage balls and races.

Assembly

Mainshaft

1. From rear of shaft, assemble second-speed gear (hub of gear toward rear of shaft).

2. Install first-second synchronizer clutch assembly onto mainshaft (sleeve taper toward rear, hub to front); together with a synchronizer ring on each side of clutch assembly so that keyways line up with clutch keys.

3. Press first-speed sleeve onto mainshaft. (A 1¾ in. ID pipe cut to convenient length makes a suitable tool).

4. Install first - speed gear (hub toward front) and press onto the rear bearing with snap - ring grooves toward front of transmission. Be sure bearing is firmly seated.

5. Choose selective fit snap-ring (.087, .090, .093 or .096 in.) and install it into groove in mainshaft behind rear bearing. Maximum clearance of snap-ring and rear face should be between zero and .005 in.

NOTE: always use new snap-ring.

6. Install third-speed gear (hub to front of transmission) and third-speed gear synchronizing ring (notches to front).

7. Install third and fourth-speed gear clutch assembly with both sleeve taper and hub toward front.

8. Install snap-ring onto mainshaft in front of third and fourth-speed clutch, with ends of snap-ring seated behind spline teeth.

9. Install rear bearing retainer. Spread snap - ring in plate, to allow ring to drop around rear bearing, and press on the end of mainshaft until snap - ring engages the groove in rear bearing.

10. Install reverse gear (shift collar to rear).

11. Install speedometer drive gear.

Countergear

1. Install roller spacer into counter gear.
2. With heavy grease to assist, install a spacer in either end of countergear, 28 roller bearings, then a spacer and 28 more rollers. Then, install another spacer. In the other end of the countergear, do the same.
3. Insert dummy shaft into counter gear.

Transmission

1. Rest case on side with cover opening toward mechanic. Install countergear tanged thrust washers in place, holding with heavy grease. Make sure tangs are in proper notches.
2. Set countergear in place. Use care not to disturb tanged washers.
3. Position transmission case so that it rests on front face.
4. Lubricate and insert countershaft in rear. Turn countershaft so flat on end of shaft is horizontal and facing bottom of case.
   NOTE: the flat of shaft must be horizontal and toward bottom to mate with rear bearing retainer when installed.
5. Align countergear with shaft in rear and hole in front of case (pushing dummy shaft out front of case) until flat of shaft is flush with rear of case. Be sure thrust washers remain in place.
6. Check end-play of counter gear (dial indicator should be used). If end-play is more than .025 in. install new thrust washer.
7. Install cage and 17 roller bearings into main drive gear. Use heavy grease to hold bearings.
8. Install main drive gear with bearings through side opening of case and into position in front bore.
9. Place gasket in position of rear bearing retainer.
10. Install fourth-speed synchronizing ring onto main drive gear (notches toward rear).
11. Position tanged thrust washer for reverse idler on machined face. Position front reverse idler gear next to thrust washer (hub facing toward rear of case).

Caution Before attempting to install mainshaft to case, slide the third-fourth synchronizer clutch sleeve forward into fourth-speed detent position.

12. Lower mainshaft assembly into case. Be sure notches on fourth-speed synchronizer ring correspond to keys in clutch assembly.
13. With guide pin in rear bearing retainer aligned with hole in rear of case, tap rear bearing retainer into position with soft hammer.
14. From rear of case, insert reverse idler gear, engaging splines with portion of front gear in case.
15. Place gasket in position on rear face of bearing retainer.
16. Install remaining flat washer on reverse idler shaft.
17. Install reverse idler shaft, roll pin, and thrust washer into gears and front boss of case. Make sure to pick up front tanged thrust washer.
18. Pull reverse shifter shaft to left side of extension and rotate shaft to bring reverse shift fork forward in extension (reverse detent position). Start extension onto transmission case, while slowly pushing in on shifter shaft to engage the shift fork with the reverse gear shift collar. Then, pilot the reverse idler shaft into the extension housing, permitting the extension to slide into the transmission case.
19. Install extension and retainer-to-case attaching bolts.
20. Push or pull reverse shifter shaft to line up grooves in the shaft with the holes in the boss and drive in the lock pin. Install shift lever.
21. Press bearing onto main drive gear (snap-ring groove in front), and into case until several main drive gear retaining nut threads are exposed.
22. Lock transmission by shifting into two gears. Install main drive gear retaining nut onto the gear shaft and draw it up tight. Be sure bearing is completely seated against shoulder. Torque retaining nut to 40 ft. lbs. and lock in place by staking into main drive gear shaft hole with punch. Do not damage shaft threads.
23. Install main drive gear bearing retainer, gasket attaching bolts and bolt lock retainers. Use a suitable seal on bolts. Tighten to 20 ft. lbs.
24. Shift mainshaft third-fourth sliding clutch sleeve into neutral position and first-second sliding clutch into second gear (forward) detent position. Shift side cover third-fourth shift lever into neutral detent and first-second shift lever into second gear detent position.
25. Install side cover, with gasket, and carefully position in place. A dowel pin provides proper alignment position. Install bolts and tighten evenly to avoid distortion. Torque to 20 ft. lbs.

Type-19

Saginaw 4-Speed

Application

Camaro, 1967-70
Chevy II, 1966-70
Chevelle, 1966-70

NOTE: repair procedures on the Saginaw four-speed transmission are typical of the Muncie four-speed. Parts, however, are not interchangeable.

Chevrolet, 1966-70
Corvette, 1967-70

Tempest, 1967-68 (6 Cyl.)
Firebird, 1969 (6 Cyl.)

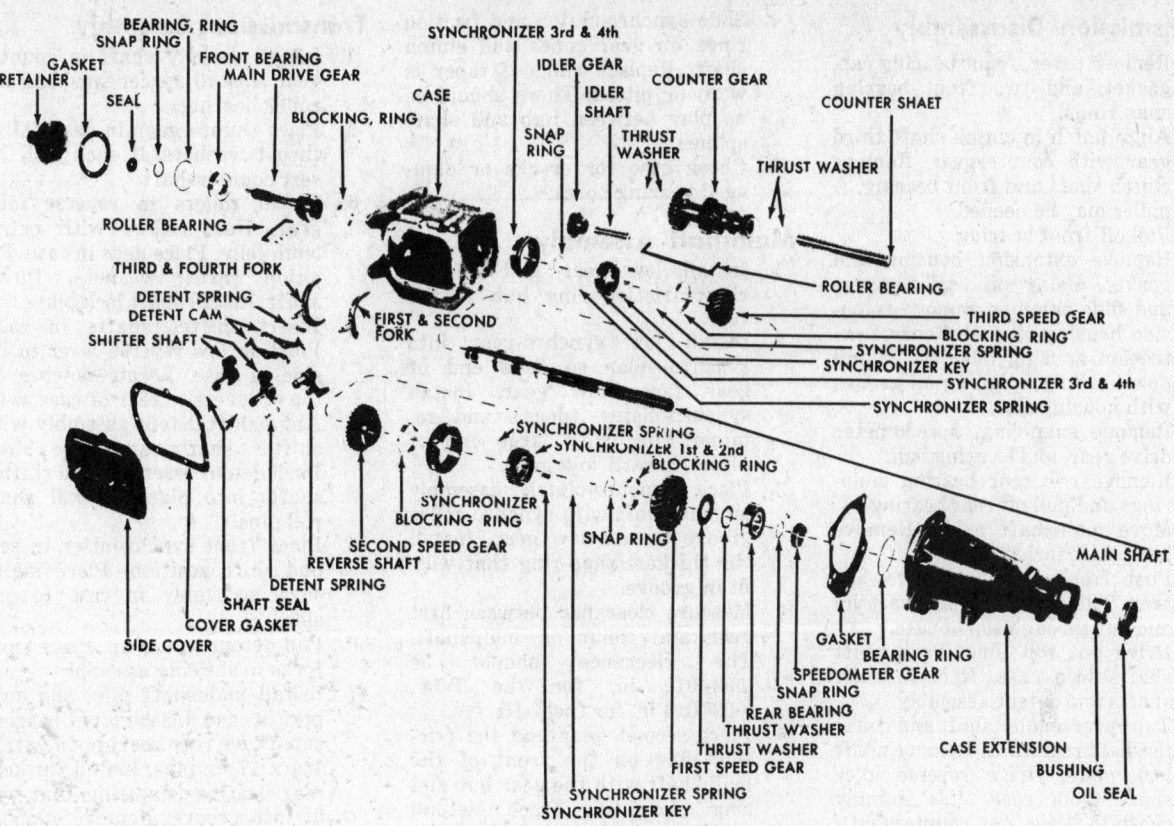

Saginaw 4-speed transmission

Type-20
Warner T-14, T-15 Fully
Synchronized 3-Speed

T-14 Application
American Motors (232), 1968-70

T-15 Application
American Motors (232, 290, 304, 360), 1968-70

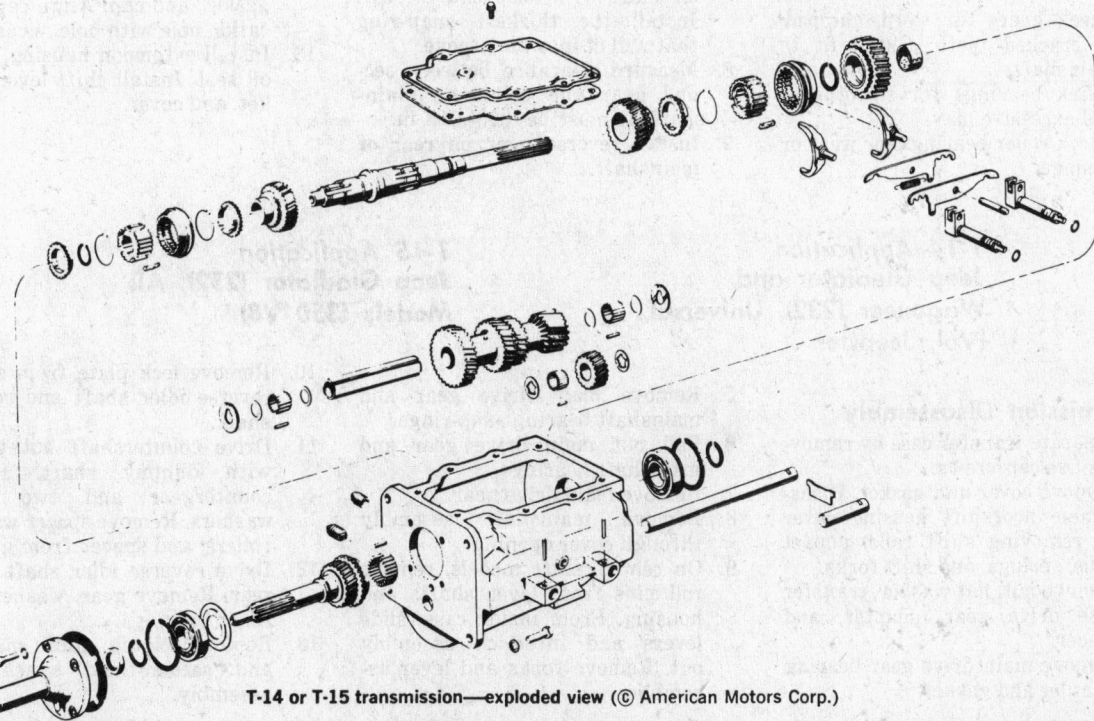

T-14 or T-15 transmission—exploded view (© American Motors Corp.)

Transmission Disassembly

1. Remove cover, front bearing cap, gasket, and two front bearing snap rings.
2. Align notch in clutch shaft third gear with countergear. Remove clutch shaft and front bearing. A puller may be needed.
3. Pull off front bearing.
4. Remove extension housing and gasket. Using oil seal remover and slide hammer, remove extension housing oil seal. Remove extension housing bushing. Install new bushing, aligning oil groove with housing slot.
5. Remove snap-ring, speedometer drive gear, and locating ball.
6. Remove two rear bearing snap-rings and pull off rear bearing.
7. Move mainshaft aside. Remove both shift forks.
8. Push front synchronizer toward rear. Tilt front of mainshaft up and out through top of case.
9. Drive out roll pins. Push shift shafts into case. Remove shift shafts and detent assembly.
10. Tap reverse idler shaft and countershaft rearward. Remove shaft lock plate. Drive reverse idler shaft from case. Use dummy shaft to drive out countershaft.

Mainshaft Disassembly

1. From front of shaft, remove front snap-ring, second-third synchro-clutch assembly, and second gear.
2. From rear of shaft, remove reverse gear, rear snap-ring, rear synchro-clutch assembly, and low gear.

Inspection

1. Check gears for worn, chipped, or cracked teeth. Check fit to mainshaft.
2. Check bearings for smoothness and excessive play.
3. Check roller bearings for wear or damage.

4. Slide synchro-clutch and friction rings on gear cones and clutch shaft. Replace rings if taper is worn or pitted. There should be no play between hub and shaft splines.
5. Check case for cracks or damaged bearing bores.

Mainshaft Assembly

1. Install low gear and friction ring; friction ring hub to the rear.
2. Install low synchro-gear into synchro-collar so deep end of gear faces low gear. Install synchro-plates (dogs) and retainer ring with large end of plates toward low gear.
3. Place synchro-clutch assembly on mainshaft with synchro-collar groove toward low gear. Install the thickest snap-ring that will fit in groove.
4. Measure clearance between first gear and collar on mainshaft. The clearance should be .003-.012 in. for the T-14; .003-.014 in. for the T-15.
5. Place second gear and the friction ring on the front of the mainshaft with the gear hub and ring forward. Place second synchro-gear into synchro-collar with deep end of gear facing rear of shaft.
6. Hold synchro-clutch assembly with one synchro-plate, or dog, in 12 o'clock position. Install tang of retainer ring into the dog at 12 o'clock and install ring clockwise. On opposite side, start with the same dog and install ring clockwise.
7. Place second synchro assembly on shaft with deep end to rear. Install the thickest snap-ring that will fit into the groove.
8. Measure clearance between second gear and collar on mainshaft. It must be .003-.018 in.
9. Install reverse gear on rear of mainshaft.

Transmission Assembly

1. Install dummy shaft in counter gear. Install spacer washers and roller bearings.
2. Place countergear in case. Align thrust washers at each end. Insert countershaft.
3. Install rollers in reverse idler gear. Hold rollers with petroleum jelly. Place gear in case. Position thrust washers. Insert shaft. Install shaft lock plate.
4. Insert shifter shafts in case. Position low-reverse lever to inside of case. Locate notches on top of levers to rear of case stud. Align shift detent assembly with shifter shafts and case stud. Push detent assembly and shifter shafts into place. Install shaft roll pins.
5. Place front synchronizer in second shift position. Place mainshaft assembly in case to one side.
6. Pull detent levers up. Place shift forks in shifting assembly.
7. Install mainshaft pilot end support in case. Install front bearing cap. Drive rear bearing on with a 1¼ x 17 in. pipe. Install thickest rear bearing snap-ring that will fit into groove. Remove support and bearing cap.
8. Install locating ball, speedometer drive gear, and snap-ring.
9. Press front bearing onto clutch shaft.
10. Place rollers in clutch shaft. Hold with petroleum jelly.
11. Place friction ring on mainshaft. Slide clutch shaft into position from front.
12. Install thickest front bearing snap-ring that will fit in groove, gasket, and cap. Align cap lubrication hole with hole in case.
13. Install extension housing. Install oil seal. Install shift levers, gasket, and cover.

T-14 Application
Jeep Gladiator and Wagoneer (232), Universal (V6), Jeepster

Transmission Disassembly

1. Separate transfer case by removing five capscrews.
2. Remove cover and gasket. Disassemble floorshift housing cover by removing shift rails, poppet balls, springs, and shift forks.
3. Remove nut, flat washer, transfer case drive gear, adapter, and spacer.
4. Remove main drive gear bearing retainer and gasket.

T-15 Application
Jeep Gladiator (232), All Models (350 V8)

5. Remove main drive gear and mainshaft bearing snap-rings.
6. Pull out main drive gear and mainshaft bearings.
7. Remove main drive gear.
8. Remove mainshaft assembly through cover opening.
9. On remote shift models, remove roll pins from lever shafts and housing. From inside case, slide levers and interlock assembly out. Remove forks and lever assemblies.

10. Remove lock plate from slots in reverse idler shaft and countershaft.
11. Drive countershaft out to rear with dummy shaft. Remove countergear and two thrust washers. Remove spacer washers, rollers, and spacer from gear.
12. Drive reverse idler shaft out to rear. Remove gear, washers, and roller bearings.
13. Remove clutch hub snap-ring and second-third synchronizer assembly.

1 Retainer screws
2 Main drive gear bearing retainer
3 Retainer gasket
4 Oil seal
5 Snap ring (small)
6 Snap ring (large)
7 Main drive gear bearing
8 Oil Retaining washer (slinger)
9 Main drive gear
10 Mainshaft pilot bearing rollers
11 Case
12 Nut
13 Flatwasher
14 Spacer

29 Countershaft front thrust washer (large)
30 Countershaft gear
31 Reverse idler gear bearing washer

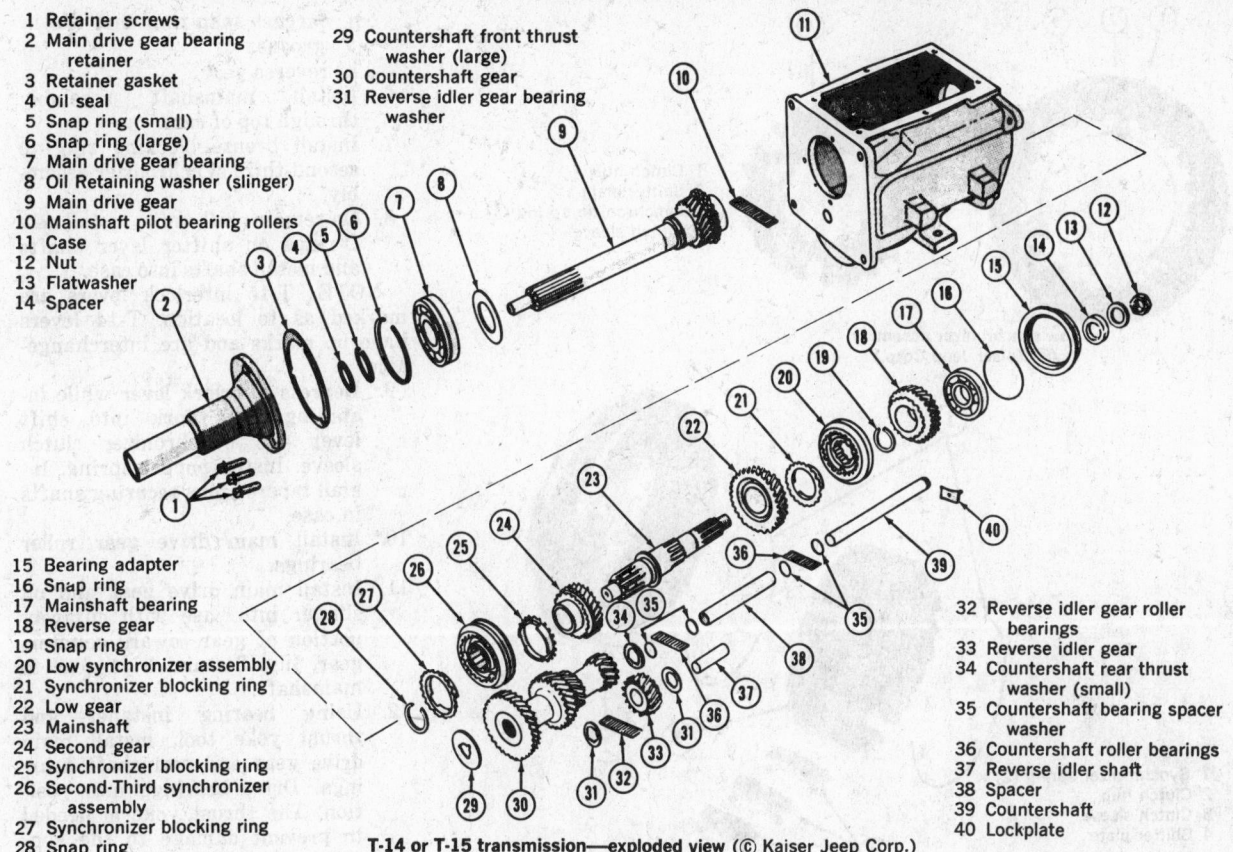

15 Bearing adapter
16 Snap ring
17 Mainshaft bearing
18 Reverse gear
19 Snap ring
20 Low synchronizer assembly
21 Synchronizer blocking ring
22 Low gear
23 Mainshaft
24 Second gear
25 Synchronizer blocking ring
26 Second-Third synchronizer assembly
27 Synchronizer blocking ring
28 Snap ring

32 Reverse idler gear roller bearings
33 Reverse idler gear
34 Countershaft rear thrust washer (small)
35 Countershaft bearing spacer washer
36 Countershaft roller bearings
37 Reverse idler shaft
38 Spacer
39 Countershaft
40 Lockplate

T-14 or T-15 transmission—exploded view (© Kaiser Jeep Corp.)

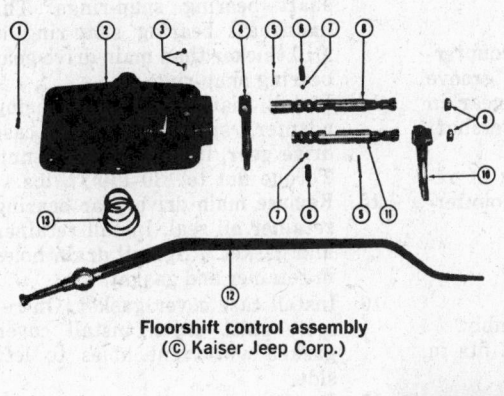

1 Control lever housing pin
2 Control housing
3 Interlock plunger and plug
4 Second-third shift fork
5 Shift fork pin
6 Poppet spring
7 Poppet ball
8 Second-third shift rail
9 Shift rail caps
10 Low-reverse shift fork
11 Low-reverse shift rail
12 Shift lever
13 Shift lever support spring

Floorshift control assembly
(© Kaiser Jeep Corp.)

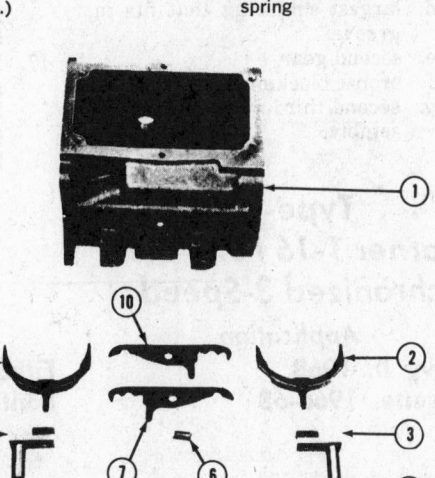

1 Case
2 Low-reverse shift fork
3 Low-reverse shift lever shaft
4 Tapered Pin
5 O-Ring
6 Poppet spring
7 Second-third interlock lever
8 Second third shift lever shaft
9 Second-third shift fork
10 Low-reverse interlock lever

Remote shift interlock assembly
(© Kaiser Jeep Corp.)

14. Remove second and reverse gears.
15. Remove clutch hub snap-ring and low synchronizer assembly.
16. Remove low gear.

Synchronizer Disassembly and Assembly

1. Remove springs. Low synchronizer has only one spring; second-third, two.
2. Mark sleeve and hub before separating.
3. Remove hub.
4. Remove three shifter plates from hub.
5. Inspect all parts for wear.
6. Assemble in reverse order of disassembly. On second-third unit, make sure that spring openings are 120 degrees from each other, with spring tension opposed.

NOTE: if a synchronizer assembly is replaced on a floor shift unit, the shift fork operating the synchronizer being replaced must have the letter A just under the shaft hole on the side opposite the pin.

Inspection

1. Wash all parts in solvent.
2. Air dry.
3. Check case bearing and shaft bores. Check for cracks or burrs.
4. Check all gears and bronze blocking rings for cracks, and chipped, worn, or cracked teeth.

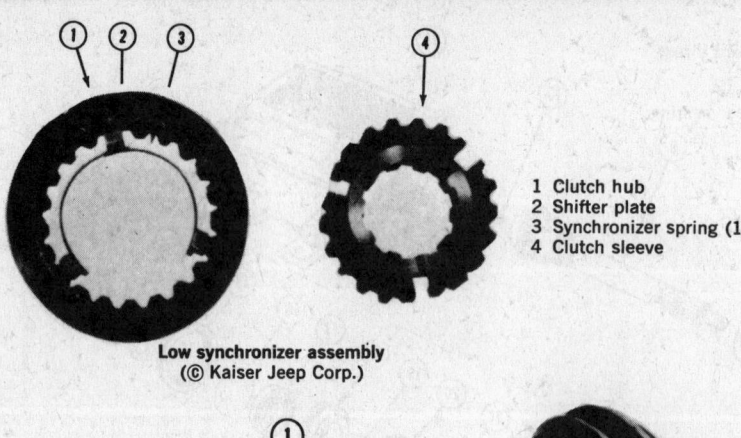

1 Clutch hub
2 Shifter plate
3 Synchronizer spring (1)
4 Clutch sleeve

Low synchronizer assembly
(© Kaiser Jeep Corp.)

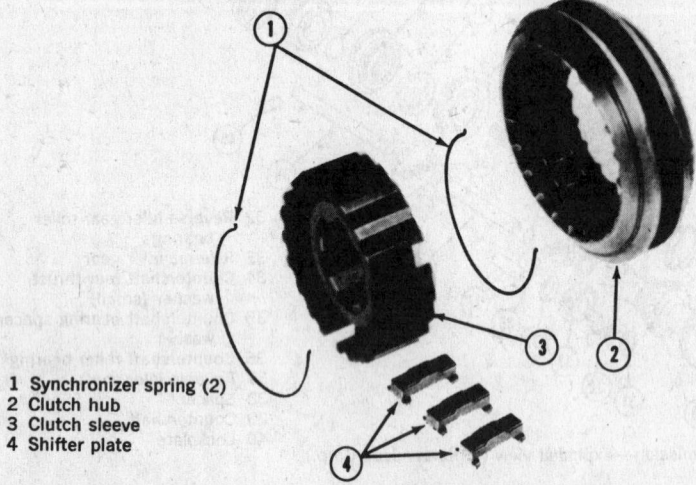

1 Synchronizer spring (2)
2 Clutch hub
3 Clutch sleeve
4 Shifter plate

Second-third synchronizer assembly
(© Kaiser Jeep Corp.)

If any gears are replaced, also replace the meshing gears.

5. Check all bearings and bushings for wear or damage.
6. Check that synchronizer sleeves slide freely on clutch hubs.

Transmission Assembly

1. Place reverse idler gear with dummy shaft, roller bearing, and thrust washers in case. Install reverse idler shaft.
2. Assemble countershaft center spacer, four bearing spacers, and bearing rollers in countershaft gear.
3. Install large countergear thrust washer in front of case. Position small thrust washer on countergear hub with lip facing groove in case. Holding countergear in position, push in countershaft from rear.
4. Install lock plate in slots of reverse idler shaft and countershaft.
5. Install to mainshaft:
   a. low gear.
   b. bronze blocking ring.
   c. low synchronizer assembly.
   d. largest snap-ring that fits in groove.
   e. second gear.
   f. bronze blocking ring.
   g. second-third synchronizer assembly.
   h. largest snap-ring that fits in groove.
   i. reverse gear.
6. Install mainshaft assembly through top of case.
7. Install bronze blocking ring to second-third synchronizer assembly.
8. On remote shift units, install new O-rings on shifter lever shafts and install shafts into case.

NOTE: T-15 interlock levers are marked as to location. T-14 levers have no marks and are interchangeable.

9. Depress interlock lever while installing shift fork into shift lever and synchronizer clutch sleeve. Install poppet spring. Install tapered pins securing shafts in case.
10. Install main drive gear roller bearings.
11. Install main drive gear and oil slinger into case with cutaway portion of gear toward countergear. Install main drive gear to mainshaft.
12. Using bearing installer and thrust yoke tool, install main drive gear and mainshaft bearings. Drive bearings into position. The thrust yoke is needed to prevent damage to the synchronizer clutch.
13. Install main drive gear and mainshaft bearing snap-rings. The mainshaft bearing snap-ring is .010 thicker than main drive gear bearing snap-ring.
14. Install mainshaft rear bearing adapter, spacer, transfer case drive gear, flat washer, and nut. Torque nut to 130-170 ft. lbs.
15. Replace main drive gear bearing retainer oil seal. Install retainer and gasket. Align oil drain holes in retainer and gasket.
16. Install case cover gasket. On remote shift units, install cover gasket with vent holes to left side.
17. Position gear train and floorshift assembly in neutral. Insert shifter forks into clutch sleeves. Install and torque bolts to 8-15 ft. lbs.

Type-21
Warner T-16 Fully Synchronized 3-Speed

Application

Camaro, 1967-68
Chevelle, 1966 (396), 1967-68
Chevrolet, 1965-67 (396-427), 1968

Chevy II, 1968
Corvette, 1966-68

Firebird, 1967 (V8)
Pontiac, 1964 (HD)

Disassembly of Unit

1. Clean outside of transmission to keep foreign matter from entering the case and to improve working conditions.
2. Remove side cover assembly and shift forks.
3. Remove extension housing attaching screws. Rotate extension clockwise to expose reverse idler gear shaft.

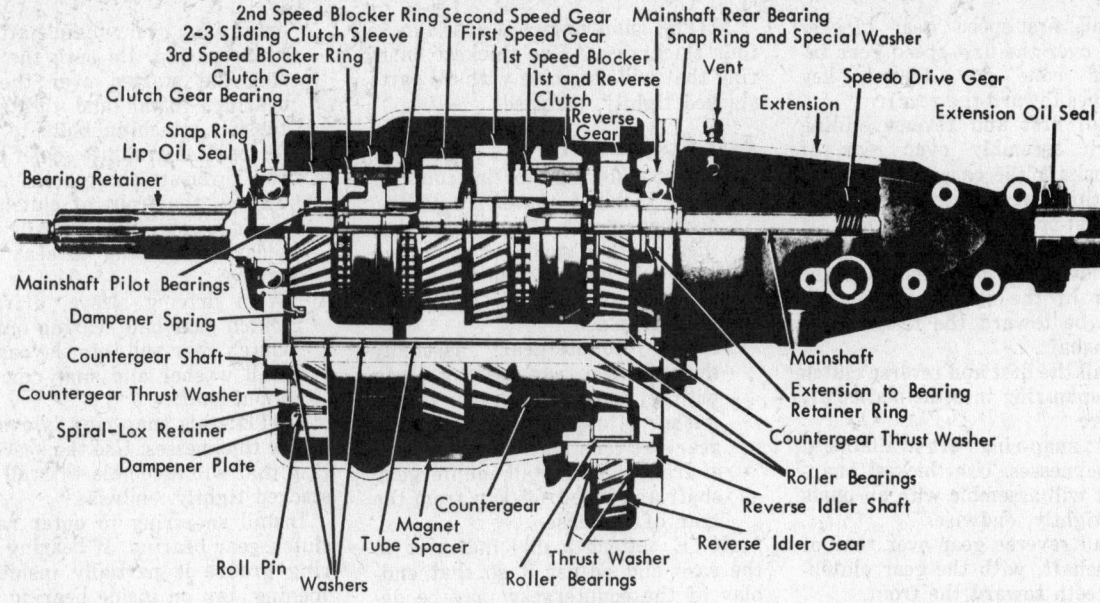

Warner T-16 transmission

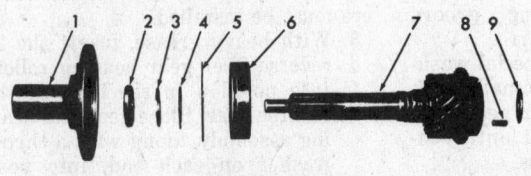

Mainshaft assembly
(© G.M. Corp.)

1 Retainer
2 Lip seal
3 Snap ring
4 Special washer
5 Snap ring
6 Clutch gear bearing
7 Clutch gear
8 Mainshaft pilot bearings
9 Bearing spacer

4. With a long drift through the side cover opening, drive the reverse idler shaft and Woodruff key through the rear of the case.
5. Rotate the extension counterclockwise to expose countergear shaft and, with a brass drift, drive shaft and Woodruff key out the rear of case.
6. With the countergear dropped to the bottom of case, remove the mainshaft and extension assembly through the rear of the case, remove the mainshaft pilot roller bearings from the clutch gear.
7. Remove snap-ring which retains mainshaft rear bearing and remove the extension from the rear bearing and mainshaft by tapping on end.
8. Remove the clutch gear bearing retainer and clutch gear bearing snap-ring and washer from the mainshaft.
9. Drive the clutch gear through its bearing into the case and remove the bearing by tapping from inside the case.
10. Remove countergear and roller bearings, both countergear thrust

washers, reverse idler gear and 25 roller bearings, and both idler gear thrust washers from the case.

Mainshaft Disassembly

1. With snap-ring pliers, remove second and third-speed sliding clutch hub snap-ring from mainshaft. Then remove clutch assembly, second-speed blocker ring and second-speed gear from front of mainshaft.
2. Remove rear bearing snap-ring from mainshaft groove.
3. With reverse gear supported on press plates, apply pressure to the rear of the mainshaft to remove reverse gear, rear bearing, special washer, snap-ring and the speed-

ometer drive gear from rear of mainshaft.
4. Remove first and reverse sliding clutch hub snap-ring from the mainshaft and remove clutch assembly, first-speed blocker ring and first-speed gear from rear of mainshaft.

Cleaning and Inspection

Clean and inspect all parts for wear or other damage. Replace if necessary.

The countergear anti-rattle plate is spring-loaded. If weak or broken, remove plate and/or spring. Replace as necessary.

Synchronizer clutch hubs and sliding sleeves are selected assemblies and should be kept together as in the original assembly. The keys and springs, however, may be replaced if worn or broken.

If bushing in rear of extension housing needs replacement, remove the seal. Then, drive bushing into extension housing. Drive new bushing into the case from the rear. Lubricate inside diameter of bushing and seal, then install new oil seal with driver.

If the lip seal in the clutch bearing retainer needs replacing, pry out the old seal. Then, drive a new seal into place until seal bottoms in its bore.

Mainshaft Assembly

1. Install first-speed gear onto rear of mainshaft, with gear clutching teeth to the rear.

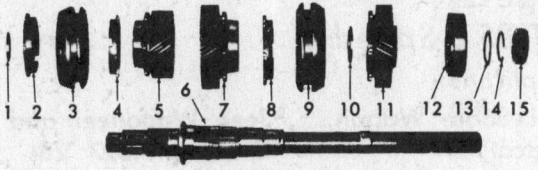

Clutch gear components
(© G.M. Corp.)

1 Snap ring
2 3rd speed blocker ring
3 2-3 clutch assembly
4 2nd speed blocker ring
5 2nd speed gear
6 Mainshaft
7 1st speed gear
8 1st speed blocker ring
9 1st reverse clutch assembly
10 Snap ring
11 Reverse gear
12 Rear bearing
13 Special washer
14 Snap ring
15 Speedo drive gear

2. Install first-speed gear blocker ring over the first-speed gear tapered cone end (clutch key notches toward the rear).
3. Install first and reverse sliding clutch assembly over rear of mainshaft (be careful to engage the three keys with the notches of first-speed blocker ring). If properly installed, the straightest side of the clutch hub and the taper of the sliding sleeve will both be toward the rear of the mainshaft.
4. Install the first and reverse clutch hub snap-ring into the mainshaft groove.

NOTE: snap-rings are available in three thicknesses. Use thickest snap-ring that will assemble with all parts stacked tightly, endwise.

5. Install reverse gear over rear of mainshaft, with the gear clutching teeth toward the front.
6. Press mainshaft rear bearing over rear of mainshaft, with its outer race snap-ring groove closest to reverse gear.
7. Install rear bearing special washer and snap-ring onto mainshaft.
8. Press the speedometer gear onto the rear of mainshaft until centered on the shaft boss.
9. Install the second-speed gear over the front of the mainshaft, with the gear clutching teeth toward the front.
10. Install the second-speed gear blocker ring over the second-speed gear tapered cone end (clutch key notches toward the front).
11. Install the second and third sliding clutch assembly over the front of the mainshaft, engaging the clutch keys with the notches of the second-speed blocker ring. If properly installed, the straightest side of the clutch hub should be toward the rear. The clutch sliding sleeve taper should be toward the front of the mainshaft.
12. Install the second and third-speed clutch hub retainer snap-ring onto the mainshaft.

NOTE: snap-rings are available in four thicknesses. Use thickest snap-ring that will assemble with all parts stacked tightly, endwise.

Assembly of Unit

1. Insert tube spacer in countergear. Install a spacer, 20 rollers, another spacer, 20 more rollers, then another spacer at each end of countergear. Use heavy grease and the dummy shaft to hold them in place.
2. Insert countergear assembly through case rear opening along with a tanged countergear thrust washer (tang away from the gear) at each end (large washer at front) and install countergear shaft and Woodruff key from the rear of the case.

NOTE: attach a dial indicator to the case and adjust it so that end-play of the countergear may be determined. If end-play greater than 0.025 in. is shown, new thrust washers must be installed.

3. With heavy grease, insert the 25 reverse idler gear bearing rollers into position in the bore of the reverse gear. Place gear and bearing assembly, along with a thrust washer on each end, into position inside the case. The beveled edge of the gear teeth face toward the front of the case.
4. Load the pilot bearing rollers into clutch gear, using grease to hold them in place, and position gear in case. Do not install clutch gear ball bearing at this time.
5. Stand case on end with clutch gear down through hole in bench. Place third-speed blocker ring over clutch gear.
6. Install mainshaft assembly from the rear of the case, picking up the spacers, pilot bearing and third-speed blocker rings.
7. Install reverse idler shaft and Woodruff key.
8. Install extension housing gasket to rear of case. Then, using snap-ring pliers, expand the extension-to-bearing snap-ring and install

extension over mainshaft and rear bearing. Be sure the snap-ring has started over the rear bearing. Install and tighten extension attaching bolts to 35-45 ft. lbs. Use graphite sealer on the two lower attaching bolts.
9. Tap on the front of clutch gear shaft to force rear bearing-to-extension snap-ring to seat in its groove.
10. With driving sleeve, drive the clutch gear ball bearing onto the clutch gear and into the case. Install washer and snap-ring onto input shaft.

NOTE: this snap-ring is available in five thicknesses. Use thickest snap-ring that will assemble with all parts stacked tightly, endwise.

Install snap-ring to outer race of clutch gear bearing. If bearing snap-ring groove is partially inside case opening, tap on inside bearing outer race with a long drift through side cover opening.

NOTE: if mainshaft does not turn freely, check clutch sliding sleeves for neutral positions and that the blocker rings are free on their gear cone surfaces.

11. Install gaskets, clutch gear bearing retainer and lip seal assembly with oil drain passages at bottom. Then, tighten attaching screws to 15-20 ft. lbs. Use graphite sealer on threads of retainer bolts.

NOTE: install two retainer-to-case gaskets (.010 and .015 in. thick) instead of the one .025 in. production gasket removed.

12. Place the shift forks in the clutch sleeve grooves with the first and reverse fork hump toward the bottom.
13. Install side cover and gasket. Torque retaining bolts to 15-20 ft. lbs.

NOTE: the two side cover-to-case retaining bolts have special oil sealing splines and must be used at these two through locations.

Type-22
Warner T-85 3-Speed
Application

American Motors, 1964, 1966 (327 V8)

Disassembly

1. With transmission drained and mounted in an adequate stand, remove the cap screws, washers, and side cover assembly of transmission.
2. Remove four cap screws and washers holding the mainshaft rear bearing retainer extension to

Buick, 1964 LeSabre Wagon, 1964-65 Wildcat

case and move the extension away from the case about 1/2 in. Then rotate retainer to expose countershaft end and lock key.
3. From front of transmission, drive countershaft to rear, using countershaft dummy.
4. Work countershaft all the way out of cluster, end spacers, rollers and roller bearing spacer.

Jeep Wagoneer and Gladiator (327 V8)

5. Leave dummy shaft in place in the countershaft gear cluster to keep the bearings and spacers in location.
6. Lower countergear cluster down into bottom of case, then remove rear bearing retainer extension, gasket, and mainshaft assembly from transmission case.
7. Remove mainshaft front bearing

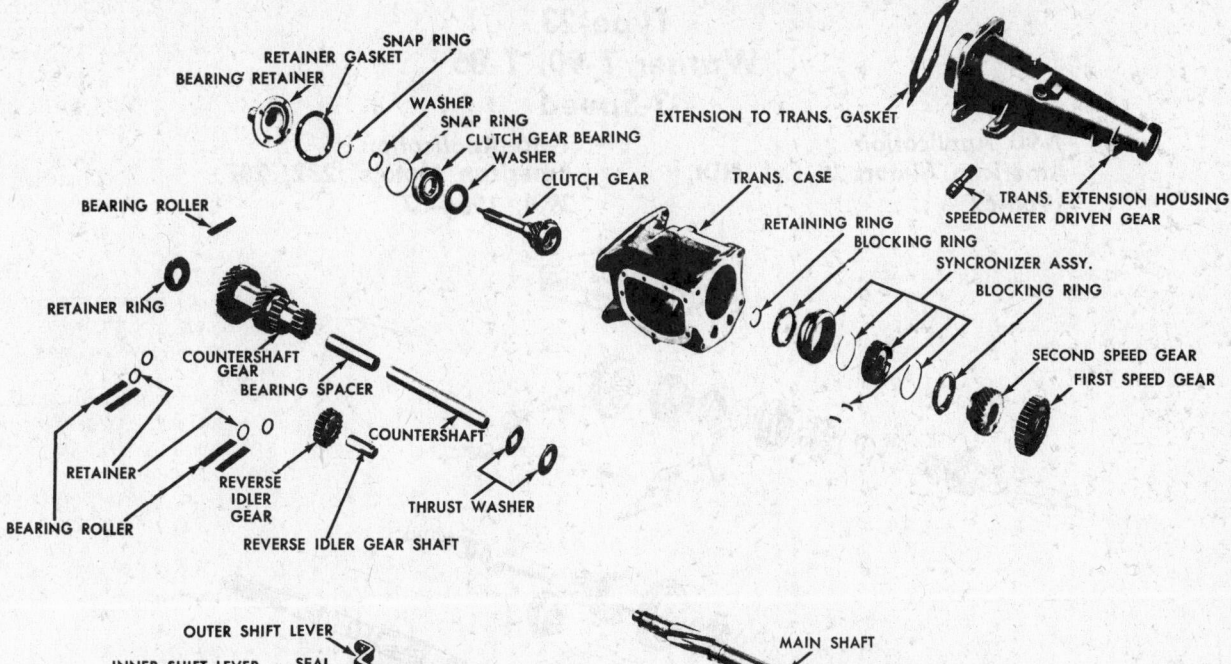

Warner T-85 transmission (© Buick Div., G.M. Corp.)

rollers from inside the main drive gear.

8. Remove bearing spacing washer from front end of mainshaft.

9. Remove attaching cap screws and washers, then the main drive gear bearing retainer from front of case.

10. Remove main drive gear bearing snap ring from front side of main drive gear bearing.

11. With transmission case up-ended (front up) press main drive gear out of bearing.

12. Remove oil retainer from main drive gear.

13. Tap main drive gear bearing out through front of case and remove main drive gear bearing snap ring from shaft.

14. Using brass drift, drive reverse idler shaft to rear of case to clear lock key.

15. Remove lock key and, from rear of case, drive idler gear shaft into

case, then remove idler gear and shaft from case.

16. Lift out countergear, dummy shaft, and washers from case.

17. Remove synchronizing ring from front side of second and third-speed clutch. Remove clutch hub retaining snap-ring from front end of mainshaft.

18. Remove second- and third-speed clutch sleeve from clutch hub, then remove clutch hub from mainshaft.

19. Remove two clutch key springs and three clutch keys from clutch hub.

20. Remove rear synchronizer ring and second-speed gear from the mainshaft.

21. Remove first and reverse sliding gear from the mainshaft.

22. Remove speedometer driven gear lock plate to extension bolt and lock washer, then remove lock plate.

23. Insert screwdriver in lock plate slot in fitting and pry fitting, gear and shaft from extension.

24. Remove mainshaft rear bearing front snap-ring from rear bearing retainer extension and tap mainshaft and rear bearing out of retainer extension.

25. Remove snap-ring from rear of speedometer drive gear.

26. Remove speedometer drive gear, detent ball and speedometer gear front snap-ring.

27. Remove mainshaft rear bearing rear snap-ring.

28. Using a press, press mainshaft rear bearing toward rear of shaft until loose and remove.

29. Remove seal from rear bearing retainer.

Assembly

1. Assemble by reversing the above procedure.

Type-23
Warner T-90, T-86
3-Speed

T-90 Application
American Motors (6 Cyl. HD),
1966-67

T-86 Application
American Motors (232, 287, 290), 1964-67

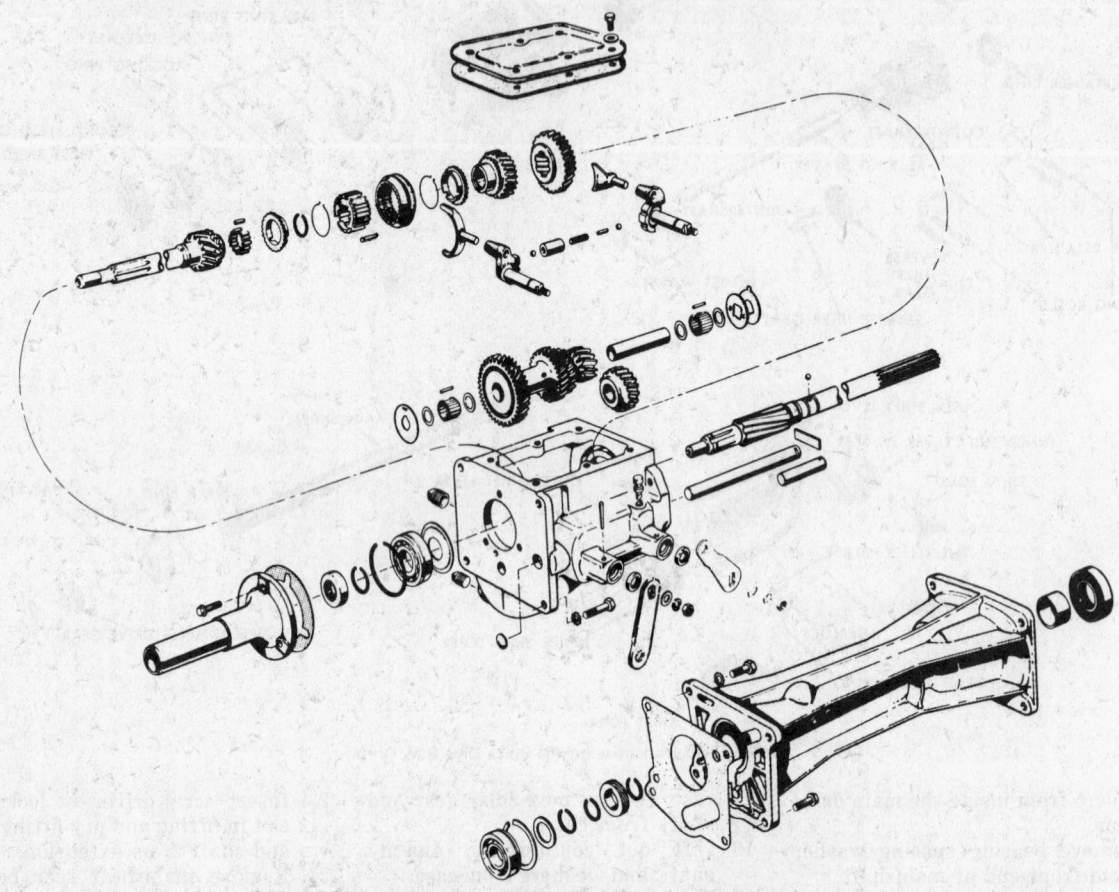

T-90 or T-86 transmission—exploded view (© American Motors Corp.)

Disassembly
1. Remove cover.
2. Remove front bearing cap, clutch shaft snap-ring and bearing lock-ring.
3. Remove the front bearing, using a bearing puller and a thrust yoke to prevent damaging synchronizer clutches.
4. Remove extension housing. Drive out seal from inside housing with oil seal remover and installer tool. Use bushing remover and installer tool to replace bushing.
5. Move mainshaft assembly back about ¾ in. Lower front end of clutch shaft, move mainshaft assembly over countergear and out of shift forks. Remove clutch shaft from front of case.
6. Check roller bearings inside rear of clutch shaft for wear or damage.
7. Remove snap-ring, speedometer drive gear, and key.
8. Remove snap-ring, synchro-clutch assembly, second gear, friction ring, and low-reverse sliding gear. Press off rear bearing.
9. Remove shifter forks.
10. Remove shaft lock plate from rear of case.
11. Drive countershaft out to rear, using dummy shaft. Lower countergear to bottom of case.
12. Drive out reverse idler shaft. Remove reverse idler gear and countergear.
13. Remove outer shift levers and shifter shaft lock pins. Remove shifter shafts and two interlock ball bearings. Remove interlock sleeve and spring. Remove shaft oil seals from case.

Inspection
1. Wash all parts in solvent.
2. Air dry.
3. Check gears for worn, cracked, or chipped teeth. Check fit to shaft. Gears should fit smoothly without excessive play between splines.
4. Check bearings for cracked races, and worn or scored balls.
5. Slide synchronizer rings on cones of second gear and clutch shaft. Replace rings if there is excessive wear or pits on the taper.
6. Check case bearing recesses for wear or scoring. Check case for cracks.

Assembly
1. Install new shift shaft oil seals.
2. Install low-reverse shift shaft, interlock sleeve, ball bearing, pin, and spring. Install second-high shift shaft and second ball bearing.
3. Shift mechanism into any gear position. With one end of interlock sleeve against shift shaft quadrant, clearance between opposite end of sleeve and quadrant on other shaft should be .001-.007 in. Interlock sleeves

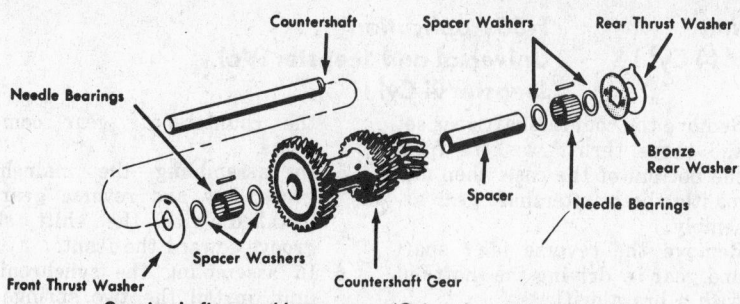

Countershaft assembly sequence (© American Motors Corp.)

are available in several sizes for adjustment.

4. Install lock pins and shift levers.
5. Install dummy shaft and bearings in countergear. Install thrust washers; the bronze front washer must index with the case.
6. Place countergear and dummy shaft in bottom of case.
7. Install reverse idler gear with chamfered side of teeth to front of case. Drive in shaft from rear.
8. Align slots in countershaft and reverse idler shaft. Position countergear and drive in countershaft. Install lock plate.
9. Install shifter forks.

10. Press rear bearing onto mainshaft. Install key, speedometer drive gear, and snap-ring.
11. Install low-reverse sliding gear on mainshaft with sliding collar to front. Gear should slide easily. Install second gear with tapered cone to front.
12. Install rear bearing snap-ring. Install the thickest mainshaft rear snap-ring which will fit into groove.
13. Install synchro-clutch assembly and thickest mainshaft front snap-ring that will fit in groove.
14. When synchro-clutch hub is pressed against snap-ring, there should be .003-.010 in. clearance

between second gear and shoulder on mainshaft.
15. Install 14 rollers in clutch shaft. Retain with light grease.
16. Install front friction ring on clutch shaft. Insert clutch shaft through top of case. Install mainshaft assembly through rear of case, moving to right to engage shifter forks in synchro-clutch collar and low-reverse sliding gear. Guide clutch shaft onto main shaft.
17. Install a new oil seal in extension housing.
18. Install extension housing.
19. Install oil slinger, concave side to rear. Drive in front bearing using thrust yoke to prevent synchronizer damage.
20. Install thickest clutch shaft snap-ring that will fit in groove.
21. Install front bearing cap with new gasket. Choose thickness of gasket to give zero clutch shaft end play.
22. Check clearance of friction rings. The clearance should be .056-.145 in.
23. Install cover and gasket.

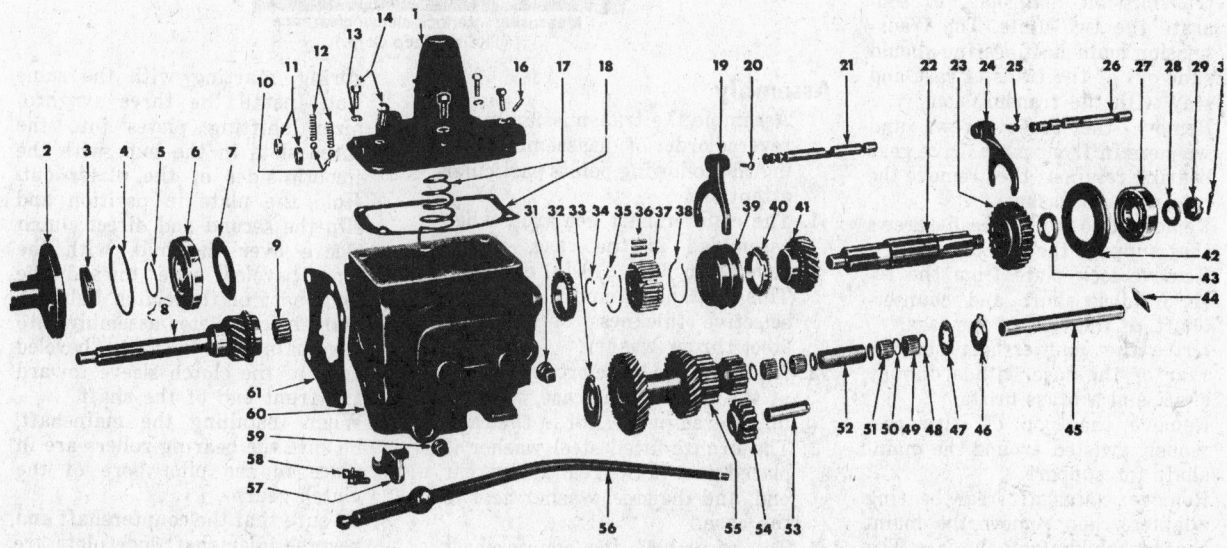

T-90, T-86 transmission, Jeep

1 Bearing retainer bolt
2 Bearing retainer
3 Bearing retainer oil seal
4 Bearing snap-ring
5 Main drive gear snap-ring
6 Main drive gear bearing
7 Front bearing washer
8 Main drive gear
9 Pilot roller bearing
10 Poppet ball
11 Shift rail cap
12 Poppet spring
13 Lockwasher
14 Shift housing bolt
15 Shift housing
16 Interlock plunger
17 Shift lever spring

18 Shift tower gasket
19 High and intermediate shift fork
20 Shift fork pin
21 High and intermediate shift rail
22 Main shaft
23 Sliding gear
24 Low and reverse shift fork
25 Shift fork pin
26 Low and reverse shift rail
27 Rear bearing
28 Main shaft washer
29 Main shaft nut
30 Cotter pin
31 Filler plug
32 Blocking ring

33 Front countershaft thrust washer
34 Clutch hub snap-ring
35 Synchronizer spring
36 Synchronizer plate
37 Clutch hub
38 Synchronizer spring
39 Clutch sleeve
40 Blocking ring
41 Second speed gear
42 Rear bearing adapter
43 Bearing spacer
44 Lock plate
45 Countershaft
46 Rear countershaft thrust washer

47 Rear countershaft thrust washer
48 Countershaft bearing washer
49 Countershaft bearing rollers
50 Countershaft bearing washer
51 Countershaft bearing rollers
52 Countershaft bearing spacer
53 Reverse gear shaft
54 Reverse idler gear
55 Countershaft gear set
56 Shift lever
57 Oil collector
58 Oil collector screw
59 Drain plug
60 Transmission case
61 Bearing retainer gasket

T-90 Application
Jeep Universal (4 Cyl.)

T-86 Application
**Universal and Jeepster (V6),
Jeepster (4 Cyl.)**

Disassembly

1. Drain lubricant and flush out the case.
2. If a transfer case is involved, remove its rear cover.
3. If a power take-off is involved, remove the shift unit which replaces the cover.
4. Remove cotter pin, nut and washer and remove the transfer case main drive gear.
5. Remove the transmission shift cover.
6. Loop a piece of wire around the mainshaft just back of second-speed gear. Twist the wire and attach one end to the right front cover screw, the other end to the left cover screw. Tighten the wire to prevent the mainshaft from pulling out of the case when the transfer case is removed. Should the mainshaft come out, the synchronizer parts will drop into the bottom of the case.
7. Remove five screws holding the transfer case to the rear face of the transmission.
8. Support the transfer case, then tap lightly on the end of the transmission mainshaft to separate the two units. The transmission mainshaft bearing should slide out of the transfer case and stay with the transmission.
9. Remove the three screws and washers in front main drive gear bearing retainer, then remove the retainer and gasket.
10. Remove two hollow-head screws that support the oil collector.
11. Remove lock plate from the reverse idler shaft and countershaft, at the rear of the case.
12. Drive the countershaft out the rear of the case with a dummy shaft and a brass drift.
13. Remove the loop of wire previously twisted around the mainshaft for support.
14. Remove mainshaft rear bearing adapter, then remove the mainshaft assembly from the case. The assembly may be removed through the case rear opening. Remove main drive gear.

15. Remove the countershaft gear set and three thrust washers from the bottom of the case, then dismantle the countershaft gear assembly.
16. Remove the reverse idler shaft and gear by driving the shaft out with a brass drift.
17. On column shift models, check clearance between ends of interlock sleeve and notched surface of each shift lever. The correct clearance is .001-.007 in. Several sizes of interlock sleeves are available for adjustment.

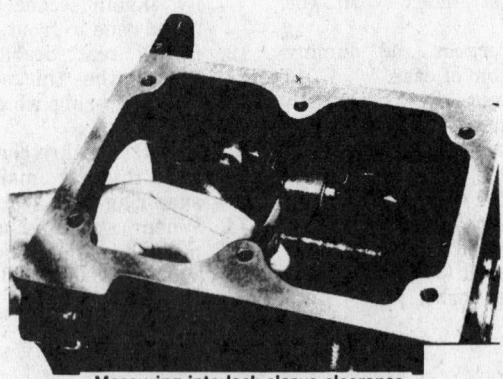
Measuring interlock sleeve clearance
(© Kaiser Jeep Corp.)

Assembly

Assemble the transmission in the reverse order of disassembly, giving the following points particular attention.

1. The countershaft gear set, when assembled in the case, should have from .012-.018 in. end-play. This clearance is controlled by the selective thickness of the rear steel thrust washer.
2. Assemble the large bronze washer at the front of the case with the lip entered in the slot in the case.
3. The bronze-faced steel washer is placed next to the gear at the rear end, and the steel washer next to the case.
4. To assemble the countershaft bearing rollers, use a dummy shaft. Use grease and a loading sleeve to facilitate reassembly of

the countershaft gear components.
5. In assembling the mainshaft gears, low and reverse gear is installed with the shift shoe groove toward the front.
6. In assembling the synchronizer unit, install the two springs in the high and intermediate clutch hub with spring tension opposed. Place the right lipped end of a spring in the slot of the hub and place the spring in the hub. Turn the hub around and make the same installation with the other spring, starting with the same slot. Install the three synchronizer shifting plates into the three slots in the hub, with smooth sides of the plates out. Hold the plate in position and slip the second and direct clutch sleeve over the hub, with the long beveled edge toward the long part of the clutch hub. Install the completed assembly onto the mainshaft with the beveled edge of the clutch sleeve toward the front end of the shaft.
7. When installing the mainshaft, be sure the bearing rollers are in place in the pilot bore of the clutch gear.
8. Be sure that the countershaft and reverse idler shaft lock plate are in position and completely recessed into the indents of the transfer case.

Type-24
Warner T-96 3-Speed

Application
**American Motors, 1964-70
(199, 232 6 Cyl.)**

Disassembly

1. Remove top cover.
2. Remove front bearing cap, clutch shaft snap-ring, and bearing lock-ring.

3. Use a bearing puller and a thrust yoke to remove front bearing.
4. Remove oil slinger.

5. Remove extension housing. Replace rear bearing oil seal and extension housing bushing if necessary.

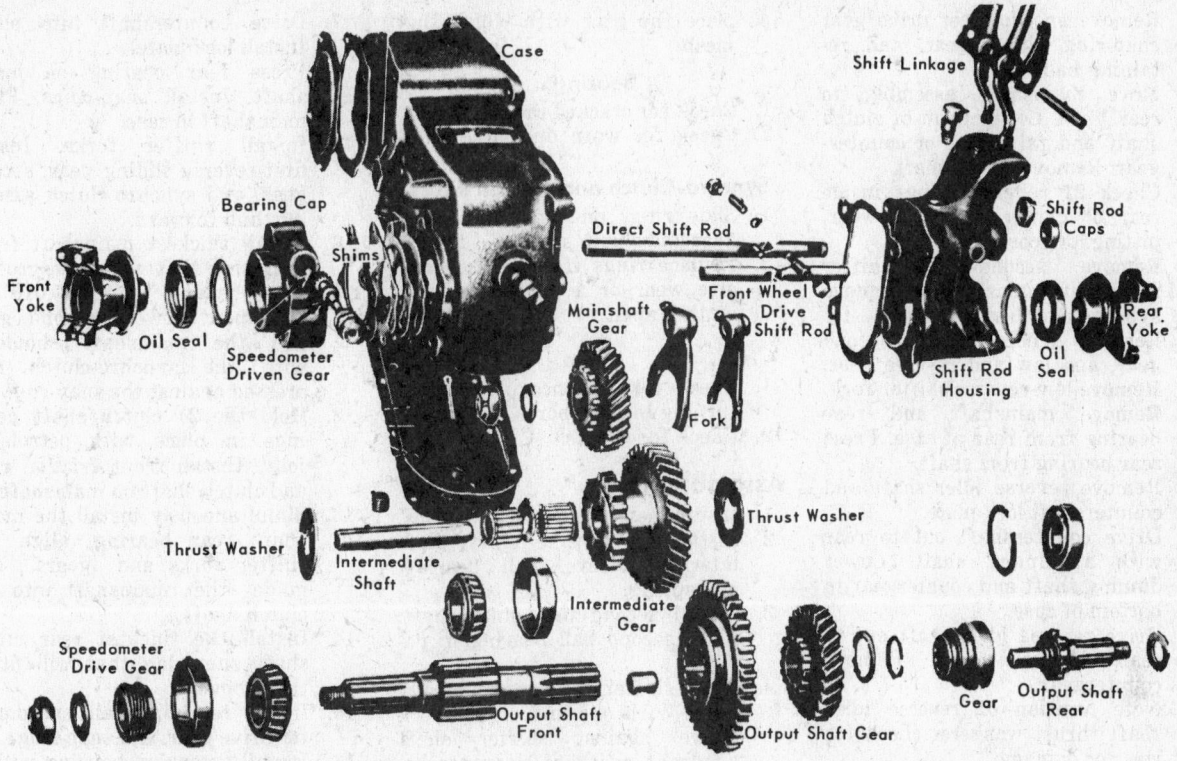

Case

Shift Linkage

Bearing Cap

Shims

Direct Shift Rod

Shift Rod Caps

Front Yoke

Oil Seal

Speedometer Driven Gear

Mainshaft Gear

Front Wheel Drive Shift Rod

Shift Rod Housing

Oil Seal

Rear Yoke

Thrust Washer

Fork

Thrust Washer

Intermediate Shaft

Intermediate Gear

Speedometer Drive Gear

Output Shaft Front

Output Shaft Gear

Gear

Output Shaft Rear

Transfer case

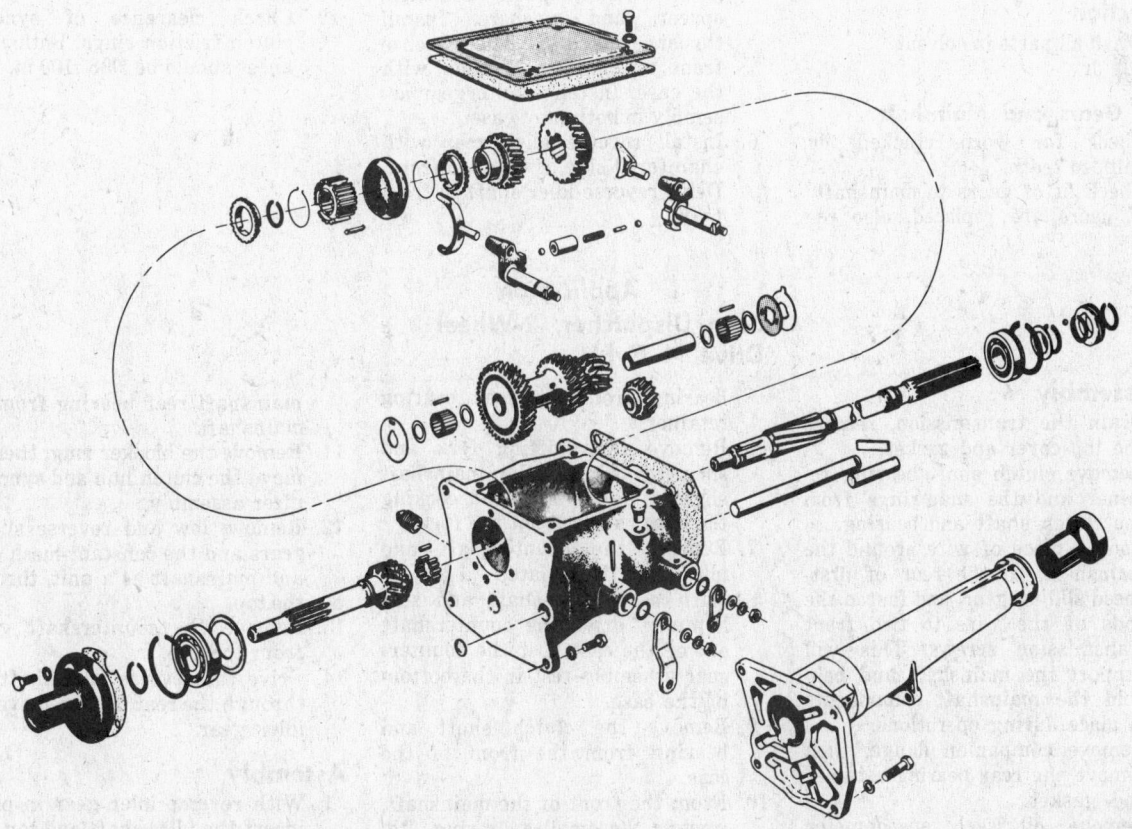

T-96 transmission—exploded view (© American Motors Corp.)

6. Remove speedometer drive gear snap-ring, drive gear, and retaining ball.
7. Move mainshaft assembly to rear ½ in. Lower front of clutch shaft and raise rear of countergear. Remove clutch shaft.
8. Check 21 roller bearings inside rear of clutch shaft for wear, pitting, or scoring.
9. Remove second-high shifter fork. Tilt mainshaft to remove synchro-clutch snap-ring.
10. Remove synchro-clutch. second gear, and low and reverse gear.
11. Remove low-reverse shifter fork.
12. Remove mainshaft and rear bearing from rear of case. Press rear bearing from shaft.
13. Remove reverse idler shaft and countershaft lock plate.
14. Drive countershaft out to rear with a dummy shaft. Lower dummy shaft and countergear to bottom of case.
15. Drive reverse idler shaft out to rear. Remove gear. Remove countergear.
16. Note position of reverse idler shaft thrust washers; check for wear or damage.
17. Remove outer shift levers and shifter shaft lock pin. Remove shifter shafts from inside case. Remove two interlock ball bearings. Remove interlock sleeve, pin, and spring. Remove shifter shaft O-rings.

Inspection
1. Wash all parts in solvent.
2. Air dry.

Gears and Mainshaft
1. Check for worn, cracked, or chipped teeth.
2. Check fit of gears to mainshaft. If gears are replaced, also re-

place the gear with which they mesh.

Bearings
1. Check for cracked races.
2. Check for worn or scored balls.

Synchro-Clutch and Friction Rings
1. Slide rings on cones of second gear and clutch shaft.
2. Replace rings if there is excessive wear or a pitted condition on the taper.

Case
1. Check for evidence of bearings turning in their bores.
2. Check for cracks.

Assembly
1. Install new shift shaft O-rings.
2. Install low-reverse shift shaft interlock sleeve, ball bearing, and spring.
3. Install second-high shift shaft. Place second ball bearing in position.
4. Place shifter mechanism in any gear. With one end of interlock sleeve against shifter shaft quadrant, measure clearance between opposite end of sleeve and the other quadrant. Clearance should be .001-.007 in. Selective lengths of interlock sleeves are available for adjustment. Install lock pins and shift levers.
5. Install dummy shaft in countergear. Install needle bearings, spacer, and washers. Install thrust washers. The bronze front washer must index with the case. Install countergear assembly in bottom of case.
6. Install reverse idler gear with chamfered side of teeth to front. Drive reverse idler shaft in from rear.

7. Drive countershaft into place. Install lock plate.
8. Press rear bearing on mainshaft. Install snap-rings. Place mainshaft in case.
9. Install shifter forks. Install first-reverse sliding gear, second gear, and synchro-clutch assembly, hub forward.
10. Install thickest mainshaft front snap-ring that will fit in groove.
11. There should be .003-.010 in. clearance between second gear and the mainshaft shoulder, with the synchro-clutch hub pressed against the snap-ring.
12. Hold the 21 clutch shaft bearings in place with petroleum jelly. Install front friction ring and clutch shaft on mainshaft.
13. Simultaneously install the mainshaft rear bearing, align the shifter forks and gears, and guide the mainshaft into the clutch shaft.
14. Install the thickest rear mainshaft snap-ring that will fit in the groove.
15. Install retaining ball, speedometer drive gear, and snap-ring.
16. Install extension housing with a new oil seal.
17. Place oil slinger on clutch shaft with concave side to rear. Install front bearing using thrust yoke. Install thickest snap-ring that will fit in the groove.
18. Install bearing cap and a new gasket.
19. Check clearance of synchro-clutch friction rings. Both clearances should be .036-.100 in.

Application
Jeep Dispatcher, 2-Wheel Drive (4 Cyl.)

Disassembly
1. Drain the transmission, remove the top cover and gasket.
2. Remove clutch shaft bearing retainer and the snap-rings from the clutch shaft and bearing.
3. Loop a piece of wire around the mainshaft, to the rear of first-speed sliding gear, and fasten the ends of the wire to two front transmission screws. This will support the mainshaft and help hold the mainshaft components in place during operations.
4. Remove companion flange, then remove the rear bearing retainer and gasket.
5. Remove oil seal, speedometer drive gear, and mainshaft rear

bearing from the rear bearing retainer.
6. Remove the shifting fork and shoe by sliding the mainshaft slightly to the rear and cocking the shaft away from the fork.
7. Remove the countershaft and idler shaft lock plate.
8. With a dummy shaft and soft hammer, drive the countershaft out of the case. Let the countergear assembly rest in the bottom of the case.
9. Remove the clutch shaft and bearing from the front of the case.
10. From the front of the mainshaft, remove the small snap-ring. Remove the large snap-ring and the

mainshaft rear bearing from the mainshaft.
11. Remove the blocker ring, then remove the clutch hub and synchronizer assembly.
12. Remove low and reverse sliding gears and the constant-mesh gear and mainshaft as a unit, through the top.
13. Remove the countershaft gears from the top.
14. Drive the reverse idler shaft out through the rear, then remove the idler gear.

Assembly
1. With reverse idler gear in place, insert the idler shaft and tap into place.

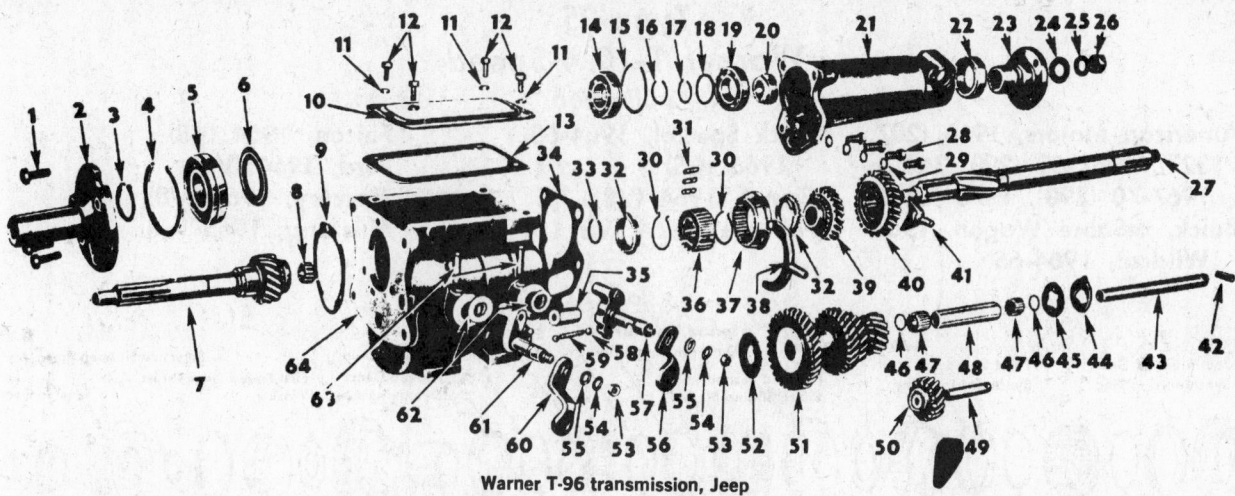

Warner T-96 transmission, Jeep

1 Main bearing retainer bolt	18 Rear bearing washer
2 Main bearing retainer	19 Rear main shaft bearing
3 Main drive gear snap-ring	20 Speedometer drive gear
4 Bearing snap-ring	21 Rear bearing retainer
5 Main drive gear bearing	22 Main shaft oil seal
6 Oil baffle	23 Coupling flange
7 Main drive gear	24 Main shaft washer
8 Pilot bearing rollers	25 Main shaft nut lockwasher
9 Bearing retainer gasket	26 Main shaft nut
10 Case cover	27 Main shaft
11 Case cover bolt gasket	28 Rear bearing retainer bolt
12 Case cover bolt	29 Retainer bolt lockwasher
13 Case cover gasket	30 Synchronizer spring
14 Rear main shaft bearing	31 Synchronizer shifting plate
15 Rear bearing snap-ring	32 Blocking ring
16 Main shaft snap-ring	33 Clutch hub snap-ring
17 Rear bearing snap-ring	34 Rear bearing retainer gasket

35 Interlock sleeve	49 Reverse idler gear shaft
36 Clutch hub	50 Reverse idler gear
37 Clutch sleeve	51 Countershaft gear
38 High and intermediate shift fork	52 Thrust washer
39 Second speed gear	53 Control lever to shaft nut
40 Low and reverse gear	54 Lever to shaft lockwasher
41 Low and reverse shift fork	55 Lever to shaft washer
42 Idler and countershaft lock plate	56 Low and reverse control lever
43 Countershaft	57 Low and reverse shift lever
44 Thrust washer	58 Poppet ball
45 Thrust washer	59 Poppet spring
46 Countershaft bearing shift spacer	60 High and intermediate control
47 Countershaft bearing rollers	61 High and intermediate shift lever
48 Countershaft bearing long spacer	62 Shift shaft oil seal
	63 Shift lever shaft pin
	64 Transmission case

2. Assemble the countershaft gear by inserting a dummy shaft and the tubular bearing spacer into the countergear. Grease both ends of the gear bore and insert 20 bearing rollers into each end of the gear. Add a roller bearing spacer washer to each end of the gear and place the assembly in the bottom of the case, large end of gear forward.

3. Install the mainshaft rear bearing onto the mainshaft with the closed side of the bearing toward the front end of the shaft. Replace the bearing snap-ring and mainshaft snap-ring.

4. Start the shaft through the rear of the case and assemble the low and reverse sliding gear on the shaft, fork groove toward the rear.

5. Assemble second-speed gear on the shaft, synchronizer teeth toward the front.

6. To assemble the synchronizer unit, install the two springs in the high and intermediate clutch hub. These springs must be installed with the spring tension opposed. Install the three synchronizer shifting plates into the three slots in the hub, with the smooth side of the plates out. With the plates in position, slip the second and third-speed clutch sleeve over the hub. Install the two blocking rings, one on each side of the hub.

7. Install shifting fork and shoe.

8. Grease the clutch shaft pilot bore and install 13 bearing rollers.

9. Install the clutch shaft main drive gear and bearing assembly, (closed side of bearing toward inside of case).

10. Enter the mainshaft bearing in the case and assemble the shaft in position.

11. Install the front bearing retainer.

12. Invert transmission, then line up the countergear bore with the countershaft bore in the case. Enter the countershaft into the case from the rear and entirely displace the dummy shaft while seating the countershaft in the case.

13. Install countershaft-idler-shaft lock plate.

14. Install rear bearing snap-ring and bearing onto the shaft.

15. Attach rear bearing retainer and gasket to the case.

16. Install speedometer drive gear onto the shaft with the shoulder of the gear front. Replace rear oil seal if necessary.

17. Install companion flange, then check operation of transmission.

18. Fill with lubricant to level, then install transmission case cover.

Type-25
Warner T-10 4-Speed
Application

American Motors, 1966 (287, 327), 1967-69 (290, 343), 1967-70 (390), 1970 (360)	**Buick Special, 1964-65, 1966 (GS)**	**Falcon, 1964 (V8)**
	Comet, 1964 (V8)	**Ford, 1964 (V8)**
Buick, LeSabre Wagon, 1964 Wildcat, 1964-65	**Fairlane, 1964 (V8)**	**Mercury, 1964 (V8)**
		Mustang, 1965 (V8)

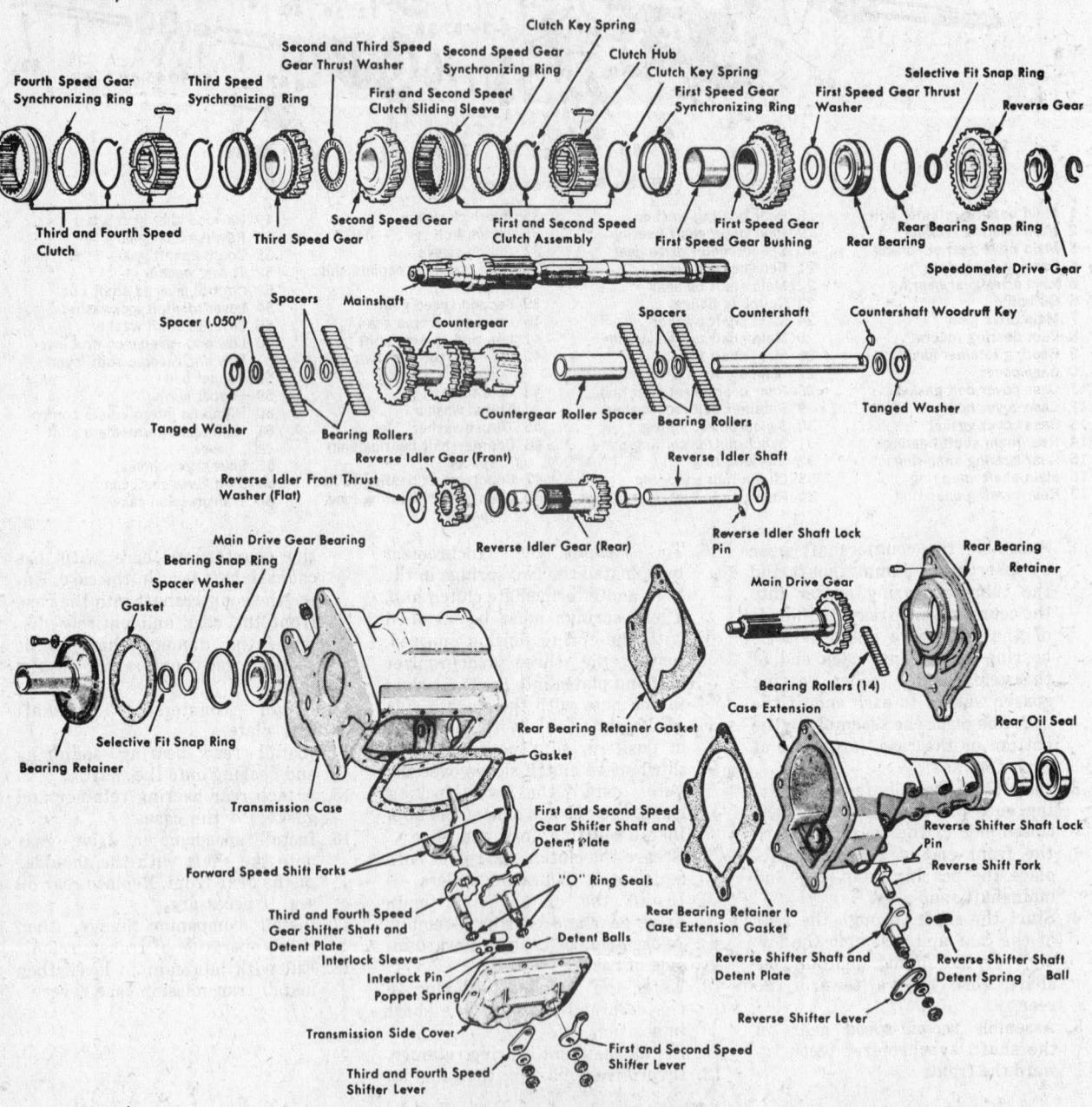

Warner T-10 transmission (General Motors application)
(© Buick Div., G.M. Corp.)

Disassembly

1. Drain transmission, mount in adequate stand. Then remove the side cover and shift controls.
2. Remove four bolts from front bearing retainer, then remove retainer and gasket.
3. Remove output shaft companion flange.
4. Drive lock pin up from reverse shifter lever boss, then pull shift-shaft out about 1/8 in. to disengage shifter fork from reverse gear.
5. Remove five bolts from the case extension and tap the extension (with soft hammer) rearward. When idler gear shaft is out as far as it will go, move extension to the left so the reverse fork clears the reverse gear. Remove extension and gasket.
6. Remove rear bearing snap-ring from mainshaft.
7. Remove case extension oil seal.
8. Remove speedometer drive gear with puller.
9. Remove the reverse gear, reverse

idler gear and tanged thrust washer.

10. Remove self-locking bolt holding the rear bearing retainer to transmission case.

11. Remove the entire mainshaft assembly.

12. Unload bearing rollers from main drive gear and remove fourth-speed synchronizer blocking ring.

13. Lift the front half of reverse idler gear and its thrust washer from the case.

14. Remove the main drive gear snap-ring and remove spacer washer.

15. With soft hammer, tap main drive gear toward rear and out of front bearing.

16. From inside the case, tap out front bearing and snap-ring.

17. From the front of the case, tap out the countershaft, using dummy shaft.

18. Then lift out the countergear assembly with both tanged washers.

19. Dismantle the countergear, consisting of 80 rollers, six .050 in. spacers and a roller tubular spacer.

20. Remove mainshaft front snap-ring and slide third and fourth-speed clutch assembly, third-speed gear and synchronizer ring, second and third-speed gear thrust bearing, second-speed gear and second-speed synchronizer ring from front of mainshaft.

21. Spread rear bearing retainer snap-ring and press mainshaft out of retainer.

22. Remove the mainshaft rear snap-ring.

23. Support first and second-speed clutch assembly and press on rear of mainshaft to remove shaft from rear bearing, first-speed gear, and synchromesh ring, first and second-speed clutch sliding sleeve and first-speed gear bushing.

Assembly

Mainshaft

1. From the rear of the mainshaft, assemble first and second-speed clutch assembly to mainshaft (sliding clutch sleeve taper toward the rear, hub to the front) and press the first-speed gear bushing onto the shaft.

2. Install first-speed gear synchronizing ring so notches in ring align with keys in hub.

3. Install first-speed gear (hub toward front) and the first-speed gear thrust washer. Be sure the grooves in the washer are facing first-speed gear.

4. Press on the rear bearing, with the snap-ring groove toward the front of the transmission. Be sure the bearing is firmly seated against the shoulder on the mainshaft.

5. Install the selective fit snap-ring onto the mainshaft behind the

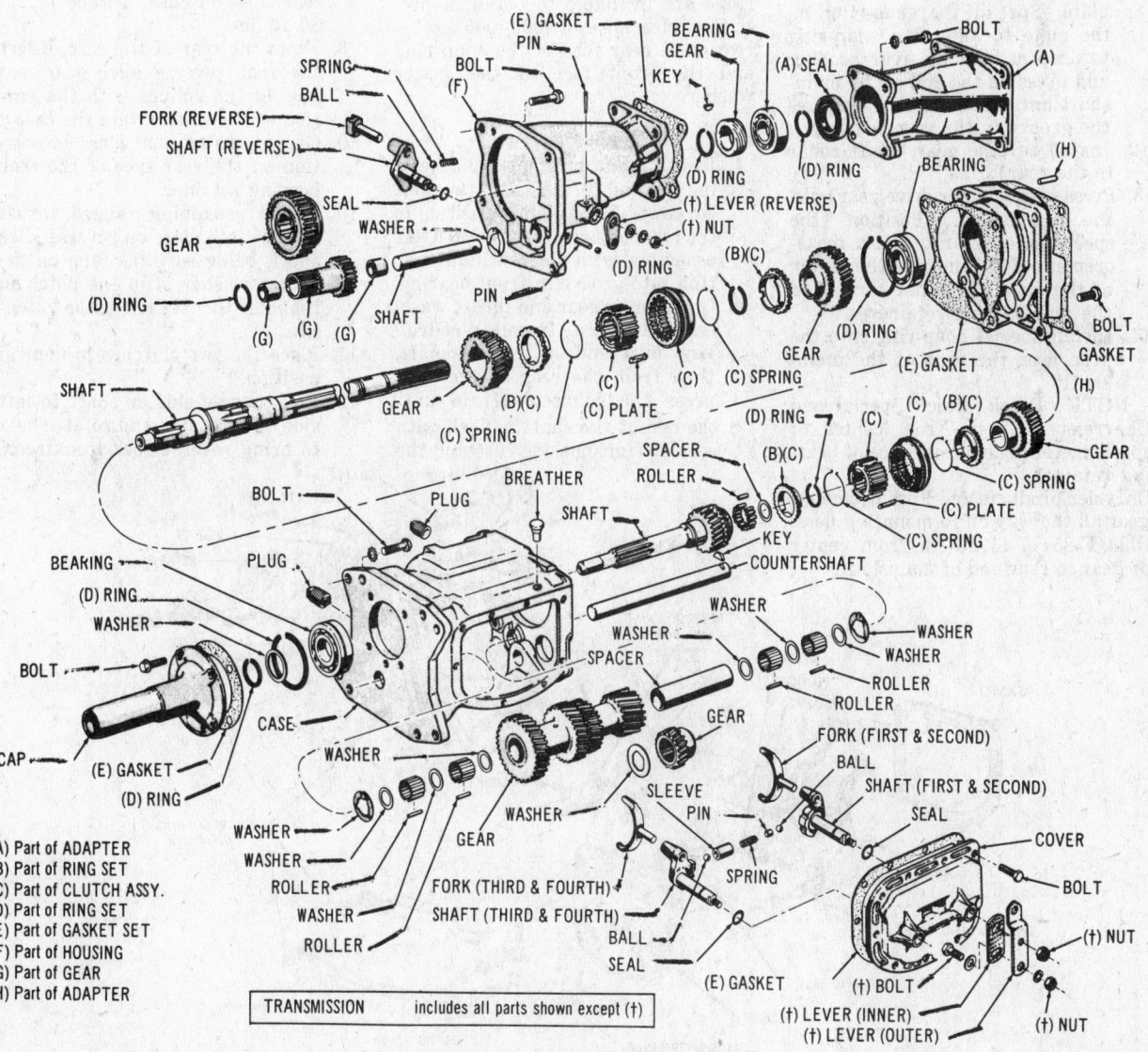

(A) Part of ADAPTER
(B) Part of RING SET
(C) Part of CLUTCH ASSY.
(D) Part of RING SET
(E) Part of GASKET SET
(F) Part of HOUSING
(G) Part of GEAR
(H) Part of ADAPTER

TRANSMISSION includes all parts shown except (†)

Warner T-10 transmission (American Motors application)

rear bearing. Use the thickest ring that will fit between the rear face of the bearing and the front face of the snap-ring.

6. From the front of the mainshaft, install the second-speed gear synchronizing ring so that notches in the ring correspond with the keys in the hub.

7. Install the second-speed gear (hub toward the back) and install the second and third-speed gear thrust bearing.

8. Install third-speed gear (hub to front) and third-speed gear synchronizing ring (notches front).

9. Install third and fourth-speed gear clutch assembly (hub and sliding sleeve) with taper front, being sure keys in the hub correspond with notches in third-speed gear synchronizing ring.

10. Install snap-ring (.086-.088 in. thickness) into groove in mainshaft, in front of the third and fourth-speed clutch assembly.

11. Install rear bearing retainer plate. Spread the snap-ring on the plate to allow the snap-ring to drop around the rear bearing and press on the end of the mainshaft until the snap-ring engages the groove in the rear bearing.

12. Install reverse gear (shift collar to the rear).

13. Press speedometer drive gear onto the mainshaft. Position the speedometer gear to get a measurement of $4\frac{1}{2}$ in. from the center of the gear to the flat surface of the rear bearing retainer.

14. Install special snap-ring into the groove at the rear of the mainshaft.

NOTE: Buick, Buick Special and Chevrolet: $4\frac{1}{2}$ in. from center of speedometer gear to rear face of bearing retainer.
Chrysler products and Ford products: gear all the way on to mounting boss. Olds F-85: 7-11/16 in. from center of gear to rear end of mainshaft.

Countergear

1. Install countergear dummy and tubular roller bearing spacer into the countergear.

2. Using heavy grease to hold the rollers, install 20 bearing rollers in either end of the countergear, two spacers, 20 more rollers, then one spacer. Install the same combination of rollers and spacers in the other end of the countergear.

3. Set the countergear assembly in the bottom of the transmission case, be sure the tanged thrust washers are in their proper position.

Main Drive Gear

1. Press bearing (snap-ring groove front) onto main drive gear until the bearing fully seats against the shoulder on the gear.

2. Install spacer washer and selective fit snap-ring in the groove in the main drive gear shaft.
NOTE: variable thickness snap-rings are available to obtain a prescribed clearance of .000-.005 in. between the rear face of the snap ring and the front face of the spacer washer.

Transmission

1. Install main drive gear and bearing assembly through the side cover opening and into position in the transmission front bore. After assembly is in place, install snap ring into groove in front bearing.

2. Lift countergear and thrust washers into place. Install Woodruff key into end of countershaft, then from the rear of the case, press the countershaft in until the end of the shaft is flush with rear of transmission case and the

dummy shaft is displaced. Endplay in the countergear must not exceed .025 in.

3. Install the 14 bearing rollers into the grease-coated end of the main drive gear.

4. Using heavy grease, position gasket on front face of rear bearing retainer. Install the fourth-speed synchronizing ring onto main drive gear with clutch key notches toward rear of transmission.

5. Position the reverse idler gear thrust washer on the machined face of the ear cast in the case for the reverse idler shaft. Position the front reverse idler gear on top of the thrust washer, hub facing toward rear of case.

6. Lower the mainshaft assembly into the case, with the notches of the fourth-speed synchronizing ring corresponding to the keys in the clutch assembly.

7. Install self-locking bolt, holding the rear bearing retainer to the transmission case. Torque to 20-30 ft. lbs.

8. From the rear of the case, insert the rear reverse idler gear, engaging the splines with the portion of the gear within the case.

9. Grease gasket, and place in position on the rear face of the rear bearing retainer.

10. Install remaining tanged thrust washer into place on reverse idler shaft, being sure the tang on the thrust washer is in the notch in the idler thrust face of the extension.

11. Place the two clutches in neutral position.

12. Pull reverse shifter shaft to left side of extension and rotate shaft to bring reverse fork to extreme

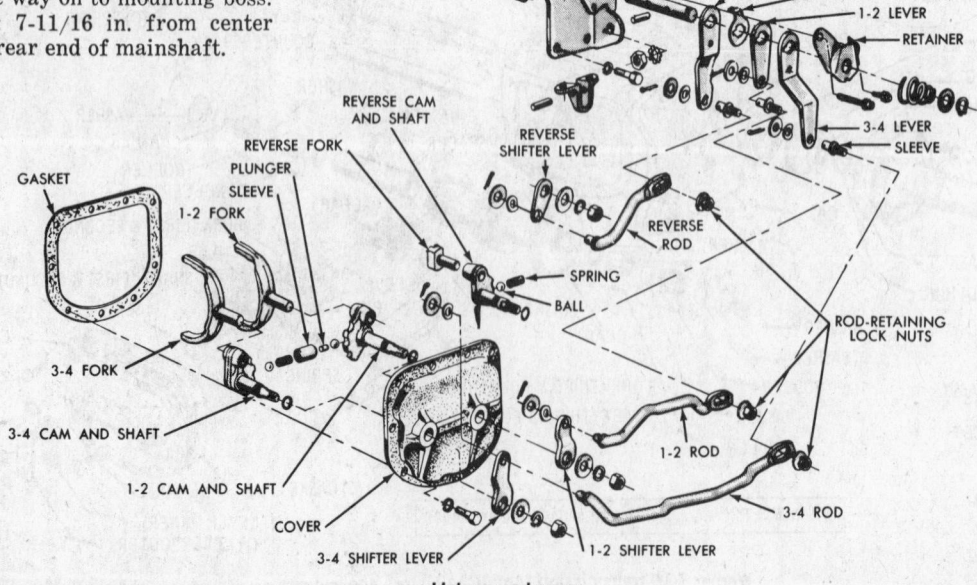

Linkage and cover

forward position in extension. Line up forward and reverse idler gears.

13. Start the extension onto the transmission case by inserting reverse idler shaft through reverse idler gears. Push in on shifter until shift fork engages reverse gear shift collar. When the fork engages, rotate the shifter shaft to move reverse gear rearward. This will allow the extension to slide onto the transmission case.

14. Install three extension and retainer to case attaching bolts and torque to 35-45 ft. lbs. Install

two extension to retainer attaching bolts and torque to 20-30 ft. lbs. Use sealer on the lower, right attaching bolt.

15. Adjust reverse shift shaft so that groove in shaft lines up with hole in boss. Drive in lock pin from top of boss.

16. Install the main drive gear bearing retainer and gasket, being sure the oil well lines up with the oil outlet hole. Install four sealer-coated attaching bolts and torque to 15-20 ft. lbs.

17. Install a shift fork into each clutch sleeve.

18. With both clutches in neutral, install side cover gasket and lower side cover into place.

19. Install attaching bolts and torque to 10-20 ft. lbs. Use sealer on the lower right bolt.

20. Install first and second, and third and fourth shift levers, lock washers and nuts.

NOTE: when used in American Motors cars equipped with torque tube drive, the tailshaft extension housing is peculiar to torque tube design. Both torque tube and open driveshaft designs are shown in the accompanying illustration.

Type-26
Warner T-18 4-Speed
Application
Jeep Gladiator and Wagoneer (232 Six and 327 V8)

This transmission is essentially the same as the Warner T-98, except for different mounting flanges.

Type-27
Warner T-89 3-Speed
Application
Jeep Gladiator and Wagoneer

This transmission is essentially the same as the Warner T-85, except for different internal gear ratios.

Type-28
Warner T-98 4-Speed
Application
Jeep Universal (4 Cyl.)

This is a Warner transmission, synchronized in second, third and fourth speeds only.

Transmission Disassembly

1. Remove transmission—transfer case adapter plate and gasket.
2. Remove transmission control housing assembly.
3. Mark the two blocking rings, third and high synchronizing gear, and third and high synchronizing sleeve. Also mark the blocking ring, low and second synchronizing gear, and low and second synchronizer sleeve.
4. Slide the low-speed gear toward the rear of the transmission case.
5. Disengage the reverse gearshift arm from the reverse idler gear and remove the reverse gearshift arm from the reverse mounting pin.
6. Move the low-speed gear back into neutral position.
7. Remove rear bearing retainer. Remove the snap-ring from the main drive pinion (clutch shaft) and the outer race of the drive pinion ball bearing.
8. Remove the main drive pinion ball bearing and oil slinger.
9. Remove the snap-ring from the outer bearing race of the transmission mainshaft ball bearing,

then, with a bearing puller, pull the bearing.
10. Separate the mainshaft assembly from the main drive pinion.
11. Lift the mainshaft assembly from the case.
12. Remove the main drive pinion from the case.
13. Remove the mainshaft pilot rollers from the drive pinion.
14. Mark relation between synchronizer gears and splines on the mainshaft.
15. Disassemble the mainshaft by removing the snap-ring holding third and high synchronizer assembly onto the mainshaft.
16. Remove the snap-ring holding second-speed synchronizer onto the mainshaft.
17. Slide second-speed synchronizer and second-speed gear from the mainshaft.
18. Remove the two remaining snap-rings, spacer, and thrust washer from the main shaft.
19. Remove the two large lock rings and push the synchronizer gear out of the sleeve.
20. If second-speed synchronizer assembly is to be disassembled, wrap the assembly in a cloth to prevent losing the lock balls and springs. Then, push the gear out of the sleeve in a direction oppo-

site the shift fork groove. Remove the cloth and lift balls, springs, and plates out of the gear.
21. Remove the lock plate for the countershaft and the reverse idler gear shaft.
22. With a pry bar in the slot of the reverse idler gear shaft, loosen the shaft. Then, slip the reverse idler shaft out of the housing and gear. Lift reverse idler gear from the case.
23. To remove the countershaft, use a dummy shaft, (1⅛ x 9.850 in.) to displace the countershaft and keep the countergear components intact. After the countershaft has been pushed entirely out of the countergear and case, remove the gear.
24. Completely disassemble the countergear assembly.
25. To disassemble the reverse idler gear assembly, remove one of the snap-rings and tap out the washer, both sets of bearing rollers, center spacer, and sleeve. Remove the remaining snap-ring.

Transfer Case Disassembly

1. Remove propeller shaft flange nut, pull the flange.
2. Remove the lower cover.

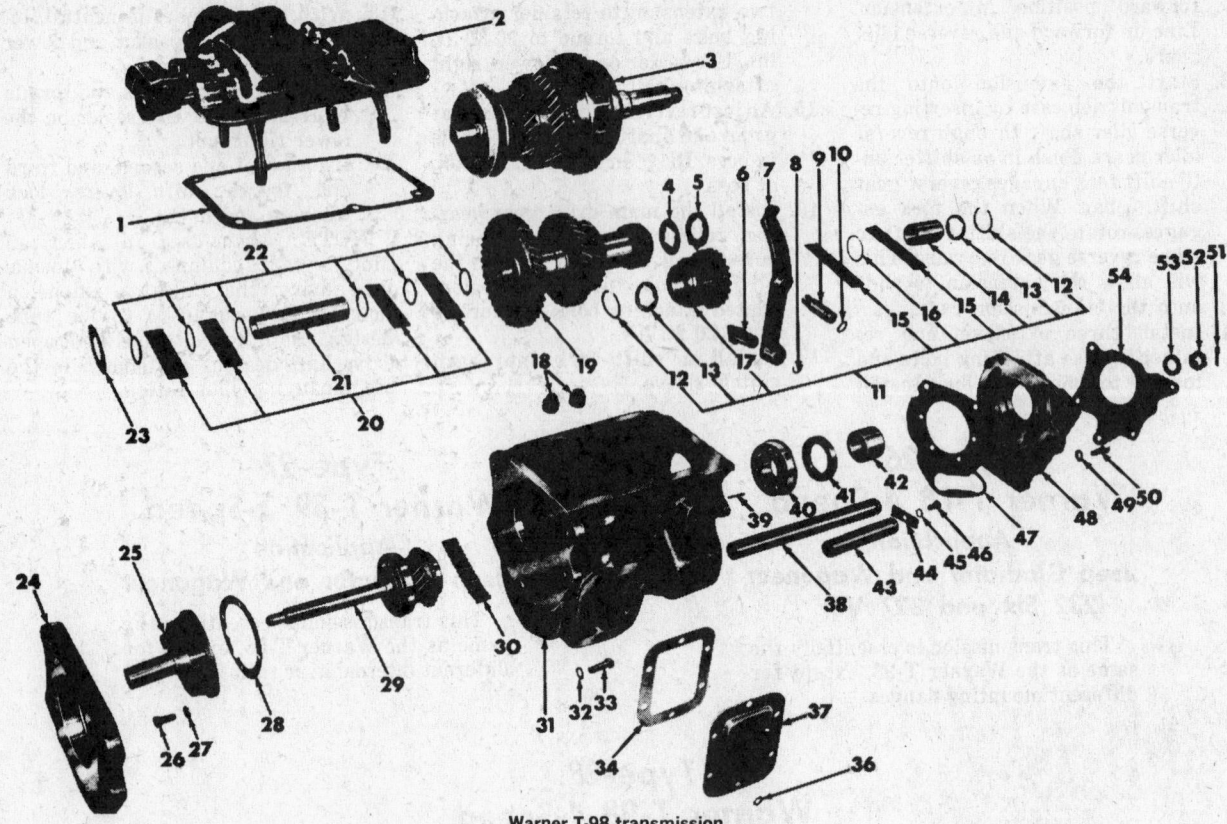

Warner T-98 transmission

1	Control housing gasket	15	Bearing rollers	28	Bearing retainer gasket	41	Oil seal
2	Control housing assembly	16	Spacer	29	Main drive gear	42	Spacer
3	Mainshaft assembly	17	Reverse idler gear	30	Bearing rollers	43	Reverse idler shaft
4	Thrust washer	18	Pipe plug	31	Transmission case	44	Lock plate
5	Thrust washer	19	Countershaft gears	32	Adapter plate lock washer	45	Lock plate lock washer
6	Reverse shifting shoe	20	Bearing rollers	33	Adapter plate bolt	46	Lock plate bolt
7	Reverse shifting arm	21	Spacer	34	Side opening cover gasket	47	Gasket
8	C washer	22	Washer	35	Side opening cover bolt	48	Adapter plate
9	Reverse shifting arm pivot	23	Thrust washer	36	Side opening cover lock	49	Adapter plate lock washer
10	O-ring	24	Bellhousing to transmission		washer	50	Adapter plate bolt
11	Reverse idler gear assembly		adapter plate	37	Side opening cover	51	Cotter key
12	Snap-ring	25	Bearing retainer	38	Countershaft	52	Nut
13	Thrust washer	26	Bearing retainer bolt	39	Shifting arm pivot taper pin	53	Washer
14	Sleeve	27	Bearing retainer lock washer	40	Bearing	54	Gasket

3. Remove lock plate screw, lock washer and lock plate.

4. Drive out the intermediate shaft through the rear of the case.

5. Remove intermediate gear, with thrust washers and needle bearings, through the bottom of the case.

6. Remove poppet plugs, springs, and balls on both sides of output bearing cap. Shift FWD to the engaged position (shaft forward).

7. Remove screws holding front output bearing cap and remove the cap as an assembly.

8. Remove the screws holding the rear output bearing cap to the speedometer gear.

9. With a rawhide hammer, drive against the front end of the mainshaft to start the rear bearing from the case. Use a special forked wedge to force the front bearing from its seat on the shaft. Loosen snap-ring and slide it forward on the shaft. Then, drive the shaft on through the case to remove the rear bearing. As the shaft is removed, the gears will remain in the case and can be removed from the bottom, as can the snap-ring and thrust washer.

10. Remove set screw in sliding gear shift fork. This will permit the shift rod to be removed.

Front Cap

1. Remove pin and attaching nut, then pull the universal joint yoke.

2. Remove the oil seal.

3. Remove set screw from shift fork and shifting rod. Clutch gear and fork can now be removed together.

4. Remove output shift shaft by pressing it through the bearing.

5. Remove bearing retaining snap-ring, then remove the bearing.

6. Remove the transfer case shift rod oil seal.

Rear Cap

1. Remove the pin, nut, and washer which hold the universal joint companion flange. Then, remove the companion flange and handbrake assembly. Remove the oil seal.

2. Remove the speedometer driven gear, the cap screw attaching the cap and the cap. There may be shims between the cap and the transfer case housing.

3. Remove speedometer driving gear.

Transfer Case Assembly

Reassembly of the transfer case is a reverse procedure to that of disassembly. When installing the shift rail oil seals into the front bearing cap, protect the seals against possible damage by the shift rail notches. When installing the intermediate gears, support the thrust washer with a

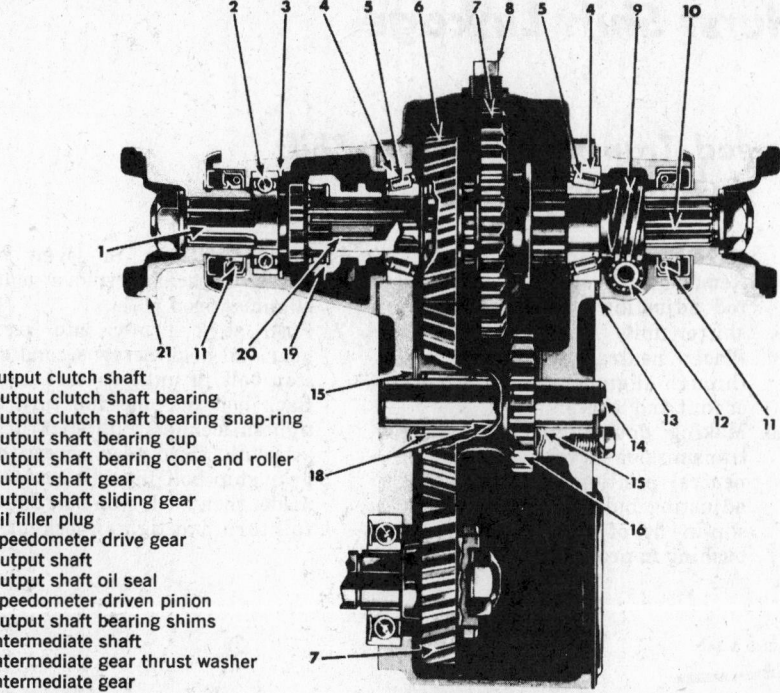

1 Output clutch shaft
2 Output clutch shaft bearing
3 Output clutch shaft bearing snap-ring
4 Output shaft bearing cup
5 Output shaft bearing cone and roller
6 Output shaft gear
7 Output shaft sliding gear
8 Oil filler plug
9 Speedometer drive gear
10 Output shaft
11 Output shaft oil seal
12 Speedometer driven pinion
13 Output shaft bearing shims
14 Intermediate shaft
15 Intermediate gear thrust washer
16 Intermediate gear
17 Main shaft gear
18 Intermediate shaft bearings
19 Output shaft clutch gear
20 Output clutch shaft pilot bushing
21 Companion flange front

Transfer case

pilot pin. This tool correctly locates the thrust washer until the gear assembly is positioned. When the rear bearing cap assembly is installed, check the end movement of the mainshaft, which determines the adjustment of the tapered roller bearings. For correct bearing adjustment, the shaft should have .004-.008 in. endplay. Adjustment is made by selective shim installation between the cap and the case. Shims of various thicknesses are available for this adjustment. Do not install rear cap oil seal until the bearings are correctly adjusted.

Transmission Assembly

Assemble in reverse order of disassembly. Pay particular attention to the following:

1. Install countershaft from rear, with bronze front thrust washer and steel backed bronze rear thrust washer installed with lugs engaged in notches in end of gear cluster. Do not seat countershaft until reverse idler gear and shaft have been installed.

2. Install reverse idler shaft until lock plate slot is adjacent to countershaft slot. Insert lock plate and tap shafts in together.

3. Assemble second speed synchronizer by installing low-second hub into low-second gear. Install retaining ring in gear.

Slide hub out of gear until holes in hub are clear of gear. Install shifter plates and springs. Push hub back into gear. Push a shifter plate toward center of gear while installing ball. Repeat for other two balls. Push hub into gear until balls snap into position.

4. Assemble third-fourth synchronizer by installing springs with tension opposed. Place right

lipped end of one spring in a hub slot. Place spring in hub. Turn hub around and repeat operation with other spring. Install three synchronizer shifting plates in hub with smooth side of plates out. Slip clutch sleeve over hub with long beveled edge toward long part of hub. Install two blocking rings.

5. Assemble mainshaft, placing threaded end up. Install snapring and thrust washer with recessed side covering snap-ring. Install bearing rollers around shaft and hold with rubber band. Install spacer. Install second gear, tapered shoulder up. Install snap-ring and blocking ring. Install second speed synchronizer and snap-ring. Install third gear, tapered shoulder to front. Install third-fourth synchronizer assembly and snapring.

6. Install main drive gear assembly and bearings in front of case and mainshaft assembly through top of case. Temporarily install bearing retainer.

7. Install mainshaft bearing snapring; press bearing into case. Remove bearing retainer. Install oil slinger and snap-ring. Press main drive gear bearing into case. Use thickest snap-ring that will fit in groove.

8. Measure space between main drive gear bearing retainer and case. Install a gasket .003-.005 in. thicker than this measurement.

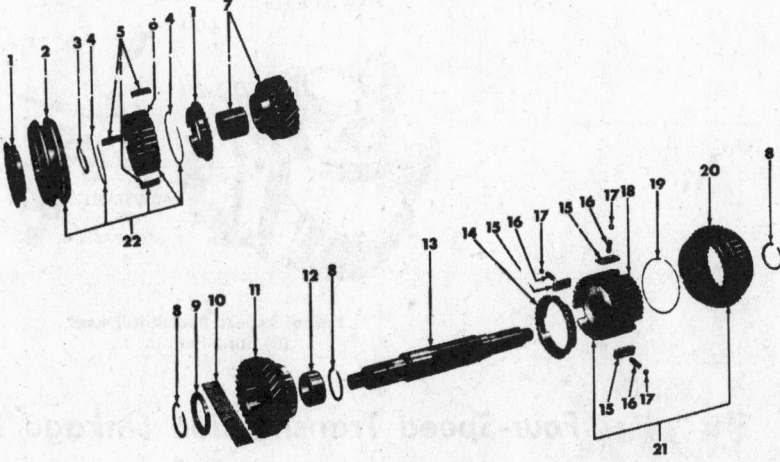

Mainshaft assembly

1 Blocking ring
2 Direct and third clutch sleeve
3 Snap-ring
4 Spring
5 Shifting plate
6 Direct and third clutch hub
7 Third speed gear assembly
8 Snap-ring
9 Thrust washer
10 Bearing rollers
11 Second speed gear
12 Spacer
13 Mainshaft
14 Blocking ring
15 Shift plate
16 Poppet spring
17 Ball
18 Low and second clutch hub
19 Retaining ring
20 Low and second speed gear
21 Second synchronizer assembly
22 Direct and third synchronizer assembly

Hurst Shift Linkage

Three-Speed Transmission Floor Shift Linkage Adjustment

The floor-mounted three-speed transmission linkage uses two shift rods and levers. Adjustment can be made with the aid of a neutral alignment rod, supplied with the floor shift linkage kit, or a 1/4 in. diameter rod.

Adjustment Procedure

1. Place shifter unit in the neutral position.
2. Back both shifter stop bolts out of shifter frame until only a few threads remain engaged.
3. Remove both shifting rods and rod adjusting buttons from the shifter unit.
4. Place neutral alignment rod through alignment holes in shifter unit and levers.
5. Making doubly sure that both transmission levers are in the neutral position, adjust the rod adjusting buttons to permit easy slip-in fit of button into nylon bushing in proper lever.
6. Fasten buttons in lever with spring clips and remove neutral alignment rod.
7. Push stick firmly into second gear and hold. Screw second gear stop bolt in until contact is felt. Back bolt out one full turn and tighten locknut. Pull stick firmly back into third gear, screw third gear stop bolt in until contact is made, then back stop bolt out one full turn and tighten locknut.

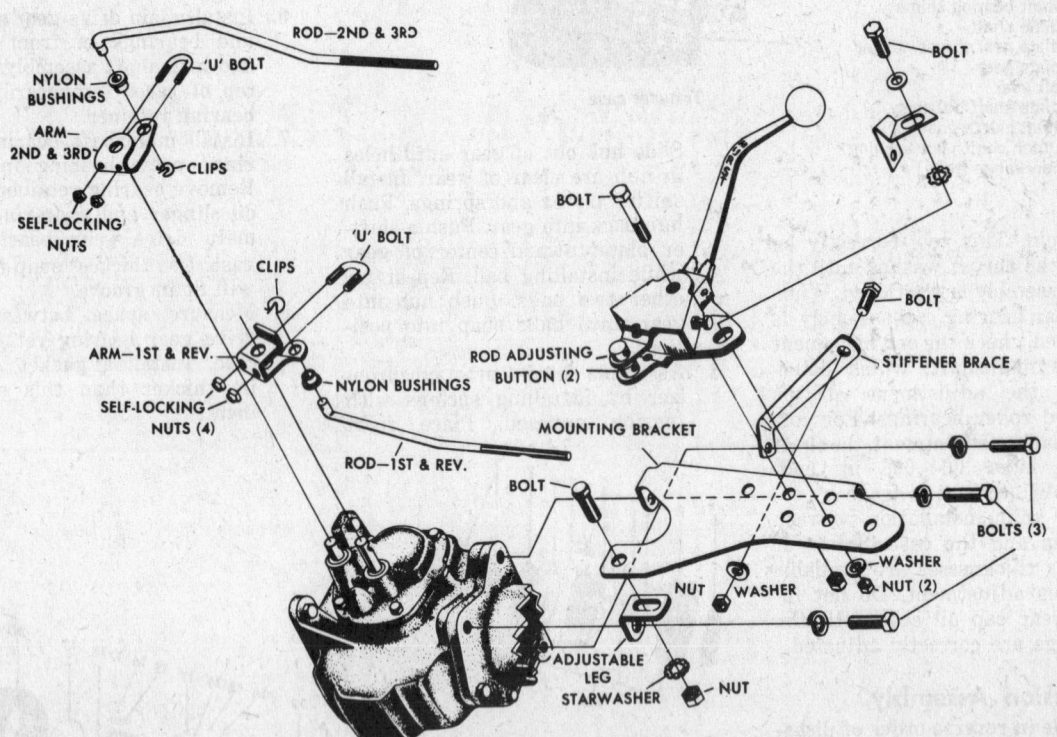

Typical 3-speed floorshift linkage
(© Hurst Perf. Inc.)

Four-Speed Transmission Linkage Adjustment

The four-speed transmission gearshift linkage uses three shift rods and levers. The adjustment can be made with the aid of a neutral alignment rod, supplied with the floor shift linkage kit, or a 1/4 in. diameter rod.

Adjustment Procedure

1. Place shifter unit in the neutral position.
2. Back both shifter stop bolts out of shifter frame until only a few threads remain engaged.
3. Remove shifting rods and rod adjusting buttons from the shifter unit.
4. Align levers with shifter frame and insert neutral alignment rod through notches in frame and holes in levers.
5. Rotate transmission arms backward and forward. The neutral position for each arm can be felt at the mid-position of full travel. Reverse arm must be moved to the end of its travel toward the front (disengaged position).
6. Adjust positions of buttons on each rod to permit easy slip-in fit of button into nylon bushing in proper lever.

NOTE: transmission arms must remain in neutral positions while align-

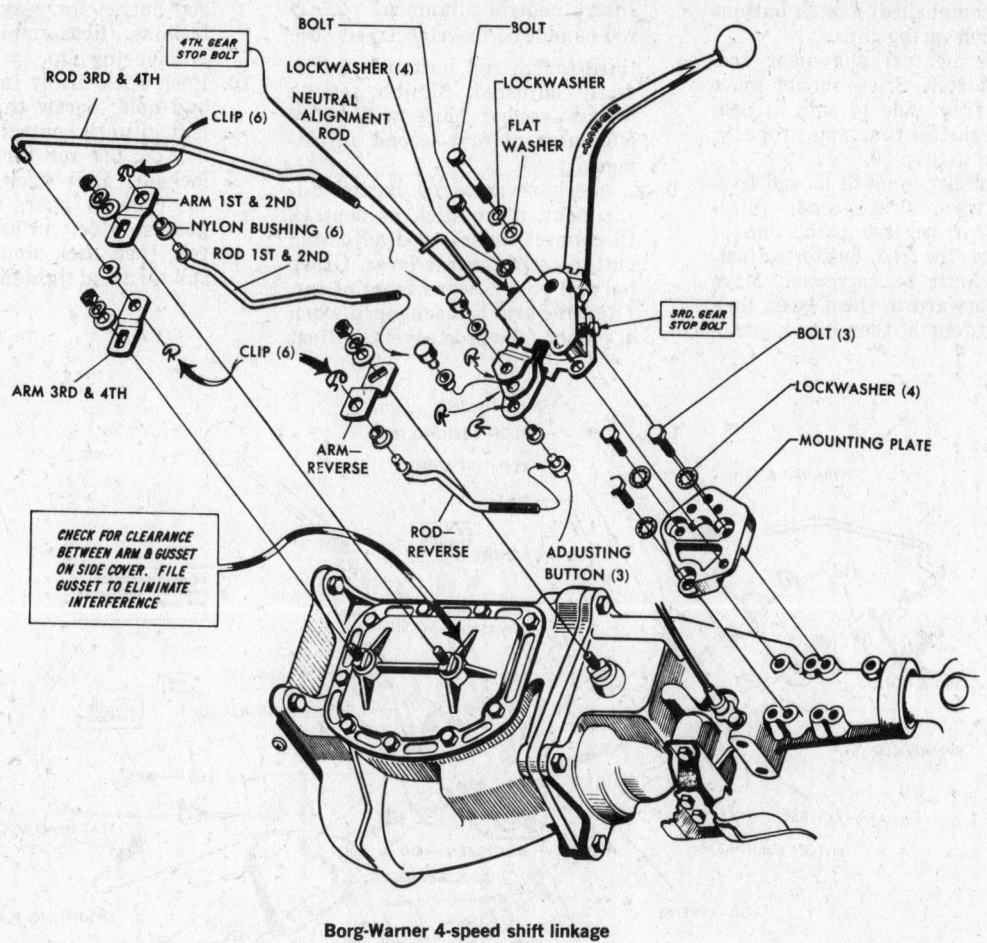

Borg-Warner 4-speed shift linkage
(© Hurst Perf. Inc.)

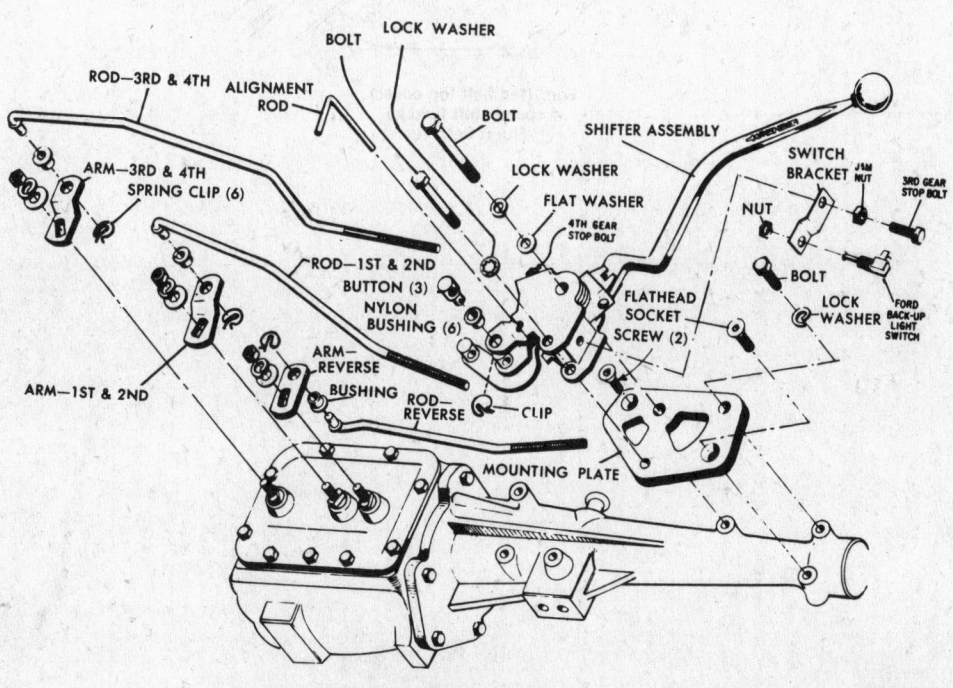

Dagenham 4-speed shift linkage
(© Hurst Perf. Inc.)

ment is accomplished. Fasten buttons in levers with spring clips.

7. Remove neutral alignment rod. Test shifter. Stick should move freely from side to side in neutral. If shifter functions properly, proceed to Step 10.

8. If the stick cannot be moved freely between first-second, third-fourth or reverse path, one or more of the rod button adjustments must be corrected. Move stick forward to third gear, then back to fourth, then into neutral.

Insert neutral alignment rod. If rod cannot be inserted freely, the third-fourth rod button is incorrectly adjusted. Similar testing of first-second shift will prove alignment of first-second adjustment.

9. To check reverse rod button adjustment, place stick at neutral. Disconnect reverse rod adjusting button from reverse lever. Grasp rod and push toward front of car. (Reverse arm is disengaged when at end of forward travel). Adjust

rod button for easy slip-in fit in bushing. Reassemble and fasten with spring clip.

10. Push stick firmly into third gear and hold. Screw third gear stop bolt in until contact is felt. Back bolt out one full turn and tighten locknut. Pull stick firmly back into fourth gear, screw fourth gear stop bolt in until contact is felt, then back stop bolt out one full turn and tighten locknut.

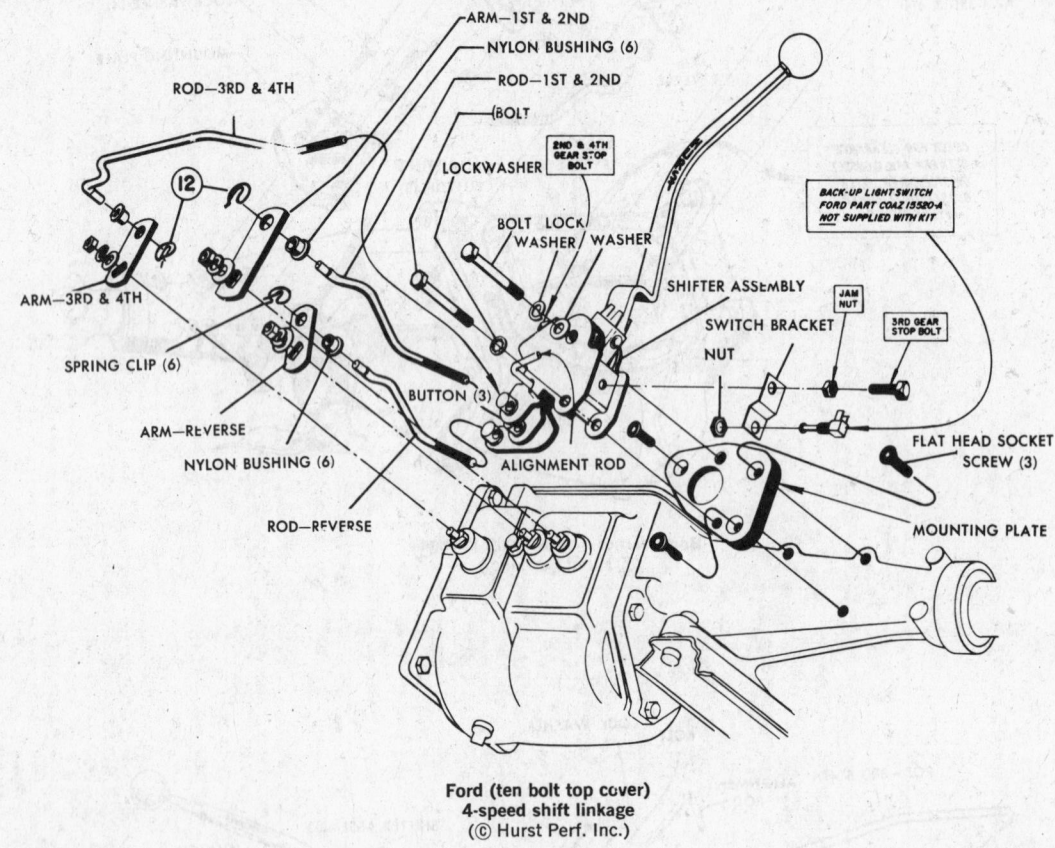

Ford (ten bolt top cover)
4-speed shift linkage
(© Hurst Perf. Inc.)

OVERDRIVE TRANSMISSION AVAILABILITY

Make and Model (No. cyl./cu. in. displ.)	1964	1965	1966	1967	1968	1969
American Motors						
Ambassador (6/232)		X	X	X	X	X
Ambassador (8/287)	X	X	X			
Ambassador (8/290)				X	X	
Ambassador (8/327)		X				
American (6/195)	X	X				
American (6/199)			X	X	X	X
American (6/232)						X
AMX (8/290)					X	
Classic (6/199)	X	X				
Classic (6/232)		X	X			
Classic (8/287)	X	X	X			
Classic (8/327)		X				
Javelin (6/232)					X	X
Marlin (6/232)			X	X		
Marlin (8/287)			X			
Rebel (6/232)				X	X	X
Rebel (8/290)					X	
Rogue (6/232)					X	
Ford Motor Company						
Fairlane (8/221)						
Fairlane (8/260)	X					
Fairlane (8/289)		X	X	X		
Falcon (6/144)	X					
Falcon (6/170)	X					
Ford 300 (6/223)						
Ford 300 (8/292)						
Ford 300 (8/352)						
Ford 300 (8/390)						
Ford (6/240)	X	X	X	X		
Ford (8/289)	X	X	X	X		
Ford (8/352)	X					
Ford (8/390)	X	X				
Galaxie (6/223)						
Galaxie (8/292)						
Galaxie (8/352)						
Galaxie (8/390)						
Galaxie 500 (6/223)						
Galaxie 500 (8/292)						
Galaxie 500 (8/352)						
Galaxie 500 (8/390)						
500 XL (8/352)						
500 XL (8/390)						
Mercury (8/390)	X	X				
Meteor (8/221)						
Meteor (8/260)						

Make and Model (No. cyl./cu. in. displ.)	1964	1965	1966	1967	1968	1969
General Motors Corporation						
Bel Air (6/250)					X	
Bel Air (8/307)					X	
Biscayne (6/250)					X	
Biscayne (8/307)					X	
Caprice (8/307)					X	
Chevelle (6/194)		X	X			
Chevelle (6/230)		X	X	X	X	
Chevelle (8/283)	X	X	X	X		
Chevelle (8/307)					X	
Chevrolet (6/230)	X	X				
Chevrolet (6/250)			X	X		
Chevrolet (8/283)	X	X	X	X		
Concours (6/230)					X	
Concours (8/307)					X	
Impala (6/250)					X	
Impala (8/307)					X	
Malibu (6/230)					X	
Malibu (8/307)					X	
Nomad (6/230)					X	
Nomad (8/307)					X	
Studebaker Corporation						
Challenger (6/170)	X					
Challenger (8/259)	X					
Challenger (8/289)	X					
Commander (6/170)	X					
Commander (6/194)		X				
Commander (8/259)	X					
Commander (8/283)		X				
Commander (8/289)	X					
Cruiser (6/194)		X				
Cruiser (8/259)	X					
Cruiser (8/283)		X				
Cruiser (8/289)	X					
Daytona (8/283)		X				
Daytona (8/289)	X					
Hawk (8/289)	X					
Lark (6/170)						
Lark (8/259)						
Lark (8/289)						

Overdrive Transmissions

Operation of the Overdrive

Mechanical Controls

An overdrive is designed so that when the instrument panel control knob is in the in position (overdrive functioning) the final drive is through a free-wheeling clutch, so that any time the car speed is greater than that of the engine free-wheeling takes place. However, if the car speed is greater than about 27 mph, the solenoid will act to lock the sun gear so that free-wheeling cannot occur.

Locking the sun gear causes the planetary pinions of the overdrive to rotate the internal ring gear at a greater speed than that of the input shaft. The increase of output over input speed is approximately 7:10. That is, for every 0.7 of a revolution of the input shaft, the output makes a complete revolution.

When the dash control is pulled out, the shift rail is moved to the rear position. This causes the shifter fork to move the sun gear into engagement with the planet carrier, thus locking the two together. As a result, the planetary gear set is locked and the output shaft rotates with the input shaft.

If the dash control is in, and the manual gear shift lever is moved to reverse, the shift rail of the overdrive is moved to the rear by a cam on the transmission reverse gear fork. Thus the overdrive is automatically locked out when in reverse.

When the overdrive is operative, but not engaged, the car will free-wheel. However, once it moves up into overdrive it will no longer free-wheel. This is a normal condition.

Where an overdrive is known to be in good condition mechanically, but does not function properly, it is a good idea to check the solenoid, using battery.

Electrical Controls

When the speed of the car reaches approximately 27 mph (the cut-in point of the governor) the governor contacts close, completing a control circuit to ground. This causes a current to flow through the relay coil, causing the relay contacts to close. The power circuit to the solenoid is completed at the relay, and current flows through the solenoid windings causing the solenoid armature to move to the in position.

The armature of the solenoid is moved by two coils, a holding coil and a traction coil. When the armature moves in, it automatically disconnects the traction coil and re-

mains in position due to the holding coil, which is lighter and takes less current.

The movement of the armature is conveyed to the pawl, which blocks the sun gear with a spring so that the pawl is not forced into engagement. When the driver eases up on the accelerator momentarily, the pawl is permitted to nudge the balk ring out of its way and engage the control plate. As soon as the pawl has engaged the plate the car is in overdrive.

The overdrive shifts down automatically when the vehicle speed drops to the cut-out speed of the governor (about 21 mph). At this point, the governor contacts open, interrupting the control circuit and so causing the relay points to open. This releases the solenoid armature and it returns to the out position. This, of course, pulls the pawl from engagement with the control ring, and the car is out of overdrive.

Downshift from overdrive to direct

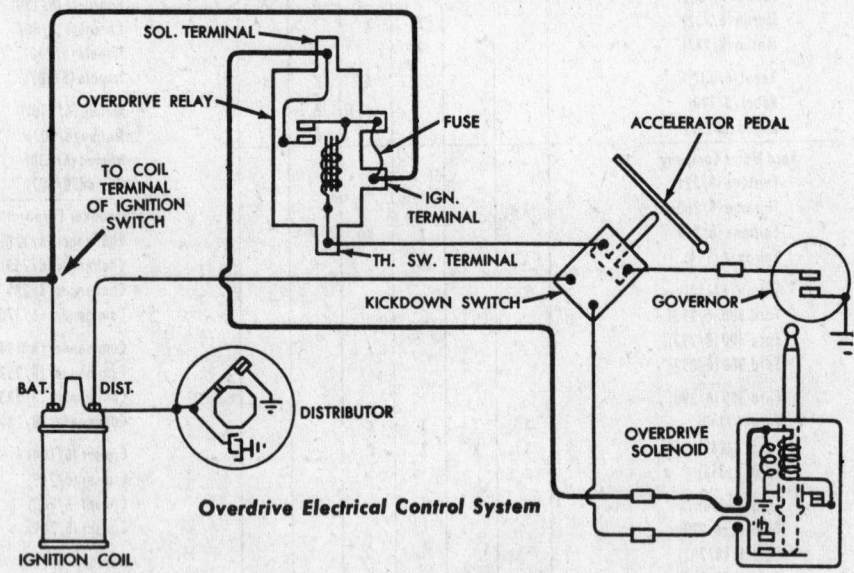

Overdrive Electrical Control System

Schematic wiring diagram of typical overdrive installation (© Ford Motor Co.)

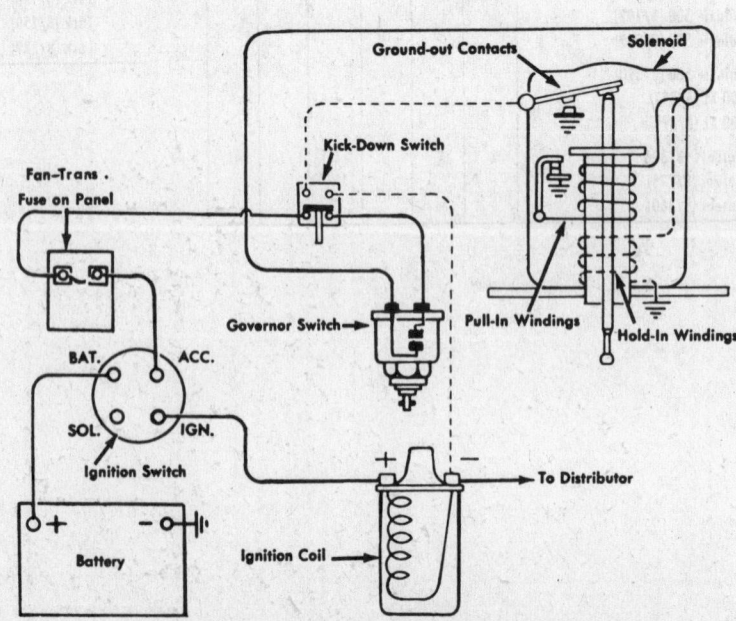

Schematic wiring diagram, American Motors. Solid line is governor solenoid circuit. Dotted line is kick-down circuit.
(© American Motors Corp.)

drive, even though the governor is calling for overdrive, is accomplished by a switch under the accelerator pedal called the kickdown switch. The switch has two functions in the circuit:

First, it opens the circuit to release the solenoid. However, the pawl cannot release itself from the control plate until the driving torque is removed.

Second, it momentarily grounds out the engine ignition circuit. This shorting out lasts only about one-half of a revolution of the crankshaft, but it is enough to release the torque on the pawl so that it can retract and so drop the car into direct drive for faster acceleration.

This ability of the kickdown switch to ground out the ignition is controlled by another set of contacts in the solenoid, which restore the ignition as soon as the solenoid armature has returned to the out position.

On some early models, there is a rail switch (reverse lock-out switch) which locks out the governor circuit when the dash control is in the out position. This rail switch is actuated by the shift rail. On later models, the movement of the shift rail accomplishes the same thing, mechanically, by moving the control plate so that the pawl cannot engage it.

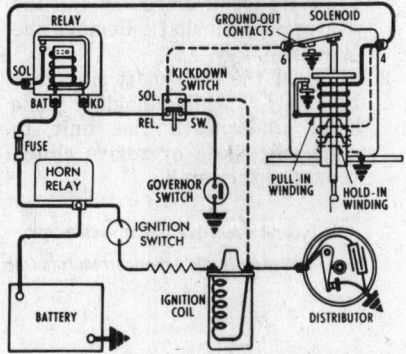

Schematic wiring diagram, Chevrolet
(© G.M. Corp.)

Troubleshooting Chart

Overdrive Does Not Engage

1. Check that the control lever on the unit moves as far, in each direction, as it should and that hand control on dash has ¼ in. free travel before lever moves.
2. Be sure that fuse on relay has not burnt out.
3. With ignition on, connect test lamp from fuse clip to ground. If lamp does not light, replace wire from switch to relay.
4. Connect test lamp between kickdown switch terminal on relay and ground. If relay does not click, replace it.

5. Disconnect wire from solenoid terminal on relay. Connect test lamp between this solenoid terminal and ground. Ground kickdown switch terminal on relay. If lamp does not light, replace the relay.
6. Disconnect the wire from the relay at the solenoid. Connect test lamp between end of wire and ground. Ground the kickdown switch terminal on the relay. If lamp does not light, replace the wire from relay to solenoid.
7. Ground kickdown switch terminal on the relay. Solenoid should click. If it does not, replace the solenoid.
8. Operate kickdown switch several times to be sure it is not sticking. Ground each terminal on the switch in turn. Solenoid should click when one terminal is grounded. If it does not, replace kickdown switch.

Overdrive Does Not Release

1. With control pushed in turn ignition key on. If a click is heard in the overdrive relay, the trouble is electrical. If no click, the trouble is mechanical and unit must be removed and overhauled.
2. Disconnect wire at governor. If overdrive then releases, replace governor.
3. Disconnect wire from governor at kickdown switch. If overdrive then releases, the wire is grounded.
4. Disconnect wire from relay and wire from governor, at the kickdown switch. If the overdrive then releases, replace the kickdown switch.
5. Disconnect wire from kickdown switch at the relay. If overdrive then releases, replace the wire; it is grounded.
6. Disconnect wire from solenoid at the relay. If overdrive then releases, replace the relay.
7. Loosen the screws holding the solenoid to the case. If the solenoid can be pulled out, without turning it, the pin which normally prevents the solenoid plunger from turning has been sheared or omitted. It will be necessary to assemble the solenoid correctly. If it can be withdrawn ½ in., without being turned, the solenoid is defective and should be replaced.

Overdrive Does Not Kick Down

1. With engine running, ground the kickdown switch terminal on the solenoid. If engine stops, replace solenoid.
2. With engine running, depress the kickdown switch and ground each of the terminals, in turn. If engine dies, all is well. If engine does not die when one terminal

is grounded, replace the kickdown switch. If engine does not die at any time, replace the wire from the switch to the coil.
3. Ground the governor terminal at the governor. This should cause the solenoid to click on. Now, operate the kickdown switch. This should cause a second click as the solenoid releases. If there is no second click, the kickdown switch is defective.

Engine Stops When Kickdown Switch is Depressed

1. With engine running disconnect wire from kickdown switch at solenoid. Operate kickdown switch. If engine does not stop, replace the solenoid. If engine does stop, replace the wire.

Car Will Not Move Unless Overdrive Is Locked Out, Acts like Slipping Clutch

1. Remove unit and overhaul free-wheel mechanism.

Stays in Overdrive When Hand Control Is Positioned to Lock Out of Overdrive

1. Check hand control. Lever on unit must move from stop to stop. If all is well there, the shift rail is binding. Remove and overhaul.

Does Not Reverse Unless Locked Out

1. Check operation of gear-shift mechanism on transmission. The shift rail in the overdrive is not being moved by reverse fork.

Servicing the Overdrive

The overdrive gets its oil supply from the regular transmission. Therefore the standard transmission of a car equipped with an overdrive requires an extra pint of lubricant. The overdrive has its own drain plug.

Some models also have a separate fill plug, through which the extra pint should be installed when transmission, plus overdrive assembly, has been drained. One must remove the transmission and overdrive assembly from the car in order to be able to do any mechanical work on the overdrive.

Overdrive Disassembly— GM Type

First, remove the lock-out switch, if present. (The lock-out switch is located toward the rear of the overdrive on the left side.) Then, turn over the assembly to permit the two steel balls under the switch to drop out.

Remove the governor assembly and the overdrive housing bolts (which hold the overdrive housing to the

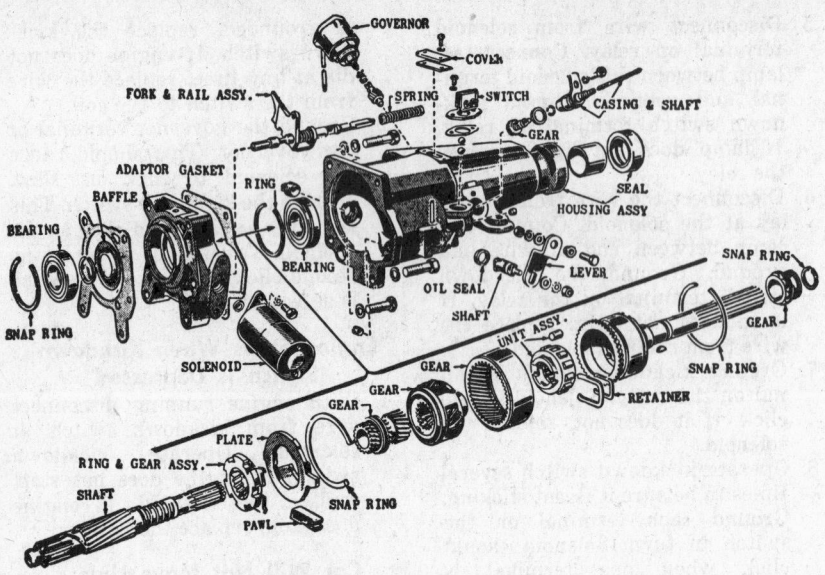

Overdrive assembly, Ford installation shown (© Ford Motor Co.)

transmission housing). Remove the shift rail pin and the cover, which is located on top of overdrive housing.

NOTE: do not remove the bolt holding the adapter to the transmission case at this time.

Pull the shift rail lever and shaft out as far as it will go. Reach down through the hole in the top of the overdrive housing and spread the snap-ring, then tap with a soft hammer on the end of the overdrive mainshaft, while pulling the housing toward the rear. This will separate the overdrive housing from the internal parts.

Now, remove the overdrive mainshaft from the assembly. Remove the clutch assembly retainers, the clutch and planetary gear assemblies and the sun gear and shift rail. These will come off as a unit, after the retainer is raised to permit their removal from the shaft.

Inspection After Disassembly

If the overdrive has a history of noisy gears, examine the sun gear, ring gear, and planetary pinions for scratches, nicks, burrs or roughness. If any are found, it will be necessary to replace the defective gears.

If the overdrive has a history of slipping in normal drive (not overdrive), examine the free-wheeling rollers and the roller cage for roughness or pits.

If the free-wheeling rollers have little depressions worn in the free-wheeling cam, it will be necessary to replace both the cam and the free-wheeling rollers.

American Motors Overdrive Units

Overdrive Shift Shaft

1. Drive out the shift shaft lock pin, then remove the shift shaft out of the case.

Governor and Solenoid

1. With a 1⅜ in. thin, open-end wrench, remove the governor from the overdrive case.
2. Remove solenoid attaching screws, then rotate solenoid one-quarter turn clockwise to release the plunger from the overdrive locking pawl.

Overdrive Case

1. Remove rear bearing snap ring and spacer washer from overdrive mainshaft.
2. Remove the four cap screws, then remove the overdrive case from the transmission case adapter housing.
NOTE: while withdrawing the overdrive case, keep tapping the end of the overdrive mainshaft to prevent the overdrive rollers from dropping out of position.

Overdrive Disassembly

1. Slide the speedometer gear and the governor drive gear from their Woodruff drive key and the overdrive mainshaft. Remove the Woodruff key.
2. Slide off the mainshaft ring-gear assembly, while holding one hand underneath the unit to catch the loose overdrive clutch rollers as they spill.

Typical overdrive transmission unit, American Motors (© American Motors Corp.)

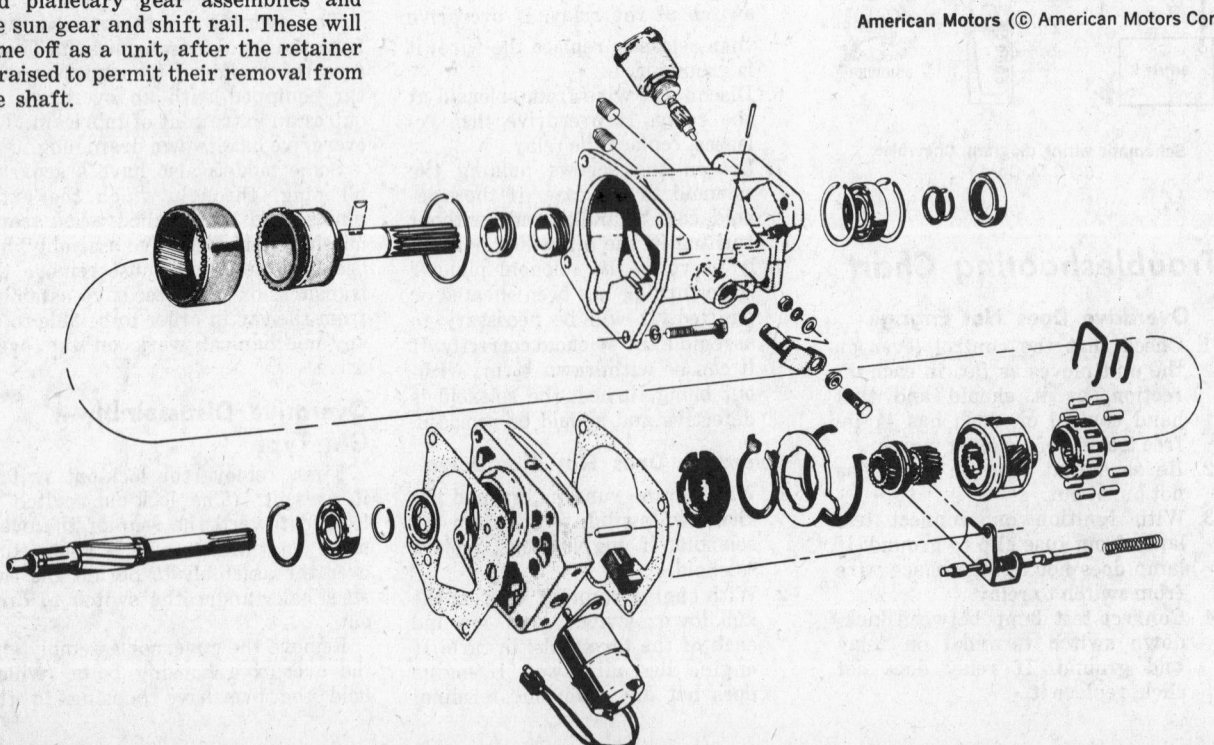

3. To separate the ring gear from the mainshaft, remove the large ring gear snap-ring.
4. To remove overdrive clutch cam, remove the cam snap-ring.
5. Remove the remaining lock ring on the mainshaft, then remove the pinion cage.
6. Remove the overdrive shifter rail and sun gear and collar from the mainshaft as a unit. When the sun gear is free of the sun gear hub, the shifter rail assembly can be separated from the sun gear shifting collar.
7. Remove the large snap-ring that holds the sun gear cover plate, hub and overdrive balk ring in place. Remove the cover plate and trough assembly and the sun gear hub assembly. The overdrive locking pawl can now be lifted out.
8. The control lever and shaft may be removed from the overdrive case, also remove the retracting spring from the overdrive case.

Overdrive Unit Assembly and Installation

1. Install the hub and balk ring assembly, chamfered side of the ring against the sun gear hub.
2. Install the locking pawl, positioning the pawl and balk ring in the locked out position, with the pawl on the step of the ring for correct installation of the solenoid.
3. Install the cover plate and trough in position and lock it in place with the large snap-ring. The snap-ring is available in various thicknesses from .0625—.0705 in. Use the thickest ring that will fit the groove.
4. Install the fork of the shifter rod into the sun gear shift collar. Then hold them together as you slide the sun gear onto the mainshaft and the shifter rod into the opening in the bearing adapter.
5. Install the pinion cage lock ring onto the mainshaft, then install the pinion cage assembly onto the mainshaft. The pinion cage pinions will mesh with the sun gear and the cage will butt up against the lock ring previously installed.

6. Position the free-wheeling cam onto the mainshaft so the counterbore of the cam slides over the machined surface of the pinion cage.
7. Install the free-wheeling cam snap-ring onto the mainshaft. This ring is available in thicknesses of .063—.073 in. Use the thickest ring which will fit the groove.
8. Replace the ring gear on the overdrive mainshaft and lock it in place with a snap-ring. This ring is available in various thicknesses from .055—.059 in. Use the thickest ring which will fit the groove.
9. Replace the free-wheeling rollers in the free-wheeling cage. A rubber band wound around the roller and cage assembly, just tight enough to hold the rollers in their retracted position, will simplify the installation of the ring gear.
10. Rotate the roller and cage assembly counterclockwise so that the rollers are at the bottom of their cams. The rubber band should hold the cage and rollers in this position. This will permit installing the overdrive mainshaft and ring gear.
11. Insert the Woodruff key into the mainshaft, then slide on the governor drive gear and the speedometer drive gear. The governor gear (small one) goes on first.
12. Install the shift shaft oil seal, then the retractor spring in the overdrive case.
13. Holding the bearing adapter to the transmission case, work the overdrive case onto the overdrive assembly. Secure the assembly with the four attaching cap screws.
14. Push the shift shaft into the case so that the operating cam will engage with the slot in the shift rod. Then install the lock pin to hold the shaft in position.
15. Insert the solenoid plunger in the opening in the bearing adapter and engage it with the notch in the locking pawl. Rotate the solenoid one-quarter turn counter-

clockwise to lock the pawl and plunger together.
16. Pull the solenoid to make sure the plunger is locked with the pawl, then install and tighten the two solenoid attaching screws.
17. Thread the governor into the overdrive case, then tighten just enough to prevent oil seepage.
18. Drive rear bearing on the overdrive mainshaft. Install rear spacer and snap-ring. This snap-ring is available in various thicknesses, from .087-.102 in. Use the thickest snap-ring that will fit the groove.
19. Install a new oil seal in the torque tube adapter, then attach the adapter to the overdrive case with the four cap screws.
20. Place the oil slinger on the clutch shaft, concave side toward the rear.
21. Install the large lock ring onto the clutch shaft bearing and place the bearing on the clutch shaft.
22. With the use of tool J-2995 and a thrust yoke J-6652, install the bearing onto the clutch shaft and into the transmission case. These are special tools. In the absence of special tools, be sure to have a helper to back-up the clutch gear with a substantial brass support (a drift or bar) to remove the possibility of damaging the synchronizer rings while driving the clutch bearing into place.
23. When the bearing is properly seated, install the spacer washer and install the clutch shaft bearing snap ring. This ring is available in various thicknesses, from .087-.102 in. Use the thickest ring which fits the groove.
24. Install the clutch shaft bearing cap, without a gasket. Then check the clearance between the bearing cap and case with a feeler gauge. Clutch shaft end-play should be zero in. Therefore, the clearance between the bearing cap and case must be accomplished by the installation of a gasket of a select thickness.
25. Install the bearing cap and new gasket then attach with four cap screws.

Drive Axles and Differential

Service Procedure Index

Make and Model		Adjust Axle Shaft End-Play	REPLACEMENT			
			Axle Shaft and/or Bearing	Axle Outer Oil Seal	Axle Inner Oil Seal	Pinion & Ring Gear and/or Pinion Bearing
Buick②	1964–70	2	4	none	13	18
Cadillac	1964–70	2	4	10		15
Eldorado	1967–70	See Car Section				
Capri (Mercury)	1970	none	8	none	8	14
Chevrolet	1964	2	4	10	none	16
	1965–70	2	5	none	13	18
Chevelle, Chevy II, Camaro	1964–70	2	5	none	13	18
Chrysler and Imperial	1964–70	①	①	①	①	①
Continental	1964–70	2	4	none	13	19
Corvair	1960–69	See Car Section				
Corvette	1964–70	none	21	21	21	21
Dodge, Challenger	1964–70	①	①	①	①	①
Fairlane, Meteor③	1964–70	2	4	none	13	19
Falcon, Comet, —6 cyl.	1964–70	2	4	none	13	20
Mustang, Cougar —8 cyl.	1964–70	2	4	none	13	19
Firebird	1967–70	2	4	none	13	18
Ford④	1964–70	2	4	none	13	19
Jeep	1964–70	3	6	11	13	18
Mercury	1964–70	2	4	none	13	19
Oldsmobile	1964–70	2	4	none	13	18
F-85, Cutlass, 442	1964–67, 70	2	4	none	13	18
	1968–69	none	5	none	13	18
Toronado	1966–70	See Car Section				
Plymouth	1964–70	①	①	①	①	①
Pontiac⑤	1964–70	2	4	none	13	15
Rambler (American Motors)	1964–70	9	9	9	13	18
Tempest	1964–70	2	4	none	13	18
Valiant, Dart, Barracuda	1964–70	①	①	①	①	①

Number in Columns Refer to Section Numbers in Text

① —All Chrysler Corporation cars are serviced according to axle type which is identified by use of the chart:

Axle	Filler Location	Cover Fastening
7¼ in.	Cover	9 Bolts
8¼ in.	Carrier, Right Side	10 Bolts
8¾ in.	Carrier, Right Side	Welded
9¼ in.		
9¾ in.	Cover	10 Bolts

Service on the 9¼ in. and 9¾ in. axles is identical.

② —1970 Buick Special and Skylark assembled in Canada: Use Section 5 in place of Section 4.

③ —1969 and 1970 Meteor with 240-IV and 302-2V engines: Use Section 5 in place of Section 4.

④ —1967 and 1968 Ford with 240-1V and 289-2V, and 1969 and 1970 Ford with 240-1V and 302-2V engines: Use Section 5 in place of Section 4.

⑤ —1970 Pontiac GTO with 455 engine, Grand Prix with 455 engine, Grand Prix with 400 engine and 4-speed transmission and 400 engine with automatic transmission and 3.55 rear drive ratio: Use Section 5 in place of Section 4.

Section Page Numbers

Axle	End-play Adjustment	Axle Shaft and/or Bearing Replacement	Outer oil Seal Replacement	Inner Oil Seal Replacement	Ring & Pinion Gear and/or Pinion Bearing Replacement
7¼ in.	2	4	none	13	17
8¼ in.	2	5	none	13	17
8¾ in.	1	7	7	12	17
9¼ in.					
9¾ in.	1	7	7	12	17

Drive Axles

Diagnosis

Noises
1. Wheels loose on drums.
2. Worn drum or axle shaft keyways (key-type axles).
3. Hub bolts loose.
4. Burned or scored wheel bearings, carrier bearings or pinion bearings.
5. Insufficient or improper lubricant.
6. Bent axle shaft, wheel or hub.
7. Improperly adjusted wheel bearings, carrier bearings or pinion bearings.
8. Excessive gear lash between drive gear and pinion.
9. Loose pinion driving flange.
10. Damaged gears.

Overheating of Unit
1. Insufficient or improper lubricant.
2. Bearing adjusted too tightly.
3. Excessively worn gears.

Loss of Lubricant
1. Lubricant level too high.
2. Improper type lubricant.
3. Clogged breather.
4. Worn oil seals, pinion or wheel bearing areas.

Excessive Backlash
1. Excessive gear wear.
2. Excessive spline wear.
3. Worn or improperly adjusted gears.

4. Loose pinion driving flange.

Limited Slip Differential— Only One Wheel Spins
1. Improper type of lubricant.
2. Clutch plates improperly assembled.
3. Clutch plates burnt.

Limited Slip Differential— Chatter on Turns
1. Improper type lubricant, not penetrating between plates, causing plates to stick together.

Section 1
Axle Shaft End-Play Adjustment

Chrysler—1964-70
Dodge—1964-70

Plymouth—1964-70
Valiant, Dart—1964-70

1. Remove wheel and brake drum on the right-hand shaft.
2. Through hole in flange, remove five nuts from retainer plate. The lock on one of the studs will have to be removed at this time.
3. Pull the assembly with tool C-3971 or equivalent.
4. Remove inner seal, using tool C-637. Install new seal, using tool C-839 or equivalent.
5. Back off adjustment on right

axle shaft assembly until inner face of adjuster is flush with inner face of retainer. Slide assembly into housing, using care not to damage seal, and install five retaining nuts. Torque to 30-35 ft. lbs.
6. With dial indicator mounted on brake support, tighten adjuster until there is zero end-play. Back off approximately four notches to get end-play of 0.013-0.023 in.

7. Hit end of left axle with nonmetallic mallet to seat right wheel bearing against adjuster, in order to get true end-play reading.
8. Install the adjuster lock under one of the nuts.
9. Recheck end-play and repeat adjustment procedure if necessary.
10. Remove indicator and reinstall drum, drum retaining clips and wheel.

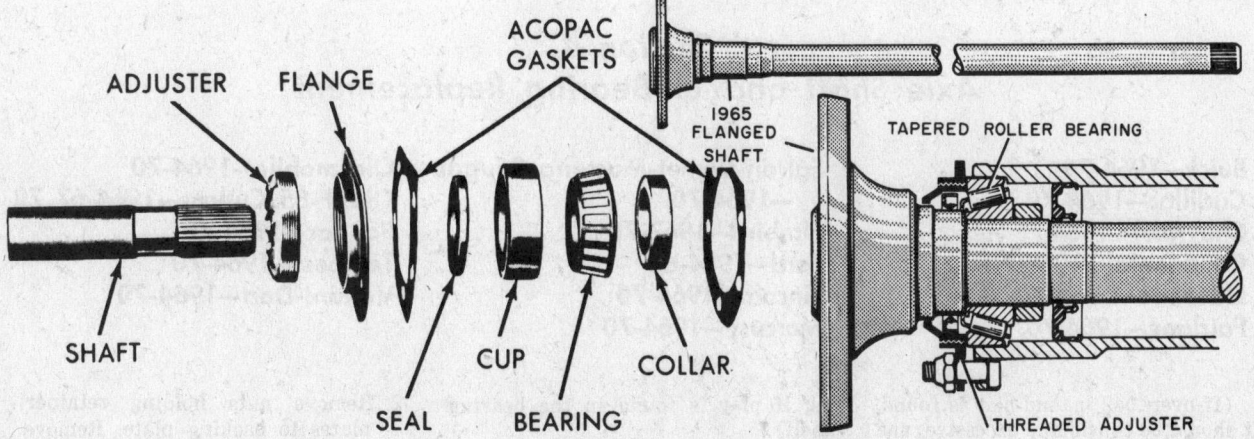

1965—Chrysler-Dodge-Plymouth axle shaft and bearing

Section 2
Axle Shaft End-Play Adjustment

Buick—1964-70
Cadillac—1964-70
Camaro—1967-70
Chevrolet—1964-70
Chevy II—1964-70
Chevelle—1964-70

Continental—1964-70
Fairlane—1964-69
Falcon-Comet—1964-69
Firebird—1967-70
Ford—1964-66
Mercury—1964-66

Mustang-Cougar—1965-69
Oldsmobile—1964-70
Pontiac—1964-70
Tempest—1964-70
Toronado—1969-70
Valiant-Dart—1964-69

On this construction, the axle bearing is held in place by pressure from the backing plate. The bearing, a radial ball bearing, is held to the axle shaft by a ring, which has been shrunken on. If end-play exists in this construction, either the bearing is loose on the shaft, the bearing retainer has moved on the shaft, or the backing plate is bent where it retains the bearing in the housing.

To correct this condition, see Axle Shaft and/or Bearing Replacement, Sections 4 and 5.

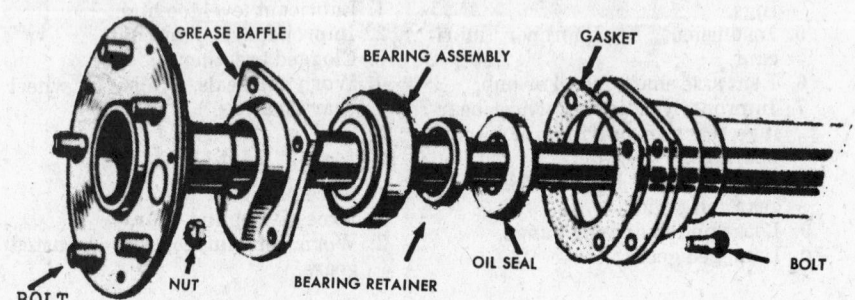

Axle shaft, bearing and oil seals

Section 3
Axle Shaft End-Play Adjustment

Jeep—1964-70

1. Remove the wheel and tire and, using puller, remove brake drum from tapered shaft. (Do not use knock-off puller, as bearing or other parts may be damaged.)
2. Disconnect the brake line from wheel cylinder.
3. Remove the backing plate.
4. To reduce the axle play, remove shims, and, to increase end-play, add shims.
5. To measure results, reinstall the backing plate and thoroughly tighten retaining bolts. A dial

gauge can be set up to ascertain exact bearing play.
6. Reconnect the brake line and bleed air from brake system.
7. Reinstall hub and drum, wheel and tire.
8. Recommended end-play is from .012 to .020 in.

On this construction, with tapered roller bearing, end-play adjustment is accomplished by adding or removing shims from behind the backing plate.

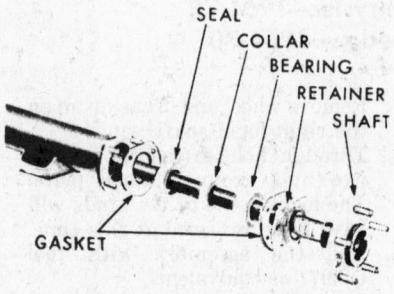

Adjusting axle shaft end-play

Section 4
Axle Shaft and/or Bearing Replacement

Buick—1964-70
Cadillac—1964-70
Chevrolet—1964
Continental—1964-70
Eldorado—1967-70
Fairlane—1964-70

Falcon-Comet-Mustang-Cougar —1964-70
Firebird—1967-70
Ford—1964-66
Lincoln—1964-70
Mercury—1964-70

Oldsmobile—1964-70
Olds F-85, Cutlass—1964-67, 70
Pontiac—1964-70
Tempest—1964-70
Valiant-Dart—1964-70

(If over .042 in. end-play is found, it should be considered excessive, and the bearing should be adjusted by shim changes, or replacement of bear-

ing if play is found in the bearing itself).
1. Remove wheel, tire and brake drum.

2. Remove nuts holding retainer plates to backing plate. Remove brake line.
3. Remove backing plate.

4. Pull out axle shaft and bearing assembly. A slide hammer-type puller is best for removal when bearing is tight in housing.
5. With chisel, nick bearing retainer in three or four places so that retainer will then slip off.
6. Press off bearing and replace with new one by pressing into position.
7. Press on new retainer.
8. Reassemble in housing, being sure bearing is completely seated.
9. Reinstall backing plate, retainer, drum and wheel and tire. Bleed brakes as required.

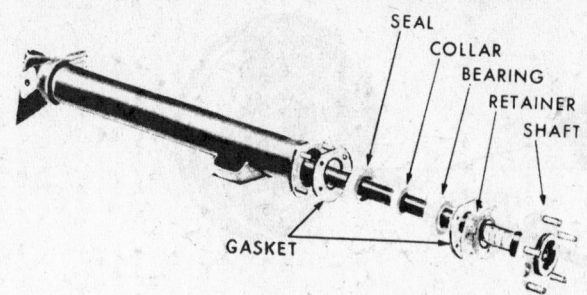

Typical rear axle shaft bearing

Section 5
Axle Shaft and/or Bearing and Seal Replacement

Chevrolet—1965-70
Chevy II—1964-70

Chevelle—1964-70
Camaro—1967-70

Plymouth—1964-70

Axle Shaft Removal
1. Remove wheel, tire assembly and brake drum.
2. Clean all dirt from cover area and

remove cover and drain lubricant.
3. Remove differential pinion shaft lock screw and shaft.
4. Push shaft toward center and re-

move C-lock from button at inner end of shaft.
5. Remove axle shaft, using care not to damage oil seal.

Axle shaft lock arrangement

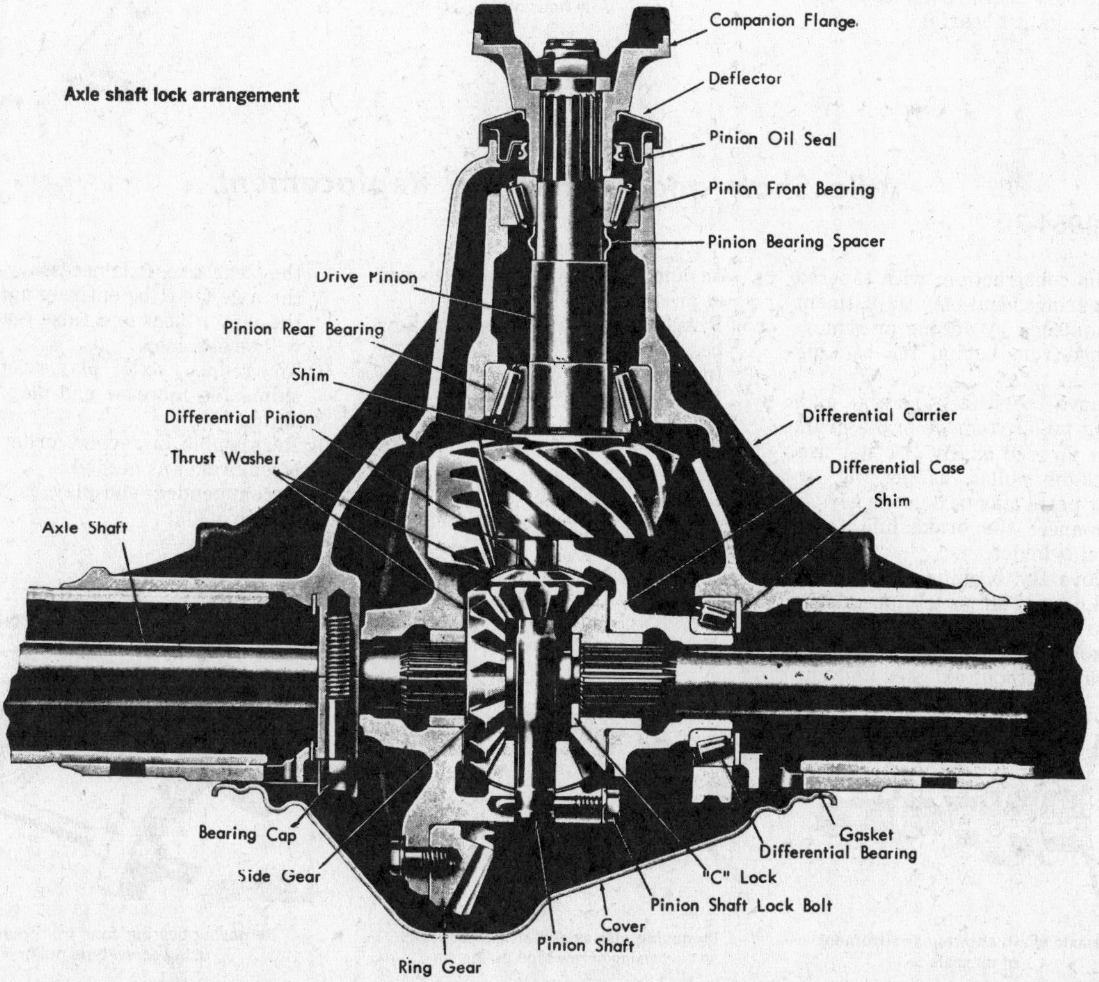

Companion Flange
Deflector
Pinion Oil Seal
Pinion Front Bearing
Pinion Bearing Spacer
Drive Pinion
Pinion Rear Bearing
Shim
Differential Pinion
Thrust Washer
Axle Shaft
Differential Carrier
Differential Case
Shim
Bearing Cap
Side Gear
Gasket
Differential Bearing
"C" Lock
Pinion Shaft Lock Bolt
Cover
Pinion Shaft
Ring Gear

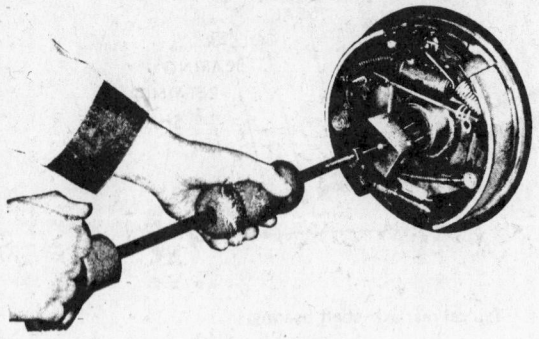

Bearing and seal removal

Bearing and seal installation

Oil Seal and/or Bearing Replacement

1. Insert tool J-8119 into bore and position behind bearing outer race.
2. With slide hammer, pull seal and bearing from housing.
3. Lubricate new bearing with wheel bearing lubricant and install. Make sure bearing completely bottoms against shoulder.
4. Pack space between seal lips with high melting point wheel bearing lubricant and place in housing. Make sure seal is completely bottomed against bearing.

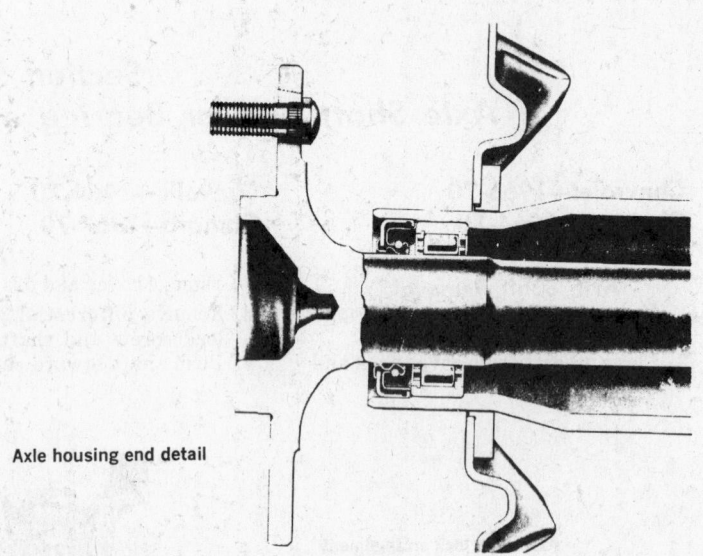

Axle housing end detail

Section 6
Axle Shaft and/or Bearing Replacement

Jeep—1964-70

On this construction, with tapered roller bearings, end-play adjustment is accomplished by adding or removing shims from behind the backing plate.

1. Remove the wheel and tire and, using puller, remove brake drum from tapered shaft. (Do not use knock-off puller, as bearing or other parts may be damaged.)
2. Disconnect the brake line from wheel cylinder.
3. Remove the backing plate. Note thickness of shims used, to aid in reassembly.
4. Remove axle shaft and bearing. (Drum, without axle key and nut

on about half-way, may be used as an inertia puller).
5. Press off old, and press on new bearing cone.
6. Install new seal, as required.
7. Place axle shaft in proper position and drive in cone. Do not use hammer and punch to locate cone, as damage may occur.
8. Check axle shaft end-play. A dial gauge can be set up to measure

the clearance. It is necessary that the axle shaft be entirely against the center block or a false reading will be obtained.
9. To reduce axle play, remove shims. To increase end-play, add shims.
10. Reassemble in reverse order and bleed brakes as needed.
11. Recommended end-play is from .012 to .020 in.

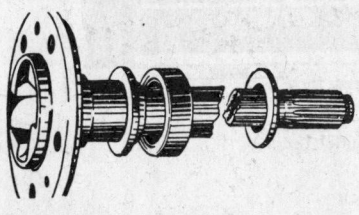

Rear axle shaft, showing arrangement of oil seals

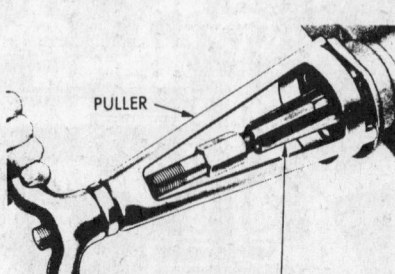

PULLER

REAR AXLE SHAFT

Removing rear axle shaft and bearing, using screw-type puller

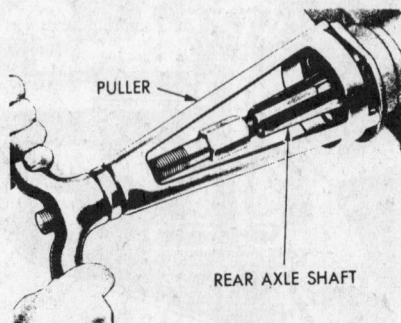

Removing bearing cone from rear axle, using screw-type puller

Section 7
Axle Shaft and/or Bearing Replacement

Chrysler—1964-70
Dodge—1964-70

Removal

1. Remove wheel and brake drum.
2. Through hole in flange, remove five nuts from the retainer plate. The right-hand shaft, with the adjuster in the retainer plate, will also have a lock on one of the studs that will be removed at this time.
3. Pull the assembly with tool C-3971 or equivalent.
4. Remove inner seal, using tool C-637. Install new seal, using tool C-839 or equivalent.

Disassembly

1. Install the shaft in vise and notch the collar with a chisel.
2. Remove the roller retainer by cutting it off at the lower edge with a chisel.

Caution To prevent the possibility of nicking the shaft or damaging the sealing surface, slide protective sleeve SP5041 over the shaft.

Plymouth—1964-70
Valiant, Dart—1964-70

3. Grind a section from the flange of the inner bearing cone.
4. Remove rollers from the section of flange which was ground away.
5. Pull retainer down as far as possible and remove by using side cutters.
6. Remove cup from axle shaft.
7. Using tool C-3971 or equivalent, remove cone.
8. Replace seal in plate.

Assembly

1. Install retainer plate onto shaft.
2. Install new bearing cone, cup, and collar, using tool C-3971 or equivalent.

Installation

1. Install new gasket (Chrysler part 2467173) on left end of housing and add the backing plate with gasket (Chrysler part 2404191).
2. Slide left axle shaft assembly into position. Use care not to damage seal.
3. Install retainer plate nuts and torque to 30-35 ft. lbs.

4. Repeat Step 1 on right side of housing.
5. Back off adjuster on right axle shaft assembly until inner face of adjuster is flush with inner face of retainer. Slide right assembly into housing, being careful not to damage seal, and install five retaining nuts. Torque to 30-35 ft. lbs.
6. With dial indicator mounted on brake support, tighten adjuster until there is zero end-play. Back off approximately four notches to get end-play of .013-.023 in.
7. Hit end of left axle with non-metallic mallet to seat right wheel bearing against adjuster, in order to get true end-play reading.
8. Install the adjuster lock under one of the nuts.
9. Recheck end-play and repeat adjustment procedure, if needed.
10. Remove indicator and reinstall drum, drum retaining clips and wheel.

Section 8
Axle Shaft and/or Bearing Replacement

Capri—1970

The Salisbury rear end of the Capri is assembled with metric bolts, therefore be sure to use the correct tool when servicing this unit.

1. Jack up car and remove wheels, brake drum retaining screw, and brake drums.
2. Remove the screws that secure the bearing retainer plate to the axle housing. These are accessi-

ble through holes in the axle shaft flange.
3. Pull the axle shaft and bearing assembly out of the axle housing.
4. Loosen the inner retaining ring by nicking it with a chisel.
5. Press on new bearing. Do not attempt to press on both bearing and retainer ring at the same time. Before assembling the retainer ring onto the shaft, the

inside of the retainer and the shaft journal should be wiped clean. These parts must not be degreased or lubricated.
6. Insert the shaft in the housing, tap home.
7. Install the four bolts and spring washers to secure bearing retainer plate; torque to 15-18 ft. lbs.
8. Replace brake drums, brake drum retaining screw, wheels.

Section 9
Axle Shaft and/or Bearing Replacement

American Motors 1964-70

On this construction, with tapered roller bearings, end-play adjustment is accomplished by adding or removing shims from behind the backing plate. The hub and axle shaft are serrated to fit together on the taper.

1. Remove rear wheels, drums, and hubs. Rear Hub Puller J-1644 or J-736 should be used to remove the hub. The use of a "knock-out" may cause damage

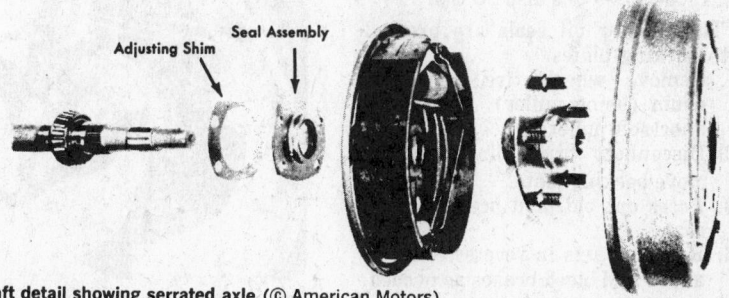

Axle shaft detail showing serrated axle (© American Motors)

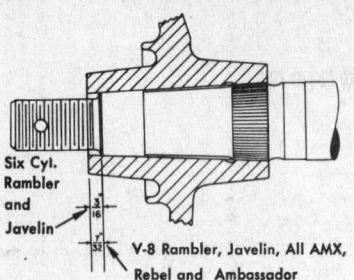

Six Cyl. Rambler and Javelin

3/16

7/32

V-8 Rambler, Javelin, All AMX, Rebel and Ambassador

American Motors hub to axle clearance
(© American Motors)

to the rear wheel bearings or the thrust block. The hub is held to the drum by three cap screws.

2. Remove brake support plate assembly, oil seal, and axle shims from axle shaft.

3. Use Axle Shaft Bearing Puller Tool J-2498 to pull the axle shaft and bearing from the axle tube. Remove inner oil seal.
4. Press off bearing.
5. Press on new bearing.
6. Apply sealer to outer surface of oil seal, install.
7. Install axle shafts, outer bearing cup, shims, oil seal retainer plate, and brake support plate. Tighten to 30-35 ft. lbs.
8. Slide hub onto axle shaft, align serrations and keyway on the hub with the axle shaft key. A new hub must be installed when a new replacement axle shaft is installed, because of the serrations, but a new hub may be put

on an old axle shaft if the serrations on the shaft are not damaged.

9. Tighten the axle shaft nut until the face of the hub is 3/16 in. from the edge of the outer taper of the shaft on six cylinder Ramblers, Hornets and Javelins, and 7/32 in. for all V-8 Ramblers, Hornets, and Javelins and all Rebels, AMX, and Ambassadors. Tighten shaft nut to 250 ft. lbs.
10. Seat bearings by striking ends of axle shafts with soft hammer. Attach dial indicator to support plate. End-play should be 0.004-0.008 in. To obtain the desired end-play, add or remove shims from the left side only.

Section 10
Axle Outer Oil Seal Replacement

Cadillac—1964-1970

On these models, an outer oil seal is included in the bearing. There is a drain plate which tends to drain away whatever lubricant gets past the bearing to the inside of the backing

Chevrolet—1964

plate. Cadillac however, uses a synthetic O-ring seal recessed into the outer diameter of the axle bearing race. Chevrolet uses an inner and outer oil seal on the bearing.

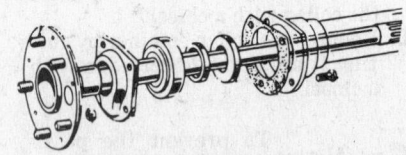

Rear axle shaft, showing arrangement of oil seals
(© G.M. Corp.)

Section 11
Axle Outer Oil Seal Replacement

Jeep—1964-1970

1. Remove wheel, tire and, using puller, remove brake drum from tapered shaft. (Do not use knock-off puller, as bearing or other parts may be damaged.)
2. Disconnect the brake line from wheel cylinder.
3. Remove backing plate.

4. These backing plate bolts also hold the oil seal, which can now be replaced. Replace shims, as removed, unless bearing is to be adjusted at same time.
5. Reinstall in reverse order of above.

Support Plate

Oil Seals

Outer oil seal   Shims

Section 12
Axle Outer Oil Seal Replacement

Chrysler—1964-1970
Dodge—1964-1970
Plymouth—1964-1970

1965-1970—See also Section 7.

These outer oil seals are pressed into backing plates.

1. Remove wheel, tire, hub and drum (using puller). Do not use knock-off puller.
2. Disconnect brake line and remove backing plate.
3. Press out old, and press in new, seal.
4. Replace parts in reverse order of above and bleed brakes as needed.

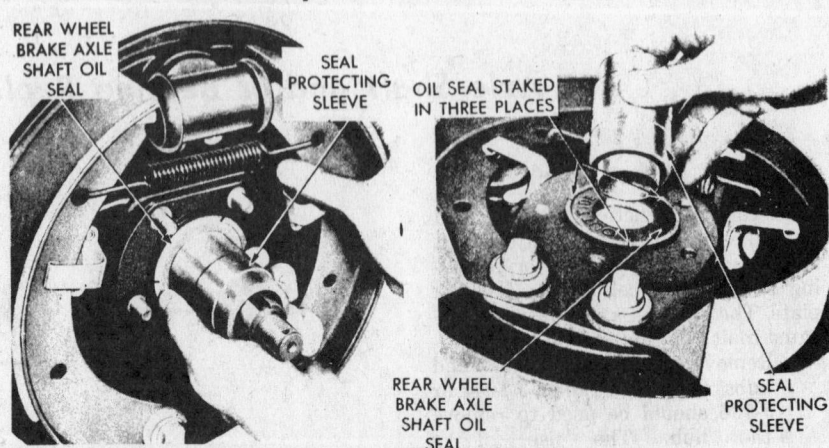

REAR WHEEL BRAKE AXLE SHAFT OIL SEAL

SEAL PROTECTING SLEEVE

OIL SEAL STAKED IN THREE PLACES

REAR WHEEL BRAKE AXLE SHAFT OIL SEAL

SEAL PROTECTING SLEEVE

Removing or installing rear brake support plate, using seal protecting sleeve

Section 13
Axle Inner Oil Seal Replacement

All Makes—All Models

(See Index For Exceptions)

The inner oil seal is seated in the inside of the bearing, in the axle tube on all American cars except those noted in index.

Remove the axle shaft as outlined in the paragraphs devoted to axle shafts for the car on which you are working. After the axle shaft, bearing and outer oil seal have been removed, reach into the housing with an inertia-type puller and pull out the inner oil seal.

The new oil seal is driven, by its rim, into the housing. Install the bearing, outer oil seal and axle shaft, replace the hub and drum assemblies and the wheels.

Section 14
Ring and Pinion and/or Pinion Bearing Replacement

Capri—1970

This unit is assembled with metric bolts, therefore be sure to use the correct tools when servicing the unit.

Differential Removal

1. Jack car and pull axle shafts. See Section 8.
2. Mark driveshaft and pinion flanges for realignment, remove four bolts and washers, the ten retaining bolts, the cover, gasket, and drain the axle.
3. Unscrew differential bearing cap screws, mark and remove caps. Using two pry bars, remove the differential.

Differential Installation

1. Place differential in housing. Position bearing caps as marked. Torque screws to 43-49 ft. lbs.
2. Position new gasket, axle case cover, secure with bolts, torque to 22-29 ft. lbs.
3. Attach driveshaft, install axle shafts. See Section 8.

Differential Reconditioning

1. Remove bearings from each side of differential assembly. Remove shims.
2. Unscrew the ring gear retaining bolts and remove the ring gear from the differential assembly.
3. Remove locking pin which secures the differential pinion shaft in the differential case, remove the pinion shaft, differential pinions, differential side gears and adjusting shims.
4. Hold drive pinion flange and remove nut, pull off pinion flange using suitable puller, remove pinion and bearing spacer from pinion.
5. Pull off large roller bearing from pinion shaft, remove spacer shims from pinion shaft.
6. Remove small taper roller bearing together with radial oil seal from axle housing.
7. Drive bearing races out from axle housing.
8. Install pinion bearing races, pulling races squarely into position.
9. To determine total side play of differential case in the housing, press the taper roller bearings on differential case without shims. Install pressure blocks into axle tubes and install differential into housing. Install bearing caps, tighten, loosen, then re-tighten finger tight. Mount dial indicator gauge on axle housing so feeler contacts side of ring gear and dial reads zero. By moving the differential, the total side play can be measured. Record this measurement. Remove differential and pressure blocks.
10. To determine the thickness of the pinion bearing spacer, use the trial and error method. Install the pinion with a selected spacer, small taper bearing, drive pinion flange, and the old self-locking nut. Tighten the nut to 72-87 ft. lbs. and rotate the pinion several times using an in. lbs. torque wrench. If the rotating torque required is too high, the spacer is too thin, and should be replaced with a thicker one. If the torque is too low, a thinner spacer should be used. Correct torque is 13-19 in. lbs. Remove the old nut. To check the spacer thickness, use the method described in *Pinion Mesh Markings, Section 26.*
11. Check that the new radial oil seal has grease between the two sealing lips and coat it with sealing compound, install it and a new self locking nut; torque to 72-87 ft. lbs.
12. Remove differential case bearings. Position shims as indicated by the total side play figure, one half of the required amount on each side, on the differential case. Press the taper roller bearings on the differential case. Insert the case in the axle housing and center. Position the bearing caps as marked. Insert the screws and torque to 43-49 ft. lbs.
13. Position the dial indicator feeler in a vertical position on one ring gear tooth and check the tooth flank backlash. If the backlash is not within 0.005-0.009 in. (0.12-0.22 mm.) the differential must be removed again. If the backlash is too large, remove shims from the ring gear face side and transfer to ring gear back side. Reverse procedure if backlash is too small. Do not increase or decrease number of shims, but only interchange between one side and the other.
14. For proper tooth contact pattern check, see *Pinion Mesh Markings, Section 26.*
15. Position a new gasket and the axle case cover on the axle case, secure with bolts and torque to 22-29 ft. lbs.
16. Install driveshaft, axles and wheels. See Section 8.

Section 15
Ring and Pinion Gear and/or Pinion Bearing Replacement

Cadillac—1970

Differential Carrier Removal

1. Support car on rear axle housing and remove wheels, drums, and axle shafts.
2. Remove two attaching bolts and lockwashers that secure differential carrier nose bumper arm.
3. Disconnect driveshaft and support out of way, loosen cover bolts and drain case, remove cover and gasket.
4. Remove bearing caps, differential adjuster nut lock and nut, differential shim, differential gear case.
5. Remove bolts on pinion, remove pinion.

Differential Carrier Installation

1. Install pinion and original pinion shims, install bolts and tighten to 30 ft. lbs.
2. Install bearing cups on side bearings. Wide bearing cup goes on left-side bearing.
3. Install differential case in housing, install differential case adjuster nut.
4. Install right-side bearing cap and two cap bolts. Do not tighten bolts. Align bearing cap threads with adjuster nut threads by pulling cap straight back on bolts and gently pushing cap into position. Tighten bolts to 50 ft. lbs.
5. Slide differential case to extreme right, tighten adjuster nut until it makes contact with right-side bearing.
6. Slide left-side bearing and cup to right, install a 0.100 in. shim between carrier and left-side bearing cup. Use a feeler gauge to determine shim thickness required.
7. Position dial indicator on carrier, with feeler against plate as shown in picture. Using spanner wrench or pin wrench, tighten

adjuster nut until a reading of 0.003-0.004 in. is obtained. Rotate gear case to check for steady reading.

8. Position dial indicator gauge against heel of ring gear tooth. Correct reading is 0.005-0.010 in. If reading is too high, increase shim, if reading is too low, decrease shim. Recheck backlash, if it cannot be obtained, reset the spread and repeat Steps 5-7.
9. Final check backlash with the method described in *Pinion Mesh Markings, Section 26*.
10. If it is necessary to change pinion shim, backlash must be rechecked again.
11. Install lock tab on adjuster nut and tighten bolt to 18 ft. lbs.
12. Install new differential carrier cover gasket, install differential cover, tighten bolts to 30 ft. lbs.
13. Install axle shafts. Install two attaching bolts and lockwashers that secure differential carrier nose bumper arm and tighten to 50 ft. lbs.
14. Install driveshaft, fill axle to proper level, then mount wheels.

Pinion Disassembly

1. Remove and discard pinion O-ring.
2. Using pinion flange holder, remove pinion nut. A torque multiplier is recommended because it will require 200+ ft. lbs. to remove the nut.
3. Remove pinion flange. Pry oil seal out of pinion retainer and discard.
4. Remove pinion outer bearing. Press outer bearing outer cup from retainer. Press out inner bearing outer cup. Remove pinion bearing collapsible spacer from pinion shaft and discard.
5. Remove pinion inner bearing.

Pinion Assembly

1. Install inner bearing on pinion

shaft. Install inner bearing cup.
2. Install outer bearing, outer bearing cup.
3. Install new collapsible spacer on pinion shaft.
4. Position pinion retainer on pinion, lubricate pinion outer bearing and install outer bearing in retainer.
5. Position oil seal on retainer, be sure it is lubricated. Install pinion nut, washer and oil seal and installer and press oil seal into retainer by tightening pinion nut. When properly installed, oil seal will protrude 1/16 in. from machined face of retainer.
6. Remove pinion nut, washer and oil seal installer from pinion.
7. Position pinion flange on pinion shaft using marks made during disassembly, install pinion nut.
8. Using care, preload pinion nut to 22-30 in. lbs. with a new bearing and 15-20 in. lbs. with an old bearing.

Differential Case Disassembly

1. Remove side bearings on differential case using puller.
2. With differential case in holder mark alignment of differential case and ring gear, loosen ring gear-to-case bolts.
3. Remove three bolts and insert pilots, remove remaining bolts, tap pilots to loosen and remove ring gear.
4. Remove differential cross-shaft, pinion gears and side gears.

Differential Case Assembly

1. Install side gears and thrust washers.
2. Install pinion gears and thrust washers.
3. Insert cross-shaft and roll pin.
4. Using guide pins and alignment marks, position ring gear on case, insert bolts and tighten to 85 ft. lbs.
5. Install side bearings.

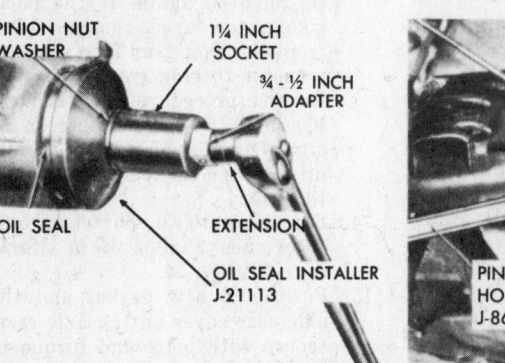

Installing pinion oil seal

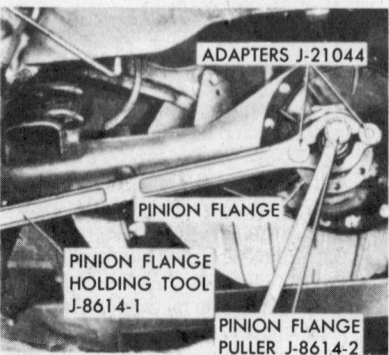

Removing pinion flange

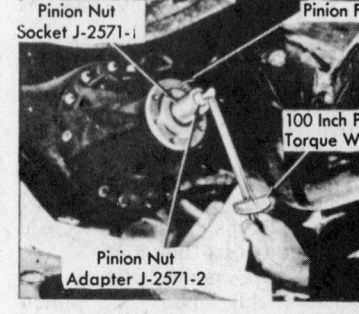

Checking preload—Cadillac pinion seal replacement
(© Cadillac Div., G.M. Corp.)

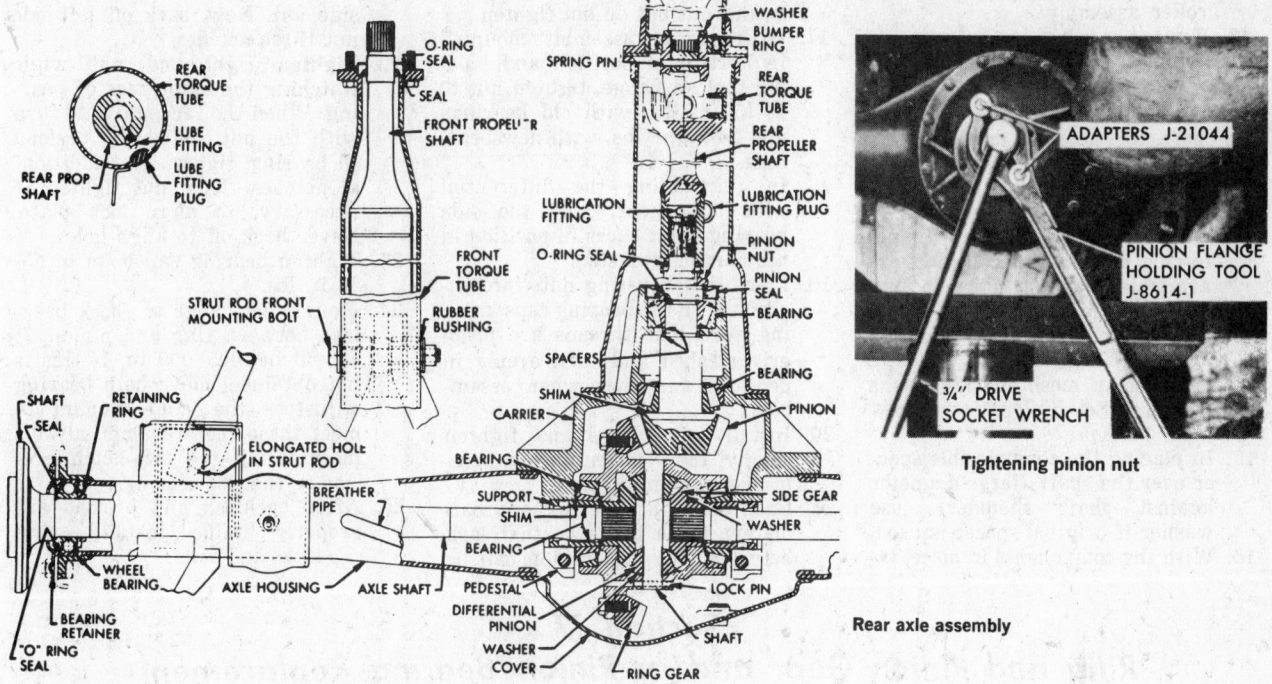

Rear axle assembly

Tightening pinion nut

Section 16
Ring and Pinion Gear and/or Pinion Bearing Replacement

Chevrolet—1964

Differential Removal

1. Support car on rear axle housing and remove axle shafts, as described in Section 4.
2. Remove the companion flange U-bolts and support propeller shaft away from work area.
3. Loosen nuts holding carrier to mounting studs and back carrier away far enough to drain lubricant. Then, remove nuts and lift carrier from housing.

Differential Reinstallation

1. Be sure that axle housing is thoroughly cleaned and install new gasket.
2. Carefully install the carrier assembly over the mounting studs and install nuts. Tighten evenly to 45-60 ft. lbs.
3. Install propeller shaft, axle shafts, drums and wheels.
4. Bring lubricant to filler plug level. Lower car to floor.

Differential Reconditioning

1. Mount differential in holding fixture.
2. Measure pinion to ring gear backlash at points 180° apart. The lowest reading should be within .005-.009 in.
3. Mark the bearing adjusting nuts, bearing caps and carrier with

Oldsmobile, F-85—1964

punch marks, one mark on one side and two marks on the opposite. These marks can then be matched in reassembly.

4. Remove bearing cap lock screws and locks.
5. Determine side bearing preload as follows:
   A. Loosen each bearing cap attaching bolt (one-quarter to one-half turn) just enough to allow movement of adjusting

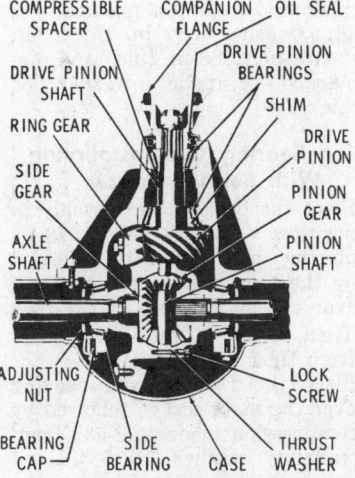

Differential assembly

Pontiac—1964

nut. Tap lightly to see if cap is loose in threads.
   B. Back off right-hand adjusting nut and watch the outer race of the bearing.
   C. If side bearing preload is correct, the outside bearing race should continue to turn until the adjusting nut is loosened two or three notches.
6. Remove the cap bolts, caps and adjusting nuts.
7. Remove the case, holding the outer races against the bearings.
   Important: do not mix side bearing outer races. They must be assembled in the original position in which they were removed.
   NOTE: the various parts can now be replaced by use of proper bearing pullers.
   In reassembly, check pinion bearing preload with in. lb. torque wrench. Preload should be 10-15 in. lbs. for old bearings and 24-32 in. lbs. for new bearings.
8. With the housing held in a horizontal position, remove the companion flange nut and washer.
9. Using a puller, remove the companion flange.
10. Remove the drive pinion, rear bearing, and spacer.
11. Remove the oil seal.

12. Remove the front inner race and roller assembly.
13. Remove rear bearing outer race.
    Thoroughly inspect all parts. Note seal faces on companion flange for possible nicks or scratches and housing bore for possible damages that could cause seal leakage.
14. In reinstalling the drive pinion assembly, proper shims must be used to locate the proper pinion-to-ring mesh. If special tools are available, the markings on pinions can be followed to obtain this mesh. Otherwise, the trial and error method may be used, using gear mesh marking as a standard. See *Pinion Mesh Markings, Section 26.*
15. In placing the compressible spacer over the shaft (large diameter against shaft shoulder), use washer if original spacer is used.
16. With the components in place, install the companion flange, washer and nut but do not tighten.
17. With carrier assembly mounted in horizontal position and holding tool on flange, tighten nut to 10-15 in. lbs. with old bearings and 24-32 in. lbs. with new bearings.
18. In reinstalling the differential with ring gear, hold the side bearing outer races in position in the carrier pedestals.
19. Place the adjusting nuts into position with the bearing caps, making sure that threads are properly meshed and turn evenly in pedestal and caps when assembled.
20. Install cap screws and tighten snugly, then loosen approximately one-quarter turn.
21. Back off right-side nut approximately three turns so that lash between ring and pinion can be removed by tightening the left-side nut. Now, back off left-side nut three notches.
22. Tighten right-hand nut while watching the outer race of bearing. When the race starts to turn with the nut, indicating preload on bearing, tighten two additional notches. Turn nut tighter, if necessary, to align lock plate. Never back off to align holes.
23. Tighten bearing cap bolts to 65-85 ft. lbs.
24. Mount dial gauge to check backlash between ring and pinion. It should be .005-.009 in. If this is not obtained, move both bearing adjusters the same amount to meet these requirements, always maintaining the two-notch preload on the side bearings.
25. After backlash and preload are properly obtained, install bearing cap locks and bolts.

Section 17
Ring and Pinion Gear and/or Pinion Bearing Replacement

Chrysler—1964-70
Dodge—1964-70

Plymouth—1964-70
Valiant, Dart—1964-70

Differential Removal

1. Raise and support car under rear axle housing.
2. Remove wheels, hubs and drums and rear axle shafts, as described in appropriate section. Drain lubricant.
3. Disconnect rear universal joint and place out of way.
4. Remove attaching bolts and remove carrier assembly to bench.

Differential Installation

1. Using new gasket, insert carrier in housing, install nuts. Torque nuts to 45 ft. lbs. Install rear universal joint.
2. Install lubricant. Reinstall axle shafts, hubs, drums, wheels and tires. Lower car to floor.

Differential Reconditioning

1. Mount carrier in holding tool, with pinion flange up.
2. Remove pinion shaft nut and washer. Remove flange with suitable puller.
3. With proper puller screwed into oil seal, pull seal from housing.
4. Rotate assembly with holding tool to allow oil slinger, shim pack and spacer (where used), to drop from the carrier.
5. Make identifying marks on the differential pedestals, adjusting nuts, and bearing caps for reassembly identification.
6. Remove the bearing lock screws and locks.
7. Remove the cap bolts, caps and bearing adjusting nuts.
8. Remove the differential assembly, being sure to keep each bearing cup with the proper bearing.
9. Remove the pinion and rear bearing cone from the carrier.
10. The bearing cone and pinion locating washer may now be removed from the shaft.
11. The bearing cups may now be removed from the carrier housing.
12. If ring gear is to be replaced it can now be done.
    NOTE: the ring gear attaching screws have left-hand threads.
    For example: with a washer .086 in. thick and a pinion marked +2, install a washer .084 in. thick; or, with a washer .086 in. thick and pinion marked —2, install a washer .088 in. thick.

Pinion Bearing Cup Installation, With Bearing Spacer

1. Place the bearing cups evenly in position. Assemble tool C-578 by placing spacer SP-2919, followed by the rear pinion bearing cone, over the main screw of the tool. Then, insert it into the carrier from the gear side.
2. Place the front pinion bearing over the main screw, followed by compression sleeve SP-535, centralizing washer SP-534, and main screw nut SP-533. Hold the compression sleeve with flange holding tool C-3281 and tighten the nut, allowing the tool to rotate as the nut is tightened in order not to damage the bearing cups. Do not remove tool after installing the cups.
   NOTE: pinions with bearing spacer require bearing preload before depth adjustment.
3. With tool installed in carrier, remove the main screw nut, centralizing washer, compression sleeve and front pinion bearing.
4. Install the pinion bearing spacer, the larger bore of spacer next to the rear bearing.
5. Position sleeve SP-1730 in the front bearing, making sure the sleeve is flush with the rear of the bearing.
6. Position the original shims, previously removed from the drive pinion shaft, over the sleeve and slide the sleeve bearing and shims over the tool main screw until the shims rest against the spacer.
7. Install the tool compression sleeve SP-535 (square end out), centralizing washer SP-534 and main screw nut SP-533. Turn carrier in stand to bring nut to top.
8. Tighten tool nut to 240 ft. lbs., using holding tool on the compression sleeve to hold the assembly in several positions to make a complete revolution while tightening. Remove holding tool and rotate the assembly several

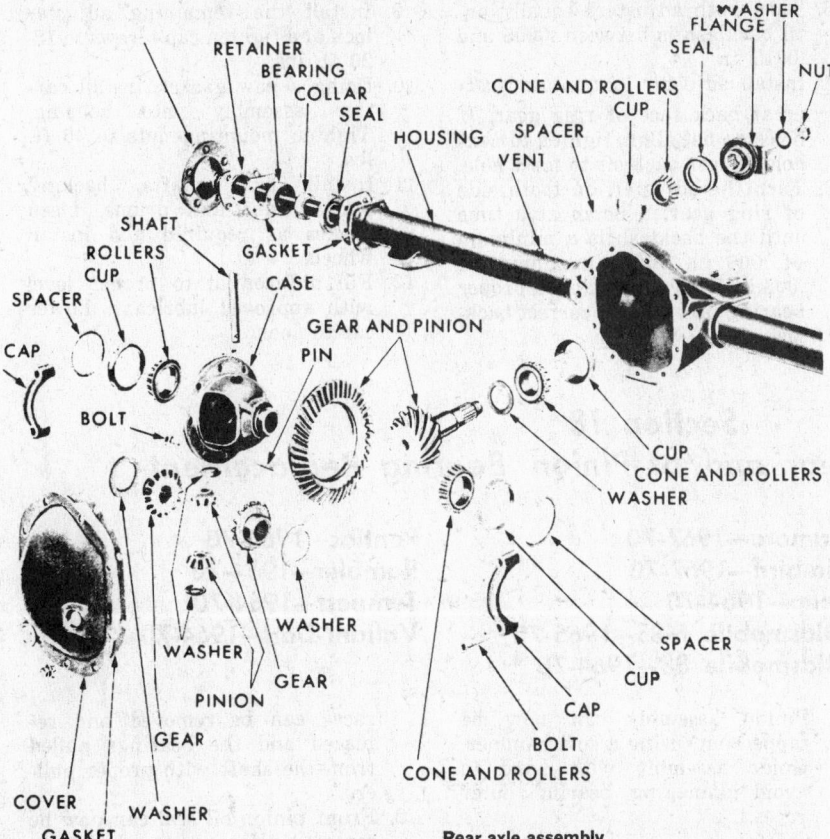

RETAINER
BEARING
COLLAR
SEAL
HOUSING
WASHER
FLANGE
SEAL
NUT
CONE AND ROLLERS
CUP
SPACER
VENT
SHAFT
ROLLERS
CUP
GASKET
CASE
SPACER
CAP
GEAR AND PINION
PIN
BOLT
CUP
CONE AND ROLLERS
WASHER
WASHER
WASHER
SPACER
GEAR
CUP
PINION
CAP
GEAR
BOLT
COVER
WASHER
CONE AND ROLLERS
GASKET

Rear axle assembly

turns, in both directions, to align the bearing rollers. Torque to 240 ft. lbs.

9. Keep nose of carrier up. With in. lb. torque wrench C-685, read the torque while the wrench is moving at least through one full rotation. Reading should be 20-30 in. lbs. A thicker shim under front bearing will reduce torque, and a thinner shim will increase torque. When proper preload torque is obtained, do not remove tool.

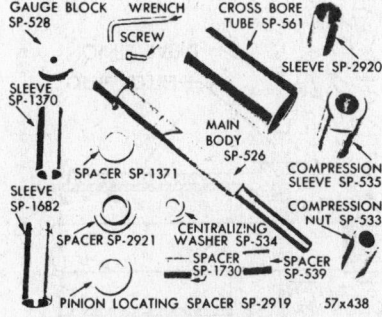

GAUGE BLOCK
SP-528
WRENCH
SCREW
CROSS BORE
TUBE SP-561
SLEEVE SP-2920
SLEEVE
SP-1370
MAIN
BODY
SP-526
SPACER SP-1371
COMPRESSION
SLEEVE SP-535
SLEEVE
SP-1682
COMPRESSION
NUT SP-533
SPACER SP-2921
CENTRALIZING
WASHER SP-534
SPACER
SP-1730
SPACER
SP-539
PINION LOCATING SPACER SP-2919
57x438

Tool set C-758

10. Obtain depth of pinion to ring gear mesh by turning assembly and holder with the pinion nut down. Install gauge block SP-528 on the end of tool, securing it with the Allen screw, flat portion next to pedestal. Position arbor

SP-561 in the bearing pedestals and hold by installing the bearing caps. A .002 in. shim should be inserted before completely tightening the caps.

11. Select a washer to insert between the gauge block and arbor.

12. Measure thickness of this washer. If it has a — marking, add this amount to the above-selected washer. If it has a + marking, subtract that amount. This will give the thickness of the washer to be installed to get the proper pinion-to-ring gear mesh.

13. Remove tool arbor, tool and bearings, shims, spacer, tool sleeve and rear bearing cone from the tool main screw.

14. With shaft end, or pinion, facing up, install the selected spacer washer onto the pinion gear shaft. The chamfer must face the pinion head.

15. Place the rear bearing cone on the pinion shaft (small side away from pinion head).

16. Install rear bearing cone onto pinion shaft with tool DD-996. An arbor press will help, with the aid of this tool.

17. Install the tubular spacer, the selected shim pack, and the front bearing in its cup in the carrier.

18. Install oil seal.

19. Insert the pinion shaft up through the carrier, then install the companion flange.

20. Remove tool and install plain washer, convex side up, and the nut.

21. Tighten nut to 240 ft. lbs.

Pinion Bearing Cup Installation, Without Bearing Spacer

Steps 1 and 2 same as above, with bearing spacer.

NOTE: pinions without bearing spacers require depth of mesh adjustment first, before bearing preload.

3. Install pinion locating spacer SP-2921 onto main body of tool C-758. Place large end of rear bearing over shoulder of spacer and insert tool into carrier.

4. Place front bearing (small end down), compression sleeve SP-535, centralizing washer SP-534, and compression nut SP-533 on main body.

5. With nose of carrier up, place flange holding tool C-3281 on compression sleeve. Allow the assembly to rotate, while tightening the nut to 25-35 in. lbs. Rotate tool several times to align bearings. Recheck torque.

6. Rotate carrier and insert gauge block SP-528 into end of the tool with the Allen screw.

7. Position arbor SP-561 in the pedestals of the carrier.

8. Hold arbor in place by installing bearing caps. A .002 in. shim should be inserted before completely tightening the caps.

9. Select washer and shims as detailed in section above covering pinion with bearing spacers, Steps 11 and 12.

10. Remove the tool with shim pack, bearing cone, pinion locating washer, and spacer from the carrier and tool.

11. Install correct washer onto pinion shaft, with chamfer facing the pinion head.

12. Place rear bearing cone on the pinion shaft (small side away from pinion head). Install rear bearing cone onto pinion shaft, with tool DD-996, and install the selected pinion shim pack.

13. Insert pinion shaft up through the carrier and install front bearing.

14. Install oil seal with lip toward bearing.

15. Install companion flange, plain washer (concave side down), and nut.

16. Tighten to 240 ft. lbs. Rotate several times to align bearing rollers and recheck torque.

Carrier Installation

1. Install the differential bearing cups onto their respective bearings and position assembly in the carrier.

2. Install the caps and bolts and tighten bolts finger tight. Be sure

all identification markings are properly aligned and positioned.

3. Tighten the adjusters enough to square the bearing cups with the bearings and eliminate end-play. Allow some backlash to remain.

4. Tighten one corresponding bolt on each cap to 85-90 ft. lbs.

5. Set dial gauge to contact outer end of ring gear tooth and take readings at 90° intervals to find the spot with the least clearance. Do not rotate ring gear after this position is found.

6. Turn both adjusters equally until backlash is between .0005 and .0015 in.

7. Install adjuster lock into adjuster at back face of ring gear. If holes do not align, tighten to next hole. Never back off to meet hole.

8. Turn the adjuster, on tooth side of ring gear, a notch at a time until the backlash is a minimum of .006 in. and a maximum of .008 in. This will establish proper bearing preload and correct backlash.

9. Install the remaining adjuster lock and tighten cap screws to 15-20 ft. lbs.

10. Using a new gasket, install carrier assembly into housing. Tighten mounting nuts to 45 ft. lbs.

11. Install axle shafts, backing plates, hubs and drums, bleed brakes as required and install wheels.

12. Fill differential to proper level with approved lubricant. Lower car to floor.

Section 18
Ring and Pinion Gear and/or Pinion Bearing Replacement

Buick—1964-70
Buick Special—1964-70
Chevrolet—1965-70
Chevelle—1964-70
Chevy II—1964-70

Camaro—1967-70
Firebird—1967-70
Jeep—1964-70
Oldsmobile F-85—1965-70
Oldsmobile 88—1964-70

Pontiac—1965-70
Rambler—1964-70
Tempest—1964-70
Valiant-Dart—1964-70

The above-captioned models are all equipped with an integral housing-type rear axle assembly. On this type axle all adjustments—pinion depth, pinion bearing preload, differential bearing preload, ring and pinion backlash—are controlled and adjusted by means of shims. There are no screw adjusters.

Differential Removal

1. Raise car under frame, allowing axle housing to hang by shock absorbers.

2. Remove wheels, brake drums and axle shafts, as described in correct section.

3. Drain lubricant. Remove rear cover from housing.
   (It would be well, at this point, to check and record backlash as an aid in reassembly).

4. Disconnect rear universal joint and wire propeller shaft to exhaust line to hold away from working area.

5. Remove companion flange nut, washer and companion flange.

6. Remove the four bearing cap bolts.

7. Reinstall the bearing caps using four 7/16-14 x 4½ in. cylinder head bolts, finger tight, as safety bolts.

8. Remove differential assembly by using a special slide hammer bolted to the ring gear or by using two pry bars behind the differential case. After removing past a certain point, the assembly will fall free onto the two safety bolts. These can now be removed and the unit taken to the bench.
NOTE: caps must be marked for proper reassembly.

9. Pinion assembly can now be tapped out, using a soft hammer. Guide assembly with hand to avoid damaging bearing outer races.

Differential Parts Replacement

1. Examine pinion shaft bearing outer races and bearings. If replacement is indicated, the outer races can be removed and replaced and the bearings pulled from the shaft with proper pullers.

2. Front pinion oil seal can now be replaced. Use proper driver to avoid damage.

3. If rear bearing is to be replaced, place original pinion shim on shaft and press bearing into

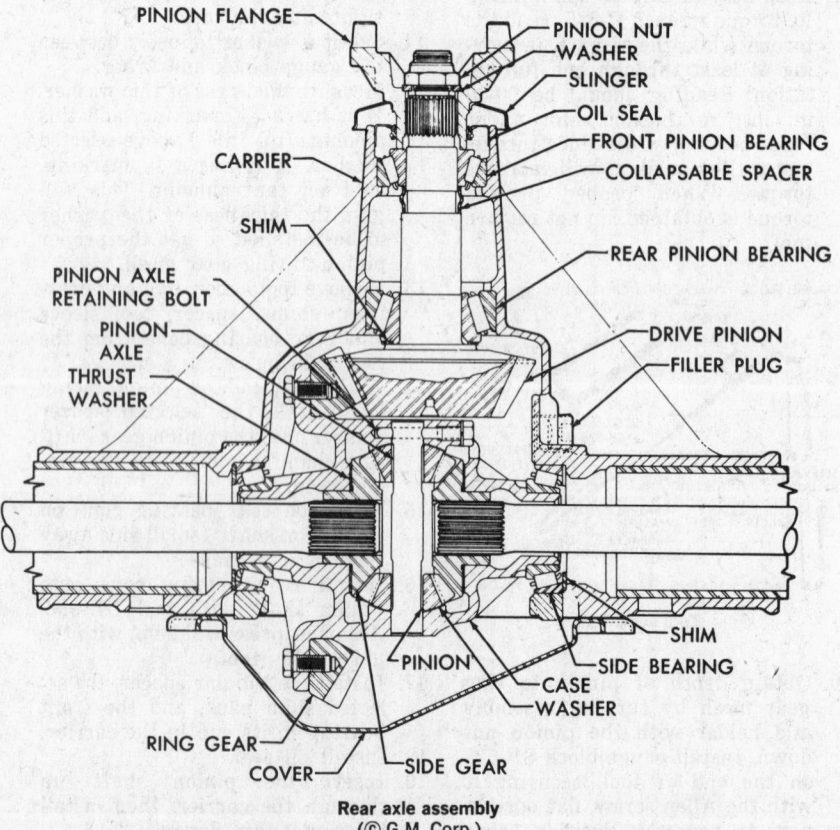

Rear axle assembly
(© G.M. Corp.)

place. (Original shim is used to check for proper pinion mesh and may have to be changed for final assembly).

4. If new bearings are used, use original preload spacer. If original bearings are used, select preload spacer and install. Place front bearing in position and press to contact preload spacer.

5. Install companion flange, washer and nut and tighten to very slight pressure. Check preload. Preload should be 20-30 in. lbs. with new bearings and 12-20 in. lbs. with original bearings.

6. Continue to tighten pinion nut in very small increments. Less than one-eighth turn is advisable. Repeatedly test for proper preload, and frequently turn the shaft in bearings to be sure proper seating is obtained.

(In differential case installation, bearing preload and side spacing must be considered. Do not use original production shims, as they are cast iron and may be damaged when being tapped into place. Service spacers are furnished in .170 in. thickness, with shims available in .002 in. increments from .040-.082 in. Make sure all bearing surfaces and shim

surfaces are absolutely clean before starting assembly).

7. Place differential case, with bearing outer races, in position in carrier.

8. Install .170 in. thick service spacer between each race and housing.

9. If new bearings are being used, add shims to both spacers to produce a total thickness on each side to correspond to the original shim thickness. If original bearings are used, increase this thickness .002 in. on each side.

10. Slip left shim between bearing and spacer, then tap right shim carefully into place.

11. As a safety precaution, install the 7/16-14 x 4½ in. head bolts into the caps.

12. Rotate differential several times to be sure bearings are properly seated. With torque wrench on ring gear attaching bolt, check bearing preload. It should be 30-40 in. lbs. with new bearings and 20-30 in. lbs. with used bearings.

13. Mount dial gauge to read gear lash at outer edge of ring gear. Check lash at three or four positions to be sure that there is not more than .002 in. runout. If more than this is found, correc-

tions should be made to such items as burrs, uneven bearing seating, distorted differential case, etc.

14. Gear lash at the minimum point should be .005-.009 in. for all new gears. If original gears are reinstalled, the measurement of backlash taken before disassembly should be maintained within ±.001 in. If lash is not within specifications, addition or removal of shims must be done to obtain desired measurements. Always keep the total shim measurement the same as when started, simply moving from one side to the other. This will maintain the bearing preload. Each .004 in. shim moved will change backlash approximately .002 in.

15. After proper backlash is obtained, remove the four safety head bolts and install the regular cap bolts. Tighten to 60-75 ft. lbs.

NOTE: at this point, it is best to make a red lead test to check contact pattern. If proper pattern is not obtained, it may be necessary to change shim on pinion shaft, thus changing pinion depth.

16. Reinstall axle shafts, drums, hubs and wheels, install proper lubricant and lower car to floor.

Section 19
Ring and Pinion Gear and/or Pinion Bearing Replacement

Continental—1964-70
Comet—8 Cyl., 1964-70

This construction uses a removable pinion bearing housing, in addition to the removable carrier assembly.

Pinion depth is controlled by shims placed between the carrier housing and the pinion bearing housing.

Differential Carrier Removal

1. Remove filler plug and drain lubricant.
2. Disconnect rear universal from companion flange and tie it to one side, out of work area.
3. Remove wheels, drums and axle shafts, as in Section 4.
4. Remove the nuts holding carrier to axle housing.
5. Remove carrier assembly and mount in holding fixture.

Differential Carrier Installation

1. Install carrier into housing and install nuts. Tighten to 30-40 ft. lbs.
2. Assemble rear universal to companion flange. Torque U-bolt nuts to 12-15 ft. lbs.

Fairlane—1964-70
Ford—1964-70

3. Install backing plates, axle shafts, drums and wheels. Where axle shafts of different lengths are used the longer shaft goes to the right side.
4. Fill the differential to the bottom of filler plug hole with approved lubricant.

Differential Carrier Reconditioning

Before disassembling, check for gear runout, parts chipping or other damage. Check and note amount of backlash.

1. Mark bearing supports and caps for aid in proper assembly.
2. Remove locks, caps and adjusting rings, then remove differential from carrier.
3. Necessary replacements can now be made.
4. Turn the carrier with pinion shaft upright and remove the bolts securing the pinion retainer assembly. Remove the O-ring.
5. Remove the pinion locating shim; measure and record its thickness.
6. Remove the pinion pilot bearing.

Lincoln—1964-70
Mercury—1964-70

Use proper tool to avoid damaging housing or bearing. When replacing pilot, use caution so as not to crack the bearing support.

7. With pinion retainer assembly mounted in holding tool, remove the pinion nut and washer. Then, remove the companion flange.
8. Remove the oil slinger and seal.
9. Press the pinion shaft out of the retainer and front bearing cone. Discard the collapsible spacer.
10. Remove the rear bearing cone.
11. Bearing cups may now be removed and replaced.
12. In reassembling the pinion shaft and bearings, install the rear bearing cone onto the pinion shaft first.
13. Place a new collapsible spacer on the pinion shaft.
14. Place the pinion gear and rear bearing cone in the retainer.
15. Install the front pinion bearing cone onto the shaft.
16. Place oil slinger on shaft.
17. Place new pre-soaked oil seal in position, coating outer edges with oil resistant sealer.

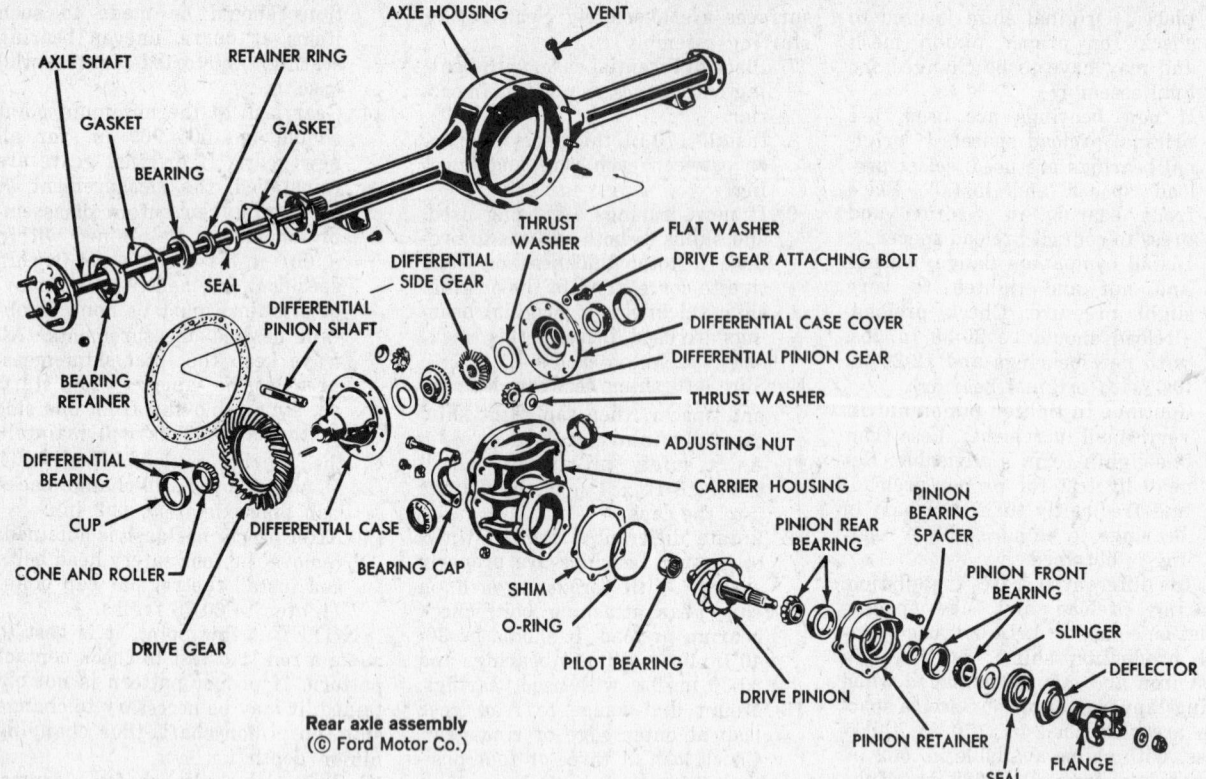

AXLE HOUSING VENT

AXLE SHAFT   RETAINER RING

GASKET   GASKET

BEARING

SEAL

THRUST WASHER

FLAT WASHER

DRIVE GEAR ATTACHING BOLT

DIFFERENTIAL SIDE GEAR

DIFFERENTIAL CASE COVER

DIFFERENTIAL PINION SHAFT

DIFFERENTIAL PINION GEAR

THRUST WASHER

BEARING RETAINER

ADJUSTING NUT

CARRIER HOUSING

DIFFERENTIAL BEARING

PINION BEARING SPACER

CUP

PINION REAR BEARING

PINION FRONT BEARING

CONE AND ROLLER

DIFFERENTIAL CASE

BEARING CAP

SLINGER

DRIVE GEAR

SHIM

O-RING

DEFLECTOR

PILOT BEARING

DRIVE PINION

PINION RETAINER

SEAL

FLANGE

Rear axle assembly
(© Ford Motor Co.)

18. Place companion flange on shaft and start nut with washer.
19. Tighten the pinion shaft nut to a maximum of 175 ft. lbs. While tightening, continually rotate the shaft to be sure the bearings are properly seated and rollers aligned.
20. Check the bearing preload. With used bearings and seal, preload should be 8-12 in. lbs. With new bearings, preload should be 17-27 in. lbs. If the torque on the pinion shaft nut is less than 175 ft. lbs , after proper preload is obtained, a new collapsible spacer must be used. Never back the nut off to reduce preload. Install a new spacer and retorque to obtain the right preload.

If new gears are being used, check the mating numbers on the pinion and ring gear. The proper shim must be selected for the drive pinion retainer. The number on the pilot end of the pinion indicates the shim thickness to be used; —1 indicates that a shim .001 in. thinner than standard is required. +1 indicates that a shim .001 in. thicker than standard is required. Markings are based on a standard shim thickness of .020 in. When replacing gear sets, these numbers should be compared with the old ones so the shims can be selected allowing for the difference between markings of the different gear sets.

If there are painted timing (index) marks on either the old or new gears, line up the matching marks when the ring gear is meshed with the pinion.
21. Lubricate the O-ring with axle lubricant and snap it into place on the pinion retainer. Do not roll into place.
22. Place the pinion gear and retainer in the carrier, with the proper shim. Be careful not to disturb the O-ring in groove. Torque bolts to 30-40 ft. lbs.
23. Coat the differential pedestal bores with lubricant so that bearing cups will move easily, and place the differential, with bearings, in the carrier. Slide along the bores until a slight backlash can be felt. Set the adjusting nuts until they just contact the bearing cups.
24. Position the caps on the carrier, checking the match marks formerly made to be sure of proper assembly. Also, check that threads are not crossed. Install the bolts and torque to 55-65 ft. lbs.
25. Loosen the cap bolts and retorque to 20 ft. lbs.
26. Loosen right-side nut, tooth side of ring gear, until it is away from the cup. Tighten left-side nut until the ring gear mates with the pinion, with no backlash, being sure the right-side nut is still loose. Tighten right-side nut two notches beyond point where it contacts the bearing. Rotate ring gear several times to be sure the bearings are properly seated.
27. Loosen the right-side nut to release preload. If there is any backlash, tighten left-side nut just enough to remove. Tighten right-side nut to obtain proper preload. This will be two and one-half to three notches tight. The backlash should be .008-.012 in. Torque the cap bolts to 55-65 ft. lbs. Always be sure that, in adjusting the nuts, final rotation is in a tightening direction. Do not back off.

At this point, it is well to check the pinion lash markings in Section 26. If no corrections are required, install assembly into axle housing, as described above.

Section 20
Ring and Pinion Gear and/or Pinion Bearing Replacement

Falcon-Comet-Mustang-Cougar—6 Cylinder, 1964-70

There are four different axle housings and three types of differential case used in this rear end construction. The design change depends upon car model and engine size, therefore, the model-year identification plate should be used to identify correct model and gear ratio when ordering replacement parts.

The rear axle assembly uses an integral-type housing. The axle shafts are held in the housing by ball bearings and a bearing retaining plate at the axle housing outer ends.

The differential assembly is mounted on two opposed tapered roller bearings. The bearings support and position the differential between two pedestals and spanner adjustable caps. Differential bearing preload and drive gear backlash is controlled by the thrust of these large pedestal nuts against the differential side bearing cups.

The drive pinion assembly is mounted on two opposed tapered roller bearings. Pinion bearing preload is adjusted by a collapsible spacer on the pinion shaft. Pinion and ring gear tooth contact is adjusted by shims between the rear bearing cone and pinion gear. Gear ratio is stamped on a metal tag attached to the differential case inspection cover. This axle is not equipped with a lubricant drain.

Rear Axle Housing Assembly Removal

1. Raise car and support on the underbody.
2. Loosen the differential carrier inspection cover, drain and discard lubricant.
3. Disconnect drive shaft at rear U-joint.
4. Disconnect rear shock absorbers at axle.
5. Remove both rear axle shafts.
6. Separate brake T from rear axle housing and separate hydraulic brake line from its retaining clip.
7. Disconnect both rear brake backing plates from the axle housing and wire them out of the way.
8. Support the axle housing on a jack, then remove spring U-bolts and plates.
9. Lower axle housing and remove it from underneath the car.

Rear Axle Housing Assembly Installation

1. Raise the axle housing into position and install spring U-bolts and plates. Torque spring U-bolt nuts to 13-20 ft. lbs.
2. Attach brake backing plates. Use new gaskets on each side of the backing plates.
3. Install axle shafts, brake drums and wheels.
4. Attach brake line T-fitting to the axle housing, and fasten the hydraulic brake line to its retainer on the housing.
5. Raise the axle housing and connect the shock absorbers.
6. Connect driveshaft at rear U-joint.

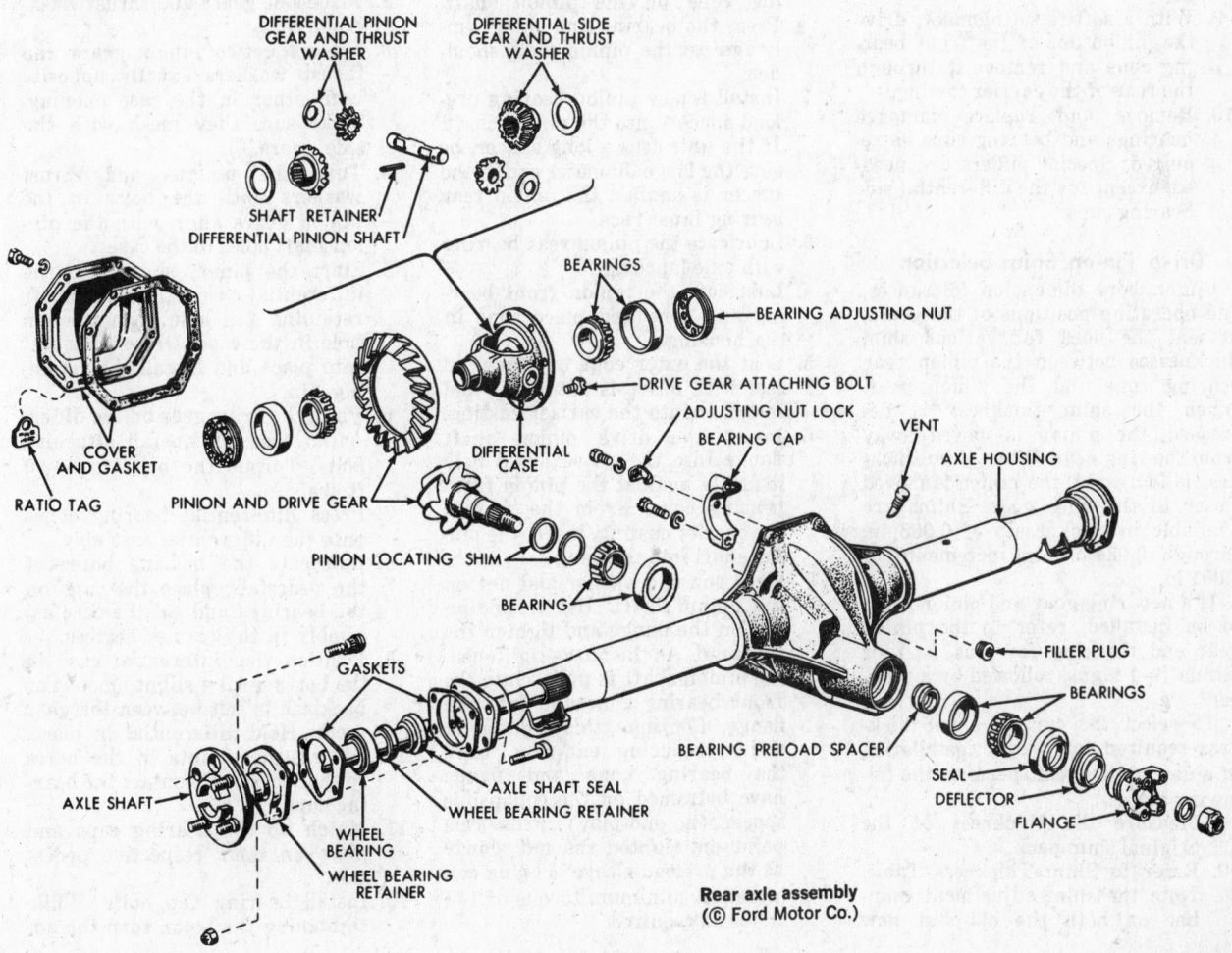

DIFFERENTIAL PINION GEAR AND THRUST WASHER — DIFFERENTIAL SIDE GEAR AND THRUST WASHER — SHAFT RETAINER — DIFFERENTIAL PINION SHAFT — BEARINGS — BEARING ADJUSTING NUT — DRIVE GEAR ATTACHING BOLT — ADJUSTING NUT LOCK — BEARING CAP — VENT — AXLE HOUSING — COVER AND GASKET — RATIO TAG — PINION AND DRIVE GEAR — DIFFERENTIAL CASE — PINION LOCATING SHIM — BEARING — FILLER PLUG — BEARINGS — BEARING PRELOAD SPACER — SEAL — DEFLECTOR — FLANGE — AXLE SHAFT — GASKETS — AXLE SHAFT SEAL — WHEEL BEARING RETAINER — WHEEL BEARING — WHEEL BEARING RETAINER

Rear axle assembly
(© Ford Motor Co.)

7. Clean inspection cover and axle housing. Apply sealer to both sides of inspection cover gasket. Daub cover attaching screw threads with anti-seize compound. Attach inspection cover and torque screws to 15-20 ft. lbs.
8. Fill axle to level with proper lubricant.

Differential Case and Drive Pinion Removal and Disassembly

With the axle assembly out of the car and the inspection cover removed:
1. Remove the differential side bearing adjusting nut-locks.
2. Mark the differential bearing caps, adjusting nuts and case for relative identity during reassembly.
3. Remove differential bearing pedestal cap bolts and caps.
4. Remove differential case and bearing cups.
5. Remove ring gear attaching bolts. Separate ring gear from differential case with a soft-faced hammer.
6. Drive out the differential pinion shaft retaining pin with a punch.
7. Use a drift to drive out the differential pinion shaft. Remove the gears and thrust washers.
8. Hold the drive pinion flange and remove pinion nut and flat washer.
9. With a soft-faced hammer, drive the pinion out of the front bearing cone and remove it through the rear of the carrier casting.
10. Remove and replace damaged bearings and bearing cups as required. Special pullers are needed, except for the differential side bearing cups.

Drive Pinion Shim Selection

Pinion bore dimension tolerances, and operating positions of the gears, present the need for various shim thicknesses between the pinion rear bearing cone and the pinion gear. When the shim thickness is decreased, the pinion is moved away from the ring gear. When shim thickness is increased, the pinion is moved closer to the ring gear. Shims are available in thicknesses of 0.008 in. through 0.024 in. in increments of 0.001 in.

If a new ring gear and pinion set is to be installed, refer to the pinion gear end marking for plus (+) or minus (−) signs, followed by a number.

To select the correct shim thickness required for proper installation of a new pinion gear, perform the following steps:
1. Measure the thickness of the original shim pack.
2. Refer to Shim Thickness Table. Note the shim adjustment number on both the old and new

Old Pinion Marking	New Pinion Marking								
	−4	−3	−2	−1	0	+1	+2	+3	+4
+4	+0.008	+0.007	+0.006	+0.005	+0.004	+0.003	+0.002	+0.001	0
+3	+0.007	+0.006	+0.005	+0.004	+0.003	+0.002	+0.001	0	−0.001
+2	+0.006	+0.005	+0.004	+0.003	+0.002	+0.001	0	−0.001	−0.002
+1	+0.005	+0.004	+0.003	+0.002	+0.001	0	−0.001	−0.002	−0.003
0	+0.004	+0.003	+0.002	+0.001	0	−0.001	−0.002	−0.003	−0.004
−1	+0.003	+0.002	+0.001	0	−0.001	−0.002	−0.003	−0.004	−0.005
−2	+0.002	+0.001	0	−0.001	−0.002	−0.003	−0.004	−0.005	−0.006
−3	+0.001	0	−0.001	−0.002	−0.003	−0.004	−0.005	−0.006	−0.007
−4	0	−0.001	−0.002	−0.003	−0.004	−0.005	−0.006	−0.007	−0.008

pinion. Observe the amount of change shown in the table under New Pinion Marking, in line with Old Pinion Marking. Note the reading. Add or subtract the reading obtained from that of the original shim pack and note the result.
3. To the above result, add or subtract the shim thickness correction indicated during the gear tooth contact pattern check. This final result is the shim thickness required for installation of the new drive pinion.

Drive Pinion Assembly and Installation

1. Place the shim and pinion bearing cone on the pinion shaft. Press the bearing and shim firmly against the pinion shaft shoulder.
2. Install a new pinion bearing preload spacer onto the pinion shaft. If the unit uses a long spacer, be sure the large diameter end of the spacer is against the pinion rear bearing inner race.
3. Lubricate the pinion rear bearing with axle lubricant.
4. Lubricate the pinion front bearing and cone and place cone in the housing.
5. Coat the outer edge of a new oil seal with oil resistant sealer and install it into the carrier casting.
6. Insert the drive pinion shaft flange into the oil seal and hold it firmly against the pinion front bearing cone. From the rear of the carrier casting, insert the pinion shaft into the flange.
7. Place the flat washer and nut on the pinion shaft. Use a holding tool on the flange and tighten the shaft nut. As the nut is tightened, the pinion shaft is pulled into the front bearing cone and into the flange. Continue this tightening action, reducing end-play, until the bearing cone and flange have bottomed on the collapsible spacer (no end-play). From this point on, tighten the nut slowly as the preload sleeve is being collapsed. A minimum torque of 140 ft. lbs. is required.

Caution If the nut is overtorqued, the pinion shaft must be removed and a new collapsible sleeve installed. Do not decrease the preload by loosening the pinion shaft nut.
8. As soon as there is preload on the bearings, turn the pinion shaft in both directions a few times to seat them.
9. Adjust pinion preload on used bearings to 10-16 in. lbs. On new bearings, 17-27 in. lbs.

Differential Case Assembly and Installation

1. Lubricate all differential parts with axle lubricant before they are installed.
2. Place side gears and thrust washers in case.
3. Place the two pinion gears and thrust washers exactly opposite each other in the case opening. Make sure they mesh with the side gears.
4. Turn the pinions and thrust washers until the holes in the pinion gears align with the pinion shaft holes in the case.
5. Start the pinion shaft into the differential case. Align the shaft retaining pin hole with the pin hole in the case. Drive the shaft into place and install the retaining pin.
6. Place the drive gear on the differential case and install attaching bolts. Torque the bolts to 40-50 ft. lbs.
7. Press differential bearing cones onto the differential assembly.
8. Lubricate the bearing bores of the pedestals, place the cups on the bearings and set the case assembly in the carrier casting.
9. Position the differential case in the bores until a slight amount of backlash is felt between the gear teeth. Hold differential in place.
10. Set adjusting nuts in the bores so that they just contact the bearing cups.
11. Match up the bearing caps and place them on their respective pedestals.
12. Install bearing cap bolts. While tightening the bolts, turn the ad-

justing nut with a spanner wrench.

13. If the adjusting nuts bind as the pedestal caps are tightened, remove the caps and inspect for cross-threading or other damage. When satisfied that the nuts are correctly seated in the pedestals and caps, tighten the cap bolts. Now, loosen the cap bolts, then torque them to 5 ft. lbs.

14. Loosen the right-hand adjusting nut until it is away from the cup. Tighten the left-hand nut until the ring gear is just fully meshed.

15. Loosen the left-hand adjusting nut one to one and one-half notches. Tighten the right-hand adjusting nut to the specified two and one-half to three notches beyond the point where it first contacts the bearing cup. As preload is applied from the right-hand side, the ring gear is moved away from the pinion. This usually results in correct backlash. Backlash should be 0.008-0.012 in.

16. Torque differential cap bolts to 40-50 ft. lbs. Measure pinion-to-ring gear backlash at several points around the ring gear in order to check runout. Runout should not exceed 0.002 in.

17. Run a gear tooth contact pattern check.

Section 21
Corvette Three-Link Rear Suspension

Rear Wheel Bearing, Spindles and Oil Seal Adjustment and/or Replacement

To measure roller bearing end-play on spindle. (Should be .001-.007 in.):

1. Raise rear of car until wheels clear ground.
2. Remove outboard axle driveshaft flange from spindle.
3. Mark camber cam and bolt in relation to bracket.
4. Loosen and turn camber bolt to force strut rods outward. This will give clearance to drop axle driveshaft.
5. Mount dial indicator on inboard surface of torque arm with pointer contacting spindle end.
6. Measure and record in and out movement of spindle.

To remove wheel spindle and support:

1. Remove wheel and tire assembly.
2. Remove spindle cotter pin, nut and washer and pull drive flange from spindle.
3. Remove brake drum or caliper and disc and pull spindle from support.
4. Pry out inner and outer seals. Remove races and rollers. Remove shims and spacers.
NOTE: bearing cups may be removed while spindle support is mounted to torque arm.
5. Disconnect brake line and remove nuts holding backing plate and spindle. Hang backing plate on frame.
6. Disconnect lower end of shock absorber. Support end of spring if necessary.
7. Remove cotter pin and nut from strut rod mounting shaft and drop strut rod. Support may now be removed.

To reassemble spindle and support:

1. If using original bearings, pack with lubricant and place in support with spacer and original shims.
2. Replace both seals. Be sure they are well seated.
3. Proceed to assemble spindle and support in sequence opposite to removal steps.
4. If using new wheel bearings, assemble with a .145 in. shim and check end-play as in Steps 1 through 6.
NOTE: shims are furnished in thicknesses from .097-.145 in. in increments of .006 in. If the .145 in. shim allows excess play, use thinner shim.
5. Now, select proper shims needed to bring the play to the specified .001-.007 in. tolerance.

Axle Driveshaft Replacement

1. Remove U-bolt clamps from inboard side gear yoke.
2. Remove bolts securing outboard flange to spindle flange.

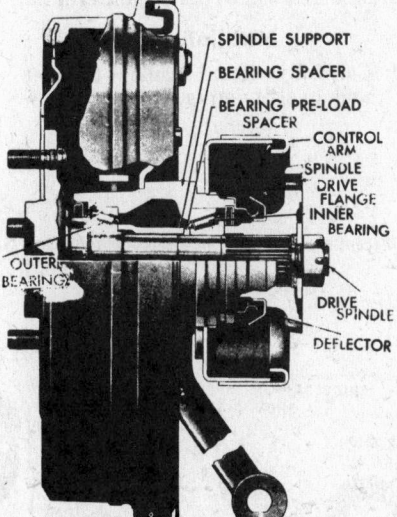

Spindle and support cross-section
(© Chevrolet Div., G.M. Corp.)

3. Pry driveshaft out at outboard end.
4. Joints may now be repaired or replaced as in the conventional cross and trunnion-type universal joint.
5. Assemble inboard trunnions into side gear yokes and install U-bolts. Torque to 14-18 ft. lbs.
6. Install outboard flange to spindle bolts and torque to 70-90 ft. lbs.

Differential Removal and/or Ring and Pinion Gear and Bearing Replacement

1. Raise rear of car and support on frame, slightly forward of control arm pivots.

Marking camber cam and bracket
(© Chevrolet Div., G.M. Corp.)

Checking spindle end-play
(© Chevrolet Div., G.M. Corp.)

2. Remove wheels and tires.
3. Place jack under spring at link bolt and raise spring until nearly flat.

4. With chain around crossmember, secure spring in this flattened position. A C-clamp should be used to prevent chain from slipping on spring.
5. Lower jack and remove link bolt.
6. Repeat Steps 3, 4 and 5 on opposite side.
7. Disconnect axle driveshafts at carrier yokes.
8. Disconnect carrier front support bracket at crossmember. Remove bolts, then remove the bracket.

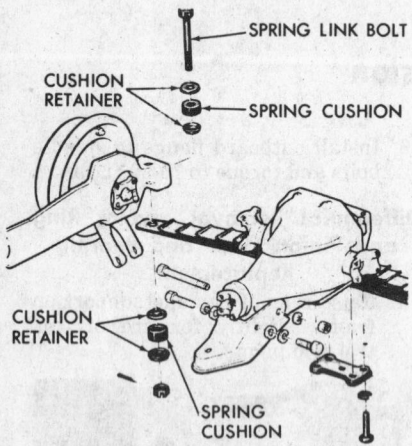

Spring mounting with link bolt
(© Chevrolet Div., G.M. Corp.)

9. Disconnect propeller shaft at transmission and companion flange.
10. Slide transmission yoke forward. Drop propeller shaft down and out toward rear.
11. Mark camber cam and bolt relative location on strut rod bracket and loosen cam bolts.

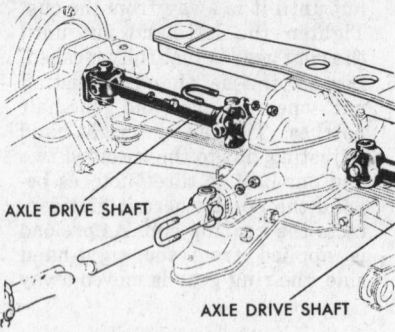

AXLE DRIVE SHAFT

AXLE DRIVE SHAFT

Axle driveshaft

12. Remove bolts securing bracket to carrier. Then, drop bracket.
13. Remove camber cam bolts and swing strut rods up and out of way.
14. Remove carrier - to - cover bolts. Loosen slowly to allow lubricant to drain.
15. With bolts out, pull carrier slightly out of cover and drop nose to clear crossmember. Work carrier down and out.

NOTE: use care not to damage gasket mounting surface in order to avoid any possible oil-leak condition.

To recondition, follow steps as described in Section 19. See chart for shim identification. Note that oil seals are used in housing at side gear yokes and should be replaced whenever disturbed in reconditioning differential.

Lubricate and tap preload shims into position using plastic hammer or brass drift, while rotating differential.

Installation

1. Carefully clean inside of cover and gasket surface. Install new gasket on cover with liberal amount of grease.

Shim Identification

Shim	Notches I.D.	Notches O.D.
.064	None	2
.066	None	3
.068	None	4
.070	1	None
.072	1	1
.074	1	2
.076	1	3
.078	1	4
.080	2	None
.082	2	1
.084	2	2
.086	2	3
.088	2	4
.090	3	None
.092	3	1
.094	3	2

2. Make two guide pins by cutting heads from two 1/4-13 x 1 1/4 in. bolts up to 1966 cars, two 1/2-13 x 1 1/4 in. bolts for 1967 to 1970 cars, and slotting unthreaded end.
3. Install these guide bolts into two holes below center, one on each side of carrier.
4. Carefully raise carrier into position, aligning the guide studs in cover.
5. Install cover bolts and tighten evenly.
6. Starting with Step 13 of previous procedure, Differential Removal, reverse sequence until Step 2 is reached. Be sure to return camber cam and bolt positions to their proper positions (Step 11).
7. Fill to plug level with hypoid lubricant and reinstall plug.

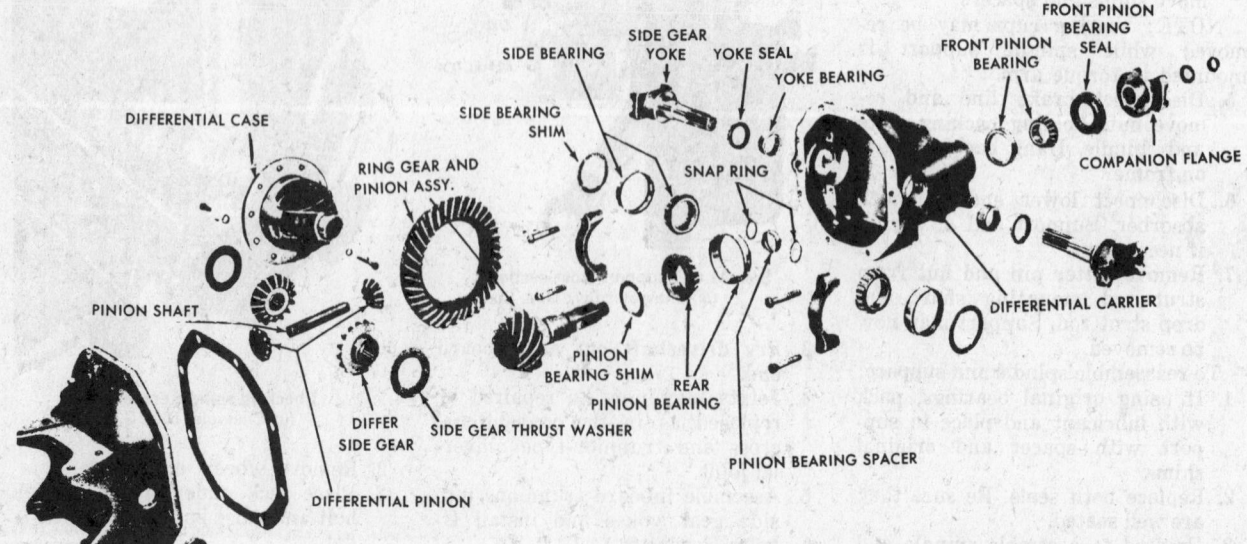

Corvette differential—exploded view (© Chevrolet Div., G.M. Corp.)

Section 22
Shim Pack Selection By Pinion Code Markings

There are several marks on most pinions, usually the part number, the number indicating a matched gear set, and the pinion depth code mark. The pinion depth code mark can be identified easily because it is usually etched rather than stamped and it is almost always etched on the rear face of the tooth itself rather than on the body or hub of the gear. In almost all cases, a plus or minus symbol is etched, rather than stamped, on the rear face of the tooth itself rather than on the body or hub of the gear. In almost all cases, a plus or minus symbol is etched on one gear tooth and the number is etched on the next tooth. If neither a plus or minus sign nor a number is found, the marking is presumed to be zero.

A plus sign indicates there is too much metal on the pinion (uses less shims.) A minus sign indicates too little metal on the gear (uses more shims.) The number states how many thousandths too much or too little.

Using the chart, start at the number between the old markings and no change in shims will be necessary to compensate for pinion depth. It may be necessary, however, to adjust the shim pack thickness to compensate for new bearings.

Example:		
Old bearing meas.	1.495 in. across	
Parallel measures	.500 the parallel	
Old bearing is	.995 in.	
New bearing meas.	1.499 in. across	
Parallel measures	.500 the parallel	
New bearing is	.999 in. wide	
New bearing	.999	
Old bearing	.995	
Difference	.004 in.	

For a pinion having different markings from the old one, refer to the chart. The chart shows that if you count toward the bottom, the shim pack is decreased; toward the top increases the shim pack.

For example, an old pinion has —5 etched on the tooth. The new pinion has +3 etched on the tooth. The shim pack has .038 in. total thickness. On the chart starting with —5 (the old pinion mark), count to +3 (the mark on the new pinion). This equals —8 blocks or .008 in. to be subtracted from the shim pack. The original pack had .038 in. Subtract-

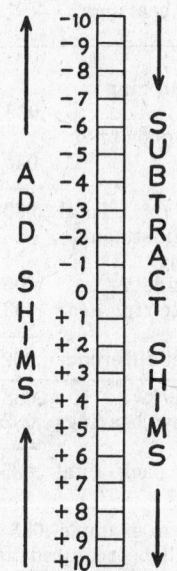

ing .008 in., new pack will have .030 in. shim thickness.

Caution   Keep in mind that the shim adjustment just made is to compensate for the manufacturing difference between the old and new pinion. It probably will be necessary to compensate further if new bearings are to be used.

Section 23
Shim Pack Selection for New Bearings

The standard manufacturing tolerance on the total width of a tapered roller bearing can be as much as .008 in. Therefore, in order to make proper allowances for possible differences between the old and new bearings, it is necessary to know the total width of both the old and the new bearing.

On the old bearing, remove the cone from the pinion and the outer race from the housing. Assemble the

bearing on the bench and place a machinist parallel bar across the outer race. Then, using micrometers, measure from the face of the inner cone to the face of the parallel bar. Subtract the thickness of the parallel bar. The difference will be the total width of the bearing. Make a note of this figure. Do the same with the new bearing and make a note of the figure. The difference between the width

of the old and new bearing will be the amount of shims which will have to be changed to accommodate the new bearing.

Remove .004 in. from the shim pack if the new bearing is thicker than the old bearing.

If the new bearing is thinner, add shims to the pack.

Section 24
Differential Side Bearing Shim Selection

When selecting shim pack thickness for differential side bearings, keep in mind that the bearings are preloaded a sufficient amount to spread the pedestals approximately .010 in. This means that the final shim packs make the differential .010 in. wider than the space where it must fit. Remove both differential side bearings and carefully measure, with a micrometer, the shim packs found behind each bearing. Make a

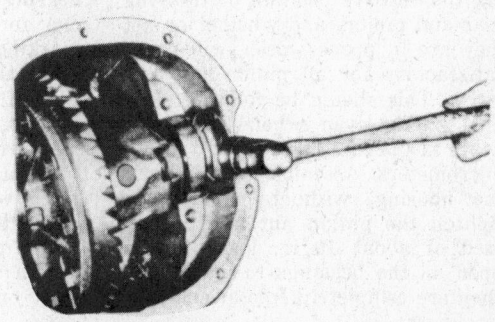

Installing differential side bearing adjusting shims

note of the thickness of each pack and also the total thickness of both packs. Refer to section Shim Pack Selection for New Bearings and alter the total shim pack to compensate for the new bearings.

Example:	
Shim pack on ring gear side	.038 in.
Shim pack opposite ring gear	.045 in.
Total	.083 in.
Required adjustment for bearings	
Ring gear side	.005 added
Side opposite ring gear	.003 removed
Net difference	.002 in.
Old shim pack	.083 in.
Add for new bearings	.002 added
New shim pack total	.085

In the above example .085 in. shim thickness will be required to set the differential up with proper preload. It remains, then, to decide how many on the ring gear side, how many on the side opposite the ring gear.

Press the bearing cones onto the differential without any shims. Make sure the bearing cones are seated firmly against the shoulder.

Slip the races over the bearing cones and install the differential into the carrier. Do not install the pedestal caps. Using two screwdrivers or putty knives, pry against the outer race of the bearing on the ring side, forcing the ring gear into mesh with the pinion. Pry until there is zero lash between the ring gear and pinion.

Caution Be sure the bearing on the side opposite the ring gear is fully entered into its race, otherwise the differential may tend to cock slightly, giving a false reading. To insure that the bearing is firmly seated, some shops install the pedestal cap finger tight on the side opposite the ring gear. This offers some resistance to the pry bars and keeps the bearing firmly seated.

With the differential pried over until there is zero lash in the gears, select, from the shim pack, a combination of shims which will just take up the space between the outer race and the shoulder in the carrier on the ring gear side. The shims are too small, of course, to fit around the race at the same time, but the selection can be rolled around in the space to be sure they fit tightly all the way around.

This group of shims will be just right for the ring gear side of the differential.

Remove the differential, pull off the bearing cone on the ring gear side and install the group of shims just selected onto the differential. Reinstall the bearing, pressing it firmly against the shim pack.

The balance of the shims go behind the bearing cone on the side opposite the ring gear. Remove that bearing, install the shims and reinstall the bearing, seating it firmly.

As stated in the first section, the differential assembly is now .010 in. wider than the space into which it must fit. If a spreader is available, spread the carrier just barely enough to slip the differential into place. If a spreader is not available, start the assembly into the carrier by cocking the outer races slightly, so that their edges will enter the carrier bearing seat. Then, press the differential into the carrier. Quite a bit of force is needed, since pressing the differential spreads the carrier something more than .010 in.

Install the pedestal caps and secure.

Section 25
Shim Combination Chart

Most bearing manufacturers supply, for each of their bearings, shims in .005 in., .007 in. and .020 in. thickness. Using combinations of the above three shims the following shim packs can be obtained:

Required thickness	Shim combinations	
.005	(1) .005	
.007	(1) .007	
.010	(2) .005	
.012	(1) .005	(1) .007
.014	(2) .007	

Required thickness	Shim combinations	
.015	(3) .005	
.017	(2) .005	(1) .007
.019	(1) .005	(2) .007
.020	(1) .020	
.021	(3) .007	
.022	(3) .005	(1) .007
.024	(2) .005	(2) .007
.025	(1) .005	(1) .020
.026	(1) .005	(3) .007
.027	(1) .007	(1) .020
.028	(4) .007	

Required thickness	Shim combinations		
.029	(3) .005	(2) .007	
.030	(2) .005	(1) .020	
.031	(2) .005	(3) .007	
.032	(1) .020	(1) .007	(1) .005
.003	(1) .005	(4) .007	
.034	(1) .020	(2) .007	
.035	(1) .020	(3) .005	

The above sample list can be continued in increments of .001 in. to any desired pack thickness.

Section 26
Pinion Mesh Markings

The following method of determining the relative position of the ring gear and pinion, and whether or not they are in proper mesh, will prove satisfactory for all pinion and ring gears. This should be followed by a final check, even when the pinion depth has been determined by special micrometers. Assemble the pinion in the housing, without preload, and tighten the pinion nut until a preload of about 10 in. lbs. is developed on the bearings to insure that they are completely free of end-play.

This, of course, is not the final bearing preload setting, but is a good one for checking the pinion mesh markings.

Install the differential assembly and adjust it to provide from .004-.008 in. backlash of the ring gear, measured at the rim of the gear.

Paint five or six of the ring gear teeth with red or white lead and, while the helper brakes the ring gear with a piece of wood, slowly turn the pinion until the ring gear makes at least one full revolution. The mesh of the pinion with the ring gear will be indicated as a mark in the red lead on the ring gear teeth. Compare this mark with the accompanying photographs. The caption on each photograph explains whether the mark indicates the pinion is too deep or too shallow, the ring gear too close or too far away.

When the marking is found to be improper, it is customary to make trial changes in increments of .005 or .007 in. If changing the shim .005-.007 in. throws the marking from too

DRIVE SIDE

MOVE PINION TOWARDS
REAR OF CAR

COAST SIDE   DRIVE SIDE

COAST SIDE

MOVE RING GEAR
CLOSER TO PINION

MOVE PINION TOWARDS
FRONT OF CAR

CORRECT MESH MARKINGS

TOE END OF TOOTH

MOVE RING GEAR
AWAY FROM PINION

HEEL END OF TOOTH

Proper tooth contact

deep to too shallow, the proper distance is about half-way between.

If, after changing this increment of shims, the mark still indicates that more must be changed, it is advisable to continue changing in the same increments, that is .005-.007 in.

While considerable time is generally required to disassemble the rear, press off the bearings, change the shims, press the bearing back on and reassemble the rear; this is still the only positive method of determining if the rear will operate quietly after it is finally installed in the vehicle.

Section 27
Improved Traction Differentials

Two general types of improved traction differentials are used. Either type accomplishes the same purpose. The direction of extra pulling power is to the rear wheel with the most traction.

The cone clutch-type drives the axle shaft through side gear, seating in the cone, the cone seating in the case. The pressure actuating these clutches is obtained either through cross pins riding up ramps in the case or through thrust springs exerting pressure on thrust blocks.

The disc-type drives the axle shaft through a series of Bellville and friction plates. The pressure is obtained by the cross pins riding up ramps in

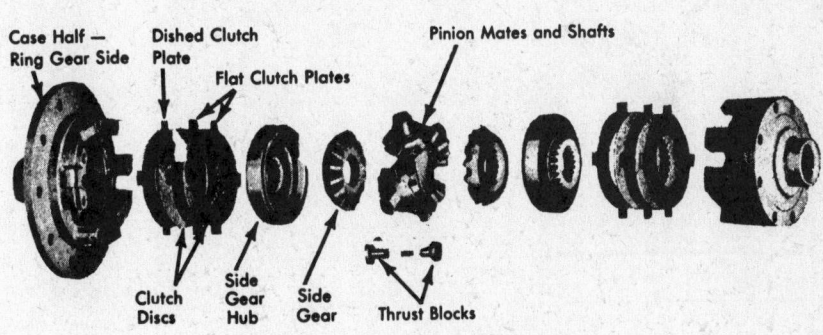

Disc type improved traction differential

Cone type improved traction differential

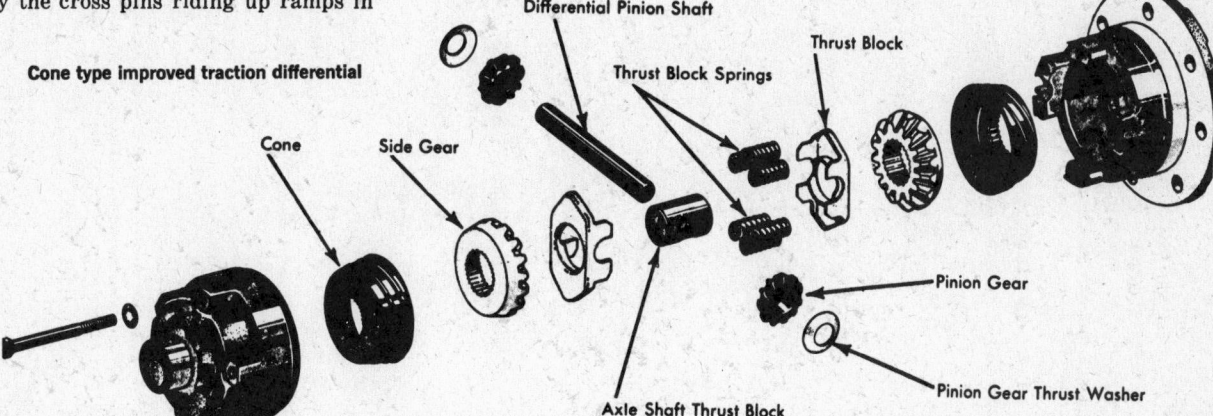

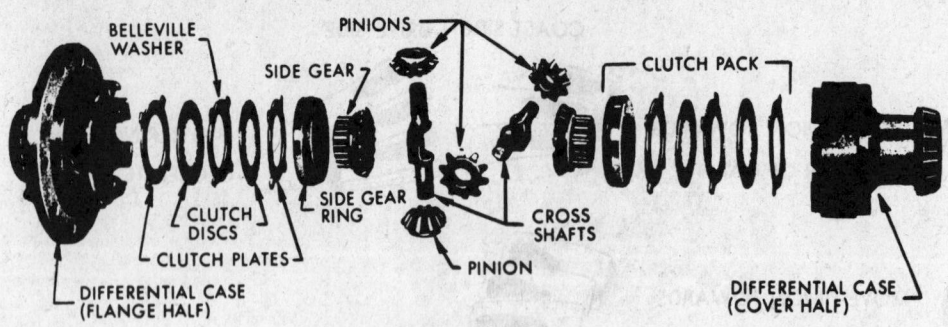

BELLEVILLE WASHER
PINIONS
SIDE GEAR
CLUTCH PACK
CLUTCH DISCS
CLUTCH PLATES
SIDE GEAR RING
CROSS SHAFTS
PINION
DIFFERENTIAL CASE (FLANGE HALF)
DIFFERENTIAL CASE (COVER HALF)

Disc type improved traction differential

the case. In this type differential, the number of plates used will vary from car to car.

In servicing these assemblies, the general procedures for pinion gears, shafts and bearings, and for ring gears, bearings and cases are followed.

For servicing the improved traction part of differential, proceed as follows:

1. Clamp case assembly in brass-protected vise and mark the case halves for proper reassembly.
2. Separate the case halves and remove the internal actuating parts.
3. Replace the required parts and reassemble in reverse of above.
4. Also, be sure to use the manufacturer's specified lubricant.

Caution Dynamic wheel balancers should not be used unless the wheel opposite the one being balanced is completely removed. With this type differential, only one driving wheel has to be in contact with the ground for the vehicle to move.

Hydraulic Brakes

Brake Type Index

Section Numbers Refer to Sections in Text

Make	Model	Section	Make	Model	Section
Buick	1964–70	7, 12, 16	Firebird	1967–70	7, 12
Buick Special	1964–70	7, 12, 16	Ford	1964–70	6
Cadillac	1964–70	7	Javelin and AMX	1968–70	4, 12
and Eldorado	1967–70	12, 16	Jeep	1964–70	8
Capri	1970	11	Maverick	1970	6
Chevrolet	1964–70	7	Mercury	1964–70	6
	1968–70	16		1967–70	12
Camaro and	1964–70	7	Montego	1968–70	6, 12
Chevelle and	1968–70	16	Mustang	1965–70	6
Chevy II				1966–70	12
Chrysler	1964–70	6	Oldsmobile	1964–70	7
	1966–70	13	and Toronado	1967–70	12, 16
Comet	1964–70	6	Olds F-85	1964–70	7, 14
	1967–70	12	Plymouth	1964–66	6
Continental	1964	6		1966–70	12, 13, 14, 15
	1967–70	12	Pontiac	1964 Front	1
Corvair	1960–69	7		1964 Rear	2
Corvette	1964	2		1965–70	7
	1964	10	Rambler		
	1965–70	12	American	1964–66	4
Cougar	1967–70	6	Ambassador	1964–69	6, 9
	1967–70	12		1970	6
Dodge and	1964–66	6	Classic and	1965–69	9
Dart and	1966–70	12, 13, 14	Ambassador	1965–70	12
Challenger	1969–70	15	Tempest	1964–70	7
Fairlane	1964–69	6	Thunderbird	1964–66	6
	1966–70	11		1965–70	12
Falcon	1964–70	6	Valiant	1964–70	2
	1967–70	11		1966–70	12, 13, 14, 15

Section Page Numbers

Section	Page	Section	Page	Section	Page	Section	Page
1	1231	5	1233	9	1237	13	1242
2	1232	6	1234	10	1238	14	1244
3	1232	7	1235	11	1238	15	1246
4	1233	8	1236	12	1240	16	1248

Wheel cylinders—1228
Master cylinders—1229
Brake bleeding—1230

Hydraulic Brakes

Brake Pressure Differential Warning System

Description

Since the introduction of dual master cylinders to the hydraulic brake system, a pressure differential warning signal has been added.

This signal consists of a warning light, usually located in the instrument cluster. The light is activated by a differential pressure switch located somewhere below the master cylinder. Both front and rear brake systems are connected to this switch assembly.

The purpose of the warning light is to signal the driver that there is trouble somewhere in the brake system. The signal indicates a hydraulic brake pressure differential of about 80 to 150 psi.

This should warn the driver that a leak or other hydraulic failure has occurred in either front, rear, or both branches of this dual braking system.

Brake Pressure Differential Warning Light Remains on

After a failure of the front or the rear branch of the hydraulic system, which has been corrected and the brakes bled, the dual brake warning light may continue to remain on due to the pressure differential valve remaining in the off-center position. (See illustration)

Caution Front wheel balancing of disc brake-equipped cars may also cause a pressure differential in the front branch of the brake system.

To centralize the pressure differential valve and turn off the warning light, a pressure difference must be created in the opposite branch of the hydraulic system from the one that was repaired or bled last.

1. Place the ignition switch in the acc or on position.
2. Loosen the differential valve assembly brake tube nut at the outlet port on the opposite side of brake system that was wheel balanced, repaired and/or bled last.
3. Depress brake pedal slowly to build line pressure until the pressure differential valve is moved to a centralized position and the warning light goes out. Then, immediately retighten the outlet port tube nut.
4. Check the fluid level in both compartments of the master cylinder and fill them to ¼-½ in. of the top, if necessary.
5. Turn ignition switch to the off position.
6. Before operating the car, check brake pedal height, firmness and general brake performance.

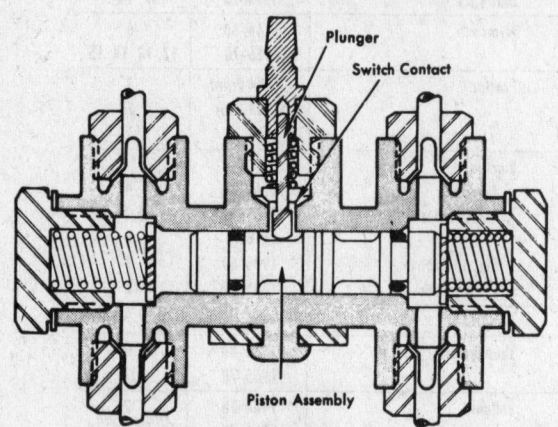

Warning light switch assembly, "on" position

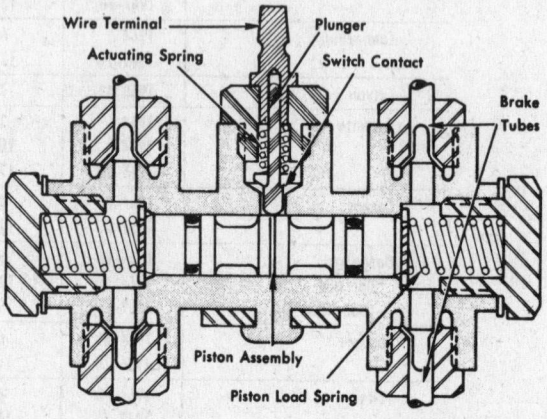

Warning light switch assembly, "off" position

Hydraulic Systems

Wheel Cylinders

There are two general types of wheel cylinders, single piston and double piston. The double piston type is divided into step bored and straight bored. Single piston wheel cylinders are used on Chrysler Lockheed brakes having two forward shoes. The purpose of step boring wheel cylinders is to get greater pressure on the forward shoe. For this reason, the larger bore is toward the forward shoe.

No fluid whatever should leak from a wheel cylinder. The slightest leak, just enough to dampen the dust boot, is too much. If a cylinder is

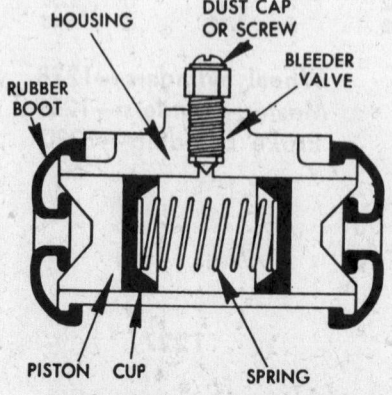

Typical straight bore wheel cylinder

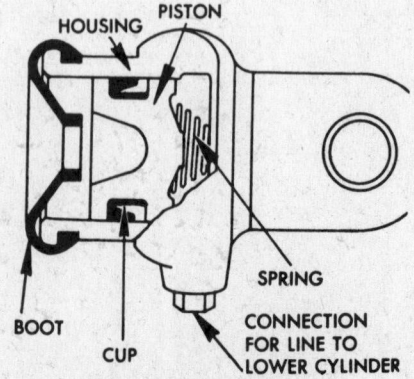

Typical single piston wheel cylinder

only suspected of leaking, it must be serviced or replaced.

Service on Wheel Cylinders

Jack up the car and remove the wheel and drum assembly from the brake to be serviced. Remove the brake shoe assemblies from the backing plate. At this point it is a good idea to thoroughly clean up the backing plate and the outside of the wheel cylinder before removing it.

Disconnect the hydraulic line. The flexible line on the front brakes is disconnected, first, where it joins to the metal line at the frame bracket. With a wrench, hold the end of the flexible line and, with another wrench, turn the compression nut on the metal line counterclockwise to disconnect. Now, using a wrench right at the wheel cylinder, take off the flexible line. Do not twist the flexible line, since this probably will damage it too much for further use. The line at the rear brakes is usually metal right up to the wheel cylinder, with the compression nut screwed right into the cylinder fitting.

After the line is disconnected, remove the mounting bolts which hold the wheel cylinder to the backing plate.

On Lockheed brakes having two forward shoes, the flexible line goes to the top cylinder only. A rigid line is used on the brake side of the backing plate to connect the top cylinder to the bottom cylinder.

Remove the boots from the cylinder, then take out the pistons and cups. If there are any scratches lengthwise on the piston outside diameter, there are very likely matching scratches on the inside of the cylinder, and it will be necessary to either replace the cylinder or have it honed.

In any case, thoroughly clean the cylinder and, holding it up to the light, examine the inside surface for roughness or pits. Examine it slowly and carefully as it is vital that the inside surface be perfectly smooth and regular. The reflected light pattern on the inside surface should be uninterrupted. If any defects are found, the cylinder must be honed or replaced.

Mount the spring, cups, pistons and boots into the wheel cylinder, dipping each part in clean brake fluid. Install a wheel cylinder clamp on the cylinder to hold the pistons in place and mount the cylinder onto the backing plate. Connect the lines and mount the brake shoes, drum and wheel.

Master Cylinders

Servicing the Conventional Master Cylinder

Remove the cylinder from the car and, before taking it apart, thoroughly clean the outside of the casting to eliminate the possibility of getting grit or dirt inside. To take the cylinder apart, first remove the boot from the cylinder, then remove the lock wire from the end, releasing the internal parts. Take out the washer, piston, cup, spring, and check valve.

Inspect the piston for scratches and pits and, if any are found, use a new one. Always use a new cup. Using a light, slowly and carefully examine the inside surface for pits and scratches. If any are found, the cylinder must be honed or replaced.

If the cylinder is the type with a partially closed bore, also examine the machined face at the front (closed) end, since this surface is the seat for the check valve and, if it is pitted or scratched, the valve will not seat properly. On cylinders with a straight-through bore, the check valve seats on the adapter fitting in the front of the cylinder. This fitting can be removed to inspect the check valve surface.

To reassemble the cylinder, first dip each part in clean brake fluid, then, install the check valve, spring, cup, piston, washer and lock ring. Except for those cylinders where the pushrod and washer are assembled under the lock ring, the boot is first attached to the pushrod and, after the cylinder is mounted on the car, the big end is attached to the cylinder and secured.

Servicing the Dual-Type Master Cylinders

Dual-type master cylinders are of two general types; those used with drum brakes, and those used with disc brakes.

The drum type contains a front and rear piston, operated in tandem. The two outlets have tubes to the front and rear wheels connected through residual pressure valves. A safety switch is mounted, in vertical position, with brake tubes connected. Failure in either line causes a warning light to operate on the dashboard.

The disc type contains a front and rear piston, operated in tandem. The two outlets have tubes to the front

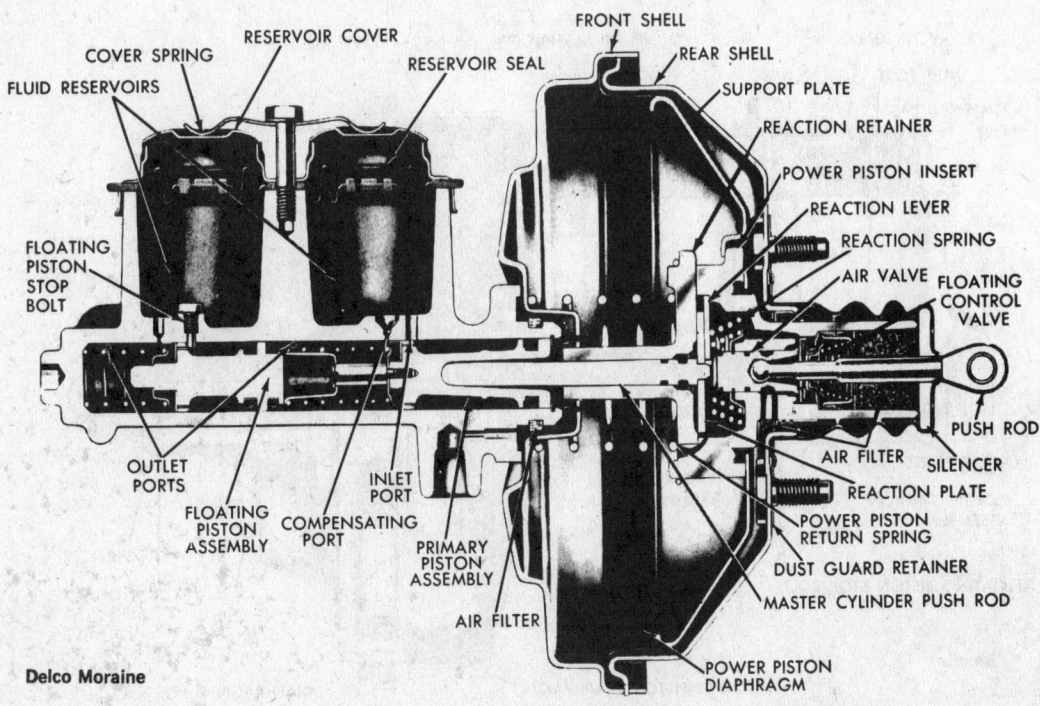

Delco Moraine

and rear wheels. Only the tube to the rear brakes is connected through residual pressure valve. A safety switch is mounted in vertical position with brake tubes connected. Failure in either line causes a warning light to operate on the dashboard.

Disassembly

1. With master cylinder removed from vehicle, thoroughly clean all external parts.
2. Remove cap retainer and remove cover and seal. Drain fluid.
3. Remove seal from cover.
4. Remove lock ring screw and retainer from open end of cylinder.
5. Remove floating piston stop bolt.
6. Remove rear piston assembly.
7. Tamp end of cylinder on wood block to remove front piston. Or, if binding, use air pressure to force out. (If air pressure is used new cups should be installed.)
8. With cylinder in vise, outlet holes up, remove tube seats. Easy-outs or small screws inserted in seat assist.
9. Remove residual valve/s and spring/s. All parts and seals should be thoroughly inspected and replaced as required. The tube seats should always be replaced.
10. Before reassembling, coat all parts with clean brake fluid and place on clean surface.
11. With cylinder in vise, outlet holes up, install residual valve/s and spring/s. Be sure springs are completely seated in bore.
12. Install new check valves so that, when properly seated, they are flush with the machined top of

outlet boss. Then, install new tube seats and seat with spare tube nut.
13. With cylinder mounted in vise, open end up, insert the components in to bore in sequence.
14. Install the retaining ring or lock screw and plate in end and install the floating piston stop bolt.
15. Install reservoir cover with new cover seal.

Bleeding Brakes

Brake fluid is, for all practical purposes, non-compressible. The inclusion of even small amounts of air will

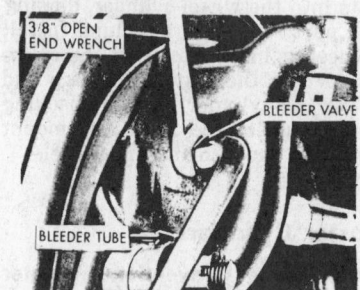

Bleeding brakes

cause the brake to feel soft, spongy and rubbery, since the fluid under pressure will compress the air.

The purpose of bleeding brakes is to expel all of the entrapped air from the hydraulic system. Cleaning and flushing the system is a much bigger job, although it includes bleeding.

Bleeding is not considered a service adjustment, since it in no way affects the actual adjustment of the brakes.

Exclusive of pressure bleeding and flushing, there are two methods in

common use to bleed brakes. The first is sometimes called the safety method, the second is generally referred to as bleeding out.

Safety Method of Bleeding Brakes

The advantages of this method are that is requires only one man and is cleaner. Proceed as follows: clean the bleeder screw at each wheel.

NOTE: Lockheed brakes, with two wheel cylinders and two forward shoes, have two bleeder screws at each front wheel, one at each wheel cylinder.

Attach a hose to the bleeder screw.

NOTE: special hoses are made for this purpose. Place the end of the hose in a jar which has been filled with fluid, preferably brake fluid, but any liquid, even water, will do. Make sure the end of the hose is down in the fluid. The idea is to prevent air from backing up into the cylinder. Now, fill the master cylinder.

Open the bleeder screw about one-quarter turn and then depress the brake pedal, slowly, all the way to the floor, hold it there for about one second, then, very slowly, let it come up. Air being ejected from the system will appear as bubbles in the bleeder jar. Repeat until no more bubbles appear in the jar while the pedal is being depressed.

Caution Refill the master cylinder when the pedal has been depressed three times.

Close the bleeder screw and remove the bleeder hose. Repeat at all bleeder screws. Do not reuse the fluid in the bleeder jar.

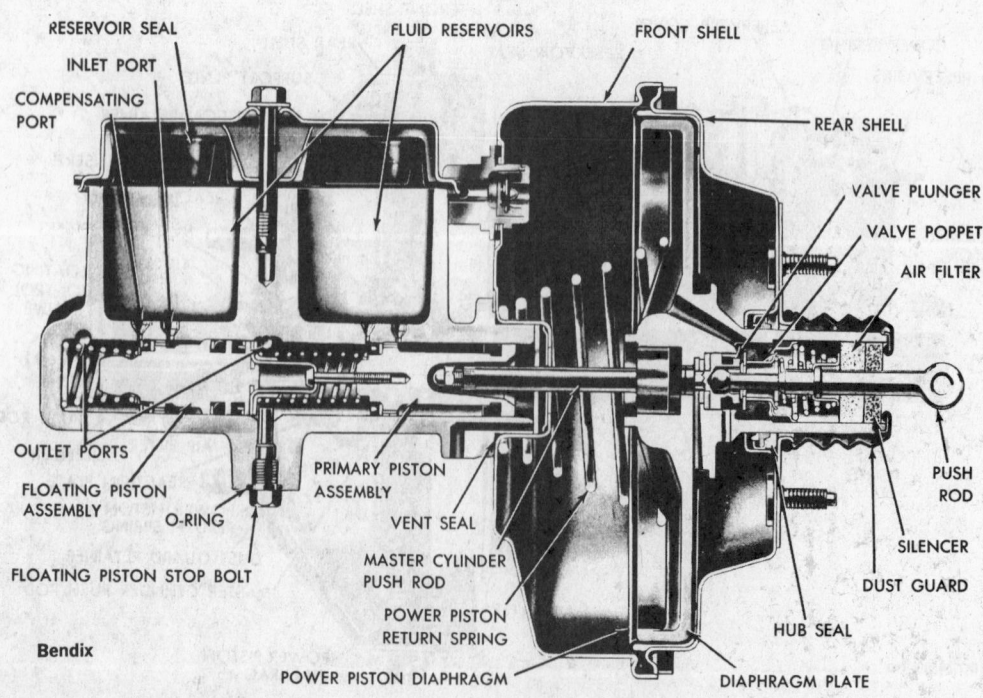

Bendix

To bleed with a pressure bleeder, be sure all dirt is away from the cylinder cap. Connect bleeder and apply pressure, as specified by equipment maker. Proceed from wheel to wheel until all air is expelled.

On dual-type master cylinders, the Moraine-type has reservoirs separated, and either section can be bled separately. On the Bendix-type, there is a vent between reservoirs. In this type, it is necessary to cap one section with a solid cap to prevent fluid and pressure loss through the cap vent hole.

Adjustment Procedures

Section 1
Bendix Brake with Eccentric Anchors

These are hydraulically actuated two-shoe brakes which have a certain amount of self-energy in either direction.

The two shoes are arranged on the backing plate so they bear against (but are not attached to) the anchor pin, which is located at the top of the backing plate. The lower end of the shoes are linked together by an adjustable sleeve, known as a star wheel. The shoes are held to the star wheel by a spring, which also acts as a detent on the star wheel.

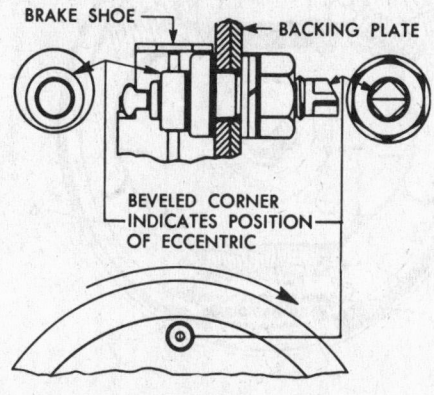

Bendix eccentric anchor

Minor or Service Adjustment

The minor or service adjustment is used to compensate for normal lining wear. Jack up the car and remove one wheel and drum assembly to inspect the brakes. If all is well, replace the wheel and drum assembly.

Remove the plug from the adjusting hole cover at the lower edge of the backing plate. Working through the

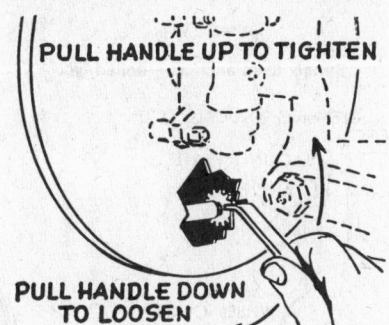

Adjusting star wheel

slot in the backing plate, expand the shoes by prying the star wheel adjuster with the special brake tool or a bent screwdriver (handle of the tool moves toward the axle to expand the shoes). Keep moving the star wheel until the shoes are very tight against the drum and it is impossible to turn the wheel. Now, retract the shoes by moving the star wheel in the opposite direction until the wheels

are free. Replace the plug in adjusting slot. Repeat at all four wheels.
Road test the car.

Major or Complete Adjustment

The major, or complete, adjustment is performed when new linings have been installed or when there is brake difficulty which cannot be overcome by a minor adjustment.

Jack up the car and remove one wheel and drum assembly to inspect the lining. If all is well, replace the drum only but leave the wheel off. Remove the other three wheels, but leave the drums in place. Notice that each drum is provided with a slot for a feeler gauge.

Remove the cover from the star wheel adjusting slot and pry the star wheel handle of the tool toward the axle, until the brakes drag very lightly. Now, loosen the jam nut on the anchor. This is the large nut located at the top of the backing plate.

Turn the drum until the feeler gauge slot is 1-1½ in. away from the heel end of the secondary shoe and insert a .010 in. feeler gauge between the lining and the drum at this point. Now, turn the eccentric anchor until the feeler gauge is just gripped. Holding the eccentric anchor at this point securely tighten the anchor jam nut.

Go back to the star wheel slot and turn the star wheel handle of the tool towards the axle until the shoes are very tight against the drum, then

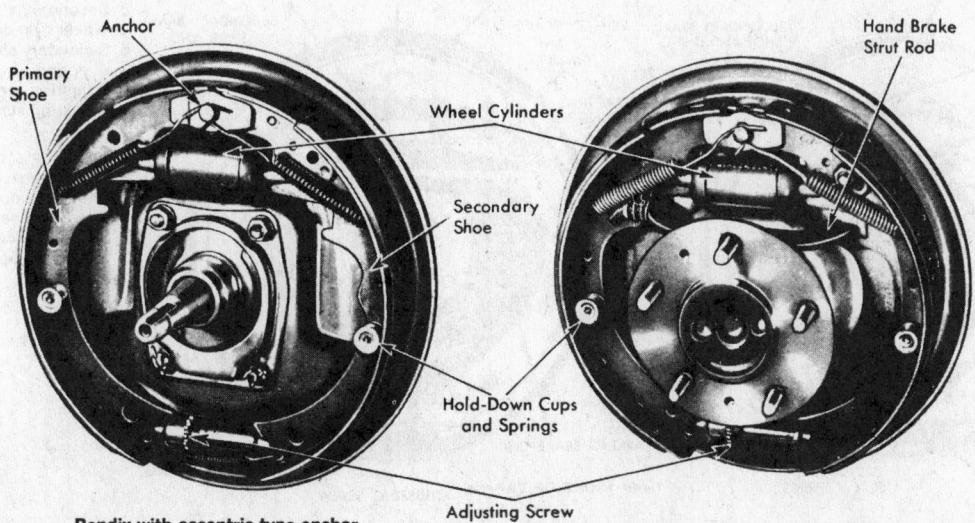

Bendix with eccentric type anchor

slowly back off until the shoes are just free. Repeat this procedure at all four wheels, then replace the wheels. Road test the car.

Handbrake Adjustments

The handbrake adjustment on these brakes is usually done in conjunction with the service brake adjustment, since the procedure for adjusting the handbrake requires that the first part of a minor service adjustment be performed on the rear wheels.

The procedure is as follows: make sure the cables are not binding in the conduit, then, remove the plug from the adjusting slot at the lower edge of the backing plate on both rear wheels.

Expand the brake shoes by prying the adjusting star wheel handle of the tool toward the axle, until the shoes are very tight against the drum.

Do this at both rear wheels, then, back off on the star wheel adjuster (handle of the tool away from the axle) until the back wheels are just free of drag.

Disconnect handbrake cable at cross-shaft or equalizer bar. Pull all slack from cables and adjust cable to proper tension (by feel) and reconnect. Test by turning both rear wheels to be sure no drag has been created.

Section 2
Bendix Brake with Fixed Anchors

These are hydraulically actuated two-shoe brakes which have a certain amount of self-energy in either direction. In appearance, the shoes look to be the same as the Bendix brakes with eccentric and sliding anchors. The difference is that, on the

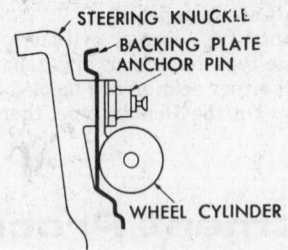

Chevrolet Front

Bendix fixed anchors—Chevrolet front

Bolted Type

Bendix fixed anchors—bolted type

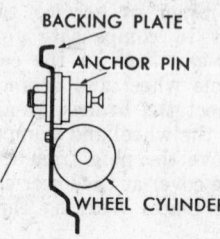

Ford Front

Bendix fixed anchors—Ford front

Swedged Type

Bendix fixed anchors— swedged type

fixed anchor brake and the self-centering anchor brake, no adjustment whatever is provided for the anchor.

Adjustments

Both minor and major adjustments are made as in Section 1, except that, because of the fixed anchor, the centering of the shoes is omitted.

NOTE: 1960 Ford left rear star wheel turns in opposite direction to adjust.

The parking brake is also adjusted as in Section 1.

Section 3
Bendix Brake with Fixed Anchors

These are the same as in Section 2 except for minor spring changes, see illustrations. Adjustments are the same as for Sections 1 and 3.

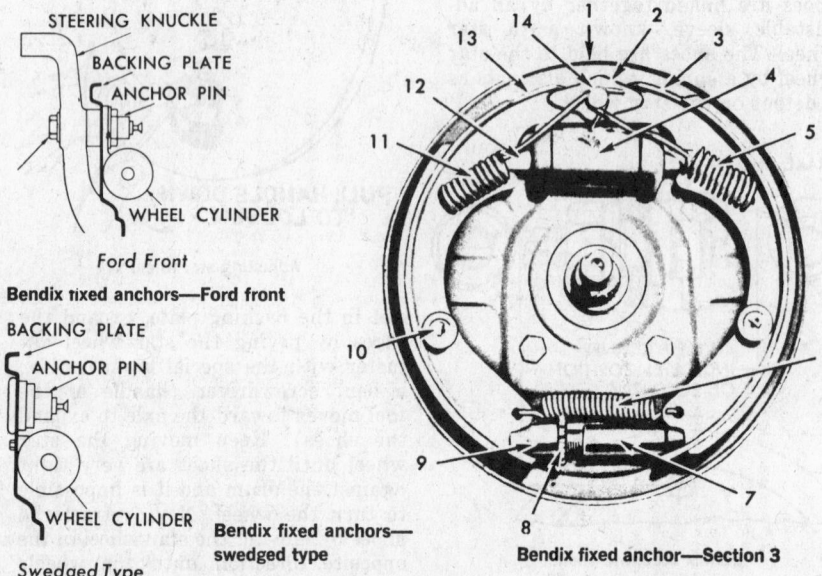

Bendix fixed anchor—Section 3

1 Flange plate
2 Shoe guide plate
3 Secondary shoe
4 Wheel cylinder
5 Secondary shoe return spring
6 Adjusting screw spring
7 Adjusting screw nut

8 Adjusting screw
9 Adjusting screw socket
10 Shoe hold down assembly
11 Primary shoe return spring
12 Wheel cylinder push rod
13 Primary shoe
14 Anchor pin

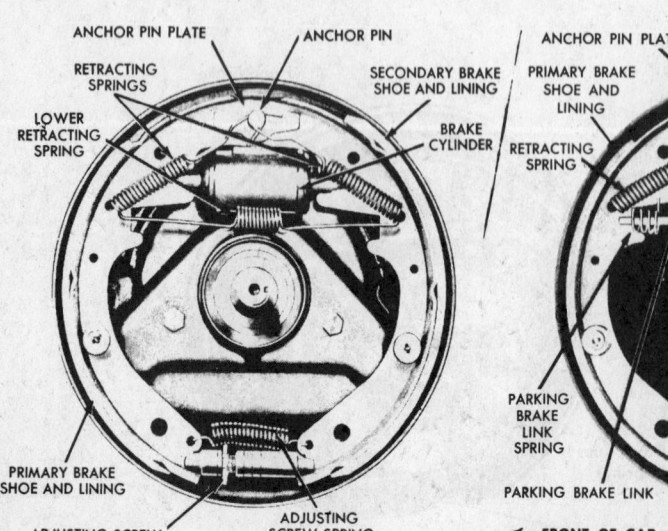

Bendix with fixed anchors

Section 4
Bendix Brake with
Fixed Anchor

These are similar to those in Section 2. Principal changes are in shoe retainers and springs, and the lack of an anchor pin plate, see illustration. Adjustments are the same as in Sections 1 and 2.

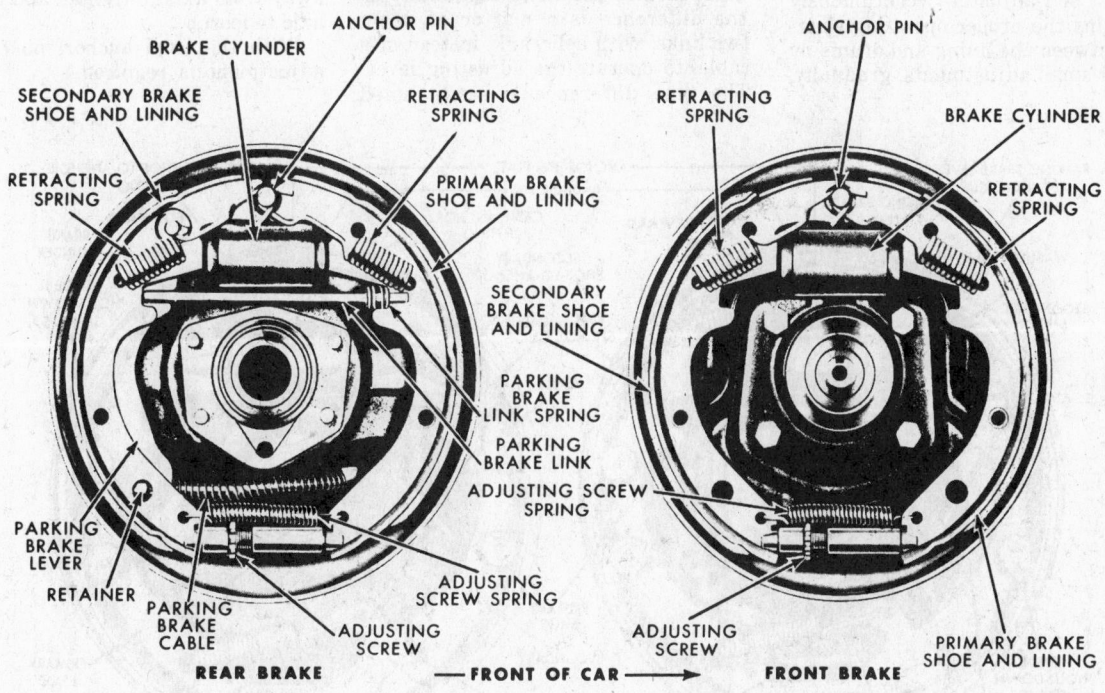

Bendix with fixed anchors

Section 5
Bendix Brake with
Fixed Anchor

These are similar to those in Section 2. Principal changes are in shoe retainers and springs.
Adjustments are the same as in Sections 1 and 2.

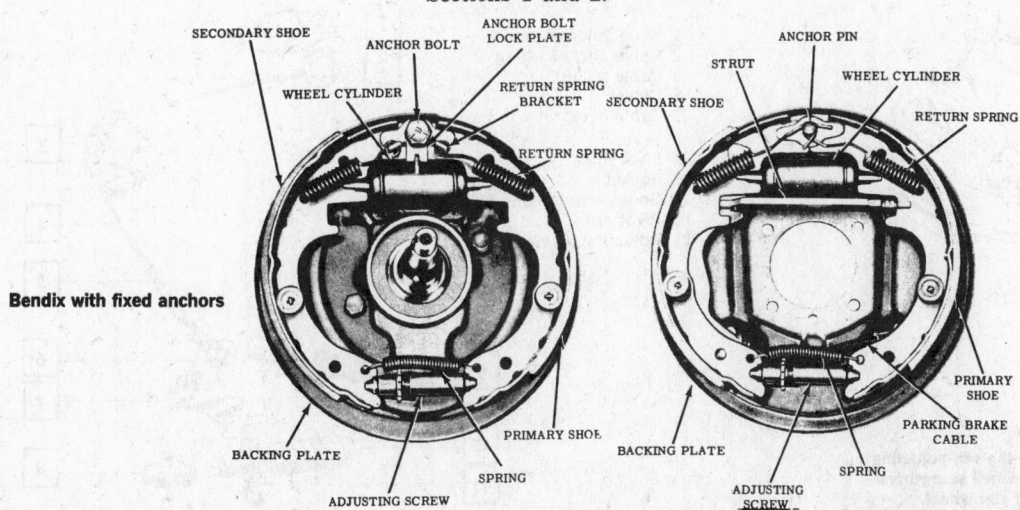

Bendix with fixed anchors

Section 6
Bendix Brake with Fixed Anchor and Self Adjuster

These brakes are very similar to the brakes common to the Bendix with fixed anchors in Section 2, but with a self-adjusting feature.

The self-adjuster continuously maintains the proper operating clearance between the lining and drums by making small adjustments, gradually,

as lining wears. When brakes are applied while backing the car, the lever engages the star wheel whenever the lining to drum clearance is too great. The star adjuster is moved one notch at a time until proper clearance is obtained.

The self-adjusting parts consist of a shoe guide plate, cable, cable guide, lever with pivot hook and a star wheel and pivot assembly, duo-servo-type. Some models use the compound-type, the difference is a matter of using two links, with bellcrank, instead of a cable, to operate the adjusting lever. The type difference is easily noted

when working with the system. The functions are the same.

Manual adjustment can be made through the backing plate slot by using a screwdriver or special adjusting tool. An ice pick or narrow screwdriver should also be inserted to raise the lever from the star wheel.

Star wheel rotation is opposite to that of non self-adjusting brakes. That is, the tool is inserted and pried away from axle to tighten and toward axle to loosen.

With the fixed anchor, no further adjustment is required.

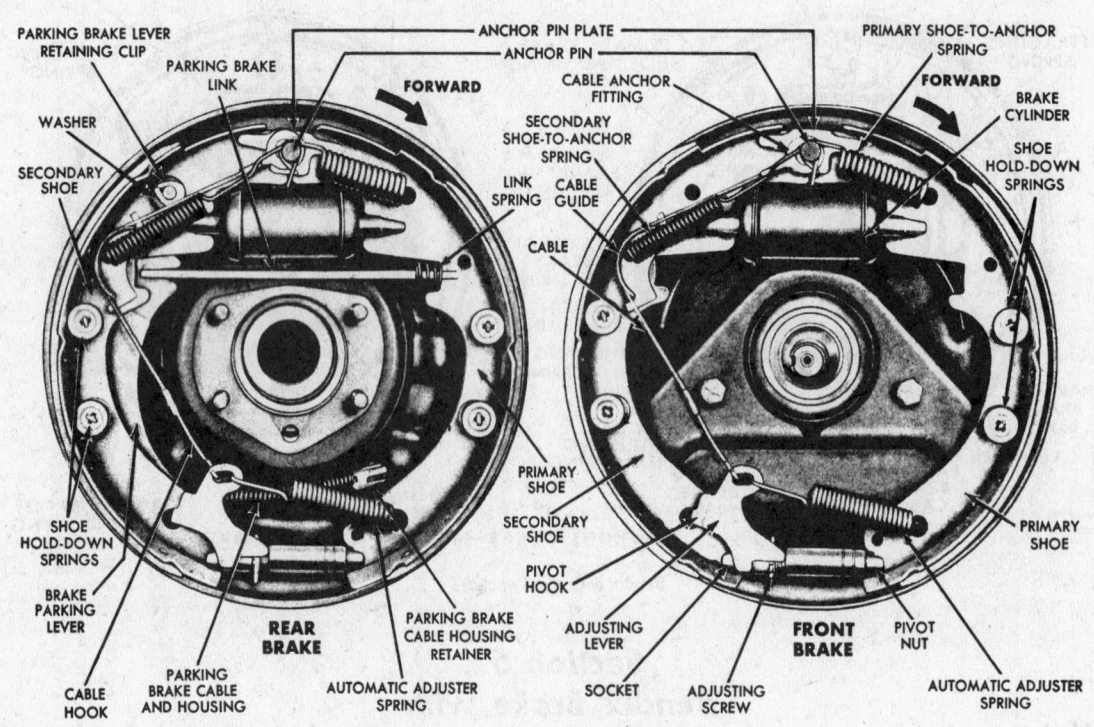

Bendix with fixed anchor—self-adjusting

Shelf-adjusting brake components

1 Shoe guide plate
2 Cable anchor fitting
3 Cable guide
4 Cable
5 Cable hook
6 Lever
7 Pivot hook
8 Socket
9 Spring—automatic adjuster
10 Pivot nut
11 Adjusting screw

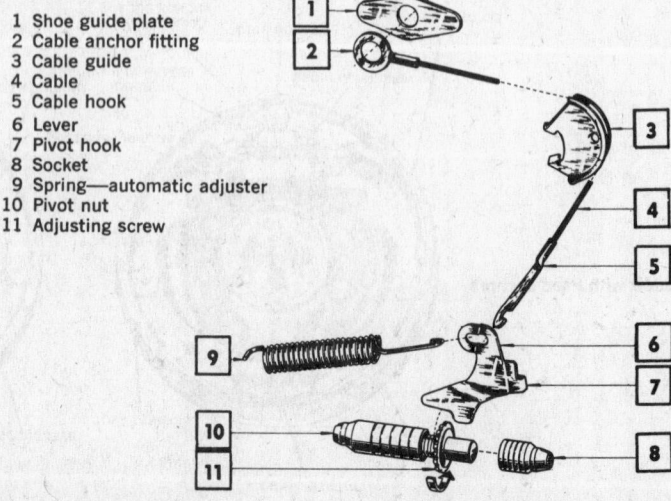

Self-adjuster method. Push the self-adjusting lever out of the way with a small screwdriver or ice pick to back off star wheel.

Section 7
Bendix Brake with Fixed Anchor and Self Adjuster

All procedures with this brake are the same as in Section 6.

The difference is structural. Self-adjusting is accomplished through an actuating link and adjusting lever, contacting the star wheel adjuster. Self-adjusters are operated when brakes are applied while moving car in reverse.

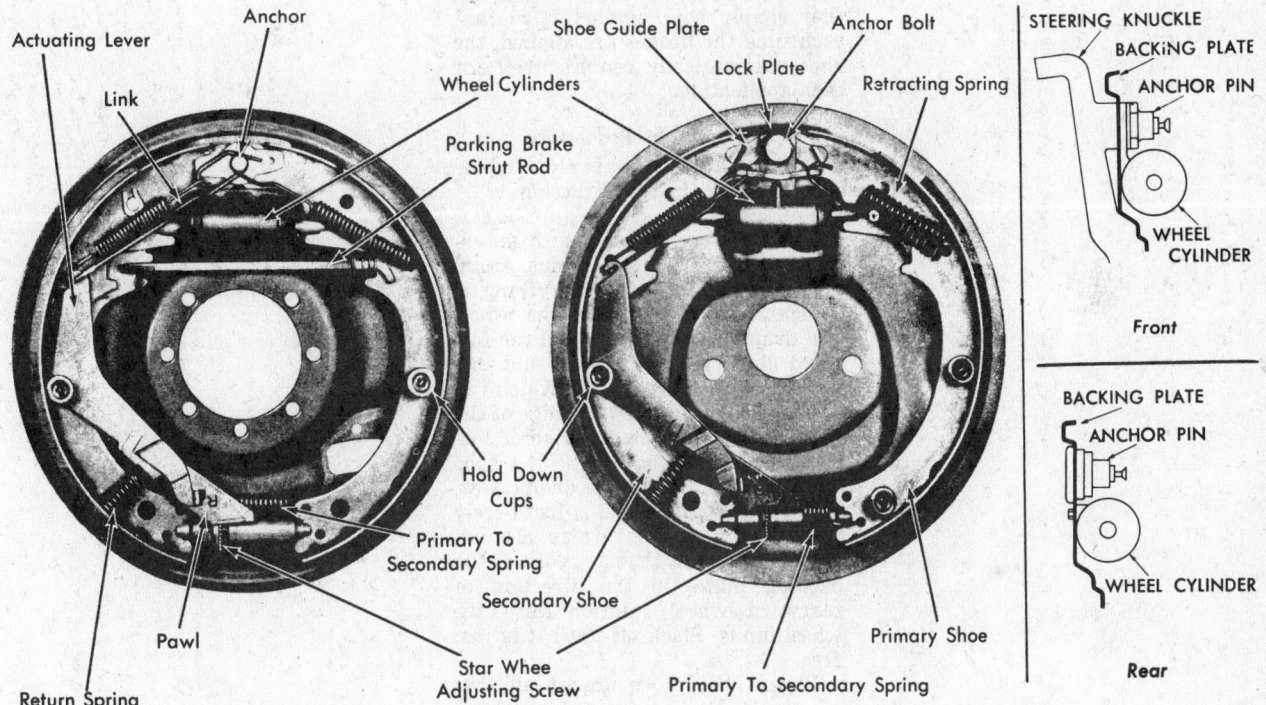

Bendix with fixed anchor—self-adjusting

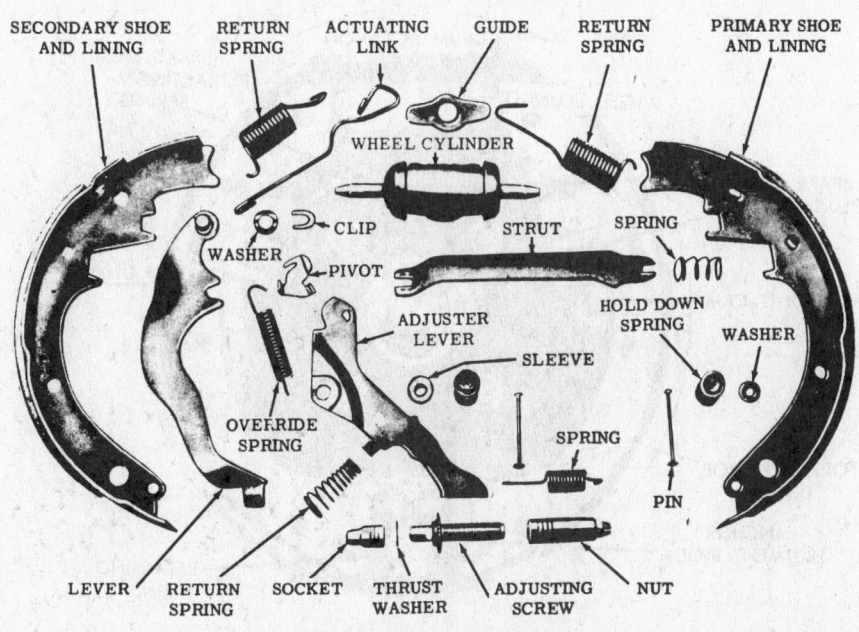

Self-adjusting brake components

Section 8
Bendix Brake with Self-Centering Shoes

This brake is completely different from, and should not be confused with, those listed above. On Bendix self-centering brakes, two shoes, hydraulically-actuated and not connected together, are used.

Instead of being connected with the usual star wheel adjuster at the bottom of the backing plate, these brakes are fitted with a thrust block against which the shoes bear. However, the shoes are not attached to this block, they simply bear against it so that each time the brakes are applied, the shoe automatically centers itself for each application.

Adjustment Procedure

Only one adjustment is provided for each shoe on this construction.

Proceed as follows: jack up the car, and remove one wheel and drum assembly and inspect the linings, drum, and wheel cylinder. If everything is to specifications, replace the wheel and drum assembly. Now, on the forward shoe, loosen the jam nut and turn the eccentric adjuster, which will be found near the forward edge of the backing plate, in the direction of forward wheel rotation until the shoe binds heavily against the drum. Now, back off until the wheel is just barely free. Then, turn the reverse shoe adjuster, located at the back edge of the backing plate, in the direction of rearward wheel rotation until the wheel binds. Slack off until it is just free.

Repeat the procedure at all four wheels.

Road test the car.

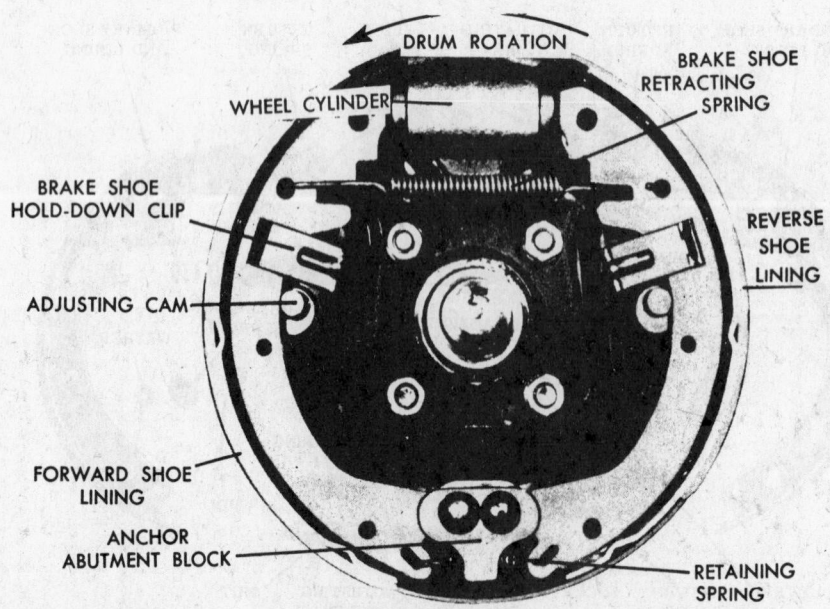

Bendix with self-centering shoes

Section 9
Wagner Brake with Self-Centering Shoes

This brake is somewhat similar to the duo-servo-type, in that, when pressure is applied, the primary shoe contacts first and the drum motion carries it against the secondary shoe and assembly, then against the keystone-shaped anchor block. The heel end of the secondary shoe now slides on the anchor block, and the assembly centers itself in the drum.

Adjustments

Jack up car. Be sure parking brake cables do not hold rear shoes from

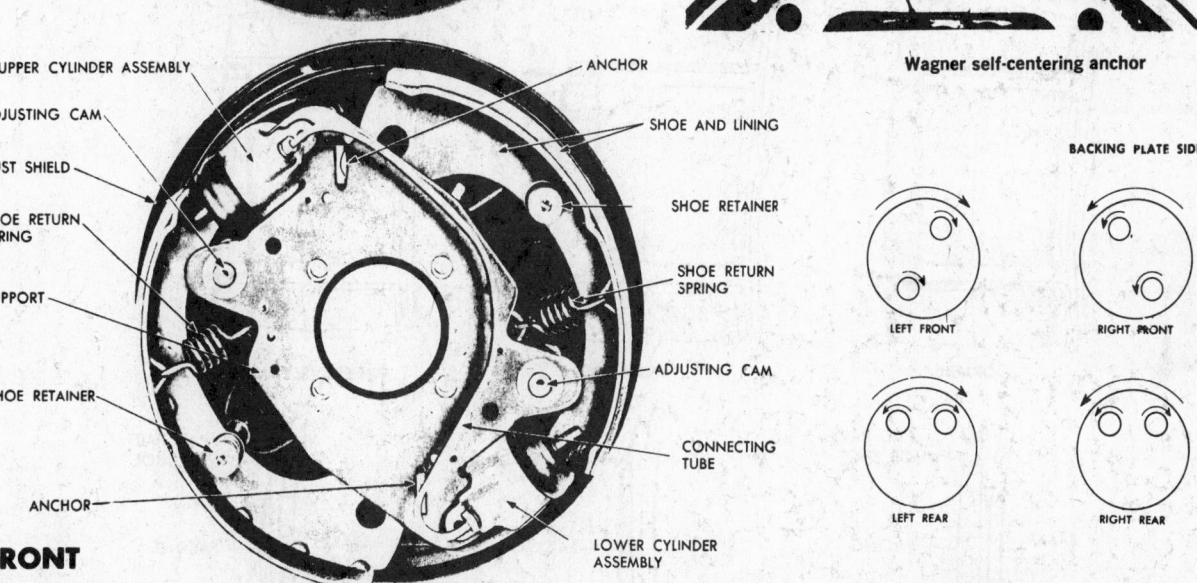

Wagner self-centering anchor

Rotation of adjustment eccentrics

anchor. Adjust the star wheel until the shoes are tight in the drum. Press the brake pedal a few times to be sure shoes have centered, then try to turn star wheel to be sure shoes are fully expanded. Back off star wheel until wheels are free.

Be sure that anchor block is free to rotate on anchor pin and that the curved side (see illustration) is toward front of car. Anchor pin must not move in backing plate.

Parking Brake Adjustment

This procedure is same as in Section 1.

DRUM ROTATION

ANCHOR PIN

PRIMARY SHOE LINING

WHEEL CYLINDER

RETAINING SPRING

STAR-WHEEL ADJUSTER

ANCHOR BLOCK (See Fig. 2)

RETRACTING SPRING

SECONDARY SHOE LINING

BRAKE SHOE HOLD-DOWN

FRONT

UPPER CYLINDER ASSEMBLY

ADJUSTING CAM

DUST SHIELD

SHOE RETURN SPRING

SUPPORT

SHOE RETAINER

ANCHOR

ANCHOR

SHOE AND LINING

SHOE RETAINER

SHOE RETURN SPRING

ADJUSTING CAM

CONNECTING TUBE

LOWER CYLINDER ASSEMBLY

REAR

CYLINDER ASSEMBLY

SHOE RETURN SPRINGS

SUPPORT

SHOE AND LINING

ADJUSTING CAMS

DUST SHIELD

SHOE RETAINERS

ANCHORS

BACKING PLATE SIDE

LEFT FRONT

RIGHT FRONT

LEFT REAR

RIGHT REAR

Chrysler total contact

Section 10
Bendix Brake with Fixed Anchor Self-Adjusting and Metallic Lining

These brakes are similar to those of Section 8 but use metallic block lining.

Adjustments are covered in Section 1.

In relining, blocks must be free of any nicks or burrs. Hands must be clean while handling. All oil or grease should be kept from lining. Do not use solvent for cleaning.

As the self-adjuster operates on the primary shoes, adjustment takes place on forward brake application.

NOTE: brake shoes with metallic lining require a specially finished brake drum (honed to a 20 micro-inch finish). Metallic brakes are not recom-

mended for replacement on vehicles having standard brake drums not honed to specifications.

The metallic shoe assembly has two layers of different material. Only the top layer is braking material. During inspection, one may easily be misled since the braking layer does not appear much different from the bonding layer. Close inspection will reveal the difference.

To prevent damage to the drum from worn shoes, replace the brake shoes whenever the thickness of the braking layer becomes less than 3/32 in. measured at the heel or toe of the pad.

Heavy duty metallic self-adjusting brake

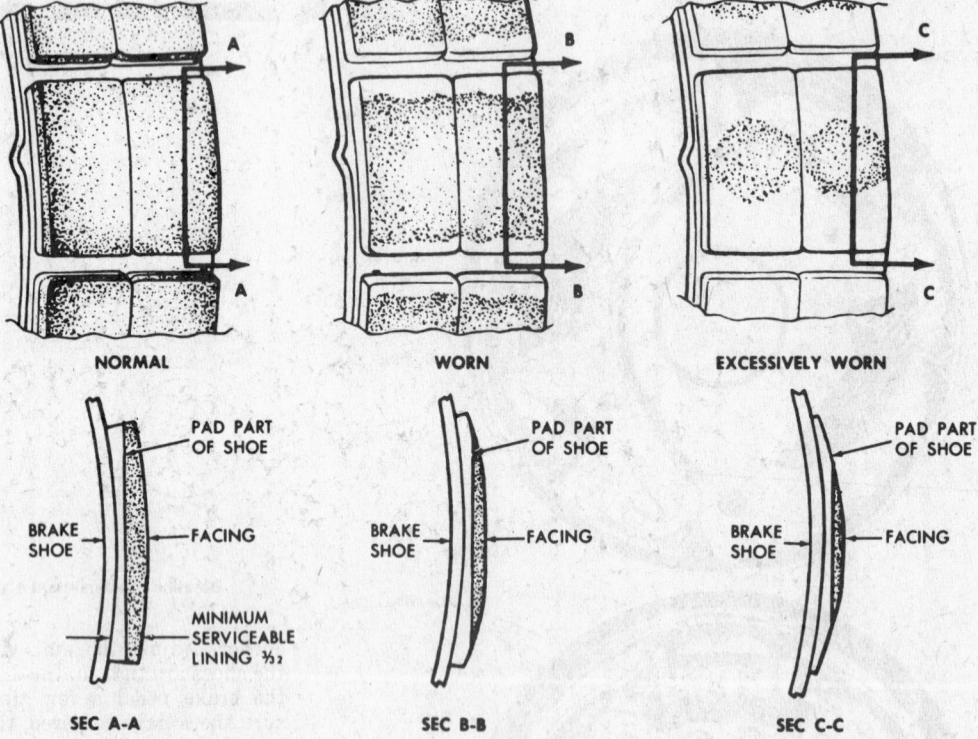

Section 11
Capri Brake System

Pressure Differential Valve Centering Procedure

Remove the dust cover at the bottom of the valve body and insert the centering tool through the opening. The piston can be moved to the center position with this tool, which can be fabricated from a small screw driver.

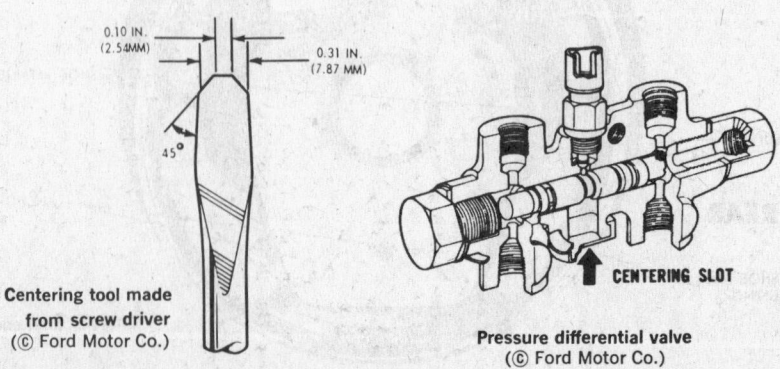

Centering tool made from screw driver
(© Ford Motor Co.)

Pressure differential valve
(© Ford Motor Co.)

Bleeding the System

To bleed the system, three screws are provided—one on each of the front disc brakes and one on the left rear drum brake. The major difference from U.S. cars is that the centering tool must hold the piston in the pressure differential valve centered. After centering the valve, proceed to bleed the brakes as normal.

Disc Brake Pad Removal and Installation

1. Jack up car and remove tires and wheels.
2. Pull out retaining pin clips, retaining pins and remove pads and shims. If necessary, use needle nosed pliers carefully.
3. Push pistons in cylinders until they seat. If the master cylinder was full, it may overflow, so remove some fluid.
4. Install new shims and pads.
5. Reinstall retaining pins and secure with retaining pin clips.
6. Replace wheels and lower car.

Piston Seal Removal and Replacement

1. Lift car and remove wheels.
2. Remove drake pads. Push brake pedal to bring pistons in contact with disc.
3. Remove brake line from union on rear of caliper and plug ends.
4. Bend lock tabs, remove the two caliper retaining bolts and remove caliper assembly.
5. Partially remove piston, remove retaining ring and remove the bellows from lower point of piston skirt. Remove piston.

6. Pull sealing bellows from ring in cylinder.
7. Repeat for other cylinder.
8. Clean metal parts in alcohol or suitable solvent. Inspect piston and cylinder bore for scoring and wear. Replace parts showing wear and all rubber parts.
9. Insert piston seal in groove in cylinder. Insert bellows to cylinder with lip outwards installed in groove in cylinder.
10. Lubricate piston with brake fluid and insert, crown first, through bellows into cylinder.
11. Install inner edge of bellows into groove in piston skirt.
12. Push piston into cylinder as far as possible. Secure bellows to piston with retaining ring.
13. Repeat for other piston.
14. Using new locking plate, replace caliper assembly and tighten bolts to 40-50 ft. lbs. Bend tabs on locking plate.
15. Attach brake line to caliper.
16. Push pistons into cylinders and insert pads and shims. Secure pads with retaining pins and clips.
17. Replace wheels and lower car, bleed system.

Checking the Runout

Disconnect the connecting rod from the steering arm at its outer end, after removing the cotter pin, castellated nut and separating the ball joint. Attach dial indicator to steering arm and check runout. If it exceeds 0.0035 in. either cure the problem if it is misalignment of hub or dirt or dust, or replace the disc.

Disc and Hub Replacement

1. Remove the caliper assembly.
2. Remove the grease cap from the end of the hub.
3. Remove cotter pin and adjusting nut retainer, adjusting nut, thrust washer and outer bearing race.
4. Remove hub and disc from spindle.
5. Separate disc from hub by bending lock tabs and removing the bolts. Discard the locking plate and bolts.
6. Clean mating surfaces of hub and disc.
7. Align mating marks, place disc on hub and install new locking plate and bolts. Tighten bolts to 30-34 ft. lbs. and bend lock tabs.
8. Replace hub and disc assembly on spindle. Install outer bearing race, thrust washer and adjusting nut.
9. Tighten nut to 27 ft. lbs. while rotating disc to insure proper seating of the bearings.
10. Slacken nut 90 degrees and install nut retainer so castellations line up with cotter pin hole in wheel spindle. Install new cotter pin.
11. Check runout.
12. Tap grease cap in place in end of hub.

Rear Brakes

The rear brakes are self-adjusting drum type brakes similar to the type used in U.S. cars, except that the self-adjustment occurs when the parking brake is applied.

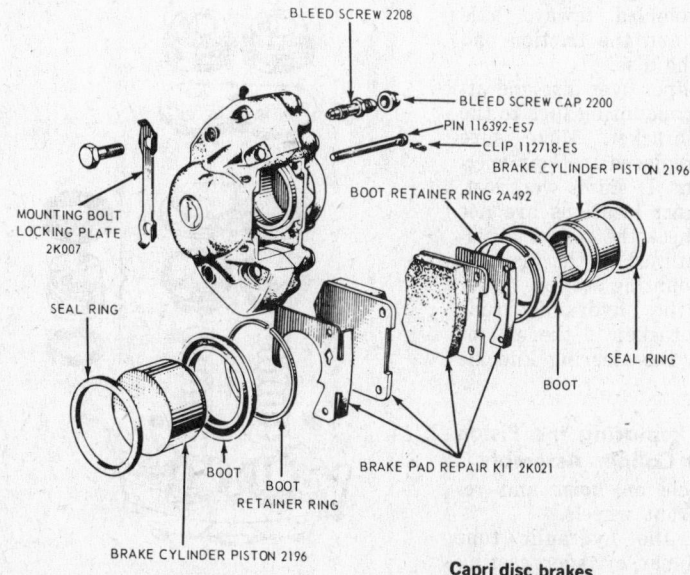

Capri disc brakes
(© Ford Motor Co.)

Capri rear brakes
(© Ford Motor Co.)

Disc Brakes – Bendix (Series "E")

Section 12

General Information

The following information must be considered as typical of the Bendix Series E caliper disc brake. Specific data on different car applications may vary slightly from that shown. In general, however, the basic information should still apply.

The caliper disc brake should be inspected periodically to determine the degree of wear of the friction pads. To inspect the pads for wear, remove the front wheels from the hubs and observe the pads through the two openings in the bridge of the caliper. When the pads have worn down to approximately ⅛ in. thickness maximum pad life has been achieved. Do not let the pads wear until the shoe plate rubs on the disc, as this will tend to score the disc surface. When maximum pad life has been obtained, the shoe plate and friction pad assembly should be replaced. It is recommended that all of the pads of the front brakes be replaced at the same time. Refer to the following procedure for replacing the shoe plate and friction pad assemblies.

Procedure for Replacing the Shoe Plate and Friction Pad Assemblies

1. Place vehicle on hoist and remove the front wheels. Work only on one brake at a time.
2. Disconnect the hydraulic tube mounting bracket from the end of the bolt in the steering knuckle arm. If possible, don't disconnect the tube or loosen any hydraulic fittings. This will eliminate the need for bleeding the system.
3. Remove the two caliper mounting bolts, using care not to loosen shims (if used) between the mounting lugs and the caliper mounting bracket. The shims must be replaced in same locations when the caliper is remounted.
4. Lift caliper assembly off disc and use a length of rod or wire to hang caliper to vehicle upper suspension. This will relieve any strain on the hydraulic connections and will also leave both free to install new friction pad assemblies.
5. Lift worn pad assemblies from the caliper.
6. Push the four pistons back until they bottom in their cylinders. Some resistance will be encountered, since the displaced fluid will have to flow back through the small compensating port lo-

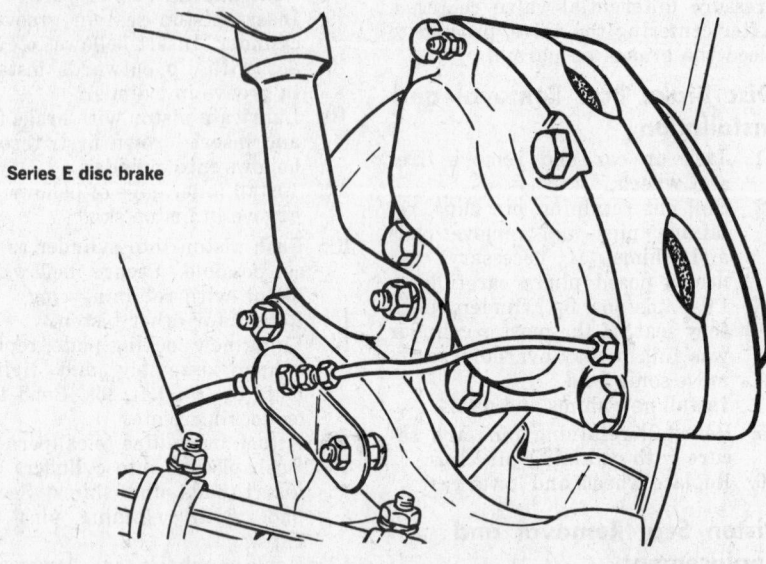

Series E disc brake

cated in the master cylinder. If the master cylinder reservoir is relatively full of fluid, some fluid may overflow through the breather port when all of the pistons have been bottomed in their cylinders. Use caution to prevent the pressure developed, while bottoming one piston, from pushing the other pistons out of their bores.

7. Install new friction pad assemblies (one at a time) into the brake caliper. The pistons must be completely bottomed to provide necessary clearance for installing the new pad assemblies and reinstalling the caliper over the disc. Flat edge of pad assembly is assembled toward axle center line and the friction pad is toward the disc.
8. Straddle caliper over disc and attach caliper mounting lugs to the mounting bracket. Make sure shims are replaced in their original locations. If shims were lost, or their exact locations are not known, recheck the caliper alignment as outlined in the installation and mounting section.
9. Reattach the hydraulic tube mounting bracket to the end of the bolt in the steering knuckle arm.

Procedure for Replacing the Piston Seals in the Caliper Assembly

1. Place vehicle on hoist and remove the front wheels.
2. Disconnect the hydraulic tube from the inner cylinder casting of the caliper. Plug the open end

of the tube to prevent loss of fluid.

3. Remove the two caliper mounting bolts, using care not to loosen the shims (if used) between the mounting lugs and the cali-

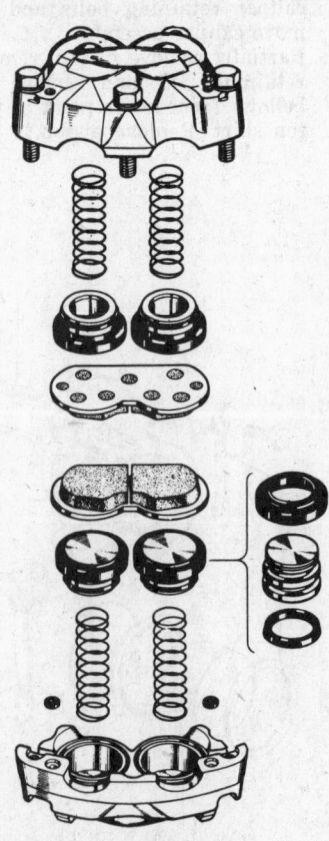

Brake caliper

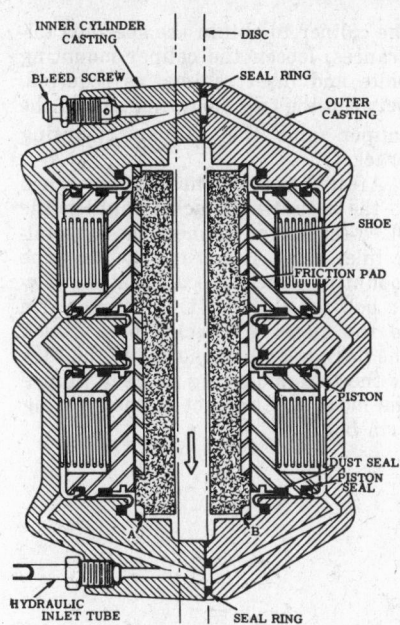

Cross-section, caliper and passages

per mounting bracket. The shims must be replaced in same locations when the caliper is re-mounted.

4. Lift the caliper assembly off the disc. Open bleed-screw and let excess fluid drain out of caliper cylinders and passages. Place caliper on clean area of work-bench and proceed with disassembly.

5. Lift the shoe plate and friction pad assemblies from the cylinder castings.

   NOTE: use care to keep friction pads clean, do not contaminate with brake fluid, grease, or oil.

6. Remove the bleed-screw from the inner cylinder casting.

7. Clamp caliper lugs securely in a vise and remove the three cylinder bolts.

8. Remove the two small seal rings from recesses in the outer cylinder casting.

9. Disengage the dust seals from grooves in the cylinder castings and remove pistons from their cylinders. Low-pressure air applied with caution to the hydraulic inlet or bleed-port will aid in removing the pistons.

10. On each piston, disengage the dust seal and piston hydraulic seal from their respective grooves. Use a small, smooth, end screwdriver or a scribing tool to lift the seals from their grooves.

11. Clean all metal parts with a commercial cleaning fluid such as Bendix Metalclene. Clean rubber parts with alcohol.

12. Inspect the rubber parts and replace with new parts when defects are found. It is recommended that the rubber parts be replaced with new parts if the

brake has been in use for a considerable length of time without service (such as 40-50,000 miles).
   A. The piston seals should not show signs of deterioration and the lip of the seal should not have any cuts or nicks.
   B. The dust seals should not have cracks due to flexing.
   C. The seal rings should not show a permanent set.

13. Inspect the seal grooves in the pistons for nicks and burrs. If possible, remove any nicks or burrs with fine emery cloth.

14. Inspect the cylinder bores for scoring. If light scoring is found, it normally can be removed with fine crocus cloth. If the scoring is deep, the cylinder casting must be replaced. Also check area near bottom of cylinder bores for corrosive deposits. Remove any deposits with fine crocus cloth, if possible.

15. Inspect inner and outer cylinder castings for any cracks, especially near the holes in the mounting lugs and the holes for the cylinder clamping bolts.

16. Lubricate the piston hydraulic and dust seals with brake fluid and reinstall seals in the piston grooves. Note that the hydraulic seal lip points toward open end of piston.

17. Lubricate the cylinder bores with coat of brake fluid and insert pistons into the cylinders. Be sure coat of brake fluid covers seal lip. Do not damage the seal lip. Insert end of a scribing tool between seal lip and edge of the cylinder and move the scribing tool around the circumference of the seal lip. This will permit the seal lip to enter the cylinder without damage.

18. Push in each piston until it bottoms in its cylinder.

19. Clamp the lugs of the inner cylinder casting in a vise. Insert the two seal rings in the recesses in the outer cylinder casting. Place the inner and outer cylinder castings together and install the three cylinder bolts. Tighten bolts securely so that the two rings seal around the fluid cross-over passages and there is metal to metal contact between the castings.

20. Install the shoe plate and friction pad assemblies in the inner and outer cylinder castings.

21. Replace the bleed-screw in the inner cylinder casting.

22. Reinstall the caliper assembly in the vehicle. Straddle the caliper over the disc and attach caliper mounting lugs to the mounting bracket. Make sure shims are replaced in their original locations. If shims were lost, recheck the caliper alignment as outlined in

the installation and mounting section.

23. Reattach the hydraulic tube to the inner cylinder casting and then bleed the brakes as outlined in the following procedure.

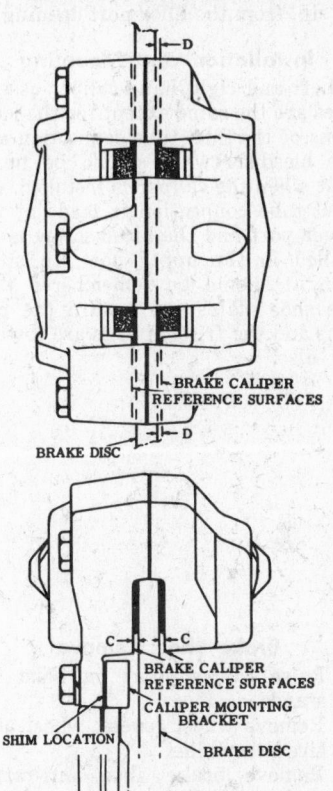

Alignment of caliper to disc

Bleeding Procedure

1. Inspect system to make sure all connections are tightened and all bleed-screws are closed.

2. Clean dirt from the filler cap of the master cylinder reservoir, then remove the filler cap.

3. Fill the master cylinder reservoir with approved brake fluid.

4. Attach a drain tube to the bleed-screw on brake located at the farthest hydraulic-line distance from the master cylinder. Immerse open end of the tube in a container having a small amount of brake fluid. Keep end of tube submerged in the fluid.

5. Open bleed-screw and have assistant operate the brake pedal through its full stroke. Tighten bleed-screw before releasing pedal. Repeat operation until fluid, free of bubbles, comes out of the drain tube. Then securely tighten the bleed-screw.

6. Repeat Steps 3, 4 and 5 at remaining wheel brakes, bleeding the shortest line last. (If work was done only on front brakes, the rear brakes normally will not have to be bled.) After the system has been properly bled, the brake pedal should feel firm.

7. Apply the brake pedal firmly and hold pressure in the system. Inspect the entire system for leaks.
8. Add fluid to the master cylinder reservoir if necessary. Fluid level should be maintained at about ½ in. from the filler port opening.

Installation and Mounting

Left and right-hand caliper assemblies are the same except for the locations of the inlet tube connection and the bleed-screw It should be noted that when the caliper is installed, the inlet tube connection is made at the lower port and the bleed-screw is installed in the upper port. In addition, it should be remembered that the shoe plates are toward the pistons and the friction pads are toward the disc.

To install the caliper, straddle the caliper over the brake disc with the disc located between the two friction pads. Attach the mounting lugs to the caliper mounting bracket using the two bolts and lockwashers. Tighten bolts and then check the alignment of the caliper with the disc. Use feeler gauges to check that the caliper is centered relative to the disc rubbing surfaces. Check spacing betweeen the caliper reference surfaces. Make this check at both ends of the caliper. Dimensions "C", should be equal, to within 0.008 in. Also inspect caliper to see if it is parallel to the disc rubber surfaces. Use feeler gauges to check dimensions "D," which should be within 0.002 in. If it is necessary to shift the caliper to obtain the specified tolerances, loosen the caliper mounting bolts and insert shims, as required, between the two mounting lugs of the caliper and the caliper mounting bracket.

After the caliper has been attached to the mounting bracket, tighten the bleed-screw and connect the hydraulic inlet tube. Attach tube to the tube mounting bracket located on the lower bolt which holds the steering arm to the steering knuckle pin. Attach the flexible hose connector to the tube at the tube mounting bracket. Secure the mounting bracket to the steering arm bolt.

Disc Brakes by Budd

Section 13

Brake Shoe Removal

1. Raise car, support on hoist or stands.
2. Remove wheel covers, wheel and tire assemblies.
3. Remove brake shoe anti-rattle spring.
4. Remove bolts that attach caliper assembly to steering knuckle and steering knuckle arm.
5. Remove caliper from disc by slowly sliding the caliper assembly up and away from the braking disc. As the friction pads start to clear the disc, insert piston compression tool C-3992 between the piston insulator pads and turn knob on tool until pistons are fully compressed.

Cleaning and Inspection

Check for piston seal leaks and for any damage of the piston boot. Wipe the cavity clean (between inner and outer caliper housings). Check piston boot for proper seating in the piston groove and under the coiled retainer in the caliper groove. If a leak is evident, it will necessitate disassembly of the caliper and the replacement of piston seals and boots.

Brake Shoe Installation

1. Slide shoe and lining assemblies into position in caliper assembly (one at a time).
2. Position shoes, then install tool C-3992 between them. Adjust tool so that it holds the shoes apart

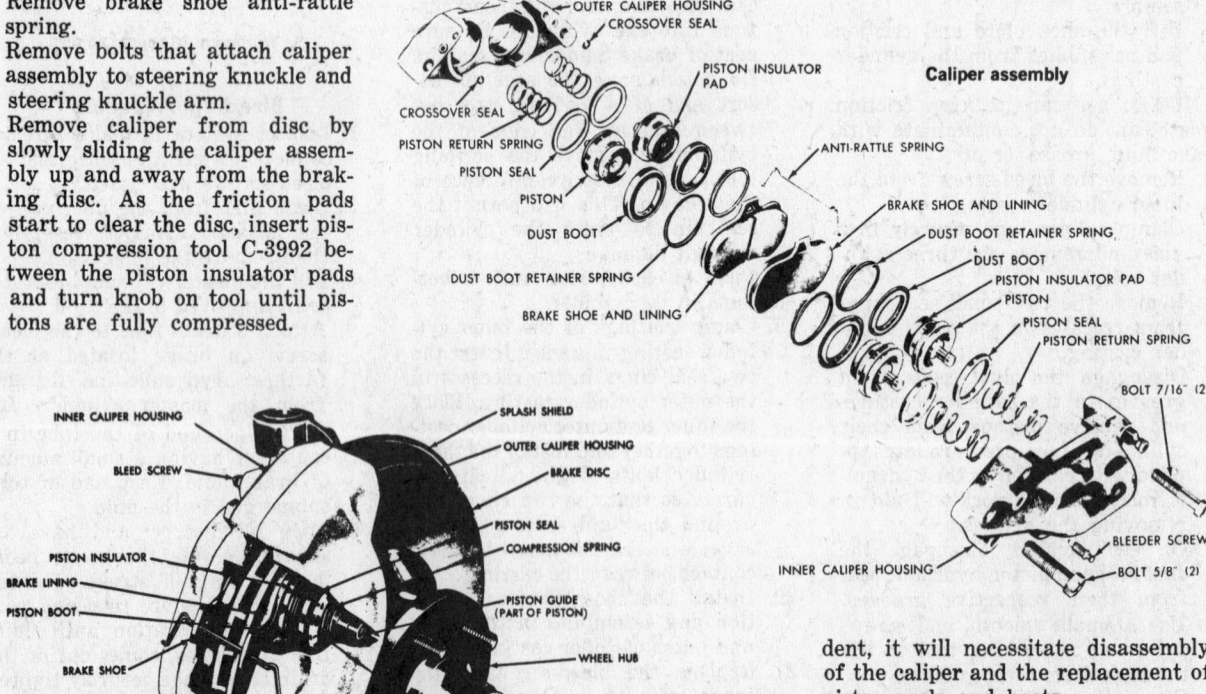

Budd disc brake

but is not exerting much force against the lining surface. The tool must be permitted to slip when contacted by the disc.

3. Slide the caliper assembly into position over the brake disc and align the mounting holes. (As the caliper is being lowered, the disc will force the tool out of the caliper.)
4. Install caliper mounting bolts and torque to 70-80 ft. lbs.
5. Install anti-rattle springs.
6. Install wheel and tire assemblies.

Caution Road test the car, making several severe stops to wear away any foreign material on the brakes and to seat the units.

Caliper Removal
1. Raise car, support on hoist or stands.
2. Remove wheel covers, wheel and tire assemblies and anti-rattle springs.
3. Disconnect brake line at caliper housing and cap the line to prevent fluid loss.
4. Remove caliper to steering knuckle attaching bolts.
5. Remove by sliding caliper assembly up and away from braking disc.

Caliper Disassembly
1. Drain caliper, then place caliper assembly in a vise. Remove piston compression tool C-3992 and the four bolts that hold the two halves of the caliper together.
2. Separate halves and remove the two cross-over seals.
3. Using a small screwdriver, pry out exposed end of piston dust boot retainer-spring and uncoil from its groove to release dust boot.
4. Using the same screwdriver, carefully work the dust boot out of the groove. Be sure to hold the piston compressed during this operation.
5. Remove piston, piston seal and dust boot from caliper. Return piston return spring.
6. Remove piston dust boot by grasping the edge and pulling it out of its groove.
7. With the fingers, roll piston seal out of piston groove. Discard the seal.
8. Remove the remaining three pistons in the same way.
9. Remove bleeder-screw from inner caliper housing.

Cleaning and Inspection
Clean all parts in alcohol, wipe or blow dry. Check dust boots for punctures and tears. If damaged, replace with new boots. Inspect for bore scores and pitting. Bores that show scoring may be honed with tool C-

3993, providing the bore diameter is not increased more than .002 in.

NOTE: when using hone C-3993, do not fail to install the hone baffle-plate before using.

Caliper Assembly
1. Clamp caliper housing in a vise. Coat cylinder bores with silicone grease provided in Seal and Boot Package P/N2585298, then install piston return spring.
2. Coat inner and outer surfaces of a new piston seal with the above silicone grease and work over piston land and down into piston groove. Use the fingers only.
3. Install dust boot on piston with lip of seat toward piston pad.
4. Install piston assembly over return spring and press down until piston bottoms in the bore.
5. With a small, blunt screwdriver, work lip of boot into groove around diameter of bore.
6. Install coil retainer spring by inserting one end in position over boot in groove. Continue to install around diameter of bore until retainer is fully seated. Be sure that boot is completely locked in position by retainer and that retainer is fully seated in groove over boot.
7. Install remaining pistons in the same way, then test pistons for smooth operation in their bores by depressing with the fingers.
8. Lightly clamp outer caliper half in vise and install new cross-over passage seals in position in recess of caliper mating surface.
9. Place mating caliper half over the one clamped in the vise and install attaching bolts. Torque bolts to 55 ft. lbs. (7/16 in. bolts) and 150 ft. lbs. (5/8 in. bolts).
10. Install brake shoes in caliper assembly and insert piston compression tool C-3992 between lining pads. Turn knob on tool until pistons are compressed. Adjust tool so that it will slip easily when contacted by disc.
11. Install bleeder-screw and tighten lightly.

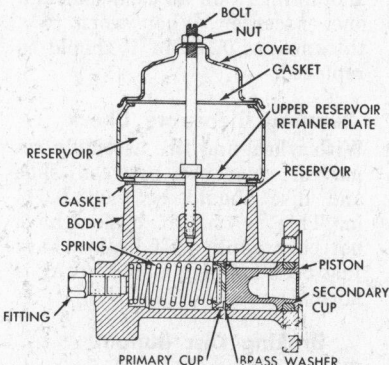

BOLT
NUT
COVER
GASKET
UPPER RESERVOIR RETAINER PLATE
RESERVOIR
RESERVOIR
GASKET
BODY
SPRING
PISTON
SECONDARY CUP
FITTING
PRIMARY CUP   BRASS WASHER

Disc brake master cylinder

Checking Fluid Level
When removing master cylinder cover, hold the center bolt with a screwdriver while backing off master cylinder cover retaining-nut.

Bleeding the System
Bleeding procedures are quite similar to current practices on drum-type brakes. However, manual bleeding of the system may require more pumping of the pedal due to the increased fluid requirements of the disc brake system. It may be necessary to check, and add fluid to, the master cylinder more frequently during bleeding. Don't get fluid on the lining or disc.

Pressure bleeding operations remain unchanged. However, the adapter cover must be kept tightly attached to the master cylinder to prevent leakage.

Caliper Installation
Before installing the caliper assembly over the brake disc, check disc for runout. Wobble or runout should not exceed .005 in.
1. Install the caliper assembly over the disc and align mounting holes. Install mounting bolts and torque to 70-80 ft. lbs. (As caliper is being lowered into place, disc will force tool out of caliper.)
2. Install anti-rattle spring.
3. Reconnect brake line at caliper housing, then follow normal bleeding procedure. Replenish brake fluid in master cylinder.
4. Install wheel and tire assembly and wheel covers.

Brake Disc and Hub Removal
1. Remove wheel cover and wheel and tire assembly.
2. Remove caliper assembly (previously described).
3. Remove grease cup, cotter pin, nut-lock, nut, washer and outer wheel bearing.
4. Pull disc and hub from wheel spindle.

Disc from Hub Removal
1. Clamp wheel hub in a vise with disc attaching bolts facing up.
2. With a prick punch, make a mark in the hub and one on the disc to index for balance upon reassembly.
3. Remove disc to hub attaching bolts.
4. Separate disc from hub.

NOTE: if wheel studs are to be replaced, use a press.

If disc is damaged, replace with a new disc. Light scoring and grooving which appear after normal use are not detrimental to brake operation. However, a heavily scored disc will impair efficiency and increase lining wear. Do not reface the braking disc.

Disc onto Hub Installation

1. Place disc on hub and align bolt holes and punch marks.
2. Install attaching bolts and torque to 55-60 ft. lbs.

Disc and Hub Assembly Installation

1. Slide the disc and hub assembly onto the spindle.
2. Install outer bearing, washer and nut.
3. Adjust wheel bearing nut to 90 in. lbs. while rotating the hub and disc.
4. Position locknut on nut with one pair of slots in line with cotter pin hole.
5. Back off the adjusting nut and lock assembly one slot, then install cotter pin. The resulting adjustment should be 0-.003 in. (no preload) end-play.
6. Install dial indicator and check for disc wobble or runout. Runout should not exceed .005 in.
7. Clean grease cup and install.
8. Install caliper assembly as previously described.
9. Install wheel and tire assembly and the wheel cover.

Master Cylinder

There is no functional difference between this unit and other master cylinders except, in this case, residual pressure is controlled elsewhere in the system. There is no residual pressure maintained in the disc brake portion of the hydraulic system. Therefore, the residual pressure valve is located in the brake line, after the brake line T. It influences the rear brakes only.

Master Cylinder Removal

1. Remove nuts that hold master cylinder to power brake unit.
2. Disconnect brake tube at master cylinder outlet, then plug hole.
3. Slide master cylinder straight out and away from power brake unit and remove from the car.

Master Cylinder Disassembly

1. Clean outside of unit to prevent contamination.
2. Remove master cylinder cover retaining-nut. Remove cover and gasket, then drain brake fluid.
3. Remove piston retainer attaching screw and remove piston.
4. Remove brass washer between piston and cup. Remove primary cup and spring.
5. Using a wrench, loosen and remove center stud and remove upper reservoir and gasket. Remove plate from the inside of the reservoir.

Master Cylinder Assembly

1. Install piston return spring in cylinder bore with closed end outward.
2. Install primary cup in cylinder bore with open end of cup over closed end of spring.
3. Install secondary cup on master cylinder piston and install piston assembly in bore with flat brass washer between piston and primary cup.
4. Install piston retainer and screw to hold piston assembly in place.
5. Position reservoir gasket on lower body, then install retaining plate in upper reservoir.
6. Place upper reservoir in position on reservoir gasket, then install center stud. Tighten to 50 in. lbs.

Master Cylinder Installation

1. Slide master cylinder over mounting studs on power brake unit, being sure the brake pushrod engages the piston. Install attaching nuts and torque to 200 in. lbs.
2. Attach brake line to master cylinder and tighten securely.
3. Refill master cylinder and reservoirs.
4. Bleed brake system as previously described.
5. Install reservoir cover and gasket. Install attaching nut and tighten.

Section 14

Disc Brakes by Kelsey-Hayes

General Information

This system uses the conventional drum and lining assemblies for the rear brakes. The front consists of discs, with calipers, shoes and linings hydraulically operated to grip the discs.

When pressure is applied, the caliper pistons move the lining against both sides of the disc. The seals are deflected by the hydraulic pressure. When pressure is released, the seals relax or retract, pulling the pistons away from the shoes and lining.

There is no hydraulic pressure to the calipers when the brakes are not applied because the fluid source at the master cylinder by-passes the residual check valve.

A proportioning valve between master cylinder and rear brakes provides balanced braking action between front and rear under a wide variety of conditions. The valve regulates pressure applied to rear brakes when high pressures are required at the front brakes. If a premature rear wheel slide is experienced, it indicates, generally, that pressure is not being properly controlled through this valve. Service should be replacement of the unit.

Shoe and Lining Wear Check

1. Remove wheel and tire assembly, caliper splash shield and shoe and lining as described later.
2. Take three micrometer readings across middle section of the shoe and lining; one at each end and one at center. When worn to a thickness of 0.180 in. it should be replaced.

Running Clearance Check

1. With wheel and tire assembly removed, clearance between shoe and disc should be 0.003-0.006 in. This is true if brakes have not been applied. If brakes have just been applied, a slight drag is normal.

Braking Disc Runout

1. Check wheel bearing adjustment. If any play is found, it is important that a careful adjustment be made.
2. Install dial gauge so that point may be mounted to contact a spot about 1 in. from edge of disc.
3. Rotate disc and read dial gauge. If reading is more than 0.0025 in. the disc should be replaced. Do not attempt to refinish disc.

Brake Shoe Removal and Installation

1. Raise vehicle on lift, or place on jack stands.
2. Remove wheel and tire assembly.
3. Remove caliper splash shield and anti-rattle spring assembly.
4. Grab shoes at each end-tab and remove by pulling outward. If a rust ring interferes, force piston back into bore by placing water-pump-type pliers at corners of shoe and caliper housing.
5. To reinstall, reverse the above. It will be necessary to push the pistons back into bores to allow for the thicker shoes. Use a flat-

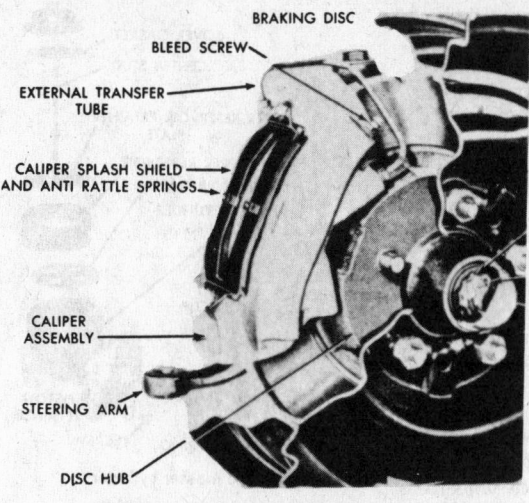

Disc brake assembly—right front

sided metal bar to push the pistons back until bottomed.

6. Slide shoe with lining into caliper with ears resting on bridges of caliper. Be sure shoe is fully seated with lining facing disc.

7. With master cylinder properly filled, pump brake pedal until a firm pedal is secured.

8. Reinstall antirattler, dust shield, wheel and tire and lower car. Road test. Then recheck fluid.

Caliper Removal

1. Raise vehicle on lift or place on jack stands.
2. Remove wheel and tire assembly.
3. Disconnect brake hose at frame

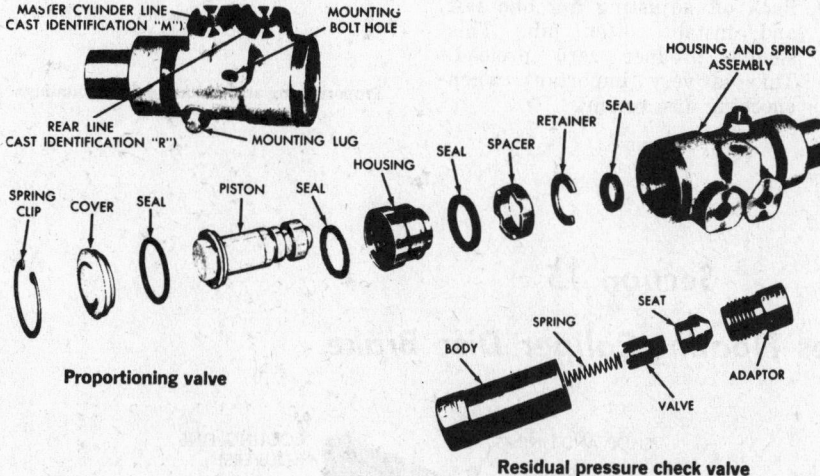

Proportioning valve

Residual pressure check valve

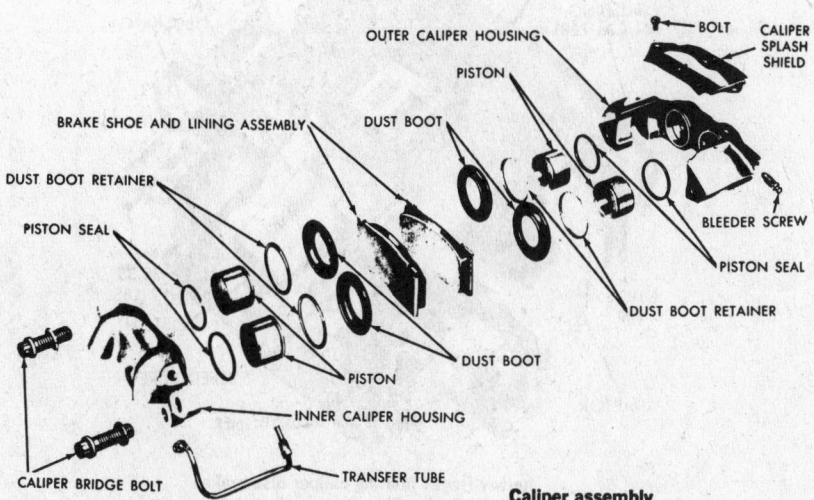

Caliper assembly

mounting bracket. Plug tube to prevent loss of fluid.

4. Remove caliper mounting bolts. If necessary to install new brake hose, scribe mark on bracket where hose enters, and position retaining clip underneath. Be sure, at reassembly, that open end of clip faces out and away from caliper.

5. Slowly slide caliper assembly up and away from disc and remove to bench.

Caliper Disassembly

1. Remove splash shield and anti-rattle spring. Remove jumper tube at caliper.
2. Mount assembly in vise with protected jaws. Remove transfer tube.
3. Remove shoe and lining assemblies.
4. Remove bridge bolts and separate caliper assemblies. In reassembly be sure these bolts are properly torqued to 70-80 ft. lbs.
5. Peel dust boot out and away from caliper housing retainer and out of piston groove. Remove remaining dust boots in same manner.
6. With tool C-3999 or equivalent, remove each of the pistons. Use care not to damage any parts by scratching. Draw piston straight out as any cocking will cause difficulty and possible damage.
7. With small pointed wooden or plastic tool, remove piston seals from piston bores. Discard seals.

Careful inspection of all parts and thorough cleaning is a must. All seals and dust boots should be replaced. Minor scratches may be cleaned up with crocus cloth. Deeper scratches in bores may be honed but not over .002 in. If more is required, replace caliper housing. Dark stains on bore walls caused by seals will do no harm. Be sure all dust, grit and other foreign matter is completely removed.

Caliper Reassembly

1. Clamp inner caliper half in vise with protected jaws, and at mounting lugs.
2. If new dust boot retainer ring is to be installed, housing should be thoroughly cleaned and a Loctite Sealant Grade H or equivalent applied to ring surface that fits in groove. Then install ring.
3. Dip new piston seals in clean brake fluid and install in caliper grooves. Gently work seals around groove with fingers until properly seated. (Be sure seals are not twisted or rolled.)
4. Coat outside of pistons with clean brake fluid and install in bore with open end out. Boot retaining groove facing out of piston.

5. Position piston squarely and with steady pressure push into proper position.

6. Install new dust boots over caliper retaining rings and into grooves.

7. Install other half on one in vise, install bridge bolts and tighten to 70-80 ft. lbs.

8. Install transfer tube and thoroughly tighten.

9. Install bleeder-screw. Do not tighten.

Caliper Installation

Before installing caliper assembly, check disc run-out as described under Braking Disc Runout.

1. Install caliper assembly over disc and line up bolt holes. Install bolts and tighten to 45-60 ft. lbs. Check that disc runs squarely and centrally in opening. There should be approximately .090-.120 in. clearance between disc and caliper. There should also be a minimum of .050 in. from either disc face to machined groove in outboard caliper.

2. Install shoe and lining assemblies between caliper and disc.

3. Install splash shield and anti-rattler.

4. Open bleeder-screw, and connect brake line. Allow caliper to fill with brake fluid; then close bleeder-screw. Be sure all air bubbles

have escaped. Refill master cylinder to proper level.

5. Pump brake pedal several times to actuate piston seals and position shoe and lining assemblies.

6. Check system for fluid leaks.

7. Install wheel and tire assembly. It is very important that wheel stud nuts be tightened to 55 ft. lbs.

8. Lower car to floor.

The vehicle should be tested with several 40 mph stops to wear away any foreign matter and to seat the unit. If this is not done a side pull may be experienced.

Braking Disc and Hub Removal and Installation

1. After jacking car, remove wheel and tire assembly.

2. Remove caliper assembly. Do not disconnect brake line. Support caliper assembly with wire hook to avoid strain on hose.

3. Remove outer wheel bearing.

4. Pull disc and hub from spindle.

5. Install in reverse of above.

6. Adjust wheel bearing, while rotating disc, to 90 in. lbs.

7. Position locknut on adjusting nut so a pair of slots line up with cotter pin hole.

8. Back off adjusting nut one slot and install cotter pin. This should produce zero preload. This is very important when checking disc runout.

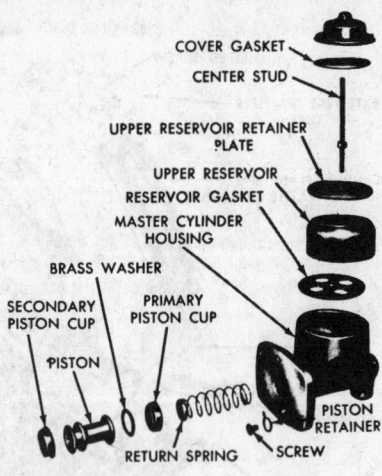

Disc brake master cylinder

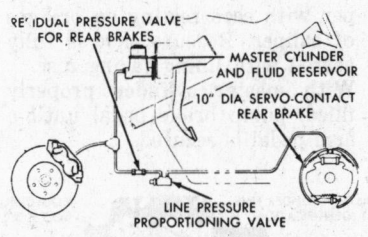

Proportioning and residual valve mounting

Section 15

Kelsey-Hayes Floating Caliper Disc Brake

General Information

This system uses a single piston to force the pads against both sides of the disc. The braking force initially forces the inboard pad against the disc, but as the pressure increases, the caliper mover inboard, sliding on the guide pins, and provides a clamping force on the disc.

A porportioning valve as used to match forces between the front and rear brakes. This results in better control under adverse conditions. Automatic adjustment of running clearance is maintained through the nature of these brakes.

Shoe and Lining Wear Check

1. Remove wheel and tire assemblies, and calipers.

2. Measure at three places with a micrometer. Replace if less than 0.180 in.

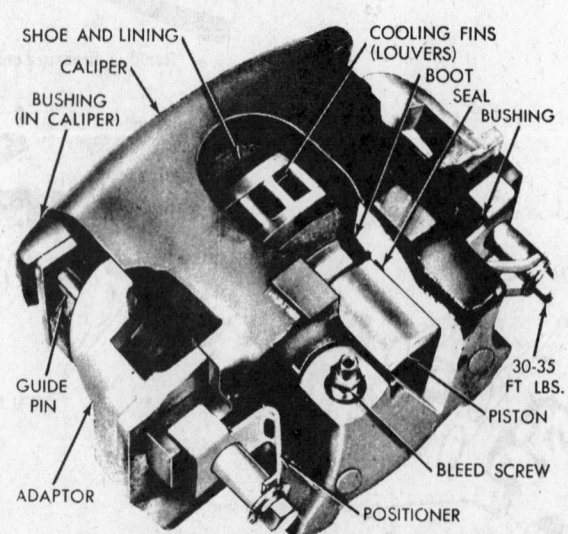

Kelsey-Hayes floating caliper disc brakes
(© Dodge Div., Chrysler Corp.)

Brake Shoe Removal and Installation

1. Raise vehicle on lift or place on jackstands.
2. Remove front wheels and tires.
3. Remove caliper guide pins, positioners that hold caliper to adaptor and anti-rattle springs.
4. Remove caliper from disc by slowly sliding caliper assembly out and away from backing disc. Be careful not to damage the brake hose.
5. Slide outboard shoe and lining assembly out of caliper. Slide inboard shoe and lining assembly out of adaptor.
6. Remove outer bushings from caliper by pressing out of bore using wooden or plastic stick. Discard bushings.
7. Slide inner bushings off guide pins and discard. Remove positioners from guide pins and discard.
8. Inspect piston for seal leaks or damaged dust boot. Replace if leaky or damaged.
9. Push piston back in bore.
10. Install new inner guide pin bushing on caliper with flanged end on inboard side. Compress flanges of outboard bushings in fingers and work into position in hole from outboard side of caliper.
11. Slide new shoe and lining assembly into position in adaptor and caliper.
12. Holding outboard lining in position, carefully slide caliper down into position in adaptor and over disc. Align guide pin

holes of adaptor, inboard and outboard shoes.
13. Install new positioners over guide pins with open ends toward outside, and arrows up. Install guide pins through bushings, caliper, adaptor, inboard and outboard shoes into outer bushings in caliper and anti-rattle spring.
14. Press in on end of guide pins and thread pins into adaptor, using extreme care not to cross threads. Tighten to 30-35 ft. lbs.
15. Pump brake pedal several times. If firm pedal is not obtained, bleed system.
16. If the system was full before changing pads, it may be necessary to remove fluid from master cylinder to complete the operation.

Caliper Removal

1. Raise car on lift or jackstands. Remove wheels and tires.
2. Disconnect front brake hose from caliper and tube at mounting bracket.
3. Remove guide pins and positioners that attach caliper to adapter. Slide caliper out and away from disc and adapter. Remove inboard and outboard shoe and lining.

Caliper Disassembly

1. Mount caliper in vise equipped with protector jaws. Do not over tighten, for distortion will result.
2. Remove dust boot.
3. Using tool C-4087, remove piston from caliper. Do not scratch,

burr or otherwise damage piston. Do not use compressed air to remove piston.
4. Use small pointed wood or plastic stick to remove seal from groove in bore. Do not use screw driver or other metal tool that will scratch cylinder bore.
5. Press outer bushing caliper using wooden stick.
6. Remove inner bushing and discard, remove bleeder screw.

Cleaning Note

Clean all parts in alcohol or suitable cleaner. Blow dry with compressed air. Inspect piston bore for scoring or pitting. Hone with tool C-4095 only, if deep scratches are present, provided bore is not increased more than 0.002 in. Otherwise replace with new caliper housing. Piston should be replaced if it shows pitting, scoring or if the plating is worn. Black stains are caused by the seal and cause no harm. Clean and handle all parts extremely carefully.

Caliper Assembly

1. Clamp caliper in vise with protector jaws. Do not tighten excessively.
2. Dip new piston seal in lubricant supplied in kit and install in groove in bore. Do not use old seal.
3. Coat new piston dust boot with lubricant, leaving a generous amount inside boot. Install in caliper by working into groove using fingers only.
4. Plug high pressure inlet to cal-

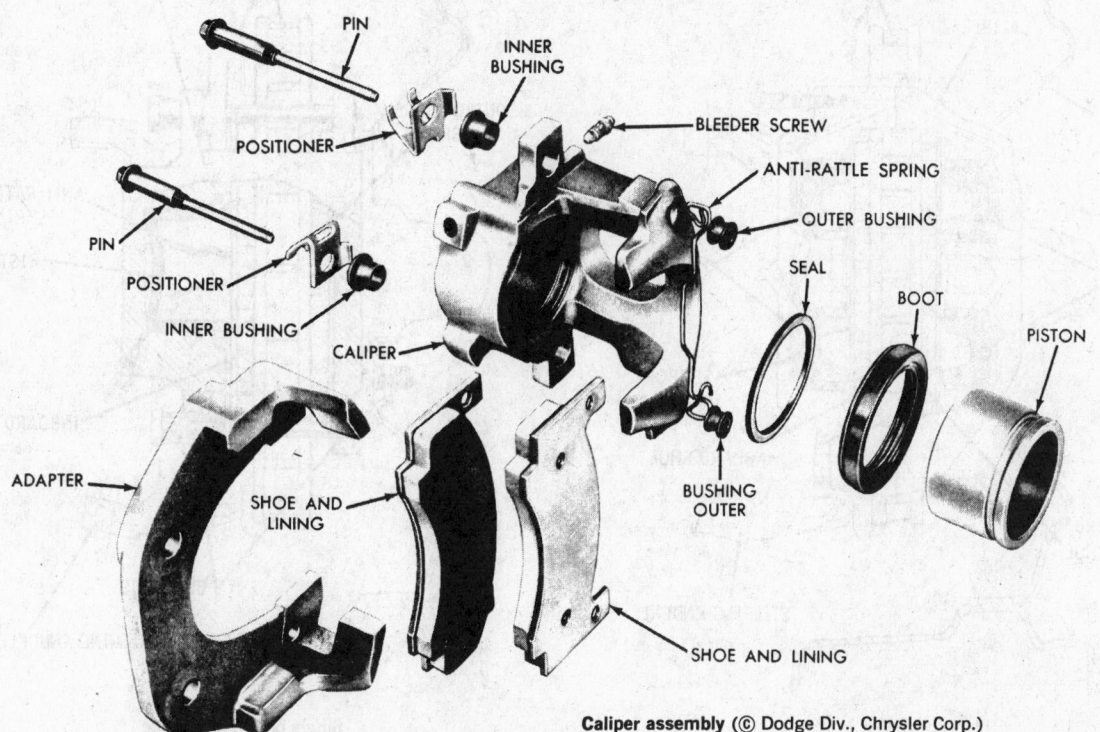

Caliper assembly (© Dodge Div., Chrysler Corp.)

iper and bleeder screw hole, then coat piston with lubricant. Spread boot and install piston. Remove plug and carefully push piston to bottom of bore.

5. Install new inner guide pin bushings in caliper with flanged end on inboard side. Compress flanges of outboard bushing with fingers and work in position in hole from outboard side. Press in until seated. Install bleeder screw.
6. Check runout and thickness before installing.

Caliper Installation

1. Inspect the lining. If any is worn or contaminated with fluid, all pads must be replaced.

2. Connect brake hose to caliper and tighten.
3. Proceed with procedure outlined in *Brake Shoe Removal and Installation* Steps 12 through 16.

Checking Disc Runout and Thickness

1. With dial indicator mounted on steering arm, check runout. This should not exceed 0.0025 in.
2. If disc is to be refinished, no more than 0.015 in. may be removed from either side, providing this does not reduce thickness below minimum thickness of 0.980 in.

Disc and Hub Removal

1. Raise vehicle and remove wheels and tires. Remove caliper assembly.
2. Remove grease cup, cotter pin, nut lock, nut, thrust washer and outer bearing.
3. Pull disc and hub off spindle.

Disc and Hub Installation

1. Slide disc and hub on spindle.
2. Install bearing, thrust washer and nut. Tighten nut to 90 ft. lbs.
3. Install lock nut on nut and back off adjusting nut one slot.
4. Install grease cup. Install caliper assembly.

Section 16

Delco Moraine, Single Cylinder Disc Brake

General Information

Operations herein can only be assumed to be typical, representative of this type brake system. Operations may vary in detail from car to car, therefore, it is recommended that the system being worked on be carefully examined for differences which might alter procedure.

Disassembly

NOTE: check brake lining any time the wheels are removed (tire rotation, etc.). Inboard lining may be checked by looking through inspection hole in caliper. A rule of thumb policy is when the thickness of the lining is about the same as the thickness of the metal shoe, all shoe and

lining assemblies should be replaced.
1. Remove ⅔ of the total capacity of the front master cylinder reservoir. Discard the fluid.
NOTE: removing the fluid prevents reservoir overflow when the single cylinder is pushed back into its bore to remove the caliper.

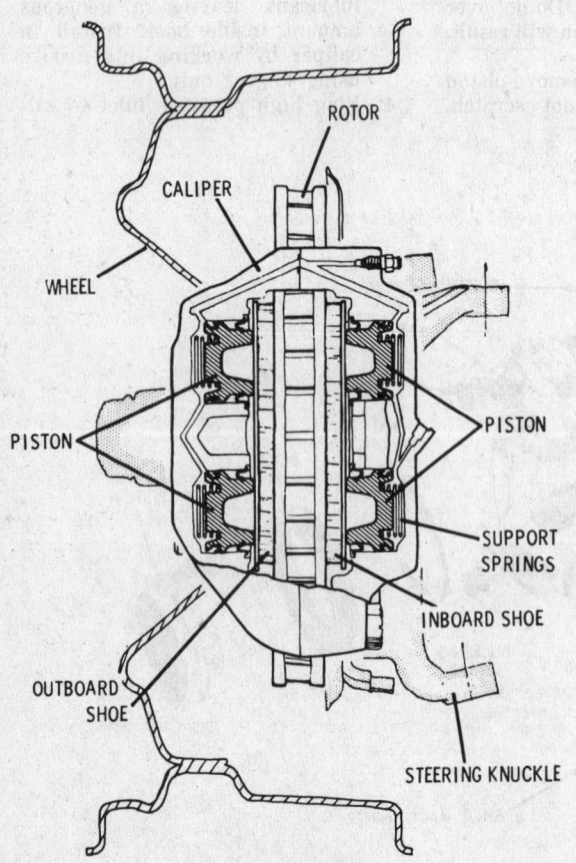

Four piston disc brake

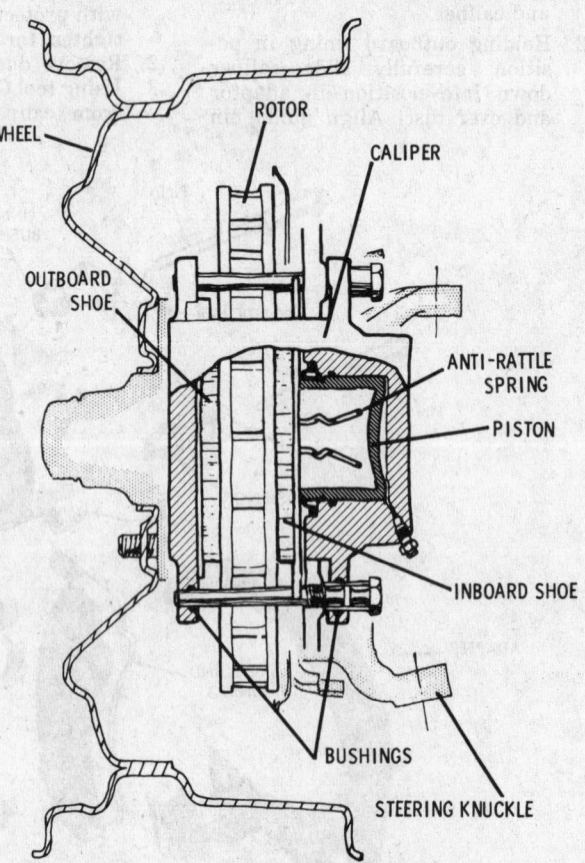

Single piston disc brake

Sliding caliper disc brake

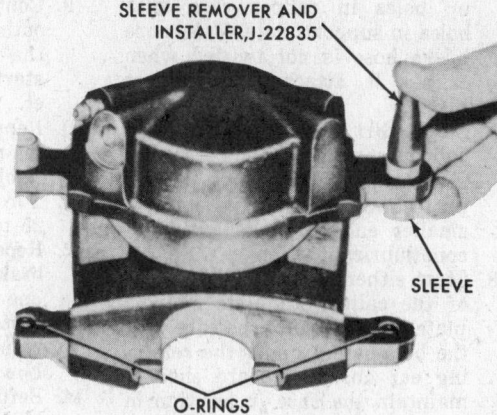

Removing sleeves

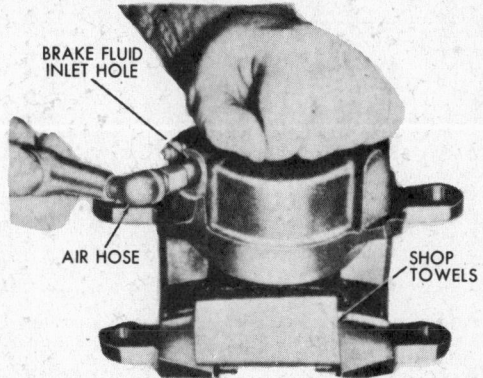

Removing piston

2. Raise car and remove front wheels.

3. Position a 7 in. C-clamp on the caliper so that the solid side rests against the back of the caliper assembly. The screw-end rests against the back of the outboard shoe.

4. Tighten C-clamp until caliper moves out far enough to push piston to bottom of cylinder. This will relieve the pressure on shoe and lining assemblies.

5. Remove C-clamp.

6. Remove two bolts that hold caliper to support plate.

NOTE: it is not necessary to disconnect brake hose when relining this type brake.

7. Slide caliper off disc.

NOTE: do not permit caliper to hang from brake hose. It may be supported by a hook-shaped wire attached to upper control arm.

8. Remove inboard shoe and spring from support plate.

9. Outboard shoe may also have disengaged itself from caliper as caliper was removed from disc. If it did not disengage itself, remove outboard shoe.

10. Push two sleeves out from inboard ears of caliper.

11. Remove four O-rings, two on inboard ears and two on outboard ears.

12. Examine piston area for fluid leaks (moisture around boot area). If there is any evidence of leakage, the caliper should be overhauled.

Caution Do not use air pressure to clean caliper because of the possibility of unseating dust boot.

13. Examine disc for scoring or other damage. If scoring is apparent, replace hub and disc assembly.

NOTE: disc only may be replaced on Fleetwood, Eldorado and Toronado. Machining of disc is not recommended.

Assembly

1. Make sure the piston is bottomed in the piston bore. Make sure the dust boot has folded back correctly as the piston moved back.

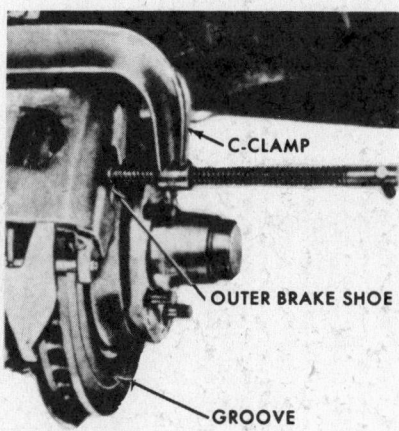

Pushing piston into bore

2. Wipe out grooves in calipers and install O-rings.

3. Lubricate new sleeves on outer surface with silicone lubricant. Push in two sleeves and position the sleeves so that end toward the shoe and lining assembly is flush with the machined surface on the ear.

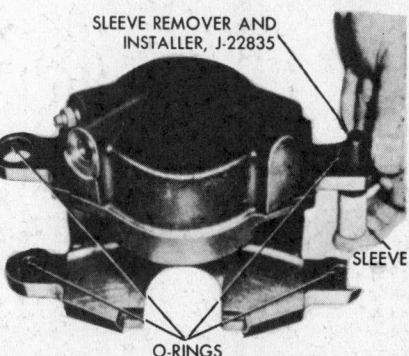

Installing sleeves

4. Place outboard shoe and lining into the caliper so that the lining faces toward the disc. The two ears at the end of the shoe fit over the ears on the caliper. The flange on the bottom of the shoe fits into the cut-out section of the caliper.

5. Position spring inside cut-out section of piston. Place flat inboard shoe and lining in support plate so that lining is toward disc.

6. Position caliper on disc and line

up holes in caliper ears with holes in support plate. Make sure brake hose is not twisted when caliper is attached to support plate.

7. Clean dirt and corrosion from caliper mounting bolts. Do not use abrasives; they may remove protective plating. Lubricate smaller ends of bolts with silicone lubricant.

8. Start either bolt into inboard ear of the caliper and into support plate. At this point, be sure that the bolt passes under the retaining ear on the inboard shoe to maintain the shoe in position in the caliper.

9. Continue the bolt on through the outboard ear of the caliper until the threads on the bolt can be started into the mounting bracket.

10. Repeat Steps 8 and 9 in placing remaining bolt into caliper assembly.

11. Torque caliper mounting bolts to 25 to 35 ft. lbs.

12. Repeat operation on other disc brake assembly. Never reline just one front brake.

13. Install wheels, tighten wheel mounting nuts to 90-120 ft. lbs. Lower car.

14. Before moving the car, pump the brake pedal two or three times to insure a firm pedal.

15. Check the master cylinder reservoirs and fill to within $1/8$ to $3/8$ in. of the top.

16. Any time the front brakes are relined the rear brakes should be checked.

Bleeding

Bleeding operation is identical to that for drum brakes, except that the metering valve must be depressed and held open. The valve stem can be taped or hand-held.

Caution   If this type system is pressure-bled, do not exceed 30 psi.

POWER BRAKE IDENTIFICATION CHART

Car	Bendix Treadle Vac	Bendix Master Vac	Bendix Dual Diaphragm	Bendix Single Diaphragm	Moraine Diaphragm	Moraine Diaphragm	Moraine Diaphragm	Midland-Ross Diaphragm	Kelsey-Hayes Bellows	Kelsey-Hayes Diaphragm	Bendix Single Diaphragm Hydrovac	Bendix Dual Diaphragm Hydrovac	Delco Moraine Tandem Diaphragm	Midland-Ross Tandem Diaphragm
	Section 1	Section 2	Section 3	Section 4	Section 5	Section 6	Section 7	Section 8	Section 9	Section 10	Section 11	Section 12	Section 13	Section 14
American Motors				1964-70										
Buick				1965-70	1964		1965-70						1970	
Buick Special							1965-70			1964				
Cadillac				1964-70			1964-70							
Capri	See Hydraulic Brake Section													
Camaro and Chevy II				1965-70			1964	1970						
Chevrolet			1970	1964-70				1970			1970	1970		
Chevelle							1964-70	1970						
Chrysler			1964-70	1964, 1967-70				1965-70		1964-66				
Continental		1964-70	1968-70					1965-67						
Corvette				1964-70			1964-70	1970						
Cougar				1967-70				1964-70		1965-66				1967-69
Dodge, Dart, Challenger	1965-70	1965-70	1967-70											
Fairlane, Falcon				1964-70						1964				
Firebird, Tempest, GTO							1964-70							
Ford			1966-70	1966-70						1964-70				
Jeep				1964-70										
Mercury, Montego				1967-70						1964-70				
Mustang, Comet				1965-70										
Plymouth			1967-70	1964-70		1965-70		1964-70		1965				1967-69
Pontiac		1964-70	1967-70		1965-70		1964-70							
Thunderbird				1967-70				1964-70						
Oldsmobile			1967-70	1964-68	1966-70		1965			1964			1970	
F-85 and Cutlass			1967-70	1964-68	1966-70		1965			1964-65				

Section Page Numbers

Section	Page	Section	Page	Section	Page
1	1252	6	1257	10	1264
2	1254	7	1259	11	1265
3	1255	8	1260	12	1267
4	1255	9	1262	13	1268
5	1256			14	1270

Power Brakes

The following items are in addition to those listed under non-power brakes. Check those items first.

Dragging Brakes

(Refer to non-power brakes first)
1. Excessive hydraulic seal friction or binding.
2. Compensator port plugged.
3. Sticking valve plunger.
4. Improper booster pushrod length.
5. Return of fluid blocked.

6. Valve sleeve not properly positioned or staked.

Grabbing Brakes

(Refer to non-power brakes first)
1. Sticking actuating valve.
2. Broken plunger stem.

Hard Pedal

(Refer to non-power brakes first)
1. Faulty vacuum check valve.

2. Collapsed or leaking vacuum hose.
3. Plugged vacuum fittings.
4. Leaking vacuum chamber.
5. Diaphragm assembly out of place.
6. Vacuum leak in forward vacuum housing.

Pedal Goes to Floor (or Excessively Low)

(Refer to non-power brakes first)
1. Shoe (s) hanging at backing plate.
2. Broken plunger stem.
3. Self-adjusters not operating.

Section 1
Bendix Treadle-Vac Power Brakes

The Treadle-Vac is a self-contained unit in which the master cylinder of the regular hydraulic system is part of the unit. It is therefore called an adjacent-type unit. It is worthy to note that this unit is an atmospherically balanced unit in which, when the brakes are not applied, there is atmospheric pressure on both sides of the piston.

Treadle-Vac Removal

1. Disconnect the hydraulic line at the hydraulic cylinder and cap so that all the fluid does not run out. Disconnect the vacuum line at the unit. Remove the brake pedal pivot pin and tip the upper end of the pedal back to expose the pushrod pivot pin under the pedal pad.

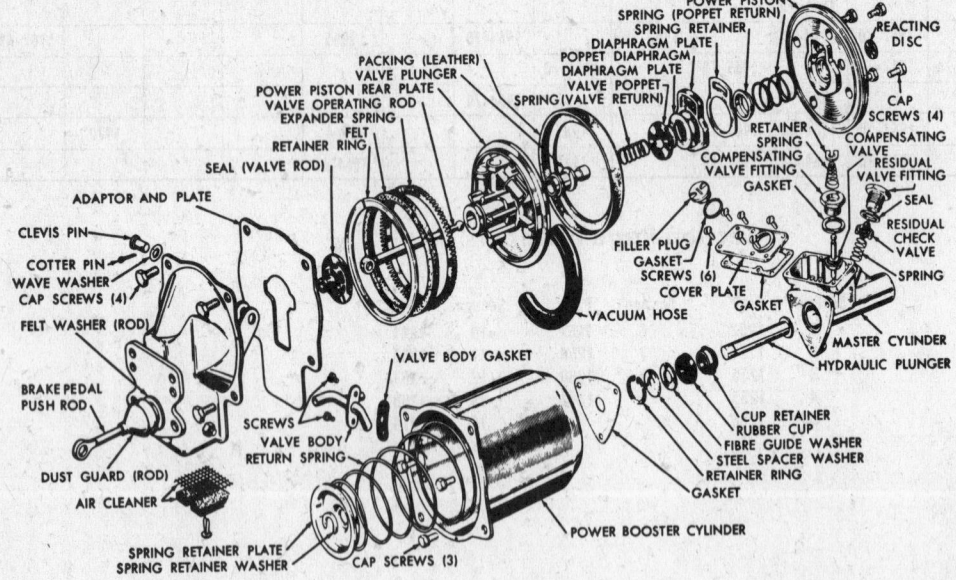

Exploded view of Treadle-Vac unit (hydraulic reaction type) (© Oldsmobile Div., G.M. Corp.)

End plate removal
(© Oldsmobile Div., G.M. Corp.)

Vacuum piston and valve removal
(© Oldsmobile Div., G.M. Corp.)

Sleeve valve and vacuum hose removal
(© Oldsmobile Div., G.M. Corp.)

2. Remove the brake pedal and the accelerator pedal. Fold back the carpet. Unfasten the steering column grommet and slide it up the column.
3. Remove steering column cover plate to toe pan screws and remove the cover plate and the power brake unit as an assembly.
4. Remove four bolts to release the unit from the plate.
5. Reverse the procedure to reinstall.

NOTE: be careful not to pull pushrod out of the plunger.

Bleeding the Treadle-Vac

Bleed the brakes in the normal manner. This unit does not have any bleed-screws. Fill to within 1/4 in. of top.

Overhaul of Treadle-Vac Unit

Repair parts for the unit come in kits. Always install all the parts supplied in the kit, regardless of whether it appears necessary.

Be sure to keep the parts clean, as any sign of grease will interfere with the operation of the rubber seals. Be sure to scribe across the vacuum cylinder and the hydraulic cylinder to facilitate assembly to the same position. Do the same thing to show assembly position of the cover and the hydraulic cylinder. Use alcohol to clean the parts and be sure that no fluid gets on the valve assembly. Inspect all parts for damage and excessive wear. If the inside of the vacuum cylinder is rusted or corroded, it can be cleaned with steel wool or emery cloth, but be sure to remove all traces of the abrasive. If it is scored it must be replaced. Replace the hydraulic plunger if it is rusted or scored.

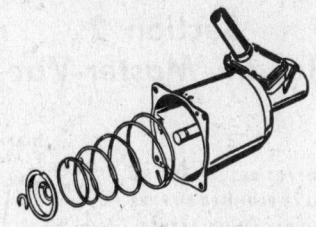

Vacuum piston return spring and retainer removal (© Oldsmobile Div., G.M. Corp.)

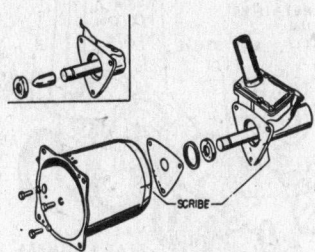

Vacuum cylinder and leather seal removal (© Oldsmobile Div., G.M. Corp.)

Residual check valve removal (© Oldsmobile Div., G.M. Corp.)

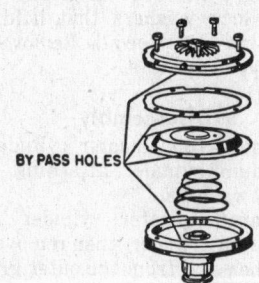

Diaphragm cover and valve spring removal (© Oldsmobile Div., G.M. Corp.)

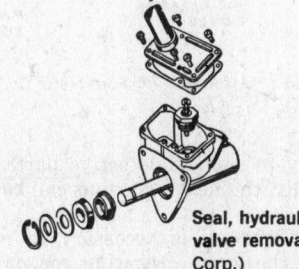

Seal, hydraulic plunger and compensating valve removal (© Oldsmobile Div., G.M. Corp.)

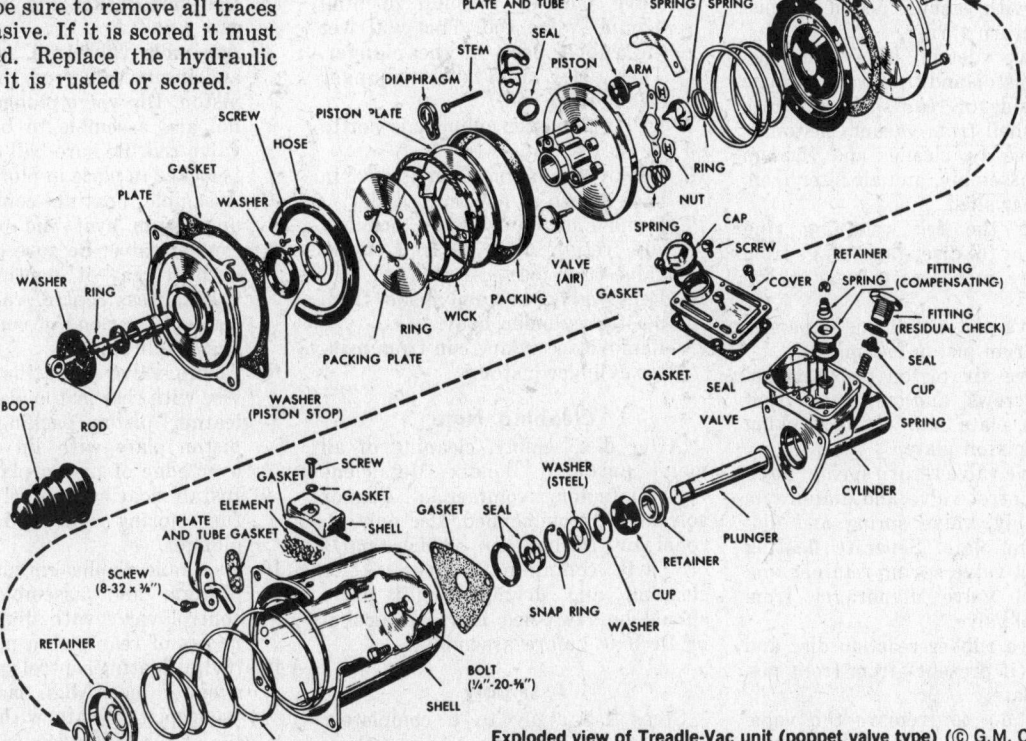

Exploded view of Treadle-Vac unit (poppet valve type) (© G.M. Corp.)

Section 2
Bendix Master-Vac

Removal

1. Disconnect clevis at brake pedal to pushrod.
2. Remove vacuum hoses from power cylinder.
3. Disconnect hydraulic line from master cylinder.
4. Remove the four attaching nuts and lock washers that hold the unit to the firewall. Remove the power brake unit.

Disassembly

1. Remove four master cylinder to vacuum cylinder attaching nuts and washers.
2. Separate master cylinder from vacuum cylinder, then remove the rubber seal from the outer groove at end of master cylinder.
3. Remove the pushrod from the power section. (Do not disturb adjusting screw.)
4. Remove pushrod boot and valve operating rod.
5. Scribe alignment marks across the rear shell and vacuum cylinder. Remove all but two of the end plate attaching screws (opposite each other). Hold down on the rear shell while removing the two remaining screws to prevent the piston return spring from expanding.
6. Scribe a mark across the face of the piston, to index the mark on the rear shell, and remove rear shell with vacuum piston and piston return spring.
7. Remove vacuum hose from vacuum piston and from vacuum tube on inside of rear shell. Separate rear shell from vacuum piston.
8. Remove air cleaner and vacuum tube assembly, and air filter from the rear shell.
9. Spring the felt retaining ring enough to disengage ring from grooves in bosses on rear piston plate.
10. Remove piston felt and expander ring from piston assembly.
11. Remove six piston plate attaching screws and separate front piston plate and piston packing from piston plate.
12. Remove valve return spring, floating control valve and diaphragm assembly, valve spring and diaphragm plate. Separate floating control valve spring-retainer and control valve diaphragm from control valve.
13. Remove rubber reaction disc and shim (if present) from front piston plate.

NOTE: do not remove the valve operating rod and valve plunger from the rear piston plate unless it is neces-

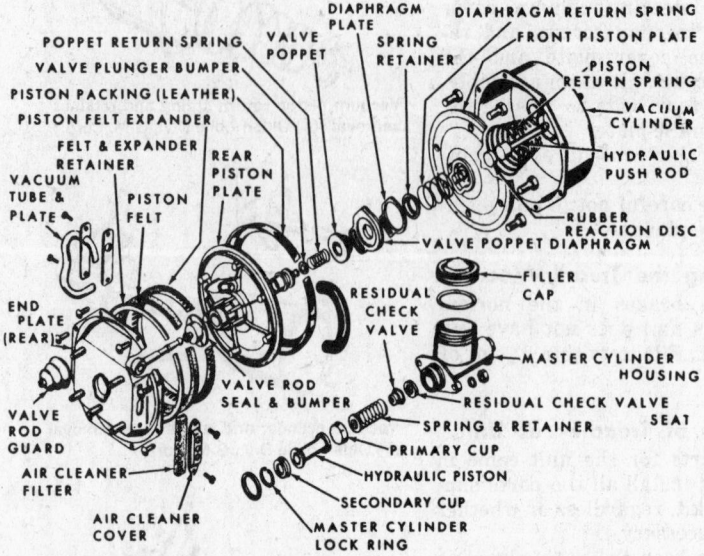

Bendix Master-Vac (© Ford Motor Co.)

sary to replace defective parts. Normally, the next two steps can be omitted.

14. When it is necessary to replace the valve operating rod or valve plunger, remove valve rod seal from groove in piston plate and pull seal over end of rod.
15. Hold piston with valve plunger side down and inject alcohol into valve plunger through opening around valve rod. This will wet the rubber lock in the plunger. Then drive or pry valve plunger off the valve rod.

NOTE: if master cylinder is not to be rebuilt, omit Steps 16-19.

16. Remove snap-ring from groove in base at end of master cylinder.
17. Remove piston assembly, primary cup, retainer spring, and check-valve from master cylinder.
18. Remove filler cap and gasket from master cylinder body.
19. Remove secondary cup from master cylinder piston.

Cleaning Note

After disassembly, cleaning of all metal parts in Bendix Metalclene or satisfactory commercial cleaner solvent is recommended. Use only alcohol or Declene on rubber parts or parts containing rubber. After cleaning and drying, metal parts should be rewashed in clean alcohol or Declene before assembly.

Assembly

Steps 1-5 apply to a completely disassembled master cylinder. Otherwise, omit these steps (1-5).

1. Coat bore of master cylinder with brake fluid.
2. Dip secondary cup in brake fluid and install on master cylinder piston.
3. Dip other piston parts in brake fluid and assemble the piston. Install piston.
4. Install snap-ring into groove of cylinder.
5. Use new gasket and install filler cap.
6. Assemble valve rod seal on rod and insert valve rod through the piston. Dip valve plunger in alcohol and assemble to ball end of valve rod. Be sure ball end of rod is locked in place in plunger.
7. Assemble floating control valve diaphragm over end of floating control valve. Be sure diaphragm is in recess of floating control valve. Press control valve spring retainer over end of control valve and diaphragm.
8. Clamp valve operating rod in a vise with rear piston plate up. Lay leather piston packing on rear piston plate with lip of leather over edge of piston plate.
9. Install floating control valve return spring over end of valve plunger.
10. Assemble diaphragm plate to diaphragm and assemble floating control valve with diaphragm in recess of rear piston plate.
11. Install floating control valve spring over retainer. Align and assemble front piston plate with rear piston plate. Center the floating control valve spring on front piston

plate and center valve plunger stem in hole of piston.

12. Holding front and rear piston plates together, loosely install six piston plate cap screws.

13. Install shim and rubber reaction disc in recess at center of front piston plate.

NOTE: a piston assembling ring (tool J-7780) is handy in assembling the piston.

14. Place the assembling tool over piston packing, turn piston assembly upside down and assemble the expander ring against inside lip of leather packing. Saturate felt with Bendix Vacuum Cylinder Oil or Delco shock absorber fluid—type A, then assemble in expander ring. Assemble retainer ring over bosses on rear piston plate. Be sure retainer is anchored in grooves of piston plate.

15. Assemble air cleaner filter over vacuum tube of air cleaner and attach air cleaner shell in position with screws.

16. Slide vacuum hose onto vacuum inlet tube of piston and align hose to lay flat against piston.

17. Wipe a coat of vacuum cylinder oil on bore of cylinder. Remove assembling ring from vacuum piston and coat leather piston packing with vacuum cylinder oil.

18. Install rear shell over end of valve operating rod and attach vacuum hose to tube end on each side of end plate.

19. Center small diameter end of piston return spring in vacuum cylinder. Center large diameter of spring on piston. Check alignment mark on piston with marks on vacuum cylinder and rear shell, compress spring and install two attaching screws at opposite sides

to hold rear shell and cylinder together. Now, install balance of screws and tighten evenly.

20. Dip small end of pushrod boot in alcohol and assemble guard over end of valve operating rod and over flange of shell.

21. Insert large end of pushrod through hole in end of vacuum cylinder and guide into hole of front piston plate.

NOTE: before going on with assembly, check the distance from the outer end of the pushrod to the master cylinder mounting surface on the vacuum cylinder. This measurement should be 1.195-1.200 in.

22. After pushrod adjustment is correct, replace rubber seal in groove on master cylinder body.

23. Assemble master cylinder to the vacuum cylinder at four studs. Replace lock washers and nut and securely tighten.

Section 3
Bendix Dual Diaphragm Type

Since 1962, with some Chrysler products, a new power brake unit is used. It features a direct pedal connection to a vacuum unit mounted on the firewall, with the master cylinder directly mounted to booster.

The booster chamber contains two diaphragms and is under constant engine vacuum. When brakes are applied, the control valve is opened to allow atmospheric pressure behind both diaphragms. This provides the power boost to the master cylinder.

This vacuum-suspended system provides reserve against fade. Pedal linkages are eliminated, no additional vacuum storage tanks are needed.

Removal

1. Remove bolt attaching pedal to pushrod.
2. Disconnect fluid line from master cylinder.
3. Disconnect vacuum line from check valve.
4. Remove four attaching nuts and washers under dash.
5. Pull booster and cylinder assembly forward from support bracket.
6. Remove four nuts and washers holding master cylinder to booster and remove cylinder.

Installation

Reverse the above procedure.

NOTE: do not attempt to disassemble the booster. It is serviced only by the dealer.

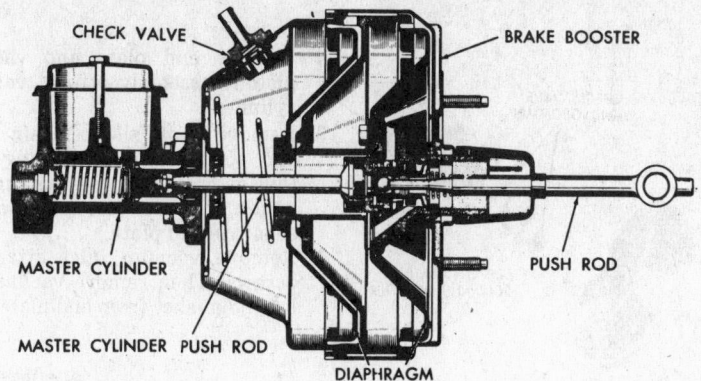

CHECK VALVE   BRAKE BOOSTER

MASTER CYLINDER   PUSH ROD

MASTER CYLINDER PUSH ROD

DIAPHRAGM

Cutaway view of brake booster and master cylinder

Section 4
Bendix Single Diaphragm

Disassembly

1. After emptying unit and thoroughly cleaning outside, mount in vise with pushrod up. Clamp on sides of master cylinder.
2. Remove rear retainer and boot and install nuts on rear shell studs. Scribe line between two

shells for reassembly. With pry bar behind nuts, rotate rear shell so that cutouts in shell line up with tangs of front shell.

Caution Loosen carefully, as it is spring loaded.

3. Lift rear shell and plate and

valve rod and plunger assembly from unit. Remove return spring.

4. Remove plate and valve body, valve rod and plunger assembly from unit. Do not remove rear shell bearing seal unless defective and new seal is on hand for replacement.

Dual master cylinder type

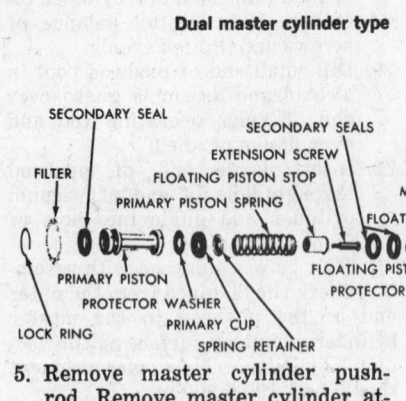

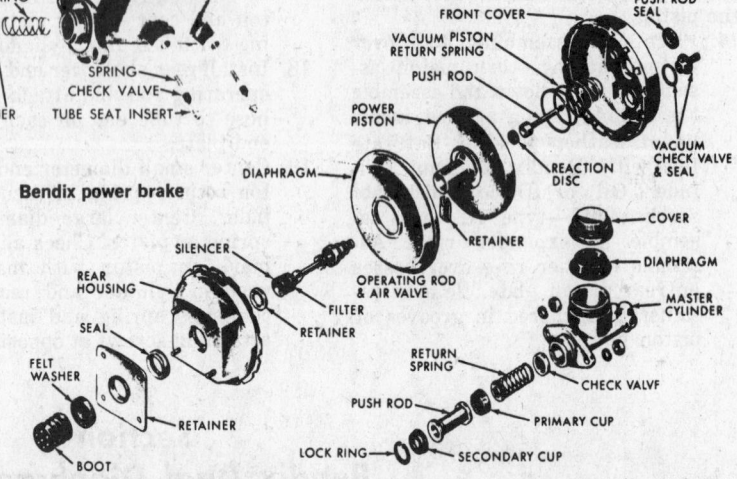

Bendix power brake

5. Remove master cylinder pushrod. Remove master cylinder attaching nuts and separate cylinder from shell.
6. Remove seal and vacuum check valve from front shell.
7. Pry off filter retainer and remove air silencer and filter from body. Do not chip plastic. Remove diaphragm from plate and valve body, then remove valve retainer key, valve rod and plunger. Press reaction disc from diaphragm plate.
8. If required, recondition master cylinder in conventional manner. Inspect all parts thoroughly, replacing any parts with scuffing, nicks, or scratches that cannot be cleaned with crocus cloth. Diaphragms that show deterioration or cracking also should be replaced.

9. Reassemble in reverse of above procedures.
   Gauging is only necessary when major structural parts have been replaced. To gauge, remove master cylinder so that pushrod is exposed. Place gauge over piston rod so that it fits between the two studs on front of housing. The gap between the cut-out of gauge and piston rod should not be more than .010 in. Proper clearance may be obtained by adjusting screw to meet the requirements.

Section 5
Moraine Diaphragm Type

Disassembly

1. Empty system and thoroughly clean outside of assembly. Be sure no oils or greases contact diaphragm units.

Major parts of power brake unit

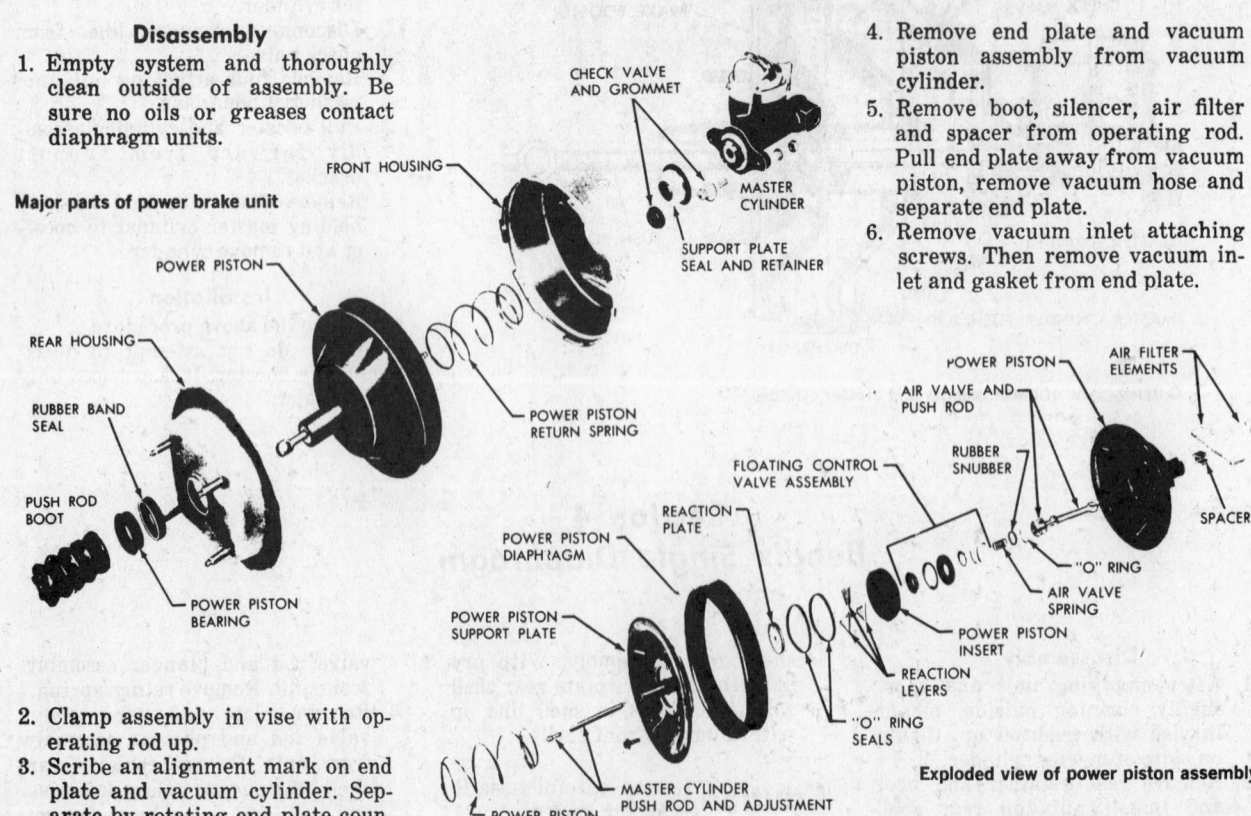

2. Clamp assembly in vise with operating rod up.
3. Scribe an alignment mark on end plate and vacuum cylinder. Separate by rotating end plate counterclockwise.

4. Remove end plate and vacuum piston assembly from vacuum cylinder.
5. Remove boot, silencer, air filter and spacer from operating rod. Pull end plate away from vacuum piston, remove vacuum hose and separate end plate.
6. Remove vacuum inlet attaching screws. Then remove vacuum inlet and gasket from end plate.

Exploded view of power piston assembly

7. Remove rubber bearing from end plate.
8. Remove vacuum piston return spring from cylinder.
9. Remove vacuum cylinder.
10. Remove support plate and diaphragm from vacuum piston assembly and separate diaphragm from support plate.
11. Remove master cylinder pushrod from plate.
12. Remove O-ring and reaction levers from vacuum piston.
13. Push air valve and operating rod from floating valve side of vacuum piston. This will dislodge floating valve from seat.
14. Rotate air valve extension and unlock it from air valve. Remove air valve return spring, floating valve return spring and floating valve.
15. Remove air valve O-rings.
16. Remove vacuum hose. Disassemble floating valve.

Thoroughly clean and inspect all parts. Check for scoring and pitting, dents and nicks. Inspect the diaphragm for nicks and deterioration and abrasion.
17. Check master cylinder parts and replace as in normal master cylinder repairs.
18. Assemble in reverse of the disassembly procedure. Keep all diaphragm parts clean and free of oil and grease. Lubricate other parts with clean hydraulic brake fluid.
19. Check pushrod adjustment.
   A. With vacuum unit assembled and master cylinder removed, position gauge over pushrod end, with legs of gauge over pushrod and resting on vacuum cylinder. Adjustment is correct if screw contacts gauge or legs of gauge are .020 in. from vacuum cylinder.
   B. If necessary rotate screw until this adjustment is attained.

Section 6
Moraine Diaphragm Type

Removal

1. Disconnect clevis at brake pedal assembly.
2. Remove vacuum hoses from power brake unit.

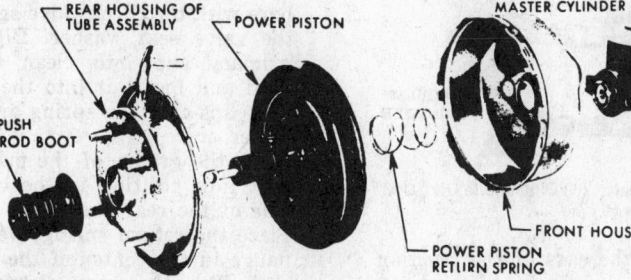

Major parts of Moraine diaphragm unit

3. Disconnect hydraulic line from master cylinder.
4. Remove the four nuts and lock washers holding the unit to the firewall. Lift out the power brake unit.
5. Remove the cotter pin, clevis pin and washer from the end of the brake pedal pushrod.
6. Remove the cotter pin, clevis pin and washer from the end of the power brake operating rod. Remove the wave washer.
7. Remove the cotter pin, then tap out the pivot pin. Remove two operating levers and the pivot collar.

Disassembly

1. Place unit in vise with pushrod up. Clamp firmly on side of master cylinder reservoir.
2. Place long wooden hammer handle in position to bear against two studs. Rotate counterclockwise to separate housings.
3. Lift rear housing and power piston assembly from unit.

4. Remove hose from rear housing tube and remove pushrod boot.
5. Remove power piston return spring from front housing.
NOTE: unless front housing or

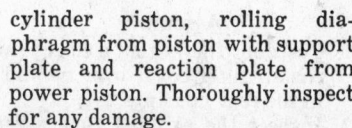

master cylinder assembly is damaged and must be replaced there is no need to remove front housing from master cylinder assembly.

Caution Protect all parts from nicks or cuts. Keep grease, oils and foreign matter from diaphragm.

6. Remove screws and lift master

cylinder piston, rolling diaphragm from piston with support plate and reaction plate from power piston. Thoroughly inspect for any damage.
7. Remove retaining ring from master cylinder piston and remove reaction plate.
8. Push master cylinder piston through support plate. Then remove O-ring, secondary seals and rubber bumper from inside piston.
9. Remove power piston assembly from reaction disc, reaction levers, air valve reaction plate, floating control valve return spring and power piston seal ring.
10. Turn power piston over and inspect hose. Replace hose if needed.
11. Remove air valve boot. Force pushrod air valve assembly through power piston, dislodging floating control valve and diaphragm assembly. If pushrod air-valve assembly is bad, replace with complete new assembly.
12. Remove floating control valve spring retainer from floating con-

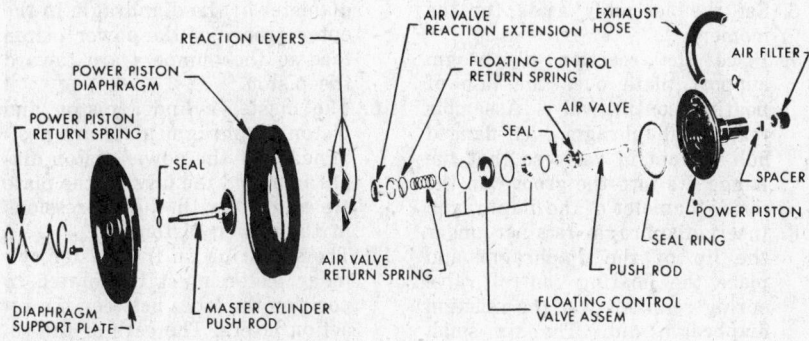

Exploded view of power piston assembly

trol valve and diaphragm. Remove reaction diaphragm retainer. Then remove reaction diaphragm and support plate from hub of floating control valve.
13. Check master cylinder parts and replace as required.
14. Carefully inspect all parts and replace as needed, reassembling in reverse of disassembly procedure. Coat all parts with clean brake fluid during assembly. This does not apply to diaphragm.

Cleaning Note

Use Declene or fresh brake fluid to clean all parts thoroughly. Soak in the fluid and use hair brush to dislodge all foreign matter. Use compressed air on all orifices and valve holes, then air dry.

Gasoline, kerosene or any other mineral cleaner will react with rubber parts. Lubricate with brake fluid only.

Assembly

1. Position the filter element and the air filter cover over two holes in the rear housing.
2. Inspect the exhaust tube assembly for damage. Insert the tube (with new gasket), into the rear housing and install two screws.
3. Insert the small rubber bumper into the counterbore of the master cylinder piston. Place two secondary seals in grooves of master cylinder piston, with lips toward the small drilled holes in the end of the piston.
4. Install the O-ring into the second groove from the counterbored end of the master cylinder piston. Lubricate with silicone grease and insert piston into the piston diaphragm plate, from the flange side.
5. Place reaction lever plate (with raised rim away from piston diaphragm plate) over piston and install snap-ring into the groove.
6. Set the assembly aside for the moment.
7. Place the reaction diaphragm support plate over the hub of floating control valve. Assemble reaction diaphragm over flanged hub of control valve so that the flange fits into the groove on the inside diameter of the diaphragm.
8. Insert diaphragm retainer under the lip of the diaphragm and place the floating control valve spring retainer over the reaction diaphragm hub. The six small rubber bumpers must be positioned in the spring retainer.
9. Place power piston in vise with the flat surface up. (Do not clamp.)
10. Insert the pushrod end of the air valve pushrod into the piston and press down into the piston to seat the air valve.

11. Wipe a light film of silicone grease on the outer diameter of reaction diaphragm.
12. Place the floating control valve and diaphragm assembly down over the air valve and press the outer edge of the reaction diaphragm into position in the power piston.
13. Place the air valve return spring inside the floating control valve to seat on the air valve. Place the floating control valve spring to seat on the flange of the spring retainer. Position the air valve reaction plate over the air valve so the low center portion rests on the inner spring (air valve return spring). The outer spring (floating control valve spring) will also be in position to engage the reaction lever plate.

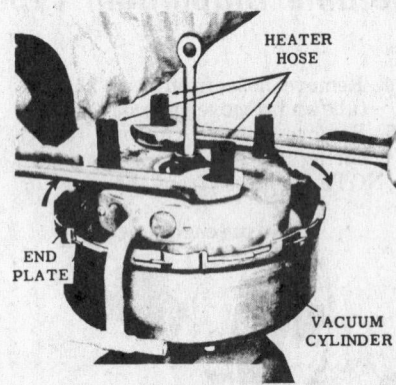

Installing end plate (© Oldsmobile Div., G.M. Corp.)

14. Position the ears of the reaction levers in the machined locations in the power piston and rest the small ends on the reaction lever plate. Insert seal-ring into the inner groove on the flat surface of the power piston face.
15. Pull the skirt of the power piston diaphragm down over the piston and place the bead on the inner diameter of the diaphragm in the outer groove of the power piston face so the bumpers are toward the piston.
16. The master cylinder piston and piston diaphragm plate are positioned over the power piston diaphragm, and the lugs on the plate are aligned with the depressions in the power piston.
17. The snap-ring on the master cylinder piston must be rotated to position the lobes between the reaction levers. The ears of the reaction levers must stay in place in the machined seats on the power piston as the diaphragm plate contacts the power piston face. The position of the bead and levers must be correct as the support plate is finally positioned and the four ¼ in. screws tightened through the support plate

and into the power piston. Torque to 65-85 in. lbs.
18. Turn power piston assembly upside down to assemble the air valve boot. Position large lip of boot in groove of power piston. Then place small diameter lip in the groove in the air valve.
19. Coat the outside diameter of the tube-like boss on power piston with rubber trim cement and allow a minute or two to air dry. Then slip the exhaust hose fully on the boss so the hose is parallel to the power piston.
20. Place the gasket on the front shell. Place the seal, master cylinder to housing, in the groove on the master cylinder body.
21. Place the front shell on a flat surface with studs up. Insert and position master cylinder body on studs and press it into the front shell. Install lockwashers and nuts, and torque to 15-20 ft. lbs.
22. Turn the assembly upside down in a vise (so that the master cylinder bore is now accessible). Insert the valve seat washer into the bore. Press check valve into open end of spring and retainer assembly, and position in the bore with the check valve against the valve seat washer. Dip the primary cup into clean brake fluid and insert it into the bore with lips over the spring and retainer.
23. Work the groove of the pushrod boot into position in the center hole of the rear shell.
24. Place the return spring over the flange in the center of the front shell. Place the power piston over the front shell and insert the master cylinder piston through the return spring into the master cylinder bore.
25. Press the power piston assembly down and hold it in position while placing the bead of the power piston diaphragm into the recess in the rim of the front shell.
26. Place the exhaust hose onto the exhaust tube of the rear shell assembly. Force the pushrod through the boot on the rear shell. Press shell down and align the locking lugs.

NOTE: the following gauging operation is necessary only after structural parts have been replaced. It is also necessary if the exact number of shims removed at disassembly are not known.

With the brake cylinder in the released position, force air into the hydraulic outlet of the master cylinder. If air passes freely through the compensating port (the smaller of the two holes in the bottom of the master cylinder reservoir), shimming can be considered satisfactory. If air does not flow, remove the master cylinder

and add shims until it does flow freely.

When the correct number of shims is in position, the master cylinder is assembled to the housing. After inserting the master cylinder piston into the bore and positioning the cylinder on the studs, the four attaching nuts should be torqued to 15-20 ft. lbs.

Installation

If linkage was removed, proceed as follows:

1. Place one operating lever on the collar. Tap in place until it seats against shoulder. Install second operating lever.
2. Place the lever and collar in position and install the pivot pin and cotter key.
3. Place one of the wave washers on the inner side of the outer operating lever (over the upper hole) and install the clevis pin, flat washer and cotter pin to the power brake assembly operating rod.
4. Place the remaining wave washer on the inner side of the outer operating lever (over the middle hole) and install the clevis pin, flat washer and cotter key to the brake pedal pushrod.
5. Mount the power brake assembly in place and install the four attaching nuts and lock washers.
6. Hook the hydraulic line to the master cylinder.
7. Attach the vacuum lines to the power unit.
8. Attach the pushrod clevis to the brake pedal assembly. Adjust pedal height by means of clevis on brake pedal pushrod at the pedal.

Section 7
Moraine Diaphragm Type

Disassembly

1. Empty assembly and thoroughly clean outside surfaces.
2. Scribe mark on top center of front and rear housings and in line with the master cylinder reservoir cover.
3. Separate the housings by rotating one counterclockwise.
4. Lift power piston and rear housing assembly from unit.
5. Remove clevis from pushrod. Remove jam nut. Remove boot from rear of housing. Remove retaining ring on pushrod that holds silencer in place and remove silencer.

der large divided locking lug and proceed to pull it from under the other two small locking lugs on power piston.
12. Remove reaction retainer, piston rod, reaction plate, reaction levers and air valve spring.
13. Remove reaction bumper and air valve spring retainer from air valve.
14. Place power piston tool (J-21524) with square shank in vise. Hold support plate and power piston with tube of piston up.
15. Pull diaphragm edges away from support plate so that steel support plate can be gripped by hand. Position assembly so that the three lugs on tool fit into three notches in piston.
16. Press down on support plate and rotate counterclockwise until support plate separates from power piston.
17. Remove rolling diaphragm from support plate.
18. Place power piston, tube down, in vise. Do not clamp tube.
19. Remove snap-ring on air valve.
20. Using power ram, press and remove air valve assembly, using 3/8 in. drive extension as a remover.
21. Removal of air valve pushrod disassembles the floating control valve, floating valve retainer, pushrod limiter washer and air filters from the power piston.
22. Remove floating control valve assembly from pushrod, as it must be replaced by new floating control valve assembly.
23. Master cylinder pushrod can now be pushed from the center of reaction retainer. Remove O-rings from groove in master cylinder piston rod.

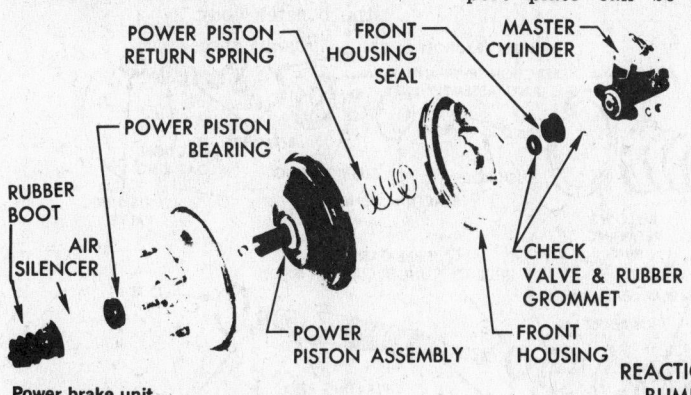

POWER PISTON RETURN SPRING — FRONT HOUSING SEAL — MASTER CYLINDER

POWER PISTON BEARING

RUBBER BOOT

AIR SILENCER

CHECK VALVE & RUBBER GROMMET

POWER PISTON ASSEMBLY — FRONT HOUSING

Power brake unit

6. Remove power piston return spring from front housing.
7. While holding master cylinder in vise, remove front housing from master cylinder.
8. Remove nuts and lock washers and remove master cylinder. Master cylinder reconditioning is in the conventional manner.
9. Remove front housing seal, then the vacuum check valve and grommet from front housing.
10. Remove silencer from neck of power piston tube.
11. Remove lock-ring from power piston by prying one end from un-

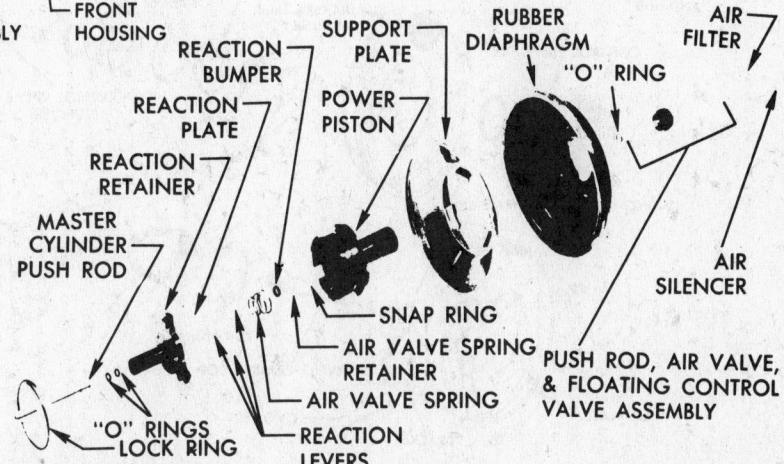

REACTION BUMPER — REACTION PLATE — REACTION RETAINER — MASTER CYLINDER PUSH ROD — SUPPORT PLATE — POWER PISTON — RUBBER DIAPHRAGM — "O" RING — AIR FILTER — AIR SILENCER — PUSH ROD, AIR VALVE, & FLOATING CONTROL VALVE ASSEMBLY

"O" RINGS LOCK RING — SNAP RING — AIR VALVE SPRING RETAINER — AIR VALVE SPRING — REACTION LEVERS

Power piston assembly

24. Carefully clean and inspect all parts, replace all seals, clean up any corrosion. Where corrosion is present, the entire system should be completely flushed and all new fluid used. Minor cleaning may be done with crocus cloth.

25. Reassemble in reverse of the above procedures.

Gauging is only necessary when major structural parts such as front or rear housing, power piston assembly, master cylinder piston or master cylinder assembly is replaced. To gauge, remove master cylinder so that master cylinder pushrod is exposed. Place gauge over piston rod so that it fits between the two studs on front of housing. The gap between the cutout of gauge and piston rod should be not more than .010 in. If corrections must be made a service adjustable piston rod and adjusting screw should be used to secure the proper clearance.

Section 8
Midland-Ross Diaphragm Type

The self-contained booster assembly is mounted on the engine side of the firewall. It is connected directly to the brake pedal. This booster is not equipped with a separate vacuum tank.

The master cylinder is attached to the forward side of the booster. The balance of the hydraulic brake system is identical to other standard service brakes.

Booster Removal

1. From inside the car, remove the horseshoe-type retainer holding the booster pushrod link onto the brake pedal pin. Then slide the link off the pin.
2. Disconnect wires from the stoplight switch.
3. Disconnect hydraulic line at master cylinder.
4. Disconnect manifold vacuum hose from the booster. If car is equipped with a transmission vacuum throttle valve hose, remove this hose.
5. Remove nuts and washers from booster bracket mounting studs on fire wall. Then pull booster and bracket assembly off the studs, sliding the pushrod link out of the hole in the panel.

Booster Repairs

1. Separate master cylinder from booster body.
2. Remove air filter cover and hub and the filter from the booster body.
3. Remove the vacuum manifold mounting bolt, manifold, gaskets and vacuum check valve from the booster body.
4. Disconnect the valve operating rod from the lever by removing its retaining clip, washers, and pivot pin.
5. Disconnect the lever from the booster end plate brackets by removing its retaining clip, washers, and pivot pin.
6. Remove two brackets from the end plate.
7. Remove the rubber boot from the valve operating rod.
8. To remove the bellows, control valve, and diaphragm assemblies, remove large C-ring that holds the rear seal adapter assembly to the booster end plate.
9. Scribe matching lines on the booster body and the end plate. Then remove the ten retaining screws. Tap the outside of the plate with a soft hammer and separate the plate from the booster body.
10. Push the bellows assembly into the vacuum chamber and remove the bellows, control valve, and diaphragm as an assembly from the booster body.

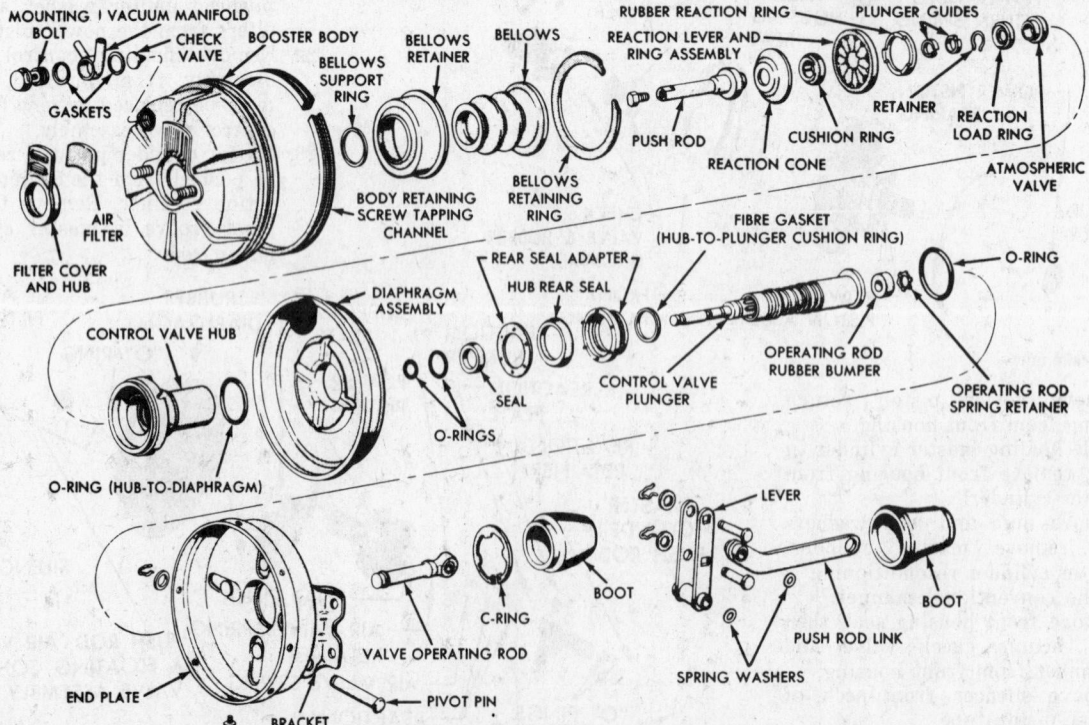

Disassembled view of booster

11. Remove the outer O-ring from the control valve hub.

12. To disassemble the bellows, pushrod, and control valve assemblies, remove the large bellows retaining ring, bellows, bellows retainer, and support ring from the diaphragm and valve assembly.

13. Remove the retainer and support ring from the bellows.

14. Remove pushrod assembly, the reaction lever and ring assembly, and the rubber reaction ring from the control valve hub.

15. Remove the reaction cone and cushion ring from the pushrod assembly. Then disassemble the reaction levers from the ring.

16. Remove the two plastic plunger guides from the control valve plunger. Then remove the retainer that holds the reaction load ring and atmospheric valve on the control valve hub.

17. Slide the reaction load ring and atmospheric valve from the control valve hub.

18. Separate the control valve hub and the plunger assembly from the diaphragm by sliding the plunger and rear seal adapter from the rear of the hub. Then remove the hub outer O-ring from the front side of the diaphragm.

19. To disassemble the control valve plunger, remove the hub rear seal adapter from the valve plunger assembly, and remove the seal from the adapter.

20. Remove the O-rings, the seal, and the fiber gaskets from the plunger.

21. If the plunger assembly needs to be replaced, hold the plunger and pull out the valve operating rod with pliers. Do not separate the operating rod and plunger unless the plunger is to be replaced.

Assembly

1. If valve operating rod was removed for replacement of plunger, install a new rubber bumper and spring retainer on the rod before installing it on the replacement plunger. Then push the rod firmly until it bottoms in the plunger.

2. Install fiber gaskets, plunger seal, and the two O-rings on the plunger assembly.

3. Install the valve hub rear seal in the adapter assembly with the sealing lip toward the rear. Then slide the adapter assembly onto the plunger with the small diameter end of the hub toward the rear.

4. To assemble the control valve, pushrod, and bellows assemblies, install the hub outer O-ring. Then install the plunger with the seal adapter and the hub on the diaphragm. To do this, hold the hub on the front side of the diaphragm and insert the plunger assembly in the hub from the rear side of the diaphragm.

5. Install atmospheric valve and then the reaction load ring onto the plunger and hub. Compress the valve spring, and install the load ring retainer into the groove of the plunger.

6. Install two plastic plunger guides into their grooves on the plunger.

7. Install rubber reaction ring into the valve hub so that the ring locating knob indexes in the notch in the hub, with the ring tips toward the front.

8. Assemble the reaction lever and ring assembly, and install the assembly into the valve hub.

9. Install the reaction cone and cushion ring on the pushrod. Then install the pushrod assembly on the valve hub so that the plunger indexes in the rod.

10. Assemble the bellows, retainer, and support ring. The ring should be positioned on the middle fold of the bellows.

11. Position the bellows assembly on the diaphragm, and secure it with the retaining ring. Make sure the retaining ring is fully seated.

12. Install the bellows, control valve, and diaphragm assemblies with a screwdriver, moving the booster body retaining screw tapping channel just enough to provide a new surface for the self-tapping attaching screws.

13. Install the diaphragm, the control valve components, and the bellows as an assembly into the booster body. (Be sure the lip of the diaphragm is evenly positioned on the retaining radius of the booster body.) Pull the front lip of the bellows through the booster body, and position it around the outer groove of the body.

14. Install O-ring in the front side of the end plate, and locate the plate on the booster body. Align the scribed lines, compress the two assemblies together with a clamp. Then install all ten self-tapping attaching screws.

15. Install the large C-ring onto the rear seal adapter at the rear side of the end plate.

Pushrod Adjustment

The pushrod has an adjusting screw to maintain the correct relationship between the control valve plunger and the master cylinder piston after the booster is completely assembled. If this screw is not properly adjusted, the brakes may drag.

To check adjustment of the screw, make a gauge to the dimensions shown. Place this gauge against the master cylinder mounting surface of

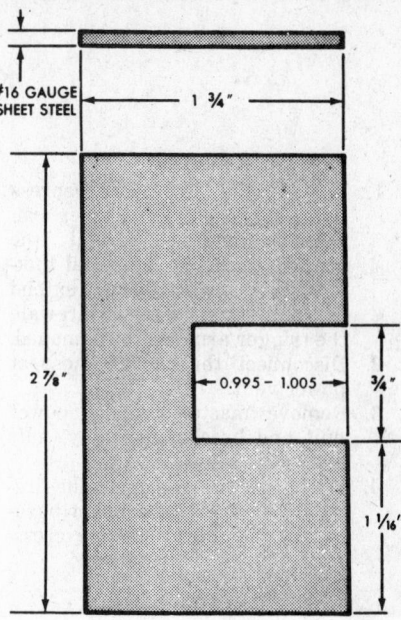

Pushrod gauge

the booster body. The pushrod screw should be adjusted so that the end of the screw just touches the inner edge of the slot in the gauge.

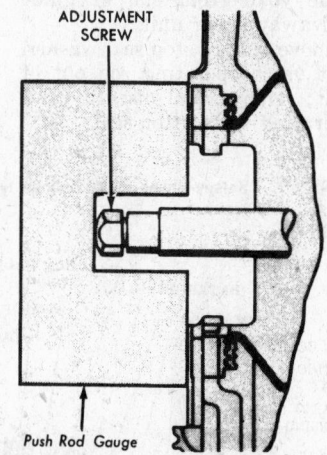

Checking pushrod screw with gauge

Booster Installation

1. Install rubber boot on the valve operating rod.

2. Position the two mounting brackets on the end plate, and install retaining nuts.

3. Connect the lever assembly to the lower end of the mounting brackets with its pivot pin. Then install the spring washer and retaining clip.

4. Connect the valve operating rod to the upper end of the lever with its pivot pin, washer, and retaining clip.

5. Install the vacuum check valve, the vacuum manifold, the two gaskets, and the mounting bolt. Torque the mounting bolt to 8-10 ft. lbs.

Section 9
Kelsey-Hayes Bellows Type

Removal

1. Use a pedal depressor. Depress the pedal to prevent trigger arm from extending beyond the bracket limits. If the pedal linkage is permitted to extend through the hole in the firewall, the trigger arm may be damaged.
2. Disconnect the vacuum hose at power vent.
3. Remove master cylinder power unit and bracket assembly, nuts and lock washers.
4. Withdraw the unit from the firewall. Don't lose the nylon bushing at the pedal linkage crosspin.

Disassembly

1. Remove mounting plate to unit attaching nuts.
2. Slide plate off and away from unit.
3. Remove mounting plate O-ring.
4. With an Allen wrench, back out two set screws enough to permit removal of yoke.
5. Slide yoke from end of guide and away from unit.
6. Remove rubber stop seal washer.
7. Lift valve operating rod out of unit; remove, and discard valve operating rod button-seal.

8. Remove nuts that attach outer mounting plate.
9. Gently pry and lift the plate straight up and away from the unit.
10. Compress bellows by hand to expose the guide bearing. Slide bearing off end of guide.
11. Remove and discard bearing seal from outside bearing.
12. Peel outer lip of bellows from around inner mounting plate.
13. Remove plate and lift out return spring and return spring retainer.
14. Remove bolts and lock washers that attach valve cover to valve. Lift off cover.
15. Remove and discard O-ring from valve cover.
16. Remove air valve spring from center of valve.
17. Remove air filter and pull the air valve out of the housing using a small bent wire, (paper clip).
18. Place valve housing, end down, on bench. Remove bellows from valve by peeling back outer lip of bellows.
19. Lift bellows up, away from the valve.
20. Remove guide to valve body

bolts and lift off guide to expose vacuum valve, valve spring and seals.
21. Remove seals, then lift out vacuum valve and retainer.
22. Remove valve housing to guide seal.
23. Insert valve housing and remove air valve seal from its groove in the valve body.

Assembly

Lubricate all seals and O-rings with silicone grease before installation.

1. Insert new air valve seal into bore of valve housing (with lips of seal facing out).
2. Position new vacuum valve in retainer.
3. Invert valve housing and install vacuum valve and retainer in housing.
4. Press down firmly on retainer to snap it in place.
5. Position new valve housing-to-guide seal in groove provided.
6. Install new vacuum valve seal in bore guide, (lip of seal toward bottom of bore).

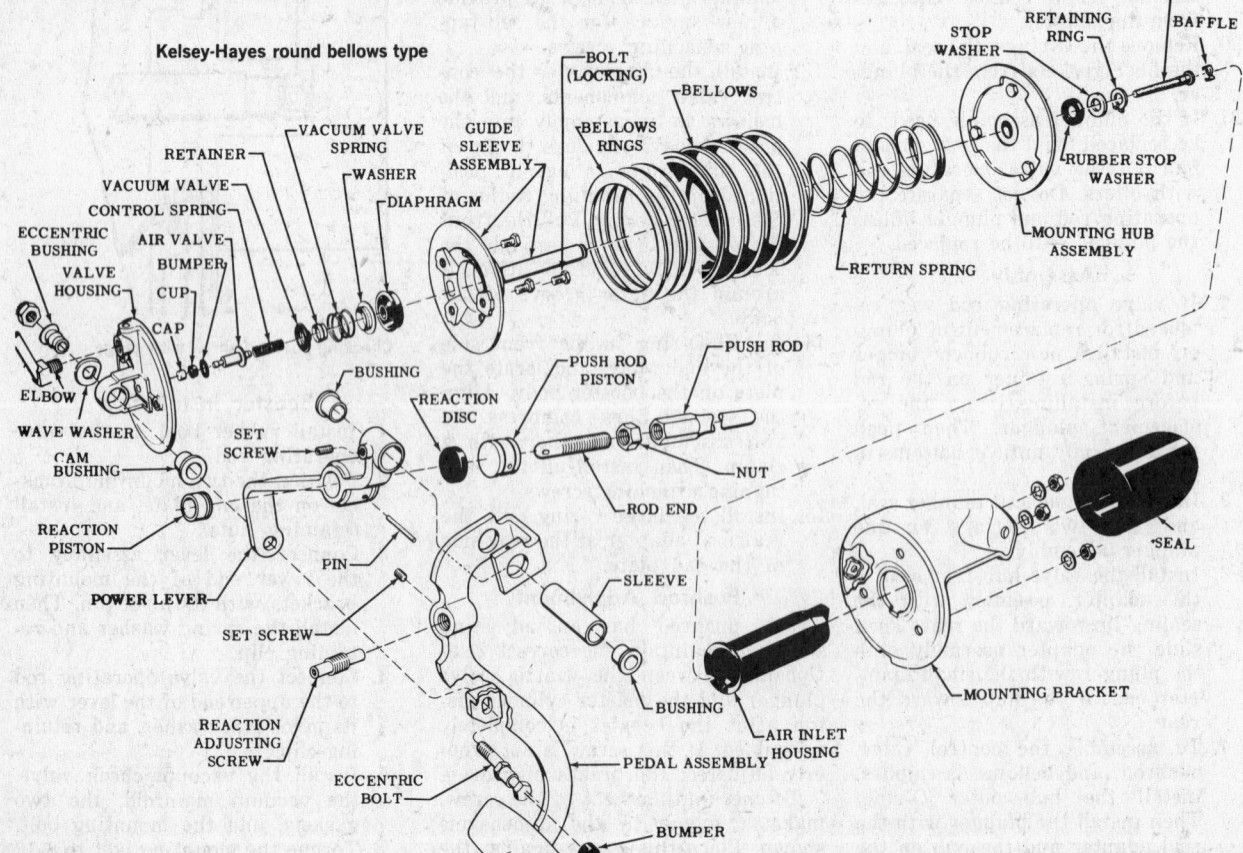

Kelsey-Hayes round bellows type

7. Install vacuum valve spring in center of valve.
8. Position guide over vacuum valve, lining up bolt in guide with bolt holes in valve body.
9. Lower guide down against valve body. Be certain that the tapered portion of valve enters seal evenly.
10. Press down on guide to seal and install bolts and lock washers. Tighten bolts evenly.
11. If new bellows is being installed, position supports in bellows.
12. Using holding fixture (made from 3 in. length of 4 in. diameter pipe) to support guide and valve assembly in upright position, install the bellows. Be sure arrows on edge of bellows and housing are aligned.
13. With assembly in holding fixture, wipe the outer surface of the air valve with silicone grease and insert the small end into bore of housing.
14. Use finger pressure to test for free movement of valve against vacuum valve spring.
15. Install air valve spring into recess in air valve and air filter.
16. Install new valve housing cover O-ring on shoulder provided on valve body hub.
17. Position valve body cover over valve housing, with notch in edge of cover matching arrow on bellows. Be sure air valve spring rests on dimple in center of cover.
18. Press cover down evenly over valve housing to seat over O-ring, install bolts, then tighten securely.

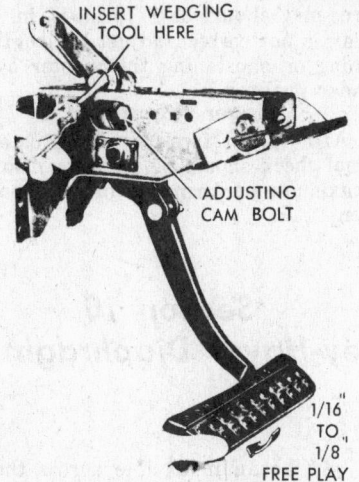

Pedal play adjustment (© Ford Motor Co.)

19. Remove assembly from holding fixture and invert unit.
20. Wipe guide lightly with silicone grease and install return spring.
21. Position spring evenly around hub of valve housing and guide.
22. Place spring retainer and inner mounting plate over spring, (be sure the arrow marks on the plate are in line with the arrow on the edge of the bellows).
23. Compress return spring and fold bellows lip over edge of plate.
24. Install new guide bearing seal in groove inside bearing bore.
25. Seat the seal snugly in the bearing.
26. Lubricate inside of bearing with silicone grease and slide it over guide, while compressing the bel-

lows. Bearing to be installed with lip of seal facing out.
27. Push bearing down over guide and into pocket of plate.
28. Release bellows and the bearing will ride up guide with plate into position.
29. Install bearing to mounting plate O-ring and lower outer mounting plate down on assembly.
30. Install nuts and draw down finger tight.
31. Slide new valve operating rod seal ring over nylon bumper on end of rod and into groove.
32. Install rod in center of guide.
33. Press on end of rod to test for free operation or movement of air and vacuum vanes. A two-step movement should be felt when rod is depressed and released fully.
34. Place new stop seal washer in position and install yoke on end of guide.
35. Compress bellows slightly and alternately tighten set screws. Hub of yoke must be down snug against shoulder of guide, with set screws aligned with tapered holes in guide.
36. Tighten mounting plate nuts.
37. Place mounting bracket in position, with long centerline of bracket at right angle to long centerline of unit section.
38. Install nuts and lock washers, then tighten securely.

Pedal Free-Play (On the Car)

A pedal free-play check should be made with no vacuum in the system. Apply the brakes several times

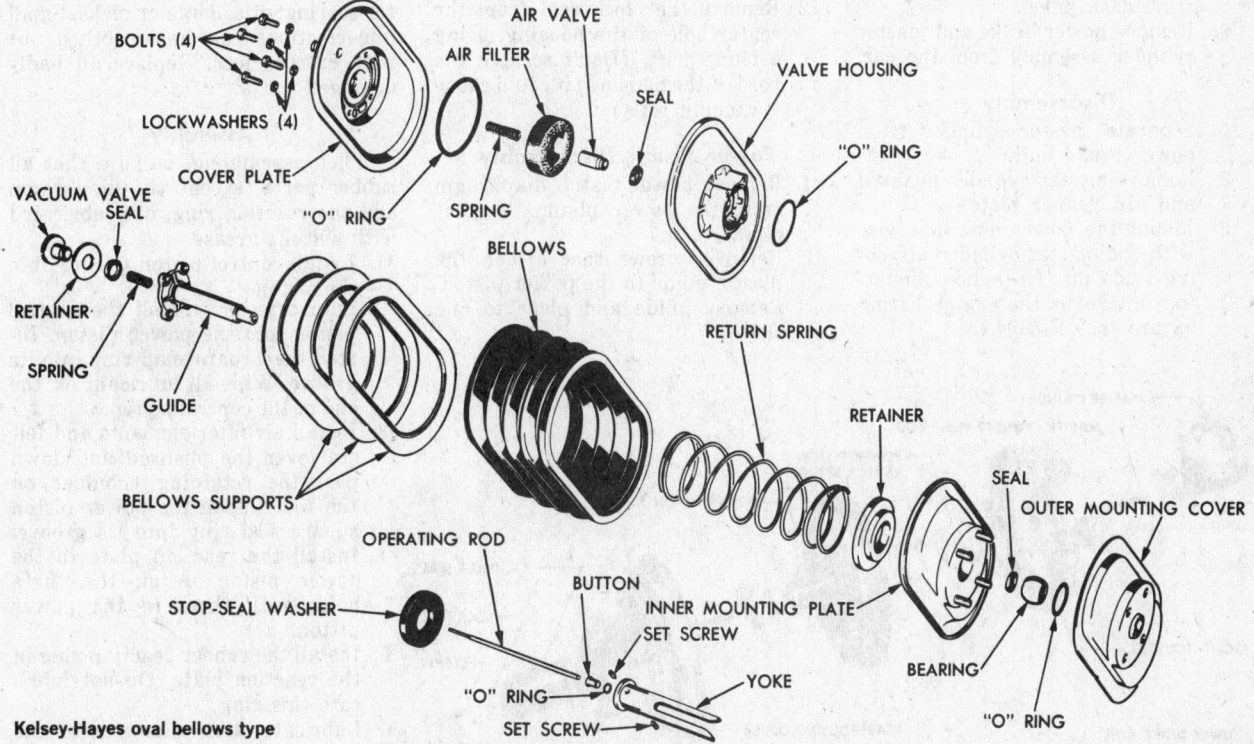

Kelsey-Hayes oval bellows type

(engine not running) to exhaust the vacuum supply. Insert a screwdriver between the trigger pivot and the rear side of the hole in the power brake lever. This will force the brake pedal and the power lever apart. Check free-play with screwdriver in this position by pressing lightly at the pad end of the brake pedal. Pedal free-play should be 1/16 in.-1/8 in. If play is not correct, adjust by lengthening or shortening the master cylinder pushrod.

Trigger Adjustment

After pedal free-play is adjusted, final check should be made to insure maximum performance of the booster.

A slight amount of rotation of the adjusting cam clockwise will speed up a slow pedal return.

A slight amount of rotation of the adjusting cam counterclockwise will eliminate a time delay during a fast application.

Section 10
Kelsey-Hayes Diaphragm Type

Identification

This power brake unit was incorporated in production on Chrysler Corporation cars during the first part of 1964.

The Kelsey-Hayes power brake unit can be identified by the twist-lock method of locking the housing and cover together, plus the white-colored vacuum check valve assembly.

Service

Removal

1. With engine off, apply brakes several times to equalize internal brake pressure.
2. Disconnect hydraulic line from master cylinder.
3. Disconnect vacuum hose from power brake check valve.
4. Disconnect power brake from brake pedal (under instrument panel).
5. Disconnect power brake unit from dash panel.
6. Remove power brake and master cylinder assembly from the car.

Disassembly

1. Separate master cylinder from power brake unit.
2. Remove master cylinder pushrod and air cleaner plate.
3. Mount the power unit in a vise with the master cylinder attaching-studs up. (A handy mounting bracket is the engine lifting fixture tool C-3804.)

4. Scribe an index line across the housing and cover for reassembly reference.
5. Pry out the housing lock. Do not damage the lock, as it must be used at assembly.
6. Remove check valve from cover by prying out of rubber grommet.
7. Place parking brake flange holding tool C-3281 over the master cylinder mounting studs.
8. Rotate the tool and cover in a counterclockwise direction. Then, separate the cover from the housing. This will expose the power piston return spring and diaphragm.
9. Lift out the power piston return spring. Remove the brake unit from the vise.
10. Remove power piston by slowly lifting the piston straight up.
11. Remove air cleaner, guide seal and seal retainer from the cover.
12. Remove the block seal from the center hole of the housing, using a blunt drift. (Don't scratch the bore of the housing; it could cause a vacuum leak.)

Power Piston Disassembly

1. Remove power piston diaphragm from the power piston. Keep it clean.
2. Remove screws that attach the plastic guide to the power piston. Remove guide and place to one side.

3. Remove the power piston square seal ring, reaction ring insert, reaction ring and reaction plate.
4. Depress operating rod slightly, then remove the Truarc snapring.
5. Remove control piston by pulling the operating rod.
6. Remove the O-ring seal from the end of the control piston.
7. Remove the filter elements and dust felt from the control piston rod.

Cleaning and Inspection

Thoroughly wash all metal parts in a suitable solvent and dry with compressed air. The power diaphragm, plastic power piston and guide should be washed in a mild soap and water solution. Blow dust and all cleaning material out of internal passages. All rubber parts should be replaced, regardless of condition. Install new air filters at assembly. Inspect all parts for scoring, pits, dents or nicks. Small imperfections can be smoothed out with crocus cloth. Replace all badly damaged parts.

Assembly

When assembling, be sure that all rubber parts, except the diaphragm and the reaction ring, are lubricated with silicone grease.

1. Install control piston O-ring onto the piston.
2. Lubricate and install the control piston into the power piston. Install the Truarc snap-ring into its groove. Wipe all lubricant off the end of the control piston.
3. Install air filter elements and felt seal over the pushrod and down past the retaining shoulder on the rod. Install the power piston square seal ring into its groove.
4. Install the reaction plate in the power piston. Align the three holes with those in the power piston.
5. Install the rubber reaction ring in the reaction plate. Do not lubricate this ring.
6. Lubricate outer diameter of the reaction insert and install in the

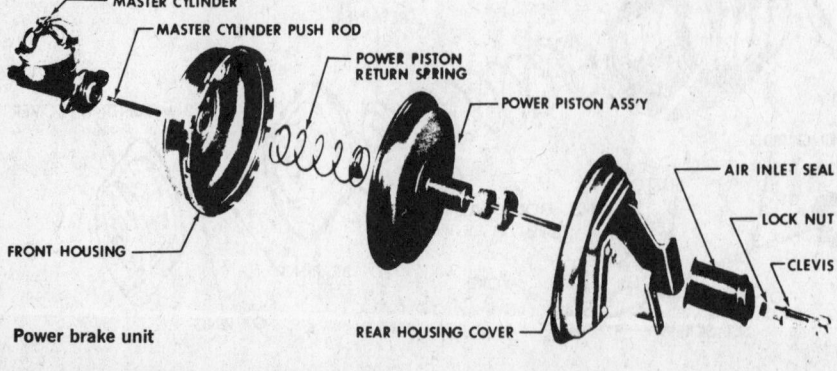

MASTER CYLINDER
MASTER CYLINDER PUSH ROD
POWER PISTON RETURN SPRING
POWER PISTON ASS'Y
AIR INLET SEAL
LOCK NUT
CLEVIS
FRONT HOUSING
REAR HOUSING COVER

Power brake unit

reaction ring.

7. Install reaction insert bumper into the guide.

8. Place guide on the power piston, align the holes with the aligning points on the power piston. Install retaining screws and torque to 80-100 in. lbs.

9. Install diaphragm on power piston; be sure that the diaphragm is correctly seated in the power piston groove.

10. With the housing blocked to prevent damage, install the block seal in the housing, using tool C-3205 or suitable substitute.

11. Install a new cover seal on the retainer and lubricate thoroughly, inside and out, with silicone grease, then install in the cover bore. Install new air filter.

12. Lubricate check valve grommet and install the vacuum check valve.

13. Mount the power unit in a vise, with master cylinder attaching studs up. (Use the same engine lifting fixture, tool C-3804, if available.)

14. Apply a light coating of silicone grease to the bead, *outer edge only*, of the power piston diaphragm.

15. Install the power piston assembly in the housing with the operating rod down.

16. Install the power piston return spring into the flange of the guide.

17. Place the cover over the return spring and press down on the cover. At the same time, pilot the guide through the seal.

18. Rotate the cover to lock it to the housing. Be sure the scribe lines are in correct index and that the diaphragm is not pinched during assembly.

19. Install the housing lock on one of the long tangs of the housing.

20. Remove the power unit from the vise and remove mounting bracket, if used.

21. Install the master cylinder pushrod and air cleaner plate, then install the master cylinder on the studs. Install attaching nuts and washers. Torque to 200 in. lbs.

Installation

1. Install firewall to power brake seal.

2. Install power brake unit onto firewall and torque the attaching nuts to 200 in. lbs.

3. Install pushrod to brake pedal attaching bolt. Torque to 30 ft. lbs.

4. Install vacuum hose onto the power brake unit.

5. Attach the hydraulic tube and fill the master cylinder. Bleed hydraulic system.

6. Adjust stop-light switch if necessary.

Testing Operations

Vacuum Leak, Released Position

With transmission in Neutral or Park, stop the engine and wait one minute. Apply the brake several times. Each application should provide less and less assist, as a result of normal depletion of reserve vacuum.

The number of applications on reserve vacuum assist will depend on how hard the pedal is applied and how far the pedal travels. If vacuum assist is not present, an air leak is indicated.

Unit Operation

After depleting the reserve vacuum, apply light foot pressure to the pedal, then start the engine. If the power system is working properly, the pedal will fall away slightly, as vacuum is restored.

Vacuum Leak, Holding Position

With transmission in neutral or park, stop the engine while holding a moderately heavy load steadily on the brake pedal. After one minute, release and apply the pedal several times. If there is no vacuum assist during this test, but the system was normal during test No. 1, there is an air leak within the unit. Some units will leak air internally if the pedal load is light. This is a normal condition.

Hydraulic Leak Test

Depress the brake pedal lightly while engine is running, maintaining constant pressure. If the pedal falls noticeably in one minute, the hydraulic system is probably leaking.

If the pedal has a spongy feel, air may be present in the hydraulic system.

Road test the car, making a brake application at about 40 mph to determine if the car stops evenly and quickly.

If system tests are satisfactory and the pedal travels to within 1 in. of the floor pan, the brake shoes probably need adjustment or replacement.

General Inspection

1. Check for free operation and return of the brake pedal. If binding exists, check all pivot points for binding, then lubricate as required.

2. Check stop-light switch for operation and possible adjustment.

3. Inspect master cylinder for proper fluid level.

4. Inspect vacuum line and connections for leaks.

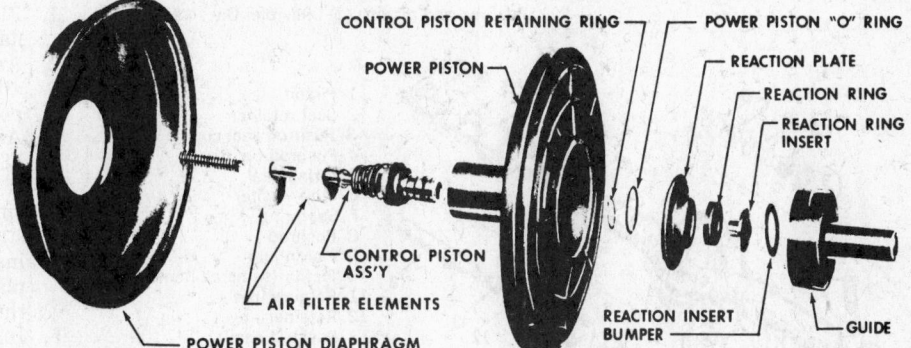

Power piston assembly

CONTROL PISTON RETAINING RING — POWER PISTON "O" RING — REACTION PLATE — REACTION RING — REACTION RING INSERT — GUIDE — REACTION INSERT BUMPER — POWER PISTON DIAPHRAGM — AIR FILTER ELEMENTS — CONTROL PISTON ASS'Y — POWER PISTON

Section 11
Bendix Single Diaphragm Hydrovac

Removal

1. Empty system and thoroughly clean outside of assembly.

2. Scribe across front and rear shells. Clamp hydraulic cylinder in vise in horizontal position.

3. Remove clamp and separate shells.

4. Turn rim of diaphragm back and

press support plate to compress diaphragm return spring.

5. Using snap-ring pliers, remove snap-ring from groove inside bore of hydraulic cylinder.

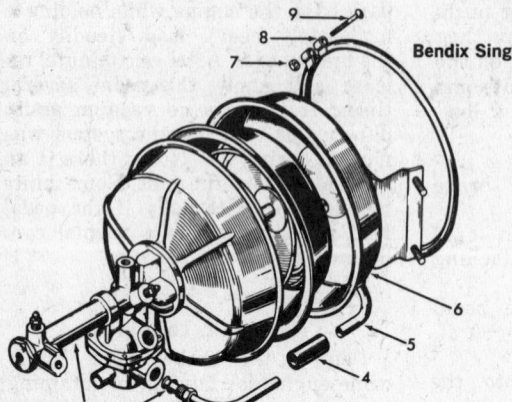

Bendix Single Diaphragm Hydrovac Power Brake
(© Chevrolet Div., G.M. Corp.)

1 Hydraulic cylinder
2 Control tube seal
3 Control tube nut
4 Vacuum hose
5 Rear shell control tube
6 Rear shell
7 Nut
8 Clamp band
9 Clamp bolt

6. Pull diaphragm assembly straight out of cylinder. Parts may be removed from assembly by removing nut from diaphragm end of assembly.

Pushrod and Piston Disassembly

1. Piston rod parts are removed from the pushrod from the piston end of the pushrod.
2. Slide retainer ring out of groove in piston and separate piston from pushrod by removing retainer pin.
3. With thin bladed screw driver remove piston cap from piston. Slide off seal retainer, pushrod seal cup, pushrod bearing, stop washer and snap ring.
4. Remove "O"-ring seal from bearing groove.
5. If necessary, remove pushrod seal from end of pushrod with pliers and discard.

washer and cup from valve piston.

4. Using screw driver, pry off plastic valve body cover.
5. Pry off plastic retainer, lift off

Control valve (© Chevrolet Div., G.M. Corp.)

1 End cap
2 Gasket
3 Hydraulic cylinder housing
4 Cup
5 Valve piston
6 Washer
7 Diaphragm
8 Washer
9 Valve body spring
10 Vacuum poppet and stem
11 Poppet spring
12 Control valve body
13 Capscrews
14 Poppet
15 Retainer
16 Valve body cover
A Pressure check valve
B Spring
C Snap-ring retainer
D Snap-ring

Pushrod and piston (© Chevrolet Div., G.M. Corp.)

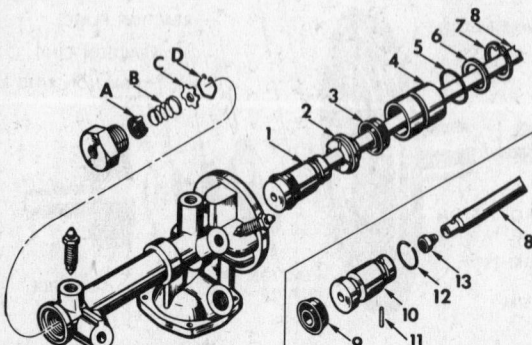

1 Piston
2 Seal retainer
3 Pushrod seal cup
4 Pushrod bearing
5 O-ring seal
6 Stop washer
7 Snap-ring
8 Pushrod
9 Piston cup
10 Piston (same as item 1)
11 Retainer pin
12 Retainer ring
13 Pushrod seal

Control Valve Disassembly

1. Scribe across flanges of control valve body and hydraulic cylinder housing.
2. Remove four cap screws and lift off valve body and spring.
3. Remove washer, diaphragm,

atmospheric poppet and then remove vacuum poppet and stem and poppet spring from valve body.

6. Using 1⅜ in. wrench, remove end cap and gasket.
7. If residual pressure check valve

is used, lift snap-ring from groove inside the cap. Remove spring retainer, spring and residual pressure check valve.

Piston and Pushrod Assembly

1. To install a new pushrod seal in pushrod, place the new seal (rubber side down) on a clean block of wood. Hold pushrod vertically with drilled end of rod resting on shaft end of seal and strike upper end with a soft mallet to seat.
2. Dip cup in brake fluid and assemble cup on piston.
3. Slide snap-ring, stop washer, bearing with "O"-ring installed in groove, pushrod seal cup, seal retainer, retainer ring and piston assembly on pushrod in order.

4. Assemble retainer pin through holes in piston and rod. Secure pin in place with retainer ring, making sure ring is seated in groove on piston.

Front Shell Assembly

1. Place new gasket in groove on flange of hydraulic cylinder.
2. Assemble front shell to cylinder, aligning cutout in shell with porting in cylinder flange.
3. Assemble capscrew, tighten to 130-230 in. lbs.
4. If diaphragm plate and return spring were disassembled, install nut and tighten to 160-200 in. lbs., then washer, diaphragm plate, concave side first, on threaded end of pushrod.
5. Install diaphragm, washer, and nut on pushrod, and tighten nuts to 160-200 in. lbs.
6. Coat hydraulic cylinder bore with clean brake fluid. Dip piston and bearing parts in brake fluid. Slide diaphragm return spring onto pushrod, large diameter of the spring first.
7. Seat spring against concave surface of diaphragm plate and align entire assembly with hy-

draulic cylinder bore.

8. Carefully insert hydraulic piston assembly, retainer, seal and bearing into cylinder bore. Press against diaphragm and plate to compress return spring and seat stop washer against bearing inside bore and install snap-ring in groove in cylinder bore.

Control Valve Assembly

1. Assemble cup facing up on control valve piston. If two cups are used install them on piston back to back. Then assemble piston, washer, and diaphragm and seat inner bed of diaphragm in piston groove.
2. Lay piston and diaphragm assembly aside. Assemble poppet return spring, vacuum poppet, and atmospheric poppet in valve body from opposite sides.
3. Snap poppet retainer over end of vacuum poppet stem.
4. Assemble cover securely in groove of valve body. Assemble spring retainer washer and spring on end of control valve piston. Insert control valve piston, diaphragm and spring assembly in valve body with spring, around bosses.
5. Press outer bead of diaphragm onto valve body groove and,

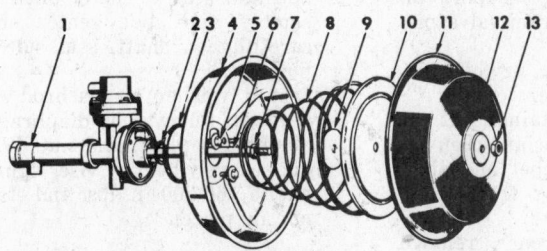

Diaphragm and front shell assembly   (© Chevrolet Div., G.M. Corp.)

1 Hydraulic cylinder
2 Snap-ring
3 Gasket
4 Front shell
5 Capscrews
6 Pushrod
7 Nut
8 Return spring
9 Washer
10 Diaphragm plate
11 Diaphragm
12 Washer
13 Nut

while holding spring compressed, dip piston and cup in clean brake fluid.

6. Align valve body to scribe marks and assemble control valve piston carefully in its bore below hydraulic cylinder bore. Secure valve body with four capscrews.
7. Tighten screws to 40-60 in. lbs. Install check valve, spring, retainer and snap ring in end fitting. Install fitting and gasket on end of hydraulic cylinder and tighten fitting to 50-85 ft. lbs. If removed, install bleed screws, tighten to 10-15 ft. lbs.

Rear Shell, Control Tube, Hose Assembly

1. Coat bead of diaphragm or

flanges of front and rear shells with talcum powder or equivalent and place rear shell on diaphragm.
2. Align scribe marks on both shells. Make certain bead of diaphragm is seated in outer flanges of both shells and hold rear shell in position while assembling clamp band over flanges. Position opening of clamp band 45 degrees off vertical center line away from mounting bracket.
3. Squeeze ends of band together and tighten nut and bolt.
4. Place new seal on control tube, and assemble hose to tube on rear shell and to control tube. Attach tube and nut to control valve body port.

Section 12
Bendix Tandem Diaphragm Hydrovac

Control Tube, Rear Shell, Center Shell Disassembly

1. Empty system and thoroughly clean outside of assembly.
2. Scribe across shells and clamp bands. Clamp hydraulic cylinder in vise in horizontal position.

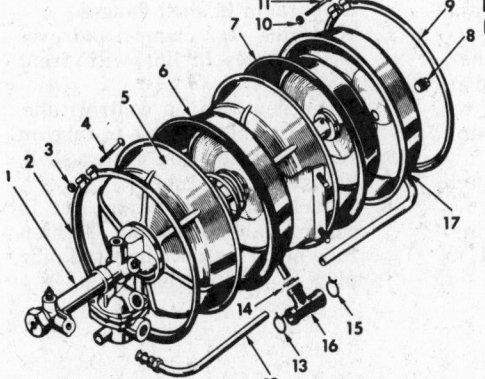

Bendix Tandem Diaphragm Type Power Brake   (© Dodge Div., Chrysler Corp.)

1 Hydraulic cylinder
2 Clamp ring
3 Nut
4 Bolt
5 Front shell
6 Center shell
7 Diaphragm
8 Pipe plug
9 Clamp ring
10 Nut
11 Bolt
12 Control tube
13 Hose clamp
14 Hose clamp
15 Hose clamp
16 Hose tee
17 Rear shell

3. Remove hose clamps, hose tee. Disconnect control tube from control valve housing.

4. Remove rear clamp band and rear shell. Remove front clamp band and center shell and diaphragm assembly.

Diaphragm Return Spring, Hydraulic Parts Disassembly

1. Use a ¾ in. diameter hose tee or a block of wood with a ¾ in. hole over the end of the pushrod and press to compress return spring. Using snap-ring pliers,

remove snap-ring from groove inside bore of end plate.
2. Using both hands, carefully pull assembly straight out of assembly. Parts may be removed from assembly by removing retainer ring and spring retainer from pushrod. To remove front shell, remove three capscrews and lift front shell and support plate off flange of cylinder.

Pushrod and Piston Disassembly

1. Piston parts are removed from the pushrod from the piston end of the pushrod.
2. Slide retainer ring out of groove in piston and separate piston from pushrod by removing retainer pin.
3. With thin bladed screw driver remove piston cap from piston. Slide off seal retainer, pushrod seal, pushrod bearing, stop washer and snap-ring.

Control Valve and Check Valve Disassembly

1. Scribe across flanges of control

valve body and hydraulic cylinder housing.
2. Remove four cap screws and lift off valve body and spring.
3. Remove washer, diaphragm, washer and cup from valve piston.
4. Using screw driver, pry off plastic valve body cover.
5. Pry off plastic retainer, lift off atmospheric poppet and then remove vacuum poppet and stem and poppet spring from valve body.
6. Using 1⅜ in. wrench, remove end cap and gasket.
7. To disassemble pressure check valve, lift snap-ring from groove inside end cap. Remove spring retainer, spring and residual pressure check valve.

Diaphragm and Center Shell Disassembly

1. Clamp nut on end of shaft in vise and remove nut, diaphragm, and washer from shaft. Carefully slide center shell assembly off shaft and remove seal from groove inside brass bearing hub with ice pick.
2. Remove shaft and diaphragm assembly from vise and insert drift punch through holes in shaft next to diaphragm. Use 1¾ in. wrench to remove nut from end of shaft.
3. Slide washer, diaphragm, diaphragm plate and washer off end of shaft. No further disassembly is possible without damage to parts.

Cleaning and Inspection Note

Thoroughly clean all metal parts in denatured alcohol or suitable cleaner. Inspect all parts not included in repair pack for excessive wear. Replace any that are worn or damaged. Inspect control valve body atmospheric valve seat. If damaged, replace housing.

Diaphragm and Center Shell Assembly

1. Slide washer, diaphragm plate, diaphragm, washer on shaft and thread nut until tight.
2. Insert drift punch through holes in shaft next to diaphragm and use 1¾ in. wrench to tighten nut securely. Stake nut to washer in two places. Press new O-ring seal into groove inside brass bearing

hub of center shell. Clamp nut on end of shaft in vise and lubricate shaft with lubriplate.
3. Lubricate inside of brass bearing hub and seal in center shell assembly and slide center shell carefully on shaft, seal side of hub last.
4. Install washer on pushrod end of shaft, followed by diaphragm, diaphragm plate, and nut. With nut securely held by vise, tighten nut to 160-200 in. lbs. and stake nut in place.

Control Valve and Check Valve Assembly

1. Assemble poppet return spring, vacuum poppet and stem and atmospheric poppet in valve body from opposite sides. Snap poppet retainer over end of vacuum poppet stem.
2. Assemble valve body cover securely in groove of body. Dip cups in clean brake fluid and assemble back to back on valve piston.
3. Align valve body to scribe mark, seat outer edge of diaphragm in groove in valve body and assemble four capscrews.
4. To assemble residual pressure check valve, install check valve in end cap, followed by spring and spring retainer, concave center diameter centered on spring. Compress spring and install snap-ring in groove inside end cap. Place new gasket on end cap and install end cap in cylinder.

Piston and Pushrod Assembly

1. To install a new pushrod seal in pushrod, place new seal (rubber side down) on a clean block of wood. Hold pushrod vertically with drilled end of rod resting on shaft end of seal and strike upper end with a soft mallet to seat.
2. Dip cup in clean brake fluid and assemble cup to piston.
3. Slide snap-ring, stop washer, bearing with O-ring installed in groove, pushrod seal cup, seal retainer, retainer ring and piston assembly on pushrod in order.
4. Assemble retainer pin through holes in piston and rod. Secure pin in place with retainer ring, making sure ring is seated in groove on piston.

Front Shell, Return Spring and Hydraulic Parts Assembly

1. Place new gasket in groove on flange of hydraulic cylinder.
2. Assemble front shell and support plate to cylinder, aligning cutouts in shell and plate with porting in cylinder flange, and assemble capscrews securely.
3. Coat hydraulic cylinder bore with clean brake fluid. If removed, install spring retainer and snap-ring on end of pushrod.
4. Dip hydraulic piston and bearing parts in brake fluid. Slide diaphragm return spring on pushrod, small diameter end first, over hydraulic piston end. Seat spring against concave surface of spring retainer and then align entire assembly.
5. Carefully insert piston, bearing and seals in bore.
6. Slide hose tee or block of wood with hole on end of pushrod and press against it to compress return spring. Using both hands seat stop washer against bearing inside bore and install snap-ring in groove in cylinder bore.

Shells and Diaphragms Assembly

1. With hydraulic cylinder held in vise with bleed screws up, guide center shell and diaphragm assembly on end of pushrod, seating rod in seat in end of rear shaft.
2. Align scribe marks on center shell and front shell and press together to seat bead of front diaphragm in flanges of shells. Dust diaphragm and flanges with talcum powder.
3. Assemble clamp band over shell flanges, opening 45 degrees from vertical center line on side away from bracket.
4. Tighten nut and bolt of band.
5. Assemble rear shell to scribe marks and seat bead of rear diaphragm in shell flanges.
6. Assemble rear clamp band over shell flanges in line with front clamp.
7. Place new seal on control tube, and assemble hose tee to tube on rear shell, tube on center shell, and to control tube.
8. Attach control tube and nut to control valve body port. Tighten tube nut. If removed, install pipe plug in port in rear shell.

Section 13
Delco-Moraine Tandem-Diaphragm Type

Disassembly

1. Empty system and clean outside

thoroughly. Be sure no oil or grease contacts the diaphragm

unit.
2. Scribe front and rear housings

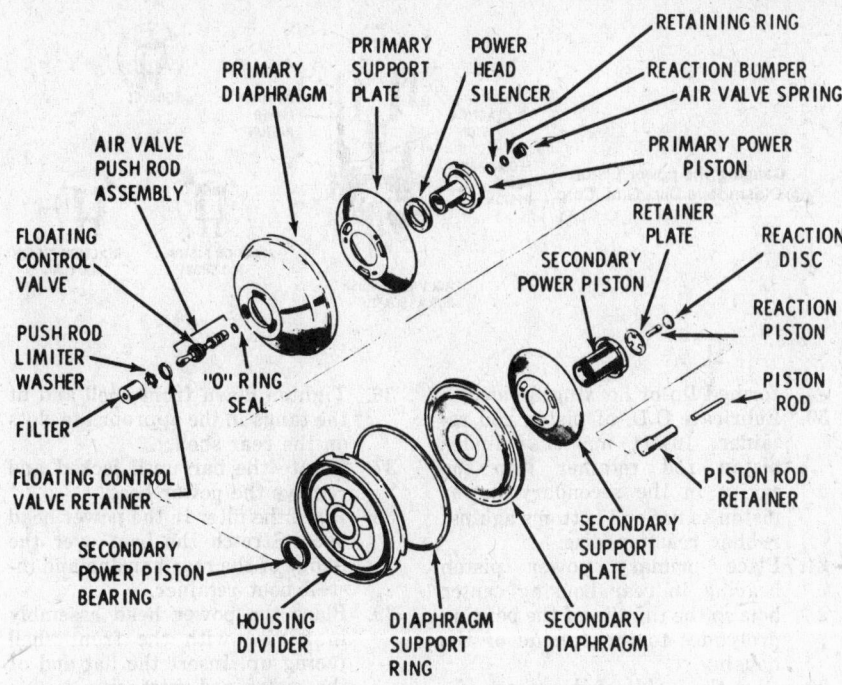

Delco-Moraine Tandem Diaphragm Power Brake (© Oldsmobile Div., G.M. Corp.)

for alignment. Remove front housing seal and master cylinder pin.
3. Using Tool J-23456 separate shells. Remove front shell, return spring retainer plate and piston rod retainer.
4. Remove dust boot retainer, boot and pushrod from rear housing. Remove felt silencers.
5. Remove power piston assembly from rear shell and primary power piston bearing from center opening.
6. Lift bead of secondary diaphragm and remove diaphragm
7. Mount Tool J-23101 in vise and position secondary power piston on tool.
8. Fold back primary diaphragm and unscrew primary power piston from secondary power piston.
9. Remove the housing divider from the secondary power piston and the bearing from the housing divider.
10. Fold back secondary diaphragm and unscrew secondary support plate from secondary power piston. Remove diaphragm from support plate.
11. Push down on end of reaction piston of secondary power piston with a pencil to remove the reaction piston.
12. Remove the air valve spring from the end of the air valve.
13. Position primary power piston on small jaws of Tool J-23101. Fold back primary diaphragm and primary support plate from primary power piston.
14. Remove primary diaphragm

from primary support plate.
15. Remove air filter and pushrod limiter washer from primary power piston.
16. Remove power head silencer from neck of power piston tube.
17. Remove rubber reaction bumper from air valve.
18. Remove snap ring from air valve.
19. Pull primary power piston to remove air valve pushrod assembly.
20. Remove O-ring seal from air valve.

Assembly

1. Discard all old rubber parts except the power diaphragms. Lubricate O-ring with silicone lubricant and place on air valve.
2. If floating control valve needs replacement, replace complete air valve pushrod assembly.
3. Lubricate floating control valve with silicone lubricant and place air valve end into tube of primary power piston. Push air valve pushrod assembly so floating control valve bottoms on tube section of primary power piston.
4. Press retainer into primary power piston with Tool J-23175.
5. Place pushrod limiter washer over pushrod and position on the floating control valve.
6. Install filter element over the pushrod eye and press into primary power piston tube.
7. Place snap-ring into groove in air valve.
8. Install rubber reaction bumper on air valve.
9. If either power piston has been replaced, proceed to step 10,

otherwise proceed to step 17.
10. To maintain correct power output, the power piston assembly must be gauged for selective fit of the reaction piston if either power piston is replaced .
11. Hand tighten the secondary power piston to the primary power piston without the air valve spring.
12. Insert reaction piston into its cavity in the secondary power piston.
13. With the secondary power piston pointing up, push on the reaction piston to insure that it is seated on the air valve.
14. Insert Tool J-23470 into secondary power piston so that the outer edges of the gauge rests on the bottom, with the two gauging levels within the smaller reaction piston cavity.
15. Move the gauge to the left or right of the nose of the reaction piston. The correct size reaction piston is indicated when the nose of the piston hits the lower level of the gauge and clears the higher level of the gauge while permitting the outer edges of the gauge to remain seated on the larger cavity of the secondary power piston. If the reaction piston is too short, both levels of the gauge will clear the nose. If the piston is too long, the gauge will not seat.
16. Repeat steps 12 through 15 until a correct size piston is found. Selective fit reaction pistons are available as follows:

98 SERIES

Color	Size
Red	1.148 in.–1.154 in.
Yellow	1.161 in.–1.167 in.
Blue	1.174 in.–1.180 in.

TORONADO SERIES

Red	1.178 in.–1.184 in.
Yellow	1.191 in.–1.197 in.
Blue	1.204 in.–1.210 in.
White	1.217 in.–1.223 in.

17. Lubricate and seat rubber reaction disc in large cavity of secondary power piston on reaction piston.
18. Unlock secondary power piston from primary power piston.
19. Assemble the primary diaphragm to the primary support plate from the side of the support plate opposite the locking tangs. Press the raised flange on the I.D. of the diaphragm through the center hole of the support plate. Be sure that the edge of the support plate center hole fits into the groove in the raised flange of the diaphragm. Lubricate diaphragm I.D. and the raised surface of the flange with

silicone lubricant.

20. Fold diaphragm away from support plate. Place primary support plate and diaphragm assembly over tube of primary power piston. Screw on.

21. Place power head silencer on tube of primary power piston.

22. Assemble secondary diaphragm, support plate and power piston as described in paragraphs 19 and 20.

23. Place secondary diaphragm support ring on secondary power piston assembly so it rests on edge of diaphragm.

24. Hold housing divider with formed lip facing down and insert the secondary bearing so the extended lip of the bearing faces up.

25. Position Tool J-23188 on threaded end of secondary power piston. Hold the housing divider with the formed lip up and press it down over the tool and onto the secondary power piston till it rests on the supporting ring. Remove Tool J-23188.

26. Pick up primary power piston assembly and position small end of air valve return spring on air valve so it contacts air valve retaining ring.

27. Position the primary power piston on the tubular portion of the secondary power piston. Seat air valve return spring on raised center of secondary piston.

28. Screw secondary power piston onto primary power piston securely.

29. Fold primary diaphragm onto support plate and pull bead over

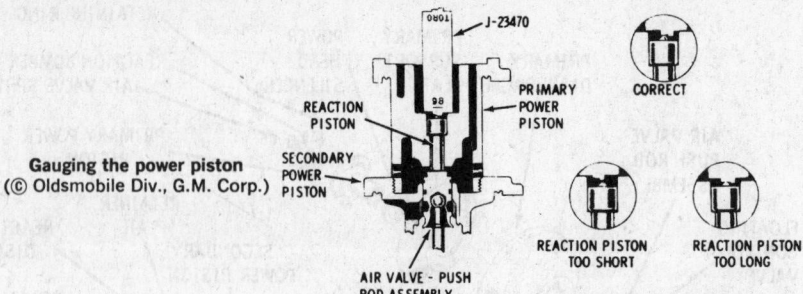

Gauging the power piston (© Oldsmobile Div., G.M. Corp.)

CORRECT

REACTION PISTON TOO SHORT

REACTION PISTON TOO LONG

PRIMARY POWER PISTON

REACTION PISTON

SECONDARY POWER PISTON

AIR VALVE - PUSH ROD ASSEMBLY

formed lip of housing divider.

30. Lubricate O.D. of piston rod retainer. Insert master cylinder piston rod retainer into the cavity in the secondary power piston so flat end bottoms against rubber reaction disc.

31. Place primary power piston bearing in rear housing center hole so the thin lip of the bearing protrudes to the outside of the housing.

32. Coat the inside of the power piston bearing with silicone lubricant and assemble primary power piston to rear shell by pressing the tube through the bearing. Press down until the housing seats against the shell.

33. Mount Tool J-23456 in vise and position rear shell in tool.

34. Place piston rod retainer plate on the end of the power piston and install power piston return spring.

35. Lower front shell over rear shell and position bar on front shell bearing.

36. Tighten down front shell and fit the tangs in the appropriate slots on the rear shell.

37. Rotate the bar until locked and remove the power head.

38. Place the filter in the power head boot. Stretch the boot over the flange of the rear housing and install boot retainer.

39. Place the power head assembly in a vise with the front shell facing up. Insert the flat end of the piston rod retainer.

40. Press on the master cylinder piston rod to check the seat.

41. Place gauge J-22647 over the piston rod in a position that allows it to be moved left and right without contacting the studs. Correct adjustment allows the lower step to contact the piston rod and the upper step to miss. Adjust the rod accordingly.

42. Lubricate the I.D. of the front housing seal and insert seal.

43. Position master cylinder on housing, install nuts and torque to 20 ft. lbs.

Section 14
Midland Ross Tandem Diaphragm

Removal

1. With engine turned off, apply brakes several times to equalize pressures.

2. Disconnect hydraulic brake lines from master cylinder and vacuum hose from check valve.

3. Under instrument panel, remove bolt connecting brake pedal link to pedal.

4. Remove unit from car.

Disassembly

1. Remove master cylinders from unit.

2. Remove filter cover and retainer, separate and remove filter.

3. Disconnect valve operating rod from power lever assembly by removing clevis.

4. Remove lower pivot and power lever, pedal link, boot retaining

plate and boot.

5. Remove rear bracket and rubber boot from valve operating rod.

6. Scribe a mark on covers and clamp band, remove clamp band.

7. Push bellows retaining lip into front cover chamber and sep-

CLAMP

BELLOWS SUPPORT RINGS

Bellows assembly (© Dodge Div., Chrysler Corp.)

arate bellows, control hub and diaphragm assemblies from covers.

8. Remove vacuum check valve rubber grommet from body. Remove rear seal from rear cover.

Cover and Diaphragms Disassembly

1. Remove the bellows clamp with a screw driver, and remove the bellows from the control hub. Remove the two bellows support rings.

2. Remove pushrod and reaction cone assembly and the reaction lever assembly from control hub.

3. Remove plastic pushrod guide, retainer and reaction cone from pushrod assembly.

4. Remove operating rod from valve plunger. Pull pushrod firmly un-

til plastic retainer shears. Remove all pieces from groove in pushrod and plunger.

5. Fold back bead of rear diaphragm, turn diaphragm plate and remove.
6. Remove O-ring from center assembly.
7. Roll ribbed bead of front diaphragm from center plate assembly retaining ring and remove center assembly.
8. Remove center seal from center plate assembly.
9. Using tool C-3984, turn control hub clockwise 1/4 turn while holding retaining plate. Separate diaphragm, retaining plate, tail stock extension, tail stock and two O-rings.

Control Hub Disassembly

1. Remove snap-ring retainer which holds plunger to control hub and separate. It may be necessary to file away the small burr from the end of the plunger before they can be separated.
2. Compress spring of plunger assembly toward rubber valve and remove spring retainer. Separate spring, washer, rubber valve from spring end and O-ring and fibre washer from other end of plunger.

Cleaning Note

Clean all metal parts in suitable solvent, and dry with compressed air. Plastic parts should be cleaned with soap and water. Replace all rubber parts and filters. Inspect and replace all parts showing wear. Lubricate all rubber parts except diaphragms and reactor ring with silicone lubricant.

Control Hub Assembly

1. Assemble O-ring and fibre washer on plunger. Insert plunger into hub and install snap-ring.
2. Assemble rubber valve onto control hub so that the two round holes in the valve index with the two projections on the hub. Assemble washer into hub so the

flat edge indexes with the flat projection of the hub.
3. Install spring and spring retaining clip over plunger by compressing spring.

Diaphragm Assembly

1. Install tail stock over plunger with flat indexing on flat on control hub.
2. Insert large O-ring in groove between tail stock and control hub. Insert small O-ring in sealing ring on rear of tail stock, install tail stock extension over tail stock and O-rings.
3. Install ribbed diaphragm to control hub retainer plate and place over tail stock extension.
4. Using tool C-3984, turn control hub assembly 1/4 turn counterclockwise to the stop.
5. Insert center seal into center plate assembly and install over tail stock extension. Roll ribbed diaphragm bead over center plate retaining ring.
6. Insert O-ring into sealing groove of rear diaphragm plate assembly and install over tail stock extension.
7. Rotate rear diaphragm plate assembly 1/4 turn counterclockwise to the stop.
8. Place groove of rear diaphragm sealing bead over mating lip of center plate assembly.
9. Install lever assembly in control hub with rubber protrusions toward hub. Be sure the levers are evenly spaced within the hub.
10. Assemble the reaction cone, retainer, and plastic pushrod guide on pushrod. Install pushrod assembly in control hub so that the pushrod indexes in the valve plunger.
11. Install two bellows support rings on bellows. Install bellows on control hub so the lip on bellows indexes in groove on hub.
12. Secure bellows to hub by assembling large bellows clamp.
13. Assemble rear seal into rear cov-

er. Position diaphragms, bellows, and control hub as assembled into rear cover. Make sure lip of seal fits over rear flange.
14. Install rubber grommet in front cover with larger diameter on outside of unit. Install vacuum manifold assembly through grommet.
15. Install front cover to rear cover, making sure that scribe lines are aligned. Pull front lip of bellows through front cover and position it evenly around hole in front cover.
16. Install clamp band over lips of front and rear cover, aligning scribe marks. It is necessary to force the covers together, requiring about 300 lbs. Engine vacuum or a vise may be used. Tighten the band.
17. Install rubber boot to valve operating rod and assemble plastic retainer to end of the rod. Insert rod into plunger so that the retainer engages groove in plunger. Install lip of boot in groove of rear seal.
18. Install brackets on rear cover.
19. Install pedal link, power lever, clevis pin, boot retaining plate and boot. Attach assembly to brackets.
20. Connect power lever to operating rod.
21. Position air filter in metal retainer and snap plastic cover on. Install assembly onto mounting studs of front cover.
22. Install master cylinders to mounting studs and torque to 100 ft. lbs.
23. Reinstall dash panel to power brake seal.
24. Position unit on dash, tighten nuts to 200 ft. lbs.
25. Lubricate the bearing surface of the bolt to connect the pedal link with the pedal linkage. Tighten to 30 ft. lbs.
26. Connect vacuum hoses and hydraulic brake lines.

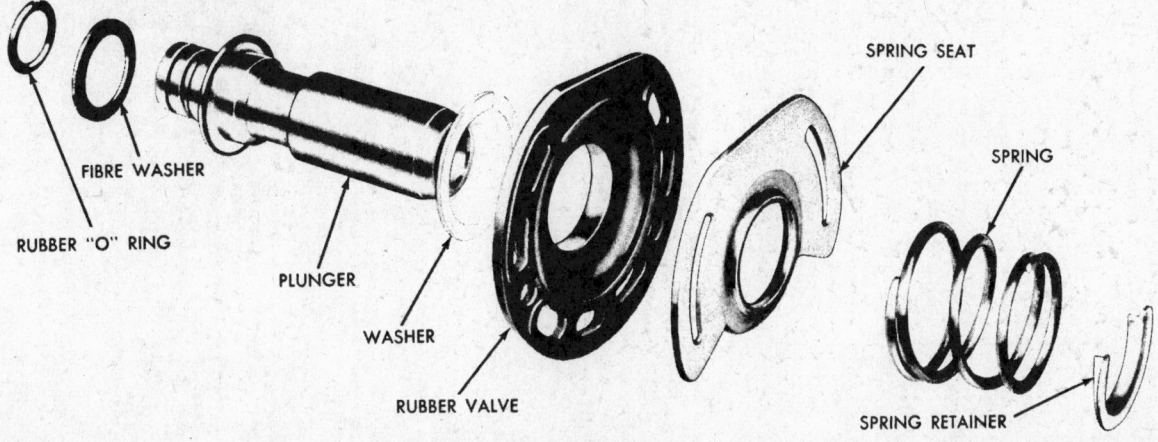

FIBRE WASHER

RUBBER "O" RING

PLUNGER

WASHER

RUBBER VALVE

SPRING SEAT

SPRING

SPRING RETAINER

Plunger assembly (© Dodge Div., Chrysler Corp.)

Manual Steering

Application Index

There are six different types of steering gears used. Each type of steering gear and the make of car it is installed in is listed below. The section numbers refer to the sections in the text that cover that particular type of steering gear.

Gear Types

A Gemmer worm and double roller tooth with screw adjusted mesh
B Ford steering gear, recirculating ball
C Saginaw steering gear, recirculating ball
D Ross cam and lever type
E Chrysler steering gear, recirculating ball
F Rack and pinion steering gear

Section Page Numbers

Section	Page	Section	Page
1	1273	5	1279
2	1273	6	1280
3	1274	7	1284
4	1276		

Gear Type	Section	Make	Year
A	2	American Motors Corp.	
		Rambler American	1964–67
		Kaiser Jeep Corp.	
		Jeepster	1967–70
		Jeepster Commando	1967–70
B	3	Ford Motor Co. and Lincoln-Mercury Division	
		All models except Capri	1964–70
C	4	General Motors Corp.	
		All models	1964–70
		American Motors Corp.	
		All models except Rambler American	1964–70
		Kaiser Jeep Corp.	
		Jeep Wagoneer	1966–70
		Jeep Panel Delivery	1966–70
D	5	Kaiser Jeep Corp.	
		Jeep Universal	1964–70
E	6	Chrysler Corp.	
		All models	1964–70
F	7	Ford Motor Co.	
		Capri	1970

Condition	Possible Cause	Correction
Hard steering	(a) Low or uneven tire pressure.	(a) Inflate tires to recommended pressures.
	(b) Insufficient lubricant in the steering gear housing or in steering linkage.	(b) Lubricate as necessary.
	(c) Steering gear shaft adjusted too tight.	(c) Adjust according to instructions.
	(d) Front wheels out of line.	(d) Align the wheels. See "Front Suspension."
	(e) Steering column misaligned.	(e) See "Steering Column—Manual Transmission."
Excessive play or looseness in the steering wheel	(a) Steering gear shaft adjusted too loose or badly worn.	(a) Replace worn parts and adjust according to instructions.
	(b) Steering linkage loose or worn.	(b) Replace worn parts. See "Front Wheel Alignment."
	(c) Front wheel bearings improperly adjusted.	(c) Adjust according to instructions.
	(d) Steering arm loose on steering gear shaft.	(d) Inspect for damage to the gear shaft and steering arm, replace parts as necessary.
	(e) Steering gear housing attaching bolts loose.	(e) Tighten attaching bolts to specifications.
	(f) Steering arms loose at steering knuckles.	(f) Tighten according to specifications.
	(g) Worn ball joints.	(g) Replace the ball joints as necessary. See "Front Suspension."
	(h) Worm-shaft bearing adjustment too loose.	(h) Adjust worm bearing pre-load according to instructions.

Section 1
Steering Gear Alignment

Before any steering gear adjustments are made, it is recommended that the front end of the car be raised and a thorough inspection be made for stiffness or lost motion in the steering gear, steering linkage and front suspension. Worn or damaged parts should be replaced, since a satisfactory adjustment of the steering gear cannot be obtained if bent or badly worn parts exist.

It is also very important that the steering gear be properly aligned in the car. Misalignment of the gear places a stress on the steering worm shaft, therefore a proper adjustment is impossible. To align the steering gear, loosen the steering gear-to-frame mounting bolts to permit the gear to align itself. Check the steering gear to frame mounting seat, if there is a gap at any of the mounting bolts, proper alignment may be obtained by placing shims where excessive gap appears. Tighten the steering gear-to-frame bolts. Alignment of the gear in the car is very important and should be done carefully so that a satisfactory, trouble-free gear adjustment may be obtained.

Section 2
Gemmer Worm and Double Roller Tooth Type With Screw Adjusted Mesh

The steering gear is of the worm and roller type with a 24 to 1 gear ratio. The cross shaft is straddle mounted with a bearing surface at the top and bottom points of the shaft mounting areas. The three tooth cross shaft roller is mounted in ball bearings. The proper lubricant used in the gear box is S.A.E. 90 Extreme Pressure Lubricant.

The external adjustments given below will remove all play from the steering gear. Before doing these adjustments, refer to Section 1 to insure that the steering gear requires adjustment.

Worm Bearing Adjustment

1. Turn the steering wheel about one full turn from straight ahead and secure it so it doesn't move.
2. Determine if there is any worm

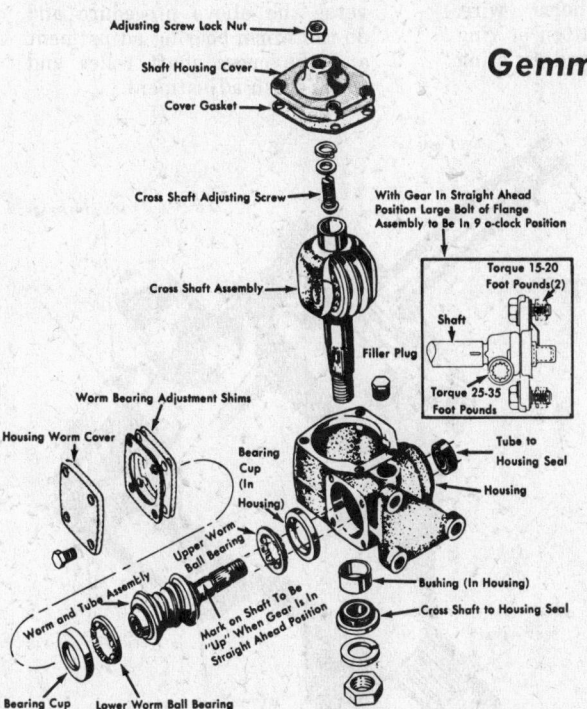

Adjusting Screw Lock Nut

Shaft Housing Cover

Cover Gasket

Cross Shaft Adjusting Screw

With Gear In Straight Ahead Position Large Bolt of Flange Assembly to Be In 9 o-clock Position

Cross Shaft Assembly

Torque 15-20 Foot Pounds (2)

Shaft

Filler Plug

Torque 25-35 Foot Pounds

Worm Bearing Adjustment Shims

Housing Worm Cover

Bearing Cup (In Housing)

Tube to Housing Seal

Housing

Upper Worm Ball Bearing

Worm and Tube Assembly

Mark on Shaft To Be "Up" When Gear Is In Straight Ahead Position

Bushing (In Housing)

Cross Shaft to Housing Seal

Bearing Cup

Lower Worm Ball Bearing

Gemmer worm and double roller tooth type steering gear
(© American Motors Corp.)

gear end-play by shaking the front wheel sideways and noting if there is any end movement that may be felt between the steering wheel hub and the steering jacket tube. (*Be sure any movement noted is not looseness in the steering jacket tube.*)

3. If end play is present, adjust the worm bearings by loosening the four cover cap screws about ⅛ in. Separate the top shim, using a knife blade, and remove it. Do not damage the remaining shims or gaskets.

4. Replace the cover and recheck the end-play again. If necessary, repeat steps 2 and 3 until the end play movement is as small as possible without tightening the steering gear too much.

NOTE: adjustment may be done with the Pitman arm disconnected. With the steering wheel turned about one full turn from straight ahead, using the spring scale tool J-544, adjust with the shims as given above until the spring scale pull is between ¼ and ⅝ ft. lbs.

Cross Shaft Roller and Worm Mesh Adjustment

1. Turn the steering wheel to the middle of its turning limits with the Pitman arm disconnected. The steering gear roller should be on the worm high spot.

2. Shake the Pitman arm sideways to determine the amount of clearance between the worm cross shaft roller. Movement of more than 1/32 in. indicates the roller and worm mesh must be adjusted.

3. Loosen the adjusting screw lock nut and tighten the external

cross shaft adjusting screw a small amount. Recheck the clearance by shaking the Pitman arm. Repeat until the clearance is correct. (*Do not overtighten.*)

NOTE: the cross shaft roller and worm mesh adjustment may be done, using spring scale tool J-544, by measuring the amount of wheel pull as the external cross shaft adjusting screw is tightened. When the spring scale pull is between ⅞ and 1⅛ ft. lbs., the adjustment is correct.

Line-up Dash Marks on Steering Shaft and Wheel

J-544

Checking pull at steering wheel rim
(© American Motors Corp.)

4. Tighten the Pitman arm attaching nut to 100-125 ft-lbs. The steering wheel nut (if loosened) should be tightened to 15-20 ft-lbs. torque.

Steering Gear Removal and Installation

1. Disconnect the horn wire. Remove the horn button or ring by pressing down and twisting counterclockwise.

2. Remove the steering wheel, using Puller Tool J-21232. Note the line up of dash marks on steering shaft and steering wheel.

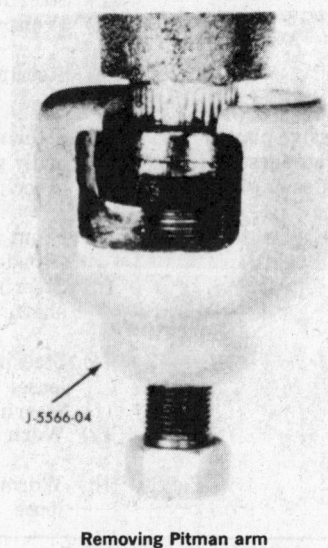

J-5566-04

Removing Pitman arm
(© American Motors Corp.)

3. Remove Pitman arm with Puller Tool J-5566-04.

4. Loosen the steering jacket tube assembly at the support plate and at the instrument panel mounting bracket. Remove the steering gear attaching bolts, and remove the steering gear from the car. On some cars, two flexible couplings are used in the steering shaft assembly.

5. To install the steering gear, reverse the above procedure and do the worm bearing adjustment and the cross shaft roller and worm mesh adjustment.

Section 3
Ford Steering Gear—Recirculating Ball Type

Steering Worm and Sector Gear Adjustments

The ball nut assembly and the sector gear must be adjusted properly to maintain a minimum amount of steering shaft end play and a minimum amount of backlash between the sector gear and the ball nut. There are only two adjustments that may be done on this steering gear and they should be done as given below:

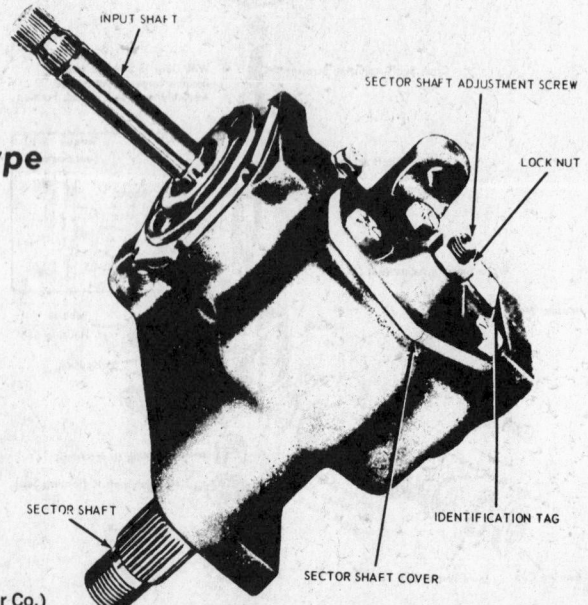

INPUT SHAFT

SECTOR SHAFT ADJUSTMENT SCREW

LOCK NUT

IDENTIFICATION TAG

SECTOR SHAFT COVER

SECTOR SHAFT

Ford manual steering gear, recirculating ball type (© Ford Motor Co.)

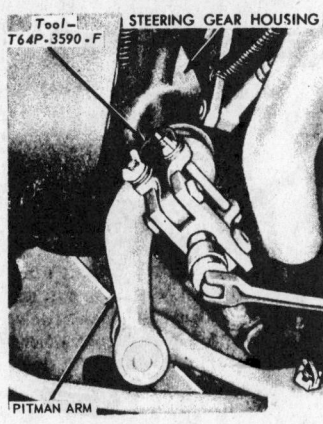

Removing Pitman arm (© Ford Motor Co.)

1. Disconnect the Pitman arm from the steering Pitman-to-idler arm rod.
2. Loosen the locknut on the sector shaft adjustment screw and turn

Steering gear adjustments (© Ford Motor Co.)

the adjusting screw counterclockwise.

3. Measure the worm bearing preload by attaching an in.-lbs. torque wrench to the steering wheel nut. With the steering wheel off center, note the reading required to rotate input shaft about 1½

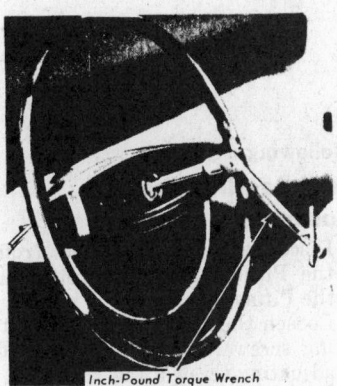

Checking steering gear pre-load
(© Ford Motor Co.)

turns either side of center. If the torque reading is not about 4-5 in-lbs., adjust the gear as given in the next step.

4. Loosen the steering shaft bearing adjuster lock nut and tighten or back off the bearing adjusting screw until the reload is within the specified limits.
5. Tighten the steering shaft bearing adjuster lock nut, and recheck the reload torque.
6. Turn the steering wheel slowly to either stop. Turn *gently* against the stop to avoid possible damage to the ball return guides. Then rotate the wheel 2¾ turns to center the ball nut.
7. Turn the sector adjusting screw clockwise until the proper torque (9-10 in-lbs.) is obtained that is necessary to rotate the worm gear past its center (high spot).
8. While holding the sector adjusting screw, tighten the sector screw adjusting locknut to the proper torque (32-40 ft-lbs.) and recheck the backlash adjustment.
9. Connect the Pitman arm to the steering arm-to-idler arm rod.

Steering Gear Removal and Installation

1. Remove the bolt(s) that holds the flex coupling to the steering shaft.
2. Remove the nut and lock washer that secures the Pitman arm to the sector shaft using Tool T64P-3590-F.
3. To gain enough clearance on some cars with standard transmissions, it may be necessary to disconnect the clutch linkage. On some cars with 8-cylinder engines, it may be necessary to lower the exhaust system.

4. Remove the steering gear-to-side rail bolts and remove the steering gear.
5. When installing the steering gear, position the steering gear and flex coupling in place and install and torque the steering gear-to-side rail bolts to 50-65 ft-lbs.
6. If clutch linkage has been disconnected, reposition and adjust it. If the exhaust system has been lowered, reinstall it to its proper position.
7. Position the Pitman arm and the sector shaft and install the attaching nut and lock washer. Tighten the nut to 150-225 ft-lbs.
8. Install the flex coupling attaching nut(s) and tighten to specification (tilt steering column, one bolt—20-37 ft-lbs.; fixed column, two bolts—10-22 ft-lbs.)

Steering Gear Disassembly and Assembly

1. Rotate the steering shaft three turns from either stop.
2. Remove the sector shaft adjusting screw locknut and the housing cover bolts and remove the sector shaft with the cover. Remove the cover from the shaft by turning the screw clockwise. Keep the shim with the screw.
3. Loosen the worm bearing adjuster nut and remove the adjuster assembly and the steering shaft upper bearing.
4. Carefully pull the steering shaft and ball nut from the housing, and remove the steering shaft lower bearing. *Do not run the ball nut to either end of the worm gear to prevent damaging the ball return guides. Disassem-*

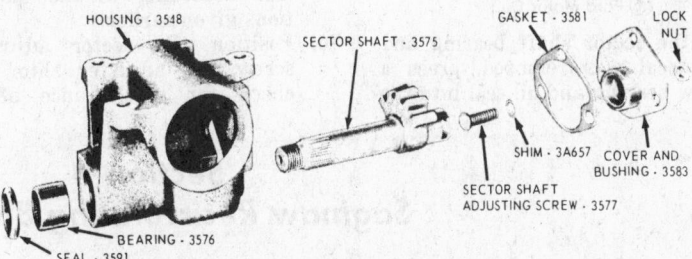

Sector shaft and housing disassembled (© Ford Motor Co.)

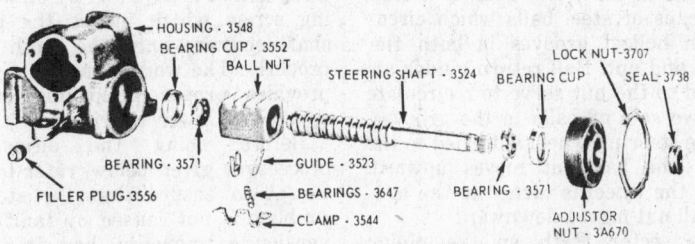

Steering shaft, ball nut, and bearings disassembled (© Ford Motor Co.)

ble the ball nut only if there are signs of binding or tightness.

5. To disassemble the ball nut, remove the ball return guide clamp and the ball return guides from the ball nut. *Keep ball nut clamp side up until ready to remove the ball bearings.*
6. Turn the ball nut over and rotate the worm shaft from side to side until all 50 balls have dropped out into a clean pan. With all balls removed, the ball nut will slide off the wormshaft.
7. Remove the upper bearing cup from the bearing adjuster and the lower cup from the housing. It may be necessary to tap the housing or the adjuster on a wooden block to jar the bearing cups loose.
8. If the inspection shows bearing damage, the sector shaft bearing and the oil seal should be pressed out.

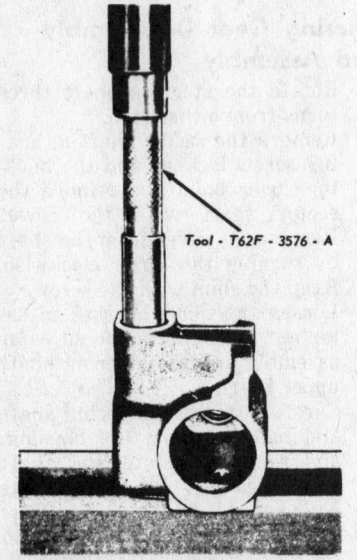

Removing oil seal and bearing
(© Ford Motor Co.)

9. If the sector shaft bearing and oil seal was removed, press a new bearing and oil seal into the housing. Do not clean, wash, or soak seals in cleaning solvent. Apply the recommended steering gear lubricant to the housing and seals.
10. Install the bearing cup in the lower end of the housing and a bearing cup in the adjuster nut.
11. Install a new seal in the bearing adjuster if the old seal was removed.
12. Insert the ball guides into the holes in the ball nut, lightly tapping them if necessary to seat them.
13. Insert 25 balls into the hole in the top of each ball guide. If necessary, rotate the shaft slightly to distribute the balls evenly in the circuit.
14. Install the ball guide clamp, tightening the screws to the proper torque. Check that the worm shaft rotates freely.
15. Coat the threads of the steering shaft bearing adjuster, the housing cover bolts, and the sector adjusting screw with a suitable oil-resistant sealing compound. Do not apply sealer to female threads and *Do not get sealer on the steering shaft bearings.*
16. Coat the worm bearings, sector shaft bearings, and gear teeth with steering gear lubricant.
17. Clamp the housing in a vise, with the sector shaft axis horizontal, and place the steering shaft lower bearing in its cup. Place the steering shaft and ball nut assemblies in the housing.
18. Position the steering shaft upper bearing on top of the worm gear and install the steering shaft bearing adjuster, adjuster nut, and the bearing cup. Leave the nut loose.
19. Adjust the worm bearing preload according to the instructions given earlier.
20. Position the sector adjusting screw and adjuster shim, and check for a clearance of not

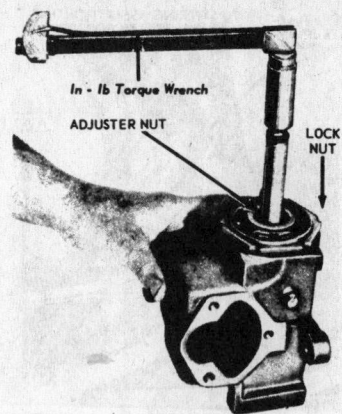

Checking steering shaft bearing pre-load
(© Ford Motor Co.)

more than 0.002 in. between the screw head and the end of the sector shaft. If the clearance exceeds 0.002 in., add enough shims to reduce the clearance to under 0.002 in. clearance.
21. Start the sector shaft adjusting screw into the housing cover. Install a new gasket on the cover.
22. Rotate the steering shaft until the ball nut teeth mesh with the sector gear teeth, tilting the housing so the ball will tip toward the housing cover opening.
23. Lubricate the sector shaft journal and install the sector shaft and cover. With the cover moved to one side, fill the gear with gear lubricant (about 0.97 lb.). Push the cover and the sector shaft into place, and install the two top housing bolts. Do not tighten the bolts until checking to see that there is some lash between the ball nut and the sector gear teeth. Hold or push the cover away from the ball nut and tighten the bolts to the proper torque (30-40 ft-lbs.).
24. Loosely install the sector shaft adjusting screw lock nut and adjust the sector shaft mesh load as given earlier. Tighten the adjusting screw lock nut.

Section 4
Saginaw Recirculating Ball Type

The steering gear is of the recirculating ball nut type. The ball nut, mounted on the worm gear, is driven by means of steel balls which circulate in helical grooves in both the worm and nut. Ball return guides attached to the nut serve to recirculate the two sets of balls in the grooves. As the steering wheel is turned to the right, the ball nut moves upward. When the wheel is turned to the left, the ball nut moves downward.

The sector teeth on the pinion shaft and the ball nut are designed so that they fit the tightest when the steering wheel is straight ahead. This mesh action is adjusted by an adjusting screw which moves the pinion shaft endwise until the teeth mesh properly. The worm bearing adjuster provides proper preloading of the upper and lower bearings.

Before doing the adjustment procedures given below, refer to Section 1 to ensure that the steering problem is not caused by faulty suspension components, bad front end alignment, etc. Then, proceed with the following adjustments.

Worm Bearing Preload Adjustment

1. Disconnect the ball stud from the Pitman arm, and retighten the Pitman arm nut.
2. Loosen the Pitman shaft adjusting screw lock nut and back off adjusting screw a few turns.
3. Attach a spring scale Tool J-544 to the steering wheel and measure the pull needed to move the steering wheel when off the high

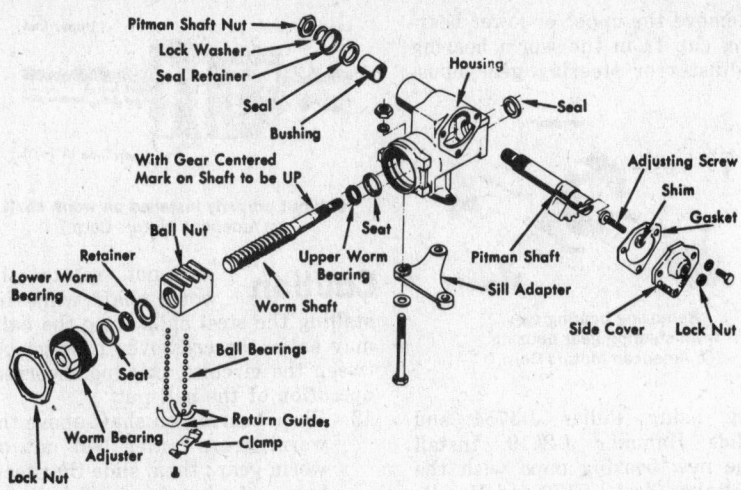

Saginaw steering gear, recirculating ball type (© American Motors Corp.)

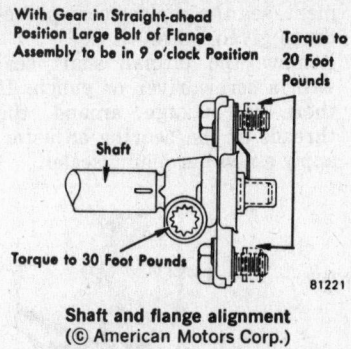

Shaft and flange alignment
(© American Motors Corp.)

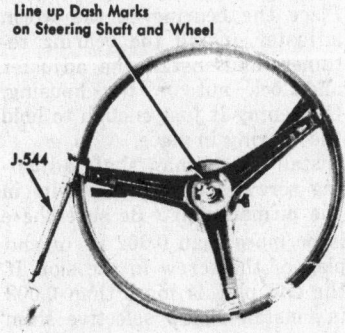

Checking pull at steering wheel rim
(© American Motors Corp.)

point. The pull should be between $1/8$ and $3/8$ ft-lb.

4. To adjust the worm bearing, loosen the worm bearing adjuster lock nut with a brass drift and turn the adjuster screw until the proper pull is obtained. When adjustment is correct, tighten the adjuster lock nut, and recheck with the spring scale again.

Sector and Ball Nut Backlash Adjustment

1. After the worm bearing preload has been adjusted correctly, loosen the Pitman shaft adjusting screw lock nut and turn the Pitman shaft adjusting screw clockwise until a pull of $3/4$ to $1 1/8$ ft-lbs. is shown on the spring scale. When the adjustment is correct, tighten the Pitman shaft adjusting screw lock nut and recheck the adjustment.
2. Turn the steering wheel to the center of its turning limits (pitman arm disconnected). If the steering wheel is removed, the mark on the steering shaft should be at top center.
3. Connect the ball stud to the Pitman arm, tightening the attach-

ing nut to the proper torque (115 ft-lbs.).

Steering Gear Removal and Installation

1. Remove the flexible coupling bolts.
2. Remove the Pitman arm, using Puller tool J-5566-04.

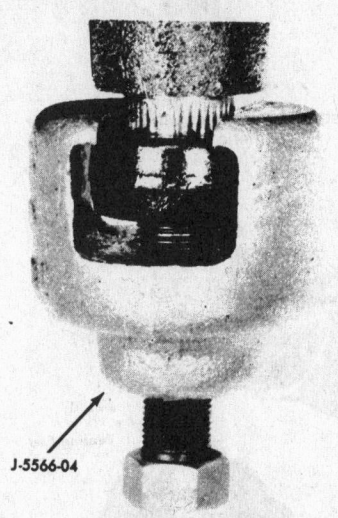

Removing Pitman arm
(© American Motors Corp.)

3. Remove the steering gear mounting screws and lower the steering gear.
4. Center the steering gear with the index mark up. Mark on shaft on flange must be aligned at assembly.
5. Insert the flexible coupling bolts to the shaft and tighten to 20 ft-lbs. torque. Tighten the pinch bolt to 30 ft-lbs. Tighten the steering gear mounting screws to 45 ft-lbs. torque, and the Pitman arm nut to 115 ft-lbs. torque. *Stake Pitman arm nut with*

a center punch to insure nut retention.

NOTE: whenever the steering gear is removed for replacement or overhaul, or the mounting bolts are loosened for any reason, the steering column *must* be realigned to the gear assembly. Slight misalignment of the steering column may cause increased steering effort and extra wear to the steering components. Refer to the section on steering columns for alignment procedures.

Steering Gear Disassembly and Assembly

1. After removing the steering gear from the car, place the steering gear assembly in a bench vise.
2. Rotate the worm shaft until it is centered with the mark facing upward. Remove three cover attaching screws and the adjusting screw lock nut. Remove the cover and gasket by turning adjusting screw clockwise through the cover.
3. Remove the adjusting screw with its shim from the slot in the end of the pitman shaft. Remove the pitman shaft from the housing being careful not to damage the seal in the housing.
4. Loosen the worm bearing adjuster lock nut with a brass drift and remove the adjuster and bearing. Remove the bearing retainer with a screwdriver.
5. Remove the worm and shaft assembly with the ball nut assembly and bearing. Remove the ball nut return guide clamp by removing screws. Remove the guides, turn ball nut over, and remove the steel balls by rotating the shaft from side to side. After all the steel balls have been removed, take the ball nut off the worm shaft.
6. Clean all parts in solvent. Inspect all bearings, bearing cups, bushings, seals, worm groove, and gear teeth for signs of wear, scoring, pitting, etc. If the pitman shaft bushings or seal, steering shaft seal, or upper and lower bearing cups need replace-

ment, see the replacement procedures given below.

7. Remove the pitman shaft seal with a screwdriver or punch. If there is leakage around the threads of the bearing adjuster, apply a non-hardening sealer.

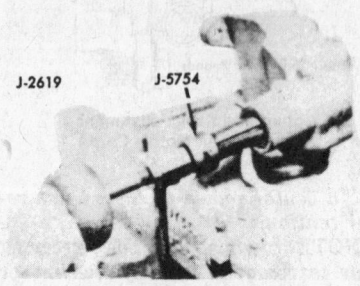

Removing Pitman shaft bushing
(© American Motors Corp.)

8. Remove faulty bushings from the pitman shaft with Puller J-5754 and Slide Hammer J-2619. Install new bushings with Tool J-7133, seating the inner end of the bushing flush with the inside surface of the housing.

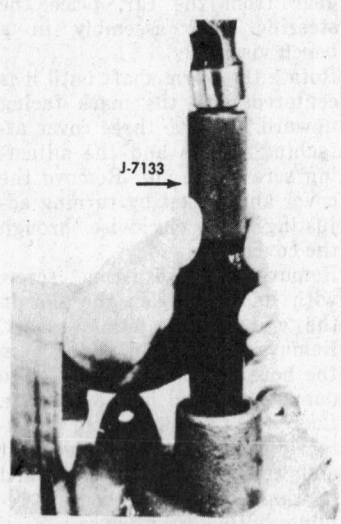

Installing Pitman shaft bushing
(© American Motors Corp.)

9. Remove the steering shaft seal with a punch or screwdriver. Tap new seal in place, using a section of tubing to seat the seal.

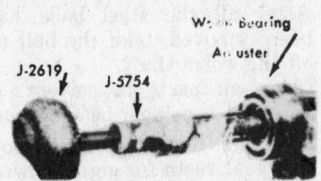

Removing bearing cup
from worm bearing adjuster
(© American Motors Corp.)

10. Remove the upper or lower bearing cup from the worm bearing adjuster or steering gear hous-

Removing bearing cup
from steering gear housing
(© American Motors Corp.)

ing using Puller J-5754 and Slide Hammer J-2619. Install the new bearing cups with the Installer Tool J-7170 and Handle J-8092 or J-8592.

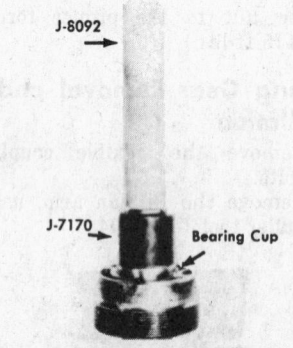

Installing bearing cup in worm adjuster
(© American Motors Corp.)

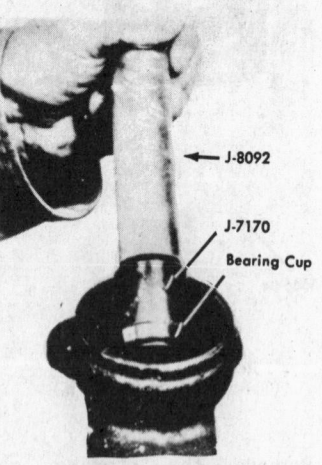

Installing bearing cup
in steering gear housing
(© American Motors Corp.)

11. Lubricate all seals, bushings, and bearings before installing into the steering gear assembly.

12. Position the ball nut on the worm shaft. Install the steel balls in the return guides and the ball nut, placing an equal number in each circuit of the ball nut. Install the return guide clamp and screws.

Ball nut properly installed on worm shaft
(© American Motors Corp.)

Caution   do not rotate the worm shaft while installing the steel balls since the balls may enter the crossover passage between the circuits, causing incorrect operation of the ball nut.

13. Place bearing on shaft above the worm gear, center ball nut on worm gear; then, slide the steering shaft, bearing, and ball nut into the housing. *Do not damage the steering shaft seal in the housing.*

14. Place the bearing in the worm adjuster, install the bearing retainer, and install the adjuster and lock nut on the housing, tightening it just enough to hold the bearing in place.

15. Install the pitman shaft adjusting screw and selective shim in the pitman shaft. Be sure there is no more than 0.002 in. of end play of the screw in the slot. If the end play is more than 0.002 in., install a new selective shim to get the proper clearance. Shims are available in four thicknesses: 0.063 in., 0.065 in., 0.067 in., and 0.069 in.

16. Install the pitman shaft and adjusting screw with the sector and ball nut positioned as shown.

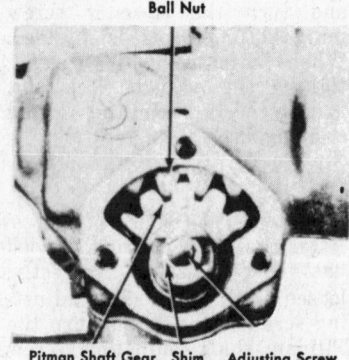

Positioning the Pitman shaft
and ball nut in housing
(© American Motors Corp.)

17. Install the cover and gasket on the adjusting screw, turning screw counterclockwise until it extends through the cover from ⅝ to ¾ in. Install the cover attaching screws.

18. Tighten the pitman shaft adjusting screw so that the teeth on the shaft and the ball nut engage

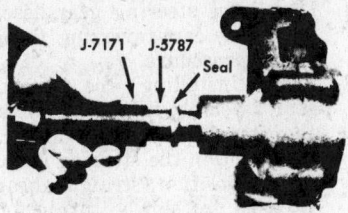

Installing Pitman shaft seal
(© American Motors Corp.)

but do not bind. Final adjustment must be made later.

19. Install the pitman shaft seal using Seal Protector J-5787 and Installer J-7171.
20. Fill steering gear with a good quality chassis lubricant. Turn the steering gear from one extreme to the other to make sure it does not bind. *Do not allow the ball nut to strike the ends of the ball races on the worm gear to avoid damaging the ball return guides.*
21. Install the steering gear as described previously. Do the final adjustments on the worm bearing preload and the sector and ball nut backlash adjustments.

Section 5
Ross Cam and Twin Lever Type

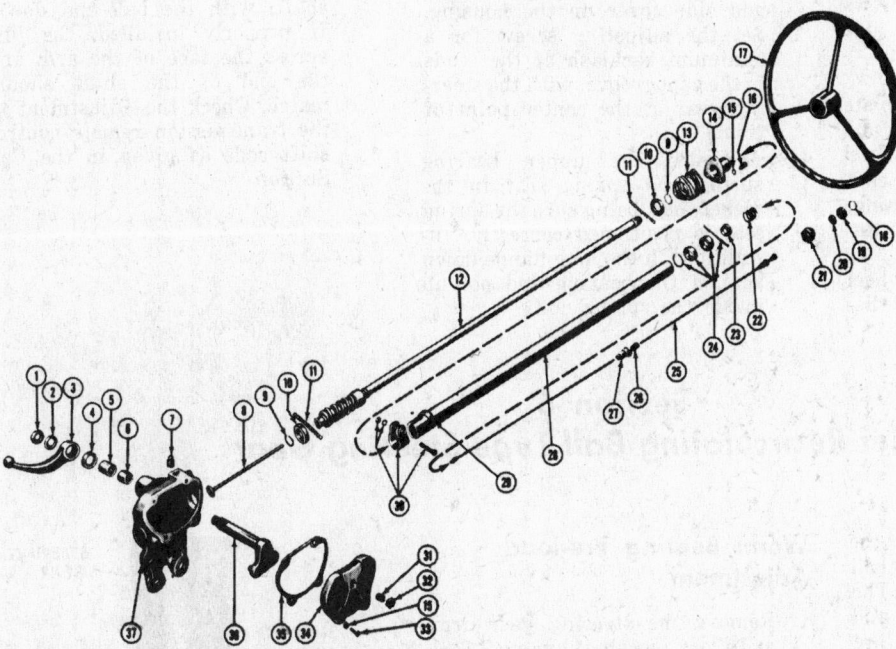

Ross steering gear, cam and lever type (© Kaiser Jeep Corp.)

1 Nut
2 Lockwasher
3 Steering gear arm
4 Lever shaft oil seal
5 Outer housing bushing
6 Inner housing bushing
7 Filler plug
8 Cover and tube
9 Ball retaining ring
10 Cup
11 Ball (steel)
12 Tube and cam
13 Shims
14 Upper cover
15 Lockwasher
16 Bolt
17 Steering wheel
18 Horn button retainer
19 Horn button
20 Horn button cap
21 Nut
22 Spring
23 Spring seat
24 Bearing
25 Horn cable
26 Horn button spring
27 Spring cup
28 Steering column
29 Oil hole cover
30 Clamp
31 Adjusting screw
32 Nut
33 Bolt
34 Side cover
35 Gasket
36 Shaft and lever
37 Housing

The cam and lever steering gear consists of a spiral cam and a cross shaft and lever assembly with two lever studs. When the steering wheel is turned, the cam moves the studs, causing the cross shaft to rotate and the steering arm to move. There are two adjustments that must be done: end play of the steering shaft, and backlash adjustment of the lever studs (tapered pins).

Steering Gear End-Play Adjustment

The steering gear is adjusted for minimum end-play by removing shims which are located between the upper cover and the steering gear housing. Before doing this adjustment, loosen the housing side cover adjusting screw to free the pins in the cam groove. Loosen the cover enough to remove one or more shims as required. As each shim is removed, retighten the cover and check the end-play. Adjustment should leave a slight drag in the steering gear but still allow it to turn easily. Replacement shims are available in three thicknesses: 0.002 in., 0.003 in., and 0.010 in. (0.0508, 0.0762, and 0.254 mm).

Steering Gear Backlash Adjustment

The steering gear lever studs (tapered pins) backlash adjustment is done by turning the adjusting screw in until a slight drag is felt through the mid-position when turning the steering wheel slowly from one stop to the other. This adjustment is done when the steering wheel is centered in the straight ahead driving position. The cam groove is cut shallow in this area to provide a close adjustment of the pins. Do not adjust the screw in any other position of the steering gear.

Steering Gear Removal and Installation

1. Remove the directional signal unit from the steering column (see section on steering columns).
2. Remove the steering column bracket attached to the instrument panel.
3. Remove the upper section of the floor pan.
4. Disconnect the shift rods from the shift levers at the lower end of the steering column.
5. Disconnect the horn wire from the lower end of the steering gear assembly.
6. Remove the steering arm from the steering gear assembly. Remove the attaching bolts from the steering gear and bring it up through the floor pan opening.
7. Installation of the steering gear assembly is the reverse of the procedure given above. Readjust the remote control shaft rods after they have been installed.

Steering Gear Disassembly and Assembly

When disassembling early production models, the line across the face of the steering arm and the end of the shaft should match. On later production models, blind splines on the lever shaft and in the steering gear arm engage, allowing correct positioning of the arm.

1. Remove the steering gear arm with a puller tool (C-3646).
2. Loosen the adjusting screw lock nut and unscrew the adjusting screw two turns.
3. Remove the side cover screws and washers. Remove the side cover and gasket.
4. Remove the lever shaft.
5. Remove the upper cover plate screws, the upper cover, and gasket. Remove the cam, wheel
6. Inspect all bearings, seals, cam grooves, etc. for wear, chipping, scoring, and other damage. Replace all worn parts.
7. Assemble all parts to the wheel tube in the reverse order of dis-assembly. Assemble the cam, wheel tube, and bearing assembly and install in the housing, seating the lower bearing ball cup correctly in the housing.
NOTE: new plastic retainer type cam bearings are available for Ross steering gears that replace, and are interchangeable with, the lock ring type cam bearings on gears equipped with early type cams.
8. With the adjusting shims installed in the housing, install the upper cover and adjust the cam bearings.
9. Install the lever shaft, gasket, and side cover on the housing. Set the adjusting screw for a minimum backlash of the studs in the cam groove, with the steering gear at the center point of travel.
10. Assemble the upper bearing spring and spring seat in the jacket tube, being sure the spring seat is positioned correctly. Install it with the long flange down against the bearing and not up inside the spring coil.

11. Install the steering gear assembly in the car, reversing the removal procedure.
12. After installing the steering gear assembly in the car, install the steering wheel temporarily and position the steering gear in the midposition (straight ahead). Do this by slowly turning the steering wheel from one stop to the other, and turning it to the middle of its turning range.
13. With the steering wheel held steady and the front wheels facing straight ahead, install the steering arm on the lever shaft, with the ball end down. If properly installed, the line across the face of the arm and the end of the shaft should match. Check the adjustment to the transmission remote control shift rods as given in the Car Section.

Section 6
Chrysler Recirculating Ball Type Steering Gear

The steering gear is similar to the Saginaw type recirculating ball steering gear in many ways. The differences are mainly in design and torque limits. The adjustments are presented briefly below. Before doing these adjustments, refer to the instructions in Section 1.

Worm Bearing Pre-load Adjustment

1. Remove the steering gear arm and lock washer, using Tool C-3646.
2. Remove the horn button or horn ring.

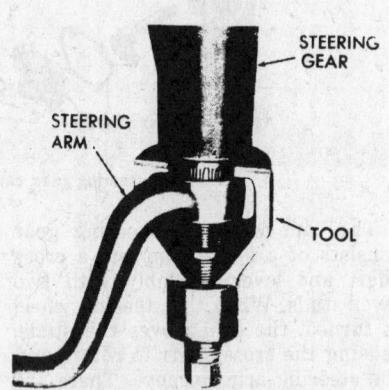

Removing steering gear arm
(© Chrysler Corp.)

3. Loosen the cross shaft adjusting screw lock nut, and back out the adjusting screw about two turns.
4. Turn the steering wheel two complete turns from the straight ahead position, and place torque wrench Tool C-3380 on the steering shaft nut.
5. Rotate the steering shaft at least one turn toward the straight ahead position while measuring the torque on the torque wrench. The torque should be between 1½ and 4½ in-lbs. to move the steering wheel. If torque is not within these limits, loosen the worm shaft bearing adjuster lock nut and turn the adjuster

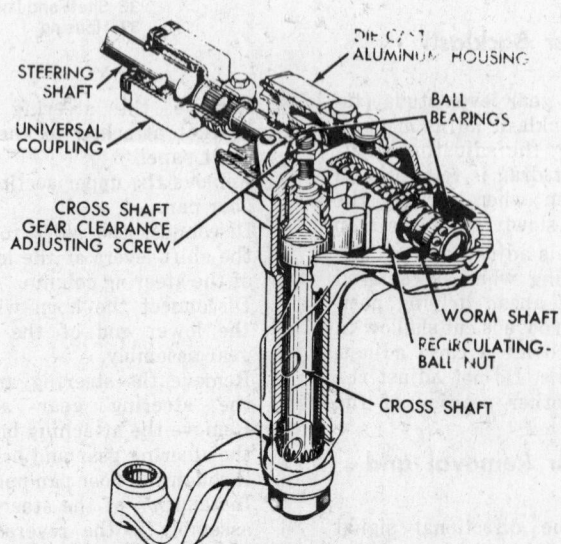

Chrysler steering gear, recirculating ball type
(© Chrysler Corp.)

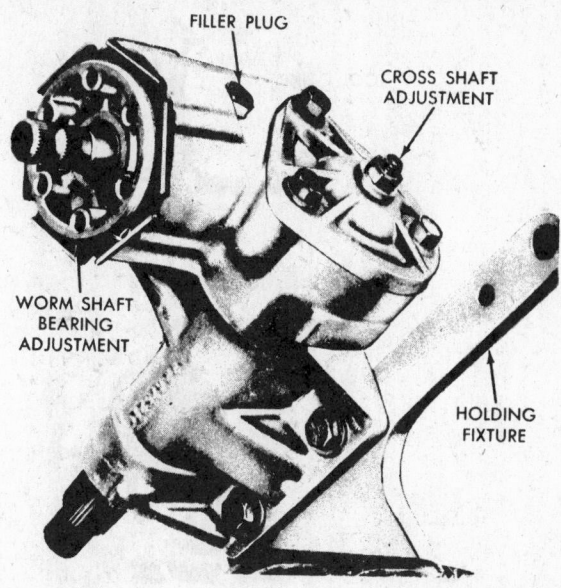

FILLER PLUG

CROSS SHAFT ADJUSTMENT

WORM SHAFT BEARING ADJUSTMENT

HOLDING FIXTURE

Steering gear adjustment locations (© Chrysler Corp.)

TOOL C-3786

CROSS SHAFT

Removing the cross shaft (© Chrysler Corp.)

clockwise to increase the pre-load or counterclockwise to decrease the pre-load. Use the adjuster wrench Tool C-3884 to do the adjustment. When the preload is correct, hold the adjuster screw steady and tighten the lock nut.

Ball Nut Rack and Sector Mesh Adjustment

1. Turn steering wheel from one stop to the other, counting the number of turns. Turn the steering wheel back exactly half way, to the center position.
2. Turn the cross shaft adjusting screw clockwise to remove all lash between ball nut rack and the sector gear teeth, then tighten adjusting screw lock nut to 35 ft-lbs. torque.
3. Turn the steering wheel about ¼ turn away from the center or high spot position. With the torque wrench Tool C-3380 on the steering wheel nut measure the torque required to turn the steering wheel through the high spot at the center position. The reading should be between 8¼ and 11¼ in-lbs. torque. This is the total of the worm shaft bearing preload and the ball nut rack and sector gear mesh load. Readjust the cross shaft adjustment screw if necessary to obtain a correct torque reading.
4. After completing the adjustments, place the front wheels in a straight ahead position, and with the steering wheel and steering gear centered, install the steering arm on cross shaft. Tighten the steering arm retaining nut to 180 ft-lbs. torque.

Steering Gear Removal and Installation

To avoid damaging the steering column, it may be necessary to remove it before taking the steering gear out of the car. This is very important on cars equipped with energy-absorbing type steering columns. See the section on Steering Columns in the Car Section for proper removal, installation, and alignment procedures.

1. After removing the steering column, remove the steering arm retaining nut and lock washer. Remove the steering arm with Tool C-3646.
2. Remove the steering gear attaching bolts, and remove the steering gear from the car.
3. To install the steering gear, position the gear assembly on the frame and install the attaching bolts and lock washers. Tighten the bolts to the proper torque (100 ft-lbs.).
4. Rotate the worm shaft by hand and center the cross shaft to the middle of its travel. Align the master serration on the cross shaft with the splines in the steering arm. Install the steering arm with its lock washer and nut, tightening it to 180 ft-lbs. torque.
5. Install and align the steering column as given in the Steering Column section.

Steering Gear Disassembly and Assembly

1. Attach the steering gear assembly to a holding fixture (Tool C-3323) and put the holding fixture in a bench vise. Thoroughly

clean the outside surface of the steer-gear assembly before disassembly.
2. Loosen the cross shaft adjusting screw lock nut, and back out the adjusting screw about two turns to relieve the mesh load between the ball nut rack and the sector gear teeth. Remove the Cross-shaft seal as given in the procedure for cross shaft seal replacement.
3. Position the steering gear worm shaft in a straight ahead position.
4. Remove the attaching bolts from the cross shaft cover and slowly remove the cross shaft while sliding arbor Tool C-3786 into the housing. Remove the lock nut from the adjusting screw and remove the screw from the cover by turning the screw clockwise. Slide the adjustment screw and its shim out of the slot in the end of the cross shaft.
5. Loosen the worm shaft bearing adjuster lock nut with a brass drift (punch) and remove the lock nut. Hold the worm shaft steady while unscrewing the adjuster, using the wrench from Tool Set C-3884. Slide the worm adjuster off the shaft.

Caution handle the adjuster carefully to avoid damaging the aluminum threads. Also, do not run the ball nut down to either end of the worm shaft to avoid damaging the ball guides.

6. Carefully remove the worm and ball nut assembly. This assembly is serviced as a complete assembly only and is not to be disassembled or the ball return guides removed or disturbed.
7. Remove the cross shaft needle bearing by placing the gear housing in an arbor press; insert Tool C-3786 in the lower end of the housing and press both bearings through the housing. The cross shaft cover assembly, including a needle bearing or bushing, is serviced as an assembly.
8. Remove the worm shaft oil seal from the worm shaft bearing ad-

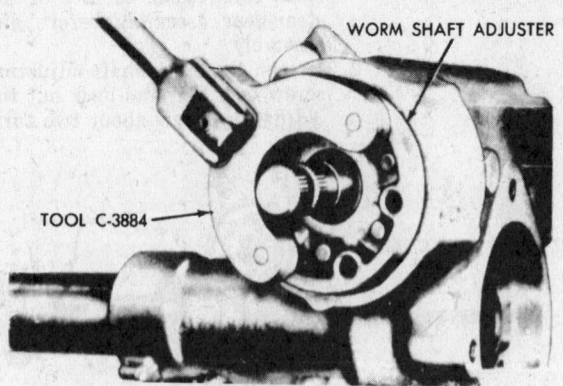

Removing the worm shaft adjuster (© Chrysler Corp.)

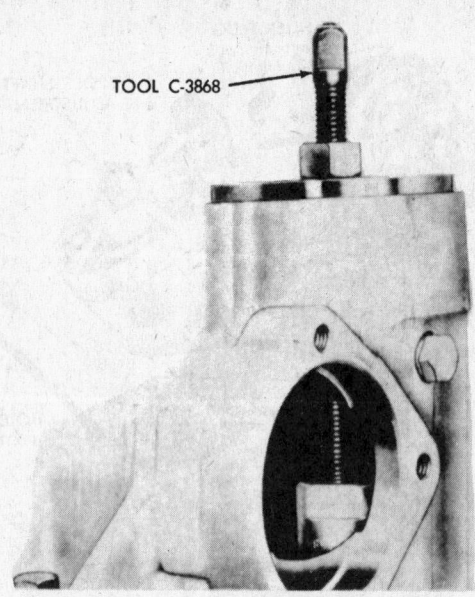

Removing the lower bearing cup
(© Chrysler Corp.)

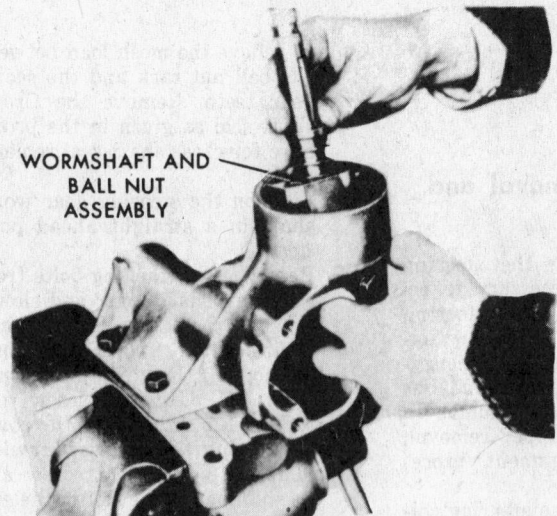

WORMSHAFT AND BALL NUT ASSEMBLY

Removing the worm shaft
and ball nut assembly
(© Chrysler Corp.)

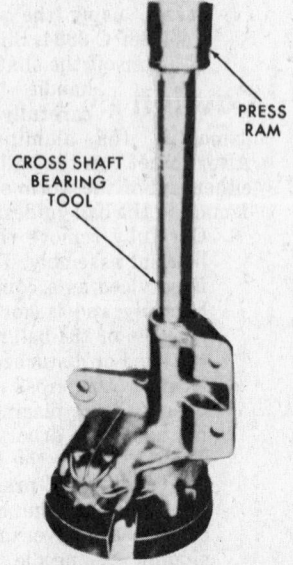

PRESS RAM

CROSS SHAFT BEARING TOOL

Removing the cross shaft inner
and outer bearings
(© Chrysler Corp.)

juster by inserting a blunt punch behind the seal and tapping alternately on each side of the seal until it is driven out of the adjuster.

9. Remove the worm shaft in the same manner as that given in step 8. *Be careful not to cock the bearing cup and distort the adjuster counter bore.*

10. Remove the lower cup if necessary by placing the locking head jaws of remover tool C-3868 behind the bearing cup and expanding the remover head by pressing down on the center plunger of the tool. Pull the bearing cup out by turning the remover screw clockwise while holding the center screw steady.

11. Wash all parts in clean solvent and dry thoroughly. Inspect all parts for wear, scoring, pitting, etc. Test operation of the worm shaft and ball nut assembly. If ball nut does not travel smoothly and freely on the worm shaft or if there is binding, replace the assembly.

NOTE: extreme care must be taken when handling the aluminum worm bearing adjuster to avoid thread damage. Also, be careful not to damage the threads in the gear housing. Always lubricate the worm bearing adjuster before screwing it into the housing.

12. Inspect the cross shaft for wear and check the fit of the shaft in the housing bearings. Inspect the fit of the shaft pilot bearing in the housing. Be sure the worm shaft is not bent or damaged.

13. Install the cross shaft outer needle bearing by placing the bearing on the end of Tool C-3786

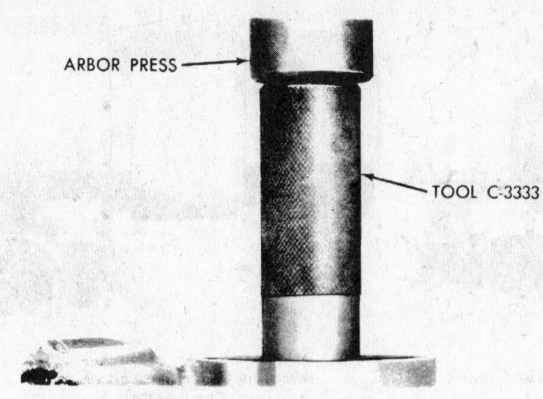

Installing the inner bearing (© Chrysler Corp.)

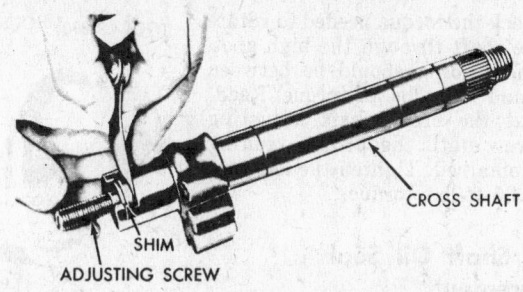

Measuring cross shaft adjusting screw end clearance
(© Chrysler Corp.)

with an adapter ring. Press the bearing into the housing about ½ in. below the end of the bore. Leave space for the new oil seal.

14. Install the inner needle bearing by placing it on Tool C-3333 and pressing it into the inside end of the housing bore flush with the inside end of the bore surface.

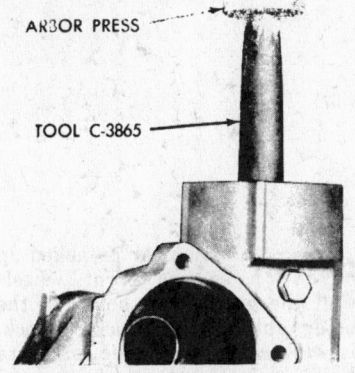

Installing the worm shaft lower bearing cup
(© Chrysler Corp.)

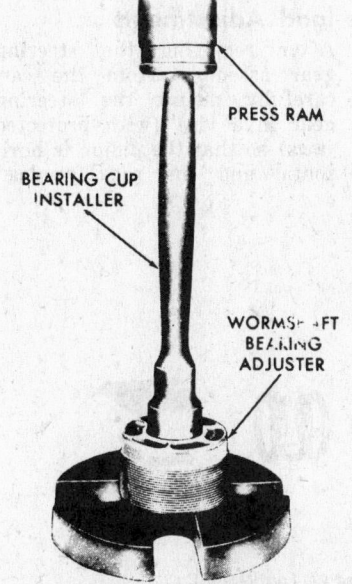

Installing the worm shaft upper bearing cup
(© Chrysler Corp.)

15. Install the worm shaft bearing cups (upper and lower) by placing them and their spacers in the adjuster nut and pressing them in place with Tool C-3865.

16. Install the worm shaft oil seal by placing the seal in the worm shaft adjuster with the metal seal retainer up. Drive the seal into place with a suitable sleeve until it is just below the end of the bore in the adjuster.

Caution Apply a coating of steering gear lubricant to all moving parts during assembly. Also, put lubricant on and around oil seal lips.

17. Clamp the holding fixture and housing in a bench vise with the bearing adjuster opening upward. Place a thrust bearing in the lower cup in the housing.

18. Hold the ball nut from turning and insert the worm shaft and ball nut assembly into the housing with the end of the worm shaft resting in the thrust bearing. Place the upper thrust bearing on the worm shaft. Thoroughly lubricate the threads on the adjuster and the threads in the housing.

19. Place a protective sleeve of tape over the splines on the worm shaft to avoid damaging the seal. Slide the adjuster assembly over the shaft.

20. Thread the adjuster into the housing and, with Tool wrench C-3884 and the splined nut set, tighten the adjuster to 10 ft-lbs. while rotating the worm shaft to seat the bearings.

21. Loosen the adjuster so no bearing pre-load exists. Then, using a torque wrench Tool C-3380, tighten the adjuster for a worm shaft bearing pre-load of 1⅛ to 4½ in-lbs. Tighten the bearing adjuster lock nut, and recheck the pre-load to be sure it didn't change.

22. Before installing the cross shaft,

pack the worm shaft cavities in the housing above and below the ball nut with steering gear lubricant. A good grade of multi-purpose lubricant may be used if steering gear lubricant is not available. *Do not use gear oil.* Pack enough lubricant into the worm cavities to cover the worm.

23. Slide the cross shaft adjusting screw and shim into the slot in the end of the shaft. Check the end clearance for no more than 0.004 in. clearance. If the clearance is not within the limit, remove old shim and install a new shim, available in three different thicknesses, to get the proper clearance.

24. Start the cross shaft and adjuster screw into the bearing in the housing cover. Using a screwdriver through the hole in the cover, turn the screw counterclockwise to pull the shaft into the cover. Install the adjusting screw lock nut, but do not tighten at this time.

25. Rotate the worm shaft to center the ball nut.

26. Place a new gasket on the housing cover and install the cross shaft and cover assembly into the steering gear housing. *Be sure to coat the cross shaft and sector teeth with steering gear lubricant before installing the cross shaft in the housing.* Allow some lash between the cross shaft sector teeth and the ball nut rack. Install and tighten the cover bolts to 25 ft-lbs. torque.

27. Place the cross shaft seal on the cross shaft with the lip of the seal facing the housing. Place the installing adapter SP-3828 from tool C-3880 against the seal with short step toward the seal. Place the nut from Tool C-3880 on the cross shaft and turn it down against the adapter, pressing the seal into the housing until the step on the adapter touches the housing. Remove the tool.

28. Turn the worm shaft about ¼ turn away from the center of the "high-spot" position. Using torque wrench C-3380 and a ¾ in. socket on the worm shaft spline,

check the torque needed to rotate the shaft through the high spot. The reading should be between 8 and 11 in-lbs. of torque. Readjust the cross shaft adjusting screw until the proper reading is obtained. Tighten the lock nut to 35 ft-lbs. torque.

Cross Shaft Oil Seal Replacement

1. Remove the steering gear arm retaining nut and lock washer. Remove the steering gear arm using Tool C-3880.
   NOTE: Tool C-3880 consists of three pieces: adapter SP3056; half rings SP-1932; and nut SP-3610.
2. Slide the threaded adapter over the end of the cross shaft and install the nut portion of the tool on the shaft. Maintain pressure on the adapter with the tool nut while turning the adapter into the oil seal until the adapter grips the seal firmly. Place two half rings and the retainer over

Removing the cross shaft oil seal
(© Chrysler Corp.)

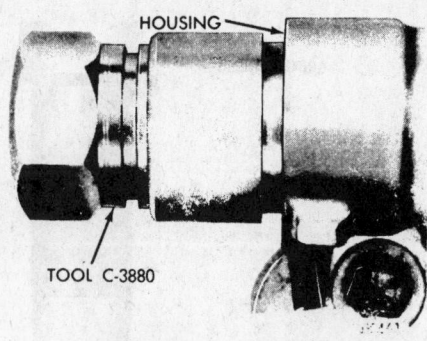

Installing the cross shaft oil seal
(© Chrysler Corp.)

both portions of the tool. Turn the tool nut counterclockwise to pull the seal from the housing.
3. Place a new oil seal onto the splines of the cross shaft with the lip of the seal facing the housing. Place the installing adapter SP-3052 from Tool C-3880 against the seal and press it in until a gap of ¼ in. exists between the adapter and

the housing.
4. Place the nut from Tool C-3880 on the cross shaft and turn it down against the adapter until the seal is pressed into the housing.
5. Remove the tool, and install the steering gear arm, lock washer, and retaining nut. Tighten the nut to 180-ft-lbs. torque.

Section 7
Rack and Pinion Gear Type

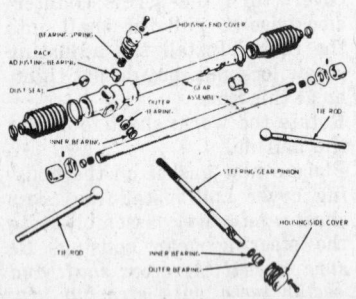

Rack and pinion steering gear disassembled
(© Ford Motor Co.)

The rack and pinion steering gear is mounted in rubber insulators on brackets attached to the front crossmember. Movement of the steering wheel is passed by the steering shaft through a universal joint and flexible coupling to the helically-toothed pinion. Rotation of the pinion causes the rack to move sideways and the connecting rods, attached to the ends of the rack, transmit this movement to the spindle arms and thus cause the road wheels to turn. The steering mechanism provides for about 3½ turns of the steering wheel from lock to lock (approximately 4.92 inches of travel of the rack).
The steering gear has three tenths (0.3) U.S. pint of oil, S.A.E. 90 Hy-

poid, installed during manufacture and normally requires no more lubrication. But if it should be necessary to add or refill the steering gear, *Do not fill completely with oil.* Leave some space to prevent a pressure buildup during use.
Due to the design and construction of the rack and pinion gear, there are only two adjustments required: a. rack damper adjustment, and b. pinion bearing pre-load adjustment. Both adjustments are done by varying the thickness of a shim pack under a cover plate. It is necessary to remove the assembly from the car to do any adjustment.

Caution When car is jacked up so the front wheels are off the ground, do not turn the steering quickly from lock to lock. This will cause hydraulic pressure to build up in the steering gear and may burst or blow off the bellows.

Rack Damper and Pinion Pre-load Adjustments

1. After removing the steering gear assembly from the car, carefully mount the steering gear in a vise (with protected jaws) so that the pinion is horizontal and the rack pre-load

Rack damper and shims disassembled (© Ford Motor Co.)

Steering pinion pre-load shims
(© Ford Motor Co.)

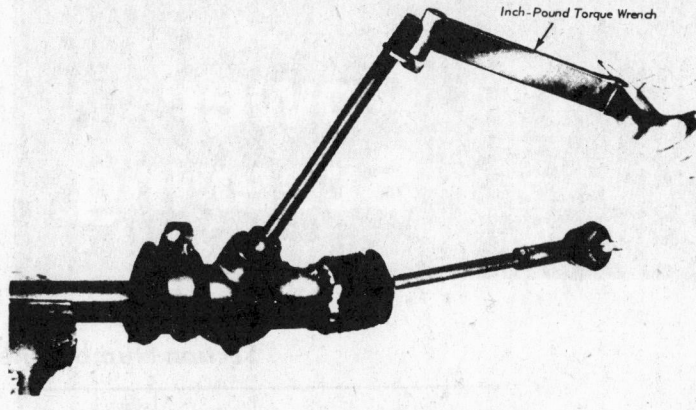

Measuring pinion turning torque (© Ford Motor Co.)

cover is facing up.

2. Remove the two cover attaching screws from the housing.

3. Lift off the cover plate, shim pack and gaskets, and remove the spring and slider.

4. Remove the two screws holding the pinion bearing pre-load cover plate to the housing and lift off the cover plate, shim pack, and gasket.

5. Replace the pinion cover plate and loosely install the attaching screws. Do not install the shim pack and gaskets.

6. Tighten the attaching screws on the cover plate until the cover plate just touches the pinion bearing. Using a feeler gauge, measure the gap between the cover plate and the housing and note the reading. Check that the screws are evenly tightened by measuring the fit between the plate and housing next to each screw.

7. Make up a shim pack (including two gaskets that sandwich the shim pack) that is 0.002 to 0.004 in. smaller than the measured gap. This shim pack will provide the proper pre-load on the pinion bearing when installed.

8. Remove the cover plate and the attaching screws and install the assembled shim pack. Reinstall the cover plate and the attaching screws, coating the threads with sealing compound, and tighten them to 6-8 ft-lbs. torque.

9. To adjust the rack damper (also called a "slipper"), it is necessary that the distance between the cover plate and the top of the rack damper be 0.0005-0.0035 in.

10. Place the rack damper on the rack bar and seat it firmly. Measure the distance from the top of the rack damper to the cover plate mounting surface of

the housing and note the reading.

11. Make up a shim pack (including two gaskets that sandwich the shim pack) that is 0.0005 to 0.0035 in. greater than the measurement just made. *It is extremely important that this measurement be accurate to ensure correct steering. Otherwise excessive wear and possible failures will result.*

12. Install the spring into the recess in the rack damper, and position the shim pack, gaskets, and cover plate over the spring. Install the attaching screws, using sealing compound on the threads, and tighten them to 6-8 ft-lbs. torque.

13. Using an in-lb. torque wrench, check the torque required to start the pinion rotating. The torque reading should be 10-18 in-lbs. If the reading is not within these limits, the adjustment is incorrect and must be redone. Check the shimming again, and inspect the gear assembly for tight bearings, broken gear teeth, lack of lubricant, etc.

Steering Gear Removal and Installation

1. With the steering wheel set straight ahead, raise the car and place jack stands under the front frame.

2. Remove the nut and bolt holding the flexible coupling to the pinion splines. Bend back the lock tabs and remove the attaching screws holding the steering gear to the mounting brackets on the crossmember. Remove the screws, locking plates, and U-clamps.

3. Remove the cotter pins and the castellated nuts holding the connecting rod ends to the spindle

arms. Using the steering ball joint separator Tool 3290-C, separate the connecting rod ends from the spindle arms. Remove the castellated nuts and pull the steering gear from the car. It will be necessary to turn one wheel to the lock position to allow the steering gear to be moved sideways enough to clear the stabilizer bar.

4. Remove the connecting rod ends and the locknuts, noting the number of turns it takes to unscrew them.

5. When installing the steering gear, replace the lock nuts and the connecting rod ends by turning them the exact number of turns it took to remove them.

6. With the steering wheel in the straight ahead position, locate the steering gear and align the mating splines on the flexible coupling and pinion shaft. Check the condition of the rubber mounting insulators and replace them if necessary.

7. Fasten the steering gear assembly to the mounting brackets on the crossmember with the U-clamps Put new locking plates under the attaching screws and tighten the screws to 15-18 ft-lbs. torque.

8. Place the connecting rods on the spindle arms. Install the castellated nuts, tighten to 18-22 ft-lbs. torque, and install new cotter pins.

9. Tighten the flexible coupling to the pinion shaft attaching bolt to 12-15 ft-lbs. torque.

10. Raise the car enough to remove the jack stands and lower the car to the ground.

11. Check the front wheel toe-in and lock angles. Check position of steering wheel spokes.

Power Steering

Application Index

Section Numbers Refer to Sections in Text

Make	Year	Gear Type	Sections
Buick	1964–70	D	2, 6
Buick Special	1964–70	D	2, 6
Cadillac and Eldorado	1964–70	D	2, 6
Chevelle	1964–70	D	2, 6
Chevrolet	1964	B	2, 5
	1965–70	D	2, 6
Camaro and Chevy II	1964–70	B	2, 5
Chrysler	1964–70	C	2, 7
Comet	1964–67	A	2, 4
Continental	1964	E	2, 8
	1965–70	F	2, 9
Corvette	1964–70	B	2, 5
Cougar	1966–70	A	2, 4
Dodge and	1964–70	C	2, 7
Dart and Challenger	1964–70	C	2, 7
Fairlane	1964–69	A	2, 4
Falcon	1964–65	A	2, 4
	1966–70	D	2, 6
Firebird	1964–70	D	2, 6
Ford	1964	A	2, 4
	1965–70	F	2, 9
	1965	D	2, 6

Make	Year	Gear Type	Sections
Javelin and AMX	1968–70	D	2, 6
Jeep "J"	1963–66	D	2, 6
Mercury	1964	A	2, 4
	1965–70	F	2, 9
	1965	D	2, 6
Montego	1968–70	A	2, 4
Mustang	1965–70	A	2, 4
Oldsmobile and Toronado	1964–70	D	2, 6
Oldsmobile Toronado	1967–70	D	2, 6
Oldsmobile F-85	1964–70	D	2, 6
Plymouth	1964–70	C	2, 7
Pontiac	1964–70	D	2, 6
Rambler and Hornet	1964–70	D	2, 6
Tempest	1964–70	D	2, 6
Thunderbird	1964	E	2, 8
	1965–70	F	2, 9
Valiant	1964–70	C	2, 7

Gear Type

A—Bendix Linkage
B—Saginaw Linkage
C—Chrysler Full-Time Constant-Control
D—Saginaw Rotary
E—Ford 1st Type Torsion Bar
F—Ford 2nd Type Torsion Bar

Section Page Numbers

Section	Title	Page
1	General	1289
2	Prelim. tests	1289
3	Oil pumps	1290
4	Bendix linkage-type	1302
5	Saginaw linkage-type	1305
6	Saginaw rotary-type	1309
7	Chrysler full time constant control	1313
8	Ford torsion bar (1st type)	1317
9	Ford torsion bar (2nd type)	1319

Diagnosis charts—1287-89

Specifications—1323-27

Diagnosis Linkage Type

Hard Steering in Both Directions
1. Loose pump drive belt.
2. Improperly lubricated linkage points.
3. Low tire air pressure.
4. Pitman arm stud too tight on valve body.
5. Power cylinder piston rod bent and/or binding in rod guide.
6. Low oil pump pressure; sticking flow control or relief valve.
7. Power cylinder seal leakage.
8. Bind in steering column.
9. Air in hydraulic system.
10. Improper gear adjustment.

Hard Steering in One Direction
1. Unequal or low tire pressure.
2. Bent power cylinder piston rod.
3. Bind in steering column.
4. Front end out of alignment.
5. Improperly adjusted spool valve.

Car Constantly Turns to One Side
1. Unequal tire pressure.
2. Pitman arm stud binding in valve body.
3. Binding in steering column.
4. Spool valve sticking or improperly adjusted.

Slow Recovery After Turns
1. Low tire pressure.
2. Pitman arm stud binding in valve body.
3. Air in hydraulic system.
4. Binding linkage and front suspension points.
5. Bent power steering cylinder rod.
6. Improper wheel alignment.
7. Improperly adjusted spool valve.

Loose Steering (May be Diagnosed as a Kickback)
1. Improperly adjusted steering gear.
2. Shock absorbers not operating properly.

Oil Leakage
1. Check all hose and gasket areas and seals at pump shaft and piston rod.

Noises
1. Loose pump drive belt.
2. Low oil level.
3. Worn pump bushings.
4. Dirt or sludge in pump.
5. Pitman arm stud loose, either end.
6. Tie rod clamps not in horizontal position at bottom of tie rods.
7. Piston rod-to-frame loose.
8. Kinks or sharp bends in hoses.
9. Air in system.
10. Dirt in system causing valves to stick.

Rotary Type

Hard Steering
1. Low or uneven tire pressure.
2. Improper gear adjustment.
3. Improper caster.
4. Improper camber.
5. Insufficient or incorrect hydraulic fluid.
6. Twisted or bent suspension parts, frame, and linkage components.
7. Tight wheel bearings.
8. Steering spindle bent.
9. Pump belt out of adjustment.
10. Pump output low.
11. Air in system.
12. Steering gear out of adjustment.
13. Valve spool out of adjustment.
14. Valve spool sticking.
15. Steering linkage binding.

Hard Steering—Straight Ahead
1. Steering adjustment too tight.
2. Steering gear shaft binding.

Hard Steering—Turning or Parking
1. Oil level low.
2. Pump pressure low.
3. Pressure loss in steering gear due to leakage past O-rings.
4. Pressure loss between valve spool and sleeve.
5. Pressure loss past piston ring or scored housing bore.

Loose Steering
1. Loose wheel bearings.
2. Loose tie rod ends or linkage.
3. Worn ball joints.
4. Worn suspension parts.
5. Fluid level low.
6. Insufficient mesh load.
7. Insufficient worm bearing preload.
8. Valve spool out of adjustment.

Erratic Steering
1. Oil or brake fluid on brake lining.
2. Out of round brake drums.
3. Improperly adjusted brakes.
4. Under-inflated tires.
5. Broken spring or other details in suspension system.
6. Improper caster.

Pull to One Side—(The Continuing Tendency of the Car to Veer in One Direction Only.)
1. Incorrect tire pressure.
2. Wheel bearings improperly adjusted.
3. Dragging brakes.
4. Improper caster or camber.
5. Sagging springs.
6. Incorrect toe-in.
7. Oil or brake fluid on brake linings.
8. Front and rear wheels out of alignment.
9. Bent suspension parts.
10. Worn shock absorbers.

Wandering—(Tendency of Car to Veer in One Direction and, Upon Correction, to Veer in the Opposite Direction.)
1. Incorrect tire pressure.
2. Incorrect or uneven caster and camber.
3. Ball joints worn.
4. Loose tie rod ends.
5. Improper toe-in.
6. Steering gear adjustments too tight or backlash excessive.
7. Bent suspension parts.

Wheel Tramp—(Excessive Vertical Motion of Wheels.)
1. Improper balance of wheels, tires and brake drums.
2. Loose tie rod ends or steering connections.
3. Incorrect tire pressure.
4. Inoperative or loose shock absorber.

Shimmy—(Oscillation of Wheels about Spindle Ball Joints.)
1. Incorrect tire pressure.
2. Worn shock absorbers.
3. Loose wheel bearings.
4. Incorrect steering gear adjustments.
5. Loose tie rod ends.
6. Eccentric or bulged tires.
7. Ball joints worn.
8. Wheel runout.
9. Incorrect toe-in.
10. Worn idler arm bushings.

Road Shocks
1. Worn or damaged shock absorbers.
2. Weak or sagging springs.
3. Improper caster.
4. Improperly adjusted steering gear or linkage.
5. Improperly inflated tires.

Tire Wear
1. Incorrect tire pressure.
2. Incorrect toe-in.
3. Dragging brakes.
4. Improper camber.
5. Bent or broken suspension parts.
6. Excess mileage without rotating tires.
7. Wheel runout.

Binding or Poor Recovery
1. Steering gear shaft binding.
2. Steering gear out of adjustment.
3. Steering linkage binding.
4. Valve spool binding due to dirt, or burred edges.
5. Valve spool out of adjustment.
6. Interference at sector shaft arm and ball stud.
7. Travel regulator stop out of adjustment.

Loss of Power Assist
1. Pump inoperative.
2. Lines damaged.
3. Power cylinder damaged.
4. Valve spool out of adjustment.

Loss of Power Assist —One Direction

1. Valve spool out of adjustment.
2. By-pass in control valve inoperative.

Noisy Pump

1. Air being drawn into pump.
2. Lines touching other parts of car.
3. Oil level low.
4. Excessive back pressure caused by obstructions in lines.
5. Excessive wear of internal parts.

Poor Return of Steering Gear to Center

1. Valve spool sticking.
2. Valve spool out of adjustment.

Steering Wheel Surge When Turning

1. Valve spool sticking.
2. Excessive internal leakage.

Chrysler Constant Control

Hard Steering

1. Improper tire inflation.
2. Improper wheel alignment.
3. Low fluid level (accompanied by pump noise).
4. Loose pump belt (accompanied by squealing noise).
5. Slipping pump belts (caused by oil on belts).
6. Binding steering leakage.
7. Low pump output.
8. Steering gear malfunction.
   A. Cross-shaft adjusted too tight.
   B. Pressure control valve stuck in closed position.
   C. External fluid leakage at:
      cross-shaft lower oil seal.
      cross-shaft adjusting screw seal.
      cross-shaft cover O-ring seal.
      valve body to gear housing O-rings.
   D. Damaged valve pivot lever.
      if pressure gauge will build up to 850 to 950 psi, check for:
      damaged cross-shaft bearings.
      dirt or chips in steering gear.
      damaged thrust bearings or excessive preload.
      binding worm and piston assembly.
   E. Excessive internal leakage. If pressure gauge will not build up to 850 to 950 psi, check for: Damaged O-rings or seals in or between the cylinder head and the column jacket support.

Poor Recovery from Turns

1. Improper tire inflation.
2. Improper wheel alignment.
3. Binding steering linkage.
4. Damaged steering tube bearings.

5. Mast jacket and gear housing not properly aligned.
6. Steering gear malfunctions.
   A. Improper cross-shaft adjustment.
   B. Pressure control valve stuck in open position.
   C. Column jacket support spanner nut loose.
   D. Damaged valve pivot lever.
   E. Improper worm thrust bearing adjustment.
   F. Burrs or nicks in reaction ring grooves in cylinder head or column jacket support.
   G. Damaged cylinder head-to-worm seal.
   H. Dirt or chips in steering gear.
   I. Damaged worm and piston assembly.

Wanders to Either Side

1. Improper tire inflation.
2. Improper wheel alignment.
3. Steering wheel off center.
4. Improper valve body adjustment.
   wanders to right—move valve body up on gear housing.
   wanders to left—move valve body down on gear housing.
5. Damaged valve pivot lever.
6. Column jacket support spanner nut loose.

Temporary Increase in Effort When Making Turn

1. Low fluid level.
2. Loose pump belt.
3. Slipping pump belts.
4. Binding steering linkage.
5. Engine idle too slow.
6. Defective pump.
7. Air in system. (bleed system)
8. Steering gear malfunction.
   A. External leakage.
   B. Improper cross-shaft adjustment.
   C. Excessive internal leakage.

Excessive Play in Steering Wheel

1. Improper cross-shaft adjustment.
2. Column jacket support spanner nut loose.
3. Improper worm thrust bearing adjustment.

Lack of Assistance

(ONE DIRECTION)
1. Fluid leaking past worm shaft cast iron seal ring or ferrule O-ring.
(BOTH DIRECTIONS)
1. Broken ring on worm piston.
2. Piston end plug loose.
3. Damaged reaction seal.
4. Slipping pump belt.
5. Low pump output.

Noises

1. Buzzing noise in Neutral, stops when steering wheel is turned—pressure control valve sticking.
2. Howling noise—low fluid level.
3. Squealing noise — loose pump belt.

Ford Torsion Bar

Jerky Steering

1. Low fluid or leaking condition, air in system.
2. Improper linkage and suspension lubrication.
3. Worn or damaged linkage and/or suspension.
4. Loose steering gear mounting.
5. Improper gear adjustment.
6. Erratic brakes.
7. Improper wheel bearing adjustment.
8. Wheels and tires out of balance.

Loose or Too Easy Steering

1. Loose or worn linkage and suspension connections.
2. Loose steering gear mounting.
3. Improper gear adjustment.
4. Improper wheel bearing adjustment.

Hard Steering and/or Loss of Power Assist

1. Low tire pressure.
2. Uneven tire tread or variation in tire sizes.
3. Low fluid, air in system.
4. Improper lubrication.
5. Obstruction in power steering oil lines.
6. Binding ball joints or linkage.
7. Insufficient pump pressure.
8. Improper gear adjustment.
9. Improper front wheel alignment.
10. Spool valve binding or out of adjustment.
11. Obstruction in gear.

Noises

1. Low fluid, air in system.
2. Front or rear spring damages or line up.
3. Improper lubrication.
4. Obstruction in lines.
5. Inoperative shock absorbers.
6. Worn, loose or damaged linkage and suspension parts.
7. Loose steering gear mountings.
8. Improper gear adjustment.
9. Improper front wheel alignment.
10. Excessive wear in gear or pump parts.

Shimmy or Wheel Tramp

1. Improper tire pressure.
2. Front or rear spring condition.
3. Low fluid or air in system.
4. Defective shock absorbers.
5. Worn or damaged linkage and or suspension parts.
6. Loose steering gear mounting.
7. Improper gear adjustment.
8. Wheels and tires out of balance.
9. Improper alignment.

Pulls to One Side or Side to Side Wander

1. Improper tire pressure.
2. Uneven tire tread or variation in tire size.

3. Uneven or overloaded vehicle.
4. Front or rear spring condition.
5. Bent spindle or spindle arm.
6. Weak shock absorber.
7. Worn or damaged linkage and/or suspension.
8. Incorrect gear adjustment.
9. Front brake condition.
10. Incorrect alignment.

Tires Squeal on Turns

1. Improper tire pressure.
2. Uneven tire tread or variation in tire size.
3. Bent spindle or spindle arm.
4. Worn or damaged linkage and/or suspension.
5. Incorrect alignment.
6. Driving into turn at too high speed.

Poor Recovery Following Turn

1. Improper tire pressure.
2. Improper lubrication.
3. Worn or damaged linkage and/or suspension.
4. Insufficient pump pressure.
5. Improper gear adjustment.
6. Improper alignment.

Section 1
General Information

The procedures for maintaining, adjusting, and repairing the power steering systems and components discussed in this chapter are to be done only after determining that the steering linkages and front suspension systems are correctly aligned and in good condition. All worn or damaged parts should be replaced before attempting to align the power steering systems. After correcting any condition that could affect the power steering, do the preliminary tests of the steering system components given in *Section 2.*

Section 2
Preliminary Tests

Lubrication

Proper lubrication of the steering linkage and the front suspension components is very important for the proper operation of the steering systems of cars equipped with power steering. Check the steering gear box for sufficient lubricant by removing the filler plug and checking the level. Add enough fluid gear oil S.A.E. 90 to bring the oil level to the filler plug hole if necessary.

Caution   Do not use a pressure gun to add fluid gear oil since the pressure will force the oil out of the steering gear box.

NOTE: the Chrysler power steering systems use the same lubricant in the steering gear box and in the power steering pump reservoir.

Air Bleeding

Air bubbles in the power steering system must be removed from the fluid. Be sure the reservoir is filled to the proper level and the fluid is warmed up to the operating temperature. Then, turn the steering wheel through its full travel three or four times until all the air bubbles are removed. *Do not hold the steering wheel against its stops.* Recheck the fluid level.

Fluid Level Check

1. Run the engine until the fluid is at the normal operating temperature. Then, turn the steering wheel through its full travel three or four times, and shut off the engine.
2. Check the fluid level in the steering reservoir. On cars built before 1968, the fluid level is checked by removing the reservoir cap and looking in the filler tube for the fluid level. On 1968 and later cars, a dipstick is provided in the filler tube that shows the proper fluid level. If the fluid level is low, add enough fluid to raise the level to the Full mark on the dipstick or filler tube. Use automatic transmission fluid, type A.

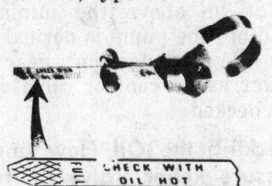

Power steering pump dipstick
(© Ford Motor Co.)

Pump Belt Check

1. Inspect the pump belt for cracks, glazing, or worn places. Using a belt tension gauge, check the belt tension for the proper range of adjustment. The amount of tension varies with the make of car and the condition of the belt. New belts (those belts used less than 15 minutes) require a higher figure. The belt deflection method of adjustment may be used only if a belt tension gauge is not available. The belt should be adjusted for a deflection of 3/8 to 1/2 in.

Fluid Leaks

Check all possible leakage points (hoses, power steering pump, or steering gear) for loss of fluid. Turn engine on and rotate the steering wheel from stop to stop several times. Tighten all loose fittings and replace any defective lines or valve seats.

Turning Effort

Check the turning effort required to turn the steering wheel after aligning the front wheels and inflating the tires to the proper pressure.

1. With the vehicle on dry pavement and the front wheel straight ahead, set the parking brake and turn the engine on.
2. After a short warm-up period of the engine, turn the steering wheel back and forth several times to warm the steering fluid.
3. Attach a spring scale to the steering wheel rim and measure the pull required to turn the steering wheel one complete revolution in each direction. The effort needed to turn the steering wheel should not exceed the limits given in the specifications.

NOTE: this test may be done with the steering wheel removed and a torque wrench applied on the steering wheel nut. See the section on manual steering for a discusion of this test.

Power Steering Pump Flow

Since the power steering pump provides all the power assist in a power steering system, the pump must operate properly at all times for the system to work. After performing all the checks given above, the power steering pump may be tested for proper flow by doing the following procedure:

1. Disconnect the pressure and return lines at the power steering pump and connect the test pressure and return lines to the pump. The test pressure and return lines are connected to a pressure gauge and two manual valves.
2. Open the two manual valves, connect a tachometer to the engine, and start the engine. Run the engine at idle speed until the reservoir fluid temperature reaches about 165-175 degrees Fahrenheit. *This temperature must be maintained during the*

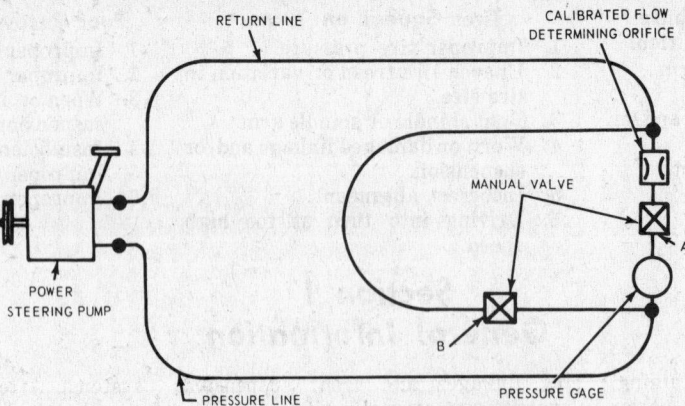

Power steering pump test circuit diagram (© Ford Motor Co.)

test. Manual valve B may be partially opened to create a back pressure of no more than 350 psi to aid the temperature rise. Reservoir fluid must be at the proper level.

3. After the engine and the reservoir fluid are sufficiently warmed up, close the manual valve B. Note the pressure gauge reading. It must be a minimum of 620 psi.

4. If the pressure reading is below the minimum acceptable pressure, the pump is defective and must be repaired. If the pressure reading is at or above the minimum value, the pump is normal. Open manual valve B and proceed to the pump fluid pressure test.

Power Steering Pump Fluid Pressure Test

1. Keep the lines and pressure gauge connected as in the Pump Flow Test.

2. With manual valve A and B opened fully, run the engine at the proper idle speed. Then, close manual valve A and manual valve B, in that order. *Do not keep both valves closed for more than 5 seconds since the fluid temperature will increase abnormally and cause unnecessary wear to the pump.*

3. With both manual valves closed, the pressure reading should be as given in the specifications. If the pressure reading is below the minimum reading, the pump is defective and must be repaired. If the pressure reading is at or above the minimum reading, the pump is normal and the power steering gear or power assist control valve must be checked.

Checking the Oil Flow and Pressure Relief Valve in the Pump Assembly

When the wheels are turned hard right, or hard left, against the stops, the oil flow and pressure relief valves come into action. If these valves are working and are not stuck there should be a slight buzzing noise. Do not hold the wheels in the extreme position for over three or four seconds because, if the pressure relief valve is not working, the pressure could get high enough to damage the system.

Power Steering Hose Inspection

1. Inspect both the input and output hoses of the power steering pump for worn spots, cracks, or signs of leakage. Replace hose if defective, being sure to reconnect the replacement hose properly. Many power steering hoses are identified as to where they are to be connected by special means, such as, fittings that will only fit on the correct pump fitting, or hoses of special lengths.

Test Driving Car to Check the Power Steering

When test driving to check power steering, drive at a speed between 15 and 20 mph. Make several turns in each direction. When a turn is completed, the front wheels should return to the straight ahead position with very little help from the driver.

If the front wheels fail to return as they should and yet the steering linkage is free, well oiled and properly adjusted, the trouble is probably due to misalignment of the power cylinder or improper adjustment of the spool valve.

Section 3
Power Steering
Oil Pumps

The power steering oil pump supplies all the power assist used in power steering systems of all designs. There are various designs of oil pumps used by the automobile manufacturers but all pumps supply power to operate the steering systems with the least effort. All power steering pumps have a reservoir tank built onto the oil pump. These pumps are driven by belts turned by pulleys on either the engine, the rear of the generator, or the front of the crankshaft.

During operation of the engine at idle speed, there is provision for the power steering pump to supply more fluid pressure. During driving speeds or when the car is moving straight ahead, less pressure is needed and the excess is relieved through a pressure relief and flow control valve. The pressure relief part of the valve is inside the flow control part and is basically the same for all pumps. The flow control valve regulates, or controls, the constant flow of fluid from the pump as it varies with the demands of the steering gear. The pressure relief valve limits the hydraulic pressure built up when the steering gear is turned against its stops.

During all pump disassembly work, make sure all work is done on a clean work surface. Clean the outside of the pump thoroughly and do not allow dirt of any kind to get inside the pump. Do not immerse the shaft oil seal in solvent.

If replacing the rotor shaft seal, be extremely careful not to scratch sealing surfaces with tools.

Pump Overhaul

Vane Type Power Steering Pump

The vane type power steering pump is used in Saginaw steering systems. The operation is basically the same as that of the roller type pumps. Centrifugal force moves a number of vanes outward against the pump ring, causing a pumping action of the fluid to the control valve.

Removal and Installation

1. Disconnect hoses at the pump, securing them in a raised position to prevent oil drainage. Cap or cover the ends of the hoses to keep dirt out.
2. Install two caps on the fittings on the pump to prevent oil drainage.

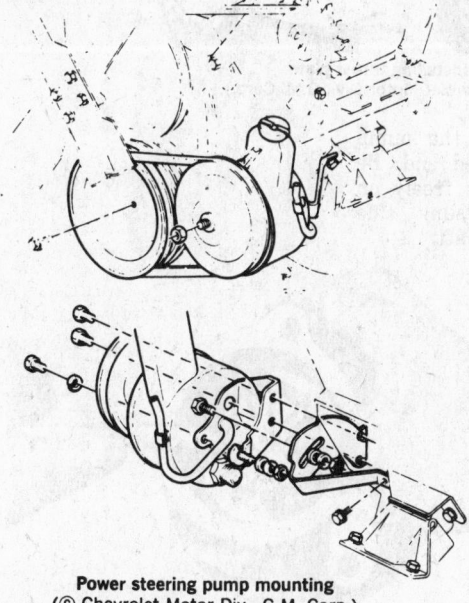

Power steering pump mounting
(© Chevrolet Motor Div., G.M. Corp.)

3. Loosen the bracket-to-pump mounting nuts, move pump toward engine slightly, and remove the pump drive belt.
4. Remove the bracket-to-pump bolts and remove the pump from the car.
5. While holding the drive pulley steady, loosen and remove the pulley attaching nut. Slide the pulley off the shaft. *Do not hammer the pulley off the shaft because the pump will be damaged.*
6. To install the pump on the car, reverse the removal procedure. Always use a new pulley nut, tightening it to 35-45 ft-lbs. torque.
7. After reconnecting the hoses to the pump, fill the reservoir with fluid and bleed the pump of air by turning the drive pulley counterclockwise (as viewed from the front) until air bubbles do not appear.

8. Install the pump drive belt over the pulley, move the pump against the belt until tight enough, and tighten the mounting bolts and nuts.
9. Bleed the air from the system as given previously.

Disassembly

1. Clean the outside of the pump in a non-toxic solvent before disassembling.
2. Mount the pump in a vise, being careful not to squeeze the front hub of the pump too much.
3. Remove the union and seal.
4. Remove the reservoir retaining studs and separate the reservoir from the housing.
5. Remove the mounting bolt O-rings and the union O-rings.

6. Remove the filter and filter cage; discard the filter element.
7. Remove the end plate retaining ring by compressing the retaining ring and then prying it out with a screwdriver. The retaining ring may be compressed by inserting a small punch in the 1/8 in. diameter hole in the housing and pushing in until the ring clears the groove.
8. Remove the end plate. The end plate is spring-loaded and should rise above the housing level. If it is stuck inside the housing, a slight rocking or gentle tapping should free the plate.
9. Remove the shaft woodruff key and tap the end of the shaft gently to free the pressure plate, pump ring, rotor assembly, and thrust plate. Remove these parts as one unit.
10. Remove the end plate O-ring. Separate the pressure plate, pump ring, rotor assembly, and thrust plate.

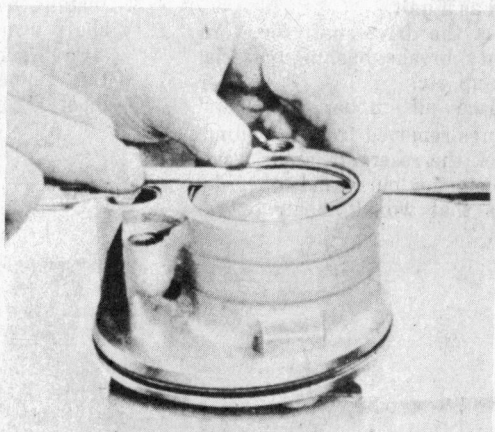

Removing end plate ring
(© Chevrolet Motor Div., G.M. Corp.)

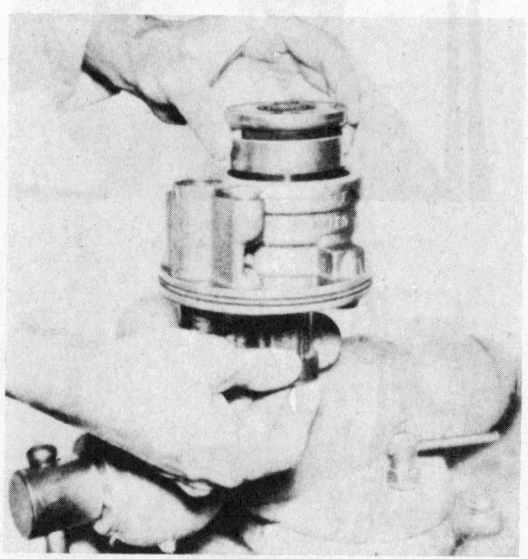

Removing impeller unit
(© Chevrolet Motor Div., G.M. Corp.)

Inspection

Clean all metal parts in a non-toxic solvent and inspect them as given below:

1. Check the flow control valve for free movement in the housing bore. If the valve is sticking, see if there is dirt or rough spots in the bore.
2. Check the cap screw in the end of the flow control valve for looseness. Tighten if necessary being careful not to damage the machined surfaces.
3. Inspect the pressure plate and the pump plate surfaces for flatness and that there are no cracks or scores in the parts. Do not mistake the normal wear marks for scoring.
4. Check the vanes in the rotor assembly for free movement and that they were installed with the radiused edge toward the pump ring.
5. If the flow control valve plunger is defective, install a new part. It is factory calibrated and supplied as a unit.
6. Check the drive shaft for worn splines, breaks, bushing material pick-up, etc.
7. Replace all rubber seals and O-rings removed from the pump.
8. Check the reservoir, studs, casting, etc. for burrs and other defects that would impair operation.

Assembly

1. Install a new shaft seal in the housing and insert the shaft at the hub end of housing, splined end entering mounting face side.
2. Install the thrust plate on the dowel pins with the ported side facing the rear of the pump housing.

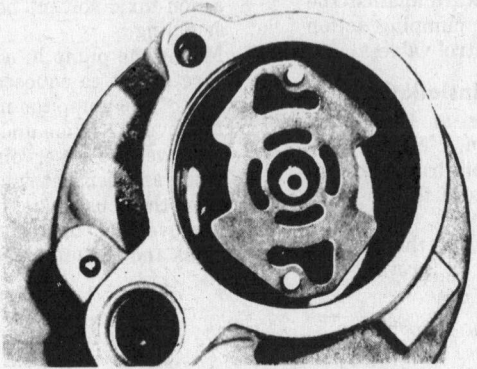
Installing thrust plate
(© Chevrolet Motor Div., G.M. Corp.)

3. Install the rotor on the pump shaft over the splined end. Be sure the rotor moves freely on the splines. Countersunk side must be toward the shaft.

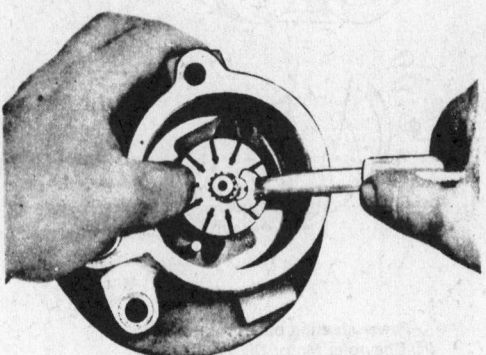
Installing shaft snap ring
(© Chevrolet Motor Div., G.M. Corp.)

4. Install the shaft retaining ring. Install the pump ring on the dowel pins with the direction of the rotation arrow to the rear of the pump housing. Rotation is clockwise as seen from the pulley.

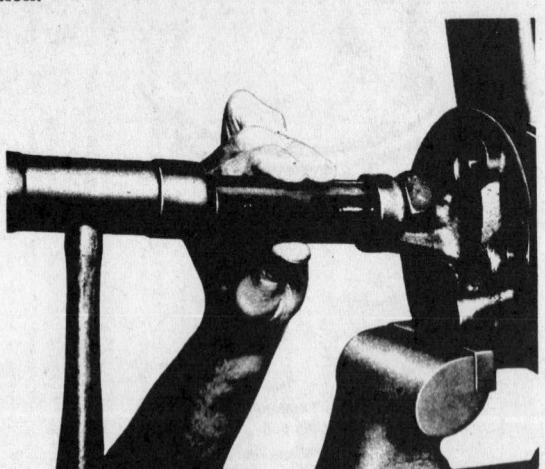
Installing shaft seal
(© Chevrolet Motor Div., G.M. Corp.)

Shaft installation
(© Chevrolet Motor Div., G.M. Corp.)

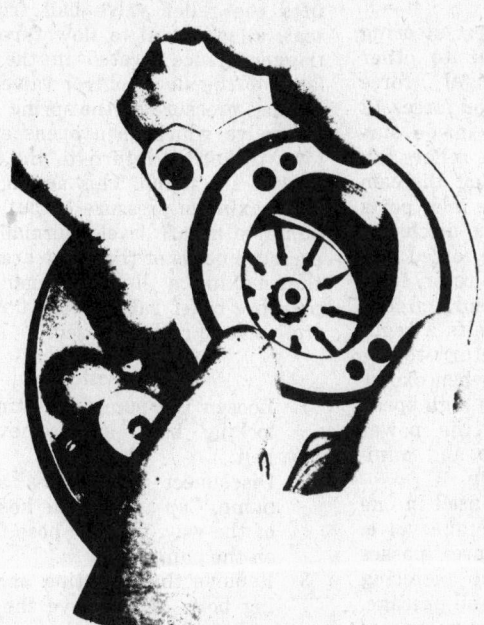

Installing pump ring
(© Chevrolet Motor Div., G.M. Corp.)

5. Install the vanes in the rotor slots with the radius edge towards the outside.

Installing vanes
(© Chevrolet Motor Div., G.M. Corp.)

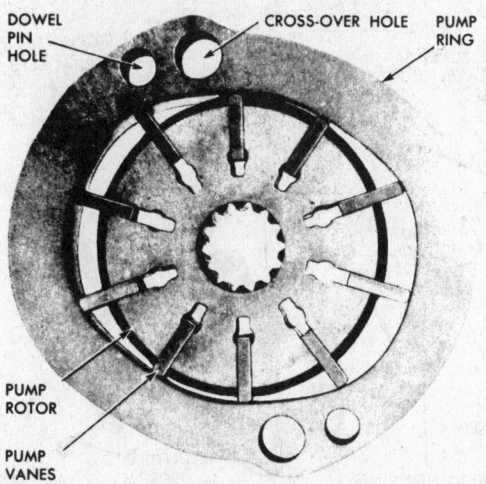

DOWEL PIN HOLE CROSS-OVER HOLE PUMP RING

PUMP ROTOR

PUMP VANES

Correct vane assembly
(© Chevrolet Motor Div., G.M. Corp.)

6. Lubricate the outside diameter and chamfer of the pressure plate with petroleum jelly so as not to damage the O-ring and install the plate on the dowel pins with the ported face toward the pump ring. Seat the pressure plate by placing a large socket on top of the plate and pushing down with the hand.
7. Install the pressure plate spring in the center groove of the plate.

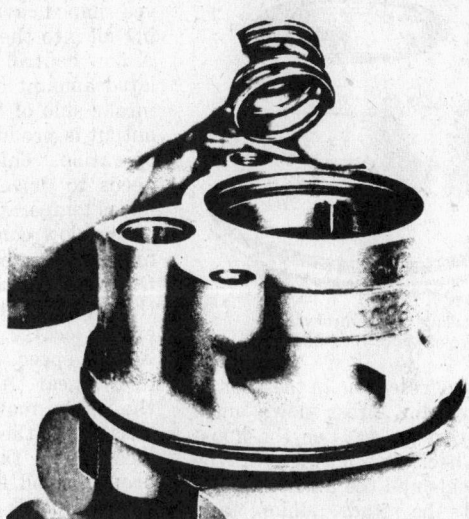

Installing pressure plate spring
(© Chevrolet Motor Div., G.M. Corp.)

8. Install the end plate O-ring. Lubricate the outside diameter and chamfer of the end plate with petroleum jelly so as not to damage the O-ring and install the end plate in the housing using an arbor press. Install the end plate retaining ring while pump is in the arbor press. Be sure the ring is in the groove and the ring gap is positioned properly.

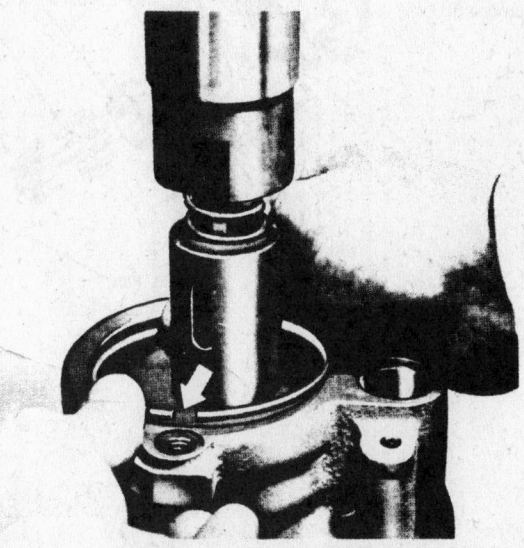

Installing end plate retaining ring
(© Chevrolet Motor Div., G.M. Corp.)

9. Install the flow control spring and plunger, hex head screw end in bore first. Install the filter cage, new filter stud seals and union seal.

Installing flow control valve
(© Chevrolet Motor Div., G.M. Corp.)

10. Place the reservoir in the normal position and press down until the reservoir seats on the housing. Check the position of the stud seals and the union seal.
11. Install the studs, union, and drive shaft woodruff key. Support the shaft on the opposite side of the key when tapping the key into place.

Roller Type Power Steering Pump

The roller type power steering pump is designed similar to other constant flow centrifugal force pumps. A star-shaped rotor forces 12 steel rollers against the inside surface of a cam ring. As the rollers follow the eccentric pattern of the cam ring, oil is drawn into the inlet ports and exhausted through the discharge ports while the rollers are forced into vee shaped cavities of the rotor, forcing oil into the high pressure circuit. A flow control valve permits a regulated amount of oil to return to the intake side of the pump when excess output is produced during high speed operation. This reduces the power needs to drive the pump and minimizes temperature build-up.

The flow control valve used in one make of pump is a two-stage valve. Oil under high pressure passes through two holes into a metering circuit located in a sealed oil passage. At low speed, about 2.7 gpm. passes to the gear. As speed increases and the valve moves, excess oil is by-passed to the inlet and the valve blocks flow through one hole. This drops the oil flow to about 1.6 gpm. at high speeds.

When steering conditions produce excessive pressure needs (such as turning the wheels against the stops), the pressure built up in the steering gear exerts force on the spring end of the flow control valve.

This end of the valve contains the pressure relief valve. High pressure lifts the relief valve ball from its seat, allowing oil to flow through a trigger orifice located in the front land of the flow control valve. This reduces pressure on the spring end of the valve which then opens and allows the oil to return to the intake side of the pump. This action limits the maximum pressure output of the pump to a safe level. Normally, the pressure needs of the pump are below the maximum limits, causing the pressure relief ball and the flow control valve to remain closed.

Removal

1. Loosen the pump mounting and locking bolts and remove the belt.
2. Disconnect both hoses at the pump. Cap and tie the hoses out of the way. Cap the hose fittings on the pump.
3. Remove the mounting and locking bolts and remove the pump and brackets from the car.

Installation

1. Position the pump and brackets on the engine and install the mounting and locking bolts.
2. Install the drive belt and adjust for the proper tension.
3. Connect the pressure and return hoses, using a new pressure hose O-ring.
4. Fill the pump reservoir to the top of the filler neck with power steering fluid.
5. Start the engine and turn the steering wheel several times from stop to stop to bleed the pump of air. Check the oil level and add fluid if necessary.

NOTE: when checking the oil level, see that the level is as follows: engine cold-bottom of filler tube; engine hot-half way up filler tube.

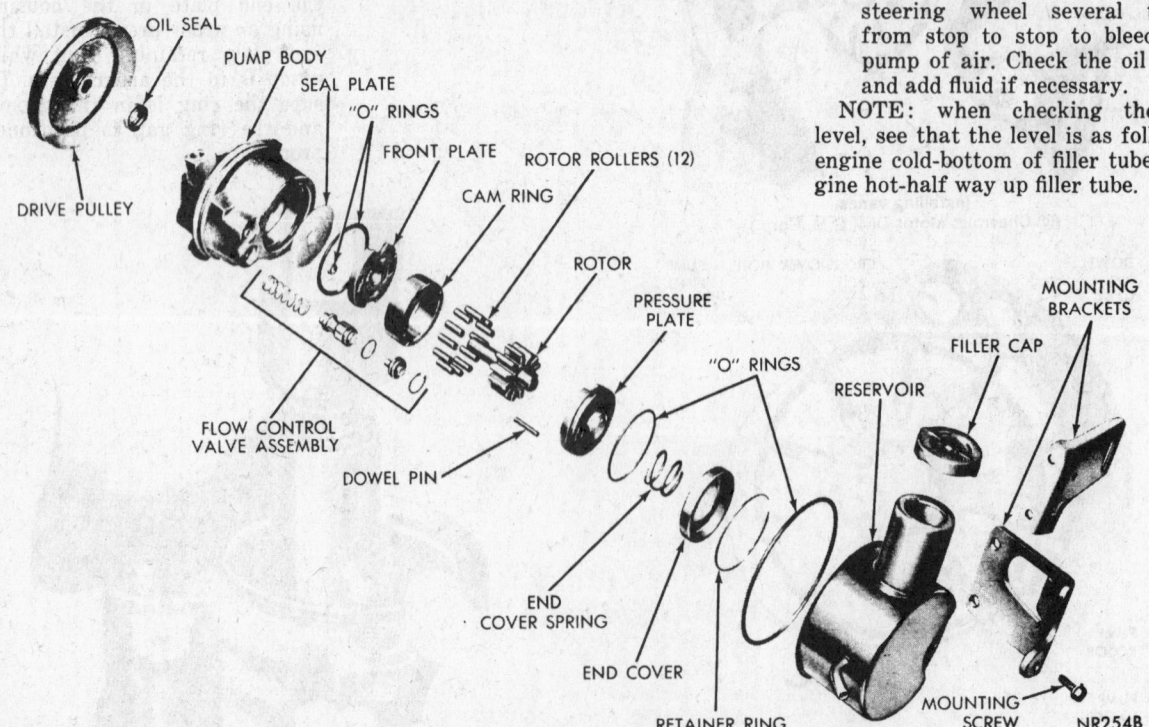

OIL SEAL
PUMP BODY
SEAL PLATE
"O" RINGS
FRONT PLATE
ROTOR ROLLERS (12)
CAM RING
DRIVE PULLEY
ROTOR
PRESSURE PLATE
"O" RINGS
MOUNTING BRACKETS
FILLER CAP
RESERVOIR
FLOW CONTROL VALVE ASSEMBLY
DOWEL PIN
END COVER SPRING
END COVER
RETAINER RING
MOUNTING SCREW
NR254B

Chrysler 1.06 power steering pump, disassembled view (© Chrysler Corp.)

Disassembly

1. Remove pump from engine, drain reservoir, and clean outside of pump. Clamp the pump in a vise at the mounting bracket.
2. Remove the drive pulley.
3. Remove the shaft seal by installing the seal remover adapter over the end of the drive shaft with the large end toward the pump. Place the seal remover

tween the ring and the housing. Remove the ring.
7. Remove the end cover and spring from the housing. It may be necessary to tap the cover gently to loosen it in the housing.
8. Remove the pump from the vise and turn the pump over so the rotating group may come out of the housing. Tap the end of the drive shaft to loosen these parts.

3. Inspect the drive shaft for excessive wear and the seal area for nicks or scoring. Replace if necessary.
4. Inspect the end plates, rollers, rotor and cam ring for nicks, burrs, or scratches. If any of the components are damaged enough to cause poor operation of the pump, all the interior parts may have to be replaced to prevent later failures.

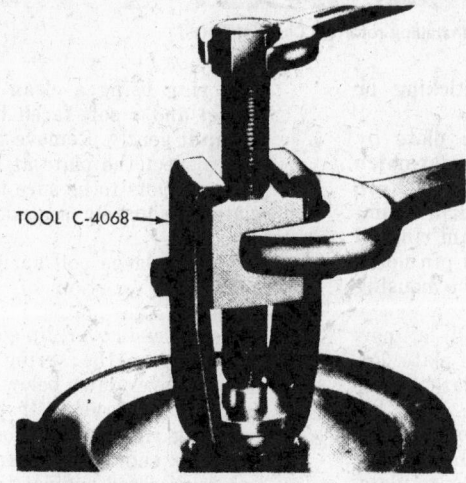

TOOL C-4068

Removing drive pulley (© Chrysler Corp.)

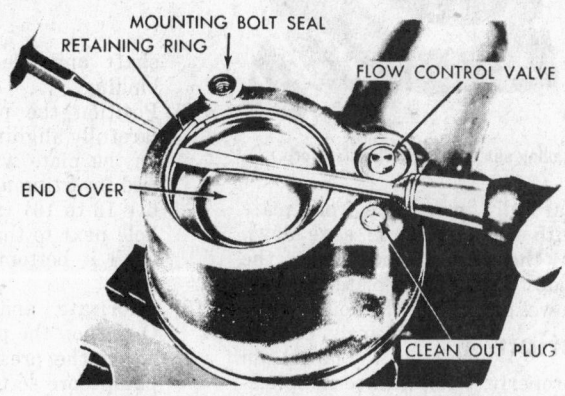

MOUNTING BOLT SEAL
RETAINING RING
FLOW CONTROL VALVE
END COVER
CLEAN OUT PLUG

Removing end cover retaining ring (© Chrysler Corp.)

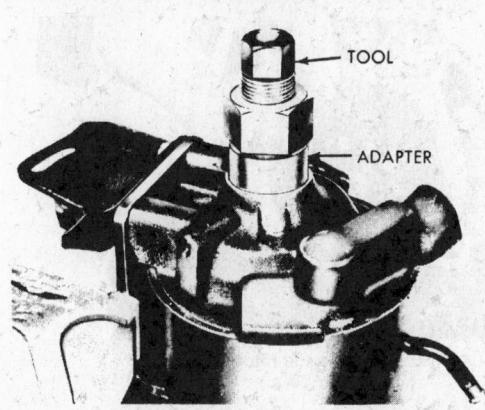

TOOL
ADAPTER

Removing shaft seal (© Chrysler Corp.)

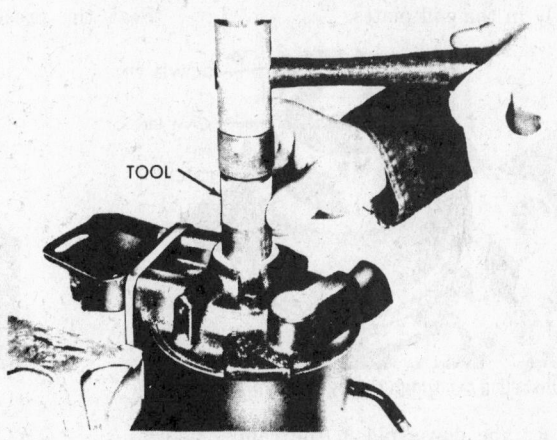

TOOL

Installing shaft seal (© Chrysler Corp.)

tool over the shaft and through the adapter. Then, screw the tapered thread well into the metal portion of the seal. Tighten the large drive nut and remove the seal.
4. Remove the pump from the vise and remove the bracket mounting bolts. Remove the bracket.
5. Remove the reservoir and place the pump in a soft-faced vise with the shaft down. Discard the mounting bolt and the reservoir O-rings.
6. Move the end cover retaining ring around until one end of the ring lines up with the hole in the pump body. Insert a small punch in the hole and push it in far enough to bend the ring so a screwdriver can be inserted be-

Lift the pump body off the rotating group. Check that the seal plate is removed from the bottom of the housing bore.
9. Discard the O-rings from the pressure plate and end cover.
10. Remove the snap ring, bore plug, flow control valve and spring from the housing. Discard the O-ring. If necessary to disassemble the flow control valve for cleaning, see the procedure for disassembly.

Inspection

1. Remove the clean out plug with an Allen wrench.
2. Wash all metal parts in clean, non-toxic solvent. Blow out all passages with compressed air and air dry all cleaned parts.

5. Inspect the pump body drive shaft bushing for excessive wear. Replace the pump body and bushing as one assembly.

Assembly

1. Install the 1/8 in. pipe clean out plug, tightening it to 80 in-lbs. torque.
2. Place the pump body on a clean flat surface and install a new shaft seal into the bore.
3. Install a new end cover O-ring into the groove in the pump bore. Be sure to lubricate the O-ring with power steering fluid before installing it.
4. Lubricate and install a new O-ring in the groove on the pump body where the reservoir fits snugly.

5. Install the brass seal plate to the bottom of the housing bore. Align the notch in the seal plate with the dowel pin hole in the housing.

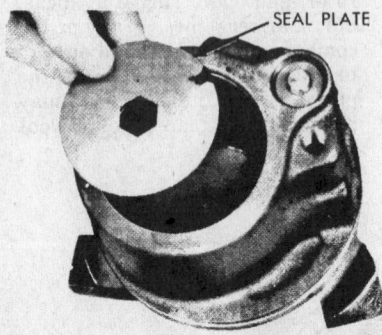

Installing seal plate (© Chrysler Corp.)

6. Carefully install the front plate with the chamfered edge down in the pump bore. Align the index notch in the plate with the dowel pin hole in the housing.

Caution Be extremely careful to align the dowel pin hole properly. Pump can be completely assembled with the dowel pin not seated properly in the dowel pin hole in the housing and positioned improperly in the end plates.

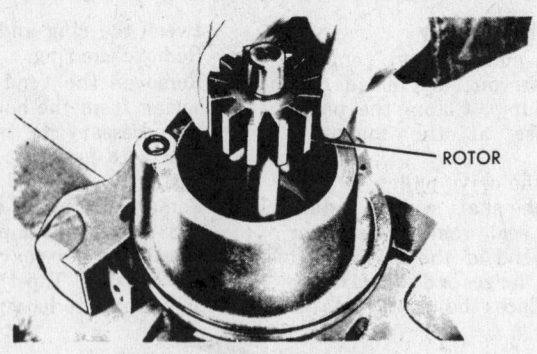

Installing rotor (© Chrysler Corp.)

shaft and are not sticking or binding.

9. Position the pressure plate by carefully aligning the index notch on the plate with the dowel pin and inserting a clean drill (number 13 to 16) in the cam ring oil hole next to the dowel pin notch until it bottoms on the housing floor.

10. Lubricate and install a new O-ring on the pressure plate. Position the pressure plate in the pump bore so that the dowel pin is in the index notch on the plate and the drill extends through the oil passage in the pressure plate. Seat the pressure plate on the

cam ring using a clean 1⅛ in. socket and a soft-faced hammer to tap it gently. Remove the drill and inspect the plate at both oil passage slots to be sure that the plate is squarely seated on the cam ring.

11. Place the large coil spring over the raised portion of the installed pressure plate.

12. Place the end cover, lip edge facing up, over the spring. Press the end cover down below the retaining ring groove. Install the retaining ring in the groove. Be sure the end cover chamfer is squarely seated against the snap ring.

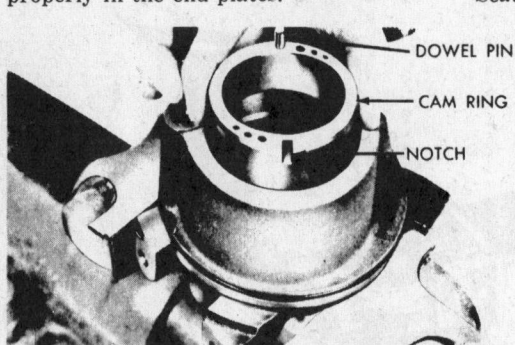

Installing cam ring (© Chrysler Corp.)

7. Place the dowel pin in the cam ring and position the cam ring inside the pump bore. Notch in the cam ring must be facing up (away from the pulley end of pump housing). If the cam ring has two notches, one machined and one cast, install the cam ring with the machined notch up. Check the amount of dowel pin extending above the cam ring surface. If more than 3/16 in. is showing, the dowel pin is not seated in the index hole in the housing.

8. Install the rotor and shaft in the cam ring and carefully install the 12 steel rollers in the cavities of the rotor. Lubricate the rotor, rollers, and the inside surface of the cam ring with power steering fluid. Rotate the shaft by hand to be sure all the rollers are seated parallel with the

Installing rollers in rotor
(© Chrysler Corp.)

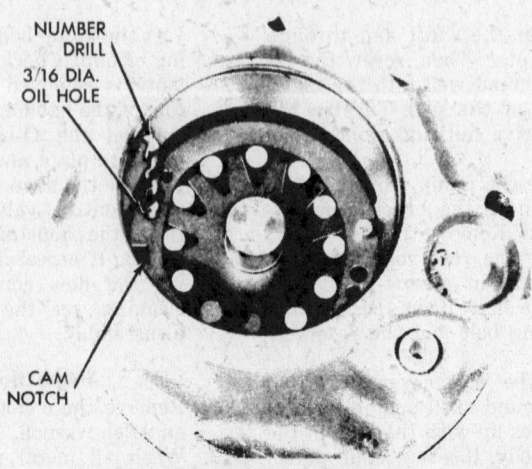

Aligning oil holes (© Chrysler Corp.)

INDEX NOTCH
DOWEL PIN
PRESSURE PLATE

Installing pressure plate (© Chrysler Corp.)

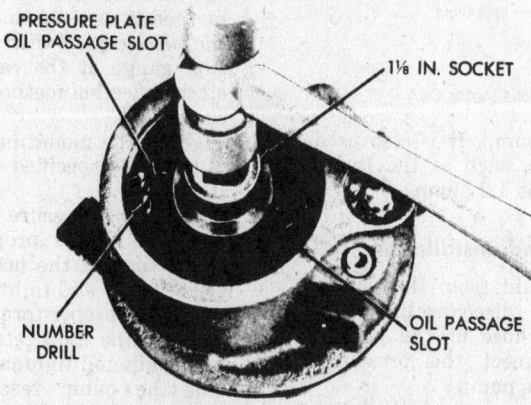

PRESSURE PLATE
OIL PASSAGE SLOT
1⅛ IN. SOCKET
NUMBER DRILL
OIL PASSAGE SLOT

Seating pressure plate (© Chrysler Corp.)

13. Replace the reservoir mounting bolt seal.
14. Lubricate the flow control valve assembly with power steering fluid and insert the valve spring and valve in the bore. Install a new O-ring on the bore plug, lubricate with fluid, and carefully install in the bore. Install the snap ring with the sharp edge up. *Do not depress the bore plug more than 1/16 in. below th snap ring groove.*
15. Place the reservoir on the pump body and visually align the mounting bolt hole. Tap the reservoir down on the pump with a plastic-faced hammer.
16. Remove the pump from the vise and install the mounting brackets with the mounting bolts on the pump. Tighten the bolts to 18 ft-lbs. torque.
17. Install the drive pulley by using the installer tool as follows. Place the pulley on the end of the shaft and thread the installer tool into the ⅜ in. threaded hole in the end of the shaft. Put the installer shaft in a vise and tighten the drive nut against the thrust bearing, pressing the pulley on the shaft until it is flush with the end of

the shaft. *Do not try to press the pulley on the shaft without the special installer tool since the pump interior will be damaged by any other installation procedure.* A small amount of drive shaft end play will be seen when the pulley is installed. This end play is necessary and will be minimized by a thin coat of oil between the rotor and the end plates when the pump is operating.
18. Install the pump assembly on the engine, install the drive belt and hoses (use new O-ring on pressure hose), and check for leaks.

Flow Control Valve Disassembly

1. After removing the pump from the engine and the reservoir from the pump, remove the snap ring and plug from the flow bore. Discard the O-ring.
2. Depress the control valve against the spring pressure and allow the valve to spring out of the bore. If the valve is stuck in the bore or it did not come out of the bore far enough, it may be necessary to tap the housing lightly to remove the valve.
3. If the valve has dirt or foreign particles on it or in its bore, the rest of the pump needs cleaning. The hoses should be flushed and the steering gear valve body reconditioned. If the valve bore is badly scored, replace the pump body and the flow control valve.
4. Remove any nicks or burrs by gently rubbing the valve with crocus cloth. Clamp the valve land in a vise with soft-jaws and remove the hex head ball seat and shims. Note the number and gauge (thickness) of the shims on the ball seat. They must be re-installed for the same shim thickness to keep the same value of relief pressure.
5. Remove the valve from the vise and remove the pressure relief ball, guide, and spring.

Flow Control Valve Assembly

1. Insert the spring, guide and pressure relief ball in the end of the flow control valve.
2. Install the hex head plug using the exact number and thickness shims that were removed. Tighten the plug to 80 in-lbs. torque.
3. Lubricate the valve with power steering fluid and insert the flow control valve spring and valve in the housing bore. Install a new O-ring on the bore plug, lubricate with fluid and carefully install into the bore. Install the snap ring. *Do not depress the bore plug more than 1/16 in. beyond the snap ring groove.*

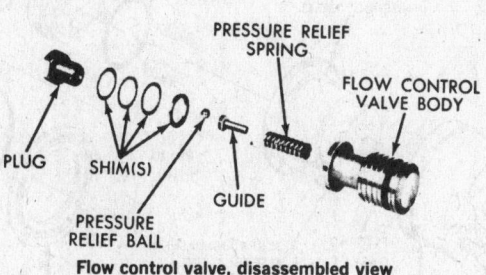

PRESSURE RELIEF SPRING
FLOW CONTROL VALVE BODY
PLUG
SHIM(S)
GUIDE
PRESSURE RELIEF BALL

Flow control valve, disassembled view (© Chrysler Corp.)

Slipper Type Power Steering Pump

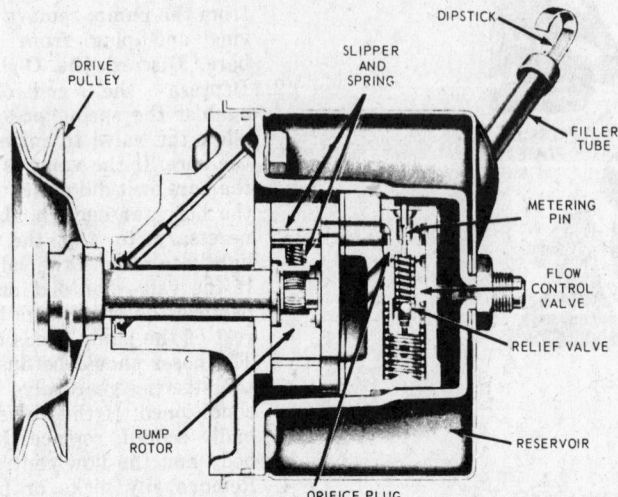

Ford-Thompson power steering pump, sectional view (© Ford Motor Co.)

The slipper type power steering pump is a belt-driven constant displacement assembly that uses a number of spring-loaded slippers in the pump rotor to force oil from the inlet side to the flow control valve. Openings in the metering pin allow a flow of about two gpm. of fluid to the steering gear before the flow control valve directs the excess fluid to the inlet side of the pump again. Maximum pressure in the pump is limited by the pressure relief valve which opens when the oil pressure exceeds the maximum pressure limits.

The slipper type power steering pump discussed in this section is used in Ford cars except the Lincoln Continental and Continental Mark III and is called the Ford-Thompson

power steering pump. It is also used in Chrysler cars, such as the Imperial, and called the 1.2 pump.

Removal and Installation

1. Drain the fluid from the pump reservoir by disconnecting the fluid return hose at the pump. Then, disconnect the pressure hose from the pump.
2. Remove the mounting bolts from the front of the pump. On eight cylinder engines, there is a nut on the rear of the pump that must be removed. After removing all the mounting bolts and nuts from the pump, move the pump inward to loosen the belt tension and remove the belt

from the pulley. Then, remove the pump from the car.
3. To reinstall the pump after maintenance or repair, position the pump on the mounting bracket and loosely install the mounting bolts and nuts. Put the drive belt over the pulley and move the pump outward against the belt until the proper belt tension is obtained. Measure the belt tension with a belt tension gauge for the proper adjustment. Only in cases where a belt tension gauge is not available should the belt deflection method be used. If the belt deflection method is used, be sure to check the belt tension with a belt tension gauge at the earliest time since deflection method is not accurate.
4. Tighten the mounting bolts and nuts to the specified torque limits.
5. Tighten the pressure hose fitting hex nut to the proper torque. Then, connect the pressure hose to the pump and tighten the hose nut to the proper torque.
6. Connect the fluid return hose to the pump and tighten the clamp.
7. Fill the pump reservoir with power steering fluid, and bleed the air bubbles from the system.
8. Check for leaks and recheck the fluid level. If necessary, add fluid to raise the level properly.

Disassembly

1. Drain as much fluid from the pump as possible after removing

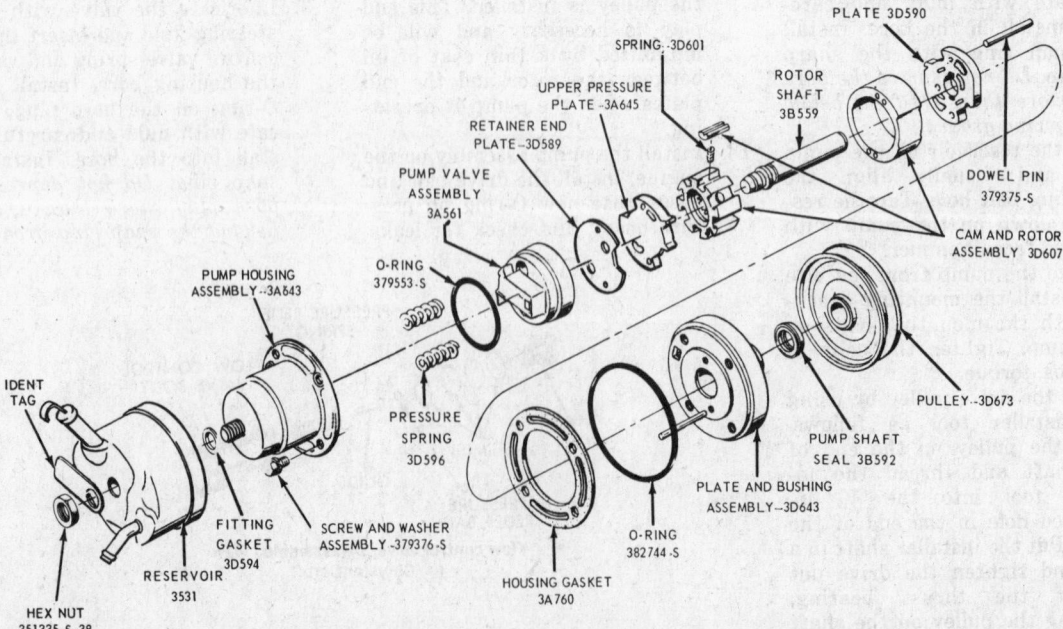

Ford-Thompson power steering pump, disassembled view (© Ford Motor Co.)

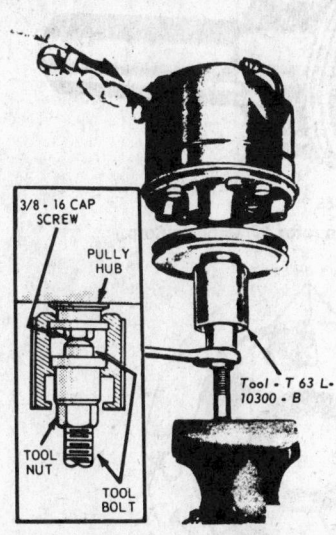

Removing drive pulley (© Ford Motor Co.)

the pump from the car.
2. Install a 3/8-16 in. capscrew in the end of the pump shaft to avoid damaging the shaft end with the pulley remover tool. Install the pulley remover tool on the pulley hub and place the pump and remover tool in a vise as shown. Hold the pump steady and turn the tool nut counterclockwise to draw the pulley off the shaft. *The pulley must be removed without in and out pressure on the pump shaft to avoid damaging the internal thrust washers.*
3. Remove the pump reservoir by installing the pump in a holding fixture with an adapter plate in a vise with the reservoir facing up.
4. Remove the outlet fitting hex nut and any other attaching parts from the reservoir case.

Removing pump reservoir (© Ford Motor Co.)

5. Invert the pump so the reservoir is now facing down. Using a wooden block, remove the reservoir by tapping around the flange until the reservoir is loose. Remove the reservoir

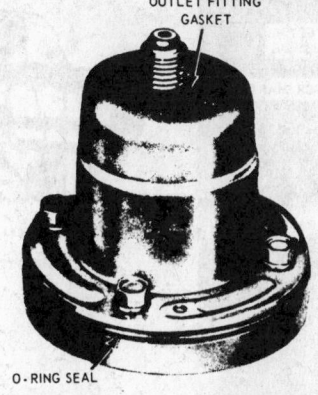

Pump gasket locations (© Ford Motor Co.)

O-ring seal and the outlet fitting gasket from the pump.
6. Again invert the pump assembly in the vise, remove the pump housing holding bolts, and remove the pump housing.
7. Remove the housing cover, the O-ring seal, and the pressure springs from inside the pump housing. Remove the pump cover gasket and discard it.
8. Remove the retainer end plate and upper pressure plate. In some pumps, the end plate and the upper pressure plate are made in one unit.
9. Remove the loose fitting dowel pin. Be careful not to bend the fixed dowel pin which remains in the housing plate assembly.
10. Remove the rotor assembly being careful not to let the slippers and springs fall out of the rotor. It may not be necessary to disassemble the rotor assembly unless the lower pressure plate, housing plate, rotor shaft and/or seal is to be replaced. However, the rotor assembly may be disassembled by removing the slippers and springs from the cam ring.

11. Clean any rust, dirt, burrs, or scoring from the pulley end of the rotor shaft before removing the shaft from the housing plate. The shaft must come out without restrictions to avoid scoring or damaging the bushing. Remove the pump rotor shaft.
12. Remove the lower pressure plate.
13. Remove the rotor shaft seal after first wrapping a piece of 0.005 in. shim stock around the shaft and pushing it into the inside of the seal until it touches the bushing. With a sharp tool, pierce the seal body and pry the seal out. *Do not damage the bushing, housing, or the shaft.* Install a new seal using the tool shown and a soft-faced hammer.
14. If the pump has a flow control valve, disassemble according to instructions given in the section on the roller type power steering pump.

Inspection

1. Wash all metal parts in clean, non-toxic solvent. Blow out all oil passages with compressed air and air dry all cleaned parts.
2. Inspect the drive shaft for excessive wear and the seal area for nicks or scoring. Replace if necessary.
3. Inspect the pressure plates, slippers, rotor, and cam ring for nicks, burrs, or scratches. If any of the parts are damaged enough to cause poor operation or binding of the pump, replace the defective part.
4. Inspect the pump body drive shaft bushing for excessive wear. Replace if necessary.

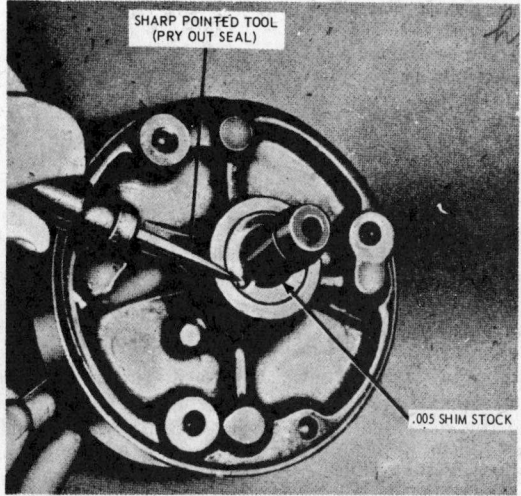

Rotor shaft seal removal (© Ford Motor Co.)

Rotor shaft seal installation (© Ford Motor Co.)

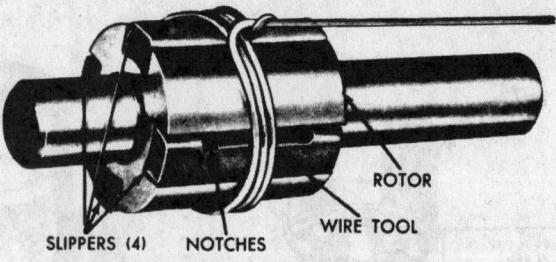

Slippers installed in rotor (© Chrysler Corp.)

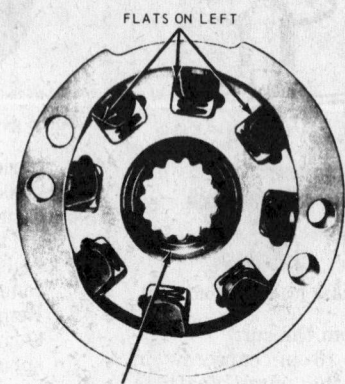

Correct slipper installation,
Ford-Thompson power steering pump
(© Ford Motor Co.)

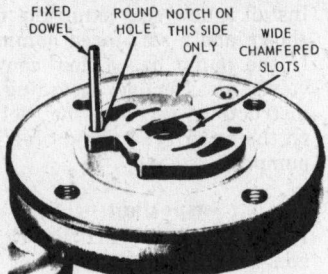

Lower pressure plate installed
(© Ford Motor Co.)

Assembly

1. With the pump assembly positioned on the adapter plate in the holding fixture, install the lower pressure plate on the anchor pin with the chamfered slots at the center hole facing up.

2. Lubricate the rotor shaft with power steering fluid and insert the shaft into the lower pressure plate and housing plate.

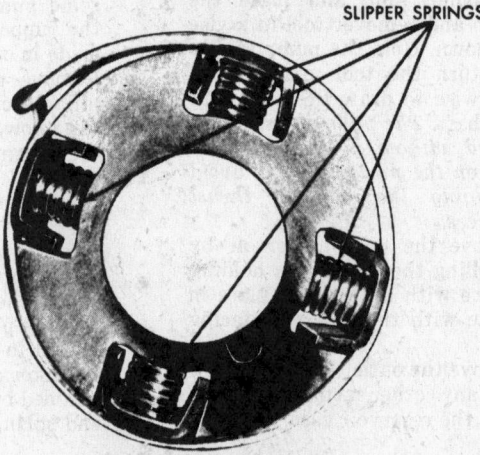

Correct slipper installation,
Chrysler 1.2 power steering pump
(© Chrysler Corp.)

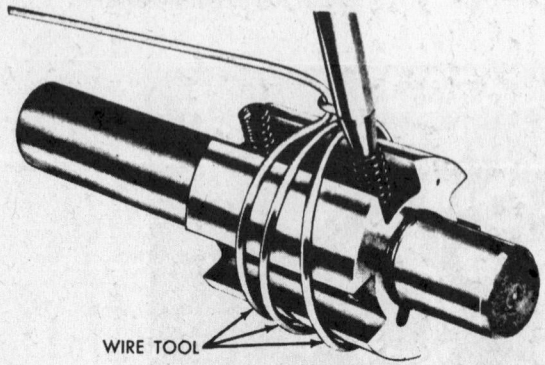

Installing slipper springs (© Chrysler Corp.)

Cam and rotor installation
(© Ford Motor Co.)

3. Assemble the rotor, slippers, and springs by wrapping a piece of wire around the rotor, installing the springs, and sliding a slipper in each groove of the rotor over the springs. Then, insert the assembly into the cam ring. Be sure the flat side of the slippers are toward the left side as shown. (Ford-Thompson power steering pump). The Chrysler power steering pump slippers are installed as shown. Be sure that the springs are installed straight and are not cocked to one side under the slippers.

4. Install the cam ring and rotor assembly on the drive shaft with the fixed dowel passing through the first hole to the left of the cam notch when the arrow on the cam outside diameter is pointing toward the lower pressure plate. If the cam and rotor

assembly does not seat properly, turn the rotor shaft slightly until the spline teeth mesh, allowing the cam and rotor to drop into position.

5. Insert the loose fitting dowel through the cam insert and lower pressure plate into the hole in the housing plate assembly. When both dowels are installed properly, they will be the same height.

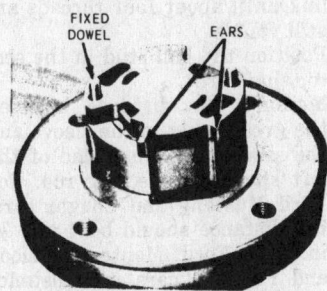

Upper pressure plate installation
(© Ford Motor Co.)

6. Install the upper pressure plate so the tapered notch is facing down against the cam insert. The fixed dowel should pass through the round dowel hole and the loose dowel through the long hole. The slot between the ears on the outside of the pressure plate should match the notch on the cam insert.

7. Install the retainer end plate so the slot on the end plate matches the notches on the upper pressure plate and the cam insert.

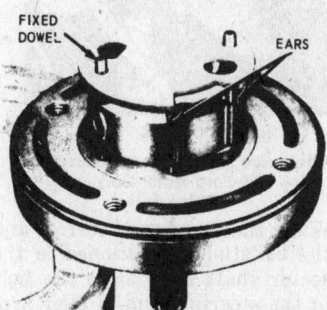

Retainer end plate installation
(© Ford Motor Co.)

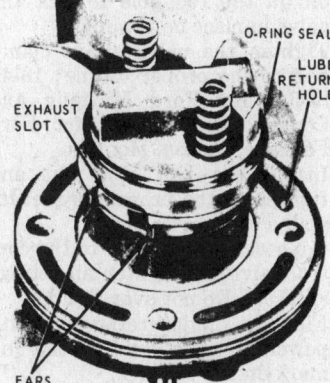

Valve and pressure spring installation
(© Ford Motor Co.)

8. Install the pump valve assembly O-ring seal on the pump valve assembly. *Do not twist the seal.*

9. Place the pump valve assembly on top of the retainer end plate with the large exhaust slot on the pump valve in line with the outside notches of the cam, upper pressure plate, and retainer end plate. All parts must be fully seated. If correctly installed, the relief valve stem will be in line with the lube return hole in the pump housing plate.

10. Put small amounts of vaseline on the pump housing plate to hold the cover gasket in place. Install the cover gasket in place.

11. Insert the pressure plate springs into the pockets in the pump valve assembly.

Tool T69P-3B586-A

Pump housing installation
(© Ford Motor Co.)

12. Install Tool T69P-3B586-A and plug the intake hole in the housing.

13. Lubricate the inside of the housing and the housing cover seal with power steering fluid. Make and install two studs for use as positioning guides, one in the bolt hole nearest the drain hole and the other in the bolt hole on the opposite side of the housing plate.

14. Align the small lube hole in the housing rim and the lube hole in the housing plate. Install the housing, using a steady, even, downward pressure. *Do not jar the pressure spring out of position.* Remove the guide studs and loosely install the housing retaining bolts finger tight. Remove the Tool T69P-3B586-A.

15. Tighten the retaining bolts evenly to 28-32 ft-lbs. until the housing flange contacts the gasket.

Checking pump rotational torque
(© Ford Motor Co.)

16. Install a 3/8-16 hex head screw into the end of the rotor shaft and put a torque wrench on it. Check the amount of torque needed to rotate the rotor shaft. If the torque is more than 15 in-lbs., loosen the retaining bolts slightly and rotate the rotor shaft. Then, retighten the retaining bolts evenly. *Do not use the pump if the shaft torque exceeds 15 in-lbs.*

17. Release the pin in the bench holding fixture and shake the pump assembly back and forth. If there is a rattle, the pressure springs have fallen out of their seats and must be reinstalled.

18. Install the reservoir O-ring seal on the housing plate without twisting it. Lubricate the seal and install the reservoir, aligning the notch in the reservoir flange with the notch in the outside edge of the pump housing plate and bushing assembly. Using only a soft-faced hammer, tap at the rear outer corners of the reservoir. Inspect the assembly to be sure the reservoir is fully seated on the housing plate.

19. Install the identification tag (if one was removed) on the outlet valve fitting. Install the outlet valve fitting nut and tighten to 43-45 ft-lbs. torque.

20. Turn the pump assembly over and install the pulley using the tool used to remove the pulley. Turn the tool nut clockwise to draw the pulley on the shaft until it is flush with the shaft end. *Do not exert inward and outward pressures on the shaft to avoid damaging the internal thrust areas. Remove the tool.*

Section 4
Bendix Linkage—Type Power Steering System

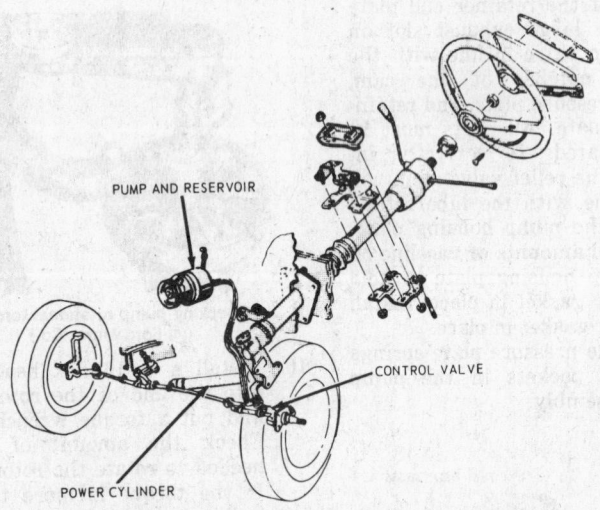

PUMP AND RESERVOIR

CONTROL VALVE

POWER CYLINDER

Bendix linkage-type power steering system (© Ford Motor Co.)

The Bendix linkage-type power steering system is a hydraulically controlled linkage-type steering system composed of an integral pump and fluid reservoir, a control valve, a power cylinder, connecting fluid lines, and the steering linkage. The hydraulic pump, which is driven by a belt turned by the engine, draws fluid from the reservoir and provides fluid pressure through hoses to the control valve and the power cylinder. There is a pressure relief valve to limit the pressures within the steering system to a safe level. After the fluid has passed from the pump to the control valve and the power cylinder, it returns to the reservoir.

The Bendix linkage-type steering system when used in Ford compact cars is called the Ford Non-Integral Power Steering System.

In-Car Adjustments

Control Valve Centering Spring Adjustment

1. Raise the car and remove the spring cap attaching screws and remove the spring cap.

Caution be very careful not to position the hoist adapters of two post hoists under the suspension and/or steering components. Place the hoist adapters under the front suspension lower arms.

2. Tighten the adjusting nut snug (about 90-100 in-lbs.); then, loosen the nut 1/4 turn (90 degrees). Do not tighten the adjusting nut too tight.
3. Place the spring cap on the valve housing. Lubricate and install the attaching screws and washers. Tighten the screws to 72-100 in-lbs. torque.

4. Lower the car and start the engine. Check the steering effort using a spring scale attached to the steering wheel rim for a torque of no more than 12 lbs.

Power Steering Control Valve Removal and Installation

1. Raise the car on a hoist. If a two post hoist is used, be sure to place the hoist adapters under the front suspension steering arms. *Do not allow the hoist adapters to contact the steering linkage.*
2. Disconnect the four fluid line fittings at the control valve and drain the fluid from the lines. Turn the front wheels back and forth to force all the fluid from the system.
3. Loosen the clamping nut and bolt at the right end of the sleeve.
4. Remove the roll pin from the steering arm-to-idler arm rod through the slot in the sleeve.

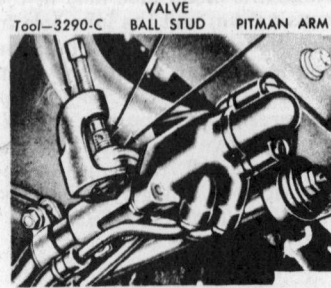

VALVE
Tool—3290-C BALL STUD PITMAN ARM

Removing control valve ball stud
(© Ford Motor Co.)

5. Remove the control valve ball stud nut.
6. Remove the ball stud from the sector shaft arm using Tool 3290-C.

7. After turning the front wheels fully to the left, unthread the control valve from the center link steering arm-to-idler arm rod.
8. To install the control valve, thread the valve on the center link until about four threads are still visible.
9. Position the ball stud in the sector shaft arm.
10. Measure the distance between the grease plug in the sleeve and the stud at the inner end of the left spindle connecting rod. For Ford Mustang and Cougar cars, the distance should be 4 and 7/8 in. For Ford Montego, Falcon and Fairlane cars, the distance should be 5 and 5/8 in. If the distance is not correct, disconnect the ball stud from the sector shaft arm and turn the valve on the center link until the correct distance is obtained.

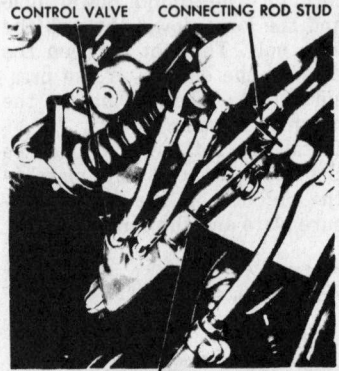

CONTROL VALVE   CONNECTING ROD STUD

LUBRICATION PLUG

Control valve installation measurements
(© Ford Motor Co.)

11. When the distance is correct and the ballstud is positioned in the sector shaft arm, align the hole in the steering arm-to-idler arm rod with the slot near the end of the valve sleeve. Install the roll pin in the rod hole to lock the valve in place on the rod.
12. Tighten the valve sleeve clamp bolt to the proper torque: 18-42 in-lbs. for Ford Mustang and Cougar; 42-60 in-lbs. for Ford Fairlane, Falcon, Montego.
13. Install the ball stud nut and tighten to the proper torque. Install a new cotter pin.
14. Connect all fluid lines to the control valve and tighten all fittings securely. Do not over-tighten.
15. Fill the fluid reservoir with power steering fluid to the full mark on the dipstick.
16. Start the engine and run it for a few minutes to warm the fluid in

the power steering system. Turn the steering wheel back and forth to the stops and check the system for leaks.

17. Increase the engine idle speed to about 1000 rpm, turn the steering wheel back and forth several times, and stop the engine. Check the control valve and hose connections for leaks.

18. Recheck the fluid level and add fluid if necessary.

19. Start the engine again, and check the position of the steering wheel when the front wheels are straight ahead. *Do not make any adjustments until toe-in is checked.*

20. With engine running, check front wheel toe-in (see Front Wheel Alignment section).

21. Check steering wheel turning effort which should be equal in both directions.

Power Steering Power Cylinder Removal and Installation

1. Disconnect the two fluid lines from the power cylinder and drain the fluid.

4. Remove the insulator sleeve and washer from the end of the power cylinder.

5. Inspect the tube fittings and seats in the power cylinder for nicks, burrs, or other damage. Replace the seats or tubes if damaged.

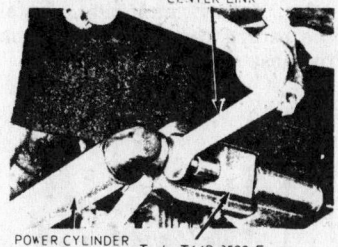

Disconnecting power cylinder stud
(© Ford Motor Co.)

6. Install the washer, sleeve and the insulator on the end of the power cylinder rod.

7. While extending the rod as far as possible, insert the rod in the bracket on the frame and then, compress the rod so the stud may be inserted in the center link. Secure the stud with the castellated nut and a new cotter pin.

warm the fluid. Turn the steering wheel back and forth to the stops to fill the system. Stop the engine.

11. Recheck the fluid level and add fluid if necessary. Check for fluid leaks.

12. Start the engine again, turn the steering wheel back and forth, and check for leaks while the engine is running.

Control Valve Disassembly and Assembly

1. Clean the outside of the control valve of dirt and fluid.

2. Remove the centering spring cap from the valve housing. The control valve should be put in a soft-faced bench vise during disassembly. Clamp the control valve around the sleeve flange only to avoid damaging the valve housing, spool, or sleeve.

3. Remove the nut from the end of the valve spool bolt. Remove the washers, spacer, centering spring, adapter, and the bushing from the bolt and valve housing.

4. Remove the two bolts holding the valve housing and the sleeve together. Separate the valve housing and the sleeve.

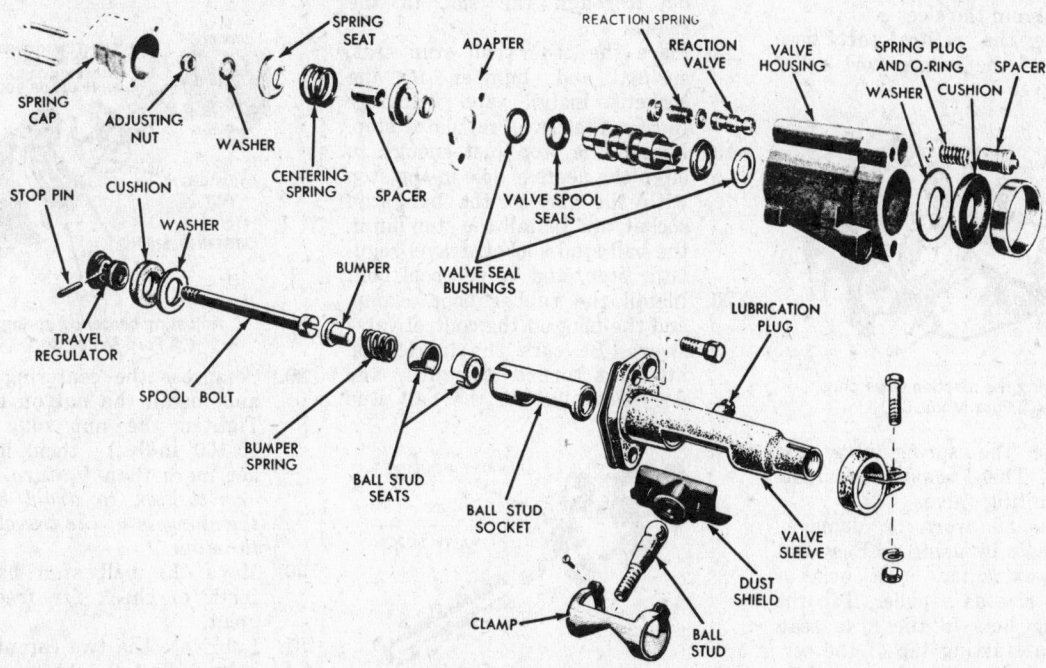

Control valve, disassembled view (© Ford Motor Co.)

2. Remove the pal nut, attaching nut, washer and the insulator from the end of the power cylinder rod. Remove the cotter pin and castellated nut holding the power cylinder stud to the center link.

3. Disconnect the power cylinder stud from the center link as shown.

8. Install the insulator, washer, nut, and a pal nut on the power cylinder rod.

9. Connect the two fluid lines to their proper ports on the power cylinder.

10. Fill the reservoir with power steering fluid to the full mark on the dipstick. Start the engine and run for a few minutes to

5. Remove the plug from the sleeve. Push the valve spool out of the centering spring end of the valve housing, and remove the seal from the spool.

6. Remove the spacer, bushing, and valve housing.

7. Drive the stop pin out of the travel regulator stop with a punch and hammer. *Pull the*

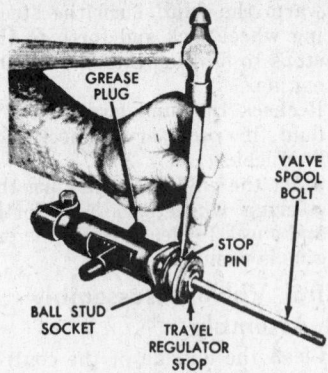

Removing the stop pin
(© Ford Motor Co.)

head of the valve spool bolt tightly against the travel regulator stop before driving the pin out of the stop.

8. Turn the travel regulator stop counterclockwise in the valve sleeve to remove the stop from the sleeve.
9. Remove the valve spool bolt, spacer, and rubber washer from the travel regulator stop.
10. Remove the rubber boot and clamp from the valve sleeve. Slide the bumper, spring, and ball stud seat out of the valve sleeve, and remove the ball stud socket from the sleeve.
11. Remove the return port hose seat and the return port relief valve.

Removing the reaction valve plug
(© Ford Motor Co.)

12. Remove the spring plug and O-ring. Then, remove the reaction limiting valve.
13. Replace all worn or damaged hose seats by using an Easy-Out screw extractor or a bolt of proper size as a puller. Tap the existing hole in the hose seat, using a starting tap of the correct size. *Remove all metal chips from the hose seat after tapping.* Place a nut and washer on a bolt of the same size as the tapped hole. The washer must be large enough to cover the hose seat port. Insert the bolt in the tapped hole and remove the hose seat by turning the nut clockwise and drawing the bolt out. Install a new hose seal in the port, and thread a bolt of the correct size in the port. Tighten

the bolt enough to bottom the seal in the port.
14. Coat all parts of the control valve assembly with power steering fluid. Seals should be coated with lubricant before installation.
15. Install the reaction limiting valve, spring, and plug. Install the return port relief valve and the hose seat.

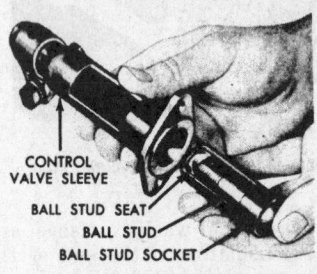

Installing ball socket, seal, and bracket
(© Ford Motor Co.)

16. Insert one of the ball stud seats (flat end first) into the ball stud socket, and insert the threaded end of the ball stud into the socket.
17. Place the socket in the control valve sleeve so that the threaded end of the ball stud can be pulled out through the slot in the sleeve.
18. Place the other ball stud seat, spring, and bumper in the socket. Install and securely tighten the travel regulator stop.
19. Loosen the stop just enough to align the nearest hole in the stop with the slot in the ball stud socket, and install the stop pin in the ball stud socket, travel regulator stop, and valve spool bolt.
20. Install the rubber boot, clamp, and the plug on the control valve sleeve. Be sure the lubrication fitting is turned on tightly and does not bind on the ball stud socket.

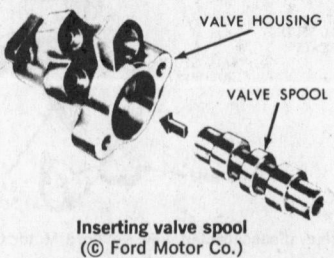

Inserting valve spool
(© Ford Motor Co.)

21. Insert the valve spool in the valve housing, rotating it while installing it.
22. Move the spool toward the centering spring end of the housing, and place the small seal bushing, and spacer in the sleeve end of the housing.
23. Press the valve spool against the inner lip of the seal and, at the

same time, guide the lip of the seal over the spool with a small screwdriver. *Do not nick or scratch the seal or the spool during installation.*
24. Place the sleeve end of the housing on a flat surface so that the seal, bushing, and spacer are at the bottom end and push down the valve spool until it stops.
25. Carefully install the spool seal and bushing in the centering spring end of the housing. Press the seal against the end of the spool, guiding the seal over the spool with a small screwdriver. *Do not nick or scratch the seal or the spool during installation.*
26. Pick up the housing, and slide the spool back and forth in the housing to check for free movement.
27. Place the valve sleeve on the housing so that the ball stud is on the same side of the housing as the ports for the two power cylinder lines. Install the two bolts in the sleeve, and torque them to the proper torque.
28. Place the adapter on the centering spring end of the housing, and install the bushing, washers, spacers, and centering spring on the valve spool bolt.

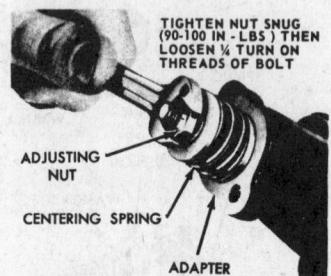

Adjusting centering spring
(© Ford Motor Co.)

29. Compress the centering spring, and install the nut on the bolt. Tighten the nut snug (about 90-100 in-lbs.); then, loosen it not more than ¼ turn. *Do not over-tighten to avoid breaking the stop pin at the travel regulator stop.*
30. Move the ball stud back and forth to check for free movement.
31. Lubricate the two cap attaching bolts. Install the centering spring cap on the valve housing, and tighten the two cap bolts to the proper torque.
32. Install the nut on the ball stud so that the valve can be put in a vise. Then, push forward on the cap end of the valve to check the valve spool for free movement.
33. Turn the valve around in the vise, and push forward on the sleeve end to check for free movement.

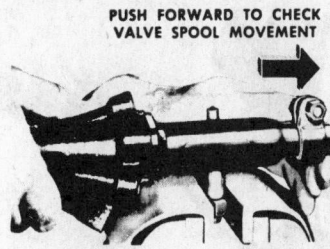

PUSH FORWARD TO CHECK
VALVE SPOOL MOVEMENT

Inspecting valve spool movement
(© Ford Motor Co.)

Power Cylinder Seal Removal and Installation

2. Pull the piston rod out all the way to remove the scraper, bushing, and seals. If the seals cannot be removed in this manner, remove them by carefully prying them out of the cylinder with a sharp pick. *Do not damage the shaft or seal seat.*
3. Coat the new seals with power steering fluid and place the parts on the piston rod which should be coated with grease or lubricant.

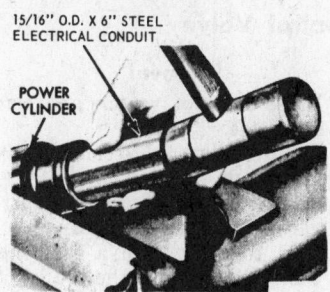

15/16" O.D. X 6" STEEL
ELECTRICAL CONDUIT

POWER
CYLINDER

Installing power cylinder seals
(© Ford Motor Co.)

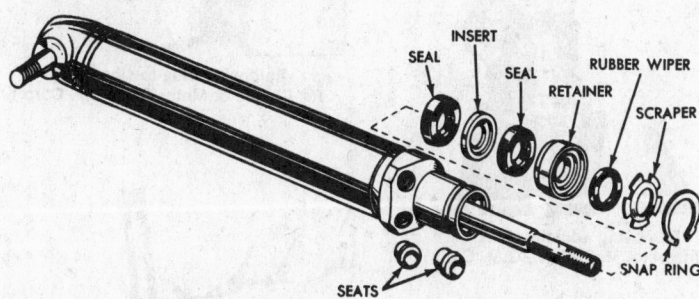

INSERT
SEAL
SEAL
RETAINER
RUBBER WIPER
SCRAPER
SNAP RING
SEATS

Power cylinder, disassembled view (© Ford Motor Co.)

1. Clamp the power cylinder in a vise, and remove the snap ring from the end of the cylinder. *Do not distort or crack the cylinder in the vise.*

4. Push the rod in all the way, and install the parts in the cylinder with a deep socket slightly smaller than the cylinder opening.

Power Steering Pump Removal and Replacement

To remove or replace the power steering pump, see the section on the slipper type pump for instructions.

Power Steering Gear Removal and Replacement

To remove and replace the steering gear, see the section on Manual Steering for instructions.

Section 5
Overhaul of Saginaw Linkage-Type

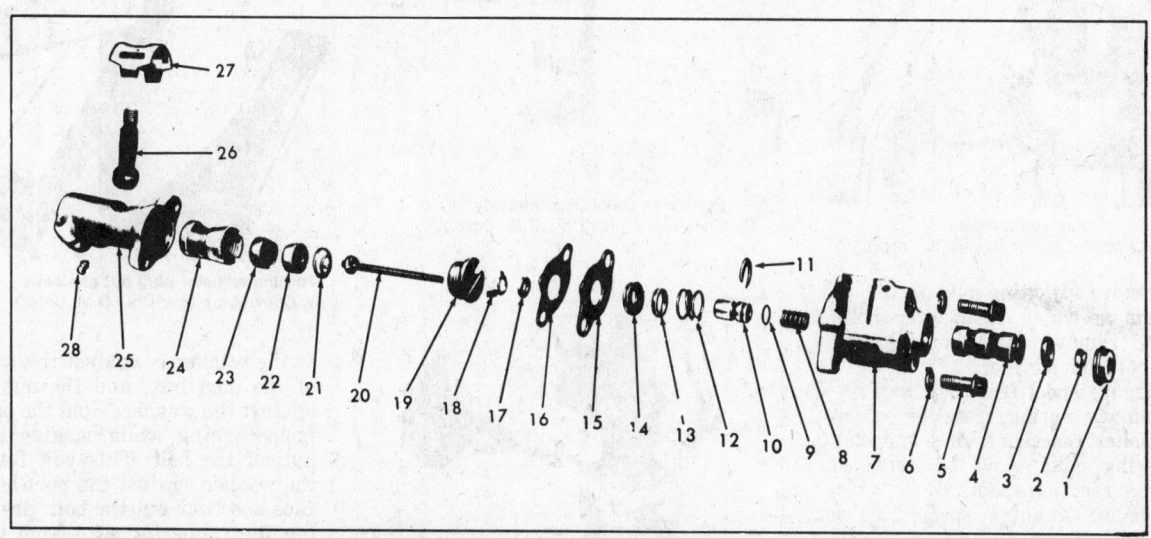

Control valve, disassembled view (© Chevrolet Motor Div., G.M. Corp.)

1 Dust cover	9 "O" ring seal	15 Annulus spacer	20 Valve shaft
2 Adjusting nut	10 Valve reaction	16 Gasket	21 Ball seat spring
3 Vee block seal	spool	17 Valve shaft	22 Ball seat
4 Valve spool	11 Spring thrust	washer	23 Ball seat
5 Valve mounting	washer	18 Plug to sleeve	24 Sleeve bearing
bolts	12 Valve spring	key	25 Adapter housing
6 Lock washer	13 Spring retainer	19 Ball adjuster	26 Ball stud
7 Valve housing	14 Annulus seal	nut	27 Dust shield
8 Valve adjustment			28 Lubrication fitting
spring			

Control Valve

Removal

1. Raise front of vehicle and place on stands.
2. Remove relay rod to control valve clamp bolt.
3. Disconnect two pump to control valve hose connections and allow fluid to drain into a container, then disconnect the valve to power cylinder hoses.
4. Remove ball stud to pitman arm retaining nut and disconnect control valve from pitman arm.
5. Turn steering gear so that pitman arm is away from valve, to allow working room, and unscrew control valve from relay rod.
6. Remove control valve from vehicle.

Disassembly

1. Place valve assembly in vise with dust cap end up and remove dust cap.

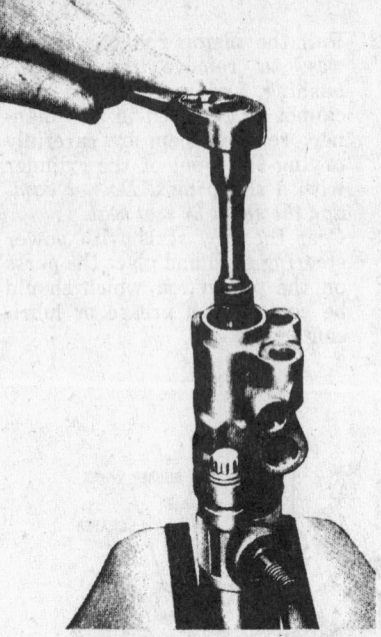

Removing adjusting nut
(© Chevrolet Motor Div., G.M. Corp.)

Removing plug-to-sleeve key
(© Chevrolet Motor Div., G.M. Corp.)

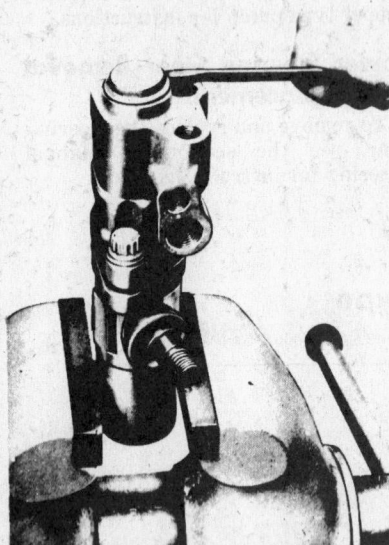

Dust cap removal
(© Chevrolet Motor Div., G.M. Corp.)

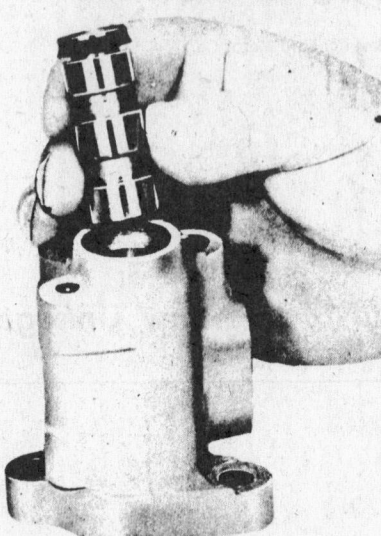

Removing spool from housing
(© Chevrolet Motor Div., G.M. Corp.)

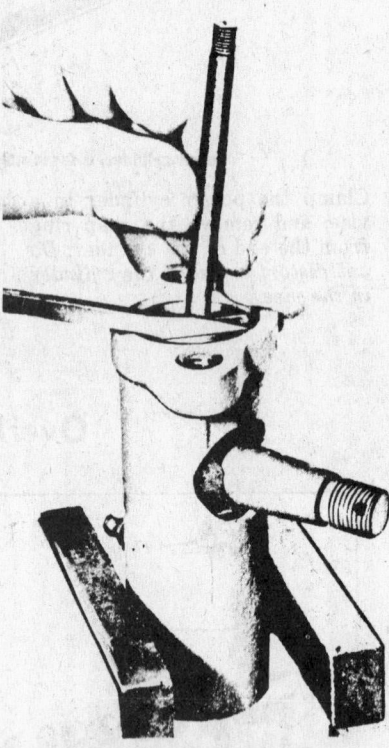

Turning adjuster plug out of sleeve
(© Chevrolet Motor Div., G.M. Corp.)

2. Remove adjusting nut.
3. Remove valve to adapter bolts and remove valve housing and spool from adapter.
4. Remove spool from housing.
5. Remove spring, reaction spool, washer, reaction spring, and seal. O-ring may now be removed from reaction spool.
6. Remove annulus spacer, valve shaft washer and plug-to-sleeve key.
7. Carefully turn adjuster plug out of sleeve. Use care not to nick the top surface.
8. If necessary to replace a connector seat, tap threads in center hole using a 5/16-18 tap. Thread a bolt with a nut and a flat washer into the tapped hole

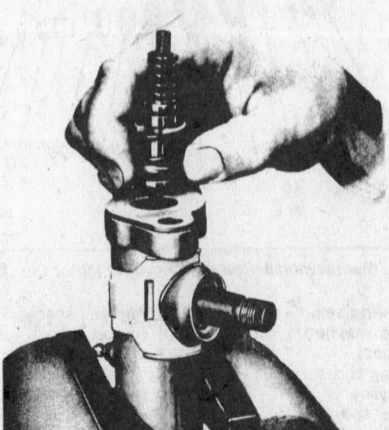

Removing valve parts from shaft
(© Chevrolet Motor Div., G.M. Corp.)

so the washer is against the face of the port boss and the nut is against the washer. Hold the bolt from turning while backing the nut off the bolt. This will force the washer against the port boss face and back out the bolt, drawing the connector seat from the top cover housing. Discard the old connector seat and clean the housing out thoroughly to remove any metal tapping chips. Drive a new connector seat against the housing seat being careful not to damage either the connector seat or the housing seat.

Removing connector seat
(© Chevrolet Motor Div., G.M. Corp.)

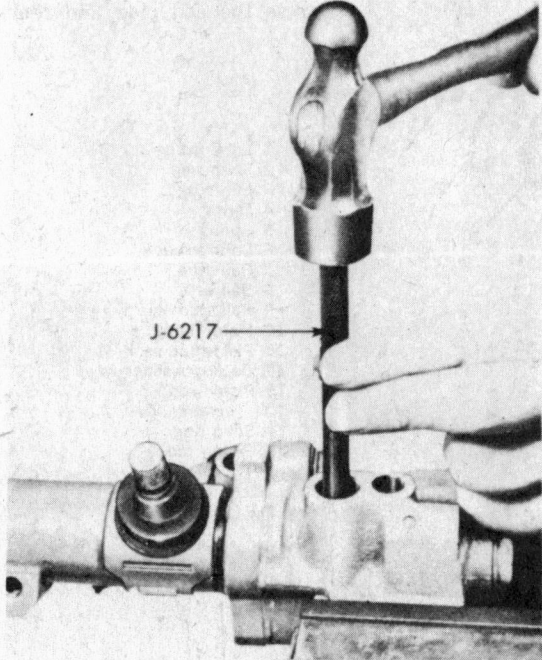

J-6217 →

Installing connector seat
(© Chevrolet Motor Div., G.M. Corp.)

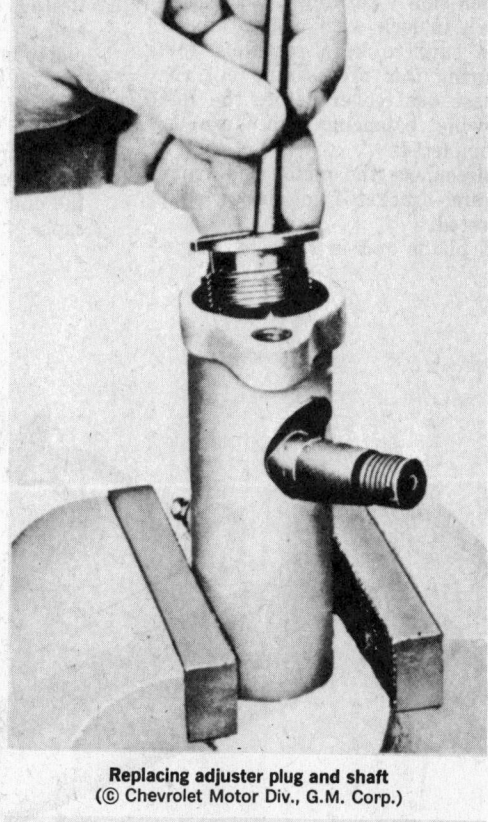

Replacing adjuster plug and shaft
(© Chevrolet Motor Div., G.M. Corp.)

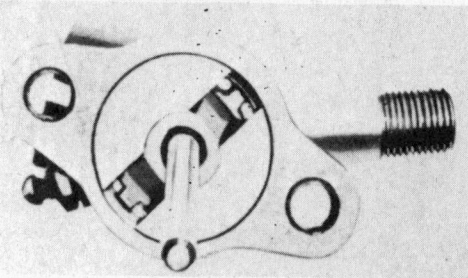

Proper key installation
(© Chevrolet Motor Div., G.M. Corp.)

9. Remove adapter from vise and turn over to allow spring and one of the two ball seats to drop out.
10. Remove ball stud with other ball seat and allow sleeve to fall free.

Inspection

1. Wash all parts in non-toxic solvent and blow dry with air.
2. Inspect all parts for scratches, burrs, distortion, excessive wear and replace all such worn or damaged parts.
3. Replace all seals and gaskets.

NOTE: Corvette valves incorporate a 55 pound centering spring which might be interchanged with Chevrolet and Chevelle and Chevy II springs. They should not be interchanged as the other springs are only 30 pounds. Corvette valves are stamped with an X on the dust cover.

Assembly

1. Replace sleeve and ball seat in adapter, then the ball stud and then the other ball seat and spring. (small end down)

2. Place adapter in vise. Put the shaft through the seat in the adjuster plug and screw adjuster plug into sleeve.
3. Turn plug in until tight, then back off until slot lines up with notches in sleeve.
4. Insert key. Be sure small tangs on end of key fit into notches in sleeve.
5. Install valve shaft washer, annulus spacer, and reaction seal (lip up), spring retainer, reaction spring and spool, then washer and adjustment spring. Install O-ring seal on reaction spool before installing spool on shaft. Install washer with chamfer up.
6. Install seal on valve spool with lip down. Then install spool, being careful not to jam spool in

housing.
7. Install housing with spool onto adapter. The side ports should be on the same side as the ball stud. Bolt the housing to the adapter.
8. Depress the valve spool and turn the locknut into the shaft about four turns. Use a clean wrench or socket.

NOTE: always use a new nut.

Installation and Balancing

1. Install the control valve on the relay rod so that control valve bottoms, then back off enough (if necessary) to install the clamp bolt. Do not back off more than two turns. There will be approximately 1/16-1/8 in. gap.
2. Tighten control valve clamping-bolt and install ball stud to pit-

man arm.

3. Reconnect the four hoses to the valve.

4. Fill system with type A fluid and bleed air by running engine, then slowly turning wheels from lock to lock with engine idling. Be sure to keep reservoir full during this process. Do not replace dust cover before the following balancing procedure is completed.

5. Disconnect the piston rod from frame bracket if not already separated.

6. If piston rod is retracted, turn

7. With valve properly balanced it should be possible to move the rod in and out manually.

8. Shut off engine and connect piston rod to frame bracket.

9. Restart engine with front wheels still off ground. If the wheels do not turn in either direction from center, the valve has been properly balanced. Correct the condition by rebalancing the valve if necessary.

10. After proper adjustment, grease the end of valve and install dust cap.

2. Check hose connection seats for damage and replace if necessary.

3. For service other than seat or seal replacement, it is necessary to replace the power cylinder.

4. The ball stud may be replaced by removing snap-ring.

Disassembly and Reassembly

1. To remove piston rod seal, remove snap-ring and pull out on rod. Remove back-up washer, piston rod scraper, and piston rod seal from rod.

2. To remove the ball stud, depress the end plug and remove

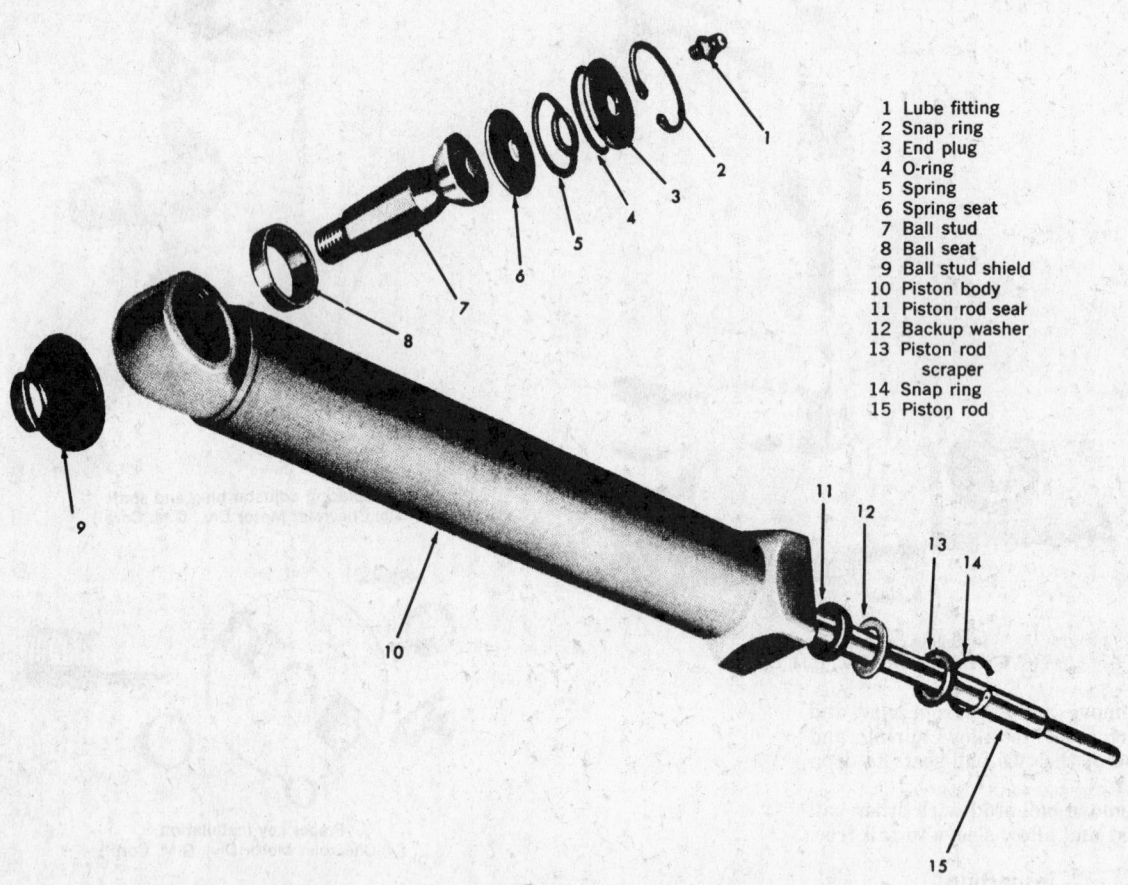

1 Lube fitting
2 Snap ring
3 End plug
4 O-ring
5 Spring
6 Spring seat
7 Ball stud
8 Ball seat
9 Ball stud shield
10 Piston body
11 Piston rod seal
12 Backup washer
13 Piston rod
   scraper
14 Snap ring
15 Piston rod

Power cylinder, disassembled view (© Chevrolet Motor Div., G.M. Corp.)

adjusting nut clockwise until rod begins to move out. Then, turn nut counterclockwise until rod just begins to move in. Now, turn the nut clockwise just exactly half the rotation needed to change the direction of piston rod movement. If piston rod is extended before starting, reverse the above to get the mid-point in piston movement.

Caution Do not turn the nut back and forth more than is absolutely necessary to balance the valve.

Power Cylinder

Removal

1. Remove the two hoses which are connected to the cylinder and drain fluid into a container.

2. Remove power cylinder from frame bracket.

3. Remove cotter pin and nut and pull stud out of relay rod.

4. Remove cylinder from vehicle.

Inspection

1. Check seals for leaks around cylinder rod. If leaks are found, replace seals.

the snap ring. Push on the end of the ball stud and the end plug, spring, spring seat and ball stud and seal may be removed. If the ball seat is to be replaced, it must be pressed out.

3. Reverse disassembly procedure. Be sure snap-ring is properly seated.

Installation

1. Install power cylinder on vehicle in reverse of removal procedure.

2. Reconnect the hydraulic lines, fill system and bleed out air as described in the installation and

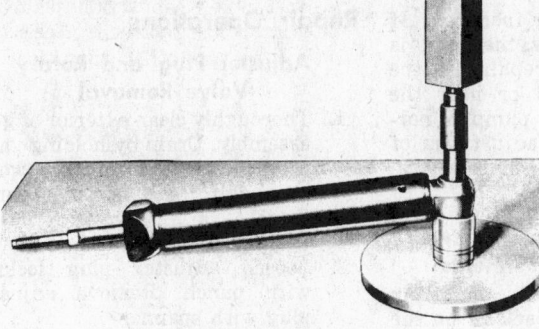

Removing ball cup
(© Chevrolet Motor Div., G.M. Corp.)

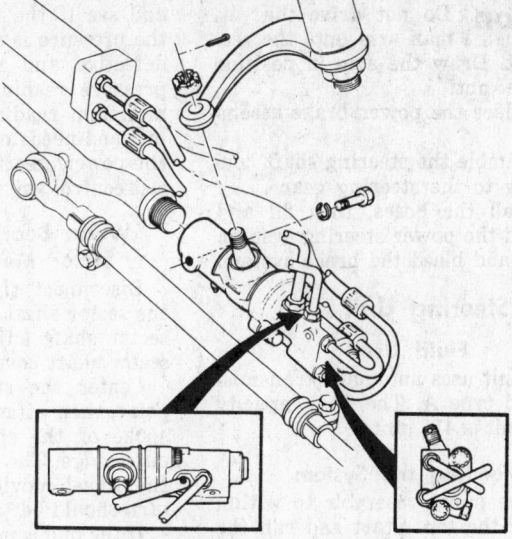

Power steering hose installation
(© Chevrolet Motor Div., G.M. Corp.)

Installing ball cup
(© Chevrolet Motor Div., G.M. Corp.)

balancing section of control valve servicing.

Power Steering Hoses

Carefully inspect the hoses. When installing hoses be sure to install in such a position as to avoid all chafing or other abuse when making sharp turns.

Section 6
Saginaw Rotary Type Power Steering

The rotary type power steering gear is designed with all components in one housing.

The power cylinder is an integral part of the gear housing. A double-acting type piston allows oil pressure to be applied to either side of the piston. The one-piece piston and power rack is meshed to the sector shaft.

The hydraulic control valve is composed of a sleeve and valve spool. The spool is held in the neutral position by the torsion bar and spool actuator. Twisting of the torsion bar moves the valve spool, allowing oil pressure to be directed to either side of the power piston, depending upon the directional rotation of the steering wheel, to give power assist.

On many cars of the General Motors Corporation, a modified version of the rotary valve power steering system provides variable ratio steering to assist the driver to steer the car easier and safer. The steering gear ratio will vary from a high ratio of about 16:1 while steering straight ahead to a lower gear ratio of about 12.4:1 while making a full turn to either side. See the specifications for the exact gear ratios.

Roller Pump Removal and Installation

Removal

Remove the reservoir cover and use a suction gun to empty the reservoir. Disconnect the hoses from the pump and tie them in a raised position to prevent oil drainage. Loosen the pump adjusting screw and remove the pump belt, then take out the retaining bolts and remove the pump and reservoir.
NOTE: on cars equipped with air conditioning the pump assembly is removed from underneath the vehicle.

Installation

Position the pump assembly and install the retaining bolts. Be sure there is clearance between the pump bracket and the engine front support bracket. Install the hoses and place the pump belt on the pulley. Adjust the belt to ½ in. deflection, then tighten the adjusting screw.

Connect the hoses to the pump assembly.

Fill the reservoir to within ½ in. of the top with automatic transmission fluid type A.

Start the engine and rotate the steering wheel several times to the right and left to expel air from the system, then recheck the oil level and install the reservoir cover.

Rotary Steering Unit Removal and Installation

Removal
1. Remove the lines from the steering gear and tie them in a raised position to prevent oil drainage.
2. Remove the power brake assembly.
3. Using a puller, remove the pitman arm from the sector shaft.
4. Take out the nuts and bolts that hold the steering shaft coupling to the steering gear.
5. Remove the bolts that secure the gear assembly to the frame side rail and remove the gear assembly from the vehicle.

Installation
1. Position the gear assembly against the frame side rail and install the retaining bolts.
2. Install the pitman arm, aligning the large spline on the sector shaft with the large groove in the arm.

Caution Do not drive the pitman arm onto the sector shaft. Draw the arm in position with the nut.

3. Replace the power brake assembly.
4. Assemble the steering shaft coupling to the steering gear.
5. Install the hoses, then fill and bleed the power steering system.
6. Fill and bleed the brake system.

Power Steering Unit

Fluid Used

This unit uses automatic transmission fluid type A. The fluid capacity for the unit is 4½ pints.

Bleeding the System

Fill the pump reservoir to within ½ in. of the top. Start and run the engine to attain normal operating temperatures. Now, turn the steering wheel through its entire travel three or four times to expel air from the system, then recheck the fluid level.

Checking Steering Effort

Run the engine to attain normal operating temperatures. With the wheels on a dry floor, hook a pull scale to the spoke of the steering wheel at the outer edge. The effort required to turn the steering wheel should be 3½-5 lbs. If the pull is not within these limits, check the hydraulic pressure.

Pressure Test

To check the hydraulic pressure, disconnect the pressure hose from the gear. Now connect the pressure gauge between the pressure hose from the pump and the steering gear housing. Run the engine to attain normal operating temperatures, then turn the wheel to a full right and a full left turn to the wheel stops.

Hold the wheel in this position only long enough to obtain an accurate reading.

The pressure gauge reading should be within the limits specified. If the pressure reading is less than the minimum pressure needed for proper operation, close the valve at the gauge and see if the reading increases. If the pressure is still low, the pump is defective and needs repair. If the pressure reading is at or near the minimum reading, the pump is normal and needs only an adjustment of the power steering gear or power assist control valve.

Worm Bearing Preload and Sector Mesh Adjustments

Disconnect the pitman arm from the sector shaft, then back off on the sector shaft adjusting screw on the sector shaft cover.

Center the steering on the high point, then attach a pull scale to the spoke of the steering wheel at the outer edge. The pull required to keep the wheel moving for one complete turn should be ½-⅔ lbs.

If the pull is not within these limits, loosen the thrust bearing locknut and tighten or back off on the valve sleeve adjuster locknut to bring the preload within limits. Tighten the thrust bearing locknut and recheck the preload.

Slowly rotate the steering wheel several times, then center the steering on the high point. Now, turn the sector shaft adjusting screw until a steering wheel pull of 1-1½ lbs. is required to move the worm through the center point. Tighten the sector shaft adjusting screw locknut and recheck the sector mesh adjustment.

Install the pitman arm and draw the arm in position with the nut.

Repair Operations

Adjuster Plug and Rotary Valve Removal

1. Thoroughly clean exterior of gear assembly. Drain by holding valve ports down and rotating worm back and forth through entire travel.
2. Place gear in vise.
3. Loosen adjuster plug locknut with punch. Remove adjuster plug with spanner.
4. Remove rotary valve assembly by grasping stub shaft and pulling it out.

Adjuster Plug Disassembly

1. Remove upper thrust bearing retainer with screwdriver. Be careful not to damage bearing bore. Discard retainer. Remove spacer, upper bearing and races.
2. Remove and discard adjuster plug O-ring.
3. Remove stub shaft seal retaining ring (Truarc pliers will help) and remove and discard dust seal.
4. Remove stub shaft seal by prying out with screwdriver and discard.
5. Examine needle bearing and, if required, remove same by pressing from thrust bearing end.
6. Inspect thrust bearing spacer, bearing rollers and races.
7. Reassemble in reverse of above.

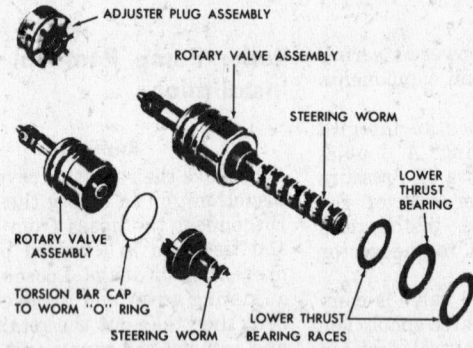

Adjuster plug and O-ring removal
(© Pontiac Div., G.M. Corp.)

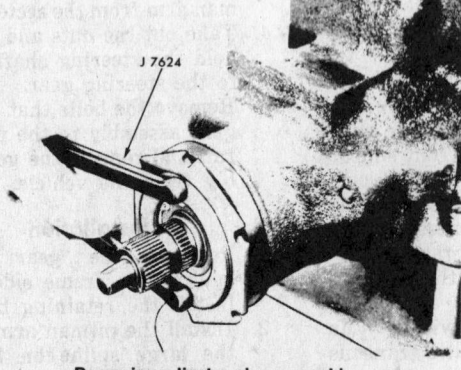

Removing adjuster plug assembly
(© Pontiac Div., G.M. Corp.)

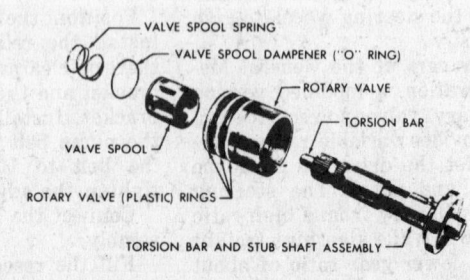

Rotary valve, disassembled view
(© Pontiac Div., G.M. Corp.)

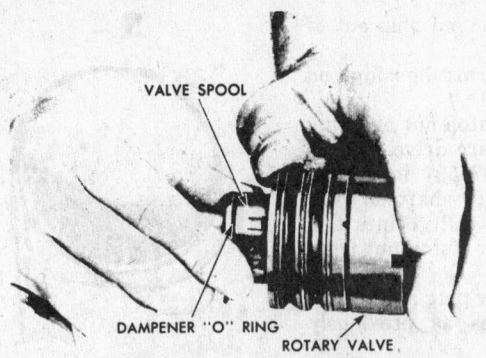

Removing valve spool from rotary valve
(© Pontiac Div., G.M. Corp.)

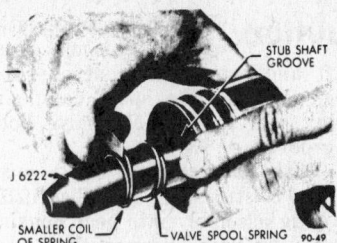

Installing valve spool spring
(© Buick Motor Div., G.M. Corp.)

Rotary Valve Disassembly

Repairs are seldom needed. Do not disassemble unless absolutely necessary. If the O-ring seal on valve spool dampener needs replacement, perform this portion of operation only.

1. Remove cap-to-worm O-ring seal and discard.
2. Remove valve spool spring by prying on small coil with small screwdriver to work spring onto bearing surface of stub shaft. Slide spring off shaft. Be careful not to damage shaft surface.
3. Remove valve spool by holding the valve assembly in one hand with the stub shaft pointing down. Insert the end of pencil or wood rod through opening in valve body cap and push spool until it is out far enough to be removed. In this procedure, rotate to prevent jamming. If spool becomes jammed it may be necessary to remove stub shaft, torsion bar and cap assembly.

Rotary Valve Reassembly

Caution All parts must be free and clear of dirt, chips, etc., before assembly and must be protected after assembly.

1. Lubricate three new back-up O-

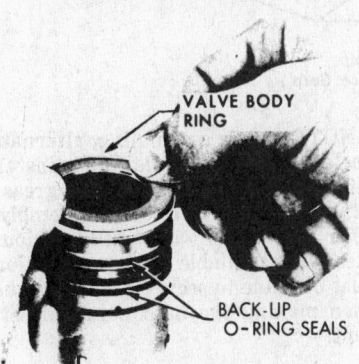

Installing valve body rings
(© Buick Motor Div., G.M. Corp.)

ring seals with automatic transmission oil and reassemble in the ring grooves of valve body. As-

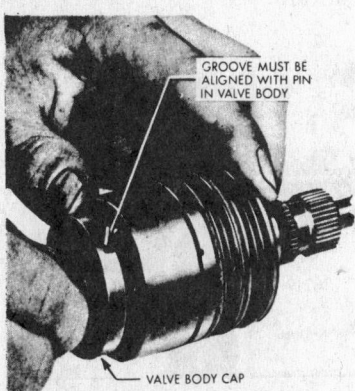

Assembling stub shaft, torsion bar, and cap assembly
(© Buick Motor Div., G.M. Corp.)

semble three new valve body rings in the grooves over the O-ring seals by carefully slipping over the valve body.

NOTE: if the valve body rings seem loose or twisted in the grooves, the heat of the oil during operation will cause them to straighten.

2. Lubricate a new dampener O-ring with automatic transmission oil and install in valve spool groove.
3. Assemble stub shaft torsion bar and cap assembly in the valve body, aligning the groove in the valve cap with the pin in the valve body. Tap lightly with soft hammer until cap is against valve body shoulder. Valve body pin must be in the cap groove. Hold parts together during the remainder of assembly.
4. Lubricate spool. With notch in spool toward valve body, slide the spool over the stub shaft. Align the notch on the spool with the spool drive pin on stub shaft and carefully engage spool in valve body bore. Push spool evenly and with slight rotating motion until spool reaches drive pin. Rotate spool slowly, with some pressure, until notch engages pin. Be sure dampener O-ring seal is evenly distributed in the spool groove.

Caution Use extreme care because spool to valve body clearance is very small. Damage is easily caused.

5. With seal protector (J-6222) over stub shaft, slide valve spool spring over stub shaft, with small diameter of spring going over shaft last. Work spring onto shaft until small coil is located in stubshaft groove.
6. Lubricate a new cap to O-ring seal and install in valve body.

Adjuster Plug and Rotary Valve Installation

1. Align narrow pin slot on valve body with valve body drive pin on the worm. Insert the valve assembly into gear housing by pressing against valve body with finger tips. Do not press on stub shaft or torsion bar. The return hole in the gear housing should be fully visible when properly assembled.

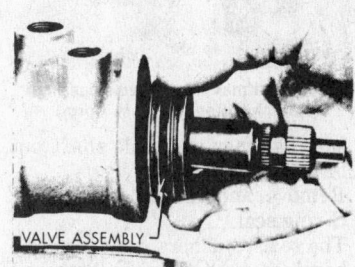

Inserting valve assembly in housing
(© Buick Motor Div., G.M. Corp.)

Valve assembly installed in housing
(© Buick Motor Div., G.M. Corp.)

Caution Do not press on stub shaft as this may cause shaft and cap to pull out of valve body, allowing the spool dampener O-ring seal to slip into valve body oil grooves.

2. With protector (J-6222) over end of stub shaft, install adjuster plug assembly into gear housing snugly with spanner, then back plug off approximately one-eighth turn. Install plug locknut but do not tighten. Adjust preload as described in the adjustment section.
3. After adjustment, tighten locknut.

Pitman Shaft Removal and Installation

1. Completely drain the gear assembly and thoroughly clean the outside.
2. Place gear in vise.
3. Rotate stub shaft until pitman shaft gear is in center position. Remove side cover retaining bolts.
4. Tap end of pitman shaft with

Removing pitman shaft needle bearings
(© Buick Motor Div., G.M. Corp.)

soft hammer and slide shaft out of housing.
5. Remove and discard side cover O-ring seal.
6. The seals, washers, retainers and bearings may now be removed and examined.
7. Examine all parts for wear or damage and replace as required.
8. Install in reverse of above. Make proper adjustment as described in adjustment section.

Rack-Piston Nut and Worm Assembly Removal

1. Completely drain the gear assembly and thoroughly clean the outside.
2. Remove pitman shaft assembly, previously described.
3. Rotate housing end plug retaining ring so that one end of ring is over hole in gear housing. Spring one end of ring so screwdriver can be inserted to lift out ring.
4. Rotate stub shaft to full left turn

position to force end plug out of housing.
5. Remove and discard housing end plug O-ring seal.
6. Remove rack-piston nut end plug with ½ in. square drive.
7. Insert Tool (J-7539) in end of worm. Turn stub shaft so that rack-piston nut will go into tool and remove rack-piston nut from gear housing.
8. Remove adjuster plug and rotary valve assemblies as previously described.
9. Remove worm and lower thrust bearing and races.
10. Remove cap to O-ring seal and discard.

Installing pitman shaft seals
(© Buick Motor Div., G.M. Corp.)

Rack-Piston Nut and Worm Disassembly and Reassembly

1. Remove and discard piston ring and back-up O-ring on rack-piston nut.
2. Remove ball guide clamp and return guide.
3. Place nut on clean cloth and remove ball retaining tool. Make sure all balls are removed.
4. Inspect all parts for wear, nicks, scoring or burrs. If worm or rack-pinion nut need replacing, both must be replaced as a matched pair.
5. In reassembling reverse the above.

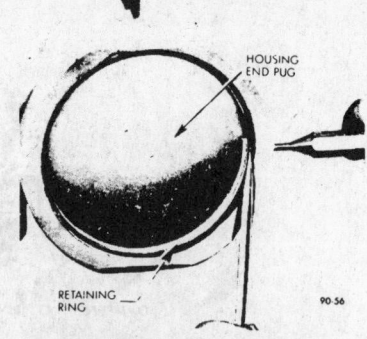

Removing housing end plug retaining ring
(© Buick Motor Div., G.M. Corp.)

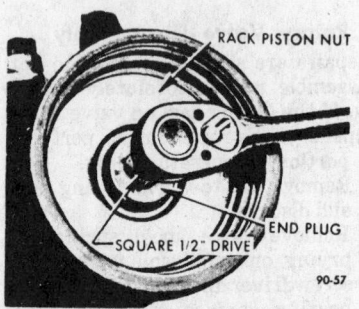

Removing rack-piston nut end plug
(© Buick Motor Div., G.M. Corp.)

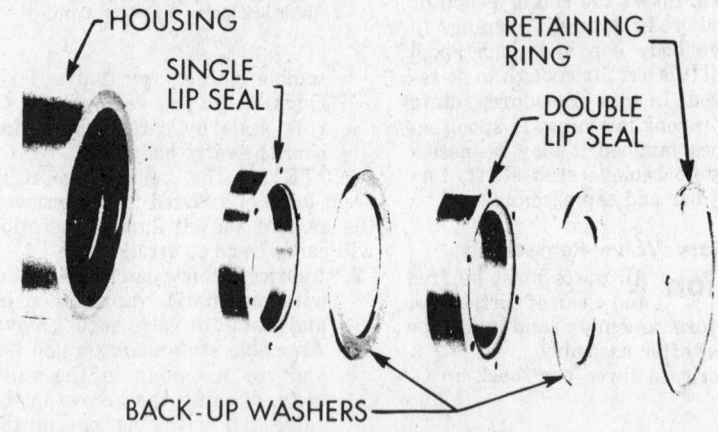

HOUSING   SINGLE LIP SEAL   RETAINING RING   DOUBLE LIP SEAL   BACK-UP WASHERS

Pitman shaft seals
(© Buick Motor Div., G.M. Corp.)

NOTE: when assembling, alternate black and white balls, and install guide and clamp. Packing with grease helps in holding during assembly. When new balls are used, various sizes are available and a selection must be made to secure proper torque when making the high point adjustment.

Rack-Pinion Nut and Worm Assembly Installation

1. Install in reverse of removal procedure.
2. In all cases use new O-ring seals.
3. Make adjustments as described in that section.

RACK-PISTON NUT

3/4" SOCKET

WORM

J 7539

90-58

Removing rack-piston nut
(© Buick Motor Div., G.M. Corp.)

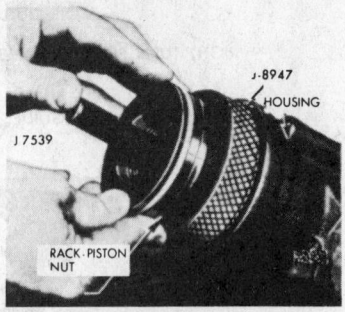

J-8947

HOUSING

J 7539

RACK-PISTON NUT

Installing rack-piston nut—
43, 44, 45000 series
(© Buick Motor Div., G.M. Corp.)

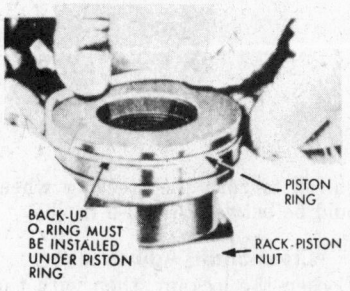

PISTON RING

BACK-UP O-RING MUST BE INSTALLED UNDER PISTON RING

RACK-PISTON NUT

Installing rack-piston ring on rack-piston nut
(© Buick Motor Div., G.M. Corp.)

PLAIN

BLACK

ROTATE WORM COUNTER CLOCKWISE

Loading balls in rack-piston nut
(© Buick Motor Div., G.M. Corp.)

J-7576

J-7539

RACK-PISTON NUT

HOUSING

Installing rack-piston nut—
46, 48, 49000 series
(© Buick Motor Div., G.M. Corp.)

TORQUE END PLUG TO 50 LB. FT.

Torquing rack-piston nut end plug
(© Buick Motor Div., G.M. Corp.)

Section 7
Chrysler Full Time Power Steering Constant Control Type

The power steering gear system for Chrysler Corporation cars, starting in 1958 is called the Constant Control type. This system consists of a hydraulic pressure pump, a power steering gear and connecting hoses.

The power steering gear housing contains a gear shaft and sector gear, a power piston with gear teeth milled into the side of the piston which is in constant mesh with the gear shaft sector teeth, a worm shaft which connects the steering wheel to the power piston through a coupling. The worm shaft is geared to the piston through recirculating ball contact.

A pivot lever is fitted into the spool valve at the upper end and into a drilled hole in the center thrust bearing race at the lower end. The center thrust bearing race is held firmly against the shoulder of the worm shaft by two thrust bearings, bearing races and an adjusting nut. The pivot lever pivots in the spacer which is held in place by the pressure plate.

When the steering wheel is turned to the left the worm shaft moves out of the power piston a few thousandths of an inch, the center thrust bearing race moves the same distance since it is clamped to the worm shaft. The race thus tips the pivot lever and moves the spool valve down, allowing oil under pressure to flow into the left-turn power chamber and force the power piston down. As the power piston moves, it rotates the cross-shaft sector gear and, through the steering linkage, turns the front wheels.

On a right turn the worm shaft moves into the power piston, the center thrust bearing race thus tips the pivot lever and moves the spool valve up, allowing oil under pressure to flow into the right power chamber and force the power piston up.

Constant Control Power Steering Unit Removal and Installation

Disconnect the battery, then remove the horn ring. Disconnect the horn wire and remove the steering wheel.

Disconnect the turn signal wires at the connectors. Remove the mast jacket support bracket and clamp at the instrument panel. Loosen the two bolts that hold the mast jacket to the gear housing, then push the mast

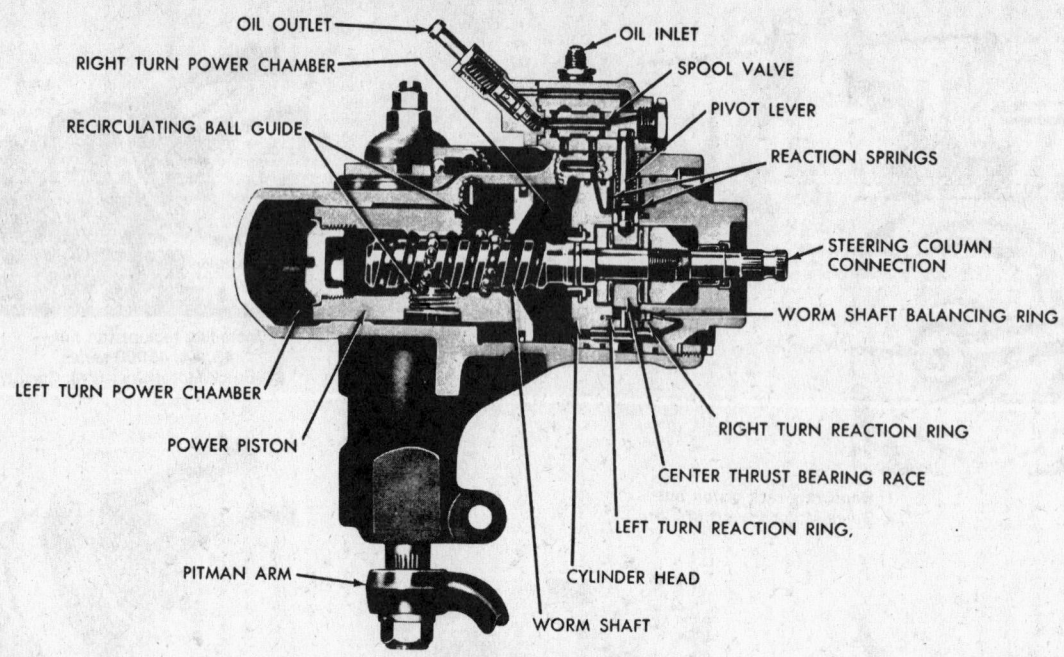

OIL OUTLET
RIGHT TURN POWER CHAMBER
RECIRCULATING BALL GUIDE
OIL INLET
SPOOL VALVE
PIVOT LEVER
REACTION SPRINGS
STEERING COLUMN CONNECTION
WORM SHAFT BALANCING RING
LEFT TURN POWER CHAMBER
POWER PISTON
RIGHT TURN REACTION RING
CENTER THRUST BEARING RACE
LEFT TURN REACTION RING,
PITMAN ARM
CYLINDER HEAD
WORM SHAFT

Chrysler power steering gear
(© Chrysler Corp.)

jacket upward to expose the coupling pin. Remove the pin.

Remove the drag link from the pitman arm.

Disconnect the pressure and return hoses at the steering gear and cap the ends of the hoses to avoid loss of fluid.

Remove the pitman arm nut and washer, then, with a pitman arm puller, remove the arm from the cross-shaft.

Insert the assembly through the engine compartment and install the gear housing to frame bolts, tightening the bolts finger tight. Align the steering tube coupling with the steering column tube and install the coupling pin, insulator and inserts.

Slide the mast jacket down in position over the gear housing and tighten the clamp bolts. Install the mast jacket support bracket and clamp at the instrument panel, but do not tighten. Connect the turn signal wires, then install the steering wheel, horn wire and horn ring.

Adjust the mast jacket so that there is 1/8 in. clearance between it an the steering wheel, then tighten the mast jacket clamp.

Now, tighten the mast jacket support bracket to the instrument panel, then tighten the gear housing to frame bolts. Install the pitman arm and drag link. Connect the pressure and return hoses, then fill the reservoir to the level mark with type A automatic transmission fluid and bleed the system.

Chrysler Constant Control Power Steering Unit Maintenance

This unit uses automatic transmission fluid type A. The fluid capacity for the unit is two quarts.

Bleeding the System

Fill the reservoir to the level mark with automatic transmission fluid type A. Start the engine and allow it to idle and come to normal operating temperature. Now turn the steering wheel several times to the right and left to expel air from the system, then recheck the oil level.

Checking Steering Effort

Raise the front wheels off the floor. Now, start the engine, then hook a spring scale to the spoke of the steering wheel at the outer edge. The effort required to turn the steering wheel should be between 5 and 9 in. lbs.

Cross-Shaft Adjustment

Loosen the locknut, then with the cross-shaft on center, tighten the cross-shaft adjusting screw until the backlash just disappears. From this position, turn the screw in 3/8 to 1/2 turn. Hold the screw in place and tighten the locknut to 50 ft-lbs. torque.

Pressure Test

Connect the pressure test hoses with the pressure gauge installed between the pump and steering gear.

Now, fill the reservoir to the level mark, then start the engine and bleed the system. Allow the engine to idle until the fluid in the reservoir is between 150° F. and 170° F. Now turn

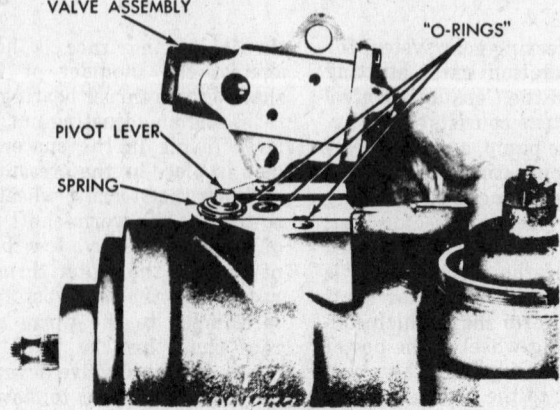

VALVE ASSEMBLY
"O-RINGS"
PIVOT LEVER
SPRING

Removing valve body assembly
(© Chrysler Corp.)

the steering wheel to the extreme right and check the pressure reading, then turn the wheel to the extreme left and check the reading again. The gauge reading should be equal in each direction. If not, it indicates excessive internal leakage in the unit.

The pressure should read between 1200-1300 psi for satisfactory power steering operation.

Repair Operations

Reconditioning

1. Drain gear by turning worm shaft from limit to limit with oil connections held downward. Thoroughly clean outside.
2. Remove valve body attaching screws and remove body and three O-rings.
3. Remove pivot lever and spring. Pry under spherical head with a screwdriver.

Caution Use care not to collapse slotted end of valve lever as this will destroy bearing tolerances of the spherical head.

4. Remove steering gear arm from sector shaft.
5. Remove snap-ring and seal back up washer.
6. Remove seal. Tool C-3880 will assist and prevent damage to relative parts.
7. Loosen gear shaft adjusting screw locknut and remove gear shaft cover nut. Tool C-3633 or equivalent is recommended.
8. Rotate wormshaft to position sector teeth at center of piston travel. Loosen power train retaining nut. (tool C-3634)
9. Insert tools C-3323 and C-3875 into housing until both tool and shaft are engaged with bearings.
10. Turn worm shaft either to full left or full right (depending on car application) to compress power train parts. Then remove power train retaining nut as mentioned above.
11. Remove housing head tang washer.
12. While holding power train completely compressed, pry on piston teeth with screwdriver, using shaft as a fulcrum, and remove complete power train.

Caution Maintain close contact between cylinder head, center race and spacer assembly and the housing head. This will eliminate the possibility of reactor rings becoming disengaged from their grooves in cylinder and housing head. It will prohibit center spacer from separating from center race and cocking in the housing. This could make it impossible to remove the power train without damaging involved parts.

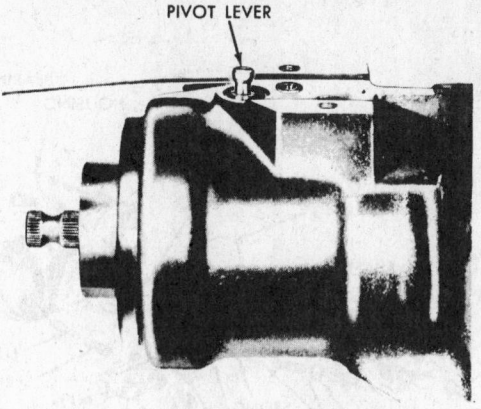

PIVOT LEVER

Removing pivot lever
(© Chrysler Corp.)

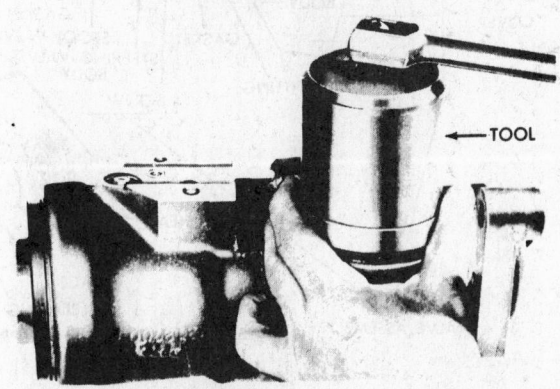

TOOL

Removing gear shaft retaining nut
(© Chrysler Corp.)

TOOL

Removing power train retaining nut
(© Chrysler Corp.)

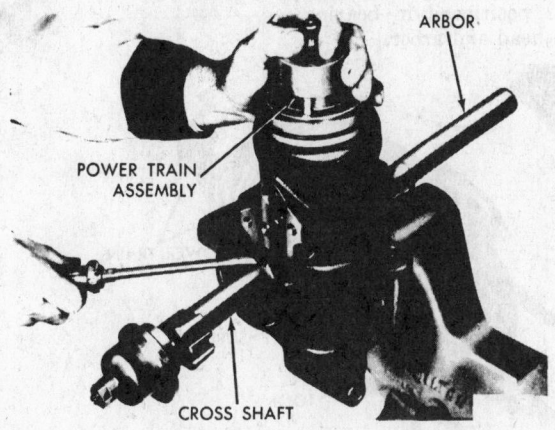

ARBOR

POWER TRAIN ASSEMBLY

CROSS SHAFT

Removing power train
(© Chrysler Corp.)

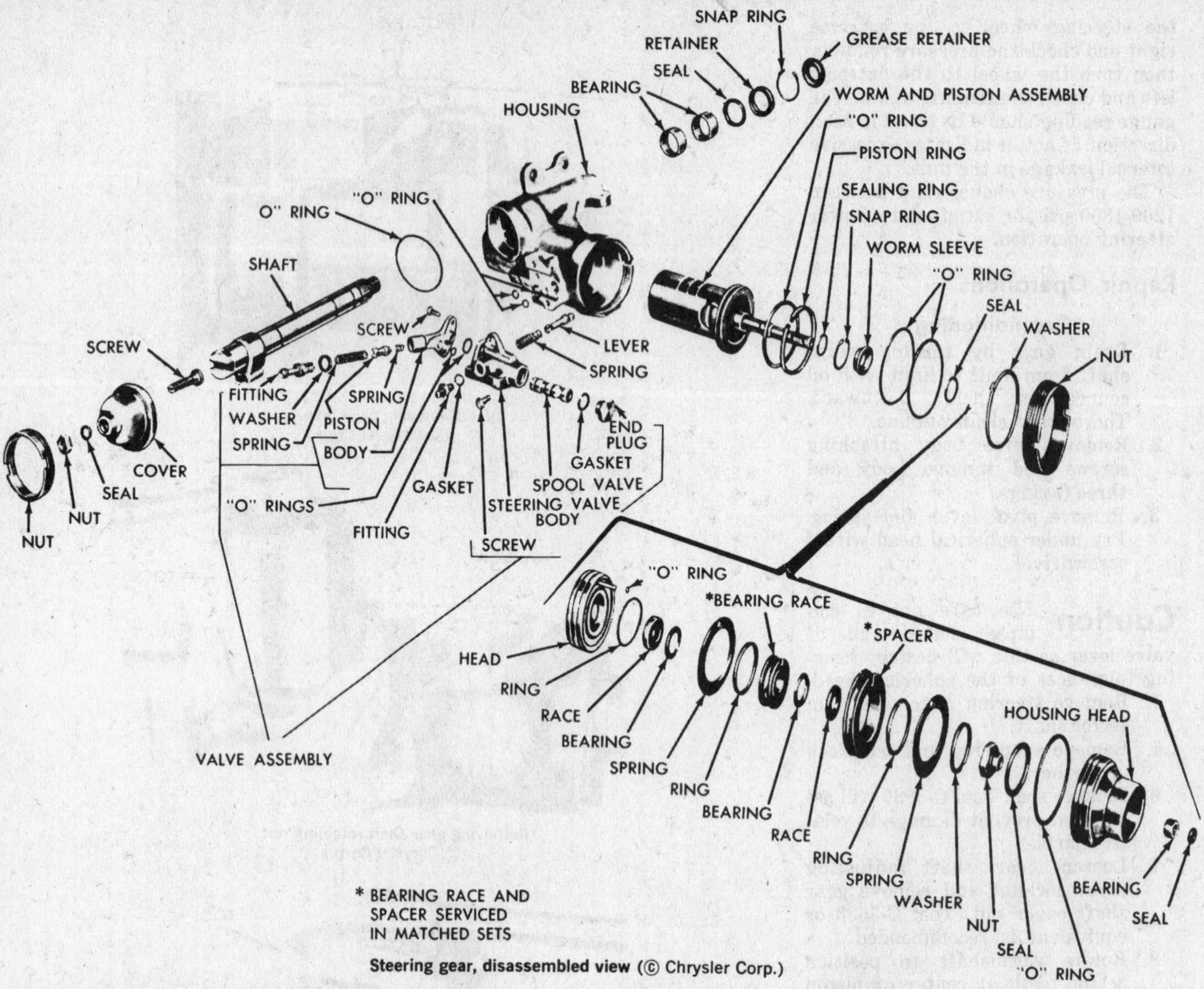

SNAP RING
RETAINER
SEAL
BEARING
GREASE RETAINER
HOUSING
WORM AND PISTON ASSEMBLY
"O" RING
PISTON RING
SEALING RING
SNAP RING
WORM SLEEVE
"O" RING
"O" RING
SEAL
WASHER
NUT

"O" RING
"O" RING
SHAFT
SCREW
SCREW
LEVER
SPRING
FITTING
SPRING
WASHER
PISTON
SPRING
BODY
COVER
SEAL
"O" RINGS
GASKET
FITTING
GASKET
SCREW
STEERING VALVE BODY
SPOOL VALVE
END PLUG
GASKET
NUT
NUT

VALVE ASSEMBLY

"O" RING
*BEARING RACE
*SPACER
HEAD
RING
RACE
BEARING
SPRING
RING
BEARING
RACE
RING
SPRING
WASHER
NUT
SEAL
"O" RING
HOUSING HEAD
BEARING
SEAL

* BEARING RACE AND
SPACER SERVICED
IN MATCHED SETS

Steering gear, disassembled view (© Chrysler Corp.)

13. Place power train in soft-jawed vise in vertical position. The worm bearing rollers will fall out. Use of arbor tool C-3999 will hold rollers when the housing is removed.

14. Raise housing head until worm-shaft oil shaft just clears the top of wormshaft and position arbor tool C-3929 on top of shaft and into seal. With arbor in position, pull up on housing head until arbor is positioned in bearing. Remove head and arbor.

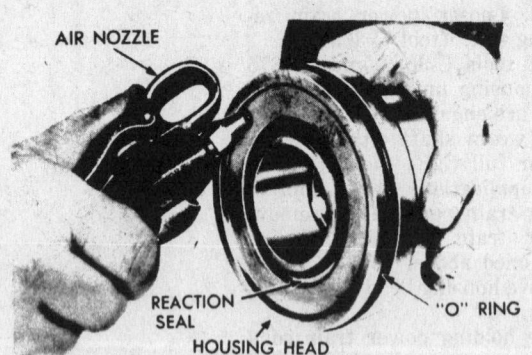

AIR NOZZLE

REACTION SEAL
"O" RING
HOUSING HEAD

Removing reaction seal from wormshaft support
(© Chrysler Corp.)

15. Remove large O-ring from housing head groove.
16. Remove reaction seal from groove in face of head with air pressure directed into ferrule chamber.
17. Remove reactor spring, reactor ring, worm balancing ring and spacer.
18. While holding wormshaft from turning, turn nut with enough force to release staked portions from knurled section and remove nut.

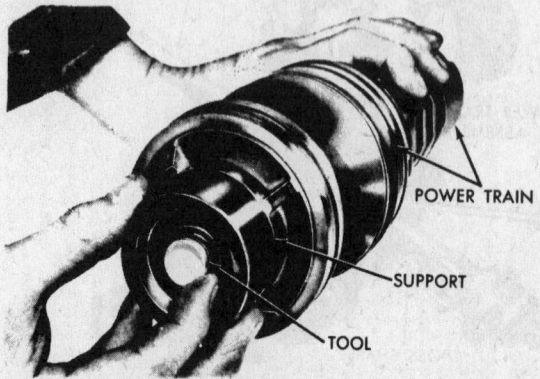

POWER TRAIN
SUPPORT
TOOL

Retaining bearing rollers with arbor tool
(© Chrysler Corp.)

19. Remove upper thrust bearing race (thin) and upper thrust bearing.
20. Remove center bearing race.
21. Remove lower thrust bearing and lower thrust bearing race (thick).
22. Remove lower reaction ring and reaction spring.
23. Remove cylinder head assembly.
24. Remove O-rings from outer grooves in head.
25. Remove reaction O-ring from groove in face of cylinder head. Use air pressure in oil hole located between O-ring grooves.
26. Remove snap-ring, sleeve and rectangular oil seal from cylinder head counterbore.
27. Test wormshaft operation. Not more than 2 in. lbs. should be required to turn it through its entire travel, and with a 15 ft. lb. side load.

NOTE: the worm and piston is serviced as a complete assembly and should not be disassembled.

28. Assemble in reverse of above, noting proper adjustments and preload requirements following.
29. When cover nut is installed, tighten to 20 ft-lbs. torque.
30. Valve mounting screws should be tightened to 200 in.-lbs. torque.
31. With hoses connected, system bled, and engine idling roughly, center valve unit until not self-steering. Tap on head of valve body attaching screws to move valve body up, and tap on end plug to move valve body down.
32. With steering gear in center, tighten gear shaft adjusting lock, loosen adjusting screw until lash just disappears.

33. With steering gear in center position or halfway from lock to lock, loosen adjusting screw until lash just disappears.
34. Continue to tighten 3/8 to 1/2 turn and tighten locknut to 50 ft. lbs.

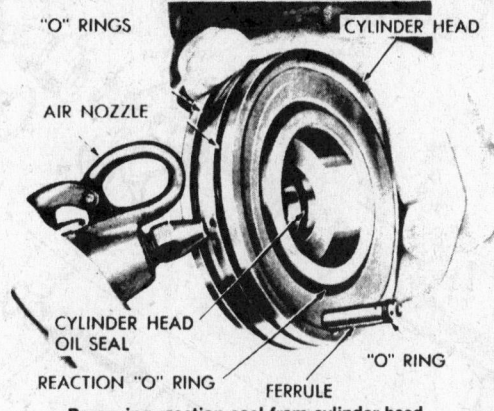

Removing reaction seal from cylinder head
(© Chrysler Corp.)

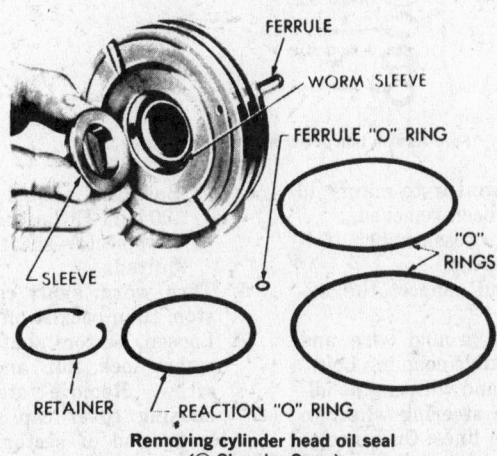

Removing cylinder head oil seal
(© Chrysler Corp.)

Section 8
Ford Torsion Bar Power Steering
(1st Type)

This is a torsion bar type with steering gear, power unit and control valve combined as one unit. On some applications the pump is driven from the front of crankshaft rather than by belt.

On Lincoln-Continental the windshield wiper is driven by oil from power steering unit and fed through the hydraulic line returning from gear to pump.

Steering Gear Removal and Installation

Removal

1. Remove pressure and return lines from gear housing. Plug openings and cap lines.
2. Disconnect horn ground wire from sleeve alignment bolt. Remove bolt securing flexible coupling to gear worm shaft.
3. Raise car. Remove splash shield, if used.
4. Where necessary, remove exhaust resonator pipe and front end crossmember.
5. Remove pitman arm from sector shaft.
6. In some cases it will be necessary to remove the transmission linkage rods from equalizer shaft and force shaft outward. This compresses spring within shaft and frees it from the inner ball joint.
7. Remove gear housing mounting bolts. In some cases fender splash shield may require loosening or removal.
8. Pull gear assembly from flexible coupling and remove from vehicle.

Installation

1. Center the gear by turning the worm shaft to either stop, then turn it back approximately two turns.
2. Center the steering wheel and insert the input shaft into the flexible coupling.
3. Secure gear to frame or body member with the attaching bolts.
4. Install pitman arm on sector shaft. Torque to 100-130 ft. lbs.
5. Position equalizer shaft and install transmission linkage rods.

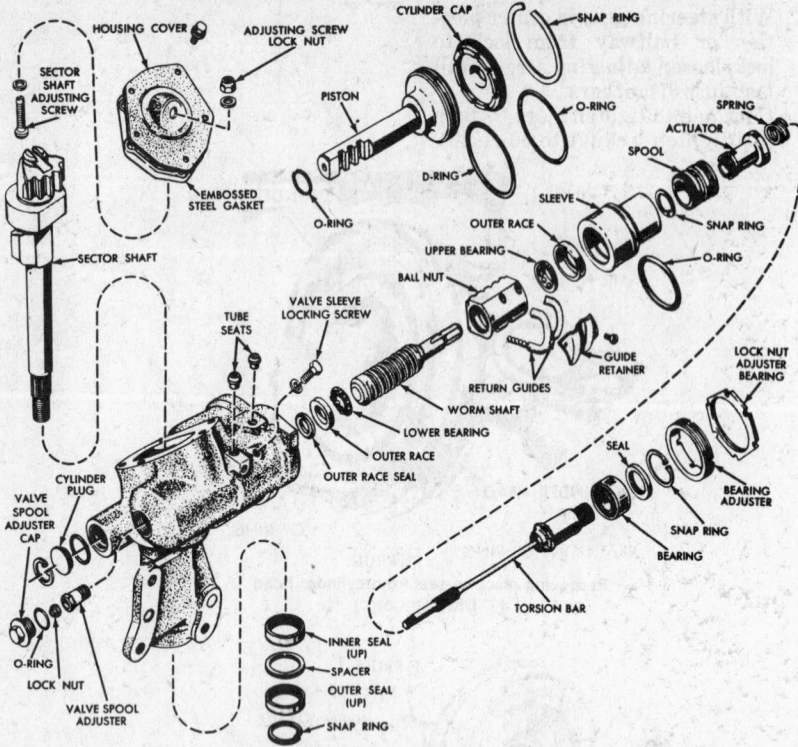

Ford torsion bar power steering (1st type)

6. Install resonator-to-manifold pipe if it has been removed.
7. Attach front crossmember if it has been removed.
8. Lower car and connect the hydraulic lines.
9. Connect horn ground wire and install the flexible coupling bolt.
10. Fill reservoir and with engine idling, cycle the steering wheel to bleed air from lines. On Lincoln-Continental with oil-operated windshield wiper, have wipers operating to be sure of complete air elimination.

Repair Operations

Disassembly

1. Drain fluid and thoroughly clean unit before any disassembly. A non-toxic solvent is recommended.
2. Mount unit in vise. An adapter is recommended to avoid any damage to unit.
3. Remove cylinder plug snap-ring and, with air, remove cylinder cap from bore.
4. Remove cylinder cap and then remove cap O-ring.
5. Check backlash between sector gear and pinion rack.
   A. Mount dial gauge with indicator shaft against machined surface of outside piston diameter.
   B. Hold sector shaft firmly and move piston by hand. Complete travel in both directions and note needle deflection.

C. Backlash should not exceed .004 in. Either new piston or new sector shaft may be required.
6. Turn worm shaft completely to stop, then back it off 1¾ turns.
7. Loosen sector shaft adjusting screw lock nut and adjusting screw. Remove steering gear housing cover cap screws. Tap lower end of sector shaft with soft hammer until shaft and cover can be removed. Slide cover to one side to remove from sector shaft and remove screw from cover.
8. Push piston out of housing. Remove piston O-ring. Remove rack bore O-ring.
9. Loosen valve sleeve alignment bolt.
10. Remove valve adjuster cap and remove O-ring from cap.
11. Remove bearing adjuster locknut and adjuster.
12. Remove torsion bar and sleeve assembly. Lightly tap lower end of bar with soft hammer.
13. Remove sector shaft oil seal retaining snap-ring. Remove outer seal, spacer and inner seal.

Torsion Bar and Sleeve Disassembly

1. Position ball nut assembly in vise. Protect the assembly with clean cloth. Remove valve spool sleeve from bar assembly and remove O-ring from sleeve.
2. Remove valve spool adjuster lock-nut from lower end of tor-

sion bar. Remove valve spool adjuster from the bar.
3. Remove torsion bar, spool, actuator, seal, bearing and race from worm shaft. If necessary, use soft hammer to tap end of torsion bar. The valve spool and actuator are spring loaded. Discard lower bearing race seal. Separate spool and actuator from torsion bar by turning spool and actuator while turning bar.
4. Remove valve spool snap-ring and remove spool from actuator.
5. Do not disassemble ball nut assembly unless there is evidence of roughness or other visible damage.
6. If ball nut is to be disassembled follow directions as covered in this type manual steering gear.

Torsion Bar and Sleeve Assembly and Steering Gear Reassembly

Reverse procedures outlined above. Use new seals, O-rings and gaskets. Following instructions giving adjustments required.

Adjustments

Worm Bearing Preload and Sector Mesh

1. With gear in car and pitman arm disconnected or with gear on bench, attach in. lb. torque wrench to steering wheel nut of worm shaft, using adapter.
2. Measure force required to move shaft approximately 20° away from stop. Reading should be 6-8 in. lbs.
3. If not within limits, loosen adjuster locknut and turn bearing adjuster to obtain proper reading, then tighten locknut. Use care that adjuster does not turn while tightening locknut.
4. Recheck preload.
5. Locate mechanical center of gear by rotating completely to right or left, to stop, then back off 1¾ turns.
6. With in. lb. torque wrench still connected, rotate gear to right stop and take reading through turning to left stop. Then take reading while returning to right stop.
7. The over-center mesh load should be 15-17 in. lbs. If there is a variation between left rotation and right rotation, use the higher figure.
8. If over-center load is not within the limits, loosen sector locknut and turn adjusting screw. Each one-sixteenth turn alters preload approximately 2 in. lbs Re-tighten lock-nut and recheck preload.

Valve Centering

1. Align the slot in the sleeve with the lock screw and the torsion bar in the housing. Be sure seals are properly seated. Be sure sleeve is bottomed.
2. With bearing adjuster and locknut installed, install spool centering wrench in spool adjuster and locate the spool so that the valley between lands can be seen through the pressure port. Lock spool adjuster with locknut. This is a preliminary adjustment.
3. With gear completely assembled and the ball nut located in center of travel, installed, and connected to reservoir, bleed system by idling engine and rotating gear back and forth.
4. Connect a 0-2000 psi gauge between pump pressure line and inlet port (upper hole). Be sure valve gauge hand valve is completely open.
5. With engine idling and repeated gear rotating from stop to stop, rotate gear with in. lb. torque wrench to the left stop to obtain 300 psi on gauge.
6. Reverse gear direction to opposite stop and again pull to secure 300 psi on gauge.
7. If more than 3 in. lbs. pull on torque wrench is required to obtain the 300 psi in one direction more than the other, loosen the valve spool adjuster locknut and rotate the adjuster to equalize the required necessary torque. Tighten the locknut before each reading is made.

Section 9
Ford Torsion Bar Power Steering
(2nd Type)

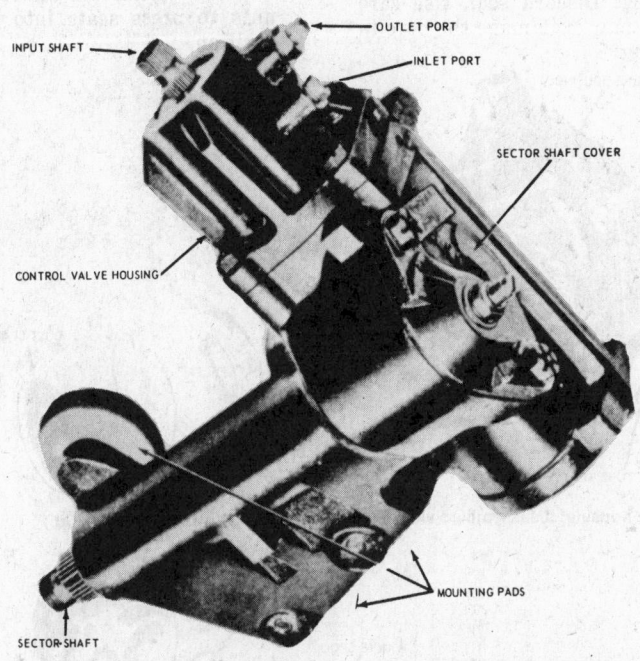

INPUT SHAFT

OUTLET PORT

INLET PORT

SECTOR SHAFT COVER

CONTROL VALVE HOUSING

MOUNTING PADS

SECTOR SHAFT

Power steering gear (© Lincoln-Mercury Div., Ford Motor Co.)

This is a torsion bar type with steering gear, power unit and control valve combined as one unit. On some applications the pump is driven from the front of crankshaft rather than by belt.

On Lincoln-Continental the windshield wiper is driven by oil from power steering unit and fed through the hydraulic line returning from gear to pump.

Steering Gear Removal and Installation

Removal

1. Disconnect pressure and return lines from gear housing. Plug openings and cap lines.
2. Disconnect ground strap from gear housing.
3. Remove flex coupling attaching bolt.
4. Remove left brace-to-torque box attaching bolt.

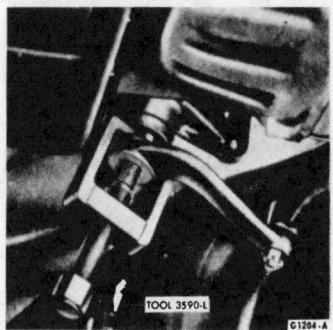

TOOL 3590-L

G-1204-A

Removing pitman arm
(© Lincoln-Mercury Div., Ford Motor Co.)

5. Loosen bolt attaching brace to side rail and swing to one side.
6. Remove pitman arm.
7. Remove exhaust pipe, (Manifold-to-resonator.)
8. Disconnect linkage rod at equalizer. Remove equalizer stud from side rail. Move shaft up and out of way.
9. Remove bolt from lower end of splash shield and move shield out of way.
10. Remove gear attaching bolts. Support gear before removing last bolt.
11. Move steering gear downward to free from flex joint and rotate assembly counterclockwise to clear side rail and engine.
12. Remove three gear mounting pads.
13. Reinstall in reverse of above.
14. Fill reservoir, idle engine and rotate steering gear from side to side. Keep fluid to level. Check for leaks. On Lincoln-Continental, wipers should be operating during the bleeding process.

Repair Operations

Disassembly

1. Remove housing mounting pads.
2. Invert steering gear and rotate

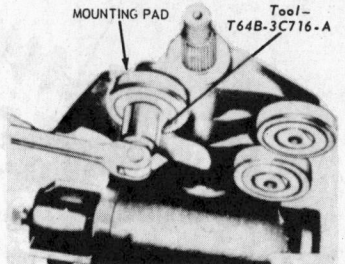

MOUNTING PAD

Tool—
T64B-3C716-A

Removing or installing
steering gear mounting pads
(© Lincoln-Mercury Div., Ford Motor Co.)

shaft back and forth to completely drain. Thoroughly clean outside of housing to prevent dirt interference while working on assembly.

3. Mount in soft-jawed vise or special holding fixture.

4. Remove Teflon locknut and brass washer from adjusting screw.

5. Turn input shaft to either lock then back 1¾ turns to center the gear.

6. Remove sector shaft cover attaching screws and identifying tag.

7. Tap lower end of sector shaft with soft hammer to loosen it, then lift cover and shaft from housing as a unit. Discard O-ring.

8. Turn cover counterclockwise and remove from adjusting screw.

9. Remove valve housing attaching bolts. Lift valve housing from steering gear housing, while holding piston to prevent it from rotating off the worm shaft. Remove housing and passage O-rings and discard.

10. With valve body and piston on end (piston end down), rotate input shaft counterclockwise out of piston. Allow ball bearings to drop into piston.

11. With cloth over open end of piston, invert to remove balls.

12. Remove ball guide clamp and guides.

13. Install valve body in holding fixture. Do not use vise. Remove locknut and retaining nut.

14. Slide worm and valve assembly out of valve housing.

15. Remove shim from valve housing bore.

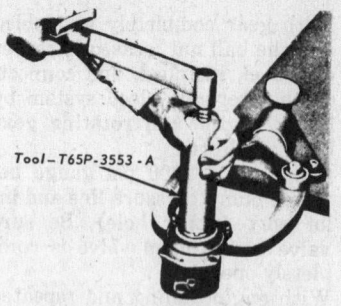

Tool—T65P-3553-A

Removing or installing lock nut
(© Lincoln-Mercury Div., Ford Motor Co.)

Valve Housing R & R

1. Remove dust seal from rear of housing. Tools T59L-100-B and T58L-101-A will assist.

2. Remove snap-ring from housing.

3. Turn fixture to invert valve housing.

4. From opposite the seal end, gently tap bearing and seal from housing. Discard seal. Use care not to mar or damage the bore in housing.

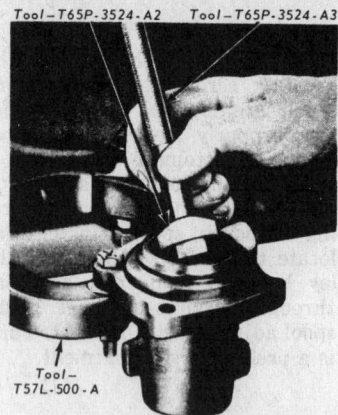

Tool—T65P-3524-A2   Tool—T65P-3524-A3

Tool—T57L-500-A

Removing bearing and oil seal
(© Lincoln-Mercury Div., Ford Motor Co.)

5. Remove inlet and outlet seats if damaged. Use an EZ-out.

6. If installing new seats, coat with vaseline and position them in housing. Install and tighten tube nuts to press seats into proper position.

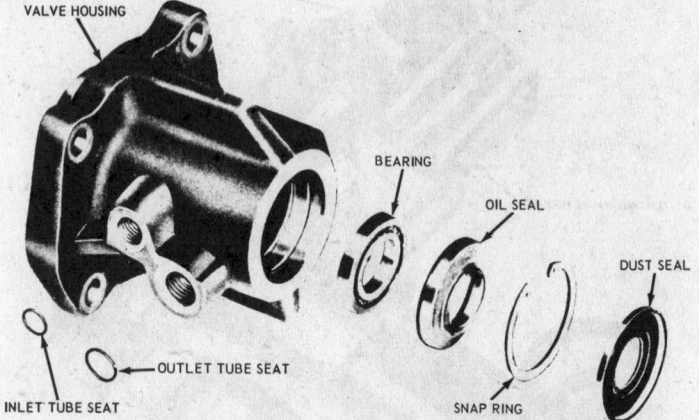

VALVE HOUSING

BEARING

OIL SEAL

DUST SEAL

OUTLET TUBE SEAT

INLET TUBE SEAT

SNAP RING

Valve housing, disassembled view (© Lincoln-Mercury Div., Ford Motor Co.)

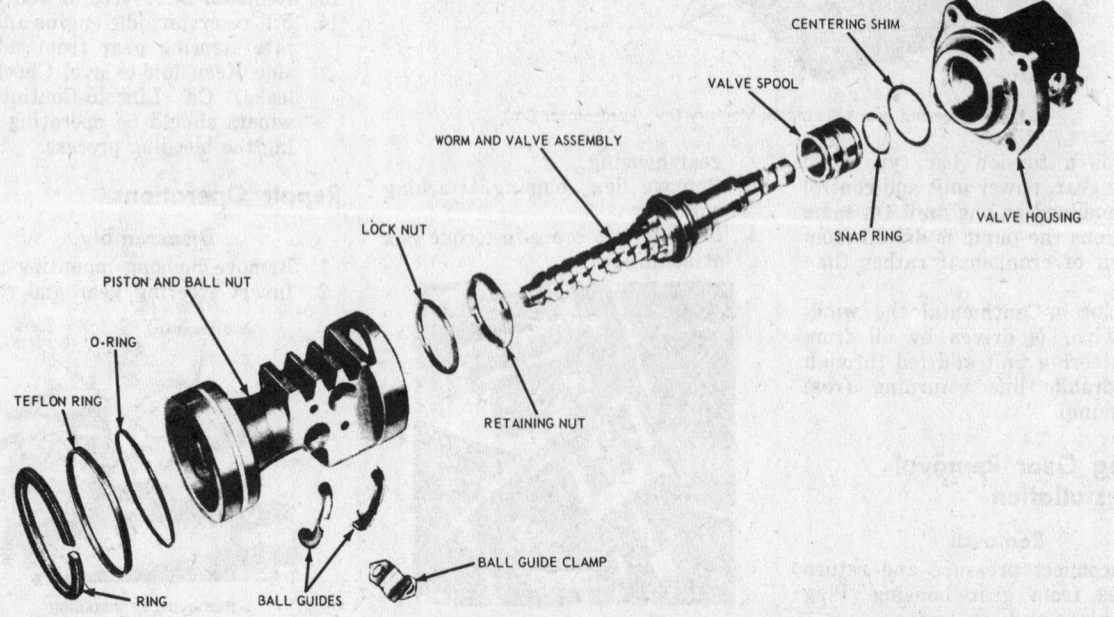

CENTERING SHIM

VALVE SPOOL

WORM AND VALVE ASSEMBLY

SNAP RING

VALVE HOUSING

LOCK NUT

PISTON AND BALL NUT

O-RING

TEFLON RING

RETAINING NUT

RING

BALL GUIDES

BALL GUIDE CLAMP

Ball nut and valve housing, disassembled view (© Lincoln-Mercury Div., Ford Motor Co.)

Installing valve housing bearing
(© Lincoln-Mercury Div., Ford Motor Co.)

7. Coat bearing and seal surfaces with vaseline.
8. Position bearing in housing and, with proper tool, see that it is thoroughly seated.
9. Lubricate new seal with gear lubricant and place in housing with metal side of seal facing outward. Drive seal into housing until outer edge of seal does not quite clear the snap-ring groove.
10. Place snap-ring in housing and gently drive ring in until it seats in its groove. This will properly locate the seal.
11. Place dust seal in housing with rubber side out and drive into place. The seal must be located behind the undercut.

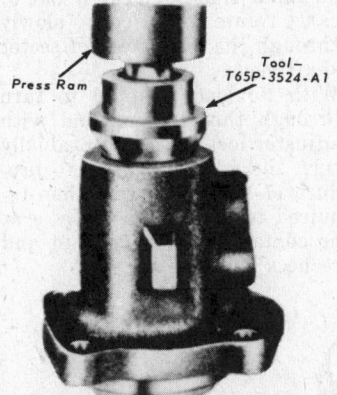

Installing oil seal in valve housing
(© Lincoln-Mercury Div., Ford Motor Co.)

Worm and Valve R & R

1. Remove snap-ring from end of actuator.
2. Slide control valve spool off of actuator.
3. Install valve spool evenly and slowly with slight oscillating motion into flanged end of housing with identification groove between the valve spool lands outward. Check for freedom of valve movement within working area. Spool should enter housing freely and fall by its own weight.
4. If spool is not free, check for

burrs and remove with a hard stone.
5. Check valve for burrs and if burrs are found, stone valve in a radial direction only. Be sure valve is entirely free.
6. Remove spool from housing.
7. Slide spool into actuator, making sure groove in spool annulus is toward worm.
8. Install snap-ring to retain spool.
9. Check clearance between spool and snap ring. It should be 0.002-0.005 in. If not within these limits, select snap-ring that will produce 0.003 in.

Piston and Ball Nut R & R

1. Remove the Teflon ring and O-ring from piston and ball nut.
2. Dip new O-ring in gear lubricant and install on piston and ball nut.
3. Install new Teflon ring, using care not to stretch more than necessary.

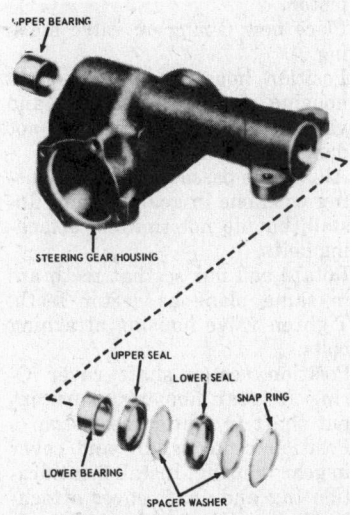

Steering gear housing, disassembled view
(© Lincoln-Mercury Div., Ford Motor Co.)

Gear Housing R & R

1. Remove snap-ring and spacer washer from lower end of housing.

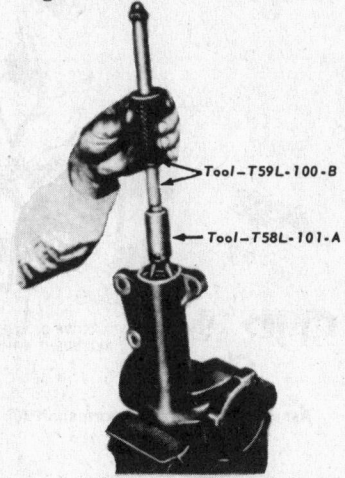

Removing outer seal
(© Lincoln-Mercury Div., Ford Motor Co.)

2. Remove outer seal from housing. Lift out spacer washer.
3. Remove upper seal in same manner as lower seal.
4. Press upper and lower bushings

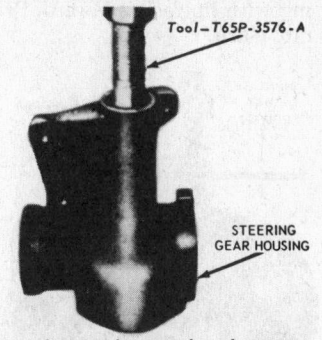

Removing steering gear housing upper and lower bushings
(© Lincoln-Mercury Div., Ford Motor Co.)

from housing if worn or defective.
5. Press new bushings into place.

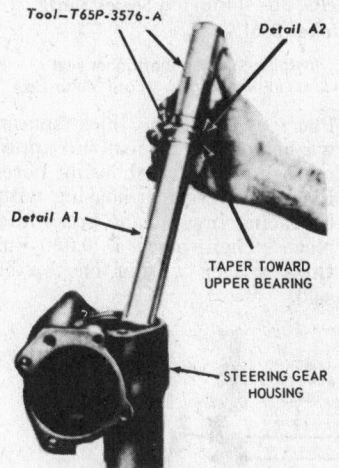

Installing upper bushing in steering gear housing
(© Lincoln-Mercury Div., Ford Motor Co.)

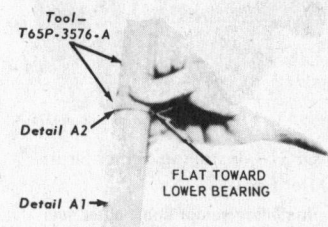

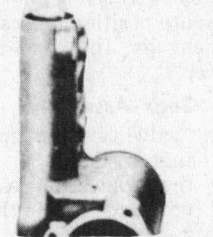

Installing lower bushing in steering gear housing
(© Lincoln-Mercury Div., Ford Motor Co.)

6. Dip both sector shaft seals in gear lubricant.
7. Apply lubricant to sector shaft seal bore of housing and position sector shaft inner seal in housing with lip facing inward. Press into place.

FIG. 30—Installing Sector Shaft Inner Seal

Installing sector shaft inner seal
(© Lincoln-Mercury Div., Ford Motor Co.)

8. Place a 0.090 in. thick spacer washer on top of seal and apply more lubricant to housing bore.
9. Place outer seal in housing with lip facing inward and press into place. Then place a 0.090 in. thick spacer washer on top of seal.

FIG. 31—Installing Sector Shaft Outer Seal

Installing sector shaft outer seal
(© Lincoln-Mercury Div., Ford Motor Co.)

10. Place snap-ring in housing and press into position to locate seals and engage the snap-ring in groove.

Gear Assembly

1. Mount valve housing in fixture with flanged end up.
2. Place the same thickness valve spool centering shim in the housing as was removed. Use only one shim.
3. Install worm and valve in housing.
4. Install retaining nut in housing and torque to 55-65 ft. lbs.
5. Install locknut and tighten to 20-30 ft. lbs.
6. Place piston on bench with ball guide holes facing up. Insert worm shaft into piston so that first groove is in alignment with hole nearest to center of piston.
7. Place ball guide in position and feed balls (27) into guide, turning worm clockwise. If all balls have not been fed into guide upon reaching stop, rotate shaft back and forth while installing balance of balls.
8. Secure guides in ball nut with the clamp.
9. Position new lube passage O-ring in counterbore of housing.
10. Apply vaseline to Teflon seal on piston.
11. Place new O-ring on valve housing.
12. Position housing spacer ring in housing and slide piston and valve into gear housing. Do not damage Teflon seal.
13. Align lube passage in valve housing with one in gear housing. Install, but do not tighten, attaching bolts.
14. Rotate ball nut so that teeth are in same plane as sector teeth. Tighten valve housing attaching bolts.
15. Position sector shaft cover O-ring in gear housing. Turn input shaft to center the piston.
16. Position sector shaft and cover in gear housing. Install identification tag and shaft cover attaching bolts and tighten. Make adjustments as described under "Adjustments."

Adjustments

Valve Spool Centering

1. Install a 0-2000 psi pressure gauge into the pressure line from pump to gear inlet.
2. Be sure valve of gauge is entirely open.
3. Fill reservoir to proper level.
4. With engine idling, cycle steering from side to side through full travel. Check and keep system full of proper oil.
5. After the above bleeding, with engine at approximately 1000 rpm and the steering wheel centered, attach an in. lb. torque wrench, by means of socket, to steering wheel nut. Apply pressure to torque wrench in either direction to produce gauge reading of 250 psi.
6. The torque reading should be similar in both directions. If more than 4 in. lbs. difference in either direction is recorded, the gear must be removed. The valve centering shim must be replaced by either a thicker or thinner shim. Use only one shim. To make this adjustment out of car use the same procedure, but the pressure reading must be made at the right and left stops, instead of either side of center.

Over Center Position Load

1. With no fluid in gear and torque wrench on steering wheel nut, or on input shaft of gear if out of car, rotate the gear slowly through the high point of sector shaft mesh.
2. With no load required to turn through this position and with adjuster locknut loose, gradually turn the adjuster screw to produce 11-12 in. lbs. more than required to turn when mesh is at no contact. Tighten locknut and recheck the torque.

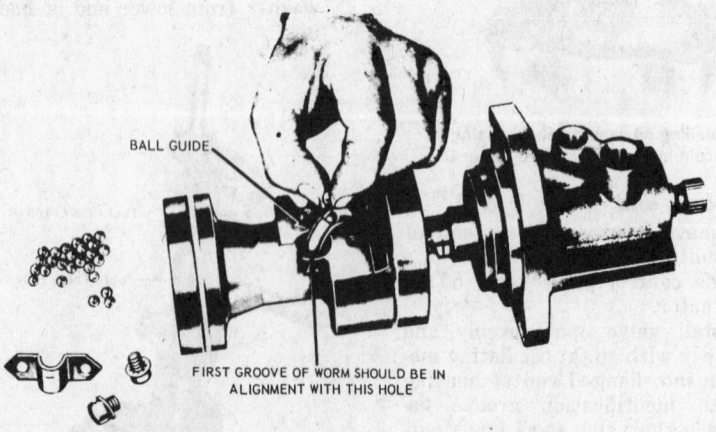

Assembling piston on wormshaft (© Lincoln-Mercury Div., Ford Motor Co.)

Steering Wheel Turning Effort

Vehicle Model	Lbs. Effort
Ford, Mercury	5.0
Cougar, Fairlane, Falcon, Montego, Mustang	6.5
Lincoln Continental, Mark III, Thunderbird	3.75

Power Steering Pump Service Specifications

Description	Ford—Thompson	Eaton
Pump Rotor Shaft End Play		.017 in. Max.— .003 in. Min.
Max. Torque Allowed to Rotate Rotor Shaft		15 In-Lb.
Stamped Housing to Plate Assy. Screw and Washer Assy.		28-32 Ft.-Lb.
Reservoir to Stamped Housing Nut		43-47 Ft.-Lb.
Cam Ring to Pressure Plate Screw		20 In-Lb.
Housing to Cover Screw		15-20 Ft.-Lb.

Ford—Thompson Power Steering Pump Torque Limits (Ft-Lbs)

Description	Ford, Mercury, Thunderbird	Cougar, Falcon, Fairlane, Montego, Mustang
Pump Rear Mounting Nut	14-18	20-30
Pressure Hose Nut	20-30	20-30
Pump Attaching Nuts	25-35	30-40
Bracket to Engine Attaching Nuts 6-Cyl.	12-16	7-10
302-351 CID	18-25	18-25
390, 427, 428, 429 CID	30-40	30-40
Bracket to Cylinder Head	25-35	25-35
Belt Adjustment Bolt	—	25-35
Rear Support Bracket Attaching Nuts—240 CID	45-60	
Bracket to Engine Adapter—170-200 CID		25-35
Engine Adapter to Engine 170-200 CID		17-25
250 CID		30-40
Engine Adapter to Pump—250 CID		25-35

Power Assist Steering Gear Specifications

Vehicle	Mustang, Cougar	Falcon, Montego, Fairlane
Model	SMB-K ①	SMA-B
Gear type	Power assist and manual ②	Power assist
Gear ratio	16:1	16:1
Turns of strg. gr. (lock to lock) ③	3¾	4
Lube type	ESW—M1C87—A	
Lube capacity (Lb.)	.55 ± .05	.87 ± .07
Worm bearing preload (in lb) ④	3-4 ②	3-4
Total center meshload (in lb) ⑤	8-9 ②	8-9
Adjustments (All models)	Adjusting screw to bottom of sector shaft T slot clearance: .000—.002	

① Production only.
② When used for improved or competition handling, worm bearing preload must be adjusted to 4-5 In Lb and total center meshload must be adjusted to 9-10 In Lb.
③ Gear only—not attached to Pitman arm.
④ Torque required to rotate input shaft at approximately 1½ turns either side of center (gear out of vehicle or Pitman arm disconnected).
⑤ Required to rotate input shaft and worm assembly past the center high point.

Integral Power Steering Gear Specifications

Description	Ford Design (XR—50)	Saginaw Design
Type	Recirculating ball torsion bar	
Ratio	17:1	17.5:1
Turns of steering wheel (lock to lock-linkage disconnected)	4	4⅛
Fluid specifications	M—2C33—F	
Fluid capacity (included in pump reservoir fill)	1.6 Pints (Approx.)	
Phosphorescent Dye Additive (for leak detection)	M99B103—A (4 Oz. per quart)	
Sector shaft end play—linkage disconnected	None	.002" Max.
Sector shaft mesh load. Total over mechanical center position. Must be① greater than worm Bearing preload torque, shown below	14 In Lb (Max) (Exc. Lincoln) 17 In Lb (Max.) (Lincoln)	14 In Lb (Max)
Worm bearing preload	2-7 In Lb	4-7 In Lb②
Clearance between valve spool & retaining ring	.0035—.0005" Preferable .002"	—
Pressure variation between right & left turn (at 250 P.S.I.)— Check efforts each side of center	4 In Lb Max. Variation	
Clearance between inner sector seal and housing	.025"	—

① 8-9 in lb (XR—50 exc. Lincoln) 11-12 in lb for Lincoln, 4—8 in lb for Saginaw Gear.
② 3 in lb in excess of valve assy. Drag total worm bearing preload and seal drag not to exceed 8 in lb.

Steering Gear Torque Limits (Ft-Lb)

Description	*Manual and Power Assist*			
	Fairlane, Falcon, Montego	Mustang Cougar	Ford (XR-50) Power Steering	Saginaw Power Steering
Sector shaft cover bolts	17-25	15-22	55-70	30-35
Mesh load adjusting screw lock nut	32-40 18-42 (In-Lb) Pwr.	32-40	25-35	20-30
Ball return guide clamp screw	42-60 (In-Lb) Man.	18-42 (In-Lb)	60-120 (In-Lb)	3-6
Preload adjuster lock nut	60-80	45-60		50-110
Valve housing to gear housing screw			35-45	
Race retaining inner nut			①	
Race lock nut			②	
Piston end cap			50-75	50-100
Set screw-rack adjustment				
Lubricant fill plug and vent	3-9 ③	3-9 ③		

① With tool T66P-3553-B - compute the torque as follows:

$$\text{Torque} = \frac{\text{Length of Torque Wrench} \times 60 \text{ Lb. Ft.}}{\text{Length of Torque Wrench} + 5.5 \text{ Inches}}$$

Example: With 13 inch torque wrench

$$\frac{13 \text{ In.} \times 60 \text{ Lb. Ft.}}{13 \text{ In.} + 5.5 \text{ In.}} = \frac{13 \times 60 \text{ Lb. Ft.}}{18.5} = 0.703 \times 60 = 42 \text{ (Lb. Ft.)}$$

② With tool T66P-3553-B - compute the torque as follows:

$$\text{Torque} = \frac{\text{Length of torque wrench} \times 25 \text{ Lb. Ft.}}{\text{Length of torque wrench} + 5.5 \text{ Inches}}$$

③ Minimum of one thread must remain exposed when installed.

Steering Linkage Torque Limits (Ft-Lbs)—Ford

Description	Cougar, Fairlane, Falcon, Montego Mustang	Ford, Mercury	Continental Mark III Thunderbird	Lincoln Continental
Cylinder mounting bracket to underbody or frame (side hole) (bottom hole)	28-35 35-43			
Power cylinder to bracket	18-24			
Power cylinder to bracket lock nut	3-5			
Steering tie rod end to spindle arm	30-40 ①	35-47 ①	35-47 ①	40-55
Idler arm mounting bracket to underbody or frame	28-35	28-35	28-35	20-30
Pitman arm to control valve (power steering)	35-47 ①			
Steering gear to side rail or frame	50-65	50-65	50-65	50-65
Pitman arm to sector shaft	150-225	150-225	150-225	150-200
Idler Arm to Pitman arm-to-idler arm rod	60-80 ①	60-80 ①	60-80 ①	40-55
Spindle connecting rod clamp to adjusting sleeve	10-14	8-13	8-13	19-26
Pitman arm to steering arm-to-idler arm rod	35-47 ①	35-47 ①	35-47 ①	40-55
Steering spindle arm connecting rod to arm-to-idler arm rod	30-40 ①	35-47 ①	35-47 ①	40-55

① Torque to low limit of specification; then, tighten the nut to the nearest cotter pin slot and insert the cotter pin.

Steering-Gear—Lincoln-Continental

Model	XR-50
Type	Recirculating ball torsion bar
Ratio	17:1
Turns of steering wheel (lock to lock) ①	3.6
Lube type ②	C1AZ-19582-A
Lube capacity—Lbs	1.6 Pints (Approx)

① Gear only—not attached to Pitman arm
② Phosphorescent dye additive for leak detection 4 oz per quart of fluid

Steering Gear Adjustments

Description	
Sector shaft mesh load total over mechanical center position —must be 11—12 in-lbs greater than the off-center torque of 2 to 7 in-lbs	14 in-lbs maximum ②
Clearance between valve spool and retaining ring	0.002—0.005 in Preferable 0.003 in
Pressure variation between right and left turns (at 250 lbs pressure)—Check efforts each side of center	4 in-lbs Maximum Variation
Clearance between inner sector seal and housing	0.025 Inch

Steering Gear and Column Torque Limits— Lincoln-Continental

Description	Ft-Lbs	Description	Ft-Lbs
Meshload adjusting screw locknut	25—35	Steering column	
Ball return guide clamp screw (in-lbs)	42—70	Steering wheel attaching nut	20—30
Gear cover to gear housing —power steering	55—65	Steering column to support bracket (instrument panel)	45—60 In-Lbs
Valve housing to gear housing—power steering	35—45	Tilt wheel column flange screws	20—30
Race retaining inner nut— power steering	①	Neutral switch actuator to shift tube (in-lbs)	18—30
Race lock nut—power steering	②	Steering shaft flex coupling bolt	25—35
Piston end cap—power steering	50—75		

① With Tool T65P-3553-A—Torque to 55—65 Ft-Lbs
   With Tool T66P-3553-B, Compute the Torque as Follows:

$$\text{Torque} = \frac{\text{Length of Torque x 60 Ft-Lbs}}{\text{Length of Torque Wrench} + 5.5 \text{ Inches}}$$

Example—With 13-inch Torque Wrench

$$\frac{13 \text{ in x 60 Ft-Lbs}}{13 \text{ In x 5.5 In}} \quad \frac{13 \text{ x 60 Ft-Lbs}}{18.5} \quad —0.704 \text{ x 60 } —42 \text{ Ft-Lbs}$$

② With Tool T65P-3553-A — Torque to 20—30 Ft-Lbs
   With Tool T66P-3553-B, Compute the Torque as Follows:

$$\text{Torque} = \frac{\text{Length of Torque Wrench x 25 Ft-Lbs}}{\text{Length of Torque Wrench} + 5.5 \text{ Inches}}$$

Power Steering Gear—Chrysler

Type	Constant control full time power
Ratio	15.7 to 1
Wheel turns—stop to stop	3 ½
Cross shaft bearings	2 needle bearings and 1 direct bearing on grey iron cover
Worm shaft thrust bearing pre-load	16-24 Ozs.
Cross shaft adjustment	Tighten adjusting screw ⅜ to ½ turn past zero back lash (center of high spot)
Fluid capacity of hydraulic system	4 pts. (3¾ imperial pts.)
Type of fluid	Power steering fluid part No. 2084329 or equivalent

Pump

Type	Constant displacement— 1.06 cu. in. per revolution
Maximum pressure	1200 to 1300 PSI
Pump output 1.06 pump	
High level	2.5 to 3.0 gpm
Low level	1.4 to 1.8 gpm
Type of fluid	Power steering—Part No. 2084329 or equivalent. Do not use Type "A" Transmission fluid

Power Steering Gear Torques

	Foot Pounds		Pounds Foot Inch
Gear housing to frame bolt	100	Steering column support nut	140
Gear shaft adjusting screw lock nut	50	Steering shaft coupling bolts	200
Gear shaft cover nut	20	Valve body attaching bolts	200
Pump inlet fitting	30	Valve body end plug	25
Steering arm nut	120	Steering wheel nut	27

Pump Torques

Location	Foot Pounds
High pressure hose fittings	13
Pump bracket bolts	23
Flow control valve plug	7
Bracket mounting bolts	30
1/8 inch pipe clean out plug	7

Chevrolet Corvette, Camaro, Chevy II

Steering Gear

Type	Semi-reversible, recirculating ball
Ratio	16:1
Overall ratio (fast)	17:1
(road)	19.6:1
(with power steering)	17:1
Linkage	
Type	Relay
Location	To rear of wheels
Tie rods	2

Provision on steering arms for adjusting for road or street, or fast steering ratio (manual steering feature only).

Steering Bolts Torques

	Foot Pounds
Steering gear to frame	25-35
Idler arm to frame	25-35
Tie rod stud nut	25-35
Steering damper to frame bracket	23-33
Steering damper to relay rod	20-28
Damper bracket to relay rod	20-28
Mast jacket spring stop	20-25
Mast jacket U-bolt	10-15
Steering wheel hub nut	35-40
Steering coupling pinch bolts	25-35
Power steering pump pulley	55-65
Power steering pump mounting nut	15-24
Pump bracket bolts	20-30
Power cylinder to relay rod	42-47
Power cylinder to frame bracket	20-26
Valve to pitman arm	42-47
Valve clamp bolt	15-22

Chevrolet Power Steering— Pump Pressure

Vehicle	Constant Ratio Steering	Variable Ratio Steering
Chevrolet		1350-1450
Chevelle		1200-1300 (L-6) 1350-1450 (V-8)
Station Wagon	900-1000	
Monte Carlo		1350-1450
Nova		1200-1300 (L-6) 1350-1450 (V-8)
Corvette	870-1000	

Chevrolet Power Steering—Gear Ratios

Vehicle	Constant Ratio				Variable Ratio			
	Regular		Fast		Regular		Fast	
	Gear	Overall*	Gear	Overall*	Gear	Overall	Gear	Overall
Chevrolet					16:1 to 12.4:1	19.3:1 to 15.5:1		
Chevelle & Monte Carlo					16:1 to 12.4:1	18.6:1 to 14.5:1		
119" W.B. Sta. Wgn.	17.5:1	21.2:1						
116" W.B. Sta. Wgn.	17.5:1	21.6:1						
Nova					16:1 to 12.4:1	18.9:1 to 14.7:1	16:1 to 12.4:1	14.25:1 to 11:1
Corvette			16:1	17.6:1				

*On Center

Chevrolet Adjustment Specifications—
Power Steering

	Chevrolet, Chevelle, Station Wagon, Monte Carlo and Nova
Ball drag	3 in. lb. max.
Thrust bearing preload	½ - 2 in. lb. in excess of valve assy. drag
Over center preload	3 - 6 in. lb. in excess of above
Total steering gear preload	14 in. lb. max.
	Corvette
Worm bearing preload	4 to 7 in. lb.
Sector lash adjustment	4 to 10 in. lb. excess of above
Total steering gear preload	14 in. lb. max.

Power Steering Pump Drive Belt Tension Limits (Lbs.)

Make	Age of Belt	
	New	Used
Chrysler Corp. All models	120	70
Ford Motor Co. All models	120-150	90-120
General Motors Corp. Pontiac	6 cyl-115 V-8 140	75 100
Cadillac	100	55-70
Oldsmobile	Adjust to proper mark on tension gauge BT-33-70M	
Buick	Adjust to proper mark on tension gauge BT-33-70M	
Chevrolet	Adjust to proper tension as shown on Tool J-7316	
American Motors Corp. All models	125-145 lbs 90-110 lbs. Use Tool J-7316 to measure tension	

Maintenance Index

Number in column is the section for the job

Car	Year	Wheel Bearings	Ball Joints		Caster	Camber
			Upper	Lower		
Buick	1964–70	2	5	5	8	8
Buick Skylark, GS	1964–70	2	4	5	8	8
Cadillac	1964–70	2	④	5	10	16
Eldorado—See Car Section	1967–70	2	4	4	10	16
Camaro	1967–70	2	4	5	8	8
Capri	1970		See Car Section			
Chevelle	1964–70	2	4	5	8	8
Chevrolet	1964	2	4	4	8	8
	1965–70	2	4	4	18	15
Chevy II	1965–67	2	4	4	18	15
	1968–70	2	4	4	8	8
Chrysler and Imperial	1964–66	2	7	7	17	17
	1967–70	2	7	④	17	17
Continental and Mark III	1964–70	2	4	4	11	11
Corvair	1960–1964	2	4	5	14	12
	1965–69	2	4	5	14	15
Corvette	1964–70	2	4	4	8	8
Dodge 880 and Dart	1964	2	7	7	17	17
	1965–70	2	7	④	17	17
Fairlane and Meteor and Montego	1964–65	2	4	4	11	11
	1966–70	2	4	4	13	13
Falcon and Comet and Cougar	1964–65	2	4	4	8	8
	1966–70	2	4	4	13	13
Ford	1964	2	4	4	8	8
	1965–70	2	4	4	11	11
Jeep, Except "J" Model ①	1964–70	2	4	None	①	①
Maverick	1970	2	4	4	13	13
Mercury	1964	2	4	4	8	8
	1965–70	2	4	4	11	11
Mustang	1965–66	2	4	4	8	8
	1967–70	2	4	4	13	13
Oldsmobile Oldsmobile Toronado—See Car Section	1964–70	2	4	5	8	8
Oldsmobile F-85	1964–70	2	4	5	8	8
Plymouth	1964–70	2	7	④	17	17
Pontiac	1964–70	2	4	4	8	8
American Motors, Except Rambler	1970	2	4	4	10	15
American	1968–69	2	None	4	10	15
	1964–67	2	None	4	17	17
Rambler American	1964–69	2	None	4	10	15
Tempest and Firebird	1964–70	2	4	5	8	8
Thunderbird	1964–66	2	5	4	13	13
	1967–70	2	4	4	11	11
Valiant and Barracuda	1964–70	2	7	④	17	17

④ —Lower ball joint replaceable with steering arm only.
① —Caster controlled by front transverse spring seat. Camber controlled by shims behind the upper control arm frame brackets.—"J" models, see car section.

Section Page Numbers

Section	Page	Section	Page	Section	Page	Section	Page	Section	Page	Section	Page
1	1332	4	1336	7	1337	10	1338	13	1339	16	1341
2	1335	5	1337	8	1338	11	1339	14	1340	17	1341
3	1335	6	1337	9	1338	12	1339	15	1340	18	1341

Diagnosis chart—1332

Suspension, Front End Alignment

Wheel Alignment

For a car to have safe steering control with a minimum of tire wear, certain established rules must be followed. These rules fix the value of planes, angles and radii relative to each other and to car and tire dimensions. Some factors of this proven plan are built-in, with no provision for adjustment; others are adjustable within limits. The entire scheme, being a relative one, depends one upon the other for factor values. It is, therefore, difficult to change some of the established settings without influencing others. This system is called steering geometry or wheel alignment and requires a complete check of all the factors involved.

Definitions of these factors and explanations of the effect each one has

the car on a straight piece of highway, may be the first indication of front end misalignment.

Check the steering wheel spoke position relative to straight-ahead car travel. If the wheel is not in a normal straight position, the condition may be corrected during toe-in adjustment.

1. Loosen the tie-rod adjusting clamps.

2. If the steering wheel position is satisfactory, shorten or lengthen both tie rods (the same amount) to the required toe-in specifications.

If, however, after correcting the toe-in value, the steering wheel position is wrong, shorten the right and lengthen the left tie-rod, or shorten the left and lengthen the right tie-rod (the same amount) to obtain the desired steering wheel position.

arm should be disconnected from the sector shaft. The steering wheel (or gear) should be turned from extreme right to extreme left to determine the halfway point in its turning scope. This will be the spot on the gear that is in action during straight-ahead driving and in which position the steering gear should be adjusted.

With the steering wheel in the straight-ahead position and the steering gear adjusted to no-lash status, reconnect the pitman arm.

To obtain front wheel to steering wheel relations, refer to steering wheel adjustment illustration.

Steering Geometry

Camber Angle

Camber is the amount that the front wheels are inclined outward at

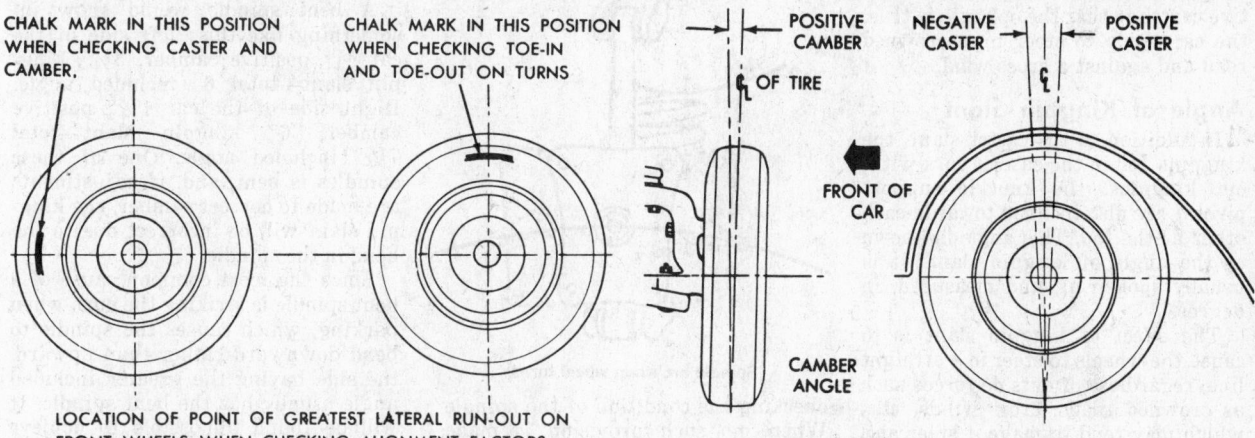

CHALK MARK IN THIS POSITION WHEN CHECKING CASTER AND CAMBER.

CHALK MARK IN THIS POSITION WHEN CHECKING TOE-IN AND TOE-OUT ON TURNS

LOCATION OF POINT OF GREATEST LATERAL RUN-OUT ON FRONT WHEELS WHEN CHECKING ALIGNMENT FACTORS

POSITIVE CAMBER

℄ OF TIRE

FRONT OF CAR

CAMBER ANGLE

NEGATIVE CASTER

POSITIVE CASTER

Wheel position for checking alignment

Caster and camber angles

on the car are given in the following paragraphs.

For adjustment data relative to each separate car and year, refer to the car section. Geometry specifications are found under front suspension of the car being serviced.

Toe-in, Relative to Steering Wheel Position

Toe-in, or straightaway alignment, is a value given the negative distance between the front extremes of the two front, or rear, wheels relative to a like measurement taken at the rear extremes of the same wheels.

This factor of alignment is blamed for more steering and tire wear troubles than any other. The careless or inexperienced mechanic will attempt all sorts of corrective measures with a turn of the tie-rod adjustments.

Always check steering wheel alignment in conjunction with and at the same time as, toe-in. In fact, the steering wheel spoke location, with

If the car has been wrecked, or indicates any evidence of steering gear or linkage disturbance, the pitman

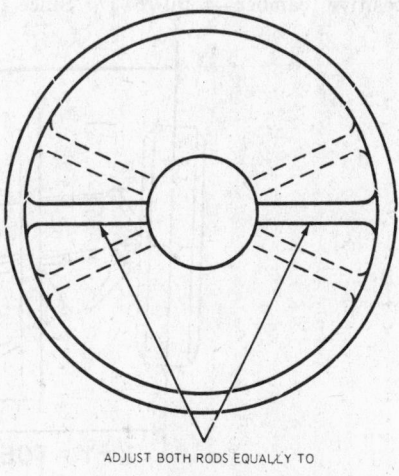

ADJUST BOTH RODS EQUALLY TO MAINTAIN NORMAL SPOKE POSITION

Steering wheel position

the top. Camber is spoken of, and measured, in degrees from the perpendicular.

The purpose of the camber angle is to take some of the load off the spindle outboard bearing.

Caster Angle

Caster is the amount that the kingpin (or in the case of cars without kingpins, the knuckle support pivots) is tilted towards the back of the car. Caster is usually spoken of, and measured, in degrees. Positive caster means that the top of the kingpin is tilted toward the back of the car. Positive caster is indicated by the sign +.

Negative caster is exactly opposite, the top of the kingpin is tilted toward the front of the car. This is generally indicated by —. Negative caster is sometimes referred to as reverse caster.

The effect of positive caster is to cause the car to steer in the direction

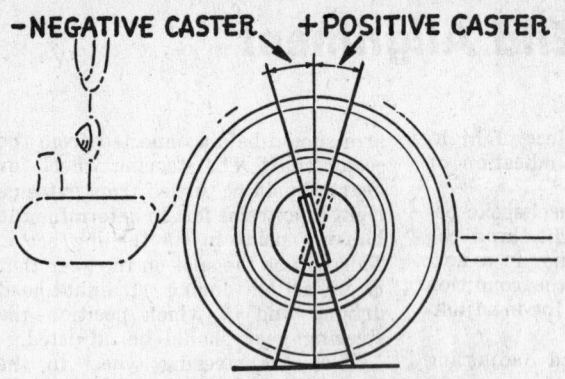

Caster angle. Note that if the pivot tilts forward, caster is negative; tilted back, caster is positive.

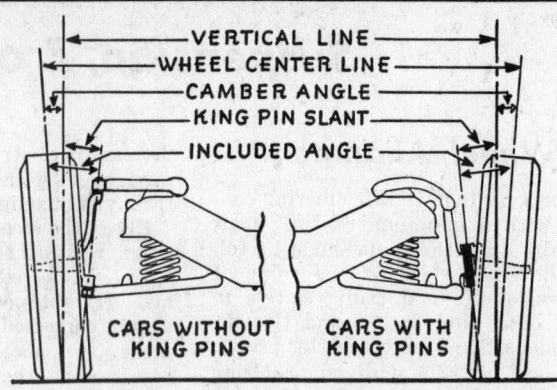

Camber, king pin slant and included angle

in which it tends to go. This is not necessarily a straight line, since cars with independently sprung front wheels usually steer easily. Positive caster in the front wheels may cause the car to steer down off a crowned road or steer in the direction of a cross wind. For this reason, many of our modern cars are arranged with negative caster so that the opposite is true, the car tends to steer up a crowned road and against a cross wind.

Angle of Kingpin Slant

In addition to the caster slant, the kingpins (or in the case of cars without kingpins, the knuckle support pivots) are also inclined towards each other at the top. This angle is known as the angle of kingpin slant. It is usually spoken of, and measured, in degrees.

The effect of kingpin slant is to cause the wheels to steer in a straight line, regardless of outside forces such as crowned roads, cross winds, etc., which may tend to make it steer at a tangent. As the spindle is moved from extreme right to extreme left it apparently rises and falls. Notice that it reaches its extreme high position when the wheels are in the straight-ahead position. Now, in actual operation, the spindle cannot rise and fall because the wheel is in constant contact with the ground. Therefore, the car itself will rise at the extreme right turn and come to its lowest position at the straight ahead point, and again rise for an extreme left turn. The weight of the car will tend to cause the wheels to come to the straight ahead position, which is the lowest position of the car itself.

Included Angle

Included angle is the name given to that angle which includes kingpin slant and camber. It is the relation between the centerline of the wheel and the centerline of the kingpin (or in the case of cars without kingpins, the knuckle support pivots). This angle is built in to the knuckle (spindle) forging and will remain

constant throughout the life of the car, unless the spindle itself is damaged.

When checking a car on the front end stand, always check kingpin slant as well as camber unless some provision is made on the stand for

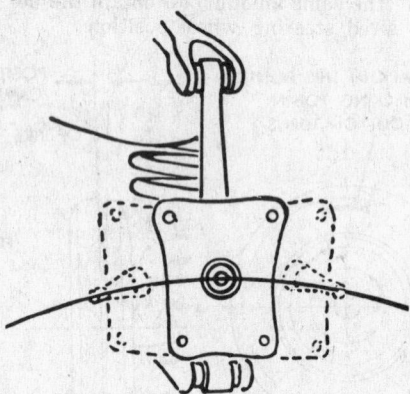

Spindle arc when wheel turns

checking the condition of the spindle. Where no such provision is made, add the kingpin slant to the camber for each side of the car. These totals should be exactly the same, regardless of how far from the norm any of the readings are.

For instance the left side of the car checks $5\frac{1}{2}°$ kingpin slant and $1°$ positive camber—total $6\frac{1}{2}°$. Since

both sides check exactly the same for the included angle, it is unlikely that both spindles are bent exactly the same amount in the same direction. It may be assumed that the spindles, in this instance, are not bent, and adjusting to correct for camber will automatically correct kingpin slant.

A bent spindle would show up something like this: left side of the car $\frac{3}{4}°$ positive camber, $5\frac{1}{4}°$ kingpin slant—total $6°$ included angle. Right side of the car $1\frac{1}{4}°$ positive camber, $6°$ kingpin slant—total $7\frac{1}{4}°$ included angle. One of these spindles is bent, and, if adjustments are made to correct camber, the kingpin slant will be incorrect due to the bend in the spindle.

Since the most common cause of a bent spindle is striking the curb when parking, which causes the spindle to bend downward rather than upward, the side having the greater included angle usually has the bent spindle. It will be found impossible to achieve good alignment and minimum tire wear unless the bent spindle is replaced.

Toe-In

Toe-in is the amount that the front wheels are closer together at the front than they are at the back. This di-

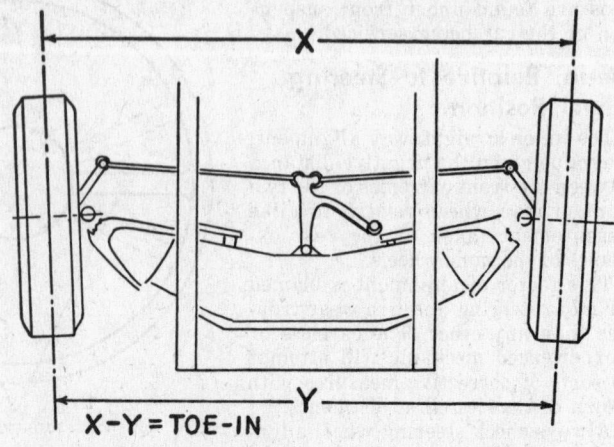

Toe-in. Note that the wheels are closer at the front than at the rear

mension is usually spoken of, and measured, in inches or fractions of inches.

Generally speaking, the wheels are toed-in because they are cambered. When a car operates with 0° camber it will be found to operate with zero toe-in. As the required camber increases, so does the toe-in. The reason for this is that the cambered wheel tends to steer in the direction in which it is cambered. Therefore, it is necessary to overcome this tendency of the wheel by compensating very slightly in the direction opposite to that which it tends to roll. Caster and camber both have an effect on toe-in, therefore toe-in is the last thing on the front end which should be corrected. Always set caster, camber and kingpin slant (included angle) before setting toe-in.

Toe-Out—Steering Radius

When a car is steered into a turn, the outside wheel of the automobile scribes a much larger circle than the inside wheel. Therefore, the outside wheel must be steered to somewhat less angle than the inside wheel. This difference in angle is sometimes called toe-out on turns.

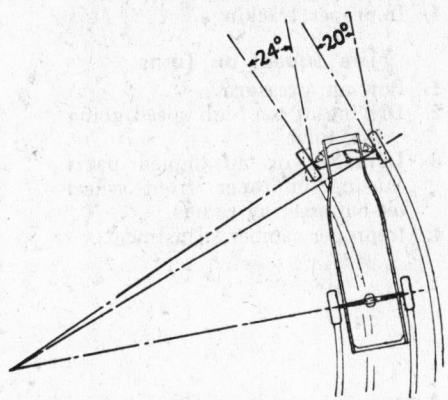

Toe-out. Inside wheel turns a greater number of degrees

The change in angle from toe-in in the straight ahead position to toe-out in the turn position is caused by the relative position of the steering arms to the kingpin and to each other.

If a line were drawn from the center of the kingpin through the center of the steering arm tie-rod attaching-hole, at each wheel these lines would be found to cross almost exactly in the center of the rear axle.

If the front end angles, including toe-in, are set correctly, and the toe-out is found to be incorrect, one or both of the steering arms are bent.

Tracking

While tracking is more a function of the rear axle and frame than it is the front axle, it is difficult to properly align the front suspension if the car does not track straight.

Tracking means that the centerline of the rear axle follows exactly the path of the centerline of the front axle when the car is moving in a straight line.

With cars which have equal tread, front and rear, the rear tires will follow in exactly the tread of the front tires when the car is moving in a straight line. However, there are many cars whose rear tread is wider than the front tread. On such cars the rear axle tread will straddle the front axle tread an equal amount on both sides when the car is moving in a straight line.

Perhaps the easiest way to check a car for tracking is to stand directly in back of it and watch it move in a straight line down the street. If the observer will stand as near to the center of the car as possible, he can observe readily, even with the difference in perspective between the front and the back wheels, whether or not they are tracking properly. If the car is observed to track incorrectly, the difficulty will be found in either the frame or the rear axle alignment.

Perhaps the easiest way to check

tracking is to park the car on a level floor, drop a plumb-line from the extreme outer edge of the front suspension lower A-frame, using the same drop point on each side of the car. Make a chalk mark where the plumb-line strikes the floor.

Do the same with the rear axle, selecting a point on the rear-axle housing for the plumb-line.

Now, measure diagonally from the left rear mark to the right front mark and from the right rear mark to the left front mark. These two diagonal measurements should be exactly the same. A ¼ in. variation would be acceptable.

If the measurements taken diagonally are not the same for both sides, measure from the right rear to the right front and from the left rear to the left front. These two measurements should also be the same, within ¼ in.

If the diagonal measurements are different, but the longitudinal measurements are the same on both sides, the frame is swayed (diamond-shaped).

However, in the event that the diagonal measurements are uneven and the longitudinal measurements are also uneven, the car is tracking incorrectly, the rear axle is misaligned.

If the diagonal measurements and the longitudinal measurements are uneven, but the car tracks correctly as observed on the street, a kneeback is indicated.

Kneeback means that one complete side of the front suspension is bent back.

This is often caused by crimping the front wheels against the curb when parking the vehicle, then starting up without straightening out the wheels.

It is possible to have caster, camber, toe-in, kingpin slant and included angle exactly correct and, if the car has a kneeback or does not track properly, have it handle very poorly.

Diagnosis

Steering Wheel Spoke Position Not Centered

1. Start with steering gear set on high-spot.
2. Check for proper toe-in.
3. Check for proper relation between lengths of each tie rod.

Front End Rides Hard

1. Improper tires.
2. Improper air pressure.
3. Shock absorbers too severe or malfunctioning.

Rides Too Soft

1. Improper air pressure.
2. Loss of spring load-rate, (weak springs).
3. Weak or leaking shock absorbers.

Car Steers to One Side At All Times

1. Incorrect caster angle.
2. Incorrect camber angle.
3. Incorrect kingpin or wheel support angle.
4. Unequal air pressure, or unequal tread.
5. Unequal or one-side brake drag.
6. Unequal shock absorber control.
7. Bent or damaged steering suspension components.
8. Uneven or weak spring condition, front or rear, causing car to sit unevenly.
9. Improper tracking.

Car Steers to One Side Only When Brakes Are Applied

1. Improper brake adjustment, shoes or anchors.
2. Grease or foreign substance on brake lining.
3. Excessive wear or bent condition in suspension components.

Car Steers Down Off Crowned Road But Normally on Flat Road

1. Excessive positive camber at one or both sides.
2. Weak or uneven shock absorber action.
3. Excessive or unequal wear in suspension components.

Car Wanders—Steers Erratically

1. Incorrect caster.
2. Improper tire pressure or unequal treads.
3. Excessive wear or damaged suspension components.
4. Power steering gear damaged, causing power operation to function abnormally.

Car Steers Hard

1. Binding steering or suspension parts.
2. Improper lubrication.
3. Improper (too large) tires.
4. Low air pressure.

Tires Cup on Outside Edge With Ripple Wear Pattern

1. Generally incorrect camber or toe-in.

Tires Wear Unevenly in Center And Faster Than Outer Edges

1. Generally too much air pressure.

Tires Wear and Scuff on Both Outer Edges, Not in Center

1. Generally low air pressure.

Uneven Outer Edge Wear—Center and Inner Edge Wear Normal

1. With adjustments normal this is generally caused by driving into turns at too high speed. Do not confuse with outer edge cupping.

Unequal Tire Wear Between Front Wheels

1. Unequal air pressure.
2. Unequal tire quality or size.
3. Bent or worn steering suspension components.
4. Improper tracking.

Tire Squeal on Turns

1. Low air pressure.
2. Driving at too high speed going into turn.
3. Damaged or misaligned parts causing improper front wheel toe-out-steering radius.
4. Improper camber adjustment.

Section 1
Suspension and Ball Joint Checks
General Procedure

In checking suspension and ball joints it is advisable to follow factory recommendations. For all practical purposes, however, the following procedures are entirely satisfactory.

Suspension System

This check is made with the ball joints loaded, so that suspension elements, other than ball joints, may be checked.

When the front spring, or torsion bar, is supported by the lower control arm, the jackstand should be located under the front crossmember or frame.

When the front spring is supported by the upper control arm, the jackstand should be located under the lower control arm.

Vehicles supported in their above respective positions may have the parts of the suspension system examined for wear.

Vertical or horizontal movement at the road wheel should not exceed the following:

up to and including 16 in. . . ¼ in.
16 in. to 18 in. . . . . ⅜"
more than 18 in. . . . . ½"

Ball Joint Check

When checking the ball joints for wear, they must free of any load, therefore, the following jacking procedure is in order.

When the front spring, or torsion bar, is supported by the lower control arm, the jackstand should be located under the lower control arm.

When the front spring is supported by the upper control arm, the jackstand should be located under the crossmember or frame.

Replace the upper ball joint, if any noticeable play is present in the joint, when the spring or torsion bar is supported by the lower arm.

Replace the lower ball joint, if any noticeable play is present in the joint, when the spring is supported by the upper control arm.

Replace the load carrying ball joint if the side-play (horizontal motion) of the tire, when rocked, exceeds manufacturers' tolerances; or if the up-and-down (vertical motion) exceeds manufacturers' tolerances. (See charts of tolerance.)

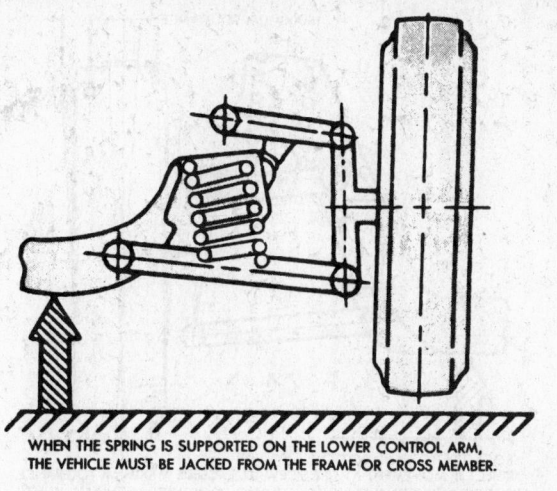

WHEN THE SPRING IS SUPPORTED ON THE LOWER CONTROL ARM, THE VEHICLE MUST BE JACKED FROM THE FRAME OR CROSS MEMBER.

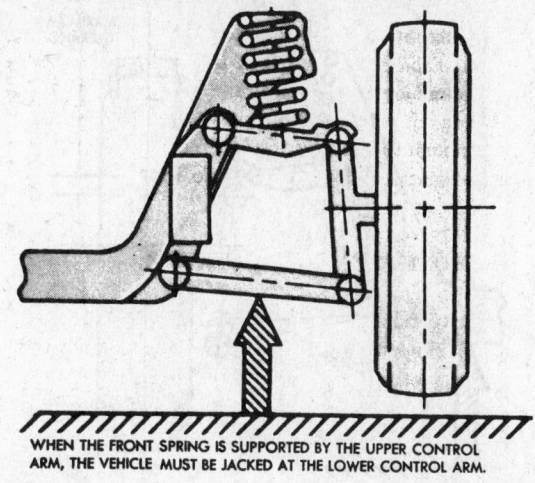

WHEN THE FRONT SPRING IS SUPPORTED BY THE UPPER CONTROL ARM, THE VEHICLE MUST BE JACKED AT THE LOWER CONTROL ARM.

Steering and suspension jacking procedure

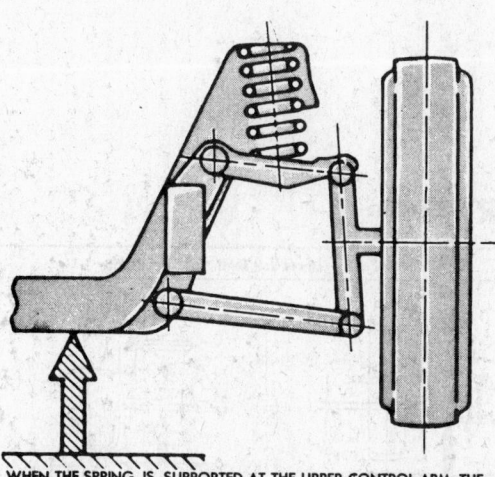

WHEN THE SPRING IS SUPPORTED AT THE UPPER CONTROL ARM, THE CAR MUST BE HOISTED AT THE FRAME.
REJECT IF LOWER JOINT IS PERCEPTIBLY LOOSE.

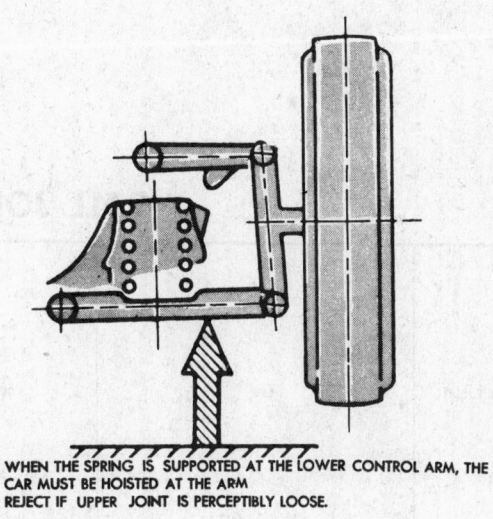

WHEN THE SPRING IS SUPPORTED AT THE LOWER CONTROL ARM, THE CAR MUST BE HOISTED AT THE ARM
REJECT IF UPPER JOINT IS PERCEPTIBLY LOOSE.

Ball joint inspection jacking procedure

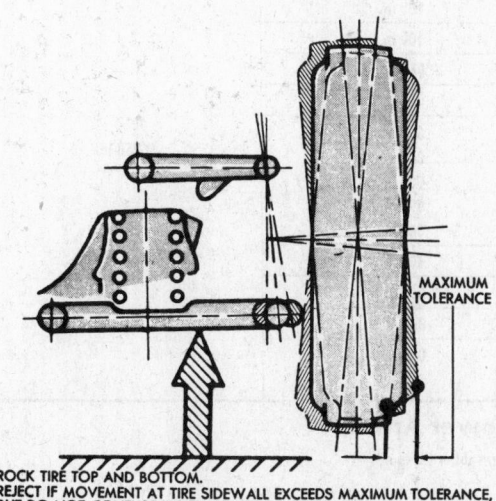

ROCK TIRE TOP AND BOTTOM.
REJECT IF MOVEMENT AT TIRE SIDEWALL EXCEEDS MAXIMUM TOLERANCE, BUT DO NOT CONFUSE WHEEL BEARING LOOSENESS WITH BALL JOINT WEAR.

MAXIMUM TOLERANCE

Check ball joint radial (side) play

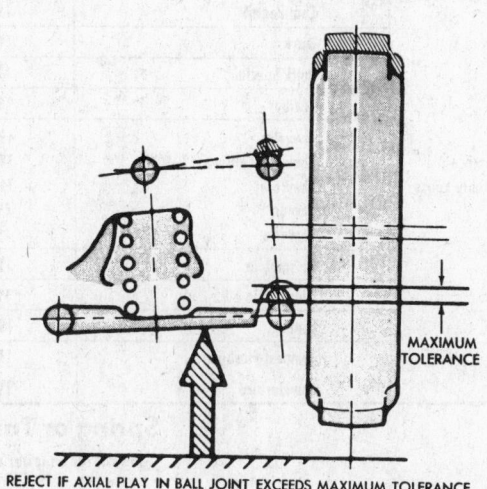

REJECT IF AXIAL PLAY IN BALL JOINT EXCEEDS MAXIMUM TOLERANCE.

MAXIMUM TOLERANCE

Check ball joint axial (up-and-down) play

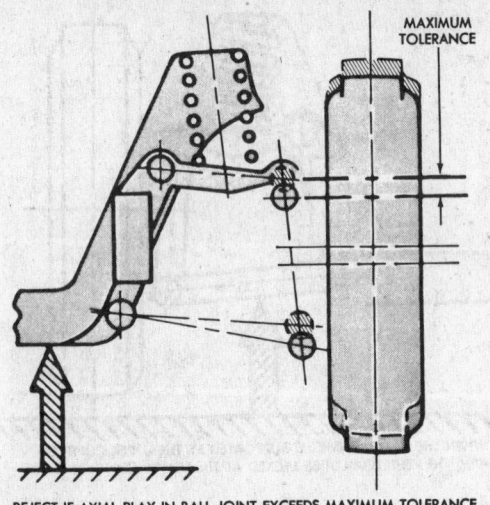

REJECT IF AXIAL PLAY IN BALL JOINT EXCEEDS MAXIMUM TOLERANCE.

Check ball joint axial (up-and-down) play

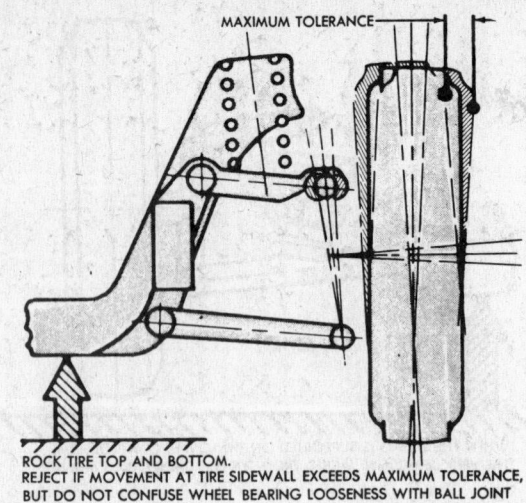

ROCK TIRE TOP AND BOTTOM.
REJECT IF MOVEMENT AT TIRE SIDEWALL EXCEEDS MAXIMUM TOLERANCE, BUT DO NOT CONFUSE WHEEL BEARING LOOSENESS WITH BALL JOINT WEAR.

Check ball joint radial (side) play

BALL JOINT CHART 1

Upper Ball Joint	Car	Year	Up-and-Down Play / Lower Ball Joint — At the Joint should not exceed the following	Side Play — At the Tire should not exceed the following
	All Chrysler Corporation	1964–67	.050 in.	
		1968–1970	.080 in.	
	Ford & Mercury	1964–70	ZERO	
	Continental	1964–70	.200 in.	
	Buick	1964–70	.100 in.	
	Buick Special	1964–70	.060 in.	
	Cadillac	1964–70	.062 in.	
Replace if Noticeably Loose	Chevelle	1964–70	.060 in.	.250 in.
	Chevy II, Camaro (67–70)	1968–70	.060 in.	
	Chevrolet	1964–70	.093 in.	
	Corvair	1960–69	.093 in.	
	Corvette	1964–70	.062 in.	
	Oldsmobile	1964–70	.125 in.	
	Oldsmobile F-85	1964–70	.060 in.	
	Pontiac	1964–70	.060 in.	
	Tempest-Firebird	1964–70	.060 in.	
	Thunderbird	1967–70	Zero	

Spring or Torsion Bar on Lower Arm

Support car on lower arm. Upper arm must clear rubber bumper.

BALL JOINT CHART 2

Lower Ball Joint	Car	Year	Up-and-Down Play Upper Ball Joint At the Joint should not exceed the following	Side Play At the Tire should not exceed the following
Replace if Noticably Loose	Fairlane-Montego, Maverick	1964–70	.200 in.	
	Falcon-Comet, Cougar, Mustang	1964–70	.200 in.	
	Thunderbird	1964–66	ZERO	.250 in.
	Chevy II	1964–67	.093 in.	
	All American Motors	1964–69 1970	No Upper Ball Joint .080 in.	.250 in.

Spring on Upper Arm
Support car on cross member or side rails.

Section 2
Wheel Bearing and Seal Replacement

Place jack under lower suspension arm. Remove hub cover and grease cap. Remove spindle nut, keyed washer and outer bearing. Slide off hub and drum. In some cases drum wear may require loosening of brake adjustment.

At this point brakes and drums should be inspected for any non-standard condition.

With hub and drum on bench, remove seal and inner bearing. Thoroughly clean all parts. Drive out inner and outer races of roller type. Use care not to mar the bearing surfaces.

Pack bearings thoroughly with approved lubricant. In replacing cups a bearing race driver is best. If a punch is used be sure it is blunt and drive parts in carefully to avoid any cocking of the bearings.

Install new grease seal in hub. Reassemble hub and drum on spindle and replace outer bearing, key washer and nut.

A common method of adjustment is to tighten to zero clearance and back off to first cotter pin castellation. Some manufacturers recommend tightening to approximately 10 to 12 ft. lbs. then backing off nut one flat or one-sixth turn. If cotter pin holes do not line up, loosen slightly.

Readjust brake if necessary and install grease cap and hub cover. Remove jack.

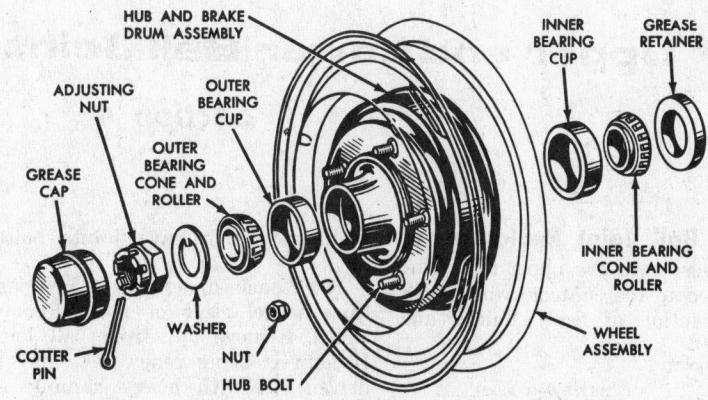

Front wheel roller bearing—typical of drum brakes

Section 3
Kingpin and Bushing Replacement

Kingpin and bushings can be placed in two general classes; (A) with bushings in knuckle, (B) with bushings in spindle. Generally, the first type will be found in General Motors cars and some Dodges.

In all cases, jack up the car and remove the hub and drum as described in the wheel bearing section.

(A) Remove backing plate-to-knuckle bolts and lift assembly, with brakes, from knuckle. Suspend it with a piece of wire to prevent damage to brake hose.

Drive out lock-pin or bolt. With sharp punch remove top welch plug.

Now drive pin and bottom welch plug down through knuckle and support.

Drive bushings from spindle and replace with new. Be sure when driving bushing that grease holes line up with those in knuckle.

Align and ream bushings to a snug running fit for the new kingpin.

Insert the kingpin through the top of spindle, support, thrust bearing (with shims to control up-and-down play) and bottom of spindle. Keep kingpin in proper rotation so that lock-pin can be entered. Replace lock-pin or bolt securely. Replace upper and lower Welch plugs.

Reinstall backing plate with steering arms and properly lubricate. Reinstall hubs, drums and wheels and remove jacks.

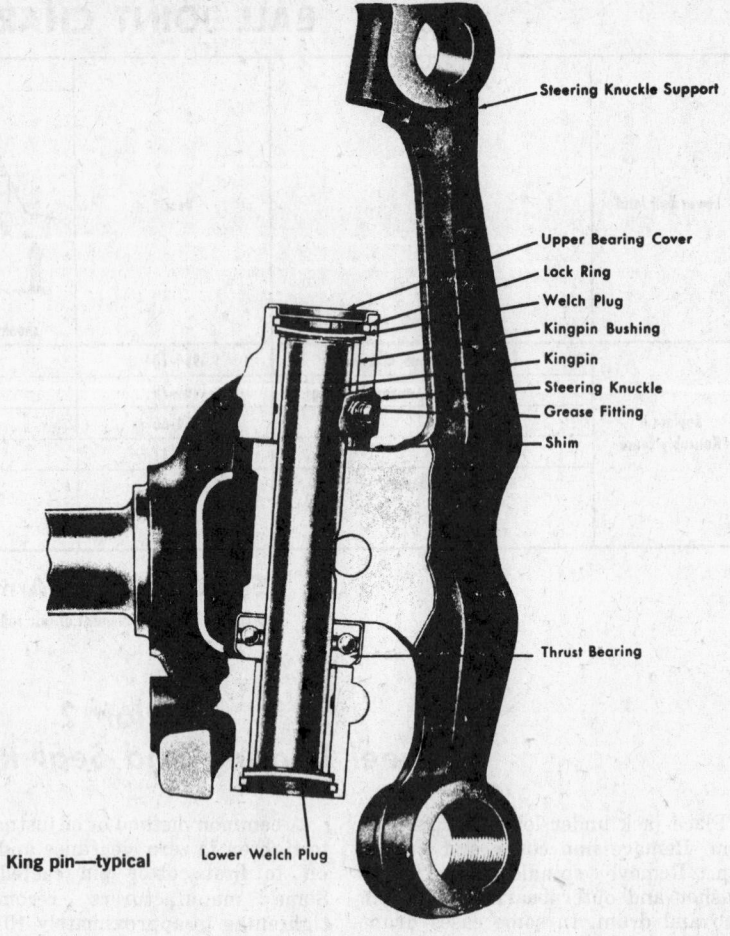

Steering Knuckle Support
Upper Bearing Cover
Lock Ring
Welch Plug
Kingpin Bushing
Kingpin
Steering Knuckle
Grease Fitting
Shim
Thrust Bearing

King pin—typical   Lower Welch Plug

Upper and Lower Ball Joint Replacement

Section 4

Upper Ball Joint Replacement

On these cars the upper ball joint is riveted to the control arm. (With the exception of early Buicks and

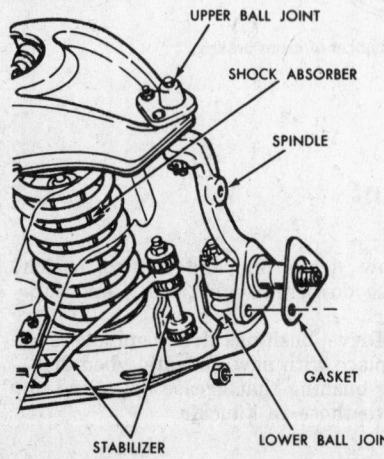

Ford ball joint

UPPER BALL JOINT
SHOCK ABSORBER
SPINDLE
GASKET
STABILIZER   LOWER BALL JOINT

Cadillacs which have joints bolted in.)

Place jack under lower arm and raise wheel clear of floor. Remove wheel. Remove nut from ball joint. If joint is being renewed it may be driven out with heavy hammer. If threads are to be saved, a spreader tool should be used.

After removing joint from knuckle support, cut off rivets at upper arm. Drilling rivet eases this job.

To replace ball joint, install in upper arm, using the special bolts furnished with the joint for the purpose. Do not use ordinary bolts.

Next, set the taper into the upper end of the knuckle support, install nut and cotter pin. Alignment should be checked.

Lower Ball Joint Replacement

This operation is the same as above, except for Corvair and Tempest.

The lower ball joint on the Tempest and early Corvair is threaded into the lower arm. The joint has a self-tapping thread. Use 1-9/16 in. socket for Tempest and 1⅝ in. for Corvair.

The lower ball joint on the Corvair, late production, is pressed in.

The Greenbriar models are riveted. Alignment should be checked.

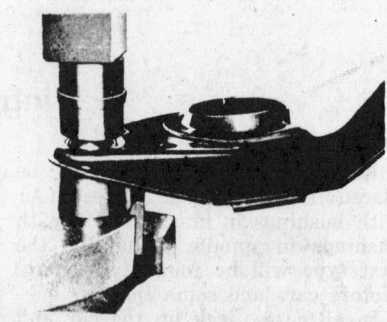

Removing Corvair ball joint from support arm

Section 5

These ball joints are pressed into support arms. To replace pressed-in units it is necessary to remove front spring and support arm.

After removing wheel and drum, loosen nut slightly at ball joint taper and hammer around area to loosen. If new ball joints are to be installed it is not necessary to protect threads.

After spring and support arm are removed, place under arbor press with suitable tool, or large socket, and press from arm.

Replace ball joint by reversing the pressing procedure.

Special tools are available that are of C-clamp-type and can be used on some cars to avoid removal of springs and support arms.

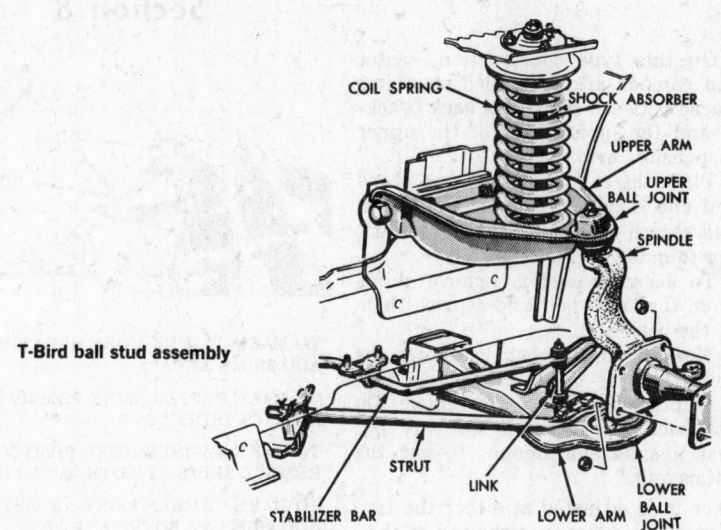

T-Bird ball stud assembly

Section 6

Proceed as in Section 4. Bolts instead of rivets are used. In assembly, use the special bolts furnished. Do not use ordinary bolts. The bolts and nuts should be tightened to 35-40 ft. lbs.

Section 7

Upper Ball Joint Replacement

On these cars the upper ball joint is threaded into the control arm.

Place jack under lower arm and release load on torsion bar. Raise wheel clear of floor. Remove wheel. Remove nut from ball joint. If joint is being renewed, it may be driven with heavy hammer. If threads are to be saved, a spreader tool should be used.

After removing joint from knuckle support, the ball joint can be unscrewed from the support arm. Tools C-3560, C-3561 or C-3714 are recommended for this operation.

In replacing the ball joint, be sure to engage the threads in the control arm squarely. Torque to 125 ft. lbs. If this torque can not be obtained, check

for bad threads in arm or on joint. Install new balloon seal over joint.

Place joint in knuckle and install nut. Torque to 55 ft. lbs. on Valiant and Dart models beginning 1963, and 100 ft. lbs. on all others. Reload and adjust torsion bar and reset height.

Lower Ball Joint Replacement

This operation is the same as replacing the upper ball joint described in above section.

NOTE: exception: Dodge, Plymouth and Valiant. On these cars the lower ball joint is integral with the steering arm and is not serviced separately. To service this unit: remove the upper arm bumper. Raise car so that front suspension is under no load. If jacks are used, a support

must be placed between jack and K-member.

Remove the wheel and drum assembly. Remove the two lower bolts holding the steering arm to the backing plate.

Remove tie-rod end from the steering arm. Do not damage seal. Remove the ball joint stud from lower control arm. A spreading tool will aid this operation.

To replace, install new seal on ball joint. Bolt the steering arm to the backing plate. Insert the ball joint into control arm and torque nut to 100 ft. lbs.

Reinstall the tie rod end. Install drum and wheel. Reload and adjust torsion bar and reset height.

Caster, Camber and Toe-in Adjustment

Section 8

On this type construction, caster and camber are controlled by shims placed between the frame back bracket and the inner shaft of the upper suspension arm.

Place the car on a front end stand and check caster, camber and toe-in and record the readings before starting to make corrections.

To decrease caster, remove shims from the front bolt and replace them at the back bolt.

Keep in mind when loosening the bolts that the upper suspension arm is helping to support the vehicle. Do not back the bolts all the way off, just a sufficient amount to get the shims out.

Toe-in is adjusted at either the tie rod itself or a sleeve at the end of the tie-rod. Loosen the clamp bolt and turn the tie-rod (or sleeve) until correct toe-in is obtained.

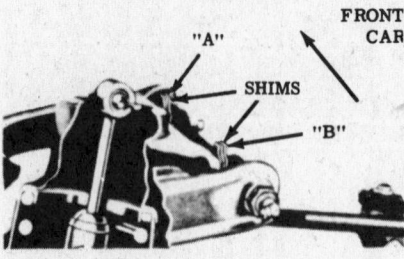

TO MAKE CASTER MORE NEGATIVE: ADD SHIMS AT "A".

TO MAKE CASTER MORE POSITIVE: REMOVE SHIMS AT "A".

TO MAKE CAMBER MORE POSITIVE: REMOVE SHIMS AT BOTH "A" & "B".

TO MAKE CAMBER MORE NEGATIVE: ADD SHIMS AT BOTH "A" & "B".

NOTE: BY ADDING OR SUBTRACTING AN EQUAL AMOUNT OF SHIMS FROM "A" AND "B", CAMBER WILL CHANGE WITHOUT AFFECTING CASTER ADJUSTMENT.

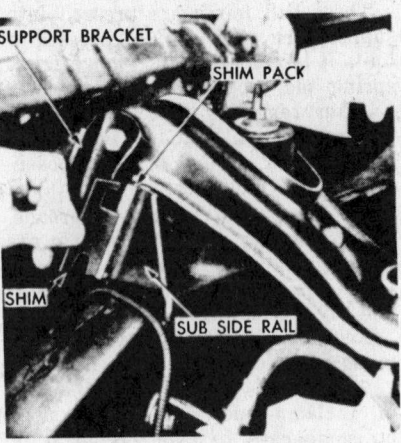

Shim pack—typical of early Chrysler products

Shim method—typical of General Motors and some Ford products

Section 9

Both caster and camber are adjusted at the eccentric pin in the outer end of the upper support arm.

Set the car up on a front end stand and check caster, camber and toe-in readings.

Make a note of each reading so as to determine which way the eccentric adjuster should be turned.

Remove the grease fitting from the front bushing and, working through the grease fitting with an Allen wrench, engage the eccentric pin at the front. Loosen the clamp bolt in the top of the knuckle support, which will leave the eccentric pin free to turn.

Now, turn the eccentric pin (which has a right-hand thread) first in the direction which tends to correct caster. The pin would be turned counterclockwise to increase caster or clockwise to decrease caster. If it is necessary to decrease caster, turn the

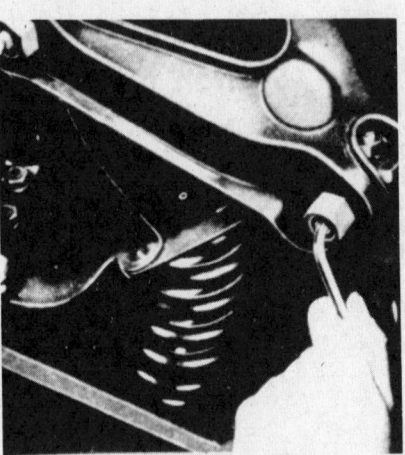

Upper, outer eccentric pin method

pin clockwise until the correct caster setting is attained and then rock it back and forth until the correct cam-

ber reading is obtained. Since both adjustments are made at the same pin, it is sometimes absolutely impossible to get exact readings. However, it will be found that with very little error in caster, the correct camber reading can be obtained.

Recheck the caster and camber readings on the front end stand and, once certain they are correct, secure the clamp bolt and reinstall the grease fitting.

Toe-in is adjusted either at the tie-rod itself or, on some models, sleeves at the ends of each of the tie rods.

Where the tie-rod itself is turned, simply loosen the clamp bolt which holds the tie-rod to the end assembly and turn the rod. In cases where sleeves are used at the end of the tie-rods, loosen the clamp bolts which prevent the sleeve from turning and turn the sleeve until the correct toe-in is obtained.

Section 10

Caster is adjusted by lengthening or shortening tie-struts at the frame crossmember. Lengthening the struts by loosening the front, and tightening the rear, nuts produces more negative caster. Reverse this for more positive caster. One turn makes approximately ½° change.

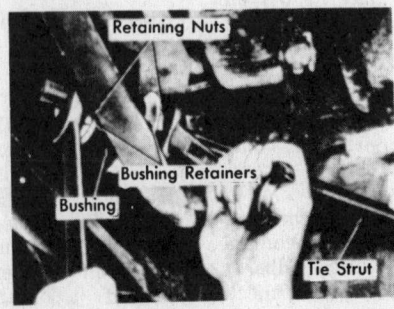

Strut rod control of caster

Section 11

On these cars, caster and camber are controlled by moving the upper control arm assembly where mounted at the frame. The arm-to-frame mounting bolts pass through slotted holes in frame or underbody. Tool number 62F3000A, or equivalent, is necessary to make these adjustments.

With the tool installed, loosen the upper arm inner shaft retaining bolts two or three turns, or enough to allow serrations to pass when turning one or both of the tool adjusting screws. Turning both in, clockwise, will increase camber. Turning both out, or counterclockwise, will reduce camber. Turning front screws in and rear screw out will give positive caster. Turning front screw out and rear screw in will give negative caster.

Caution Loosen or tighten tool screws not more than two turns, alternately, to avoid collapsing sheet metal. Adjust caster first. One complete screw turn will give about 1/4° change in setting. After adjustments have been made, tighten inner shaft retaining-bolts to 150-160 ft. lbs. and recheck the caster and camber to be sure no accidental changes have been made.

Section 12

This is a camber adjustment only; caster adjustment is made at another point.

This is the first adjustment to be performed on the front suspension. The camber adjustment is made by means of shims between the upper control arm inner shaft and the front crossmember. Although shims can be changed at either the front or rear attachment, it is important that the shimming be done equally, so as to have no effect on caster. Adding shims at both front and rear of support shaft will decrease positive camber. The procedure for adjustment is to loosen the upper support shaft-to-crossmember bolts, add or remove shims (equally) as required and retighten the bolts. Opposing sides should not vary more than 1/4°.

Section 13

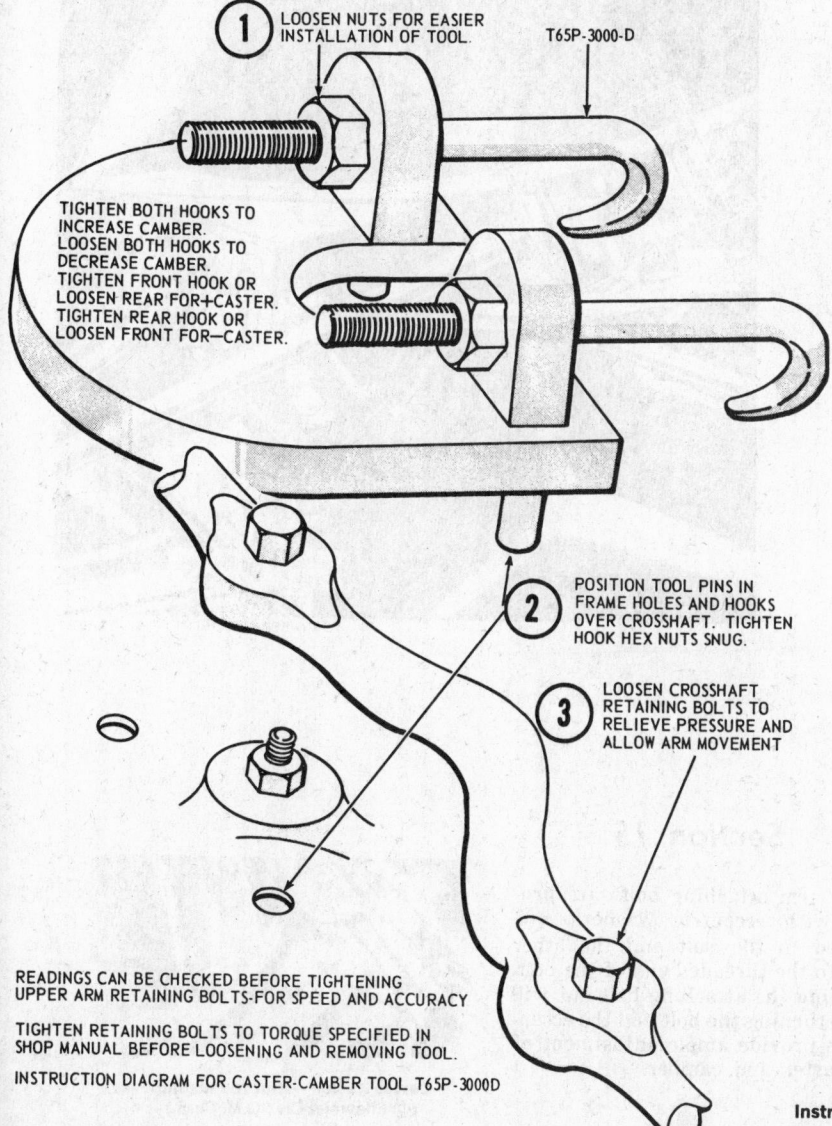

① LOOSEN NUTS FOR EASIER INSTALLATION OF TOOL.

T65P-3000-D

TIGHTEN BOTH HOOKS TO INCREASE CAMBER. LOOSEN BOTH HOOKS TO DECREASE CAMBER. TIGHTEN FRONT HOOK OR LOOSEN REAR FOR+CASTER. TIGHTEN REAR HOOK OR LOOSEN FRONT FOR—CASTER.

② POSITION TOOL PINS IN FRAME HOLES AND HOOKS OVER CROSSHAFT. TIGHTEN HOOK HEX NUTS SNUG.

③ LOOSEN CROSSHAFT RETAINING BOLTS TO RELIEVE PRESSURE AND ALLOW ARM MOVEMENT

READINGS CAN BE CHECKED BEFORE TIGHTENING UPPER ARM RETAINING BOLTS-FOR SPEED AND ACCURACY

TIGHTEN RETAINING BOLTS TO TORQUE SPECIFIED IN SHOP MANUAL BEFORE LOOSENING AND REMOVING TOOL.

INSTRUCTION DIAGRAM FOR CASTER-CAMBER TOOL T65P-3000D

Adjust the camber by removing or installing shims between the pivot bracket of the front suspension lower arm and the mounting bracket on the underbody in the engine compartment. The removal of shims between the mounting and pivot brackets will move the lower ball joints inward. The installation of shims between the mounting and pivot brackets will move the lower ball joint outward. Camber adjusting shims are available in several standard shim thicknesses. A 1/16 in. change of shim thickness will change the camber angle 1/3°. The total shim stack thickness should not exceed 11/16 in.

The caster adjustment is made by repositioning the strut on the lower arm. Adjust the caster by loosening the rearward washers, nuts and bolts. Lift the strut so that the strut serrations will be free from the serrations on the lower arm. Lengthen the distance between the strut forward mount and the side of the lower arm to decrease the caster angle. Decrease the distance between the strut forward mount and the side of the lower arm to increase the caster angle. Tighten the rearward nuts that retain the strut to the lower arm. Check the caster, camber, and toe-in alignment for the correct settings listed in the specifications.

Check the steering wheel spoke position when the front wheels are in the straight-ahead position. If the spokes are not in their normal position, they can be properly adjusted while toe-in is being adjusted.

Instruction diagram—caster-camber tool no. T65P-3000-D

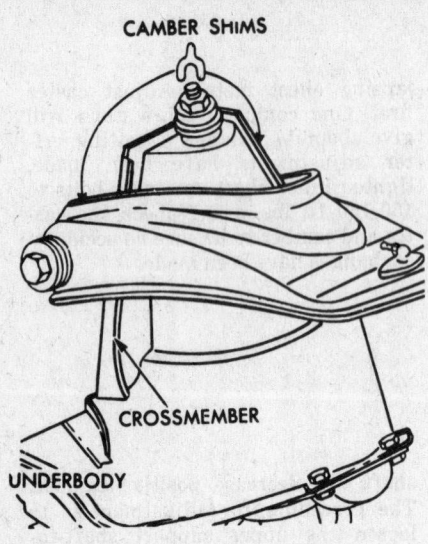

CAMBER SHIMS

CROSSMEMBER

UNDERBODY

Corvair camber—shim method
(© Chevrolet Div., G.M. Corp.)

1. Loosen the two clamp bolts on each spindle connecting rod sleeve.
2. Adjust toe-in. If the steering wheel spokes are in their normal position, lengthen or shorten both rods equally to obtain correct toe-in. If the steering wheel spokes are not in their normal position, make the necessary rod adjustments to obtain correct toe-in and steering wheel spoke alignment.
3. Recheck toe-in and steering wheel spoke alignment. If toe-in is correct, and the steering wheel spokes are still not in their normal position, turn both connecting rod sleeves upward or downward the same number of turns to move the steering wheel spokes.
4. When toe-in and steering wheel spoke alignment are both correct, torque the clamp bolts on both connecting rod sleeves to specifications.

Section 14

This is a caster adjustment only; camber adjustment is made at another point.

Caster is adjusted by turning two nuts at the rear of the strut rod. Lengthening this rod by turning the nuts increases caster, shortening this rod by turning the nuts decreases caster. Caster should be to specifications and be within ¼° of opposing side.

CASTER ADJUSTMENT   STRUT

NEGATIVE CASTER

POSITIVE CASTER

NEGATIVE CAMBER   POSITIVE CAMBER

CAMBER ADJUSTMENT

Caster and camber adjustments (see Sections 14 and 15)

Section 15

The lower control arm attaching bolts are provided with eccentric cams, or washers. Loosening the lower control arm pivot bolt and rotating the eccentrics will move the lower control arm in or out. This will alter the camber.

NOTE: as in the case of American Motors (except American), the lower control arm attaching bolts are provided with eccentric washers, one attached to the bolt and the other keyed to the threaded end of the bolt. Loosening the attaching bolt nut will permit turning the bolt and the eccentrics to provide ample adjustment of both caster and camber.

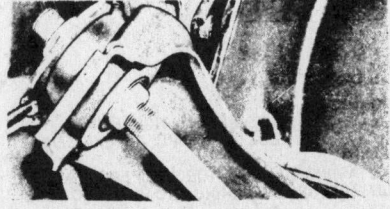

Corvair caster—strut rod method
(© Chevrolet Div., G.M. Corp.)

Section 16

This is a camber adjustment only; caster adjustment is made at another point.

Camber change is made by turning the camber eccentric at the upper ball joint. This eccentric is between the ball stud and the upper knuckle. Loosen ball stud nut and tap up to free eccentric enough to rotate it in its mounting. Keep the stud to the rear of the eccentric to hold the steering arm angle within limits.

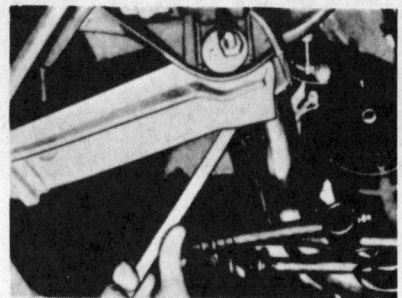

Lower arm eccentric mounting
(Chevy II shown)

Section 17

The upper control arm attaching bolts are fitted through eccentric washers, one welded to head and one keyed to threaded end.

To set caster or camber, loosen nuts and rotate with washers.

Turning both bolts equal amounts will adjust camber. Turning either front or rear bolt, one more than the other, will adjust caster.

Always recheck caster, camber and toe-in when one is moved.

After caster and camber readings are set to specifications, torque the cam bolt nuts to 65 ft. lbs.

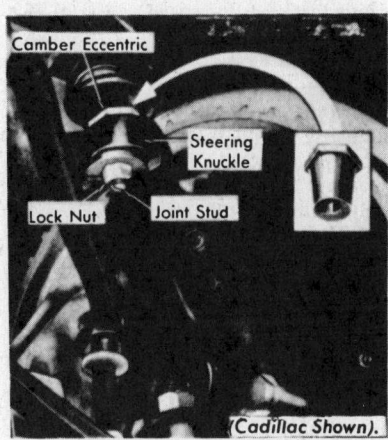

Upper ball joint camber adjustment
(© Cadillac Div., G.M. Corp.)

Section 18

This is a caster adjustment only; camber adjustment is made at another point.

Caster is adjusted by turning two nuts at the front of the strut rod. Lengthening this rod by turning the nuts increase caster, shortening this rod by turning the nuts decreases caster. Caster should be to specifications and be within 1/4° of opposing side.

Ford ball joint

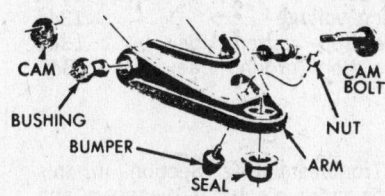

Typical of late Chrysler products

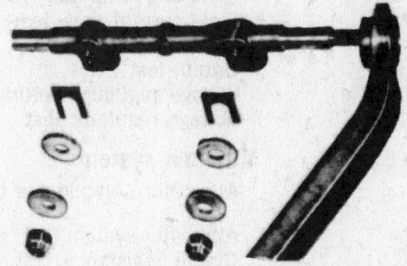

Lower arm shim control (Rambler American)
(© American Motors Corp.)

Basic Electrical Diagnosis

Index

NOTE: See Troubleshooting Section in this manual for more electrical diagnosis and battery tests.

Basic Electrical Diagnosis

To satisfy the growing trend toward organized engine diagnosis and tune-up, the following gauge and meter hook-ups, as well as diagnosis procedures are covered. The most sophisticated tune-up and diagnostic facilities are no more than a complex of the basic gauges and meters in common, everyday use. Therefore, to understand gauge and meter hook-ups, their applications and procedures, it is to be equipped with the know-how to perform the most exacting diagnosis.

Testing the Battery

Specific Gravity Test— Hydrometer

Before attempting any electrical checks, it is important that the condition of the battery be known.

The battery is an electrochemical warehouse for the energy and current requirements of the various electrical components of the vehicle. This all-important unit serves three different purposes. It supplies current for the starting motor, ignition system, lights, accessories, etc. It helps control voltage in the entire electrical system, and it furnishes current to the system when electrical demands are greater than the output of the generator.

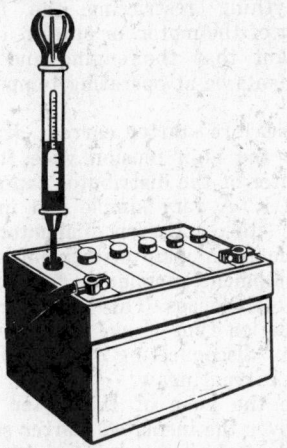

While not technically exact, a practical measurement of the chemical condition of the battery is indicated by measuring the specific gravity of the acid (electrolyte) contained in each cell. The electrolyte in a fully charged battery is usually between 1.260 and 1.280 times as heavy as pure water at the same temperature. Variations in the specific gravity readings for a fully charged battery may differ. Therefore, it is most important that all battery cells produce an equal reading.

As a battery discharges, a chemical change takes place within each cell. The sulphate factor of the electrolyte combines chemically with the battery plates, reducing the weight of the electrolyte. A reading of the specific gravity of the acid, or electrolyte, of any partially charged battery will therefore be less than that taken in a fully charged one.

The hydrometer is the instrument in general use for determining the specific gravity of liquids. The battery hydrometer is readily available from many sources, including your local auto replacement parts house. The following chart gives an indication of specific gravity value, related to battery charge condition.

Specific Gravity Reading	Charged Condition
1.260-1.280	Fully charged
1.230-1.250	Three-quarter charged
1.200-1.220	One-half charged
1.170-1.190	One-quarter charged
1.140-1.160	Just about flat
1.110-1.130	All the way down

Testing Battery Polarity

Battery polarity is very important, especially since the introduction of AC generators. Permanent damage to the diodes of alternators (AC generators) will result from reversing polarity.

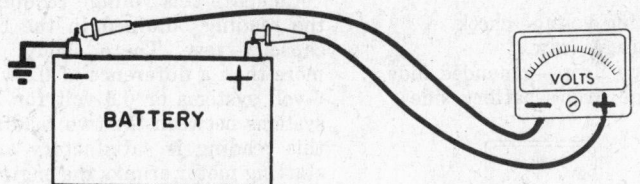

To determine battery polarity, turn the voltmeter selector to the high reading scale. Connect voltmeter leads to the battery posts. If the gauge needle moves in the correct direction, the positive lead of the meter is on the positive (+) post of the battery. If the gauge needle moves in the wrong direction, polarity is reversed.

Know Your Instruments
Ohmmeter

An ohmmeter is used to measure electrical resistance in a unit or circuit. The ohmmeter has a self-contained power supply. In use, it is connected across (or in parallel with) the terminals of the unit being tested.

Ammeter

An ammeter is used to measure current (amount of electricity) flowing through a unit, or circuit. Ammeters are always connected in the line (in series) with the unit or circuit being tested.

Voltmeter

A voltmeter is used to measure voltage (electrical pressure) pushing the current through a unit, or circuit. The meter is connected across

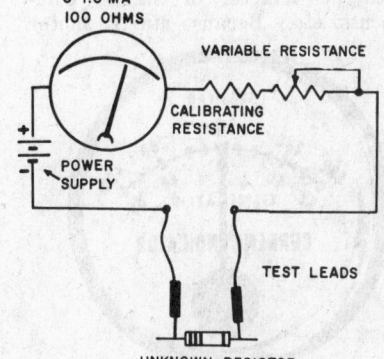

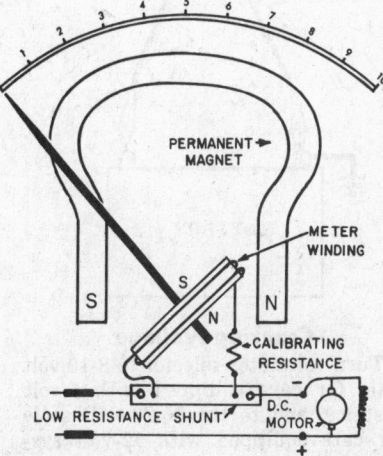

the terminals of the unit being tested. The meter reading will be the difference in pressure (voltage drop) between the two sides of the unit.

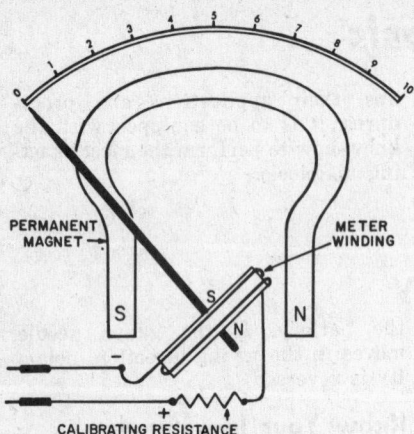

Testing Battery Capacity

It is not conclusive enough that a specific gravity test be made of the chemical condition of the battery by hydrometer. The electrical aspects must also be known. The battery must have capacity. It must be capable of maintaining a useful voltage, even while delivering the heavy load required in cranking the engine.

Capacity Test

While there are many very good testers available for the specific purpose of testing batteries, a satisfactory battery capacity test can be made with a simple voltmeter.

With engine at operating temperature, proceed as follows:

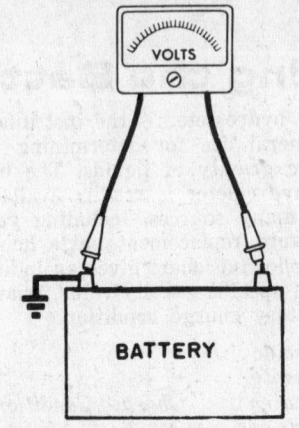

With prods (two ice picks will do) forced into the center of each battery post, and the voltmeter set to the high voltage scale (16-20 volts), connect voltmeter test leads to the battery post prods. Turn on the lights, heater, and accessories. Remove the center (high tension) wire from the distributor cap and ground it. Crank the engine for 20 second periods, three or four times. Wait about one-half minute between cranking sessions to prevent overheating the starter. During these cranking tests, the voltage should not drop below 5 volts for a 6-volt battery or 10 volts for a 12-volt battery. Even at temperatures below freezing, the voltmeter should not register below 9.4 volts on a 12-volt system.

Testing the Starting Motor

Extreme load conditions of the starter motor and its circuit require that the battery be kept in a fully charged condition and that starter circuit resistance be kept at a maximum.

Testing the Starter Circuit

The starter circuit should be divided and tested in four separate phases:

1. Cranking voltage check.
2. Amperage draw.
3. Voltage drop—grounded side.
4. Voltage drop—battery side.

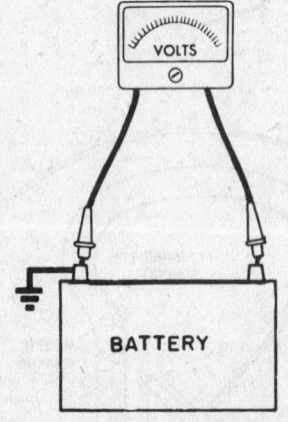

Cranking Voltage

Turn voltmeter selector to 8-10 volt scale for cars equipped with 6-volt systems, and to the 16-20 volt scale for cars equipped with 12-volt systems.

Connect voltmeter leads to prods tapped into the battery posts (observe polarity and reverse meter leads if necessary). Remove the high tension wire from the distributor cap and ground it to prevent starting. Now, close the starting switch. Observe both voltmeter reading and cranking speed. The cranking speed should be even, and at a satisfactory rate of speed, with a voltmeter reading of 4.8 volts or more for 6-volt systems, and at least 9.6 volts for 12-volt systems.

Compare this voltage reading with the reading obtained in the battery capacity test. There should be no more than a difference of 0.3 volt for 6-volt systems or 0.6 volt for 12-volt systems between the two readings. If this reading is satisfactory and the starting motor cranks the engine with sufficient speed, the starting motor circuit is in good condition.

Amperage Draw

The amount of current the starter motor draws is usually (but not always) associated with the physical or mechanical aspects of cranking the engine. (Mechanical trouble in the engine, frozen or worn starter parts, misaligned starter or starter components, etc.) Because starter motor

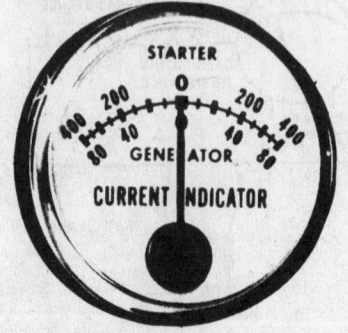

amperage draw is directly influenced by anything restricting the free turning of the motor, or starter, it is important that the engine and all components be at operating temperatures.

To measure starter current draw, remove the high tension wire from the center of the distributor cap and ground it. A very simple and inexpensive starter current indicator is available from your instrument and test equipment supplier or from the larger tool houses. This indicator is an induction type gauge and shows, without disconnecting any wires, starter current draw.

Place the yoke of the meter directly over the insulated starter supply cable (cable must be straight for a minimum of 2 in.). Close the starter switch for about 20 seconds. Watch the meter dial and record the average reading. If the indicator swings in the wrong direction, reverse the position of the meter. On 6-volt systems, normal draw for small to medium size engines is 150 to 225 amperes. Larger and high compression engines may draw as much as 400 amperes. On 12-volt systems, the current draw should be about one half the amount registered for the 6-volt system.

More exacting, but complex equipment is available from many name brand manufacturers. This equipment consists of a combination voltmeter, ammeter, and carbon pile rheostat. When using this equipment, follow the equipment manufacturer's procedures and recommendations.

High amperage and lazy performance would suggest, an excessively

stiff engine, friction in the starter or starter drive, grounded starter field or armature.

Normal amperage and lazy performance suggest, high resistance, possibly poor connections somewhere in the starter circuit.

Low amperage and lazy or no performance suggest, battery condition poor or bad cables or connections along the line.

Voltage Drop—Grounded Side

With a voltmeter on the 3-volt scale, without disconnecting any wires, connect one test lead of the voltmeter to a prod secured in the grounded battery post. The other test lead is connected to a cleaned, bare metal portion of the starter motor housing. Close the starter switch and note the voltmeter reading. If the reading is the same as battery reading, the ground circuit is open somewhere between the battery and the starter. In many cases

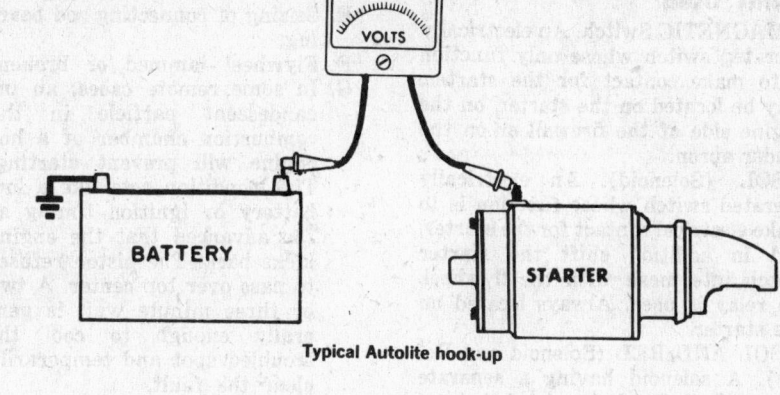

Typical Autolite hook-up

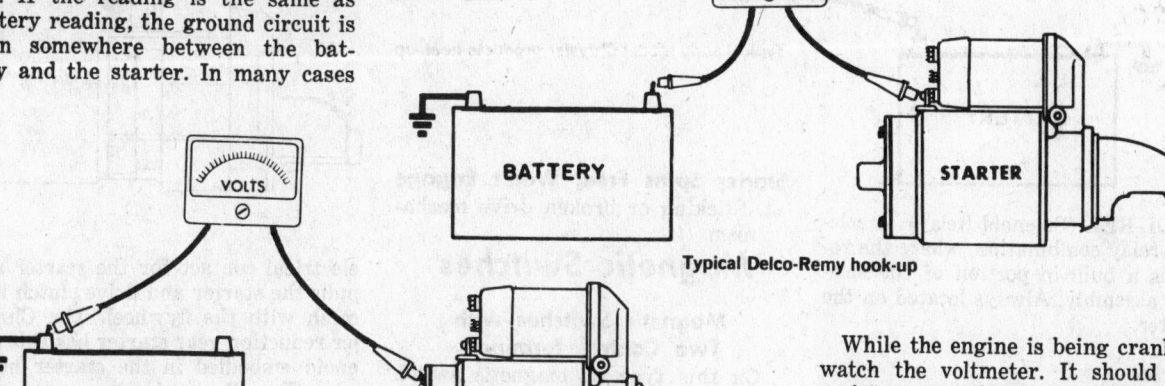

Typical Delco-Remy hook-up

the reading will be very small. The reading shown will indicate voltage drop (loss) between battery ground post and starter housing. The drop should not exceed 0.2 volt for a 6-volt system or 0.4 volt for a 12-volt system. If the voltage drop is above the specified amount, the next step is to isolate and correct the cause. It can be a bad cable or connection anywhere in the battery-to-starter ground circuit. A check of this type should progress along the various points of possible trouble, between the battery ground post and the starter motor housing, until the trouble spot has been located.

NOTE: due to the design of the Chrysler reduction gear starter, testing is limited to measuring voltage drop to starter cable connection.

Voltage Drop—Battery Side

Bad starter cranking may result from poor connections or faulty components of the battery or hot phase of the starter motor circuit. To check this phase of the circuit, without disconnecting any wires, connect one

lead of a voltmeter to a prod secured in the hot post of the battery and the other voltmeter lead to the terminal of the starting motor. The meter should be set to the 16-20 volt scale. Before closing the starter switch, the voltmeter reading will be that of the battery. After closing the starter switch, change the selector on the voltmeter to the 3-volt scale. With a jumper wire between the relay battery terminal and the relay starter switch terminal, crank the engine. If the starting motor cranks the engine, the relay (solenoid) is operating.

While the engine is being cranked, watch the voltmeter. It should not register more than 0.5 volt. If more than this, check each part of the circuit for voltage drop to isolate the trouble, (high resistance).

Without disturbing the voltmeter-to-battery hook-up, move the free voltmeter lead to the battery terminal of the relay (solenoid), and crank the engine. The voltmeter should show no more than 0.1 volt.

If this reading is correct, move the same voltmeter lead to the starting motor terminal of the relay (solenoid). While the engine is being cranked, the voltmeter should show no more than 0.3 volt. If it does, the trouble lies in the relay.

If the reading is correct, the trouble is in the cable or connections between the relay and the starting motor.

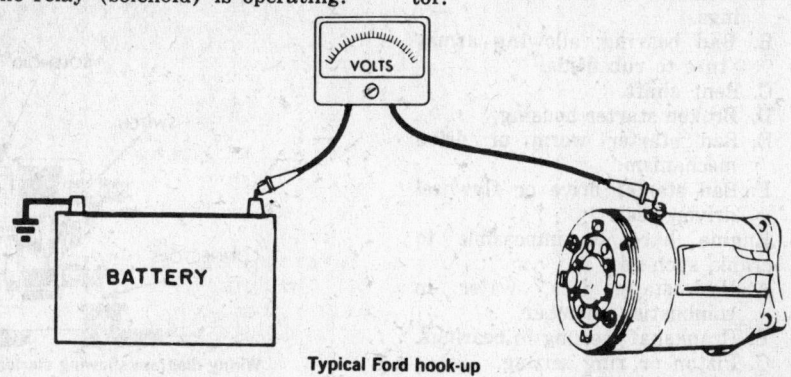

Typical Ford hook-up

Terms Used

MAGNETIC Switch. An electrically operated switch whose only function is to make contact for the starter. May be located on the starter, on the engine side of the firewall or on the fender apron.

SOL. (Solenoid). An electrically operated switch whose function is to make electrical contact for the starter, and in addition shift the starter clutch into mesh with the flywheel. No relay is used. Always located on the starter.

SOL AND REL (Solenoid and Relay). A solenoid having a separate relay, the relay being located some distance from the solenoid, usually on the fender apron or the engine side of the firewall. The solenoid is always located on the starter.

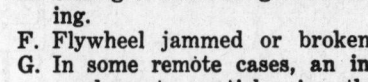
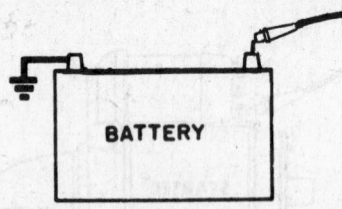

SOL-REL (Solenoid-Relay). A solenoid-relay combination, where the relay is a built-in portion of the solenoid assembly. Always located on the starter.

Starter

Starter Won't Crank the Engine

1. Dead battery.
2. Open starter circuit, such as:
   A. Broken or loose battery cables.
   B. Inoperative starter motor solenoid.
   C. Broken or loose wire from starter switch to solenoid.
   D. Poor solenoid or starter ground.
   E. Bad starter switch, (ignition, dash button or carburetor).
3. Defective starter internal circuit, such as:
   A. Dirty or burnt commutator.
   B. Stuck, worn or broken brushes.
   C. Open or shorted armature.
   D. Open or grounded fields.
4. Starter motor mechanical faults, such as:
   A. Jammed armature end bearings.
   B. Bad bearing, allowing armature to rub fields.
   C. Bent shaft.
   D. Broken starter housing.
   E. Bad starter worm or drive mechanism.
   F. Bad starter drive or flywheel driven gear.
5. Engine hard or impossible to crank, such as:
   A. Hydrostatic lock, water in combustion chamber.
   B. Crankshaft seizing in bearings.
   C. Piston or ring seizing.

D. Bent or broken connecting rod.
E. Seizing of connecting rod bearing.
F. Flywheel jammed or broken.
G. In some remote cases, an incandescent particle in the combustion chamber of a hot engine will prevent starting. This condition acts like a low battery or ignition timing so far advanced that the engine kicks back. The piston refuses to pass over top center. A two or three minute wait is generally enough to cool the troubled spot and temporarily clear the fault.

Typical early model Chrysler products hook-up

Starter Spins Free, Won't Engage

1. Sticking or broken drive mechanism.

Magnetic Switches

Magnetic Switches with Two Control Terminals

On this type of magnetic switch current is supplied from the ignition switch or transmission neutral button to one of the magnetic switch control terminals. The other control terminal is connected to the transmission neutral safety switch (on the transmission) where it is grounded.

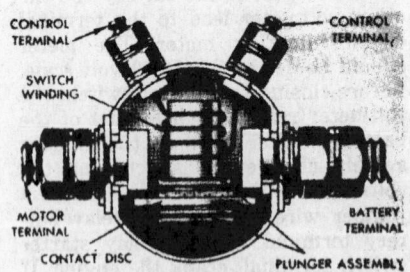

Schematic diagram of a magnetic switch with two control terminals

Magnetic Switches with Ignition Resistor By-Pass Terminals

Used with 12-volt systems. All normally use a magnetic switch with a single control terminal. The second terminal is an ignition resistor by-pass terminal. To trouble-shoot this type of switch, ignore the ignition terminal and check according to the instructions given for magnetic switches with internal grounds. (Except Studebaker.)

Solenoids Without Relays

This type of starter solenoid is always mounted on the starter. Makes

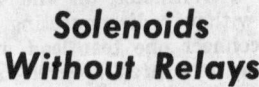
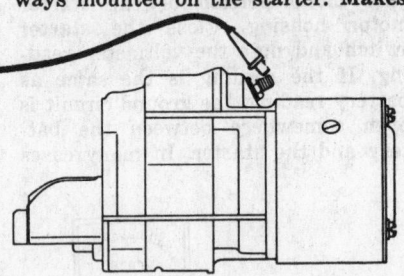

electrical contact for the starter and pulls the starter and drive clutch into mesh with the flywheel. The Chrysler reduction gear starter has this solenoid embodied in the starter housing. (See illustration.)

There is only one control terminal on the solenoid. Except for Chevrolet, all of the 12-volt models are fitted with an ignition resistor by-pass terminal on the solenoid.

The ignition by-pass terminal is usually marked IGN.

Solenoids With Separate Relays

The solenoid itself is always mounted on the starter. In addition to making contact for the starter, it also pulls the starter drive clutch gear into mesh with the flywheel. A single control terminal is used on the solenoid itself. On V8 Buicks with

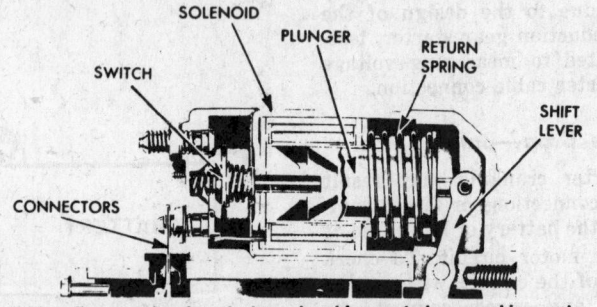

Wiring diagram showing starter solenoid mounted on cranking motor

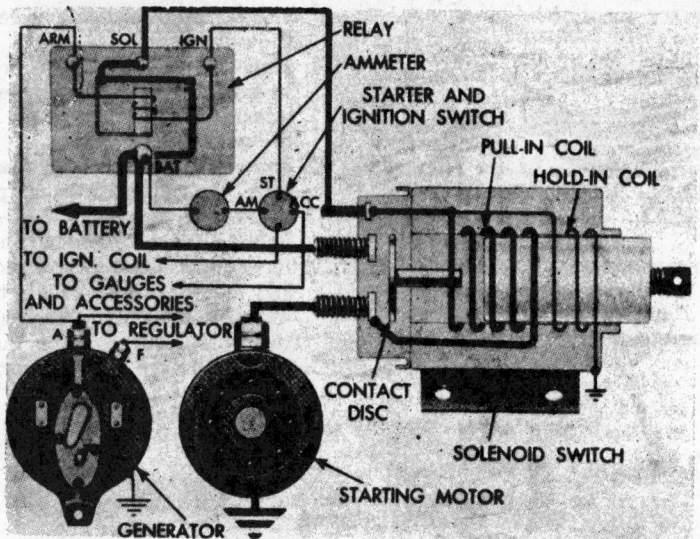

Pictorial drawing of solenoid with a separate relay

12-volt systems, there is also an ignition resistor by-pass terminal on the solenoid.

Solenoids With Built-In Relays

These units are always mounted on the starter and are connected, through linkage, to the starter drive clutch. The relay portion is a square box built into and integral with the front end of the solenoid assembly.

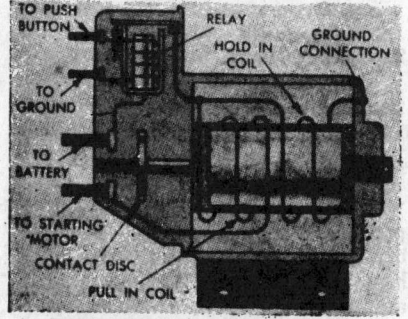

Pictorial drawing of solenoid with a built-in relay

Neutral Safety Switches

The purpose of the neutral safety switch is to prevent the starter from cranking the engine except when the transmission is in neutral or park.

On Chrysler, Dodge and Plymouth the neutral safety switch is located on the transmission. It serves to ground the solenoid or magnetic switch, whichever is used.

On all other cars the neutral safety switch is located either at the bottom of the steering column, where it contacts the shift mechanism, or on the steering column, underneath the dash.

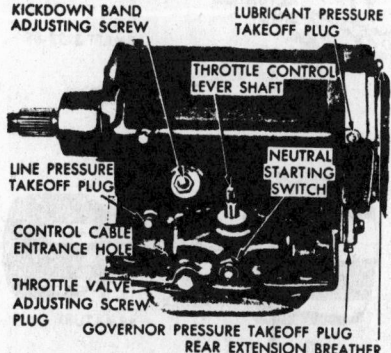

Location of the neutral safety switch on Chrysler, Dodge, Dart, Plymouth and Valiant

On most cars, the neutral safety switch and the back-up light switch are combined into a single switch mechanism.

See the car sections and/or the automatic transmission section for specific details.

Troubleshooting Neutral Safety Switches Quick Test

If the starter fails to function and the neutral safety switch is to be checked, a jumper can be placed across its terminals. If the starter then functions the safety switch is defective.

In the case of Chrysler, Dodge and Plymouth, the neutral safety switch has one wire. This wire must be grounded for testing purposes. If the starter works with the wire grounded, the switch is defective.

Neutral Safety Switch—Back-Up Light Switch

When the neutral safety switch is built in combination with the back-up light switch, the easiest way to tell which terminals are for the back-up lights is to take a jumper and cross every pair of wires. The pair of wires which light the back-up lamps should be ignored when testing the neutral safety switch. Once the back-up light wires have been located, jump the other pair of wires to test the neutral safety switch. If the starter functions only when the jumper is placed across these two wires, the neutral safety switch is defective or requires adjustment.

Adjustment of neutral safety switches for each of the automatic transmissions is given in the automatic transmission section of this manual.

Reduction-Gear Starting Motor

(Chrysler Corporation Products)

This starter weighs 15 pounds and has 15% fewer parts. The housing is die-cast aluminum.

A 3.5 to 1 reduction, combined with the starter to ring gear ratio, results in a total gear reduction of about 45 to 1.

NOTE: the high-pitched sound is caused by the higher starter speed.

The positive shift solenoid is enclosed in the starter housing and is energized through the ignition switch. When ignition switch is turned to start, the solenoid plunger engages drive gear through a shifting fork. At the completion of travel, the plunger closes a switch to revolve the starter.

The tension of the spring-type shifting prevents a butt-tooth lock up and motor will not start before total shift.

An overrunning clutch prevents motor damage if key is held on after engine starts.

No lubrication is required due to Oilite bearings.

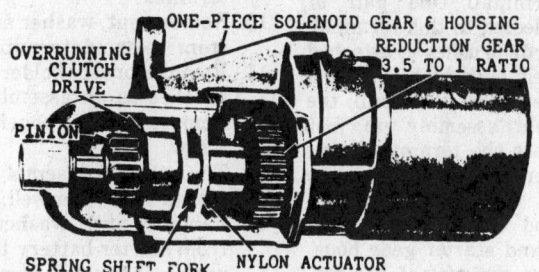

Reduction gear starting motor (reduction gear side)

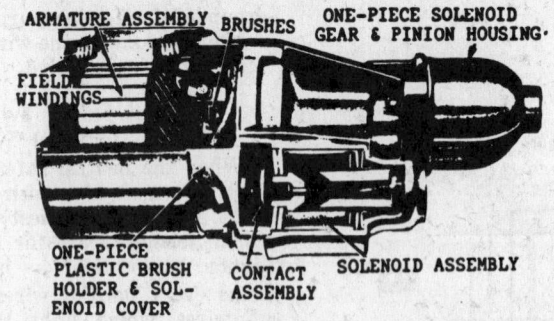

Reduction gear starting motor (solenoid side)

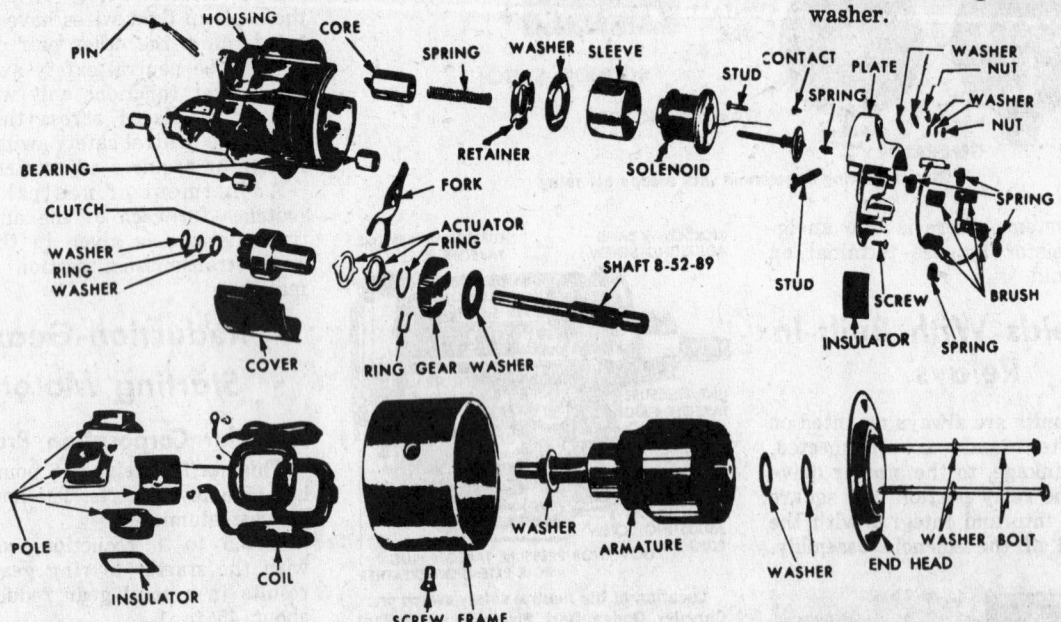

Reduction gear starter motor

Disassembly

1. Support assembly in a vise equipped with soft jaws. Do not clamp. Care must be used not to distort or damage the die cast aluminum.
2. Remove the thru-bolts and the end housing.
3. Carefully pull the armature up and out of the gear housing, and the starter frame and field assembly. Remove the steel and fiber thrust washer.

NOTE: on eight cylinder engines the starting motors have the wire of the shunt field coil soldered to the brush terminal. The six cylinder engines have the four coils in series and do not have a wire soldered to the brush terminal. One pair of brushes is connected to this terminal. The other pair of brushes is attached to the series field coils by means of a terminal screw. Carefully pull the frame and field assembly up just enough to expose the terminal screw and the solder connection of the shunt field at the brush terminal. Place two wood blocks between the starter frame and starter gear housing to facilitate removal of the terminal screw and unsoldering of the shunt field wire at the brush terminal.

4. Support the brush terminal with a finger behind terminal and remove screw.
5. On eight cylinder engine starters unsolder the shunt field coil lead from the brush terminal and housing.
6. The brush holder plate with terminal, contact and brushes is serviced as an assembly.
7. Clean all old sealer from around plate and housing.
8. Remove the brush holder attaching screw.
9. On the shunt type, unsolder the solenoid winding from the brush terminal.
10. Remove nut, washer and insulator from solenoid terminal.
11. Remove brush holder plate with brushes as an assembly.
12. Remove gear housing ground screw.
13. The solenoid assembly can be removed from the well.
14. Remove nut, washer and seal from starter battery terminal and remove terminal from plate.
15. Remove solenoid contact and plunger from solenoid and remove the coil sleeve.
16. Remove the solenoid return spring, coil retaining washer, retainer and the dust cover from the gear housing.
17. Release the snap-ring that locates the driven gear on pinion shaft.
18. Release front retaining ring.
19. Push pinion shaft toward the rear and remove snap-ring, thrust washers, clutch and pinion, and two shift fork nylon actuators.
20. Remove driven gear and friction washer.
21. Pull shifting fork forward and remove moving core.
22. Remove fork retainer pin and shifting fork assembly. The gear housing with bushings is serviced as an assembly.

To Reassemble: reverse the above procedures. At last portion of assembly, after gear housing ground screw has been securely tightened, clean area at joint between brush holder plate to field frame and housing mating joint. Apply a bead of brush plate sealer around four sides of joint. (A sealer such as MoPar 2421847 is suggested.) Be sure the joints are thoroughly sealed.

Ford Positive Engagement Starter

This starting motor is a series-parallel wound, four pole, four brush unit. It is equipped with an over-running clutch drive pinion, which is engaged with the flywheel ring gear by an actuating lever, operated by a movable pole piece. This pole piece is hinged to the starter frame and can drop into position through an opening in the frame.

Three conventional field coils are located at three pole piece positions.

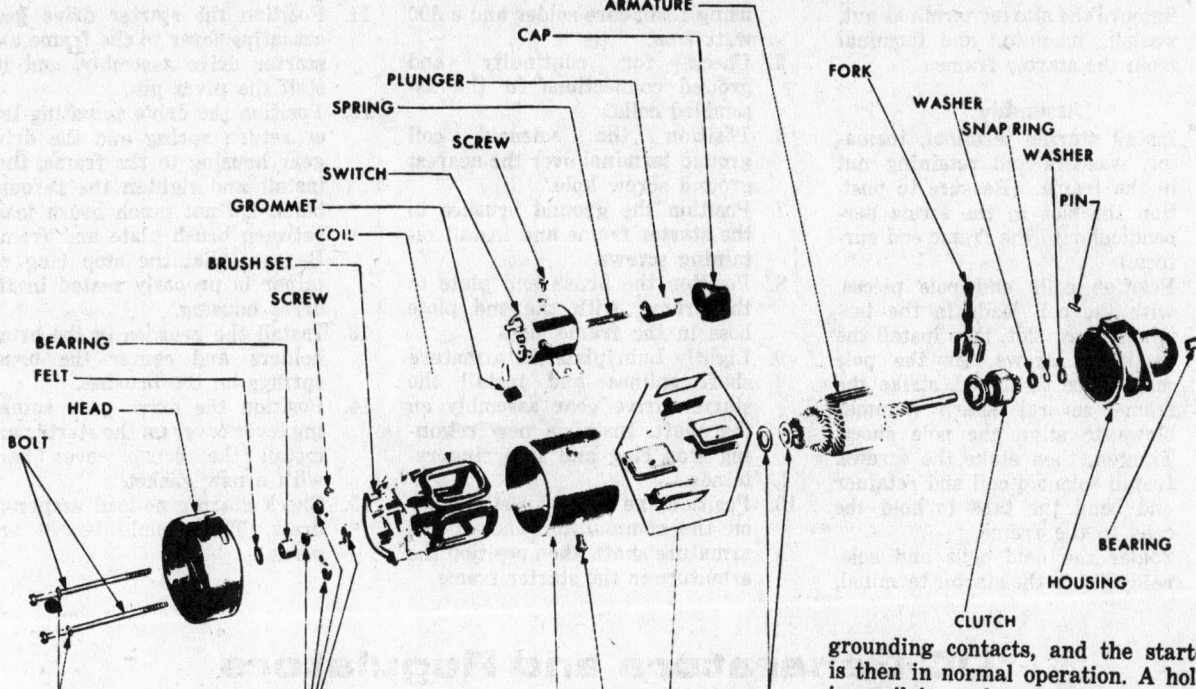

ARMATURE
CAP
PLUNGER
SPRING
SCREW
SWITCH
GROMMET
COIL
BRUSH SET
SCREW
BEARING
FELT
HEAD
BOLT
WASHER
BRUSH SPRING

FORK
WASHER
SNAP RING
WASHER
PIN
BEARING
HOUSING
CLUTCH

WASHER
POLE PIECE
SCREW
FRAME

Conventional Chrysler starter motor

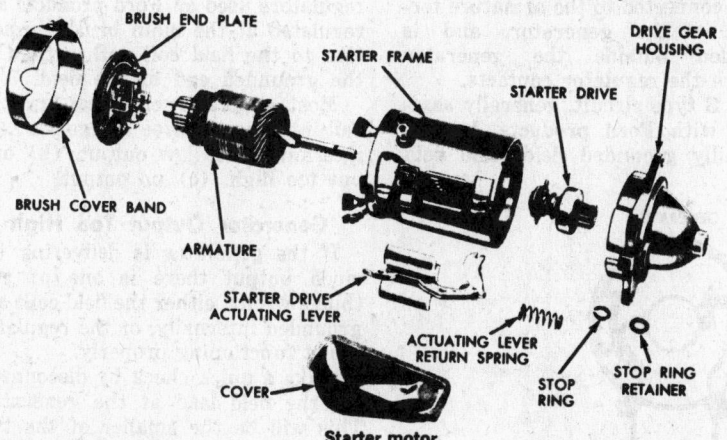

BRUSH END PLATE
BRUSH COVER BAND
ARMATURE
STARTER DRIVE ACTUATING LEVER
COVER
ACTUATING LEVER RETURN SPRING
STOP RING
STOP RING RETAINER
STARTER FRAME
STARTER DRIVE
DRIVE GEAR HOUSING

Starter motor

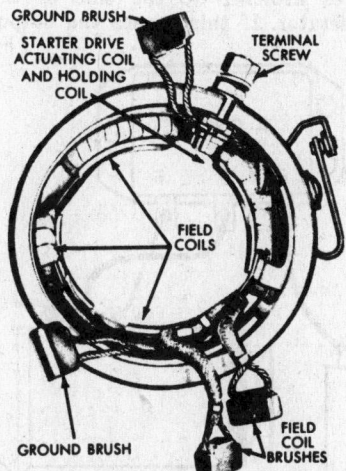

GROUND BRUSH
STARTER DRIVE ACTUATING COIL AND HOLDING COIL
TERMINAL SCREW
FIELD COILS
GROUND BRUSH
FIELD COIL BRUSHES

Field coil assembly

grounding contacts, and the starter is then in normal operation. A holding coil is used to hold the movable pole shoe in the fully seated position during the engine cranking operation.

Cars equipped with automatic transmissions have a starter neutral switch circuit control. This is to prevent operation of the starter if the selector lever is not in neutral or park.

Reconditioning Procedure

Disassembly

1. Remove brush cover band and starter drive gear actuating lever cover. Observe the brush lead locations for reassembly, then remove the brushes from their holders.
2. Remove the thru-bolts, starter drive gear housing and the drive gear actuating lever return spring.
3. Remove the pivot pin retaining the starter gear actuating lever and remove the lever and the armature.
4. Remove the stop ring retainer. Remove and discard the stop ring holding the drive gear to the armature shaft; then remove the drive gear assembly.
5. Remove the brush end plate.
6. Remove the two screws holding the ground brushes to the frame.
7. On the field coil that operates the starter drive gear actuating lever, bend the tab up on the field retainer and remove the field coil retainer.
8. Remove the three coil retaining screws. Unsolder the field coil leads from the terminal screw, then remove the pole shoes and coils from the frame (use a 300 watt iron).

The fourth field coil is designed to serve also as an engaging coil and a hold-in coil for the operation of the drive pinion.

When the ignition switch is turned to the start position, the starter relay is energized and current flows from the battery to the starter motor terminal. This prime surge of current first flows through the starter engaging coil, creating a very strong magnetic field. This magnetism draws the movable pole piece down toward the starter frame, which then causes the lever attached to it to move the starter pinion into engagement with the flywheel ring gear.

When the movable pole shoe is fully seated, it opens the field coil,

9. Remove the starter terminal nut, washer, insulator and terminal from the starter frame.

Assembly

1. Install starter terminal, insulator, washers and retaining nut in the frame. (Be sure to position the slot in the screw perpendicular to the frame end surface.)
2. Position coils and pole pieces, with the coil leads in the terminal screw slot, then install the retaining screws. As the pole screws are tightened, strike the frame several sharp hammer blows to align the pole shoes. Tighten, then stake the screws.
3. Install solenoid coil and retainer and bend the tabs to hold the coils to the frame.
4. Solder the field coils and solenoid wire to the starter terminal, using rosin-core solder and a 300 watt iron.
5. Check for continuity and ground connections in the assembled coils.
6. Position the solenoid coil ground terminal over the nearest ground screw hole.
7. Position the ground brushes to the starter frame and install retaining screws.
8. Position the brush end plate to the frame, with the end plate boss in the frame slot.
9. Lightly Lubriplate the armature shaft splines and install the starter drive gear assembly on the shaft. Install a new retaining stop ring and stop ring retainer.
10. Position the fiber thrust washer on the commutator end of the armature shaft, then position the armature in the starter frame.
11. Position the starter drive gear actuating lever to the frame and starter drive assembly, and install the pivot pin.
12. Position the drive actuating lever return spring and the drive gear housing to the frame, then install and tighten the through bolts. Do not pinch brush leads between brush plate and frame. Be sure that the stop ring retainer is properly seated in the drive housing.
13. Install the brushes in the brush holders and center the brush springs on the brushes.
14. Position the drive gear actuating lever cover on the starter and install the brush cover band with a new gasket.
15. Check starter no-load amperage draw. This should be 70 amperes.

DC Generators and Regulators

Testing DC Generator Systems

The generator system consists of the generator, generator regulator (voltage relay, current relay, and cut out relay) battery, charge indicator, and all the wires and other parts involved in the charging circuit.

Basically there are two types of DC charging systems in current use. These are different only in the way in which the field circuit of the generator is connected and grounded.

In the A type circuit, generally associated with General Motors and Chrysler products, the field circuit is externally grounded. The field circuit is connected to the armature terminal in the generator, and is grounded outside the generator through the regulator contacts.

The B type circuit, generally associated with Ford products, has an internally grounded field. The field circuit is connected to the armature circuit in the regulator, and is grounded inside the generator.

Generator Trouble-Shooting

NOTE: all DC generators with regulators used on Ford products are regulated at the main brush connection to the field coil rather than at the grounded end of the field.

Most generator electrical troubles fall into one of three categories: (a) intermittent or low output. (b) output too high. (c) no output.

Generator Output Too High

If the generator is delivering too much output there is one of two things wrong, either the field coils are grounded internally, or the regulator is not functioning properly.

Make a quick check by disconnecting the field lead at the generator. This will be the smaller of the two wires attached to the side of the generator. If this causes the output

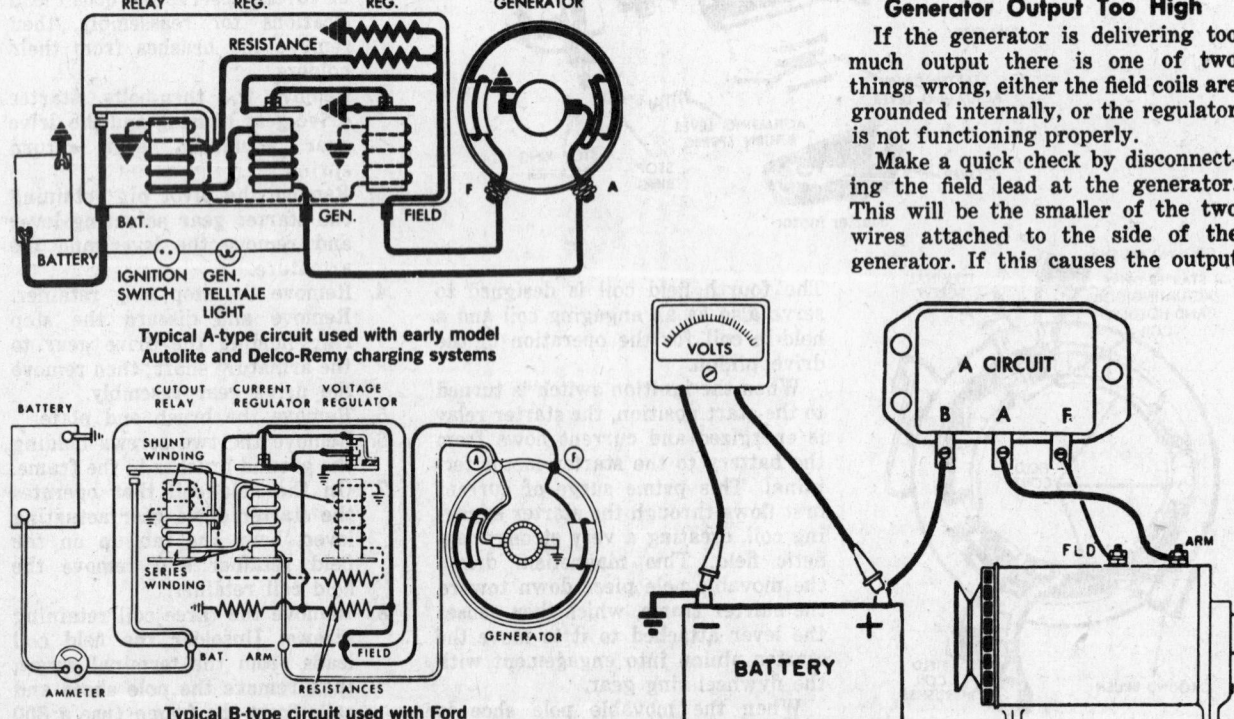

Typical A-type circuit used with early model Autolite and Delco-Remy charging systems

Typical B-type circuit used with Ford products charging systems

to drop to a very low value, then the regulator is at fault or the field wire is grounded somewhere in its harness. If, after disconnecting the field, the generator output remains too high, there is a ground in the field circuit inside of the generator, and the field coil should be replaced.

Generator Output Too Low

To check the voltage; if no generator output is indicated proceed as follows:

1. Hook up voltmeter between the two battery posts (use prods).
2. Run the engine at 1500-1800 rpm and observe voltmeter reading. An increase in reading, beyond battery voltage indicates that the charging system is working.
3. Compare this reading with specifications. If voltage output is less than specifications, trouble may be in the generator or regulator.

Caution When DC generators are used with Delco-Remy regulators having double contact voltage control, do not ground the field lead, as this may cause the high voltage points in the regulators to burn.

Test Instruments Required

Commercial test sets are made especially for testing regulators and generators conveniently and easily; however, if a test set is not available, the instruments necessary to test a generator, cutout relay, voltage or current regulator are: a voltmeter

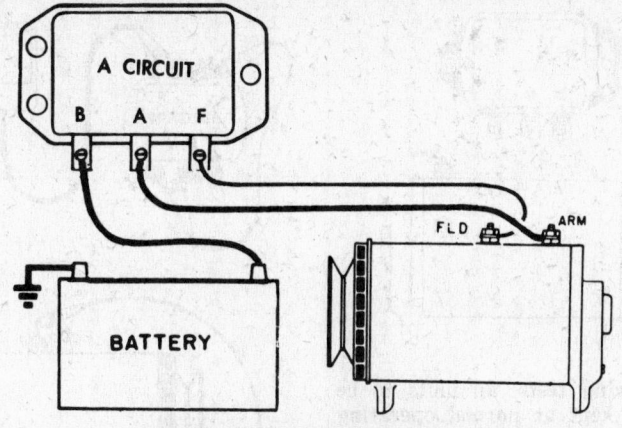

Connect jumper wire to battery (B) terminal of regulator. Momentarily, touch the other end of jumper to the armature terminal of the generator or regulator.

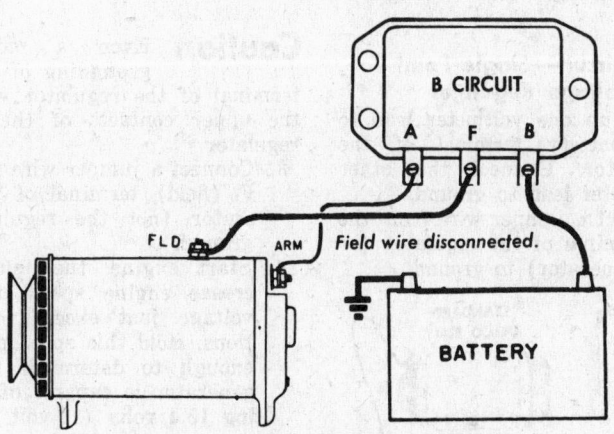

Remove field wire from regulator. Momentarily flash the disconnected wire to "Bat." terminal of regulator.

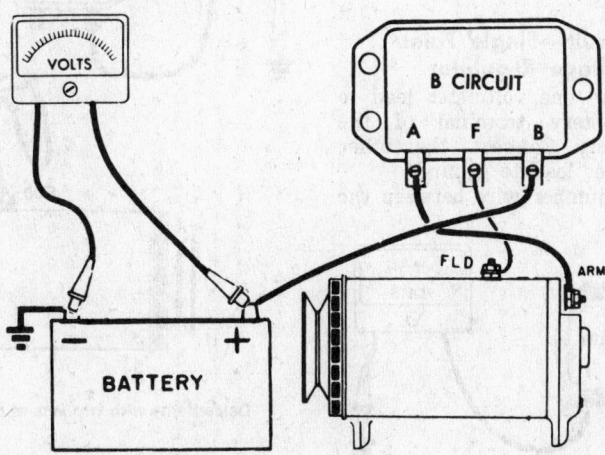

with a full scale reading from zero to 10 volts (for 6-volt systems) or zero to 20 volts (for 12-volt systems), an ammeter with a full scale reading from minus 5 amp. to 40 amp. positive, a 3/4 ohm variable resistance capable of carrying 50 amp. (for 6-volt systems) or a 1 1/2 ohm variable resistance (for 12-volt systems). It is well to have also a 1 1/2 ohm 1000 watt variable resistor to test batteries.

Polarizing the Generator

Because the DC generator will build up voltage that will cause current to flow in either direction, depending upon the polarity of residual magnetism of the pole shoes, the DC generator must be polarized. Any time a DC generator has been repaired or replaced on a car, the generator must be polarized before starting the engine.

Generators of the A circuitry type are polarized by connecting a jumper lead from the battery terminal of the regulator to the armature terminal of the regulator or generator. A touch of the jumper lead is all that is necessary to polarize the generator.

Generators of the B circuitry type are polarized by disconnecting the field lead at the regulator and momentarily flashing this lead to the battery terminal of the regulator.

The Regulator

The generator regulator is the next step in diagnosing a faulty charging system. Manufacturers' recommendations are that, when making any of

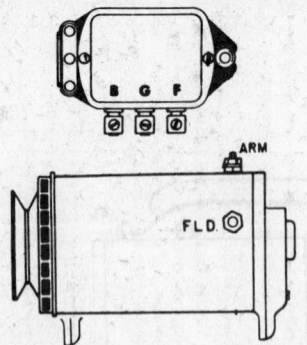

the following tests, all units to be tested be kept at normal operating temperatures. This usually takes about 15 minutes of engine running. Readings should all be taken with the regulator cover in place. It does make a difference, due to magnetic influence.

A Circuit—Single Point Voltage Regulator

1. Hook up one voltmeter lead to the battery terminal of the regulator. Connect the other voltmeter lead to ground.
2. Connect a jumper wire from the F terminal of the regulator (or the generator) to ground.

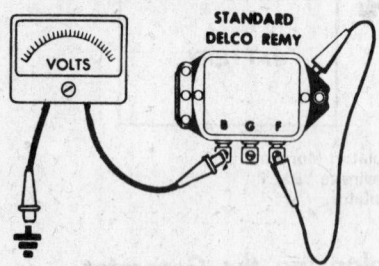

3. Start the engine, then slowly increase engine speed until the voltage just exceeds specifications. Hold this speed only long enough to determine that the generator is capable of producing 7.8 volts (6-volt system), or at least 15.4 volts (12-volt system).

Caution Do not increase engine speed beyond 1 volt of specifications.

If generator voltage output is adequate, repair or replace regulator.

If generator voltage output is under specifications, repair or replace the generator.

A Circuit—Dual Point Voltage Regulator

1. Hook up one voltmeter lead to the battery terminal of the regulator. Connect the other voltmeter lead to ground.
2. With this type regulator, the F (field) wire must be disconnected from the generator or regulator.

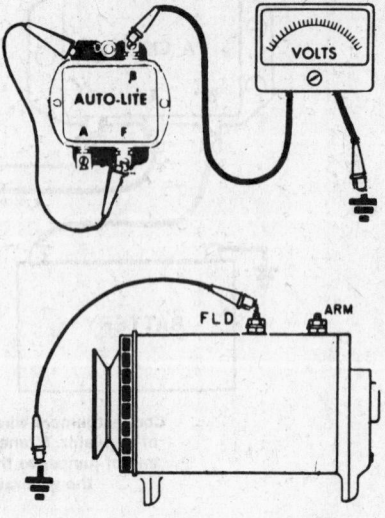

Caution Even a momentary grounding of the field terminal of the regulator will burn the upper contacts of the voltage regulator.

3. Connect a jumper wire from the F (field) terminal of the generator (not the regulator) to ground.
4. Start engine, then slowly increase engine speed until the voltage just exceeds specifications. Hold this speed only long enough to determine that the generator is capable of producing 15.4 volts (12-volt system).

Caution Do not increase engine speed beyond 1 volt of specifications.

B Circuit—Single Point Voltage Regulator

1. Hook up one voltmeter lead to the battery terminal of the regulator. Connect the other voltmeter lead to ground.
2. Run a jumper wire between the

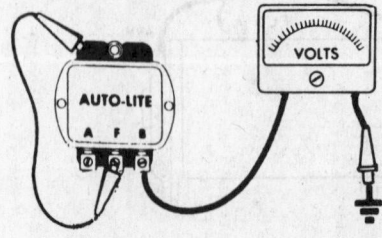

A (armature) and F (field) terminals of either the regulator or generator.

3. Start engine and slowly increase engine speed until the voltage just exceeds specifications. Hold this speed only long enough to determine that the generator is capable of producing 7.8 volts (6-volt system), or at least 15.4 volts (12-volt system).

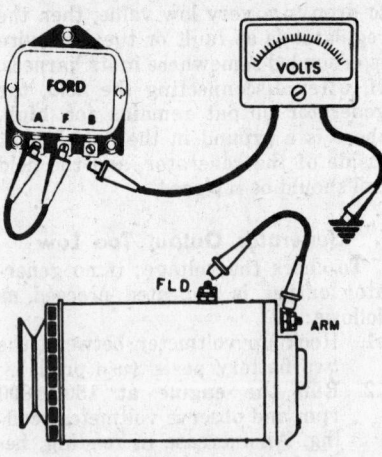

Caution Do not increase engine speed beyond 1 volt of specifications.

B Circuit—Dual Point Voltage Regulator

1. Hook up one voltmeter lead to the battery terminal of the regulator. Connect the other voltmeter lead to ground.
2. Disconnect the F (field) wire from the regulator.
3. Hook up a jumper wire from the A (armature) terminal of the regulator or generator to the wire just disconnected from the regulator.

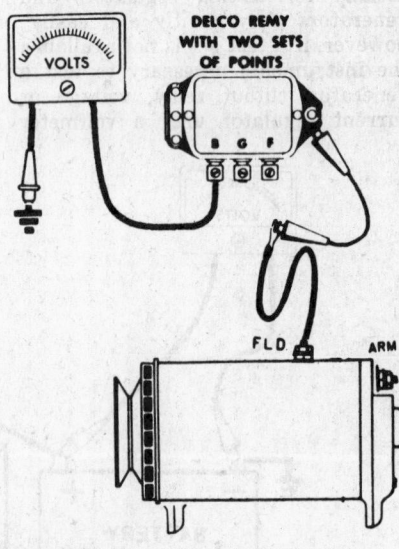

Delco-Remy with two sets of points

Caution Even a momentary energizing of the regulator field terminal will cause burning and permanent damage to the voltage regulator upper points. As a precautionary measure, because it is sometimes difficult to identify single point from dual point regulators, follow the procedure for the dual set unit.

4. Start engine, then slowly increase engine speed until voltage

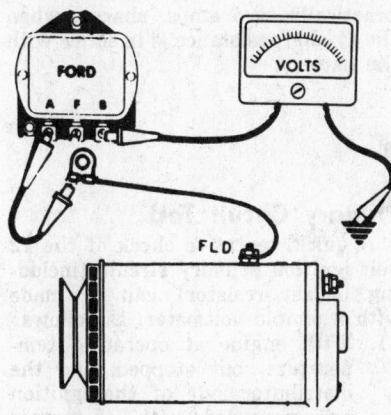

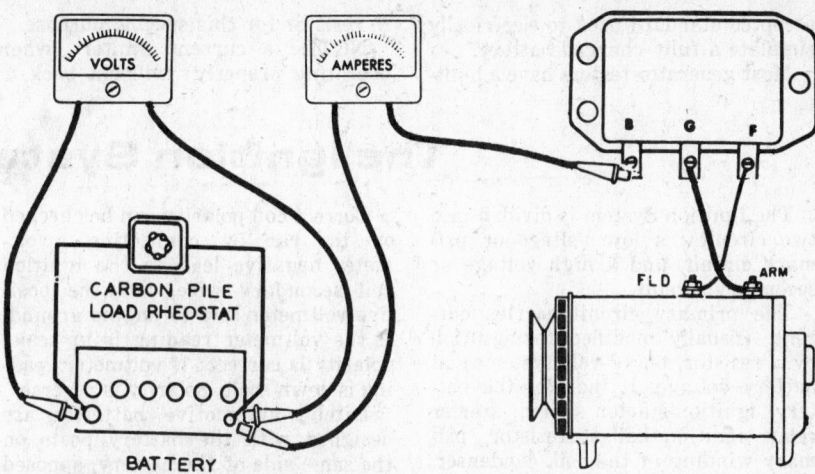

just exceeds specifications. Hold this speed only long enough to determine that the generator is capable of producing 15.4 volts (12-volt system.)

Caution Do not increase engine speed beyond 1 volt of specifications.

Adjusting the Voltage Regulator

Most regulator troubles are caused by oxidized or burned regulating, or current limiting points. These points, like distributor breaker points, have a limited life span. Average point oscillating frequency of the voltage regulating unit may be as much as 70,000-75,000 cycles per hour.

Experience dictates that both customer and mechanic are best served by the following practice. Aside from simple voltage regulator adjustment, any malfunctions of the regulator (voltage regulator, current regulator, or cut-out) are best handled by renewing the entire regulating unit. Remember that most regulating units that give trouble (aside from voltage control) are usually fatigued and are approaching the end of their dependability span. All things considered, it is impractical to guarantee continued satisfactory service.

Before attempting any adjustment of the regulator, two tests should be made; both at generator and regulator operating temperatures. These tests are to establish the voltage and amperage output of the system under practical conditions—a partially charged battery, and a battery which is fully charged.

Bear in mind that the voltage output should remain constant. The amperage reading, however, will be influenced by the charge condition of the battery. A low battery requiring greater generator output (higher amp. reading) than a battery of full charge.

Connect the voltmeter and ammeter as illustrated.

Test 1

Discharge the battery (if it is fully charged) by using a carbon pile, or turn on headlights, heater, radio, etc. Disconnect the distributor cap center wire and ground it. Operate the starter to partially discharge the battery.

Reconnect the distributor cap center wire, then start the engine. Slowly accelerate and watch the meters. Note highest reading on both voltmeter and ammeter.

This is the maximum charging potential of this generator-regulator setting.

Test 2

With the same meter hook-up as in Test 1, run the engine at about 1500-2000 rpm, long enough to be reasonably sure that the battery is fully charged; or, place a 1/4 ohm resistor in series with the voltage regulator battery wire. This will influence the charging system to the same degree as that of a fully charged battery.

NOTE: the 1/4 ohm resistor is the

VOLTAGE ADJUSTMENT
(BEND SPRING HANGER DOWN TO INCREASE SETTING)

CURRENT ADJUSTMENT
(BEND SPRING HANGER DOWN TO INCREASE SETTING)

AUTO-LITE ➡

CUT-OUT ADJUSTMENT
(BEND SPRING HANGER DOWN TO INCREASE SETTING)

CURRENT ADJUSTMENT
(TURN SCREW CLOCK WISE TO INCREASE SETTING)

CUT-OUT ADJUSTMENT
(TURN SCREW CLOCK WISE TO INCREASE SETTING)

DELCO-REMY ➡

VOLTAGE ADJUSTMENT
(TURN SCREW CLOCK WISE TO INCREASE SETTING)

VOLTAGE ADJUSTMENT
(BEND SPRING REST UP TO INCREASE SETTING)

CURRENT ADJUSTMENT
(BEND SPRING REST UP TO INCREASE SETTING)

FORD ➡

CUT-OUT ADJUSTMENT
(BEND SPRING REST UP TO INCREASE SETTING)

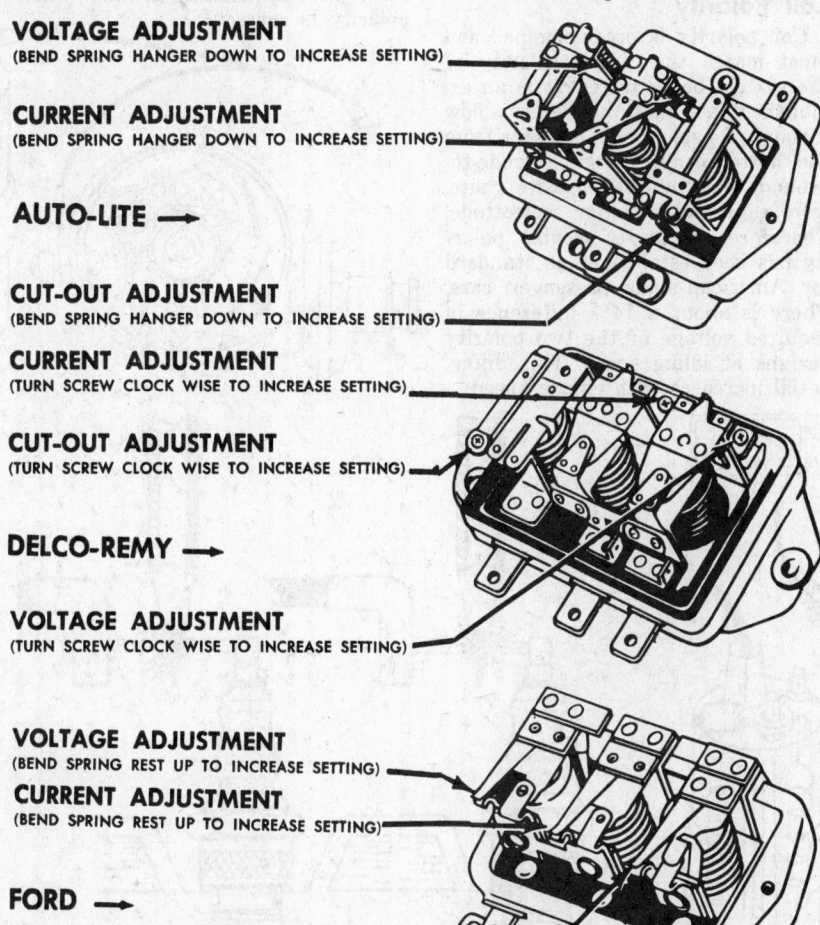

accepted standard used to electrically simulate a fully charged battery.

Most generator testers have a built-in resistor for this specific purpose.

NOTE: a current limiter (when operating properly) will cut back to practically zero amps. charge, when the 1/4 ohm resistance is in series with the battery.

The Ignition System

The Ignition System is divided into two circuits; a low voltage or primary circuit, and a high voltage or secondary circuit.

The primary circuit carries current, (usually modified for ignition by a resistor, on 12-volt systems) at battery voltage. It includes the battery, ignition-starter switch, starter relay, ignition ballast resistor, primary winding of the coil, condenser, contact points, and ground.

The secondary circuit begins with the ignition coil. Secondary voltage is a product of the coil and emerges from the secondary terminal and flows through a cable to the distributor cap. It is distributed by the rotor, through the distributor cap and cables, to the spark plugs, and to ground.

Caution Secondary circuit pressure could reach as high as 30,000 volts.

Coil Polarity

Coil polarity is predetermined and must match the circuit polarity of the system being tested. It is an established fact that the electron flow through the spark plug is better from the hotter center plug electrode-to-ground than by the opposite route, from ground-to-center electrode. Therefore, negative ground polarity has been established as standard for American-made passenger cars. There is about a 14% difference in required voltage of the two polarity designs at idling speed. This differential increases with engine speed.

Correct coil polarity can be checked on the car by connecting a voltmeter negative lead to the ignition coil secondary wire, and the positive voltmeter lead to engine ground. If the voltmeter reading is up-scale, polarity is correct: if voltmeter reading is down-scale, polarity is reversed.

Lately, automotive batteries are designed with the battery posts on the same side of the battery, opposed to the earlier diagonal post design. Therefore, terminal size and cable length will discourage improper battery installation. This results in the battery and distributor terminals of the coil being the most likely points of possible reversal of polarity.

Another tentative, but less precise, method is to hold a regular carbon-cored wooden lead pencil in the gap between a disconnected spark plug wire and ground. It is possible to observe the direction of spark flow, from wire-to-pencil-to-ground when polarity is correct.

Primary Circuit Test

A quick, tentative check of the 12 volt ignition primary circuit (including ballast resistor) can be made with a simple voltmeter, as follows:

1. With engine at operating temperature, but stopped, and the distributor side of the ignition coil grounded with a jumper wire, hook up a voltmeter between the ignition coil (switch side) and a good ground.
2. Jiggle the ignition switch (switch on) and watch the meter. An unstable needle will indicate a defective ignition switch.
3. With ignition switch on (engine stopped) the voltmeter

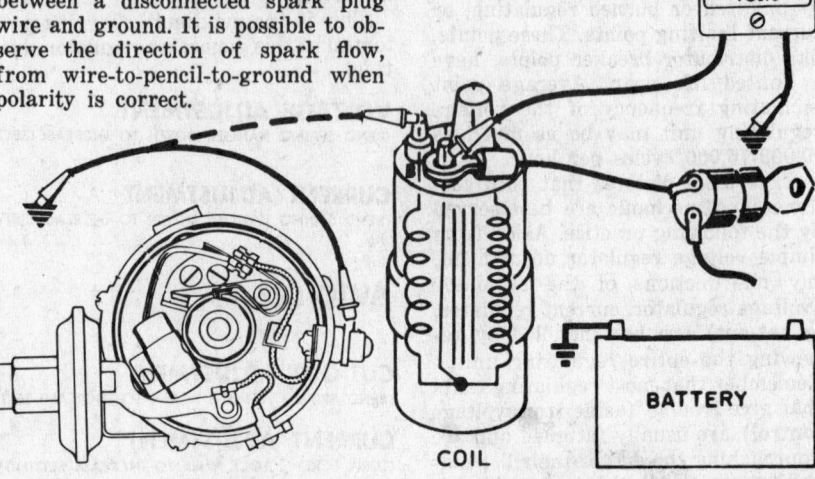

should read 5.5 to 7 volts for 12-volt systems.
4. Crank the engine. Voltmeter should read at least 9 volts during cranking period.
5. Now remove the jumper wire from the coil. Start the engine. Voltmeter should read from 9.0 volts to 11.5 volts (depending upon generator output) while running.

Coil Resistance, with Ohmmeter —Primary Circuit

To check ignition coil resistance, primary side, switch ohmmeter to

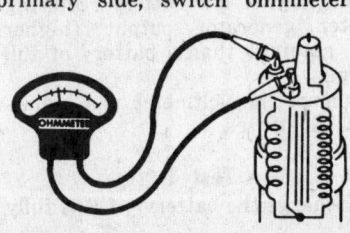

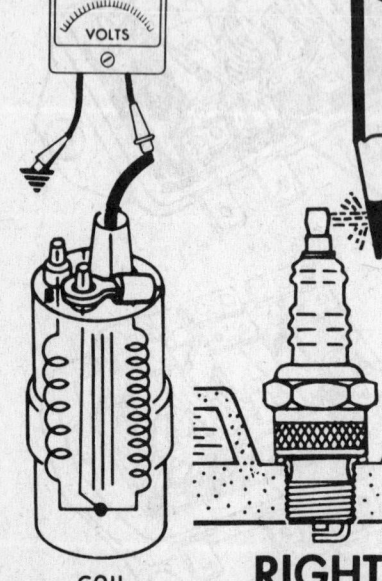
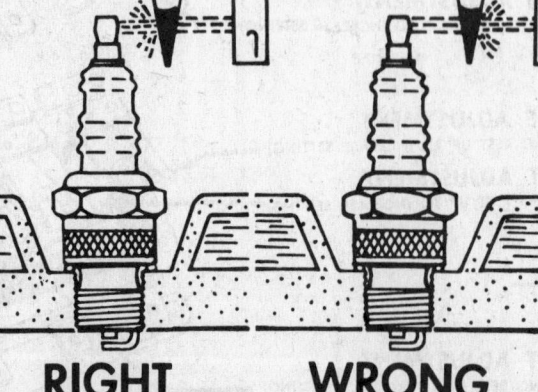

COIL   **RIGHT**   **WRONG**

low scale. Connect the ohmmeter leads across the primary terminals of the coil and read the low ohms scale.

Coils requiring ballast resistors should read about 1.0 ohm resistance. 12-volt coils, not requiring external ballast resistors, should read about 4.0 ohms resistance.

Coil Resistance, with Ohmmeter —Secondary Circuit

To check ignition coil resistance, secondary side, switch ohmmeter to high scale. Connect one test lead to the distributor cap end of the coil secondary cable. Connect the other test lead to the distributor terminal of the coil. A coil in satisfactory condition should show between 4 K and 8 K on the scale. Some special coils (Mallory, etc.) may show a resistance as high as 13 K. If the reading is

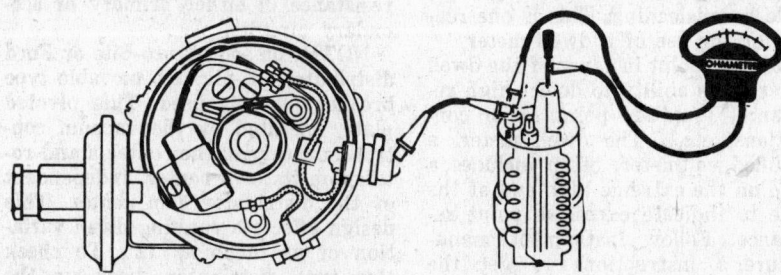

much lower than 4 K, the coil probably has shorted secondary turns. If the reading is extremely high (40 K or more) the secondary winding is either open, there is a bad connection at the coil terminal, or resistance is high in the cable.

If both primary and secondary windings of the coil test good, but the ignition system is still unsatisfactory, check the system further.

Ballast Resistor

Some sort of ballast resistor is used with most American-made cars equipped with 12-volt ignition systems. This resistance may be built into the ignition coil, or it may be a special wire of specific resistance, comprising the primary ignition circuit.

To provide a greater safety margin of sufficient voltage for high speed operation, a special ignition coil is used with whatever type of ballast resistance is used. Other reasons for

ballast resistance are to limit to a safe maximum the primary current flow through the coil and through the distributor contact points. This helps protect the contact points at slow engine speed when they are closed for a longer period of time. The resistor also protects against excessive build-up of primary current when the ignition switch is on with the engine stopped and ignition points closed.

On some systems, the resistor is removed from the ignition circuit during engine cranking, then with the ignition connected directly to battery voltage. This keeps ignition voltage as high as possible while cranking. The by-pass type system can have the by-pass factor built into the ignition switch, or it may be part of the starter solenoid.

Primary Circuit—Distributor Side

With the voltmeter on the 16-20 volt scale, connect one voltmeter lead to ground. Connect the other voltmeter lead to the distributor side of the coil. Remove the high tension wire from the coil and ground it. Close ignition switch and slowly bump the engine to open and close the points. When the distributor con-

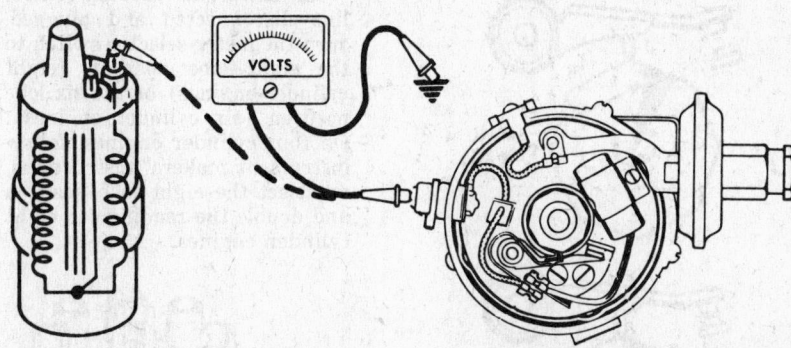

tacts make and break during cranking, the voltmeter reading should be from one-third to one-half battery voltage. Normally, with engine stopped and with points open, the reading will be the same as battery voltage. Furthermore, with the engine stopped and the points closed and in good condition, the reading will be close to zero.

If while cranking, the voltmeter reading remains zero or close to it, the trouble may be one or more of the following:

A. No current at distributor. Disconnect the distributor primary wire from the top of the coil. Now, take a voltmeter reading from the distributor terminal of the coil to make sure that the current is going through the circuit.

B. Points are not opening because of mechanical (points or cam) failure or maladjustment. Dual points in parallel, one set not opening.

C. The movable point, the stud at the primary distributor wire termi-

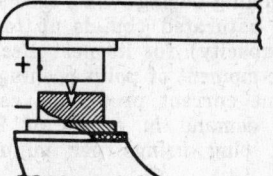

Pitting caused by low capacity

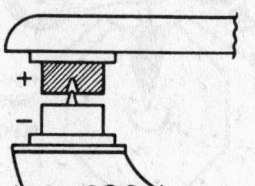

Pitting caused by high capacity

nal, or the pigtail wire is grounded.

D. The condenser has a dead short. An ohmmeter check of the condenser will show this condition. Connect one test lead of an ohmmeter to the body of the condenser and the other test lead to the pigtail. If the meter shows the slightest reading, the condenser is shorted. With a few exceptions, a visual inspection of the distributor contact points will generally indicate the condition of the

NOTE: If just one side opens up, the resistance will be doubled.

SHORTED

OPEN

Ballast resistor

condenser. An open, or shorted condenser will not function. A condenser of too great capacity will cause metal to transfer from the movable distributor point to the stationary point. This will cause a pit on the movable point. An under-capacity condenser, causes metal to leave the stationary point and build up on the movable point.

Any excessive resistance in either the primary or secondary circuits will upset the sensitive balance of the ignition system and cause the ignition points to pit.

Ignition Point Dwell

It is very important that point dwell be adjusted to exact specifications before any attempt is made to time the engine.

Point dwell (cam angle) is the degree value for the closed attitude of ignition points for each make-and-break period of a distributor cycle. It is that phase of ignition system functioning during which the coil becomes saturated (builds up to voltage capacity) for its next discharge at the moment of point opening.

Some current production car engines demand in excess of 23,000 spark plug firings per minute to

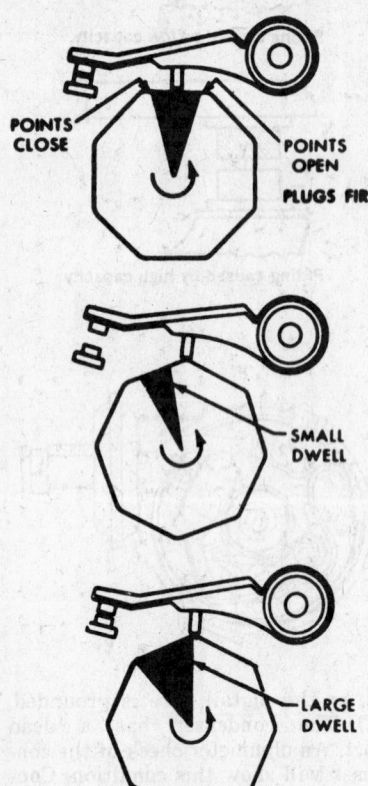

POINTS CLOSE — POINTS OPEN — PLUGS FIRE

SMALL DWELL

LARGE DWELL

fire their cylinders. This places a tremendous demand upon the ignition system, particularly the aspect of coil build up (saturation) and discharge time.

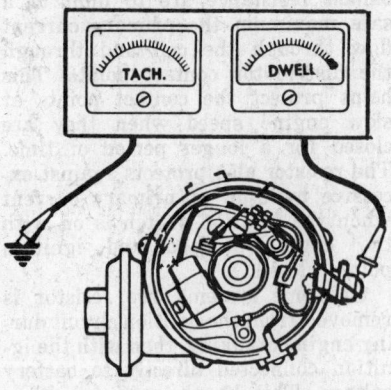

TACH. DWELL

While it is true that ignition points can be adjusted by using a thickness gauge, the results, even when using new points, are sometimes inconclusive. Point gap is incidental to particular distributor cam shape and could be misleading. This is one reason for the use of a dwell meter.

Another point in favor of the dwell meter is its ability to detect high resistance, (oxidized points, poor connections, etc.). The dwell meter, a modified voltmeter, often includes a band on the extreme high end of the scale to indicate excessive point resistance. Follow instrument manufacturer's instructions to get the most out of your particular equipment.

The most informative procedure is to use both methods (dwell meter and point gap) then compare the two. Many times the comparison is surprising, and leads to the location of previously unnoticed distributor troubles.

Using the Dwell Meter

1. With distributor vacuum control line disconnected and plugged, turn the meter selector switch to the eight lobe position (eight cylinder engines) or the six lobe position (six cylinder engines). On four cylinder engines, follow instrument makers' instructions, or select the eight lobe position and double the reading for eight cylinder engines.

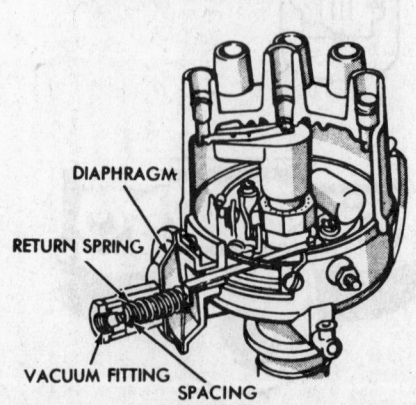

DIAPHRAGM

RETURN SPRING

VACUUM FITTING — SPACING WASHER

2. Connect one tach-dwell meter lead to the coil terminal of the distributor and the other meter lead to ground.
3. Start the engine and operate it at idle speed. Note reading on dwell meter. On eight cylinder engines, (single contacts) dwell should read 26°-32°. Double contacts should read 26°-32° (each set), or 34°-40° combined. Six cylinder engines should show a dwell of 36°-45° and four cylinder engines, a dwell in the area of 40°. These are tentative figures and cover a wide latitude. It is therefore, urgent that manufacturers' specifications be followed without exception.

An excessive variation in dwell, (over 3°) as engine speed is increased usually indicates a worn distributor shaft, bushing, breaker plate or high resistance in either primary or secondary circuits.

NOTE: on some Auto-Lite or Ford distributors, a pivoted, movable type breaker plate is used. This pivoted plate, operated by the vacuum control unit, carries the contacts and rotates on its own center, independent of the distributor cam center. This design affects a running dwell variation of as much as 12°. To check this type distributor, hook up the distributor vacuum control line. Increase engine rpm and observe dwell changes at various engine rpm and throttle attitudes.

NOTE: experience dictates that all distributor adjustments are best performed with the use of a good off-car distributor tester.

Dwell information at idle speed is given for each engine in Tune-up Specifications Table of all of the cars in the Car Section of this manual.

Ignition Timing

Ignition timing is a term applied to the relationship of piston travel and moment of spark in a gas engine.

Due to the many variables involved, such as compression ratio, temperature, humidity, elevation,

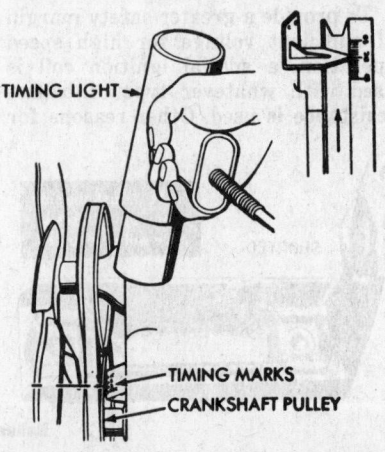

TIMING LIGHT

TIMING MARKS
CRANKSHAFT PULLEY

fuel octane value, engine condition, work load, etc. published timing data must be considered approximate; some tolerance permitted.

Ignition timing consists of basic (prime) timing and dynamic (variable) timing.

It is very important that point dwell be correct before setting timing.

Basic Timing

Basic timing can be checked quite accurately by using one of the many timing lights, (strobe-flashers) available. A timing light, when properly connected to No. 1 spark wire, (or the exact opposite cylinder in firing sequence of any multiple cylinder, four stroke cycle automotive engine) will indicate the moment of ignition for that cylinder. Index markings may be on either the rotating member of the crankshaft, (vibration damper or flywheel) with the pointer stationary, or the index may be on an engine stationary member, with the pointer or scribe mark rotating.

NOTE: because ignition timing is directly affected by distributor contact spacing, points should be adjusted to specifications before timing is attempted.

1. Unless otherwise stated by the manufacturer, the distributor control vacuum line should be disconnected, and plugged, to prevent fuel induction disturbance.
2. Hook up the timing light, (power or otherwise) according to the equipment instructions.
3. With engine at operating temperature and adjusted to function smoothly, run engine at low idle. Use a tachometer and be sure the rpm is below the speed of governor advance influence.
4. Shine the timing light on the indexing area (balancer or flywheel) and note the degree value indicated by the pointer.
5. Rotate the distributor body one way or the other until the pointer appears to correspond with the index value published.

Caution
Power timing (on the road adjusting for ping) cannot be tolerated, especially on engines equipped with exhaust emission (air pollution) control devices.

Possible Indications and Causes
If The Flash Is Intermittent:
A. The test light is defective.
B. The test light has a bad connection.
C. Distributor points are bad or badly out of adjustment.
D. Distributor grounding is poor.
E. Distributor cap is cracked or tracking.

F. Spark plug gaps too small.
G. Broken or badly worn rotor.

If Pointer Appears To Move On The Index Scale (Unfixed At Constant Engine Speed):
A. Distributor governor weights loose or with broken springs.
B. Distributor shaft or bushing worn.
C. Rotor loose or broken.
D. Distributor base plate loose.
E. Cam lobes worn.

Dynamic Timing

To accurately check and calibrate dynamic timing through all attitudes of engine operation, more sophisticated equipment than the common strobe-light is needed. A distributor tester, an oscillograph, or one of the more complex timing lights equipped with an advance value index is needed.

It is possible, however, to determine to some degree, the action of both governor advance and vacuum control mechanisms with a tachometer and a timing light.

NOTE: before checking dynamic timing with a timing light, extend the index graduations on the timing member involved, by about 30°. This should be done with chalk or white paint, in increments of 5°, on the rotating member, whether that member carries the index or the pointer. Some measurements and extreme care will be necessary in making this extension.

Governor Control
1. Repeat Steps 1 through 4 of basic timing procedures.
2. By watching the timing light flash on the timing index, determine the exact engine rpm that starts the distributor to advance. Compare this with published specifications.
3. It is equally important that distributor advance progresses steadily with engine speed. It is just as important that a decrease in engine speed will smoothly and gradually return the index pointer to its original position.
4. After checking the indications against specifications, turn the engine off and make corrections, if necessary.

NOTE: dynamic timing cannot be *accurately* checked using the above method; therefore it is recommended that no attempt be made to modify advance curves (especially on exhaust emission equipped engines) unless the proper distributor test facilities are available.

Vacuum Control
Vacuum control action can be observed and evaluated by using a tachometer and a timing light.

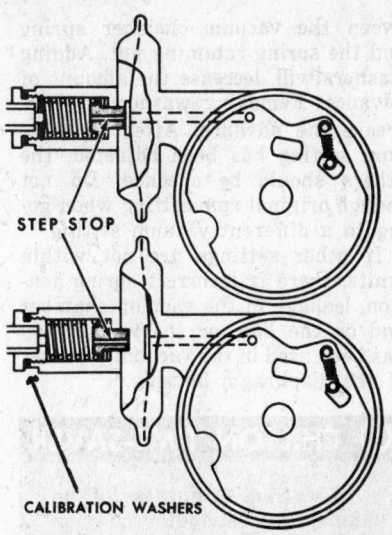

STEP STOP

CALIBRATION WASHERS

This type of spark control, whether used as the only means of control or used in conjunction with a governor type mechanism, operates through a spring loaded vacuum chamber. This chamber is attached to the side of the distributor, then, through linkage, to the breaker plate (or pick-up assembly of transistorized magnetic pulse distributors).

Carefully metered vacuum is piped to one side of the spring loaded diaphragm of the vacuum unit. Vacuum controlled timing is then the result of differential (vacuum-spring) pressures.

In the case of vacuum-only controls, (Ford Loadmatic, etc.) metering is more critical; therefore, a manometer or a very accurately calibrated vacuum gauge is required, in conjunction with the tachometer and timing light.

To Check Vacuum Control:
1. Hook up a tachometer and the timing light in the conventional manner.
2. Connect a good vacuum gauge or a manometer into the vacuum line between the carburetor and distributor.
3. With engine at operating temperature and adjusted to function smoothly, run engine at low idle.
4. Shine timing light on the indexing area and observe the vacuum reading and timing light index relationship.
5. Compare these readings with published vacuum advance data.

Indications and Causes
If timing is not within degree range at specified vacuum reading, a faulty vacuum or mechanical control mechanism, or Loadmatic control mechanism is defective.

If all parts are good, adjustments in control valve can be made by changing the calibration washers be-

tween the vacuum chamber spring and the spring retaining nut. Adding washers will decrease the amount of advance. Removing washers will increase the advance. After one vacuum setting has been adjusted, the others should be checked. Do not change original rpm setting when going to a different vacuum setting.

If other settings are not within limits, there is incorrect spring tension, leakage in the vacuum chamber and/or the line, or the wrong stop has been used in the vacuum chamber of the diaphragm housing.

⏻ CHILTON TIME-SAVER

If extreme tightness of an aluminum distributor body in the crankcase of the six cylinder engine is experienced, the following method of removal may help: play the discharge of a carbon dioxide fire extinguisher directly into the distributor body for about five seconds. The extreme cold of the CO_2 gas should shrink the distributor body and pilot enough to permit removal.

After cleaning the oxide from the distributor pilot and obtaining a free fit in the crankcase distributor bore, grease the distributor body pilot, install new O rings and reinstall distributor into crankcase.

Transistor Ignition

There are two basic types of transistorized ignition systems; those with breaker points, and those without. Both types furnish more uniform electrical performance throughout the speed range of the engine than does the conventional system, and they operate with a negligible amount of maintenance.

Caution Because we are dealing with transistors and diodes, it is advisable to discontinue some of the previous methods of ignition circuit testing. Heat, shock, or reverse polarity may cause extensive damage to the components of the system.

Breaker Point Type

Trouble Checks

1. Check for the existence and approximate quality of spark by pulling the high-tension wire out of the distributor cap. With ignition switch on and the loose end of this wire held about 1/4 in. from the engine block, crank the engine. This should produce a good, strong spark.
2. Reconnect the high-tension coil wire to the distributor cap and make a similar test at the spark plugs. A good spark at the plugs indicates the trouble to be other than ignition output failure.
3. If the spark is weak or non-ex-

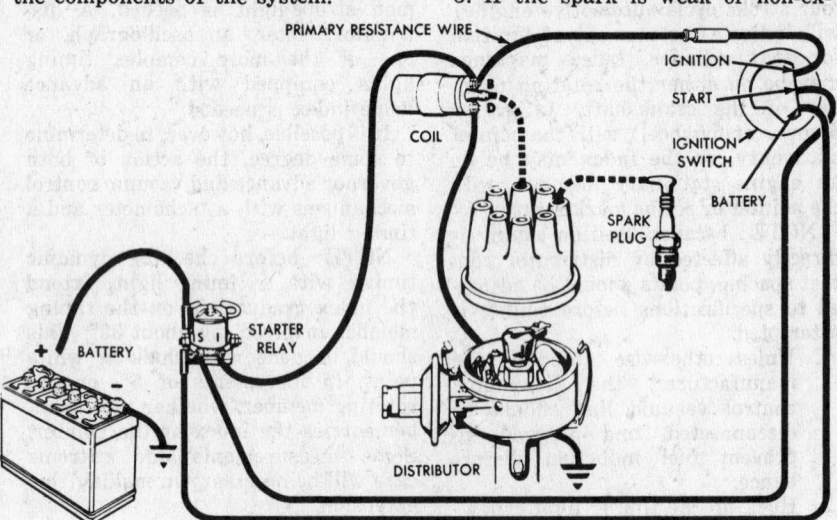

Breaker point type distributor circuit

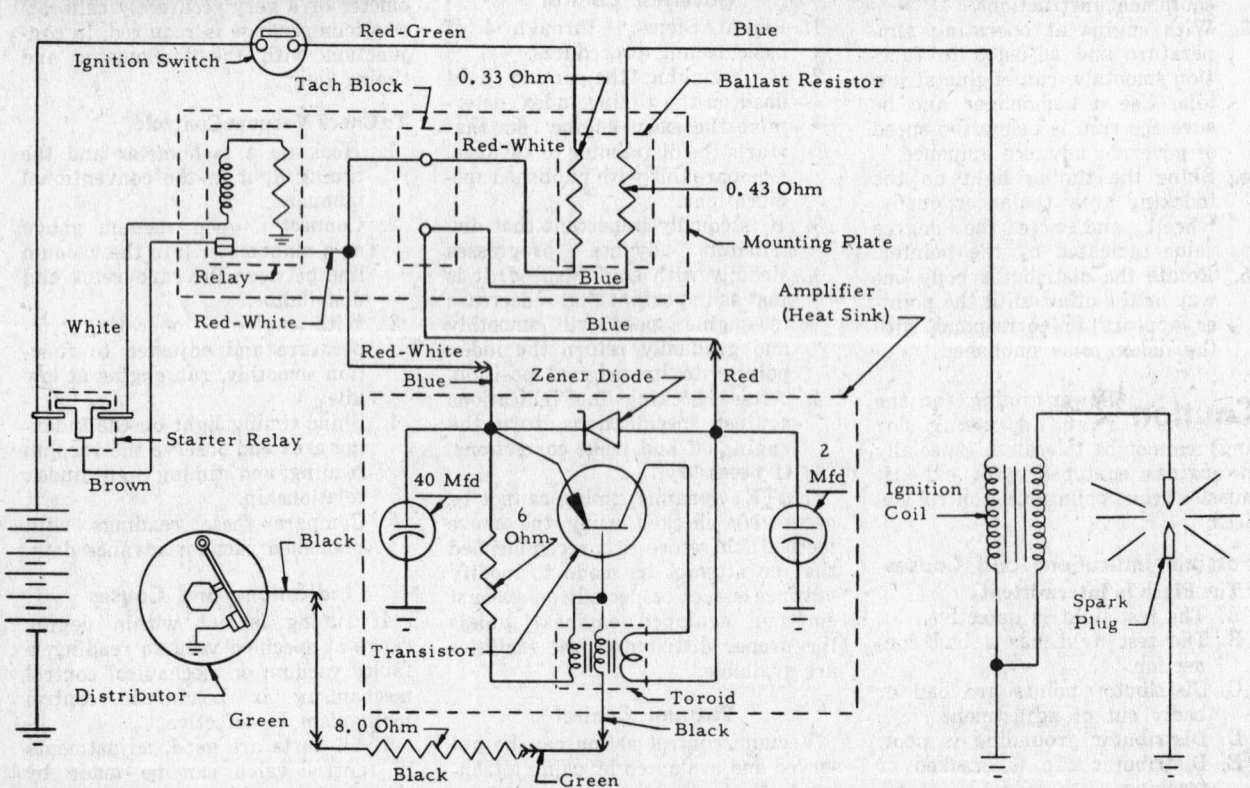

Breaker point type schematic circuit

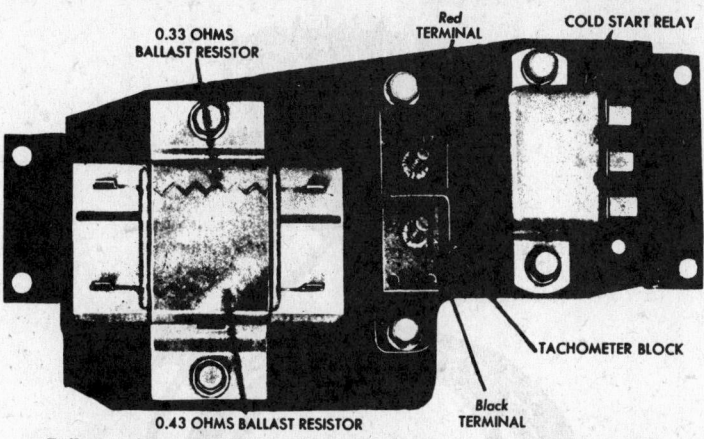

Ballast resistors, tachometer block and cold start relay—cover removed

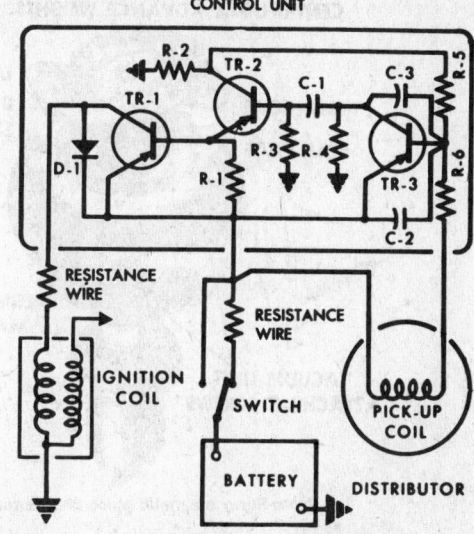

Pulse distributor schematic circuit

istent, remove the cover from the mounting plate and connect a dwell meter to the tachometer block. Hook up the red lead to the terminal in the red area and the black lead to black.

4. With ignition on, crank engine and note meter reading. A dwell angle of less than 45° (for an eight lobe cam) indicates that the transistor is working properly.

5. A dwell meter reading of zero indicates that the breaker points are dirty or not closing and should be replaced or adjusted.
   A dwell meter reading of 45° (8 lobe cam) indicates the following:
   A. Ignition system is not being supplied with current.
   B. Breaker points are not opening.
   C. Transistor is defective.

6. Disconnect the bullet connector from the distributor lead and again crank the engine. A meter reading of zero indicates trouble in the breaker points. A meter reading of 45° (eight lobe cam) indicates power source or transistor trouble. To determine which, connect a voltmeter, or test light to the red/green lead terminal of the ballast resistor and crank the engine. A reading of 45° or steady light indicates transistor failure. Replace the transistor amplifier assembly. Absence of any indication shows an open circuit between the battery and the transistor.

7. A weak spark in Steps 1 or 2 indicates a weak ignition coil. Turn ignition off, replace coil, then repeat Step 1. Do not attempt to test coil, because its low impedance will cause inaccurate readings.

8. To jump the ignition switch from under the hood, disconnect the ballast resistor-to-ignition switch wire (red with green chaser) at the ballast resistor. Then, con-

nect a jumper from the positive battery terminal to the vacated ballast resistor blade terminal. If the resistor is not open, the ignition system will now be supplied with current. Check this provision before going further.

9. The only locations from which to get troubleshooting information are the tachometer block terminals, the ballast resistor terminals and the distributor primary bullet connector. By connecting a voltmeter and an ammeter at these points, with a well charged battery and the brown wire disconnected from the cold start relay, cranking amperages and voltages may be obtained that should correspond with the following chart.

Caution

Do not pierce insulation material to obtain meter readings.

Magnetic Pulse Type

The Delco-Remy, magnetic pulse, fully transistorized ignition system uses a magnetic pulse distributor— no breaker points. This system switches power electronically rather than with ignition contact points. Instead of the familiar cam and breaker plate assembly, this distributor uses a rotating iron timer core and a magnetic pickup assembly. The magnetic pick-up assembly consists of a bearing plate on which are sandwiched a ceramic ring-type permanent magnet, two pole pieces and a pick-up coil. The pole pieces are

doughnut shaped steel plates with accurately spaced internal teeth, one tooth for each cylinder of the engine.

A critically important part is the iron timer core. It has a number of equally spaced projections or vanes and is attached to, and rotates with, the distributor shaft.

The transistor control unit, the switchbox of the system, is mounted in an aluminum case and contains three transistors, a zener diode, a condenser and five small resistors. The zener diode is a circuit protection device. Remaining components control and switch ignition-coil current electronically; there are no moving parts in the control unit.

The ignition coil is of standard design except for a special winding. The external primary resistor is a ceramic type, similar to those used on various conventional systems.

The magnetic pulse distributor provides a triggering pulse or signal for the transistor control unit. Within the distributor, a magnetic field is produced through the internal teeth of the upper and lower pole pieces by the permanent magnet between them. As the vanes of the iron timer core on the distributor shaft pass near the pole teeth as the shaft rotates, the magnetic field alternately builds up and collapses. Thus, a voltage pulse is induced in the pick-up coil each time a vane of the iron core passes a tooth on the pole pieces.

Each voltage pulse is conducted to the transistor control unit where it

Reference	Cranking rpm		Idle (700 rpm)		Normal (1500 rpm)	
	Volts	Amps	Volts	Amps	Volts	Amps
A—Battery	10.4	—	13.2	—	14.7	—
B—Emitter	9.5	4.6	13.0	5.6	14.0	12.5
C—Emitter	8.0	4.6	11.2	5.6	10.0	12.5
D—Collector	4.0	4.2	4.6	5.0	9.5	12.0
E—Collector	2.0	4.2	2.2	5.0	4.5	12.0
F—Base	4.0	0.4	5.8	0.6	3.5	0.5

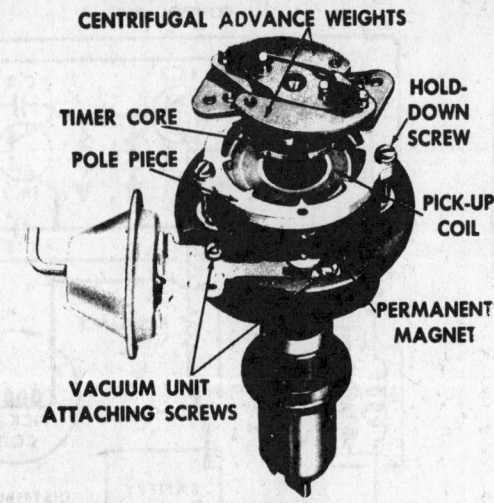

CENTRIFUGAL ADVANCE WEIGHTS

TIMER CORE

POLE PIECE

HOLD-DOWN SCREW

PICK-UP COIL

PERMANENT MAGNET

VACUUM UNIT ATTACHING SCREWS

Delco-Remy magnetic pulse distributor

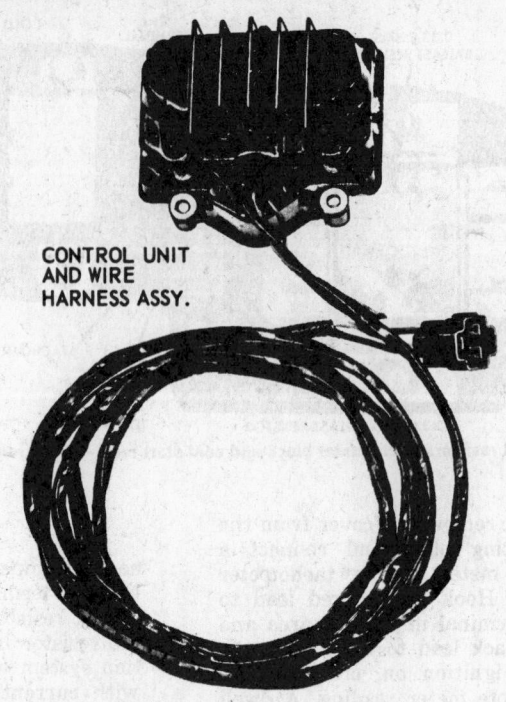

CONTROL UNIT AND WIRE HARNESS ASSY.

Control unit

turns on the triggering transistor, causing it to turn off the switching transistor. This action interrupts the current flow through the ignition coil primary winding, causing the coil to fire the spark plug. The switching transistor then automatically returns to an on condition, permitting coil current to build up for the next firing.

The rotating iron timer core is designed with multiple vanes, four in the case of eight cylinder engines, and three in the case of six cylinders.

Diagnosis

This diagnosis guide has been taken from material furnished by Pontiac Division, General Motors Corporation. It can be applied, however, to other transistorized ignition systems of the same basic type.

Cautions

1. Don't use 18 volts or 24 volts for emergency starting.
2. Never crank engine with coil high-tension lead or more than three spark plug leads disconnected.
3. Don't short circuit between coil positive terminal and ground.
4. On any repair that necessitates replacement of control unit or ignition resistor, perform complete charging system check before releasing the unit. Basic cause of trouble may be high or uncontrolled charging rate.

Intermittent Miss

1. Check pick-up coil connector to make sure that wire harness leads and pick-up coil leads are tight in connector bodies.
2. Check for loose connections throughout the system.
3. Engine roughness or miss indicates a short or open in the distributor pick-up coil. Connect an ohmmeter between two distributor pick-up coil terminals in

connector body. Resistance should read 300-400 ohms. If resistance is infinite, coil is open; if resistance is low, coil is shorted. In either case, the pick-up coil must be replaced.
4. Connect an ohmmeter from either of the terminals to the distributor housing. The reading should be infinite. If not, the wiring is grounded.

Diagnosis

Hard or No Start

Disconnect any one spark plug lead and crank the engine while holding the lead about ¼ in. from ground. Caution: Do not perform this test by using the coil or distributor secondary wire—damage to the system will result.

If Spark Occurs

Reconnect spark plug. Problem is not in ignition system.

If No Spark Occurs

1. Reconnect spark plug.
2. Check ignition coil primary by connecting an ohmmeter between the two terminals. An infinite reading indicates an open primary.
3. Check ignition coil secondary by connecting an ohmmeter between high-voltage center tower and either primary terminal. If the reading is infinite, the secondary is open.
4. Check ground resistor with a low-range ohmmeter. Turn ignition switch off and connect leads

across resistor terminals. The reading should be .43 ohms.
5. Check by-pass resistor with a low-range ohmmeter. Reading should be .68 ohms with ignition switch off.
6. Check continuity of system as a whole, using a low-range (0-20 volt) voltmeter, in the following manner:

STEP 1
A. Connect the voltmeter between ignition coil positive terminal and ground.
B. Turn on ignition switch and observe reading. Voltage should be 7-9 volts.
C. If reading is 11-12 volts, there is an open circuit between this point and ground. The circuit includes the coil primary winding, ballast resistor and connecting wiring.
D. If reading is zero, there is an open circuit between this point and battery. Go on to Step 2.

STEP 2
A. Connect voltmeter between other ballast resistor and ground. Observe reading with ignition switch on.
B. If the reading is 11-12 volts, there is an open circuit between this ballast resistor and the ignition coil. The circuit consists of the control unit and associated wiring. If wiring is OK, the control unit must be replaced.

AC Generators

Is It the Alternator or the Voltage Regulator?

The first step in diagnosing troubles of the charging system, is to identify the source of failure. Does the fault lie in the alternator or the regulator? The next move depends upon preference or necessity; either repair or replace the offending unit.

It is just as easy to separate an alternator, electrically, from the AC regulator as it is to separate its counterpart, the DC generator from its regulator.

AC generator output is controlled by the amount of current supplied to the field circuit of the system.

Unlike the DC generator, an AC generator is capable of producing substantial current at idle speed. Higher maximum output is also a possibility. This presents a potential danger when testing. As a precaution, a field rheostat should be used in the field circuit when making the following isolation test. The field rheostat permits positive control of the amount of current allowed to pass through the field circuit during the isolation test. Unregulated alternator capacity could ruin the unit.

NOTE: most manufacturers of precision gauges offer special test connectors, in sets, that will adapt to the leads and connections of any AC charging system.

Caution Before attempting the isolation test, disconnect the field wire from the regulator. Failure to take this precaution can cause instant burning and permanent damage to the regulator.

Isolation Test

(By-passing the Regulator)

1. Connect voltmeter leads to two prods driven into the battery posts.
2. Disconnect field wire from the FLD terminal of the voltage regulator.
3. Connect one lead of a field rheostat to the undisturbed IGN terminal of the regulator, and the other field rheostat lead to the wire that was removed from the FLD terminal of the regulator.
4. With field rheostat turned to the low side of the scale, (high resistance) start the engine and adjust throttle to about 2,000 rpm.
5. Slowly move field rheostat control knob to decrease resistance, (allowing more current to flow through the field circuit) until voltmeter reading slightly exceeds manufacturers' specifications.

NOTE: under load conditions, observe the alternator for arcing or any other evidence of malfunction.

6. If alternator performs satisfactorily, repair or replace the regulator. Conversely, if the voltmeter reading is zero, or below specifications, repair or replace the alternator.

Alternator Test Plans

The following is a procedure pattern for testing the various alternators and their control systems.

There are certain precautionary measures that apply to alternator tests in general. These items are listed in detail to avoid repetition when testing each make of alternator, and to encourage a habit of good test procedure.

1. Check alternator drive belt for condition and tension.
2. Disconnect battery cables, check physical, chemical, and electrical condition of battery.
3. Be absolutely sure of polarity before connecting any battery in the circuit. Reversed polarity will ruin the diodes.
4. Never use a battery charger to start the engine.
5. Disconnect both battery cables when making a battery recharge hook-up.

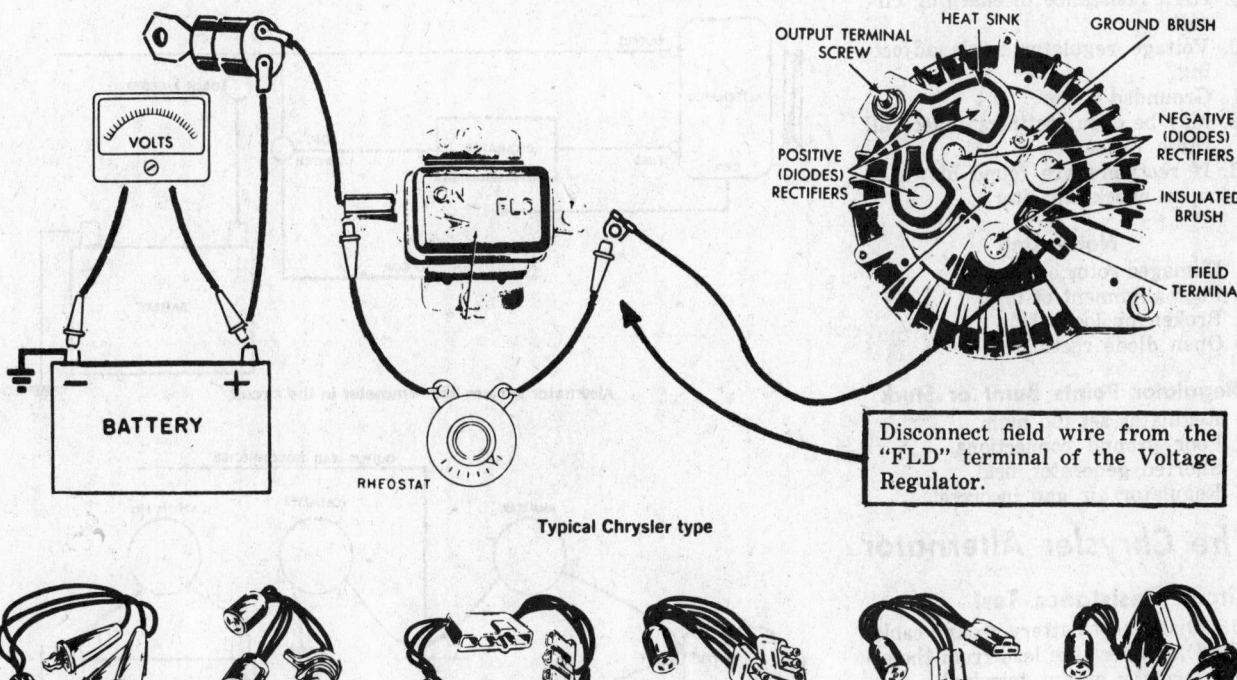

Typical Chrysler type

Disconnect field wire from the "FLD" terminal of the Voltage Regulator.

Left to right: Chrysler adapter; Delcotron adapter; Delcotron adapter; Motorola adapter; Motorola adapter; FoMoCo adapter

6. Be sure of polarity hook-up when using a booster battery for starting.
7. Never ground the alternator output or battery terminal.
8. Never ground the field circuit between alternator and regulator.
9. Never run any alternator on an open circuit with the field energized.
10. Never try to polarize an alternator.
11. Do not attempt to motor an alternator.
12. The regulator cover must be in place when taking voltage limiter readings.
13. The ignition switch must be in off position when removing or installing the regulator cover.
14. Use insulated tools only to make adjustments to the regulator.
15. When making engine idle speed adjustments, always consider potential load factors that influence engine rpm. To compensate for electrical load, switch on the lights, radio, heater, air conditioner, etc.

Diagnosis

Low or No Charging

1. Blown fuse.
2. Broken or loose fan belt.
3. Voltage regulator not working.
4. Brushes sticking.
5. Slip ring dirty.
6. Open circuit.
7. Bad wiring connections.
8. Bad diode rectifier.
9. High resistance in charging circuit.
10. Voltage regulator needs adjusting.
11. Grounded stator.
12. May be open rectifiers (check all three phases).
13. If rectifiers are found blown or open, check capacitor.

Noisy Unit

1. Damaged rotor bearings.
2. Poor alignment of unit.
3. Broken or loose belt.
4. Open diode rectifiers.

Regulator Points Burnt or Stuck

1. Regulator set too high.
2. Poor ground connections.
3. Shorted generator field.
4. Regulator air gap incorrect.

The Chrysler Alternator

Circuit Resistance Test

1. Disconnect battery ground cable.
2. Disconnect the lead from the alternator output terminal.
3. Hook up an ammeter as follows:
   A. Connect the positive lead to the alternator output terminal.
   B. Connect the negative lead to

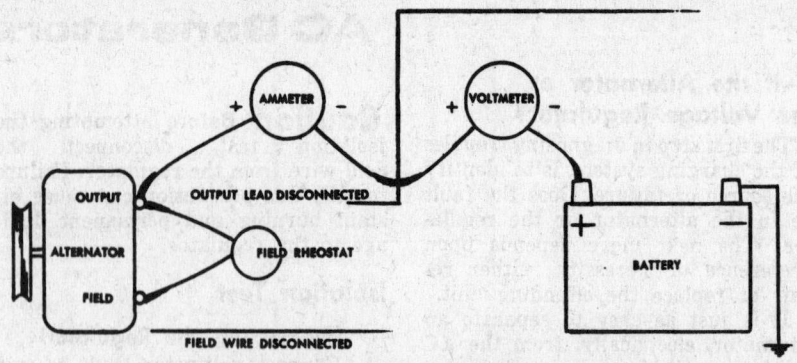

Plan to check charging circuit resistance

the lead just disconnected from the alternator output terminal.
4. Hook up voltmeter as follows:
   A. Connect the positive voltmeter lead to the lead just disconnected from the alternator output terminal.
   B. Connect the negative voltmeter lead to the positive battery post.
5. Disconnect the lead from the alternator field terminal.
6. With field rheostat set to the open position, connect rheostat leads to the alternator field terminal and the alternator output terminal.
7. Hook up a tachometer to the engine.
8. Connect the battery ground cable.
9. Start the engine and set speed to recommended rpm.

10. Slowly adjust the field rheostat toward the closed position until the ammeter registers 10 amps.
11. The voltmeter reading will now show the voltage drop in the charging circuit. There should not be more than 0.2 volt drop.
12. Adjust field rheostat to the open position.
13. If the voltage drop exceeds 0.2 volt, stop the engine, clean and tighten all circuit connections, then repeat the test.

Current Output Test

1. The ammeter and field rheostat hook up should remain the same as for the circuit resistance check.
2. Connect the voltmeter negative lead to the battery negative post.
3. With carbon pile load control set in the off position, connect one lead to each battery post.

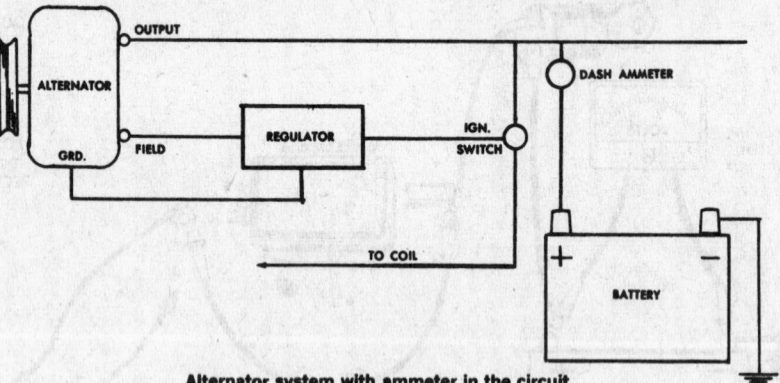

Alternator system with ammeter in the circuit

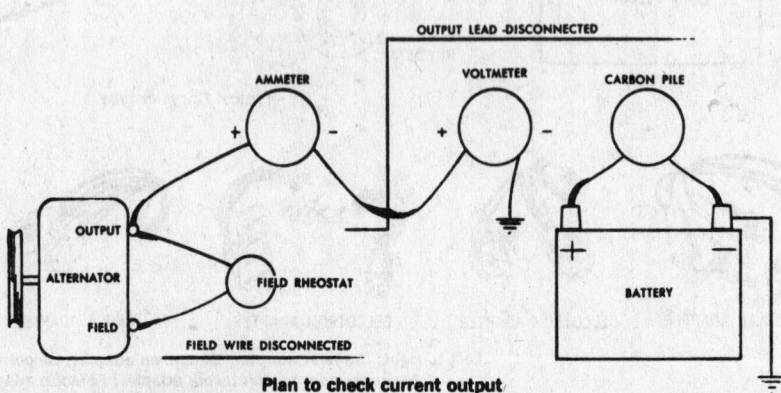

Plan to check current output

4. Start the engine and adjust speed to recommended rpm.

5. Slowly move the field rheostat control toward the closed position. Note voltmeter and ammeter readings. Maintain a 15 volt reading by adjusting the carbon pile control. With field rheostat closed, note ammeter reading.

6. Now, open the field rheostat and turn the carbon pile control to off position.

7. Compare ammeter reading with manufacturer's specifications. The reading should be no less than specified.

8. If below specifications:
   A. Check drive belt tension, adjust if necessary, then repeat the test.
   B. If results are still below specifications, internal trouble is indicated. Make a field current draw test.

9. If the readings are satisfactory, the next test, field current draw will be unnecessary.

Field Current Draw Test

Caution Disconnect battery ground at the battery, then:

1. With the output lead connected to the alternator output terminal, connect an ammeter positive lead to the alternator output terminal.

2. Connect the ammeter negative lead to a field rheostat lead.

3. Turn the field rheostat control to the open position.

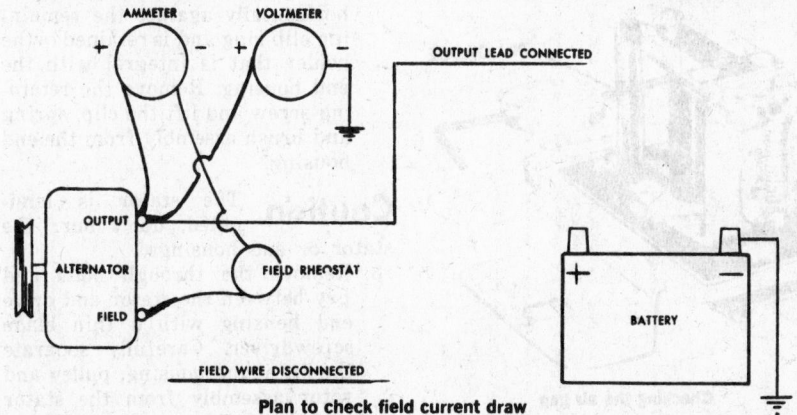

Plan to check field current draw

4. Now, connect the remaining field rheostat lead to the alternator field terminal.

5. Hook up the voltmeter leads by connecting the positive lead to the alternator output terminal and the negative lead to ground.

6. The ground cable may now be reconnected to the battery negative post.

7. Turn the field rheostat control to the closed position, then check the ammeter and voltmeter readings.

8. Turn the field rheostat control to the open position, then compare the ammeter and voltmeter readings with manufacturers' specifications.

9. If readings are below specifications, the trouble probably lies in the brushes or slip rings.

10. If readings are above specifications, look for trouble in the rotor field windings.

11. If readings are as specified, but the alternator did not deliver its rated output, look for trouble in the stator or diodes.

12. Replace the alternator.

Voltage Regulator Setting

1. With the battery ground cable disconnected at the battery and the lead disconnected from the alternator output terminal, hook up an ammeter as follows:
   A. Connect the ammeter positive lead to the alternator output terminal.

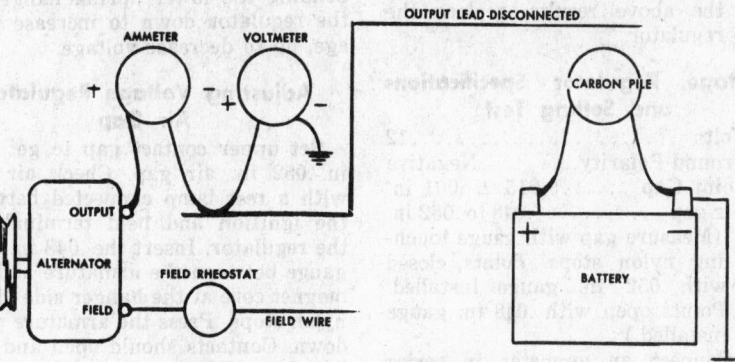

Plan to check voltage regulator setting

   B. Connect the ammeter negative lead to the lead just disconnected from the alternator output terminal.

2. Hook up a voltmeter as follows:
   A. Connect the voltmeter positive lead to the lead just disconnected from the alternator output terminal.
   B. Connect the negative lead to ground.

3. Disconnect lead from alternator field terminal.

4. Hook up a field rheostat (with control in open position) to the alternator field terminal and to the lead which was just disconnected from this field terminal.

5. Connect a tachometer to the engine.

6. Hook up the ground cable to the battery.

7. Hook up a carbon pile (load control turned to the off position) between the battery posts.

8. With engine operating at 1500 rpm, adjust field rheostat control to the closed position and watch the voltmeter and ammeter.

Caution Voltmeter reading must not exceed 15 volts. Continue to run the engine at this setting until the regulator has reached operating temperature.

9. With field rheostat in fully closed position, drop engine speed to recommended rpm for testing.

10. If the charging rate is above rec-

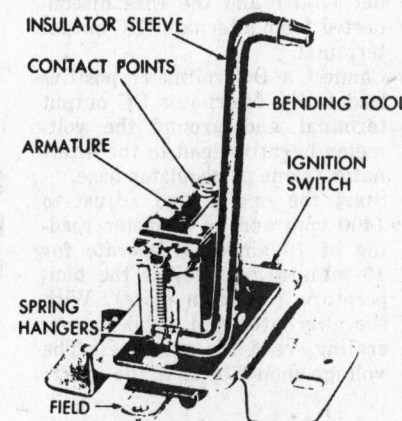

Adjusting spring tension

ommendations, the battery is probably in an undercharged condition. Either recharge the battery or replace it with a battery in fully charged condition.

11. If the charging rate is below 15 amps, regulate the carbon pile load to obtain a 15 amp rate.

12. With the voltage regulator temperature at running specifications, turn the carbon pile load control off and increase engine

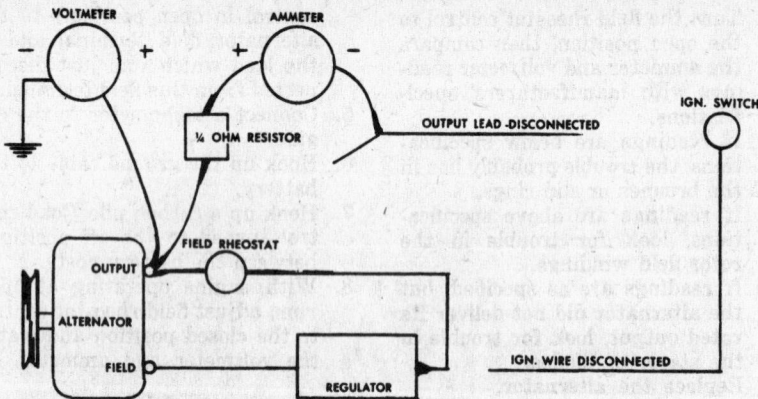

Plan to check voltage regulator with a ¼ ohm resistor in the circuit

rpm to manufacturer's specifications. The ammeter should read 5 amps. or less and voltage should increase 0.2 volt to 0.7 volt. This indicates that both sets of points are working, within specified limits.

13. If the regulator fails to produce the above results, replace the regulator.

Voltage Regulator Specifications and Setting Test

Volts 12
Ground Polarity . . . . . . .Negative
Point Gap . . . . . . .015 ± .001 in.
Air gap . . . . .048 to .052 in.
(Measure gap with gauge touching nylon stops. Points closed with .052 in. gauge installed. Points open with .048 in. gauge installed.)

1. Connect an ammeter in series with the alternator output terminal (Bat.) and the wire disconnected from alternator DC output terminal.

2. Connect a DC voltmeter positive lead to the alternator DC output terminal and ground the voltmeter negative lead to the alternator frame or regulator base.

3. Start the engine and adjust to 1400 rpm and an ammeter reading of 10 amperes. Operate for 15 minutes to stabilize the temperature (cover in place). With the alternator and regulator operating, read the voltmeter. The voltage should be as in the chart:

Operating Voltage (°F)—Surrounding Temperature

Temp. in degrees	20°	40°	60°	80°	100°	120°	140°
Minimum setting	13.9 to	13.82 to	13.74 to	13.65 to	13.56 to	13.48 to	13.40 to
Maximum setting	14.5	14.42	14.34	14.25	14.16	14.08	14.0

Temperature checked about 2 in. from the cover. There will be a slightly higher voltage at higher engine speeds. However, this must not exceed 0.7 volts at any temperature range.

Caution Do not short circuit between the spring hanger and base or spring. Use an insulated bending tool. Grounding the spring hanger will damage the regulator (burn the contact spring).

Adjust the voltage if required by bending the lower spring hanger of the regulator down to increase voltage, up to decrease voltage.

Adjusting Voltage Regulator Air Gap

Set upper contact gap to get .048 in.-.052 in. air gap. Check air gap with a test lamp connected between the ignition and field terminals on the regulator. Insert the .048 in. wire gauge between the armature and the magnet core at the hanger side of the nylon stops. Press the armature plate down. Contacts should open and test light should go out. Insert the .052 in.

PIN GAUGE

Checking the air gap

gauge in the same position and depress the armature. Upper contacts should be closed and test lamp should remain lighted. Bend upper contact support as necessary to maintain the air gap adjustment.

Adjusting Lower Contact Clearance

Set lower contact gap to .015 in. (± .001 in.) by bending the lower contact arm. Press the armature down so that the armature rests on the nylon stop and recheck the contact gap.

Note: press down on the armature and not on the contact spring.

Specifications

Rotation . . . .clockwise at drive end
Voltage 12
Standard output (Amperes) 35
Heavy duty output 40
Brushes (number) 2
Field coil draw . . . . . .2.38 amperes minimum to 2.75 amperes maximum at 12-volts or 2.97 amperes minimum to 3.43 amperes maximum at 15 volts at 70° F alternator operating temperature at 750 alternator rpm.
Rectifier resistance . .4 to 10 ohms in the forward direction. Essentially none in reverse direction
Condenser capacity . . .5 microfarad

Disassembly

To prevent damage to the brush assemblies they should be removed before proceeding with the disassembly of the alternator. The insulated brush is mounted in a plastic holder that positions the brush vertically against one of the slip rings.

1. Remove the retaining screw, flat washer, nylon washer and field terminal and carefully lift the plastic holder containing the spring and brush assembly from the end housing.

2. The ground brush is positioned horizontally against the remaining slip ring and is retained in the holder that is integral with the end housing. Remove the retaining screw and lift the clip, spring and brush assembly from the end housing.

Caution The stator is laminated, don't burr the stator or end housings.

3. Remove the through bolts and pry between the stator and drive end housing with a thin blade screwdriver. Carefully separate the drive end housing, pulley and rotor assembly from the stator and rectifier housing assembly.

4. The pulley is an interference fit on the rotor shaft. Remove with puller, tool C-3615 and special adapters, tool SP-3002.

5. Remove the three nuts and washers and, while supporting the end frame, tap the rotor shaft with a plastic hammer and separate the rotor and end housing.

6. The drive end ball bearing is an interference fit with the rotor shaft. Remove the bearing with puller, tool C-3615 and adapters as follows:

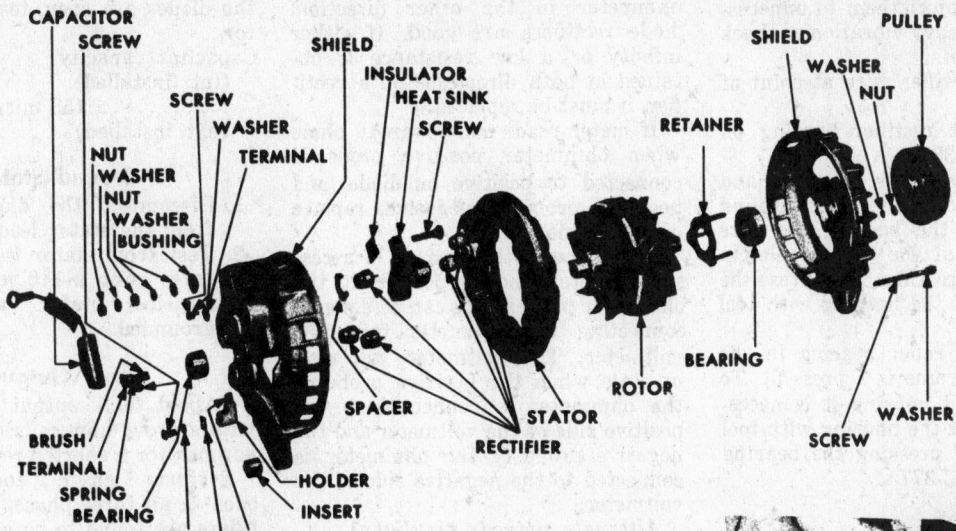

CAPACITOR SCREW — SCREW — NUT — WASHER — NUT — WASHER — BUSHING — TERMINAL — WASHER — SHIELD — INSULATOR HEAT SINK — SCREW — RETAINER — SHIELD — WASHER — NUT — PULLEY

BRUSH — TERMINAL — SPRING — BEARING — WASHER — BRUSH — TERMINAL — HOLDER — INSERT — SPACER — RECTIFIER — STATOR — ROTOR — BEARING — WASHER — SCREW

Alternator

BRUSH ASSEMBLY

Insulator brush

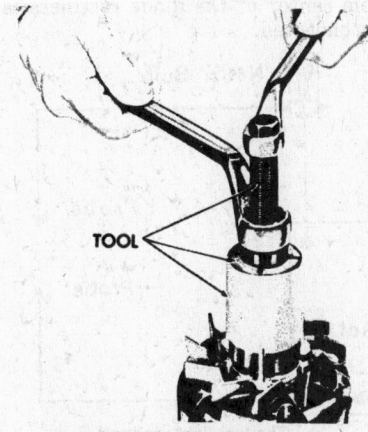

TOOL

Pulling bearing from rotor shaft

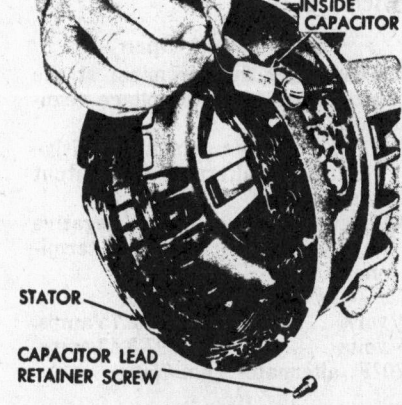

INSIDE CAPACITOR — STATOR — CAPACITOR LEAD RETAINER SCREW

Inside capacitor

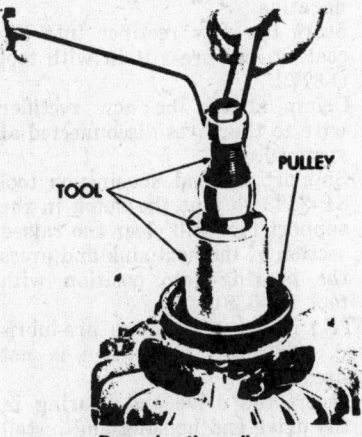

TOOL — PULLEY

Removing the pulley

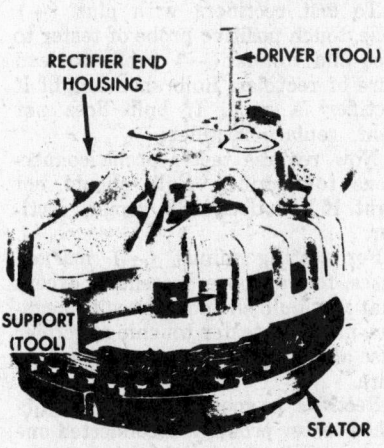

RECTIFIER END HOUSING — DRIVER (TOOL) — SUPPORT (TOOL) — STATOR

Removing a rectifier

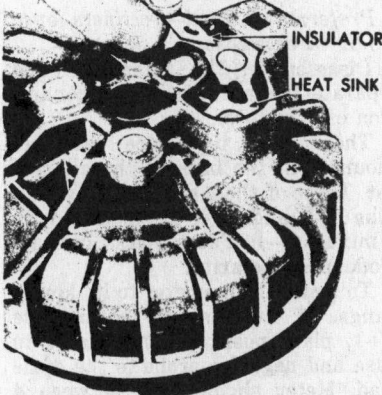

INSULATOR — HEAT SINK

Heat sink and insulator

A. Position the center screw of puller on rotor shaft.
B. Place the thin lower end of the adapters SP-3375 under the bearing equally spaced and the upper end of the adapters around the center screw.
C. Hold adapters and center screw in position with the sleeve.

Caution Tool sleeve must bottom on bearings, otherwise adapters may be damaged.

D. Turning center screw clockwise while holding the body of the tool will draw the bearing from the shaft.

NOTE: further dismantling of the rotor is not advisable, as the remain-

der of the rotor assembly is not serviced separately.

7. Remove the DC output terminal nuts and washers and remove terminal screw and inside capacitor (on units so equipped).

NOTE: the heat sink is also held in place by the terminal screw.

8. Remove the insulator.

NOTE: three positive rectifiers are pressed into the heat sink and three negative rectifiers in the end housing. When removing the rectifiers, it is necessary to support the end housing and/or heat sink to prevent damage to these castings. Another caution is in order relative to the diode recti-

fiers. Don't subject them to unnecessary jolting. Heavy vibration or shock may ruin them.

    A. Cut rectifier wire at point of crimp.

    B. Support rectifier housing on tool C-3771.

NOTE: this tool is cut away and slotted to fit over the wires and around the bosses in the housing. Be sure that the bore of the tool completely surrounds the rectifier, then press the rectifier out of the housing with tool SP-3380.

NOTE: the roller bearing in the rectifier end frame is a press fit. To protect the end housing it is necessary to support the housing with tool SP-3383 when pressing the bearing out with tool C-3770.

Bench Tests

Field Circuit Open

1. Disconnect field terminal at the alternator and the voltage regulator.
2. Connect a DC ammeter positive lead to the alternator DC output terminal.
3. Connect the ammeter negative lead to the alternator field terminal.

Field Current Draw

12 volts ....2.38-2.75 amps.
15 volts ....2.97-3.43 amps.
(70°F, alternator rpm 750)

Testing Silicon Diode Rectifiers With Ohmmeter

Preferred method—rectifiers open in all three phases.

Disassemble the alternator and separate the wires at the Y-connection of the stator.

There are six diode rectifiers mounted in the back of the alternator. Three of them are marked with a plus (+), and three are marked with a minus (—). These marks indicate diode case polarity.

To test, set ohmmeter to its lowest range. If case is marked positive (+), place positive meter probe to case and negative probe to the diode lead. Meter should read between 4 and 10 ohms. Now, reverse leads of ohmmeter, connecting negative meter probe to positive case and positive meter probe to wire of rectifier. Set meter on a high range. Meter needle should move very little, if any (infinite reading). Do this to all three positive diode rectifiers.

The three with minus (—) marks on their cases are checked the same way as above. Only now the negative ohmmeter probe is connected to the case for a reading of 4 to 10 ohms. Reverse leads as above for the other part to test.

If a reading of 4 to 10 ohms is obtained in one direction and no reading (infinity) is read on the

ohmmeter in the other direction, diode rectifiers are good. If either infinity or a low resistance is obtained in both directions on a rectifier, it must be replaced.

If meter reads more than 10 ohms when ohmmeter positive probe is connected to positive on diode, and negative probe to negative, replace diode rectifier.

NOTE: with this test, it is necessary to determine the polarity of the ohmmeter probes. This can be done by connecting the ohmmeter to a DC voltmeter. The voltmeter will read up-scale when the positive probe of the ohmmeter is connected to the positive side of the voltmeter and the negative probe of the ohmmeter is connected to the negative side of the voltmeter.

Alternate method—test light.

Make-up a tester as shown in sketch. Refer to first paragraph of the preferred method. Be sure lead from center of the diode rectifiers is disconnected.

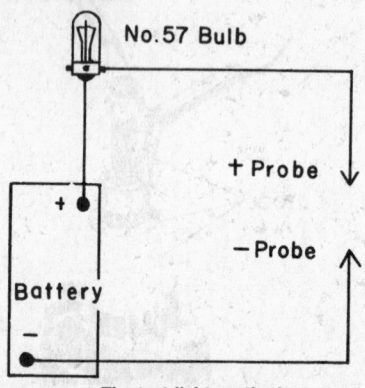

The test light method

To test rectifiers with plus (+) case, touch positive probe of tester to case and minus (—) probe to lead wire of rectifier. Bulb should light if rectifier is good. If bulb does not light, replace rectifier.

Now reverse tester probe connections to rectifier. Bulb should not light. If bulb does light, replace rectifier.

For testing minus (—) marked cases, follow above procedure, except that now bulb should light with negative probe of tester touching rectifier case and positive probe touching lead wire.

Rectifier is good if the bulb lights when tester probes are connected one way, and does not light when tester connections are reversed.

Rectifier must be replaced if the bulb does not light either way tester is connected. Also, replace rectifier if bulb lights for both ways tester is connected to the diode rectifier.

NOTE: the usual cause of an open or blown diode or rectifier is a defective capacitor or a battery that has been installed in reverse polarity. If the battery is installed properly and

the diodes are open, test the capacitor.

Capacitor capacity:
  (int. installed)
    ....158 microfarad, min.
  (ext. installed) .... .5 microfarad

Ground Stator

1. Disconnect the diode rectifiers from the stator leads.
2. Test from stator leads to stator core, using a 110-volt test lamp. Replace stator assembly if grounded.

Low Output

(About 50% output accompanied with a growl-hum caused by a shorted phase or a shorted rectifier.)

Perform Steps 1, 2 and 3 (rectifier open in all three phases). If the rectifiers are found to be within specifications, replace the stator assembly.

Current Output Too High (No Control) Caused by Open Rectifier or Open Phase

Perform Steps 1, 2 and 3 (rectifier open in all three phases). If the rectifier tests satisfactorily, inspect the stator connections before replacing the stator.

Assembly

1. Support the heat sink or rectifier end housing on circular plate, tool SP-3377.
2. Check rectifier identification to be sure the correct rectifier is being used. The part numbers are stamped on the case of the rectifier. They are also marked, red for positive and black for negative.
3. Start the new rectifier into the casting and press it in with tool C-3772.
4. Crimp attach the new rectifier wire to the wires disconnected at removal.
5. Support the end housing on tool SP-3383 so that the notch in the support tool will clear the raised section of the heat sink and press the bearing into position with tool SP-3381.

NOTE: new bearings are pre-lubricated, additional lubrication is not required.

6. Insert the drive end bearing in the drive end housing and install the bearing plate, washers and nuts to hold the bearing in place.
7. Position the bearing and drive end housing on the rotor shaft and while supporting the base of the rotor shaft, press the bearing and housing in position on the rotor shaft with an arbor press and tool C-3769.

Caution Be careful that there is no cocking of the bearing at installation; or damage

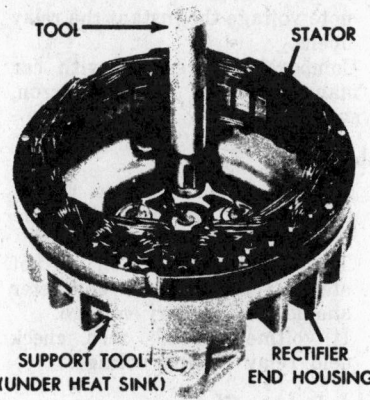

TOOL — STATOR

SUPPORT TOOL (UNDER HEAT SINK) — RECTIFIER END HOUSING

Installing diode rectifier

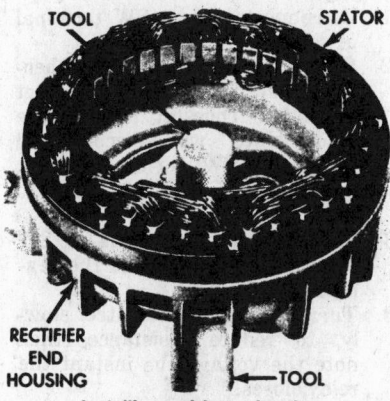

TOOL — STATOR

RECTIFIER END HOUSING — TOOL

Installing end frame bearing

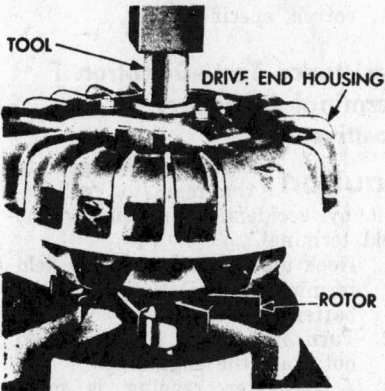

TOOL — DRIVE END HOUSING

ROTOR

Installing drive end frame and bearing on rotor shaft

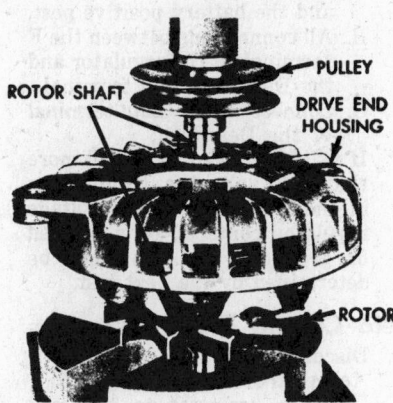

ROTOR SHAFT — PULLEY

DRIVE END HOUSING

ROTOR

Installing the pulley

will result. Press the bearing on the rotor shaft until the bearing contacts the shoulder on the rotor shaft.

8. Install pulley on rotor shaft. Shaft of rotor must be supported so that all pressing force is on the pulley hub and rotor shaft.

NOTE: Do not exceed 6,800 lbs. pressure. Pulley hub should just contact bearing inner race.

9. Some alternators will be found to have the capacitor mounted internally. Be sure the heat sink insulator is in place.

10. Install the output terminal screw with the capacitor attached through the heat sink and end housing.

11. Install insulating washers, lockwashers and locknuts.

12. Make sure the heat sink and insulator are in place and tighten the locknut.

13. Position the stator on the rectifier end housing. Be sure that all of the rectifier connectors and phase leads are free of interference with the rotor fan blades and that the capacitor (internally mounted) lead has clearance.

14. Position the rotor assembly in the rectifier end housing. Align the through bolt holes in the stator with both end housings.

15. Enter stator shaft in the rectifier end housing bearing, compress stator and both end housings manually and install through bolts, washers and nuts.

16. Install the insulated brush and terminal attaching screw.

17. Install the ground screw and attaching screw (see illustration).

18. Rotate pulley slowly to be sure the rotor fan blades do not hit the rectifier and stator connectors.

Testing the Alternator System (In the Car)

With the battery fully charged and in good condition and the engine at normal operating temperature, proceed with tests as follows:

Caution Never ground the field circuit between the alternator and the regulator as this will result in damage to the voltage regulator.

Charging Circuit Resistance Test

1. Connect a DC ammeter which will read a minimum of 40 amperes in series with the alternator DC output terminal and the DC output lead which was disconnected from the alternator.

2. Connect a DC voltmeter positive lead to the DC output lead that was disconnected from the alternator, and the voltmeter negative lead to the battery positive post.

NOTE: A voltage drop test across

each connection will locate any bad connections.

3. Start the engine and adjust the speed to produce 10 amperes from the alternator. The voltage reading should not exceed .2 volt. If there is higher voltage drop, clean and tighten all connections in the charging circuit and recheck charging circuit resistance.

Delcotron By Delco-Remy

Although the unit is built to give long periods of trouble-free service, a regular inspection procedure should be followed to enjoy the maximum life that is built into each generator.

Inspection

Frequency of inspection is determined by exposure to adverse operating conditions. High speed operation, high temperatures, dust and dirt all increase wear on brushes, slip rings and bearings.

At regular periods, inspect terminals for corrosion and loose connections, and the wiring for frayed insulation. Check mounting bolts for tightness, and the belt for alignment, proper tension and wear. Because of higher load capacity and the higher inertia of the heavy rotor used in AC generators, belt tension is more critical. Tension should be adjusted according to the vehicle manufacturer's recommendations. In most cases this will be slightly greater than for DC generators.

Noisy Generator

Noise in a Delcotron may be caused by worn or dirty bearings, loose mounting or a loose drive pulley. In the self-rectifying generator, noise may come from a shorted or open diode.

Indicator or Ammeter

There are two types of regulators used with this system.

A. A three-unit regulator, containing a voltage regulator, a field relay, and an indicator lamp relay.

B. A two-unit regulator, containing a voltage regulator, and a field relay. In this instance, the indicator lamp relay is omitted. The two unit regulator is used in circuits equipped with an ammeter. There is no L terminal on the regulator.

Except for the indicator lamp relay test, all other tests are conducted in the same way.

Indicator Lamp Relay Test

1. Turn ignition switch on; do not start the engine. If indicator does not burn, check for burned out bulb.

2. If light does burn, but stays on

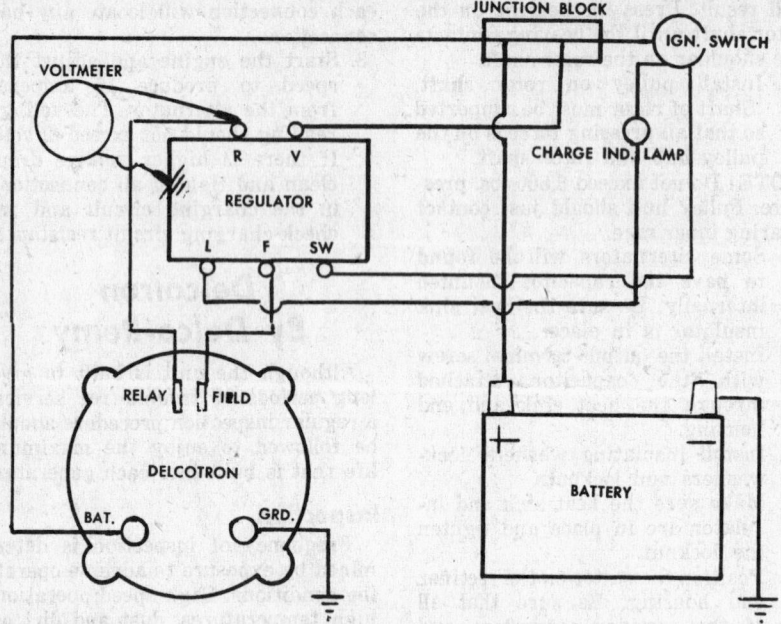

Plan to check indicator lamp relay

after the engine is started and the generator is in operation:

A. Connect voltmeter leads to the R terminal and a ground on the regulator.

B. If voltmeter reading is more than 5 volts, the indicator lamp relay is defective.

C. If voltmeter reading is 5 volts or less, trouble lies elsewhere in the system.

Indicator Lamp Relay Opening Voltage Test

1. With a 50-75 ohm field rheostat connected to the R and V terminals of the regulator, turn rheostat control to the open position.
2. Connect a voltmeter to the R and ground terminal on the regulator.
3. Turn the field rheostat control slowly, to reduce resistance, then

note voltage the instant the relay opens.

4. Compare this reading with car manufacturers', or Delcotron, specifications.

Field Relay Test

1. Connect a voltmeter to the F terminal and the ground on the regulator.
2. Turn ignition switch on; do not start the engine. Voltmeter should read battery voltage.
3. If voltmeter reads zero, check field relay closing voltage.

Field Relay Closing Voltage Test

1. Disconnect lead at SW terminal of regulator.
2. Connect a 50-75 ohm field rheostat to the SW terminal. Connect the other rheostat lead to the lead which was disconnected from the SW terminal. Then, turn the field rheostat control to the open position.
3. Connect voltmeter leads to SW terminal and ground of the regulator.
4. Turn field rheostat control slowly, to reduce resistance, then note the voltage the instant the relay closes.
5. Compare this voltmeter reading with car manufacturers', or Delcotron, specifications.

Resistance Test—Delcotron F Terminal to Battery Positive Post

Caution Serious damage to the regulator will result by accidentally grounding the field terminal.

1. Hook up a voltmeter to the field terminal of the Delcotron and the battery, positive post.
2. Turn on ignition switch but do not start the engine.
3. If voltmeter reading is more than 0.3 volt, clean and tighten:
   A. All connections between the V terminal on the regulator and the battery positive post.
   B. All connections between the F terminal on the regulator and the Delcotron field terminal.
   C. Connections at Bat. terminal of the Delcotron.
4. If voltmeter reading is still more than 0.3 volt, the trouble lies in the field relay or the voltage regulator contacts, or a shorted or grounded field. This can be determined by the next test.

Field Current Draw Test

1. Disconnect the lead from the F terminal of the regulator.
2. Hook an ammeter to the lead just disconnected and to the F terminal of the regulator.

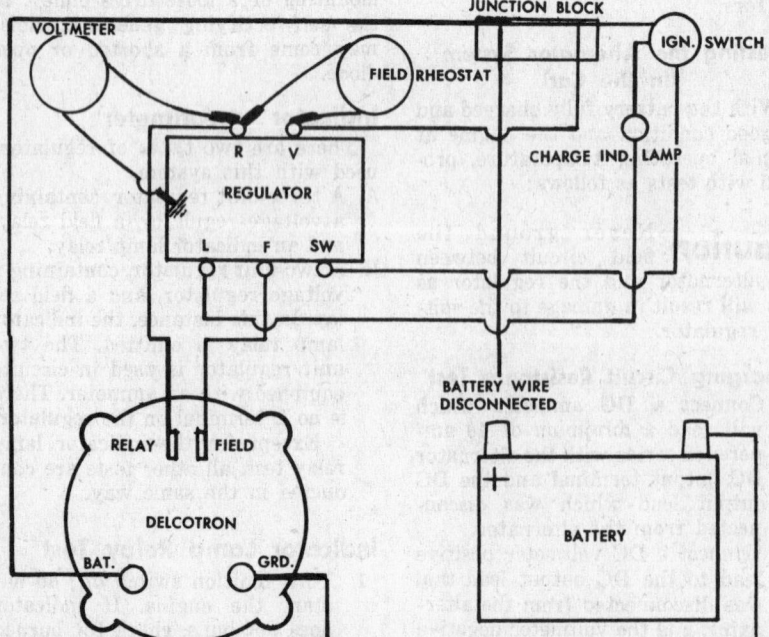

Plan to check indicator lamp relay opening voltage

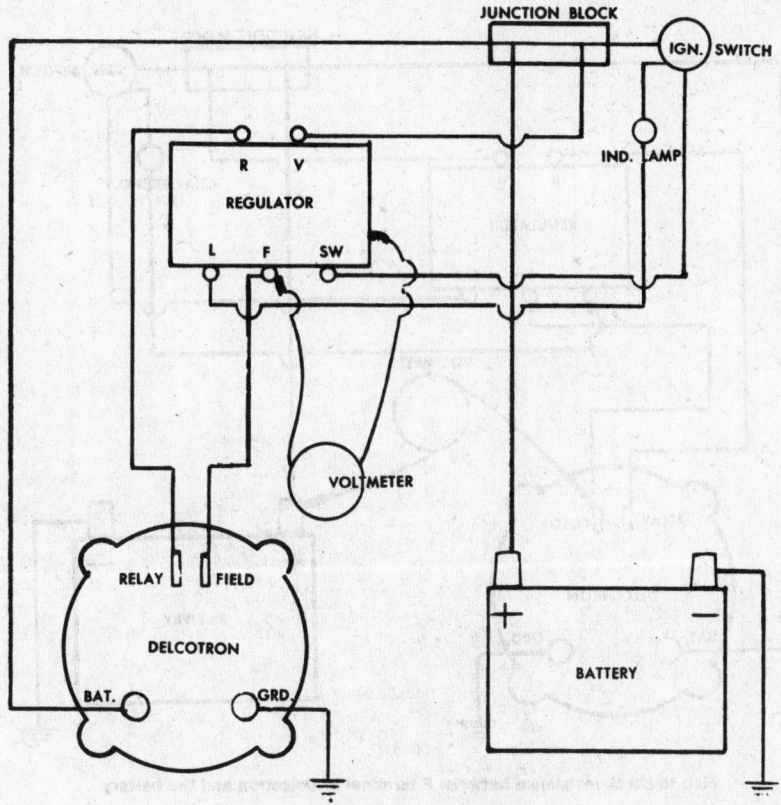

Plan to check Delcotron field relay

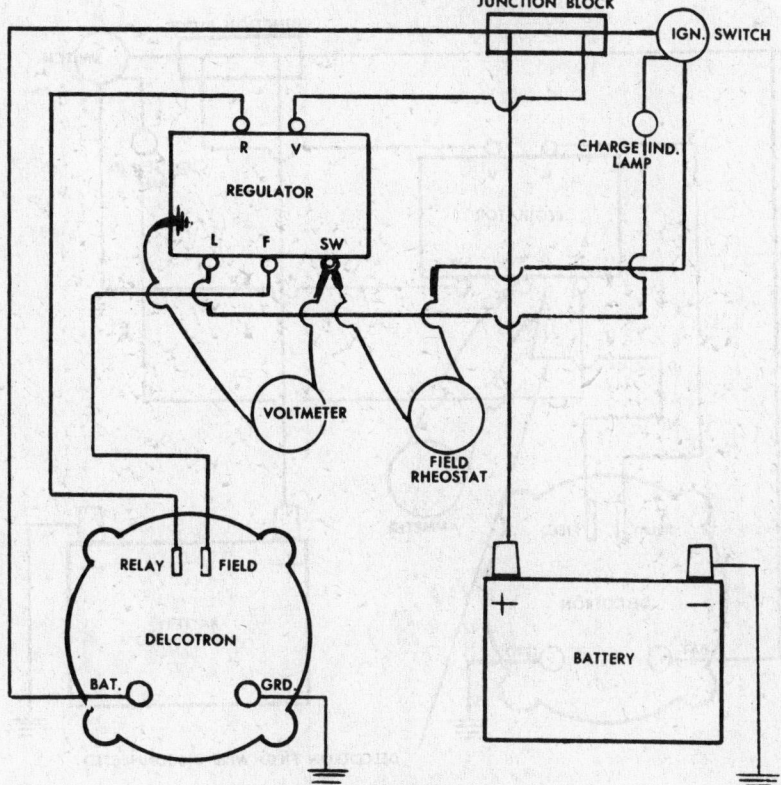

Plan for checking field relay closing voltage

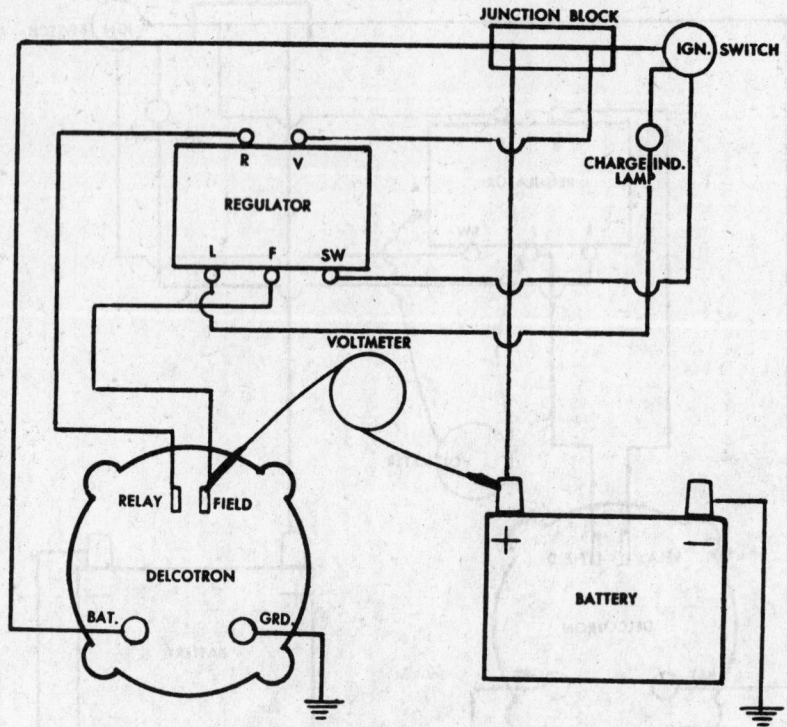

Plan to check resistance between F terminal of Delcotron and the battery

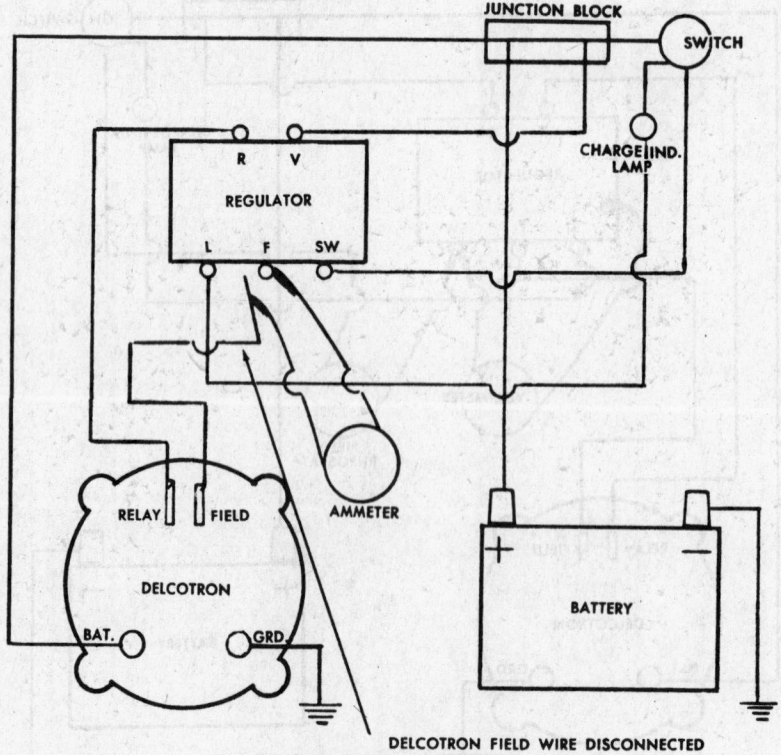

DELCOTRON FIELD WIRE DISCONNECTED

Plan to check field current draw

3. Turn ignition switch on and note the ammeter reading.

4. Compare this reading with car manufacturers', or Delcotron, specifications.

Delcotron Output Test

1. Disconnect lead from Bat. terminal of Delcotron.
2. Hook an ammeter to the lead just disconnected, and to the Bat. terminal of the Delcotron.
3. Hook up the voltmeter leads to the Bat. terminal and ground terminal of the Delcotron.
4. Disconnect the lead from the F terminal of the regulator.
5. Hook up a field rheostat to the V terminal of the regulator and to the lead that was disconnected from the F terminal of the regulator. Rotate the field rheostat control knob to the open position.
6. With a carbon pile load control hooked up to the battery posts, start the engine and set engine rpm to manufacturers' specifications.
7. Slowly turn field rheostat control knob toward closed position. Note ammeter and voltmeter readings.

Caution Be careful not to exceed the recommended regulator voltage setting. This is controlled by the carbon pile load.

8. Compare ammeter and voltmeter readings with specifications.

Voltage Regulator Setting Test

1. Disconnect the lead from Bat. terminal of regulator.
2. Connect ammeter leads and a ¼ ohm resistor (25 watts or more) in series with the Bat. terminal of the Delcotron and the lead which was disconnected from this terminal.
3. Disconnect lead from the V terminal of the regulator, then connect a jumper lead from the V terminal on the regulator to the battery terminal on the Delcotron.
4. Disconnect the lead from the F terminal of the regulator and connect a field rheostat (25 watt-25 ohm or more) to the F terminal of the regulator and the lead which was disconnected from this terminal. Adjust the field rheostat to the open position.
5. Hook the voltmeter leads to the V terminal and the ground of the regulator.
6. Connect the carbon pile load control to each battery post.
7. Start the engine and set speed to 1500 rpm.
8. Slowly turn the field rheostat control to the closed position. Be sure that voltage reading does

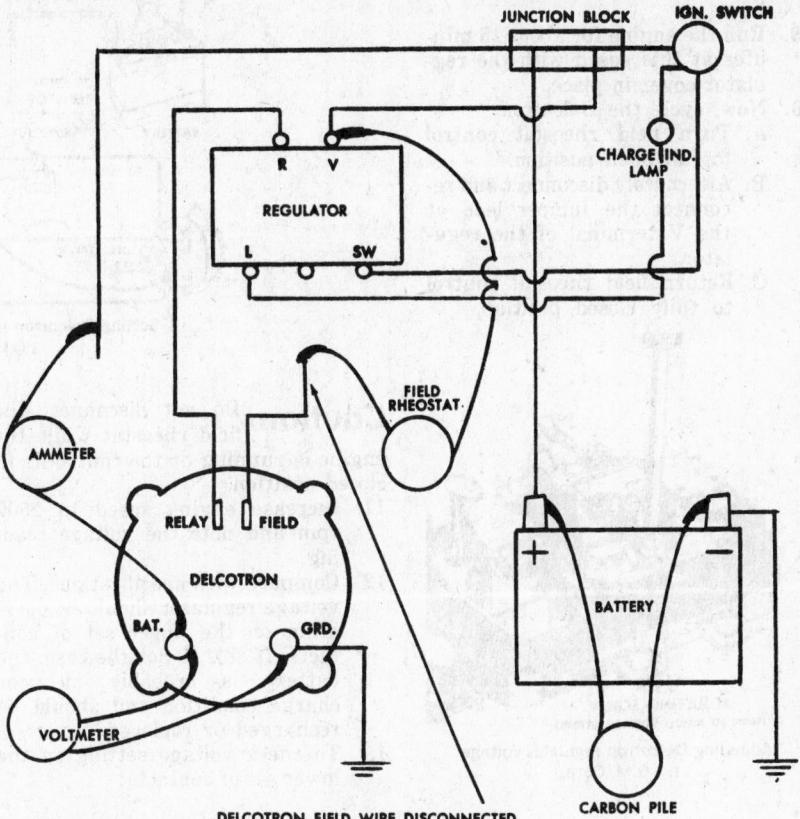

DELCOTRON FIELD WIRE DISCONNECTED

Plan to check Delcotron output

Plan to check voltage regulator setting

not exceed the recommended setting.

9. Run the engine for about 15 minutes at this speed with the regulator cover in place.

10. Now, cycle the Delcotron.
   A. Turn field rheostat control to full open position.
   B. Alternately disconnect and reconnect the jumper lead at the V terminal of the regulator.
   C. Return field rheostat control to fully closed position.

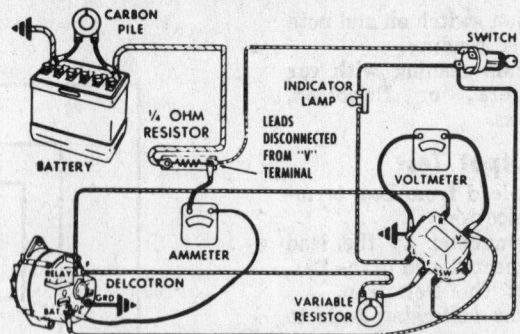

Setting Delcotron regulator (double contact)
(© G.M. Corp.)

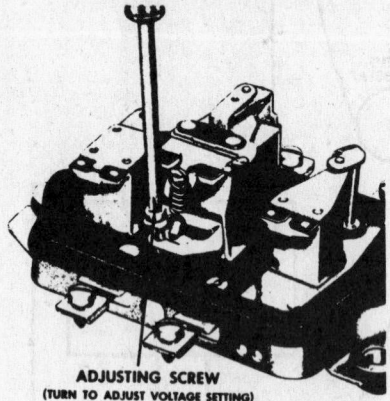

ADJUSTING SCREW
(TURN TO ADJUST VOLTAGE SETTING)

Adjusting Delcotron regulator voltage
(© G.M. Corp.)

Caution Do not disconnect the field rheostat while the engine is running or the control is in closed position.

11. Increase engine speed to 2500 rpm and note the voltage reading.

12. Compare with specifications. The voltage regulator should be operating on the upper set of contacts. If this is not the case, the battery is probably in poor charge condition and should be recharged or replaced.

13. To check voltage setting on the lower set of contacts:

A. Have the engine running at 2500 rpm, then slowly increase resistance in the field rheostat until the regulator starts to operate on the lower set of contacts.

B. Note voltage reading and compare with manufacturers' specifications.

C. If increasing the resistance with the field rheostat fails to cause the regulator to operate on the lower set of points, turn the field rheostat control to fully closed position, then turn the carbon pile load control to lightly

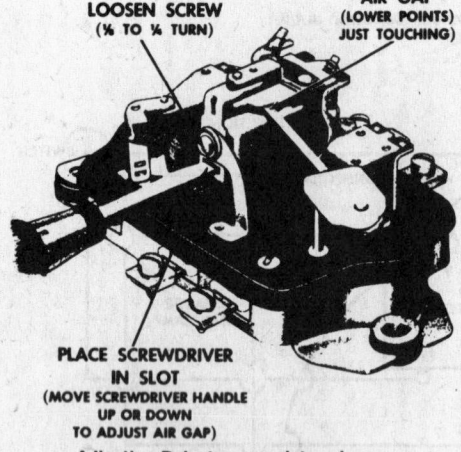

LOOSEN SCREW
(⅛ TO ¼ TURN)

AIR GAP
(LOWER POINTS)
JUST TOUCHING)

PLACE SCREWDRIVER IN SLOT
(MOVE SCREWDRIVER HANDLE UP OR DOWN TO ADJUST AIR GAP)

Adjusting Delcotron regulator air gap
(© G.M. Corp.)

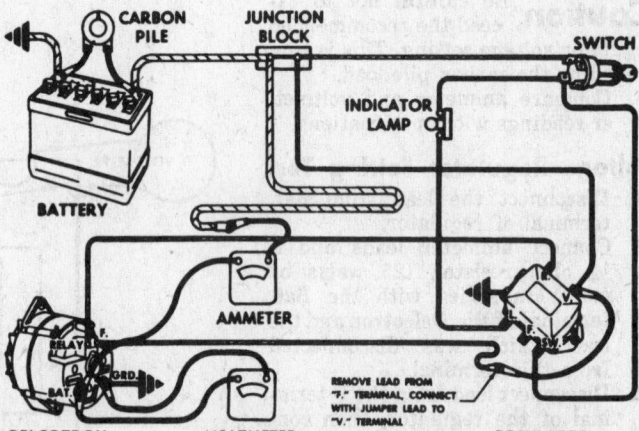

Checking Delcotron regulator output

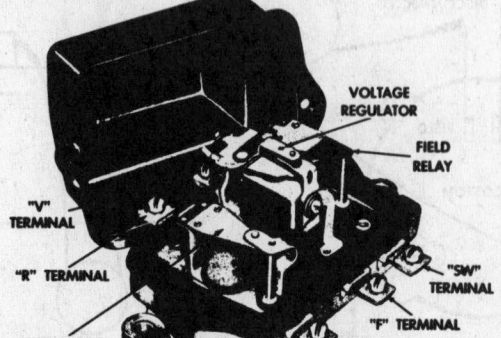

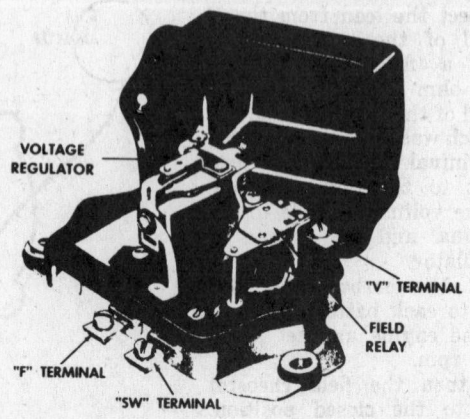

Three-unit and two-unit Delcotron regulators (© G.M. Corp.)

load the battery. Now, adjust the field rheostat control to cause the regulator to operate on the lower set of points.

D. Note reading and compare it with the manufacturers' specifications.

Disassembly

After long periods of operation, or at engine overhaul, the generator should be removed, disassembled and cleaned. The unit consists of four

ring end frame and the positive case diodes are mounted into the insulated heat sink. Diodes with a negative case have positive polarity leads, whereas positive case diodes have negative polarity leads. To further identify these diodes, the negative case diodes have right-hand threads and the positive case diodes left-hand threads.

Diodes can be checked for shorts or opens with an ohmmeter. To check for shorts, connect the negative lead of the ohmmeter to the

can be done by connecting the ohmmeter to a DC voltmeter. The voltmeter will read up scale when the positive lead of the ohmmeter is connected to the positive side of the voltmeter and the ohmmeter negative lead is connected to the negative side of the voltmeter.

To check for an open circuit, connect the negative lead of the ohmmeter to the positive lead of the diode and the positive lead of the ohmmeter to the negative case of the diode. Reverse the connections for

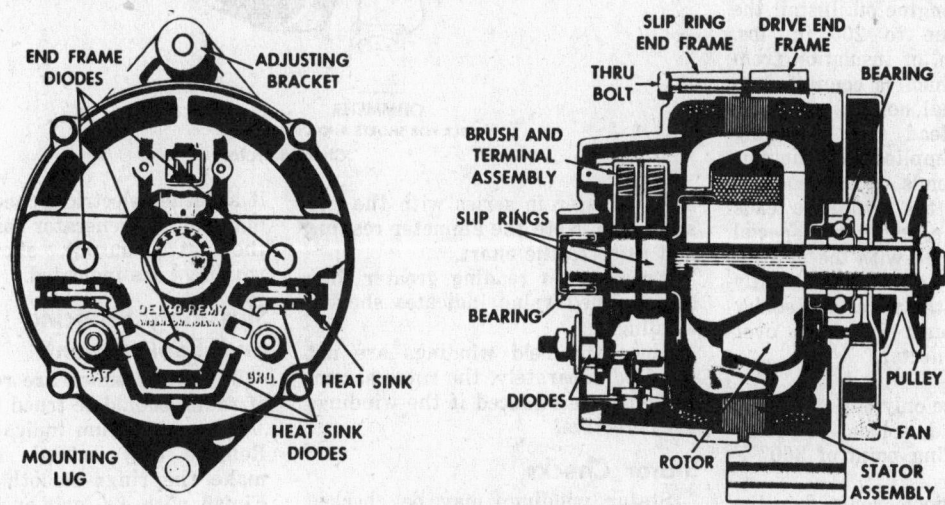

Cross-section of Delcotron

main components—the two end frames, the stator and the rotor.
1. Remove four thru-bolts.
2. Separate drive end frame and rotor from stator assembly by prying with screwdriver. Note that separation is between stator frame and drive end frame.
3. Place tape over slip ring end frame bearing to seal dirt.
4. Lightly clamp rotor in vise to remove shaft nut.

Caution Do not distort rotor by tightening vise too much.

5. After nut removal, take off washer, pulley, fan and collar.
6. Separate drive end frame from rotor shaft.

Diode Tests

There are six silicon diodes mounted on the slip ring end frame assembly. Three of these diodes are mounted in the end frame, and three are mounted into a bracket, or heat sink, which is attached to, but insulated from, the slip ring end frame. The Bat. terminal on the generator is attached to the insulated heat sink.

All diodes are marked with either a + or − on the hexagonal head, to identify the polarity of the case. On a generator to be used with a negative ground system, the negative case diodes are mounted into the slip

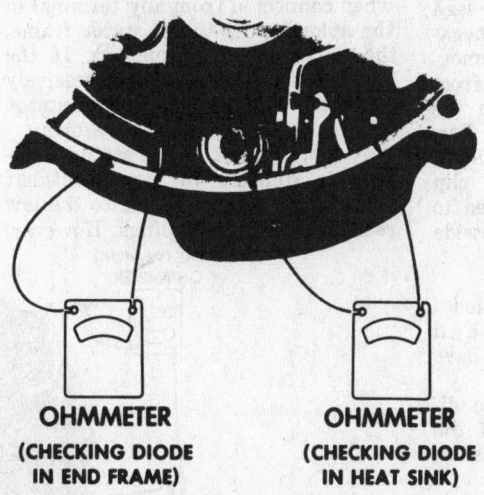

OHMMETER
(CHECKING DIODE
IN END FRAME)

OHMMETER
(CHECKING DIODE
IN HEAT SINK)

Checking diodes

negative case of the diode. Connect the positive lead of the ohmmeter to the positive lead of the diode. Reverse the connections for positive case diodes. Ohmmeter readings may vary considerably when checking diodes, but if the reading is below 300 ohms, the diode most likely is faulty and should be replaced. Push and pull on the lead to show up loose connections. Use an ohmmeter scale on which the 300 ohm value can be accurately read.

NOTE: the polarity of the ohmmeter leads must be determined. This

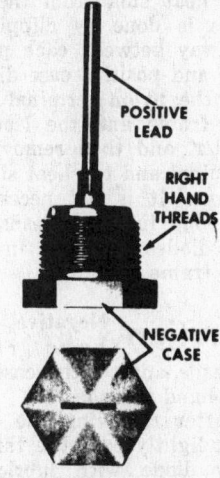

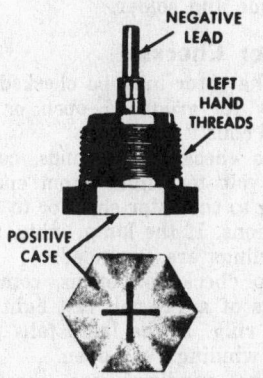

Diode rectifiers

positive case diodes. If the ohmmeter reads an infinite resistance (all the way), the diode is open and must be replaced.

Diode Replacement

To replace a diode that is mounted in the outside frame, use diagonal cutters to clip the leads on each side of the diode lead. Leave about ½ in. of lead length on each side of the diode lead to match the replacement diode. Remove the defective diode.

Before installing the new diode, lightly coat the threads with silicon grease or light engine oil. Install the diode and torque to 200 in. lbs. Strip about ½ in. of insulation from the long lead which is connected to the nylon terminal holder, and place over the long lead the insulating sleeve which is supplied with the new diode. Join the ends of the leads of the new diode to the respective leads in the end frame, using the special connectors provided with the new diode. Crimp the connectors tightly over the ends of the wires and solder securely. Then push the sleeve over the soldered connector.

Caution Use only 60% tin, 40% lead solder, or other solder with melting point of 360° F or above.

To replace a diode mounted in the heat sink, it is necessary to remove the heat sink from the end frame. This is done by clipping the lead midway between each pair of negative and positive case diodes, removing the nylon terminal holder from the frame and the leads from the holder, and then removing the Bat. terminal and the heat sink attaching screws. It is not necessary to clip the long insulated leads attached to the diodes mounted in the outside end frame.

Caution Negative case diodes have right-hand threads, and positive case diodes have left-hand threads.

After removing the defective diode, lightly coat the threads of the new diode with lubricant, install, torque and solder.

Rotor Checks

The rotor may be checked electrically for grounded, open, or shorted field coils.

To check for grounds, connect a 100-volt test light from either slip ring to the rotor shaft or to the laminations. If the lamp lights, the field windings are grounded.

To check for opens, connect the leads of a 110-volt test light to each slip ring. If the lamp fails to light, the windings are open.

The windings are checked for short-circuits by connecting a battery

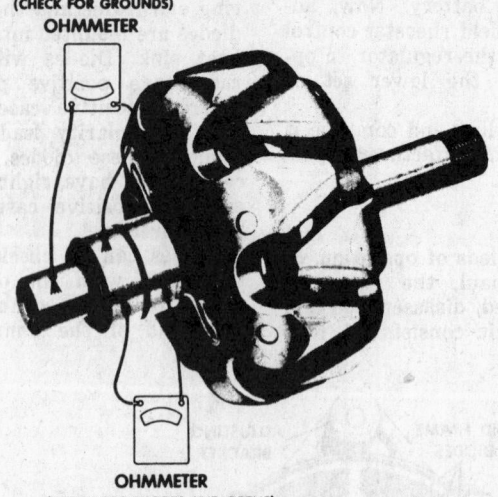

(CHECK FOR GROUNDS)
OHMMETER

OHMMETER
(CHECK FOR SHORTS AND OPENS)

Checking rotor

and ammeter in series with the two slip rings. Note the ammeter reading and refer to the chart.

An ammeter reading greater than the specified value indicates shorted windings.

Since the field windings are not serviced separately, the rotor assembly must be replaced if the windings are defective.

Stator Checks

Stator windings may be checked for grounded, open, or shorted windings. If a 110-volt test lamp lights when connected from any terminal in the nylon holder to the stator frame, the windings are grounded. If the lamp fails to light when successively connected between each pair of stator terminals, the windings are open.

A short circuit in the stator windings is difficult to locate without laboratory equipment, due to the low resistance of the windings. However,

if all other electrical checks are normal and the generator fails to supply the rated output, shorted stator windings are indicated.

Slip Ring Servicing and Replacement

Slip rings which are rough or out of round should be trued in a lathe to .001 in. maximum indicator reading. Remove only enough material to make the rings smooth and round. Finish with 400 grit or finer polishing cloth and blow away all dust.

Slip rings which must be replaced can be removed from the shaft with a gear puller, after the leads have been unsoldered. The new assembly should be pressed on with a sleeve which just fits over the shaft; this will apply all the pressure to the inner slip ring collar and prevent damage to the outer slip ring. Only pure tin solder should be used when reconnecting field leads.

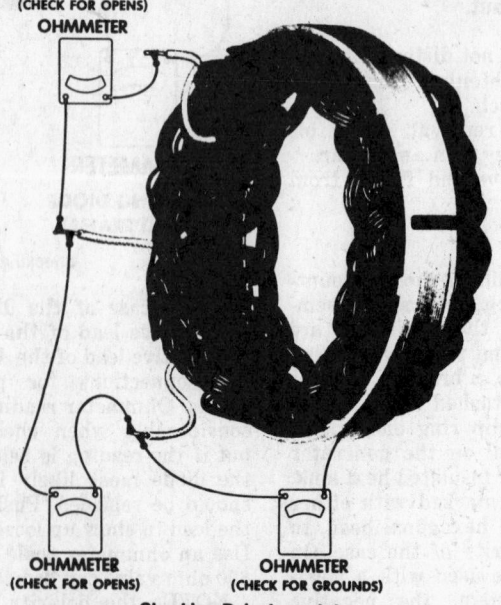

(CHECK FOR OPENS)
OHMMETER

OHMMETER
(CHECK FOR OPENS)

OHMMETER
(CHECK FOR GROUNDS)

Checking Delcotron stator
(© G.M. Corp.)

Brush Replacement

The extent of brush wear can be determined by comparison with a new brush. If brushes are one-half worn, they should be replaced.

1. Remove brush holder assembly from end frame by removing two holder assembly screws.
2. Place springs and brushes in the holder and insert straight wire or pin into holes at bottom of holder to retain brushes.
3. Attach holder assembly onto end frame.

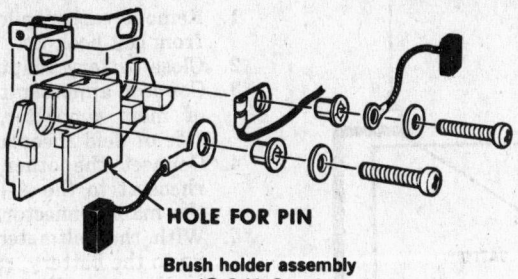

HOLE FOR PIN

Brush holder assembly
(© G.M. Corp.)

Reassembly

Reassembly is the reverse of disassembly.

The Ford System

To Make All Following Tests

In addition to the regular precautionary measures:

1. Disconnect both battery cables, clean the terminal posts and clamps, then reconnect the ground cable only.
2. Install a special battery make-and-break adapter switch between battery, positive (+) post, and battery cable. Close the adapter switch.

Field Relay Test

1. Withdraw regulator main connector and remove regulator cover.
2. Connect field rheostat leads to the S terminal of the regulator and the A+ terminal of the regulator. Place rheostat control in open position.
3. Connect one voltmeter lead to the regulator body (ground). Run the other voltmeter lead to the A+ regulator terminal.
4. Turn ignition switch on. The reading should be the same as existing battery voltage.
5. Remove the voltmeter connection from the regulator A+ terminal and connect it to the S terminal.
6. Now, slowly turn the rheostat control knob until field relay contacts close. Note voltage reading at this instant. This is relay closing voltage.
7. Repeat Step 6 several times. If closing voltage is not as specified by the manufacturer, or the contacts are not operating properly, replace the regulator.

Alternator Output on the Car

1. Connect test ammeter leads to the battery adapter switch. Open and close switch, and check polarity.

NOTE: when the battery adapter switch is opened, current will flow through the test ammeter. Never operate the alternator with an open circuit.

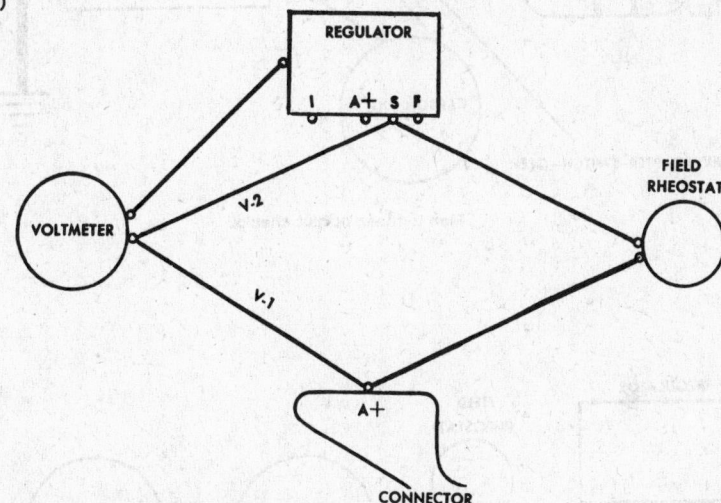

Plan to make field relay checks

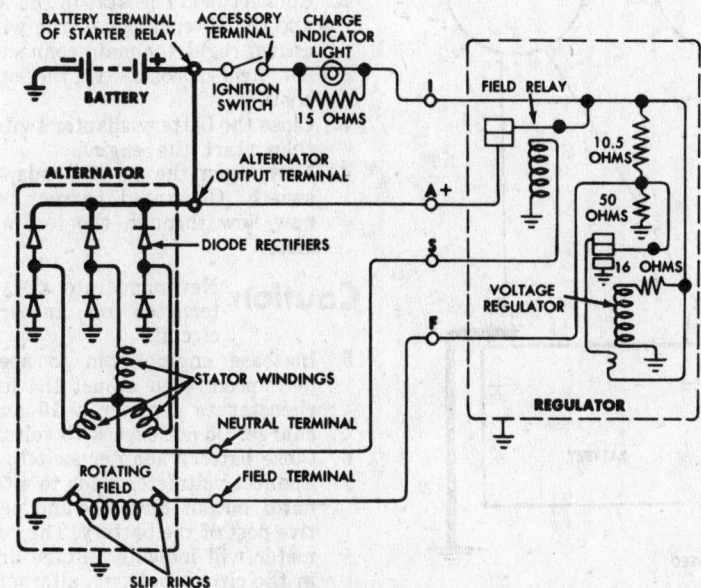

Ford alternator schematic
(© Ford Motor Co.)

2. Withdraw main connector from the regulator. Connect the field rheostat leads to the A+ and F terminals of the main connector. Set field rheostat to open position.
3. Connect carbon pile load control leads to the battery posts.
4. Connect voltmeter lads to their polar respective battery posts.
5. Throw the battery adapter switch to the closed position, then start the engine.
6. Open the battery adapter switch. Current being generated will now flow through the test ammeter.
7. Increase engine speed to specified rpm, then adjust the field rheostat until the voltmeter registers exactly 15 volts.
8. Read the ammeter, then add 5 amps. to this reading, to determine the total alternator output. The addition of 5 amps. to the meter reading is to compensate

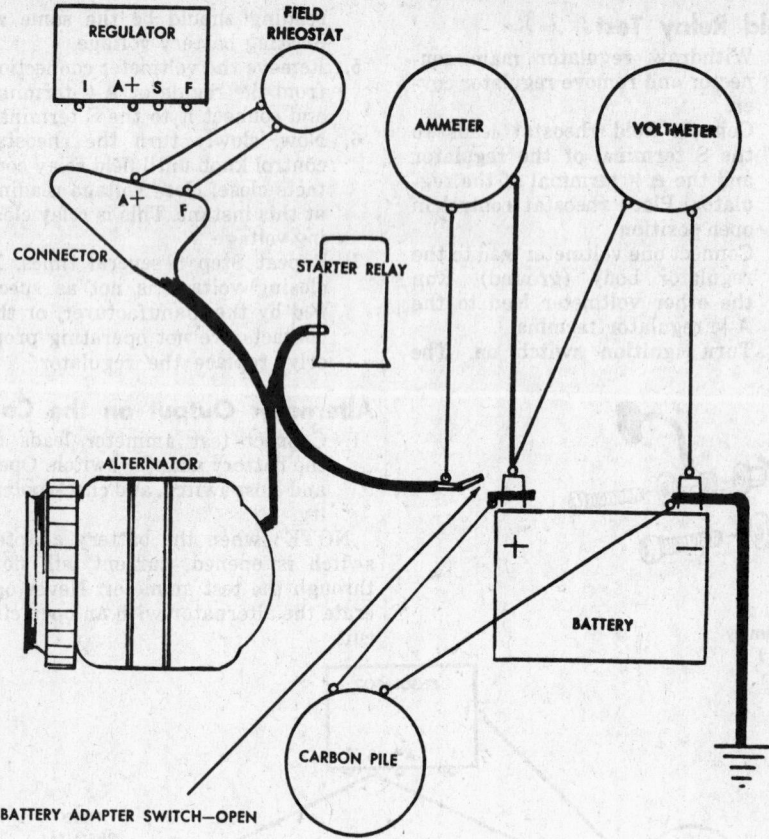

Plan to make output checks

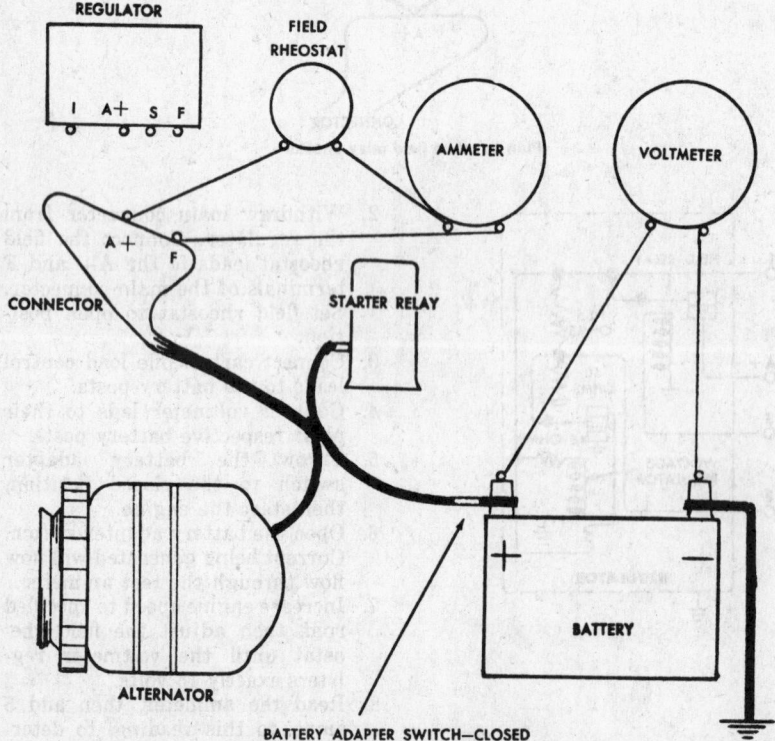

Plan to make field current draw check

for the current used by the ignition, and the alternator field circuits.

NOTE: maximum current output may not be possible if the battery is fully charged. When this situation exists, turn the field rheostat toward the resistance out position, holding the voltage at 15 volts by applying a load with the carbon pile load control.

If the rated output is not produced, go on to the next test.

Field Current Draw Test

1. Remove carbon pile load control from last hook-up.
2. Close battery adapter switch.
3. Connect ammeter to F terminal of main connector, and to one side of field rheostat.
4. Connect the other side of the rheostat to the A+ terminal of the main connector.
5. With the voltmeter still hooked up to the battery, run the engine at the prescribed rpm.
6. Adjust field rheostat to obtain voltmeter reading of 15 volts.
7. Now, read the ammeter. This reading is field current draw.
8. Reading should comply with specifications. If values do not agree, alternator reconditioning or replacement is in order.
   A zero reading indicates an open circuit.
   A low reading indicates trouble in the field circuit; could be slip rings or brushes.
   A high reading indicates shorted or grounded turns in the fields.

Charging Circuit Resistance Test

1. Hook the ammeter into the circuit across the battery adapter switch.
2. Connect field rheostat to the A+ and the F terminals of the withdrawn regulator main connector. Set field rheostat to the open position.
3. Close the battery adapter switch, then start the engine.
4. Now, open the battery adapter switch. Generated current will now flow through the test ammeter.

Caution Never operate the alternator on an open circuit.

5. Increase engine rpm to specified speed, then adjust the field rheostat to register a 10 amp. charge. Do not exceed 15 volts.
6. Close battery adapter switch.
7. Connect voltmeter leads to alternator output terminal and positive post of the battery. The voltmeter will indicate voltage drop in the circuit between alternator and battery. It should not exceed 0.3 volt.

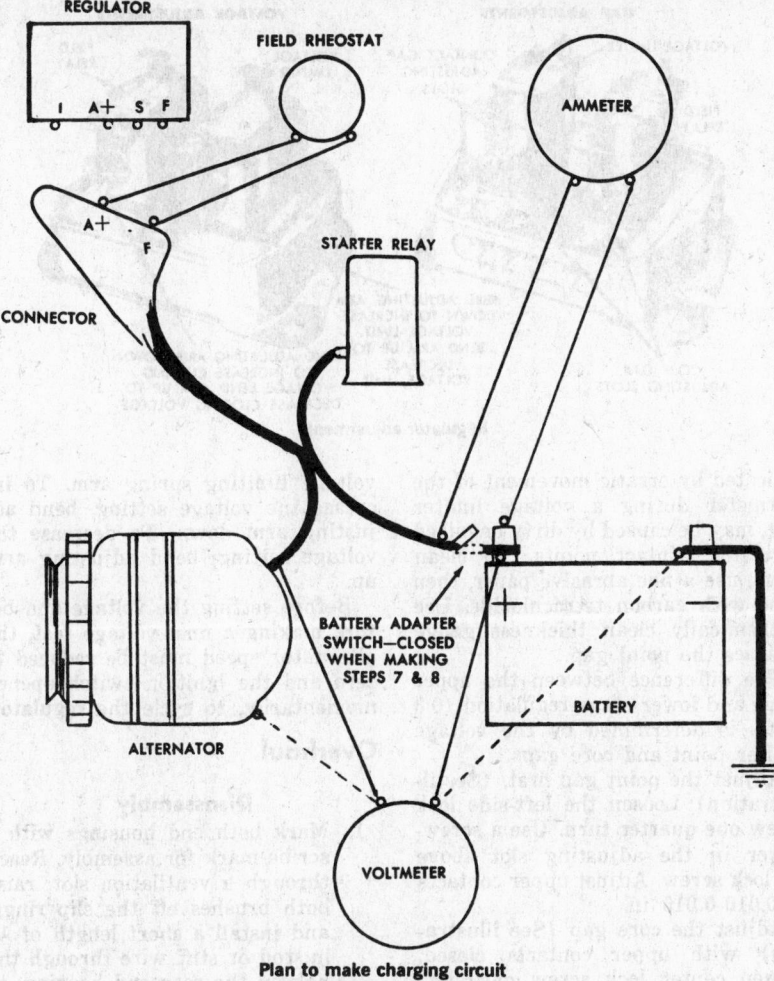

Plan to make charging circuit resistance check

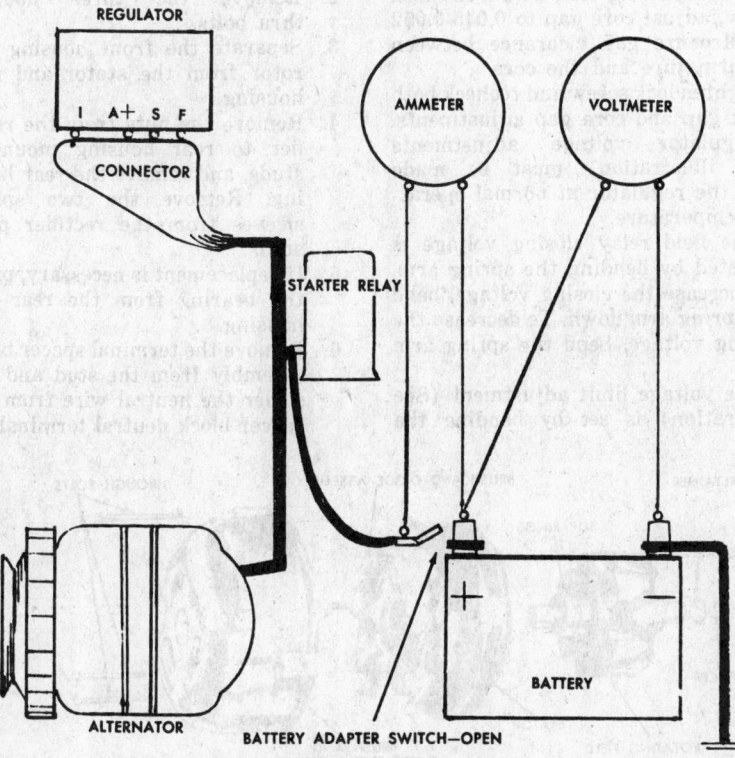

Plan to make voltage limiter check

8. Now, change voltmeter leads to the alternator frame and the battery negative post. This voltmeter reading will show voltage drop in the circuit between alternator frame and the battery, negative post. It should not exceed .05 volt.

If voltage drop in Steps 7 and 8 is in excess of specifications, stop engine, clear and tighten connections and repeat the test.

Voltage Limiter Test

NOTE: all limiter (regulator) tests and calibrations must be made at operating temperature, with the regulator cover in place.

1. With the regulator main connector plugged into the regulator, hook the test ammeter to the battery positive terminal and to the battery adapter switch (switch open).
2. Connect voltmeter to both positive and negative posts of battery.
3. Close the battery adapter switch then start the engine.
4. Open the battery adapter switch. Now, start the engine and run it at specified rpm for about five minutes.
5. If test ammeter registers more than 10 amps. stop the engine, disconnect battery cables and replace, or charge, the battery.
6. Then, reconnect cables to fully charged battery, and run the engine at specified rpm for about five minutes.
7. Check voltmeter reading against specifications. With battery charged and voltage setting as specified, the ammeter should read less than 2 amps.
8. If results do not agree with specifications, replace the regulator.

Diode Tests

To test positive diodes, make connections as shown. Contact the probe to each diode lead.

To test negative diodes, make connections as shown. Follow the same procedure as for positive diodes.

Good diodes will be indicated as 2 amps. or more and readings alike within two graduations.

Field Circuit Test

Current draw should show within specifications. If there is little or no current flow, the field or brushes have a high resistance or are open. A current flow much higher than specifications indicates shorted or grounded turns or brush leads touching. If the test indicates the field to be shorted or open and the field brush assembly or slip rings are not at fault, the entire rotor must be replaced.

Regulator Bench Adjustments

Erratic operation of the regulator,

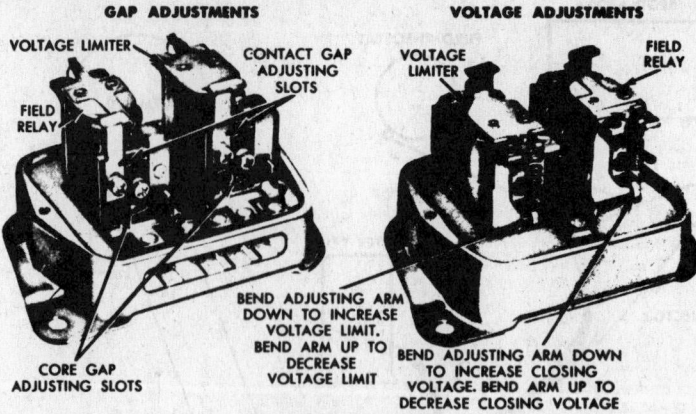

Regulator adjustments

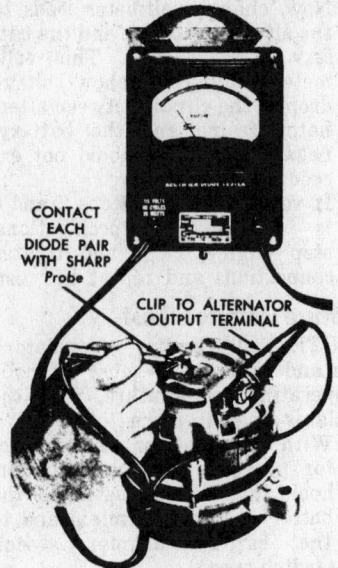

Positive diode test

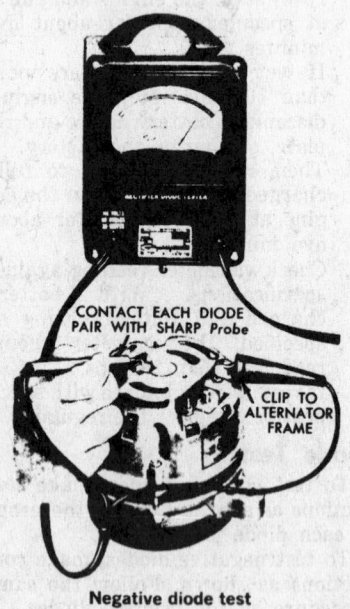

Negative diode test

indicated by erratic movement of the voltmeter during a voltage limiter test, may be caused by dirty or pitted regulator contact points. To clean them, use a fine abrasive paper, then wipe with carbon tetrachloride. Use a chemically clean thickness gauge to space the point gap.

The difference between the upper stage and lower stage regulation (0.3 volt), is determined by the voltage limiter point and core gaps.

Adjust the point gap first. (See illustration) Loosen the left-side lock screw one-quarter turn. Use a screwdriver in the adjusting slot above the lock screw. Adjust upper contacts to 0.010-0.015 in.

Adjust the core gap (See illustration) with upper contacts closed. Loosen center lock screw one-quarter turn. With a screwdriver inserted in the adjusting slot, below the lock screw, adjust core gap to 0.045-0.052 in. Measure gap clearance between the armature and the core.

Tighten lock screw and recheck both point gap and core gap adjustments.

Regulator voltage adjustments (See illustration) must be made with the regulator at normal operating temperature.

The field relay closing voltage is adjusted by bending the spring arm. To increase the closing voltage, bend the spring arm down. To decrease the closing voltage, bend the spring arm up.

The voltage limit adjustment (See illustration) is set by bending the voltage limiting spring arm. To increase the voltage setting, bend adjusting arm down. To decrease the voltage setting, bend adjusting arm up.

Before setting the voltage and before making a final voltage test, the alternator speed must be reduced to zero and the ignition switch opened momentarily, to cycle the regulator.

Overhaul

Disassembly

1. Mark both end housings with a scribe mark for assembly. Reach through a ventilation slot, raise both brushes off the slip rings, and install a short length of ⅛ in. rod or stiff wire through the hole in the rear end housing, to hold the brushes off the slip rings
2. Remove the three housing thru bolts.
3. Separate the front housing and rotor from the stator and rear housing.
4. Remove the nuts from the rectifier to rear housing mounting studs, and remove the rear housing. Remove the two spacer sleeves from the rectifier plate studs.
5. If replacement is necessary, press the bearing from the rear end housing.
6. Remove the terminal spacer block assembly from the stud and unsolder the neutral wire from the spacer block neutral terminal.

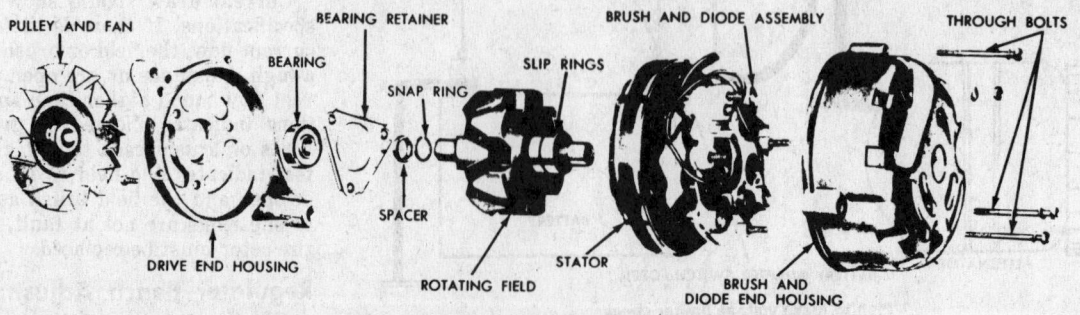

Alternator components

7. If the brushes are being replaced, straighten the field brush terminal blade locking tabs and remove the terminal blade from the terminal spacer block. Remove the brushes and holders from the assembly.

8. If either diode plate is being replaced, carefully unsolder the leads from the diodes.

Caution Use only a 100-watt soldering iron. Leave the soldering iron in contact with the diode terminals only long enough to remove the wires. Use pliers as temporary heat sinks in order to protect the diodes.

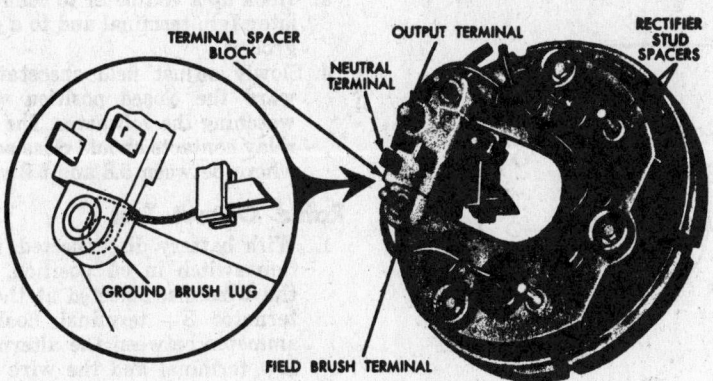

Stator, heat sink and terminal spacer assembly

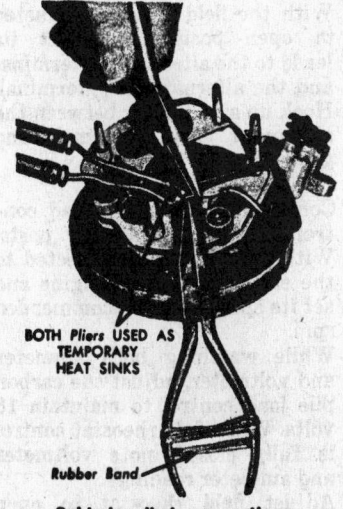

Soldering diode connections

9. Remove the three insulated diode plate screws and the insulators, and separate the diode plates.

10. Remove the drive pulley. On alternator pulleys with threaded holes in the outer end of the pulley, use a standard puller for removal.

11. Remove the three screws that hold the front bearing retainer, and remove the front housing.

12. If the bearing is being replaced, pull, using a bearing puller.

Remove the bearing retainer and spacer. It will not be necessary to remove the stop ring, unless it has been damaged.

Cleaning and Inspection

1. The rotor, stator, diode rectifier assemblies, and bearings are not to be cleaned with solvent. These parts are to be wiped off with a clean cloth. Cleaning solvent may cause damage to the electrical parts or contaminate the bearing internal lubricant. Wash all other parts in solvent and dry them.

2. Rotate the front bearing on the driveshaft. Check for any scraping noise, looseness or roughness that will indicate that the bearing is excessively worn. As the bearing is being rotated, look for excessive lubricant leakage. If any of these conditions exist, replace the bearing.

3. Place the rear end housing on the slip ring end of the shaft and rotate the bearing on the shaft. Make a similar check for noise, looseness or roughness. Inspect the rollers and cage for damage. Replace the bearing if these conditions exist, or if the lubricant is missing or contaminated.

4. Check both the front and rear housings for cracks. Check the front housing for stripped threads in the mounting holes. Replace defective housings.

5. Pulleys that have been removed and installed several times may have to be replaced because of increased bore diameter. A pulley is not suitable for reuse if more than one-quarter of the shaft length will enter the pulley bore with light pressure. Replace any pulley that is bent.

6. Check all wire leads on both the stator and rotor assemblies for loose soldered connections, and for burned insulation. Solder all poor connections. Replace parts that show burned insulation.

7. Check the slip rings for damaged insulation and runout. If the slip rings are more than 0.0005 in. out of round, take a light cut (minimum diameter limit 1.2 in.) from the face of the rings to true them. If the slip rings are badly damaged, the entire rotor will have to be replaced, as they are serviced as a complete assembly.

8. Replace any parts that are burned or cracked. Replace brushes that are worn to less than 0.350 in. in length. Replace the brush spring if it has less than 7-12 oz. tension.

Assembly

1. If the stop ring on the driveshaft was damaged, install a new stop ring. Push the new ring onto the shaft and into the groove.

2. Position the front bearing spacer on the driveshaft against the stop-ring, and position the bearing retainer on the shaft with the flat surface of the retainer outward.

3. Putting pressure on the inner race only, press the new bearing onto the shaft until it contacts the spacer.

4. Place the front housing over the shaft, with the bearing positioned in the front housing cavity. Install the bearing retainer mounting screws.

5. Press the pulley onto the shaft until the hub just touches the inner race of the front bearing.

6. If a new diode plate is being installed, mount the two plates together so that they are insulated from each other (insulating spacer between the plates are cupped insulator under the screw head). Do not tighten the screws at this time. Solder the wire leads to the diodes in as short a time as possible to avoid heat damage. Use only a 100-watt iron.

7. Insert the new field brush terminal blade into the slot in the terminal spacer block, with the brush pig-tail extended toward the brush holder pivots.

8. Install the brush holders and brush spring onto the terminal spacer block, then position the brushes in the brush holders.

9. Solder the neutral wire to the neutral terminal. Position the terminal spacer block assembly on the rectifier plate mounting studs, with the ground brush lug over the mounting stud farthest from the output terminal. Tighten the diode plate assembly screws.

10. Place the spacers on the rectifier mounting studs farthest from the terminal block.

11. Install the rear bearing so that

the open end of the bearing is flush with the inner surface of the housing boss. Allow space under the outer end of the bearing during installation.

12. Place the rectifier plate and stator assembly in the rear end housing and mount the rectifier plates to the housing.

13. Retract the brushes and insert a short length of 1/8 in. rod or stiff wire through the hole in the rear end housing to hold the brushes in the retracted position.

14. Wipe clean the rear bearing surface of the rotor shaft.

15. Position the rear housing and stator assembly over the rotor

and align the scribe marks made during disassembly, install the housing through bolts. Remove the brush retracting rod.

The Leece-Neville System

Field and Lamp Relay Test

1. Disconnect the battery wire from B terminal of the regulator.
2. Disconnect the N wire from the regulator.
3. With the field rheostat in the open position, connect its leads to the disconnected battery wire

and the N terminal of the regulator.

4. Hook up a voltmeter to the regulator N terminal and to a good ground.
5. Slowly adjust the field rheostat toward the closed position. The field relay contacts should close somewhere between 1.9 and 2.4 volts.

Load Relay Test

1. Disconnect the battery wire from B terminal of the regulator.
2. With field rheostat in the open position, connect its leads to the disconnected battery wire and the regulator Ign. terminal.
3. Hook up a voltmeter to the regulator Ign. terminal and to a good ground.
4. Slowly adjust field rheostat toward the closed position while watching the voltmeter. The load relay contacts should close somewhere between 5.8 and 6.2 volts.

Rated Output Test

1. With battery disconnected, ignition switch in off position, and the wire disconnected at the alternator B+ terminal, hook an ammeter between the alternator B+ terminal and the wire that was just disconnected from this terminal.
2. Disconnect wire from the alternator F terminal.
3. With the field rheostat adjusted to open position, connect its leads to the alternator F terminal and the alternator B+ terminal.
4. Hook up a voltmeter between the alternator B+ terminal and ground.
5. Reconnect battery cables.
6. Connect a carbon pile load control between the battery posts.
7. With a tachometer connected to the engine, start the engine and set its speed at the recommended rpm.
8. While watching both ammeter and voltmeter, adjust the carbon pile load control to maintain 15 volts. When field rheostat control is fully closed, note voltmeter and ammeter readings.
9. Adjust field rheostat to open position and carbon pile control to off.
10. Compare readings with specifications. Readings should be at least equal to rated output specified by the manufacturer. If output complies with specifications, proceed to the voltage regulator test. If readings are below rated output, it indicates possible internal troubles.

Field Current Draw Test

1. Disconnect battery ground cable.
2. Disconnect carbon pile, volt-

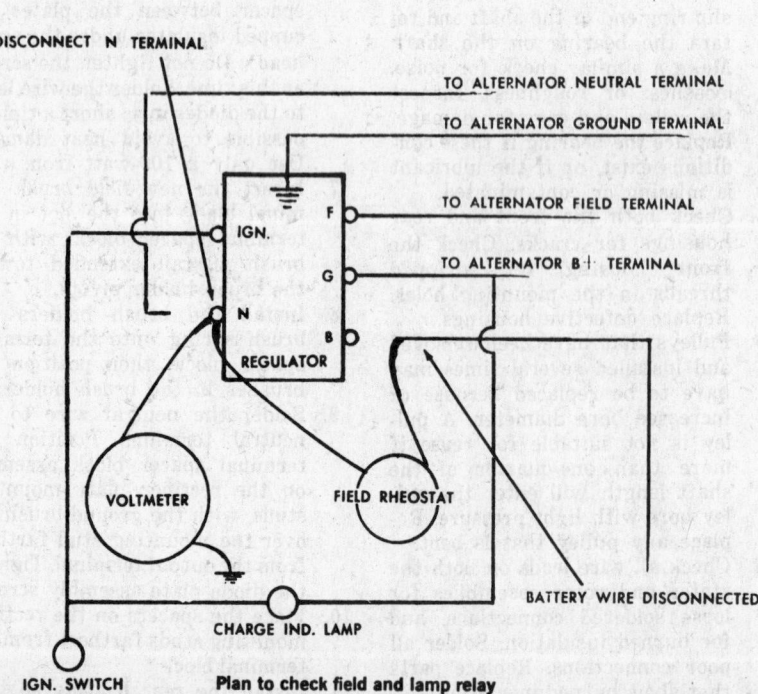

Wiring plan of Leece-Neville 6000 negative ground system, ammeter equipped

DISCONNECT N TERMINAL

TO ALTERNATOR NEUTRAL TERMINAL

TO ALTERNATOR GROUND TERMINAL

TO ALTERNATOR FIELD TERMINAL

TO ALTERNATOR B+ TERMINAL

IGN.

G

N

B

REGULATOR

VOLTMETER

FIELD RHEOSTAT

BATTERY WIRE DISCONNECTED

CHARGE IND. LAMP

IGN. SWITCH   **Plan to check field and lamp relay**

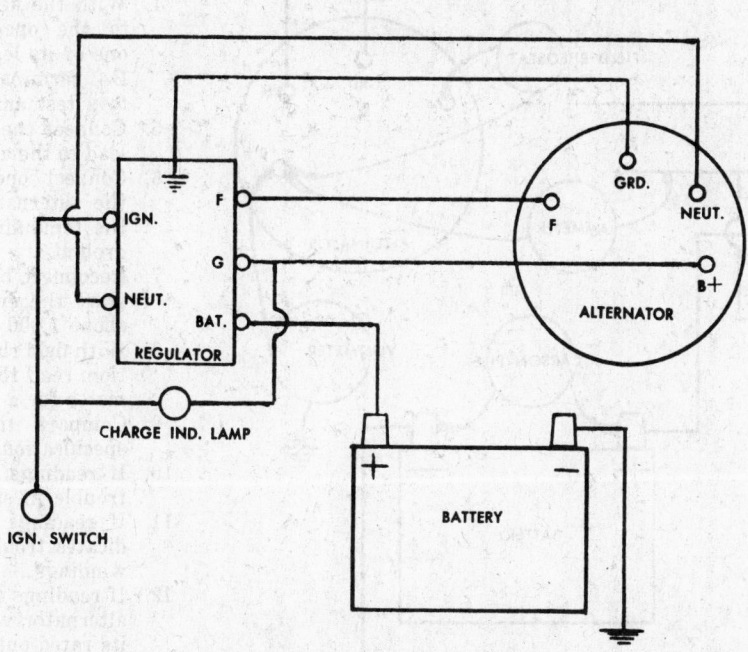

Wiring plan of Leece-Neville 6000 negative ground system, charge indicator light equipped

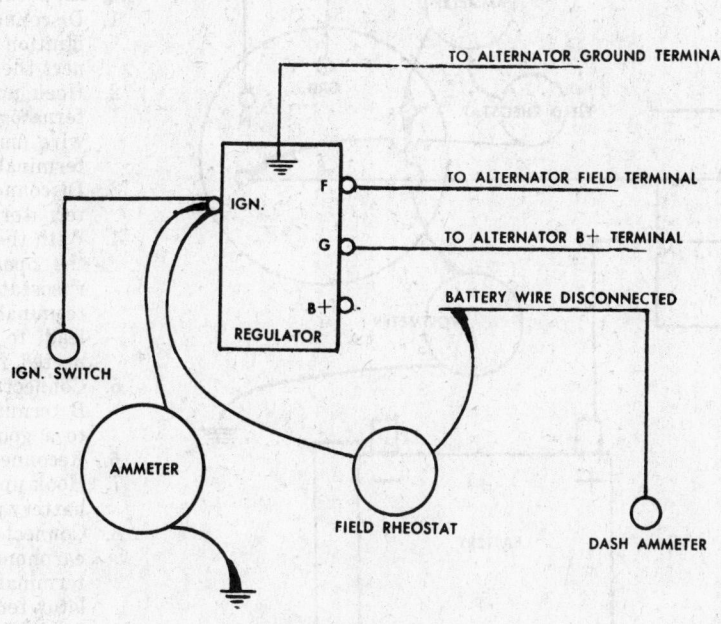

Plan to check load relay

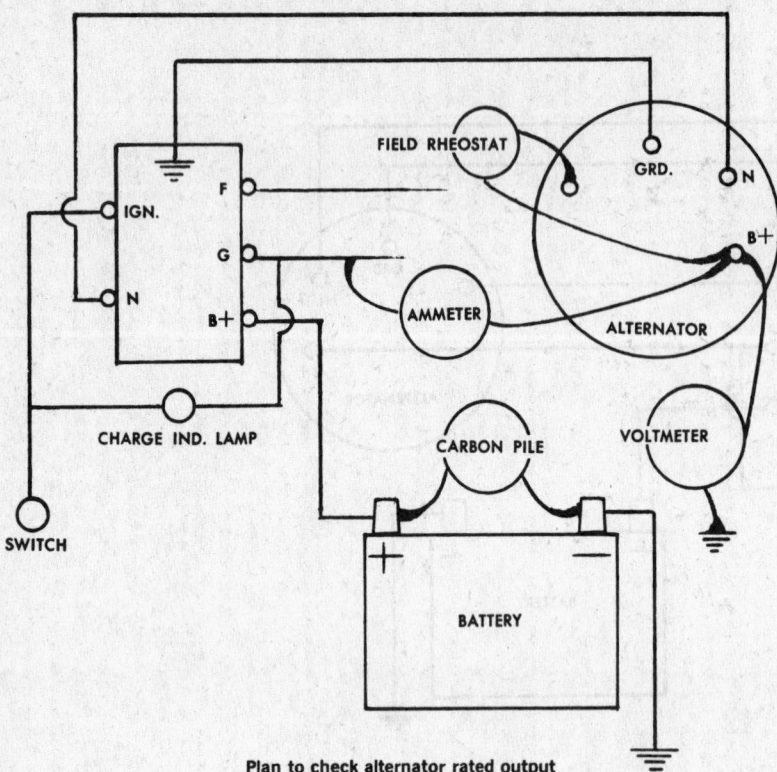

Plan to check alternator rated output

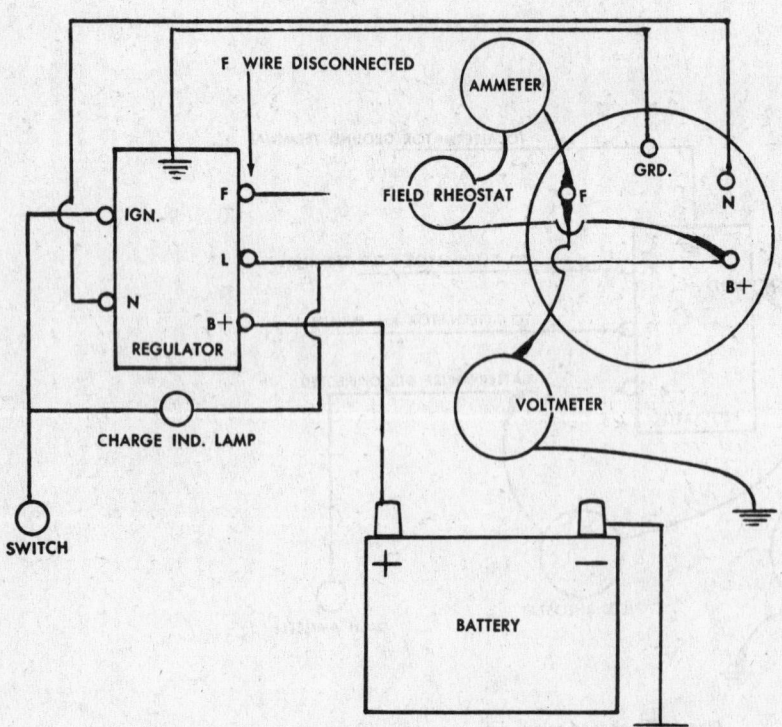

Plan to check alternator field current draw

meter, ammeter and field rheostat from the previous check.

3. Reconnect the regular circuit wire to the alternator B+ terminal.

4. With the field rheostat adjusted to the open position, connect one of its leads to the alternator B+ terminal and the other lead to a test ammeter lead.

5. Connect the remaining ammeter lead to the alternator F terminal.

6. Connect one voltmeter lead to the alternator F terminal and the remaining voltmeter lead to ground.

7. Reconnect battery ground cable, start the engine, and run it at about 1,000 rpm.

8. With field rheostat in closed position, read the ammeter and voltmeter for a very brief period.

9. Compare these readings with specifications.

10. If readings are low, it indicates trouble in slip rings or brushes.

11. If readings are too high, it indicates trouble in the rotor field windings.

12. If readings are as specified on an alternator which did not deliver its rated output, look for trouble in the stator or diodes.

Voltage Regulator Test

NOTE: manufacturers' recommended settings are established with the regulator cover off. Generally, a decrease of 0.2 to 0.3 volt will result when the cover is installed and the regulator brought to normal operating temperature.

1. Disconnect battery, be sure the ignition switch is off, and disconnect the alternator B+ terminal.

2. Hook up an ammeter to the alternator B+ terminal and to the wire just disconnected from this terminal.

3. Disconnect the wire from the alternator F terminal.

4. With the field rheostat turned to the open position, connect one rheostat lead to the alternator F terminal and the other rheostat lead to the wire just disconnected from the F terminal.

5. Connect voltmeter leads to the B terminal of the regulator and to a good ground.

6. Reconnect battery cables.

7. Hook up a carbon pile across the battery posts.

8. Connect a pair of 1,000 ohm + earphones to the regulator F terminal and the L (gen.) regulator terminal.

9. Start the engine and gradually increase its speed while listening for a vibrating sound in the earphones. This indicates the operation of the regulator top contacts. Note voltmeter reading, compared with manufacturers' specifications.

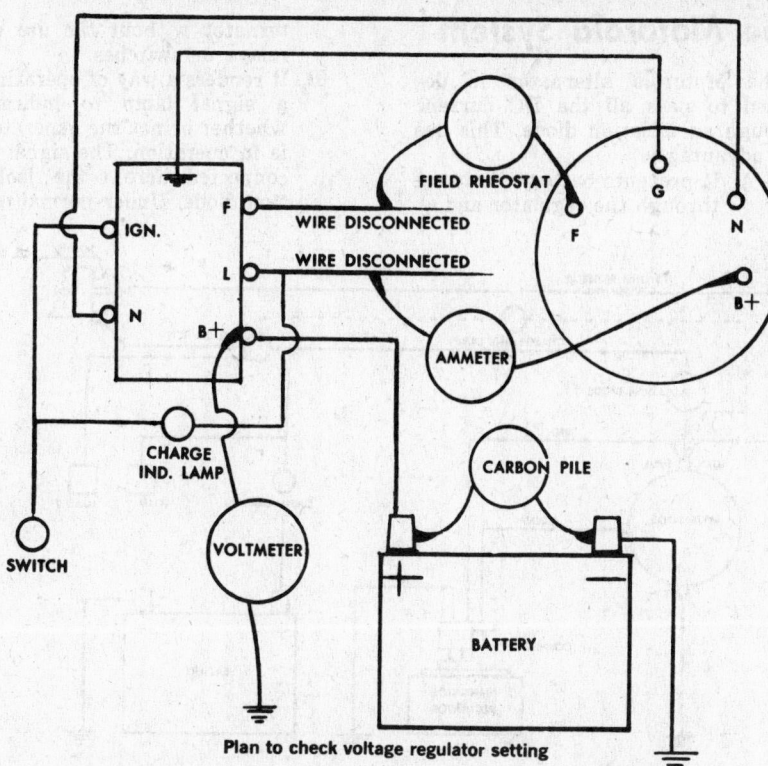

Plan to check voltage regulator setting

10. Continue to increase engine rpm until earphones cease to pick up vibration. Voltage should read maximum just before vibrating sound stops. Compare with manufacturers' specifications. The silencing of the vibration indicates transfer of action to the lower contacts.

11. Resume a gradual increase in engine rpm until the vibrating sound starts again. This sound is produced by the lower contacts. Compare voltmeter reading with specifications.

Final Regulator Setting

This is a final setting, to be made at operating temperature through all variations of speed and vibration.

1. Run the engine about 15 minutes, or enough to develop normal operating temperature at the regulator.
2. Set engine speed at 1,750 rpm.
3. Adjust carbon pile load to produce a 25 amp. charging rate.
4. Open and close the field rheostat four or five times.
5. Turn carbon pile load to full off.

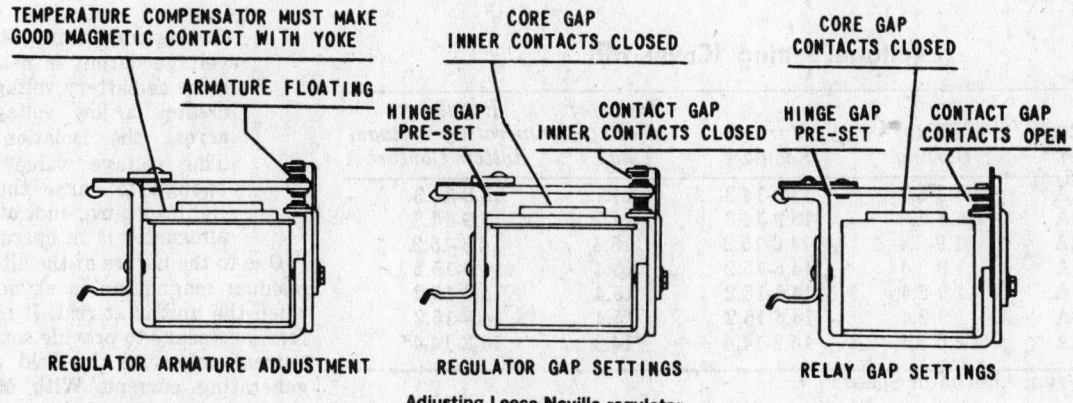

Adjusting Leece-Neville regulator

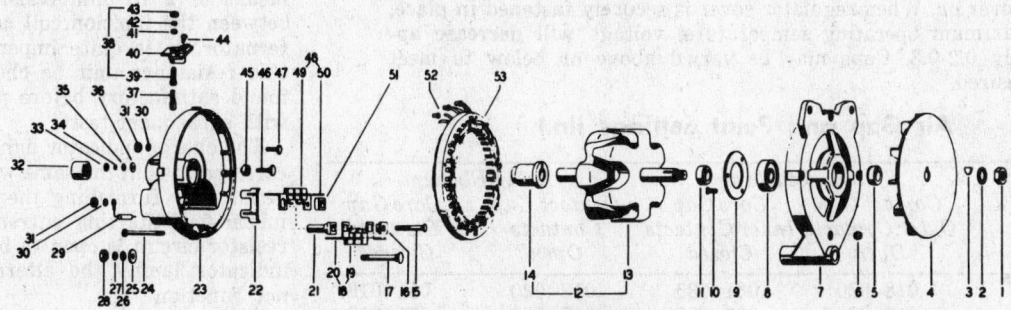

Exploded view Leece-Neville alternator

1 Rotor shaft nut	14 Slip ring	27 Washer
2 Shaft nut washer	15 Screw terminal	28 Hex nut
3 Woodruff key	16 Bushing insulator	29 Capacitor assembly
4 Fan assembly	17 Terminal screw	30 Washer
5 Fan spacer	18 Rectifier assembly	31 Hex nut
6 Bolt nut	19 Rectifier	32 Roller bearing
7 Drive end frame	20 Rectifier	33 Nut
8 Ball bearing	21 Insulator bushing	34 Lockwasher
9 Ball bearing retainer plate	22 Support-rectifier	35 Guard washer
10 Bearing retainer plate screws	23 Slip ring end housing	36 Insulation washer
11 Spacer	24 Thru bolt	37 Brush and spring assembly
12 Rotor and slip ring assembly	25 Guard washer	38 Brush holder assembly
13 Rotor	26 Washer	39 Screw

40 Brush holder only
41 Guard washer
42 Lockwasher
43 Nut
44 Screw
45 Insulator bushing
46 Terminal screw
47 Terminal screw
48 Rectifier assembly
49 Rectifier
50 Rectifier mount
51 Spacer insulator
52 Eyelet
53 Stator assembly

6. With field rheostat fully closed, note the voltmeter reading and compare it with regulator operating specifications.

7. If regulator does not operate within specified limits, replace it.

Disassembly

1. Remove two brush-holder assembly screws and remove holder with brushes.

2. Remove shaft nut, preventing pulley from slipping by gripping a fan belt around pulley and in vise. Remove pulley.

3. Remove four thru-bolts and carefully pry two end housings apart.

4. Place tape over slip ring end housing bearing to keep out dirt.

5. Remove fan, key and spacer and with puller remove drive end housing.

6. Bearings from both housings can be removed after taking off retainer at drive end or tape from slip ring end.

7. To remove individual rectifier, loosen solder around cell while pressing it out. A piece of asbestos or fiber will help concentrate heat.

The Motorola System

The Motorola alternator is designed to pass all the DC current through an isolation diode. This has two advantages.

A. It prevents battery discharge through the regulator and alternator without the use of relays or switches.

B. It renders a way of operating a signal lamp to indicate whether or not the generator is in operation. The signal is connected across the isolation diode. Under normal operating conditions, alternator voltage output is nearly the same as battery voltage. This creates a low voltage flow across the isolation diode. The voltage value is not enough to cause the signal light to burn, indicating the alternator is in operation.

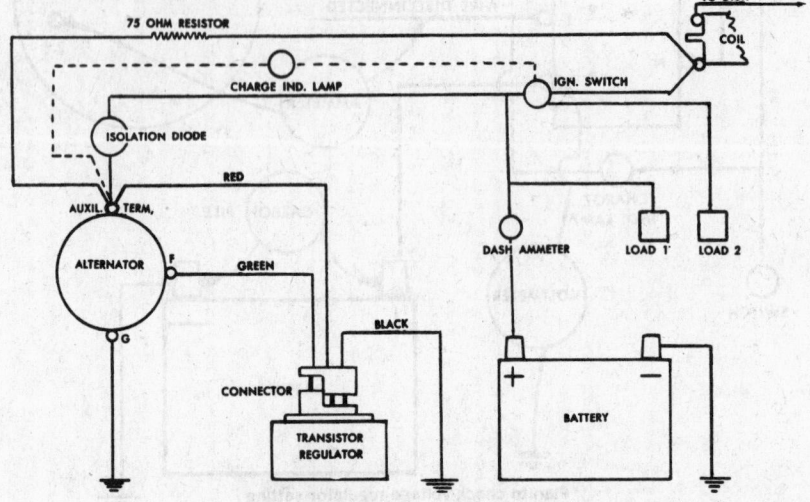

Motorola alternator system charging circuit

Due to the nature of the alternator, residual magnetism is at near zero when the unit is at rest. It is, therefore, necessary to provide some small current to excite the field prior to generating current. With Motorola, this priming current is supplied by means of a 75 ohm resistance unit between the ignition coil and the alternator. It is quite important that this resistance unit be checked and found satisfactory before proceeding with subsequent tests.

The charge indicator light on some cars operates in the same way as this resistor by furnishing the necessary initial field starting current. If this resistor circuit is open (a burned out indicator lamp) the alternator will not function.

The regulator is a sealed unit and should require no adjustment. It is therefore, recommended that nonfunctioning regulators be replaced.

Alternator System Test

1. Be sure that the battery is disconnected and the ignition switch is off.

2. Disconnect the wire from the alternator output terminal.

3. Connect the positive lead of a

Voltage Setting (Cover off)

Regulator Number	Relay Closing	Starts to Regulate	Transfer Voltage (max.)	Regulator Operating Voltage, Bottom Contact
3532 RA	5.8-6.2	13.9-14.3	14.7	13.9-14.3
3533 RA	5.8-6.2	13.9-14.3	14.7	13.9-14.3
3628 RA	1.9-2.4	14.6-15.2	15.4	14.6-15.2
3629 RA	1.9-2.4	14.6-15.2	15.4	14.6-15.2
3630 RA	1.9-2.4	14.6-15.2	15.4	14.6-15.2
3631 RA	1.9-2.4	14.6-15.2	15.4	14.6-15.2
3687 RA	2.0-2.1	14.2-14.4	14.8	14.2-14.4*

* 3000 rpm Alternator Speed

NOTE: voltage regulator settings will be approximately ½ volt lower with the cover on. When regulator cover is securely fastened in place, and at maximum operating temperature, voltage will decrease approximately 0.2-0.3. Gaps may be varied above or below to meet settings desired.

Air Gap and Point Settings (in.)

Regulator Number	Voltage Regulator Contact Gap Outer Contacts Open	Core Gap Inner Contacts Closed	Relay Element Contact Gap Contacts Open	Core Gap Contacts Closed*
3532 RA	.018-.020	.031-.039	.018-.020	.011-.013
3533 RA	.018-.020	.031-.039	.018-.020	.011-.013
3628 RA	.018-.020	.030-.035	.018-.020	.005-.007
3629 RA	.018-.020	.030-.035	.018-.020	.005-.007
3630 RA	.018-.020	.030-.035	.018-.020	.005-.007
3631 RA	.018-.020	.030-.035	.018-.020	.005-.007
3687 RA	.018-.020	.035-.043	.018-.020	.009-.011

* Settings are made with contacts closed and armature held to core.

CAUTION: when checking these regulators for maximum output do not use a jumper from F terminal to ground, or F terminal to G terminal. Doing so will result in burned jumpers, or fused contacts.

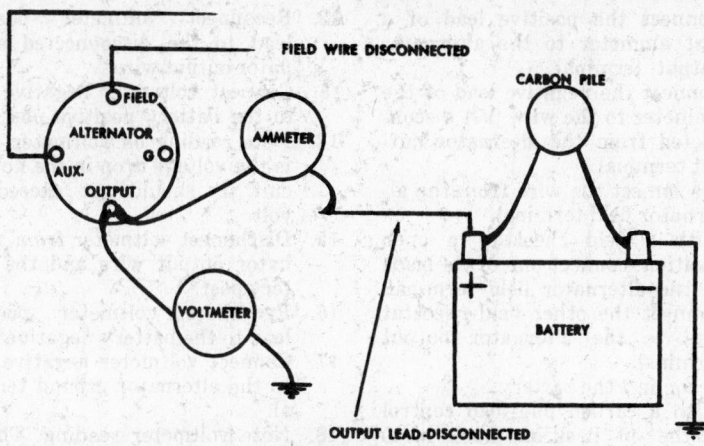

Plan for making alternator system checks

test ammeter to the alternator output terminal and the negative lead of the test ammeter to the wire that was just disconnected from this terminal.

4. Connect the positive lead of a voltmeter to the alternator output terminal and the negative lead of the voltmeter to ground.
5. Reconnect the battery.
6. Connect a carbon pile load control unit (control in off position) to both battery posts.
7. Start engine and run at fast idle for as long as is required to bring it to normal operating temperature.
8. Gradually increase rpm and adjust carbon pile control until current output is reached.
9. Operation is normal if the system develops within 5 amps. of the rated output and voltage registers 14 ± 1 volts.
10. If voltmeter registers over 15 volts, at rated output, replace the regulator.
11. If system is satisfactory at low engine rpm but output is not enough at higher rpm, check for slipping of the drive belt.
12. If current output is low and the output terminal voltage is more than 13 volts, check the isolation diode.

13. If the current output and the output terminal voltage are both low, check the alternator rated output.
14. If current output at specified terminal voltage is produced, check resistance of the charging circuit.

Alternator Output Test

This check is of the alternator only.

1. The ammeter, voltmeter and carbon pile test instruments remain connected into the circuit as with the last check.

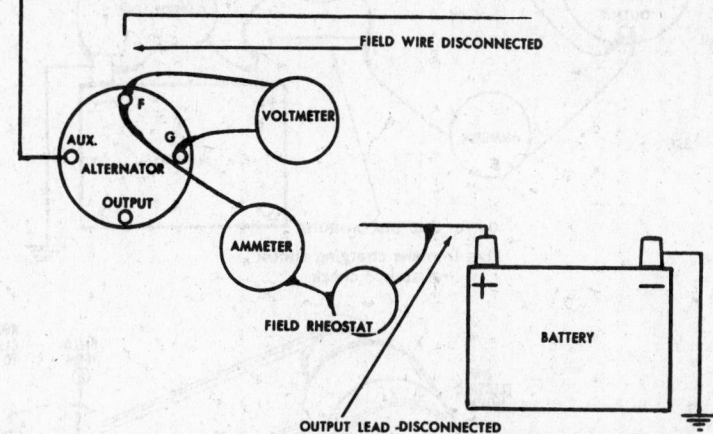

Plan to make field current draw check

2. Disconnect the battery, also the wire from the alternator field terminal.
3. With a field rheostat, adjusted to open position, connect its leads to the alternator field terminal and to the alternator output terminal.
4. Reconnect the battery.
5. Start and run the engine at 1750 rpm.
6. Gradually adjust field rheostat toward the closed position while watching the voltmeter and ammeter (do not exceed 15 volts). Control this voltage with the carbon pile load adjustment. When field rheostat has reached the closed position, note ammeter reading.
7. Adjust field rheostat to the open position and the carbon pile load control off.
8. Compare notes taken in Step 6 with manufacturers' specifications.
9. If readings are low, check drive belt for proper tension.

Field Current Draw Test

1. With battery disconnected, disconnect the wires from the alternator output terminal and the alternator field terminal.
2. With a field rheostat in the open position, connect its leads

to the disconnected alternator output wire and to the positive lead of the test ammeter.
3. Connect the negative ammeter lead to the alternator field terminal.
4. Connect the positive voltmeter lead to the alternator field terminal.
5. Connect the negative voltmeter lead to the alternator ground terminal.
6. Reconnect the battery.
7. Start and run the engine at fast idle.
8. Adjust field rheostat to closed position, then note the voltmeter and ammeter readings.

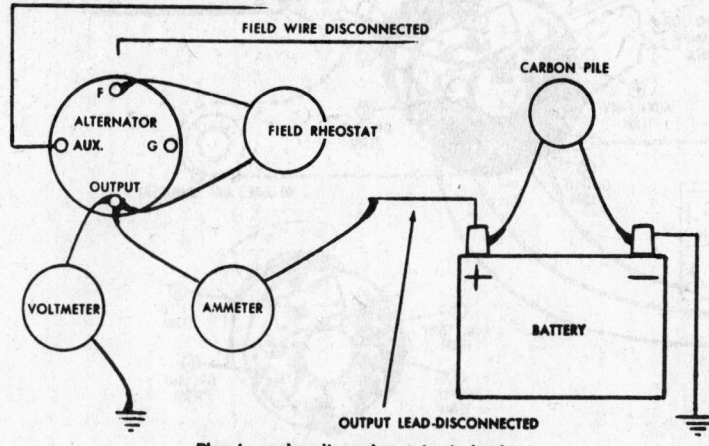

Plan to make alternator output check

9. Adjust field rheostat control to the open position.
10. Compare the readings obtained in Step 8 with manufacturers' specifications.
11. If readings are zero, there is an indication of trouble in the field coil, or the connections between field coil and slip ring.
12. If readings are low, there is probable trouble in the slip rings or brushes.
13. If readings are high, the field coil is probably shorted.
14. If readings are normal, on an alternator which failed to produce its rated output, the probable cause lies in the stator or diodes. Replace the alternator in this case.

Charging Circuit Resistance Test

1. With battery disconnected and the ignition switch off, disconnect the wire from the alternator output terminal.

2. Connect the positive lead of a test ammeter to the alternator output terminal.
3. Connect the negative lead of the ammeter to the wire just disconnected from the alternator output terminal.
4. Disconnect the wire from the alternator field terminal.
5. With a field rheostat in open position, connect one of its leads to the alternator field terminal.
6. Connect the other field rheostat lead to the alternator output terminal.
7. Reconnect the battery.
8. With a carbon pile load control in the off position, hook it to both battery posts.
9. Start and set to 1500-1700 rpm.
10. Adjust field rheostat control toward the closed position enough to get a charging rate of 10 amps. Use the carbon pile load control to prevent exceeding 15 volts. Hold at 15 volts.
11. Disconnect voltmeter from the alternator output terminal and the ground.

12. Reconnect voltmeter positive lead to the disconnected alternator output wire.
13. Connect voltmeter negative lead to the battery positive post.
14. Note reading on voltmeter. This is the voltage drop in the hot circuit. It should not exceed 0.3 volt.
15. Disconnect voltmeter from alternator output wire and the battery post.
16. Reconnect voltmeter positive lead to the battery negative post.
17. Connect voltmeter negative lead to the alternator ground terminal.
18. Note voltmeter reading. This is the voltage drop in the ground circuit. This reading should not exceed 0.15 volt.

Bench Tests

Alternator System

1. Mount alternator to be tested according to typical plan. (See illustration.) Alternator must turn clockwise as viewed from pulley end.
2. Connect regulator to alternator and mount regulator solidly on heat sink.
3. Connect ammeter, voltmeter, indicator lamp, battery and carbon pile load.

NOTE: the No. 57X lamp is used to excite field of alternator to develop an output. The resistor wire, or any 75 ohm, 3 watt resistor may be used.

Optional Alternator Test

This test excludes the regulator from the circuit and can be conducted if the regulator is suspected and a known good regulator is not available.

1. Use the same procedure as described in previous paragraph

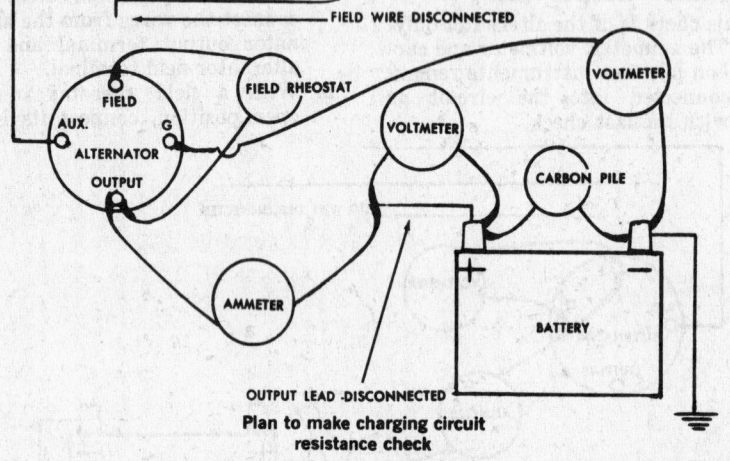

Plan to make charging circuit resistance check

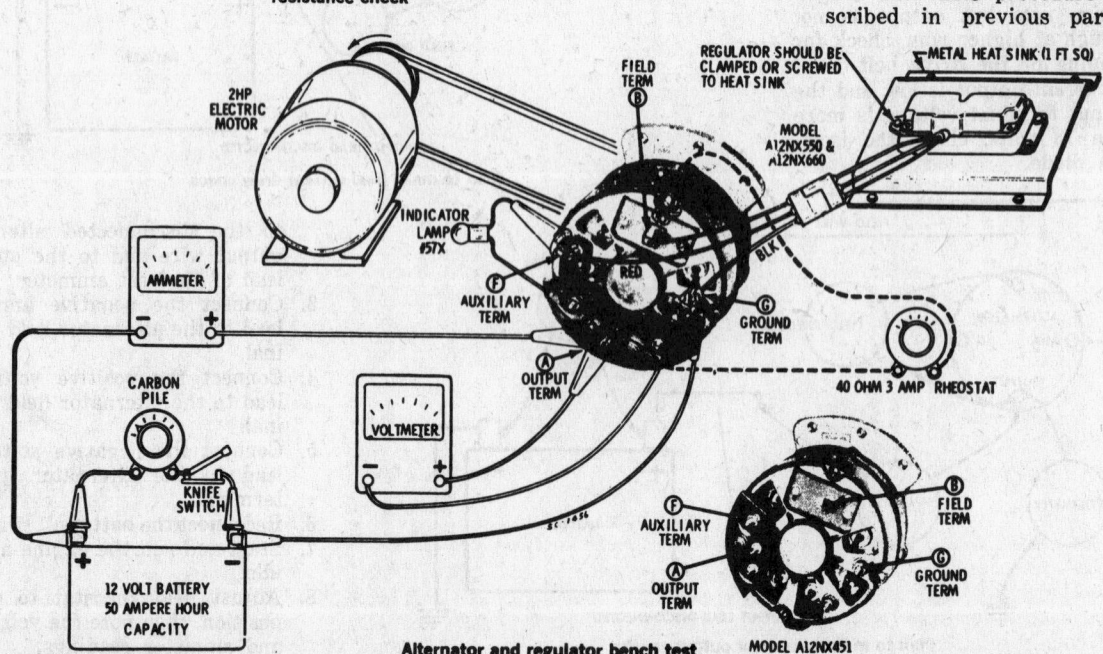

Alternator and regulator bench test   MODEL A12NX451

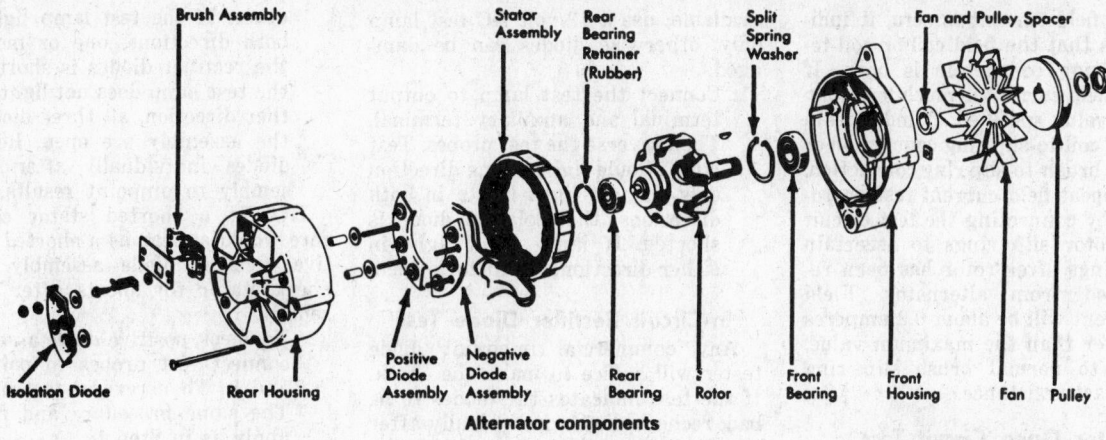

Alternator components

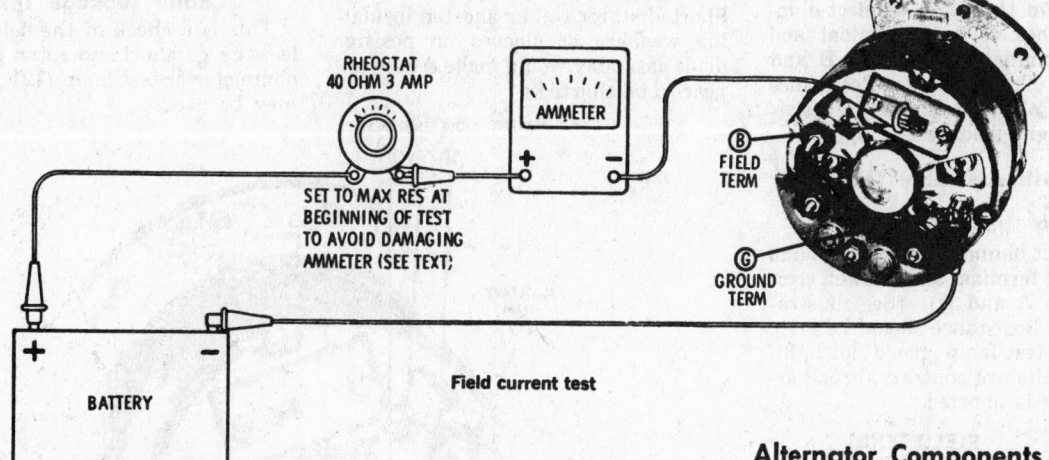

RHEOSTAT
40 OHM 3 AMP

AMMETER

SET TO MAX RES AT
BEGINNING OF TEST
TO AVOID DAMAGING
AMMETER (SEE TEXT)

B FIELD TERM

G GROUND TERM

BATTERY

Field current test

with the exception of the next two steps.

2. Disconnect voltage regulator plug and the field terminal wire.

3. Connect a 40 ohm (3 amp.) rheostat from alternator field terminal to output terminal. Set rheostat to maximum resistance.

4. Start electric drive motor. Increase load with carbon pile and at the same time increase field current by reduction of rheostat resistance.

> Do not allow output terminal voltage to exceed 15 volts.

5. Continue to increase the load and decrease rheostat resistance while holding output voltage below 15 volts until rheostat resistance is zero. At this point, maximum field current is being applied.

Alternator is working if the rated output is available. If rated current output is not available, check alternator.

Regulator Test

This test should be made with a known good alternator and a fully charged battery.

1. Set carbon pile for minimum load (maximum resistance) to prevent

discharging the battery when alternator is not running.

2. Start electric drive motor (alternator rpm should be 3000 to 4000).

3. Increase load with carbon pile until a 10 amp. output is reached. Allow a 15 minute temperature stabilization period. Regulator is working properly if the voltage at output terminal is according to specifications and the alternator is supplying a 10 amp. load.

Isolation Diode Circuit Test

Excessive leakage through the isolation diode will discharge the battery. The rate of discharge depends upon the degree of leakage. Normal and tolerable leakage is less than .001 amperes.

1. To check isolation diode leakage, connect the regulator to the battery. (See illustration.) Do not operate the alternator.

2. Measure the voltage from auxiliary terminal F to ground terminal G. The voltage appearing at the auxiliary terminal should not exceed 0.1 volt. Voltage greater than this indicates leakage through the isolation diode. Check the isolation diode with a commercial diode tester, or with a 12-volt DC test lamp.

Alternator Components Test

Rotor (Field Coil) Test

This test checks the condition of the field coil for open or shorted field coil turns, badly worn or sticky brushes and open connections.

1. Construct a field current test circuit. (See illustration).

2. Connect test circuit to field terminal and ground terminal (test point B and G).

NOTE: set rheostat to maximum resistance (40 ohms) before making connections

3. Slowly reduce resistance of rheostat to zero. With full battery voltage applied to field coil (battery must be fully charged), the field current should be 1.2 to 1.7 amps. Turn rotor by hand while taking reading to indicate value of brush contact. However, a slight fluctuation of current (.2 amps.) is normal.

4. If current is not within limits, inspect brushes and slip rings for dirt, bad brushes and connections. Check brush assemblies for shorts and continuity.

Reinstall repaired or known good brush assembly and repeat the test. If field current is above the maximum value, it indicates that the field coil is shorted or rotor-to-field coil has shorted turns.

If field current is zero, it indicates that the field coil or coil-to-slip ring connection is open. If the field current is much less than the value specified, it indicates a poor coil-to-slip ring connection or poor brush-to-slip ring connection.

Repeat field current test on rotor by connecting the test circuit to rotor slip rings to ascertain findings after rotor has been removed from alternator. Field current will be about 0.2 amperes higher than the maximum value, due to normal brush slip ring contact resistance.

Rotor Open Circuit Test

An ohmmeter may be used to check continuity of the rotor. Connect ohmmeter probes to field terminal and ground terminal (test points B and G. See illustration). Resistance should be about 6 ohms. If resistance is high, field coil is open.

Brush Assembly Test

Insulation

1. Connect ohmmeter or a test lamp to field terminal and bracket (test points A and D). (See illustration.) Resistance should be high, or the test lamp should not light. If results are contrary, brush assembly is shorted.

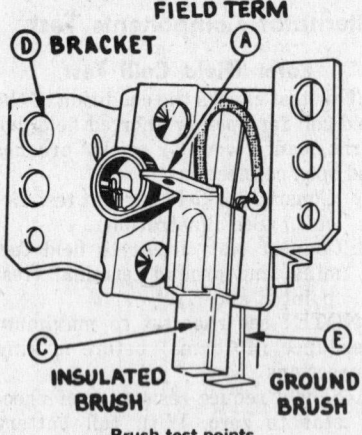

FIELD TERM
D BRACKET A
C INSULATED BRUSH
E GROUND BRUSH

Brush test points

Continuity

2. Connect an ohmmeter to field terminal and brush (points A and C). Resistance should be zero. Move brush and brush lead wire to make sure that the brush lead wire connections are not intermittent. Resistance should not vary when brush and lead wire are being moved around.
3. Connect ohmmeter to bracket and grounded brush, points E and D. Resistance should be zero. Repeat same test on brush lead wire as in Step 2.

Isolation Diode Test

If a commercial diode tester is not available, use a 12-volt DC test lamp only, otherwise diodes can be damaged.

1. Connect the test lamp to output terminal and auxiliary terminal. Then reverse the test probes. Test lamp should light in one direction only. If test light lights in both directions, the isolation diode is shorted. If lamp won't light in either direction, the diode is open.

In-Circuit Rectifier Diode Test

Any commercial in-circuit diode tester will suffice to make the check. If the test indicates the diodes to be bad, recheck diodes individually after the diode assemblies have been disconnected from the stator assembly. Shorted stator coil or shorted insulating washers or sleeves on positive diode assembly would make diodes appear to be shorted.

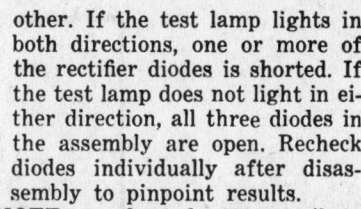

POSITIVE DIODE TERMINALS

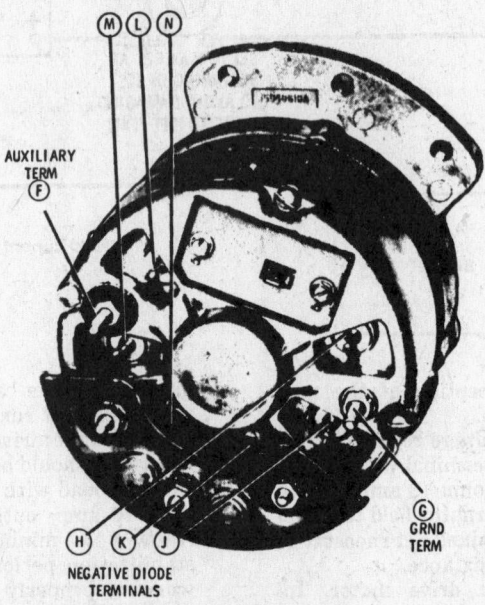

AUXILIARY TERM F

NEGATIVE DIODE TERMINALS

In-circuit diode test points

1. To check negative diode assembly, connect tester to test points G and H. (See illustration.) Then, successively check between test points G and J and G and K.
2. To check positive diode assembly, connect tester to test points F and L. Then successively check between test points F and M and F and N.

Rectifier Diode Test, With Lamp

When an in-circuit diode tester is not available, a 12-volt DC test lamp may be used to indicate a shorted diode. The lamp, however, will not show an open condition unless all three diodes of either assembly are open.

1. Connect test light probes to points G and H, then reverse test probes. The lamp should light in one direction but not in the

other. If the test lamp lights in both directions, one or more of the rectifier diodes is shorted. If the test lamp does not light in either direction, all three diodes in the assembly are open. Recheck diodes individually after disassembly to pinpoint results.

NOTE: a shorted stator coil to core would appear as a shorted negative rectifier diode assembly. Also check stator for shorts after disassembly.

2. To check positive diode assembly, connect test probes to points F and L. Then reverse test probes. The same procedure and results apply as in Step 1.

Rotor Leakage Test

This is a check of the field coil for leakage or shorts to rotor poles. An ohmmeter or test lamp (12V or 120V) may be used.

1. Remove the brush assembly to gain access to rotor slip rings.
2. Connect ohmmeter or test lamp probes to one of the slip rings and the ground terminal, points G and P. (See illustration.) Ohmmeter resistance should be infinite or test lamp should not light. If condition is contrary, leakage or a short exists between field coil and rotor.

Repeat test after rotor has been removed from the alternator to pinpoint findings.

In-Circuit Stator Leak Test

When making this test, consideration must be given to the rectifier diodes that are connected to the stator winding. The negative diode assembly will conduct in one direction when properly polarized. A shorted diode in the negative recti-

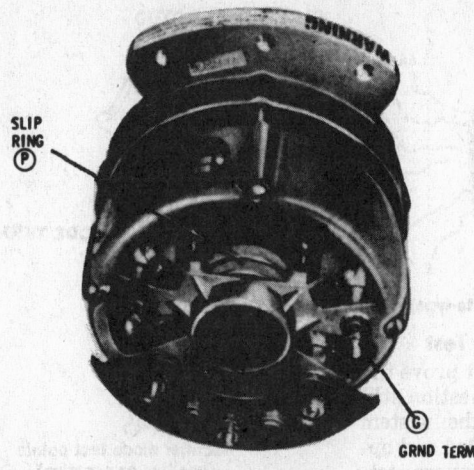

Rotor leakage test points

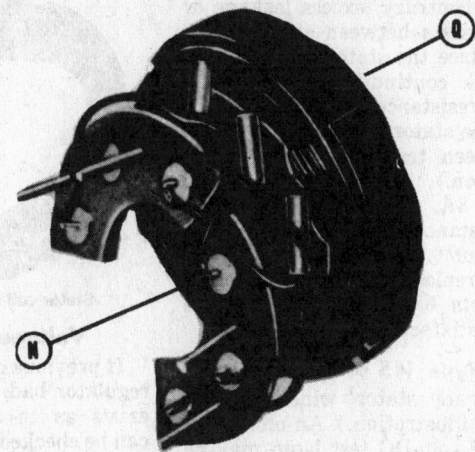

Stator leakage test points

In-circuit stator leakage test points

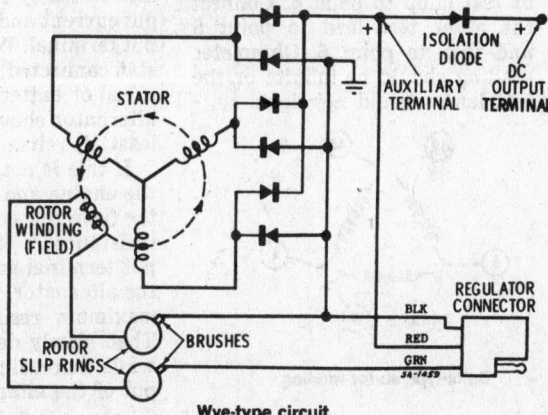

Wye-type circuit

fier-diode assembly will make the stator appear to be shorted. Therefore, the rectifier-diode plate assembly and stator must be checked individually after the alternator has been disassembled if the problem has been narrowed down to the stator.

1. Connect a 12-volt DC test lamp to a diode terminal of the negative diode assembly and ground terminal, points G and H. (See illustration.)
2. Reverse test probes. The test lamp should light in one direction but not in the other direction. If lamp does not light in either direction, it indicates that all three rectifiers in the negative diode assembly are open. If the lamp lights in both directions, the stator winding is shorted to stator or one of the negative rectifier diodes is shorted.

Out-of-Circuit Stator Leak Test

Disassemble alternator and remove the rectifier-diode plates and stator as an assembly. (See illustration.)

An ohmmeter or 12-volt DC test lamp may be used.

1. Connect ohmmeter or test lamp probes to one of the rectifier diode terminals and to the stator test points N and Q.

Resistance reading should be infinite (or the test light should not light). If result is contrary, high leakage or a short exists between stator winding and stator. In either case, stator should be replaced.

Stator Coil Leak and Continuity Test

This check is for shorts or leakage between stator coil windings. The 30 and 40 amp. alternators use a Wye-Type winding. (See illustration.) 45 ampere models use a Delta type winding. (See illustration.)

Wye Type (30 and 40 Amp. Models)

1. Separate winding ends (See illustration.) An ohmmeter or 12-volt DC test lamp may be used.
2. Connect one lead of the ohmmeter or test lamp to point 1. Connect the other test lead to point 2 and then to point 3. Ohmmeter reading should be infinite or the test lamp should not light.
3. Connect test leads to points 2 and 3. Ohmmeter reading should be infinite, or the test lamp should not light.

In test 2 or test 3, if the test re-

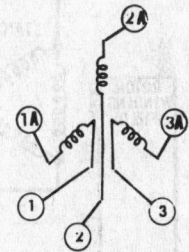

Wye-type stator winding

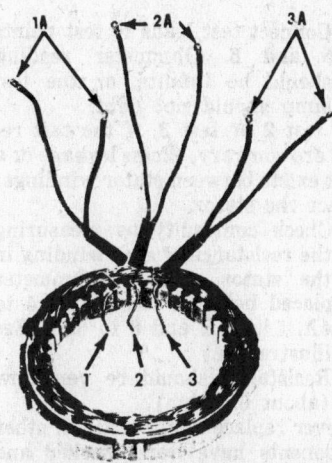

Stator coil test points (Wye-type)

sults are contrary, excess leakage or a short exists between stator windings; replace the stator.

4. Check continuity by measuring the resistance of each winding in the stator with an ohmmeter, between test points (See illustration), 1 to 1A, 2 to 2A and 3 to 3A.
   Resistance should be very low (about 0.1 ohm).

Never replace stator until all other components have been checked and proven satisfactory.

Delta Type (45 Amp. Models)

1. Separate stator winding ends. (See illustration.) An ohmmeter or 12-volt DC test lamp may be used.
2. Connect one lead of the ohmmeter or test lamp to point 4. Connect the other test lead to point 5 and then to point 6. Ohmmeter reading should be infinite or the test lamp should not light.

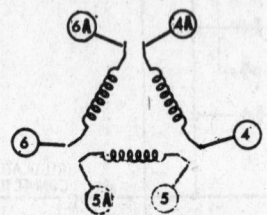

Delta-type stator winding

3. Connect test leads to test points 5 and 6. Ohmmeter reading should be infinite, or the test lamp should not light.
   In test 2 or test 3, if the test results are contrary, excess leakage or a short exists between stator windings; replace the stator.
4. Check continuity by measuring the resistance of each winding in the stator with an ohmmeter placed between test points 4 to 4A, 5 to 5A and 6 to 6A. (See illustration.)
   Resistance should be very low (about 0.1 ohm).

Never replace stator until all other components have been checked and proven satisfactory.

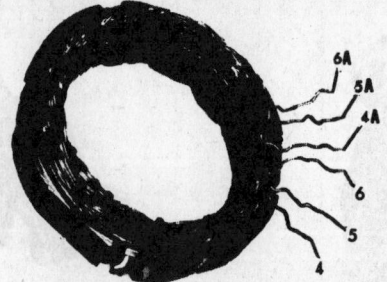

Stator coil test points (Delta-type)

Voltage Regulator Test

If previous checks did not prove the regulator bad, but some question still exists as to reliability, the system can be checked while installed and operating in the car. A volt-ampere tester can be used. The system should be first tested by taking a reading of output current and the voltage at the output terminal. With a carbon pile rheostat connected across the output terminal or battery, rated current of the alternator should be available with at least 13 volts.

If this is not obtainable, shut down the engine and disconnect the regulator from the system. Insert a variable resistance in series between the output terminal and the field terminal of the alternator. Start the engine, with maximum resistance in the circuit. Then slowly decrease field resistance while increasing the load on the output of the alternator.

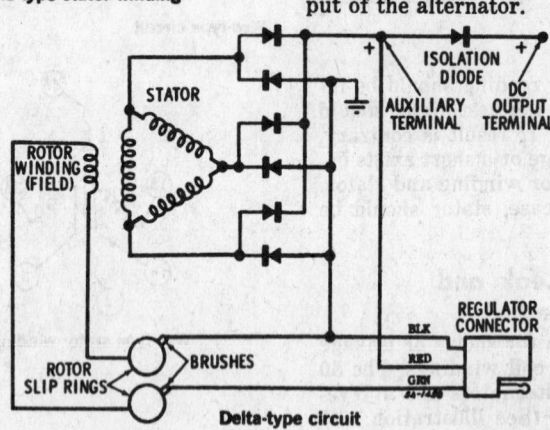

Delta-type circuit

Caution Observe voltage at output terminal and be sure to decrease the field resistance and increase the output load without exceeding 15 volts at the output terminal.

Continue to decrease field resistance and increase output load until there is no resistance in the field circuit. If rated current output is now obtained with 13 volts at the output terminal, the voltage regulator is bad. If rated current is not available under these conditions, the alternator is bad and must be removed for repairs or replacement.

Out-of-Circuit Rectifier Diode Test

If a commercial diode tester is not

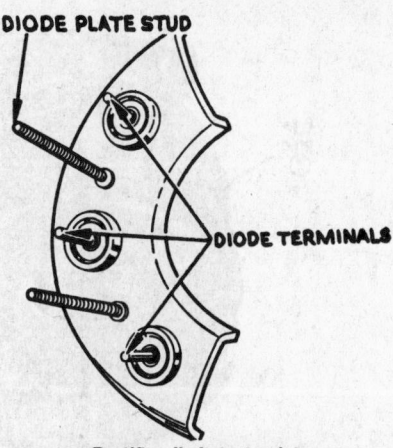

Rectifier diode test points
(positive or negative)

available, check the diodes with a 12-volt DC test lamp only.

Caution When unsoldering the stator wires from the rectifier diode assembly, provide a heat sink to the diode terminal with a pair of longnosed pliers.

1. Connect test lamp probes to diode terminal and diode plate stud (see illustration), then reverse test lamp probes. The test light should light in one direction but not in the other.
   If the test lamp lights in both directions, the diode is shorted. If test lamp does not light in either direction, the diode is open.
   Test the remaining diodes of the assembly in the same manner. Replace entire assembly if one of the diodes is found to be bad.

The Prestolite System

The Prestolite AC generator regulator with a circuit breaker incorporates a polarity sensing feature for the purpose of isolating generator output from battery potential when the system is out of operation. This feature prevents damage to the components of the charging system in the event of battery polarity reversal.

Circuit Resistance Test

1. With the battery ground cable disconnected at the battery, and the lead disconnected from the generator output A terminal, hook up an ammeter as follows:
   A. Connect an ammeter, positive lead to the generator output A terminal.
   B. Connect the ammeter, negative lead to the lead that was just removed from the generator A terminal.
2. Disconnect the lead from the generator, field F terminal.
3. With a field rheostat control turned to open position, connect

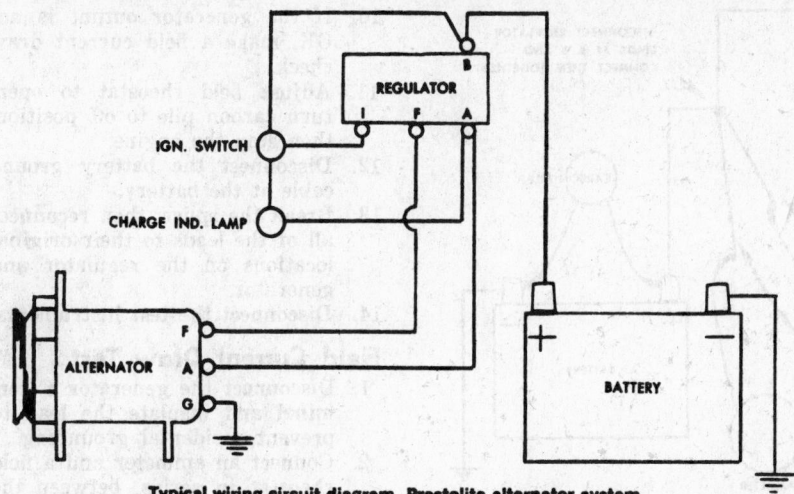

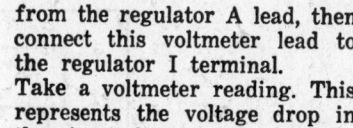

Typical wiring circuit diagram, Prestolite alternator system

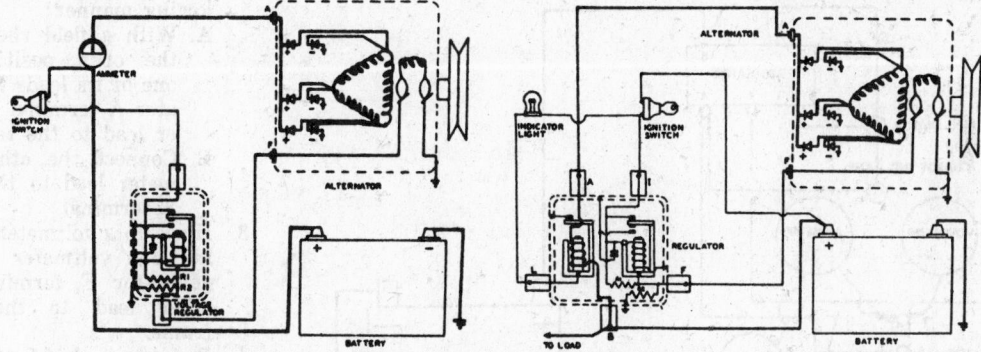

Single and two-unit Prestolite circuits

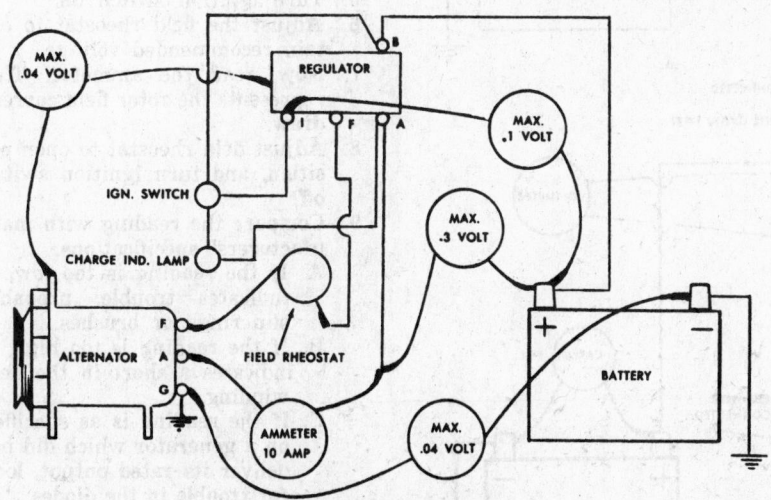

Plan to make charging circuit resistance check

from the regulator A lead, then connect this voltmeter lead to the regulator I terminal.

11. Take a voltmeter reading. This represents the voltage drop in the circuit between the battery, positive post and the regulator, I terminal. The drop must not exceed 0.1 volt.

12. Disconnect both voltmeter leads, then connect one voltmeter lead to the battery negative post. Connect the other voltmeter lead to the housing of the generator.

13. Take a voltmeter reading. This represents voltage drop in the ground circuit between the battery, negative post and the generator. The drop must not exceed 0.04 volt.

14. Disconnect the voltmeter lead from the battery negative post and connect this same voltmeter lead to a good ground on the regulator.

15. Take a voltmeter reading. This represents voltage drop in the ground circuit between generator and regulator. The drop must not exceed 0.04 volt.

16. Turn field rheostat control to open position, then stop the engine.

Alternator Output Test

With the battery ground cable disconnected at the battery, and the ignition switch in off position, proceed as follows:

1. Disconnect the leads from regulator terminals A, F and B. Then connect the leads all together and temporarily insulate them.

2. Hook up a voltmeter to the generator output A terminal and the generator frame.

3. Connect an ammeter between the generator output A terminal and spliced connection of the leads removed from the regulator A, F and B terminals.

4. Reconnect the battery ground cable.

5. With a carbon pile adjusted to the open position, connect it to both battery posts. Then start the engine and adjust the speed to 1750 rpm.

one lead of this rheostat to the generator F terminal and the other lead to the lead which was just disconnected from this F terminal.

4. Hook up an engine tachometer.

5. Connect battery ground cable.

6. Start engine and set speed to 850 rpm.

7. Adjust field rheostat toward closed position enough to effect a 10 amp. charging rate.

8. Connect a voltmeter to the battery positive post and to the lead which was disconnected from the generator output A terminal.

9. Take a voltmeter reading. This represents the voltage drop in the circuit between the battery positive post and the regulator A terminal. The drop must not exceed 0.3 volt.

10. Disconnect the voltmeter lead

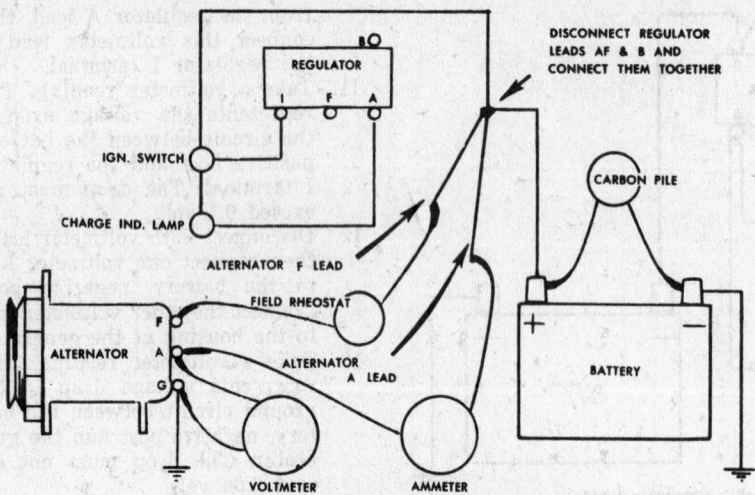

Plan to make alternator output test

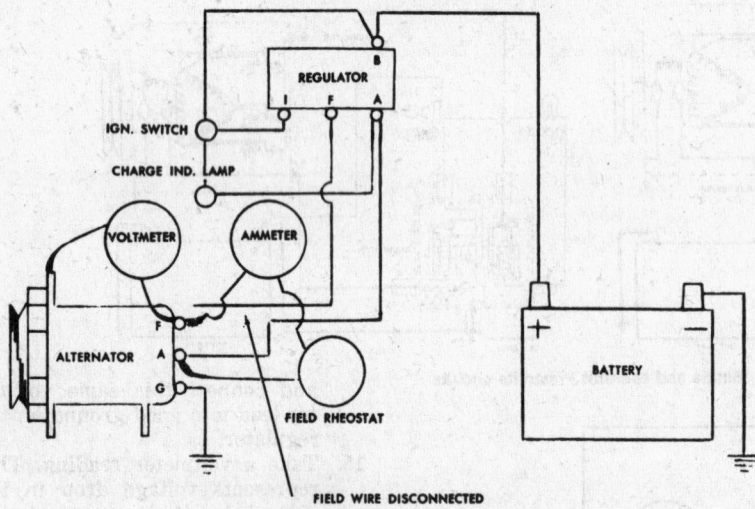

Plan to make field current draw test

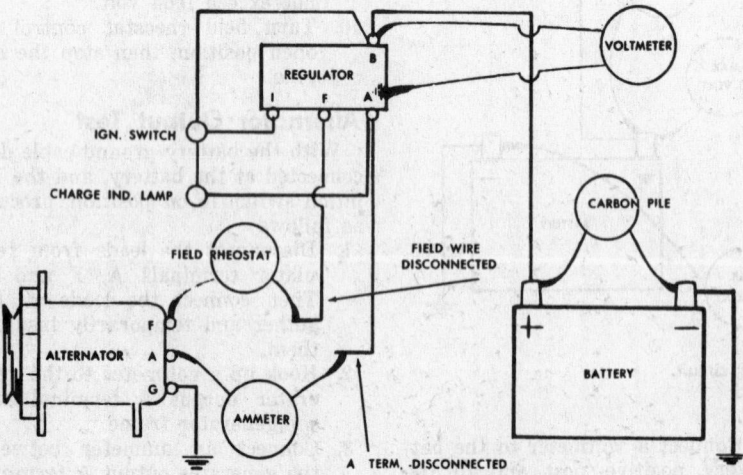

Plan to make voltage regulator check

10. If the generator output is not OK, make a field current draw check.
11. Adjust field rheostat to open, turn carbon pile to off position, then stop the engine.
12. Disconnect the battery ground cable at the battery.
13. Break the splice, then reconnect all of the leads to their original locations on the regulator and generator.
14. Disconnect the test instruments.

Field Current Draw Test

1. Disconnect the generator F terminal and insulate the lead to prevent accidental grounding.
2. Connect an ammeter and a field rheostat in series, between the generator F terminal and the generator A terminal in the following manner:
   A. With a field rheostat set in the open position, connect one of its leads to the generator A terminal and the other lead to the test ammeter.
   B. Connect the other test ammeter lead to the generator F terminal.
3. Hook up a voltmeter by connecting one voltmeter lead to the generator F terminal and the other lead to the generator frame.
4. Reconnect the battery ground cable.
5. Turn ignition switch on.
6. Adjust the field rheostat to obtain recommended voltage.
7. Now, read the ammeter. This represents the rotor field current draw.
8. Adjust field rheostat to open position, and turn ignition switch off.
9. Compare the reading with manufacturers' specifications.
   A. If the reading is too low, it indicates trouble, probably slip rings or brushes.
   B. If the reading is too high, it indicates a short in the field winding.
   C. If the reading is as specified on a generator which did not deliver its rated output, look for trouble in the diodes.

Voltage Regulator Setting

1. With the battery ground cable disconnected at the battery and the ignition switch in off position, hook up an ammeter to the generator A terminal and to the lead just disconnected from this A terminal.
2. Connect a field rheostat to the generator F terminal and to the lead just disconnected from this F terminal. Adjust rheostat to open position.
3. Connect a voltmeter to the reg-

6. With a field rheostat connected to the generator F terminal and spliced connection of the leads removed from the regulator A, F and B terminals, adjust the rheostat toward closed position.
7. Hold a voltage of 14.2 with the carbon pile load control.

8. When the field rheostat control is all the way closed, the ammeter reading should agree with manufacturers' specifications, with a tolerance of ± 2 amps.
9. If the generator output has checked out as good, proceed to the voltage regulator check.

ulator B terminal and a good ground on the frame of the regulator.

4. Connect the battery ground cable.

5. Hook up a carbon pile (load control turned to the off position) between the battery posts.

6. Start the engine and while operating at 850 rpm, adjust the field rheostat to obtain a 10 amp. charge. Be careful not to exceed factory voltage recommendation. Run engine long enough to develop normal regulator operating temperature.

7. Open, then fully close the field rheostat to cycle the system.

8. Adjust the carbon pile load to obtain a 10 amp. generator output while reading the voltmeter. This reading is on the upper contacts.

9. Turn carbon pile to off position and increase engine rpm to 1750, then read the voltmeter. This voltage should increase and the amperage decrease if the regulator lower point set is working. This is known as spread.

10. If the battery is in fully charged condition, the spread may be checked in the following manner.

A. Adjust load control to produce 15 amps. of charge, then read the voltmeter. This reading is on the upper set of points.

B. Adjust load control to off position, wait about one-half minute, then read the voltmeter. This reading is on the lower set of points. The difference between the two readings is spread.

C. Spread should be 0.2 to 0.5 volts higher when operating on lower grounding contact. Operation on lower contacts must not exceed 14.7 volts. If necessary, adjust regulator so as not to exceed 14.7 volts.

11. If regulator does not operate within specifications, renew the regulator.

Disassembly

1. Remove thru bolts and tap ends lightly with plastic hammer to separate ends from rotor.

2. Remove nuts and washers from negative rectifier brackets, and the nuts, washers and insulator bushings from the positive rectifier brackets, and separate the slip ring end head.

3. Remove insulated brush, gripping brass terminal on brush head with pliers and pulling from field terminal insulator.

4. Remove screw that attaches ground brush. Do not lose brush springs.

5. To remove pulley nut, grip pulley with fan belt and vise to hold while breaking nut loose. Then remove pulley with suitable puller.

6. To remove drive end head, remove key, fan and spacer. Then, using suitable puller remove head.

7. Remove retaining plate and press out drive end head bearing.

8. The slip end head bearing can be removed with a puller.

9. The rectifiers can now be pressed out and in. In cutting and crimping leads keep as close to the sleeve as possible.

Regulators

The voltage regulator is a dual contact type. Field F terminal and ignition I terminal are insulated from base. A regulating resistor is connected between these terminals.

To adjust, remove the cover, loosen stationary contact support screw and move support up or down. Use following specification tables for settings.

Alternator

	Model No. ALC 5001	Model No. ALD 5001	Model No. ALD 5005
Rotation—Clockwise, Drive End	Yes	Yes	Yes
Voltage	12	12	12
Rated Output (Amps.)	45	40	40
Brushes (No.)	2	2	2
Rotor Field Coil Draw (At 70°F) At 10 Volts	2.34—2.43 Amps.	2.32—2.42 Amps.	2.32—2.42 Amps.
Current Output—4200 rpm (Engine speed 1750 rpm)	42 Amps. 14.2 Volts	38 Amps. 14.2 Volts	38 Amps. 14.2 Volts
Ground Polarity	Negative	Negative	Negative
Control	VR	VR	VR

Voltage Regulator

	Model No. VBS 6201A	Model No. VBT 6201A	Model No. VBS 6201A-1
Volts	12	12	12
Ground Polarity	Negative	Negative	Negative
Point Gap (in.)	.010	.010	.010
Air Gap (in.)	.034-.038	.034-.038	.034-.038
Relay Air Gap (in.)	.015-.020	——	.015-.020
Relay Opening (Amps.)	2.0-2.5	——	2.0-2.5
Relay Closing (Amps.)	1.5 Max.	——	1.5 Max.
Operating Voltage (Upper Contact) 50°F 80°F 110°F 140°F Tolerance 14.3 14.1 13.0 13.7 ±.2		—All Models—	
Voltage Spread (Between upper and lower contact operation)	.1-.3 Volts	.1-.3 Volts	.1-.3 Volts
Capacitor (Color band indicates ground end)	.1 MFD	.1 MFD	.1 MFD
Resistance			
R^1	10 Ohms.	10 Ohms.	10 Ohms.
R^2	20 Ohms.	20 Ohms.	20 Ohms.
VR Winding	52-58 Ohms.	52-58 Ohms.	52-58 Ohms.
1600-1800 rpm Alternator Speed, 15 minute warm-up, 10 Amp. load		—All Models—	
3400-3600 rpm Alternator Speed, 15 Amp. load Upper contacts—2-7 Amp. Charge rate—lower grounding contacts		—All Models—	
All tests to be made with a fully charged battery		—All Models—	

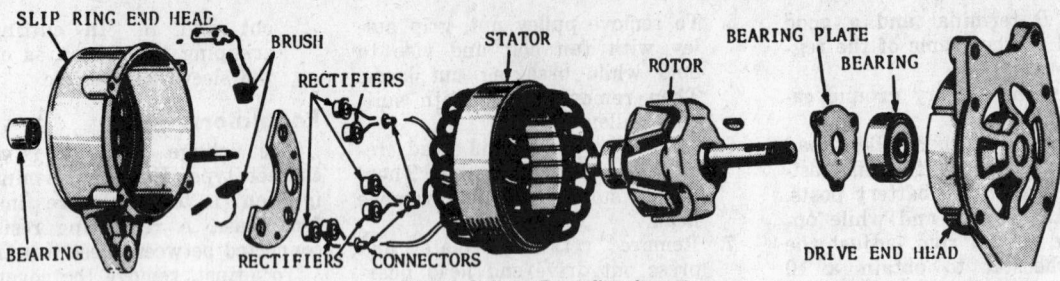

SLIP RING END HEAD BRUSH STATOR BEARING PLATE BEARING

RECTIFIERS ROTOR

BEARING RECTIFIERS CONNECTORS DRIVE END HEAD

Exploded view—Prestolite alternator

C.S.I. AC Generator

The C.S.I. system is an integrated AC generating system containing a built-in voltage regulator. Removal and displacement is essentially the same as for the standard AC generator. Specialized service procedures are as follows:

Diode Trio Initial Testing

1. Before removing this unit, easily identified in the illustration, connect an ohmmeter between the brush lead clip and the end frame. The lowest reading scale should be used for this test.

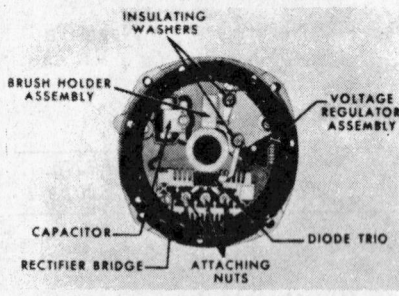

INSULATING WASHERS

BRUSH HOLDER ASSEMBLY

VOLTAGE REGULATOR ASSEMBLY

CAPACITOR DIODE TRIO

RECTIFIER BRIDGE ATTACHING NUTS

Slip ring end frame—C.S.I. AC generator
(© Pontiac Motor Div., G.M. Corp.)

2. After taking a reading, reverse the lead connections. If the meter reads zero, the brush lead clip is probably grounded, due to omission of the insulating sleeve or insulating washer.

Diode Trio Removal

1. Remove the three nuts which secure the stator.
2. Remove stator.
3. Remove the screw which secures the diode trio lead clip, then remove diode trio.
NOTE: The position of the insulating washer on the screw is critical; make sure it is returned to the same position on reassembly.

Diode Trio Testing

1. Connect an ohmmeter, on lowest range, between the single brush connector and one stator lead connector (see illustration).
2. Observe the reading, then reverse the meter leads. Repeat this test with each of the other two stator lead connectors. The readings on each of these tests

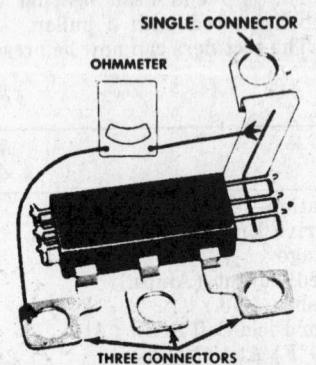

SINGLE CONNECTOR

OHMMETER

THREE CONNECTORS

Testing diode trio—C.S.I. AC generator
(© Pontiac Motor Div., G.M. Corp.)

should NOT be identical, there should be one low and one high reading for each test. If this is not the case, replace the diode trio.
CAUTION: Do not use high voltage on the diode trio.

Rectifier Bridge Testing

1. Connect an ohmmeter between the heat sink (ground) and the base of one of the three terminals (see illustration). Then, reverse the meter leads and take a reading. If both readings are identical, the bridge is defective and must be replaced.

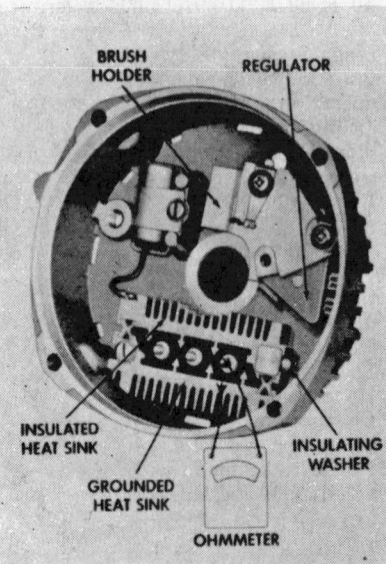

BRUSH HOLDER REGULATOR

INSULATED HEAT SINK

GROUNDED HEAT SINK

INSULATING WASHER

OHMMETER

**Testing rectifier bridge diodes—
C.S.I. AC generator**
(© Pontiac Motor Div., G.M. Corp.)

2. Repeat this test with the remaining two terminals, then between the INSULATED heat sink (as opposed to the GROUNDED heat sink in previous test) and each of the three terminals. As before, if any two readings are identical, on reversing the meter leads, the rectifier bridge must be replaced.

Rectifier Bridge Removal

1. Remove the attaching screw and the BAT. terminal screw.
2. Disconnect the condenser lead.
3. Remove the rectifier bridge.
NOTE: The insulator between the insulated heat sink and the end frame is extremely important to the operation of the unit. It must be replaced in exactly the same position on reassembly.

Brush and/or Voltage Regulator R & R

1. Remove two brush holder screws and stator lead to strap nut and washer, brush holder screws and one of the diode trio lead strap attaching screws.
NOTE: The insulating washers must be replaced in the same position on reassembly.
2. Remove brush holder and brushes. The voltage regulator may also be removed at this time, if desired.

Brush holder—C.S.I. generator
(© Pontiac Motor Div., G.M. Corp.)

3. Brushes and brush springs must be free of corrosion and must be undamaged and completely free of oil or grease.
4. Insert spring and brushes into holder, noting whether they slide freely without binding. Insert wooden or plastic toothpick into bottom hole in holder to retain brushes.

NOTE: The brush holder is serviced as a unit; individual parts are not available.

5. Reassemble in reverse order of disassembly.

Voltage Regulator Testing

NOTE: The voltage regulator must be tested with the C.S.I. unit still in place.

1. Disconnect battery ground strap.
2. Connect an ammeter in series with the BAT. terminal of the generator and the lead removed from that terminal.
3. Reconnect battery ground strap, then turn on all accessories to place a load on the system.
4. Connect a carbon pile across the battery terminals.
5. Ground the field winding by inserting an insulated screwdriver into the test hole in the alternator frame and depressing the tab. Do not push the tab into the generator more than 1 in. (See illustration.)
6. Run the engine at moderate rpm, equivalent to 30-40 mph in high gear, and adjust the carbon pile to obtain maximum current output.
7. If the output is within 10% of the rated output of the alternator

**Grounding tab for voltage regulator test—
C.S.I. AC generator**
(© Pontiac Motor Div., G.M. Corp.)

and the system does not charge properly, the voltage regulator is defective and must be replaced.

Generator—C.S.I.
Amps.—cold rating 55
Output @ rpm .... 30 @ 2000
Output @ rpm .... 55 @ 5000
Field current draw 4.0-4.5
Regulator—C.S.I.
Model 1116368
            (Pontiac parts no.)
Normal range ..... 13.5-16.0 volts

Slip Ring End Frame Bearing and Seal R & R

1. With stator removed, press out bearing and seal, using a socket or similar tool that fits inside the end frame housing. Press from outside to inside, supporting the frame inside with a hollow cylinder (large, deep socket) to allow the seal and bearing to pass.
2. The bearings are sealed for life and permanently lubricated. If a bearing is dry, do not attempt to repack it, as it will throw off the grease and contaminate the inside of the generator.
3. Using a flat plate, press the new bearing from the outside toward the inside. A large vise is a handy press, but care must be exercised so that end frame is not distorted or cracked. Again, use a deep socket to support the inside of the end frame.
4. From inside the end frame, insert seal and press flush with housing.
5. Install stator and reconnect leads.

Carburetors

Carter, Ball and Ball Carburetors

YEAR	MODEL OR TYPE	FLOAT LEVEL (IN.)		FLOAT DROP (IN.)		Pump Travel Setting (IN.)	CHOKE SETTING		ON THE CAR ADJUSTMENTS			
		Prim.	Sec.	Prim.	Sec.		Unloader (IN.)	Housing	Idle Screw Turns Open	Idle Speed (RPM)	Fast Idle Speed (RPM)	Dashpot Plunger Clearance (IN.)
Buick—Buick Special												
1964	AFB-3633S, 35S, 65S	7/32	7/32	3/4	3/4	7/16	7/32	INDEX	1/2-3/4	D-500	600	.000
1964	AFB-3634S	7/32	7/32	3/4	3/4	1/2	7/32	INDEX	1/2-2	D-500	600	.000
1964	AFB-3645S	7/32	7/32	3/4	3/4	1/2	7/32	INDEX	1/2-3/4	D-500	600	.000
1964	AFB-3646S	9/32	9/32	3/4	3/4	1/2	7/32	INDEX	1/2-2	D-500	600	.000
1965	AFB-3921S, 22S, 23S	7/32	7/32	3/4	3/4	7/16	7/32	INDEX	1	D-500	600	.000
1965	AFB-3924S, Rear Syn.	7/32	7/32	3/4	3/4	1/2	7/32	INDEX	1	D-500	600	...
1965	AFB-3925S, Rear Aut.	9/32	9/32	3/4	3/4	1/2	7/32	INDEX	1	D-500	600	...
1966	AFB-4053S, 4054S	7/32	7/32	3/4	3/4	7/16	.215	INDEX	3/4	D-500	600	...
1966	AFB-4055S	7/32	7/32	3/4	3/4	7/16	.160	1-RICH	1	D-500	600	...
1966	AFB-4056S	7/32	7/32	3/4	3/4	7/16	.160	INDEX	1	D-500	600	...
1966	AFB-4059S	7/32	7/32	3/4	3/4	7/16	.215	INDEX	3/4	D-500	600	...
1966	AFB-4060S	7/32	7/32	3/4	3/4	1/2	.215	INDEX	3/4	D-500	600	...
1966	AFB-4061S	7/32	7/32	3/4	3/4	7/16	.160	1-RICH	1	D-500	600	...
1966	AFB-4179S, 80S, 81S	7/32	7/32	3/4	3/4	1/2	.215	INDEX	3/4	D-500	600	...
1967	AFB-4331S, Aut.	1 13/32	1 13/32	3/4	3/4	7/16	.160	1-RICH	1	D-550	600	...
1967	AFB-4332S, Man.	1 13/32	1 13/32	3/4	3/4	7/16	.160	INDEX	1	D-550	600	...
1967	AFB-4344S, A.I.R.	1 13/32	1 13/32	3/4	3/4	17/32	.160	2-RICH	1	D-600	600	...

*—With screw on bottom step of cam.
N—Neutral.
D—Drive.

Cadillac

YEAR	MODEL OR TYPE	Prim.	Sec.	Prim.	Sec.	Pump Travel	Unloader	Housing	Idle Screw	Idle Speed	Fast Idle	Dashpot
1964	AFB-3655S	3/8	3/8	15/16	15/16	15/32	9/32	INDEX	2 1/2	D-480	1725	①
1965	AFB-3903S	3/8	3/8	15/16	15/16	15/32	9/32	INDEX	2 1/2	D-480	1725	①
1966	AFB-4168S, 4169S, 4171S	3/8	3/8	15/16	15/16	5/16	9/32	INDEX	2 1/2	D-480	1725	①

①—By trial.
D—Drive.

Chevelle

YEAR	MODEL OR TYPE	Prim.	Sec.	Prim.	Sec.	Pump Travel	Unloader	Housing	Idle Screw	Idle Speed	Fast Idle	Dashpot
1964	WCFB—ALL	7/32	1/4	3/4	3/4	1/2	3/16	INDEX	1/2-2	N-475	1750	.060
1965	WCFB—ALL	7/32	1/4	3/4	3/4	1/2	3/16	INDEX	1/2-2	N-475	1750	.060
1966	6 Cyl.-YF-4079S, 4080S	1/2	...	1 3/16	...	...	.260	...	1 1/2	...	...	...
1966	V8-AVS-4027S, 4028S	1 15/32	...	2	...	11/32	.170	...	1 1/2	...	2200	...
1967	6 Cyl.-YF	7/32	...	1 3/16	...	...	.250	...	...	...	...	...

N—Neutral.

Chevrolet

YEAR	MODEL OR TYPE	Prim.	Sec.	Prim.	Sec.	Pump Travel	Unloader	Housing	Idle Screw	Idle Speed	Fast Idle	Dashpot
1964	WCFB	7/32	1/4	3/4	3/4	1/2	3/16	INDEX	1 1/2	D-500	1750	.060
1965	WCFB	7/32	1/4	3/4	3/4	1/2	3/16	INDEX	1 1/2	D-500	1750	...
1966	WCFB	7/32	1/4	3/4	3/4	1/2	3/16	INDEX	1 1/2	D-500	1750	...
1964	AFB-3720S-21S	7/32	7/32	3/4	3/4	1/2	1/4	1-LEAN	1 1/2	D-500	1750	.060
1964	AFB-3499S	7/32	7/32	3/4	3/4	1/2	1/4	2-LEAN	1 1/2	N-750	1750	.060
1964	AFB-3361S-62S	7/32	7/32	3/4	3/4	1/2	1/4	2-RICH	1 1/2	N-750	1750	.060
1965	AFB-3720SA-21SA	7/32	7/32	3/4	3/4	1/2	1/4	1-LEAN	1 1/2	N-750	1750	...
1965	AFB-3499S	7/32	7/32	3/4	3/4	1/2	1/4	2-RICH	1 1/2	N-750	1750	...
1966	6 Cyl.-YF-4079S, 4080S	1/2	...	1 3/16	...	...	.260	...	1 1/2	500	...	...
1966	V8-AVS-4027S, 4028S	1 15/32	...	2	...	11/32	.170	...	...	...	2200	...
1967	6 Cyl.-YF	7/32	...	1 3/16	...	...	.250	...	1 1/2	...	...	...

①—Set pump arm parallel4

YEAR	MODEL OR TYPE	FLOAT LEVEL (IN.)		FLOAT DROP (IN.)		Pump Travel Setting (IN.)	CHOKE SETTING		ON THE CAR ADJUSTMENTS			
		Prim.	Sec.	Prim.	Sec.		Unloader (IN.)	Housing	Idle Screw Turns Open	Idle Speed (RPM)	Fast Idle Speed (RPM)	Dashpot Plunger Clearance (IN.)
Chevy II												
1964	YF	7/16	...	1 3/16	...	...	...	...	1 1/2	500	...	...
1965	YF	7/16	...	1 3/16	...	...	...	...	1 1/2	500	...	...
1966	4 Cyl.-VF-3379S, 3402S	1/2	...	1 3/16	...	...	...	...	...	...	...	...
1966	6 Cyl.-VF-4079S, 4080S	1/2	...	1 3/16	...	...	.260	...	1 1/2	...	...	...
1966	V8-AVS-4027S, 4028S	1 15/32	...	2	...	11/32	.170	...	...	...	2200	...
1967	4 Cyl.-YF	7/32	...	1 3/16	...	...	.250	...	...	...	...	...
1967	6 Cyl.-YF	7/32	...	1 3/16	...	...	.260	...	1 1/2	...	...	...
Chrysler Imperial												
1964	BBD-3685S	1/4	...	...	...	①	1/4	2-RICH	3/4	500	700	...
1964	AFB-3614S	7/32	7/32	3/4	3/4	7/16	3/8	INDEX	1-2	500	700	...
1964	AFB-3505S	9/32	7/32	1/2	1/2	7/16	...	...	1-2	500	1400	...
1964	AFB-3615S-3644S	7/32	7/32	3/4	3/4	7/16	3/8	2-RICH	1-2	500	700	...
1965	AFB-3855S, 56S	7/32	7/32	3/4	3/4	7/16	3/8	2-RICH	1-2	500	700	...
1965	AFB-3858S	7/32	7/32	3/4	3/4	7/16	3/8	2-RICH	1-2	500	700	...
1965	AFB-3859S, 60S	7/32	7/32	3/4	3/4	7/16	3/8	INDEX	1-2	550	700	...
1965	AFB-3871S	7/32	7/32	3/4	3/4	7/16	3/8	2-RICH	1-2	500	700	...
1965	BBD-3849S, 50S	5/16	...	...	...	①	1/4	2-RICH	3/4	500	650	...
1966	AFB-4130S, 4131S	7/32	7/32	3/4	3/4	7/16	3/8	2-RICH	1-2	500	700	...
1966	AFB-4132S, 33S, 4136S, 37S	7/32	7/32	3/4	3/4	7/16	3/8	INDEX	...	650	1500	...
1966	BBD-4125S	5/16	...	...	...	①	1/4	2-RICH	1	500	700	...
1966	BBD-4126S	5/16	...	...	...	①	1/4	2-RICH	1	600	1600	...
1967	AFB-4298S, 99S	7/32	7/32	3/4	3/4	7/16	3/8	2-RICH	1-2	500	700	...
1967	AFB-4309S, 10S, w/cap.	5/16	5/16	3/4	3/4	7/16	5/16	INDEX	1-2	650	1400	...
1967	AFB-4326S, 27S	7/32	7/32	3/4	3/4	7/16	3/8	INDEX	1-2	650	750	...
1967	AFB-4311S, 12S, w/cap.	5/16	5/16	3/4	3/4	7/16	5/16	INDEX	1-2	650	1400	...
1967	AFB-4328S, 29S, w/cap.	5/16	5/16	3/4	3/4	7/16	3/8	INDEX	1-2	650	1400	...
1967	BBD-4296S, 97S	5/16	...	...	...	...	1/4	2-RICH	1 1/2	550‡	700†	...
1968	BBD-4422S-Std. Tr.	5/16	...	...	...	29/32	1/4	2-RICH	1 1/2	650	1600	...
1968	BBD-4423S-Aut. Tr.	5/16	...	...	...	29/32	1/4	2-RICH	1 1/2	600	1600	...
1968	AVS-4426S-Std. 383	5/16	...	1/2	...	7/16	1/4	INDEX	1-2	650	1600	...
1968	AVS-4429S-Auto. 400	5/16	...	1/2	...	7/16	1/4	INDEX	1-2	650	1400	...
1969	AVS-4616S-383-Aut	5/16	...	1/2	...	7/16	1/4	INDEX	1-2	650	1700	...
1969	AVS-4711S-383 Std.	5/16	...	1/2	...	7/16	1/4	INDEX	1-2	700	1700	...
1969	AVS-4638S-Aut. AC.	5/16	...	1/2	...	7/16	1/4	INDEX	1-2	650	1700	...
1969	AVS-4617S-4618S-440	7/32	...	1/2	...	7/16	1/4	INDEX	1-2	700	1700	...
1969	AVS-4640S-440-Aut. AC	7/32	...	1/2	...	7/16	1/4	INDEX	1-2	650	1700	...
1969	BBD-4613S-383 Std.	5/16	...	...	...	1	1/4	2-RICH	1 1/2	700	1600	...
1969	BBD-4614S-383-Aut.	5/16	...	...	...	1	1/4	2-RICH	1 1/2	600	1600	...
1970	BBD-4725S-383, Std. w. CAS	5/16	...	...	...	1	1/4	2-RICH	...	750	1700	...
1970	BBD-4726S-383, Auto. w. CAS, w/o AC	5/16	...	...	...	1	1/4	2-RICH	...	650	1700	...
1970	BBD-4894S-383, Auto. w. CAS, w. AC	5/16	...	...	...	1	1/4	2-RICH	...	650	1700	...
1970	BBD-4727S-383, Std. w. ECS	5/16	...	...	...	1	1/4	2-RICH	...	750	1700	...
1970	BBD-4728S-383, Auto. w. ECS	5/16	...	...	...	1	1/4	2-RICH	...	650	1700	...
1970	AVS-4736S-383, Auto. w/o AC, w. CAS	5/16	...	1/2	...	7/16	1/4	2-RICH	...	700	1700	...
1970	AVS-4732S-383, Auto. w. AC, w. CAS	5/16	...	1/2	...	7/16	1/4	2-RICH	...	700	1700	...
1970	AVS-4734S-383, Auto. w. ECS	5/16	...	1/2	...	7/16	1/4	2-RICH	...	700	1700	...
1970	AVS-4737S-440, Std. w. CAS	7/32	...	1/2	...	7/16	1/4	2-RICH	...	900	2000	...
1970	AVS-4738S-440, Auto. w. CAS, w/o AC	7/32	...	1/2	...	7/16	1/4	2-RICH	...	800	1800	...
1970	AVS-4741S-440, Auto w. AC, w. CAS	7/32	...	1/2	...	7/16	1/4	2-RICH	...	800	1800	...
1970	AVS-4739S-440, Std. w. ECS	7/32	...	1/2	...	7/16	1/4	2-RICH	...	900	2000	...
1970	AVS-4740S-440, Auto. w. ECS	7/32	...	1/2	...	7/16	1/4	2-RICH	...	800	1800	...

① —Set pump arm parallel.
(‡)—Man. trans. w/cap. 650, auto.-600.
(†)—Man. trans. w/cap. 1700, auto.-1400.

CAS—Cleaner Air System
ECS—Evaporation Control System
CAP—Cleaner Air Package

Corvair

YEAR	MODEL OR TYPE	Prim.	Sec.	Prim.	Sec.		Unloader	Housing	Idle Screw Turns Open	Idle Speed	Fast Idle Speed	Dashpot
1963	YH—ALL	5/8	...	2 1/16	...	...	7/16	1-LEAN	...	850	①	...
1964	YH-3311S	5/8	...	2 1/16	...	...	7/16	1-LEAN	3/4	850	①	...
1965-66	YH-4141SA	5/8	...	2 1/16	...	...	7/16	1-LEAN	3/4	850	①	...

① —.030 in. clearance at choke valve—set on highest step.

YEAR	MODEL OR TYPE	FLOAT LEVEL (IN.)		FLOAT DROP (IN.)		Pump Travel Setting (IN.)	CHOKE SETTING		ON THE CAR ADJUSTMENTS			
		Prim.	Sec.	Prim.	Sec.		Unloader (IN.)	Housing	Idle Screw Turns Open	Idle Speed (RPM)	Fast Idle Speed (RPM)	Dashpot Plunger Clearance (IN.)
Dodge												
1964	BBS-3675S, 76S, 81S	1/4	...	...	...	.060	3/16	2-RICH	1	500	700	...
1964	BBS-3677S, 78S	1/4	...	...	...	.060	3/16	2-RICH	1	500	700	...
1964	BBS-3679S, 80S	1/4	...	...	...	.060	3/16	2-RICH	1	500	700	...
1964	BBD-3682S, 83S	1/4	...	...	...	1/16	1/4	INDEX	1	500	700	...
1964	BBD-3684S	5/16	...	...	...	Ⓐ	1/4	2-RICH	3/4	500	700	...
1964	AFB-3611S	7/32	7/32	3/4	3/4	7/16	3/8	INDEX	1 1/2	500	700	...
1965	BBD-3849S	5/16	...	...	...	Ⓐ	1/4	2-RICH	3/4	500	600	...
1965	BBD-3850S	5/16	...	...	...	Ⓐ	1/4	2-RICH	3/4	500	700	...
1965	BBS-3833S, 34S, 35S, 36S	1/4	...	...	...	1/16	3/16	2-RICH	1	550	700	...
1965	BBS-3837S, 38S, 39S, 40S	1/4	...	...	...	1/16	3/16	2-RICH	1	550	700	...
1965	AFB-3853S, 54S	7/32	7/32	3/4	3/4	1/16	1/32	2-RICH	1-2	650	700	...
1965	AFB-3855S, 56S	7/32	7/32	3/4	3/4	1/16	3/8	2-RICH	1-2	500	700	...
1965	AFB-3859S, 60S-HP	7/32	7/32	3/4	3/4	7/16	3/8	INDEX	1-2	550	700	...
1966	BBD-4125S, 4127S	5/16	...	...	...	Ⓐ	1/4	2-RICH	1	500	700	...
1966	BBD-4113S, 4114S	1/4	...	...	...	.060	1/4	2-RICH	1	500	700	...
1966	BBS-4099S	1/4	...	...	...	.060	3/16	2-RICH	1-2	550	700	...
1966	AFB-4119S	7/32	7/32	3/4	3/4	7/16	7/32	2-RICH	1-2	600	625	...
1966	AFB-4130S, 4131S	7/32	7/32	3/4	3/4	1/16	3/8	2-RICH	1-2	500	700	...
1966	BBD-4113S, 4114S	1/4	...	...	...	.060	1/4	2-RICH	1	500	700	...
1966	BBD-4115S	1/4	...	...	...	...	1/4	INDEX	2	700	1400	...
1966	BBD-4116S	1/4	...	...	...	...	1/4	INDEX	2	650	1500	...
1966	BBD-4125S, 4127S	5/16	...	...	...	Ⓐ	1/4	2-RICH	1	500	700	...
1966	BBD-4126S, 4128S	5/16	...	...	...	Ⓐ	1/4	2-RICH	1	600	1500	...
1966	BBS-4099S, 4100S, 4103S, 04S	1/4	...	...	...	.060	3/16	2-RICH	1-2	550	700	...
1966	BBS-4101S, 02S, 4105S, 06S	1/4	...	...	...	.060	3/16	2-RICH	1-2	650	1500	...
1966	AFB-4130S, 4131S	7/32	7/32	3/4	3/4	7/16	3/8	2-RICH	1 1/2	500	700	...
1966	AFB-4132S, 33S, 4136S, 37S	7/32	7/32	3/4	3/4	7/16	3/8	INDEX	1 1/2	600	1500	...
1966	AFB-4119S	7/32	7/32	3/4	3/4	7/16	7/32	2-RICH	1 1/2	600	625	...
1966	AFB-4120S	7/32	7/32	3/4	3/4	7/16	7/32	2-RICH	1 1/2	600	700	...
1966	AFB-4121S, 4122S	7/32	7/32	3/4	3/4	7/16	7/32	INDEX	1 1/2	650	1600	...
1967	BBS-4286S, 87S	1/4	...	...	...	.060	3/16	2-RICH	1-2	550	700	...
1967	BBS-4302S, cap.	1/4	...	...	...	.060	3/16	2-RICH	1-2	700①	1550①	...
1967	BBD-4113SA, 4114SA	1/4	...	...	...	.060	1/4	2-RICH	1	500	700	...
1967	BBD-4115SA, 16SA, cap.	1/4	...	...	...	...	1/4	INDEX	2	700②	1400②	...
1967	BBD-4296S, 97S	5/16	...	...	...	29/32	1/4	2-RICH	1 1/2	550	700	...
1967	BBD-4306S, 07S, cap.	5/16	...	...	...	Ⓐ	1/4	2-RICH	1 1/2	650③	1700③	...
1967	AFB-4298S, 99S	7/32	7/32	3/4	3/4	7/16	1/4	2-RICH	1-2	500	700	...
1967	AFB-4309S, 10S, cap.	5/16	5/16	3/4	3/4	7/16	5/16	INDEX	1-2	650	1400	...
1967	AFB-4326S, 27S	7/32	7/32	3/4	3/4	7/16	3/8	INDEX	1-2	650	750	...
1967	AFB-4311S, 12S	5/16	5/16	3/4	3/4	7/16	3/16	INDEX	1-2	650	1400	...
1968	BBS-4414S-Std. Tr.	1/4	...	...	...	.060	3/16	2-RICH	1-2	N-700	1550	...
1968	BBS-4415S-Aut. Tr.	1/4	...	...	...	.060	3/16	2-RICH	1-2	N-650	1700	...
1968	BBD-4416S-Std. Tr.	1/4	...	...	...	...	1/4	2-RICH	2	700	1400	...
1968	BBD-4417S-Aut. Tr.	1/4	...	...	...	...	1/4	2-RICH	2	650	1700	...
1968	BBD-4420S-Std. Tr.	1/4	...	...	...	...	1/4	2-RICH	1	650③	1300③	...
1968	BBD-4422S, 23S	5/16	...	...	...	29/32	1/4	2-RICH	1 1/2	650	1600	...
1968	AVS-4426S, 4401S	5/16	...	1/2	...	7/16	1/4	INDEX	1-2	650	1600	...
1968	AVS-4429S	5/16	...	1/2	...	7/16	1/4	INDEX	1-2	650	1400	...
1969	AVS-4615S-383-Std.	5/16	...	1/2	...	7/16	1/4	INDEX	1-2	700	1700	...
1969	AVS-4616S-4638S-383-A.T.	5/16	...	1/2	...	7/16	1/4	INDEX	1-2	650	1700	...
1969	AVS-4711S-383-Std.	5/16	...	1/2	...	7/16	1/4	INDEX	1-2	700	1700	...
1969	AVS-4611S-340-Std.	7/32	...	1/2	...	7/16	1/4	INDEX	1-2	750	1700	...
1969	AVS-4612S, 4639S, 340-Aut.	7/32	...	1/2	...	7/16	1/4	INDEX	1-2	700	1700	...
1969	AVS-4617S-440-Std.	7/32	...	1/2	...	7/16	1/4	INDEX	1-2	700	1700	...
1969	AVS-4618S, 4640S-440-A.T.	7/32	...	1/2	...	7/16	1/4	INDEX	1-2	650	1700	...
1969	BBS-4601S-170-M.T.	1/4	...	...	...	...	3/16	2-RICH	1-2	750	1600	...
1969	BBS-4602S-170-Aut.	1/4	...	...	...	...	3/16	2-RICH	1-2	750	1800	...
1969	BBD-4605S-273-Std.	1/4	...	...	...	...	1/4	INDEX	2	700	1500	...
1969	BBD-4606S-273-A.T.	1/4	...	...	...	...	1/4	INDEX	2	650	1600	...
1969	BBD-4607S-318-Std.	1/4	...	...	...	...	1/4	INDEX	1	700	1300	...
1969	BBD-4608S-318-A.T.	1/4	...	...	...	...	1/4	INDEX	1	650	1700	...
1970	BBS-4715S-198, Std. w. CAS.	1/4	...	...	...	...	3/16	2-RICH	...	750④	1800	...
1970	BBS-4716S-198, Auto. w. CAS	1/4	...	...	...	...	3/16	2-RICH	...	750④	1800	...
1970	BBS-4717S-198, man. w. ECS	1/4	...	...	...	...	3/16	2-RICH	...	740④	1800	...
1970	BBS-4718S-198, Auto. w. ECS	1/4	...	...	...	...	3/16	2-RICH	...	750④	1800	...
1970	BBD-4721S-318, Std. w. CAS	1/4	...	...	...	...	1/4	INDEX	...	750	1600	...

YEAR	MODEL OR TYPE	FLOAT LEVEL (IN.)		FLOAT DROP (IN.)		Pump Travel Setting (IN.)	CHOKE SETTING		ON THE CAR ADJUSTMENTS			
		Prim.	Sec.	Prim.	Sec.		Unloader (IN.)	Housing	Idle Screw Turns Open	Idle Speed (RPM)	Fast Idle Speed (RPM)	Dashpot Plunger Clearance (IN.)
Dodge												
1970	BBD-4722S-318, Auto. w/o AC, w. CAS	1/4	...	...	...	...	1/4	INDEX	...	700	2000	...
1970	BBD-4895S-318, Auto. w. AC, w. CAS	1/4	...	...	...	...	1/4	INDEX	...	700	2000	...
1970	BBD-4723S-318, Std. w. ECS	1/4	...	...	...	...	1/4	INDEX	...	750	1600	...
1970	BBD-4724S-318, Auto. w. ECS	1/4	...	...	...	...	1/4	INDEX	...	700	2000	...
1970	BBD-4725S-383, Std. w. CAS	5/16	...	...	...	1	1/4	2-RICH	...	750	1700	...
1970	BBD-4726S-383, Auto. w. CAS, w/o AC	5/16	...	...	...	1	1/4	2-RICH	...	650	1700	...
1970	BBD-4894S-383, Auto. w. AC, w. CAS	5/16	...	...	...	1	1/4	2-RICH	...	650	1700	...
1970	BBD-4725S-383, Std. w. ECS	5/16	...	...	...	1	1/4	2-RICH	...	750	1700	...
1970	BBD-4728S-383, Auto. w. ECS	5/16	...	...	...	1	1/4	2-RICH	...	650	1700	...
1970	AVS-4933S-340, Std. w. CAS	7/32	...	1/2	...	7/16	1/4	INDEX	...	950	2000	...
1970	AVS-4934S-340, Auto. w/o AC, w. CAS	7/32	...	1/2	...	7/16	1/4	INDEX	...	900	2000	...
1970	AVS-4935S-340, Auto. w. AC, w. CAS	7/32	...	1/2	...	7/16	1/4	INDEX	...	900	2000	...
1970	AVS-4936S-340, Std. w. ECS	7/32	...	1/2	...	7/16	1/4	INDEX	...	950	2000	...
1970	AVS-4937S-340, Auto. w. ECS	7/32	...	1/2	...	7/16	1/4	INDEX	...	900	2000	...
1970	AVS-4736S-383, Auto. w/o AC, w. CAS	5/16	...	1/2	...	7/16	1/4	2-RICH	...	700	1700	...
1970	AVS-4732S-383, Auto. w. AC, w. CAS	5/16	...	1/2	...	7/16	1/4	2-RICH	...	700	1700	...
1970	AVS-4734S-383, Auto. w. ECS	5/16	...	1/2	...	7/16	1/4	2-RICH	...	700	1700	...
1970	AVS-4737S-440, Std. w. CAS	7/32	...	1/2	...	7/16	1/4	2-RICH	...	900	2000	...
1970	AVS-4738S-440, Auto. w/o AC, w. CAS	7/32	...	1/2	...	7/16	1/4	INDEX	...	800	1800	...
1970	AVS-4741S-440, Auto. w. AC, w. CAS	7/32	...	1/2	...	7/16	1/4	INDEX	...	800	1800	...
1970	AVS-4739S-440, Std. w. ECS	7/32	...	1/2	...	7/16	1/4	INDEX	...	900	2000	...
1970	AVS-4740S-440, Auto. w. ECS	7/32	...	1/2	...	7/16	1/4	INDEX	...	800	1800	...
1970	AFB-4742S-426, Hemi Std. or Auto.	7/32	...	3/4	...	7/16	...	...	...	900	...	...
1970	AFB-4745S-426, Hemi Std.	7/32	...	3/4	...	7/16	1/4	2-RICH	1-2	900	2000	...
1970	AFB-4746S-426, Hemi Auto.	7/32	...	3/4	...	7/16	1/4	2-RICH	1-2	900	2000	...

① —Auto. trans. 650, 1900.　　③ —Auto. trans. 600, 1400.　　④ —Set pump arm parallel.　　CAS—Cleaner Air System

② —Auto. trans. 650, 1500.　　④ —With headlights and AC on.　　CAP—Cleaner Air Package　　ECS—Evaporation Control System

YEAR	MODEL OR TYPE	Prim.	Sec.	Prim.	Sec.	Pump Travel	Unloader	Housing	Idle Screw	Idle Speed	Fast Idle	Dashpot
Jeep												
	YF-2467S	9/32	...	1 1/4	...	...	NONE	NONE	1-2 1/2	...	550	...
	WCD-2204S	3/16	...	NONE	...	11/32	1/8	INDEX	1-2	...	...	...
	YF-951S-A	5/16	...	NONE	...	NONE	NONE	NONE	3/4-1 3/4	...	...	...
—	596S, 636 SA, 4 Cyl.	3/8	...	...	...	17/64	...	...	...	...	...	...
—	YF-938SD, 4 Cyl.	5/16	...	...	...	...	...	...	3/4-1 3/4	...	...	...
—	YF-2392S, 4 Cyl.	9/32	...	...	...	...	...	...	1/2-2	...	...	...
—	YF-4002S, F-4 Cyl. w/o Emission control	17/64	...	...	...	...	...	...	...	650–700	...	...
—	YF-4366S, F-4 Cyl. w. Emission control	17/64	...	...	...	...	...	...	...	650–700	...	...
—	RBS-4016S-232(6)	21/32	...	...	...	7/16	7/64	RICH	...	550	...	...
—	RBS-4163S-232(6)	21/32	...	...	...	7/16	...	...	...	550	...	...
—	RBS-232(6) w. Emission control	21/32	...	...	...	...	...	1-RICH	...	650–700	...	1/8

YEAR	MODEL OR TYPE	Prim.	Sec.	Prim.	Sec.	Pump Travel	Unloader	Housing	Idle Screw	Idle Speed	Fast Idle	Dashpot
Lincoln-Continental												
1964	AFB	3/16	3/16	23/32	23/32	17/32	1/8	1-RICH	1 1/2	D-465	700	...
1965	AFB	3/16	3/16	23/32	23/32	17/32	1/8	1-RICH	1 1/2	D-465	700	...
1966	AFB	3/16	3/16	23/32	23/32	17/32	1/8	1-RICH	1 1/2	D-465	700	...
1967-68	C7VF-A w. AC, Auto:, w. Emission control	3/16	...	23/32	...	①	.096	1-RICH	...	450	...	.437 ②
1967-68	C7VF-B w/o AC, Auto., w. Emission control	3/16	...	23/32	...	①	.096	1-RICH	...	450	...	.437 ②
1967-68	C7VF-C w. AC, Auto., w/o Emission control	3/16	...	23/32	...	①	.096	1-RICH	...	450	...	.437 ②
1967-68	C7VF-D w/o AC, Auto., w/o Emission control	3/16	...	23/32	...	①	.096	1-RICH	...	450	...	.437 ②

① —Accelerator pump position—inner hole.　　② —At wide open throttle.

YEAR	MODEL OR TYPE	Prim.	Sec.	Prim.	Sec.	Pump Travel	Unloader	Housing	Idle Screw	Idle Speed	Fast Idle	Dashpot
Plymouth												
1964	BBS-3679S, 3680S	1/4	...	...	...	.060	3/16	2-RICH	1	550	700	...
1964	BBD-3682S, 3683S	1/4	...	...	...	...	1/4	INDEX	1	500	700	...
1964	BBD-3684S	1/4	...	...	...	①	1/4	2-RICH	3/4	500	700	...
1964	AFB-3611S-383 cu. in.	7/32	7/32	3/4	3/4	7/16	5/8	INDEX	1 1/2	500	700	...
1964	AFB-3705SA-426 cu. in.	7/32	7/32	3/4	3/4	17/32	...	...	1 1/2	900	...	...

Plymouth, continued

YEAR	MODEL OR TYPE	FLOAT LEVEL (IN.)		FLOAT DROP (IN.)		Pump Travel Setting (IN.)	CHOKE SETTING		ON THE CAR ADJUSTMENTS			
		Prim.	Sec.	Prim.	Sec.		Unloader (IN.)	Housing	Idle Screw Turns Open	Idle Speed (RPM)	Fast Idle Speed (RPM)	Dashpot Plunger Clearance (IN.)
1965	BBS-3839S, 40S	1/4	...	...	...	...	3/16	2-RICH	1	550	700	...
1965	BBS-3837S, 38S	1/4	...	...	...	...	3/16	2-RICH	1	550	700	...
1965	BBS-3833S, 34S	1/4	...	...	...	...	3/16	2-RICH	1	550	700	...
1965	BBD-3847S, 48S	1/4	...	...	...	...	1/4	INDEX	1	500	700	...
1965	AFB-3853S, 54S	7/32	7/32	3/4	3/4	7/16	7/32	INDEX	1-2	600	625	...
1965	AFB-3855S, 56S	7/32	7/32	3/4	3/4	7/16	3/8	2-RICH	1-2	500	700	...
1965	AFB-3859S, 60S	7/32	7/32	3/4	3/4	7/16	3/8	INDEX	1-2	550	700	...
1966	BBD-4113S, 4114S	1/4	...	...	...	.060Ⓐ	1/4	2-RICH	2	500	700	...
1966	BBD-4115S, 4116S	1/4	...	...	...	...	1/4	INDEX	2	700	1400	...
1966	BBD-4125S, 4127S	5/16	...	...	...	①	1/4	2-RICH	1	500	700	...
1966	BBD-4126S	5/16	...	...	...	①	1/4	2-RICH	1	650	1600	...
1966	BBD-4128S	5/16	...	...	...	①	1/4	2-RICH	1	600	1400	...
1966	AFB-4130S, 4131S	7/32	7/32	3/4	3/4	7/16	3/8	INDEX	1½	500	700	...
1966	AFB-4132S, 33S, 4136S, 37S	7/32	7/32	3/4	3/4	7/16	3/8	INDEX	1½	650	1500	...
1966	AFB-4119S, 4120S	7/32	7/32	3/4	3/4	7/16	3/8	2-RICH	1½	...	700	...
1966	AFB-4121S, 4122S	7/32	7/32	3/4	3/4	7/16	7/32	INDEX	1½	700	1500	...
1967	BBD-4113SA, 14SA	1/4	...	...	...	.060Ⓐ	1/4	2-RICH	1	500	700	...
1967	BBD-4115SA, M.T. cap	1/4	...	...	...	...	1/4	INDEX	2	700	1400	...
1967	BBD-4296S, 97S	5/16	...	...	...	1/16	1/4	2-RICH	1½	550	700	...
1967	BBD-4306S, M.T. cap	5/16	...	...	...	.050Ⓐ	1/4	2-RICH	1	650	1700	...
1967	AFB-4294S, 95	7/32	7/32	3/4	3/4	7/16	7/32	2-RICH	1-2	600	650	...
1967	AFB-4304S, 05S	5/16	5/16	3/4	3/4	7/16	7/32	INDEX	1-2	650	1550	...
1967	AFB-4298S, 99S	7/32	7/32	3/4	3/4	7/16	3/8	2-RICH	1-2	500	700	...
1967	AFB-4309S, 10S, cap.	5/16	5/16	3/4	3/4	7/16	5/16	INDEX	1-2	650	1400	...
1967	AFB-4326S, 27S	7/32	7/32	3/4	3/4	7/16	5/16	INDEX	1-2	650	750	...
1967	AFB-4311S, 12S, cap.	5/16	5/16	3/4	3/4	7/16	7/16	INDEX	1-2	650	1400	...
1967	AFB-4328S, 29S, cap.	5/16	5/16	3/4	3/4	7/16	3/8	INDEX	1-2	650	1500	...
1967	AFB-4139S, 4324S, Front	5/16	5/16	3/4	3/4	7/16	...	...	1-2	750	...	...
1967	AFB-4343S, 4325S, Rear	7/32	7/32	3/4	3/4	7/16	1/4	1-RICH	1-2	750	1500	...
1968	BBS-4414S-Std. Tr.	1/4	...	...	...	.060Ⓐ	3/16	2-RICH	1-2	N-700	1550	...
1968	BBS-4415S-Aut. Tr.	1/4	...	...	...	.060Ⓐ	3/16	2-RICH	1-2	N-650	1700	...
1968	BBD-4416S-Std. Tr.	1/4	...	...	...	...	1/4	2-RICH	2	700	1400	...
1968	BBD-4417S-Aut. Tr.	1/4	...	...	...	...	1/4	2-RICH	2	650	1700	...
1968	BBD-4420S-Std. Tr.	1/4	...	...	...	...	1/4	2-RICH	1	650	1300	...
1968	BBD-4422S, 23S	5/16	...	...	...	29/32	1/4	2-RICH	1½	650	1600	...
1968	AVS-4426S, 4401S	5/16	...	1/2	...	7/16	1/4	INDEX	1-2	650	1600	...
1968	AVS-4429S	5/16	...	1/2	...	7/16	1/4	INDEX	1-2	650	1400	...
1969	BBD-4605S, 273, Std.	1/4	...	...	...	...	1/4	INDEX	2	700	1500	...
1969	BBD-4606S, 273, Aut.	1/4	...	...	...	...	1/4	INDEX	2	650	1600	...
1969	BBD-4607S, 318, Std.	1/4	...	...	...	...	1/4	INDEX	1	700	1300	...
1969	BBD-4608S, 318, Aut.	1/4	...	...	...	...	1/4	INDEX	1	650	1700	...
1969	BBD-4613S, 383, Std.	5/16	...	...	...	1	1/4	2-RICH	1½	700	1600	...
1969	BBD-4614S, 383, Aut.	5/16	...	...	...	1	1/4	2-RICH	1½	600	1600	...
1969	AVS-4611, 340, Std.	7/32	...	1/2	...	7/16	1/4	INDEX	1-2	750	1700	...
1969	AVS-4612, 340, Aut.	7/32	...	1/2	...	7/16	1/4	INDEX	1-2	700	1700	...
1969	AVS-4615, 383, Std.	5/16	...	1/2	...	7/16	1/4	INDEX	1-2	700	1700	...
1969	AVS-4616, 383, Aut.	5/16	...	1/2	...	7/16	1/4	INDEX	1-2	650	1700	...
1969	AVS-4617, 440, Std.	7/32	...	1/2	...	7/16	1/4	INDEX	1-2	700	1700	...
1969	AVS-4618, 440, Aut.	7/32	...	1/2	...	7/16	1/4	INDEX	1-2	650	1700	...
1969	AVS-4638, 340, Aut.	5/16	...	1/2	...	7/16	1/4	INDEX	1-2	650	1700	...
1969	AVS-4639, 340, Aut.	7/32	...	1/2	...	7/16	1/4	INDEX	1-2	700	1700	...
1969	AVS-4640, 440, Aut.	7/32	...	1/2	...	7/16	1/4	INDEX	1-2	650	1700	...
1969	AVS-4682, 383, Aut.	5/16	...	1/2	...	7/16	1/4	INDEX	...	650	1700	...
1969	AVS-4711, 383, Std.	5/16	...	1/2	...	7/16	1/4	INDEX	...	700	1700	...
1969	AFB-4619S, 426, Front	7/32	...	3/4	...	7/16	...	...	...	750	...	...
1969	AFB-4620S, 426, Rear	7/32	...	3/4	...	7/16	1/4	2-RICH	1-2	750	2000	...
1969	AFB-4621S, 426, Rear	7/32	...	3/4	...	7/16	1/4	2-RICH	1-2	750	2000	...
1970	BBS-4715S-198, Std. w. CAS.	1/4	...	...	...	...	3/16	2-RICH	...	750②	1800	...
1970	BBS-4716S-198, Auto. w. CAS	1/4	...	...	...	...	3/16	2-RICH	...	750②	1800	...
1970	BBS-4717S-198, Man. w. ECS	1/4	...	...	...	...	3/16	2-RICH	...	750②	1800	...
1970	BBS-4718S-198, Auto. w. ECS	1/4	...	...	...	...	3/16	2-RICH	...	750②	1800	...
1970	BBD-4721S-318, Std. w. CAS	1/4	...	...	...	...	1/4	INDEX	...	750	1600	...
970	BBD-4722S-318, Auto. w/o AC, w. CAS.	1/4	...	...	...	...	1/4	INDEX	...	700	2000	...
970	BBD-4895S-318, Auto. w. AC, w. CAS	1/4	...	...	...	...	1/4	INDEX	...	700	2000	...
970	BBD-4723S-318, Std. w. ECS.	1/4	...	...	...	...	1/4	INDEX	...	750	1600	...
1970	BBD-4724S-318, Auto. w. ECS	1/4	...	...	...	...	1/4	INDEX	...	700	2000	...

Plymouth, continued

YEAR	MODEL OR TYPE	FLOAT LEVEL (IN.)		FLOAT DROP (IN.)		Pump Travel Setting (IN.)	CHOKE SETTING		ON THE CAR ADJUSTMENTS			
		Prim.	Sec.	Prim.	Sec.		Unloader (IN.)	Housing	Idle Screw Turns Open	Idle Speed (RPM)	Fast Idle Speed (RPM)	Dashpot Plunger Clearance (IN.)
1970	BBD-4725S-383, Std. w. CAS	5/16	...	...	...	1	1/4	2-RICH	...	750	1700	...
1970	BBD-4726-383, Auto. w. CAS, w/o AC	5/16	...	...	...	1	1/4	2-RICH	...	650	1700	...
1970	BBD-4494S-383, Auto. w. AC, w. CAS	5/16	...	...	...	1	1/4	2-RICH	...	650	1700	...
1970	BBD-4727S-383, Std. w. ECS	5/16	...	...	...	1	1/4	2-RICH	...	750	1700	...
1970	BBD-4728S-383, Auto. w. ECS	5/16	...	...	...	1	1/4	2-RICH	...	650	1700	...
1970	AVS-4933S-340, Std. w. CAS	7/32	...	1/2	...	7/16	1/4	INDEX	...	950	2000	...
1970	AVS-4934S-340, Auto. w/o AC, w. CAS	7/32	...	1/2	...	7/16	1/4	INDEX	...	900	2000	...
1970	AVS-4935S-340, Auto. w. AC, w. CAS	7/32	...	1/2	...	7/16	1/4	INDEX	...	900	2000	...
1970	AVS-4936S-340, Std. w. ECS	7/32	...	1/2	...	7/16	1/4	INDEX	...	950	2000	...
1970	AVS-4937S-340, Auto. w. ECS	7/32	...	1/2	...	7/16	1/4	INDEX	...	900	2000	...
1970	AVS-4736S-383, Auto. w/o AC, w. CAS	5/16	...	1/2	...	7/16	1/4	2-RICH	...	700	1700	...
1970	AVS-4732S-383, Auto. w. AC, w. CAS	5/16	...	1/2	...	7/16	1/4	2-RICH	...	700	1700	...
1970	AVS-4734S-383, Auto. w. ECS	5/16	...	1/2	...	7/16	1/4	2-RICH	...	700	1700	...
1970	AVS-4737S-440, Std. w. CAS	7/32	...	1/2	...	7/16	1/4	2-RICH	...	900	2000	...
1970	AVS-4738S-440, Auto. w/o AC, w. CAS	7/32	...	1/2	...	7/16	1/4	INDEX	...	800	1800	...
1970	AVS-4741S-440, Auto. w. AC, w. CAS	7/32	...	1/2	...	7/16	1/4	INDEX	...	800	1800	...
1970	AVS-4739S-440 Std. w. ECS	7/32	...	1/2	...	7/16	1/4	INDEX	...	900	2000	...
1970	AVS-4740S-440, Auto. w. ECS	7/32	...	1/2	...	7/16	1/4	INDEX	...	800	1800	...
1970	AFB-4742S-426, Hemi Std. or Auto.	7/32	...	3/4	...	7/16	...	...	...	900	...	...
1970	AFB-4745S-426, Hemi Std.	7/32	...	3/4	...	7/16	1/4	2-RICH	1-2	900	2000	...
1970	AFB-4746S-426, Hemi Auto.	7/32	...	3/4	...	7/16	1/4	2-RICH	1-2	900	2000	...

① —Set pump arm parallel.
② —With headlights and AC on.
④ —Pump travel is relative to bowl vent adjustment. To adjust, back off the idle speed screw. Place pump stroke rod in medium stroke hole in throttle lever, be sure the bowl vent clip on the pump stem is in the center of the three notches. With throttle valves tightly closed insert a drill of the proper size between the bowl vent valve and air horn. Adjustment is made by bending the pump operating rod.

CAP—Cleaner Air Package.
CAS—Cleaner Air System.
ECS—Evaporation Control System.

Pontiac

YEAR	MODEL OR TYPE	FLOAT LEVEL (IN.)		FLOAT DROP (IN.)		Pump Travel Setting (IN.)	CHOKE SETTING		ON THE CAR ADJUSTMENTS			
		Prim.	Sec.	Prim.	Sec.		Unloader (IN.)	Housing	Idle Screw Turns Open	Idle Speed (RPM)	Fast Idle Speed (RPM)	Dashpot Plunger Clearance (IN.)
1964	AFB—All	21/64	21/64	23/32	23/32	.300	.150	1-RICH	1 1/2	N-550	2500	...
1965	AFB-3896S	5/16	5/16	23/32	3/16	...	5/32	1-RICH	1 1/2	N-550	2500	...
1965	AFB-3895S, 3898S	3/8	3/8	23/32	5/16	...	5/32	1-RICH	1 1/2	N-550	2500	...
1966	AFB-4030S	1/4	1/4	23/32	23/32	5/16	5/32	1-RICH	1 1/2	...	2500	...
1966	AFB-4033S, 4041S	5/16	5/16	23/32	23/32	5/16	5/32	1-RICH	1 1/2	...	2500	...
1966	AFB-4034S	1/4	1/4	23/32	23/32	5/16	5/32	1-RICH	1 1/2	...	2800	...
1966	AFB-4035S	3/8	3/8	23/32	23/32	5/16	5/32	1-RICH	1 1/2	...	2500	...
1967	AFB-4242S	5/16	5/16	23/32	23/32	3/8	5/32	1-RICH	...	500	2800	...
1967	AFB-4243S	3/8	3/8	23/32	23/32	3/8	5/32	1-RICH	...	600	2500	...
1967	AFB-4244S	1/4	1/4	23/32	23/32	3/8	5/32	1-RICH	...	600	2500	...
1967	AFB-4245S	5/16	5/16	23/32	23/32	3/8	5/32	1-RICH	...	...	...	...

Rambler—American Motors

YEAR	MODEL OR TYPE	FLOAT LEVEL (IN.)		FLOAT DROP (IN.)		Pump Travel Setting (IN.)	CHOKE SETTING		ON THE CAR ADJUSTMENTS			
		Prim.	Sec.	Prim.	Sec.		Unloader (IN.)	Housing	Idle Screw Turns Open	Idle Speed (RPM)	Fast Idle Speed (RPM)	Dashpot Plunger Clearance (IN.)
1964	RBS-L, Hd.	15/32	...	...	...	.0292	7/64	INDEX	1/4-1/4	N-500	1800	1/8
1964	RBS-OHV	15/32	...	...	...	.0292	7/64	INDEX	1/4-1/4	N-500	1800	1/8
1964	WCD-OHV	1/8	...	5/8	...	.033	3/16	INDEX	1/2-2	N-500	1800	1/8
1965	RBS-3708S, 3766S	15/32	...	...	...	.0292	7/64	INDEX	1/4-1 1/4	N-500	1800	1/8
1965	WCD-3706S	1/8	...	5/8	...	.033	3/16	INDEX	1/4-1 1/4	N-500	1800	1/8
1966-67	RBS-3882S	15/32	...	...	...	.029	1/8	1-RICH	1/4-1 1/4	N-600	2000	1/8
1966	WCD-4191S	1/8	...	5/8	...	.033	3/16	INDEX	1/4-1 1/4	N-500	1800	1/8
1967	WCD-3888S	1/4	...	...	...	.025	3/16	INDEX	1/4-1 1/4	N-600	1900	...
1967	WCD-4365S, Aut.	7/32	...	...	...	.022	5/16	2-RICH	1/4-1 1/4	N-600	1350	...
1968	RBS	15/32	...	...	...	.029	1/8	1-RICH	1/4-1 1/4	N-1 1/4	...	...
1968	WCD	1/4	...	...	...	.025	3/16	INDEX	1/4-1 1/4	N-600		...
1968	WCD, Aut. Tr.	7/32	...	...	...	.022	5/16	2-RICH	1/4-1 1/4	N-500		...
1968	WCD-4410S	7/32	...	...	...	①	3/16	INDEX	1	N-600	2000	...
1968	WCD-4537S	7/32	...	...	...	①	3/16	INDEX	1	N-525	2000	...
1968	AFB-4467S	5/16	...	2	...	15/64	5/32	2-RICH	1	650	2000	...
1968	AFB-4468S	5/16	...	2	...	17/64	5/32	INDEX	1	550	2000	...
1968	AFB-4469S	5/16	...	2	...	17/64	5/32	2-RICH	1/2	650	2000	...

YEAR	MODEL OR TYPE	FLOAT LEVEL (IN.) Prim.	FLOAT LEVEL (IN.) Sec.	FLOAT DROP (IN.) Prim.	FLOAT DROP (IN.) Sec.	Pump Travel Setting (IN.)	CHOKE SETTING Unloader (IN.)	CHOKE SETTING Housing	Idle Screw Turns Open	Idle Speed (RPM)	Fast Idle Speed (RPM)	Dashpot Plunger Clearance (IN.)

American Motors—Rambler, continued

YEAR	MODEL OR TYPE	FLOAT LEVEL Prim.	FLOAT LEVEL Sec.	FLOAT DROP Prim.	FLOAT DROP Sec.	Pump Travel Setting	Unloader	Housing	Idle Screw Turns Open	Idle Speed	Fast Idle Speed	Dashpot Clearance
1969	AFB-4660S	11/32	…	2	…	21/64	5/32	2-RICH	2	…	2000	…
1969	AFB-4661S	11/32	…	2	…	21/64	5/32	INDEX	2	…	2000	…
1969	AFB-4662S	11/32	…	2	…	21/64	5/32	INDEX	2	…	2000	…
1969	AFB-4663S	11/32	…	2	…	21/64	5/32	INDEX	2	…	2000	…
1969	AFB-4664S	11/32	…	2	…	21/64	5/32	INDEX	2	…	2000	…
1969	AFB-4665S	11/32	…	2	…	21/64	5/32	INDEX	2	…	2000	…
1969	WCD-4667S, 68S	7/32	…	…	…	…	3/16	INDEX	1 1/2	…	…	…
1970	YF-4768S-199, 232, Std.	29/64	…	1 1/4	…	…	21/64	INDEX	…	②	2300	7/64
1970	YF-4767S, Auto.-199, 232 ①	29/64	…	1 1/4	…	…	19/64	INDEX	…	②	2300	…
1970	YF-4978S-232, Auto. ④	29/64	…	1 1/4	…	…	19/64	1-RICH	…	②	2300	…
1970	YF-4770S-232, Std.	29/64	…	1 1/4	…	…	19/64	INDEX	…	②	2300	3/32
1970	YF-4769S-232, Auto.	29/64	…	1 1/4	…	…	19/64	INDEX	…	②	2300	…
1970	WCD-4950S-232, Std.	7/32	…	…	…	…	3/16	INDEX	…	②	2000	3/32
1970	WCD-4816S-232, Auto.	7/32	…	…	…	…	3/16	INDEX	…	②	2000	3/32
1970	WCD-4817S-232, Std.-Auto.	7/32	…	…	…	…	3/16	INDEX	…	②	2000	3/32

① —Set pump arm parallel.
② —Refer to engine compartment decal for idle speed.
③ —Prior to engine code 209E22.
④ —Effective with engine code 209E22.
N—Neutral.

Tempest

YEAR	MODEL OR TYPE	FLOAT LEVEL Prim.	FLOAT LEVEL Sec.	FLOAT DROP Prim.	FLOAT DROP Sec.	Pump Travel Setting	Unloader	Housing	Idle Screw Turns Open	Idle Speed	Fast Idle Speed	Dashpot Clearance
1964	AFB-ALL	5/16	5/16	23/32	23/32	.30	.150	1-RICH	1 1/2	N-550	2500	…
1965	AFB-3895S, 3899S	3/8	3/8	23/32	23/32	.30	.150	1-RICH	1 1/2	…	2500	…
1965	AFB-3896S, 3900S	5/16	5/16	23/32	23/32	.30	.150	1-RICH	1 1/2	N-550	2500	…
1966	AFB-4030S	1/4	1/4	23/32	23/32	.375	5/32	1-RICH	1 1/2	…	2500	…
1966	AFB-4031S, 4033S	5/16	5/16	23/32	23/32	.375	5/32	1-RICH	1 1/2	…	2500	…
1966	AFB-4034S, 4041S	5/16	5/16	23/32	23/32	.375	5/32	1-RICH	1 1/2	…	2800	…
1966	AFB-4035S	3/8	3/8	23/32	23/32	.375	5/32	1-RICH	1 1/2	…	2500	…
1966	AFB-4036S	1/4	1/4	23/32	23/32	.375	5/32	1-RICH	1 1/2	…	2800	…
1967	AFB-4242S	5/16	5/16	23/32	23/32	.375	5/32	1-RICH	1 1/2	500	2800	…
1967	AFB-4243S	3/8	3/8	23/32	23/32	.375	5/32	1-RICH	1 1/2	…	…	…

Valiant—Dart—Barracuda—Challenger

YEAR	MODEL OR TYPE	FLOAT LEVEL Prim.	FLOAT LEVEL Sec.	FLOAT DROP Prim.	FLOAT DROP Sec.	Pump Travel Setting	Unloader	Housing	Idle Screw Turns Open	Idle Speed	Fast Idle Speed	Dashpot Clearance
1964	BBS-3675S, 76S	1/4	…	…	…	.060	3/16	2-RICH	1	550	700	…
1964	BBS-3677S, 78S	1/4	…	…	…	.060	3/16	2-RICH	1	550	700	…
1964	BBD-3767S, 68S	1/4	…	…	…	1/16	1/4	INDEX	1	500	700	…
1965	BBS-3833S, 34S	1/4	…	…	…	.060	3/16	2-RICH	1	550	700	…
1965	BBS-3839S, 40S	1/4	…	…	…	.060	3/16	2-RICH	1	550	700	…
1965	BBS-3837S, 38S	1/4	…	…	…	.060	3/16	2-RICH	1	550	700	…
1965	BBD-3843S, 44S	1/4	…	…	…	1/16	3/16	2-RICH	1	550	700	…
1966	BBS-4099S	1/4	…	…	…	.060	3/16	2-RICH	1-2	550	700	…
1966	BBD-4113S, 14S	1/4	…	…	…	.060	3/16	2-RICH	1-2	550	700	…
1966	AFB-4119S, 20S	3/32	…	3/4	…	…	1/32	2-RICH	1-2	500	700	…
1966	BBS-4099S, 4100S, 4103S, 04S	1/4	…	…	…	.060	7/16	2-RICH	1-2	550	700	…
1966	BBS-4101S, 02S, 4105S, 06S	1/4	…	…	…	.060	3/16	2-RICH	1-2	700	1500	…
1966	BBD-4113S, 4114S	1/4	…	…	…	.060	1/4	2-RICH	1	500	700	…
1966	BBD-4115S, 4116S	1/4	…	…	…	…	1/4	INDEX	1-2	700	1400	…
1966	AFB-4119S, 4120S	7/32	7/32	3/4	3/4	7/16	1/32	2-RICH	1 1/2	…	625	…
1966	AFB-4121S, 4122S	7/32	7/32	3/4	3/4	7/16	1/32	INDEX	1 1/2	…	1500	…
1967	BBD-4113SA, 14SA	1/4	…	…	…	.060	1/4	2-RICH	1	500	700	…
1967	BBD-4115SA, 16SA	1/4	…	…	…	…	1/4	INDEX	2	700	1400	…
1967	BBS-4286S, 87S	1/4	…	…	…	.060	3/16	2-RICH	1-2	700	…	…
1967	BBS-4302S, 03S	1/4	…	…	…	.060	3/16	2-RICH	1-2	700	…	…
1967	AFB-4294S, 95S	7/32	7/32	3/4	3/4	7/16	1/32	2-RICH	1-2	600	650	…
1967	AFB-4304S, 05S, cap.	5/16	5/16	3/4	3/4	7/16	1/32	INDEX	1-2	650	1550	…
1968	BBS-4414S	1/4	…	…	…	.060	3/16	2-RICH	1-2	N-700	1550	…
1968	BBS-4415S	1/4	…	…	…	.060	3/16	2-RICH	1-2	N-650	1700	…
1968	BBD-4416S	1/4	…	…	…	…	1/4	2-RICH	2	700	1400	…
1968	BBD-4417S	1/4	…	…	…	…	1/4	2-RICH	2	650	1700	…
1968	BBD-4420S	1/4	…	…	…	…	1/4	2-RICH	1	650	1300	…
1968	BBD-4422S, 23S	5/16	…	1/2	…	29/32	1/4	2-RICH	1 1/2	650	1600	…
1968	AVS-4426S, 4401S	5/16	…	1/2	…	7/16	1/4	INDEX	1-2	650	1600	…
1968	AVS-4429S	5/16	…	1/2	…	7/16	1/4	INDEX	1-2	650	1400	…

Valiant—Dart—Barracuda—Challenger

YEAR	MODEL OR TYPE	FLOAT LEVEL (IN.)		FLOAT DROP (IN.)		Pump Travel Setting (IN.)	CHOKE SETTING		ON THE CAR ADJUSTMENTS			
		Prim.	Sec.	Prim.	Sec.		Unloader (IN.)	Housing	Idle Screw Turns Open	Idle Speed (RPM)	Fast Idle Speed (RPM)	Dashpot Plunger Clearance (IN.)
1969	BBS-4601S-170 Std.	1/4	...	...	...	...	3/16	2-RICH	1-2	750	1600	...
1969	BBS-4602-170 Aut.	1/4	...	...	...	...	3/16	2-RICH	1-2	750	1800	...
1969	BBD-4605S-273 Std.	1/4	...	...	...	...	1/4	INDEX	2	700	1500	...
1969	BBD-4606S, 273 Aut.	1/4	...	...	...	...	1/4	INDEX	2	650	1600	...
1969	BBD-4607S, 318 Std.	1/4	...	...	...	...	1/4	INDEX	1	700	1300	...
1969	BBD-4608S, 318 Aut.	1/4	...	...	...	...	1/4	INDEX	1	650	1700	...
1969	BBD-4613S, 383 Std.	5/16	...	...	...	1	1/4	2-RICH	1½	700	1600	...
1969	BBD-4614S, 383 Aut.	5/16	...	...	...	1	1/4	2-RICH	1½	600	1600	...
1969	AVS-4611S, 340 Std.	7/32	...	1/2	...	7/16	1/4	INDEX	1-2	750	1700	...
1969	AVS-4612S, 4639S, 340 Aut.	7/32	...	1/2	...	7/16	1/4	INDEX	1-2	700	1700	...
1969	AVS-4615S, 383 Std.	5/16	...	1/2	...	7/16	1/4	INDEX	1-2	700	1700	...
1969	AVS-4616S, 4638S, 383 Aut.	5/16	...	1/2	...	7/16	1/4	INDEX	1-2	650	1700	...
1969	AVS-4711S, 383 Std.	5/16	...	1/2	...	7/16	1/4	INDEX	...	700	1700	...
1969	AVS-4682S, 383 Aut.	5/16	...	1/2	...	7/16	1/4	INDEX	...	650	1700	...
1970	BBS-4715S-198, Std. w. CAS	1/4	...	...	...	...	3/16	2-RICH	...	750①	1800	...
1970	BBS-4716S-198, Auto. w. CAS	1/4	...	...	...	...	3/16	2-RICH	...	750①	1800	...
1970	BBS-4717S-198, Man. w. ECS	1/4	...	...	...	...	3/16	2-RICH	...	750①	1800	...
1970	BBS-4718S-198, Auto. w. ECS	1/4	...	...	...	...	3/16	2-RICH	...	750①	1800	...
1970	BBD-4721S-318, Std. w. CAS	1/4	...	...	...	...	1/4	INDEX	...	750	1600	...
1970	BBD-4722S-318, Auto. w/o AC. w. CAS	1/4	...	...	...	...	1/4	INDEX	...	700	2000	...
1970	BBD-4895S-318, Auto. w. AC, w. CAS	1/4	...	...	...	...	1/4	INDEX	...	700	2000	...
1970	BBS-4715S-198, Std. w. CAS	1/4	...	...	...	...	1/4	INDEX	...	750	1600	...
1970	BBD-4724S-318, Auto. w. ECS	1/4	...	...	...	...	1/4	INDEX	...	700	2000	...
1970	BBD-4725S-383, Std. w. CAS	5/16	...	...	...	1	1/4	2-RICH	...	750	1700	...
1970	BBD-4726S-383, Auto. w. CAS, w/o AC	5/16	...	...	...	1	1/4	2-RICH	...	650	1700	...
1970	BBD-4894S-383, Auto. w. AC, w. CAS	5/16	...	...	...	1	1/4	2-RICH	...	650	1700	...
1970	BBD-4727S-383, Std. w. ECS	5/16	...	...	...	1	1/4	2-RICH	...	750	1700	...
1970	BBD-4728S-383, Auto. w. ECS	5/16	...	...	...	1	1/4	2-RICH	...	650	1700	...
1970	AVS-4933S-340, Std. w. CAS	7/32	...	1/2	...	7/16	1/4	INDEX	...	950	2000	...
1970	AVS-4934S-340, Auto. w/o AC, w. CAS	7/32	...	1/2	...	7/16	1/4	INDEX	...	900	2000	...
1970	AVS-4935S-340, Auto. w. AC, w. CAS	7/32	...	1/2	...	7/16	1/4	INDEX	...	900	2000	...
1970	AVS-4936S-340, Std. w. ECS	7/32	...	1/2	...	7/16	1/4	INDEX	...	950	2000	...
1970	AVS-4937S-340, Auto. w. ECS	7/32	...	1/2	...	7/16	1/4	INDEX	...	900	2000	...
1970	AVS-4736S-383, Auto. w/o AC, w. CAS	5/16	...	1/2	...	7/16	1/4	2-RICH	...	700	1700	...
1970	AVS-4732S-383, Auto. w. AC, w. CAS	5/16	...	1/2	...	7/16	1/4	2-RICH	...	700	1700	...
1970	AVS-4734S-383, Auto. w. ECS	5/16	...	1/2	...	7/16	1/4	2-RICH	...	700	1700	...
1970	AVS-4737S-440, Std. w. CAS	7/32	...	1/2	...	7/16	1/4	2-RICH	...	900	2000	...
1970	AVS-4738S-440, Auto. w/o AC, w. CAS	7/32	...	1/2	...	7/16	1/4	INDEX	...	800	1800	...
1970	AVS-4741S-440, Auto. w. AC, w. CAS	7/32	...	1/2	...	7/16	1/4	INDEX	...	800	1800	...
1970	AVS-4739S-440, Std. w. ECS	7/32	...	1/2	...	7/16	1/4	INDEX	...	900	2000	...
1970	AVS-4740S-440, Auto. w. ECS	7/32	...	1/2	...	7/16	1/4	INDEX	...	800	1800	...
1970	AFB-4742S-426, Hemi Std. or Auto.	7/32	...	3/4	...	7/16	...	...	...	900	...	...
1970	AFB-4745S-426, Hemi Std.	7/32	...	3/4	...	7/16	1/4	2-RICH	1-2	900	2000	...
1970	AFB-4746S-426, Hemi Auto.	7/32	...	3/4	...	7/16	1/4	2-RICH	1-2	900	2000	...

① —With headlights and AC on.

SINGLE BARREL—YH TYPE

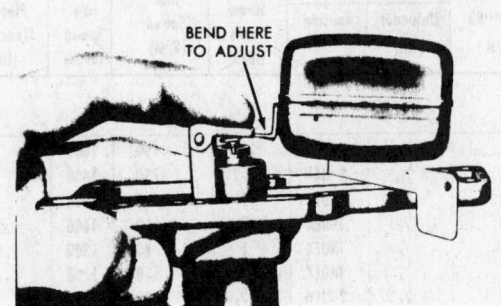

BEND HERE
TO ADJUST

Adjusting float level

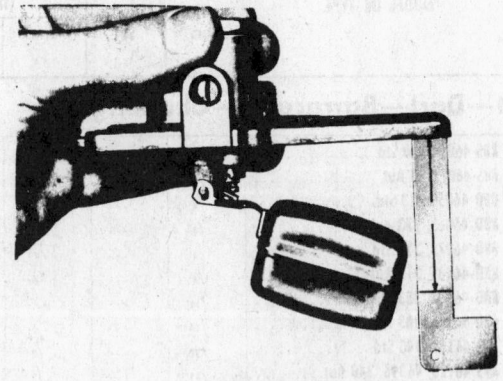

Measuring float drop

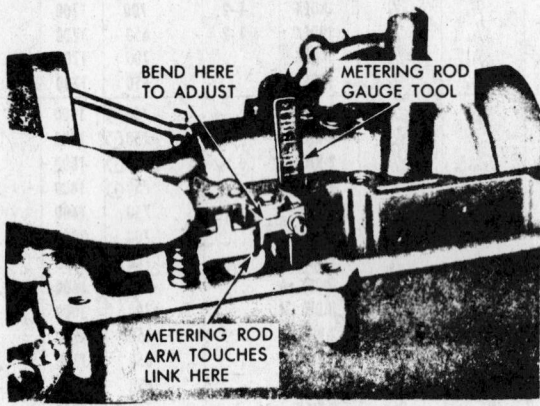

BEND HERE
TO ADJUST

METERING ROD
GAUGE TOOL

METERING ROD
ARM TOUCHES
LINK HERE

Adjusting metering rod arm

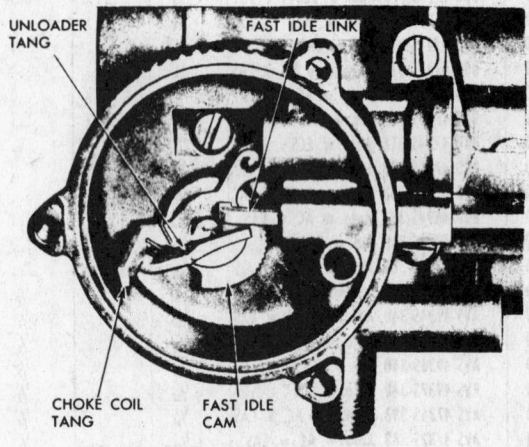

UNLOADER
TANG

FAST IDLE LINK

CHOKE COIL
TANG

FAST IDLE
CAM

Choke housing and fast idle linkage

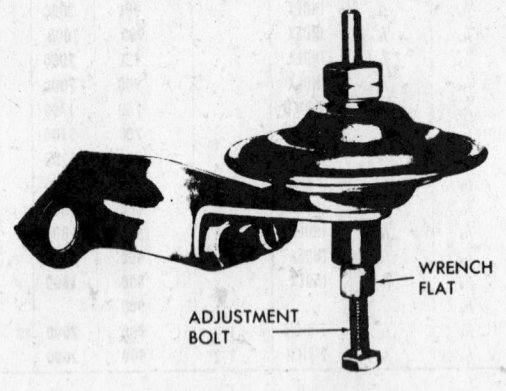

WRENCH
FLAT

ADJUSTMENT
BOLT

Throttle return check valve

OIL FEED

OIL DRAIN

CHOKE CLEAN
AIR TUBES

CHOKE HEAT
PASSAGE

Carburetor fittings and passages

SINGLE BARREL—BBS TYPE

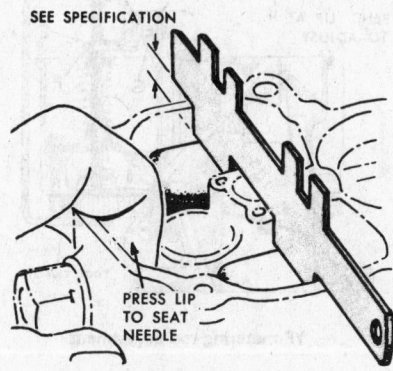

BBS float adjustment

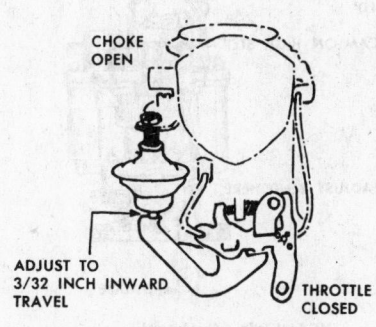

BBS dashpot adjustment

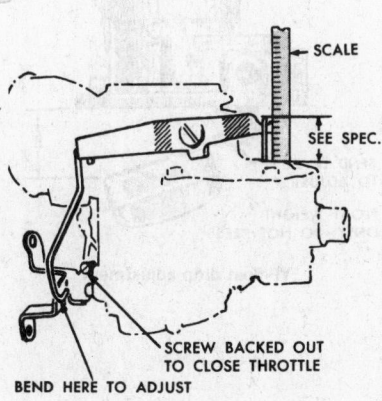

BBS fast idle adjustment

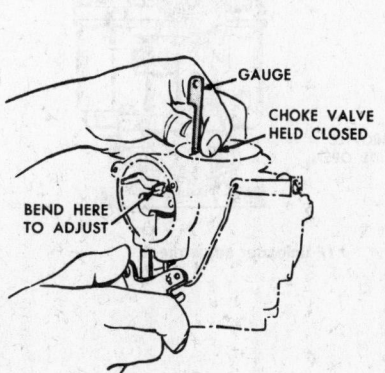

BBS pump adjustment

BBS unloader adjustment

SINGLE BARREL—YF TYPE

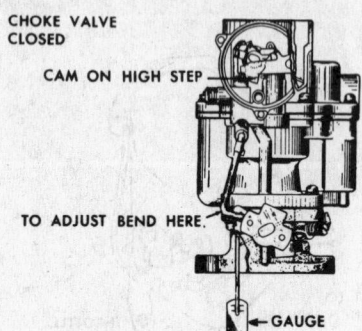

CHOKE VALVE CLOSED

CAM ON HIGH STEP

TO ADJUST BEND HERE

←GAUGE

YF fast idle adjustment

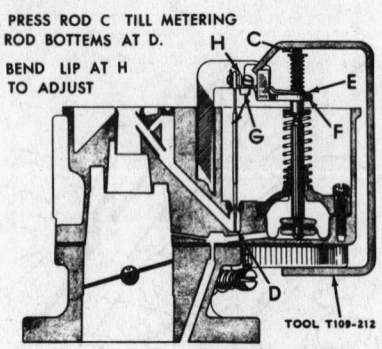

PRESS ROD C TILL METERING ROD BOTTEMS AT D.

BEND LIP AT H TO ADJUST

H   C

E

G   F

D

TOOL T109-212

YF metering rod adjustment

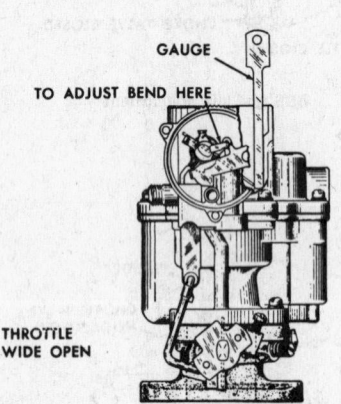

GAUGE

TO ADJUST BEND HERE

THROTTLE WIDE OPEN

YF unloader adjustment

GASKET REMOVED

A

GAUGE

YF float level adjustment

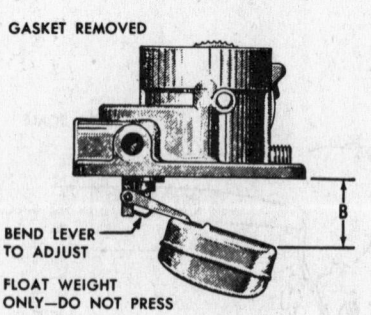

GASKET REMOVED

B

BEND LEVER TO ADJUST

FLOAT WEIGHT ONLY—DO NOT PRESS

YF float drop adjustment

SINGLE BARREL—AS TYPE

AS float adjustment (early models)

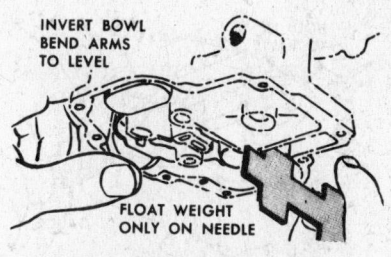

AS float adjustment (late models)

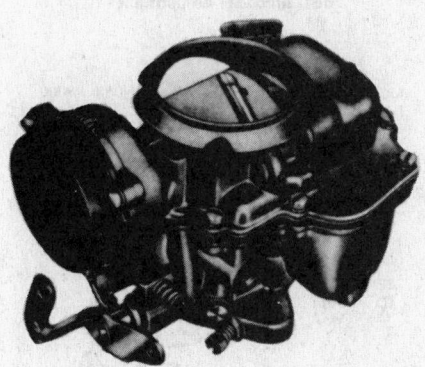

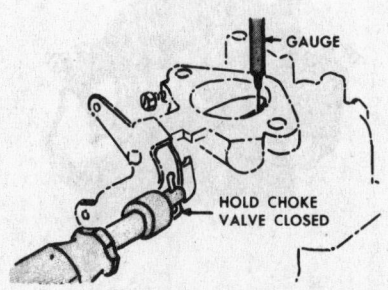

AS fast idle adjustment

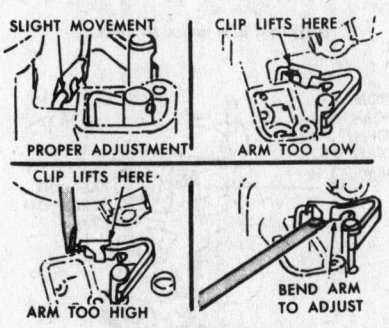

AS metering rod adjustment

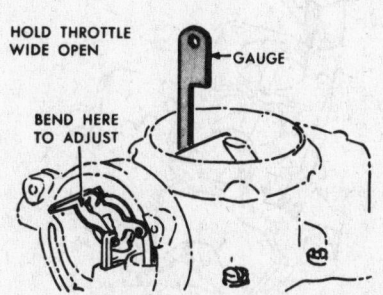

AS unloader adjustment

TWO BARREL—BBD TYPE

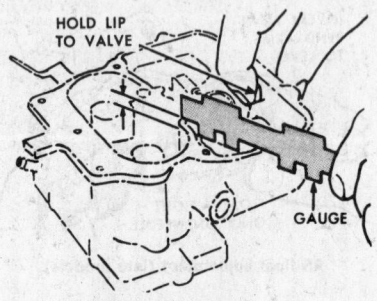

BBD float adjustment

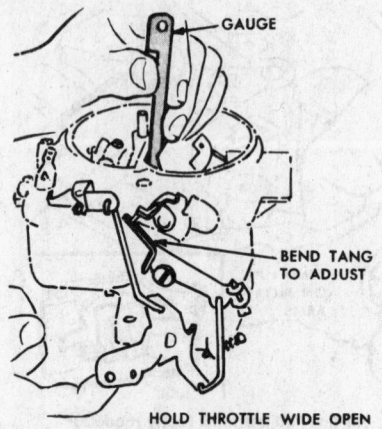

BBD unloader adjustment

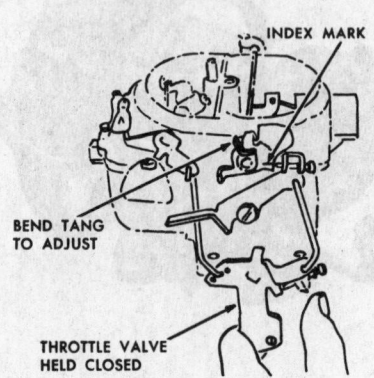

BBD fast idle index alignment

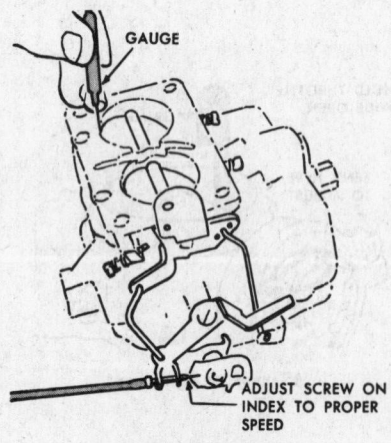

BBD fast idle adjustment

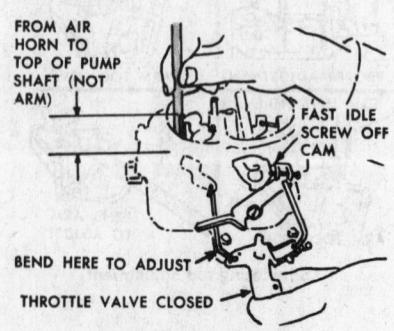

BBD pump adjustment

FOUR BARREL—WCFB TYPE

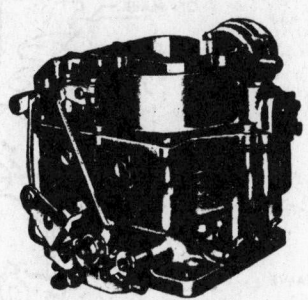

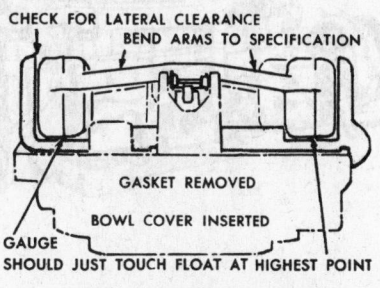

CHECK FOR LATERAL CLEARANCE
BEND ARMS TO SPECIFICATION

GASKET REMOVED

BOWL COVER INSERTED

GAUGE

SHOULD JUST TOUCH FLOAT AT HIGHEST POINT

WCFB float level adjustment

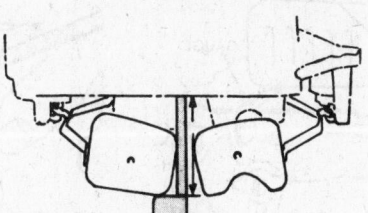

WCFB float drop adjustment

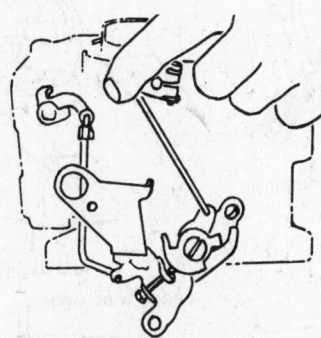

WCFB fast idle adjustment

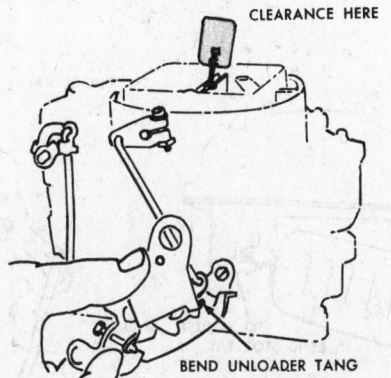

CLEARANCE HERE

BEND UNLOADER TANG

WCFB unloader adjustment

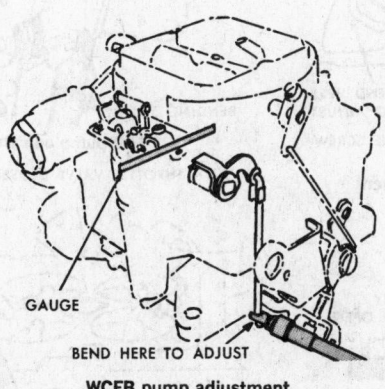

GAUGE

BEND HERE TO ADJUST

WCFB pump adjustment

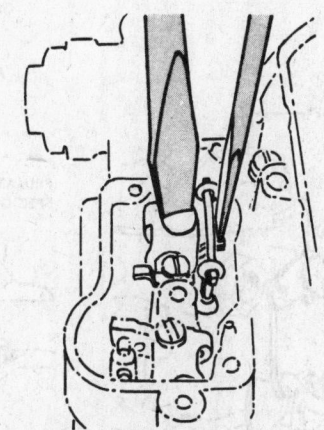

WCFB metering rod adjustment

FOUR BARREL—AFB TYPE

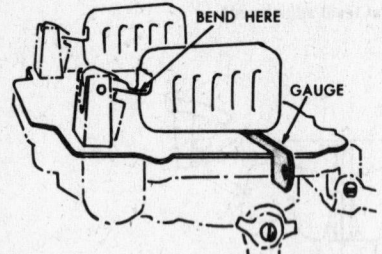

AFB float level adjustment

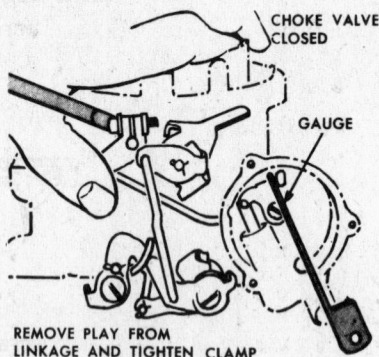

CHOKE VALVE CLOSED

GAUGE

REMOVE PLAY FROM LINKAGE AND TIGHTEN CLAMP

AFB choke piston lever adjustment

CHOKE VALVE CLOSED ON GAUGE WITH SCREW ON MARK

AFB fast idle throttle adjustment

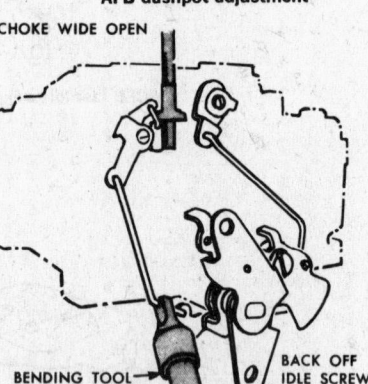

BEND HERE TO ADJUST

AFB dashpot adjustment

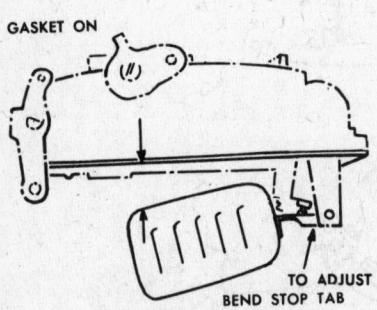

GASKET ON

TO ADJUST BEND STOP TAB

AFB float drop adjustment

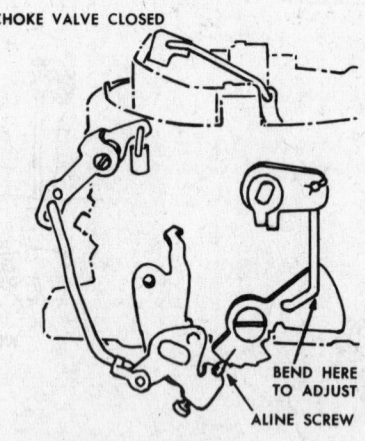
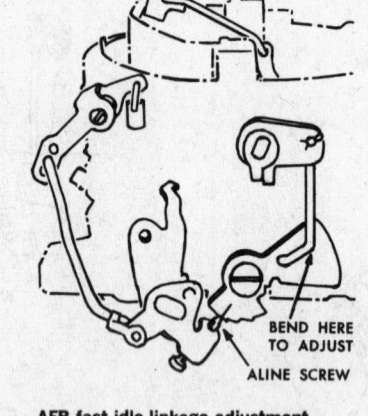

CHOKE VALVE CLOSED

BEND HERE TO ADJUST

ALINE SCREW

AFB fast idle linkage adjustment

CHOKE WIDE OPEN

BENDING TOOL

BACK OFF IDLE SCREW

AFB pump adjustment

WITH THROTTLE VALVE CLOSED

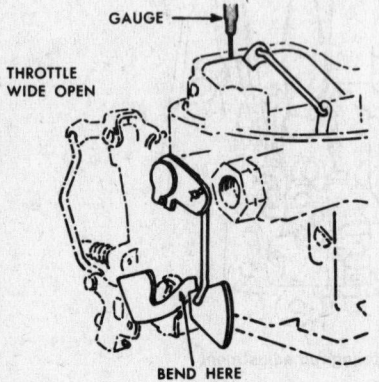

GAUGE

THROTTLE WIDE OPEN

BEND HERE

AFB unloader adjustment

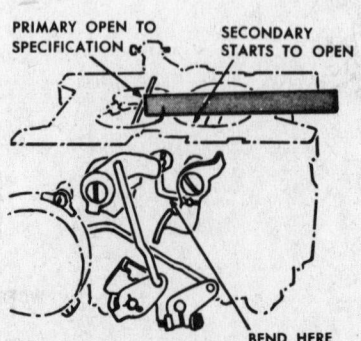

PRIMARY OPEN TO SPECIFICATION

SECONDARY STARTS TO OPEN

BEND HERE

AFB secondary throttle adjustment

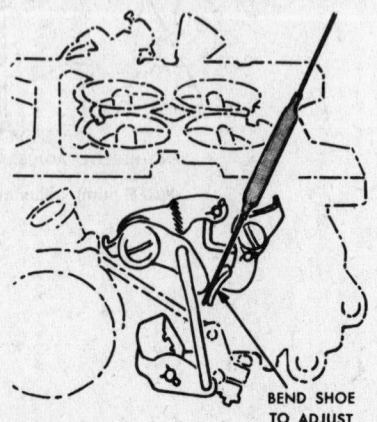

BEND SHOE TO ADJUST

FOUR BARREL – AVS TYPE

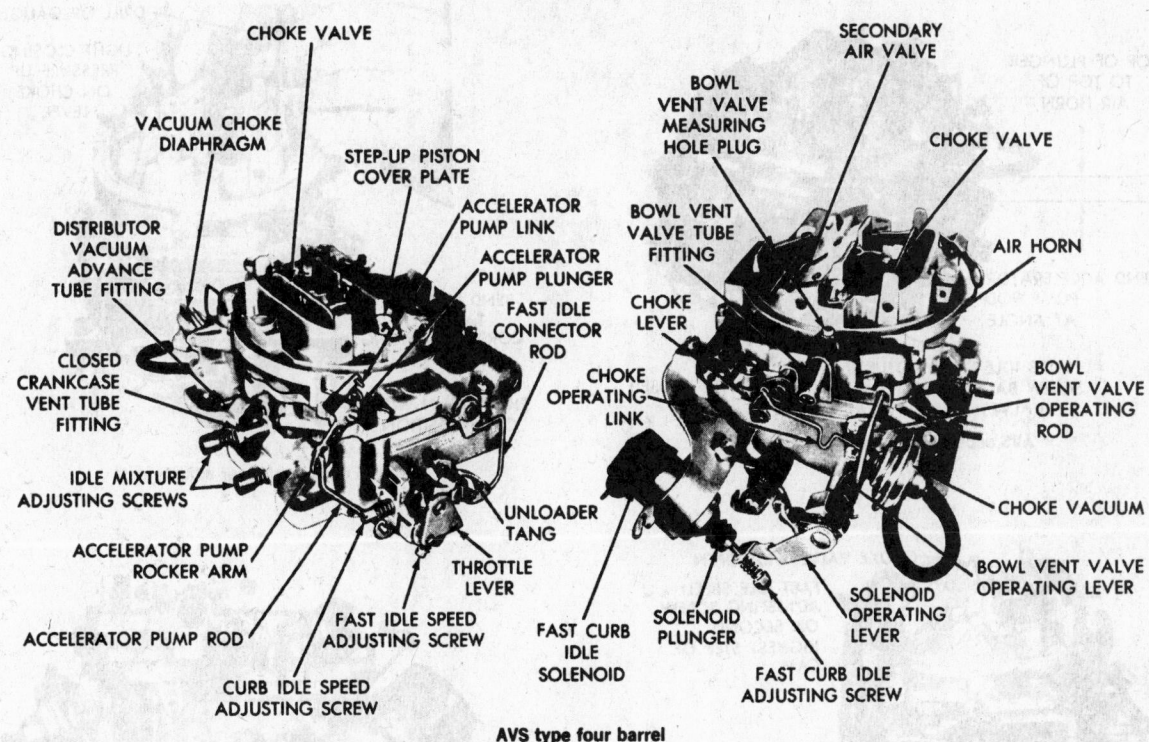

CHOKE VALVE

VACUUM CHOKE DIAPHRAGM

STEP-UP PISTON COVER PLATE

ACCELERATOR PUMP LINK

ACCELERATOR PUMP PLUNGER

DISTRIBUTOR VACUUM ADVANCE TUBE FITTING

FAST IDLE CONNECTOR ROD

CLOSED CRANKCASE VENT TUBE FITTING

IDLE MIXTURE ADJUSTING SCREWS

ACCELERATOR PUMP ROCKER ARM

UNLOADER TANG

ACCELERATOR PUMP ROD

THROTTLE LEVER

FAST IDLE SPEED ADJUSTING SCREW

CURB IDLE SPEED ADJUSTING SCREW

SECONDARY AIR VALVE

BOWL VENT VALVE MEASURING HOLE PLUG

CHOKE VALVE

BOWL VENT VALVE TUBE FITTING

AIR HORN

CHOKE LEVER

CHOKE OPERATING LINK

BOWL VENT VALVE OPERATING ROD

CHOKE VACUUM

BOWL VENT VALVE OPERATING LEVER

FAST CURB IDLE SOLENOID

SOLENOID PLUNGER

SOLENOID OPERATING LEVER

FAST CURB IDLE ADJUSTING SCREW

AVS type four barrel

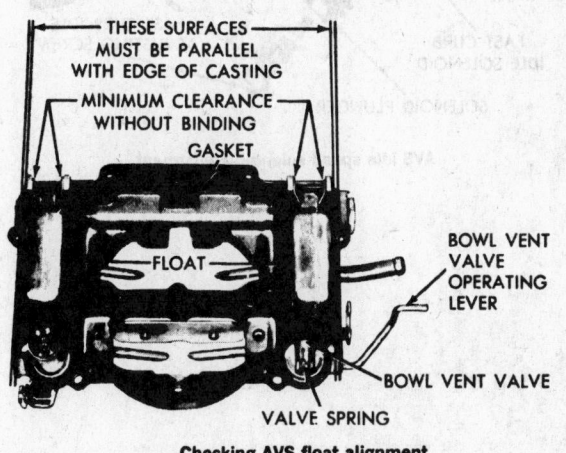

THESE SURFACES MUST BE PARALLEL WITH EDGE OF CASTING

MINIMUM CLEARANCE WITHOUT BINDING GASKET

FLOAT

BOWL VENT VALVE OPERATING LEVER

BOWL VENT VALVE

VALVE SPRING

Checking AVS float alignment

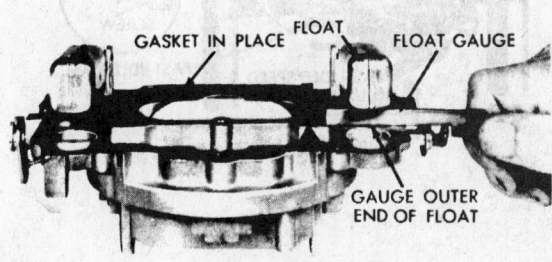

GASKET IN PLACE

FLOAT

FLOAT GAUGE

GAUGE OUTER END OF FLOAT

Checking AVS float height

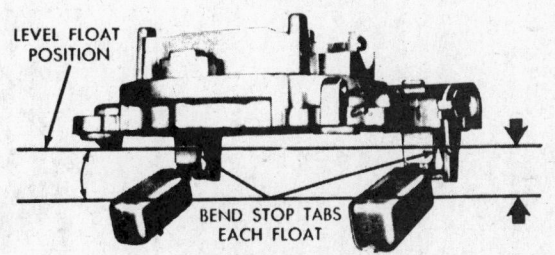

LEVEL FLOAT POSITION

BEND STOP TABS EACH FLOAT

Checking AVS float drop

FOUR BARREL— AVS TYPE

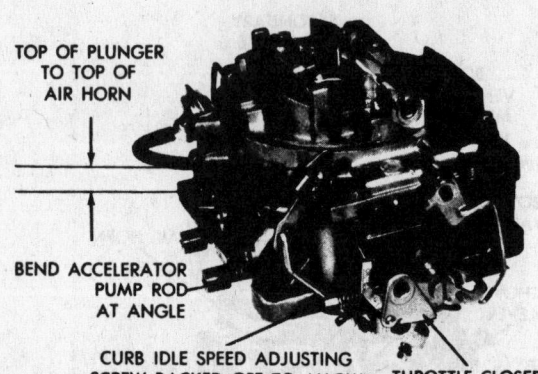

TOP OF PLUNGER TO TOP OF AIR HORN

BEND ACCELERATOR PUMP ROD AT ANGLE

CURB IDLE SPEED ADJUSTING SCREW BACKED OFF TO ALLOW VALVES TO FULLY SEAT

THROTTLE CLOSED

AVS accelerator pump adjustment

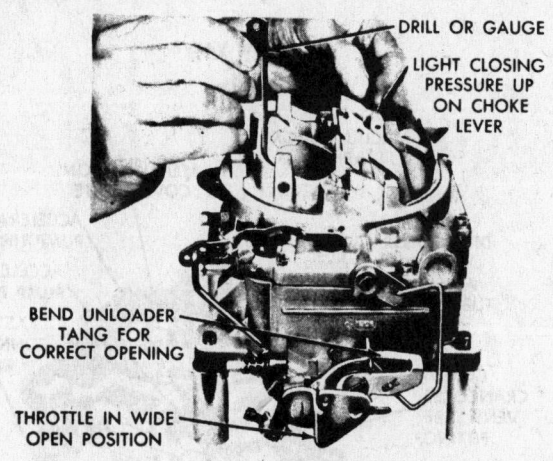

DRILL OR GAUGE

LIGHT CLOSING PRESSURE UP ON CHOKE LEVER

BEND UNLOADER TANG FOR CORRECT OPENING

THROTTLE IN WIDE OPEN POSITION

AVS choke unloader adjustment

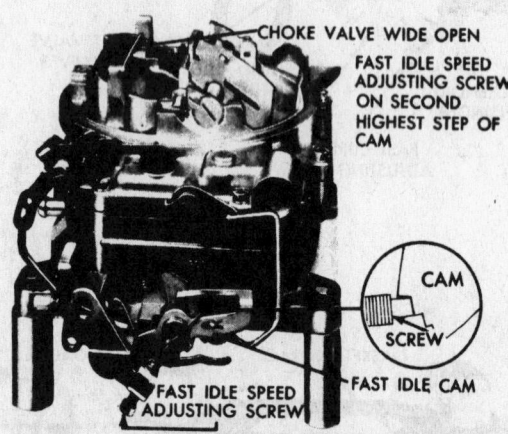

CHOKE VALVE WIDE OPEN

FAST IDLE SPEED ADJUSTING SCREW ON SECOND HIGHEST STEP OF CAM

CAM

SCREW

FAST IDLE CAM

FAST IDLE SPEED ADJUSTING SCREW

AVS fast idle adjustment

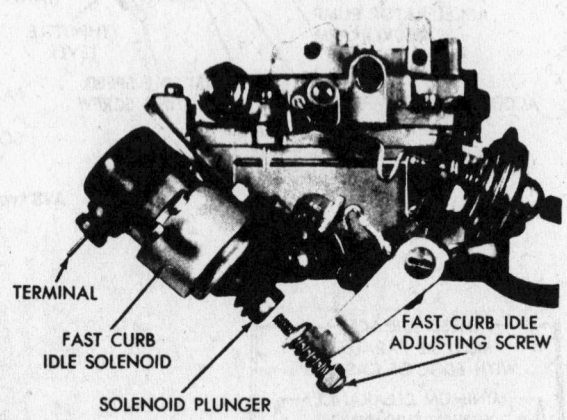

TERMINAL

FAST CURB IDLE SOLENOID

SOLENOID PLUNGER

FAST CURB IDLE ADJUSTING SCREW

AVS idle speed solenoid adjustment

TWO BARREL—WGD TYPE

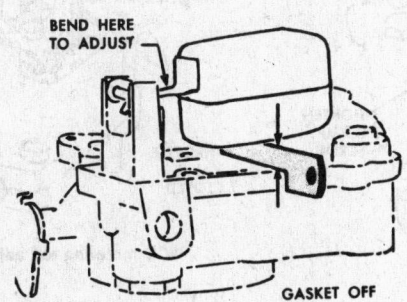

WGD float adjustment

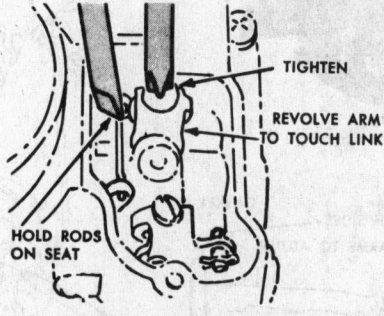

WGD metering rod adjustment

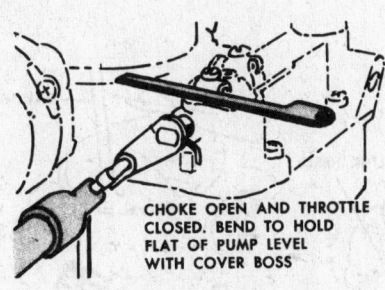

WGD pump adjustment

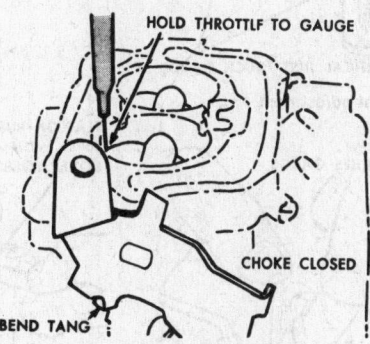

WGD fast idle adjustment

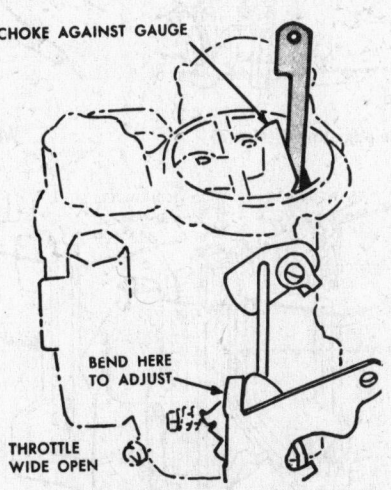

WGD unloader adjustment

TWO BARREL—WCD TYPE

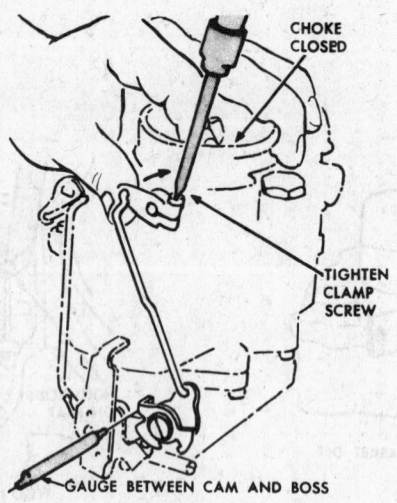

CHOKE CLOSED

TIGHTEN CLAMP SCREW

GAUGE BETWEEN CAM AND BOSS

WCD fast idle adjustment (step 1)

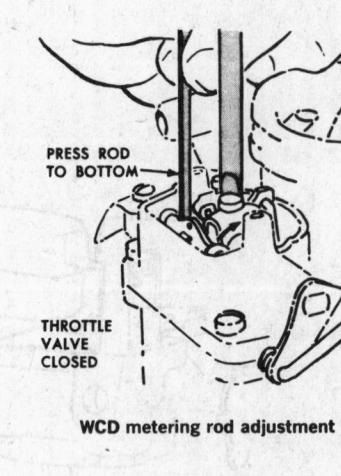

PRESS ROD TO BOTTOM

THROTTLE VALVE CLOSED

WCD metering rod adjustment

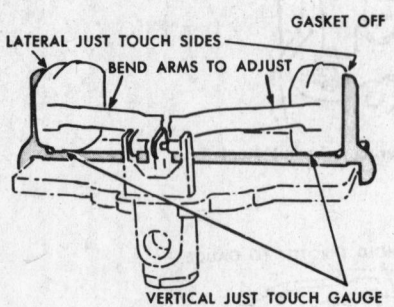
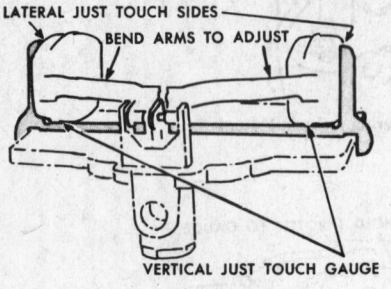

GASKET OFF

LATERAL JUST TOUCH SIDES

BEND ARMS TO ADJUST

VERTICAL JUST TOUCH GAUGE

WCD float adjustment

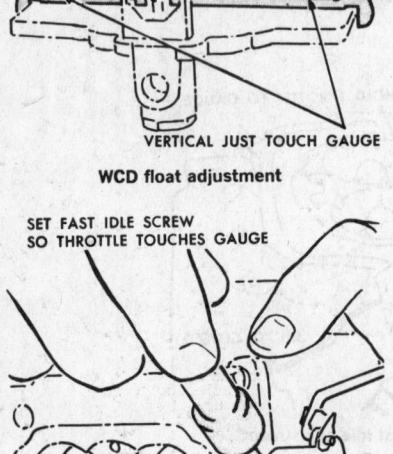

SET FAST IDLE SCREW SO THROTTLE TOUCHES GAUGE

CHOKE CLOSED

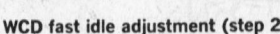

WCD fast idle adjustment (step 2)

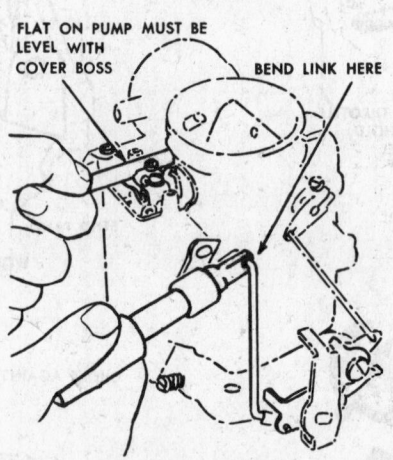

FLAT ON PUMP MUST BE LEVEL WITH COVER BOSS

BEND LINK HERE

WCD pump adjustment

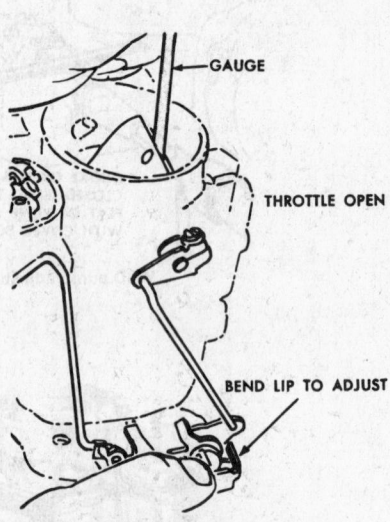

GAUGE

THROTTLE OPEN

BEND LIP TO ADJUST

WCD unloader adjustment

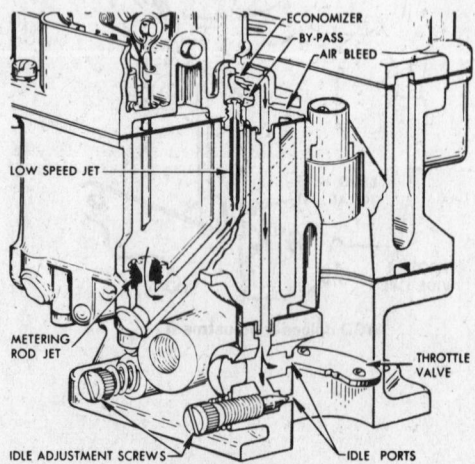

ECONOMIZER
BY-PASS
AIR BLEED

LOW SPEED JET

METERING ROD JET

IDLE ADJUSTMENT SCREWS

IDLE PORTS

THROTTLE VALVE

Low speed system

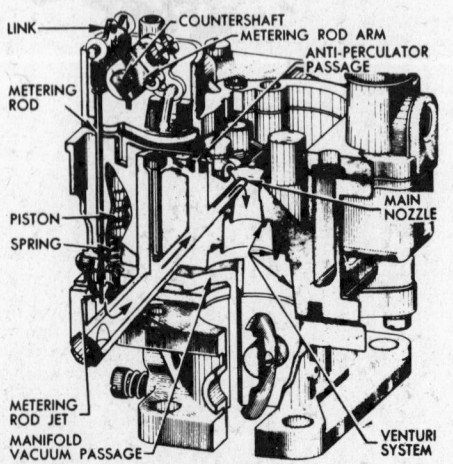

LINK
COUNTERSHAFT
METERING ROD ARM
ANTI-PERCULATOR PASSAGE

METERING ROD

PISTON

SPRING

MAIN NOZZLE

METERING ROD JET
MANIFOLD VACUUM PASSAGE

VENTURI SYSTEM

High speed and power systems

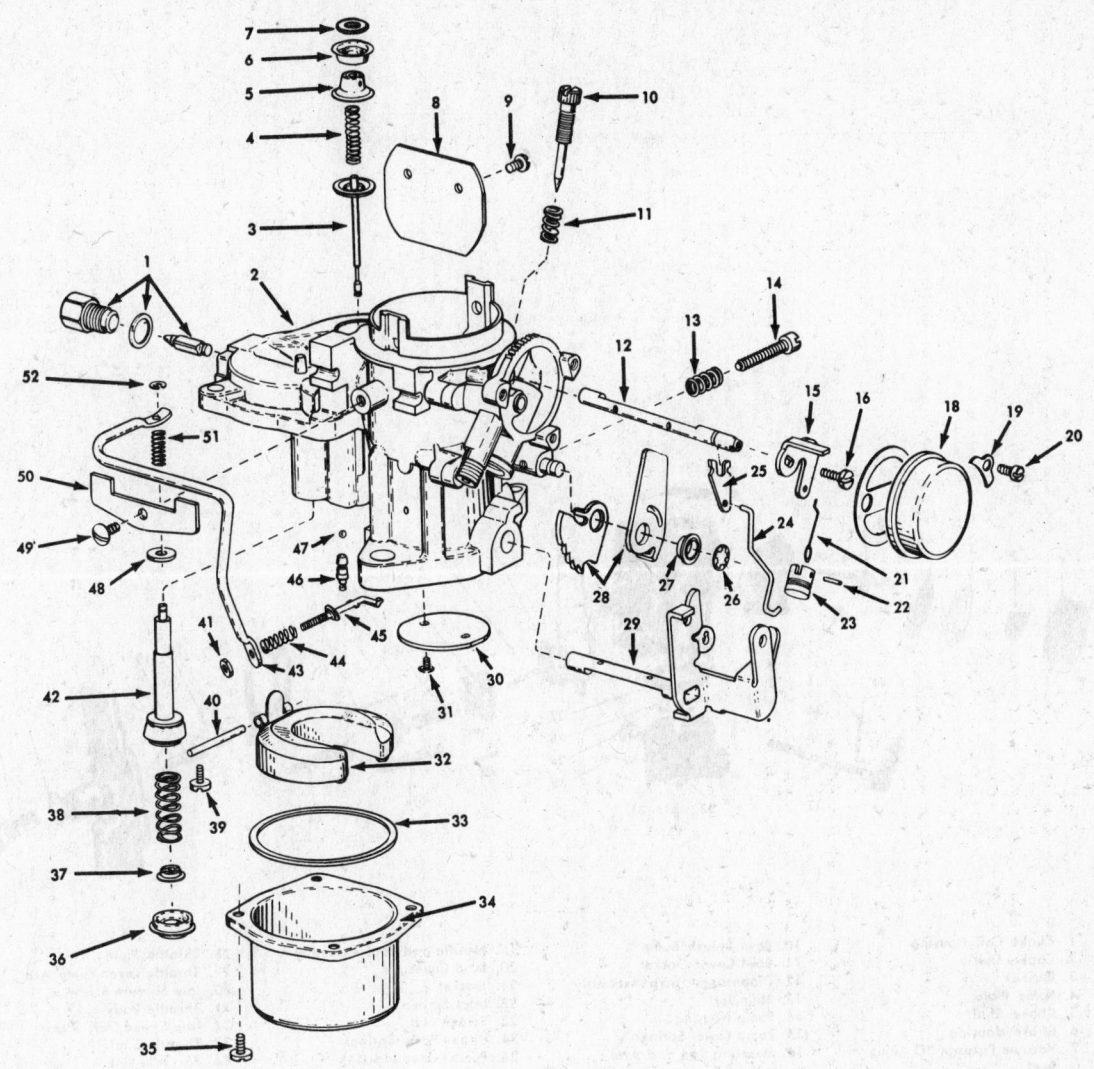

1 Needle and seat assembly
2 Main body
3 Step-up rod and diaphragm assembly
4 Step-up diaphragm spring
5 Diaphragm retainer
6 Diaphragm cover
7 Cover washer
8 Choke valve
9 Choke valve attaching screw (2)
10 Idle adjustment screw
11 Idle adjustment screw spring
12 Choke shaft
13 Throttle lever adjustment screw
   spring
14 Throttle lever adjustment screw

15 Choke piston lever
16 Choke piston lever attaching
   screw
17 Coil housing gasket
18 Thermostatic coil and housing
   assembly
19 Coil housing retainer (3)
20 Coil housing attaching screw (3)
21 Choke piston link (wire)
22 Choke piston pin
23 Choke piston
24 Fast idle connector rod
25 Choke lever
26 Fast idle cam retainer
27 Fast idle cam collar washer
28 Fast idle cam

29 Throttle shaft and lever assembly
30 Throttle valve
31 Throttle valve attaching screw
   (2)
32 Float and lever assembly
33 Bowl gasket
34 Bowl
35 Bowl attaching screw (4)
36 Pump cover
37 Pump intake check ball retainer
38 Lower pump spring
39 Float pin attaching screw (2)
40 Float lever pin
41 Pump adjustment nut
42 Pump plunger assembly
43 Pump arm
44 Pump delayer spring
45 Pump connector link
46 Pump inlet ball seat
47 Pump discharge check ball
48 Plunger shaft washer
49 Pump arm retainer attaching
   screw
50 Pump arm retainer
51 Upper arm spring
52 Plunger shaft retainer

Disassembled view, Carter RBS

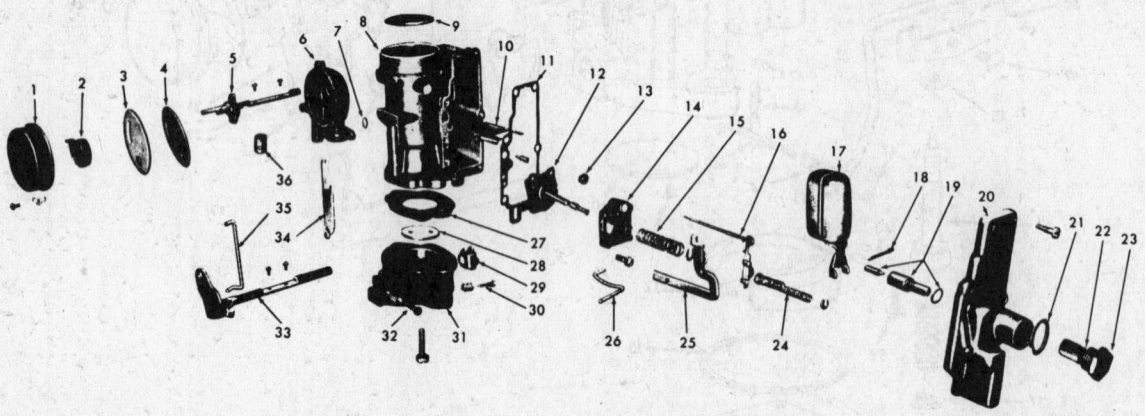

1. Choke Coil Housing
2. Choke Coil
3. Gasket
4. Baffle Plate
5. Choke Shaft
6. Choke Housing
7. Vacuum Passage "O" Ring Seal
8. Caburetor Body
9. Choke Plate
10. Bowl Splash Baffle
11. Bowl Cover Gasket
12. Diaphragm Pump Assembly
13. Main Jet
14. Pump Housing
15. Pump Lower Spring
16. Metering Rod and Arm Assembly
17. Float
18. Hinge Pin
19. Needle and Seat Assembly
20. Bowl Cover
21. Gasket
22. Inlet Screen
23. Screen Nut
24. Upper Pump Spring
25. Pump Actuating Link
26. Connector Link
27. Gasket
28. Throttle Plate
29. Throttle Lever Pump Arm
30. Idle Mixture Screw
31. Throttle Body
32. Idle Speed (Air) Screw
33. Throttle Shaft
34. Fast Idle Link
35. Fast Idle Connector Link
36. Choke Piston

Carter YH

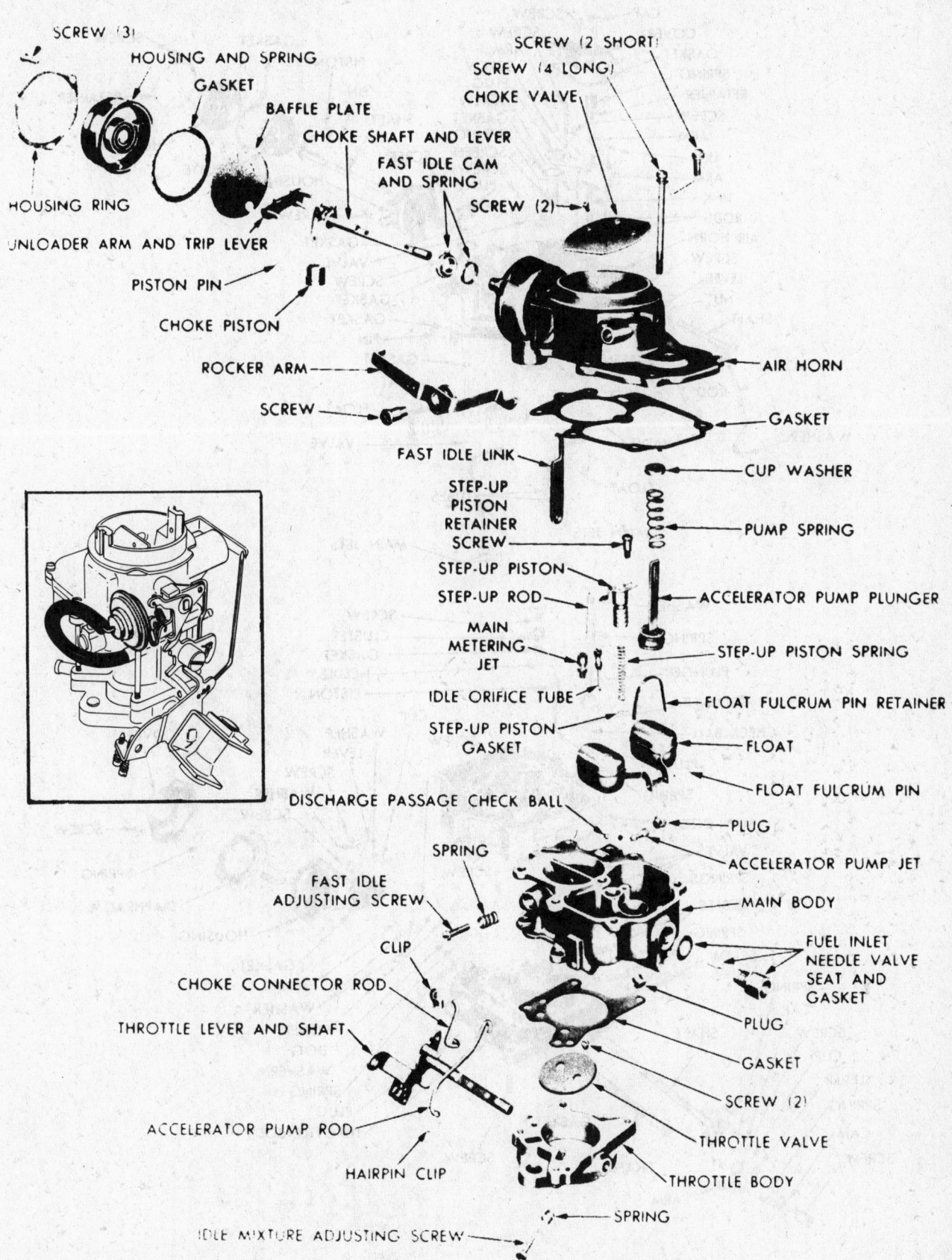

SCREW (3)
HOUSING AND SPRING
GASKET
BAFFLE PLATE
CHOKE SHAFT AND LEVER
FAST IDLE CAM AND SPRING
SCREW (2 SHORT)
SCREW (4 LONG)
CHOKE VALVE
SCREW (2)
HOUSING RING
UNLOADER ARM AND TRIP LEVER
PISTON PIN
CHOKE PISTON
ROCKER ARM
SCREW
AIR HORN
GASKET
FAST IDLE LINK
CUP WASHER
STEP-UP PISTON RETAINER SCREW
PUMP SPRING
STEP-UP PISTON
STEP-UP ROD
ACCELERATOR PUMP PLUNGER
MAIN METERING JET
STEP-UP PISTON SPRING
IDLE ORIFICE TUBE
FLOAT FULCRUM PIN RETAINER
STEP-UP PISTON GASKET
FLOAT
FLOAT FULCRUM PIN
DISCHARGE PASSAGE CHECK BALL
PLUG
ACCELERATOR PUMP JET
SPRING
MAIN BODY
FAST IDLE ADJUSTING SCREW
FUEL INLET NEEDLE VALVE SEAT AND GASKET
CLIP
CHOKE CONNECTOR ROD
PLUG
THROTTLE LEVER AND SHAFT
GASKET
SCREW (2)
ACCELERATOR PUMP ROD
THROTTLE VALVE
HAIRPIN CLIP
THROTTLE BODY
SPRING
IDLE MIXTURE ADJUSTING SCREW

Carter BBS single barrel (typical)

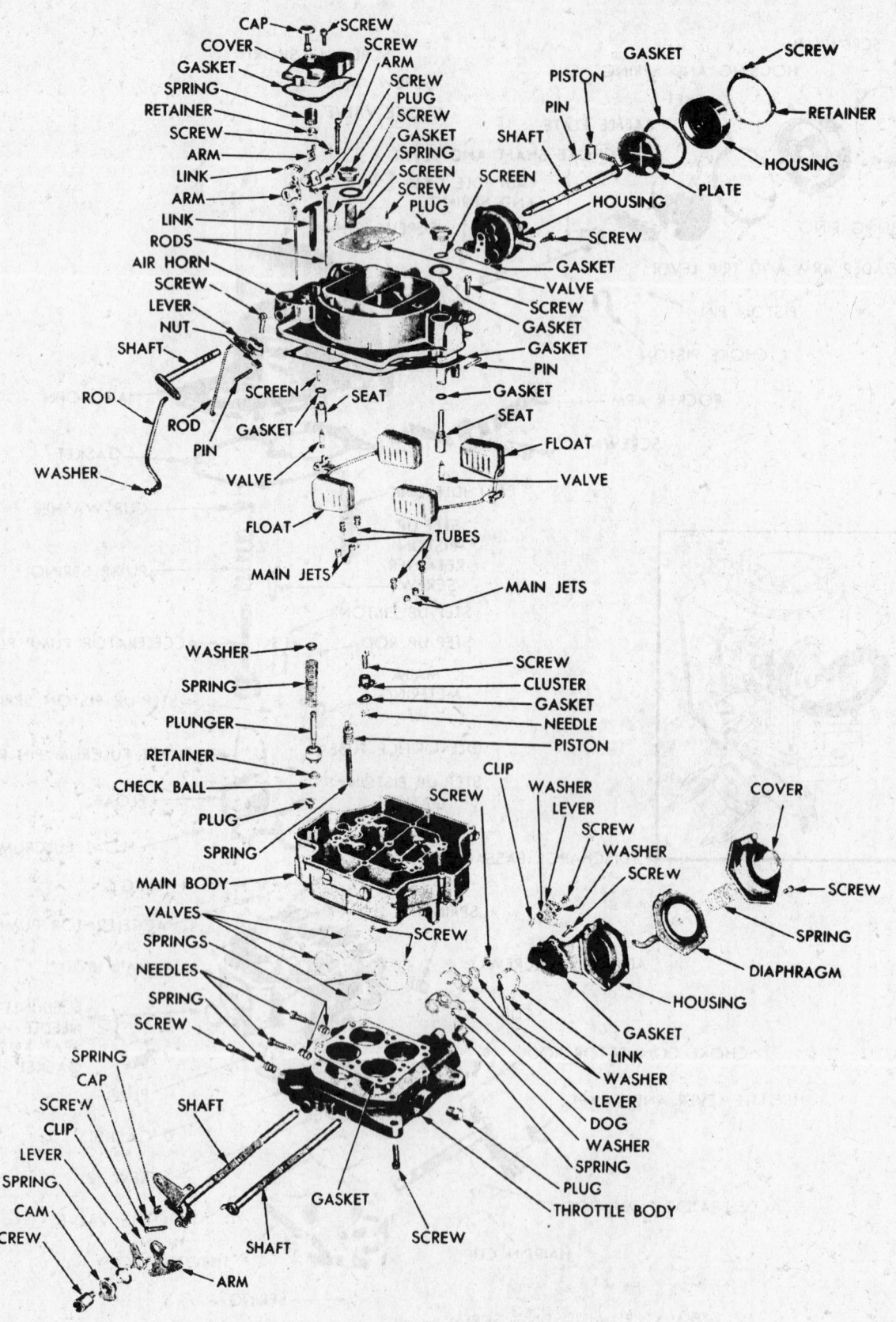

Carter WCFB four-barrel (typical)

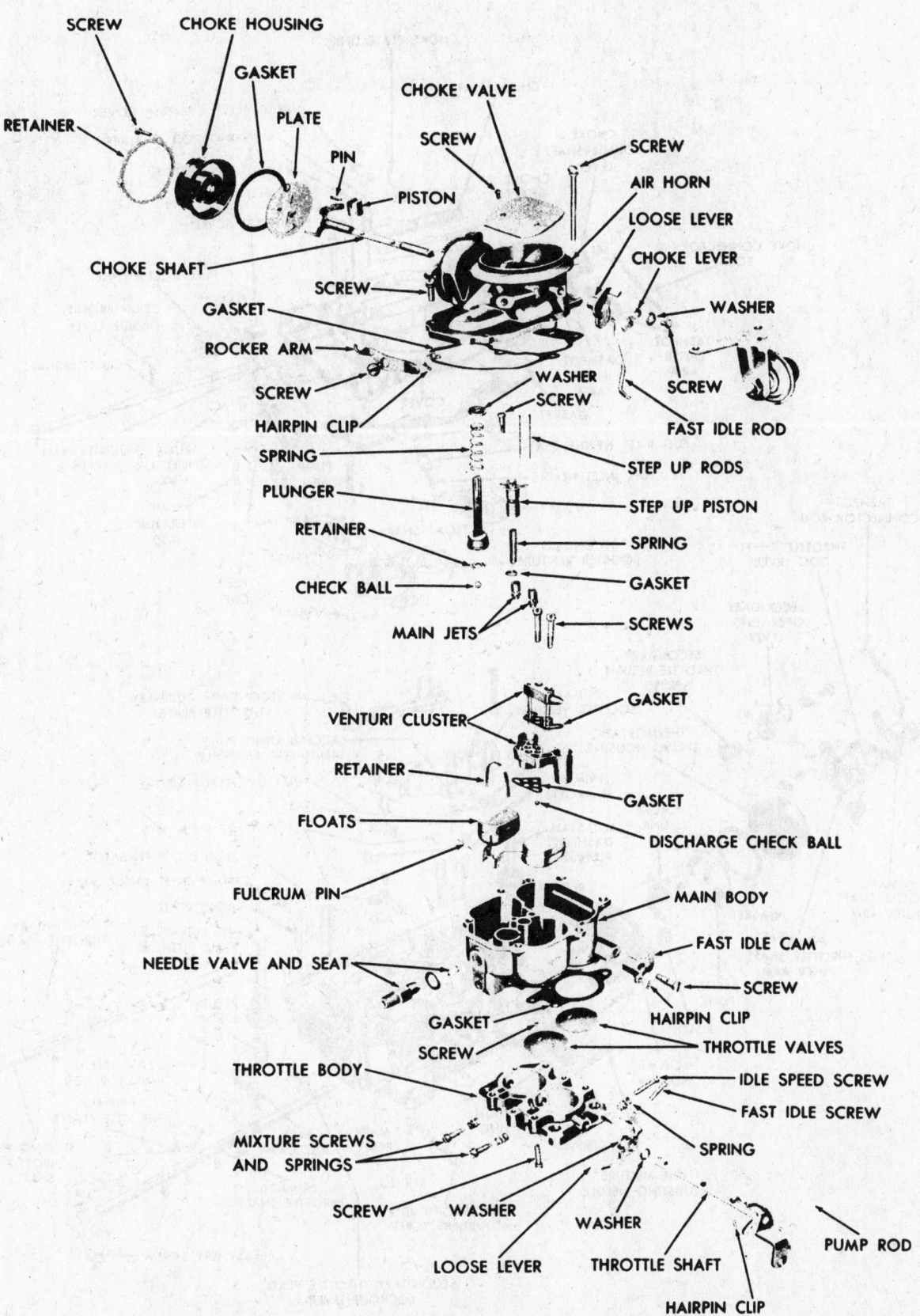

Carter BBD two-barrel (typical)

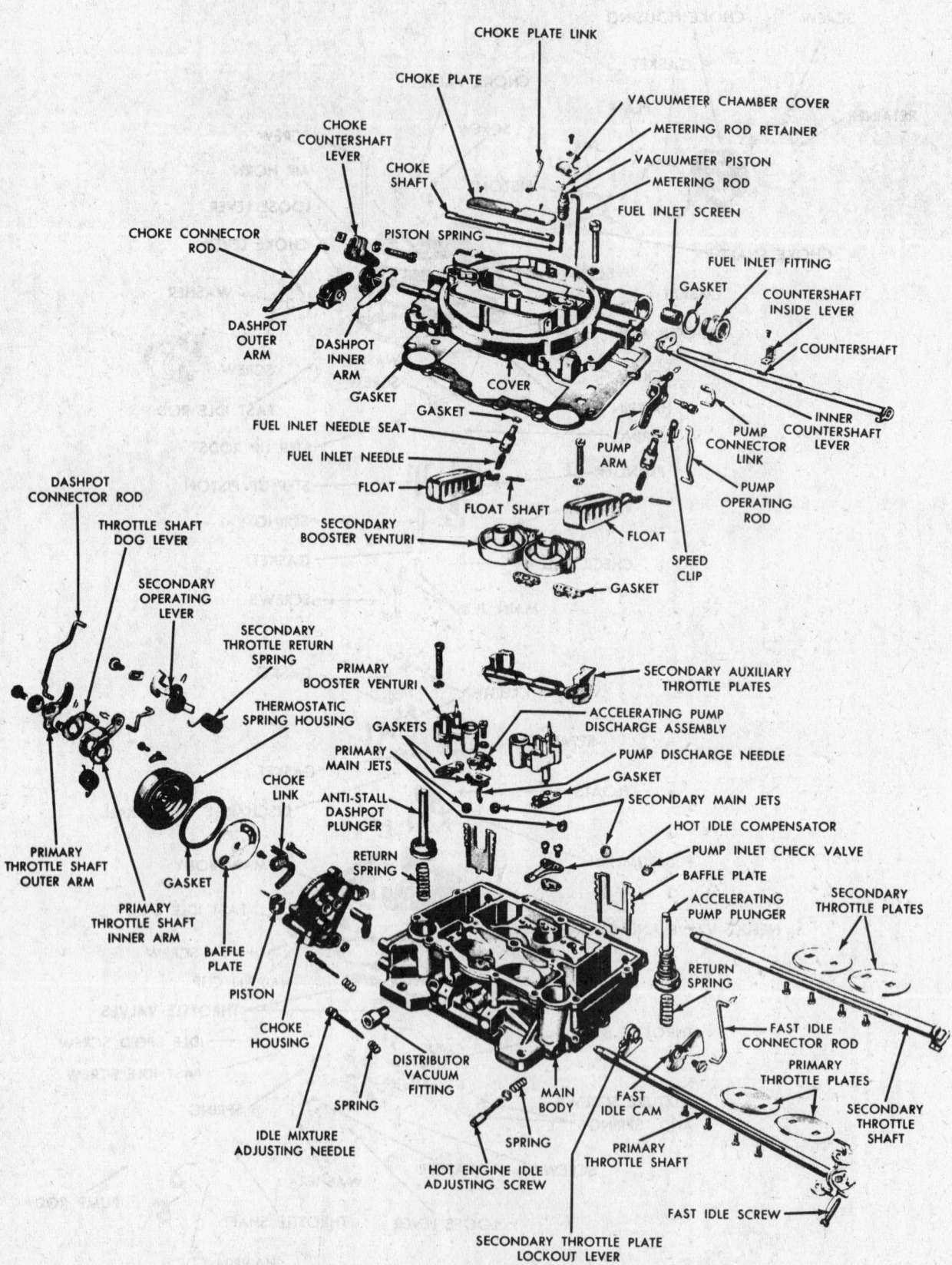

CHOKE PLATE LINK

CHOKE PLATE

VACUUMETER CHAMBER COVER

METERING ROD RETAINER

CHOKE COUNTERSHAFT LEVER

CHOKE SHAFT

VACUUMETER PISTON

METERING ROD

FUEL INLET SCREEN

CHOKE CONNECTOR ROD

PISTON SPRING

FUEL INLET FITTING

GASKET

COUNTERSHAFT INSIDE LEVER

COUNTERSHAFT

DASHPOT OUTER ARM

DASHPOT INNER ARM

INNER COUNTERSHAFT LEVER

GASKET

COVER

GASKET

PUMP CONNECTOR LINK

FUEL INLET NEEDLE SEAT

PUMP ARM

PUMP OPERATING ROD

FUEL INLET NEEDLE

DASHPOT CONNECTOR ROD

FLOAT

FLOAT SHAFT

FLOAT

THROTTLE SHAFT DOG LEVER

SECONDARY BOOSTER VENTURI

SPEED CLIP

SECONDARY OPERATING LEVER

GASKET

SECONDARY THROTTLE RETURN SPRING

PRIMARY BOOSTER VENTURI

SECONDARY AUXILIARY THROTTLE PLATES

THERMOSTATIC SPRING HOUSING

GASKETS

ACCELERATING PUMP DISCHARGE ASSEMBLY

PRIMARY MAIN JETS

PUMP DISCHARGE NEEDLE

CHOKE LINK

GASKET

ANTI-STALL DASHPOT PLUNGER

SECONDARY MAIN JETS

HOT IDLE COMPENSATOR

RETURN SPRING

PUMP INLET CHECK VALVE

PRIMARY THROTTLE SHAFT OUTER ARM

GASKET

BAFFLE PLATE

SECONDARY THROTTLE PLATES

PRIMARY THROTTLE SHAFT INNER ARM

ACCELERATING PUMP PLUNGER

BAFFLE PLATE

RETURN SPRING

PISTON

FAST IDLE CONNECTOR ROD

CHOKE HOUSING

PRIMARY THROTTLE PLATES

DISTRIBUTOR VACUUM FITTING

SPRING

MAIN BODY

FAST IDLE CAM

SECONDARY THROTTLE SHAFT

IDLE MIXTURE ADJUSTING NEEDLE

SPRING

PRIMARY THROTTLE SHAFT

HOT ENGINE IDLE ADJUSTING SCREW

FAST IDLE SCREW

SECONDARY THROTTLE PLATE LOCKOUT LEVER

Carter AFB four-barrel (typical)

Ford, Autolite Carburetors

American Motors

YEAR	MODEL OR TYPE	Float Level (IN.)	Fuel Level (IN.)	Fast Idle Cam (IN.)	Pump Travel Setting	Choke Setting	ON THE CAR ADJUSTMENT: Idle Screw Turns Open	Idle Speed (RPM)	Fast Idle Speed (RPM)	Dashpot Plunger Clearance (IN.)
1968	6200-8HA2-290, Auto.	3/8	...	#3 hole①	5/64	INDEX	1 1/2	550②	1600	3/32
1968	6200-8HM2-290, Std.	3/8	...	#3 hole①	/64	INDEX	1 1/2	650	1600	3/32
1968	6200-8ZA2-343, Auto.	3/8	...	#3 hole①	5/64	INDEX	1 1/2	550②	1600	3/32
1969	6200-9HM2-290, Std.	1/2	...	#3 hole①	5/64	INDEX	...	③	1600	9/64
1969	6200-9HA2-290, Auto.	1/2	...	#3 hole①	5/64	INDEX	...	③	1600	9/64
1969	6200-9ZA2-343, Auto.	1/2	...	#3 hole①	5/64	INDEX	...	③	1600	9/64
1970	2100-0DM2-304, Std.	3/8	...	#3 hole①	13/64	INDEX	...	③	1600	1/8
1970	2100-0DA2-304, Auto.	3/8	...	#3 hole①	13/64	2-RICH	...	③	1600	1/8
1970	2100-0RA2-360, Auto.	3/8	...	#3 hole①	13/64	1-RICH	...	③	1600	1/8
1970	4300-0WM4-360, 390 Std.	13/16	...	center hole	19/64	2-RICH	...	③	1600	1/16
1970	4300-0WA4-360, 390, Auto.	13/16	...	center hole	19/64	2-RICH	...	③	1600	1/8

① —Inboard hole on pump lever.
② —In drive.
③ —See decal for idle speed.

Fairlane & Montego

YEAR	MODEL OR TYPE	Float Level (IN.)	Fuel Level (IN.)	Fast Idle Cam (IN.)	Pump Travel Setting	Choke Setting	Idle Screw Turns Open	Idle Speed (RPM)	Fast Idle Speed (RPM)	Dashpot Plunger Clearance (IN.)
1964	170-6 Cyl.-Std.	1	...	...	3/16	...	1-1 1/2	500	...	...
1964	170-6 Cyl.-Aut.	1	...	...	3/16	...	1-1 1/2	500	...	3 1/2
1964	200-6 Cyl.-Aut.-1 Bbl.	1	...	...	3/16	...	1-1 1/2	525	...	3 1/2
1964	260-V8-Std.-2 Bbl.	1/2	3/4	...	①	2-RICH	1-1 1/2	500	1300	...
1964	260-V8-Aut.-2 Bbl.	1/2	3/4	3/32	①	2-RICH	1-1 1/2	525	1600	5/64
1964	289-V8-Std.-2 Bbl.	5/8	7/8	3/32	①	2-RICH	1-1 1/2	525	1300	...
1964	289-V8-Aut.-2 Bbl.	5/8	7/8	3/32	①	2-RICH	1-1 1/2	525	1600	...
1964	289-V8-Std.-4 Bbl.	21/32	29/32	1/16	①	3-LEAN	1-1 1/2	750	1800	5/64
1965	200-6 Cyl.-Aut.-1 Bbl.	1 3/32	...	...	3/16	INDEX	1 1/2	525	1500	3 1/2
1965	289-V8-Std.-2Bbl.	15/32	7/8	.130⑤	①	INDEX	1 1/2	600	1400	...
1965	289-V8-Aut.-2 Bbl.	15/32	7/8	.120⑤	①	2-RICH	1 1/2	500	1600	5/64
1965	289-V8-Std.-4 Bbl.	29/64	29/32	.125⑤	①	INDEX	1 1/4	575	1400	...
1965	289-V8-Aut.-4 Bbl.	29/64	29/32	.109⑤	①	2-RICH	1 1/4	500	1600	5/64
1966	200-6 Cyl.-Std.-1 Bbl.	1 3/32	...	...	3/16	1-LEAN	1-1 1/2	575	1400	...
1966	200-6 Cyl.-Aut.-1 Bbl.	1 3/32	...	...	3/16	INDEX	1-1 1/2	525	1500	3 1/2
1966	289-V8-Aut.-2 Bbl.	31/64	7/8	.130⑤	①	2-RICH	1-1 1/2	500	1600	5/64
1966	289-V8-Std.-4 Bbl.	17/32	29/32	.120⑤	①	2-RICH	1-1 1/2	575	1400	...
1966	289-V8-Aut.-4 Bbl.	9/16	15/16	.109⑤	①	2-RICH	1-1 1/2	500	1600	5/64
1966	390-V8-Std.-2 Bbl.	31/64	7/8	.200⑤	①	INDEX	1-1 1/2	575	1300	...
1966	390-V8-Aut.-2 Bbl.	31/64	7/8	.180⑤	①	INDEX	1-1 1/2	500	1400	5/64
1966	390-V8-Std.-4 Bbl.	17/32	29/32	.160⑤	①	2-RICH	1-1 1/2	575	1200	...
1966	390-V8-Aut.-4 Bbl.	17/32	29/32	.140⑤	①	1-RICH	1-1 1/2	500	1300	5/64
1967	200-6 Cyl.-Std.-1 Bbl.	1 3/32	...	...	3/16	1-LEAN	1-1 1/2	575	1300	...
1967	200-6 Cyl.-Aut.-1 Bbl.	1 3/32	...	...	3/16	INDEX	1-1 1/2	500	1500	3 1/2
1967	200-6 Cyl.-Std.-1 Bbl.-Ex. Em.	1 3/32	...	...	3/16	2-LEAN	1-1 1/2	700	1400	2
1967	200-6 Cyl.-Aut.-1 Bbl.-Ex. Em.	1 3/32	...	...	3/16	INDEX	1-1 1/2	500	1500	2
1967	289-V8-Std.-2 Bbl.	3/8	3/4	7/64	①	INDEX	1-1 1/2	575	1400	...
1967	289-V8-Std.-2 Bbl.-Ex. Em.	17/32	29/32	7/64	⑥	INDEX	1-1 1/2	625	1400	.125
1967	289-V8-Aut.-2 Bbl.	17/32	29/32	7/64	①	2-RICH	1-1 1/2	475	1600	.075
1967	289-V8-Aut.-2 Bbl.-Ex. Em.	17/32	29/32	7/64	⑥	2-RICH	1-1 1/2	550	1600	.125
1967	289-V8-Std.-4 Bbl.	25/32	...	.100	#1	INDEX	1-1 1/2	600	1400	...
1967	289-V8-Std.-4 Bbl.	25/32	...	.100	#1	INDEX	1-1 1/2	625	1400	1/8
1967	289-V8-Aut.-4 Bbl.	25/32	...	.100	#1	INDEX	1-1 1/2	525	1600	...
1967	289-V8-Aut.-4 Bbl.-Ex. Em.	25/32	...	.100	#1	INDEX	1-1 1/2	550	1600	1/8
1967	390-V8-Std.-2 Bbl.	31/64	7/8	.160	①	INDEX	1-1 1/2	575	1300	...
1967	390-V8-Std.-2 Bbl.-Ex. Em.	17/32	29/32	.170	⑥	INDEX	1-1 1/2	625	1300	.095
1967	390-V8-Aut.-2 Bbl.	3/8	3/4	.150	①	2-RICH	1-1 1/2	475	1400	...
1967	390-V8-Aut.-2 Bbl.-Ex. Em.	17/37	29/22	.150	⑥	INDEX	1-1 1/2	550	1500	.095
1967	390-V8-Std.-4 Bbl.	25/32	...	.100	#1	INDEX	1-1 1/2	575	1200	...
1967	390-V8-Std.-4 Bbl.-Ex. Em.	25/32	...	.100	#1	INDEX	1-1 1/2	625	1300	1/8

YEAR	MODEL OR TYPE	Float Level (IN.)	Fuel Level (IN.)	Fast Idle Cam (IN.)	Pump Travel Setting	Choke Setting	ON THE CAR ADJUSTMENT: Idle Screw Turns Open	Idle Speed (RPM)	Fast Idle Speed (RPM)	Dashpot Plunger Clearance (IN.)

Fairlane & Montego, continued

YEAR	MODEL OR TYPE	Float Level (IN.)	Fuel Level (IN.)	Fast Idle Cam (IN.)	Pump Travel Setting	Choke Setting	Idle Screw Turns Open	Idle Speed (RPM)	Fast Idle Speed (RPM)	Dashpot Plunger Clearance (IN.)
1968	200-6 Cyl.-Std.-1 Bbl.-Therm	1 3/32	...	...	(8)	2-LEAN	1-1 1/2	700	1400	(9)
1968	200-6 Cyl.-Aut.-1 Bbl.-Imco	1 3/32	...	...	(8)	1-LEAN	1-1 1/2	550	1500	(9)
1968	289-V8-Std.-2 Bbl.-Therm	3/8	3/4	.110	(7)	INDEX	1-1 1/2	625	1200	.125
1968	289-V8-Aut.-2 Bbl. Imco	3/8	3/4	.120	(7)	1-LEAN	1-1 1/2	550	1400	.125
1968	302-V8-Std.-2 Bbl.-Therm	3/8	3/4	...	(7)	INDEX	1-1 1/2	625	1200	.125
1968	302-V8-Aut.-2 Bbl.-Imco	3/8	3/4	.120	(7)	1-LEAN	1-1 1/2	550	1400	.125
1968	390-V8-Std.-2 Bbl.-Therm	...	7/8	.170	(1)	INDEX	1-1 1/2	625	1300	.125
1968	390-V8-Aut.-2 Bbl.-Imco	31/64	7/8	...	(1)	INDEX	1-1 1/2	550	1500	.125
1969	250-6 Cyl. Aut. 1 Bbl. C90F-A	1 3/32	...	...	.190	3-LEAN	...	550	1600	.080
1969	250-6 Cyl. Std. 1 Bbl. C90F-B	1 3/32	...	...	.190	1-LEAN	...	700	1400	.080
1969	250-6 cyl. aut. 1 Bbl. C90F-K	1 3/32	...	...	.90	3-LEAN	...	550	1600	(10)
1969	250-6 Cyl. Std. 1 Bbl. C90F-J	1 3/32	...	...	.190	1-LEAN	...	700	1400	(10)
1969	302-V8-Std.-2 Bbl.-C8HF-BD	3/8	3/4	.110	(1)	2-RICH	...	650	1400	1/8
1969	302-V8-Aut.-2 Bbl.-C9ZF-G	3/8	3/4	.110	(7)	INDEX	...	550	1600	1/8
1969	351-V8-Std.-2 Bbl.-C9ZF-A	9/16	15/16	.130	(1)	1-RICH	...	650	1300	7/64
1969	351-V8-Aut.-2 Bbl.-C90F-C	31/64	7/8	.100	(1)	2-RICH	...	550	1600	...
1969	351-V8-Std.-4 Bbl.-C9ZF-C	13/16	...	.130	#2	2-LEAN	...	675	1250	3/32
1969	351-V8-Aut.-4 Bbl.-C90F-D	13/16	...	.100	#2	1-LEAN	...	575	1400	...
1969	390-V8-Std.-4 Bbl.-C9ZF-E	13/16	...	.210	#3	INDEX	...	700	1300	1/8
1969	390-V8-Aut.-4 Bbl.-C90F-E	13/16	...	.230	#3	1-LEAN	...	550	1400	...
1970	2100—DOAFC-302, Std.	7/16	13/16	.130	#3 (11)	1-RICH	...	800/500 (12)	1400	(10)
1970	2100—DOAFD-302, Auto.	7/16	13/16	.130	#2 (11)	1-RICH	...	575	1500	1/8
1970	2100—DOAFV-302, Auto w. AC	7/16	13/16	.130	#2 (11)	1-RICH	...	600/500 (12)	1500	(10)
1970	2100—DOOFK—351C, Std.	7/16	13/16	.190	#4 (11)	INDEX	...	700/500 (12)	1500	(10)
1970	2100—DOOFL-351C, Auto.	7/16	13/16	.130	#3 (11)	1-RICH	...	600	1500	1/8
1970	2100—DOOFM-351C, Auto, w. AC	7/16	13/16	.130	#3 (11)	1-RICH	...	600/500 (12)	1500	(10)
1970	4300—DOOFZ, AB-351C, Std.	.79-.85	...	.160	#2	INDEX	...	800/500 (12)	1250	(10)
1970	4300—DOOFY, AC-351C, Auto.	.79-.85	...	.180	#2	INDEX	...	600	1400	.080
1970	4300—DOOFAA, AD-351C, Auto. w. AC	.79-.85	...	.180	#2	INDEX	...	600/500 (12)	1400	(10)
1970	4300—DOAFL, AB, AL-429, Std.	25/32	...	.220	#2	INDEX	...	700	1400	.070
1970	4300—DOAF, AG, AM-429, Auto	25/32	...	.170	#2	INDEX	...	600	1300	.070

(1)—Install pump operating rod in the inside hole in pump link and no. 3 hole in pump over travel lever for climatic conditions.
(2)—Install pump rod in top hole for winter and second hole from bottom for summer.
(3)—Position fast idle screw on second stop of fast idle cam.
(4)—Engine hot with fast idle screw on high step of cam.
(5)—Clearance between choke plate and air horn.
(6)—Install pump operating rod in the outside hole in pump link and no. 3 hole.
(7)—Install pump operating rod in the inside hole in pump link and no. 2 hole in pump over travel lever.
(8)—With throttle plate closed, install 3/32 in. diam. pin in Hi position, .090 clearance between pin and pump cover.
(9)—Two turns in after contact.
(10)—Solenoid equipped.
(11)—Rod position—overtravel lever.
(12)—Higher speed with solenoid energized.

Falcon—Comet—Mustang—Cougar

YEAR	MODEL OR TYPE	Float Level (IN.)	Fuel Level (IN.)	Fast Idle Cam (IN.)	Pump Travel Setting	Choke Setting	Idle Screw Turns Open	Idle Speed (RPM)	Fast Idle Speed (RPM)	Dashpot Plunger Clearance (IN.)
1964	144-6 Cyl.-Falcon-Std.	1	...	...	(2)	NONE	1-1 1/2	525	...	...
1964	144-6 Cyl.-Falcon-Aut.	1	...	...	(2)	NONE	1-1 1/2	525	...	3 1/2
1964	170-6 Cyl.-Falcon-Std.	1	...	...	(2)	NONE	1-1 1/2	525	...	...
1964	170-6 Cyl.-Falcon-Aut.	1	...	...	(2)	NONE	1-1 1/2	525	...	3 1/2
1964	170-6 Cyl.-Comet-Std.	1	...	(1)	(2)	INDEX	1-1 1/2	525	1300	...
1964	170-6 Cyl.-Comet-Aut.	1	...	(1)	(2)	INDEX	1-1 1/2	525	1500	3 1/2
1964	200-6 Cyl.-Falcon-Aut.	1	...	...	...	NONE	1-1 1/2	525	...	3 1/2
1964	200-6 Cyl.-Mustang-Aut.	1	...	...	...	NONE	1-1 1/2	525	...	3 1/2
1964	200-6 Cyl.-Comet-Aut.	1	...	...	...	INDEX	1-1 1/2	525	1500	3 1/2
1964	200-6 Cyl.-Mustang-Aut.	1	...	...	...	INDEX	1-1 1/2	525	1500	3 1/2
1964	260-V8-Std.-2 Bbl.	1/2	3/4	3/32 (4)	...	2-RICH	1-1 1/2	600	1300	...
1964	260-V8-Aut.-2 Bbl.	1/2	3/4	3/32 (4)	...	2-RICH	1-1 1/2	500	1600	5/64
1964	289-V8-Comet-Std.-4 Bbl.	21/32	29/32	1/16 (4)	(2)	1-LEAN	1-1 1/2	600	1300	...
1964	289-V8-Mustang-Std.-4 Bbl.	21/32	29/32	1/16 (4)	(2)	1-LEAN	1-1 1/2	600	1300	...
1964	289-V8-Comet-Aut.-4 Bbl.	21/32	29/32	1/16 (4)	(2)	3-LEAN	1-1 1/2	500	1500	5/64
1964	289-V8-Mustang-Aut.-4 Bbl.	21/32	29/32	1/16 (4)	(2)	3-LEAN	1-1 1/2	500	1500	5/64
1965	170-6 Cyl.-1 Bbl.-Std.	1 3/32	...	(1)	3/16	INDEX	1 1/2	600	1300	...
1965	170-6 Cyl.-1 Bbl.-Aut.	1 3/32	...	(1)	3/16	INDEX	1 1/2	525	1500	3 1/2
1965	170-6 Cyl.-1 Bbl.-Std.	1 3/32	...	(1)	3/16	2-LEAN	1 1/2	600	1300	...
1965	170-6 Cyl.-1 Bbl.-Aut.	1 3/32	...	(1)	3/16	INDEX	1 1/2	525	1500	3 1/2
1965	200-6 Cyl.-1 Bbl.-Std.	1 3/32	...	(1)	3/16	1-LEAN	1 1/2	600	1300	...
1965	200-6 Cyl.-1 Bbl.-Aut.	1 3/32	...	(1)	3/16	INDEX	1 1/2	525	1500	3 1/2
1965	260-V8-2 Bbl.-Std.	19/64	3/4	.125 (4)	#4	2-LEAN	1 1/2	600	1400	...
1965	260-V8-2 Bbl.-Aut.	19/64	3/4	.078 (4)	#3	2-RICH	1 1/2	500	1600	5/64
1965	289-V8-2 Bbl.-Std.	15/32	7/8	.130 (4)	#4	INDEX	1 1/2	600	1400	...

Falcon—Comet—Mustang—Cougar, continued

YEAR	MODEL OR TYPE	Float Level (IN.)	Fuel Level (IN.)	Fast Idle Cam (IN.)	Pump Travel Setting	Choke Setting	ON THE CAR ADJUSTMENT: Idle Screw Turns Open	Idle Speed (RPM)	Fast Idle Speed (RPM)	Dashpot Plunger Clearance (IN.)
1965	289-V8-2 Bbl.-Aut.	15/32	7/8	.120④	#3	2-RICH	1-1/2	500	1600	5/64
1965	289-V8-225 Hp.-4 Bbl.-Std.	29/64	29/32	.125④	#3	INDEX	1-1/4	600	1400	...
1965	289-V8-225 Hp.-4 Bbl.-Aut.	29/64	29/32	.120④	#3	2-RICH	1-1/4	500	1600	5/64
1965	289-V8-271 Hp.-4 Bbl.-Std.	29/64	7/8	.200④	#3	INDEX	1-1/4	775	1800	...
1965	289-V8-271 Hp.-4 Bbl.-Std.	15/32	7/8	.160④	#3	...	1-1/4	775	1400	...
1965	289-V8-271 Hp.-4 Bbl.-Aut.	29/64	7/8	.140④	...	INDEX	1-1/4	675	1800	5/64
1966	170-6 Cyl.-Std.-1 Bbl.	1 3/32	...	...	3/16	2-LEAN	1-1-1/2	575	1400	3-1/2
1966	170-6 Cyl.-Aut.-1 Bbl.	1 3/32	...	...	3/16	INDEX	1-1-1/2	525	1500	3-1/2
1966	200-6 Cyl.-Std.-1 Bbl.	1 3/32	...	...	3/16	1-LEAN	1-1-1/2	575	1400	...
1966	200-6 Cyl.-Aut.-1 Bbl.	1 3/32	...	...	3/16	INDEX	1-1-1/2	525	1500	3-1/2
1966	289-V8-Aut.-2 Bbl.	31/64	7/8	.130④	#3	2-RICH	1-1-1/2	500	1600	5/64
1966	289-V8-Std.-4 Bbl.	17/32	29/32	.120④	#3	2-RICH	1-1-1/2	575	1400	...
1966	289-V8-Aut.-4 Bbl.	9/16	15/16	.120④	#3	2-RICH	1-1-1/2	500	1600	5/64
1966	390-V8-Std.-2 Bbl.	31/64	7/8	.200④	#3	INDEX	1-1-1/2	575	1300	...
1966	390-V8-Aut.-2 Bbl.	31/64	7/8	.180④	#3	INDEX	1-1-1/2	500	1400	5/64
1966	390-V8-Std.-4 Bbl.	17/32	29/32	.160④	#3	2-RICH	1-1-1/2	575	1200	...
1966	390-V8-Aut.-4 Bbl.	17/32	29/32	.140④	#3	1-RICH	1-1-1/2	500	1300	5/64
1967	170-6 Cyl.-Std.-1 Bbl.	1 3/32	...	...	.190	2-LEAN	1-1-1/2	575	1300	...
1967	170-6 Cyl.-Std.-1 Bbl.-Ex. Em.	1 3/32	...	...	.190	2-LEAN	1-1-1/2	700	1400	2
1967	170-6 Cyl.-Aut.-1 Bbl.	1 3/32	...	...	.190	INDEX	1-1-1/2	500	1500	3-1/2
1967	170-6 Cyl.-Aut.-1 Bbl.-Ex. Em.	1 3/32	...	...	.190	INDEX	1-1-1/2	550	1500	2
1967	200-6 Cyl.-Std.-1 Bbl.	1 3/32	...	...	.190	1-LEAN	1-1-1/2	575	1300	...
1967	200-6 Cyl.-Std.-1 Bbl.-Ex. Em.	1 3/32	...	...	.190	2-LEAN	1-1-1/2	700	1400	2
1967	200-6 Cyl.-Aut.-1 Bbl.	1 3/32	...	...	.190	INDEX	1-1-1/2	500	1500	3-1/2
1967	200-6 Cyl.-Aut.-1-Bbl.-Ex. Em.	1 3/32	...	...	.190	INDEX	1-1-1/2	500	1500	2
1967	289-V8-Std.-2 Bbl.	3/8	3/4	.110	#3	INDEX	1-1-1/2	575	1400	...
1967	289-V8-Std.-2 Bbl.-Ex. Em.	17/32	29/32	.110	#3	INDEX	1-1-1/2	625	1400	.125
1967	289-V8-Aut.-2 Bbl.	17/32	29/32	.110	#3	2-RICH	1-1-1/2	475	1600	.075
1967	289-V8-Aut.-2 Bbl.-Ex. Em.	17/32	29/32	.110	#3	2-RICH	1-1-1/2	550	1600	.125
1967	289-V8-Std.-4 Bbl.	25/32	...	.100	...	INDEX	1-1-1/2	...	...	...
1967	289-V8-Std.-4 Bbl.-Ex. Em.	25/32	...	.100	#1	INDEX	1-1-1/2	625	1400	1/8
1967	289-V8-Aut.-4 Bbl.	25/32	...	.100	#1	INDEX	1-1-1/2	525	1600	...
1967	289-V8-Aut.-4 Bbl.-Ex. Em.	25/32	...	.100	#1	INDEX	1-1-1/2	550	1600	1/8
1967	390-V8-Std.-2 Bbl.	31/64	7/8	.160	#3	INDEX	1-1-1/2	575	1300	...
1967	390-V8-Std.-2 Bbl.-Ex. Em.	17/32	29/32	.170	#4	INDEX	1-1-1/2	625	1300	.095
1967	390-V8-Aut.-2 Bbl.	3/8	3/4	.150	#3	2-RICH	1-1-1/2	475	1400	...
1967	390-V8-Aut.-2 Bbl.-Ex. Em.	17/32	29/32	.150	#4	INDEX	1-1-1/2	550	1500	.095
1967	390-V8-Std.-4 Bbl.	25/32	...	.100	#1	INDEX	1-1-1/2	575	1200	...
1967	390-V8-Std.-4 Bbl.-Ex. Em.	25/32	...	.100	#1	INDEX	1-1-1/2	625	1300	1/8
1967	390-V8-Aut.-4 Bbl.	25/32	...	.100	#1	INDEX	1-1-1/2	475	1400	...
1967	390-V8-Aut.-4 Bbl.-Ex. Em.	25/32	...	.100	#1	INDEX	1-1-1/2	550	1500	1/8
1968	200-6 Cyl.-Std.-1 Bbl.-Therm	1 3/32	...	...	⑤	2-LEAN	1-1-1/2	700	1400	⑥
1968	200-6 Cyl.-Aut.-1 Bbl.-Imco	1 3/32	...	...	⑤	1-LEAN	1-1-1/2	550	1500	⑥
1968	289-V8-Std.-2 Bbl.-Therm	3/8	3/4	.110	...	INDEX	1-1-1/2	625	1200	.125
1968	289-V8-Aut.-2 Bbl.-Imco	3/8	3/4	.120	...	1-LEAN	1-1-1/2	550	1400	.125
1968	302-V8-Std.-4 Bbl.-Therm	13/16	...	.090	7/16	INDEX	1-1-1/2	625	1250	.062
1968	302-V8-Aut.-4 Bbl.-Imco	13/16	...	.100	7/16	2-RICH	1-1-1/2	550	1400	.093
1969	200-6 Cyl.-Std.-1 Bbl., C8DF-B	1 3/32	...	...	.150	3-LEAN	...	750	1400	3
1969	200-6 Cyl.-Aut.-1 Bbl., C8OF-B	1 3/32	...	...	.190	1-LEAN	...	550	1500	2
1969	250-6 Cyl.-Aut.-1 Bbl., C9OF-A	1 3/32	...	...	.190	3-LEAN	...	550	1600	.080
1969	250-6 Cyl.-Aut.-1 Bbl., C9OF-K	1 3/32	...	...	.190	3-LEAN	...	550	1600	⑩
1969	250-6 Cyl.-Std.-1 Bbl., C9OF-B	1 3/32	...	...	.190	1-LEAN	...	700	1400	...
1969	250-6 Cyl.-Std.-1 Bbl., C9OF-J	1 3/32	...	...	.190	1-LEAN	...	700	1400	⑩
1969	302-V8-Std.-2 Bbl., C8AF-BD	3/8	3/4	.110	①	2-RICH	...	650	1400	1/8
1969	302-V8-Aut.-2 Bbl., C9AF-A	3/8	3/4	.110	⑦	INDEX	...	550	1600	1/8
1969	302-V8-Aut.-2 Bbl., C9ZF-G	3/8	3/4	.110	⑦	INDEX	...	550	1600	1/8
1969	351-V8-Std.-2 Bbl., C9ZF-A	9/16	15/16	.130	①	1-RICH	...	650	1300	7/64
1969	351-V8-Aut.-2 Bbl., C9ZF-B	31/64	7/8	.100	①	2-RICH	...	550	1600	...
1969	351-V8-Std.-4 Bbl., C9ZF-C	13/16	...	.130	#2	2-LEAN	...	675	1250	3/32
1969	351-V8-Aut.-4 Bbl., C9ZF-D	13/16	...	.100	#2	1-LEAN	...	575	1400	...
1969	390-V8-Std.-4 Bbl., C9ZF-E	13/16	...	.210	#3	INDEX	...	700	1300	1/8
1969	390-V8-Aut.-4 Bbl., C9ZF-F	13/16	...	.230	#3	1-LEAN	...	550	1400	...
1969	170-6 Cyl.-1 Bbl., C8DF-G	7/32	...	...	...	INDEX	...	...	...	...
1969	170-6 Cyl.-1 Bbl., C8DF-H	7/32	...	...	...	1-LEAN	...	...	...	...
1970	2100—DOAFC-302, Std.	7/16	13/16	.130	#3⑨	1-RICH	...	800/500 ⑩	1400	⑧
1970	2100—DOAFD-302, Auto.	7/16	13/16	.130	#2⑨	1-RICH	...	575	1500	1/8

Falcon—Comet—Mustang—Cougar, continued

YEAR	MODEL OR TYPE	Float Level (IN.)	Fuel Level (IN.)	Fast Idle Cam (IN.)	Pump Travel Setting	Choke Setting	ON THE CAR ADJUSTMENT: Idle Screw Turns Open	Idle Speed (RPM)	Fast Idle Speed (RPM)	Dashpot Plunger Clearance (IN.)
1970	2100—DOAFV-302, Auto w. AC	7/16	13/16	.130	#2⑨	1-RICH	...	600/500 ⑩	1500	⑧
1970	2100—DOAFE-351W, Std.	7/16	13/16	.190	#3⑨	2-LEAN	...	700/500 ⑩	1300	⑧
1970	2100—DOAFF-351W, Auto.	7/16	13/16	.170	#4⑨	2-LEAN	...	575	1600	1/8
1970	2100—DOAFV-351W Auto, w. AC	7/16	13/16	.170	#4⑨	2-LEAN	...	600/500 ⑩	1600	⑧
1970	4300—DOOFZ, AB-351C, Std.	.79-.85	...	.160	#2	INDEX	...	800/500 ⑩	1250	⑧
1970	4300—DOOFY, AC-351C, Auto.	.79-.85	...	.180	#2	INDEX	...	.600	1400	.080
1970	4300—DOOFAA, AD-351C, Auto. w. AC	.79-.85	...	.180	#2	INDEX	...	600/500 ⑩	1400	⑧

①—Set fast idle screw on high step of cam. Adjust screw to recommended rpm.
②—Install pump rod in top hole for winter and second hole from bottom for summer.
③—Clearance between top side of fast idle cam and cast stop on back of choke housing.
④—Clearance between choke plate and air horn wall.
⑤—Install 7/32 diam. roll pin in Hi position.
⑥—Two turns after contact.
⑦—Install pump operating rod in the inside hole in pump link and #2 hole in pump over-travel lever.
⑧—Solenoid equipped.
⑨—Rod position—overtravel lever.
⑩—Higher speed with solenoid energized.

Ford—Thunderbird

YEAR	MODEL OR TYPE	Float Level (IN.)	Fuel Level (IN.)	Fast Idle Cam (IN.)	Pump Travel Setting	Choke Setting	ON THE CAR ADJUSTMENT: Idle Screw Turns Open	Idle Speed (RPM)	Fast Idle Speed (RPM)	Dashpot Plunger Clearance (IN.)
1964	1-Bbl. 223-6 Cyl.-Std.	1	...	...	③	...	1-1½	525	...	...
1964	1-Bbl. 223-6 Cyl.-Aut.	1	...	...	③	...	1-1½	525	...	3½
1964	2-Bbl. 289-V8-Std.	5/8	7/8	3/32 ②	③	2-RICH	1-1½	575	1300	...
1964	2-Bbl. 289-V8-Aut.	5/8	7/8	3/32 ②	③	2-RICH	1-1½	475	1600	5/64
1964	4-Bbl. 352-V8-Std.	21/32	29/32	1/16 ②	③	1-LEAN	1-1½	575	1300	...
1964	4-Bbl. 352-V8-Aut.	21/32	29/32	1/16 ②	③	3-LEAN	1-1½	475	1500	5/64
1964	4-Bbl. 390-V8-Std.	21/32	29/32	1/16 ②	③	1-RICH	1-1½	575	1300	...
1964	4-Bbl. 390-V8-Aut.	21/32	29/32	1/16 ②	③	1-LEAN	1-1½	475	1500	5/64
1965	1-Bbl. 240-6 Cyl.-Std.	1/32	...	...	③	INDEX	1-1½	525	1300	3½
1965	1-Bbl. 240-6 Cyl.-Aut.	1/32	...	...	③	INDEX	1-1½	525	1500	...
1965	2-Bbl. 289-V8 Std.	15/32	7/8	.130 ②	③	INDEX	1-1½	600	1400	...
1965	2-Bbl. 289-V8-Aut.	15/32	7/8	.120 ②	③	2-RICH	1-1½	500	1600	5/64
1965	4-Bbl. 352-V8-Std.	15/32	7/8	.130 ②	③	INDEX	1-1½	600	1300	5/16
1965	4-Bbl. 352-V8-Aut.	15/32	7/8	.120 ②	③	INDEX	1-1½	500	1500	5/16
1965	2-Bbl. 390-V8-Std.	15/32	7/8	.140 ②	③	2-RICH	1-1½	600	1300	...
1965	2-Bbl. 390-V8-Aut.	15/32	7/8	.150 ②	③	2-RICH	1-1½	500	1500	5/64
1965	4-Bbl. 390-V8-Std.	15/32	7/8	.130 ②	③	INDEX	1-1½	600	1500	5/16
1965	4-Bbl. 390-V8-Aut.	15/32	7/8	.120 ②	③	INDEX	1-1½	500	1500	5/16
1965	4-Bbl. 390-V8-T Bird	15/32	7/8	.125 ②	...	INDEX	1½	500	1500	3/32
1966	240-6 Cyl.-Std.-1 Bbl.	1 3/32	...	.200 ②	③	INDEX	1-1½	525	1500	...
1966	240-6 Cyl.-Aut.-1 Bbl.	1 3/32	...	.200 ②	③	INDEX	1-1½	525	1600	3½
1966	289-V8-Std.-2 Bbl.	31/64	7/8	.140 ②	③	INDEX	1-1½	575	1400	...
1966	289-V8-Aut.-2 Bbl.	31/64	7/8	.130 ②	③	2-RICH	1-1½	500	1600	5/64
1966	352-V8-Aut.-4 Bbl.	17/32	29/32	.140 ②	③	INDEX	1-1½	500	1500	5/64
1966	390-V8-Std.-2 Bbl.	31/64	7/8	.200 ②	③	INDEX	1-1½	500	1400	...
1966	390-V8-Aut.-2 Bbl.	31/64	7/8	.180 ②	③	INDEX	1-1½	500	1400	5/64
1966	390-V8-Std.-4 Bbl.	17/32	29/32	.160 ②	③	2-RICH	1-1½	575	1200	...
1966	390-V8-Aut.-4 Bbl.	17/32	29/32	.140 ②	③	1-RICH	1-1½	500	1300	5/64
1966	428-V8-Std.-4 Bbl.	17/32	29/32	.160 ⑦	③	2-RICH	1-1½	575	1200	...
1966	428-V8-Aut.-4 Bbl.	17/32	29/32	.140 ②	③	1-RICH	1-1½	500	1300	5/64
1967	240-6 Cyl.-Std.-1 Bbl.	1 3/32	...	...	③	INDEX	1-1½	525	1500	6
1967	240-6 Cyl.-Aut.-1 Bbl.	1 3/32	...	...	③	INDEX	1-1½	500	1600	6
1967	240-6 Cyl.-Std.-1 Bbl.-Ex. Em.	1 3/32	...	...	③	INDEX	1-1½	600	1500	2
1967	240-6 Cyl.-Aut.-1 Bbl.-Ex. Em.	1 3/32	...	...	③	1-LEAN	1-1½	500	1600	2
1967	289-V8-Std.-2 Bbl.	31/64	7/8	.130	③	INDEX	1-1½	575	1400	...
1967	289-V8-Aut.-2 Bbl.	31/64	7/8	.110	③	2-RICH	1-1½	475	1600	...
1967	289-V8-Std.-2 Bbl.-Ex. Em.	17/32	29/32	.110	③	INDEX	1-1½	625	1400	.125
1967	289-V8-Aut.-2 Bbl.-Ex. Em.	17/32	29/32	.110	③	2-RICH	1-1½	550	1600	.110
1967	390-V8-Std.-2 Bbl.	31/64	7/8	.160	③	INDEX	1-1½	575	1300	...
1967	390-V8-Aut.-2 Bbl.	3/8	3/4	.150	③	2-RICH	1-1½	475	1400	...
1967	390-V8-Std.-2 Bbl.-Ex.Em.	17/32	29/32	.170	③	INDEX	1-1½	625	1300	.095
1967	390-V8-Aut.-2 Bbl.-Ex. Em.	17/32	29/32	.150	③	INDEX	1-1½	550	1500	.095
1967	390-V8-Std.-4 Bbl.	25/32	...	.100	#1 hole	INDEX	1-1½	575	1200	...
1967	390-V8-Aut.-4 Bbl.	25/32	...	.100	#1 hole	INDEX	1-1½	475	1400	...
1967	390-V8-Std.-4 Bbl.-Ex. Em.	25/32	...	.100	#1 hole	INDEX	1-1½	625	1300	1/8
1967	390-V8-Aut.-4 Bbl.-Ex.Em.	25/32	...	.100	#1 hole	INDEX	1-1½	550	1500	1/8
1967	428-V8-Std.-4 Bbl.	25/32	...	.100	#2 hole	INDEX	1-1½	575	1200	...
1967	428-V8-Aut.-4 Bbl.	25/32	...	.090	#2 hole	INDEX	1-1½	475	1200	...
1967	428-V8-Std.-4 Bbl.-Ex. Em.	25/32	...	.090	#2 hole	INDEX	1-1½	475	1200	...
1967	428-V8-Aut.-4 Bbl.-Ex.Em.	25/32	...	.100	#1 hole	INDEX	1-1½	550	1500	1/8

YEAR	MODEL OR TYPE	Float Level (IN.)	Fuel Level (IN.)	Fast Idle Cam (IN.)	Pump Travel Setting	Choke Setting	ON THE CAR ADJUSTMENT: Idle Screw Turns Open	Idle Speed (RPM)	Fast Idle Speed (RPM)	Dashpot Plunger Clearance (IN.)

Ford—Thunderbird, continued

YEAR	MODEL OR TYPE	Float Level (IN.)	Fuel Level (IN.)	Fast Idle Cam (IN.)	Pump Travel Setting	Choke Setting	Idle Screw Turns Open	Idle Speed (RPM)	Fast Idle Speed (RPM)	Dashpot Plunger Clearance (IN.)
1968	240-6 Cyl.-Aut.-1 Bbl.-Imco	1 3/32	...	...	Ⓐ	3-LEAN	1-1 1/2	500	1600	.080
1968	302-V8-Std.-2 Bbl.-Therm	3/8	3/4	...	③	INDEX	1-1 1/2	625	1200	.125
1968	302-V8-Aut.-2 Bbl.-Imco	3/8	3/4	.120	③	1-LEAN	1-1 1/2	550	1400	.125
1968	390-V8-Std.-2 Bbl.-Therm	31/64	7/8	.170	③	INDEX	1-1 1/2	625	1300	.125
1968	390-V8-Aut.-2 Bbl.-Imco	3/8	3/4	.170	③	INDEX	1-1 1/2	550	1500	.125
1968	390-V8-Std.-4 Bbl.-Therm	25/32	...	.100	#3 hole	1-RICH	1-1 1/2	625	1300	.093
1968	390-V8-Aut.-4 Bbl.-Imco	25/32	...	.100	#3 hole	2-RICH	1-1 1/2	550	1400	.093
1968	428-V8-Std.-4 Bbl.-Therm	25/32	...	.100	#3 hole	1-RICH	1-1 1/2	625	1300	.093
1968	428-V8-Aut.-4 Bbl.-Imco	25/32	...	.100	#3 hole	2-RICH	1-1 1/2	550	1400	.093
1969	240-6 cyl.-Aut.-1 Bbl.-C8AF-E	1 3/32	...	...	.190	3-LEAN	...	500	1600	.080
1969	302-V8-Std.-2 Bbl.-C8AF-BD	3/8	3/4	.110	①	2-RICH	...	650	1400	1/8
1969	302-V8-Aut.-2 Bbl.-C9AF-A	3/8	3/4	.110	④	INDEX	...	550	1600	1/8
1969	390-V8-Std.-2 Bbl.-C9AF-B	3/64	7/8	.170	①	1-RICH	...	650	1300	1/8
1969	390-V8-Aut.-2 Bbl.-C9AF-C	31/64	7/8	.100	①	2-RICH	...	550	1500	1/8
1969	429-V8-Aut.-2 Bbl.-C9AF-J	31/64	7/8	.100	①	2-RICH	...	550	1500	1/8
1969	429-V8-Std.-4 Bbl.-C9AF-G	25/32	...	.220	#2	INDEX	...	650	1200	3/32
1969	429-V8-Aut.-4 Bbl.-C9AF-R	25/32	...	.160	#2	1-RICH	...	575	1300	...
1969	429-V8-Aut.-4 Bbl.-C8SF-H	25/32	...	.160	#2	1-RICH	...	550	1300	3/32
1970	2100—DOAFC-302, Std.	7/16	13/16	.130	#3⑥	1-RICH	...	800/500⑦	1400	⑤
1970	2100—DOAFD-302, Auto	7/16	13/16	.130	#2⑥	1-RICH	...	575	1500	1/8
1970	2100—DOAFV-302, Auto w. AC	7/16	13/16	.130	#2⑥	1-RICH	...	600/500⑦	1500	⑤
1970	2100—DOAFE-351W, Std.	7/16	13/16	.190	#3⑥	2-LEAN	...	700/500⑦	1300	⑤
1970	2100—DOAFF-351W, Auto	7/16	13/16	.170	#4⑥	2-LEAN	...	575	1600	1/8
1970	2100—DOAFV-351W Auto, w. AC	7/16	13/16	.170	#4⑥	2-LEAN	...	600/500⑦	1600	⑤
1970	2100—DOAFY-390, Std.	7/16	13/16	.170	#3⑥	1-RICH	...	750/500⑦	1400	⑤
1970	2100—DOAFZ-390, Auto	7/16	13/16	.170	#3⑥	...	...	575	1500	1/8
1970	2100—DOAFAA-390, Auto	7/16	13/16	.160	#3⑥	...	...	600/500⑦	1500	⑤
1970	2100—DOAFJ-429, Auto	7/16	13/16	.160	#3⑥	...	...	590	1400	1/8
1970	2100—DOAFT-429, Auto, w. AC	7/16	13/16	.160	#3⑥	...	...	600/500⑦	1400	⑤
1970	4300—DOAFM, AD, AJ-428, Auto	1.00	...	.120	#3	2-RICH	...	600	1600	.080
1970	4300—DOAFR, AE, AK—428 Auto w. AC	1.00	...	.120	#3	2-RICH	...	600/500⑦	1600	⑤
1970	4300—DOAFL, AB, AL-429, Std.	25/32	...	.220	#2	INDEX	...	700	1400	.070
1970	4300—DOAF, AG, AM-429, Auto	25/32	...	.170	#2	INDEX	...	600	1300	.070
1970	4300—DOSF, A, D, E-429, Auto	25/32	...	.170	#2	INDEX	...	600	1300	.070

① —Clearance between top side of fast idle cam and cast stop on back of choke housing.

② —Clearance between choke plate and air horn wall.

③ —With link in inboard hole in pump lever, place overtravel lever in long stroke hole. For extremely cold weather intermediate for moderate, and short stroke for warm weather.

④ —Install pump operating rod in the inside hole in pump link and No. 2 hole in pump overtravel lever.

⑤ —Solenoid equipped.

⑥ —Rod position—overtravel lever.

⑦ —Higher speed with solenoid energized.

Ⓐ —Install 3/32 diam. Roll pin in Hi position.

JEEP

YEAR	MODEL OR TYPE	Float Level (IN.)	Fuel Level (IN.)	Fast Idle Cam (IN.)	Pump Travel Setting	Choke Setting	Idle Screw Turns Open	Idle Speed (RPM)	Fast Idle Speed (RPM)	Dashpot Plunger Clearance (IN.)
1970	4300—OWM4-390, Std.	13/16	...	...	center hole	2-RICH	...	①	1600	1/16
1970	4300—OWA4-390, Auto.	13/16	...	...	center hole	2-RICH	...	①	1600	1/8

① —See decal for idle speed.

Mercury—Lincoln

YEAR	MODEL OR TYPE	Float Level (IN.)	Fuel Level (IN.)	Fast Idle Cam (IN.)	Pump Travel Setting	Choke Setting	Idle Screw Turns Open	Idle Speed (RPM)	Fast Idle Speed (RPM)	Dashpot Plunger Clearance (IN.)
1964	2-Bbl.-V8-390-250 Hp.-Std.	21/32	29/32	1/16 ③	②	1-LEAN	1-1 1/2	575	1300	NONE
1964	2-Bbl.-V8-390-250 Hp.-Aut.	21/32	29/32	1/16 ③	②	1-LEAN	1-1 1/2	475	1500	5/64
1964	2-Bbl.-V8-390-266 Hp.-Aut.	5/8	7/8	9/64 ③	②	2-LEAN	1-1 1/2	475	1500	5/64
1964	4-Bbl.-V8-390-Std.	21/32	29/32	1/16 ③	②	1-RICH	1-1 1/2	575	1300	NONE
1964	4-Bbl.-V8-390-Aut.	21/32	29/32	1/16 ③	②	1-LEAN	1-1 1/2	475	1500	5/64
1965	2-Bbl.-V8-390-Std.	15/32	7/8	.140 ③	②	2-RICH	1-1 1/2	600	1300	...
1965	2-Bbl.-V8-390-Aut.	15/32	7/8	.150 ③	②	2-RICH	1-1 1/2	500	1500	5/64
1965	4-Bbl.-V8-390-Std.	15/32	7/8	.120 ③	②	INDEX	1-1 1/2	600	1500	5/16
1965	4-Bbl.-V8-390-Aut.	15/32	7/8	.120 ③	②	INDEX	1-1 1/2	500	1500	5/64
1966	390-V8-Std.-2 Bbl.	31/64	7/8	.200 ③	②	INDEX	1-1 1/2	500	1400	...
1966	390-V8-Aut.-2 Bbl.	31/64	7/8	.180 ③	②	INDEX	1-1 1/2	500	1400	5/64
1966	390-V8-Std.-4 Bbl.	17/32	29/32	.160 ③	②	2-RICH	1-1 1/2	575	1200	...
1966	390-V8-Aut.-4 Bbl.	17/32	29/32	.140 ③	②	1-RICH	1-1 1/2	500	1300	5/64
1966	410-V8-Std.-4 Bbl.	17/32	29/32	.160 ③	②	2-RICH	1-1 1/2	575	1200	...
1966	410-V8-Aut.-4 Bbl.	17/32	29/32	.140 ③	②	1-RICH	1-1 1/2	500	1300	5/64
1966	428-V8-Std.-4 Bbl.	17/32	29/32	.160 ③	②	2-RICH	1-1 1/2	575	1200	...
1966	428-V8-Aut.-4 Bbl.	17/32	29/32	.140 ③	②	1-RICH	1-1 1/2	500	1300	5/64

YEAR	MODEL OR TYPE	Float Level (IN.)	Fuel Level (IN.)	Fast Idle Cam (IN.)	Pump Travel Setting	Choke Setting	ON THE CAR ADJUSTMENT			
							Idle Screw Turns Open	Idle Speed (RPM)	Fast Idle Speed (RPM)	Dashpot Plunger Clearance (IN.)

Mercury—Lincoln, continued

YEAR	MODEL OR TYPE	Float Level (IN.)	Fuel Level (IN.)	Fast Idle Cam (IN.)	Pump Travel Setting	Choke Setting	Idle Screw Turns Open	Idle Speed (RPM)	Fast Idle Speed (RPM)	Dashpot Plunger Clearance (IN.)
1967	390-V8-Std.-2 Bbl.	31/64	7/8	.160	②	INDEX	1-1½	575	1300	...
1967	390-V8-Aut.-2 Bbl.	3/8	3/4	.150	②	2-RICH	1-1½	475	1400	...
1967	390-V8-Std.-2 Bbl.-Ex. Em.	17/32	29/32	.170	②	INDEX	1-1½	625	1300	.095
1967	390-V8-Aut.-2 Bbl.-Ex. Em.	17/32	29/32	.150	②	INDEX	1-1½	550	1500	.095
1967	390-V8-Std.-4 Bbl.	25/32	...	.100	#1 hole	INDEX	1-1½	575	1200	...
1967	390-V8-Aut.-4 Bbl.	25/32	...	.100	#1 hole	INDEX	1-1½	475	1400	...
1967	390-V8-Std.-4 Bbl.-Ex. Em.	25/32	...	.100	#1 hole	INDEX	1-1½	625	1300	1/8
1967	390-V8-Aut.-4 Bbl.-Ex. Em.	25/32	...	.100	#1 hole	INDEX	1-1½	550	1500	1/8
1967	410-V8-Std.-4 Bbl.	25/32	...	.100	#2 hole	INDEX	1-1½	575	1200	...
1967	410-V8-Aut.-4 Bbl.	25/32	...	.090	#2 hole	INDEX	1-1½	475	1200	...
1967	410-V8-Std.-4 Bbl.-Ex. Em.	25/32	...	.100	#1 hole	INDEX	1-1½	625	1300	1/8
1967	410-V8-Aut.-4 Bbl.-Ex. Em.	25/32	...	.100	#1 hole	INDEX	1-1½	550	1500	1/8
1967	428-V8-Std.-4 Bbl.	25/32	...	.100	#2 hole	INDEX	1-1½	575	1200	...
1967	428-V8-Aut.-4 Bbl.	25/32	...	.090	#2 hole	INDEX	1-1½	475	1200	...
1967	428-V8-Std.-4 Bbl.-Ex. Em.	25/32	...	.090	#2 hole	INDEX	1-1½	475	1200	...
1967	428-V8-Aut.-4 Bbl.-Ex. Em.	25/32	...	.100	#1 hole	INDEX	1-1½	550	1500	1/8
1968	390-V8-Std.-2 Bbl.-Therm	31/64	7/8	.170	#3 hole	INDEX	1-1½	625	1300	.125
1968	390-V8-Aut.-2 Bbl.-Imco	3/8	3/4	.170	#3 hole	INDEX	1-1½	550	1500	.125
1968	390-V8-Aut.-2 Bbl.-Imco	31/64	7/8	.100	#3 hole	INDEX	1-1½	550	1500	.125
1968	390-V8-Std.-4 Bbl.-Therm	25/32	...	.100	#3 hole	1-RICH	1-1½	625	1300	.093
1968	390-V8-Aut.-4 Bbl.-Imco	25/32	...	.100	#3 hole	2-RICH	1-1½	550	1400	.093
1968	428-V8-Std.-4 Bbl.-Therm	25/32	...	.100	#3 hole	1-RICH	1-1½	625	1300	.093
1968	428-V8-Aut.-4 Bbl.-Imco	25/32	...	.100	#3 hole	2-RICH	1-1½	550	1400	.093
1969	390-V8-std.-2 Bbl.-C9AF-B	31/64	7/8	.170	①	1-RICH	...	650	1300	1/8
1969	390-V8-Aut.-2 Bbl.-C9AF-C	31/64	7/8	.100	①	2-RICH	...	550	1500	1/8
1969	390-V8-Aut.-2 Bbl.-C9MF-A	31/64	7/8	.120	①	2-RICH	...	550	1500	1/8
1969	429-V8-Aut.-2 Bbl. C9AF-J	31/64	7/8	.100	①	2-RICH	...	550	1500	1/8
1969	429-V8-aut.-4 Bbl. C9AF-R	25/32	...	.160	#2	1-RICH	...	575	1300	...
1970	2100—DOAFC-302, Std.	7/16	13/16	.130	#3⑤	1-RICH	...	800/500 ⑥	1400	④
1970	2100—DOAFD-302, Auto.	7/16	13/16	.130	#2⑤	1-RICH	...	575	1500	1/8
1970	2100—DOAFV-302, Auto w. AC	7/16	13/16	.130	#2⑤	1-RICH	...	600/500 ⑥	1500	④
1970	2100—DOAFE-351W, Std.	7/16	13/16	.190	#3⑤	2-LEAN	...	700/500 ⑥	1300	④
1970	2100—DOAFF-351W, Auto.	7/16	13/16	.170	#4⑤	2-LEAN	...	575	1600	1/8
1970	2100—DOAFV-351W, Auto, w. AC	7/16	13/16	.170	#4⑤	2-LEAN	...	600/500 ⑥	1600	④
1970	2100—DOAFY-390, Std.	7/16	13/16	.170	#3	1-RICH	...	750/500 ⑥	1400	④
1970	2100—DOAFZ-390, Auto	7/16	13/16	.160	#3⑤	...	...	575	1500	1/8
1970	2100—DOAFAA-390, Auto	7/16	13/16	.160	#3⑤	...	...	600/500 ⑥	1500	④
1970	2100—DOAFJ-429, Auto	7/16	13/16	.160	#4⑤	...	...	590	1400	1/8
1970	2100—DOAFT-429, Auto, w. AC	7/16	13/16	.160	#3⑤	...	...	600/500 ⑥	1400	④
1970	4300—DOAFM, AD, AJ-428, Auto	1.00	...	.120	#3	2-RICH	...	600	1600	.080
1970	4300—DOAFR, AE, AK-428, Auto w. AC	1.00	...	.120	#3	2-RICH	...	600/500 ⑥	1600	④
1970	4300—DOAFL, AB, AL-429, Std.	25/32	...	.220	#2	INDEX	...	700	1400	.070
1970	4300—DOAF, AG, AM-429, Auto	25/32	...	.170	#2	INDEX	...	600	1300	.070
1970	4300—DOVF, A, B, C-460, Auto.	25/32	...	.170	#2	1-RICH	...	600	1250	.100

① —Clearance between top side of fast idle cam and cast stop on back of choke housing.

② —With link in inboard hole in pump lever, place over travel lever in longest stroke hole. For extremely cold weather, intermediate for moderate, and short stroke for warm weather.

③ —Clearance between choke plate and air horn wall.

④ —Solenoid equipped.

⑤ —Rod position—overtravel lever.

⑥ —Higher speed with solenoid energized.

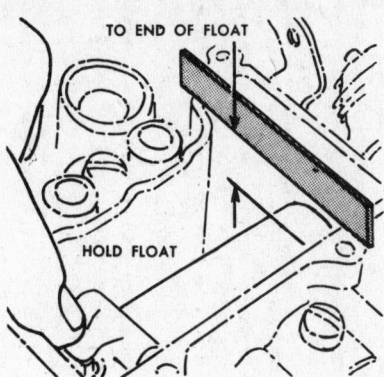

Ford two- and four-barrel float setting

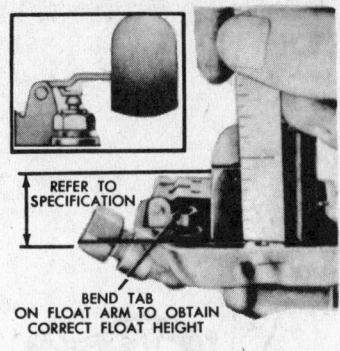

Ford single-barrel float adjustment

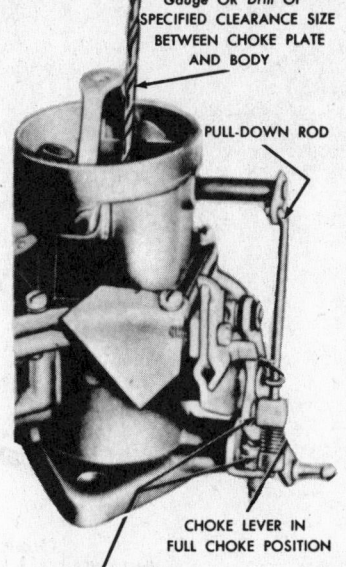

Ford single-barrel manual choke pull-down adjustment

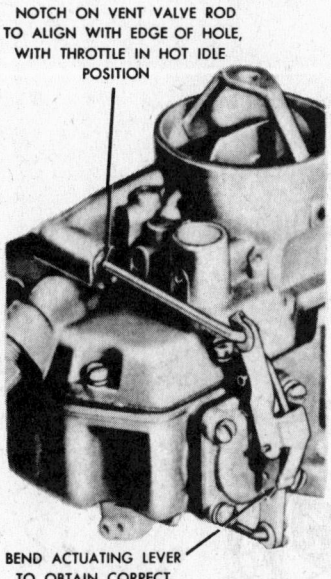

Ford single-barrel float bowl vent valve adjustment

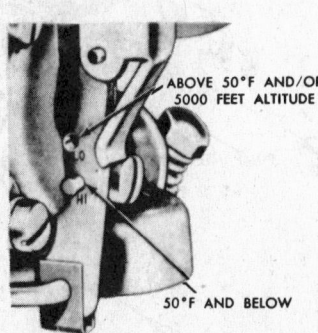

Ford single-barrel idle fuel mixture adjustment

Ford single-barrel accelerating pump lever adjustment

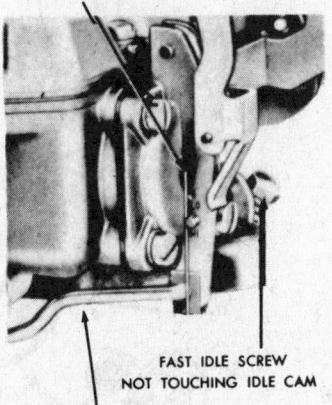

Ford single-barrel accelerating pump adjustment

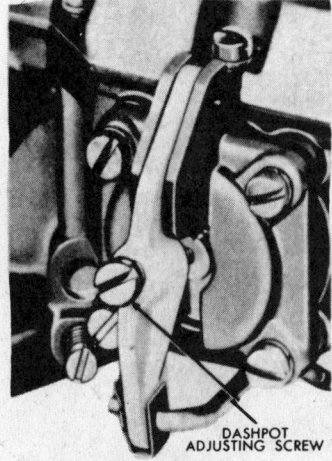

Anti-stall dashpot adjustment

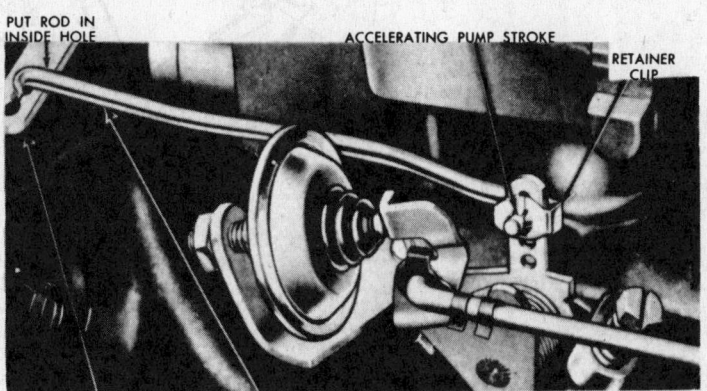

Accelerating pump adjustment

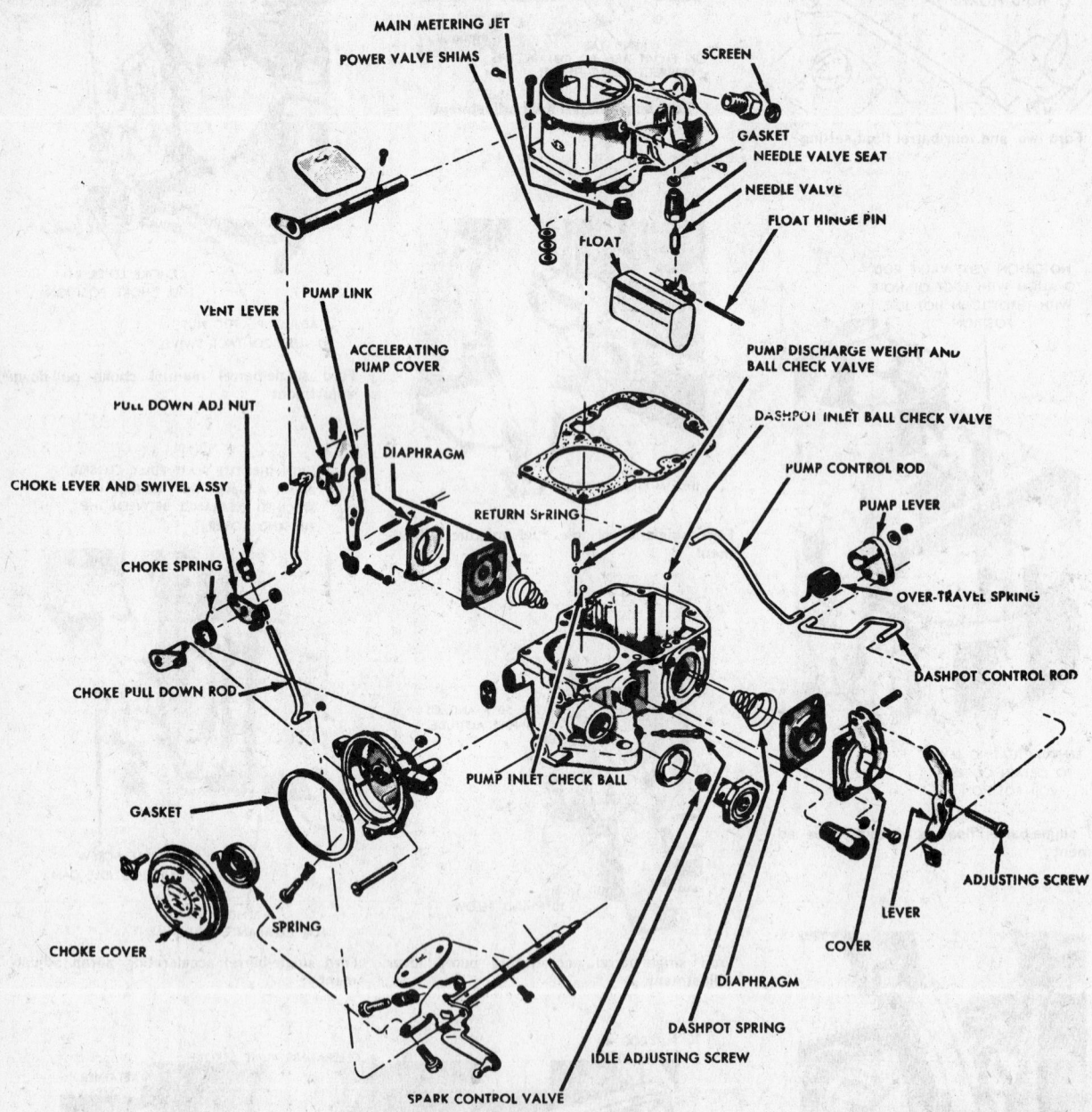

MAIN METERING JET
POWER VALVE SHIMS
SCREEN
GASKET
NEEDLE VALVE SEAT
NEEDLE VALVE
FLOAT HINGE PIN
FLOAT
PUMP LINK
VENT LEVER
ACCELERATING PUMP COVER
PUMP DISCHARGE WEIGHT AND BALL CHECK VALVE
PULL DOWN ADJ NUT
DASHPOT INLET BALL CHECK VALVE
DIAPHRAGM
PUMP CONTROL ROD
CHOKE LEVER AND SWIVEL ASSY
PUMP LEVER
RETURN SPRING
OVER-TRAVEL SPRING
CHOKE SPRING
DASHPOT CONTROL ROD
CHOKE PULL DOWN ROD
PUMP INLET CHECK BALL
ADJUSTING SCREW
GASKET
LEVER
COVER
CHOKE COVER
SPRING
DIAPHRAGM
DASHPOT SPRING
IDLE ADJUSTING SCREW
SPARK CONTROL VALVE

Ford single-barrel (typical)

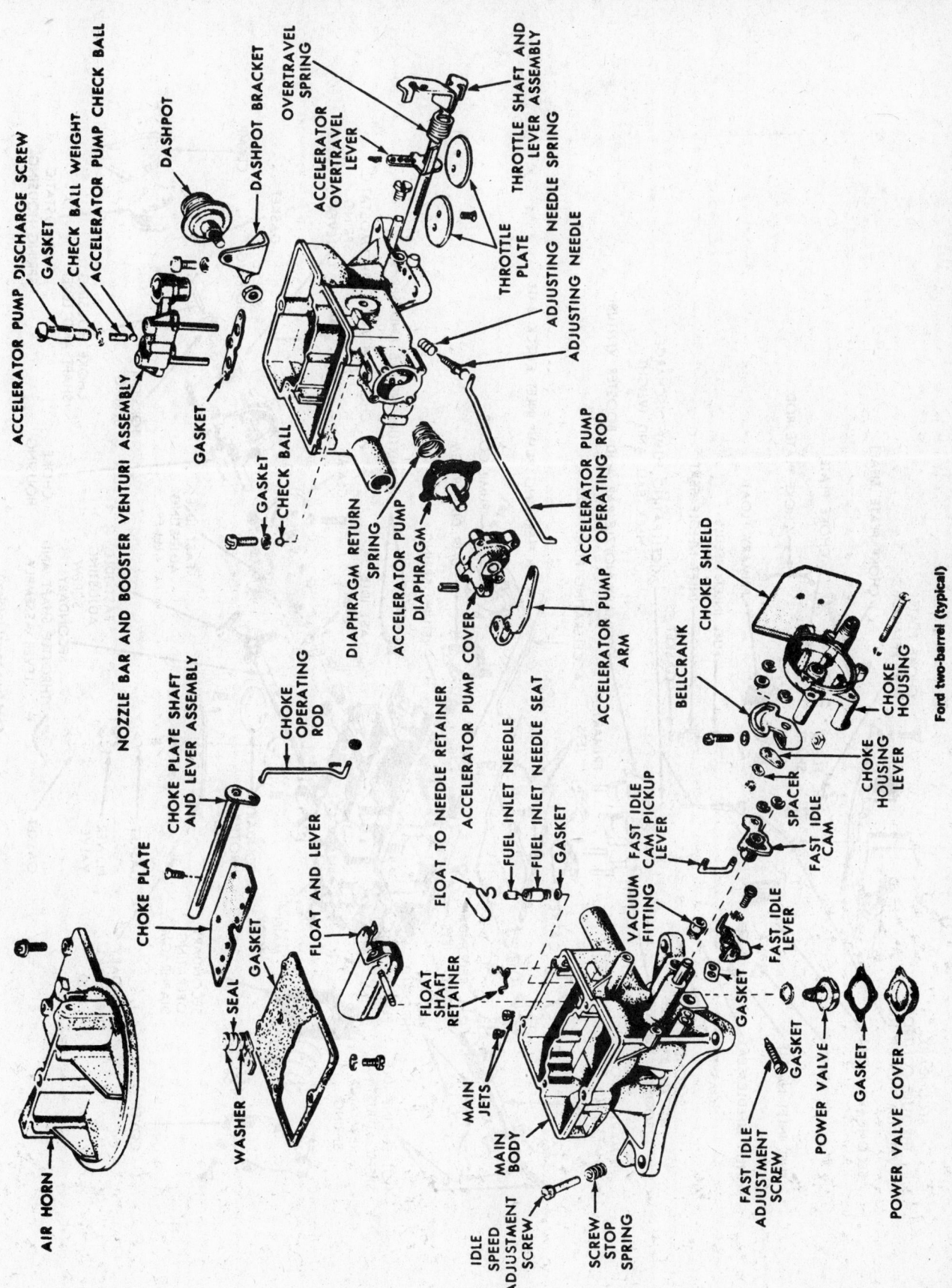

ACCELERATOR PUMP DISCHARGE SCREW
GASKET
CHECK BALL WEIGHT
ACCELERATOR PUMP CHECK BALL
DASHPOT
DASHPOT BRACKET
OVERTRAVEL SPRING
ACCELERATOR PUMP OVERTRAVEL LEVER
THROTTLE SHAFT AND LEVER ASSEMBLY
THROTTLE PLATE
ADJUSTING NEEDLE SPRING
ADJUSTING NEEDLE
NOZZLE BAR AND BOOSTER VENTURI ASSEMBLY
GASKET
GASKET
CHECK BALL
DIAPHRAGM RETURN SPRING
ACCELERATOR PUMP DIAPHRAGM
ACCELERATOR PUMP COVER
ACCELERATOR PUMP ARM
ACCELERATOR PUMP OPERATING ROD
CHOKE SHIELD
BELLCRANK
CHOKE HOUSING
CHOKE HOUSING LEVER
SPACER
CHOKE PLATE
CHOKE PLATE SHAFT AND LEVER ASSEMBLY
CHOKE OPERATING ROD
FLOAT AND LEVER
FLOAT TO NEEDLE RETAINER
FUEL INLET NEEDLE
FUEL INLET NEEDLE SEAT
GASKET
VACUUM FITTING
FAST IDLE CAM PICKUP LEVER
FAST IDLE CAM
FAST IDLE LEVER
AIR HORN
WASHER
SEAL
GASKET
FLOAT SHAFT RETAINER
MAIN JETS
MAIN BODY
IDLE SPEED ADJUSTMENT SCREW
SCREW STOP SPRING
FAST IDLE ADJUSTMENT SCREW
GASKET
POWER VALVE
GASKET
GASKET
POWER VALVE COVER

Ford two-barrel (typical)

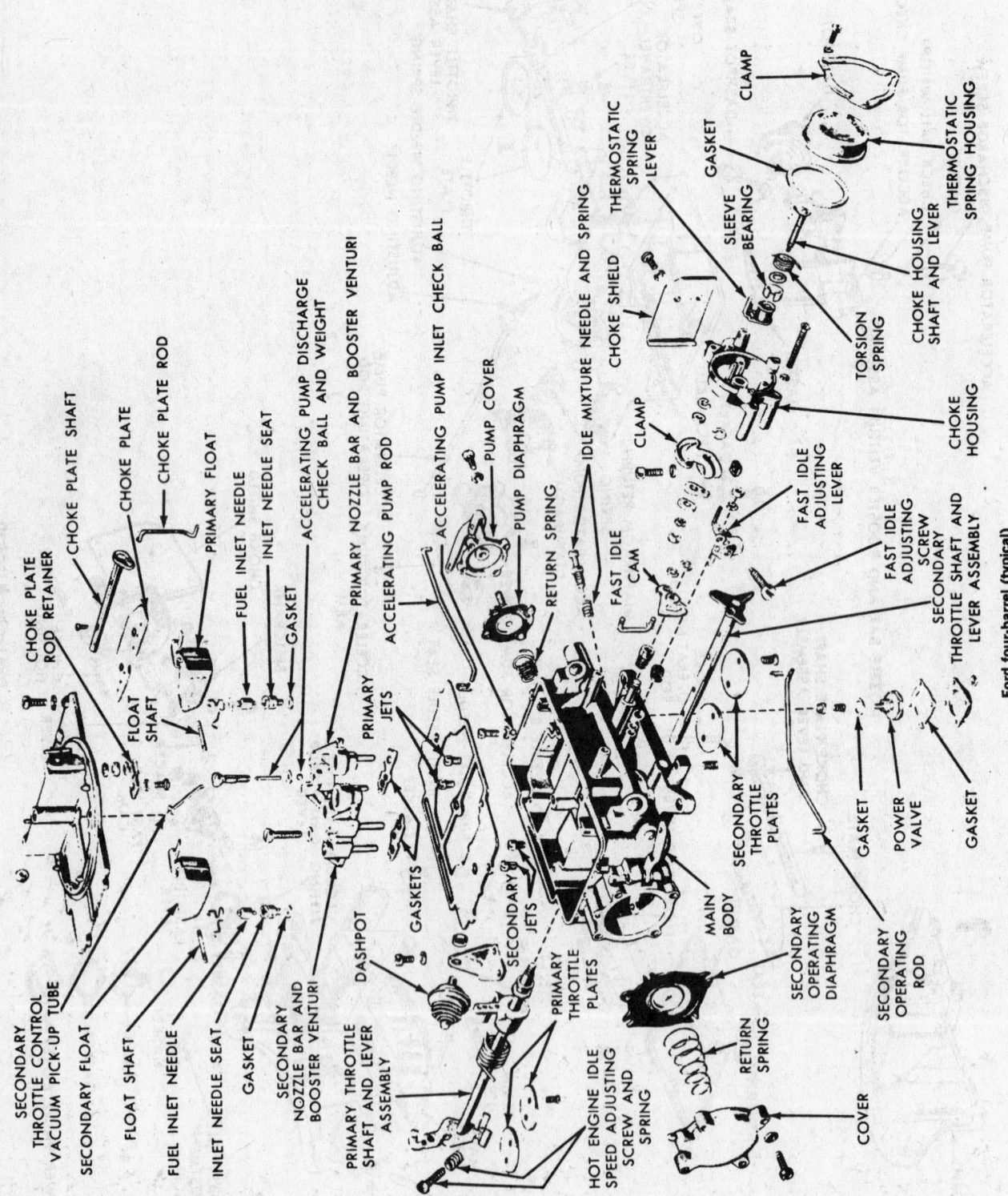

Ford four-barrel (typical)

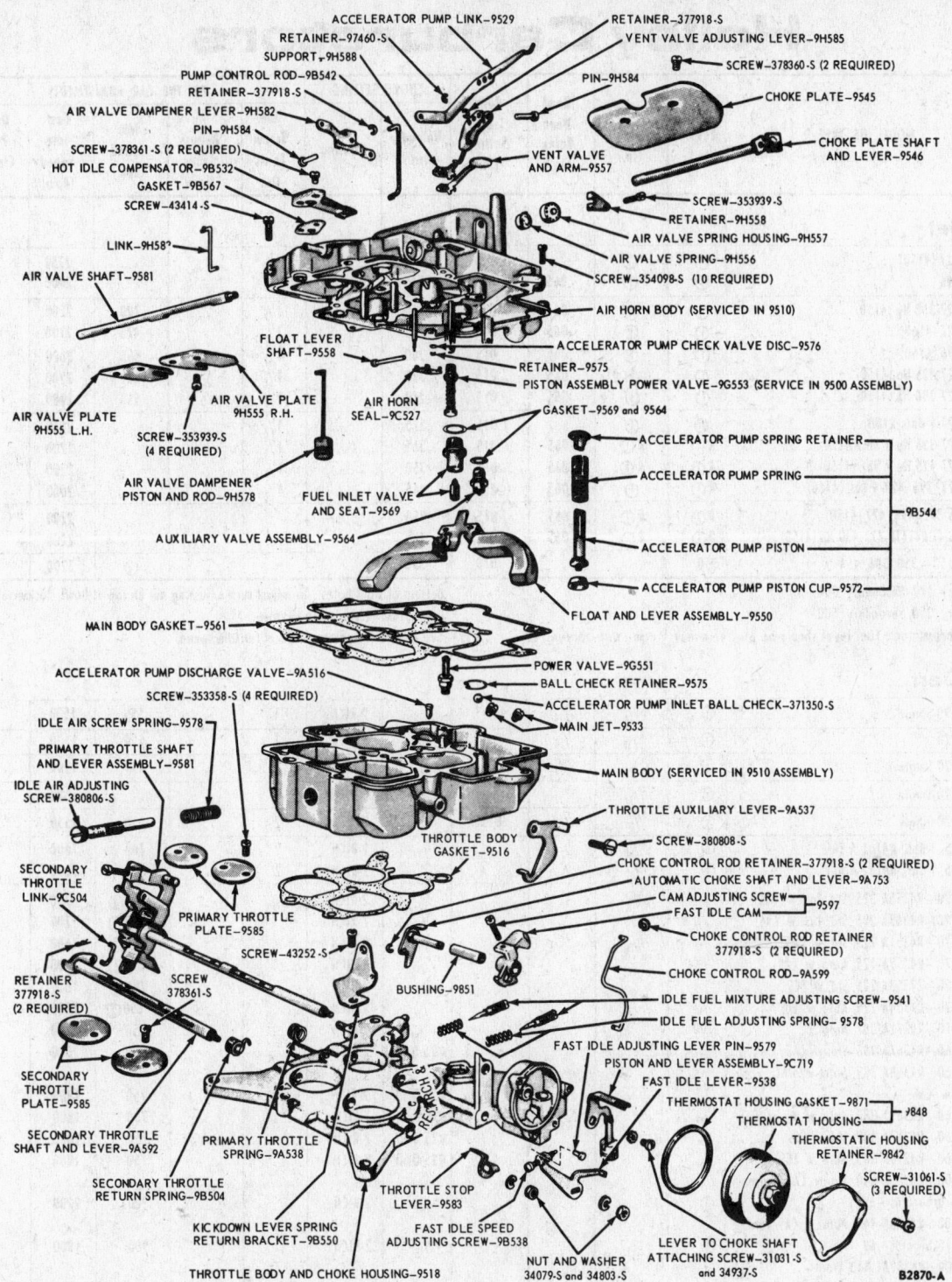

ACCELERATOR PUMP LINK-9529
RETAINER-97460-S
SUPPORT-9H588
PUMP CONTROL ROD-9B542
RETAINER-377918-S
AIR VALVE DAMPENER LEVER-9H582
PIN-9H584
SCREW-378361-S (2 REQUIRED)
HOT IDLE COMPENSATOR-9B532
GASKET-9B567
SCREW-43414-S
LINK-9H58?
AIR VALVE SHAFT-9581
FLOAT LEVER SHAFT-9558
AIR VALVE PLATE 9H555 L.H.
AIR VALVE PLATE 9H555 R.H.
SCREW-353939-S (4 REQUIRED)
AIR HORN SEAL-9C527
AIR VALVE DAMPENER PISTON AND ROD-9H578
FUEL INLET VALVE AND SEAT-9569
AUXILIARY VALVE ASSEMBLY-9564

RETAINER-377918-S
VENT VALVE ADJUSTING LEVER-9H585
SCREW-378360-S (2 REQUIRED)
PIN-9H584
CHOKE PLATE-9545
CHOKE PLATE SHAFT AND LEVER-9546
VENT VALVE AND ARM-9557
SCREW-353939-S
RETAINER-9H558
AIR VALVE SPRING HOUSING-9H557
AIR VALVE SPRING-9H556
SCREW-354098-S (10 REQUIRED)
AIR HORN BODY (SERVICED IN 9510)
ACCELERATOR PUMP CHECK VALVE DISC-9576
RETAINER-9575
PISTON ASSEMBLY POWER VALVE-9G553 (SERVICE IN 9500 ASSEMBLY)
GASKET-9569 and 9564

ACCELERATOR PUMP SPRING RETAINER
ACCELERATOR PUMP SPRING
ACCELERATOR PUMP PISTON
ACCELERATOR PUMP PISTON CUP-9572
9B544

FLOAT AND LEVER ASSEMBLY-9550

MAIN BODY GASKET-9561
ACCELERATOR PUMP DISCHARGE VALVE-9A516
SCREW-353358-S (4 REQUIRED)
IDLE AIR SCREW SPRING-9578
PRIMARY THROTTLE SHAFT AND LEVER ASSEMBLY-9581
IDLE AIR ADJUSTING SCREW-380806-S
SECONDARY THROTTLE LINK-9C504
PRIMARY THROTTLE PLATE-9585
SCREW-43252-S
RETAINER 377918-S (2 REQUIRED)
SCREW 378361-S
SECONDARY THROTTLE PLATE-9585
SECONDARY THROTTLE SHAFT AND LEVER-9A592
SECONDARY THROTTLE RETURN SPRING-9B504
PRIMARY THROTTLE SPRING-9A538
KICKDOWN LEVER SPRING RETURN BRACKET-9B550
THROTTLE BODY AND CHOKE HOUSING-9518

POWER VALVE-9G551
BALL CHECK RETAINER-9575
ACCELERATOR PUMP INLET BALL CHECK-371350-S
MAIN JET-9533
MAIN BODY (SERVICED IN 9510 ASSEMBLY)
THROTTLE AUXILIARY LEVER-9A537
SCREW-380808-S
CHOKE CONTROL ROD RETAINER-377918-S (2 REQUIRED)
AUTOMATIC CHOKE SHAFT AND LEVER-9A753
CAM ADJUSTING SCREW
FAST IDLE CAM
9597
CHOKE CONTROL ROD RETAINER 377918-S (2 REQUIRED)
CHOKE CONTROL ROD-9A599
IDLE FUEL MIXTURE ADJUSTING SCREW-9541
IDLE FUEL ADJUSTING SPRING-9578
FAST IDLE ADJUSTING LEVER PIN-9579
PISTON AND LEVER ASSEMBLY-9C719
FAST IDLE LEVER-9538
THERMOSTAT HOUSING GASKET-9871
THERMOSTAT HOUSING
9848
THERMOSTATIC HOUSING RETAINER-9842
SCREW-31061-S (3 REQUIRED)

THROTTLE BODY GASKET-9516
BUSHING-9851
THROTTLE STOP LEVER-9583
FAST IDLE SPEED ADJUSTING SCREW-9B538
NUT AND WASHER 34079-S and 34803-S
LEVER TO CHOKE SHAFT ATTACHING SCREW-31031-S and 34937-S

B2870-A

Carburetor assembly, four-barrel 4300 (1967)

Holley Carburetors

Chevrolet

YEAR	MODEL OR TYPE	Float Level (In.)	Fuel Level (In.)	Bowl Vent Valve (In.)	Pump Travel Setting (In.)	CHOKE SETTING: Unloader (In.)	CHOKE SETTING: Housing	ON THE CAR ADJUSTMENTS: Idle Screw Turns Out	Air Bypass Turns Open	Idle Speed (Rpm)	Fast Idle Speed (Rpm)	Dashpot Plunger Clearance (In.)
1964-65	327 (4150)	①	①	...	.015	.375	...	1 1/2	...	...	2250	...
1965	396	①	①	.065	.015	.275	...	1 1/2	...	...	2000	...
1966	327-350 Hp (4150)	①	①	.065	.015	.260	...	1 1/2	...	700	2200	...
1966	327 (4160)	①	①	.065	.015	.260	...	1 1/2	...	475	2200	...
1966	396 (4160)	①	①	.065	.015	.260	...	1 1/2	...	550	2000	...
1966	427-425 Hp (4150)	①	①	.065	.015	.350	...	1 1/2	...	800	2200	...
1966	427-390 Hp (4160)	①	①	.065	.015	.260	...	1 1/2	...	550	2000	...
1967	427-2 Bbl. (2100)	①	①	...	.015	.275	...	1 1/2	...	...	2200	...
1967	327-325 Hp-4 Bbl. (4150)	A①	A①	.065	.015	.265	...	1 1/2	...	...	2200	...
1967	427-425 Hp-4 Bbl. (4150)	A①	A①	.065	.015	.350	...	1 1/2	...	...	2200	...
1967	327, 396, 427-4 Bbl. (4160)	A①	①	.065	.015	.265	...	1 1/2	...	...	2000	...
1968-69	V8-302, 396, 427 (4150)	B①	B①	.065	.015	.350	...	...	...	...	2200	...
1968-69	V8-CORVETTE-427-430 Hp (4150)	B①	B①	.065	.015	.350	...	...	...	...	2200	...
1970	4150—350, 396, 454	.350	...	...	.015	.350	...	...	...	②	2200	...

A—Primary .170, Secondary .300.

B—Primary .350, Secondary .500.

①—Float adjustment: Fuel level should be plus or minus 1/32 in. with threads at bottom of sight holes. To adjust turn adjusting nut on top of bowl clockwise, to lower, counterclockwise to raise.

②—See engine compartment decal for idle speed.

Dodge-Dart

YEAR	MODEL OR TYPE	Float Level (In.)	Fuel Level (In.)	Bowl Vent Valve (In.)	Pump Travel Setting (In.)	CHOKE SETTING: Unloader (In.)	CHOKE SETTING: Housing	ON THE CAR ADJUSTMENTS: Idle Screw Turns Out	Air Bypass Turns Open	Idle Speed (Rpm)	Fast Idle Speed (Rpm)	Dashpot Plunger Clearance (In.)
1964	1920-Single	①	11/16	1/16	...	...	2-RICH	1	...	550	1500	...
1965	1920	①	27/32	1/16	...	...	2-RICH	1	...	550	1500	...
1966	1920-Single	①	27/32	3/32	...	...	2-RICH	2	...	550	1500	...
1967	1920-Single	①	27/32	3/32	...	...	2-RICH	2	...	...	1500	...
1968	1920-Single	①	27/32	3/32	.015	...	2-RICH	2	...	650	1550	...
1969	225, 1 Bbl., R4161-A Std.	①	27/32	3/32	...	9/32	2-RICH	2	...	700	1600	...
1969	225, 1 Bbl., R4162-A Aut.	①	27/32	3/32	...	9/32	2-RICH	2	...	650	1800	...
1970	1920—R4355A-225, Std. Taxi w. CAS	27/32	...	...	...	9/32	2-RICH	...	...	650②	1700	...
1970	1920—R4363A-225, Std. Taxi w. CAS	27/32	...	...	...	9/32	2-RICH	...	...	650②	1700	...
1970	1920—R4351A-225, Std. w. CAS	27/32	...	...	...	9/32	2-RICH	...	...	700②	1600	...
1970	1920—R4352A-225, Auto. w. CAS	27/32	...	...	...	9/32	2-RICH	...	...	650②	1800	...
1970	1920—R4353A-225, Std. w. ECS	27/32	...	...	...	9/32	2-RICH	...	...	700②	1600	...
1970	1920—R4354A-225, Auto. w. ECS	27/32	...	...	...	9/32	2-RICH	...	...	650②	1800	...
1970	2210—R4371A-383, Auto.	.200	...	...	...	11/64	2-RICH	...	...	650	1700	...
1970	4160—R4367A-383, Std. w. CAS	15/64	...	...	...	#25 Drill	2-RICH	...	...	750	2000	...
1970	4160—R4368A-383, Auto. w/o AC, w. CAS	15/64	...	...	...	#25 Drill	2-RICH	...	...	750	1800	...
1970	4160—R4369A-383, Auto. w. AC, w. CAS	15/64	...	...	...	#25 Drill	2-RICH	...	...	750	1800	...
1970	4160—R4217A-383, Std. w. ECS	15/64	...	...	...	#25 Drill	2-RICH	...	...	750	2000	...
1970	4160—R4218A-383, Auto. w. ECS	15/64	...	...	...	#25 Drill	2-RICH	...	...	750	1800	...
1970	2300—R4375A-440, Std. w. CAS, center carburetor—6V	③	...	...	...	5/32	2-RICH	...	...	900	2200	...
1970	2300—R4376A-440, Auto. w. CAS, center carburetor—6V	③	...	...	...	5/32	2-RICH	...	...	900	1800	...
1970	2300—R4382AF-440, front carburetor—6V	③	...	...	...	...	...	...	...	...	...	...
1970	2300—R4383AR-440, rear carburetor—6V	③	...	...	...	...	...	...	...	...	...	...
1970	2300—R4374A-440, Std. w. ECS, center carburetor—6V	③	...	...	...	5/32	2-RICH	...	...	900	2200	...
1970	2300—R4144A-440, Auto. w. ECS, center carburetor—6V	③	...	...	...	5/32	2-RICH	...	...	900	1800	...
1970	2300—R4175AF-440, front carburetor-6V	③	...	...	...	...	...	...	...	...	...	...
1970	2300—R4365AR-440, rear carburetor-6V	③	...	...	...	...	...	...	...	...	...	...
1970	4160—R4366A-440, Auto. w. CAS	15/64	...	...	...	#25 Drill	2-RICH	...	...	650	1600	...
1970	4160—R4360A-440, Auto. w. ECS	15/64	...	...	...	#25 Drill	2-RICH	...	...	650	1600	...

①—Use gauge C-3903.

②—With headlights and AC on.

③—Fuel level should be at bottom edge of sight plug hole.

CAS—Cleaner Air System

ECS—Evaporation control system

Ford—T-Bird

YEAR	MODEL OR TYPE	Float Level (In.)	Fuel Level (In.)	Bowl Vent Valve (In.)	Pump Travel Setting (In.)	CHOKE SETTING Unloader (In.)	CHOKE SETTING Housing	ON THE CAR ADJUSTMENTS Idle Screw Turns Out	Air Bypass Turns Open	Idle Speed (Rpm)	Fast Idle Speed (Rpm)	Dashpot Plunger Clearance (In.)
1964	2-Barrel	Ⓐ	①	...	...	...	INDEX	1-1½	...	485	1800	.075
1964	4-Barrel	Ⓐ	①	...	...	...	INDEX	1-1½	...	...	1800	...
1965	4-Barrel (Single)	Ⓐ	①	...	.013	...	INDEX	1-1½	...	750	1800	...
1965	4-Barrel-Frt.	Ⓐ	①	...	...	...	INDEX	1-1½	1/16	750	1800	...
1965	4-Barrel-Rear	Ⓐ	①	...	...	...	...	1-1½	...	750	...	...
1966	4-Barrel (Single)	Ⓐ	①	...	.013	...	INDEX	1-1½	...	750	1800	...
1966	4-Barrel-Frt.	Ⓐ	①	...	...	...	3-LEAN	1-1½	1/16	750	1800	...
1966	4-Barrel-Rear	Ⓐ	①	...	...	...	...	1-1½	...	750	...	...
1967	V8-Frt.	Ⓐ	①	...	...	...	INDEX	1½	1/16	750	...	...
1967	V8-Rear	Ⓐ	①	...	...	...	...	1½	...	750	...	...
1969	4150C-428 CJ Std.	...	...	...	...	...	1-RICH	...	...	700	1350	.100
1969	4150C-428 CJ Auto.	...	...	...	...	...	2-RICH	...	...	600	1500	.100
1970	4150C-302 Boss	...	...	...	...	...	...	...	...	800/850 ②	...	...
1970	4150C-428 CJ Std.	...	...	...	.015③	...	...	...	...	725/500 ②	1900	.140
1970	4150C-428 CJ Auto.	...	...	...	.015③	...	...	...	...	675/500 ②	2100	.200
1970	4150C-429 SCJ, Auto.	...	...	...	...	...	2-RICH	...	...	650/500 ②	2400	Solenoid
1970	4150C-429 SCJ Std.	...	...	...	...	...	2-RICH	...	...	700/500 ②	2200	...
1970	4150C-429 Boss	...	...	...	...	...	...	...	...	700/500 ②	2200	Solenoid

①—Bottom of sight plug.
Ⓐ—Invert fuel bowl, turn adjusting nut until base of float is parallel with floor of bowl.
②—Higher rpm—solenoid energized.
③—Min. spring adjustment.

Jeep

YEAR	MODEL OR TYPE	Float Level (In.)	Fuel Level (In.)	Bowl Vent Valve (In.)	Pump Travel Setting (In.)	CHOKE SETTING Unloader (In.)	CHOKE SETTING Housing	ON THE CAR ADJUSTMENTS Idle Screw Turns Out	Air Bypass Turns Open	Idle Speed (Rpm)	Fast Idle Speed (Rpm)	Dashpot Plunger Clearance (In.)
	Single-2300	...	Ⓐ	...	...	...	INDEX	1	...	550		...
	Single-1920	...	Ⓐ	...	...	...	INDEX	1	...	550		...
	2415	...	Ⓐ	...	...	...	INDEX	1	...	550		...
	2500	...	Ⓐ	...	...	...	INDEX	1	...	550		...
	2209-327 Std.	9/16	...	...	#2 Hole	3/16	INDEX	...	...	550	1800	5/32
	2209-327 Auto.	9/16	...	...	#2 Hole	3/16	INDEX	...	...	500	1800	5/32
	4160-327 Std.	①	...	...	...	.180	1-RICH	...	...	550	1800-2000	.090-.120
	4160-327 Auto.	①	...	...	...	.180	1-RICH	...	...	500	1800-2000	.090-.120
	2209-327 w. Emission control	9/16	...	...	...	3/16	INDEX	½-1½	...	650/700	1800	5/32
	4160-327 w. Emission control	①	...	...	#1 Hole	.180	1-RICH	½-1½	...	650/700	1600-1800	3/32

Ⓐ—Engine running at idle speed—Fuel should be within 1/16 in. of the bottom line of the check plug port.
①—Bottom of sight plug opening.

Mercury

YEAR	MODEL OR TYPE	Float Level (In.)	Fuel Level (In.)	Bowl Vent Valve (In.)	Pump Travel Setting (In.)	CHOKE SETTING Unloader (In.)	CHOKE SETTING Housing	ON THE CAR ADJUSTMENTS Idle Screw Turns Out	Air Bypass Turns Open	Idle Speed (Rpm)	Fast Idle Speed (Rpm)	Dashpot Plunger Clearance (In.)
1964	4-Barrel	Ⓐ	①	...	.013	...	INDEX	1-1½	...	...	1800	...
1965	4-Barrel	Ⓐ	①	...	.015	...	INDEX	1-1½	...	750	1800	...
1966	4-Barrel	Ⓐ	①	...	.015	...	INDEX	1-1½	...	750	1800	...
1967	4-Barrel	Ⓐ	①	...	.015	...	2-RICH	1-1½	...	525	1800	...
1969	4150C-428 CJ Std.	...	...	...	...	...	1-RICH	...	...	700	1350	.100
1969	4150C—428 CJ Auto.	...	...	...	...	...	2-RICH	...	...	600	1500	.100
1970	4150C—302 Boss	...	...	...	...	...	...	...	...	800/850 ②	...	...
1970	4150C—428 CJ Std.	...	...	...	.015③	...	...	...	...	725/500 ②	1900	.140
1970	4150C—428 CJ Auto.	...	...	...	.015③	...	...	...	...	675/500 ②	2100	.200
1970	4150C—429 SCJ Auto.	...	...	...	...	...	2-RICH	...	...	650/500 ②	2400	Solenoid
1970	4150C—429 SCJ Std.	...	...	...	...	...	2-RICH	...	...	700/500 ②	2200	...
1970	4150C—429 Boss	...	...	...	...	...	...	...	...	700/500 ②	2200	Solenoid

①—Bottom of sight plug.
Ⓐ—Invert fuel bowl, turn adj. nut until base of float is parallel with floor of bowl.
②—Higher rpm—solenoid energized.
③—Min spring adjustment.

Plymouth—Barracuda—Challenger

YEAR	MODEL OR TYPE	Float Level (In.)	Fuel Level (In.)	Bowl Vent Valve (In.)	Pump Travel Setting (In.)	CHOKE SETTING Unloader (In.)	CHOKE SETTING Housing	ON THE CAR ADJUSTMENTS Idle Screw Turns Out	Air Bypass Turns Open	Idle Speed (Rpm)	Fast Idle Speed (Rpm)	Dashpot Plunger Clearance (In.)
1964	1920-Single	①	11/16	1/16	...	...	2-RICH	1	...	550	1500	...
1965	1920-Single	①	27/32	1/16	...	...	2-RICH	1	...	550	700	...
1966	1920-Single	①	27/32	3/32	...	...	2-RICH	2	...	550	1500	...
1967	1920-Single	①	27/32	3/32	...	...	2-RICH	2	...	550	700	...
1968	4160—R3918A-440, Auto.	7/64	...	...	...	5/32	2-RICH	...	...	550	700	...
1969	1920—R4165A-225, Auto.-Std.	Gauge	...	...	...	9/32	2-RICH	2	...	650②	1700	...
1969	1920—R4161A-225, Std.	Gauge	...	...	...	9/32	2-RICH	2	...	700②	1600	...
1969	1920—R4162A-225, Auto.	Gauge	...	...	...	9/32	2-RICH	2	...	650②	1800	...
1969	1920—R4163A-225, Std.	Gauge	...	...	...	9/32	2-RICH	2	...	700②	1600	...

YEAR	MODEL OR TYPE	Float Level (In.)	Fuel Level (In.)	Bowl Vent Valve (In.)	Pump Travel Setting (In.)	CHOKE SETTING		ON THE CAR ADJUSTMENTS				
						Unloader (In.)	Housing	Idle Screw Turns Out	Air Bypass Turns Open	Idle Speed (Rpm)	Fast Idle Speed (Rpm)	Dashpot Plunger Clearance (In.)

Plymouth—Barracuda—Challenger, continued

YEAR	MODEL OR TYPE	Float Level (In.)	Fuel Level (In.)	Bowl Vent Valve (In.)	Pump Travel Setting (In.)	Unloader (In.)	Housing	Idle Screw Turns Out	Air Bypass Turns Open	Idle Speed (Rpm)	Fast Idle Speed (Rpm)	Dashpot Plunger Clearance (In.)
1969	1920—R4164A-225, Auto.	Gauge	...	...	...	9/32	2-RICH	2	...	650②	1800	...
1969	4160-R4166A-440, Auto.	15/64	...	...	...	5/32	2-RICH	...	...	600	1400	...
1970	1920-R4355A-225, Std. Taxi w. CAS	27/32	...	...	...	9/32	2-RICH	...	...	650②	1700	...
1970	1920-R4363A-225, Std. Taxi w. CAS	27/32	...	...	...	9/32	2-RICH	...	...	650②	1700	...
1970	1920-R4351A-225, Std. w. CAS	27/32	...	...	...	9/32	2-RICH	...	...	700②	1600	...
1970	1920-R4352A-225, Auto. w. CAS	27/32	...	...	...	9/32	2-RICH	...	...	650②	1800	...
1970	1920-R4353A-225, Std. w. ECS	27/32	...	...	...	9/32	2-RICH	...	...	700②	1600	...
1970	1920-R4354A-225, Auto. w. ECS	27/32	...	...	...	9/32	2-RICH	...	...	650②	1800	...
1970	2210-R4371A-383, Auto.	.200	...	...	9/16	11/64	2-RICH	...	...	650	1700	...
1970	4160-R4367A-383, Std. w. CAS	15/64	...	...	...	#25 Drill	2-RICH	...	...	750	2000	...
1970	4160-R4368A-383, Auto. w/o AC, w. CAS	15/64	...	...	...	#25 Drill	2-RICH	...	...	750	1800	...
1970	4160-R4369A-383, Auto. w. AC, w. CAS	15/64	...	...	...	#25 Drill	2-RICH	...	...	750	1800	...
1970	4160-R4217A-383, Std. w. ECS	15/64	...	...	...	#25 Drill	2-RICH	...	...	750	2000	...
1970	4160-R4218A-383, Auto. w. ECS	15/64	...	...	...	#25 Drill	2-RICH	...	...	750	1800	...
1970	2300-R4375A-440, Std. w. CAS, center carburetor—6V	③	...	...	...	5/32	2-RICH	...	...	900	2200	...
1970	2300-R4376A-440, Auto. w. CAS, center carburetor—6V	③	...	...	...	5/32	2-RICH	...	...	900	1800	...
1970	2300-R4382AF-440, front carburetor—6V	③	...	...	...	...	...	...	...	...	...	...
1970	2300-R4383AR-440, rear carburetor—6V	③	...	...	...	...	...	...	...	...	...	...
1970	2300-R4374A-440, Std. w. ECS, center carburetor—6V	③	...	...	...	5/32	2-RICH	...	...	900	2200	...
1970	2300-R4144A-440, Auto. w. ECS, center carburetor—6V	③	...	...	...	5/32	2-RICH	...	...	900	1800	...
1970	2300-R4175AF-440, front carburetor—6V	③	...	...	...	...	...	...	...	...	...	...
1970	2300-R4365AR-440, rear carburetor—6V	③	...	...	...	...	...	...	...	...	...	...
1970	4160-R4366A-440, Auto. w. CAS	15/64	...	...	...	#25 Drill	2-RICH	...	...	650	1600	...
1970	4160-R4360A-440, Auto. w. ECS	15/64	...	...	...	#25 Drill	2-RICH	...	...	650	1600	...

① —Use gauge C-3903. ③ —Fuel level should be at bottom edge of sight plug hole.
② —With headlights on.

Rambler

YEAR	MODEL OR TYPE	Float Level (In.)	Fuel Level (In.)	Bowl Vent Valve (In.)	Pump Travel Setting (In.)	Unloader (In.)	Housing	Idle Screw Turns Out	Air Bypass Turns Open	Idle Speed (Rpm)	Fast Idle Speed (Rpm)	Dashpot Plunger Clearance (In.)
1964	Single-1909	⑥	②	...	...	5/16	INDEX	1	...	550	1700	...
1964	2-Bbl.-2300	⑤	②	.060	③	3/16	1-RICH	1	...	550	1700	3/32 ①
1964	4-Bbl.-4150	⑤	②	.060	③	3/16	1-LEAN	1	...	550	1700	7/64 ①
1965	Single-1909	5/16	②	...	...	9/32	INDEX	1/4-2 3/4	...	550	1700	9/64
1965	2-Bbl.-2300	11/32 ④	1/2 ⑦	.060	③	.180	INDEX	1	...	550	1700	3/32
1965	4-Bbl.-4150	⑤	②	.060	③	.180	1-LEAN	1	...	550	1700	3/32
1966	6 Cyl.-199-1 Bbl.-Std.	5/16	...	1/16	...	15/64	1-RICH	3/4-1 1/4	...	...	1500	...
1966	6 Cyl.-199-1 Bbl.-Aut.	5/16	...	1/16	...	15/64	INDEX	3/4-1 1/4	...	...	1600	5/32
1966	6 Cyl.-232-1 Bbl.-Std.	5/16	...	1/16	...	15/64	1-RICH	3/4-1 1/4	...	...	1400	...
1966	6 Cyl.-232-1 Bbl.-Aut.	5/16	...	1/16	...	15/64	INDEX	3/4-1 1/4	...	...	1600	5/32
1966	V8-287-2 Bbl.	11/32 ④	1/2 ⑦	5/64	③	3/16	INDEX	1/2-1 1/2	...	...	1800	5/32
1966	V8-327-4 Bbl.	⑤	②	.060	③	.180	1-LEAN	1/2-1 1/2	...	...	2000	3/32
1967	6 Cyl.-Std.	5/16	...	1/6	...	15/64	⑧	3/4-1 1/4	...	...	1500	...
1967	V8-2 Bbl.	11/32 ④	1/2 ⑦	5/64	③	3/16	1-LEAN	1/2-1 1/2	...	...	1800	5/32
1967	V8-4 Bbl.	⑤	②	.060	③	.180	1-LEAN	1/2-1 1/2	...	...	2000	3/32
1968	6 Cyl.-Std.	5/16	...	1/16	...	15/64	1-RICH	1	...	600	1600	...
1968	6 Cyl.-Aut.	5/16	...	1/16	...	15/64	1-RICH	1	...	525	1600	...

①—At curb idle.
②—To bottom of sight hole.
③—Tighten screw 1/4 turn beyond touch point.
④—With bowl cover inverted, measure distance between float and bowl cover.
⑤—Parallel with floor of bowl.
⑥—Use gauge J-10231.
⑦—Measure from fuel level to top of bowl.
⑧—199—INDEX, 232—1-LEAN.

YEAR	MODEL OR TYPE	Float Level (IN.)	Fuel Level (IN.)	Fast Idle Cam (IN.)	Pump Travel Setting	Choke Setting	ON THE CAR ADJUSTMENT			
							Idle Screw Turns Open	Idle Speed (RPM)	Fast Idle Speed (RPM)	Dashpot Plunger Clearance (IN.)

Valiant

YEAR	MODEL OR TYPE	Float Level	Fuel Level	Fast Idle Cam	Pump Travel	Choke Setting	Idle Screw Turns Open	Idle Speed	Fast Idle Speed	Dashpot Plunger Clearance
1964	1920-Single	①	$^{11}/_{16}$	$^{1}/_{16}$	...	...	2-RICH	1	...	550	1500	...
1965	1920-Single	①	$^{27}/_{32}$	$^{1}/_{16}$	...	...	2-RICH	1	...	550	700	...
1966	1920-Single	①	$^{27}/_{32}$	$^{3}/_{32}$	...	$^{3}/_{16}$	2-RICH	1-2	...	550	700	...
1967	1920-Single	①	$^{27}/_{32}$	$^{3}/_{32}$	...	$^{3}/_{16}$	2-RICH	1	...	550	700	...
1969	1920-R4165A-225, Auto.-Std.	Gauge	...	...	...	$^{9}/_{32}$	2-RICH	2	...	650②	1700	...
1969	1920-R4161A-225, Std.	Gauge	...	...	...	$^{9}/_{32}$	2-RICH	2	...	700②	1600	...
1969	1920-R4162A-225, Auto.	Gauge	...	...	...	$^{9}/_{32}$	2-RICH	2	...	650②	1800	...
1969	1920-R4163A-225, Std.	Gauge	...	...	...	$^{9}/_{32}$	2-RICH	2	...	700②	1600	...
1969	1920-R4164A-225, Auto.	Gauge	...	...	...	$^{9}/_{32}$	2-RICH	2	...	650②	1800	...
1970	1920-R4363A-225, Std. Taxi w. CAS	$^{27}/_{32}$	...	...	...	$^{9}/_{32}$	2-RICH	...	...	650②	1700	...
1970	1920-R4351A-225, Std. w. CAS	$^{27}/_{32}$	...	...	...	$^{9}/_{32}$	2-RICH	...	...	700②	1600	...
1970	1920-R4352A-225, Auto. w. CAS	$^{27}/_{32}$	...	...	...	$^{9}/_{32}$	2-RICH	...	...	650②	1800	...
1970	1920-R4353A-225, Std. w. ECS	$^{27}/_{32}$	...	...	...	$^{9}/_{32}$	2-RICH	...	...	700②	1600	...
1970	1920-R4354A-225, Auto. w. ECS	$^{27}/_{32}$	...	...	...	$^{9}/_{32}$	2-RICH	...	...	650②	1800	...

① —Use gauge C-3903.
② —With headlights on.

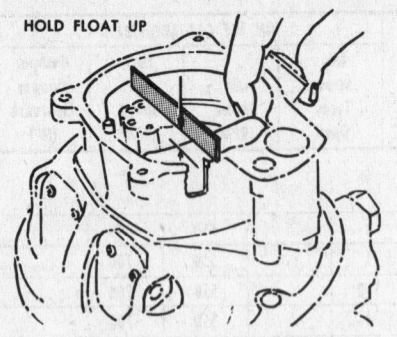

Holley four-barrel float setting

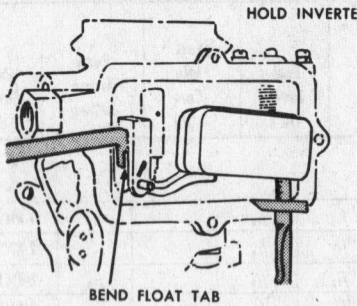

BEND FLOAT TAB

Holley single-barrel float drop (© Ford Motor Co.)

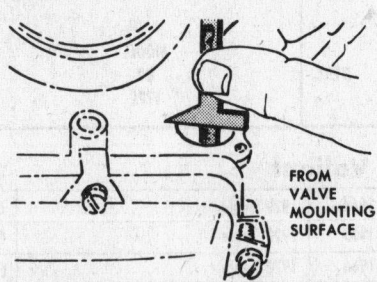

FROM VALVE MOUNTING SURFACE

Holley single-barrel fuel level check (© Ford Motor Co.)

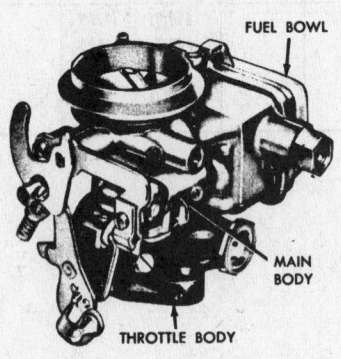

FUEL BOWL

MAIN BODY

THROTTLE BODY

Holley single-barrel carburetor (© Ford Motor Co.)

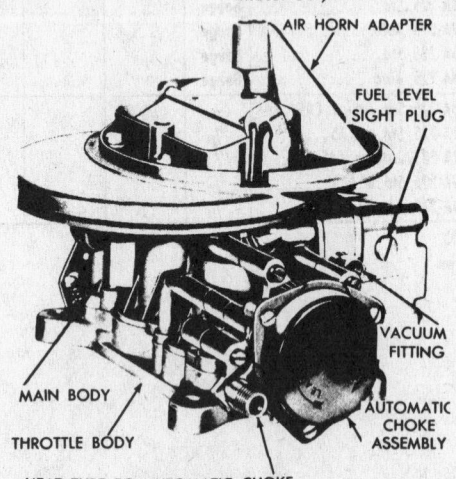

AIR HORN ADAPTER

FUEL LEVEL SIGHT PLUG

VACUUM FITTING

AUTOMATIC CHOKE ASSEMBLY

MAIN BODY

THROTTLE BODY

HEAT TUBE TO AUTOMATIC CHOKE

Holley two-barrel carburetor

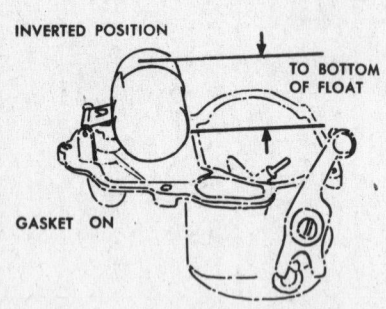

INVERTED POSITION

TO BOTTOM OF FLOAT

GASKET ON

Holley two-barrel float setting (© Ford Motor Co.)

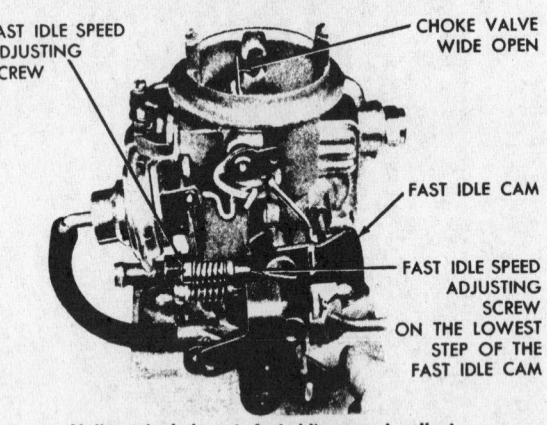

FAST IDLE SPEED ADJUSTING SCREW

CHOKE VALVE WIDE OPEN

FAST IDLE CAM

FAST IDLE SPEED ADJUSTING SCREW ON THE LOWEST STEP OF THE FAST IDLE CAM

Holley single-barrel fast idle speed adjustment (on the car)

SCALE

SURFACE OF CARBURETOR

SURFACE OF FUEL

Holley single-barrel wet fuel level measurement

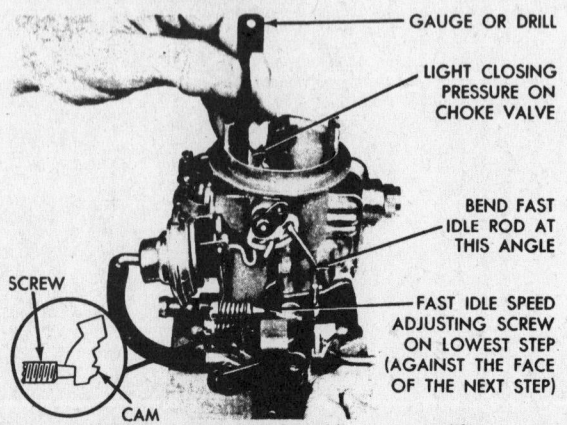

GAUGE OR DRILL

LIGHT CLOSING PRESSURE ON CHOKE VALVE

BEND FAST IDLE ROD AT THIS ANGLE

FAST IDLE SPEED ADJUSTING SCREW ON LOWEST STEP (AGAINST THE FACE OF THE NEXT STEP)

SCREW

CAM

Holley single-barrel fast idle cam position adjustment

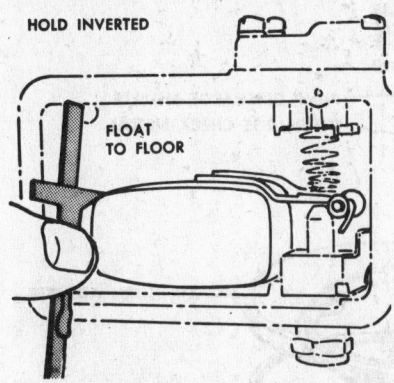

Holley two- and four-barrel float adjustment (© Ford Motor Co.)

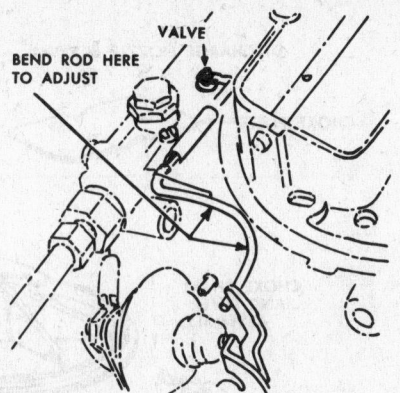

Holley dual bowl vent adjustment (© Ford Motor Co.)

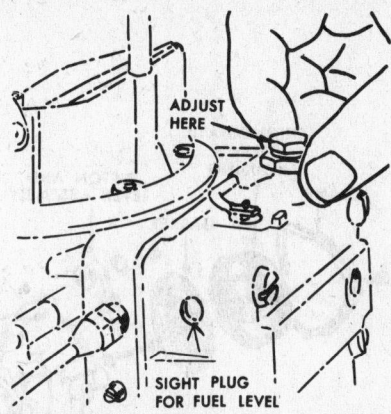

Holley two- and four-barrel fuel level adjustment (© Ford Motor Co.)

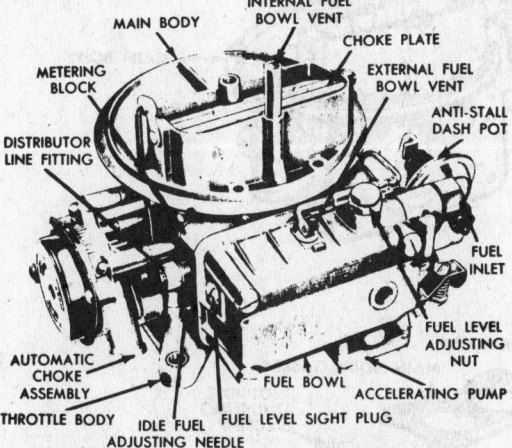

Holley dual carburetor (© Ford Motor Co.)

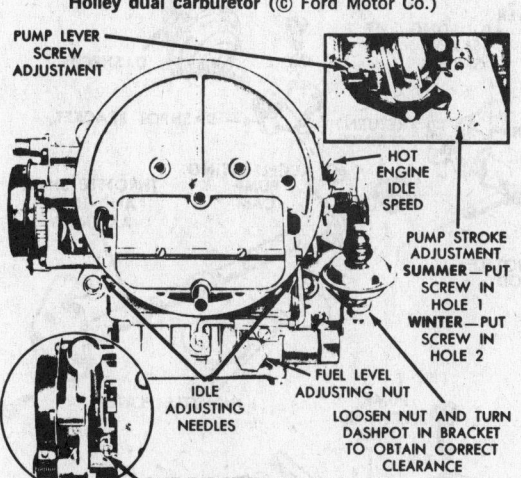

Holley dual carburetor idle adjustment (© Ford Motor Co.)

Holley single-barrel float setting

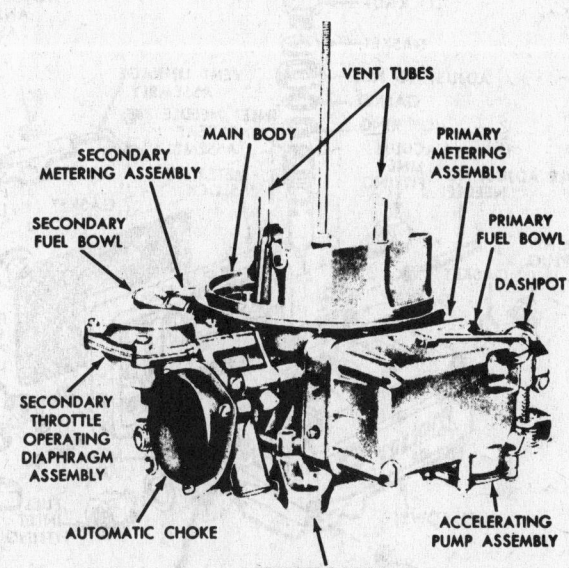

Holley four-barrel carburetor (© Ford Motor Co.)

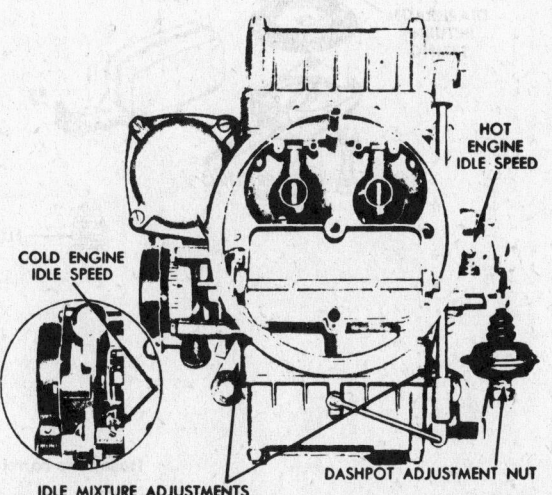

Holley four-barrel idle adjustment (© Ford Motor Co.)

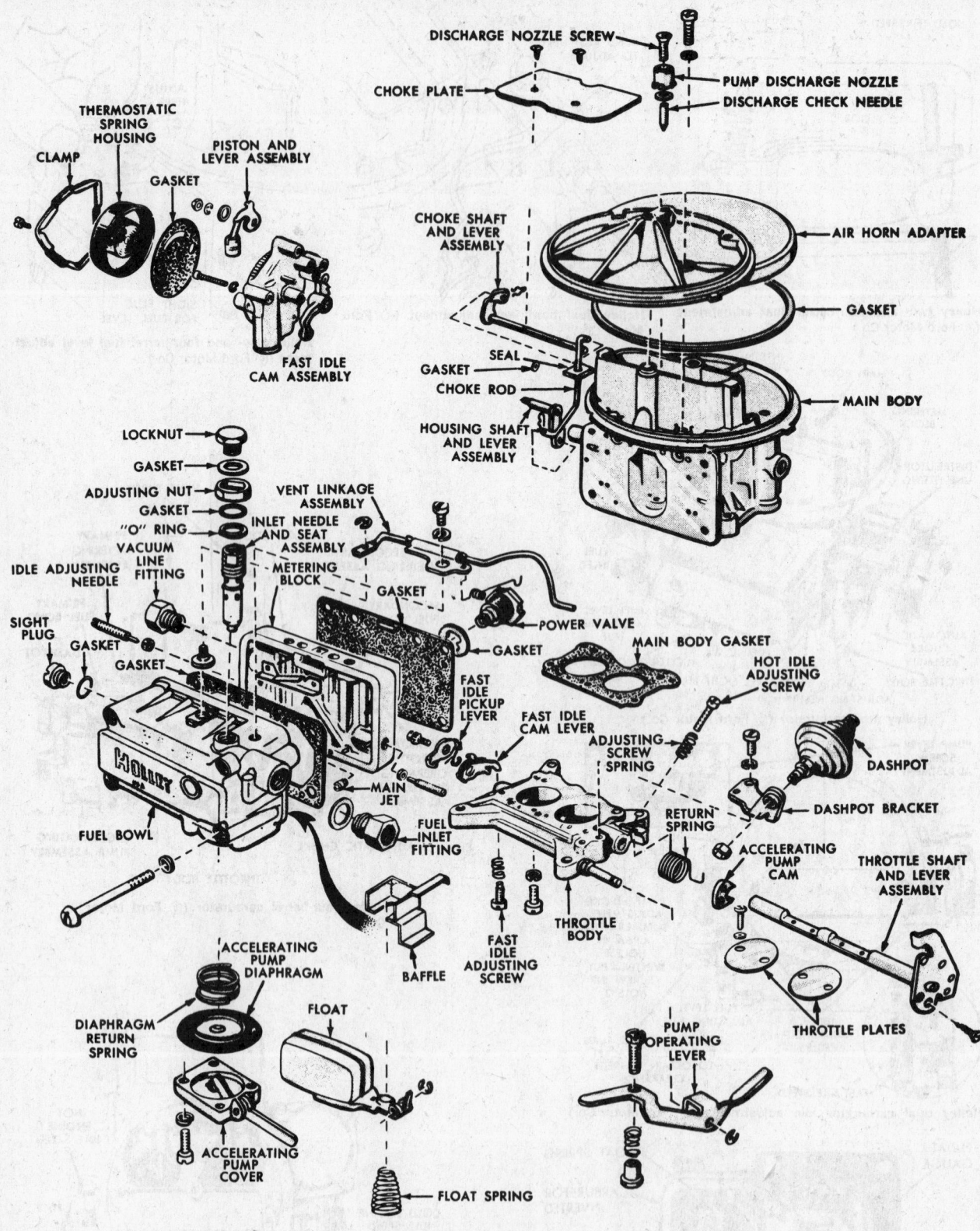

Holley two-barrel (typical) (© Ford Motor Co.)

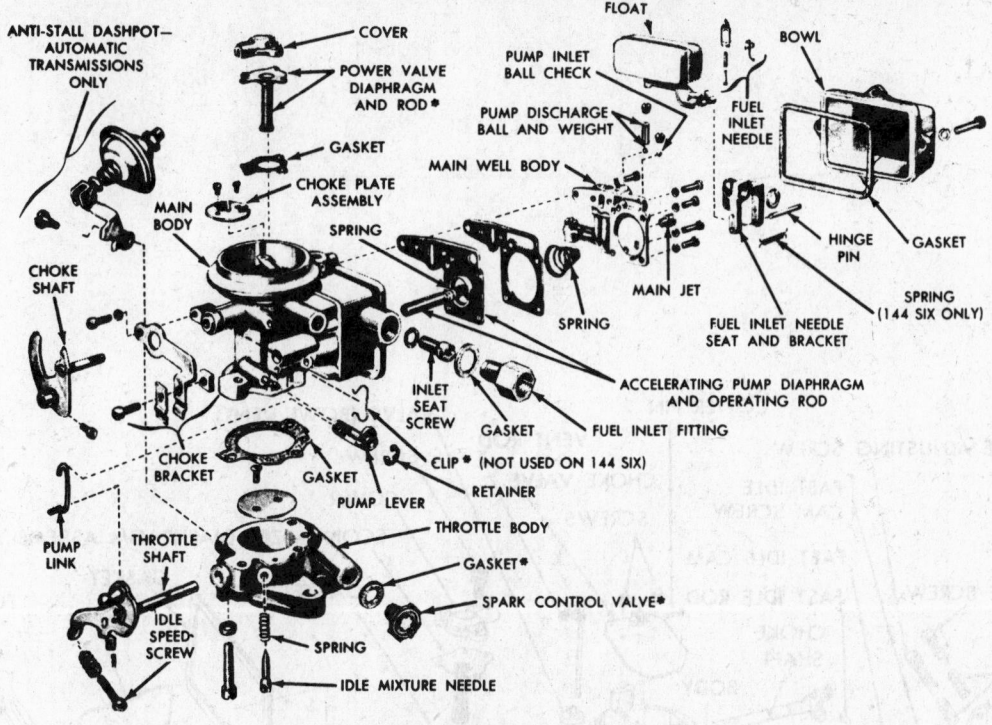

Holley single-barrel (typical) (© Ford Motor Co.)

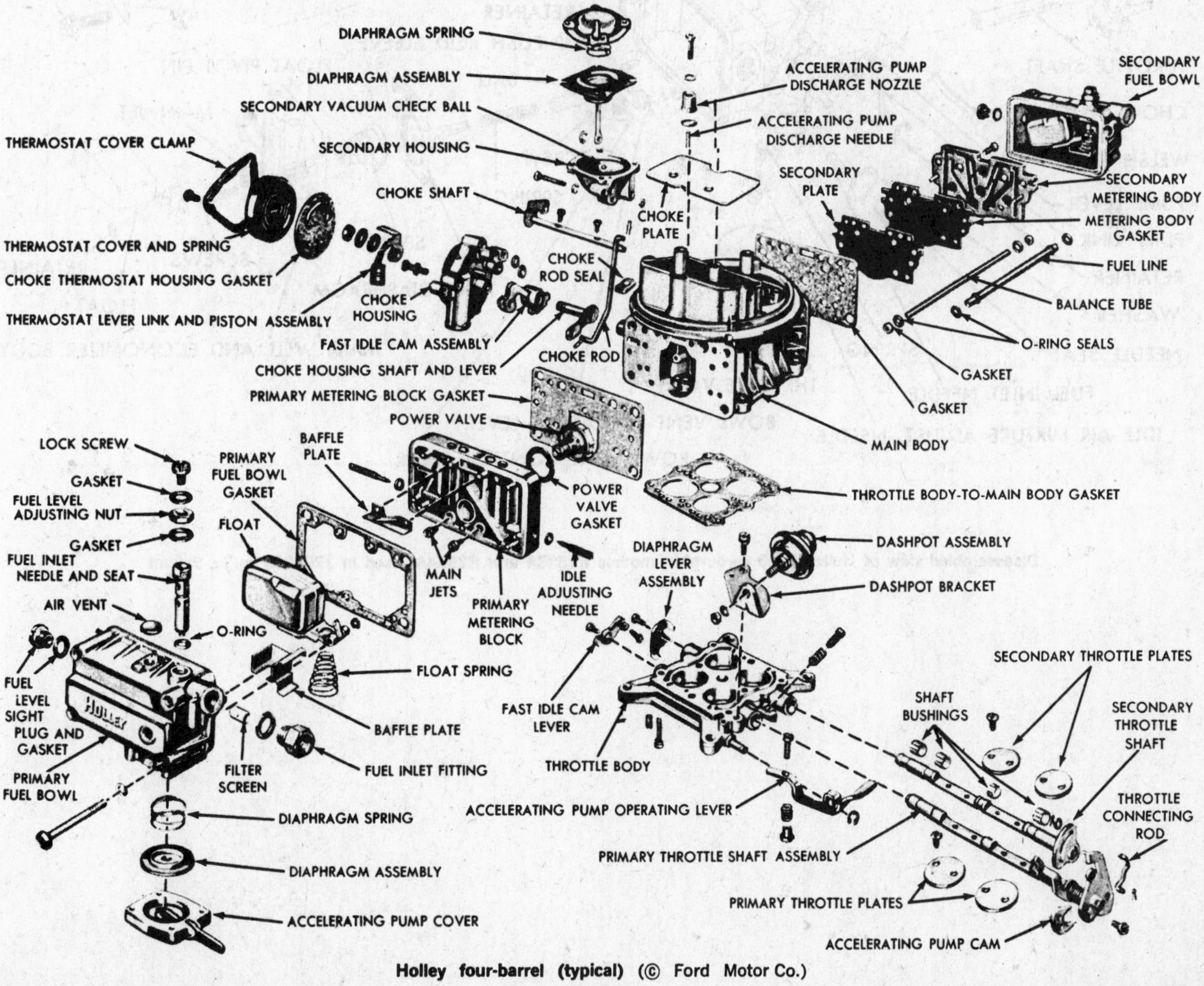

Holley four-barrel (typical) (© Ford Motor Co.)

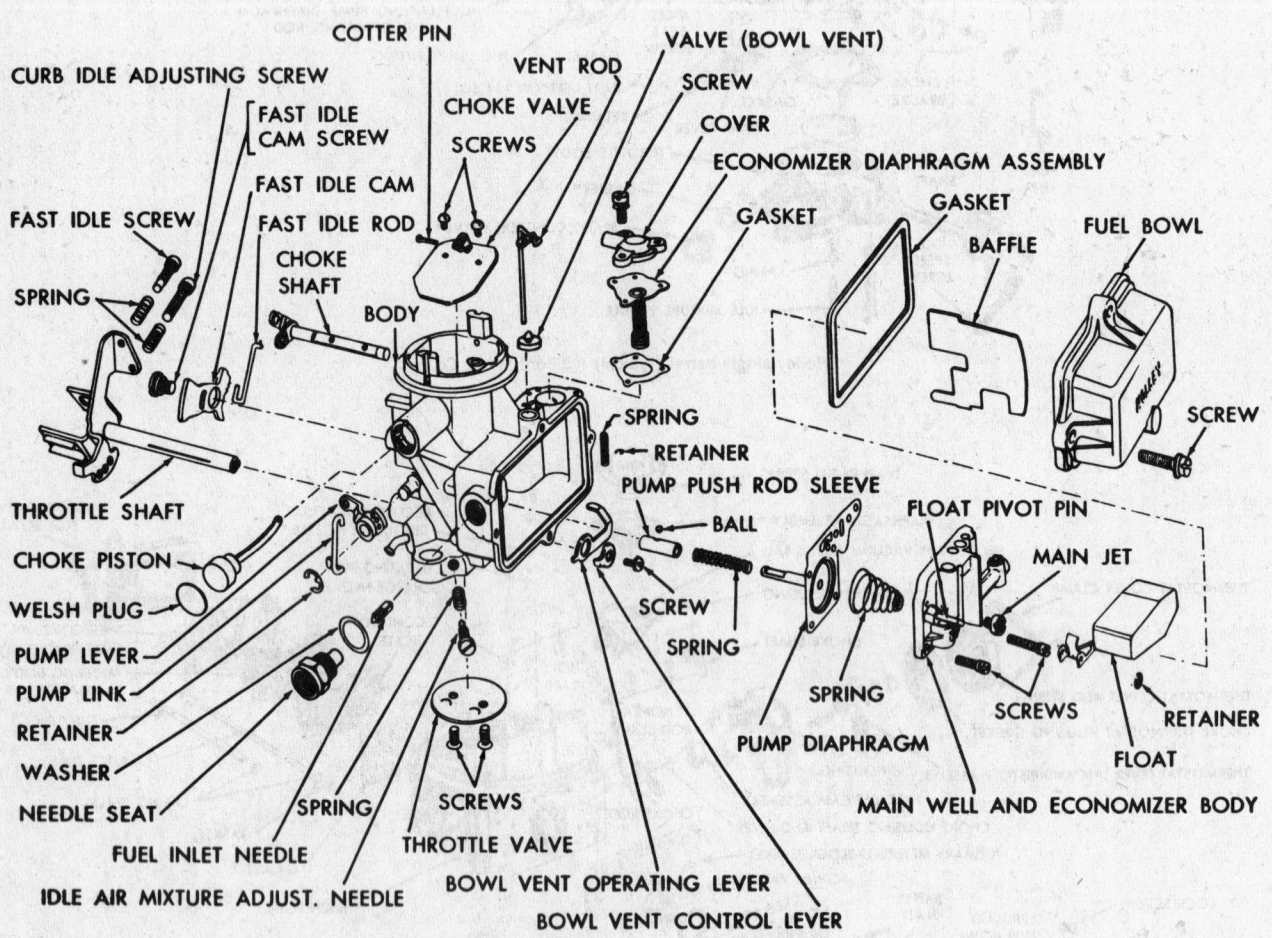

COTTER PIN

CURB IDLE ADJUSTING SCREW

VENT ROD

VALVE (BOWL VENT)

SCREW

CHOKE VALVE

COVER

FAST IDLE CAM SCREW

SCREWS

ECONOMIZER DIAPHRAGM ASSEMBLY

FAST IDLE CAM

GASKET

GASKET

FAST IDLE SCREW

FAST IDLE ROD

BAFFLE

FUEL BOWL

CHOKE SHAFT

SPRING

BODY

SCREW

THROTTLE SHAFT

SPRING

RETAINER

PUMP PUSH ROD SLEEVE

FLOAT PIVOT PIN

CHOKE PISTON

BALL

MAIN JET

WELSH PLUG

SCREW

PUMP LEVER

SPRING

PUMP LINK

SPRING

RETAINER

PUMP DIAPHRAGM

SCREWS

WASHER

RETAINER

NEEDLE SEAT

MAIN WELL AND ECONOMIZER BODY

FLOAT

SPRING

SCREWS

FUEL INLET NEEDLE

THROTTLE VALVE

IDLE AIR MIXTURE ADJUST. NEEDLE

BOWL VENT OPERATING LEVER

BOWL VENT CONTROL LEVER

Disassembled view of Holley 1920 carburetor, models R2533A and R2534A, used in 170-225 cu. in. Valiant

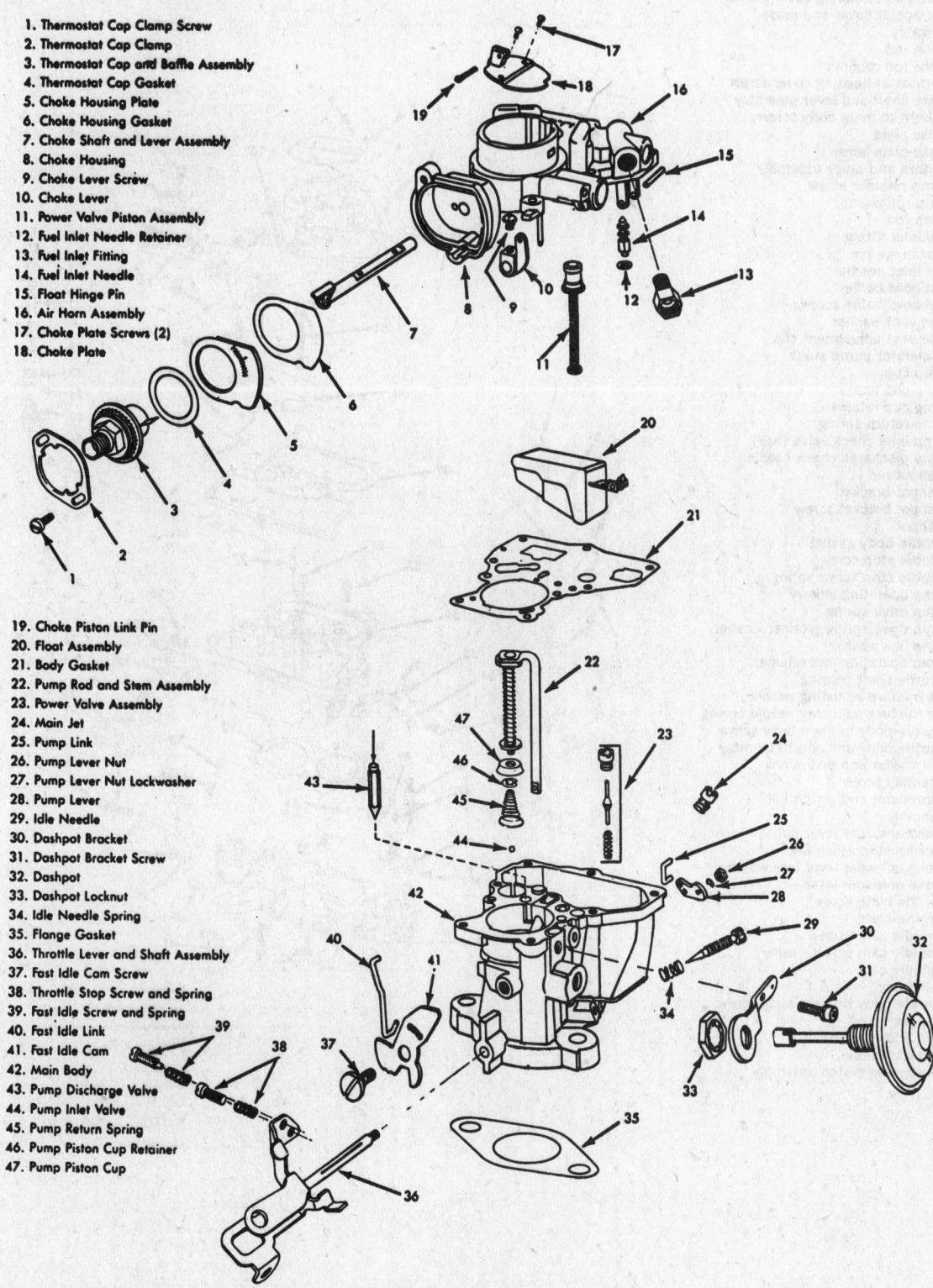

1. Thermostat Cap Clamp Screw
2. Thermostat Cap Clamp
3. Thermostat Cap and Baffle Assembly
4. Thermostat Cap Gasket
5. Choke Housing Plate
6. Choke Housing Gasket
7. Choke Shaft and Lever Assembly
8. Choke Housing
9. Choke Lever Screw
10. Choke Lever
11. Power Valve Piston Assembly
12. Fuel Inlet Needle Retainer
13. Fuel Inlet Fitting
14. Fuel Inlet Needle
15. Float Hinge Pin
16. Air Horn Assembly
17. Choke Plate Screws (2)
18. Choke Plate

19. Choke Piston Link Pin
20. Float Assembly
21. Body Gasket
22. Pump Rod and Stem Assembly
23. Power Valve Assembly
24. Main Jet
25. Pump Link
26. Pump Lever Nut
27. Pump Lever Nut Lockwasher
28. Pump Lever
29. Idle Needle
30. Dashpot Bracket
31. Dashpot Bracket Screw
32. Dashpot
33. Dashpot Locknut
34. Idle Needle Spring
35. Flange Gasket
36. Throttle Lever and Shaft Assembly
37. Fast Idle Cam Screw
38. Throttle Stop Screw and Spring
39. Fast Idle Screw and Spring
40. Fast Idle Link
41. Fast Idle Cam
42. Main Body
43. Pump Discharge Valve
44. Pump Inlet Valve
45. Pump Return Spring
46. Pump Piston Cup Retainer
47. Pump Piston Cup

Disassembled view of Holley 1909 carburetor, 1964 Rambler

1 Thermostat housing cover gasket
2 Thermostat cover and guide assembly
3 Choke rod
4 Choke rod retainer
5 Thermostat housing cover screw
6 Choke shaft and lever assembly
7 Air horn to main body screw
8 Choke plate
9 Choke plate screw
10 Air horn and plugs assembly
11 Clamp retainer screw
12 Pump rod clamp
13 Pump rod
14 Fuel inlet fitting
15 Float hinge pin
16 Fuel inlet needle
17 Fuel bowl baffle
18 Fuel bowl baffle screw
19 Bowl vent washer
20 Bowl vent adjustment clip
21 Accelerator pump shaft
22 Pump cup
23 Pump cup liner
24 Pump cup retainer
25 Pump return spring
26 Pump inlet check valve (ball)
27 Pump discharge check needle
28 Dashpot nut
29 Dashpot bracket
30 Dashpot bracket screw
31 Dashpot
32 Throttle body gasket
33 Throttle stop screw
34 Throttle stop screw spring
35 Pump operating link
36 Pump drive spring
37 Pump drive spring retainer washer
38 Pump link washer
39 Pump operating link retainer
40 Throttle shaft bearing
41 Idle mixture adjusting needle
42 Idle mixture adjusting needle spring
43 Throttle body to main body screw
44 Throttle body and shaft assembly
45 Thermostat and piston link assembly screw
46 Thermostat and piston link assembly
47 Choke unloader lever nut
48 Choke piston assembly
49 Choke unloader lever lock washer
50 Choke unloader lever
51 Throttle plate screw
52 Throttle plate
53 Fast idle cam screw
54 Fast idle cam screw washer
55 Fast idle cam
56 Main body
57 Throttle body to main body screw
58 Main jet
59 Main body gasket
60 Float assembly
61 Power valve piston assembly

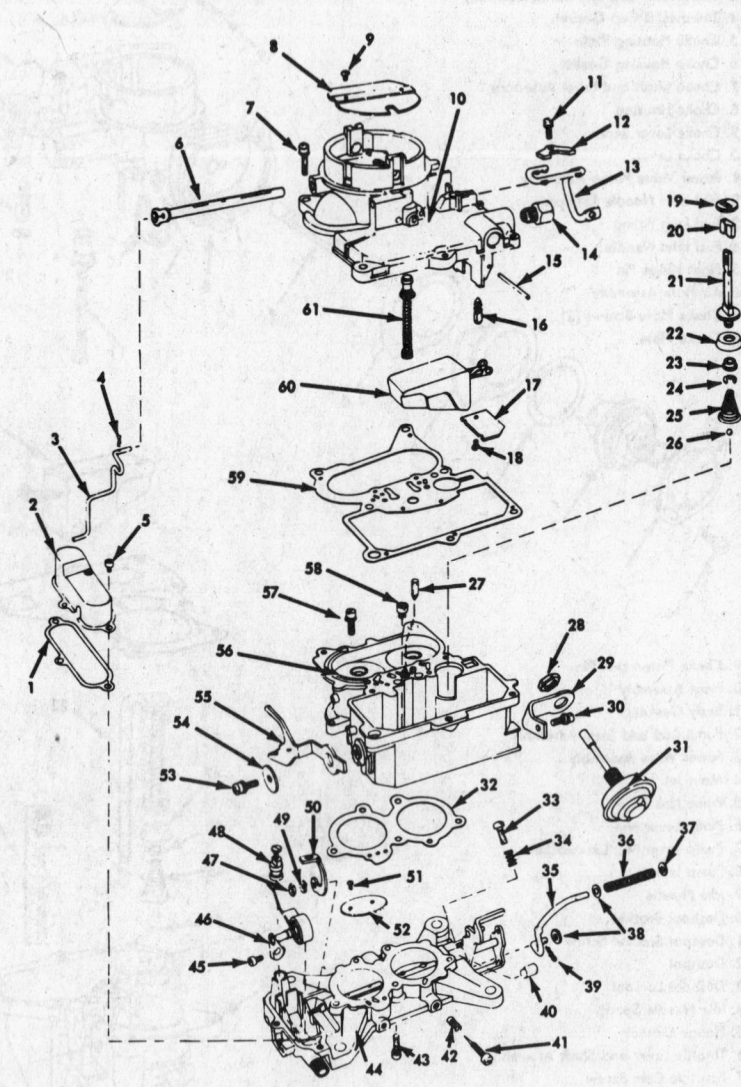

Disassembled view of Holley 2209 carburetor, 1965 Rambler

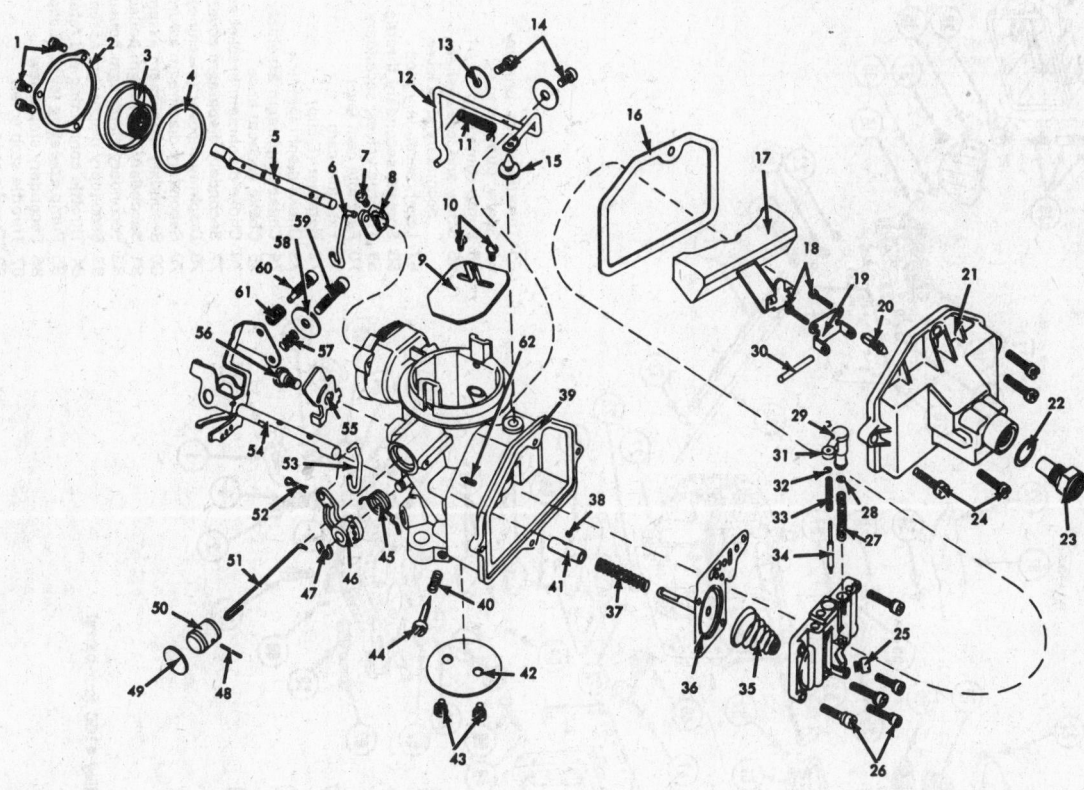

1 Choke thermostat cover screws	21 Float bowl	42 Throttle plate
2 Choke thermostat cover retainer	22 Fuel inlet fitting gasket	43 Throttle plate screws
3 Choke thermostat cover assembly	23 Fuel inlet fitting	44 Idle mixture adjustment needle
4 Choke thermostat cover gasket	24 Float bowl screws	45 Pump return spring
5 Choke shaft assembly	25 Main jet	46 Pump operating cam lever
6 Fast idle connector rod	26 Metering block screws	47 Pump operating cam lever retainer
7 Choke lever screw	27 Power valve piston spring	48 Choke piston pin
8 Choke lever	28 Power valve piston spring washer	49 Choke piston plug
9 Choke plate	29 Power valve retainer	50 Choke piston
10 Choke plate screws	30 Float shaft pin	51 Choke piston link
11 Bowl vent rod spring	31 Power valve piston	52 Pump link retainer
12 Bowl vent rod	32 Power valve spring washer	53 Pump line
13 Vent rod retainer washer	33 Power valve spring	54 Throttle shaft assembly
14 Vent rod retainer screws	34 Power valve	55 Fast idle cam
15 Bowl vent valve	35 Pump return spring	56 Fast idle cam screw
16 Fuel bowl gasket	36 Pump diaphragm assembly	57 Throttle stop screw spring
17 Float assembly	37 Pump override spring	58 Vent rod operating washer
18 Float shaft bracket screws	38 Pump push rod sleeve ball	59 Throttle stop screw
19 Float shaft retaining bracket	39 Main body	60 Fast idle adjusting screw
20 Fuel inlet needle	40 Idle mixture needle spring	61 Fast idle adjusting screw spring
	41 Pump push rod sleeve	62 Choke piston stop screw

Disassembled view of Holley 1931 carburetor, 1966 Rambler

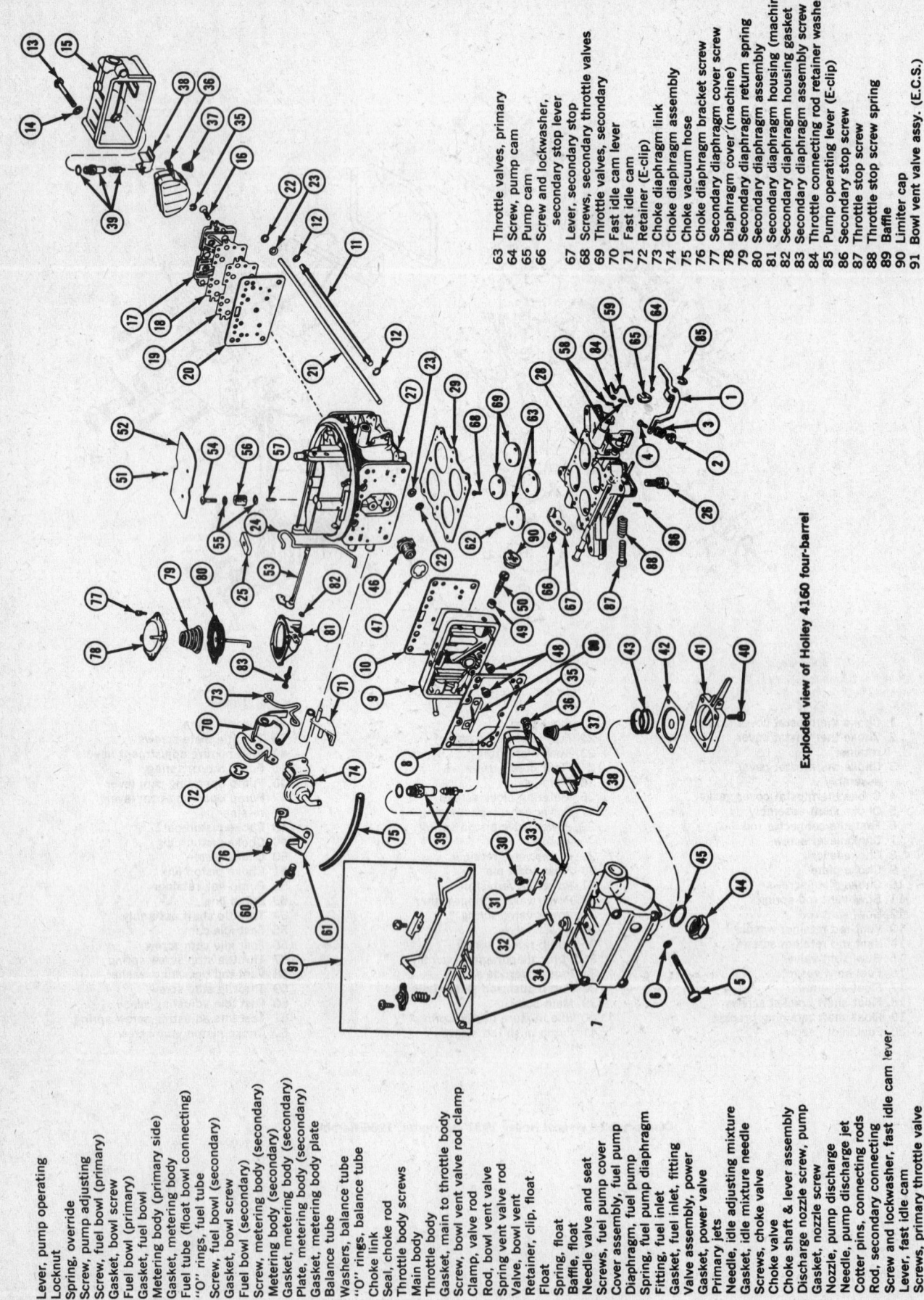

Exploded view of Holley 4160 four-barrel

1 Lever, pump operating
2 Locknut
3 Spring, override
4 Screw, pump adjusting
5 Screw, fuel bowl (primary)
6 Gasket, bowl screw
7 Fuel bowl (primary)
8 Gasket, fuel bowl
9 Metering body (primary side)
10 Gasket, metering body
11 Fuel tube (float bowl connecting)
12 "O" rings, fuel tube
13 Screw, fuel bowl (secondary)
14 Gasket, bowl screw
15 Fuel bowl (secondary)
16 Metering body (secondary)
17 Metering body (secondary)
18 Gasket, metering body (secondary)
19 Plate, metering body (secondary)
20 Gasket, metering body plate
21 Balance tube
22 Washers, balance tube
23 "O" rings, balance tube
24 Choke link
25 Seal, choke rod
26 Throttle body screws
27 Main body
28 Throttle body
29 Gasket, main to throttle body
30 Screws, fuel pump cover
31 Clamp, valve rod
32 Rod, bowl vent valve
33 Spring vent valve rod
34 Valve, bowl vent
35 Retainer, clip, float
36 Float
37 Spring, float
38 Baffle, float
39 Needle valve and seat
40 Cover assembly, fuel pump
41 Cover assembly, fuel pump
42 Diaphragm, fuel pump
43 Spring, fuel pump diaphragm
44 Fitting, fuel inlet
45 Gasket, fuel inlet, fitting
46 Valve assembly, power
47 Gasket, power valve
48 Primary jets
49 Needle, idle adjusting mixture
50 Needle, idle mixture needle
51 Screws, choke valve
52 Choke valve
53 Choke shaft & lever assembly
54 Discharge nozzle screw, pump
55 Gasket, nozzle screw
56 Nozzle, pump discharge
57 Needle, pump discharge jet
58 Cotter pins, connecting rods
59 Rod, secondary connecting
60 Screw and lockwasher, fast idle cam lever
61 Lever, fast idle cam
62 Screws, primary throttle valve

63 Throttle valves, primary
64 Screw, pump cam
65 Pump cam
66 Screw and lockwasher, secondary stop lever
67 Lever, secondary stop
68 Screws, secondary throttle valves
69 Throttle valves, secondary
70 Fast idle cam lever
71 Fast idle cam
72 Retainer (E-clip)
73 Choke diaphragm link
74 Choke diaphragm assembly
75 Choke vacuum hose
76 Choke diaphragm bracket screw
77 Secondary diaphragm cover screw
78 Diaphragm cover (machine)
79 Secondary diaphragm return spring
80 Secondary diaphragm assembly
81 Secondary diaphragm housing (machine)
82 Secondary diaphragm housing gasket
83 Secondary diaphragm assembly screw
84 Throttle connecting rod retainer washer
85 Pump operating lever (E-clip)
86 Secondary stop screw
87 Throttle stop screw
88 Throttle stop screw spring
89 Baffle
90 Limiter cap
91 Bowl vent valve assy. (E.C.S.)

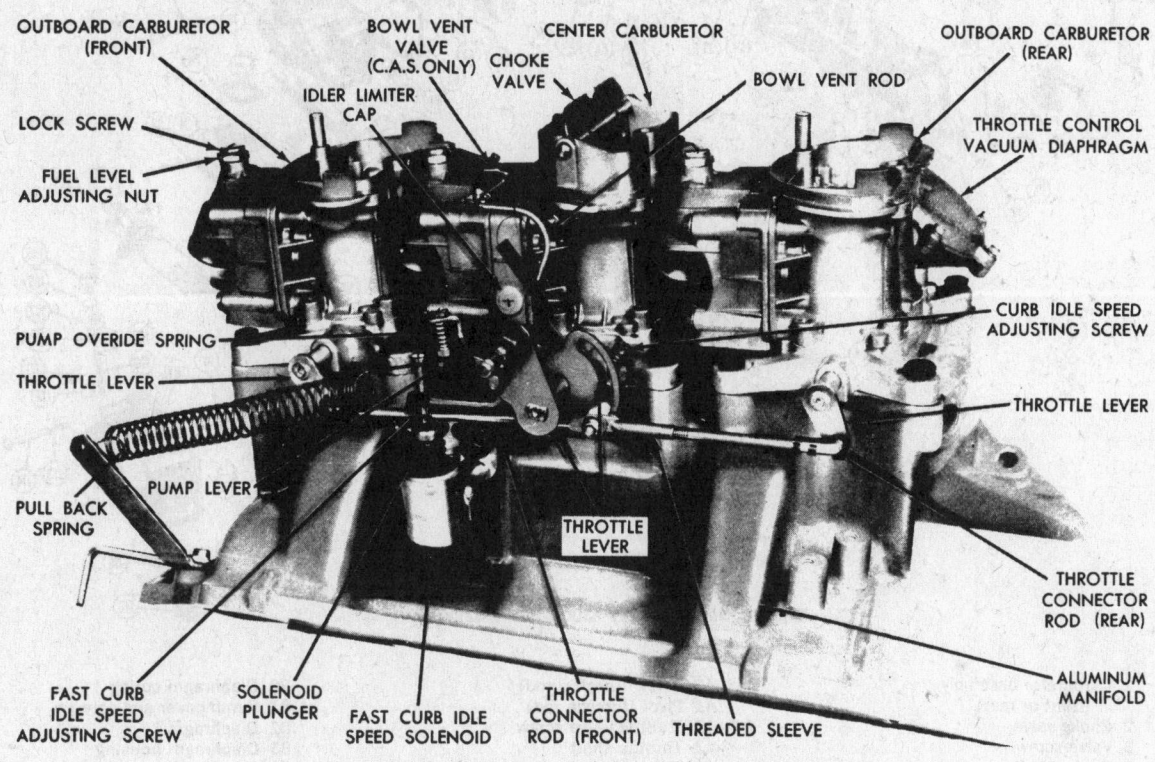

Triple installation of Holley 2300 two-barrel carburetors (Dodge 440 Six-Pack)

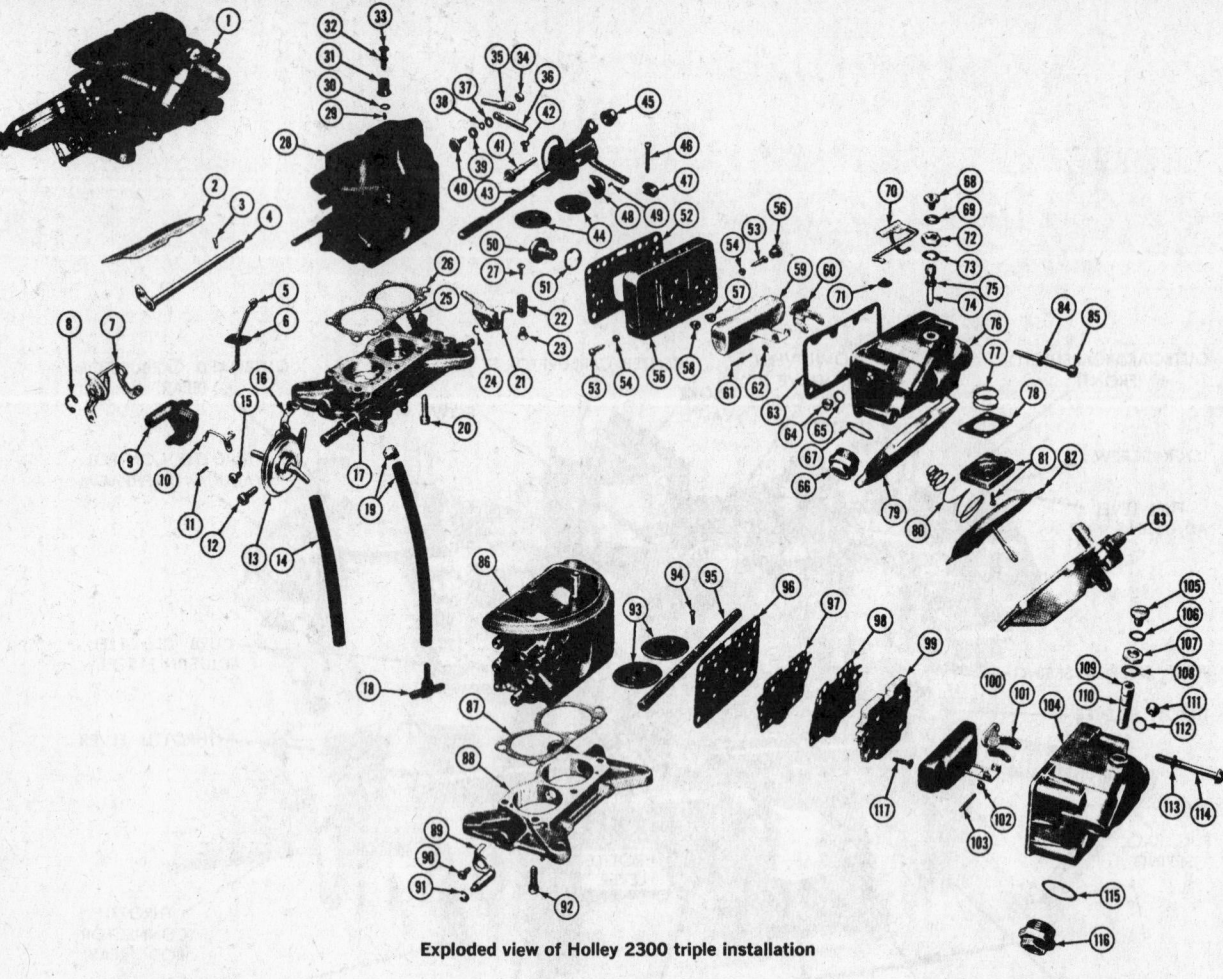

Exploded view of Holley 2300 triple installation

1	Carburetor assembly (front or rear)
2	Choke valve
3	Valve screw
4	Choke shaft
5	Operating rod
6	Seal
7	Choke control lever
8	"E" clip
9	Fast idle cam
10	Link
11	Link pin
12	Screw, diaphragm bracket
13	Choke diaphragm
14	Vacuum tube
15	Screw
16	Fast idle cam lever
17	Throttle body (center carb.)
18	Plastic "T"
19	Tube and clamp
20	Screw, throttle body
21	Lever
22	Spring
23	Nut, adjusting
24	Spacer
25	Curb idle screw and spring
26	Gasket
27	Screw Accelerator Pump Adj.
28	Main body (center carb.)
29	Discharge check needle
30	Gasket
31	Cluster, discharge
32	Gasket
33	Cluster screw
34	Nut
35	Sleeve (rear)
36	Sleeve (front)
37	Washer, throttle rod sleeve
38	Spacer, throttle rod sleeve
39	Washer, throttle rod sleeve
40	Screw (throttle rod)
41	Pivot (throttle rod)
42	Throttle valve screw
43	Throttle shaft
44	Throttle valves
45	Insulator
46	Adjusting screw (solenoid)
47	Nut (adjusting screw)
48	Cam, pump
49	Pump cam screw
50	Power valve
51	Power valve gasket
52	Gasket
53	Idle mixture screw
54	Idle mixture screw gasket
55	Metering body
56	Limiter cap (plastic)
57	Metering jet
58	Metering jet
59	Float
60	Float hinge bracket
61	Fulcrum pin
62	Float spring
63	Gasket
64	Sight plug
65	Sight plug gasket
66	Fuel inlet fitting
67	Fuel inlet fitting gasket
68	Lock screw
69	Gasket
70	Bowl vent rod
71	Bowl vent valve
72	Adjusting nut
73	Gasket, nut
74	Fuel inlet needle seat
75	Seat "O" ring
76	Fuel bowl
77	Spring, pump
78	Diaphragm, pump
79	Diaphragm cover
80	Diaphragm spring
81	Pump cover and screw
82	Diaphragm
83	Diaphragm housing
84	Seal (bowl screw)
85	Bowl screw
86	Main body (front or rear)
87	Gasket
88	Throttle body
89	Lever
90	Screw
91	"E" clip
92	Screw, throttle body
93	Throttle valves
94	Valve screw
95	Throttle shaft
96	Gasket
97	Plate
98	Gasket
99	Metering body
100	Float
101	Float hinge bracket
102	Float spring
103	Fulcrum pin
104	Fuel bowl
105	Lock screw
106	Gasket
107	Adjusting nut
108	Gasket
109	Inlet needle seat
110	Seat "O" ring
111	Sight plug
112	Gasket
113	Seal (bowl screw)
114	Bowl screw
115	Gasket
116	Fuel inlet fitting
117	Clutch head screw (metering body)

Rochester Carburetors

Buick

YEAR	MODEL OR TYPE	FLOAT LEVEL (IN.) Prim. (IN.)	Sec. (IN.)	FLOAT DROP (IN.) Prim. (IN.)	Sec. (IN.)	Pump Travel Setting (IN.)	CHOKE SETTING Unloader (IN.)	Housing	ON THE CAR ADJUSTMENTS Idle Screw Turns Open	Idle Speed (RPM)	Fast Idle Speed (RPM)	Dashpot Plunger Clearance (IN.)
1964	2GC—7024046	1/2	...	1 29/32	...	1 11/32	.085	2-RICH	1	500	...	...
1964	2GC—7024047	1/2	...	1 29/32	...	1 11/32	.085	INDEX	1	500	...	...
1964	4GC—7024040	19/32	1 11/32	1 7/16	1 3/16	1 1/64	.120	INDEX	1 1/2	500	...	...
1964	4GC—7024044	19/32	1 21/64	1 7/32	1 3/16	7/8	.120	2-RICH	1 1/2	500	...	...
1964	4GC—7024045	19/32	1 21/64	1 7/32	1 3/16	1	.120	INDEX	1 1/2	500	...	...
1965	2GC—7025046/7	19/32	...	1 29/32	...	1 11/32	.136	INDEX	1	500	...	...
1965	4GC—7025040	1 13/32	1 13/32	1 1/16	1 1/16	1 1/64	.120	INDEX	1 1/2	500	...	...
1966	2GC—7026046	1/2	...	1 27/32	...	1 5/32	.140	INDEX	2	550	...	...
1966	2GC—7026047	1/2	...	1 27/32	...	1 11/32	.140	INDEX	2	550	...	...
1966	2GC—7036048	1/2	...	1 27/32	...	1 11/32	.140	INDEX	1	550	...	...
1966	4GC—7026040	1 1/16	1 1/16	1 1/16	1 1/16	1	.120	INDEX	1 1/2	500	...	...
1966	4MC—7026240/1	1/4	...	...	...	9/32	.325	INDEX	1 1/2	500	...	...
1966	4MC—7036240/1	1/4	...	...	...	9/32	.325	INDEX	1 1/2	500	...	...
1967	2GC—7027044, 46	15/32	...	1 1/16	...	31/32	.140	INDEX	2	550	...	...
1967	2GC—7027045, 49, Calif.	15/32	...	1 1/16	...	31/32	.140	INDEX	2	600	...	...
1967	4MV—7027146, 48	7/32	...	...	...	9/32	.325	...	1	600	...	.030 (B)
1967	4MV—7027147, 49, Calif.	7/32	...	...	...	9/32	.325	...	1	600	...	.030 (B)
1967	4MV—7027240	9/32	...	...	...	9/32	.325	...	1	600	...	.030 (B)
1967	4MV—7027241, Calif.	9/32	...	...	...	13/32	.325	...	1	600	...	.030 (B)
1968	2GV—7028140, 41	15/32	...	1 7/32	...	1 11/32	.140	(A)	2	See Decal	...	.030
1968	4MV—7028244, 45	5/16	...	...	...	9/32	.325	(A)	2	See Decal	...	.030
1968	4MV—7028240, 242	13/32	...	...	...	9/32	.325	(A)	2	See Decal	...	.030
1968	4MV—7028243	15/32	...	...	...	9/32	.325	(A)	2	See Decal	...	.030
1969	2GV—7029140, 41	15/32	...	1 7/32	...	1 11/32	.140	...	...	...	...	...
1969	4MV—7029244, 45	5/16	...	...	...	13/32	.325	...	2	620	...	...
1969	4MV—7029240, 43	3/8	...	...	...	13/32	.325	...	2	620	...	...
1969	4MV—7029241	5/16	...	...	...	13/32	.325	...	2	720	...	...
1970	2GV-7040143-350, Std.	15/32	...	1 7/8	...	1 15/32	.200	...	...	700	...	...
1970	2GV-7040142-350, Auto.	15/32	...	1 7/8	...	1 13/32	.180	...	...	600 (C)	...	...
1970	2GV-7040446-350 Auto., CAL	15/32	...	1 7/8	...	1 13/32	.180	...	...	600 (C)	...	...
1970	4MV-7040245-350 Std.	5/16	...	...	...	13/32	.335	...	...	700 (C)	720 (C)	.030
1970	4MV-7040244-350 Auto.	5/16	...	...	...	13/32	.335	...	...	600 (C)	650 (C)	.030
1970	4MV-7040243-455 Std.	3/8	...	...	...	9/32	.335	...	...	700 (C)	720 (C)	.030
1970	4MV-7040240-455 Auto.	3/8	...	...	...	9/32	.335	...	...	600 (C)	650 (C)	.030
1970	4MV-7040247-455 Riviera	3/8	...	...	...	9/32	.325	...	...	600 (C)	650 (C)	.030

(A)—Remove upper end of choke rod, close choke valve, place rod in notch. Bend rod to adjust. Place rod in std. hole.
(B)—Vacuum break setting—.230.
(C)—In Drive.
CAL—California

Buick Special

YEAR	MODEL OR TYPE	FLOAT LEVEL Prim. (IN.)	Sec. (IN.)	FLOAT DROP Prim. (IN.)	Sec. (IN.)	Pump Travel Setting (IN.)	CHOKE Unloader (IN.)	Housing	Idle Screw Turns Open	Idle Speed (RPM)	Fast Idle Speed (RPM)	Dashpot Plunger Clearance (IN.)
1964	BC—7024148/9	1 7/32	...	1 3/4	...	...	.230	INDEX	1 3/4	550	...	...
1964	2GC—7024046	1/2	...	1 29/32	...	1 11/32	.085	2-RICH	1	550	...	...
1964	2GC—7024047	1/2	...	1 29/32	...	1 11/32	.085	INDEX	1	550	...	...
1964	4GC—7024044/5	1 21/64	1 3/8	1 19/32	1 3/16	7/8	.120	2-RICH	1 1/2	550	...	...
1965	BC—7025148/9	1 7/32	...	1 3/4	...	...	.230	INDEX	1 3/4	550	...	...
1965	2GC—7025046/7	19/32	...	1 29/32	...	1 11/32	.136	INDEX	1	550	...	...
1966	2GC—7026144	1/2	...	1 27/32	...	1 5/32	.140	INDEX	2	550	...	...
1966	2GC—7026145	1/2	...	1 27/32	...	1 11/32	.140	INDEX	2	550	...	...
1966	2GC—7036144, Calif.	1/2	...	1 27/32	...	1 1/16	.140	INDEX	2	550	...	...
1966	2GC—7026046	1/2	...	1 27/32	...	1 11/32	.140	INDEX	2	550	...	...
1966	2GC—7026047	1/2	...	1 27/32	...	1 11/32	.140	INDEX	2	550	...	...
1966	2GC—7036046, Calif.	1/2	...	1 27/32	...	1 5/32	.140	INDEX	2	550	...	...
1966	2GC—7036048, Calif.	1/2	...	1 27/32	...	1 5/32	.140	INDEX	1	550	...	...
1966	4GC—7026040	1 1/16	1 1/16	1 1/16	1 1/16	1	.120	INDEX	1 1/2	500	...	...

Buick Special, continued

YEAR	MODEL OR TYPE	FLOAT LEVEL (IN.) Prim. (IN.)	FLOAT LEVEL (IN.) Sec. (IN.)	FLOAT DROP (IN.) Prim. (IN.)	FLOAT DROP (IN.) Sec. (IN.)	Pump Travel Setting (IN.)	CHOKE SETTING Unloader (IN.)	CHOKE SETTING Housing	ON THE CAR ADJUSTMENTS Idle Screw Turns Open	Idle Speed (RPM)	Fast Idle Speed (RPM)	Dashpot Plunger Clearance (IN.)
1967	2GC—7027040	1/2	...	1 9/32	...	1 1/16	.140	INDEX	2	550	...	...
1967	2GC—7027042	1/2	...	1 9/32	...	1 5/32	.140	INDEX	2	550	...	...
1967	2GC—7027041, Calif.	1/2	...	1 9/32	...	1 1/16	.140	INDEX	2	600	...	...
1967	2GC—7027044/46	15/32	...	1 9/16	...	1 5/32	.140	INDEX	2	550	...	...
1967	2GC—7027045/49, Calif.	15/32	...	1 9/16	...	1 5/32	.140	INDEX	2	600	...	...
1968	MV—7028014	5/16	...	...	...	...	.350	...	...	650	...	...
1968	MV—7028047	5/16	...	...	...	...	.325	...	...	700	...	...
1968	2GV—7028140/41	15/32	...	1 9/32	...	1 11/32	.140	...	...	...	...	...
1968	4MV—7028244/45	5/16	...	...	...	13/32	.325	...	2	700	...	.030
1968	4MV—7028242	3/8	...	...	...	9/32	.325	...	2	700	...	.030
1968	4MV—7028243	7/16	...	...	...	11/32	.325	...	2	700	...	.030
1969	MV—7029014, A.T.	1/4	...	...	...	...	.170	...	2	620	...	...
1969	MV—7029047, M.T	9/32	...	...	...	...	.190	...	2	720	...	...
1969	2GV—7029140/41	15/32	...	1 7/32	...	1 11/32	.140	...	...	...	...	...
1969	4MV—7029244/45	5/16	...	...	...	13/32	.325	...	2	...	...	.030
1969	4MV—7029242/43	3/8	...	...	...	13/32	.325	...	2	620	...	...
1970	2GV-7040143-350, Std.	15/32	...	1 7/8	...	1 15/32	.200	...	...	700	...	...
1970	2GV-7040142-350, Auto.	15/32	...	1 7/8	...	1 13/32	.180	...	...	600Ⓐ	...	...
1970	2GV-7040446-350 Auto., CAL	15/32	...	1 7/8	...	1 13/32	.180	...	...	600Ⓐ	...	...
1970	4MV-7040245-350 Std.	5/16	...	...	...	13/32	.335	...	...	700Ⓐ	720Ⓐ	.030
1970	4MV-7040244-350, Auto.	5/16	...	...	...	13/32	.335	...	...	600Ⓐ	650Ⓐ	.030
1970	4MV-7040243-455, Std.	3/8	...	...	...	9/32	.335	...	...	700Ⓐ	720Ⓐ	.030
1970	4MV-7040240-455, Auto.	3/8	...	...	...	9/32	.335	...	...	600Ⓐ	650Ⓐ	.030
1970	4MV-7040246-455, Stage 1	5/16	...	...	...	9/32	.335	...	...	600Ⓐ	650Ⓐ	.030
1970	MV-7040015-250, Std.	1/4	...	...	...	...	.350	...	...	750	900	...
1970	MV-7040014-250, Auto.	1/4	...	...	...	...	.350	...	...	600Ⓐ	650	...

Ⓐ—In drive.
CAL—California

Cadillac

YEAR	MODEL OR TYPE	FLOAT LEVEL Prim. (IN.)	FLOAT LEVEL Sec. (IN.)	FLOAT DROP Prim. (IN.)	FLOAT DROP Sec. (IN.)	Pump Travel Setting (IN.)	Choke Unloader (IN.)	Choke Housing	Idle Screw Turns Open	Idle Speed (RPM)	Fast Idle Speed (RPM)	Dashpot Plunger Clearance (IN.)
1964	4GC	1 7/16	1 3/8	1 1/2	1 5/32	27/32	.125	INDEX	1 1/2–2 1/2	480	1700	①
1965	4GC	1 7/16	1 3/8	1 1/2	1 1/16	13/16	.125	INDEX	1 1/2–2 1/2	480	1700	①
1966	4GC	1 7/16	1 3/8	1 1/2	1 1/16	13/16	.125	INDEX	1 1/2–2 1/2	480	1700	①
1967	4MV	7/32	...	...	...	11/32	.312	...	...	500	1725	...
1967	4MV, Calif.	7/32	...	...	...	11/32	.300	...	...	550	1700	...
1968	4MV-Std.	1/4	...	...	...	11/32	.312	②	1	See Decal		.030
1968	4MV-Eldorado	11/32	...	...	...	11/32	.312	②	1	See Decal		.030
1969	4MV-Std.	1/4	...	...	...	11/32	.300	②	...	...	1750	.030
1969	4MV-Eldorado	3/8	...	...	...	11/32	.300	②	...	...	1750	.030
1970	4MV	.240	...	...	...	.344	.300	INDEX	...	600③	1900–1950④	.030

①—By trial.
②—Remove upper end of choke rod, close choke valve, place rod in notch. Bend rod to adjust, place rod in std. hole.
③—In drive—AC idle speed-up—900- 950 rpm.
④—AC off.

Chevelle—Camaro—Monte Carlo

YEAR	MODEL OR TYPE	FLOAT LEVEL Prim. (IN.)	FLOAT LEVEL Sec. (IN.)	FLOAT DROP Prim. (IN.)	FLOAT DROP Sec. (IN.)	Pump Travel Setting (IN.)	Choke Unloader (IN.)	Choke Housing	Idle Screw Turns Open	Idle Speed (RPM)	Fast Idle Speed (RPM)	Dashpot Plunger Clearance (IN.)
1964	BC—ALL	1 9/32	...	...	...	...	.350	①	1 1/2	500	.060	...
1964	2GC—ALL	1 23/64	...	...	...	57/64	.360	INDEX	1 1/2	500	...	...
1964	4GC—ALL	1 33/64	1 33/64	2 1/4	2 1/4	1 1/16	.235	INDEX	1 1/2	475	...	...
1965	BV	1 9/32	...	1 3/4	...	...	.350	...	1 1/2	500	Ⓐ	...
1965	2GV	1 23/64	...	1 29/32	...	57/64	13/64	...	1 1/2	500	Ⓐ	...
1965	4GC	1 33/64	1 37/64	2 1/4	2 1/4	1 1/16	.235	INDEX	1 1/2	500	Ⓐ	...
1966	BV	1 9/32	...	1 3/4	...	...	.350	...	1 1/2	475	Ⓐ	...
1966	2GV	3/4	...	1 3/4	...	1 1/8	.215	...	1 1/2	475	...	...
1966	4GC	1 17/32	1 19/32	2 1/4	2 1/4	1 1/16	.250	INDEX	1 1/2	475	Ⓐ	...
1967	BV	1 9/32	...	1 3/4	...	...	.350	...	1 1/2	...	...	...
1967	2GV	3/4	...	1 3/4	...	1 1/8	.215	...	1 1/2	...	...	...
1967	4MV	9/32	...	...	...	13/32	.260	...	...	500	2200	...
1967	4MV	9/32	...	...	...	13/32	.300	...	...	...	2000	...
1968	MV-Aut.	9/32	...	...	...	...	.350	...	1 1/2	...	2400	...
1968	MV-Std.	9/32	...	...	...	...	.350	...	1 1/2	...	2400	...
1968	2GV	3/4	...	1 3/4	...	1 1/8	.200	...	...	...	...	...
1968	4MV-Aut.	3/16	...	...	...	9/32	.300	...	2	...	2400	...
1968	4MV-Std.	3/16	...	...	...	9/32	.300	...	2	...	2400	...

Chevelle—Camaro—Monte Carlo, continued

YEAR	MODEL OR TYPE	FLOAT LEVEL (IN.) Prim. (IN.)	Sec. (IN.)	FLOAT DROP (IN.) Prim. (IN.)	Sec. (IN.)	Pump Travel Setting (IN.)	CHOKE SETTING Unloader (IN.)	Housing	ON THE CAR ADJUSTMENTS Idle Screw Turns Open	Idle Speed (RPM)	Fast Idle Speed (RPM)	Dashpot Plunger Clearance (IN.)
1969	MV-Aut.	1/4	...	...	...	...	.350	...	...	...	2400	...
1969	MV-Std.	1/4	...	...	...	...	.350	...	...	...	2400	...
1969	MV-Aut. w/AC	1/4	...	...	...	...	.350	...	...	...	2400	...
1969	2GV-Std.	27/32	...	1 3/4	...	1 1/8	.215	...	1	...	...	...
1969	2GV-Std. w/AC	27/32	...	1 3/4	...	1 1/8	.215	...	1	...	2000	...
1969	2GV-Aut.	27/32	...	1 3/4	...	1 1/8	.215	...	1	...	2000	...
1969	2GV-Aut. w/AC	27/32	...	1 3/4	...	1 1/8	.215	...	1	...	2000	...
1969	4MV-Aut.	7/32	...	...	...	5/16	.450	...	2	...	2400	.015
1969	4MV-Std.	7/32	...	...	...	5/16	.450	...	2	...	2400	.015
1969	4MV-Aut.	1/4	...	...	...	5/16	.450	...	2	...	2400	.015
1969	4MV-Std.	1/4	...	...	...	5/16	.450	...	2	...	2400	.015
1970	MV-230, Auto.	1/4	...	...	...	...	.350	...	...	②	2400	...
1970	MV-230, Std.	1/4	...	...	...	...	.350	...	...	②	2400	...
1970	MV-250, Auto.	1/4	...	...	...	...	.350	...	...	②	2400	...
1970	MV-250, Std.	1/4	...	...	...	...	.350	...	...	②	2400	...
1970	2GV-1.25-307, Auto.	27/32	...	1 3/4	...	1 1/8	.215	...	...	②	2200–2400	...
1970	2GV-1.25-307, Auto. w. AC	27/32	...	1 3/4	...	1 1/8	.215	...	...	②	2200–2400	...
1970	2GV-1.25-307, Std.	27/32	...	1 3/4	...	1 1/8	.160	...	...	②	2200–2400	...
1970	2GV-1.25-307, Std. w. AC	27/32	...	1 3/4	...	1 1/8	.225	...	...	②	2200–2400	...
1970	2GV-1.50-350, Auto.	23/32	...	1 3/8	...	1 17/32	.325	...	...	②	2200–2400	...
1970	2GV-1.50-350, Auto. w. AC	23/32	...	1 3/8	...	1 17/32	.325	...	...	②	2200–2400	...
1970	2GV-1.50-350, Std.	23/32	...	1 3/8	...	1 17/32	.275	...	...	②	2200–2400	...
1970	2GV-1.50-350, Std. w. AC	23/32	...	1 3/8	...	1 17/32	.275	...	...	②	2200–2400	...
1970	2GV-1.50-400, Auto.	23/32	...	1 3/8	...	1 17/32	.325	...	...	②	2200–2400	...
1970	2GV-1.50-400, Auto. w. AC	23/32	...	1 3/8	...	1 17/32	.325	...	...	②	2200–2400	...
1970	2GV-1.50-400, Std.	23/32	...	1 3/8	...	1 17/32	.325	...	...	②	2200–2400	...
1970	2GV-1.50-400, Std. w. AC	23/32	...	1 3/8	...	1 17/32	.325	...	...	②	2200–2400	...
1970	4MV-350, Auto., 300 hp.	1/4	...	...	...	5/16	.450	...	...	②	2400	.020
1970	4MV-350, Std., 300 hp.	1/4	...	...	...	5/16	.450	...	...	②	2400	.020
1970	4MV-350, Std., 325 hp.	1/4	...	...	...	5/16	.450	...	...	②	2400	.015
1970	4MV-396, 325 hp., 400, 330 hp., Auto.	1/4	...	...	...	5/16	.450	...	...	②	2400	.020
1970	4MV-396, 325 hp., 400, 330 hp., Std.	1/4	...	...	...	5/16	.450	...	...	②	2400	.020
1970	4MV-396, 350 hp., Auto.	1/4	...	...	...	5/16	.450	...	...	②	2400	.020
1970	4MV-396, 350 hp., Std.	1/4	...	...	...	5/16	.450	...	...	②	2400	.020
1970	4MV-454, 345-360-390 hp., Auto.	1/4	...	...	...	5/16	.450	...	...	②	2400	.020
1970	4MV-454, 360-390 hp., Auto.	1/4	...	...	...	5/16	.450	...	...	②	2400	.020
1970	4MV-454, 360-390 hp., Std.	1/4	...	...	...	5/16	.450	...	...	②	2400	.020

Ⓐ—Screw on second step with gauge.   ①—Std. trans.—1-RICH; auto. trans.—2-RICH.   ②—See engine compartment decal for idle speed.

Chevrolet

YEAR	MODEL OR TYPE	FLOAT LEVEL (IN.) Prim. (IN.)	Sec. (IN.)	FLOAT DROP (IN.) Prim. (IN.)	Sec. (IN.)	Pump Travel Setting (IN.)	CHOKE SETTING Unloader (IN.)	Housing	ON THE CAR ADJUSTMENTS Idle Screw Turns Open	Idle Speed (RPM)	Fast Idle Speed (RPM)	Dashpot Plunger Clearance (IN.)
1964	BC	1 1/8	...	1 3/4	...	...	.350	...	1 1/2	475	...	...
1964	2GV	1 23/64	...	1 29/32	...	57/64	.360	Ⓑ	1 1/2	475	...	...
1964	4GC—327 cu. in.	1 33/64	1 23/64	2 1/4	2 1/4	1 1/16	.235	INDEX	1 1/2	475	...	②
1964	4GC-409 cu. in.	13/16	1 1/2	1 1/2	1 1/8	1 3/32	.130	INDEX	1 1/2	475	...	②
1965	BV	1 7/32	...	1 3/4	...	...	.350	...	1 1/2	475	Ⓐ	...
1965	2GV	1 23/64	...	1 29/32	...	57/64	.208	...	1 1/2	475	Ⓐ	...
1965	4GC—327 cu. in.	1 33/64	1 37/64	2 1/4	2 1/4	1 1/16	.235	INDEX	1 1/2	475	Ⓐ	...
1965	4GC-409 cu in.	13/16	1/2	1 1/2	1 1/8	1 3/32	.130	INDEX	1 1/2	475	Ⓐ	...
1966	BV	1 7/32	...	1 3/4	...	...	.350	...	1 1/2	475	Ⓒ	...
1966	2GV	3/4	...	1 3/4	...	1 1/8	.215	...	1 1/2	475	Ⓒ	...
1966	4GC	1 17/32	1 19/32	2 1/4	2 1/4	1 1/16	.250	INDEX	1 1/2	475	...	...
1966	4MV	7/32	...	...	...	13/32	.260	...	1 1/2	...	2200	...
1966	4MV	9/32	...	...	...	13/32	.300	...	1 1/2	...	2000	...
1967	BV	1 7/32	...	1 3/4	...	...	.350	...	1 1/2	500	Ⓒ	...
1967	2GV	3/4	...	1 3/4	...	1 1/8	.215	...	1 1/2	550	Ⓒ	...
1967	4MV	9/32	...	...	...	13/32	.260	...	...	500	2200	...
1967	4MV	9/32	...	...	...	13/32	.300	...	...	...	2000	...
1968	MV—Aut.	9/32	...	...	...	...	.350	...	1 1/2	...	2400	...
1968	MV—Std.	9/32	...	...	...	...	.350	...	1 1/2	...	2400	...
1968	2GV	3/4	...	1 3/4	...	1 1/8	.200	...	...	...	...	...
1968	4MV—Aut.	9/32	...	...	...	9/32	.260	...	2	...	2400	...
1968	4MV—Std.	9/32	...	...	...	9/32	.300	...	2	...	2400	...
1968	4MV—Aut.	3/16	...	...	...	9/32	.300	...	2	...	2400	...
1968	4MV—Std.	3/16	...	...	...	9/32	.300	...	2	...	2400	...

YEAR	MODEL OR TYPE	FLOAT LEVEL (IN.)		FLOAT DROP (IN.)		Pump Travel Setting (IN.)	CHOKE SETTING		ON THE CAR ADJUSTMENTS			
		Prim. (IN.)	Sec. (IN.)	Prim. (IN.)	Sec. (IN.)		Unloader (IN.)	Housing	Idle Screw Turns Open	Idle Speed (RPM)	Fast Idle Speed (RPM)	Dashpot Plunger Clearance (IN.)
Chevrolet, continued												
1969	MV—Aut.	1/4	...	...	...	...	.350	...	...	...	2400	...
1969	MV—Std.	1/4	...	...	...	...	.350	...	...	...	2400	...
1969	2GV—Aut. 327	3/4	...	1 3/4	...	1 13/32	.275	...	1	...	2000	...
1969	2GV—Std. 327	3/4	...	1 3/4	...	1 13/32	.275	...	1	...	2000	...
1969	2GV—Aut. 396	3/4	...	1 3/4	...	1 13/32	.275	...	1	...	2000	...
1969	2GV—Std. 396	3/4	...	1 3/4	...	1 13/32	.275	...	1	...	2000	...
1969	4MV—Aut. 350	7/32	...	...	...	5/16	.450	...	2	...	2400	.015
1969	4MV—Std. 350	7/32	...	...	...	5/16	.450	...	2	...	2400	.015
1969	4MV—Aut. 427	1/4	...	...	...	5/16	.450	...	2	...	2400	.015
1969	4MV—Std. 427	1/4	...	...	...	5/16	.450	...	2	...	2400	.015
1970	MV-230, Auto.	1/4	...	...	...	...	.350	...	...	③	2400	...
1970	MV-230, Std.	1/4	...	...	...	...	.350	...	...	③	2400	...
1970	MV-250, Auto.	1/4	...	...	...	...	.350	...	...	③	2400	...
1970	MV-250, Std.	1/4	...	...	...	...	.350	...	...	③	2400	...
1970	2GV-1.25-307, Auto.	27/32	...	1 3/4	...	1 1/8	.215	...	...	③	2200-2400	...
1970	2GV-1.25-307, Auto. w. AC	27/32	...	1 3/4	...	1 1/8	.215	...	...	③	2200-2400	...
1970	2GV-1.25-307, Std.	27/32	...	1 3/4	...	1 1/8	.160	...	...	③	2200-2400	...
1970	2GV-1.25-307, Std. w. AC	27/32	...	1 3/4	...	1 1/8	.225	...	...	③	2200-2400	...
1970	2GV-1.50-350, Auto.	23/32	...	1 3/8	...	1 17/32	.325	...	...	③	2200-2400	...
1970	2GV-1.50-350, Auto. w. AC	23/32	...	1 3/8	...	1 17/32	.325	...	...	③	2200-2400	...
1970	2GV-1.50-350, Std.	23/32	...	1 3/8	...	1 17/32	.275	...	...	③	2200-2400	...
1970	2GV-1.50-350, Std. w. AC	23/32	...	1 3/8	...	1 37/32	.275	...	...	③	2200-2400	...
1970	2GV-1.50-4.00, Auto.	23/32	...	1 3/8	...	1 17/32	.325	...	...	③	2200-2400	...
1970	2GV-1.50-4.00, Auto. w. AC	23/32	...	1 3/8	...	1 17/32	.325	...	...	③	2200-2400	...
1970	2GV-1.50-400, Std.	23/32	...	1 3/8	...	1 17/32	.325	...	...	③	2200-2400	...
1970	2GV-1.50-400, Std. w. AC	23/32	...	1 3/8	...	1 17/32	.325	...	...	③	2200-2400	...
1970	4MV-350, Auto., 300 h.p.	1/4	...	...	...	5/16	.450	...	...	③	2400	.020
1970	4MV-350, Std., 300 h.p.	1/4	...	...	...	5/16	.450	...	...	③	2400	.020
1970	4MV-350, Std., 325 h.p.	1/4	...	...	...	5/16	.450	...	...	③	2400	.015
1970	4MV-454, 345-360-390 h.p., Auto.	1/4	Sec.	...	...	5/16	.450	...	...	③	2400	.020
1970	4MV-454, 360-390, h.p., Auto.	1/4	...	...	...	5/16	.450	...	...	③	2400	.020
1970	4MV-454, 360-390 h.p., Std.	1/4	...	...	...	5/16	450	...	...	③	2400	.020

①—Std. trans. 1-rich; automatic trans. 2-rich.   ②—By trial.   ③—See engine compartment for idle speed.   ④—Screw on 2nd step with gauge.
⑤—Std. Trans.—1-LEAN; auto. trans.—INDEX.

Chevy II

YEAR	MODEL OR TYPE	FLOAT LEVEL (IN.)		FLOAT DROP (IN.)		Pump Travel Setting (IN.)	CHOKE SETTING		ON THE CAR ADJUSTMENTS			
		Prim. (IN.)	Sec. (IN.)	Prim. (IN.)	Sec. (IN.)		Unloader (IN.)	Housing	Idle Screw Turns Open	Idle Speed (RPM)	Fast Idle Speed (RPM)	Dashpot Plunger Clearance (IN.)
1964	6 Cyl.-BV—ALL	1 9/32	...	1 3/4	...	...	.350	...	1 1/2	500	②	...
1964	V-8-2GV—ALL	1 23/64	...	1 29/32	...	57/64	.203	①	1 1/2	475	...	...
1965	6 Cyl.-BV	1 9/32	...	1 3/4	...	...	.350	...	1 1/2	500	②	...
1965	V8-2GV	1 23/64	...	1 29/32	...	57/64	.203	...	1 1/2	500	...	...
1965	V8-4GC	1 32/64	1 37/64	2 1/4	2 1/4	1 1/16	.235	INDEX	1 1/2	...	...	...
1966	6 Cyl.-BV	1 9/32	...	1 3/4	...	...	.350	...	1 1/2	475	②	...
1966	V8-2GV	3/4	...	1 3/4	...	1 1/8	.215	...	1 1/2	475	...	...
1966	V8-4GC	1 17/32	1 19/32	2 1/4	2 1/4	1 1/16	.250	INDEX	1 1/2	...	...	...
1967	BV	1 1/32	...	1 3/4	...	...	.350	...	1 1/2	500	③	...
1967	2GV	3/4	...	1 3/4	...	1 1/8	.215	...	1 1/2	500	...	...
1967	4MV-Aut.	9/32	...	...	...	13/32	.260	...	...	500	2200	...
1967	4MV-Std.	9/32	...	...	...	13/32	.300	...	...	...	2000	...
1968	M	9/32	...	...	...	...	...	...	1 1/2	...	2400	...
1968	MV-Aut.	9/32	...	...	...	...	.350	...	1 1/2	...	2400	...
1968	MV-Std.	9/32	...	...	...	...	.350	...	1 1/2	...	2400	...
1968	2GV	3/4	...	1 3/4	...	1 1/8	.200	...	...	...	...	...
1968	4MV	9/32	...	...	...	9/32	.300	...	2	...	2400	...

YEAR	MODEL OR TYPE	FLOAT LEVEL (IN.)		FLOAT DROP (IN.)		Pump Travel Setting (IN.)	CHOKE SETTING		ON THE CAR ADJUSTMENTS			
		Prim. (IN.)	Sec. (IN.)	Prim. (IN.)	Sec. (IN.)		Unloader (IN.)	Housing	Idle Screw Turns Open	Idle Speed (RPM)	Fast Idle Speed (RPM)	Dashpot Plunger Clearance (IN.)

Chevy II, continued

YEAR	MODEL OR TYPE	Prim.	Sec.	Prim.	Sec.	Pump	Unloader	Housing	Idle Screw	Idle Speed	Fast Idle	Dashpot
1969	M............................	1/4	...	...	...	...	...	...	...	...	2400	...
1969	MV-Aut........................	1/4	...	...	...	.350	...	...	...	...	2400	...
1969	MV-Std........................	1/4	...	...	...	.350	...	...	...	...	2400	...
1969	2GV-Std.	27/32	...	1 3/4	...	1 1/8	.215	...	1	...	...	...
1969	2GV-Aut.	27/32	...	1 3/4	...	1 1/8	.215	...	1	...	2000	...
1969	4MV-Std. 350	7/32	...	...	...	5/16	.450	...	2	...	2400	.015
1969	4MV-Aut. 350	7/32	...	...	...	5/16	.450	...	2	...	2400	.015
1969	4MV-Std. 396	1/4	...	...	...	5/16	.450	...	2	...	2400	.015
1969	4MV-Aut. 396	1/4	...	...	...	5/16	.450	...	2	...	2400	.015
1970	M-153, (4 Cyl.)	1/4	...	...	...	...	...	...	...	④	2400	...
1970	MV-230, Auto.	1/4	...	...	...	...	.350	...	...	④	2400	...
1970	MV-230, Std.	1/4	...	...	...	...	.350	...	...	④	2400	...
1970	MV-250, Auto.	1/4	...	...	...	...	.350	...	...	④	2400	...
1970	MV-250, Std.	1/4	...	...	...	...	.350	...	...	④	2400	...
1970	2GV-1.25-307, Auto.	27/32	...	1 3/4	...	1 1/8	.215	...	...	④	2200–2400	...
1970	2GV-1.25-307, Auto. w. AC ...	27/32	...	1 3/4	...	1 1/8	.215	...	...	④	2200–2400	...
1970	2GV-1.25-307, Std.	27/32	...	1 3/4	...	1 1/8	.160	...	...	④	2200–2400	...
1970	2GV-1.25-307, Std. w. AC	27/32	...	1 3/4	...	1 1/8	.225	...	...	④	2200–2400	...
1970	2GV-1.50-350, Auto.	23/32	...	1 3/8	...	1 17/32	.325	...	...	④	2200–2400	...
1970	2GV-1.50-350, Auto. w. AC ...	23/32	...	1 3/8	...	1 17/32	.325	...	...	④	2200–2400	...
1970	2GV-1.50-350, Std.	23/32	...	1 3/8	...	1 17/32	.275	...	...	④	2200–2400	...
1970	2GV-1.50-350, Std. w. AC	23/32	...	1 3/8	...	1 17/32	.275	...	...	④	2200–2400	...

① —Std. trans. 1-rich, auto. trans. index.
② —With .060 clearance at choke valve—Set on next to high step.
③ —With .090 clearance at choke valve—Set on next to high step.
④ —See engine compartment decal for idle speed.

Corvair

YEAR	MODEL OR TYPE	Prim.	Sec.	Prim.	Sec.	Pump	Unloader	Housing	Idle Screw	Idle Speed	Fast Idle	Dashpot
1960	H-7015300 Std.	1 13/64	...	1 3/4	...	...	.25	INDEX	2	450–500	...	...
1960	H-7013600, Auto.	1 13/64	...	1 3/4	...	...	.25	INDEX	2	475–500	...	...
1961	H-7019100, 7019101	1 13/64	...	1 3/4	...	...	...	MANUAL	...	500	...	...
1962	H-7023100, 7023101	1 13/64	...	1 3/4	...	INDEX	.25	③	1 1/2	①	...	...
1963	H—7023100-1-2	1 13/64	...	1 3/4	①	INDEX	.250	INDEX	1 1/2	500	①	...
1963	RC	3/8	...	...	...	INDEX	.325	INDEX	...	850	1500	...
1964	H—7024022-3-4	1 5/64	...	1 3/4	①	INDEX	.312	INDEX	1 1/2	500	①	...
1965	HV—7025023-4-5	1 1/16	...	1 1/2	①	INDEX	.312	INDEX	1 1/2	625	①	...
1966	HV	1 1/16	...	1 9/16	①	INDEX	.325	INDEX	1 1/2	625	①	...
1967	HV	1 1/16	...	1 9/16	...	INDEX	.325	...	1 1/2	500 ②	...	...
1968	HV	1 1/16	...	1 13/16	...	①	.500	...	1 1/2	...	...	...
1969	HV	1 1/16	...	1 13/16	...	①	.500	...	1 1/2	...	...	...

① —See Corvair car section for balancing and linkage adjustments.
② —500 rpm—95 HP, 650 rpm—110 HP.
③ —Two turns up from free entry in lever.

Jeep

YEAR	MODEL OR TYPE	Prim.	Sec.	Prim.	Sec.	Pump	Unloader	Housing	Idle Screw	Idle Speed	Fast Idle	Dashpot
—	2G-7026082-V6, Std. w/o Emission control, Universal	1 3/32	...	1 1/8	...	1 5/32	...	...	3 in	650–700	...	...
—	2G-7027082-V6, Std. w. Emission control, Universal	1 5/32	...	1 1/8	...	1 5/32	...	...	3 in	650–700	...	...
—	2G-7026089-V6, Auto., Jeepster	1 5/32	...	1 1/8	...	1 32	.136	INDEX	3 in.	...	...	...
—	2GV-7028088-350, Wagoneer, Gladiator	1 5/32	...	1 3/4	...	1 11/32	.140	STD. HOLE	3 in	650–700	...	...

YEAR	MODEL OR TYPE	FLOAT LEVEL (IN.) Prim. (IN.)	FLOAT LEVEL (IN.) Sec. (IN.)	FLOAT DROP (IN.) Prim. (IN.)	FLOAT DROP (IN.) Sec. (IN.)	Pump Travel Setting (IN.)	CHOKE SETTING Unloader (IN.)	CHOKE SETTING Housing	ON THE CAR ADJUSTMENTS Idle Screw Turns Open	ON THE CAR ADJUSTMENTS Idle Speed (RPM)	ON THE CAR ADJUSTMENTS Fast Idle Speed (RPM)	ON THE CAR ADJUSTMENTS Dashpot Plunger Clearance (IN.)
Oldsmobile												
1964	2GC	1/2	...	1 13/16	...	1 7/16	.150	1-LEAN	1 1/2	550	2000	.050
1964	4GC	9/32	1 3/8	1 1/2	1 5/16	1	.115	INDEX	1 1/2	550	1600	.020
1965	2GC—330 Eng. Syn.	3/4	...	1 7/8	...	1 7/16	.160	INDEX	1 1/2	525	1800	...
1965	2GC—330 Eng. Aut.	19/32	...	1 7/8	...	1 7/16	.160	INDEX	1 1/2	550	...	...
1965	4GC—425 Eng.	1 15/32	1 3/8	1 1/2	1 9/32	1	.120	1-RICH	1 1/2	550	1000	...
1965	4GC—425 Eng.	1 15/32	1 3/8	1 1/16	1 1/16	1	.120	1-RICH	1 1/2	550	1000	...
1966	2GC—7036159	3/4	...	1 7/8	...	1 7/16	.160	INDEX	1 1/2	600	900	.075
1966	2GC—7036053	19/32	...	1 7/8	...	1 7/16	.160	INDEX	1 1/2	...	...	...
1966	4MV—7036254	11/32	...	...	...	9/32	.300	Up-Notch	...	550	700	.075
1966	4MV—7036250	...	...	...	...	...	...	...	...	...	...	...
1967	2GC	19/32	...	1 9/16	...	1 7/16	.160	INDEX	1 1/2	...	...	...
1967	2GC	1/2	...	1 9/16	...	1 7/16	.160	INDEX	1 1/2	600	900	...
1967	4MV	1/4	...	...	...	5/16	.325	R-Notch	1 1/2	550	700	...
1968	2GC	9/16	...	1 3/8	...	1 7/16	.160	1-LEAN	...	...	...	...
1968	2GC	9/16	...	1 3/8	...	1 7/16	.160	INDEX	...	...	...	...
1968	4MV	1/4	...	...	...	5/16	.200	CENTER	2	...	...	...
1969	2GC	9/16	...	1 3/8	...	1 7/16	.170	INDEX	...	...	...	...
1969	2GC—350	9/16	...	1 3/8	...	1 7/16	.170	1-LEAN	...	...	...	...
1969	4MV	1/4	...	...	...	5/16	.200	CENTER	2	700	...	...
1970	2GC-7040154-455 (L-33), Auto.	9/16	...	1 3/8	...	1 3/8	.140	INDEX	...	575	900	.170
1970	2GC-7040155-350, 455, Std.	9/16	...	1 3/8	...	1 3/8	.140	1-LEAN	...	②	900	.170
1970	2GC-7040156-350, Auto.	9/16	...	1 3/8	...	1 3/8	.140	INDEX	...	575	900	.170
1970	2GC-7040159-455, Auto.	9/16	...	1 3/8	...	1 3/8	.140	INDEX	...	575	900	.170
1970	4MC-7040250-350	1/4	...	...	...	INNER HOLE	.200	INDEX	...	①	①	.060
1970	4MC-7040251-455	1/4	...	...	...	INNER HOLE	.200	INDEX	...	①	①	.060
1970	4MC-7040252-455 Toronado	1/4	...	...	...	INNER HOLE	.200	1-RICH	...	①	①	.060

① —See decal for idle speeds.

② —350 cu. in.–750 rpm, 455 cu. in.–675 rpm.

Oldsmobile F-85

YEAR	MODEL OR TYPE	FLOAT LEVEL Prim.	FLOAT LEVEL Sec.	FLOAT DROP Prim.	FLOAT DROP Sec.	Pump Travel Setting	Unloader	Housing	Idle Screw Turns Open	Idle Speed	Fast Idle Speed	Dashpot Plunger Clearance
1964	2GC—V-8	25/32	...	1 29/32	...	1 3/16	.260	INDEX*	1 1/2	550	1900	.020
1964	2GC—V-6 Std.	1 17/64	...	1 29/32	...	3/4	.175	INDEX	1	525	650	...
1964	2GC—V-6 Aut.	1 17/64	...	1 29/32	...	3/4	.175	1-RICH	1	525	650	...
1964	4GC	1 7/16 ■	1 3/8 ■	1 3/8	1 1/8	1 1/64	.187	2-RICH	1 1/2	550	1600	...
1965	BC—V-6	1 9/32	...	1 7/8	...	...	.300	INDEX	1 1/2	550	...	...
1965	2GC	3/4	...	1 7/8	...	1 7/16	.160	INDEX	1 1/2	550	...	...
1965	4GC	1 15/32	1 3/8	1 1/2	1 3/16	1	.120	1-RICH	1 1/2	550	1000	...
1966	BV	1 9/32	...	1 3/4	...	...	.350	...	1 1/2	550	...	...
1966	2GC—7036159	3/4	...	1 7/8	...	1 7/16	.160	INDEX	1 1/2	600	900	.075
1966	4MV	11/32	...	...	...	9/32	.300	Up-Notch	...	...	...	...
1967	BV	1 9/32	...	1 3/4	...	...	.350	...	1 1/2	500	...	...
1967	2GC	19/32	...	1 9/16	...	1 7/16	.160	INDEX	550	700	...	...
1967	2GC	1/2	...	1 9/16	...	1 7/16	.160	INDEX	...	575	900	...
1967	2GC	19/32	...	1 9/16	...	1 7/16	.160	1-LEAN	...	...	...	...
1967	4MV	1/4	...	...	...	5/16	.325	R-Notch	...	550	700	...
1968	2GC	9/16	...	1 3/8	...	1 7/16	.160	1-LEAN	...	...	...	...
1968	2GC	9/16	...	1 3/8	...	1 7/16	.160	INDEX	...	...	...	...
1968	4MV	1/4	...	...	...	5/16	.200	CENTER	2	...	...	...
1969	MV	5/16	...	...	...	...	.350	...	...	...	...	...
1969	2GC	9/16	...	1 3/8	...	1 7/16	.170	INDEX	...	...	...	...
1969	2GC, 350	9/16	...	1 3/8	...	1 7/16	.170	1-LEAN	...	...	...	...
1969	4MV	1/4	...	...	...	5/16	.200	CENTER	2	700	...	...
1970	MV-7040014-250, Auto.	1/4	...	...	...	...	.350	...	...	①	①	...
1970	MV-7040017-250, Std.	1/4	...	...	...	...	.350	...	...	①	①	...
1970	2GC-7040154-455 (L-33), Auto.	9/16	...	1 3/8	...	1 3/8	.140	INDEX	...	575	900	.170
1970	2GC-7040155-350, 455, Std.	9/16	...	1 3/8	...	1 3/8	.140	1-LEAN	...	②	900	.170
1970	2GC-7040156-350, Auto.	9/16	...	1 3/8	...	1 3/8	.140	INDEX	...	575	900	.170
1970	2GC-7040159-455, Auto.	9/16	...	1 3/8	...	1 3/8	.140	INDEX	...	575	900	.170

YEAR	MODEL OR TYPE	FLOAT LEVEL (IN.)		FLOAT DROP (IN.)		Pump Travel Setting (IN.)	CHOKE SETTING		ON THE CAR ADJUSTMENTS			
		Prim. (IN.)	Sec. (IN.)	Prim. (IN.)	Sec. (IN.)		Unloader (IN.)	Housing	Idle Screw Turns Open	Idle Speed (RPM)	Fast Idle Speed (RPM)	Dashpot Plunger Clearance (IN.)

Oldsmobile F-85, continued

YEAR	MODEL OR TYPE	Prim.	Sec.	Prim.	Sec.	Pump	Unloader	Housing	Idle Screw	Idle Speed	Fast Idle	Dashpot
1970	4MC-7040250-350	1/4	...			INNER HOLE	.200	INDEX	...	①	①	.060
1970	4MC-7040251-455	1/4	...			INNER HOLE	.200	INDEX	...	①	①	.060
1970	4MC-7040253-442 Model, Std.	1/4	...			INNER HOLE	.200	INDEX	...	①	①	.060
1970	4MC-7040255-350 W-31	1/4	...			INNER HOLE	.200	INDEX	...	①	①	.060
1970	4MC-7040256-W-30, 442, Std.	1/4	...			INNER HOLE	.200	INDEX	...	①	①	.060
1970	4MC-7040257-442 Model, Auto.	1/4	...			INNER HOLE	.200	INDEX	...	①	①	.060
1970	4MC-7040257-455 W-33, Auto.	1/4	...			INNER HOLE	.200	INDEX	...	①	①	.060
1970	4MC-7040258-442 W-30, Auto.	1/4	...			INNER HOLE	.200	INDEX	...	①	①	.060

*—1 Notch Lean on synchromesh.
■—Heel measurement shown, Toe 1/2 in. Primary, 3/8 in. Secondary.
①—See decal for idle speeds.
②—350 cu. in.—750 rpm, 455 cu. in.—675 rpm.

Pontiac

YEAR	MODEL OR TYPE	Prim.	Sec.	Prim.	Sec.	Pump	Unloader	Housing	Idle Screw	Idle Speed	Fast Idle	Dashpot
1964	2GC●	5/8	...	1 3/4	...	1 21/64	.160	INDEX	1 1/2	500	...	.064
1964	2GC▲	11/16	...	1 3/4	...	1 1/8	.160	INDEX	1 1/2	500	...	.094
1964	2GC—①*	23/32	...	1 3/4	...	1 1/8	.160	INDEX	1 1/2	500	.056②	...
1964	2GC—②■	23/32	...	1 3/4	...	55/64	...	...	...	...	...	...
1965	2GC	19/32	...	1 3/4	...	1 11/32	.160	INDEX	1 1/2	550		
1965	2GC—Frt. & Rear	21/32	...	1 3/4	...	27/32	...	...	1 1/2			...
1965	2GC—Center	11/16	...	1 3/4	...	1 1/8	.160	INDEX	1 1/2	550		.055
1966	2GC	19/32	...	1 3/4	...	1 21/64	.160	INDEX	1 1/2	550		
1966	2GC—Frt. & Rear	23/32	...	1 3/4	...	27/32	...	...	1 1/2			
1966	2GC—Center	23/32	...	1 3/4	...	1 21/64	.160	INDEX	1 1/2			
1967	2GC-7027066—M.T.	9/16	...	1 7/16	...	1 11/32	.160	INDEX	...	600		
1967	2GC-7027060—A.T.	9/16	...	1 7/16	...	1 11/32	.160	INDEX	...	500		
1967	4MV-7027263—M.T.	3/16	...		...	9/32	.325	Cent-Notch	...	600		
1967	4MV-7027262—A.T.	3/16	...		...	9/32	.325	Cent-Notch	...	500		
1968	2GC-7028060 A.T.	9/16	...	1 3/4	...	1 11/32	.180	...	...	...	...	...
1968	2GV-7028066 M.T.	9/16	...	1 3/4	...	1 11/32	.180	...	...	...	...	...
1968	4MV-7028262-64	1/4	...	...	...	9/32	.300	...	2	...	...	...
1968	4MV-7028263-65	1/4	...	...	...	9/32	.300	...	2	...	...	...
1969	2GV-7029060 A.T.	9/16	...	1 3/4	...	1 11/32	.180	...	...	...	...	...
1969	2GV-7028066 M.T.	9/16	...	1 3/4	...	1 11/32	.180	...	...	...	...	...
1969	4MV-7029262	9/32	...	...	...	9/32	.300	Cent-Notch	2	...	2300	.030
1969	4MV-7029263 M.T.	9/32	...	...	...	9/32	.300	Cent-Notch	2	...	2300	.030
1969	4MV-7029268 A.T.	9/32	...	...	...	9/32	.300	Cent-Notch	2	...	2300	.030
1970	2GV—7040060-400, Auto.	11/16	...	1 3/4	...	1 11/32	.180	...	...	①	①	...
1970	2GV—7040460-400, Auto.-cal.	11/16	...	1 3/4	...	1 11/32	.180	...	...	①	①	...
1970	2GV—7040461-400, Auto. w/AC-cal.	11/16	...	1 3/4	...	1 11/32	.180	...	...	①	①	...
1970	2GV—7040062-350, Auto.	9/16	...	1 3/4	...	1 11/32	.180	...	...	①	①	...
1970	2GV—7040462-350, Auto-cal.	9/16	...	1 3/4	...	1 11/32	.180	...	...	①	①	...
1970	2GV—7040462-350, Auto. w/AC-cal.	9/16	...	1 3/4	...	1 11/32	.180	...	...	①	①	...
1970	2GV—7040064-400 Altitude, Auto.	11/16	...	1 3/4	...	1 11/32	.180	...	...	①	①	...
1970	2GV—7040066-400, Std.	11/16	...	1 3/4	...	1 11/32	.180	...	...	①	①	...
1970	2GV—7040466-400, Std.-cal.	11/16	...	1 3/4	...	1 11/32	.180	...	...	①	①	...
1970	2GV—7040071-350, Std.	9/16	...	1 3/4	...	1 11/32	.180	...	...	①	①	...
1970	2GV—7040471-350, Std.-cal.	9/16	...	1 3/4	...	1 11/32	.180	...	...	①	①	...
1970	2GV—7040072-350 Altitude, Std.	9/16	...	1 3/4	...	1 11/32	.180	...	...	①	①	...
1970	4MV—7040262-455, Auto-Bonnev. sm. valve	9/32	...	...	...	...	...	CENTER NOTCH	...	①	①	.025
1970	4MV—7040562-455, Auto-Bonnev. sm. valve-cal.	9/32	...	...	...	...	...	CENTER NOTCH	...	①	①	.025
1970	4MV—7040263-400 Std. exc. Ram-Air	9/32	...	...	...	...	...	CENTER NOTCH	...	①	①	.025
1970	4MV—7040563-400, Std.-Exc. Ram-Air-cal.	9/32	...	...	...	...	...	CENTER NOTCH	...	①	①	.025

YEAR	MODEL OR TYPE	FLOAT LEVEL (IN.)		FLOAT DROP (IN.)		Pump Travel Setting (IN.)	CHOKE SETTING		ON THE CAR ADJUSTMENTS			
		Prim. (IN.)	Sec. (IN.)	Prim. (IN.)	Sec. (IN.)		Unloader (IN.)	Housing	Idle Screw Turns Open	Idle Speed (RPM)	Fast Idle Speed (RPM)	Dashpot Plunger Clearance (IN.)
Pontiac, continued												
1970	4MV—7040264-400, Aut.-exc. Ram-Air	9/32	...				...	CENTER NOTCH	...	①	①	.025
1970	4MV—7040564-400, Auto-exc. Ram-Air-cal	9/32	...				...	CENTER NOTCH	...	①	①	.025
1970	4MV—7040267-455, Std.	9/32						CENTER NOTCH		①	①	.025
1970	4MV—7040567-455, Std.-cal	9/32						CENTER NOTCH		①	①	.025
1970	4MV—7040567-455, Std.-cal	9/32						CENTER NOTCH		①	①	.025
1970	4MV—7040268-455, Auto.	9/32						CENTER NOTCH		①	①	.025
1970	4MV—7040568-455, Auto-cal	9/32						CENTER NOTCH		①	①	.025
1970	4MV—7040274-455, Altitude, Auto.	9/32	...					CENTER NOTCH	...	①	①	.025

Ⓐ—Set cam on #2 step.
●—2GC—With -11/16 throat.
▲—2GC—With 1 7/17 throat.
✱—2GC—① Is center carburetor on three 2-barrel application.
■—2GC—① Are the front and rear carburetors with this application.
①—See decal for idle speeds.
CAL—California

Tempest—Firebird

YEAR	MODEL OR TYPE	FLOAT LEVEL Prim.	FLOAT LEVEL Sec.	FLOAT DROP Prim.	FLOAT DROP Sec.	Pump Travel Setting	Unloader	Housing	Idle Screw Turns Open	Idle Speed	Fast Idle Speed	Dashpot
1964	BV	1 9/32	...	1 1/4	...	.040	.230	...	1 1/2	475	...	...
1964	2GC	5/8	...	1 1/4	...	1 11/32	.160	INDEX	1 1/2	500		
1964	4GC	1 21/64	1 21/64	1 1/4	1 1/4	15/16	.152	INDEX	1 1/2	700★	①	
1965	BV-7025167	1 9/32	...	1 1/8	...	.040	.230	...	1 1/2	...	...	...
1965	2GC-7025062, 326 Eng.	5/8	...	1 1/4	...	1 11/32	.160	INDEX	1 1/2			
1965	2GC—389 Eng., Frt.	23/32	...	1 1/4	...				1 1/2			
1965	2GC—389 Eng., Cent.	23/32	...	1 1/4	...	1 11/32		INDEX	1 1/2			
1965	2GC—389 Eng., Rear	23/32	...	1 1/4	...				1 1/2			
1966	BV	1 9/32	...	1 1/8	...	.040	.230		1 1/2			
1966	2GC	5/8	...	1 1/8	...	1 11/32	.160	INDEX	1 1/2	600	900	.075
1966	4MV	11/32	...	...		9/32	.300	Up-Notch	...	550	700	.075
1966	BV—6 Cyl, 230 Eng.	1 5/16	...	1 1/8	...	.040	.230		1 1/2			
1966	4MV—6 Cyl.	7/32	...	...		9/32	.300		1 1/2			
1966	2GC—389 Eng., Frt.	21/32	...	1 1/4	...	27/32			1 1/2			
1966	2GC—389 Eng., Cent.	19/32	...	1 1/4	...	1 21/64	.160		1 1/2			
1966	2GC—389 Eng., Rear	21/32	...	1 1/4	...	27/32			1 1/2			
1966	2GC—326 Eng.	19/32	...	1 1/4	...	1 21/64	.160	INDEX	1 1/2			
1967	BV—6 Cyl., 230 Eng.	1 9/32	...	1 1/8	...	.040	.230		1 1/2	...	...	...
1967	GC—7027062, 071	9/16	...	1 1/16	...	1 11/32	.160	INDEX	1 1/2			
1967	GC	9/16	...	1 1/16	...	1 11/32	.160	INDEX	1 1/2			
1967	4MV—7027260-261, M.T.	7/32	...	...		9/32	.325	Cent-Notch	...	600	2600	
1967	4MV—7027260-261, A.T.	7/32	...	...		9/32	.325	Cent-Notch	...	500	2800	
1967	4MV—7027262-263	9/16	...	...		9/32	.325	Cent-Notch	...	550	2500	
1968	BV-7028168	1 11/32	...	1 1/8	...	...	.230	...				
1969	MV-7029165-67, M.T.	9/32	...	...		...	.450	...		...	2400	...
1969	MV-7029166-68, A.T.	9/32	...	...		...	.450	...			2800	...
1969	2GV-7029062, A.T.	9/16	...	1 1/4	...	1 11/32	.180					
1969	4MV	9/32	...	...		9/32	.300	Cent-Notch	2	...	2300	.030
1969	4MV—Ram Air	9/32	...	...		1/4	.300	Cent-Notch	2	...	2800	.030
1970	MV-7040014-250, Auto.	1/4	...	...		...	.350	...		①	①	...
1970	MV-7040017-250, Std.	1/4	...	...		...	.350	...		①	①	...
1970	2GV-7040060-400, Auto.	11/16	...	1 1/4	...	1 11/32	.180			①	①	
1970	2GV-7040460-400, Auto.-cal.	11/16	...	1 1/4	...	1 11/32	.180			①	①	
1970	2GV-7040461-400, Auto. w/AC-cal	11/16	...	1 1/4	...	1 11/32	.180			①	①	
1970	2GV-7040062-350, Auto.	9/16	...	1 1/4	...	1 11/32	.180			①	①	
1970	2GV-7040462-350, Auto-cal.	9/16	...	1 1/4	...	1 11/32	.180			①	①	
1970	2GV-7040463-350, Auto. w/ac-cal	9/16	...	1 1/4	...	1 11/32	.180			①	①	
1970	2GV-7040064-400 Altitude, Auto.	11/16	...	1 1/4	...	1 11/32	.180			①	①	
1970	2GV-7040066-400, Std.	11/16	...	1 1/4	...	1 11/32	.180			①	①	
1970	2GV-7040466-400, Std.-cal.	11/16	...	1 1/4	...	1 11/32	.180			①	①	
1970	2GV-7040071-350, Std.	9/16	...	1 1/4	...	1 11/32	.180			①	①	
1970	2GV-7040471-350, Std.-cal	9/16	...	1 1/4	...	1 11/32	.180			①	①	
1970	2GV-7040072-350 Atltitude, std.	9/16	...	1 1/4	...	1 11/32	.180			①	①	...
1970	4MV-7040263-400 Std.-exc. Ram Air	9/32	...	...		...	...	CENTER NOTCH	...	①	①	.025

Tempest—Firebird, continued

YEAR	MODEL OR TYPE	FLOAT LEVEL (IN.)		FLOAT DROP (IN.)		Pump Travel Setting (IN.)	CHOKE SETTING		ON THE CAR ADJUSTMENTS			
		Prim. (IN.)	Sec. (IN.)	Prim. (IN.)	Sec. (IN.)		Unloader (IN.)	Housing	Idle Screw Turns Open	Idle Speed (RPM)	Fast Idle Speed (RPM)	Dashpot Plunger Clearance (IN.)
1970	4MV—7040263-400 Std. exc. Ram-Air-cal.	9/32	...	...	...	...	...	CENTER NOTCH	...	①	①	.025
1970	4MV—7040264-400, Auto.-exc. Ram-Air	9/32	...	...	...	...	...	CENTER NOTCH	...	①	①	.025
1970	4MV—7040564-400, Auto.-exc. Ram-Air-cal.	9/32	...	...	...	...	...	CENTER NOTCH	...	①	①	.025
1970	4MV—7040267-455, Std.	9/32	...	...	...	...	...	CENTER NOTCH	...	①	①	.025
1970	4MV—7040567-455, Std.-cal.	9/32	...	...	...	...	...	CENTER NOTCH	...	①	①	.025
1970	4MV—7040268-455, Auto...........	9/32	...	...	...	...	...	CENTER NOTCH	...	①	①	.025
1970	4MV—7040568-455, Auto-cal.	9/32	...	...	...	...	...	CENTER NOTCH	...	①	①	.025
1970	4MV—7040270-400, Auto, Ram Air III, IV	9/32	...	...	...	...	...	CENTER NOTCH	...	①	①	.025
1970	4MV—7040570-400, Auto, Ram-Air III, IV-cal.	9/32	...	...	...	...	...	CENTER NOTCH	...	①	①	.025
1970	4MV—7040273—400, Std. Ram-Air III, IV	9/32	...	...	...	...	...	CENTER NOTCH	...	①	①	.025
1970	4MV—7040274-455 Altitude, Auto....	9/32	...	...	...	...	...	Cent-Notch		①	①	.025

①—Auto. trans. 3000; std. trans. 2800.
★—Auto. trans. 600 "DR"; std. trans. 700 Neutral.

CAL—California
ⓓ —See decal for idle speeds.

MODEL R—RC

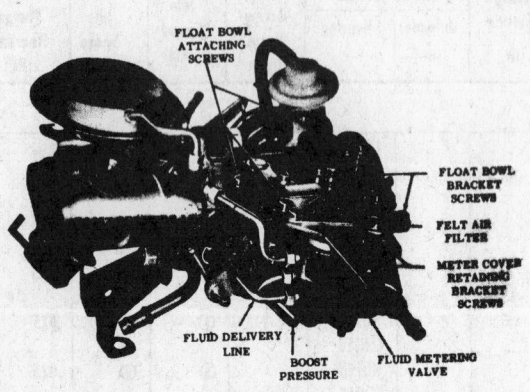

FLOAT BOWL
ATTACHING
SCREWS

FLOAT BOWL
BRACKET
SCREWS

FELT AIR
FILTER

METER COVER
RETAINING
BRACKET
SCREWS

FLUID DELIVERY
LINE

BOOST
PRESSURE

FLUID METERING
VALVE

(© G.M. Corp.)

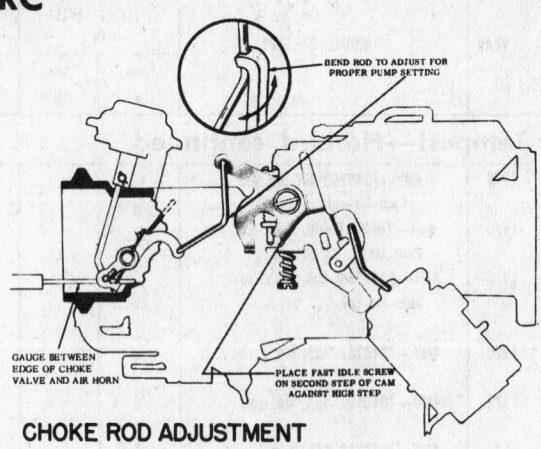

BEND ROD TO ADJUST FOR
PROPER PUMP SETTING

GAUGE BETWEEN
EDGE OF CHOKE
VALVE AND AIR HORN

PLACE FAST IDLE SCREW
ON SECOND STEP OF CAM
AGAINST HIGH STEP

CHOKE ROD ADJUSTMENT

Choke rod adjustment (© G.M. Corp.)

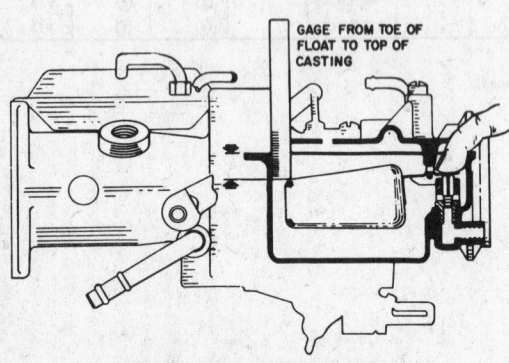

GAGE FROM TOE OF
FLOAT TO TOP OF
CASTING

FLOAT LEVEL ADJUSTMENT

Float level adjustment (© G.M. Corp.)

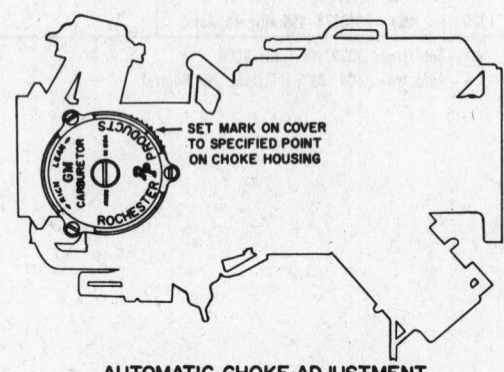

SET MARK ON COVER
TO SPECIFIED POINT
ON CHOKE HOUSING

GM
CARBURETOR

AUTOMATIC CHOKE ADJUSTMENT

Automatic choke adjustment (© G.M. Corp.)

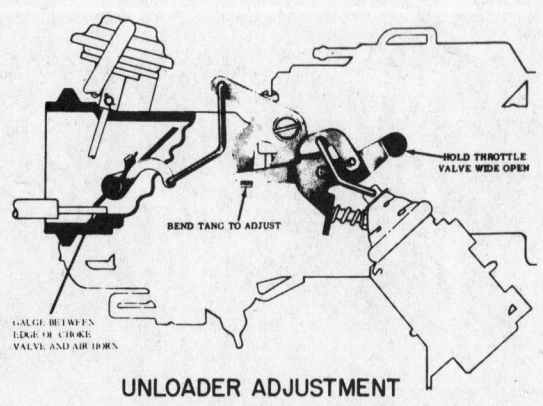

HOLD THROTTLE
VALVE WIDE OPEN

BEND TANG TO ADJUST

GAUGE BETWEEN
EDGE OF CHOKE
VALVE AND AIR HORN

UNLOADER ADJUSTMENT

Unloader adjustment (© G.M. Corp.)

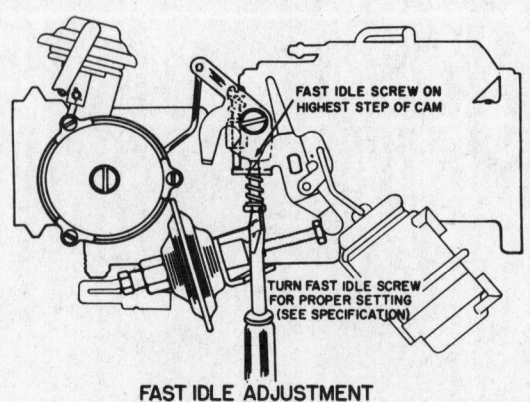

FAST IDLE SCREW ON
HIGHEST STEP OF CAM

TURN FAST IDLE SCREW
FOR PROPER SETTING
(SEE SPECIFICATION)

FAST IDLE ADJUSTMENT

Fast idle adjustment (© G.M. Corp.)

SINGLE BARREL—BC TYPE

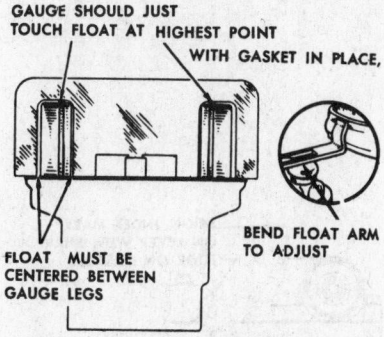

BC float level adjustment (© G.M. Corp.)

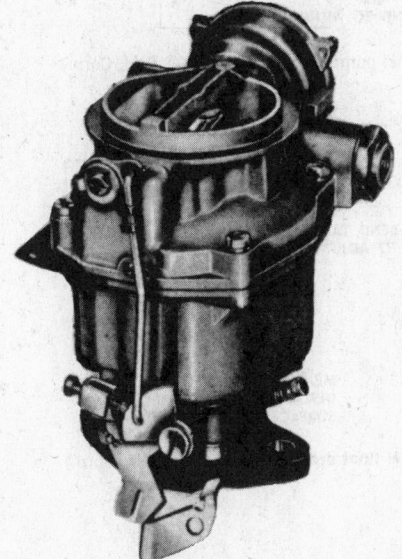

(© G.M. Corp.)

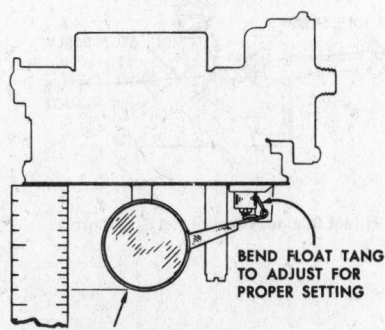

BC float drop adjustment (© G.M. Corp.)

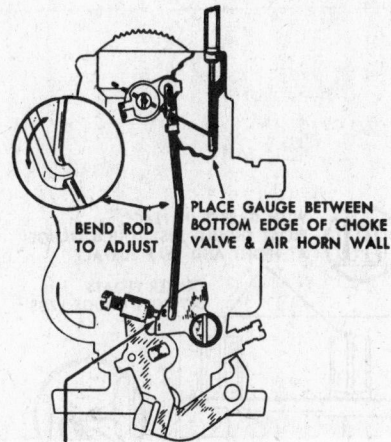

BC choke rod setting

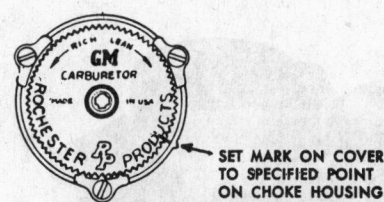

BC automatic choke adjustment

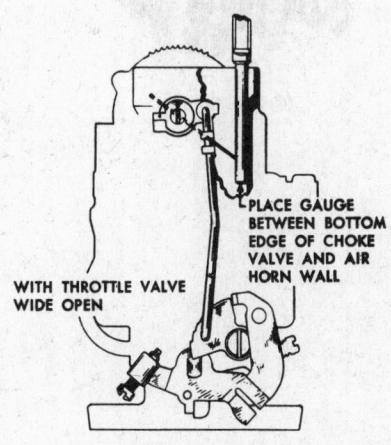

BC unloader setting (© G.M. Corp.)

SINGLE BARREL—H TYPE

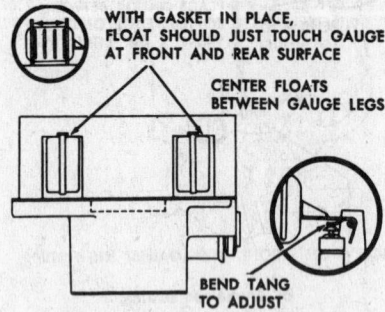

WITH GASKET IN PLACE, FLOAT SHOULD JUST TOUCH GAUGE AT FRONT AND REAR SURFACE

CENTER FLOATS BETWEEN GAUGE LEGS

BEND TANG TO ADJUST

H float level adjustment (© G.M. Corp.)

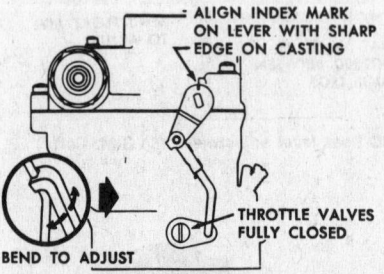

ALIGN INDEX MARK ON LEVER WITH SHARP EDGE ON CASTING

THROTTLE VALVES FULLY CLOSED

BEND TO ADJUST

H pump rod adjustment (© G.M. Corp.)

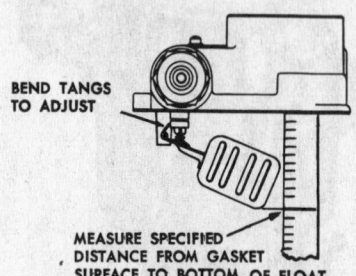

BEND TANGS TO ADJUST

MEASURE SPECIFIED DISTANCE FROM GASKET SURFACE TO BOTTOM OF FLOAT

H float drop adjustment (© G.M. Corp.)

(© G.M. Corp.)

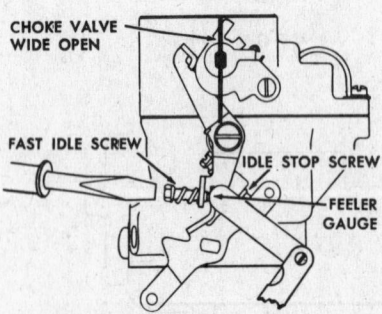

CHOKE VALVE WIDE OPEN

FAST IDLE SCREW

IDLE STOP SCREW

FEELER GAUGE

H fast idle adjustment (© G.M. Corp.)

FOUR BARREL—4GC TYPE

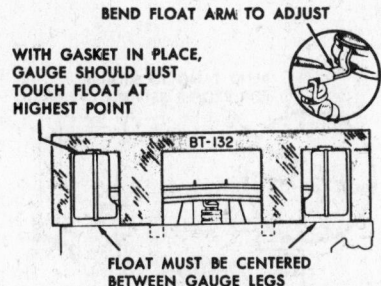

BEND FLOAT ARM TO ADJUST

WITH GASKET IN PLACE, GAUGE SHOULD JUST TOUCH FLOAT AT HIGHEST POINT

BT-132

FLOAT MUST BE CENTERED BETWEEN GAUGE LEGS

4GC float level adjustment

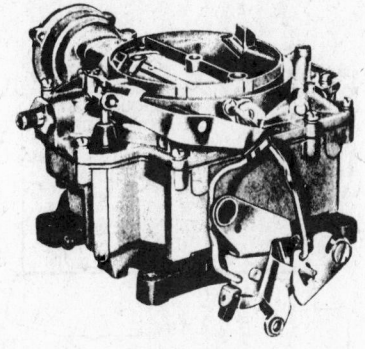

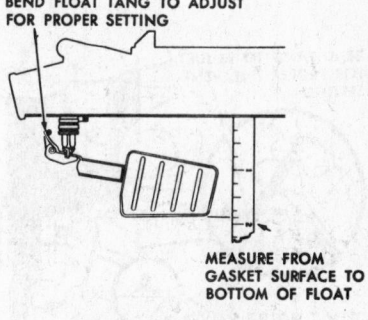

BEND FLOAT TANG TO ADJUST FOR PROPER SETTING

MEASURE FROM GASKET SURFACE TO BOTTOM OF FLOAT

4GC float drop adjustment

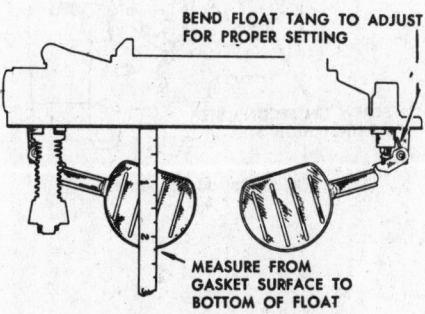

BEND FLOAT TANG TO ADJUST FOR PROPER SETTING

MEASURE FROM GASKET SURFACE TO BOTTOM OF FLOAT

4GC float drop adjustment

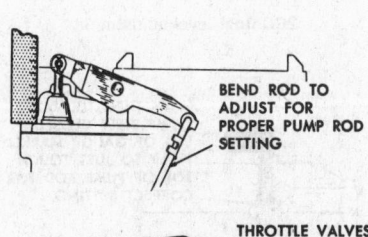

BEND ROD TO ADJUST FOR PROPER PUMP ROD SETTING

THROTTLE VALVES FULLY CLOSED

4GC pump rod setting

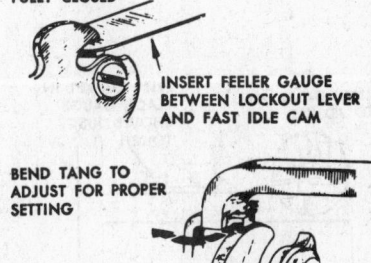

CHOKE VALVE FULLY CLOSED

INSERT FEELER GAUGE BETWEEN LOCKOUT LEVER AND FAST IDLE CAM

BEND TANG TO ADJUST FOR PROPER SETTING

4GC secondary lockout adjustment

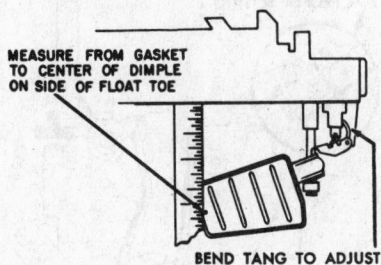

MEASURE FROM GASKET TO CENTER OF DIMPLE ON SIDE OF FLOAT TOE

BEND TANG TO ADJUST

4GC float adjustment

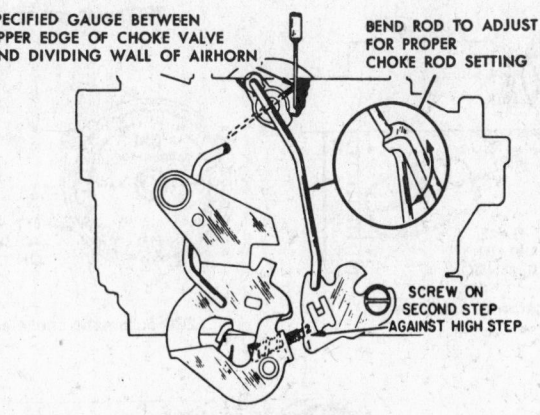

SPECIFIED GAUGE BETWEEN UPPER EDGE OF CHOKE VALVE AND DIVIDING WALL OF AIRHORN

BEND ROD TO ADJUST FOR PROPER CHOKE ROD SETTING

SCREW ON SECOND STEP AGAINST HIGH STEP

4GC choke rod adjustment

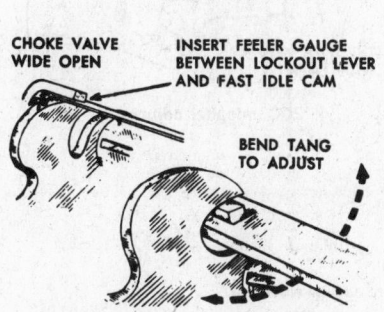

CHOKE VALVE WIDE OPEN

INSERT FEELER GAUGE BETWEEN LOCKOUT LEVER AND FAST IDLE CAM

BEND TANG TO ADJUST

4GC secondary contour adjustment

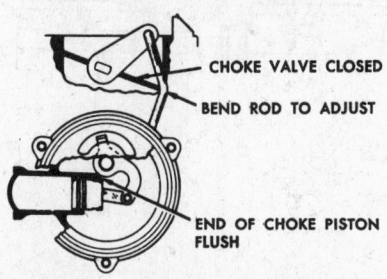

CHOKE VALVE CLOSED

BEND ROD TO ADJUST

END OF CHOKE PISTON FLUSH

4GC intermediate choke rod adjustment

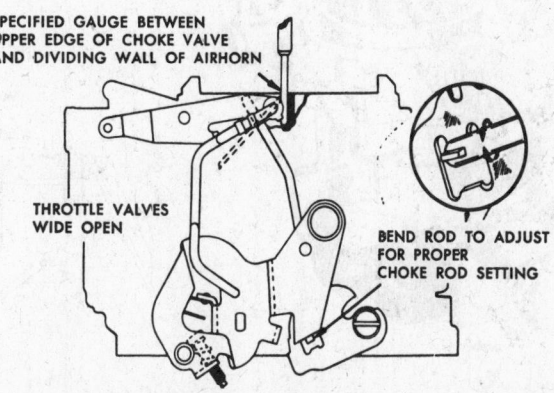

SPECIFIED GAUGE BETWEEN UPPER EDGE OF CHOKE VALVE AND DIVIDING WALL OF AIRHORN

THROTTLE VALVES WIDE OPEN

BEND ROD TO ADJUST FOR PROPER CHOKE ROD SETTING

4GC unloader adjustment

TWO BARREL—2GC TYPE

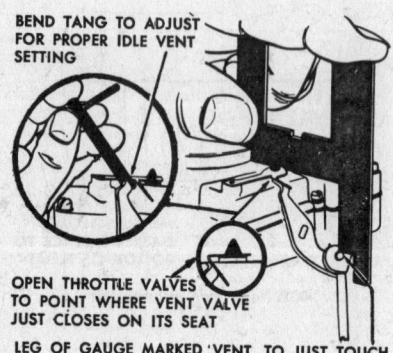

BEND TANG TO ADJUST FOR PROPER IDLE VENT SETTING

OPEN THROTTLE VALVES TO POINT WHERE VENT VALVE JUST CLOSES ON ITS SEAT

LEG OF GAUGE MARKED 'VENT TO JUST TOUCH

2GC idle vent adjustment

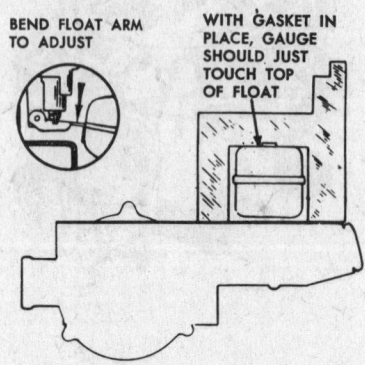

BEND FLOAT ARM TO ADJUST

WITH GASKET IN PLACE, GAUGE SHOULD JUST TOUCH TOP OF FLOAT

2GC float level adjustment

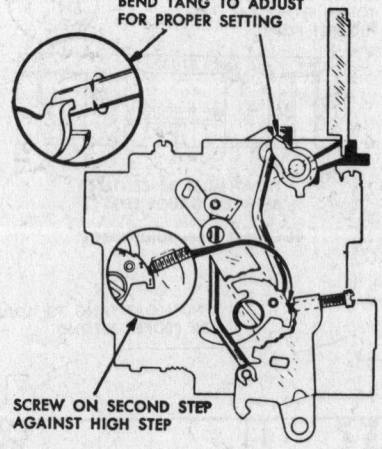

BEND TANG TO ADJUST FOR PROPER SETTING

SCREW ON SECOND STEP AGAINST HIGH STEP

2GC choke rod adjustment

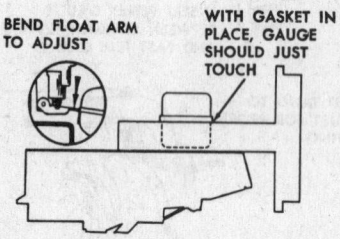

BEND FLOAT ARM TO ADJUST

WITH GASKET IN PLACE, GAUGE SHOULD JUST TOUCH

2GC float level adjustment

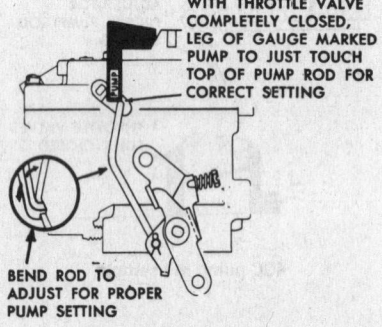

WITH THROTTLE VALVE COMPLETELY CLOSED, LEG OF GAUGE MARKED PUMP TO JUST TOUCH TOP OF PUMP ROD FOR CORRECT SETTING

BEND ROD TO ADJUST FOR PROPER PUMP SETTING

2GC pump rod adjustment

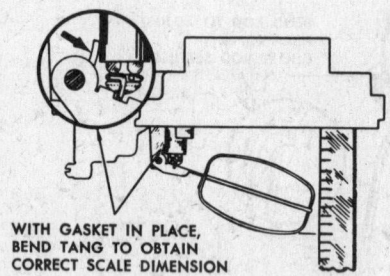

WITH GASKET IN PLACE, BEND TANG TO OBTAIN CORRECT SCALE DIMENSION

2GC float drop adjustment

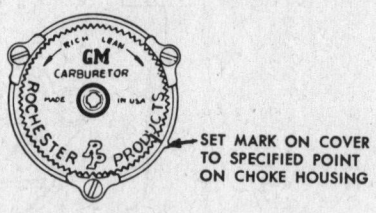

RICH LEAN
GM CARBURETOR
MADE IN USA
ROCHESTER PRODUCTS

SET MARK ON COVER TO SPECIFIED POINT ON CHOKE HOUSING

2GC automatic choke adjustment

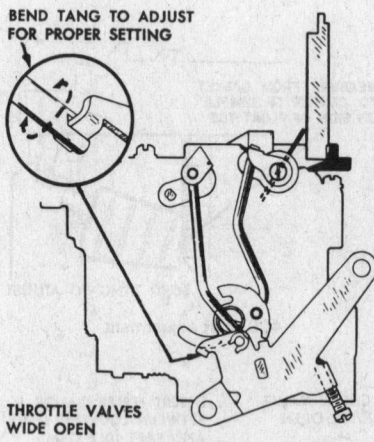

BEND TANG TO ADJUST FOR PROPER SETTING

THROTTLE VALVES WIDE OPEN

2GC unloader adjustment

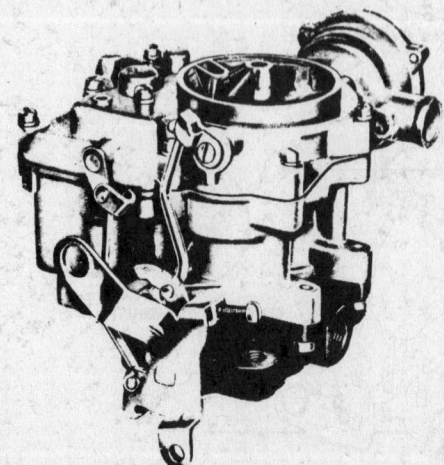

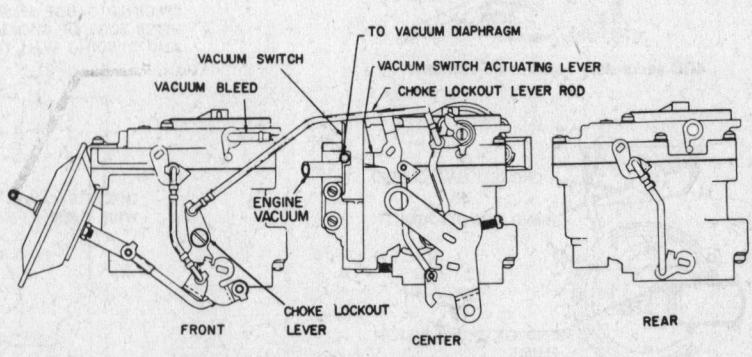

TO VACUUM DIAPHRAGM

VACUUM SWITCH

VACUUM SWITCH ACTUATING LEVER

VACUUM BLEED

CHOKE LOCKOUT LEVER ROD

ENGINE VACUUM

CHOKE LOCKOUT LEVER

FRONT   CENTER   REAR

Typical three two-barrel carburetor installation

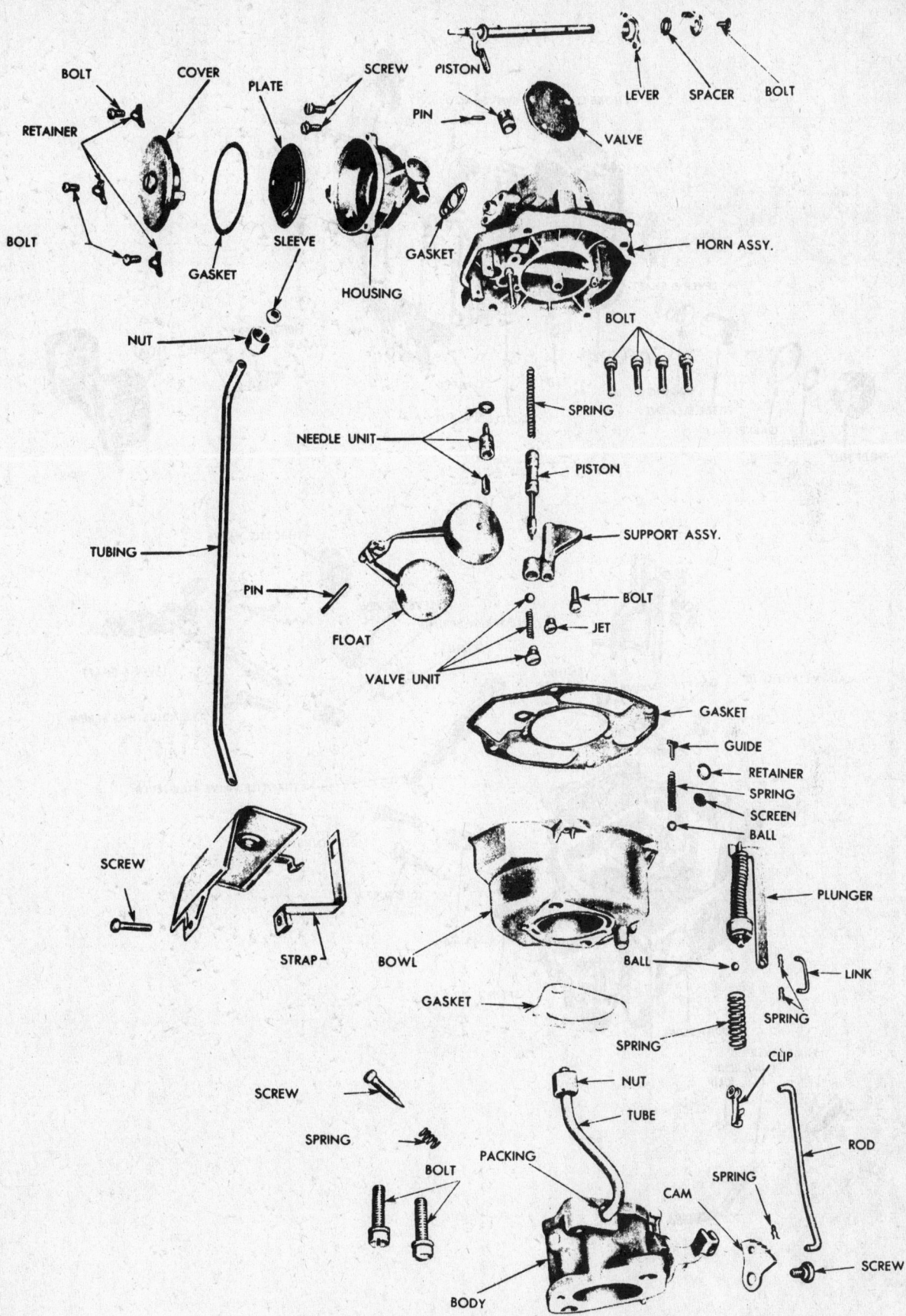

Rochester BC single-barrel (© Chevrolet Div., G.M. Corp.)

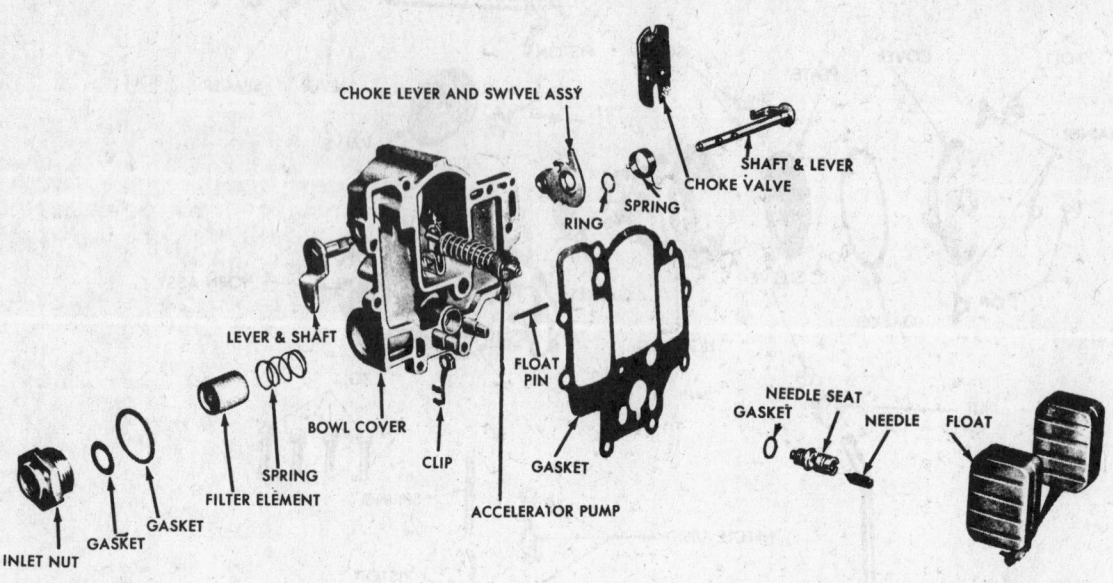

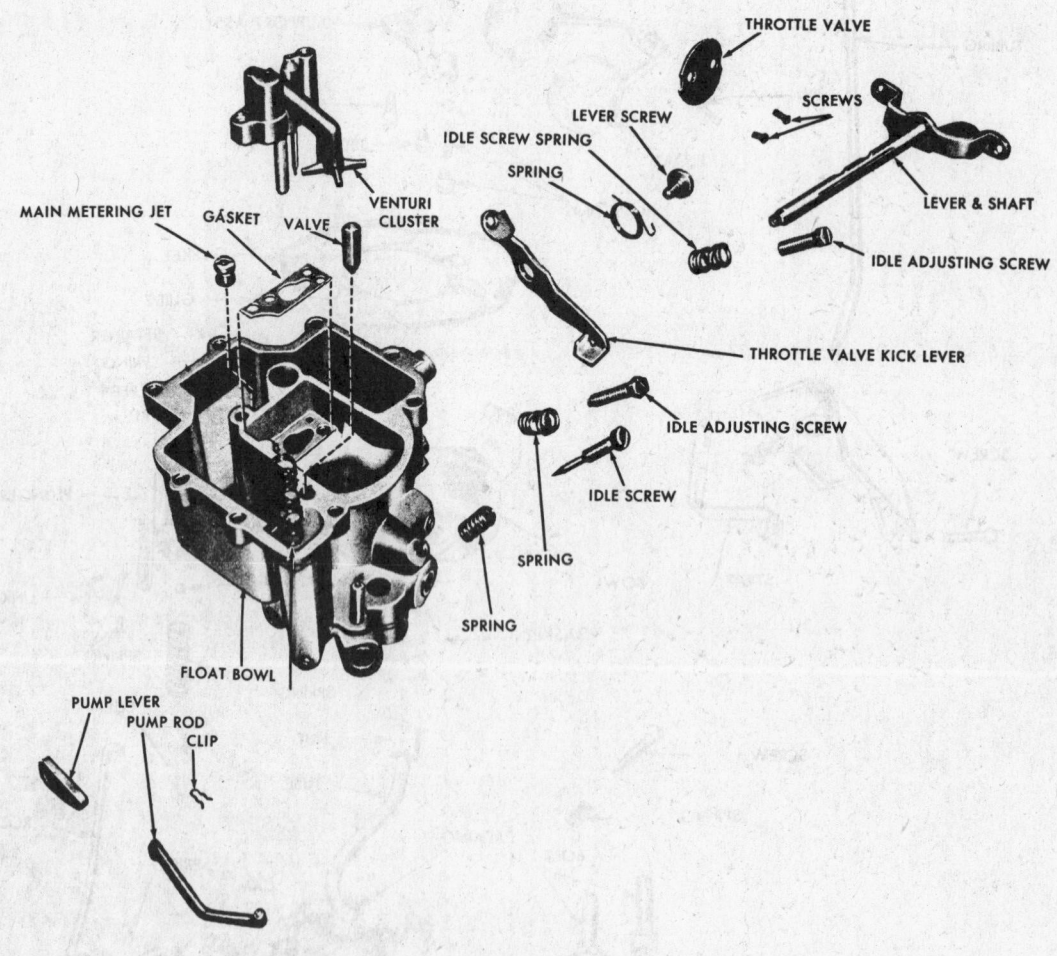

Rochester Model H single-barrel (Corvair) (© G.M. Corp.)

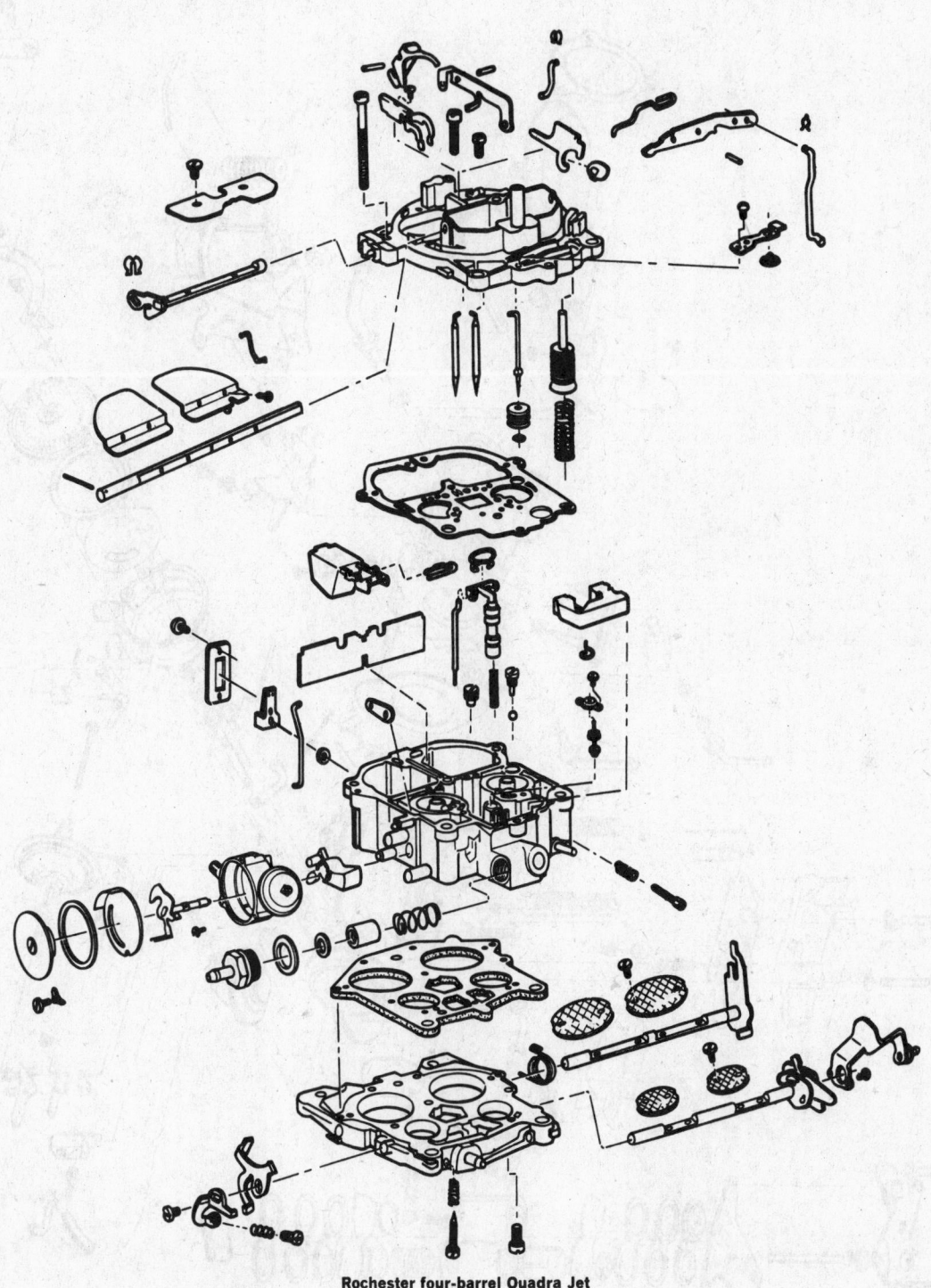

Rochester four-barrel Quadra Jet

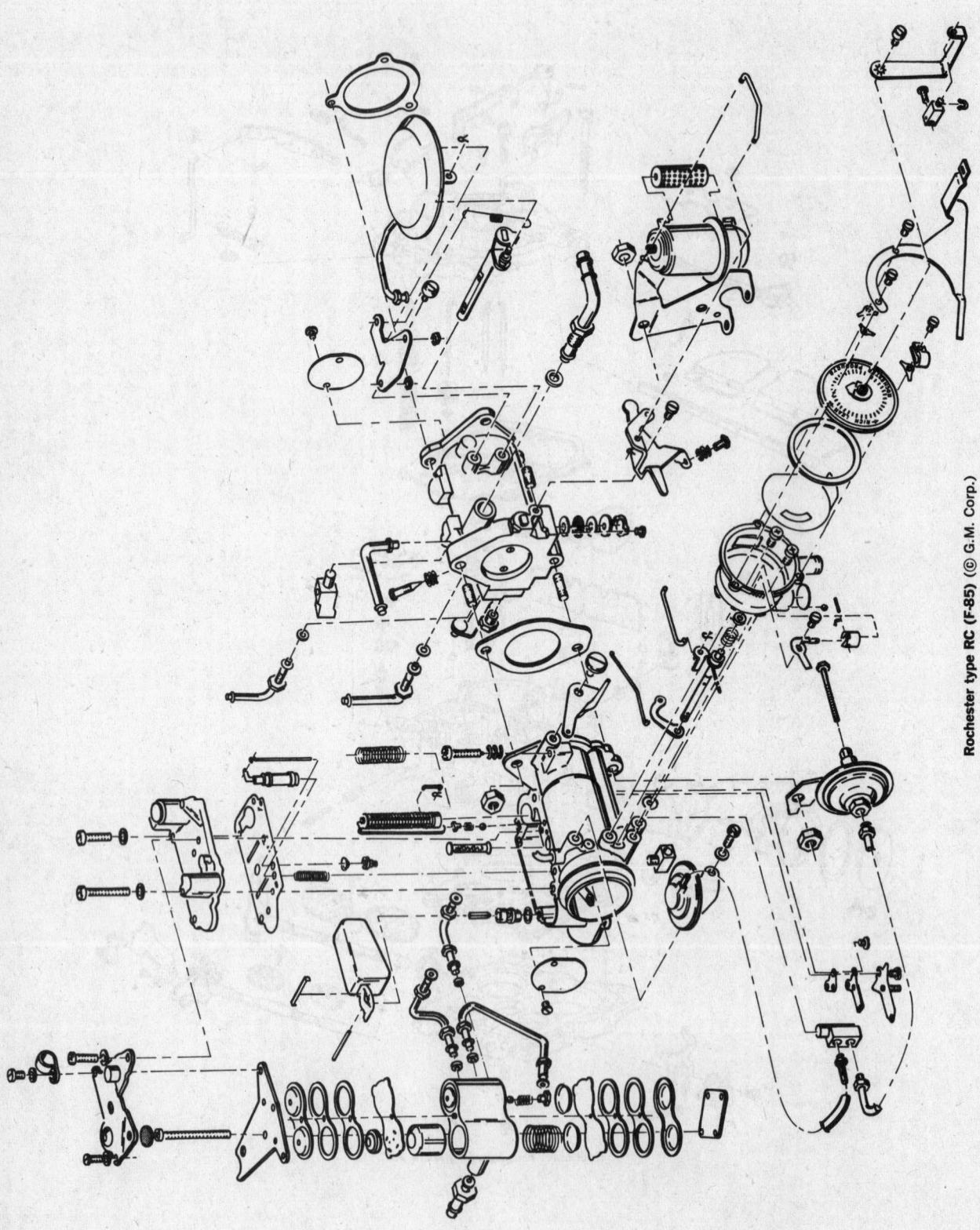

Rochester type RC (F-85) (© G.M. Corp.)

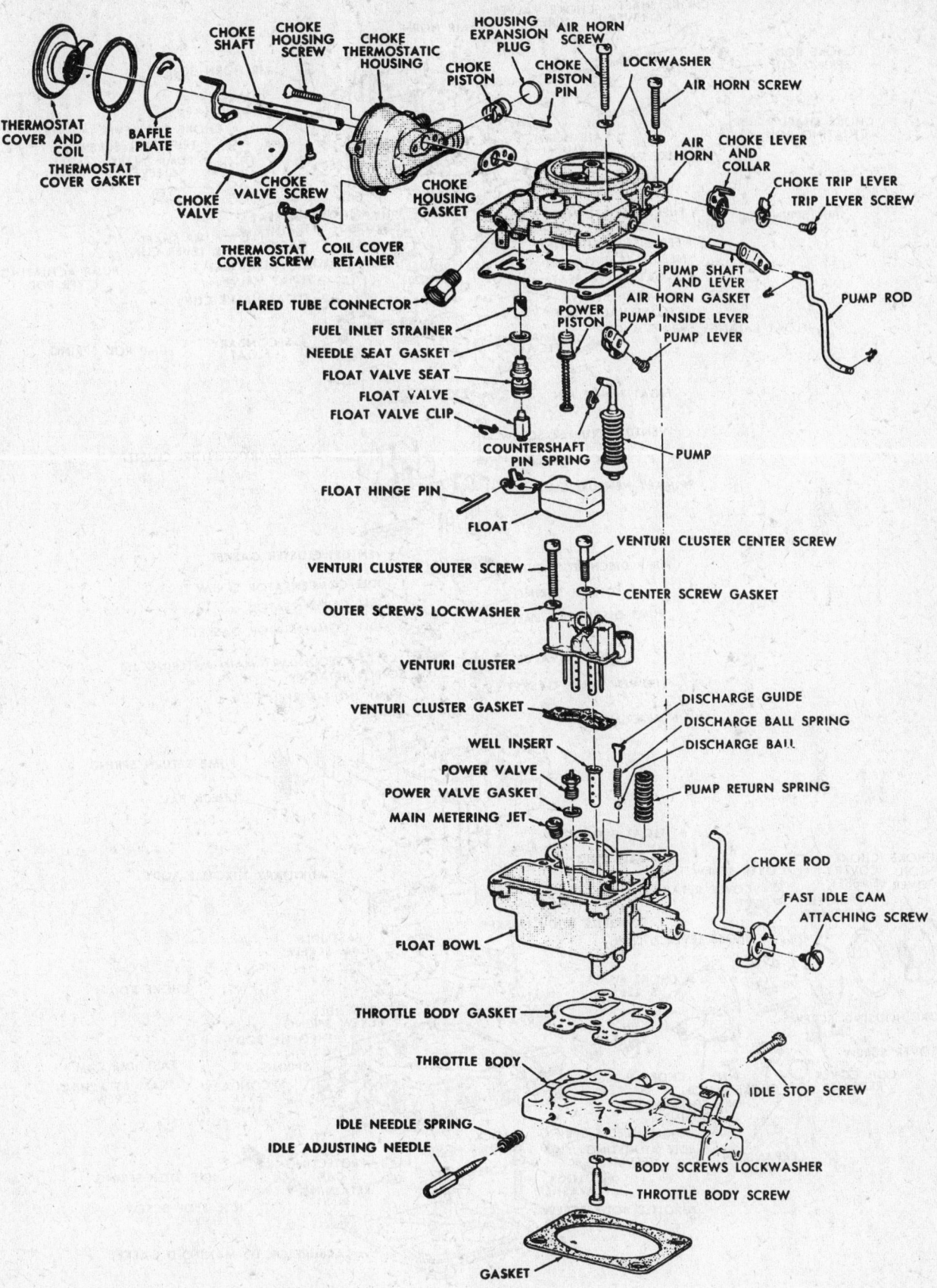

Rochester 2GC two-barrel (© Buick Div., G.M. Corp.)

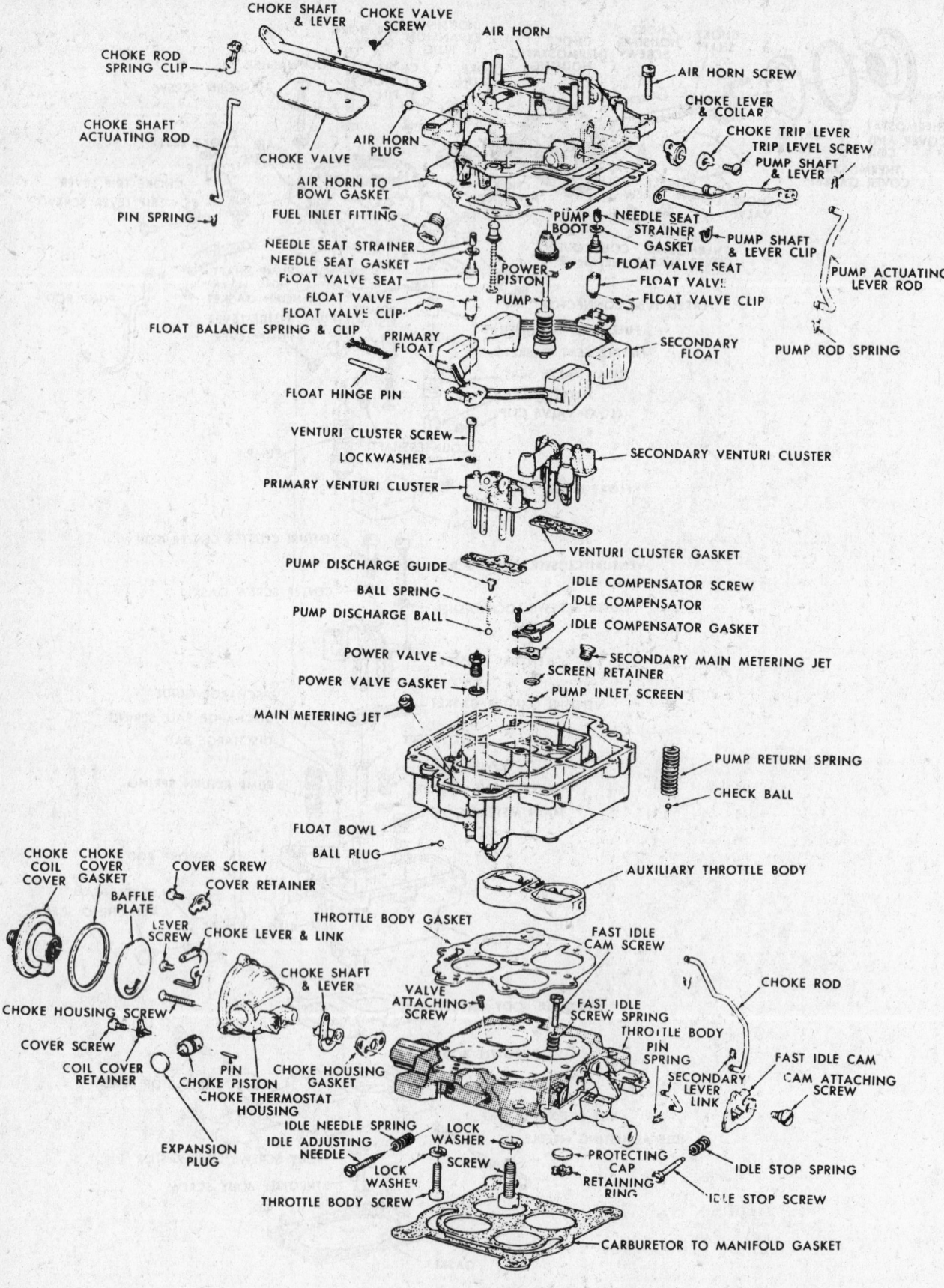

Rochester 4GC four-barrel (© Buick Div., G.M. Corp.)

4MV

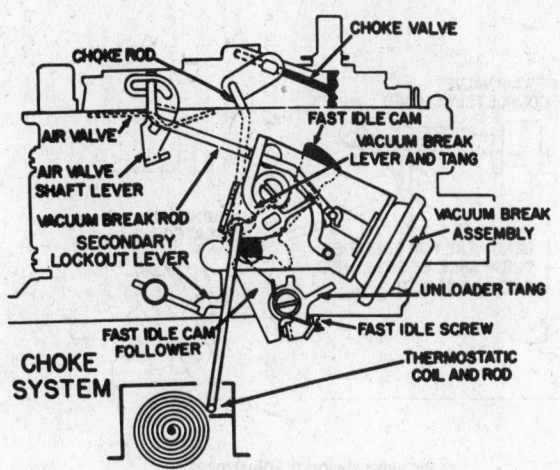

Choke system 327-350 cu. in. engine

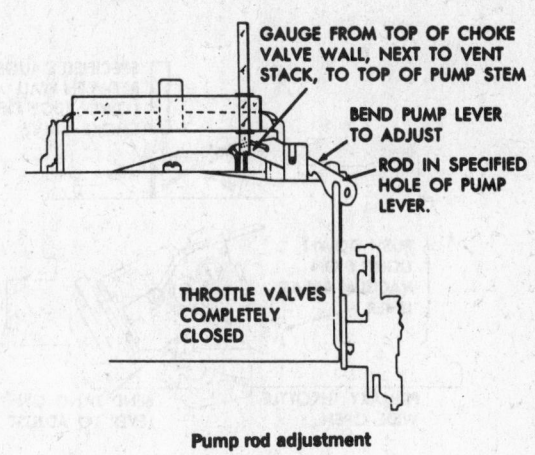

Pump rod adjustment

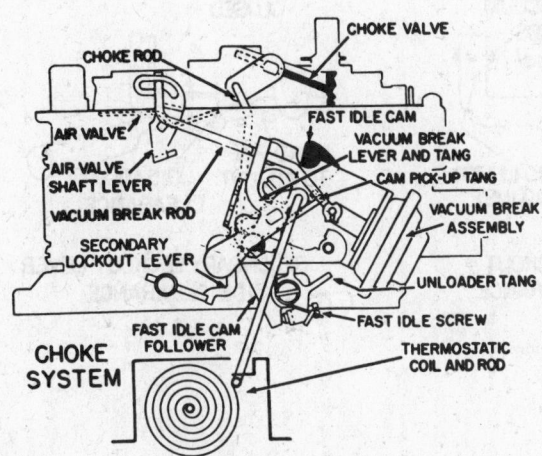

Choke system 396-427 cu. in. engine

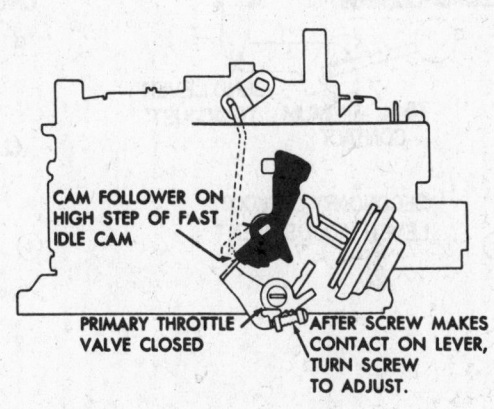

Fast idle adjustment

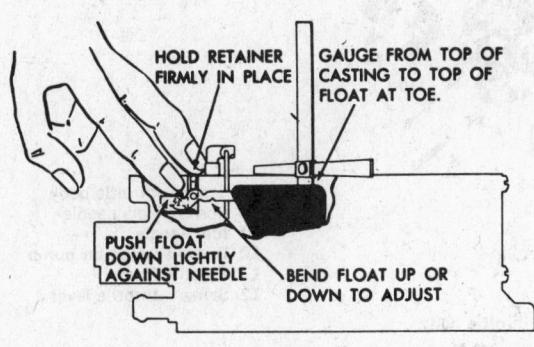

Float level adjustment

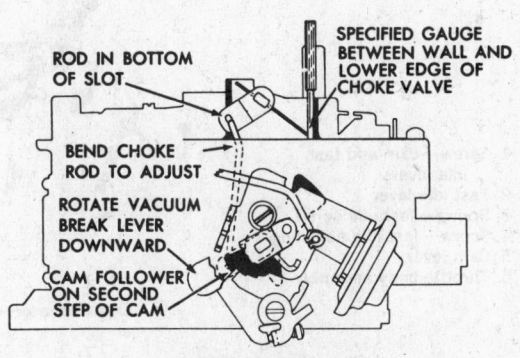

Choke rod adjustment

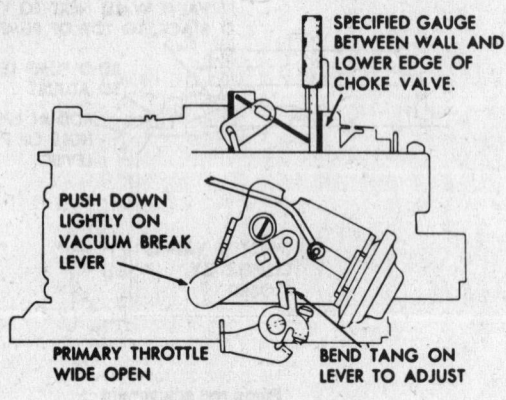

SPECIFIED GAUGE BETWEEN WALL AND LOWER EDGE OF CHOKE VALVE.

PUSH DOWN LIGHTLY ON VACUUM BREAK LEVER

PRIMARY THROTTLE WIDE OPEN

BEND TANG ON LEVER TO ADJUST

Unloader adjustment

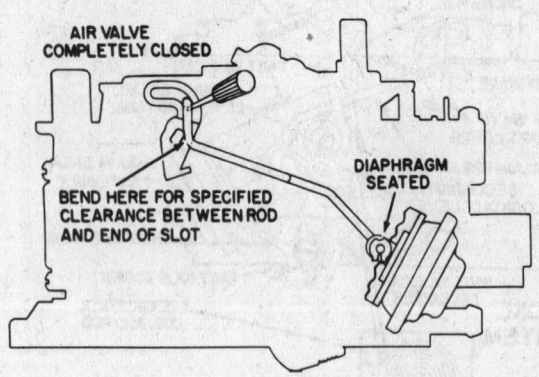

AIR VALVE COMPLETELY CLOSED

DIAPHRAGM SEATED

BEND HERE FOR SPECIFIED CLEARANCE BETWEEN ROD AND END OF SLOT

Air valve dashpot adjustment

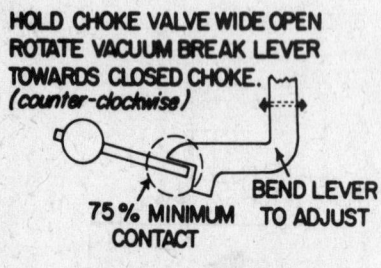

HOLD CHOKE VALVE WIDE OPEN ROTATE VACUUM BREAK LEVER TOWARDS CLOSED CHOKE. *(counter-clockwise)*

75% MINIMUM CONTACT

BEND LEVER TO ADJUST

SECONDARY LOCKOUT LEVER ADJUSTMENT

①

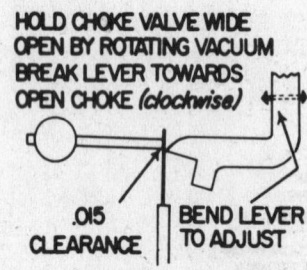

HOLD CHOKE VALVE WIDE OPEN BY ROTATING VACUUM BREAK LEVER TOWARDS OPEN CHOKE *(clockwise)*

.015 CLEARANCE

BEND LEVER TO ADJUST

SECONDARY LOCKOUT OPENING CLEARANCE

②

Secondary lockout adjustment

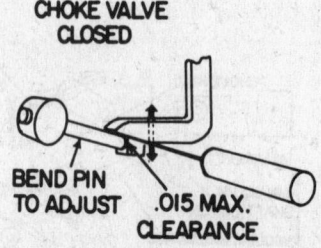

CHOKE VALVE CLOSED

BEND PIN TO ADJUST

.015 MAX. CLEARANCE

SECONDARY LOCKOUT LEVER SIDE CLEARANCE

③

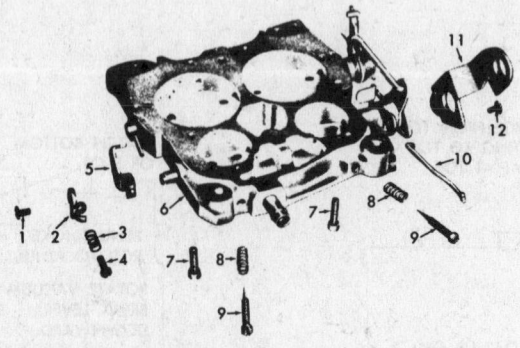

1 Screw—cam and fast idle levers
2 Fast idle lever
3 Spring—fast idle screw
4 Screw—fast idle adjusting
5 Cam lever
6 Throttle body assembly

7 Screw—throttle body
8 Spring—idle needle
9 Idle needle
10 Rod—accelerator pump
11 Throttle lever
12 Screw—throttle lever

Exploded view—4MV Quadrajet throttle body

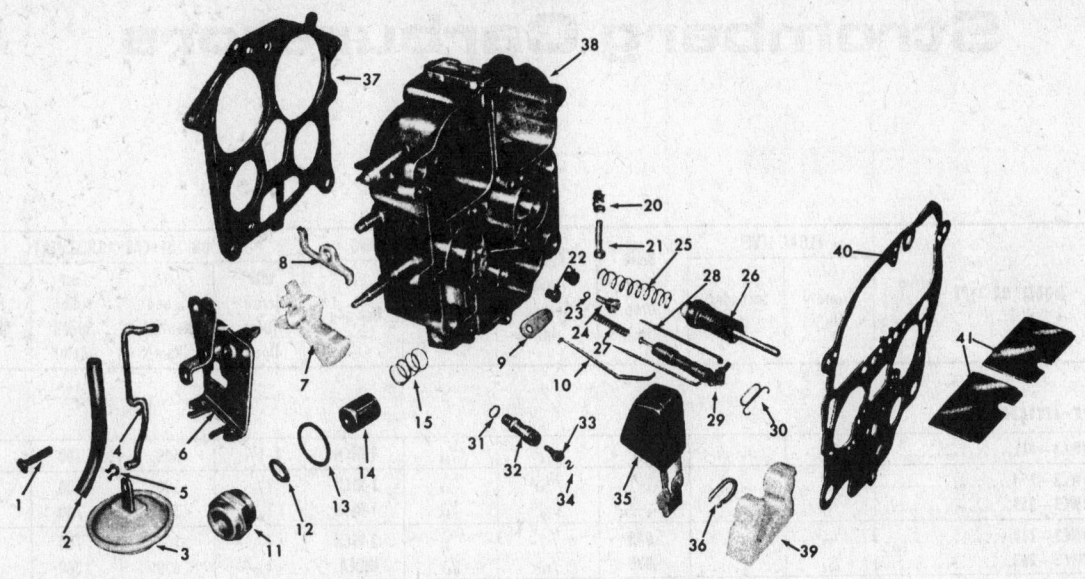

Exploded view—4MV Quadrajet float bowl

1 Screw—choke control	14 Filter—fuel inlet	30 Spring—metering rod primary
2 Hose—vacuum break	15 Spring—fuel filter	31 Gasket—float needle seat
3 Vacuum break	20 Spring—idle speed screw	32 Seat—float needle
4 Link—vacuum break	21 Screw—idle speed	33 Needle—float
5 Clip—Vacuum break rod	22 Jet—primary	34 Pull clip—float needle
6 Bracket assembly—	23 Ball—pump discharge	35 Float assembly
choke control	24 Retainer—pump discharge	36 Hinge pin—float assembly
7 Cam—fast idle	ball	37 Gasket—throttle body
8 Lever—secondary lock out	25 Spring—pump return	38 Float bowl assembly
9 Lever—choke intermediate	26 Pump assembly	39 Insert—float bowl
10 Rod—choke	27 Spring—power piston	40 Gasket—air horn
11 Nut—fuel inlet	28 Metering rod—primary	41 Baffle—float bowl
12 Gasket—fuel filter	29 Power piston assembly—	
13 Gasket—fuel inlet nut	primary	

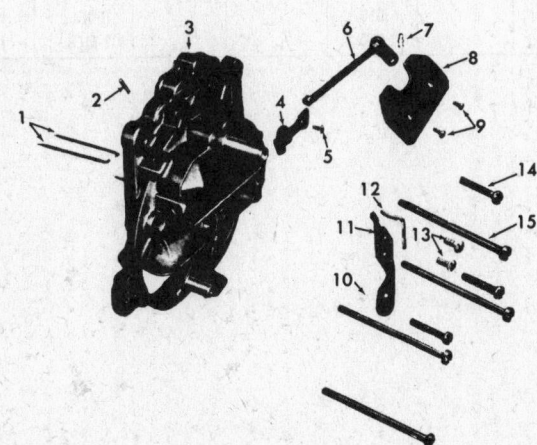

1 Metering rod—secondary	9 Screw—choke valve
2 Roll pin—pump lever	10 Clip—pump rod
3 Air horn assembly	11 Lever—pump actuating
4 Idle vent valve	12 Lever—idle vent valve
5 Screw—idle vent valve	13 Screw—air horn
6 Choke shaft and lever	14 (Counter sunk)
assembly	screw—air horn (short)
7 Clip—choke rod	15 Screw—air horn (long)
8 Choke valve	

Exploded view—4MV Quadrajet air horn

Stromberg Carburetors

YEAR	MODEL OR TYPE	FLOAT LEVEL Primary (In.)	FLOAT LEVEL Secondary (In.)	Bowl Vent Drop (In.)	Pump Travel Setting (In.)	CHOKE SETTING Unloader (In.)	CHOKE SETTING Housing	ON THE CAR ADJUSTMENTS Idle Screw Turns Open	ON THE CAR ADJUSTMENTS Idle Speed Neutral (Rpm)	ON THE CAR ADJUSTMENTS Fast Idle Speed (Rpm)	ON THE CAR ADJUSTMENTS Dashpot Plunger Clearance (In.)
Chrysler-Imperial											
1964	WWC3—ALL	1/8	...	5/64	7/16	15/64	1-RICH	1-1½	500	700	...
1965	WWC3—254	5/32	...	1/16-3/32	11/32	15/64	1-RICH	1½	500	700	...
1965	WWC3—255	5/32	...	1/16-3/32	7/16	15/64	1-RICH	½	500	700	...
1966	WWC3—262	5/32	...	.040	7/16	15/64	2-RICH	1½	500	700	...
1966	WWC3—263	5/32	...	.020	7/16	15/64	INDEX	1½	600	1300	...
1967	WWC3	5/32	...	.030	7/16	15/64	2-RICH	1½	...	1300	...
Dodge-Dart											
1964	WW3—ALL	7/32	...	5/64	5/64	15/64	INDEX	1¼	500	700	...
1965	WW3—248, 49, 50, 51	7/32	...	5/64	...	5/16	INDEX	1¼	500	700	...
1965	WWC3—254	5/32	...	1/16-3/32	11/32	15/64	1-RICH	1½	500	700	...
1965	WWC3—255	5/32	...	1/16-3/32	7/16	15/64	1-RICH	1½	500	700	...
1966	WW3—258, 259	7/32	...	.060	...	5/16	2-RICH	1¼	500	700	...
1966	WW3—260, 261	7/32	...	.050	...	5/16	INDEX	1½	650	1450	...
1966	WWC3—262	5/32	...	.040	7/16	15/64	2-RICH	1½	500	700	...
1966	WWC3—263	5/32	...	.020	7/16	15/64	INDEX	1½	600	1300	...
1967	WW3	7/32	...	.060	...	5/16	2-RICH	1¼	500	700	...
1967	WW3	7/32	...	.050	...	5/16	INDEX	1½	650	1450	...
1967	WWC3	7/32	...	.040	7/16	15/64	2-RICH	1½	...	...	...
Plymouth											
1964	WW3—239, 240	7/32	...	5/64	5/64	15/64	INDEX	1¼	500	700	...
1965	WW3—248, 49, 50, 51	7/32	...	5/64	5/64	5/16	INDEX	1¼	500	700	...
1965	WWC3—254	5/32	...	1/16-3/32	11/32	15/64	1-RICH	1½	500	700	...
1965	WWC3—255	5/32	...	1/16-3/32	7/16	15/64	1-RICH	1½	500	700	...
1966	WW3—258, 259	7/32	...	.060	...	5/16	2-RICH	1¼	500	700	...
1966	WW3—260, 261	7/32	...	.050	...	5/16	INDEX	1½	650	1450	...
1966	WWC3—262	5/32	...	.040	7/16	15/64	2-RICH	1½	500	700	...
1966	WWC3—263	5/32	...	.020	7/16	15/64	INDEX	1½	600	1300	...
1967	WW3	7/32	...	.060	...	5/16	2-RICH	1¼	500	700	...
1967	WW3	7/32	...	.050	...	5/16	INDEX	1½	650	1450	...
1967	WWC3	5/32	...	.040	7/16	15/64	2-RICH	1½	...	...	...

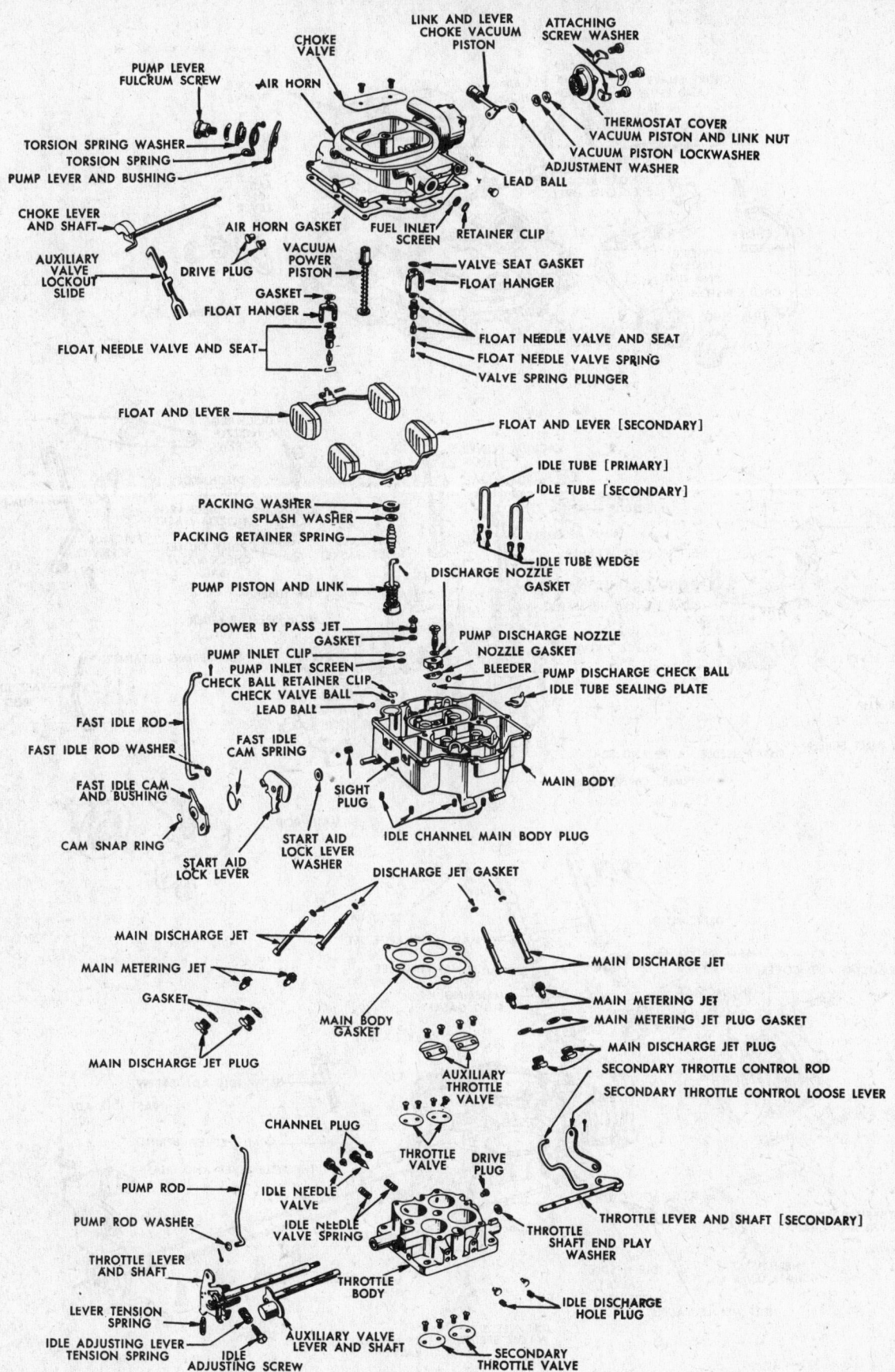

Stromberg WA4 four-barrel

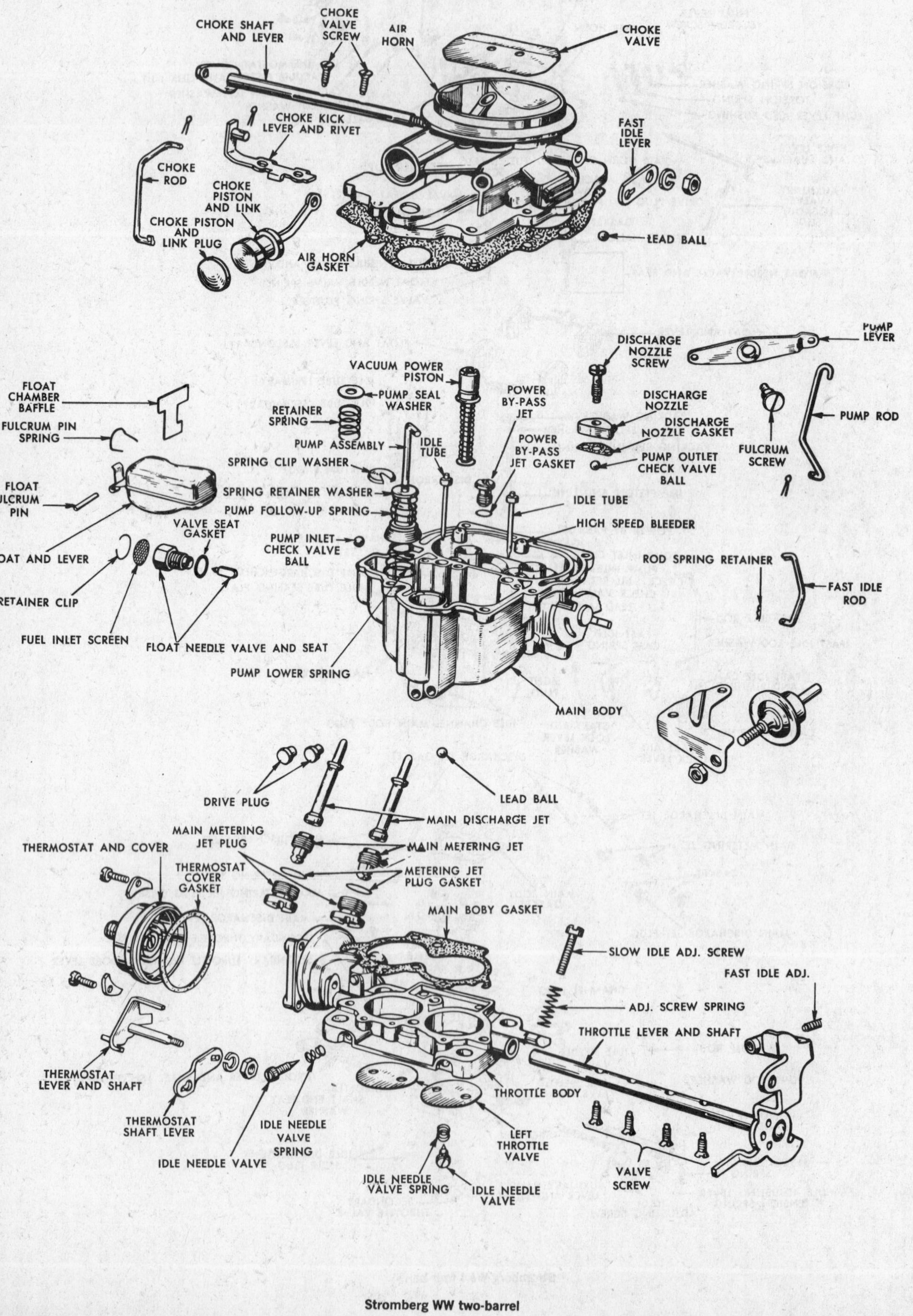

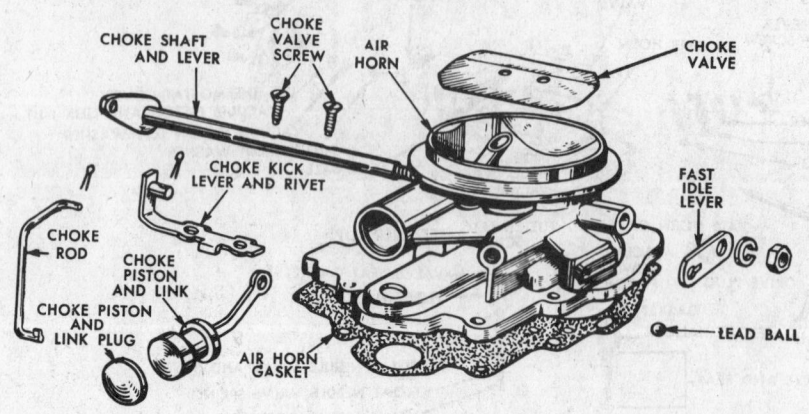

CHOKE SHAFT AND LEVER
CHOKE VALVE SCREW
AIR HORN
CHOKE VALVE
CHOKE KICK LEVER AND RIVET
FAST IDLE LEVER
CHOKE ROD
CHOKE PISTON AND LINK
CHOKE PISTON AND LINK PLUG
AIR HORN GASKET
LEAD BALL

VACUUM POWER PISTON
PUMP SEAL WASHER
POWER BY-PASS JET
DISCHARGE NOZZLE SCREW
PUMP LEVER
FLOAT CHAMBER BAFFLE
RETAINER SPRING
DISCHARGE NOZZLE
FULCRUM PIN SPRING
PUMP ASSEMBLY
IDLE TUBE
POWER BY-PASS JET GASKET
DISCHARGE NOZZLE GASKET
PUMP ROD
SPRING CLIP WASHER
FLOAT FULCRUM PIN
SPRING RETAINER WASHER
PUMP OUTLET CHECK VALVE BALL
FULCRUM SCREW
VALVE SEAT GASKET
PUMP FOLLOW-UP SPRING
FLOAT AND LEVER
PUMP INLET CHECK VALVE BALL
IDLE TUBE
RETAINER CLIP
HIGH SPEED BLEEDER
ROD SPRING RETAINER
FUEL INLET SCREEN
FAST IDLE ROD
FLOAT NEEDLE VALVE AND SEAT
PUMP LOWER SPRING
MAIN BODY

DRIVE PLUG
LEAD BALL
MAIN DISCHARGE JET
MAIN METERING JET PLUG
MAIN METERING JET
THERMOSTAT AND COVER
THERMOSTAT COVER GASKET
METERING JET PLUG GASKET
MAIN BOBY GASKET
SLOW IDLE ADJ. SCREW
FAST IDLE ADJ.
ADJ. SCREW SPRING
THROTTLE LEVER AND SHAFT
THERMOSTAT LEVER AND SHAFT
THERMOSTAT SHAFT LEVER
IDLE NEEDLE VALVE SPRING
THROTTLE BODY
LEFT THROTTLE VALVE
IDLE NEEDLE VALVE
IDLE NEEDLE VALVE SPRING
IDLE NEEDLE VALVE
VALVE SCREW

Stromberg WW two-barrel

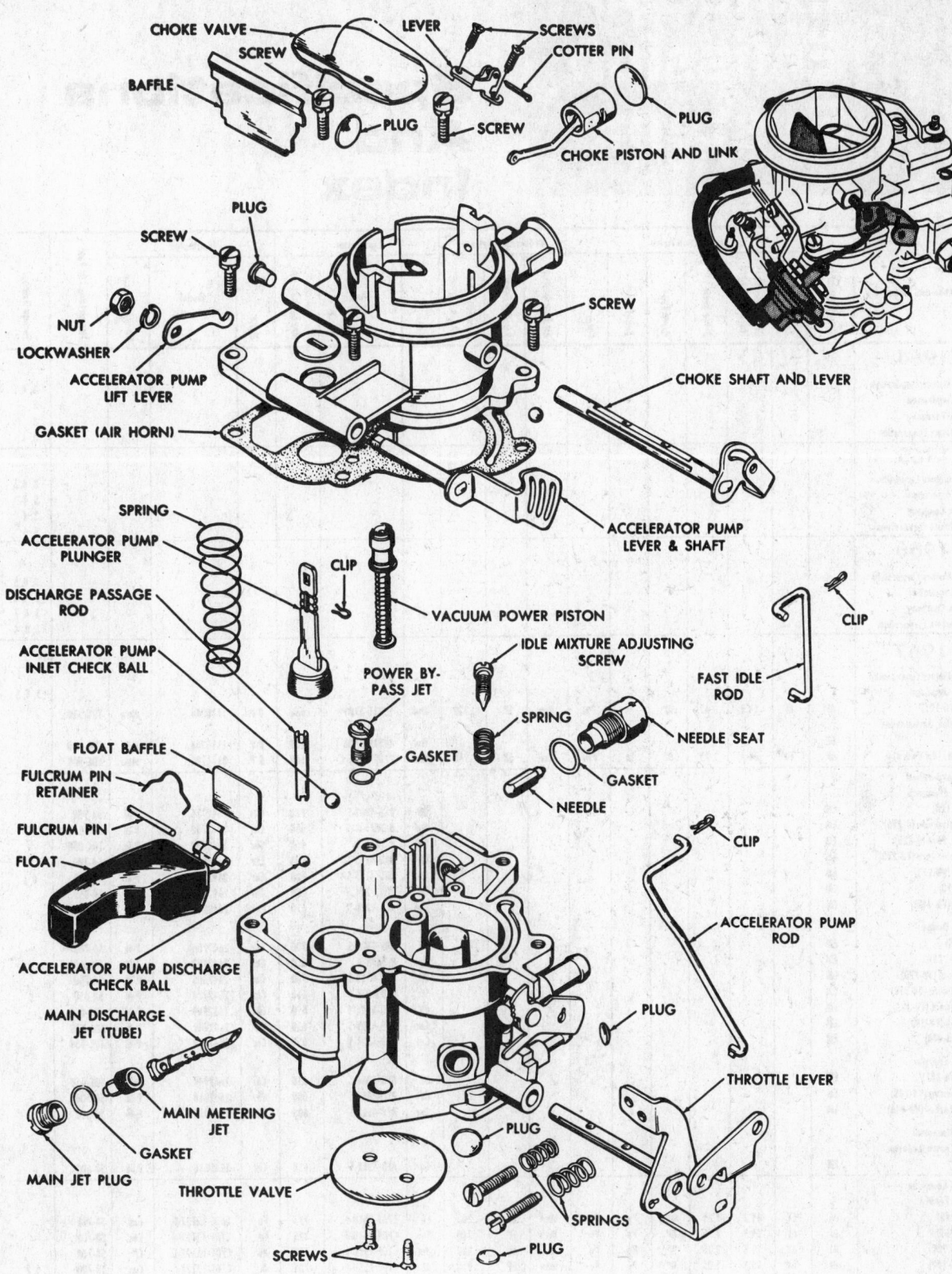

Stromberg WA3 single-barrel

Exhaust Emission Systems

Specifications And Index

Make and Model	Type System	Air Injection Pump						Air Injection System			Carburetor			Distributor			Timing (deg.) @ rpm	Section Number
		Type Pump	Displacement (cu. in. per rev.)	Drive Ratio	Drive Type	Relief Valve	Filter	Air Distribution	Point of Entry	Injection Tube I.D. (in.)	Make	Model	Idle Speed (rpm)	Make	Model	Vacuum Source		
1964																		
American Motors Corporation																		3, 4, 5
Chrysler Corporation																		3
Ford Motor Company																		3
General Motors Corporation																		3, 4
1965																		
American Motors Corporation																		3, 4, 5
Chrysler Corporation																		3, 4
Ford Motor Company																		3, 4
General Motors Corporation																		3, 4
1966																		
American Motors Corporation																		3, 4, 5
Chrysler Corporation																		3, 4, 6
Ford Motor Company																		3, 4, 7
General Motors Corporation																		3, 8, 4
1967																		
American Motors Corporation																		3, 4, 5
Rambler																		
American (6-199)	AI	EV	19.3	.65	BP	In	Ppr	Man	EP	.222	Hol	1931C-3709	600	D-P	1110366	Man	TDC-500	
Rebel, Marlin, Ambassador (6-232)	EM										Hol	1931C-3708-1	600	D-P	1110366	Man	TDC-500	
Ambassador (Air-Guard)	AI	EV	19.3	1.25	BP	In	Ppr	Man	EP	.285	Hol	2209-3484-1	600	D-P	1111106	Man	TDC-500	
Chrysler Corporation																		
Plymouth																		
Valiant (6-170)	EM										BB	BBS-4302-S	700	Chr	2642786	P-M	5A-700	
Valiant, Barracuda (6-225)	EM										Hol	R-3671-A	650	Chr	2642792	P-M	TDC-650	
Belvedere, Fury (6-225)	EM										Hol	R-3673-A	650	Chr	2642792	P-M	TDC-650	
Valiant, Belvedere (V8-273)	EM										BB	BBD-4115-SA	700	Chr	2642805	P-M	5A-700	
Barracuda (V8-273)	EM										BB	BBD-4115-SA	700	Chr	2642805	P-M	5B-700	4, 6
Fury (V8-318)	EM										Str	WW3-274	650	Chr	2642724	P-M	5A-650	
Belvedere (V8-440)	EM										Car	AFB-4328-S	650	Chr	2642819	P-M	TDC-650	
Dodge																		
Dart (6-170)	EM										BB	BBS-4302-S	700	Chr	2642786	P-M	5A-700	
Coronet (6-225)	EM										Hol	R-3673-A	650	Chr	2642792	P-M	TDC-650	
Dart, Coronet (V8-273)	EM										BB	BBD-4115-SA	700	Chr	2642805	P-M	5A-700	
Charger, Polara (V8-318)	EM										Str	WW3-274	650	Chr	2642724	P-M	5A-650	
Polara, Monaco (V8-383)	EM										BB	BBD-4306-S	650	Chr	2642949	P-M	TDC-650	
Monaco 500 (V8-383)	EM										Car	AFB-4309-S	650	Chr	2642949	P-M	TDC-650	
Coronet (V8-440)	EM										Car	AFB-4328-S	650	Chr	2642819	P-M	TDC-650	
Chrysler																		
Newport (V8-383)	EM										BB	BBD-4306-S	650	Chr	2642949	P-M	TDC-650	
Town & Country (V8-383)	EM										BB	BBD-4307-S	600	Chr	2642810	P-M	5B-600	
300, New Yorker (V8-440)	EM										Car	AFB-4312-S	600	Chr	2642816	P-M	5B-600	
Imperial																		
Imperial, Crown, LeBaron (V8-440)	EM										Car	AFB-4312-S	600	Chr	2642816	P-M	5B-600	
Ford Motor Company																		
Ford																		
Falcon (6-170)	AI	EV	19.3	1.25	BP	Pp	Po	Man	EP	.260	Fo	C7DF-9510-J	775	Fo	C6DF-12127-D	Cas	5B-700	
Falcon (6-200)	AI	EV	19.3	1.25	BP	Pp	Po	Man	EP	.260	Fo	C7DF-9510-J	725	Fo	C7DF-12127-C	Cas	5B-700	
Mustang (6-200)	AI	EV	19.3	1.25	BP	Pp	Po	Man	EP	.260	Hol	C7OF-9510-J	650	AL	C7DF-12127-C	CP	5B-700	
Fairlane (6-200)	AI	EV	19.3	1.25	BP	Pp	Po	Man	EP	.250	AL	C7OF-9510-N	725	AL	C7DF-12127-C	Cas	5B-700	4, 7
Ford (6-240)	AI	EV	19.3	1.25	BP	Pp	Po	Man	EP	.260	AL	C7AF-9510-AA	600	AL	C6AF-12127-AC	C	TDC-600	
Falcon (V8-289)	AI	EV	19.3	1.30	BP	Pp	Po	Man	EP	.260	AL	C7DF-9510-G	650	Fo		Cas	1½ B-625	
Mustang (V8-289)	AI	EV	19.3	1.25	BP	Pp	Po	Man	EP	.260	AL	C7DF-9510-G	650	AL	C70F-12127-D	CP	1½ B-625	
Fairlane (V8-289)	AI	EV	19.3	1.30	BP	Pp	Po	CH	EP	.250	AL	C7DF-9510-G	650	AL	C70F-12127-D	Cas	1½ B-62	
Ford (V8-289)	AI	EV	19.3	1.25	BP	Pp	Po	Man	EP	.260	AL	C7AF-9510-S	600	AL	C7OF-12127-D	C	TDC-600	
Thunderbird (V8-390)	AI	EV	19.3	1.33	BP	Pp	Po	Man	EP	.260	AL	C7AF-9510-AY	550	AL	C7MF-12127-H	Tsu	6B-525	

Make and Model	Type System	Type Pump	Displacement (cu. in. per rev.)	Drive Ratio	Drive Type	Relief Valve	Filter	Air Distribution	Point of Entry	Injection Tube I.D. (in.)	Make	Model	Idle Speed (rpm)	Make	Model	Vacuum Source	Timing (deg.) @ rpm	Section Number
Mercury																		
Comet (6-200)	AI	EV	19.3	1.25	BP	Pp	Pe	Man	EP	.260	AL	C7DF-9510-J	650	AL	C7DF-12127-C	Cas	5B-600	
Comet (V8-289)	AI	EV	19.3	1.30	BP	Pp	Pe	CH	EP	.260	AL	C7DF-9510-U	650	AL	C70F-12127-D	Cas	1B-500	
Cougar (V8-289)	AI	EV	19.3	1.25	BP	Pp	Pe	CH	EP	.250	AL	C7DF-9510-G	650	AL	C70F-12127-D	Cas	1½ B-650	
Comet (V8-390)	AI	EV	19.3	1.33	BP	Pp	Pe	Man	EP	.250	Fo	C70F-9510-C	625	Fo	C70F-12127-F	Cas	6B-625	
Mercury (V8-390)	AI	EV	19.3	1.25	BP	Pp	Pe	Man	EP	.260	AL	C7AF-9510-BF	625	AL	C7AF-12127-Z	Cas	6B-610	
Mercury (V8-410)	AI	EV	19.3	1.25	BP	Pp	Pe	Man	EP	.260	AL	C7AF-9510-AG	625	AL	C7SF-12127-B	Cas	6B-610	
Mercury (V8-428)	AI	EV	19.3	1.25	BP	Pp	Pe	Man	EP	.260	AL	C7AF-9510-AG	625	AL	C7AF-12127-AC	Cas	6B-610	
Continental																		
(V8-462)	AI	EV	19.3	1.25	BP	Pp	Pe	CH	EP	.250	Car	C6VF-9510-F	500	AL	C6VF-12127-A	Cas	10B-500	
General Motors Corporation																		
Chevrolet																		
Chevy II (4-153)	NA																	
Corvair (6-164)	AI	EV	19.3	1.25	BP	P	N	Man	EP	.256	Roc	7036014	700	DR	1110369	C	TDC-700	
Chevy II (6-194)	AI	EV	19.3	1.25	BP	P	N	CH	EP	.257	Car	3909405	700	DR	1110388	C	2B-700	
Camaro, Chevelle (6-230)	AI	EV	19.3	1.25	BP	P	N	CH	EP	.257	Car	3905975	700	DR	1110387	C	4B-700	
Chevrolet (6-250)	AI	EV	19.3	1.25	BP	P	N	CH	EP	.257	Car	3905975	700	DR	1110351	C	4B-700	
Chevy II, Chevelle, Chevrolet (V8-283)	AI	EV	19.3	1.25	BP	P	N	Man	EP	.257	Roc	7037101	700	DR	1111256	C	TDC-700	
Camaro (V8-327)	AI	EV	19.3	1.25	BP	P	N	Man	EP	.257	Roc	7037101	700	DR	1111101	C	2B-700	
Corvette (V8-327)	AI	EV	19.3	1.25	BP	P	N	Man	EP	.257	Hol	3906635	700	DR	1111194	C	6B-700	
Chevelle (V8-396)	AI	EV	19.3	1.25	BP	P	N	Man	EP	.257	Roc	7037211	700	DR	1111169	C	4B-700	
Pontiac																		
Firebird, Tempest (6-230)	AI	EV	19.3	.95	BP	Sld	Cac	Man	EP	.256	Roc		700	DR	1110391	C	TDC-700	3, 4, 8
Firebird, Tempest (V8-326)	AI	EV	19.3	1.20	BP	Sld	Cac	CH	EP	.250	R-C		700	DR	1111199	C	6B-700	
Tempest (V8-400)	EM										R-C	7027263	700	DR	1111261	C	6B-700	
Catalina, Executive, Bonneville (V8-400)	AI	EV	19.3	1.20	BP	Sld	Cac	CH	EP	.250	Car	AFB-4245-S	700	DR	1111254	C	6B-700	
Grand Prix (V8-400)	AI	EV	19.3	1.20	BP	Sld	Cac	CH	E?	.250	Car	AFB-4245-S	700	DR	1111244	C	6B-700	
2 + 2 (V8-428)	AI	EV	19.3	1.20	BP	Sld	Cac	CH	EP	.250	R-C	7037263	700	DR	1111237	C	6B-700	
Buick																		
Special (V6-225)	AI	EV	19.3	1.25	BP	P	Cac	Man	EP	.252	Roc	2GC	550	DR	1110342	C	5B-650	
Special (V8-300)	AI	EV	19.3	1.25	BP	P	Cac	Man	EP	.252	Roc	2GC	650	DR	1111147	C	2½ B-550	
Special, Sportwagon (V8-340)	AI	EV	19.3	1.25	BP	P	Cac	Man	EP	.252	Roc	2GC	650	DR	1111144	C	2½ B-550	
LeSabre (V8-340)	AI	EV	19.3	1.25	BP	P	Cac	Man	EP	.250	Roc	2GC	650	DR	1111159	C	2½ B-550	
Special (V8-400); Wildcat, Electra, Riviera (V8-430)	AI	EV	19.3	1.25	BP	P	Cac	CH	EP	.250	Roc	4MV	600	DR	1111149	C	2½ B-550	
Oldsmobile																		
F-85 (6-250)	AI	EV	19.3	1.25	BP	Psl	Pe	Man	EP	.257	Car	YF	600	DR	1110351			
F-85 (V8-330)	AI	EV	19.3	1.25	BP	Psl	Pe	Man	EP	.257	Roc	2GC	600	DR	1111029			
Vista-Cruiser, Delmont 88 (V8-330)	AI	EV	19.3	1.25	BP	Psl	Pe	Man	EP	.257	Roc	2GC	600	DR	1111048			
F-85 Supreme (V8-330)	AI	EV	19.3	1.25	BP	Psl	Pe	Man	EP	.257	Roc	4MV	600	DR	1111048			
F-85 (V8-400)	AI	EV	19.3	1.25	BP	Psl	Pe	Man	EP	.257	Roc	4MV	600	DR	1111188			
Delta 88, Delmont 88 (V8-425)	AI	EV	19.3	1.25	BP	Psl	Pe	Man	EP	.257	Roc	2GC	600	DR	1111042			
98, Toronado (V8-425)	AI	EV	19.3	1.25	BP	Psl	Pe	Man	EP	.257	Roc	4MV	600	DR	1111151			
Cadillac																		
Calais, DeVille, Fleetwood (V8-429)	AI		19.3	1.50	BP	Pp		Man	CH	.244	Roc	4MV	550	DR	1111262	C	5B-550	
Fleetwood Eldorado (V8-429)	AI		19.3	1.50	BP	Pp		CH	Man	.313	R-C	4MV	550	DR	1111262	C	5B-550	
1968																		
American Motors Corporation																		4, 5
Rambler																		
American (6-199)	EM										Hol	1931-3966-A	600	DR	1110444	Man	TDC-600	
Rogue, Rebel, Javelin, Ambassador (6-232)	EM										Hol	1931-3968-A	600	DR	1110444	Man	TDC-600	
Rebel, AMX, Ambassador (Air Guard)	AI	EV	19.3	1.25	BP	In	Cen	SHM	EP	.285	AM	8HM2	650	DR	1111106	Man	TDC-650	
Chrysler Corporation																		
Plymouth																		
Valiant, Signet (6-170)	EM										BB	RBS-4414-S	700	Chr	2875199	P-M	5A-700	
Barracuda (6-225)	EM										Hol	R-3919-A	650	Chr	2875364	P-M	TDC-650	
Belvedere, Satellite, Fury (6-225)	EM										Hol	R-3921-A	650	Chr	2875364	P-M	TDC-650	
Valiant, Signet, Belvedere, Satellite (V8-273)	EM										BB	BBD-4416-S	700	Chr	2875334	P-M	5A-700	
Barracuda, Satellite, Fury, VIP, Suburban (V8-318)	EM										BB	BBD-4420-S	650	Chr	2875342	P-M	5A-650	4, 6
Barracuda (V8-340)	EM										Car	AVS-4424-S	700	Pres	2875086		TDC-700	
Road Runner (V8-383)	EM										Car	AVS-4426-S	650	Chr	2875356		TDC-650	
GTX (V8-440)	EM										Car	AVS-4428-S	650	Chr	2875102		TDC-650	
Dodge																		
Dart (6-170)	EM										BB	RBS-4414-S	700	Chr	2875199	P-M	5A-700	
Coronet (6-225)	EM										Hol	R-3921-A	650	Chr	2875364	P-M	TDC-650	
Dart, Coronet (V8-273)	EM										BB	BBD-4416-S	700	Chr	2875334	P-M	5A-700	
Coronet, Charger, Polara (V8-318)	EM										BB	BBD-4420-S	650	Chr	2875342	P-M	5A-650	
Dart (V8-340)	EM										Car	AVS-4424-S	700	Pres	2875086		TDC-700	
Monaco (V8-383)	EM										BB	BBD-4422-S	650	Chr	2875352		TDC-650	
Coronet, Charger (V8-440)	EM										Car	AVS-4428-S	650	Chr	2875102		TDC-650	

Make and Model	Type System	Air Injection Pump: Type Pump	Displacement (cu. in. per rev.)	Drive Ratio	Drive Type	Relief Valve	Air Injection System: Filter	Air Distribution	Point of Entry	Injection Tube I.D. (in.)	Carburetor: Make	Model	Idle Speed (rpm)	Distributor: Make	Model	Vacuum Source	Timing (deg.) @ rpm	Section Number
Chrysler																		
Newport (V8-383)	EM										BB	BBD-4422-S	650	Chr	2875352	P-M	TDC-650	
Town & Country (V8-383)	EM										BB	BBD-4423-S	600	Chr	2875354	P-M	7½ B-600	
300, New Yorker (V8-440)	EM										Hol	R-3918-A	600	Chr	2875362	P-M	7½ B-600	
Imperial																		
Crown, LeBaron (V8-440)	EM										Hol	R-3918-A	600	Chr	2875362	P-M	7½ B-600	
Ford Motor Company																		
Ford																		
Falcon (6-170)	AI	VP	19.3	1.00	BP	Po	Cen	Man	EP	.260	Car	C8DF-9510-A	775	AL	C8DF-12127-A	C-M	6B-775	
Falcon (6-200)	AI	VP	19.3	1.25	BP	Po	Cen	Man	EP	.260	AL	C8OF-9510-A	725	AL	C8DF-12127-C	C-M	6B-725	
Mustang (6-200)	AI	VP	19.3	1.25	BP	Po	Cen	Man	EP	.260	AL	C8OF-9510-A	725	AL	C8DF-12127-C	C-M	6B-725	
Fairlane, Torino (6-200)	AI	VP	19.3	1.25	BP	Po	Cen	Man	EP	.260	AL	C8OF-9510-A	725	AL	C8DF-12127-E	C-M	6B-725	
Ford, Galaxie, XL (6-240)	AI	VP	19.3	1.25	BP	Po	Cen	Man	EP	.260	Car	C8AF-9510-V	525	AL	C8AF-12127-A	CP	6B-525	
Falcon (V8-289)	AI	VP	19.3	1.30	BP	Po	Cen	CH	EP	.260	AL	C8AF-9510-AK	650	AL	C8TF-12127-F	C-M	6B-650	
Mustang (V8-289)	AI	VP	19.3	1.30	BP	Po	Cen	CH	EP	.260	AL	C8AF-9510-AK	650	AL	C8TF-12127-F	C-M	6B-650	
Fairlane, Torino (V8-302)	AI	VP	19.3	1.30	BP	Po	Cen	Man	EP	.260	AL	C8AF-9510-AK	650	AL	C8AF-12127-E	C-M	6B-650	
Ford, Galaxie, XL, LTD (V8-302)	AI	VP	19.3	1.25	BP	Po	Cen	Man	EP	.260	AL	C8AF-9510-AK	650	AL	C8AF-12127-E	CP	6B-650	
Thunderbird (V8-390)	EM										AL	C8AF-9510-B	550	AL	C7AF-12127-AC	C	6B-550	
Mercury																		
Comet, Montego (6-200)	AI	VP	19.3	1.25	BP	Po	Cen	Man	EP	.260	AL	C8OF-9510-A	725	AL	C8DF-12127-C	C-M	6B-725	
Cyclone, Montego (V8-302)	AI	VP	19.3	1.25	BP	Po	Cen	Man	EP	.260	AL	C8AF-9510-AK	650	AL	C8AF-12127-E	C-M	6B-650	
Cougar, XR-7 (V8-302)	AI	VP	19.3	1.30	BP	Po	Cen	Man	EP	.260	AL	C8AF-9510-AK	650	AL	C8AF-12127-E	C-M	6B-650	4, 7
Monterey, Montclair, Commuter, Colony Park (V8-390)	AI	VP	19.3	1.25	BP	Po	Cen	Man	EP	.260	AL	C8AF-9510-M	650	AL	C8AF-12127-M	C	6B-650	
Parklane, Brougham, Marquis (V8-390)	EM										AL	C8AF-9510-B	550	AL	C7AF-12127-AC	C	6B-550	
Parklane S-55 (V8-428)	EM										AL	C8AF-9510-B	550	AL	C8AF-12127-Y	C	6B-550	
Continental																		
(V8-462)	EM										Car	C8VF-9510-E	550	AL	C8VF-12127-F	C	10B-550	
General Motors Corporation																		
Chevrolet																		
Chevy II (4-153)	AI	VP	19.3		BP	DV	Cen	CH	EP	.257	Roc	7028009	750	DR	1110447	C	TDC-750	
Corvair, Monza (6-164)	AI	VP	19.3	1.15	BP	DV	Cen	Man	EP	.256	Roc	7028005	700	DR	1110434	C	6B-700	
Camaro, Chevy II, Chevelle, Malibu, Nomad, Concours (6-230)	AI	VP	19.3	1.15	BP	DV	Cen	CH	EP	.257	Roc	7028017	700	DR	1110436	C	TDC-700	
Biscayne, Bel Air, Impala (6-250)	AI	VP	19.3	1.15	BP	DV	Cen	CH	EP	.257	Roc	7028017.	700	DR	1110439	C	TDC-700	
Chevy II, Chevelle, Malibu, Nomad, Concours, Biscayne, Bel Air, Impala, Caprice (V8-307)	AI	VP	19.3	1.15	BP	DV	Cen	Man	EP	.267	Roc	7028101	700	DR	1111257	C	2B-700	
Camaro (V8-327)	AI	VP	19.2	1.15	BP	DV	Cen	Man	EP	.257	Roc	7028101	700	DR	1111440	C	2A-700	
Corvette (V8-327)	AI	VP	19.3	1.15	BP	DV	Cen	Man	EP	.257	Roc	7028207	700	DR	1111194	C	4B-700	
Chevelle SS-396 (V8-396)	AI	VP	19.3	1.15	BP	P	Cen	Man	EP	.257	Roc	7028211	700	DR	1111169	C	4B-700	
Pontiac																		
Tempest, LeMans, Safari, Firebird (6-250)											Roc	7028065	700	DR	1110430		TDC-700	4, 8
Tempest, LeMans, Safari, Firebird (V8-350)											Roc	7028071	700	DR	1111281		9B-700	
GTO (V8-400)											Roc	7028263	850	DR	1111449		9B-850	
Catalina, Executive (V8-400)											Roc	7028066	800	DR	1111272		9B-800	
Bonneville, Grand Prix (V8-400)											Roc	7028263	850	DR	1111448		9B-850	
Buick																		
Special, Skylark (6-250)	EM										Roc	MV	700	DR	1110439	C	TDC-700	
Special, Skylark, Sportwagon, LeSabre (V8-350)	EM										Roc	2GV	700	DR	1111330	Man	TDC-700	
GS-350 (V8-350)	EM										Roc	4MV	700	DR		Man	TDC-550	
GS-400 (V8-400)	EM										Roc	4MV	700	DR	1111285	Man	TDC-550	
Wildcat, Electra, Riviera (V8-430)	EM										Roc	4MV	550	DR	1111285	Man	TDC-550	
Oldsmobile																		
F-85, Cutlass (6-250)	EM										Roc	1BV	650	DR	1110351	C	4B-500	
F-85, Cutlass, Supreme, Vista-Cruiser, Delmont (V8-350)	EM										Roc	2GC	650	DR	1111286	C	5B-850	
4-4-2 (V8-400)	EM										Roc	4MV	725	DR	1111287	C	TDC-850	
Delta (V8-455)	EM										Roc	2GC	650	DR	1111288	C	5B-850	
Ninety-Eight (V8-455)	EM										Roc	4MV	550	DR	1111289	C	10B-850	
Toronado (V8-455)	EM										Roc	4MV	550	DR	1111289	C	7½ B-850	
Cadillac																		
Calais, DeVille, Fleetwood, Eldorado (V8-472)	AI		19.3	1.20	BP	N	Cen	Man	CH	.579	Roc	4MV	550	DR	1111239	C	5B-550	
1969																		
American Motors Corporation																		
Rambler																		
Rambler (6-199)	EM										Car	RBS-4633S	600	DR	1110444	Man	TDC-600	4, 5
Rambler (6-232)	EM										Car	RBS-4631S	600	DR	1110444	Man	TDC-600	
Rebel, Javelin, Ambassador (6-232)	EM										Car	WCD-4667S	600	DR	1110444	Man	TDC-600	
AMX (Air-Guard)	AI	EV	19.3	1.25	Be	In	Cen	SHM	EP	.285	Car	AFB-4660S	650	DR	1111198	Man	TDC-650	

Make and Model	Type System	Air Injection Pump Type Pump	Displacement (cu. in. per rev.)	Drive Ratio	Drive Type	Relief Valve	Filter	Air Injection System Air Distribution	Point of Entry	Injection Tube I.D. (in.)	Carburetor Make	Model	Idle Speed (rpm)	Distributor Make	Model	Vacuum Source	Timing (deg.) @ rpm	Section Number
Chrysler Corporation																		
Plymouth																		
Valiant (6-170)	EM										Car	BBS-4601S		Chr	2875813	CP	5A-	
Barracuda (6-225)	EM										Hol	R-416A		Chr	2875822	CP	TDC-	
Belvedere, Satellite, Fury (6-225)	EM										Hol	R-4163A	700	Chr	2875822	CP	TDC-700	
Satellite, Fury, VIP (V8-318)	EM										Car	BBD-4607S	700	Chr	2875796	CP	TDC-700	
Road Runner (V8-383)	EM										Car	AVS-4615S	700	Pres	2875715	CP	TDC-700	
GTX (V8-440)	EM										Car	AVS-4617S	700	Pres	2875772	CP	TDC-700	4, 6
Dodge																		
Dart (6-170)	EM										Car	BBS-4601S	750	Chr	2875813	CP	5A-750	
Coronet, Charger (6-225)	EM										Hol	R-4163A	700	Chr	2875822	CP	TDC-700	
Coronet, Polara (V8-318)	EM										Car	BBD-4607S	700	Chr	2875796	CP	TDC-700	
GTS (V8-340)	EM										Car	AVS-4611S	750	Pres	2875782	CP	TDC-750	
Super Bee (V8-383)	EM										Car	AVS-4615S	700	Pres	2875715	CP	TDC-700	
Monaco (V8-383)	EM										Car	BBD-4613S	700	Chr	2875742	CP	TDC-700	
Coronet, Charger (V8-440)	EM										Car	AVS-4617S	700	Pres	2875772	CP	TDC-700	
Chrysler																		
Newport (V8-383)	EM										Car	BBD-4613S	700	Chr	2875742	CP	TDC-700	
300, New Yorker (V8-440)	EM										Hol	R-4166A	600	Chr	2875764	CP	7½ B-600	
Imperial																		
Crown, LeBaron (V8-440)	EM										Hol	R-4166A	600	Chr	2875764	CP	7½ B-600	
Ford Motor Company																		
Ford																		
Falcon (6-170)	EM										Car	C8DF-9510-H	750	AL	C9DF-12127-B		6B-750	
Futura, Mustang (6-200)	EM										AL	C9DF-9510-B	750	AL	C8DF-12127-C		6B-750	
Custom, Galaxie 500, XL (6-240)	EM										Car	C8AF-9510-BF	775	AL	C8AF-12127-A		6B-775	4, 7
Mustang, Fairlane, Torino, Cobra (6-250)	EM										AL	C9OF-9510-B	700	AL	C9OF-12127-R		6B-700	
LTD (V8-302)	EM										AL	C8AF-9510-BD	650	AL	C8AF-12127-E		6B-650	
Mustang (V8-351)	EM										AL	C9ZF-9510-A	650	AL	C9OF-12127-M		6B-650	
Thunderbird (V8-429)	EM										AL	C8SF-9510-H	550	AL	C8VF-12127-A		6B-550	
Mercury																		
Comet, Montego, Cyclone (6-250)	EM										AL	C9OF-9510-B	700	AL	C9OF-12127-R		6B-700	
Cougar (V8-351)	EM										AL	C9ZF-9510-A	650	AL	C9OF-12127-M		6B-650	
Monterey, Marauder, Marquis (V8-390)	EM										AL	C9AF-9510-B	650	AL	C9AF-12127-J		6B-650	
Brougham (V8-429)	EM										AL	C9AF-9510-J	550	AL	C8VF-12127-A		6B-550	
Lincoln																		
Continental, Mark III (V8-460)	EM										AL	C8VF-9510-J	550	AL	C8VF-12127-G		10B-550	
General Motors Corporation																		
Chevrolet																		
Chevy Nova (4-153)	AI	AV	19.3	1.15	Bo	DV	Con	CN	EP	.257	Roc	7029008	750	DR	1110457	C	TDC-750	
Corvair (6-164)	AI	AV	19.3	1.15	Bo	DV	Con	Man	EP	.256	Roc	7028005	700	DR	1110452	C	6B-700	
Camaro, Chevelle (6-230)	AI	AV	19.3	1.15	Bo	DV	Con	CN	EP	.257	Roc	7029017	700	DR	1110459	C	TDC-700	
Biscayne, Bel Air, Impala (6-250)	AI	AV	19.3	1.15	Bo	DV	Con	CN	EP	.257	Roc	7029017	700	DR	1110464	C	TDC-700	
Caprice (V8-327)	AI	AV	19.3	1.15	Bo	DV	Con	CN	EP	.257	Roc	7029127	700	DR	1111483	C	2A-700	
Corvette (V8-350)	AI	AV	19.3	1.15	Bo	DV	Con	Man	EP	.257	Roc	7029203	700	DR	1111490	C	4B-700	4, 8
Pontiac																		
Tempest, Firebird (6-250)											Roc	7029165	500	DR	1110475		TDC-500	
GTO, Grand Prix (V8-400)											Roc	7029263	1000	DR	1111952		9B-1000	
Catalina, Executive (V8-400)											Roc	7028066	1000	DR	1111952		9B-1000	
Bonneville (V8-428)											Roc	7029262	650	DR	1111960		9B-650	
Buick																		
Special, Skylark (6-250)	EM										Roc	MV	700	DR	1110439	C	TDC-700	
Skylark, LeSabre (V8-350)	EM										Roc	2GV	700	DR	1111938	Man	TDC-700	
GS (V8-350)	EM										Roc	4MV	700	DR	1111334	Man	TDC-550	
Wildcat, Electra, Riviera (V8-430)	EM										Roc	4MV	550	DR	1111335	Man	TDC-550	
Oldsmobile																		
F-85, Cutlass (6-250)	EM										Roc	1BV	750	DR	1110463	Por	TDC-700	
Cutlass, Supreme (V8-350)	EM										Roc	4MV	625	DR	1111930	Por	8B-850	
Delta (V8-350)	EM										Roc	2GC	625	DR	1111961	Por	6B-850	
4-4-2 (V8-400)	EM										Roc	4MV	750	DR	1111932	Por	2B-850	
Delta Custom, Royale (V8-455)	EM										Roc	2GC	625	DR	1111934	Por	6B-850	
Ninety-Eight, Toronado (V8-455)	EM										Roc	4MV	550	DR	1111935	Por	8B-850	
Cadillac																		
Calais, DeVille, Fleetwood, Eldorado (V8-472)	AI		19.3	1.20	Bo	H	Con	Man	CN	.579	Roc	4MV	550	DR	1111239	Car	5B-550	
1970																		
American Motors Corporation																		
Hornet (6-199)	EM										Car	YF-4768S	600	DR	1110481	Man	TC-650	
Hornet, Javelin, Rebel, Ambassador (6-232)	EM										Car	YF-4768S	600	DR	1110481	Man	TC-650	
Ambassador (V8-304)	AI	EV	19.3	1.25	Bo	In	Con	SNM	EP	.285	AM	ODM-2	650	DR	1112018	Man	TC-900	
AMX (V8-360)	AI	EV	19.3	1.25	Bo	In	Con	SNM	EP	.285	AM	ORA-2	650	DR	1111987	Man	TC-800	4, 5

Make and Model	Type System	Type Pump	Displacement (cu. in. per rev.)	Drive Ratio	Drive Type	Relief Valve	Filter	Air Distribution	Point of Entry	Injection Tube I.D. (in.)	Carb Make	Carb Model	Idle Speed (rpm)	Dist Make	Dist Model	Vacuum Source	Timing (deg.) @ rpm	Section Number
Chrysler Corporation																		
Plymouth																		
Duster, Valiant (6-198)	EM										Car	BBS-4715S[1]		Chr	3438237	CP		
Barracuda, Belvedere, Satellite, Fury (6-225)	EM										Hol	R-4351A[1]	700	Chr	2875822	CP	TC-700	
Satellite, Fury (V8-318)	EM										Car	BBD-4721S[1]	750	Chr	3438255	CP	TC-750	
Duster (V8-340)	EM										Car	AVS-4933S[1]		Chr	3438325	CP		
Cuda, Road Runner (V8-383)	EM										Hol	R-4367A[1]	700	Chr	3438233	CP	TC-700	
Fury (V8-440)	EM										Hol	R-4366A[1]	650	Chr	3438219	CP		4, 6
Dodge																		
Dart, Swinger (6-198)	EM										Car	BBS-4715S[1]	750	Chr	3438237	CP	TC-750	
Challenger, Coronet, Charger (6-225)	EM										Hol	R-4351A[1]	700	Chr	2875822	CP	TC-700	
Coronet, Polara (V8-318)	EM										Car	BBD-4721S[1]	750	Chr	3438222	CP	TC-750	
Dart, Swinger (V8-340)	EM										Car	AVS-4933S[1]	900	Chr	3438317	CP		
Challenger, Coronet, Monaco (V8-383)	EM										Hol	R-4367A[1]	700	Chr	3438231	CP	TC-700	
Coronet, Charger (V8-440)	EM										Car	AUS-4737S[1]	900	Chr	3438222	CP	TC-900	
Chrysler																		
Newport (V8-383)	EM										Hol	R-4370A[1]	750	Chr	3438231	CP	TC-750	
300, New Yorker (V8-440)	EM										Hol	R-4366A[1]	650	Chr	3438219	CP	7½B-650	
Imperial																		
Crown, LeBaron (V8-440)	EM										Hol	R-4366A[1]	650	Chr	3438219	CP	7½B-650	
Ford Motor Company																		
Ford																		
Maverick (6-170)	EM										Car	D0DF-R-9510	750	AL	C9DF-B-12127	C	6B-750	
Falcon, Futura, Mustang (6-200)	EM										Car	D0DF-M-9510	750	AL	D0DF-C-12127	C	6B-750	
Custom, Galaxie 500 (6-240)	EM										Car	D0AF-A-9510	850	AL	C8AF-A-12127		6B-850	
Fairlane, Torino (6-250)	EM										Car	D0ZF-E-9510	675	AL	D0OF-A-12127	C	6B-675	
Torino (V8-302)	EM										AL	D0AF-C-9510	800	AL	D0AF-Y-12127	C	6B-800	
Mustang, Ford (V8-351)	EM										AL	D0AF-9510-E[2]	800[3]	AL	D0AF-H-12127[4]	C	6B-800	
Torino (V8-429)	EM										AL	D0AF-9510-AL	700	AL	FAC-12029-A		6B-700	
Thunderbird (V8-429)	EM										AL	D0SF-E-9510	600	AL	D0AF-12127-Z		6B-600	4, 7
Mercury																		
Montego (6-250)	EM										Car	D0ZF-C-9510	750	AL	D0OF-A-12127	C	6B-750	
Cougar (V8-302)	EM										AL	D0AF-C-9510	700	AL	D0AF-Y-12127	C	6B-700	
Cougar, Cyclone (V8-351)	EM										AL	D0AF-E-9510	800	AL	D0AF-H-12127		6B-800	
Monterey, Marauder, Marquis (V8-390)	EM										AL	D0AF-Y-9510	750	AL	C8AF-12127M		6B-750	
Cyclone (V8-429)	EM										AL	D0AF-9510-AL	700	AL	C9AF-Y-12127	C	6B-700	
Lincoln																		
Continental, Mark III (V8-460)	EM	PD	19.3	1.25	Be	In	Con	SM	EP	.250	AL	D0VF-C-9510	590	AL	D0VF-12127B		10B-600	
General Motors Corporation																		
Chevrolet																		
Nova (4-153)	AI	AV	19.3	1.15	Be	DV	Con	CH	EP	.257	Roc	7040008	750	DR	1110457	C	TC-750	
Camaro (6-230)	AI	AV	19.3	1.15	Be	DV	Con	CH	EP	.257	Roc	7029017	700	DR	1110459	C	TC-700	
Malibu, Biscayne, Bel Air, Impala (6-250)	EM										Roc	7040017	750	DR	1110463	C	TC-750	
Monte Carlo, Caprice, Corvette (V8-350)	EM										Roc	7040113	750	DR	1112001	C	TC-750	
Pontiac																		
Tempest, LeMans, Firebird (6-250)	EM										Roc	7040017	700	DR	1110463		TC-700	
Catalina (V8-350)											Roc	7040071	800	DR	1112008		9B-800	
GTO, Executive, Grand Prix (V8-400)											Roc	7040263	950	DR	1111176		9B-950	
Bonneville (V8-455)											Roc	7040267	950	DR	1112012		9B-950	4, 8
Buick																		
Skylark (6-250); Skylark 350 (V8-350)	EM										Roc	1MV	700	DR	1110439		6B-600	
LeSabre (V8-350)	EM										Roc	2GV	700	DR	1111964	Man	6B-600	
GS (V8-350)	EM										Roc	4MV	700	DR	1111986	Man	6B-600	
GS, LeSabre, Wildcat, Electra, Riviera (V8-455)	EM										Roc	4MV	700	DR	1111984	Man	6B-600	
Oldsmobile																		
F-85, Cutlass (6-250)	EM										Roc	1BV	750	DR	1115184	Por	TC-750	
Cutlass, Delta (V8-350)	EM										Roc	2GC	750	DR	1111976	Por	10-1100	
4-4-2, Delta (V8-455)	EM										Roc	4MC	700	DR	1111982	Por	12B-1100	
Ninety Eight, Toronado (V8-455)	EM										Roc	4MC	575[5]	DR	1111981	Por	8B-1100	
Cadillac																		
Calais, DeVille, Fleetwood (V8-472)	EM										Roc	4MV	600	DR	1111939	C	7½B-600	
Eldorado (V8-500)	EM										Roc	4MV	600	DR	1111939	C	7½B-600	

1 — Excluding the state of California.
2 — Autolite D0OF-K on 351-2V Cleveland engine.
3 — Idle speed (RPM) 750 on 351-2V Cleveland engine.
4 — Autolite D0OF-T-12127 on 351-2V Cleveland engine.
5 — Toronado 600 RPM.
For Volkswagen and Capri, see the car section.
A—After top dead center
AI—Air injection
AL—Autolite
AM—American Motors

AV—Semi-articulated vane
B—Before top dead center
Be—Belt
BB—Ball and Ball
BP—Belt and pulley
C—Carburetor
Cac—Carburetor air cleaner
Car—Carter
Cas—Carburetor air stream
Cen—Centrifugal
CH—Cylinder head
Chr—Chrysler

C-M—Carburetor air stream and intake manifold
CP—Carburetor port
D-P—Delco-Remy or Prestolite
DR—Delco-Remy
DV—Diverter valve
EM—Engine modifications
EP—Exhaust ports
EV—Eccentric vane
Fa—Ford
Hol—Holley
In—Integral
Man—Manifold

N—None
NA—Not available
P—Pressure
PD—Positive displacement
Pe—Polyurethane element, replaceable
P-M—Port in carburetor or manifold
Po—Poppet
Por—Ported
Pp—Pressure sensitive poppet
Ppr—Pleated Paper, replaceable
Pres—Prestolite
Psl—Poppet, spring-loaded

R-C—Rochester or Carter
Roc—Rochester
SHM—Separate header manifold
SM—Supply manifold at rear of engine to integral gallery in head
Sld—Spring-loaded disc
Str—Stromberg
TDC, TC—Top dead center
Tsu—Thermal sensing unit
VP—Vane pump

Exhaust Emission Systems

In the past few years, development of exhaust emission controls has been stimulated by a growing federal and public concern about air pollution. Regulations are currently in effect prescribing the maximum allowable emission of unburned hydrocarbons (HC) and poisonous carbon monoxide (CO) in engine exhaust. Emissions from engine crankcase and fuel supply are also restricted. In future years, further regulations will be enforced concerning emission of oxides of nitrogen (NO_x).

To comply with government regulations concerning vehicle exhaust, many engine alterations had to be made. All of these changes are intended to effect more complete burning of the fuel. The result should be a substantial reduction in unburned hydrocarbons and other noxious fumes in terminal exhaust. This is the prime reason for modification.

The highly complex nature of the subject matter prevents any simple solution to the case. Rather, it requires that more sophisticated devices be incorporated into the system to achieve tighter controls over terminal exhaust levels.

To provide best vehicle operation with lowest emission of harmful by-products, the car manufacturer's specifications and maintenance recommendations must be followed to the letter.

To ensure the high state of control needed to minimize terminal exhaust problems and still preserve safe and satisfactory performance levels, the following engine adjustment and control factors must be maintained:

A. Engine idle speed.
B. Ignition timing.
C. Engine idle air-fuel ratio.
D. Vacuum advance control valve (where used).
E. Air injector pump and circuit (where that type system is used).

Because this information is both vital and relative to engine and engine application rather than to engine alone, pertinent tune up data will be found posted in a conspicuous place in the engine compartment of all American built passenger cars beginning with 1968.

Many service procedures, tests, and adjustments have to be changed from those of previous years, particularly in the areas of fuel induction and ignition. To maintain the required results, the fuel-air ratio must not be overly rich and the engine must be kept in good operating condition.

Emission Systems Diagnosis Guide

Symptoms—Rough Idle
Fuel System
1. Leak in vacuum system (lines, connections, diaphragms, etc.).
2. Engine idle too low.
3. Idle fuel mixture incorrect.
4. Air leaks at manifold, carburetor, etc.
5. Power valve leaking fuel.
6. Idle fuel system air bleeds or fuel passages restricted.
7. Secondary throttle not closing (4-V carburetor).
8. Float setting incorrect.
9. Hot and cold air intake system stuck in Heat On position.

Ignition System
1. Defective or poorly adjusted points.
2. Poorly functioning spark plugs.
3. Incorrect ignition timing.
4. Insufficient secondary voltage at plug wires.

Exhaust System
1. Exhaust control valve inoperative (if so equipped).

Engine
1. Leak in vacuum system (lines, connections, diaphragms, etc.).
2. Air leaks at manifold, carburetor, etc.
3. Poor compression (head gasket, exhaust or intake valve leaks, failure of rings, cracked or broken piston, etc.).
4. Inoperative crankcase ventilator (PCV) valve, or restricted tubing.
5. Improperly adjusted valve tappets.
6. Worn camshaft lobes.
7. Incorrect valve timing.

Overheating at Idle
With engine operating, check warning light, or temperature gauge. Trouble could develop in the distributor vacuum control valve (temperature sensing valve).

Engine Stalls
Fuel System
1. Idle speed too low.
2. Idle mixture out of adjustment.
3. Carburetor fast idle too low.
4. Choke out of calibration or adjustment.
5. Choke mechanism binding.
6. Float level too high.
7. Interference with fuel flow.
8. Bad fuel pump.
9. Obstructed fuel lines or tank vent.
10. Carburetor icing.
11. Vapor lock.
12. Inoperative carburetor dashpot.

Engine Noises
Thermactor Pump
The thermactor pump, like any air pump, may produce a detectable sound. This noise should not, however, be of a level audible to anyone within the passenger compartment.

A new air pump, until broken in, may produce a slight squealing or chirping sound.

If pump noise is objectionable, check the following:
1. Drive belt alignment and tension.
2. Loose mounting.
3. Hoses disconnected or leaking.
4. Interference of hoses and body.
5. Defective centrifugal filter fan.
6. Defective relief valve.
7. Improper pressure-setting plug, broken spring, or plug missing. If, after checking the above noise factors, the trouble still persists, replace or repair the pump.

Ignition
1. Ignition, initial timing, too far advanced.
2. Poor engine mounting.
3. Leaking cylinder head gasket.
4. Crankcase ventilator circuit inoperative (stuck open or closed).
5. Poor compression, valves, tappet clearance, or piston and ring assemblies.
6. Worn camshaft lobes.
7. Incorrect valve timing.

Exhaust System
1. Air by-pass valve vacuum line collapsed, plugged, disconnected or leaking, causing backfire.
2. Malfunctioning air by-pass valve, causing backfire.
3. Malfunctioning distributor vacuum advance control valve, causing backfire.
4. Exhaust system leak.

General

The following systems, or variations thereof, are used to ventilate the crankcase of four cycle engines. The many designs range from a simple road draft tube, to multiple and complicated gas emission types, some of which include special ignition and

induction systems. These breathing devices tend to reduce crankcase oil contamination which may result in engine sludging and wear. The prime purpose of the more complex designs is to reduce the discharge of poisonous unburned hydrocarbons into the air we breathe.

No matter what the name, how simple or involved the ventilating system, they all need attention to a varying degree, depending upon design and the amount of work they perform.

Section 1
Road Draft Tube

The only service required for this system is a periodic cleaning of crankcase air intake filters and road draft tubes.

The breathing filter should be thoroughly cleaned with a solvent, then saturated with light oil and blown clean at 6,000 mile intervals (or less, if conditions warrant it).

The road draft tube requires less frequent attention. Driving and engine conditions determine the service intervals to prevent sludge formations in the tube or oil separator.

If there is any evidence of crankcase pressure, (leaking seals, engine sludging, etc.) the tube should be checked as a possible source of trouble. This is always a good precaution on all engine reconditioning jobs.

Type 1 Tests

Symptoms

Symptoms of crankcase ventilation failure of the Type 1 system is heavy engine sludging, excessive oil filter tube smoking, poor rocker arm and tappet lubrication, fouled spark plugs, main bearing and engine seals leaking, etc.

Test

A simple and tentative test of the ventilating system may be made by the following checks:
1. Make sure that the oil filter cap and filter (if so equipped) is clean and clear of obstruction.
2. Admit small quantities of compressed air into the oil filler tube, while noting the ability of the draft tube to exhaust this air. The engine should be at operating temperature for this test.

Section 2
Air Cleaner Type

Air cleaner type crankcase ventilating systems do not use a control valve. Air entering the carburetor, through the air cleaner, picks up

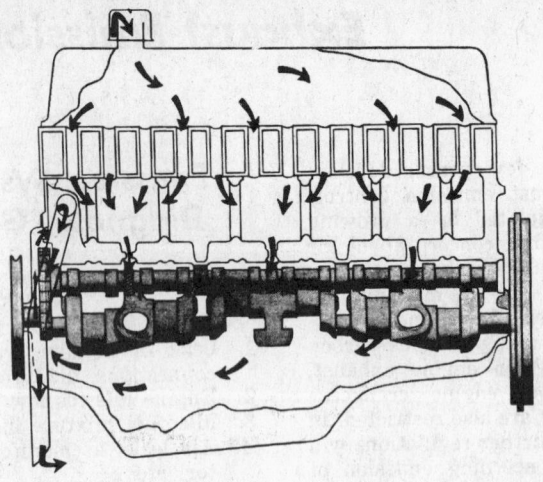

Road draft type of crankcase ventilation

crankcase gases by way of an emission tube. Emission flow is controlled by the amount of air allowed to enter the crankcase via the crankcase breather and cap. The breather cap usually contains a filter.

Service to this type emission system should focus upon the two points of filtration: the crankcase air inlet, and the carburetor air cleaner.

Service intervals should not exceed 4,000 miles.

Air cleaner type system

Type 2 Tests

Symptoms

Symptoms of crankcase ventilating failure of the Type 2 system are the same as with Type 1.

Test

1. Make sure that the oil filler cap and cap filter (if so equipped) are clean.
2. Clean or replace the carburetor air cleaner.
3. Bring engine to operating temperature.

4. With the engine not running, blow small quantities of compressed air into the oil filler Note the ability of the crankcase-to-carburetor air cleaner tube to exhaust these oil fumes and air. A clean system will permit free flow of air.

Section 3
Flow Control (PCV) Valve

Outside air entering the engine via the breather cap passes through the crankcase and is admitted, along with poisonous and otherwise harmful by-pass gases, to the intake manifold for combustion in the engine. Because these gases (air and unburned hydrocarbons) are admitted

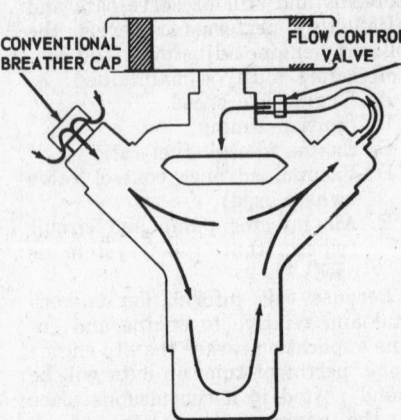

HOSE TO AIR CLEANER   CONVENTIONAL BREATHER CAP   FLOW CONTROL VALVE

Flow control valve type system

to the induction system on the engine side of the carburetor throttle plate, this flow must be controlled. A valve (many are spring loaded) regulates the flow. During idle (low manifold pressure) intake manifold vacuum is high. The high vacuum overcomes plunger weight (or valve spring tension) and seats the plunger

in the valve. With the plunger seated, all of the ventilating air passes through a calibrated orifice in the valve plunger. This results in minimum crankcase ventilation.

As engine speed increases, and manifold vacuum decreases, the plunger weight (or valve spring) unseats the control valve plunger. This increases crankcase air flow and results in maximum ventilation.

Service to this unit is confined to the air breather inlet and the regulator valve. In some cases the valve assembly does not come apart. A faulty valve must be repaired or replaced. Service periods on this type of system depend mainly upon driving and engine conditions; however, service intervals should not exceed 6,000 miles.

Type 3 Tests

Symptoms

Because Type 3 enters the induction system on the engine side of the carburetor throttle plate, fuel-air ratio is involved. A plugged or obstructed ventilating system reveals itself in the same manner as Types 1 and 2, sludging, etc.

An open, or poorly seated, regulator valve upsets the carburetor fuel-air mixture. Therefore, excess crankcase fumes entering the intake manifold at idle speed cause an uneven engine and stalling. The degree of imbalance depends upon the amount of crankcase fumes allowed to leak past the poorly seated regulator valve.

Gauge Test

There are various devices available for testing this equipment, mainly by reading positive or negative crankcase pressures.

Alternate Test

Failure of the Type 3 system falls into two general categories: the system is obstructed (reduced, or no crankcase ventilation) or, there is too much ventilation (regulator valve not seated at engine idle).

The first condition, (not enough ventilation) builds up crankcase pressure and results in sludging and other failures peculiar to Types-1 and 2. An obstructed condition may be detected in the system by observing its ability to pass compressed air and crankcase fumes through the system, (including the regulator valve).

The second condition, too much ventilation, (regulator valve not seating at idle) usually manifests itself in the form of rough engine idle and stalling. This situation is quite common. It may be very misleading because the fuel-air ratio is unbalanced, and resembles a case of intake manifold leak, or carburetor trouble.

To test the second condition:
1. Bring engine to operating temperature.
2. Hook up a tachometer to the idling engine and record the rpm.
3. Disconnect the regulator valve from the line to the intake manifold.
4. Make a plug (see note below) with a metered hole of the same value as that of the disconnected regulator valve. Then insert the plug in the end of the disconnected line to the intake manifold.
5. Compare engine rpm with that which was recorded with the regulator valve in place. Any difference (±20 rpm) indicates a poorly seated regulator valve.

NOTE: make a tapered plug of hard rubber, plastic, hard wood, metal, etc., that will fit into the PCV hose end (or connection) to the intake manifold. Drill a hole, lengthwise, through the tapered plug, the same size as the hole in the end of the regulator valve plunger.

There are various orifice sizes used, depending upon the cubic inch displacement of the engine involved. The following is a general engine size-to-valve plunger orifice chart.

Engine size, cu. in.	Orifice size,	Approximate
90-140	.056-.075 in.	1/16 in.
141-200	.074-.088 in.	5/64 in.
201-275	.087-.099 in.	3/32 in.
276-350	.098-.112 in.	7/64 in.
351- up	.111-.125 in.	1/8 in.

When properly inserted, this tool will admit the same amount of air into the induction system as the standard equipment regulator valve, in good operating condition. Five test plugs (of various orifice sizes) are enough to test all popular American made passenger car engines for regulator valve condition at engine idle.

Section 4
Fully Closed
Ventilating Systems

This system breathes filtered air through a tube running from the carburetor air cleaner to the crankcase (a closed cap is used on oil filler tube). The emission side of the system is of the flow control valve type or a variation thereof. A flame ar-

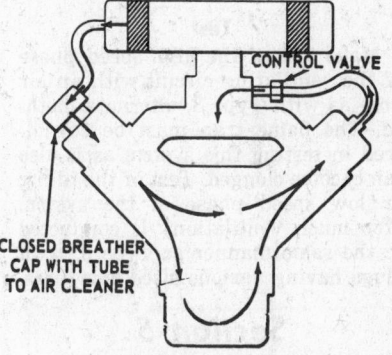

Closed ventilation system

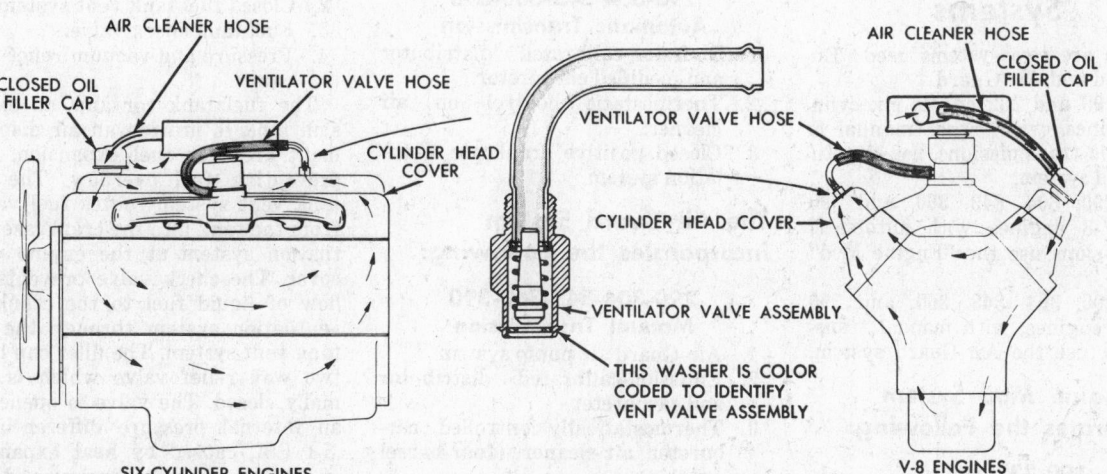

SIX-CYLINDER ENGINES   V-8 ENGINES

Fully closed ventilation system
(© Chrysler Corp.)

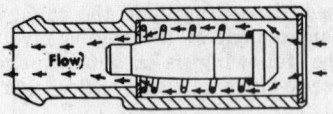

PCV valve (typical)

restor may be incorporated into the inlet side of the system. Being a fully closed ventilating system, don't forget, when checking engine oil, to replace the dip stick all the way, to completely seal the dip stick opening.

Service to the fully closed system should not exceed 6,000 mile intervals.

Type 4 Tests

Symptoms

This system is fully closed and depends upon the carburetor air cleaner for crankcase inlet breathing. While the engine is idling, or at low throttle operation, system failures are potentially the same as Type 3. However, at part throttle or high speed operation, any restriction of air flow into the system (dirty air cleaner or flame trap) may cause negative crankcase pressures. This condition may tend to cause premature oil failure and sludging.

Test

Make test of the high speed phase of the ventilating circuit with an air hose as with Type 3, alternate method. The flame trap must be considered in testing this system as it also can become clogged. Test of the idling or low speed phase of the system (too much ventilation) is conducted in the same manner as Type 3, with plugs having various sized apertures.

Section 5
American Motors
Systems

There are two systems used: Engine-Mod and Air-Guard.

The 199 and 232 cu. in. six cylinder engines, with either manual or automatic transmissions, use the Engine-Mod system.

The 290, 304, 343, 360, and 390 C.I.D. V-8 engines with automatic transmissions use the "Engine Mod" system.

The 290, 304, 343, 360, and 390 CID V8 engines, with manual transmissions, use the Air-Guard system.

The Engine Mod System Incorporates the Following:

199-232 cu. in.
1. Low-quench combustion chamber.

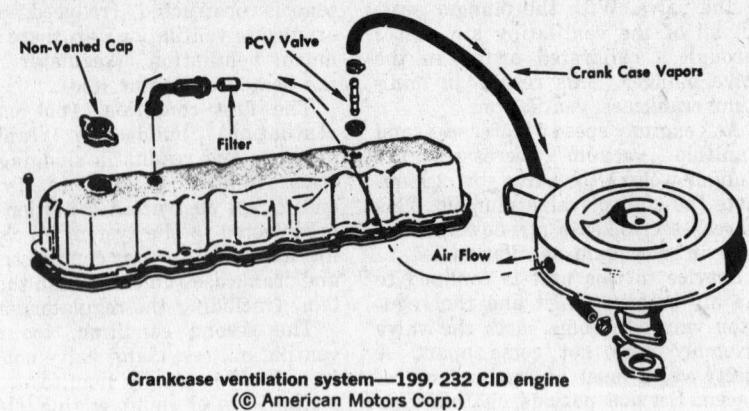

Crankcase ventilation system—199, 232 CID engine
(© American Motors Corp.)

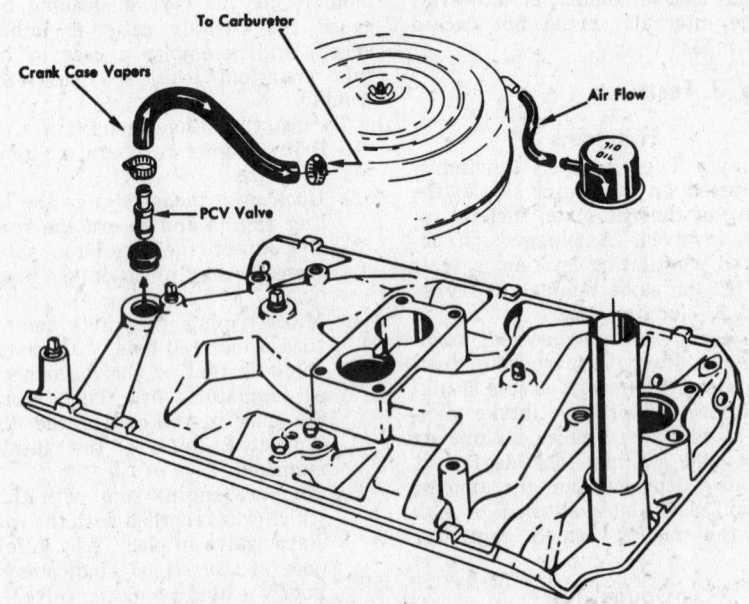

Crankcase ventilation system—290, 343, 390 CID engine
(© American Motors Corp.)

2. Emission-calibrated distributor and modified carburetor.
3. Closed positive crankcase ventilation system.

290-304-343-360-390 Automatic Transmission
1. Emission-calibrated distributor and modified carburetor.
2. Thermostatic control on air cleaner.
3. Closed positive crankcase ventilation system.

The Air-Guard System Incorporates the Following:

290-304-343-360-390 Manual Transmission
1. Air-Guard air pump system.
2. Emission-calibrated distributor and carburetor.
3. Thermostatically-controlled carburetor air-cleaner (four-barrel only).
4. Closed positive crankcase ventilation system.

1970 Models

All 1970 vehicles sold in California have an evaporative emission control system incorporating the following:
1. Fuel expansion tank.
2. Closed fuel tank vent system.
3. Fuel tank check valve.
4. Pressure and vacuum relief filler cap.

The fuel tank contains an expansion tank to provide an air displacement area for fuel expansion, thus preventing fuel overflow. The fuel tank vent system routes fuel vapors from the tank into the crankcase ventilation system at the engine valve cover. The check valve prevents the flow of liquid fuel to the crankcase ventilation system through the fuel tank vent system. The filler cap has a two way relief valve which is normally closed. The valve is opened by an internal pressure differential of .5-1 PSI, caused by heat expansion, or by an external pressure differential of .25-.5 PSI, caused by fuel usage.

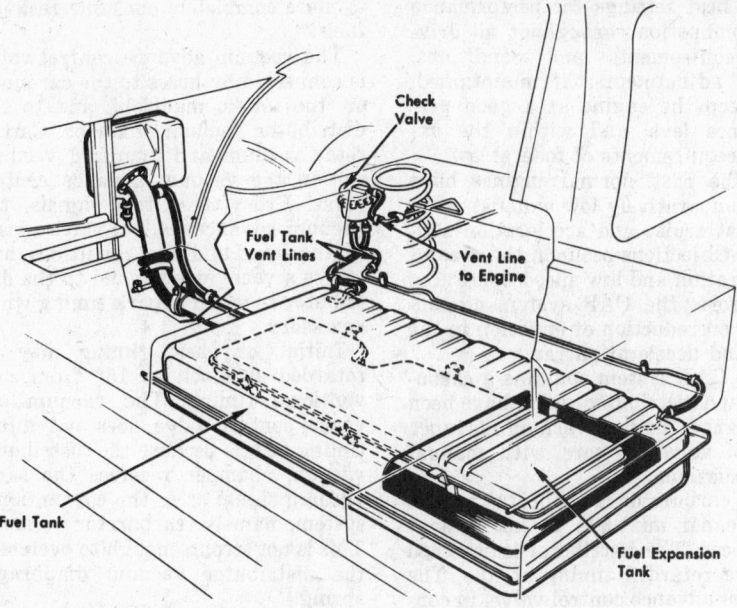

Evaporative emission control system (© American Motors Corp.)

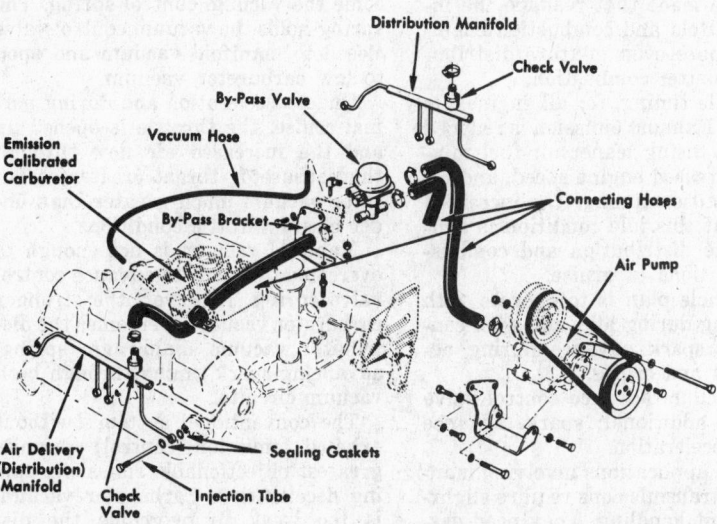

"Air-Guard" System

Air Pump

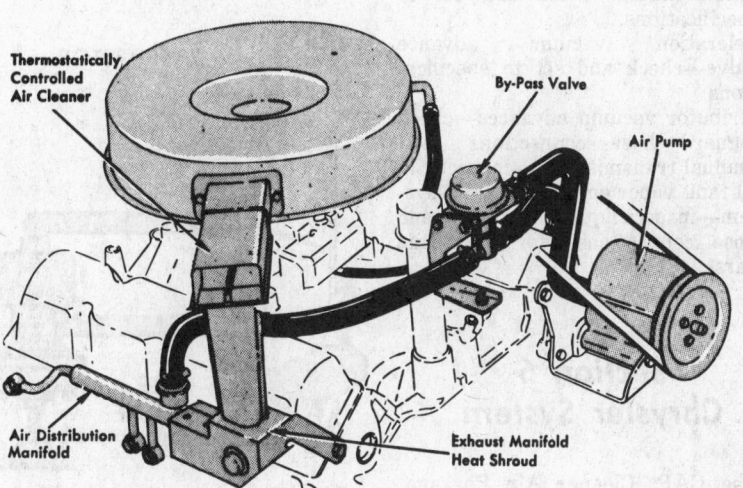

Air-Guard system—290, 343, 390 CID with manual transmission

A deceleration valve has been added to 199, 232, and 390 engines with manual transmission. The function of this valve is to provide full distributor vacuum advance on deceleration.

Idle Speed and Mixture Adjustments

In common practice with other American car manufacturers, American Motors Corporation cars have pertinent tune-up data particular to specific engine-transmission requirements posted in a conspicuous location in the engine compartment.

The following adjustment procedures must be performed exactly as outlined to obtain correct idle speed and mixture settings.

1. On carburetors without plastic idle mixture limiter cap(s), turn mixture screw(s) clockwise until seated lightly, then turn screw(s) counterclockwise two turns for a starting point.
2. Start engine and run until operating temperature is reached. Do not idle engine over three minutes at a time. If adjustment is not completed in three minutes, run engine at 2000 rpm for one minute, then return to idle and finish adjustment.
3. If equipped with Air-Guard system, stop engine and disconnect bypass valve air inlet hose. Restart engine.
4. Adjust idle speed to specified rpm.
5. On carburetors without plastic idle mixture limiter cap(s), turn mixture screw(s) counterclockwise (richer) until engine slows. Turn both screws equally. Turn screw(s) clockwise (leaner) until rpm increases, then continue turning clockwise until rpm decreases. Now turn screw(s) counterclockwise (richer) until

highest possible idle rpm is obtained.

6. On carburetors with idle mixture limiter cap(s), start from full rich stop(s) and turn screw(s) clockwise (leaner) until rpm drops. Turn both screws equally. Now turn screw(s) counterclockwise (richer) until highest possible idle rpm is obtained. If a satisfactory adjustment cannot be obtained, remove idle limiter cap(s) and adjust as in steps 1 and 5. Install new idle limiter cap(s) over idle mixture screw(s) with ear(s) positioned against full rich stop(s). Be careful not to disturb setting while installing cap(s).

7. If idle speed changed by more than 30 rpm during adjustment, reset idle speed and repeat adjustment. Adjust final idle speed.

8. Stop engine. Connect bypass valve air inlet hose.

Services

Required Each 12,000 Miles

Engine oil filler cap (filter type)—clean

Heat valve (exhaust manfiold)—inspect and lubricate

Drive belts (condition and tension)—inspect and correct

Carburetor air cleaner element—clean every 12,000 miles; replace every 24,000 miles

PCV valve—replace

PCV filter (6 cylinder)—clean

Carburetor—inspect and adjust choke, adjust idle speed and mixture to specifications

Spark plugs—inspect, clean and re-gap (replace if required)

Ignition points, coil and spark plug wires—inspect (replace if required)

Distributor cam lubricator—rotate at 12-36-60,000 miles; replace at 24-48-72,000 miles

Ignition timing—check and set to specifications.

Deceleration vacuum advance valve—check and set to specifications

Distributor vacuum advances—check

Air-guard hose connections (V-8 manual transmission)—inspect

Fuel tank vapor emission control system—inspect liquid check valve and hose connections (for California cars)

Section 6
Chrysler System

The CAP, Cleaner Air Package, system is designed to constantly control carburetion and ignition timing at the best settings for performance and combustion throughout all driving requirements and conditions. These adjustments, if maintained, will keep the engine at a good performance level and within the exhaust requirements of federal law.

In the past, normal engines have had comparatively low exhaust emissions at cruise and acceleration attitudes. Objections occur in the area of deceleration and low speed operation. Therefore, the CAP system concentrates on reduction of emission in the idle, and deceleration ranges.

The CAP system contains a carburetor and distributor which have been redesigned, and a vacuum advance control valve on cars with manual transmissions.

The carburetor is calibrated to provide leaner mixtures at idle and at low speed. The distributor is designed to give retarded timing at idle. The vacuum advance control valve, in conjunction with the distributor, provides advanced timing during deceleration.

A number of dimensional changes have been made that reshape the intake manifold and combustion chamber for more even mixture distribution and better combustion.

CAP idle timing, for all engines, is retarded. Exhaust emission is reduced at idle by using leaner air-fuel mixtures, increased engine speed, and retarded ignition timing. The increased air flow at this idle condition is similar to the distribution and combustion conditions of cruise.

The whole plan is to operate with late timing during idle, and with conventional spark advance during acceleration and cruise.

The vacuum advance control valve provides additional spark advance during deceleration.

Engine applications involving manual shift transmissions require slightly different handling. They need earlier ignition of the fuel mixture to accomplish efficient combustion and to more completely consume residual fuels.

The vacuum advance control valve is connected by hoses to the carburetor, the intake manifold, and to the distributor vacuum chamber. Carburetor vacuum and manifold vacuum act on the vacuum advance control valve. From these two signals, the vacuum advance control valve senses engine speed and load conditions, and relays a vacuum command to the distributor to modify spark timing when necessary.

Initial, or basic, timing may be retarded as much as 15° from conventional timing. The vacuum advance control valve does not affect timing at idle because the distributor vacuum chamber receives the same vacuum signal as in the conventional system, namely, carburetor vacuum. This is not strong enough to overcome the distributor vacuum diaphragm spring.

Manifold vacuum acts on the vacuum control valve diaphragm. However, it is not strong enough to overcome the vacuum control spring. The spring holds the vacuum control valve closed to manifold vacuum and open to low carburetor vacuum.

Under acceleration and during normal cruise, the throttle is opened up and the increased air flow through the carburetor throat creates carburetor vacuum much greater than under closed throttle conditions.

Manifold vacuum is not enough to overcome the vacuum advance control valve spring. However, the stronger carburetor vacuum overcomes the distributor vacuum diaphragm spring, advancing spark timing through both vacuum circuits.

The conventional system (without exhaust emission control) permits greatest objectionable emissions during deceleration. Carburetor vacuum is too weak to overcome the distributor vacuum advance diaphragm spring.

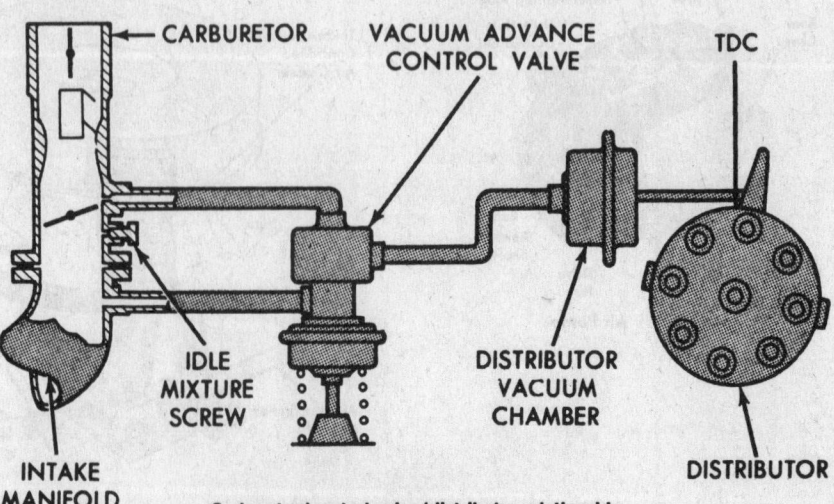

Carburetor/control valve/distributor relationship

Vacuum advance control valve

Manifold vacuum is at its strongest during deceleration. Therefore, the CAP system, when equipped with a vacuum advance control valve, uses manifold vacuum instead of carburetor vacuum to control spark timing. To summarize: during deceleration, manifold vacuum is strong enough to overcome the vacuum advance control valve spring and the distributor vacuum diaphragm spring, moving the spark timing to maximum advance.

1970 Models

The CAP system has been improved to CAS, Cleaner Air System. This system utilizes:
1. Heated air intake system.
2. Modified carburetor.
3. Lower compression ratio.
4. Solenoid retarded distributor.

The heated air system is used on all engines except 340, 426 Hemi, 440 Six-Pack, and those with the fresh air scoop option. When air temperature is below 10° F, intake air flows through a manifold heat stove into the air cleaner. When air temperature is between 10 and 100°F, intake air is a mixture of heated and unheated air. The volume of heated and unheated air is regulated by a thermostat, a vacuum diaphragm, and an

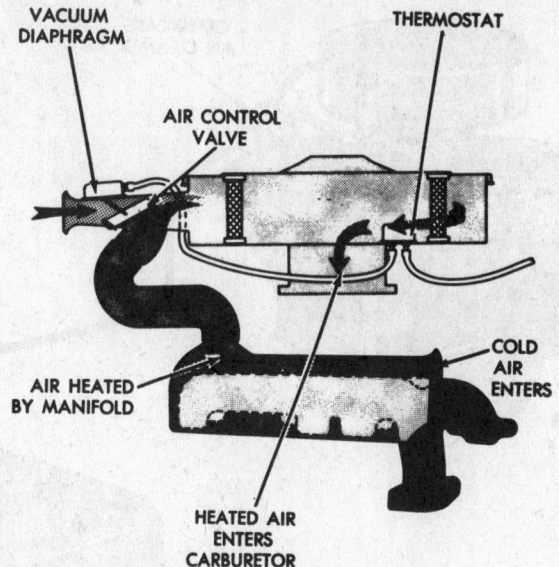

Heated air intake system
(© Chrysler Corp.)

air control valve in the air cleaner housing. The thermostat regulates the air control valve opening. Under conditions of greatly reduced vacuum, such as a sudden burst of acceleration, the vacuum diaphragm overrides the thermostat to open the air control valve fully to unheated air.

Carburetors have leaner mixtures and external idle mixture limiting devices. The automatic choke has been redesigned to release more quickly on engine warmup. An electrical solenoid throttle stop is used on 340, 440, 440 Six-Pack, and 426 Hemi engines. These engines use high idle speeds to achieve acceptable emission levels. The solenoid throttle stop de-

energizes when the ignition is switched off, allowing the throttle blades to close more completely. This prevents running on.

318, 383, and 440 (except 440 Six-Pack) engines have lower compression ratios to reduce hydrocarbon emissions.

All 383 and 440 (except 440 Six-Pack) engines have a solenoid in the distributor advance mechanism to retard ignition timing when the throttle is closed. This system is inactivated during cold starting. Timing must be set at closed throttle to give full retard.

All 1970 vehicles sold in California have an Evaporation Control System,

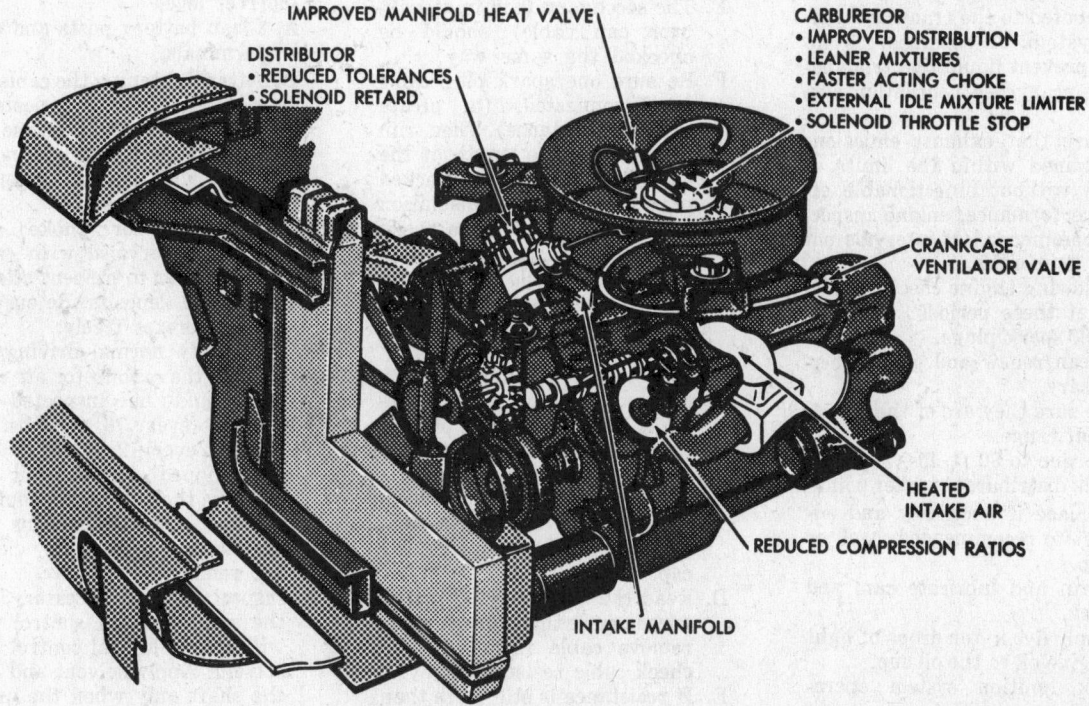

Cleaner air system features (© Chrysler Corp.)

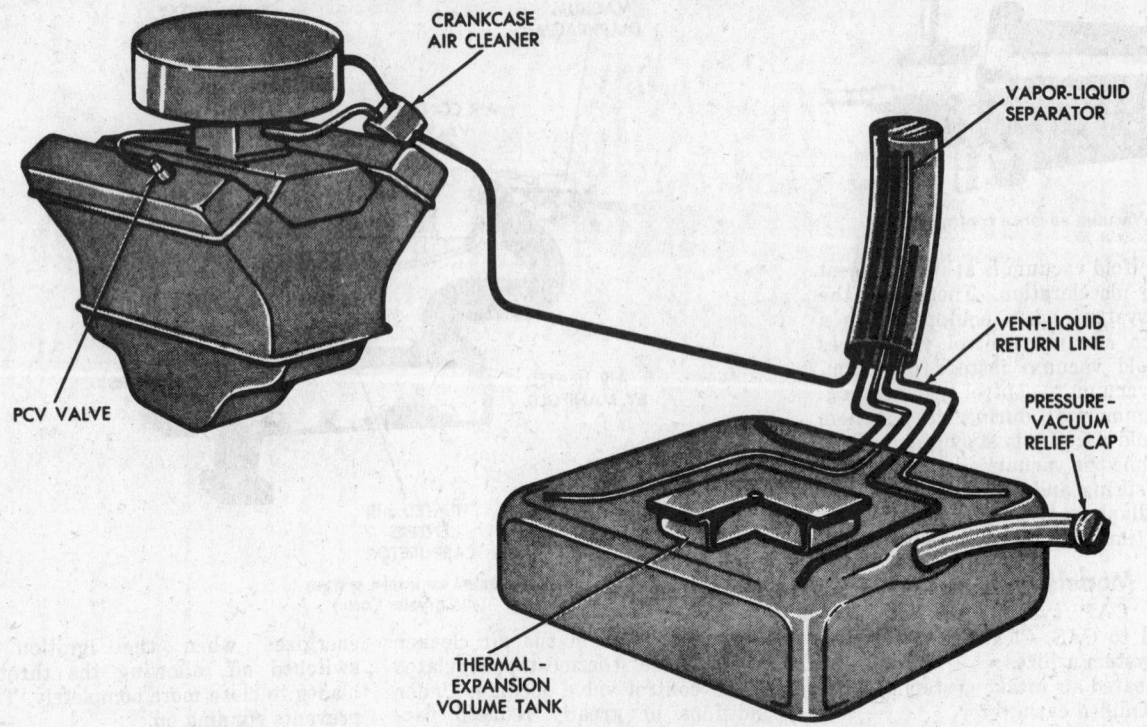

CRANKCASE
AIR CLEANER

VAPOR-LIQUID
SEPARATOR

VENT-LIQUID
RETURN LINE

PCV VALVE

PRESSURE -
VACUUM
RELIEF CAP

THERMAL -
EXPANSION
VOLUME TANK

Evaporation control system (© Chrysler Corp.)

ECS, to reduce evaporation losses from the fuel system. The system has an expansion tank in the main fuel tank. This prevents spillage due to expansion of warm fuel. A special filler cap with a two way relief valve is used. An internal pressure differential of .5-1 PSI, caused by thermal expansion, opens the valve, as does an external pressure differential of .25-5 PSI, caused by fuel usage. Fuel vapors from the carburetor and fuel tank are routed to the crankcase ventilation system. A separator is installed to prevent liquid fuel from entering the crankcase ventilation system.

To insure that exhaust emissions are maintained within the limits of legislation without objectionable effects on performance, engine inspection is recommended at intervals not to exceed 12 months.

The following engine checks should be made at these periods:

1. Check spark plugs.
   A. Clean/renew and gap as necessary.
   B. Be sure they are of the proper heat range.
   C. Torque to 30 ft. lbs.
2. Check distributor breaker points.
   A. Replace if necessary and adjust to recommended dwell or gap.
   B. Clean and lubricate cam and wick.
   C. Apply five to ten drops of light engine oil to the oil cup.
3. Check ignition system operation.

A. Connect one end of test probe to a good ground.
B. Disconnect secondary ignition cable at spark plug end.
C. Insulate secondary cable end from ground.
D. With engine idling, move test probe along entire length of cable. If cable insulation is breaking down, there will be a spark jump from the point of trouble to the probe.
E. The secondary (coil to distributor cap cable) should be checked the same way.
F. Be sure one spark plug cable is disconnected, (to create greater resistance). Then, run probe along the length of the coil secondary cable. Cracked or otherwise poor insulation should be obvious by a spark jump at the point of probe. Bad cables should be replaced.
4. Resistance of secondary cables should also be checked. Use an ohmmeter for this purpose.
   A. Remove cable from spark plug and install proper adapter between cable and spark plug.
   B. Remove cap from distributor, (secondary cables attached).
   C. Connect ohmmeter between plug adapter and the corresponding electrode inside the cap.
   D. Read the ohmmeter. If resistance is more than 30,000 ohms, remove cable from cap and check cable resistance only.
   E. If resistance is still more than 30,000 ohms, replace cable. If

resistance is much less than 30,000 ohms, clean cap wire towers or renew the cap.
5. Inspect ignition coil for indications of oil leakage. Any indication of oil leakage justifies replacing the coil.
6. Test battery with a hydrometer. Specific gravity should be 1.220 or more, with temperature corrections. Add mineral-free water if required, to bring fluid up to correct level.
   A. Clean battery posts and cable terminals.
   B. After tightening the cable post clamps, coat battery posts and clamps with light grease.
   C. Reseal any breaks around the posts with rubber cement to retard corrosion.
7. The carburetor choke shaft should be serviced with carburetor cleaner to prevent sticking from gum deposits. Be sure the choke operates freely.
8. To satisfy normal driving conditions, the carburetor air cleaner should be inspected and cleaned every 6 months. Replace it every 2 years.
   If equipped with an oil bath cleaner, the oil level should be checked every 6 months, and should be thoroughly cleaned and refilled once a year.
9. Inspect, and if necessary, free the manifold heat control valve with manifold heat control valve solvent. Apply solvent and work the shaft only when the unit is cold.

10. A very critical and much neglected item of engine performance is the crankcase emission control valve.

The crankcase emission control valve depends entirely upon the fine balance of spring pressure opposed to manifold vacuum. This balance can very easily be upset by small deposits of varnish or sludge. A closed valve induces crankcase pressure and engine sludge. A poorly seated (open) valve upsets engine idle and results in poor low speed operation.

Inspect and service the closed crankcase ventilating system every 6 months and replace the control valve every 12 months.

A. A tentative check of PCV valve operation is to idle the engine.

B. Disconnect the ventilator valve assembly from the rocker cover. You should be able to hear the valve plunger, when shaking the valve.

C. If the valve is open, a hissing sound will be heard from the valve opening and vacuum can be felt with the finger.

D. If the valve is closed, or the line is obstructed anywhere between the valve and the intake manifold, crankcase pressure will develop. Check for plugging of the hose, freedom of the connection at the carburetor base or a valve stuck closed. Free vacuum passages. Renew the PCV valve if necessary.

E. Another approach is through the use of a tachometer hookup while pinching the PCV hose. If pinching the hose has no effect on engine idle rpm, the system is plugged or the PCV valve is not opening.
If pinching the hose smooths out the idle, the PCV valve is not seating.

11. To adjust curb idle, have the engine warm and running, with a tachometer hooked up. Adjust idle rpm to factory specifications.

A. The air cleaner must be installed, transmission in neutral, and headlights (six cylinder cars only) on high beam. If car is equipped with air conditioner, turn it off.
On cars equipped with a dash pot, adjust the stem so it does not contact throttle lever.

12. Check ignition timing.

A. Disconnect the distributor vacuum line and plug the opening to prevent vacuum leak.

B. Have air cleaner installed, transmission in neutral and headlights (six cylinder cars only) on high beam.

C. Set ignition basic timing to factory specifications.

D. On 1970 models, check distributor solenoid for proper operation by disconnecting wire at carburetor. Timing should advance 5½ degrees and engine speed should increase.

E. Unplug and reconnect vacuum line. Basic timing should remain at factory recommendations.

13. To adjust air/fuel ratio, use a combustion analyzer and tachometer.

A. Warm up the engine thoroughly.

B. Transmission in Neutral, and headlights, (six cylinder cars only) on high beam and engine running at prescribed rpm.

C. Disconnect hose between distributor vacuum control assembly (manual transmission) and intake manifold, at control valve end. Plug end of hose.

D. Insert probe of combustion analyzer into tail pipe as far as possible (minimum of two feet). On dual exhaust cars, use left side tail pipe. If garage exhaust system is used, a plenum chamber or other means must be used to reduce the vacuum of the exhaust system to ½ in. water or less.

14. The air/fuel ratio specified for CAP is 14:1 or higher (air cleaner installed). The ratio for CAS should be 14.2:1.

NOTE: when using the combustion analyzer to adjust air/fuel ratio, do not rotate mixture screws more than 1/16 turn at a time. The ratio must be changed in very small increments. The higher the reading, the leaner the mixture. Idle mixture adjustment limiting caps must not be removed unless idle mixture is being set with a combustion analyzer. Maintain correct idle rpm during mixture adjustments.

A. Adjust mixture screw 1/16 turn richer (counterclockwise) then wait ten seconds for meter to stabilize.

B. Repeat Step A until meter shows a definite increase in richness (less than 14.0).

C. Then adjust mixing screw 1/16 turn leaner (clockwise) until the ratio reads factory recommendations or higher.

D. If idle speed changes as mixture screw is adjusted, readjust independently to obtain smooth idle, providing you stay within the recommended air/fuel ratio range.

F. Unplug and reconnect distributor vacuum control hose. If idle speed changes, adjust

control valve setting.

NOTE: if air cleaner was removed to make mixture screw adjustments, recheck air/fuel reading after installing cleaner. Readjust to specified value or higher.

15. To check setting of distributor vacuum control valve:

A. Warm up engine.

B. Connect a tachometer.

C. Connect a vacuum gauge into the distributor vacuum line. Use a t-fitting with the same inside diameter as the line.

D. If the carburetor is equipped with a dashpot, adjust so that it does not contact the throttle lever at idle speed.

E. Disconnect and plug the hose that connects the distributor vacuum control valve to the intake manifold.

F. Remove the distributor vacuum hose at the distributor and plug the hose. Distributor vacuum must be 0-6 in. Hg. with the engine idling. If reading is higher than 6 in. Hg., recheck idle speed, timing, and air/fuel ratio.

16. To check operation of distributor vacuum control valve:

A. Remove vacuum connection plugs and reconnect vacuum hoses.

B. Remove the distributor vacuum valve cover.

C. Run engine speed up to 2000 rpm and hold for about 5 seconds.

D. Release the throttle. Distributor vacuum should increase to about 15 in. Hg. for a minimum of 1 second, distributor vacuum must then fall below 6 inches within 3 seconds after throttle is released.

17. To adjust vacuum control valve. If it requires less than 1 second or more than 3 seconds to obtain the correct vacuum reading:

A. Turn the adjusting screw counterclockwise to increase the time distributor vacuum remains above 6 in. Hg., or clockwise to decrease the time distributor vacuum remains above 6 in. Hg. One turn of the adjusting screw changes the valve setting about ½ in. Hg. If the valve cannot be adjusted to specifications, it must be replaced.

B. Install vacuum control valve cover and reset carburetor dashpot.

18. A carburetor dashpot is a common name given to a device used to slow down the closing of the throttle when the accelerator pedal is suddenly released. The purpose of the dashpot is to prevent engine stalling.

To adjust the dashpot:

A. With engine idling at recommended rpm and air/fuel mixture properly adjusted, connect a tachometer to the engine.

B. Postion throttle lever so that the actuating tab on the lever contacts the stem of the dashpot, but does not depress it.

C. If adjustment is necessary, loosen the locknut that holds the dashpot to the bracket. Rotate the unit to obtain the desired setting. Tighten the locknut.

D. Recheck distributor vacuum control valve.

E. If distributor vacuum does not fall below 6 in. of Hg. within 4 seconds after throttle has been released, readjust or replace dashpot.

19. Road test the car, noting performance in all driving ranges. If deficiencies exist, they should be corrected before the car is released to the owner.

All adjustments should be in the interest of better performance and economy. However, any changes must comply with CAP specifications and anti-smog legislation.

Caution If contrary or confusing tune-up data is encountered, refer to the decal containing information pertinent to the specific engine-vehicle application. This decal is posted in a conspicuous location in the engine compartment of each vehicle, beginning with 1968 production.

THIS VEHICLE HAS
CHRYSLER CLEANER AIR PACKAGE
TO REDUCE EXHAUST EMISSIONS
REGULAR SERVICE IS REQUIRED TO
MAINTAIN LOWEST EXHAUST EMISSIONS

28898 **43**

IDLE
SETTINGS
RPM = 600 IN NEUTRAL
TIMING = TDC
MIXTURE = 14.0 A./F. RATIO OR HIGHER

Data plate

Section 7
The Ford System

The Ford Motor Company employs two methods of exhaust emission control, the IMCO (improved combustion) and the Thermactor (air injection) types. Both of these types are used extensively throughout their family of engines. The Thermactor type can be readily identified by the belt driven air pump.

The IMCO system produces an acceptable level of emissions by use of carburetor and distributor modifications; no air pump is necessary.

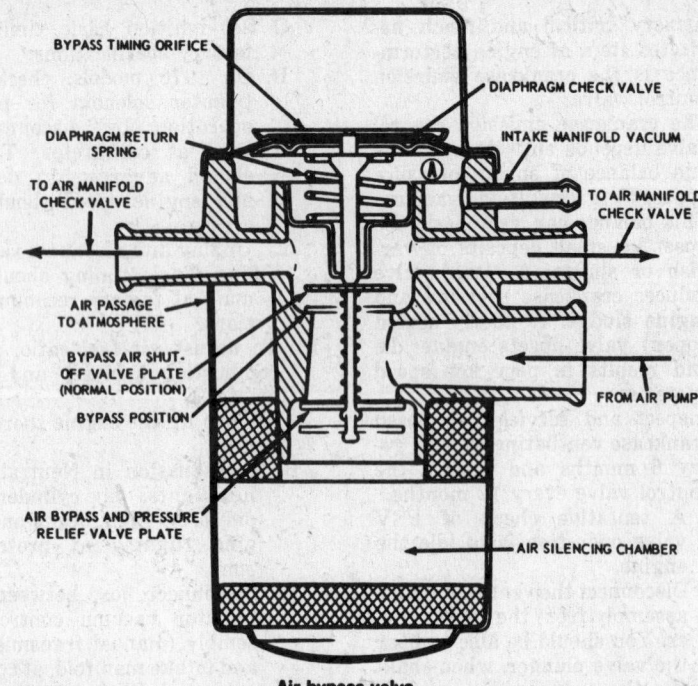

Air bypass valve

With this system, the exhaust gas emissions are reduced in the combustion system rather than by burning them in the exhaust manifolds.

The Thermactor method of exhaust emission control provides injected fresh air into the hot exhaust stream through the engine exhaust ports. At this point, the fresh air mixes with the hot exhaust gases, extending the burning period of these gases. This induced burning lowers the hydrocarbon and carbon monoxide content of exhaust by converting some of it into harmless carbon dioxide and water.

The Thermactor system consists of the following major parts:

1. Air supply pump.
2. Air by-pass valve.
3. Check valves.
4. Internal or external air manifold/s.
5. Air supply tubes (external air manifolds only).

Air for the Thermactor system is cleaned by a centrifugal filter fan mounted on the air pump driveshaft. An element-type air cleaner is not required. A pressure relief valve is installed in the pump housing to prevent air pump outlet pressure from becoming excessive. The pressure setting of the relief valve is controlled by a replaceable plastic plug. Air pump bearings are sealed and lubricated with a lifetime lubricant. Rotor vane and bearing clearances are established during initial assembly. No adjustments are required.

Air supply from the pump is controlled by the air bypass valve. During engine deceleration, the air bypass valve opens and air delivery to the cylinder head ports is momentarily diverted to the atmosphere.

A check valve is incorporated in the air inlet side of the air manifold/s to prevent exhaust backflow into the air pump and air bypass valve during the air bypass cycle. This valve also is utilized in the case of drive belt or pump failure.

Pertinent tune-up data such as engine idle speed, ignition timing, engine idle air/fuel ratio, etc. will be found posted on a decal, in a conspicuous location in the engine compartment of all cars beginning with 1968.

The car manufacturer installs the necessary equipment to control vehicle air pollutants; it is the responsibility of the owner to see that this equipment remains effective by maintaining it to the manufacturers' recommendations. Even a clogged air cleaner may not seem important to the unknowing car owner, but it is most detrimental to proper performance, fuel economy and emission control.

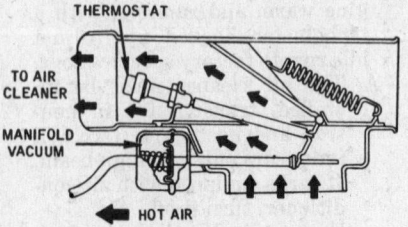

Duct and valve assembly in heat-on position—warm-up

Crankcase Ventilator System

Crankcase ventilator systems are covered earlier in this Unit Repair Section and must be maintained in good repair to complement either the IMCO or Thermactor systems of exhaust emission control.

Ventilator System Test

1. Replace the crankcase ventilator regulator valve with a known good valve of identical design.
2. Start the engine and compare idle conditions with the prior idle conditions. If engine idle is satisfactory (with new valve installed) trouble exists in the old valve and it should be replaced.
3. If loping or rough idle still persists with the good regulator valve in place, the trouble is elsewhere. Check crankcase ventilator system for restriction at the intake manifold or carburetor spacer.
4. If the system is not restricted, explore further the aspects of tune-up and the various elements of exhaust emission control.

AC Ventilator System Test Method

This tester (AC) is operated by engine vacuum through the engine

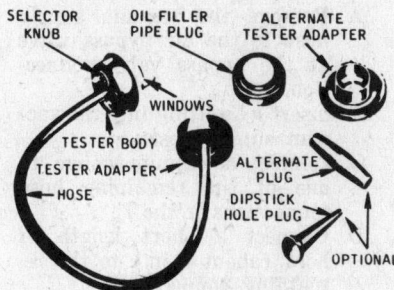

AC crankcase ventilation system tester

oil filler tube opening.

1. With engine operating at normal temperature, connect one end of hose to tester body, the other end to the tester adapter.
2. Use plug supplied with the tester to close the opening in the dipstick tube.
3. Insert tester adapter into the filler cap opening and turn the selector knob to No. 2 for the six cylinder 240 cu.in. and the V8 engines. Turn the knob to No. 4 for the six cylinder 170 and 200 cu. in. engines.
4. Disconnect the air inlet hose at the oil filler tube or oil filler cap and plug the tube (or cap) and hose openings.

5. With plugs secure, and tester tube free of kinks, hold tester body upright and note color in the tester windows. See chart for probable causes of trouble in the crankcase ventilation system.
6. Clean or replace the troublesome components. Then, repeat tests.

Thermactor Diagnosis and Tests

Prior to performing any extensive tests of the Thermactor system:

1. Be sure that a problem exists.
2. Determine that the engine, as a basic unit, is functioning properly by performing the following:

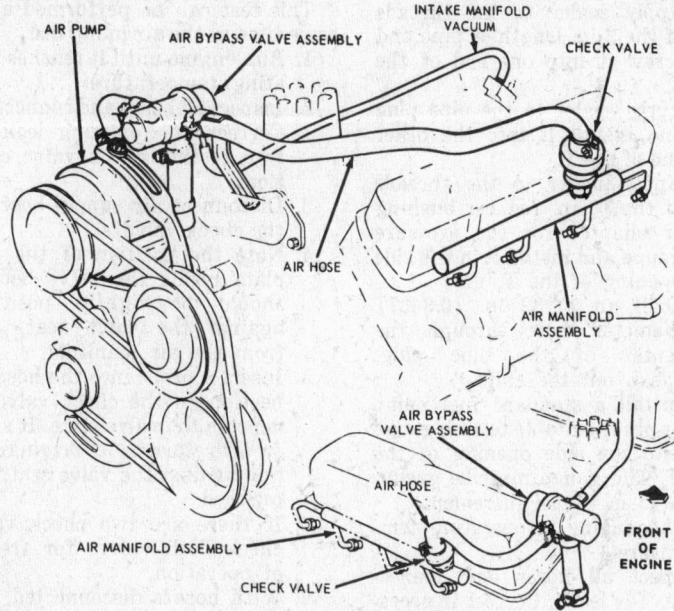

Thermactor installation—V8

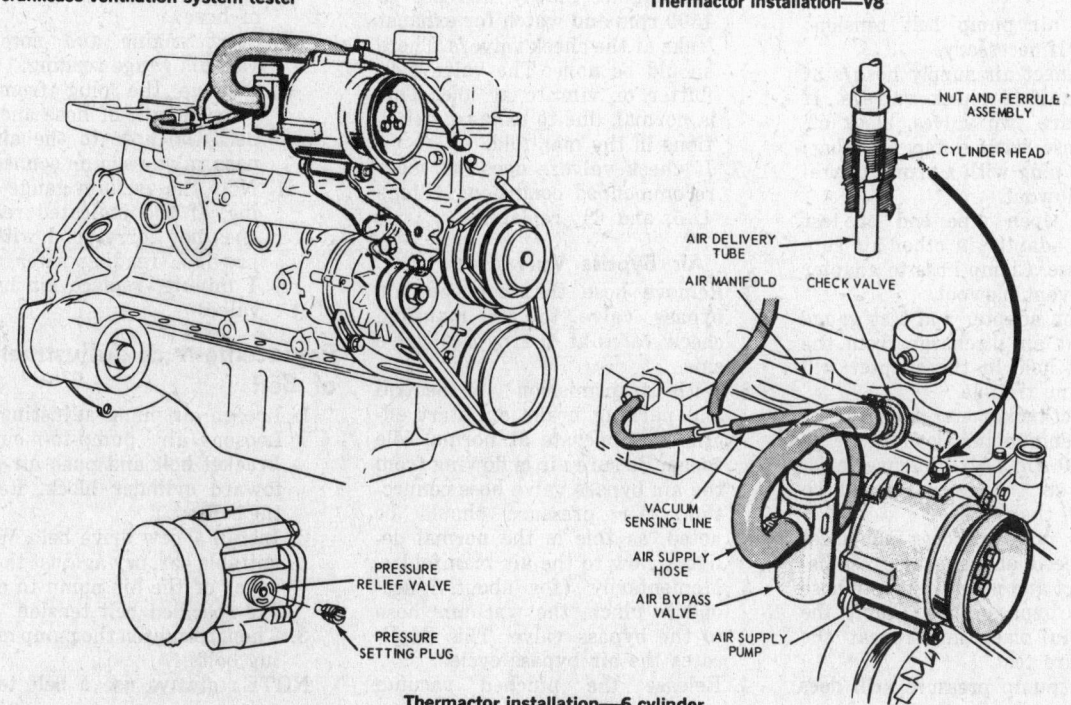

Thermactor installation—6 cylinder

disconnect the air bypass valve vacuum sensing line at the intake manifold. Plug the manifold connection to prevent leakage.

3. Normal engine diagnosis procedures can then be performed.

Air Pump Test

1. To test the air supply pump, a test gauge adapter must be made. Make the adapter as follows:
   A. Obtain a ½ in. pipe T.
   B. A 2 in. length of ½ in. galvanized pipe, threaded at one end only.
   C. A ½" pipe plug.
   D. A ½ in. reducer bushing or suitable gauge adapter.
   E. Apply sealer to the threads of the 2 in. length of pipe and screw it into one end of the T.
   F. Apply sealer to the pipe plug and install it into the other end of the T.
   G. Apply sealer to the threads of the ½ in. reducer bushing or adapter for the pressure gauge and install it in the side opening of the T.
   H. Drill an 11/32 in. (0.3437) diameter hole through the center of the pipe plug. Clean out the chips.
   I. Install a standard fuel pump or other suitable testing gauge into the side opening of the T. The gauge must be graduated in ¼ psi increments.
2. Bring engine to operating temperature.
3. Inspect all hoses and connections for leaks. Correct if necessary.
4. Check air pump belt tension; adjust if necessary.
5. Disconnect air supply hose/s at air manifold check valve/s. If there are two valves, block off one hose with a tapered plug; secure plug with a clamp to prevent blowout.
6. Insert open pipe end of test gauge adapter in other air supply hose. Clamp hose to adapter to prevent blowout.
7. Position adapter and test gauge so that air discharge from the drilled hole in the adapter will cause no trouble.
8. Connect tachometer to engine.
9. Start engine and slowly increase to 1500 rpm. Air pressure registered on the gauge should be greater than 1 psi.
10. If air pressure does not meet or exceed above pressures disconnect and plug air supply hose at the bypass valve. Clamp the plug in place and repeat the pressure test.
11. If air pump pressure still does

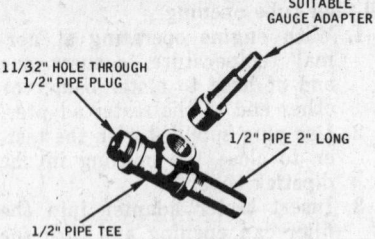

Air supply pump test gauge adapter

not meet minimum requirements, install a new air pump and repeat the pump test.

12. Replace the air pump as determined by the result of this test.

Check Valve Test

This test can be performed at the same time as the air pump test.

1. Run engine until it reaches operating temperature.
2. Inspect all hoses and connections. Correct any existing leaks before testing check valve operation.
3. Disconnect air supply hose/s at the check valve/s.
4. Note the position of the valve plate inside the valve body. It should be lightly positioned against the valve seat, away from the air manifold.
5. Insert a probe into the hose connection on the check valve and depress the valve plate. It should freely return to its original position against the valve seat when released.
   If there are two check valves, check both valves for freedom of operation.
6. With hose/s disconnected, start the engine. Slowly increase to 1500 rpm and watch for exhaust leaks at the check valve/s. There should be none. The valve may flutter or vibrate at idle. This is normal, due to exhaust pulsations in the manifold.
7. If check valve/s does not meet recommended conditions (Steps 4, 5, and 6), replace.

Air Bypass Valve Tests

1. Remove hose that connects air bypass valve to air manifold check valve at the bypass valve side.
2. With transmission in neutral and parking brake on, start engine and operate at normal idle speed. Be sure air is flowing from the air bypass valve hose connection. Air pressure should be noted, as this is the normal delivery flow to the air manifold/s.
3. Momentarily (for about 5 seconds) pinch the vacuum hose to the bypass valve. This duplicates the air bypass cycle.
4. Release the pinched vacuum

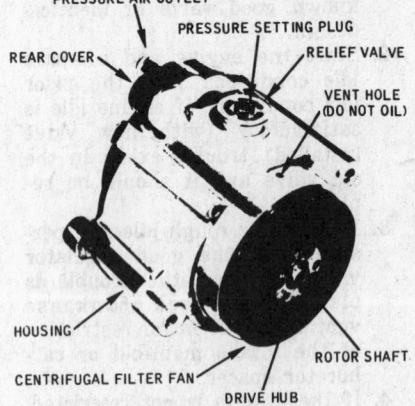

Thermactor air pump

hose. Air flow through the air bypass valve should diminish or stop for a short period. The length of time required to resume normal flow cannot be established. Variables in engine vacuum and the length of time the vacuum line is pinched off are determining factors.

5. Check the by-pass valve for diaphragm leakage by performing the following check:
   A. Remove the vacuum supply hose to the air bypass valve at the bypass valve connection.
   B. Insert a T-fitting into the vacuum supply hose.
   C. Connect a vacuum gauge to one of the remaining hose connections on the T.
   D. Connect a short length of hose (about 3 in.) to the remaining connection.
   E. Insert a suitable plug in the open end of the short length of hose.
   F. Start engine and note the vacuum gauge reading.
   G. Remove the plug from the short length of hose and connect the hose to the air bypass valve vacuum connection.
   H. Note the vacuum gauge reading. If the indicated reading does not correspond with the previous reading after about 1 minute, replace air bypass valve.

Replacement or Adjustment of Belt

1. Loosen air pump adjusting bolt. Loosen air pump-to-mounting bracket bolt and push air pump toward cylinder block. Remove the belt.
2. Install a new drive belt. With a suitable bar, pry against the rear cover of the air pump to obtain the specified belt tension.
3. Then, retighten the pump mounting bolts.

NOTE: always use a belt tension

gauge (T63L-8620-A) to check belt tension, and follow tool manufacturers' instructions and specifications. Any belt that has operated for 10 minutes can be considered a used belt. Adjust accordingly.

Air Bypass Valve Replacement

1. Disconnect air and vacuum hoses at the air bypass valve body.
2. Position the air by-pass valve, and connect the respective hoses.

Check Valve Replacement

1. Disconnect air supply hose at the valve. (Use a 1¼ in. crowfoot wrench, the valve has a standard, right-hand pipe thread).
2. Clean the threads on the air manifold adapter (air supply tube on 289 or 302 V8 engine) with a wire brush. Do not blow compressed air through the check valve in either direction.
3. Install check valve and tighten.
4. Connect air supply hose.

Air Manifold (Except 289 or 302 V8)

Removal

1. Disconnect air supply hose at check valve, position the hose out of the way and remove the valve.
2. Loosen all of the air manifold-to-cylinder head tube coupling nuts (compression fittings).

Cleaning and Inspection

Inspect air manifold for damaged threads and fittings and for leaking connections. Repair or replace as required.

Clean manifold and associated parts with kerosene. Do not dry with compressed air.

Installation

1. Position the air manifold/s on the cylinder head. Be sure all of the tube coupling nuts are aligned with the cylinder head.
2. Screw each coupling nut into the cylinder head, 1-2 threads. Tighten the tube coupling nuts.
3. Install the check valve and torque it to specifications.
4. Connect air supply hose to the check valve.

Air Supply Tube (289 V8)

Removal

1. Disconnect air supply hose at the check valve and position hose out of the way.
2. Remove check valve.
3. Remove air supply tube bolt and seal washer.
4. Carefully remove air supply tube and seal washer from the cylinder head.

Cleaning and Inspection

Inspect air supply tube for evidence of leaking threads or seal surfaces. Examine the attaching bolt head, seal washers and supply tube surface for leaks. Inspect the attaching bolt and cylinder head threads for damage.

Clean air supply tube, seal washers and bolt with kerosene. Do not dry with compressed air.

Installation

1. Install seal washer and air supply tube on cylinder head. Be sure it is positioned in the same manner as before removal.
2. Install seal washer and mounting bolt. Torque to specifications.
3. Install check valve; torque to specifications.
4. Connect air supply hose to the check valve.

Air Nozzle Replacement (Except 289 or 302 V8)

Normally, air nozzles should be replaced during cylinder head reconditioning. A nozzle may be replaced, however, without removing the cylinder head, by removing the air manifold and using a hooked tool.

Clean nozzles with kerosene and a stiff bristle brush. Inspect air nozzles for eroded tips.

Air Pump Filter Fan Replacement

1. Loosen air pump as described in earlier procedures.
2. Remove drive pulley attaching bolts and pull the pulley off the air pump shaft.
3. Pry the outer disc loose, then remove the centrifugal filter fan. Care must be used to prevent foreign matter from entering the air intake hole, especially if the fan breaks during removal. Do not attempt to remove the metal drive hub.
4. Install the new filter fan by drawing it into position with the pulley bolts.

NOTE: some 1966-67 air pumps have air filters with replaceable non-cleanable elements.

Air Pump

Removal

1. Disconnect air inlet and outlet hoses at the air pump.
2. Loosen pump belt tension adjuster.
3. Disengage drive belt.
4. Remove mounting bolt and air pump.

Installation

1. Position air pump on the mounting bracket and install the mounting bolt.
2. Place drive belt in pulleys and attach the adjusting arm to the air pump.
3. Adjust drive belt tension to specifications and tighten adjusting arm and mounting bolts.
4. Connect air inlet and outlet hoses to the air pump.

Air Pump Relief Valve

Replacement

Do not disassemble the air pump on the car to replace relief valve, but remove the pump from the engine.

1. Position tool T66L-9A486-D on the air pump and remove the relief valve with the aid of a slide hammer (T59L-100-B).
2. Position relief valve on the pump housing and hold tool T66L-9A486-B in position.
3. Use a hammer to lightly tap the tool until the relief valve is seated.

Relief Valve Pressure-Setting Plug Replacement

1. Compress locking tabs inward (together) and remove the plastic pressure-setting plug.
2. Before installing the new plug, be sure that the plug is the correct one. The plugs are color-coded.
3. Insert the plug in the relief valve hole and push in until it snaps into place.

Fuel System

Hot and Cold Air Intake

All, except high performance and police interceptor engines use a hot and cold intake system which regulates the temperature of carburetor intake air. This system improves engine performance during warm-up. It reduces the tendency of the carburetor to ice before reaching normal operating temperature.

The heat factor of carburetor intake air is controlled by a valve plate and a vacuum override built into a duct attached to the air cleaner. The exhaust manifold shroud tube is attached to a shroud over the exhaust manifold, which is used as a heat source. The duct has an opening at the outer end to permit entry of cooler air from the engine compartment. A thermostatic bulb within the duct and the vacuum override motor attached to the duct and connected to the thermostat lever provides the means to balance the air temperature for various engine operating conditions.

During warm-up, the thermostat is in the retracted position. The valve plate is held in the heat-on position (up) by the valve plate spring, shutting off the air from the engine compartment. All air is then drawn from the shroud around the exhaust manifold.

During cold acceleration, additional intake air is provided by the vacuum motor control. Decrease in intake manifold vacuum during acceleration permits the vacuum motor to override the thermostat control. This opens the system to both engine compartment air and heated air from the exhaust manifold shroud.

As the temperature of the air passing the thermostatic bulb increases, the thermostat starts to expand and forces the valve plate down. This allows cooler air from the engine compartment to enter the air cleaner. When the air reaches maximum temperature, the valve plate will be in heat-off position (down) so that only engine compartment air is being used.

The 390 GT V8 engine is equipped with a similar duct and valve assembly. It operates in the same manner as on other engines, except that the vacuum override is not used. Instead, a vacuum motor is installed on the outer edge of the air cleaner. When manifold vacuum is low, during engine loading or high-speed operation, a spring in the vacuum motor opens the motor valve plate in the air cleaner. This provides the optimum air supply for greater volumetric efficiency needed for full-power operation.

Carburetor Adjustments

Since 1968 the carburetors on all Ford Motor Company cars are equipped with idle fuel mixture adjustment limiters. These limiters control the maximum idle fuel richness and help prevent unqualified persons from making overly rich idle adjustments. There are two kinds of idle limiters: external and internal.

The external-type plastic idle limiter cap is installed on the head of the idle fuel mixture adjusting screw/s.

Any adjustment made on carburetors having this type of limiter must be within the range of the idle adjusting limiter.

NOTE: under no circumstances should the idle adjusting limiters or the limiter stops on the carburetor be mutilated or deformed to render the limiter inoperative.

The power valve cover must be installed with the limiter stops on the

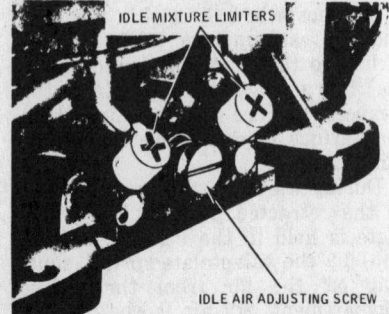

Idle air adjusting screw and mixture limiters
(Carter 4-V)

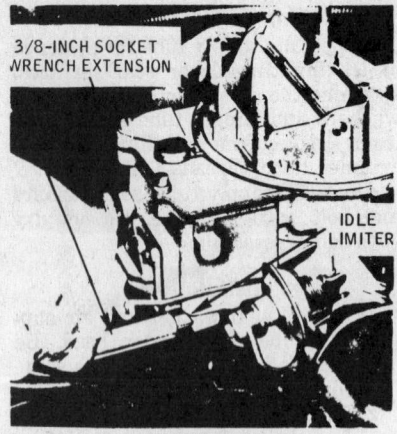

Installing idle limiters

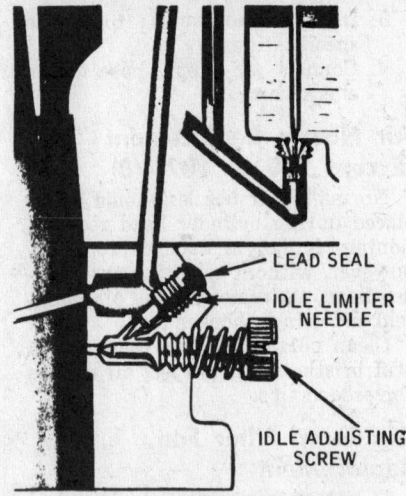

Idler adjusting screw and limiter
(Carter 1-V)

cover in position to provide a positive stop for the tabs on the idle adjusting limiters. A satisfactory idle should be obtainable within the range of the idle adjusting limiters, if all other engine systems are operating within specifications.

The internal needle-type limiter is located in the idle channel and is not externally visible. It is capped by a lead seal. The limiter is installed and sealed during manufacture.

Following are the normal procedures necessary to properly adjust engine idle and fuel mixture. Specific operations should be followed in the proper sequence whenever the idle speed or idle fuel adjustments are made.

Normal Idle Fuel Settings (Engine Off)

1. Set idle fuel mixture screw/s and limiter cap/s to the full counterclockwise position.
   On carburetors without limiter caps, establish an initial idle mixture screw setting by turning the screw inward until it is lightly seated; then turning it outward one to one and one-half turns. Do not turn the screw/s tightly against their seats. Dam-

age may result which will require renewal of the screw/s.
2. Back off the idle speed adjusting screw until the throttle plate/s seat in the throttle bore/s.
3. Be sure the dashpot or solenoid throttle positioner (if so equipped) is not interfering with the throttle lever. It may be necessary to loosen the dashpot or solenoid to allow the throttle plate to seat.
4. Turn the idle speed adjusting screw (except Thunderbird and Lincoln) inward until it just makes contact with the screw stop on the throttle shaft and lever assembly. Then, turn the screw outward one and one-half turns to establish a preliminary idle speed adjustment.
   On Lincoln and Thunderbird, turn the idle air adjusting screw inward until it lightly seats, then turn the screw outward three and one-half turns.
5. Set parking brake before making the idle mixture and speed adjustments. On a car with a parking brake (vacuum) release, remove the vacuum line from the power unit of the vacuum release parking brake assembly. Plug vacuum line, then set parking brake. The vacuum power unit must be deactivated to keep the parking brake engaged when the engine is operated with transmission in Drive.

Normal Idle Fuel Settings (Engine Running)

1. Engine temperature must be at stable operating level before any attempt is made at idle adjustments.
2. Check initial ignition timing and distributor advance and retard.
NOTE: use an accurate tachometer when checking initial timing, idle fuel mixture and speed as outlined in a later paragraph (initial ignition timing).
3. On manual shift transmissions, the idle setting must be made only when the transmission is in Neutral. On automatic transmissions, the idle setting is made with the selector lever in Drive, except as noted when using an exhaust analyzer.
4. Be sure the choke plate is in fully open position.
5. On carburetors equipped with a hot idle compensator, or where the idle compensator is in the crankcase ventilation hose, be sure the compensator is seated to allow for proper idle adjustment.
6. Turn headlights on high beam, to place the alternator under load, in order to properly adjust engine idle speed.

7. Final idle speed is adjusted with the air conditioner turned off.

8. Adjust engine curb idle to specifications. On Thunderbird and Lincoln, readjust idle air by-pass screw, as required, to correct idle speed; tachometer reading to be taken with air cleaner in place. Adjust solenoid throttle positioner to obtain correct idle speed.

9. Turn idle mixture screw/s inward to obtain the smoothest idle possible within the range of idler limiter/s. Turn idle mixture screws inward an equal amount. Check for idle smoothness only with air cleaner installed.

Additional Idle Speed and Mixture Procedures

If a satisfactory idle is not obtained after following the above procedure, additional checks may be necessary.

1. Check the following items and, if necessary, correct.
   A. Vacuum leaks.
   B. Ignition system wiring.
   C. Spark plugs.
   D. Distributor point condition and dwell.
   E. Initial ignition timing.
2. If idle condition is not satisfactory after above items have been checked, perform the following mechanical checks.
   A. Fuel level and fuel bowl vent.
   B. Crankcase ventilation system.
   C. Valve lash (mechanical tappets) and dry tappet clearance, (hydraulic tappets) using the collapsed tappet method for checking.
   D. Engine compression.
3. After going through the above checks, there may still be unsatisfactory performance. The cause may be a lean idle fuel mixture. If this condition is encountered, check the air-fuel ratio with the aid of an exhaust analyzer. With it, adjust the ratio to specifications.

Exhaust Gas Analyzer

The use of the gas analyzer is recommended only after the Normal Fuel Setting Procedures and Additional Idle Speed and Mixture Procedures have been performed, and engine idle is still unsatisfactory.

Except for laboratory equipment, all brand name gas analyzers in common use depend upon thermal conductivity to register.

A typical procedure is as follows:
1. Connect a gas analyzer to the car.
2. If the car is equipped with a Thermactor type exhaust emission system, disconnect the air

injector pump supply hose at the pump or check valve/s. Do not adjust for the drop in engine idle speed, which occurs when the air supply hose is disconnected. However, note the amount of rpm drop for use in Step 4.

3. Observe gas analyzer reading.

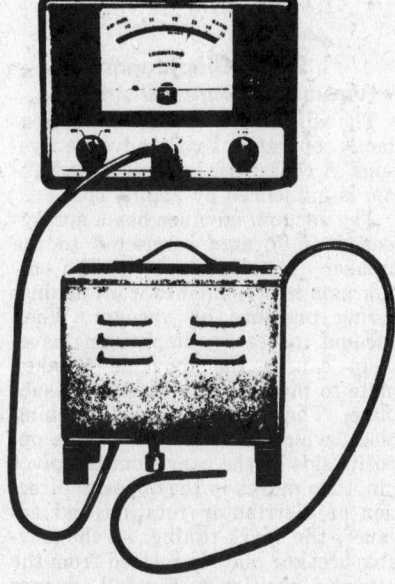

Exhaust analyzer

The value must be taken with air cleaner installed. Refer to specifications for the proper air-fuel ratio.

4. Turn idle mixture adjusting screw/s as required, within the range of the idle limiter, until the required ratio is obtained. On multi-venturi carburetors, turn screws an equal amount. Correct for any changes in engine idle speed immediately, as soon as the idle screw/s are turned. Refer to the drop in engine rpm obtained when the Thermactor air pump hose/s were disconnected in Step 2, then correct idle speed to the rpm noted. Allow at least 10 seconds following each idle mixture screw adjustment for the analyzer reading to properly respond and stabilize.

NOTE: thermal conductivity exhaust gas analyzers indicate a false, rich reading if the air-fuel mixture is extremely lean. To check for this condition, partially hand choke the carburetor, or quickly open and close the throttle a few times to momentarily enrich the mixture. The meter will reflect the false rich condition, then will deflect in the lean direction as the rich condition subsides. The meter will gradually stabilize to a richer reading as the abnormally lean air-fuel ratio is regained.

Cars equipped with automatic transmissions must be in Neutral

while above is being done.

5. If air-fuel ratio is to specifications, and the various engine systems are functioning properly, no further adjustments should be made.

If air-fuel ratio is not to specifications, as indicated by the analyzer, it may be corrected by altering the controlled limits of the carburetor idle fuel system. Refer to the following procedures:

Removal and Installation of Idle Limiter Caps

1. Remove the plastic limiter caps by cutting them with side cutters and a knife. After the cut is made, pry the limiter apart.
2. Adjust carburetor air-fuel ratio, using the exhaust analyzer.
3. When air-fuel ratio is within limits, install a colored plastic service limiter cap.

   When installing the cap, use care not to disturb the idle mixture. Position the cap so that it is in the maximum counterclockwise position with the tab of the limiter against the stop on the carburetor. The idle mixture adjusting screw will then be at the maximum allowable outward, or rich, setting.

   Install service cap by pushing with the thumb or with a 3/8 in. socket extension.
4. Recheck the air-fuel ratio with air cleaner installed, using the exhaust gas analyzer.

Removal and Installation of Lead Seal and Readjustment of Idle Limiter

1. Remove the lead seal covering the idle limiting needle in the throttle body. Pick it out with an ice pick.
2. With idle adjusting needle at maximum rich setting, slowly back out the idle limiter, one-sixteenth turn at a time, until the specified air-fuel ratio reading is obtained on the exhaust analyzer.
3. After obtaining the specified air-fuel ratio reading, install a new lead seal over the idle limiter. Drive the seal into the hole with a small punch until the lead contacts the head of the screw. After the idle limiter has been reset, and the air-fuel ratio and idle conditions are found to be satisfactory, stamp or scribe the letter R on the carburetor identification tag, just above the name Autolite, to indicate that the carburetor has been reworked.

Ignition System

Emission control features in the ignition system are:

1. Dual-diaphragm vacuum advance distributor.
2. Distributor vacuum control valve (coolant temperature sensing).
3. Distributor vacuum advance control valve (deceleration).
4. Distributor modulator.

All distributors are equipped with both vacuum and centrifugal advance units. Vacuum advance governs the ignition timing (spark advance) at low engine speeds (rpm) or low engine loading. The centrifugal advance, in combination with the vacuum advance, controls the ignition timing at higher engine speeds and heavy engine loading to provide the correct ignition timing for maximum engine performance. A dual-diaphragm advance is used on some engines to provide additional ignition timing retardation during engine idle operation.

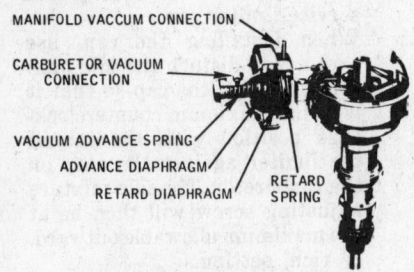

MANIFOLD VACUUM CONNECTION
CARBURETOR VACUUM CONNECTION
VACUUM ADVANCE SPRING
ADVANCE DIAPHRAGM
RETARD DIAPHRAGM
RETARD SPRING

Distributor with dual-diaphragm vacuum advance

The distributor vacuum control valve advances the spark timing under conditions of prolonged idling, thus preventing overheating. The deceleration valve provides advanced timing on deceleration. The distributor modulator prevents spark advance below a set vehicle speed on either acceleration or deceleration.

Single-Diaphragm Vacuum Advance Distributor

The distributor has two independently operated spark advance systems. A centrifugal advance mechanism is influenced by engine rpm.

The vacuum advance has a spring-loaded diaphragm connected to the breaker plate assembly. The diaphragm is activated to work against spring pressure by vacuum. When vacuum increases, diaphragm movement causes the movable breaker plate to pivot on the stationary subplate. The breaker point rubbing block, which is positioned on the opposite side of the cam from the pivot pin, then moves in the opposite direction of distributor rotation and advances the spark timing. As the movable breaker plate is rotated from the retarded position to the full-advance position, the point dwell decreases slightly. This is due to the breaker

point rubbing block and the cam rotating on different axes.

Dual-Diaphragm Vacuum Advance Distributor

On dual-diaphragm distributors, the centrifugal advance unit is the same as on single-diaphragm units. The dual-diaphragm unit consists of two independently operating diaphragms. The advance (primary) diaphragm uses carburetor vacuum to advance ignition timing. The retard (secondary) diaphragm is activated by intake manifold vacuum, which provides additional timing retardation during closed throttle deceleration and idle.

The advance diaphragm (outer) is coupled to the movable breaker plate in a manner similar to that in single-diaphragm distributors. An increase in vacuum moves the diaphragm against the advance diaphragm spring. This action causes the movable breaker plate to oppose distributor rotation; therefore, ignition timing is advanced. This advance is calibrated to occur during normal road-load operation, but not during deceleration or idle.

When intake manifold vacuum is applied to the retard diaphragm (inner), it moves in toward the distributor. This permits the diaphragm

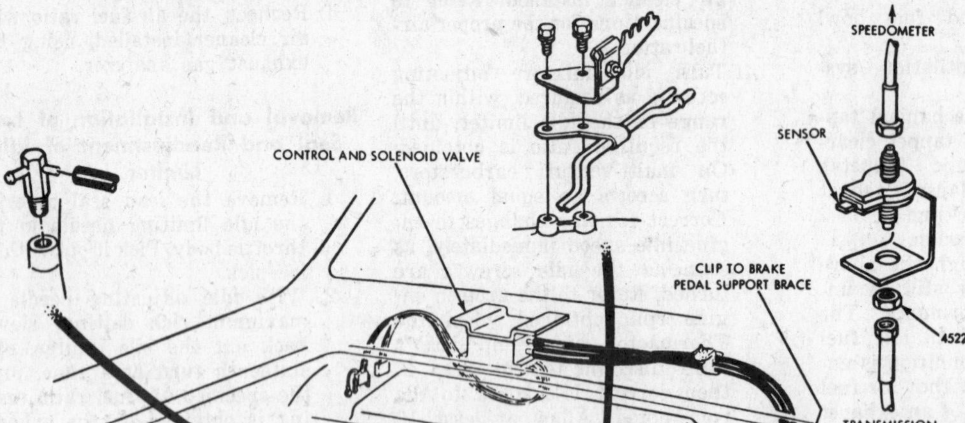

SPEEDOMETER
SPEED CONTROL SENSOR
SENSOR
9E720 ASSY.
45222-S2
TRANSMISSION

CONTROL AND SOLENOID VALVE
CLIP TO BRAKE PEDAL SUPPORT BRACE

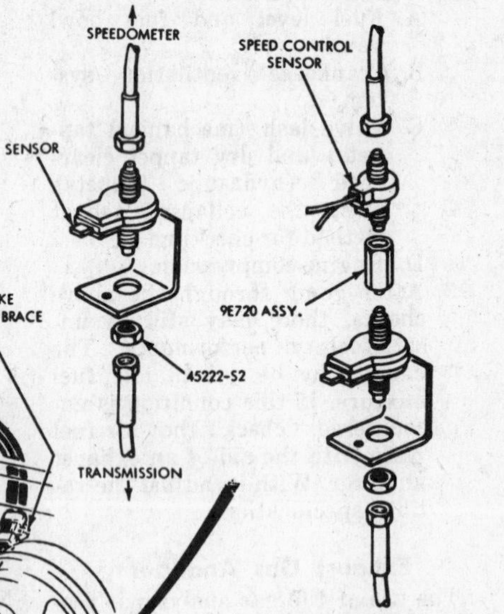

INSTALLATION FOR DISTRIBUTOR MODULATOR AND SPEED CONTROL

Typical distributor modulator installation (© Ford Motor Company)

spring to move the breaker plate in the same direction as distributor rotation. This retardation phase of ignition timing occurs during engine idle or deceleration, except when a distributor modulator is installed in the vacuum supply line.

Distributor Modulator.

The distributor modulator ensures close control of distributor spark advance during acceleration and deceleration. Modulator components are a speed sensor, a thermal switch, an electronic control amplifier, and a three way solenoid valve. All these units operate together to control vacuum to the distributor. The control amplifier and solenoid valve are a single unit mounted inside the passenger compartment. The speed sensor is connected to the speedometer cable. The thermal switch is mounted near either front door hinge pillar on the outside of the cowl panel. The modulator prevents distributor vacuum spark advance until a speed of 21-31 mph is reached when accelerating, and cuts out vacuum advance when a speed of 25-15 mph is reached on deceleration.

Distributor Vacuum Control Valve

The distributor vacuum control valve (temperature-sensing valve), is incorporated into the distributor vacuum advance supply line in order to provide advanced timing during prolonged idling. This valve is installed in the coolant outlet housing in order to sense engine coolant temperatures. Normally, the valve connects two ports: regular source vacuum at the carburetor and the distributor port. During periods of prolonged idle, if engine temperature rises, the valve closes the regular source (intake manifold vacuum) and connects the distributor port to the alternate source vacuum port. On certain engines, the latter port is capped with a filter. The advanced timing causes an immediate increase in engine speed, which will be maintained until engine temperature returns to normal.

Distributor Vacuum Deceleration Control Valve

On some engines, a distributor vacuum advance control valve (deceleration valve) is incorporated into the

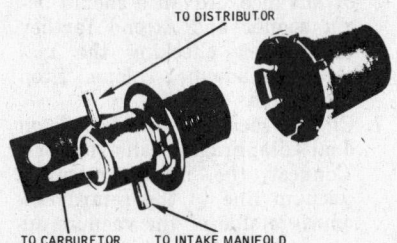

Distributor vacuum advance control valve

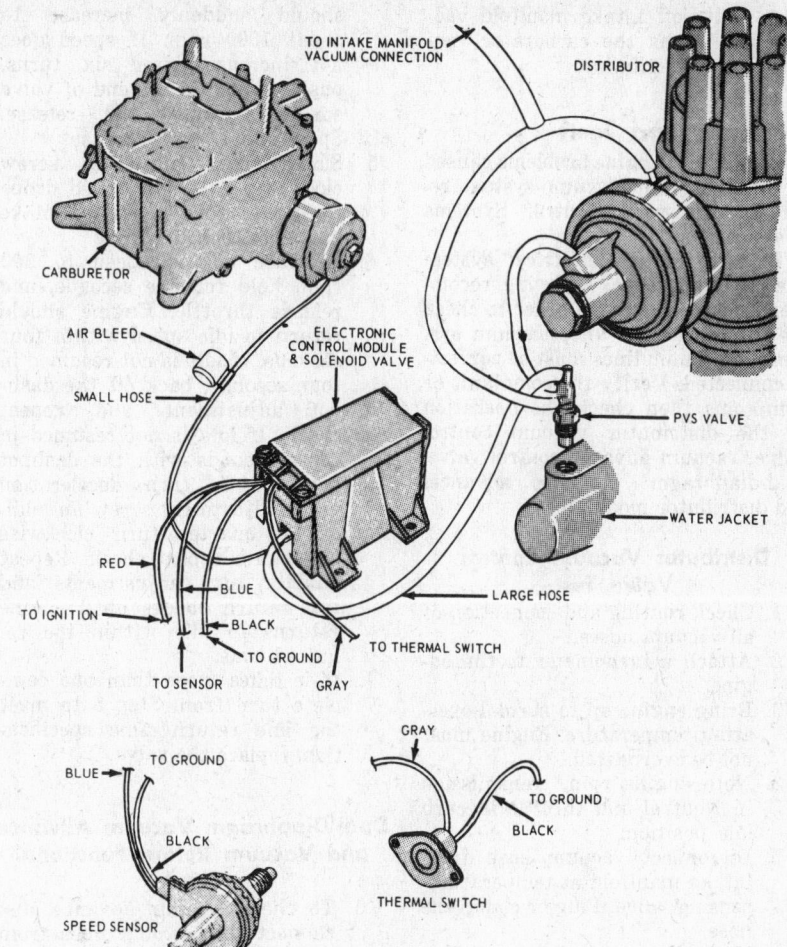

Distributor modulator details (© Ford Motor Company)

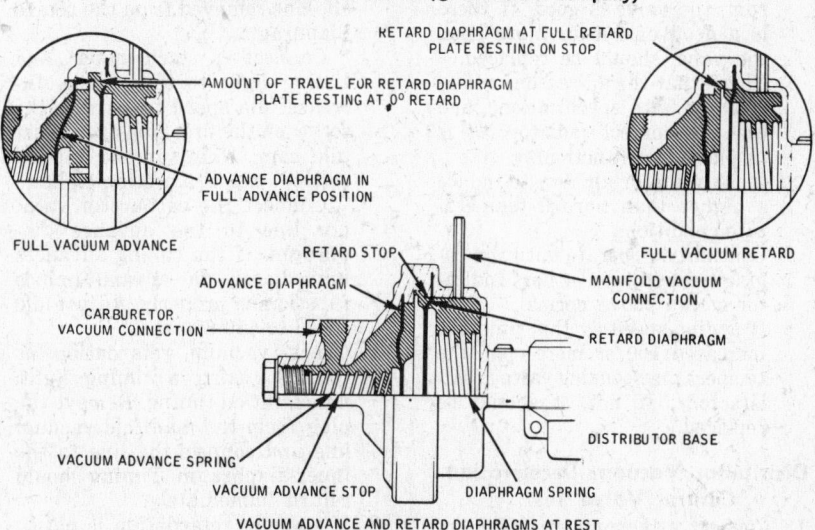

Dual diaphragm vacuum advance mechanism

distributor vacuum system to provide additional control of ignition timing. This device is used in conjunction with the dual-diaphragm vacuum advance unit. Normally, the advance diaphragm (outer) is connected to a vacuum port on the carburetor. During deceleration, when intake manifold vacuum rises, the deceleration valve closes off the carburetor vacuum and provides direct intake manifold vacuum to the distributor advance diaphram (outer). This permits maximum ignition advance to prevent after-burning or popping in the engine exhaust system. When the vehicle slows down and the engine is operating at idle, the deceleration

valve shuts off intake manifold vacuum and opens the carburetor vacuum to the distributor.

Diagnosis and Tests

To pinpoint engine problems caused by the distributor vacuum system, refer to Emission Control Systems Diagnosis Guide.

To test ignition electrical system use an oscilloscope, following recommended procedures. In order to check the distributor advance vacuum system the vacuum lines must be correctly connected. Verify the complaint or symptom; then check the operation of the distributor vacuum control valve, vacuum advance control valve, dual-diaphragm vacuum advance, and distributor modulator.

Distributor Vacuum Control Valve Test

1. Check routing and connection of all vacuum hoses.
2. Attach a tachometer to the engine.
3. Bring engine up to normal operating temperature. Engine must not be overheated.
4. Note engine rpm, transmission in Neutral and throttle in curb idle position.
5. Disconnect vacuum hose from intake manifold at temperature-sensing valve. Plug or clamp the hose.
6. Note idle rpm with hose disconnected. If there is no change in rpm, the valve is good. If there is a drop of 100 or more rpm, the valve should be replaced.
7. Check that the all-season cooling mixture is to specifications, and that the correct radiator cap is in place and functioning.
8. Block radiator air flow to induce a higher than normal temperature condition.
9. Continue to operate until the engine temperature or heat indicator shows above normal.
   If engine speed by this time has increased 100 or more rpm, the temperature-sensing valve is satisfactory. If not, it should be replaced.

Distributor Vacuum Deceleration Control Valve Test

1. Connect a tachometer to the engine and bring engine to normal operating temperature.
2. Check idle speed and set to specifications with headlights on high beam if necessary.
3. Turn off headlights and note idle rpm.
4. Remove plastic cover from valve. Slowly turn adjusting screw counterclockwise without pressing in. After five and no more than six turns, idle speed

should suddenly increase to about 1000 rpm. If speed does not increase after six turns, push inward on the end of valve spring retainer and release. Speed should now increase.

5. Slowly turn adjusting screw clockwise until idle speed drops to speed noted in Step 3. Make one more turn clockwise.
6. Increase engine speed to 200 rpm, hold for five seconds, and release throttle. Engine should return to idle speed within four seconds. If idle is not resumed in four seconds, back off the dashpot adjustment and repeat check. If idle is not resumed in three seconds with the dashpot backed off, turn deceleration valve adjustment screw an additional quarter turn clockwise and again repeat check. Repeat quarter turn adjustments and idle return checks until engine returns to idle within the required time.
7. If it takes more than one complete turn from Step 5 to meet the idle return time specification, replace the valve.

Dual-Diaphragm Vacuum Advance and Vacuum Retard Functional Check

1. To check vacuum advance, disconnect the vacuum lines from both the advance (outer) and retard (inner) diaphragms. Plug the line removed from the retard diaphragm.
   Connect a tachometer and timing light to the engine. Increase idle speed by setting the screw on the first step of the fast idle cam. Note ignition timing setting, using a timing light.
   Connect the carburetor vacuum line to the advance diaphragm. If the timing advances immediately, the advance unit is functioning properly. Adjust idle speed to 550-600 rpm.
2. Check vacuum retardation as follows: using a timing light, note ignition timing. Remove the plug from the manifold vacuum line and connect the line to the inner diaphragm. Timing should retard immediately.
3. If vacuum retardation is not to specifications, replace the dual-diaphragm advance unit. If the advance (vacuum) does not function properly, calibrate the unit on a distributor test stand, following the procedures for dual advance distributors. If the advance part of the unit cannot be calibrated, or if either diaphragm is leaking, replace the dual-diaphragm vacuum advance unit.

Diaphragm Leak Test Dual-Diaphragm Distributors

To check the diaphragm for leakage, place distributor on a test stand. Without connecting the vacuum line to the distributor, adjust the vacuum pressure of the tester to maximum position. Hold a finger on the open end of the tester vacuum hose. Note maximum reading obtained. Do not exceed 25 in. Hg.

If reading is 25 in. Hg. or less, connect tester vacuum line to the vacuum fitting of the advance diaphragm. Without changing any adjustments, the maximum gauge reading should not be less than it was above. If it is less, the diaphragm is leaking and should be replaced.

Repeat the above procedure with the vacuum line connected to the retard diaphragm.

Adjustments

The procedure for setting ignition timing includes functional checks for the advance and retard positions of the diaphragm assembly.

Initial Ignition Timing

1. Clean and mark timing pointer and index marks. Be sure the distributor vacuum lines are properly connected.
2. Disconnect the vacuum line (single-diaphragm distributors) or vacuum lines (dual-diaphragm distributors), and plug the disconnected vacuum line/s.
3. Connect a timing light to No. 1 cylinder plug wire. Hook up a tachometer.
4. Start engine, reduce idle speed to 600 rpm, and adjust initial timing to specifications.
5. Check mechanical (centrifugal) advance for proper operation. Start engine and accelerate to about 2000 rpm. If timing advances, the centrifugal advance is functioning. Note the engine speed when advance begins and how far (in degrees) it advances. Stop the engine.
6. Unplug the carburetor vacuum line and connect it to the distributor vacuum advance unit (outer diaphragm on dual-diaphragm distributors). Start engine and set idle to about 2000 rpm. Note engine speed when advance begins and the degrees of advance. Advance should begin sooner and extend farther than when checking the mechanical advance alone. Stop the engine.
7. Check vacuum retardation on dual-diaphragm distributors. Connect the intake manifold vacuum line to the retard diaphragm side of the vacuum advance. Operate engine at normal idle speed. Initial timing

should retard to zero (TDC). On some engines it will go as low as 6° ATDC.

8. If the vacuum advance or vacuum retard phases (dual-diaphragm distributors) are not working properly (refer to Steps 6 and 7 above), remove the distributor and check it on a distributor tester.

Replace the dual diaphragm unit if the advance portion cannot be calibrated, the retard portion is not to specifications, or either diaphragm is leaking.

9. Unplug vacuum line/s and connect it to the distributor vacuum advance assembly. Remove timing light and tachometer.

Evaporative Emission Control System—1970 Models

Due to regulations regarding gasoline vapor emission in the State of California, all Ford Motor Company automobiles shipped into that state are equipped with an evaporative emission control system for 1970.

The system uses:
1. A non-vented filler cap
2. A fill vent and a vapor vent, both located inside the fuel tank.
3. A fuel expansion tank or vapor separator.
4. A three-way control valve.
5. A carbon canister.

The fill vent and the vapor vent both deliver fuel vapors to the vapor separator. The vapor separator returns any liquid fuel to the fuel tank and vapor to the carbon canister through the three-way valve. The three way valve also relieves vacuum within the fuel tank, caused by fuel consumption or thermal contraction, and pressure, caused by thermal expansion or a blocked vapor line. The carbon canister stores fuel vapors

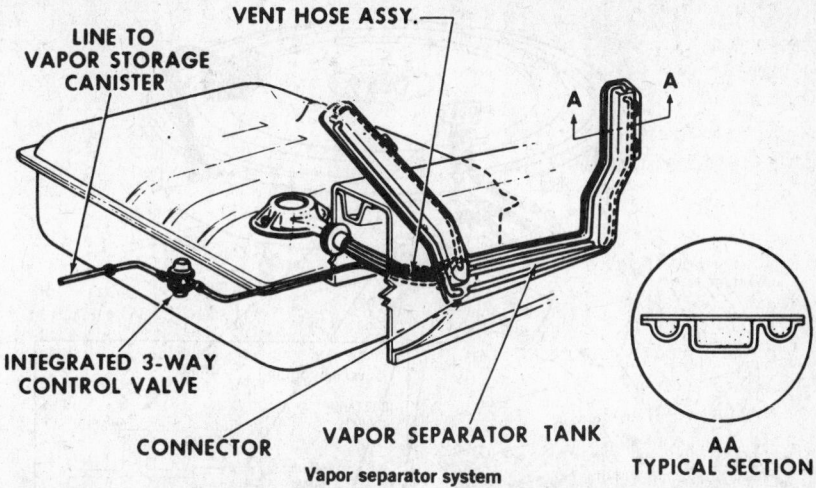

VENT HOSE ASSY.

LINE TO VAPOR STORAGE CANISTER

A A

INTEGRATED 3-WAY CONTROL VALVE

CONNECTOR

VAPOR SEPARATOR TANK

AA TYPICAL SECTION

Vapor separator system

until they are drawn into the engine air cleaner and burned. On vehicles not equipped with a carbon canister, fuel vapor is discharged into the crankcase ventilation system through the PCV valve.

Section 8
General Motors System

Description

The General Motors Corporation has two basic systems of exhaust emission control to satisfy the needs of their various car division requirements. These two systems are the Controlled Combustion System (CCS) and the Air Injection Reaction (AIR) system. These two systems are easily recognized and identifiable by the presence or absence of an air injector pump.

The Controlled Combustion System (CCS) consists of an air cleaner assembly which includes a temperature

sensor, vacuum motor, control damper assembly and the necessary connecting vacuum hoses.

The Air Injection Reactor (AIR) system consists of a belt driven air pump which forces air through manifolds or special heads into the area of the exhaust valves.

Caution

Never attempt to operate a vehicle equipped with the AIR system with the pump drive disconnected.

Controlled Combustion System (CCS)

The Controlled Combustion System (CCS) functions as follows: During engine warm-up with engine compartment temperatures 85°F or below, the temperature sensor is closed. This allows engine vacuum to be directed to the vacuum motor, closing the damper to outside air. With the damper closed, the cool air will flow through the openings at the ends of the shroud, where it is heated. The heated air then flows up through the

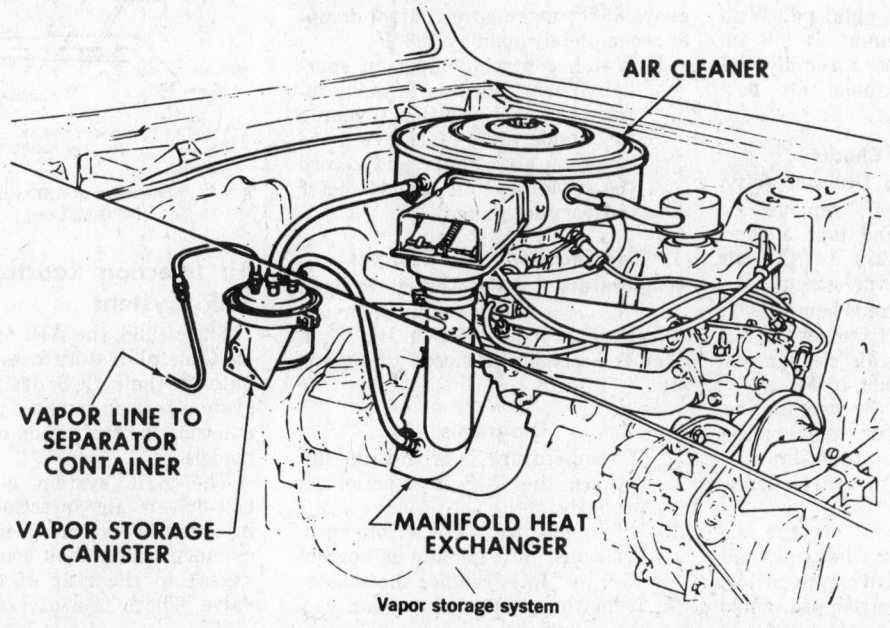

AIR CLEANER

VAPOR LINE TO SEPARATOR CONTAINER

VAPOR STORAGE CANISTER

MANIFOLD HEAT EXCHANGER

Vapor storage system

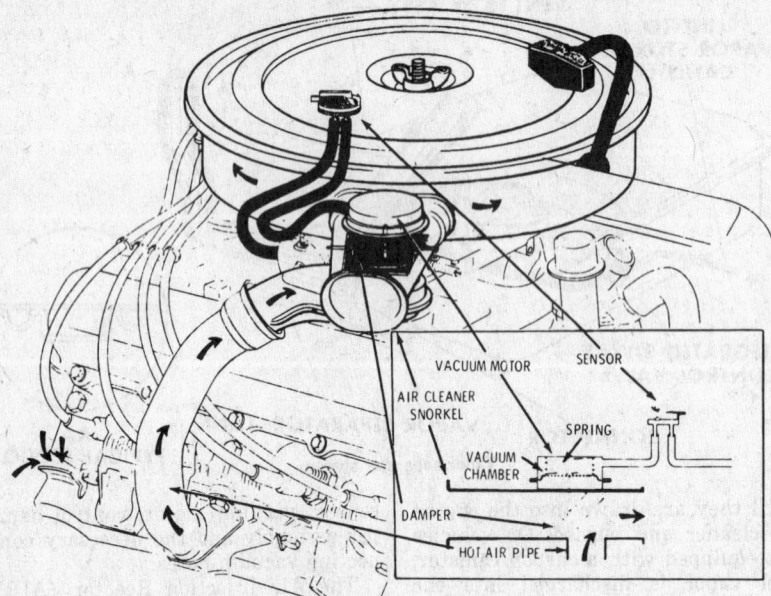

Controlled combustion system

hot air pipe and adapter into the air cleaner. As the temperature inside the cleaner reaches about 100°F, the sensor bleeds off vacuum to the vacuum motor, causing the control damper to open, permitting underhood air to be admitted and mixed with the heated air from the heat shroud. This mixture tends to keep the carburetor inlet air temperature at about 100°F.

Under full throttle or below 6-8 in. Hg. the vacuum motor will no longer hold the valve open to hot air. The hot air pipe is closed off, allowing only outside air to enter the air cleaner.

The vacuum motor and control damper assembly in the left snorkel of some special option engines have no sensor and are controlled only by intake manifold vacuum. On these models, the snorkel remains closed until full throttle is obtained. With intake manifold vacuum at 6-8 in. Hg. the control damper assembly will open, allowing maximum air flow into the air cleaner.

Functional Checks
Vacuum Motor and Damper Assy.
1. With engine off, remove air cleaner cover and tape a thermometer (J-5421) to the air cleaner next to the sensor.
   NOTE: if temperature is below 80°F, continue to Step 2. If temperature is above 80°F, remove air cleaner and allow to cool to at least 75°F.
2. Install a T into the vacuum line, at vacuum motor, and connect a vacuum gauge into the line.
3. With engine off, control damper will be open.
4. Install the cover on the air cleaner, without wing nut, and start engine. With engine at idle speed, the control damper should

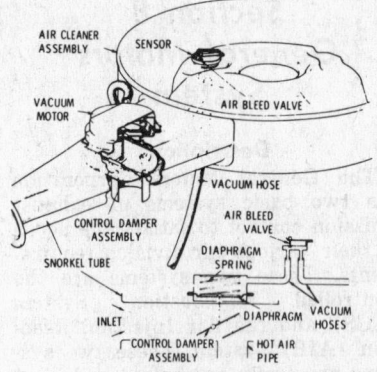

Operation (regulating mode)

close, unless surrounding temperature is above 85°F.
NOTE: if surrounding temperature is above 85°F, make sure control damper is completely open by 128°F.
5. Watch control damper in snorkel. When it begins to move toward the open position (opening to outside air), quickly remove cover on air cleaner and record the readings on thermometer and vacuum gauge.

Specifications for Damper
Temperature: 85°F—starts to open
            128°F—fully open
Vacuum at motor: 5-9 in. Hg. when damper assembly is closed to outside air.

Diagnosis
1. If temperature is within specifications, the CCS is functioning properly.
2. If temperature is not within specifications, but vacuum is correct (5-9 in. Hg.), replace the sensor.
3. If neither temperature nor vac-

uum are within specifications, replace vacuum motor.

Vacuum Motor Replacement
1. Remove air cleaner.
2. Disconnect vacuum hose from motor.
3. Drill out the two spot welds with a 1/16 in. drill, then, enlarge the holes as necessary.
4. Remove motor retaining strap.
5. Lift up motor, and unhook linkage at control damper assembly.
6. Drill a 7/64 in. hole in the snorkel tube at center of motor retaining strap in illustration.
7. Insert vacuum motor linkage into control damper assembly.
8. Use the motor retaining strap and sheet metal screw, provided in the motor service replacement package, to hold the retaining strap and motor to the snorkel tube.
9. Make sure that the screw does not interfere with operation of damper.
10. Connect vacuum hose to motor and install air cleaner.

Sensor Replacement
1. Remove air cleaner.
2. Detach hoses at sensor.
3. Pry up tabs on sensor retaining clips. Remove clips and retainers. Note position of sensor for reinstallation, then remove sensor.
4. Install by reversing removal procedure.

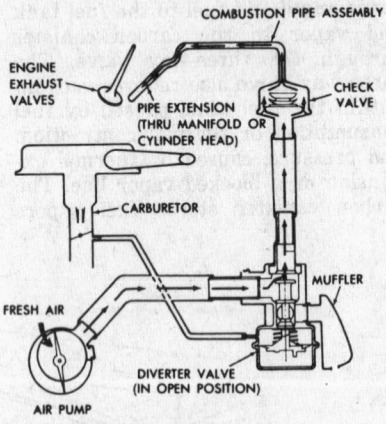

Schematic of AIR system
(© G.M. Corp.)

Air Injection Reactor (AIR) System

Since 1968, the AIR system is used on General Motors cars as an alternate to their CCS design. The AIR system is used to satisfy the exhaust emission control needs of certain car models.

The AIR system consists of a belt-driven air injection pump, air injection tubes (or specially designed cylinder heads that contain air passages) to the rear of each exhaust valve. There is a diverter valve and

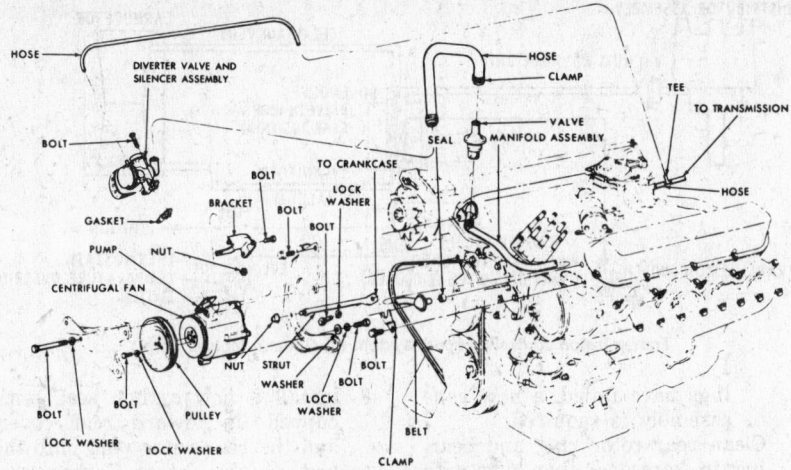

Air injection reactor system

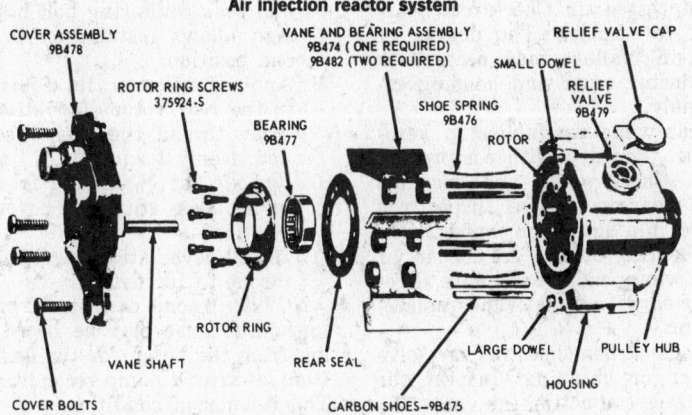

Air injection reactor pump

silencer on the air pump to control pressures within the system, and a check valve to protect the hoses and pump from hot gases.

NOTE: do not attempt to operate vehicle with drive belt disconnected.

Air Pump

Intake air is admitted through a centrifugal fan at the front of the pump. There, dirt and other materials are separated from the air by centrifugal force. Air is routed from the pump to the air injection manifold, or the specially cast cylinder head/s, through flexible tubes.

The only serviceable component of the air pump is the centrifugal fan. Do not consider the pump to be defective if it squeaks when turned by hand. Do not lubricate the pump.

NOTE: When steam, or solvent cleaning the engine, mask off the centrifugal filter fan to prevent contamination.

Drive Belt

1. Inspect drive belt for wear or deterioration. Replace if necessary.
2. Check belt tension and adjust, if below 75 lbs., using a belt tension gauge.
3. A used belt should read 75 ±5 lbs. and a new belt 125 ±5 lbs.

Caution Do not pry on pump housing.

Tube and Hose

Hoses of the AIR system are of special composition, able to withstand high temperature. Use no substitute material.

When replacing or renewing hoses or tubes, route them according to original plan.

Check Valve

The check valve should be inspected whenever the pump outlet tube is disconnected from the check valve, or whenever check valve failure is suspected.

To test the check valve, blow through the valve, then attempt to suck back through it. Flow should be in only one direction (toward the tube).

Air Injection Tubes or Nozzles

There are different methods of air injection into the exhaust system (ducts and injection nozzles cast into the head/s, nozzles screwed into the head/s, and nozzles screwed into the exhaust manifold/s at each exhaust port).

Regardless of the method used, the air discharge orifice must be clear and show no indication of burning. The

orifices or nozzles should be examined whenever the exhaust manifold or head is removed for any purpose. Clean or renew nozzles as necessary.

Air Injection Manifold/s, Where Used

The air injection manifold/s is/are held in place by clamps. These clamps must be installed correctly to insure proper sealing. Each feeder tube of the manifold is sealed by a packing gland which fits on the tube below the flange.

Diverter Valve and Check Valve

The diverter valve is attached, or adjacent, to the side of the pump. Its purpose is to sense manifold vacuum through a fitting at the carburetor. During sudden deceleration, vacuum increases, causing the valve to open, allowing air from the injector pump to pass through the valve and silencer to the atmosphere. Approximate duration of the valve opening is 2 seconds. This valve also controls pressure within the system by diverting excessive pump output to the atmosphere, through the silencer.

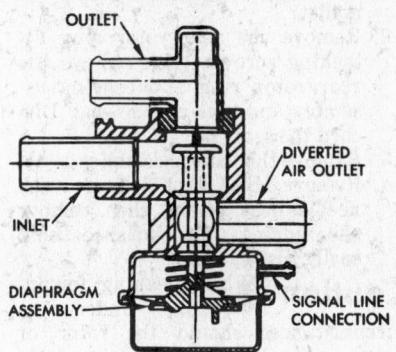

VALVE IN OPEN POSITION

Diverter valve

In most cases, the air injection manifold is fitted with a screw-on check valve near the right cylinder head. This valve has a one-way diaphragm which prevents hot exhaust gases from backing up into the hose and pump, causing damage. This is to protect the system in the event of pump belt failure, abnormally high exhaust system pressure, or air delivery hose breakage.

Air Injection Pump Reconditioning

Caution The following reconditioning procedure is submitted for the convenience of shops having adequate facilities, and is not a General Motors Corporation common practice or recommended policy.

The vane-type air injection pump as used with the AIR system consists of parts as illustrated in the accompanying exploded view.

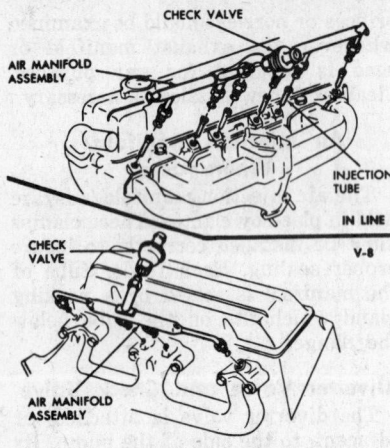

Air injection tubes

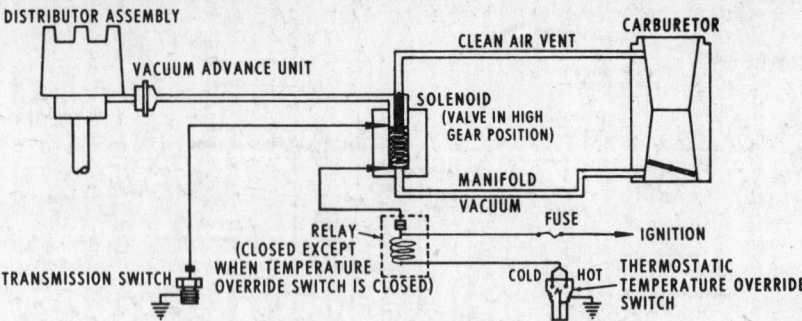

Transmission controlled spark system (© General Motors Corp.)

NOTE: before disassembly, rotate the drive hub quickly, first three-quarter turn counterclockwise, then one-quarter turn clockwise. Roughness may be felt which could indicate vane bearing failure.

1. Place the drive hub in a vise and remove the cover assembly by tapping straight up with a plastic mallet.
2. Remove the rear rotor ring attaching screws, then remove the rear rotor ring and bearing assembly and rear carbon seal. Discard the seal.
3. Remove the vane assemblies.
4. Remove the vane shoes with needle nose pliers, then remove shoe springs. Carbon shoes chip easily; use caution.

Caution When cleaning for inspection, under no circumstances should the rotor or cover assemblies be immersed in cleaning solvent.

Cleaning and Inspection

1. Blow carbon dust from the housing and rotor assembly and cover assembly with compressed air.
2. Wipe stubborn deposits from the components with a cloth saturated with kerosene. Remaining deposits that do not respond to petroleum solvents may be removed with denatured alcohol.
3. Inspect inside of housing for vane or rotor wear. Excessive wear justifies replacement of housing and rotor.
4. Inspect rear cover as follows:
   A. Check rear bearing inner race for grooves or scratches. This damage indicates probable rear bearing failure; the bearing and cover assembly must be replaced.
   B. Inspect the vane pivot pin. This pin is precision ground. Grooves, scratches, or discoloration indicate probable vane bearing failure. If bear-

ings have failed, a new vane assembly is required.
5. Clean rear rotor ring and bearing in kerosene, then dry with compressed air. Check rear bearing for failure. The degree of bearing failure may necessitate replacing rotor and housing assembly.
6. Clean vane assemblies in kerosene, then dry with compressed air. Check vanes and bearings for wear or damage. Inspect the pivot pin again for confirmation of bearing failure. All corners of the vanes must be square. Broken edges require vane replacement.
7. Check carbon shoes for excessive wear or chipping. Inspect the surface contacting the vane. The shoes should show small grooves from action. This condition is normal. However, chipped or broken shoes must be replaced.

Rear Bearing Replacement and Pump Assembly

1. With bearing remover J-21058 and J-21844, pull the rear bearing out of rotor ring.
2. With the same tools (J-21085 and J-21844), pull the rear bearing into the rotor ring until bearing is 1/32 in. beneath the rotor ring surface. Pull on the lettered end of the bearing.
3. Lubricate the vane bearings and rear bearing with special air injection pump grease or its equivalent; wipe away any excess.
4. Position vanes on 3/8 x 4 in. rod.
5. Position rotor with one vane slot at stripper, then support the housing in a vise by the drive hub and install the vanes (with rod).
6. Install one carbon shoe on each side of every vane. Shoes must be installed with bearing surface against vane and with rounded edge of contact toward outside diameter of rotor.
7. Install shoe springs into the deepest of the two shoe slots. The arc of the spring should be installed toward the shoe. Push shoe spring flush with or below rotor surface.

8. Install a new carbon seal, with cupped lip toward rear cover, and the rear rotor ring onto the end of the rotor. Carbon seal and rear rotor ring bolt hole design allows installation in only one position.
9. Apply Loctite or its equivalent to the rear rotor ring attaching screw thread, then install screws and evenly torque to 37 in. lbs.
10. Remove the 3/8 in. rod from the vanes, then start the cover into position.
11. Install cover attaching bolts and torque to 10 ft. lbs.

NOTE: in some cases a chirping or squeaking noise may be heard coming from the pump for the first 5 or 10 miles after a pump reconditioning. This is a normal condition.

Transmission Control Spark (TCS)

On most 1970 models, distributor vacuum advance has been eliminated in the lower forward gears. A solenoid vacuum switch is energized in the lower gears by a switch at the transmission. The solenoid shuts off the vacuum supply to the distributor, venting the vacuum advance unit to the carburetor to prevent air locking of the advance unit at an advanced position. TCS is overridden by a temperature switch for cold engine operation. Some engines also have a hot override switch. Vacuum advance is provided only in high gear, except for four speed transmissions, which have vacuum advance in third and fourth gears.

TCS Troubleshooting

If full vacuum is found in all gears, the cause may be:
1. Fuse blown.
2. Wire disconnected at solenoid.
3. Wire disconnected at transmission switch.
4. Transmission switch failed.
5. Temperature override switch energized—check by disconnecting electrical lead.
6. Solenoid failed.

If no vacuum is found in high gear, the cause may be:
1. Clean air line and distributor

vacuum line reversed at solenoid.
2. Foreign material in solenoid.
3. Plunger return spring broken.
4. Distributor or manifold vacuum hose broken or disconnected.
5. Transmission switch or wire shorted to ground.

Evaporative Emission Control System (ECS)

ECS is installed on all 1970 models sold in California. The function of this system is to prevent fuel vapors from escaping to the atmosphere. A special non-vented filler cap is used. This cap is equipped with a two way valve to relieve pressure or vacuum within the fuel tank. A liquid-vapor separator is mounted in front of the gasoline tank. The separator prevents liquid fuel from passing to the vapor storage canister in the engine compartment. The canister contains a filter which must be cleaned every 12,000 miles. From the canister, vapors are vented to the crankcase ventilation system and/or to the carburetor air cleaner intake.

Canister and/or Canister Filter Removal

1. Raise vehicle on hoist.
2. Note installed position of hoses on canister.
3. Disconnect hoses from top of canister.
4. Loosen clamps and remove canister.
5. If replacing filter, remove bottom of canister and pull out filter.

Canister and/or Canister Filter Inspection

1. Check hose connection openings. Assure that they are open.
2. On four barrel carburetor models, check operation of purge valve by applying vacuum to the valve. A good valve will hold vacuum.

Canister and/or Canister Filter Installation

1. Install new filter.
2. Assemble bottom of canister to canister body.
3. Install canister and tighten clamp bolts.
4. Connect hoses to top of canister

in same position as in Step 3 above.

Canister Purge Valve Disassembly

1. Disconnect lines at valve.
2. Snap off valve cap (Slowly remove cap as diaphragm is under spring tension). Remove diaphragm, spring retainer and spring.
3. Replace parts as necessary. Check orifice openings.

Canister Purge Valve Assembly

1. Install spring, spring retainer, diaphragm and cap.
2. Connect lines to valve.

Separator Removal

1. Raise vehicle on hoist.
2. Disconnect lines from separator.
3. Remove retaining screw and remove separator.

Separator Installation

1. Install separator and its retaining screw.
2. Connect lines to separator.
3. Lower vehicle and remove from hoist.

Dash Gauges And Indicators

Gauge Type Index

Numbers in Columns Refer to Section Numbers in Text

Make and Model	1964	1965	1966	1967	1968	1969	1970
American Motors							
AMX	...	...	...	...	2, 4	2, 4	2, 4
Gremlin	...	...	...	...	...	...	2, 4
Hornet	...	...	...	...	...	...	2, 4
Javelin	...	...	...	...	2, 4	2, 4	2, 4
Rambler	2, 4	2, 4	2, 4	2, 4	2, 4	2, 4	2, 4
Chrysler Corporation							
Chrysler	2, 4	2, 4	2, 4, 5	2, 4	2, 4	2, 4, 5	2, 4, 5
Challenger	...	...	...	...	...	...	2, 5
Dart, Barracuda	2, 4	2, 4, 5	2, 4, 5	2, 4, 5	2, 4, 5	2, 4, 5	2, 4, 5
Dodge	2, 4, 5	2, 4, 5	2, 4, 5	2, 4, 5	2, 4, 5	2, 4, 5	2, 4, 5
Imperial	2, 4	2, 4	2, 4	2, 4, 5	2, 4, 5	2, 4, 5	2, 4, 5
Plymouth	2, 4, 5	2, 4, 5	2, 4, 5	2, 4, 5	2, 4, 5	2, 4, 5	2, 4, 5
Valiant	2, 4, 5	2, 4, 5	2, 4, 5	2, 4, 5	2, 4, 5	2, 4, 5	2, 4, 5
Ford Motor Company							
Capri	...	...	...	...	...	...	2,4
Comet	2, 5	2, 5	2, 5	2, 5	...	...	...
Cougar	...	...	...	2, 4, 5	2, 4, 5	2, 5	2, 4, 5
Fairlane-Torino	2, 4	2, 4	2, 4	2, 4	2, 4	2, 4	2, 4
Falcon	2, 4	2, 4	2, 4	2, 4	2, 4	2, 4	2, 4
Ford	2, 4	2, 4	2, 4	2, 4	2, 4	2, 4	2, 4
Lincoln Continental	2, 4	2, 4	2, 4	2, 4	2, 4	2, 4	2, 5
Mark III	...	...	...	...	2, 5	2, 5	2, 5
Maverick	...	...	...	...	...	...	2, 4

Make and Model	1964	1965	1966	1967	1968	1969	1970
Mercury	2, 5	2, 5	2, 5	2, 4, 5	2, 4	2, 4	2, 4
Montego	...	...	...	...	2, 5	2, 4	2, 4, 5
Mustang	...	2, 4	2, 4	2, 4, 5	2, 4, 5	2, 4, 5	2, 4, 5
Thunderbird	2, 5	2, 5	2, 5	2, 5	2, 4, 5	2, 5	2, 5
General Motors							
Buick	3, 4	3, 4	3, 4, 5	3, 4, 5	3, 4, 5	3, 4	3, 4
Buick Special	3, 4	3, 4	3, 4	3, 4	3, 4	3, 4	3, 4, 5
Cadillac	3, 4	3, 4	3, 4	3, 4	3, 4	3, 4	3, 4
Cadillac Eldorado	3, 4	3, 4	3, 4	3, 4	3, 4	3, 4	3, 4
Camaro	...	...	...	3, 4, 5	3, 4, 5	3, 4, 5	...
Chevelle	3, 4, 5	3, 4, 5	3, 4, 5	3, 4, 5	3, 4, 5	3, 4, 5	3, 4
Chevrolet	3, 4	3, 4, 5	3, 4, 5	3, 4, 5	3, 4, 5	3, 4	3, 4
Chevy II, Nova	3, 4	3, 4	3, 4	3, 4	3, 4, 5	3, 4, 5	3, 4, 5
Corvair	3, 4 (1960–64)	3, 4	3, 4	3, 4	3, 4	3, 4	...
Corvette	3, 4, 5	3, 4, 5	3, 4, 5	3, 4, 5	3, 4, 5	3, 4, 5	3, 4, 5
Oldsmobile	3, 4	3, 4	3, 4	3, 4	3, 4	3, 4	3, 4
Oldsmobile F85	3, 4	3, 4	3, 4	3, 4	3, 4, 5	3, 4, 5	3, 4, 5
Oldsmobile Toronado	...	...	3, 4	3, 4	3, 4	3, 4	3, 4
Pontiac	3, 4, 5	3, 4, 5	3, 4, 5	3, 4	3, 4	3, 4	3, 4
Pontiac Firebird	...	...	...	3, 4, 5	3, 4, 5	3, 4, 5	3, 4, 5
Pontiac Tempest, GTO	3, 4, 5	3, 4, 5	3, 4, 5	3, 4, 5	3, 4, 5	3, 4, 5	3, 4, 5

Section Page Numbers

Dash Gauges and Indicators

There are various systems used to indicate values of heat, pressure, current flow and fuel supply. The following are the more popular systems used.

Bourdon Tube

This gauge consists of a flattened tube bent to form a curve. The curve tends to straighten under internal pressure (caused by captivated gas in the case of a heat indicator, or the engine oil pressure if used as an oil gauge). The curved tube is geared or linked to an indicator needle which may be read on a calibrated scale.

Bi-Metallic or Thermal

This gauge is activated by the difference in the expansion factors of a bi-metal bar. A sending unit consisting of a variable resistance conductor influences current flow to a voltage limiter, or directly to a heating element coiled around a bi-metal bar in the gauge.

Magnetic

In this system, the indicator needle is moved by changing the balance between the magnetic pull of two coils built in the gauge. When the ignition switch is in the off position, the pointer may rest any place on the gauge dial. Balance is controlled by the action of a sending unit or a tank unit containing a rheostat, the value of which varies with temperature, pressure or movement of a float arm.

Warning Lights

This system is quite popular and may be used to indicate heat, low pressure or as a battery discharge indicator. General Motors uses a two-light temperature indicator version of this unit in some models.

Section 1
Bourdon Tube

Oil Pressure

The gauge is the pressure expansion type and is activated by oil pressure developed by the oil pump, acting directly on the mechanism of the

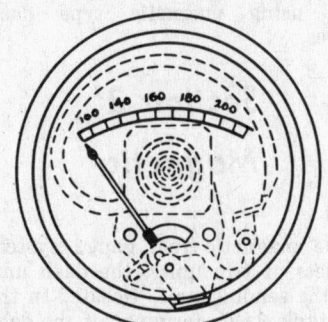

gauge. The gauge is connected by a small tube to the main oil passage in the engine oiling system. This design registers the full pressure of the oil pump.

Testing

The best way to test this type gauge is by comparison with a gauge of known accuracy. The hook-up can be made by using a T-fitting where the gauge tube taps into the engine oil pressure system. Connect the test gauge and the car gauge to the T-fitting and read both gauges through the entire range of engine speeds.

Be sure there is enough engine oil of correct viscosity before making the test.

A gauge pointer that flutters is usually an indication that oil has entered the gauge oil tube. The tube should contain trapped air to cushion the pulsations of the oil pump and relief valve. Oil can work up into the gauge line as a result of a gauge or tube leak or improper installation. To correct this condition, renew the unit or correct the leak; then, with the gauge line disconnected at both ends, blow the line clear. Connect line at gauge first and then at the engine.

If the gauge reads too low or reads no pressure, test for a possible obstruction by disconnecting the line at the gauge. Hold the end of the line over an empty container, then start the engine. After a few bubbles, oil should flow steadily. If flow is satisfactory at this point, the gauge is bad and should be renewed. If the flow is unsatisfactory at this point, a process of further elimination must be continued to determine the point of pressure failure. When reconnecting the gauge tube, always clear the tube, then hook up the gauge end first.

Temperature

The heat indicator is a vapor pressure activated, completely sealed unit. It makes use of heat expansion which creates a pressure move to a pointer on a calibrated gauge face. The entire unit consists of a Bourdon tube connected by a capillary tube to a vapor bulb. The vapor bulb is inserted by a plug into the engine cooling system at some point in the cylinder block or head. The heat of the engine coolant causes the sealed-in liquid in the bulb to expand in proportion to temperature rise. This vapor pressure influences the Bourdon tube in the gauge and positions the pointer on the gauge dial.

When installing a new unit, a loop must be carefully formed in the tube somewhere between the engine and the car body to compensate for vibration and other movement. The tube

must never be sharply bent or permitted to chafe by rubbing.

Testing

The unit is not serviceable and testing is best accomplished by comparison with another gauge of known accuracy.

Section 2
Bi-Metal

Fuel

Except for some of the older model cars, it is safe to assume that these bi-metal or thermal type gauges operate on the principle of constant applied voltage and are sensitive only to changes originating at the sending unit. Constant voltage is provided through the use of a voltage regulator contained within one of the gauges comprising the gauge system, or by an independent voltage limiter mounted close to the dash gauge and in the gauge circuit.

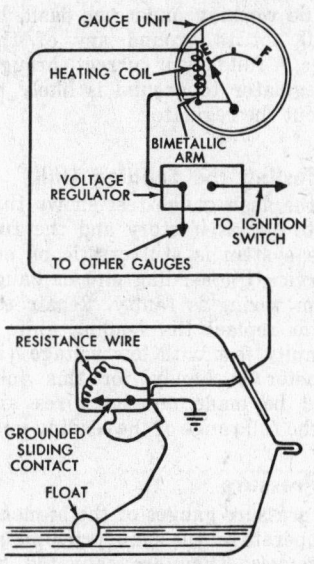

Fuel gauge circuit

The fuel gauge system consists of a sending unit, located in the fuel tank, and a registering unit mounted in the instrument cluster. The sending unit is a rheostat that varies its resistance depending on the amount of fuel in the tank.

Testing the Dash Gauge

Caution Gauge systems using constant voltage regulators should not be grounded while testing. An excess of 5 volts is likely to burn out the unit.

To safely test this type of voltage regulated system:

1. Have ignition switch in off position.
2. Connect terminals of two, series connected, flashlight batteries to

the terminals of the gauge to be tested.

3. The 3 volts should cause the gauge to read approximately full scale. If the gauge is inaccurate or does not register, replace it. If the gauge unit is still erroneous, the sending

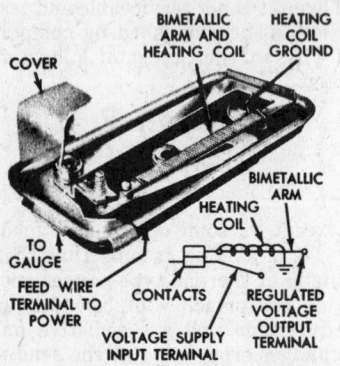

Constant voltage regulator

unit, or the wiring to the sending unit is at fault.

If both fuel gauge and temperature gauge are in error, in the same manner, the constant voltage regulator is probably at fault.

While working under the dash, be careful not to ground any of the gauges. A full flow of current through the regulator to ground is likely to burn out the regulator.

Testing the Sending Unit

If the dash gauge test shows that unit to be satisfactory and the fuel gauge system is still erratic or out of service, the sending unit or gauge system wiring is faulty. Repair the wire or replace the sending unit. A continuity test with low voltage (an ohmmeter is handy for this job) should be made on the wires and over the full range of the sending unit.

Oil Pressure

Oil pressure gauges of the bi-metal type operate on the same principle as gas gauges. They are activated by temperature and the difference in the expansion factors of two bonded metals in a bi-metal bar.

The pressure sending unit consists of a pressure-activated variable resistor. This sealed unit is usually screwed into the crankcase to become part of the engine oil pressure circuit. As pressure is applied to one side of a diaphragm, linkage advances

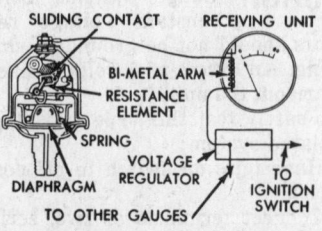

Meter-type oil gauge circuit

a contact arm across the coils of a resistor. This action reduces resistance in the gauge circuit, thus increasing current flow and heat to the bi-metal arm in the gauge. The gauge is calibrated to read oil pressure in psi.

Testing the Dash Unit

In testing the dash gauge and constant voltage regulator, use the same procedure and exercise the same caution as outlined in testing the bi-metal or thermal fuel gauge.

Testing the Sending Unit

The most dependable way to test this type oil pressure sending unit is by replacement and comparison with a new one or a substitute of known accuracy.

Temperature

The temperature gauge consists of a sending unit, mounted in the cylinder head or block, and a remote resistor unit (temperature gauge) mounted on the instrument panel. The principle of operation is essentially the same as the bi-metallic fuel

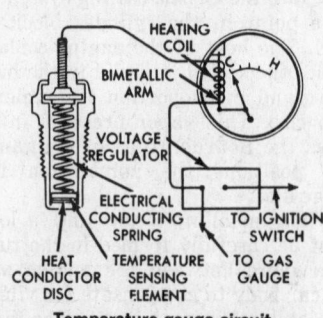

Temperature gauge circuit

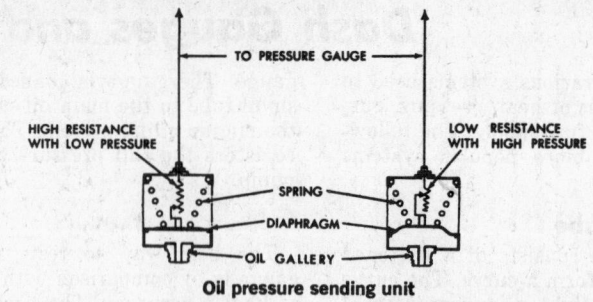

Oil pressure sending unit

King-Seeley resistance-type oil pressure gauge

gauge, the exception being that the resistance of the sending unit is influenced by engine temperature instead of tank fuel level, as with the fuel gauge.

The temperature sending unit is constructed with a coil spring and sensing disc. Current passing through this coil encounters increased resistance, proportional to an increase in temperature. The dash gauge, which is calibrated in heat units, registers this resistance change.

Testing the Dash Gauge

In testing the dash gauge and constant voltage regulator, use the same procedure and exercise the same caution as outlined in testing the bi-metal or thermal fuel gauge.

Testing the Sending Unit

There are various and complex methods of testing these units. However, the most practical way is by replacement and comparison with a new one of known accuracy.

NOTE: this sending unit, while similar in appearance, is not interchangeable with the unit used in systems using magnetic type dash gauges.

Section 3

Magnetic

Fuel

The magnetic fuel gauge system consists of two units, the dash unit and the sending unit, (located in the fuel tank.) One terminal of the dash

unit is connected to the ignition switch so that the system is active only when the ignition switch is turned on. With the ignition switch turned off, the pointer may come to rest at any position on the dial of the dash unit.

The gauge pointer is moved by changing the magnetic pull of two coils in the unit. The magnetic pull is controlled by the action of the tank unit which contains a variable rheostat, the value of which varies with movement of a float and arm.

When the ignition switch is on and the tank unit arm is in the full position, the current flow to ground is through the resistor, battery coil and the ground coil. Because the ground coil has more windings than the battery coil, it builds up a stronger magnetic field and the pointer is pulled to the full position.

When the tank unit arm is in the empty position, the current flow is through the resistor, the battery coil and the wire to ground at the tank unit. The pointer is thus pulled to the empty position. The resistor in series with the battery coil is for a balance of resistance between the two coils in the dash unit, because the ground coil has more windings.

Testing the Dash Gauge

Disconnect the wire from the tank unit. Using a new tank unit of known accuracy, clip a test wire from the body of the test unit to ground. Clip another test wire from the connector of the test unit to the tank unit wire. With ignition on, moving the float arm through its entire range, the dash gauge should respond proportionally.

Testing the Tank Unit

If tests indicate that the trouble lies in the tank unit, remove the unit and check for mechanical failure. The unit may have either a loaded float or something binding.

An electrical check for circuit continuity may be made throughout the unit's range.

Temperature

The temperature gauge system consists of a magnetic dash unit and a resistance-type sending unit screwed into the water jacket of the cylinder head or the engine block.

The dash unit has two magnetic poles. One of the windings is connected to the ignition switch and ground. This electromagnet exerts a steady pull to hold the gauge pointer to the left or "cold" position whenever the ignition switch is turned on.

The other winding in the dash unit connects to a ground through the engine sending unit. This electromagnet exerts a steady pull on the gauge pointer toward the right, or "hot"

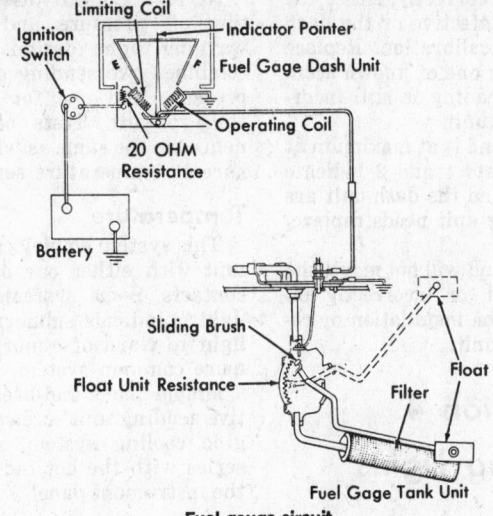

Fuel gauge circuit

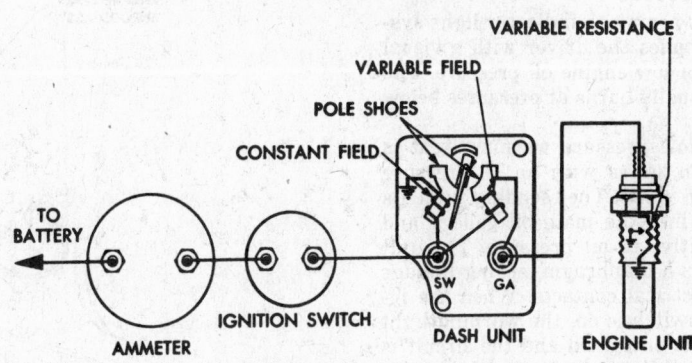

Water temperature gauge

side of the gauge. The strength of this pull is dependent upon the current allowed to pass through the engine unit (sending unit) resistor.

The sending unit, located in the engine cooling system, contains a flat disc (thermistor) that changes resistance as its temperature varies.

NOTE: this sending unit, while similar in appearance, is different and is not interchangeable with the unit used in systems using bi-metal or thermal dash gauges. The resistance of the thermistor disc is maximum when the temperature is cold and minimum when hot. The decrease in resistance allows more current to flow through the electromagnet connected to the engine unit. The resulting increase in magnetic pull causes the gauge pointer to move to the right, or "hot" side.

Tests

1. Disconnect the wire at the sending unit and turn on the ignition switch. The gauge hand should stay against the left side stop pin.
2. Ground the wire disconnected from the sending unit. With the ignition switch still on, the gauge hand should swing across the dial to the right stop pin.

Corrective Measures

If the gauge hand does not stay to the left, either the wire is grounded between the dash unit and the engine unit or the dash unit is defective.

Test further by disconnecting the sending unit wire at the gauge. Turn on the ignition switch. If the gauge hand stays on the left-hand stop pin, replace the disconnected wire. If the gauge hand still moves, replace the gauge.

If the gauge hand does not swing across the dial, there is an open circuit in the wire between the sending unit and gauge, the gauge is defective, or current is not reaching the dash gauge.

Test further by grounding the sending unit terminal of the dash gauge and turn on the ignition switch. If the gauge hand now moves, replace the disconnected wire. If the gauge hand does not move, connect a test lamp from the dash unit. If the test lamp does not light, test the wire between the ignition switch and the dash unit by connecting a test lamp to the accessory terminal at the ignition switch and ground. The test lamp should burn.

If the gauge hand operates correctly, but the gauge does not indi-

cate temperature correctly, either the sending unit is defective or the dash gauge is out of calibration. Replace sending unit with one of known accuracy. If gauge reading is still incorrect, replace dash unit.

If the gauge hand is at maximum at all times, and tests 1 and 2 indicate that the wiring and the dash unit are good, the sending unit needs replacement.

If the gauge hand will not move, the dash unit is bad, or incorrectly installed. Correct the installation or replace the gauge unit.

Section 4
Warning Lights

Oil Pressure

The warning or indicator light system supplies the driver with a visual signal of low engine oil pressure. The light usually burns at pressures below 5 psi.

The low pressure warning light is wired in series with an oil pressure sending unit. The sending unit is tapped into the main oil gallery and is sensitive to oil pressure. The unit contains a diaphragm, spring linkage and electrical contacts. When the ignition switch is on, the warning light circuit is energized and the circuit is completed through the closed contacts in the sending unit. When the engine starts, oil pressure will compress the diaphragm, opening the contact points and breaking the circuit.

Tests

Connect indicator bulb in series with sending unit. Apply air or oil pressure to sending unit, controlling the pressure with a control valve and positive indicating gauge. With less than 5 psi the light should be on. Over 5 psi, the light should go out. The system may tolerate a variation of 1 psi.

NOTE: Corvair uses a combination oil pressure and temperature warning light design. This design combines two sending units (one for pressure and one for heat) in the same circuit. Tests of the components are the same as with other pressure or temperature sending units.

Temperature

This system employs a heat sending unit with either one or two sets of contacts. Some systems use a green light to indicate subnormal, and a red light to warn of abnormal heat. The more common system, however, uses a simple make-and-break heat-sensitive sending unit screwed into the engine cooling system, and wired in series with the hot indicator light in the instrument panel.

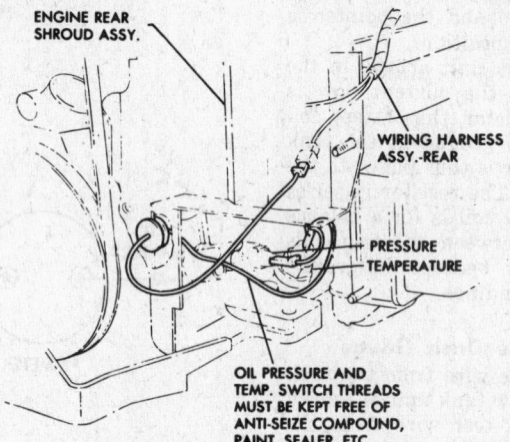

Oil pressure and temperature senders

The two-light system uses a bi-metal element mounted between two signal circuits. Normal operating temperature (somewhere between 120°F. and 250°F.) will cause the bi-metal bar to assume a position of no contact between the low and the high temperature circuit. When the ignition switch is turned on, with a cold engine, the cold (green) circuit is complete. If the engine becomes hot enough to move the bi-metal bar so that it touches the contacts of the hot circuit, the hot (red) light comes on.

This hot signal indicates that temperatures are in the area of 250°F. in the sealed cooling system.

Tests

Tests of this signal system are best performed by a check of continuity and by replacement with components of known accuracy.

Charge Indicator

A light is used to indicate general generator operation. When generator output is below battery potential, a red light is shown. When generator output is above battery potential, other factors (wiring, voltage regulator, etc.) being normal, the light is out.

The charge indicator bulb is connected to the charging circuit, obtaining its ground through the voltage regulator. When the generator output rises above battery potential, the current flow from the generator reverses the direction of flow in the charging circuit and the light goes out. At this time, the charge indicator light no longer is grounded through the voltage regulator.

When an alternator is used, it is necessary to supply a small amount of excitation current to the alternator field, due to the small amount of residual magnetism. Current can be

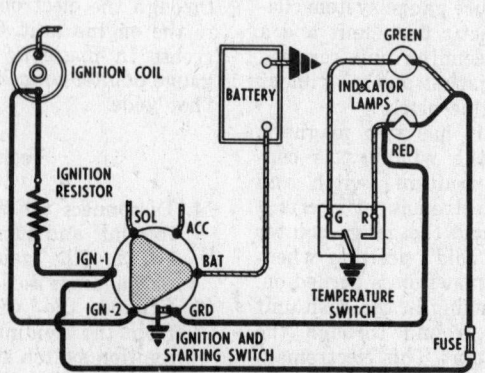

Temperature indicator circuit

Battery   12 Volt Test Light

Restriction Orifice Should Be Contained in Gauge Unit to Prevent Fluctuations of Gauge Needle

Oil Pressure Sending Unit or Pressure Switch

Air Pressure Control Valve

(Never Allow Over 150 Lbs.)

Source of Air Pressure

Oil pressure warning light test procedure

supplied through the indicator light to the regulator terminal on the alternator. This current has a value of about .25 amperes and will cause the indicator light to come on when the ignition switch is turned on.

As the alternator starts to generate, an output voltage will develop at the regulator terminal. When this voltage exceeds the battery voltage, the isolation diode will pass current to the battery and system. The indicator light is in parallel with the isolation diode. As current flows through the isolation diode, about 1 volt is developed across the diode; therefore, 1 volt will be developed across the indicator light. This is not enough to light the indicator, thus, the light will go out when the alternator is supplying sufficient current.

Caution In some instances, a glow of the indicator light will be noticed at night while cruising. This can exist in a system that is operating normally and with a fully charged battery. The condition, while of no great importance, is feedback, probably caused by an overload of equipment and accessories.

If the situation is objectionable, a resistor (20 to 25 ohms) may be installed in the circuit on the generator side of the indicator light.

Section 5
Ammeters

The automotive ammeter is a gauge or meter used to indicate direction and relative value of current flow. This type of charge indicator is usually equipped with a dampening device to reduce pointer fluctuation during current surge from the voltage regulator.

The meter will show charge when the battery is being charged and discharge when the battery is being discharged. It merely gives an indication of the state of charge of the battery, since it shows a relatively high charging rate when the battery is low, and a low charging rate when the battery is near full charge. Just after cranking the engine, the meter will show a generous swing toward the charge side for a short time, if lights and accessories are turned off. As the energy spent in cranking is restored to the battery, the pointer will gradually move back toward center but should stay on the charge side. If the battery is low, however, the indicator will show a high charging rate for an indeterminate length of time.

The charge indicator does not show the charging rate of the generator, since energy supplied by the generator to electrical units, other than the

battery and horns, does not pass through it.

At speeds above 30-35 mph, with all lights and accessories on, the indicator should show a reading somewhere on the charge side, depending on the state of the battery. Above this speed, the indicator should never show a discharge reading; if it does, the generator and regulator should be tested.

Aluminum Thread Repairs

Aluminum is the most abundant metallic element in existence. In its pure state it is silvery-white, very light in weight and has extremely good heat and electrical conductivity. These properties make it very desirable for use in the automobile industry.

Aluminum, when alloyed for engine blocks, transmission case construction and certain other applications, has one outstanding deficiency: enough thread strength to hold attaching and mounting bolts.

Any of the repair shops or garages having maintenance facilities for automobiles, trucks or power equipment need a method for dealing with aluminum. In some cases, an entire cylinder block, crankcase or transmission case may have to be scrapped because of one hole in which stripped threads can't be repaired. Or, a unit may be put out of service for a long time while the threads are repaired by drilling oversize and retapping or by welding and redrilling.

Permanent repairs to stripped sparkplug holes, gear cases, cylinder blocks, intake and exhaust manifolds or virtually any point can now be repaired quickly and on the spot. This is accomplished by the use of stainless steel wire thread inserts.

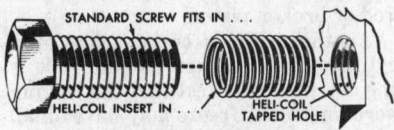

Heli-Coil installation
(© Chrysler Corp.)

Wire screw thread inserts are precision formed coils of various sizes. They are tailored to accept a bolt or stud of the original size. When inserted into a hole, the diamond shaped wire forms nominal size internal threads that are stronger and more durable than the original aluminum threads. Greater thread load is possible and wear resistance is improved.

For convenience, inserts are packaged in kits of the most popular sizes. Each kit contains inserts of a given thread size with a tap and inserting tool, a separate T-handle to drive the inserting tools and a pressure

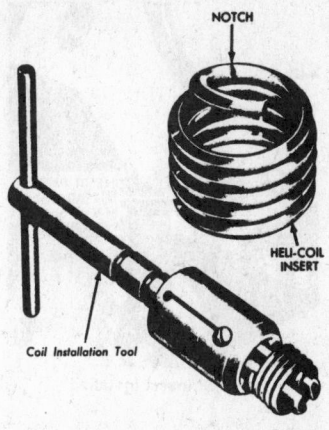

plate for easy starting of inserts. Sizes are available from 8-32 through ⅝-11.

To repair damaged threads the following procedure should be carefully followed:

1. Drill damaged threads out of the stripped hole. Use the same drill size as the bolt thread outside diameter. Use a 5/16 in. drill for a 5/16-18 thread, for example.

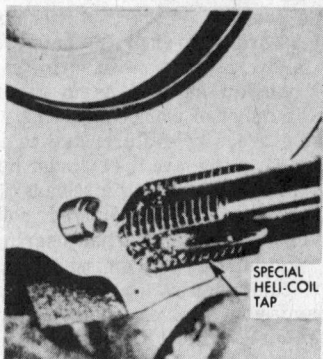

Tapping new thread with special tap supplied with inserts

2. Select the correct special tap supplied with the kit. The tap is marked for the size and thread desired. As an example, the special tap marked 5/16-18 will not cut the same thread as a standard 5/16-18 tap. It will, however, form a thread large enough to suit the coiled insert.

3. Select the correct thread inserting tool. These tools are marked for the hole and thread size to be re-

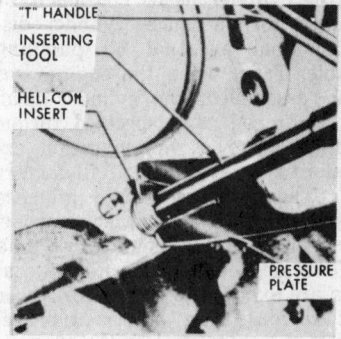

Installing insert

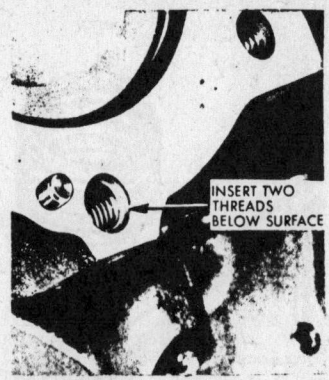

Heli-Coil insert installed

Heli-Coil Insert			Drill	Tap	Insert. Tool	Extracting Tool
Thread Size	Part No.	Insert Length (In.)	Size	Part No.	Part No.	Part No.
1/2-20	1185-4	3/8	17/64 (.266)	4 CPB	528-4N	1227-6
5/16-18	1185-5	15/32	Q (.332)	5 CPB	528-5N	1227-6
3/8-16	1185-6	9/16	X (.397)	6 CPB	528-6N	1227-6
7/16-14	1185-7	21/32	29/64 (.453)	7 CPB	528-7N	1227-16
1/2-13	1185-8	3/4	33/64 (.516)	8 CPB	528-8N	1227-16

stored. Place the thread insert on the tool and adjust the sleeve to the proper length for the insert being used.

4. Press the insert against the face of the tapped hole while turning the tool clockwise. This will wind the insert into the drilled hole. Continue this action until the insert is one-half turn below the face of the work.

5. Reaching through the thread insert, bend the insert tang straight up and down until it breaks off at the notch.

6. If an error has been made and the results are unsatisfactory, the insert can be removed with the extractor tool. Place the extractor tool in the insert so that the blade contacts the top turn of the coil, one-quarter to one-half turn from the end. Tap the tool solidly with a hammer. This will cause the blade to cut into the insert. Press downward on the tool and turn it counterclockwise until the insert is removed.

The Torque Wrench And Its Application

The torque wrench is one of the least understood, therefore, the most neglected items in the auto serviceman's tool kit. A careful mechanic with a torque wrench far outclasses his fellow craftsmen in quality of workmanship as well as in the reduced number of work comebacks.

Some areas of failure, due to improper torquing, are the cylinder head bolts or nuts, bearing preloads (as needed with certain tapered roller bearings), engine journal bearings, and automatic transmissions and their components. In the case of transmission valve bodies, many valves are subject to binding and malfunction in improperly torqued valve body housings.

The most common area of neglect, and consequent trouble, is engine cylinders and cylinder heads.

As an example: it is not unusual to have a complaint of excessive oil consumption developing in an engine having no previous history of using oil. The trouble occurs at about the same time the cylinder head is disturbed for one reason or another.

Incredible as it may seem, the shape of each cylinder is subject to stress and some distortion by tightening the cylinder head. Too much, or unequal, torque causes cylinder distortion and eventual oil pumping trouble. True, when the engine was being assembled, the cylinder head was tightened to a specified torque reading. However, age, vibration, expansion, contraction, etc., all factors of age, heat, and seasoning, influence the value of the original torque. These changes are not always to a uniform degree. Some fastener locations are exposed to greater or lesser heat factors than others and are, therefore, subject to varying changes in stress. These differences alter the original torque reading and change the cylinder head-to-cylinder block pressure in varying degrees.

Whether or not the difference in head-to-block pressure permits the cylinder head gasket to leak or causes it to be blown, depends entirely upon the degree of gasket compression, pressure change and torque variation that has taken place between the head and block. A blown cylinder head gasket may result. Cylinder compression pressure may be lost to the atmosphere, it may be lost between adjacent cylinders, or the leak can be into the water jacketing and cooling system. The latter type leak may cause very serious trouble because of hydraulically locked cylinders due to coolant leaks into the cylinders while the engine is stopped, then it being trapped on the compression stroke of starting. In many cases, a cylinder locked in this manner results in a bent rod or broken piston.

Another serious consequence of a cylinder head gasket coolant leak is the detrimental effect that the leaking permanent antifreeze may have on all of the engine's internal parts. Under proper circumstances this condition could cause irreparable engine damage.

Whether or not the head gasket holds, does not alter the fact that the roundness of the cylinders is influenced by torque applied to the head. While the original tightening of the head, distorted the cylinders at the time the engine was built, piston and ring assemblies remain almost unaffected.

During the break-in period, the rings more or less assume the approximate shape of the cylinders, regardless of distortion. This wear pattern is conducive to satisfactory oil control and compression sealing until disturbed.

Any deviation from the original cylinder head torque application changes the form of cylinder distortion and, changes the piston ring-to-cylinder wall contact or wear pattern. This situation creates a potential blow-by and oil consumption problem. A condition of this kind is quite common, although the cause is seldom recognized as being related to improper cylinder head tightening.

To summarize: always retorque the cylinder head in the correct sequence, to factory specifications.

Bearing preloads are also measured with a torque wrench and, where applicable, are quite critical. Bearing preloads are required in the areas of the differential drive pinion, and at the differential case pedestal mounting.

Another location which may benefit from an established torque value is the front wheel tapered roller bearing adjustment.

As an example, one manufacturer's suggested wheel bearing adjustment is as follows: rotate wheel and drum while torquing the wheel bearing adjusting nut to 17-25 ft. lbs. torque. Back off the adjusting nut one-half turn and retighten to 10-15 in.lbs. while rotating the drum and wheel. Selectively position nut retainer on adjusting nut so that a set of slots are in line with cotter pin hole. Adjusting nut should not be rotated in this operation. Lock the adjusting nut and retainer with cotter pin so that cotter pin end does not interfere with seating of wheel static collector in spindle hole.

Journal bearings, both engine connecting rods and main bearings, require a specific torque. These specifications are supplied by the engine manufacturer and are included in the

torque specifications tables of each car make in this manual. The prime reason for torque in the case of journal bearings, is to assure positive security without distortion.

The automatic transmission valve body, assembly and attachment is another area critical to torque and possible distortion. By the very nature of most valve body designs, (sliding spring-loaded piston-type valves within a labyrinth of oil passages and bores) the slightest distortion may render one or more phases of transmission operation inoperative. Torque on the valve body fastener bolts is usually given in inch-pounds of torque.

A torque wrench is based upon the law of leverage—force times distance equals the twisting energy about a point. Founded on this principle, problems of torque, relative to torque wrenches, may be solved. Lever length is the distance from the center of the driven point or fastener being tightened to the axis of the torque wrench handle pivot. This is the point where the drive load should be applied.

Torque is twist; the standard unit of measure is foot-pounds, or inch-pounds in some cases of low or extremely critical torque. If the lever length is measured in feet and the force in pounds, torque developed should be quoted in foot-pounds. If the lever length is measured in inches and the force in pounds, torque developed should be quoted as inch-pounds.

Example:

Foot-pounds x 12 = inch-pounds
Inch-pounds ÷ 12 = foot-pounds

In some tight spots of poor accessibility, an adapter may be used. It may, or may not, add length to the effective leverage of the torque wrench. If it does add length, the total length of the assembly is the sum of the length of the adapter, plus the lever length of the torque wrench. (See the illustration.) Sometimes, damaged threads or rust interference may cause an incorrect torque reading. To overcome the difficulty of obtaining a correct reading in a case of damaged or rusted threads, simply add the torque resistance of the untightened nut or bolt to quoted specifications. Be

sure that the rundown resistance is measured near the last rotation, as burr or other interference may have become modified at this point.

In reaching a final correct torque reading, it is quite possible to experience a seizing manifested by a cracking noise at the nut or bolt. Should this set or seizing take place, it is necessary to break the set by rotating the fastener in a loosening direction, then again revert to tightening with a steady pull on the torque wrench. During this smooth tightening procedure, the torque reading should be taken.

Cylinder Head Tightening Sequence

1. Tighten all head fasteners to one-third specified torque.
2. Apply a second tightening two-thirds final torque.
3. Retighten all fasteners to full torque, in proper sequence.
4. Retighten to full torque again in reverse sequence to insure against any untorqued fasteners.
5. After running engine, retorque to final specifications and in normal sequence.

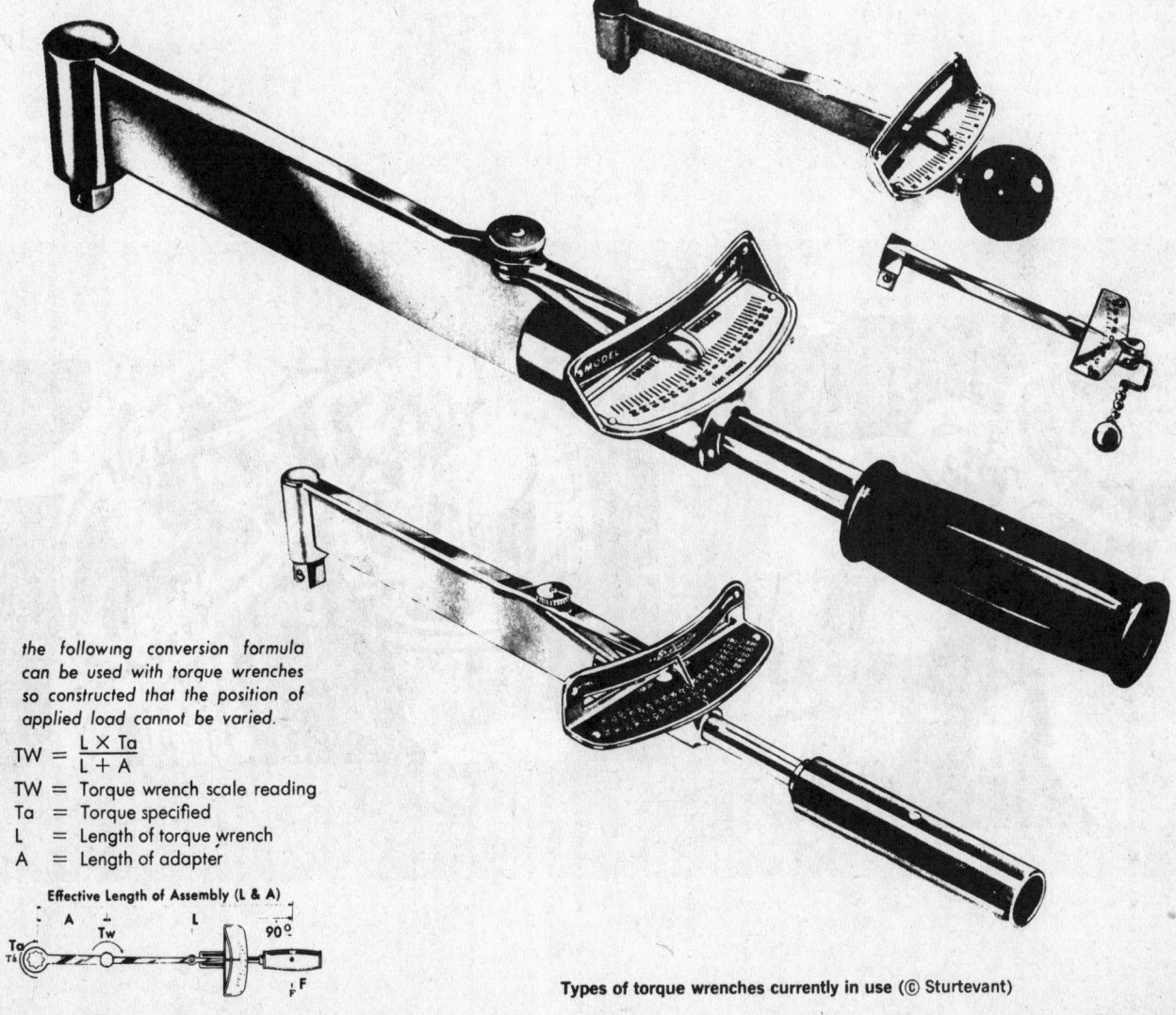

the following conversion formula can be used with torque wrenches so constructed that the position of applied load cannot be varied.

$$TW = \frac{L \times Ta}{L + A}$$

TW = Torque wrench scale reading
Ta = Torque specified
L = Length of torque wrench
A = Length of adapter

Effective Length of Assembly (L & A)

Types of torque wrenches currently in use (© Sturtevant)

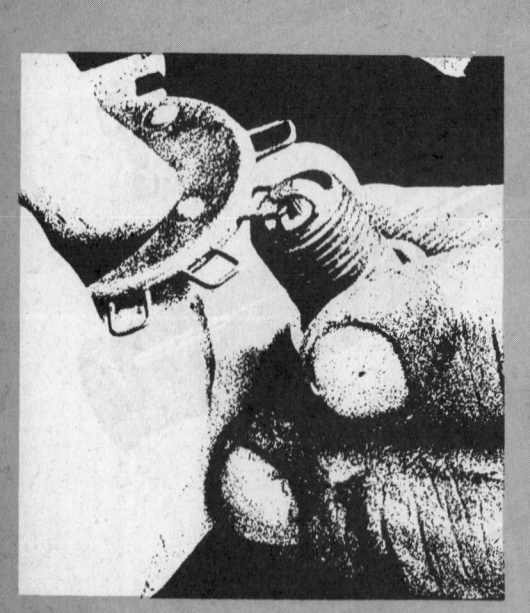

Troubleshooting Section

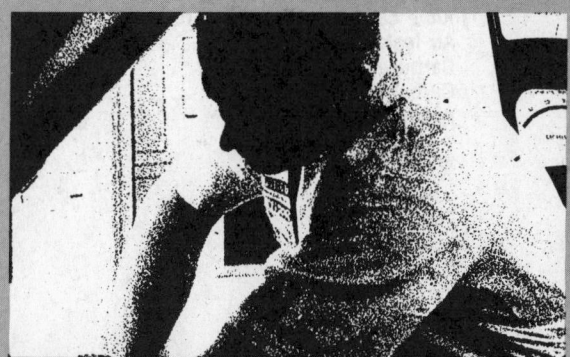

Troubleshooting and Tune-up

Index

IMPORTANT: Many of the diagnostic tests mentioned in this section can be found, in more complete form, in the section entitled Basic Electrical Diagnosis in this manual. This, in particular, applies to ignition system tests and starter motor tests and service.

Diagnosis

Diagnosing engine problems requires a combination of knowledge, mechanical skill and good test equipment. However, testing equipment is only as good as the skill of the person operating it.

Engine trouble can be separated into four general groups:
1. Won't start
2. Performs poorly
3. Improper lubrication
4. Noise

Won't Start

Assuming the engine will crank, there are four reasons why an engine won't start:
1. The combustion chamber is not receiving adequate spark.
2. The proper fuel-air mixture is not reaching the cylinders.
3. Poor compression.
4. Timing is incorrect.

Some of the above reasons overlap in scope. However, a thorough probe into all four factors should uncover the scientific fault.

Weak or No Spark To Combustion Chambers

1. Burnt or poorly gapped distributor points.
2. Short or open condenser.
3. Short or open coil.
4. Poor primary circuit from ignition switch to points in distributor.
5. Neutral safety switch bad or out of adjustment.
6. Low or dead battery.
7. On overdrive equipped cars, bad kickdown switch.
8. High starter draw.
9. Defective ignition switch.
10. High resistance anywhere in primary or secondary ignition circuit.
11. Moisture and dirt on distributor cap, ignition wires, and plug porcelains.
12. Cracked distributor cap.
13. Broken rotor.
14. Reversed coil polarity.

Improper Fuel—Air Mixture To Cylinders

1. Fuel tank empty or line to fuel pump blocked or open. Many car fuel pumps are now equipped with a replaceable filter in the tank.
2. Fuel tank not vented to atmosphere.
3. Weak or completely bad fuel pump. (Make fuel pump volume and pressure test.)
4. Fuel stoppage at carburetor filter or float level valve.

5. Leaking carburetor float.
6. Poorly adjusted float level.
7. Air, (vacuum) leak at carburetor or intake manifold, (beware of faulty double action fuel pumps and vacuum operated accessories).
8. Poor compression, (low vacuum reading).
9. Inferior fuel.
10. Obstructed or dirty carburetor air cleaner.
11. Leaky manifold heat riser.
12. Automatic choke stuck or out of adjustment.

NOTE: a choke thermostat spring that has fatigued needs replacement, not mere adjustment.

13. Engine valves not seating properly, burnt, sticky or out of adjustment.
14. Hydraulic valve lifter trouble (improper oil pressure, worn or stuck tappets).
15. Camshaft wear or breakage.
16. Valve timing incorrect.
17. Manifold heat control valve not working.
18. Vaporlock, under certain conditions.
19. Low engine operating temperature (below 170° F).
20. High engine operating temperature (above 190° F).

Poor Compression

1. Poorly seating exhaust or intake valves.
2. Worn or sludged-up rings.
3. Blown head gasket.
4. Valve timing incorrect.
5. Cracked or broken piston.
6. Beware of short connecting rods in cars equipped with dual engine options. It is mechanically possible to install the connecting rods of the shorter stroked engine into the bores and onto the crankshaft of the higher displacement engine. This error will shorten the piston stroke and result in low compression.

Timing Incorrect

1. Poorly gapped distributor points.
2. Distributor static timing incorrect.
3. Distributor governor advance, faulty.
4. Distributor, vacuum control, defective.
5. Distribution of secondary current to spark plugs, out of firing sequence.
6. Ignition timing not coincident with cylinder timing.
7. Valve timing incorrect.

Performs Poorly

The purpose of this manual is to point out, in concise form, the best and quickest route to factory performance standards. Special performance coverage is not possible here.

1. Distributor trouble, points, condenser, timing, etc.
2. Weak coil.
3. Bad or incorrect spark plugs or wiring.
4. Carburetor out of adjustment or broken.
5. Leaky intake manifold or carburetor gasket.
6. Leaky vacuum operated units or accessories.
7. Poor fuel—compression ratio—timing balance.
8. Faulty fuel supply.
9. Poor compression, valves, rings, gasket, etc.
10. Exhaust back pressure—clogged muffler or tail pipe.
11. Clogged air filter or inlet.
12. Cooling system failure.
13. Leaky heat riser or stuck heat control valve.

Improper Lubrication

1. Low oil level.
2. Very high oil level (causing foaming).
3. Clogged oil pump screen.
4. Oil pump sucking air, loose pump mount inside crankcase.
5. Dirty oil.
6. Clogged or saturated oil filter.
7. Oil contamination, (such as water, acid, gum, abrasives, antifreeze, etc.)
8. Improper viscosity or quality of oil.
9. By-pass valve stuck.

NOTE: all name-brand oils are good, and engine manufacturers' recommendations should be followed. However, extreme service exposure may alter engine requirements necessitating more frequent oil changes, heavy duty oil, use of detergents, etc.

Noise

As long as combustion and reciprocating engines are used to furnish power, a certain degree of noise will be present. The best we can do is to keep the noise level as low as possible.

To catalogue all of the causes of noise would require coverage larger than one volume will permit. We therefore will limit our list to mechanical causes most common to the automotive gasoline engine.

1. Valve Noise.
   A. Sticky valves.
   B. Stuck hydraulic tappets.
   C. Dry hydraulic tappets.
   D. Bent valve stem.
   E. Warped valve head.
   F. Broken valve spring.
   G. Tappet out of adjustment.
   H. Rocker arm loose.
   I. Pushrod bent or worn.
   J. Rocker shaft worn or loose.
   K. Valve or tappet guide worn.
2. Piston Noise.
   A. Collapsed or broken piston.
   B. Scored piston or cylinder wall.

C. Broken piston rings.
D. Carbon on piston and cylinder head making physical contact.
E. Top cylinder ridge striking top piston ring.
F. Loose or broken piston pin.

3. Crankshaft Bearing Noise.
A. Loose connecting rod bearing.
B. Loose main bearing.
C. Bent connecting rod, (sounds like bearing).
D. Loose flywheel or converter, (sounds like bearing).
E. Loose crankshaft damper or pulley, (sounds like bearing).
F. Crankshaft end-play, (noticeable when actuating clutch).
G. Crankshaft misalignment.

4. Detonation.
Detonation is spontaneous combustion within the cylinder. It is caused by an imbalance of compression ratio, heat, fuel value, and timing. Detonation can be annoying, wasteful and very destructive to engine parts.

The following factors or combinations should be checked as contributing elements.
A. Cooling system temperature.
B. Spark plug appearance.
C. Ignition timing index.
D. Fuel octane rating.
E. Incandescence within the combustion area caused by carbon particles, sharp edges, burnt spark plugs, etc.
F. Lean carburetor mixture.
G. Stuck manifold heat-control valve.

Professional Approach

The Oscilloscope

It is a time consuming job even to remove spark plugs from some of the V8 engines. Therefore, it is good logic to test engine components while performing their normal duties under various conditions. The oscilloscope will do this.

While this tool is no cure-all, it does produce a television screen picture of the situation as it exists within the running engine.

This picture can be translated into functions of the engine that are familiar to the mechanic.

A good oscilloscope can determine:
1. Compression balance.
2. Condition of spark plugs.
3. Condition of distributor points.
4. Ignition coil or condenser problems.
5. Bad ignition wiring.
6. Distributor point dwell.
7. Cracked distributor cap.
8. Worn or broken rotor.
9. Worn distributor points.
10. Bad secondary wires and terminals.
11. Reversed polarity.
It is not possible here to outline the various oscilloscope hook-ups. Follow the manufacturer's instructions.

The following operations depend largely upon the results of the oscilloscope picture.

Compression Check

Due to the added torque of modern engine starting systems, plus changes in engine camshaft design, it is no longer practical to determine cylinder compression balance by listening to the starter while cranking the engine. For positive cranking pressure readings, use a cylinder compression gauge.

Because engine design and other factors influence compression, a pressure test of all of the cylinders in any one engine will furnish a yardstick of comparison for that particular engine only. Therefore there can be no table established for compression pressure standards.

1. Run the engine until normal temperature has been reached.
2. Carefully remove spark plug wires.
3. Blow foreign matter from spark plug wells. Then loosen plugs one turn.
4. Replace plug wires, start engine and snap throttle open once or twice (this should clear the engine cylinders of dislodged carbon particles).
5. Stop engine and remove plug wires and spark plugs.
6. Remove carburetor air cleaner or cleaners and (depending upon intake manifold design) block

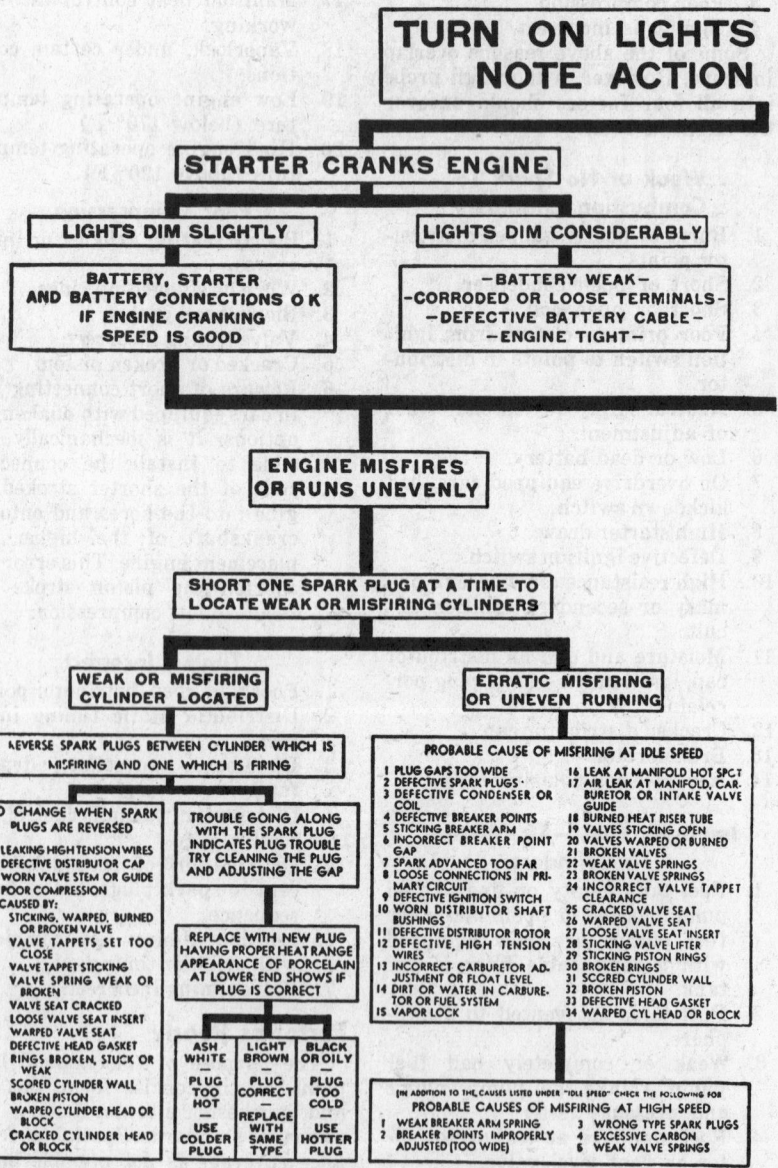

throttle valve or valves wide open.

7. Hook up starter remote control cable and switch. Then, using a compression gauge in the cylinder, crank the engine through about four compression strokes. Record the highest gauge pressure for that cylinder.

8. Proceed with remaining cylinders in the same manner.

9. Some variation in cylinders is to be expected; however, the difference should not exceed 15 lbs.

Poor Compression

In the event of poor compression in two adjacent cylinders of a multicylinder engine, the probable cause is a blown head gasket between them. This condition can generally be identified by an identical pressure reading for the two cylinders. The condition can also be spotted by a popping back through the carburetor.

Low compression may also be the result of a leaky head gasket at a point between the compression chamber and a coolant passage of the water jacket. This will usually affect individual cylinders and will probably manifest itself as excess water in the crankcase. A rusty dipstick or water droplets in the oil filler cap is also a good indication of coolant entering the crankcase. A bad coolant leak into the combustion chamber may stop the crankshaft from turning. The coolant hydraulically prevents completion of the compression stroke in the bad cylinder. Any further attempt at cranking the engine, under these conditions, can cause severe damage.

To locate a compression leak, if the cooling system is not involved, pour a tablespoon full of SAE 30 oil into the combustion chamber. Crank the engine to spread the oil. If the gauge reading is 10 psi. or more higher than the original reading, bad rings are a probability. If the compression pressure has not been improved, the trouble could be valves. However, one other remote possibility would be a crack or hole in the piston that has been temporarily sealed by oil.

Compression Troubles—Causes

Generally, compression troubles show up in the form of hard starting, oil consumption, or poor engine performance, (especially at idle). There is also the possibility of backfiring at the carburetor and afterburning in the muffler.

The best way to check compression pressures is with a compression gauge inserted into the combustion chamber at the spark plug hole. Crank the engine at least four revolutions, then record the reading. All cylinders should read within 20 lbs. of specified cranking pressures and be within 20 lbs. of each other.

1. Worn or sludge choked rings.
2. Broken rings or pistons.
3. Bent or wrong connecting rod or piston.
4. Blown or leaking head gasket.
5. Hydro-locked engine.
6. Leaky cylinder block, or heads.
7. Poorly seated valves.
8. Incorrect valve timing.
9. Broken timing chain or gears.
10. Worn or broken camshaft.
11. Insufficient valve train lash.

Compression Troubles—Corrections

1. Worn or sludge choked rings do, in extreme cases, leak compression so badly that starting is impossible. This is a condition of gradual development, however, and not likely to happen overnight. When ring condition has deteriorated to this degree, oil consumption, smoky exhaust, plug fouling and general poor performance is evident. To isolate a ring condition from other causes of poor compression, take a compression reading in the regular manner and record the results. Then, squirt about a tablespoon full of SAE 30 oil into all the cylinders and take a second compression test and record these readings. An appreciable increase in pressure reading will indicate probable ring failure or sludging. In either case, this evidence warrants removal of at least one piston assembly for visual inspection.

If, upon examination, rings or piston ring grooves are worn, renew the

SHOOTING
LACK OF POWER

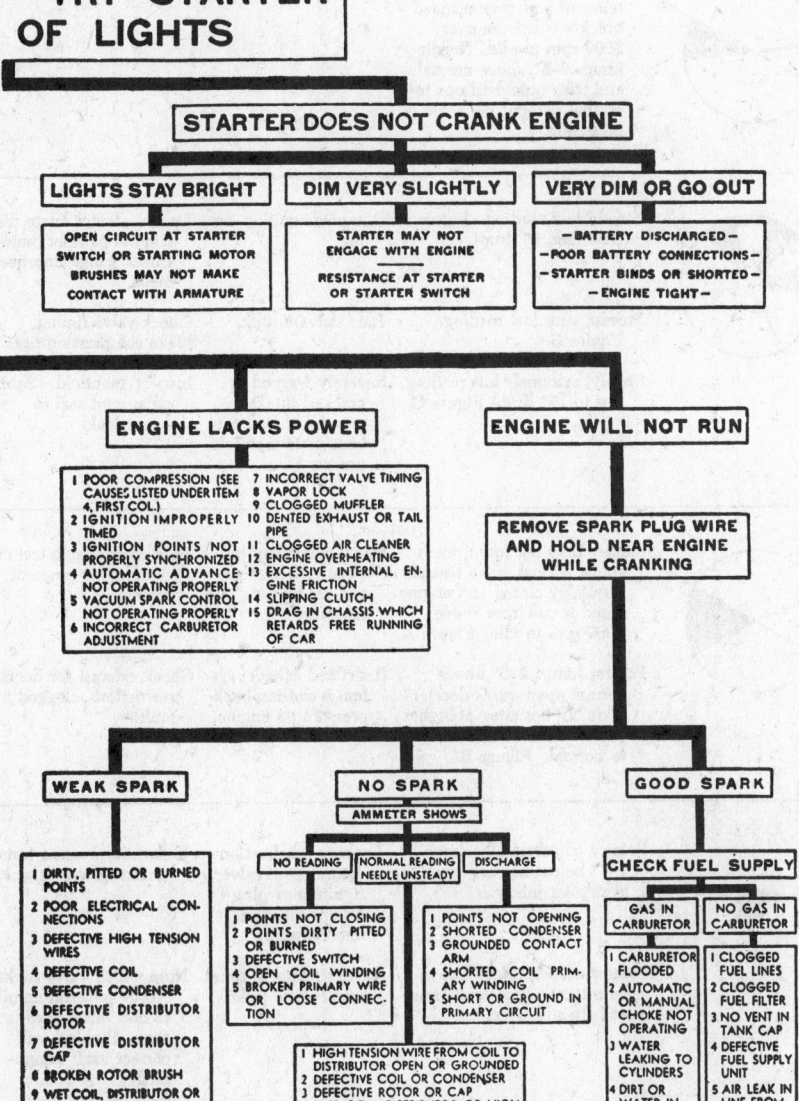

necessary parts. If rings and grooves are merely choked with carbon and sludge, carefully clean out the piston grooves and drainback holes, then install new, correctly sized rings. Never reinstall the old rings.

2. Broken rings or pistons cause poor compression in varying degrees, proportional to the amount of breakage. It is possible, but not likely, to have breakage in all of the cylinders of a multi-cylinder engine. Breakage in all, or most, of the cylinders could result in hard starting and poor performance in addition to oil consumption and other difficulties. However, if the breakage is confined to one or two cylinders of a six or eight cylinder engine, poor idle, stalling and an uneven flow of power will be the result. To isolate the trouble, check for the miss (electrically). Remove the spark plugs from suspicious cylinders and check for fouling, then take a compression reading on all of the cylinders. All cylinders should read within 20 psi of each other. The piston will have to be removed from the low compression cylinder to make any kind of correction. At this time, (piston removed) a visual inspection should pinpoint the trouble—whether it is rings or piston. Renew the offending part.

3. A bent rod is a short rod, but a wrong rod or piston can be either too short or too long.

A bent connecting rod will reduce (proportional to the degree of the bend) the total compression potential of that cylinder. A bent rod, unless it is severe enough to interfere with crankshaft rotation, may cause a noise, but should not result in hard starting. It may, however, due to low compression pressure, cause rough idling and poor performance.

The presence of wrong parts, (piston or connecting rod) in an engine that is in service is rather remote but entirely possible.

If the wrong piston or rod results in the piston extending too far above normal travel, interference would occur and something would surely lock up or break.

In the case of a short rod, a piston designed for lower compression ratio, or a piston built with less than proper length above the piston pin, lower compression pressures will result in that cylinder than in other normal cylinders of the same engine. If a compression gauge reading indicates this condition, remove the cylinder head/heads and compare the piston top position with that of its opposing cylinder. With the exception of the 90°, V6 (Buick, Olds, Tempest, etc.) the comparison cylinder is the one directly opposed in firing sequence to the one in question. If the piston in the bad cylinder is at a lower point of stroke than its comparison piston, remove the piston and rod assembly for visual inspection. Also determine the cause. Coolant in the combustion area at the time of cranking is the usual cause of a bent rod.

4. A blown or leaky head gasket may manifest itself in more than one way. It may or may not result in extensive engine damage. At the first sign of an engine exhaust or coolant leak, locate and correct the trouble.

An engine cylinder exhaust leak, except in the case of severe piston and ring damage (where the exhaust leak is into the crankcase) is usually at the cylinder head gasket. This leak may be an external one, and it is generally audible and fairly easy to locate with an oil can or by listening.

The more difficult ones are those which connect adjacent cylinders, or those which leak exhaust gases into the cooling system, and coolant into the cylinders.

When a cylinder head gasket is open between adjacent cylinders, there are symptoms of crossfiring and loss of power with poor idle and bad low speed engine performance. There is also a possibility of backfiring at the carburetor. Do not operate a car, while in this condition, without the carburetor air cleaner. The air cleaner is a pretty good flametrap in preventing fire. Proof of this type cylinder head leak is indicated by identical low compression readings in adjacent cylinders.

When the cylinder head gasket is open from the cylinder compression

	Vacuum Reading	Possible Reasons	Next Test
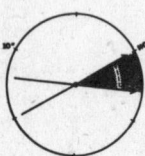	Steady gauge reading 18–20″ at all speeds. Throttle is released and engine speed quickly cuts from over 2000 rpm to idle. Needle jumps 2–5″ above normal and then quickly drops to normal without pause or hesitation.	Normal engine performance.	Vacuum okay. Go to compression tests.
	Steady low reading. Figure A. (less than 2″ drop)	Retarded ignition timing.	Loosen clamp, rotate distributor to reset timing. Check gauge improvement.
	Steady, very low reading. Figure B.	Late valve timing.	Check valve timing. Make compression tests.
	Steady, extremely low reading. (up to 15″ drop) Figure C.	Severely warped or cracked intake manifold. Bad carburetor-to-manifold gasket.	Inspect manifold. Squirt oil around seal to detect leak.
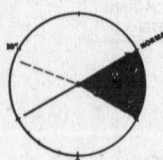	Pointer does not jump much above normal when throttle is quickly closed and engine speed is cut from above 2000 rpm to idle. Figure A.	Piston rings may be worn or defective and are blowing into crankcase.	Take compression test of cylinders to pinpoint trouble.
	Pointer jumps 2–5″ above normal upon quick deceleration but hesitates at higher pressure before returning to normal. Figure B.	Restricted exhaust system is causing back-pressure on engine.	Check exhaust for dents, restrictions, clogged muffler.
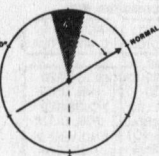	Pointer (rhythmically drops 1–7″ below normal vacuum at regular intervals.	Leaking combustion chamber or valve; (ignition or plug failure involving one cylinder).	Make compression tests; (make ignition check).
	Pointer drops rapidly but intermittently (not every time) and then recovers.	Valve sticking at times won't close tight.	Note which valve sticks. Apply penetrating oil to one valve guide at a time. Problem will correct itself temporarily.

chamber to the cooling system, one or more symptoms may be present. There may or may not (depending upon the extent of leak) be an engine miss due to a wet spark plug tip. There may be indication of water in the oil pan. This should cause droplets of condensation to form on the oil dipstick and the underside of the oil fill tube cap. This sort of leak can cause extensive engine damage, especially to connecting rods and pistons. There is also the threat of antifreeze damage to other internal engine parts.

If the leak is very small, (a temporary condition) there may be a tendency of the coolant to show foam or bubbles in the expansion tank of the radiator. This is usually the first indication of leak between the combustion area and the cooling system and should be corrected immediately.

The most acute stage of head gasket coolant leak is hydraulic lock. This is an advanced condition brought on by coolant leaking into the cylinder while the engine is stopped (usually overnight). Due to the incompressibility of water, the piston is locked in the cylinder on its compression stroke. Result, a bent rod or worse.

To identify cylinder head leak and isolate it requires some simple deduction.

First, identify the leak as an exhaust leak (external or between cylinders). If it is this type leak, the compression gauge, plus some common sense, will localize it. Remove the head, plane or otherwise resurface it, and reinstall it with a new head gasket. Torque head to manufacturer's specifications.

Second, if the leak is into the cooling system, a cooling system (radiator) pressure gauge should reveal the trouble. This gauge, installed in place of the radiator pressure cap and pumped up to about 20 psi., should retain this pressure for about 5 minutes.

NOTE: leaving the gauge on for extended periods may be misleading, because the engine cools and pressures drop due to natural contraction.

An appreciable drop in cooling system pressure indicates a leak somewhere in the cooling system. A visual inspection of spark plugs is a good barometer of presence of water in a cylinder.

Remove the cylinder head, and depending upon the degree of leak and how long the leak has existed, do operations necessary to restore the engine to normal.

NOTE: if permanent antifreeze has circulated through the lubrication system of the engine for any length of time, damage to internal engine parts could be extensive.

While the following procedure is no panacea, it will remove the residue resulting from ethylene glycol leaking into the engine. Pour a solution of 60% engine oil and 40% butyl cellusolve in the crankcase. Run the engine at idle for ½ hour. Drain the oil pan, then, flush by running the engine at idle with a mixture of 60% engine oil and 40% kerosene. Drain and refill crankcase with appropriate viscosity engine oil.

5. A hydro-locked engine mistakenly diagnosed as a bad starter, or a weak battery, is likely to produce a bent connecting rod or worse. The cause of engine lock-up should be thoroughly investigated before further efforts to crank the engine are attempted.

6. Leaky cylinder block or heads is a rather rare condition, but one that does occur now and then. It may be due to porosity, a crack, or the result of overboring a cylinder. There is also the possibility of steel sleeve-to-cylinder head chafing of the head gasket on aluminum engines. This action is slow in developing, but may cause a leak at the head gasket. A combustion chamber containing very little, or no carbon, is a pretty good indication of water having been present in the cylinder.

7. Poorly seated valves may be the result of insufficient valve lash, weak or broken valve springs, loose or cracked valve seat inserts, sticking valves, worn valve stems or guides, and warped or burned valve heads. Excessively high oil pressure has occasionally prevented valves from seating in hydraulic tappet-equipped engines.

Vacuum Reading	Possible Reasons	Next Test
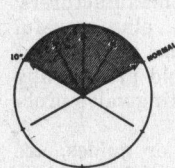 Pointer wavers rapidly between 10-20″ at idle becoming worse with higher rpm.	Weak or broken spring causing valve to close slowly.	Remove valve covers; check condition of springs.
Wavers irregularly at idle; fluctuates rapidly in smaller range at higher rpm.	Manifold leak at intake port—upsets and reduces cylinder draft.	Squirt oil around manifold; check vacuum increase when oil fills leak. Replace faulty gasket.
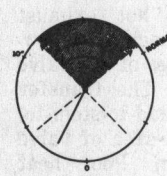 Drifts at idle; stabilizes at higher engine rpm.	Burnt valve; combustion chamber leak.	Make compression tests.
 Wavers irregularly in one range despite engine speed.	Unbalanced carburetion; improper spark plug gap, ignition timing; poor valve seating.	Adjust carburetors; check plug gap; check distributor and advance spark; make compression tests.
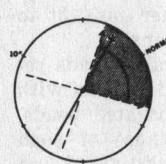 Vacuum averages lower than normal at idle, needle fluctuates almost 3″ on both sides of normal.	Worn valve guides admitting air—upsetting carburetion.	Squirt oil on guide seals. Check vacuum improvement.

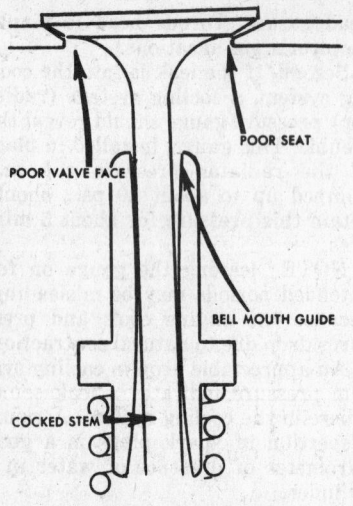

Valve stem and seat wear

Aside from noise, malfunctioning of any part of the valve train will influence intake manifold pressure, and engine performance. This is readily detectable through the use of a vacuum gauge. See Unit Repair Section.

Valve seat angles

A compression gauge test on the engine should isolate the troublesome cylinder, or cylinders. See Unit Repair Section.

Insufficient valve lash is a contributing factor in causing burned valves because it reduces the dwell time of the valve face contact with the valve seat. Heat transfer of the hot valve head to the comparatively cool valve seat is reduced, causing valve temperatures to mount beyond their heat tolerances. This results in valve head warping and burning. In addition, reduced valve lash changes the intended valve timing pattern of opening, closing, and duration. See section on vacuum gauge for correct method of valve lash adjustment.

Weak or broken valve springs may, or may not, cause noise. The condition and degree of valve spring failure controls the amount of influence this situation has on engine performance. To detect some cases of weak or broken springs, a visual inspection of the valve springs may be required.

Loose or cracked valve seat inserts act the same as poorly seated valves from any other cause. They can be detected with a vacuum gauge and located with a compression gauge; then identified by visual inspection.

A loose or damaged valve seat insert may be replaced by removing the old seat with a special valve seat insert remover or, by drilling two holes in the old insert (on opposite sides). Be very careful to use a drill of smaller diameter than the width of the seat insert, and not to drill all the way through the insert. Split the seat into two halves with a sharp chisel. Loose seats, in many cases, may be lifted out without drilling or chiseling.

After the old seat has been removed, thoroughly clean the seat counterbore of head or block (depending upon engine design). The new insert should measure from .002 to .004 in. larger O.D. than the I.D. of the counterbore, both measurements taken at room temperature. It may be necessary to enlarge the counterbore to accommodate an oversize valve seat insert. Special equipment is necessary for this operation.

When satisfied that the counterbore is clean and the correct size for the replacement insert, (.002 to .004 in. less than the O.D. of the insert) pack the insert in dry ice. Permit insert to remain in dry ice for 10 minutes. It will now drive into the counterbore with an adequate flat-surfaced driver.

These inserts are exceptionally hard and require a high speed reseating grinder to obtain a satisfactory contact angle for the valve face.

Sticking valves manifest themselves by clicking (valve tap), engine miss, or both. Valve stem-to-guide clearance has been reduced by build-up of varnish and hard carbon to the extent of preventing the valve spring from returning the valve to its seat. As varnish and carbon (both good insulators) start to form on the valve stem and in the guide, heat transfer is reduced and valve operating temperatures rise. As temperatures go up, valve seats burn and temperatures rise even higher. This is a vicious cycle, repairable only by grinding valves and cleaning carbon, and replacing necessary parts.

Advanced engine design, better temperature control, and improved lubricating oils have greatly reduced the sticking valve problem. However close observance to manufacturers' valve specifications, plus careful workmanship and the renewal of valve seats, should prevent any recurrence of the sticking valve problem.

Worn valve stems or guides, are very closely related to sticking valves and require the same corrective measures.

Worn stems or guides permit misalignment of the valve face with its seat. This, in addition to affecting compression, permits hot exhaust gases and carbon to reach and insulate the bearing surfaces of the valve stem from the guide. The transfer of heat from valve head-to-stem-to-guide is the most vital area of valve cooling in the system. Poor heat transfer results in a ruined exhaust valve.

In the case of worn intake valve stems, or guides, loosely fitting valve guides permit poor seat alignment, plus the passage of excess oil into the combustion chambers. This condition contributes to oil consumption, carbon and varnish formation and early oil control ring failure due to gum and carbon. If valve stem-to-guide clearance is greater than specifications, renew the guides; the valves too, if the stems are worn.

Cylinder heads with guides cast integrally with the head, can be reamed to accommodate valves with oversize stems. In either case, fit to factory recommendations.

Warped or burned valve heads require replacement, as do valves with stretched stems or separated heads.

In any event, the cause of such condition should be investigated and corrected to prevent repetition of the same failure.

Typical valve train
(© American Motors Corp.)

Valve heads that have pulled off their stems, or stems which have stretched, are both good proof of excessive valve spring stiffness.

8. Seriously incorrect valve timing is unusual except in instances of careless installation of gears or chain at the time of replacement.

There is, however, a possibility of chain jumping when there is extreme wear in chain and sprockets. Very old engines of three point chain suspension are also subject to this fault if proper chain tension is neglected. There is also the possibility of tim-

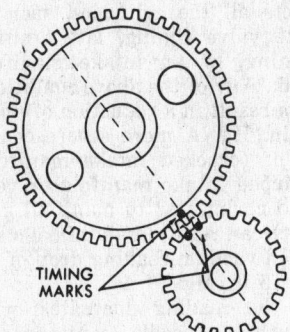

Timing mark alignment—gear

ing drive belt jumping on the Pontiac Tempest 6 cylinder overhead cam engine, if the drive belt is out of adjustment.

This ingenious design of belt drive for the camshaft and the distributor-fuel pump-oil pump assembly is three point suspended and critical to tension adjustment. An engine which is out of valve timing, to any great degree will not run. Any attempt to crank it may damage valves and/or pistons due to interference (overhead valve engines). An engine which is out of time by 180° will not only fail to start, but may catch fire if an attempt is made to start it.

A jumped chain, depending upon the number of links and sprocket teeth involved, will affect engine performance in varying degrees from sluggish response to complete failure and possible damage.

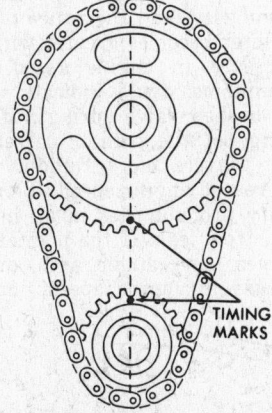

Timing mark alignment—chain

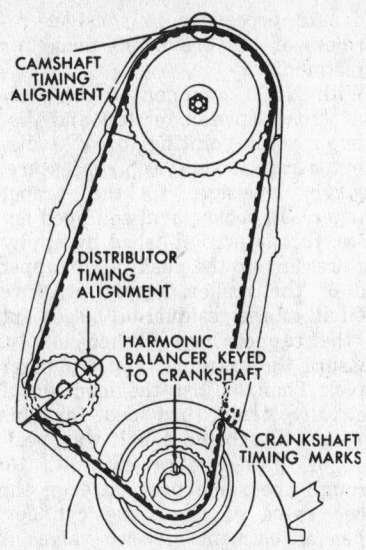

Timing mark alignment—OHC belt

Because the distributor is usually driven by the camshaft (Corvair, Tempest ohc-6 cylinder, and Volkswagen are exceptions) valve timing is relative to distributor timing. A camshaft or crankshaft out of time by only one tooth, or link, will influence distributor timing. In fact, a quick check for a jumped chain is to match spark timing with the crankshaft timing index. A compression check should reflect any change in valve timing by an overall low reading. A vacuum gauge reading will also show lows in inches of mercury.

Regardless of cause, (careless installation of chain or gears, or jumped and out of register valve timing) the timing case cover must be removed to re-establish correct timing.

NOTE: after establishing correct valve timing, take a compression reading to affirm, or rule out, the possibility of valve or other valve-to-piston interference damage.

9. A broken timing chain, or gears, provide obvious evidence. Except for Corvair, Tempest ohc-6 cylinders and Volkswagen, the distributor rotor will not rotate when cranking the engine. The above three cars drive their distributors from the crankshaft. Their drive is independent of camshaft action.

A positive check for broken timing chain or gears involves removal of rocker covers and watching for valve train response to crankshaft rotation. No reaction indicates a broken drive chain, broken sprocket, broken gear, or a broken camshaft. In any case, it requires removal of the timing case cover for visual inspection. Replace necessary parts, then take a compression test to check the possibility of valve-to-piston interference damage.

10. A worn camshaft and a broken camshaft are worlds apart in symp-

toms. A worn camshaft presents a problem in diagnosis, depending upon the degree of wear. It produces varying results on a compression gauge and hard to recognize values on a vacuum gauge.

The simplest and most positive check is to measure intake and exhaust valve lift of all of the cylinders with a dial indicator. Then, compare these readings with each other, and with the specifications given in the Valve Specifications Charts of the car involved. These charts are near the beginning of each car coverage, in the Car Section.

A worn camshaft can usually be attributed to one of two things, poor lubrication or excessively heavy valve springs. In either case, renew the camshaft, (and possibly the tappets) weigh valve springs, and, if necessary, replace them with springs of standard pressure and length. And finally, clean the oil supply area and use high quality oil of correct viscosity.

A broken camshaft will prevent the engine from running. It will react according to the location of fracture. It is most likely to break at or just behind the gear or sprocket. In this case it will produce symptoms of a broken chain or gears.

If the shaft is broken anywhere else along its length, the valve train servicing cylinders beyond the break will be dead. Valve action beyond the break will stop, while action ahead of the break will respond to crankshaft rotation (assuming there is no valve head-to-piston interference to prevent the crank from turning). At any rate, a new camshaft is in order.

NOTE: it is very likely that valve head or piston damage has resulted from engine out-of-time crankshaft rotation. Take a compression test of all of the cylinders.

11. Insufficient valve train lash, or not enough tappet clearance, is a very common and costly condition that goes undetected for long periods. It goes unnoticed because there are no accompanying noises as there are with excess tappet clearance.

The prime damage resulting from too little tappet clearance is valve burning.

Tappet clearances less than specifications, tend to hold the valve off of its seat for greater periods of time. This upsets the predetermined valve-to-valve seat heat transfer time, resulting in valve failure.

If tappet clearances are extremely small, valve expansion may prevent the valve from seating at all.

Another aspect of less-than-standard tappet clearances is the effect it has on valve time planning. Engine designers, to achieve certain results, have built into the breathing pattern of the engine, a particular opening

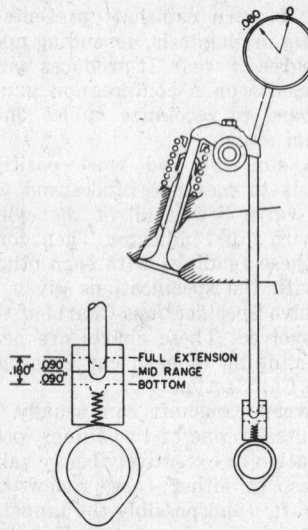

Measuring hydraulic tappet clearances

and closing time for both intake and exhaust valves (valve timing). There is also a period (in degrees of crankshaft rotation) during which both intake and exhaust valves are open (valve overlap). This timing design is planned with specific tappet clearances in mind. Any deviation from established tappet clearances (valve lash) upsets the entire timing pattern and reduces engine performance.

Adjust mechanical tappets to factory specifications according to hot or cold recommendations. If specifications call for a hot adjustment, run the engine long enough to stabilize engine temperatures. Also, if components of the valve train have been disturbed in any way, it is good practice to run the engine long enough to settle mechanical clearances to normal running fit before final adjustment. Adjust hydraulic tappet systems at that point of the valve train designed for this purpose.

Hydraulic tappet systems are designed to operate at zero lash. All valve train lash is taken up hydraulically by tappet design, and the valve functions with no clearance.

All hydraulic tappets have a built-in plunger travel-limit, through which extremes in valve train variables are compensated. An ideal adjustment of a tappet is, therefore, midway in its plunger dry travel limit. This allows (depending upon design) about .060-.080 in. above or below the center of plunger travel for automatic hydraulic positioning of the plunger. While these are typical tolerances, they are safe and illustrate the latitude of hydraulic plunger limits.

There are factory recommended procedures for adjusting hydraulic tappet clearances. See Valve System in each car coverage, in the Car Section.

However, the following is a typical and safe procedure to consider regardless of the provisions made for adjustment:

With No. 1 cylinder on compression stroke (piston on top and distributor rotor pointing to No. 1 distributor cap tower) apply pressure, through leverage, to the tappet plunger. On rocker arm equipped engines, this is accomplished by applying pressure to the pushrod or tappet end of the rocker. Apply pressure until all of the residual oil bleeds out of the tappet body. When empty, measure the extreme tappet plunger travel. Then, adjust the member of the valve train, provided for this purpose, to achieve a half-way position of plunger travel limit. Go through the same procedure on the other valve of the same cylinder. When adjustment of both valves is complete for No. 1 cylinder, proceed to the next cylinder in firing order, and repeat the operation just completed.

This method is more time consuming than some factory recommended practices, but it is a safe and typical procedure that works.

Vacuum Gauge

The vacuum gauge can be a very useful instrument in the hands of an experienced mechanic. It does, however, have its limitations. To know these limits and to use the vacuum gauge as intended is a vital part of engine diagnosis.

Due to atmospheric variables, engine design (even with engines of the same basic series) and many other factors, it is impractical to establish a fixed vacuum gauge reading for any make or model engine. Still, a vacuum reading, in inches of mercury (in. Hg.), is a very reliable index of performance value for comparison purposes.

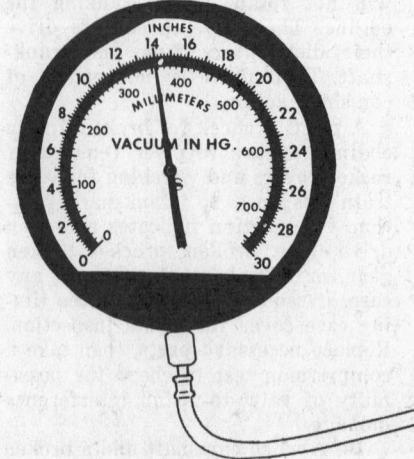

Vacuum gauge

Primary Engine Analysis

Most engines are equipped with some sort of plug or connection in the intake manifold. Remove the plug or fitting and replace it with a suitable vacuum gauge adapter. This connection must be at some point on the intake manifold, on the engine side of the carburetor throttle plate, (the underside of a downdraft carburetor). Connect a tachometer to the engine, and set the idle speed to specifications, in neutral. While running the engine, observe the following:

1. If the reading is low but steady, it indicates a condition that affects all the cylinders, such as late valve timing, late ignition timing, or an intake manifold leak. Any of the above conditions can result in a reduction of up to 2 in. Hg. A more severe condition, (cracked intake manifold, warped intake manifold or very bad carburetor to manifold gasket) can result in an intake manifold vacuum reading drop of up to 15 in. Hg.

2. If the reading fluctuates, with rhythm, (needle continues to waver with a regular pattern) it indicates trouble in one area. A spark plug may be fouled, a valve seat burned, a head gasket blown, a distributor cap cracked or tracking, or any number of things that would upset normal combustion in one cylinder.

3. A wavering and irregular gauge needle may be caused by any of the conditions listed in the above paragraph, but with the conditions existing in more than one cylinder. Other items that may cause an irregular gauge needle are poorly adjusted carburetor, high carburetor float level, wear in distributor shaft or bushing, faulty distributor advance control mechanism, excessively advanced or retarded spark timing, dual ignition points unsynchronized, wide spark plug gaps or poorly seated intake valves.

4. If the pointer wavers with irregularity, and the range of the pointer becomes greater with an increase in engine speed, the symptoms would indicate weak or broken valve springs. If the range of the pointer decreases, but wavers more rapidly with increased engine speed, the difficulty probably lies in an intake manifold leak. If the pointer becomes more stable with an increase in engine speed, indica-

tion would be toward trouble in the carburetor or ignition.

5. Normal response to sudden throttle opening and closing is a wide sweep of the gauge needle. If this does not occur, there must be leakage in the intake system. If failure of manifold gaskets or valve seats has been explored, poorly sealing piston rings are a possibility.

6. If the distributor timing is advanced to record the highest steady vacuum gauge reading, then retarded to reduce the gauge reading ½ in. Hg., and if the gauge reading is holding steady within the normal range for this engine, read the spark setting with a timing light. Ignition timing should index to specifications, plus or minus 2°.

Specific Conditions

Low Vacuum Gauge Reading

If the gauge needle registers a low reading, a correction may be effected by an ignition timing adjustment. In the event that this does not correct the situation, explore the possibilities of a leak in the intake system by the following procedure:

1. Remove the carburetor air cleaner.

2. Disconnect the primary lead from the ignition switch side of the ignition coil and, if using an auxiliary starter switch, have the ignition switch in the on position before bumping the starter. On some cars, failure to observe this precaution can cause damage to the grounding circuit in the ignition switch.

3. Crank the engine for about ten revolutions, while the hand is held tightly over the carburetor air horn. A vacuum gauge reading during the cranking period should register 13 in. Hg. or better. Individual engine design characteristics prevent the establishment of exact values for this test. If the reading is low, the next step should narrow down the source of the leak.

4. Remove the carburetor and very thoroughly seal off the carburetor opening in the intake manifold. Repeat the same cranking procedure as in the previous step and watch the gauge reading. If the reading is higher than in the previous test, the leak is in the carburetor, or at the carburetor to manifold mounting-flange. Some of the carburetor vacuum leaks could be in the distributor vacuum control line, the throttle shaft, a cracked or broken carburetor body or mounting flange, or a bad mounting gasket.

5. Another means of detecting vac-

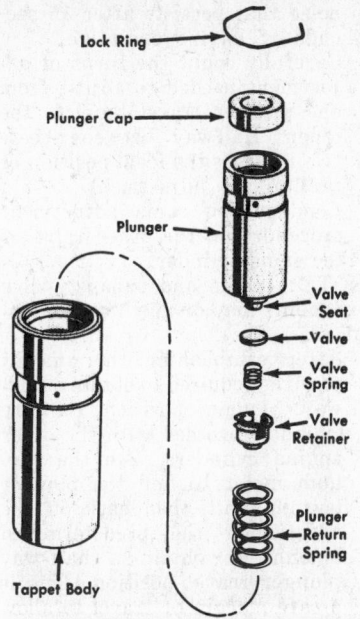

Lock Ring

Plunger Cap

Plunger

Valve Seat

Valve

Valve Spring

Valve Retainer

Plunger Return Spring

Tappet Body

Typical hydraulic lifter
(© American Motors Corp.)

uum leaks is by squirting oil at the suspected point and watching the vacuum gauge needle. The gauge needle should rise when oil is applied to the spot.

6. Exhaust restrictions (back pressure) can also cause a low vacuum gauge reading. The engine must breathe freely through both intake and exhaust systems to produce maximum power. Restriction anywhere in the exhaust system (tailpipe, muffler, resonator, exhaust pipe, exhaust manifold, manifold heat control valve or in the exhaust valve ports) can cause back pressure, especially at high engine speed. This will show as lower than normal vacuum gauge reading, at sustained, higher, engine rpm.

Unsteady Gauge Reading

If the gauge needle wavers throughout a wide range, the trouble could be a blown head gasket between adjacent cylinders, ignition cross-fire, caused by poor plug wire insulation or induction firing of cylinders, or a cracked distributor cap. The trouble could be one or more weak or broken valve springs, spark plugs that are old or fouled from oil or a rich mixture or poorly gapped. Sticky valves, worn valve guides and poorly seating valves can also cause an unsteady needle. To isolate the trouble and make the diagnosis more conclusive it is suggested that precise and related gauges and instruments be used.

Vacuum Meter Route To Hydraulic Tappet Adjustment

The hydraulic valve lifter automatically adjusts to valve train clear-

ances throughout the entire range of engine temperatures.

Unlike the solid tappet, hydraulic lifters compensate for valve lash and operate at zero lift clearance. This is accomplished by using engine oil held by a plunger and check valve within a hollow tappet body. The tappet body (with the exception of designs peculiar to six cylinder Tempest and some others) is directly responsive to camshaft lobe contour. The tappet plunger adjusts hydraulically to the clearance needs of the valve train to maintain zero lash at all times; through all operating conditions and temperature changes. There is a two-fold advantage here. The first is quiet operation. The second is greater valve port openings without increasing cam lobe lift. Because there are no clearance losses with hydraulic tappets through valve train lash, valve timing controls, (opening, closing, and overlap valves) are not influenced by tappet clearance, as with solid lifters.

Hydraulic tappets, however, all have one thing in common; they must be adjusted or their components selected to maintain certain mechanical limits. Zero tappet clearance is then adjusted hydraulically within the confines of these plunger travel extremes.

If valve train linkage is, in effect, too long, the tappet plunger will be forced against the bottom of its tappet body. This condition will prevent the engine valve from seating, causing engine miss, valve burning, extended valve overlap, and valve or piston damage.

The opposite of this condition, is pushrods or other valve train components which are, in effect, too short. This situation, if beyond the mechanical limits of tappet plunger travel, demands more automatic take-up than plunger travel allows, defeating the purpose of the system. This causes excessive valve train noise, poor engine performance due to late valve opening and early valve closing (diminishing overlap) and reduced engine breathing.

Because of tappet plunger travel limits, and the fact that the length of the tappet assembly is in a state of constant change, valve train components must be held within close dimensions, or some provision made for adjustment, so that an attitude of the plunger can be established somewhere near mid-way of its possible travel limits. This is a safety precaution to insure the greatest latitude of automatic lash adjustment. Some engine designs use selective length pushrods to establish this safety factor. These rods are of various lengths, graduated in increments of .060 in. Other designs control initial valve train length right in the

tappet assembly by using selective tappet components. Still others have adjustments provided for in either the rocker arm, (pushrod end) or at the rocker arm pedestal mount. In both of the later designs, (adjustable types) primary adjustment specifications are supplied by the car manufacturer. Due to difference in thread sizes of the adjusting elements and rocker-to-valve lift ratios, it is impractical to generalize and assume a fixed number of adjusting screw or pedestal nut turns needed to locate a safe mid-way position for plunger travel within the tappet body.

In the event that specifications are not available for hydraulic valve tappet adjustments on any specific engine, the following vacuum meter method can be safely employed.

1. Hook up a vacuum gauge to the intake manifold in the usual way, so that it is in plain view while adjusting the valves.
2. Remove the rocker cover. On V8 engines, select the rocker cover on the most accessible bank of cylinders.
3. With a strip of aluminum foil, (heavy cardboard will do) fashion an oil splash deflector to protect the engine and everything in the immediate vicinity against oil drench.
4. With engine running at normal operating temperature, reduce valve lash at the adjusting point, (in quarter turn stages) while watching the vacuum meter.
5. Continue with these quarter turn stages until the tappet plunger travel limit has been reached. The limit will have been reached when the hydraulic compensating factor of the tappet can no longer accommodate the adjustment. This will show on the vacuum meter as a cylinder miss.
6. From this point, turn the adjustment counterclockwise, in quarter turn increments, until the opposite extreme is reached in tappet plunger travel. This end of travel limit is indicated by tappet

noise that persists after 15 seconds of engine operation.
7. Carefully count the turns of adjustment needed to adjust from one plunger travel limit to the other. Halfway between these two limits is the ideal position of preliminary adjustment.
8. Conduct the same adjustment procedure on the other valve of the same cylinder.
NOTE: intake and exhaust valve adjustments may or may not be identical.
9. After establishing the amount of turns required to obtain a midway plunger position for one cylinder, proceed with the other engine cylinders. Use the vacuum meter to find the plunger bottom limit, then back off the adjustment the predetermined amount to obtain a half-way plunger travel position. This is a safe margin of travel to allow for hydraulic tappet action.

Battery, Cranking

Whenever electrical trouble develops it is wise to make a quick check of battery and cables to be sure that the source of current is adequate. This check will also shed some light on the engine cranking system.

1. Switch on the headlights.
2. With lights shining on a wall or other reflective surface, close the starter switch.
3. When the starter switch is closed, one of three things will normally happen: (A.) lights will remain bright and the engine will crank at normal speed; (B.) lights will dim considerably; or go out; or (C.) lights will not be influenced, and no cranking will take place.
(A.) The first situation indicates a normal electrical condition of the battery,

cables and connections, and the cranking system.
(B.) The second condition indicates a poor connection in the cranking circuit between battery and starter motor, or, the battery is only partially charged. Or, there is a condition in the starting motor, or in the engine itself which causes an excessive draw on battery current. A low battery will manifest itself in a clattering sound at the starter solenoid (when the solenoid is energized). Test battery with a hydrometer and give it a high rate discharge test to isolate the trouble. If battery is found to be OK, check the cranking motor.
(C.) The third condition indicates an open circuit somewhere in the cranking motor, starter switch, solenoid or wiring.

Cranking Troubles—Causes

Caution Cars equipped with manual shift transmissions can be push-started. However, some automatic transmission-equipped cars (due to no-rear-pump-design) cannot be push-started. In these cases a fully charged battery must be installed or jumper cables used for cranking current. Be very careful not to reverse polarity especially on systems using alternating current generators. Permanent damage may be done to the diodes of the alternator.

Starter Troubles—Causes

1. Starter won't respond at all.
2. Engine cranks slowly.
3. Engine won't crank, accompanied by clacking noise.
4. Starter turns but engine won't crank.
5. Starter cranks engine but is noisy.

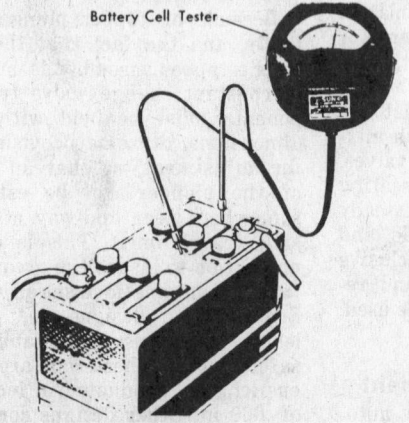

Checking battery cell voltage

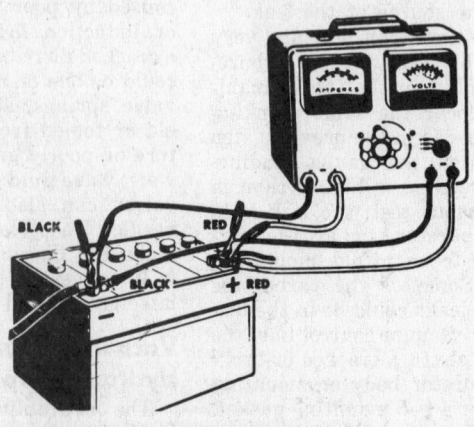

Battery high rate discharge test

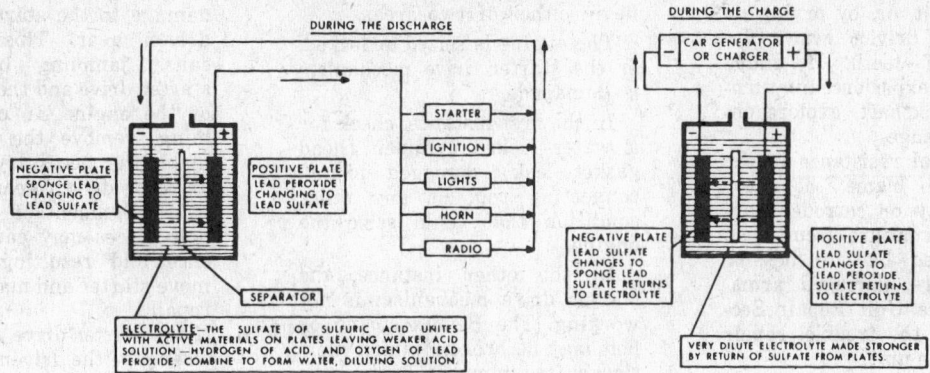

DURING THE DISCHARGE

STARTER
IGNITION
LIGHTS
HORN
RADIO

NEGATIVE PLATE SPONGE LEAD CHANGING TO LEAD SULFATE

POSITIVE PLATE LEAD PEROXIDE CHANGING TO LEAD SULFATE

SEPARATOR

ELECTROLYTE—THE SULFATE OF SULFURIC ACID UNITES WITH ACTIVE MATERIALS ON PLATES LEAVING WEAKER ACID SOLUTION—HYDROGEN OF ACID, AND OXYGEN OF LEAD PEROXIDE, COMBINE TO FORM WATER, DILUTING SOLUTION.

DURING THE CHARGE

CAR GENERATOR OR CHARGER

NEGATIVE PLATE LEAD SULFATE CHANGES TO SPONGE LEAD SULFATE RETURNS TO ELECTROLYTE

POSITIVE PLATE LEAD SULFATE CHANGES TO LEAD PEROXIDE SULFATE RETURNS TO ELECTROLYTE

VERY DILUTE ELECTROLYTE MADE STRONGER BY RETURN OF SULFATE FROM PLATES.

Schematic diagram of battery electro-chemical action

6. Starter drive will not disengage flywheel.

Starter Troubles—Corrections

1. Starter won't respond, may be caused by any one or more of the items listed under Battery Troubles—Causes. In addition, there may be an open circuit in the cranking system. This can be in the ignition-starter switch, the starter solenoid, any of the cranking circuit wires, or in the starter motor proper. To correct the situation, turn on headlamps and close the starter switch. If lights dim or go off, check electrical supply and battery connections; charge battery, or tighten connections. If lights are not affected, the trouble may lie in an open circuit. It may just be a faulty, or maladjusted, automatic transmission starter safety switch. Locate the trouble by bridging the various switches and wiring components until a good circuit is established and the starter motor turns.

Caution When bridging the heavy wire connections at the starter or solenoid, use a battery cable, or its equivalent as a preventative measure against burning the hands.

By process of elimination, the trouble spot can be isolated and the component repaired or replaced.

2. Slow cranking may be the result of mechanical resistance of the engine (heavy oil, seized engine parts, low temperatures, bent parts, etc.). There is also a possibility of high friction or mechanical interference within the starter motor or drive unit. A third possibility is electrical resistance in the starter.

If slow cranking is the result of mechanical resistance in the engine, check engine oil for viscosity. Drain heavy oil and refill crank case to level with oil appropriate to temperature conditions.

If slow cranking is caused by seizing of engine parts, or bent parts causing unusual friction, undue engine resistance may be felt by removing all of the spark plugs which eliminates the effects of compression. Then, try to turn the crankshaft by the fan belt, or a socket and L-handle applied to the vibration damper

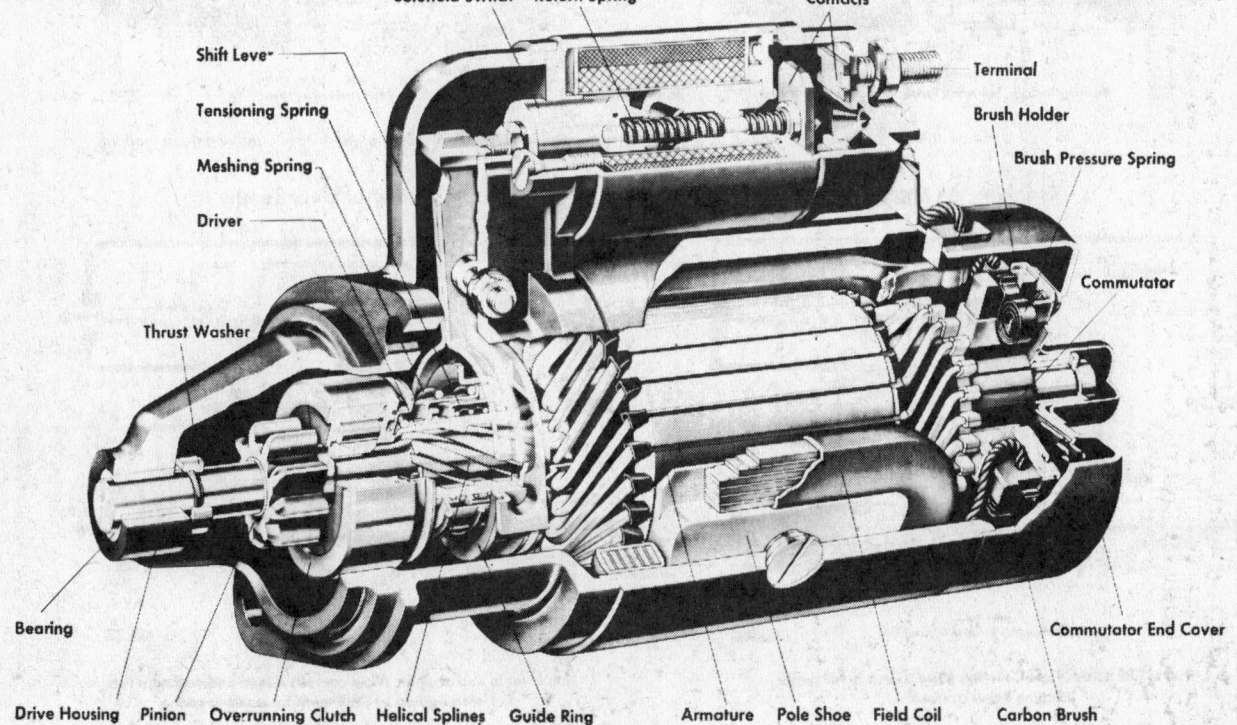

Solenoid Switch · Return Spring · Contacts

Shift Lever · Terminal

Tensioning Spring · Brush Holder

Meshing Spring · Brush Pressure Spring

Driver · Commutator

Thrust Washer

Bearing · Commutator End Cover

Drive Housing · Pinion · Overrunning Clutch · Helical Splines · Guide Ring · Armature · Pole Shoe · Field Coil · Carbon Brush

Typical starter motor
(© Bosch)

attaching bolt, or, by prying on the starter driven gear (flywheel gear) teeth. If undue resistance is experienced in turning the crankshaft, explore engine for damage.

If electrical resistance in the starter is to blame, the cause may be dirty or corroded connections in armature circuit, field coils damaged, bad brushes, or brush holders, damaged armature, etc. See Unit Repair Section. Locate the trouble, repair or replace the unit.

3. If the engine won't crank and closing the starting circuit causes a clacking sound at the starter solenoid, the trouble probably lies in battery voltage being below that required to overcome starter solenoid switch spring tension. Refer to Battery Troubles—Causes.

4. If the starter turns but the engine won't crank, the trouble can lie in either of two areas.

The engine is seized or locked, or the starter drive mechanism is damaged.

In the first instance, check for a water locked cylinder (head gasket leak) damaged piston, connecting rod, or any other condition that could seize the engine.

In the other instance, the starter drive mechanism is not working. The Bendix spring or bolt may be broken, pinion inertia shaft gummy or bent, drive housing broken, pinion broken, starter driven gear missing teeth, armature shaft broken.

In the overrunning clutch type of drive, the overrunning clutch may be worn.

5. Where the starter cranks the engine, accompanied by noise, the engine itself may be damaged. In addition, there may be damage to the starter drive, or driven gear. This may even cause jamming between the starter drive and the driven gear of the engine. In case of jamming, remove the starter and make the necessary repairs to drive or driven gears.

Misalignment of armature or pole pieces may cause interference and resulting noise. Remove starter and make necessary repairs.

6. If the starter drive gear will not disengage the driven gear, either the solenoid switch is energized, the solenoid switch is sticking closed, or there is mechanical interference with the releasing mechanism. Correct the trouble by making electrical checks of the ignition-starter switch and the solenoid. If this check does not reveal the trouble, remove the starter and correct the mechanical difficulty.

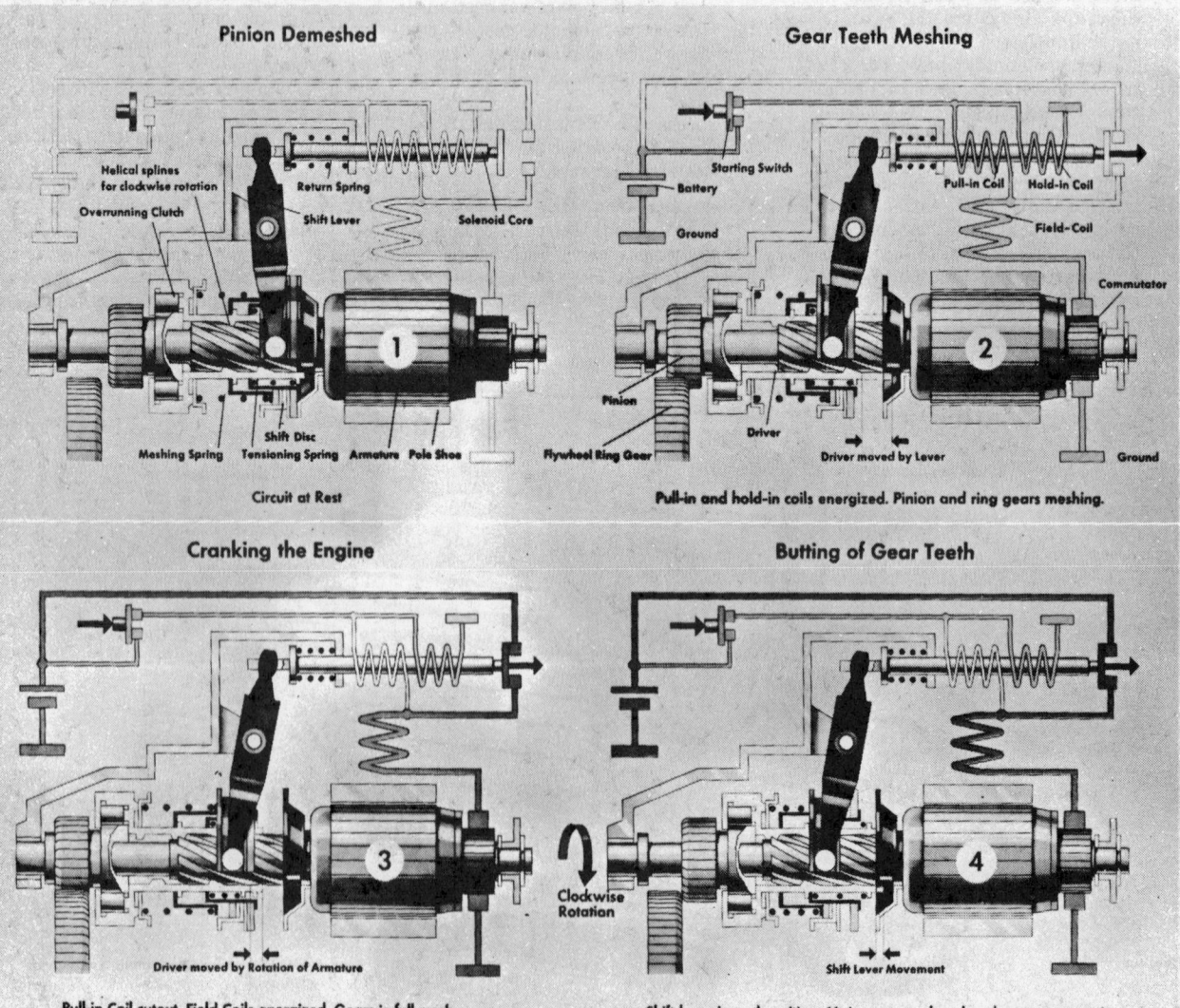

Typical starter motor operation
(© Bosch)

Car Battery

Selection of Battery

The modern car battery (with very few exceptions) is of 12 volt potential. It is a lead-acid unit of various ampere hours capacity, depending upon the required work load (radio, air conditioning, electrical windows, tail gate, telephone, etc.). Every new car model seems to put more load on the reservoir of electrical supply—the battery.

Batteries come in different forms as specified and designed by the car manufacturer to service a car, and are peculiar to the car's electrical needs. The battery originally supplied with the car is generally adequate. The time of real concern is when the battery is to be replaced, or when making extra demands upon the electrical system.

The prime purpose of the battery is to supply a source of energy for cranking the car engine. It also provides the necessary power for the ignition system. A battery can (for a limited time) supply adequate current to satisfy electrical demands during periods when requirements exceed generator output.

Replacing a Battery

The convenient and most popular way to store new batteries is in a dry state. They are given a charge (with special equipment) at the time of manufacture. A dry charged unit will hold this charge almost indefinitely, in the absence of moisture.

Before deciding on a battery which will fulfill your needs, first, consider some of the essentials that may put the replacement battery in a different category from that of the unit originally supplied with the vehicle. By the time the original battery has outlived its usefulness, resistance in the wiring circuits has probably built up considerably. The starter will have lost some of its zip. The ignition system is probably less efficient, and there is the likelihood that electrical accessories have been added.

All of the above reasons are justification for choosing a battery of greater capacity than the one supplied by the manufacturer.

Preparation

After ample study of electrical needs has been done and a selection made, place the new battery on a bench or work table. Never activate a battery while in the car. Remove vent caps from all the cells.

Fill each cell carefully, using sulfuric acid and distilled water (electrolyte) to a strength of 1.250-1.265 specific gravity to about ⅜ in. above the top of the separators, or to indicated level mark.

Place a battery type thermometer in one of the center cells. Check specific gravity of the electrolyte with a battery hydrometer. The battery temperature must be above 80°F, and specific gravity must be above 1.250 prior to installing the battery.

In charging 12 volt batteries, set charging rate at 35 amperes (6 volt batteries at 70 amperes) until electrolyte has reached 80° F and electrolyte gravity is 1.250 or higher. Lower charging rates may also be used to obtain 80° F and 1.250 specific gravity. When charging, do not exceed 125° F. Normally, 10-15 minutes charging will be enough. However, colder climates may take a little longer.

When the battery is removed from the charger, re-level, if necessary, with electrolyte, and replace the vent plugs.

When installing, make sure that both ends of the battery cables are clean and securely tightened when attachment is made.

Start engine and see that the generator is charging with lights and all accessories on.

Caution Be careful not to install or charge the battery in reverse. Damage to battery and generator can result especially if the car is equipped with an alternator or transistorized radio.

The electrical state of the battery should be checked again in about a year.

Caution Since electrolyte is extremely corrosive to metals and many other materials, do not discard by pouring into sinks or drains. If battery acid is spilled on battery during filling or charging, or on bench or clothing, immediately flush off with generous amounts of water and baking soda or ammonia.

Battery Troubles—Causes

1. Battery too small for the job (accessories, etc.).
2. Tired battery (worn out).
3. Corroded battery connections.
4. Generator not charging.
5. Generator charging rate low.
6. Regulator out of operation.
7. Regulator out of adjustment.
8. Regulator has poor ground.
9. Alternator inoperative.
10. Loose generator or alternator drive belt.
11. Constant drain of current due to short circuit, etc.

Battery Troubles—Corrections

1. Battery capacity may be less than requirements demand. Additional accessories, too frequent use of starter, low operational speeds, etc. require a greater source of electrical supply. Install a larger capacity battery.
2. Either age or abuse is the usual cause of a tired battery. No amount of charging will offer more than temporary relief. Install a new battery of proper capacity.
3. Corroded battery posts and connections result from chemical reaction of dissimilar metals to battery electrolyte. Excessive corrosion at a battery post is usually an indication of the failure of a seal between the post and the battery cover. Clean post and cable clamp, seal post-to-battery cover with rubber cement or other plastic material,

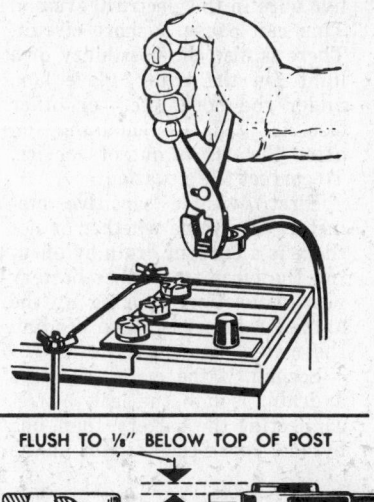

FLUSH TO ⅛" BELOW TOP OF POST

RING

Battery terminal connections
(© G.M. Corp.)

then coat post with petroleum jelly, install cable clamp and tighten.
4. Generator not charging can be caused by many things within the generator, in which case the unit should be removed for bench work. See Unit Repair Section.
5. Low generator charging rate may be caused by a loose drive belt, loose or poor battery post connections, high resistance in charging circuit or a poor or improperly adjusted regulator. See Unit Repair Section.
6. Regulator may be out of operation because of burned points in the regulator or any open circuit in the control system. See Unit Repair Section.
7. Regulator out of adjustment. See Unit Repair Section.
8. A poor ground at the regulator can become involved. A possible cause of trouble is a poor regulator ground in any of the externally grounded field circuits (A circuits). In the heavy-duty, (B

circuits) the field is grounded within the generator. For detailed information in this area, refer to Unit Repair Section.

9. The alternator may be inoperative because of damaged diodes, poor internal connections, open, grounded, or shorted field circuit, grounded or shorted stator windings. In any event, the unit should be removed to the test bench. See Unit Repair Section.

10. A loose generator (AC or DC) drive belt will cause low, or partial charging. Correct by adjusting drive belt.

11. A constant drain of current from the battery may be caused by a frayed bit of insulation on any live wire in the electrical system. This can cause a short circuit. There is also the possibility of a light (in the trunk, glove box, under the hood, etc.) or other electric gadget remaining on after the car is out of service. To correct the situation:

First, with a sensitive ammeter, determine whether or not there is a current drain by opening the circuit at either battery post connection, hooking up the ammeter in series, then reading the meter.

Second, if the meter registers a drain, isolate the leak by reconnecting the battery, then, one by one, check each circuit at the fuse block. This is a tedious but unavoidable procedure and consists of removing each fuse and testing that circuit with the prods of an ammeter (in series). The circuit which activates the meter is the guilty one; identify the troublesome spot by elimination. Correct the trouble by correcting the short or replacing the switch or other electrical component.

In the event that the fuse block does not furnish evidence of the trouble, check the circuits which are protected with circuit breakers, (headlamps, parking lamps, seat and window controls, etc.).

Ignition System

1. Select the most accessible spark plug and remove its distributor wire.
2. Hold this exposed distributor wire top 1/4 in. from the spark plug terminal.
3. Have someone crank the engine, (ignition switch on). A good, fat spark should bridge the 1/4 in. gap, once every two revolutions of the crankshaft, (four-stroke-cycle engines, or once every revolution for two-stroke-cycle engines). If current is reaching this point, read Fuel System Troubles.

Ignition System Troubles —Causes

In general, hard starting and engine failure problems can be traced to some phase of the electrical system (battery, points, distributor, coil, etc.).

If the electrical supply and cranking systems have been checked out and approved or corrected, the following electrical causes for hard starting and engine failure should be checked.

1. Burned or damaged distributor points.
2. Points out of adjustment.
3. Defective condenser.
4. Shorted or open primary circuit.
5. Faulty ballast resistor (12 volt systems).
6. Ignition coil failure.
7. Poor distributor ground.
8. Ignition switch failure.
9. Neutral safety switch out of adjustment.
10. Bad kickdown switch on overdrive equipped cars.
11. Moisture on distributor cap (inside or out).
12. Cracked or carbon tracked distributor cap.
13. Failure of spark plugs.
14. Failure of plug or distributor wires.

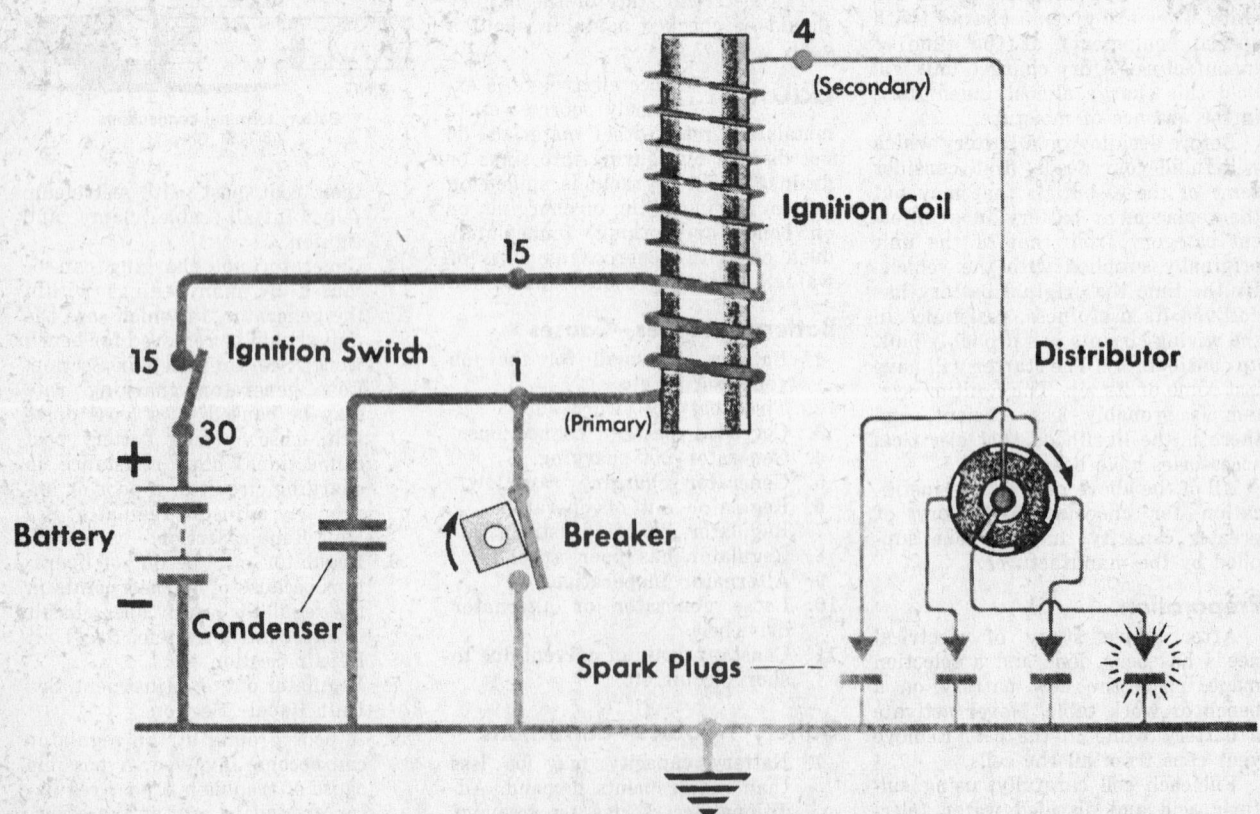

Typical ignition wiring schematic
(© Bosch)

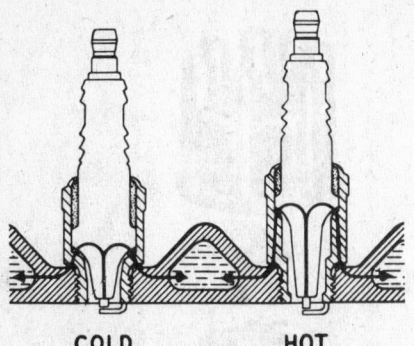

COLD   **HOT**

Heat travel in hot and cold spark plugs

15. Induction firing of cylinders.
16. Distributor rotor damage.
17. Incorrect distributor firing sequence.

Ignition System Troubles —Corrections

1. A visual inspection of the points will reveal their physical condition and give a fair indication of their electrical value in the primary circuit. Appearance of the points will also furnish a fair index to the condition of the condenser. Burned or otherwise damaged ignition points may cause hard starting and very poor engine performance. Do not file or otherwise recondition points. It is poor economy. Renew the set. If points show burning, renew the condenser also.

NOTE: always reset ignition timing after replacing ignition points. Point gap, or dwell time, is directly related to ignition timing.

2. A visual inspection of point gap will indicate (within the scope of starting or normal engine performance) whether or not trouble lies in points out of adjustment. If necessary, adjust points, then set distributor timing.

3. The best indication of condenser failure (short of an electrical test with a condenser tester), is to examine the contact surfaces of the distributor points. Burned or cindered points indicate a faulty condenser. Renew the condenser and points.

4. A shorted or open primary cir-

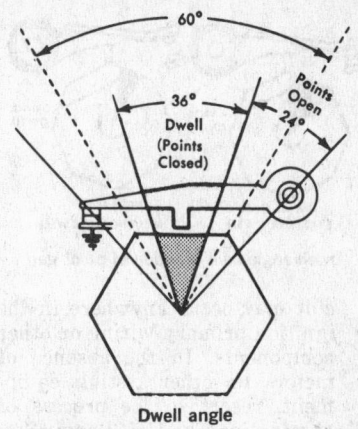

Dwell angle

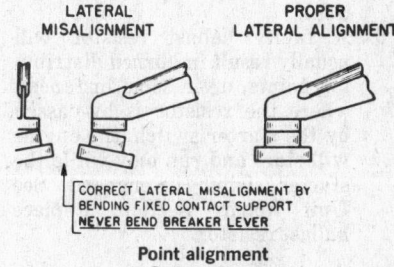

Point alignment
(© G.M. Corp.)

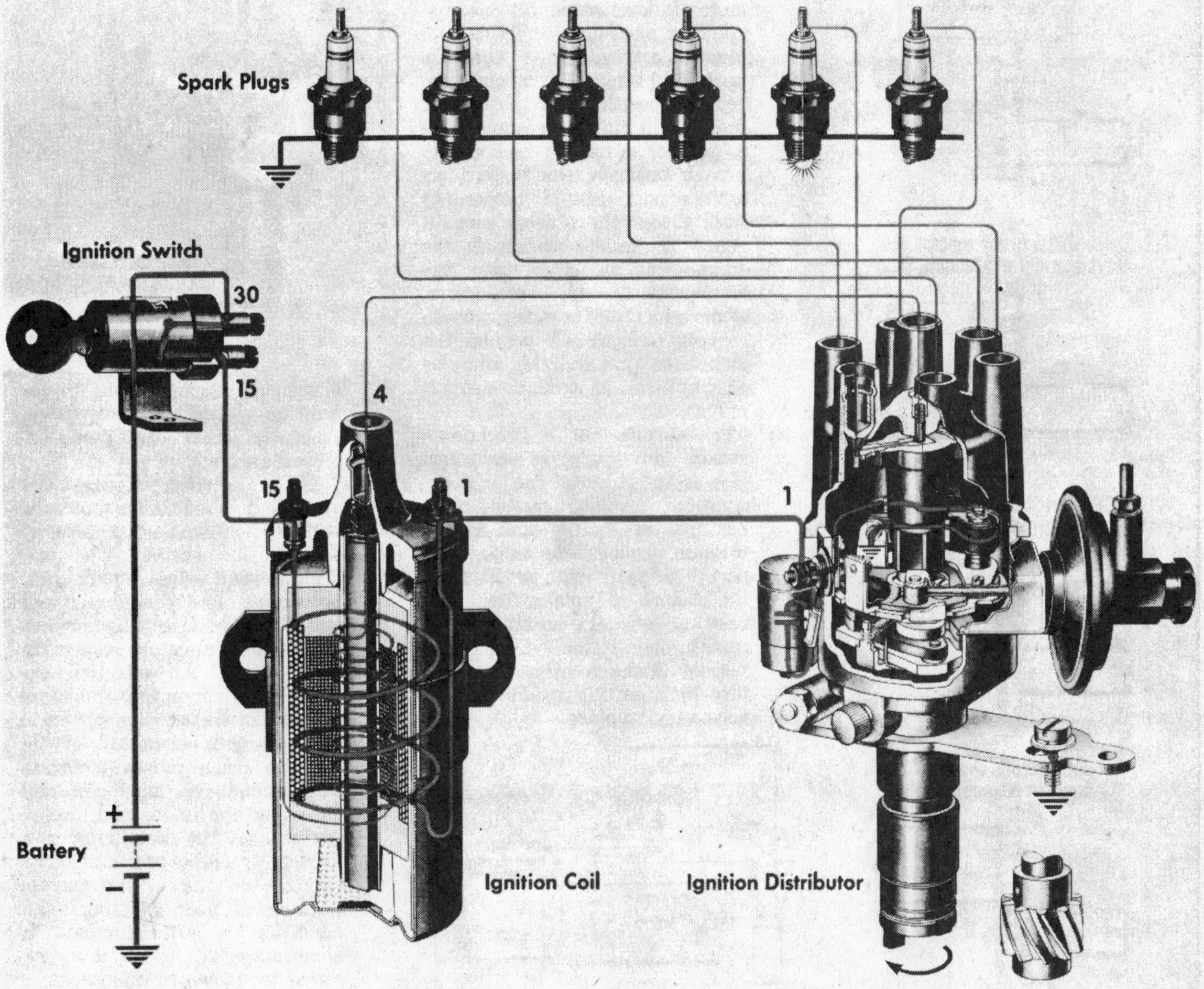

Spark Plugs

Ignition Switch

30

15

4

15   1

1

Battery

Ignition Coil   **Ignition Distributor**

Typical ignition system, showing major components
(© Bosch)

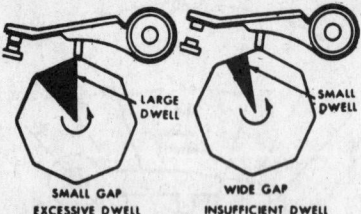

Dwell angle as a function of point gap

cuit may occur anywhere in the ignition primary wiring or other components. In the absence of meters or other testing equipment, resort to the process of elimination. See Unit Repair Section.

5. A faulty ballast resistor will usually result in burned distributor points, or in some instances, where the resistor is by-passed by the starter switch, the engine will start and run only while the starter switch is engaged. See Unit Repair Section. Replace ballast resistor.

Primary Winding without Current No Magnetic Field

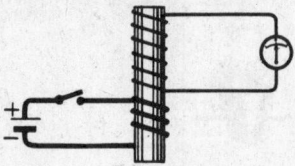

Cutting-in Process Development of Magnetic Field

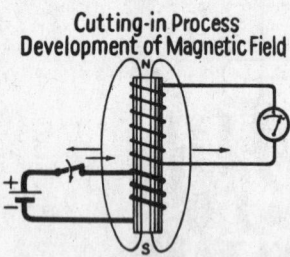

DC Current in Primary Winding Constant Magnetic Field

Interrupting Process Collapse of Magnetic Field

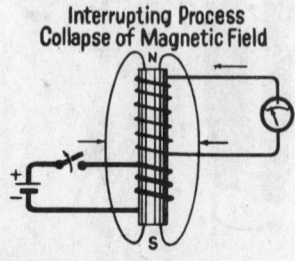

Ignition coil operation

6. Ignition coil failure is possible, but improbable, due to quality control and product improvement. However, an open or a shorted coil is a possibility. Coil deterioration is usually a gradual process and occurs over a period of time. This dying process manifests itself by hard starting, cylinder miss under heavy load, and engine miss at high rpm. To be sure, see Unit Repair Section.

7. A poor distributor ground may cause hard starting and poor performance, identical to ignition coil failure. In case of suspected distributor ground trouble, install a permanent jumper wire from distributor to ground.

8. Ignition switch failure may show up in the starter circuit or the primary ignition circuit. Some circuit designs involve the ballast resistor. See Unit Repair Section.

9. Neutral safety switches, on automatic transmission equipped cars, are more of a nuisance than a threat in hard starting or poor performance cases. If the starter motor fails to respond to normal starting procedures, (lights and other electrical units function normally) try all transmission selector positions. If the starter motor still fails to respond, use a jumper wire on the safety switch (transmission in park, or neutral position). If the jumper wire closes the starter circuit, renew the safety switch. If the jumper has no effect upon the starter, the trouble is elsewhere. If movement of the safety switch corrects the trouble, adjust the switch so that starting may be accomplished in park or neutral only.

10. The purpose of a kickdown switch on overdrive equipped cars is to interrupt the primary ignition circuit, momentarily cutting the engine and transmission torque long enough to permit a gear change. Due to the nature of this switch, the ignition primary circuit could remain open if the switch is defective. Make a circuit continuity check on this switch and, if necessary, replace.

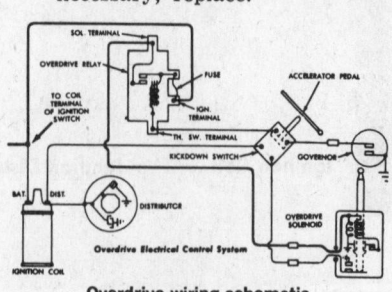

Overdrive wiring schematic
(© Ford Motor Co.)

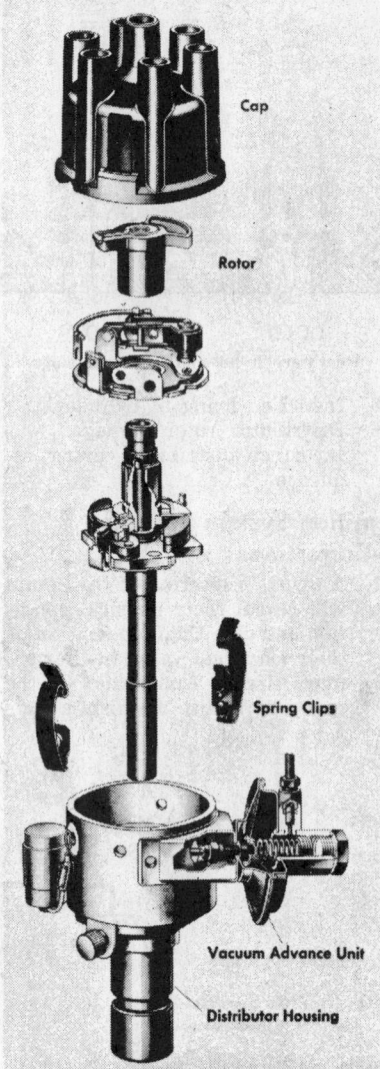

Typical distributor
(© Bosch)

11. Moisture is a detriment to the ignition system in many ways, especially when the parts involved are dirty.

Under certain atmospheric and heat conditions, moisture will form on various surfaces of the ignition system. The coil, high tension wires, spark plug porcelains, and distributor cap are all parts subject to short circuits due to moisture, especially when dirty. An effective approach to correcting moisture troubles of the ignition system is to thoroughly clean all of the ignition circuit wires (primary and secondary), plus coil surfaces and the inside and outside surfaces of the distributor cap.

12. Carbon tracking and cracks inside the distributor cap may be a cause of hard starting. This condition is quite common to some materials, but it is aggravated by moisture. If cracks or tracking are present, renew the distributor cap.

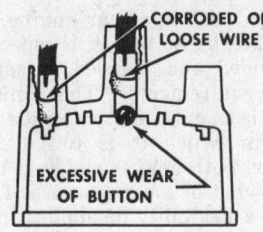

CORRODED OR LOOSE WIRE

EXCESSIVE WEAR OF BUTTON

Distributor cap malfunctions

Spark Plug Torque

Thread size	Cast-Iron Heads	Aluminum Heads
10 mm.	14	11
14 mm.	30	27
18 mm.	34*	32
7/8 in.—18	37	35

* 17 ft. lbs. for tapered plugs using no gaskets.

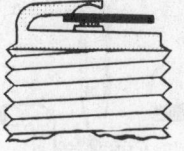 

WRONG   RIGHT

Gapping spark plugs

13. Spark plug failure may be the result of many things, from breakage to plain fatigue. However, hard starting problems resulting from plug failure are usually caused by a short in the secondary circuit. This short may be the result of high resistance at the spark plug electrodes (plug gap too wide) or badly fouled plug tip. Or, it may be a cracked or dirt-shorted porcelain.

A spark plug causing poor engine performance usually manifests itself by causing the cylinder to misfire under heavy engine load (open throttle at low speed). While an oscilloscope will quickly identify spark plug difficulties, a visual check of the plugs is generally enough to reveal any existing trouble. If at all in doubt, replace with a correctly gapped new plug of the proper heat range.

14. Spark plug and distributor wires, like other components of the secondary ignition circuit (coil, heavy coil wire, distributor cap, rotor, plug wires, and spark plugs) carry very high voltage. This high pressure (voltage) must be contained by good dielectric insulation (or adequate dry air space). This means that the wire insulation must not be cracked, contaminated or damaged.

Especially subject to damage due to rough handling is the non-metallic type of secondary voltage distributor and plug cables.

Splashed   Wet fouled (oil or gas)   Core bridged   Gap bridged

Scavenger deposits   Overheating   Chipped insulator   Mechanical damage (preignition or detonation)

Reversed coil polarity   Cold fouled   Normal

Spark plug damage

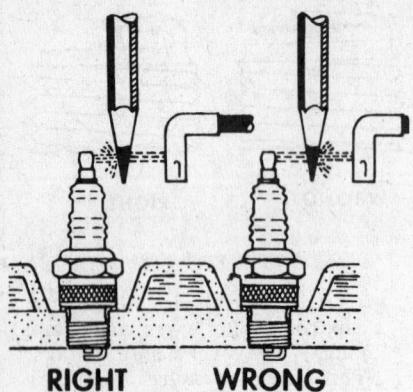

RIGHT   WRONG

Coil polarity determined by watching spark. Spark jumps from wire to pencil to plug if correct.

Instead of a wire core, these cables have a core of carbon-impregnated string calibrated to a predetermined resistance.

Bending or other rough handling of this cable fractures the carbon core and increases resistance to the point of hard starting and engine miss.

15. Induction firing is a crossfiring of cylinders of consecutive firing order. Spark plugs are fired out of time by induced voltage. This condition may cause backfiring, engine roughness, detonation and possible engine damage.

Any time high voltage is being conducted by cable, a magnetic

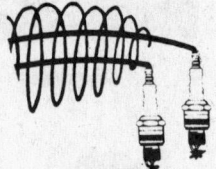

Induction firing

field is present around that cable. If another plug cable is sufficiently close, and parallel, to the first cable, enough induced voltage may be absorbed by the second to fire both cylinders at the same time. If the induction-fired plug is the next in firing order, timing is extremely overadvanced in that one cylinder.

To ensure against induction firing of consecutively firing cylinders, separate the two critical cables as much as possible. Cross

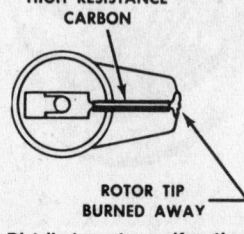

HIGH RESISTANCE CARBON

ROTOR TIP BURNED AWAY

Distributor rotor malfunctions

the cables if you must, but don't run them parallel.

16. Distributor rotor failure is a possibility. Any physical damage should be obvious upon inspection; the center, or collector brush, may be worn or broken; or the distributor rotor finger broken or burned to the extent of causing high secondary voltage resistance. Rotor fit on the distributor shaft should be snug and positive.

A quick check for a grounded rotor is to remove the center wire from the distributor cap and hold it about ¼ in. from the distributor rotor, while someone cranks the engine ignition on. If a spark occurs across this gap, the rotor is grounded. If at all in doubt about the rotor, replace it.

17. Incorrect distributor firing sequence is probably caused by disturbance of the original cylinder timing. This is most likely to occur as the outcome of some maintenance operation involving the distributor cap or spark plugs. The effect could be, engine miss, backfiring (with a possibility of fire) and failure of the engine to run. While it is technically possible to time the ignition to any engine cylinder that happens to be at the top of its compression stroke, (the V6 Buick and some others are exceptions) it is wise to stick to the original wiring plan. The spark plug cable length, predetermined cable resistance, distributor vacuum control line position, etc., are all reasons why the original design should be followed. See Firing Order illustrations near the beginning of each car in the Car Section.

Fuel System

Fuel Value

Any material used to produce heat or power by combustion may be called fuel. However, for our purpose, and to satisfy the demands of the popular American made automobile, all fuels except those familiar to the general public will be excluded from this treatment. Diesel oil and the exotic fuels of competition do not apply here. They belong, and serve their purpose in an entirely different field. Commercial fuel is known to all of us as gasoline.

Automotive fuel is referred to as gasoline. However, weight for weight, air is about 15 times as necessary as gasoline in supplying a useful combustible mixture for the average

American passenger car engine.

Gasoline, as we buy it, is especially compounded and blended to satisfy many requirements. The finished product is as carefully planned as the engine in which it is to be used. However, both engine and fuel must be somewhat of a compromise. There must be a flexibility of application in matching the fuel to the engine and to its work load.

Among the many fuel qualities essential to good performance and long engine life, and the one which is most optional, is the octane rating (speed of burning).

Speed of burning is controlled by many things, some of which are inherent in the crude from which the gasoline is made, some controlled by refining (cracking, mixing, etc.) and some, by the addition of varying quantities of antiknock compounds.

Tetraethyl lead, benzol, alcohol, etc. are some of the substances used to control speed of burning and establish octane values. The burning characteristics of the fuel must be compatible with engine design, torque requirements, and the compression ratio of the engine used.

Uncontrolled burning (detonation) can be defined as an explosion of the last parts of the burning mixture and is usually, but not always, audible as a ping or knock. Detonation is not always the result of a poor selection of fuels; other factors contribute. They may be combustion chamber deposits, cooling system inadequacies, lean fuel mixtures, ignition timing and controls, etc. Regardless of cause, detonation can produce extensive and permanent engine damage in short order. When most high compression engines are adjusted to factory specifications, and to develop their maximum power and fuel economy, they require a fuel of 90 to 100+ octane value. Also, if the chosen fuel satisfies a properly tuned engine, nothing can be gained by using a fuel of higher octane number.

The octane number of a fuel is established, with some variations in the laboratory, by using a reference fuel and a test engine. Usual test procedure has been to run a fuel of unknown antiknock value in a specific single-cylinder test engine, and to record the intensity of knock. Then, run a fuel in the same test engine, consisting of known amounts of iso-octane and normal heptane that will produce the same resistance to detonation as the fuel being tested. The percentage of iso-octane comprising the reference fuel now assumes a meaning. It represents an octane number; one that can be used to identify the burning characteristics of the fuel being tested.

For fuels with antiknock qualities above 100, the reference fuel used is 100% iso-octane, plus a known

amount of antiknock additive. As an example: the antiknock value of a certain premium fuel, using tetraethyl lead as a knock deterrent, might be expressed as 100 + 1.5. This would indicate that the gasoline has the same antiknock characteristics as pure iso-octane to which 1.5 milliliters of tetraethyl lead had been added to each gallon of iso-octane. Above 100 octane, values also employ a formula by which the milliliters of antiknock additive are computed and transposed into octane numbers above 100, such as 105, 110, 115, and so on. This becomes highly technical and is beyond the scope of this manual.

To summarize: premium fuels are the slower burning fuels of higher octane number and are generally better suited to the needs of high compression engines. However, to gain the full mechanical and economic advantage of this higher priced fuel, ignition timing must be adjusted accordingly. The slower burning aspects of premium fuel will usually tolerate earlier ignition timing, other things being equal.

Vaporlock

A most difficult problem of fuel and the automobile fuel system is vaporlock. This has been greatly reduced as our knowledge of refining and fuel technology in general has increased.

The function of an ideal fuel system is that of delivering a continuous and uninterrupted supply of reliable fuel to the engine at the pressure required and in whatever quantity needed to satisfy the various engine and driver demands.

One aim in fuel system design is to avoid any route or trapping of the fuel that may contribute to the retention of heat. This could cause bubbles or vapor to form in areas of the fuel system under the various conditions of acceleration, hot weather and high altitude driving. Most of these problems have been solved by a combination of engine fuel induction design and by fuel technology, plus geographic and seasonal planning. However, the chance of vaporlock still exists in some areas of the country, at certain times of the year—generally in the spring and fall.

The petroleum industry, anticipating seasonal fuel requirements, attempts to satisfy these atmospheric needs by adjusting their fuel compounding formula to the humidity and heat of summer as encountered in such places as Saint Louis and New Orleans. The winter needs of Duluth or International Falls, Minnesota, call for an entirely different compound—high in volatility for easy starting and efficient burning during cold weather starting and engine warm-up. If the same fuel were to be used throughout the year for all the pressure, humidity, and elevation differences in the United States, there would be constant combustion trouble.

Because of the pressure drop in the fuel line, on the suction side of the fuel pump, there is a tendency for fuel to boil or bubble under sub-normal boiling temperatures. Also, the volume of the fuel in the vapor state

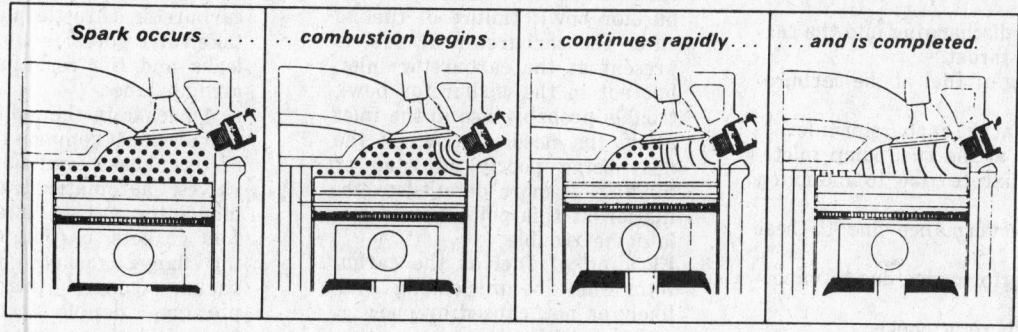

Spark occurs..... ...combustion begins.. ...continues rapidly... ...and is completed.

Normal combustion

Ignited by hot deposit.. ..regular ignition spark... ..ignites remaining fuel. ..flame fronts collide.

Preignition

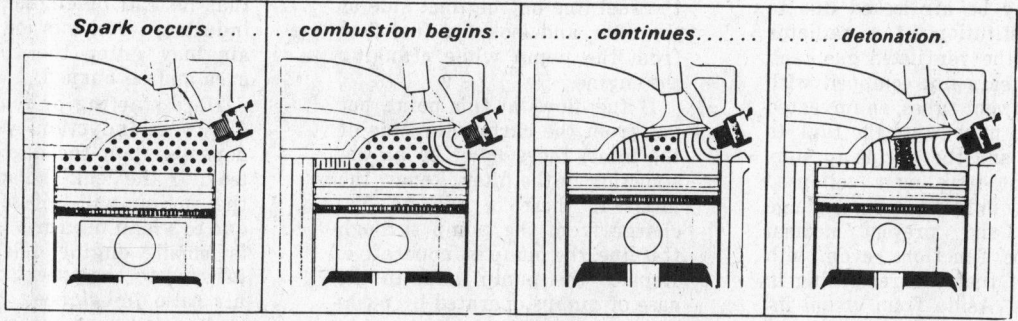

Spark occurs... ...combustion begins.. ......continues...... ......detonation.

Detonation ("knock-ping")

is over 1,000 times its volume in the liquid state. This aggravates the matter still more by reducing the weight of the fuel being delivered. When occupied by vapor instead of liquid fuel, the pump is in a state of vaporlock. When this condition exists, the pump is incapable of delivering a continuous flow of fuel to the carburetor. This, of course, interrupts engine performance and in extreme cases stops the engine completely. The carburetor cannot successfully meter a mixture of vapor and fuel. Therefore, lean mixtures, rough operation, and engine stalling result.

Fuel System Troubles—Causes

If the electrical supply system, the cranking system, and the ignition systems have been eliminated as cause for hard starting or poor engine performance, fuel is the next area to explore.

Following are some of the difficulties encountered with the fuel system, including induction.

1. No fuel at the fuel tank standpipe.
2. No fuel discharging into the carburetor throat.
3. Flooding of fuel at the carburetor.
4. No fuel at the carburetor inlet.
5. No fuel at the fuel pump inlet.
6. Poor mixture due to induction leak.
7. Poor performance due to heat riser.
8. Poor performance due to vaporlock.
9. The carburetor choke.
10. Carburetor icing.

Fuel System Troubles —Corrections

1. The first thing to determine is whether or not there is fuel in the fuel tank. The gas gauge could be inoperative; there may be an obstruction in the gas line at the fuel pick-up, (standpipe) or at the fuel tank filter. There is a past history of obstruction and leakage failure of the flex-line that connects the gas tank line to the fuel pump. The gas tank bottom may be bent from striking a road obstruction, or the tank may be air locked due to the substitution of a radiator cap for the ventilated gas tank cap. On cars not equipped with gas tank vent tubes, an unvented tank cap may cause the tank to collapse and the engine to stop.

   A bent tank or a collapsed tank may influence the gas gauge reading and prevent normal standpipe functions, even with plenty of fuel still remaining in the tank. Aside from visual inspection, blow a reverse flow of

air into the tank from the disconnected fuel line and listen at the open end of the fill-pipe for escaping air. No sound of escaping air indicates obstruction some place in the line or tank. Clear sound of escaping air indicates no fuel at the standpipe opening in the tank. A gurgling sound indicates the presence of fuel, (or other liquid) in the tank.

2. The next step, after being sure of a supply of fuel in the tank, is to check for fuel discharge into the intake manifold, at the carburetor throat.

   To do this, remove the carburetor air cleaner (downdraft carburetors) and while looking down the carburetor throat snap the throttle open once or twice. A spurt of fuel into the throat area will indicate the presence of fuel in the carburetor. No fuel at this point could suggest failure of the accelerator pump to discharge fuel. The more likely cause, however, is an empty carburetor bowl; failure of fuel to enter the carburetor. If fuel is present at the carburetor inlet, but not in the carburetor bowl, trouble probably lies in the inlet filter, the needle valve, or the float being poorly adjusted or binding. A more detailed examination at this point should isolate the trouble.

3. Flooding of fuel at the carburetor may be misleading and, likely as not, caused by poor ignition, bad timing or some other malfunction, aside from the fuel system.

   Carburetor-induced flooding is generally caused by a loaded float, a cracked carburetor body, a loose carburetor plug or jet, or a faulty needle valve or seat. In any event (aside from needle and seat trouble), carburetor work is required.

4. If, by disconnecting the feed line at the carburetor inlet, there is no fuel flow while cranking the engine, the fuel line filter is clogged, or the fuel pump is not working.

   To check the filter, disconnect the fuel line on the inlet side of the filter and look for fuel flow from the pump while cranking the engine.

   If fuel flows at this point, but not from the carburetor side of the filter, there must be an obstruction in the filter. Renew the filter. If there is still no discharge from the pump through this line, the pump is inoperative. Replace the pump, and, in the case of pumps operated by pushrods, renew the rods.

NOTE: an increasing number of fuel pumps are sealed at manufacture and are not to be serviced. Check pump construction before attempting to rebuild the unit.

5. The only other reason for lack of fuel supply should lie in the fuel tank or lines which were covered at the beginning of this treatment under Fuel System —Troubles.

6. An induction system leak at the intake manifold may cause hard starting, poor performance, burned valves, detonation, and, in some engine designs, oil consumption.

   Unless supercharged, the intake manifold is subject to negative air pressure (vacuum); directly related to, and influenced, by design, atmospheric conditions, and attitude of operation. These variables prevent establishing any fixed set of values for vacuum meter readings covering all engines. However, all of the induction system, from carburetor throttle plate to intake valve ports, is subject to air leaks and the resulting fuel-to-air imbalance.

   Air leaks in the induction system are most common in the area of crankcase emission control valves, automatic transmission modulator valves, intake manifold gaskets, carburetor mounting flange gaskets, distributor vacuum control lines or diaphragm, double acting fuel pumps, vacuum operated windshield wipers, power brakes, or cracked intake manifolds.

   In the event of a crankcase emission control valve not seating, (a very common fault), a double acting fuel pump diaphragm leaking, or an intake manifold gasket leak, (352, 390 Ford, and some others), engine oil fumes may be inducted into the combustion chambers. This situation may result in engine oil consumption in addition to a vacuum leak.

   The same is true in the case of an automatic transmission modulator valve that is leaking, except that instead of engine oil being inducted, transmission oil and air may enter the combustion area and be burned.

   Hard starting or rough idling due to an induction system leak depends upon the degree of the leak. If the leak is bad enough, the engine will refuse to start due to a lean mixture. If the leak is small, engine idle will be affected to the extent of fuel-to-air ratio imbalance.

   A pretty good barometer of an

air leak is the carburetor idle mixture adjustment necessary to obtain a smooth running engine. An air leak anywhere in the induction system will require a reduction of air at the fuel-to-air mixture adjusting screw to produce good idle. If a smooth idle can only be had by closing, or nearly closing off the idle air adjustment, it is logical to assume that there is an air leak somewhere in the induction system.

If it has been established that there is an induction system leak, the next step is to locate it.

A rule-of-thumb process of elimination is to check out each possibility of leak, one at a time, while watching for a change in engine idle speed. When the source of trouble has been temporarily disconnected from the induction system (as an example: a bad power brake supply line) and the manifold connection capped, there should be a noticeable change in engine rpm.

If all of the potential trouble-makers associated with the intake manifold check out OK, the leak is probably in the intake manifold itself, or at the intake manifold or carburetor flange gaskets. If the suspected leak is at an accessible spot, a squirt of oil in that area will be drawn into the induction system through the opening. This action will affect manifold pressure and fuel mixture, resulting in a change in engine idle speed. Subsequent applications of oil to the intake manifold and carburetor flange gasket areas should help locate the troubled spot if it does exist at these attaching points. The same method of squirt-and-observe may be used to explore suspicious looking cracks in the intake manifold. When the leak is located, repair or replace the troublesome unit.

7. Poor performance due to heat riser failure is rather remote except in the area of the heat control valve.

The purpose of the heat riser is to assist in vaporizing the atomized fuel-air mixture from the carburetor.

The heat riser is a portion of the exhaust system separate from, but immediately adjacent to, the intake manifold; it is generally a part of the exhaust manifold. This heat riser is equipped with a spring and counterweight loaded exhaust flow control valve. The purpose of this valve is to direct exhaust gases around the walls of the intake manifold to help vaporize fuel while the engine is below optimum combustion temperatures; then, as the engine warms up, to divert the excessive heat of exhaust gases away from the intake manifold.

There is a remote possibility of exhaust gases being admitted into the intake manifold at this spot through a cracked housing or (in some cases) a bad gasket.

The most common failure in the heat riser area is with the manifold heat control valve. The valve may remain open (counterweight in the down position) reducing exhaust influence on the induction gas temperatures. This situation would tend to reduce engine efficiency and build up crankcase dilution and sludge. Free the control valve and, if necessary, renew the control valve spring.

8. Vaporlock is a partial or complete stoppage of fuel flow caused by fuel vaporizing in the liquid fuel area of the system.

Partial vaporlock affects performance by restricting the flow of liquid gas at, or to, the carburetor. This condition is first noticeable at idle speeds (rough idle and stalling). An engine operating with vaporlock may respond to acceleration with the same degree of hesitation as that resulting from a carburetor with a bad accelerator pump, or it may tend to ping due to lean mixture.

Complete vaporlock will stop the engine and make it very difficult to restart until the engine and the fuel system have cooled sufficiently to reduce the vapor in the system.

To correct vaporlock, be sure to use a premium fuel. Check the antipercolator passages in the carburetor. Check fuel pump pressure (low pump pressure will aggravate vaporlock). Be sure fuel lines are insulated from the exhaust system and that a generous amount of air flow is routed to cool the lines. An insulator block may be necessary between carburetor and intake manifold.

Vaporlock may occur more frequently at high altitudes or in the event of a sudden warm spell in early spring, before the refineries have changed to summer type fuels.

Percolation is another condition caused by boiling, this time in the carburetor. In this case, usually at idle or after a hard run. Due to expansion, liquid gasoline is dribbled from the carburetor jets into the intake manifold, causing a flooded engine. An engine which has stalled due to flooding will usually respond to about ten seconds of wide open throttle cranking.

9. The carburetor choke is designed to change the ratio of gasoline-to-air through the progression of heat changes in the engine. Its prime purpose is to aid in starting a cold engine by supplying a rich fuel mixture.

The simple choke uses a flue-type plate, mounted slightly off-center in the carburetor throat, above the throttle plate. It is controlled manually by the operator. However, the automatic choke has been almost entirely accepted by the motoring public, because a properly functioning automatic choke does a better job of constantly adjusting the air-to-fuel ratio from one temperature extreme of the engine to the other.

A sensing device is designed and calibrated to react, by expansion or contraction, to the slightest warmup temperature change, and to compensate for the change in choke setting. The same degree of surveillance and manual adjustment by using a hand operated choke is impossible. To check for hard starting due to malfunctioning of the choke, (manual or automatic) make sure that the choke plate is entirely closing off the carburetor throat when choke is applied (cold engine).

To check closing of the automatic choke, remove the air cleaner, then open throttle once, all the way. The unloader tang in the automatic choke mechanism should now permit the choke plate to close. If it does not, linkage is binding or out of adjustment. Choke adjustment should very closely comply with manufacturers' specifications. Index marks (usually scribed on the choke thermostat housing) are for this purpose.

However, adjustment of the thermostat housing will usually satisfy minor requirements, but only renewal of the thermostat spring can entirely replace the calibration originally built into the part.

In the absence of marks or specifications, a rule-of-thumb automatic choke adjustment is:

First, make sure that the engine is cold and that choke plate and linkage are free.

Second, adjust the tension on the thermostat spring to just close the choke plate in the carburetor throat.

Third, rotate thermostat cover

in the direction of choke closing, (rich) about 10°. On Chrysler and some General Motors products, using the well-type choke thermostat, loosen the mounting post locknut and adjust index marks to specifications, or bend choke connecting link to effect the same choke attitude as previously stated (about two scribe marks on the calibration plate).

10. Carburetor icing is a phenomenon familiar to temperate and frigid climates. It can, however, happen under certain atmospheric conditions, almost anywhere.

 Air at 40°F may support almost ½ lb. of water per 1000 cu. ft. This is not much water, but it is enough to cause icing.

The moisture laden air enters the carburetor throat and becomes part of the atomized mixture. As the fuel and water laden air emerge from the carburetor venturi, gas expansion and consequent refrigeration (heat absorption from surrounding area) take place.

If the air is humid, is moderately low in temperature, and the fuel is high in volatility, ice may form on the throttle plate. As fuel evaporates, it adds to the cooling effect of expansion and the throttle plate temperature is quickly lowered to below freezing. Air-borne water particles are promptly frozen and accumulate on the throttle plate. Any formation of ice at this point will interfere with normal air flow and the mechanical movement of the throttle valve itself. This may cause stalling and erratic idle, with poor low speed performance.

Recently, the larger oil companies are treating their products with ice-retarding additives to prevent this condition. If confronted with carburetor icing, be sure that your fuel supply is winter blended for this purpose. Anhydrous compounds (under various trade names) are available from most gas stations for the absorption of water from fuel tanks. This additive should also help in a case of suspected icing.

ANTI-FREEZE INFORMATION

Freezing and Boiling Points of Solutions According to Percentage of Alcohol or Ethylene Glycol

Freezing Point of Solution	Alcohol Volume %	Alcohol Solution Boils at	Ethylene Glycol Volume %	Ethylene Glycol Solution Boils at
20°F.	12	196°F.	16	216°F.
10°F.	20	189°F.	25	218°F.
0°F.	27	184°F.	33	220°F.
−10°F.	32	181°F.	39	222°F.
−20°F.	38	178°F.	44	224°F.
−30°F.	42	176°F.	48	225°F.

Note: above boiling points are at sea level. For every 1,000 feet of altitude, boiling points are approximately 2°F. lower than those shown. For every pound of pressure exerted by the pressure cap, the boiling points are approximately 3°F. higher than those shown.

To Increase the Freezing Protection of Anti-Freeze Solutions Already Installed

Cooling System Capacity Quarts	Number of Quarts of ALCOHOL Anti-Freeze Required to Increase Protection													
	From +20°F. to					From +10°F. to					From 0°F. to			
	0°	−10°	−20°	−30°	−40°	0°	−10°	−20°	−30°	−40°	−10°	−20°	−30°	−40°
10	2	2 3/4	3 1/2	4	4 1/2	1	2	2 1/3	3 1/4	3 3/4	1	1 3/4	2 1/2	3
12	2 1/2	3 1/4	4	4 3/4	5 1/4	1 1/4	2 1/4	3	3 3/4	4 1/2	1 1/4	2	2 3/4	3 1/2
14	3	4	4 3/4	5 1/2	6	1 1/2	2 1/2	3 1/2	4 1/2	5	1 1/4	2 1/2	3 1/4	4
16	3 1/4	4 1/2	5 1/2	6 1/4	7	1 3/4	3	4	5	5 3/4	1 1/2	2 3/4	3 3/4	4 3/4
18	3 3/4	5	6	7	7 3/4	2	3 1/4	4 1/2	5 3/4	6 1/2	1 3/4	3	4 1/4	5 1/4
20	4	5 1/2	6 3/4	7 3/4	8 3/4	2	3 1/4	5	6 1/4	7 1/4	1 3/4	3 1/2	4 3/4	5 3/4
22	4 1/2	6	7 1/2	8 1/2	9 1/2	2 1/4	4	5 1/2	6 3/4	8	2	3 3/4	5 1/4	6 1/2
24	5	6 3/4	8	9 1/4	10 1/2	2 1/2	4 1/2	6	7 1/2	8 3/4	2 1/4	4	5 1/2	7
26	5 1/4	7 1/4	8 3/4	10	11 1/4	2 3/4	4 3/4	6 1/2	8	9 1/2	2 1/2	4 1/2	6	7 1/2
28	5 3/4	7 3/4	9 1/2	11	12	3	5 1/4	7	8 3/4	10 1/4	2 1/2	4 3/4	6 1/2	8
30	6	8 1/4	10	11 3/4	13	3	5 1/2	7 1/2	9 1/4	10 3/4	2 3/4	5	7	8 3/4

Test radiator solution with proper tester. Determine from the table the number of quarts of solution to be drawn off from a full cooling system and replace with concentrated anti-freeze, to give the desired increased protection. For example, to increase protection of a 22-quart cooling system containing Alcohol anti-freeze, from +10°F. to −20°F. will require the replacement of 5 1/2 quarts of solution with concentrated anti-freeze.

Cooling System Capacity Quarts	Number of Quarts of ETHYLENE GLYCOL Anti-Freeze Required to Increase Protection													
	From +20°F. to					From +10°F. to					From 0°F. to			
	0°	−10°	−20°	−30°	−40°	0°	−10°	−20°	−30°	−40°	−10°	−20°	−30°	−40°
10	1 3/4	2 1/4	3	3 1/2	3 3/4	3/4	1 1/2	2 1/4	2 3/4	3 1/4	3/4	1 1/2	2	2 1/2
12	2	2 3/4	3 1/2	4	4 1/2	1	1 3/4	2 1/2	3 1/4	3 3/4	1	1 3/4	2 1/2	3 1/4
14	2 1/4	3 1/4	4	4 3/4	5 1/2	1 1/4	2	3	3 3/4	4 1/2	1	2	3	3 1/2
16	2 1/2	3 1/2	4 1/2	5 1/4	6	1 1/4	2 1/2	3 1/2	4 1/4	5 1/4	1 1/4	2 1/4	3 1/4	4
18	3	4	5	6	7	1 1/2	2 3/4	4	5	5 3/4	1 1/2	2 1/2	3 3/4	4 3/4
20	3 1/4	4 1/2	5 3/4	6 3/4	7 1/2	1 3/4	3	4 1/4	5 1/2	6 1/2	1 1/2	2 3/4	4 1/4	5 1/4
22	3 1/2	5	6 1/4	7 1/4	8 1/4	1 3/4	3 1/4	4 3/4	6	7 1/4	1 3/4	3 1/4	4 1/2	5 1/2
24	4	5 1/2	7	8	9	2	3 1/2	5	6 1/2	7 1/2	1 3/4	3 1/2	5	6
26	4 1/4	6	7 1/2	8 3/4	10	2	4	5 1/2	7	8 1/4	2	3 3/4	5 1/2	6 3/4
28	4 1/2	6 1/4	8	9 1/2	10 1/2	2 1/4	4 1/4	6	7 1/2	9	2	4	5 3/4	7 1/4
30	5	6 3/4	8 1/2	10	11 1/2	2 1/2	4 1/2	6 1/2	8	9 1/2	2 1/4	4 1/4	6 1/4	7 3/4

Test radiator solution with proper hydrometer. Determine from the table the number of quarts of solution to be drawn off from a full cooling system and replace with undiluted anti-freeze, to give the desired increased protection. For example, to increase protection of a 22-quart cooling system containing Ethylene Glycol (permanent type) anti-freeze, from +20°F. to −20°F. will require the replacement of 6 1/4 quarts of solution with undiluted anti-freeze.

ANTI-FREEZE CHART

Temperatures Shown in Degrees Fahrenheit
+32 is Freezing

Quarts of ALCOHOL Needed for Protection to Temperatures Shown Below

Cooling System Capacity Quarts	1	2	3	4	5	6	7	8	9	10	11	12	13
10	+23°	+11°	−5°	−27°									
11	+25	+13	0	−18	−40°								
12		+15	+3	−12	−31								
13		+17	+7	−7	−23								
14		+19	+9	−3	−17	−34°							
15		+20	+11	+1	−12	−27							
16		+21	+13	+3	−8	−21	−36°						
17		+22	+16	+6	−4	−16	−29						
18		+23	+17	+8	−1	−12	−25	−38°					
19		+24	+17	+9	+2	−8	−21	−32					
20			+18	+11	+4	−5	−16	−27	−39°				
21			+19	+12	+5	−3	−12	−22	−34				
22			+20	+14	+7	0	−9	−18	−29	−40°			
23			+21	+15	+8	+2	−7	−15	−25	−36°			
24			+21	+16	+10	+4	−4	−12	−21	−31			
25			+22	+17	+11	+6	−2	−9	−18	−27	−37°		
26			+22	+17	+12	+7	+1	−7	−14	−23	−32		
27			+23	+18	+13	+8	+3	−5	−12	−20	−28	−39°	
28			+23	+19	+14	+9	+4	−3	−9	−17	−25	−34	
29			+24	+19	+15	+10	+6	−1	−7	−15	−22	−30	−39°
30			+24	+20	+16	+11	+7	+1	−5	−12	−19	−27	−35

+ Figures are above Zero, but below Freezing.

− Figures are below Zero. Also below Freezing.

Quarts of ETHYLENE GLYCOL Needed for Protection to Temperatures Shown Below

Cooling System Capacity Quarts	1	2	3	4	5	6	7	8	9	10	11	12	13	14
10	+24°	+16°	+4°	−12°	−34°	−62°								
11	+25	+18	+8	−6	−23	−47								
12	+26	+19	+10	0	−15	−34	−57°							
13	+27	+21	+13	+3	−9	−25	−45							
14			+15	+6	−5	−18	−34							
15			+16	+8	0	−12	−26							
16			+17	+10	+2	−8	−19	−34	−52°					
17			+18	+12	+5	−4	−14	−27	−42					
18			+19	+14	+7	0	−10	−21	−34	−50°				
19			+20	+15	+9	+2	−7	−16	−28	−42				
20				+16	+10	+4	−3	−12	−22	−34	−48°			
21				+17	+12	+6	0	−9	−17	−28	−41			
22				+18	+13	+8	+2	−6	−14	−23	−34	−47°		
23				+19	+14	+9	+4	−3	−10	−19	−29	−40		
24				+19	+15	+10	+5	0	−8	−15	−23	−34	−46°	
25				+20	+16	+12	+7	+1	−5	−12	−20	−29	−40	−50°
26				+17	+13	+8	+3	−3	−9	−16	−25	−34	−44	
27				+18	+14	+9	+5	−1	−7	−13	−21	−29	−39	
28				+18	+15	+10	+6	+1	−5	−11	−18	−25	−34	
29				+19	+16	+12	+7	+2	−3	−8	−15	−22	−29	
30				+20	+17	+13	+8	+4	−1	−6	−12	−18	−25	

For capacities over 30 quarts divide true capacity by 3. Find quarts Anti-Freeze for the ⅓ and multiply by 3 for quarts to add.